Catalogue of Mean UBV Data on Stars

Jean-Claude Mermilliod   Monique Mermilliod

Catalogue of Mean UBV Data on Stars

Springer-Verlag

New York Berlin Heidelberg London Paris
Tokyo Hong Kong Barcelona Budapest

Jean-Claude Mermilliod
Institut d'Astronomie
Université de Lausanne
CH 1290 Chavannes-des-Bois
Switzerland

Monique Mermilliod
Institut d'Astronomie
Université de Lausanne
CH 1290 Chavannes-des-Bois
Switzerland

Library of Congress Cataloging-in-Publication Data
Mermilliod, Jean-Claude.
    Catalogue of mean UBV data on stars / Jean-Claude Mermilliod,
  Monique Mermilliod
        p.   cm.
    Includes bibliographical references.
    ISBN 0-387-94355-2. — ISBN 3-540-94355-2
    1. Stars — Spectra — Classification — Catalogs.   2. UBV system
  (Astronomy) — Catalogs.   I. Mermilliod, Monique.   II. Title.
  Q5881.M47   1994
  523.8′22′0212 — dc20   94-28224

Printed on acid-free paper.

Production managed by Hal Henglein; manufacturing supervised by Jacqui Ashri.
Photocomposed copy prepared from the authors' TeX file.
Printed and bound by Braun-Brumfield, Ann Arbor, MI.
Printed in the United States of America.

9 8 7 6 5 4 3 2 1

ISBN 0-387-94355-2 Springer-Verlag New York Berlin Heidelberg
ISBN 3-540-94355-2 Springer-Verlag Berlin Heidelberg New York

Contents

Introduction

The fundamental paper by Johnson and Morgan describing the standard stars, the first cluster observations, and the properties of the UBV system was published in 1953 (Johnson & Morgan 1953). We celebrated in 1993 the 40th anniversary of this photometric system, which became used all around the world. Although some justified criticisms were raised against the definition of the U filter or the original reduction procedure of the $U - B$ index, the UBV photometric system contains two of the most important parameters, namely the apparent V magnitude and the $B - V$ colour index. They are very much used, but no recent catalogue of mean values has been prepared for several years, although the number of published measurements has nearly doubled. The previous catalogue (Nicolet 1978) collected the UBV data published prior to 1976 for 53845 stars. The data collections have been continuously updated, and the latest catalogue of original data has been described by Mermilliod (1987a). He also analysed the content and produced lists of erroneous or suspect measurements (Mermilliod 1987b, 1987c).

The publication of the present catalogue has been delayed much beyond our wishes due to the underestimation of the time necessary to collect positions, many of which were not previously known, for several thousand stars. The collaboration of our group in the preparation of the Hipparcos Input Catalogue required the updating of several collections of photometric data and also took a lot of manpower that could not be used to complete the UBV catalogue. Furthermore, the referee who reviewed this project proposed to include spectral types, while Prof. C. Jaschek and B. Hauck suggested that full references to the bibliography should be added. Due to the elapsing time, we added to the 1986 catalogue of original data (Mermilliod 1987a) all those published until the end of 1989, and in January 1993, we decided to include also the data from the three years 1990, 1991 and 1992 which meanwhile had been compiled.

Cape BV (Arp 1958) system, Eggen's UBV_E (Eggen & Sandage 1960), the $UBVRI$ in the Johnson (Mendoza 1963), Kron (Kron & Smith 1951), Cousins (1980), Kunkel & Rydgren (1979), Moffet & Barnes (1979), and Neckel and Chini (1980) systems. The references allow one to distinguish the original system in which the UBV data were obtained. The relation is given in Table I.

As discussed in Mermilliod (1987a), numerous checks have been performed to improve the reliability of the data. The preparation of the Hipparcos Input Catalogue required the confrontation of photometric data, not only within the UBV system, but also with other systems (Geneva 7-colours, $uvby$, Walraven). These comparisons provided further checks of the correctness of the V magnitude and $B - V$ colour index. Still, 27803 (27%) stars have data from one observation published in one source only and therefore no internal check could be performed for these stars. This value is, however, a little overestimated because, in many cases, the real number of observations was not published and has been set to 1 by default.

Table I

Reference coding of the photometric systems

First	-	Last	Photometric system	
1	-	999	UBV	(1966 - 1983)
1001	-	1423	UBV	WPC (1953 - 1966)
1460	-	1784	UBV	(1983 - 1992)
2001	-	2042	Cape BV	Nicolet's cat.
3001	-	3080	UBV_E	Eggen
4001	-	4003	UBViyz	Jennens & Helfer
5002	-	5013	$UBVRI$	Kunkel & Rydgren
6001	-	6011	$UBVRI$	Moffet & Barnes
7002	-	7010	$UBVRI$	Neckel & Chini
8001	-	8114	$UBVRI$	Johnson

1 Source of the UBV Data

The main source of UBV data used to compute mean values is the compilation made by Mermilliod (1987a) and the data compiled at the Institute of Astronomy of Lausanne University until the end of 1992. The catalogue contains the photometric data in the original UBV system (Johnson & Morgan 1953), as well as the V, $B - V$, and $U - B$ parameters from other systems that use the same magnitude and index definitions: the

2 Computation of Mean Values

As shown in the analysis presented by Mermilliod (1987b, 1987c), rather large erratic errors may appear in the UBV system. This means that a star having several agreeing sources of data may present discrepant values with differences larger than 0.2 mag in V and larger than 0.1 in $B - V$ or $U - B$. It would be difficult to take into account all these suspect data individually, and a new scheme to compute mean values has been developed, which gives automatically a low weight to the discrepant data. A two-step iterative procedure has been adopted: the first step consists of a simple weighted mean, the weight being the number of measurements to the 2/3 power. The exponent used gives slightly less weight to observations with a number of measurements much larger than usually found. This is also the unique step for stars with two sources of measurements only.

$$\overline{X} = \frac{\sum_i X_i \, n_i^{2/3}}{\sum_i n_i^{2/3}}$$

where X represents V, $(B - V)$, or $(U - B)$, and n is the number of measurements. The second step uses the differences $E_i = X_i - \overline{X}$ of the individual values (X_i) to the mean $(\overline{X})$ to compute new weighted mean values. The weight (p_i) has the form:

$$p_i = \frac{\alpha \, \sqrt{2} \, \varepsilon_X \, n_i^{2/3}}{\sqrt{E_i^2 + \varepsilon_X^2}}$$

where

$$\alpha = \frac{N}{\sum_i n_i^{2/3}}$$

is a normalisation factor, so that the sum of the weights is close to the sum of sources when the error (E_i) is close to the mode of the errors (ε_X). The weighted mean values are simply computed with the formula

$$\overline{X} = \frac{\sum_i p_i \, X_i}{\sum_i p_i}$$

The values of ε_X have been estimated from the distribution of the differences for stars with two sources as presented in Table XI of Mermilliod (1987b). The adopted values for ε_V, $\varepsilon_{(B-V)}$, $\varepsilon_{(U-B)}$, are equal to 0.01, 0.0075, and 0.011, respectively. This means that the expression

$$\frac{\sqrt{2} \, \varepsilon_X}{\sqrt{E_i^2 + \varepsilon_X^2}} \sim 1$$

for values of E_i close to ε_X and p_i is close to unity for the majority of the stars. The dispersion is computed in the usual way:

$$\sigma_X = \sqrt{\frac{\sum_i (X_i - \overline{X})^2 \, p_i}{\sum_i p_i}}$$

As desired, this procedure gives a lower weight to discrepant values, and the computed mean value is closer to that defined by the majority of the sources. However, data from older publications, e.g. those included in the Washington photoelectric catalogue (WPC) (Blanco et al. 1968), may be less precise, like the $U - B$ colours of references WPC 18 and 34. Therefore, the weights for these

publications with references between 1001 and 1425 (as taken from the WPC) have been multiplied by 0.5. Half weight has also been given to the reference 15 of the present catalogue, which contains partly new data, but most of them are combined with values taken from the literature. Thus, the combination of both 0.5 weights gives a full weight to most of these data.

Mean values of variable stars have been computed in the same manner. Due to the procedure used to include them in the catalogue (only the brightest value of the published light curve or list of observations has been taken), the final dispersion may be artificially small and not representative of the real amplitude of variation. For a number of stars with large deviant data which could alter too much the computation of mean values (see Mermilliod 1987c), the number of measurements has been set "manually" to 0. Thus, the reference is kept but the values are not taken into account in the computation of the mean value.

3   Identifications

3.1   Principles

The Lausanne identifiers (LID, Mermilliod 1978), used for the computer version of the photometric catalogues, have been translated to restore the original identifications and in order the HD(E) and DM numbers. However, BS (=HR) numbers are also very useful, as is the information on duplicity (WDS) or variability (GCVS and variable star names). The files developed to improve the coherence of the identifiers in our photometric database, which give lists of codes asssociated to several commonly used catalogues, have been used in the reverse sense to add cross-identifications concerning mainly three catalogues: LSS, NLS, and Giclas.

As already described in previous papers (Mermilliod 1978, 1987a), the problem of stellar identification in a catalogue is a serious one. It is clear that it is not always possible to retain the originally published identification because too many identifiers exist for each star, and most of them are generally used by the astronomers. It is highly unfortunate that no consensus has yet emerged from common practice to define primary identifiers to be used. Furthermore, we feel that it is absurd to identify a $V = 6.29$ magnitude star as an "A" labelled on a map!

Accordingly, one of our main contributions in building the UBV catalogue is to collect the data under a sole designation for each star. The amount of work needed to carry out such a project is undoubtedly one of the most important parts of the whole chain. As an example, several hundred stars have been identified in the Durchmusterung catalogues, mostly by comparing the published and DM maps. The identification principles continuously maintained during this work are recalled below.

3.2 Field stars

3.2.1 HD, HDE, DM, and other catalogues or lists

The basic identifiers remain the HD and HDE numbers. DM identifications following the HD convention are used for fainter stars. If one of these two identifiers is not available, an attempt has first been made to relate the stars, as far as possible, to the main catalogues collecting the same kind of stars. We have tried to follow the acronyms proposed by Fernandez et al. (1983) for many commonly used designations. Those used are listed in Table 6. However, we had to find a solution for many faint field stars or small lists and sequences for which acronyms have not been, or will never be, formed. The solution adopted here was to designate anonymous stars by the paper reference and the star number used in that paper: AAS35,353 # 5.

3.2.2 Stars in sequences by constellation

Several papers published one or several sequences grouped by constellations. The faint stars that do not appear in the HD or DM catalogues were designated by the constellation name, the number of the sequence, and the star name, like: Cas sq 1 # 18. The reference for the sequence numbering and identification is given in Table 7. Stars for which we have found coordinates are recorded in the main catalogue (Table 1), while the fainter ones (without coordinates) are listed in the second part (Table 2).

3.2.3 Stars in multiple tables

Sometimes, papers present the data in several tables corresponding to sequences observed in various fields. We have retained the table number to differentiate between the stars with identical numbers: AJ79,1294 T4# 7.

3.2.4 Sequences around NGC clusters or galaxies

A number of calibrating sequences around NGC or IC globular clusters or galaxies have been published. The syntax retained has the form: NGC 45 sq1 10. The references for the sequences are given in Table 8. We have systematically transformed the letter designation into numbers: upper case (A to Z) become 1 to 26, while lower case (a to z) become 27 to 52. Greek letters would start at 53. We have identified the brighter stars in the HD and DM catalogues, and most coordi-

nates given here come from the GSC. The fainter stars without coordinates are listed in Table 2.

3.2.5 Sequences around active galaxies or quasars

There exist also a large number of short sequences around active galaxies or quasars. We have designated the sequence stars by the name of the galaxy or quasar followed by the star number within the sequence. We have also systematically transformed the letters into their numeric values. The references for the sequences are given in Table 9.

3.3 Cluster stars

Cluster stars offer an even bigger challenge to reduce the various numbering systems used into a coherent one. Once a basic reference has been adopted for a cluster, all other numbering systems are transformed to the adopted one, and a cross-identification table is built to keep track of the transformations. Such tables have been described by Mermilliod (1979) and have been considerably developed for the database for stars in open clusters (Mermilliod 1988, 1992). The adopted numbering system, as given by the references given in Table 10, was used first, but was often extended either by simply continuing the sequential numeration or by adding a constant to the non-common star numbers of any new numbering system. This is indicated in Table 10 by the presence of a second reference and an indication of the value of the constant added, for instance: (no + 1000). For globular clusters, the problem is more difficult due to the complex style often used to identify the stars. As a rule, published identifications have been retained, because there are seldom two or more sources of photoelectric data.

The catalogue gives in addition to the cluster identifications the HD and DM numbers when they exist, from the compilation of Mermilliod (1986) and later development done for the database. Because the cluster numbering occupies the cross-identification field, further cross-references, with variable-, double-, or luminous star names, are given in Table 5. We have used the old-fashioned way of writing the cluster and star names, because the formalism proposed by the IAU would have been difficult to use here.

3.4 Variable stars

The variable star designation has been added in the cross-identification field to facilitate the use of the catalogue. This is based on a compilation of the variable star names made at Lausanne based on the tape version of the General Catalogue of Variable Stars distributed by the Strasbourg Data Center and the successive numbering lists (57th to 70th) published in the Information Bulletin of Variable Stars. It may be incomplete for "anonymous" stars. The presence of variable star names in the cross-identification field implies that the star is variable and the remark "V" has not been added. However, if the field is already filled and a variable star name exists, a "V" has been added and the variable star name can be found in Table 5. Finally, if a star has been declared variable by the observer, but has not yet received a variable star designation, the remark "V" has been indicated.

3.5   Double stars

Double stars remain an important source of trouble, mainly because indication of the diaphragm used or simply the component observed is often lacking. Unless otherwise indicated in the literature, we have assumed that double stars with separation less than 10 seconds of arc have not been separated, and the photometry refers to the joined luminosity and colours. A strict criterion is difficult to apply since the limiting separation below which double stars are observed together has changed with time due to instrument improvement. We have given an individual entry to each component (A, B, ...) if they have been measured separately and an additional one if the magnitude and colours also exist for the combined systems (AB, AP, ABC, BC, CD, and so on). The multiplicity is given in the remark column. If the WDS identification could not be written in the main catalogue (Table 1) because the field is already occupied, it is given in Table 5. It may also happen that a star has been seen double at the telescope, or that the author noted that a companion is included in the diaphragm. Therefore, a remark "AB" has been added, but no WDS number corresponds to it.

4   Coordinates

A very large effort has been devoted to collecting available positions from the literature to add coordinates to the catalogue. Positions for the HD, HDE, and many DM stars have been provided by the Strasbourg Data Center thanks to the courtesy of Dr. Daniel Egret. Positions for fainter stars have been taken directly from the Durchmusterung catalogues and precessed to 1950. Data for various kinds of stars have been collected from the relevant catalogues or lists published. However, many thousands of stars remained without coordinates after this first compilation because they were identified on charts only. Thanks to the recent availability of the Guide Star Catalogue (GSC) (Lasker et al. 1988) and the superb program developed by Luc Weber (Geneva Observatory), it has been possible to obtain coordinates for more than 5000 stars brighter than about 14.5, depend-ing on the GSC local limit. The coordinates were found by comparing visually the published charts with that of the GSC plotted on the screen. However, a number of charts were found to be completely unusable due to inappropriate scale, incompleteness, and so on. Therefore, some stars that should normally have had coordinates do not.

The coordinates were then used to perform more cross-identifications with the Durchmusterung catalogues, which finally became quite easy and reliable with the program developed by Bernard Pernier (Geneva Observatory). In addition, the coordinates made it possible to find identical stars still recorded under different names. This mostly applies to anonymous stars. Nothing could be done for stars fainter than the limit of the GSC, and about 4000 additional stars do not yet have any position. They are listed in Table 2, together with their mean *UBV* data. The solution of the problem of coordinates will still require large efforts. It may be worth making this effort because faint stars are potentially interesting for the magnitude scale calibration of Schmidt plates.

Stars in clusters offered also several problems because few positions were available. All coordinates contained in the database for stars in open clusters (Mermilliod 1988, 1992) were used, but this was not sufficient. Therefore, rectangular (x,y) positions in mm or arbitrary units were transformed to equatorial coordinates with a second order astrometric program, on the basis of at least 6 standard stars. The lists of standard stars were often enlarged with positions found in the GSC.

The GSC provided most of the coordinates for small clusters and standard sequences for the larger ones. In total, more than 5000 stars have been cross-identified with the GSC fields. Finally, cluster maps have been measured with a Hipad Plus digitizing tablet (Houston Instruments) and the x,y positions have been transformed to equatorial coordinates. For most clusters, the residuals of the transformation were less than 1 second of arc, which is quite sufficient for the precision used in the present catalogue. However, a number of fainter cluster stars remain for which no coordinates could be determined, and they are presented with the central coordinates of the cluster. Central coordinates only are also given for globular cluster stars because of the lack of time and standards, except for NGC 104 (47 Tuc) and NGC 362. Tucholke (1992a, 1992b) published coordinates for these two clusters just when this catalogue approached completion and was so kind as to send us a computer readable copy which allowed us to take them into account.

The coordinates for the equinox 1950 are given as frequently as possible to the second of time and to the tenth of minute of arc in declination. When the last digit is missing (e.g. for HD-type positions) the fraction of minute has been multiplied by 60 and recorded as seconds. In this case, no decimals are given in the declination. Less precise data can therefore be identified in this manner. The equinox 1950 has been chosen because it is still the most abundant for the catalogues or publications consulted in preparing the present compilation.

5   Spectral Types

Spectral types have been taken from the Michigan catalogues vol. I to IV. (Houk & Cowley 1975, Houk 1978, Houk 1982, Houck & Smith-Moore 1988). Further spectral classifications were taken from Jaschek's catalogue of selected types (Jaschek 1978) or from Kennedy's (1983) compilation, as distributed by the Strasbourg Data Center. The spectral types for the bright stars have been improved with those published in the Bright Star Catalogue (Hoffleit & Jaschek 1982) when the luminosity class was missing. However, small differences in subtypes or luminosity classes may remain. For stars in open clusters, the data come from the database for stars in open clusters (Mermilliod 1988, 1992). When no MK types were available, we have looked for one-dimensional types, from the HD catalogue or the Spectral Durchmusterung, or the types quoted in the publications where the UBV data were found. Many spectral classifications for proper motion stars were taken from Bidelman & Lee (1975), Lee (1984), and Stephenson (1986).

6   Errors

Although we spent a lot of time in checking the data included in this catalogue, there certainly remain undetected errors. They may result from the observation of a wrong star, typographical mistakes in the publications, or errors made when entering the data in the computer. One example lately solved is provided by the star CpD -62° 2675. Reference 1684 observed He3 775, which is identified to CpD -62° 2675 by Henize (1976), and gives blue colours, while reference 1696 uses the CoD number (-62° 638) and finds red colours. The GSC field shows clearly two distinct stars separated by 1' 32" in declination, with about the same magnitude. But both the CoD and CpD contain one star only, the same, because the coordinates agree closely. We have used the small difference in the GSC magnitude to tentatively attach the red colours to the CpD star (southern component at 12^h 15 35, -62° 43.9 1950) and the blue ones to He3 775 (northern component, at 12^h 15 30, -62° 42.4, 1950). Obvious problems will be seen when comparing the UBV colours and the spectral types. The spectral types have been taken from existing compilations or catalogues, and it happens that they more or less strongly disagree with UBV colours. A number of problems have been solved, but in most cases the solution is not straighforward, because our catalogue correctly reflects the data found in the literature. We shall appreciate receiving comments or suggestions to correct the errors and improve the data collection.

7   Catalogue Content

7.1   Table 1

The content of the successive columns is the following:

- (1) HD, HDE numbers
- (2) DM identifications according to the HD convention
- (3) Miscellanous identifications. The references for the acronyms used are given in Table 6.
- (4) A ⋆ means that further cross-identifications are given in Table 5.
- (5) Remarks on variability (V) or duplicity (ABC..)
- (6) Equatorial coordinates for the equinox 1950
- (7) Number of sources used in the computation of the mean value
- (8) Mean V magnitude and σ over the mean
- (9) Mean $B - V$ index and σ over the mean
- (10) Mean $U - B$ index and σ over the mean
- (11) Total number of observations, by summing up the individual sources
- (12) Spectral types
- (13) Keys to the bibliographic references. The references are listed in Table 3. When not enough space was available in the main table, the key list is continued in Table 4. This is indicated by an * following the end of the reference list.

7.2   Table 2

This table has the same format as Table 1, except that it contains stars for which coordinates were not available at the time of the catalogue preparation.

7.3   Table 3

Table 3 contains the full reference to the publications in which UBV data were found. It gives the authors' names, the journal references and titles.

7.4 Table 4

Because of the limited length of the key reference field in the main catalogue (Table 1), additional reference keys are given in Table 4. This table is entered by the star name, instead of the coordinates. It contains first the DM, then the HD, HDE, and cluster stars. The few field stars are listed at the end of the table. Indication of the components of a double or multiple system is given when necessary to attribute the references to the right object.

7.5 Table 5

Table 5 contains additional cross-identifications of the WDS, GCVS, LS, LSS, and Giclas catalogues when more than one identification exists for any star. As in Table 4, the stars are listed by name order: DM, HD, HDE, cluster and field stars. It contains also a few spectral types when they were too long to fit in the reserved place in Table 1.

7.6 Table 6

Table 6 lists the references for the acronyms used in Table 1. We tried to follow the precepts of Fernandez et al. (1983).

7.7 Tables 7, 8, 9

These three tables have the same format and give the references for the numbering of stars in various regions or around extragalactic objects. Table 7 refers to stellar sequences by constellation, Table 8 to sequences around NGC and IC globular clusters and galaxies, and Table 9 contains the references for photoelectric sequences around quasars and active galaxies.

7.8 Table 10

Table 10 gives the references of the adopted numbering systems for NGC, IC, and anonymous open and globular clusters. When an indication like (no + 1000) is written, it means that 1000 has been added to the star numbers of the paper referenced that are not common with those of the first paper (see, for example, NGC 2539). If the numbers used in the catalogue are larger than those used in the quoted references, it means that the numbering has been extended. The definitions are recorded in cross-identification tables, which can be obtained from the authors, in printed form or by E-mail (mermio@scsun.unige.ch).

8 Statistics

The stellar content of the present catalogue is illustrated by a few statistics or diagrams.

8.1 Magnitude distribution

The distribution of the V magnitude is shown in Fig. 1. The number of stars per interval of magnitude increases regularly up to mag 10 and then decreases. The inner, hatched histogram represents the contribution of stars in clusters.

8.2 Number of sources

We have also counted the number of stars having 1, 2, 3, ..., n independent sources of data and show the results in Table II. The number of stars having n data sources has a strong maximum at $n = 1$, and decreases readily from 3 to 20. For one third of the stars having one data source (27803), the observation relies on one observation only.

Table II

Number of data sources

N_S	N_{star}	N_S	N_{star}	N_S	N_{star}	N_S	N_{star}
1	75917	6	911	11	100	16	19
2	14601	7	586	12	49	17	21
3	5814	8	338	13	46	18	11
4	2735	9	221	14	28	19	6
5	1489	10	156	15	23	>20	37

8.3 Distribution of σ over V, $B-V$, and $U-B$

For the stars with at least 3 sources of data, the distribution of the sigmas over the mean has been plotted (Fig. 2). All three distributions have a pronounced maximum in the interval 0.005 - 0.010, but the width extent toward larger values is different. The distribution of $\sigma(B-V)$ is the narrowest, with a large fraction of sigmas smaller than .020 mag. The distribution of $\sigma(U-B)$ is larger and reaches 0.040 or even 0.050 mag, which is also normal, due to the problems on this colour index. The distribution of $\sigma(V)$ is related not only to the precision of the photometry but also to the variability of the stars. It is also sensitive to small errors in the zero point calibration. It is therefore not surprising to note an extended right wing toward higher sigmas.

8.4 UBV diagrams

The $(V, B - V)$ and $(U - B, B - V)$ planes have been plotted for the whole catalogue. The former diagram (Fig. 3) shows clearly the clump red giants around $(B - V) = 1.05$ and red supergiants, around $(B - V) = 1.8$. The blue stars around $(B - V) = -0.20$ and $V = 13$ to 15 are blue supergiants from the Magellanic Clouds. The latter diagram (Fig. 4) is very densely populated. It shows standard features, such as the very reddened blue stars extending nearly to $(B - V) = 2$, or the M dwarfs, well separated from the giant sequence.

8.5 Galactic distribution

The distribution of the catalogue stars in galactic coordinates is presented separately for the stars brighter than $V = 10$ (Fig. 5) and the fainter ones (Fig. 6). For the brighter stars, the galactic plane, the Gould Belt, and the galactic poles are very well depicted. For the fainter stars, the galactic plane, the nearby clusters, galactic poles, and Magellanic Clouds are well marked. In addition, a number of special areas with a larger number of stars show up.

9   Acknowledgments

This catalogue is based on data collected over 20 years of collaboration between the Institute of Astronomy of the University of Lausanne and the Strasbourg Data Center. This long-term work has been made possible by continuous support from the Swiss National Fund for Scientific Research (FNRS). The data were first entered in the computer by Ms. C. Corbat (on punched cards) and then by Mr. D. Lechaire. We express our gratitude to Dr. D. Egret (Strasbourg), who provided us with the positions contained in Simbad, Dr. J.-L. Halbwachs and M. Wenger (Strasbourg), for furnishing spectral types, also from Simbad, Dr. H.-J. Tucholke for the coordinates in the globular clusters NGC 104 and 362, and Dr. W. Warren for signaling problems and typographic errors and for providing additional HDE cross-identifications. Mrs. J. Rosvick contributed very efficiently to the task of extracting coordinates from the GSC catalogue in many fields, and especially for the sequences around globular clusters and galaxies. Dr. S. Berthet contributed to the preparation of the catalogue of spectral types. We are also thankful to Dr. P. Bartholdi (Geneva Observatory) for the design of the method to compute the average values and generous help with the preparation of the printed catalogue, Dr. D. Mégevand for his advice on the table preparation with LaTeX, Dr. M. Grenon (Geneva Observatory) for helpful discussion and criticism of the data, B. Pernier for help in identifying stars in the Durchmusterung Catalogues, and L. Weber for the interactive program to use the GSC catalogue on workstations.

References

Arp H.C. 1958, AJ 63, 118

Bidelman W.P., Lee S.-G. 1975, AJ 80, 239

Blanco V.M., Demers S., Douglass G.G., FitzGerald M.P. 1968, Publ. U. S. Naval Obs. XXI

Cousins A.W.J. 1980, SAAO Circ. 1, 234

Eggen O.J., Sandage A.R. 1960, MNRAS 120, 79

Fernandez A., Lortet M.-C., Spite F. 1983, AAS 52 no 4

Henize K.G. 1976, ApJS 30, 491

Hoffleit D., Jaschek C. 1982, The Bright Star Catalogue (fourth revised version) Yale Univ. Obs. (New Haven)

Houk N. 1978, University of Michigan catalogue of two-dimensional spectral types for the HD stars. Vol. 2

Houk N. 1982, University of Michigan catalogue of two-dimensional spectral types for the HD stars. Vol. 3

Houk N., Cowley A.P. 1975, University of Michigan catalogue of two-dimensional spectral types for the HD stars. Vol. 1

Houk N., Smith-Moore M. 1988, University of Michigan catalogue of two-dimensional spectral types for the HD stars. Vol. 4

Jaschek M. 1978, Bull. Inform. CDS 15, 121

Johnson H.L., Morgan W.W. 1953, ApJ 117, 313

Kennedy P.M. 1983, MK classification catalogue (Mt Stromlo Obs.)

Kron G.E., Smith J.L. 1951, ApJ 113, 324

Kunkel W.E., Rydgren A.E. 1979, AJ 84, 633

Lasker, B.M., Jenkner, H., and Russell, J.L. 1988, in IAU Symposium 133, Mapping the Sky – Past Heritage and Future Directions, ed. S. Debarbat, J.A. Eddy, H.K. Eichhorn, and A.R. Upgren (Dordrecht: Reidel), p.229

Lee S.-G. 1984, AJ 89, 702

Mendoza E.E. 1963, Tonantzintla Tacubaya 3, 137

Mermilliod J.-C. 1978, Inform. Bull. CDS 14, 32

Mermilliod J.-C. 1979, A&AS 36, 163

Mermilliod J.-C. 1986, A&AS 63, 293

Mermilliod J.-C. 1987a, A&AS 71, 413

Mermilliod J.-C. 1987b, A&AS 71, 119

Mermilliod J.-C. 1987c, Bull. Inform. CDS 32, 37

Mermilliod J.-C. 1988, Bull. Inform. CDS 35, 77

Mermilliod J.-C. 1992, Bull. Inform. CDS 40, 115

Moffet T.J., Barnes III T.G. 1979, PASP 91, 180

Neckel T., Chini R. 1980, A&AS 39, 411

Nicolet B. 1978, A&AS 34, 1

Sanduleak N. 1969, Contr. CTIO no 89

Stephenson C.B. 1986, AJ 92, 139

Tucholke H.-J. 1992a, A&AS 93, 293

Tucholke H.-J. 1992b, A&AS 93, 311

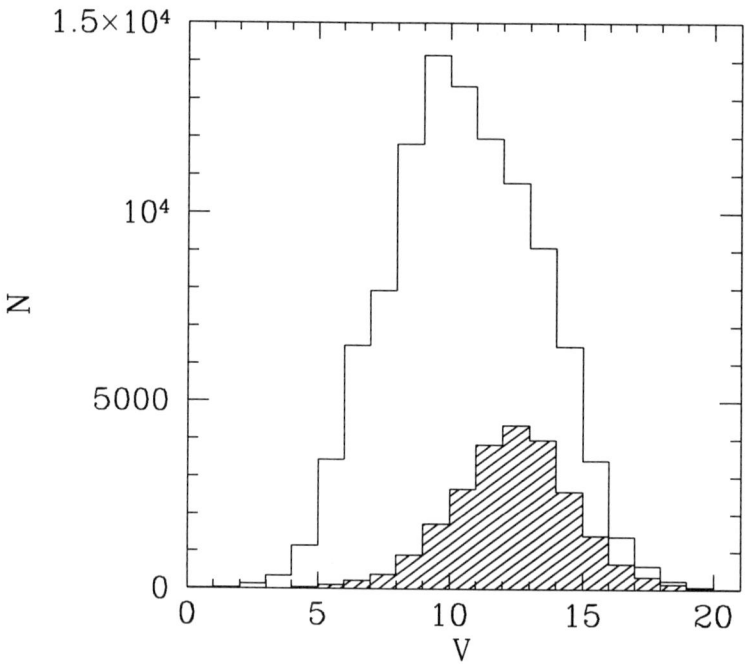

Figure 1: Distribution of the V magnitude for the stars in the catalogue. The dashed histogram represents the contribution of stars in clusters.

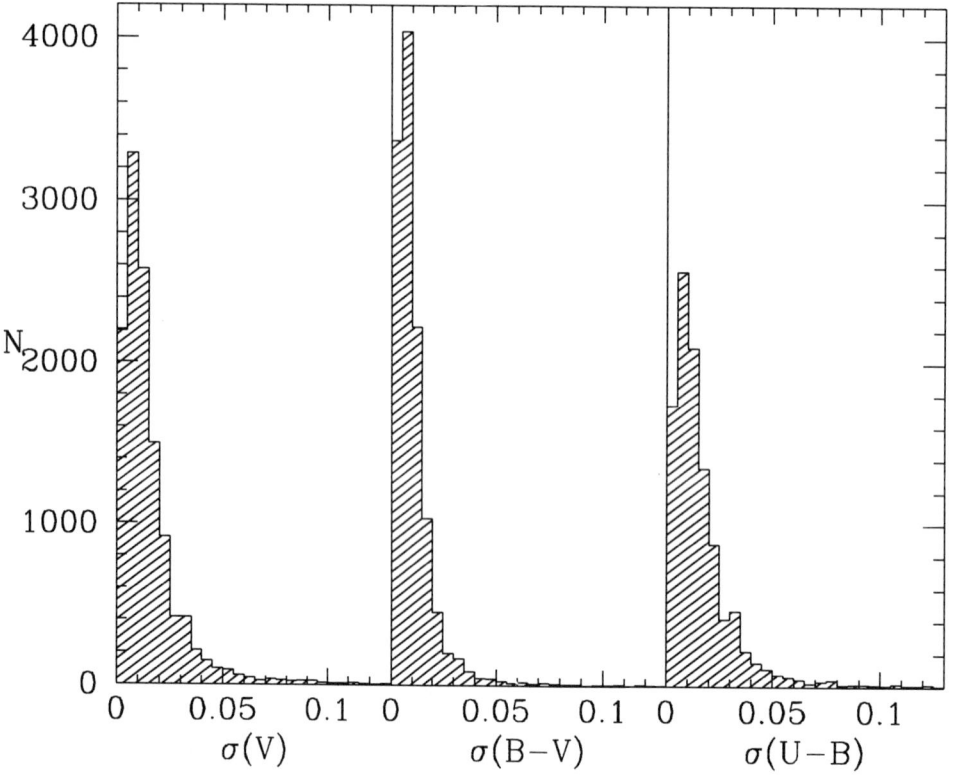

Figure 2: Distribution of the errors on the average magnitudes and colours.

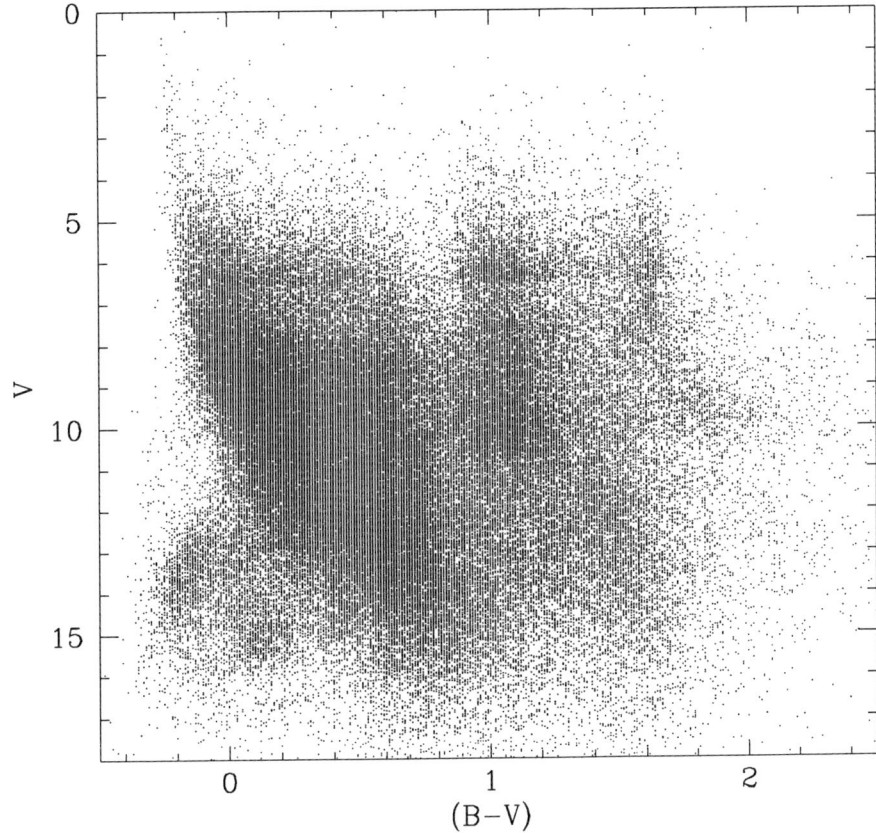

Figure 3: Colour-magnitude diagram for all stars in the catalogue.

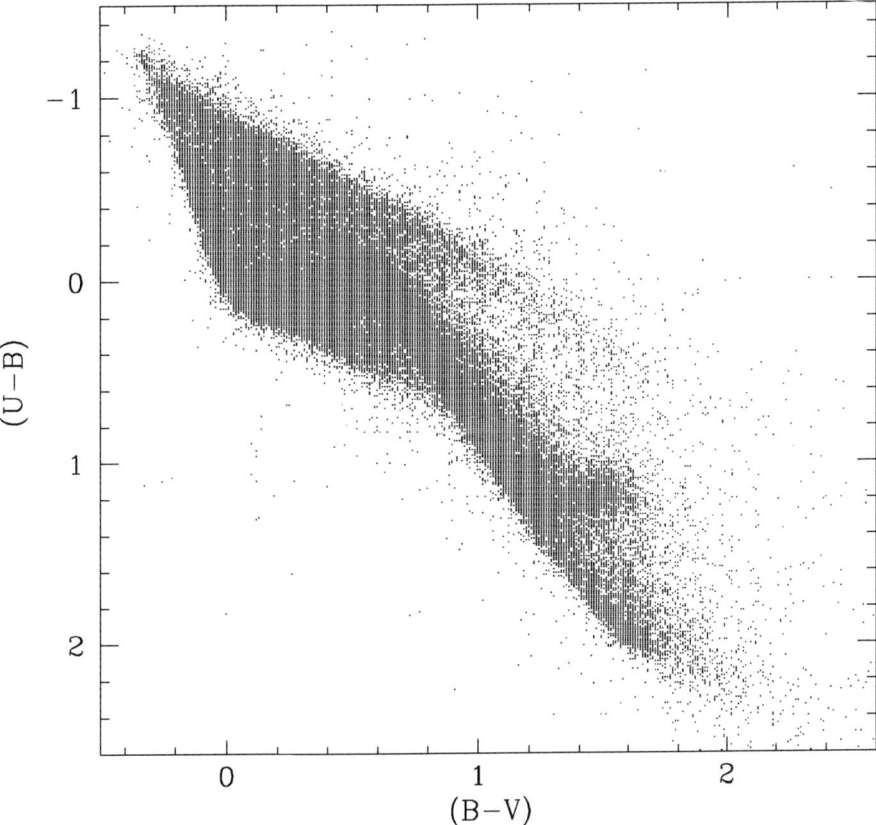

Figure 4: Colour-colour diagram for all stars in the catalogue.

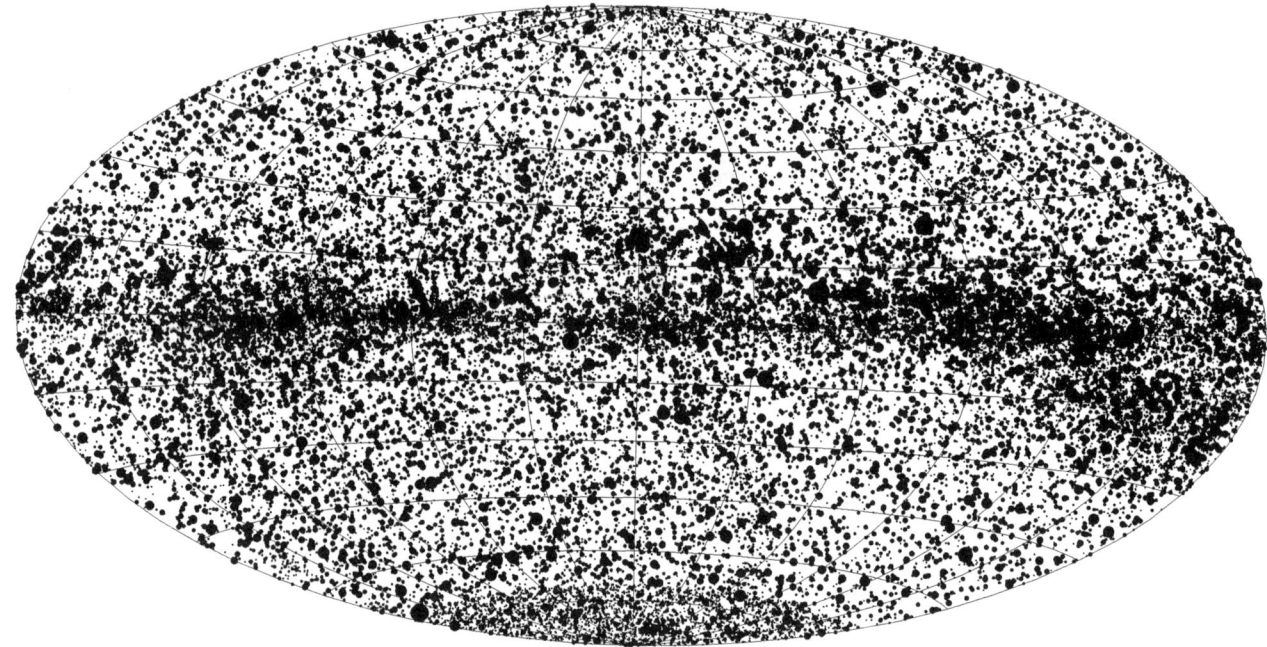

Figure 5: Galactic distribution of the stars brighter than $V = 10.0$.

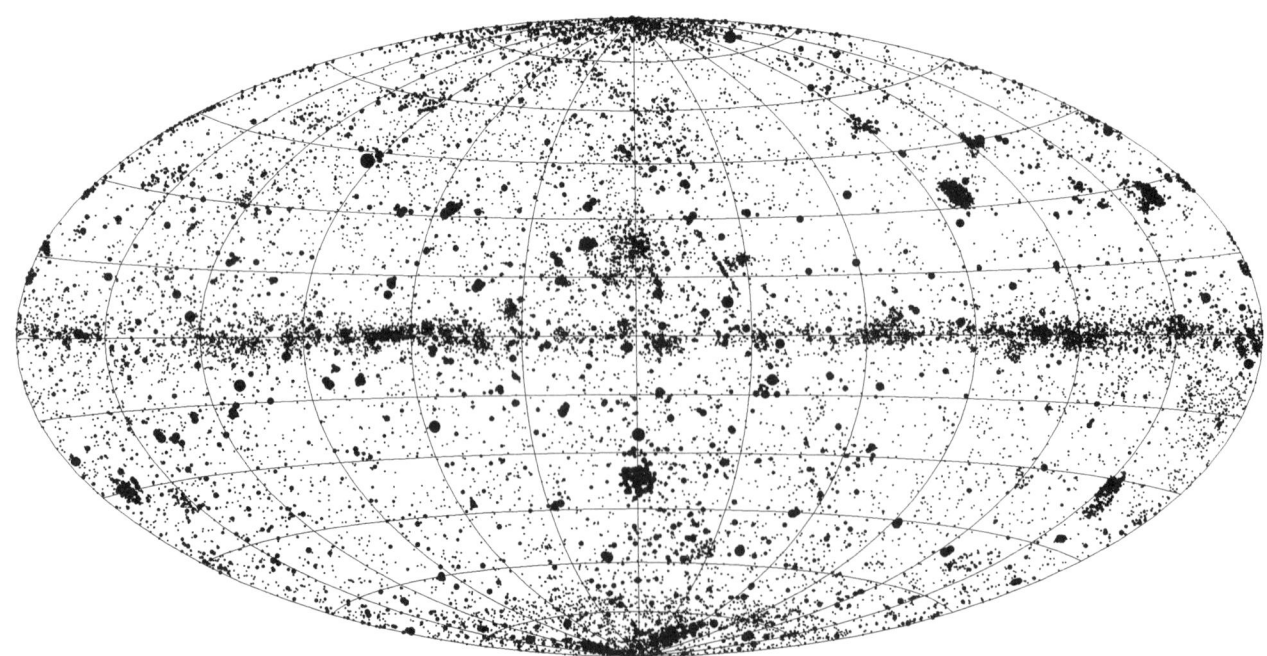

Figure 6: Galactic distribution of the stars fainter than $V = 10.0$.

The Tables

Table 1   13

Table 1: The Catalogue

HD	DM	Other Id	N Rem	α_{1950}	δ_{1950}	S	V	σ_V	B–V	σ_{B-V}	U–B	σ_{U-B}	n	Spectrum	References
		LP 704 - 96		0 00 00	−08 42.9	1	12.33		1.14		1.14		1		1696
225009	+65 01987	HR 9094	⋆ A	0 00 01	+65 49.2	4	5.87	.009	1.08	.009	0.91	.005	4	G8 III	1,15,150,1028
	−29 18936	BI 1 - 46		0 00 01	−28 57.3	1	10.71		1.26		1.32		4	K2	1675
225010	+65 01988	IDS23575N6532	B	0 00 04	+65 49.3	3	7.33	.012	0.08	.009	0.06	.012	5	A2 V	1,150,1028
225015	−9 06310	IDS23575S0903	AB	0 00 05	−08 46.0	1	9.05		0.47		-0.11		7	G0	176
	+19 05193			0 00 06	+19 38.7	1	10.15		0.41		0.00		2		1375
	−29 18938	BI 1 - 47		0 00 06	−29 21.4	1	10.23		0.38		-0.07		6	A7	1675
		G 267 - 18		0 00 06	−34 30.0	1	14.90		0.47		-0.44		2	DC	3079
	−30 19793	BI 1 - 48		0 00 08	−29 44.2	1	12.15		0.60		0.04		2		1675
240479	+59 02819	IDS23576N5924	A	0 00 09	+59 41.0	1	10.22		0.19		-0.02		2	A7	401
240480	+59 02819	IDS23576N5924	A	0 00 10	+59 40.9	1	9.38		1.36		1.22		1	K7	401
240480	+59 02819	IDS23576N5924	AB	0 00 10	+59 40.9	1	8.96		0.84		0.36		1	K7	401
225031	−22 06251			0 00 11	−22 10.1	1	9.30		0.72		0.32		2	G8/K0 V	1375
225023	+35 05159	IDS23576N3516	A	0 00 12	+35 32.2	1	7.52		0.18		0.09		2	A0	1601
	−30 19794	BI 1 - 49		0 00 13	−30 04.6	1	11.56		0.93		0.71		3		1675
225041	−25 16760	RU Scl		0 00 14	−25 13.4	2	9.44	.020	0.08	.005	0.06	.020	2	F0/2 V	668,700
225040	−24 17956	IDS23577S2426	AB	0 00 15	−24 09.2	2	9.62	.010	-0.06	.005	-0.17	.005	9	B9 V	152,1068
	−31 19569	BI 1 - 50		0 00 17	−30 36.8	1	11.86		0.71		0.19		2	K5	1675
		WLS 0-20 # 10		0 00 18	−20 12.8	1	11.27		0.61		0.04		2		1375
225047	−30 19795	BI 1 - 51	AB	0 00 22	−30 25.7	4	8.43	.005	0.02	.016	0.01	.016	13	A0 Vb	1068,1675,2012,3004
225050	−49 14325	JL 150		0 00 23	−49 10.7	1	10.15		0.50		-0.03		2	G0	132
	+60 02660			0 00 24	+61 05.1	1	9.66		1.85				2		369
225045	−20 06703	HR 9095		0 00 24	−20 19.5	2	6.24	.005	0.53	.000	0.07	.010	7	F6 V	15,1071
	+55 03074	G 217 - 29		0 00 25	+56 26.9	2	9.75	.008	0.94	.015	0.74	.018	3	K3	1658,7010
225062	−61 06792			0 00 28	−60 58.2	1	7.81		1.03		0.87		11	K0 III	1704
225064	+51 03769			0 00 31	+52 11.2	1	8.14		0.88		0.51		2	K0	1566
	+31 05024			0 00 32	+31 35.5	1	8.74		0.54		0.00		2	F8	1733
		LS I +59 026		0 00 32	+59 57.2	1	12.34		0.35		-0.44		2		41
	−30 19796	BI 1 - 52		0 00 33	−29 38.2	1	11.64		0.60		0.11		2		1675
225069	−24 17960	HR 9096		0 00 34	−24 25.4	1	6.45		1.18				4	K1 III	2007
		CS 29517 # 44		0 00 35	−14 41.1	1	14.69		0.05		0.17		1		1736
	−30 17797	BI 1 - 53		0 00 36	−30 27.9	1	11.32		0.56		-0.05		4	G3	1675
225076	−25 16766			0 00 37	−24 52.0	1	7.96		0.42				4	F2/3 V	2012
225073	+16 05034			0 00 39	+17 16.5	1	6.57		1.12		1.10		2	K0	1648
		CS 22957 # 31		0 00 39	−03 06.0	1	14.55		0.46		-0.16		1		1736
225077	−31 19571	BI 1 - 134		0 00 39	−31 29.4	3	8.87	.014	0.08	.011	0.04	.025	13	A1mA2V	1068,2012,3004
240483	+59 02821			0 00 41	+59 40.9	1	9.64		0.17		0.04		2	B9	401
	−30 19799	BI 1 - 55		0 00 41	−29 36.5	1	10.67		0.62		-0.01		2	G4	1675
	−31 19572	BI 1 - 54		0 00 41	−30 35.2	1	10.35		0.43		-0.06		2	F1	1675
	+21 05017			0 00 43	+22 09.8	1	9.73		0.93		0.54		2	G0	1375
		Steph 2199		0 00 45	+04 24.5	1	12.04		1.52		1.26		2	M2	1746
	−30 19800	BI 1 - 56		0 00 45	−30 06.6	1	11.30		0.57		-0.04		4	K5	1675
	−30 19801	BI 1 - 57		0 00 46	−30 18.1	1	10.38		0.42		-0.10		2	F1	1675
		CS 29503 # 12		0 00 48	−23 55.7	1	13.30		-0.25		-0.95		1		1736
		PHL 2580		0 00 48	−23 57.0	2	13.30	.042	-0.24	.025	-0.92	.049	6	sdB	286,1036
	−30 19802	BI 1 - 58	AB	0 00 49	−30 05.8	1	10.62		0.50		-0.07		2	F5	1675
225094	+62 02356	HR 9097, V639 Cas		0 00 51	+63 21.8	5	6.24	.011	0.33	.010	-0.54	.004	12	B3 Iae	15,154,1012,1025,8031
225100	−36 16149			0 00 51	−36 14.6	2	9.16	.020	0.24	.010	0.08	.015	10	A2 III	152,1068
	+61 02583			0 00 52	+62 04.8	2	8.69	.000	0.12	.000			5	B8 V	1025,1118
225099	−29 18941	BI 1 - 59	⋆ AB	0 00 52	−29 31.7	2	9.43	.005	0.42	.001	-0.05	.011	9	F3 V	702,1675
225102	−47 14782			0 00 52	−47 06.7	1	11.14		0.04		0.05		4	A0/1 V	152
225095	+54 03103			0 00 53	+55 16.4	1	7.91		-0.02		-0.74		3	B2 IVne	1212
	−31 19574	BI 1 - 60		0 00 53	−30 35.8	1	10.61		0.44		-0.10		2		1675
		WLS 0-20 # 5		0 00 54	−17 34.3	1	10.95		0.63		0.13		2		1375
		LP 644 - 28		0 00 55	−03 32.4	1	12.09		0.45		-0.17		2		1696
225107	+11 05090			0 00 56	+12 04.7	1	7.46		0.29		0.03		2	A5	1733
	+60 02662			0 00 56	+61 22.3	1	8.43		1.32				2	K2	1025
		LS I +60 075		0 00 57	+60 42.3	1	10.76		0.48		-0.36		2		41
		LS I +60 074		0 00 57	+60 55.5	1	10.99		0.26		-0.53		2		41
		CS 22957 # 36		0 00 58	−05 01.0	1	14.43		0.42		-0.17		1		1736
		G 266 - 32		0 00 58	−17 00.8	1	14.67		-0.05		-0.94		1		3062
225111	−30 19803	BI 1 - 61		0 00 58	−30 00.2	2	10.20	.030	0.37	.002	-0.03	.011	7	A5 (III)	702,1675
	−30 19804	BI 1 - 62		0 00 59	−30 00.9	1	10.44		0.69		0.23		5		1675
		BI 1 - 64		0 01 00	−30 32.4	1	12.50		0.77		0.28		2		1675
	−31 19576	BI 1 - 63		0 01 00	−30 45.4	2	10.62	.025	0.45	.018	-0.08	.032	4	F7	1675,3004
225119	−29 18945	BI 1 - 65		0 01 04	−28 41.7	3	8.21	.015	-0.12	.015	-0.35	.023	13	Ap Si	220,1068,1675
		HA 140 # 84		0 01 04	−28 58.5	1	11.96		0.67		0.14		9		1499
	−30 19805	BI 1 - 67		0 01 04	−30 19.1	1	11.42		0.53		-0.05		2	G0	1675
		HA 140 # 85		0 01 05	−28 54.2	1	12.22		0.72		0.20		9		1499
225118	−29 18946	BI 1 - 66	⋆	0 01 07	−28 40.3	3	8.25	.009	0.78	.010	0.37	.035	8	G8 V	79,1675,2034
		LS I +60 076		0 01 10	+60 43.4	1	11.52		0.48		-0.31		2		41
225132	−18 06417	HR 9098		0 01 11	−17 36.9	11	4.55	.008	-0.05	.005	-0.07	.031	182	B9 IVn	3,9,15,26,244,1020*
	−30 19807	BI 1 - 68		0 01 15	−30 21.5	1	11.04		0.54		-0.07		2	F8	1675
225136	+65 01993	HR 9099		0 01 16	+66 26.0	2	6.30	.005	1.66	.010	1.94		6	M4 III	369,3001
240487	+59 02823	IDS23587N6007	AB	0 01 18	+60 23.9	1	9.41		0.11				3	B5	1025
225155	−29 18943	BI 1 - 69		0 01 19	−28 40.3	1	7.67		0.72		0.28		5	G5	1675
225146	+60 02663	LS I +60 077		0 01 22	+60 49.5	3	8.59	.004	0.37	.020	-0.64	.000	6	B0 Ibp	41,1012,1025
		GD 578		0 01 24	−31 04.5	1	12.94		0.15		0.08		1		3061

HD	DM	Other Id	N Rem	α₁₉₅₀	δ₁₉₅₀	S	V	σ_V	B–V	σ_B-V	U–B	σ_U-B	n	Spectrum	References
		Smethells 150		0 01 24	−42 50.	1	11.55		1.26				1		1494
225157	−45 15227			0 01 24	−45 34.0	1	8.53		0.23				4	A6 V	2012
	+48 04234			0 01 25	+49 24.9	1	9.05		0.35				1	F0	21
	+60 02664	LS I +60 078	⋆ A	0 01 26	+60 35.6	1	10.62		0.37		-0.48		2		41
	+60 02664	LS I +60 078	⋆ AB	0 01 26	+60 35.6	1	9.94		0.35		-0.54		2		401
	+60 02664	IDS23589N6017	B	0 01 26	+60 35.6	1	10.65		0.33		-0.56		2		41
	−31 19579	BI 1 - 70		0 01 26	−30 36.0	1	11.12		0.56		-0.07		2		1675
225160	+61 02585	LS I +61 121	⋆ B	0 01 28	+61 56.6	5	8.20	.009	0.27	.010	-0.73	.009	15	O8 e	41,1011,1012,1025,1775
		LP 988 - 105		0 01 29	−43 43.8	1	12.74		1.29				1		3061
		UU Cet		0 01 31	−17 16.5	3	11.63	.048	0.22	.053	0.03	.032	3	F1:	597,699,700
	−60 07712	AO Tuc		0 01 34	−59 45.8	1	10.87		0.16		0.10		1	A2	700
225185	−10 06219	IDS23590S1042		0 01 37	−10 25.4	1	9.84		0.68		0.22		3	G5	1696
225180	+61 02586	HR 9100	⋆	0 01 38	+62 00.6	3	5.87	.008	0.30	.013	0.29		16	A1 III	41,1025,1118
225187	−30 19809	BI 1 - 71	⋆ V	0 01 38	−30 24.8	3	7.13	.028	-0.09	.015	-0.31	.013	8	B8 V	1068,1675,3004
240489	+59 02824	LS I +59 027		0 01 40	+59 52.6	1	8.82		0.08		-0.76		3	B0	41
225179	+64 01893			0 01 40	+64 52.5	1	7.59		1.68				2	K5	369
	−30 19810	BI 1 - 72		0 01 40	−30 19.7	1	10.62		1.11		0.92		6		1675
		NGC 7822 - 7	⋆	0 01 41	+66 04.2	2	10.67	.019	1.42	.047	0.45	.033	4		405,1285
225186	−18 06419			0 01 42	−17 42.2	1	9.07		0.16				4	A1/2 IV/V	2012
	−30 19811	BI 1 - 73		0 01 42	−30 02.2	1	10.81		0.56		0.03		5	G4	1675
		LS I +55 004		0 01 43	+55 35.8	1	10.76		0.08		-0.73		2		41
	−26 16890	G 266 - 35		0 01 43	−26 11.5	1	8.60		0.75		0.27		1	G5	79
	−30 19812	BI 1 - 74		0 01 45	−30 06.1	1	10.52		0.45		-0.11		2	F1	1675
225194	+12 05061	G 30 - 39		0 01 46	+12 40.7	1	8.62		0.64		0.17		1	G5	333,1620
225190	+53 03280			0 01 46	+53 59.8	3	7.80	.054	-0.09	.007	-0.57	.007	6	B3 V	399,555,963
		LS I +62 062		0 01 46	+62 47.1	1	11.46		0.57		-0.17		2		41
225197	−17 06868	HR 9101		0 01 46	−16 48.4	3	5.78	.015	1.08	.013	1.05	.006	8	K2 III	58,2035,3016
		G 30 - 40		0 01 47	+14 08.2	1	11.19		0.96		0.68		1		333,1620
225200	−29 18950	BI 1 - 75		0 01 47	−29 32.8	4	6.39	.006	0.01	.005	-0.05		15	A0 IV shell	15,1068,2006,2012
		G 217 - 30	A	0 01 50	+57 47.4	2	10.47	.016	0.78	.007	0.23	.018	5		1658,7010
		G 217 - 30	B	0 01 50	+57 47.4	1	13.13		1.66				3		7010
		LS I +62 063		0 01 50	+62 46.5	1	11.26		0.40		-0.29		2		41
225206	−30 19813	BI 1 - 76		0 01 50	−29 39.6	4	7.76	.016	-0.06	.011	-0.16	.024	12	B9.5Vp	1068,1675,2012,3004
		LS I +60 079		0 01 51	+60 13.1	1	11.32		0.40		-0.44		2		41
225208	−47 14785			0 01 51	−47 20.8	1	9.80		0.86				4	K0 V	2034
	−29 18951	BI 1 - 77		0 01 53	−28 59.9	1	10.90		0.48		-0.05		4		1675
		LS I +62 064		0 01 54	+62 43.0	1	12.73		0.44		-0.27		1		41
		LS I +60 080		0 01 56	+60 58.1	1	10.97		0.35		-0.41		2		41
	−30 19814	BI 1 - 79		0 01 56	−30 31.2	1	11.68		0.61		0.05		2		1675
	−30 19814	BI 1 - 78		0 01 56	−30 31.2	1	12.09		0.34		0.06		2		1675
225212	−11 06194	HR 9103		0 01 57	−10 47.3	2	4.95	.010	1.62	.005	1.85		9	K3 Ib	2035,8100
	−30 19815	BI 1 - 80		0 01 59	−29 55.8	1	11.04		0.51		-0.05		2		1675
		PHL 645		0 02 00	−24 41.9	1	13.84		-0.25		-1.17		2	sdO	3064
225233	−73 02346	HR 9106		0 02 00	−73 10.5	3	7.31	.004	0.44	.004	0.00	.010	17	F5 V	15,1075,2038
225218	+41 04933	HR 9105	⋆ AB	0 02 02	+41 48.8	1	6.12		0.15		0.14		2	B9 III	1733
225217	+42 04827	SU And	A	0 02 02	+43 16.4	2	8.20	.015	2.51	.065	4.49	.360	2	C5 II	109,414
225217	+42 04827		B	0 02 02	+43 16.4	1	12.77		0.38		0.01		1		414
	−30 19816	BI 1 - 81		0 02 03	−30 13.4	1	11.24		0.99		0.77		2		1675
225216	+66 01679	HR 9104		0 02 05	+66 53.3	1	5.67		1.07		0.93		3	K1 III	1355
		LP 988 - 102		0 02 05	−40 57.8	1	12.84		1.63		1.30		2		3078
		LS I +61 122		0 02 06	+61 13.3	1	11.30		0.26		-0.56		2		41
	+60 02666			0 02 06	+61 26.9	1	9.55		0.48				2	F4 V	1025
225224	−8 06240			0 02 06	−08 27.8	1	8.71		1.43		1.76		1	K2	1746
		BI 1 - 82		0 02 06	−30 31.0	1	12.41		0.57		0.05		2		1675
		LS I +59 028		0 02 07	+59 57.4	1	10.15		1.01		0.78		2		41
		LS I +60 081		0 02 07	+60 18.8	1	10.90		0.36		-0.45		2		41
	−29 18954	BI 1 - 84		0 02 07	−29 27.0	1	11.19		0.47		-0.07		2	F6	1675
		LS I +60 082		0 02 08	+60 31.4	1	12.20		0.40		-0.31		2		41
225248	−29 18953	BI 1 - 83		0 02 08	−29 07.8	1	8.43		1.30		1.50		4	K0	1675
		WLS 0 35 # 9		0 02 10	+35 14.3	1	13.10		0.78		0.20		2		1375
		G 266 - 36		0 02 10	−23 36.6	1	13.61		1.45		1.25		1		3061
225253	−72 02800	HR 9108		0 02 10	−71 42.9	3	5.58	.003	-0.11	.007	-0.41	.005	20	B8 IV/V	15,1075,2038
	+37 04930	IDS23596N3737	AB	0 02 12	+37 53.8	1	9.39		0.74		0.28		2	G5	3016
	+67 01598	NGC 7822 - 9		0 02 13	+67 54.6	2	9.13	.015	0.75	.045	-0.34	.005	4	O9	25,1285
	−30 19817	BI 1 - 85		0 02 14	−30 18.0	1	11.14		0.54		-0.08		2	F9	1675
225239	+33 04828	HR 9107		0 02 16	+34 22.8	5	6.10	.013	0.63	.004	0.08	.012	10	G2 V	15,22,908,1003,3026
		IX Cas		0 02 16	+49 57.3	2	11.21	.022	0.56	.038	0.41		2	F7	851,1399
225264	−30 19818	BI 1 - 86		0 02 18	−29 54.7	4	8.28	.008	0.01	.011	-0.03	.022	13	A0 Vas	1068,1675,2012,3004
	−30 19819	BI 1 - 87		0 02 18	−30 03.7	1	10.82		0.71		0.24		2		1675
225257	+57 02855	IDS23598N5758	AB	0 02 19	+58 15.2	1	6.60		0.03		-0.57		3	B1 V +B6 V	399
		CS 22876 # 40		0 02 19	−34 30.3	2	15.08	.010	0.41	.008	-0.20	.009	3		1580,1736
225273	+63 02103			0 02 20	+64 30.2	1	8.14		0.44				2	F5	1118
225282	−30 19820	BI 1 - 88		0 02 20	−30 32.1	4	8.32	.030	0.04	.006	0.05	.014	15	A1 Va	152,1068,1675,3004
225276	+25 05068	HR 9109	⋆ A	0 02 21	+26 22.2	1	6.25		1.40		1.59		1	K4 IIIb	3016
225276	+25 05068	IDS23598N2606	B	0 02 21	+26 22.2	1	10.72		1.02		0.79		2		3016
236293	+50 04233			0 02 22	+50 54.1	1	8.12		1.24		1.10		2	K2	1566
	−30 19821	BI 1 - 89		0 02 23	−29 50.1	1	11.55		0.59		0.06		2	F8	1675
	−30 19823	BI 1 - 91		0 02 24	−30 26.1	1	11.47		0.58		-0.02		2		1675
	−30 19822	BI 1 - 90		0 02 26	−30 33.6	1	9.85		1.07		0.96		5	gG8	1675
225292	+26 04744			0 02 27	+27 23.8	2	6.47	.010	0.96	.010	0.68	.010	6	G8 II	985,1501

Table 1   15

HD	DM	Other Id	N	Rem	α_{1950}	δ_{1950}	S	V	σ_V	B–V	σ_{B-V}	U–B	σ_{U-B}	n	Spectrum	References
225291	+44 04549	IDS23598N4507	A		0 02 27	+45 23.7	1	7.47		0.55				1	F8 V	680
225291	+44 04549	IDS23598N4507	AB		0 02 27	+45 23.7	1	7.31		0.58		-0.11		2	F8 V	680
225291	+44 04549	IDS23598N4507	B		0 02 27	+45 23.7	1	9.88		0.71				1		680
225299	-70 03038				0 02 27	-70 29.3	1	8.13		0.71				4	G5 V	2012
225213	-37 15492				0 02 28	-37 36.2	6	8.56	.027	1.45	.008	0.99	.042	14	M2 V	258,1311,2012,2017*
225297	-36 16162	G 267 - 24			0 02 29	-36 17.7	1	7.73		0.54		0.01		1	G0 V	79
	+44 04548	G 171 - 39			0 02 30	+45 30.7	2	9.94	.039	1.49	.010	1.24	.049	3	M2	680,3072
225289	+60 02667	HR 9110, V567 Cas			0 02 30	+61 02.1	4	5.78	.015	-0.08	.006	-0.37	.028	8	B8 p HgMn	1025,1079,1118,3016
6	-1 04525	HR 2			0 02 30	-00 46.8	3	6.30	.007	1.10	.000	1.01	.009	19	K0	15,1061,3047
	-29 18957	BI 1 - 92			0 02 34	-29 29.6	1	11.33		0.44		-0.05		2		1675
4	+29 05059	IDS00000N2946	A		0 02 35	+30 03.1	1	7.78		0.43		0.07		1	F0	3016
4	+29 05059	IDS00000N2946	B		0 02 35	+30 03.1	1	9.38		0.51		0.06		1	F0	3016
3	+44 04550	HR 1	⋆ A		0 02 35	+44 57.1	2	6.71	.001	0.06	.004	0.07	.004	4	A1 Vn	695,1733
		LP 49 - 275			0 02 36	+72 57.0	1	14.33		-0.07		-0.91		3		940
		CS 29503 # 4			0 02 36	-26 48.5	1	16.08		-0.23		-1.00		1		1736
		LB 3127			0 02 36	-62 54.	1	12.66		0.46		-0.03		2		45
17	+34 05061				0 02 38	+35 17.6	1	6.87		0.16		0.10		2	A2	1336
	-30 19825	BI 1 - 93			0 02 43	-29 57.8	1	11.48		0.61		0.08		2	F8	1675
	-31 19590	BI 1 - 94			0 02 44	-30 35.6	1	12.32		0.46		-0.10		2		1675
	-30 19826	BI 1 - 95			0 02 45	-29 47.0	1	11.43		0.68		0.16		2	F7	1675
26	+7 05128				0 02 47	+08 30.6	6	8.21	.009	1.06	.014	0.51	.019	14	G4 V:p	993,1003,1319,1775,3077,8005
28	-6 06357	HR 3, BC Psc	AB		0 02 47	-05 59.2	8	4.61	.013	1.04	.006	0.88	.013	24	K1 III	15,58,1075,1425,2012*
31	-22 06257				0 02 47	-21 35.6	3	8.37	.010	-0.06	.000	-0.16	.020	11	B9 V	1068,2012,3007
	+62 02362				0 02 48	+63 30.7	1	9.74		0.20				2	B7 III	1025
55	-68 03597				0 02 48	-68 06.2	2	8.49	.035	1.07	.015	0.87		8	K5 V	2012,3072
		G 266 - 37			0 02 53	-26 15.2	1	12.29		1.11		0.99		1		3061
	-30 19828	BI 1 - 98			0 02 54	-30 23.5	1	11.33		0.70		0.18		1	G9	1675
50	-30 19827	BI 1 - 96			0 02 54	-30 34.1	3	9.73	.008	0.28	.005	0.06	.011	9	A8 V	702,1675,3004
		BI 1 - 97			0 02 55	-29 45.4	1	12.46		0.66		-0.01		2		1675
	-30 19829	BI 1 - 99			0 02 56	-30 10.1	1	11.26		0.55		-0.01		2	F7	1675
66	-43 15584				0 02 56	-43 10.0	1	7.46		1.62		1.77		7	M3 III	3007
60	+16 05041				0 02 57	+16 49.3	2	8.81	.015	1.03	.002	0.74		4	K0	882,3040
		LS I +60 083			0 02 58	+60 33.2	1	12.15		0.44		-0.42		2		41
	+63 02105	LS I +63 037			0 02 58	+63 32.3	1	9.18		0.68				2	A9 II	1025
	-30 19830	BI 1 - 100			0 03 00	-30 19.0	1	12.07		0.73		0.25		4		1675
	-30 19831	BI 1 - 101			0 03 01	-30 03.3	1	12.05		1.02		0.78		2		1675
		G 130 - 41			0 03 02	+28 54.2	2	13.23	.000	0.82	.015	0.25	.015	2		333,1620,1658
73	+42 04831				0 03 02	+43 07.4	2	8.47	.014	-0.18	.000	-0.90	.009	4	B1.5IV	399,963
38	+45 04408	G 171 - 40	⋆ A		0 03 02	+45 32.1	1	9.01		1.44		1.29		2	K2	680
38	+45 04408	G 171 - 40	⋆ AB		0 03 02	+45 32.1	5	8.23	.010	1.44	.016	1.23	.013	18	K2 +K2	680,938,1013,1197,3072
38	+45 04408	IDS00002N4516	B		0 03 02	+45 32.1	2	9.04	.054	1.43	.019	1.22	.019	3	K2	680,3003
	+87 00220				0 03 04	+87 36.7	1	8.94		1.07		0.81		3	K0	1733
	-30 19832	BI 1 - 103			0 03 04	-30 11.9	1	11.27		0.98		0.70		4	K1	1675
		CS 22876 # 30			0 03 04	-36 58.2	1	12.99		-0.05		-0.21		1		1736
240496	+57 02862	LS I +58 010			0 03 05	+58 13.0	1	9.86		0.42		0.32		2	A5	41
	-30 19833	BI 1 - 102			0 03 05	-29 45.0	1	12.00		0.96		0.61		2		1675
		JL 155			0 03 06	-57 01.	1	15.40		0.72		-0.79		1		832
87	+12 05063	HR 4			0 03 08	+13 07.1	1	5.51		0.90				2	G5 III	71
86	+16 05042				0 03 08	+16 34.1	1	8.54		1.16				1	K0	882
		LS I +62 065			0 03 09	+62 17.7	1	11.11		0.28		-0.45		2		41
91	-30 19834	BI 1 - 104			0 03 10	-30 14.3	3	9.93	.010	0.32	.009	0.00	.006	8	F0 V	702,1675,3004
		Steph 2			0 03 13	-08 23.0	1	12.44		1.45		1.23		1	M0	1746
	-29 18960	BI 1 - 105			0 03 14	-29 15.1	1	10.48		1.00		0.71		2	dK0	1675
		LS I +60 084			0 03 19	+60 45.5	1	11.26		1.53		0.59		2		41
	-29 18961	BI 1 - 108			0 03 20	-29 12.6	1	11.45		0.56		0.06		1	F6	1675
105	-42 16587				0 03 20	-42 01.8	2	7.51	.005	0.60	.000			8	G0 V	1075,2013
101	+17 05035				0 03 21	+17 57.5	1	7.45		0.56		0.04		2	F8	1648
	-30 19837	BI 1 - 107			0 03 21	-30 21.4	1	10.67		0.50		-0.04		2	F7	1675
113	+17 05036	G 30 - 42	⋆ AB		0 03 22	+17 48.0	2	8.58	.000	0.92	.010	0.66		3	K0	333,882,1620
104	-31 19594	BI 1 - 106			0 03 22	-30 38.3	2	8.01	.000	0.99	.005	0.75	.025	9	G5 III	702,1675
117	-31 19595	BI 1 - 109	⋆ AB		0 03 24	-30 36.4	2	9.16	.085	0.51	.046	0.03	.038	7	F4 V	702,1675
	+60 02668	LS I +60 085	⋆ AB		0 03 25	+60 35.3	3	8.96	.011	0.46	.013	-0.47	.005	6	B1 III	41,1012,1025
		CS 29503 # 11			0 03 25	-24 27.3	1	13.91		0.00		0.02		1		1736
	-30 19838	BI 1 - 110			0 03 25	-29 56.2	1	11.65		0.61		0.04		2	G4	1675
114	+15 04937				0 03 27	+16 10.4	1	7.87		0.93				1	G5	882
108	+62 02363	LS I +63 039			0 03 27	+63 24.1	3	7.39	.021	0.17	.017	-0.79	.000	6	O6 fpe	1011,1012,1025
110	+39 05219				0 03 29	+40 08.4	1	6.63		0.90		0.54		1	G5	583
		CS 29503 # 8			0 03 29	-24 53.9	1	14.38		0.04		0.12		1		1736
		BPM 2			0 03 30	-83 39.	1	13.67		0.62		0.00		1		3065
		LS I +62 066			0 03 33	+62 56.7	1	12.14		0.31		-0.44		2		41
		LS I +61 123			0 03 34	+61 45.1	1	10.69		0.83		0.51		2		41
		LS I +59 029			0 03 36	+59 34.5	1	11.32		0.55		-0.37		2		41
141	-29 18964	BI 1 - 111	⋆ A		0 03 36	-29 25.8	4	7.90	.005	-0.03	.004	-0.09	.025	16	A0 IVn	1068,1675,2012,3004
123	+57 02865	HR 5, V640 Cas	⋆ AB		0 03 38	+58 09.5	2	5.98	.015	0.69	.014	0.20	.013	9	G5 V	292,938,1118,3030
134	+62 02364				0 03 39	+63 06.7	2	8.79	.000	0.19	.005			5	A7 V	1025,1118
		CS 29503 # 14			0 03 41	-23 05.7	1	14.66		0.09		0.17		1		1736
	-30 19839	BI 1 - 112			0 03 41	-29 59.0	1	12.22		0.75		0.38		2	F8	1675
	-30 19840	BI 1 - 113			0 03 41	-30 23.0	1	11.72		0.63		0.08		2		1675
		LP 464 - 7			0 03 42	+15 13.0	1	13.32		1.50		1.23		1		1696
	+60 02669				0 03 42	+60 47.9	2	7.70	.027	1.90	.004			5	M2	369,1025

HD	DM	Other Id	N Rem	α_{1950}	δ_{1950}	S	V	σ_V	B–V	σ_{B-V}	U–B	σ_{U-B}	n	Spectrum	References
		Smethells 153		0 03 42	−66 07.	1	12.17		1.55				1		1494
135	+59 02828			0 03 43	+60 10.7	2	7.95	.030	0.24	.030			4	F0	1025,1118
142	−49 14337	HR 6	★AB	0 03 44	−49 21.2	5	5.70	.015	0.52	.007	0.02	.004	17	G1 IV	15,258,1075,2012,3077
160	−64 04404	IDS00012S6448	AB	0 03 45	−64 31.2	1	7.84		0.46				4	F3 IV	2012
		3C 2 # 1		0 03 46	−00 20.0	1	12.02		1.50		1.88		1		327
	−30 19841	BI 1 - 114		0 03 48	−30 01.0	1	11.10		1.03		0.84		2	G0	1675
		CS 22876 # 34		0 03 48	−35 33.9	1	13.73		0.40		0.05		1		1736
		LS I +62 067		0 03 50	+62 13.9	1	10.58		0.23		-0.65		2		41
144	+63 02107	HR 7		0 03 50	+63 55.1	3	5.57	.025	-0.02	.013	-0.22	.015	6	B9 IIIe	1079,1118,3023
	−30 00001	BI 1 - 115		0 03 51	−30 08.8	1	10.37		1.21		1.31		4	F8	1675
157	−20 06720			0 03 52	−20 20.7	5	10.67	.015	-0.03	.004	-0.09	.011	26	B9 V	152,1026,1068,1775,3007
	−30 00002	BI 1 - 116		0 03 54	−30 34.5	1	11.61		0.62		0.08		2	K3	1675
		G 31 - 25		0 03 57	+05 32.8	2	10.94	.030	1.00	.015	0.90		4	K1	202,333,1620
		3C 2 # 3		0 03 57	−00 16.6	1	11.75		0.74		0.29		1		327
168	+12 05066			0 03 58	+12 33.2	1	8.61		1.22				2	K0	882
		G 30 - 43		0 03 59	+10 08.5	1	12.41		1.24		1.14		1		333,1620
236301	+58 02700			0 04 00	+58 36.8	1	9.26		0.42		0.21		2	A5	401
163	+62 02367			0 04 00	+62 56.8	2	8.50	.005	0.34	.000			5	F2 V	1025,1118
166	+28 04704	HR 8	★A	0 04 01	+28 44.7	6	6.10	.019	0.75	.003	0.32	.014	32	K0 V	15,1013,1197,1355,1758,3026
	−31 00001	BI 1 - 117		0 04 06	−30 55.8	1	11.13		0.68		0.16		4	G4	1675
		BI 1 - 118		0 04 07	−30 38.0	1	12.54		0.75		0.34		7		1675
		3C 2 # 2		0 04 10	−00 17.3	1	11.99		1.09		0.85		1		327
	+59 02829	LS I +60 086		0 04 12	+60 19.4	2	9.85	.009	0.39	.005	-0.64	.005	4	B0 IV:pen	41,1012
		G 158 - 27		0 04 12	−07 47.9	2	13.76	.034	1.98	.015	1.75	.128	8		419,3078
		PHL 666		0 04 12	−27 38.	1	13.93		-0.25		-1.16		2	sdO	3064
		Steph 4		0 04 13	+17 34.6	1	11.99		1.31		1.20		1	K5	1746
186	+43 04635			0 04 13	+44 20.1	1	9.02		-0.09		-0.71		1	F5	963
		LS I +59 030		0 04 13	+59 34.9	1	10.97		0.35		-0.61		2		41
	+61 02593			0 04 13	+62 05.7	1	9.98		0.22				3	B6 III	1025
202	−21 06537	G 266 - 42		0 04 13	−21 21.9	3	9.80	.044	0.58	.005	0.00	.004	6	G2/3 V	79,1696,3062
		LS I +62 068		0 04 16	+62 01.6	1	11.48		0.24		-0.52		2		41
203	−23 00004	HR 9		0 04 17	−23 23.1	3	6.17	.004	0.39	.005	0.02		8	F2 IV	15,79,2027
	−31 00003	BI 1 - 119		0 04 17	−30 40.0	1	11.82		0.66		0.00		2		1675
	−31 00004	BI 1 - 120		0 04 20	−30 36.2	1	10.86		0.42		-0.08		2	F8	1675
208	−4 06025	G 158 - 28		0 04 21	−03 54.1	2	8.20	.015	0.57	.030	-0.02	.005	2	G0	1658,3056
	+27 04677	G 130 - 42		0 04 23	+27 57.2	1	10.25		0.74		0.28		1	K0	333,1620
212	−31 00006	BI 1 - 121		0 04 23	−30 52.9	3	10.23	.009	0.40	.003	-0.05	.009	7	F2 V	702,1675,3004
205	+63 02108			0 04 24	+64 27.4	1	8.76		0.32				2	A5	70
228	−4 00005			0 04 24	−44 15.8	1	8.54		0.22		0.14		6	G8 III	1673
	−31 00007	BI 1 - 122		0 04 28	−30 38.0	1	11.75		0.62		0.07		2	G6	1675
225	−19 06570			0 04 29	−19 06.0	1	8.23		0.48		0.04		2	F5	1375
	−30 00003	BI 1 - 123		0 04 31	−29 49.2	1	11.49		1.24		1.13		5		1675
		TonS 142		0 04 32	−33 45.4	2	14.65	.028	0.19	.019	-0.65	.039	4		1736,3064
236305	+58 02703		V	0 04 33	+59 30.5	1	9.32		1.82				2	F0	369
		G 266 - 43		0 04 33	−25 41.4	1	14.95		1.62				1		3061
		LS I +60 087		0 04 34	+60 02.4	3	10.88	.031	0.55	.000	0.14	.018	5	B8 II:	41,1012,1215
		TOU28,33 T9# 104		0 04 34	+62 52.7	1	9.68		0.32				2		1025
		G 266 - 45		0 04 34	−27 49.8	1	14.06		1.56		1.07		1		3061
235	−31 00010	BI 1 - 124		0 04 35	−30 50.4	4	8.66	.012	0.06	.011	0.06	.013	15	A0 Vbn	1068,1675,2012,3040
	+57 02870	LS I +58 011		0 04 42	+58 03.6	2	10.02	.055	0.50	.009	0.01	.017	3	B8 Iab	41,1215
		LS I +61 124		0 04 42	+61 37.2	1	11.05		0.48		-0.24		2		41
250	+20 05430	G 131 - 23		0 04 44	+20 41.4	1	9.14		0.74		0.28		1	G5	333,1620
255	−7 06153			0 04 45	−07 24.3	1	10.40		0.05		0.06		4	A0	152
256	−18 06428	HR 10		0 04 45	−17 39.9	4	6.18	.005	0.13	.008	0.08	.021	10	A6 Vn	15,79,252,2012
276	−77 00001			0 04 46	−77 00.5	1	7.54		0.41				4	F2 V	2012
249	+25 05073			0 04 47	+26 10.4	1	7.38		1.00		0.75		2	K1 IV	3026
		G 130 - 43		0 04 48	+28 58.5	2	14.19	.009	1.49	.009	1.40	.000	6		203,3078
268	−26 00008	G 266 - 46		0 04 49	−25 38.0	1	7.05		0.45		-0.01		1	F5 V	79
		G 267 - 33		0 04 49	−29 52.2	1	15.47		1.53				1		3061
264	+16 00001			0 04 55	+17 01.5	1	8.50		1.59				1	K0	882
	+62 00001	LS I +62 069		0 04 56	+62 48.1	2	10.37	.019	0.31	.019	-0.65	.014	4	B2 IV-V:pe	41,1012
		G 266 - 47		0 04 56	−26 19.4	1	14.85		1.60				1		3061
236307	+59 00001			0 04 57	+60 27.9	1	9.79		0.22		0.16		2	A0	401
		GD 2		0 04 58	+33 00.8	2	13.84	.014	-0.28	.005	-1.19	.014	5	DA	1727,3060
	+61 00001			0 04 58	+61 38.7	1	8.52		1.90				3	M0	369
283	−24 00008	G 266 - 49		0 04 58	−24 05.9	2	8.69	.000	0.80	.000	0.34		8	K0 V	158,2012
	+62 00002			0 04 59	+62 51.7	1	9.15		0.21				2		1025
	−30 00006	BI 1 - 125		0 05 00	−30 34.0	1	10.59		0.44		-0.08		6	F8	1675
291	+38 00002			0 05 05	+38 45.4	2	8.02	.000	0.43	.005	0.00	.005	5	F5 V	105,1733
		CS 22876 # 32		0 05 05	−35 48.0	2	12.83	.000	0.40	.007	-0.26	.009	4		1580,1736
306	−39 00010			0 05 05	−38 57.9	1	9.44		0.08		0.02		4	A0 V	152
294	+0 00003			0 05 06	+00 49.0	1	8.25		0.17		0.14		3	A0	379
		LS I +58 012		0 05 08	+58 58.8	1	12.23		0.64		-0.25		2		41
	+62 00003	IDS00026N6248	AB	0 05 08	+63 04.8	1	8.59		0.42		0.10		2	F5 I	401
315	−3 00002	HR 11		0 05 10	−02 49.6	10	6.44	.013	-0.14	.008	-0.46	.008	89	B8 IIIp Si	10,15,79,220,989,1061*
319	−23 00013	HR 12	★AB	0 05 14	−22 47.2	3	5.93	.004	0.14	.000	0.06		11	A1 V	15,1068,2012
313	+37 00003			0 05 16	+38 26.6	1	7.66		0.16		0.10		2	A3	1336
		CS 29503 # 27		0 05 18	−26 47.6	1	14.39		0.04		0.12		1		1736
		JL 161		0 05 18	−49 18.	1	14.83		0.51		-0.04		1		832
		G 267 - 34		0 05 22	−30 15.4	1	13.14		1.50				1		3061

Table 1   17

HD	DM	Other Id	N Rem	α_{1950}	δ_{1950}	S	V	σ_V	B–V	σ_{B-V}	U–B	σ_{U-B}	n	Spectrum	References
245	+85 00412	G 265 - 1		0 05 23	+86 30.6	1	8.37		0.65		-0.07		2	G2	1003
	+62 00004	LS I +63 044		0 05 24	+63 04.1	1	8.82		0.62				2	F2 II	1025
		PHL 6304		0 05 24	-23 57.	1	14.53		-0.27		-1.03		1		286
	+43 00004			0 05 25	+44 17.8	1	9.16		0.38		0.13		1	F0	963
		G 30 - 45		0 05 26	+07 43.9	1	13.08		1.52				1		906
331	+36 00002			0 05 27	+36 59.6	1	9.05		0.22		0.15		2	A3	1336
343	-30 00010	BI 1 - 126		0 05 29	-30 18.6	2	9.82	.040	0.31	.005	0.02	.004	4	F0 Vp	1675,3004
344	-34 00017	HR 13		0 05 31	-33 48.5	1	5.68		1.12				4	K1 III	2032
	-30 00012	BI 1 - 127		0 05 32	-29 41.8	1	9.61		0.96		0.62		3	dG9	1675
	+14 00003	G 30 - 46		0 05 38	+14 43.9	1	11.01		0.89		0.69		1	K2	333,1620
352	-3 00003	HR 14, AP Psc	AB	0 05 38	-02 43.6	5	6.10	.030	1.37	.012	1.13	.032	23	K2 III	10,15,58,1061,3016
	-38 00011	G 267 - 35		0 05 40	-38 33.5	1	9.99		0.56		-0.04		1	G0	79
350	+33 00005			0 05 42	+34 13.0	1	8.78		0.31		0.13		2	A2	1336
		BPM 46039		0 05 42	-25 53.	1	13.14		1.21		1.12		1		3061
	-29 00015	G 266 - 52		0 05 43	-28 48.4	2	11.71	.025	0.55	.002	-0.11	.003	4		1696,3061
	+61 00005			0 05 44	+62 06.3	2	8.96	.010	0.12	.005			4	B6 V	1025,1118
360	-9 00005	HR 16		0 05 44	-09 06.1	4	5.99	.005	1.04	.004	0.83	.004	11	K0	15,79,253,1061
		G 31 - 26		0 05 45	-05 31.5	3	12.12	.035	0.60	.018	-0.14	.025	4		202,1620,1696
358	+28 00004	HR 15, α And	⋆ A	0 05 48	+28 48.9	7	2.07	.038	-0.11	.012	-0.45	.022	23	B8 IVp	15,1034,1049,1118*
		MCC 351		0 05 53	+17 08.7	2	10.76	.019	1.50	.020			3	K7	1017,1619
		CS 22876 # 31		0 05 53	-36 25.9	1	14.18		0.27		0.08		1		1736
394	-30 00013	BI 1 - 128		0 05 54	-30 20.8	1	9.51		0.48		0.03		5	G6 IV	1675
371	+62 00005			0 05 55	+62 55.6	2	6.41	.005	1.03	.015	0.71		6	G3 II	1025,1355
395	-44 00012			0 05 55	-44 33.1	1	9.85		0.16		0.11		4	A2 IV/V	152
	-26 00021	BPM 46043		0 05 56	-26 21.9	1	10.49		0.46		-0.01		1		3061
		G 31 - 27		0 05 57	+02 56.0	1	13.95		1.52				2		202
		LS I +59 031		0 05 57	+59 50.4	1	12.33		0.57		-0.25		2		41
		G 266 - 53		0 05 59	-20 40.5	1	11.45		0.96		0.79		1		3061
	-30 00014	BI 1 - 129		0 05 59	-30 00.3	1	9.81		1.38		1.64		1	G0	1675
402	-18 00003	HR 18		0 06 00	-17 51.3	3	6.07	.012	1.65	.012	2.00	.018	15	M0 III	15,1071,3007
400	+35 00008	HR 17		0 06 06	+36 21.0	3	6.17	.079	0.49	.017	-0.07	.030	10	F7 V	254,1067,3037
		JL 162		0 06 06	-46 36.	1	16.37		0.42		-1.11		1		132
	+54 00003	LS I +55 005		0 06 09	+55 23.4	1	10.47		0.15		-0.69		2		41
		GD 3		0 06 10	+44 44.2	3	12.55	.034	0.09	.046	-0.03	.023	10	sdAp	272,308,979
405	+13 00003			0 06 11	+13 51.9	2	7.81	.044	1.13	.012	1.02		3	K0	882,3040
416	+36 00004	IDS00037N3640	AB	0 06 16	+36 56.2	1	8.89		0.31		0.07		2	A5	1336
417	+24 00003	HR 19		0 06 17	+25 11.0	3	6.23	.005	0.97	.005	0.73	.010	9	K0 III	15,1003,1355
415	+37 00007			0 06 19	+37 44.8	1	8.97		0.30		0.19		2	A2	1336
437	-22 00007			0 06 21	-22 27.5	1	7.12		1.64		1.90		4	M2 III	3007
		LS I +57 002		0 06 24	+57 09.3	1	11.39		0.20		-0.54		2		41
		BPM 46048		0 06 24	-27 24.	1	14.12		1.43		1.14		1		3061
		G 266 - 55		0 06 26	-25 23.9	1	12.09		1.38		1.24		1		3061
		G 266 - 56		0 06 26	-25 28.3	1	13.56		1.16		0.86		1		3061
448	+17 00007	HR 22		0 06 27	+17 56.0	1	5.53		1.04				2	G9 III	71
		G 130 - 46		0 06 27	+27 22.3	1	11.67		1.26		1.12		2	M1	7010
455	-30 00019	BI 1 - 130		0 06 28	-30 12.6	1	10.07		0.46		-0.08		4	F0	1675
	-27 00016	G 266 - 57		0 06 29	-27 24.1	2	11.76	.054	1.50	.019	1.15		3	M0	1705,3061
432	+58 00003	HR 21, β Cas	⋆ A	0 06 30	+58 52.4	10	2.27	.008	0.34		0.10	.012	42	F2 III-IV	1,15,22,1004,1077*
469	-54 00019	HR 23	⋆ AB	0 06 31	-54 16.8	4	6.32	.010	0.74	.000	0.38		15	G4 IV	15,258,1075,2012
431	+78 00001	HR 20	⋆ AB	0 06 32	+79 26.2	2	6.01	.000	0.20	.009	0.10		4	A6 IV+A8 IV	1381,3030
		LS I +60 088		0 06 36	+60 37.9	1	12.55		0.42		-0.23		2		41
443	+64 00003			0 06 37	+64 47.5	1	7.01		0.92		0.67		5	K0	1355
		G 130 - 47		0 06 39	+25 00.0	2	11.48	.028	1.35	.037	1.20	.066	4		333,1620,3024
		CS 29503 # 21		0 06 39	-24 06.7	1	15.28		0.10		0.17		1		1736
471	+24 00004	G 130 - 48	⋆ A	0 06 40	+25 00.4	2	7.78	.009	0.67	.014	0.15	.005	4	G0	1620,3024
		G 30 - 48		0 06 43	+08 44.8	2	13.74	.022	1.38	.022	1.16		6		1663,3078
		LS I +59 032		0 06 43	+59 51.1	1	12.19		0.40		-0.43		2		41
		LS I +61 125		0 06 43	+61 08.8	1	11.28		0.80		0.12		2		41
		G 267 - 39		0 06 43	-30 51.6	1	14.92		0.77		0.17		1		3061
483	+16 00003			0 06 45	+17 15.4	1	7.05		0.64		0.18		2	G2 III	1648
470	+57 00009			0 06 46	+58 23.3	2	8.24	.000	0.08	.000	-0.41	.005	6	B5 Vn	206,401
		LS I +62 070		0 06 47	+62 30.6	1	11.31		0.40		-0.62		2		41
		LP 524 - 25		0 06 48	+03 33.8	1	11.53		1.38		1.26		1	K7	1696
		G 30 - 49		0 06 48	+06 24.1	2	12.37	.039	1.01	.000	0.81		3		202,333,1620
		LS I +60 089		0 06 48	+60 16.2	1	11.55		0.19		-0.62		2		41
		BPM 46055		0 06 48	-26 50.	1	14.31		0.87		0.37		1		3061
493	-28 00016	HR 24	⋆ AB	0 06 48	-28 16.0	3	5.40	.004	0.42	.005	0.06		8	F3 V	404,2009,2024
489	+18 00003	G 131 - 29	⋆ AB	0 06 53	+18 50.2	3	7.94	.000	0.66	.000	0.17	.010	3	G3 V	333,1620,3026
496	-46 00018	HR 25		0 06 53	-46 01.4	6	3.87	.012	1.02	.011	0.84	.007	23	K0 III	15,1075,2012,4001*
491	+0 00008			0 06 54	+00 57.9	1	7.28		1.22		1.31		3	G5	379
	+61 00008	KN Cas		0 06 58	+62 23.4	1	9.49		1.72		0.58		2	M1 Ibpe	8032
		CS 22876 # 33		0 06 59	-35 42.7	1	13.37		0.05		0.02		1		1736
		G 130 - 49		0 07 01	+30 52.2	1	16.85		0.25		-0.70		3		3060
	+25 00009	IDS00044N2608	AB	0 07 02	+26 24.2	1	8.83		1.16		1.01		2	K0	1733
		LS I +60 090	⋆ A	0 07 02	+60 23.1	1	11.96		0.31		-0.49		2		41
		LS I +60 090	⋆ B	0 07 02	+60 23.1	1	12.31		0.26		-0.47		2		41
		LS I +62 071		0 07 02	+62 50.4	1	11.56		0.25		-0.58		2		41
		G 130 - 50		0 07 05	+30 52.3	1	13.21		1.41		1.11		3		3016
	+9 00008			0 07 07	+10 14.2	1	9.50		0.49		-0.03		3	F8	196
508	+61 00009			0 07 07	+62 18.5	2	8.26	.009	0.34	.005	0.15		4	A9 V	401,1025

HD	DM	Other Id	N Rem	α_{1950}	δ_{1950}	S	V	σ_V	B–V	σ_{B-V}	U–B	σ_{U-B}	n	Spectrum	References
536	−38 00023	IDS00046S3752	AB	0 07 10	−37 35.6	1	8.34		0.43				4	F3/5 V	2012
535	−36 00024	IDS00047S3649	AB	0 07 11	−36 32.2	1	9.85		0.25		0.01		5	A5 II/III	1068
		CS 22876 # 41		0 07 12	−34 16.0	1	14.10		0.21		0.04		1		1736
		CS 22876 # 38		0 07 13	−34 55.9	1	14.19		0.14		0.16		1		1736
		LP 524 - 76		0 07 16	+04 31.6	1	12.70		0.66		0.06		2		1696
		BD Cas		0 07 16	+61 14.0	1	10.84		1.50				1		592
530	+10 00007	IDS00047N1052	A	0 07 17	+11 08.3	1	8.37		0.41		0.02		3	F5	196
545	−3 00009			0 07 21	−02 50.4	1	6.85		1.61		1.98		2	K2	3016
540	+48 00019			0 07 23	+49 21.1	1	8.80		−0.05		−0.34		1	A0	963
236323	+58 00009			0 07 23	+59 23.5	1	8.93		0.17				2	A0	1118
		LS I +60 091		0 07 23	+60 15.5	1	11.00		0.25		−0.63		2		41
565	−62 00009			0 07 23	−62 34.5	1	6.34		0.17		0.15		7	A6 V	1628
562	−26 00035			0 07 26	−26 09.2	2	7.65	.000	0.14	.005	0.04		8	A2 V	1068,2012
560	+10 00008	HR 26	⋆A	0 07 28	+10 52.1	1	5.56		−0.07		−0.21		2	B9 Vn	542
560	+10 00008	HR 26	⋆AB	0 07 28	+10 52.1	5	5.53	.010	−0.07	.005	−0.21	.019	15	B9 V +G5 Ve	228,542,1049,1079,1319
560	+10 00008	IDS00049N1035	B	0 07 28	+10 52.1	1	9.44		0.34		−0.14		1	G5 Ve	542
		LS I +60 092		0 07 30	+60 15.9	1	12.20		0.30		−0.52		2		41
		LP 464 - 397		0 07 36	+10 47.0	1	11.42		1.09		1.02		1		1696
		BPM 46063		0 07 36	−29 07.	1	12.50		1.28		1.16		1		3061
		LS I +59 033		0 07 38	+59 37.0	1	11.53		0.71		−0.24		2		41
575	−20 00009			0 07 38	−20 15.7	1	9.47		0.52		0.03		2	F7 V	1375
		LS I +61 126		0 07 40	+61 47.7	1	10.74		0.62		−0.25		2		41
		PHL 703		0 07 42	−26 28.9	3	12.86	.003	−0.15	.023	−1.16	.015	7	sdO	286,1036,3064
571	+45 00017	HR 27		0 07 43	+45 47.6	8	5.04	.004	0.40	.009	0.25	.011	25	F2 II	15,39,253,254,1119*
		CS 22876 # 37		0 07 44	−35 03.4	1	15.00		0.42		−0.21		1		1736
587	−6 00011	HR 29		0 07 45	−05 31.6	2	5.83	.005	0.98	.000	0.75	.005	7	K1 III	15,1061
585	+17 00009			0 07 46	+18 22.3	1	9.25		1.12				3	K0	882
	−22 00022			0 07 46	−22 03.7	1	10.27		0.51		0.01		2		1375
636	−82 00004	HR 30		0 07 49	−82 30.1	3	5.27	.007	1.05	.000	0.93	.005	14	K1/2 III	15,1075,2038
	+61 00012			0 07 51	+61 38.5	1	9.33		1.12				2	K0	1025
584	+56 00011	HR 28		0 07 52	+56 53.2	1			−0.08		−0.42		1	B7 IV	1079
		LS I +62 072		0 07 54	+62 21.0	1	11.67		0.24		−0.41		2		41
593	+58 00011	LS I +59 034		0 07 56	+59 23.7	3	6.70	.009	0.01	.016	−0.75	.010	18	B1 V	41,1012,3016
		WLS 0 20 # 8		0 07 57	+19 43.7	1	11.62		−0.23		−0.90		2		1375
613	+32 00011			0 07 57	+32 51.1	2	6.83	.004	1.43	.010	1.67	.015	5	K4 III	1003,1355
594	+57 00018			0 07 59	+58 28.2	2	8.14	.010	0.11	.010	−0.46	.019	6	B3 IV	206,401
		JL 163		0 08 00	−50 32.	1	12.88		−0.17		−0.94		1	sdB:	132
611	+59 00005	IDS00054N6015	A	0 08 03	+60 31.7	2	8.23	.025	1.12	.015			3	G2 Ib-II	1025,6009
236327	+57 00019			0 08 04	+58 28.1	2	8.68	.014	0.10	.009	−0.44	.009	5	B9	206,401
627	+57 00022	IDS00055N5813	AB	0 08 08	+58 29.5	2	8.29	.005	0.15	.009	−0.29	.005	5	B7 V	206,401
	+62 00011			0 08 08	+62 53.7	1	9.42		0.27				2	B5 V	1025
645	−13 00013	HR 31		0 08 09	−12 51.5	5	5.84	.015	1.00	.005	0.84	.022	16	K0 III	15,252,2013,2029,3077
		G 267 - 44		0 08 11	−29 14.9	1	13.32		0.98		0.80		1		3061
		PHL 712		0 08 12	−20 54.	1	14.96		0.07		0.04		1		3064
661	−73 00004	HR 32	⋆AB	0 08 13	−73 30.2	4	6.63	.004	0.37	.004	0.05	.010	23	F3/F5 IV	15,1075,2012,2038
	+59 00006			0 08 14	+60 09.9	1	9.57		0.18				2	B8	1025
657	−39 00027			0 08 14	−39 02.8	1	8.67		0.59		0.11		1	G2 V	79
	+62 00012			0 08 16	+63 07.9	1	8.13		1.73				3	K2	1025
	+64 00007	LS I +64 017		0 08 16	+64 41.1	1	9.64		0.56		−0.48		2	B5	405
232123	+51 00016			0 08 21	+52 09.9	1	8.89		0.50		0.07		2	F5	1375
		G 31 - 29		0 08 30	+04 55.8	2	11.55	.034	1.40	.019	1.17		3	M0:	202,333,1620
236330	+57 00024	IDS00059N5722	A	0 08 30	+57 38.4	1	10.60		0.93		0.02		2	A0	405
		BPM 46072		0 08 30	−26 52.	1	13.78		1.47		1.26		1		3061
	−6 00015	G 31 - 30		0 08 31	−06 03.8	4	10.83	.039	1.41	.027	1.37	.190	6	M2	202,1619,1620,1658
684	+9 00012			0 08 40	+10 21.5	1	8.39		1.33		1.40		3	K0	196
	+62 00014			0 08 40	+62 56.7	1	9.32		0.38				2	F0 III	1025
679	+59 00008	IDS00061N5924	AB	0 08 41	+59 41.1	1	8.89		0.19				2	A2	1118
685	+7 00013			0 08 42	+07 40.2	1	7.68		1.17		1.10		2	K2	1733
		G 243 - 16		0 08 42	+58 04.4	1	9.50		1.25		1.07		3	K7	7010
		BPM 16421		0 08 42	−58 07.	1	14.20		0.48				2		3065
693	−16 00017	HR 33		0 08 43	−15 44.5	3	4.90	.009	0.49	.002	−0.01		12	F7 V	3,15,2024
		G 266 - 60	AB	0 08 44	−21 00.1	1	11.82		0.45		−0.19		3	F4	1696
691	+29 00014			0 08 46	+30 10.3	1	7.96		0.75		0.34		2	K0 V	3026
	+61 00013			0 08 46	+61 46.4	1	9.72		0.32				2	A0	1025
		CS 29503 # 29		0 08 47	−26 43.3	1	13.62		0.07		0.14		1		1736
705	−33 00035		A	0 08 47	−33 04.7	1	8.58		0.70		0.17		1	G6 V	3061
705	−33 00035		B	0 08 47	−33 04.7	1	9.45		0.34		0.08		1		3061
704	−33 00036			0 08 48	−32 40.9	2	8.44	.019	0.12	.005	0.07		9	A3 V	1068,2012
702	−9 00022			0 08 51	−08 38.3	1	9.81		0.51				2	G0	1594
	+35 00018			0 08 54	+35 39.2	1	9.33		0.94		0.57		2	K0	1601
701	−3 00013			0 08 54	−02 53.8	1	9.82		0.53		−0.06		1	K2	979
		BPM 46078		0 08 54	−28 27.	1	11.91		0.74		0.25		1		3061
	+17 00011			0 08 55	+18 01.5	1	10.52		0.61		0.09		2		1375
	+63 00008	IDS00063N6324	AB	0 08 55	+63 39.4	1	9.98		0.22				4	B3 III	1025
		LS I +60 093		0 08 58	+60 41.5	1	11.91		0.40		−0.30		2		41
	+62 00016			0 08 58	+62 38.6	1	9.73		0.25				5	B8 V	1025
	+62 00015			0 08 58	+63 09.9	2	9.21	.004	0.14	.004			6	A0 V	70,1025
698	+57 00028	LS I +57 003		0 08 59	+57 56.0	3	7.10	.011	0.17	.019	−0.42	.011	8	B5 II:	41,1012,3016
714	+18 00011			0 09 01	+18 37.9	2	7.66	.015	1.03	.019	1.02		3	K0	882,3077
720	−28 00026	HR 34	⋆A	0 09 02	−28 04.7	2	5.42	.005	1.35	.005	1.46		6	K5 III	2009,3005

Table 1   19

HD	DM	Other Id	N	Rem	α_{1950}	δ_{1950}	S	V	σ_V	B–V	σ_{B-V}	U–B	σ_{U-B}	n	Spectrum	References
741	−59 00008				0 09 03	−59 11.3	2	8.34	.005	0.62	.000	0.13		5	G3 V	258,2033
	−12 00013				0 09 04	−11 38.8	1	10.95		0.65		0.11		std		1347
725	+56 00016	LS I +56 005			0 09 06	+56 59.6	3	7.09	.027	0.63	.008	0.58	.019	5	F5 Ib-II	41,1355,6009
		RY Psc			0 09 07	−02 01.5	2	11.92	.102	0.22	.016	0.20	.075	2	F2.5	597,699
	−39 00031	G 267 - 45			0 09 08	−39 30.3	2	9.91	.000	1.07	.010	0.96		7	K5 V	2033,3079
		LS I +62 073			0 09 11	+62 45.6	1	11.57		0.45		-0.42		2		41
	−32 00037	G 267 - 47			0 09 11	−32 34.8	1	10.70		0.88		0.59		2		3061
739	−35 00042	HR 35			0 09 12	−35 24.8	3	5.24	.007	0.44	.005			9	F4 V	15,2012,2024
232130	+50 00026				0 09 13	+50 42.7	1	9.45		0.26		0.08		2	A0	1566
	+37 00019				0 09 17	+37 36.8	1	10.62		1.02		0.71		2	K0	1375
		G 267 - 48			0 09 18	−31 30.0	1	15.47		1.57				1		3061
783	−80 00003				0 09 21	−80 27.1	1	7.20		0.92		0.63		3	G8 III	1730
743	+47 00021	HR 36			0 09 22	+47 52.5	1	6.16		1.45				2	gK4	71
		Feige 1			0 09 24	+11 49.0	1	13.59		0.04		0.07		3		308
		WLS 24 50 # 6			0 09 24	+50 32.5	1	11.82		0.59		0.03		2		1375
		BPM 46086			0 09 24	−23 33.	1	13.97		0.49		-0.23		1		3061
770	−42 00032				0 09 25	−42 27.0	1	6.53		1.04		0.84		6	K0 III	1628
		G 267 - 49			0 09 26	−29 49.5	1	15.36		1.67				1		3061
		LP 464 - 429			0 09 27	+11 20.2	1	12.13		0.91		0.65		1		1696
		CS 22876 # 42			0 09 28	−34 16.5	1	13.12		0.36		-0.21		1		1736
761	+52 00019	IDS00069N5304		AB	0 09 30	+53 20.7	2	6.87	.044	0.25	.000	0.02	.000	3	A7 Vn +F2 V	292,3030
		LS I +61 127			0 09 32	+61 06.7	1	12.57		0.52		-0.36		2		41
		LP 936 - 164			0 09 32	−38 28.7	1	12.08		1.00		0.82		1		1696
		BPM 46090			0 09 36	−32 11.	1	14.45		0.70		0.22		1		3061
787	−18 00014	HR 37			0 09 37	−18 13.0	4	5.26	.014	1.47	.007	1.63	.005	11	K4 III	15,1071,2024,3016
		G 217 - 37			0 09 39	+50 09.0	2	14.37	.009	0.43	.019	-0.41	.019	4		940,1658
784	+21 00010				0 09 41	+22 16.7	1	7.67		1.77		1.93		3	M1	1648
774	+62 00020				0 09 42	+62 36.9	1	7.49		1.72				3	K2	1025
	+84 00002				0 09 45	+84 41.2	1	10.44		0.09		0.01		6	A0	1219
	+13 00013	G 30 - 52			0 09 55	+14 17.2	2	8.59	.000	0.81	.000	0.27	.000	3	K0	927,1620
	+59 00011				0 09 56	+60 12.1	1	9.56		1.84				2		369
812	−22 00023	G 266 - 64			0 09 57	−22 20.6	1	8.74		0.71		0.21		1	K0 V +G	79
		G 131 - 35			0 09 58	+21 26.3	1	11.68		1.46		1.24		3	K6	7010
	+59 00012	LS I +59 035			0 10 00	+59 56.9	1	10.38		1.02		0.70		2	G0 II	41
823	−31 00052	G 267 - 50			0 10 01	−31 08.1	1	8.67		0.75		0.31		1	G8/K0 V	3061
236338	+55 00018				0 10 02	+55 43.7	1	9.27		0.52				26	G0	6011
	+4 00015				0 10 03	+05 28.4	1	9.40		1.19		1.16		2	K0	1375
822	−14 00023				0 10 04	−13 53.3	1	8.13		0.03		0.05		4	A1 V	152
834	−27 00037	G 266 - 65			0 10 06	−27 08.0	5	7.96	.009	0.95	.006	0.71	.010	24	K1 IV	79,1075,1775,2033,3061
860	−80 00004				0 10 06	−80 28.6	1	9.10		1.01		0.69		3	G8 II/IV	1730
	+62 00022				0 10 10	+63 03.7	1	9.42		0.31				2	A6 V	1025
		WLS 0 20 # 11			0 10 11	+22 17.4	1	10.23		0.25		0.19		2		1375
		BPM 46098			0 10 12	−23 38.	1	14.29		0.50		-0.20		1		3061
		JL 166			0 10 12	−46 08.	1	15.23		-0.23		-1.11		2	sdOB	132
831	+14 00011				0 10 14	+15 16.9	1	8.63		1.10				2	K0	882
829	+36 00012	HR 38			0 10 14	+37 24.9	2	6.72	.005	-0.12	.010	-0.71	.009	4	B2 V	154,1223
	−32 00048	BPM 46099			0 10 14	−32 23.9	1	10.67		0.83		0.47		1		3061
		G 267 - 52			0 10 17	−33 09.2	1	13.30		1.34		1.12		1		3061
		BPM 46101			0 10 18	−26 27.	1	11.79		1.06		0.93		1		3061
856	−32 00049				0 10 18	−32 10.6	1	8.78		0.73		0.32		1	G3/5 V	3061
		LB 3130			0 10 18	−72 12.	1	12.70		0.01		-0.01		3		45
841	+61 00015				0 10 21	+62 27.0	1	7.73		1.13				3	F2 V	1025
842	+55 00021	LS I +55 006			0 10 22	+55 34.9	1	7.95		0.51		0.49		3	A9 I	41
	−75 00012				0 10 22	−75 00.2	1	11.34		0.47		0.04		2		1730
		JL 167			0 10 24	−53 37.	1	13.28		0.55		-0.01		1		832
	−29 00040				0 10 26	−29 29.6	1	9.42		0.83		0.49		1		3061
		G 241 - 76		⋆ AB	0 10 27	+69 03.0	1	12.45		1.59		1.27		3	M6	7010
863	+34 00013				0 10 28	+35 00.9	2	7.71	.000	0.95	.000	0.67	.005	6	G5	196,583
	+63 00010				0 10 31	+64 04.3	1	9.11		0.50				2	F5	1118
	+4 00019				0 10 32	+05 21.0	1	10.59		1.20				1	dM0	1632
	+19 00020				0 10 32	+20 06.1	1	9.00		1.36		1.18		1	M0	3003
861	+61 00016				0 10 32	+61 45.8	1	6.64		0.19				2	Am	1025
874	+16 00011				0 10 35	+16 38.7	2	6.53	.025	1.05	.005	0.91	.010	4	K1 III	1648,3016
		BPM 46105			0 10 36	−22 49.	1	12.51		0.61		-0.02		1		3061
886	+14 00014	HR 39, γ Peg		⋆ A	0 10 39	+14 54.3	15	2.83	.014	-0.23	.011	-0.86	.013	53	B2 IV	15,154,1006,1020*
	+62 00023	LS I +62 074			0 10 41	+62 58.0	1	9.90		0.28		-0.59		2	B2 III	41
	+64 00013				0 10 41	+65 20.1	1	10.29		0.57		0.19		4	B2.5V	206
890	−17 00017	IDS00081S1743		A	0 10 41	−17 27.8	1	7.53		1.24		1.32		2	K1/2 III	1375
		LB 3131			0 10 42	−62 09.	1	12.33		0.44		-0.01		2		45
		LS I +62 075			0 10 43	+62 19.7	1	11.36		0.41		-0.21		2		41
901	−27 00038				0 10 43	−26 36.2	1	7.44		1.28		1.35		3	K1/2 III	1657
902	−38 00040				0 10 43	−38 06.1	1	7.19		1.53				4	K3/4 III	2012
		LS I +58 013			0 10 45	+58 56.5	1	10.76		0.46		-0.26		2		41
		LS I +60 094			0 10 45	+60 56.0	1	9.95		0.59		-0.45		2		41
	−28 00042	BPM 46107			0 10 47	−28 03.1	1	11.04		0.84		0.53		1		3061
	+18 00015	G 32 - 2			0 10 48	+18 47.8	1	9.86		1.19		1.17		1	K7	333,1620
895	+26 00013	HR 40		⋆ AB	0 10 48	+26 42.6	1	6.30		0.65		0.33		3	G3 III+F2IV	3016
895	+26 00013	IDS00082N2626		C	0 10 48	+26 42.6	1	10.37		0.63		0.09		3		3016
		LB 3132			0 10 48	−60 47.	1	12.48		0.52		-0.04		2		45
894	+67 00007				0 10 52	+67 46.9	1	8.35		0.36		-0.01		2	F0	25

HD	DM	Other Id	N	Rem	α_{1950}	δ_{1950}	S	V	σ_V	B–V	σ_{B-V}	U–B	σ_{U-B}	n	Spectrum	References
	+63 00012	LS I +63 052			0 10 54	+63 53.2	1	9.76		0.55		-0.45		2	O9 Ib	1011
		BPM 46108			0 10 54	-29 45.	1	13.45		1.00		0.66		1		3061
	-0 00017	IDS00084S0040		A	0 10 55	+01 28.0	1	10.65		0.72				2	K2	202
905	+40 00029	HR 41			0 10 55	+40 45.6	1	5.72		0.31		-0.02		2	F0 IV	253
	+63 00013				0 10 55	+63 42.3	2	8.88	.005	0.50	.005			5	F6 V	1025,1118
925	-46 00037				0 10 56	-46 32.4	1	10.27		0.56		0.06		1	F7/G0 (V)	832
904	+46 00026				0 10 59	+46 54.9	1	7.98		0.18		0.11		2	A2	1601
916	+35 00027				0 11 00	+35 53.6	1	8.18		0.07		0.03		2	A0	1336
923	-30 00048				0 11 02	-29 51.2	3	8.60	.017	0.18	.005	0.16	.024	12	A5 IV	355,1068,2012
	-23 00049				0 11 03	-23 23.2	1	10.74		0.48		-0.03		3		1554
		G 31 - 35			0 11 04	+00 02.8	2	15.33	.014	0.22	.014	-0.55	.019	4		316,3028
		NGC 45 sq1 10			0 11 05	-23 22.3	1	14.02		0.55		-0.06		4		1554
232137	+51 00025				0 11 06	+52 22.9	1	8.58		0.11		0.11		2	A0	1566
		NE # 7			0 11 06	+67 37.	1	13.42		2.64		1.35		1		25
232138	+53 00024				0 11 07	+54 11.2	1	8.72		-0.05		-0.85		1	B3	963
929	+17 00017	IDS00086N1805		AB	0 11 08	+18 21.5	2	8.08	.040	1.06	.015	0.80		4	G5	882,3040
		LP 764 - 105			0 11 08	-14 55.4	1	13.57		0.54		-0.18		2		1696
		G 266 - 68			0 11 08	-28 34.0	1	11.45		1.16		1.18		1		3061
		G 266 - 67			0 11 08	-28 34.7	1	11.07		0.94		0.75		1		3016
	-23 00051				0 11 09	-23 24.1	1	11.80		1.03		0.77		3		1554
942	-26 00056	HR 42			0 11 10	-26 17.9	3	5.94	.004	1.54	.005			11	K5 III	15,1075,2032
943	-26 00057	HR 43			0 11 12	-26 33.8	2	6.20	.088	1.49	.005	1.65		6	K4 III	2009,3005
957	-49 00035				0 11 15	-48 57.7	1	6.96		0.18		0.13		7	A5 IV	1628
936	+59 00015				0 11 16	+59 43.3	1	6.88		1.12		0.88		4	G8 II	1355
236346	+58 00016	LS I +59 036			0 11 17	+59 09.0	1	8.86		0.89		0.68		2	F0 I	41
		NE # 6			0 11 18	+67 37.	1	11.71		1.55		1.32		1		25
		LP 880 - 691			0 11 18	-32 44.	1	13.22		0.84				1		3061
	-80 00006				0 11 19	-80 10.8	1	10.51		0.47		0.00		3		1730
	+72 00010				0 11 20	+73 06.8	1	8.25		1.82		1.96		3	M2	1733
		NGC 45 sq1 5			0 11 22	-23 28.8	1	13.03		0.59		0.03		4		1554
	-39 00041				0 11 22	-38 36.8	3	11.61	.016	0.57	.004	0.09	.004	15		831,937,1554
955	-18 00025				0 11 23	-17 49.4	2	7.37	.030	-0.17	.015	-0.64	.025	6	B3/5 V	1068,3007
		WLS 0 35 # 8			0 11 24	+34 51.8	1	13.90		0.75		0.09		2		1375
		LS I +61 128			0 11 24	+61 58.7	1	11.83		0.59		-0.31		2		41
		NGC 45 sq1 6			0 11 24	-23 19.6	1	13.27		0.88		0.59		4		1554
		NGC 45 sq1 7			0 11 24	-23 31.7	1	13.52		0.55		-0.02		4		1554
		NGC 55 sq1 8			0 11 24	-39 20.1	2	13.59	.005	0.68	.000	0.13	.000	10		831,1554
		NGC 55 sq1 9			0 11 24	-39 24.9	2	13.63	.005	0.98	.005	0.78	.000	10		831,1554
		L 50 - 73			0 11 24	-72 06.	1	15.31		0.37		-0.45		1		3062
952	+32 00021	HR 44			0 11 26	+32 55.7	2	6.25		0.00	.015	-0.03	.020	8	A1 V	985,1049
962	+59 00016				0 11 26	+60 26.5	3	7.71	.004	0.70	.028	0.41		6	F2 V	401,1025,1118
		GR 2			0 11 26	-29 46.3	2	13.91	.054	1.54	.010	1.21		3		481,3061
1032	-85 00002	HR 47			0 11 28	-85 16.3	3	5.77	.004	1.69	.018	2.09	.005	16	M0/1 III	15,1075,2038
		SE # 7			0 11 30	+66 03.	1	10.54		0.82		0.38		2		25
967	-12 00020	G 158 - 44			0 11 30	-11 35.2	2	8.36	.000	0.65	.025	0.02	.035	2	G5	79,3056
		BPM 46120			0 11 30	-22 44.	1	13.44		1.00		0.87		1		3061
		NGC 45 sq1 4			0 11 31	-23 38.7	1	12.23		0.81		0.49		4		1554
987	-75 00015	IDS00091S7515		A	0 11 32	-74 57.9	1	8.75		0.74		0.25		2	G6 V	1730
961	+65 00021				0 11 33	+66 03.9	1	8.21		0.22		0.18		2	A3	25
	-23 00055				0 11 34	-23 29.1	1	9.93		1.16		1.11		3	K0	1554
977	+12 00013				0 11 37	+13 16.5	1	8.62		1.07				2	G5	882
		LS I +62 076			0 11 39	+62 37.5	1	11.94		0.35		-0.37		2		41
236348	+59 00017	IDS00090N5947		AB	0 11 40	+60 03.8	1	8.51		0.73		0.22		3	G5	1386
		LS I +62 077			0 11 40	+62 56.4	1	11.73		0.63		-0.41		2		41
		S012 # 4			0 11 40	-80 09.7	1	11.15		1.31				2		1730
	-6 00027	G 31 - 36			0 11 41	-05 51.5	3	10.91	.015	0.80	.011	0.41	.020	4		202,1620,1658
	+61 00017				0 11 43	+62 23.0	1	9.78		0.16				2	B3 V	1025
1004	-55 00035				0 11 46	-55 20.8	1	6.66		0.46		-0.03		5	F5 IV/V	1628
		SE # 4			0 11 48	+66 15.	1	12.58		0.82		0.41		1		25
		SE # 8			0 11 48	+66 19.	1	13.26		1.00		0.89		1		25
		SE # 2			0 11 48	+66 21.	1	12.01		0.98		0.60		1		25
		SE # 1			0 11 48	+66 23.	1	10.70		0.64		0.12		2		25
		CS 29503 # 36			0 11 48	-23 15.3	1	14.24		0.15		0.09		1		1736
1000	-21 00017	G 266 - 71			0 11 49	-21 28.3	1	6.90		0.47		0.05		1	F7 V	79
1002	-28 00047	G 266 - 70			0 11 49	-27 48.4	1	6.46		0.64		0.17		1	G5 V	3061
1024	-72 00014				0 11 51	-71 44.7	1	9.34		0.40		-0.02		1	F3 V	744
	-14 00027				0 11 52	-13 36.3	1	10.92		0.11		0.10		3	A2	1026
1014	-8 00026	HR 46, AD Cet		★AB	0 11 54	-08 03.5	5	5.13	.009	1.58	.026	1.82	.010	31	M3 III	15,1075,1256,2024,3051
		G 267 - 56			0 11 54	-33 15.1	1	13.62		1.35		1.04		2		3061
		BPM 46125			0 11 54	-33 30.	1	12.09		1.02		0.85		1		3061
		G 30 - 53			0 11 55	+07 31.9	1	10.94		1.24		1.23		1	K5	333,1620
1017	-19 00019				0 11 55	-19 22.6	1	9.31		0.33		0.03		6	A9 V	1068
		NGC 45 sq1 11			0 11 55	-23 28.6	1	14.07		0.65		0.15		4		1554
		NGC 45 sq1 9			0 11 55	-23 28.6	1	14.00		0.52		0.00		5		1554
		NGC 45 sq1 8			0 11 58	-23 33.0	1	13.96		0.79		0.27		4		1554
		LS I +62 078			0 11 59	+62 41.5	1	11.61		0.39		-0.29		3		41
	+67 00010				0 12 00	+67 43.0	1	10.74		0.74		0.32		2		25
		NE # 2			0 12 00	+67 48.	1	12.61		0.85		0.23		2		25
1013	+19 00027	HR 45			0 12 01	+19 55.7	17	4.80	.006	1.57	.007	1.91	.019	126	M2 III	1,3,15,116,1077,1118*
1009	+63 00015				0 12 01	+64 17.3	1	8.41		0.17				3	B9 Vmn?	1118

Table 1   21

HD	DM	Other Id	N	Rem	α_{1950}	δ_{1950}	S	V	σ_V	B–V	σ_{B-V}	U–B	σ_{U-B}	n	Spectrum	References
		LS I +62 079			0 12 03	+62 16.0	1	10.66		0.29		-0.57		2		41
		NGC 55 sq1 10			0 12 03	-39 39.8	2	14.26	.005	0.92	.005	0.71	.005	9		831,1554
1010	+61 00018				0 12 06	+61 55.1	1	8.49		0.48				3	F6 V	1025
1038	-19 00021	HR 48, AE Cet			0 12 06	-19 12.6	7	4.43	.022	1.65	.012	1.98	.013	60	M1 III	3,15,1024,1075,2024*
236351	+55 00028				0 12 07	+55 55.8	2	9.14	.014	0.21	.031			27	A3	851,6011
236352	+57 00038				0 12 07	+58 20.9	1	9.28		1.70				1	K7	592
1037	-15 00032				0 12 07	-15 04.9	1	6.64		1.03		0.82		4	G8 III/IV	3040
1026	+62 00029	IDS00094N6217		A	0 12 08	+62 33.8	1	8.04		0.18		0.05		2	A0	401
1026	+62 00029	IDS00094N6217		B	0 12 08	+62 33.8	1	8.61		0.13		0.09		1	A0	401
		LS I +62 080			0 12 12	+62 31.8	1	11.48		0.36		-0.46		2		41
		SE # 5			0 12 12	+66 08.	1	10.98		0.69		0.53		2		25
	-39 00043	UY Scl			0 12 14	-39 31.5	1	11.53		0.91		0.57		1		937
	+65 00024				0 12 15	+66 20.9	1	8.78		2.22				3	K2	369
		LS I +62 081			0 12 16	+62 33.5	1	11.12		0.40		-0.35		2		41
		NGC 55 sq2 11			0 12 16	-39 29.9	1	12.08		0.96		0.66		1		937
1348	-89 00001				0 12 17	-88 38.5	2	7.25	.009	0.09	.004	0.03	.054	14	B9.5IV	1628,1704
		BPM 46129			0 12 18	-25 41.	1	11.92		0.93		0.60		1		3061
1048	+21 00013	HR 49			0 12 20	+22 00.4	2	6.26		-0.02	.005	0.04	.068	5	A1p	1049,3050
236355	+55 00029				0 12 20	+56 03.7	1	9.16		1.67				1	K7	851
1064	-10 00030	HR 51			0 12 21	-09 50.8	2	5.74	.005	-0.08	.000			7	B9 V	15,2012
	-0 00024				0 12 23	+00 21.1	1	9.87		0.89				2		202
1061	+8 00019	HR 50, UU Psc	⋆	A	0 12 24	+08 32.6	1	6.01		0.32		0.04		4	F0 IV	3024
1061	+8 00019	HR 50, UU Psc	⋆	AB	0 12 24	+08 32.6	2	5.78	.005	0.32	.005	0.02	.015	7	F0 IV +A7	15,1256
1061	+8 00019	IDS00098N0816		B	0 12 24	+08 32.6	1	7.62		0.40		-0.03		4	A7	3024
1060	+15 00022				0 12 24	+15 57.0	1	8.53		1.05				2	G5	882
		BPM 46132			0 12 24	-22 39.	1	12.61		1.25		1.12		1		3061
		BPM 46133			0 12 24	-25 04.	1	11.83		0.93		0.61		1		3061
	+57 00041				0 12 25	+57 57.8	1	9.44		1.12				1		592
1057	+60 00013				0 12 26	+61 29.4	1	8.44		1.33				3	K2 II	1025
		LS I +60 095			0 12 27	+60 41.7	1	12.36		0.50		-0.11		2		41
1089	-35 00065	HR 54			0 12 27	-35 10.9	1	6.17		1.34				4	K3 III	2032
1069	+59 00019				0 12 29	+60 29.7	1	7.67		1.43				3	K2 I	1025
1074	+32 00026				0 12 30	+32 45.2	1	7.35		1.02		0.73		2	K0	1375
	+57 00042	SY Cas			0 12 30	+58 08.8	2	9.42	.006	0.79	.000	0.56		2	F5	592,1399
		NE # 4			0 12 30	+67 44.	1	11.67		0.65		0.42		2		25
		G 266 - 75			0 12 30	-27 19.5	1	14.01		1.41		1.28		1		3061
1075	+30 00026	HR 52			0 12 31	+31 15.5	1	6.34		1.58		1.86		2	K5	1733
1087	-28 00051				0 12 31	-27 46.8	1	9.13		0.99		0.75		2	K1 IV	3061
1070	+58 00018	LS I +59 037	⋆	A	0 12 32	+59 30.1	3	8.01	.017	0.59	.005	0.48	.038	7	A5 II:	41,450,1012
1070	+58 00018	IDS00099N5913		B	0 12 32	+59 30.1	1	11.59		0.38		0.26		2		450
1088	-34 00056				0 12 32	-33 51.7	1	9.09		0.53		0.04		1	G0 V	3061
1086	-18 00030				0 12 33	-17 32.2	3	9.81	.015	0.19	.005	0.15	.017	14	A2 III	152,1068,1775
1083	+26 00023	HR 53	⋆	A	0 12 35	+27 00.3	1	6.35		-0.02		0.02		3	A1 Vn	1049
		G 217 - 41			0 12 35	+52 47.8	1	10.41		0.90		0.52		2	K3	7010
		NE # 5			0 12 36	+67 43.	1	11.26		0.73		0.26		2		25
		BPM 46141			0 12 36	-30 15.	1	14.01		0.77		0.22		1		3061
		G 130 - 59			0 12 37	+24 38.3	1	14.48		0.92		0.44		3	dK	3029
236356	+59 00021				0 12 37	+59 51.8	1	9.29		1.85				2	K7	369
		LS I +61 129			0 12 37	+61 58.6	1	12.70		0.31		-0.18		2		41
	+58 00019	LS I +58 014			0 12 39	+58 50.5	1	11.38		0.72		-0.31		2		41
		LS I +61 130			0 12 40	+61 58.4	1	12.55		0.43		-0.29		2		41
	+61 00019				0 12 40	+62 25.1	1	9.12		0.58				2	F2 III	1025
	-25 00068				0 12 42	-25 26.0	1	10.82		0.33		0.04		5	A5	1068
		G 243 - 19			0 12 44	+64 00.0	1	10.66		1.09		0.87		2	K3	7010
1116	-40 00040	IDS00102S4039		AB	0 12 45	-40 22.4	1	7.39		0.98				4	G8/K1 III	2012
		JL 171			0 12 48	-47 16.	1	17.40		-0.60		-1.20		1		132
	+63 00018	LS I +63 057			0 12 49	+63 52.2	2	10.26	.015	0.52	.005	-0.47	.015	3	O7	405,483
1098	+61 00020				0 12 53	+61 50.0	1	8.21		0.98				3	G5	1025
1112	-4 00017				0 12 54	-03 55.9	1	9.09		-0.06		-0.21		3	B9	1026
		G 158 - 50			0 12 54	-16 24.4	1	11.53		1.74		1.26		1		3079
		BPM 46144			0 12 54	-22 11.	1	13.09		1.43		1.25		1		3061
	-25 00069				0 12 54	-24 53.4	1	12.15		0.95		0.56		2		3061
		BPM 46145			0 12 54	-26 58.	1	12.59		0.69		0.15		1		3061
		BPM 46146			0 12 54	-32 35.	1	11.45		0.75		0.25		2		3061
1122	+12 00014				0 12 58	+13 08.5	1	7.93		1.60				2	K2	882
		NGC 55 sq1 6			0 12 59	-39 39.2	2	12.77	.000	0.22	.000	0.15	.000	10		831,1554
		G 267 - 58			0 13 00	-35 28.4	1	14.62		1.58		1.08		1		3079
		LB 3133			0 13 00	-62 56.	1	13.62		0.38		-0.04		2		45
1136	-25 00070	IDS00105S2454		AB	0 13 02	-24 37.4	1	10.02		1.06		1.20		1	G1 V	8084
		GR 3			0 13 03	-15 38.0	1	14.36		1.47		1.16		2		481
	-26 00067	G 266 - 77			0 13 03	-26 32.5	1	12.07		1.22		1.09		1		3061
		GR 4			0 13 06	-30 02.7	1	14.31		1.57		0.82		1		481
1128	+58 00022				0 13 08	+59 09.8	3	7.90	.011	0.05	.008	-0.03	.025	4	B9	598,653,1118
		DO 58			0 13 11	+06 30.9	1	10.78		1.68				3		1532
	+59 00022	MU Cas			0 13 11	+60 09.2	1	10.81		0.30		-0.26		6	A0	1768
		PHL 756			0 13 12	-26 02.	1	16.58		-0.15		-1.28		1		286
		G 30 - 55			0 13 13	+13 16.4	1	12.61		1.67				1		1705
1142	+60 00016				0 13 14	+60 43.4	2	6.45	.000	0.84	.033	0.47		4	G8 III	401,1025
		SZ Scl			0 13 14	-31 22.6	1	12.30		0.68		0.08		2		3064
1151	+1 00025				0 13 18	+02 25.5	1	8.38		0.33		0.14		2	A5	1375

HD	DM	Other Id	N	Rem	α_{1950}	δ_{1950}	S	V	σ_V	B–V	σ_{B-V}	U–B	σ_{U-B}	n	Spectrum	References
1141 +76 00005		HR 55		⋆ AB	0 13 22	+76 40.4	1	6.35		-0.07		-0.27		1	B8 V +B9 V	1079
		BPM 46154			0 13 24	-21 35.	1	13.22		1.20		0.90		1		3061
		CS 29503 # 46			0 13 25	-25 41.7	1	12.08		0.18		0.16		1		1736
1168 +13 00026					0 13 27	+14 25.8	1	8.42		1.32				2	K2	882
1171 -23 00070		G 266 - 78			0 13 27	-22 41.9	2	9.77	.005	0.73	.007	0.20	.039	3	G8 V	79,3061
		BPM 46156			0 13 30	-24 44.	1	13.26		1.47		1.17		1		3061
		BPM 46157			0 13 30	-26 01.	1	11.84		1.04		0.87		1		3061
		LB 3134			0 13 30	-62 25.	1	12.52		0.06		0.05		2		45
	+47 00043				0 13 31	+47 36.9	1	8.87		1.67		1.89		2	M0	1375
236362 +55 00036					0 13 31	+55 35.4	1	9.01		0.42		0.00		2	F5	1502
1166 +61 00021					0 13 31	+61 36.9	1	8.39		1.23				3	G8 III	1025
		G 267 - 61			0 13 33	-31 15.4	1	13.77		1.49		1.18		1		3061
		LS I +59 038			0 13 36	+59 16.1	1	11.82		0.72		-0.34		2		41
		BPM 46161			0 13 36	-24 41.	1	13.28		1.49		1.26		1		3061
		G 32 - 6			0 13 37	+19 35.6	2	12.33	.018	1.44	.028	1.03	.077	5		3078,7009
1187 -32 00072		HR 57		⋆ A	0 13 37	-31 43.4	3	5.66	.004	1.35	.000	1.50		9	K5 III	15,2012,3009
		G 32 - 7			0 13 38	+19 35.7	2	13.27	.015	1.52	.056	1.25	.121	5		3078,7009
		LP 824 - 274			0 13 38	-24 40.7	1	14.08		0.56		-0.23		1		3062
		CS 29503 # 47			0 13 38	-26 05.6	1	14.80		0.10		0.09		1		1736
		LS I +61 131			0 13 39	+61 38.8	1	12.07		0.29		-0.48		2		41
1221 -76 00019		HR 58			0 13 39	-76 11.4	3	6.46	.014	0.98	.006	0.73	.005	21	G8/K0 III	15,1075,2038
1196 -28 00061		G 266 - 79			0 13 40	-28 03.3	1	9.39		0.65		0.09		1	G5 V	3061
		PHL 761			0 13 42	-24 06.	1	15.38		-0.05		-0.87		2	DA:	3064
1185 +42 00041		HR 56		⋆ A	0 13 43	+43 19.0	1	6.15		0.05		0.03		2	A2 Vp	3016
1185 +42 00041		IDS00111N4303		B	0 13 43	+43 19.0	1	10.21		0.70		0.19		2	A0	3016
		NGC 55 sq1 4			0 13 44	-39 28.3	2	11.93	.005	0.55	.005	0.00	.005	10		831,1554
	+15 00028	G 30 - 56			0 13 45	+16 22.2	4	9.74	.008	0.56	.006	0.02	.033	6	G0 V	333,927,1003,1620,1696
		AO 1024			0 13 45	+43 49.4	1	10.92		0.50		0.00		1	F4 V	1748
		LS I +58 015			0 13 47	+58 48.2	1	10.44		0.44		-0.51		2		41
1207 -34 00067					0 13 49	-33 44.8	1	9.53		0.79		0.38		1	G8/K0 V	3061
		AO 1025			0 13 50	+43 26.1	1	10.61		0.54		-0.03		1	F5 V	1748
		G 267 - 63			0 13 50	-31 44.4	1	13.05		1.55		1.08		1		3061
		LS I +60 096			0 13 53	+60 19.9	1	10.77		0.42		-0.40		1		41
		NGC 55 sq2 3			0 13 53	-39 47.9	1	13.72		0.87				2		937
	-47 00051	Smethells 15			0 13 53	-46 59.9	1	11.05		1.26				1	K5	1494
1202 +28 00029		IDS00113N2844		A	0 13 56	+29 00.9	1	8.23		1.02		0.67		2	K0	3032
1202 +28 00029		IDS00113N2844		BC	0 13 56	+29 00.9	1	9.70		0.24		0.10		2	A0	3032
1201 +61 00022		LS I +62 082			0 13 58	+62 16.4	2	8.68	.005	0.60	.028	0.36		4	A0 III	41,1025
	-40 00042				0 13 58	-39 45.2	1	9.80		0.37		-0.03		4	F5	937
1227 +7 00027		HR 59			0 14 00	+07 57.8	4	6.11	.011	0.92	.005	0.65	.009	12	G8 II-III	15,1256,1355,3016
1213 +15 00030					0 14 00	+15 33.5	1	8.32		0.38		-0.01		3	A5	1117
	-40 00043				0 14 00	-39 37.1	3	10.20	.008	1.03	.000	0.79	.014	14		831,937,1554
		LS I +61 132			0 14 01	+61 29.0	1	11.41		0.40		-0.48		2		41
1233 -33 00071					0 14 01	-33 02.3	1	7.33		1.20		1.13		6	K0 III	1673
1249 -46 00054					0 14 01	-45 53.3	1	9.56		0.43		-0.02		1	F5 V	832
1247 -39 00051					0 14 03	-39 06.5	1	9.75		0.45		-0.08		3	F3 V	937
1228 +1 00028					0 14 06	+01 34.4	1	7.05		1.60		1.72		8	M5 III	3040
1223 +35 00035		IDS00115N3604		AB	0 14 06	+36 21.1	1	6.99		0.05		-0.01		2	A0	1336
1212 +41 00028					0 14 06	+41 34.3	1	8.64		0.59		0.05		1	F8	3016
1211 +46 00043					0 14 06	+47 00.1	1	8.33		-0.02		-0.32		2	B9	1601
		BPM 46166			0 14 06	-33 39.	1	13.59		1.36		1.23		1		3061
		Smethells 158			0 14 06	-50 33.	1	12.39		1.51				1		1494
	+20 00017	G 131 - 45			0 14 07	+20 34.7	1	9.98		0.95		0.65		1	K3	333,1620
1224 +35 00034		IDS00115N3556		AB	0 14 07	+36 12.8	2	7.74	.020	0.47	.005	-0.07	.025	4	F6 V +F9 V	292,3030
		NGC 55 sq1 7			0 14 07	-39 27.3	2	12.81	.005	0.64	.000	0.01	.000	11		831,1554
	+61 00024	LS I +61 133			0 14 08	+61 52.7	1	10.46		0.26		-0.47		2	B3	41
		LS I +62 083			0 14 08	+62 37.3	1	10.76		0.22		-0.56		1		41
		NGC 55 sq2 5			0 14 08	-39 07.4	1	13.21		0.71		0.06		2		937
1256 -21 00024		HR 61			0 14 10	-20 29.3	4	6.49	.017	-0.11	.019	-0.47	.011	11	B6 III/IV	1068,1079,2009,3007
	-40 00044				0 14 10	-39 46.2	1	9.51		1.09		0.96		4	K0	937
1222 +46 00044					0 14 11	+46 58.0	1	9.21		0.00		-0.21		2		1601
		AO 1026			0 14 12	+43 57.7	1	10.32		1.00		0.64		1	K0 V	1748
1257 -30 00063					0 14 12	-30 01.6	1	9.72		0.56		-0.07		1	F7 V	3061
1239 +60 00021		HR 60		⋆ A	0 14 15	+61 15.3	2	5.73	.010	0.91	.025	0.59		3	G8 III	1118,3016
1239 +60 00021		IDS00116N6059		B	0 14 15	+61 15.3	1	10.21		1.48		1.26		2		3016
1238 +62 00038					0 14 15	+63 16.5	2	7.86	.005	0.08	.005			4	A0	1025,1118
		NGC 55 sq2 7			0 14 15	-39 35.7	1	11.84		1.14				3		937
		NGC 55 sq2 8			0 14 16	-39 40.5	1	13.44		0.75				2		937
		G 130 - 60			0 14 18	+24 04.2	1	13.66		1.54		1.21		2	dM3	3016
		SE # 3			0 14 18	+66 20.	1	11.29		0.90		0.73		1		25
		BPM 46168			0 14 18	-22 21.	1	13.51		1.32				1		3061
1254 +22 00034					0 14 19	+22 58.7	1	6.92		1.06		0.87		2	K0 III-IV	1648
		NGC 55 sq1 5			0 14 20	-39 26.5	3	12.37	.000	0.56	.011	-0.06	.011	14		831,937,1554
		G 31 - 39			0 14 22	+04 51.2	2	13.80	.013	1.62	.000			6		906,940
		G 266 - 80			0 14 22	-25 45.7	1	13.84		1.18		1.05		2		3061
		NGC 55 sq2 9			0 14 23	-39 30.4	1	12.74		0.79		0.50		2		937
1273 -53 00042					0 14 24	-52 55.9	2	6.84	.000	0.64	.000	0.05		8	G2 V	158,1075
	+40 00045	G 171 - 44			0 14 25	+40 40.1	2	8.99	.026	1.37	.026	1.18		3	M0	1017,3072
1280 +37 00034		HR 63			0 14 28	+38 24.2	12	4.61	.005	0.06	.005	0.04	.008	62	A2 V	1,15,374,1006,1049*
1279 +47 00050		HR 62			0 14 30	+47 40.2	2	5.89		-0.09	.000	-0.44	.000	9	B7 III	1079,1267

Table 1   23

HD	DM	Other Id	N	Rem	α_{1950}	δ_{1950}	S	V	σ_V	B−V	σ_{B-V}	U−B	σ_{U-B}	n	Spectrum	References
		LS I +60 097			0 14 32	+60 35.8	1	12.11		0.44		-0.29		2		41
1287	+62 00040				0 14 36	+62 32.9	3	8.55	.008	0.09	.011	0.07		6	A0 V	401,1025,1118
1324	−79 00007	HR 64			0 14 37	−79 03.5	3	6.76	.003	0.45	.005	0.07	.010	16	F5 V	15,1075,2038
	−19 00027	G 266 - 81			0 14 39	−18 35.4	1	10.68		0.63		0.04		1		79
	+18 00024	G 32 - 9			0 14 40	+18 36.4	1	10.51		1.31		1.20		1	K5	1620
		G 130 - 61			0 14 41	+28 54.0	1	11.52		1.50				1	M0	1746
1302	−14 00038				0 14 41	−14 23.1	1	9.53		-0.09		-0.22		3	B8/9 V	3007
		LS I +62 084			0 14 42	+62 53.0	1	11.54		0.52		-0.36		2		41
1320	−44 00052				0 14 45	−44 07.8	3	7.96	.009	0.65	.005	0.09	.000	11	G5 V	158,2012,3077
	−39 00054				0 14 46	−39 17.8	1	10.14		0.97		0.67		3		937
1309	+15 00035	IDS00122N1557		A	0 14 47	+16 14.0	1	8.78		0.63		0.14		5	F5	3016
1309	+15 00035	IDS00122N1557		B	0 14 47	+16 14.0	1	9.80		0.67		0.18		5	G0	3016
		BPM 46175			0 14 48	−25 15.	1	11.62		0.67		0.17		1		3061
		G 30 - 57			0 14 49	+11 08.9	1	11.58		1.26		1.14		1	K5	333,1620
1317	+8 00024	IDS00122N0819		AB	0 14 50	+08 35.9	1	7.16		0.67		0.14		3	F4 V	3024
1317	+8 00024	IDS00122N0819		C	0 14 51	+08 35.9	1	7.79		0.43		-0.09		2	F4 V	3024
		AO 1027			0 14 51	+44 00.4	1	10.47		0.96		0.64		1	G5 III	1748
		G 158 - 51			0 14 54	−09 01.1	1	12.49		1.57		1.25		1		801
		Smethells 159			0 14 54	−47 43.	1	11.36		1.22				1		1494
1315	+35 00040				0 14 56	+36 16.4	2	7.58	.005	0.95	.000	0.68	.005	6	G5	196,583
		G 266 - 82			0 14 56	−22 19.2	1	14.67		1.60		1.25		1		3061
1343	−19 00030	HR 66			0 15 00	−19 19.7	2	6.44	.005	0.38	.005			7	F3 V	15,2012
1372	−72 00023	AQ Tuc			0 15 02	−72 11.6	1	9.93		0.29		0.18		1	F3/5	744
1337	+50 00046	HR 65, AO Cas			0 15 03	+51 09.3	3	6.02	.115	-0.05	.080	-0.99	.015	6	O9 IIInn	15,154,1011
	−25 00084	BPM 46177			0 15 04	−25 33.2	1	10.51		0.61		0.09		2	G0	3061
		Smethells 161			0 15 06	−64 40.	1	11.30		1.31				1		1494
		G 158 - 53			0 15 07	−11 02.3	1	14.00		1.47		1.38		3		3078
		LS I +61 134			0 15 08	+61 27.4	1	10.77		0.78		0.62		2		41
	−9 00040	G 158 - 52			0 15 08	−08 57.8	3	10.99	.003	1.42	.025	1.20		4	K7	801,1017,1619
		HA 116 # 99			0 15 08	−14 10.3	1	12.13		0.73		0.34		9		1499
1334	+58 00024				0 15 09	+58 46.8	3	7.71	.018	0.00	.009	-0.58	.010	4	B2.5V	598,653,1118
	−53 00046				0 15 09	−53 04.9	1	10.38		0.99		0.87		1		1696
1349	+50 00047				0 15 12	+50 47.8	1	8.66		0.25		0.09		2	A5	1566
	+67 00016				0 15 12	+67 45.5	1	10.34		0.90		0.32		1	A0	1215
1367	+0 00028	HR 67			0 15 13	+01 24.6	3	6.18	.017	0.94	.005	0.72	.014	9	K0 II	15,1256,3016
1352	+15 00039				0 15 14	+16 03.2	1	7.22		0.47		-0.01		2	F6 V	3016
		LP 764 - 107			0 15 14	−14 53.1	1	10.69		0.87		0.57		1		1696
1365	+7 00032				0 15 18	+07 35.3	1	7.52		0.31		0.15		2	A5	1375
		BPM 46182			0 15 18	−29 39.	1	15.27		0.58		-0.15		1		3061
1368	−0 00035	G 31 - 41			0 15 20	+00 05.9	7	8.86	.015	0.54	.008	-0.03	.023	18	F9 V	202,258,333,1003,1620*
	+39 00051	G 171 - 46			0 15 20	+39 57.1	1	9.97		0.82		0.29		2	G9	7010
1399	−72 00024				0 15 20	−72 16.2	1	10.35		1.09		1.10		1	K0	744
1375	+11 00034				0 15 21	+12 29.6	1	6.53		0.96		0.71		2	G8 II	1733
		G 266 - 83			0 15 22	−18 11.4	1	13.20		1.58				1		3061
		G 131 - 50			0 15 24	+20 40.8	1	11.79		1.47				1	M3	906
		G 131 - 51			0 15 24	+20 40.9	1	11.09		1.41				1	K7	906
1388	−14 00042				0 15 25	−13 44.0	2	6.52	.024	0.60	.005	0.09	.015	3	G2 V	79,3026
		HA 68 # 280			0 15 30	+15 37.7	1	11.26		0.88		0.51		8		1499
1374	+45 00052				0 15 30	+45 56.2	1	6.82		1.31		1.47		3	K2 III	1501
1326	+43 00044	GX + GQ And	⋆	AB	0 15 31	+43 44.4	10	8.09	.017	1.57	.014	1.22	.026	29	M1 V	1,22,680,1118,1197*
1326	+43 00044	IDS00127N4327		C	0 15 31	+43 44.4	6	11.06	.030	1.79	.012	1.39	.010	15		1,680,694,3078,8006,8105
1384	+51 00039				0 15 32	+52 08.1	1	8.10		1.03		0.99		2	G5	1566
		HA 116 # 180			0 15 33	−14 14.0	1	12.25		0.43		-0.03		9		1499
		AO 1028			0 15 34	+43 54.5	1	11.27		1.09		0.83		1	G8 III	1748
1383	+60 00025	LS I +61 135			0 15 35	+61 27.0	2	7.63	.000	0.27	.005	-0.67	.005	4	B1 II	41,1012
		BPM 46188			0 15 36	−21 52.	1	13.71		0.82		0.32		1		3061
		G 267 - 66			0 15 39	−31 32.7	1	13.22		0.54		-0.19		1		3061
		MCC 356			0 15 42	+09 55.5	2	10.90	.011	1.49	.005			3	M1	1619,1705
1404	+35 00044	HR 68			0 15 42	+36 30.5	8	4.52	.011	0.06	.010	0.07	.009	31	A2 V	15,1007,1013,1049*
1419	+10 00025	HR 69			0 15 43	+10 55.7	1	6.05		1.03				2	K0 III	71
		HA 68 # 216			0 15 43	+15 32.6	1	12.41		0.58		-0.03		9		1499
		AO 1029			0 15 43	+43 44.7	1	8.09		1.57		1.22		1	M2 V	1748
		AO 1030			0 15 44	+43 29.1	1	10.76		0.49		-0.05		1	F5 V	1748
	+60 00026				0 15 44	+60 37.5	1	9.01		2.18				3	M0 III	369
1421	−2 00034				0 15 45	−02 17.6	1	8.88		0.48				1	K2	366
1431	−21 00031	IDS00132S2142		AB	0 15 45	−21 25.0	3	6.68	.009	-0.01	.012	-0.26	.027	10	B9/9.5IV	1068,2012,3007
1400	+61 00032				0 15 46	+61 55.5	1	6.96		1.54		1.67		5	K7 I	1355
1403	+37 00040				0 15 47	+38 27.5	1	9.05		0.00		-0.11		2	A0	1336
	−27 00074	BPM 46192			0 15 49	−27 20.4	1	10.48		0.62		0.10		2		3061
		LS I +62 085			0 15 56	+62 57.3	1	12.18		0.43		-0.21		2		41
	+5 00032				0 15 58	+05 49.0	1	10.17		1.58				3	M3 III:	1532
	−31 00099				0 15 59	−30 53.7	1	12.75		0.14		0.18		1	A0	966
		BPM 46195			0 16 00	−30 09.	1	11.96		0.81		0.41		1		3061
1439	+30 00035	HR 71			0 16 01	+31 14.4	2	5.87	.01	-0.01	.005	-0.01	.015	5	A0 IV	603,1049
1438	+42 00048	HR 70	⋆	AB	0 16 03	+43 30.8	1	6.11		-0.08		-0.36		1	B8 V	1079
		CS 22882 # 2			0 16 04	−30 53.8	1	12.75		0.14		0.20		1		1736
		G 267 - 68			0 16 04	−31 55.5	1	15.93		0.95		0.57		1		3061
1450	+15 00043	IDS00135N1526		ÀB	0 16 07	+15 42.8	1	8.64		0.55		-0.08		1	F5	1117
1461	−8 00038	HR 72			0 16 07	−08 19.7	5	6.45	.007	0.68	.000	0.29	.005	25	G0 V	15,1071,1075,2018,2029
1474	−30 00079				0 16 11	−30 14.1	1	8.38		0.51		-0.01		2	G0 V	3061

HD	DM	Other Id	N Rem	α_{1950}	δ_{1950}	S	V	σ_V	B–V	σ_{B-V}	U–B	σ_{U-B}	n	Spectrum	References
1483	−43 00064	HR 73		0 16 13	−43 30.8	4	6.32	.009	1.21	.009			18	K2 III	15,1075,2013,2028
	+62 00049	LS I +63 061		0 16 14	+63 17.7	1	8.94		0.41		-0.52		2	B0.5III	1012
1457	+59 00030	LS I +60 098		0 16 18	+60 03.5	3	7.82	.010	0.61	.016	0.57		5	F0 I	41,1118,6009
		G 266 - 86		0 16 20	−25 13.9	1	14.80		1.49		1.11		1		3061
		PHL 786		0 16 24	−32 12.	2	14.45	.025	-0.17	.000	-1.09	.020	6		286,3064
1470	+36 00029			0 16 26	+36 33.3	1	9.10		0.24		0.05		2	A5	1336
1480	+21 00024			0 16 27	+22 05.5	1	8.74		1.16		1.03		2	K0	1648
		AO 1031		0 16 29	+43 56.4	1	10.06		1.61		2.09		1	M2 III	1748
		AO 1032		0 16 31	+43 51.5	1	11.47		0.49		-0.02		1	F6 V	1748
1479	+58 00028	V377 Cas	★AB	0 16 31	+59 25.7	2	7.82	.014	0.34	.005	0.06		4	F0	401,1118
		G 31 - 42		0 16 32	+04 04.5	1	14.29		1.71		0.12		1		906
	+20 00024	G 131 - 53		0 16 32	+20 34.3	1	9.38		0.62		0.12		1	G0	333,1620
	−10 00047	G 158 - 59		0 16 33	−10 14.3	1	9.94		1.33		1.16		1	K7 V	3072
		GR 5		0 16 33	−24 01.4	1	15.12		1.43		0.88		1		481
	+64 00028			0 16 35	+64 52.8	1	9.28		1.97				3		369
1486	+58 00030	TV Cas		0 16 36	+58 51.7	3	7.25	.025	0.07	.021	-0.05	.025	4	B9 V	598,653,1118
1502	+13 00034			0 16 42	+13 46.6	2	8.35	.005	0.93	.010	0.69		4	K0	882,1648
		BPM 46202		0 16 42	−22 36.	1	11.48		0.75		0.22		1		3061
		TonS 149		0 16 42	−25 28.9	2	12.70	.009	0.01	.009	0.15	.013	3		1036,3064
		G 266 - 87		0 16 44	−20 11.1	1	12.50		1.55				1		3061
1515	−29 00071			0 16 44	−28 46.0	1	9.44		0.55		-0.04		1	G0 V	3061
		SB 126		0 16 46	−30 18.5	3	14.53	.000	0.01	.005	0.10	.014	4	A0	966,1736,3064
	−72 00028			0 16 46	−71 45.6	3	10.61	.006	0.47	.005	-0.06	.005	12		366,522,1582,1696
	−29 00074			0 16 47	−29 27.9	1	10.40		0.49		-0.05		1	F5	3061
		LS I +59 039		0 16 48	+59 45.7	1	10.78		0.50		-0.18		2		41
		PHL 789		0 16 48	−32 16.	1	15.74		-0.33		-0.99		1		286
		NGC 104 - 3201		0 16 48	−72 20.2	2	13.80	.015	0.88	.005	0.51	.010	8		529,1582
1522	−9 00048	HR 74	★A	0 16 53	−09 06.1	8	3.54	.020	1.21	.012	1.24	.025	27	K2 III	3,15,1075,2012,3024*
1522	−9 00048	IDS00143S0923	B	0 16 53	−09 06.1	1	12.87		0.78		0.33		3		3024
		PHL 2856		0 16 54	−22 06.	1	15.31		0.22		-0.58		1	DA:	3064
		AO 1033		0 16 55	+43 36.8	1	11.35		0.00		-0.26		1	B9 V	1748
1541	−38 00068			0 16 55	−38 02.2	2	9.72	.034	0.27	.000	0.14	.005	9	A5 III(m)	152,1068
		NGC 104 - 2201		0 16 56	−72 24.8	2	13.99	.000	0.77	.010	0.45	.010	8		529,1582
		NGC 104 - 3202		0 16 57	−72 19.9	1	14.66		0.57		0.03		3		529
		BPM 46204		0 17 00	−21 13.	1	12.47		1.16		1.07		1		3061
		NGC 104 - 3204		0 17 00	−72 18.5	2	14.62	.005	0.94	.010	0.69	.073	9		529,1582
		G 158 - 61		0 17 02	−10 53.2	1	14.74		1.05		0.44		2		3062
		GR 6		0 17 02	−29 17.5	2	14.20	.010	1.51	.005	1.19	.010	2		481,3061
1527	+39 00056	HR 75		0 17 03	+40 27.1	2	6.35	.013	1.18	.001	1.10		5	K1 III	71,1501
		NGC 104 - 3203		0 17 04	−72 18.7	2	14.11	.005	0.79	.000	0.36	.000	7		529,1582
		LS I +62 086		0 17 06	+62 06.9	1	10.70		0.16		-0.57		2		41
		G 266 - 89		0 17 07	−28 25.8	1	14.27		1.52		1.11		1		3078
232158	+49 00049			0 17 09	+50 28.4	1	8.27		1.22		1.04		2	K2	1601
1552	+42 00055			0 17 14	+42 36.2	1	8.65		0.30		0.09		2	F2 III	1733
	−31 00112	G 267 - 74		0 17 14	−31 11.2	2	12.14	.015	0.43	.025	-0.19	.005	5		1696,3061
1546	+43 00053	VX And		0 17 15	+44 25.9	1	8.51		4.43				1	C8	109
		BPM 46211		0 17 18	−22 53.	1	12.98		1.45		1.19		1		3061
1563	+15 00047			0 17 21	+15 58.4	1	6.60		0.98		0.89		2	K0	1117
1544	+61 00038	LS I +61 136		0 17 22	+61 47.3	2	8.13	.012	0.15	.008	-0.77	.004	8	B0.5III (n)	41,1012
1562	+37 00042	IDS00148N3741	A	0 17 23	+37 57.2	2	6.99	.000	0.61	.010	0.04	.015	4	G0	979,3026
		G 171 - 50		0 17 24	+42 27.2	1	13.07		0.49		-0.22		2		1658
		BPM 46216		0 17 24	−26 31.	1	14.60		1.28		1.05		1		3061
1561	+48 00079	HR 76		0 17 25	+48 35.3	1	6.54		0.03		0.04		2	A0 Vs	1733
		LB 433		0 17 26	+13 35.7	2	15.22	.000	-0.12	.000	-0.97	.000	2		1298,3016
1589	−25 00096			0 17 27	−25 23.9	1	7.48		1.12		1.08		1	K1 III	3061
		NGC 104 - 2302		0 17 27	−72 23.0	1	14.11		1.00		0.80		3		529
1581	−65 00013	HR 77		0 17 29	−65 10.1	7	4.23	.009	0.57	.006	0.02	.007	38	G0 V	15,1075,1311,2038*
1588	−18 00041			0 17 31	−17 58.7	1	6.72		1.28				4	K1 III/IV	2009
	+61 00039	LS I +62 087		0 17 33	+62 11.2	2	8.54	.014	0.31	.009	-0.80	.033	4	B0.5IV	41,1012
1596	−33 00091			0 17 33	−32 37.1	1	10.00		0.80		0.37		1	G8 III/IV	3061
		GR 7		0 17 36	−17 20.3	1	11.69		1.51		1.18		3	M1	481
	−26 00095	LP 824 - 401		0 17 37	−25 54.1	1	10.61		0.85		0.54		1		3061
	+61 00040	LS I +62 088		0 17 41	+62 07.1	2	9.54	.005	0.48	.005	-0.50	.005	4	B2 I-IIp	41,1012
		G 266 - 90		0 17 41	−26 59.4	1	14.21		0.79		0.36		1		3061
		NGC 104 - 3302		0 17 43	−72 17.2	2	15.04	.010	0.86	.020	0.64	.030	8		529,1582
		LS I +58 016		0 17 44	+58 32.5	1	12.54		0.43		-0.16		2		41
		G 32 - 12		0 17 46	+17 40.5	1	13.82		0.89		0.47		2		1658
1606	+30 00042	HR 78		0 17 47	+30 39.5	3	5.90		-0.10	.009	-0.47	.023	7	B7 V	252,1049,1079
236382	+59 00036	LS I +60 099		0 17 47	+60 00.5	1	8.81		0.26		-0.60		2	B0	41
		SB 133		0 17 47	−29 04.2	3	14.48	.013	0.05	.009	0.12	.019	6	A0 pec	295,966,1736
1620	−33 00095	G 267 - 76	★AB	0 17 47	−33 02.7	2	8.51	.000	0.72	.002	0.18	.000	3	G6 V	79,3061
		CS 22882 # 5		0 17 48	−29 47.2	2	14.28	.010	0.60	.014	-0.07	.001	3		1580,1736
1619	−25 00099			0 17 49	−24 58.8	4	8.64	.010	0.35	.015	0.13	.020	12	A1 mA7-F3	79,355,1068,2012
1601	+48 00084			0 17 51	+48 41.5	1	6.47		0.53		0.24		2	G0	1733
1614	+44 00062	IDS00152N4457	AB	0 17 53	+45 13.9	1	7.01		0.54		0.06		2	F5	1601
1604	+35 00052			0 17 55	+36 10.2	1	8.93		0.01		-0.15		2	A0	1336
		CS 22882 # 1		0 17 55	−31 55.7	2	14.81	.010	0.40	.007	0.04	.009	3		1580,1736
		NGC 104 - 2401		0 17 55	−72 22.7	1	15.56		0.85				3		529
		3C 9 # 2		0 17 56	+15 23.7	1	14.40		0.80		0.29		2		327
		3C 9 # 1		0 17 58	+15 24.8	1	13.28		1.02		0.77		3		327

Table 1 25

HD	DM	Other Id	N Rem	α_{1950}	δ_{1950}	S	V	σ_V	B−V	σ_{B-V}	U−B	σ_{U-B}	n	Spectrum	References
1613	+61 00043			0 17 59	+61 36.1	1	6.87		2.02		2.13		3	M2 II:	8100
		NGC 104 - 2308		0 17 59	−72 32.6	1	13.89		0.89		0.39		1		522
		NGC 104 - 2310		0 18 00	−72 34.5	1	14.56		1.03		0.62		2		522
1635	+7 00036	HR 80		0 18 01	+07 54.8	3	5.36	.005	1.34	.000	1.55	.011	9	gK3	15,1256,1355
	+4 00037	AJ71,719 R410		0 18 02	+04 40.2	1	9.68		0.86		0.42		2		1375
		WLS 20 5 # 10		0 18 03	+04 40.1	1	10.56		0.38		0.20		2		1375
		NGC 104 - 3306		0 18 03	−72 11.7	1	14.11		0.83		0.38		1		366
		SB 136		0 18 06	−21 54.	1	12.24		0.15		0.15		2	A3	295
		NGC 104 - 3407		0 18 07	−72 17.9	2	12.96	.000	1.22	.005	1.16	.025	7		522,529
1632	+32 00045	HR 79		0 18 08	+32 38.0	1	5.79		1.60				2	K5 III	71
1633	+25 00037			0 18 09	+26 14.2	1	7.37		1.40		1.69		1	K5 III	979
1686	−77 00010			0 18 09	−77 30.0	1	7.87		0.55				4	F6 V	2012
		LS I +55 008	⋆ V	0 18 10	+55 26.1	1	12.79		0.07		−0.79		2		41
		LS I +59 040		0 18 10	+59 33.9	1	12.31		0.51		−0.35		2		41
1624	+66 00019	IDS00154N6707	AB	0 18 10	+67 23.4	1	7.91		0.81		0.38		2	G5	3030
		NGC 104 - 3305		0 18 10	−72 12.6	1	13.05		1.20		1.10		2		522
		NGC 104 - 3312		0 18 12	−72 07.7	1	14.13		0.86		0.26		1		522
		NGC 104 - 3307		0 18 12	−72 11.2	1	14.23		0.96		0.49		1		522
1655	−26 00102			0 18 13	−25 44.6	1	8.34		1.48				6	G5 III	955
1642	+17 00033			0 18 14	+18 12.3	3	7.64	.015	0.93	.012	0.61	.032	7	G5	882,1648,3040
		NGC 104 - 3416		0 18 14	−72 14.5	2	14.02	.009	0.79	.009	0.35	.047	4		366,522
		NGC 104 - 3415		0 18 16	−72 14.7	2	14.02	.015	0.88	.015	0.34	.010	3		366,522
1641	+32 00048	IDS00157N3226	AB	0 18 17	+32 42.1	1	6.85		0.45		−0.02		1	F5	695
	−29 00081	G 267 - 78		0 18 17	−29 09.5	1	11.04		1.00		0.81		2		3061
1663	+10 00032	HR 81	⋆ AB	0 18 19	+10 42.0	1	6.56		0.04		−0.07		2	A0 V + A2	1733
		G 266 - 94		0 18 19	−25 50.6	1	12.61		1.43		1.16		1		3061
		Steph 25		0 18 21	+17 21.6	1	11.23		1.46		1.18		2	M0	1746
1685	−70 00012	HR 83		0 18 21	−69 54.1	5	5.50	.004	−0.04	.005	−0.13	.011	22	B9 V	15,611,1075,2016,2038
		G 267 - 79		0 18 23	−28 30.7	1	15.17		1.49				1		3061
		NGC 104 - 3414		0 18 23	−72 14.9	2	14.01	.006	0.87	.014	0.46	.000	7		366,939
		NGC 104 - 2309		0 18 23	−72 33.8	1	12.81		0.63		0.01		1		522
		G 266 - 95		0 18 24	−28 03.0	1	14.68		1.45		1.10		1		3061
1683	−39 00072			0 18 25	−39 31.0	1	7.14		0.50		0.01		6	F6 V	1628
		NGC 104 - 3410		0 18 25	−72 19.4	1	13.18		1.17		1.05		2		522
		NGC 104 - 3409		0 18 25	−72 19.9	1	14.34		0.83		0.28		3		529
1707	−72 00031	IDS00161S7232	AB	0 18 27	−72 15.3	5	8.30	.009	1.15	.008	1.12	.005	32	K0 IV	366,522,529,1582,1704
	+56 00045			0 18 28	+57 21.1	1	9.14		1.80				2		369
1671	+37 00045	HR 82		0 18 29	+37 41.5	4	5.19	.011	0.42	.012	0.04	.009	9	F5 III	253,254,1118,3053
		NGC 104 - 3501		0 18 30	−72 21.3	2	12.12	.010	1.47	.000	1.70	.035	7		522,529
1677	+35 00053			0 18 31	+35 36.1	1	7.37		0.24		0.09		2	A0	1336
		NGC 104 - 3412		0 18 31	−72 16.9	2	14.06	.025	0.80	.010	0.38	.015	2		366,522
	+19 00048	G 131 - 55		0 18 32	+20 18.7	1	10.06		0.96		0.73		1	K3	333,1620
		NGC 104 - 3505		0 18 32	−72 20.5	1	15.39		0.73		0.47		3		529
		NGC 104 - 2507		0 18 32	−72 24.2	1	14.08		0.82		0.36		1		522
1658	+66 00020	IDS00158N6627	AB	0 18 35	+66 43.7	1	7.23		0.01		−0.10		2	B8 V +A1 V	3016
1658	+66 00020	IDS00158N6627	C	0 18 35	+66 43.7	1	10.16		0.45		−0.06		2		3016
		BPM 46230		0 18 36	−20 46.	1	12.57		0.58		−0.11		1		3061
1692	−26 00104			0 18 36	−26 32.3	1	9.36		0.51		−0.06		1	G0 V	3061
		NGC 104 - 3309		0 18 36	−72 10.0	1	11.80		0.54		0.05		5		522
	+59 00038	MZ Cas		0 18 37	+59 40.0	1	9.13		2.49				2	M2 Iab	369
		NGC 104 - 2311		0 18 37	−72 34.5	1	14.00		0.72		0.06		2		522
		NGC 104 - 2416		0 18 38	−72 30.2	2	12.66	.000	1.32	.005	1.38	.030	7		522,1582
		GD 603		0 18 42	−33 59.1	1	14.67		−0.05		−0.82		1		3061
		NGC 104 - 1304		0 18 42	−72 36.6	1	14.45		0.97		0.56		1		522
		NGC 104 - 1201		0 18 42	−72 38.3	1	12.85		1.30		1.40		2		522
		LS I +61 137		0 18 43	+61 27.6	1	11.42		0.56		−0.38		2		41
		NGC 104 - 3503		0 18 44	−72 21.5	1	13.96		0.77		0.37		2		522
1721	−36 00104			0 18 45	−36 04.5	2	7.14	.010	1.58	.005	1.90		8	M0 III	2012,3040
		NGC 104 - 1301		0 18 46	−72 35.6	1	12.60		1.20		1.20		2		522
		BPM 16078		0 18 48	−51 57.	1	14.30		0.60		−0.20		3		3065
		LS I +61 138		0 18 50	+61 10.9	1	12.21		0.65		−0.22		1		41
	−46 00076			0 18 51	−46 00.1	3	10.40	.018	1.49	.013	1.20	.005	6	M0	158,1705,3079
	+67 00026			0 18 52	+67 38.0	1	9.13		2.10				2		369
		NGC 104 - 1305		0 18 52	−72 36.6	1	14.33		0.63		0.03		2		522
		NGC 104 - 1105		0 18 53	−72 42.9	1	13.69		1.10		0.85		3		522
		LS I +62 089		0 18 54	+62 19.7	1	11.78		0.36		−0.44		2		41
1697	+60 00037			0 18 56	+61 24.9	2	7.26	.005	0.51	.014	−0.01		4	F8 V	401,1118
		CS 22882 # 8		0 18 56	−29 07.6	2	13.97	.000	0.38	.000	−0.16	.011	2		966,1736
1715	+35 00055			0 19 00	+35 56.4	1	8.09		0.45		−0.09		2	A5	1336
1714	+35 00056			0 19 00	+35 58.5	1	8.50		0.35		0.13		2	A0	1336
	+60 00038	NGC 103 - 60		0 19 00	+60 47.4	2	9.08	.000	1.12	.005	0.92	.005	6	G8 III	260,523
		NGC 104 - 2426		0 19 00	−72 31.7	1	12.10		1.52		1.75		2		522
1737	−29 00086	HR 84		0 19 01	−29 15.5	4	5.17	.009	1.01	.004	0.84		20	K0 III	3,15,1075,2032
		LS I +58 017		0 19 02	+58 03.8	1	10.86		0.45		−0.27		2		41
	+60 00038	NGC 103 - 61		0 19 02	+60 48.1	2	8.94	.010	1.41	.005	1.23	.010	6	G8 III	260,523
	+63 00033	LS I +64 021		0 19 04	+64 19.1	1	9.44		0.57		−0.34		2	B1 V	1012
1748	−33 00108	G 267 - 83	⋆ ABC	0 19 04	−33 20.0	1	9.69		1.03		0.76		1	K2	3061
		TonS 153		0 19 06	−22 51.	1	15.57		−0.06		0.08		1		3064
		LS I +61 139		0 19 08	+61 28.8	1	11.82		0.38		−0.45		4		41
	+60 00039	LS I +61 140		0 19 09	+61 28.4	3	9.46	.000	0.26	.020	−0.69	.005	6	O9 V	41,1011,1012

HD	DM	Other Id	N	Rem	α_{1950}	δ_{1950}	S	V	σ_V	B–V	σ_{B-V}	U–B	σ_{U-B}	n	Spectrum	References
		NGC 104 - 3516			0 19 09	−72 14.5	1	13.20		1.18		0.88		2		522
		NGC 104 - 2511			0 19 09	−72 25.7	1	13.84		0.84		0.45		2		522
		NGC 104 - 2429			0 19 10	−72 32.9	1	14.61		0.98		0.63		1		522
		NGC 104 - 1402			0 19 10	−72 33.9	1	13.39		0.59		0.03		2		522
		NGC 104 - 2508			0 19 11	−72 24.5	2	13.25	.010	0.77	.005	0.35	.010	9		522,1582
		NGC 104 - 3517			0 19 14	−72 14.9	1	14.90		-0.15		-0.45		2		522
		LS I +62 090			0 19 15	+62 40.2	1	11.15		0.20		-0.44		2		41
1760	−20 00050	HR 85, T Cet			0 19 15	−20 20.1	2	5.33	.266	1.81	.020	1.66	.155	3	M5 IIe	58,3076
		NGC 104 - 1207			0 19 15	−72 42.7	2	14.08	.010	0.84	.010	0.40		5		522,1582
	+60 00040	LS I +61 142			0 19 16	+61 14.7	1	10.67		0.29		-0.58		2	K0 III	41
		NGC 104 - 1407			0 19 16	−72 35.0	1	13.78		1.07		0.78		2		522
		NGC 104 - 2428			0 19 17	−72 32.1	1	13.23		1.16				1		522
1743	+61 00048	LS I +61 141			0 19 18	+61 54.5	1	8.33		0.14		-0.74		18	B0.2 IV	41
		PS 3			0 19 18	−32 59.	1	13.07		0.32		0.12		1		295
		NGC 104 - 1406			0 19 18	−72 34.3	1	13.45		1.11		0.94		2		522
		NGC 104 - 2601			0 19 19	−72 25.5	1	13.83		1.06		0.82		2		522
		NGC 104 - 3512			0 19 20	−72 16.6	2	11.81	.020	1.65	.025	1.96		4		522,529
		NGC 104 - 2602			0 19 20	−72 25.1	1	13.00		1.09		0.86		3		522
1766	−23 00111	IDS00168S2334		ABC	0 19 21	−23 17.0	2	7.49	.147	0.61	.050	0.10	.049	4	G1 V	938,3061
		NGC 104 - 1208			0 19 21	−72 43.2	1	14.32		0.75		0.12		3		522
		NGC 104 - 1306			0 19 22	−72 36.9	1	15.49		0.90				3		529
1801	−78 00009	HR 87			0 19 22	−77 42.2	3	5.96	.007	1.40	.000	1.40	.000	14	K3 III	15,1075,2038
		NGC 104 - 3519			0 19 24	−72 13.6	1	14.40		1.02		0.70		1		522
		NGC 104 - 1408			0 19 24	−72 35.1	1	13.53		1.09		0.88		2		522
		NGC 104 - 1205			0 19 25	−72 41.5	1	11.85		1.61		1.94		2		522
		NGC 104 - 1309			0 19 26	−72 39.9	1	13.20		1.04		0.80		3		522
		GD 605			0 19 27	−24 42.0	2	14.43	.025	-0.33	.002	-1.26	.040	5		286,3064
		BPM 46239			0 19 27	−31 41.0	1	11.16		1.49		1.09		1		3061
		LP 465 - 2			0 19 28	+12 17.1	1	13.70		1.46		1.07		1		1696
		Steph 29			0 19 28	+21 05.8	1	10.00		0.94		0.73		1	K5	1746
		LS I +61 143			0 19 28	+61 28.0	1	12.39		0.27		-0.49		2		41
	+61 00049	G 243 - 21			0 19 28	+61 44.9	2	9.66	.005	0.66	.005	0.14		3	G8	1118,1658
		NGC 104 - 3518			0 19 28	−72 13.8	1	14.11		0.82		0.43		1		522
		CS 22882 # 6			0 19 29	−29 46.1	2	14.13	.009	0.42	.008	-0.16	.008	4		1580,1736
1779	−27 00098	G 266 - 97			0 19 32	−26 59.2	3	8.93	.023	0.67	.009	0.13	.010	7	G5 V	79,2033,3077
	−25 00112	G 266 - 98			0 19 33	−25 13.1	2	9.43	.010	0.70	.005	0.16	.025	2	G5	79,3061
1780	−39 00079	G 267 - 84			0 19 33	−38 58.5	1	9.30		0.75		0.29		1	G8 IV/V	79
	−25 00113	IDS00171S2511		A	0 19 34	−24 55.5	1	12.30		1.09		0.98		1		3061
		NGC 104 - 2603			0 19 35	−72 27.7	1	12.90		1.27		1.21		2		522
		BPM 46243			0 19 36	−22 51.	1	12.06		0.86		0.56		1		3061
	+60 00042	LS I +61 144			0 19 37	+61 22.7	1	10.17		0.24		-0.50		2	B5	41
1778	+52 00052				0 19 38	+53 19.4	1	8.16		0.42		0.10		1	F3 II	8100
1790	−28 00102				0 19 38	−27 40.6	1	9.59		0.51		-0.02		1	F6 V	3061
		NGC 104 - 2605			0 19 38	−72 25.3	1	12.50		1.22		1.23		2		522
		NGC 104 - 1206			0 19 38	−72 41.2	1	14.13		0.83		0.32		2		522
1769	+65 00043				0 19 39	+65 54.3	1	8.59		0.50		-0.04		2	F5	1733
		Steph 31			0 19 40	+21 06.6	1	10.74		1.54		1.90		3	M0	1746
		LS I +61 145			0 19 41	+61 33.0	1	11.22		0.17		-0.64		3		41
	+32 00053				0 19 42	+32 47.5	1	8.71		1.71		1.91		2	M0	1733
		LP 765 - 71			0 19 44	−18 03.5	1	13.29		1.14		1.05		1		1696
1794	+34 00042				0 19 45	+35 15.6	1	7.02		1.61		1.93		2	K5	1625
	+37 00053				0 19 46	+38 06.9	1	10.01		0.99		0.77		1		1722
		NGC 104 - 2525			0 19 47	−72 29.9	1	12.43		1.29		1.25		2		522
		G 171 - 54			0 19 48	+47 30.6	1	11.21		1.29		1.23		3	M0	1723
	+5 00043	V Psc		⋆	0 19 49	+06 23.7	1	11.22		1.33				2	M1 V	202
		NGC 104 - 2604			0 19 49	−72 28.4	1	13.07		1.04		0.75		2		522
		NGC 104 - 1429			0 19 49	−72 34.3	1	14.10		0.86				2		529
1796	+12 00025	HR 86		⋆ A	0 19 50	+13 12.3	1	6.24		1.23				2	K3 III	71
		NGC 104 - 1310			0 19 51	−72 38.9	1	14.01		0.72				3		529
1815	−27 00101	G 266 - 100			0 19 52	−27 18.3	4	8.32	.010	0.89	.005	0.62	.007	11	K1 V	79,158,2018,3077
		BPM 46249			0 19 54	−34 19.	1	13.29		0.47		-0.26		1		3061
1813	−24 00113				0 19 55	−24 32.6	1	8.95		1.58		1.82		4	M2/3 (III)	3007
		AO 838			0 19 59	+38 21.0	1	10.74		1.60		1.92		1		1722
		NGC 104 - 1510			0 19 59	−72 34.7	2	12.15	.005	1.44	.010	1.57		5		522,529
		NGC 104 - 1412			0 19 59	−72 38.2	2	11.66	.020	0.45	.000	-0.03		7		529,1582
		G 130 - 65			0 20 02	+23 38.0	2	11.62	.000	0.43	.000	-0.25	.000	3	sdF8	333,927,1620
		G 131 - 58			0 20 03	+16 46.0	1	11.61		0.86		0.46		2		333,1620
		G 267 - 87			0 20 03	−31 27.2	1	13.83		1.47		1.17		1		3061
	−28 00107	G 266 - 101			0 20 04	−27 51.3	1	11.59		0.72		0.07		1		3061
		NGC 104 - 4502			0 20 04	−72 11.8	1	13.28		0.96		0.61		2		522
		NGC 104 - 1505			0 20 04	−72 32.3	3	12.15	.013	1.50	.012	1.76	.070	10		522,529,1582
		NGC 104 - 1506			0 20 04	−72 32.9	2	13.26	.010	1.16	.010	0.84		4		522,529
		NGC 104 - 1311			0 20 04	−72 39.0	1	14.16		0.72				3		529
	+23 00046				0 20 05	+23 38.0	1	9.30		1.25		1.17		2	K0	1064
1811	+53 00053				0 20 05	+54 22.1	1	7.14		0.89		0.50		2	G5	1566
		NGC 104 - 4503			0 20 05	−72 11.3	1	12.39		1.22		1.12		2		522
		NGC 104 - 1411			0 20 06	−72 35.6	1	13.94		0.87		0.52		2		522
		G 267 - 89			0 20 07	−31 27.1	1	14.30		1.50				1		3061
1810	+61 00050	LS I +61 146		⋆ AB	0 20 08	+61 57.9	2	8.17	.014	0.06	.014	-0.82	.019	8	B0 IV	41,399
		LS I +61 147			0 20 09	+61 57.9	1	11.11		0.02		-0.52		1		41

Table 1   27

HD	DM	Other Id	N	Rem	α_{1950}	δ_{1950}	S	V	σ_V	B–V	σ_{B-V}	U–B	σ_{U-B}	n	Spectrum	References
		LS I +60 100		A	0 20 10	+60 55.4	1	11.74		0.22		-0.42		2		41
		LS I +60 100		B	0 20 10	+60 55.4	1	12.04		0.24		-0.36		2		41
1828	-10 00057				0 20 10	-09 54.0	1	8.19		0.44		-0.04		38	F8	978
		NGC 104 - 4506			0 20 11	-72 10.0	1	13.54		1.11		0.88		2		522
1822	+33 00033				0 20 12	+34 26.0	1	9.43		0.17		0.09		2	A0	1336
		BPM 16085			0 20 12	-55 57.	1	14.48		0.58		-0.14		2		3065
1833	-10 00058	BD Cet			0 20 14	-09 30.5	1	7.97		1.14		0.93		1	G5	1641
		NGC 104 - 1312			0 20 14	-72 40.1	1	14.58		0.98		0.49		2		522
		NGC 104 - 1313			0 20 14	-72 41.2	1	14.44		1.00		0.47		2		522
1834	-10 00060				0 20 16	-09 33.3	1	8.62		1.45		1.72		49	K0	978
		NGC 104 - 1519			0 20 16	-72 35.6	1	13.89		0.88		0.39		1		522
		AO 840			0 20 18	+38 32.3	1	10.60		0.58		0.07		1		1722
1835	-13 00060	HR 88, BE Cet	⋆	A	0 20 18	-12 29.2	5	6.39	.006	0.66	.003	0.23	.038	22	G3 V	15,79,1738,2030,3077
		BPM 46255			0 20 18	-22 48.	1	12.61		0.82		0.43		1		3061
1831	+37 00054				0 20 19	+38 28.6	2	6.72	.026	1.61	.009	1.81	.014	6	M1	814,1722
		NGC 104 - 3730			0 20 19	-72 18.8	1	11.94		1.62		1.84		2		1468
		NGC 104 - 1414			0 20 20	-72 37.4	1	14.24		1.06		0.67		1		522
1832	+21 00033	G 131 - 59			0 20 23	+22 06.1	1	7.60		0.63		0.13		1	F8	333,1620
		NGC 104 - 1605			0 20 23	-72 24.5	1	12.36		1.45		1.60		2		522
		BPM 46256			0 20 24	-23 50.	1	12.44		0.82		0.27		1		3061
		NGC 104 - 1518			0 20 25	-72 35.2	1	13.06		1.08		0.75		2		522
		NGC 104 - 1513			0 20 26	-72 32.6	1	12.41		1.32		1.32		3		522
	+28 00051				0 20 29	+28 55.5	1	9.55		0.94				1	G5	1213
		NGC 104 - 4312			0 20 29	-72 03.1	1	12.70		1.30		1.22		1		522
		NGC 104 - 4509			0 20 30	-72 09.2	1	12.99		1.21		0.93		2		522
		AO 841			0 20 31	+38 10.3	1	10.89		0.16		0.13		1		1722
1845	+54 00048	T Cas			0 20 31	+55 30.9	2	9.11	.700	1.77	.195	0.31	.145	2	M4	8022,8027
	-53 00077				0 20 31	-52 47.8	2	9.83	.015	0.85	.005	0.44		6	K0	2012,3077
1879	-16 00058				0 20 32	-16 13.2	3	6.42	.016	1.59	.012	1.90	.034	15	M0 III	1003,1024,3040
1844	+59 00045				0 20 33	+59 57.5	1	7.62		0.30				2	A0	1118
		NGC 104 - 1315			0 20 33	-72 40.5	1	14.07		0.85		0.37		1		522
		G 217 - 46			0 20 34	+60 56.2	1	11.07		1.31		1.27		2	K7	7010
		WLS 20 5 # 7			0 20 35	+02 38.2	1	11.28		0.53		0.06		2		1375
232174	+50 00061				0 20 35	+50 55.0	1	8.56		0.98				53	G5	6011
		NGC 104 - 1316			0 20 35	-72 39.9	1	13.17		1.19		0.97		2		522
		G 242 - 51			0 20 36	+76 54.7	2	11.33	.005	1.45	.017	1.19	.019	5		419,7009
	+6 00031				0 20 37	+07 27.7	1	9.74		1.15		0.99		2	K0	1375
1875	+35 00063				0 20 37	+36 24.5	1	9.82		0.15		0.12		2	A0	1336
		NGC 104 - 2728			0 20 37	-72 23.8	1	14.18		0.82		0.37		2		350
		NGC 104 - 1516			0 20 37	-72 34.4	1	13.94		1.04		0.70		2		522
		LS I +62 091			0 20 39	+62 45.2	1	10.97		0.46		-0.34		1		41
		NGC 104 - 3741			0 20 39	-72 18.1	1	14.66		0.96		0.67		2		350
1874	+51 00059				0 20 40	+52 26.1	1	8.29		0.99		0.73		2	K0	1566
	+30 00050				0 20 41	+31 04.3	1	10.15		1.26		1.16		1		272
1873	+54 00049				0 20 41	+55 31.5	1	8.13		0.06		-0.20		2	B9	1566
		G 31 - 46			0 20 43	-00 06.9	2	9.92	.000	0.91	.010	0.62	.025	2		1620,1658
1909	-31 00138	HR 89, AV Scl			0 20 43	-31 18.8	3	6.55	.012	-0.06	.022	-0.33	.020	10	B9 IV	1068,2032,3007
1912	-51 00088				0 20 44	-50 49.5	1	7.63		1.46		1.75		4	K4 III	1673
		NGC 104 - 1419			0 20 44	-72 36.9	1	14.19		0.97		0.72		2		522
		NGC 104 - 3743			0 20 45	-72 17.6	1	14.58		0.82		0.48		3		350
		NGC 104 - 2743			0 20 45	-72 26.1	2	14.13	.015	0.80	.010	0.34	.034	3		343,350
		NGC 104 - 1704			0 20 45	-72 26.9	1	13.00		1.25				2		1582
		NGC 104 - 1320			0 20 45	-72 41.5	1	13.54		1.13		0.87		2		522
1910	-33 00118				0 20 46	-33 26.8	1	8.75		1.07		0.90		1	K3/4 V	3061
		CS 22882 # 9			0 20 47	-28 48.2	2	15.49	.005	0.41	.008	-0.20	.009	3		1580,1736
		BPM 46260			0 20 48	-22 51.	1	13.04		0.41		-0.10		1		3061
		NGC 104 - 4409			0 20 48	-72 05.0	1	13.95		0.79		0.29		1		522
		NGC 104 - 2742			0 20 48	-72 26.0	1	13.04		1.08		0.75		2		343
		NGC 104 - 1521			0 20 48	-72 35.2	2	14.00	.002	0.79	.008	0.32	.004	7		366,939
		NGC 104 - 1322			0 20 49	-72 42.4	1	14.57		0.92		0.39		2		522
		AO 842			0 20 50	+38 25.8	1	11.38		1.52		1.90		1		1722
		NGC 104 - 1523			0 20 51	-72 42.7	2	13.94	.002	0.80	.011	0.39	.000	7		366,939
		G 130 - 66			0 20 52	+24 01.5	1	14.28		1.59		1.21		4		203
1923	-30 00101				0 20 52	-30 07.4	1	6.82		1.56		1.83		4	M2 III	3007
		NGC 104 - 2741			0 20 52	-72 25.8	1	13.18		1.03				3		343
		AO 843			0 20 53	+38 19.3	1	11.66		1.05		0.72		1		1722
		NGC 104 - 1522			0 20 53	-72 34.7	1	13.56		1.09		0.87		2		522
1905	+37 00057	IDS00182N3802		A	0 20 54	+38 18.2	1	9.88		0.22		0.14		2	A3	1336
	+62 00068				0 20 54	+63 05.4	1	9.61		0.58		-0.38		2	B1 II	1012
	-27 00103	G 266 - 102			0 20 54	-26 38.2	1	11.65		0.92		0.71		1		3061
		NGC 104 - 4715			0 20 54	-72 17.1	1	11.23		1.58		1.59		1		1468
		NGC 104 - 1319			0 20 54	-72 40.7	1	13.08		0.93		0.60		2		522
		CS 22882 # 36			0 20 55	-31 29.5	2	14.52	.010	0.38	.007	-0.20	.010	3		1580,1736
	-23 00127	BPM 46263			0 20 57	-23 30.6	1	11.03		0.83		0.50		1		3061
		NGC 104 - 1526			0 20 57	-72 35.8	3	14.00	.012	0.85	.009	0.31	.022	9		366,522,939
	+61 00056	LS I +62 092			0 20 58	+62 02.3	1	10.04		0.33		-0.58		2	B0 Iab:	41
		LS I +59 041			0 20 59	+59 34.3	1	11.48		0.45		-0.42		2		41
	+59 00047				0 20 59	+60 06.0	1	8.99		0.51				2	F8	1118
		WLS 24 50 # 5			0 21 00	+52 17.4	1	13.88		0.66		0.06		2		1375
		LP 825 - 528			0 21 00	-26 49.	1	13.55		1.36		1.15		1		3061

HD	DM	Other Id	N Rem	α_{1950}	δ_{1950}	S	V	σ_V	B–V	σ_{B-V}	U–B	σ_{U-B}	n	Spectrum	References
		NGC 104 - 4414		0 21 01	−72 06.3	2	14.51	.110	0.94	.000	0.51	.060	4		343,522
		NGC 104 - 4601		0 21 02	−72 10.6	1	12.29		1.32		1.42		2		522
1918	+44 00076			0 21 03	+44 48.7	2	7.60	.004	0.98	.012	0.66	.008	8	G9 III	1003,1775
		G 31 - 47		0 21 03	−04 02.3	2	10.66	.015	1.11	.019	1.04		3		202,1620
		NGC 104 - 1528		0 21 04	−72 36.3	2	14.08	.006	0.81	.006	0.28	.021	7		366,939
	+61 00059	LS I +61 148		0 21 05	+61 39.5	1	10.02		0.21		-0.49		2	B2	41
		G 158 - 70		0 21 05	−07 22.8	1	12.90		0.92		0.50		2		3062
		NGC 104 - 4411		0 21 05	−72 05.2	2	13.64	.465	1.01	.030	0.87		2		343,522
		NGC 104 - 1524		0 21 05	−72 35.3	3	14.09	.009	0.86	.007	0.36	.018	10		366,522,939
	+28 00054	SW And		0 21 06	+29 07.5	2	9.23	.079	0.22	.026	0.19	.000	2	A7 III	668,699
		AO 844		0 21 07	+38 13.5	1	11.73		1.55		1.70		1		1722
236398	+59 00048			0 21 07	+60 15.5	1	8.98		0.22				2	A0	1118
	−25 00123	G 266 - 103		0 21 07	−24 55.6	2	10.01	.010	0.58	.002	-0.02	.020	2	G0	79,3061
1946	−30 00104			0 21 07	−29 45.8	1	9.24		0.59		0.06		1	F7/8 V	3061
1917	+60 00045			0 21 08	+60 41.2	1	6.58		0.36				2	F2	1118
		G 267 - 93		0 21 08	−34 55.8	1	14.59		1.24		0.82		1		3061
		NGC 104 - 1735		0 21 09	−72 28.3	2	13.32	.025	1.06	.025	0.91	.005	4		343,350
		NGC 104 - 1529		0 21 09	−72 36.5	1	13.90		0.86		0.31		2		522
		LS I +62 093		0 21 10	+62 00.4	1	11.84		0.39		-0.46		2		41
		NGC 104 - 1727		0 21 10	−72 27.2	1	14.59		0.92		0.65		2		350
	+59 00049	LS I +60 101		0 21 12	+60 14.2	1	10.95		0.41		-0.43		3		41
		NGC 104 - 1733		0 21 12	−72 28.1	1	14.36		0.96		0.66		2		350
1941	+28 00055	IDS00186N2857	B	0 21 14	+29 13.1	1	8.48		0.58		0.09		2		3032
1901	+77 00009			0 21 14	+77 33.4	1	8.40		0.62		0.13		2	F5	1375
		LP 705 - 20		0 21 15	−14 21.3	1	11.95		0.50		-0.19		2		1696
		G 266 - 104		0 21 15	−21 53.0	1	12.04		1.02		0.99		1		3061
1942	+28 00056	IDS00186N2857	A	0 21 16	+29 13.5	2	8.25	.040	0.54	.012	0.29	.203	11		1320,3016
		NGC 103 - 41		0 21 16	+61 02.9	1	13.13		1.55		1.43		1		260
	+61 00061	LS I +62 094		0 21 17	+62 10.5	1	11.19		0.21		-0.59		2		41
		NGC 104 - 4602		0 21 17	−72 11.0	1	13.07		1.19		0.87		1		522
		NGC 104 - 1424		0 21 17	−72 39.2	1	15.41		0.85				3		529
1930	+60 00047	NGC 103 - 53		0 21 18	+61 13.6	3	8.52	.011	0.76	.005	0.39	.005	8	G5	260,523,1118
		BPM 46272		0 21 18	−22 07.	1	14.72		0.45		-0.18		1		3061
		NGC 103 - 43	⋆ V	0 21 19	+61 03.9	2	11.47	.044	1.10	.029	0.73	.039	3		41,260
		NGC 104 - 1324		0 21 20	−72 40.7	1	14.08		0.83		0.35		2		522
		NGC 104 - 1604		0 21 22	−72 33.7	1	13.01		1.21		1.22		3		522
	+17 00044	G 32 - 16		0 21 23	+17 50.9	1	9.68		0.88		0.48		1	K1	333,1620
1967	+37 00058	HR 90, R And	⋆ AB	0 21 23	+38 18.0	7	7.39	.227	1.97	.031	1.24	.112	10	S6.6 e	15,635,814,3001,8015*
	+61 00064	LS I +62 095		0 21 23	+62 20.8	1	9.51		0.29		-0.57		2		41
		CS 22882 # 11		0 21 23	−28 10.3	2	15.63	.000	0.00	.000	0.04	.011	2		966,1736
		NGC 104 - 1422		0 21 23	−72 39.4	2	13.17	.030	0.67	.015	0.31		8		529,1582
		NGC 103 - 46		0 21 24	+61 06.3	1	12.35		1.98		2.37		2		260
		BPM 46274		0 21 24	−21 32.	1	12.09		0.89		0.51		1		3061
		BPM 46275		0 21 24	−21 32.	1	13.98		1.16		0.85		1		3061
1952	+43 00072			0 21 25	+43 59.3	1	6.66		0.41		0.17		3	F5 V	1501
232178	+50 00064			0 21 25	+50 37.0	1	8.85		1.04				53	G5	6011
		NGC 103 - 45		0 21 25	+61 05.7	1	13.13		2.27		2.40		2		260
1989	−31 00143			0 21 25	−31 32.9	1	9.26		0.97		0.71		1	K1 IV/V	3061
		NGC 103 - 39	⋆	0 21 26	+61 02.0	2	10.69	.050	0.40	.020	-0.33	.010	4		41,260
		NGC 104 - 4415		0 21 26	−72 05.2	3	12.35	.015	1.06	.005	0.80	.031	6		343,366,522
		NGC 104 - 1603		0 21 26	−72 33.3	1	11.89		1.58		1.87		3		522
		NGC 103 - 40		0 21 28	+61 02.1	1	11.13		0.51		0.32		2		260
1939	+63 00038			0 21 30	+64 01.4	1	8.57		0.25				2	B9	1118
		BPM 46278		0 21 30	−20 13.	1	12.57		1.41		1.09		1		3061
		L 170 - 39		0 21 30	−56 37.	1	12.12		1.07		0.83		1		3062
		NGC 104 - 4419		0 21 30	−72 04.5	2	15.74	.010	0.52	.000	0.03	.010	4		343,522
		NGC 104 - 1602		0 21 30	−72 32.7	1	13.24		1.04		0.72		3		522
1987	−10 00065	S Cet		0 21 31	−09 36.3	1	9.87		1.61		0.99		1	M4	975
1966	+53 00064			0 21 32	+53 32.8	1	7.52		0.46		-0.01		2	F5	1566
1950	+59 00051			0 21 32	+60 10.6	1	7.58		0.15				2	B9	1118
1976	+51 00062	HR 91	⋆ AB	0 21 33	+51 44.6	5	5.57	.015	-0.12	.009	-0.60	.010	13	B5 IV	15,154,1012,1223,1267
		NGC 104 - 4737		0 21 33	−72 14.2	3	12.97	.010	1.24	.015	1.08	.045	9		343,350,3036
		NGC 104 - 1425		0 21 33	−72 38.5	2	14.08	.020	1.02	.015	0.74		5		522,529
		LS I +56 006		0 21 34	+56 27.2	1	10.90		0.29		-0.61		2		41
		NGC 103 - 47		0 21 34	+61 05.7	1	10.88		1.39		1.30		2		260
		NGC 104 - 4515		0 21 34	−72 07.1	1	14.41		0.99		0.55		1		343
1999	−38 00105			0 21 35	−37 41.3	2	8.31	.005	-0.12	.005	-0.46	.010	6	B8 II	1068,3007
		NGC 104 - 4417		0 21 35	−72 05.3	3	14.10	.021	0.83	.029	0.35	.028	3		343,366,522
		NGC 104 - 4739		0 21 36	−72 14.2	2	14.66	.000	0.95	.040	0.66	.020	5		350,3036
		NGC 104 - 1531		0 21 36	−72 34.6	3	14.08	.025	0.83	.008	0.44	.025	6		366,522,529
		NGC 103 - 48		0 21 37	+61 06.4	1	10.89		0.48		0.28		2		260
	+62 00071			0 21 37	+63 26.5	1	9.23		0.42				2	F8	1118
	+61 00067	NGC 103 - 65		0 21 39	+61 33.8	3	8.78	.013	0.35	.012	0.13	.004	7	A0 Ib	41,260,523
		NGC 104 - 4418		0 21 40	−72 03.8	4	12.15	.013	1.46	.005	1.65	.026	2		343,366,522,528
		NGC 104 - 4740		0 21 40	−72 14.2	1	14.10		0.76		0.45		2		3036
2007	−17 00054			0 21 41	−17 02.8	3	9.48	.009	0.21	.004	0.17	.010	21	A3 III	830,1068,1783
		NGC 104 - 4745		0 21 41	−72 15.2	1	14.27		0.62		0.15		2		3036
		Smethells 165		0 21 42	−62 28.	1	11.33		1.39				1		1494
		NGC 104 - 4741	V	0 21 42	−72 14.1	3	11.92	.025	1.64	.009	1.92	.010	8		350,1468,3036
		G 31 - 48		0 21 43	−03 09.1	2	12.71	.024	1.32	.029	1.05		3		202,1620

Table 1

29

HD	DM	Other Id	N	Rem	α_{1950}	δ_{1950}	S	V	σ_V	B–V	σ_{B-V}	U–B	σ_{U-B}	n	Spectrum	References
		NGC 103 - 50			0 21 44	+61 07.4	1	12.37		0.27		-0.26		2		260
		NGC 104 - 4743			0 21 44	-72 14.5	3	14.37	.039	0.96	.021	0.61	.087	6		343,350,3036
	-73 00021	NGC 104 - 1533			0 21 45	-72 35.6	3	10.08	.005	0.71	.008	0.20	.005	16	K0	366,522,529
		NGC 104 - 1427			0 21 45	-72 38.8	1	15.11		1.00				3		529
1996	+25 00046				0 21 46	+26 07.1	1	8.07		0.93				3	K1 III	20
		NGC 104 - 4603			0 21 46	-72 10.8	1	12.13		1.41		1.59		1		522
		NGC 104 - 1765			0 21 46	-72 28.1	1	14.17		0.82		0.44		2		350
		LS I +61 152			0 21 47	+61 26.1	1	11.18		0.27		-0.48		2		41
		L 170 - 38			0 21 48	-56 35.	1	11.08		1.08		0.95		1		3062
		NGC 104 - 1761			0 21 48	-72 28.7	1	14.52		0.90		0.71		2		350
2041	-72 00034	IDS00196S7249		AB	0 21 48	-72 31.8	3	8.88	.011	0.48	.007	0.00		10	F6 IV/V	366,447,522
	+37 00059				0 21 49	+38 04.5	1	9.48		0.43		-0.03		1		1722
2026	-29 00106				0 21 49	-29 15.5	3	8.12	.012	0.14	.000	0.14	.014	13	A1 V	355,1068,2012
2014	-25 00126	G 266 - 106			0 21 50	-25 07.4	1	9.60		0.89		0.64		1	K2 V	3061
		NGC 104 - 4728			0 21 50	-72 12.4	3	11.70	.025	1.53	.009	1.69	.026	10		343,1468,3036
		NGC 104 - 1535			0 21 50	-72 36.3	1	15.43		0.88				3		529
		NGC 104 - 4729			0 21 51	-72 13.1	3	12.07	.019	1.53	.017	1.78	.012	12		343,350,3036
		NGC 104 - 4751			0 21 51	-72 13.8	1	12.04		1.54		1.86		2		3036
1994	+53 00066	NQ Cas			0 21 52	+54 01.0	1	9.71		1.81				1	C4.5	1238
2025	-27 00108	G 266 - 107			0 21 53	-27 18.3	3	7.93	.027	0.94	.015	0.74	.131	9	K2 V	258,2012,3061
		NGC 103 - 51			0 21 54	+61 07.2	1	11.50		1.29		1.15		2		260
		NGC 104 - 5705			0 21 54	-72 13.6	1	13.86		1.11				3		343
		NGC 104 - 13			0 21 54	-72 21.	1	12.49		1.42		1.61		5		343
		NGC 104 - 15			0 21 54	-72 21.	2	13.80	.050	1.06	.018	0.73	.025	4		350,3036
		NGC 104 - 16			0 21 54	-72 21.	2	14.61	.010	0.94	.034	0.59	.010	3		350,3036
		NGC 104 - 17			0 21 54	-72 21.	1	14.26		0.99		0.64		2		3036
		NGC 104 - 41			0 21 54	-72 21.	1	13.30		1.13		0.89		2		350
		NGC 104 - 45			0 21 54	-72 21.	1	13.42		1.08		0.81		2		350
		NGC 104 - 58			0 21 54	-72 21.	1	14.51		0.92		0.57		2		350
		NGC 104 - 61			0 21 54	-72 21.	1	14.47		0.98		0.67		3		350
		NGC 104 - 63			0 21 54	-72 21.	2	13.60	.015	1.08	.013	0.85	.036	11		350,3036
		NGC 104 - 66			0 21 54	-72 21.	1	12.58		1.29		1.22		2		350
		NGC 104 - 69			0 21 54	-72 21.	1	14.11		0.89		0.54		3		350
		NGC 104 - 89			0 21 54	-72 21.	1	14.36		0.94		0.74		2		350
		NGC 104 - 90			0 21 54	-72 21.	1	13.20		1.06		0.79		2		350
		NGC 104 - 92			0 21 54	-72 21.	1	12.84		1.24		1.21		3		3036
		NGC 104 - 100			0 21 54	-72 21.	1	14.15		0.80		0.39		2		350
		NGC 104 - 125			0 21 54	-72 21.	1	13.07		1.21		1.17		2		350
		NGC 104 - 130			0 21 54	-72 21.	1	14.50		0.94		0.55		3		350
		NGC 104 - 133			0 21 54	-72 21.	1	14.63		0.97		0.54		3		350
		NGC 104 - 139			0 21 54	-72 21.	1	12.67		1.32		1.19		4		343
		NGC 104 - 162			0 21 54	-72 21.	1	13.46		1.13		1.01		std		350
		NGC 104 - 167			0 21 54	-72 21.	1	13.88		1.03		0.77		2		343
		NGC 104 - 174			0 21 54	-72 21.	1	14.18		0.81		0.38		2		350
		NGC 104 - 182			0 21 54	-72 21.	1	14.28		1.07		0.90		3		350
		NGC 104 - 206			0 21 54	-72 21.	1	11.83		1.61		1.93		1		1468
		NGC 104 - 207			0 21 54	-72 21.	1	14.86		0.91		0.60		2		350
		NGC 104 - 212			0 21 54	-72 21.	1	11.88		1.52		1.62		3		350
		NGC 104 - 238			0 21 54	-72 21.	1	14.43		1.01		0.65		2		350
		NGC 104 - 244			0 21 54	-72 21.	1	14.04		0.87		0.47		2		350
		NGC 104 - 263			0 21 54	-72 21.	1	14.29		0.85		0.46		2		350
		NGC 104 - 265			0 21 54	-72 21.	1	13.95		0.90		0.54		2		350
		NGC 104 - 271			0 21 54	-72 21.	1	12.56		1.24		1.21		2		350
		NGC 104 - 275			0 21 54	-72 21.	1	14.93		0.92		0.66		2		350
		NGC 104 - 277			0 21 54	-72 21.	1	12.30		1.49		1.82		2		350
		NGC 104 - 284			0 21 54	-72 21.	1	12.11		1.47		1.64		2		350
		NGC 104 - 285			0 21 54	-72 21.	1	14.09		0.85		0.38		2		350
		NGC 104 - 286			0 21 54	-72 21.	1	14.18		0.73		0.46		2		350
		NGC 104 - 287			0 21 54	-72 21.	1	13.29		1.11				2		343
		NGC 104 - 290			0 21 54	-72 21.	1	14.45		0.93		0.52		2		350
		NGC 104 - 301			0 21 54	-72 21.	2	10.19	.019	0.76	.005	0.37		6		447,1468
		NGC 104 - 383			0 21 54	-72 21.	1	11.83		1.50		1.28		1		1468
		NGC 104 - 413			0 21 54	-72 21.	1	11.63		1.66		1.87		1		1468
		NGC 104 - 415			0 21 54	-72 21.	1	11.83		1.61		1.87		1		1468
		NGC 104 - 523		V	0 21 54	-72 21.	1	11.78		1.57		1.54		1		1468
		NGC 104 - 1421			0 21 54	-72 21.	3	12.04	.083	1.70	.019	1.75	.000	11		522,529,1582
		NGC 104 - 2303			0 21 54	-72 21.	1	10.60		0.46		-0.04		4		529
		NGC 104 - 2304			0 21 54	-72 21.	1	14.45		0.92				4		529
		NGC 104 - 3515			0 21 54	-72 21.	1	14.05		0.76		0.35		1		522
		NGC 104 - 5701			0 21 54	-72 21.	2	9.07	.024	0.75	.012	0.29		5		447,1468
		NGC 104 - 6407			0 21 54	-72 21.	2	12.81	.015	1.16	.005	1.10	.010	3		343,522
		NGC 104 - 6415			0 21 54	-72 21.	2	14.08	.000	0.86	.006	0.46	.030	7		366,939
		NGC 104 - 9002			0 21 54	-72 21.	1	16.12		0.75		0.20		3		529
		NGC 104 - 9004			0 21 54	-72 21.	1	17.51		0.54				2		529
		NGC 104 - 9005			0 21 54	-72 21.	1	17.18		0.61				1		529
		NGC 104 - 9007			0 21 54	-72 21.	1	17.48		0.53				1		529
		NGC 104 - 9008			0 21 54	-72 21.	1	16.66		0.74				3		529
		NGC 104 - 9009			0 21 54	-72 21.	1	18.18		0.44				2		529
		NGC 104 - 9010			0 21 54	-72 21.	1	16.98		0.76				1		529
		NGC 104 - 9011			0 21 54	-72 21.	1	15.80		0.66				2		529

HD	DM	Other Id	N Rem	α_{1950}	δ_{1950}	S	V	σ_V	B–V	σ_{B-V}	U–B	σ_{U-B}	n	Spectrum	References
		NGC 104 - 9015		0 21 54	−72 21.	1	15.47		0.85		0.39		3		529
		NGC 104 - 9017		0 21 54	−72 21.	1	17.48		0.78				1		529
		NGC 104 - 9018		0 21 54	−72 21.	1	17.54		0.62				1		529
		NGC 104 - 9019		0 21 54	−72 21.	1	18.24		0.66				2		529
		NGC 104 - 9020		0 21 54	−72 21.	1	17.96		0.41				1		529
		NGC 104 - 9021		0 21 54	−72 21.	1	17.93		0.60				1		529
		NGC 104 - 9022		0 21 54	−72 21.	1	18.84		0.74				2		529
		NGC 104 - 9024		0 21 54	−72 21.	1	17.12		0.58		0.01		1		529
		NGC 104 - 9025		0 21 54	−72 21.	1	16.18		0.65				4		529
		NGC 104 - 9030		0 21 54	−72 21.	1	16.08		0.86				4		529
		NGC 104 - 9032		0 21 54	−72 21.	1	16.36		0.71				4		529
		NGC 104 - 9035		0 21 54	−72 21.	1	18.10		0.57				1		529
		NGC 104 - 9042		0 21 54	−72 21.	1	18.92		0.75				3		529
		NGC 104 - 9043		0 21 54	−72 21.	1	17.75		0.65				1		529
		NGC 104 - 9045		0 21 54	−72 21.	1	17.01		1.64				1		529
		NGC 104 - 9051		0 21 54	−72 21.	1	11.85		1.59				3		529
		NGC 104 - 9052		0 21 54	−72 21.	1	12.98		1.23				4		529
		NGC 104 - 9066		0 21 54	−72 21.	1	17.32		0.57				1		529
		NGC 104 - 9067		0 21 54	−72 21.	1	17.66		0.61				2		529
		NGC 104 - 9068		0 21 54	−72 21.	1	18.40		0.71				1		529
		NGC 104 - 9072		0 21 54	−72 21.	1	17.48		1.24				1		529
		NGC 104 - 9073		0 21 54	−72 21.	1	16.66		0.78				2		529
		NGC 104 - 9074		0 21 54	−72 21.	1	17.96		0.60				2		529
		NGC 104 - 9075		0 21 54	−72 21.	1	18.26		0.53				1		529
		NGC 104 - 9076		0 21 54	−72 21.	1	17.28		1.02				1		529
		NGC 104 - 9077		0 21 54	−72 21.	1	16.44		0.91				2		529
		NGC 104 - 9087		0 21 54	−72 21.	1	15.89		0.63				2		529
		NGC 104 - 9088		0 21 54	−72 21.	1	17.07		1.79				1		529
		NGC 104 - 9089		0 21 54	−72 21.	1	16.51		0.50				2		529
		NGC 104 - 9091		0 21 54	−72 21.	1	18.39		0.59				2		529
		NGC 104 - 9092		0 21 54	−72 21.	1	17.75		0.62				2		529
		NGC 104 - 9093		0 21 54	−72 21.	1	17.01		0.76				2		529
		NGC 104 - 9200		0 21 54	−72 21.	1	18.80		0.67				1		529
		NGC 104 - 9201		0 21 54	−72 21.	1	18.66		0.71				1		529
		NGC 104 - 9203		0 21 54	−72 21.	1	18.94		0.64				1		529
		NGC 104 - 9204		0 21 54	−72 21.	1	19.05		0.85				1		529
		NGC 104 - 9206		0 21 54	−72 21.	1	19.00		0.78				1		529
		NGC 104 - 9208		0 21 54	−72 21.	1	18.68		0.68				1		529
		NGC 104 - 9502	V	0 21 54	−72 21.	1	9.98		1.47		0.88		1		1468
		NGC 104 - 9503	V	0 21 54	−72 21.	1	11.39		1.52		1.08		1		1468
		NGC 104 - 9505	V	0 21 54	−72 21.	1	11.69		1.65		1.84		1		1468
		NGC 104 - 9507	V	0 21 54	−72 21.	1	11.81		1.71		2.12		1		1468
		NGC 104 - 9511	V	0 21 54	−72 21.	1	12.06		1.67		1.61		1		1468
		NGC 104 - 9513	V	0 21 54	−72 21.	1	12.28		1.65		1.86		1		1468
		NGC 104 - 9516	V	0 21 54	−72 21.	1	11.79		1.64		1.86		1		1468
		NGC 104 - 9518	V	0 21 54	−72 21.	1	11.87		1.68		2.16		1		1468
		NGC 104 - 9528	V	0 21 54	−72 21.	1	11.99		1.64		1.75		1		1468
		NGC 104 - 5401		0 21 55	−72 03.2	3	13.87	.041	0.91	.013	0.50	.078	3		343,366,522
		NGC 104 - 5703		0 21 55	−72 14.5	2	12.53	.000	1.29	.010	1.27	.020	5		350,3036
		NGC 104 - 5702		0 21 55	−72 15.1	1	13.50		1.09		0.90		2		3036
		NGC 104 - 8501		0 21 55	−72 36.9	2	14.11	.019	0.83	.023	0.37		4		366,529
		G 130 - 68		0 21 56	+29 45.7	1	14.54		1.67				4		940
2023	−3 00049	HR 94		0 21 56	−02 29.7	4	6.06	.012	1.22	.008	1.21	.007	27	K1 III	15,361,1061,3016
		NGC 104 - 8503		0 21 56	−72 35.3	1	15.51		0.85				3		529
		LS I +62 096		0 21 58	+62 08.4	2	10.45	.024	0.41	.029	-0.29	.000	7		41,523
		NGC 104 - 8704		0 21 58	−72 27.9	3	11.92	.010	1.57	.007	1.99	.040	5		350,1468,3036
		NGC 104 - 8502		0 21 58	−72 36.7	1	13.91		0.88				3		529
2037	−27 00110			0 21 59	−27 11.6	2	8.35	.009	0.28	.005	-0.02		8	A5 IV	1068,2012
		NGC 104 - 5502		0 21 59	−72 08.5	3	14.09	.021	0.82	.025	0.40	.007	5		343,366,939
		NGC 103 - 37		0 22 00	+60 59.5	1	14.06		1.25		1.08		1		260
		LS I +55 009		0 22 01	+55 30.5	1	11.67		0.15		-0.62		2		41
2011	+61 00069	HR 93		0 22 01	+61 33.3	2	5.40		0.00	.010	-0.17		3	B9 III	1079,1118
		NGC 104 - 8705		0 22 01	−72 28.1	1	12.29		1.41		1.71		2		350
		NGC 104 - 8603		0 22 01	−72 34.0	1	14.07		0.80		0.27		1		366
		NGC 104 - 5704		0 22 02	−72 14.8	1	14.06		0.72		0.32		2		350
2072	−72 00037			0 22 02	−72 15.4	1	9.09		0.76				1	G5	366
		NGC 104 - 8706		0 22 02	−72 27.8	1	14.10		0.81		0.32		2		3036
2035	+13 00046			0 22 03	+14 02.3	1	6.46		1.19		1.12		2	K0	1648
	−25 00128			0 22 04	−24 48.1	1	10.02		0.59		0.10		2	G0	3061
		BPM 46283		0 22 06	−22 12.	1	13.15		0.81		0.27		1		3061
		WLS 100 75 # 6		0 22 09	+75 00.8	1	10.52		0.50		0.15		2		1375
	+61 00070			0 22 10	+62 05.0	1	8.71		1.36		1.44		5	K2 III	523
		NGC 104 - 5402		0 22 10	−72 03.1	1	14.58		0.65		0.15		1		343
		LS I +60 102		0 22 13	+60 05.1	1	10.80		0.45		-0.02		3		41
		NGC 103 - 54		0 22 13	+61 11.7	1	11.70		1.15		0.97		1		260
		NGC 104 - 5403		0 22 13	−72 03.2	1	14.62		0.91		0.39		1		343
		NGC 104 - 5404		0 22 13	−72 05.8	3	12.81	.013	1.08	.008	0.81	.015	13		343,366,522
		NGC 104 - 5717		0 22 14	−72 15.5	2	13.16	.000	1.16	.000	1.00	.005	3		350,3036
		NGC 104 - 8507		0 22 14	−72 36.2	1	14.07		0.79		0.37		2		522
		NGC 103 - 16	⋆ V	0 22 15	+61 02.1	1	13.93		0.45		0.01		1		260

Table 1   31

HD	DM	Other Id	N	Rem	α_{1950}	δ_{1950}	S	V	σ_V	B–V	σ_{B-V}	U–B	σ_{U-B}	n	Spectrum	References
		NGC 104 - 5405			0 22 15	−72 05.6	2	15.75	.040	0.88	.000	0.32	.040	2		343,366
		NGC 103 - 17			0 22 16	+61 01.7	1	14.38		0.47		0.09		1		260
2070	−51 00095				0 22 17	−51 19.0	4	6.81	.009	0.60	.004	0.12		15	G0 V	1020,1075,1311,2012
		NGC 104 - 8528			0 22 17	−72 33.9	1	13.72		0.92		0.48		1		522
		NGC 103 - 15			0 22 18	+61 02.6	1	13.86		0.37		-0.14		1		260
2068	−24 00128			A	0 22 18	−23 43.8	2	8.48	.019	0.43	.005	-0.04		6	F3 V	2012,3061
2068	−24 00128			B	0 22 18	−23 43.8	1	9.98		0.59		0.06		1		3061
		BPM 46289			0 22 18	−26 49.	1	15.32		1.30		1.21		1		3061
		NGC 103 - 14			0 22 19	+61 03.0	1	13.04		0.36		-0.14		1		260
	+44 00085	G 171 - 57			0 22 21	+45 24.8	1	9.61		1.06		1.00		1	K2	1658
2032	+62 00073				0 22 21	+62 55.7	1	8.70		0.58		0.56		2	A2	523
		NGC 104 - 8508			0 22 21	−72 35.9	1	13.80		0.87		0.44		2		522
		Kron 3 sq # 1			0 22 21	−73 04.1	1	11.28		0.66		0.18		4		1669
2055	+49 00072				0 22 22	+50 27.1	1	8.29		0.15		0.11		1	A0	1462
2054	+52 00061	HR 96			0 22 23	+52 46.2	2	5.76		-0.07	.004	-0.31	.004	4	B9 IV	985,1079
		NGC 104 - 8727			0 22 23	−72 28.0	1	13.92		0.80		0.38		2		350
		NGC 103 - 55			0 22 24	+61 09.7	1	11.92		0.37		0.21		1		260
		BPM 46291			0 22 24	−27 40.	1	11.12		0.55		-0.05		2		3061
2057	+47 00088	G 171 - 58			0 22 25	+47 46.2	1	7.52		0.59		0.15		2	F8	1733
	+63 00041	LS I +64 023			0 22 25	+64 23.5	1	10.46		0.81		-0.27		2	B0 III	1012
	+61 00071	LS I +62 097			0 22 26	+62 20.7	1	10.20	.005	0.20	.015	-0.60	.015	5		41,523
2077	−5 00060				0 22 26	−04 47.7	2	8.63	.005	0.11	.005	0.18	.040	10	A0	152,1775
		NGC 104 - 8726			0 22 26	−72 28.3	1	14.14		0.84		0.40		2		350
2064	+29 00062				0 22 27	+29 34.2	1	8.34		0.46		0.04		11	F5	1371
2043	+61 00072				0 22 27	+62 16.0	1	9.17		0.26		-0.07		3	A2	523
		NGC 104 - 8403			0 22 27	−72 38.0	1	15.82		0.68				2		529
		NGC 103 - 20			0 22 29	+61 00.5	1	16.91		0.99		0.39		1		260
		BPM 46295			0 22 30	−33 09.	1	12.73		1.38		1.20		1		3061
		NGC 103 - 19			0 22 31	+61 01.3	1	16.72		0.67		0.48		1		260
2098	−31 00154	G 267 - 98			0 22 32	−30 58.3	2	7.56	.010	0.62	.005	0.16	.005	2	G2 V	79,3061
		NGC 104 - 8729			0 22 32	−72 27.6	1	14.18		0.88		0.41		2		350
		NGC 103 - 22			0 22 34	+61 00.0	1	14.95		0.65		0.46		1		260
		G 217 - 48			0 22 35	+60 48.0	1	12.84		1.00		0.70		2		1658
		NGC 104 - 5406			0 22 35	−72 04.3	4	12.82	.008	1.29	.010	1.24	.031	6		343,366,522,528
		NGC 103 - 36			0 22 36	+60 52.8	1	14.60		1.99		2.18		1		260
		BPM 46298			0 22 36	−25 59.	1	12.90		0.64		0.05		2		3061
	+5 00048				0 22 37	+06 00.9	1	9.88		0.62		0.05		2	F8	1375
		NGC 104 - 5508			0 22 38	−72 08.6	1	14.15		0.87		0.35		5		1582
		LS I +60 103			0 22 39	+60 49.8	1	10.96		0.26		-0.42		3		41
		NGC 104 - 5723			0 22 39	−72 15.3	1	14.19		0.79		0.35		2		350
2084	+29 00064				0 22 40	+29 49.1	1	8.34		0.96		0.74		1	G8 II	8100
		NGC 103 - 11			0 22 40	+61 03.7	1	16.52		0.99		0.50		1		260
		NGC 103 - 26			0 22 41	+61 01.4	1	16.23		0.69		0.26		1		260
		BPM 46301			0 22 42	−25 15.	1	13.01		0.64		0.14		1		3061
		NGC 104 - 5510			0 22 42	−72 08.0	4	14.36	.021	0.72	.025	0.23	.034	9		343,366,522,1582
		NGC 103 - 8			0 22 44	+61 03.7	1	15.81		0.57		0.39		1		260
	+63 00043	LS I +64 024			0 22 44	+64 27.7	1	10.73		0.76		-0.30		2	B0.5V:n	1012
		NGC 103 - 56			0 22 45	+61 10.3	1	11.07		0.33		0.23		3		260
		Kron 3 sq # 2			0 22 45	−73 03.3	1	12.39		0.91		0.51		4		1669
		NGC 103 - 7			0 22 46	+61 03.3	1	14.94		0.57		0.13		1		260
		NGC 103 - 23			0 22 47	+61 00.5	1	13.96		0.85		0.32		1		260
		GR 8			0 22 47	−16 40.9	1	13.52		1.39		1.18		2		481
		NGC 104 - 5514			0 22 47	−72 06.9	1	14.15		1.00		0.63		5		1582
		NGC 104 - 5511			0 22 47	−72 07.9	1	15.10		0.96		0.53		5		1582
		NGC 104 - 5731			0 22 47	−72 16.5	1	14.22		0.83		0.44		2		350
		NGC 103 - 34			0 22 49	+60 56.3	1	9.91		1.19		0.95		std		260
		NGC 103 - 28			0 22 49	+61 01.8	1	14.54		0.36		-0.01		1		260
		AS Cas			0 22 49	+63 57.2	2	11.76	.046	1.17	.051	0.83		2		1462,1772
		NGC 104 - 8406			0 22 49	−72 37.2	1	12.37		1.43		1.57		2		522
2114	+1 00057	HR 97		⋆ A	0 22 50	+01 39.8	3	5.76	.007	0.86	.000	0.57	.008	8	G5 III	15,1256,1355
		CS 29497 # 7			0 22 50	−25 01.4	1	14.85		0.20		0.04		1		1736
		NGC 104 - 5411			0 22 50	−72 06.3	2	14.08	.002	0.88	.008	0.43	.004	8		366,939
		NGC 103 - 6			0 22 51	+61 03.5	1	16.28		0.81		0.41		1		260
		LS I +62 098			0 22 51	+62 08.2	1	11.76	.005	0.31	.010	-0.40	.020	6		41,523
		NGC 103 - 35			0 22 52	+60 53.0	1	11.74		0.22		-0.37		1		260
	+61 00073				0 22 52	+62 20.0	1	10.45		0.36		0.21		3		523
		NGC 104 - 5512			0 22 52	−72 08.1	2	15.12	.015	0.90	.030	0.68		10		529,1582
		NGC 104 - 5740			0 22 52	−72 14.8	1	15.06		0.87		0.68		2		350
		NGC 104 - 6702			0 22 53	−72 17.1	2	15.14	.019	0.86	.024	0.52	.161	3		350,3036
		NGC 103 - 2			0 22 54	+61 02.4	1	17.27		1.09		0.80		1		260
		NGC 103 - 57			0 22 54	+61 10.5	1	11.36		1.67		1.84		1		260
		BPM 46303			0 22 54	−22 58.	1	14.01		1.13		1.10		1		3061
		BPM 46304			0 22 54	−32 59.	1	11.19		1.37		1.18		1		3061
		NGC 104 - 5513			0 22 54	−72 08.5	3	14.16	.018	1.00	.021	0.74	.070	11		343,529,1582
		NGC 103 - 1			0 22 55	+61 02.8	1	13.27		0.39		0.00		1		260
		NGC 104 - 5412			0 22 55	−72 06.5	4	14.55	.023	0.95	.013	0.47	.032	6		343,366,522,529
		NGC 104 - 6701			0 22 55	−72 17.3	1	11.67		1.55		2.19		1		1468
2083	+71 00016				0 22 56	+71 31.8	1	6.89		0.03		-0.83		2	B1 V	1012
		LS I +62 099			0 22 57	+62 30.1	2	11.40	.010	0.51	.005	-0.32	.000	5		41,523
2131	−35 00125	G 267 - 101			0 22 57	−35 06.5	2	8.85	.005	0.51	.000	-0.04	.015	3	F7 V	79,3061

HD	DM	Other Id	N Rem	α_{1950}	δ_{1950}	S	V	σ_V	B–V	σ_{B-V}	U–B	σ_{U-B}	n	Spectrum	References
		NGC 104 - 8601		0 22 57	−72 32.2	1	14.12		0.80		0.38		1		522
2126	+33 00039			0 22 58	+33 50.4	2	8.23	.004	1.03	.004	0.76	.008	9	K0 III-IV	1003,1775
		NGC 104 - 5739		0 22 58	−72 15.1	1	12.40		1.41				3		529
		Kron 3 sq # 4		0 22 58	−73 04.2	1	14.63		0.84				4		1669
2102	+57 00077			0 22 59	+57 41.2	1	7.51		1.68				2	M1	369
		NGC 104 - 6703		0 22 59	−72 16.9	2	14.19	.054	0.82	.007	0.36	.000	3		350,3036
		NGC 103 - 33		0 23 00	+60 58.5	2	12.64	.010	0.26	.005	-0.26	.020	5		41,260
2141	−16 00065	G 266 - 110		0 23 00	−16 24.6	2	8.91	.008	0.65	.008	0.13	.017	7	G3 V	79,1775
		NGC 104 - 8417		0 23 01	−72 36.6	1	13.71		0.80		0.34		2		522
		Kron 3 sq # 6		0 23 03	−73 05.7	1	15.42		0.82				4		1669
		NGC 104 - 5408		0 23 04	−72 04.3	1	15.01	.075	0.92	.010	0.50	.035	2		343,366
		NGC 104 - 6717		0 23 04	−72 18.2	1	12.96		1.20		1.14		5		343
		Kron 3 sq # 5		0 23 05	−73 06.8	1	15.14		0.96				4		1669
2111	+63 00045			0 23 06	+63 34.6	2	8.79	.005	0.53	.005	0.04		3	F8	1118,1462
		BPM 46306		0 23 06	−22 49.	1	12.33		0.88		0.54		1		3061
		NGC 104 - 5409		0 23 06	−72 04.0	3	13.79	.021	1.10	.005	0.85	.036	3		343,366,522
		NGC 104 - 5520		0 23 06	−72 07.6	3	15.54	.061	0.89	.013	0.37	.060	3		343,366,522
2140	+6 00043			0 23 07	+07 24.9	1	6.78		1.30		1.51		2	K0	3077
		NGC 103 - 31		0 23 09	+60 58.8	1	14.85		1.65		1.29		1		260
		LS I +62 100		0 23 09	+62 15.0	2	11.14	.000	0.27	.005	-0.49	.005	5		41,523
	+61 00074	LS I +62 101		0 23 09	+62 16.4	4	9.56	.009	0.32	.000	-0.66	.008	9	O9 Vnn	41,523,1011,1012
	+62 00074			0 23 09	+62 52.4	1	10.57		0.34		0.07		4		523
2151	−77 00016	HR 98		0 23 09	−77 32.1	8	2.80	.005	0.62	.005	0.11	.006	21	G1 IV	15,1020,1034,1075*
		NGC 103 - 30		0 23 11	+60 59.4	1	15.68		1.06		1.05		1		260
2178	−22 00065			0 23 11	−21 54.5	2	7.64	.005	0.06	.000	0.05		8	A0/1 V	1068,2012
2181	−50 00076			0 23 11	−50 28.2	1	10.44		0.10		0.12		4	A1 V	152
	−72 00040			0 23 11	−71 44.5	1	11.84		0.52				1		1409
		NGC 104 - 5627		0 23 11	−72 11.2	4	12.48	.035	1.31	.011	1.32	.005	5		350,447,528,529
		NGC 103 - 32		0 23 13	+60 58.5	1	12.29		0.58		0.21		2		260
2155	+34 00052			0 23 14	+34 54.9	1	8.64		0.08		0.07		2		1336
		NGC 104 - 5521		0 23 14	−72 07.8	3	15.64	.015	0.85	.037	0.31	.044	3		343,366,522
		NGC 104 - 6742		0 23 14	−72 20.4	2	14.15	.000	0.77	.002	0.36	.038	4		350,3036
		NGC 104 - 7719		0 23 14	−72 24.4	1	14.61		0.95		0.69		3		350
2154	+34 00053			0 23 15	+35 01.8	1	8.03		0.15		0.00		2	A0	1336
		NGC 104 - 5414		0 23 15	−72 05.6	2	14.91	.015	0.90	.010	0.42	.029	3		343,522
		NGC 103 - 29		0 23 16	+61 02.7	2	11.09	.022	0.15	.004	-0.65	.013	12		41,260
	−15 00067			0 23 16	−14 45.2	1	11.42		0.96		0.75		1		1696
		NGC 104 - 8602		0 23 16	−72 31.7	1	12.84		1.25		1.20		1		522
2243	−75 00037			0 23 16	−74 40.9	1	8.74		1.08		0.95		2	G8 III	1117
	+62 00075			0 23 17	+63 02.3	1	9.76		0.58		0.37		4		523
		LS I +62 102		0 23 18	+62 05.8	2	10.75	.015	0.15	.005	-0.61	.020	5		41,523
	+61 00075			0 23 18	+62 22.0	1	10.57		0.56		0.00		4		523
		LS I +62 103		0 23 18	+62 59.6	2	12.26	.005	0.45	.010	-0.32	.025	7		41,523
		NGC 104 - 6743		0 23 18	−72 20.8	1	14.61		0.95		0.73		2		350
		NGC 104 - 6744		0 23 19	−72 20.9	1	13.95		1.03		0.93		2		350
	+61 00076			0 23 20	+62 13.1	1	9.66		0.50		0.05		3	F2	523
	+30 00057			0 23 21	+31 01.2	1	9.60		-0.11		-0.76		1	B9	963
		NGC 121 sq1 5		0 23 21	−71 54.6	1	11.81		0.89				1		1409
2216	−50 00078			0 23 22	−50 00.3	1	9.82		0.47		-0.05		1	F5 V	832
		Kron 3 sq # 3		0 23 22	−73 06.8	1	13.56		0.52		-0.05		4		1669
		Cas sq 1 # 18		0 23 23	+62 11.3	1	13.34		0.37		-0.20		5		523
		NGC 104 - 5522		0 23 23	−72 08.0	2	15.35	.020	0.41	.005	0.06	.045	4		343,522
		NGC 104 - 6718		0 23 23	−72 17.9	1	15.45		0.93		0.52		2		350
2189	+35 00069			0 23 24	+35 32.6	1	7.51		0.14		0.06		2	A3	1336
		LS I +62 104		0 23 24	+62 43.6	2	12.01		0.30	.030	-0.46	.030	7		41,523
		L 506 - 26		0 23 24	−31 28.8	1	12.22		1.04		0.89		1		1696
		NGC 104 - 6747		0 23 24	−72 20.7	2	12.03	.019	1.58	.029	1.73	.083	3		350,1468
		NGC 104 - 7718		0 23 24	−72 24.4	1	14.84		0.92		0.59		2		350
		NGC 104 - 8517		0 23 24	−72 35.0	1	12.62		1.29		1.27		3		522
2225	−31 00161			0 23 25	−31 28.5	1	10.02		1.52		1.63		4	M3 III	3007
		NGC 104 - 5525		0 23 26	−72 10.5	3	14.51	.022	0.93	.024	0.65	.005	4		343,366,529
2191	+19 00064			0 23 27	+19 52.2	1	6.68		-0.04		-0.07		3	A0 III	1501
		Cas sq 1 # 19		0 23 27	+62 24.9	1	14.22		0.32		0.04		5		523
		G 267 - 103		0 23 27	−31 27.5	1	12.20		1.02		0.83		2		3061
2170	+55 00072	IDS00208N5614	AB	0 23 28	+56 30.2	1	6.67		0.87		0.45		4	G5 III	1355
		Cas sq 1 # 20		0 23 28	+62 12.3	1	14.13		0.81		-0.24		5		523
		G 158 - 77		0 23 28	−10 53.5	1	13.52		1.37		1.17		1	sdM1	3062
2222	−28 00121			0 23 28	−28 33.0	1	8.90		0.68		0.23		2	G5 V	3061
2252	−71 00013			0 23 28	−71 30.1	1	10.21		0.48				1	F5	1409
	+62 00077			0 23 30	+62 52.2	1	10.29		0.59		0.17		2		523
		G 158 - 78		0 23 30	−10 54.4	1	16.22		0.36		-0.45		1	DA	3062
		Cas sq 1 # 22		0 23 31	+62 39.3	1	10.78		0.33		0.14		2		523
		NGC 104 - 7726	V	0 23 33	−72 23.5	3	11.80	.007	1.62	.004	1.88	.033	7		350,1468,3036
		NGC 104 - 8518		0 23 33	−72 37.5	1	13.11		1.10		0.81		2		522
		NGC 104 - 6752		0 23 34	−72 20.4	1	14.69		0.98		0.60		2		350
		NGC 104 - 8519		0 23 34	−72 35.1	1	14.06		0.80		0.34		2		522
	+61 00077			0 23 34	+62 29.0	2	9.58	.004	0.20	.004	-0.64	.013	5	B1 IV	523,1012
		CS 29527 # 8		0 23 36	−21 04.0	1	14.60		0.67		0.14		1		1736
		NGC 104 - 6605		0 23 36	−72 14.8	2	14.97	.005	-0.08	.010	-0.52	.029	3		343,366
		Florsch 13		0 23 36	−74 28.6	2	12.01	.041	0.51	.006	0.00	.013	13		767,1499

Table 1   33

HD	DM	Other Id	N Rem	α_{1950}	δ_{1950}	S	V	σ_V	B–V	σ_{B-V}	U–B	σ_{U-B}	n	Spectrum	References
2207	+50 00072	TU Cas		0 23 37	+51 00.2	5	7.18	.011	0.35	.012	0.32	.058	5	F3 II	1399,1462,1767,1772,6011
	+61 00079	LS I +61 154		0 23 37	+61 59.1	1	9.52		0.70		0.49		2	A1 Ib	41
		Cas sq 1 # 25		0 23 37	+62 25.6	1	14.01		0.28		0.11		5		523
2206	+61 00080			0 23 39	+62 05.8	1	8.65		0.40		-0.05		3	F0	523
		NGC 104 - 5645		0 23 40	-72 13.3	2	13.70	.041	1.09	.036	0.94		5		528,529
		HA 200 # 1173		0 23 40	-74 25.2	1	12.34		0.58		0.09		9		1499
	+61 00081	LS I +61 155		0 23 41	+61 57.5	1	10.22		0.65		0.31		2	A0 Ib	41
	+62 00079	LS I +63 073		0 23 42	+63 09.9	2	9.22	.000	0.41	.000	-0.56	.000	4	O9.5IV	1011,1012
		NGC 104 - 5527		0 23 42	-72 08.7	4	13.59	.023	1.08	.005	0.84	.021	9		343,366,522,529
		LS I +59 042		0 23 43	+59 44.8	1	12.29		0.67		-0.03		2		41
		Cas sq 1 # 24		0 23 43	+62 29.1	1	12.27		0.36		0.20		3		523
		NGC 121 sq1 14		0 23 43	-71 52.0	1	13.44		0.75				1		1409
		G 1 - 3		0 23 45	+01 07.0	1	13.81		1.18				2		202
2262	-44 00101	HR 100		0 23 45	-43 57.4	6	3.93	.010	0.17	.003	0.10	.007	23	A6 Vn	15,1075,2012,3023*
		NGC 121 sq1 6		0 23 46	-71 45.7	1	11.82		0.64			std			1409
		NGC 104 - 5417		0 23 46	-72 06.1	4	14.10	.020	0.85	.014	0.36	.020	12		343,366,522,939
	+60 00051	LS I +61 156		0 23 47	+61 08.2	1	9.23		0.62		0.23		2	A2 Ib	1012
	+62 00080	LS I +62 105		0 23 48	+62 55.7	2	11.38	.015	0.53	.010	-0.27	.020	6		41,523
		SB 171		0 23 48	-23 17.	1	14.60		-0.17		-0.56		2		3064
		BPM 46314		0 23 48	-31 21.	1	12.87		0.74		0.19		1		3061
2261	-42 00116	HR 99	AB	0 23 49	-42 34.6	8	2.39	.013	1.09	.005	0.89	.011	32	K0 IIIb	3,15,1020,1034,1075*
		Cas sq 1 # 27		0 23 50	+62 56.3	1	10.95		1.35		1.21		5		523
2268	-19 00058	V1171 Cyg		0 23 50	-18 58.2	1	6.25		1.61		1.89		4	M1 III	3007
	-32 00128	CS22882 # 35		0 23 50	-32 14.1	2	12.40	.000	-0.02	.000	0.01	.012	2	A0	966,1736
	+63 00048	LS I +64 025	★ AB	0 23 51	+64 09.3	1	9.12		0.88		-0.19		2	B1 III:enn	1012
		CS 29527 # 17		0 23 51	-19 07.1	1	15.04		0.04		0.10		1		1736
		Cas sq 1 # 29		0 23 52	+62 36.5	1	11.21		0.27		-0.17		4		523
2169	+79 00010			0 23 52	+79 46.5	2	6.56	.015	-0.10	.000	-0.33	.000	5	B9	985,1733
		NGC 104 - 5419		0 23 52	-72 07.4	4	14.03	.015	0.87	.010	0.35	.040	11		343,366,522,939
		NGC 104 - 5529		0 23 53	-72 10.0	7	11.90	.019	1.59	.009	1.85	.028	32		343,350,366,447,522*
		G 1 - 1		0 23 54	+05 07.9	2	14.83	.058	0.92	.019	0.63		3		202,1658
		NGC 104 - 5418		0 23 54	-72 06.0	1	15.33		0.87		0.31		2		343
		NGC 104 - 6614		0 23 56	-72 15.0	2	13.77	.023	1.03	.033	0.99		4		528,529
	+61 00082	IDS00212N6211	AB	0 23 57	+62 27.2	1	9.97		0.16		-0.45		5	B8	523
		NGC 104 - 8520		0 23 57	-72 34.1	1	13.71		0.78		0.40		2		522
	+29 00071	G 130 - 71		0 24 01	+29 55.5	1	10.05		1.00		0.93		1	K0	333,1620
	+61 00083			0 24 01	+62 26.3	1	10.34		0.22		-0.01		5	B9	523
		LP 645 - 25		0 24 02	-03 05.7	1	12.12		0.44		-0.24		2		1696
		Cas sq 1 # 31		0 24 03	+62 54.3	1	12.29		1.38		1.29		6		523
		LTT 17100		0 24 03	+69 52.5	2	10.51	.033	1.48	.005	1.21	.019	4		196,801
2273	-0 00063	HR 101		0 24 03	-00 19.6	2	6.18	.005	0.91	.000	0.55	.000	7	G8 III-IV	15,1256
		NGC 104 - 5421		0 24 03	-72 06.8	2	14.78	.000	0.99	.000	0.38	.000	2		343,522
		AA26,31 # 101		0 24 06	+62 51.	1	12.61		0.43		-0.24		2		483
		NGC 104 - 6616		0 24 06	-72 15.6	1	13.38		1.11		0.89		1		528
		Cas sq 1 # 33		0 24 07	+62 55.2	1	11.82		0.80		0.47		2		523
	+61 00084			0 24 09	+62 09.6	1	10.52		0.61		0.11		4		523
		Cas sq 1 # 35		0 24 09	+62 30.5	1	11.53		0.40		0.28		4		523
	-71 00015			0 24 09	-71 33.2	1	10.66		1.06				1		1409
		NGC 104 - 5308		0 24 09	-72 03.0	3	14.09	.012	1.03	.021	0.73	.027	3		343,366,522
		Cas sq 1 # 34		0 24 10	+62 34.0	1	11.98		0.58		0.13		3		523
		NGC 121 sq1 15		0 24 10	-71 51.9	1	14.38		0.73				3		1409
		NGC 121 sq1 8		0 24 11	-71 56.6	1	12.46		0.71				1		1409
		NGC 104 - 7504		0 24 12	-72 14.2	1	14.03		0.86		0.47		1		522
		NGC 104 - 5530		0 24 12	-72 14.5	1	13.14		1.06		0.73		3		522
		NGC 104 - 6513		0 24 12	-72 28.8	1	15.08		0.85				3		529
	+60 00052	NGC 103 - 59		0 24 13	+60 53.7	2	8.70	.015	1.13	.005	0.94	.005	9	G5	260,523
236416	+59 00056			0 24 14	+60 07.5	1	8.92		0.64				27	G5	6011
		NGC 104 - 6509		0 24 14	-72 39.4	1	13.61		1.12				1		529
		Cas sq 1 # 38		0 24 15	+62 54.1	1	11.99		0.72		0.25		3		523
		NGC 104 - 5309		0 24 16	-72 04.3	3	12.20	.007	1.48	.009	1.60	.040	7		343,366,522
	+60 00053	LS I +60 105		0 24 17	+60 46.7	1	9.52		0.14		-0.74		2	B2	41
	-23 00156			0 24 17	-23 13.6	1	10.68		0.38		-0.09		1	G0	3061
	-26 04343			0 24 17	-25 57.1	2	10.17	.014	-0.11	.011	-0.55	.006	6		540,976
		L 170 - 27		0 24 17	-55 41.6	1	15.14		0.17		-0.67		2		3065
		NGC 104 - 5422		0 24 17	-72 06.8	3	12.49	.013	1.39	.012	1.50	.014	8		343,366,522
		Cas sq 1 # 37		0 24 18	+62 55.2	1	12.66		0.56		0.21		5		523
		NGC 104 - 6502		0 24 18	-72 12.5	4	14.05	.018	0.79	.016	0.35	.008	11		343,366,522,529,939
	-33 00143	IDS00217S3253	A	0 24 19	-32 36.7	2	12.09	.029	1.00	.015	0.59	.024	6	K3:	1773,3062
		NGC 104 - 6501		0 24 20	-72 11.8	3	13.72	.008	1.07	.020	0.81	.040	5		343,366,522
		NGC 104 - 7502		0 24 20	-72 30.8	1	13.94		1.05		0.71		2		522
		G 1 - 2		0 24 21	+07 48.9	2	13.40	.015	1.54	.019	1.10		3		202,333,1620
		NGC 104 - 7601		0 24 21	-72 26.5	1	13.79		0.88		0.48		1		522
		NGC 104 - 7507		0 24 21	-72 28.0	1	13.80		1.07		0.73		2		522
		NGC 104 - 5427		0 24 22	-72 10.2	3	12.96	.024	1.24	.017	1.20	.022	4		343,522,528
2303	+17 00052			0 24 24	+17 37.6	1	8.71		0.54		0.11		2	F8	1733
		BPM 46318		0 24 24	-21 53.	1	13.05		0.68		0.10		1		3061
	-26 00134	G 266 - 114		0 24 24	-26 25.1	2	11.26	.190	0.95	.018	0.78	.028	2		1696,3061
		NGC 104 - 6511		0 24 24	-72 14.0	1	14.06		0.83				2		529
		NGC 104 - 7525		0 24 24	-72 32.5	1	13.90		0.65		0.02		2		522
	-26 00135			0 24 25	-26 04.8	1	10.22		0.55		-0.04		1	F5	3061

HD	DM	Other Id	N Rem	α_{1950}	δ_{1950}	S	V	σ_V	B–V	σ_{B-V}	U–B	σ_{U-B}	n	Spectrum	References
		NGC 121 sq1 10		0 24 25	−71 48.8	1	12.96		1.19				1		1409
		NGC 121 sq1 9		0 24 25	−71 56.8	1	12.70		0.56				1		1409
		NGC 121 sq1 12		0 24 26	−71 46.8	1	13.08		0.87				3		1409
2302	+24 00052			0 24 27	+24 45.9	2	6.66	.010	0.47	.000	0.05	.005	6	F7 IV	985,1501
		LS I +64 026		0 24 27	+64 25.8	2	11.20	.165	0.94	.055	−0.12	.010	4	O9.5V	342,7005
		WLS 24 50 # 7		0 24 28	+47 51.9	1	13.41		0.49		0.07		2		1375
	+37 00067			0 24 29	+38 25.1	1	9.76		0.32		0.06		1		1722
		NGC 104 - 6514		0 24 32	−72 14.9	2	14.34	.020	0.96	.020	0.56	.010	2		343,366
2339	−41 00102			0 24 34	−40 46.1	1	10.21		0.05		0.06		4	A0 V	152
2315	+24 00053			0 24 35	+25 18.1	1	7.49		1.19		1.26		3	K3 III	833
	+62 00081	LS I +62 106		0 24 35	+62 47.1	3	11.05	.005	0.35	.010	−0.51	.005	6	B2 III	41,523,1012
2333	−8 00065	IDS00220S0826	AB	0 24 35	−08 09.2	1	8.16		0.73		0.23		2	G5 V +K0 V	3030
		NGC 104 - 5310		0 24 35	−72 04.0	2	13.40	.010	0.64	.000	0.16	.020	4		343,522
		LS I +61 157		0 24 37	+61 24.1	1	11.57		0.23		−0.08		2		41
		Cas sq 1 # 40		0 24 38	+62 19.0	1	13.46		0.37		−0.26		5		523
		NGC 104 - 6504		0 24 38	−72 12.4	1	12.26		0.69		0.26		4		522
2349	−33 00148			0 24 41	−32 42.5	1	8.65		0.93		0.63		1	K0 IV	3061
2311	+61 00086			0 24 43	+62 27.4	1	8.96		0.11		0.07		4	A0	523
2386	−71 00016			0 24 43	−71 30.9	1	9.06		0.90				1	G2/5 III	1409
		NGC 104 - 6505		0 24 43	−72 12.7	1	15.69		0.76		0.34		2		343
		NGC 104 - 6531		0 24 43	−72 18.9	1	13.12		0.87				1		447
2363	−26 00138	HR 102		0 24 44	−25 49.4	1	5.98		1.03		0.88		2	K0 III	58
	−70 00019			0 24 44	−70 31.7	1	10.29		0.98		0.83		1		1696
		NGC 104 - 7503		0 24 44	−72 30.2	1	13.21		1.10		0.78		3		522
2343	+30 00060			0 24 45	+30 37.0	1	8.20		1.04		0.78		2	K1 III	1733
		Cas sq 1 # 41		0 24 45	+62 29.8	1	13.08		1.44		1.05		6		523
		NGC 104 - 8415		0 24 45	−72 34.7	1	14.07		0.84		0.36		1		522
2367	−37 00134			0 24 47	−36 40.9	1	9.08		1.55		1.67		4	M3 III	3007
		NGC 104 - 5312		0 24 47	−72 05.8	3	12.19	.023	1.50	.009	1.77	.029	6		343,366,522
	+60 00056	NGC 103 - 58		0 24 48	+61 06.7	3	9.30	.017	0.50	.009	0.03	.005	8	F8	260,523,1118
		NGC 104 - 8416		0 24 48	−72 34.9	1	13.43		1.15		1.09		1		522
		Cas sq 1 # 43		0 24 49	+62 29.8	1	12.55		0.73		0.26		5		523
	−35 00136			0 24 49	−34 57.8	1	10.46		0.64		0.18		1	G5	3061
		Cas sq 1 # 44		0 24 50	+62 24.3	1	12.62		0.98		0.63		6		523
		G 267 - 109		0 24 50	−31 22.5	1	12.42		1.38				1		3061
		NGC 104 - 7510		0 24 50	−72 27.1	1	13.12		1.13		0.95		3		522
		Cas sq 1 # 47		0 24 52	+62 27.2	1	13.09		0.28		−0.19		4		523
2342	+36 00056	AQ And		0 24 53	+35 18.5	1	7.92		3.79				1	N0	109
		Cas sq 1 # 46		0 24 53	+62 29.8	1	12.11		1.29		1.00		4		523
2329	+57 00085			0 24 54	+58 16.6	2	7.45	.012	0.08	.004	−0.50	.008	7	B3 V	1012,1775
		Cas sq 1 # 48		0 24 54	+62 24.3	1	11.90		2.06		2.17		6		523
2358	+15 00059	IDS00223N1528	A	0 24 55	+15 44.9	2	6.45	.005	0.23	.005	0.14	.000	4	A5	252,3016
2359	+13 00052			0 24 56	+14 09.0	2	8.45	.080	1.59	.012	1.99		4	K2	882,3040
		Cas sq 1 # 49		0 24 58	+62 23.6	1	10.97		1.49		1.44		6		523
		Cas sq 1 # 45		0 24 58	+62 30.1	1	11.77		1.38		1.26		6		523
2373	+6 00052			0 24 59	+06 57.5	1	8.83		0.09		−0.05		2	A0	1026
2383	−34 00147			0 24 59	−34 34.2	1	9.39		0.53		0.01		1	F6 V	3061
		SB 179		0 25 00	−32 01.	1	13.15		0.31		−0.02		1	A2:pec	295
		BPM 46325		0 25 00	−35 36.	1	10.67		0.72		0.23		1		3061
		NGC 104 - 6518		0 25 00	−72 15.9	2	13.94	.045	1.02	.025	0.72	.040	2		343,366
2381	−32 00136			0 25 01	−32 06.3	1	7.70		0.38		−0.04		1	F2 V	3061
	+61 00087	LS I +61 158		0 25 02	+61 59.6	1	9.50		0.56		0.33		2	A0 Ib	41
		Cas sq 1 # 50		0 25 04	+62 17.8	1	10.63		1.27		1.10		6		523
		NGC 104 - 7407		0 25 04	−72 30.9	1	14.08		0.85		0.40		2		522
	+4 00059	G 1 - 4		0 25 06	+04 33.8	2	10.68	.009	0.88	.005	0.56		4	K2	202,333,1620
		G 32 - 21		0 25 06	+09 19.6	1	14.66		1.57		1.10		1		1773
		G 1 - 5		0 25 06	+09 19.8	1	14.11		1.55		1.07		2		1773
2395	−20 00067			0 25 06	−20 24.7	2	6.78	.014	0.26	.028	0.13		8	A9 V	1068,2012
		Florsch 21		0 25 06	−72 57.6	1	11.47		0.43		−0.16		3		767
236419	+59 00061	LS I +60 106		0 25 07	+60 20.9	2	9.15	.004	0.31	.000	−0.51	.000	11	B2 III	41,1012
		Cas sq 1 # 52		0 25 07	+62 24.5	1	13.14		0.43		0.24		4		523
		LS I +63 074		0 25 07	+63 49.4	1	10.87		0.85		0.23		2	B7 (III):	1012
	−73 00018			0 25 07	−72 57.2	1	11.36		0.49		−0.17		2		1696
		Cas sq 1 # 51		0 25 08	+62 39.2	1	11.70		0.66		0.53		4		523
		NGC 121 sq1 16		0 25 08	−71 45.8	1	14.76		0.68				4		1409
2368	+52 00073			0 25 09	+52 40.7	1	7.78		0.45		−0.06		2	F5	1566
		NGC 104 - 7409		0 25 09	−72 29.4	1	14.10		1.02		0.67		2		522
2391	+12 00036			0 25 10	+13 25.7	1	8.73		1.21				3	K2	882
		BPM 46328		0 25 12	−19 55.	1	11.36		0.91		0.67		1		3061
	+62 00082			0 25 13	+62 56.6	1	10.37		0.35		0.27		3		523
2404	−33 00151	G 267 - 110		0 25 13	−32 39.8	2	9.02	.005	0.75	.000	0.21	.025	2	G6 V	79,3061
		NGC 104 - 6402		0 25 13	−72 10.6	4	14.04	.051	0.84	.034	0.41	.043	6		343,366,522,939
		Cas sq 1 # 54		0 25 14	+62 53.2	1	12.75		1.52		1.26		6		523
2415	−30 00127			0 25 15	−29 47.4	3	11.11	.016	0.02	.012	0.00	.006	22	B9/A0 V	152,1068,1775
		G 31 - 52		0 25 16	−06 45.1	2	11.59	.054	1.22	.000	1.14		3	K4	202,3079
		NGC 104 - 7408		0 25 16	−72 30.2	1	14.51		0.75		0.25		2		522
		LS I +60 107		0 25 18	+60 49.6	1	12.81		0.31		−0.54		3		41
		NGC 104 - 6408		0 25 18	−72 14.9	1	12.81		1.29		1.12		1		522
		Cas sq 1 # 55		0 25 19	+62 52.4	1	13.21		0.54		0.01		5		523
		NGC 104 - 6301		0 25 20	−72 08.0	3	14.28	.008	1.00	.013	0.60	.016	5		343,366,522

Table 1 35

HD	DM	Other Id	N Rem	α_{1950}	δ_{1950}	S	V	σ_V	B–V	σ_{B-V}	U–B	σ_{U-B}	n	Spectrum	References
		LS I +61 159		0 25 21	+61 25.1	1	11.67		0.32		-0.50		2		41
		Cas sq 1 # 56		0 25 21	+62 29.9	1	12.12		0.53		0.52		4		523
		LS I +63 075		0 25 21	+63 51.3	1	11.11		0.99		-0.20		2	B1 V	1012
2410	+18 00051			0 25 22	+19 14.3	1	6.38		1.00		0.78		3	K0	1648
		Cas sq 1 # 57		0 25 22	+62 11.7	1	10.80		2.09		2.35		6		523
		NGC 121 sq1 13		0 25 22	-71 49.3	1	13.42		0.64				4		1409
		CS 29527 # 16		0 25 23	-19 14.3	1	13.98		0.07		0.17		1		1736
		BPM 46331		0 25 24	-30 38.	1	12.60		0.94		0.59		2		3061
		BPM 46332		0 25 24	-32 21.	1	13.95		0.63		-0.02		1		3061
	+62 00083			0 25 25	+62 43.0	1	10.42		0.42		0.18		4		523
2411	+17 00055	HR 103, TV Psc		0 25 26	+17 37.0	2	4.92	.196	1.65	.005	1.80	.033	7	M3 III	3051,8032
	+55 00081	LS I +56 007		0 25 26	+56 07.2	2	10.07	.000	0.07	.000	-0.65	.009	5	B1.5V:e:nn	41,1012
2429	-33 00152	HR 105, η Scl		0 25 27	-33 17.0	5	4.82	.021	1.64	.005	1.80	.004	20	M4 III	15,1075,2012,3007,8015
		NGC 104 - 6418		0 25 27	-72 20.3	1	13.32		1.08				1		447
2438	-12 00072	AG Cet		0 25 28	-11 56.1	1	7.25		1.42				4	M5 III	2009
2466	-72 00041			0 25 28	-72 09.5	5	8.50	.015	0.50	.013	0.00	.000	28	F8 IV/V	366,447,522,529,1582
		NGC 104 - 6419		0 25 28	-72 21.1	1	14.57		0.82				1		447
		Cas sq 1 # 58		0 25 30	+62 44.1	1	12.48		0.54		0.34		2		523
		BPM 46335		0 25 30	-28 17.	1	13.05		0.86		0.56		1		3061
		Cas sq 1 # 63		0 25 31	+62 20.4	1	10.88		1.21		1.03		4		523
2421	+43 00092	HR 104		0 25 32	+44 07.1	2	5.19	.014	0.04	.006	-0.01		4	A2 Vs	1363,3023
		Cas sq 1 # 60		0 25 34	+62 42.6	1	11.58		0.27		-0.46		2		523
		Cas sq 1 # 64		0 25 36	+62 08.9	1	12.39		0.28		-0.29		3		523
2436	+15 00063	HR 106		0 25 37	+16 10.1	1	6.06		1.58		2.00		2	K5 III	252
		Cas sq 1 # 62		0 25 37	+62 30.0	1	12.63		0.28		-0.27		4		523
2434	+34 00059			0 25 39	+35 01.3	1	8.88		0.30		0.19		2	A5	1336
		Cas sq 1 # 61		0 25 39	+62 34.8	1	12.32		1.03		0.50		5		523
		Cas sq 1 # 68		0 25 40	+62 30.2	1	10.91		1.35		1.25		5		523
		Cas sq 1 # 66		0 25 40	+62 50.4	1	10.74		0.47		0.04		2		523
		BPM 46340		0 25 42	-22 27.	1	13.21		0.75		0.22		1		3061
		Cas sq 1 # 65		0 25 43	+62 55.9	1	10.91		1.37		1.34		4		523
		BD +61 00090a		0 25 44	+62 26.8	2	10.19	.009	0.18	.000	-0.69	.014	8		41,523
	+61 00090	LS I +62 107		0 25 44	+62 26.8	1	10.02		1.31		1.31		6		523
2505	-72 00042			0 25 44	-72 24.4	4	9.56	.015	1.12	.017	1.06	.008	14	K0	366,447,522,1582
2454	+9 00047	HR 107		0 25 45	+09 55.0	6	6.04	.012	0.43	.005	-0.06	.019	28	F6 Va	15,253,254,361,1256,3013
		Cas sq 1 # 73		0 25 45	+62 14.1	1	12.57		0.29		-0.32		4		523
		Cas sq 1 # 67		0 25 45	+62 30.9	1	13.53		0.33		-0.16		6		523
		GR 9		0 25 45	-32 44.7	1	15.16		1.66		0.81		2		481
236422	+56 00063			0 25 46	+56 52.8	1	9.31		1.48		1.63		1		1748
		Cas sq 1 # 72		0 25 46	+62 17.6	1	13.17		0.71		0.23		5		523
		Cas sq 1 # 70		0 25 48	+62 52.8	1	12.41		2.07		2.22		6		523
		Cas sq 1 # 71		0 25 49	+62 28.6	1	11.28		0.62		0.09		3		523
		G 242 - 56		0 25 49	+80 38.1	1	11.37		1.15		1.00		2	K5	7010
2453	+31 00059	GR And		0 25 50	+32 09.7	4	6.90	.015	0.08	.013	0.04	.006	11	A2p	220,603,1063,1202
2475	-21 00057	HR 108	★ AB	0 25 51	-20 36.6	7	6.41	.010	0.59	.013	0.02	.044	24	G3 V	15,176,1075,2018,2024*
		NGC 104 - 7420		0 25 52	-72 21.6	1	11.73		0.52				1		447
		Cas sq 1 # 78		0 25 53	+62 17.5	1	10.77		1.86		1.82		6		523
		BPM 46341		0 25 54	-25 53.	1	12.83		0.40		-0.22		1		3061
		BPM 46342		0 25 54	-29 20.	1	12.19		0.88		0.38		1		3061
		Cas sq 1 # 79		0 25 55	+62 13.4	1	10.58		1.23		1.11		3		523
		Cas sq 1 # 75		0 25 55	+62 34.4	1	12.12		0.41		-0.27		6		523
236423	+59 00063			0 25 56	+60 19.8	1	9.00		0.47				2	F5	1118
	+61 00091	LS I +62 108		0 25 56	+62 22.8	2	10.71	.024	0.14	.019	-0.64	.005	6		41,523
		Cas sq 1 # 76		0 25 56	+62 25.3	1	12.23		0.22		-0.49		2		523
		NGC 104 - 6421		0 25 56	-72 21.3	1	14.01		0.82				1		447
2461	+38 00054	IDS00233N3832	AB	0 25 57	+38 48.8	2	8.34	.000	0.96	.000	0.68	.005	5	K0 III	1501,1601
		Cas sq 1 # 80		0 25 57	+62 12.8	1	14.57		0.36		-0.12		4		523
		NGC 121 sq1 11		0 25 57	-71 58.5	1	13.04		0.53				1		1409
2472	+32 00069			0 25 58	+33 21.5	1	7.47		1.47		1.71		2	K0	1625
		G 217 - 52		0 25 59	+57 39.4	1	14.45		0.85		0.36		2		1658
2451	+61 00092	LS I +62 109		0 25 59	+62 13.8	3	8.74	.008	0.19	.004	-0.72	.007	14	B0.5IV	41,523,1012
2478	-30 00132			0 25 59	-30 23.3	1	9.30		0.55		0.04		2	G0 V	3061
2490	-40 00093	HR 109		0 25 59	-40 11.5	6	5.42	.005	1.56	.005	1.90	.000	22	M0 III	15,1075,2013,2028*
		Hiltner 33		0 26 00	+62 48.	2	10.49	.000	0.38	.009	-0.61	.009	4	B0(V)p(e)	523,1012
2488	-12 00075	IDS00235S1147	A	0 26 00	-11 30.8	1	6.87		0.53		0.08		1	F8 V	1738
		Cas sq 1 # 82		0 26 01	+62 31.1	1	11.44		0.38		0.23		2		523
		Steph 41		0 26 01	-02 56.4	1	11.31		1.36		1.25		1	K7	1746
		Cas sq 1 # 85		0 26 02	+62 09.2	1	13.10		1.07		0.46		5		523
		Cas sq 1 # 81		0 26 02	+62 45.4	1	12.44		1.94		2.05		6		523
		Cas sq 1 # 84		0 26 03	+62 09.8	1	12.53		0.65		0.18		6		523
	-17 00063			0 26 04	-16 30.0	1	9.63		1.14		1.17		1	K5 V	3072
		LP 705 - 38		0 26 06	-11 11.5	1	12.34		1.00		0.84		1		1696
	-28 00138	G 267 - 113		0 26 06	-28 33.7	1	11.76		0.81		0.40		1		3061
		NGC 104 - 7320		0 26 07	-72 25.6	1	11.92		1.76		1.91		1		522
2469	+56 00065			0 26 08	+56 56.0	1	7.28		0.92		0.55		1	G8 III-I	1748
		G 31 - 53		0 26 08	-06 55.9	2	12.16	.058	1.49	.005	1.05		3		202,3079
		NGC 104 - 7327		0 26 08	-72 22.8	2	14.00	.002	0.78	.003	0.33	.017	7		366,939
		NGC 104 - 7313		0 26 08	-72 29.4	1	14.13		0.81		0.37		1		366
		Cas sq 1 # 92		0 26 09	+62 09.1	1	13.43		1.83				5		523
		Cas sq 1 # 91		0 26 09	+62 13.7	1	12.78		0.64		0.15		6		523

HD	DM	Other Id	N Rem	α_{1950}	δ_{1950}	S	V	σ_V	B–V	σ_{B-V}	U–B	σ_{U-B}	n	Spectrum	References
		Cas sq 1 # 89		0 26 09	+62 19.3	1	11.82		1.94		2.10		6		523
		Cas sq 1 # 87		0 26 09	+62 44.8	1	11.79		1.38		1.16		6		523
		Cas sq 1 # 86		0 26 09	+62 45.9	1	13.37		0.62		0.29		6		523
		Cas sq 1 # 90		0 26 10	+62 18.2	1	11.83		0.72		0.33		5		523
		NGC 104 - 7204		0 26 11	−72 37.4	2	14.05	.002	0.82	.002	0.40	.030	7		366,939
		Cas sq 1 # 88		0 26 12	+62 23.7	1	11.70		0.28		0.16		3		523
		CS 29527 # 19		0 26 12	−18 40.5	1	12.25		-0.05		-0.17		1		1736
		NGC 104 - 7311		0 26 12	−72 29.7	1	14.08		0.82		0.33		1		366
	+60 00057	LS I +60 108		0 26 14	+60 37.7	1	10.06		0.87		0.74		3	A7 II	41
		NGC 104 - 6209		0 26 14	−72 11.9	1	12.48		0.79				1		447
		Cas sq 1 # 94		0 26 15	+62 08.8	1	13.44		1.45				6		523
		WD0026+136		0 26 16	+13 38.1	1	15.91		-0.12		-1.10		2	DA1	1727
		LS I +62 110		0 26 16	+62 08.4	2	10.28	.005	0.26	.000	-0.41	.005	4		41,523
		MtW 44 # 9		0 26 17	+30 26.6	1	15.12		0.61		-0.01		4		397
2507	+36 00066	HR 110		0 26 17	+36 37.4	1	6.26		0.92				2	G5 III	71
		LS I +56 008		0 26 17	+56 24.7	1	10.75		0.10		-0.63		2		41
		Cas sq 1 # 96		0 26 17	+62 07.9	1	11.40		0.58		0.34		2		523
2529	−51 00113	HR 111		0 26 18	−50 48.5	1	6.26		1.09				4	K0 III	2009
		MtW 44 # 18		0 26 19	+30 23.6	1	10.88		1.43		1.58		4		397
		MtW 44 # 16		0 26 19	+30 24.6	1	13.05		0.57		0.02		4		397
		MtW 44 # 17		0 26 19	+30 25.5	1	16.07		0.42		-0.05		4		397
	+60 00058	LS I +60 109		0 26 19	+60 42.7	1	10.09		0.24		-0.67		2	B2	41
		Cas sq 1 # 93		0 26 21	+62 50.4	1	9.94		2.13		2.17		6		523
2527	−25 00155	IDS00238S2512	AB	0 26 21	−24 54.8	1	7.13		0.31				4	A9 V	2012
		Cas sq 1 # 100		0 26 22	+62 09.0	1	10.88		1.36		1.14		6		523
		NGC 104 - 6311		0 26 22	−72 16.3	1	14.08		0.85		0.45		1		366
		Cas sq 1 # 97		0 26 24	+62 49.4	1	11.44		1.29		1.07		6		523
		G 31 - 54		0 26 24	−02 43.6	1	13.25		1.54				2		202
		NGC 104 - 7326		0 26 24	−72 23.0	1	13.06		1.07				1		447
2525	−2 00059			0 26 25	−01 49.0	1	10.11		0.61		0.12		2	G0	1696
2547	−49 00099			0 26 25	−49 22.5	1	10.14		0.42		-0.01		1	F2 V	832
		HA 44 # 28		0 26 26	+30 06.5	1	11.33		0.74		0.22		9		1499
		Cas sq 1 # 99		0 26 26	+62 10.2	1	12.07		0.33		-0.17		3		523
		Cas sq 1 # 98		0 26 27	+62 38.6	1	11.42		1.18		1.14		6		523
		G 267 - 114		0 26 28	−31 25.8	1	14.10		0.71		0.00		2		3061
		NGC 104 - 7208		0 26 28	−72 33.8	2	14.04	.030	0.82	.040	0.37	.010	2		366,522
		NGC 188 - 5014		0 26 29	+84 58.1	1	12.27		1.47		1.62		1		4002
		Cas sq 1 # 102		0 26 30	+62 08.1	1	11.63		0.28		-0.40		3		523
2588	−71 00018	IDS01458N1232	AB	0 26 30	−71 34.1	1	10.27		0.65				1	G0	1409
		LS I +60 110		0 26 31	+60 15.7	1	10.70		0.54		0.13		3		41
		SB 191		0 26 32	−23 56.6	3	14.04	.017	0.05	.026	0.19	.022	4	A0	295,1736,3064
2537	+28 00071			0 26 33	+29 16.3	1	9.02		0.19		0.15		11	A2	1371
		Cas sq 1 # 101		0 26 33	+62 31.2	1	11.50		0.56		0.06		2		523
2555	−28 00141			0 26 33	−27 40.1	1	9.24		0.87		0.50		1	K1 III	3061
2558	−50 00096			0 26 33	−50 19.7	1	9.78		0.19		0.12		4	A4 V	152
		BPM 46351		0 26 36	−23 11.	1	14.15		0.88		0.38		1		3061
		NGC 104 - 7212		0 26 36	−72 31.3	2	14.11	.000	0.86	.030	0.37	.035	2		366,522
2553	−12 00077			0 26 38	−11 52.5	1	7.34		1.62		1.95		3	K5 III	1657
		CS 22882 # 31		0 26 38	−30 39.1	2	15.27	.000	-0.16	.000	-0.52	.014	2		966,1736
236424	+56 00066			0 26 39	+56 40.2	1	8.31		0.99		0.48		1		1748
		LS I +62 111		0 26 39	+62 06.1	2	11.08	.005	0.34	.014	-0.40	.024	7		41,523
		CS 29527 # 15		0 26 40	−19 26.7	1	14.24		0.37		-0.21		1		1736
		Cas sq 1 # 103		0 26 41	+62 24.6	1	11.73		1.12		0.83		5		523
2552	+28 00072			0 26 42	+28 33.1	1	7.31		1.47				4	K3 III	20
		BPM 46352		0 26 42	−23 25.	1	13.78		0.51		-0.01		1		3061
		TonS 159		0 26 42	−30 39.	1	15.17		-0.11		-0.61		2		3064
		BPM 46353		0 26 42	−30 44.	1	11.66		1.19		1.05		2		3061
		BPM 46354		0 26 42	−31 13.	1	15.11		0.59		-0.07		1		3061
		NGC 104 - 7216		0 26 43	−72 29.8	1	14.07		0.86		0.39		1		366
2551	+29 00085			0 26 44	+30 27.5	1	8.81		0.38		-0.01		4	A5	397
2582	−25 00162	G 266 - 119		0 26 44	−25 12.3	3	9.08	.008	0.68	.000	0.24	.017	8	G3 V	79,1775,3061
		NGC 104 - 7219		0 26 44	−72 27.0	3	14.01	.008	0.89	.015	0.40	.015	8		366,522,939
		Cas sq 1 # 104		0 26 45	+62 30.7	1	10.73		0.28		0.04		2		523
		CS 22882 # 32		0 26 45	−31 40.0	2	15.41	.000	0.08	.000	0.23	.010	2		966,1736
2585	−38 00138	T Scl		0 26 45	−38 11.1	1	10.91		1.59		1.03		1	K2 (III)	975
		SB 192		0 26 46	−31 42.3	3	14.21	.014	0.07	.010	0.15	.015	5		295,966,1736
2535	+61 00094			0 26 47	+61 47.3	1	7.00		1.06		1.02		3	K2 III	37
		AJ84,127 # 2240		0 26 48	+38 26.	1	11.03		1.14		1.09		1		1510
		MCC 359		0 26 48	+38 27.0	1	10.68		1.08				1	K8	1017
		BPM 46357		0 26 48	−32 45.	1	13.77		1.53				1		3061
		SB 194		0 26 48	−33 07.	1	13.42		0.19		0.10		1	A0	295
	+61 00095			0 26 49	+62 27.7	1	9.62		0.51		0.01		3		523
2584	−34 00159			0 26 49	−34 22.7	1	9.42		0.61		0.12		1	G1 V	3061
		NGC 129 - 173		0 26 50	+59 52.1	3	11.14	.011	0.28	.021	0.17	.012	4		49,88,1756
	−3 00056	G 31 - 55		0 26 50	−02 36.5	3	10.65	.035	0.56	.007	-0.13	.005	5		202,1620,1696
		LP 465 - 26		0 26 51	+12 36.7	1	11.54		0.94		0.81		1		1696
		NGC 104 - 7220		0 26 53	−72 26.7	1	12.53		1.27		1.27		1		522
		NGC 104 - 7217		0 26 53	−72 29.4	1	13.70		1.11		0.90		1		522
2597	−46 00127			0 26 54	−45 57.2	1	8.70		0.27		0.00		4	A7/8 V	152
		NGC 129 - 191		0 26 55	+59 54.5	2	15.07	.060	0.65	.020	0.60	.170	2		88,1756

Table 1

HD	DM	Other Id	N Rem	α_{1950}	δ_{1950}	S	V	σ_V	B–V	σ_{B-V}	U–B	σ_{U-B}	n	Spectrum	References
		NGC 129 - 181		0 26 55	+59 54.9	2	13.94	.010	1.57	.020	1.31		2		88,1756
		Cas sq 1 # 109		0 26 56	+62 05.8	1	12.02		0.33		-0.37		3		523
2520	+74 00014			0 26 56	+74 58.0	3	8.41	.005	0.62	.002	0.05	.023	5	K0	979,3026,8112
		NGC 129 - 189		0 26 57	+59 53.0	1	14.77		1.69				1		88
		LS I +62 112		0 26 58	+62 24.8	2	11.57	.010	0.24	.005	-0.56	.024	7		41,523
		HA 44 # 113		0 26 59	+30 06.7	1	11.71		1.21		1.03		9		1499
		Cas sq 1 # 108		0 26 59	+62 47.4	1	10.63		0.35		-0.34		2		523
	+62 00089	LS I +62 113		0 26 59	+62 47.5	1	10.15		0.49		-0.56		2	B0 V	41
		G 266 - 120		0 26 59	-20 05.1	1	11.35		0.80		0.43		1		3061
		0027 259		0 27 00	+25 53.9	1	11.22		0.03		-0.08		2		1727
		NGC 129 - 193		0 27 00	+59 53.8	1	15.27		0.84		0.85		1		88
		Cas sq 1 # 107		0 27 00	+62 52.8	1	11.60		0.56		0.11		2		523
		NGC 188 - 5015		0 27 00	+84 47.5	1	10.91		1.63		1.78		1		4002
2559	+56 00067			0 27 01	+56 45.3	1	7.50		0.03		-0.34		1	B9	1748
		NGC 129 - 176		0 27 02	+59 55.7	2	12.80	.010	0.63	.010	0.11	.045	4		88,1756
2577	+60 00060			0 27 03	+60 54.3	1	9.10		0.28				2	A2	1118
	+19 00075	IDS00245N1945	AB	0 27 04	+20 01.4	1	10.11		0.57		0.05		3	F2	3032
		NGC 129 - 182		0 27 04	+59 56.6	2	14.12	.107	0.71	.097	0.34	.078	3		88,1756
		CS 29497 # 12		0 27 05	-23 27.6	1	10.74		0.39		-0.03		1		1736
2613	-23 00173	IDS00246S2344	AB	0 27 05	-23 27.7	1	10.28		0.32		0.02		5	A7 III	1068
		NGC 129 - 125		0 27 06	+59 55.9	3	11.78	.009	0.43	.016	-0.14	.005	6		49,88,1756
2615	-33 00163	G 267 - 15		0 27 06	-32 50.9	1	7.61		0.43		-0.03		1	F5 V	79
		GD 615		0 27 07	-32 47.1	1	12.22		0.43		-0.22		1		3061
		NGC 129 - 121		0 27 08	+59 54.6	1	12.15		0.43		0.00		2	B7 V	88
		LS I +60 111		0 27 08	+60 04.6	1	11.33		0.63		-0.38		3		41
	+62 00090			0 27 09	+62 46.6	1	10.32		0.34		0.22		3		523
236429	+59 00065	NGC 129 - 200	⋆ V	0 27 10	+59 56.2	6	8.72	.020	1.08	.026	0.75	.038	6	G1 Ib-II	41,49,851,1399,1772,6011
		LS I +62 114	⋆ AB	0 27 10	+62 33.8	2	12.33	.010	0.60	.000	-0.35	.010	7		41,523
		NGC 129 - 111		0 27 11	+60 01.1	2	10.91	.009	0.51	.005	0.32	.005	4		49,1756
		NGC 129 - 115		0 27 12	+59 52.8	2	11.43	.010	0.74	.005	0.25	.005	3		49,1756
		SB 198		0 27 12	-20 57.	1	13.09		0.22		0.09		2	A5	295
		G 31 - 56		0 27 13	+00 44.6	2	12.34	.029	1.04	.019	0.75		3		202,1620
		NGC 129 - 195		0 27 13	+59 51.4	1	16.40		1.17				1		88
		NGC 129 - 113		0 27 13	+59 54.8	2	13.82	.044	0.60	.005	0.34	.044	3		49,1756
		LS I +62 115		0 27 13	+62 33.3	2	12.46	.005	0.63	.005	-0.34	.005	6		41,523
		NGC 129 - 186		0 27 17	+59 49.8	2	14.62	.034	0.70	.029	0.50	.083	3		88,1756
		Cas sq 1 # 116		0 27 17	+62 46.0	1	13.40		0.48		-0.11		5		523
236433	+59 00067	NGC 129 - 170		0 27 18	+59 52.5	6	8.87	.008	0.95	.018	0.67	.009	47	F5 Ib	41,49,88,851,1756,6011
		Florsch 33		0 27 18	-72 57.6	1	11.16		0.52		-0.05		3		767
		Cas sq 1 # 117		0 27 19	+62 30.2	1	11.49		0.24		-0.41		2		523
		NGC 129 - 174		0 27 20	+59 53.6	3	12.58	.024	0.46	.021	0.08	.015	4		49,88,1756
		NGC 129 - 105		0 27 20	+59 55.5	3	11.14	.004	0.38	.022	-0.16	.007	5	B9 V	49,88,1756
		Cas sq 1 # 119		0 27 20	+62 29.5	1	11.15		1.14		0.84		4		523
	+62 00091			0 27 20	+62 57.0	1	10.16		0.48		0.00		3		523
2630	-15 00084	HR 115		0 27 20	-15 08.4	3	6.14	.019	0.38	.004	-0.02		9	F2/3 V	15,2012,3053
2632	-32 00154	HR 116		0 27 20	-32 23.5	1	6.57		1.34				4	K2/3 III	2006
		NGC 129 - 178		0 27 21	+59 54.5	3	12.87	.009	0.40	.019	-0.14	.019	3		49,88,1756
		Cas sq 1 # 118		0 27 21	+62 36.1	1	11.50		0.46		0.41		2		523
2629	-1 00052			0 27 21	-01 23.6	1	7.48		0.34		-0.07		3	F3 V	3037
		NGC 129 - 177		0 27 22	+59 49.9	2	12.86	.019	0.56	.029	-0.05	.005	3		88,1756
		Cas sq 1 # 121		0 27 22	+62 20.2	1	11.55		1.33		1.10		2		523
2534	+80 00010			0 27 22	+81 05.9	1	8.11		0.15		0.09		5	A0	1219
	+53 00079			0 27 23	+54 10.7	1	10.10		1.08		1.02		1		3072
		NGC 129 - 190		0 27 23	+59 50.6	1	14.92		0.80				1		88
		Cas sq 1 # 120		0 27 23	+62 26.7	1	11.14		0.21		-0.22		2		523
		NGC 129 - 192		0 27 24	+59 56.5	2	15.24	.015	0.71	.010	0.45	.000	2		88,1756
		NGC 129 - 185		0 27 24	+59 57.0	2	14.33	.000	0.69	.065	0.43	.115	2		88,1756
2642	-31 00186			0 27 24	-31 19.1	1	9.83		0.67		0.12		1	G2/3 V	3061
2610	+63 00053			0 27 25	+64 28.4	1	7.27		-0.02				4	A0	1118
		Cas sq 1 # 122		0 27 26	+62 18.8	1	11.19		0.25		-0.59		2		523
2643	-32 00156	IDS00250S3250	AB	0 27 26	-32 33.0	1	8.25		1.03		0.91		1	K1 III	3061
		NGC 129 - 194		0 27 27	+59 50.0	1	15.96		1.38				1		88
		NGC 129 - 179		0 27 27	+59 50.7	3	13.12	.017	0.57	.016	0.04	.027	4		49,88,1756
		NGC 129 - 96		0 27 27	+59 51.7	2	12.27	.028	0.50	.005	-0.07	.014	4		49,1756
2620	+51 00087			0 27 28	+52 14.4	1	9.09		0.10		0.08		2	A0	1566
		G 266 - 122		0 27 28	-17 14.2	1	13.84		1.52		1.10		3		481
		G 266 - 123		0 27 28	-24 55.0	2	12.66	.035	1.49	.020	1.12		2		1696,3061
2628	+28 00075	HR 114, GN And	⋆ AB	0 27 29	+29 28.6	3	5.22	.014	0.26	.008	0.08	.005	6	A7 III	39,1118,3058
2637	-4 00054	HR 117	⋆ AB	0 27 29	-04 10.4	4	5.72	.013	1.54	.014	1.88	.018	13	M0 III	15,938,1256,3016
2641	-30 00138			0 27 29	-30 30.4	2	9.52	.005	0.13	.014	0.10		8	A0 V	1068,2012
		NGC 129 - 187		0 27 30	+59 52.4	1	14.61		0.90		0.18		1		88
2626	+59 00068	NGC 129 - 201	⋆ AB	0 27 32	+59 42.1	4	5.94	.004	0.01	.007	-0.36	.003	13	B9 IIIn	15,49,1079,1118
236436	+59 00068	NGC 129 - 171	⋆ C	0 27 32	+59 54.5	4	9.27	.008	0.46	.009	-0.14	.014	22	F6 V	49,88,1655,1756
		NGC 129 - 184		0 27 33	+59 53.6	3	14.12	.047	0.55	.031	0.33	.013	5		49,88,1756
		NGC 129 - 172		0 27 36	+59 53.6	3	10.89	.013	0.44	.013	-0.08	.005	7	B6 III-IV	49,88,1756
		Cas sq 1 # 123		0 27 36	+62 45.2	1	10.65		0.52		0.42		2		523
2619	+64 00052	LS I +64 027		0 27 36	+64 59.8	1	8.31		0.57		-0.44		2	B0.5III	1012
		BPM 46364		0 27 36	-20 09.	1	11.23		0.79		0.22		1		3061
		NGC 129 - 175		0 27 37	+59 54.6	2	12.76	.000	0.52	.044	0.15	.010	3		88,1756
	-31 00188	G 267 - 119		0 27 38	-31 06.8	1	11.50		0.97		0.79		1		3061

HD	DM	Other Id	N	Rem	α_{1950}	δ_{1950}	S	V	σ_V	B–V	σ_{B-V}	U–B	σ_{U-B}	n	Spectrum	References
236440	+56 00069				0 27 40	+56 59.4	1	8.52		1.21		1.11		1		1748
		NGC 129 - 63			0 27 40	+59 56.8	1	11.87		0.39		-0.07		2		88
		NGC 129 - 61			0 27 40	+59 57.5	1	11.71		0.34		-0.17		2	B5 V	88
2589	+76 00010	HR 112			0 27 40	+76 44.6	4	6.22	.004	0.85	.004	0.52	.035	18	K0 IV	15,252,1003,1258
2698	-60 00038				0 27 40	-60 04.2	1	8.13		0.97		0.69		9	G8 III	1704
	+61 00100				0 27 41	+62 17.5	1	10.68		1.21		0.90		4		523
	+61 00099				0 27 41	+62 21.7	1	10.64		0.29		-0.50		6		523
		NGC 129 - 180			0 27 42	+59 49.6	1	13.71		0.59		0.29		1		88
		AA26,31 # 102			0 27 42	+62 10.	1	12.28		0.32		-0.31		3		483
		PS 9			0 27 42	-29 31.	1	12.22		0.41		-0.19		2		295
		BPM 46365			0 27 42	-29 59.	1	13.36		0.92		0.68		1		3061
	-76 00052				0 27 42	-76 03.2	1	10.46		0.93		0.69		1		1696
	+6 00058	G 1 - 6			0 27 43	+06 54.2	2	10.68	.000	1.07	.000	0.97		3	K2	202,333,1620
2670	-19 00067	G 266 - 124			0 27 43	-19 28.0	1	8.54		0.62		0.05		1	G3 V	3061
		NGC 129 - 48			0 27 44	+60 00.8	2	9.63	.005	0.44	.009	-0.04	.005	4	F6 V	49,1756
		Cas sq 1 # 124			0 27 44	+62 52.8	1	10.35		1.25		1.02		2		523
		TPHE A			0 27 44	-46 47.9	1	14.65		0.79		0.38		12		1764
		NGC 129 - 188			0 27 46	+59 52.9	1	14.74		1.59		0.80		1		88
		NGC 129 - 202		★	0 27 46	+60 02.5	3	11.71	.022	0.31	.025	-0.46	.022	7		41,49,1756
		NGC 129 - 183			0 27 47	+59 56.7	1	14.01		0.46		0.28		1		88
		NGC 129 - 47			0 27 51	+59 52.2	2	13.71	.020	0.46	.015	0.17	.030	2		49,1756
		G 242 - 59			0 27 51	+73 05.6	1	10.63		1.36		1.30		3		7010
		NGC 188 - 5020			0 27 51	+84 48.9	1	10.91		1.63		1.78		1		4002
		CS 22882 # 30			0 27 51	-30 37.4	2	14.82	.005	0.38	.007	-0.18	.010	3		1580,1736
		TPHE B			0 27 51	-46 44.5	1	12.33		0.41		0.16		17		1764
2654	+61 00101				0 27 52	+62 04.7	3	7.37	.011	-0.02	.005	-0.68	.010	6	B2 V	401,523,1118
		TPHE C			0 27 52	-46 49.1	1	14.38		-0.30		-1.22		23		1764
	+21 00055	G 131 - 71			0 27 53	+22 29.7	1	9.22		0.93		0.68		1	K5	333,1620
2696	-24 00179	HR 118			0 27 53	-24 03.8	5	5.18	.007	0.13	.005			11	A3 V	15,1020,2012,2016,2024
		CS 22882 # 12			0 27 53	-27 52.0	2	15.25	.010	0.38	.007	-0.19	.010	3		1580,1736
		TPHE D			0 27 53	-46 47.7	1	13.12		1.55		1.87		23		1764
		AJ89,1229 # 16			0 27 54	+04 03.8	1	12.25		1.41				3		1532
		NGC 129 - 42			0 27 54	+59 52.6	2	14.49	.035	0.62	.020	0.46	.065	2		49,1756
		TPHE E			0 27 54	-46 41.2	1	11.63		0.44		-0.10		8		1764
		LB 1558			0 27 54	-46 49.	1	14.41		-0.32		-1.20		1		832
2676	+35 00083				0 27 55	+35 53.6	1	9.17		0.20		0.10		2	A5	1336
		SB 202			0 27 55	-28 25.6	3	14.25	.019	0.17	.010	0.10	.029	5	A0	295,966,1736
		LS I +62 116			0 27 56	+62 00.9	2	11.37	.010	0.33	.010	-0.42	.015	4		41,523
2690	+2 00058				0 27 57	+02 59.4	1	8.40		1.19		1.11		2	K2	1375
2665	+56 00070				0 27 58	+56 47.4	2	7.71	.010	0.79	.002	0.13	.019	3	G5 III	1748,3016
2719	-19 00068	G 266 - 125	B		0 28 00	-18 52.1	1	11.20		0.91		0.58		1		3061
		G 32 - 26			0 28 01	+14 25.9	2	13.24	.010	0.70	.005	0.05	.025	2		333,1620,1658
2688	+31 00067	IDS00253N3136	A		0 28 01	+31 51.6	1	7.51		0.46		0.07		1	F2	695
		NGC 188 - 5021			0 28 01	+84 53.1	1	12.64		1.24		1.37		1		4002
2724	-41 00116	HR 119, BB Phe			0 28 01	-41 13.0	2	6.18	.005	0.34	.005			7	F2 III	15,2012
2726	-48 00102	HR 120			0 28 01	-48 29.4	2	5.68	.005	0.36	.005			7	F2 V	15,2012
		NGC 129 - 31			0 28 02	+59 52.3	1	14.33		0.89		0.33		1		49
		NGC 129 - 24			0 28 03	+59 56.7	2	11.77	.005	0.38	.000	-0.14	.015	2	B6 V	49,88
		Cas sq 1 # 129			0 28 03	+62 40.1	1	12.42		0.44		-0.05		5		523
2664	+63 00055				0 28 03	+63 47.3	2	8.12	.010	0.47	.000	0.14		3	F8	1118,1462
		NGC 129 - 22			0 28 04	+59 55.6	1	12.99		0.42		0.04		1		49
		LS I +62 117		★ AB	0 28 05	+62 40.0	2	11.70	.005	0.44	.000	-0.29	.000	5		41,523
2717	-10 00087				0 28 05	-09 40.1	1	9.97		0.17		0.13		4	A0	152
	+18 00062	G 32 - 27			0 28 07	+18 47.6	1	10.70		0.92		0.73		2	K0	333,1620
2719	-19 00068	G 266 - 126	A		0 28 07	-18 53.8	1	7.54		0.49		0.00		1	F5 V	3061
		NGC 129 - 16			0 28 09	+59 54.8	1	12.26		0.39		0.10		1		49
	+61 00103				0 28 09	+62 04.9	1	10.15		0.56		0.08		3		523
	+61 00102				0 28 09	+62 21.8	1	9.67		1.62		1.57		4		523
		NGC 129 - 204			0 28 10	+59 50.7	1	14.08		0.49		0.33		1		49
2713	+27 00072				0 28 11	+27 50.1	1	8.80		0.39				3	F2 IV	20
		NGC 129 - 13			0 28 11	+59 55.2	1	12.74		0.83		0.30		1		49
232206	+54 00086				0 28 14	+55 30.9	1	8.33		1.32		1.39		2	K2	1566
2732	+28 00080				0 28 15	+29 15.1	1	8.17		0.97				2	K1 III	20
2710	+56 00071				0 28 16	+57 05.3	1	7.69		0.11		0.12		1	A0	1748
		NGC 129 - 203			0 28 16	+59 50.7	1	11.95		0.46		0.25		1		49
	+58 00069	LS I +59 045			0 28 17	+59 13.5	1	9.79		0.49		-0.33		2		41
		NGC 129 - 205			0 28 23	+59 53.1	1	14.96		0.58		0.14		1		49
		WLS 24 50 # 9			0 28 24	+50 13.2	1	12.33		0.72		0.22		2		1375
2711	+55 00094				0 28 24	+55 43.5	1	8.83		0.04		-0.32		3	A0	1502
		AA26,31 # 103			0 28 24	+62 52.	1	13.40		0.49		-0.31		2		483
	+62 00093	IDS00256N6248	AB		0 28 24	+63 04.5	1	9.59		0.36		-0.25		2	B3	523
		G 217 - 54			0 28 25	+55 18.4	1	10.57		0.86		0.55		1	K0	1658
		TPHE F			0 28 25	-46 50.1	1	12.47		0.86		0.53		3		1764
	+61 00104				0 28 26	+62 04.1	1	9.47		0.19		-0.43		3	B5	523
236445	+56 00072				0 28 28	+56 58.5	1	8.70		0.21		-0.04		1		1748
2760	-10 00089	IDS00259S1038	A		0 28 28	-10 21.6	1	6.86		0.20		0.10		4	A3	3016
	+61 00105	LS I +62 118			0 28 29	+62 09.1	4	9.29	.012	0.27	.005	-0.69	.011	9	O9 V	41,523,1011,1012
	+59 00069		V		0 28 30	+59 38.1	1	10.10		1.90				2		369
		AA26,31 # 104			0 28 30	+62 06.	1	11.97		0.36		-0.29		4		483
	+62 00095	IDS00253N6223	AB		0 28 30	+63 01.9	1	10.07		0.49		-0.01		2		523

Table 1

39

HD	DM	Other Id	N Rem	α_{1950}	δ_{1950}	S	V	σ_V	B–V	σ_{B-V}	U–B	σ_{U-B}	n	Spectrum	References
		BPM 46369		0 28 30	−31 33.	1	15.00		1.35		1.14		1		3061
	+62 00096			0 28 31	+63 03.9	1	9.75		0.34		0.22		6		523
2729	+65 00067	HR 121		0 28 31	+66 14.6	1	6.17		-0.10		-0.48		2	B6 V	154
	−26 00154			0 28 31	−26 27.1	1	10.38		0.62		0.12		1	G5	3061
2761	−10 00090	IDS00259S1038	BC	0 28 33	−10 21.6	1	8.43		0.62		0.09		4	G0	3032
2762	−32 00163	G 267 - 120		0 28 33	−31 54.8	2	9.56	.010	0.55	.005	-0.08	.010	2	F8/G0 V	79,3061
		LS I +60 113		0 28 34	+60 24.7	1	11.63		0.26		-0.50		3		41
236446	+59 00070	NGC 129 - 165	⋆ A	0 28 36	+59 58.8	2	8.71	.000	1.79	.039	2.12	.034	3	M0 Ib	49,88
		SB 208		0 28 36	−30 57.	1	14.28		0.06		0.20		2		295
		G 130 - 73		0 28 40	+33 21.0	1	11.58		1.39		1.26		1	K5	333,1620
2756	+38 00063			0 28 40	+38 44.1	1	8.49		0.18		0.08		3	A3 V	833
		TPHE G		0 28 40	−46 39.3	1	10.44		1.55		1.92		3		1764
	+62 00097			0 28 41	+62 58.4	1	10.08		1.22		0.98		6		523
236448	+59 00071	NGC 129 - 196		0 28 42	+59 53.8	1	9.36		0.18		-0.21		1	A0 V	49
		NGC 129 - 168		0 28 42	+59 57.3	2	12.36	.005	0.36	.010	-0.09	.005	2		49,88
2766	+38 00064			0 28 44	+38 42.3	2	7.33		0.11		0.17		2	A3 V	105
2745	+59 00072	NGC 129 - 197		0 28 44	+59 47.8	3	8.90	.014	0.21	.023	0.16	.020	4	A0 V	49,851,1118
2767	+32 00080	HR 122	⋆ A	0 28 46	+33 18.4	1	5.87		1.14				2	K1 III	71
2796	−17 00070			0 28 46	−17 04.2	4	8.50	.007	0.75	.011	0.25	.055	8	FW	742,955,979,1594
		PS 12		0 28 46	−23 31.9	2	14.56	.045	0.03	.039	0.05	.040	2		295,1736
2799	−36 00166			0 28 46	−35 45.1	2	10.99	.029	0.05	.010	0.08	.024	10	B9/A0 V	152,1068
		AA26,31 # 105		0 28 48	+62 17.	1	12.38		0.32		-0.31		4		483
236449	+59 00073	NGC 129 - 164		0 28 49	+59 58.8	2	8.57	.000	1.95	.034	2.15		3	K2 Ib	49,88
		NGC 129 - 169		0 28 51	+59 57.2	1	13.54		0.76		0.25		1		88
		LS I +62 119		0 28 52	+62 24.2	2	11.75	.000	0.29	.015	-0.51	.015	5		41,523
		NGC 188 - 5026		0 28 52	+84 53.8	1	12.95		1.10		0.92		1		4002
		NGC 129 - 166		0 28 54	+59 59.2	2	9.97	.040	0.16	.015	-0.16	.000	2	B9 V	49,88
		SB 210		0 28 54	−28 14.1	2	12.94	.000	0.02	.000	0.07	.011	3	A0	966,1736
2774	+52 00092	HR 124		0 28 56	+52 33.8	3	5.59	.010	1.15	.005	1.13	.015	9	K2 III	15,1003,1355
236453	+56 00074			0 28 56	+56 37.7	1	8.87		0.19		0.22		1		1748
		CS 29497 # 17		0 28 56	−24 14.0	1	14.16		0.26		-0.09		1		1736
		NGC 129 - 167		0 28 59	+59 57.5	1	12.36		1.95		2.32		1		88
		LS I +60 114		0 28 59	+60 17.5	1	10.72		0.38		-0.26		2		41
		LS I +62 120		0 28 59	+62 36.9	2	11.46	.005	0.84	.029	-0.18	.005	7		41,523
2806	+15 00073			0 29 00	+15 44.7	1	6.86		1.21		1.30		2	K0	3040
	+17 00064			0 29 00	+17 56.6	1	9.54		0.46		-0.04		2	F5	1375
2772	+53 00082	HR 123	⋆ AB	0 29 00	+54 14.8	5	4.75	.054	-0.10	.004	-0.34	.020	10	B7.5V+B8.5V	15,1079,1363,3023,8015
		LS I +62 121		0 29 00	+62 24.5	2	10.98	.020	0.33	.005	-0.49	.010	5		41,523
		SB 212		0 29 00	−23 13.	1	13.32		0.00		0.02		2	A0	295
		SB 211		0 29 00	−24 14.	1	14.14		0.22		-0.04		2		3064
		Ki 14- 133		0 29 01	+62 52.6	2	11.77	.005	0.67	.005	0.11	.029	6		41,523
		GR 10		0 29 01	−35 46.1	1	14.28		1.49		1.18		2		481
2821	−37 00165			0 29 01	−37 28.6	1	8.08		1.17		0.88		4	G5 III	119
2834	−49 00115	HR 125	⋆ A	0 29 01	−49 04.8	5	4.76	.008	0.02	.004	0.04	.000	16	A0 V	15,1068,1075,2012,3023
		G 158 - 95		0 29 02	−06 08.3	2	12.75	.000	1.64	.005	1.20	.019	6		316,3078
		CS 29497 # 22		0 29 02	−25 11.1	1	13.42		-0.02		-0.05		1		1736
2831	−26 00157			0 29 05	−26 18.0	1	9.83		0.68		0.23		1	G2 V	3061
2830	−2 00069			0 29 07	−02 04.2	1	7.03		0.08		0.08		2	A0	1375
2789	+66 00035			0 29 08	+66 53.1	1	8.22		0.39		-0.40		3	B3 Vne	1212
2814	+36 00077	IDS00265N3625	A	0 29 09	+36 41.5	1	8.20		0.88		0.43		2	G0	3016
2814	+36 00077	IDS00265N3625	B	0 29 09	+36 41.5	1	11.46		0.69		0.05		2		3016
2814	+36 00077	IDS00265N3625	C	0 29 09	+36 41.5	1	12.65		0.87		0.32		3		3016
2814	+36 00077	IDS00265N3625	D	0 29 09	+36 41.5	1	10.78		0.41		0.04		2		3016
	−24 00187	BPM 46374		0 29 11	−23 40.9	1	10.61		0.67		-0.01		3		3061
	+63 00059			0 29 12	+63 34.3	1	10.22		0.62				2		1118
	+66 00034	V547 Cas	⋆ AB	0 29 12	+66 57.9	2	10.31	.025	1.50	.045	1.16		5	M3	196,3078
		PG0029+024		0 29 16	+02 21.9	1	15.27		0.36		-0.18		2		1764
2884	−63 00050	HR 126	⋆ A	0 29 16	−63 14.0	5	4.36	.008	-0.06	.009	-0.16	.014	11	B9 V	15,1034,1075,3023,8029
2885	−63 00050	HR 127	⋆ CD	0 29 17	−63 14.5	5	4.54	.008	0.15	.006	0.02	.007	11	A2 V +A7 V	15,1034,1075,3030,8029
2840	+27 00080	IDS00266N2759	B	0 29 18	+28 14.0	2	8.87	.015	0.33	.010			5	A2	20,6009
236455	+56 00075			0 29 18	+56 49.9	1	9.35		0.14		0.01		1		1748
2863	−31 00197			0 29 18	−31 11.6	1	10.37		0.64		0.12		1	G6 (III)	3061
		LB 1559		0 29 18	−47 41.	1	12.36		-0.26		-0.93		3	sdB	45
2839	+27 00080	IDS00266N2759	A	0 29 19	+28 14.7	2	8.84	.000	1.05	.000			5	K1 III	20,6009
2838	+29 00094			0 29 19	+30 14.8	1	8.48		1.44		1.70		11	K2	1371
	+4 00067			0 29 20	+04 53.0	1	10.22		0.49		-0.03		2	F5	1375
2857	−6 00086			0 29 21	−05 32.5	2	9.99	.015	0.20	.010	0.23	.120	2	A2	979,3016
2861	−28 00148	IDS00269S2823	A	0 29 21	−28 06.3	2	9.60	.005	0.76	.000	0.26	.019	3	G6/8 V	79,3064
	+62 00099	Ki 14 - 58	⋆	0 29 22	+62 55.9	2	11.02	.029	0.40	.010	-0.45	.010	6	Be	41,523
	+29 00095			0 29 24	+30 15.7	1	10.21		1.30				1	R0	1238
	+61 00108			0 29 24	+62 05.5	1	9.60		1.33		1.27		6		523
		G 267 - 122		0 29 32	−32 01.0	1	15.97		1.33				1		3061
2851	+57 00098	IDS00268N5747	AB	0 29 33	+58 03.8	1	7.10		0.37		0.02		3	F2 V	1501
2866	+34 00073			0 29 34	+34 43.1	1	6.73		0.17		0.09		2	A5 IV	1336
2880	−5 00077	IDS00270S0544	AB	0 29 34	−05 27.3	3	8.54	.018	0.84	.004	0.47	.017	10	K0 V +K1 V	176,214,292,3032
2893	−13 00089			0 29 35	−12 34.2	1	8.27		0.58		0.02		2	G1 V	1375
2892	+0 00071			0 29 38	+00 54.7	3	9.36	.009	1.32	.005	1.40	.006	25	K5	989,1375,1729
	+6 00063			0 29 41	+06 38.9	1	9.96		0.27		0.16		3	A0	1026
		LS I +62 124		0 29 42	+62 47.5	2	11.59	.005	0.42	.010	-0.40	.010	6		41,523
		BPM 46379		0 29 42	−30 08.	1	12.98		1.23		0.88		2		3061

HD	DM	Other Id	Rem	α₁₉₅₀	δ₁₉₅₀	S	V	σ$_V$	B–V	σ$_{B-V}$	U–B	σ$_{U-B}$	n	Spectrum	References
2888	+42 00099	HR 128		0 29 44	+43 13.1	2	6.68	.019	-0.01	.000	-0.20	.000	7	A1 Vn	1733,3016
		Cas sq 1 # 147		0 29 47	+62 13.2	1	11.71		0.43		0.25		4		523
		CS 29527 # 22		0 29 47	−19 42.5	1	11.88		0.02		0.08		1		1736
		L 405 - 60		0 29 48	+18 11.5	1	11.25		0.68		0.01		2		1696
		BPM 46384		0 29 48	−29 02.	1	12.15		0.94		0.61		2		3064
2913	+6 00064	HR 132	★A	0 29 49	+06 40.8	3	5.67	.008	-0.01	.005	-0.18	.005	12	B9.5V	15,1256,3032
2913	+6 00065	IDS00272N0624	B	0 29 49	+06 40.8	1	9.55		0.71		0.18		5		3016
2874	+50 00093			0 29 49	+50 49.7	1	8.30		0.25		0.13		2	A2	1566
	+61 00109			0 29 49	+62 12.8	1	10.65		0.30		0.13		4	B9	523
	+62 00101	LS I +62 125		0 29 49	+62 59.6	2	9.65	.005	0.42	.000	-0.30	.000	4	B2	41,523
	−38 00157			0 29 51	−38 04.5	1	12.99		0.00		0.10		2	A0	295
		LS I +62 126		0 29 52	+62 48.6	2	11.34	.020	0.34	.010	-0.39	.015	4		41,523
2916	−25 00185			0 29 52	−24 55.4	1	7.27		0.37		-0.03		1	F3 III/IV	3061
		LS I +63 079		0 29 53	+63 01.8	1	11.09		0.52		-0.23		2	B2 V	1012
		BPM 46385		0 29 54	−20 54.	1	13.08		0.53		-0.17		1		3061
		NGC 152 sq1 26		0 29 54	−73 24.9	1	13.51		0.65		0.13		3		959
2911	+17 00067			0 29 55	+18 31.2	1	6.81		1.54		1.92		3	M0 III	1501
		DZ And		0 29 55	+25 44.7	1	10.22		1.23		1.29		1		842
2909	+27 00082	G 69 - 1		0 29 55	+27 55.3	1	8.72		0.65		0.08		1	G5	333,1620
		LS I +62 127		0 29 55	+62 22.8	2	10.64	.010	0.38	.005	-0.43	.020	4		41,523
2924	+26 00076	HR 133		0 29 56	+27 18.3	1	6.70		0.07		0.10		2	A2 IV	1733
2910	+19 00079	HR 131	★A	0 29 58	+20 01.2	2	5.37	.008	1.07	.010	0.96	.007	5	K0 III	58,3016
		KUV 00300-1810		0 29 59	−18 09.9	1	16.36		0.14		0.74		1	DA	1708
		NGC 152 sq1 25		0 29 59	−73 24.8	1	13.16		0.53		0.02		3		959
		BPM 46387		0 30 00	−31 01.	1	11.12		0.63		0.07		2		3061
2925	+22 00079	IDS00274N2238	AB	0 30 01	+22 55.0	3	6.84	.012	0.92	.002	0.59	.007	9	G8 III-IV	1003,1355,3077
2901	+53 00090			0 30 01	+53 50.7	3	6.92	.034	1.23	.011	1.27	.022	6	K2 III	979,1355,3016
2907	+56 00078			0 30 05	+56 54.5	1	8.20		0.20		0.07		1	A2	1748
		BPM 46389		0 30 06	−31 08.	1	14.00		1.48		1.24		2		3061
2905	+62 00102	HR 130, κ Cas		0 30 08	+62 39.4	14	4.16	.011	0.13	.011	-0.79	.014	59	B1 Iae	1,15,41,154,369,1012*
		G 267 - 123		0 30 08	−31 08.9	2	11.85	.044	1.16	.010	1.13	.024	3		1696,3061
	+61 00111			0 30 09	+62 10.3	1	9.72		0.52		-0.01		6		523
2942	+27 00084	HR 134	★A	0 30 10	+28 00.3	1	6.30		1.00		0.73		1	G8 II	3016
2942	+27 00084	HR 134	★AB	0 30 10	+28 00.3	1	6.26		0.99		0.80		1	G8 II	8100
2942	+27 00084	IDS00276N2744	B	0 30 10	+28 00.3	1	11.26		0.83		0.41		2		3016
2933	+35 00090	IDS00275N3518	AB	0 30 11	+35 34.9	1	7.94		-0.02		-0.24		2	A0	1336
	+43 00100	G 171 - 62		0 30 11	+44 26.5	2	10.29	.001	0.79	.001	0.41	.010	3	K0	1658,7010
		Cas sq 1 # 151		0 30 11	+62 11.1	1	10.86		0.33		-0.45		6		523
		BPM 46390		0 30 12	−27 14.	1	11.24		0.73		0.27		2		3064
2960	−34 00176			0 30 13	−34 34.9	1	9.31		1.60		1.99		4	K5/M0	3007
	+41 00080			0 30 14	+41 58.9	1	9.90		-0.13		-0.75		1	B5	963
		CS 22170 # 3		0 30 14	−10 03.0	1	13.71		-0.04		-0.10		1		1736
	−25 00189	BPM 46392		0 30 15	−25 33.9	1	10.55		0.61		-0.01		2		3064
		BPM 46383		0 30 18	−21 37.	1	11.20		0.81		0.38		1		3061
		G 172 - 4		0 30 19	+44 27.7	1	16.59		0.30		-0.50		1	DA	3060
2904	+70 00024			0 30 19	+70 42.4	2	6.40	.019	-0.02	.015	-0.08	.024	6	A0 Vn	985,1733
		G 132 - 3		0 30 20	+41 43.6	1	11.24		1.48		0.84		1	M0	1620
2928	+61 00113	LS I +62 129		0 30 22	+62 15.7	3	8.62	.007	0.61	.007	0.08	.007	7	A0 Iab	41,523,1012
2952	+54 00101	HR 135		0 30 23	+54 37.2	1	5.93		1.04		0.84		2	K0 III	3016
		BPM 46395		0 30 24	−31 31.	1	12.09		0.77		0.29		1		3061
3003	−63 00052	HR 136	★AB	0 30 28	−63 18.4	3	5.09	.017	0.04	.008	0.01	.005	6	A0 V	258,2035,3023
		CS 22179 # 7		0 30 31	−04 29.4	1	14.03		0.25		0.06		1		1736
	+63 00061			0 30 32	+63 43.6	1	9.65		0.91		-0.17		2	B1 Iab	1012
		CS 22882 # 13		0 30 34	−28 05.1	1	14.24		0.53		-0.17		1		1736
3000	−29 00149			0 30 34	−29 21.0	1	9.36		0.76		0.36		3	G5/6 V	3064
		SB 221		0 30 35	−21 27.8	2	14.01	.019	0.11	.015	0.14	.016	3	A0	295,1736
		LP 405 - 61		0 30 36	+18 54.5	1	11.66		1.17		1.18		1		1696
3002	−34 00181			0 30 36	−34 03.1	1	9.57		0.24		0.10		6	A5 IV/V	1068
	−24 00201			0 30 41	−24 28.3	1	10.55		0.78		0.37		1	K0	3061
		BPM 46399		0 30 42	−21 47.	1	14.30		1.10		1.00		1		3061
2974	+59 00076			0 30 44	+60 16.4	2	7.90	.009	-0.01	.005	-0.21		4	B9	401,1118
		LS I +62 130		0 30 48	+62 53.5	2	11.49	.015	0.43	.010	-0.39	.015	4		41,523
		SB 225		0 30 54	−28 51.0	3	14.30	.058	0.08	.027	0.18	.050	5	A0:pec	295,966,1736
236464	+57 00102		V	0 31 00	+57 56.4	1	8.71		1.69					M2	369
	+62 00105	LS I +62 131		0 31 00	+62 51.1	2	8.80	.000	1.27	.020	1.04	.055	5	F8	41,523
		AA124,216 # 1		0 31 00	−44 41.	1	11.72		0.46		-0.08		1		1478
		AA124,216 # 2		0 31 00	−44 41.	1	14.74		1.55		1.12		1		1478
		AA124,216 # 3		0 31 00	−44 41.	1	16.77		1.04		1.00		1		1478
		AA124,216 # 4		0 31 00	−44 41.	1	16.03		0.62		-0.08		1		1478
		AA124,216 # 5		0 31 00	−44 41.	1	14.41		0.57		-0.05		1		1478
		LB 3144		0 31 00	−76 44.	1	10.95		0.07		0.01		4		45
	+62 00106			0 31 01	+62 56.3	1	10.76		0.29		-0.13		3		523
3029	+19 00083			0 31 02	+20 09.5	1	7.84		0.27		0.08		2	A3	1375
		WLS 40-15 # 6		0 31 02	−14 32.0	1	11.68		0.65		0.14		2		1375
	−16 00090	RX Cet		0 31 07	−15 45.7	3	11.03	.083	0.22	.054	0.05	.055	3	A7	597,699,700
	−22 00092			0 31 08	−21 37.6	1	10.67		0.99		0.56		2	G5	3061
3042	+6 00068			0 31 10	+07 24.7	1	9.12		0.67		0.08		2	G0	1375
		KUV 00312-1837		0 31 10	−18 37.1	1	16.66		-0.04		-1.06		1	DB	1708
3060	−35 00168			0 31 10	−35 26.6	1	8.43		0.94		0.64		4	G6 III	119
	−15 00098			0 31 11	−14 33.9	1	9.95		0.98		0.81		1		79

Table 1   41

HD	DM	Other Id	N Rem	α_{1950}	δ_{1950}	S	V	σ_V	B−V	σ_{B-V}	U−B	σ_{U-B}	n	Spectrum	References
3059	−30 00156	HR 138		0 31 13	−29 50.0	4	5.55	.005	1.26	.010	1.23		13	K1 III	15,1075,2032,3005
232218	+52 00104			0 31 15	+52 49.5	1	8.92		0.52				3	G5	6009
3074	−35 00170	G 267 - 128	★ AB	0 31 17	−35 16.2	5	6.40	.013	0.62	.004	0.15	.004	12	F8/G0 V	79,938,1311,2012,3061
3112	−71 00020	HR 139, θ Tuc		0 31 17	−71 32.5	3	6.12	.003	0.24	.003	0.19	.010	22	A7 IV	15,1075,2038
	−19 00078	G 266 - 129		0 31 18	−18 38.6	1	10.77		1.09		1.05		1		3061
3039	+59 00078			0 31 19	+59 37.4	1	7.96		0.53				2	G0	1118
3085	−40 00123			0 31 22	−40 12.0	1	7.42		−0.03		−0.05		1	B9 V	3007
		G 158 - 100		0 31 23	−12 24.5	3	14.89	.028	0.67	.004	−0.07	.031	12		538,940,1658,3056
		NGC 152 sq1 2		0 31 23	−73 22.0	1	14.51		0.99		−0.70		3		959
	−28 00154	G 266 - 130		0 31 24	−27 53.4	1	10.09		0.70		0.19		2	G0	3064
	+61 00117			0 31 27	+62 21.3	1	10.41		0.64		0.20		4		523
3038	+65 00070	HR 137		0 31 29	+66 28.5	1			−0.08		−0.33		1	B9 III	1079
		GD 620		0 31 29	−26 26.1	1	13.69		0.40		−0.27		2		3064
		NGC 152 sq1 4		0 31 29	−73 24.3	1	14.84		0.76		0.32		4		959
		CS 29527 # 25		0 31 30	−21 01.1	1	14.86		0.05		0.15		1		1736
		BPM 46408		0 31 30	−25 50.	1	13.70		1.03		0.96		2		3064
		TonS 163		0 31 30	−27 23.	1	14.18		−0.26		−1.20		2		3064
		BPM 46410		0 31 30	−32 46.	1	14.39		1.38				1		3061
		BPM 46411		0 31 30	−36 27.	1	13.08		0.50		−0.25		2		3061
		LP 28 - 158		0 31 36	+70 55.	1	13.51		1.65				4		940
		G 266 - 132		0 31 36	−23 20.0	1	13.56		1.49				1		3061
		BPM 46414		0 31 36	−27 35.	1	12.78		1.04		0.90		3		3064
		SB 231		0 31 36	−28 47.2	3	13.19	.005	−0.01	.000	−0.01	.013	5	A0 pec	295,966,1736
		Florsch 61		0 31 36	−72 41.4	1	12.31		0.80		0.41		4	G7	767
3101	−32 00183	IDS00292S3250	AB	0 31 37	−32 33.7	2	8.04	.030	1.51	.012	1.61	.017	10	M3 III	1673,3007
3037	+68 00034			0 31 38	+69 09.5	1	8.39		1.29		1.24		4	K0 III	206
		CS 22882 # 29		0 31 38	−31 08.5	2	14.78	.010	0.39	.007	−0.20	.009	3		1580,1736
	+21 00064			0 31 39	+22 08.3	1	9.62		1.35				1	M9	1238
3067	+62 00107	IDS00289N6221	AB	0 31 41	+62 37.7	1	7.67		0.48				2	F5	1118
		NGC 152 sq1 1		0 31 41	−73 22.4	1	12.72		0.54		−0.03		8		959
236466	+56 00084	LS I +57 004		0 31 42	+57 15.9	2	8.77	.030	1.02	.015	9.77		5	G5	41,6009
		BPM 46415		0 31 42	−22 48.	1	13.85		1.45		1.25		1		3061
3108	−33 00192	BPM 46416	★ AB	0 31 42	−33 03.3	1	10.32		0.98		0.68		2	K0	3061
3079	+47 00138	G 171 - 64		0 31 43	+47 38.4	2	7.38	.009	0.54	.005	0.02	.009	4	F8	196,1658
3068	+61 00118			0 31 43	+62 18.9	2	8.06	.022	0.51	.013	0.01		5	F7 V	523,1118
	+61 00119			0 31 45	+62 20.8	1	8.87		0.65		0.16		3	F8	523
3049	+71 00023			0 31 45	+72 28.4	1	9.43		0.20		0.18		2	A0	1375
	−49 00129			0 31 46	−49 22.1	1	10.22		0.48		−0.01		2		832
3119	−36 00189	G 267 - 130		0 31 47	−36 32.4	1	9.84		0.55		−0.06		1	F6/7 V	3061
		BPM 46418		0 31 48	−26 13.	1	12.81		0.91		0.54		4		3064
		BPM 46419		0 31 48	−31 17.	1	14.29		0.62		0.00		1		3061
		BPM 46420		0 31 54	−25 20.	1	13.53		0.64		−0.04		2		3064
3125	−5 00083	IDS00294S0506	AB	0 31 57	−04 49.3	1	7.01		0.73		0.30		4	G0	3024
3125	−5 00083	IDS00294S0506	C	0 31 57	−04 49.3	1	9.27		0.57		0.07		4	G0	3024
	+61 00120			0 31 58	+62 17.7	1	9.89		0.40		0.01		6		523
		BPM 46421		0 32 00	−32 58.	1	14.56		0.49		−0.14		1		3061
	−17 00081			0 32 05	−17 23.7	1	10.08		0.44		−0.09		2	F5	1375
3158	−53 00117	HR 140		0 32 05	−52 39.0	5	5.57	.003	0.47	.005	−0.02	.005	16	F6 V	15,158,612,1075,2035
		CS 22170 # 2		0 32 06	−10 45.4	1	14.08		0.09		0.16		1		1736
		BPM 46422		0 32 06	−25 17.	1	12.91		1.25		1.08		2		3064
	+25 00077	G 69 - 3		0 32 07	+25 47.5	1	9.48		1.00		0.76		1	K0	333,1620
	+61 00122	LS I +62 132	★ A	0 32 08	+62 10.4	1	10.73		0.23		−0.56		2	B2 V:pe	41
	+61 00122	LS I +62 132	★ AB	0 32 08	+62 10.4	1	10.39		0.32		−0.59		2		1012
3175	−63 00056			0 32 08	−63 20.2	1	9.28		−0.17		−0.61		1	B4 V	55
	−38 00173	G 267 - 132		0 32 10	−37 42.9	1	12.23		0.47		−0.23		2		1696
		G 266 - 134		0 32 11	−25 11.9	2	12.61	.015	0.67	.005	0.08	.002	3		1696,3064
		BPM 46423		0 32 12	−21 25.	1	13.95		0.70		0.02		1		3061
	+62 00108	LS I +62 133		0 32 13	+62 42.4	2	10.87	.015	0.17	.010	−0.53	.020	4		41,523
3151	+22 00086			0 32 17	+22 37.7	2	8.90	.014	0.37	.010	−0.06	.009	4	F2	1375,1733
3123	+62 00109	LS I +62 134		0 32 18	+62 42.9	3	7.25	.017	0.59	.016	0.53	.030	5	F0	41,523,6009
		BPM 46425		0 32 18	−29 25.	1	12.72		0.60		−0.02		3		3064
3166	+12 00057	HR 141		0 32 20	+13 05.8	1	6.39		1.14		1.10		2	K0	1733
3130	+60 00070			0 32 22	+61 29.9	1	8.83		0.57				2	F8	1118
	+38 00072	IDS00297N3837	A	0 32 23	+38 53.6	1	8.46		1.27		1.26		3	K0	833
		NGC 152 sq1 3		0 32 23	−73 26.1	1	11.17		0.58		0.06		7		959
		LP 825 - 448		0 32 24	−25 36.0	2	12.30	.020	0.71	.010	0.16	.060	2		1696,3061
3149	+51 00103			0 32 27	+51 33.9	1	7.38		0.59		0.07		1	F8	1462
236469	+58 00077	LS I +59 046		0 32 27	+59 02.2	1	9.13		0.92		0.58		2	F4 I	41
3165	+36 00087	IDS00298N3617	A	0 32 29	+36 33.4	1	6.60		1.45		1.64		1	G8 III-IV	3016
3165	+36 00087	IDS00298N3617	B	0 32 29	+36 33.4	1	9.19		0.97		0.66		1	K0	3016
3180	−25 00205			0 32 29	−24 41.7	1	9.62		0.43		−0.14		1	F3 V	3061
		BPM 46430		0 32 30	−25 53.	1	11.75		0.82		0.43		2		3064
		BPM 46431		0 32 30	−32 19.	1	13.76		1.32		1.30		2		3061
3187	−26 00171			0 32 33	−26 24.1	2	7.62	.005	0.98	.000	0.71		5	G8 III	79,2012
3188	−27 00166			0 32 33	−27 01.2	1	9.51		0.61		0.09		1	G3 V	3061
3221	−62 00047			0 32 35	−62 11.4	1	10.30		1.05		0.71		1	K5 V	3072
		BPM 46429	★	0 32 36	−22 53.3	1	13.78		0.70		0.15		1		3061
		LP 825 - 458		0 32 36	−23 34.	1	14.81		0.73		0.28		1		3061
3147	+67 00056			0 32 39	+67 39.1	1	6.91		2.05		2.34		5	K2 Ib-II	1355
3196	−4 00062	HR 142	★ AB	0 32 40	−03 52.1	6	5.20	.007	0.56	.011	0.08	.005	29	F8 V	15,1075,1197,1256*

HD	DM	Other Id	N Rem	α_{1950}	δ_{1950}	S	V	σ_V	B–V	σ_{B-V}	U–B	σ_{U-B}	n	Spectrum	References
		CS 22179 # 6		0 32 40	−04 31.5	1	13.14		0.26		0.10		1		1736
		BPM 46435		0 32 42	−22 53.	1	11.84		0.47		−0.12		1		3061
3222	−64 00051			0 32 42	−63 57.8	3	8.57	.017	0.85	.014	0.53	.020	11	K2 V	1097,2018,3078
		G 266 - 135		0 32 45	−17 35.5	1	14.94		0.23		−0.58		1		3061
		BPM 46437		0 32 48	−29 10.	1	13.06		0.89		0.45		2		3064
		GR 13		0 32 49	−19 29.7	1	14.67		1.52		1.04		2		481
3203	+34 00079			0 32 50	+35 23.0	1	7.71		0.07		−0.01		2	A0	1336
3215	−7 00085			0 32 50	−06 39.4	1	9.74		0.17		0.17		4	A0	152
		CS 29527 # 35		0 32 53	−21 17.3	1	13.17		0.04		0.09		1		1736
3202	+35 00098			0 32 54	+36 18.2	1	8.56		0.10		0.05		2	A0	1336
		SB 235		0 32 54	−21 17.	1			0.03		0.18		3	A0	295
3278	−60 00041			0 32 58	−60 05.4	1	9.57		0.82		0.46		2	G8 IV/V	1730
3212	+33 00075	IDS00303N3356	AB	0 32 59	+34 12.3	1	8.09		0.30		0.06		2	A2	1336
3191	+60 00071	LS I +61 160		0 32 59	+61 11.1	2	8.58	.004	0.45	.004	−0.46	.008	7	B1 IV:nn	41,1012
3229	−1 00068	HR 143		0 32 59	−00 46.8	3	5.93	.008	0.44	.002	−0.03	.008	9	F5 IV	15,1061,3016
		G 1 - 7		0 33 02	+01 36.9	3	15.56	.038	0.19	.041	−0.67	.009	6		202,419,3028
		CS 29527 # 42		0 33 02	−19 31.6	1	15.42		0.13		0.15		1		1736
3225	+35 00099			0 33 04	+35 58.3	1	8.91		0.20		0.12		2	A5	1336
		G 172 - 11		0 33 04	+52 24.8	1	12.58		1.54				1		906
3244	−26 00173			0 33 04	−25 40.5	1	8.23		0.28				4	A7 V	2012
		G 1 - 8		0 33 05	+01 09.6	1	13.53		1.34				2		202
		G 217 - 59		0 33 05	+52 25.3	1	15.08		1.71				1		906
		LS I +60 115		0 33 05	+60 14.1	1	12.12		0.78		−0.26		3		41
		L 87 - 87		0 33 06	−68 55.	1	12.10		1.01		0.89		1		1696
		SK 1		0 33 07	−74 25.	2	12.65	.085	0.01	.030	−0.11	.005	7		416,573
3256	−15 00105			0 33 08	−15 00.6	1	8.58		0.09		0.11		4	A1 V	152
		LP 825 - 474		0 33 12	−23 04.	1	15.33		1.60				1		3061
3303	−55 00117	HR 148		0 33 14	−55 05.8	3	6.06	.004	1.00	.004			11	K0 IV	15,1075,2009
		G 32 - 36		0 33 18	+12 19.4	1	11.26		1.31		1.25		1	K7	333,1620
		Steph 58		0 33 18	−05 00.9	1	12.07		1.49		1.26		1	M0	1746
		BPM 46440		0 33 18	−21 49.	1	13.93		0.68		0.03		1		3061
3302	−48 00121	HR 147		0 33 18	−48 16.5	2	5.51	.000	0.45	.003	−0.01		6	F7 V	58,2009
3268	+12 00059	HR 145		0 33 19	+12 56.1	2	6.38	.055	0.50	.010	−0.04	.035	4	F7 V	254,3037
		CS 29527 # 45		0 33 20	−18 13.6	1	14.08		0.27		−0.15		1		1736
		LP 765 - 73		0 33 20	−20 11.2	2	12.29	.010	0.97	.010	0.78	.005	2		1696,3061
3240	+53 00102	HR 144		0 33 21	+53 53.6	2	5.07	.014	−0.11	.001	−0.37		4	B7 III	154,1363
3284	+6 00075			0 33 22	+07 20.0	1	8.44		0.17		0.07		2	A2	1733
3266	+29 00105	G 69 - 4	★ A	0 33 22	+29 43.4	1	8.43		0.66		0.10		4	G2 V	3026
3266	+29 00105	G 69 - 4	★ AB	0 33 22	+29 43.4	2	7.96	.020	0.68	.015	0.10	.020	3	G2 V	1003,1658
3266	+29 00105	IDS00307N2927	B	0 33 22	+29 43.4	1	9.26		0.69		0.14		4		3026
	−22 00099			0 33 23	−21 53.6	1	10.61		0.71		0.16		2	F5	3061
		CS 22882 # 22		0 33 23	−30 16.7	2	14.99	.000	0.10	.000	0.20	.010	2		966,1736
		BPM 46442		0 33 24	−22 45.	1	12.18		0.46		−0.08		2		3061
3249	+62 00116			0 33 29	+62 56.8	2	8.52	.019	0.10	.009	−0.36		4	A0	401,1118
	−10 00109			0 33 29	−09 47.0	1	11.25		1.41				1		1746
3264	+47 00145			0 33 30	+48 16.9	1	7.50		−0.05		−0.63		1	B2 V	1423
3311	−19 00085			0 33 30	−19 02.3	3	9.02	.011	0.15	.005	0.11	.000	13	A1 mA3-A5	355,1068,2012
3325	−15 00109	HR 150		0 33 32	−15 14.9	3	6.45	.005	1.06	.005	0.91	.011	8	K0 III	15,79,1071
3283	+59 00084	HR 146	★	0 33 36	+60 03.1	3	5.78	.007	0.28	.017	0.27	.040	6	A4 III	41,1118,8100
3337	−21 00070	IDS00311S2119	AB	0 33 37	−21 02.0	1	7.85		0.43		−0.04		1	F3 V	3061
3326	−23 00220	HR 151, BG Cet		0 33 38	−23 07.0	4	6.05	.009	0.30	.004			16	A5mA5-F0	15,1020,2012,2012
3359	−49 00138	IDS00318S4941	B	0 33 40	−49 24.1	1	8.37		0.78				4	G8 IV	2012
3322	+26 00091	HR 149, PY And		0 33 41	+26 58.8	3	6.46		−0.08	.010	−0.42	.023	8	B8 IIIp Hg	1049,1079,3033
		LP 705 - 100		0 33 41	−08 57.1	1	12.12		0.67		0.07		2		1696
3375	−60 00042			0 33 44	−59 59.5	1	6.91		0.36		−0.03		7	F2 V	1628
		LS I +59 047		0 33 45	+59 58.5	1	10.55		0.19		−0.61		2		41
3395	−73 00039			0 33 47	−73 29.8	4	7.84	.017	0.93	.012	0.58	.007	9	G5 III	14,1408,1425,3021
3365	−38 00185	G 267 - 137	★ AB	0 33 51	−38 34.1	1	8.36		0.64		0.13		1	G2/3 V	79
		CS 22882 # 25		0 33 54	−31 12.5	2	14.91	.000	0.03	.000	0.13	.011	2		966,1736
232227	+53 00104			0 33 57	+54 21.5	1	10.55		0.15		0.01		2	A0 V	1003
		LS I +59 048		0 33 57	+59 25.0	1	10.87		0.73		−0.44		2		41
	−26 00179	RT Scl		0 33 59	−25 56.9	1	10.20		0.34		−0.07		3	A7	3064
	−29 00169			0 33 59	−29 33.5	1	10.43		0.71		0.26		1	G5	3061
		BPM 46449		0 34 00	−24 53.	1	13.49		0.86		0.44		1		3061
		G 131 - 78		0 34 03	+18 58.5	1	15.00		0.85		0.16		2		1658
3346	+43 00113	HR 152		0 34 03	+44 12.8	2	5.10	.025	1.58	.015	1.89	.064	8	K5 III	1355,3016
		SB 246		0 34 06	−22 42.	2	13.67	.015	0.16	.009	0.16	.003	3	A0	295,1736
	+61 00131			0 34 07	+62 16.1	1	9.56		0.53				2	F2	1118
3360	+53 00105	HR 153		0 34 10	+53 37.3	7	3.66	.017	−0.20	.006	−0.85	.024	24	B2 IV	15,154,1119,1203,1223*
		AAS35,353 # 5		0 34 10	−73 36.3	1	12.78		0.68		0.20		4	G4	767
3379	+14 00076	HR 155, AG Psc		0 34 11	+14 57.4	2	5.87	.021	−0.15	.009	−0.64	.047	3	B2.5IV	154,1034
	−24 00225	BPM 46445		0 34 11	−23 16.0	1	11.09		0.65		0.15		1		3061
3369	+32 00101	HR 154	★ A	0 34 12	+33 26.7	8	4.35	.010	−0.13	.011	−0.56	.008	28	B5 V	15,154,603,1084,1203*
3369	+32 00102	IDS00315N3310	B	0 34 12	+33 26.7	2	8.62	.004	0.20	.004	0.09	.004	9	F0 V	1084,3024
3369	+32 00101	IDS00315N3310	C	0 34 12	+33 26.7	1	13.01		0.53		−0.08		1		3024
	−23 00225	BPM 46451		0 34 12	−23 17.	1	10.26		0.70		0.04		1		3061
		S112 # 3		0 34 12	−60 08.5	1	11.01		0.56		0.02		2		1730
		L 122 - 4		0 34 12	−60 13.	1	13.84		0.26		−0.61		2		3060
3405	−49 00141	IDS00318S4941	AB	0 34 14	−49 24.3	1	6.78		0.64				4	G0/1 V	2012
		LP 765 - 50		0 34 16	−19 06.7	1	14.42		0.69		−0.20		2		3061

Table 1   43

HD	DM	Other Id	N	Rem	α_{1950}	δ_{1950}	S	V	σ_V	B–V	σ_{B-V}	U–B	σ_{U-B}	n	Spectrum	References
		CS 22882 # 24			0 34 16	−30 41.2	2	14.46	.009	0.40	.007	-0.18	.009	4		1580,1736
	+60 00073	LS I +61 161			0 34 17	+61 05.1	2	9.65	.005	0.57	.000	-0.43	.005	4	B1 Ib	41,1012
3419	−60 00043				0 34 17	−60 09.4	1	9.56		0.42		-0.03		2	F5 V	1730
		G 69 - 6			0 34 22	+28 53.7	1	13.26		1.56		1.15		1		333,1620
3403	−25 00222	G 266 - 138			0 34 22	−24 46.1	2	8.18	.020	0.64	.005	0.10	.025	2	G8/K0 V	79,3061
		LS I +62 135			0 34 23	+62 33.8	1	10.53		0.28		-0.58		2		41
3413	−4 00064				0 34 24	−03 40.5	1	7.88		1.18		1.19		3	K0	1657
		CS 29527 # 39			0 34 24	−20 13.3	1	14.55		0.13		0.17		1		1736
3444	−65 00058	HR 160			0 34 26	−65 24.0	3	6.41	.003	1.26	.000	1.38	.000	22	K2/3 III	15,1075,2038
3411	+23 00084	HR 156			0 34 29	+23 44.4	2	6.17	.010	1.18	.010	1.27	.010	5	K2 III	1501,1733
	−25 00224				0 34 29	−25 14.0	1	11.75		0.75		0.29		1		3064
		WLS 40 0 # 10			0 34 30	−00 22.3	1	11.84		0.54		0.00		2		1375
		BPM 46455			0 34 30	−25 31.	1	11.98		0.77		0.33		2		3064
3412	+14 00078				0 34 32	+14 52.0	1	8.11		0.56		0.11		2	F5	1648
		AAS35,353 # 6			0 34 36	−73 33.9	1	13.60		-0.07		-0.43		3	B6	767
3366	+72 00035	IDS00316N7221	A		0 34 38	+72 37.2	2	6.95	.000	-0.01	.005	-0.63	.020	7	B3	399,555
	−20 00095	G 268 - 10			0 34 38	−19 44.7	1	10.92		0.81		0.31		1		3061
3421	+34 00086	HR 157			0 34 40	+35 07.5	2	5.42	.073	0.88	.005	0.45	.034	3	G2.5 IIa	252,8100
3431	+39 00138				0 34 44	+40 03.5	1	6.86		0.02		0.01		2	A0	401
		AAS35,353 # 7			0 34 45	−73 24.7	1	14.25		-0.06		-0.43		3	B6	767
3443	−25 00225	HR 159		⋆ AB	0 34 47	−25 02.5	7	5.57	.010	0.72	.009	0.20	.009	21	K1 V +G	15,214,678,1197,2012*
		LP 825 - 507			0 34 48	−22 36.	1	16.08		1.51						3061
		BPM 46458			0 34 48	−29 55.	1	14.28		1.26		0.95		1		3064
		BPM 46459			0 34 48	−32 15.	1	12.74		0.95		0.64		1		3061
		BPM 46462			0 34 48	−34 53.	1	13.67		1.46		1.08		1		3061
3408	+66 00043				0 34 52	+67 22.5	1	7.40		1.68				3	K5	369
3460	−37 00205	G 267 - 141		⋆ AB	0 34 54	−37 33.7	3	6.99	.009	0.72	.009	0.17	.005	10	G8 V	1311,2012,3061
3457	+2 00080	HR 161			0 34 56	+02 51.7	5	6.39	.009	1.34	.009	1.52	.000	15	K4 III	15,1003,1071,1355,3040
		G 131 - 80			0 34 56	+17 34.0	1	11.98		1.18		1.10		1	K4-5	333,1620
		G 266 - 141			0 34 56	−21 10.1	1	14.50		0.47		-0.54		4		3060
		CS 22882 # 26			0 34 58	−32 02.8	2	14.43	.010	0.44	.009	-0.20	.007	3		1580,1736
3488	−55 00130	HR 162			0 34 59	−54 40.1	3	6.40	.008	0.99	.005			11	K0 II/III	15,1075,2035
		SAO 4166			0 35 00	+79 37.0	1	10.55		0.60		0.07		2		1375
		BPM 46464			0 35 00	−34 21.	1	12.68		0.87		0.52		2		3061
3447	+41 00096				0 35 01	+42 15.0	1	7.92		0.45		0.04		2	F5	1375
		G 69 - 7			0 35 02	+23 22.9	1	12.04		1.37		1.20		1	M0	333,1620
		CS 22882 # 21			0 35 05	−30 07.5	2	13.84	.010	0.39	.007	-0.20	.009	3		1580,1736
		WD0035+124			0 35 08	+12 29.2	1	16.38		-0.23		-1.19		2	DA1	1727
		LP 705 - 68			0 35 09	−09 42.9	1	11.52		0.82		0.45		1		1696
	+25 00090				0 35 12	+25 33.4	1	8.52		0.89		0.61		1	G8 III	3077
		BPM 46465			0 35 12	−21 15.	1	12.73		0.65		0.00		1		3061
		BPM 46468			0 35 12	−34 55.	1	12.40		1.51		1.06		1		3061
		Florsch 90			0 35 12	−72 50.3	1	12.35		0.42		-0.09		3	F2	767
		AAS35,353 # 9			0 35 14	−72 54.4	1	13.42		0.80		0.39		3	G9	767
		CS 29527 # 38			0 35 15	−20 24.7	1	14.41		0.25		-0.08		1		1736
		CS 22179 # 14			0 35 18	−06 54.3	1	14.55		0.19		0.18		1		1736
		BPM 46466			0 35 18	−28 08.	1	12.73		1.20		0.96		3		3064
		BPM 46467			0 35 18	−31 59.	1	13.02		1.27		1.12		2		3061
3517	−56 00121				0 35 19	−55 56.1	1	8.68		1.46		1.20		6	M6 III	3007
		GD 630			0 35 25	−22 46.1	1	13.40		0.65		-0.04		1		3061
3503	+1 00108				0 35 26	+02 29.3	2	7.61	.005	1.09	.015	0.90	.010	3	K0	1355,1375
		G 218 - 5			0 35 28	+52 03.5	1	10.48		1.42		1.27		2	M0	7010
		BPM 46470			0 35 30	−20 36.	1	13.67		0.71		0.08		1		3061
		BPM 46472			0 35 30	−21 07.	1	13.65		1.00		0.72		1		3061
		G 266 - 143			0 35 30	−25 11.4	1	12.78		1.40		1.18		1		3064
		BPM 46475			0 35 30	−26 04.	1	12.00		1.25		1.17		1		3064
		BPM 46476			0 35 30	−32 07.	1	13.71		0.68		0.03		2		3061
	−37 00211	G 267 - 142			0 35 30	−36 41.9	1	10.77		0.77		0.25		1	K0	3061
3512	−1 00075	IDS00330S0103	A		0 35 31	−00 46.7	3	6.66	.007	1.29	.024	1.38	.005	7	K2 III	1003,1311,3040
3512	−1 00075	IDS00330S0103	B		0 35 31	−00 46.7	1	13.53		0.64		0.15		4		3024
3601	−76 00068				0 35 31	−75 52.9	1	7.97		0.21		0.09		3	A2 mA3-A7	1117
3513	−9 00119				0 35 32	−09 07.4	1	10.15		0.12		0.17		4	A0	152
		G 158 - 104			0 35 33	−07 21.4	2	13.49	.000	0.89	.005	0.39	.027	3		1658,3062
		BPM 46478			0 35 34	−26 26.1	1	12.58		0.80		0.36		1		3061
	−26 00192				0 35 34	−26 26.1	1	10.54		0.54		0.00		2	F8	3064
		LS I +66 007			0 35 35	+66 48.8	1	10.20		0.88		-0.15		2		405
		GR 14			0 35 37	−30 48.7	1	14.30		1.44		1.20		2		481
		GR 15			0 35 40	−22 33.6	1	15.11		1.46		1.27		1		481
3489	+59 00091				0 35 41	+60 03.0	2	6.77	.020	1.73	.030	1.79	.005	5	K3 Ib-II	1355,8100
		CS 22882 # 27			0 35 43	−32 04.4	2	15.11	.000	0.42	.000	-0.26	.012	2		966,1736
	+46 00127	G 172 - 16			0 35 45	+47 21.5	1	10.90		0.58		-0.07		1	F8	1658
3520	+37 00106				0 35 46	+37 44.7	1	8.32		0.49		0.00		2	F5	1375
	+60 00075				0 35 49	+61 27.5	1	9.53		1.90				2	K7 III	369
		G 266 - 144			0 35 49	−18 58.9	1	11.71		1.09		0.81		1		3061
	−15 00115				0 35 50	−15 16.4	5	10.87	.007	-0.20	.008	-0.81	.022	37	B2	830,989,1026,1729,1783
		LS I +59 049			0 35 51	+59 47.1	1	11.96		0.73		-0.17		2		41
3560	−32 00217				0 35 53	−32 18.6	1	8.81		0.51		0.03		1	F7 V	3061
3546	+28 00103	HR 163			0 35 54	+29 02.4	7	4.36	.015	0.87	.004	0.47	.005	38	G5 III	15,1003,1363,3016*
3440	+81 00013	HR 158			0 35 54	+82 13.1	1	6.40		0.55		-0.01		2	F8	1375
		BPM 46481			0 35 54	−26 40.	1	13.24		0.50		-0.15		2		3064

HD	DM	Other Id	N Rem	α_{1950}	δ_{1950}	S	V	σ_V	B–V	σ_{B-V}	U–B	σ_{U-B}	n	Spectrum	References
		LB 3147		0 35 54	−61 21.	1	12.09		0.07		0.14		2		45
		Steph 63		0 35 55	+08 41.1	1	12.34		1.07		1.03		1	K4	1746
		NGC 188 - 3149		0 35 56	+85 00.9	1	15.30		0.82		0.42		1		217
3519	+59 00092			0 35 57	+59 33.1	2	6.73	.009	-0.02	.018	-0.41		5	B8 II	1118,1501
	+1 00111			0 35 58	+02 23.6	1	10.65		0.16		0.06		2	A2	1026
3556	+5 00085			0 35 59	+05 44.9	1	8.77		0.59		0.10		2	G0 V	1003
3567	−9 00122	G 270 - 23		0 36 00	−08 34.6	9	9.25	.010	0.46	.007	-0.15	.006	18	F5 V	516,742,1003,1064*
3571	−31 00241			0 36 00	−31 30.1	1	10.47		0.77		0.35		1	G5/8 (III)	3061
		AAS35,353 # 10		0 36 00	−73 12.2	1	14.25		-0.06		-0.40		3	B6	767
3580	−21 00084	BB Cet		0 36 02	−20 34.3	2	6.75	.005	-0.12	.005	-0.52	.045	6	Ap Si	1068,3007
		NGC 188 - 3002		0 36 03	+84 55.7	1	15.60		0.57		0.03		1		217
		AAS35,353 # 11		0 36 04	−73 06.6	1	13.61		0.75		0.35		3	G8	767
		BPM 46485		0 36 06	−24 31.	1	12.71		1.35		1.28		2		3061
		CS 29527 # 52		0 36 07	−22 07.5	1	15.35		0.12		0.20		1		1736
	+30 00089	G 69 - 8	⋆ A	0 36 08	+30 44.7	1	9.04		0.63		0.04		1	G0	333,1620
		G 69 - 9		0 36 10	+32 45.9	2	11.03	.019	1.13	.003	1.01	.020	4	K5	1620,1723
		SB 257		0 36 10	−29 14.9	3	14.01	.005	0.14	.005	0.17	.009	5	A0:pec	295,966,1736
		LP 825 - 615		0 36 12	−25 18.	1	13.33		0.92		0.57		1	M4	3061
3659	−73 00041			0 36 12	−73 28.5	1	9.89		0.51				1	F0/F2V:	2001
	+73 00026	G 242 - 61		0 36 13	+74 28.5	1	9.24		0.68		0.14		2		3026
		SB 259		0 36 13	−24 43.4	2	13.95	.000	0.05	.010	0.20	.044	3	A0	295,1736
		G 69 - 10		0 36 14	+30 20.5	1	11.06		1.54		1.20		1	M1	3078
		CS 29527 # 53		0 36 16	−22 21.9	1	15.21		0.06		0.20		1		1736
3606	−27 00187	IDS00338S2743	AB	0 36 17	−27 26.8	1	9.34		0.61		0.04		1	G3 V	3061
3604	−19 00092			0 36 18	−18 45.7	4	9.59	.023	-0.01	.009	-0.07	.012	15	A0 V	152,1026,1068,3007
		NGC 188 - 3160		0 36 19	+84 56.0	1	17.58		1.06				1		217
		NGC 188 - 3159		0 36 20	+84 54.0	1	16.94		0.91				1		217
		NGC 188 - 3004		0 36 21	+84 55.1	1	11.66		1.14		1.08		7		217
3574	+48 00192	HR 164	⋆ A	0 36 23	+49 04.8	1	5.43		1.64				2	K7 III	71
	−25 00235	IDS00339S2539	B	0 36 23	−25 22.9	1	10.01		0.17		0.07		1	A2	3061
		CS 22179 # 13		0 36 24	−06 05.8	1	14.85		0.04		0.10		1		1736
		LP 825 - 617		0 36 24	−25 33.	1	14.06		0.88		0.44		1		3061
		BPM 46487		0 36 24	−27 47.	1	12.87		0.90		0.65		3		3064
3689	−74 00053			0 36 24	−74 14.2	7	7.42	.018	0.48	.009	-0.04	.019	23	F6 V	14,1117,1408,1425*
3622	−26 00196	IDS00339S2609	AB	0 36 25	−25 52.2	1	7.77		0.22				4	A5 V	2012
		WLS 48 40 # 6		0 36 26	+39 00.7	1	11.84		0.40		-0.02		2		1375
		CS 22170 # 20		0 36 26	−12 00.8	1	14.91		0.38		-0.17		1		1736
3619	−13 00108			0 36 26	−12 35.3	1	9.99		0.12		0.11		4	A1 V	152
		LS I +61 162		0 36 28	+61 57.6	1	11.90		0.42		-0.50		2		41
3657	−57 00132			0 36 28	−56 56.4	1	9.25		0.45		-0.16		3	G8/K0III/IV	3037
		CS 29527 # 47		0 36 30	−19 21.6	1	14.79		0.15		0.11		1		1736
		BPM 46488		0 36 30	−20 55.	1	13.13		1.17		0.99		1		3061
3646	−43 00167			0 36 30	−43 34.2	2	7.66	.001	0.96	.004	0.78	.019	9	G8/K0 III	1117,1673
		NGC 188 - 3001		0 36 32	+84 56.1	1	15.52		0.69		0.21		2		217
3630	−22 00109			0 36 33	−21 33.1	1	9.31		0.61		0.03		1	G2/3 V	3061
3628	+2 00084	G 1 - 9		0 36 36	+02 51.3	5	7.34	.020	0.63	.010	0.12	.005	10	G2 V	22,202,333,1003,1620,3079
		BPM 46490		0 36 36	−26 55.	1	14.20		0.91		0.81		2		3064
		BPM 46491		0 36 36	−28 33.	1	13.32		0.85		0.30		3		3064
		NGC 188 - 3162		0 36 37	+84 55.5	1	17.18		0.95				1		217
		NGC 188 - 3145		0 36 37	+85 02.4	1	14.97		0.69		0.12		1		217
3627	+30 00091	HR 165	⋆ A	0 36 39	+30 35.3	7	3.27	.029	1.28	.013	1.49	.015	17	K3 III	15,1119,1355,1363*
3627	+30 00091	IDS00340N3019	B	0 36 39	+30 35.3	1	12.44		1.47				1	M2	3016
		AAS35,353 # 12		0 36 39	−73 51.1	1	14.21		-0.17		-0.74		2	B2	767
3719	−73 00042	HR 169		0 36 44	−73 24.7	11	6.85	.016	0.12	.008	0.11	.007	80	A1 mA3-A7	14,15,978,1075,1408*
3651	+20 00085	HR 166	⋆ A	0 36 45	+20 58.9	12	5.88	.019	0.85	.008	0.57	.009	60	K0 V	1,3,15,22,1077,1118*
3625	+51 00120			0 36 45	+52 17.4	1	8.61		0.46		-0.01		2	F8	1566
3616	+60 00078	IDS00339N6109	AB	0 36 45	+61 25.6	1	8.48		0.09				2	A0	1118
		BPM 46493	A	0 36 48	−29 23.	1	13.27		1.21		1.20		1		3064
		BPM 46493	B	0 36 48	−29 23.	1	14.28		1.35				1		3064
		BPM 46496		0 36 48	−30 04.	1	11.77		0.68		-0.01		2		3064
		NGC 188 - 3163		0 36 53	+84 55.5	1	17.41		0.75				1		217
		BPM 46495		0 36 54	−26 32.	1	14.08		1.01		0.74		2		3064
		LS I +60 117		0 36 56	+60 41.0	3	10.00	.015	0.35	.015	-0.60	.005	7	O9 IV	41,1011,1012
3637	+62 00130			0 36 56	+62 57.2	1	7.74		0.50				3	F6 IV	1118
		NGC 188 - 3161		0 36 58	+84 56.5	1	16.98		0.90				1		217
3677	−25 00242			0 36 58	−24 47.9	1	9.49		0.62		0.03		1	G2/3 V	3061
		AAS35,353 # 13		0 36 58	−72 25.6	1	13.27		0.49		-0.04		3	F5	767
3738	−72 00057			0 37 02	−72 33.4	1	8.48		0.29		0.06		7	A9 V	1408
		CS 22179 # 11		0 37 03	−04 52.1	1	13.10		0.24		0.15		1		1736
	+9 00073	G 1 - 10		0 37 04	+10 23.0	2	10.55	.005	1.15	.015	0.97		2	K5:	1620,1746
	−27 00193			0 37 04	−27 07.9	1	11.68		1.16		1.14		2		3061
3647	+61 00144			0 37 05	+62 11.8	1	9.11		0.27				3	A0	1118
		NGC 188 - 3155		0 37 05	+84 59.4	1	16.10		0.74		0.26		1		217
		G 266 - 147		0 37 08	−27 08.0	1	11.68		1.19		1.14		1		3064
	+84 00011	NGC 188 - 3138		0 37 12	+85 04.4	1	9.78		1.52		1.73		5	K5	217
		G 1 - 11		0 37 13	−00 00.4	1	13.51		1.55				2		202
	−27 00194	G 266 - 148		0 37 13	−26 43.9	1	10.23		1.29		1.20		4		3064
	+59 00096	LS I +59 050		0 37 14	+59 40.5	1	9.96		0.74		0.53		2	A4 Ib	41
	+70 00035	G 242 - 62	⋆ A	0 37 14	+70 45.1	2	9.91	.015	0.60	.005	-0.02	.000	3	G5	1620,1658
		NGC 188 - 1		0 37 14	+85 06.4	1	13.04		1.02		0.74		1		4002

Table 1   45

HD	DM	Other Id	N Rem	α_{1950}	δ_{1950}	S	V	σ_V	B–V	σ_{B-V}	U–B	σ_{U-B}	n	Spectrum	References
		GD 8		0 37 15	+31 16.1	1	14.66		-0.22		-1.20		1	DAwk	3060
	-27 00195	BPM 46509		0 37 16	-27 15.9	1	10.82		0.76		0.47		2		3064
3690	+20 00087	HR 167	⋆ A	0 37 17	+21 09.9	1	5.36		1.16		1.05		1	K0 III	3016
3690	+20 00087	HR 167	⋆ AB	0 37 17	+21 09.9	1	5.40		1.09		0.91		1	K0 III+F3 V	4001
3690	+20 00087	IDS00347N2053	B	0 37 17	+21 09.9	1	8.67		0.40		-0.03		2	F3 V	3016
3673	+59 00097			0 37 18	+59 55.6	1	8.55		0.08				2	A0	1118
		BPM 46507		0 37 18	-20 08.	1	10.98		0.71		0.21		1		3061
		BPM 46508		0 37 18	-25 01.	1	12.22		1.01		0.78		1		3061
3715	-15 00118			0 37 19	-14 34.4	2	9.47	.015	0.44	.010	-0.07		3	F3 V	742,1594
		NY Cas		0 37 20	+58 19.5	1	13.09		0.75				1		1772
3737	-42 00209			0 37 21	-41 56.0	2	8.44	.005	0.57	.009			8	F6/7 V	1075,2013
3722	-25 00245		A	0 37 23	-25 18.4	1	8.76		0.70		0.13		4	G6 V	3064
3722	-25 00245		B	0 37 23	-25 18.4	1	13.74		1.39		1.24		1		3061
3736	-38 00205			0 37 26	-37 41.0	2	8.58	.009	0.19	.005	0.14		8	A6 V	1068,2012
3681	+58 00088			0 37 27	+59 11.4	1	7.04		1.11		1.05		3	K0 IIIp	37
		G 266 - 150		0 37 28	-28 18.5	1	12.70		0.68		-0.01		3		3064
		NGC 188 - 2046		0 37 29	+84 59.2	2	13.34	.008	0.55	.013	0.15	.008	7		217,217
3750	-45 00201	HR 171		0 37 29	-45 04.3	4	6.00	.005	1.14	.005	1.09		17	K1 III	15,1075,1117,2028
		BPM 46517		0 37 30	-27 32.	1	13.02		0.69		-0.03		2		3064
3735	-34 00224	HR 170, Z Scl		0 37 30	-34 14.1	4	6.68	.008	0.52	.006	-0.03	.016	11	F7 V	58,1311,2035,3061
3699	+58 00089			0 37 32	+59 14.4	1	7.26		1.92				2	K5 III	369
		AAS35,353 # 14		0 37 32	-72 43.0	1	13.52		0.50		-0.04		3	F5	767
	-44 00170			0 37 33	-44 31.7	1	11.39		1.51				1	M0.5	1705
		NGC 188 - 2028		0 37 35	+84 57.6	1	15.86		0.70		0.23		1		217
		BD -9 00129b		0 37 36	-09 12.0	1	9.59		0.40		-0.05		6		1732
3748	-9 00129			0 37 37	-09 12.3	1	9.86		0.13		-0.41		6	A5	1732
3712	+55 00139	HR 168, α Cas	⋆ A	0 37 39	+56 15.8	6	2.22	.010	1.17	.007	1.13	.008	30	K0 II-III	15,1118,1119,3016*
3743	+23 00092	IDS00350N2330	A	0 37 41	+23 46.8	1	7.21		0.19		0.14		2	A3	3024
3743	+23 00092	IDS00350N2330	B	0 37 41	+23 46.8	1	9.51		0.34		0.06		3	A3	3024
3743	+23 00092	IDS00350N2330	C	0 37 41	+23 46.8	1	11.10		0.56		0.09		1		3016
3761	-29 00190			0 37 41	-28 50.6	1	9.94		0.60		0.07		2	G3 V	3064
		BPM 70297		0 37 42	-19 42.	1	12.06		0.94		0.60		1		3061
		G 1 - 12		0 37 47	+07 06.1	2	13.31	.005	0.82	.005	0.32	.010	2		333,1620,1658
		WLS 40 20 # 10		0 37 47	+19 50.4	1	12.91		1.09		0.90		2		1375
	+63 00070	LS I +63 090		0 37 48	+63 38.5	1	9.17		0.72		-0.37		2	B0 Ib	1012
		WLS 40 0 # 7		0 37 49	-02 11.8	1	11.70		0.39		-0.04		2		1375
3724	+59 00098			0 37 50	+60 03.0	1	8.69		0.49				2	F5	1118
		NGC 188 - 2026		0 37 52	+84 57.0	1	13.68		0.95		0.65		1		4002
	+84 00012	NGC 188 - 2029		0 37 52	+84 57.8	1	9.55		0.50		0.03		7	M0 III	217
		BPM 46523		0 37 54	-21 18.	1	11.71		0.64		-0.01		1		3061
3782	-21 00090			0 37 56	-21 16.1	1	9.77		0.82		0.39		1	G8/K0 IV/V	3061
3794	-17 00109	HR 172	⋆ AB	0 37 58	-16 47.4	4	6.49	.005	0.92	.004	0.53	.000	15	G8 III	15,79,1071,1075
		BPM 46527		0 38 00	-29 42.	1	12.56		0.77		0.26		2		3064
		BPM 46529		0 38 00	-32 59.	1	12.52		0.75		0.28		1		3061
		LP 937 - 87		0 38 00	-33 49.0	2	12.37	.010	0.94	.007	0.72	.047	2		1696,3061
3795	-24 00263	HR 173	⋆	0 38 02	-24 04.4	6	6.14	.005	0.71	.011	0.21	.003	17	G3 V	15,79,1075,1311,2007,3079
3808	-27 00200	G 266 - 151		0 38 03	-27 32.9	2	8.79	.009	0.72	.012	0.18	.023	4	G5 V	79,3064
3765	+39 00154	G 132 - 15		0 38 04	+39 55.3	6	7.36	.008	0.94	.006	0.73	.020	19	K2 V	22,333,1003,1197,1620*
3810	-30 00194			0 38 05	-30 08.6	2	8.72	.050	0.63	.015	0.09	.030	2	G3/5 V	79,3061
	+66 00052	LS I +66 008		0 38 06	+66 34.3	1	10.77		0.74		0.19		1	A0 Iab:	1215
		NGC 188 - 2027	⋆ V	0 38 06	+84 57.2	1	15.30		0.68		0.22		1	G3 V	217
		LB 1566		0 38 06	-55 18.	2	13.09	.014	-0.30	.009	-1.17	.018	5	sdO	45,832
3823	-60 00046	HR 176		0 38 06	-59 44.1	5	5.89	.009	0.56	.007	0.01	.005	21	F6/8 V	15,1311,2012,2012,3079
		LS I +63 092		0 38 09	+63 08.9	1	11.59		0.90		-0.07		3		483
3807	-5 00101	HR 174	⋆ A	0 38 10	-04 37.6	3	5.92	.018	1.09	.010	0.87	.022	10	K0 III	15,1061,3016
		NGC 188 - 2025		0 38 11	+84 56.8	1	15.57		0.92		0.49		1		217
		SB 272		0 38 12	-26 03.	1	13.04		0.25		0.07		1	A5	295
3777	+56 00108			0 38 14	+56 53.8	1	7.54		0.11		0.18		2	Am	8100
		LS I +60 118		0 38 15	+60 35.4	1	12.29		0.44		-0.46		3		41
3821	-8 00117	IDS00357S0747	A	0 38 15	-07 30.3	1	7.02		0.62		0.08		2	G0	3016
3821	-8 00117	IDS00357S0747	AB	0 38 15	-07 30.3	1	6.96		0.65		0.11		4	G0	158
3821	-8 00117	IDS00357S0747	B	0 38 15	-07 30.3	1	10.02		1.22		1.08		2		3016
3836	-34 00230	G 267 - 155		0 38 16	-34 34.3	2	9.85	.010	0.59	.002	0.10	.029	3	G3 V	79,3061
	+62 00133	LS I +62 136		0 38 18	+62 39.7	1	10.14		0.76		0.26		2	A0 II	41
		BPM 46533		0 38 18	-26 32.	1	14.79		0.84		0.36		2		3064
		BPM 46535		0 38 18	-28 20.	1	12.39		0.86		0.41		2		3064
		CS 29497 # 30		0 38 19	-24 24.0	1	12.65		0.32		-0.14		1		1736
		NGC 188 - 2		0 38 20	+85 06.6	2	14.85	.037	0.78	.009	0.36	.005	4		217,217
		CS 22170 # 9		0 38 21	-10 36.0	1	14.00		0.04		0.13		1		1736
3948	-81 00009			0 38 23	-80 55.9	1	7.90		0.74				4	G2 IV/V	2012
3817	+38 00090	HR 175		0 38 24	+39 11.1	6	5.32	.012	0.89	.007	0.60	.004	25	G8 III	15,1007,1013,1355*
	-23 00254	G 266 - 154		0 38 24	-23 16.7	2	10.37	.010	0.73	.002	0.27	.015	2	G5	79,3061
3844	-28 00196	G 266 - 153		0 38 24	-27 57.3	1	9.79		0.66		0.10		3	G3/5 V	3064
	+18 00092	G 32 - 40		0 38 29	+18 58.9	1	9.92		0.93		0.75		1	K3	333,1620
3827	+38 00091			0 38 29	+39 19.8	3	8.01	.008	-0.24	.007	-1.01	.006	18	B0.7 V((n))	399,401,1775
	+49 00160			0 38 31	+49 55.1	1	10.28		1.11		0.88		2		1375
3867	+40 00180	IDS00361S4105	AB	0 38 31	-40 48.3	1	8.31		0.10		0.05		2	A1 V	152
	-52 00078			0 38 31	-51 38.9	2	11.28	.000	0.50	.005	-0.25	.029	5	sdG8	1097,1696
		NGC 188 - 3137		0 38 33	+85 03.3	2	12.72	.030	0.56	.005	0.03	.000	4		217,217
3861	+8 00094			0 38 37	+09 04.9	1	6.55		0.49		0.01		3	F8 V	3016

HD	DM	Other Id	N	Rem	α_{1950}	δ_{1950}	S	V	σ_V	B−V	σ_{B-V}	U−B	σ_{U-B}	n	Spectrum	References
	−20 00117	G 266 - 155			0 38 40	−20 08.2	1	11.32		0.82				1		3061
		SB 276			0 38 40	−26 13.0	2	13.84	.019	0.13	.002	0.17	.007	3	A0	295,1736
3877	−27 00203	IDS00362S2723		AB	0 38 43	−27 06.7	1	9.99		0.57		−0.03		1	G0 V	3061
	+40 00143				0 38 46	+41 05.6	1	9.96		1.32		1.45		4		894
		G 69 - 12			0 38 48	+30 59.7	1	11.64		1.36		1.24		1	M2	333,1620
		NGC 188 - 3136			0 38 48	+85 05.9	2	12.28	.008	0.63	.004	0.09	.000	7		217,217
	−25 00259	BPM 46543			0 38 48	−25 16.0	1	10.40		0.57		0.05		3	G0	3061
		BPM 46546			0 38 48	−27 30.	1	14.08		1.16		1.04		1		3064
		BPM 46547			0 38 48	−31 38.	1	12.78		0.53		−0.13		2		3061
		BPM 46548			0 38 48	−36 06.	1	15.39		0.61		−0.02		1		3061
		CS 22170 # 13			0 38 49	−08 39.8	1	13.21		−0.02		−0.10		1		1736
3885	−20 00118				0 38 49	−20 08.2	3	9.81	.025	−0.09	.007	−0.29	.019	11	Ap Si	152,1026,1068
	−13 00116				0 38 53	−13 27.5	1	10.73		1.32				2	K5	1619
		AAS35,353 # 16			0 38 55	−72 52.0	1	14.62		−0.02		0.03		4	A0	767
		NGC 188 - 2067			0 38 56	+85 02.1	1	14.93		0.75		0.23		1		217
3883	+23 00094	HR 178		⋆ A	0 38 57	+24 21.3	2	6.04	.000	0.26	.000	0.21	.010	6	A7 m	3058,8071
3908	−13 00117				0 38 57	−12 57.9	1	8.81		1.07		0.90		2	G8 III/IV	1375
		G 266 - 157			0 38 58	−22 37.3	1	14.60		0.65		−0.32		4		3061
3919	−46 00180	HR 180			0 38 58	−46 21.6	5	4.59	.003	0.97	.005	0.72	.004	22	G8 III	15,1075,2012,8015,8029
3909	−21 00091				0 39 00	−20 34.4	1	7.68		1.57		1.89		3	K4 III	1657
		PASP80,437 # 2			0 39 00	−26 37.	1	12.25		1.12		0.58		1		622
3856	+65 00083	HR 177			0 39 03	+65 52.4	1	5.83		1.04		0.87		5	G9 III-IV	1355
	−24 00272	G 266 - 158			0 39 03	−24 15.9	1	11.57		0.99		0.75		1		1696
	−34 00239	G 267 - 156			0 39 06	−33 53.8	2	10.60	.050	1.44	.000	1.17	.000	5		481,3062
3892	+38 00094				0 39 08	+38 37.3	3	7.64	.010	−0.05	.005	−0.26	.022	13	B9 V	401,833,1775
		NGC 188 - 2043			0 39 08	+84 58.5	1	15.87		0.73		0.20		1		217
		Steph 69			0 39 10	−09 35.5	1	12.98		1.54		1.30		2	M0	1746
		G 32 - 41			0 39 12	+18 03.4	1	10.99		1.27		1.25		1	K7	333,1620
		CS 22179 # 21			0 39 13	−03 16.9	1	14.41		0.14		0.19		1		1736
		NGC 188 - 14			0 39 14	+84 49.7	2	10.76	.005	0.62	.000	0.21	.005	3		217,217
		Florsch 109			0 39 15	−72 40.1	1	11.58		0.54		0.06		3	F7	767
3901	+49 00164	HR 179			0 39 16	+50 14.3	6	4.80	.015	−0.11	.007	−0.62	.026	16	B2.5V	15,154,1119,1223,1363,8015
		NGC 188 - 3018			0 39 17	+84 53.1	3	11.39	.022	1.56	.019	1.82	.039	4	K4 IIIb	217,387,4002
		BPM 46554			0 39 18	−27 14.	1	14.00		0.88		0.43		3		3064
		BPM 46557			0 39 18	−31 31.	1	12.77		1.48		1.21		1		3061
		WLS 40-15 # 10			0 39 20	−15 16.4	1	11.86		0.80		0.38		2		1375
		G 267 - 157			0 39 20	−35 37.3	1	13.38		1.30		1.10		4		3079
	−34 00241				0 39 21	−33 56.1	2	11.22	.004	0.48	.012	−0.08	.011	11		1698,1765
		NGC 188 - 2047			0 39 23	+84 59.6	1	13.47		0.66		0.09		1		217
		NGC 188 - 4			0 39 24	+85 04.	2	13.29	.010	0.34	.020	0.16	.005	2		217,217
		NGC 188 - 6			0 39 24	+85 04.	2	12.10	.005	0.53	.009	0.04	.014	5		217,217
		NGC 188 - 1201			0 39 24	+85 04.	1	12.25		0.62		0.06		1		217
		NGC 188 - 200			0 39 24	+85 04.	1	14.89		0.78				1		217
	+40 00144				0 39 25	+41 06.9	1	10.35		1.14		1.05		4		894
		WLS 40 20 # 7			0 39 27	+17 50.8	1	12.62		0.70		−0.01		2		1375
	+64 00071				0 39 28	+65 26.9	1	9.63		1.99				2		369
		CS 29527 # 59			0 39 28	−21 14.6	1	14.77		−0.02		0.05		1		1736
		G 69 - 13			0 39 29	+24 40.6	1	12.87		1.36		1.19		1		333,1620
		LS I +63 094			0 39 29	+63 46.6	1	10.75		0.87		−0.26		2	B0 pe(III)?	1012
3953	−10 00136			V	0 39 29	−09 55.3	1	8.83		1.51		1.34		20	M2	1732
		PG0039+049			0 39 30	+04 53.3	1	12.88		−0.02		−0.87		3	sdB	1764
		BPM 30295			0 39 30	−47 46.	1	14.49		0.49		−0.22		6		3065
3980	−57 00143	HR 183, ξ Phe		⋆ AB	0 39 30	−56 46.6	5	5.69	.010	0.20	.009			19	Ap SrEuCr	15,1075,2013,2024,2029
		NGC 188 - 1059			0 39 33	+84 58.7	1	13.95		1.01		0.69		1		217
		AAS35,353 # 17			0 39 34	−73 03.4	1	13.33		0.03		−0.01		3	A0	767
3979	−43 00190				0 39 35	−43 23.9	1	7.06		0.36		0.09		2	G8 II/III	1117
4229	−86 00009				0 39 36	−85 58.5	2	6.79	.000	1.26	.000	1.45		5	K3 III	258,1075
3924	+57 00132	HR 181			0 39 38	+58 28.8	1			−0.01		−0.12		1	B9.5III	1079
	−13 00120				0 39 38	−12 40.2	1	10.59		0.81		0.46		1		79
	−16 00117				0 39 38	−15 38.1	1	10.47		0.29		0.22		4	A2	1068
		AAS35,353 # 18			0 39 38	−72 38.6	1	14.26		0.48		−0.14		3	F0	767
		NGC 188 - 2012			0 39 39	+84 56.4	1	16.23		0.72		0.24		1		217
		G 269 - 19			0 39 40	−34 57.0	1	15.26		0.87		0.57		1		3061
		LS I +62 137			0 39 42	+62 58.4	1	11.66		0.79		0.08		2		41
		SB 283			0 39 42	−33 17.	1	14.54		0.08		0.21		2		295
		AJ89,1229 # 23			0 39 48	+07 50.4	1	14.32		1.52				2		1532
3972	+3 00093	IDS00372N0337		AB	0 39 49	+03 53.6	3	7.59	.015	0.51	.011	0.02	.008	9	F6 V +G7 V	176,292,938
3970	+26 00108				0 39 49	+27 22.2	1	7.20		0.34		0.19		3	F2 II	1501
3950	+51 00133	V486 Cas			0 39 49	+52 03.8	1	6.91		0.12		−0.68		2	B1 III	1003
		NGC 188 - 1071			0 39 50	+84 57.1	1	15.09		0.67		0.14		1		217
3949	+61 00152				0 39 51	+62 29.5	1	7.79		0.15				2	A0	1118
		NGC 188 - 2009			0 39 51	+84 55.3	1	15.31		0.74		0.19		1		217
		MN83,95 # 101			0 39 51	−73 50.9	1	11.17		1.32				1		591
		WLS 40 20 # 5			0 39 52	+22 24.8	1	11.87		0.42		0.00		2		1375
3940	+63 00081	LS I +64 031			0 39 52	+64 01.0	2	7.27	.004	0.72	.009	0.05		5	A1 Ia	1012,8031
		G 267 - 158			0 39 53	−36 59.5	1	12.74		1.56				1		3061
		TonS 166			0 39 54	−26 33.	1	11.18		0.56		−0.31		4		1401
		SB 285			0 39 54	−28 50.	1	13.62		0.18		0.12		2	A3	295
3999	−32 00254				0 39 54	−32 11.4	2	9.34	.028	0.12	.009	0.10	.009	8	A2 V	152,1068
		NGC 188 - 1069			0 39 55	+84 57.7	3	12.32	.038	1.30	.011	1.37	.020	8	K3 IIIab	217,387,4002

Table 1   47

HD	DM	Other Id	N	Rem	α₁₉₅₀	δ₁₉₅₀	S	V	σ_V	B–V	σ_B-V	U–B	σ_U-B	n	Spectrum	References
		CS 22170 # 21			0 39 55	−11 53.4	1	14.16		0.40		-0.20		1		1736
	−33 00239	G 267 - 150			0 39 55	−33 02.9	2	10.98	.019	0.43	.002	-0.18	.005	3		1696,3061
		CS 22170 # 14			0 39 57	−08 04.7	1	12.59		0.13		0.10		1		1736
4011	−34 00245				0 39 57	−34 15.9	1	9.62		0.34		0.12		4	A9 V	1068
		G 267 - 159		A	0 39 58	−33 19.0	2	13.08	.005	0.47	.005	-0.25	.010	4		1696,3061
		G 267 - 159		B	0 39 58	−33 19.0	1	14.03		0.46		-0.20		1		3061
		HV 821			0 39 58	−74 00.	3	11.69	.016	0.95	.050	0.61	.155	3	F8 Ia	689,1511,3075
		BPM 46560			0 40 00	−28 10.	1	12.48		1.35		1.19		2		3064
		Smethells 172			0 40 00	−44 40.	1	11.17		1.24				1		1494
		R 1			0 40 00	−74 00.	1	11.68		0.84				5		8033
		LS I +60 119			0 40 02	+60 11.8	1	11.05		0.43		-0.14		2		41
		CS 22179 # 19			0 40 02	−05 16.9	1	13.87		0.17		0.20		1		1736
	+34 00106	FF And			0 40 03	+35 15.6	2	10.38	.000	1.44	.032	1.08		4	M3	196,1017
3989	+45 00181				0 40 03	+45 39.4	3	7.06	.008	1.52	.000	1.79	.017	15	K5 III	1003,1355,4001
3983	+51 00134				0 40 03	+52 09.5	1	8.44		0.14		-0.26		2	A0	1375
	+61 00153	LS I +61 163			0 40 03	+61 57.7	2	9.31	.009	0.85	.004	0.34	.004	6	A0 Ib	41,1012
		NGC 188 - 2055			0 40 03	+85 00.3	1	15.38		0.65		0.18		1		217
	−57 00144	Smethells 17			0 40 03	−57 09.0	1	10.45		1.16				1	K3	1494
	−18 00113				0 40 04	−17 34.4	1	11.00		0.62		0.05		2		1375
		LS I +64 032			0 40 05	+64 38.7	1	11.29		0.63		-0.38		2	B0.5 IV	1012
		NGC 188 - 3022			0 40 09	+84 52.7	1	15.28		0.69		0.22		1		217
		LP 705 - 89			0 40 12	−12 55.7	1	12.39		0.54		-0.13		2		1696
4053	−36 00241				0 40 12	−36 17.8	2	7.07	.015	1.61	.005	1.99		8	M1 III	2035,3007
		NGC 188 - 3125			0 40 13	+85 05.8	1	16.32		0.76		0.29		1		217
		NGC 188 - 1066			0 40 14	+84 57.7	1	14.97		0.88		0.42		1		217
	−17 00117	G 268 - 30			0 40 14	−17 11.4	1	10.33		0.75		0.34		1		79
	+60 00085	NGC 225 - 8			0 40 15	+61 30.9	1	10.13		0.51		0.44		1	A9 III	49
4014	+15 00106				0 40 16	+16 23.4	1	6.54		1.18		1.21		2	K0	1648
		G 69 - 14			0 40 17	+22 23.1	1	11.55		1.44		1.18		1	M1	1620
		NGC 188 - 1050			0 40 18	+84 59.0	1	13.86		0.55		0.03		3		217
		LP 881 - 311			0 40 18	−28 13.	1	16.00		1.80				1		3061
		BPM 46566			0 40 18	−32 44.	1	13.67		1.06		0.84		1		3061
		NGC 188 - 1065			0 40 19	+84 57.9	1	15.35		0.68		0.13		1		217
		CS 29509 # 3			0 40 19	−31 23.0	1	15.91		0.18		0.16		1		1736
4065	−39 00175	HR 185		⋆ AB	0 40 19	−38 44.2	6	6.05	.008	-0.03	.008	-0.13	.005	22	B9.5V	15,1020,1068,2012*
		AE And			0 40 20	+41 32.8	1	17.00		0.10		-0.81		1		8066
4089	−66 00047	HR 187			0 40 21	−65 44.5	4	5.38	.003	0.49	.005	0.00	.009	24	F6 V	15,1075,2038,3026
	+60 00086	NGC 225 - 4			0 40 22	+61 29.2	2	9.71	.012	1.67	.006	1.09		35	K2 III	49,1655
	+61 00154	V594 Cas			0 40 22	+61 38.2	2	10.57	.038	0.56	.021	-0.28	.030	3	Be	41,206,1283
		NGC 188 - 1056			0 40 25	+84 58.9	1	16.41		0.82		0.26		1		217
	−57 00145				0 40 25	−57 13.2	1	10.41		0.96		0.72		1	K0	1696
		NGC 188 - 1116			0 40 26	+84 56.2	1	13.84		0.95		0.48		1		217
		G 1 - 13			0 40 27	+06 50.9	1	15.23		0.90		0.37		2		1658
4063	−23 00267				0 40 27	−22 53.7	1	10.03		1.01		0.84		1	K1/2 (IV)	3061
4084	−56 00133				0 40 27	−56 07.1	2	8.62	.010	1.17	.005	1.01	.029	3	K1p Ba	565,3048
4088	−60 00048	HR 186			0 40 28	−60 32.2	5	5.98	.005	1.32	.000	1.45		20	K5 III	15,1075,2020,2027,3005
		LS I +60 120			0 40 29	+60 00.2	1	11.57		0.61		-0.25		2		41
		NGC 225 - 17			0 40 29	+61 32.4	1	11.48		0.35		0.08		1	A2	49
4004	+63 00083	LS I +64 034			0 40 29	+64 29.3	2	10.20	.025	0.37	.000	-0.25	.015	4	WN5	1012,1359
	+60 00087	NGC 225 - 9			0 40 30	+61 29.5	2	10.34	.016	0.55	.010	0.07	.013	30	A0 III	49,1655
	+71 00031	G 242 - 65			0 40 30	+71 54.3	5	10.20	.019	0.40	.017	-0.19	.013	13	F5	1003,1064,1620,1658,3016
		G 1 - 14			0 40 30	−00 43.7	2	12.94	.063	1.23	.005	1.16		3		202,1620
		CS 22179 # 22			0 40 30	−03 21.0	1	15.32		0.15		0.15		1		1736
		CS 22170 # 15			0 40 30	−08 31.6	1	14.00		0.31		0.11		1		1736
		CS 22170 # 16			0 40 30	−09 03.0	1	13.95		-0.13		-0.55		1		1736
		AzV 1a			0 40 32	−73 33.2	1	13.96		-0.07		-0.90		1	B3	767
		NGC 188 - 1047			0 40 33	+85 00.6	2	14.93	.005	0.72	.000	0.19	.019	3		217,217
	−38 00222				0 40 33	−38 24.0	1	10.46		-0.22		-0.92		17	B8	1732
4080	−35 00220				0 40 34	−35 35.3	1	10.05		0.69		0.22		1	G8 (IV)(+G)	3061
		NGC 188 - 1057			0 40 36	+84 58.5	1	13.59		1.14		0.92		1	K0 IVa	217
		AAS53,255 # 8			0 40 38	−72 51.	1	13.07		1.80				1		1686
		NGC 188 - 1063			0 40 39	+84 57.8	1	15.85		0.72		0.10		1		217
		G 266 - 160			0 40 40	−20 58.6	1	12.54		1.42				1		3061
4058	+46 00146	HR 184			0 40 40	+46 45.1	2	4.96	.020	0.18	.000	0.09		6	A5 V	1118,3058
		NGC 225 - 18			0 40 41	+61 37.3	1	12.04		0.40		0.28		1	A5	49
		NGC 188 - 3026			0 40 41	+84 51.2	1	14.12		0.91		0.36		1		4002
		NGC 188 - 2212			0 40 41	+84 55.6	1	17.16		0.92		0.77		1		217
		NGC 188 - 2073			0 40 43	+85 02.1	1	15.03		0.69		0.17		1	G2 V	217
	+60 00089	NGC 225 - 13			0 40 44	+61 27.3	1	10.92		0.60		0.06		1	G0	49
		NGC 225 - 28			0 40 45	+61 32.4	1	13.87		1.05		0.62		1		49
		NGC 225 - 15			0 40 46	+61 29.7	1	11.02		0.94		0.54		1	G8	49
		NGC 188 - 1062			0 40 46	+84 57.7	1	14.81		0.78		0.22		1		217
		NGC 188 - 2072			0 40 47	+85 01.6	1	12.48		1.31		1.36		3	K1 IIIab	387
		AF And			0 40 49	+40 55.9	2	16.08	.030	0.15	.035	-0.84	.015	2		8066,8067
		NGC 188 - 1061			0 40 49	+84 57.6	1	14.10		1.10		0.86		1		217
	−29 00206	BPM 46572			0 40 49	−28 42.7	1	10.76		0.93		0.66		2		3064
	+33 00099	G 69 - 15		⋆ A	0 40 52	+33 34.6	2	8.73	.025	1.13	.000	1.10	.005	2	K5	333,1620,3072
	+60 00090	NGC 225 - 14			0 40 52	+61 24.1	1	11.02		0.63		0.20		1	F8	49
		NGC 188 - 1176			0 40 52	+85 00.9	1	18.50		0.90				1		217
4096	−1 00088	IDS00383S0126		B	0 40 52	−01 09.1	1	10.86		1.03		0.86		1	F7 V	3016

HD	DM	Other Id	N Rem	α_{1950}	δ_{1950}	S	V	σ_V	B–V	σ_{B-V}	U–B	σ_{U-B}	n	Spectrum	References
4096	−1 00088	G 70 - 4	⋆ A	0 40 52	−01 09.9	4	9.23	.034	0.66	.009	0.18	.005	5	G5	202,927,1620,3016
4112	−35 00222			0 40 52	−35 08.4	1	10.12		0.80		0.47		2	K1/2 V	3061
		WLS 40 0 # 5		0 40 54	+01 47.8	1	12.16		0.58		-0.02		2		1375
	+46 00147			0 40 54	+47 29.4	1	9.60		0.65		0.15		2	G5	1375
		LB 3152		0 40 54	−73 03.	1	12.27		0.15		0.13		2		45
		LS I +60 121		0 40 55	+60 07.9	1	11.65		0.50		-0.36		2		41
		NGC 225 - 12		0 40 55	+61 30.8	1	10.89		0.23		0.04		1	A0 V	49
	+61 00157	NGC 225 - 10		0 40 55	+61 33.7	1	10.63		0.28		-0.02		1	A1 V	49
		NGC 188 - 1048		0 40 55	+84 59.0	2	14.88	.015	0.76	.029	0.27	.015	3		217,217
		G 69 - 16		0 40 56	+28 11.0	1	14.53		1.58		0.79		3		316
		NGC 225 - 22		0 40 56	+61 26.7	1	12.56		0.79		0.24		1		49
	+60 00091	NGC 225 - 5		0 40 56	+61 27.9	2	9.87	.001	0.96	.003	0.40	.002	37	G6 III	49,1655
		NGC 225 - 20		0 40 56	+61 29.0	1	12.46		0.81		0.26		1		49
4076	+60 00092			0 40 57	+61 00.1	1	9.04		0.37				3	F0	70
		AAS35,353 # 19		0 40 57	−73 03.9	1	12.74		0.54		0.04		1	F7	767
		NGC 188 - 1060		0 40 58	+84 58.2	1	15.17		0.67		0.12		1		217
		NGC 188 - 1117		0 40 58	+84 58.4	1	17.35		0.92		0.80		1		217
		AN298,163 # 203		0 40 58	−73 54.5	1	12.47		0.93		0.50		2		590
		NGC 188 - 1058		0 40 59	+84 58.7	1	16.77		0.68		-0.05		1		217
4122	−12 00125			0 40 59	−12 10.6	1	9.20		0.28		0.08		4	A8 IV/V	152
		NGC 188 - 1169		0 41 00	+84 55.9	1	17.59		0.97		0.73		1		217
		NGC 188 - 1177		0 41 00	+84 58.9	1	18.57		1.20				2		217
		NGC 188 - 1045		0 41 00	+85 00.4	1	15.00		0.72		0.14		2		217
		BPM 46578		0 41 00	−26 10.	1	11.03		0.57		-0.04		4		3064
		AzV 1		0 41 01	−73 26.8	3	12.66	.018	0.12	.011	0.17	.027	10	A0 I	573,733,767
		AN298,163 # 201		0 41 01	−73 52.2	1	13.34		0.52		0.32		2		590
	+23 00097	G 69 - 17		0 41 02	+23 36.8	1	10.97		1.32		1.21		1	M0	333,1620
4138	−44 00183			0 41 02	−44 23.8	1	8.04		1.51		1.76		4	K4 III	1117
	+41 00119	CC And		0 41 03	+42 00.5	1	9.31		0.31		0.11		1	F3 IV-V	597
		LS I +61 165		0 41 03	+61 49.1	1	10.90		0.67		-0.31		3		41
		NGC 188 - 2211		0 41 03	+84 55.4	1	17.38		0.95		0.66		1		217
	−26 00231			0 41 03	−26 08.2	1	11.71		0.70		1.20		1		3061
		AN298,163 # 202		0 41 04	−73 53.9	1	11.91		1.10		0.92		3		590
4128	−18 00115	HR 188		0 41 05	−18 15.6	14	2.04	.013	1.02	.006	0.87	.009	214	K0 III	3,9,15,26,198,1020*
		G 266 - 161		0 41 05	−24 09.1	2	12.20	.005	1.23	.027	1.12	.054	3	K4	1696,3061
		G 266 - 162		0 41 06	−24 09.4	1	14.24		1.47		1.15		2	K4	3061
	+61 00158	NGC 225 - 11		0 41 07	+61 32.7	1	10.67		0.27		0.12		1	A1 V	49
4150	−58 00042	HR 191	⋆ A	0 41 07	−57 44.2	8	4.36	.004	0.00	.004	-0.01	.007	26	A0 IV	15,200,1075,2012,2020*
		NGC 188 - 1178		0 41 09	+84 57.5	1	19.16		1.84				2		217
4129	−25 00277			0 41 09	−25 00.2	1	9.65		0.68		0.09		1	G5 V	3061
		AN298,163 # 205		0 41 09	−73 49.2	1	11.53		0.67		0.08		3		590
4119	+32 00122			0 41 10	+33 02.2	2	6.92	.000	0.30	.000	0.04	.010	5	A8 V	1501,1625
		NGC 188 - 1107		0 41 10	+84 56.9	1	15.95		1.39		1.11		1		217
		NGC 225 - 27		0 41 11	+61 26.4	1	13.65		0.89		0.33		1		49
		AAS35,353 # 20		0 41 11	−73 24.5	1	13.60		-0.02		-0.29		2	B9	767
		NGC 188 - 1044		0 41 12	+85 00.9	1	16.36		0.75		0.29		1		217
		BPM 46580		0 41 12	−27 38.	1	12.84		1.00		0.54		2		3064
		BPM 46581		0 41 12	−33 51.	1	12.02		0.50		-0.02		1		3061
		LS I +64 035		0 41 13	+64 11.3	1	11.38		0.59		-0.34		2	B1 p(e)(V)	1012
		NGC 225 - 24		0 41 14	+61 25.9	1	13.03		0.60		0.09		1		49
4118	+41 00120			0 41 15	+42 27.5	1			0.06		0.04		std		1231
		NGC 225 - 31		0 41 15	+61 28.4	1	15.43		1.55		0.64		1		49
		NGC 225 - 23		0 41 15	+61 29.1	1	12.97		0.92		0.32		1		49
		NGC 188 - 1118		0 41 15	+84 58.5	1	16.98		1.06		0.76		1		217
4148	−31 00266			0 41 15	−30 42.6	1	8.94		0.46		-0.10		1	F3 V	3061
		NGC 225 - 16		0 41 16	+61 34.6	1	11.42		0.29		0.21		1	A3 V	49
		NGC 188 - 2076		0 41 16	+85 03.5	1	12.46		1.21		1.19		3	K0 IIIb	387
		CS 22170 # 24		0 41 16	−11 38.7	1	15.06		0.12		0.19		1		1736
		NGC 225 - 30		0 41 17	+61 33.0	1	14.37		1.07		0.51		1	F5	49
		NGC 188 - 1179		0 41 18	+84 57.2	1	18.27		1.12				2		217
		NGC 188 - 3029		0 41 19	+84 50.4	1	15.38		0.68		0.22		1		217
		NGC 188 - 1106		0 41 19	+84 56.7	2	11.79	.024	0.53	.005	-0.01	.029	6		217,217
		NGC 188 - 1043		0 41 19	+85 00.7	1	14.93		0.62		0.13		2		217
4145	−12 00126	HR 190		0 41 19	−12 17.0	3	6.01	.009	1.11	.005	1.09	.007	7	K1 III	58,79,2009
		NGC 225 - 19		0 41 20	+61 34.3	1	12.35		0.49		0.26		1	A7	49
		NGC 188 - 1110		0 41 20	+84 56.4	1	16.86		0.80		0.53		1		217
		CS 29527 # 61		0 41 20	−21 02.1	1	13.77		0.02		0.02		1		1736
		CS 29497 # 38		0 41 20	−27 00.2	1	15.45		0.11		0.18		1		1736
		AzV 2		0 41 21	−73 39.5	2	11.97	.019	0.04	.005	-0.70	.005	9	B6 Ia0:e	425,573
		NGC 225 - 26		0 41 22	+61 38.9	1	13.17		0.81		0.54		1		49
		NGC 188 - 1049		0 41 22	+84 59.0	2	15.65	.010	0.72	.020	0.18	.000	2		217,217
		AN298,163 # 207		0 41 22	−73 46.6	1	12.62		0.62		-0.02		1		590
		AN298,163 # 208		0 41 23	−73 53.5	1	11.87		1.08		0.86		1		590
		NGC 225 - 25		0 41 24	+61 33.3	1	13.15		0.79		0.16		1	F6	49
		LB 1571		0 41 24	−51 53.	1	15.73		-0.27		-1.08		1		832
	+59 00109	LS I +60 122		0 41 25	+60 00.3	1	10.11		0.38		-0.51		2	B1	41
		NGC 188 - 1168		0 41 25	+84 56.4	1	17.76		0.93		0.60		1		217
4156	−8 00128	G 270 - 49		0 41 25	−07 48.3	4	9.39	.019	0.59	.004	0.02	.008	8	G0	516,742,1064,3077
4158	−21 00099			0 41 25	−20 40.4	2	9.54	.027	0.29	.014	0.00	.014	7	A1/2	152,1068
		NGC 225 - 29		0 41 28	+61 30.0	1	13.93		0.79		0.33		1		49

Table 1 49

HD	DM	Other Id	N	Rem	α_{1950}	δ_{1950}	S	V	σ_V	B–V	σ_{B-V}	U–B	σ_{U-B}	n	Spectrum	References
		NGC 188 - 1109			0 41 28	+84 56.5	1	16.22		0.76		0.22		2		217
4157	−21 00100				0 41 28	−20 31.3	1	9.63		0.02		-0.04		4	B9.5V(n)	1068
4159	−21 00101				0 41 28	−21 21.2	1	9.89		1.06		0.96		2	K0	3061
		BPM 46584			0 41 30	−23 35.	1	13.18		1.15		1.12		1		3061
		WLS 40 0 # 9			0 41 31	+00 14.2	1	12.39		0.63		0.12		2		1375
		NGC 188 - 1075			0 41 31	+84 58.1	2	13.88	.009	1.11	.019	0.95	.080	4		217,387
		NGC 188 - 2080			0 41 32	+85 03.1	1	15.69		0.69		0.15		1		217
4133	+53 00131				0 41 33	+53 52.7	1	7.54		0.03		-0.03		2	A0	1566
		G 268 - 32			0 41 33	−14 11.6	1	12.11		0.47		-0.27		2		1696
	+60 00094	NGC 225 - 2			0 41 34	+61 30.4	1	9.64		0.20		-0.14		1	B9 IV	49
		NGC 188 - 1112			0 41 34	+84 56.0	1	14.40		0.69		0.13		2		217
		NGC 188 - 1108			0 41 34	+84 56.6	2	15.15	.005	0.73	.005	0.25	.005	2		217,217
		NGC 188 - 1134			0 41 34	+84 57.6	1	17.60		1.06				1		217
		NGC 188 - 1076			0 41 35	+84 58.2	1	14.94		0.68		0.15		1		217
4075	+75 00036	G 242 - 67			0 41 36	+75 40.0	2	7.19	.000	0.76	.010	0.39	.010	7	G5	1355,3026
4142	+47 00181	HR 189			0 41 39	+47 35.4	5	5.64	.024	-0.12	.014	-0.56	.002	14	B5 V	15,154,1003,1119,1203
4188	−11 00128	HR 194			0 41 40	−10 52.9	7	4.76	.011	1.01	.009	0.84	.007	25	K0 III	15,58,1075,1425,2012*
	+60 00095	NGC 225 - 7			0 41 41	+61 28.7	2	10.15	.022	0.18	.005	-0.07	.011	3	A0 IV	49,1655
		G 32 - 44			0 41 42	+12 20.7	1	12.78		1.49		1.02		1		333,1620
		NGC 188 - 1171			0 41 42	+84 59.5	1	18.69		1.02				1		217
4189	−31 00271			A	0 41 42	−30 41.4	2	9.03	.005	0.53	.002	0.00	.015	3	F8 V	79,3061
4189	−31 00271			B	0 41 42	−30 41.4	1	14.15		0.41		-0.24		1		3061
		BPM 46587			0 41 42	−36 48.	1	12.43		1.25		1.10		1		3061
	+61 00162	NGC 225 - 3			0 41 43	+61 37.6	1	9.67		0.17		-0.15		1	A0 IV	49
		LS I +63 097			0 41 43	+63 38.1	1	10.60		0.72		-0.43		2	B1 pe(IV-V)	1012
		NGC 188 - 1113			0 41 43	+84 55.8	2	15.97	.000	0.78	.005	0.29	.000	2		217,217
		LS I +60 123			0 41 44	+60 43.6	1	11.90		0.63		-0.20		2		41
	+61 00163	NGC 225 - 1			0 41 44	+61 32.3	2	9.28	.007	0.14	.005	-0.21	.005	29	B8 III	49,1655
		NGC 188 - 1041			0 41 46	+85 00.6	1	15.23		0.73		0.19		2		217
4211	−39 00181	HR 195		⋆	0 41 47	−38 41.8	4	5.88	.028	1.14	.023	1.18		11	K1 III	58,2018,2032,3025
	+63 00087	LS I +64 036			0 41 48	+64 03.8	1	9.87		0.52		-0.47		2	B0.5IV	1012
		NGC 188 - 1111			0 41 48	+84 56.4	1	15.21		0.68		0.16		1		217
	+61 00164	NGC 225 - 6			0 41 49	+61 36.1	2	10.00	.005	0.15	.000	-0.01	.014	20	A0 V	49,1655
		G 69 - 18			0 41 50	+29 38.3	2	13.29	.005	0.99	.009	0.45	.014	4		333,419,1620
		NGC 225 - 21			0 41 50	+61 34.5	1	12.55		1.02				1		49
		NGC 188 - 1135			0 41 50	+84 57.6	1	16.97		0.93		0.45		1		217
		AAS35,353 # 21			0 41 50	−73 01.0	1	12.81		0.41		0.01		1	F3	767
	+64 00076	LS I +64 037			0 41 51	+64 59.5	1	9.10		0.59		-0.39		2	B1 Ib	1012
		NGC 188 - 1042			0 41 51	+85 00.8	1	15.06		0.97		0.66		2		217
		NGC 188 - 5			0 41 51	+85 07.4	2	12.89	.009	0.69	.005	0.12	.027	5		217,217
		G 70 - 5			0 41 51	−00 53.9	2	13.52	.022	0.76	.007	0.11	.022	6		1620,1658
		G 270 - 52			0 41 51	−06 20.6	1	13.38		0.69		-0.03		2		1658
		AzV 2b			0 41 52	−73 45.3	1	13.49		-0.14		-0.98		2	B0	767
4174	+39 00167	EG And			0 41 53	+40 24.4	2	7.20	.020	1.67	.035	1.49	.125	2	M2 e	1591,1753
		G 268 - 35			0 41 53	−26 19.2	1	13.93		1.29		1.10		2		3064
4209	−30 00213				0 41 54	−30 23.1	1	9.34		0.64		0.08		1	G3 V	3061
		AAS35,353 # 22			0 41 55	−73 35.0	1	14.22		-0.07		-0.45		3	B6	767
4180	+47 00183	HR 193, ø Cas		⋆ A	0 41 56	+48 00.7	9	4.57	.046	-0.07	.009	-0.51	.016	23	B5 IIIe	15,154,379,1119,1203*
		AA131,200 # 2		A	0 41 57	+55 30.0	1	14.59		1.12		0.30		2		7005
		AA131,200 # 2		B	0 41 57	+55 30.0	1	14.00		1.24		0.37		2		7005
		NGC 188 - 2177			0 41 57	+84 54.9	1	16.03		0.74		0.27		1		217
		NGC 188 - 1105			0 41 57	+84 56.9	2	12.32	.015	1.19	.020	1.11	.035	6		217,387
		CS 22179 # 20			0 41 57	−04 40.2	1	13.99		0.16		0.17		1		1736
		AN298,163 # 212			0 41 57	−73 53.2	1	13.38		0.60		0.49		2		590
		CS 22179 # 18			0 41 58	−06 34.5	1	15.87		0.32		0.17		1		1736
	−19 00109				0 41 58	−19 14.5	1	10.23		1.47		1.23		1	K7	3061
4208	−27 00223	G 267 - 162			0 41 58	−26 47.5	3	7.78	.019	0.66	.009	0.16	.005	7	G5 V	79,2033,3064
		NGC 188 - 1104			0 41 59	+84 57.5	1	15.39		0.69		0.11		3		217
		NGC 188 - 1020			0 41 59	+84 59.0	2	14.25	.055	0.91	.031	0.50	.011	5		217,4002
		NGC 188 - 3117			0 41 59	+85 04.3	1	12.46		0.70		0.17		2		217
		NGC 188 - 1033			0 42 00	+85 00.1	1	14.53		0.76		0.26		2		217
		NGC 188 - 1034			0 42 00	+85 00.4	1	15.83		0.73		0.22		1		217
		BPM 46592			0 42 00	−31 33.	1	13.73		1.47		1.26		1		3061
		NGC 188 - 1103			0 42 02	+84 57.3	1	15.14		0.62		0.10		2		217
	+40 00154				0 42 03	+41 02.4	1	9.15		1.00		0.75		4	G5	894
		NGC 188 - 1124			0 42 03	+84 58.7	1	17.13		1.02		0.60		1		217
		LP 646 - 4			0 42 03	−06 13.2	1	12.02		0.57		-0.12		2		1696
4251	−50 00181				0 42 03	−49 58.8	1	7.69		0.03		0.06		4	A0/1 V	152
		NGC 188 - 1032			0 42 04	+85 00.2	1	14.40		0.64		0.14		2		217
		NGC 188 - 1081			0 42 05	+84 58.2	1	15.80		0.76		0.16		2		217
4227	−36 00249	IDS00397S3639		AB	0 42 05	−36 22.9	1	9.59		0.72		0.15		1	K1/2 +(G)	3061
		NGC 188 - 1102			0 42 06	+84 57.1	1	14.50		0.60		0.10		2		217
	−19 00111	G 268 - 37			0 42 07	−19 13.4	4	10.76	.005	1.42	.036	1.12	.025	11	K7	158,1619,1705,3061
4305	−73 00044				0 42 07	−72 38.2	5	9.76	.005	0.45	.016	-0.12	.000	14	F5 V	1408,1425,2001,2010,2013
		LS I +64 038			0 42 08	+64 07.2	1	10.69		0.49		-0.46		2	B1 IV p	1012
		G 267 - 163			0 42 09	−27 39.1	1	14.78		0.61		-0.19		1		3061
		NGC 188 - 1080			0 42 10	+84 58.6	1	15.35		0.72		0.13		2		217
		NGC 188 - 2223			0 42 10	+85 01.8	1	17.52		1.08		0.70		2		217
		AN298,163 # 213			0 42 11	−73 47.9	1	14.84		0.41		-0.76		2		590
		NGC 188 - 1017			0 42 12	+84 58.7	1	14.99		0.69		0.17		2		217

HD	DM	Other Id	N Rem	α_{1950}	δ_{1950}	S	V	σ_V	B–V	σ_{B-V}	U–B	σ_{U-B}	n	Spectrum	References
		NGC 188 - 2088		0 42 12	+85 03.2	1	12.99		1.22				2	K1 IIIb	387
		LS I +63 100		0 42 13	+63 41.6	1	11.22		0.65		-0.31		2	B2 III	1012
	+63 00089	LS I +64 039	★ A	0 42 13	+64 07.1	1	9.50		0.60		-0.41		2	B1 Ib	1012
		G 1 - 19		0 42 14	-00 23.1	1	12.96		1.09		0.71		1		1620
		SB 298		0 42 14	-27 27.3	2	13.06	.005	0.07	.003	0.16	.000	3	A2	295,1736
		NGC 188 - 1101		0 42 15	+84 56.9	1	15.00		0.68		0.07		2		217
4247	-22 00127	HR 197		0 42 16	-22 16.8	4	5.23	.004	0.34	.005			13	F0 V	15,1020,2012,2024
		NGC 188 - 1082		0 42 17	+84 58.3	2	16.74	.024	0.81	.010			3		217,217
		LS I +63 101		0 42 18	+63 26.3	1	11.05		0.70		-0.28		2	B1 III	1012
4161	+74 00027	HR 192, YZ Cas	★ A	0 42 18	+74 42.9	1	5.66		0.05		0.07		2	A2 IV	3024
4161	+74 00027	IDS00390N7426	B	0 42 18	+74 42.9	1	11.23		0.94		0.64		4		3024
		NGC 188 - 3033		0 42 18	+84 52.2	1	15.53		0.66		0.17		1		217
		NGC 188 - 1173		0 42 18	+84 56.2	1	18.62		0.82				1		217
4309	-74 00060			0 42 18	-74 32.3	2	7.59	.019	0.50	.009			8	F6 IV/V	1075,2013
		NGC 188 - 1100		0 42 19	+84 55.8	1	15.23		0.70		0.21		1		217
		NGC 188 - 1175		0 42 19	+84 57.9	1	19.96		1.56				1		217
		NGC 188 - 2233		0 42 20	+85 02.6	1	18.05		1.12				1		217
4294	-63 00072	HR 199	★ AB	0 42 22	-62 46.3	3	6.06	.012	0.44	.005	0.11		9	F5 III/IV	15,404,2029
4041	+84 00013			0 42 24	+84 46.6	2	8.88	.035	0.12	.015	0.11	.065	2	A2	217,217
		NGC 188 - 1031		0 42 24	+85 00.6	1	16.24		0.78		0.30		1		217
		NGC 188 - 2224		0 42 24	+85 02.3	1	17.60		1.20		0.89		2		217
4222	+54 00143	HR 196	★ AB	0 42 25	+54 56.9	1	5.42		0.04		0.05		2	A2 Vs	3023
		CS 29509 # 10		0 42 26	-29 30.7	1	14.60		0.14		0.09		1		1736
		GR 16		0 42 27	-17 12.0	1	14.71		1.48		1.27		2		481
		NGC 188 - 1016		0 42 28	+84 58.9	1	15.91		0.84		0.38		1		217
		NGC 188 - 2234		0 42 28	+85 01.6	1	18.00		1.20				1		217
4272	-12 00128			0 42 29	-12 25.0	1	7.24		1.06		0.86		3	K0 III	1657
4275	-23 00282			0 42 29	-23 08.2	1	10.00		0.57		-0.01		2	G2/3 V	3061
		NGC 188 - 1083		0 42 30	+84 58.3	1	16.74		0.81		0.55		1		217
		SB 302		0 42 30	-29 31.	1	14.16		0.13		0.10		2	A0	295
4256	+1 00131	G 1 - 20		0 42 31	+01 31.2	4	8.05	.013	0.99	.011	0.90	.021	10	K2 V	22,202,333,1620,3031
4214	+64 00077			0 42 31	+64 49.2	1	8.56		0.94				2	G5	70
		NGC 188 - 1170		0 42 31	+84 58.0	1	18.95		1.41				1		217
		NGC 188 - 1015		0 42 31	+84 58.8	1	14.81		0.82		0.33		2		217
4308	-66 00053			0 42 32	-65 54.8	3	6.54	.005	0.65	.005	0.11	.009	12	G5 V	1311,2018,3079
4269	+22 00113			0 42 33	+23 19.0	1	7.40		-0.02		-0.05		2	B8	1733
		NGC 188 - 1091		0 42 34	+84 57.4	1	14.84		0.71		0.14		1		217
4293	-43 00207	HR 198		0 42 35	-42 56.9	2	5.93	.005	0.29	.005			7	F0 V	15,2012
		LS I +63 102		0 42 36	+63 04.8	1	10.50		0.66		-0.31		2	B0.5 V	1012
4271	-0 00109	IDS00401S0018	A	0 42 36	-00 01.2	1	7.00		0.54		0.01		2	F8	3024
4271	-0 00109	IDS00401S0018	B	0 42 36	-00 01.2	1	14.56		1.66		1.17				3024
		NGC 188 - 1099		0 42 37	+84 56.0	1	14.94		0.73		0.26		1		217
		AzV 3		0 42 37	-73 14.4	1	14.12		-0.11		-0.51		3	B0	733
4360	-74 00062	AN298,163 # 216		0 42 39	-73 46.5	1	10.12		0.46		-0.09		2	F0	590
		NGC 188 - 1097		0 42 41	+84 56.5	1	14.98		0.97		0.63		1		217
		NGC 188 - 1085		0 42 41	+84 58.0	1	14.57		1.04		0.76		1		217
		NGC 188 - 1014		0 42 41	+84 59.2	1	15.45		0.72		0.13		1		217
		G 1 - 21		0 42 42	+01 24.2	1	13.33		1.18		1.63		1		1658
		G 70 - 9		0 42 42	+01 24.2	1	13.76		1.15		1.56		1		1658
4253	+57 00138	LS I +57 005		0 42 42	+57 45.9	1	8.38		0.20		-0.17		2	B9 II	8100
4304	-54 00166	HR 202		0 42 42	-53 59.3	4	6.15	.004	0.53	.003	-0.08		15	F7 V	15,612,1075,2027
	+13 00099	G 32 - 46	★ A	0 42 43	+14 05.4	1	9.79		0.76		0.28		1	G8	333,1620
	+16 00071	G 33 - 8		0 42 43	+16 42.7	1	9.97		0.96		0.76		1	K3	333,1620
		NGC 188 - 1093		0 42 43	+84 57.0	1	14.53		0.70		0.15		1		217
		NGC 188 - 1003		0 42 44	+84 58.7	1	13.96		0.48		0.07		2		217
	-24 00298			0 42 44	-24 21.6	1	10.06		0.81		0.46		1	K0	3061
		NGC 188 - 1092		0 42 47	+84 57.3	1	15.39		0.69		0.13		1		217
		NGC 188 - 1001		0 42 49	+84 58.5	2	11.95	.059	1.13	.005	0.85	.030	5	G8 IIIb	217,387
4301	-5 00120	HR 201		0 42 51	-04 54.2	3	6.15	.008	1.61	.010	2.01	.010	11	M0 III	15,1256,3016
	+63 00092	LS I +64 041		0 42 52	+64 07.8	1	10.68		0.57		-0.35		2	B2 III:n	1012
3959	+86 00009			0 42 52	+86 40.0	1	8.65		0.17		0.08		3	A3	1219
4266	+55 00157	LS I +56 009		0 42 53	+56 30.1	1	6.95		0.41				3	F2 I	6009
		NGC 188 - 1174		0 42 53	+84 57.2	1	18.70		1.27				1		217
4316	-44 00194			0 42 53	-43 50.1	1	9.70		0.42		0.02		3	F3 V	1117
		CS 29527 # 55		0 42 54	-22 02.6	1	12.30		0.15		0.09		1		1736
		BPM 46599		0 42 54	-29 46.	1	12.40		0.99		0.80		2		3064
4306	-10 00155			0 42 55	-09 49.1	2	9.01	.005	0.74	.005	0.05		6	G0	742,1594
		WLS 40-15 # 9		0 42 55	-14 56.7	1	12.39		0.77		0.26		2		1375
		G 33 - 9		0 42 56	+16 42.0	2	10.60	.000	0.57	.000	-0.06	.000	3	G0	333,927,1620
		NGC 188 - 1094		0 42 57	+84 56.9	1	16.39		0.79		0.22		1		217
4307	-13 00128	HR 203	★ A	0 42 58	-13 09.1	5	6.14	.015	0.60	.008	0.07	.017	13	G2 V	79,742,2009,3026,7009
		NGC 188 - 1096		0 43 01	+84 56.6	1	16.56		0.86		0.33		1		217
		NGC 188 - 1172		0 43 01	+84 58.2	1	19.03		1.30				2		217
		NGC 188 - 1004		0 43 02	+84 58.5	1	16.28		0.83		0.27		1		217
		MN201,73 # 1		0 43 02	-34 08.6	1	14.44		1.50				1		1495
		NGC 188 - 1090		0 43 05	+84 57.5	1	15.69		0.75		0.24		1		217
4329	-29 00213			0 43 05	-28 57.5	2	10.10	.009	0.15	.019	0.12	.019	8	A0 V	152,1068
		SK 4		0 43 05	-74 15.	1	13.89		-0.21		-1.02		3		573
4312	+25 00112			0 43 06	+25 54.0	1	7.44		1.52		1.90		2	K5 II	8100
		NGC 188 - 2093		0 43 06	+85 02.3	1	14.94		0.81		0.40		1		217

Table 1 51

HD	DM	Other Id	Rem	α_{1950}	δ_{1950}	S	V	σ_V	B–V	σ_{B-V}	U–B	σ_{U-B}	n	Spectrum	References
		LS I +62 138		0 43 07	+62 57.9	1	11.78		0.48		−0.24		2		41
		NGC 188 - 1005		0 43 07	+84 58.4	1	14.79		0.64		0.34		1	K0 III-IV	217
		NGC 188 - 1167		0 43 07	+84 59.0	1	18.42		0.84				1		217
4327	−21 00106	IDS00406S2127	AB	0 43 08	−21 10.9	3	9.53	.010	0.14	.012	0.09	.029	9	A2 IV	937,1068,2012
		CS 29527 # 67		0 43 10	−19 03.2	1	14.62		0.25		−0.13		1		1736
		AAS53,255 # 9		0 43 10	−73 21.	1	12.65		2.14				1		1686
		AAS35,353 # 23		0 43 10	−73 26.3	1	13.03		1.02		0.63		2	K0	767
4296	+60 00099			0 43 11	+60 52.6	1	8.81		0.14				2	A0	1118
		NGC 188 - 1006		0 43 11	+84 58.4	1	14.85		0.95		0.61		1		217
		AzV 4		0 43 11	−72 58.2	2	13.84	.033	0.09	.005	−0.76	.010	7	B0	573,733
4338	−17 00132	HR 206	★ AB	0 43 12	−16 41.8	2	6.46	.005	0.32	.005			7	F0 IV	15,2012
		Smethells 173		0 43 12	−51 54.	1	11.91		1.48				1		1494
		G 268 - 42		0 43 14	−21 51.0	1	13.86		0.77		−0.02		1		3061
		NGC 188 - 1095		0 43 15	+84 56.7	1	16.05		0.72		0.26		1		217
		CS 29527 # 56		0 43 15	−21 46.1	1	11.46		0.18		0.16		1		1736
		MN201,73 # 2		0 43 15	−34 58.4	1	11.97		0.54				1		1495
		G 1 - 22		0 43 16	+02 54.6	1	14.30		1.29				2		202
		AzV 5		0 43 16	−73 39.5	1	13.90		−0.07		−0.91		7	B1	425
		NGC 188 - 1009		0 43 17	+84 58.9	1	15.48		0.84		0.38		2		217
4322	+40 00158			0 43 18	+40 32.2	1	7.58		0.48		−0.01		2	F5	1601
		BPM 46604		0 43 18	−27 34.	1	14.25		0.72		0.14		3		3064
		PHL 6605		0 43 18	−31 40.	1	11.65		−0.02		−0.02		2		3064
		L 219 - 23		0 43 18	−51 00.	1	15.14		1.02		0.92		1		3062
		Florsch 166		0 43 18	−73 32.5	1	11.34		0.51		0.00		3	F5	767
4321	+54 00148	HR 204		0 43 22	+55 01.9	1	6.51		0.17		0.15		2	A2 III	1733
		CS 22942 # 3		0 43 22	−24 49.2	1	15.12		0.42		−0.16		1		1736
4335	+44 00160	HR 205		0 43 24	+44 35.3	3	6.05	.000	−0.07	.010	−0.27	.011	7	B9.5IIIp	220,252,1079
		NGC 188 - 1010		0 43 24	+84 59.3	1	14.90		0.76		0.22		2		217
		BPM 46606		0 43 24	−34 22.	1	12.17		0.83		0.38		1		3061
		BPM 46607		0 43 24	−35 26.	1	13.46		0.62		0.00		1		3061
		G 172 - 21		0 43 25	+44 07.3	1	12.40		0.77		0.24		2		1658
		G 267 - 166		0 43 25	−33 35.0	1	12.66		0.70		0.10		1		3061
4378	−42 00249	IDS00410S4227	A	0 43 25	−42 10.8	1	8.48		1.17		1.14		2	K5/M0 V	3008
4378	−42 00249	IDS00410S4227	AB	0 43 25	−42 10.8	1	7.88		1.26				4	K5/M0 V	2012
4378	−42 00249	IDS00410S4227	B	0 43 25	−42 10.8	1	9.27		1.27		1.18		2	K7	3008
4391	−48 00176	HR 209	★ AB	0 43 25	−47 49.6	4	5.80	.006	0.64	.000			18	G1 V	15,1075,2018,2028
		NGC 188 - 1122		0 43 26	+84 58.6	1	17.50		1.00		0.62		1		217
		AN298,163 # 101		0 43 26	−71 38.3	1	12.58		0.61		0.15		2		590
		LS I +61 166		0 43 27	+61 22.8	1	11.25		0.68		−0.34		2		41
		LS I +64 044		0 43 27	+64 06.5	1	10.50		0.53		−0.36		2	B2 III	1012
		NGC 188 - 1008		0 43 28	+84 58.9	1	16.43		0.72		0.25		1		217
4295	+68 00049	HR 200		0 43 29	+69 03.1	1	6.39		0.41		−0.04		2	F3 V	1733
		AzV 6		0 43 29	−73 31.7	1	13.46		0.03		−0.90		4	B0 Ia	425
	−12 00130	G 270 - 61		0 43 30	−12 00.9	1	11.04		1.02		0.91		1		1696
4376	−24 00309			0 43 30	−24 26.1	1	9.67		0.62		0.11		1	G3 V	3061
		BPM 46610		0 43 30	−34 30.	1	13.85		0.95		0.69		1		3061
4374	−6 00131	G 270 - 62	★	0 43 34	−05 55.0	2	9.17	.025	0.64	.010	0.10	.015	3	G0	1064,1658
		NGC 188 - 2185		0 43 35	+84 55.3	1	15.41		0.70		0.18		1		217
4389	−26 00245			0 43 35	−25 39.2	1	8.80		0.98		0.66		3	K0 III	937
		BPM 46608		0 43 36	−22 25.	1	11.80		1.04		0.94		2		3061
	−26 00244			0 43 36	−26 30.5	1	10.47		1.05		0.90		1	K2	4003
4399	−29 00215			0 43 38	−29 06.0	1	9.64		0.28		0.03		3	A9 V	1068
		AAS35,353 # 25		0 43 38	−73 24.8	1	13.25		0.76		0.27		2	G8	767
		NGC 253 sq1 11		0 43 39	−25 48.3	1	12.31		0.77		0.29		3		937
	−27 00240			0 43 40	−27 18.4	1	9.83		1.12		1.08		1	K0	4003
		AAS35,353 # 26		0 43 40	−73 37.4	1	14.14		−0.05		−0.50		2	B7	767
		LS I +64 045		0 43 41	+64 11.1	1	11.04		0.56		−0.26		2	B3 III	1012
		NGC 188 - 1166		0 43 42	+84 58.3	1	17.96		1.26				1		217
		NGC 247 sq2 16		0 43 42	−20 54.3	1	15.00		0.84				2		1554
4396	−21 00107			0 43 42	−20 56.2	2	9.36	.005	0.63	.005	0.08	.015	8	G3 V	937,1554
4457	−68 00027			0 43 42	−68 08.8	1	9.78		0.90		0.64		1	K2 V	3072
		AzV 7		0 43 42	−73 20.9	1	14.57		−0.26		−1.01		4	B0	425
4398	−23 00293	HR 210	★	0 43 43	−22 47.7	4	5.49	.013	0.98	.002	0.69		12	G3 V	15,2013,2029,3061
4547	−80 00012			0 43 43	−79 56.2	1	8.77		1.42		1.60		4	K1/2 [III]	1700
		AzV 8		0 43 45	−73 15.6	1	13.53				−0.48		4	B8	425
		CS 22942 # 4		0 43 46	−24 15.7	3	14.62	.017	0.16	.010	0.07	.007	3		966,1736,1736
4428	−44 00198			0 43 46	−44 22.6	1	8.64		0.64				4	G3 V	2012
		AzV 9		0 43 46	−73 30.5	2	13.02	.034	−0.05	.025	−0.73	.015	8	B2	425,573
4362	+58 00101	HR 207		0 43 47	+59 18.1	6	6.40	.019	1.08	.011	0.76	.018	9	G0 Ib	1118,1218,1355,4001*
4349	+61 00168			0 43 47	+62 16.9	1	8.51		0.31				2	A2	1118
4414	−26 00247			0 43 48	−25 48.6	3	9.08	.009	0.29	.005	0.00	.005	8	F0 V	378,937,2012
		BPM 46612		0 43 48	−27 10.	1	14.45		0.52		−0.22		3		3064
		BPM 46613		0 43 48	−34 55.	1	12.09		1.15		0.97		2		3061
		NGC 188 - 1165		0 43 49	+84 58.2	1	17.85		0.77				1		217
		NGC 247 sq1 5		0 43 50	−21 15.6	2	11.14	.023	0.65	.005	0.13	.019	4		937,1554
		AzV 9a		0 43 52	−72 59.0	1	15.16		−0.20		−0.86		4	B0	767
		NGC 247 sq1 2		0 43 54	−21 01.7	2	11.21	.015	1.31	.010	1.33	.019	6		937,1554
		NGC 253 sq1 14		0 43 54	−25 41.9	1	13.53		0.59		0.16		1		937
	+33 00106	G 69 - 21		0 43 55	+33 33.4	3	10.34	.004	0.67	.009	0.13	.012	4	G3	1620,1658,7010
4408	+14 00111	HR 211		0 43 56	+15 12.2	2	5.34	.029	1.62	.022	1.76		7	M4 IIIa	71,3001

HD	DM	Other Id	N	Rem	α_{1950}	δ_{1950}	S	V	σ_V	B–V	σ_{B-V}	U–B	σ_{U-B}	n	Spectrum	References
	−26 00248				0 43 59	−25 38.9	1	11.24		0.81		0.37		3		937
		AzV 10			0 43 59	−73 56.2	2	12.65	.010	-0.05	.005	-0.74	.005	6	B3 Ia	425,573
		BPM 46616			0 44 00	−25 54.	1	13.46		1.30		1.16		2		3061
		WLS 40 20 # 9			0 44 01	+20 37.6	1	11.93		0.49		-0.17		2		1375
		LS I +61 167			0 44 02	+61 52.3	1	12.10		0.60		-0.34		2		41
		G 1 - 23			0 44 03	+02 31.0	2	13.53	.034	1.02	.029	0.90		3		202,1620
	−25 00298				0 44 03	−25 08.5	1	9.99		1.10		0.97		2	K2	937
		AzV 11			0 44 03	−73 32.6	1	13.56		-0.10		-0.78		3	B2	425
4452	−29 00219				0 44 05	−28 49.6	1	9.21		0.64		0.09		1	G3/5 V	3061
4455	−33 00293				0 44 05	−33 25.0	1	8.97		0.47		-0.05		1	F5 V	3061
		NGC 188 - 2231			0 44 06	+84 58.0	1	17.71		1.10		0.90		1		217
		NGC 247 sq2 14			0 44 06	−20 57.2	1	13.83		0.73				2		1554
		BPM 46615			0 44 06	−23 20.	1	11.42		0.76		0.25		3		3061
		NGC 253 sq2 4			0 44 06	−25 43.9	1	11.65		0.64		0.28		8		1554
		LS I +61 168			0 44 08	+61 45.6	1	11.84		0.51		-0.30		2		41
		CS 22942 # 2			0 44 08	−24 59.4	2	13.89	.009	0.58	.013	-0.11	.002	4		1580,1736
	+42 00170	G 172 - 23			0 44 10	+43 23.5	1	9.76		0.88		0.51		2	K2	7010
		NGC 247 sq2 13			0 44 10	−21 07.9	1	13.47		0.77		0.43		2		1554
		BPM 46618			0 44 12	−24 12.	1	12.02		0.57		-0.02		1		3061
		SB 317			0 44 12	−28 13.	1	13.32		0.29		0.13		1		295
		AzV 12			0 44 12	−73 22.7	2	13.22	.004	-0.17	.008	-0.98	.008	7	B2	425,573
		NGC 188 - 2121			0 44 14	+84 58.6	1	16.26		0.71		0.30		1		217
		NGC 188 - 2202			0 44 15	+84 57.4	1	17.21		0.84		0.59		1		217
4485	−34 00280				0 44 17	−33 41.5	1	10.53		0.09		0.01		4	A0 IV	1068
		LP 646 - 11			0 44 21	−07 37.5	1	13.04		0.88		0.52		1		1696
	−25 00299				0 44 22	−25 07.5	1	10.52		0.42		-0.04		2		937
4610	−79 00016			V	0 44 22	−79 05.7	1	9.88		0.27		0.19		2	A6 V	1700
4382	+74 00029	HR 208			0 44 23	+74 34.5	2	5.39		-0.06	.009	-0.51	.060	3	B8 III	1079,3016
	−19 00117				0 44 23	−18 47.6	2	10.59	.005	0.61	.007	0.07	.058	3		1696,3061
		NGC 247 sq2 15			0 44 23	−21 06.1	1	14.25		0.68		0.22		2		1554
		MN201,73 # 5			0 44 23	−32 48.4	1	13.04		0.11				1		1495
236526	+55 00163				0 44 24	+55 42.4	1	9.38		0.49		0.24		2	F8	1502
		GR 17			0 44 24	−27 07.9	1	14.84		1.46		1.02		2		481
		BPM 46621			0 44 24	−31 56.	1	14.06		0.73		0.04		1		3061
		BPM 46622			0 44 24	−34 16.	1	11.69		1.12		0.90		1		3061
4482	+11 00096	HR 213			0 44 25	+11 42.1	1	5.50		0.97				2	G8 II	71
	−25 00300				0 44 25	−25 30.8	2	11.60	.000	0.82	.005	0.48	.000	5		937,1554
		G 268 - 46			0 44 27	−26 24.7	1	13.98		1.64		1.24		1		3064
		AAS53,255 # 10			0 44 28	−72 57.	1	12.86		1.98				1		1686
		BPM 46623			0 44 30	−25 09.	2	12.03	.030	0.95	.007	0.69	.015	4		937,3064
4460	+47 00201				0 44 31	+47 32.0	3	8.41	.011	-0.16	.005	-0.89	.019	7	B1.5V ((n))	399,401,555
	−12 00134				0 44 33	−12 09.0	4	11.77	.005	-0.30	.009	-1.25	.046	31	Op	989,1026,1729,1732
		AzV 13			0 44 33	−73 18.6	2	14.50	.032	-0.08	.018	-0.96	.005	5	B1:	733,767
4442	+60 00105				0 44 34	+61 13.1	1	8.68		0.42				2		1118
		SB 322			0 44 34	−23 36.5	3	12.93	.045	0.17	.010	0.09	.009	5	A5	295,966,1736
4490	+18 00101	HR 214, XX Psc			0 44 35	+19 18.3	1	6.13		0.27		0.20		8	F0 Vn	3
4479	+38 00112				0 44 36	+38 45.4	1	7.53		1.30		1.33		2	K0 III	105
	−34 00283				0 44 39	−34 27.2	1	13.50		0.44		-0.20		1		3060
4502	+23 00106	HR 215, ζ And		⋆ A	0 44 41	+23 59.7	6	4.05	.023	1.11	.011	0.90	.026	16	K1 IIe	15,1355,1363,3016*
4530	−29 00221				0 44 41	−29 24.5	1	9.16		0.44		-0.03		1	F7 V	3061
		L 50 - 14			0 44 42	−70 22.	1	12.28		0.71		0.15		2		1696
4599	−72 00065	MN83,95 # 201			0 44 42	−72 29.6	8	10.31	.016	0.19	.007	0.12	.020	23	A2 III/IV	343,366,522,591,1254*
		AzV 14			0 44 42	−73 22.3	2	13.74	.037	-0.20	.009	-0.97	.009	12	B0	425,573
		G 1 - 24			0 44 43	+03 36.7	1	13.79		1.56				2		202
4609	−72 00066				0 44 45	−72 34.4	7	10.77	.013	0.43	.007	-0.01	.006	23	F5	343,366,522,1254,2001*
		NGC 188 - 3108			0 44 46	+85 05.0	1	15.51		0.73		0.20		1		217
		NGC 253 sq2 6			0 44 46	−25 46.4	1	13.89		0.51		-0.01		4		1554
		MN201,73 # 6			0 44 47	−30 08.5	1	14.04		0.10				1		1495
4526	+5 00104	HR 216			0 44 48	+06 28.1	3	5.98	.007	0.94	.000	0.70	.004	9	G8 III	15,1061,1355
		Klemola 15			0 44 48	+09 43.	1	10.27		-0.24		-0.93		3		1026
		NGC 247 sq2 11			0 44 50	−20 46.7	1	12.65		0.76		0.32		3		1554
4440	+71 00037	HR 212			0 44 51	+72 24.1	1	5.86		1.01		0.82		4	K0 IV	1355
		NGC 247 sq1 6			0 44 52	−20 52.2	1	13.70		0.72		0.22		2		937
4555	−26 00252				0 44 53	−25 39.2	2	9.30	.005	0.62	.019	0.15	.015	9	G3 V	937,1554
		Florsch 185			0 44 53	−73 17.5	1	11.40		1.03		0.96		1	K4	767
4539		PHL 830			0 44 54	+09 42.	2	10.31	.010	-0.22	.005	-0.88	.016	10	sdB	963,1732
		AzV 15			0 44 54	−73 41.2	2	13.18	.015	-0.21	.005	-1.01	.010	6	O9 Iab	733,835
		NGC 247 sq1 3			0 44 56	−20 52.8	2	10.79	.015	0.85	.000	0.46	.000	7		937,1554
		MN201,73 # 7			0 44 56	−29 58.2	1	12.36		0.55				1		1495
		AN298,163 # 107			0 44 56	−71 36.6	1	11.23		1.20		1.21		3		590
	+13 00108	G 32 - 49			0 44 59	+14 22.2	1	10.94		0.89		0.58		1	G9	333,1620
		NGC 253 sq2 18			0 45 00	−25 22.9	1	12.60		0.62		0.10		3		1554
4551	+24 00115				0 45 01	+25 15.7	1	7.88		0.04		0.00		3	A0	833
	−30 00233				0 45 01	−29 57.9	1	11.42		0.66				1		1495
		NGC 188 - 2112			0 45 03	+84 59.8	1	16.12		0.73		0.28		1		217
	−26 00253				0 45 03	−25 44.6	2	10.80	.005	0.53	.019	0.10	.019	9		937,1554
4550	+25 00118				0 45 04	+26 01.1	1	6.89		1.09		0.94		2	K0 III	3016
		AN298,163 # 108			0 45 04	−71 48.6	1	12.65		0.72		0.15		2		590
4513	+59 00119				0 45 05	+60 03.1	2	8.74	.005	0.48	.020			3	F2	592,1118
		AzV 16			0 45 06	−73 24.8	3	13.12	.025	0.11	.017	-0.80	.023	5	B1	425,573,1513

Table 1

53

HD	DM	Other Id	N Rem	α_{1950}	δ_{1950}	S	V	σ_V	B–V	σ_{B-V}	U–B	σ_{U-B}	n	Spectrum	References
4572	−26 00254			0 45 08	−25 40.1	2	8.91	.010	0.56	.015	0.06	.024	9	G0 V	937,1554
4597	−37 00273	G 267 - 171	★	0 45 08	−37 12.5	3	7.83	.015	0.54	.000	−0.04	.014	9	F7/8 V	79,158,2033
		MN201,73 # 8		0 45 08	−38 18.4	1	12.93		0.73				1		1495
		NGC 247 sq2 12		0 45 09	−21 03.3	1	12.74		0.60		−0.02		2		1554
		NGC 188 - 2131		0 45 10	+84 57.8	1	16.35		0.76		0.34		1		217
		G 269 - 34		0 45 10	−27 50.7	1	12.78		1.17		1.08		1		3064
		NGC 247 sq2 10		0 45 11	−21 09.9	1	12.58		0.88		0.59		2		1554
		Steph 82		0 45 12	+05 45.5	1	11.63		1.44		1.22		1	K7	1746
		LS I +62 139		0 45 12	+62 43.1	2	11.40		0.48	.000	−0.53	.000	4	O7	1011,1012
		BPM 46625		0 45 12	−27 38.	1	12.22		0.44		−0.22				3064
	−29 00222			0 45 12	−29 17.8	2	10.52	.000	1.34	.000	1.49		2		1495,4003
		AJ63,118 # 17		0 45 12	−72 27.	5	10.62	.025	1.20	.007	1.21	.012	18		343,366,522,2001,3021
		HV 824		0 45 12	−72 57.8	2	12.04	.010	0.63	.027	0.38		2		1511,3075
		AJ74,1000 M31# 15		0 45 14	+41 31.5	1	10.54		1.09				3		209
4585	−18 00127	HR 218		0 45 14	−18 20.1	2	5.70	.001	1.30	.001	1.46		6	K3 III	58,2007
		AzV 17		0 45 14	−73 22.3	1	13.76		−0.09		−0.79		9	B1	425
4568	+20 00105	HR 217		0 45 15	+20 39.1	1	6.56		0.49		0.01		2	F8 V	254
4651	−73 00047			0 45 15	−73 05.9	1	9.78		0.46		0.00		4	F3 Vp	1408
		AzV 17a		0 45 15	−73 22.7	1	13.37		−0.07		−0.85		1	B2	767
		NGC 253 sq2 8		0 45 17	−25 46.1	1	14.26		0.64		0.25		5		1554
		MN201,73 # 9		0 45 17	−37 48.4	1	14.41		0.67				2		1495
		MN83,95 # 202		0 45 17	−72 27.7	1	10.61		1.21				1		591
		NGC 188 - 2113		0 45 18	+84 59.8	1	16.45		0.81		0.40		1		217
		NGC 247 sq2 9		0 45 18	−21 04.2	1	12.37		0.55		0.02		3		1554
		NGC 247 sq2 7		0 45 19	−21 19.0	1	11.94		0.57				3		1554
		NGC 253 sq1 6		0 45 19	−25 50.9	1	12.66		0.46		−0.14		3		937
		AJ74,1000 M31# 14		0 45 20	+41 23.8	1	11.03		1.12				3		209
		HV 1430		0 45 20	−73 12.7	1	13.97		0.81				1		1511
		NGC 247 sq1 14		0 45 22	−21 02.4	2	11.97	.000	0.83	.014	0.40	.014	4		937,1554
		MN83,95 # 203		0 45 22	−72 25.0	1	12.14		0.98				1		591
		CS 22170 # 32		0 45 24	−10 32.7	1	15.54		0.09		0.20		1		1736
		BPM 46631		0 45 24	−33 10.	2	12.96	.034	0.35	.039	−0.15	.019	3		295,3061
		AzV 18		0 45 24	−73 22.8	3	12.46	.008	0.03	.004	−0.80	.019	5	B1 Ia0	425,573,1513
4607	−10 00164	G 270 - 71		0 45 25	−09 37.3	1	7.73		0.51		−0.01		4	F5	158
	−26 00257			0 45 25	−25 52.6	1	10.29		1.07		0.75		3	K0	937
	−33 00302	G 267 - 172		0 45 25	−33 02.3	1	10.51		0.92		0.44		1		3061
		AzV 19		0 45 25	−73 06.3	2	13.00	.000	0.01	.015	−0.16	.000	6	B6 Iab	425,835
4581	+34 00117			0 45 26	+35 27.9	1	8.66		1.27		1.44		2	K2	1625
		MN201,73 # 10		0 45 26	−38 38.7	1	15.45		0.68				1		1495
		AJ74,1000 M31# 6		0 45 28	+41 06.1	1	10.35		1.01				3		209
4560	+63 00094			0 45 30	+64 26.9	1	10.82		2.74		2.64		3	A2	8032
		NGC 188 - 2115		0 45 30	+85 00.2	1	15.85		0.73		0.26		1		217
		NGC 247 sq1 12		0 45 30	−20 59.8	2	11.78	.000	0.75	.000	0.33	.005	4		937,1554
		NGC 247 sq1 13		0 45 30	−21 00.1	1	14.23		0.74				1		937
		NGC 247 sq2 6		0 45 30	−21 08.0	1	11.82		0.60		−0.04		3		1554
		BPM 46628		0 45 30	−25 14.	1	12.88		1.30		1.30		2		3064
		BPM 46629		0 45 30	−26 56.	1	14.75		1.63		1.13		1		3064
4623	−30 00240	IDS00431S2953	AB	0 45 30	−29 37.0	1	7.58		0.32				4	F0 IV	2012
		MN201,73 # 11		0 45 31	−38 48.4	1	15.23		0.74				1		1495
		MN201,73 # 12		0 45 32	−38 48.7	1	15.82		0.77				1		1495
	+40 00163			0 45 33	+41 19.5	1	9.33		1.58				3	K5	209
		CS 22179 # 28		0 45 33	−04 07.3	1	14.91		0.04		−0.04		1		1736
4622	−22 00134	HR 220		0 45 33	−21 59.7	5	5.57	.010	−0.06	.004	−0.11	.007	15	B9 V	15,1068,1075,2024,3007
4604	+42 00174			0 45 35	+43 24.7	1	9.26		0.08		0.13		26	A3	1655
		G 32 - 50		0 45 36	+16 24.1	1	12.25		1.32		1.25		1	K7	333,1620
4631	−25 00306	G 268 - 48	★ AB	0 45 36	−25 21.1	4	8.89	.015	0.77	.008	0.43	.016	10	G8 V	79,937,1554,3064
		BPM 46635		0 45 36	−28 32.	1	14.08		1.48		1.50		2		3064
		BPM 46636		0 45 36	−29 51.	1	12.10		0.75		0.05		2		3064
		MN201,73 # 13		0 45 36	−38 28.3	1	13.73		0.74				2		1495
		MN137,55 # 233		0 45 38	−73 19.5	1	14.91		−0.09		−0.73		3		967
4579	+63 00095			0 45 40	+64 24.9	1	9.10		0.51				3	F2	1118
		AzV 20		0 45 40	−73 17.8	4	12.11	.022	0.30	.019	−0.11	.040	11	B8:Ia	35,425,835,967
	−27 00246			0 45 41	−27 11.2	1	10.82		0.60		−0.05		2		3064
4627	+6 00105	HR 221		0 45 42	+07 01.6	3	5.92	.005	1.10	.007	1.00	.008	8	G8 III	15,1061,1355
		AzV 21		0 45 43	−73 27.3	2	14.13	.015	−0.14	.015	−0.92	.005	9	B1	425,573
4628	+4 00123	HR 222	★ A	0 45 45	+05 01.4	10	5.74	.013	0.89	.008	0.59	.010	40	K2 V	15,22,1013,1061,1355,1509*
		AAS53,255 # 16		0 45 47	−73 21.	1	12.85		2.21				1	M1 Ia	1686
	+60 00111			0 45 48	+60 46.0	1	9.54		0.68				2		70
		NGC 188 - 2116		0 45 48	+84 58.5	1	15.97		0.70		0.21		1		217
		LP 766 - 26		0 45 48	−15 11.5	1	11.63		0.70		0.21		2		1696
		AN298,163 # 111		0 45 50	−71 47.3	1	12.27		0.82		0.51		3		590
		CS 22942 # 5		0 45 51	−23 47.9	2	15.53	.000	0.11	.000	0.29	.009	2		966,1736
		AzV 22		0 45 51	−73 24.0	2	12.24	.023	−0.07	.041	−0.66	.000	5	B2 Ia	425,835
		AzV 23		0 45 51	−73 39.1	2	12.25	.005	0.08	.009	−0.65	.014	4	B3 Ia	733,835
	−14 00142	G 270 - 73		0 45 52	−14 23.1	2	10.86	.005	0.49	.005	−0.10	.015	3		79,1696
	−32 00305	IDS00434S3208	AB	0 45 52	−31 51.7	1	10.33		0.60		0.06		1		3061
		MN137,55 # 220		0 45 52	−73 13.4	1	14.88		1.34				4		967
4499	+81 00018		V	0 45 54	+81 41.7	1	7.44		1.11				9	G5	1778
		AzV 24		0 45 54	−73 18.8	2	13.83	.063	−0.16	.019	−0.85	.024	6	O9	733,835
		NGC 188 - 3049		0 45 56	+84 54.9	1	15.35		0.66		0.20		1		217

HD DM	Other Id	N	Rem	α_{1950}	δ_{1950}	S	V	σ_V	B–V	σ_{B-V}	U–B	σ_{U-B}	n	Spectrum	References
4613 +64 00083	LS I +65 006			0 45 58	+65 16.7	1	8.79		0.58		-0.35		2	B1 II	1012
	AzV 25			0 45 58	-73 28.7	1	13.19		-0.09		-0.56		4	B5 Iab	425
4636 +50 00147	HR 223			0 45 59	+50 41.8	3	4.90	.009	-0.11	.000	-0.43	.000	5	B9 III	1079,1363,3023
	CS 22170 # 27			0 45 60	-11 49.1	1	13.09		0.38		0.03		1		1736
4681 -46 00214				0 46 00	-45 59.9	1	10.68		-0.04		-0.11		4	B9 IV	152
	AzV 26			0 46 01	-73 24.6	3	12.52	.018	-0.19	.009	-0.98	.015	9	B0 Ia	425,573,1513
4725 -73 00048				0 46 02	-73 08.4	1	10.81		0.46		0.02		4	F2 V	1408
	AAS50,119 # 104			0 46 03	-73 13.1	1	15.26		-0.04				3		967
	AzV 27			0 46 03	-73 30.6	2	12.19	.020	0.13	.010	0.01	.010	8	A2 Iab	733,835
4614 +57 00150	HR 219		⋆ A	0 46 04	+57 33.1	2	3.45	.000	0.57	.005	0.03	.010	5	G3 V	22,3078
4614 +57 00150	HR 219		⋆ AB	0 46 04	+57 33.1	14	3.44	.007	0.57	.006	0.01	.013	181	G3 V +K7 V	1,15,61,71,254,1077*
4614 +57 00150	IDS00430N5717	B		0 46 04	+57 33.1	2	7.51	.005	1.39	.005	1.03		4	K7 V	22,3078
	NGC 188 - 3057			0 46 04	+84 55.8	1	15.76		0.71		0.14		1		217
	NGC 253 sq2 5			0 46 04	-25 29.8	1	12.62		0.86		0.64		5		1554
4656 +6 00107	HR 224		⋆ A	0 46 05	+07 18.8	11	4.43	.008	1.50	.008	1.85	.015	37	K5 III	3,15,245,1061,1355*
	NGC 253 sq2 7			0 46 05	-25 33.8	1	14.08		0.56		0.05		3		1554
	G 69 - 24			0 46 06	+26 45.0	1	12.38		1.52		1.20		1	M4	333,1620
	Hiltner 65			0 46 06	+63 54.9	1	11.15		0.51		-0.26		2	B3 IV	1012
	AzV 28			0 46 06	-73 37.6	1	13.42		-0.08		-0.80		3	B0	733
	LP 766 - 92			0 46 07	-17 09.7	1	12.87		0.44		-0.19		2		1696
	G 70 - 14			0 46 08	-01 39.4	3	10.89	.013	0.59	.013	-0.05	.013	6		927,1620,1658
	AzV 29			0 46 08	-72 47.9	1	14.71		0.01		-0.91		3	B0	733
	NGC 188 - 3096			0 46 09	+85 03.8	1	15.43		0.66		0.10		1		217
4670 +17 00106				0 46 10	+18 02.5	1	7.94		-0.02		-0.16		2	B9	1648
	AJ74,1000 M31# 12			0 46 11	+40 37.5	1	11.09		1.25				3		209
	AJ74,1000 M31# 5			0 46 12	+40 56.1	1	11.48		0.84				3		209
	AzV 30			0 46 13	-72 46.3	1	14.21		0.00		-0.19		4	B5	733
	G 1 - 26			0 46 14	+04 33.1	1	14.79		1.25				2		202
4691 -29 00225				0 46 15	-28 46.0	2	6.78	.000	0.34	.004	-0.02	.002	6	F0 V	1628,3061
4669 +40 00165				0 46 16	+40 51.6	1	7.69		1.16				3	K0	209
	BPM 46644			0 46 18	-33 35.	1	14.59		0.93				1		3061
4709 -54 00180				0 46 18	-53 36.3	1	8.30		1.17		0.89		6	K1/2 III	1673
	MN137,55 # 169			0 46 18	-73 12.8	1	14.99		0.65		0.12		4		967
	AzV 31			0 46 18	-73 22.9	2	12.49	.034	0.29	.015	-0.05	.010	6	B8:Ia	425,835
4647 +56 00131			V	0 46 19	+56 48.2	1	7.07		1.77				3	M2 III	369
	NGC 247 sq1 7			0 46 20	-20 49.4	2	10.26	.019	0.59	.002	-0.05	.037	4		937,3061
4676 +16 00076	HR 225		⋆ A	0 46 21	+16 40.3	2	5.08	.004	0.50	.004	0.00		5	F8 V	1363,3026
	AJ74,1000 M31# 3			0 46 21	+40 50.9	1	12.04		1.09				3		209
	AzV 32			0 46 21	-73 00.1	1	14.20		-0.14		-0.91		6	B1	425
	LS I +63 106			0 46 23	+63 35.8	1	10.71		0.66		-0.27		2	B2 V:pe	1012
	Case *M # 23			0 46 24	+64 29.	2	10.67	.050	2.85	.080			2	M2 Iab	138,148
	Steph 83			0 46 26	+42 38.6	1	10.33		1.08		1.05		1	K7	1746
	G 243 - 41			0 46 26	+67 40.8	2	10.42	.001	1.02	.010	0.92	.015	3	K3	1658,5010
	NGC 247 sq1 8			0 46 28	-20 54.8	1	11.76		0.65		0.16		3		937
	AzV 33			0 46 28	-73 55.7	2	13.54	.014	-0.15	.009	-0.72	.000	4	B3 Iab	425,835
4685 +40 00167				0 46 30	+40 48.6	1	7.12		1.05				3	G5	209
4635 +69 00045	G 242 - 70		⋆ AB	0 46 30	+70 10.5	2	7.76	.005	0.90	.005	0.68	.000	5	K0	1355,1658
	BPM 46645			0 46 30	-21 33.	1	13.80		0.74		0.14		2		3061
	G 1 - 27			0 46 31	+05 09.0	8	12.37	.016	0.55	.012	0.02	.015	19		1,202,1281,1620,1663,1698*
	NGC 247 sq1 9			0 46 31	-20 58.1	1	10.58		1.21		1.10		3		937
	AzV 34			0 46 31	-73 40.0	1	13.81		0.20		-0.05		3	B2:	733
4738 -49 00210				0 46 32	-48 52.0	1	8.14		0.35				4	F0 IV/V	2012
	AzV 35			0 46 32	-73 07.4	1	14.13		0.22		-0.18		4	B0:	733
	AAS35,353 # 28			0 46 32	-73 31.2	1	13.14		-0.02		-0.66		3	B5	767
4840 -80 00013				0 46 32	-79 48.1	1	9.72		1.17		1.02		2	G5	1700
-30 00248				0 46 34	-29 47.7	1	10.69		0.33		0.07		1	A7	378
	AzV 36			0 46 34	-73 00.0	1	14.02		-0.03		-1.00		5	B1	425
	AAS50,119 # 106			0 46 35	-73 16.5	1	15.80		-0.11				2		967
-36 00273	G 267 - 175			0 46 36	-35 59.6	1	11.27		0.73		0.07		1		3061
	AAS50,119 # 107			0 46 36	-73 16.5	1	15.35		0.43				2		967
	SK 22			0 46 36	-73 23.	1	13.37		-0.06		-1.00		1		835
4737 -47 00229	HR 229			0 46 38	-46 58.2	4	6.27	.008	0.90	.005			18	G8 III	15,1075,2013,2028
	AzV 37			0 46 38	-73 19.7	2	12.86	.023	-0.03	.014	-0.52	.005	5	B5 Iab	733,835
	AzV 38			0 46 39	-73 16.9	3	12.85	.015	0.13	.007	-0.24	.004	5	A0 Ia	425,655,835
4674 +59 00125	IDS00437N6013	AB		0 46 41	+60 29.2	1	8.02		0.50				2	F5	1118
	NGC 247 sq1 10			0 46 41	-21 10.4	1	10.13		0.64		0.10		1		937
	AzV 39			0 46 41	-73 43.1	2	12.97	.027	0.03	.032	-0.45	.005	5	B7 Iab	733,835
	NGC 188 - 3070			0 46 42	+84 57.8	1	15.42		0.71		0.23		1		217
	MN201,73 # 14			0 46 42	-35 18.6	1	13.09		1.49				1		1495
	Smethells 174			0 46 42	-50 26.	1	10.75		1.43				1		1494
4751 -43 00226				0 46 43	-42 50.1	1	7.60		-0.12		-0.34		2	B8 V	3007
4720 -2 00111				0 46 44	-01 53.9	1	8.73		0.33		0.02		2	A3	1375
	NGC 188 - 3094			0 46 45	+85 03.6	1	13.27		1.34		1.32		3		387
	AJ74,1000 M31# 11			0 46 46	+40 42.8	1	12.17		1.16				3		209
	NGC 247 sq1 11			0 46 46	-21 09.1	1	14.05		0.65		0.09		1		937
4732 -24 00345	HR 228		⋆ AB	0 46 46	-24 24.5	2	5.90	.002	0.95	.000	0.72		6	K0 III	58,2009
	AzV 40			0 46 46	-72 27.8	1	14.42		-0.05		-0.95		5	B1	425
4722 -19 00127				0 46 47	-19 16.8	1	9.91		0.56		0.02		1	F6 V	3061
	AzV 41			0 46 47	-73 09.2	1	14.56		-0.10		-0.84		3	B2	425
-28 00251				0 46 48	-27 47.4	1	11.45		0.26		0.06		1	A5	378

Table 1

55

HD	DM	Other Id	N Rem	α_{1950}	δ_{1950}	S	V	σ_V	B–V	σ_{B-V}	U–B	σ_{U-B}	n	Spectrum	References
	−56 00154			0 46 48	−56 22.2	1	10.20		−0.11		−0.75		1	B5	832
	+42 00182			0 46 49	+43 22.6	1	9.08		0.33		0.04		26	F5	1655
		LS I +60 124		0 46 49	+60 42.8	1	9.89		0.40		−0.51		2	B2:V:pnne	1012
	−31 00300			0 46 49	−30 54.1	1	10.09		0.76		0.28		2	G2	3061
4694	+63 00097	LS I +64 049		0 46 50	+64 21.9	1	8.50		0.72		−0.16		2	B3 Ia	1012
		AzV 42		0 46 51	−73 14.6	2	13.50	.025	−0.04	.016	−0.69	.025	8	B5	425,655
4729	+2 00109			0 46 52	+02 36.2	1	9.33		0.27		0.06		2	A2	1375
	+40 00169			0 46 52	+40 34.1	1	9.63		1.22				3		209
		AAS50,119 # 108		0 46 52	−73 19.5	1	15.50		0.00				2		967
4815	−75 00064	HR 236		0 46 52	−75 11.7	5	5.06	.007	1.37	.005	1.67	.012	21	K4 III	15,611,1075,2038,3053
	−32 00318	G 267 - 176		0 46 53	−32 24.6	1	10.51		0.55		−0.07		1		3061
		MN201,73 # 15		0 46 54	−26 08.3	1	12.00		0.64				1		1495
4730	−14 00145	HR 227	⋆ AB	0 46 55	−13 49.9	2	5.59	.005	1.31	.010	1.54		6	K3 III	2009,3077
4747	−23 00315	G 268 - 53		0 46 57	−23 29.2	4	7.16	.009	0.78	.009	0.31	.020	10	G8/K0 V	79,2024,2033,3077
		CS 22170 # 40		0 46 60	−08 13.1	1	13.86		0.48		−0.14		1		1736
		BPM 46653		0 47 00	−23 19.	1	14.38		0.60		−0.05		1		3061
		AzV 43		0 47 00	−73 02.6	1	14.08		−0.13		−1.00		7	B1	425
		MN137,55 # 148		0 47 00	−73 21.3	1	12.63		0.91		0.50		6		967
		AzV 44		0 47 01	−73 41.2	1	13.97		−0.34		−1.00		3	O9	425
		XY Cas		0 47 02	+59 50.3	2	9.63	.013	0.97	.010	0.65		2	G4	592,1399
		CS 22170 # 36		0 47 02	−09 07.1	1	13.56		0.42		−0.19		1		1736
4727	+40 00171	HR 226		0 47 03	+40 48.4	11	4.53	.010	−0.15	.007	−0.58	.005	33	B5 V+F8 V	1,15,154,637,985,1006*
	+40 00170			0 47 03	+40 55.2	1	10.04		1.16				3		209
		AzV 46		0 47 03	−72 17.3	2	13.35	.066	−0.16	.023	−0.89	.014	4	B1	425,835
		AzV 45		0 47 03	−73 38.4	1	14.15		0.04		−0.28		5	B8	425
		AzV 47		0 47 05	−73 42.2	1	13.38		−0.26		−0.99		3	B0	425
4772	−24 00347	HR 232		0 47 06	−23 38.0	4	6.27	.013	0.13	.005	0.13		19	A3 IV	15,1068,2012,2012
		Smethells 175		0 47 06	−54 52.	1	9.49		1.19				1		1494
4744	+29 00141	IDS00444N2954	A	0 47 10	+30 10.7	2	7.61	.005	1.06	.005	0.78	.010	3	G8 IV	1003,3077
	−19 00129	G 268 - 54	⋆ AB	0 47 10	−18 56.7	2	9.78	.005	0.74	.005	0.20	.015	2	G5	79,3061
4757	+26 00131	HR 230	⋆ AB	0 47 11	+27 26.3	1	5.56		0.36		0.07		3	F4 III	1733
4758	+26 00131	HR 231	⋆ AB	0 47 12	+27 26.3	2	5.57	.005	0.37	.005	0.06		5	F5 III	3026,6009
		BPM 46657		0 47 12	−26 27.	1	14.63		1.36		1.12		2		3064
		BPM 46658		0 47 12	−27 32.	1	11.88		0.93		0.61		2		3064
4781	−8 00145			0 47 13	−08 07.3	1	7.83		1.31		1.36		3	K0	1657
4666	+77 00027			0 47 15	+77 40.9	1	6.74		0.27		0.12		2	A5	1375
	+60 00114	LS I +60 125		0 47 16	+60 37.9	1	9.87		0.50		−0.49		2	B2 III:pe	1012
4862	−74 00067			0 47 17	−73 38.3	4	11.02	.025	−0.04	.017	−0.71	.000	13	B3 I	425,573,1277,8033
236547	+59 00128			0 47 18	+59 51.6	1	8.92		1.70				2	M0	70
		L 123 - 30		0 47 18	−61 17.7	2	12.16	.025	1.46	.005	1.05		2		1494,3078
		LB 3162		0 47 18	−61 40.	1	12.05		0.35		0.03		3		45
	+62 00156			0 47 19	+62 51.4	1	8.72		0.99				3	G5	70
		BPM 46660		0 47 24	−27 20.	1	14.99		0.58		−0.10		2		3064
		AAS53,255 # 28		0 47 24	−72 58.	1	13.51		1.79				1		1686
		CS 22942 # 9		0 47 26	−23 29.5	2	14.76	.000	0.28	.013	−0.04	.004	2		966,1736
		AzV 49		0 47 26	−73 39.8	1	13.92		0.01		−0.19		9	A0	425
		AzV 50		0 47 28	−73 08.9	2	13.10	.010	−0.14	.025	−0.92	.005	5	B1	425,573
	−27 00260	G 268 - 55		0 47 29	−27 06.2	1	10.11		0.63		0.04		1		3061
4778	+44 00176	HR 234, GO And		0 47 30	+44 43.8	2	6.12	.015	0.02	.002	−0.04	.000	9	A0p Cr	1202,3033
4768	+58 00119	LS I +59 055		0 47 30	+59 24.0	1	7.57		0.38		−0.34		2	B5 Ib	1012
		BPM 46659		0 47 30	−24 08.	1	13.68		0.71				1		3061
		BPM 46662		0 47 30	−26 41.	1	14.55		0.55		−0.35		3		3064
		G 268 - 56		0 47 32	−18 51.9	1	11.97		1.36		1.24		1		3061
4777	+49 00215	IDS00448N5005	AB	0 47 34	+50 21.5	1	7.82		0.47		−0.05		2	F5 V +F7 V	292
		G 70 - 17		0 47 34	−04 42.1	2	11.26	.024	0.86	.015	0.50		3		202,1620
4838	−55 00164			0 47 34	−54 52.0	1	9.58		1.17		1.08		1	K4 V	3072
	+61 00175	LS I +62 141		0 47 35	+62 05.6	1	9.59		0.76		−0.10		2	B3 II	1012
		BPM 46664		0 47 36	−28 23.	1	13.48		0.85		0.34		2		3064
4813	−11 00153	HR 235		0 47 37	−10 54.8	9	5.19	.019	0.51	.011	0.00	.012	41	F7 IV-V	3,15,1007,1013,1075*
		CS 22170 # 31		0 47 37	−11 06.2	1	13.98		0.14		0.15		1		1736
4893	−74 00068			0 47 37	−73 45.0	1	8.45		1.40		1.64		7	K2 III	1408
4775	+63 00099	HR 233	⋆	0 47 40	+63 58.5	2	5.36	.030	0.51	.015	0.14		6	B9.5 +	1118,3026
	+40 00173			0 47 41	+41 27.5	1	8.83		1.21				3	K2	209
4850	−47 00233			0 47 41	−47 33.9	1	9.60		0.07		0.12		4	A0 V	152
4822	−6 00145			0 47 42	−06 28.0	1	9.49		0.57		0.01		2	G0	1064
		G 267 - 177		0 47 42	−32 02.1	1	12.31		1.15		0.93		1		3061
4849	−44 00216	HR 239, AZ Phe		0 47 43	−43 40.0	3	6.48	.008	0.29	.008	0.12		8	A9/F0 III	15,612,2012
		G 69 - 26		0 47 44	+23 40.6	2	14.87	.000	1.50	.008	1.29	.004	7		316,3079
		MN137,55 # 87		0 47 44	−73 18.0	1	14.87		0.37				3		967
		AzV 51		0 47 45	−73 07.5	1	14.05		−0.14		−0.86		2	B0	425
		AAS19,271 # 121	AB	0 47 45	−73 47.3	1	13.90		−0.14		−0.97		3		416
	−31 00305	G 267 - 178		0 47 47	−30 52.6	2	11.66	.019	0.44	.000	−0.16	.009	4		1696,3061
		BPM 46666		0 47 48	−25 20.	1	11.04		1.04		1.02		3		3064
		BPM 16274		0 47 48	−52 25.	1	14.20		−0.01		−0.79		2	DA	3065
	−32 00327	G 267 - 179		0 47 49	−31 36.2	2	10.41	.015	0.96	.015	0.77	.000	3		79,3061
		AzV 52		0 47 49	−73 58.5	2	13.75	.029	−0.13	.034	−0.97	.039	3	B1	425,835
	+41 00146			0 47 50	+41 55.1	1	9.66		1.26				3	K2	209
		CS 29509 # 27		0 47 50	−31 16.3	1	12.44		0.31		−0.12		1		1736
		LP 406 - 19		0 47 54	+17 51.1	1	12.87		1.00		0.71		1		1696
232289	+50 00159	G 172 - 24		0 47 54	+51 06.7	1	9.64		0.62		0.06		1	G5	1658

HD	DM	Other Id	N	Rem	α_{1950}	δ_{1950}	S	V	σ_V	B–V	σ_{B-V}	U–B	σ_{U-B}	n	Spectrum	References
	+73 00038				0 47 54	+74 13.8	2	9.14	.084	1.21	.005	1.09	.042	4	K0	979,3016
	−28 00258			V	0 47 54	−27 42.6	2	9.62	.019	0.88	.007	0.42	.022	3	G0	3061,4003
4857	−35 00278	IDS00455S3502		AB	0 47 55	−34 45.7	1	10.00		0.78		0.40		1	K0 V	3061
4835	+9 00097				0 47 56	+10 08.4	1	8.61		0.28		0.09		3	A2	1733
		CS 22179 # 26			0 47 56	−03 23.5	1	13.76		0.09		0.16		1		1736
		WLS 48 40 # 10			0 47 57	+39 33.0	1	11.27		1.04		0.69		2		1375
		CS 22183 # 10			0 47 57	−03 23.6	1	13.74		0.09		0.17		1		1736
4831	+24 00123				0 47 58	+25 18.8	1	7.20		0.94		0.65		2	G8 III	1625
		AzV 53			0 47 58	−73 09.0	1	12.96		0.16		−0.22		3	A0 Ia	425
		Steph 85			0 47 59	+12 39.4	1	12.34		1.50		1.22		1	M0	1746
		AzV 54			0 48 00	−73 53.9	1	14.10		−0.13		−0.67		8	B3	425
236550	+59 00128				0 48 01	+60 04.0	1	8.92		1.70				2	K7	369
		AzV 55			0 48 01	−73 34.1	1	13.40		−0.07		−0.31		7	B5 Ib	425
		POSS 295 # 2			0 48 03	+29 40.6	1	13.43		0.53				1		1739
4741	+77 00029				0 48 03	+78 21.2	1	8.47		0.78		0.33		3	G5	3026
		AzV 56			0 48 04	−73 12.0	5	11.16	.007	0.00	.011	−0.78	.015	16	B2 Ia	425,573,655,967,1513
4818	+50 00161	HR 238, V526 Cas			0 48 05	+51 14.2	2	6.37	.015	0.29	.005	0.11	.010	3	F2 IV	39,1733
	+41 00148				0 48 06	+41 39.2	1	8.90		1.48				3	K5	209
		AzV 57			0 48 06	−72 40.8	1	14.24		−0.19		−0.91		4	B2	733
		MN201,73 # 16			0 48 07	−35 38.6	1	13.33		0.21				1		1495
		AJ74,1000 M31# 7			0 48 08	+40 53.8	1	11.55		1.33				3		209
4810	+63 00101				0 48 08	+64 03.2	2	8.40	.009	0.14	.000	0.07		4	A2	401,1118
		MN137,55 # 24			0 48 08	−73 16.5	1	14.63		−0.11		−0.51		3		967
	−30 00253				0 48 09	−30 14.3	1	11.46		0.18		0.09		1	A2	378
		AzV 58			0 48 11	−73 08.1	1	14.38		−0.22		−0.92		4	B2	733
		AzV 59			0 48 11	−73 27.9	1	13.40		0.06		−0.45		1	A0	425
		BPM 46670			0 48 12	−20 00.	1	12.35		0.72		0.12		1		3061
		AzV 61			0 48 13	−72 27.7	2	13.71	.037	−0.23	.009	−1.09	.023	4	O5 V	425,835
		G 1 - 28			0 48 14	+10 05.4	1	10.50		0.78		0.36		1	G9	333,1620
		NGC 188 - 5215			0 48 14	+84 54.8	1	13.44		1.15		1.07		1		4002
		AzV 62			0 48 14	−73 11.4	1	14.34		0.06		−0.72		3	B3	425
		AzV 63			0 48 14	−73 26.5	1	13.48		0.07		−0.52		5	A0	425
		AzV 60			0 48 14	−73 56.3	2	12.89	.098	−0.05	.028	−0.70	.009	4	B2 Ia	425,835
		LP 406 - 20			0 48 16	+15 32.1	1	12.66		0.91		0.66		1		1696
		WLS 48 40 # 7			0 48 16	+38 00.4	1	11.30		0.45		−0.01		2		1375
		AJ74,1000 M31# 9			0 48 16	+40 48.5	1	12.31		0.90				3		209
4817	+61 00178	HR 237			0 48 16	+61 32.0	6	6.07	.023	1.88	.031	1.84	.042	12	K5 Ib-II	138,252,1103,1355*
4891	−24 00358				0 48 17	−23 56.5	2	8.32	.009	0.64	.000	0.09		5	G1 V	2012,3061
		BPM 46675			0 48 18	−26 39.	1	11.50		0.62		0.05		2		3064
		AzV 64			0 48 19	−73 40.4	1	13.92		−0.14		−0.93		3	B1	425
		MN137,55 # 23			0 48 20	−73 16.5	1	12.55		0.97		0.85		6		967
		AJ63,118 # 25			0 48 20	−73 24.1	1	11.00		0.13				3		2001
		AzV 66			0 48 20	−73 32.7	1	13.48		−0.15		−0.89		3	B0	425
	+73 00039	G 242 - 71			0 48 21	+74 12.1	2	9.77	.000	0.49	.002	−0.12	.005	3	sdF8	979,1658
4976		AzV 65			0 48 21	−73 24.0	6	11.02	.031	0.13	.016	−0.59	.009	28	B6 I e	425,573,1277,1408*
4969	−72 00068				0 48 22	−71 45.9	4	9.54	.007	0.55	.013	0.06	.015	11	F7 V	1408,1425,2001,2010
4841	+62 00160	LS I +63 110			0 48 23	+63 30.6	1	6.86		0.57		−0.30		2	B5 Ia	1012,8100
		BPM 46676			0 48 24	−27 51.	1	15.42		0.61		−0.12		2		3064
		BPM 46677			0 48 24	−29 35.	1	13.06		0.93		0.56		2		3064
		AzV 67			0 48 24	−72 48.8	1	13.66		−0.16		−0.96		2	B0	425
4828	+66 00066				0 48 25	+67 18.5	1	8.57		2.04				3	K5	369
4919	−51 00209	HR 242, ρ Phe			0 48 25	−51 15.6	6	5.21	.010	0.36	.006	0.13	.000	20	F3 III	15,258,612,1075,2013,2029
		AzV 67a			0 48 27	−72 59.9	1	14.11		0.29		0.21		1	F5	767
		BSD 8 # 1391			0 48 28	+61 49.2	1	11.43		0.52		0.15		3	A0	1185
		AzV 68			0 48 28	−72 40.2	1	14.65		−0.11		−0.96		4	B2	733
4899	+1 00149				0 48 30	+02 28.4	1	7.60		0.61		0.16		2	G0	1375
4854	+60 00117				0 48 30	+60 51.9	1	9.23		0.17				3	B9 V	1118
		BPM 46678			0 48 30	−26 00.	1	13.75		0.48		−0.18		3		3064
		BPM 46679			0 48 30	−27 33.	1	12.80		0.75		0.24		3		3064
		AzV 70			0 48 31	−72 54.4	2	12.39	.005	−0.16	.005	−0.96	.015	5	B0 Ia	425,573
		AzV 69			0 48 31	−73 09.7	2	13.36	.018	−0.22	.005	−1.02	.009	5	B0	425,835
		BSD 8 # 1390			0 48 34	+61 19.6	1	9.95		1.23				1	G6	1103
		BSD 8 # 1392			0 48 35	+61 50.8	1	10.85		0.36		0.27		3	B9	1185
		CS 22170 # 37			0 48 35	−08 40.2	1	15.10		0.51		−0.04		1		1736
		MN201,73 # 18			0 48 35	−31 18.7	1	14.04		0.83				1		1495
4906	+18 00111	G 32 - 53			0 48 36	+18 31.3	3	8.75	.025	0.77	.015	0.23	.028	4	G0	680,927,1620
		CS 22170 # 28			0 48 36	−11 24.8	1	11.75		0.19		0.05		1		1736
		BPM 46680			0 48 36	−26 22.	1	13.67		0.73		0.01		2		3064
		BPM 46682			0 48 36	−31 08.	1	11.76		0.50		−0.10		1		3061
		AzV 71			0 48 37	−72 28.4	3	13.58	.010	0.01	.015	−1.00	.009	5	B0	425,835,1513
		CS 22942 # 12			0 48 39	−25 48.0	1	14.14		0.29		−0.06		1		1736
		AzV 72			0 48 40	−72 25.3	2	12.95	.005	−0.06	.009	−0.55	.037	4	B6 Iab	425,835
4881	+50 00164	HR 241			0 48 41	+51 18.0	1	6.22		0.03		−0.05		3	B9.5 V	1733
		HV 829			0 48 41	−73 00.2	3	11.71	.024	0.69	.038	0.46	.045	3	G0:I	689,1511,3075
		AzV 73			0 48 41	−73 19.5	1	14.08		−0.17		−0.91		2	B0	733
		BPM 46685			0 48 42	−26 24.	1	11.28		1.12		1.06		2		3064
		AzV 74			0 48 42	−73 01.4	1	11.74		0.72		0.51		2	G0:I	35
5069	−79 00019				0 48 42	−79 29.2	1	9.96		0.50		0.08		4	F5/7 [IV]	1700
4913	+17 00112	G 32 - 54			0 48 43	+18 28.3	3	9.22	.012	1.20	.004	1.16	.014	4	K2	333,680,1620,3072
4928	+2 00118	HR 243			0 48 44	+03 06.9	4	6.36	.005	1.07	.000	0.89	.008	10	K0 III	15,1256,1355,3002

Table 1   57

HD	DM	Other Id	N	Rem	α_{1950}	δ_{1950}	S	V	σ_V	B−V	σ_{B-V}	U−B	σ_{U-B}	n	Spectrum	References
		AJ74,1000 M31# 8			0 48 44	+40 46.2	1	10.96		1.15				3		209
4902	+40 00177				0 48 44	+40 57.6	1	7.29		-0.04		-0.16		2	A0	401
	+60 00119				0 48 45	+60 35.3	1	10.95		0.30		0.18		3	B9	1185
		AzV 75			0 48 46	-73 08.8	2	12.79	.005	-0.16	.005	-0.99	.005	5	O9	733,835
		MN137,55 # 9			0 48 46	-73 14.1	1	13.17		0.19		0.28		5		967
		AzV 77			0 48 47	-73 04.0	1	13.91		-0.22		-0.93		3	B0	425
232292	+50 00165				0 48 48	+50 55.7	1	8.12		0.99		0.67		2	K0	1566
	+60 00121				0 48 48	+61 21.4	1	9.70		0.31				1	A0	1103
		BPM 46686			0 48 48	-31 48.	1	13.65		0.60		-0.10		1		3061
		BPM 46687			0 48 48	-34 32.	1	13.30		0.98		0.75		1		3061
		BPM 16285			0 48 48	-54 29.	1	15.29		-0.02		-0.78		1	DA	3065
5030	-74 00069				0 48 48	-73 45.2	5	11.18	.021	0.11	.015	-0.51	.010	27	A0 Ia	425,573,1277,1408,8033
		LB 3164			0 48 48	-74 48.	1	12.89		0.08		0.01		3		45
		CS 22170 # 34			0 48 49	-10 11.7	1	14.30		0.10		-0.29		1		1736
		BSD 8 # 1395			0 48 52	+61 34.0	1	11.20		0.40		0.25		3	A0	1185
		BSD 8 # 1396			0 48 53	+61 43.1	1	10.53		0.36		-0.20		3	B5	1185
		LB 1580			0 48 54	-45 25.	1	13.09		0.59		-0.05		2		45
		AzV 79			0 48 54	-73 06.2	1	13.71		0.23		0.25		3	F2	425
5045	-74 00070				0 48 54	-73 44.7	5	11.04	.028	-0.02	.020	-0.89	.008	26	B3 Ia	425,573,1277,1408,8033
	-28 00262				0 48 56	-27 39.5	1	10.41		1.08		0.92		1	K0	4003
		AzV 80			0 48 57	-73 03.9	1	13.33		-0.14		-0.92		6	B1	425
4935	+11 00106	IDS00464N1214		AB	0 48 58	+12 30.8	1	6.72		0.33		0.01		3		196
		AzV 81			0 48 59	-73 43.3	2	13.28	.005	-0.11	.015	-0.86	.000	2	WN4.5+O5	425,835
4965	-3 00113				0 49 00	-03 24.9	2	7.26	.011	0.12	.008	0.10	.008	7	A0	252,1509
		BPM 46690			0 49 00	-25 52.	1	11.65		0.57		-0.01		2		3064
		BPM 46691			0 49 00	-30 55.	1	12.13		0.93		0.74		1		3061
		AAS53,255 # 36			0 49 00	-72 59.	1	12.71		1.75				1		1686
	+11 00108				0 49 02	+12 03.7	1	9.44		1.04		0.79		3	K0	196
		AzV 82			0 49 02	-72 58.9	1	14.13		-0.22		-1.02		3	B2	733
		CS 22170 # 38			0 49 03	-08 30.6	1	13.34		0.27		0.12		1		1736
		BSD 8 # 816			0 49 04	+61 15.9	1	11.28		0.94		0.67		3	Mc	1185
4972	-15 00156	G 268 - 59			0 49 04	-15 09.5	1	9.78		1.09		1.00		1	K2/3 V	3072
4967	-23 00332	G 268 - 60		★ A	0 49 04	-23 10.7	3	8.96	.038	1.28	.009	1.15	.039	7	K5 V	680,2013,3073
4973	-24 00365				0 49 04	-23 51.1	1	7.93		1.02				4	K0 III	2012
		AzV 83		AB	0 49 05	-72 58.4	1	13.38		-0.19		-0.90		4	B1	425
		BPM 46692			0 49 06	-32 16.	1	12.48		0.84		0.40		1		3061
		BSD 8 # 282			0 49 09	+60 06.5	1	12.17		0.34		0.04		3	A0	1185
		BSD 8 # 814			0 49 09	+60 21.7	1	11.66		0.41		-0.01		3	A0	1185
		BSD 8 # 809			0 49 09	+60 24.0	1	11.71		0.62		0.29		3	B5	1185
		G 132 - 36			0 49 10	+37 35.0	1	15.41		1.68				3		419
	-25 00326				0 49 10	-24 55.0	1	10.44		0.61		0.10		1	G0	3061
		AzV 84			0 49 10	-72 24.5	1	14.69		-0.20		-0.97		6	B0	425
		CS 29509 # 26			0 49 12	-31 30.6	1	14.98		0.11		0.05		1		1736
		BPM 46694			0 49 12	-34 08.	1	11.73		0.55		-0.05		1		3061
4970	+15 00126				0 49 13	+16 26.7	1	7.86		1.09		0.98		3	G5	1648
4950	+46 00183				0 49 13	+47 19.4	1	8.83		0.00		-0.21		2	A0	401
	+63 00102	LS I +64 052			0 49 13	+64 24.7	1	10.00		0.57		-0.41		2	B1 II	1012
		GR 18			0 49 14	-25 02.7	1	14.18		1.47		1.15		1		481
		CS 22942 # 8			0 49 15	-23 05.6	2	15.51	.000	0.13	.000	0.09	.009	2		966,1736
		AzV 85			0 49 15	-73 09.3	1	13.75		-0.07		-0.99		1	B0	425
4931	+59 00132	LS I +59 056			0 49 16	+59 49.1	4	8.68	.018	0.38	.021	0.26		7	B8 V	592,1103,1118,1185
4988	-17 00150				0 49 17	-17 20.7	1	9.88		0.42		-0.03		2	F3 V	1375
		AzV 86			0 49 17	-72 16.6	2	12.80	.035	-0.15	.015	-0.97	.005	5	B1	425,835
		AA9,95 # 104			0 49 17	-72 59.6	1	11.31		1.82		1.52		4	M5 III-I	35
		BPM 46693			0 49 18	-21 21.	1	13.34		0.76		0.32		1		3061
		Hiltner 74			0 49 19	+64 16.9	1	11.31		0.45		-0.39		2	B2 V	1012
		PHL 6772			0 49 19	-32 10.4	3	13.64	.030	0.10	.016	0.15	.006	5	A0	295,1736,3064
	-52 00112				0 49 19	-51 38.6	1	11.27		0.49		-0.26		2	G8 VI	3077
		AAS53,255 # 39			0 49 19	-72 59.	1	11.37		1.88				1	M5III-I	1686
		G 69 - 27			0 49 20	+20 18.7	1	11.53		1.44		1.22		1	K7	333,1620
		BSD 8 # 284			0 49 21	+59 56.1	1	12.41		0.49		0.40		3	A0	1185
		BSD 8 # 291			0 49 21	+60 07.6	1	11.28		0.35		0.16		3	B7	1185
		AzV 87		AB	0 49 21	-72 56.4	1	13.90		-0.38		-0.94		3	B0	425
	-22 00281				0 49 22	-22 33.4	1	10.96		0.63		0.13		2		1375
		AzV 88			0 49 23	-73 04.8	1	13.11		0.15		0.22		2	F2	425
		BPM 46696			0 49 24	-21 11.	1	12.94		0.49		-0.12		1		3061
	-27 00272				0 49 25	-26 41.0	1	9.83		0.39		-0.05		4		366
		AzV 91			0 49 25	-73 11.5	2	12.59	.023	0.02	.005	-0.47	.005	4	B8 Ia	425,835
		AzV 90			0 49 25	-73 14.5	2	12.57	.014	0.08	.000	0.14	.036	5	A5 Ib	733,835
		AzV 89			0 49 25	-73 32.1	1	14.47		-0.28		-0.90		2	B0	425
		G 1 - 30			0 49 26	+06 41.4	2	12.18	.015	0.52	.005	-0.15	.020	2	sdF8	333,1620,1696
		BSD 8 # 283			0 49 27	+59 26.6	1	12.13		0.30		0.23		3	B5	1185
		BSD 8 # 293			0 49 27	+60 01.6	1	12.88		0.23		-0.17		3	B7	1185
4710	+84 00014				0 49 27	+85 25.9	1	8.08		1.07		0.88		2	K0	1502
		BSD 8 # 823			0 49 28	+60 52.3	1	11.13		0.69		0.45		3	B8	1185
		BSD 8 # 824			0 49 28	+60 54.9	1	10.50		0.51				1	F2 p	1103
		AzV 92			0 49 29	-72 34.3	1	13.66		-0.12		-0.71		6	B2 Iab	425
		WLS 40 0 # 8			0 49 30	+00 10.5	1	11.25		0.62		0.10		2		1375
5024	-31 00319				0 49 30	-31 13.8	1	9.21		0.37				4	A9 V	2012
		BSD 8 # 2			0 49 32	+59 15.0	1	12.45		0.28		-0.16		3	B6 p	1185

HD	DM	Other Id	N	Rem	α_{1950}	δ_{1950}	S	V	σ_V	B–V	σ_{B-V}	U–B	σ_{U-B}	n	Spectrum	References
5042	−44 00226	HR 245	★	AB	0 49 32	−43 58.8	2	6.89	.005	0.36	.005			7	F2/3 V	15,2012
		BSD 8 # 296			0 49 33	+59 59.8	1	12.52		0.48				3	B7	1185
		BSD 8 # 285			0 49 33	+60 03.2	1	9.70		0.36		−0.11		3	B7	1185
	+41 00150				0 49 34	+42 03.3	1	9.15		1.26				3	K0	209
		AzV 96			0 49 34	−72 23.6	2	12.68	.069	−0.14	.005	−0.91	.015	5	B1 Ia	425,835
		AzV 93			0 49 34	−72 57.7	1	14.13		−0.13		−0.81		2	B0:	35
		AzV 94			0 49 34	−73 05.9	1	13.99		−0.34		−0.89		3	B1	425
4985	+45 00223				0 49 35	+45 32.0	1	8.43		0.00		0.01		2	A0	401
	−28 00266				0 49 35	−28 33.7	1	11.53		0.99				1		1495
		AzV 95			0 49 35	−73 00.4	1	13.91		−0.30		−1.03		3	O9	425
		BPM 46700			0 49 36	−27 44.	1	14.02		0.61		−0.10		2		3064
		WLS 100 0 # 6			0 49 37	+00 15.3	1	10.91		0.79		0.26		2		1375
5041	−14 00152	IDS00471S1346		AB	0 49 37	−13 30.1	2	9.43	.019	0.90	.007	0.57	.010	6	K1 V +K3 V	214,3016
		WLS 100 75 # 10			0 49 38	+75 28.1	1	10.15		0.49		0.06		2		1375
		AzV 98			0 49 38	−72 39.2	2	11.47	.020	0.05	.010	−0.40	.005	7	B9 Ia	573,733
		AzV 99			0 49 38	−73 04.1	1	13.01		−0.09		−0.71		7	B6 Ia-Iab	425
		AzV 97			0 49 38	−73 31.9	1	13.30		−0.12		−0.78		5	B2 Iab	425
		BSD 8 # 289			0 49 39	+59 25.8	1	12.06		0.32		−0.05		3	B5	1185
		BSD 8 # 299			0 49 39	+59 55.4	1	11.68		0.25		−0.22		3	B9	1185
	+19 00138				0 49 40	+20 06.3	1	10.06		0.54		0.06		2		1375
		CS 22170 # 29			0 49 40	−11 20.9	1	13.08		0.17		0.07		1		1736
5116	−73 00050				0 49 40	−72 57.0	1	9.67		1.24		1.34		5	G8/K1	35
		AAS53,255 # 42			0 49 40	−75 14.	1	13.16		1.54				1		1686
		CS 22166 # 4			0 49 41	−11 20.9	1	13.08		0.16		0.08		1		1736
		AzV 100			0 49 41	−73 14.1	1	14.29		−0.19		−0.75		2	B3	425
5007	+24 00128	IDS00470N2515		AB	0 49 42	+25 30.5	2	7.67	.005	1.18	.000	1.19	.000	5	K1 III	833,1625
4978	+60 00122				0 49 42	+61 26.8	1	9.27		0.24				2	B9 V	1118
5061	−36 00296				0 49 43	−36 01.2	2	8.65	.000	0.02	.014	0.00		7	A0 V	1068,2012
	+59 00133				0 49 44	+60 20.6	1	9.77		0.37		0.23		3	B5	1185
	−11 00162				0 49 44	−10 56.2	6	11.19	.009	−0.08	.011	−1.06	.038	36	sdO+G	963,989,1026,1729,1732,6009
5058	−23 00334	IDS00473S2309		AB	0 49 46	−22 53.2	1	7.13		0.53		0.02		1	F8/G1 V	79
		AzV 101			0 49 46	−73 32.3	3	12.13	.011	0.07	.006	−0.47	.008	10	B5 Ia	733,767,835
		BPM 46701			0 49 48	−27 08.	1	12.56		0.99		0.82		2		3064
		BPM 46702			0 49 48	−27 20.	1	14.16		0.69		0.17		2		3064
		AzV 102			0 49 49	−72 48.4	1	14.29		−0.20		−0.82		3	B2	425
		AzV 103			0 49 50	−73 02.2	1	13.36		−0.12		−0.83		3	B1 Iab	425
		BSD 8 # 826			0 49 52	+60 23.9	1	11.70		0.40		−0.24		3	B6	1185
		BSD 8 # 828			0 49 52	+60 39.6	1	9.94		0.23		−0.04		3	B8	1185
5067	−27 00277				0 49 52	−26 40.2	4	8.72	.425	0.98	.009	0.75	.000	10	K0 III	366,421,1495,4003
		AA9,95 # 107			0 49 52	−72 56.4	1	12.95		0.09		0.12		2	G0 V	35
		AzV 104			0 49 52	−73 04.2	1	13.13		−0.22		−0.93		3	B1 Ia	425
	+40 00183				0 49 53	+41 15.2	1	9.57		1.05				3	K0	209
5005	+55 00191	LS I +56 013	★	ABC	0 49 53	+56 21.4	2	7.76	.000	0.09	.000	−0.83	.000	6	O6.5V	1011,1209
		BSD 8 # 1412			0 49 53	+61 40.1	1	11.20		0.41		0.20		3	A0 p	1185
4994	+63 00105	IDS00469N6323		AB	0 49 54	+63 39.2	1	7.99		0.36				4	F0	1118
		BPM 46703			0 49 54	−26 47.	1	13.35		0.68		0.09		3		3064
		AzV 105			0 49 55	−72 44.2	2	12.23	.010	0.02	.015	−0.24	.010	6	A0 Iab	733,835
	+39 00197				0 49 56	+40 30.9	1	10.06		1.54				3		209
	+40 00184				0 49 57	+40 40.1	1	8.54		1.06				3	K0	209
		BSD 8 # 309			0 49 57	+60 05.7	1	10.35		0.22		0.00		3	B8	1185
		AzV 106			0 49 57	−72 53.5	1	14.32		−0.11		−0.85		3	B1	425
	+60 00123				0 49 58	+61 00.2	1	9.63		0.21		−0.20		3	A0	1185
		POSS 295 # 39			0 49 59	+31 39.3	1	17.56		1.66				1		1739
		BSD 8 # 1411			0 49 59	+61 33.1	1	10.18		0.30		0.15		3	B8	1185
		AzV 107			0 49 59	−72 52.0	1	13.63		0.31		0.31		3	F5	425
		G 268 - 63			0 50 00	−26 06.5	2	12.53	.005	0.51	.010	−0.24	.010	4		1696,3064
		BPM 46705			0 50 00	−26 32.	1	11.98		0.89		0.53		2		3064
		BPM 46706			0 50 00	−28 15.	1	12.69		0.54		−0.26		2		3064
		BPM 46707			0 50 00	−29 03.	1	12.89		0.71		0.19		3		3064
		BSD 8 # 316			0 50 03	+59 52.5	1	12.36		0.29		−0.23		3	A0 p	1185
4853	+82 00020	HR 240			0 50 03	+83 26.2	2	5.61	.015	0.09	.000	0.10	.005	5	A4 V	985,1733
		BSD 8 # 831			0 50 04	+60 23.2	1	10.01		0.18		−0.09		3	B6	1185
5015	+60 00124	HR 244	★	A	0 50 04	+60 51.0	8	4.80	.012	0.54	.008	0.09	.030	26	F8 V	15,351,1103,1118,3026*
5078	−21 00130				0 50 04	−21 16.0	1	7.85		0.65		0.22		1	G3/5 V	3061
	−29 00247				0 50 04	−29 24.8	1	10.03		1.08		0.94		1	K2	4003
		AzV 109			0 50 04	−72 55.5	1	13.73		−0.13		−0.85		3	B1 Ib	425
5065	+39 00198				0 50 06	+39 58.4	1	6.78		0.56		0.07		2	G0	1601
		BPM 70406			0 50 06	−18 58.	1	13.50		0.82		0.46		1		3061
		Smethells 177			0 50 06	−41 31.	1	11.92		1.49				1		1494
		AzV 110		AB	0 50 07	−73 00.3	3	12.13	.023	0.08	.004	−0.33	.007	6	A0 Ia	35,425,835
5074	+18 00114				0 50 08	+18 41.5	1	8.34		1.44		1.62		2	K2	1648
5066	+37 00159	HR 246			0 50 08	+38 16.6	1			0.02		0.10		4	A2 V	1049
		NGC 288 - 35			0 50 08	−26 45.7	1	15.95		−0.06		0.03		3		366
		AAS35,353 # 29			0 50 08	−72 47.6	1	13.46		−0.21		−0.99		4	B1	767
5031	+60 00126				0 50 09	+61 20.8	2	9.02	.015	0.46	.015	−0.05		5	A0 V	1118,1185
5096	−13 00146				0 50 09	−13 12.5	1	7.90		0.38		0.00		2	F2 V	1375
		NGC 188 - 10			0 50 10	+84 56.0	2	10.90	.019	1.41	.015	1.50	.024	4		387,4002
		AzV 111			0 50 10	−72 48.2	1	13.84		−0.16		−0.82		4	B1	425
		AAS35,353 # 30			0 50 10	−73 05.1	1	12.74		0.57		0.00		4	F0	767
		BSD 8 # 1415			0 50 11	+61 18.1	1	10.70		0.34		−0.01		3	A0	1185

Table 1   59

HD	DM	Other Id	N	Rem	α_{1950}	δ_{1950}	S	V	σ_V	B–V	σ_{B-V}	U–B	σ_{U-B}	n	Spectrum	References
		AzV 112			0 50 11	−72 49.5	1	14.15		−0.17		−1.01		4	B0	425
		AzV 110a			0 50 11	−73 38.7	1	13.85		−0.14		−0.90		5	B1	767
		NGC 288 - 82			0 50 12	−26 52.	1	14.30		0.65		−0.06		5		1638
		NGC 288 - 84			0 50 12	−26 52.	1	15.41		0.85		0.45		1		366
		NGC 288 - 90			0 50 12	−26 52.	1	14.14		0.59		−0.10		1		366
		NGC 288 - 103			0 50 12	−26 52.	1	14.73		0.99		0.44		1		366
		NGC 288 - 111			0 50 12	−26 52.	1	15.34		0.16				2		1523
		NGC 288 - 114			0 50 12	−26 52.	1	14.59		0.70		0.14		1		366
		NGC 288 - 115			0 50 12	−26 52.	1	14.85		0.99		0.28		1		366
		NGC 288 - 153			0 50 12	−26 52.	1	16.15		−0.09				2		1523
		NGC 288 - 156			0 50 12	−26 52.	1	14.44		0.99		0.53		1		366
		NGC 288 - 157			0 50 12	−26 52.	1	15.77		0.00				2		1523
		NGC 288 - 162			0 50 12	−26 52.	1	15.92		0.00		−0.03		1		366
		NGC 288 - 163			0 50 12	−26 52.	2	15.61	.015	0.87	.020	0.26		2		366,1523
		NGC 288 - 213			0 50 12	−26 52.	2	13.03	.000	1.37	.005	1.41		3		366,1523
		NGC 288 - 216			0 50 12	−26 52.	1	15.36		0.18				2		1523
		NGC 288 - 503			0 50 12	−26 52.	3	13.13	.035	0.97	.021	0.69		4		366,421,1523
		NGC 288 - 504			0 50 12	−26 52.	2	15.27	.000	0.54	.049	−0.04		3		366,421
		NGC 288 - 505			0 50 12	−26 52.	1	16.30		0.96		0.45		3		366
		NGC 288 - 507			0 50 12	−26 52.	2	13.94	.005	1.32	.000	1.08		2		366,1523
		NGC 288 - 508			0 50 12	−26 52.	1	14.50		0.68		0.00		1		366
		NGC 288 - 514			0 50 12	−26 52.	2	16.37	.035	−0.06	.010	−0.33		5		366,1523
		NGC 288 - 525			0 50 12	−26 52.	3	11.68	.009	0.62	.010	0.04	.009	8		366,1523,1638
		NGC 288 - 545			0 50 12	−26 52.	2	12.84	.005	0.86	.020	0.62		2		366,1523
		NGC 288 - 546			0 50 12	−26 52.	1	17.80		0.63				3		366
		BPM 46712			0 50 12	−28 12.	1	12.08		0.60		−0.05		2		3064
		AAS35,353 # 31			0 50 12	−73 37.0	1	13.74		−0.17		−0.66		3	B3	767
5072	+38 00129				0 50 13	+38 46.2	1	7.77		0.55		−0.03		2	F7 IV	105
5098	−24 00376	HR 247		⋆ A	0 50 13	−24 16.7	1	5.48		1.27		1.21		2	K1 III	3005
5098	−24 00376	IDS00478S2433		AB	0 50 13	−24 16.7	1	5.46		1.24				4	K1 III	2006
		SB 353			0 50 13	−26 10.7	2	14.61	.000	0.19	.000	0.18	.008	2		966,1736
		CS 22170 # 30			0 50 14	−11 17.4	1	14.26		−0.04		−0.13		1		1736
		CS 22166 # 5			0 50 14	−11 17.6	1	14.26		−0.02		−0.13		1		1736
		AzV 113			0 50 14	−72 40.3	1	14.28		−0.12		−0.96		3	B1	733
		AA9,95 # 109			0 50 14	−73 02.3	2	12.87	.000	0.10	.000	0.16	.050	6	A1 V	35,767
		NGC 288 - 30			0 50 15	−26 45.9	1	16.87		−0.13		−0.55		3		366
		NGC 288 - 28			0 50 16	−26 45.9	1	16.98		0.70				2		366
		NGC 288 - 27			0 50 16	−26 46.5	2	14.95	.030	0.92	.025	0.53		4		366,1523
		AzV 114			0 50 16	−72 55.6	1	14.93		−0.19		−1.02		4	B0	425
		POSS 295 # 40			0 50 17	+31 27.6	1	18.47		1.39				2		1739
5106	−19 00140				0 50 17	−18 50.8	1	8.70		0.65		0.09		1	G2 V	3061
		AzV 115			0 50 17	−72 28.4	1	13.57		−0.18		−0.98		6	B1	425
5084	+38 00131				0 50 18	+38 53.2	1	8.23		0.33		0.04		2	F3 V	105
		BPM 46714			0 50 18	−27 45.	1	12.15		1.09		0.76		3		3064
		BPM 46715			0 50 18	−29 35.	1	12.06		1.05		0.88		2		3064
		BPM 46716			0 50 18	−34 14.	1	10.71		0.94		0.65		1		3061
5093	+17 00113	IDS00477N1734		AB	0 50 19	+17 50.1	1	8.96		1.01		0.71		2	K0	1375
		NGC 288 - 25			0 50 19	−26 46.3	1	16.62		0.75		0.13		2		366
		NGC 288 - 70			0 50 20	−26 43.6	1	14.04		1.21		0.72		1		366
		NGC 288 - 24			0 50 20	−26 46.3	2	15.92	.010	−0.04	.005	−0.10		3		366,1523
		NGC 288 sq1 11			0 50 20	−27 03.0	1	14.34		0.92				2		421
		NGC 288 - 45			0 50 21	−26 44.3	1	12.49		0.50		−0.08		2		366
		NGC 288 - 21			0 50 21	−26 46.1	1	15.88		0.91		0.34		1		366
		NGC 288 - 22			0 50 21	−26 46.4	2	14.28	.005	0.68	.005	−0.01		2		366,1523
		NGC 288 - 23			0 50 21	−26 46.8	2	12.77	.013	0.72	.009	0.22	.017	6		366,1638
		AzV 116			0 50 21	−72 24.2	2	13.53	.058	−0.19	.010	−1.02	.024	3	B0	425,835
5210	−75 00066				0 50 22	−75 08.5	2	8.72	.015	0.59	.010	0.08		2	G1/2 V	1642,1738
	+58 00130				0 50 23	+58 44.4	1	10.68		0.20		0.03		3	B9	1185
		BSD 8 # 841			0 50 23	+61 15.0	1	11.15		0.36		0.20		3	B9	1185
		CS 22183 # 11			0 50 23	−03 09.0	1	12.55		0.13		0.18		1		1736
		NGC 288 - 46			0 50 23	−26 44.9	1	15.41		0.90		0.33		2		366
		NGC 288 sq1 10			0 50 23	−27 02.9	1	14.25		0.53				2		421
		NGC 288 - 47			0 50 24	−26 44.6	1	17.22		0.78				2		366
		NGC 288 - 20			0 50 25	−26 45.7	1	15.51		0.74		0.22		1		366
5135	−52 00116				0 50 25	−51 49.3	1	7.88		0.72				4	G3 IV/V	2012
236567	+58 00131				0 50 26	+58 53.3	1	9.95		0.16		0.01		3	A0	1185
	−26 00283				0 50 26	−26 29.9	1	11.68		0.99				1		1495
		NGC 288 - 19			0 50 26	−26 46.1	2	14.66	.034	0.98	.039	0.59		3		366,1523
		CS 29509 # 23			0 50 26	−33 00.4	1	13.51		−0.03		−0.17		1		1736
		SB 354			0 50 26	−33 00.4	2	13.52	.010	−0.05	.014	−0.15	.027	2	A0	295,1736
5071	+59 00134				0 50 27	+60 23.7	2	7.78	.025	0.11	.005	0.08		6	A0 V	1118,1185
5112	−1 00114	HR 248			0 50 27	−01 24.9	10	4.76	.009	1.57	.009	1.91	.017	39	M0 III	3,15,58,975,1075,1425*
		NGC 288 - 15			0 50 28	−26 46.6	1	16.50		1.07				2		366
		AA9,95 # 110			0 50 29	−73 02.4	1	12.59		0.53		0.02		3	F6 V	35
		AzV 117			0 50 29	−73 34.7	1	13.88		−0.01		−0.59		4	A0	425
		BPM 46717			0 50 30	−25 22.	1	13.37		1.26		1.02		1		3061
		NGC 288 - 51			0 50 30	−26 45.0	2	15.21	.005	1.48	.005	1.46		3		366,1523
		NGC 288 - 52			0 50 30	−26 45.2	1	13.36		0.66		0.05		1		366
		AzV 118			0 50 30	−72 24.9	2	13.11	.075	−0.06	.050	−1.03	.000	4	B1	425,835
	+40 00185				0 50 31	+40 54.7	1	10.44		1.09				3		209

HD	DM	Other Id	N	Rem	α_{1950}	δ_{1950}	S	V	σ_V	B–V	σ_{B-V}	U–B	σ_{U-B}	n	Spectrum	References
5190	−70 00037	HR 252		⋆ AB	0 50 31	−69 46.5	4	**6.21**	.003	**0.55**	.004	**0.07**	.008	23	F7 IV/V	15,611,1075,2038
		AzV 119			0 50 31	−73 26.9	1	13.99		−0.29		−0.90		3	B1	425
		LP 766 - 37			0 50 32	−19 00.0	1	13.77		0.94		0.64		1		3061
		NGC 288 - 54			0 50 32	−26 45.4	2	**15.19**	.010	**0.82**	.083	0.46		3		366,1523
		CS 29509 # 31			0 50 32	−30 10.9	1	13.97		0.14		0.17		1		1736
		CS 29509 # 37			0 50 33	−28 17.1	1	15.49		0.07		0.14		1		1736
		AzV 120			0 50 33	−72 25.3	1	14.43		−0.22		−1.04		4	O9 V	425
	+18 00115	G 32 - 56			0 50 34	+18 52.6	1	9.53		0.80		0.38		1	G5	333,1620
5133	−31 00325	G 269 - 49			0 50 34	−30 37.7	6	**7.15**	.015	**0.94**	.011	**0.68**	.008	26	K2 V	79,1075,2013,2024*
		NGC 288 - 7			0 50 35	−26 45.8	2	**13.87**	.005	**1.13**	.005	0.83		3		366,1523
		NGC 288 - 2			0 50 35	−26 46.7	2	**15.96**	.034	**−0.02**	.058	0.11		3		366,1523
5134	−33 00330	G 269 - 50			0 50 35	−32 55.9	2	**9.12**	.015	**0.57**	.010	**0.05**	.005	2	G0 V	79,3061
		POSS 295 # 17			0 50 36	+31 30.1	1	15.02		1.55				1		1739
		NGC 288 - 6			0 50 36	−26 45.8	1	17.12		−0.11		−0.71		3		366
	−30 00265				0 50 36	−30 07.6	1	11.80		0.68				1		1495
		NGC 288 - 1			0 50 37	−26 46.7	2	**14.01**	.015	**1.45**	.010	1.37		2		366,1523
	+63 00108	LS I +64 053			0 50 38	+64 29.5	1	10.67		0.68		−0.21		2	B3 III:	1012
		WLS 40-15 # 8			0 50 38	−15 00.4	1	10.61		0.62		0.12		2		1375
		NGC 288 - 55			0 50 38	−26 45.0	1	14.94		1.61				1		366
		NGC 288 - 4			0 50 38	−26 45.8	1	15.01		0.99		0.46		1		366
		AzV 121			0 50 38	−72 55.9	1	11.31		0.70		0.60		3	G0	35
5132	−18 00143				0 50 39	−17 55.4	1	7.64		0.32				4	A5/7mA5-F3	2012
		HA 92 # 309			0 50 40	+00 29.8	1	13.84		0.51		−0.02		1		1764
		POSS 295 # 41			0 50 40	+32 06.8	1	18.96		1.65				4		1739
		POSS 295 # 1			0 50 40	+33 27.4	1	13.42		0.83				2		1739
		MN201,73 # 24			0 50 40	−31 08.5	1	11.93		0.78				1		1495
		AzV 122			0 50 40	−72 57.1	1	12.70		0.03		−0.56		2	B6 Ia	35
		BSD 8 # 1421			0 50 41	+61 42.7	1	10.13		0.30		0.08		3	B7	1185
		AA9,95 # 113			0 50 41	−73 01.5	1	12.90		1.07		−0.24		3		35
		HA 92 # 235			0 50 42	+00 20.0	1	10.60		1.64		1.98		2		1764
		WLS 100 75 # 5			0 50 42	+77 15.8	1	11.20		0.49		0.10		2		1375
		AzV 135			0 50 42	−72 23.0	1	13.98		−0.26		−0.98		4	B0	425
		AzV 124			0 50 42	−72 27.1	1	13.50		−0.06		−1.05		5	B0	425
		AzV 123			0 50 42	−73 07.2	1	13.22		−0.03		−0.45		4	B8 Iab	425
5118	+36 00148	HR 249			0 50 43	+37 08.9	2	**6.07**	.007	**1.14**	.001	1.13		7	K2 III	71,1501
5143	+3 00120	IDS00480N0333		AB	0 50 45	+03 48.9	1	7.21		0.35		0.03		2	F0	3016
5156	−25 00338	HR 251		⋆ AB	0 50 45	−25 02.9	2	**6.44**	.005	**0.44**	.000	0.01		6	F3 V	404,2032
		AzV 125			0 50 45	−73 35.0	2	**12.55**	.009	**0.04**	.018	**−0.66**	.009	5	B3 Ia	733,835
		BSD 8 # 333			0 50 46	+60 10.6	1	11.32		0.21		−0.37		3	B6	1185
	+84 00016	NGC 188 - 8			0 50 46	+85 01.9	1	8.76		0.93		0.64		2	G5	4002
	−2 00119				0 50 46	−02 09.4	1	9.91		0.60		0.05		2		1375
		AzV 126			0 50 46	−72 55.6	1	13.47		−0.02		−0.90		2	B0 Ia	35
	−38 00288				0 50 47	−38 00.2	1	12.04		0.65		0.17		2		1554
		BPM 46720			0 50 48	−26 52.	2	**12.45**	.009	**0.86**	.002	**0.45**	.004	11		1638,3064
		BPM 46721			0 50 48	−31 20.	1	11.45		0.85		0.45		1		3061
		AzV 127			0 50 48	−73 29.3	1	12.52		0.26		0.27		3	F2	425
		AzV 128			0 50 49	−72 27.8	1	14.34		−0.22		−1.03		4	B0	425
5175	−44 00229				0 50 50	−44 02.5	1	9.71		0.38		−0.22		2	A5 (Ibw)	1696
5207	−60 00058				0 50 50	−59 56.4	1	9.61		0.36		0.06		3	A3 mA8-F3	1097
		AAS35,353 # 33			0 50 50	−72 57.3	1	13.10		0.79		0.31		4	G7	767
		1Zw 051+12 # 4			0 50 51	+12 22.5	1	14.56		1.22		1.18		1		327
	−23 00339				0 50 51	−23 13.1	1	9.98		0.90		0.60		2	K0	3061
5173	−39 00215	IDS00485S3915		AB	0 50 51	−38 59.2	1	8.34		0.38				4	F0 V	2012
		AzV 129			0 50 52	−72 37.2	1	14.19		−0.12		−0.78		4	B1	733
5128	+51 00179	HR 250		⋆ AB	0 50 54	+52 25.1	1	6.27		0.19		0.15		3	A5 m	8071
	−27 00284				0 50 54	−26 48.9	1	10.28		0.56		0.03		3		366
		BPM 46724			0 50 54	−28 36.	1	14.40		0.82		0.20		2		3064
		GD 659			0 50 54	−33 16.2	2	**13.36**	.000	**−0.23**	.015	**−1.11**	.040	2		295,622
		NGC 288 sq2 7			0 50 55	−26 50.8	1	13.92		1.02				5		1638
		AzV 130			0 50 55	−73 21.3	1	14.13		−0.13		−0.80		3	B0	425
		G 269 - 52			0 50 56	−33 01.7	1	12.67		1.39		1.25		2		481
		AzV 131			0 50 56	−73 01.4	1	12.61		0.07		−0.46		3	B8 Ia	35
5110	+65 00106				0 50 57	+66 09.9	1	7.05		0.40		0.02		3	F0	379
	−29 00253				0 50 58	−29 34.8	1	11.61		0.72				1		1495
		AzV 132		AB	0 50 58	−73 28.2	1	13.63		0.09		−0.75		3	B2:	425
		GR 19			0 50 59	−15 11.7	1	11.83		1.40		1.24		2		481
		AzV 133			0 50 59	−72 53.0	1	13.91		−0.31		−0.92		3	B0	425
	+59 00136	LS I +59 057			0 51 01	+59 44.7	1	10.11		0.19		−0.45		3	B5	1185
		CS 22166 # 1			0 51 01	−14 42.8	1	12.70		−0.01		−0.02		1		1736
		AzV 134			0 51 01	−73 20.5	1	12.77		0.11		0.22		3	F2	425
	−38 00289				0 51 02	−37 57.6	1	10.69		0.60		0.05		4		1554
		G 172 - 28			0 51 03	+45 40.4	1	11.10		1.43		1.25		2	M2	7010
		BSD 8 # 337			0 51 04	+59 56.1	1	9.99		0.29		0.26		3	B9	1185
		NGC 288 sq1 5			0 51 04	−26 42.0	1	13.01		0.61				5		421
5205	−39 00217				0 51 04	−39 20.3	1	8.60		0.36				4	F0 IV/V	2012
5109	+68 00058	IDS00479N6830		A	0 51 05	+68 46.6	1	9.73		0.55		0.07		4	G0	7009
5109	+68 00058	IDS00479N6830		AB	0 51 05	+68 46.6	2	**9.45**	.290	**0.60**	.025	0.09		4	G0	1619,1625
	−38 00290				0 51 05	−37 41.7	1	9.62		0.69				4	G5	1554
		AA9,95 # 117			0 51 05	−72 59.6	1	14.06		−0.11		−0.86		2	B2 II	35
		BPM 70420			0 51 06	−18 59.	1	12.70		0.55		−0.13		2		3061

Table 1 61

HD	DM	Other Id	N	Rem	α_{1950}	δ_{1950}	S	V	σ_V	B–V	σ_{B-V}	U–B	σ_{U-B}	n	Spectrum	References
		AA9,95 # 120			0 51 06	−72 51.2	1	11.08		0.77		0.41		2	G9 V	35
		AA9,95 # 118			0 51 06	−72 59.2	1	13.26		-0.06		-0.87		2	B2 Iab	35
5109	+68 00058	IDS00479N6830		B	0 51 07	+68 46.6	1	10.40		0.63		0.16		4	G0	7009
5185	−22 00150	G 268 - 65			0 51 07	−22 28.4	3	8.92	.009	0.72	.016	0.25	.004	6	G6 V	79,1731,3061
5277	−73 00052				0 51 07	−73 23.3	5	10.95	.025	0.13	.008	-0.18	.017	28	A0 Ia	425,573,1277,1408,8033
5186	−24 00378	G 268 - 66			0 51 08	−24 17.9	1	8.67		0.65		0.10		1	G5 V	79
		AzV 137		AB	0 51 08	−73 00.3	2	12.38	.045	-0.08	.020	-0.78	.010	6	B4 Ia	35,573
		AzV 138			0 51 08	−73 04.6	1	14.28		0.00		-0.97		3	B0	425
		1Zw 051+12 # 1			0 51 10	+12 23.5	1	13.72		0.70		0.17		1		327
236570	+59 00137				0 51 10	+60 18.3	1	9.33		0.14		0.24		3	B5	1185
	+60 00127				0 51 10	+61 07.1	1	9.51		0.25		-0.30		3	B8	1185
		G 269 - 53			0 51 10	−29 43.7	1	12.74		1.34		1.10		3		3064
		CS 29509 # 25			0 51 10	−32 07.9	1	15.87		0.09		0.18		1		1736
		AzV 139			0 51 11	−73 29.3	1	14.14		-0.11		-0.77		3	B2	425
		BPM 46735			0 51 12	−31 21.	1	12.40		0.63		0.07		2		3061
		BPM 46736			0 51 12	−32 04.	1	11.94		0.80		0.34		1		3061
		Florsch 264			0 51 12	−71 57.9	1	13.45		-0.08		-0.33		2	B8	767
		HA 92 # 322			0 51 13	+00 31.3	1	12.68		0.53		0.00		1		1764
5217	−27 00285				0 51 14	−26 43.5	3	9.36	.348	0.97	.008	0.73	.010	8	K1 III	366,421,3064
		BPM 46738			0 51 14	−26 45.2	2	12.87	.005	0.59	.002	-0.09		6		421,3064
		NGC 288 sq1 7			0 51 14	−26 47.8	1	13.56		0.65				5		421
	−38 00291				0 51 14	−37 44.5	1	11.05		0.87				4		1554
		AzV 140			0 51 14	−71 54.7	2	11.87	.040	0.48	.010	0.36	.000	4	F3 Iab	425,573
		1Zw 051+12 # 3			0 51 15	+12 17.5	1	13.32		0.84		0.52		1		327
		POSS 295 # 37			0 51 15	+29 45.7	1	17.39		1.63				1		1739
232310	+52 00194				0 51 15	+53 21.7	1	9.37		0.01		-0.30		2	B9	1566
		BSD 8 # 16			0 51 15	+58 48.8	1	10.66		0.27		-0.08		3	B9	1185
	−14 00163				0 51 15	−14 28.8	1	10.46		0.84		0.58		1		79
		AAS35,353 # 35			0 51 15	−73 26.6	1	12.44		0.67		-0.02		4	G0	767
		NGC 300 sq2 6			0 51 16	−37 54.8	1	13.10		0.57				3		1554
		BPM 46732			0 51 18	−23 40.	1	14.11		0.56		-0.14		1		3061
		BPM 46739			0 51 18	−32 27.	1	12.47		1.01		0.85		1		3061
5229	−38 00293				0 51 18	−38 14.5	2	8.75	.000	1.21	.005	1.21	.005	8	K2 III	910,1554
		NGC 288 sq1 3			0 51 20	−26 38.5	1	12.65		0.46				3		421
		POSS 295 # 9			0 51 21	+27 39.5	1	14.02		1.25				2		1739
	+41 00158				0 51 21	+41 50.2	1	10.06		1.03				3	K0	209
		GR 20			0 51 21	−25 16.3	1	14.22		1.57		1.20		1		481
5291	−73 00053				0 51 21	−72 54.4	6	10.85	.035	0.04	.010	-0.64	.002	27	B6 Ia	35,573,1277,1408,1571,8033
		BSD 8 # 342			0 51 22	+59 45.3	1	12.12		0.40		0.29		3	A0	1185
		BSD 8 # 855			0 51 22	+60 25.9	1	10.89		0.32		-0.18		3	B6	1185
5302	−73 00054				0 51 23	−73 22.9	1	9.98		1.53				3	K0	1408
		BPM 46741			0 51 24	−26 04.	1	13.94		0.42		-0.19		1		3061
5227	−29 00255				0 51 24	−28 48.9	1	8.55		1.34		1.52		1	K2/3 III	4003
		GR 21			0 51 24	−35 16.0	1	15.21		1.67		1.22		2		481
5213	+21 00116				0 51 25	+22 11.7	1	8.52		0.19		0.14		2	A3	1375
		NGC 288 sq1 9			0 51 25	−27 05.9	1	14.22		0.65				2		421
		AzV 141			0 51 25	−72 49.9	1	14.50		-0.03		-0.63		4	B3:	425
		AAS35,353 # 36			0 51 26	−72 40.7	1	13.40		-0.13		-0.65		4	B3	767
5303	−75 00068	CF Tuc			0 51 27	−74 55.4	4	7.61	.013	0.71	.009	0.17	.011	21	G2/5 V	1641,1642,1738,2033
		NGC 300 sq2 5			0 51 28	−37 57.3	1	12.96		1.20				2		1554
5171	+60 00128				0 51 29	+60 56.8	1	9.46		0.18				3	A2	1118
		BSD 8 # 1430			0 51 30	+62 00.9	1	11.49		0.46		0.30		3	A0	1185
		BPM 46743			0 51 30	−21 16.	1	12.98		0.87		0.60		1		3061
		BPM 46744		A	0 51 30	−26 50.	1	13.76		1.08		0.80		2		3064
		BPM 46744		B	0 51 30	−26 50.	1	14.34		0.92		0.52		2		3064
		BPM 46745			0 51 30	−26 59.	1	12.32		0.89		0.54		2		3064
		AzV 142			0 51 31	−72 38.2	1	12.89		0.20		0.30		5	F5	425
		MN201,73 # 25			0 51 32	−31 58.2	1	15.12		1.50				1		1495
5223	+23 00123				0 51 33	+23 47.8	2	8.33	.000	1.43	.000			2	R3p	1238,3025
		NGC 288 sq1 8			0 51 33	−26 48.2	1	13.74		1.10				4		421
	−28 00280				0 51 33	−28 23.0	2	10.37	.010	1.16	.014	1.15		2		1495,4003
5276	−63 00083	HR 257, BQ Tuc			0 51 33	−63 08.6	2	5.70	.000	1.57	.000	1.77		5	M4 III	2035,3002
		AAS35,353 # 37			0 51 33	−73 31.7	1	12.85		0.17		-0.03		2	A5	767
		Steph 90			0 51 35	+04 20.2	1	11.20		1.55		1.91		1	M0	1746
		BPM 46748			0 51 36	−28 07.	1	12.42		0.58		-0.05		3		3064
		BPM 46749			0 51 36	−32 02.	1	11.62		0.79		0.40		1		3061
	+67 00077				0 51 37	+67 38.3	1	9.29		0.54		-0.01		2	B5	1733
	−33 00336	BPM 46750			0 51 37	−33 31.6	1	11.07		1.04		0.88		1		3061
5248	−18 00147	G 268 - 68			0 51 40	−18 18.1	1	8.94		0.81		0.52		2	K0 V	3061
	−29 00259				0 51 40	−29 00.9	1	11.06		0.21		0.11		1	A5	378
	+63 00109				0 51 41	+64 15.3	1	10.17		0.31				2		1118
		HA 92 # 245			0 51 42	+00 23.6	1	13.82		1.42		1.19		8		1764
		AzV 143			0 51 42	−72 54.6	1	14.12		-0.19		-1.02		5	B0	425
5236	+24 00136				0 51 43	+25 26.7	2	8.41	.005	0.34	.005	0.06	.010	5	F0	833,1625
	+60 00129				0 51 45	+61 14.6	2	9.24	.021	0.42	.017			3	F4 V	1103,1118
	−20 00160	G 268 - 70			0 51 45	−19 56.3	1	11.12		0.97		0.73		1		3061
5268	−9 00181	HR 255			0 51 46	−09 00.7	1	6.16	.005	0.92	.005	0.54	.000	9	G5 IV	15,1061,3077
		CS 22942 # 11			0 51 46	−25 20.8	2	12.93	.009	0.65	.015	-0.02	.001	4		1580,1736
	−29 00260				0 51 46	−28 48.9	1	11.36		1.02		0.82		1		4003
		BSD 8 # 865			0 51 47	+60 24.2	1	11.60		0.35		0.01		3	B7	1185

HD	DM	Other Id	N Rem	α_{1950}	δ_{1950}	S	V	σ_V	B–V	σ_{B-V}	U–B	σ_{U-B}	n	Spectrum	References
		GD 660		0 51 47	−20 07.9	1	13.63		0.48		−0.29		1		3061
	−24 00385			0 51 47	−23 50.3	2	10.86	.010	0.58	.002	−0.01	.030	4		1696,3061
		AAS35,353 # 38		0 51 47	−72 26.3	1	13.09		−0.05		−0.30		4	B8	767
		Hiltner 76		0 51 48	+63 14.0	1	11.21		0.36		−0.42		2	B2 V	1012
		BPM 46753		0 51 48	−27 16.	1	14.90		1.35		1.14		2		3064
		NGC 300 sq1 16		0 51 48	−38 11.2	1	19.06		1.64				1		910
		AAS35,353 # 39		0 51 48	−72 51.0	1	14.02		0.05		0.09		3	A0	767
	−37 00322			0 51 49	−36 40.7	1	9.92		0.55		−0.09		2	F8	3061
		AzV 144		0 51 50	−73 12.5	1	14.06		−0.18		−0.94		4	B0	425
		MN201,73 # 26		0 51 51	−27 38.1	1	11.92		1.41				1		1495
		AzV 145		0 51 51	−72 53.9	1	13.35		−0.11		−0.89		3	B1	425
5211	+59 00141			0 51 52	+59 48.4	2	9.02	.003	0.47	.004			5	F4 III	70,1103
		AzV 146		0 51 52	−73 26.2	1	13.67		−0.06		−0.69		3	B3	425
		PHL 6850		0 51 54	−27 11.9	3	13.05	.023	0.08	.010	0.16	.013	5	A0	295,966,1736
		BPM 46755		0 51 54	−35 04.	1	14.00		1.48		1.13		1		3061
		POSS 295 # 10		0 51 55	+31 35.6	1	14.27		0.74				1		1739
		NGC 300 sq1 14		0 51 55	−38 10.6	1	17.90		1.22		1.26		4		910
5267	+18 00122	HR 254	⋆ AB	0 51 56	+18 55.1	2	5.79		−0.02	.010	−0.14	.000	5	A0 V +A4 V	1049,1733
		HA 92 # 248		0 51 57	+00 24.0	1	15.35		1.13		1.29		2		1764
		AzV 148		0 51 57	−72 58.7	1	14.28		−0.23		−1.00		4	B0	425
		AzV 147		0 51 57	−73 21.7	1	13.92		−0.10		−0.89		3	B1	425
		BSD 8 # 870		0 51 59	+60 32.7	1	10.26		0.33		−0.20		3	G0 p	1185
5233	+60 00130			0 51 59	+61 23.5	2	8.41	.008	0.06	.008			3	B3 V	1103,1118
		HA 92 # 249		0 52 00	+00 24.8	1	14.32		0.70		0.24		8		1764
		BPM 46754		0 52 00	−20 25.	1	12.51		1.30		1.18		1		3061
		BPM 46756		0 52 00	−27 14.	1	13.78		1.16		1.02		2		3064
		NGC 300 sq1 13		0 52 00	−38 11.2	1	17.58		0.53		−0.16		4		910
		POSS 295 # 38		0 52 01	+31 42.5	1	17.40		1.55				2		1739
5234	+58 00134	HR 253	⋆ A	0 52 01	+58 42.2	7	4.82	.012	1.22	.008	1.25	.013	19	K2 III	15,1103,1118,1355*
	+60 00131			0 52 01	+61 21.4	1	9.48		0.25				1	A0	1103
		HV 834		0 52 01	−72 33.	3	11.91	.051	0.60	.019	0.51	.075	3	F8 Ia	689,1511,3075
5370	−74 00072			0 52 01	−74 08.4	1	8.49		0.51				1	F6 IV/V	1642
		HA 92 # 250		0 52 03	+00 22.7	1	13.18		0.81		0.48		9	K0	1764
		BSD 8 # 25		0 52 03	+58 48.0	1	11.19		0.36		0.30		3	A0	1185
		CS 22183 # 6		0 52 04	−04 33.7	1	14.33		0.09		0.07		1		1736
232316	+50 00179			0 52 05	+50 56.0	1	8.65		0.04		−0.17		2	A0	1566
		AAS35,353 # 40		0 52 05	−72 51.1	1	14.06		−0.13		−0.67		6	B3	767
		AAS53,255 # 64		0 52 06	−71 59.	1	12.53		1.95				1		1686
		NGC 300 sq1 12		0 52 07	−38 12.5	1	16.04		0.59		0.04		4		910
	−38 00298			0 52 07	−38 16.3	2	11.17	.000	1.03	.000	0.95	.000	21		910,1554
5244	+63 00112			0 52 09	+64 16.5	1	7.59		0.23				2	A3	1118
5301	−32 00360			0 52 09	−31 54.9	1	9.22		0.75		0.28		1	G5 IV/V	3061
		HA 92 # 330		0 52 10	+00 27.2	1	15.07		0.57		−0.12		1		1764
	−38 00299			0 52 10	−38 08.3	2	11.24	.000	0.57	.000	0.07	.000	23		910,1554
		NGC 300 sq1 11		0 52 10	−38 11.0	1	15.26		0.97		0.56		7		910
		AzV 149		0 52 10	−73 04.6	1	13.96		−0.06		−0.86		2	B2	425
	+64 00093	LS I +64 054		0 52 12	+64 58.5	1	10.18		0.45		−0.33		2	B2 III	1012
		AAS35,353 # 41		0 52 12	−72 30.1	1	12.98		−0.09		−0.59		4	B5	767
		AzV 150		0 52 13	−72 44.8	1	12.72		−0.05		−0.53		3	B6 Ia	425
		HA 92 # 252		0 52 14	+00 23.1	1	14.93		0.52		−0.14		18		1764
5273	+47 00242	HR 256		0 52 14	+48 24.5	2	6.29	.020	1.69	.010	2.00	.005	5	M2.5 IIIa	1733,3001
		BSD 8 # 352		0 52 16	+59 41.1	1	11.45		0.19		−0.32		3	B5 p	1185
		SB 363		0 52 16	−28 30.1	2	13.73	.058	0.19	.000	0.16	.010	3	A3:	295,1736
		AzV 151		0 52 16	−73 02.1	2	12.28	.024	−0.05	.010	−0.75	.000	6	B6 Ia	573,733
5286	+22 00146	HR 258	⋆ AB	0 52 17	+23 21.5	5	5.46	.008	1.01	.010	0.90	.004	13	G6 IV+K6 IV	1355,1381,3030,4001,8032
	−32 00361			0 52 17	−32 15.4	1	10.09		0.56		−0.05		1	G0	3061
		HA 92 # 253		0 52 18	+00 24.1	1	14.09		1.13		0.95		17		1764
5294	+23 00125	IDS00497N2332	A	0 52 19	+23 49.9	1	7.38		0.65		0.10		3	G5	196
		AAS35,353 # 42		0 52 19	−72 02.0	1	14.40		0.03		0.09		3	A0	767
		AzV 152		0 52 19	−72 47.9	3	11.87	.007	0.17	.004	0.20	.024	7	A3 Iab	573,733,1571
		AzV 153		0 52 19	−72 53.5	1	13.58		0.00		−0.46		7	B6 Ib	425
		POSS 295 # 35		0 52 20	+29 27.0	1	17.37		1.48				2		1739
		NGC 300 sq1 17		0 52 20	−38 08.1	1	19.98		0.82		0.60		2		910
		NGC 300 sq1 15		0 52 21	−38 13.0	1	19.06		0.98				1		910
		AAS35,353 # 43		0 52 21	−72 54.5	1	12.69		0.56		0.07		2	F8	767
5309	+12 00108			0 52 23	+12 34.6	1	8.22		1.17		1.13		3	K0	196
		MtW 92 # 67		0 52 24	+00 27.8	1	12.50		0.66		0.24		5		397
		BPM 46758	A	0 52 24	−27 36.	1	13.75		1.35		1.24		1		3064
		BPM 46758	B	0 52 24	−27 36.	1	13.42		0.50		−0.05		1		3061
		AAS53,255 # 66		0 52 24	−71 53.	1	10.76		1.53				1		1686
		AzV 154		0 52 25	−72 57.8	1	13.55		0.41		−0.88		3	B0:	425
		HA 92 # 335		0 52 26	+00 28.0	1	12.52		0.67		0.21		1	G2	1764
5292	+41 00164			0 52 26	+42 24.1	1	8.82		0.00		−0.06		2	A0	1375
		CS 22166 # 9		0 52 26	−14 42.3	1	15.11		−0.21		−1.04		1		1736
5319	+0 00142			0 52 27	+00 31.2	6	8.05	.008	0.99	.009	0.82	.013	47	K0 III	281,397,989,1117,1729,6004
		MtW 92 # 77		0 52 29	+00 27.0	1	13.60		0.84		0.50		4		397
		HA 92 # 339		0 52 29	+00 27.9	1	15.58		0.45		−0.18		8		1764
		MtW 92 # 76		0 52 29	+00 27.9	1	15.47		0.50		−0.18		4		397
		WLS 48 40 # 9		0 52 29	+40 03.2	1	13.29		0.62		0.12		2		1375
		NGC 300 sq1 9		0 52 29	−38 07.7	1	14.43		0.86		0.57		4		910

Table 1 63

HD	DM	Other Id	N	Rem	α_{1950}	δ_{1950}	S	V	σ_V	B–V	σ_{B-V}	U–B	σ_{U-B}	n	Spectrum	References
		BPM 46759			0 52 30	−22 12.	1	12.89		1.12		1.05		1		3061
		BPM 46760			0 52 30	−25 37.	1	14.04		0.79		0.22		2		3064
		BPM 46761			0 52 30	−27 40.	1	12.86		1.14		1.06		2		3064
		BSD 8 # 1450			0 52 31	+62 03.2	1	10.67		0.40		0.26		3	A0	1185
		AzV 155			0 52 31	−72 59.2	1	14.34		−0.08		−0.16		4	B0:	733
	+35 00167				0 52 32	+35 46.9	1	8.89		1.58		1.93		3	K5	1601
	−23 00349				0 52 32	−22 58.6	1	9.59		1.93				2		369
5316	+23 00126	HR 259			0 52 34	+24 17.2	2	6.22	.013	1.62	.007	1.75		7	M4 IIIab	71,3001
		BSD 8 # 356			0 52 34	+59 31.9	1	11.03		0.34		0.29		3	B9	1185
	+63 00113				0 52 34	+64 01.5	1	10.15		0.28				4		1118
		AzV 156			0 52 34	−72 34.0	1	14.17		−0.08		−0.14		3	A0	425
		BSD 8 # 880			0 52 35	+60 28.4	1	11.11		0.31		−0.39		3	B5 p	1185
		CS 22166 # 8			0 52 35	−13 43.5	1	14.20		0.40		−0.18		1		1736
		HA 92 # 188			0 52 36	+00 07.0	1	14.75		1.05		0.75		6		1764
		HA 92 # 342			0 52 36	+00 27.0	7	11.62	.010	0.44	.007	−0.04	.008	97	F8	281,397,975,989,1729*
		POSS 295 # 28			0 52 36	+29 13.7	1	16.73		1.31				1		1739
		BPM 46763			0 52 36	−28 22.	1	12.21		0.66		0.17		2		3064
		BPM 46764			0 52 36	−36 20.	1	12.20		0.94		0.62		1		3061
		AzV 157			0 52 38	−72 33.3	1	14.33		−0.11		−0.94		3	B0	425
		AzV 158			0 52 38	−72 35.2	1	14.06		−0.18		−0.95		3	B1	425
	+0 00143	G 1 - 31			0 52 39	+00 39.8	3	10.61	.015	1.14	.008	1.12	.017	6	K2	202,1620,1764
		BSD 8 # 29			0 52 39	+58 41.7	1	10.96		0.27		0.24		3	A0	1185
		HA 92 # 409			0 52 40	+00 39.9	1	10.63		1.14		1.14		3	K3	1764
		AAS35,353 # 44			0 52 40	−72 39.8	1	13.33		0.13		−0.54		3	B8	767
		HA 92 # 410			0 52 41	+00 45.6	1	14.98		0.40		−0.13		13		1764
		BSD 8 # 360			0 52 41	+59 56.2	1	11.69		0.26		−0.20		3	A0	1185
		AAS35,353 # 45			0 52 41	−72 53.7	1	12.99		0.05		0.09		6	A2	767
		HA 92 # 412			0 52 42	+00 45.6	1	15.04		0.46		−0.15		13		1764
		NGC 300 sq1 8			0 52 42	−38 07.5	1	14.18		0.56		−0.06		4		910
		AzV 159			0 52 43	−72 32.1	1	13.56		0.32		0.39		4	F2	425
		POSS 295 # 16			0 52 45	+31 33.0	1	14.98		1.11				2		1739
		Hiltner 78			0 52 47	+63 43.0	1	11.06		0.64		−0.25		2	B2 IV	1012
		HA 92 # 259			0 52 48	+00 24.3	1	15.00		0.64		0.11		1		1764
		HA 92 # 345			0 52 50	+00 34.9	1	15.22		0.75		0.12		1		1764
	−22 00160				0 52 50	−22 18.3	1	10.36		0.86		0.55		1	K2	3061
5365	−22 00161				0 52 51	−22 12.2	1	8.14		1.09		1.05		4	K1 III/IV	1731
		HA 92 # 347			0 52 52	+00 34.6	1	15.75		0.54		−0.10		2		1764
5334	+46 00201				0 52 53	+47 07.6	1	8.76		−0.02		−0.17		2	A0	401
		AAS35,353 # 46			0 52 53	−72 34.2	1	14.29		0.06		0.05		3	A0	767
		HA 92 # 260			0 52 55	+00 20.9	1	15.07		1.16		1.12		4		1764
		BSD 8 # 1452			0 52 55	+61 39.8	1	10.58		0.32		0.04		3	B9	1185
5364	−8 00164				0 52 55	−08 07.7	1	9.85		0.25		0.13		4	A0	152
		NGC 300 sq1 7			0 52 55	−38 09.1	1	13.11		0.93		0.63		4		910
		HA 92 # 348			0 52 56	+00 28.4	2	12.11	.000	0.59	.003	0.05	.003	13	G0	281,1764
		AAS35,353 # 47			0 52 56	−73 19.2	1	11.92		0.38		−0.04		4	F2	767
5363	−7 00142				0 52 57	−06 41.1	2	9.57	.005	0.28	.003	0.19	.006	5	A0	152,1736
		HA 92 # 417			0 52 58	+00 36.9	1	15.92		0.48		−0.19		3		1764
		AAS53,255 # 71			0 52 59	−70 22.	1	12.42		1.64				2		1686
5344	+46 00202				0 53 00	+46 36.3	1	7.99		−0.02		−0.30		2	A0	401
		NGC 300 sq1 10			0 53 00	−37 47.9	1	14.76		0.70		0.09		5		910
		AAS35,353 # 48			0 53 00	−73 03.8	1	13.22		−0.01		−0.44		4	B8	767
		AzV 160			0 53 00	−73 08.1	1	13.99		−0.22		−0.83		4	B0	425
	−27 00294				0 53 01	−26 52.0	1	9.96		1.01		0.87		1	K0	4003
		LP 706 - 62			0 53 02	−08 27.8	1	11.63		0.67		0.09		2		1696
	−26 00296				0 53 02	−26 08.5	1	11.99		1.46		1.28		1		3064
		AG Tuc			0 53 02	−66 58.5	1	12.28		0.16		0.10		1		700
		AAS35,353 # 49			0 53 04	−72 48.1	1	13.40		0.03		−0.12		4	A0	767
		G 1 - 32			0 53 05	+04 12.6	2	11.80	.005	1.34	.019	1.17		3	K5	202,1620
5326	+65 00110				0 53 05	+66 08.6	1	8.33		0.08		−0.46		2	A0	401
5403	−38 00312				0 53 05	−37 47.6	2	8.67	.000	0.70	.000	0.20	.000	11	G5 V	910,1554
		HA 92 # 263			0 53 06	+00 20.1	4	11.78	.004	1.05	.005	0.85	.006	77	G5	281,989,1729,1764
		G 268 - 76	A		0 53 06	−26 16.6	2	13.90	.045	1.51	.010	1.08	.010	4		481,3064
		G 268 - 76	B		0 53 06	−26 16.6	1	14.03		0.46		−0.13		1		3061
		BPM 46771			0 53 06	−27 22.	1	14.40		0.91		0.55		3		3064
236577	+58 00135				0 53 09	+58 43.7	1	9.47		1.15				1	K0	1103
		SB 366			0 53 09	−23 44.7	3	12.26	.005	0.20	.010	0.08	.024	5	A5	295,966,1736
	−38 00313				0 53 09	−37 43.9	2	10.79	.000	0.58	.000	0.07	.000	14		910,1554
5457	−70 00040	HR 270			0 53 09	−69 47.8	3	5.45	.004	1.10	.005	1.01	.005	16	K2 III	15,1075,2038
		L 220 - 80			0 53 10	−52 06.5	2	12.36	.020	1.50	.026			2		1494,1705
5342	+60 00135				0 53 11	+60 59.0	3	8.02	.016	0.12	.013	−0.16		7	B8 II	1103,1118,1185
5384	−8 00167	HR 263			0 53 11	−07 37.0	5	5.85	.005	1.52	.006	1.93	.006	19	K5 III	15,1071,1075,1425,3077
5374	+38 00146				0 53 12	+38 53.9	2	8.94	.010	0.05	.010	0.02	.000	6	A0 V	833,1601
	+60 00134	LS I +61 175			0 53 12	+61 29.5	1	10.59		0.73		−0.32		3	O7	1185
		BPM 46775			0 53 12	−32 31.	1	11.87		0.76		0.30		1		3061
		POSS 295 # 36			0 53 14	+29 31.8	1	17.39		1.50				2		1739
5343	+57 00172				0 53 14	+57 43.6	1	6.21		1.37				2	K3 III	71
		AzV 161			0 53 14	−72 25.8	2	11.78	.005	0.03	.010	−0.41	.005	4	A0 Ia	425,573
5382	+26 00151	HR 262			0 53 17	+26 56.3	4	6.09	.009	0.13	.010	0.10	.011	10	A5 IV	985,1049,1716,1733
		BSD 8 # 368			0 53 17	+59 58.6	1	12.51		0.28		0.24		3	A0	1185
		MN201,73 # 28			0 53 17	−36 08.1	1	13.22		0.64				1		1495

HD	DM	Other Id	N Rem	α_{1950}	δ_{1950}	S	V	σ_V	B–V	σ_{B-V}	U–B	σ_{U-B}	n	Spectrum	References
		BSD 8 # 894		0 53 18	+60 38.5	1	10.71		0.67		0.10		3	B5	1185
		AzV 162		0 53 18	−73 12.4	1	13.77		−0.02		−1.06		9	B0	425
5333	+69 00054			0 53 19	+70 11.0	1	8.51		0.26		−0.51		3	A2	1733
		G 268 - 78		0 53 19	−18 30.4	1	14.93		0.61		−0.12		2		3061
5424	−28 00287			0 53 19	−28 09.8	3	9.48	.014	1.14	.000	0.80	.002	4	G8 II	565,1495,3048
		HA 92 # 497		0 53 20	+00 55.5	1	13.64		0.73		0.26		1		1764
		G 270 - 98		0 53 20	−11 44.0	1	15.26		0.35		−0.50		1	DA	3060
5499	−74 00074			0 53 20	−74 34.5	3	6.69	.003	0.99	.002	0.82		55	K1 IV	978,2033,3040
		AAS35,353 # 50		0 53 21	−72 45.6	1	13.33		−0.10		−0.55		3	B6	767
5372	+51 00188			0 53 22	+52 13.3	1	7.52		0.64		0.23		2	G5	1566
		AAS35,353 # 51		0 53 22	−72 26.0	1	13.61		−0.15		−0.64		4	B3	767
		AzV 163		0 53 22	−73 02.7	1	14.18		−0.17		−0.97		4	B1	425
		HA 92 # 498		0 53 23	+00 54.4	1	14.41		1.01		0.79		1		1764
236579	+58 00136			0 53 23	+58 46.8	1	8.13		0.97				1	K0	1103
		BSD 8 # 370		0 53 23	+59 56.2	1	12.00		0.34		0.21		3	A0	1185
5425	−30 00277	G 269 - 58		0 53 23	−29 56.9	1	9.48		1.18		1.14		2	K4 III	3072
		HA 92 # 500		0 53 24	+00 54.2	1	15.84		1.00		0.21		1		1764
		HA 92 # 425		0 53 25	+00 36.7	1	13.94		1.19		1.17		19		1764
		HA 92 # 426		0 53 26	+00 36.7	1	14.47		0.73		0.18		4		1764
		HA 92 # 501		0 53 26	+00 54.6	1	12.96		0.61		0.07		1		1764
	−26 00298			0 53 26	−25 43.3	1	8.58		3.26		1.35		1		109
		POSS 295 # 11		0 53 28	+31 39.1	1	14.28		1.37				2		1739
		AAS35,353 # 52		0 53 28	−72 50.5	1	14.27		−0.15		−0.66		3	B3	767
		AAS35,353 # 53		0 53 29	−72 42.4	1	13.39		−0.10		−0.25		3	B8	767
		BPM 46780		0 53 30	−27 02.	1	14.45		0.64		0.00		3		3064
5445	−28 00288	HR 268		0 53 30	−28 02.8	2	6.12	.020	1.66	.005	1.97		8	M0 III	2032,3007
		G 1 - 33		0 53 31	+10 34.0	1	12.55		1.35		1.06		1		333,1620
5358	+63 00116			0 53 31	+64 30.4	1	9.01		0.15				2	A2	1118
5437	−12 00162	HR 267		0 53 31	−11 32.2	4	5.32	.021	1.50	.014	1.78		14	K4 III	15,1075,2035,3016
	−29 00267			0 53 31	−29 34.9	1	10.83		0.90		0.52		1		4003
		HA 92 # 355		0 53 32	+00 34.6	1	14.97		1.16		1.20		7		1764
5418	+13 00127			0 53 32	+13 40.9	1	6.44		0.98		0.72		2	G8 II	1648
	−28 00289			0 53 32	−27 43.1	1	11.08		0.75		0.22		1		4003
		HA 92 # 427		0 53 33	+00 44.1	1	14.95		0.81		0.35		1		1764
		AzV 164		0 53 33	−72 57.5	1	14.16		−0.26		−0.86		4	B0	425
		HA 92 # 502		0 53 34	+00 48.2	1	11.81		0.49		−0.09		1	G0	1764
		SB 371, UV Scl		0 53 36	−26 40.	2	13.55	.045	0.32	.020	0.09	.045	4	A3:pec	295,561
		G 269 - 59		0 53 36	−27 01.1	1	15.48		1.64				1		3061
		BPM 46782		0 53 36	−35 52.	1	13.80		1.44		1.12		2		3061
		L 123 - 45		0 53 36	−61 58.	2	11.83	.015	0.54	.015	−0.16	.020	4		1696,3060
		Florsch 311		0 53 36	−72 03.6	1	13.26		0.07		0.05		3	A1	767
		AzV 165		0 53 36	−72 42.2	1	12.79		−0.02		−0.36		5	B7 Iab	425
		AAS35,353 # 54		0 53 36	−72 49.5	1	13.47		0.04		−0.01		4	A0	767
		AAS35,353 # 55		0 53 37	−73 05.5	1	13.74		0.04		−0.15		4	B9	767
		AzV 166		0 53 37	−73 30.4	2	12.45	.005	0.03	.005	−0.34	.005	5	B9 Iab	425,835
		GR 61		0 53 39	−06 26.5	1	14.79		−0.09		−0.99		1		3060
	−47 00277	Smethells 17		0 53 39	−47 24.9	1	11.06		1.30				1		1494
		AzV 167		0 53 39	−72 38.0	1	14.14		0.21		0.22		7	F2	425
		AAS35,353 # 57		0 53 39	−72 39.8	1	14.21		−0.12		−0.53		5	B5	767
5395	+58 00138	HR 265		0 53 40	+58 54.7	8	4.63	.008	0.96	.007	0.68	.015	23	G8 III-IV	15,1103,1118,1355*
5394	+59 00144	HR 264, γ Cas	⋆ AB	0 53 40	+60 26.8	2	2.48	.109	−0.15	.051	−1.07	.024	19	B0 IV:e	15,154,1119,1212,1223*
5357	+67 00081	HR 261		0 53 40	+68 30.4	1	6.39		0.37		−0.04		1	F0 m	254
		CS 22942 # 22		0 53 40	−23 09.5	2	15.81	.000	0.03	.000	0.29	.010	2		966,1736
		CS 22942 # 21		0 53 40	−23 34.1	2	14.34	.000	0.37	.000	0.09	.011	2		966,1736
		HA 92 # 430		0 53 42	+00 37.0	1	14.44		0.57		−0.04		18		1764
		BSD 8 # 905		0 53 42	+60 43.7	1	11.04		0.40		−0.30		3	B5	1185
5351	+68 00060	G 242 - 73		0 53 42	+68 46.6	3	9.13	.011	1.03	.008	0.89	.013	5	K4 V	1003,1658,3072
5454	−23 00358			0 53 42	−22 45.6	1	9.79		0.44		−0.01		4	F3 V	1731
		BPM 46783		0 53 42	−26 45.	1	13.82		0.58		−0.10		3		3064
	−27 00300			0 53 42	−26 57.7	1	10.84		1.17		1.11		1		4003
		MN201,73 # 29		0 53 43	−31 28.2	1	12.25		0.42				2		1495
		AzV 167a		0 53 43	−73 00.7	1	14.60		−0.15		−0.90		4	B1	767
	+29 00155			0 53 44	+29 59.7	1	9.43		1.06		0.82		2	K0	1733
5409	+59 00147			0 53 44	+59 45.8	2	7.85	.015	0.04	.022			7	B9 V	1103,1118
	−36 00326	G 269 - 60		0 53 44	−35 39.4	2	11.17	.000	0.46	.002	−0.13	.004	6		79,3061
		AAS53,255 # 77		0 53 44	−72 52.	1	12.80		1.86				1	M0 Iab	1686
5408	+59 00146	HR 266	⋆ AB	0 53 45	+60 05.6	4	5.55	.000	−0.07	.000	−0.31	.015	8	B8 V +B9 V	1079,1103,1118,3016
		BSD 8 # 37		0 53 46	+59 00.8	1	10.92		0.22		0.20		3	B8	1185
5393	+62 00172			0 53 46	+62 59.4	1	9.04		0.03				2	A0	1118
		AzV 168		0 53 46	−73 04.9	1	13.99		−0.08		−0.76		4	B2	425
5392	+63 00117	LS I +64 055		0 53 48	+64 15.8	1	7.01		0.99				2	F4 I	1118
		BPM 46785		0 53 48	−27 54.	1	13.50		0.81		0.09		4		3064
		SB 373		0 53 48	−33 13.	1	13.71		0.20		0.19		1	A0:	295
		CS 22166 # 12		0 53 50	−11 33.0	1	14.73		0.19		0.10		1		1736
	−28 00291	G 269 - 61		0 53 51	−28 24.6	1	10.96		0.96		0.72		3		3061
	+0 00145			0 53 52	+01 26.6	1	10.58		0.05		−0.01		3	B9	1026
	−28 00292			0 53 52	−27 44.1	1	10.97		1.37				1		4003
		HA 92 # 276		0 53 53	+00 25.6	2	12.04	.002	0.63	.000	0.07	.006	24		281,1764
		CS 22942 # 17		0 53 54	−26 53.4	1	15.94		−0.02		−0.14		1		1736
		MN201,73 # 30		0 53 54	−32 48.1	1	14.43		1.44				1		1495

Table 1   65

HD	DM	Other Id	N	Rem	α_{1950}	δ_{1950}	S	V	σ_V	B–V	σ_{B-V}	U–B	σ_{U-B}	n	Spectrum	References
		HV 873			0 53 54	−71 05.	1	12.95		1.08		0.77		1	F8:I	689
5497	−36 00327				0 53 57	−35 53.0	2	9.05	.014	0.22	.000	0.03		7	A4 V	1068,2012
		AAS53,255 # 79			0 53 57	−72 52.	1	12.88		1.72				1		1686
		AzV 169			0 53 57	−73 01.1	1	13.90		-0.13		-0.86		4	B1	425
		G 32 - 59			0 53 58	+17 11.6	2	13.71	.009	1.64	.009	1.31	.009	4		419,3079
5448	+37 00175	HR 269		⋆A	0 53 58	+38 13.7	5	3.86	.007	0.13	.010	0.15	.007	12	A5 V	15,1049,1363,3016,8015
	+79 00022	G 242 - 74		⋆A	0 53 58	+79 50.4	1	9.78		0.54		-0.10		1	G0	1658
5496	−31 00362				0 53 58	−31 27.3	3	10.58	.012	-0.06	.007	-0.18	.004	9	B9 V	152,1068,3007
		POSS 295 # 22			0 53 59	+30 42.9	1	16.11		1.72				2		1739
		BPM 46789			0 54 00	−28 24.	1	10.95		0.99		0.69		2		3064
		PHL 6918			0 54 00	−31 28.	1	10.60		-0.06		-0.18		2	A0 V	3064
		AAS35,353 # 58			0 54 00	−72 44.7	1	13.28		0.07		0.06		3	A0	767
	−29 00270				0 54 01	−28 55.6	1	10.25		0.86		0.46		1	G5	4003
5487	−20 00167				0 54 02	−19 42.9	2	8.08	.014	0.09	.005	0.11		8	A2 IV/V	1068,2012
		AzV 170			0 54 02	−73 33.6	1	14.09		-0.23		-1.02		8	B0	425
5429	+60 00136				0 54 03	+61 17.8	2	8.82	.020	0.24	.040	-0.10		5	A0	1118,1185
	−27 00303				0 54 04	−27 29.9	1	10.35		1.23		1.37		1	G0	4003
	−83 00016				0 54 04	−82 56.4	1	8.43		1.63		1.64		10	M0	1704
	−24 00405	G 268 - 80			0 54 05	−24 28.5	1	11.23		1.22		1.18		1	K5	3061
		AAS35,353 # 59			0 54 05	−72 47.1	1	13.03		0.01		-0.19		3	A0	767
		AzV 171			0 54 05	−72 55.6	3	13.29	.031	0.00	.010	-0.35	.010	11	B6 Iab	425,835,967
		G 2 - 6			0 54 06	+01 21.0	1	13.71		1.51				2		202
		Hiltner 79			0 54 06	+63 00.0	1	11.67		0.30		-0.47		2	B1 V	1012
		CS 22942 # 18			0 54 06	−26 15.9	2	15.23	.000	0.23	.000	-0.03	.008	2		966,1736
	−38 00318				0 54 08	−38 05.9	2	10.15	.000	0.47	.000	-0.04	.000	9	F5	910,1554
5495	+1 00168				0 54 09	+01 36.3	1	9.24		0.58		0.05		10	K0	1371
		AAS35,353 # 60			0 54 09	−72 27.9	1	13.52		0.02		-0.05		6	A0	767
	−24 00407				0 54 10	−24 18.4	1	9.94		0.86		0.49		1	K0	3061
		AzV 172			0 54 10	−72 25.1	1	13.37		0.06		-0.16		5	B8	425
		BSD 8 # 912			0 54 12	+60 29.7	1	11.51		0.35		-0.11		3	B6	1185
		AAS50,119 # 202			0 54 12	−72 56.7	1	12.02		1.40		1.16		8		967
		HA 92 # 282			0 54 13	+00 22.3	2	12.96	.004	0.32	.001	-0.04	.001	28		281,1764
		AAS35,353 # 61			0 54 14	−71 54.1	1	13.43		0.51		0.01		3	F8	767
		POSS 295 # 20			0 54 16	+29 47.6	1	16.10		1.57				2		1739
		CS 29509 # 40			0 54 16	−29 56.4	1	14.91		0.20		0.05		1		1736
		AAS35,353 # 62			0 54 16	−72 45.7	1	14.30		0.01		-0.11		3	A0	767
		HA 92 # 507			0 54 17	+00 49.7	3	11.34	.009	0.95	.011	0.68	.004	15	G0	281,1764,6004
		HA 92 # 508			0 54 17	+00 53.4	3	11.67	.004	0.54	.008	-0.04	.003	16	F5	281,1764,6004
5505	+0 00146				0 54 17	+01 24.7	3	9.00	.004	1.08	.006	0.96	.003	29	K5	989,1371,1729
		BPM 46793			0 54 18	−27 22.	1	14.03		0.67		-0.01		2		3064
		BPM 46794			0 54 18	−29 41.	1	11.53		0.86		0.56		3		3064
	−66 00071				0 54 18	−65 42.7	2	10.10	.025	0.96	.025	0.71	.030	2	K5 V	1696,3072
		HA 92 # 364			0 54 19	+00 27.8	2	11.67	.001	0.60	.003	-0.03	.003	15	G0	281,1764
		HA 92 # 433			0 54 20	+00 44.5	3	11.66	.008	0.67	.012	0.12	.005	16	G2	281,1764,6004
		NGC 330 sq1 9			0 54 20	−72 45.3	2	12.36	.015	0.59	.000	0.09	.000	7		1464,1571,1669
5524	−26 00303				0 54 23	−25 38.0	3	7.19	.010	0.13	.012	0.07	.030	9	A2/3 V	561,1068,2012
5458	+61 00185				0 54 24	+62 17.9	1	8.99		0.25				3	B5p	1118
		BPM 46795			0 54 24	−29 12.	1	13.20		1.24		1.19		1		3064
	+64 00106	LS I +64 056			0 54 28	+64 35.4	1	10.34		0.69		-0.21		2	B1 V	1012
		CS 22183 # 12			0 54 28	−03 33.1	1	13.40		-0.01		-0.08		1		1736
		BSD 8 # 395			0 54 30	+60 14.7	1	11.12		0.18		-0.08		3	B9	1185
		BPM 46798			0 54 30	−21 41.	1	13.10		1.09		1.00		1		3061
		BPM 46799			0 54 30	−24 56.	1	11.75		0.82		0.40		1		3061
5546	−30 00283				0 54 30	−30 01.9	1	10.22		0.22		0.09		3	A5 IV/V	1068
		Florsch 323			0 54 30	−70 55.7	1	12.88		0.54		0.06		3	F7	767
		AzV 173			0 54 31	−73 10.7	2	12.76	.000	-0.07	.010	-0.81	.015	5	B3 Ia	425,835
5516	+22 00153	HR 271		⋆A	0 54 32	+23 08.9	7	4.41	.011	0.94	.005	0.69	.005	27	G8 III-IV	15,1007,1013,1355*
		NGC 330 - 1			0 54 33	−72 45.	1	13.37		0.04		-0.42		2		1571
		NGC 330 - 8			0 54 33	−72 45.	1	13.19		-0.08		-0.80		2		1571
		NGC 330 - 9			0 54 33	−72 45.	1	12.49		0.15		0.12		3		1571
		NGC 330 - 14			0 54 33	−72 45.	1	12.95		-0.10		-0.72		2		1571
		NGC 330 - 1001			0 54 33	−72 45.	1	15.60		0.50		0.70		3		1571
		NGC 330 - 1035			0 54 33	−72 45.	1	17.07		-0.17				2		1571
		NGC 330 - 1078			0 54 33	−72 45.	1	13.39		0.78		0.34		2		1571
		NGC 330 - 1201			0 54 33	−72 45.	2	13.36	.010	0.07	.010	-0.03	.015	6		1571,1669
		NGC 330 - 1203			0 54 33	−72 45.	2	14.01	.005	1.28	.020	1.15		8		1571,1669
		NGC 330 - 1208			0 54 33	−72 45.	1	15.35		-0.18		-0.69		1		1571
		NGC 330 - 1209			0 54 33	−72 45.	1	15.35		-0.11				1		1571
		NGC 330 - 1211			0 54 33	−72 45.	1	13.81		0.03		-0.20		2		1571
		NGC 330 - 1219			0 54 33	−72 45.	1	14.28		0.16		0.03		1		1571
		NGC 330 - 1222			0 54 33	−72 45.	1	13.92		0.03		0.06		2		1571
		NGC 330 - 1224			0 54 33	−72 45.	1	13.52		1.87				1		1571
		NGC 330 - 1243			0 54 33	−72 45.	1	13.50		0.26		0.29		2		1571
	+0 00147				0 54 34	+00 50.8	2	9.98	.007	1.07	.002	1.03		8	G5	281,6004
		POSS 295 # 27			0 54 39	+31 29.6	1	16.67		1.60				1		1739
5501	+59 00154	LS I +59 058			0 54 39	+59 58.1	1	8.74		0.36		-0.05		3	A3 II	1185
5543	+0 00148				0 54 40	+01 05.7	2	8.26	.022	0.27	.004	0.05	.009	12	A3	1117,1371
		G 70 - 23			0 54 40	−03 47.5	2	12.58	.063	1.24	.010	1.33		3		202,1620
		BSD 8 # 400			0 54 42	+60 07.6	1	11.67		0.23		0.20		3	A0	1185
		BSD 8 # 401			0 54 42	+60 14.5	1	10.60		0.22		-0.17		3	B6 p	1185

HD	DM	Other Id	N Rem	α_{1950}	δ_{1950}	S	V	σ_V	B–V	σ_{B-V}	U–B	σ_{U-B}	n	Spectrum	References
		PHL 912		0 54 42	−22 38.	1	14.34		0.08		−0.94		1		3064
		TonS 180 # 4		0 54 42	−22 38.	1	15.29		1.15		0.66		4		1687
		BPM 46801		0 54 42	−30 44.	1	13.79		1.28		1.12		1		3061
		Florsch 330		0 54 42	−70 31.7	1	12.63		0.52		−0.08		3		767
		HA 92 # 288		0 54 43	+00 20.6	4	11.63	.004	0.86	.005	0.47	.006	70	G3	281,989,1729,1764
		LS I +61 176		0 54 43	+61 36.6	1	10.60		0.72		−0.37		2	B1 pe(IV-V)	1012
		MN201,73 # 31		0 54 46	−36 28.3	1	12.51		1.10				1		1495
		TonS 180 # 3		0 54 47	−22 35.8	1	14.14		0.65		0.06		4		1687
		MN201,73 # 32		0 54 47	−29 18.3	1	11.78		1.54				2		1495
	−2 00129	G 70 - 24		0 54 48	−02 05.0	5	9.46	.021	0.86	.010	0.49	.016	11	K0	158,196,202,1620,1705
5526	+45 00237	HR 272		0 54 49	+45 34.2	1	6.12		1.02		0.89		2	K2 III	252
		WLS 112 60 # 6		0 54 49	+59 24.6	1	11.32		1.25		0.87		2		1375
	+66 00075	IDS00517N6710	A	0 54 51	+67 25.7	1	8.44		2.08				3	M0	369
		CS 22942 # 19		0 54 52	−25 42.3	1	12.71		0.86		0.13		1		1736
	+17 00129			0 54 53	+18 20.8	1	9.55		1.18		1.32		1	K0	1746
		BSD 8 # 405		0 54 54	+60 15.7	1	11.20		0.13		−0.53		3	B6 p	1185
		BPM 46804		0 54 54	−31 37.	1	11.56		0.69		0.04		2		3061
		POSS 295 # 29		0 54 55	+30 16.9	1	16.80		1.63				3		1739
5513	+59 00156		V	0 54 55	+59 50.1	2	9.04	.025	0.24	.030	0.19		7	A2 V	1118,1185
		AzV 174		0 54 55	−72 17.8	2	12.43	.005	0.23	.005	0.28	.010	6	A7 Iab	425,835
		G 1 - 35		0 54 56	+10 19.2	1	11.67		0.72		0.18		1	G5	333,1620
		CS 29509 # 41		0 54 56	−30 08.3	1	14.68		0.06		0.13		1		1736
		AzV 175		0 54 56	−72 52.7	2	13.63	.033	−0.20	.056	−0.80	.005	4	B1	425,835
		AzV 176		0 54 57	−72 41.3	1	13.93		−0.10		−0.84		3	B1	733
		BSD 8 # 403		0 54 59	+59 42.1	1	11.10		0.29		0.14		3	A0	1185
		TonS 180 # 1		0 54 59	−22 41.4	1	9.21		0.39		−0.03		4		1687
		BPM 46803		0 55 00	−21 50.	1	11.72		0.90		0.66		2		3061
		BPM 46805		0 55 00	−27 24.	1	12.48		0.94		0.58		2		3064
5618	−40 00209			0 55 00	−39 47.8	2	8.86	.014	0.16	.005	0.11		7	A3 V	1068,2012
		Smethells 180		0 55 00	−51 52.	1	10.77		1.30				1		1494
		AzV 177		0 55 01	−72 19.7	1	14.60		−0.22		−1.11		4	O5 V	425
		AJ63,118 # 35		0 55 01	−72 46.5	1	12.35		0.50				4		2001
		BSD 8 # 1481		0 55 02	+62 02.1	1	11.09		0.69		−0.12		3	B2	1185
5601	−11 00177	BC Cet		0 55 02	−10 44.8	1	7.64		−0.06		−0.18		4	A0	152
		AzV 183		0 55 02	−71 55.4	2	13.04	.019	−0.09	.009	−0.50	.009	4	B5 Iab	425,835
	+59 00157			0 55 03	+59 36.6	1	10.36		0.21		−0.23		3	B7 V	1185
5678	−73 00057	MN83,95 # 401		0 55 03	−73 15.8	1	10.45		0.53				1	F2	591
236589	+55 00215	LS I +56 017		0 55 04	+56 09.7	1	9.32		0.15		−0.69		2	B1 II	1012
5588	+0 00149			0 55 05	+01 30.9	1	6.96		1.19		1.18		2	G5	1117
		IRC +30 019	AB	0 55 05	+28 47.3	1	13.82		2.47				2		8019
		MN201,73 # 33		0 55 05	−29 18.4	1	12.10		0.57				1		1495
		MN83,95 # 408		0 55 05	−73 26.5	1	12.76		0.68				1		591
		TonS 180 # 2		0 55 06	−22 41.2	1	12.74		0.56		−0.01		4		1687
		AzV 178		0 55 06	−72 44.9	1	14.38		−0.19		−0.80		3	B2	425
5575	+28 00157	HR 274		0 55 07	+28 43.3	1	5.42		1.08				2	K0	71
		TonS 180 # 5		0 55 07	−22 41.3	1	14.62		0.69		0.12		4		1687
5600	+1 00176			0 55 08	+01 49.4	1	7.99		0.46		0.01		6	F8	1371
		AzV 179		0 55 09	−72 12.4	1	14.76		0.01		−0.93		4	B0	733
5553	+55 00216	LS I +55 010		0 55 10	+55 33.3	2	8.14	.005	0.02	.014	−0.76	.019	4	B1.5III-IV	399,555
5617	−19 00147	IDS00527S1932	AB	0 55 10	−19 16.1	2	6.92	.014	0.08	.005	0.05		7	A2 V	1068,2012
5633	−62 00075			0 55 10	−62 31.0	4	9.49	.026	1.31	.018	1.25		8	M0 V	1494,1705,2018,3078
		MN83,95 # 406		0 55 10	−73 26.9	1	11.96		0.48				1		591
	+44 00205			0 55 11	+45 05.0	1	9.48		0.35		0.02		1	F0	1462
	−44 00254			0 55 11	−43 49.2	1	11.04		0.50		−0.19		2		1696
5552	+61 00187	LS I +61 177		0 55 13	+61 39.7	2	9.07	.010	0.70	.010	−0.29	.000	5	B1 Ia	1012,1185
		AzV 180		0 55 13	−72 40.4	2	13.16	.042	0.00	.009	−0.70	.009	4	B5 Ia	425,835
		MN201,73 # 35		0 55 15	−33 08.6	1	13.89		1.50				1		1495
		MN201,73 # 34		0 55 15	−35 18.3	1	11.82		1.03				1		1495
		AzV 181		0 55 15	−72 33.8	1	13.83		−0.01		−0.28		7	A0	425
		AAS50,119 # 204		0 55 15	−72 46.4	1	14.90		−0.15		−0.98		5		967
		AAS50,119 # 203		0 55 15	−72 47.5	1	13.42		0.16		0.21		8		967
5631	−31 00376			0 55 16	−30 45.5	1	8.43		1.33		1.44		1	K2 III	3061
5612	+12 00119	HR 276		0 55 17	+13 25.6	4	6.32	.004	0.90	.008	0.58	.008	19	G8 III	15,1007,1013,8015
5536	+63 00118			0 55 17	+63 50.0	1	8.89		0.18				2	A0	1118
5624	−2 00131			0 55 17	−02 02.2	1	6.51		1.04		0.81		2	K1 III	1375
		TonS 180 # 6		0 55 17	−22 44.6	1	13.80		0.55		−0.01		4		1687
		BSD 8 # 413		0 55 18	+60 01.0	1	11.46		0.22		−0.10		3	B6	1185
5630	−26 00311			0 55 18	−26 29.6	1	9.98		0.34		0.05		2	F2 V	561
		AzV 182		0 55 18	−72 24.3	1	14.33		−0.14		−1.09		4	B0	425
5700	−70 00042			0 55 19	−69 53.5	1	8.77		0.00		0.01		4	A1 V	1408
5551	+62 00175	LS I +63 116		0 55 20	+63 26.6	1	7.71		0.62		−0.36		2	B1.5Ib	1012
5550	+65 00115	HR 273		0 55 20	+66 04.9	1			−0.02		−0.14		1	A0 III	1079
		AAS35,353 # 65		0 55 21	−72 25.1	1	13.29		0.04		−0.02		6	A1	767
5597	+38 00148			0 55 23	+39 12.4	1	6.52		1.16		1.15		2	K1 III	105
	−83 00018			0 55 23	−82 53.2	1	9.62		0.71		0.18		1	G5	1696
	+58 00142	LS I +59 059		0 55 24	+59 23.2	1	9.71		0.34		−0.62		3	B0	1185
		BPM 46810		0 55 24	−32 38.	1	13.14		0.85		0.55		1		3061
		MN83,95 # 302		0 55 24	−72 03.3	1	12.54		0.96				1		591
		POSS 295 # 31		0 55 25	+29 37.8	1	17.00		1.70				1		1739
	−28 00298			0 55 25	−28 13.1	1	10.72		1.01		0.84		1		4003

Table 1   67

HD	DM	Other Id	N Rem	α_{1950}	δ_{1950}	S	V	σ_V	B–V	σ_{B-V}	U–B	σ_{U-B}	n	Spectrum	References
5644 −28 00299				0 55 27	−28 19.9	1	9.17		1.52		1.86		1	K2 III	54
	−30 00290			0 55 28	−30 16.0	1	10.79		0.63		−0.04		1		378
5608 +33 00140		HR 275		0 55 29	+33 40.9	1	5.98		1.00				2	K0	71
		AAS35,353 # 66		0 55 29	−72 44.9	1	13.18		0.02		−0.15		3	B7	767
		AzV 184		0 55 31	−72 38.6	1	14.18		−0.08		−0.47		4	B8	425
		MN83,95 # 414		0 55 31	−73 23.6	1	14.51		0.03				1		591
5660 −23 00367				0 55 33	−22 51.9	1	8.48		1.53		1.80		4	K4/5 III	1731
		CS 22183 # 13		0 55 36	−02 38.1	1	14.42		0.21		0.12		1		1736
		POSS 295 # 15		0 55 37	+31 48.5	1	14.82		1.30				1		1739
		MN83,95 # 301		0 55 37	−72 03.0	1	11.01		0.93				1		591
5595 +61 00188				0 55 38	+61 50.4	1	9.30		0.25		−0.04		3	B8 V	1185
5641 +20 00131		HR 277	⋆ AB	0 55 39	+21 08.1	2	6.46		0.07	.015	0.07	.034	5	A2 V	1049,1733
5659 −16 00162		IDS00532S1613	A	0 55 41	−15 57.1	1	7.79		0.43		−0.03		2	dF6	3016
5659 −16 00162		IDS00532S1613	AB	0 55 41	−15 57.1	1	7.07		0.46		−0.04		1	dF6 + dF7	258
5659 −16 00162		IDS00532S1613	B	0 55 41	−15 57.2	1	7.85		0.44		−0.04		2	dF7	3016
		AzV 185		0 55 41	−72 17.7	1	13.28		0.04		−0.45		3	B8:	425
		BPM 70468		0 55 42	−18 20.	1	12.27		0.93		0.32		1		3061
		BPM 46815		0 55 42	−32 01.	1	13.08		0.62		0.05		1		3061
		Florsch 348		0 55 42	−70 28.9	1	13.01		0.77		0.33		3		767
		AAS50,119 # 205		0 55 42	−72 56.2	1	12.83		0.42		0.04		8		967
5774		U Tuc		0 55 42	−75 16.2	1	9.65		1.22		0.70		1	Me	975
		BSD 8 # 937		0 55 43	+61 10.4	1	12.35		0.40		0.25		3	B9	1185
		G 269 - 62	A	0 55 43	−26 11.4	1	13.73		1.17				1		3061
		G 269 - 62	B	0 55 43	−26 11.4	1	13.57		0.74				1		3064
5654 +6 00135				0 55 44	+06 34.5	1	6.74		1.08		0.94		2	K0 III	1733
		MN201,73 # 36		0 55 45	−34 28.3	1	13.23		0.77				1		1495
		AzV 186		0 55 46	−72 49.3	2	14.02	.057	−0.25	.073	−1.03	.012	10	O9	425,967
5676 −26 00314				0 55 47	−26 08.8	1	7.89		1.11		1.05		3	K1 III	1657
		AzV 187		0 55 47	−71 36.1	2	12.07	.015	−0.12	.015	−0.75	.000	6	B2.5Ia	425,573
		Dril 4		0 55 48	−29 08.	1	12.04		0.58		−0.01		2	G0 V	561
		AzV 188		0 55 48	−72 09.9	1	14.15		−0.18		−0.91		3	B1	425
		AAS50,119 # 207		0 55 48	−72 48.6	1	14.64		0.14		−0.99		5		967
	−24 00414			0 55 49	−24 10.2	2	12.35	.000	−0.03	.000	−0.08	.012	2	A0	966,1736
		BSD 8 # 1487		0 55 51	+62 09.4	1	10.70		0.47		0.31		3	B9	1185
		SK 67		0 55 51	−72 32.	1	12.24		0.08		−0.54		2		835
5638 +46 00215				0 55 52	+46 46.0	1	6.85		−0.11		−0.58		2	B3 V	401
5684 −27 00308				0 55 52	−27 21.7	1	8.36		1.26		1.45		1	K2 III	54
		AzV 189		0 55 52	−72 45.0	1	14.51		−0.23		−0.87		3	B1	425
		AzV 190		0 55 53	−72 43.0	1	13.52		−0.01		−0.31		4	B6 Iab	425
		CS 22166 # 16		0 55 54	−15 03.3	1	12.73		0.61		−0.03		1		1736
		AzV 191		0 55 55	−72 29.2	1	13.63		−0.16		−0.92		4	B1	425
		AzV 192		0 55 56	−72 38.0	1	14.58		−0.32		−0.99		4	O9	425
		AzV 193		0 55 56	−72 40.7	1	15.40		−0.16		−0.94		3	B1	425
		SB 389		0 55 57	−24 31.9	2	14.38	.000	0.18	.000	0.07	.008	2	A0	966,1736
	−28 00302	G 269 - 63		0 55 58	−28 07.3	2	11.77	.000	1.57	.012	1.13		4	M4	1705,3078
5580 +73 00047				0 55 60	+73 43.1	1	8.26		1.12		0.92		3	K0	1733
		CS 22166 # 18		0 55 60	−15 25.0	1	13.72		0.38		0.08		1		1736
		BPM 46817		0 56 00	−22 27.	1	13.74		1.06		0.93		1		3061
	−30 00296			0 56 00	−30 20.6	1	11.95		1.30				1		3061
		POSS 295 # 26		0 56 02	+31 49.9	1	16.46		1.63				1		1739
		AzV 194	AB	0 56 04	−72 51.7	1	13.95		−0.22		−0.96		3	B0	425
		AzV 195		0 56 04	−72 56.4	1	13.53		−0.02		−0.28		3	B5 Ib	425
5738 −49 00268				0 56 05	−48 50.6	1	8.88		0.00				4	A0/1 V	2012
5649 +61 00189				0 56 11	+61 38.5	2	8.68	.000	0.63	.001			3	G0 V	1103,1118
		G 242 - 75		0 56 11	+69 12.8	1	10.05		0.79		0.39		1	K0	1658
		G 1 - 37		0 56 12	+01 07.0	2	11.06	.039	1.02	.000	0.99		3		202,1620
5737 −30 00297		HR 280		0 56 12	−29 37.6	12	4.31	.009	−0.16	.007	−0.52	.013	136	B7 IIIp	3,9,15,200,561,747*
	+58 00145	IDS00532N5857	AB	0 56 13	+59 13.2	1	9.76		0.26		0.09		3	A0	1185
		AzV 196		0 56 13	−72 43.8	1	13.93		−0.13		−0.84		3	B1	425
5720 −6 00176				0 56 14	−06 09.1	1	6.57		1.14					K0 III	2009
5722 −12 00173		HR 279		0 56 14	−11 39.0	3	5.62	.012	0.94	.014	0.67	.008	8	G7 III	58,2035,3016
5736 −27 00309				0 56 14	−27 05.1	1	9.79		0.60		0.03		1	G3 V	3061
	−38 00327	G 269 - 65		0 56 14	−37 36.9	2	11.97	.024	0.45	.007	−0.17	.002	3		1696,3061
		G 69 - 36		0 56 15	+29 33.1	1	11.27		0.93		0.68		3	K3	333,1620
5690 +38 00154				0 56 15	+38 39.9	1	8.50		0.04		0.04		4	F0 V	833
5534 +79 00024				0 56 15	+80 16.6	1	6.69		0.37		0.08		3	F2	985
5735 −20 00174				0 56 16	−19 54.1	1	7.12		1.60		1.80		7	M1 III	3040
		MN83,95 # 407		0 56 16	−73 21.1	1	12.12		1.17				1		591
		AAS50,119 # 208		0 56 17	−72 51.5	1	15.41		−0.20		−0.86		3		967
		LS I +62 147		0 56 18	+62 46.0	1	10.49		0.49		−0.37		2	B2 III	1012
		BPM 46827		0 56 18	−28 22.	1	13.92		0.44		−0.23		2		3064
		G 269 - 66		0 56 18	−30 49.3	1	12.40		1.27				1		3061
5771 −61 00050		HR 281		0 56 18	−60 58.0	2	6.22	.005	0.10	.000			7	A2 V	15,2012
		G 1 - 38		0 56 19	+04 32.6	2	13.23	.049	1.25	.015	1.06		3		202,1620
		BSD 8 # 1491		0 56 20	+61 24.8	1	10.63		0.51		−0.23		3	B5 p	1185
		G 269 - 67		0 56 20	−31 43.2	1	14.10		1.45				1		3061
		AzV 203		0 56 20	−71 42.3	1	14.14		−0.23		−0.98		3	B0	425
		AAS50,119 # 209		0 56 20	−72 58.2	1	14.20		−0.05		−0.64		5		967
		AzV 197		0 56 21	−72 34.0	1	11.86		0.45		0.35		3	F5	425
		AzV 198		0 56 21	−72 35.7	1	13.41		0.37		0.34		3	F5	425

HD	DM	Other Id	N Rem	α_{1950}	δ_{1950}	S	V	σ_V	B–V	σ_{B-V}	U–B	σ_{U-B}	n	Spectrum	References
		AzV 199	AB	0 56 21	−72 51.8	1	13.41		0.18		−0.08		3	A3 Ib	425
		MN83,95 # 403		0 56 23	−73 21.3	1	11.53		1.06				1		591
		BPM 46830		0 56 24	−29 20.	1	13.57		0.96		0.58		3		3064
		AAS35,353 # 68		0 56 24	−72 23.2	1	13.18		0.00		−0.17		6	B9	767
		BSD 8 # 945		0 56 25	+60 34.7	1	11.62		0.31		−0.31		3	B3	1185
5744	−6 00177			0 56 25	−05 40.7	3	10.66	.012	−0.02	.003	−0.17	.010	7	A0	152,1026,1736
		MN83,95 # 405		0 56 25	−73 19.7	1	11.91		0.72				1		591
		AzV 208		0 56 26	−71 59.4	1	14.10		0.01		−1.03		3	O9	425
	+27 00157			0 56 27	+27 33.2	1	9.64		0.60		0.05		4	G0	625
		G 269 - 68		0 56 27	−28 30.1	2	13.89	.010	1.59	.034	1.17		3		481,3064
		AzV 200		0 56 27	−72 54.6	3	12.14	.032	0.07	.007	−0.51	.005	6	A0	425,573,1571
		G 269 - 69		0 56 28	−28 58.0	1	13.52		1.55		1.12		2		3064
		AzV 201		0 56 29	−72 27.1	1	14.04		−0.14		−0.93		4	B0	425
		BPM 46831		0 56 30	−19 53.	1	11.46		0.73		0.07		1		3061
		BSD 8 # 946		0 56 31	+60 16.9	1	10.08		1.14				1	gG6 p	1103
		BSD 8 # 1495		0 56 32	+61 19.4	1	12.14		0.48		−0.21		3	B4	1185
		AzV 207		0 56 33	−72 10.6	1	14.37		−0.22		−1.05		3	O7 V	425
		AzV 202		0 56 33	−72 23.6	1	14.33		−0.18		−0.95		4	B0	425
	+65 00116			0 56 35	+66 00.1	1	8.90		1.83				3	M0	369
		AAS50,119 # 210		0 56 35	−72 46.9	1	15.15		0.08		−0.16		5		967
5717	+46 00220			0 56 36	+46 34.3	1	8.48		0.02		−0.20		2	B9	401
		JL 219		0 56 36	−56 51.	1	15.03		0.23		−0.58		1		832
		MN83,95 # 404		0 56 36	−73 17.4	1	11.75		0.52				1		591
		BSD 8 # 949		0 56 37	+60 16.6	2	10.38	.029	0.17	.008	−0.48		4	B6 p	1103,1185
5769	−30 00299			0 56 37	−29 40.3	3	9.31	.010	0.20	.004	0.08	.005	8	A4 V	561,1068,2012
5766	−24 00420			0 56 38	−23 57.6	1	9.50		0.64		0.08		1	F5/6 V	3061
5702	+61 00191			0 56 39	+61 59.4	2	8.81	.000	0.49	.002			3	F7 V	1103,1118
		SK 70		0 56 39	−72 05.	2	12.63	.009	0.59	.009	0.13	.014	4	G0	767,835
5689	+62 00178	LS I +63 117		0 56 40	+63 20.3	2	9.13	.000	0.34	.000	−0.68	.000	4	O6 V	1011,1012
		LS I +64 058		0 56 41	+64 23.5	2	10.99	.000	0.72	.000	−0.20	.005	4	B2 III	1012,1012
		AzV 204		0 56 41	−72 51.4	1	14.38		−0.12		−0.93		4	B1:	425
		LP 882 - 179		0 56 42	−27 58.	1	16.20		1.45				1		3061
		AzV 205		0 56 42	−72 37.7	2	12.30	.008	0.12	.000	−0.17	.000	7	A2 Ia	425,835
		DQ And		0 56 44	+45 08.2	2	11.23	.045	0.50	.005	0.29		2		592,1462
		AzV 206		0 56 44	−72 00.9	2	13.44	.005	−0.22	.005	−0.96	.000	4	B0	425,835
5781	−1 00124	IDS00542S0113	AB	0 56 47	−00 56.6	1	7.62		0.51		0.01		4	F5	176
		AAS50,119 # 211		0 56 47	−72 48.1	1	15.61		−0.07		−0.61		3		967
		BPM 46834		0 56 48	−22 17.	1	14.02		0.91		0.64		1		3061
5780	+0 00159	IDS00543N0015	A	0 56 49	+00 30.6	2	7.65	.020	1.44	.008	1.67	.008	8	K5 II-III	1003,3077
		AAS50,119 # 212		0 56 49	−72 57.4	1	14.62		−0.04		−0.43		5		967
		Psc sq 1 # 2		0 56 50	+27 38.2	1	12.50		0.74		0.14		4		625
		LS I +63 118		0 56 50	+63 33.1	1	11.72		0.53		−0.34		2	B1 V	1012
		AAS53,255 # 96		0 56 51	−72 49.7	2	13.67	.041	1.84	.016			7		967,1686
		GD 670		0 56 53	−35 49.9	1	13.94		0.71		−0.04		1		3061
		CS 22166 # 21		0 56 54	−11 46.3	1	15.10		0.08		0.18		1		1736
5816	−36 00357			0 56 55	−35 45.6	1	9.30		0.28				4	F0 V	2012
		AzV 210		0 56 55	−72 32.5	2	12.67	.005	−0.04	.015	−0.78	.024	3	B3 Ia	425,835
		AzV 209		0 56 55	−72 41.1	1	14.46		−0.17		−0.97		4	B0	425
	+60 00140			0 56 56	+60 52.7	1	9.36		1.15				2		70
	+44 00212	IDS00541N4437	AB	0 56 58	+45 06.5	1	10.57		0.54				1	F8	592
5747	+59 00161			0 56 59	+60 14.5	1	7.07		0.96				1	G8 II	1103
		AAS50,119 # 214		0 56 59	−72 57.9	1	15.26		−0.16		−0.85		3		967
		Psc sq 1 # 4		0 57 00	+27 36.6	1	11.45		0.56		−0.01		4		625
		BPM 70483		0 57 00	−18 58.	1	13.34		1.33		1.22		1		3061
		BPM 46837		0 57 00	−32 07.	1	11.20		0.63		0.00		1		3061
		BPM 46839		0 57 00	−36 18.	1	11.77		1.26		1.18		1		3061
5805	−1 00125	IDS00545S0144	AB	0 57 01	−01 27.7	1	8.66		0.69		0.28		5	G5	176
		AzV 211		0 57 01	−72 42.4	3	11.51	.010	0.10	.007	−0.45	.005	11	A0 Ia	425,573,1571
		POSS 295 # 13		0 57 02	+31 42.8	1	14.31		1.51				2		1739
		G 269 - 70		0 57 03	−26 47.3	1	15.95		0.47		−0.43		2		3064
5824	−32 00395			0 57 03	−32 14.1	1	9.70		0.29		0.03		4	A9 V	3064
5825	−35 00336			0 57 03	−34 54.4	2	7.66	.000	0.97	.005	0.51	.039	3	G8p Ba	565,3048
		W Psc		0 57 04	+27 40.5	1	10.50		1.53		1.15		1	M2e	625
	−14 00181	G 268 - 88		0 57 06	−14 15.1	1	10.59		1.08		0.94		1	K4 V	3072
		AAS35,353 # 70		0 57 06	−72 01.0	1	12.43		0.49		0.04		3	F5	767
		MN83,95 # 402		0 57 06	−73 17.1	1	11.24		0.48				1		591
5764	+47 00272			0 57 07	+47 45.0	1	7.15		−0.12		−0.65		2	B8	401
5715	+70 00065	HR 278		0 57 08	+70 42.8	1	6.39		0.13		0.11		2	A4 IV	3050
		AAS50,119 # 215		0 57 08	−72 49.1	1	14.27		1.51				5		967
	−29 00286	G 269 - 72		0 57 11	−28 59.5	1	12.58		0.59		−0.12		5		3064
5788	+43 00193	HR 282	★ AB	0 57 12	+44 26.5	1	5.70		−0.01		−0.06		2	B9.5V	1733
		BPM 46840		0 57 12	−22 19.	1	12.04		0.63		0.07		1		3061
		BPM 46842		0 57 12	−32 37.	1	11.69		0.41		−0.15		1		3061
		BPM 46843		0 57 12	−33 26.	1	12.40		0.85		0.45		2		3061
		AzV 213		0 57 12	−72 12.7	2	12.09	.025	0.13	.000	−0.03	.000	5	A2 Iab	425,573
		AAS50,119 # 216		0 57 12	−72 55.1	1	15.55		−0.14		−0.74		3		967
		AzV 212		0 57 12	−72 59.8	1	14.39		−0.17		−0.87		3	B1	425
5789	+43 00193	HR 283	★ AB	0 57 13	+44 26.7	1			−0.01		−0.05		1	B9.5Vn	1079
5820	+5 00131	HR 284, WW Psc	★ A	0 57 14	+06 12.8	4	6.11	.017	1.67	.007	1.96	.042	10	M2 III	15,1256,1733,3055
		AzV 214		0 57 14	−72 29.4	1	13.34		0.07		−0.72		3	B3 Iab	733

Table 1   69

HD	DM	Other Id	N Rem	α_{1950}	δ_{1950}	S	V	σ_V	B–V	σ_{B-V}	U–B	σ_{U-B}	n	Spectrum	References
		AAS50,119 # 217		0 57 14	−72 56.8	1	11.81		1.61		1.98		8		967
		AzV 215		0 57 15	−72 48.2	3	12.77	.006	−0.12	.015	−0.94	.025	12	B1 Ia	425,835,967
		AzV 214a		0 57 15	−73 03.6	1	14.91		−0.17		−0.93		8	B1	767
	−24 00426			0 57 16	−23 58.1	1	10.06		0.82		0.43		1	G5	3061
	−27 00317			0 57 16	−27 01.5	2	11.09	.039	0.19	.024	0.12	.019	3	A3	378,561
		AAS35,353 # 71		0 57 18	−72 53.8	1	14.30		0.18		0.38		3	A5	767
	+62 00180			0 57 19	+63 17.1	1	9.30		2.07				2	K5	369
		AzV 216		0 57 20	−73 00.7	2	14.14	.073	−0.19	.041	−0.91	.001	9	B0	425,967
	−26 00320			0 57 21	−26 23.1	1	11.89		0.58				1		1495
		AzV 217		0 57 21	−72 35.0	1	14.59		−0.20		−0.89		4	B2:	425
		CS 22183 # 17		0 57 23	−03 28.6	1	14.27		0.06		0.14		1		1736
		AzV 218		0 57 24	−72 35.8	1	13.80		−0.13		−0.92		3	B1	425
		BSD 8 # 441		0 57 25	+60 14.5	1	11.66		0.25		0.21		3	B7	1185
5868	−26 00321	IDS00550S2624	AB	0 57 25	−26 08.1	2	8.54	.005	0.50	.016	−0.06		5	F5 V	2012,3061
5776	+62 00181	LS I +62 148		0 57 26	+62 45.6	1	8.05		0.52		0.11		2	A0 Ib	1012
		AzV 219		0 57 26	−72 32.7	1	14.50		−0.21		−0.89		2	B2:	425
		Psc sq 1 # 3		0 57 27	+27 38.1	1	11.98		0.62		0.06		4		625
		MN201,73 # 40		0 57 27	−32 58.1	1	15.10		1.11				1		1495
		Florsch 379		0 57 27	−72 42.5	1	11.67		0.61		0.16		5	G0	767
	−0 00154			0 57 29	+00 06.7	1	9.51		1.01		0.65		2	K0	1375
5865	−22 00172	IDS00550S2245	A	0 57 29	−22 28.8	1	10.32		0.63		0.07		1	G2/5 (V)	3061
5865	−22 00172	IDS00550S2245	B	0 57 29	−22 28.8	1	10.88		0.75		0.33		1		3061
		CS 22942 # 24		0 57 29	−23 46.8	2	14.15	.010	0.40	.007	−0.20	.009	2		1580,1736
		AzV 220		0 57 29	−72 21.9	1	14.50		−0.22		−1.08		4	O9	425
		G 33 - 28		0 57 30	+15 53.7	1	11.24		1.19		1.16		1	K3	333,1620
5797	+59 00163	V551 Cas		0 57 30	+60 10.5	4	8.47	.020	0.26	.030	0.19	.005	13	Am	220,1103,1118,1185
		Be 62 - 1		0 57 30	+63 40.5	1	12.28		1.70		1.43		3		903
		AzV 221		0 57 30	−72 47.7	2	13.46	.005	−0.11	.014	−0.75	.005	4	B0 Ia	425,835
	+58 00150			0 57 31	+59 02.3	1	10.16		0.25		0.12		3	B9	1185
	−27 00318			0 57 32	−26 53.8	1	10.94		0.80		0.37		1		4003
		AzV 222		0 57 32	−72 37.1	1	13.20		−0.07		−0.75		3	B3 Iab	425
		WLS 112 60 # 12		0 57 33	+58 18.4	1	11.01		0.20		0.12		2		1375
		BSD 8 # 1502		0 57 33	+61 31.4	1	11.92		0.56		−0.24		3	B0	1185
		G 1 - 39		0 57 34	+04 27.9	2	10.52	.024	1.05	.005	1.01		3	K3	202,1620
		AzV 223		0 57 34	−72 55.1	1	13.67		−0.20		−0.90		3	B0	425
		Be 62 - 2		0 57 35	+63 40.4	1	13.58		1.08		0.43		4		903
		AzV 224		0 57 35	−72 20.8	1	14.22		−0.14		−0.99		4	B0	425
5857	+17 00135	IDS00550N1740	AB	0 57 36	+17 55.9	3	7.72	.679	0.95	.184	0.49	.444	8	G5	196,1619,7008
		POSS 295 # 18		0 57 37	+30 33.4	1	16.03		1.45				2		1739
		AzV 225		0 57 37	−72 45.9	1	14.05		−0.01		−0.46		5	B8	425
		Be 62 - 13		0 57 40	+63 40.4	1	12.52		0.83		−0.07		4		903
		AzV 226		0 57 40	−72 33.3	1	14.42		−0.29		−1.07		3	B0	425
		CS 22183 # 14		0 57 42	−02 33.6	1	14.21		0.03		0.08		1		1736
	−27 00319			0 57 42	−27 01.8	1	11.27		1.07		0.91		1		4003
		BSD 8 # 970		0 57 44	+60 48.9	1	11.68		0.41		0.25		3	B9	1185
		AzV 227		0 57 44	−72 58.6	1	14.76		−0.18		−1.03		4	B1	425
		AzV 228		0 57 44	−73 01.3	1	14.86		−0.14		−1.05		4	B1	425
5854	+37 00190			0 57 45	+37 31.3	1	7.19		0.13		0.13		2	A0	1375
5679	+81 00025	U Cep	★ A	0 57 45	+81 36.4	1	6.83		−0.05		−0.40		2	G8 III	3024
5679	+81 00025	IDS00534N8120	B	0 57 45	+81 36.4	1	11.83		0.73		0.29		3		3024
		AAS35,353 # 73		0 57 45	−72 14.5	1	13.98		−0.12		−0.40		4	B6	767
		Be 62 - 3		0 57 47	+63 38.6	1	13.53		0.83		0.05		4		903
5980		AzV 229		0 57 47	−72 26.0	4	11.80	.084	−0.21	.027	−0.99	.015	16	Wp	425,573,1277,8033
6005	−74 00079			0 57 47	−74 01.1	1	9.33		0.53		0.04		4	F6 V	1408
5871	+20 00139			0 57 48	+20 58.8	1	6.82		1.39		1.61		2	K2	1648
		BPM 46847	A	0 57 48	−28 15.	1	14.27		1.47		1.12		1		3064
		BPM 46847	B	0 57 48	−28 15.	1	13.55		0.74		0.16		2		3061
		BPM 46848		0 57 48	−30 57.	1	13.19		1.20		0.98		1		3061
		AzV 230		0 57 49	−72 17.2	2	12.73	.020	−0.02	.040	−0.79	.010	5	B2	425,835
		AzV 231		0 57 49	−72 36.4	1	14.20		0.00		−0.32		3	A0	425
5902	−21 00149			0 57 51	−20 32.7	1	9.05		0.52		−0.07		1	G0 V	3061
	−26 00323	G 268 - 90		0 57 51	−25 53.0	2	9.96	.018	1.11	.014	1.03	.018	5	K2	79,3064
		AzV 232		0 57 52	−72 26.9	2	12.36	.005	−0.20	.005	−1.00	.005	8	O9 Ia0-Ia	425,573
		AAS53,255 # 100		0 57 52	−72 32.	1	13.13		1.87				1	M0 Ia-Iab	1686
	−14 00184	G 270 - 111		0 57 53	−13 52.0	1	10.10		1.03		0.89		1	K3 V	3072
		Be 62 - 9		0 57 54	+63 41.0	1	12.50		0.67		−0.18		2		903
		AzV 233		0 57 56	−72 59.0	1	14.65		−0.21		−0.88		3	B0	425
		BSD 8 # 1505		0 57 57	+61 45.1	1	11.81		0.68		0.14		3	F8	1185
		CS 22942 # 28		0 57 60	−25 33.9	1	13.54		0.32		0.10		1		1736
		SB 404		0 58 00	−25 33.9	2	13.55	.005	0.31	.005	0.06	.019	3	A pec	561,966
		BPM 46850		0 58 00	−33 18.	1	12.57		0.80		0.33		1		3061
	+58 00151			0 58 01	+58 59.0	1	9.97		0.30		0.27		3	B7	1185
		BSD 8 # 448		0 58 01	+59 48.9	1	10.81		0.32		0.23		3	B9	1185
	+62 00183	LS I +63 123		0 58 01	+63 10.3	1	8.96		0.69		−0.24		2	B1 V:p	1012
5922	−27 00322			0 58 01	−27 10.6	1	10.05		0.62		0.02		1	G3/5 V	3061
5882	+50 00198			0 58 02	+50 36.6	2	7.68	.015	−0.05	.000	−0.59	.015	4	B2.5Vn	401,1733
	+63 00124	Be 62 - 8		0 58 02	+63 39.4	2	11.10	.000	0.55	.000	0.02	.009	6	B1 V	903,1012
5851	+59 00166	IDS00550N5949	AB	0 58 03	+60 05.4	1	8.11		0.48				1	F5 IV	1103
		Be 62 - 6		0 58 03	+63 41.1	1	10.91		0.61		−0.27		3		903
		AzV 239		0 58 03	−72 02.6	1	13.67		−0.13		−0.74		4	B2	425

HD	DM	Other Id	N Rem	α_{1950}	δ_{1950}	S	V	σ_V	B–V	σ_{B-V}	U–B	σ_{U-B}	n	Spectrum	References
		AzV 234		0 58 03	−72 20.4	2	12.98	.005	−0.12	.009	−0.72	.000	4	B2 Iab	425,835
		AAS53,255 # 102		0 58 04	−72 37.	1	12.87		1.60				1		1686
		Be 62 - 10		0 58 06	+63 42.1	1	13.97		0.71		0.10		3		903
	+65 00121	LS I +65 008		0 58 06	+65 38.8	1	10.54		1.03		0.52		1	B9	1215
		SB 406, RM Cet		0 58 06	−16 12.	2	11.90	.075	0.08	.010	0.09	.035	2	A5 pec	668,700
5919	−2 00140			0 58 07	−01 55.7	1	6.88		1.60		1.72		5	M4 III	3007
		AzV 235		0 58 07	−73 01.0	4	12.18	.021	−0.14	.032	−0.95	.007	13	B1 Ia	425,573,1277,1408
		AJ63,118 # 34		0 58 07	−73 01.1	1	12.04		−0.08				1		2001
5934	−26 00326			0 58 08	−26 01.1	1	7.34		1.03				4	F8 III/IV	2012
		BSD 8 # 1508		0 58 09	+61 44.1	1	10.94		0.48		0.35		3	A0	1185
		Be 62 - 12		0 58 11	+63 39.9	2	11.16	.013	0.65	.004	−0.21	.004	6		903,1012
		Be 62 - 4		0 58 11	+63 41.5	1	12.18		0.62		−0.29		3		903
		AzV 236		0 58 11	−72 59.4	1	14.70		−0.17		−1.00		4	B0	425
	−27 00323			0 58 12	−26 54.3	1	10.11		0.76		0.28		1	K0	3061
		BPM 46853		0 58 12	−27 58.	1	13.92		1.52		1.00		1		3061
		G 243 - 50		0 58 13	+61 06.3	2	10.95	.168	1.55	.040	1.29		6		1625,1723
		Be 62 - 5		0 58 13	+63 42.7	1	13.75		0.70		0.25		3		903
		AzV 237		0 58 13	−72 35.2	2	12.58	.014	0.03	.014	−0.40	.005	4	B8 Ia	425,835
		G 269 - 74		0 58 14	−37 42.1	1	14.30		0.46		−0.25		2		3060
	+70 00068			0 58 15	+71 26.8	1	9.96		1.46				6		1197
		AzV 244		0 58 15	−71 56.9	1	14.41		−0.20		−1.06		3	B0	425
		AzV 238		0 58 15	−72 29.7	1	13.77		−0.22		−1.02		5	O9	425
		Be 62 - 14		0 58 16	+63 37.0	1	13.39		0.65		−0.07		3		903
	−28 00310			0 58 16	−27 54.2	2	10.92	.025	1.13	.004	1.07		2		1495,4003
		POSS 295 # 33		0 58 18	+30 07.2	1	17.23		1.45				2		1739
		BSD 8 # 75		0 58 18	+58 53.1	1	11.64		0.38		0.32		3	A0	1185
		Be 62 - 11		0 58 18	+63 40.5	1	11.98		0.66		−0.30		3		903
		LS I +63 125		0 58 18	+63 40.5	1	10.92		0.61		−0.29		2	B1 V	1012
5917	+28 00166			0 58 19	+28 43.8	1	8.60		0.69		0.28		1	G8 III	979
5890	+60 00143	IDS00553N6030	AB	0 58 19	+60 46.5	2	8.77	.002	0.43	.002			4	F5 V	70,1103
	−27 00325			0 58 19	−26 51.5	2	11.10	.000	1.15	.017	1.07		2		1495,4003
		AAS35,353 # 74		0 58 19	−72 45.3	1	13.52		0.03		−0.11		3	A0	767
	+61 00194	IDS00552N6157	A	0 58 21	+62 13.0	1	10.71		0.52		0.36		3	B8	1185
		AzV 250		0 58 21	−72 11.8	2	12.82	.005	−0.02	.000	−0.13	.009	4	B9 Ib	425,835
		AzV 240		0 58 21	−72 40.0	1	14.03		−0.14		−0.72		4	B3	425
	+41 00186			0 58 22	+42 05.0	1	10.15		0.49		0.00		2		1375
		G 70 - 28		0 58 22	−04 43.4	2	13.27	.044	1.72	.010	1.27		3		202,3078
5962	−27 00326			0 58 22	−27 29.0	1	9.47		1.42		1.70		1	K3 III	54
		AAS53,255 # 105		0 58 22	−72 36.	1	12.98		1.91				1		1686
		AAS53,255 # 104		0 58 23	−73 00.	1	13.45		1.56				1		1686
		BSD 8 # 77		0 58 24	+58 50.7	1	11.18		0.29		0.25		3	B9	1185
		BPM 46856		0 58 24	−26 33.	1	14.54		1.12		0.95		1		3064
		BPM 46857		0 58 24	−29 07.	1	14.37		0.99		0.80		1		3061
	−31 00400			0 58 24	−31 01.6	1	10.33		0.47		−0.11		1	F8	3061
6045	−70 00044			0 58 24	−70 26.8	1	8.00		0.43		−0.03		1	F3/5 V	1408
		AzV 241		0 58 24	−73 11.5	1	14.44		0.07		−0.44		4	B8	425
		BSD 8 # 981		0 58 26	+60 37.2	1	12.40		0.30		−0.23		3	B0	1185
5976	−25 00397			0 58 26	−24 41.6	1	9.57		0.56		0.00		1	F8/G0 V	3061
		Sculptor sq # 4		0 58 26	−34 01.0	1	14.21		1.07				2		527
	+15 00150	G 33 - 30		0 58 27	+16 03.3	3	10.66	.008	0.54	.004	−0.17	.004	4	sdF7	333,927,1620,1696
5916	+44 00215			0 58 27	+45 11.0	3	6.85	.004	0.90	.005	0.45	.012	7	G8 III-IV	1003,1355,1462
	+45 00253			0 58 27	+45 58.7	1	9.36		−0.02		−0.41		2	A0	401
	−37 00366	BPM 46860		0 58 27	−36 43.5	1	10.89		0.91		0.61		1		3061
		LP 938 - 61		0 58 27	−36 43.6	1	12.02		1.15		1.23		1		1696
		AAS53,255 # 106		0 58 27	−72 02.	1	13.80		1.84				1		1686
		AzV 245		0 58 27	−72 10.7	1	13.34		−0.08		−0.50		3	B5 Iab	733
		AzV 242		0 58 27	−72 30.1	3	12.08	.025	−0.12	.010	−0.91	.010	6	B1.5Ia	425,573,1513
		BSD 8 # 1512		0 58 28	+61 52.0	1	11.34		0.61		0.23		3	A0	1185
		BSD 8 # 1513		0 58 28	+61 59.5	1	11.96		0.53		−0.09		3	B5	1185
		SB 408		0 58 28	−28 21.6	3	14.22	.012	0.21	.015	0.06	.025	4	A0	295,561,1736
		AAS53,255 # 107		0 58 28	−72 22.	1	13.36		1.59				2		1686
		AzV 243		0 58 28	−73 03.4	2	13.89	.019	−0.22	.000	−0.99	.015	3	B0	425,835
		GD 9		0 58 29	−04 27.5	1	15.38		0.03		−0.75		1	DA	3060
		Sculptor sq # 1		0 58 31	−33 57.5	1	10.09		0.57				3		527
		POSS 295 # 21		0 58 34	+30 28.0	1	16.11		1.62				4		1739
		BSD 8 # 1515		0 58 34	+62 12.0	1	11.02		0.56		0.20		3	B7	1185
		G 14 - 55	A	0 58 35	−29 31.6	2	11.37	.041	1.50	.012	1.16		3	M2	1705,1764
		LP 416 - 94	A	0 58 35	−29 31.6	1	10.99		1.25				2		920
		G 14 - 55	B	0 58 35	−29 31.6	1	13.69		1.66				1		1705
		LP 416 - 94	B	0 58 35	−29 31.6	1	13.61		0.66				1		920
		Sculptor sq # 2		0 58 35	−34 05.6	1	12.40		0.56				1		527
		BPM 46859		0 58 36	−31 44.	1	13.04		0.49		−0.17		1		3061
		BSD 8 # 459		0 58 37	+59 33.4	1	12.08		0.40		0.28		3	A0	1185
		AzV 246		0 58 37	−71 45.5	2	14.54	.010	−0.19	.034	−0.96	.015	9	B3	733,767
		BSD 8 # 1514		0 58 39	+61 23.8	1	10.85		0.44		−0.31		3	B4	1185
		AzV 248		0 58 39	−71 45.7	2	14.82	.020	−0.23	.015	−0.97	.010	10	B0:	733,767
		AzV 247		0 58 39	−71 49.6	1	14.31		−0.25		−1.06		4	O9	425
		AAS53,255 # 109		0 58 39	−71 58.	1	13.25		1.96				1		1686
		AAS35,353 # 75		0 58 39	−72 49.4	1	12.58		0.38		−0.04		5	F1	767
		LP 586 - 41		0 58 40	+02 37.6	1	12.08		1.05		0.93		1		1696

Table 1   71

HD	DM	Other Id	N	Rem	α_{1950}	δ_{1950}	S	V	σ_V	B–V	σ_{B-V}	U–B	σ_{U-B}	n	Spectrum	References
6090		BT Tuc			0 58 41	−73 00.7	1	10.58		0.74				1	G5	2001
		Be 62 - 7			0 58 42	+63 41.1	1	13.81		0.62		−0.12		3		903
	−38 00342	G 269 - 75			0 58 42	−38 18.3	1	10.09		0.57		−0.10		1		79
		AzV 249			0 58 42	−72 57.5	1	13.75		−0.19		−0.91		3	B0	425
		G 33 - 31			0 58 44	+14 58.9	2	12.56	.000	0.87	.005	0.33	.015	2		333,1620,1658
		AzV 251			0 58 44	−72 46.9	1	14.75		−0.27		−1.11		5	O9 Vn	425
		AzV 255		AB	0 58 45	−71 46.1	2	12.80	.015	−0.20	.010	−0.99	.015	5	B0	425,573
		AzV 254			0 58 45	−71 47.2	2	11.63	.015	0.14	.000	0.11	.010	8	A3 Ia	425,573
	+60 00144				0 58 48	+60 55.8	1	10.21		0.42		0.17		3	A3 III	1185
		BPM 46861			0 58 48	−26 54.	1	14.22		0.55		−0.19		1		3061
		AzV 252			0 58 51	−72 27.1	2	13.07	.000	−0.12	.024	−0.78	.010	3	B2 Ia	425,835
		AAS53,255 # 110			0 58 52	−72 22.	1	13.68		1.66				1		1686
		G 242 - 76			0 58 53	+71 25.0	1	10.06		1.47		1.15		1		3078
		SB 410			0 58 54	−33 58.9	3	12.60	.025	−0.21	.033	−1.05	.013	4		295,1036,3064
		Sculptor sq # 3			0 58 55	−33 58.9	1	12.66		−0.22				1		527
		HV 11423			0 58 57	−71 53.6	1	12.52		1.84				2	M0 Iab	5007
	−33 00378	BPM 46864			0 58 58	−32 43.7	1	10.77		0.62		0.03		1		3061
6055	−39 00260	HR 288			0 58 58	−39 11.2	3	5.59	.008	1.18	.000			11	K1 III	15,1075,2032
		AAS53,255 # 111			0 58 58	−72 14.	1	12.64		1.83				1		1686
		AzV 253			0 58 58	−73 06.9	2	13.33	.005	−0.09	.005	−0.51	.000	4	B6 Iab	425,835
		AJ63,118 # 33			0 58 59	−71 48.6	2	11.59	.009	0.14	.003	0.06		4		522,2001
		BPM 46863		A	0 59 00	−29 38.	1	13.79		1.52		1.25		1		3064
		BPM 46863		B	0 59 00	−29 38.	1	13.98		0.95		0.50		1		3061
		BPM 46865			0 59 00	−34 54.	1	11.62		1.08		0.96		2		3061
6009	+24 00163				0 59 01	+25 01.4	1	6.71		0.78		0.39		2	G8 IV	1625
		G 218 - 20			0 59 03	+53 54.8	1	15.12		1.84				1		906
		AzV 256			0 59 03	−72 46.6	1	14.62		−0.12		−1.04		4	B0	425
5817	+81 00027				0 59 04	+81 50.0	1	8.45		0.60		0.01		3	G5	196
		BPM 70494			0 59 04	−19 40.	1	10.52		0.70		0.11		1		3061
6054	−28 00314				0 59 05	−27 43.2	2	9.19	.005	1.51	.010	1.49		2	K5/M0	54,1495
		MN201,73 # 44			0 59 05	−36 48.4	1	14.89		1.44				1		1495
		AAS53,255 # 112			0 59 05	−72 52.	1	12.98		1.72				1		1686
		AzV 257			0 59 06	−72 40.0	2	12.78	.010	−0.07	.000	−0.71	.019	6	B3 Iab	425,573
		AzV 258			0 59 06	−72 45.9	2	13.75	.000	0.03	.005	−1.10		5	O9	425,1513
	+58 00154				0 59 07	+58 43.2	1	10.92		0.22		−0.39		3	B6	1185
		AzV 259			0 59 07	−71 50.3	1	14.68		−0.15		−0.93		4	B0	425
		AAS35,353 # 76			0 59 07	−72 29.3	1	13.65		−0.05		−0.36		2	B8	767
		AAS53,255 # 114			0 59 09	−71 54.	1	11.90		1.83				1	M0 Iab	1686
		BSD 8 # 1520			0 59 10	+61 38.6	1	10.69		0.66		0.07		3	B6	1185
6037	−17 00180				0 59 10	−16 32.0	1	6.47		1.18				4	K2/3 III	2009
		MN201,73 # 45			0 59 10	−35 58.1	1	15.40		1.42				1		1495
		AAS53,255 # 115			0 59 10	−72 27.	1	12.13		1.94				1	M0 Ia	1686
		WLS 100 0 # 5			0 59 11	+02 26.7	1	12.30		0.68		0.21		2		1375
		PS 44			0 59 12	−23 32.	1	12.15		0.36		−0.07		2		295
		MN201,73 # 47			0 59 12	−35 28.2	1	14.16		1.44				1		1495
		AzV 260			0 59 12	−72 29.7	2	13.27	.009	−0.14	.042	−0.82	.005	4	B1 Iab	425,835
6018	+48 00322				0 59 13	+49 12.0	1	8.30		0.03		−0.08		2	A0	401
		MN201,73 # 48			0 59 13	−35 58.6	1	13.59		1.16				3		1495
6107	−61 00058				0 59 16	−61 07.8	1	6.90		0.65				4	G1/2 V	2013
		AzV 264			0 59 16	−72 15.8	2	12.36	.005	−0.14	.010	−0.90	.005	5	B0 Ia0-Ia	425,573
6172	−73 00059				0 59 16	−72 58.0	7	7.68	.017	1.33	.009	1.47	.009	45	K2/3 III	14,1408,1425,1704,2001*
	−7 00160				0 59 17	−06 34.8	1	10.85		0.31		0.26		1		1736
		AAS53,255 # 116			0 59 17	−73 08.	1	11.94		1.78				1	K5 Ia	1686
		BPM 46869			0 59 18	−36 20.	1	14.05		0.62		0.00		2		3061
		POSS 295 # 25			0 59 19	+31 25.2	1	16.36		1.56				1		1739
		BSD 8 # 469			0 59 19	+59 35.7	1	11.53		0.51		0.37		3	B7	1185
6067	−20 00183	G 268 - 96			0 59 19	−19 41.2	2	8.23	.020	0.62	.005	0.14	.020	2	G5 V	79,3061
		BSD 8 # 471			0 59 20	+60 09.1	1	11.75		0.32		0.18		3	A0	1185
		AzV 261			0 59 20	−72 46.9	1	14.06		−0.13		−0.84		4	B0	425
		BSD 8 # 1000			0 59 21	+60 33.4	1	11.64		0.28		0.22		3	B9	1185
	+62 00186	LS I +62 149			0 59 21	+62 42.9	2	9.84	.024	0.35	.004	−0.51	.004	8	B1 V	1012,1775
		MN201,73 # 50			0 59 22	−35 48.1	1	14.35		0.66				2		1495
		AAS53,255 # 117			0 59 22	−72 30.	1	12.96		1.59				1		1686
6028	+50 00202	HR 287			0 59 23	+50 46.0	2	6.48	.005	0.12	.000	0.16	.005	4	A3 V	1501,1733
		AAS35,353 # 77			0 59 23	−72 12.4	1	13.09		0.87		0.40		4	G9	767
	+61 00195	V388 Cas		A	0 59 24	+62 04.5	5	9.56	.014	1.50	.007	1.21	.005	19	K5 V	1,1118,1197,1775,3072
6088	−26 00334				0 59 24	−26 09.4	1	9.75		0.35				1	F0 V	1495
		Dril 11			0 59 24	−26 49.	1	14.15		0.62		−0.01		2	G2	561
		BPM 46868			0 59 24	−31 13.	1	12.67		0.72		0.12		1		3061
		Florsch 411			0 59 24	−71 26.5	1	13.32		0.55		−0.06		3	F7	767
6064	+1 00191				0 59 25	+02 15.4	1	7.58		0.48		−0.02		2	F6 V	3026
		BSD 8 # 94			0 59 25	+58 44.0	1	10.31		0.33		0.21		3	B9	1185
	+60 00145				0 59 25	+61 24.8	1	10.12		1.23				1	G8 V	851
		AzV 262			0 59 25	−72 55.6	1	13.04		−0.20		−0.95		5	B0	425
		AAS35,353 # 78			0 59 26	−72 42.5	1	14.05		0.12		0.14		2	A3	767
6017	+60 00146	IDS00564N6103		AB	0 59 27	+61 19.2	2	8.84	.000	0.56	.014	0.08		12	F5 III	1118,1371
	−10 00216	G 270 - 118			0 59 27	−10 08.9	3	10.48	.037	1.24	.022	1.11	.010	7	K5 V	158,1705,3072
		MN201,73 # 51			0 59 27	−36 08.1	1	14.48		0.72				1		1495
		AzV 263			0 59 27	−72 29.1	2	12.82	.040	−0.01	.010	−0.31	.050	5	B6 Ib	425,835
		LP 586 - 48			0 59 28	+01 57.6	1	10.52		0.89		0.63		2		1696

HD	DM	Other Id	N Rem	α₁₉₅₀	δ₁₉₅₀	S	V	σ_V	B–V	σ_B–V	U–B	σ_U–B	n	Spectrum	References
		CS 22942 # 29		0 59 28	−27 12.4	2	14.83	.010	0.40	.007	-0.25	.009	2		1580,1736
		AAS53,255 # 118		0 59 28	−72 19.	1	13.16		1.63				1		1686
		AAS53,255 # 119		0 59 28	−72 21.	1	12.92		1.45				1		1686
		GR 22		0 59 29	−26 01.9	2	11.87	.030	1.24	.010	1.14		4		481,3061
6193	−72 00075		AB	0 59 29	−71 48.4	5	9.76	.012	0.35	.018	0.01	.039	10	A9 V	522,1408,1425,2001,3021
		AAS53,255 # 120		0 59 29	−72 54.	1	13.89		1.70				1		1686
		MN201,73 # 52		0 59 30	−30 58.5	1	12.34		0.61				1		1495
		MN201,73 # 53		0 59 30	−32 48.3	1	13.46		0.31				1		1495
		CS 29509 # 49		0 59 31	−31 38.2	1	15.10		0.01		0.07		1		1736
		AzV 266		0 59 31	−72 43.6	2	12.62	.015	-0.13	.005	-0.90	.024	6	B1 Ia	425,573
		AzV 265		0 59 31	−73 01.7	1	14.56		-0.21		-0.85		4	B2	425
		BSD 8 # 475		0 59 32	+60 11.3	1	12.74		0.27		-0.01		3	A0	1185
		BSD 8 # 1003		0 59 32	+60 16.7	1	12.09		0.29		0.04		3	B6	1185
		G 1 - 41		0 59 33	+01 04.8	1	13.82		1.20				2		202
	+60 00147			0 59 33	+60 31.4	1	10.30		0.24		-0.10		3	B8 V	1185
		AAS35,353 # 80		0 59 33	−72 32.2	1	12.66		0.09		0.15		4	A2	767
6027	+58 00155			0 59 34	+59 01.1	2	7.84	.021	1.48	.072	1.75		5	K2 V	1619,7008
	−26 00336			0 59 34	−25 52.6	1	11.61		0.77				1		1495
5996	+68 00067	G 243 - 53		0 59 35	+68 57.6	3	7.67	.000	0.75	.005	0.27	.020	4	G5	908,1658,3026
	−30 00314			0 59 35	−29 47.0	1	12.85		0.29		-0.02		3	A5	561
		NGC 362 sq2 3		0 59 35	−71 07.1	1	10.94		1.13		1.05		1		1461
		CS 22183 # 28		0 59 36	−07 18.7	1	14.00		0.14		0.20		1		1736
		AzV 267		0 59 36	−72 22.7	1	14.92		-0.24		-1.10		3	O8 Vn	425
		AzV 268		0 59 37	−72 28.8	2	13.18	.056	-0.14	.009	-0.86	.005	4	B0 Ia	733,835
		AzV 269		0 59 37	−72 48.7	2	11.43	.025	0.13	.005	0.06	.015	9	A5 Ia	425,573
		AzV 270		0 59 38	−72 33.6	2	11.42	.000	0.03	.005	-0.41	.015	7	A0 Ia	425,573
	−35 00347			0 59 39	−34 59.9	1	9.87		0.76		0.37		1	G8	3061
		NGC 362 sq2 35		0 59 40	−71 01.9	1	13.10		1.07		1.12		2		1461
		AAS53,255 # 121		0 59 40	−72 29.	1	13.21		1.95				1		1686
		BSD 8 # 1533		0 59 41	+62 08.1	1	12.20		0.51		-0.29		3	B0	1185
		AzV 271		0 59 41	−72 33.4	1	13.46		-0.16		-0.92		3	B0 Iab	425
6094	+5 00138	IDS00571N0520	AB	0 59 42	+05 35.9	1	8.82		0.52		-0.05		2	G0	3016
	+60 00149			0 59 43	+60 35.5	1	9.83		0.22		-0.45		3	B4 III	1185
6026	+61 00196			0 59 43	+62 20.4	1	7.20		0.95		0.58		2	K0	1375
		AzV 272		0 59 44	−72 36.3	1	14.52		-0.15		-0.62		3	B2	425
	+59 00168			0 59 45	+60 24.2	1	10.89		0.67		0.12		3	B8 V	1185
		Hiltner 96		0 59 46	+63 43.9	1	11.03		0.57		-0.19		2	B3 III	1012
6047	+64 00116			0 59 46	+64 34.1	1	8.31		0.48				2	F8	1118
		AAS53,255 # 122		0 59 46	−72 21.	1	12.87		1.68				2		1686
		AAS53,255 # 123		0 59 46	−72 23.	1	13.27		1.63				1		1686
		LP 586 - 52		0 59 47	+01 43.2	1	11.70		0.97		0.79		1		1696
6048	+59 00169			0 59 47	+59 51.5	1	8.74		0.23		-0.24		3	B8 II	1185
6101	+4 00158	G 1 - 43		0 59 48	+04 47.4	2	8.19	.049	1.00	.005	0.86		3	K2	202,1620
	+58 00156			0 59 48	+59 05.2	1	11.28		0.31		0.16		3	B6	1185
		BPM 46871		0 59 48	−29 08.	1	13.72		1.27				1		3061
		AzV 273		0 59 48	−72 23.2	2	12.17	.010	0.06	.005	0.07	.020	5	A1 Ib	425,573
		WLS 48 40 # 8		0 59 49	+40 03.4	1	11.35		0.43		-0.01		2		1375
	+60 00151			0 59 49	+60 49.9	1	9.25		1.20				4	G8 III	70
6138	−31 00408			0 59 50	−31 04.0	1	9.53		1.04		0.83		1	G8/K0 (III)	54
6139	−32 00406			0 59 50	−31 53.5	1	8.79		0.51				4	F6/7 V	2012
	−11 00192			0 59 51	−10 41.0	2	10.04	.021	1.35	.016	1.22		4	K7 V	1619,3072
		GR 23		0 59 51	−17 17.5	1	13.81		1.32		1.17		3		481
		AzV 274		0 59 51	−72 39.4	1	14.02		-0.07		-1.08		4	B0	425
		AAS35,353 # 81		0 59 51	−73 12.7	1	14.41		-0.01		-0.08		2	B9	767
6222	−72 00076			0 59 52	−71 49.1	7	7.76	.029	1.11	.009	1.05	.016	28	K1 III	343,366,522,1408,1425*
		AzV 275	V	0 59 53	−73 12.5	2	13.67	.180	-0.25	.000	-1.00	.015	2	B1	733,835
		LB 3171		0 59 54	−60 36.	1	11.29		0.44		0.00		2		45
		AzV 276	V	0 59 54	−72 25.4	1	14.31		-0.01		-0.74		1	B0	733
		AzV 277		0 59 54	−72 30.8	1	14.07		-0.15		-0.85		3	B2	425
6192	−57 00220	HR 295		0 59 55	−57 16.3	4	6.10	.004	0.94	.005			18	G8 III	15,1075,2013,2029
		AzV 278	AB	0 59 55	−72 57.9	1	14.23		-0.18		-0.74		3	B0	425
		AzV 279		0 59 56	−72 19.2	1	14.20		-0.25		-1.03		3	O9	425
		BSD 8 # 1018		0 59 57	+60 36.7	1	10.30		0.23		-0.30		3	B6	1185
	−31 00410			0 59 57	−30 52.4	2	9.57	.030	1.30	.010	1.47	.005	2	K3 III	54,3061
6084	+51 00216	IDS00570N5116	AB	0 59 58	+51 31.8	1	6.83		0.02		-0.51		2	B5	401
6073	+60 00153			0 59 58	+61 30.4	1	8.21		1.19		0.89		4	G5 II	1371
		AzV 284		0 59 58	−71 56.2	1	14.22		-0.19		-0.90		3	B1	425
		AzV 281		0 59 58	−72 11.7	1	14.37		-0.07		-1.07		3	B0	425
		G 172 - 34		0 59 59	+46 46.8	1	10.80		1.41		1.33		3	K7	1723
		G 172 - 35		0 59 59	+46 47.0	1	12.76		1.51		1.08		5		1723
		G 1 - 44		1 00 00	+03 34.7	1	13.08		1.51		1.38		1		1620
6156	−22 00183	G 268 - 101	★ AB	1 00 00	−21 52.7	2	7.91	.010	0.80	.000	0.41	.005	2	K0 V	79,3061
		AzV 280		1 00 00	−72 18.6	1	14.66		-0.26		-0.94		1	B0	425
6166	−31 00411			1 00 01	−30 39.6	1	9.15		1.07		0.92		1	G8 IV	54
6178	−32 00410	HR 293		1 00 03	−31 49.2	5	5.50	.012	0.08	.008	0.13		17	A2 V	15,1020,1068,2012,2012
6118	+31 00168	HR 291		1 00 04	+31 32.2	3	5.50	.005	-0.05	.005	-0.18	.017	7	B9.5V	1049,1079,3023
6153	+3 00147			1 00 04	+04 08.7	1	8.62		0.62		0.01		2	G0	3016
6116	+40 00209	HR 290	★ A	1 00 05	+41 04.6	6	5.96	.019	0.16	.009	0.10	.008	22	A5 m	15,1007,1013,3052*
	+60 00154			1 00 05	+60 52.1	1	9.17		0.47				2	F6 V	70
		NGC 362 sq2 11		1 00 05	−71 04.1	1	12.98		1.10		0.92		2		1461

Table 1

73

HD	DM	Other Id	N Rem	α_{1950}	δ_{1950}	S	V	σ_V	B–V	σ_{B-V}	U–B	σ_{U-B}	n	Spectrum	References
		AAS53,255 # 124		1 00 05	−72 54.	1	13.12		1.41				1		1686
		AAS53,255 # 125		1 00 07	−73 48.	1	15.48		1.39				1		1686
	+1 00194	IDS00576N0141	AB	1 00 08	+01 57.9	1	9.52		0.77		0.28		3	G9 V	3030
6114	+46 00243	HR 289	⋆ AB	1 00 08	+47 06.5	2	6.46	.010	0.24	.005	0.07	.002	7	A9 V	254,3030
	+61 00195	G 243 - 55	⋆ B	1 00 08	+62 05.8	2	13.65	.005	1.68	.000	0.79	.000	6		1,3072
		NGC 362 sq2 39		1 00 09	−71 01.3	1	14.07		0.63		0.03		2		1461
		G 33 - 34		1 00 10	+19 56.2	1	14.18		0.90		0.39		2		1658
		AAS35,353 # 82		1 00 10	−71 59.6	1	14.16		-0.10		-0.50		3	B5	767
		AAS53,255 # 126		1 00 10	−72 22.	1	13.12		0.94				1		1686
	+59 00172			1 00 11	+59 56.7	2	9.61	.002	0.49	.005			3	F8 V	70,1103
	−23 00382			1 00 11	−23 19.8	1	10.45		0.81		0.38		1	G5	3061
		AzV 282		1 00 11	−72 29.3	1	14.83		-0.25		-1.13		2	O7 V	425
6189	−25 00410	G 268 - 102		1 00 13	−25 34.6	1	9.85		0.93		0.65		2	K2 V	3061
6197	−35 00350			1 00 13	−34 38.3	1	9.02		0.96		0.51		2	G8 IV	3061
		CS 22953 # 3		1 00 15	−61 59.8	2	13.71	.009	0.69	.016	-0.04	.002	4		1580,1736
		BSD 8 # 1543		1 00 16	+61 33.7	1	12.38		0.59		-0.12		3	B6	1185
6208	−48 00251			1 00 16	−48 28.2	1	9.36		0.07		0.07		4	A0 V	152
		NGC 362 sq2 17		1 00 16	−71 06.6	1	14.44		0.03		0.16		1		1461
		AAS53,255 # 127		1 00 16	−72 08.	1	12.99		1.95				1		1686
		AAS53,255 # 128		1 00 16	−72 16.	1	12.84		1.60				1	K2 I	1686
		AAS35,353 # 83		1 00 16	−72 30.5	1	14.78		-0.14		-0.55		3	B5	767
		AzV 283		1 00 16	−73 01.4	1	13.82		-0.12		-0.70		3	B3	425
6176	+13 00153			1 00 18	+14 22.9	1	9.54		0.39		-0.03		1	F5	1776
		AzV 286		1 00 18	−72 20.4	3	12.32	.029	0.15	.011	0.30	.029	8	A5 Ib	425,573,1277
		AzV 287	AB	1 00 18	−72 28.8	2	12.85	.009	0.00	.000	-0.92	.019	4	B0 Ia	425,835
		AAS53,255 # 129		1 00 18	−72 56.	1	13.76		1.79				1		1686
		BSD 8 # 107		1 00 19	+58 52.7	1	10.09		0.20		0.12		3	B8	1185
		L 87 - 68		1 00 19	−67 54.9	1	14.66		1.36		1.01		3		3079
		AzV 285		1 00 19	−72 48.7	1	13.96		-0.07		-1.02		3	O9	425
6186	+7 00153	HR 294		1 00 21	+07 37.3	8	4.27	.011	0.96	.009	0.69	.009	33	K0 III	3,15,1061,1355,1363*
		CS 22183 # 24		1 00 21	−05 01.0	1	11.93		0.05		0.06		1		1736
		AzV 290		1 00 21	−72 18.5	1	13.93		-0.15		-0.83		3	B1	425
		BSD 8 # 1028		1 00 22	+61 15.5	1	11.00		0.39		-0.27		3	B6	1185
		AzV 289		1 00 22	−72 51.7	2	12.43	.000	-0.14	.010	-0.94	.020	5	B1 Ia0	425,573
		AzV 288		1 00 22	−73 11.8	1	14.60		-0.25		-0.84		3	B0	425
6185	+15 00154			1 00 23	+15 52.1	1	8.56		1.04		0.66		2	K0	1648
6111	+61 00198			1 00 23	+61 40.2	1	8.85		0.55				3	F8 V	1118
	−28 00317			1 00 23	−27 54.3	1	11.58		0.47				1		1495
		AAS53,255 # 130		1 00 23	−72 47.	1	13.48		1.92				1		1686
		BPM 46880		1 00 24	−24 56.	1	11.43		0.60		0.08		1		3061
		AzV 293		1 00 26	−71 46.1	1	14.19		-0.05		-0.94		3	B1	425
		AzV 294		1 00 26	−71 57.3	1	14.64		-0.14		-1.13		4	O9	425
		AzV 291		1 00 26	−72 31.6	2	14.82	.035	-0.14	.000	-0.55	.015	7	B1	733,767
		AzV 292		1 00 26	−72 35.2	2	13.11	.010	-0.08	.010	-0.56	.015	6	B3 Iab	425,573
	+60 00156			1 00 27	+60 56.4	1	9.01		1.96				3	K5 V	369
6129	+62 00191	IDS00573N6309	AB	1 00 28	+63 25.2	1	8.14		0.56				2	G0	1118
		BSD 8 # 1546		1 00 29	+62 02.8	1	10.68		0.42		-0.04		3	B7	1185
		AAS53,255 # 131		1 00 29	−72 25.	1	13.42		1.74				1		1686
		AAS53,255 # 132		1 00 29	−72 42.	1	13.31		1.42				1		1686
6234	−37 00378	G 269 - 81		1 00 30	−37 35.0	2	8.52	.005	0.58	.000	0.01	.010	3	G5 V	79,3061
		BPM 16715		1 00 30	−58 10.	1	14.25		0.32		-0.22		3		3065
		LB 3172		1 00 30	−62 55.	1	12.48		0.04		0.06		2		45
		AzV 296		1 00 30	−72 29.4	1	14.38		-0.23		-1.11		3	O7.5Vn	425
		AzV 295		1 00 30	−73 14.5	1	15.16		-0.18		-0.96		4	B0	425
		BSD 8 # 112		1 00 31	+59 04.4	1	11.54		0.43		0.29		3	B9	1185
6130	+60 00157	HR 292	⋆ AB	1 00 31	+60 48.4	2	5.92	.001	0.49	.002			4	F0 II	1103,1118
6203	−5 00177	HR 296	⋆ A	1 00 31	−05 06.2	4	5.44	.021	1.11	.012	1.04	.011	10	K0 III-IV	15,1003,1256,3016
6218	−28 00318			1 00 31	−27 53.1	1	9.86		0.77				3	K1 (III) +F	1495
		AzV 297		1 00 31	−72 16.5	1	12.18		-0.06		-0.56		4	B7 Ia	425
		G 33 - 35		1 00 31	+19 49.8	1	11.40		1.61				1	M2	1746
		BSD 8 # 497		1 00 32	+59 56.7	1	11.00		0.29		0.14		3	B8 p	1185
		AzV 298		1 00 33	−72 19.0	2	12.51	.009	0.03	.042	-0.12		6	A0 Iab	425,1277
6245	−47 00313	HR 299		1 00 34	−46 40.0	3	5.37	.013	0.90	.000			11	G8 III	15,1075,2035
		BSD 8 # 1547		1 00 35	+62 08.2	1	10.63		0.39		0.00		3	B7	1185
		NGC 362 - 1011		1 00 36	−71 07.	1	16.16		0.73		0.48		1		1461
		NGC 362 - 1012		1 00 36	−71 07.	1	15.49		0.62		0.16		1		1461
		NGC 362 - 1014		1 00 36	−71 07.	1	16.70		0.88		0.27		1		1461
		NGC 362 - 1064		1 00 36	−71 07.	1	16.06		1.39				1		1461
		NGC 362 - 2009		1 00 36	−71 07.	1	15.25		0.21		0.34		2		1461
		NGC 362 - 2063		1 00 36	−71 07.	1	15.14		0.24		0.21		2		1461
		NGC 362 - 2066		1 00 36	−71 07.	1	16.04		0.19		-0.04		2		1461
		NGC 362 - 2068		1 00 36	−71 07.	1	14.76		0.67		0.01		1		1461
		NGC 362 - 2069		1 00 36	−71 07.	1	14.77		0.54		-0.06		1		1461
		NGC 362 - 3006		1 00 36	−71 07.	1	15.04		0.63		-0.07		1		1461
		NGC 362 - 3010		1 00 36	−71 07.	1	15.47		0.60		0.24		1		1461
		NGC 362 - 3014		1 00 36	−71 07.	1	13.48		1.17		0.73		1		1461
		NGC 362 - 3032		1 00 36	−71 07.	1	15.52		0.67		-0.01		1		1461
		NGC 362 - 3070		1 00 36	−71 07.	1	12.77		1.38		1.37		1		1461
		NGC 362 - 4103		1 00 36	−71 07.	1	14.56		1.01		0.61		1		1461
		NGC 362 - 4104		1 00 36	−71 07.	1	13.64		1.14		0.83		1		1461

HD	DM	Other Id	N	Rem	α_{1950}	δ_{1950}	S	V	σ_V	B–V	σ_{B-V}	U–B	σ_{U-B}	n	Spectrum	References
		NGC 362 - 1002			1 00 36	−71 07.6	1	14.22		1.06		0.70		1		1461
		AzV 301			1 00 36	−72 15.9	1	14.24		-0.15		-0.91		3	B0	425
		AzV 300			1 00 36	−72 27.4	1	14.46		-0.19		-1.05		3	B0	425
		AzV 299			1 00 36	−72 38.3	1	14.61		-0.05		-1.06		3	O9	425
		CS 22942 # 37			1 00 38	−23 34.6	2	14.15	.000	0.10	.000	0.22	.010	2		966,1736
		NGC 362 - 1001			1 00 39	−71 07.5	1	15.11		0.73		0.15		1		1461
		NGC 362 sq2 14			1 00 41	−71 05.8	1	13.42		1.14				8		1677
		AAS53,255 # 133			1 00 41	−72 18.	1	13.29		1.63				1		1686
		AzV 302			1 00 42	−72 38.2	1	14.35		-0.25		-0.87		2	B0	425
	−28 00320				1 00 43	−27 47.3	1	11.52		0.52				2		1495
		NGC 362 sq2 33			1 00 43	−70 58.9	1	12.96		0.74		0.23		2		1461
		NGC 362 sq2 37			1 00 43	−71 01.4	1	13.75		1.18		1.01		2		1461
		AzV 303			1 00 43	−72 16.4	1	12.81		-0.13		-0.83		3	B1 Iab	733
		NGC 362 - 1004			1 00 44	−71 08.2	1	15.00		0.82		0.34		2		1461
		AzV 304			1 00 45	−72 55.3	1	14.98		-0.20		-0.98		6	B0.5V	425
		NGC 362 - 1013			1 00 46	−71 10.3	1	15.79		0.67				2		1461
6006	+81 00030				1 00 48	+81 41.6	1	7.89		0.02		-0.08		3	A0	1219
6255	−27 00341				1 00 48	−27 22.6	1	8.94		1.61		1.93		1	K5/M0 III	54
		PHL 7081			1 00 48	−30 00.	1	11.24		-0.02		-0.23		6	A0	3064
6311	−66 00080	HR 304, CC Tuc			1 00 48	−65 43.5	3	6.25	.019	1.63	.004	1.90	.000	24	M2 III	15,1075,2038
		Florsch 437			1 00 48	−71 31.0	1	12.45		0.53		-0.03		3	F6	767
		AAS53,255 # 134			1 00 48	−73 11.	1	12.85		1.85				1		1686
6201	+43 00206				1 00 50	+43 47.9	1	8.76		-0.15		-0.96		2	F5	401
	−30 00324				1 00 50	−29 59.5	1	11.19		-0.01		-0.33		1	A0	378
		NGC 362 - 1006			1 00 50	−71 07.3	1	15.50		0.17		0.18		3		1461
		AzV 305			1 00 50	−72 32.9	1	12.86		0.28		0.35		4	F5	425
		BSD 8 # 501			1 00 51	+60 15.2	1	12.65		0.41		0.09		3	B8	1185
	+61 00199				1 00 51	+62 02.0	1	9.51		0.32		0.18		3	A0 II	1185
6254	−26 00343				1 00 51	−26 26.9	2	8.01	.005	0.99	.000	0.69		6	G8 IV	2012,3077
		NGC 362 - 1007			1 00 52	−71 08.1	1	14.53		0.84		0.45		2		1461
		NGC 362 - 1010			1 00 53	−71 08.9	1	15.57		0.68		0.20		1		1461
		NGC 362 - 1017			1 00 53	−71 09.5	1	14.98		1.08		0.88		2		1461
6268	−28 00322				1 00 54	−28 08.9	5	8.11	.041	0.85	.007	0.26	.010	22	G0	54,955,2013,2033,3077
		SB 418			1 00 54	−29 28.	1	13.23		0.36		-0.03		2	A5	561
6269	−30 00325	HR 300			1 00 54	−29 47.6	5	6.28	.010	0.94	.009	0.58	.010	11	G5 IV	15,54,79,2012,3061
		Florsch 444			1 00 54	−71 32.5	1	13.41		0.07		0.16		3	A2	767
		AzV 306			1 00 54	−71 55.8	1	15.29		-0.20		-0.85		4	B0	425
6229	+22 00170				1 00 55	+23 30.0	1	8.60		0.71		0.16		1	G5 III	3077
		NGC 362 sq2 32			1 00 55	−70 57.2	1	12.29		0.69		0.15		3		1461
		NGC 362 - 4084			1 00 55	−71 06.7	1	12.94		1.50		1.60		2		3036
		AzV 307			1 00 55	−72 55.8	1	14.13		-0.25		-0.95		4	B0	425
		G 70 - 31			1 00 56	+04 48.7	1	11.82		0.74		0.30		1		1620
6182	+61 00200	LS I +61 180			1 00 57	+61 32.6	2	8.25	.015	0.48	.010	-0.47	.010	5	B1 Ibp	1012,1371
		NGC 362 - 1019			1 00 57	−71 09.5	1	15.58		1.00		0.67		2		1461
		BSD 8 # 1043			1 00 58	+60 43.2	1	10.77		0.49		0.15		3	B7	1185
		AAS53,255 # 135			1 00 59	−72 32.	1	12.71		1.75				1	K5-M0 I	1686
6226	+46 00245				1 01 00	+47 22.5	2	6.81	.005	-0.04	.005	-0.59	.005	5	B2 IV-V	399,401
6290	−41 00260				1 01 00	−41 17.4	2	7.00	.004	1.60	.006	1.95	.000	12	M1 III	1673,3007
		BSD 8 # 117			1 01 01	+58 52.8	1	12.19		0.40		0.30		3	B9	1185
		NGC 362 - 4100			1 01 01	−71 04.9	1	12.63		1.66		1.77		2		3036
		BSD 8 # 503			1 01 02	+59 37.8	1	11.19		0.35		0.09		3	B8	1185
	−26 00345	G 268 - 103			1 01 02	−26 05.5	1	11.16		0.92		0.63		1		3061
		NGC 362 - 4018			1 01 02	−71 03.8	1	13.79		1.17		1.11		1		1461
		NGC 362 - 1023			1 01 03	−71 08.3	1	13.25		1.17		1.25		1		3036
		AzV 309			1 01 03	−72 41.3	1	13.37		-0.15		-0.88		3	B0	425
		AAS35,353 # 86			1 01 03	−72 53.4	1	13.63		-0.02		-0.22		6	B9	767
		BSD 8 # 1044			1 01 04	+60 47.0	1	12.66		0.66				3	B0	1185
		AAS35,353 # 87			1 01 04	−72 35.2	1	13.75		-0.09		-0.13		2	B7	767
		AzV 308			1 01 04	−73 07.0	1	14.10		-0.03		-0.95		4	O9	425
6211	+51 00220	HR 298			1 01 05	+52 14.1	1	5.99		1.47				2	K2	71
		BSD 8 # 1554			1 01 05	+61 28.0	1	10.53		0.56		0.10		3	B8	1185
		AAS53,255 # 136			1 01 05	−72 18.	1	13.01		1.59				1		1686
		AzV 310			1 01 05	−72 23.5	2	11.63	.015	0.29	.005	0.29	.050	6	F0 Ia	35,573
		AAS53,255 # 137			1 01 05	−72 27.	1	13.31		1.32				1		1686
		Dril 47			1 01 06	−28 30.	1	11.87		0.95		0.76		2	G8 III	561
	−29 00311				1 01 06	−28 55.0	1	10.79		0.67		0.10		1		1099
		AzV 311			1 01 06	−72 12.3	1	14.55		-0.12		-0.97		3	B1	425
		BSD 8 # 118			1 01 07	+58 50.2	1	10.84		0.55		0.36		3	B6	1185
		NGC 362 - 4091			1 01 08	−71 06.7	1	13.36		1.08		0.96		3		1461
		AzV 313			1 01 08	−72 16.6	1	14.48		-0.24		-0.91		3	B1	425
		AzV 312			1 01 08	−72 28.2	1	13.62		-0.28		-0.92		4	B0	425
		BSD 8 # 508			1 01 09	+60 02.6	1	12.76		0.36		0.04		3	B6	1185
		LS I +63 130			1 01 09	+63 30.3	1	10.04		0.65		-0.27		2	B1 IV	1012
		NGC 362 - 4098			1 01 09	−71 06.5	1	15.73		0.74		-0.13		1		1461
		BSD 8 # 1045			1 01 10	+60 46.4	1	12.27		0.67				3	A0	1185
		AzV 314			1 01 11	−72 32.8	1	12.90		-0.12		-0.64		6	B2 Iab	425
		AAS35,353 # 88			1 01 12	−72 30.3	1	14.11		-0.12		-0.66		3	B5	767
	+36 00187				1 01 13	+36 33.9	1	7.92		1.12		1.01		2	G5	1601
6210	+60 00158	HR 297			1 01 13	+61 18.6	3	5.84	.003	0.55	.009	0.11		5	F7 IV	351,1103,1118
		AzV 317			1 01 13	−72 04.4	2	12.89	.005	-0.20	.000	-0.99	.005	6	B0 Ia	733,835

Table 1   75

HD	DM	Other Id	N	Rem	α_{1950}	δ_{1950}	S	V	σ_V	B–V	σ_{B-V}	U–B	σ_{U-B}	n	Spectrum	References
	−72 00077				1 01 13	−72 26.5	4	10.90	.011	0.08	.014	-0.46	.005	14	B8 Ia	35,573,1277,8033
		AzV 316			1 01 13	−73 03.0	1	14.09		-0.20		-0.83		4	B1 V	425
6288	+0 00174	HR 301		⋆A	1 01 14	+01 06.0	1	6.11		0.25		0.06		2	F1 V	3024
6288	+0 00174	HR 301		⋆AB	1 01 14	+01 06.0	1	6.03	.005	0.28	.005	0.11	.015	7	F1 V	15,1256
6288	+0 00174	IDS00587N0050		B	1 01 14	+01 06.0	1	9.51		0.82		0.44		2		3024
		G 1 - 45			1 01 14	+04 48.3	3	14.03	.070	0.23	.085	-0.54	.035	5	DA	419,1620,3028
		CS 22166 # 60			1 01 14	−15 32.9	1	14.23		0.42		-0.17		1		1736
		POSS 295 # 34			1 01 15	+30 31.3	1	17.31		1.56				2		1739
		NGC 362 - 4027			1 01 15	−71 02.4	1	16.03		0.83		0.39		1		1461
6262	+37 00199				1 01 16	+38 25.2	1	7.08		1.69		1.86		2	M3 III	1601
6334	−60 00072	IDS00592S6038		AB	1 01 16	−60 21.9	2	6.81	.005	0.48	.005	0.03		6	F6/8 III/IV	404,2012
		NGC 362 - 1044			1 01 17	−71 05.5	1	13.32		1.28		1.36		2		1461
		NGC 362 - 4049			1 01 17	−71 08.0	1	13.67		1.05		0.99		1		1461
		NGC 362 - 4047			1 01 17	−71 12.0	1	12.78		1.45		1.47		1		1461
		AzV 318			1 01 17	−72 25.9	1	13.59		-0.21		-0.86		4	B0 Ib	425
		AAS53,255 # 138			1 01 17	−72 40.	1	12.90		1.79				1		1686
		POSS 295 # 32			1 01 18	+30 04.5	1	17.21		1.57				1		1739
		TonS 186			1 01 18	−25 35.	1	13.91	.005	0.25	.015	0.07	.025	4	A5	561,3064
		PHL 966			1 01 18	−27 08.	1	15.32		-0.32		-1.16		1		3064
		NGC 362 - 1066			1 01 18	−71 05.2	2	13.27	.042	0.93	.005	0.56	.052	4		1461,3036
		NGC 362 - 4011			1 01 18	−71 07.6	1	13.75		1.14		0.90		1		1461
		AzV 319			1 01 18	−72 48.7	1	13.22		0.06		-0.40		3	A0	425
		AzV 320		AB	1 01 19	−72 17.1	1	12.94		-0.17		-0.78		1	B2 Ia	35
6274	+25 00160				1 01 20	+26 17.1	1	8.92		0.54		0.02		3	F7 V	3026
		G 268 - 104			1 01 20	−17 16.0	1	13.90		1.58		1.16		2		481
		AzV 321			1 01 20	−72 24.1	1	13.88		-0.21		-1.05		5	O9	425
		G 70 - 33			1 01 21	−04 07.3	4	12.34	.024	0.56	.012	-0.16	.039	6		202,1620,1696,3056
		MN201,73 # 58			1 01 21	−35 48.3	1	12.54		0.63				1		1495
		POSS 295 # 30			1 01 22	+29 39.3	1	16.94		0.89				1		1739
6249	+57 00188				1 01 22	+57 42.8	1	7.56		0.12		0.02		2	B9	401
		NGC 362 sq2 1			1 01 22	−70 52.0	1	10.43		0.90		0.63		1		1461
		NGC 362 sq2 38			1 01 22	−71 12.2	1	13.81		1.12		0.84		2		1461
	−24 00469				1 01 23	−24 17.1	2	12.77	.000	0.09	.000	0.15	.009	4	A0	966,1736
		AzV 322			1 01 23	−72 41.6	2	13.85	.037	-0.10	.005	-0.96	.009	4	B0	35,425
		AAS35,353 # 89			1 01 23	−72 56.0	1	14.14		-0.11		-0.62		4	B5	767
6286	+25 00161	BE Psc		⋆A	1 01 24	+26 19.1	1	8.26		0.96		0.46		2	G2 V	3016
		BPM 46893			1 01 24	−26 26.	1	11.76		0.63		0.08		3		3061
		MN201,73 # 59			1 01 24	−35 28.0	1	12.85		0.54				1		1495
	−46 00293				1 01 24	−46 02.3	1	11.63		1.11		0.86		1		3078
6406	−72 00078				1 01 24	−72 21.4	1	9.63		0.46		-0.01		6	F3/5 IV/V	35
		AzV 323			1 01 24	−72 46.2	1	13.24		0.26		0.29		3	F2	425
6446	−78 00028				1 01 24	−77 49.1	2	7.20	.025	1.39	.010	1.37		8	K2 IIp	2012,3077
6259	+51 00221				1 01 26	+52 11.7	1	8.51		0.54		0.02		2	G0	1566
		NGC 362 sq2 36			1 01 26	−71 01.5	1	13.73		1.17		0.81		2		1461
		AzV 324		AB	1 01 26	−72 27.3	1	12.93		-0.08		-0.58		3	B5 Iab	35
		AzV 325			1 01 27	−72 12.9	1	14.70		-0.22		-0.88		3	B1	425
		AAS53,255 # 139			1 01 29	−72 12.	1	13.11		1.73				1		1686
		Dril 15			1 01 30	−27 08.	1	13.53		0.69		0.21		1	G3	561
		MN201,73 # 60			1 01 30	−31 58.5	1	13.18		0.61				1		1495
		LB 3175			1 01 30	−63 32.	1	12.89		0.03		0.03		2		45
		AAS53,255 # 140			1 01 30	−72 45.	1	12.52		1.84				1	M6 III:	1686
		AAS53,255 # 141			1 01 30	−72 50.	1	12.88		1.75				1		1686
5848	+85 00019	HR 285			1 01 31	+85 59.4	6	4.23	.026	1.21	.012	1.31	.017	13	K2 II-III	15,1025,1363,1462*
		NGC 362 - 5243			1 01 31	−71 09.2	1	11.88		0.88		0.51		2		1461
		AAS35,353 # 90			1 01 31	−72 28.6	1	13.04		0.09		0.15		3	A3	767
		CS 22942 # 35			1 01 32	−24 08.8	2	14.55	.009	0.63	.014	-0.07	.000	4		1580,1736
		AzV 327		AB	1 01 32	−72 18.2	2	13.09	.200	-0.20	.019	-1.03	.029	6	B0 I	35,767
		Steph 112			1 01 33	+10 35.4	1	12.03		1.49		1.17		1	M0	1746
		AzV 326			1 01 33	−72 41.9	1	13.93		-0.16		-0.84		3	B0	425
	+59 00177				1 01 34	+60 14.3	1	9.97		0.23		-0.32		3	B2 III	1185
		BSD 8 # 1050			1 01 34	+60 50.5	1	10.63		0.35		0.06		3	B7	1185
6322	−19 00168				1 01 34	−18 50.0	3	9.26	.026	-0.05	.005	-0.10	.012	9	B9 V	1068,2012,3064
	−28 00329				1 01 35	−28 13.3	2	10.00	.020	0.98	.000	0.79		2	K0 III	54,1495
		AAS53,255 # 142			1 01 35	−72 25.	1	13.10		1.86				1		1686
		AAS53,255 # 143			1 01 35	−72 35.	1	13.00		1.60				1	M6 III:	1686
		SB 423			1 01 36	−33 55.	1	13.83		0.06		0.10		1	A0	295
		R 29			1 01 36	−72 23.	1	12.74		1.18				3		1277
		NGC 362 - 2049			1 01 37	−71 10.0	1	14.01		1.08		0.85		2		1461
6431	−73 00060				1 01 37	−72 38.1	1	9.74		1.03		0.78		4	K0	35
		NGC 362 - 2045			1 01 38	−71 11.5	1	15.45		0.56		-0.09		1		1461
	−15 00192	G 268 - 107			1 01 39	−14 51.5	1	10.59		0.77		0.32		2		1696
6340	−35 00361				1 01 39	−34 56.7	2	9.00	.018	0.10	.014	0.03		7	A2 V	1068,2033
	−35 00362	BPM 46901			1 01 39	−35 26.4	3	10.58	.004	0.78	.012	0.33	.008	8		79,1696,3061
		NGC 362 - 3063			1 01 40	−71 04.8	2	12.75	.034	1.61	.010	1.76	.029	3		1461,3036
		NGC 362 sq1 2			1 01 41	−71 18.9	2	11.12	.000	1.08	.005	0.95	.010	8		489,1677
		AAS53,255 # 144			1 01 41	−72 23.	1	12.53		1.10				1		1686
6339	−30 00329				1 01 42	−29 38.6	1	9.85		0.49		0.03		1	F5 V	1099
	−35 00360	G 269 - 87			1 01 42	−34 56.4	5	10.25	.006	0.77	.011	0.18	.013	16	G5 IV	79,1311,1696,2018,3061
		NGC 362 sq2 30			1 01 42	−70 50.8	1	10.77		0.52		0.08		1		1461
		NGC 362 - 6431			1 01 42	−71 03.9	1	12.73		1.40		1.12		2		1461

HD	DM	Other Id	N	Rem	α_{1950}	δ_{1950}	S	V	σ_V	B–V	σ_{B-V}	U–B	σ_{U-B}	n	Spectrum	References
		AAS53,255 # 145			1 01 42	−72 56.	1	13.09		1.59				1		1686
6301	+28 00174	HR 303			1 01 43	+29 23.5	3	6.20	.014	0.42	.005	−0.05	.024	9	F7 IV-V	71,254,272
		NGC 362 - 2047			1 01 43	−71 10.0	1	14.17		0.97		0.71		2		1461
		AzV 328			1 01 43	−72 42.4	1	13.73		−0.22		−0.94		3	B1 Iab	425
	+60 00159				1 01 45	+60 55.2	1	10.13		0.34		−0.05		3	B6 III	1185
		Feige 11			1 01 46	+03 57.6	4	12.06	.006	−0.24	.006	−0.98	.007	100	sdB	281,989,1026,1729,1764
		RX And			1 01 46	+41 01.8	1	10.77		0.00		−0.69		1		698
		AzV 329			1 01 46	−72 18.6	1	13.79		−0.17		−1.00		4	B0	425
6314	+39 00249	HR 305			1 01 47	+39 43.4	1	6.74		0.30		0.02		2	F0 Vn	254
		AzV 330			1 01 47	−72 12.6	1	13.65		−0.05		−1.10		3	O9	425
		POSS 295 # 12			1 01 48	+33 04.5	1	14.31		1.55				2		1739
		BPM 46898			1 01 48	−26 46.	1	14.42		0.95		0.52		1		3061
		Dril 16			1 01 48	−27 00.	1	11.78		0.52		−0.02		1	G0 V	561
		BPM 46899			1 01 48	−28 09.	1	11.85		1.28		1.20		1		3061
		Florsch 461			1 01 48	−70 53.1	1	12.57		0.62		−0.01		2		767
		AzV 331			1 01 48	−72 18.5	2	13.17	.015	0.10	.011	0.12		5	A5	733,1277
		AzV 332			1 01 48	−72 22.7	3	12.38	.024	−0.25	.032	−1.03	.000	9	WN3 + O7 Ia	425,573,1277
		G 69 - 45			1 01 49	+25 51.0	1	10.03		1.27		1.17		3	K5	333,1620
		NGC 362 - 3037			1 01 49	−71 06.3	1	13.19		1.22		0.98		1		1461
		NGC 362 - 2041			1 01 49	−71 10.0	1	16.33		1.60				1		1461
		NGC 362 - 2043			1 01 49	−71 11.5	1	13.73		1.05		0.68		2		1461
		AzV 333			1 01 50	−73 09.8	1	13.69		−0.06		−0.46		4	B9	425
6300	+50 00212	HR 302			1 01 51	+50 44.5	3	6.53	.013	−0.08	.010	−0.58	.001	7	B3 V	154,1119,1223
		AzV 335			1 01 52	−72 45.4	1	14.89		−0.17		−0.87		3	B1	425
		AzV 334			1 01 52	−73 13.0	1	13.81		−0.21		−0.91		6	B0	425
		NGC 362 sq2 9			1 01 53	−70 53.0	1	12.47		0.59		−0.01		1		1461
		NGC 362 - 3044			1 01 53	−71 04.2	1	12.85		1.49		1.64		3		3036
		AAS53,255 # 146			1 01 53	−72 19.	1	12.41		1.61				1		1686
		AAS35,353 # 91			1 01 53	−72 52.7	1	13.77		−0.04		−0.50		2	B5	767
6348	−3 00146	G 70 - 35			1 01 54	−02 38.0	6	9.16	.023	0.80	.000	0.36	.005	11	G5	202,516,742,1064,1620,1658
6364	−27 00345				1 01 54	−27 25.4	3	9.63	.012	0.29	.004	0.04	.005	6	A5/7 III	561,1068,1495
6365	−30 00330				1 01 54	−30 17.4	2	9.84	.052	0.24	.009	0.02	.014	4	A3 III/IV	1068,1099
		NGC 362 - 3011			1 01 54	−71 07.0	1	12.63		1.57		1.48		2		1461
		AAS53,255 # 147			1 01 54	−72 41.	1	13.81		1.32				1		1686
		AzV 336			1 01 54	−72 47.6	1	15.00		−0.18		−0.91		3	B0	425
		AAS53,255 # 148			1 01 55	−73 08.	1	11.90		1.99				1		1686
		NGC 362 - 2020			1 01 57	−71 08.2	1	12.88		1.48		1.60		2		3036
		AAS35,353 # 93			1 01 58	−72 55.6	1	13.59		−0.06		−0.56		3	B7	767
6378	−26 00348	G 268 - 109			1 01 59	−25 52.1	4	9.77	.029	1.19	.016	1.17	.030	12	K4/5 V	158,1495,2033,3061
		NGC 362 - 3022			1 01 59	−71 05.3	1	15.80		0.99		0.20		1		1461
		AAS53,255 # 149			1 01 59	−72 15.	1	13.21		1.36				1		1686
		AAS35,353 # 94			1 01 59	−72 33.0	1	13.33		0.18		0.11		2	A5	767
		NGC 362 - 3039			1 01 62	−71 05.7	2	12.64	.047	1.58	.005	1.79	.122	4		1461,3036
	+62 00193				1 02 01	+63 17.9	1	10.03		0.91				4		70
		NGC 362 - 2071			1 02 02	−71 11.0	1	15.49		0.79		0.64		1		1461
		NGC 362 - 3001			1 02 05	−71 06.9	1	14.42		0.84		0.13		1		1461
		AzV 337			1 02 06	−72 19.9	1	12.78		−0.11		−0.76		4	B1 Ia	733
		AzV 338			1 02 06	−72 31.5	2	12.61	.004	0.00	.009	−0.21	.004	6	A0 Iab	35,835
		MN201,73 # 62			1 02 07	−30 08.3	1	13.73		1.50				1		1495
		NGC 362 - 3008			1 02 09	−71 05.5	1	15.90		0.60		−0.05		1		1461
		AAS35,353 # 95			1 02 09	−72 48.4	1	13.01		0.04		0.05		3	A0	767
6413	−48 00259				1 02 10	−48 12.5	3	6.83	.021	1.60	.009	1.94		12	M1 III	2012,2013,3040
		NGC 362 - 3007			1 02 10	−71 05.2	1	15.17		0.93		0.39		1		1461
		AAS35,353 # 96			1 02 10	−72 37.7	1	13.87		0.00		−0.46		4	B9	767
		CS 22942 # 31			1 02 11	−26 47.8	2	14.98	.000	0.08	.000	0.15	.010	2		966,1736
6403	−34 00411	IDS00598S3404		AB	1 02 11	−33 48.0	1	6.43		1.09				4	K0 III	2033
		NGC 362 - 2078			1 02 11	−71 09.0	1	14.72		1.10		0.28		2		1461
		AAS53,255 # 150			1 02 11	−72 18.	1	13.10		1.59				1	M7:V:	1686
		AzV 339a			1 02 11	−72 20.3	1	12.80		0.21		0.29		2	F5	767
		AzV 339			1 02 11	−73 08.4	2	12.65	.005	−0.10	.010	−0.73	.005	5	B3 Iab	425,835
		CS 29514 # 3			1 02 12	−26 47.9	1	15.01		0.09		0.14				1736
		Dril 18			1 02 12	−27 16.	1	12.21		0.76		0.40		1	G9 IV	561
		BPM 46903			1 02 12	−28 04.	1	14.06		1.41		1.01		1		3061
		AzV 340			1 02 12	−72 23.5	1	12.71		−0.06		−0.86		4	B2 Ia	733
6312	+62 00194	G 243 - 58		★ AB	1 02 14	+63 27.4	1	8.29		0.56				2	G0	1118
6402	−31 00426				1 02 14	−30 45.4	1	8.24		0.52				4	F6 V	2012
		AzV 341			1 02 14	−72 44.1	1	15.02		−0.19		−0.98		3	B1	425
6320	+63 00134				1 02 15	+63 56.7	1	8.68		0.58				3	G5	1118
6386	+4 00172	HR 307			1 02 17	+05 23.3	3	6.01	.007	1.53	.008	1.82	.014	23	gK5	15,1256,8040
		BPM 46905			1 02 18	−33 32.	1	11.81		0.76		0.42		1		3061
		Smethells 182			1 02 18	−57 55.	1	11.45		1.31				1		1494
		AzV 342		AB	1 02 19	−72 18.8	1	12.59		−0.03		−0.62		3	B1 Iab	35
	−31 00427	BPM 46904			1 02 20	−30 58.6	1	10.94		0.68		0.16		2		3061
		BSD 8 # 537			1 02 21	+59 16.3	1	11.58		0.26		−0.44		3	B5	1185
		NGC 362 - 3004			1 02 21	−71 05.3	2	13.67	.000	1.22	.005	1.02	.015	6		1461,1677
		NGC 362 - 2074			1 02 21	−71 08.5	1	14.62		0.39		0.52		1		1461
		AAS35,353 # 98			1 02 21	−72 10.3	2	12.71	.015	0.42	.015	−0.05	.005	5	F2	35,767
		AzV 343		AB	1 02 21	−72 58.1	2	13.05	.000	−0.07	.028	−0.49	.005	4	B4 Iab	425,835
6434	−40 00239				1 02 22	−39 44.9	2	7.72	.005	0.60	.005	−0.01		5	G2/3 V	742,2012
		AAS35,353 # 97			1 02 22	−72 40.9	1	13.35		0.69		0.16		2	G9	767

Table 1   77

HD	DM	Other Id	N	Rem	α_{1950}	δ_{1950}	S	V	σ_V	B–V	σ_{B-V}	U–B	σ_{U-B}	n	Spectrum	References
		G 268 - 110			1 02 23	−18 23.8	1	14.47		1.87				5		940
		AAS53,255 # 151			1 02 23	−72 21.	1	13.07		1.54				1	M6 III	1686
		AzV 344			1 02 24	−72 12.6	1	14.72		−0.15		−1.07		3	B0	425
		NGC 362 - 3005			1 02 25	−71 04.9	1	15.24		0.98		0.46		1		1461
		AA9,95 # 224			1 02 25	−72 24.5	1	11.26		1.54		1.84		4	K5 III-I	35
		AzV 345	A		1 02 26	−72 49.2	1	13.56		−0.12		−0.88		1	B2	35
		AzV 345	AB		1 02 26	−72 49.2	1	13.23		−0.04		−0.90		1	B2	835
6397	+14 00163	HR 308		⋆ A	1 02 27	+14 40.7	1	5.70		0.40		−0.08		5	F4 II-III	254
		AzV 345a			1 02 28	−72 20.8	1	13.66		−0.23		−1.07		2	O9	767
	−29 00319				1 02 29	−29 25.9	1	12.43		0.59		0.00		1		1099
		AAS53,255 # 152			1 02 29	−72 20.	1	13.35		1.16				1	K5-M0 I	1686
		BPM 46907			1 02 30	−33 37.	1	10.91		0.58		−0.05		1		3061
		NGC 362 - 2080			1 02 30	−71 08.3	1	15.53		0.88		0.08		1		1461
		AJ63,118 # 19			1 02 30	−72 29.0	2	10.71	.015	0.48	.001	-0.03		23		35,2001
		NGC 362 sq2 16			1 02 32	−71 11.7	1	14.18		0.66		−0.05		1		1461
		LP 526 - 61			1 02 33	+09 01.6	1	12.82		1.38		1.25		1		1696
		NGC 362 sq2 10			1 02 33	−72 02.3	1	12.58		1.08		0.95		3		1461
		AzV 345b			1 02 33	−72 22.1	1	14.30		−0.21		−0.94		3	B2	767
		AzV 346			1 02 34	−72 37.0	1	13.72		−0.17		−0.71		3	B3	425
6426	−14 00203				1 02 35	−14 01.6	1	7.29		1.08		0.96		3	K0 III	1657
		AzV 347			1 02 35	−72 08.0	2	12.10	.024	−0.01	.005	-0.45	.010	6	A0 Ia	425,573
		AAS53,255 # 153			1 02 35	−72 15.	1	13.49		1.60				1		1686
	+9 00124				1 02 36	+10 00.4	1	10.61		−0.01		−0.01		3	A0	1026
		Dril 19			1 02 36	−25 58.	1	14.63		0.22		0.01		1	A2	561
		Dril 20			1 02 36	−28 57.	1	13.42		1.31		1.30		2	G8	561
		Florsch 473			1 02 36	−74 05.1	1	13.04		0.32		0.22		3	F5	767
		AAS53,255 # 154			1 02 37	−73 06.	1	13.15		1.85				2	M0 Ia-Iab	1686
	+59 00178				1 02 38	+60 11.4	1	10.33		0.33		0.26		3	A0 V	1185
6343	+65 00129				1 02 38	+65 42.2	2	7.26	.010	0.16	.000	-0.24	.000	5	B8	379,1212
		AzV 348			1 02 40	−72 57.5	1	14.35		−0.19		−0.85		4	B1	425
		G 172 - 38			1 02 41	+49 11.3	1	11.08		0.76		0.30		1		1658
		VW Cas			1 02 41	+61 29.2	3	10.45	.028	1.08	.033	0.69		3	F9	851,1399,1772
		AAS53,255 # 155			1 02 41	−72 14.	1	13.33		1.75				1		1686
		AAS53,255 # 156			1 02 41	−72 18.	1	13.73		1.43				1		1686
6382	+60 00162				1 02 42	+60 30.3	4	8.29	.015	0.15	.017	0.12		8	A3 V	1103,1118,1185,1379
6537	−73 00061				1 02 42	−72 40.2	1	10.49		0.44		0.02		9	F2	35
		AAS35,353 # 100			1 02 43	−72 58.8	1	13.79		−0.08		−0.66		4	B5	767
		AzV 349			1 02 44	−72 12.7	1	14.53		−0.24		−1.02		3	O9	425
		NGC 362 sq2 31			1 02 45	−71 06.2	1	11.48		0.59		0.09		2		1461
		AzV 351			1 02 45	−72 25.2	1	13.58		−0.15		−0.95		2	B0	35
6536	−72 00079				1 02 45	−72 31.2	7	8.42	.015	1.00	.007	0.79	.008	51	G8/K0 III	35,1408,1425,2001*
		AzV 350			1 02 45	−72 56.8	1	14.06		−0.14		−0.90		3	B0	425
		BSD 8 # 547			1 02 46	+60 13.6	1	11.18		0.26		0.18		3	B8	1185
6451	−20 00191	IDS01003S2023	AB		1 02 46	−20 07.3	3	8.57	.013	0.24	.005	0.15	.004	12	A2/3mA8-A9	355,1068,2012
		NGC 362 sq1 1			1 02 46	−71 14.3	3	10.90	.004	0.85	.004	0.52	.020	11		489,1461,1677
		AJ63,118 # 26			1 02 47	−72 29.4	2	11.13	.036	1.15	.020	1.15		5		35,2001
		G 69 - 47			1 02 48	+28 13.6	3	14.81	.007	1.87	.033	1.51	.042	5		203,906,3078
	+59 00179				1 02 48	+60 23.7	2	9.99	.009	0.19	.009	0.06		4	A1 V	1103,1185
	−5 00184	G 70 - 36			1 02 48	−05 27.6	2	11.19	.005	1.10	.010	1.05		3	K4.5	202,1620
		AAS53,255 # 157			1 02 49	−73 01.	1	12.30		1.58				1		1686
		MN201,73 # 64			1 02 50	−28 58.2	1	14.86		1.64				1		1495
6440	+14 00167	IDS01002N1452	A		1 02 51	+15 07.5	1	9.17		1.07		0.86		1	K2 V	680
6440	+14 00167	IDS01002N1452	AB		1 02 51	+15 07.5	3	8.64	.035	1.18	.058	1.01	.009	5	K2 V	196,680,3072
6440	+14 00167	IDS01002N1452	B		1 02 51	+15 07.5	1	9.92		1.38		1.27		1	G5	680
		BSD 8 # 550			1 02 51	+59 21.9	1	11.51		0.34		0.18		3	B9	1185
		NGC 362 sq1 3			1 02 51	−71 12.4	3	11.58	.005	0.61	.004	0.07	.010	11		489,1461,1677
		AAS35,353 # 101			1 02 52	−72 48.8	1	12.81		0.07		0.16		3	A2	767
		CS 22166 # 26			1 02 53	−14 15.3	1	16.04		−0.08		−0.85		1		1736
		POSS 295 # 23			1 02 54	+29 36.5	1	16.23		1.53				1		1739
		AAS53,255 # 158			1 02 54	−72 21.	1	13.18		1.99				1		1686
		AzV 352			1 02 55	−72 37.7	1	13.11		0.04		−0.07		3	A0 Ib	35
		AzV 353			1 02 55	−72 43.7	1	13.99		−0.22		−0.84		3	B0	425
		AAS53,255 # 159			1 02 55	−73 08.	1	13.77		1.65				1		1686
6461	−13 00196				1 02 56	−13 10.3	2	7.65	.015	0.78	.010	0.33		3	G3 V	742,1594
		NGC 362 sq2 34			1 02 57	−71 00.1	1	12.98		0.58		0.01		2		1461
		CS 22166 # 30			1 02 58	−12 13.6	1	13.58		0.42		−0.20		1		1736
		AzV 356			1 02 58	−72 20.7	1	13.56		−0.10		−0.94		4	B1 Iab	425
		AzV 354			1 02 58	−72 42.2	1	14.74		−0.19		−0.86		4	B1	425
		AzV 355			1 02 58	−73 04.0	1	14.33		−0.25		−0.85		4	B0	425
6456	+20 00156	HR 310		⋆ A	1 03 00	+21 12.4	2	5.34		−0.02	.015	−0.06	.005	6	A1 Vn	1049,1084
		Dril 21			1 03 00	−26 58.	1	13.56		0.61		0.01		1	G5	561
		AAS53,255 # 160			1 03 00	−72 17.	1	12.33		1.97				1	K5-M0 I	1686
		AzV 357			1 03 00	−72 29.9	1	13.01		0.00		−0.12		5	A0 Ib	35
		AAS35,353 # 102			1 03 00	−73 16.0	1	14.51		−0.08		−0.57		3	B5	767
6457	+20 00157	HR 311		⋆ B	1 03 01	+21 11.9	1	5.56		−0.06		−0.17		3	A0 Vn	1084
6417	+56 00191				1 03 01	+57 29.3	2	7.12	.010	0.03	.015	-0.56	.005	5	B3 III	399,401
6408	+59 00180				1 03 01	+60 02.6	1	8.93		0.26		0.19		5	A2 V	1185
6492	−39 00284				1 03 01	−39 05.1	2	9.17	.023	0.28	.005	0.10	.009	7	A9 V	152,1068
		AAS53,255 # 161			1 03 01	−72 55.	1	13.70		1.65				1		1686
		AzV 358			1 03 02	−72 16.2	1	14.23		−0.04		−1.03		3	B0	425

HD	DM	Other Id	N	Rem	α_{1950}	δ_{1950}	S	V	σ_V	B–V	σ_{B-V}	U–B	σ_{U-B}	n	Spectrum	References
		AzV 358a			1 03 05	−72 20.5	1	14.10		−0.19		−0.91		3	B0	767
6482	−10 00229	HR 315			1 03 06	−10 14.8	6	6.11	.016	1.01	.005	0.86	.004	20	K0 III	15,1007,1013,2007*
		Florsch 483			1 03 06	−71 32.5	1	13.01		0.45		−0.12		2	F3	767
	−72 00080	AJ63,118 # 24			1 03 08	−72 24.0	2	11.05	.039	−0.15	.017	−0.71		8		35,2001
		BSD 8 # 556			1 03 10	+60 06.0	1	12.24		0.47		−0.16		3	B4	1185
		BSD 8 # 1075			1 03 10	+60 34.6	1	10.92		0.37		0.24		3	B8	1185
	−29 00324				1 03 10	−28 59.6	1	11.75		0.85				1		1495
6416	+61 00206	HR 309			1 03 12	+62 29.7	2	6.53	.009	0.21	.014	0.05		4	A5 Vn	1118,3016
		SB 435			1 03 12	−24 14.	1	13.38		−0.03		−0.11		2	A0	295
		BPM 46913			1 03 12	−26 20.	1	13.96		0.70		0.09		1		3061
		JL 225			1 03 12	−54 21.	2	14.34	.000	0.55	.055	−0.07	.010	2		132,832
		NGC 362 sq1 9			1 03 12	−71 06.8	3	14.59	.007	0.79	.011	0.29	.008	9		489,1461,1677
		AzV 361			1 03 12	−72 16.3	1	14.29		−0.14		−0.90		3	B0	425
		AzV 359			1 03 12	−72 42.0	1	14.73		−0.20		−0.98		1	B1	425
		AzV 362			1 03 13	−72 22.3	4	11.36	.012	−0.02	.008	−0.76	.000	15	B3 Ia0	35,573,1277,2001
		AzV 360			1 03 13	−72 38.1	1	14.46		−0.26		−0.97		3	B1	35
6479	+4 00175	HR 313	⋆	A	1 03 14	+04 38.6	5	6.35	.035	0.38	.009	−0.06	.027	15	F4 V	15,254,1071,1084,3016
	−28 00340	G 268 - 115		A	1 03 14	−27 46.8	1	12.03		0.45		−0.24		5		3061
	−28 00340			B	1 03 14	−27 46.8	1	13.22		1.09		0.90		3		3061
	−30 00336				1 03 14	−29 40.6	1	11.34		0.62		0.10		1		1099
	−15 00198				1 03 15	−14 56.5	1	11.66		0.79		0.40		2		1696
		AzV 363			1 03 15	−72 38.9	1	14.61		−0.14		−0.96		1	B0	425
6480	+4 00176	HR 314	⋆	B	1 03 16	+04 38.6	5	7.26	.041	0.48	.008	−0.05	.037	15	F8 V	15,254,1071,1084,3016
		BSD 8 # 559			1 03 16	+60 10.3	1	10.63		0.47		−0.28		3	B5 p	1185
		AzV 364			1 03 16	−72 15.3	1	14.13		−0.18		−1.04		3	B0	425
		AzV 367			1 03 16	−72 24.6	4	11.23	.024	0.07	.015	−0.58	.000	11	B6 Ia0	35,573,1277,2001
		CS 29514 # 8			1 03 17	−24 14.7	1	13.40		0.02		0.00		1		1736
		AzV 365			1 03 17	−72 43.2	1	14.54		−0.27		−1.07		3	B0 II	425
		BSD 8 # 1578			1 03 18	+61 58.8	1	11.56		1.17		0.32		3	B0	1185
		LP 882 - 275			1 03 18	−31 26.	1	15.52		1.68				1		3061
6609	−72 00082				1 03 18	−72 09.0	1	10.20		0.52		−0.01		3	F7/G0	35
		AzV 369		AB	1 03 18	−72 18.6	3	11.08	.029	0.58	.000	0.47	.015	11	F4 Ia	35,573,2001
		AzV 368			1 03 18	−72 33.5	1	14.39		−0.20		−0.98		1	B0	425
		AzV 366			1 03 18	−73 02.7	2	13.54	.019	−0.13	.015	−0.80	.019	3	B2	425,835
	+62 00197				1 03 19	+63 21.4	1	9.62		0.49				3	F8	1118
		AzV 370			1 03 19	−72 38.5	1	15.10		−0.23		−1.02		3	B0 V:	425
		G 269 - 91			1 03 20	−29 06.6	1	13.70		1.50				1		3061
		AzV 371			1 03 20	−73 05.5	1	13.91		−0.12		−0.70		1	B0 Ia	425
6519	−32 00429				1 03 21	−32 13.1	1	9.02		0.61				4	G2/3 V	2012
		NGC 362 sq1 7			1 03 21	−71 06.6	3	13.81	.017	0.65	.009	0.05	.018	10		489,1461,1677
		BSD 8 # 1080			1 03 22	+60 16.6	1	11.98		0.40		−0.19		3	B5	1185
		BSD 8 # 1079			1 03 22	+60 20.7	1	11.76		0.50		0.01		3	B7	1185
6504	−24 00480				1 03 22	−23 51.7	1	8.27		0.50		0.00		1	F5 V	3061
		NGC 362 sq1 6			1 03 22	−71 12.8	2	12.86	.009	0.96	.019	0.60	.028	11		489,1677
		AzV 372			1 03 22	−73 02.8	2	12.63	.005	−0.18	.005	−0.96	.005	5	B0	425,835
		AJ63,118 # 21			1 03 23	−72 24.9	2	10.85	.027	1.23	.057	1.09		5		35,2001
		AzV 373			1 03 23	−72 55.9	2	12.19	.024	−0.09	.000	−0.83	.010	6	B2 Ia	425,573
		PHL 7147			1 03 24	−20 05.	1	11.02		−0.01		−0.05		2	A2	3064
6517	−26 00356	IDS01010S2650		A	1 03 24	−26 34.4	1	8.95		0.63		0.07		6	G0 V	3061
6517	−26 00356	IDS01010S2650		AB	1 03 24	−26 34.4	1	8.91		0.64				4	G4 VI	2012
6517	−26 00356	IDS01010S2650		B	1 03 24	−26 34.4	1	11.52		1.00		0.81		6		3061
6517	−26 00356	IDS01010S2650		C	1 03 24	−26 34.4	1	9.23		0.36		−0.09		2		3061
		BPM 46917			1 03 24	−27 54.	1	15.39		0.18		−0.61		1		3062
		BSD 141 # 144			1 03 24	−30 40.	1	12.04		0.93		0.71		4		1099
		AAS53,255 # 163			1 03 24	−72 31.	1	13.03		1.48				1		1686
6515	−22 00193				1 03 25	−21 50.0	1	8.48		0.37				4	A9 V	2012
		AAS53,255 # 164			1 03 25	−73 12.	1	13.27		1.64				1		1686
6476	+31 00180	IDS01007N3139		A	1 03 26	+31 54.9	2	6.27	.005	1.31	.005	1.45	.005	5	K3 III	1501,1733
6447	+63 00139	IDS01002N6352		AB	1 03 26	+64 07.6	1	7.52		0.57				2	G0	1118
	+56 00194				1 03 27	+57 18.0	1	9.81		1.93				2	M0	369
		AzV 374			1 03 27	−72 42.9	1	13.11		−0.17		−0.87		3	B1 Iab	425
		AzV 375			1 03 27	−72 52.2	1	13.91		−0.09		−0.84		3	B2 II	425
	+63 00138				1 03 28	+63 51.8	1	8.90		1.51				3	K0	1118
		BSD 8 # 1088			1 03 29	+60 59.7	1	11.27		0.49		0.19		3	B8	1185
		G 269 - 93			1 03 29	−27 53.0	1	14.10		0.60		0.05		3		3060
6500	+0 00181	G 2 - 19			1 03 30	+01 26.5	2	9.45	.010	0.59	.005	0.00		3		202,1620
		POSS 295 # 4			1 03 30	+29 21.1	1	13.70		0.90				3		1739
		BPM 46925			1 03 30	−33 20.	1	11.66		0.45		−0.06		2		3061
6623	−72 00083				1 03 30	−72 00.1	4	7.37	.010	1.09	.005	0.97	.009	10	K0 III	35,258,1075,3077
	−29 00325				1 03 31	−29 09.7	1	10.93		1.03		0.61		1		1099
6532	−27 00355	AP Scl			1 03 32	−26 59.8	3	8.44	.007	0.15	.005	0.07	.010	9	Ap SrCrEu	561,1068,2012
	−30 00338				1 03 32	−30 05.7	1	11.68		1.16				1		1495
		AzV 376			1 03 32	−72 40.6	1	13.54		−0.07		−0.84		5	B1	35
	−30 00339				1 03 33	−29 44.2	1	11.68		1.18		0.98		1		1099
		AzV 378			1 03 33	−72 21.5	1	13.88		−0.24		−1.07		4	B0	425
		AzV 377			1 03 34	−73 04.3	1	14.78		−0.34		−1.20		3	O9	425
6530	−10 00230	HR 317			1 03 35	−10 06.4	2	5.57	.005	0.01	.000			7	A1 V	15,2027
	−23 00395	G 268 - 114			1 03 35	−22 43.1	1	10.21		1.08		1.03		1	K5	3061
		WLS 100 75 # 7			1 03 36	+72 40.8	1	11.35		1.25		0.99		2		1375
		BPM 46921			1 03 36	−27 49.	1	12.05		0.43		−0.20		1		3061

Table 1 79

HD	DM	Other Id	N Rem	α_{1950}	δ_{1950}	S	V	σ_V	B–V	σ_{B-V}	U–B	σ_{U-B}	n	Spectrum	References
		BPM 46922		1 03 36	−30 03.	1	11.73		0.58		0.00		1		3061
		BPM 46924		1 03 36	−32 27.	1	12.45		0.70		0.14		1		3061
		HA 141 # 276		1 03 37	−30 04.6	1	12.31		0.64		-0.02		1		1099
		AzV 379		1 03 38	−72 59.4	1	13.42		0.03		0.07		4	A3 Ib	733
		BSD 8 # 569		1 03 39	+59 24.7	1	11.75		0.53		0.36		3	B4	1185
6560	−32 00430			1 03 39	−32 07.3	1	8.10		1.09				4	K0 III	2012
		WLS 100 0 # 9		1 03 40	−01 07.5	1	11.42		0.67		0.22		2		1375
6559	−24 00484	HR 320		1 03 42	−24 15.6	2	6.14	.007	1.08	.004	1.00		6	K1 III	58,2009
		Dril 23		1 03 42	−27 33.	1	12.35		0.53		-0.02		1	F8	561
	−30 00341			1 03 42	−29 40.9	1	11.98		0.70		0.24		1		1099
	+55 00242			1 03 43	+55 52.3	1	9.81		1.87				2		369
6655	−73 00063			1 03 43	−72 49.2	4	8.04	.004	0.55	.011	0.01	.010	13	F8 V	1408,1425,2001,2033
	+63 00137	G 243 - 60		1 03 44	+63 40.2	6	8.99	.019	1.30	.008	1.22	.013	16	K7 V	1003,1013,1017,1197*
		CS 29518 # 6		1 03 44	−31 22.9	1	14.79		0.21		0.02		1		1736
		POSS 295 # 24		1 03 45	+30 01.1	1	16.32		1.72				2		1739
	−29 00326			1 03 46	−28 57.7	1	10.37		0.98		0.62		1	G8 III	54
6475	+59 00181			1 03 48	+59 35.5	2	6.87	.001	0.00	.002			4	A2 V	1103,1118
6474	+63 00141	V487 Cas		1 03 48	+63 30.4	2	7.61	.010	1.62	.030	1.52		7	G0 Ia	1118,1355
6662	−71 00037			1 03 48	−71 12.0	1	8.17		1.16		1.02		4	G8 III	1461
		AAS53,255 # 165		1 03 49	−72 55.	1	13.51		1.75				1		1686
		AAS53,255 # 166		1 03 50	−73 12.	1	13.31		1.42				1		1686
6595	−47 00324	HR 322	★ AB	1 03 51	−46 59.2	7	3.30	.009	0.89	.006	0.57	.007	20	G8 III	15,278,1020,1075,2012*
		AzV 381		1 03 52	−72 34.0	1	14.33		-0.30		-0.97		1	B0	425
		AzV 380		1 03 52	−73 19.8	1	13.56		-0.12		-0.91		1	B1	835
		G 70 - 37		1 03 53	−02 26.6	1	13.00		1.41				2		202
		AzV 383	AB	1 03 53	−72 44.6	1	14.06		-0.25		-1.02		4	O9	425
		CS 22166 # 29		1 03 54	−12 42.2	1	14.52		0.00		0.04		1		1736
		Dril 24		1 03 54	−26 25.	1	11.14		0.58		0.04		1	G0 IV	561
		LB 3176		1 03 54	−64 21.	1	11.61		0.09		0.13		2		45
		NGC 362 sq1 8		1 03 54	−71 09.9	2	14.18	.010	0.79	.000	0.29	.000	6		489,1677
		NGC 362 sq1 4		1 03 54	−71 11.0	3	12.34	.000	0.53	.010	-0.03	.015	11		489,1461,1677
		AAS53,255 # 167		1 03 54	−72 33.	1	13.25		1.44				1		1686
		AzV 384		1 03 54	−72 49.5	2	12.61	.000	-0.02	.020	-0.35	.015	6	A0 Ia	425,573
		AzV 382		1 03 54	−73 04.2	2	11.41	.005	0.05	.010	-0.43	.005	7	B8 Ia	425,573
		Steph 115		1 03 56	+11 36.0	1	11.65		1.37		1.25		1	K7	1746
6557	+12 00135	HR 319		1 03 56	+12 41.3	1	6.12		0.96				2	G8 III	71
		AzV 385		1 03 56	−72 33.7	1	13.86		-0.24		-0.83		2	B1 Iab	425
6497	+56 00196	HR 316		1 03 57	+56 40.2	5	6.42	.009	1.19	.006	1.25	.017	11	K2 III	15,1003,1355,3071,4001
	+59 00182			1 03 57	+60 23.2	1	10.85		0.15				1	B9 V	1103
		G 243 - 61		1 03 57	+69 36.7	1	10.58		1.01		0.82		2	K5	7010
6569	−15 00200	G 270 - 141		1 03 57	−14 33.7	1	9.53		0.90		0.60		1	K1 V	3072
6594	−35 00374			1 03 57	−35 03.9	1	8.10		0.61				4	G2 V	2011
		AzV 386		1 03 57	−72 44.7	1	14.12		-0.24		-0.90		3	B0	425
		BSD 8 # 580		1 03 58	+60 04.6	1	11.52		0.38		0.16		3	A0	1185
		BSD 8 # 1597		1 04 00	+61 26.3	1	12.01		0.34		-0.25		3	B6	1185
		Dril 25		1 04 00	−27 31.	1	12.97		0.62		0.02		1	G5 V	561
6554	+34 00183			1 04 01	+34 52.7	1	7.53		0.10		0.10		2	A2	1625
	−37 00402	G 269 - 98		1 04 01	−37 00.7	2	10.13	.019	1.04	.010	0.92	.005	3		79,3061
	−29 00327	IDS01017S2934	AB	1 04 04	−29 18.0	2	10.81	.015	0.45	.003	-0.13	.061	3	F6 VI	1099,1730
	−30 00344			1 04 04	−29 37.0	1	10.37		0.53		0.02		1		1099
		POSS 295 # 3		1 04 05	+30 42.9	1	13.62		0.65				2		1739
		BSD 8 # 1099		1 04 05	+60 17.1	1	11.95		0.27		-0.41		3	B5 p	1185
		BSD 8 # 1097		1 04 05	+60 17.6	1	12.70		0.48		0.38		3	F6	1185
		AzV 387		1 04 05	−72 32.0	1	13.90		-0.29		-1.00		3	O9	425
		AzV 388		1 04 05	−72 45.4	1	14.12		-0.21		-1.03		3	B0	733
		LP 826 - 539		1 04 06	−25 32.	1	15.07		0.71		0.33		1		3061
6619	−36 00417	HR 323, AW Scl		1 04 06	−35 55.7	7	6.61	.015	0.14	.004	0.14	.028	30	A1 V	15,216,355,1068,2032*
	−9 00219	G 270 - 144		1 04 08	−09 22.0	1	10.73		0.91		0.67		1		1696
		NGC 362 sq2 6		1 04 09	−71 13.0	1	11.45		1.01		0.81		2		1461
		AzV 389		1 04 09	−72 30.6	1	13.59		-0.18		-0.99		3	B0	425
6540	+52 00262	HR 318	★ A	1 04 10	+53 13.9	1	6.29		1.55		1.65		2	K0	1733
		AzV 390		1 04 10	−72 02.3	2	13.15	.030	0.04	.010	-0.66	.035	4	B1:	35,573
		BSD 8 # 1601		1 04 12	+61 26.9	1	11.27		0.26		0.18		3	B8	1185
		BSD 8 # 1600		1 04 12	+61 28.0	1	12.45		0.43		0.34		3	B7	1185
6592		Z Cet		1 04 12	−01 44.9	1	10.85		1.70		0.75		1	M4	975
		NGC 362 sq2 2		1 04 13	−71 17.3	1	10.45		1.46		1.76		1		1461
		AAS35,353 # 104		1 04 13	−72 36.6	1	13.25		-0.08		-0.31		4	B8	767
		AzV 391		1 04 13	−73 02.7	1	14.69		-0.14		-0.68		5	B2	425
	−34 00423	BPM 46931		1 04 15	−33 36.4	1	10.84		0.57		0.00		1		3061
6616	−16 00177			1 04 16	−15 46.6	1	7.20		1.00		0.74		6	G8/K0 III	1628
		GD 678		1 04 16	−25 14.8	2	14.01	.000	0.37	.000	-0.14	.000	4		3061,3061
		BSD 8 # 1105		1 04 17	+60 22.2	1	10.84		0.45		-0.38		3	B0	1185
6564	+48 00337			1 04 18	+49 17.3	1	6.82		-0.04		-0.28		2	A0	401
		POSS 295 # 19		1 04 19	+29 07.5	1	16.06		1.66				1		1739
		WLS 112 60 # 10		1 04 19	+59 25.8	1	12.28		0.53		0.26		2		1375
6586	+37 00210	IDS01015N3807	AB	1 04 20	+38 23.0	1	8.86		0.27		0.10		2	F8 V +F8 V	3030
6586	+37 00210	IDS01015N3807	C	1 04 20	+38 23.0	3	7.57	.000	0.51	.010	0.05	.005	8	A5	292,1381,3030
		AzV 392		1 04 20	−71 35.1	1	12.57		0.03		0.03		2	A3 Ib	425
		AzV 393		1 04 22	−72 35.7	2	11.47	.035	-0.03	.005	-0.85	.010	7	B1.5Ia:	35,573
		LP 526 - 66		1 04 23	+05 38.3	1	12.09		0.91		0.68		1		1696

HD	DM	Other Id	N Rem	α_{1950}	δ_{1950}	S	V	σ_V	B–V	σ_{B-V}	U–B	σ_{U-B}	n	Spectrum	References
	+15 00162			1 04 23	+15 37.8	1	10.38		0.59		0.13		1	G5	1776
6628	−23 00403			1 04 23	−23 07.3	1	7.68		0.86		0.32		2	G5 V	3061
	−30 00345			1 04 23	−30 25.4	1	11.34		0.40		-0.01		1	A0	1099
		CS 29514 # 6		1 04 24	−25 17.8	1	14.20		0.31		0.00		1		1736
		Dril 26		1 04 24	−25 18.	1	14.15		0.42		-0.09		1	A7	561
		L 171 - 54		1 04 24	−58 22.	1	12.74		0.50		-0.23		2		1696
	+60 00165			1 04 25	+61 02.2	1	10.61		0.47		-0.14		3		1185
6538	+62 00203			1 04 25	+63 25.6	3	7.67	.007	0.99	.003			30	G5	61,71,1118
6626	−5 00190			1 04 26	−05 10.4	1	8.02		0.09		0.11		4	A0	152
		AAS53,255 # 168		1 04 26	−73 08.	1	13.19		1.88				1		1686
6685	−61 00070			1 04 27	−61 25.2	1	8.16		1.12		1.04		9	K1 III	1704
		AzV 394		1 04 27	−72 39.4	1	13.29		-0.22		-0.82		3	B1	425
		BSD 8 # 1114		1 04 29	+60 47.3	1	12.43		0.45		-0.25		3	B0	1185
		AzV 396		1 04 29	−72 29.5	1	14.23		-0.23		-1.03		4	B0	425
		AzV 395		1 04 30	−73 05.5	1	15.00		-0.05		-1.05		3	O9	425
		GD 679		1 04 31	−33 36.5	3	13.52	.029	-0.28	.010	-1.13	.026	4		295,1036,3064
		AzV 397		1 04 31	−72 34.0	1	13.81		-0.07		-0.99		2	O9	35
12697	−89 00011			1 04 32	−89 26.4	1	9.69		0.72		0.32		1	G2 V	1117
		AzV 398		1 04 34	−72 12.0	1	13.85		0.09		-0.77		6	B2	425
		AzV 399		1 04 34	−72 40.1	2	12.38	.015	-0.01	.000	-0.23	.010	5	B7 Ib	425,573
		POSS 295 # 7		1 04 35	+28 49.2	1	13.94		1.23				2		1739
		PS 48		1 04 35	−28 23.8	2	13.19	.015	0.22	.004	0.03	.020	3	A0	561,1736
6673	−51 00273			1 04 35	−51 15.4	3	8.84	.016	0.93	.011	0.65	.040	11	K2 V	158,2012,3077
		SB 449		1 04 36	−28 34.	1	13.15		0.24		0.09		1	A0	295
	−29 00331			1 04 36	−29 11.7	1	12.38		0.58		0.02		1		1099
		BPM 30551	⋆V	1 04 36	−46 26.	1	15.42		0.17		-0.50		1	DA	832
		Florsch 509		1 04 36	−72 53.6	1	12.96		0.09		0.09		3	A2	767
		AzV 400		1 04 37	−72 28.9	1	14.61		-0.11		-0.86		4	B2	425
		AAS53,255 # 169		1 04 37	−72 31.	1	13.35		1.20				1		1686
6670	−30 00348			1 04 38	−29 52.9	3	9.36	.014	0.33	.009	-0.02	.024	9	A9 V	561,1068,2012
	−33 00407	BPM 46934		1 04 38	−32 45.0	2	10.72	.034	0.56	.005	-0.04	.029	3		1696,3061
		BSD 8 # 1121		1 04 41	+60 22.0	1	12.66		0.44		0.03		3	B8	1185
		BSD 8 # 1120		1 04 41	+60 46.0	1	11.75		0.55		0.05		3	B5	1185
6580	+61 00211			1 04 42	+62 23.2	1	7.83		1.95				2	M1	369
		L 292 - 41	⋆	1 04 42	−46 25.	2	15.30	.040	0.28	.010	-0.58	.000	2	DA	832,3065
		AzV 401		1 04 42	−72 51.4	1	13.20		0.19		0.31		3	F2	425
6581	+61 00212			1 04 44	+62 04.1	3	8.55	.013	0.10	.014	-0.11		7	B8 III	70,1103,1371
		HA 141 # 182		1 04 44	−29 41.4	1	13.40		0.86				1		1099
		AzV 402	AB	1 04 44	−72 43.7	2	13.37	.005	-0.11	.005	-0.98	.000	5	B0	35,573
		BSD 8 # 1124		1 04 47	+60 16.2	1	12.23		0.62		0.17		3	B2	1185
6668	−24 00496	HR 325		1 04 47	−24 15.8	5	6.36	.007	0.23	.015	0.09	.004	16	A3 III	15,58,1020,1068,2012
		G 132 - 56		1 04 48	+39 09.4	1	11.20		1.05		0.90		1	K3	1620
		GR 25		1 04 48	−17 45.4	1	14.85		1.47		1.30		1		481
		CS 22183 # 29		1 04 49	−06 02.6	1	13.87		0.19		0.26		1		1736
		AzV 403		1 04 49	−73 00.1	1	14.87		-0.20		-1.03		3	B0	425
	−33 00408	G 269 - 104		1 04 50	−32 41.9	1	10.76		1.38		1.22		1		3061
		NGC 381 - 24		1 04 51	+61 18.4	1	11.59		0.42		0.28		1		1670
6473	+79 00029	HR 312		1 04 51	+79 44.7	1	6.23		0.94		0.58		2	K0	1733
	−30 00350			1 04 51	−30 11.2	2	11.19	.005	0.95	.000	0.62	.005	2		79,3061
	+59 00186	LS I +60 141		1 04 53	+60 17.3	1	10.11		0.32		-0.45		3	B5 II	1185
		BSD 8 # 1127		1 04 53	+60 38.6	1	12.04		0.53		0.26		3	B7 p	1185
		BPM 46938		1 04 54	−36 56.	1	10.53		0.88		0.50		1		3061
		AzV 410		1 04 54	−72 17.8	1	13.30		-0.23		-0.90		4	B1 Ia-lab	425
6582	+54 00223	HR 321	⋆A	1 04 55	+54 40.6	13	5.16	.015	0.69	.007	0.10	.009	62	G5 Vp	1,15,22,792,908,1003*
		NGC 381 - 38		1 04 55	+61 14.0	1	12.94		0.48		0.32		1		1670
		G 268 - 119		1 04 55	−17 53.6	1	11.29		1.20		1.16		1	K4-5	3061
6660	+22 00176	G 34 - 2	⋆	1 04 56	+22 41.7	3	8.41	.012	1.13	.005	1.07	.008	6	K4 V	22,333,1620,3072
		AzV 404		1 04 56	−72 38.1	2	12.22	.025	-0.12	.000	-0.79	.000	5	B2 Ia-lab	425,573
		G 132 - 57		1 04 57	+33 56.2	2	13.35	.004	1.59	.004	1.18		6		419,3078
6645	+46 00266	IDS01021N4618	AB	1 04 58	+46 35.1	1	7.41		1.19		1.11		1	K2p	8100
		NGC 381 - 46		1 04 58	+61 17.0	1	13.91		0.50		0.36		1		1670
		CS 29518 # 2		1 04 58	−32 33.9	1	15.30		0.10		0.14		1		1736
		NGC 381 - 48		1 04 59	+61 19.0	1	12.76		0.39		-0.21		1		1670
		NGC 381 - 49		1 05 00	+61 19.8	1	13.79		0.55		0.29		1		1670
	−35 00381	IDS01027S3524	AB	1 05 00	−35 07.5	1	11.02		1.04		0.93		1	K0	3061
6782	−71 00039			1 05 00	−71 34.3	1	9.54		0.61		0.08		3	G0/2 V	1408
6692	−19 00180			1 05 01	−19 24.6	1	9.25		0.70		0.27		1	G6 V	3061
		AzV 405		1 05 01	−72 12.0	1	14.19		-0.16		-0.93		4	B2	733
		AAS53,255 # 170		1 05 01	−72 33.	1	13.47		1.45				1		1686
		G 268 - 122		1 05 03	−15 46.4	2	12.26	.078	1.41	.005	1.19	.019	3		481,3061
		AzV 407		1 05 03	−72 26.8	1	14.31		-0.20		-1.01		4	B0	425
		AzV 406		1 05 03	−72 48.5	1	14.85		-0.17		-0.93		3	B1	425
		NGC 381 - 55		1 05 04	+61 18.5	1	14.98		0.66		0.44		1		1670
		NGC 381 - 58		1 05 04	+61 18.5	1	13.38		0.60		0.41		1		1670
		CS 22166 # 32		1 05 04	−12 27.0	1	13.25		0.12		0.15		1		1736
		BSD 8 # 1134		1 05 05	+60 18.5	1	12.45		0.41		-0.15		3	B5	1185
		BSD 8 # 1136		1 05 06	+60 46.4	1	12.08		0.41		-0.33		3	B0	1185
		AAS35,353 # 106		1 05 07	−73 14.0	1	14.63		-0.15		-0.69		3	B3	767
		BSD 8 # 1619		1 05 07	+61 34.8	1	12.39		0.46		0.30		3	F2	1185
		AAS53,255 # 171		1 05 07	−72 45.	1	12.83		1.50				1		1686

Table 1   81

HD	DM	Other Id	N	Rem	α_{1950}	δ_{1950}	S	V	σ_V	B–V	σ_{B-V}	U–B	σ_{U-B}	n	Spectrum	References
		AzV 408			1 05 07	−72 53.6	1	13.66		−0.15		−0.82		4	B1	425
6658	+43 00234	HR 324			1 05 08	+43 40.6	3	5.04	.014	0.11	.005	0.14	.000	7	A3 m	1363,3058,8071
	+62 00207	HS Cas			1 05 08	+63 19.2	2	9.72	.109	2.52	.015	2.19		5	M4 Ia	369,8032
		AzV 409a			1 05 08	−72 13.1	1	14.28		−0.04		−0.89		4	B2	767
		AzV 409			1 05 09	−73 26.3	2	13.51	.005	−0.01	.009	−1.01	.014	4	B0	425,835
6664	+38 00194				1 05 10	+38 59.2	2	7.86	.044	0.62	.024	0.05	.024	3	G1 V	105,8100
		G 243 - 62			1 05 10	+63 15.0	2	11.52	.000	0.87	.009	0.45	.009	4		538,1658
		NGC 381 - 78			1 05 11	+61 19.9	1	10.95		0.35		0.30		1		1670
6724	−29 00334				1 05 11	−29 33.2	4	9.29	.017	0.40	.011	0.01	.016	12	F0 V	54,1068,1099,2012
		Dril 30			1 05 12	−26 35.	1	12.85		0.61		0.08		1	G2 V:	561
6725	−32 00439	IDS01028S3212	AB		1 05 12	−31 55.9	1	9.62		0.68		0.19		1	G3 V	3061
		G 2 - 21			1 05 13	+12 37.0	1	12.17		1.47		1.19		1	M0:	333,1620
		NGC 381 - 98			1 05 13	+61 16.4	1	12.71		0.47		−0.18		1		1670
		NGC 381 - 93			1 05 13	+61 18.6	1	12.28		0.42		0.29		1		1670
		NGC 381 - 91			1 05 13	+61 20.0	1	12.96		0.46		0.59		1		1670
		NGC 381 - 85			1 05 13	+61 21.4	1	16.03		0.56		0.24		1		1670
		NGC 381 - 87			1 05 13	+61 22.4	1	11.48		0.39		0.27		1		1670
6723	−29 00335				1 05 13	−28 58.3	5	9.07	.024	0.29	.009	0.02	.008	13	A8 V	561,1068,1099,1730,2012
		AAS53,255 # 172			1 05 13	−72 32.	1	12.02		1.71				1		1686
6678	+42 00240				1 05 14	+42 34.9	1	7.71		0.05		0.04		2	A0	1601
		NGC 381 - 99			1 05 14	+61 17.6	1	14.07		0.67		0.29		1		1670
		NGC 381 - 97			1 05 14	+61 19.2	1	13.48		0.55		0.39		1		1670
6722	−28 00348				1 05 14	−27 59.3	2	7.97	.009	1.08	.005	0.86		5	G8 III	79,2018
6680	+31 00185	HR 327			1 05 15	+31 44.7	4	6.25	.005	0.39	.010	−0.02	.022	10	F5 IV	15,254,1003,3026
6706	−10 00238	HR 329			1 05 15	−10 03.2	2	5.75	.059	0.43	.009	0.00		5	F7 IV	254,2007
		HA 141 # 141			1 05 15	−29 28.1	1	14.23		1.03				1		1099
		HA 141 # 253			1 05 15	−30 01.9	1	13.64		0.99		0.35		1		1099
6695	+19 00185	HR 328			1 05 16	+20 28.4	1			0.13		0.13		3	A3 V	1049
		NGC 381 - 104			1 05 16	+61 19.0	1	13.24		0.48		0.36		1		1670
		NGC 381 - 96			1 05 16	+61 24.7	1	11.64		0.37		0.28		1		1670
		NGC 381 - 90			1 05 16	+61 27.5	1	11.67		0.37		0.30		1		1670
6708	−16 00183	G 268 - 123			1 05 16	−16 13.0	2	9.27	.008	0.58	.008	−0.03	.013	7	G1 V	79,1775
6720	−20 00203				1 05 16	−19 34.6	1	7.95		0.77		0.31		1	G8 V	3061
		NGC 381 - 105			1 05 17	+61 22.0	1	12.54		0.41		0.28		1		1670
6718	−9 00221	G 270 - 148			1 05 17	−08 30.0	1	8.44		0.65		0.17		1	G0	1658
6688	+44 00240				1 05 18	+44 31.8	1	7.47		−0.09		−0.39		2	B8 III	401
		NGC 381 - 111			1 05 18	+61 16.2	1	15.78		1.11		0.49		1		1670
	−27 00363				1 05 18	−26 41.6	1	10.51		0.49		−0.11		1	F5	3061
		LB 1591			1 05 18	−50 27.	1	12.90		0.02		0.05		2		45
		Smethells 183			1 05 18	−63 41.	1	11.43		1.54				1		1494
		NGC 381 - 118			1 05 19	+61 16.7	1	12.88		1.63		1.64		1		1670
		BSD 8 # 1628			1 05 19	+61 17.7	1	12.79		0.52		0.34		3	B0	1185
		NGC 381 - 116			1 05 19	+61 18.7	1	15.03		0.86		0.22		1		1670
		NGC 381 - 109			1 05 19	+61 20.6	1	14.69		1.01		0.50		1		1670
		NGC 381 - 103			1 05 19	+61 26.0	1	11.39		0.34		0.25		1		1670
		NGC 381 - 120			1 05 20	+61 17.6	1	12.75		0.50		0.29		1		1670
6793	−62 00089	HR 332			1 05 20	−62 02.5	3	5.35	.017	0.89	.005			11	G5 III	15,1075,2007
		NGC 381 - 123			1 05 23	+61 18.8	1	15.14		0.81		0.33		1		1670
		NGC 381 - 122			1 05 23	+61 20.5	1	13.04		0.51		0.33		1		1670
6741	−27 00365				1 05 23	−27 15.2	1	8.85		1.04		0.89		2	G8 III	561
		NGC 381 - 127			1 05 24	+61 17.3	1	14.32		0.45		0.31		1		1670
	−29 00336				1 05 24	−28 49.6	1	9.88		1.03		0.91		1	K0 III	54
6827					1 05 24	−72 05.	1	10.39		0.84		0.48		3	G5	35
6734	+1 00212				1 05 25	+01 43.9	4	6.46	.009	0.86	.009	0.47	.007	14	K0 IV	1003,1355,1509,3077
		NGC 381 - 131			1 05 25	+61 16.3	1	16.01		0.97		0.14		1		1670
		BSD 8 # 1630			1 05 25	+61 43.7	1	10.91		0.45		−0.29		3	B5	1185
		NGC 381 - 138			1 05 26	+61 15.3	1	15.11		1.41		1.08		1		1670
7070	−84 00020				1 05 26	−83 51.5	1	7.40		1.08		0.88		7	K0 III	1628
		NGC 381 - 142			1 05 27	+61 16.5	1	12.28		0.74		0.18		1		1670
		AzV 411			1 05 27	−72 29.2	1	13.83		−0.26		−1.06		8	O8.5V	425
		NGC 381 - 147			1 05 28	+61 19.0	1	13.96		0.36		−0.03		1		1670
		NGC 381 - 133			1 05 28	+61 23.6	1	12.84		1.25		0.84		1		1670
		HA 141 # 142			1 05 28	−29 25.2	1	14.50		0.96				1		1099
6676	+57 00200	HR 326			1 05 29	+57 59.8	4	5.79	.013	0.00	.014	−0.27	.004	7	B8 V	252,1079,1103,1379
6715	+21 00150	G 34 - 3			1 05 30	+21 42.6	1	7.65		0.66		0.18		1	G5	333,1620
		NGC 381 - 151			1 05 30	+61 17.1	1	15.07		0.55		0.32		1		1670
		AzV 412			1 05 30	−72 43.6	1	12.97		−0.09		−0.50		4	B6 Iab	425
		NGC 381 - 159			1 05 31	+61 15.4	1	14.82		1.06		0.54		1		1670
		NGC 381 - 155			1 05 31	+61 18.5	1	11.97		0.41		0.29		1		1670
6663	+62 00209				1 05 31	+63 25.4	1	8.19		0.50				5	G0	70
	−26 00364	G 268 - 126			1 05 31	−25 56.4	1	11.30		0.97		0.77		1		3061
		HA 141 # 213			1 05 31	−29 45.6	1	14.08		0.67				1		1099
		HA 141 # 214			1 05 31	−29 47.7	1	13.39		0.64				1		1099
6767	−42 00391	HR 331	⋆	AB	1 05 31	−41 45.2	9	5.21	.008	0.16	.006	0.08	.008	46	A3 IV/V	15,116,278,977,977*
		AAS35,353 # 107			1 05 33	−71 42.3	1	12.48		0.47		−0.07		4	F4	767
		NGC 381 - 164			1 05 34	+61 19.6	1	13.64		0.63		0.28		1		1670
		CS 22166 # 34			1 05 34	−12 27.6	1	14.47		0.15		0.16		1		1736
	+60 00168	IDS01025N6113	AB		1 05 35	+61 29.6	1	10.49		0.37		0.24		3	B8 V	1185
		BSD 8 # 1148			1 05 36	+60 53.7	1	12.19		0.35		0.00		3	B5 p	1185
		BSD 8 # 1147			1 05 37	+61 11.5	1	11.53		0.42		−0.16		3	B5	1185

HD	DM	Other Id	N	Rem	α_{1950}	δ_{1950}	S	V	σ_V	B–V	σ_{B-V}	U–B	σ_{U-B}	n	Spectrum	References
		NGC 381 - 165			1 05 37	+61 24.9	1	12.80		0.60		0.13		1		1670
		AzV 413			1 05 37	−72 36.5	1	14.99		−0.19		−1.10		3	O9	425
		CS 22953 # 5			1 05 40	−61 41.3	2	13.63	.009	0.60	.013	−0.08	.001	4		1580,1736
		Florsch 523			1 05 42	−70 33.5	1	11.30		1.01		0.87		3		767
		AzV 414			1 05 44	−72 42.5	1	13.07		−0.07		−0.39		4	B8	425
6761	+9 00132				1 05 45	+09 38.5	1	6.70		1.64		2.00		3	M1	1733
6884	−73 00064				1 05 46	−72 44.0	5	10.54	.047	0.10	.012	−0.57	.010	51	B9 Iae	35,573,835,1277,8033
6763	+4 00190	HR 330		⋆ A	1 05 47	+05 23.1	9	5.52	.014	0.33	.007	−0.01	.011	39	F0 III-IV	15,253,254,1003,1007*
6675	+68 00074	LS I +69 003			1 05 47	+69 25.2	2	6.91	.009	0.29	.013	−0.64		5	B0.5III	1012,8031
		BPM 46942			1 05 48	−26 14.	1	13.82		0.60		0.00		1		3061
		R 40			1 05 48	−72 44.	1	12.83		0.41				3		1277
	+60 00169	OX Cas			1 05 51	+61 12.5	3	9.94	.018	0.28	.008	−0.54	.027	9	B1 V	1012,1185,1768
		AzV 416			1 05 52	−73 02.2	1	13.69		−0.02		−0.35		4	B9	425
		G 269 - 106			1 05 53	−29 04.3	1	13.45		1.70				1		3079
		AzV 417			1 05 53	−72 37.8	1	13.93		0.38		0.37		2	F5:	425
		LS I +62 150			1 05 54	+62 31.5	1	10.42		0.77		−0.31		2	B0 III	1012
		BPM 30559			1 05 54	−48 30.	1	15.04		0.48		−0.20		1		3065
		AzV 418			1 05 54	−72 34.2	2	14.56	.029	−0.16	.019	−0.67	.015	6	B2:	733,767
	−29 00338				1 05 55	−29 25.5	1	12.04		0.94		0.71		1		1099
		AzV 424			1 05 55	−72 23.5	1	13.11		−0.22		−0.95		4	O9 Ia	425
		G 1 - 48			1 05 56	+05 56.8	1	12.86		1.51		1.49		1		1620
		AzV 419			1 05 56	−72 31.3	1	13.73		−0.09		−0.85		3	B2	425
		AAS35,353 # 109			1 05 56	−72 56.7	1	14.10		−0.20		−0.73		6	B3	767
		CS 22166 # 39			1 05 59	−13 39.4	1	14.62		−0.07		−0.28		1		1736
		AzV 420			1 05 59	−72 33.6	2	13.14	.005	−0.23	.009	−0.96	.009	4	B2 Iab	425,573
	+59 00190				1 06 00	+60 18.0	1	10.28		0.28		0.11		3	B9 III	1185
		AzV 421			1 06 00	−72 38.8	1	14.55		−0.24		−0.87		4	B1	425
	+16 00120	G 33 - 40			1 06 01	+17 00.0	3	10.52	.025	1.28	.014	1.25	.022	5	K6	1658,1746,7008
		CS 22166 # 36			1 06 01	−12 43.9	1	14.39		0.10		0.17		1		1736
		CS 22166 # 37			1 06 01	−13 11.7	1	14.94		0.32		−0.17		1		1736
		AAS53,255 # 173			1 06 02	−72 47.	1	12.69		1.65				1		1686
6756	+45 00283				1 06 03	+45 40.6	2	8.09	.009	−0.08	.005	−0.51	.014	8	B8	401,1775
	+62 00212	LS I +63 136			1 06 03	+63 24.9	1	9.62		0.79		−0.17		2	B0.5V	1012
6806	−24 00503				1 06 03	−24 30.3	1	8.21		0.55		0.03		1	F7 V	79
		HA 141 # 186			1 06 03	−29 39.8	1	14.21		0.90		0.46		1		1099
236633	+59 00191	LS I +60 142			1 06 04	+60 21.7	2	9.11	.040	0.40	.010	−0.56	.025	5	B0.5III	1012,1185
6805	−10 00240	HR 334		⋆ A	1 06 04	−10 26.8	7	3.44	.013	1.16	.005	1.19	.008	23	K2 III	15,1075,2012,3016*
	−22 00206				1 06 05	−22 23.9	1	10.77		0.38		0.00		2	A5	3061
	−33 00417				1 06 05	−32 59.5	1	12.22		−0.24		−0.89		1	DA:	1036
		BSD 141 # 205			1 06 06	−29 27.	1	11.67		1.18		0.98		4		1099
		TonS 192			1 06 06	−33 00.9	1	12.23		−0.23		−0.98		2		3064
6870	−62 00090	BS Tuc			1 06 06	−62 08.3	2	7.50	.058	0.24	.019	−0.05		6	A2 IIwp	2012,3026
		Florsch 535			1 06 06	−70 55.4	1	12.99		0.46		−0.14		3		767
		AzV 422			1 06 06	−72 31.7	1	14.16		−0.22		−0.74		6	B2	425
		CS 29514 # 9			1 06 08	−22 58.5	1	14.75		0.15		0.09		1		1736
		AAS53,255 # 174			1 06 08	−72 43.	1	12.72		1.76				1		1686
		AzV 423			1 06 09	−73 06.9	2	13.29	.014	−0.20	.009	−0.99	.000	4	B0 Ia	425,573
		BSD 8 # 164			1 06 10	+58 58.1	1	11.20		0.42		0.24		3	A0	1185
6931	−72 00084				1 06 11	−72 16.0	1	10.92		0.20		0.08		2	A3/5	1408
		AzV 425			1 06 11	−72 44.1	1	14.72		−0.15		−0.85		2	B0	425
		BSD 8 # 1164			1 06 12	+60 46.7	1	10.91		0.36		0.15		3	B9	1185
		PHL 1000			1 06 12	−27 06.9	3	12.54	.015	−0.14	.010	−0.59	.012	7	A0	561,1036,3064
		TonS 193			1 06 12	−33 56.	1	15.26		0.08		−0.50		1		3064
		AzV 426			1 06 12	−72 49.5	1	14.16		−0.24		−1.03		4	B0 III	425
		CS 29514 # 2			1 06 13	−27 09.2	1	12.58		−0.14		−0.56		1		1736
	+55 00249				1 06 16	+55 48.1	1	9.25		1.72				2		369
6772	+51 00236				1 06 17	+52 20.2	1	9.00		0.28		0.11		2	A2	1566
6882	−55 00241	HR 338, ζ Phe		⋆ ABC	1 06 17	−55 30.8	6	3.92	.026	−0.09	.007	−0.41	.013	18	B6 V+B9 V	15,200,1075,2012,3023,8029
		AzV 427		AB	1 06 17	−72 42.6	1	13.94		−0.14		−0.85		5	B0	425
		CS 29514 # 10			1 06 18	−22 52.9	1	14.36		0.34		−0.19		1		1736
		Dril 33			1 06 18	−27 44.	1	13.32		0.91		0.63		2	K0	561
6815	+8 00177				1 06 19	+09 27.8	1	7.30		−0.06		−0.16		16	B9	8040
		AAS53,255 # 175			1 06 19	−72 29.	1	12.41		1.84				1		1686
		G 33 - 42			1 06 20	+19 09.6	1	11.36		1.00		0.94		1	K3	333,1620
6552	+82 00030				1 06 20	+83 11.5	1	8.83		0.13		0.08		5	A2	1219
	−29 00344				1 06 20	−28 54.4	1	11.00		0.54		0.02		1		1099
		AAS53,255 # 176			1 06 20	−72 40.	1	12.58		1.70				1		1686
6855	−34 00439				1 06 21	−34 34.8	1	9.41		0.38				4	F0 V	2012
6869	−47 00333	IDS01041S4712		AB	1 06 21	−46 56.1	4	6.88	.014	0.29	.012	0.06	.011	14	A9 V	116,278,1075,2011
	−29 00345				1 06 22	−28 43.1	1	10.04		1.31		1.46		1	K2 III	54
		AAS35,353 # 111			1 06 23	−72 36.3	1	13.56		−0.22		−0.86		4	B1	767
		AzV 428			1 06 23	−72 56.9	2	12.58	.009	0.02	.004	−0.23	.009	6	B8 Iab	425,835
		NGC 381 - 207			1 06 24	+61 07.2	1	9.38		0.22		−0.50		1		1670
		CS 22166 # 38			1 06 25	−13 25.9	1	14.94		0.06		0.16		1		1736
		AzV 429			1 06 26	−72 16.8	1	14.75		−0.21		−1.02		4	B0	425
		G 268 - 128			1 06 27	−15 13.8	1	13.78		1.58		1.06		3		481
6755	+60 00170	G 243 - 63			1 06 29	+61 16.8	4	7.71	.014	0.72	.005	0.08	.005	6	F8 V	70,792,1003,1658
6868	−30 00360	IDS01041S3009		AB	1 06 29	−29 53.3	1	8.06		0.49				1	F5 V	1099
		SB 462			1 06 30	−21 47.	1	13.36		0.07		0.20		2	A0	295
6795	+56 00207				1 06 31	+57 05.1	1	6.64		0.44		0.02		3	F3 V	1501

Table 1 83

HD	DM	Other Id	N Rem	α_{1950}	δ_{1950}	S	V	σ_V	B–V	σ_{B-V}	U–B	σ_{U-B}	n	Spectrum	References
		AzV 430		1 06 31	−72 57.3	1	14.20		0.03		−0.01		5	A0	425
		AzV 431		1 06 32	−72 24.6	2	12.99	.016	0.11	.004	0.17	.004	9	A2 Ib	425,835
		CS 22183 # 31		1 06 33	−04 59.4	1	13.62		0.62		−0.04		1		1736
	−22 00207			1 06 33	−21 48.2	1	10.40		0.68		0.18		2	G5	3061
		GD 11		1 06 34	+37 16.8	1	15.25		−0.23		−1.05		1	DAs	3060
	−20 00208			1 06 34	−20 29.4	1	9.80		0.55		0.03		1	G5	3061
6811	+46 00275	HR 335	⋆ AB	1 06 35	+46 58.5	4	4.25	.008	−0.07	.002	−0.34	.000	11	B6 IV +B9 V	15,1079,1363,8015
6880	−31 00455	G 269 - 109		1 06 35	−31 11.7	3	9.10	.007	0.78	.008	0.34	.029	9	G8/K0 V	158,2013,3061
		AzV 432		1 06 35	−72 35.5	1	14.21		−0.22		−0.92		2	B0	425
		TonS 195		1 06 36	−33 25.9	2	12.08	.054	−0.12	.024	−0.69	.015	3		832,3064
		AzV 433	AB	1 06 36	−72 31.9	1	14.38		−0.16		−1.02		4	B0	425
		AzV 434		1 06 37	−71 57.4	1	12.67		0.40		0.32		3	F5	425
6997	−73 00065			1 06 37	−73 16.8	3	9.07	.018	0.90	.015	0.57	.015	8	G8 III	1408,1425,2001
		BSD 8 # 1647		1 06 38	+61 54.7	1	11.53		0.48		0.37		3	B8	1185
	+56 00208			1 06 39	+56 47.8	1	9.49		2.02				2		369
		AAS53,255 # 178		1 06 39	−73 03.	1	12.92		1.67				1		1686
		AAS35,353 # 112		1 06 40	−72 52.5	1	14.55		−0.15		−0.68		4	B3	767
6848	+21 00156			1 06 42	+22 18.7	1	8.38		0.20		0.10		2	A3	1648
		PASP80,437 # 6		1 06 42	−28 39.	1	13.29		0.86		0.43		1		622
	−33 00421			1 06 42	−33 23.8	1	12.10		−0.21		−0.62		1	B5	1036
		AzV 435		1 06 44	−72 15.8	1	14.13		−0.07		−1.01		5	B0	425
6895	−34 00444			1 06 45	−33 52.5	1	8.81		0.64				4	G0 V	2012
		G 268 - 130		1 06 47	−24 57.4	1	12.38		1.53		1.28		1		3061
		BSD 8 # 1178		1 06 48	+60 34.5	1	13.00		0.51				3	B8	1185
		BPM 46950		1 06 48	−32 04.	1	11.52		0.62		0.02		1		3061
6908	−36 00437			1 06 49	−36 24.5	1	9.17		0.80		0.38		1	G8 V	3061
		HA 141 # 34		1 06 50	−28 59.4	1	14.32		0.80		0.21		1		1099
6833	+53 00236			1 06 51	+54 28.3	2	6.74	.010	1.17	.005	0.91	.019	7	G8 III	1003,1355
		AzV 436		1 06 53	−72 39.4	1	14.17		−0.15		−1.00		3	B0	35
		BPM 46951		1 06 54	−20 09.	1	12.49		0.90		0.66		1		3061
		Dril 34		1 06 54	−28 39.	1	12.42		2.26		5.55		2	C	561
6860	+34 00198	HR 337	⋆ A	1 06 55	+35 21.4	7	2.05	.015	1.58	.010	1.97	.020	23	M0 IIIa	15,1118,1119,3016*
7008	−71 00040			1 06 56	−70 56.0	1	8.38		1.14		1.04		4	K0 III	727
		AAS35,353 # 113		1 06 56	−72 45.3	1	14.29		−0.05		−0.31		3	B8	767
		AJ63,118 # 27		1 06 56	−73 10.5	2	11.18	.029	0.71	.000	0.27		7		35,2001
		BSD 8 # 1182		1 07 00	+60 30.9	1	11.50		0.45		−0.18		3	B6 p	1185
6832	+61 00218	IDS01039N6116	AB	1 07 00	+61 32.6	3	8.29	.009	0.06	.015	−0.29		12	B9 III	70,1103,1371
	+58 00179			1 07 01	+58 46.7	2	10.05	.022	0.40	.012	0.34		4	B9	1103,1185
6823	+62 00214			1 07 01	+63 18.2	1	9.29		0.30				2	A0	1118
6822	+63 00146	IDS01038N6407	AB	1 07 01	+64 22.8	1	8.20		0.14				3	A0	1118
6843	+51 00239	IDS01041N5130	AP	1 07 02	+51 46.0	1	7.94		0.77		0.43		1	F5	3054
6843	+51 00239	IDS01041N5130	B	1 07 02	+51 46.0	1	10.94		0.56		0.11		1		3054
6843	+51 00239	IDS01041N5130	C	1 07 02	+51 46.0	1	12.22		0.62		0.26		1		3032
6319	+86 00017	HR 306		1 07 03	+86 52.9	2	6.24	.021	1.12	.011	0.99		83	gK2	252,1258
		POSS 52 # 1		1 07 04	+65 10.7	1	15.10		1.71				1		1739
		AzV 437		1 07 04	−73 29.2	1	13.54		−0.07		−0.98		3	B1	425
	+61 00219	V554 Cas		1 07 05	+62 15.7	1	9.46		2.09				2	M1 ep+B	369
		SB 466		1 07 06	−22 57.	1	12.88		0.28		0.04		1	A3:	295
		SB 467		1 07 06	−26 55.	1	13.36		0.41		−0.21		2	A pec	561
	−30 00364			1 07 06	−29 40.7	1	12.10		0.90		0.60		1		1099
		BSD 8 # 1184		1 07 07	+60 57.7	1	12.33		0.39		−0.24		3	B5	1185
		GR 27		1 07 07	−20 44.6	1	13.74		1.52		1.17		1		481
6903	+18 00153	HR 339		1 07 09	+19 23.5	1	5.55		0.69				2	G0 III	71
6893	+33 00181			1 07 10	+33 36.9	2	6.99	.000	0.51	.005	0.10	.005	5	F8 V	1501,1625
		BSD 8 # 174		1 07 10	+58 44.7	2	11.30	.008	0.50	.009	0.33		4	B9	1103,1185
		BSD 8 # 621		1 07 11	+59 42.1	1	10.96		0.45		−0.29		3	B6	1185
		G 270 - 159		1 07 11	−03 55.6	1	11.12		0.86		0.53		1		1658
		AAS35,353 # 114		1 07 11	−71 50.1	1	14.48		−0.01		0.03		5	A0	767
		L 221 - 49	⋆	1 07 12	−53 12.	1	15.04		0.38		−0.09		3		3065
		AAS53,255 # 180		1 07 14	−72 25.	1	13.70		1.55				1		1686
		AAS53,255 # 181		1 07 14	−72 39.	1	12.47		1.75				1		1686
	+60 00174			1 07 15	+60 54.1	1	10.09		0.23		0.11		3	B7 V	1185
6829	+68 00077	HR 336	⋆ A	1 07 15	+68 30.8	1	5.29		−0.02		−0.06		1	A0 Vnn	3023
	−11 00218	G 270 - 160		1 07 15	−10 48.7	1	10.53		0.71		0.27		2		1696
		AzV 438		1 07 15	−72 53.5	2	12.95	.023	0.01	.009	−0.25	.005	4	A0 Iab	425,835
6841	+63 00147			1 07 17	+63 54.6	3	7.29	.003	0.22	.003			41	A3	61,71,1118
6958	−40 00256	IDS01050S4041	AB	1 07 17	−40 24.5	1	8.41		0.12		0.07		4	A2 IV/V	152
6996	−57 00251	IDS01052S5708	AB	1 07 17	−56 51.7	2	7.13	.005	0.46	.000	−0.04		6	F5 IV/V	2012,3037
		AzV 439		1 07 17	−72 05.8	1	14.81		−0.24		−1.08		5	O9.5III	425
		GR 28		1 07 19	−14 36.4	1	14.15		1.31		1.10		2		481
6871	+58 00181			1 07 20	+58 41.9	1	8.23		0.19				1	A3	1103
6890	+52 00272			1 07 22	+53 28.0	1	8.28		0.88		0.55		2	G5	1566
		AzV 440		1 07 22	−72 08.7	1	14.64		−0.20		−1.07		4	B0	425
		POSS 295 # 6		1 07 24	+31 23.8	1	13.91		0.51				5		1739
		BPM 46952		1 07 24	−20 29.	1	14.50		0.42		−0.28		1		3061
		BPM 46953		1 07 24	−21 36.	1	13.38		0.58		−0.04		2		3061
		BPM 46954		1 07 24	−23 29.	1	13.09		0.60		−0.10		1		3061
		BSD 8 # 1192		1 07 25	+61 12.2	1	10.50		0.28		−0.12		3	B6 p	1185
	+59 00194			1 07 26	+59 47.1	1	10.83		0.35		0.32		3	B9 V	1185
		AzV 441		1 07 26	−72 42.0	1	13.89		−0.27		−0.97		5	B0	425

HD	DM	Other Id	N Rem	α_{1950}	δ_{1950}	S	V	σ_V	B−V	σ_{B-V}	U−B	σ_{U-B}	n	Spectrum	References
		GD 12		1 07 27	+26 45.1	1	14.93		0.09		-0.64		1		3060
6920	+41 00219	HR 340		1 07 28	+41 49.0	2	5.66	.009	0.59	.004	0.12		5	F7 V	1119,1501
		AJ63,118 # 29		1 07 28	−73 09.5	2	11.30	.029	0.52	.006	0.02		7		35,2001
		AzV 442		1 07 28	−73 18.5	3	11.34	.014	0.13	.003	0.05	.010	7	A3 Ia	425,573,2001
	+59 00195			1 07 30	+59 46.1	1	10.43		0.18		0.02		3	A0 V	1185
	−7 00185			1 07 30	−07 19.8	1	9.76		0.62		0.04		2	G0	1696
		AAS33,107 # 2		1 07 30	−70 57.	1	12.32		0.60		0.12		4		127
		BSD 8 # 1196		1 07 31	+60 40.9	1	10.94		0.30		-0.32		3	B9	1185
6966	+14 00175	HR 344		1 07 32	+15 24.5	1	6.06		1.50				2	M0 III	71
		AAS53,255 # 182		1 07 33	−72 54.	1	9.54		1.66				1		1686
		AzV 444		1 07 34	−72 37.6	1	14.53		-0.11		-1.08		3	O9	425
7099	−73 00066			1 07 34	−72 48.1	5	10.96	.012	-0.06	.014	-0.83	.014	24	B3 Ia	425,573,1277,1408,8033
6953	+24 00186	HR 341		1 07 36	+25 11.6	3	5.80	.002	1.47	.004	1.82	.000	8	K5 III	71,833,3001
6918	+50 00228	IDS01047N5029	AP	1 07 36	+50 44.8	2	6.88	.020	0.43	.017	-0.03	.010	4	F5	1566,3032
6918	+50 00228	IDS01047N5029	B	1 07 36	+50 44.8	1	10.94		0.36		0.21		2		3016
6918	+50 00228	IDS01047N5029	C	1 07 36	+50 44.8	1	10.43		0.18		0.03		1		3032
6918	+50 00228	IDS01047N5029	D	1 07 36	+50 44.8	1	12.87		1.13		0.69		1		3032
6918	+50 00228	IDS01047N5029	E	1 07 36	+50 44.8	1	12.14		0.43		0.21		1		3032
		AzV 447		1 07 38	−72 31.8	1	14.20		-0.23		-1.00		4	B1	425
		G 270 - 162		1 07 39	−04 50.3	1	11.62		0.95		0.79		1		1658
7006	−26 00377			1 07 39	−26 27.6	1	7.34		1.13		1.10		7	K0 III	1628
236640	+58 00183			1 07 40	+59 26.3	1	10.10		0.26		-0.02		3	A2	1185
6976	−9 00227	HR 346		1 07 41	−09 10.3	4	6.39	.007	1.03	.008	0.88	.019	10	K0 III	15,79,1061,3016
		SB 469		1 07 42	−28 15.	1	13.03		0.36		-0.14		2	A pec	561
		LP 882 - 326		1 07 42	−31 23.	1	15.92		1.70				1		3061
	+58 00184			1 07 43	+58 40.7	1	9.93		1.21				1	G0	1103
	−11 00220	G 270 - 163		1 07 45	−10 37.1	6	9.20	.009	0.57	.005	-0.01	.018	8	G0	79,516,742,1064,1594,3002
7041	−57 00252			1 07 45	−56 37.3	2	9.03	.009	0.80	.000	0.32		4	G6/8wF8IV/V	742,1594
		AzV 445		1 07 45	−72 41.2	2	12.78	.000	-0.09	.014	-0.64	.005	5	B3 Iab	425,573
		AAS53,255 # 184		1 07 45	−73 01.	1	13.27		1.61				1		1686
	−29 00355			1 07 47	−29 30.4	1	10.90		0.74		0.29		1		1099
	+58 00187			1 07 48	+58 45.1	2	10.37	.017	0.33	.023	0.18		4	B8	1103,1185
236642	+58 00185			1 07 48	+59 16.2	2	8.64	.000	1.80	.000			4	M0	70,369
		BPM 46956		1 07 48	−24 50.	1	13.03		1.55		1.30		1		3061
	+58 00186			1 07 49	+58 57.9	1	10.87		0.34		0.15		3	B9	1185
7040	−45 00382			1 07 50	−44 45.7	3	9.45	.011	0.37	.008	-0.05	.011	14	F2 V	116,863,2011
		CS 22953 # 16		1 07 50	−59 00.9	2	14.68	.010	0.60	.014	-0.10	.001	3		1580,1736
6990	+14 00176			1 07 51	+15 24.1	2	9.30	.074	1.70	.022	2.09		5	K5	882,3040
6798	+78 00034	HR 333		1 07 52	+79 24.5	2	5.62	.020	0.01	.005	0.03	.000	5	A3 V	985,1733
	+58 00188			1 07 53	+59 05.8	1	10.60		0.35		-0.28		3	B7	1185
		BSD 8 # 640		1 07 53	+59 22.6	1	11.35		0.38		-0.37		3	B4 p	1185
		BPM 46957		1 07 54	−26 35.	1	12.61		1.54				1		3061
		AAS33,107 # 3		1 07 54	−70 57.	1	11.46		0.52		0.02		5		127
6948	+55 00255			1 07 55	+56 30.5	1	7.26		-0.03		-0.32		2	B9	401
	+59 00198			1 07 56	+59 45.4	1	9.78		0.23		-0.02		3	B8 III	1185
		AzV 446		1 07 56	−73 25.4	1	14.71		-0.30		-1.11		5	O8	425
7014	+1 00221	HR 347		1 07 59	+02 10.8	5	5.94	.011	1.51	.008	1.86	.030	23	K4 III	15,252,361,1071,1375
7126	−71 00044			1 07 59	−70 54.6	1	9.65		0.44		-0.05		5	F5 V	727
		BSD 8 # 641		1 08 00	+59 49.8	1	11.30		0.31		0.20		3	B9	1185
		BSD 8 # 1208		1 08 01	+60 50.7	1	11.96		0.37		-0.26		3	B5	1185
		GR 30		1 08 01	−15 26.2	1	14.22		1.60		1.19		1		481
6961	+54 00236	HR 343	⋆ A	1 08 03	+54 53.1	13	4.33	.005	0.17	.002	0.13	.006	112	A7 V	1,15,985,1006,1363,1502*
		G 2 - 25		1 08 04	+12 50.6	1	13.39		0.98		0.70		1		333,1620
7082	−58 00081	HR 350		1 08 05	−57 57.5	1	6.41		0.89				4	G6 II/III	2035
7022	+10 00136	G 1 - 50		1 08 06	+10 44.0	2	9.07	.024	0.76	.005	0.38		3	G5	202,333,1620
		PHL 3361		1 08 06	−26 22.	3	13.16	.020	-0.02	.008	0.01	.046	4	A0	295,561,3064
		LP 938 - 116		1 08 06	−38 06.4	1	13.29		1.54		1.18		1		1696
		Florsch 569		1 08 06	−73 34.1	1	13.15		0.07		0.11		3	A2	767
		PHL 7200		1 08 07	−25 56.7	2	14.29	.010	0.01	.002	0.05	.030	3	A0	561,1736
		CS 29514 # 16		1 08 07	−26 20.8	1	13.19		0.00		0.03		1		1736
6960	+63 00149	HR 342		1 08 10	+63 56.3	5	5.55	.011	-0.07	.004	-0.13	.005	46	B9.5V	61,71,1079,1118,3023
7059	−28 00359			1 08 10	−28 02.7	1	8.87		0.39		-0.01		2	F2 V	561
		Florsch 571		1 08 12	−71 55.6	1	12.66		0.74		0.18		4	G1	767
		LP 767 - 76		1 08 13	−18 15.8	1	14.13		1.50		1.23		2		481
7011	+45 00291			1 08 16	+45 55.1	1	6.60		1.60		1.89		2	K5	1601
7047	+8 00183	G 1 - 51	⋆ A	1 08 17	+09 17.7	3	7.21	.011	0.57	.005	0.09	.007	19	G0	202,333,1620,8040
		AzV 448		1 08 17	−72 46.2	1	14.89		-0.25		-1.10		3	O8.5Vn	425
		AAS35,353 # 117		1 08 19	−72 34.6	1	14.13		-0.15		-0.75		4	B3	767
		V 459 Cas		1 08 20	+60 52.9	1	10.34		0.26		0.24		5	B9	1768
		AzV 449		1 08 20	−72 23.7	2	12.81	.005	-0.08	.037	-0.45	.019	4	B8 Iab	425,835
7034	+30 00181	HR 349		1 08 21	+31 09.6	2	5.15	.004	0.25	.008	0.23		6	F0 V	1363,1501
7019	+36 00201	HR 348	⋆ AB	1 08 21	+37 27.5	2			-0.10	.000	-0.40	.014	5	B7 III-IV	1049,1079
	+61 00220	LS I +62 151		1 08 22	+62 27.8	1	9.68		0.92		0.13		2	B7 Ib	1012
7057	+16 00123			1 08 24	+16 30.7	1	8.20		1.22				1	K2	882
		BPM 46962		1 08 24	−20 07.	1	13.39		0.80		0.31		1		3061
6972	+64 00127	HR 345, RU Cas		1 08 25	+64 45.2	3	5.56	.026	-0.09	.007	-0.29	.000	6	B9 IV	70,1079,3023
7095	−43 00342			1 08 26	−43 03.0	2	9.88	.011	0.32	.015	0.02	.020	14	A9 V	116,863,2011
		CS 22166 # 41		1 08 27	−13 58.7	1	14.36		0.32		-0.15		1		1736
7010	+59 00199			1 08 28	+60 14.4	1	8.15		1.05				1	K0 IV	1103
		BSD 8 # 646		1 08 30	+59 55.2	1	12.08		0.35		0.26		3	B8	1185

Table 1   85

HD	DM	Other Id	N	Rem	α_{1950}	δ_{1950}	S	V	σ_V	B–V	σ_{B-V}	U–B	σ_{U-B}	n	Spectrum	References
7124	−50 00323				1 08 30	−49 53.5	1	9.52		0.23		0.09		4	A4 V	152
		CS 29518 # 15			1 08 33	−27 58.0	1	15.41		0.10		0.16		1		1736
7187	−73 00067	IDS01071S7329	AB		1 08 33	−73 13.4	8	7.13	.016	0.08	.014	0.06	.012	30	A1/2 V	14,35,1408,1425,2001*
	−68 00041				1 08 34	−67 43.1	2	9.79	.005	1.55	.010	1.18	.019	6		158,3079
	+61 00221				1 08 35	+61 45.0	1	9.70		0.27		−0.01		3	B8	1185
		BPM 46963			1 08 36	−25 38.	1	12.32		0.83		0.41		1		3061
		BPM 46964			1 08 36	−32 20.	1	13.17		0.79		0.30		1		3061
7076	+13 00175				1 08 37	+14 25.6	1	7.95	.010	0.99	.024	0.72		3	K0	882,1648
	+60 00177				1 08 38	+60 54.1	2	9.54	.002	0.49	.002			3	F5 V	1103,1118
		BSD 8 # 1676			1 08 39	+62 01.2	1	12.54		0.58		−0.13		3	B0	1185
		AzV 451			1 08 39	−72 38.2	1	14.15		−0.23		−1.07		4	O9	425
	+79 00033				1 08 41	+79 50.0	1	10.15		1.33		1.17		2	K0	1375
		AzV 450			1 08 41	−71 53.8	1	14.85		−0.02		−0.13		4	A0	425
		AAS35,353 # 118			1 08 44	−73 04.9	1	14.15		−0.05		−0.17		5	B9	767
		GR 32			1 08 45	−14 21.5	1	14.66		1.39		1.38		1		481
7087	+20 00172	HR 351			1 08 46	+20 46.2	6	4.66	.007	1.02	.004	0.82	.006	21	K0 III	15,1355,1363,3016*
7136	−32 00469				1 08 48	−32 30.8	1	8.13		0.93				4	G3 III	2012
7222	−73 00068				1 08 49	−73 01.0	1	10.62		0.08		0.06		2	A1 IV/V	1408
		PHL 7202			1 08 51	−21 54.8	4	13.11	.045	0.00	.005	0.06	.010	6	A0	295,966,1736,3064
7107	+9 00138				1 08 52	+10 01.6	1	6.50		0.90		0.58		3	G8 III	985
	+60 00179	IDS01058N6039	AB		1 08 53	+60 55.4	1	8.90		0.42				4	F0 V	70
	−31 00475				1 08 53	−31 10.9	1	10.36		0.62		0.00		1	G0	3061
7106	+29 00190	HR 352			1 08 54	+29 49.5	6	4.51	.005	1.09	.006	1.01	.008	15	K0 IIIb	15,1118,1355,3016*
		GR 34			1 08 55	−30 53.5	1	14.44		1.48		1.21		1		481
		LS I +62 152			1 09 00	+62 02.9	1	10.86		0.62		−0.40		3		1185
		BPM 46967			1 09 00	−23 16.	1	12.63		1.12		0.89		1		3061
7151	−43 00346			V	1 09 00	−42 56.8	1	9.40		1.48				4	F6 V	1075
		Florsch 585			1 09 00	−71 05.9	1	12.61		0.57		−0.01		3	F7	767
	+60 00180	LS I +61 182			1 09 01	+61 03.6	1	9.32		0.25		−0.68		2	B0:pe	1012
		GR 35			1 09 02	−34 24.2	1	14.91		1.46		1.21		1		481
		AzV 453			1 09 02	−71 57.9	1	12.61		−0.06		−0.34		4	A0 Iab	425
		AzV 452			1 09 02	−73 32.7	2	13.56	.005	−0.07	.005	−0.35	.023	5	B9 Ib	425,835
	+60 00181				1 09 04	+61 28.7	1	10.33		0.33		−0.16		3	B7	1185
	−30 00373				1 09 04	−29 45.4	1	9.72		1.07		0.94		2	K0 III	54
		GR 36			1 09 05	−28 23.5	1	13.63		1.40		1.26		2		481
		BPM 46971			1 09 06	−26 40.	1	11.28		0.70		−0.02		4		3061
		BPM 46972			1 09 06	−26 40.	1	13.17		0.47		−0.23		2		3061
7132	+13 00176				1 09 07	+13 36.7	1	8.57		0.35		0.07		2	F0	1648
7147	−3 00161	HR 353			1 09 11	−02 31.0	3	5.93	.025	1.40	.000	1.68	.020	8	K4 III	15,1003,1256
		BPM 46975			1 09 12	−27 32.	1	13.50		0.95		0.70		1		3061
7165	−31 00480				1 09 12	−30 50.0	1	9.09		1.24		1.18		1	K2 III	54
		L 293 - 10			1 09 12	−45 10.	1	13.71		0.61		−0.09		2		1696
		BSD 8 # 662			1 09 13	+60 08.1	1	12.10		0.53		−0.12		3	B0	1185
	−33 00436				1 09 13	−32 52.6	1	9.80		0.69		0.17		1		3061
	−0 00182				1 09 14	−00 21.2	1	10.38		0.73		0.25		2	G5	1375
7083	+63 00156				1 09 15	+64 21.4	2	8.03	.009	0.05	.005	−0.23		4	A0	401,1118
		CS 22174 # 8			1 09 16	−10 55.9	1	13.62		0.28		−0.05		1		1736
		AzV 454			1 09 16	−72 59.0	2	13.63	.095	−0.21	.036	−0.98	.005	5	B2	425,835
	+60 00183				1 09 19	+61 02.8	1	10.66		0.29		0.19		3	A1 V	1185
7172	−26 00384				1 09 19	−26 13.5	1	8.79		0.65		0.05		1	K0 III +(G)	3061
		AzV 455			1 09 20	−72 58.5	1	14.88		−0.29		−1.01		3	B0	425
7103	+61 00223	LS I +61 183			1 09 21	+61 37.2	1	8.35		0.53		−0.28		2	B3 Ib	1012
		BSD 8 # 1686			1 09 22	+62 06.9	1	12.35		0.76				3		1185
7185	−30 00376				1 09 23	−30 12.0	1	9.23		1.20		1.08		2	K2 III	54
		0109+22 # 1			1 09 24	+22 28.	1	11.58		0.56		0.08		1		1482
		0109+22 # 2			1 09 24	+22 28.	1	13.37		0.67		0.14		1		1482
		0109+22 # 3			1 09 24	+22 28.	1	14.15		0.75		−0.13		1		1482
		0109+22 # 4			1 09 24	+22 28.	1	14.26		0.81		0.41		1		1482
		0109+22 # 5			1 09 24	+22 28.	1	15.14		0.62		0.02		1		1482
		0109+22 # 6			1 09 24	+22 28.	1	15.47		0.55		−0.03		1		1482
		0109+22 # 7			1 09 24	+22 28.	1	15.75		0.57		0.03		1		1482
		0109+22 # 8			1 09 24	+22 28.	1	15.74		0.87		0.74		1		1482
		0109+22 # 9			1 09 24	+22 28.	1	12.37		0.74		0.20		1		1482
7104	+61 00224				1 09 24	+61 30.6	2	8.84	.002	0.08	.000			3	B8 V	70,1103
		AzV 456			1 09 26	−72 58.8	2	12.88	.000	0.10	.000	−0.74	.000	4	B0	425,835
7184	−27 00389				1 09 27	−26 37.3	2	9.89	.010	0.21	.005	0.07	.000	5	A1/2 III/IV	561,1068
		CS 22174 # 1			1 09 28	−12 10.6	1	14.39		0.56		0.00		1		1736
		BSD 8 # 196			1 09 29	+58 58.7	1	12.59		0.44		0.36		3	B8	1185
236655	+59 00203	LS I +60 146			1 09 29	+60 02.9	1	9.06		0.49		0.23		3	B8 Ib	1185
7203	−35 00416				1 09 30	−35 03.7	1	8.46		0.60				4	G0 V	2012
		UZ Cas			1 09 31	+60 56.8	1	10.94		0.90				1	G2	1772
		AzV 457			1 09 31	−73 34.1	1	14.13		−0.09		−1.07		5	O9	425
		CS 22946 # 4			1 09 32	−20 42.0	2	15.46	.000	0.02	.000	0.17	.011	2		966,1736
		G 271 - 11			1 09 34	+00 52.9	1	11.45		0.59		0.06		1		1658
7194	−5 00207				1 09 37	−04 45.5	2	8.89	.005	0.06	.005	0.10	.023	20	A0	152,1775
		AzV 459			1 09 39	−72 43.3	1	13.76		−0.10		−0.69		3	B3 Ib	425
7158	+44 00261	HR 355			1 09 40	+45 04.3	2	6.11	.000	1.64	.005	2.01	.012	7	M1 III	1733,3001
		AzV 458			1 09 41	−72 13.3	1	14.34		−0.06		−1.04		5	O9	425
7181	+31 00196	IDS01074N3133	B		1 09 44	+31 56.8	2	8.30	.013	0.20	.004	0.09	.013	5	B9	904,3032
		BSD 8 # 1244			1 09 44	+61 03.2	1	11.98		0.39				3	B5	1185

HD	DM	Other Id	N Rem	α_{1950}	δ_{1950}	S	V	σ_V	B–V	σ_{B-V}	U–B	σ_{U-B}	n	Spectrum	References
		BSD 8 # 1698		1 09 45	+61 36.5	1	10.93		0.30		-0.03		3	B9	1185
		G 269 - 119		1 09 46	-32 42.4	1	12.19		0.72		0.12		1		3061
	+59 00205	LS I +59 064		1 09 47	+59 39.3	1	10.56		0.29		-0.49		3	B1 V	1185
7235	-38 00418			1 09 47	-38 12.5	1	8.62		1.54		1.67		4	M2/3 III	3007
236657	+55 00266			1 09 48	+55 40.4	1	8.59		0.20		0.12		2	A0	1502
		CS 29514 # 15		1 09 48	-26 29.4	1	13.24		-0.29		-0.98		1		1736
		PHL 1003		1 09 48	-26 31.	4	13.15	.013	-0.24	.008	-1.04	.010	11	DA:	295,561,1036,3064
7220	-26 00387			1 09 48	-26 34.4	1	8.41		1.12				4	K1 III	2012
		AzV 460		1 09 52	-72 14.5	1	14.12		-0.01		-0.99		6	O9	425
	+60 00185			1 09 53	+61 08.8	1	10.00		0.22		0.00		3	B7 V	1185
		SB 486		1 09 54	-21 44.	1	12.97		0.08		0.11		2	A0	295
		LB 1595		1 09 54	-51 13.	1	14.89		-0.25		-1.17		1		832
		AzV 461		1 09 54	-72 25.7	1	14.66		-0.31		-1.03		4	B0	733
7261	-45 00398			1 09 55	-44 45.8	1	9.50		0.54				4	F7 V	2011
		AAS35,353 # 120		1 09 56	-72 42.5	1	14.06		-0.20		-0.75		5	B2	767
		AzV 462		1 09 56	-72 47.3	2	12.57	.009	-0.16	.014	-0.88	.009	4	B1 Ia	425,835
7208	+0 00197	G 2 - 28		1 09 57	+00 43.1	2	8.70	.010	0.93	.015	0.72		3	K0	202,333,1620
7157	+60 00186	HR 354	★ AB	1 09 58	+61 26.5	3	6.40	.018	0.01	.006	-0.28	.009	4	B9 V	401,1079,1103
		CS 22946 # 3		1 09 58	-20 58.5	2	14.81	.010	0.50	.011	-0.18	.005	2		1580,1736
	+61 00226			1 09 59	+62 06.9	1	9.23		1.28				1	K2	1103
		BPM 46979		1 10 00	-27 55.	1	13.23		1.10		1.17		1		3061
		G 268 - 135		1 10 01	-17 16.2	5	12.06	.021	1.84	.018	1.33	.048	14		481,940,1619,1705,3078
		CS 29514 # 11		1 10 02	-23 05.0	1	12.52		0.10		0.16		1		1736
7259	-31 00484	HR 358		1 10 02	-31 04.0	3	6.51	.006	0.48	.000	0.03		8	F5 IV	15,54,2012
7269	-38 00419			1 10 02	-38 30.8	1	9.02		0.52		0.02		1	F7 V	858
	-23 00441			1 10 03	-22 48.4	1	9.91		0.50		-0.08		1	F8	3061
		POSS 295 # 8		1 10 04	+32 20.8	1	13.96		1.27				2		1739
		BSD 8 # 1701		1 10 04	+62 05.4	1	9.99		0.49				1	A2 p	1103
		AzV 464		1 10 05	-73 28.2	2	13.60	.059	-0.08	.009	-1.06	.000	5	O9	425,835
7215	+31 00197	IDS01074N3133	A	1 10 06	+31 48.6	2	7.00	.021	0.04	.005	-0.01	.020	5	A0 V	904,3032
		GR 37		1 10 07	-17 34.8	1	14.06		1.34		1.03		1		481
		LP 467 - 20		1 10 08	+10 19.0	1	11.93		1.12		1.10		1		1696
		AzV 466		1 10 09	-72 36.1	1	13.91		0.04		-0.14		3	B8 Ib	733
7231	+15 00175			1 10 10	+16 30.0	1	8.27		1.00				2	K0	882
		AzV 463		1 10 12	-72 23.4	2	12.16	.023	0.08	.000	0.12	.014	5	A2 Ib	425,835
7205	+40 00248	G 132 - 62		1 10 13	+41 23.4	1	7.27		0.77		0.31		1	G5	333,1620
7229	+29 00195	HR 356	★ A	1 10 14	+29 48.0	1	6.21		1.01		0.75		3	G9 III	542
7229	+29 00195	HR 356	★ AB	1 10 14	+29 48.0	2	6.19	.004	1.00	.006	0.73		3	G9 III+G1 V	71,542
7229	+29 00195	IDS01075N2932	B	1 10 14	+29 48.0	1	10.25		0.79		0.29		3		542
		BSD 8 # 1264		1 10 14	+60 22.2	1	11.49		0.43		-0.34		3	B4	1185
		BSD 8 # 1705		1 10 16	+62 08.4	1	10.55		0.63				1	G1	1103
7268	-7 00196			1 10 17	-07 02.9	1	6.60		0.94		0.75		3	G8 III	3016
7280	-29 00376	AN Scl	★ AB	1 10 18	-29 26.7	1	8.80		0.89		0.49		1	K0 III	3061
		JL 234		1 10 18	-56 30.	1	15.68		0.15		-1.26		1		132
7226	+38 00214			1 10 20	+38 48.4	1	8.30		1.24		1.32		3	K2 III	833
		BSD 8 # 1265		1 10 20	+60 41.6	1	11.81		0.33		0.24		3	B9	1185
		AzV 465		1 10 21	-73 00.0	1	15.06		-0.30		-1.04		3	B0	425
		BSD 8 # 1708		1 10 22	+62 07.4	1	10.97		0.62				1	F8	1103
7279	-25 00480			1 10 22	-25 30.0	1	9.59		1.26		1.15		1	K4 V	3061
		BPM 46982		1 10 24	-25 01.	1	13.45		1.50		1.35		1		3061
		AzV 467	AB	1 10 25	-73 00.1	1	14.55		-0.18		-1.00		5	B0	425
		BSD 8 # 1268		1 10 26	+60 37.7	1	10.84		0.23		0.02		3	B9	1185
7311	-35 00420			1 10 27	-35 28.2	1	6.96		1.05		0.86		4	K0 III	1628
7312	-38 00420	HR 359, AI Scl		1 10 27	-38 07.3	3	5.91	.009	0.29	.009	0.12		8	F0 III	15,858,2012
7254	+33 00187			1 10 28	+33 50.0	1	6.75		-0.15		-0.47		2	B8	401
		BPM 46983		1 10 30	-26 02.	1	14.02		1.58				1		3061
		BPM 46986		1 10 30	-33 14.	1	12.79		0.95		0.65		2		3061
		SB 490		1 10 31	-18 13.5	2	13.45	.000	0.09	.000	0.09	.010	2	A0	966,1736
236664	+58 00195	LS I +58 025		1 10 34	+58 50.1	2	9.99	.030	0.23	.020	-0.62	.059	5	B0.5V	1012,1185
7288	+13 00180			1 10 35	+14 16.5	1	9.60		1.19		1.02		2	K0	3040
		BPM 46984		1 10 36	-24 48.	1	14.62		0.50		-0.15		1		3061
		Dril 42		1 10 36	-25 20.	1	12.65		0.40		-0.11		2	F3	561
7323	-36 00460			1 10 36	-36 00.6	2	7.83	.000	0.10	.018	0.05		7	A0 V	1068,2012
		BSD 8 # 695		1 10 37	+59 32.6	1	10.87		0.30		-0.46		3	B5	1185
		CS 22174 # 2		1 10 37	-12 09.2	1	13.01		0.09		0.08		1		1736
		WLS 120 5 # 6		1 10 38	+04 43.8	1	11.01		0.48		0.00		2		1375
		AzV 468		1 10 38	-72 56.8	1	15.14		-0.27		-1.11		3	O8.5V	425
	+80 00032			1 10 40	+81 12.9	1	9.85		0.01		-0.70		3	B5	1219
		AzV 469		1 10 40	-72 43.6	2	13.21	.014	-0.22	.005	-1.00	.009	4	B1 Ia	425,835
		BSD 8 # 1709		1 10 41	+62 09.4	1	10.83		0.89				1	G2	1103
		GR 38		1 10 44	-34 54.2	1	15.00		1.67				1		3061
7320	-2 00181	G 270 - 174		1 10 46	-02 07.6	5	8.95	.009	0.69	.004	0.13	.003	10	G0	516,742,1064,1375,1658
	-2 00180	G 70 - 48		1 10 47	-01 59.4	2	10.06	.005	0.97	.019	0.81		3	K0	202,1620
7442	-74 00088			1 10 47	-74 10.4	1	7.20		0.59		0.12		6	F8/G0 V	1628
		LB 3186		1 10 48	-61 08.	1	11.22		-0.05		-0.21		2		45
		Florsch 615		1 10 48	-73 33.6	1	12.95		0.11		0.10		4	A2	767
		BSD 8 # 696		1 10 49	+59 53.5	1	10.91		0.37		-0.16		3	B9	1185
	-34 00464			1 10 49	-34 11.7	1	10.59		0.60		0.11		1		1730
		CS 22946 # 2		1 10 50	-21 52.0	2	14.59	.015	0.35	.009	-0.25	.006	2		1580,1736
7252	+60 00188	LS I +60 147		1 10 53	+60 37.1	2	7.13	.011	0.07	.019	-0.74		3	B1 V	1012,1103

Table 1   87

HD	DM	Other Id	N Rem	α_{1950}	δ_{1950}	S	V	σ_V	B–V	σ_{B-V}	U–B	σ_{U-B}	n	Spectrum	References
	−34 00465			1 10 53	−34 02.2	1	10.25		0.34		0.04		2	A5	1730
5914	+88 00004	HR 286		1 10 57	+88 45.3	5	6.46	.010	0.11	.006	0.09	.015	93	A3 V	15,252,1258,1462,1733
		AzV 470		1 10 59	−71 24.9	1	13.14		0.01		−0.13		3	B9 Ib	425
		BPM 70605		1 11 00	−19 27.	1	12.56		0.69		0.10		2		3061
		BPM 46989		1 11 00	−23 10.	1	14.16		1.58		0.96		1		3061
		BPM 46990		1 11 00	−26 16.	1	14.18		0.83		0.22		1		3061
		LP 883 - 78		1 11 00	−31 53.5	1	13.63		1.48		1.11		2		481
		POSS 52 # 2		1 11 01	+66 44.3	1	15.76		1.59				1		1739
7318	+23 00158	HR 360	⋆ A	1 11 02	+24 19.2	1	4.65		1.06		0.89		2	K0 III	542
7318	+23 00158	HR 360	⋆ AB	1 11 02	+24 19.2	8	4.66	.014	1.04	.009	0.86	.021	19	K0 III	15,542,1355,1363,3016*
7318	+23 00158	IDS01083N2403	B	1 11 02	+24 19.2	1	9.11		0.92		0.25		2	K0	542
236669	+55 00271			1 11 02	+55 44.9	1	7.98		1.47		1.65		2		1733
7357	−19 00200			1 11 02	−19 25.8	1	9.70		0.62		0.06		1	G5 V(w)	3061
	+60 00190			1 11 03	+60 41.5	1	9.75		0.28		−0.12		3	B8	1185
7251	+63 00161			1 11 03	+64 21.7	1	9.15		0.51				2	F8	1118
7307	+26 00199	RT Psc		1 11 04	+26 52.1	1	7.93		1.53		1.56		1	M1	3076
	+24 00191			1 11 06	+24 41.8	1	9.98		0.23		0.13		1	A0	1307
		LP 883 - 332		1 11 06	−32 24.5	1	13.12		0.56		−0.13		2		1696
		LP 883 - 77		1 11 06	−32 24.5	1	13.14		0.58		−0.15		2		3061
7344	+6 00174	HR 361	⋆ A	1 11 07	+07 18.7	1	5.24		0.27		0.09		4	A7 IV	3024
7344	+6 00174	IDS01085N0703	ABC	1 11 07	+07 18.7	2	4.85	.005	0.33	.005	0.07	.015	7	A7 IV+F7 V	15,1256
7345	+6 00175	HR 362	⋆ BC	1 11 08	+07 18.9	2	6.29	.025	0.49	.010	0.02	.030	7	F7 V	254,3053
		AAS35,353 # 121		1 11 12	−73 16.7	1	14.26		−0.05		−0.42		3	B8	767
		CS 22174 # 5		1 11 14	−11 32.4	1	13.88		−0.02		0.00		1		1736
		WLS 112 60 # 5		1 11 15	+61 48.4	1	10.97		0.77		0.22		2		1375
		BSD 8 # 1717		1 11 16	+61 26.1	1	11.41		0.35		−0.19		3	B3	1185
	+60 00191	LS I +61 184		1 11 17	+61 04.2	1	9.94		0.61		−0.14		2	B2 IV	1012
		E1 - X		1 11 18	−44 29.	1	12.04		0.61				4		1075
		Florsch 624		1 11 18	−71 10.2	1	12.28		0.52		−0.01		2	F6	767
	+15 00176			1 11 19	+16 13.9	2	9.84	.015	1.22	.005	1.20	.010	2	K5	1776,3072
	+60 00192			1 11 19	+60 37.6	1	9.34		0.24		−0.23		3	B5	1185
7351	+27 00196	HR 363		1 11 20	+28 16.0	2	6.42	.005	1.68	.015	1.90	.025	11	M2	1501,3055
		BSD 8 # 704		1 11 20	+59 52.8	1	12.22		0.48		0.31		3	A0	1185
7413	−34 00467			1 11 21	−33 59.7	1	9.53		0.50		0.05		2	F7/8 V	1730
	−72 00087	AJ63,118 # 22		1 11 21	−71 52.1	1	10.87		0.46				5		2001
7343	+34 00213			1 11 22	+34 45.5	1	8.70		1.26		1.27		2	K0	1625
7352	+25 00197			1 11 23	+25 33.8	1	8.35		0.55		0.03		3	G0 V	833
7402	−31 00497	IDS01090S3139	AB	1 11 23	−31 22.9	2	7.74	.009	1.08	.005	0.93		5	G8 III/IV	54,2012
		AzV 472		1 11 23	−72 59.9	2	12.65	.005	−0.13	.014	−0.83	.000	4	B2 Ia	425,835
		POSS 295 # 14		1 11 24	+28 25.9	1	14.39		1.16				3		1739
		BPM 46995		1 11 24	−26 56.	1	14.30		0.41		−0.10		1		3061
		AzV 471		1 11 24	−73 31.4	2	13.65	.005	−0.07	.005	−0.86	.000	5	B2 Ib	573,733
7385	−8 00214	IDS01089S0809	A	1 11 25	−07 53.2	1	7.55		0.99		0.79		3	G5	1657
7374	+15 00177	HR 364		1 11 28	+15 52.2	4	5.96	.020	−0.08	.007	−0.40	.013	9	B8 III	252,1049,1079,3016
		BSD 8 # 1719		1 11 28	+61 22.3	1	11.18		0.36		0.35		3	A0	1185
		CS 22953 # 23		1 11 28	−61 29.4	2	13.70	.005	0.52	.011	−0.22	.004	3		1580,1736
		BSD 8 # 705		1 11 32	+60 15.7	1	11.93		0.75		0.19		3	B5	1185
7519	−72 00088			1 11 32	−71 53.9	3	9.74	.000	0.43	.011	−0.07	.005	8	F3/5 V	1408,1425,2001
		CS 29518 # 29		1 11 33	−30 02.2	1	12.25		0.15		0.05		1		1736
		AJ63,118 # 32		1 11 33	−71 50.5	1	11.53		0.53				5		2001
236673	+59 00210			1 11 34	+60 19.6	1	9.55		0.94				3	K2	70
7331	+60 00193	IDS01084N6025	AB	1 11 34	+60 40.5	2	7.25	.016	0.46	.016	−0.04		3	F6 V +F6 V	1103,3030
	+6 00179			1 11 35	+07 02.9	1	8.95		1.79		2.05		6	M0	3049
7428	−36 00470			1 11 35	−36 27.0	1	8.36		1.10				4	K1 III	2012
		SB 503		1 11 36	−26 40.	1	11.53		0.24		−0.03		2	A5	561
7330	+61 00229			1 11 39	+62 25.0	2	8.49	.008	0.07	.009			3	A0	70,1103
		Steph 127	AB	1 11 39	−09 02.7	1	12.26		1.20		1.26		1	K7	1746
7424	−17 00219	G 268 - 138		1 11 39	−16 41.4	4	10.08	.022	0.72	.004	0.03	.012	5	G8 V	79,516,1064,3077
		AzV 474		1 11 39	−71 51.8	1	14.35		−0.09		−0.34		6	B8	425
	−36 00472			1 11 40	−36 07.8	1	10.06		0.65		0.21		1	G5	3061
		BSD 8 # 1721		1 11 41	+62 05.7	1	10.30		0.42		0.08		3	B8	1185
	+7 00188			1 11 42	+07 34.3	1	9.93		0.37		0.11		2	A5	1375
	+23 00159	RU Psc		1 11 42	+24 09.1	1	9.94		0.22		0.14		1	A2	668
	+59 00214	LS I +60 149		1 11 42	+60 13.7	1	10.05		0.63		−0.16		3	B4	1185
		LP 883 - 82		1 11 42	−28 46.	1	14.70		1.62				1		3061
		AzV 473		1 11 43	−73 36.1	1	12.81		0.59		0.43		3	F5	35
7393	+18 00163			1 11 44	+18 51.8	1	7.96		1.00		0.81		2	G5	1648
		BPM 46996		1 11 48	−25 20.	1	12.29		0.80		0.23		2		3061
		G 269 - 125		1 11 49	−37 12.7	1	13.85		1.66		1.34		1		3061
		BSD 8 # 720		1 11 50	+59 52.4	1	11.97		0.51		−0.23		3		1185
		BSD 8 # 1304		1 11 51	+60 51.5	1	11.02		0.42		−0.36		3	B2	1185
7438	−8 00215	G 270 - 176	⋆ B	1 11 51	−08 10.7	2	7.85	.022	0.78	.002	0.35	.002	4		58,3077
		BSD 8 # 1725		1 11 52	+61 24.6	1	11.69		0.56		0.23		3	B7 p	1185
7392	+38 00220			1 11 53	+39 11.1	1	7.72		0.96		0.63		2	K0 III	105
7439	−8 00216	HR 366	⋆ A	1 11 53	−08 11.5	7	5.13	.029	0.45	.016	−0.06	.010	22	F5 V	15,258,1197,1256,2012*
		SB 505		1 11 54	−21 18.	1	12.16		0.21		0.12		1	A5	295
		BPM 46997		1 11 54	−28 56.	1	11.83		0.59		0.09		1		3061
	+59 00216			1 11 55	+60 13.8	1	9.49		0.24		−0.10		3	B7	1185
		LP 707 - 37		1 11 55	−10 32.3	1	12.25		1.20		1.16		1		1696
		AA9,95 # 305		1 11 55	−73 37.7	1	13.30		1.10		0.79		3	G7 III	35

HD	DM	Other Id	N Rem	α_{1950}	δ_{1950}	S	V	σ_V	B−V	σ_{B-V}	U−B	σ_{U-B}	n	Spectrum	References
		G 172 - 45		1 11 56	+42 53.3	1	9.98		0.62		0.07		2	G5	7010
		BSD 8 # 718		1 11 56	+60 14.7	1	10.31		0.50		-0.04		3	B9	1185
		POSS 295 # 5		1 11 58	+33 12.3	1	13.86		0.59				2		1739
7370	+60 00194			1 11 59	+60 36.0	2	8.86	.000	0.13	.001			3	B8 II	1103,1118
	−31 00505			1 11 59	−30 43.7	1	9.97		0.99		0.73		1	G8 III	54
		LS I +58 026		1 12 00	+58 02.3	1	11.24		0.19		-0.14		1		1215
7238	+79 00036	HR 357		1 12 00	+79 38.7	2	6.27	.021	0.42	.001	-0.09		3	F5 Vs	71,254
		JL 236		1 12 00	−53 00.	1	13.45		-0.28		-1.04		1	sdB	132
	+59 00217			1 12 01	+59 53.3	1	9.23		1.12				2	F6 Iab:	70
		BSD 8 # 723		1 12 02	+60 15.0	1	12.00		0.50		-0.09		3	B5	1185
		CS 22174 # 3		1 12 03	−12 09.7	1	14.85		-0.01		0.02		1		1736
		POSS 52 # 3		1 12 05	+65 26.2	1	16.78		1.63				2		1739
7446	+6 00181	HR 367		1 12 06	+06 43.9	4	6.03	.007	1.08	.000	1.02	.004	10	K0	15,1061,1355,1738
7583	−74 00090			1 12 07	−73 36.0	6	10.21	.019	0.13	.005	-0.34	.006	67	A0 Ia/0	35,573,835,1277,1408,8033
		BSD 8 # 727		1 12 08	+59 51.3	1	11.04		0.40		-0.15		3	B7	1185
		CS 22946 # 20		1 12 10	−18 46.7	2	14.88	.000	0.13	.000	0.19	.009	2		966,1736
7487	−32 00484			1 12 10	−31 36.4	1	8.76		0.54				4	F7 V	2012
7498	−42 00432			1 12 11	−42 09.5	2	8.47	.000	0.56	.000	0.06		8	G0 V	158,2033
	+59 00219			1 12 13	+60 08.4	1	9.88		0.37		-0.08		3	B8	1185
7516	−51 00313			1 12 13	−50 40.8	2	7.34	.000	1.13	.002	1.10		11	K1 III	1673,2012
		G 269 - 127		1 12 15	−30 29.6	1	15.48		1.13		0.86		1		3061
7476	−1 00162	HR 368	★ A	1 12 16	−01 14.4	3	5.69	.008	0.43	.004	-0.02	.015	10	F5 IV-V	15,1061,3016
		GR 40		1 12 17	−16 41.7	1	14.36		1.49		1.28		1		481
		AzV 476		1 12 17	−73 33.2	1	13.55		-0.07		-0.88		2	B2 II	35
		AzV 477		1 12 18	−73 23.7	2	13.14	.019	-0.17	.000	-0.82	.005	4	B2 Iab	425,835
7526	−48 00313			1 12 22	−48 31.1	1	10.02		1.32		1.39		3	C1,2	864
		BPM 46999		1 12 24	−23 09.	1	13.65		1.45		1.28		1		3061
		BSD 8 # 1318		1 12 27	+60 51.6	1	11.97		0.56		0.17		3	B8	1185
7415	+63 00164			1 12 28	+64 03.7	1	8.81		0.23				2	A0	1118
7533	−45 00412			1 12 28	−45 31.9	3	8.90	.005	0.43	.000	-0.01	.016	16	F3 V	977,1075,2011
		BPM 47000		1 12 30	−20 43.	1	13.03		1.00		0.82		2		3061
		Dril 44		1 12 30	−28 14.	1	14.00		0.29		-0.04		2	A2	561
7432	+58 00203	IDS01094N5846	AB	1 12 31	+59 01.7	1	9.62		0.16				2	A2 V	1119
7634	−72 00089			1 12 31	−71 58.7	3	9.79	.005	1.04	.014	0.89	.025	9	G8/K0	1408,1425,2001
		BPM 47002		1 12 36	−31 26.	1	12.59		0.79		0.30		1		3061
		Smethells 184		1 12 36	−54 12.	1	11.09		1.48				1		1494
		BSD 8 # 732		1 12 44	+59 41.1	1	12.24		0.69		0.18		3		1185
		AW Cas		1 12 49	+61 21.4	1	11.75		1.26				1		1772
7458	+61 00233			1 12 51	+61 38.7	1	7.30		0.38				1	F0 V	1103
7621	−65 00125			1 12 51	−65 01.4	3	9.03	.005	0.76	.006	0.38		9	K0 IV/V	158,1705,2033
		AA9,95 # 312		1 12 53	−73 39.0	1	10.72		0.71		0.15		4	G0 V	35
7568	−28 00383			1 12 56	−28 31.2	1	9.40		0.94		0.70		1	K0 IV	54
7570	−46 00346	HR 370		1 12 56	−45 47.9	9	4.96	.009	0.57	.006	0.10	.006	29	G0 V	15,116,278,977,1020*
7581	−45 00415			1 12 57	−44 53.1	4	8.36	.009	0.49	.003	0.06	.008	18	F5 IV/V	58,977,1075,2011
		AzV 478		1 12 57	−73 28.5	2	11.55	.010	0.10	.000	0.03	.010	7	A0 Ia	35,573
	+59 00221			1 12 58	+59 34.3	1	10.14		0.40		0.22		3	A0	1185
		BPM 47004		1 13 00	−25 28.	1	13.86		0.94		0.60		1		3061
		BPM 47005		1 13 00	−25 41.	1	13.09		0.90		0.55		1		3061
		BPM 47008		1 13 00	−33 08.	1	11.11		0.61		0.07		1		3061
	−30 00403			1 13 01	−30 32.8	1	11.00		0.55		0.01		2		3061
		G 268 - 141		1 13 05	−15 51.7	1	13.17		1.46		1.19		2		481
	−26 00411	IDS01107S2621	AB	1 13 05	−26 04.7	1	11.02		0.50		-0.05		2	F5	3061
7595	−28 00384			1 13 07	−28 31.7	1	9.72		1.20		1.14		1	K0/1 III	54
		BSD 8 # 1742		1 13 11	+61 42.5	1	10.25		0.45		-0.32		3	B5 p	1185
		BPM 47009		1 13 12	−21 19.	1	12.39		0.99		0.76		1		3061
		BPM 47013		1 13 12	−32 52.	1	11.83		0.85		0.44		1		3061
7340	+80 00034			1 13 14	+81 17.8	1	7.61		0.97		0.67		3	G5	1733
7563	+15 00181			1 13 16	+15 55.6	2	7.70	.015	0.10	.010	0.10	.005	3	A2 III	1648,1776
7580	−10 00263			1 13 17	−10 03.1	1	8.31		0.17		0.06		4	A0	152
		BPM 47011		1 13 18	−23 17.	1	13.36		0.93		0.72		1		3061
		BPM 47012		1 13 18	−26 27.	1	12.20		0.62		-0.06		1		3061
7693	−69 00044		★ CD	1 13 19	−69 05.1	1	7.22		1.00				3	K2 V	2024
		AA9,95 # 314		1 13 20	−73 25.9	1	12.22		0.54		-0.01		3	F7 V	35
7561	+25 00205	Z Psc		1 13 21	+25 30.3	1	6.84		2.62		3.39		1	N0	109
		BSD 8 # 736		1 13 21	+59 56.4	1	10.74		0.45		-0.42		3	B3	1185
		CS 29514 # 25		1 13 23	−24 22.6	1	14.69		0.16		-0.41		1		1736
		BPM 47014		1 13 24	−24 34.	1	13.47		0.90		0.55		3		3061
		BPM 47015		1 13 24	−28 47.	1	12.12		1.31		0.93		1		3061
		LP 883 - 120		1 13 24	−29 26.	1	13.03		1.05		0.91		1		3061
		AzV 479		1 13 26	−73 36.0	2	12.46	.027	-0.15	.005	-0.98	.005	5	B1 Ia	35,573
7546	+47 00357	HR 369		1 13 27	+47 49.1	2	6.61		-0.05	.004	-0.34	.019	5	B9 IIIp Si	220,1079
		LP 707 - 43		1 13 27	−11 26.3	1	13.03		0.69		0.14		2		1696
		AzV 480		1 13 27	−72 37.4	1	14.48		-0.03		-1.11		4	O9	425
	−10 00264	G 270 - 183		1 13 30	−09 45.5	1	10.89		0.72		0.14		2		1696
7629	−24 00548			1 13 30	−24 14.2	1	7.13		0.30				4	A9 IV	2011
7602	+16 00129			1 13 31	+17 22.3	3	8.24	.016	1.40	.005	1.69	.010	5	K2	882,979,1648
7578	+32 00223	HR 371		1 13 31	+32 51.1	1	6.02		1.15				2	K1 III	71
		CS 22174 # 10		1 13 31	−10 09.2	1	14.69		0.11		0.21		1		1736
		AzV 481		1 13 31	−72 57.8	1	14.91		-0.24		-1.02		4	O9	425
7755	−73 00071			1 13 31	−73 31.3	1	9.61		1.24		1.30		7	K0/1 III	35

Table 1   89

HD	DM	Other Id	N	Rem	α_{1950}	δ_{1950}	S	V	σ_V	B–V	σ_{B-V}	U–B	σ_{U-B}	n	Spectrum	References
7604	+12 00155	IDS01109N1312	A		1 13 32	+13 28.3	1	8.47		1.42				3	K0	882
	−31 00517				1 13 32	−30 58.4	1	10.30		0.67		0.14		1	G5	3061
7603	+16 00130				1 13 33	+16 33.9	2	7.68	.005	1.57	.000	1.82		3	K2 V	882,3077
		BSD 8 # 1330			1 13 34	+60 37.8	1	10.87		0.25		0.02		2	B9	1185
7616	+15 00182				1 13 36	+15 54.2	1	7.91		0.48		−0.04		1	F2	1776
		G 172 - 46			1 13 37	+51 35.8	1	11.15		0.91		0.56		1		1658
	−26 00414				1 13 38	−25 57.5	1	11.99		0.29		−0.01		1	A7	561
		BSD 8 # 1331			1 13 40	+60 30.4	1	11.17		0.32		0.19		3	B8	1185
		G 34 - 15			1 13 41	+24 04.1	1	15.05		1.86		1.36		4		203
		BPM 47017			1 13 42	−26 45.	1	15.34		0.48		−0.11		1		3061
		Dril 46			1 13 42	−26 48.	1	12.46		0.43		−0.13		2	F3	561
7652	−24 00549				1 13 43	−24 09.9	2	10.03	.023	0.12	.005	0.10	.014	7	A1 IV/V	152,1068
7615	+22 00204				1 13 45	+23 19.6	2	6.69	.001	0.04	.006	0.04	.014	6	A0	1307,1499
		LP 2 - 450			1 13 47	+83 53.8	1	14.76		1.81				6		1663
7676	−34 00483	VV Scl			1 13 48	−34 24.8	1	8.40		0.20				4	Ap SrCrEu	2012
7675	−32 00492				1 13 50	−31 39.4	2	8.69	.005	1.01	.000	0.79		5	G8 III	54,2012
		AzV 482			1 13 51	−73 35.8	1	14.48		−0.25		−1.07		5	O9	425
		BSD 8 # 1334			1 13 52	+60 30.0	1	10.82		0.35		0.14		3	B6	1185
7650	+8 00200	G 33 - 48			1 13 53	+09 21.7	1	8.75		0.58		−0.02		1	G0	333,1620
		G 69 - 62			1 13 54	+25 04.1	3	10.08	.009	1.36	.021	1.22	.013	4	K4-5	333,1017,1620,3016
7640	+17 00176	G 34 - 16			1 13 55	+18 15.5	1	8.46		0.60		0.07		1	G5	333,1620
		CS 22174 # 11			1 13 57	−09 19.2	1	14.04		0.05		0.13		1		1736
7674	−14 00244				1 13 58	−13 48.7	1	8.47		1.63		1.90		4	M2 (III)	3007
7706	−42 00446				1 13 58	−42 16.4	9	6.58	.006	1.20	.002	1.27	.009	131	K2 III	116,278,861,863,977*
7586	+59 00223				1 14 00	+60 00.7	3	9.25	.031	0.22	.020	0.26		7	A0 V	70,1103,1185
7702	−28 00387				1 14 00	−28 04.6	1	9.55		1.35		1.50		1	K1 III/IV	54
7660	+1 00238				1 14 01	+02 28.3	1	7.54		0.36		0.02		2	F0	1375
		CS 29514 # 24			1 14 02	−24 32.7	1	15.21		0.15		0.10		1		1736
7672	−3 00172	HR 373, AY Cet		⋆ A	1 14 04	−02 45.8	3	5.42	.022	0.90	.005	0.45	.010	9	G5 IIIe	15,1061,3016
7788	−69 00045	HR 377		⋆ AB	1 14 05	−69 08.5	5	4.86	.005	0.48	.008	0.01	.009	22	F5 V	15,1075,2024,2038,3026
		BPM 47023			1 14 06	−23 43.	1	13.85		0.98		0.75		1		3061
		AzV 483			1 14 06	−73 35.6	2	11.88	.025	−0.10	.015	−0.86	.005	8	B1.5Ia	35,573
	+59 00224				1 14 07	+60 07.4	2	10.10	.007	0.30	.022	0.21		4	B7	1103,1185
7659	+20 00186				1 14 08	+20 47.4	1	7.03		0.41		−0.01		3	F6 V	1501
		CS 29514 # 20			1 14 09	−27 11.3	1	14.77		0.06		0.15		1		1736
7647	+44 00271	HR 372			1 14 10	+44 38.4	3	6.12	.014	1.59	.013	1.98	.016	5	K5 III	1601,1716,1733
		BPM 47024			1 14 12	−20 22.	1	13.00		0.63		0.01		1		3061
		SB 516			1 14 12	−27 10.	2	14.75	.042	0.09	.009	0.25	.042	4	A1	561,3064
236683	+58 00210				1 14 13	+58 37.5	1	9.80		0.27		−0.51		3	B0	1185
7637	+55 00279				1 14 14	+55 39.3	1	7.66		1.56		1.62		2	K2	1502
236685	+55 00278				1 14 14	+56 09.1	1	8.13		1.66				2	K5	369
		AzV 484			1 14 16	−73 39.5	1	14.02		−0.25		−1.06		4	O9	425
7636	+56 00240	LS I +57 010			1 14 18	+57 22.1	1	6.61		0.14		−0.69		3	B2 IIIne	1212
7471	+80 00035	IDS01097N8020	AB		1 14 18	+80 35.9	1	7.17		0.38		0.04		2	F0	3032
7471	+80 00035	IDS01097N8020	D		1 14 18	+80 35.9	1	11.64		0.61		0.07		2		3032
7471	+80 00035	IDS01097N8020	E		1 14 18	+80 35.9	1	15.58		1.08				1		3032
		BPM 47026			1 14 18	−27 21.	1	13.67		0.50		−0.11		1		3061
		BPM 47027			1 14 18	−28 50.	1	12.86		0.51		−0.09		1		3061
7700	+6 00189		V		1 14 19	+06 32.9	1	8.99		0.70		0.19		1	G5	1768
		CS 29518 # 26			1 14 22	−32 08.6	1	11.85		0.22		0.04		1		1736
		G 269 - 131			1 14 22	−33 48.1	1	13.93		0.72		0.04		1		3061
	+84 00019	RU Cep			1 14 24	+84 52.2	1	8.70		1.78		1.95		1	M0 III	793
		CS 29518 # 28			1 14 24	−31 20.8	1	15.00		0.09		0.18		1		1736
7669	+38 00229				1 14 25	+39 13.0	1	6.69		0.03		0.06		2	A0 V	105
7727	−3 00174				1 14 25	−02 32.4	2	6.52	.005	0.57	.005	0.04	.017	3	G0 V	258,3026
		AzV 485			1 14 25	−72 39.7	1	14.91		0.02		−0.12		4	B3	733
		NGC 457 - 295			1 14 26	+58 05.6	1	12.67		0.48		0.36		1	A0	63
		G 269 - 132			1 14 27	−29 01.7	1	12.62		0.56		−0.08		1		3060
		LP 883 - 128			1 14 27	−29 01.7	1	13.46		0.52		−0.27		3		1696
7752	−32 00499				1 14 27	−31 40.3	1	8.68		1.11		1.05		1	K0 III	54
		AzV 486			1 14 29	−73 36.5	2	12.14	.025	−0.18	.025	−0.95	.005	9	B0 Iab-Ib	35,573
		BPM 47029			1 14 30	−32 04.	1	13.86		0.70		0.23		1		3061
		AzV 487			1 14 30	−73 34.8	2	12.68	.025	−0.17	.005	−0.95	.000	5	B0 Ia	35,573
	+60 00200				1 14 31	+61 28.7	2	9.70	.022	0.19	.036	−0.20		4	B5	1103,1185
		CS 29514 # 28			1 14 35	−22 55.3	1	15.15		0.24		−0.04		1		1736
		XX And			1 14 36	+38 41.2	2	10.22	.049	0.19	.015	0.11	.042	2	A2	668,699
		SB 519			1 14 36	−27 15.	2	13.41	.005	0.03	.005	0.14	.035	4	A0	295,561
		SB 520			1 14 36	−28 01.8	3	14.58	.018	0.05	.005	0.16	.029	4	A0	561,1736,3064
		BPM 47030			1 14 36	−31 58.	1	11.44		0.56		0.01		1		3061
		AzV 488			1 14 36	−73 37.1	2	11.90	.005	−0.13	.000	−0.97	.000	9	B0 Ia	35,573
7724	+30 00196	HR 374			1 14 37	+31 28.9	1	6.36		1.14		1.12		2	K0	1733
7666	+55 00281				1 14 39	+56 22.4	1	6.66		1.02		0.90		1	K1 III-IV	1716
7787	−44 00360				1 14 39	−44 17.9	2	9.53	.011	1.07	.006	0.92		9	G8/K0 (III)	863,1075
7505	+80 00036	IDS01097N8020	C		1 14 40	+80 37.9	1	6.65		0.03		−0.03		1	A0	3032
		Dril 51			1 14 42	−27 56.	1	12.44		0.37		−0.04		2	F0	561
7746	+13 00191				1 14 43	+13 39.1	1	8.50		1.68		1.42		3	M2	3040
	+39 00294	IDS01119N3946	AB		1 14 44	+40 01.9	1	9.50		0.58		−0.04		3	F0	3016
7745	+20 00192	IDS01121N2034	A		1 14 46	+20 49.3	1	8.72		0.54		0.08		2	G5	3016
7745	+20 00192	IDS01121N2034	B		1 14 46	+20 49.3	1	10.38		0.61		0.09		4		3016
7722	+43 00262	IDS01119N4406	A		1 14 47	+44 22.1	1	6.76		1.57		1.86		2	K5	1601

HD	DM	Other Id	N	Rem	α_{1950}	δ_{1950}	S	V	σ_V	B–V	σ_{B-V}	U–B	σ_{U-B}	n	Spectrum	References
7795	−43 00371				1 14 47	−42 47.7	10	7.86	.005	−0.09	.007	−0.37	.005	133	B9 III/IV	116,278,474,863,977*
7858	−68 00047				1 14 47	−67 41.7	1	6.92		0.37		−0.05		5	F2 IV/V	1628
		BSD 8 # 1766			1 14 48	+61 41.6	1	11.27		0.27		0.15		3	B8	1185
7710	+48 00392	IDS01119N4829		AB	1 14 49	+48 44.7	1	7.14		0.07		−0.05		2	A0	3024
7710	+48 00392	IDS01119N4829		C	1 14 49	+48 44.7	1	9.00		0.31		0.12		2	A0	3024
		Steph 136			1 14 51	+04 58.4	1	11.89		1.05		0.90		1	K6	1746
7783	−16 00213				1 14 51	−15 53.5	1	9.42		0.66		0.25		1	G3 V	79
		CS 29518 # 24			1 14 52	−32 42.8	1	14.52		0.21		−0.05		1		1736
7681	+60 00201	V538 Cas			1 14 53	+61 27.3	1	7.76		1.76				1	K5	1103
7772	+13 00192				1 14 54	+13 58.8	2	7.18	.015	1.32	.015	1.44		4	K0 IV	882,1648
	+63 00167	LS I +64 062			1 14 54	+64 02.6	1	10.35		0.53		−0.36		2	B2 III	1012
		BPM 47034			1 14 54	−33 06.	1	11.67		0.63		0.14		1		3061
		AJ89,1229 # 38			1 14 55	+09 37.6	1	13.74		1.48				2		1532
	−36 00491	G 269 - 133			1 14 57	−35 58.6	1	11.42		1.53		1.19		1		3061
7817	−35 00441				1 14 59	−34 55.6	1	8.21		0.54				4	F8 V	2012
7791	+12 00159				1 15 04	+13 17.7	1	8.55		1.01				3	K0	882
	+58 00213				1 15 04	+59 23.1	1	10.50		0.38		0.20		3	B6	1185
7808	−16 00214	G 268 - 146			1 15 05	−15 45.6	6	9.76	.020	0.99	.023	0.75	.045	15	K3 V	1619,1705,1774,2034*
		BPM 47036			1 15 06	−33 02.	1	13.22		0.96		0.71		1		3061
		NGC 457 - 233			1 15 09	+58 04.3	1	12.26		0.42		0.34		1	A0	63
	+57 00238	NGC 457 - 232			1 15 10	+58 06.0	1	10.55		0.21		−0.12		2	A0	1182
7720	+61 00240	LS I +61 188			1 15 12	+61 37.8	2	8.84	.025	0.59	.010	−0.13	.030	5	B5 II	1012,1185
7804	+2 00185	HR 378			1 15 13	+03 21.1	7	5.16	.011	0.07	.004	0.10	.035	28	A3 V	3,15,1007,1013,1061*
7758	+46 00319	HR 376			1 15 13	+47 09.4	2	6.26	.005	1.36	.003	1.45		5	K1 III	71,1501
7841	−40 00294				1 15 14	−40 09.8	1	8.20		1.18		1.22		2	K2 III	1730
		BPM 47040			1 15 18	−31 46.	1	12.45		1.52		1.26		1		3061
		NGC 457 - 236			1 15 19	+57 55.1	1	11.07		0.73		0.33		1	gG8	1182
		NGC 457 - 199			1 15 19	+58 02.8	1	12.55		2.02		2.37		1		63
		G 33 - 49			1 15 20	+15 55.0	3	13.83	.015	0.12	.017	−0.79	.017	5		203,1620,3060
7916	−67 00081	HR 380		★ AB	1 15 20	−66 39.7	3	6.23	.003	0.05	.000	−0.02	.000	16	A0 V	15,1075,2038
236689	+57 00240	NGC 457 - 198		★ AB	1 15 24	+58 06.7	2	9.47	.005	0.28	.005	−0.49	.010	4	B1.5Vpe	1012,1182
		BPM 47038			1 15 24	−25 41.	1	12.16		0.98		0.77		2		3061
		NGC 457 - 273			1 15 25	+57 51.8	1	11.37		1.35		0.96		1		1182
		AzV 489			1 15 25	−73 24.4	1	14.59		−0.22		−1.06		4	O9	425
		NGC 457 - 240			1 15 27	+57 52.7	1	11.00		1.64		1.76		1	gK5	1182
7835	+13 00195				1 15 29	+14 07.2	1	9.31		0.90				3	G5	882
		BSD 8 # 1359			1 15 29	+60 39.8	1	11.60		0.48		0.07		3	B3	1185
		NGC 457 - 173			1 15 30	+57 56.8	1	12.64		0.67		0.24		1	F5	63
7886	−46 00357				1 15 33	−46 26.1	4	8.00	.005	1.01	.007	0.80	.005	18	G8 III/IV	58,977,1075,2011
		NGC 457 - 266			1 15 36	+58 11.2	1	11.70		1.21		0.95		1	gG8	63
		GR 42			1 15 36	−35 10.2	1	14.58		1.52		1.14		1		481
	−20 00233				1 15 37	−19 54.5	1	9.96		0.62		0.03		1	G0	3061
		NGC 457 - 149			1 15 39	+58 01.5	1	12.49		0.39		−0.23		2	A2	1182
	+60 00203				1 15 39	+61 18.5	1	9.44		0.20		−0.17		3	B9 V	1185
7875	−24 00562				1 15 41	−24 00.3	1	9.75		0.30		0.13		3	A2(m)A7-A8	1068
7876	−25 00515				1 15 41	−24 48.0	1	10.06		0.20		0.07		3	A1 IV	1068
7847	+17 00183				1 15 42	+18 18.9	1	7.82		0.96				3	G5	882
		BPM 47044			1 15 42	−20 23.	1	14.28		0.50				1		3061
		L 293 - 94			1 15 42	−48 25.	2	11.55	.010	1.44	.009			2		1494,1705
		NGC 457 - 151			1 15 43	+57 58.9	1	12.75		0.33		−0.31		2	A0	1182
	+57 00243	NGC 457 - 153			1 15 44	+57 56.3	2	9.48	.005	0.34	.080	−0.56	.010	4	B0 IVe	1012,1182
7898	−34 00494				1 15 44	−34 24.0	1	7.74		0.26				4	A9 IV	2012
		AzV 491			1 15 44	−73 24.7	1	14.81		−0.25		−0.98		5	O9	425
		AzV 490			1 15 44	−73 42.2	2	13.26	.059	−0.17	.009	−1.00	.005	10	B1 Iab	358,425
	+16 00133				1 15 45	+17 06.9	1	9.39		1.06		0.87		1	K0	1776
	−28 00395				1 15 45	−28 14.9	1	10.47		1.01		0.88		1	K5 V	3061
	+60 00204	IDS01126N6021		AB	1 15 47	+60 36.7	1	9.91		0.25		0.02		3	B7	1185
		G 270 - 188			1 15 47	−13 09.5	1	11.79		1.47				1		1705
	−40 00298				1 15 47	−40 15.7	1	11.13		0.71		0.27		2		1730
		BSD 8 # 1779			1 15 48	+61 36.5	1	11.45		0.71		0.00		3	B0	1185
		BPM 70637			1 15 48	−19 26.	1	13.99		0.75		0.20		1		3061
7897	−33 00478				1 15 49	−32 55.4	1	9.53		0.94		0.65		1	G8 IV	3061
7796	+59 00228				1 15 50	+60 16.3	1	9.45		0.18		0.09		3	A0	1185
7910	−46 00359				1 15 50	−45 36.8	2	8.94	.006	1.30	.003	1.40		12	K2 III	977,1075
	+57 00245	NGC 457 - 275			1 15 51	+57 50.2	1	9.85		0.27		−0.52		1	B3	1182
		NGC 457 - 154			1 15 52	+57 55.9	1	11.19		0.32		−0.48		2	B8	1182
		NGC 457 - 124			1 15 53	+58 03.6	2	11.94	.005	0.29	.005	−0.43	.000	3	B5	49,1182
		BSD 8 # 1366			1 15 53	+60 42.7	1	12.20		0.31		−0.14		3	B5	1185
		NGC 457 - 84			1 15 55	+58 02.0	1	12.23		0.29		−0.43		2	B8	1182
7896	−24 00563				1 15 55	−23 45.7	1	7.95		0.81		0.38		1	G6/8 IV	3061
7853	+36 00220	HR 379		★ AB	1 15 56	+37 07.4	1	6.46		0.23		0.14		5	A5 m	8071
7909	−33 00479				1 15 56	−33 24.0	1	7.67		1.01				4	G8 III	2012
		AzV 492			1 15 56	−73 27.7	1	12.68		−0.11		−0.98		2	B1 Iab	425
		AzV 493			1 15 56	−73 33.5	1	14.17		−0.04		−1.12		3	B0	425
		NGC 457 - 274			1 15 57	+57 50.1	1	11.80		1.36		1.25		1		63
		NGC 457 - 122			1 15 57	+58 03.9	2	12.08	.000	0.36	.040	−0.42	.030	4	B8	49,1182
	+57 00246	NGC 457 - 128			1 15 57	+57 58.5	2	9.63	.005	0.36	.005	−0.54	.000	4	B6 Ve	49,1182
		BSD 8 # 1369			1 15 59	+60 35.6	1	12.67		0.39		−0.20		3		1185
	−40 00302				1 15 59	−40 17.3	1	10.15		0.73		0.24		2		1730
		NGC 457 - 85			1 16 00	+58 01.7	2	10.70	.075	0.29	.005	−0.45	.005	3	B5	49,1182

Table 1 91

HD	DM	Other Id	N Rem	α₁₉₅₀	δ₁₉₅₀	S	V	σ_V	B–V	σ_B–V	U–B	σ_U–B	n	Spectrum	References
		SB 532		1 16 00	−33 23.	1	13.65		0.08		0.21		2	A0	295
		WLS 120 5 # 10		1 16 01	+04 10.2	1	11.11		0.61		0.11		2		1375
	+57 00247	NGC 457 - 120		1 16 01	+58 05.2	2	9.94	.010	0.20	.005	-0.70	.005	3	O9.5IV	49,1182
7908	−23 00477			1 16 01	−23 16.5	2	7.27	.000	0.28	.000	0.10		6	A9 V	2012,3013
		NGC 457 - 87		1 16 02	+58 00.3	2	12.82	.025	0.34	.005	-0.28	.005	3	A0	49,1182
		NGC 457 - 119		1 16 02	+58 04.8	2	12.25	.015	0.26	.005	-0.44	.000	3	B8	49,1182
7829	+60 00205	LS I +66 017		1 16 02	+61 25.9	2	8.41	.015	0.42	.008			4	F3 V	70,1103
		NGC 457 - 89		1 16 03	+57 59.9	1	13.58		0.72		0.20		1	F5	1182
		CS 22953 # 38		1 16 03	−59 09.5	2	14.98	.010	0.44	.009	-0.20	.007	3		1580,1736
		NGC 457 - 91		1 16 05	+57 57.8	2	11.30	.005	0.30	.020	-0.51	.015	4	B5	49,1182
7732	+76 00040	HR 375		1 16 06	+77 18.4	1	6.31		0.92				2	G5 III	71
7895	−1 00167	G 70 - 51	★ A	1 16 06	−01 07.6	8	7.99	.015	0.82	.011	0.43	.021	26	K1 V	22,202,258,1620,1705,1775*
7895	−1 00167	IDS01135S0123	C	1 16 06	−01 07.6	1	12.29		1.02		0.75		4		3062
7895	−1 00167	G 70 - 50	★ B	1 16 06	−01 08.1	4	10.68	.030	1.41	.038	1.22	.045	8	M0	202,1705,3024,7009
		BPM 47048		1 16 06	−26 04.	1	10.72		0.69		0.17		1		3061
		BPM 47051		1 16 06	−30 52.	1	12.02		0.82		0.38		1		3061
		BPM 47052		1 16 06	−34 10.	1	12.38		0.56		-0.03		1		3061
		NGC 457 - 46		1 16 07	+58 00.4	1	13.70		0.37		-0.46		1	A2	1182
		NGC 457 - 43		1 16 08	+58 01.0	2	12.96	.005	0.32	.009	-0.28	.005	2		49,1182
		NGC 457 - 41		1 16 08	+58 01.9	1	13.32		0.35		-0.20		1		1182
		NGC 457 - 78		1 16 08	+58 04.3	2	11.36	.070	0.28	.000	-0.43	.010	3	B5	49,1182
7931	−29 00411			1 16 10	−28 59.7	1	7.89		1.03				4	K0 IV	2012
7972	−44 00367			1 16 11	−44 27.6	3	8.61	.007	0.44	.001	-0.04	.011	16	F5 V	977,1075,2011
		NGC 457 - 92		1 16 12	+57 57.6	1	14.24		0.34		-0.47		1		1182
		NGC 457 - 44		1 16 12	+58 00.6	2	13.33	.000	0.38	.009	-0.03	.033	2		49,1182
		NGC 457 - 42		1 16 12	+58 01.0	2	13.19	.005	0.33	.014	-0.28	.005	2	A0	49,1182
	+57 00248	NGC 457 - 38		1 16 12	+58 02.8	2	10.11	.005	0.21	.005	-0.01	.010	3	A0	49,1182
		BPM 47050		1 16 12	−20 12.	1	10.88		0.51		-0.04		1		3061
7953	−30 00422			1 16 12	−30 26.8	1	8.92		1.25		1.35		1	K1 III	54
		NGC 457 - 49		1 16 13	+57 59.4	2	12.40	.000	0.24	.005	-0.43	.005	4	B8	49,1182
	+57 00249	NGC 457 - 37		1 16 13	+58 02.6	2	9.92	.079	0.29	.000	-0.45	.000	3	B1 V	49,1182
		NGC 457 - 77		1 16 13	+58 03.6	1	13.19		0.56		0.36		1		1182
		NGC 457 - 94		1 16 14	+57 57.0	1	13.69		0.35		-0.14		1	A0	1182
7952	−28 00398			1 16 14	−27 46.7	1	9.62		1.09		0.92		1	G8 III	54
7950	−26 00424			1 16 15	−26 09.7	2	8.70	.000	0.80	.000	0.41		5	G5 IV/V	2012,3061
		NGC 457 - 191		1 16 16	+58 10.6	1	12.38		0.49		0.42		1	A0	63
	+58 00218			1 16 16	+58 42.7	1	10.00		1.98				2		369
		NGC 457 - 48		1 16 17	+58 00.0	1	12.77		0.32		-0.19		1	B8	1182
		NGC 457 - 12		1 16 17	+58 00.5	2	12.45	.009	0.33	.038	-0.42	.033	2	B8	49,1182
		NGC 457 - 35		1 16 17	+58 02.4	2	12.23	.000	0.32	.010	-0.41	.005	3	B5	49,1182
		BPM 47058		1 16 18	−32 17.	1	11.82		0.93		0.68		1		3061
7842	+66 00103			1 16 20	+66 56.4	1	9.25		0.19		0.09		2	A0	1733
		NGC 457 - 34		1 16 21	+58 02.3	2	11.22	.009	0.25	.014	-0.45	.009	2	B5	49,1182
		NGC 457 - 76		1 16 21	+58 04.1	1	13.58		0.32		-0.18		1		1182
		NGC 457 - 50		1 16 22	+57 58.4	1	13.16		0.38		-0.20		1		1182
		NGC 457 - 16		1 16 23	+58 00.0	1	13.19		0.18		-0.25		1		49
		NGC 457 - 6		1 16 23	+58 01.7	1	11.35		0.30		-0.38		2		1182
		NGC 457 - 7		1 16 23	+58 01.8	1	11.04		0.27		-0.44		2		1182
7969	−27 00430	HR 380	★ AB	1 16 23	−26 45.1	4	8.32	.010	0.80	.003	0.42	.007	11	G8 IV	79,1775,2012,3061
		NGC 457 - 13		1 16 24	+58 00.3	1	10.81		0.26		-0.52		1	B5	49
	+57 00252	NGC 457 - 19		1 16 25	+57 59.6	3	9.51	.005	0.26	.005	-0.54	.012	4	B1 IV	49,1012,1182
		NGC 457 - 117		1 16 26	+58 06.0	1	12.66		0.31		-0.32		2	B8	1182
		NGC 457 - 118		1 16 26	+58 06.8	1	11.68		0.30		0.16		2		1182
7859	+60 00206	IDS01132N6103	AB	1 16 26	+61 19.0	2	8.78	.007	0.55	.003			3	F5	70,1103
		NGC 457 - 130		1 16 27	+57 55.9	1	12.46		1.68		1.78		1	gG8	63
		NGC 457 - 96		1 16 27	+57 57.4	1	13.07		0.68		0.14		1		1182
		NGC 457 - 17		1 16 27	+58 00.5	2	13.03	.009	0.33	.014	-0.30	.005	2		49,1182
		NGC 457 - 75		1 16 28	+58 05.4	1	14.31		0.38		0.01		1	A0	49
		Steph 137		1 16 29	+48 38.4	1	10.96		1.75		2.14		2	M0	1746
		G 245 - 21		1 16 29	+73 55.2	1	9.96		0.63		0.08		2	G5	7010
		Klemola 22		1 16 30	−01 16.	1	11.57		0.13		0.11		2		1026
		BPM 47059		1 16 30	−30 12.	1	11.85		0.68		0.19		1		3061
		BPM 47061		1 16 30	−33 15.	1	12.93		0.60		-0.01		1		3061
		NGC 457 - 31		1 16 31	+58 02.9	1	13.47		0.35		-0.05		1	A2	1182
		NGC 457 - 72		1 16 31	+58 04.1	2	12.38	.005	0.71	.015	0.16	.005	2	F5	1182,1182
7983	−9 00256	G 271 - 34	★ A	1 16 31	−09 11.8	9	8.90	.009	0.59	.007	-0.02	.010	16	G2 V	22,258,742,908,1003,1594*
7968	−13 00236			1 16 31	−12 54.4	1	10.19		-0.02		-0.04		4	B9 V	152
8001	−43 00386			1 16 31	−43 35.8	4	6.76	.006	1.48	.001	1.84	.013	27	K3 III	863,977,1075,2011
		AzV 494		1 16 31	−72 59.7	2	14.62	.072	-0.06	.027	-0.76	.005	5	B2	733,767
		NGC 457 - 74		1 16 32	+58 05.2	1	14.23		0.32		-0.04		1	A0	49
		NGC 457 - 190		1 16 32	+58 11.0	1	11.36		1.80		1.93		1		63
		NGC 457 - 21		1 16 33	+58 00.4	1	13.28		0.33		-0.17		1	A0	1182
		NGC 457 - 279		1 16 34	+57 49.0	1	11.88		1.34		1.10		1	gK0	63
	+57 00254	NGC 457 - 54	★ D	1 16 34	+57 59.1	3	10.19	.013	0.29	.011	-0.47	.008	7	B3	49,450,1182
7946	+13 00198			1 16 35	+13 37.8	1	8.74		1.09				1	K0	882
		NGC 457 - 30		1 16 35	+58 03.5	1	14.39		0.33		-0.05		1		1182
		NGC 457 - 71		1 16 35	+58 04.8	1	11.64		0.26		-0.49		2	B5	1182
		BSD 8 # 1375		1 16 35	+60 22.1	1	10.06		0.29		-0.41		3	B3	1185
8000	−30 00426			1 16 35	−30 13.4	1	9.41		0.75		0.14		1	G6 V	3061
7852	+68 00091	LS I +68 011		1 16 36	+68 50.2	1	8.84		0.34		-0.42		3	B2 III	555

HD	DM	Other Id	N	Rem	α_{1950}	δ_{1950}	S	V	σ_V	B–V	σ_{B-V}	U–B	σ_{U-B}	n	Spectrum	References
7985	−15 00244				1 16 36	−15 06.4	2	9.46	.005	0.43	.000	-0.09		3	F2/3 V	742,1594
7999	−29 00414				1 16 36	−29 27.1	1	8.98		0.63		0.11		1	G2 V	3061
		NGC 457 - 100		⋆ E	1 16 37	+57 56.7	3	10.62	.008	0.28	.005	-0.52	.008	7	B5	49,450,1182
		NGC 457 - 116			1 16 37	+58 05.6	1	12.46		0.32		-0.35		2	B8	1182
8259	−82 00022				1 16 39	−81 48.2	2	7.92		1.14		0.98		9	G8 III	1704
8024	−46 00362				1 16 40	−46 24.9	2	9.09	.000	0.44	.000	-0.06		8	F3 V	1075,2011
		NGC 457 - 60			1 16 41	+58 00.2	2	12.20	.005	0.28	.005	-0.35	.005	3	B8	49,1182
7902	+57 00257	NGC 457 - 131		⋆ C	1 16 42	+57 56.8	5	6.99	.023	0.41	.009	-0.38	.004	27	B6 Ib	49,450,1012,1084,1182
8113	−74 00092				1 16 42	−74 25.0	2	8.79	.000	1.09	.004	0.96	.027	5	K0 III	727,1408
7980	+17 00184				1 16 43	+17 52.6	2	7.64	.000	1.02	.017	0.88		3	K0	882,3040
7964	+26 00220	HR 383			1 16 43	+27 00.1	5	4.75	.033	0.03	.005	0.11	.010	9	A3 V	15,1049,1363,3023,8015
		NGC 457 - 23			1 16 43	+58 01.1	1	13.31		0.35		-0.28		1		1182
	+79 00038	G 242 - 78			1 16 43	+79 53.3	2	9.66	.009	1.32	.047	1.27		3	K8	1017,3072
8023	−45 00433				1 16 43	−45 12.0	4	8.92	.009	1.08	.002	0.97	.014	20	K0 III	58,863,977,2011
236697	+57 00258	NGC 457 - 25		⋆ V	1 16 44	+58 02.8	3	8.62	.015	2.13	.018	2.37	.070	5	M0.5Ib	138,1182,8032
7981	+14 00202				1 16 45	+14 31.8	1	8.43		1.07				2	K0 IV	882
		LP 883 - 180			1 16 45	−30 41.0	1	14.68		0.54		-0.07		3		3062
		NGC 457 - 64			1 16 46	+58 00.6	1	12.32		0.31		-0.24		2	A0	1182
		POSS 52 # 4			1 16 46	+66 45.4	1	17.58		1.57				3		1739
		CS 22946 # 11			1 16 46	−19 39.7	2	13.94	.005	0.33	.001	-0.14	.002	2		966,1736
		NGC 457 - 62			1 16 47	+58 00.2	2	11.82	.005	0.29	.015	-0.35	.020	4	B8	49,1182
		PS 60			1 16 48	−23 09.	1	14.38		0.00		0.02		2		295
		NGC 457 - 24			1 16 49	+58 01.8	1	14.45		0.31		0.00		1		49
		NGC 457 - 115			1 16 49	+58 05.4	1	13.74		0.78		0.21		1	F8	1182
		CS 29514 # 30			1 16 49	−23 10.0	1	14.40		-0.01		0.05		1		1736
		NGC 457 - 70			1 16 50	+58 03.1	1	13.33		0.30		-0.22		1	A0	1182
		NGC 457 - 135		⋆ B	1 16 52	+57 57.4	1	12.72		0.29		-0.34		1		1182
		NGC 457 - 69			1 16 52	+58 03.0	1	13.65		0.32		-0.16		1	A0	1182
		NGC 457 - 102			1 16 53	+57 59.3	1	11.79		0.28		-0.48		3		1182
		NGC 457 - 104			1 16 53	+57 59.7	1	13.41		0.29		-0.29		1		1182
		NGC 457 - 67			1 16 53	+58 01.6	1	12.36		0.29		-0.31		2		1182
		NGC 457 - 114			1 16 53	+58 05.3	1	13.69		0.61		0.47		1	A2	1182
		JL 238			1 16 54	−47 26.	1	15.71		-0.21		-0.96		1		832
		G 2 - 34			1 16 55	+13 56.1	1	13.18		1.54				1		1532
7927	+57 00260	NGC 457 - 136		⋆ A	1 16 55	+57 58.2	13	4.99	.016	0.68	.007	0.47	.028	81	F0 Ia	1,15,49,254,450,1077*
		NGC 457 - 106			1 16 55	+57 59.7	1	13.18		0.35		-0.14		1		1182
		AM Tuc			1 16 55	−68 10.7	1	11.41		0.22		0.08		1	A9.5:	700
		NGC 457 - 105			1 16 56	+57 59.4	1	11.87		0.31		-0.41		3		1182
		NGC 457 - 107			1 16 57	+58 00.0	2	12.95	.014	0.73	.028	0.21	.014	2		49,1182
		NGC 457 - 113			1 16 58	+58 05.2	1	12.22		0.29		-0.42		2	B8	1182
8033	−23 00483				1 16 58	−23 22.2	1	9.17		0.31				4	F0 V	2012
8050	−47 00385				1 16 58	−46 53.5	4	8.76	.007	0.45	.005	-0.04	.014	20	F5 V	116,977,1075,2011
8040	−35 00452	IDS01147S3501	A		1 17 00	−34 45.2	1	7.85		0.47		-0.02		4	F5 V	1731
8049	−44 00370	IDS01148S4409	AB		1 17 02	−43 53.3	3	8.71	.022	0.89	.006	0.48	.016	16	K1/2 V	977,1075,2011
		NGC 457 - 140			1 17 03	+58 00.4	2	12.85	.005	0.31	.009	-0.25	.009	2	A2	49,1182
8149	−73 00073				1 17 03	−72 59.0	1	9.63		0.43		-0.05		3	F2 V	35
8032	−6 00251	IDS01146S0551	AB		1 17 05	−05 35.3	2	8.40	.029	0.56	.005	0.06		7	F8 V +G0 V	176,3016
8038	−25 00524				1 17 06	−25 12.7	1	8.42		0.70		0.24		1	G5 V	3061
		NGC 457 - 108			1 17 07	+58 02.7	1	14.51		0.85		0.25		1		49
		WLS 112 60 # 9			1 17 07	+59 45.5	1	10.89		0.49		0.43		2		1375
		AzV 495			1 17 07	−73 38.1	1	13.89		-0.02		-0.37		2	A0	425
8095	−57 00284				1 17 10	−57 23.2	1	9.58		0.99				1	G6/K0III/IV	1642
		AzV 496			1 17 10	−73 46.8	1	12.72		0.02		-0.23		3	A1 Iab	425
		G 132 - 71			1 17 11	+38 43.6	1	11.34		1.32				2		1625
		NGC 457 - 248			1 17 12	+57 49.3	1	12.17		0.67		0.20		1	F0	63
		SB 537			1 17 12	−26 57.	1	12.96		0.28		0.10		1	A7	295
		LB 3193			1 17 12	−62 11.	1	12.69		-0.09		-0.58		2		45
	−45 00436				1 17 14	−45 22.3	2	9.90	.009	1.33	.005	1.50		8		863,1075
	+57 00263	NGC 457 - 143			1 17 15	+58 02.9	1	9.93		0.23		-0.06		2	A0	1182
8036	−1 00171	HR 385		⋆ ABC	1 17 15	−00 46.3	2	5.86	.005	0.64	.000	0.31	.005	7	G8 III+A7 V	15,1256
	+15 00192				1 17 17	+15 36.2	1	10.35		0.53		0.05		1	F5	1776
8008	+38 00241				1 17 17	+38 49.9	2	8.39	.005	0.86	.000	0.52	.010	5	G7 III	1501,1601
		CoD -22 00457			1 17 22	−22 30.1	1	10.14		0.67		0.12		2		3061
		BPM 47072			1 17 24	−22 37.	1	11.25		0.58		-0.03		1		3061
8073	−28 00407				1 17 24	−28 29.7	1	8.10		1.11				4	G8 III	2012
8106	−47 00389	AK Phe			1 17 25	−47 33.3	1	7.40		1.64				4	M4 III	2011
		BSD 8 # 1800			1 17 26	+62 09.0	1	10.78		0.29		0.03		3	B9	1185
		LS I +62 153			1 17 27	+62 15.0	1	10.22		0.50		-0.47		2	B0 V:	1012
	−23 00488				1 17 27	−23 12.1	1	10.64		0.35		-0.08		2	A5	3061
	+57 00266	NGC 457 - 217			1 17 29	+57 59.8	1	10.22		1.23		0.94		1	gK2	1182
8190	−71 00058				1 17 29	−71 22.8	1	8.21		0.52		0.04		4	G0 V	727
		NGC 457 - 283			1 17 30	+57 54.4	1	12.36		0.68		0.31		1	F2	63
	+0 00218				1 17 32	+01 30.1	1	10.17		0.50		-0.06		4	F8	1260
		NGC 457 - 215			1 17 32	+57 57.4	1	11.93		0.41		-0.35		1	B8	1182
		AzV 497			1 17 32	−73 32.6	1	13.83		-0.14		-0.76		4	B2 II:	425
		G 219 - 5			1 17 33	+57 03.7	1	10.36		1.34		1.16		1		3003
		AAS33,107 # 7			1 17 36	−71 20.	1	11.56		0.65		0.22		4		727
		TonS 208			1 17 37	−30 28.5	2	15.08	.005	0.04	.007	0.09	.021	3		1736,3064
8105	−32 00517				1 17 37	−31 38.9	2	9.14	.009	0.55	.002	0.00		5	G1 V	2012,3061
		G 269 - 140		⋆ V	1 17 37	−33 41.1	1	15.25		1.52				1		3061

Table 1   93

HD	DM	Other Id	N Rem	α_{1950}	δ_{1950}	S	V	σ_V	B–V	σ_{B-V}	U–B	σ_{U-B}	n	Spectrum	References
		LP 467 - 45		1 17 38	+12 13.1	1	12.29		0.50		-0.09		2		1696
8013	+60 00209			1 17 38	+60 41.2	2	7.53	.000	-0.02	.000			3	A0	1103,1118
	+57 00265	NGC 457 - 185		1 17 39	+58 04.3	1	9.89		0.24		-0.13		2	B8	1182
		LS I +62 154		1 17 42	+62 19.7	1	10.85		0.52		-0.45		2	B1 V	1012
		BPM 47077		1 17 42	-22 40.	1	13.86		0.50		-0.24		1		3061
8104	-24 00578			1 17 42	-24 26.4	1	9.50		0.52		0.02		1	F6 V	3061
	+61 00252	LS I +61 190		1 17 43	+61 41.5	1	8.96		0.93		0.23		3	A1 Ia	1185
8003	+63 00176	HR 384	★ A	1 17 43	+64 23.8	1	6.34		0.09				2	A2 Vnn	1118
		CS 29514 # 29		1 17 43	-22 45.5	1	12.17		0.23		0.15		1		1736
		SB 545		1 17 48	-22 29.	1	14.24		0.57		0.03		1	A0	295
7924	+75 00058			1 17 50	+76 27.0	1	7.11		0.82		0.45		2	K0	1375
8147	-46 00369			1 17 50	-45 43.3	2	9.65	.014	0.09	.002	0.10		10	A1 V	863,2011
		NGC 457 - 222		1 17 52	+58 07.2	1	11.26		1.32		1.09		1	gK2	63
7925	+75 00059	HR 381		1 17 52	+75 58.7	2	6.38	.000	0.28	.000	0.07	.005	5	F0 IVn	985,1733
		CS 22174 # 20		1 17 57	-09 03.0	1	15.06		0.40		-0.21		1		1736
8121	-11 00248	HR 388		1 17 59	-11 30.0	2	6.15	.000	1.10	.003	1.03		6	K1 III	58,2009
8130	-36 00510			1 17 59	-36 30.3	3	7.45	.018	0.05	.005	0.03	.015	9	A0 V	79,1068,2012
8111	+12 00164			1 18 00	+13 12.4	1	9.07		1.10				3	K0	882
		BPM 47083		1 18 00	-28 39.	1	10.83		0.59		0.00		1		3061
8178	-58 00095			1 18 00	-57 36.7	1	7.42		0.44				4	F3 V	1075
8110	+14 00204			1 18 01	+15 26.1	2	7.26	.009	0.87	.019	0.54		4	G8 IV	882,1648
8144	-29 00422			1 18 01	-29 14.9	1	7.37		1.08				4	K0 III	2012
		BSD 8 # 1812		1 18 02	+62 06.6	1	10.67		0.70		0.19		3	B9	1185
8120	-4 00185	HR 387		1 18 02	-03 30.5	2	6.22	.005	1.01	.005	0.85	.005	7	K1 III	15,1256
8129	-20 00249	G 268 - 155		1 18 04	-20 12.5	2	7.59	.010	0.70	.000	0.22	.005	2	G5/6 V	79,3061
8145	-30 00434			1 18 04	-29 51.7	1	8.45		0.32				4	F0 V	2012
		NGC 457 - 258		1 18 06	+58 11.9	1	12.32		0.84		0.43		1	G5	63
		BPM 47085		1 18 06	-28 00.	1	14.28		0.68		0.06		1		3061
		SS Psc		1 18 10	+21 28.5	1	10.78		0.25		0.21		1	F0	597
8109	+20 00201			1 18 11	+21 20.6	2	8.82	.000	0.48	.009	-0.05	.009	6	F5	1307,1386
8100	+37 00259	G 132 - 73		1 18 11	+37 46.3	1	7.87		0.61		0.09		1	G0	333,1620
8188	-49 00377	IDS0116IS4912	A	1 18 12	-48 56.3	1	7.57		0.54				4	F6 V	2012
8117	+24 00198			1 18 19	+24 53.9	1	7.90		0.48		-0.04		2	F5	1625
8271	-72 00092			1 18 19	-72 03.9	1	8.76		0.95		0.58		3	G8 II/IV	727
8126	+27 00215	HR 389		1 18 21	+28 28.7	1	5.23		1.39				2	gK5	71
8287	-72 00093			1 18 21	-72 06.1	1	9.50		0.56		0.02		3	F8/G2	727
		AzV 498		1 18 28	-72 56.7	1	12.79		0.01		-0.34		1	B8 Iab	767
8224	-57 00292	IDS01165S5752	AB	1 18 29	-57 36.1	1	7.00		0.54				4	F7 V	2012
		BPM 47086		1 18 30	-22 23.	1	12.31		0.73		0.22		1		3061
		BPM 47087		1 18 30	-24 09.	1	14.92		1.40		0.99		1		3061
8151	+21 00178			1 18 32	+22 06.8	1	7.01		1.56		1.94		3	K4 III	1501
	+4 00227			1 18 34	+05 21.3	1	9.79		1.52				1	K7	1632
8108	+60 00214			1 18 37	+61 15.1	1	8.72		0.40				2	F5	70
	+30 00206	G 69 - 65	★ A	1 18 39	+31 04.9	1	8.50	.009	0.92	.009	0.70	.005	4	K2 V	22,333,1620
8187	+10 00168	IDS01160N1101	AB	1 18 41	+11 16.5	1	6.90		0.37		0.14		3	F0	3016
		LB 3195		1 18 42	-64 52.	1	13.00		0.07		0.09		2		45
		G 34 - 22		1 18 44	+24 04.1	1	10.70		1.42		1.27		2	M2	7010
		AzV 499		1 18 44	-72 47.7	2	13.25	.005	-0.14	.015	-0.64	.029	3	B2 Iab-Ib	573,767
		AzV 500		1 18 44	-73 07.1	1	13.44		-0.23		-0.82		2	B1	425
8186	+13 00202			1 18 46	+13 39.7	1	8.98		1.00				2	K0	882
	-26 00426	G 268 - 151		1 18 47	-25 28.9	2	11.82	.005	0.53	.041	-0.11	.007	3		1696,3061
		BH Tuc	A	1 18 48	-73 10.0	1	13.57		0.43				1		1511
		BH Tuc	B	1 18 48	-73 10.0	1	14.28		1.01				1		1511
8160	+49 00358			1 18 51	+49 51.6	1	6.54		1.13		0.72		1	G5 Ib	1716
		BPM 47089		1 18 54	-26 41.	1	12.09		1.15		1.04		2		3061
8353	-73 00074			1 18 55	-73 00.7	1	7.48		0.33		0.04		3	A9/F0 IV	35
		AzV 501		1 18 57	-73 20.5	1	13.16		-0.04		-0.49		4	B6:I	425
		GR 45		1 19 00	-31 22.8	1	14.52		1.53		1.07		1		481
		LS I +64 067		1 19 05	+64 19.6	1	11.08		0.68		-0.35		2	B1: II:	1012
		Steph 144		1 19 06	-06 43.5	1	11.98		1.47		1.22		1	M0	1746
	-26 00442			1 19 06	-26 16.2	1	10.38		0.21		0.07		3	A0	1068
		POSS 52 # 5		1 19 08	+65 38.5	1	18.32		1.74				3		1739
	-42 00469			1 19 10	-41 54.3	3	10.14	.021	1.40	.023	1.10	.015	7	K5	158,1705,3078
		WLS 112 60 # 7		1 19 11	+58 14.5	1	11.66		0.47		-0.32		2		1375
		CS 22946 # 17		1 19 11	-17 41.6	1	14.85		0.42		-0.19		1		1736
8296	-50 00368			1 19 12	-50 00.2	1	6.75		0.44		0.03		7	F3/5 IV	1628
		CS 22174 # 23		1 19 13	-10 27.9	1	14.00		0.06		0.16		1		1736
	-28 00413	G 269 - 143		1 19 15	-28 21.0	2	11.96	.030	1.01	.027	0.91		2		1696,3061
8209	+42 00288			1 19 17	+43 19.4	1	6.67		-0.14		-0.62		1	B5	963
		LS I +61 192		1 19 17	+61 17.3	1	10.76		0.85		-0.26		2	B pe	1012
8065	+77 00049	HR 386		1 19 18	+78 27.9	1	6.07		0.39		-0.09		3	A0 Iab	196
		BPM 47091		1 19 18	-26 26.	1	13.96		1.44		1.24		1		3061
8305	-45 00444			1 19 21	-45 24.0	3	7.44	.011	0.37	.003	0.01	.010	10	F2 V	278,1075,2011
8207	+44 00287	HR 390		1 19 23	+45 16.0	5	4.87	.022	1.08	.005	1.00	.019	15	K0 III-IV	15,1355,1363,3016,8015
		BPM 47093		1 19 24	-21 31.	1	11.25		1.11		1.01		1		3061
		BPM 47098		1 19 30	-24 08.	1	13.56		0.65		0.09		1		3061
		TonS 210		1 19 30	-28 37.	1	14.83		0.22		-1.01		2		3064
		AAS35,353 # 124		1 19 31	-73 07.7	1	13.76		0.22		0.34		2	F3	767
8267	+13 00204			1 19 34	+14 04.8	2	8.04	.005	1.02	.028	0.80		4	G5	882,1648
8262	+17 00197			1 19 34	+18 25.3	3	6.97	.009	0.62	.005	0.10	.013	7	G3 V	22,1003,3026

HD	DM	Other Id	N Rem	α_{1950}	δ_{1950}	S	V	σ_V	B–V	σ_{B-V}	U–B	σ_{U-B}	n	Spectrum	References
		AzV 502		1 19 39	−73 29.4	1	14.14		−0.11		−0.92		4	B1	425
		WLS 120 5 # 5		1 19 40	+07 21.4	1	12.31		0.56		0.04		2		1375
		LP 767 - 77		1 19 41	−20 14.6	1	13.11		0.61		−0.05		2		1696
8275	+16 00139			1 19 42	+16 33.9	1	7.01		1.08		0.97		2	G5	1648
		BPM 47104		1 19 42	−32 22.	1	12.07		0.70		0.23		1		3061
8340	−45 00447	IDS01175S4511	AB	1 19 42	−44 55.3	2	9.75	.009	0.60	.002	0.11		10	G2 V	863,2011
		LB 1598		1 19 48	−55 47.	1	11.95		−0.22		−1.12		1		832
	+36 00231			1 19 49	+37 11.9	1	10.30		0.80		0.43		7		1775
8325	−18 00217			1 19 49	−17 53.7	2	10.26	.015	0.10	.000	0.13	.020	9	A0 V	152,1775
8362	−45 00448			1 19 51	−44 41.2	8	9.46	.008	1.43	.004	1.68	.010	87	K3 (III)	365,863,1460,1628,1657,1673*
		CS 22946 # 16		1 19 52	−17 27.5	2	15.83	.000	−0.08	.000	−0.01	.012	2		966,1736
		CS 29518 # 41		1 19 52	−29 14.0	1	15.05		0.11		−0.97		1		1736
8323	−8 00237			1 19 54	−07 32.9	4	9.54	.011	−0.09	.005	−0.43	.006	21	A0	152,830,1026,1783
	+12 00168	G 2 - 36	★ AB	1 19 56	+12 29.4	3	9.46	.014	1.03	.022	0.86	.002	5	K5	202,1620,3060
8310	+12 00169			1 19 57	+12 50.5	2	8.96	.010	1.07	.000	0.90		4	K2	882,3077
8338	−11 00255			1 19 58	−10 52.0	1	8.49		0.20		0.11		4	A0	152
8479	−73 00075			1 19 58	−73 17.7	1	8.40		1.04		0.77		3	K0 III	35
8261	+60 00221			1 20 00	+61 27.6	1	8.45		0.11				2	A0	70
		CS 29514 # 41		1 20 00	−23 51.8	1	12.32		0.10		0.09		1		1736
	−33 00501	G 269 - 148		1 20 00	−33 28.2	1	10.33		1.26		1.29		1		3061
8335	−1 00179	HR 393		1 20 01	−00 42.6	2	6.48	.005	1.08	.000	1.02	.005	7	gK0	15,1256
8382	−46 00383	IDS01179S4645	AB	1 20 01	−46 29.0	4	8.44	.007	0.21	.012	0.12	.021	14	A3 III/IV	116,278,1075,2011
8334	+0 00223	HR 392		1 20 02	+01 28.0	5	6.19	.004	1.51	.008	1.85	.012	15	K5 IIIab	15,1061,1355,1509,3040
8321	+16 00141	IDS01174N1640	A	1 20 02	+16 56.1	1	7.15		0.47		−0.05		1	F5 V	1776
		Wolf 63		1 20 03	+12 44.8	1	13.25		0.90		0.56		2		1696
8350	−19 00229	HR 394	★ AB	1 20 05	−19 20.5	1	6.35		0.48				4	F5/6 V	2009
		AzV 502a		1 20 07	−72 41.4	1	15.07		−0.23		−0.85		2	B1	767
	−73 00076			1 20 07	−73 14.1	1	9.52		0.61		0.10		3	G5	35
8272	+57 00274	HR 391	★ AB	1 20 09	+57 53.0	1	6.36		0.42				2	F4 V	1733
		LP 767 - 43		1 20 10	−20 02.1	1	12.01		0.86		0.55		2		1696
8381	−35 00472			1 20 10	−34 55.5	1	7.56		1.04		0.86		4	K1 III	1731
8391	−44 00388			1 20 12	−43 51.8	6	7.02	.006	0.33	.003	0.03	.005	27	F0/2 IV	278,977,977,1075,2011,3026
8436	−59 00091			1 20 15	−59 23.2	2	7.53	.004	0.84	.004	0.48	.013	5	G5 IIIp	1117,1700
		AzV 503		1 20 15	−73 12.1	1	14.74		−0.26		−1.12		2	B0	425
8358	−0 00210	BI Cet, G 71 - 3		1 20 17	+00 27.3	2	7.43	.385	0.72	.000	0.19	.007	12	G0	333,1620,1738
8521	−74 00094			1 20 17	−73 55.6	1	9.29		0.51		0.07		4	F5 V	727
		CS 22946 # 15		1 20 18	−17 41.7	2	15.59	.000	−0.04	.000	0.14	.012	2		966,1736
		BPM 47113		1 20 18	−25 26.	1	12.30		0.94		0.62		2		3061
		BPM 47114		1 20 18	−29 47.	1	13.48		1.14		0.85		1		3061
8357	+6 00211		V	1 20 20	+07 09.3	2	7.29	.013	0.86	.004	0.38	.009	16	G5	1733,1738
8380	−22 00237			1 20 20	−21 41.2	1	8.09		0.42				4	Fm δ Del	2012
8390	−35 00474			1 20 20	−35 15.1	1	8.14		1.04		0.85		4	K0 III/IV	1731
8435	−57 00296	BC Phe		1 20 21	−56 59.5	2	8.73	.039	0.71	.067	0.30		2	G6/8III/IVe	1641,1642
	+64 00156	LS I +65 011		1 20 22	+65 22.1	1	9.54		0.61		−0.36		2	B0.5III	1012
		CS 29518 # 39		1 20 25	−28 40.2	1	14.26		0.31		−0.17		1		1736
8346	+35 00260			1 20 26	+36 15.2	1	6.80		0.03		−0.05		1	A0 V	1716
	−45 00450			1 20 26	−45 34.7	1	10.88		0.64		0.14		8		1704
	−45 00450			1 20 26	−45 34.7	8	10.88	.005	0.63	.012	0.15	.006	37		116,365,863,1460,1628*
8369	+15 00201			1 20 28	+16 02.5	2	8.32	.040	1.38	.035	1.55		4	K2 III	882,3040
8519	−70 00064	IDS01190S7014	AB	1 20 28	−69 58.8	2	7.25	.004	0.42	.000	0.04		6	F3 V	727,1414
		AzV 504		1 20 28	−73 01.5	2	11.91	.005	−0.04	.009	−0.47	.005	4	B9 Ia	425,573
		SB 563		1 20 30	−30 10.	1	14.29		0.06		0.02		1		3064
8463	−61 00094			1 20 30	−61 35.5	1	7.23		1.63		1.94		7	M2 III	1704
8389	−13 00249	G 272 - 8	★ B	1 20 31	−13 13.1	1	10.30		1.38		1.26		4	K6	3024
		CS 29514 # 38		1 20 31	−26 33.2	1	14.01		0.39		0.05		1		1736
		AzV 505		1 20 32	−73 52.8	1	13.52		−0.14		−0.95		2	B2	425
8389	−13 00249	G 272 - 9	★ A	1 20 33	−13 13.6	3	7.85	.006	0.90	.002	0.70	.014	6	K0 IV	258,1705,3024
8406	−17 00247			1 20 35	−16 44.7	1	7.92		0.65		0.15		1	G5 V	79
		LP 707 - 118		1 20 38	−10 22.4	1	12.92		1.56		1.14		1		1696
8405	−0 00212	IDS01181N0003	A	1 20 40	+00 18.9	1	9.14		0.96		0.64		1	K2/3 V	1738
8388	+19 00226	HR 397		1 20 43	+20 12.5	1	5.97		1.71				2	K5	71
	−25 00551			1 20 45	−25 15.5	1	10.63		0.60		0.08		1	F8	3061
	+4 00235			1 20 46	+05 14.9	1	9.97		0.76		0.30		2	G5	1375
8373	+47 00398			1 20 46	+48 07.0	1	7.63		1.51		1.72		2	K2	1733
8375	+33 00220	HR 396		1 20 47	+33 59.0	1	6.28		0.83		0.48		4	G8 IV	1355
8559	−72 00095			1 20 47	−71 51.0	1	9.73		0.10		0.89		3	G8/K0 III	727
8374	+36 00237	HR 395		1 20 48	+37 27.3	5	5.59	.015	0.27	.011	0.11	.019	17	A1 m	985,1049,1199,3058,8071
		E1 - W		1 20 48	−44 26.	1	11.84		0.76				4		1075
	+15 00203			1 20 53	+16 10.4	1	9.09		1.28		1.28		1	K0	1776
	−19 00233	G 272 - 11		1 20 53	−18 52.6	2	10.62	.025	0.73	.015	0.18	.030	5		1696,3062
		JL 243		1 20 54	−48 04.	1	13.80		−0.28		−1.21		1		832
		AzV 506		1 20 54	−73 42.3	1	13.64		−0.28		−0.94		2	B0	425
8447	−18 00222			1 20 55	−18 11.7	2	7.06	.025	1.62	.002	1.91	.020	8	M2 III	1628,3007
	+63 00180	LS I +63 144		1 20 57	+63 41.4	2	9.99	.024	1.04	.005	0.30	.005	3	A0:Ia	1012,1215
	+62 00245	LS I +62 156		1 21 00	+62 34.0	1	10.15		0.90		−0.14		2	B1:V:pe	1012
		LP 883 - 248		1 21 00	−27 00.	1	15.19		1.67				1		3061
		BPM 47118		1 21 00	−27 14.	1	11.86		0.85		0.56		1		3061
8462	−27 00463			1 21 00	−27 28.5	1	9.78		0.77		0.36		2	G8/K0 V	3061
8472	−25 00554			1 21 02	−25 20.7	3	10.25	.018	−0.09	.006	−0.36	.012	12	B8 IV/V	152,1068,1775
8488	−43 00412			1 21 02	−43 24.6	2	9.07	.000	0.47	.000	−0.10		8	F5 V	1075,2011

Table 1 95

HD	DM	Other Id	N Rem	α_{1950}	δ_{1950}	S	V	σ_V	B–V	σ_{B-V}	U–B	σ_{U-B}	n	Spectrum	References
	−31 00561	G 269 - 149		1 21 05	−31 00.8	2	11.65	.058	1.51	.005	1.17	.029	3		481,3061
		Hiltner 123		1 21 06	+62 42.9	1	10.78		0.72		−0.22		2	B1:V:	1012
		MN182,283 # 123		1 21 06	+62 43.	1	17.10		1.41		0.64		1		705
		BPM 47117		1 21 06	−23 04.	1	11.33		0.82		0.32		1		3061
8501	−45 00456			1 21 06	−44 55.9	8	9.85	.007	0.74	.004	0.27	.006	87	G8 IV/V	365,863,1460,1628,1657,1673*
8430	+25 00232			1 21 07	+25 39.9	2	8.32	.010	0.41	.005	0.01	.010	5	F2	833,1625
8442	+17 00200			1 21 09	+17 33.5	1	6.57		0.44		−0.03		1	F0	1776
8398	+50 00267			1 21 09	+50 54.8	1	7.60		0.28		0.14		1	A2	1566
		CS 22946 # 12		1 21 11	−19 41.6	1	14.65		0.46		−0.19		1		1736
8498	−31 00562	HR 400		1 21 11	−31 12.3	4	5.84	.009	1.61	.004	2.05		13	M0 III	15,1075,2032,3035
8487	−25 00555	IDS01188S2453	AB	1 21 12	−24 36.8	1	6.65		0.24				4	A9 V	2012
	+62 00246	LS I +62 157		1 21 15	+62 31.7	1	8.74		0.91		0.01		2	B5 Ia	1012
8499	−35 00482			1 21 16	−35 34.8	2	10.14	.014	0.05	.014	0.04	.027	7	A0 V	152,1068
8484	−7 00222			1 21 19	−07 06.2	1	8.96		1.53		1.86		3	K2	3016
		G 244 - 26		1 21 20	+64 20.4	1	10.76		0.97		0.79		2	K2	7010
8468	+14 00213			1 21 21	+15 02.7	2	8.92	.020	0.99	.005	0.72		4	G8 V	882,1648
8441	+42 00293	HN And		1 21 23	+42 52.9	2	6.71		0.00	.018	0.03	.010	9	A2p	1063,1202
		E1 - b		1 21 24	−44 42.	1	13.14		0.54				4		1075
		G 133 - 8		1 21 25	+40 08.3	1	17.10		0.82		0.21		1	DC	782
	−8 00242	G 271 - 57		1 21 29	−07 58.2	1	11.17		0.77		0.34		1		1658
8344	+73 00071			1 21 31	+73 51.3	1	7.76		0.29		0.29		2	A0	1733
8512	−8 00244	HR 402	⋆ A	1 21 31	−08 26.5	10	3.60	.013	1.06	.004	0.93	.010	32	K0 IIIb	3,15,58,1075,1509*
8511	−8 00243	HR 401, AV Cet	⋆ A	1 21 32	−08 16.0	2	6.20	.005	0.24	.005	0.08	.010	7	F0 V	15,1061
	−66 00097			1 21 33	−66 26.2	1	9.88		1.01		0.86		1		1696
	+12 00172	G 34 - 27		1 21 36	+12 38.8	2	9.52	.050	1.02	.000	0.90	.010	2	K3	333,1620,1696
8497	+26 00231	IDS01189N2704	A	1 21 38	+27 19.2	1	8.74		0.70		0.20		2	G5	3032
8497	+26 00231	IDS01189N2704	B	1 21 38	+27 19.2	1	12.55		1.09		0.92		2		3032
8497	+26 00231	IDS01189N2704	C	1 21 38	+27 19.2	1	11.87		0.54		0.03		1		3032
8481	+38 00255	IDS01188N3830	A	1 21 38	+38 45.9	1	7.85		1.01		0.84		3	K1 III	1501
		CS 22946 # 14		1 21 39	−18 07.0	2	14.02	.000	0.39	.001	−0.20	.006	2		1580,1736
8544	−36 00538			1 21 39	−35 59.0	1	8.72		1.23		1.33		4	K2 III	1731
8510	+13 00207			1 21 40	+13 38.4	1	8.87		0.73				2	G5	882
		FAIRALL 9 # 1		1 21 40	−59 09.0	1	12.44		0.55		−0.04		5		1687
	−72 00096			1 21 40	−71 49.0	1	11.20		0.65		0.17		4		727
		BPM 47120		1 21 42	−32 43.	1	13.39		1.50				1		3061
		CS 22946 # 10		1 21 44	−20 15.0	2	15.60	.000	0.22	.000	0.17	.008	2		966,1736
	−25 00559			1 21 45	−25 06.2	1	10.59		0.46		−0.05		1	F2	3061
		FAIRALL 9 # 5		1 21 45	−59 10.2	1	14.16		0.65		0.09		5		1687
		G 133 - 10		1 21 46	+34 05.5	1	11.56		0.84				1	K0	1620
8700	−74 00096			1 21 46	−73 50.6	1	9.59		0.02		0.06		3	Ap Si	727
8557	−22 00240			1 21 47	−22 12.1	1	8.96		0.71		0.21		1	G5 V	3061
		MCC 374		1 21 49	+12 35.2	1	11.33		1.31				2	K7	1619
	+60 00232	LS I +60 156		1 21 50	+60 32.4	1	9.97		0.27		−0.49		2	B2 III	1012
8556	−7 00223	HR 404	⋆ AB	1 21 50	−07 10.5	5	5.91	.007	0.41	.012	−0.06	.017	18	F3 V +F4 V	15,292,1061,2029,3030
		FAIRALL 9 # 3		1 21 51	−59 03.	1	15.35		0.60		0.00		5		1687
8508	+35 00265			1 21 52	+35 48.3	1	7.87		1.02		0.76		2	G5	1601
236722	+59 00247			1 21 52	+60 17.9	1	8.51		0.12				3	B8	70
8617	−57 00302			1 21 52	−56 51.9	2	8.98	.008	1.13	.003	1.02		46	K0 III	978,1642
8581	−32 00548	G 269 - 150		1 21 53	−32 04.3	2	6.85	.000	0.58	.000	0.08	.005	2	F8 V	79,3061
		E1 - p		1 21 54	−44 39.3	1	14.78		0.80		0.38		4		1460
		AH Hyi		1 21 55	−74 37.9	1	14.76		0.61				1		1511
8464	+60 00234			1 21 58	+61 27.0	1	8.84		0.22				4	A0	70
8507	+46 00349			1 21 59	+46 54.5	1	8.05		1.03		0.75		2	G5 II	1601
	−27 00470	G 269 - 151		1 21 59	−26 51.4	1	11.43		1.12		1.00		1		3061
		AJ89,1229 # 48		1 22 01	+06 56.6	1	14.11		1.52				2		1532
8424	+70 00102	HR 398		1 22 03	+70 43.2	1	6.36		−0.02		−0.09		2	A0 Vnn	1733
8542	+19 00230			1 22 04	+19 53.9	1	8.14		0.65		0.16		2	G0	1648
8364	+76 00042			1 22 06	+77 25.2	1	7.73		0.49		0.09		2	F5	1375
		CS 29518 # 53		1 22 08	−28 19.7	1	12.31		0.25		0.10		1		1736
		FAIRALL 9 # 2		1 22 08	−59 05.4	1	13.34		0.76		0.37		10		1687
8528	+44 00296	IDS01192N4505	AB	1 22 09	+45 20.9	1	8.20		0.34		0.05		2	F0	3030
8553	+17 00202	G 33 - 56		1 22 10	+18 14.6	3	8.49	.020	0.87	.023	0.58	.000	6	K0	196,333,882,1620
8492	+62 00248			1 22 10	+63 20.0	1	9.04		0.48				2	F4 V	1118
8603	−25 00563	IDS01198S2550	AB	1 22 10	−25 34.8	2	8.83	.028	0.16	.009	0.08		8	A4 IV	1068,2012
	−32 00551	G 269 - 154		1 22 10	−32 04.1	1	10.86		1.01		0.81		1		3061
	−73 00077			1 22 10	−72 41.6	1	9.81		0.90		0.55		3	G5 III	35
236724	+56 00269			1 22 12	+57 22.6	1	8.36		1.93				2	K7	369
8589	−16 00237	HR 405	⋆	1 22 13	−15 55.2	2	6.16	.019	0.92	.005	0.54		6	G8 III/IV	252,2035
8599	−3 00195	HR 406		1 22 16	−03 06.5	2	6.14	.005	0.96	.000	0.72	.005	3	G8 III	15,1256
8561	+34 00243			1 22 17	+35 28.4	1	7.66		1.07		0.92		3	G5	1625
		BPM 47125		1 22 18	−31 24.	1	12.17		1.00		0.84		1		3061
		E1 - Z		1 22 18	−44 54.	1	12.42		0.68				4		1075
		FAIRALL 9 # 4		1 22 18	−59 04.9	1	14.68		0.71		0.27		7		1687
		G 33 - 57		1 22 22	+20 19.2	1	11.82		1.19		1.12		1	K3	333,1620
8491	+67 00123	HR 399	⋆ AB	1 22 22	+67 52.2	7	4.72	.020	1.05	.009	0.94	.008	17	K0 III	15,37,542,1118,1355*
8491	+67 00123	IDS01189N6736	CD	1 22 22	+67 52.2	1	9.18		0.66		0.34		2		542
		E1 - 48		1 22 22	−44 31.	1	11.64		0.60		−0.04		4		116
8663	−46 00391		V	1 22 25	−45 58.9	3	7.75	.006	1.48	.007	1.76	.023	18	K5/M0 III	977,1673,2011
		FAIRALL 9 # 6		1 22 25	−58 59.5	1	13.54		0.82		0.48		3		1687
		G 72 - 6		1 22 26	+27 25.5	2	12.89	.041	0.66	.054	−0.06	.005	5		308,1620

HD	DM	Other Id	N Rem	α₁₉₅₀	δ₁₉₅₀	S	V	σV	B–V	σB-V	U–B	σU-B	n	Spectrum	References
	−26 00466	G 269 - 155	⋆ AB	1 22 26	−26 29.8	2	10.44	.045	1.32	.005	1.27	.005	2		1696,3061
8637	−26 00467			1 22 27	−26 31.2	1	9.77		0.80		0.28		1	G6/8 V	79
8638	−28 00433	G 269 - 156		1 22 27	−28 05.7	4	8.30	.010	0.69	.006	0.15	.015	18	G6 V(w)	158,1075,1775,3077
8651	−42 00493	HR 408		1 22 28	−41 45.1	8	5.42	.006	1.04	.005	0.86	.005	40	K0 III	15,116,278,977,1075*
8755	−72 00097			1 22 28	−72 16.1	1	9.58		0.12		0.96		4	G8 III	727
8614	+2 00205	IDS01198N0231	A	1 22 29	+02 48.7	1	8.63		0.50		−0.01		2	F8	1375
8627	−6 00270	IDS01200S0628	AB	1 22 29	−06 12.4	1	6.74		0.21		0.12		2	A0	3016
		E1 - h		1 22 30	−44 49.5	2	13.83	.001	0.73	.018	0.23		8		1075,1460
		AAS33,107 # 16		1 22 30	−72 21.	1	12.31		0.89		0.48		4		727
8538	+59 00248	HR 403, δ Cas	⋆ A	1 22 31	+59 58.6	16	2.68	.003	0.13	.003	0.12	.006	188	A5 V	1,15,985,1006,1077,1118*
236726	+55 00320			1 22 32	+55 42.9	1	9.54		0.13		0.00		2	A0	1502
8681	−45 00463	HR 411		1 22 32	−44 47.3	11	6.26	.009	1.14	.007	1.08	.010	62	K1 II	15,116,278,365,977*
		G 269 - 157		1 22 33	−33 44.0	1	12.98		1.43		1.27		1		3061
		G 269 - 158		1 22 36	−33 44.0	1	14.87		1.50				1		3061
		E1 - d		1 22 36	−44 38.	1	13.34		0.86				4		1075
	−1 00184	G 70 - 57		1 22 37	−01 18.9	3	9.46	.027	1.03	.020	0.94	.025	4	K5	202,1620,1658
8626	+15 00208			1 22 38	+15 59.8	1	7.03		1.59				1	K5 III	882
8680	−43 00421			1 22 38	−43 01.5	1	8.93		1.61		1.88		4	M3	3007
		E1 - i		1 22 38	−44 54.1	2	13.82	.085	0.88	.050	0.61		8		1075,1460
8609	+31 00247	G 72 - 7		1 22 41	+31 43.3	1	8.96		0.76		0.27		1	G0	333,1620
8783	−72 00098			1 22 42	−72 35.1	3	7.80	.008	0.14	.010	0.17	.004	32	Ap SrEuCr	35,727,1408
	−33 00514	G 269 - 159		1 22 43	−33 06.8	1	9.81		1.45		1.15		2		3061
	−45 00464			1 22 43	−45 11.9	3	11.63	.056	0.48	.011	−0.01	.013	12		116,863,1075
		LP 883 - 299		1 22 45	−30 00.0	1	12.81		1.42		1.27		1		1696
	+62 00249	LS I +62 162		1 22 47	+62 46.4	2	10.04	.000	0.70	.000	−0.34	.000	4	O9.5V	1011,1012
		S080 # 2		1 22 47	−64 53.0	1	10.32		0.67		0.27		2		1730
8781	−71 00061	BG Hyi		1 22 47	−70 56.2	1	8.05		0.33		0.07		4	F0 II/III	727
		E1 - Y		1 22 48	−44 41.	1	12.40		0.55				4		1075
		E1 - 49		1 22 48	−44 54.0	5	11.62	.018	0.57	.024	−0.01	.025	18		116,1075,1460,1628,2021
8782	−72 00099			1 22 48	−71 46.7	1	8.72		0.52		0.00		3	F8 V	727
		SK 170		1 22 49	−74 12.1	1	13.57		−0.12		−0.64		3		733
		G 33 - 58		1 22 50	+11 53.9	2	10.28	.020	1.06	.000	0.98	.020	2		333,1620,3072
8634	+22 00226	HR 407		1 22 51	+23 15.1	2	6.20	.018	0.42	.002	−0.02		6	F5 III	71,254
		CS 22174 # 28		1 22 54	−11 46.3	1	15.15		0.07		0.12		1		1736
		CS 22174 # 34		1 22 55	−09 51.9	1	13.43		0.08		0.19		1		1736
		SK 171		1 22 55	−74 12.5	1	13.62		0.00		0.02		3		733
		CS 29518 # 47		1 22 56	−31 51.0	1	13.93		0.21		0.02		1		1736
		G 2 - 37		1 22 58	+09 30.2	1	13.11		1.69				2		1532
		BPM 47134		1 23 00	−39 24.	1	12.70		0.75		0.30		1		3061
		G 269 - 160		1 23 02	−26 16.2	1	14.94		0.40		−0.55		6		3061
236730	+58 00238			1 23 05	+58 41.0	1	9.06		1.84				2	K7	369
8729	−46 00394	AL Phe		1 23 05	−46 11.2	1	8.70		1.75				4	M3/4	1075
8735	−46 00395			1 23 06	−46 12.1	1	9.51		0.31				4	A3	1075
	−63 00115			1 23 06	−62 44.7	1	9.91		0.79		0.29		2	G5	1696
8717	−29 00454			1 23 07	−29 02.5	2	8.34	.018	0.18	.009	0.07		7	A3 V	1068,2012
8716	−27 00478			1 23 09	−26 43.1	1	8.95		0.24				4	A8 V	2012
8705	−15 00266	HR 412		1 23 10	−14 51.5	5	4.90	.010	1.23	.008	1.27	.009	23	K2.5 IIIb	3,1075,2035,3052
		E1 - a		1 23 10	−44 42.6	2	12.67	.002	0.62	.007	0.02		8		1075,1460
	+16 00148			1 23 11	+16 40.8	1	9.17		0.65		0.15		1	F8	1776
		SB 573		1 23 11	−21 01.1	2	12.79	.000	0.21	.000	0.20	.008	2	A0	966,1736
	+15 00209			1 23 12	+15 33.4	1	9.51		0.59		0.06		1	F8	1776
		L 3 - 2		1 23 12	−84 17.	1	11.00		0.51		−0.16		2		1696
		CS 22953 # 37		1 23 14	−59 31.6	2	13.63	.010	0.37	.006	−0.21	.010	3		1580,1736
		G 269 - 161		1 23 16	−29 26.7	2	13.99	.054	1.47	.002	1.07	.037	3		481,3061
	+12 00176	IDS01206N1244	AB	1 23 17	+13 00.3	1	10.63		1.19				1	K7	1632
8673	+33 00228	HR 410		1 23 17	+34 19.3	3	6.33	.008	0.47	.010	0.00	.012	7	F6 V	254,272,3026
8703	+9 00167	IDS01207N0953	A	1 23 20	+10 08.8	1	7.21		0.05		0.04		2	A0	1733
		LF 5 # 153		1 23 20	+59 46.	1	10.73		0.22		−0.61		1	B2 IV	367
	−10 00306			1 23 20	−10 10.4	1	9.10		0.56		−0.09		1		3077
8671	+42 00302	HR 409		1 23 22	+43 11.9	2	5.96	.016	0.50	.012	0.01		4	F7 V	71,272
8810	−65 00130	HR 420		1 23 22	−64 37.7	4	5.91	.006	1.56	.008	1.91	.000	19	K5 III	15,1075,1637,2038
8743	−38 00491			1 23 23	−38 26.5	1	8.85		0.17		0.10		4	A3 V	152
	−47 00422			1 23 23	−47 29.2	2	9.35	.010	1.28	.001	1.41		8	K2	863,2011
		G 269 - 162		1 23 25	−25 52.2	1	14.35		1.52		1.16		1		3061
8787	−60 00112	IDS01216S6001	B	1 23 26	−59 45.5	1	9.40		0.51		0.02		3	F2 IV	1700
8787	−60 00112	IDS01216S6001	A	1 23 27	−59 45.5	2	7.05	.004	0.38	.004	0.12	.012	13	F2 IV	1117,1700
		BPM 47136		1 23 30	−30 38.	1	14.03		0.57		−0.20		2		3061
8723	+18 00187	HR 413		1 23 33	+18 54.8	2	5.38	.024	0.38	.000	−0.11	.019	7	F2 V:	254,3026
		CS 22174 # 32		1 23 35	−11 12.6	1	14.14		0.17		0.19		1		1736
8724	+16 00149			1 23 36	+16 52.1	2	8.30	.000	0.99	.019	0.40		4	G5	882,3077
8733	+19 00238			1 23 41	+19 48.7	1	6.44		0.94		0.66		3	K0 III	1501
	+60 00238			1 23 41	+61 10.0	1	7.73		0.93		0.76		3	K2	1625
	−32 00558	G 269 - 163	⋆ AB	1 23 45	−32 21.3	1	10.74		0.82		0.38		1		3061
		BPM 47138		1 23 48	−31 01.	1	11.46		0.55		−0.06		1		3061
		SK 172		1 23 48	−73 53.2	1	13.45		−0.02		−0.38		2		733
8739	+17 00206			1 23 49	+17 56.6	1	8.57		1.23				1	K0	882
8821	−48 00367	IDS01217S4825	A	1 23 49	−48 09.4	1	7.86		0.75				1	K0 V	173
8844	−59 00096			1 23 51	−59 35.7	2	8.58	.029	0.95	.020	0.54	.020	5	G6 III	1117,1700
8904	−72 00100			1 23 51	−72 18.8	1	8.98		1.17		1.11		5	K1/2 III	727
8748	+13 00212			1 23 52	+13 42.0	1	8.39		1.30				1	K2	882

Table 1   97

HD	DM	Other Id	N	Rem	α_{1950}	δ_{1950}	S	V	σ_V	B–V	σ_{B-V}	U–B	σ_{U-B}	n	Spectrum	References
8779	−1 00189	HR 416			1 23 54	−00 39.5	2	6.40	.005	1.24	.000	1.27	.000	7	K0 IV	15,1256
		BPM 47140			1 23 54	−30 49.	1	12.58		0.89		0.57		2		3061
8747	+26 00239				1 23 58	+26 59.3	1	6.76		1.05		0.86		2	K0 III	1625
8763	+18 00189	HR 414			1 23 59	+18 58.9	3	5.49	.005	1.11	.005	1.05	.014	7	K1 III	252,1355,3002
		LF 5 # 152			1 24 00	+59 48.	1	11.30		0.28		−0.48		1	B2 V	367
		LP 467 - 65			1 24 02	+11 07.5	1	11.64		1.40		1.02		1	K6	1696
	+59 00250	LS I +59 070			1 24 02	+59 51.4	1	10.44		0.25		−0.51		1	B2 Ve	367
		CS 22174 # 27			1 24 04	−11 59.8	1	12.77		0.22		0.09		1		1736
8843	−45 00472				1 24 07	−44 57.1	2	9.63	.005	0.83	.003	0.38		10	G8 (IV)	863,2011
	−7 00230				1 24 08	−06 35.8	1	11.14		0.00		0.00		3		1026
8701	+65 00164				1 24 09	+65 49.1	1	6.98		1.92		2.19		3	K2 II:p	1355
236740	+59 00251	LS I +60 158		⋆ A	1 24 10	+60 01.5	3	7.88	.012	0.46	.004	−0.38	.009	5	B3 Ia	367,405,1012
8832	−30 00472				1 24 10	−30 32.1	1	7.46		1.20		1.15		7	K1 III	1673
		SK 175			1 24 10	−72 54.1	1	13.80		−0.18		−0.78		3		733
8833	−33 00523				1 24 11	−32 48.6	1	9.28		0.67		0.13		1	G0 V	858
		SK 174			1 24 12	−73 37.0	1	13.68		−0.18		−0.97		3		733
8792	+15 00212				1 24 15	+16 08.9	1	8.47		1.35				1	K2	882
8774	+33 00234	HR 415			1 24 15	+34 07.1	2	6.28	.010	0.45	.000	0.00	.000	7	F7 IVb vs	254,3026
8830	−24 00615	BPM 47141			1 24 15	−24 32.9	1	10.15		0.96		0.75		1	K2	3061
	+29 00238				1 24 16	+30 21.0	1	10.22		1.17				1		209
		G 2 - 38			1 24 17	+11 45.3	3	11.37	.009	0.51	.005	−0.20	.013	3		333,927,1620,1696
8803	+2 00211	HR 419		⋆ A	1 24 18	+03 16.6	1	6.70		−0.06		−0.19		1	B9 V	542
8803	+2 00211	HR 419		⋆ AB	1 24 18	+03 16.6	4	6.58	.008	−0.04	.005	−0.19	.011	10	B9 V +A8 V	15,542,1061,1079
8803	+2 00211	IDS01217N0301		B	1 24 18	+03 16.6	1	9.38		0.19		0.00		1	A8 V	542
	+77 00052				1 24 19	+77 48.6	1	10.23		0.46		0.15		2		1375
8791	+24 00212				1 24 20	+25 10.9	2	7.15	.035	1.55	.005	1.86		3	K3 II	20,8100
8829	−13 00262	HR 421			1 24 23	−13 19.0	3	5.51	.042	0.32	.012	−0.01	.005	7	F0 V	254,2007,3016
8736	+62 00253				1 24 24	+62 30.9	1	8.88		0.20				3	B9 V	1118
		SK 173			1 24 24	−73 23.	1	13.72		−0.18		−0.89		3		733
8828	−0 00229	IDS01218S0040		A	1 24 27	−00 24.7	3	7.94	.022	0.74	.008	0.31	.005	4	G5	202,258,1620
	+59 00252	LS I +60 159			1 24 29	+60 07.1	1	10.43		0.36		−0.40		1	B2 V	367
8859	−30 00475	G 269 - 165			1 24 29	−29 43.9	1	8.43		0.74		0.24		2	G5 V	3061
8801	+40 00289	HR 418		⋆ A	1 24 32	+40 50.5	1	6.46		0.27		0.03		5	AmA7-F0	8071
8784	+51 00308				1 24 32	+51 33.0	1	8.07		1.07		0.76		2	K0	1566
		NGC 559 - 1159			1 24 32	+63 02.6	1	10.88		0.74		0.35		4		8096
		AJ74,1000 M33# 8			1 24 35	+30 20.1	1	10.86		0.92				1		209
8799	+44 00307	HR 417		⋆ AB	1 24 39	+45 09.0	5	4.83	.007	0.42	.000	−0.01	.018	18	F4 IV	15,1118,3026,8015,8040
	+58 00241	LS I +58 039			1 24 40	+58 58.6	2	9.95	.015	0.27	.005	−0.58	.005	3	B1 V	367,1012
8879	−33 00525	HR 423, R Scl		⋆ A	1 24 40	−32 48.1	6	5.86	.095	4.08	.117	7.40		16	C6 II	15,109,864,3038,8015,8022
8768	+62 00254	NGC 559 - 1168		⋆	1 24 41	+63 01.4	3	8.70	.005	0.43	.019	−0.58	.014	8	O9.5IV	1011,1012,8096
		CS 29518 # 49			1 24 41	−29 57.4	1	14.43		0.19		0.06		1		1736
	+22 00234				1 24 43	+22 59.1	1	9.81		1.14		1.10		2	K7	1746
8769	+60 00240				1 24 43	+61 04.0	1	9.27		0.14				3	B9 V	1118
		NGC 559 - 1176			1 24 43	+63 02.2	1	13.48		0.53		−0.16		4		8096
		NGC 559 - 1181			1 24 46	+63 01.8	1	13.98		0.72		−0.21		4		8096
8936	−59 00098				1 24 48	−59 24.6	1	9.64		0.36		0.05		2	A9 IV/V	1117
		NGC 559 - 1189			1 24 50	+63 01.3	1	11.98		0.71		0.10		4		8096
8856	+16 00153				1 24 52	+16 31.7	2	8.57	.045	1.45	.007	1.67		4	K2	882,3040
8837	+39 00334	HR 422			1 24 52	+40 04.6	2			−0.04	.000	−0.18	.005	5	A0 III	1049,1079
		NGC 559 - 1901			1 24 52	+63 05.2	1	15.71		1.65		0.19		5		8096
		GR 47			1 24 55	−28 58.1	1	13.44		1.42		1.24		2		481
232440	+50 00280				1 24 57	+50 44.6	1	9.44		0.30		0.22		2	A2	1566
		G 2 - 39			1 24 58	+13 00.1	1	12.56		1.25		1.15		1		333,1620
		AJ74,1000 M33# 2			1 24 59	+30 03.6	1	11.88		1.13				1		209
		LP 883 - 372			1 25 00	−30 19.	1	13.69		1.50				1		3061
		LB 3209			1 25 00	−66 01.	1	11.78		−0.03		−0.09		2		45
		NGC 559 - 1902			1 25 01	+63 05.6	1	14.65		1.69		1.11		5		8096
		NGC 559 - 148			1 25 03	+63 03.2	1	14.77		0.64		0.62		3		8096
8875	+4 00251	IDS01225N0450		AB	1 25 04	+05 05.7	3	7.01	.016	0.62	.005	0.16	.021	8	G0 V	176,1381,3077
		NGC 559 - 1212			1 25 05	+63 06.6	1	10.88		0.68		0.20		3		8096
		fld I # 35			1 25 06	+58 04.4	1	11.37		1.26		1.00		2		1755
		NGC 559 - 1213			1 25 06	+63 02.6	1	14.12		0.65		0.16		4		8096
8912	−26 00480				1 25 06	−26 11.9	1	9.17		0.86		0.51		2	K0 V	3061
		NGC 559 - 145			1 25 08	+63 04.6	1	15.14		0.88		0.43		5		8096
8874	+24 00217				1 25 09	+25 25.4	1	8.41		0.40		−0.05		3	F5	833
	−25 00584	LP 827 - 229			1 25 12	−25 12.0	2	10.39	.010	0.89	.002	0.64	.007	3		1696,3061
		fld I # 34			1 25 13	+58 04.0	1	13.37		1.46		1.24		2		1755
8978	−57 00315				1 25 14	−56 41.4	1	8.68		0.46		−0.05		44	F5 V	978
		NGC 559 - 144			1 25 15	+63 05.3	1	14.15		0.96		0.43		5		8096
8932	−24 00624				1 25 16	−24 16.5	1	9.12		0.12		0.03		1	A2 V	1068
8921	−11 00272	HR 425		⋆ AB	1 25 17	−11 09.6	2	6.13	.001	1.32	.001	1.36		6	K0	58,2007
		NGC 559 - 143			1 25 18	+63 05.3	1	13.67		1.63		1.37		3		8096
		TonS 217			1 25 18	−23 39.	1	15.37		−0.03		−1.06		3		3064
8910	+15 00215				1 25 20	+15 43.8	1	7.95		1.13		1.02		2	K0 III	3077
		NGC 559 - 155			1 25 20	+63 06.9	1	12.17		0.82		0.27		4		8096
9020	−65 00132				1 25 20	−64 50.4	1	9.32		1.06		0.86		2	G8 III/IV	1730
9068	−73 00080	IDS01241S7316		AB	1 25 20	−73 00.6	2	8.69	.009	0.46	.002	−0.02	.018	8	F5/6 V	727,1408
8962	−44 00407				1 25 21	−44 11.5	2	9.84		0.58		0.05		10	G0 V	863,2011
8963	−46 00406				1 25 21	−46 06.3	6	6.96	.007	1.49	.006	1.84	.026	23	K4/5 III	116,474,977,1075,2011,2024
	+29 00242				1 25 22	+29 58.6	1	10.64		1.07				1		209

HD	DM	Other Id	N	Rem	α_{1950}	δ_{1950}	S	V	σ_V	B–V	σ_{B-V}	U–B	σ_{U-B}	n	Spectrum	References
		WLS 130 80 # 5			1 25 22	+82 28.2	1	12.04		0.35		0.11		2		1375
		S080 # 3			1 25 22	−64 31.7	1	11.06		0.87		0.48		2		1730
8930	−2 00220	G 271 - 69			1 25 23	−02 14.9	1	8.59		0.65		0.13		1	G5	1658
		NGC 559 - 151			1 25 25	+63 02.9	1	11.78		0.58		−0.11		3		8096
		NGC 559 - 154			1 25 27	+63 01.1	1	14.65		1.69		1.27		3		8096
		NGC 559 - 142			1 25 27	+63 04.7	1	13.85		0.78		0.46		3		8096
		NGC 559 - 141			1 25 27	+63 06.1	1	10.57		0.62		0.38		3		8096
		CS 22946 # 9			1 25 27	−20 17.0	2	14.70	.000	0.28	.000	0.03	.009	2		966,1736
8977	−46 00407				1 25 27	−46 24.6	6	7.70	.010	0.09	.006	0.09	.004	29	A0 V	116,278,863,1075,2011,2024
		fld I # 27			1 25 30	+58 11.2	1	12.55		1.54		1.44				1755
		NGC 559 - 120			1 25 32	+63 00.8	1	13.91		0.69		0.67		4		8096
		fld I # 26			1 25 33	+58 11.3	1	14.41		0.97		0.77		2		1755
		fld I # 36			1 25 35	+58 17.4	1	11.05		1.27		1.05		2		1755
		SK 176			1 25 36	−73 22.8	1	13.77		−0.22		−0.87		3		733
		SK 177			1 25 36	−73 24.0	1	13.63		−0.27		−1.00		3		733
		AJ74,1000 M33# 3			1 25 37	+30 10.6	1	11.31		1.17				1		209
8907	+41 00283				1 25 37	+42 00.6	1	6.65		0.49		−0.05		1	F8 V	1716
8942	+13 00216	IDS01230N1357		A	1 25 39	+14 13.1	1	8.69		0.95				1	G5	882
8975	−27 00495	IDS01234S2738		A	1 25 41	−27 22.1	1	7.43		1.01		0.78		1	G8 III	3061
8941	+16 00154				1 25 42	+16 49.3	1	6.60		0.55		0.06		1	F8 IV-V	1776
8949	+7 00213	HR 426	⋆	A	1 25 45	+07 42.2	5	6.21	.009	1.11	.006	1.08	.003	15	K1 III	15,253,1061,1355,3024
9040	−59 00099				1 25 45	−59 31.8	1	9.16		0.58		0.02		2	G0 V	1700
8983	−18 00235				1 25 46	−18 19.2	2	8.69	.000	0.18	.005	0.17		5	A3 III	79,2012
8985	−28 00451				1 25 46	−28 18.6	1	7.73		0.52		−0.01		1	F6/7 V	3061
	+16 00155				1 25 49	+16 38.5	1	9.02		1.09		0.91		1	K0	1776
		NGC 559 - 85			1 25 49	+63 05.0	2	10.55	.045	0.42	.050	0.19	.000	5		100,8096
8956	+7 00214	IDS01231N0727		B	1 25 50	+07 42.0	1	8.04		0.53		0.02		4	F8 IV-V	3024
	+30 00228	G 72 - 9			1 25 54	+31 21.1	1	9.24		1.00		0.83		1	K2	333,1620
		NGC 559 - 23			1 25 54	+63 02.4	1	12.78		1.46				2		100
		NGC 559 - 84			1 25 54	+63 05.0	1	13.52		0.61		0.44		3		100
8982	−9 00283	G 271 - 70			1 25 54	−08 38.3	1	9.28		0.55		0.04		1	F8	1658
		CS 22174 # 30			1 25 54	−11 39.8	1	14.52		0.73		0.18		1		1736
		BPM 47156			1 25 54	−23 35.	1	12.56		1.47				1		3061
		SK 178			1 25 54	−73 25.1	1	13.73		−0.18		−0.73		3		733
		NGC 559 - 20			1 25 57	+63 02.9	1	13.72		0.62		0.50		7		100
		NGC 559 - 24			1 25 58	+63 02.1	1	13.74		1.63		1.28		2		4002
9029	−45 00483				1 25 58	−45 20.1	3	9.44	.008	0.73	.002	0.32		8	G5 V	58,2011,2024
		fld I # 13			1 25 59	+58 10.1	1	12.46		0.57		0.45		2		1755
8906	+59 00258	LS I +59 071			1 25 59	+59 46.7	5	7.11	.004	0.74	.005	0.67		91	F3 Ib	61,71,1118,1355,6009
		fld I # 25			1 25 60	+58 21.5	1	13.36		0.39		0.31		2		1755
		NGC 559 - 27			1 26 00	+63 01.5	1	12.03		2.10				3		100
		NGC 559 - 19			1 26 00	+63 02.8	2	14.09	.044	0.70	.034	0.41	.151	9		100,8096
		LP 707 - 107			1 26 00	−12 57.6	1	11.46		0.65		0.09		2		1696
		BPM 47157			1 26 00	−32 28.	1	12.79		1.38				1		3061
		fld I # 24			1 26 01	+58 20.9	1	11.87		0.40		0.34		2		1755
	+61 00270				1 26 01	+61 30.2	1	9.24		2.10				3	M0	369
		fld I # 23			1 26 02	+58 20.5	1	11.79		1.37		1.24		2		1755
		fld I # 40			1 26 03	+58 07.1	1	14.26		0.56		0.32		2		1755
9026	−32 00575				1 26 03	−31 45.3	1	8.10		0.35				4	F2 III/IV	2012
		AJ74,1000 M33# 6			1 26 07	+30 15.6	1	12.42		0.94				1		209
		NGC 559 - 5			1 26 08	+63 02.7	1	13.43		1.65		1.39		1		4002
	+58 00247	LS I +58 040			1 26 09	+58 50.5	1	10.14		0.28		−0.53		1	B1	367
9027	−33 00533				1 26 09	−33 16.4	4	9.32	.019	0.14	.008	0.10	.014	14	A3 V	152,858,1068,1775
		fld I # 14			1 26 10	+58 05.0	1	13.65		0.74		0.74		2		1755
		NGC 559 - 14			1 26 10	+63 03.1	1	11.52		0.63		−0.06		2		100
		CS 22174 # 40			1 26 10	−07 39.6	1	13.18		0.22		0.00		1		1736
		NGC 559 - 32			1 26 11	+63 02.1	1	10.97		0.50		0.03		3		100
		Wolf 71			1 26 12	+12 20.9	1	11.56		0.75		0.30		2		1696
9053	−43 00449	γ Phe			1 26 12	−43 34.4	9	3.41	.016	1.57	.003	1.85	.012	28	M0-IIIa	3,15,58,278,1075,2011*
		JL 247			1 26 12	−53 16.	1	14.11		0.86		0.61		1		832
		L 222 - 53	⋆		1 26 12	−53 16.	1	14.48		0.01		−0.73		2		3065
		NGC 559 - 47			1 26 13	+63 04.6	2	13.11	.128	0.66	.079	0.55	.015	8		100,8096
		NGC 559 - 105			1 26 13	+63 06.6	1	12.83		1.98				2		100
		fld I # 12			1 26 14	+58 15.0	1	12.33		0.34		0.27		2		1755
		fld I # 11			1 26 15	+58 13.6	1	11.29		0.92		0.54		2		1755
		NGC 559 - 33			1 26 15	+63 02.3	1	11.75		0.78		0.39		3		100
		NGC 559 - 45			1 26 15	+63 04.1	1	14.87		1.02				6		100
		CS 29504 # 8			1 26 15	−35 46.1	1	14.17		0.26		0.02		1		1736
		fld I # 28			1 26 17	+57 59.3	1	14.79		1.35		0.91		2		1755
		NGC 559 - 34			1 26 17	+63 02.1	3	13.12	.052	1.84	.057	1.21	.344	9		100,4002,8096
		NGC 559 - 35			1 26 18	+63 02.4	1	13.15		1.86		1.54		1		4002
8997	+20 00226	G 34 - 32	⋆	A	1 26 20	+21 28.0	3	7.73	.011	0.97	.008	0.69	.010	7	K2 V	22,333,1355,1620
		CS 22174 # 29			1 26 20	−11 42.6	1	13.18		0.32		0.09		1		1736
		NGC 559 - 37			1 26 21	+63 02.9	2	13.68	.039	0.50	.030	−0.01	.079	8		100,8096
9163	−72 00101				1 26 21	−71 57.8	1	7.26		0.97		0.72		3	G8/K0 III	727
		fld I # 37			1 26 22	+58 21.2	1	14.06		0.51		0.34		2		1755
		NGC 559 - 39			1 26 22	+63 03.2	2	13.74	.015	1.47	.088	1.24		3		100,4002
		NGC 559 - 41			1 26 23	+63 04.0	1	13.54	.005	1.62	.208	1.73		3		100,4002
		BPM 47159			1 26 24	−23 57.	1	13.49		0.68		0.10		1		3061
	+61 00271				1 26 25	+62 09.2	1	9.88		0.21		0.08		2	A0 V	1375

Table 1   99

HD	DM	Other Id	N Rem	α_{1950}	δ_{1950}	S	V	σ_V	B–V	σ_{B-V}	U–B	σ_{U-B}	n	Spectrum	References
		NGC 559 - 76		1 26 25	+63 04.1	1	14.50		0.85				6		100
9063	−25 00597			1 26 26	−25 03.4	1	7.02		0.25				4	A8 V	2012
9065	−34 00576	HR 431, WZ Scl		1 26 27	−34 01.3	6	6.57	.016	0.31	.008	0.07	.017	17	A9/F0 V	15,258,624,858,2012,2012
9050	−14 00286			1 26 28	−14 17.5	1	8.96		0.26		0.10		4	Ap EuSr	152
		fld I # 39		1 26 29	+58 15.7	1	11.19		0.51		0.19		2		1755
		BPM 47162		1 26 30	−32 52.	1	12.76		1.10		1.00		1		3061
		BPM 47163		1 26 30	−33 39.	1	13.70		1.52		1.30		1		3061
		AJ74,1000 M33# 7		1 26 31	+30 29.5	1	11.86		0.97				1		209
9074	−24 00635			1 26 31	−23 39.3	1	9.03		0.69		0.22		1	G5 V	3061
		NGC 559 - 1266		1 26 32	+63 04.0	1	14.49		0.90		0.60		3		8096
8965	+59 00260	LS I +59 072		1 26 34	+59 59.6	5	7.28	.004	0.03	.005	-0.82	.000	94	B0.5V	61,71,367,1012,1118
	+29 00249			1 26 35	+30 06.4	1	9.40		1.00				1	K5	209
9085	−37 00564			1 26 35	−37 05.5	1	7.47		1.62		1.95		4	M2 III	3007
9084	−32 00577			1 26 37	−32 17.8	1	8.88		0.55		0.00		1	F8/G0 V	3061
8992	+58 00249	LS I +58 041		1 26 39	+58 30.3	1	7.79		0.89				2	F6 Ib	70
	−63 00118			1 26 39	−63 19.9	1	10.21		0.72		0.29		1	G3	1696
9210	−73 00081			1 26 40	−73 31.7	1	8.83		1.28		1.30		3	K1 III	727
		SU Psc		1 26 42	+19 21.4	1	10.95		-0.08		-0.61		3	B5	3064
	+62 00258	LS I +63 149	⋆ AB	1 26 42	+63 19.5	1	9.87		0.67		-0.25		2	B1 IV	1012
	+57 00301			1 26 44	+58 05.7	1	10.45		0.72		0.22		2		1755
9093	−33 00538			1 26 45	−33 36.5	2	9.14	.045	1.06	.075	0.76	.045	6	G8 III	624,858
9248	−75 00088			1 26 45	−75 18.1	1	7.57		1.68		1.99		7	M2 III	1704
		G 2 - 40		1 26 46	+10 07.9	1	14.85		0.27		-0.50		1		3060
		AAS33,107 # 24		1 26 48	−71 54.	1	11.91		0.60		0.06		4		727
		fld I # 29		1 26 54	+58 01.8	1	14.06		0.91		0.51		2		1755
		JL 249		1 26 54	−51 46.	1	16.01		-0.25		-1.30		1		132
8991	+63 00193			1 26 56	+63 36.1	1	8.03		0.07				3	A0 V	1118
		fld I # 9		1 26 57	+58 16.1	1	12.30		0.57		0.12		2		1755
		AJ74,1000 M33# 5		1 26 59	+30 12.4	1	11.12		1.00				1		209
9081	+16 00156			1 27 00	+16 50.4	1	8.13		0.68				2	G5	882
	+60 00248	LS I +61 198		1 27 00	+61 05.0	1	10.08		0.31		-0.49		1	B2 V	367
9022	+59 00261			1 27 01	+59 31.4	1	6.90		1.44				2	K3 III	1118
9070	+30 00233	IDS01241N3029	A	1 27 02	+30 45.0	1	7.95		0.71		0.26		2	G5	3026
		fld I # 10		1 27 02	+58 18.5	1	13.77		0.70		0.27		2		1755
		G 2 - 41		1 27 06	+08 11.2	1	10.48		1.08		0.94		1	K3	333,1620
9057	+46 00370	HR 430		1 27 06	+46 45.0	4	5.27	.005	1.00	.005	0.82	.002	11	K0 III	37,1355,3016,4001
		fld I # 2		1 27 06	+58 03.7	1	14.41		0.80		0.10		2		1755
		SK 179		1 27 06	−73 01.5	2	13.07	.015	-0.10	.015	-0.63	.000	5		733,835
		AJ89,1229 # 58		1 27 07	+02 58.9	1	11.37		1.60				1		1532
9133	−33 00541	XX Scl		1 27 09	−33 34.7	2	8.86	.019	0.24	.019	0.13	.015	3	A7 V	624,858
9100	+17 00210	HR 432, VX Psc		1 27 11	+18 05.9	1			0.14		0.14		3	A4 IV	1049
		fld I # 1		1 27 12	+58 03.1	1	13.81		1.13		0.96		2		1755
9132	−22 00254	HR 433	⋆ A	1 27 12	−21 53.2	5	5.12	.007	0.02	.006	0.07	.023	18	A0 V	3,15,1068,2012,3023
	+61 00277	LS I +62 165	V	1 27 13	+62 28.1	1	9.63		0.58		-0.39		2	B0 IV:nn	1012
9266	−72 00102			1 27 15	−72 12.0	1	9.07		0.29		0.07		5	F0 III	727
	−8 00256	G 271 - 75		1 27 16	−07 45.5	1	10.74		0.83		0.54		1		1658
9160	−43 00456			1 27 17	−43 24.8	3	8.30	.005	1.04	.001	0.80	.002	16	G8 III	977,1075,2011
		fld I # 21		1 27 18	+58 20.9	1	13.32		0.30		0.22		2		1755
	+60 00250	LS I +61 199		1 27 19	+61 07.1	1	10.31		0.61		-0.24		1	B1 V	367
9030	+65 00175	HR 428		1 27 21	+65 50.4	1	6.14		0.08		0.05		2	A2 Vs	3016
9151	−26 00491			1 27 21	−25 52.5	1	6.37		1.14		1.09		4	K1 III	1628
		CS 29504 # 13		1 27 22	−33 46.4	1	15.56		0.13		0.16		1		1736
		fld I # 8		1 27 24	+58 15.8	1	11.24		0.38		0.31		2		1755
9184	−47 00440	HR 435, AW Phe		1 27 24	−47 00.9	6	6.30	.008	1.65	.007	1.85	.009	25	M2 III	15,278,1075,2011,2038,3007
9183	−45 00488			1 27 25	−45 04.0	3	9.34	.009	0.94	.028	0.76	.025	13	K0 IV	116,863,2011
		fld I # 4		1 27 26	+58 04.6	1	11.42		0.62		0.09		2		1755
9021	+69 00102	HR 427		1 27 28	+70 00.5	2	5.82	.000	0.46	.005	-0.13		3	F6 V	254,1118
		fld I # 16		1 27 32	+58 00.7	1	12.27		1.35		1.16		2		1755
		fld I # 22		1 27 33	+58 24.2	1	12.46		0.69		0.29		2		1755
9138	+5 00194	HR 434	⋆ A	1 27 34	+05 53.2	10	4.84	.014	1.37	.008	1.55	.019	36	K4 III	15,333,1003,1256,1355*
		BPM 47178		1 27 36	−31 11.	1	14.51		0.62		0.10		2		3061
		LB 3211		1 27 36	−62 08.	1	12.03		0.18		0.16		2		45
		SK 180		1 27 36	−72 59.7	2	12.82	.015	-0.03	.015	-0.19	.020	5		733,835
236762	+58 00252	LS I +59 073		1 27 45	+59 28.8	2	9.58	.005	0.20	.005	-0.57	.005	3	B1.5III	367,1012
		fld I # 3		1 27 46	+58 05.8	1	13.42		0.49		0.37		2		1755
		fld I # 38		1 27 48	+58 22.6	1	14.28		0.51		0.47		2		1755
		BPM 47179		1 27 48	−26 00.	1	11.73		1.45		1.16		1		3061
	+51 00318	G 173 - 2	⋆ A	1 27 50	+52 29.1	1	10.05		0.86		0.58		1	K2	1658
		CS 22174 # 42		1 27 50	−10 00.4	1	12.46		0.00		-0.01		1		1736
		CS 22180 # 1		1 27 50	−10 00.4	1	12.45		0.00		-0.01		1		1736
9172	+25 00258			1 27 53	+25 39.4	1	7.25		0.38		-0.06		2	F0	105
		fld I # 7		1 27 53	+58 15.8	1	11.16		1.43		1.38		2		1755
9105	+62 00259	LS I +63 150		1 27 54	+63 05.4	1	7.47		0.55		-0.25		2	B5 Iab	1012
		BPM 47182		1 27 54	−31 28.	1	13.70		1.23		1.16		2		3061
		LP 587 - 69		1 27 56	+00 51.1	1	12.53		1.33		1.19		1		1696
	+6 00232			1 27 57	+06 48.0	1	9.76		1.29				1	K7	1632
9228	−26 00502	HR 436	⋆ AB	1 28 02	−26 27.9	4	5.94	.010	1.33	.009	1.41		13	K2 III	15,1075,2032,3005
		SK 183		1 28 02	−73 47.3	2	13.85	.033	-0.23	.000	-1.04	.014	4		733,835
		fld I # 6		1 28 05	+58 15.2	1	12.34		0.46		0.39		2		1755
		fld I # 42		1 28 06	+58 02.2	1	13.76		0.37		0.22		2		1755

HD	DM	Other Id	N Rem	α_{1950}	δ_{1950}	S	V	σ_V	B–V	σ_{B-V}	U–B	σ_{U-B}	n	Spectrum	References
9246	−29 00478	G 274 - 52		1 28 06	−29 31.2	1	9.44		0.85		0.46		1	K0 V	3061
		R 47		1 28 06	−72 58.	1	11.56		0.38				2		1277
		AAS33,107 # 42		1 28 06	−73 17.	1	11.65		1.10		−0.44		11		727
		SK 182		1 28 06	−73 18.	2	11.51	.085	1.15	.000	−0.27	.030	6	B5:I+KI	573,733
9202	+14 00226	AN Psc		1 28 07	+14 45.9	1	8.20		1.55		1.57		1	M2	3040
	+47 00435	G 172 - 58		1 28 07	+47 45.5	1	10.17		0.43		−0.17		1	sdF7	1658
9177	+38 00275			1 28 08	+39 14.2	2	7.59	.005	−0.06	.005	−0.24	.020	4	B9 V	105,1733
		fld I # 5		1 28 08	+58 15.2	1	13.88		0.73		0.16		2		1755
9245	−26 00503			1 28 09	−25 39.0	2	9.18	.019	0.60	.000	0.10	.012	3	G2 V	79,3061
		WLS 120 5 # 8		1 28 10	+05 05.0	1	12.00		0.52		0.01		2		1375
		fld I # 20		1 28 10	+58 18.1	1	14.85		0.42		0.31		2		1755
9136	+60 00253			1 28 12	+61 17.7	1	7.65		0.14				2	A1 V	1118
	−73 00083			1 28 12	−73 13.9	1	10.77		0.40		−0.04		5		727
		SK 181	V	1 28 12	−75 58.6	3	11.67	.220	0.39	.009	0.16	.040	10	F2 Iae	416,573,733
		G 3 - 4		1 28 13	+15 18.6	1	10.65		1.11		1.05		1	K5	1658
	+59 00266	LS I +59 074		1 28 14	+59 30.6	1	10.03		0.24		−0.59		1	B1 V	367
9146	+60 00254			1 28 16	+60 38.7	3	8.13	.003	1.59	.002			27	K3 II	61,71,1118
		fld I # 30		1 28 17	+58 11.1	1	13.85		0.66		0.32		2		1755
		CS 22174 # 43		1 28 17	−10 41.4	1	14.01		0.11		0.17		1		1736
		CS 22180 # 3		1 28 17	−10 41.4	1	13.97		0.12		0.17		1		1736
9380	−73 00084			1 28 17	−73 16.1	1	9.59		0.47		0.02		5	F5 IV/V	727
		fld I # 41		1 28 20	+58 03.0	1	12.93		0.94		0.49		2		1755
		fld I # 17		1 28 23	+58 16.9	1	13.36		0.75		0.22		2		1755
		fld I # 18		1 28 25	+58 17.5	1	12.10		1.39		1.12		2		1755
236768	+58 00256	LS I +59 075		1 28 26	+59 07.3	2	9.48	.005	0.28	.010	−0.54	.010	3	B1:V:pnn	367,1012
		MtW 45 # 9		1 28 27	+30 23.4	1	15.38		0.44				1		1500
		fld I # 15		1 28 28	+58 00.2	1	12.44		0.60		0.19		2		1755
9095	+72 00076			1 28 28	+72 39.7	1	8.17		0.68		0.17		2	G0	1375
		fld I # 19		1 28 29	+58 17.8	1	12.93		0.70		0.18		2		1755
		LF 5 # 151		1 28 30	+60 18.	1	10.84		0.22		−0.48		1	B2 Ve	367
9224	+28 00249			1 28 31	+29 09.4	2	7.27	.050	0.62	.017	0.11		4	G0 V	20,3026
		fld I # 33		1 28 31	+58 05.2	1	13.03		1.04		0.80		2		1755
9237	+6 00233			1 28 32	+07 26.3	1	9.54		0.39		−0.02		2	F5	1375
9236	+16 00164			1 28 32	+16 53.7	1	7.55		0.36		−0.02		1	F0	1776
9290	−32 00589			1 28 32	−31 43.0	1	8.97		1.07				4	K1 III	2033
		MtW 45 # 24		1 28 35	+30 22.6	1	14.27		0.73				2		1500
9252	+16 00165			1 28 37	+17 18.5	1	8.70		0.45		−0.06		1	F8	1776
		MtW 45 # 27		1 28 37	+30 18.9	1	12.32		0.74		0.26		2		1500
		MtW 45 # 28		1 28 37	+30 25.2	1	15.17		0.65				1		1500
		MtW 45 # 31		1 28 38	+30 26.6	1	15.10		0.74				1		1500
	+61 00284	IM Cas		1 28 38	+62 04.3	1	8.17		2.49				3	M3	369
		MtW 45 # 34		1 28 39	+30 22.3	1	15.19		0.94				3		1500
		MtW 45 # 40		1 28 41	+30 19.1	1	16.37		0.72				1		1500
		MtW 45 # 43		1 28 42	+30 19.7	1	13.36		0.65		0.00		3		1500
		MtW 45 # 46		1 28 44	+30 27.6	1	15.92		0.43				1		1500
		MtW 45 # 49		1 28 45	+30 29.9	1	15.44		0.92				1		1500
		MtW 45 # 48		1 28 45	+30 30.4	1	13.79		1.02				1		1500
9166	+67 00133			1 28 45	+68 09.2	3	6.76	.000	1.23	.000	1.42	.013	5	K3 III	1003,1355,3009
		MtW 45 # 51		1 28 46	+30 21.7	1	16.54		0.84				2		1500
236771	+58 00257			1 28 46	+59 12.5	1	9.31		0.22				4	A0	70
9270	+14 00231	HR 437	★ AB	1 28 48	+15 05.3	17	3.61	.006	0.97	.003	0.73	.009	135	G7 IIIa	1,3,15,985,1006,1020*
		MtW 45 # 55		1 28 48	+30 27.4	1	15.68		1.52				1		1500
		MtW 45 # 59		1 28 50	+30 28.0	1	14.79		0.73				1		1500
9200	+62 00262			1 28 50	+63 21.8	1	7.85		0.13				3	A1 V	1118
		SK 184		1 28 50	−73 33.3	2	13.71	.009	−0.01	.005	−0.74	.009	4		733,835
		MtW 45 # 72		1 28 52	+30 24.7	1	13.23		0.96		0.45		1		1500
		MtW 45 # 67		1 28 52	+30 28.3	1	13.09		0.45		−0.08		3		1500
		MtW 45 # 73		1 28 53	+30 27.2	1	10.16		0.91		0.52		9		1500
		AAS33,107 # 43		1 28 54	−73 15.	1	12.63		0.59		0.01		6		727
		MtW 45 # 77		1 28 55	+30 30.4	1	15.06		0.73				1		1500
9379	−60 00118			1 28 56	−59 55.0	2	7.93	.005	0.60	.000	0.05		5	G0 V	258,1075
		MtW 45 # 83		1 28 58	+30 29.3	1	16.29		0.89				1		1500
		fld I # 31		1 28 58	+58 10.8	1	14.06		0.72		0.19		2		1755
		MtW 45 # 87		1 28 60	+30 27.2	1	13.78		0.59		0.03		3		1500
		fld I # 32		1 29 01	+58 11.2	1	12.98		1.31		1.05		2		1755
9361	−47 00451	IDS01269S4654	AB	1 29 02	−46 39.0	1	9.40		0.53				4	F7 V	2011
		MtW 45 # 95		1 29 04	+30 26.1	1	14.32		0.63				2		1500
		MtW 45 # 97		1 29 05	+30 28.6	1	15.75		0.62				2		1500
	−26 00511			1 29 06	−25 43.9	1	10.32		0.81		0.40		1	G5	3061
		G 2 - 42		1 29 07	+03 29.7	1	13.36		1.47		1.22		1		1658
		MtW 45 # 98		1 29 07	+30 28.6	1	16.23		1.21				2		1500
9489	−73 00085			1 29 07	−73 25.9	1	8.20		1.17		1.15		4	K1 III	727
		MtW 45 # 99		1 29 09	+30 17.3	1	16.93		0.90				1		1500
		MtW 45 # 103		1 29 09	+30 19.5	1	14.66		0.73		0.30		3		1500
9336	−19 00262	IDS01268S1932	A	1 29 09	−19 16.8	1	6.82		0.26				4	A4mA4-A9	2012
		MtW 45 # 106		1 29 10	+30 17.8	1	13.40		0.79		0.30		2		1500
		MtW 45 # 105		1 29 10	+30 25.7	1	15.12		0.66				2		1500
9257	+57 00315			1 29 10	+57 42.5	1	8.57		0.14		−0.04		2	A0	1375
9362	−49 00425	HR 440		1 29 10	−49 19.9	8	3.94	.006	0.99	.005	0.70	.010	21	K0 IIIb	15,278,1020,1075,1117*
		MtW 45 # 108		1 29 11	+30 19.0	1	15.67		0.50				2		1500

Table 1   101

HD	DM	Other Id	N	Rem	α_{1950}	δ_{1950}	S	V	σ_V	B–V	σ_{B-V}	U–B	σ_{U-B}	n	Spectrum	References
9349	−30 00504				1 29 11	−30 14.5	1	6.62		0.89				4	G6/8 III	2012
		MtW 45 # 109			1 29 12	+30 22.3	1	14.97		1.01				1		1500
	+60 00261	LS I +60 161			1 29 12	+60 52.4	3	8.63	.000	0.31	.000	−0.67	.000	5	O7	367,1011,1012
		JL 250			1 29 12	−45 45.	1	15.48		0.56		−0.35		1		832
		MtW 45 # 115			1 29 13	+30 15.6	1	10.74		0.55		0.09		3		1500
		MtW 45 # 110			1 29 13	+30 24.7	1	15.78		0.55				2		1500
		MtW 45 # 113			1 29 13	+30 30.4	1	15.69		0.51		−0.01		1		1500
9313	+15 00227	BF Psc			1 29 14	+15 47.5	1	7.82		0.93				2	G5	882
		MtW 45 # 117			1 29 14	+30 23.5	1	15.71		0.72				2		1500
		MtW 45 # 119			1 29 14	+30 23.8	1	16.30		0.60				2		1500
		MtW 45 # 120			1 29 14	+30 25.5	1	16.09		0.63				2		1500
		CS 22180 # 2			1 29 14	−10 20.9	1	13.03		0.19		0.18		1		1736
		SK 185			1 29 14	−73 35.8	2	12.53	.010	0.04	.005	0.03	.005	5		573,733
	+61 00285	LS I +61 201			1 29 15	+61 42.8	2	9.41	.010	0.42	.000	−0.51	.025	3	B0.5III	367,1012
		MtW 45 # 125			1 29 16	+30 16.6	1	13.61		0.49		0.06		2		1500
9298	+34 00265	HR 438			1 29 16	+34 32.6	3	6.39		−0.11	.017	−0.42	.006	7	B7 IIIp	252,1049,1079
		MtW 45 # 126			1 29 17	+30 17.9	1	15.00		0.82		0.31		3		1500
		G 72 - 12			1 29 17	+34 18.2	2	10.82	.010	0.83	.005	0.46	.020	2	K0	1620,1658
		MtW 45 # 127			1 29 18	+30 18.9	1	14.79		0.63		0.05		3		1500
9250	+62 00264	V636 Cas			1 29 18	+63 20.2	1	7.21		1.42		1.16		2	G0 Ib	1355
		AAS33,107 # 44			1 29 18	−73 18.	1	11.07		1.32		1.38		5		727
		SK 186			1 29 19	−73 18.3	3	12.83	.005	−0.12	.004	−0.79	.004	8		416,733,835
		MtW 45 # 131			1 29 20	+30 29.5	1	14.40		0.59		0.04		2		1500
9312	+16 00167	IDS01267N1626		A	1 29 21	+16 41.6	1	6.81		0.94		0.66		2	G5	3016
		MtW 45 # 133			1 29 21	+30 22.7	1	14.34		0.51		−0.01		2		1500
		MtW 45 # 132			1 29 21	+30 29.7	1	11.38		0.72		0.17		2		1500
9256	+61 00286				1 29 21	+61 36.3	2	8.87	.001	0.09	.000			4	B7 V	1103,1118
		MtW 45 # 135			1 29 22	+30 25.8	1	16.91		1.69				1		1500
9404	−45 00500				1 29 22	−44 54.7	5	7.86	.006	0.40	.006	−0.03	.004	16	F3 V	116,278,861,1075,2011
		MtW 45 # 141			1 29 23	+30 16.9	1	15.77		1.58				2		1500
		MtW 45 # 140			1 29 23	+30 27.8	1	17.56		0.80				1		1500
		MtW 45 # 138			1 29 23	+30 29.6	1	12.86		0.62		−0.05		2		1500
	−28 00473				1 29 23	−28 22.7	1	10.76		0.24		0.09		3	A2	1068
9403	−44 00423				1 29 23	−44 06.2	5	8.18	.010	0.52	.001	0.03	.005	24	F8 V	116,278,863,1075,2011
9377	−30 00506	HR 441			1 29 25	−30 32.4	1	5.80	.017	1.07	.005			11	K0 III	15,1075,2035
		MtW 45 # 145			1 29 26	+30 26.8	1	16.43		0.60				2		1500
232459	+50 00299				1 29 26	+50 34.1	1	8.96		0.56		0.07		2	G0	1566
	−73 00086				1 29 26	−73 17.2	1	11.14		0.60		0.10		5		727
		SK 187			1 29 26	−73 36.7	2	13.19	.023	−0.20	.005	−1.01	.005	4		733,835
		MtW 45 # 147			1 29 28	+30 20.1	1	13.50		0.63		0.13		4		1500
		WLS 120 5 # 7			1 29 29	+03 54.9	1	11.11		0.46		−0.01		2		1375
		MtW 45 # 152			1 29 29	+30 23.6	1	17.02		0.53				2		1500
		MtW 45 # 151			1 29 29	+30 29.6	1	11.98		0.66		0.12		2		1500
		MtW 45 # 154			1 29 30	+30 17.3	1	17.08		0.94				2		1500
9414	−46 00420	HR 443		AB	1 29 32	−45 50.0	9	6.16	.006	0.06	.001	0.08	.005	47	A1 V	15,278,977,977,1075*
		NGC 581 - 15			1 29 33	+60 24.3	1	13.46		0.30		−0.14		1		49
9356	+0 00249	RR Cet			1 29 34	+01 05.1	1	9.15		0.18		0.13		1	A0	668
		LS I +59 077			1 29 34	+59 11.0	1	10.72		0.35		−0.39		1	B2 V	367
9477	−65 00135				1 29 35	−65 22.6	1	7.24		0.90				4	G8/K0 IV	2013
9438	−54 00342	IDS01276S5353		AB	1 29 36	−53 37.6	1	7.78		0.45				2	F5/6 V	173
		NGC 581 - 16			1 29 37	+60 24.2	1	13.18		0.30		−0.33		1		719
9569	−75 00093				1 29 37	−75 29.4	1	8.95		1.01		0.84		1	G8/K0 III	3077
	−21 00244				1 29 38	−21 27.8	1	10.76		0.64		0.00		1		79
		SK 188			1 29 38	−73 38.9	2	12.88	.014	−0.16	.009	−1.04	.028	4		733,835
		NGC 581 - 44			1 29 40	+60 24.0	1	14.62		0.34		0.12		1		719
		NGC 581 - 45			1 29 41	+60 24.4	1	14.37		0.67		−0.06		1		719
		NGC 581 - 46			1 29 42	+60 24.2	1	14.51		0.41		0.25		1		719
		NGC 581 - 33			1 29 43	+60 22.1	1	14.04		0.75		0.45		1		719
		NGC 581 - 34			1 29 43	+60 22.3	1	14.37		0.57		0.49		1		719
9400	−17 00277				1 29 43	−17 07.0	1	9.39		0.16		0.11		4	A4 V	152
9468	−60 00119				1 29 43	−59 51.0	1	7.98		0.47				4	F5 V	1075
9387	+0 00251				1 29 44	+01 18.9	1	8.13		0.24		0.07		6	A3	1320
		NGC 581 - 47			1 29 45	+60 23.8	1	13.42		0.28		−0.09		1		719
9411	−24 00651				1 29 45	−23 54.2	2	7.24	.000	0.28	.005			5	A8/9 IV/V	2012,3002
		NGC 581 - 65			1 29 46	+60 25.0	1	15.13		0.35		0.29		1		719
9399	−15 00279				1 29 46	−14 44.9	1	8.17		0.10		0.09		4	A1 V	152
9303	+59 00269				1 29 50	+60 11.6	2	7.65	.003	0.00	.001			41	A0	61,1118
		NGC 581 - 66			1 29 50	+60 26.0	1	13.80		0.28		−0.22		1		719
		NGC 581 - 43			1 29 51	+60 22.0	2	13.48	.010	0.36	.010	−0.05	.015	4		49,719
		NGC 581 - 49			1 29 51	+60 24.0	3	11.75	.017	0.24	.025	−0.35	.012	4		49,305,719
		NGC 581 - 64			1 29 51	+60 25.1	1	13.19		0.17		−0.25		1		719
		NGC 581 - 67			1 29 51	+60 26.8	1	14.52		0.46		−0.28		1		719
		NGC 581 - 35			1 29 52	+60 20.9	2	10.45	.000	0.24	.000	−0.34	.005	2	B2 V	49,719
		NGC 581 - 62			1 29 52	+60 24.7	1	14.33		0.77		0.30		1		719
9331	+52 00371	G 272 - 59			1 29 53	+52 46.6	1	8.40		0.71		0.33		1	G5	1658
		NGC 581 - 40			1 29 53	+60 21.7	1	12.38		0.20		−0.28		1		719
		NGC 581 - 61			1 29 53	+60 24.9	1	13.96		0.58		0.27		1		719
		NGC 581 - 69			1 29 53	+60 26.2	1	14.45		0.29		0.10		1		719
		NGC 581 - 68			1 29 53	+60 26.3	1	13.64		0.34		−0.07		1		719
		NGC 581 - 36			1 29 54	+60 20.6	1	14.08		0.37		−0.06		1		719

HD	DM	Other Id	N	Rem	α_{1950}	δ_{1950}	S	V	σ_V	B–V	σ_{B-V}	U–B	σ_{U-B}	n	Spectrum	References
		NGC 581 - 38			1 29 54	+60 21.0	1	15.17		0.34		0.29		1		719
		NGC 581 - 178		⋆ A	1 29 54	+60 25.3	3	9.98	.081	0.19	.065	-0.58	.057	5	B3 IV	305,450,719
		NGC 581 - 178		⋆ AB	1 29 54	+60 25.3	1	9.62		0.28		-0.58		1		8084
9311	+59 00271	NGC 581 - 176		⋆ A	1 29 54	+60 25.8	6	7.23	.029	0.29	.009	-0.40	.011	12	B5 Iab	49,305,450,719,1119,8084
		NGC 581 - 37			1 29 55	+60 21.3	1	14.64		0.12		0.38		1		719
		NGC 581 - 57			1 29 55	+60 23.7	1	13.85		0.39		0.11		1		719
		NGC 581 - 58			1 29 55	+60 23.9	1	14.45		0.44		0.10		1		719
		NGC 581 - 56			1 29 56	+60 23.6	1	12.30		0.20		-0.24		1		719
		NGC 581 - 59			1 29 56	+60 24.2	1	11.45		0.21		-0.46		1		719
		NGC 581 - 179		⋆ C	1 29 56	+60 25.3	3	11.81	.027	0.19	.014	-0.48	.016	5		305,450,719
		NGC 581 - 39			1 29 57	+60 21.6	1	13.67		0.35		0.03		1		719
		NGC 581 - 42			1 29 57	+60 22.5	3	11.24	.013	0.23	.005	-0.50	.024	4	B2 V	49,305,719
		NGC 581 - 60			1 29 57	+60 24.7	1	13.85		1.26		1.44		2		305
		G 34 - 35			1 29 59	+20 44.0	1	12.28		1.51		1.14		1		333,1620
		NGC 581 - 52			1 29 59	+60 23.0	2	13.21	.040	0.26	.040	-0.12	.035	2		49,719
9451	-27 00522	IDS01276S2704		A	1 29 59	-26 48.5	1	8.02		0.21		0.09		2	A5 V	3077
9451	-27 00522	IDS01276S2704		AB	1 29 59	-26 48.5	1	7.66		0.30				4	A5 V	2012
9451	-27 00523	IDS01276S2704		B	1 29 59	-26 48.5	1	9.17		0.54		-0.01		2		3077
		NGC 581 - 77			1 30 00	+60 24.3	1	13.66		0.66		0.28		1		719
		NGC 581 - 70		⋆ D	1 30 00	+60 25.6	3	11.75	.029	0.19	.015	-0.50	.009	6	B7	305,450,719
		NGC 581 - 83			1 30 00	+60 26.2	1	15.05		0.39		0.30		1		719
		LF 5 # 158			1 30 00	+61 19.	1	10.70		0.47		-0.42		1	B1 V	367
		NGC 581 - 180			1 30 01	+60 21.7	1	14.43		0.56		0.16		1		719
		NGC 581 - 41			1 30 01	+60 22.3	2	12.38	.040	0.11	.010	-0.23	.085	2		49,719
	-31 00622				1 30 01	-30 56.2	2	10.79	.015	0.63	.000	-0.09	.034	3	d:G8	742,3077
		NGC 581 - 53			1 30 02	+60 22.8	1	14.46		0.20		0.12		1		719
		NGC 581 - 73		⋆ E	1 30 02	+60 24.7	4	10.57	.016	0.19	.014	-0.53	.022	7	B2 V	49,305,450,719
		NGC 581 - 166			1 30 03	+60 20.5	2	13.58	.010	0.76	.010	0.25	.005	2		49,719
		NGC 581 - 54			1 30 03	+60 23.1	2	11.07	.019	0.23	.010	-0.50	.034	3		305,719
		NGC 581 - 74			1 30 03	+60 24.5	1	13.57		0.21		-0.26		1		719
		SK 190			1 30 03	-73 36.1	2	13.60	.009	-0.23	.009	-1.00	.014	4		733,835
		NGC 581 - 76			1 30 04	+60 24.3	2	11.44	.049	0.20	.029	-0.31	.088	3	B9	305,719
		NGC 581 - 75			1 30 04	+60 24.5	1	13.85		0.36		-0.08		1		719
		NGC 581 - 175			1 30 04	+60 25.1	2	12.34	.065	0.19	.035	-0.31	.065	2		305,719
		CoD -22 00526			1 30 04	-22 09.3	1	11.21		1.50		1.20		1		3078
		NGC 581 - 80			1 30 06	+60 25.6	1	14.20		0.42		0.18		1		719
		NGC 581 - 126			1 30 07	+60 22.8	1	14.91		0.38		0.32		1		719
		NGC 581 - 125			1 30 07	+60 23.1	1	13.40		0.11		-0.28		1		719
		NGC 581 - 123			1 30 07	+60 23.6	1	13.12		0.16		-0.12		1		719
		NGC 581 - 81			1 30 07	+60 26.4	2	13.27	.005	0.46	.020	0.40	.015	2		49,719
	+59 00273	NGC 581 - 127		⋆ B	1 30 08	+60 22.4	5	9.08	.024	0.21	.007	-0.56	.016	8	B2 III	49,305,367,719,1012
	+59 00274	NGC 581 - 124		⋆ A	1 30 09	+60 23.4	4	8.49	.038	2.08	.020	2.24	.072	6	M0.5Ib	138,305,719,8032
		NGC 581 - 79			1 30 09	+60 25.8	1	13.29		0.37		-0.12		1		719
9352	+57 00320	HR 439			1 30 10	+58 04.3	1	5.70		1.52		1.15		3	K0 Ib+B9 V	1355
		NGC 581 - 122			1 30 10	+60 27.3	2	11.00	.005	0.19	.019	-0.51	.039	3	B5	305,719
		LF 5 # 157			1 30 10	+61 04.	1	10.79		0.27		-0.49		1	B2 IV	367
		LS I +59 079			1 30 11	+59 14.7	1	10.68		0.35		-0.41		1	B1 Vnne	367
9463	-19 00266				1 30 11	-19 14.0	1	7.58		1.51		1.77		3	K4 III	1657
		NGC 581 - 121			1 30 12	+60 24.1	1	13.70		0.24		-0.18		1		719
		NGC 581 - 78			1 30 12	+60 24.7	2	13.45	.097	0.29	.015	-0.09	.019	3		305,719
		NGC 581 - 86			1 30 12	+60 27.5	1	10.81		1.93		1.95		1		49
		JL 251			1 30 12	-49 49.	1	14.24		-0.24		-1.21		1		832
9430	+22 00245	G 34 - 36			1 30 13	+23 26.5	2	9.04	.005	0.62	.005	0.06	.015	3	G5	1620,1658
		NGC 581 - 128			1 30 13	+60 22.8	1	12.24		0.20		-0.30		1		719
		NGC 581 - 149			1 30 14	+60 21.2	1	12.50		0.24		-0.16		1		719
		NGC 581 - 150			1 30 15	+60 21.1	1	14.92		0.42		0.19		1		719
		NGC 581 - 119			1 30 15	+60 24.8	1	15.07		0.36		0.10		1		719
9366	+54 00315				1 30 16	+54 41.3	1	6.90		1.92		2.14		2	K3 Ib	1566
		NGC 581 - 151			1 30 16	+60 20.7	1	14.67		0.24		-0.12		1		719
		NGC 581 - 129			1 30 16	+60 23.6	1	13.96		0.30		0.13		1		719
		NGC 581 - 148			1 30 17	+60 22.5	1	14.72		0.29		0.00		1		719
		NGC 581 - 117			1 30 17	+60 24.3	1	14.68		0.30		0.26		1		719
		NGC 581 - 147			1 30 18	+60 21.4	1	13.35		0.38		0.07		1		719
		NGC 581 - 116			1 30 18	+60 24.6	1	14.52		0.50		0.43		1		719
		MCC 92			1 30 18	+77 49.0	1	10.64		1.21				1	M0	1017
		NGC 581 - 130			1 30 19	+60 23.3	1	13.21		0.26		-0.28		1		719
		NGC 581 - 114			1 30 20	+60 24.4	1	13.22		0.64		0.41		1		719
		NGC 581 - 87			1 30 20	+60 26.9	1	11.35		0.26		-0.39		1	B3 V	49
		NGC 581 - 131			1 30 21	+60 23.8	1	13.68		0.73		0.25		1		719
9487	-23 00556				1 30 21	-22 40.2	3	8.66	.015	0.26	.000	0.10	.009	13	A3II/III(m)	355,1068,2012
9545	-59 00105				1 30 21	-58 38.4	1	9.18		0.08		0.08		1	A0 V	861
		NGC 581 - 145			1 30 22	+60 22.1	1	13.28		0.27		-0.18		1		719
		VES 596			1 30 22	+61 35.8	1	12.44		0.64		-0.25		2		690
9485	-17 00280				1 30 23	-16 38.7	1	9.11		0.04		0.07		4	A0 V	152
		NGC 581 - 158			1 30 25	+60 18.3	1	12.76		0.24		-0.17		1		49
9365	+59 00276	NGC 581 - 144		⋆ C	1 30 25	+60 22.0	2	8.20	.030	0.31	.005	0.03	.020	2	F0 V	49,305
		NGC 581 - 134			1 30 25	+60 22.6	1	14.41		0.34		0.13		1		719
		NGC 581 - 177			1 30 26	+60 21.2	1	13.21		0.26				1		49
		NGC 581 - 133			1 30 27	+60 23.2	1	13.41		1.11		0.97		1		719
		NGC 581 - 132			1 30 27	+60 23.3	1	14.80		0.35		0.25		1		719

Table 1   103

HD	DM	Other Id	N Rem	α_{1950}	δ_{1950}	S	V	σ_V	B–V	σ_{B-V}	U–B	σ_{U-B}	n	Spectrum	References
9497	−21 00249			1 30 28	−21 20.6	2	8.63	.026	1.52	.000	1.63	.043	6	M3 III	1375,3007
9528	−50 00410		V	1 30 29	−49 47.0	1	7.70		0.64		0.11		3	G1/2 IV/V	1730
9446	+28 00253			1 30 31	+29 00.6	1	8.35		0.68		0.14		2	G5 V	3026
9665	−75 00096			1 30 31	−74 52.1	1	8.63		0.52		0.05		3	F8 V	727
9383	+61 00291			1 30 33	+61 53.6	2	7.90	.001	0.14	.001			4	A2 V	1103,1118
		VES 597		1 30 34	+61 38.0	1	11.16		0.59		−0.45		2		690
9484	−9 00298	HR 444		1 30 34	−09 16.3	3	6.58	.003	−0.04	.000	−0.09	.006	24	A0 III	15,361,1415
	−43 00470			1 30 34	−43 15.5	1	10.71		1.15				3		1730
9544	−50 00411	HR 447		1 30 34	−49 59.0	2	6.27	.005	0.46	.000			7	F5 V	15,2012
9542	−45 00512			1 30 36	−44 59.7	2	8.32	.005	1.45	.005	1.76		12	K4 III	977,2011
9443	+40 00315			1 30 38	+40 38.7	1	7.43		0.33		−0.01		2	A5	1601
		CS 29504 # 18		1 30 38	−33 10.8	1	13.70		0.37		−0.22		1		1736
9408	+58 00260	HR 442		1 30 39	+58 58.6	5	4.70	.016	0.99	.010	0.75	.010	11	G9 IIIb	15,1118,3016,4001,8015
		NGC 581 - 111		1 30 41	+60 24.7	1	11.84		0.25		−0.39		1		49
9525	−37 00589	HR 445		1 30 42	−37 07.3	2	5.50	.005	1.02	.000			7	KO III	15,2029
		CS 22180 # 4		1 30 43	−10 44.7	1	14.07		0.25		0.09		1		1736
9512	−13 00283	G 272 - 40		1 30 47	−12 58.0	1	9.37		0.74		0.30		1	G8 V	79
9483	+29 00260			1 30 48	+30 08.0	2	8.10	.005	0.20	.005	0.11	.000	9	A3	627,1733
	−25 00636	G 274 - 62		1 30 48	−25 10.1	1	10.06		1.17		1.08		2	K6	3061
		LP 827 - 281		1 30 48	−25 20.	1	11.74		0.78		0.24		1		3061
		CS 22180 # 11		1 30 50	−11 22.1	1	14.43		0.24		0.08		1		1736
9540	−24 00658	G 274 - 63	★ A	1 30 53	−24 25.9	4	6.96	.003	0.76	.005	0.33	.005	9	K0 V	79,258,1518,2033
9540	−24 00658	IDS01286S2442	B	1 30 53	−24 25.9	1	12.77		1.30		1.15		3		1518
		LS I +61 202		1 30 54	+61 26.9	1	10.84		0.71		−0.11		3	B1 V	367
9407	+68 00113	G 243 - 70		1 30 55	+68 41.5	2	6.53	.000	0.68	.005	0.24	.005	3	G6 V	1003,1355
	+36 00276			1 30 58	+37 04.7	1	8.75		1.24		1.38		2	K2	1601
		NGC 581 - 152		1 30 83	+60 19.6	1	13.76		0.30		−0.08		1		49
232464	+51 00330	G 172 - 60		1 31 01	+52 24.7	1	9.55		0.73		0.29		1	G7	1658
9500	+34 00270			1 31 02	+35 21.1	1	7.00		1.61		1.87		3	M4 III	1501
	+75 00065	G 245 - 28	★ A	1 31 02	+75 44.5	1	10.06		0.86		0.54		1	G5	1620
		WLS 140-20 # 6		1 31 03	−19 29.7	1	11.51		1.36		1.19		2		1375
		G 72 - 15	A	1 31 05	+26 59.0	1	11.03		1.20		1.10		2	K5	7010
	−50 00413			1 31 05	−50 01.8	1	11.39		0.48		0.01		2		1730
	−74 00107			1 31 09	−74 33.5	1	9.76		1.03		0.88		3	K0	727
9535	+18 00207			1 31 11	+18 31.5	1	8.71		1.15				1	K0	1245
9562	−7 00256	HR 448		1 31 12	−07 16.8	5	5.76	.007	0.64	.003	0.21	.004	20	G2 IV	15,1061,2013,2030,3077
	−75 00099			1 31 12	−74 54.2	1	10.43		0.51		0.01		4	F5	727
	+47 00451	G 172 - 61	★ AB	1 31 13	+48 29.2	1	11.00		0.70		0.10		1	K0	1658
9575	−12 00286	IDS01288S1244	AB	1 31 14	−12 28.1	1	8.51		0.64		0.10		2	G3 V	3030
		CS 22180 # 36		1 31 17	−09 22.8	1	14.95		0.42		−0.25		1		1736
9619	−44 00434			1 31 17	−44 09.5	4	7.82	.005	0.83	.001	0.48	.003	20	K0/1 V	116,977,1075,2011
9531	+36 00277	HR 446, KK And		1 31 23	+36 58.9	2			−0.07	.005	−0.30	.000	4	B9 IV	1049,1079
		LS I +59 081		1 31 27	+59 52.5	1	10.80		0.20		−0.57		1	B2 V	367
9550	+38 00291			1 31 29	+38 35.4	1	8.26		0.32		0.10		3	A3	833
	−11 00298	G 271 - 95		1 31 30	−10 40.1	1	9.96		0.96		0.79		1		1696
236791	+58 00263			1 31 34	+59 09.7	1	8.81		1.90				3	M3 III	369
	+33 00256	IDS01288N3409	C	1 31 35	+34 25.4	2	9.56	.000	0.51	.007	0.00	.005	8	F8	196,3016
		CS 22180 # 14		1 31 35	−12 57.3	1	13.58		0.43		−0.26		1		1736
	+33 00257	IDS01288N3409	AB	1 31 38	+34 24.3	1	9.66		0.75		0.32		5	G0	3016
	−12 00290			1 31 38	−11 37.6	1	10.98		0.28		0.06		3		1026
		G 2 - 44		1 31 39	+04 54.0	1	13.52		1.54				2		202
9642	−23 00567			1 31 44	−23 02.0	1	9.07		1.54		1.59		4	M3 (III)	3007
9662	−44 00437			1 31 44	−44 36.3	4	8.61	.015	1.10	.006	0.99	.010	22	K1 III	116,863,977,2011
		LS I +61 203		1 31 47	+61 22.8	1	10.75		0.62		−0.37		1	B0 V	367
9652	−33 00565	G 274 - 65		1 31 48	−32 49.2	1	9.38		0.64		0.05		1	G3 V	79
9694	−45 00516			1 31 52	−44 39.2	3	8.54	.010	1.08	.006	0.94	.003	18	K0 III	863,977,2011
9661	−30 00520			1 31 54	−30 28.7	1	9.84		0.62		−0.01		2	G3 V	3061
		LP 468 - 115		1 31 56	+14 31.7	1	12.45		0.76		0.29		2		1696
9673	−28 00489			1 31 58	−27 37.1	2	7.89	.024	0.23	.000	0.06		9	A5 V	1068,2011
		Tr 1 - 1078		1 31 59	+61 01.8	1	13.09		0.59		0.48		1		541
9546	+62 00274	IDS01286N6234	A	1 31 59	+62 49.6	2	6.74	.032	0.98	.061	0.78		6	K1 V	1619,7008
		CS 22180 # 5		1 31 59	−10 31.7	1	14.02		0.50		−0.16		1		1736
	+28 00258	RS Tri	★ A	1 32 00	+29 20.0	1	10.27		0.29		0.16		1	A5	627
	+28 00258	RS Tri	★ AB	1 32 00	+29 20.0	1	10.11		0.31		0.14		1		627
	+28 00258	IDS01292N2905	B	1 32 00	+29 20.0	1	11.98		0.66		0.20		3		627
	+32 00270			1 32 00	+32 40.5	1	10.29		−0.16		−0.80		1		963
9734	−56 00318	IDS01301S5602	AB	1 32 00	−55 46.6	1	9.91		1.08		0.96		1	K3/5 V	3072
9659	−18 00257			1 32 01	−17 59.4	1	10.22		0.19		0.14		4	A5/7 IV	152
		Tr 1 - 1085		1 32 02	+61 03.4	1	15.03		0.40		0.05		1		541
		Tr 1 - 1088		1 32 03	+61 01.9	1	13.18		0.78		0.29		1		541
		Tr 1 - 1090		1 32 04	+61 00.9	1	13.47		0.75		0.27		1		541
9692	−28 00490			1 32 04	−28 29.5	1	6.87		1.66		2.01		4	M0 III	3007
9640	+17 00224	HR 450		1 32 06	+18 12.4	2	5.92	.024	1.50	.022	1.80		4	M2 IIIab	71,3016
		Tr 1 - 1097		1 32 06	+61 04.1	1	13.44		0.47		−0.20		1		541
9616	+32 00272			1 32 07	+32 51.9	1	6.74		0.63		0.18		3	G2 IV	1501
		Tr 1 - 1112		1 32 08	+61 01.4	1	13.72		0.41		−0.06		1		541
		Tr 1 - 1115		1 32 08	+61 01.4	1	13.98		0.44		−0.13		1		541
9648	+5 00212			1 32 09	+06 19.9	1	7.98		0.47				1	F8 IV-V	3037
		Tr 1 - 1118		1 32 09	+61 00.9	1	14.79		0.52		0.17		1		541
		Tr 1 - 1117		1 32 09	+61 03.5	1	14.52		0.45		0.06		1		541

HD	DM	Other Id	N Rem	α_{1950}	δ_{1950}	S	V	σ_V	B–V	σ_{B-V}	U–B	σ_{U-B}	n	Spectrum	References
	−67 00099			1 32 09	−67 25.9	1	10.17		0.86		0.57		1	K0	1696
		LF 5 # 156		1 32 10	+61 08.	1	10.76		0.35		-0.36		1	B2 Vnne	367
		VES 598		1 32 11	+63 43.5	1	12.91		0.89		-0.11		2		690
9672	−16 00265	HR 451		1 32 11	−15 55.9	2	5.62	.010	0.08	.005	0.05		6	A1 V	252,2035
		Tr 1 - 1125		1 32 12	+60 59.7	1	13.61		0.40		0.13		1		541
		Tr 1 - 1126		1 32 12	+61 02.7	1	14.88		0.36		0.13		1		541
		Tr 1 - 1130		1 32 13	+61 01.9	1	14.14		0.48		-0.04		1		541
		Tr 1 - 1127		1 32 13	+61 02.1	1	14.46		0.53		0.26		1		541
9670	+0 00256	G 2 - 46		1 32 14	+00 41.6	3	6.92	.011	0.52	.009	-0.02	.000	5	F8 V	333,1003,1620,3037
9655	+13 00238			1 32 14	+14 07.8	1	8.06		1.03		0.76		2	G5	1648
9638	+28 00259			1 32 14	+28 50.7	2	7.82	.049	1.15	.005	1.15		6	K2 II	20,8100
9733	−46 00433	IDS01301S4612	AB	1 32 14	−45 56.9	4	6.91	.008	0.90	.002	0.48	.002	20	G8 IV	278,863,1075,2012
9604	+52 00382			1 32 15	+53 05.4	1	6.88		0.00		-0.28		3	B8	985
		Tr 1 - 1137		1 32 15	+61 01.5	1	12.71		0.39		-0.22		1		541
		Tr 1 - 1143		1 32 15	+61 01.5	1	10.67		0.30		-0.42		1		541
		Tr 1 - 1138		1 32 15	+61 02.0	1	14.18		0.38		0.15		1		541
		Tr 1 - 1140		1 32 15	+61 02.2	1	14.19		0.24		0.10		1		541
		Tr 1 - 1146		1 32 16	+61 00.7	1	12.09		0.30		-0.27		1		541
		Tr 1 - 1145		1 32 16	+61 01.1	1	14.50		0.25		-0.12		1		541
	+60 00274			1 32 16	+61 01.1	1	10.66		0.33		-0.41		2	B3 III	1012
		Tr 1 - 1144		1 32 16	+61 03.5	1	13.87		0.43		-0.12		1		541
9570	+63 00206	IDS01288N6338	AB	1 32 16	+63 53.7	1	8.36		0.31				2	A2	1118
		Tr 1 - 1153		1 32 17	+61 01.6	1	12.02		0.41		-0.48		1		541
		Tr 1 - 1150		1 32 17	+61 01.7	1	11.30		0.33		-0.36		1		541
		Tr 1 - 1152		1 32 17	+61 02.4	1	13.55		0.42		0.03		1		541
		Tr 1 - 1156		1 32 18	+61 00.3	1	14.24		0.43		0.07		1		541
		Tr 1 - 1157		1 32 18	+61 00.7	1	15.11		0.21		-0.08		1		541
		Tr 1 - 1158		1 32 18	+61 01.9	1	9.97		0.32		-0.20		1		541
		Tr 1 - 1163		1 32 19	+61 01.1	1	15.33		0.62		0.32		1		541
		Tr 1 - 1164		1 32 19	+61 02.1	1	11.48		0.30		-0.36		1		541
	−75 00101			1 32 19	−74 49.6	1	11.23		0.91		0.60		4		727
		Tr 1 - 1170		1 32 20	+61 01.0	1	11.39		0.26		-0.42		1		541
		Tr 1 - 1173		1 32 21	+61 01.9	1	12.54		0.36		-0.24		1		541
		Tr 1 - 1178		1 32 23	+61 03.3	1	15.04		0.34		0.19		1		541
		Tr 1 - 1184		1 32 24	+60 59.6	1	13.58		0.74		0.31		1		541
		Tr 1 - 1181		1 32 24	+61 00.0	1	14.86		0.44		0.23		1		541
		Tr 1 - 1183		1 32 24	+61 00.1	1	15.26		0.34		0.16		1		541
		Tr 1 - 1191		1 32 24	+61 01.2	1	12.76		0.40		-0.01		1		541
		Tr 1 - 1193		1 32 24	+61 01.2	1	12.37		0.31		-0.38		1		541
		Tr 1 - 1187		1 32 24	+61 02.4	1	13.92		0.51		-0.10		1		541
		Tr 1 - 1182		1 32 24	+61 02.6	1	14.28		0.45		0.11		1		541
		Tr 1 - 1188		1 32 25	+61 02.9	1	13.86		0.91		0.38		1		541
		Tr 1 - 1208		1 32 29	+61 03.1	1	14.76		0.36		0.13		1		541
		Tr 1 - 1223		1 32 31	+61 00.2	1	14.35		0.65		0.17		1		541
		CS 22180 # 7		1 32 32	−11 03.0	1	13.34		0.22		0.13		1		1736
	+62 00275	LS I +63 152		1 32 34	+63 23.1	1	9.82		0.84		-0.12		2	B3:III	1012
9742	−32 00613	HR 453		1 32 34	−32 08.8	1	6.12		1.11				4	K0/1 III	2032
	+4 00275	G 2 - 47		1 32 37	+05 23.0	2	10.76	.015	0.75	.010	0.21		4	G5	202,333,1620
		Tr 1 - 1246		1 32 37	+61 01.7	1	14.07		0.42		-0.09		1		541
		Tr 1 - 1248		1 32 38	+61 00.4	1	13.99		0.86		0.20		1		541
		G 133 - 20		1 32 40	+41 47.2	3	10.97	.008	1.29	.012	1.21	.030	4	M0	1017,1510,7010
9770	−30 00529	IDS01304S3026	AB	1 32 42	−30 10.0	2	7.09	.000	0.93	.007	0.57		6	K1 V	2034,3061
9809	−50 00424			1 32 44	−50 03.6	2	8.21	.004	1.56	.000	1.88	.040	9	M0/1 III	1673,1730
9634	+59 00285			1 32 45	+59 57.5	3	7.90	.004	1.72	.004			38	K5 III	61,71,1118
9783	−30 00530	IDS01304S3026	D	1 32 46	−30 07.9	4	8.77		0.56				4	F8 V	2034
9714	+27 00248			1 32 50	+28 00.9	1	6.81		1.03				3	K1 III	20
	−34 00614	G 274 - 68		1 32 50	−33 54.7	1	11.90		0.50		-0.14		2		3062
	+63 00207			1 32 51	+64 20.5	1	10.25		1.41		1.47		4		155
		CS 22180 # 6		1 32 51	−10 48.6	1	14.20		0.11		0.19		1		1736
9582	+71 00089	Cr 463 - 75		1 32 52	+71 42.3	1	8.86		0.53		0.00		2	F8	428
9712	+40 00328	HR 452		1 32 55	+40 49.3	1	6.38		1.11		1.00		6	K1 III	1355
236798	+56 00300			1 32 56	+57 13.9	1	8.47		0.45				43	F5	6011
		WLS 112 60 # 8		1 32 57	+59 12.3	1	10.56		0.61		0.06		2		1375
9666	+58 00266			1 32 57	+59 12.3	2	7.71	.000	0.45	.004	0.03		5	F5 III	70,1375
	−40 00390			1 32 57	−40 05.4	1	11.18		0.82		0.43		1		1696
9782	−14 00299	G 272 - 48		1 32 58	−13 38.2	1	7.16		0.60		0.09		1	G0 V	79
		LF 5 # 144		1 33 00	+59 53.	1	10.72		0.28		-0.50		1	B3 V	367
9709	+46 00393			1 33 01	+46 51.6	1	7.07		-0.05		-0.43		2	B9	1601
	+60 00279	LS I +60 165		1 33 03	+60 43.5	3	9.08	.008	0.19	.004	-0.61	.013	6	B2 II	171,367,1012
	+63 00209			1 33 05	+64 12.0	1	8.96		0.57		0.14		6	B5	155
9766	+13 00240	HR 455		1 33 06	+14 24.4	3	6.22		-0.04	.003	-0.16	.025	11	B9.5III	3,1049,1079
		CS 22180 # 8		1 33 10	−11 14.0	1	14.29		0.41		-0.21		1		1736
10062	−80 00028			1 33 10	−80 11.0	1	7.26		0.12		0.11		4	A1 IV/V	1628
236799	+56 00301			1 33 11	+57 24.0	2	8.46	.002	0.55	.005	0.14		51	F5	1655,6011
9780	+16 00176	HR 457		1 33 12	+17 10.7	1	5.93		0.24		0.08		2	F0 IV	1733
9612	+73 00081	HR 449		1 33 13	+74 02.8	1			0.03		-0.17			B9 V	1079
	−26 00539	G 274 - 69		1 33 13	−26 24.2	2	11.44	.005	0.50	.002	-0.15	.020	5		1696,3062
	+61 00299	LS I +62 172		1 33 18	+62 01.8	1	9.77		0.52		-0.37		2	B8	1012
10042	−79 00040	HR 467	★ AV	1 33 18	−78 45.5	3	6.10	.004	0.97	.000	0.71	.005	11	K0 III	15,1075,2038
9831	−18 00265			1 33 19	−17 33.5	1	8.25		1.08		0.89		2	K0 III	1375

Table 1   105

HD	DM	Other Id	N Rem	α_{1950}	δ_{1950}	S	V	σ_V	B–V	σ_{B-V}	U–B	σ_{U-B}	n	Spectrum	References
		LS I +61 205		1 33 20	+61 25.7	1	10.40		0.99		-0.01		1	B2 V	367
9695	+62 00277			1 33 20	+63 09.6	1	7.62		0.09				3	B8 III	1118
	-34 00619			1 33 21	-33 54.1	1	11.92		0.48		-0.15		2		1696
9746	+47 00460	HR 454, OP And		1 33 23	+48 28.1	1	5.92		1.21				2	gK1	71
9896	-58 00123	HR 460		1 33 23	-58 23.7	4	6.01	.005	0.38	.005			18	F3 V	15,1075,2020,2027
9847	-18 00266	G 272 - 51	★ A	1 33 27	-17 47.1	4	7.17	.026	0.70	.018	0.30	.013	11	G6 IV	79,1311,2034,3026
9875	-40 00394			1 33 27	-40 05.6	1	8.00		1.58		1.93		4	M2 III	3007
	+60 00282	LS I +61 207		1 33 28	+61 10.6	1	10.63		0.42		-0.48		1	B1 III	367
236800	+59 00286	LS I +59 082		1 33 31	+59 41.3	1	9.56		0.30		-0.55		2	B1 III:n	1012
	+62 00278	LS I +63 153		1 33 33	+63 28.4	1	9.85		0.65		-0.29		2	B0.5III	1012
9856	-16 00270	HR 459		1 33 33	-15 39.3	2	5.41	.015	1.23	.015	1.20		7	K2 IIa	2009,3016
9828	+23 00211			1 33 39	+23 57.7	1	8.75		1.30		1.26		3	K0	1502
9895	-40 00395			1 33 40	-40 12.1	1	6.43		0.41		-0.07		4	F3/5 V	1628
10029	-75 00104			1 33 41	-74 55.8	1	8.99		1.11		0.94		3	K0 III	727
		NGC 609 - 1		1 33 42	+64 18.	1	15.00		2.22				2		155
		NGC 609 - 2		1 33 42	+64 18.	1	14.93		2.24				2		155
		NGC 609 - 3		1 33 42	+64 18.	1	15.24		2.00				2		155
		NGC 609 - 5		1 33 42	+64 18.	1	15.00		1.01				2		155
		NGC 609 - 6		1 33 42	+64 18.	1	15.56		0.99				2		155
		NGC 609 - 8		1 33 42	+64 18.	1	14.85		1.77				2		155
		NGC 609 - 9		1 33 42	+64 18.	1	14.45		1.00				2		155
		3C 47 # 1		1 33 43	+20 45.1	1	12.57		0.82		0.35		3		327
		G 271 - 106		1 33 45	-11 35.8	3	14.16	.021	0.17	.016	-0.52	.024	6	DA	538,1736,3060
		LP 939 - 99		1 33 45	-36 41.4	1	12.66		0.41		-0.20		3		1696
9894	-37 00603			1 33 45	-36 42.5	1	7.65		1.62		1.86		4	M2 III	3007
		G 2 - 49		1 33 47	+05 23.1	1	13.82		1.35		1.14		1		333,1620
9906	-30 00540	HR 462	★ AB	1 33 50	-30 09.8	2	5.68	.005	0.34	.005			7	F2 V	15,2011
9862	+3 00219	IDS01312N0317	AB	1 33 51	+03 31.9	1	8.64		0.30		0.16		2	Am	1026
9826	+40 00332	HR 458	★ A	1 33 51	+41 09.4	14	4.09	.013	0.54	.007	0.06	.000	49	F8 V	1,15,1004,1077,1197,1258*
9918	-37 00605	IDS01317S3748	B	1 33 55	-37 31.6	1	9.77		0.28		0.16		4	A2(m)A7-F0	152
		CS 29504 # 28		1 33 57	-34 21.4	1	14.11		0.13		0.16		1		1736
	+57 00342	RW Cas		1 33 58	+57 30.3	2	8.59	.010	0.87	.040			2	G2	934,6011
9932	-44 00455			1 34 01	-44 20.7	2	9.69	.016	0.25	.001	0.08		11	A7 V	863,2012
		CS 22180 # 34		1 34 02	-09 38.6	1	14.31		0.42		-0.24		1		1736
		CS 29504 # 29		1 34 04	-34 07.7	1	14.84		0.25		-0.02		1		1736
9841	+47 00465	IDS01310N4812	A	1 34 05	+48 27.4	1	8.78		0.17		0.06		1	A0	1722
	+71 00090	Cr 463 - 73		1 34 07	+71 56.6	1	10.00		1.72		1.82		2		428
		G 35 - 1		1 34 10	+24 29.9	2	12.60	.010	0.64	.005	-0.04	.020	2		333,1620,1658
		G 271 - 108		1 34 11	-02 28.1	1	11.96		1.01		0.87		1		1658
	+59 00289			1 34 12	+60 21.2	1	9.87		0.45		0.26		3	Am	171
9821	+59 00290			1 34 13	+59 37.5	1	8.20		0.11				2	B9	70
		G 2 - 50		1 34 14	+04 12.4	3	11.35	.015	0.51	.009	-0.16	.000	6		202,1620,1696
9811	+64 00202	LS I +64 072		1 34 16	+64 29.1	1	6.48		0.80				1	A6 Iab	6009
		Steph 173		1 34 17	+13 14.2	1	13.20		1.44		1.23		1	K5	1746
		CS 22180 # 17		1 34 22	-12 16.1	1	13.25		0.12		0.18		1		1736
9930	-15 00289			1 34 22	-14 38.4	1	8.65		0.08		0.12		4	A1 V	152
9931	-22 00266	IDS00444S2157	AB	1 34 22	-21 42.2	1	8.91		0.46		-0.01		1	G0 V	3061
9931	-22 00266	IDS00444S2157	C	1 34 22	-21 42.2	1	10.18		0.63		0.14		1		3061
	-61 00122			1 34 22	-61 19.7	1	10.10		0.53		-0.16		3	G0	1097
		G 3 - 22		1 34 26	+38 30.7	1	12.49		0.91		0.62		1		333,1620
	+62 00281	V539 Cas		1 34 26	+62 32.6	1	8.54		2.43				4	M2	369
9919	+11 00205	HR 463		1 34 27	+11 53.2	4	5.56	.029	0.35	.009	0.02	.029	18	F0 V	3,254,3037,8071
9852	+61 00303			1 34 27	+61 36.4	1	7.93		1.46		1.43		3	K1 III:p	37
	+62 00282			1 34 28	+62 39.4	1	9.62		0.58		0.32		5	F0 II	171
9774	+72 00086	HR 456	★ A	1 34 28	+72 47.2	2	5.26	.014	0.96	.000	0.72		4	G8 IIIa	1118,3016
9774	+72 00086	IDS01305N7232	B	1 34 28	+72 47.2	1	11.98		0.82		0.40		1		3016
	+29 00270			1 34 29	+30 25.4	1	9.27		1.01				1	G5	209
		LP 648 - 21		1 34 29	-07 26.8	1	13.10		0.63		-0.09		3		1696
		AJ74,1000 M33# 11		1 34 33	+30 05.2	1	11.62		1.10				1		209
10003	-45 00531			1 34 35	-45 03.6	3	8.56	.007	1.18	.006	1.20	.000	18	K1/2 III	863,977,2011
		3C 48 # 4(C')		1 34 37	+32 53.4	2	12.89	.005	0.55	.005	0.15	.010	3		157,327
		3C 48 # 10(F')		1 34 38	+32 54.9	2	14.40	.024	0.72	.000	0.21	.019	3		157,327
9939	+24 00239	G 34 - 39		1 34 39	+24 55.0	1	6.97		0.90		0.67		1	K0 IV	333,1620
10102	-73 00092			1 34 39	-72 59.7	1	8.90		1.15		1.08		4	K0 III	727
	-44 00457			1 34 41	-43 52.8	1	11.68		0.50		-0.15		3		1696
9958	-1 00219			1 34 43	-00 36.2	1	7.09		0.67		0.25		5	G0	1628
		3C 48 # 7(E)		1 34 47	+32 52.3	2	13.52	.003	0.50	.000	0.00	.000	23		157,327
9878	+61 00304			1 34 47	+62 05.9	1	6.70	.013	-0.02	.009	-0.25		6	B7 V	1118,1501
10001	-25 00660			1 34 49	-25 16.2	1	8.23		0.46				4	F3 V	2012
10002	-30 00549	G 274 - 74		1 34 49	-29 38.8	2	8.13	.000	0.85	.010	0.41		6	K1 V	2012,3008
9900	+57 00349	HR 461		1 34 50	+57 43.4	6	5.55	.009	1.38	.008	1.40	.024	24	G5 II	15,1007,1013,3016*
	+59 00293	LS I +60 166		1 34 50	+60 06.4	1	10.45		0.21		-0.54		1	B2 IV	367
		G 272 - 55		1 34 50	-17 42.5	1	15.20		0.22		-0.62		3		3062
		G 245 - 31		1 34 51	+69 22.9	1	14.49		0.76		0.05		2		1658
10123	-73 00093			1 34 51	-73 05.2	1	8.88		1.14		1.09		3	K1 III	727
	-26 00556			1 34 52	-26 23.6	1	10.38		1.34		1.27		1		79
		3C 48 # 9(F)		1 34 54	+32 53.6	2	14.54	.000	0.66	.000	0.07	.025	23		157,327
9927	+47 00467	HR 464		1 34 55	+48 22.5	9	3.57	.008	1.28	.006	1.45	.007	42	K3 III	1,15,1355,1363,3052*
	+60 00289	LS I +61 210		1 34 55	+61 06.2	2	10.18	.010	0.51	.005	-0.37	.010	3	B2 II-III	367,1012
10052	-58 00126	HR 468		1 34 55	-58 31.5	1	6.18		1.61				4	M3 III	2035

HD	DM	Other Id	N	Rem	α_{1950}	δ_{1950}	S	V	σ_V	B–V	σ_{B-V}	U–B	σ_{U-B}	n	Spectrum	References
		Cr 463 - 72			1 34 59	+71 48.9	1	11.30		0.53		-0.04		2		428
9935	+45 00396	IDS01320N4529	AB		1 35 00	+45 44.7	1	8.61		0.09		0.04		1	A1 V	695
236810	+59 00296	LS I +60 167			1 35 04	+60 18.5	2	8.69	.020	0.26	.005	-0.55	.010	3	B2 III	367,1012
		G 72 - 18			1 35 05	+32 06.0	1	13.25		0.76		0.10		1		1658
	+47 00469				1 35 05	+48 06.7	1	10.00		0.60		0.06		1	G0	1722
		G 219 - 15	AB		1 35 06	+56 58.9	1	11.24		1.41		1.14		3		7010
9911	+62 00284	IDS01318N6312	AB		1 35 07	+63 28.4	1	8.71		0.87				2	K0 V	1118
10061	-55 00331				1 35 07	-54 48.7	1	9.45		1.41		1.67		1	K2/3 V	3072
9984	+25 00269	IDS01324N2524	AB		1 35 08	+25 39.0	1	8.47		0.98				2	G8 III	20
	+40 00334				1 35 08	+40 55.7	1	8.45		1.20		1.40		3	K5 V	7008
10009	-10 00343	HR 466	★ AB		1 35 08	-09 39.6	5	6.23	.004	0.53	.000	-0.01	.010	22	F7 V	15,1071,1075,1425,2012
10038	-40 00404				1 35 08	-40 25.9	2	8.13	.015	0.27	.010	0.13		9	A2mA5-F0	355,2012
		AAS6,117 # 44			1 35 12	+62 15.	1	9.71		2.16		2.20		4	K5	171
		G 3 - 9			1 35 17	+10 04.7	2	12.19	.005	0.81	.005	0.25	.010	2		333,1620,1658
		LS I +59 084			1 35 17	+59 33.5	1	10.84		0.37		-0.47		1	B0 V	367
10069	-46 00449				1 35 19	-45 55.9	1	9.27		0.49				4	F6 IV/V	2011
9997	+38 00305				1 35 24	+39 06.1	1	8.60		0.02		0.03		2	A0 V	105
		G 271 - 115			1 35 27	-05 14.8	4	12.84	.003	0.35	.006	-0.51	.007	10		1281,1698,1705,3060
232485	+51 00348				1 35 28	+52 10.9	1	8.49		1.33		1.42		2	K2	1566
10077	-47 00489				1 35 28	-47 22.0	1	8.27		0.32				4	F2 III/IV	2012
9996	+44 00341	HR 465, GY And			1 35 30	+45 08.8	4	6.39	.035	-0.02	.072	-0.11	.013	15	B9p CrEu	196,220,1063,1202
	+17 00235	G 3 - 10			1 35 31	+17 34.7	1	9.42		0.81		0.45		1	G5	333,1620
		AO 821			1 35 32	+48 13.1	1	10.02		0.14		0.08		1		1722
232486	+51 00349				1 35 32	+52 15.9	1	9.64		0.40		0.17		2	A5	1566
10015	+28 00271	G 72 - 20			1 35 33	+29 19.0	5	8.65	.010	0.85	.008	0.56	.009	25	K0 IV-V	333,1003,1620,1775,3077,4001
		MCC 379			1 35 33	+44 20.0	1	10.74		1.14				1	K4	1017
	+61 00305	LS I +61 211			1 35 35	+61 32.3	2	10.37	.000	0.54	.005	-0.29	.015	3	B2 V	367,1012
9995	+48 00493				1 35 37	+48 39.1	1	9.04		0.05		-0.32		1	A2	1722
		LS I +59 085			1 35 37	+59 36.4	1	10.94		0.34		-0.50		1	B2 V	367
9974		LS I +57 024			1 35 38	+57 54.1	2	10.69	.000	0.02	.000	-0.86	.000	4	WN4	1011,1012
10059	-7 00269				1 35 40	-06 54.5	2	8.83	.005	0.15	.010	0.12	.005	10	A0	152,355
9964	+63 00214				1 35 42	+63 50.5	1	9.45		0.36				2	A3	1118
10101	-47 00490				1 35 42	-47 25.9	1	7.58	.005	1.02	.001	0.72	.013	16	G8 III	977,1075,2011
9973	+60 00296	LS I +60 168			1 35 44	+60 49.5	3	6.88	.012	0.86	.012	0.74		6	F5 Iab	1118,1355,6009
9972	+60 00297				1 35 46	+61 22.5	1	8.92		0.27				3	A8 V	1118
		AJ74,1000 M33# 13			1 35 47	+30 16.8	1	12.30		0.75				1		209
		AO 823			1 35 47	+48 25.9	1	10.47		1.21		1.03		1		1722
		Cr 463 - 70			1 35 48	+71 42.8	1	11.27		0.27		0.22		2		428
10144	-57 00334	HR 472			1 35 51	-57 29.4	7	0.45	.022	-0.16	.016	-0.63	.035	17	B3 Vpe	15,200,1034,1637,2012*
10045	+20 00264				1 35 53	+21 08.7	2	6.65	.000	1.43	.010	1.64	.005	5	K3 III	985,1501
		VES 599			1 35 53	+60 28.5	1	12.46		0.28		-0.35		2		8107
10121	-46 00453				1 35 58	-46 20.3	4	6.97	.007	1.16	.004	1.27	.006	18	K2 III	977,1075,2011,2024
	+43 00339				1 35 59	+43 42.8	1	9.63		1.54		2.09		4	K7 V	7009
236815	+59 00297	LS I +60 169			1 35 59	+60 09.3	1	8.52		0.22		-0.69		2	B0.5III	1012
		AJ74,1000 M33# 12			1 36 02	+30 10.3	1	10.47		1.17				1		209
		Cr 463 - 71			1 36 08	+71 40.5	1	11.31		0.28		0.21		2		428
9899	+77 00058	IDS01316N7728	A		1 36 10	+77 43.0	1	6.74		-0.03		-0.13		3	B9	1502
		VES 600			1 36 12	+58 23.8	1	12.60		0.73		0.36		2		690
10162	-48 00430				1 36 13	-48 11.3	1	7.26		0.34				4	F0 IV	2012
10142	-37 00620	HR 471			1 36 14	-36 46.8	3	5.94	.009	1.05	.000	0.90		9	K0 IV	15,2027,3077
10097	+4 00287				1 36 17	+04 52.8	3	9.14	.001	0.68	.014	0.28	.007	16	G5 V	830,1003,1783
	+29 00278				1 36 20	+30 19.5	1	10.19		1.12				1		209
10072	+43 00343	HR 469			1 36 20	+44 08.0	3	5.00	.014	0.88	.004	0.56	.005	5	G8 III	1355,1363,3016
	+56 00311				1 36 21	+57 27.5	1	9.70		1.31				1		934
	+57 00359	LS I +57 025			1 36 22	+57 34.6	1	9.91		0.26		-0.54		2	B1:V:	1012
10149	-32 00639				1 36 22	-32 25.4	1	7.45		0.73		0.38		5	G8 III +(F)	1673
10167	-43 00503				1 36 23	-43 10.9	5	6.66	.002	0.33	.001	-0.01	.006	26	F0 V	278,977,977,1075,2011
10095	+27 00261				1 36 26	+27 30.2	1	7.07		1.20				3	K3 III	20
10063	+55 00375	LS I +55 014			1 36 28	+55 31.9	1	7.39		0.25		-0.33		2	B8 Iab	1012
		Cr 463 - 69			1 36 28	+72 00.4	1	11.25		0.25		0.16		2		428
10148	-22 00272	HR 473			1 36 28	-21 31.8	1	5.56		0.34				4	F0 V	2009
10161	-25 00670	HR 474	★ A		1 36 29	-25 16.5	3	6.69	.018	-0.08	.004	-0.17		10	B9 V	15,1068,2012
		LF 5 # 140			1 36 30	+59 18.	1	10.67		0.29		-0.51		1	B0 V	367
236817	+58 00272				1 36 30	+59 20.0	1	8.71		0.63				4	G0 V	70
		G 272 - 61	★ AB		1 36 31	-18 12.5	4	12.46	.134	1.83	.027	1.10	.004	14	M3e	316,1705,3078,8039
10086	+45 00404				1 36 33	+45 37.7	2	6.62	.001	0.66	.021	0.24	.031	5	G5 IV	695,7008
		CS 22953 # 48			1 36 33	-59 15.2	2	14.19	.010	0.38	.007	-0.17	.010	3		1580,1736
10135	+13 00255				1 36 34	+14 02.0	2	6.76	.005	1.12	.000	1.00	.015	4	K0	1648,3077
10190	-45 00542	IDS01345S4506	AB		1 36 36	-44 50.9	4	8.11	.010	0.48	.003	-0.02	.012	19	F6 V	977,1075,1117,2011
19602	-22 00552				1 36 37	-21 52.6	1	10.53		0.57		-0.03		2	F3/F5V	1696
10166	-27 00555	G 274 - 77			1 36 37	-27 30.4	2	9.36	.005	0.79	.005	0.35		5	K0 V	79,2033
		Cr 463 - 78			1 36 44	+72 00.7	1	11.44		1.29		1.36		2		428
	+47 00476				1 36 45	+48 28.7	1	9.32		-0.03		-0.75		1	B5	1722
236818	+57 00362				1 36 45	+57 42.5	1	8.99		0.17				1	B8	934
		GD 420			1 36 45	+76 53.8	1	14.85		0.06		-0.63		1	DA	782
10126	+27 00262	G 72 - 21			1 36 51	+27 51.4	6	7.73	.013	0.73	.003	0.28	.014	19	G8 V	22,333,1003,1355,1620*
10241	-54 00358	HR 479	★ AB		1 36 52	-53 41.5	3	6.83	.004	0.44	.005	0.09		9	F5 IV/V	15,404,2012
	+62 00287				1 36 53	+62 51.7	1	9.05		0.33				2	B8 V	1118
		Cr 463 - 37			1 36 54	+71 38.6	1	11.34		0.65		0.09		2		428
	+48 00499				1 36 55	+48 33.0	1	10.12		0.57		0.02		1		1722

Table 1

HD	DM	Other Id	N	Rem	α_{1950}	δ_{1950}	S	V	σ_V	B–V	σ_{B-V}	U–B	σ_{U-B}	n	Spectrum	References
10186	−18 00279	IDS01345S1818	AB		1 36 56	−18 02.8	1	7.62		0.28				4	F0 IV/V	2012
10209	−29 00542				1 36 58	−29 16.6	1	7.41		0.33				4	F0 III	2012
10164	+15 00245	HR 475			1 36 59	+16 09.2	1	5.97		1.12				2	K2 III	71
		Cr 463 - 31			1 37 00	+71 57.2	1	10.34		0.45		−0.13		2		428
10110	+53 00363	HR 470			1 37 01	+53 36.9	2	6.38	.008	1.62	.016	1.98	.001	6	K5 III	595,1733
10240	−49 00464				1 37 06	−49 01.7	1	7.30		0.08				4	A1 V	2012
10156	+38 00316	IDS01342N3828	AB		1 37 07	+38 43.0	1	7.62		0.41		−0.04		3	F5	833
	+60 00301				1 37 08	+61 23.0	1	9.15		0.81				3	G5 V	70
10238	−42 00576				1 37 09	−41 44.1	1	10.32		−0.12		−0.49		4	B7/8 V	152
10107	+58 00273				1 37 11	+58 48.0	1	6.98		−0.04				2	B9	1118
	−46 00465				1 37 13	−46 02.7	1	10.74		0.32		0.33		1		7
10154	+50 00323				1 37 15	+50 39.9	1	8.89		0.39		0.16		2	A2	1566
		LS I +62 179			1 37 15	+62 20.2	1	10.58		0.62		−0.27		2	B3: II:	1012
		CS 22180 # 33			1 37 17	−10 01.2	1	14.67		0.08		0.14		1		1736
		Steph 178			1 37 18	+00 06.2	1	11.31		1.30		1.22		1	K7	1746
	+71 00092	Cr 463 - 36			1 37 20	+71 43.2	1	10.35		0.14		−0.03		2		428
	−70 00086	G 158 - 102			1 37 20	−69 58.2	1	10.82		0.70		0.17		1		1658
10125	+63 00218	LS I +63 156			1 37 22	+63 55.2	2	8.22	.000	0.31	.000	−0.65	.000	4	O9.5Ib	1011,1012
10226	−10 00349	G 271 - 125			1 37 25	−10 13.6	1	7.85		0.61		0.14		1	F8	79
10083	+71 00093	Cr 463 - 30			1 37 28	+71 59.7	1	8.91		0.27		0.10		2	A0	428
		CS 29504 # 36			1 37 28	−34 57.8	1	15.33		−0.02		−0.71		1		1736
	+63 00219				1 37 30	+63 32.5	1	10.07		0.47		0.23		2	B5	1625
10254	−23 00620				1 37 32	−23 10.0	2	7.26	.005	1.65	.005	1.98		8	M2 III	2012,3007
		G 3 - 11			1 37 34	+09 15.0	2	13.52	.015	0.73	.002	0.07	.015	3		1658,1696
10205	+39 00378	HR 477		⋆ A	1 37 37	+40 19.5	5	4.95	.009	−0.09	.007	−0.39	.015	11	B8 III	15,1079,1363,3024,8015
10205	+39 00378	IDS01347N4004	B		1 37 37	+40 19.5	1	11.54		0.65		0.05		4		3024
	−25 00676	G 274 - 80			1 37 37	−24 46.9	1	11.98		0.96		0.75		1	A0	1696
	−47 00502				1 37 37	−46 45.6	2	9.86	.005	1.16	.001	1.14		5	M0	158,1705
10204	+42 00345	HR 476			1 37 39	+43 02.7	4	5.62	.007	0.21	.001	0.14	.011	6	A5 III	39,254,695,3016
	+1 00299				1 37 40	+01 35.9	1	10.20		1.17				2	K8	1619
		LS I +60 171			1 37 41	+60 19.9	1	11.48		0.33		−0.43		1		666
		G 133 - 28			1 37 43	+41 40.9	2	11.06	.030	1.16	.009	1.12	.009	3	K5	1620,5010
		Cr 463 - 33			1 37 46	+71 50.5	1	11.18		1.98		2.61		2		428
		CS 22180 # 32			1 37 49	−09 49.1	1	14.56		−0.02		−0.04		1		1736
		Cr 463 - 38			1 37 51	+71 36.1	1	12.49		0.66		0.27		2		428
10145	+66 00145	G 244 - 31			1 37 52	+66 39.7	5	7.70	.013	0.69	.003	0.21	.012	13	G5 V	22,1003,1658,1775,3026
10278	−11 00318	G 271 - 126			1 37 53	−10 41.8	1	9.14		0.71		0.25		1	G5	79
		Cr 463 - 39			1 37 54	+71 31.7	1	11.30		0.29		0.15		2		428
10360	−56 00329	HR 486		⋆ B	1 37 54	−56 26.8	1	5.96		0.88		0.60		1	K0 V	3077
10361	−56 00329	HR 487		⋆ A	1 37 54	−56 26.9	1	5.81		0.85		0.55		1	K5 V	3077
10361	−56 00329	IDS01360S5642	AB		1 37 54	−56 26.9	4	5.06	.005	0.88	.000	0.56		11	K5 V+K0 V	15,258,2012,8015
		Steph 179			1 37 55	+49 05.1	1	11.59		1.38		1.30		1	K7	1746
	−35 00583				1 37 55	−35 27.1	1	10.31		0.78		0.34		1	G5	79
10386	−59 00117				1 37 56	−59 30.9	1	9.37		1.58		1.94		3	C3,2	864
10262	+8 00258				1 37 57	+08 30.5	1	6.33		0.40		−0.05		18	F2	696
		Cr 463 - 32			1 38 00	+71 50.6	1	12.10		1.24		1.55		2		428
		Steph 182			1 38 00	−15 49.5	1	11.56		1.39		1.29		1	K7	1746
10317	−10 00355	G 272 - 73			1 38 10	−10 05.1	1	9.11		0.71		0.21		1	G5	79
10342	−20 00309				1 38 15	−20 21.0	1	9.43		0.10		0.12		4	A0 V	152
	+60 00306				1 38 16	+60 59.6	1	9.87		1.18				35	G5 III	823
10385	−45 00547	IDS01362S4533	AB		1 38 19	−45 17.3	1	8.58		0.45		−0.02		2	F3 V	1117
10339	−13 00306				1 38 20	−12 40.6	1	9.91		0.15		0.12		4	A1 V	152
10356	−32 00655				1 38 20	−31 58.1	1	8.56		0.59		0.06		1	G0 V	79
236827	+58 00274	LS I +59 086			1 38 21	+59 20.3	1	9.51		0.29		−0.45		1	B0	367
	+60 00307	LS I +61 212			1 38 21	+61 04.4	2	10.52	.035	0.55	.010	−0.38	.020	2	B2 Ve	367,666
10309	+22 00257				1 38 22	+22 46.4	1	7.43		1.01		0.73		2	G8 III-IV	1648
10296	+27 00270				1 38 22	+28 13.8	1	8.11		1.08				2	K1 III	20
		Cr 463 - 68			1 38 28	+71 15.5	1	10.29		0.62		0.12		2		428
10354	−7 00277				1 38 30	−06 58.7	2	10.19	.014	0.12	.005	0.09	.023	7	A0	152,1026
10308	+25 00276	HR 484		⋆ A	1 38 31	+25 29.6	1	6.18		0.44		0.00		2	F2 III	105
		NGC 654 - 255			1 38 31	+61 34.4	1	12.22		0.83		0.48		2		981
		NGC 654 - 257			1 38 32	+61 41.8	1	13.65		0.76		−0.09		2		981
10424	−45 00549				1 38 32	−44 45.4	1	10.97		0.33		−0.01		1	A3/5 II	1117
	+61 00310	NGC 654 - 259		⋆	1 38 35	+61 32.7	3	9.29	.033	0.61	.014	−0.21	.025	4	B2	666,873,981
		NGC 654 - 260			1 38 35	+61 42.9	1	13.47		0.74		0.05		2		981
		NGC 654 - 261			1 38 36	+61 30.6	1	13.81		0.63		0.06		2		981
	+62 00292				1 38 36	+63 21.5	1	10.40		0.55		−0.41		2	B1:pe	1012
10221	+67 00149	HR 478, V557 Cas			1 38 36	+67 47.5	3	5.58	.019	−0.07	.017	−0.27	.044	10	A0p SiSr	1063,1118,1202
		CS 29504 # 35			1 38 36	−33 40.8	1	14.70		0.03		0.12		1		1736
10260	+60 00308	HR 481			1 38 38	+60 47.2	2	6.71		−0.03	.010	−0.42		3	B8 IIIp(Si)	1079,1118
		NGC 654 - 263			1 38 43	+61 32.8	1	13.97		0.62		−0.01		2		981
10307	+41 00328	HR 483			1 38 44	+42 21.8	17	4.95	.017	0.62	.008	0.11	.010	58	G2 V	1,15,22,254,1004,1077*
	+61 00311				1 38 44	+62 25.1	1	9.46		0.49		−0.08		3	F8	171
		Cr 463 - 40			1 38 45	+71 29.0	1	11.30		0.61		0.27		2		428
		BPM 16571			1 38 48	−55 58.4	1	14.86		−0.12		−1.06		2	DB	3065
10348	+29 00286	HR 485			1 38 49	+29 47.7	6	5.98	.010	1.01	.006	0.84	.015	16	K0 III	15,252,1007,1013,4001,8015
	+71 00095	Cr 463 - 35			1 38 49	+71 48.5	1	10.72		0.20		0.05		2		428
10433	−35 00594				1 38 49	−34 50.6	1	9.25		0.13		0.16		4	A3 IV/V	152
10380	+4 00293	HR 489			1 38 50	+05 14.1	10	4.44	.007	1.36	.007	1.56	.013	26	K3 IIIb	15,1075,1355,1363*
10349	+26 00281	G 72 - 25			1 38 51	+27 25.6	1	8.68		0.73		0.29		1	G5	333,1620

HD	DM	Other Id	N Rem	α_{1950}	δ_{1950}	S	V	σ_V	B–V	σ_{B-V}	U–B	σ_{U-B}	n	Spectrum	References
	+60 00310	AZ Cas		1 38 51	+61 10.1	2	9.26	.014	1.72	.000	0.40	.023	13	M Ib +Be	588,8032
		VES 601		1 38 53	+59 05.5	1	10.45		0.36		−0.30		2		690
		NGC 654 - 277		1 38 53	+61 39.3	1	13.97		0.84		0.32		2		981
		AJ89,1229 # 75		1 38 57	+11 22.2	1	11.25		1.62				3		1532
10293	+57 00370	HR 482	⋆ A	1 38 58	+58 22.6	1			−0.02		−0.46		1	B8 III	1079
	+58 00278	LS I +59 087		1 38 59	+59 17.8	1	10.40		0.26		−0.58		1	B2 III	367
	−16 00288			1 39 00	−15 51.7	1	10.82		0.53		−0.01		2		1696
10250	+69 00114	HR 480		1 39 01	+70 22.3	1	5.18		−0.04				2	B9 V	1118
		NGC 654 - 285		1 39 04	+61 28.1	1	13.27		1.42		1.16		2		981
		NGC 654 - 286		1 39 05	+61 39.7	1	14.25		0.67		0.04		2		981
10608	−75 00111			1 39 05	−75 14.2	1	8.09		0.45		0.08		3	F6 IV/V	727
		NGC 654 - 287		1 39 07	+61 35.2	1	13.12		2.23		1.96		2		981
		NGC 654 - 289	⋆	1 39 08	+61 32.8	3	10.95	.026	0.69	.017	−0.13	.047	4	B2 V	981,367,666
		G 133 - 29		1 39 09	+37 08.5	1	14.10		1.47		0.97		4		316
10800	−83 00027	HR 512		1 39 09	−83 13.8	4	5.86	.003	0.62	.010	0.09	.004	23	G1/2 V	15,1075,2038,3077
10390	+34 00297	HR 490		1 39 10	+34 59.6	4	5.59	.095	−0.08	.008	−0.20	.010	8	B9 IV-V	985,1049,1079,3023
10388	+38 00326	OQ And		1 39 11	+39 09.6	1	7.68		1.66		1.99		6	M3 III	1501
		NGC 654 - 290		1 39 11	+61 32.5	1	14.11		0.73		−0.05		2		981
10304	+63 00224			1 39 13	+63 38.4	1	7.84		1.12		0.92		2	K0	1625
		CS 22180 # 39		1 39 13	−08 24.6	1	14.55		0.39		−0.15		1		1736
	−68 00074			1 39 13	−68 16.7	1	8.34		0.56		−0.07		2		1773
232506	+51 00373			1 39 14	+52 08.2	1	9.41		0.00		−0.51		2	A0	1566
		NGC 654 - 293		1 39 15	+61 33.5	1	13.41		0.66		0.33		2		981
	+16 00188			1 39 16	+16 48.6	2	10.32	.021	1.17	.005			3	K4-5	1017,1619
10481	−38 00584	HR 494		1 39 16	−38 23.2	3	6.17	.008	0.42	.004	0.01		8	F2 V	15,79,2012
10453	−12 00315	HR 492	⋆ AB	1 39 17	−11 34.3	2	5.74	.000	0.44	.000	−0.03		5	F5 V+F7 V	79,2007
10442	+1 00306			1 39 18	+02 27.1	1	9.08		0.56		−0.09		2	G5	3079
	+60 00311	LS I +60 172		1 39 21	+60 35.7	3	10.02	.050	0.30	.018	−0.43	.027	6	B2 III:nn	666,1012,8107
10443	+1 00305	G 71 - 27	⋆ A	1 39 22	+01 39.2	1	8.88		0.54		−0.04		1	G0	333,1620
		Cr 463 - 34		1 39 24	+71 46.4	1	11.73		0.66		0.18		2		428
		Cr 463 - 74		1 39 26	+71 47.3	1	11.62		0.48		0.36		2		428
		Cr 463 - 67		1 39 27	+71 14.4	1	11.19		0.41		0.20		2		428
10513	−46 00473			1 39 28	−45 40.3	5	9.30	.016	0.69	.007	0.26	.010	15	G5 V	116,158,1705,2011,3077
		L 223 - 66		1 39 30	−52 51.	1	14.53		1.51				1		3062
10362	+60 00312	HR 488		1 39 32	+61 10.2	2	6.34		0.01	.005	−0.46		3	B7 II	1079,1118
		CS 22180 # 31		1 39 33	−10 15.9	1	14.87		0.12		0.08		1		1736
		NGC 654 - 302	⋆	1 39 37	+61 33.5	2	11.58	.068	0.77	.015	−0.06	.083	3		666,981
		LS I +61 219		1 39 38	+61 27.0	1	11.49		0.64		−0.27		1		666
10510	−19 00292			1 39 38	−19 26.6	1	7.69		0.40		0.03		2	F3 V	1375
	+62 00296	LS I +63 164		1 39 39	+63 21.2	1	9.84		0.54		−0.33		2	B1 IV	1012
10555	−53 00350			1 39 39	−53 11.1	1	8.36		0.96				4	G8 IV	2013
	+61 00312	LS I +62 181		1 39 40	+62 12.5	2	9.22	.000	0.53	.000	−0.39	.009	6	B1 III	171,1012
10375	+61 00313			1 39 42	+62 23.5	1	8.25		0.15				3	B9 V	1118
10553	−50 00461	HR 501		1 39 42	−50 17.4	7	6.63	.008	0.13	.005	0.12	.004	37	A3 V	15,278,977,977,1075*
10489	−11 00323	G 271 - 133		1 39 44	−11 21.3	1	9.00		0.61		0.07		1	G0	79
10477	+15 00251			1 39 45	+15 31.6	1	7.55		1.13		1.10		2	K0	1648
10417	+58 00281			1 39 45	+58 43.5	1	8.45		0.02				2	A2	70
	+61 00314	NGC 654 - 307		1 39 45	+61 50.6	2	9.45	.019	0.12	.029	−0.26	.039	3	B7 V	873,981
236835	+55 00388	V595 Cas		1 39 46	+56 15.7	4	8.85	.065	2.00	.012	2.09		8	M2 Ib:	138,148,369,8032
10476	+19 00279	HR 493	⋆ A	1 39 47	+20 01.6	15	5.24	.012	0.84	.008	0.50	.011	88	K1 V	1,3,15,22,116,801*
		NGC 654 - 308		1 39 48	+61 42.2	1	13.08		0.72		0.31		2		981
10509	−13 00311			1 39 48	−13 27.7	1	9.73		1.18		1.13		1	F3/5 V	79
10519	−18 00287	G 272 - 81	⋆ A	1 39 48	−18 08.4	6	7.46	.025	0.62	.007	0.03	.014	14	G2/3 V	22,79,742,1311,2034,3026
10607	−68 00077	IDS01383S6810	A	1 39 48	−67 55.3	3	8.32	.009	0.56	.004	−0.10	.005	7	G0wF5	258,2033,3077
		L 88 - 69		1 39 48	−67 56.0	1	12.45		1.21		1.01		2		1773
10538	−37 00650	HR 498		1 39 51	−37 05.0	3	5.71	.003	−0.01	.000			27	A0 V	15,2012,8033
	−38 00585			1 39 51	−38 11.5	1	10.02		1.05		0.65		1	K0	565
	+62 00297	LS I +63 165		1 39 53	+63 21.1	2	9.16	.013	0.64	.004	−0.36	.000	5	B1 Ib	171,1012
10537	−32 00666	HR 497		1 39 53	−32 34.7	3	5.25	.007	1.05	.006	0.79		25	K1 II/III	2032,3005,8033
		CS 29504 # 34		1 39 53	−32 49.1	1	15.53		0.11		0.21		1		1736
		NGC 654 - 310		1 39 54	+61 28.3	1	13.50		0.72		−0.13		2		981
10560	−41 00466			1 39 54	−41 22.2	1	8.67		0.16				4	A1 V	2012
10425	+59 00307	HR 491	⋆ AB	1 39 55	+60 18.0	2	5.78		−0.02	.005	−0.33		3	B8 IIIn	1079,1118
		JL 256		1 40 00	−45 59.	1	15.60		−0.35		−1.04		3		132
		L 294 - 111		1 40 00	−46 55.	1	13.62		0.86		0.39		3		1696
		LS I +60 173		1 40 01	+60 44.6	2	10.94	.020	0.34	.005	−0.43	.030	2	B2 V	367,666
		Cr 463 - 18		1 40 01	+71 23.4	1	10.35		0.26		0.04		2		428
		NGC 654 - 312		1 40 02	+61 30.6	1	14.12		0.72		0.43		2		981
		G 34 - 45		1 40 03	+22 22.2	2	13.12	.005	0.91	.010	0.44	.015	2		333,1620,1658
		NGC 654 - 313		1 40 03	+61 30.4	1	11.65		2.07		2.05		2		981
	+60 00317	NGC 663 - 573		1 40 04	+61 17.5	2	9.60	.005	0.41	.014	0.22		20	F1 V	171,823
10615	−61 00130	HR 505		1 40 05	−61 02.4	3	5.71	.004	1.26	.005			12	K2/3 III	15,1075,2035
11025	−85 00017	HR 525		1 40 06	−85 01.4	3	5.67	.007	0.94	.000	0.66	.000	18	G8 III	15,1075,2038
		LS I +60 174		1 40 07	+60 44.2	2	11.23	.035	0.36	.040	−0.48	.000	2	B1 P He	367,666
		LS I +60 175		1 40 09	+60 39.8	1	11.04		0.29		−0.51		1		666
		NGC 654 - 21		1 40 12	+61 38.2	3	12.34	.043	0.50	.017	0.28	.060	5		49,981,1046
10436	+63 00229	G 244 - 33		1 40 12	+63 34.8	7	8.42	.016	1.21	.008	1.13	.011	21	K5 V	22,171,1003,1013,1197*
10550	−4 00260	HR 500		1 40 12	−03 56.5	4	4.98	.003	1.38	.005	1.57	.016	16	K3 II-III	15,1256,3016,8100
10486	+44 00354	HR 495		1 40 13	+45 04.3	4	6.33	.008	1.02	.013	0.93	.026	11	K2 IV	252,1355,3035,8032
		LS I +60 176		1 40 13	+60 14.8	2	10.68	.010	0.50	.015	−0.42		2	B1 II	367,666

Table 1

HD	DM	Other Id	N	Rem	α₁₉₅₀	δ₁₉₅₀	S	V	σ_V	B–V	σ_B–V	U–B	σ_U–B	n	Spectrum	References
		G 3 - 13			1 40 14	+16 53.3	1	12.51		0.92		0.40		1	K3	333,1620
		Cr 463 - 45			1 40 14	+71 31.9	1	11.76		1.25		0.81		2		428
		LP 588 - 38			1 40 15	−02 37.6	1	12.35		0.64		0.02		2		1696
	+61 00315	NGC 654 - 2			1 40 16	+61 35.0	5	9.55	.031	0.88	.014	0.35	.020	19	A2 Ib	49,206,666,981,1046
	−43 00523				1 40 18	−43 22.9	1	10.78		0.72		0.18		1	F8	1696
		NGC 654 - 6			1 40 19	+61 36.6	2	13.99	.052	0.74	.047	-0.07	.009	4		981,1046
		NGC 654 - 18			1 40 19	+61 37.8	3	12.59	.065	0.69	.008	-0.14	.035	5		49,981,1046
		NGC 654 - 5			1 40 20	+61 36.6	2	13.21	.019	0.65	.005	-0.13	.028	4		981,1046
		NGC 654 - 23			1 40 20	+61 38.7	3	11.54	.031	1.23	.029	0.79	.026	6	K2 IV:	49,981,1046
		NGC 654 - 25			1 40 22	+61 39.6	3	15.09	.079	1.02	.035	0.16	.187	4		49,981,1046
		NGC 654 - 24			1 40 23	+61 39.2	2	13.92	.034	0.97	.017	0.14	.034	3		981,1046
		NGC 654 - 3			1 40 24	+61 36.1	3	11.45	.031	0.70	.019	-0.19	.028	6	B1.5IV	49,981,1046
		NGC 654 - 26			1 40 24	+61 39.3	1	13.05		0.80		-0.03		2		981
	+27 00273	G 35 - 8			1 40 25	+27 35.5	2	10.40	.000	1.46	.000	1.26		2	K7	1620,1746
		NGC 654 - 30			1 40 25	+61 38.3	1	14.43		0.65		0.01		2		981
		NGC 654 - 29			1 40 25	+61 38.9	1	13.32		0.78		0.11		2		981
		NGC 654 - 27			1 40 25	+61 39.2	1	13.23		0.78		-0.01		2		981
		NGC 654 - 10			1 40 26	+61 37.0	2	14.33	.009	0.77	.042	0.03	.014	4		981,1046
		NGC 654 - 31			1 40 26	+61 38.1	1	13.61		0.61		0.02		2		981
10497	+52 00420	LS I +52 001	★	AB	1 40 27	+52 38.1	2	6.74	.005	0.43	.005	0.41		5	F2 II	1501,6009
		NGC 654 - 8			1 40 27	+61 36.7	3	14.82	.314	1.32	.156	0.14	.151	5		49,981,1046
		NGC 654 - 105			1 40 28	+61 33.5	1	13.34		0.65		-0.17		2		981
		NGC 654 - 32			1 40 28	+61 38.6	1	13.18		0.73		-0.03		2		981
		NGC 654 - 11			1 40 30	+61 37.6	2	11.72	.013	0.64	.013	-0.20	.021	3	B1 V	981,1046
		NGC 654 - 33			1 40 30	+61 38.8	2	12.20	.030	0.68	.013	-0.21	.038	3	B2 IV	981,1046
		NGC 654 - 34			1 40 30	+61 39.0	1	13.22		0.90		0.10		2		981
10516	+49 00444	HR 496, φ Per			1 40 31	+50 26.3	9	4.06	.012	-0.04	.007	-0.93	.012	24	B2 Vep	15,154,379,1212,1223*
10474	+59 00309				1 40 31	+60 11.1	1	7.80		0.20				2	A3 V	1118
		NGC 654 - 69			1 40 32	+61 40.3	3	12.86	.128	0.93	.017	0.02	.044	6		49,981,1046
		NGC 654 - 38			1 40 33	+61 38.8	1	14.29		0.70		0.18		2		981
		NGC 654 - 70			1 40 33	+61 40.0	3	14.38	.031	0.90	.028	0.22	.017	5		49,981,1046
		Cr 463 - 44			1 40 33	+71 30.7	1	11.50		0.31		0.24		2		428
10613	−38 00589				1 40 33	−38 15.4	1	9.59		1.00				2	G8 IIIp	565
		NGC 654 - 49			1 40 34	+61 37.0	2	13.63	.075	0.77	.005	-0.08	.005	4		981,1046
		NGC 654 - 37			1 40 34	+61 39.0	1	13.34		0.80		-0.04		2		981
		NGC 654 - 71			1 40 34	+61 40.5	2	14.10	.005	0.79	.014	0.15	.066	4		981,1046
10647	−54 00365	HR 506			1 40 34	−53 59.4	3	5.54	.019	0.53	.026	0.00	.005	7	F8 V	58,2007,3037
		NGC 654 - 115			1 40 35	+61 38.3	1	12.13		0.73		0.16		2		981
		NGC 654 - 321			1 40 35	+61 38.4	1	12.09		0.79		0.06		2		981
		NGC 654 - 39			1 40 35	+61 38.9	1	13.26		0.74		-0.02		2		981
		NGC 654 - 68			1 40 35	+61 40.2	3	12.56	.032	0.94	.011	0.02	.046	6		49,981,1046
		Cr 463 - 23			1 40 35	+71 10.7	1	11.87		0.69		0.41		2		428
		NGC 654 - 112			1 40 36	+61 38.	2	14.33	.075	0.99	.000	0.39		7		981,1046
		NGC 654 - 322			1 40 36	+61 38.3	1	13.49		0.70		0.32		2		981
		NGC 654 - 40			1 40 37	+61 38.9	1	13.30		0.81		0.13		2		981
		Cr 463 - 20			1 40 37	+71 26.0	1	12.39		0.43		0.42		2		428
10747	−76 00126				1 40 37	−75 54.9	1	8.16		-0.14		-0.66		1	B2 V	55
		Steph 187			1 40 38	+33 15.8	1	12.59		1.36		1.20		1	K5	1746
		NGC 654 - 42			1 40 38	+61 38.5	2	13.50	.034	0.79	.034	0.07	.013	3		981,1046
236843	+58 00284				1 40 39	+58 34.0	1	9.07		0.94				2	G5 III	70
10485	+60 00318	NGC 663 - 390			1 40 39	+61 04.3	1	8.70		0.14				3	A3 V	70
		NGC 654 - 43			1 40 39	+61 38.3	2	12.74	.020	0.73	.015	-0.17	.015	5		981,1046
		G 271 - 137			1 40 39	−06 16.8	1	12.80		0.57		-0.06		2		1658
		NGC 654 - 50			1 40 40	+61 37.0	2	14.61	.123	0.50	.165	0.03	.014	4		981,1046
		NGC 654 - 51			1 40 40	+61 37.2	1	14.76		0.77		0.13		2		1046
		WLS 140-20 # 5			1 40 40	−17 54.5	1	12.29		0.71		0.12		2		1375
		NGC 654 - 44			1 40 41	+61 38.1	3	14.08	.026	0.86	.040	0.28	.026	5		49,981,1046
		NGC 654 - 41			1 40 41	+61 38.7	3	10.64	.030	0.85	.019	-0.09	.039	5	B2 II	666,981,1046
		NGC 654 - 113			1 40 43	+61 38.6	2	14.61	.068	0.85	.026	0.10		3		981,1046
		SK 192			1 40 43	−73 19.9	2	13.60	.005	-0.19	.000	-0.89	.009	4		733,835
		NGC 654 - 111			1 40 44	+61 35.0	2	15.06	.038	0.66	.038	0.22		3		981,1046
10494	+61 00316	NGC 654 - 1			1 40 44	+61 35.9	5	7.31	.018	1.23	.011	0.94	.036	20	F5 Ia	49,981,1046,1118,1355
		NGC 654 - 114			1 40 44	+61 36.5	2	14.70	.009	0.73	.038	0.33	.024	4		49,1046
	+64 00228				1 40 44	+65 15.7	1	8.33		1.91				4	K5	369
		G 3 - 14			1 40 45	+04 04.8	4	10.94	.019	1.53	.023	1.27		3	M2	1017,1620,1658,1705
		NGC 654 - 54			1 40 45	+61 37.7	2	13.48	.009	0.72	.009	-0.12	.026	3		981,1046
		NGC 654 - 63			1 40 45	+61 38.4	2	14.03	.035	0.76	.005	-0.01	.020	3		49,1046
		Cr 463 - 29			1 40 45	+71 44.6	1	10.84		0.71		0.22		2		428
10546	+48 00518				1 40 46	+49 24.3	1	7.44		0.00		-0.31		2	B9	401
		NGC 654 - 53			1 40 46	+61 37.4	3	11.39	.021	0.61	.005	-0.22	.068	13	B1 V	49,981,1046
		NGC 654 - 62			1 40 47	+61 39.0	1	13.93		0.70		0.10		2		981
		NGC 659 - 109			1 40 48	+60 27.	1	11.74		0.69		-0.27		1		690
		NGC 654 - 56			1 40 48	+61 37.3	3	14.53	.025	0.78	.027	0.26	.060	6		981,981,1046
		NGC 654 - 61			1 40 48	+61 38.6	2	14.46	.026	0.73	.004	0.28	.171	3		981,1046
		SK 191			1 40 48	−74 05.7	3	11.86	.012	-0.04	.008	-0.84	.007	9	B1.5IA	573,733,1412
		NGC 654 - 60			1 40 50	+61 38.2	3	12.93	.033	0.67	.011	-0.21	.007	8		49,981,1046
		NGC 654 - 324			1 40 51	+61 33.9	1	14.20		0.82		0.47		2		981
		NGC 654 - 98			1 40 51	+61 36.6	3	14.02	.033	0.64	.008	0.08	.058	6		49,981,1046
		Steph 188			1 40 52	+34 43.2	1	11.17		1.37		1.20		1	K7	1746
	+55 00393	LS I +55 016			1 40 52	+55 54.7	1	10.50		0.03		-0.70		2	B1 V	1012

HD	DM	Other Id	N	Rem	α₁₉₅₀	δ₁₉₅₀	S	V	σ_V	B–V	σ_B–V	U–B	σ_U–B	n	Spectrum	References
	+70 00125	Cr 463 - 17			1 40 52	+71 26.4	1	10.56		0.06		0.00		2		428
		NGC 654 - 104			1 40 53	+61 35.5	3	12.31	.039	0.57	.009	-0.25	.033	6		49,981,1046
		NGC 654 - 97			1 40 54	+61 36.6	3	11.83	.027	0.56	.009	-0.26	.064	8	B1.5V	49,981,1046
		L 223 - 10			1 40 54	-50 15.	1	14.46		0.83		0.22		3		3062
	+4 00302	G 3 - 16			1 40 55	+04 36.5	3	10.48	.014	0.43	.016	-0.23	.009	7	sdF5	308,979,1696
		NGC 654 - 96			1 40 56	+61 37.2	4	14.03	.034	0.71	.034	0.16	.058	9		49,981,981,1046
		NGC 654 - 110			1 40 57	+61 35.9	1	15.69		1.11		0.36		1		1046
10588	+31 00301	HR 503			1 40 58	+31 56.4	1	6.30		0.89		0.53		2	G8 III-I	1733
10543	+56 00330	HR 499		⋆ AB	1 40 59	+57 17.2	3	6.19	.012	0.12	.011	0.05	.012	10	A3 V	595,1501,1733
		VES 605			1 41 00	+58 50.4	1	12.46		0.36		-0.51		2		8107
		LS I +60 177			1 41 00	+60 22.2	2	10.66	.005	0.50	.005	-0.32	.035	2	B2 V	367,666
		NGC 654 - 92			1 41 00	+61 38.4	1	13.35		0.83		0.12		2		981
		NGC 654 - 94			1 41 01	+61 37.8	1	14.90		0.90		0.07		1		49
	+70 00126	Cr 463 - 21			1 41 01	+71 09.8	1	10.08		1.40		1.47		2		428
		Cr 463 - 22			1 41 01	+71 10.5	1	11.40		0.28		0.22		2		428
		Cr 463 - 61			1 41 01	+71 39.8	1	11.17		0.37		0.37		2		428
		Cr 463 - 43			1 41 03	+71 29.6	1	11.84		0.76		0.31		2		428
10601	+32 00307				1 41 04	+32 29.3	1	8.47		0.42		-0.05		2	G0	1733
		NGC 659 - 111		⋆	1 41 04	+60 24.8	2	10.77	.214	0.56	.015	-0.44	.034	3		666,8107
10483	+71 00100	Cr 463 - 1			1 41 07	+71 38.4	1	8.20		1.48		1.22		2	K0	428
10577	+47 00491				1 41 08	+47 57.6	1	7.02		0.02		-0.21		2	B9	401
10687	-55 00340	IDS01393S5522		AB	1 41 08	-55 06.5	1	8.75		0.46		-0.02		2	F5 V	1730
10565	+52 00424				1 41 09	+52 56.3	1	8.28		0.17		0.18		2	A2	1733
		NGC 659 - 193			1 41 09	+60 25.7	1	12.70		0.61		-0.34		2		8107
		Cr 463 - 27			1 41 10	+71 45.8	1	11.35		1.37		1.22		2		428
		NGC 654 - 103			1 41 11	+61 34.2	2	13.51	.015	0.68	.005	-0.07	.020	5		981,1046
		NGC 654 - 102			1 41 12	+61 34.8	3	11.21	.011	0.55	.008	-0.25	.041	6	B2 IV	666,981,1046
	+71 00101	Cr 463 - 55			1 41 12	+72 01.3	1	10.02		1.31		1.09		2		428
		G 173 - 16			1 41 14	+50 08.7	1	10.62		0.47		-0.14		4	sdF5	7010
		NGC 654 - 329			1 41 14	+61 24.2	1	12.36		0.83		0.06		2		981
10564	+59 00311				1 41 15	+59 50.3	1	9.06		0.13				2	A0	70
		NGC 654 - 88			1 41 16	+61 38.5	2	14.04	.099	0.78	.052	-0.14	.066	4		981,1046
		NGC 654 - 101			1 41 17	+61 34.8	3	12.22	.019	0.56	.005	-0.24	.064	6	B1 V	666,981,1046
		Cr 463 - 28			1 41 17	+71 45.0	1	10.37		0.52		-0.02		2		428
		Cr 463 - 66			1 41 18	+71 13.0	1	11.10		0.72		0.18		2		428
10859	-79 00044	HR 516		⋆ AB	1 41 18	-79 24.0	3	6.31	.007	0.95	.005	0.69	.000	18	G6/8 III	15,1075,2038
		NGC 654 - 85			1 41 21	+61 40.2	2	10.87	.009	1.37	.014	1.29	.108	4		981,1046
		LS I +61 226			1 41 22	+61 23.4	1	12.33		0.73				1		666
		Cr 463 - 16			1 41 22	+71 24.2	1	11.50		0.25		0.22		2		428
		LP 528 - 161			1 41 23	+05 48.4	1	12.67		0.57		-0.22		3		1696
10597	+45 00432	HR 504			1 41 23	+45 53.4	1	6.31		1.43		1.70		2	K5 III	1733
		Cr 463 - 42			1 41 23	+71 30.9	1	12.06		0.78		0.25		2		428
10653	-8 00302				1 41 23	-07 43.8	1	7.73		0.05		0.08		4	A0	152
	+43 00353	G 133 - 35			1 41 24	+44 12.5	2	10.18	.005	0.68	.005	0.14	.020	2	G0	1620,1658
		NGC 654 - 333			1 41 24	+61 28.2	1	13.96		0.81		0.06		2		981
10658	-5 00309	HR 507			1 41 24	-05 01.0	3	6.20	.018	1.53	.005	1.89	.000	11	K0	15,1256,2006
		PHL 1126			1 41 24	-24 20.	2	11.75	.024	-0.31	.016	-1.15	.004	4		1036,3060
	-55 00341	IDS01395S5528		AB	1 41 24	-55 13.0	1	10.52		0.41		-0.04		2	F2	1730
10582	+59 00312				1 41 25	+59 30.4	1	8.19		0.13				2	A2	70
		NGC 654 - 86			1 41 25	+61 39.0	1	13.94		0.71		-0.02		2		981
10720	-58 00138			A	1 41 25	-58 15.5	1	9.29		0.89		0.41		2	K1/2 Vp	3072
10720	-58 00138			B	1 41 25	-58 15.5	1	11.87		1.23		1.14		1		3072
		NGC 654 - 84			1 41 27	+61 39.8	2	13.70	.104	0.81	.005	0.40	.019	4		981,1046
10738	-61 00133				1 41 27	-61 36.8	1	8.26		1.07		0.96		10	K1 III	1704
10587	+56 00334	HR 502			1 41 28	+56 50.3	2	6.24	.004	0.05	.010	0.07	.008	5	A2 V	595,1733
	+71 00102	Cr 463 - 26			1 41 28	+71 57.3	1	9.29		0.31		0.05		2	B8	428
	+56 00335				1 41 30	+57 27.1	1	10.28		1.76				2		369
		LS I +64 077			1 41 30	+64 09.6	1	11.45		0.42		0.38		1		1215
		UV0141-24			1 41 30	-24 18.0	1	11.73		-0.29		-1.11		15		1732
		UV0141-24N			1 41 30	-24 18.0	1	10.67		0.46		-0.03		9		1732
		NGC 654 - 83			1 41 31	+61 40.5	2	11.58	.034	0.67	.015	0.29	.064	5		981,1046
		Cr 463 - 2			1 41 31	+71 28.8	1	9.17		0.16		0.02		2		428
		NGC 654 - 82			1 41 32	+61 41.1	2	13.21	.122	0.77	.000	0.26	.029	5		981,1046
	+71 00103	Cr 463 - 25			1 41 32	+72 01.5	1	9.84		0.30		0.14		2		428
		Cr 463 - 41			1 41 33	+71 31.6	1	11.21		1.05		0.96		2		428
	+60 00322	LS I +60 179			1 41 34	+60 33.3	3	9.69	.019	0.54	.003	-0.35	.020	5	B2 III	367,666,1012
		Cr 463 - 19			1 41 35	+71 23.2	1	12.45		0.40		0.39		2		428
		Cr 463 - 60			1 41 35	+71 38.5	1	12.38		0.67		0.10		2		428
		NGC 654 - 337			1 41 36	+61 51.1	1	12.14		2.08		2.43		2		981
	+62 00300	LS I +62 184			1 41 36	+62 34.8	2	9.93	.030	0.31	.015	-0.63	.079	3	B1 V:pen	180,1012
		L 88 - 59			1 41 36	-67 32.	2	13.87	.030	0.44	.010	-0.42	.020	2		782,3078
236854	+58 00290				1 41 39	+59 07.0	1	9.10		0.16				2	A0	70
		NGC 654 - 80			1 41 39	+61 42.2	2	10.91	.014	1.44	.000	1.36	.123	4		981,1046
10771	-67 00114				1 41 39	-66 55.6	1	9.38		0.90		0.54		4	G5/6 V	119
236857	+56 00338	IDS01384N5637		AB	1 41 42	+56 52.0	1	9.38		0.88		0.60		2	G5	3016
		NGC 659 - 129		V	1 41 42	+60 24.4	1	12.43		0.28		-0.35		2		8107
		NGC 654 - 339			1 41 42	+61 38.5	1	14.19		0.78		0.01		2		981
		G 72 - 30			1 41 44	+24 20.9	1	14.41		0.79		0.12		3		1658
	+13 00270				1 41 45	+13 47.7	1	8.64		1.11		0.94		2	K0	1648
		NGC 654 - 341			1 41 45	+61 39.3	1	12.45		0.60		-0.02		2		981

Table 1   111

HD	DM	Other Id	N	Rem	α_{1950}	δ_{1950}	S	V	σ_V	B–V	σ_{B-V}	U–B	σ_{U-B}	n	Spectrum	References
10700	−16 00295	HR 509		⋆ A	1 41 45	−16 12.0	18	3.50	.010	0.73	.007	0.22	.012	89	G8 V	1,3,15,22,58,1006*
		AAS6,117 # 50			1 41 48	+62 30.	1	10.62		1.06		0.70		3		171
		LS I +60 181			1 41 53	+60 43.1	1	11.53		0.39		-0.33		1		666
10563	+70 00128	Cr 463 - 3		⋆ A	1 41 54	+71 26.5	1	9.14		1.73		1.99		2	A0	428
	+61 00319	NGC 654 - 348			1 41 55	+61 29.7	2	9.86	.015	0.18	.019	-0.02	.044	3	B8 V	873,981
10636	+53 00379				1 41 56	+53 42.1	1	9.85		1.75				1	R6	1238
	+60 00326	LS I +60 182			1 41 57	+60 30.9	1	9.81		0.49		-0.32		1		666
		Cr 463 - 82			1 41 58	+71 26.7	1	10.23		0.17		-0.12		2		428
10682	+25 00288				1 42 00	+25 32.8	2	7.85	.005	0.36	.010	0.01	.015	6	A3	833,1625
		Cr 463 - 15			1 42 00	+71 25.7	1	12.78		0.29		0.29		2		428
10718	−21 00288	G 272 - 87			1 42 00	−20 50.6	3	8.11	.023	0.64	.000	0.12	.010	6	G3 V	79,2034,3026
		Cr 463 - 14			1 42 02	+71 23.8	1	12.61		1.30		1.19		2		428
		NGC 654 - 350			1 42 03	+61 39.4	1	11.95		0.41		0.45		2		981
		G 219 - 17			1 42 04	+53 48.5	1	15.20		1.64				1		906
		Cr 463 - 13			1 42 05	+71 22.2	1	11.27		0.14		0.18		2		428
		Cr 463 - 65			1 42 06	+71 14.4	1	11.64		-0.05		0.10		2		428
		Cr 463 - 64			1 42 09	+71 13.5	1	11.91		0.66		0.11		2		428
		NGC 663 - 169			1 42 10	+60 55.1	1	13.54		0.78		-0.03		1		49
		NGC 663 - 210			1 42 11	+60 52.9	1	12.10		0.69		-0.32		2		8107
10697	+19 00282	HR 508			1 42 12	+19 50.0	2	6.26	.000	0.72	.005	0.27	.010	10	G5 IV	1067,3077
	+60 00327	NGC 663 - 319			1 42 12	+60 47.4	1	8.17		2.24				1	K0 III	369
		CS 22180 # 30			1 42 12	−10 36.1	1	13.78		0.65		0.01		1		1736
		G 173 - 18			1 42 13	+46 17.0	1	11.43		1.43		1.15		6		1723
		NGC 663 - 130			1 42 13	+60 57.9	1	12.44		0.72		-0.21		2		8107
		NGC 654 - 357			1 42 14	+61 39.6	1	11.72		2.01		1.61		2		981
		VES 612			1 42 16	+60 07.8	1	12.41		0.64		0.32		2		8107
10779	−50 00476	IDS01403S5036		A	1 42 16	−50 21.0	1	8.76		0.22		0.07		4	A5 V	152
		G 3 - 17			1 42 18	+16 06.0	2	14.09	.019	1.71	.019			4		940,1705
		Cr 463 - 11			1 42 18	+71 17.4	1	9.80		0.33		0.11		2		428
		G 72 - 31			1 42 19	+31 17.9	1	14.80		0.19		-0.56		3		538
	+70 00131	Cr 463 - 12			1 42 19	+71 18.6	1	10.06		0.19		0.03		2		428
	+57 00382	G 219 - 20			1 42 20	+57 36.1	2	9.52	.005	0.60	.000	0.02	.002	3	G2	1658,7010
		NGC 654 - 361			1 42 20	+61 26.1	1	13.46		0.76		0.04		2		981
10595	+71 00104	Cr 463 - 24			1 42 22	+72 00.1	1	8.69		0.50		0.27		2	G0	428
		L 88 - 58			1 42 24	−67 31.	1	14.44		0.98		0.57		2		3062
236865	+56 00347				1 42 26	+56 38.5	1	8.40		2.20				3	K5	369
10694	+44 00360				1 42 27	+44 33.4	1	8.77		0.38		-0.07		2	F2 III	1601
	+60 00329	NGC 663 - 162			1 42 27	+61 04.2	1	10.01		0.45		-0.56		1	B3	666
10664	+62 00301				1 42 27	+62 41.2	1	8.94		0.26				2	B9 V	1119
		Cr 463 - 56			1 42 27	+71 48.8	1	11.29		0.43		0.22		2		428
	+60 00331	NGC 663 - 86			1 42 29	+60 59.0	3	8.94	.043	0.86	.012	0.07	.005	5	B8 Iab	49,1012,1380
		Cr 463 - 5			1 42 29	+71 31.5	1	10.37		0.21		0.02		2		428
		G 34 - 48			1 42 31	+23 02.8	2	13.10	.025	1.45	.000	1.20	.000	6	sdM3	316,3016
		G 34 - 49			1 42 32	+23 02.9	1	17.41		-0.05		-0.62		2	DC	3060
10663	+63 00236	IDS01390N6319		AB	1 42 32	+63 33.9	1	8.69		0.60				3	G2 V	1118
		NGC 663 - 83			1 42 33	+61 01.0	1	14.79		0.78		0.30		1		49
	+2 00263	G 71 - 33			1 42 35	+03 15.3	2	10.63	.000	0.49	.010	-0.16	.015	2	sdF6	1658,1696
		Cr 463 - 58			1 42 35	+71 31.6	1	11.86		0.20		0.24		2		428
	+60 00333	NGC 663 - 54		⋆ AB	1 42 38	+60 58.7	2	8.92	.005	0.80	.019	-0.11	.014	4	B5 Ib	1012,8107
		NGC 663 - 126			1 42 38	+61 03.4	1	10.68		0.68		-0.15		1	B5 III	49
		NGC 654 - 368		⋆	1 42 38	+61 46.5	3	10.56	.020	0.66	.000	-0.26	.021	5		367,981,1012
	+60 00335	NGC 663 - 323		V	1 42 39	+60 44.6	2	9.05	.088	2.36	.015			3	M2 III	138,369
		Cr 463 - 47			1 42 40	+71 37.1	1	10.69		0.31		0.19		2		428
	−25 00701	G 274 - 89			1 42 40	−25 31.7	1	10.26		0.76		0.35		1	G5	79
		NGC 663 - 135			1 42 41	+60 54.3	1	11.85		0.57		-0.30		1		49
	+60 00336	NGC 663 - 307			1 42 43	+61 11.5	1	9.09		0.78		0.20		2	B9 Iab	1012
	−30 00599	SV Scl			1 42 44	−30 19.1	1	11.14		0.15		0.10		1	A0	700
10761	+8 00273	HR 510			1 42 45	+08 54.4	8	4.26	.006	0.96	.005	0.72	.019	19	K0 III	15,1256,1355,1363*
10742	+26 00290				1 42 45	+26 52.2	1	8.39		1.00		0.74		2	K2	1733
		Cr 463 - 63			1 42 45	+71 13.6	1	12.76		0.78		0.60		2		428
232522	+54 00372	LS I +55 018		⋆ AB	1 42 46	+55 04.9	2	8.39	.243	0.18	.168	-0.79	.045	5	B1 II	1012,1212
		NGC 663 - 140			1 42 46	+60 52.8	1	10.89		0.47		-0.33		1		49
		NGC 663 - 636			1 42 47	+60 55.2	1	15.48		0.60		0.38		1		49
		Cr 463 - 46			1 42 47	+71 33.6	1	11.35		0.31		0.18		2		428
	−31 00721				1 42 47	−31 10.6	1	11.00		0.58		-0.02		1		79
	+60 00337	NGC 663 - 558			1 42 49	+61 21.0	1	9.90		0.86				1	A0 II	666
10785	−16 00301	G 272 - 89			1 42 50	−16 08.6	2	8.51	.005	0.61	.005	0.07	.005	5	G1/2 V	258,3077
		WLS 140-20 # 9			1 42 51	−20 30.0	1	11.62		0.50		-0.07		2		1375
		NGC 663 - 106			1 42 52	+60 55.0	1	13.29		0.65		-0.07		1		49
		NGC 663 - 10			1 42 53	+60 59.4	1	12.10		0.61		-0.22		2	B8	8107
232524	+54 00373	LS I +54 001			1 42 55	+54 30.6	1	9.29		-0.12		-0.90		2	B2	401
		NGC 663 - 18			1 42 55	+60 58.5	1	11.58		0.60		-0.20		1	B6	49
		NGC 663 - 19			1 42 56	+60 58.0	1	13.61		0.61		-0.03		1		49
		Cr 463 - 62			1 42 56	+71 13.2	1	10.83		2.10		2.58		2		428
	+60 00339	NGC 663 - 44		⋆ A	1 42 57	+61 00.5	5	8.48	.019	0.65	.013	-0.19	.010	8	B6 Iab	49,171,367,1012,1380
10662	+71 00105	Cr 463 - 4			1 42 57	+71 30.6	1	8.49		1.34		1.07		2	K5	428
		VES 618			1 42 58	+60 17.5	1	12.21		0.66		0.15		2		690
		NGC 663 - 61			1 42 58	+60 55.6	1	13.52		0.72		-0.17		1		49
		LF 5 # 150			1 43 00	+60 34.	1	10.86		0.52		-0.34		1	B2 V	367
		NGC 663 - 141			1 43 00	+60 52.7	2	10.66	.005	0.64	.010	-0.37	.024	3	B9 Vne	49,8107

HD	DM	Other Id	N Rem	α_{1950}	δ_{1950}	S	V	σ_V	B–V	σ_{B-V}	U–B	σ_{U-B}	n	Spectrum	References
	+60 00340	NGC 663 - 107		1 43 00	+60 55.0	1	10.16		0.60		-0.24		1	B8 Ib	49
		Cr 463 - 57		1 43 00	+71 45.5	1	11.70		0.41		0.40		2		428
		NGC 663 - 15		1 43 01	+60 58.6	1	13.90		0.76		0.00		1		49
	+72 00094	G 245 - 32		1 43 01	+73 13.4	3	9.94	.008	0.41	.008	-0.21	.010	7	sdF2:	516,1064,1658,3045
		Cr 463 - 48		1 43 02	+71 40.0	1	10.53		2.43		1.68		2		428
		NGC 663 - 62		1 43 03	+60 55.5	1	12.69		0.67		-0.09		1		49
		NGC 663 - 29		1 43 03	+60 57.1	1	13.30		0.64		-0.02		1		49
		NGC 663 - 6	V	1 43 03	+60 59.5	1	12.23		0.67		-0.20		3	B8	8107
10783	+7 00275	UZ Psc		1 43 04	+08 18.6	3	6.56	.012	-0.05	.012	-0.17	.015	11	A2p	696,1063,1202
	+19 00284	G 94 - 8		1 43 06	+20 03.6	1	9.12		0.78		0.40		1	K0	333,1620
	+60 00341	NGC 663 - 144		1 43 06	+60 52.2	1	11.37		0.54		-0.24		1		49
	+60 00344	NGC 663 - 110		1 43 07	+60 54.1	2	10.16	.065	0.56	.015	-0.22	.025	2	B2 V	367,666
	+60 00343	NGC 663 - 40	★ AB	1 43 07	+61 00.8	1	9.27		0.59		-0.28		2	B6 Ia	1012
		G 72 - 34		1 43 09	+35 40.1	2	12.98	.002	0.86	.007	0.35	.022	4		1620,1658
		NGC 663 - 66		1 43 10	+60 56.6	1	11.58		0.54		-0.31		1	B6	49
		LS I +59 091		1 43 13	+59 17.6	1	11.52		0.49		-0.40		2		8107
		NGC 663 - 30		1 43 14	+60 58.3	1	11.06		0.60		-0.32		1	B4	690
		NGC 663 - 115		1 43 15	+60 54.9	1	13.13		0.64		-0.10		1		49
	+60 00345	NGC 663 - 376		1 43 18	+61 15.2	3	9.76	.012	0.58	.004	-0.45	.000	4	B9 Iab	367,592,1012
10830	-25 00704	HR 514	★ AB	1 43 18	-25 18.1	4	5.29	.011	0.40	.010	0.01	.004	11	F2 IV +G9 V	15,404,938,2012
		SK 193		1 43 18	-74 56.	4	11.57	.006	0.18	.007	-0.35	.013	13	B9 Ia0:ep	573,733,1277,1412
		VES 623		1 43 19	+60 57.0	1	12.60		0.55		-0.13		2		8107
		NGC 663 - 68		1 43 20	+60 56.7	1	12.55		0.56		-0.18		1		49
	+60 00346			1 43 22	+61 20.1	1	9.90		1.24				1	G8 III	592
		CS 22180 # 40		1 43 22	-08 20.0	1	14.52		0.33		-0.17		1		1736
236869	+58 00296			1 43 23	+59 25.0	1	8.82		0.72				2	G5 V	70
		Cr 463 - 59		1 43 24	+71 32.4	1	11.94		0.36		0.21		2		428
	+62 00304	LS I +62 185		1 43 29	+62 49.9	1	10.25		0.47		-0.32		2	B3 IIIn	180,1012
10824	-6 00336	HR 513		1 43 29	-05 59.0	4	5.33	.005	1.52	.007	1.88	.000	18	K4 III	15,1071,1075,3016
10756	+59 00318	LS I +60 187		1 43 30	+60 25.2	2	7.53	.005	0.44	.002	-0.29	.014	4	B8 Ia	666,1012
		AAS6,117 # 101		1 43 30	+64 12.	1	10.92		0.39		0.28		2		171
		NGC 663 - 121		1 43 32	+60 57.5	1	13.44		0.67		0.02		1		49
		NGC 663 - 194		1 43 32	+61 06.5	2	11.33	.005	0.84	.029	-0.39		3		666,8107
236871	+59 00319			1 43 34	+60 07.4	2	8.81	.052	2.26	.009			4	M2 III	138,369
		NGC 663 - 120		1 43 34	+60 57.2	1	10.89		0.76		0.38		2	gG0	8107
10837	-7 00291			1 43 35	-06 41.6	1	9.23		0.16		0.14		4	A0	152
		AJ89,1229 # 79		1 43 42	+11 46.8	1	13.53		1.40				2		1532
	+60 00347	NGC 663 - 147		1 43 42	+60 56.9	2	9.46	.020	0.82	.010	0.19	.030	2	B7 II	49,367
10863	-27 00595	HR 517		1 43 42	-27 35.9	2	6.38	.005	0.37	.005			7	F2 Vn	15,2012
	+60 00348	BY Cas		1 43 45	+61 10.4	2	10.15	.018	1.17	.025	0.75		2	F5	592,1399
		LS I +60 190		1 43 46	+60 22.3	1	11.26		0.36		-0.27		1		666
		VES 626		1 43 46	+60 25.3	1	11.61		0.39		-0.19		2		690
		LF 5 # 142		1 43 50	+60 03.	1	10.76		0.26		-0.58		1	B1 v	367
10845	+16 00196	HR 515, VY Psc		1 43 52	+17 09.8	2	6.58	.023	0.25	.000	0.15	.014	5	A9 III	39,254
	+60 00351	NGC 663 - 221		1 43 53	+60 52.9	3	9.07	.031	0.64	.009	-0.22	.015	4	B6 Iab	49,367,1119
		NGC 663 - 230		1 43 54	+61 06.1	2	10.81	.020	0.53	.025	-0.25	.030	2		367,666
		Cr 463 - 49		1 43 54	+71 39.6	1	11.57		0.58		0.34		2		428
		LS I +60 192		1 43 55	+60 32.8	1	10.95		0.68		-0.34		1	B0 III	367
		G 94 - 9		1 43 56	+21 39.8	2	15.04	.010	0.24	.010	-0.64	.010	3		1620,3060
		VES 627		1 43 56	+59 18.6	1	11.22		0.37		-0.13		2		8107
10844	+25 00295			1 43 57	+25 40.2	1	8.13		0.62		0.08		2	F8	105
10806	+57 00392	IDS01406N5731	A	1 43 57	+57 46.3	1	6.75		1.10		0.85		2	G9 Ib	1733
10853	+11 00231			1 43 58	+12 09.8	2	8.90	.005	1.04	.005	0.91		2	K5	1705,3072
		NGC 663 - 222		1 43 59	+60 53.8	2	11.41	.084	0.63	.037	-0.37		4	A3	666,8107
		VES 629		1 44 00	+61 33.5	1	13.67		1.30		0.11		2		8107
		AAS6,117 # 52		1 44 00	+63 18.	1	11.62		0.69		0.50		3	A0	171
10780	+63 00238	HR 511	★ A	1 44 06	+63 36.4	8	5.63	.011	0.80	.010	0.40	.009	25	K0 V	1,15,22,1197,1355*
	+61 00326			1 44 08	+62 22.9	2	9.10	.018	1.13	.022	0.83		5	K0	171,1119
10934	-51 00419	HR 519		1 44 08	-51 03.9	6	5.48	.010	1.61	.012	1.95	.005	19	M2 III	15,678,1075,2035,3055,8015
		NGC 663 - 224		1 44 11	+61 03.3	1	12.42		0.70		-0.19		2		8107
		MWC 17	V	1 44 12	+60 27.	1	11.66		0.42		-0.20		1		1591
10939	-54 00377	HR 520		1 44 12	-53 46.3	3	5.04	.006	0.04	.006	0.06		10	A1 V	15,1637,2012
		Cr 463 - 50		1 44 14	+71 41.2	1	10.95		0.30		0.03		2		428
	-30 00607	G 274 - 92		1 44 16	-29 50.5	2	10.12	.000	0.62	.015	0.05	.005	2	G2	79,1696
		SK 194		1 44 17	-74 46.5	3	11.74	.012	0.02	.003	-0.54	.005	9	B9 Ia	573,733,1412
		LB 3227		1 44 18	-66 05.	1	13.08		0.10		-0.65		1		832
10909	-24 00751	UV For		1 44 21	-24 15.9	1	8.05		0.98		0.64		1	K0 IV	1641
11068	-74 00126			1 44 27	-73 48.2	1	7.64		0.33		0.06		3	F0 IV/V	727
10894	+10 00241			1 44 30	+10 35.7	2	7.04	.004	-0.02	.004	-0.09	.010	20	B9	696,1733
		Cr 463 - 9		1 44 31	+71 28.3	1	10.63		0.26		0.03		2		428
10948	-40 00445	IDS01424S4027	AB	1 44 32	-40 11.7	2	8.48	.088	0.25	.010	0.10		6	A8 V	2012,8100
10842	+61 00327			1 44 38	+61 35.1	2	8.84	.005	0.18	.023	0.12		7	B9 V	171,1119
		Cr 463 - 51		1 44 38	+71 36.7	1	11.82		0.30		0.30		2		428
		LF 5 # 149		1 44 40	+60 26.	1	10.37		0.36		-0.41		2	B2 IV	367
10874	+45 00447	HR 518		1 44 43	+45 58.9	2	6.31	.014	0.44	.003	0.00		4	F6 V	71,272
		Cr 463 - 52		1 44 43	+71 36.2	1	11.00		0.68		0.29		2		428
		AAS6,117 # 55		1 44 48	+63 23.	1	11.51		0.56		0.34		4		171
10881	+46 00449			1 44 50	+46 55.7	1	8.25		0.25		0.16		2	A2	1601
		LF 5 # 138		1 44 50	+58 37.	1	11.16		0.25		-0.52		1	B2 V	367
10820	+70 00133	Cr 463 - 10		1 44 51	+71 19.7	1	8.77		0.21		-0.04		2	A0	428

Table 1   113

HD	DM	Other Id	N	Rem	α_{1950}	δ_{1950}	S	V	σ_V	B–V	σ_{B-V}	U–B	σ_{U-B}	n	Spectrum	References
		Cr 463 - 53			1 44 51	+71 36.5	1	11.53		0.48		0.25		2		428
		LS I +61 238			1 44 55	+61 00.9	2	10.85	.010	0.57	.000	-0.26	.040	2	B2 Ve	367,666
10871	+59 00324	IDS01416N5957	AB		1 45 02	+60 11.6	1	8.24		0.35				2	A9 V	1025
	+71 00106	Cr 463 - 6			1 45 02	+71 30.0	1	9.34		0.23		0.15		2	B8	428
10955	-0 00274				1 45 02	-00 05.6	1	7.78		1.05		0.88		3	G5	1657
	+60 00352	NGC 663 - 625			1 45 08	+60 39.0	1	9.95		0.47				2	A9	1119
10967	-3 00262	G 271 - 154			1 45 08	-03 29.3	1	8.21		0.51		-0.02		1	F8	1658
		LS I +60 194			1 45 10	+60 16.8	1	11.55		0.33		-0.31		1		666
11022	-42 00633	HR 524			1 45 10	-42 00.6	2	6.18	.002	1.54	.000	1.91		6	K5/M0 III	58,2035
	+70 00134	Cr 463 - 8			1 45 11	+71 26.8	1	9.83		1.71		1.86		2		428
10898	+57 00399	LS I +58 049			1 45 13	+58 12.6	1	7.40		0.35		-0.52		2	B2 Ib	1012
10892	+61 00329	IDS01418N6120	A		1 45 13	+61 35.5	1	8.57		0.34				3	A4 V	1119
10892	+61 00330	IDS01418N6120	B		1 45 13	+61 35.5	1	8.94		0.93				3		1119
		G 35 - 13			1 45 14	+25 17.4	1	11.41		1.31		1.13		1	K7	333,1620
	-13 00321	G 272 - 95			1 45 15	-13 01.7	1	9.76		1.13		1.11		1	K5 V	3072
10998	-21 00300				1 45 16	-21 05.6	1	6.55		1.39				4	K3 III	2012
		CS 29504 # 45			1 45 17	-34 22.3	1	13.92		0.16		0.17		1		1736
11020	-27 00605	G 274 - 94			1 45 20	-26 59.7	3	8.98	.010	0.80	.007	0.40	.034	7	K0 V	79,2012,3008
		VES 631			1 45 22	+60 06.6	1	11.78		0.53		-0.12		2		690
	+71 00107	Cr 463 - 7			1 45 22	+71 29.5	1	9.58		0.24		-0.09		2	B8	428
	-75 00121				1 45 22	-75 02.9	1	10.31		0.48		0.02		4	F8	727
10897	+59 00325				1 45 23	+60 11.9	2	9.60	.000	0.26	.079			3	A0 V	851,1119
		Cr 463 - 76			1 45 23	+71 15.5	1	9.95		0.52		-0.05		2		428
		LB 3229			1 45 24	-51 48.	1	13.55		-0.25		-1.20		1		832
10982	+16 00203	HR 522			1 45 28	+16 42.4	3	5.86	.004	-0.04	.005	-0.14	.008	7	B9.5V	985,1049,1716
		G 173 - 22			1 45 32	+54 47.1	1	14.05		1.20		1.03		1		1658
		WLS 130 80 # 9			1 45 33	+80 16.6	1	10.69		0.27		0.13		2		1375
10942	+53 00391				1 45 35	+53 31.8	1	9.23		0.00		-0.32		2	A0	401
		Cr 463 - 77			1 45 36	+71 15.2	1	11.72		0.30		0.30		2		428
11050	-37 00687	HR 528			1 45 37	-37 24.5	4	6.32	.006	1.02	.005			18	K0 III	15,1075,2013,2029
10975	+37 00372	HR 521			1 45 41	+37 42.3	3	5.94	.008	0.97	.008	0.71	.010	5	K0 III	15,1003,1355
11038	-22 00297				1 45 43	-22 28.4	3	9.46	.019	0.54	.005	-0.03	.010	6	G1 V	742,1097,1594
		G 72 - 37			1 45 45	+32 51.0	1	14.45		1.59				4		538
	+59 00327				1 45 47	+59 49.2	1	10.40		0.71				1	G2 III	851
11007	+31 00316	HR 523			1 45 49	+32 26.3	2	5.78	.025	0.56	.020	0.00	.030	4	F8 V	254,3037
	+61 00332				1 45 49	+61 41.1	1	9.90		0.31				4	B6 V	1119
11037	+2 00270	HR 527			1 45 50	+03 26.2	3	5.91	.005	0.97	.000	0.74	.011	10	G9 III	15,1256,1355
		VES 632			1 45 51	+62 54.3	1	12.32		0.62		-0.15		2		8107
		Cr 463 - 54			1 45 51	+71 34.0	1	10.31		0.29		-0.09		2		428
232536	+53 00392				1 45 54	+53 55.1	1	10.14		-0.12		-0.82		2	F0	401
		LS I +63 167			1 45 54	+63 58.6	1	12.35		0.32		-0.48		2		180
11234	-75 00123				1 45 54	-74 59.4	1	7.34		0.82		0.38		3	G8 IV	727
	+59 00328				1 45 55	+60 19.0	1	9.82		0.17				2	B5 V	1119
		VES 633			1 45 58	+57 36.4	1	10.82		0.13		-0.31		2		8107
10964	+58 00305				1 45 58	+58 31.2	1	8.45		0.09				3	A0	70
	-74 00130				1 45 59	-73 49.9	1	10.46		0.60		0.04		4		727
	+63 00247	LS I +64 080			1 46 01	+64 04.7	3	9.74	.017	0.32	.013	-0.65	.021	8	B0 III	171,180,1012
	+60 00356				1 46 03	+60 29.8	2	10.55	.009	0.29	.018	0.19		8	B9 V	171,1119
11073	-21 00306				1 46 03	-20 53.5	1	8.56		0.36				4	F0 V	2012
	+44 00375				1 46 04	+45 02.5	1	10.19		0.13		0.11		2		1375
11074	-22 00299				1 46 05	-22 28.2	1	7.98		0.96		0.59		2	G8/K0III/IV	3008
	+60 00355				1 46 06	+61 16.4	1	9.75		0.36				2	A0 V	1119
		AAS6,117 # 58			1 46 06	+63 18.	1	11.80		0.43		0.18		3	A0	171
10980	+58 00306				1 46 08	+59 13.7	1	9.06		0.12				3	A0	70
10972	+60 00357				1 46 08	+60 46.5	1	8.49		0.10				2	B9 V	1119
		LS I +63 168			1 46 08	+63 00.7	1	11.73		0.44		-0.39		1		180
11031	+47 00508	HR 526	★ AB		1 46 09	+47 38.9	1	5.82		0.29		0.07		1	A3 V	3016
11031	+47 00508	IDS01430N4724	C		1 46 09	+47 38.9	1	9.41		0.64		0.06		1	A2	3016
		G 3 - 20			1 46 10	+17 06.2	1	12.49		1.34		1.10		1		333,1620
	+60 00358				1 46 10	+60 48.1	1	9.33		0.37				4	B5	1119
236885	+56 00357	IDS01429N5615	AB		1 46 11	+56 30.0	1	8.83		0.57		0.05		3	F8	3016
236885	+56 00357	IDS01429N5615	C		1 46 11	+56 30.0	1	9.93		0.56		0.03		3		3016
11112	-42 00638				1 46 12	-41 44.7	2	7.13	.005	0.64	.005	0.20		5	G3 V	258,2012
	+0 00293				1 46 13	+00 47.8	1	9.78		0.49		0.00		2	F4	271
	+61 00333				1 46 13	+61 32.7	1	10.00		0.27				3	B9	1119
11027	+41 00353				1 46 14	+41 44.5	1	8.68		0.35		0.04		4	F2 V	1733
11100	-26 00642				1 46 17	-26 30.1	1	7.09		0.28				4	F0 V	2012
		S052 # 3			1 46 17	-70 04.9	1	11.19		0.55		0.05		2		1730
		LS I +61 240			1 46 19	+61 13.7	1	11.09		0.54				1		666
		CS 22180 # 27			1 46 19	-10 03.3	1	13.69		0.16		0.10		1		1736
11221	-70 00093				1 46 19	-70 07.9	1	8.09		1.11		0.99		2	K0 III	1730
	+83 00038				1 46 20	+84 05.0	1	8.90		0.52		0.04		3	F8	1733
		LS I +63 169			1 46 22	+63 48.5	1	12.45		1.03		-0.28		1		180
11107	-24 00772				1 46 23	-24 07.1	1	7.90		0.47		0.09		54	F3/5 IV	978
		AAS6,117 # 59			1 46 24	+61 53.	1	12.87		0.74		0.01		3		171
		PHL 3802			1 46 24	-26 52.	1	12.31		-0.03		-0.95		2	d:B	3060
	+12 00237				1 46 28	+13 18.3	2	10.42	.029	0.05	.005	-0.02	.015	3	A0	1026,1298
11004	+61 00334				1 46 28	+61 54.2	1	8.07		0.87				2	F7 V	1025
		AJ89,1229 # 80			1 46 29	+06 09.2	1	12.78		1.49				2		1532
		G 133 - 45			1 46 31	+43 31.4	2	11.79	.015	0.54	.000	-0.13	.024	3	sdG0	1620,1658

HD	DM	Other Id	N Rem	α_{1950}	δ_{1950}	S	V	σ_V	B–V	σ_{B-V}	U–B	σ_{U-B}	n	Spectrum	References
	−21 00311	TW Cet	⋆ AB	1 46 32	−21 08.2	1	10.43		0.73		0.16		6	G5	3064
	+70 00135	Cr 463 - 79		1 46 37	+71 17.3	1	9.73		0.57		0.09		2	F8	428
11079	+25 00305			1 46 38	+26 13.5	1	6.89		-0.08		-0.51		1	B8 V	1716
		LF 5 # 148		1 46 40	+60 46.	1	10.89		0.49		-0.34		1	B1 Vn	367
		AAS6,117 # 60		1 46 42	+61 56.	1	12.53		0.82		0.68		4		171
	+61 00335			1 46 44	+61 56.2	1	9.91		0.51				2	A1 V	1119
11131	−11 00351	IDS01447S1111	B	1 46 56	−10 57.0	1	6.75		0.61		0.12		2	G0	3077
		LP 828 - 28		1 46 57	−25 44.2	1	13.33		1.02		0.82		1		1696
11301	−70 00094			1 46 60	−69 58.7	1	8.33		1.14		1.11		3	K2 III	1730
		LS I +61 242		1 47 02	+61 25.1	1	11.50		0.41				1		666
11183	−31 00753	HR 532		1 47 04	−31 19.2	1	6.36		1.22				4	K2/3 III	2007
		VES 635		1 47 05	+63 27.2	1	13.62		0.91		0.36		1		690
	+58 00310	LS I +58 050		1 47 06	+59 00.0	1	10.17		0.25		-0.53		2	B1 V	1012
	+61 00336			1 47 06	+61 57.6	1	10.56		0.34		0.22		5		171
11105	+37 00379			1 47 08	+38 18.8	1	8.62		0.37		-0.04		2	F0	1601
11171	−11 00352	HR 531	⋆ A	1 47 08	−10 56.0	12	4.66	.014	0.33	.011	0.04	.022	48	F3 III	3,15,254,1007,1013*
	−20 00345	VY Cet		1 47 10	−19 52.6	1	11.02		0.69		0.20		1		1612
		VES 636		1 47 13	+61 00.6	1	12.50		1.10		-0.04		2		8107
11094	+53 00398	TT Per		1 47 14	+53 29.7	1	8.21		1.57		1.07		1	M5 III	3001
		LP 468 - 232		1 47 18	+14 17.0	1	9.74		0.75		0.30		2		1696
		AAS6,117 # 62		1 47 18	+62 12.	1	10.11		1.70		1.95		2	M0	171
11154	+21 00243	HR 530	⋆ AB	1 47 22	+22 01.7	5	5.85	.014	0.74	.008	0.49	.011	14	K1 III+A6 V	15,938,1007,8015,8023
	+60 00361	IDS01439N6035	A	1 47 22	+60 50.1	1	10.11		0.48				2	F5 V	1119
		GD 421		1 47 22	+67 24.6	1	14.42		-0.25		-1.05		1	DAwk	782
11168	+13 00286	IDS01447N1351	A	1 47 24	+14 06.2	1	8.49		0.63		0.16		2	G0	1648
236890	+55 00423	G 173 - 25		1 47 25	+55 40.4	1	8.96		0.78		0.42		1	K0	1658
	+60 00362	LS I +61 243		1 47 26	+61 06.2	1	9.58		0.64		-0.29		2	B2 II-III	1012
	+59 00333	LS I +59 093	⋆ A	1 47 35	+59 45.7	1	10.56		0.31		-0.49		1	B2 IV	367
11092	+64 00243	IDS01440N6422	A	1 47 38	+64 36.5	3	6.56	.011	2.07	.032	2.38	.084	10	K5 Ib-IIa	138,1084,1355
		CS 22180 # 26		1 47 39	−09 52.4	1	12.26		0.04		0.02		1		1736
11262	−39 00553	HR 535	⋆ A	1 47 39	−38 39.3	3	6.37	.009	0.51	.011	-0.05		9	F6 V	15,2012,3037
		VV Cas		1 47 41	+59 38.5	2	10.28	.004	0.94	.003	0.62		2	F6	851,1399
11092	+64 00244	IDS01440N6422	B	1 47 41	+64 36.9	1	9.17		0.23		0.07		3	A0 III	1084
11249	−29 00602			1 47 41	−29 17.3	1	7.18		1.60		1.92		3	K5 III	1657
	+59 00334	LS I +60 195		1 47 42	+60 11.4	2	10.38	.015	0.26	.045	-0.62	.000	2	B0 V	367,666
		G 94 - 17		1 47 44	+18 02.9	1	10.79		1.38		1.27		1	M0	333,1620
11151	+51 00416	HR 529		1 47 44	+51 41.3	1	5.92		0.42		0.00		2	F5 V	254
	+60 00365	LS I +60 196		1 47 48	+60 49.8	1	10.21		0.71		0.14		1	B6 Ib	666
		LS I +63 170		1 47 49	+63 09.1	1	10.94		0.40		-0.50		1		180
11126	+59 00336	IDS01444N5951	AB	1 47 52	+60 06.4	2	7.98	.014	0.13	.000			4	B8 V	70,1119
11276	−25 00730			1 47 52	−24 47.2	1	8.49		1.04		0.82		41	K0 III	978
11188	+46 00463			1 47 55	+47 10.4	1	7.28		-0.02		-0.29		2	B8	1375
11202	+35 00350			1 47 57	+35 35.9	2	7.11	.005	0.57	.000	0.12	.005	6	F8 III	1501,1601
11259	−6 00348	G 271 - 161		1 47 57	−06 01.9	1	9.60		0.64		0.14		1	G0	1658
		AAS6,117 # 63		1 48 00	+61 52.	1	10.77		0.27		-0.19		3	B8	171
11288	−24 00781	IDS01457S2416	AB	1 48 00	−24 01.1	1	8.44		0.88		0.48		40	G8 IV	978
8890	+88 00007	IDS01226N8846	B	1 48 01	+89 01.5	1	8.20		0.49		0.16		1		1397
	−10 00388	G 271 - 162		1 48 02	−09 36.1	3	10.35	.014	0.43	.014	-0.19	.000	11	F3	742,1594,1658
		IC 166 - 61		1 48 04	+61 35.2	1	12.13		0.73		0.18		4		85
11163	+60 00366			1 48 05	+60 40.7	2	8.75	.027	0.42	.004			5	A5 III	70,1119
	+37 00386	NGC 752 - 437	⋆ AB	1 48 06	+37 31.7	1	8.54		0.99		0.67		1	K0	4002
11187	+54 00393			1 48 09	+54 40.7	1	7.93		0.26		0.19		4	A0p	1202
11162	+61 00338	IC 166 - 60		1 48 09	+61 35.0	2	9.19	.018	0.34	.012	0.23		std	B9 V	85,1119
11245	+25 00311	IDS01454N2521	A	1 48 10	+25 36.3	1	8.08		0.19		0.12		2	A0	105
		LF 5 # 139		1 48 10	+58 51.	1	10.70		0.25		-0.60		1	B2 V	367
		LS I +62 187		1 48 12	+62 55.5	1	11.67		0.31		-0.49		2		180
	+62 00313			1 48 12	+63 12.2	1	9.96		0.25				3	B8 V	1119
		SK 195		1 48 12	−74 11.	2	13.27	.005	-0.02	.023	-0.21	.014	4		733,835
11604	−80 00035	HR 550	⋆ A	1 48 12	−80 25.4	3	6.06	.007	0.34	.004	0.05	.015	13	F0 IV	15,1075,2038
11257	+10 00252	HR 534		1 48 13	+10 47.8	5	5.95	.014	0.30	.003	-0.05	.026	21	F2 Vw	15,254,1007,1013,8015
		G 71 - 38		1 48 15	+03 14.9	1	12.54		1.41		1.11		1		1696
		G 71 - 45		1 48 18	+03 17.2	1	14.99		1.61				4		1663
11394	−65 00152			1 48 18	−64 38.0	1	7.40		1.26				4	K1 III	2012
11236	+38 00365			1 48 19	+38 35.3	1	7.91		-0.03		-0.16		3	A0	833
11332	−48 00487	HR 537		1 48 20	−48 03.9	5	6.14	.008	1.00	.011	0.79		19	K0 III	15,1075,2018,2029,3077
11161	+65 00209	G 244 - 38	⋆ A	1 48 24	+66 12.2	1	9.03		0.62		0.08		1	G0	1658
	−50 00509			1 48 24	−49 49.7	1	11.89		0.43		-0.20		3		1696
		SK 196		1 48 26	−74 15.5	3	12.07	.029	-0.01	.009	-0.58	.009	8	B8 Ia	573,733,1412
11214	+57 00407	IDS01451N5808	AB	1 48 27	+58 24.0	1	9.04		0.14				2	A0	70
		G 3 - 24		1 48 30	+09 59.1	1	12.69		1.42		1.22		1		333,1620
		AAS6,117 # 64		1 48 30	+63 13.	1	11.93		0.62		0.31		4	B9	171
11213	+59 00337			1 48 31	+60 12.9	2	8.79	.014	0.14	.005			6	A1 V	70,1119
		IC 166 - 65		1 48 34	+61 36.8	1	12.12		2.10				1		85
	+62 00314			1 48 34	+63 13.1	2	9.91	.009	0.25	.014	-0.12		7	B8 V	171,1119
	+54 00395	LS I +55 020		1 48 35	+55 13.4	1	9.91		0.10		-0.79		2	B0 IV:p	1012
11241	+54 00396	HR 533, V436 Per		1 48 41	+54 54.0	4	5.51	.010	-0.18	.008	-0.83	.009	13	B1.5V	154,1203,1223,1234
		IC 166 - 68		1 48 41	+61 39.8	1	11.74		1.17		0.89		3		85
	+70 00138	Cr 463 - 80		1 48 41	+71 15.7	1	10.01		1.45		2.07		2		428
11251	+50 00378			1 48 42	+50 57.1	1	9.09		-0.03		-0.42		2	B9	1566
11331	−15 00326			1 48 44	−14 38.5	1	9.27		0.11		0.14		4	A2 V	152

Table 1   115

HD	DM	Other Id	N Rem	α_1950	δ_1950	S	V	σ_V	B–V	σ_B–V	U–B	σ_U–B	n	Spectrum	References
	+2 00280			1 48 45	+03 05.1	1	11.13		1.55				1	M1	1632
11309	+17 00273			1 48 48	+18 15.7	1	8.30		0.55		0.13		2	G0	1648
		IC 166 - 69		1 48 48	+61 40.1	1	12.17		0.56		0.35		3		85
		IC 166 - 70	V	1 48 49	+61 39.7	1	13.73		0.73		0.52		1		85
8890	+88 00008	HR 424, α UMi	⋆ A	1 48 49	+89 01.7	4	2.01	.018	0.60	.010	0.38	.008	15	F7:Ib-II	15,1363,1462,8015
236894	+57 00409	LS I +58 051		1 48 50	+58 11.3	3	9.37	.005	0.19	.000	-0.76	.005	5	O8 V	367,1011,1012
		LS I +60 198		1 48 50	+60 43.6	1	10.54		1.26		0.68		1	B8 Iab	1215
	+61 00339			1 48 50	+61 48.2	2	9.91	.004	0.25	.013	0.08		8	B7 III	171,1119
		IC 166 - 71		1 48 51	+61 32.8	1	12.90		0.79		0.49		1		85
	+70 00139	Cr 463 - 81		1 48 51	+71 15.6	1	9.78		1.85		2.26		2		428
		VES 637		1 48 54	+60 48.5	1	12.11		1.04		-0.14		1		8107
		BPM 16628		1 48 54	-58 48.	1	14.65		0.56		-0.08		1		3065
		LP 588 - 69		1 48 56	+02 27.9	1	11.22		0.49		-0.21		2		1696
		GD 279		1 48 56	+46 45.2	1	12.44		0.17		-0.65		5	DAsp	308
	+60 00368	LS I +60 199		1 48 57	+60 35.2	2	10.64	.015	0.59	.015	-0.35	.000	3	B1:III:	367,1012
11370	-29 00613			1 48 57	-28 49.6	1	9.58		0.20		0.10		4	A2/3 IV	152
11291	+50 00379	HR 536		1 48 58	+50 32.8	2	5.78		-0.08	.009	-0.32	.009	3	B9p HgMn	379,1079
11413	-50 00514	HR 541, BD Phe		1 48 58	-50 27.2	2	5.93	.005	0.14	.000			7	A1 V	15,2012
11353	-11 00359	HR 539	⋆ A	1 48 59	-10 34.9	10	3.73	.010	1.14	.007	1.07	.009	37	K2 III	3,15,58,1075,1425*
11379	-31 00767			1 48 59	-31 09.0	1	7.20		0.47		0.00		1	F5 V	79
		IC 166 - 1		1 49 00	+61 35.	1	16.64		1.73				1		85
		IC 166 - 2		1 49 00	+61 35.	1	15.65		0.66		0.44		1		85
		IC 166 - 5		1 49 00	+61 35.	1	16.69		0.80				1		85
		IC 166 - 35		1 49 00	+61 35.	1	15.98		0.87		0.46		1		85
		IC 166 - 52		1 49 00	+61 35.	1	16.15		0.87		0.04		1		85
		IC 166 - 62		1 49 00	+61 35.	1	16.24		1.20				2		85
		IC 166 - 63		1 49 00	+61 35.	1	14.03		2.09				1		85
		IC 166 - 64		1 49 00	+61 35.	1	14.83		0.82		0.52		2		85
		IC 166 - 72		1 49 00	+61 35.	1	14.73		1.56		1.24		1		85
		IC 166 - 73		1 49 00	+61 35.	1	15.38		0.74		0.39		2		85
		IC 166 - 74		1 49 00	+61 35.	1	16.40		1.67				2		85
	-37 00713			1 49 00	-37 36.9	2	10.52	.005	0.72	.015	0.17	.005	3		79,1696
		G 159 - 4		1 49 01	-02 25.6	1	15.28		1.17		0.65		2		3062
		G 219 - 27		1 49 04	+57 03.4	1	13.44		0.92		0.62		2	F0	1658
		G 245 - 34		1 49 04	+73 43.2	1	10.35		1.13		1.05		3	K3	7010
11378	-19 00326			1 49 06	-19 31.2	1	9.01		0.16		0.13		4	A4 III	152
232552	+54 00398	LS I +55 021		1 49 08	+55 05.1	1	8.03		0.37		-0.74		2	B0pe	1012
		IC 166 - 66		1 49 12	+61 38.4	1	13.07		0.44		0.09		1		85
11397	-17 00332	G 272 - 107		1 49 16	-16 33.6	4	8.95	.015	0.69	.008	0.11	.005	8	G6 IV/V	79,1003,1097,3077
11422	-42 00650			1 49 16	-42 10.5	1	9.98		0.09		-0.01		4	A1 V	152
11423	-43 00569			1 49 16	-43 37.0	1	11.46		0.85		0.54		1	F2/F3V	1696
	-10 00389			1 49 17	-10 02.1	1	9.81		0.93		0.56		2	K5	1375
		G 271 - 168		1 49 19	-11 02.5	1	11.80		1.51				1		1705
236896	+59 00338			1 49 22	+60 13.2	1	9.90		0.33				3	A0 Ib	1119
11569	-72 00120			1 49 24	-72 11.8	2	9.02	.000	0.37	.000	-0.15		4	Fw pec	742,1594
11348	+36 00332			1 49 25	+37 11.1	1	6.89		0.43		-0.01		2	F8	1601
	+58 00315	LS I +59 094		1 49 27	+59 19.7	1	9.99		0.19		-0.60		1	B1 V	367
		IC 166 - 67		1 49 28	+61 38.7	1	11.52		1.57		1.50		3		85
		G 3 - 26		1 49 29	+17 41.1	1	11.37		1.29		1.27		1	K6-7	1658
	+60 00369	LS I +60 200		1 49 29	+60 36.5	1	9.67		0.67		-0.26		1	B2 II	367
		LS I +61 245		1 49 29	+61 52.7	2	10.87	.017	0.39	.009	-0.44	.026	6		180,367
	+61 00341	LS I +61 244		1 49 31	+61 54.8	1	11.03		0.42		-0.46		1	B2 V	367
11317	+58 00316			1 49 36	+58 51.4	1	8.84		0.11				3	A2	70
		AAS6,117 # 67		1 49 36	+62 47.	1	11.11		0.54		0.38		3	A2	171
		AAS6,117 # 68		1 49 36	+63 24.	1	11.84		0.78		0.59		4	A0	171
	-15 00330	G 272 - 110		1 49 36	-15 09.2	1	10.39		0.83		0.54		2		1696
		SK 197		1 49 36	-74 21.	2	13.38	.009	-0.02	.005	-0.96	.033	4		733,835
	+35 00355	G 72 - 39		1 49 37	+36 06.2	2	10.09	.002	0.93	.008	0.64	.015	3	K2	1620,7010
11335	+50 00381	HR 538		1 49 38	+51 13.7	1	6.30		0.05		0.08		2	A3 V	1733
	+64 00250			1 49 38	+64 49.6	1	10.47		0.24		0.11		4		171
		G 159 - 5		1 49 38	-03 03.0	2	10.91	.030	0.76	.000	0.30	.005	2		1658,3062
		Steph 205		1 49 39	-21 17.4	1	11.72		1.20		1.17		1	K7	1746
	+61 00342	LS I +62 188	⋆ A	1 49 42	+62 18.0	3	9.53	.015	0.44	.000	-0.53	.000	13	B0.5II	171,180,1012
	+0 00301			1 49 44	+00 38.2	3	10.70	.007	0.44	.006	0.00	.000	18		281,975,6004
		VES 638		1 49 45	+55 19.9	1	10.11		0.40		0.19		2		690
	+63 00253	LS I +63 172		1 49 47	+63 56.6	2	9.35	.024	0.43	.014	-0.54	.038	4	B0 III	180,1012
	-4 00290	G 159 - 6		1 49 48	-03 41.2	1	10.48		0.95		0.80		2		3062
	-0 00286			1 49 50	+00 17.7	1	11.07		0.63		0.14		8	G2	281
	-0 00287			1 49 53	+00 17.5	1	11.24		0.65		0.07		9	G2	281
	+54 00404	LS I +54 003		1 49 53	+54 52.7	1	9.77		0.21		-0.62		2	B1 III	1012
	+62 00316	LS I +63 173	⋆ AB	1 49 56	+63 10.8	1	10.88		0.28		-0.56		2		180
		LP 884 - 50		1 49 58	-29 05.0	1	12.19		0.77		0.31		1		1696
11481	-33 00647			1 49 58	-32 47.2	1	7.96		0.10				4	A2/3 V	2012
11490	-36 00707			1 49 59	-36 29.3	1	9.22		0.30				4	A7 IV	2012
11480	-25 00741			1 50 01	-25 17.5	1	8.57		0.30				4	A3mA8-F0	2012
11512	-47 00567			1 50 01	-46 53.3	1	8.04		0.95		0.69		5	G8 III	1673
11360	+59 00342			1 50 02	+60 21.2	1	9.17		0.32				2	A5 V	1119
11457	+0 00302			1 50 03	+01 00.7	1	8.52	.001	0.38	.005	-0.03	.001	22	A5	281,1509,5006,6004
		HA 93 # 555		1 50 06	+00 57.9	1	10.69		0.79		0.25		2	G2 p	281
11374	+60 00372			1 50 07	+60 44.7	3	8.23	.007	0.42	.018	-0.04		6	F4 V	70,171,1025

HD	DM	Other Id	N Rem	α_{1950}	δ_{1950}	S	V	σ_V	B–V	σ_{B-V}	U–B	σ_{U-B}	n	Spectrum	References
	+60 00373	LS I +60 201		1 50 07	+60 59.5	1	9.70		0.86		0.71		4	A2 II	171
		VES 640		1 50 08	+60 39.8	1	11.93		0.67		0.19		2		690
11150	+80 00058			1 50 12	+80 39.9	1	7.06		1.22		1.07		2	K0	1502
11443	+28 00312	HR 544	⋆ A	1 50 13	+29 20.2	7	3.42	.012	0.49	.009	0.04	.018	24	F6 IV	15,254,1118,3053,6001*
		SK 189		1 50 13	−73 04.2	3	12.46	.028	0.04	.023	-0.59	.005	8		416,573,733
11551	−48 00497			1 50 15	−47 45.5	1	8.33		0.10				4	A1 V	2012
11428	+40 00394	HR 543	⋆ A	1 50 17	+40 29.0	1	5.40		1.32		1.41		2	gK1	1355
	−27 00646	IDS01480S2704	A	1 50 17	−26 49.0	1	9.84		0.49				4	F2	2013
	+63 00255			1 50 18	+64 24.5	1	9.37		1.94				4		369
	−27 00647	IDS01480S2704	B	1 50 18	−26 50.5	1	9.92		0.56				4	F5	2013
		SK 198		1 50 18	−74 10.	3	12.67	.010	0.15	.008	0.00	.037	8		733,835,1412
	+59 00343	LS I +60 202		1 50 19	+60 16.8	1	10.62		0.49		-0.29		1	B2 Vnne	367
	+61 00344			1 50 20	+61 45.4	2	9.15	.018	0.41	.023	0.12		7	A7 V	171,1119
11373	+65 00210			1 50 22	+65 55.9	1	8.48		1.01		0.97		2	G5	1502
		HA 93 # 12		1 50 22	−00 00.9	1	11.18		0.42		0.01		2	F4	281
11507	−23 00693	G 272 - 114		1 50 25	−22 40.9	3	8.91	.020	1.43	.012	1.24	.020	10	K5/M0 V	2034,3079,7008
		GD 19		1 50 26	+09 34.8	1	14.22		0.54		-0.19		1		3060
	+61 00345	LS I +61 246		1 50 28	+61 56.7	1	11.16		0.34		-0.61		3		180
11522	−17 00336	HR 547, BK Cet		1 50 28	−17 10.5	2	5.79	.005	0.28	.005			7	F0 IIIn	15,2027
11408	+54 00408	HR 540		1 50 29	+55 21.1	3	6.45	.007	0.19	.009	0.10	.010	9	A5 m	595,1716,8071
		Steph 207		1 50 29	−07 58.6	1	11.15		1.55		1.91		1	M0	1746
		G 71 -B5	A	1 50 31	+08 56.7	1	14.25		1.43		1.09		2		3016
		G 71 -B5	B	1 50 31	+08 56.7	1	17.16		0.21		-0.81		1		3060
	+58 00317			1 50 32	+59 18.3	1	9.75		1.04				2	K5 V	70
11401	+59 00344		V	1 50 32	+59 54.5	1	7.89		1.99				3	M3 III	369
	+60 00375	WX Cas		1 50 33	+60 51.9	1	9.90		2.53				3	M2 Iab-Ib	369
11505	−2 00311	G 71 - 40		1 50 34	−01 34.1	7	7.43	.011	0.64	.004	0.07	.016	10	G0	258,516,742,1620,1658*
		G 159 - 8		1 50 35	−03 04.0	1	12.31		0.97		0.75		1		3062
	−13 00342	IDS01481S1252	AB	1 50 35	−12 37.0	1	9.34		0.31		0.08		1	F0	966
		HA 93 # 484		1 50 36	+00 45.6	1	12.26		0.50		-0.03		13	G3	281
		CS 22171 # 1		1 50 36	−12 35.4	1	13.41		0.51		-0.07		1		1736
	+61 00347	LS I +62 189		1 50 38	+62 18.5	1	10.66		0.53		-0.26		3		180
11474	+39 00427			1 50 39	+39 38.2	1	7.73		0.54		0.05		2	F8	1601
	+59 00345			1 50 39	+60 08.1	1	10.07		0.26				2	A3 III	1119
	+62 00319		AB	1 50 39	+62 29.4	1	9.32		0.42				2	F5 III	1119
		VES 642		1 50 42	+57 36.7	1	11.87		0.27		-0.07		2		8107
11451	+49 00486			1 50 43	+49 41.8	1	6.85		1.35		1.55		1	K3 III	1716
11532	−0 00288	HA 93 # 101	A	1 50 44	+00 07.6	2	9.73	.007	0.65	.000	0.14		11	G2p	281,6004
11425	+59 00346	LS I +59 095		1 50 45	+59 44.2	1	9.22		0.74				2	Am	1119
11562	−18 00322			1 50 45	−17 38.2	1	9.27		0.61		0.11		2	G1 V	1375
11415	+62 00320	HR 542		1 50 46	+63 25.5	9	3.37	.009	-0.16	.007	-0.59	.014	23	B3 III	15,154,369,1118,1193*
11573	−33 00651			1 50 46	−33 01.4	1	8.03		0.24				4	A5 III/IV	2012
11532	−0 00288	HA 93 # 103	B	1 50 47	+00 08.5	6	8.84	.041	1.17	.005	1.15	.004	50	G5 V	281,975,989,1509,5006,6004
11502	+18 00243	HR 546, γ Ari	⋆ AB	1 50 47	+19 03.0	4	3.88	.000	-0.04	.008	-0.12	.025	8	A1p Si	15,1049,1363,8015
		WLS 140-20 # 8		1 50 47	−20 06.8	1	12.34		0.84		0.39		2		1375
		VES 643		1 50 48	+57 04.1	1	13.00		0.44		0.18		1		690
11582	−34 00722	G 274 - 112		1 50 48	−34 32.2	2	9.57	.000	0.67	.014	0.02		4	G5/6 (V)w	79,1594
236904	+58 00320			1 50 50	+58 47.6	1	9.45		0.39				2	F2	70
11581	−33 00652	G 274 - 114		1 50 50	−33 33.5	1	8.20		0.53				4	F7 V	2012
		WLS 200 25 # 12		1 50 53	+26 57.7	1	11.09		0.76		0.24		2		1375
		G 219 - 28	AB	1 50 55	+57 09.0	1	10.33		0.79		0.42		2	K3	7010
		HA 93 # 488		1 50 57	+00 46.1	1	12.16		0.57		0.04		1	G2	281
11559	+2 00290	HR 549		1 50 58	+02 56.5	7	4.61	.012	0.94	.003	0.72	.010	23	K0 III	15,1075,1355,1363*
11463	+59 00348	LS I +59 097		1 51 04	+59 38.9	2	8.21	.027	0.20	.013			5	B9 IV	70,1119
11547	+25 00319			1 51 06	+25 31.9	1	7.49		0.39		0.08		2	F0	1375
		TZ Tuc		1 51 07	−70 09.4	1	9.27		1.37				2	M3Iabe	5007
		HA 93 # 395		1 51 09	+00 42.9	1	11.64		0.67		0.09		7	G2	281
		SK 199		1 51 09	−74 29.1	2	13.41	.037	-0.02	.033	-0.16	.005	4		733,835
11471	+61 00348			1 51 10	+61 44.8	3	8.75	.012	0.07	.025	-0.04		9	B9 V	70,171,1119
11608	−21 00333			1 51 10	−21 29.7	1	9.28		0.98		0.85		2	K1/2 V	1375
11472	+59 00350	IDS01478N5926	AB	1 51 11	+59 41.2	1	8.32	.004	0.46	.000			5	F3 Vp	70,1119
	−25 00745	G 274 - 117		1 51 12	−25 17.1	1	11.36		1.01		0.93		1		1696
	+19 00302			1 51 13	+20 28.1	1	9.99		-0.10		-0.70		1		963
236905	+59 00351			1 51 13	+59 35.3	1	8.54		1.19				2	K0 III	70
11643	−39 00573	HR 554		1 51 14	−38 50.4	2	6.10	.005	1.12	.000	1.18		7	K1 II	2032,8100
11592	+9 00235			1 51 19	+10 22.4	1	6.78		0.46		-0.05		2	F5 V	1003
11519	+53 00416			1 51 23	+53 38.8	1	7.96		0.96		0.62		2	G5	1566
		AAS6,117 # 74		1 51 24	+62 31.	1	10.89		1.39		1.14		3	K0	171
		VES 644		1 51 24	+66 52.2	1	12.11		1.19		0.46		1		690
11530	+50 00390			1 51 28	+50 51.9	1	9.32		0.05		-0.33		2	A2	1566
	−0 00290	IDS01489S0019	AB	1 51 29	−00 04.0	1	10.18		0.68		0.15		2	G3p	281
		VES 645	V	1 51 32	+62 10.8	1	12.51		1.11		-0.13		2		8107
		VES 646		1 51 32	+62 32.8	1	14.46		2.08		0.63		1		690
		G 244 - 43		1 51 33	+65 23.2	1	12.20		1.42		1.33		3	M0	7010
11544	+55 00437			1 51 35	+56 20.2	2	6.81	.005	1.16	.005	0.93		6	G2 Ib	1355,6009
		LS I +62 192		1 51 36	+62 56.9	1	11.02		0.32		-0.39		2		180
		G 71 - 41		1 51 38	+01 46.7	1	15.00		0.16		-0.59		1	DA	3062
11517	+59 00354	IDS01482N5929	AB	1 51 39	+59 43.7	2	8.12	.028	0.13	.005			4	A0 V	70,1119
11695	−46 00552	HR 555, ψ Phe		1 51 39	−46 32.8	7	4.40	.011	1.59	.007	1.71	.005	32	M4 III	15,1075,2013,2029*
	+0 00304			1 51 41	+00 31.2	2	11.99	.012	0.59	.004	0.05	.008	17	G3 V	271,281

Table 1   117

HD	DM	Other Id	N	Rem	α_{1950}	δ_{1950}	S	V	σ_V	B–V	σ_{B-V}	U–B	σ_{U-B}	n	Spectrum	References
	+37 00404	NGC 752 - 400		⋆A	1 51 41	+37 57.1	1	8.91		1.02		0.80		1	G8 III	4002
11543	+58 00325				1 51 42	+59 15.8	1	8.14		0.23				3	Am	70
11640	+8 00292				1 51 44	+08 32.1	2	6.58	.045	1.71	.010	2.01	.020	5	M1	985,1733
		HA 93 # 405			1 51 45	+00 40.6	1	12.19		0.50		-0.03		9	G0 p	281
11554	+57 00425	LS I +57 028			1 51 48	+57 39.2	1	9.55		0.27		-0.58		2	B1 Vpe	1012
		AAS6,117 # 75			1 51 48	+63 13.	1	12.37		0.47		0.18		3	A0	171
	-33 00658				1 51 48	-32 47.2	1	10.42		0.62		0.04		2		1696
11542	+62 00322				1 51 49	+63 24.7	1	8.43		0.16				3	A2	1118
		HA 93 # 503			1 51 52	+00 44.8	1	12.60		0.65		0.12		13	G5	281
11636	+20 00306	HR 553			1 51 52	+20 33.9	12	2.65	.013	0.14	.006	0.13	.030	59	A5 V	1,3,15,1004,1006,1034*
		HA 93 # 30			1 51 52	-00 04.6	1	11.47		0.55		0.00		2	G2	281
11613	+39 00434	HR 551			1 51 53	+40 27.5	2	6.24	.002	1.26	.008	1.38		5	K2 III	71,1501
		LS I +62 193			1 51 53	+62 58.8	1	12.07		0.33		-0.36		1		180
11676	-12 00350				1 51 54	-12 05.8	1	8.14		1.18		1.19		2	K1 IV	1375
11683	-16 00328	G 272 - 120		⋆A	1 51 56	-15 58.0	1	9.15		0.90		0.57		1	K2 V	79
		SK 201			1 51 57	-74 28.0	2	13.01	.000	0.07	.020	0.05	.015	5		733,835
	-23 00702	G 272 - 121			1 51 58	-23 03.9	1	11.26		0.73		0.26		2		1696
11624	+36 00346	HR 552			1 52 00	+36 53.0	1	6.26		1.17				2	K0	71
		PHL 8054			1 52 00	-07 00.0	4	12.41	.006	-0.01	.004	0.00	.014	82	A0	281,989,1026,1764
		HA 93 # 407			1 52 02	+00 39.1	2	11.96	.005	0.87	.008	0.58	.007	22	G3	281,1764
		HA 93 # 317			1 52 04	+00 28.3	5	11.55	.003	0.49	.007	-0.05	.010	71	F6	271,281,989,1729,1764
11529	+67 00169	HR 548			1 52 05	+68 26.4	3	4.97	.015	-0.09	.004	-0.42	.005	6	B8 III	1079,1118,3023
11673	+13 00296				1 52 06	+13 30.7	1	7.45		1.00		0.68		2	G5	1648
		SK 200			1 52 06	-74 46.	2	13.40	.014	-0.15	.009	-0.85	.019	4		733,835
		LP 769 - 2			1 52 09	-18 41.7	1	12.16		0.47		-0.22		2		1696
	+37 00407	NGC 752 - 1			1 52 14	+37 35.6	3	9.49	.014	0.96	.001	0.65	.008	4	K0 III	1100,3067,4002
	+59 00355				1 52 14	+59 59.9	1	9.76		0.27				2	B0.5V	1119
	+0 00307				1 52 16	+00 32.4	7	9.56	.015	0.45	.003	-0.03	.009	59	F5	271,281,975,989,1729*
11606	+58 00331	LS I +59 098			1 52 16	+59 01.7	2	7.02	.010	0.06	.010	-0.86		5	B2 Vne	1118,1212
	+37 00408	NGC 752 - 3			1 52 17	+37 35.8	1	9.55		0.99		0.72		1	K0 III	1100
		G 71 - 42			1 52 17	-00 15.8	1	13.85		1.47				1		3062
11710	-6 00361				1 52 18	-05 54.1	1	10.30		0.03		0.01		2		1026
11699	-7 00319	V743 Cen			1 52 18	-06 50.2	1	8.56		0.30				1	F5	1497
11753	-43 00583	HR 558			1 52 18	-42 44.5	6	5.11	.005	-0.06	.003	-0.14	.007	25	A3 V	15,200,1075,2012,3023,8015
11605	+61 00352				1 52 20	+62 07.1	2	7.58	.010	0.12	.010			6	B8 V	1118,1119
		LS I +62 194			1 52 21	+62 03.3	1	10.91		0.46		-0.38		3		180
11623	+55 00439				1 52 24	+55 41.6	1	8.06		1.68				3	K5	369
		HA 93 # 35			1 52 24	-00 01.1	1	11.75		0.53		0.02		2		281
		VES 647			1 52 26	+60 22.9	1	11.47		0.60		-0.19		2		8107
11708	-0 00292	IDS01499S0017	AB		1 52 26	-00 02.3	1	9.46		0.44		0.02		2	F8	281
11809	-58 00166				1 52 26	-58 09.9	1	7.31		1.50		1.82		5	M0 III	1673
		SK 202			1 52 26	-74 09.3	3	12.31	.007	-0.08	.009	-0.71	.013	8		573,733,1412
	+0 00309				1 52 27	+00 38.2	1	11.92		0.75		0.25		17		281
		SK 204			1 52 27	-74 11.3	2	13.43	.019	-0.04	.014	-0.34	.023	4		733,835
	-0 00293				1 52 28	+00 25.6	3	9.79	.003	0.52	.003	-0.02	.011	30	F8	281,989,1729
		HA 93 # 333			1 52 30	+00 31.0	5	12.01	.008	0.84	.008	0.45	.014	73	G5	271,281,989,1729,1764
	+36 00348	NGC 752 - 10			1 52 30	+37 19.4	1	10.12		0.42		0.02		2	F3.5IV	1100
	-28 00595	G 274 - 120			1 52 30	-27 42.6	4	11.52	.011	0.56	.007	-0.22	.014	18		1696,1698,1765,3062
	+37 00410	NGC 752 - 12			1 52 31	+37 35.8	2	9.96	.030	0.47	.004	0.04	.004	3	F4.5IV	1100,3067
11622	+60 00380				1 52 33	+61 26.0	1	9.11		0.36				2	A6 V	1025
		LS I +61 247			1 52 33	+61 44.3	1	11.76		0.39		-0.41		2		180
		G 35 - 19			1 52 34	+22 52.1	1	11.21		1.15		1.05		1	K7	1620
		Ua Tri			1 52 36	+33 31.5	2	12.03	.154	0.14	.014	0.13	.049	2	F1	668,699
11632	+59 00356	IDS01491N6005	AB		1 52 36	+60 20.2	1	8.79		0.57				2	G0	70
		SK 203			1 52 38	-74 07.3	2	12.60	.005	0.05	.005	0.07	.010	6		573,733
		VES 648			1 52 40	+55 50.5	1	11.21		0.23		-0.16		2		690
		LF 5 # 141			1 52 40	+59 56.	1	10.85		0.39		-0.40		1	B2 Iv	367
	+37 00412	NGC 752 - 24			1 52 41	+37 38.2	2	8.93	.019	1.01	.006	0.77	.002	2	K0 III	1100,4002
11689	+42 00397				1 52 42	+42 36.0	1	7.54		0.41		0.04		2	F2	1375
	-0 00294				1 52 43	+00 20.9	5	9.40	.006	0.85	.003	0.43	.007	48	G2	271,281,989,1729,6004
	+36 00350	NGC 752 - 27			1 52 44	+37 23.2	2	9.16	.015	1.01	.000	0.76	.003	3	K0 V	1100,4002
		HA 93 # 422			1 52 45	+00 44.5	1	12.11		0.60		0.08		2	G0	281
	+55 00441	LS I +56 023			1 52 45	+56 18.6	1	9.68		0.18		-0.75		2	B1 V:pe	1012
	+60 00382			V	1 52 45	+60 42.7	1	10.13		2.17				4	M3	369
11729	+16 00217				1 52 48	+16 56.7	1	8.34		1.54		1.19		2	M2	1648
	+36 00352	NGC 752 - 32			1 52 48	+37 14.0	1	9.96		0.32		0.08		2	F0 V	1100
	+36 00351	NGC 752 - 34			1 52 49	+37 27.8	1	9.85		0.47		0.07		1	F2 II	1100
		HA 93 # 424			1 52 51	+00 42.1	5	11.62	.003	1.08	.003	0.95	.003	69	K2	281,989,1729,1764,5006
		SK 205			1 52 51	-74 18.3	3	12.66	.013	-0.07	.009	-0.63	.008	7		733,835,1412
11669	+60 00383	IDS01494N6047	AB		1 52 52	+61 02.2	2	7.43	.025	0.09	.010			6	B7 V	1118,1119
	+59 00357	LS I +60 204			1 52 53	+60 19.2	2	9.88	.079	0.49	.015	-0.58	.045	3	B0.5IV	367,1012
		VES 649			1 52 54	+62 48.5	1	13.43		0.46		-0.24		2		8107
11727	+36 00354	HR 556		⋆B	1 52 56	+37 02.0	1	5.89		1.63				2	M0 III	71
11720	+37 00415	NGC 752 - 39			1 52 56	+37 37.3	2	8.09	.014	0.97	.002	0.68	.002	2	K0 III	1100,4002
		HA 93 # 521			1 52 57	+00 51.0	1	12.06		0.57		0.04		2	G3	271
	+36 00353	NGC 752 - 41			1 52 57	+37 13.9	1	9.80		0.49		0.07		2	F6 IV	1100
11719	+42 00399				1 52 57	+42 49.0	1	7.31		1.40		1.65		3	K4 III	1501
11808	-25 00758				1 52 58	-25 36.7	1	8.52		0.20				4	A2 IV/V	2012
		SK 206			1 52 58	-74 32.1	2	13.07	.014	-0.01	.009	-0.30	.028	4		733,835
		AAS6,117 # 103			1 53 00	+64 51.	1	11.60		0.54		0.47		4		171

HD	DM	Other Id	N	Rem	α_{1950}	δ_{1950}	S	V	σ_V	B–V	σ_{B-V}	U–B	σ_{U-B}	n	Spectrum	References
		HA 93 # 244			1 53 01	+00 24.2	1	11.23		0.86		0.50		3	G9	271
		G 71 - 43			1 53 03	+02 38.5	1	12.04		1.46				2		1532
11763	+22 00284	HR 559, RR Ari			1 53 03	+23 20.0	2	5.75	.010	1.19	.003	1.04		4	K1 III	71,3016
		LP 649 - 1			1 53 03	−04 34.0	1	12.48		0.80		0.27		2		1696
		NGC 752 - 55			1 53 05	+37 44.7	1	11.40		0.44		−0.02		1		1100
		HA 93 # 429			1 53 06	+00 35.1	1	12.48		0.68		0.10		2	G3	271
	+9 00241				1 53 10	+09 48.3	1	10.44		1.71				3	K5	1532
		VES 650			1 53 10	+59 06.4	1	11.34		0.37		−0.39		2		8107
		VES 651			1 53 10	+60 02.8	1	11.56		0.79		−0.38		2		690
		HA 93 # 526			1 53 11	+00 51.5	1	10.87		0.62		0.13		3	G0	271
11749	+36 00355	HR 557		⋆ A	1 53 11	+37 00.4	5	5.69	.014	1.06	.002	0.91	.005	9	K0 III	15,1003,1355,3024,4001
11749	+36 00355	IDS01502N3646	P		1 53 11	+37 00.4	1	11.93		1.10		0.90		3		3024
		NGC 752 - 58			1 53 11	+37 25.2	1	10.46		0.40		0.02		1	F2mF4IV	1100
	+37 00416	NGC 752 - 61		⋆ A	1 53 12	+37 30.3	2	10.05	.017	0.38	.009	0.06	.009	3	F3 IV	1100,3067
		VES 652			1 53 12	+63 04.1	1	12.20		0.46		−0.33		3		8107
		AAS6,117 # 104			1 53 12	+64 53.	1	11.16		0.44		0.21		2		171
	+34 00335				1 53 13	+34 44.5	1	9.26		0.46		0.01		3	F2	1625
		NGC 752 - 62		⋆ B	1 53 13	+37 30.5	2	11.23	.021	0.42	.002	−0.02	.011	3	F3 V	1100,3067
	+37 00417	NGC 752 - 64			1 53 15	+37 32.4	2	10.57	.026	0.36	.011	0.06	.011	3	F3 IV	1100,3067
11782	+22 00285				1 53 16	+23 03.4	1	8.85		0.51		0.02		2	G5	1375
		NGC 752 - 66			1 53 16	+37 43.6	2	10.91	.013	0.44	.006	−0.02	.009	3	F4 IV-V	1100,3067
		NGC 752 - 68			1 53 17	+37 22.9	1	10.73		1.56		1.84		1		1100
	+36 00356	NGC 752 - 69			1 53 17	+37 24.0	1	10.04		0.51		0.10		1	F5.5III-IV	1100
11803	+1 00347	HR 560		⋆ AB	1 53 19	+01 36.2	6	6.01	.009	0.55	.011	0.03	.010	16	F7 V+G0 V	15,292,938,1061,1381,3077
	+36 00357	NGC 752 - 74			1 53 20	+37 23.0	1	10.73		0.42		0.01		1	F3 V	1100
	+37 00418	NGC 752 - 75			1 53 20	+37 43.4	3	8.96	.027	1.01	.002	0.76	.016	4	G9 III	1100,3067,4002
	+62 00325				1 53 22	+62 44.8	2	8.70	.004	1.40	.004	1.30		5	G8 V	171,1025
	+36 00358	NGC 752 - 77			1 53 23	+37 21.5	3	9.36	.024	1.03	.008	0.77	.031	4	K0 III	1100,3067,4002
11828	−27 00665				1 53 23	−26 44.3	1	9.27		0.65		0.12		1	G5 V	79
	+0 00311	IDS01509N0018	AB		1 53 24	+00 31.7	3	10.26	.000	0.46	.005	−0.06	.000	17		271,281,6004
		NGC 752 - 78			1 53 24	+37 32.6	1	11.56		0.67		0.18		1		1100
11716	+60 00386				1 53 24	+61 00.1	1	8.21		1.22				2	K0 III	1025
		BD +0 00311a			1 53 25	+00 32.3	2	10.94	.014	0.59	.005	0.09	.009	8	F6	271,281
		VES 653			1 53 28	+63 15.0	1	10.93		1.34		1.24		3		8107
		CS 22171 # 2			1 53 31	−12 42.9	2	13.36	.010	0.18	.006	0.16	.009	2		966,1736
11734	+59 00360				1 53 32	+59 45.2	2	8.78	.020	0.30	.020			6	B2 IV	1119,1119
	+60 00387				1 53 32	+61 00.0	1	9.91		0.33				2	Am	1025
		LS I +62 195			1 53 33	+62 49.4	1	11.06		0.32		−0.61		2		180
		NGC 752 - 88			1 53 34	+37 42.1	2	11.78	.026	0.48	.006	−0.02	.000	3	F5 V	1100,3067
	+43 00395	G 133 - 53			1 53 34	+43 40.2	1	9.59		0.92		0.66		1	K2	333,1620
	−1 00263	TX Cet			1 53 34	−00 57.6	1	10.85		0.36		0.08		3	A2	3064
	+45 00487				1 53 36	+46 16.6	1	8.74		1.33		1.46		2	K2	1733
11745	+59 00361				1 53 38	+60 14.2	1	9.83		0.25				4	B8 V	1119
	+64 00267				1 53 38	+64 58.2	1	10.52		0.48		0.41		2		171
11977	−68 00101	HR 570			1 53 40	−67 53.6	4	4.69	.011	0.95	.005	0.64	.000	21	G8 III/IV	15,1075,2038,8029
	+37 00419	NGC 752 - 96			1 53 41	+37 37.1	2	10.40	.026	0.38	.000	0.07	.004	3	F2 IV	1100,3067
		NGC 752 - 100			1 53 42	+37 45.2	1	13.19		0.42		−0.02		2		1100
		NGC 752 - 99			1 53 42	+37 45.3	1	11.93		0.28		0.14		2	F2	1100
		NGC 752 - 103			1 53 45	+37 46.6	2	10.99	.020	1.17	.002	0.98	.007	3		1100,4002
	+62 00326				1 53 45	+62 57.6	1	9.92		0.39		0.22		5	A0	171
11878	−36 00737				1 53 46	−36 29.4	2	7.93	.000	0.24	.005	0.14		5	A9 V	79,2012
	+60 00389	LS I +60 206			1 53 48	+60 53.2	3	9.27	.004	0.58	.010	−0.31	.020	8	B9 Ib	367,1012,1119
		NGC 752 - 105			1 53 49	+37 09.9	2	10.29	.038	0.42	.002	0.08	.009	3	F3 IVnp	1100,3067
	+63 00261	LS I +63 175			1 53 50	+63 47.9	3	9.67	.032	0.47	.016	−0.68	.018	5	Bpenn	180,690,1012
11812	+37 00421	NGC 752 - 108			1 53 51	+37 46.7	1	9.14		0.45		0.07		2	F4 III-IV	1100
		LP 529 - 2			1 53 52	+06 51.6	1	12.87		1.33		1.13		1		1696
11811	+37 00422	NGC 752 - 110			1 53 52	+37 47.3	1	8.94		0.84		0.45		2	G3 IV	1100
	+61 00358				1 53 52	+62 00.7	2	10.07	.009	0.35	.018	0.10		7	A0 V	171,1119
11744	+64 00268	LS I +65 014			1 53 53	+65 10.0	2	7.80	.005	0.37	.020	−0.34	.030	5	B2 III	171,399
	+37 00423	NGC 752 - 117			1 53 54	+37 37.5	2	10.27	.021	0.43	.004	0.07	.011	3	F3.5IVn	1100,3067
		CS 22171 # 8			1 53 55	−09 56.1	1	14.73		0.14		0.14		1		1736
		NGC 752 - 120			1 53 56	+37 09.2	1	12.05		0.59		0.05		1		1100
11772	+60 00390				1 53 56	+61 01.3	1	9.54		0.22				3	B8 V	1119
	−27 00668	G 274 - 123			1 53 56	−27 19.8	1	11.75		0.98		0.84		1		1696
		NGC 752 - 123			1 53 57	+37 33.4	1	11.20		0.41		−0.01		2	F3 V	3067
	+36 00361	NGC 752 - 126			1 53 58	+37 25.2	1	10.08		0.43		0.04		2	F3.5IV	1100
	−20 00370	G 272 - 124			1 53 58	−19 57.8	1	10.27		0.64		0.06		1		79
		NGC 752 - 129			1 53 59	+37 08.7	1	10.86		0.38		0.00		1	F2 V	1100
	+63 00262				1 53 59	+64 12.8	1	9.94		0.37		0.09		3		171
		NGC 752 - 130			1 54 00	+37 19.5	1	12.95		0.48		−0.02		1		1100
		NGC 752 - 132			1 54 01	+37 19.9	1	12.54		0.46		−0.05		1		1100
11937	−52 00241	HR 566		⋆ AB	1 54 01	−51 51.4	6	3.69	.009	0.85	.004	0.46	.012	30	G8 IIIb	15,678,1075,2012,3079,8029
	+3 00301				1 54 03	+30 46.8	1	9.79		1.04				1	dK7	1632
		NGC 752 - 134			1 54 03	+37 19.5	1	11.85		1.17		1.13		1		1100
		NGC 752 - 135			1 54 04	+37 38.5	2	11.23	.017	0.45	.002	0.00	.011	3	F0 V	1100,3067
		NGC 752 - 139			1 54 04	+37 41.1	2	11.78	.032	0.46	.004	−0.03	.008	4		1100,3067
	+37 00424	NGC 752 - 137			1 54 04	+37 53.4	1	8.94		1.02		0.78		1	G8 III	4002
		NGC 752 - 141			1 54 05	+37 20.0	1	13.57		0.52		−0.02		1		1100
		AAS6,117 # 79			1 54 06	+63 07.	1	12.08		0.49		−0.25		3	B2	171
	+62 00328				1 54 07	+62 50.3	1	9.40		0.40				2	B6 III	1119

Table 1   119

HD	DM	Other Id	N Rem	α_{1950}	δ_{1950}	S	V	σ_V	B–V	σ_{B-V}	U–B	σ_{U-B}	n	Spectrum	References
11800	+59 00363			1 54 10	+59 58.5	1	7.79		2.04				5	K5 Ib	1119
11995	−61 00157	HR 571		1 54 11	−61 06.4	2	6.05	.005	0.38	.005			7	F0 IV/V	15,2012
		NGC 752 - 152		1 54 12	+37 38.7	1	14.10		0.63		0.12		2		3067
		CS 22171 # 5		1 54 12	−11 48.4	1	13.76		0.28		0.03		1		1736
	−14 00363	G 272 - 125		1 54 12	−14 25.3	4	9.67	.012	0.65	.013	0.04	.004	7	G5 V	79,742,1003,1064,1594
		NGC 752 - 157		1 54 13	+37 25.8	1	12.56		0.60		-0.04		1		1100
		NGC 752 - 163		1 54 13	+37 27.3	1	12.30		0.49		-0.08		2		1100
		NGC 752 - 162		1 54 13	+37 39.6	1	12.93		0.43		-0.04		2		3067
		NGC 752 - 161		1 54 13	+37 41.5	1	11.90		0.60		0.02		1		1100
	+37 00425	NGC 752 - 159		1 54 13	+37 44.8	1	9.40		0.47		0.05		1	F0 III	1100
		NGC 752 - 164		1 54 14	+37 24.7	1	11.88		1.08		0.86		2		1100
		G 3 - 28		1 54 15	+11 25.2	1	10.56		0.77		0.35		1	K0	333,1620
		WLS 200 45 # 10		1 54 16	+44 54.5	1	10.70		0.24		0.16		2		1375
	+24 00283			1 54 17	+24 29.5	1	8.87		1.08		1.03		2	K0	1733
	+36 00363	NGC 752 - 171		1 54 19	+37 11.5	1	10.16		0.44		0.05		1	F3 IV	1100
		NGC 752 - 170		1 54 19	+37 26.2	1	13.07		0.52		-0.02		2		1100
		NGC 752 - 172		1 54 19	+37 38.7	1	14.09		0.58		0.00		2		3067
		LS I +59 099		1 54 19	+59 43.9	1	9.96		0.43		-0.40		2	B1 V	1012
		NGC 752 - 175		1 54 20	+37 24.6	1	12.89		1.03		0.87		2		1100
		NGC 752 - 174		1 54 20	+37 38.0	1	14.05		0.77		0.24		2		3067
11930	−23 00721	HR 565		1 54 20	−22 46.2	2	4.89	.015	1.43	.000	1.67		7	K3 III	2035,3053
		NGC 752 - 178		1 54 22	+37 28.0	1	13.87		0.66		0.12		2		1100
	+37 00427	NGC 752 - 177		1 54 22	+37 37.1	2	10.19	.021	0.47	.004	0.07	.017	3	F4.5IVn	1100,3067
		LS I +61 248		1 54 22	+61 45.7	1	11.87		0.43		-0.22		1		180
11821	+60 00392			1 54 24	+60 57.0	1	7.95		1.44				2	G8 V	1025
		LS I +61 249		1 54 24	+61 41.9	2	11.45	.170	0.47	.197	-0.33	.045	5	B1 V:	180,1012
		AAS6,117 # 80		1 54 24	+63 11.	1	11.19		0.70		0.31		3	B9	171
		NGC 752 - 185		1 54 25	+37 37.6	1	12.21		0.54		-0.02		1		1100
	+36 00364	NGC 752 - 187		1 54 27	+37 28.7	1	10.41		0.42		0.00		2	F3 V	1100
	+37 00428	NGC 752 - 186		1 54 27	+37 37.1	1	10.19		0.94		0.59		1	K0 III-IV	1100
		Steph 211		1 54 27	+47 27.5	1	11.90		1.50		1.16		1	M0r	1746
		NGC 752 - 189		1 54 28	+37 24.7	2	11.29	.019	0.42	.002	-0.02	.007	4		1100,3067
	+60 00393	LS I +61 250		1 54 28	+61 19.9	2	10.68	.020	0.42	.020	-0.45	.015	3	B2pe	367,1012
		G 3 - 29		1 54 29	+12 01.6	2	14.84	.014	1.44	.005	1.01	.005	5		419,3079
	+36 00365	NGC 752 - 193		1 54 30	+37 09.4	2	10.22	.033	0.38	.005	0.10	.009	2	F2 m F4IVp	1100,3067
	+41 00379	G 133 - 57		1 54 31	+41 50.1	1	9.01		0.72		0.30		1	G0	333,1620
	+60 00395			1 54 31	+61 27.3	2	8.50	.033	1.21	.012	1.10		14	G5 V	171,1119
	+62 00329			1 54 31	+62 47.3	1	9.48		1.95		1.72		4		171
	+37 00430	NGC 752 - 196		1 54 32	+37 40.4	2	10.27	.017	0.42	.002	0.06	.011	3	F3.5IVn	1100,3067
		NGC 752 - 197		1 54 33	+37 39.1	2	11.64	.034	0.44	.006	-0.02	.015	3		1100,3067
11831	+59 00364	LS I +60 207		1 54 33	+60 08.9	3	8.07	.017	0.86	.009	0.35	.013	9	A2 Ia	171,1012,1119
		LP 940 - 51		1 54 36	−32 49.4	1	12.98		0.48		-0.22		3		1696
11909	+17 00289	HR 563		1 54 37	+17 34.5	4	5.11	.010	0.92	.002	0.70	.006	13	K1 Vp	1355,1363,3016,4001
	+37 00431	NGC 752 - 205		1 54 37	+37 30.6	2	9.89	.010	0.41	.005	0.03	.015	3	F3 IV-Vp	1100,3067
		NGC 752 - 190		1 54 38	+37 26.3	1	13.68		0.47		-0.07		2		1100
11837	+57 00434			1 54 38	+57 43.2	1	8.75		0.18				2	B9 IV	1119
	+36 00367	NGC 752 - 209		1 54 39	+37 14.9	2	9.76	.042	0.06	.009	-0.01	.000	2	B9.5V	1100,3067
	+36 00368	NGC 752 - 208		1 54 39	+37 25.0	2	8.96	.014	1.08	.003	0.85	.013	2	K1 IV	1100,4002
	+37 00433	NGC 752 - 206		1 54 39	+37 34.4	2	10.05	.021	0.47	.000	0.07	.013	2	F4.5IV	1100,3067
		LS I +59 100		1 54 39	+59 05.9	1	10.41		0.32		-0.55		1	B1 V	367
	+37 00432	NGC 752 - 213		1 54 40	+37 31.6	2	9.05	.028	1.00	.008	0.75	.005	4	G9 III	1100,4002
	+61 00359			1 54 40	+61 49.4	1	10.22		0.34				3	B8 V	1119
		NGC 752 - 214		1 54 41	+37 37.8	1	10.44		0.38		0.04		1	F2 IV	1100
12087	−69 00087			1 54 41	−68 38.1	1	7.10		0.91				4	G6 III	2018
11964	−10 00403	G 272 - 126	★ A	1 54 43	−10 29.0	5	6.42	.009	0.82	.011	0.46	.017	16	G5	79,1075,1311,1518,3062
11964	−10 00403	G 272 - 127	★ B	1 54 43	−10 29.5	2	11.21	.014	1.39	.021	1.18	.000	4	K5	1518,3062
11885	+36 00370	NGC 752 - 215		1 54 44	+37 25.6	2	7.15	.019	1.20	.002	1.22	.014	2	K1 III	1100,4002
	+44 00391			1 54 44	+45 23.8	1	9.33		-0.10		-0.67		1	B8	963
		NGC 752 - 217		1 54 45	+37 37.1	2	10.46	.026	0.42	.004	0.07	.015	3	F2.5IVn	1100,3067
	+68 00138	G 244 - 44		1 54 45	+68 46.9	5	9.30	.012	0.56	.023	-0.02	.009	6	G3	516,979,1620,1658,8112
	+37 00434	NGC 752 - 218		1 54 46	+37 44.7	2	10.07	.009	0.45	.004	0.04	.000	5	F4 IV	1100,3067
	+58 00340			1 54 46	+59 22.4	1	9.56		1.91				3		369
		NGC 752 - 220		1 54 47	+37 24.8	3	9.61	.008	0.99	.005	0.73	.031	5	K0 V	1100,3067,4002
	+37 00435	NGC 752 - 219	★ V	1 54 47	+37 49.9	2	10.49	.017	0.41	.000	0.02	.009	6	F3 IV-Vnnp	1100,3067
12003	−43 00599			1 54 47	−43 25.3	1	7.87		0.23				4	A4 V	2012
		NGC 752 - 222		1 54 48	+37 32.9	2	10.96	.014	0.39	.007	0.02	.014	4	F2.5IV	1100,3067
11894	+44 00392			1 54 49	+45 21.4	1	8.14		0.25		0.04		2	A5	1601
11860	+58 00341			1 54 49	+59 23.0	3	6.65	.013	0.08	.015	0.03		8	A1 IV	401,1118,1119
		LS I +63 176		1 54 49	+63 11.7	1	11.01		0.61		-0.26		4		180
		LS I +61 251		1 54 50	+61 39.2	1	11.14		0.74		-0.30		3		180
	+37 00438	NGC 752 - 225		1 54 51	+37 34.6	1	10.15		0.55		0.06		1	F5 V	1100
11884	+46 00485			1 54 51	+46 51.1	1	6.45		1.22		1.18		2	K0	1601
		NGC 744 - 22		1 54 51	+55 14.2	1	13.22		0.48		0.34		1		49
11859	+59 00366			1 54 51	+59 48.2	3	8.63	.019	0.15	.007			8	B9 III	70,1119,1119
11858	+59 00365			1 54 51	+59 57.7	1	9.32		0.21				3	A0 IV	1119
	+29 00336			1 54 52	+30 18.7	1	8.74		0.57		-0.10		3	K0	3077
12058	−60 00167			1 54 52	−60 28.4	2	8.62	.029	1.15	.017	1.11	.005	6	K5 V	158,3072
11928	+27 00310	HR 564		1 54 53	+27 33.7	2	5.84	.014	1.58	.013	1.89		5	M2 III	71,3001
11905	+40 00407	HR 562		1 54 54	+41 27.1	1			-0.06		-0.40		1	B8 III	1079
		NGC 744 - 25		1 54 54	+55 14.6	1	13.86		0.68		0.45		1		49

HD	DM	Other Id	N	Rem	α_{1950}	δ_{1950}	S	V	σ_V	B–V	σ_{B-V}	U–B	σ_{U-B}	n	Spectrum	References
11866	+57 00435				1 54 54	+57 37.5	1	7.94		0.07				3	A2 V	1119
		NGC 752 - 234			1 54 58	+37 35.4	2	10.71	.026	0.43	.002	0.02	.013	3	F2mF4IV-V	1100,3067
		NGC 752 - 235			1 54 59	+37 33.8	1	11.43		0.48		-0.04		2	F5	1100
11857	+60 00398	HR 561			1 54 59	+61 27.3	3	6.02	.015	-0.04	.007	-0.41		6	B5 III	1079,1118,1119
		NGC 752 - 237			1 55 00	+37 26.9	2	12.32	.004	0.60	.011	0.05	.017	3		1100,3067
	+37 00439	NGC 752 - 238			1 55 00	+37 40.3	2	9.96	.005	0.45	.002	0.04	.000	4	F4.5IV	1100,3067
	+36 00372	NGC 752 - 240			1 55 01	+37 26.1	1	9.74		0.52		0.01		1	F5 V	1100
	+54 00425	NGC 744 - 2			1 55 01	+55 13.5	1	10.44		0.34		0.14		1	A1 V	49
		NGC 744 - 23			1 55 01	+55 14.7	1	13.81		0.80		0.28		1		49
236915	+58 00342				1 55 01	+59 01.6	1	8.30		2.20				1	K7	138
11983	−8 00349				1 55 02	−07 46.8	4	8.19	.007	1.51	.003	1.87	.017	35	K0	989,1375,1509,1729
		NGC 744 - 17			1 55 03	+55 15.5	1	12.67		0.37		0.09		1		49
11875	+57 00437				1 55 03	+57 52.2	1	8.41		0.35				2	F5	1025
	+36 00374	NGC 752 - 246			1 55 04	+37 15.1	1	10.09		1.00		0.84		1	K0 V	1100
		NGC 744 - 3			1 55 04	+55 15.3	1	10.62		0.45		0.18		1	A0 V	49
		NGC 744 - 19			1 55 05	+55 14.7	1	12.83		0.52		0.08		1		49
11865	+60 00400				1 55 05	+61 18.3	1	7.44		1.28				2	G8 III	1025
		NGC 744 - 6			1 55 06	+55 12.4	1	11.20		0.68		0.19		1	G2 V	49
		NGC 744 - 30			1 55 06	+55 15.5	1	14.91		0.76		0.42		1		49
12042	−52 00242	HR 573			1 55 06	−52 00.8	7	6.10	.005	0.48	.007	-0.06	.006	25	F6/7 V	15,158,258,2012,2012*
		NGC 744 - 14			1 55 07	+55 17.0	1	12.12		0.34		0.09		1		49
	+59 00367	LS I +60 208			1 55 07	+60 16.3	3	9.76	.014	0.53	.005	-0.53	.014	5	O9.5Ib	367,1011,1012
		VES 656			1 55 07	+62 35.7	1	12.58		0.60		-0.12		3		8107
12033	−38 00660				1 55 07	−37 44.4	1	7.91		1.23				4	K1 III	2012
11973	+22 00288	HR 569		⋆ A	1 55 08	+23 21.2	7	4.79	.012	0.29	.008	0.09	.004	17	F0 V	15,39,1084,1118,1193*
11973	+22 00289	IDS01524N2306		B	1 55 08	+23 21.2	2	7.37	.039	0.58	.005	0.07	.010	5	G0	1084,3024
		NGC 744 - 32			1 55 08	+55 14.4	1	15.16		0.76				1		49
		NGC 744 - 27			1 55 08	+55 15.2	1	14.24		0.62		0.48		1		49
		LS I +60 209			1 55 08	+60 23.2	2	10.13	.030	0.70	.045	-0.36	.000	3	B1 pe(V)	367,1012
	+36 00373	NGC 752 - 254			1 55 09	+37 25.4	2	10.93	.017	0.38	.004	0.05	.021	3	F2 IV-V	1100,3067
		NGC 744 - 15			1 55 09	+55 15.0	1	12.27		0.49		0.21		1		49
		VES 657			1 55 09	+65 40.3	1	12.95		0.92		0.04		2		690
11997	−5 00353	G 159 - 16			1 55 09	−05 28.4	2	9.24	.005	0.96	.007	0.77	.024	4	G5	1064,3062
11961	+30 00310				1 55 11	+30 53.5	1	7.08		1.43		1.10		2	M2	1625
		VES 658			1 55 11	+54 58.4	1	12.41		0.65		-0.05		2		690
12055	−47 00597	HR 574			1 55 11	−47 37.7	5	4.82	.009	0.87	.005	0.51	.005	22	G8 III	15,1075,2012,8015,8029
		NGC 744 - 4			1 55 12	+55 14.1	1	10.95		0.37		0.03		1	A0 V	49
		NGC 752 - 259			1 55 13	+37 25.0	2	11.40	.009	0.43	.006	-0.02	.006	3	F2	1100,3067
		NGC 752 - 261			1 55 14	+37 18.1	1	11.14		0.49		-0.03		1	F5.5IV-V	1100
		NGC 752 - 263			1 55 14	+37 20.1	1	10.91		0.38		0.02		1	F2 IVp	1100
		NGC 744 - 28			1 55 14	+55 14.4	1	14.28		0.63		0.39		1		49
		NGC 744 - 13			1 55 16	+55 17.7	1	11.93		0.39		0.16		1		49
	+60 00401				1 55 16	+61 09.3	1	9.55		0.18				4	A3 V	1119
		NGC 744 - 12			1 55 17	+55 14.3	1	11.92		0.39		0.12		1		49
		NGC 752 - 266			1 55 18	+37 23.7	2	11.23	.013	0.44	.006	-0.01	.013	3	F5 V	1100,3067
		NGC 752 - 267			1 55 18	+37 24.3	1	13.22		0.51		-0.04		1		1100
		NGC 744 - 33			1 55 19	+55 13.1	1	15.77		0.83		0.25		1		49
		NGC 744 - 24			1 55 19	+55 13.5	1	13.84		0.71		0.13		1		49
	+37 00440	NGC 752 - 271			1 55 20	+37 33.6	1	9.12		0.34		-0.04		1	F0 V	1100
		NGC 752 - 275			1 55 21	+37 22.1	1	12.19		0.55		0.01		1		1100
		NGC 744 - 10			1 55 21	+55 09.1	1	11.73		0.30		0.09		1		49
		NGC 744 - 29			1 55 21	+55 15.5	1	14.30		1.01		0.58		1		49
11949	+48 00576	HR 568			1 55 22	+48 57.7	1	5.69		1.01				2	K0 IV	71
		NGC 744 - 16			1 55 22	+55 11.3	1	12.44		0.33		0.14		1		49
		NGC 744 - 7			1 55 22	+55 11.8	1	11.40		0.25		0.02		1	A0 V	49
		NGC 744 - 34			1 55 22	+55 13.1	1	16.25		0.78		0.25		1		49
		NGC 744 - 8			1 55 23	+55 11.9	1	11.40		0.27		-0.02		1	A0 V	49
		CS 22171 # 9			1 55 23	−09 01.1	2	13.37	.000	0.41	.009	-0.17	.006	2		966,1736
		NGC 744 - 21			1 55 24	+55 15.5	1	12.93		0.46		0.21		1		49
11920	+56 00397				1 55 24	+57 02.4	1	9.00		0.11				3	A0 V	1119
12021	−2 00329	IDS01529S0233		C	1 55 24	−02 20.6	4	8.87	.004	-0.09	.006	-0.39	.017	35	A0	989,1026,1509,1729
		NGC 752 - 282			1 55 25	+37 37.5	1	13.38		0.91		0.47		1		3067
		LS I +59 101			1 55 25	+59 31.2	1	10.57		0.35		-0.51		1	B1 V	367
		NGC 744 - 11			1 55 26	+55 14.1	1	11.75		0.38		0.05		1		49
		VES 659			1 55 26	+64 33.1	1	12.72		1.64		1.26		3		690
		NGC 752 - 290			1 55 28	+37 38.0	1	13.15		0.53		-0.02		1		3067
		NGC 752 - 291			1 55 29	+37 37.5	1	12.14		1.09		0.75		1		3067
		NGC 744 - 20			1 55 29	+55 14.3	1	12.84		0.56		0.31		1		49
11948	+54 00429	NGC 744 - 1			1 55 30	+55 20.3	1	7.86		0.21		0.17		1	F0p	49
	+37 00441	NGC 752 - 295			1 55 31	+37 37.1	3	9.31	.011	0.96	.004	0.65	.007	4	G8 III	1100,3067,4002
		VES 660			1 55 31	+64 09.8	1	13.24		0.84		-0.27		3		690
	+59 00369				1 55 32	+59 55.0	1	9.94		0.29				2	B8 V	1119
11918	+60 00403				1 55 32	+61 18.0	2	8.07	.000	0.17	.020			6	B9 IV	1118,1119
		VES 661			1 55 33	+57 12.4	1	11.63		0.26		-0.15		3		8107
	+37 00442				1 55 36	+38 19.6	2	10.01	.011	-0.28	.000	-1.16	.004	15		843,963
	+58 00343	LS I +58 053			1 55 36	+58 42.0	1	9.64		0.36		-0.41		2	B2 III	1012
12037	−5 00357				1 55 36	−04 32.7	2	8.73	.004	0.33	.000	0.13	.004	8	Am	355,1026
		NGC 744 - 18			1 55 37	+55 12.5	1	12.82		0.43		0.13		1		49
		NGC 744 - 31			1 55 37	+55 14.1	1	15.06		1.27		0.99		1		49
		VES 662			1 55 37	+64 36.4	1	13.01		0.49		-0.18		3		8107

Table 1

121

HD	DM	Other Id	N Rem	α_{1950}	δ_{1950}	S	V	σ_V	B–V	σ_{B-V}	U–B	σ_{U-B}	n	Spectrum	References
12068	−29 00664	G 274 - 126	⋆ A	1 55 37	−29 05.4	1	8.26		0.70		0.21		1	G3 V	79
	+37 00444	NGC 752 - 300		1 55 38	+37 30.6	1	9.58		0.41		0.02		1	F3 III-IV	1100
		NGC 744 - 26		1 55 38	+55 12.7	1	14.02		0.54		0.36		1		49
	+59 00370			1 55 40	+59 41.0	1	9.88		0.27				3	A4 V	1119
		LF 5 # 145		1 55 40	+60 19.	1	11.03		0.61		-0.12		1	B2 V	367
12085	−34 00752	G 274 - 129		1 55 41	−33 39.1	1	9.38		0.67		0.14		1	G6 V	79
	+37 00445	NGC 752 - 306		1 55 44	+37 39.7	1	10.68		0.38		0.14		2	F2 III	1100
11947	+60 00404			1 55 44	+60 53.8	1	9.09		0.24				3	B8 V	1119
	+5 00267			1 55 47	+05 52.4	1	10.58		0.66		0.10		2		1696
12084	−29 00665			1 55 48	−29 03.5	1	8.34		0.80		0.07		1	G0 V	79
11960	+59 00371			1 55 50	+60 13.1	1	8.83		0.25				3	A3 III	1119
	+62 00332	LS I +62 198	⋆ A	1 55 50	+62 39.8	2	8.95	.025	1.96	.005	2.18		6	M2	171,369
	+62 00332	IDS01522N6225	B	1 55 50	+62 39.8	1	10.44		0.31		-0.59		1		180
	−5 00358	G 159 - 17		1 55 51	−05 25.2	1	10.69		0.97		0.71		1		3062
	−26 00709	G 274 - 125		1 55 52	−25 44.4	2	12.09	.015	0.54	.002	-0.21	.012	4		1696,3060
12027	+36 00378	NGC 752 - 309	⋆ AB	1 55 53	+37 26.6	2	8.24	.005	0.22	.000	0.12	.005	13	A3 III	1100,3067
		CS 22171 # 4		1 55 53	−12 12.7	2	14.30	.005	0.07	.013	0.12	.012	2		966,1736
	+37 00448	NGC 752 - 311		1 55 54	+37 34.3	3	9.06	.012	1.04	.008	0.81	.012	15	G9 III	1100,3067,4002
12363	−78 00042	HR 593		1 55 54	−78 35.6	3	6.14	.007	0.43	.005	0.00	.010	16	F5/6 IV/V	15,1075,2038
11946	+63 00265	HR 567	⋆ A	1 55 55	+64 22.8	2	5.27	.010	0.00	.011	-0.05		3	A0 Vn	1363,3023
	+62 00333			1 55 57	+63 22.2	1	9.71		0.31		0.18		3		171
		NGC 744 - 5		1 56 00	+55 11.2	1	11.08		1.09		0.92		1		49
11959	+62 00334			1 56 05	+63 04.3	1	8.94		0.23				2	A3 V	1119
12103	−16 00341	G 272 - 129		1 56 05	−15 39.9	1	9.17		0.70		0.13		1	G6 V	79
		NGC 744 - 9		1 56 07	+55 11.5	1	11.41		0.14		0.15		1		49
		G 35 - 20		1 56 08	+23 33.3	1	12.77		1.17		1.00		1		333,1620
11996	+59 00373			1 56 09	+59 45.9	2	7.29	.005	1.79	.005			7	K4 III	1025,1119
	+59 00372			1 56 09	+60 00.6	1	9.30		2.28				1	K5-M0 Ib	138
		VES 663		1 56 10	+58 32.3	1	12.50		1.36		1.14		2		690
12051	+32 00360	G 72 - 50	⋆ A	1 56 11	+32 58.3	2	7.14	.015	0.77	.002	0.42	.005	3	G5	333,1620,3026
236920	+55 00466			1 56 12	+55 28.4	1	7.82		0.92		0.58		2	G5	1502
		LP 769 - 82		1 56 13	−15 29.5	1	12.02		0.64		0.06		2		1696
12135	−33 00682	HR 576	⋆ AB	1 56 14	−33 18.6	3	6.35	.008	1.01	.010			11	K0 III	15,1075,2032
12025	+54 00431	U Per		1 56 15	+54 34.8	1	9.04		1.53		0.66		1	M4	817
		AAS6,117 # 86		1 56 18	+63 17.	1	10.82		0.34		-0.06		3	B9	171
	−5 00359			1 56 18	−05 02.7	3	10.46	.000	0.07	.000	0.00	.009	16	A0	830,1026,1783
12158	−43 00604			1 56 20	−43 10.4	1	10.26		0.13		0.12		4	A1/2 V	152
12014	+58 00345			1 56 22	+58 55.1	1	7.58		1.97				4	K0 Ib	1119
	−23 00736	G 272 - 131		1 56 23	−23 08.4	1	11.96		0.34		-0.19		1	DA:	1696
12362	−75 00140			1 56 26	−75 36.6	1	7.47		1.38		1.55		9	K3 III	1704
236923	+59 00374	LS I +59 103		1 56 28	+59 28.8	1	9.68		0.40		-0.45		2	B1 V	1012
	+61 00361			1 56 29	+61 30.5	2	9.12	.000	0.53	.004			5	F5 V	70,1025
		G 72 - 51		1 56 31	+36 00.1	1	13.22		1.03		0.80		1		1620
	+61 00362	LS I +62 199		1 56 31	+62 06.0	2	10.32	.018	0.40	.022	-0.50	.027	5	B2 III	180,1012
	+2 00305			1 56 32	+03 15.1	1	10.94		1.53				1	dM1	1632
12270	−66 00123	HR 584		1 56 32	−65 40.0	3	6.35	.011	0.91	.005	0.62	.000	16	G8 II/III	15,1075,2038
	+64 00275			1 56 33	+64 42.1	1	10.26		0.33		0.01		2		171
		POSS 53 # 3		1 56 35	+67 31.4	1	15.89		1.21				2		1739
		G 159 - 18		1 56 36	+03 16.6	1	10.92		1.51		0.98		1	M2	3062
		Steph 219		1 56 39	+48 22.8	1	11.90		1.77		2.24		1	K7	1746
12100	+40 00415			1 56 41	+41 06.5	1	7.48		0.89		0.54		2	G5	1601
12153	−10 00412			1 56 41	−10 25.6	1	8.71		0.44		-0.06		2	G0	1375
12167	−26 00711	G 274 - 132		1 56 41	−26 07.2	1	9.58		0.58		-0.03		1	G0 V	79
	−69 00088			1 56 43	−69 31.9	1	10.43		0.34		-0.03		2		3060
12140	+11 00261	HR 578		1 56 45	+12 03.2	2	6.07		0.19	.000	0.09	.005	5	A6 V	985,1049
12060	+59 00375			1 56 45	+59 39.9	2	8.17	.018	0.49	.004			5	F7 IV	70,1025
	+60 00408			1 56 45	+60 59.2	1	10.01		0.26				3	A2 V	1119
		LP 884 - 73		1 56 45	−30 21.1	1	11.38		0.47		-0.15		2		1696
	+12 00263			1 56 46	+13 00.4	1	10.40		0.53		-0.03		2	F8	234
12061	+58 00347			1 56 46	+58 28.6	2	8.56	.004	0.48	.000			5	F5	70,1025
12139	+20 00322	HR 577	⋆ D	1 56 49	+20 49.0	1	5.87		1.03				2	K0 III-IV	71
		Steph 220		1 56 54	+48 03.4	1	11.97		1.23		1.26		1	K6	1746
		G 159 - 19		1 56 57	−04 00.2	1	12.36		1.30		1.20		2		3062
	+60 00409			1 57 00	+61 08.2	2	10.37	.014	0.36	.018	0.25		7	A0:V	171,1119
	+64 00277	LS I +64 081		1 57 01	+64 56.8	1	9.75		0.62		0.12		3	A1 Iab	171
12206	−27 00688			1 57 02	−26 40.5	1	6.79		0.02				4	A0 V	2012
12088	+58 00350	G 220 - 5		1 57 03	+59 24.7	1	9.08		0.73				2	G5	1025
		Hiltner 197		1 57 06	+62 41.0	1	10.64		0.41		-0.39		2	B2 V	1012
		POSS 53 # 2		1 57 06	+67 40.1	1	15.23		1.45				1		1739
12287	−57 00383	IDS01554S5722	A	1 57 07	−57 06.8	1	8.28		0.92		0.58		2	K0 IV	3062
12099	+57 00446			1 57 08	+57 37.9	1	9.61		0.17				2	A3 V	1119
12287	−57 00385	IDS01554S5722	B	1 57 10	−57 07.2	1	9.10		0.52		0.02		2		3062
	+60 00411	LS I +60 210		1 57 12	+60 34.0	1	9.17		0.89				2	A0 Ib	1119
		LS I +62 200		1 57 12	+62 47.9	3	10.59	.033	0.48	.029	-0.35	.029	8	B1 V	171,180,1012
12080	+64 00278			1 57 12	+64 28.8	1	8.36		0.24		0.16		3	A0	171
12311	−62 00162	HR 591		1 57 12	−61 48.8	10	2.85	.013	0.28	.009	0.16	.035	32	F0 V	3,15,418,1020,1034*
	+58 00351	LS I +58 055		1 57 13	+58 44.3	1	10.00		0.46		-0.42		2	B1 III	1012
12112	+59 00376			1 57 17	+59 43.1	3	6.72	.017	0.14	.011			8	A3 IV	1118,1119,1119
		01H57M20S		1 57 20	+06 24.9	1	12.82		1.48		1.14		1		1696
		G 73 - 11		1 57 20	+13 20.3	1	13.34		1.58		1.13		1		333,1620

HD	DM	Other Id	N Rem	α_{1950}	δ_{1950}	S	V	σ_V	B–V	σ_{B-V}	U–B	σ_{U-B}	n	Spectrum	References
	+60 00412			1 57 20	+60 46.3	1	8.44		1.72				2	K0 IV	1025
		LS I +63 177		1 57 20	+63 03.9	1	10.39		0.81		0.31		2		180
	+62 00336			1 57 21	+62 53.4	1	9.92		0.32				2		1119
		LS I +63 178		1 57 21	+63 15.7	1	11.08		0.87		-0.24		1		180
12122	+59 00378	IDS01539N5956	AB	1 57 22	+60 10.8	1	8.94		0.22				4	B7 III	1119
12255	−21 00356	HR 583		1 57 25	−21 04.0	3	5.41	.005	1.63	.014	1.97	.019	12	M1 III	1024,2035,3007
		an L1159-16		1 57 26	+12 48.8	2	11.73	.005	0.52	.005	0.00	.010	5		234,235
		G 245 - 40		1 57 26	+73 18.1	1	14.12		1.90				1		906
		G 3 - 33	⋆ V	1 57 28	+12 49.9	7	12.28	.018	1.80	.023	1.34	.024	20		234,235,316,694,1705*
12256	−27 00692			1 57 28	−27 10.9	1	9.63		0.83		0.51		1	K1 IV	79
12275	−35 00693			1 57 28	−34 41.0	1	8.85		0.05		0.06		4	A1 V	152
		LS I +62 201		1 57 30	+62 09.6	3	11.17	.017	0.52	.020	-0.38	.032	8	B2: V:	171,180,1012
		VES 664		1 57 32	+58 31.4	1	11.15		0.60		0.13		2		690
12235	+2 00311	HR 582		1 57 33	+02 51.5	3	5.87	.004	0.62	.005	0.19	.005	9	G2 IV	15,1061,3077
	+35 00386			1 57 34	+35 32.9	1	8.93		1.03		0.75		2	G5	1601
12296	−42 00684	HR 588		1 57 35	−42 16.3	3	5.56	.007	1.06	.000			11	K1 III	15,2013,2029
236924	+58 00352			1 57 36	+59 09.0	1	8.40		1.90				3	K7	369
12212	+27 00317			1 57 39	+27 32.6	1	8.67		0.38		-0.02		7	F0	627
12274	−21 00358	HR 585		1 57 39	−21 19.2	7	3.99	.023	1.57	.011	1.90	.020	24	M0.5 III	15,1024,1075,2012*
12211	+27 00318	X Tri	⋆ AB	1 57 42	+27 38.8	2	8.87	.008	0.30	.005	0.09	.010	2	A2	627,627
12150	+57 00451	LS I +57 031	⋆ A	1 57 43	+57 57.7	3	8.62	.020	0.26	.013	-0.53	.000	5	B2 IV	1110,1119,8023
11917	+80 00061			1 57 45	+80 45.6	1	8.94		0.03		-0.11		3		1219
12161	+59 00380	IDS01543N6002	AB	1 57 48	+60 16.2	1	7.88		0.27				2	A8 III	1025
12111	+70 00153	HR 575	⋆ AB	1 57 49	+70 40.0	4	4.52	.029	0.16	.005	0.04	.019	10	A2 V +F2 V	15,1118,3023,8015
12005	+77 00073	HR 572		1 57 49	+77 40.5	1	6.04		1.14				2	K0	71
	+61 00364			1 57 52	+62 00.4	1	10.33		0.27				2	B3	1119
12292	−9 00380	HR 587, AR Cet	⋆ A	1 57 58	−08 45.9	3	5.49	.039	1.51	.016	1.34	.023	8	M3 III	15,1061,3001
12184	+59 00381			1 58 01	+59 33.7	2	8.27	.014	0.24	.000			6	A2 V	70,1119
12193	+59 00382			1 58 02	+59 43.9	3	7.60	.037	0.18	.021	0.07		10	A2 V	70,379,1119
12194	+57 00455			1 58 05	+57 53.9	1	9.61		0.21				2	A2 V	1119
		POSS 53 # 4		1 58 05	+68 00.4	1	16.68		0.96				2		1739
		LS I +61 252		1 58 06	+61 37.3	2	10.66	.103	0.45	.036	-0.52	.040	5	B1 V:	180,1012
12246	+34 00354	IDS01552N3449	A	1 58 08	+35 03.8	1	8.17		0.41		-0.08		2	F3 V	1733
		VES 665		1 58 09	+60 13.0	1	12.10		1.72		0.00		3		690
	+60 00416	LS I +60 213		1 58 10	+60 48.7	1	9.56		0.67		-0.33		2	B0.5III	1012
12192	+61 00365	LS I +61 254		1 58 11	+61 30.7	2	8.83	.009	0.28	.000	-0.47		5	B5 V	180,1119
12183	+64 00281			1 58 14	+64 34.3	1	9.08		0.33		0.35		3	A2	171
		G 3 - 34		1 58 15	+14 46.3	1	10.60		1.28		1.31		1	K6	333,1620
		G 71 - 48		1 58 17	+03 46.8	2	11.58	.015	0.73	.007	0.16	.008	4		1696,3062
12345	−13 00364	G 272 - 138		1 58 22	−13 07.1	1	8.76		0.75		0.29		1	G8 III	79
12477	−66 00125	HR 600		1 58 22	−66 18.5	3	6.09	.003	1.18	.004	1.21	.000	20	K1 III	15,1075,2038
12208	+61 00366	V598 Cas		1 58 23	+61 39.9	3	7.48	.026	1.64	.049	2.01		12	K5 V	369,1025,7008
		VES 666		1 58 24	+62 47.7	1	13.89		1.02		-0.20		3		8107
12369	−25 00789			1 58 24	−24 39.7	2	7.04	.005	1.43	.010	1.63		6	K4 III	2012,3077
12387	−41 00556			1 58 25	−40 58.0	1	7.36		0.66				4	G3 V	2012
12315	+15 00292	IDS01557N1605	AB	1 58 26	+16 19.5	1	7.64		0.42		-0.02		2	F2	1648
	+62 00338	LS I +62 202	⋆ AB	1 58 33	+62 36.1	2	9.22	.009	0.21	.000	-0.66		4	B3 II	180,1119
12243	+59 00385			1 58 34	+60 03.2	2	7.97	.020	0.10	.020	-0.16		6	B8 V	401,1119
	+61 00367			1 58 34	+62 06.3	1	10.02		0.46				3	A5 III	1119
		AAS6,117 # 90		1 58 36	+63 12.	1	10.99		0.45		0.30		4	B9	171
		CS 22171 # 13		1 58 37	−10 15.9	2	13.76	.000	0.20	.001	-0.03	.003	3		966,1736
12384	−27 00700			1 58 39	−26 38.4	1	9.93		0.06		0.07		4	A0 V	152
12394	−25 00791			1 58 40	−24 45.1	1	9.32		0.00		0.04		4	B9.5V	152
12173	+73 00108	HR 579	⋆ A	1 58 41	+73 36.6	1	6.15		0.15		0.13		3	A5 III	1733
12280	+51 00468			1 58 42	+52 24.9	1	8.20		0.44		0.00		2	F5	1566
		WLS 200 25 # 10		1 58 44	+24 11.5	1	10.57		0.55		0.08		2		1375
		CS 22171 # 12		1 58 44	−09 57.6	1	15.58		0.21		0.10		1		1736
12354	+22 00296			1 58 46	+23 09.3	1	6.60		0.32		0.02		3	F2 IV	1501
		LS I +59 105		1 58 48	+59 23.4	1	10.90		0.43		-0.15		3		8107
		CS 22171 # 16		1 58 56	−11 55.8	1	13.08		0.38		-0.20		1		1736
12303	+53 00439	HR 590		1 58 57	+54 14.8	5	5.01	.029	-0.08	.009	-0.31	.013	11	B8 III	15,1079,1363,3023,8015
	+60 00214	LS I +60 214		1 58 57	+60 45.8	1	11.57		0.71		-0.13		3		8107
12438	−30 00703	HR 594		1 59 01	−30 14.5	4	5.34	.007	0.88	.008	0.46	.005	14	G8 III	15,1311,2012,4001
12302	+58 00356	LS I +59 107		1 59 05	+59 26.9	3	8.02	.017	0.26	.011	-0.49	.012	7	B1:V:pe	379,1012,1212
	+59 00387	LS I +59 106		1 59 05	+59 50.4	1	9.59		0.63		-0.30		2	B3 II	1012
	+12 00269			1 59 06	+12 47.5	2	10.16	.014	1.19	.006			2	dK7	1619,1632
	+59 00388	LS I +59 108		1 59 06	+59 44.1	3	9.61	.009	0.60	.000	-0.24	.004	5	B3 II	64,1012,1083
12323	+54 00441	LS I +55 022		1 59 07	+55 23.0	4	8.90	.000	-0.10	.064	-0.93	.000	7	O9 V	64,1011,1012,1083
12279	+64 00282	HR 586		1 59 07	+64 39.7	2	5.99	.010	0.03	.000	0.02	.004	5	A1 Vn	401,595
12216	+71 00117	HR 580		1 59 07	+72 10.8	4	3.96	.022	0.00	.012	0.04	.004	10	A2 V	15,1118,3023,8015
236928	+59 00389	LS I +60 215		1 59 10	+60 01.0	1	9.07		1.01				5	F0 Ib	1119
12463	−37 00777			1 59 10	−37 19.2	1	10.13		0.78		0.25		1	G8	79
12376	+36 00391	IDS01562N3614	AB	1 59 14	+36 28.6	3	8.22	.012	0.83	.009	0.38	.005	10	G9 V +K0 V	292,1381,3016
12402	+27 00320			1 59 16	+28 09.6	2	6.56	.005	1.01	.005	0.79	.084	4	K1 III	1501,3040
12301	+63 00274	HR 589	⋆	1 59 17	+64 09.0	5	5.59	.011	0.37	.014	-0.28	.009	10	B8 Ib	15,64,180,1012,8031
12413	+21 00270			1 59 19	+21 51.8	1	7.19		0.46		-0.01		2	F8 V	1733
12342	+56 00409			1 59 20	+57 04.2	2	8.71	.024	0.05	.015	-0.30		5	B7 IV	401,1119
12341	+57 00461			1 59 22	+57 46.8	1	8.64		0.23				2	A2 IV	1119
		G 71 - 49		1 59 23	+01 39.7	1	14.72		0.94		0.58		2		3062
12447	+2 00317	HR 596	⋆ AB	1 59 27	+02 31.4	10	3.81	.012	0.02	.010	-0.09	.012	31	A0p SiSr	15,176,938,1063,1075*

Table 1

123

HD	DM	Other Id	N Rem	α_{1950}	δ_{1950}	S	V	σ_V	B−V	σ_{B-V}	U−B	σ_{U-B}	n	Spectrum	References
12340 +60 00420				1 59 27	+60 49.5	1	9.28		0.20				2	B9 III	1119
	−37 00779			1 59 28	−36 58.3	1	9.93		0.76		0.23		1		79
	+3 00275	G 71 - 50	★ A	1 59 30	+03 42.7	2	10.62	.000	1.22	.030			2	dK7	1625,1632
		LP 469 - 66		1 59 33	+10 10.0	1	11.69		0.94		0.72		1		1696
		WLS 200 45 # 5		1 59 33	+47 23.4	1	11.67		0.26		0.17		2		1375
		LP 469 - 67		1 59 34	+10 05.8	1	15.61		2.02				5		1663
	−29 00692			1 59 35	−29 25.2	1	11.02		0.96		0.81		1		1696
		WLS 200 25 # 7		1 59 37	+22 37.4	1	11.80		0.59		0.06		2		1375
		WLS 200-10 # 7		1 59 37	−12 07.2	1	12.10		0.66		0.23		2		1375
12365 +60 00423				1 59 38	+60 27.8	3	7.49	.012	0.08	.017	-0.10		6	A0 V	401,1118,1119
		G 159 - 28		1 59 39	+00 19.3	1	13.13		1.13		0.98		1		3062
12524 −45 00659		HR 602		1 59 42	−44 57.2	5	5.14	.005	1.49	.000	1.82	.007	18	K5 III	15,1075,2012,3005,8015
12455 +14 00328				1 59 46	+15 15.3	1	8.03		1.12		0.95		2	K0	1648
12401 +54 00444		XX Per	★ A	1 59 47	+54 59.5	3	8.17	.059	2.13	.019	1.36	.005	7	M4 Ib +B	369,8032,8049
12380 +57 00466				1 59 47	+57 59.0	1	8.54		1.17				2	G5 II	1119
12726 −76 00156				1 59 51	−76 00.0	1	9.16		1.05		0.83		1	G3/5 IV	742
		CS 22171 # 21		1 59 52	−09 17.8	1	12.38		0.20		0.15		1		1736
12479 +12 00271		HR 601		1 59 53	+13 14.2	3	5.97	.013	1.59	.004	1.83		17	M2 III	71,3001,6002
12400 +58 00359				1 59 57	+58 35.5	1	8.78		0.27				2	Am	1025
12443 +45 00523				2 00 00	+45 32.2	1	8.32		0.31		0.18		2	A3	1601
	+61 00370	LS I +61 256		2 00 00	+61 49.6	4	10.08	.020	0.68	.025	-0.41	.049	8	O9 V	64,180,1011,1012
		LS I +63 180		2 00 01	+63 26.0	2	11.25	.219	0.74	.005	-0.40	.015	3		180,8107
12471 +32 00369		HR 599	★ AB	2 00 02	+33 02.6	6	5.50	.003	0.03	.006	0.06	.012	17	A2 V	15,1007,1013,1049*
12230 +76 00063		HR 581	★ A	2 00 02	+77 02.6	2	5.33	.059	0.32	.007	0.01	.045	5	F0 Vn	985,3016
		WLS 200 45 # 7		2 00 04	+43 12.2	1	10.88		0.29		0.19		2		1375
		TOU27,47 T13# 118		2 00 04	+59 53.6	1	10.23		0.27				4		1119
		G 71 - 53		2 00 05	+01 05.1	2	13.66	.029	0.73	.015	0.07	.002	3		1658,3056
12399 +63 00277				2 00 06	+63 59.9	1	7.49		1.79		1.81		6	G5 Ia	1355
12423 +59 00394				2 00 08	+59 31.3	1	8.41		0.40				3	F0 II	1119
		G 3 - 36		2 00 09	+05 28.4	3	12.27	.040	1.45	.005	1.21	.009	8	K5	940,1774,3078
		LP 649 - 27		2 00 11	−08 21.5	1	11.61		0.82		0.36		2		1696
12563 −30 00714		HR 606		2 00 14	−29 54.3	4	6.42	.009	0.14	.004	0.11	.030	8	A3 III	15,79,2030,8071
		02H00M23S		2 00 23	+08 37.7	1	12.04		0.50		-0.14		2		1696
232588 +54 00448				2 00 23	+54 51.8	2	8.63	.000	0.07	.000	-0.71	.000	3	B1.5III	64,1012
		L 88 - 8		2 00 24	−65 13.	1	10.42		0.87		0.51		1		1696
12442 +58 00360				2 00 27	+59 19.4	2	8.79	.014	0.19	.024			6	A1 V	70,1119
12585 −30 00715				2 00 29	−30 03.3	1	9.20		0.67		0.19		2	G3 V	58
12599 −30 00716		IDS01583S3049	A	2 00 30	−30 33.7	1	8.06		1.43		1.59		3	K2/3 III	58
	+61 00371	LS I +62 203		2 00 32	+62 09.4	2	11.02	.052	0.41	.033	-0.27	.014	4	B3:II:pe	180,1012
		LS I +61 257		2 00 33	+61 49.6	1	10.57		0.66		-0.28		3		180
12596 −24 00872				2 00 33	−24 07.6	1	6.31		1.34		1.42		5	K2 III	1628
12583 −16 00356		HR 608		2 00 34	−15 32.8	2	5.86	.001	0.97	.003	0.74		6	G3 IV	58,2009
		WLS 200 25 # 5		2 00 35	+27 05.2	1	10.63		0.57		0.03		2		1375
		SK 207		2 00 36	−74 33.	2	13.91	.000	-0.22	.009	-0.94	.028	4		733,835
236935 +57 00469		LS I +58 056		2 00 38	+58 14.7	1	9.33		0.40		-0.45		2	B1:V:en	1012
12573 −0 00307		HR 607		2 00 38	−00 06.7	4	5.42	.017	0.14	.009	0.15	.024	12	A5 III	15,252,1071,8071
		CS 22171 # 18		2 00 38	−12 43.3	2	14.57	.015	0.05	.030	-0.79	.031	2		1708,1736
12617 −41 00568				2 00 40	−41 11.4	1	9.06		1.02		0.88		2	K3 V	3072
		CS 22171 # 22		2 00 43	−08 27.6	1	12.88		0.08		0.19		1		1736
12559 +17 00306				2 00 44	+18 05.3	1	8.42		0.09		0.04		2	A2	1648
12482 +59 00397				2 00 44	+60 00.3	3	7.40	.027	0.50	.013			7	F6 IV	70,1025,1025
12558 +25 00341		HR 605	★ AB	2 00 49	+25 41.7	1	5.63		0.54		0.02		1	F8 IV +F9 V	292
12533 +41 00395		HR 603	★ A	2 00 49	+42 05.4	1	2.26		1.37		1.58		1	K3-IIb	150
12533 +41 00395		IDS01578N4151	ABC	2 00 49	+42 05.4	6	2.11	.009	1.21	.024	0.92	.000	9	K3-IIb	15,1363,3034,8003*
		G 3 - 38		2 00 50	+12 20.8	1	11.54		1.34		1.28		1	K7	333,1620
12534 +41 00395		HR 604	★ BC	2 00 50	+42 05.5	1	4.84		0.03		-0.12		1	B8 V +A0 V	150
		VES 675		2 00 50	+58 53.2	1	12.05		0.72		0.26		3		8107
12494 +57 00471				2 00 53	+57 46.5	1	8.15		0.98				2	G8 IV	1119
12468 +64 00285		HR 598		2 00 53	+64 51.9	2	6.52	.002	-0.01	.003	-0.04	.009	5	A0 V	595,1733
		JL 278		2 00 54	−51 21.	1	17.10		0.40		-0.40		1		132
12608 −5 00381				2 00 55	−04 34.2	1	7.59		1.00		0.69		3	G5	1657
17579 −89 00010				2 00 55	−89 04.4	1	10.46		1.06		0.90		5	K5	826
12518 +51 00483				2 00 56	+51 43.7	1	6.66		-0.03		-0.31		2	B8	401
12594 +17 00307		HR 609		2 00 58	+18 00.8	1	6.22		1.42				2	K5	71
		MCC 385		2 00 59	−05 08.8	1	11.11		1.41				2	K7	1619
		G 159 - 30		2 00 59	−05 09.0	1	11.22		1.42		1.15		1	K7	3062
		WLS 200-10 # 5		2 00 59	−07 38.6	1	11.60		0.64		0.04		2		1375
	−21 00368	G 272 - 145		2 01 01	−21 27.8	1	11.21		1.46				1		1746
12707 −58 00184				2 01 02	−57 50.1	1	8.86		-0.13		-0.54		3	B6 V	1097
		LP 829 - 2		2 01 04	−21 46.4	1	12.45		0.80		0.34		2		1696
12510 +58 00362				2 01 09	+59 12.6	1	9.95		0.28				4	A0	1119
		VES 676		2 01 09	+62 50.1	1	13.05		0.89		-0.34		3		8107
12642 −4 00324		HR 611		2 01 09	−04 20.5	3	5.61	.009	1.59	.007	1.97	.000	14	K5 I:	15,1071,1075
236938 +59 00399				2 01 10	+60 11.9	1	8.66		0.99				2	G2 V	1025
		LB 3238		2 01 12	−51 30.	1	14.68		-0.24		-0.84		2		132
	+49 00542			2 01 13	+50 00.9	1	8.91		1.53		0.86		4		1723
12641 −1 00285		HR 610	★ A	2 01 14	−00 34.8	4	5.94	.018	0.88	.005	0.52	.021	13	G5 II-III+	15,1071,3077,8100
12641 −1 00285		IDS01587S0049	B	2 01 14	−00 34.8	1	10.77		0.74		0.32		2	G5 V	3077
12530 +56 00416				2 01 15	+56 50.9	2	8.74	.028	0.18	.014	0.15		4	A3 V	401,1119
12529 +59 00400		IDS01577N6002	AB	2 01 15	+60 16.0	1	8.22		0.61				2	F8 V	1025

HD	DM	Other Id	N Rem	α_{1950}	δ_{1950}	S	V	σ_V	B–V	σ_{B-V}	U–B	σ_{U-B}	n	Spectrum	References
12655	−19 00369			2 01 16	−18 51.6	1	8.98		0.00		0.01		4	B9 V	152
12509	+63 00281	LS I +64 085	⋆ AB	2 01 17	+64 08.8	3	7.10	.015	0.33	.025	-0.56	.020	5	B1 III	64,180,1012
		G 71 - 55		2 01 18	−00 46.5	3	10.76	.018	0.48	.007	-0.16	.012	5		1620,1696,3056
12853	−75 00144			2 01 21	−74 41.1	1	6.88		1.63		1.69		3	M5 III	1628
12544	+56 00417	IDS01580N5632	AB	2 01 23	+56 46.3	2	9.28	.025	0.11	.010	-0.63		4	B9	401,1119
	+61 00375	LS I +61 258		2 01 23	+61 50.9	4	9.56	.022	0.53	.029	-0.51	.064	8	B0.5IV	64,180,1012,1119
236940	+55 00489	LS I +56 027		2 01 24	+56 01.4	1	9.61		0.14		-0.74		2	B7	401
12556	+56 00418			2 01 26	+56 31.5	1	9.88		0.08				3	A0	1119
12638	+25 00343			2 01 31	+25 40.9	1	7.14		1.04		0.77		2	G8 III	105
12685	−18 00356			2 01 31	−17 45.2	1	7.01		1.61		1.90		3	M0/1 III	1657
		G 173 - 35		2 01 33	+49 21.7	1	15.22		1.52				1		906
12569	+59 00401			2 01 33	+60 06.5	2	7.51	.009	0.57	.013			5	F7 IV	70,1025
12581	+57 00476	LS I +58 058		2 01 35	+58 08.9	1	9.05		0.09				4	B9 V	1119
12591	+54 00453			2 01 40	+55 23.0	1	6.52		0.30		0.13		3	A9 V	1501
		CI Per		2 01 40	+56 53.5	1	12.35		0.70				1		1772
12724	−41 00574			2 01 40	−40 42.6	1	8.69		0.72		0.24		1	G8 V	258
		SK 208		2 01 42	−74 33.	2	13.45	.000	0.06	.014	0.14	.033	4		733,835
12637	+38 00402			2 01 44	+39 11.8	2	8.24	.000	0.37	.000	0.04	.010	5	F3 III	833,1733
12661	+24 00298	G 72 - 54		2 01 45	+25 10.7	1	7.50		0.74		0.28		2	K0	333,1620
12567	+63 00287	LS I +64 086		2 01 47	+64 02.9	4	8.30	.007	0.38	.015	-0.57	.010	15	B0.5III	64,171,180,1012
12590	+61 00377			2 01 52	+61 32.9	1	8.84		0.19				3	A2 IV	70
12759	−46 00604	IDS01599S4554	AB	2 01 55	−45 39.2	3	7.30	.007	0.69	.012	0.22		7	G3 V	173,258,2012
		WLS 200-10 # 10		2 01 57	−09 45.9	1	11.37		0.90		0.39		2		1375
	+58 00368			2 01 58	+59 05.0	1	9.66		0.26				2		1119
		LP 769 - 86		2 02 01	−15 55.1	1	12.10		0.48		0.01		2		1696
12756	−31 00837			2 02 05	−30 38.8	1	9.92		0.48		-0.02		2	F5 V	58
		G 94 - 29		2 02 07	+19 47.5	1	11.66		1.27		1.17		1	K3	333,1620
12624	+57 00477			2 02 08	+57 31.5	1	9.61		0.07				2	B8 V	1119
	+22 00301	G 35 - 24		2 02 09	+22 34.2	1	9.51		0.89		0.64		1	K1	333,1620
		WLS 200 25 # 9		2 02 09	+25 23.3	1	11.17		1.08		0.87		2		1375
		LP 469 - 98		2 02 12	+10 47.0	1	12.54		0.60		-0.14		2		1696
		AAS6,117 # 93		2 02 12	+63 13.	1	11.98		0.59		0.38		4	A0	171
12746	−18 00357			2 02 15	−18 15.5	1	9.44		-0.01		-0.18		4	B9 V	152
12767	−29 00706	HR 612, ν For		2 02 15	−29 32.2	4	4.69	.005	-0.17	.005	-0.51	.008	14	B9.5p (Si)	15,1075,2012,8015
	+60 00428			2 02 16	+61 26.1	2	9.46	.018	0.48	.009	0.22		5	A7 V	171,1025
12623	+62 00345			2 02 19	+62 56.1	2	7.53	.022	1.17	.009	1.11		5	K0 III	37,1025
12650	+59 00403			2 02 25	+60 02.4	1	8.32		1.40				3	G2 II	1119
13255	−83 00035			2 02 25	−82 44.9	1	8.09		0.35		-0.05		7	F0 IV/V	1628
		LS I +63 181		2 02 26	+63 07.9	2	10.26	.020	0.45	.005	-0.61	.005	4	B0.5: pe	1012,1012
		CS 22171 # 25		2 02 32	−08 11.3	1	14.90		0.08		0.14		1		1736
12659	+63 00290			2 02 34	+64 22.3	1	8.15		0.14		-0.09		3	A0	171
	−18 00359	G 272 - 148		2 02 38	−17 51.0	9	10.19	.012	1.51	.009	1.16	.030	42	dM0	116,830,1006,1075,1197*
		CS 22171 # 23		2 02 39	−08 01.8	2	15.35	.015	0.00	.004	0.10	.024	2		966,1736
		AAS6,117 # 94		2 02 42	+61 23.	1	11.22		0.41		0.28		3	A0	171
		SK 209		2 02 42	−75 12.	2	13.23	.014	-0.01	.005	-0.18	.023	4		733,835
		LP 409 - 26		2 02 47	+16 02.0	1	12.92		1.11		1.03		1		1696
12709	+56 00424			2 02 49	+57 04.1	2	7.96	.015	0.10	.005	-0.47		5	B4 IV	401,1119
12836	−38 00697			2 02 49	−38 01.8	1	10.04		0.55		-0.11		2	F8 V	1696
	+11 00275			2 02 50	+11 56.6	2	10.02	.015	1.01	.011	0.84		3		1619,3072
12878	−51 00510	IDS02010S5057	AB	2 02 52	−50 42.5	1	11.41		0.96		0.60		2	K0 III	1730
		G 272 - 151		2 02 53	−21 03.5	1	13.08		0.68		0.04		2		1696
	+62 00346			2 02 54	+62 31.4	1	10.09		0.27				2		1119
	+58 00372			2 02 55	+59 02.8	1	10.39		0.35		-0.40		2	B3 III	1012
12708	+60 00430			2 02 55	+60 48.4	1	8.68		0.28				3	B9 V	70
	+30 00331	G 72 - 56		2 02 56	+30 32.9	1	9.83		1.17		1.02		1	K6	333,1620
12740	+48 00600			2 02 58	+48 55.1	3	7.94	.007	-0.05	.010	-0.82	.019	6	B1.5II	399,401,555
12794	+13 00329			2 02 59	+14 19.7	1	9.46		0.98				1	G5	882
12793	+16 00237			2 02 59	+17 24.7	1	8.20		0.43		0.00		2	F8	1648
12835	−29 00713			2 03 01	−28 47.8	2	7.34	.001	1.48	.003	1.85		7	K4 III	58,2012
12727	+56 00425	LS I +56 028	⋆ A	2 03 02	+56 48.5	4	9.03	.018	0.08	.004	-0.74	.015	13	B2 III	64,1012,1119,1775
		VES 679	V	2 03 03	+59 25.6	1	11.71		0.46		-0.07		2		8107
12467	+80 00064	HR 597		2 03 07	+81 03.5	2	6.05	.005	0.11	.005	0.07	.010	6	A1 V	985,1733
	−28 00657	G 274 - 145	⋆ A	2 03 07	−28 18.9	1	10.88		1.40				1		1705
		G 274 - 146		2 03 09	−28 17.7	1	12.77		1.52		1.12		2		3056
	−28 00657	G 274 - 146	⋆ B	2 03 09	−28 17.7	1	12.77		1.42				1		1705
12904	−43 00635			2 03 13	−42 57.2	1	8.79		0.37				4	F3 III/IV	2012
12873	−24 00889	G 274 - 147	⋆ AB	2 03 18	−24 37.0	2	8.94	.014	0.80	.002	0.31		5	G8/K0 V	2034,3072
12750	+57 00479			2 03 19	+57 53.9	3	9.04	.007	0.60	.013	0.06		33	G5	934,1025,6011
12891	−30 00728			2 03 19	−30 32.7	1	9.63		0.40		-0.04		3	F0 V	58
	+73 00115	G 245 - 44		2 03 20	+73 46.6	2	9.88	.000	0.71	.018	0.26	.005	3	G3	1658,7010
12889	−24 00891	G 274 - 148	⋆ AB	2 03 22	−24 36.8	2	8.55	.023	0.89	.014	0.49		5	K1 V	2034,3072
12890	−28 00659			2 03 22	−28 21.0	1	8.22		1.57		1.76		4	M3 III	3007
		CS 22171 # 26		2 03 26	−09 16.4	1	14.01		0.10		0.17		1		1736
		Wolf 119		2 03 28	+03 14.4	1	13.01		1.14		1.11		1		1696
		VES 710		2 03 28	+64 07.3	1	11.86		0.64		0.08		1		1543
12831	+19 00324			2 03 30	+20 21.3	1	7.55		0.56		0.12		2	F8	1648
		VES 711		2 03 30	+64 36.5	1	12.90		0.54		-0.13		3		1543
		WLS 200 45 # 9		2 03 32	+45 00.5	1	11.07		0.99		0.69		2		1375
12872	+7 00324	HR 614, WZ Psc		2 03 34	+08 00.6	3	6.30	.004	1.63	.014	1.85	.019	13	M2 III:	15,1071,3042
12871	+13 00331			2 03 40	+13 51.8	1	9.29		0.69				1		882

Table 1

HD	DM	Other Id	N	Rem	α_{1950}	δ_{1950}	S	V	σ_V	B–V	σ_{B-V}	U–B	σ_{U-B}	n	Spectrum	References
236947	+58 00373				2 03 41	+58 33.0	5	8.64	.025	2.19	.011	2.43	.223	9	M2 Iab-Ib	138,369,1083,8016,8023
	+61 00381				2 03 42	+62 15.8	1	9.30		0.44				2	A5 V	1025
12762	+64 00291	V558 Cas			2 03 42	+65 14.7	1	8.70		0.72		0.21		5		171
		AAS6,117 # 115			2 03 42	+65 17.	1	12.26		0.80		0.76		5		171
12823	+46 00515				2 03 45	+46 59.6	1	8.24		0.17		0.11		2	A0	1601
12869	+21 00279	HR 613			2 03 46	+22 24.7	5	5.03	.008	0.12	.011	0.13	.008	25	A2 m	1049,1199,1363,3058,8071
		G 134 - 1			2 03 48	+44 57.2	2	10.24	.031	1.49	.000	1.24		4	M0	1746,7010
12868	+25 00348				2 03 52	+25 35.6	1	7.26		0.18		0.16		3	A3	833
	+64 00292				2 03 52	+65 14.9	1	10.72		0.60		0.27		5		171
12860	+33 00362				2 03 53	+33 59.4	1	9.37		0.09		0.06		2	A0	401
12923	−0 00318	HR 616			2 03 55	−00 12.2	3	6.29	.022	0.89	.010	0.60	.038	9	K0	15,252,1071
12959	−29 00718				2 03 56	−29 27.8	2	9.23	.009	0.28	.002	0.04	.001	9	A8/9 V	58,1775
12899	+13 00333				2 03 59	+14 20.8	1	8.19		0.32		0.08		2	F0	1648
12885	+25 00349	HR 615		★ AB	2 03 59	+25 28.0	2			-0.03	.009	-0.26	.000	5	B9 IV-Vn	1049,1079
12810	+58 00375				2 03 59	+58 33.9	2	8.89	.025	0.67	.070	0.11		3	G5	934,1025
	+39 00469	G 74 - 1			2 04 00	+39 31.9	1	11.00		1.27		1.22		1		1620
12761	+69 00136				2 04 02	+69 38.6	1	8.28		1.45		1.58		3	G5	1733
12986	−40 00544				2 04 09	−39 46.6	1	7.05		1.68		1.98		4	M3 III	1673
12843	+56 00427				2 04 16	+56 35.6	1	9.14		0.23				3	A3 IV	1119
236948	+57 00489	VX Per			2 04 18	+58 12.4	5	9.05	.095	1.11	.056	0.94	.249	6	G0	934,1083,1399,1772,6011
		CS 22171 # 27			2 04 20	−10 18.7	1	14.41		0.10		0.16		1		1736
12929	+22 00306	HR 617			2 04 21	+23 13.6	17	2.01	.012	1.15	.007	1.12	.009	157	K2 IIIab	1,3,15,247,637,1004*
12819	+62 00349	IDS02007N6240	A		2 04 21	+62 54.8	2	8.24	.000	0.17	.013	0.16		6	A3 V	171,1025
12883	+50 00455				2 04 23	+50 51.6	1	8.32		0.07		-0.06		2	A0	1566
12855	+57 00490				2 04 23	+58 01.7	4	8.43	.002	1.27	.006	1.30	.009	43	K0	1025,1083,1655,6011
	−30 00733	G 274 - 151			2 04 23	−29 54.0	1	11.06		0.57		-0.09		2		1696
12867	+57 00492	LS I +57 033			2 04 25	+57 28.5	3	9.41	.005	0.15	.005	-0.67	.000	6	B1 V	64,1012,1119
12854	+58 00377				2 04 25	+59 15.3	1	9.48		0.28				4	A2 V	1119
12856	+56 00429	LS I +56 029			2 04 26	+56 52.1	5	8.54	.009	0.23	.007	-0.77	.004	10	B0pe	1012,1083,1110,1212,8023
	+60 00435	LS I +61 262			2 04 28	+61 09.9	2	9.68	.000	0.47	.000	-0.38	.000	3	B2 III	64,1012
		VES 678			2 04 47	+62 38.6	1	13.61		0.65		-0.13		3		8107
12800	+70 00163	G 244 - 50			2 04 47	+71 19.1	2	6.56	.015	0.53	.005	-0.03	.019	3	F8	1658,3026
	+61 00382	LS I +61 263			2 04 48	+61 32.6	1	10.70		0.83		-0.14		2	B1:V:	1012
		SK 210			2 04 48	−75 01.	2	13.57	.005	-0.13	.005	-0.73	.019	4		733,835
	+76 00069				2 04 53	+77 20.8	1	9.70		0.63		0.14		4		1723
12983	+34 00376				2 04 54	+35 19.0	1	7.54		0.32		0.00		2	F0	1625
12906	+58 00378				2 04 54	+59 22.0	1	9.58		0.31				4	A1 V	1119
13058	−30 00737				2 04 54	−29 46.9	1	9.61		0.59		0.12		2	G0 V	58
		G 3 - 40			2 04 55	+13 40.8	1	12.51		1.46		1.25		1		333,1620
12920	+59 00413				2 04 57	+59 43.9	1	9.46		0.32				3	A1 V	1119
12882	+64 00295	LS I +64 088			2 04 57	+64 48.0	2	7.58	.075	0.35	.025	-0.54	.005	3	B6 Ia	180,1012,8100
12928	+58 00379				2 04 59	+58 37.7	2	7.83	.020	0.20	.000	-0.22		5	B8 III	401,1119
13107	−48 00557				2 04 59	−48 16.5	1	9.31		0.27		0.14		4	A2mA3-F2	152
		G 159 - 35			2 05 00	−00 49.8	1	10.49		1.23		1.23		2	K7	3077
13018	+17 00315				2 05 01	+17 47.5	1	6.62		0.22		0.11		2	A3	1648
13043	−1 00293	IDS02024S0105	A		2 05 02	−00 51.0	5	6.89	.011	0.62	.003	0.15	.004	10	G2 V	258,742,908,1003,3077
13043	−1 00293	IDS02024S0105	B		2 05 02	−00 51.0	1	10.52		1.24		1.23		2	K7	3016
	+62 00352		AB		2 05 05	+62 27.5	1	10.06		0.26				2	A9 III	1119
12953	+57 00494	HR 618, V472 Per			2 05 10	+58 11.2	11	5.69	.021	0.61	.007	-0.01	.009	45	A1 Iae	15,1012,1077,1083*
13083	−29 00728				2 05 11	−29 08.0	1	7.68		1.07		0.94		3	K1 III	58
		TOU27,47 T13# 153			2 05 12	+62 50.0	1	9.85		0.18				2		1119
12952	+58 00380				2 05 17	+58 51.3	1	9.86		0.23				3	A0 V	1119
12964	+57 00495	IDS02018N5803	A		2 05 18	+58 17.8	1	9.42		0.14				3	B6 V	1119
13057	+15 00305				2 05 20	+15 34.1	3	7.07	.041	1.64	.021	1.96	.010	5	K0	882,1648,3040
		G 94 - 33			2 05 20	+17 25.1	1	14.41		0.79		0.05		2		1658
13106	−29 00730				2 05 21	−29 34.3	1	9.23		0.33		-0.01		2	F0 V	58
12971	+59 00415				2 05 24	+59 57.5	2	7.97	.000	0.13	.005			6	A3p	1025,1119
13013	+43 00431	HR 619			2 05 25	+44 13.4	1	6.38		0.96		0.63		2	G8 III-IV	1733
13072	+10 00292				2 05 27	+10 57.2	1	7.77		1.62		1.92		2	K5	1733
13041	+37 00486	HR 620			2 05 28	+37 37.4	5	4.81	.019	0.12	.006	0.15	.010	14	A5 IV-V	15,1049,1363,3016,8015
	+27 00335	G 72 - 59		★ B	2 05 30	+28 04.3	1	10.55		0.88		0.57		1	A0	1620
	+27 00335	G 72 - 58		★ A	2 05 30	+28 04.6	1	9.99		0.79		0.33		1	F5	1620
	+55 00521				2 05 31	+55 57.3	1	10.56		0.06		-0.29		2	B6e	1543
	+55 00522				2 05 31	+56 06.4	1	10.87		0.26		-0.47		2		1543
		CS 22171 # 37			2 05 32	−09 17.9	1	14.93		0.35		-0.19		1		1736
12993	+57 00498	LS I +57 035			2 05 33	+57 41.7	6	8.96	.020	0.20	.008	-0.80	.008	12	O6.5V	64,1011,1012,1083*
		G 134 - 3			2 05 34	+44 51.7	1	10.32		0.63		0.05		1	G6	1620
12994	+56 00431				2 05 34	+56 48.0	2	8.47	.034	0.01	.015	-0.36		5	B7 IV	401,1119
	−27 00737	SS For			2 05 36	−27 06.1	2	9.49	.005	0.09	.015	0.06	.040	2	F0	668,700
	+30 00338	G 72 - 60			2 05 37	+31 09.2	3	10.28	.004	0.56	.000	-0.10	.004	5	G0	516,1620,1658
		G 173 - 39			2 05 38	+49 12.9	1	12.47		1.54				1	M5:h	1746
13130	−12 00395				2 05 42	−12 07.0	1	8.21		1.10		0.92		2	K0 III	1375
	+2 00335	G 73 - 28			2 05 43	+02 49.5	2	10.01	.020	1.05	.000	0.97		2	dK7	1620,1632
13157	−31 00856				2 05 43	−30 44.6	1	9.82		0.14		0.14		4	A0 IV	152
13071	+38 00416				2 05 50	+39 06.7	1	7.64		0.08		0.05		3	A1 V	833
13037	+57 00499				2 05 55	+57 44.5	1	8.43		0.16				3	A3 V	1119
		G 35 - 29			2 05 56	+25 00.1	3	13.22	.010	-0.04	.007	-0.85	.009	7		203,1620,3060
	−19 00391				2 05 56	−19 00.3	1	11.23		1.05		0.92		1	K5	1696
13167	−25 00830				2 05 56	−24 55.9	1	8.34		0.71				4	G3 V	2012
		L 89 - 27			2 05 56	−66 48.7	1	11.55		1.50		1.08		1		3073

HD	DM	Other Id	N	Rem	α_{1950}	δ_{1950}	S	V	σ_V	B–V	σ_{B-V}	U–B	σ_{U-B}	n	Spectrum	References
	+57 00500				2 05 57	+58 13.5	1	10.17		0.26		-0.37		2		1083
13051	+56 00432	V351 Per	6		2 05 58	+56 45.3	6	8.70	.014	0.13	.007	-0.72	.004	12	B1 IV:	1012,1083,1110,1119*
13022	+58 00383	LS I +58 063	1		2 05 58	+58 32.8	1	8.76		0.32				3	O9.5Ia	1119
13182	−29 00738		1		2 05 58	−28 58.4	1	8.71		0.99		0.74		3	K0 III	58
13078	+40 00442	BX And		⋆ A	2 05 59	+40 33.5	1	8.91		0.44		0.08		1	F2 V	627
13078	+40 00442	IDS02029N4019		B	2 05 59	+40 33.5	1	10.86		0.68		0.28		12		627
13038	+57 00501				2 06 00	+57 43.5	1	8.52		0.18				2	A5 V	1119
13165	−19 00393				2 06 01	−18 42.2	1	9.76		0.15		0.03		4	A0 V	152
		St 2 - 109a			2 06 04	+59 23.1	1	9.94		1.37		1.06		1		11
		St 2 - 139a			2 06 04	+59 23.1	1	10.20		0.75		-0.33		1		11
		St 2 - 89a			2 06 04	+59 23.1	1	12.26		0.78		0.34		2		11
13036	+58 00384	St 2 - 1			2 06 04	+59 23.1	2	8.55	.000	0.53	.000	-0.41		6	B0 Ib	1012,1119
13050	+57 00502				2 06 05	+57 51.9	1	8.52		0.34				3	A7 III	1119
		G 159 - 40			2 06 06	−00 35.7	2	10.93	.045	1.26	.010	1.15	.005	2	K5	1696,3062
	+66 00187				2 06 15	+66 54.4	1	9.51		1.95				3		369
13067	+57 00503				2 06 18	+57 48.6	1	9.55		0.20				2	A2 V	1119
13233	−29 00743				2 06 20	−28 47.2	1	9.23		1.25		1.44		3	K2 III	58
		Steph 238			2 06 23	+15 55.6	1	11.66		1.37		1.28		1	K7	1746
13215	−18 00374	HR 625			2 06 23	−18 00.9	2	6.08	.019	1.64	.024	2.02		6	M1 III	2007,3055
13232	−26 00767				2 06 23	−25 58.7	2	8.88	.010	0.29	.019	0.12		14	A0mA7-F2	355,2012
		SK 211			2 06 24	−74 54.	1	13.31		-0.11		-0.57		3		733
		G 3 - 41			2 06 27	+14 21.6	1	13.40		1.54		1.23		1		1658
		CS 22171 # 36			2 06 27	−09 26.7	1	11.66		0.25		0.08		1		1736
		L 296 - 102			2 06 30	−49 28.	1	14.21		0.55		-0.10		2		3060
13088	+58 00388	St 2 - 2			2 06 32	+58 54.9	2	8.15	.029	0.20	.000	0.19		5	A1 V	11,1119
	+62 00356				2 06 33	+62 56.5	1	10.01		1.15				1		592
13174	+25 00355	HR 623		⋆ A	2 06 34	+25 42.3	5	4.99	.012	0.33	.008	0.16	.023	9	F2 III	39,105,254,1363,3026
13161	+34 00381	HR 622			2 06 34	+34 45.1	8	3.00	.004	0.14	.010	0.10	.016	20	A5 III	1,15,1077,1355,1363*
236950	+58 00389				2 06 34	+58 42.1	1	8.31		1.68				2	K7	1025
		G 159 - 42			2 06 34	−02 36.6	1	13.54		1.25		1.07		1		3062
13138	+41 00412				2 06 35	+41 46.3	1	7.75		-0.06		-0.36		2	B9	401
		St 2 - 4			2 06 37	+59 18.4	1	10.01		0.33		0.29		2	A0	11
13201	+16 00247	HR 624			2 06 38	+16 59.4	2	6.41	.000	0.42	.005	-0.06	.000	4	F5 V	1648,1733
		St 2 - 3			2 06 38	+58 59.7	1	10.82		0.44		0.19		3	A5	11
13125	+53 00459				2 06 42	+54 05.7	1	8.19		1.57		1.98		2	K5	1566
12881	+78 00071	IDS02014N7913		B	2 06 42	+79 27.5	1	7.17		0.30		0.08		4	Am	1084
		St 2 - 5			2 06 43	+59 12.6	1	11.29		0.35		0.21		2	A0	11
	+59 00420				2 06 43	+59 55.2	1	9.27		1.47				1	K5 I + A	138
	+63 00300	LS I +63 182			2 06 44	+63 51.4	1	9.69		0.55		-0.55		2	Bpe	1012
		V395 Cas			2 06 45	+63 03.7	1	10.46		1.07				1		592
13137	+53 00460	HR 621			2 06 46	+53 36.5	2	6.30	.012	0.95	.002	0.68		5	G8 III	71,1501
	−17 00400				2 06 46	−16 34.9	1	10.91		1.41				1	K7	1746
13260	−23 00799				2 06 46	−23 13.6	1	7.23		1.46		1.70		3	K3 III	1657
13136	+55 00529	KK Per			2 06 48	+56 19.4	6	7.74	.034	2.23	.013	2.38	.044	9	M2 Ib	369,1083,1110,1355*
		St 2 - 6			2 06 48	+59 01.4	2	9.56	.000	0.31	.007	0.13		7	B8	11,1119
		G 73 - 31			2 06 49	+05 26.6	1	11.39		1.31		1.23		1	K7	1620
13122	+59 00422	St 2 - 7			2 06 50	+59 44.7	3	6.66	.023	0.33	.005	0.12		7	F5 II	11,1118,1119,8100
13111	+62 00357				2 06 54	+63 21.8	1	8.05		0.30		0.14		2	A0	171
		SK 212			2 06 54	−74 35.	1	13.22		-0.01		-0.20		3		733
		St 2 - 8			2 06 56	+58 57.7	1	10.52		0.26		-0.01		3	A0	11
		VES 716			2 06 56	+63 26.0	1	12.20		0.82		-0.35		2		1543
		CS 22171 # 31			2 06 57	−10 48.4	1	13.93		0.39		-0.20		1		1736
		St 2 - 9			2 06 59	+59 21.6	1	10.52		0.41		0.33		2	A2	11
	+59 00423	St 2 - 10			2 07 01	+59 34.7	1	10.65		0.46		0.32		2	A2	11
12927	+79 00063	IDS02014N7913		A	2 07 02	+79 27.4	1	6.47		0.24		0.07		4	A3	1084
13287	−21 00383				2 07 03	−20 50.7	1	9.68		1.35		1.60		1	K2/3 (III)	742
13248	+12 00292				2 07 04	+12 56.4	1	7.72		0.24		0.16		2	B9	1733
		G 74 - 4			2 07 07	+35 11.9	1	13.77		1.50		1.02		1		1620
13306	−29 00751				2 07 08	−29 10.3	1	9.50		0.53		0.02		3	F8/G0 V	58
13336	−44 00632	HR 632			2 07 10	−43 45.1	2	5.84	.005	1.20	.005	1.08		6	K1 III	2007,3005
13318	−31 00865				2 07 13	−30 43.5	1	8.96		0.45		0.00		2	F5 V	58
13209	+52 00528				2 07 17	+52 58.8	1	8.20		0.05		0.00		2	A0	401
13186	+57 00506				2 07 17	+58 12.6	1	9.48		1.10				2	K0	1025
13305	−24 00921	HR 630			2 07 17	−24 34.9	1	6.48		0.29				4	F0 IV/V	2035
	−39 00634				2 07 17	−39 13.9	1	10.91		0.68		0.10		2		1696
16477	−88 00022	IDS02332S8850		AB	2 07 18	−88 36.0	1	7.94		0.08		-0.10		2	B9/9.5V	1117
13247	+32 00390	IDS02044N3254		A	2 07 19	+33 07.8	1	7.76		0.01		-0.34		2	A0	401
	+62 00359				2 07 19	+63 03.6	1	9.18		1.86				1		592
13274	+14 00349				2 07 21	+15 15.4	2	8.72	.019	0.77	.007	0.20		3	G0	882,3040
		VES 717			2 07 22	+61 21.0	1	13.29		1.01		0.31		1		1543
	+61 00386				2 07 23	+61 40.4	1	8.89		1.36		1.33		17	K0	848
		AAS42,335 T3# 2			2 07 24	+61 40.	1	11.91		0.78		0.25		9		848
	+59 00427	St 2 - 11			2 07 26	+59 43.6	1	10.08		0.41		0.32		2	A0	11
	+29 00366	G 74 - 5		⋆ B	2 07 29	+29 34.5	5	8.77	.012	0.58	.002	-0.09	.010	9	F8 V	979,1620,1658,3026,8112
13335	−29 00752				2 07 29	−29 14.6	1	7.44		1.59		1.97		2	K5 III	58
13349	−29 00753				2 07 31	−28 50.5	1	8.93		0.63		0.12		2	G1/2 V	58
13296	+14 00350				2 07 32	+14 47.2	1	8.25		1.21				2	K2	882
13185	+63 00301				2 07 33	+63 54.7	1	8.79		0.22		0.11		2	A0	171
13208	+58 00391	St 2 - 13			2 07 35	+59 00.4	2	9.45	.000	0.26	.010	0.20		5	A1 V	11,1119
		CS 22171 # 34			2 07 35	−10 14.4	1	13.29		0.41		-0.23		1		1736

Table 1

HD	DM	Other Id	N	Rem	α_{1950}	δ_{1950}	S	V	σ_V	B–V	σ_{B-V}	U–B	σ_{U-B}	n	Spectrum	References
13207	+59 00429	St 2 - 11a			2 07 36	+59 44.8	2	8.23	.064	1.23	.015	0.96		5	G9 III	11,8092
	+59 00428	St 2 - 12			2 07 36	+59 50.4	1	10.17		0.38		0.30		2	A0 V	11
13196	+61 00387	IDS02040N6153	AB		2 07 40	+62 06.8	1	8.26		0.42		0.26		15	F0	848
		St 2 - 14			2 07 41	+59 19.6	1	10.90		0.55		0.35		2	A2	11
		AAS42,335 T3# 3			2 07 42	+61 41.	1	11.19		0.70		0.19		13		848
		POSS 53 # 1			2 07 43	+67 34.5	1	14.28		1.17				1		1739
13314	+14 00352				2 07 46	+14 55.0	1	8.19		1.66				2	K0	882
13359	−21 00384				2 07 47	−21 00.0	2	9.70	.000	0.71	.000	0.20		4	G1/2 V	742,1594
		AAS42,335 T3# 5			2 07 48	+61 53.	1	13.17		0.65		0.38		6		848
13294	+38 00425	IDS02048N3834	AB		2 07 50	+38 48.3	2	5.64		-0.01	.005	-0.10	.015	6	B9 V	105,1049
		G 4 - 1			2 07 51	+14 41.4	1	13.67		1.09		0.92		1		1658
13325	+18 00277	HR 631			2 07 51	+19 15.9	2	5.70	.001	1.64	.005	1.91		7	M3 IIIab	71,3001
13294	+38 00425	HR 628		⋆ A	2 07 51	+38 48.5	1	6.11		-0.07		-0.17		1	A1 Vn	105
		WLS 200-10 # 8			2 07 51	−09 49.2	1	11.84		0.47		-0.03		2		1375
13376	−30 00755				2 07 51	−30 13.9	1	8.86		0.55		0.00		3	F7 V	58
		St 2 - 15			2 07 52	+59 07.8	1	11.16		0.47		-0.27		2	B2	11
		AAS42,335 T3# 6			2 07 54	+61 47.	1	10.65		1.09		0.71		15		848
13385	−27 00746				2 07 58	−26 56.5	1	10.17		0.03		0.03		4	A0 V	152
13410	−37 00819				2 07 58	−37 37.0	1	8.49		0.51				4	F5 V	2012
13267	+56 00438	HR 627		⋆ AB	2 07 59	+57 24.6	9	6.35	.023	0.33	.005	-0.43	.013	23	B5 Ia	15,154,1012,1083,1110*
		AAS42,335 T3# 7			2 08 00	+61 53.	1	12.92		0.78		0.32		5		848
13268	+55 00534	LS I +55 026			2 08 03	+55 55.4	5	8.18	.005	0.13	.005	-0.83	.000	9	O8 Vnn	64,401,1011,1012,1083
13423	−44 00638	HR 636			2 08 05	−44 03.1	4	6.32	.005	0.90	.000			18	G8 III	15,1075,2013,2029
		LP 529 - 56			2 08 06	+09 10.4	1	12.04		1.25		1.29		1		1696
13266	+57 00509				2 08 07	+58 03.0	1	9.69		0.40				2	F2	1025
13256	+60 00447	LS I +60 222			2 08 07	+60 28.7	2	8.63	.013	1.18	.013	-0.02	.009	5	B1 Ia	171,1012
13357	+13 00343	IDS02054N1313	AB		2 08 09	+13 26.9	1	7.65		0.70				1	G5	882
		CS 22171 # 32			2 08 11	−10 15.1	1	14.53		0.46		-0.22		1		1736
		G 74 - 7			2 08 13	+39 41.5	4	14.52	.022	0.34	.029	-0.46	.022	12	DA	308,316,1620,3078
		G 133 - 71			2 08 13	+43 52.8	1	11.95		1.39		1.13		1		1620
13098	+77 00076				2 08 18	+77 40.3	1	8.50		0.16		0.10		2	A2	1375
13323	+46 00532				2 08 20	+46 58.5	1	8.06		-0.03		-0.38		2	B8	1601
13406	−13 00400				2 08 20	−13 09.7	1	7.42		0.32		0.12		4	A3mA7-F3	152
13364	+22 00312				2 08 21	+22 58.9	1	8.30		0.42		-0.04		2	F5 IV	1375
13363	+25 00362	HR 633			2 08 21	+25 42.1	2	6.01	.010	1.36	.006	1.55		5	K3 III	71,1501
		VES 719			2 08 22	+58 14.4	1	11.15		0.32		-0.17		1		1543
		St 2 - 16			2 08 23	+58 43.0	1	10.70		0.50		0.27		2	A5	11
13445	−51 00532	HR 637			2 08 25	−51 04.1	5	6.12	.010	0.82	.004	0.45	.004	11	K1 V	15,258,2012,2017,3078
13435	−28 00694		A		2 08 27	−28 27.3	4	7.06	.004	1.01	.004	0.87	.009	16	K1 III	258,1075,1775,3008
13372	+30 00347	HR 634			2 08 29	+31 17.5	4	6.24	.005	0.11	.006	0.12	.019	21	A1 m	985,1049,1199,1501
13433	−16 00389				2 08 33	−15 33.0	1	8.19		0.08		0.07		4	A0 V	152
	+45 00565	G 134 - 9			2 08 35	+45 31.7	1	9.76		0.93		0.72		1	K2	1620
	+45 00564	G 134 - 10			2 08 35	+45 41.4	1	9.65		0.89		0.62		1	K1	333,1620
13331	+56 00443				2 08 37	+57 04.7	1	8.79		0.04		-0.32		2	B8 III	401
13322	+57 00511				2 08 38	+58 06.0	1	9.59		0.35				2	A0	1119
	+8 00335	G 4 - 2			2 08 40	+09 23.3	2	10.68	.005	0.74	.000	0.18	.019	3	K0	333,1620,1696
13442	−15 00380				2 08 40	−15 18.3	2	6.71	.024	0.08	.015	0.08	.005	6	A2 IV/V	152,252
		HIC 10266			2 08 41	+56 02.3	1	11.13		1.28		1.16		5		1723
	+59 00430	St 2 - 19			2 08 41	+59 51.6	1	10.19		0.42		0.34		2	A2 V	11
13421	+7 00347	HR 635			2 08 42	+08 20.2	3	5.62	.004	0.56	.000	0.12	.008	9	G0 IV	15,1256,3077
13321	+58 00392	St 2 - 17			2 08 42	+58 48.1	1	8.76		1.90		2.15		2	K2	11
13222	+73 00121	HR 626			2 08 42	+73 47.7	1	6.29		0.91				2	G8 III	71
13470	−30 00761				2 08 42	−30 31.6	1	9.89		0.51		-0.07		3	F6 V	58
		St 2 - 18			2 08 43	+59 03.3	1	11.25		0.51		0.24		2	B8	11
13570	−61 00188				2 08 47	−61 19.9	1	7.80		1.59		1.93		8	M0 III	1704
13578	−64 00156	IDS02074S6450	A		2 08 47	−64 35.4	1	7.49		0.62				4	G3 IV	2012
13338	+57 00512	LS I +57 041			2 08 48	+57 42.4	3	9.05	.047	0.26	.009	-0.55	.000	5	B1 V	64,1012,1119
		SK 213			2 08 48	−74 38.4	1	13.68		-0.15		-0.83		3		733
	+53 00471	LS I +54 006			2 08 49	+54 24.2	1	10.02		0.00		-0.75		2	B5	401
		St 2 - 20			2 08 49	+59 19.0	1	12.66		0.75		0.24		4		11
12918	+82 00051				2 08 52	+83 19.7	1	6.43		0.96		0.74		3	K0	985
	+58 00393	St 2 - 21			2 08 53	+59 23.8	1	10.18		0.38		0.30		2	B9 V	11
		St 2 - 22		⋆	2 08 54	+59 40.0	3	11.01	.020	0.96	.010	-0.13	.000	6		11,1011,1012
		AAS42,335 T3# 8			2 08 54	+62 09.	1	10.29		0.34		-0.18		12		848
13500	−30 00766				2 08 54	−30 27.8	1	9.33		0.48		-0.03		3	F5/6 V	58
13456	−10 00447	HR 638			2 08 55	−10 17.1	3	6.00	.004	0.40	.005	-0.03		9	F5 V	15,2030,3053
	+61 00389				2 08 58	+61 55.6	1	9.60		0.49		0.46		16	F0	848
13370	+56 00445				2 09 00	+57 04.0	1	9.03		0.10				3	B9.5V	1119
		St 2 - 24			2 09 00	+58 55.5	1	11.38		0.54		0.21		2	A7	11
		St 2 - 23			2 09 00	+58 56.4	1	11.37		0.53		-0.30		2	B6	11
		AAS42,335 T3# 10			2 09 00	+61 57.	1	11.61		0.72		0.22		12		848
13512	−29 00761				2 09 00	−29 24.6	1	9.25		1.22		1.33		2	K2 III	58
13468	−2 00375	HR 639			2 09 03	−02 03.6	3	5.93	.013	0.97	.005	0.71	.012	9	K0	15,1256,3016
	+57 00513	LS I +57 044			2 09 05	+57 51.8	2	9.50	.000	0.28	.005	-0.56	.000	3	B1 III	64,1012
		WLS 200-10 # 11			2 09 06	−07 23.0	1	11.39		0.50		-0.05		2		1375
	+59 04310	St 2 - 25			2 09 08	+59 27.5	1	10.35		0.40		0.33		3	A0	11
		St 2 - 26			2 09 10	+58 59.4	1	12.55		0.71		0.00		2	B3	11
	+57 00515	LS I +57 045			2 09 12	+57 27.5	1	9.69		0.37		-0.63		2	B2pe	1012
		St 2 - 27			2 09 13	+59 20.6	1	10.57		0.52		0.39		3	A0	11
	+61 00390	IDS02055N6154	A		2 09 13	+62 08.4	1	9.75		0.49		0.09		14	F5	848

HD	DM	Other Id	N Rem	α_{1950}	δ_{1950}	S	V	σ_V	B−V	σ_{B-V}	U−B	σ_{U-B}	n	Spectrum	References
		St 2 - 29		2 09 16	+59 27.2	1	11.42		0.60		0.29		3	F0	11
	+26 00369			2 09 18	+26 55.3	1	10.18		1.29		1.26		2		1375
13379	+58 00395	St 2 - 28		2 09 19	+59 11.6	1	8.83		0.38		0.29		std	A0 V	11
13391	+56 00446	IDS02059N5644	B	2 09 20	+56 59.5	1	8.72		0.57				2	G1 V	1025
		St 2 - 30		2 09 20	+58 55.8	1	11.60		0.60		0.19		2	F0	11
13588	−47 00663			2 09 21	−46 49.2	1	7.92		0.18		0.13		4	A1mA1-F0	152
13402	+58 00396	St 2 - 31		2 09 22	+59 18.3	6	8.07	.010	0.59	.012	-0.42	.001	19	B0 Ib	11,1012,1083,1110*
		L 224 - 36		2 09 24	−52 10.	1	13.54		0.56		-0.20		2		1696
13403	+56 00449	IDS02059N5644	A	2 09 25	+56 58.4	4	7.00	.013	0.65	.012	0.11	.020	6	G3 V	979,1025,3026,8112
13480	+29 00371	HR 642, TZ Tri	★ AB	2 09 28	+30 04.2	2	4.95	.005	0.79	.003	0.31		5	G5 III+F5 V	938,1363
		St 2 - 33		2 09 29	+59 35.0	1	10.80		0.63		0.36		2	A2	11
		St 2 - 32		2 09 32	+59 23.3	1	12.04		0.94		0.38		4	F8	11
13412	+58 00397	St 2 - 34		2 09 34	+58 33.9	2	7.98	.030	0.26	.000	0.10		3	A9 III(m?)	11,1119
		G 3 - 43		2 09 35	+11 17.0	1	13.65		1.56		1.19		1		333,1620
13420	+57 00517	LS V +58 001		2 09 37	+58 19.3	2	9.61	.005	0.26	.005	-0.52		4	B4 III	1083,1119
		LP 709 - 53		2 09 37	−14 14.5	1	11.49		0.43		-0.19		2		1696
		St 2 - 35		2 09 38	+59 25.8	1	10.66		0.47		0.36		3	A2	11
13428	+58 00398	St 2 - 36		2 09 43	+59 05.2	1	9.87		0.36		0.28		2	A1 IV	11
		St 2 - 37		2 09 44	+59 40.2	1	11.34		0.62		0.37		2	A2	11
		CS 22175 # 2		2 09 45	−11 29.1	1	15.16		0.06		0.14		1		1736
13522	+23 00297	HR 644		2 09 47	+23 56.0	1	5.96		1.37				2	K0	71
13452	+53 00474	IDS02064N5344	AB	2 09 48	+53 59.1	1	8.06		0.01		-0.22		2	A0	401
	+59 00433	St 2 - 39		2 09 48	+59 35.0	1	9.97		0.51		0.40		2	A1 V	11
13599	−29 00767			2 09 48	−29 11.5	2	8.45	.000	1.32	.006	1.50	.013	9	K3 III	58,1775
13507	+39 00496			2 09 50	+40 26.1	1	7.23		0.67		0.18		9	G0	627
		WLS 200 25 # 8		2 09 51	+25 19.6	1	11.73		0.60		0.05		2		1375
		St 2 - 38		2 09 52	+59 25.5	1	12.22		0.76		0.26		3	F2	11
	+65 00237		V	2 09 53	+65 43.2	1	9.95		2.30				3	M2	369
	+2 00348	G 73 - 35		2 09 54	+03 22.8	8	10.05	.023	1.43	.010	1.10	.024	38	dM3	116,694,830,1006,1075,1705*
13437	+58 00399	St 2 - 43		2 09 54	+58 57.8	3	7.59	.007	1.22	.017	0.96		8	G5 II	11,1119,8092
13436	+59 00434	St 2 - 40		2 09 54	+59 43.3	2	8.38	.020	0.35	.000	0.02		3	F0	11,1119
		St 2 - 42		2 09 55	+59 13.5	1	10.14		0.51		0.41		2	A0	11
13450	+59 00435	St 2 - 41		2 09 57	+59 53.4	1	9.24		0.43		0.36		2	A0 V	11
		St 2 - 44		2 09 58	+59 09.0	1	10.54		0.44		0.32		3	B8	11
		LP 885 - 29		2 09 59	−27 58.1	1	11.72		1.01		0.94		2		1696
13627	−29 00770			2 09 59	−29 13.8	1	9.36		0.51		0.00		2	F6 V	58
13567	+13 00351			2 10 00	+13 41.1	1	7.10		1.15		1.00		2	K0	1648
13555	+20 00348	HR 646		2 10 00	+20 58.6	2	5.28	.005	0.43	.010	-0.05	.015	4	F5 V	254,3037
		G 134 - 14		2 10 00	+46 02.8	2	10.26	.019	1.17	.029	1.14	.012	4		1620,7010
13666	−47 00664			2 10 00	−47 24.3	1	6.82		1.40		1.52		7	K2/3 III	1628
236954	+58 00400	St 2 - 45	★	2 10 03	+58 56.2	2	9.40	.000	0.64	.004	-0.22	.004	5	B3 Ib-II	11,1012
13520	+43 00447	HR 643		2 10 04	+43 59.9	3	4.83	.009	1.48	.005	1.73	.010	7	K3.5 III	1355,1363,3053
		St 2 - 46		2 10 05	+58 48.0	1	10.92		0.49		0.26		2	A8	11
13566	+19 00332			2 10 07	+19 35.2	1	7.91		1.09		0.90		2	K0	1648
13494	+55 00543	V352 Per		2 10 08	+56 20.2	2	9.30	.000	0.18	.000	-0.65	.000	3	B1 III	64,1012
		St 2 - 47		2 10 08	+59 04.1	1	12.05		0.64		0.46		2	A5	11
13476	+57 00519	HR 641	★	2 10 09	+58 19.6	7	6.43	.020	0.60	.013	0.22	.028	19	A3 Iab	15,1012,1083,1110*
		St 2 - 48		2 10 09	+59 19.1	1	12.11		0.91		0.36		2		11
13464	+58 00401	St 2 - 49		2 10 10	+59 23.1	2	9.41	.009	0.59	.011	0.55		5	A0 V	11,1119
	+54 00487			2 10 11	+55 13.6	1	9.94		1.84				3		369
	+57 00520	LS I +58 067		2 10 11	+58 12.2	2	9.62	.000	0.39	.000	-0.51	.000	3	B1 II	64,1012
	+59 00436	St 2 - 50		2 10 12	+59 44.5	1	9.96		0.83		0.64		2	cB5	11
13612	−3 00336	HR 650	★ A	2 10 14	−02 37.6	2	5.67	.005	0.56	.005	0.07	.005	6	F8 V	1355,3077
13612	−3 00336	HR 650	★ AB	2 10 14	−02 37.6	3	5.51	.005	0.58	.000	0.08	.005	9	F8 V	15,1061,3053
13612	−3 00335	IDS02077S0252	B	2 10 14	−02 37.6	1	7.74		0.68		0.22		4	G4	3077
13565	+29 00374			2 10 15	+30 19.5	1	7.81		1.01		0.75		1	K0 II	8100
13506	+56 00452			2 10 15	+57 16.8	1	9.37		0.11				3	B9 V	1119
13530	+50 00481	HR 645	★ A	2 10 16	+50 50.1	6	5.31	.009	0.93	.005	0.62	.007	11	G8 III:	15,37,1003,1355,3016,4001
	+59 00437	St 2 - 51		2 10 17	+59 31.4	1	10.38		0.47		0.38		2	A0 V	11
		AAS42,335 T3# 12		2 10 18	+61 44.	1	11.35		0.74		-0.13		10		848
13597	+13 00352			2 10 19	+14 00.5	1	8.66		1.09				1	K0	882
13596	+14 00357	HR 648		2 10 19	+15 02.8	6	5.70	.013	1.55	.006	1.93	.029	19	M0 III	15,252,1003,3077,8003,8015
		St 2 - 52		2 10 19	+59 08.6	1	11.66		1.71		1.70		2		11
13652	−26 00795			2 10 19	−26 33.4	1	7.92		1.08				4	K0 III	2012
		TZ Per		2 10 20	+58 08.8	1	13.62		0.28		-0.58		1		698
13505	+58 00402	St 2 - 53		2 10 20	+58 30.5	2	8.57	.019	0.33	.005	0.07		4	A9 III	11,1119
13611	+8 00345	HR 649	★	2 10 21	+08 36.8	13	4.36	.009	0.89	.005	0.60	.009	52	G6 II-III	15,58,244,1071,1355*
13448	+67 00189			2 10 22	+67 53.6	1	8.09		0.28		0.16		2	F0	1733
13928	−77 00095			2 10 22	−76 51.5	1	6.66		0.34		0.07		6	F0 II/III	1628
		St 2 - 55		2 10 23	+59 12.8	1	11.76		0.90		0.38		2		11
		St 2 - 54		2 10 23	+59 19.8	1	10.95		0.59		0.49		2	B8	11
13564	+40 00462			2 10 24	+40 33.4	1	7.37		0.15		0.11		2	A3	1601
13518	+58 00403	St 2 - 56		2 10 26	+59 10.1	2	8.54	.019	0.38	.009	0.32		4	A0 V	11,1119
		VES 720		2 10 26	+59 21.9	1	12.46		0.85		0.00		1		1543
13504	+61 00392			2 10 26	+61 27.1	1	8.21		0.46		0.00		23	F6 III	848
		G 159 - 47		2 10 27	+00 47.1	1	15.04		0.96		0.51		2		3062
		L 728 - 16		2 10 27	−17 55.4	1	11.12		1.47				1		1746
		LP 769 - 90		2 10 27	−17 55.4	2	11.09	.005	1.48	.010	1.20		5	M0	158,1705
13544	+53 00480	V353 Per		2 10 28	+53 40.9	3	8.89	.018	-0.01	.000	-0.82	.005	4	B0.5IV	64,555,1012
	+58 00404	St 2 - 57		2 10 28	+59 24.0	1	10.33		0.58		0.45		3	A0	11

Table 1   129

HD	DM	Other Id	N	Rem	α_{1950}	δ_{1950}	S	V	σ_V	B–V	σ_{B-V}	U–B	σ_{U-B}	n	Spectrum	References
13610	+24 00322				2 10 31	+25 08.8	2	8.68	.019	0.55	.007	-0.07		6	F8 IV	20,3016
232618	+52 00542	LS I +52 002			2 10 31	+52 53.4	1	9.15		-0.10		-0.86		2	B2	401
	+58 00405	St 2 - 58			2 10 31	+59 26.0	1	10.00		0.49		0.39		2	A0	11
	+59 00438	St 2 - 59			2 10 32	+59 29.7	1	9.96		0.46		0.36		2	A0 V	11
13474	+65 00239	HR 640		⋆	2 10 32	+66 17.5	2	6.07	.000	0.63	.005	0.33	.005	6	G0 II-III+	985,1733
		St 2 - 60			2 10 33	+58 45.5	1	11.38		0.30		0.09		2	B9	11
13542		St 2 - 65			2 10 37	+58 53.2	1	10.15		0.44		0.35		2	A0	11
13543	+57 00521				2 10 38	+57 39.2	1	8.46		1.18				3	K0 III	1119
236955	+59 00439	St 2 - 62			2 10 41	+59 40.2	1	9.40		0.51		0.40		3	A0 V	11
13692	-21 00396	HR 651			2 10 41	-21 14.1	2	5.86	.002	1.01	.000	0.78		6	G9 III	58,2035
13709	-31 00882	HR 652			2 10 42	-30 57.5	3	5.27	.004	-0.02	.000	-0.06		8	B9 V	15,258,2012
13561	+55 00547	LS I +56 036			2 10 43	+56 16.0	1	8.83		0.09		-0.77		2	B0.5Vp	1012
		St 2 - 64			2 10 46	+59 04.3	1	10.10		0.46		0.39		2	A0	11
	+59 00442	St 2 - 61			2 10 47	+59 36.2	1	9.78		0.48		0.36		3	A0	11
13554	+59 00443	St 2 - 63			2 10 47	+59 51.3	2	9.20	.009	0.38	.007	0.05		4	A5 IV	11,1119
13594	+46 00536	HR 647		⋆ AB	2 10 49	+47 15.1	7	6.05	.014	0.40	.013	-0.07	.006	19	F4 V	15,254,292,938,1007*
13649	+24 00325				2 10 50	+25 21.8	2	7.09	.005	0.46	.005	0.01	.015	5	F7 V	833,1625
	+61 00393				2 10 50	+61 54.1	1	9.95		0.36		0.19		12	A0 V	848
	-21 00397				2 10 50	-21 25.6	1	9.88		1.36		1.26		1	K7 V	3072
		St 2 - 66		⋆	2 10 51	+59 31.2	2	11.06	.015	0.70	.000	-0.22	.019	3	B1	11,1543
	+54 00490	LS I +54 007			2 10 54	+54 49.6	2	9.52	.000	0.11	.000	-0.67	.000	3	B1 V	64,1012
		AAS42,335 T3# 15			2 10 54	+61 13.	1	11.73		0.54		0.29		7		848
13741	-36 00828				2 10 54	-35 49.9	1	9.77		0.24		0.06		4	A8 V	152
13730	-23 00822				2 10 56	-23 06.3	1	7.21		0.97				4	K0 III	2012
	+11 00299	G 4 - 6			2 10 58	+11 42.6	2	10.60	.000	0.56	.000	-0.10	.000	4	F8	516,1620
	+58 00406	St 2 - 68			2 10 59	+58 55.8	1	10.43		0.45		0.36		2	A0	11
		St 2 - 67			2 10 59	+59 34.4	1	12.21		0.71		0.14		3		11
		St 2 - 69			2 11 00	+59 11.9	1	10.67		0.57		0.45		2	A2	11
		St 2 - 70			2 11 00	+59 16.6	1	11.73		0.77		0.19		2		11
		AAS42,335 T3# 16			2 11 00	+61 14.	1	11.17		0.43		0.02		10		848
13808	-54 00432				2 11 00	-53 58.7	1	8.44		0.87		0.54		1	K2 V	3072
13580	+58 00407	St 2 - 72			2 11 01	+58 48.2	1	10.16		0.36		0.29		2	A0	11
		G 3 - 44			2 11 03	+15 45.5	1	13.16		1.07		0.75		6		940
		CS 22175 # 1			2 11 04	-11 51.6	1	12.95		0.08		0.17		1		1736
		St 2 - 71			2 11 05	+59 37.3	1	12.40		1.13		0.64		2		11
		G 159 - 48			2 11 06	-00 19.2	2	12.17	.010	0.44	.010	-0.21	.015	3		1696,3056
13721	-10 00453				2 11 06	-09 59.1	2	8.55	.000	0.29	.003	0.03	.009	22	F0 V	1003,1775
		L 296 - 91			2 11 06	-48 52.	1	12.53		0.57		-0.14		2		1696
13621	+54 00494	LS I +55 028		⋆ AB	2 11 07	+55 05.1	4	8.10	.012	0.06	.000	-0.78	.004	8	B1 V	1012,1083,1110,8023
		St 2 - 73			2 11 07	+58 58.6	1	10.68		0.53		0.41		2	A0	11
13591	+58 00408	St 2 - 74			2 11 07	+59 09.7	2	8.66	.018	0.41	.009	0.36		5	A0 IV	11,1119
13907	-71 00110	IDS02103S7125	A		2 11 07	-71 11.1	1	7.10		1.00		0.76		52	G8/K0 III	978
		VES 723			2 11 10	+62 12.3	1	12.93		1.22		0.88		1		1543
13752	-29 00784				2 11 11	-29 12.2	1	9.22		0.08		0.10		4	A0 V	152
		AAS42,335 T3# 18			2 11 12	+64 45.	1	10.82		0.57		0.17		10		848
	+60 00453				2 11 13	+61 16.1	1	10.19		0.77		0.42		12		848
13606	+58 00409	St 2 - 75			2 11 16	+59 24.7	2	9.01	.009	0.50	.005	0.40		4	B9 V	11,1119
		POSS 53 # 5			2 11 18	+67 31.9	1	17.45		1.39				2		1739
		LB 3241			2 11 18	-49 59.0	2	12.74	.025	-0.30	.015	-1.05	.070	4		45,132
13633	+57 00522	IDS02079N5802	A		2 11 24	+58 15.5	2	7.85	.014	0.17	.000	-0.24		4	B7 III	401,1119
13633	+57 00522	IDS02079N5802	B		2 11 24	+58 15.5	1	9.46		0.21		-0.13		1	B9 V	401
		St 2 - 103			2 11 24	+59 02.	1	14.70		0.92		0.41		1		11
		St 2 - 113			2 11 24	+59 02.	1	14.82		1.48				1		11
13590	+63 00310	LS I +63 185			2 11 24	+63 47.5	1	7.90		0.35		-0.41		2	B2 III	1012
		JL 286			2 11 24	-50 19.	1	14.31		-0.22		-0.82		2	sdB	132
		St 2 - 79			2 11 27	+59 10.1	1	12.00		0.69		-0.09		2	cA	11
		St 2 - 76			2 11 27	+59 34.8	1	11.32		0.64		0.41		2	A2	11
13661	+53 00486	LS I +54 008			2 11 28	+54 17.9	1	7.79		-0.01		-0.69		2	B2 IV-Ve	401
13631	+59 00445	St 2 - 77			2 11 28	+59 49.1	1	9.83		0.40		0.27		2	A1 II:	11
13776	-24 00964				2 11 29	-24 12.4	1	9.50		0.29		0.04		4	A3/5 III	152
13807	-30 00787				2 11 29	-30 07.1	2	10.21	.010	0.00	.000	-0.06	.000	12	B9.5V	152,1775
	+59 00444	St 2 - 78			2 11 30	+59 53.4	1	10.18		0.55		0.41		2	A0	11
13632	+58 00410	St 2 - 80			2 11 31	+59 22.1	1	9.54		0.49		0.38		2	A0 V	11
13643	+58 00412	St 2 - 81		⋆ AB	2 11 32	+59 16.3	1	9.03		0.45		0.37		2	B9 V	11
13656	+58 00413	St 2 - 83			2 11 34	+58 52.5	2	9.23	.005	0.44	.009	0.37		4	A0 V	11,1119
		LP 589 - 57			2 11 35	+01 19.3	1	9.99		0.99		0.76		1	K4	1696
13669	+55 00552				2 11 35	+55 33.6	1	7.90		0.35		-0.41		3	B3 IV-V	1212
13579	+67 00191	G 244 - 56		⋆ AB	2 11 35	+67 26.6	3	7.22	.032	0.91	.005	0.74	.030	8	K2 V	22,1067,1355,1658
		MCC 388			2 11 36	+01 19.2	1	10.85		0.97				2	K4	1619
		St 2 - 84			2 11 36	+58 31.7	1	10.60		0.39		0.26		2	A0	11
13654		St 2 - 82			2 11 36	+59 06.9	1	9.87		0.56		0.41		2	A0	11
		SK 214			2 11 36	-73 44.	1	13.43		-0.02		-0.14		3		733
13659	+56 00462	LS I +56 038			2 11 37	+56 41.6	4	8.65	.004	0.55	.024	-0.41	.000	7	B1 Ib	64,1012,1083,1119
	-32 00828				2 11 40	-32 16.0	2	10.31	.000	1.50	.005	1.10		5	M2.5	158,1705
13658	+57 00524				2 11 41	+57 54.6	4	8.92	.019	2.31	.000	2.67	.060	6	M1 Iab	138,1083,8016,8023
13655	+58 00414	St 2 - 82a			2 11 41	+59 06.2	2	8.87	.025	1.41	.010	1.12		5	K2	11,8092
		St 2 - 86			2 11 41	+59 19.9	1	11.40		0.71		-0.11		2	B5	11
		St 2 - 85			2 11 42	+59 07.6	1	14.18		1.48		1.08		1		11
		St 2 - 87			2 11 42	+59 35.0	1	9.98		0.48		0.38		2	A0	11
13747	+28 00374				2 11 44	+28 27.6	1	6.35		0.96		0.74		4	K1 III	985

HD	DM	Other Id	N Rem	α_{1950}	δ_{1950}	S	V	σ_V	B–V	σ_{B-V}	U–B	σ_{U-B}	n	Spectrum	References
		St 2 - 88		2 11 44	+59 30.5	1	10.37		0.55		0.44		2	A1	11
		St 2 - 90		2 11 45	+59 08.2	1	11.96		0.80		-0.06		2		11
13838	−42 00758			2 11 45	−42 00.5	1	9.54		0.24		0.10		4	A6 V	152
		St 2 - 104		2 11 48	+59 06.6	1	12.82		1.00		0.64		1		11
13677	+58 00415	St 2 - 89		2 11 49	+58 55.8	1	9.79		0.53		0.42		2	A2 V	11
		VES 725		2 11 49	+62 18.8	1	13.07		0.79		-0.34		1		1543
		St 2 - 98		2 11 50	+59 07.9	1	13.69		0.74		0.08		1		11
		St 2 - 91		2 11 52	+59 08.2	1	12.33		0.85		0.37		2		11
13676	+59 00447	St 2 - 99		2 11 52	+59 42.1	2	9.11	.014	0.53	.009	0.44		4	A3 V	11,1119
	+49 00602			2 11 54	+50 05.7	1	8.90		0.83		0.41		3		1723
		St 2 - 92		2 11 55	+58 58.2	1	11.41		0.79		0.48		1	A2	11
13689	+58 00417	St 2 - 93		2 11 56	+59 00.5	1	8.41		0.33		0.24		2	A7 V	11
13688	+58 00416	St 2 - 94		2 11 56	+59 09.2	1	8.73		0.43		0.36		2	A0 V	11
13687	+58 00418	St 2 - 96		2 11 59	+59 18.4	2	8.53	.022	0.46	.016	0.38		5	A1 V	11,1119
13717	+54 00500	LS I +55 030		2 12 00	+55 21.8	1	7.86		0.11		-0.18		2	A0 III	401
	+58 00419	St 2 - 95		2 12 00	+59 15.1	1	9.66		0.45		0.35		2	A2	11
		St 2 - 102		2 12 00	+59 31.2	1	10.71		0.60		0.51		2	A1	11
		AAS42,335 T3# 19		2 12 00	+62 00.	1	13.13		0.56		0.38		7		848
		AAS42,335 T3# 21		2 12 00	+62 01.	1	12.11		0.90		0.59		10		848
13699	+58 00420	St 2 - 100		2 12 02	+58 52.1	2	9.50	.010	0.55	.005	0.46		5	A0 V	11,1119
	+61 00394			2 12 02	+62 00.1	1	9.55		0.53		0.04		12	F5	848
		St 2 - 101		2 12 04	+58 37.4	1	10.52		0.44		0.31		2	A2	11
		St 2 - 97		2 12 04	+59 16.0	1	10.81		0.58		0.44		2		11
	−1 00306	G 159 - 50		2 12 04	−01 26.0	4	9.08	.016	0.58	.003	-0.07	.009	8	G1 V	22,742,1003,1696
	+60 00454			2 12 05	+61 22.4	1	10.29		0.37		-0.03		9		848
		LP 649 - 74		2 12 05	−07 43.1	1	11.60		0.53		-0.10		2		1696
13738	+51 00527			2 12 07	+52 16.7	1	6.94		1.44		1.63		2	K3.5III	1566
13716	+57 00525	LS I +57 048		2 12 07	+57 31.9	3	8.26	.013	0.32	.005	-0.59	.000	5	B0.5III	64,1012,1119
		St 2 - 107		2 12 08	+59 19.1	1	10.01		0.53		0.43		2	A1	11
13686	+62 00369			2 12 09	+63 00.3	1	7.01		1.87		2.08		4	K3 Ib	1355
13715	+59 00448	St 2 - 105		2 12 10	+59 48.3	2	9.19	.009	0.40	.007	0.30		4	A0 V	11,1119
13889	−42 00761			2 12 10	−41 56.6	1	9.57		0.55				3	F8 IV	1594
		G 74 - 10		2 12 11	+32 10.0	2	12.59	.022	0.81	.015	0.22	.007	3		1620,1658
		WLS 200 45 # 8		2 12 11	+44 21.9	1	11.63		0.31		0.14		2		1375
236957	+59 00449	St 2 - 106		2 12 12	+59 49.5	1	9.69		0.38		0.29		2	A0 V	11
		St 2 - 108		2 12 13	+59 21.8	1	11.11		0.64		0.49		2	A2	11
		G 35 - 35		2 12 14	+17 11.5	1	14.39		1.62		1.08		3		316
13787	+38 00442			2 12 16	+38 56.2	1	7.28		0.97		0.66		3	K0 III	833
		St 2 - 109		2 12 17	+59 09.8	1	11.84		0.71		0.33		2		11
		St 2 - 111		2 12 17	+59 25.8	1	12.51		0.76		0.01		2		11
13826	+11 00305	V Ari		2 12 18	+12 00.4	3	8.47	.075	2.06	.026	2.31	.085	10	C5 II	109,1238,3038
13745	+55 00554	V354 Per		2 12 18	+55 45.9	4	7.86	.019	0.17	.007	-0.78	.004	8	B0 III	1012,1083,1110,8023
13735	+58 00421	St 2 - 110		2 12 19	+59 10.2	2	8.91	.014	0.42	.005	0.33		4	A0 IV	11,1119
		RV Ari		2 12 22	+17 50.5	1	11.85		0.29		0.16		1		668
		G 159 - 51		2 12 24	−03 13.2	1	16.30		1.13		0.80		1		3062
13744	+57 00526	LS I +58 069		2 12 25	+58 03.7	5	7.59	.014	0.74	.012	0.18	.000	9	A0 Iab	1012,1083,1110,1119,8023
13949	−50 00638			2 12 26	−50 05.6	1	9.72		1.29				4	K5 V	2033
13734	+59 00450			2 12 27	+60 23.9	1	9.14		1.63		1.75		4	F7 V	1723
		LS I +56 040		2 12 28	+56 46.5	1	10.59		0.36		-0.41		4		1401
		St 2 - 112		2 12 31	+59 52.9	1	11.60		0.69		0.40		2	A8	11
13940	−41 00621	HR 659		2 12 31	−41 23.9	3	5.91	.005	0.97	.005			14	G8 III	15,1075,2029
13825	+23 00303			2 12 32	+24 02.5	2	6.81	.000	0.69	.000	0.28	.00	8	G8 IV	1355,3026
		G 133 - 79		2 12 32	+42 09.9	1	12.20		1.07		1.03		1		1620
13758	+57 00527	V355 Per		2 12 32	+57 30.8	3	9.06	.016	0.33	.005	-0.53	.000	5	B1 V	64,1012,1119
13836	+26 00373	G 94 - 48		2 12 34	+27 07.6	3	8.11	.010	0.70	.010	0.25	.024	5	G8 V	333,1620,1733,3026
		VES 726		2 12 34	+55 54.0	1	12.77		0.46		-0.24		1		1543
		VES 727		2 12 34	+59 02.9	1	11.85		1.00		0.10		1		1543
		St 2 - 114		2 12 34	+59 20.4	1	10.44		0.56		0.46		2	A0	11
	+59 00451	St 2 - 115		2 12 34	+59 27.2	3	9.30	.005	0.69	.000	-0.28	.015	5	OB	11,64,1012
	+58 00423	St 2 - 116		2 12 35	+59 24.9	1	10.30		0.50		0.40		2	A0	11
		LP 589 - 61		2 12 35	−02 23.8	1	11.05		0.65		0.03		2		1696
		St 2 - 118		2 12 38	+59 29.2	1	12.04		0.79		0.27		2		11
		CS 22175 # 6		2 12 38	−09 30.0	1	14.12		0.14		-0.76		1		1736
	+56 00465		AB	2 12 40	+57 23.8	1	9.41		0.88				2		1119
13772	+57 00528			2 12 40	+58 10.6	1	8.52		0.20				2	B6 IV	1119
13725	+66 00198			2 12 40	+67 03.1	2	7.05	.054	1.88	.029	2.15	.083	6	K4 II	1355,8100
		G 73 - 37		2 12 42	+07 16.0	2	11.90	.019	1.47	.005			3	M1	1532,1746
		G 134 - 20		2 12 42	+46 13.6	1	11.15		1.29		1.07		5	K7	7010
13771	+59 00452	St 2 - 117		2 12 42	+59 49.7	2	9.29	.000	0.45	.009	0.39		4	A1 V	11,1119
13818	+47 00590	HR 653		2 12 43	+47 34.8	1	6.36		1.05		0.86		2	G9 III-IV	1733
	+58 00424	St 2 - 120		2 12 45	+59 06.5	1	9.27		1.03		0.90		2	K0	11
		St 2 - 121		2 12 46	+58 49.4	1	10.73		0.69		0.46		3	A5	11
		St 2 - 119		2 12 46	+59 21.4	1	12.15		0.71		0.06		2		11
13917	−12 00418			2 12 46	−12 17.1	1	7.80		0.12		0.10		4	A2 V	152
	−11 00427	RV Cet		2 12 49	−11 02.0	3	10.60	.086	0.27	.022	0.07	.005	3	F7.5	597,699,700
13884	+15 00321			2 12 51	+15 57.8	2	7.97	.029	1.32	.010	1.41		3	K5	882,3040
13872	+24 00329	HR 657	⋆ A	2 12 52	+24 48.7	2	5.60	.005	0.49	.010	0.00	.005	6	F6 V	254,3037
13871	+25 00373	Hyades vB 157	⋆	2 12 54	+25 33.1	2	5.79	.005	0.44	.008	0.03	.016	13	F5 V	15,254,1127,3016,8015
	−12 00419	RW Cet		2 12 56	−12 26.5	1	10.09		0.34		0.02		5	A5	3064
13869	+32 00409	HR 655		2 12 58	+33 07.7	2	5.28		0.00	.005	-0.03	.000	5	A0 V	1049,3023

HD	DM	Other Id	N	Rem	α_{1950}	δ_{1950}	S	V	σ_V	B–V	σ_{B-V}	U–B	σ_{U-B}	n	Spectrum	References
		St 2 - 123			2 12 58	+58 44.1	1	11.12		0.46		0.22		2	A0	11
13945	−23 00840				2 12 58	−23 30.8	1	8.07		0.74				4	G6 IV	2012
	+58 00425	St 2 - 122			2 12 59	+59 15.5	1	9.55		0.52		0.43		2	A0 V	11
13783	+64 00312	G 244 - 59			2 12 59	+64 43.5	5	8.30	.010	0.67	.007	0.10	.014	8	G8 V	979,1003,1658,3026,8112
		St 2 - 124			2 13 00	+58 41.4	1	11.14		0.57		0.38		2	A5	11
		St 2 - 126			2 13 00	+59 29.1	1	12.43		0.82		0.31		2		11
	+58 00426	St 2 - 125			2 13 01	+59 03.8	1	10.12		0.62		0.51		2	B9	11
13936	−10 00460	HR 658			2 13 01	−09 41.9	4	6.55	.004	-0.01	.005	-0.06	.009	16	A0 V	15,1061,1509,2006
		POSS 154 # 8			2 13 03	+48 06.1	1	18.59		2.08				3		1739
		AA131,200 # 3			2 13 03	+55 09.1	1	14.52		0.57		-0.23		2		7005
13832	+54 00505				2 13 05	+55 11.7	1	10.02		0.04		-0.57		2	B9	401
	+59 00453	St 2 - 127			2 13 05	+59 30.5	1	10.05		0.56		0.46		2	A1	11
		CS 22175 # 3			2 13 05	−10 54.4	1	13.77		0.14		0.17		1		1736
		AAS42,335 T3# 23			2 13 06	+61 42.	1	11.57		0.73		0.26		11		848
		SK 215			2 13 06	−74 46.	3	12.18	.017	-0.06	.004	-0.69	.005	7	B6 Ia	573,733,1412
13831	+56 00469	V473 Per			2 13 09	+56 30.4	5	8.26	.011	0.10	.007	-0.81	.000	9	B0 IIIp	1012,1083,1110,1119,8023
13824	+57 00529				2 13 09	+57 58.9	1	8.79		0.39				2	F2 IV	1025
14003	−31 00902				2 13 11	−31 07.1	1	10.39		0.08		0.09		4	A0 V	152
13914	+15 00322				2 13 12	+15 35.3	1	8.07		0.46		0.03		2	F8	1648
		St 2 - 129			2 13 12	+58 53.5	1	11.60		0.63		0.37		6	A5	11
		St 2 - 128			2 13 12	+59 32.3	1	11.11		0.60		-0.17		2	B3	11
14141	−68 00126	HR 667			2 13 12	−68 04.5	3	5.56	.011	1.54	.004	1.84	.000	20	M1 III	15,1075,2038
		GD 25			2 13 13	+39 37.6	1	14.54		0.23		-0.61		1		3060
13959	+5 00307	IDS02106N0610		AB	2 13 16	+06 23.7	6	9.03	.027	1.09	.021	0.96	.005	10	K4 V +K4 V	196,680,1017,1381*
13841	+56 00470	NGC 869 - 3		★ B	2 13 16	+56 47.9	6	7.38	.013	0.23	.010	-0.65	.005	10	B2 Ib	64,1012,1083,1110*
		NGC 869 - 7			2 13 17	+56 46.5	1	12.08		0.23		-0.36		2		1083
	+47 00593				2 13 18	+47 26.9	1	9.40		0.32		0.15		2	A2	1375
13867	+49 00614				2 13 19	+49 35.3	1	7.57		0.02		-0.36		2	B5 Ve:	401
13893	+42 00482				2 13 21	+42 49.0	1	8.83		0.20		0.09		2	A2	1375
13854	+56 00471	NGC 869 - 16		★ A	2 13 21	+56 49.4	8	6.47	.017	0.28	.003	-0.66	.021	21	B1 Iab	15,154,1012,1083,1110*
14036	−34 00840				2 13 22	−33 59.8	1	8.66		1.18				4	K0/1 III	2033
14057	−41 00626				2 13 24	−41 18.1	1	6.97		1.62				4	M1/2 III	2012
14001	−18 00394	IDS02111S1842		AB	2 13 25	−18 28.2	3	7.93	.021	1.04	.024	0.80	.034	11	K3 V	176,938,3072
13866	+56 00475	V357 Per			2 13 27	+56 29.3	6	7.48	.025	0.18	.014	-0.64	.010	10	B2 Ib	1012,1083,1110,1119*
	+56 00473	NGC 869 - 49		★ V	2 13 27	+56 53.9	2	9.07	.005	0.25	.010	-0.64	.015	4	B1 III	1012,1110
	+58 00427	St 2 - 130			2 13 27	+59 03.1	1	10.17		0.47		0.32		2	A0	11
		LS I +58 070			2 13 29	+58 03.8	1	10.22		0.55		-0.37		2	B0.5 V	1012
		St 2 - 131			2 13 30	+59 06.9	1	10.82		0.50		0.39		2	A0	11
		AAS42,335 T3# 24			2 13 30	+61 07.	1	12.35		0.84		0.28		9		848
13830	+62 00374				2 13 30	+62 56.2	1	8.24		0.52				3	F6 IV	1118
		St 2 - 132			2 13 31	+59 14.0	1	10.49		0.50		0.32		3	A2	11
	+57 00530	PP Per			2 13 32	+58 25.1	1	9.19		2.42				1	M0 Iab-Ib	138
	+57 00531				2 13 37	+58 24.3	1	8.04		0.44				2	F5 V	1025
13890	+56 00478	V358 Per			2 13 38	+56 32.3	5	8.51	.023	0.19	.007	-0.64	.004	9	B1 III	1012,1083,1110,1119,8023
		St 2 - 133			2 13 38	+59 14.5	1	12.20		0.75		0.26		2		11
		St 2 - 134			2 13 40	+59 13.3	1	10.72		0.51		0.32		2	A1	11
		POSS 154 # 4			2 13 41	+46 51.3	1	16.65		1.25				2		1739
13864	+60 00457	IDS02100N6053		AB	2 13 41	+61 07.3	2	8.23	.003	0.61	.003	0.08	.007	17	G1 V +G4 V	292,848
13997	+11 00309	G 4 - 9			2 13 45	+12 09.1	1	7.99		0.79		0.44		1	G5	333,1620
13900	+56 00479	NGC 869 - 146			2 13 45	+56 40.0	3	9.18	.007	0.17	.005	-0.66	.005	7	B1 III	1012,1110,1119
		G 134 - 22			2 13 47	+42 44.5	2	16.21	.019	0.72	.005	0.00	.019	6	DC	538,3060
		LP 469 - 206			2 13 48	+13 21.7	1	15.79		1.98				4		1663
		LS I +58 071			2 13 48	+58 47.5	1	10.15		0.84		-0.30		2	B pe	1012
		St 2 - 135		★	2 13 48	+59 20.0	2	10.58	.005	0.65	.005	-0.28	.005	4	OB	11,1012
		St 2 - 136			2 13 49	+59 29.3	1	10.89		0.58		-0.04		2	B3	11
	+29 00385				2 13 51	+29 43.2	1	8.68		1.71		2.05		2	K7	1733
13899	+58 00428	St 2 - 137			2 13 52	+59 08.2	2	8.76	.019	0.34	.000	0.31		4	A0 V	11,1119
		AAS42,335 T3# 26			2 13 54	+61 06.	1	12.83		0.81		0.20		7		848
		JL 288			2 13 54	−48 27.	1	17.70		0.50		-0.80		1		132
14111	−39 00662				2 13 55	−38 41.6	1	9.77		0.30		0.11		4	A0mA9-F2	152
		POSS 154 # 7			2 13 56	+46 51.3	1	18.41		1.81				3		1739
13910	+56 00480	NGC 869 - 197			2 13 56	+57 08.1	1	8.11		0.09				3	B9.5V	1119
14085	−23 00852	IDS02116S2330		AB	2 13 56	−23 16.4	1	8.39		0.61				4	G8 IV	2012
	+58 00430	St 2 - 140			2 13 57	+58 52.9	1	10.19		0.69		0.40		2	A8	11
		St 2 - 139			2 13 58	+59 12.3	1	12.65		0.80		0.26		2		11
13909	+58 00431	St 2 - 138			2 13 58	+59 18.3	2	8.65	.014	0.34	.000	0.28		4	A0 V	11,1119
		LP 589 - 68			2 13 59	+00 21.9	1	11.12		1.05		0.95		1		1696
13974	+33 00395	HR 660		★ A	2 14 00	+33 59.8	11	4.86	.010	0.61	.006	0.02	.008	39	G0 V	1,15,22,254,1004,1077*
		SK 216			2 14 00	−74 11.9	2	12.80	.000	0.00	.010	-0.32	.010	5		733,835
		NGC 869 - 230			2 14 01	+56 52.9	4	13.65	.015	-0.10	.008	-0.91	.004	16		1110,1281,1727,3028
13929	+57 00533				2 14 01	+57 47.4	1	7.45		0.25				2	Am	1119
		St 2 - 141			2 14 01	+58 52.5	1	10.96		0.52		0.33		2	A5	11
14070	+5 00309				2 14 03	+06 17.9	1	9.44		0.90				1	K0	1632
	+26 00377	G 94 - 49			2 14 04	+27 11.1	1	10.40		0.66		0.04		2	G0	333,1620
	+56 00482	NGC 869 - 260			2 14 08	+56 58.1	1	9.36		0.30		-0.60		2	B1 IIIp	1012
		G 36 - 4			2 14 09	+28 36.3	1	14.02		1.13		0.79		1		1658
		GD 26			2 14 10	+38 36.4	1	13.55		0.50		-0.14		2	GD:	3060
13971	+55 00565	LS V +55 002			2 14 14	+55 37.2	1	9.20		0.13		-0.50		2	B9	1502
	+56 00484	NGC 869 - 309		★ V	2 14 14	+56 40.1	1	9.62		0.32		-0.70		2	B1 IIIe	1012
		NGC 869 - 311			2 14 15	+56 51.5	1	12.54		0.42		-0.28		2		1083

Table 1

131

HD	DM	Other Id	N Rem	α_{1950}	δ_{1950}	S	V	σ_V	B–V	σ_{B-V}	U–B	σ_{U-B}	n	Spectrum	References
13970	+55 00564	V438 Per	⋆AB	2 14 16	+56 24.6	2	8.30	.010	0.14	.000	-0.65		5	B5 Ib	401,1119
14056	+20 00363	G 4 - 10		2 14 17	+21 20.1	3	9.04	.004	0.62	.000	0.01	.004	5	G5	1620,1658,3077
14083	+16 00266			2 14 19	+17 05.1	1	8.58		0.46		0.01		2	G0	1648
		NGC 869 - 341		2 14 19	+56 51.4	1	12.64		0.31		-0.36		2		1083
13969	+56 00485	NGC 869 - 339		2 14 19	+56 51.6	5	8.84	.012	0.30	.010	-0.61	.008	10	B1 IV	1012,1083,1110,1119,8023
		St 2 - 143		2 14 19	+58 52.0	1	10.96		0.51		0.27		2	A5	11
14067	+23 00307	HR 665		2 14 20	+23 32.3	2	6.53	.020	1.03	.007	0.84	.015	4	G9 III	1733,3016
14055	+33 00397	HR 664		2 14 20	+33 37.0	7	4.01	.005	0.02	.002	0.02	.008	18	A1 Vnn	15,1049,1203,1363*
13967	+58 00434	St 2 - 144		2 14 20	+58 40.0	2	9.83	.014	0.30	.000	0.25		4	A1 V	11,1119
13966	+58 00433	St 2 - 142		2 14 22	+59 08.8	2	9.35	.005	0.36	.002	0.30		4	A1 V	11,1119
13995	+55 00566			2 14 23	+55 38.3	1	7.89		0.14		0.81		2	K0	1502
	-22 00769			2 14 23	-22 28.0	1	10.63		0.67		0.10		2		1696
		G 94 - 52		2 14 24	+26 03.3	1	11.00		1.10		1.05		5	K4:	7010
14319	-72 00163			2 14 25	-71 45.1	1	9.12		1.06				1	K1 III	1642
13982	+57 00535	HR 661		2 14 26	+57 40.2	2	5.75	.015	1.17	.005			4	K3 III	1118,1119
14287	-68 00128	HR 678		2 14 26	-67 58.7	3	5.67	.007	1.31	.005	1.48	.000	17	K2 III	15,1075,2038
13968	+57 00534			2 14 28	+58 08.8	1	9.41		0.23				2	B9 V	1119
	+58 00435	St 2 - 145		2 14 28	+59 11.9	1	10.03		0.46		0.30		2	A0	11
14129	-7 00393	HR 666		2 14 29	-06 39.1	6	5.51	.008	0.96	.005	0.78	.016	35	G8 III	3,15,418,1061,3016,8100
14014	+55 00567	LS I +56 043	⋆AB	2 14 30	+56 00.1	2	8.75	.000	0.14	.000	-0.66	.000	3	B0.5V	64,1012
		CS 22175 # 10		2 14 30	-08 03.8	1	11.99		0.11		0.10		1		1736
14082	+28 00382	IDS02116N2817	A	2 14 31	+28 30.9	1	7.02		0.50		-0.03		4	F5 V	3026
14082	+28 00382	IDS02116N2817	B	2 14 31	+28 30.9	1	7.77		0.62		0.08		4	G2 V	3026
		VES 728		2 14 31	+56 37.6	1	11.85		0.53		-0.38		1		1543
13994	+56 00486	HR 662	⋆A	2 14 32	+57 17.2	2	5.99	.010	1.08	.025	0.79		5	G8 II-III	1084,1119
13994	+56 00486	IDS02110N5703	BC	2 14 32	+57 17.2	1	9.60		0.27		0.16		3		1084
		G 36 - 7		2 14 33	+26 53.2	1	16.58		1.60		0.95		3		3016
		G 35 - 37		2 14 33	+26 54.5	2	10.93	.000	1.06	.013	0.91	.057	6	K3	333,1620,3016
14229	-56 00408			2 14 33	-55 43.5	2	9.52	.005	0.46	.005	-0.05		4	F6/7wF2 V	742,1594
	+59 00456	St 2 - 146		2 14 36	+59 34.3	3	9.87	.000	0.55	.000	-0.41	.005	4	B0.5V	11,64,1012
		NGC 869 - 581		2 14 41	+56 47.2	1	15.04		0.56		-0.19		2		1083
14012	+56 00487	IDS02110N5703		2 14 41	+57 15.6	1	9.14		0.35				2	A0	1119
14228	-52 00285	HR 674	⋆A	2 14 43	-51 44.6	8	3.55	.010	-0.12	.005	-0.37	.013	26	B8 V-IV	3,15,1020,1034,1068*
14011	+58 00436	St 2 - 148		2 14 44	+59 12.6	2	9.27	.005	0.30	.000	0.25		4	A0 V	11,1119
14026	+57 00536			2 14 45	+57 35.2	1	8.57		0.45				2	F7 IV	1025
14025	+58 00437	St 2 - 149		2 14 46	+59 08.8	2	8.45	.028	0.32	.000	0.30		4	A1 V	11,1119
	+56 00493	NGC 869 - 566		2 14 47	+56 37.2	1	9.62		0.19		-0.66		2	B1 Vpe	1012
		St 2 - 147		2 14 49	+59 28.5	1	11.42		0.50		0.34		2	B8	11
14039	+55 00570	G 220 - 7		2 14 52	+56 20.0	1	8.27		0.92		0.70		3	K1 V	196
14053	+56 00498	NGC 869 - 612	⋆AB	2 14 52	+56 46.8	4	8.41	.013	0.25	.004	-0.62	.000	7	B1 II	1012,1083,1110,8023
		NGC 869 - 622		2 14 53	+56 56.3	1	10.91		0.23		-0.53		1	B7	1541
		NGC 869 - 620		2 14 53	+56 58.5	1	13.01		0.37		-0.16		2		1083
		VES 730		2 14 53	+62 31.7	1	13.60		0.89		-0.03		2		1543
14010	+63 00315	LS I +64 090		2 14 53	+64 11.7	2	7.12	.009	0.61	.004	-0.10		5	B9 Ia	1012,8031,8100
14052	+56 00500	NGC 869 - 662		2 14 56	+56 58.7	4	8.18	.009	0.31	.012	-0.59	.004	7	B1 Ib	1012,1083,1110,8023
		St 2 - 150		2 14 56	+59 06.5	1	10.73		0.47		0.28		2	A2	11
		NGC 869 - 681		2 14 57	+56 46.0	1	12.72		0.28		-0.35		2		1083
14104	+44 00456	IDS02117N4445	A	2 14 58	+44 58.8	1	8.12		1.65		1.89		2	K2	1733
	+56 00501	NGC 869 - 692		2 14 58	+56 55.2	1	9.45		0.26		-0.67		1	B4	1541
		NGC 869 - 689		2 14 58	+56 59.3	1	13.21		0.48		0.32		2		1083
14147	+17 00339			2 14 59	+18 13.4	1	7.43		0.30		0.03		2	F0	1648
		CS 22175 # 7		2 14 59	-09 14.6	1	13.43		0.62		-0.01		1		1736
14050	+57 00537			2 15 00	+58 21.5	1	9.56		0.24				2	B5 II	1119
	+56 00502	NGC 869 - 717		2 15 01	+56 58.8	1	9.28		0.30		-0.56		2	B1 V	1110
	+56 00504	NGC 869 - 782		2 15 09	+56 52.2	1	9.45		0.27		-0.55		2	B3	1110
		G 4 - 11		2 15 10	+17 32.3	1	13.65		1.63		1.25		1		1696
14092	+56 00507	LS V +56 020		2 15 11	+56 31.9	2	9.23	.000	0.23	.000	-0.59	.000	3	B1 V	64,1012
		NGC 869 - 800		2 15 11	+56 56.6	1	12.25		0.36		-0.28		2		1110
14247	-36 00859			2 15 13	-36 12.9	1	6.70		0.98				4	K0 III	2007
14204	-2 00389			2 15 15	-02 16.3	1	7.68		1.16		1.11		3	K0	1657
	+56 00511	NGC 869 - 847		2 15 16	+56 50.2	1	9.11		0.38		-0.52		2	B3 III	1110
	+56 00510	NGC 869 - 843		2 15 16	+56 54.3	1	9.32		0.31		-0.53		2	B1.5V	1110
		NGC 869 - 837		2 15 16	+56 55.6	1	14.08		0.49		0.16		2		1110
		St 2 - 151		2 15 16	+59 15.1	1	10.88		0.49		0.32		2	A2	11
		VES 731		2 15 16	+59 36.9	1	12.40		0.54		-0.35		1		1543
		AO 1045		2 15 17	+02 51.3	1	11.41		0.98		0.84		1	K3 V	1748
	+56 00514	NGC 869 - 867		2 15 18	+56 51.7	1	10.51		0.62		0.12		2		1110
	+56 00513	NGC 869 - 864		2 15 18	+56 52.0	2	9.93	.010	0.24	.065	-0.49	.000	3	B2 Vn	1110,8023
		NGC 869 - 869		2 15 18	+56 53.0	1	11.96		0.37		-0.20		2		1110
		NGC 869 - 859		2 15 18	+56 53.5	1	10.78		0.32		-0.39		1	B2 Vn	1541
		NGC 869 - 885		2 15 19	+56 55.3	1	15.76		0.75				2		1110
14191	+19 00340	HR 669		2 15 20	+19 40.3	2	5.62		0.01	.000	0.03	.010	7	A1 Vn	252,1049
		NGC 869 - 911		2 15 21	+56 53.0	1	11.46		0.32		-0.45		2	B2 V	1541
	+56 00512	NGC 869 - 899	⋆V	2 15 21	+57 11.5	5	9.26	.038	2.46	.018	2.64	.012	12	M4 Ib	369,1083,1110,8016,8023
	+56 00515	NGC 869 - 922		2 15 22	+56 54.6	1	9.54		0.33		-0.54		1	B0.5Vn	1541
		NGC 869 - 919		2 15 22	+56 58.4	1	11.08		0.39		-0.25		1		1541
14223	-3 00347			2 15 22	-03 00.9	1	9.49		0.58		0.12		2	G0	1748
		NGC 869 - 926		2 15 23	+56 53.9	1	11.76		0.43				1	B3 V	1541
	+56 00516	NGC 869 - 929		2 15 23	+56 55.7	1	10.36		0.29		-0.46		1	B2 V	1541
		NGC 869 - 920		2 15 23	+57 12.1	1	15.16		0.83		0.53		2		1083

Table 1   133

HD	DM	Other Id	N Rem	α_{1950}	δ_{1950}	S	V	σ_V	B–V	σ_{B-V}	U–B	σ_{U-B}	n	Spectrum	References
14103 +58 00438		St 2 - 152		2 15 23	+58 58.1	1	8.90		0.50		-0.04		2	F8	11
		NGC 869 - 935		2 15 24	+56 54.1	1	14.02		0.45		0.01		1		1110
	+56 00517	NGC 869 - 936		2 15 24	+56 55.3	1	10.50		0.27		-0.51		1	B1.5V	1541
14214	+1 00410	HR 672		2 15 25	+01 31.3	6	5.59	.022	0.59	.019	0.10	.015	15	G0.5IVb	15,22,254,1003,1061,3077
		NGC 869 - 950		2 15 25	+56 57.0	2	11.27	.015	0.36	.000	-0.40	.000	3	B2 V	1110,8023
	-3 00348			2 15 25	-03 03.6	1	10.95		0.66		0.15		1		1748
		NGC 869 - 960		2 15 26	+56 53.7	1	13.69		0.43		0.00		2		1110
		NGC 869 - 963		2 15 26	+56 54.5	1	11.03		0.33		-0.52		1	B2 IV	1541
		NGC 869 - 977		2 15 27	+56 54.4	1	11.02		0.37		-0.42		1	B2 V	1541
	+56 00519	NGC 869 - 980		2 15 27	+56 54.9	1	9.65		0.35		-0.46		1	B1.5V	1541
	+56 00518	NGC 869 - 978		2 15 27	+56 55.6	2	10.61	.015	0.37	.005	-0.44	.000	3	B2 V	1110,8023
		NGC 869 - 974		2 15 27	+56 58.5	1	15.43		0.70		0.78		2		1110
		NGC 869 - 982		2 15 28	+56 53.7	1	13.87		0.43		0.13		2		1110
		NGC 869 - 991		2 15 28	+56 54.6	1	11.32		0.42				1	B2 V	1541
	+56 00520	NGC 869 - 992		2 15 28	+56 54.9	1	9.85		0.35		-0.49		1	B1 Vn	1541
		NGC 869 - 983		2 15 28	+57 11.7	1	10.65		1.27		1.26		2		1083
		LP 885 - 44		2 15 28	-32 18.3	1	11.47		0.86		0.51		1		1696
		NGC 869 - 1000		2 15 29	+56 53.3	1	13.23		0.41		0.02		2		1110
		NGC 869 - 1004		2 15 29	+56 55.3	1	10.86		0.38		-0.44		1	B2 V	1541
	+56 00521	NGC 869 - 1015		2 15 30	+56 56.9	1	10.57		0.28		-0.10		2	B8 V	1110
		St 2 - 153		2 15 30	+59 05.5	1	10.08		0.37		0.29		2	A1	11
14189 +39 00517		IDS02124N3949	A	2 15 31	+40 02.9	1	7.24		0.43		-0.03		2	F2	3054
14189 +39 00517		IDS02124N3949	B	2 15 31	+40 02.9	1	8.30		0.40		-0.04		2	F2	3054
		NGC 869 - 1041		2 15 31	+56 54.5	1	10.90		0.41		-0.42		1	B2 V	1541
		St 2 - 154		2 15 31	+58 53.2	1	10.63		0.47		0.41		2	A2	11
		NGC 869 - 1040		2 15 32	+56 58.5	1	14.50		0.68		0.40		3		1110
14134 +56 00522		NGC 869 - 1057	★ A	2 15 33	+56 54.3	8	6.55	.015	0.46	.017	-0.37	.003	13	B3 Iab	1012,1083,1110,1119*
	+56 00524	NGC 869 - 1078		2 15 35	+56 53.8	2	9.74	.005	0.34	.000	-0.49	.000	3	B1 Vn	1110,8023
		NGC 869 - 1080		2 15 35	+56 55.1	1	10.97		0.40		-0.39		1	B2 Vn	1541
		NGC 869 - 1085		2 15 35	+56 55.3	1	10.27		0.47		-0.47		1	B1.5V	1541
		St 2 - 155		2 15 35	+58 50.7	1	10.85		0.52		0.37		2	A5	11
		AO 1048		2 15 36	+02 55.1	1	11.66		0.73		0.27		1	G2 V	1748
		NGC 869 - 1094		2 15 36	+57 11.4	1	14.12		0.87		0.27		2		1083
		BPM 16781		2 15 36	-50 54.	1	15.44		0.54		-0.21		2		3065
		NGC 869 - 1109		2 15 37	+56 53.8	1	10.90		0.38				2	B2.5V	250
	+56 00525	NGC 869 - 1116		2 15 37	+56 54.3	1	9.29		0.36				2	B0.5V	250
	+56 00526	NGC 869 - 1133		2 15 38	+56 53.8	1	8.98		0.41				2	B0.5Vn	250
		NGC 869 - 1119		2 15 38	+57 01.5	1	12.57		1.40		1.23		2		1110
	+56 00527	NGC 869 - 1132		2 15 39	+56 54.0	1	8.46		0.39				2	B0.5V	250
		St 2 - 157		2 15 39	+58 49.1	1	11.30		0.55		0.32		4	A5	11
		NGC 869 - 1150		2 15 41	+56 58.4	1	13.23		0.98		0.48		4		1110
		St 2 - 156		2 15 41	+59 18.3	1	10.72		0.43		0.27		2	A2	11
	+56 00529	NGC 869 - 1161		2 15 42	+56 53.9	1	10.21		0.38				2	B1.5Vn	250
14143 +56 00530		NGC 869 - 1162	★	2 15 42	+56 56.4	8	6.65	.022	0.50	.013	-0.45	.000	14	B2 Ia	1012,1083,1110,1119*
		NGC 869 - 1181		2 15 43	+56 54.4	1	12.65		0.40		-0.23		3	B9	1110
		NGC 869 - 1166		2 15 43	+56 58.8	2	13.12	.000	0.54	.000	-0.10	.000	5		1083,1110
		NGC 869 - 1182		2 15 44	+56 58.5	1	13.62		0.52		0.16		2		1083
	+59 00461	St 2 - 158		2 15 44	+59 29.8	3	10.10	.009	0.50	.009	-0.39	.005	4	B1 II	11,64,1012
		NGC 869 - 1189		2 15 45	+56 54.9	1	13.90		0.92		0.72		3		1110
		NGC 869 - 1187		2 15 45	+56 57.2	3	10.82	.005	0.39	.000	-0.40	.000	5	B2 IV	1083,1110,8023
14222 +21 00321				2 15 46	+21 40.1	1	7.79		0.46		0.08		2	F0	1648
		NGC 869 - 1202		2 15 46	+56 53.7	2	12.12	.000	0.44	.000	-0.22	.000	5		1083,1110
14142 +58 00439		T Per		2 15 46	+58 43.9	6	8.52	.037	2.34	.008	2.59	.070	12	M2 Iab:	369,1083,1110,8007*
		NGC 869 - 1211		2 15 48	+57 11.2	2	12.58	.000	0.60	.000	0.32	.000	4		1083,1110
		LP 941 - 39		2 15 49	-35 50.4	1	11.67		1.45		1.24		1		3062
		NGC 869 - 1222		2 15 50	+56 56.8	1	15.50		0.80		0.12		2		1083
14243 +13 00368				2 15 51	+13 43.2	1	8.50		1.32				1	K0	882
		LP 245 - 14		2 15 54	+34 59.4	1	11.75		1.49		1.14		3		1723
		St 2 - 159		2 15 54	+59 37.6	1	10.79		0.51		0.41		2	B9	11
14174 +52 00564		IDS02125N5311	AB	2 15 55	+53 24.6	1	9.10		0.06		-0.25		2	A0	1566
14162 +56 00535		NGC 869 - 1268		2 15 56	+56 54.5	1	9.37		0.38		-0.49		2	B0.5V	1012
14282 -7 00397		G 159 - 55		2 15 56	-06 50.1	2	8.41	.015	0.59	.015	0.00	.025	2	G0	258,3062
		AO 1049		2 15 57	+03 05.0	1	10.97		1.50		1.68		1	K4 III	1748
14213 +45 00589		HR 671		2 15 57	+46 14.6	1	6.22		0.15		0.12		2	A4 V	1733
14161 +58 00440		St 2 - 162		2 15 58	+59 04.7	2	8.21	.024	0.30	.005	0.29		4	A1 V	11,1119
14212 +46 00552		HR 670		2 16 02	+47 09.0	2	5.32	.020	0.00	.004			3	A1 V	1363,3023
14173 +59 00466		St 2 - 160	★ B	2 16 02	+59 47.0	3	7.18	.026	1.00	.027	0.70	.052	4	G5 II	11,1119,8100
	-3 00351			2 16 02	-03 20.1	1	11.35		0.71		0.21		1		1748
	-3 00350			2 16 02	-03 32.7	1	11.11		1.15		1.11		1		1748
14252 +27 00360		HR 675	★ A	2 16 03	+28 24.8	2	5.30		0.04	.002	0.06	.007	7	A2 V	1049,3050
14172 +59 00467		St 2 - 161	★ A	2 16 05	+59 48.0	2	6.94	.020	0.22	.005	0.20		3	A2 V	11,1119
	-36 00866			2 16 05	-36 33.7	1	10.03		0.99		0.83		1	K2 V	3072
14221 +48 00648		HR 673		2 16 06	+48 43.5	1	6.42		0.41		-0.02		2	F4 V	1733
	+58 00441	St 2 - 163		2 16 06	+59 13.4	1	10.23		0.47		0.25		2	A1	11
14262 +22 00329		HR 676		2 16 08	+22 56.3	1	6.46		0.34		0.08		3	A1 V+F3 III	253
14184 +57 00541				2 16 08	+57 26.6	1	9.10		0.31				2	A9 III	1119
	+56 00541	LS V +56 025		2 16 10	+56 29.2	1	10.26		0.12		-0.64		2		401
		VES 734		2 16 14	+63 23.7	1	13.51		0.79		0.16		1		1543
14220 +51 00548				2 16 15	+52 19.9	2	7.10	.084	-0.04	.010	-0.58	.035	5	B2 V	401,1501
14183 +58 00442		St 2 - 164		2 16 15	+59 17.9	2	8.15	.030	0.21	.005	0.17		3	A0 V	11,1119

HD	DM	Other Id	N	Rem	α_{1950}	δ_{1950}	S	V	σ_V	B–V	σ_{B-V}	U–B	σ_{U-B}	n	Spectrum	References
14171	+63 00320	HR 668			2 16 21	+64 06.5	2	6.60	.000	-0.02	.005	-0.09	.008	5	B9.5V	401,595
14305	+19 00342				2 16 22	+19 27.6	1	6.83		0.54		0.03		2	F8	1648
		St 2 - 165			2 16 22	+59 11.8	1	10.93		0.41		0.32		2	A0	11
		LP 770 - 71			2 16 23	-19 45.7	1	12.97		0.69		-0.02		2		1696
14209	+58 00443	St 2 - 166			2 16 24	+58 50.5	1	9.73		0.28		0.24		2	A1 V	11
	+54 00520				2 16 26	+55 18.9	1	9.80		2.02				3		369
14272	+39 00521	HR 677			2 16 32	+39 36.4	2			-0.09	.005	-0.37	.019	4	B8 V	1049,1079
14231	+55 00581				2 16 32	+56 21.8	1	8.79		1.07		0.75		1	G5	401
	+59 00469	St 2 - 167			2 16 38	+59 38.6	1	9.92		0.41		0.34		2	A1 V	11
		CS 22175 # 11			2 16 41	-07 59.6	1	14.14		0.20		0.06		1		1736
		NGC 869 - 1575			2 16 42	+56 52.8	1	12.12		1.95		1.89		2		1083
14333	+24 00335				2 16 43	+25 17.0	1	8.03		0.06		0.00		3	A0	833
14250	+56 00545	NGC 869 - 1586		⋆V	2 16 44	+56 52.2	4	8.96	.010	0.32	.000	-0.54	.000	7	B1 IV	1012,1083,1110,8023
14242	+58 00445	V605 Cas			2 16 44	+59 26.6	5	8.37	.037	2.43	.003	2.70	.116	9	M2 Iab	138,196,1083,8016,8023
14412	-26 00828	HR 683			2 16 44	-26 10.9	5	6.35	.011	0.72	.006	0.17	.010	16	G8 V	15,678,742,2012,3077
	+59 00470	St 2 - 168			2 16 47	+59 34.5	1	9.81		0.38		0.31		2	A2 V	11
14386	-3 00353	HR 681, o Cet		⋆AP	2 16 49	-03 12.2	7	4.95	.244	1.55	.045	0.50	.323	7	M7 IIIe	15,814,975,8003,8005*
		NGC 869 - 1626			2 16 52	+56 47.3	1	13.82		1.24		0.86		2		1083
		NGC 869 - 1628			2 16 52	+56 51.7	1	13.56		0.50		-0.82		2		1083
		NGC 884 - 1641			2 16 55	+56 46.7	1	15.47		0.72		0.48		2		1083
		AO 1053			2 16 57	+02 55.4	1	11.79		0.80		0.44		1		1748
14270	+56 00547	NGC 884 - 1655		⋆V	2 16 57	+56 45.9	7	7.87	.041	2.28	.010	2.52	.042	12	M3 Iab	369,1083,1110,1355*
14411	-3 00355	IDS02143S0326	C		2 16 57	-03 11.7	3	9.33	.000	1.46	.008	1.71	.024	15	M1	244,1748,1775
		VES 737			2 16 58	+62 47.8	1	11.86		0.42		-0.14		1		1543
		LP 941 - 40			2 16 59	-37 00.9	1	11.62		1.50		1.17		1		3078
14302	+55 00587	LS I +56 045			2 17 03	+56 06.1	1	8.57		0.26		-0.56		2	B1 II-III	1012
		CS 22175 # 13			2 17 04	-07 46.1	1	12.93		0.37		-0.22		1		1736
14373	+29 00392	HR 680			2 17 08	+29 57.6	1	6.49		1.23		1.24		2	K0	1733
14417	-5 00438	HR 684			2 17 10	-04 34.5	3	6.49	.005	0.08	.000	0.09	.005	14	A3 V	15,1071,2012
14410	+15 00327				2 17 12	+16 00.4	2	9.00	.015	1.26	.005	1.14		3		882,1648
		LP 769 - 78			2 17 12	-18 21.9	1	12.62		1.23		1.07		1		1696
		POSS 154 # 2			2 17 13	+46 54.3	1	15.22		1.47				2		1739
14322	+55 00588	LS I +55 032			2 17 13	+55 40.8	7	6.81	.032	0.31	.011	-0.32	.026	21	B8 Ib	1012,1077,1083,1110*
	+75 00093				2 17 14	+75 46.0	1	8.52		1.50		0.92		2		1723
14518	-54 00440				2 17 17	-54 10.0	1	7.91		1.35		1.50		5	K3 III	1673
14311					2 17 18	+58 40.	1	9.28		1.14		0.82		2	K0	1723
	+56 00549	LS V +57 012			2 17 19	+57 13.7	1	9.82		0.33		-0.50		2	B1 V	1012
		VES 738			2 17 20	+59 51.3	1	12.81		0.74		0.19		1		1543
14643	-72 00166	BQ Hyi		⋆AB	2 17 20	-71 41.8	2	8.15	.036	0.85	.000	0.39		2	G1 IVp	1641,1642
14321	+56 00550	NGC 884 - 1781			2 17 21	+56 41.8	1	9.22		0.30		-0.54		2	B1 IV	1110
14331	+55 00590	LS I +55 033			2 17 22	+55 35.8	4	8.43	.017	0.17	.004	-0.76	.004	7	B0 III	1012,1083,1110,8023
		NGC 884 - 1789			2 17 23	+56 55.6	1	14.22		0.61		0.40		2		1083
14509	-42 00785	HR 686			2 17 25	-42 04.7	3	6.36	.007	1.16	.005			11	K2 III	15,2013,2029
		AO 1054			2 17 27	+02 33.9	1	10.65		0.50		-0.03		1	F6 V	1748
14426	+16 00277				2 17 27	+17 24.3	1	9.32		0.73				1	G5	882
14372	+46 00557	HR 679			2 17 27	+47 04.9	1	6.10		-0.08		-0.48		2	B5 V	154
14330	+56 00551	NGC 884 - 1818		⋆V	2 17 27	+56 55.8	6	7.96	.026	2.26	.007	2.54	.018	9	M1 Iab	369,1083,1110,8007*
		VES 739			2 17 27	+60 04.3	1	12.66		0.77		-0.03		1		1543
14438	+13 00371				2 17 30	+14 04.2	2	8.01	.020	1.22	.015	1.21		4	K0	882,1648
		G 74 - 16	AB		2 17 30	+34 00.9	1	12.32		1.42		1.34		2		7010
		POSS 154 # 6			2 17 31	+47 18.6	1	17.77		1.67				3		1739
		NGC 884 - 1852			2 17 32	+56 55.2	1	14.17		0.82		0.16		2		1083
	-3 00356				2 17 32	-03 01.4	1	10.64		0.55		-0.04		1		244
14459	-1 00317				2 17 33	-00 41.3	1	8.10		0.75		0.29		1	G0	258
14473	-4 00379				2 17 34	-03 45.8	1	8.06		0.50		-0.02		4	F5	244
		NGC 884 - 1876			2 17 35	+56 38.0	1	14.77		1.26		0.45		2		1083
14345	+59 00472				2 17 36	+59 26.8	1	8.62		1.63				23	K2	6011
14357	+56 00555	NGC 884 - 1899			2 17 38	+56 38.2	4	8.52	.012	0.32	.004	-0.53	.004	7	B1.5II	1012,1083,1110,8023
14328	+61 00401				2 17 38	+61 51.0	1	9.53		0.28		0.21		3	A0 V	848
14392	+49 00640	HR 682			2 17 39	+49 55.4	2	5.55		-0.09	.029	-0.40	.017	5	B9p Si	1063,3033
		VES 740			2 17 40	+57 50.8	1	11.11		0.46		-1.62		1		1543
		CS 22175 # 12			2 17 40	-08 21.9	1	12.81		0.25		0.06		1		1736
		NGC 884 - 1910			2 17 41	+56 38.3	1	13.38		0.72		0.18		2		1083
14526	-36 00874	IDS02156S3554	AB		2 17 42	-35 40.5	3	8.22	.025	0.91	.019	0.63	.024	10	K2 V	214,938,2012
	-3 00357				2 17 44	-03 06.1	2	10.72	.165	0.28	.010	0.01	.010	2		244,1748
	+58 00451	LS I +59 129	A		2 17 48	+59 12.7	1	10.07		0.52		-0.36		2	B1 III:	1012
	+58 00451		B		2 17 48	+59 12.7	1	10.36		0.35		0.22		2		1012
	+52 00570	G 173 - 52	A		2 17 51	+53 19.9	1	10.38		1.24		1.24		2		7010
	+52 00570		B		2 17 51	+53 19.9	1	14.70		0.68				3		7010
	+52 00570		C		2 17 51	+53 19.9	1	14.26		0.47				3		7010
	+52 00570		D		2 17 51	+53 19.9	1	14.94		0.74		0.14		2		7010
	+52 00570		E		2 17 51	+53 19.9	1	14.86		0.70		0.23		2		7010
	+13 00374	G 73 - 44			2 17 55	+13 26.5	1	8.97		0.79		0.31		1	K0	333,1620
		NGC 884 - 2005			2 17 55	+56 53.7	1	13.56		0.41		0.02		2		1110
		CS 22175 # 34			2 17 55	-10 51.9	1	12.60		0.37		-0.09		1		1736
		G 74 - 18			2 17 57	+36 39.4	1	14.22		1.70		1.50		4		203
		NGC 884 - 2008			2 17 57	+57 09.9	1	9.45		0.32		-0.65		2		1110
14549	-40 00589				2 18 00	-39 45.2	1	9.62		0.29		0.09		4	A2mA6-F0	152
	-21 00418				2 18 03	-20 46.5	1	10.77		0.86		0.51		2		742
	+79 00067				2 18 06	+80 13.6	1	10.45		0.17		0.16		2	A0	1375

Table 1   135

HD	DM	Other Id	N Rem	α_{1950}	δ_{1950}	S	V	σ_V	B–V	σ_{B-V}	U–B	σ_{U-B}	n	Spectrum	References
14436	+50 00530			2 18 08	+51 03.7	1	8.20		0.09		0.02		2	A0 V	401
14404	+57 00550	PR Per		2 18 08	+57 38.1	9	7.90	.048	2.30	.011	2.61	.051	15	M2 Ib	138,196,369,1083,1110*
		G 73 - 45		2 18 10	+02 45.0	1	14.78		1.69				1		906
	+56 00563	NGC 884 - 2088	⋆ V	2 18 11	+56 53.8	1	9.45		0.32		-0.65		2	B1.5Vne	1110
		NGC 884 - 2084		2 18 11	+57 09.6	1	12.09		0.73		0.40		2		1083
		NGC 884 - 2094		2 18 12	+56 55.2	1	11.90		0.30		-0.51		1		1541
		NGC 884 - 2114		2 18 14	+56 53.8	1	11.04		0.32		-0.48		1	B2 V	1541
		CS 22175 # 33		2 18 15	-10 45.2	1	14.76		0.33		0.02		1		1736
		NGC 884 - 2133		2 18 16	+56 54.1	1	12.21		0.33		-0.38		2		1110
		NGC 884 - 2126		2 18 16	+57 00.0	1	13.60		0.60		0.21		2		1083
	+67 00200			2 18 16	+67 33.3	1	8.83		0.92		0.56		2	G5	1375
		NGC 884 - 2139		2 18 17	+56 54.7	2	11.37	.010	0.28	.005	-0.49	.000	3	B2 V	1110,8023
		NGC 884 - 2134		2 18 17	+56 55.9	1	14.93		0.95		0.29		1		1110
14422	+56 00565	NGC 884 - 2138	⋆ V	2 18 17	+57 09.5	4	9.00	.083	0.50	.020	-0.64	.028	7	B1 IIIpe	1012,1083,1110,8023
14641	-56 00413	HR 688	⋆ A	2 18 17	-56 10.4	2	5.81	.000	1.55	.001	1.95		6	K5 III	58,2035
		NGC 884 - 2152		2 18 18	+56 40.9	1	13.21		1.23		0.40		2		1083
		NGC 884 - 2147		2 18 18	+56 55.1	1	14.34		0.49		0.14		3		1110
		NGC 884 - 2144		2 18 18	+57 02.2	1	12.33		0.54		-0.13		2		1083
		NGC 884 - 2155	AB	2 18 19	+56 56.5	1	15.33		0.60		0.22		1		1110
		NGC 884 - 2150		2 18 19	+57 09.0	1	14.71		0.57		0.21		2		1083
14434	+56 00567	NGC 884 - 2172		2 18 20	+56 40.6	6	8.49	.011	0.16	.000	-0.79	.005	11	O6	1011,1012,1083,1110*
		NGC 884 - 2167		2 18 20	+56 54.4	1	13.36		0.41		-0.13		2		1110
	+56 00566	NGC 884 - 2165	⋆ V	2 18 20	+56 56.3	1	9.86		0.40		-0.62		2	B1 Vne	1110
		NGC 884 - 2170		2 18 21	+56 55.1	1	15.23		0.64		0.43		2		1110
	+58 00453	LS V +59 008		2 18 21	+59 05.3	1	10.53		0.45		-0.45		2	B1 V	1012
	+20 00381	IDS02156N2108	A	2 18 22	+21 21.2	1	10.06		0.83		0.52		3	G8 V	1723
		NGC 884 - 2181		2 18 22	+56 39.8	1	15.14		0.50		0.18		2		1083
14433	+56 00568	NGC 884 - 2178		2 18 22	+57 00.9	7	6.40	.019	0.57	.008	0.03	.010	12	A1 Ia	1012,1083,1110,8007*
		NGC 884 - 2185		2 18 23	+56 52.3	2	10.92	.000	0.32	.005	-0.25	.000	3	B2 Vn	1110,8023
		NGC 884 - 2188		2 18 24	+56 55.1	2	14.99	.000	0.71	.000	0.35	.000	4		1083,1110
		NGC 884 - 2189		2 18 24	+56 59.1	1	12.26		0.43		-0.30		2		1083
		NGC 884 - 2200		2 18 25	+56 54.2	1	12.67		0.36		-0.25		2		1110
		NGC 884 - 2203		2 18 25	+56 55.1	2	14.61	.000	0.50	.000	0.05	.000	4		1083,1110
		NGC 884 - 2196		2 18 25	+56 55.8	1	11.57		0.32		-0.43		1	B1.5V	1110
		NGC 884 - 2194		2 18 25	+56 56.7	1	13.45		0.42		-0.13		2		1110
		NGC 884 - 2232		2 18 28	+56 53.8	2	11.09	.015	0.23	.015	-0.54	.000	3	B1 V	1110,8023
	+56 00571	NGC 884 - 2235		2 18 28	+56 54.6	2	9.34	.015	0.33	.000	-0.55	.000	3		1110,8023
14443	+56 00570	NGC 884 - 2227		2 18 28	+56 55.0	5	8.04	.011	0.34	.004	-0.55	.005	8	B2 II	138,1012,1083,1110,8023
		NGC 884 - 2229		2 18 28	+56 55.9	1	11.40		0.29		-0.47		1	B2 Vn	1541
14574	-27 00812			2 18 28	-27 01.4	1	10.72		0.16		0.08		4	A1/2 V	152
		NGC 884 - 2241		2 18 29	+56 53.4	1	13.58		0.43		0.02		2		1110
		NGC 884 - 2250		2 18 30	+56 51.4	1	15.10		0.56		0.29		2		1110
		NGC 884 - 2251		2 18 30	+56 54.4	2	11.55	.005	0.33	.005	-0.34	.000	3	B3 V	1110,8023
	+56 00572	NGC 884 - 2246		2 18 30	+56 54.8	3	9.89	.014	0.32	.000	-0.56	.000	5	B1.5V	1083,1110,8023
		NGC 884 - 2242		2 18 30	+56 55.7	1	10.98		0.38		-0.60		1	B1 Vnp	1541
	+70 00169	G 244 - 60		2 18 30	+70 57.1	3	8.95	.018	0.88	.010	0.63	.007	6	K2 V	22,1003,1658
		NGC 884 - 2253		2 18 31	+56 53.5	1	12.71		0.41		-0.10		2		1110
		NGC 884 - 2259		2 18 31	+56 54.9	1	13.44		0.36		-0.18		2		1083
		NGC 884 - 2258		2 18 31	+56 55.1	1	13.93		0.42		0.08		2		1083
		NGC 884 - 2255		2 18 31	+56 55.5	1	10.67		0.30		-0.49		1	B2 Vn	1541
		NGC 884 - 2269		2 18 32	+56 53.6	1	13.19		0.39		-0.19		2		1110
		NGC 884 - 2262		2 18 32	+56 57.0	2	10.49	.099	0.39	.045	-0.54	.055	3	B2 Vn	1110,1541
14442	+58 00455	LS I +59 130	A	2 18 32	+59 19.3	6	9.21	.010	0.41	.000	-0.62	.010	12	O6 ef	64,450,1011,1012,1083,1209
14442	+58 00455		B	2 18 32	+59 19.3	1	11.94		0.47		-0.21		3		450
		NGC 884 - 2270		2 18 33	+56 54.2	1	14.18		0.42		0.08		1		1110
	+56 00573	NGC 884 - 2284		2 18 34	+56 51.7	1	9.66		0.40		-0.59		2	B1 Vne	1110
14469	+55 00597	SU Per		2 18 35	+56 22.6	5	7.70	.115	2.17	.013	2.24	.068	12	M3 Iab:	369,1083,1110,8016,8023
	+56 00574	NGC 884 - 2296		2 18 35	+56 53.0	1	8.53		0.30		-0.56		2	B1 III	1110
	+56 00575	NGC 884 - 2299		2 18 36	+56 53.8	1	9.08		0.29		-0.57		1	B0.5IV	1110
	+66 00205			2 18 36	+66 50.7	1	9.35		0.52		0.17		2	B9 Ib	1215
	+56 00576	NGC 884 - 2311		2 18 37	+56 53.4	1	9.38		0.30		-0.54		2	B2 III	1110
		NGC 884 - 2330		2 18 39	+56 53.2	1	11.42		0.24		-0.39		1		1110
14629	-39 00684			2 18 39	-39 15.7	1	8.74		1.03		0.83		1	K3/4 V	3072
		NGC 884 - 2349		2 18 42	+56 52.6	2	12.81	.000	0.48	.093	0.12	.317	3		1083,1110
14476	+56 00577	NGC 884 - 2361		2 18 44	+57 02.6	1	8.75		0.38		-0.51		2	B2 III	1110
	+56 00578	NGC 884 - 2371		2 18 45	+56 53.8	1	9.25		0.32		-0.55		2	B2 III	1110
		POSS 154 # 5		2 18 48	+47 06.2	1	17.57		1.68				3		1739
14489	+55 00598	HR 685, V474 Per	⋆ A	2 18 51	+55 37.1	7	5.18	.009	0.57	.004	-0.11	.000	13	A2 Ia	15,64,851,1012,1083*
14488	+56 00583	NGC 884 - 2417	⋆ V	2 18 51	+56 52.9	5	8.49	.178	2.26	.014	2.27	.015	11	M4 Iab	369,1083,1110,8016,8023
		NGC 884 - 2421		2 18 53	+56 52.0	1	13.17		0.72		0.32		2		1083
		NGC 884 - 2426		2 18 53	+56 53.2	1	13.69		0.49		0.06		2		1083
	+47 00612	G 173 - 53		2 18 57	+47 39.1	2	9.39	.033	1.50	.009	1.18		2	M2	1017,3072
14501	+57 00551	LS V +57 017		2 18 57	+57 55.6	1	9.42		0.40		-0.32		2	B3 V:n	1012
		LP 710 - 86		2 18 59	-10 30.5	1	13.50		1.02		0.92		1		1696
	+56 00586	NGC 884 - 2488		2 19 03	+56 51.3	1	9.94		0.25		-0.59		2	B1 V	1012
14596	+15 00329			2 19 05	+15 56.1	2	7.91	.030	1.07	.013	0.74		4	K0	882,3040
14627	-3 00363			2 19 05	-03 11.3	1	9.00		0.48		-0.05		4	F8	244
		NGC 884 - 2515		2 19 07	+56 51.2	1	13.41		0.37		-0.16		2		1083
		TW Hyi		2 19 08	-73 47.5	1	12.56		0.16		0.16		1		700
		Steph 258	AB	2 19 09	+47 15.7	1	11.48		1.34		1.04		1	M0	1746

HD	DM	Other Id	N Rem	α_{1950}	δ_{1950}	S	V	σ_V	B–V	σ_{B-V}	U–B	σ_{U-B}	n	Spectrum	References
14679	−31 00942			2 19 10	−30 46.3	1	9.28		0.65				4	G5 V	2033
14520	+56 00588	NGC 884 - 2541	⋆ ABC	2 19 11	+56 51.5	5	9.15	.023	0.33	.005	-0.55	.006	9	B1.5III	401,1012,1083,1110,8023
14680	−31 00943			2 19 11	−31 09.8	3	8.81	.008	0.93	.004	0.65	.010	10	K2/3 V	158,2012,3072
		G 75 - 4		2 19 12	−04 21.5	1	13.22		1.39		1.11		1		3062
14529	+55 00599			2 19 14	+55 38.9	1	9.84		0.13		-0.14		2	A0	1083
14635	−7 00410	G 75 - 5		2 19 14	−07 06.5	1	8.98		1.08		0.96		1	K4:	1620
14528	+57 00552	S Per	⋆ A	2 19 15	+58 21.6	5	9.51	.426	2.72	.037	2.57	.103	10	M3 Iae	369,1083,1110,8016,8023
		NGC 884 - 2573		2 19 16	+56 51.8	1	14.80		0.59		0.37		2		1083
	+56 00590	NGC 884 - 2572		2 19 16	+56 55.6	1	10.02		0.34		-0.51		2		1110
	+56 00589	V360 Per		2 19 16	+57 17.1	2	9.46	.000	0.41	.000	-0.48	.000	3	B1 III	64,1012
		NGC 884 - 2575		2 19 17	+57 01.2	1	13.98		0.51		0.03		2		1083
14535	+56 00591	NGC 884 - 2589		2 19 20	+57 01.1	6	7.45	.014	0.71	.008	0.12	.008	10	A2 Ia	1012,1083,1110,8007*
14519	+62 00385			2 19 21	+63 22.5	1	7.79		1.18		1.12		2	K1 III	37
14702	−37 00889			2 19 21	−37 32.8	1	10.61		0.25		0.13		4	A9 V	152
14652	−0 00355	HR 689		2 19 23	+00 10.1	4	5.30	.021	1.65	.001	1.94	.044	17	M2 III	3,15,1061,1355
14624	+25 00388			2 19 24	+26 02.6	1	8.46		0.71		0.27		2	G5 V	3026
		NGC 884 - 2607		2 19 25	+57 08.7	1	15.18		1.14		0.64		2		1083
		NGC 884 - 2613		2 19 26	+57 08.0	1	14.98		0.97		0.29		2		1083
14542	+56 00593	NGC 884 - 2621		2 19 27	+57 09.6	4	6.98	.017	0.62	.005	-0.14	.004	7	B8 Ia	1012,1083,1110,8023
		NGC 884 - 2631		2 19 28	+56 59.6	1	12.56		0.42		-0.27		2		1083
	+64 00324			2 19 28	+64 50.8	1	10.38		0.74		0.17		2	M2	1375
		NGC 884 - 2639		2 19 29	+57 02.1	1	13.45		0.70		0.29		2		1083
	+60 00470	DN Cas		2 19 29	+60 36.2	4	9.88	.004	0.70	.000	-0.36	.000	7	O8 V	64,1011,1012,1083
		NGC 884 - 2641		2 19 30	+57 10.3	1	13.94		0.55		0.09		2		1083
14744	−35 00812			2 19 31	−35 03.8	1	9.51		0.90		0.69		1	K3 V	3072
14691	−11 00448	HR 692		2 19 35	−11 00.2	2	5.43	.032	0.37	.018	0.00		5	F0 V	254,2007
	+56 00595	NGC 884 - 2691	⋆ V	2 19 38	+56 58.3	8	8.16	.036	2.19	.038	2.31	.068	13	M0 Iab	138,196,369,1083,1110*
		LS I +57 052		2 19 39	+57 56.2	1	10.56		0.41		-0.44		2	B1 V	1012
14690	−1 00322	HR 691		2 19 39	−01 06.7	3	5.42	.008	0.31	.005	0.10	.018	8	F0 Vn	15,39,1256
		NGC 884 - 2700		2 19 40	+56 59.1	1	14.68		0.91		0.41		2		1083
	+53 00516			2 19 41	+53 46.8	1	9.97		1.90				3		369
	−4 00384	G 75 - 6	⋆ AB	2 19 41	−04 07.3	1	10.34		0.89		0.40		2		3062
14622	+40 00500	HR 687	⋆ A	2 19 43	+41 10.2	4	5.83	.015	0.27	.009	0.04	.030	7	F0 III-IV	39,254,272,3037
14622	+40 00500	IDS02166N4056	B	2 19 43	+41 10.2	1	10.49		0.50		0.03		4		450
14728	−18 00409	HR 693		2 19 44	−17 53.3	3	5.87	.008	1.23	.008	1.32	.025	8	K2 III	252,2006,3077
		LS V +57 024		2 19 45	+57 17.8	1	9.91		0.56		-0.52		4		1401
		NGC 884 - 2733		2 19 46	+56 58.3	1	13.10		0.57		0.10		2		1083
14633	+40 00501	IDS02166N4056	A	2 19 47	+41 15.2	5	7.46	.012	-0.20	.013	-1.05	.028	19	O8.5V	1,450,1011,1118,8040
14688	+16 00281			2 19 48	+16 38.6	2	6.80	.010	0.15	.000	0.13	.010	5	A2 V	1733,3016
		MWC 448		2 19 48	+57 05.	1	10.88		0.67		-0.44		4		1401
		G 4 - 16		2 19 49	+15 17.5	2	12.64	.007	0.76	.000	0.11	.017	3		1620,1658
		NGC 884 - 2752		2 19 49	+56 59.5	1	12.62		0.51		-0.20		2		1083
14580	+56 00597	NGC 884 - 2758	⋆ V	2 19 50	+56 59.1	7	8.45	.032	2.31	.024	2.51	.040	10	M0 Iab	138,369,1083,1110*
		VES 741		2 19 50	+59 28.4	1	12.03		1.00		0.02		1		1543
	+58 00458	LS I +58 074		2 19 52	+58 44.2	1	9.80		0.54		-0.59		2	B1pe	1012
	+55 00604			2 19 53	+55 53.3	1	9.97		1.24		1.20		2		1723
14687	+17 00353	G 4 - 17		2 19 54	+18 10.9	2	8.83	.015	0.91	.005	0.65	.044	3	K0	333,1620,3040
14804	−47 00713			2 19 56	−46 49.4	1	10.14		-0.08		-0.23		4	B9 V	152
	−30 00842			2 19 59	−30 29.1	1	11.43		1.06		0.93		1		1696
14605	+55 00605	V361 Per		2 20 03	+56 20.9	1	9.34		0.27		-0.75		2	B0.5Vpe	1012
	+48 00658			2 20 05	+48 48.3	1	8.78		-0.11		-0.84		2	B2	401
	+48 00661			2 20 09	+48 49.5	1	9.42		0.17		0.10		1	A0	401
14834	−50 00681			2 20 09	−49 44.9	1	7.36		0.89		0.51		6	G6 III	1673
14685	+37 00544			2 20 13	+38 01.5	1	7.14		-0.02		-0.10		2	A0	1733
14739	+16 00283			2 20 14	+17 22.3	1	7.47		0.17		0.12		2	A2	1648
14832	−43 00724	HR 698		2 20 14	−43 25.7	3	6.30	.004	0.99	.004			11	K0 III	15,1075,2032
14802	−24 01038	HR 695		2 20 15	−24 02.6	5	5.19	.007	0.60	.004	0.12	.000	17	G2 V	15,678,2012,3077,8015
14662	+54 00535	HR 690, V440 Per	⋆ AB	2 20 22	+55 08.3	3	6.27	.039	0.86	.022	0.65	.014	6	F7 Ib	138,851,1355
14788	−15 00416			2 20 22	−15 07.7	1	7.67		0.03		0.07		4	A1/2 IV	152
		GD 27		2 20 23	+48 03.2	1	14.83		0.20		-0.95		1		3060
		POSS 154 # 1		2 20 23	+48 03.2	1	15.12		-0.12				2		1739
		LS I +58 075		2 20 25	+58 05.8	1	9.44		0.43		-0.57		2	B0 IV:nn	1012
14735	+34 00425			2 20 27	+35 12.9	1	6.76		1.18		1.04		2	G5	1625
14617	+62 00387			2 20 27	+63 18.6	1	7.54		1.58		1.75		2	K2 II-III	37
14684	+50 00541			2 20 28	+50 40.6	1	8.68		-0.01		-0.39		2	A0	401
14632	+62 00388			2 20 28	+62 49.0	1	7.65		0.16		0.10		2	A0 V	401
14772	+14 00387	G 73 - 50	⋆ B	2 20 32	+15 11.3	1	9.36		0.68		0.18		1	G5	333,1620
14786	+14 00389	G 73 - 51	⋆ A	2 20 35	+15 11.4	1	8.86		0.61		0.05		1	G5	333,1620
14830	−19 00444	HR 697		2 20 36	−18 34.8	1	6.22		0.94				4	G8/K0 III	2032
14683	+55 00607			2 20 40	+56 22.1	1	10.73		0.62				1	A0 V	1543
	−12 00445			2 20 40	−11 58.1	1	11.63		0.52		-0.03		5		1732
14784	+21 00332		A	2 20 44	+22 13.9	1	8.30		0.61		0.10		3	G0	3016
14784	+21 00332		B	2 20 44	+22 13.9	1	15.93		0.07		-0.71		3	Dan	3060
14829	−11 00452			2 20 44	−10 54.4	1	10.29		0.02		0.06		4	A0	152
		G 4 - 19		2 20 47	+17 52.4	1	11.67		0.61		0.00		1		1620
236960	+58 00461	LS I +59 131		2 20 50	+59 00.2	2	9.75	.010	0.46	.005	-0.51		3	B0.5III	934,1012
		PG0220+132B		2 20 51	+13 14.5	1	14.22		0.94		0.32		2		1764
15008	−69 00113	HR 705		2 20 51	−68 53.2	5	4.08	.004	0.03	.003	0.05	.000	24	A3 V	15,1075,1075,2038,3023
14707	+57 00558	LS I +58 077		2 20 52	+58 04.5	2	9.89	.000	0.55	.000	-0.42	.000	3	B0.5III	64,1012
14840	−6 00470			2 20 53	−06 25.1	1	6.99		0.30		0.05		4	F0	244

Table 1   137

HD	DM	Other Id	N Rem	α_{1950}	δ_{1950}	S	V	σ_V	B–V	σ_{B-V}	U–B	σ_{U-B}	n	Spectrum	References
14723 +53 00521				2 20 54	+53 32.9	1	8.60		0.04		-0.31		2	B8	1566
		PG0220+132		2 20 55	+13 14.1	1	14.76		-0.13		-0.92		3	sdB	1764
14800 +32 00433		IDS02180N3303	AB	2 20 55	+33 16.9	1	7.38		0.12		0.08		2	A0	3016
14800 +32 00433		IDS02180N3303	C	2 20 55	+33 16.9	1	10.71		0.86		0.50		1	G0	3016
14800 +32 00433		IDS02180N3303	D	2 20 55	+33 16.9	1	10.19		0.79		0.32		2	G5	3016
+40 00503		IDS02178N4110	AB	2 20 55	+41 23.2	1	10.74		0.83		0.51		1		8084
		PG0220+132A		2 20 56	+13 13.9	1	15.77		0.78		-0.34		2		1764
+59 00480		IDS02173N5932	AB	2 21 00	+59 45.7	1	10.49		2.05				4	M3	369
		IC 1795 - 203		2 21 00	+61 40.	1	11.36		0.93		-0.06		4		341
		IC 1795 - 212		2 21 00	+61 40.	1	13.85		1.01		0.37		1		341
		IC 1795 - 217		2 21 00	+61 40.	1	13.42		0.77		0.28		2		341
		IC 1795 - 238		2 21 00	+61 40.	1	11.15		0.45		0.12		2		341
		IC 1795 - 262		2 21 00	+61 40.	1	10.79		0.47		0.33		2		341
14882 -30 00852		IDS02189S3019	AB	2 21 03	-30 05.7	1	6.95		0.56				2	G1 V	173
14890 -38 00797		HR 700		2 21 03	-37 48.1	4	6.52	.015	1.61	.012	1.94		13	K2 III	15,1075,2007,3005
14799 +39 00472				2 21 05	+39 07.2	1	7.83		-0.06		-0.22		1	A0 V	105
14770 +49 00649		HR 694		2 21 05	+49 46.9	2	5.19	.009	0.97	.004	0.75	.004	5	G8 III	1355,3016
15066 -72 00176				2 21 06	-71 53.9	2	8.27	.006	1.19	.000	1.24		42	K2 III	978,1642
14857 -4 00390				2 21 08	-03 36.2	1	8.42		1.63		1.90		4	M1	244
14943 -51 00571		HR 701		2 21 09	-51 19.2	2	5.91	.005	0.22	.000			7	A5 V	15,2012
+56 00607		LS V +57 029		2 21 10	+57 07.7	1	10.58		0.62		-0.43		4		1401
		L 174 - 13		2 21 12	-57 06.	1	13.60		0.65		0.02		2		1696
		CS 22175 # 30		2 21 17	-10 44.3	1	13.76		0.41		-0.17		1		1736
		MCC 222		2 21 23	+04 38.8	2	10.35	.002	1.21	.004			3	K4-5	1619,1632
+57 00564				2 21 23	+58 09.6	1	10.31		2.01				1		369
		POSS 154 # 3		2 21 24	+46 37.0	1	16.43		1.43				2		1739
+5 00332				2 21 25	+06 17.3	1	9.42		1.69				3	M0	1532
		VES 743		2 21 33	+57 59.3	1	12.68		0.75		0.07		1		1543
+39 00539		G 74 - 22		2 21 35	+39 55.5	1	9.86		1.04		0.99		1	K2	1620
14500 +79 00069				2 21 38	+80 23.6	1	8.42		0.12		0.13		3	A0	1502
		IC 1795 - 30		2 21 41	+61 53.2	1	14.86		0.90		0.62		1		504
		IC 1795 - 25		2 21 41	+61 55.3	1	14.08		0.72		0.55		1		504
14827 +54 00539		IDS02182N5448	A	2 21 43	+54 01.7	6	7.63	.006	0.04	.007	-0.21	.007	96	B9 V	401,851,1566,1601,1648,1733
14818 +55 00612		HR 696	⋆	2 21 43	+56 23.1	11	6.25	.016	0.30	.007	-0.61	.009	43	B2 Iae	15,154,369,1012,1077*
+37 00548				2 21 44	+37 33.9	2	9.74	.000	-0.10	.005	-0.50	.014	9	A0	401,1775
		LS V +57 031		2 21 44	+57 41.0	1	11.02		0.65		-0.35		1		1543
14795 +59 00484				2 21 45	+59 46.8	1	7.68		0.00		-0.41		2	B7 V	401
14875 +28 00409				2 21 46	+29 01.1	1	7.09		1.28		1.46		2	K3 III	1625
14965 -29 00859				2 21 46	-29 39.1	1	8.88		0.07		0.09		4	A1 V	152
14826 +56 00609		V441 Per	A	2 21 47	+57 12.7	8	8.26	.030	2.33	.028	2.45	.036	16	M2 Iab	138,196,369,1083,1110*
14826 +56 00609			B	2 21 47	+57 12.7	1	10.92		0.41		-0.44		3		196
		LP 885 - 55		2 21 47	-27 28.9	1	13.98		0.94		0.67		1		1696
		IC 1795 - 40		2 21 49	+61 48.5	1	13.25		1.08		0.07		4		341
		G 4 - 20		2 21 50	+14 35.4	1	14.15		1.06		0.93		1		1658
14855 +44 00483				2 21 50	+45 25.4	1	7.42		0.90		0.59		2	G5	1601
		IC 1795 - 32		2 21 50	+61 51.7	1	13.71		1.26		0.35		2		341
		MCC 5		2 21 52	+49 34.1	2	11.15	.020	1.29	.006	1.23		3	K5	1017,1726,3003
14938 -3 00372				2 21 52	-03 19.7	1	7.19		0.51		-0.04		4	F5	244
+4 00391				2 21 53	+04 43.3	1	10.28		1.56				1	M1	1632
14874 +29 00406				2 21 54	+30 25.4	1	8.16		0.67		0.08		2	G0 V	3016
14817 +60 00472		V559 Cas	⋆ AB	2 21 54	+61 19.5	2	6.99	.015	0.22	.007	-0.14	.010	5	B8 V +A0 V	1689,3030
+60 00473				2 21 58	+60 49.2	1	9.76		0.52		0.07		3	F5 V	1689
+55 00614				2 21 59	+56 20.7	1	10.52		0.49		0.34		2		1083
14694 +73 00136				2 22 01	+74 21.5	1	8.04		1.84		2.13		3	K5	1733
		IC 1795 - 51		2 22 03	+61 55.9	2	13.39	.032	0.83	.005	0.23	.149	5		341,504
14988 -26 00857		HR 703		2 22 05	-26 04.4	1	6.44		1.32				4	K3 III	2007
-35 00828				2 22 07	-35 19.7	1	11.41		1.54		1.16		1		1696
14951 +9 00316		HR 702		2 22 08	+10 23.1	2	5.46		-0.10	.000	-0.49	.008	4	B7 IV	154,1203
14918 +24 00344		IDS02193N2502	AB	2 22 10	+25 15.7	2	7.78	.025	0.82	.005	0.49		6	G5 III	20,833
14893 +36 00478				2 22 12	+36 53.6	1	7.36		-0.02		-0.11		2	B9	401
+58 00466				2 22 15	+59 01.5	1	9.94		1.34				1		934
14872 +49 00656		HR 699	⋆ A	2 22 16	+50 03.2	7	4.71	.013	1.53	.009	1.87	.029	14	K4+III	15,369,1355,1363,3016*
		IC 1795 - 55		2 22 20	+61 51.4	1	14.38		0.93		0.35		1		341
		IC 1795 - 50		2 22 23	+61 57.0	1	14.15		1.00		0.60		1		504
14949 +27 00374				2 22 25	+27 27.5	1	7.69		1.11		1.00		2	K2 II	8100
		LS I +58 079		2 22 25	+58 40.0	1	11.30		0.57		-0.43		1		1543
		LS I +59 133		2 22 25	+59 46.3	1	10.66		0.77		-0.26		2	B0 IV:	1012
15005 -4 00394		S0421	A	2 22 25	-04 07.0	1	6.89		1.02		0.82		4	K0	244
14871 +55 00616		DM Per		2 22 26	+55 52.7	1	7.95		0.10		-0.41		2	B6 V	401
		IC 1795 - 69		2 22 27	+61 56.7	1	15.38		1.00		0.65		1		504
15004 -3 00374		HR 704		2 22 27	-03 00.3	3	6.35	.017	-0.01	.005	-0.08	.044	12	A0 III	15,814,1256
		IC 1795 - 91		2 22 28	+61 47.1	1	13.80		0.85		0.46		3		341
14948 +31 00418				2 22 29	+32 10.9	1	7.18		1.48		1.71		3	K2	1733
		IC 1795 - 68		2 22 30	+61 57.2	1	16.05		0.50		0.40		1		504
		IC 1795 - 77		2 22 31	+61 53.0	1	13.24		0.86		0.23		4		341
		IC 1795 - 75		2 22 33	+61 53.3	1	14.08		0.86		0.44		2		341
15064 -41 00681		HR 706		2 22 33	-41 04.1	3	6.17	.004	0.66	.005			11	G2 V	15,2013,2029
15248 -74 00194		HR 715		2 22 33	-73 52.3	3	6.00	.003	1.09	.000	1.05	.005	21	K1 III	15,1075,2038
		Steph 265		2 22 35	+24 27.2	1	12.23		1.48		1.19		1	M0	1746
		IC 1795 - 66		2 22 37	+62 00.9	1	13.87		0.40		0.23		2		341

HD	DM	Other Id	N	Rem	α_{1950}	δ_{1950}	S	V	σ_V	B–V	σ_{B-V}	U–B	σ_{U-B}	n	Spectrum	References
		IC 1795 - 64			2 22 37	+62 03.4	1	11.85		0.47		0.18		4		341
	+28 00413				2 22 43	+28 31.9	2	9.07	.000	0.96	.000	0.73		2	K8	1017,3072
14899	+56 00621	LS I +57 056			2 22 44	+57 00.2	4	7.39	.010	0.44	.000	-0.11	.004	7	B8 Ib	1012,1083,1110,8023
		IC 1795 - 78			2 22 44	+61 53.0	1	14.24		0.99		0.45		1		341
		IC 1795 - 65			2 22 45	+62 01.7	1	13.52		1.03		0.40		2		341
	+61 00411	IC 1795 - 89			2 22 47	+61 47.3	4	10.17	.011	0.98	.011	-0.09	.020	26	O8	341,504,1011,1012
		IC 1795 - 63			2 22 47	+62 05.1	1	11.72		1.62		0.78		1		341
	+63 00333				2 22 48	+63 48.6	1	8.52		1.99				3	K5	369
		IC 1795 - 82			2 22 50	+61 50.0	1	13.03		0.84		0.11		8		341
		IC 1795 - 80			2 22 50	+61 52.7	1	13.38		1.14		0.58		2		341
		IC 1795 - 71			2 22 50	+61 56.2	2	11.82	.019	0.81	.009	0.38	.062	11		341,504
15014	+18 00303				2 22 52	+19 19.8	1	7.94		1.02		0.85		2	K0	1648
14914	+59 00486				2 22 52	+59 26.1	1	7.00		0.96		0.83		4	K0 III-IV	7008
		IC 1795 - 100			2 22 55	+62 03.1	1	13.66		0.84		0.18		2		341
		IC 1795 - 105			2 22 57	+61 59.7	2	12.93	.103	0.90	.028	0.32	.145	4		341,504
15028	+19 00355				2 22 59	+20 03.4	1	7.85		0.79		0.39		2	G5	1648
	+54 00544	LS I +54 010			2 22 59	+54 48.6	1	9.73		0.18		-0.28		1	B8 Iab	1215
		IC 1795 - 111			2 22 59	+61 46.9	1	13.93		0.98		0.05		2		341
		IC 1795 - 114			2 23 04	+61 40.9	1	13.27		1.00		-0.14		3		341
		IC 1795 - 106			2 23 04	+61 58.8	2	12.81	.102	0.89	.024	0.31	.200	3		341,504
		G 94 - 66			2 23 07	+17 44.4	1	13.87		0.88		0.23		2		1658
14947	+58 00467	LS I +58 080			2 23 08	+58 39.1	5	8.00	.022	0.45	.008	-0.60	.004	10	O6 e	64,1011,1012,1083,1209
		IC 1795 - 103			2 23 08	+62 01.4	1	12.99		0.65		0.45		2		341
14956	+57 00568	V475 Per			2 23 10	+57 27.3	5	7.20	.017	0.73	.014	-0.29	.004	10	B2 Ia	1012,1077,1083,1110,8023
		G 75 - 9			2 23 10	-01 27.7	1	13.60		1.54		1.06		1		3062
		G 73 - 56			2 23 11	+05 40.5	1	12.49		0.78		0.21		4		1696
		LS I +57 058			2 23 13	+57 36.6	1	10.49		0.43		-0.41		2	B2 IV	1012
		IC 1795 - 104			2 23 13	+62 01.2	1	11.72		0.71		0.25		3		341
14863	+71 00140				2 23 15	+71 54.6	1	7.77		0.07		-0.40		3	B8	555
		IC 1795 - 117			2 23 16	+62 02.1	1	13.37		1.40		1.22		1		341
		IC 1795 - 122			2 23 17	+61 56.9	1	13.37		1.22		0.35		2		341
15120	-24 01074				2 23 17	-24 11.8	1	8.23		1.60		1.79		4	M2/3 III	3007
	+60 00477				2 23 18	+61 06.0	1	9.90		0.35		0.25		4	A3 V	1689
		CS 22175 # 20			2 23 18	-09 06.0	1	14.08		0.37		-0.18		1		1736
15096	+5 00336	G 73 - 57			2 23 23	+05 33.2	2	7.95	.000	0.80	.014	0.32	.005	4	G5	196,333,1620
		IC 1795 - 124			2 23 23	+61 52.9	1	13.64		0.65		0.23		6		341
	+61 00412	IC 1795 - 120			2 23 24	+61 58.6	3	8.98	.011	0.39	.013	-0.16	.008	16	B6 V	341,504,1689
	-42 00816				2 23 25	-42 14.9	1	12.19		0.50		-0.26		2		3060
		IC 1795 - 130			2 23 26	+61 38.2	1	14.48		1.42		0.30		1		504
15084	+18 00305				2 23 27	+18 40.9	1	8.18		0.53		0.01		2	F7 V	1648
15105	-0 00361	R Cet			2 23 29	-00 24.2	1	8.02		1.31		1.18		1	M4	975
15233	-60 00199	HR 714			2 23 30	-60 32.1	3	5.35	.005	0.39	.010	0.06		9	F2 III	15,2012,3053
15130	-12 00451	HR 708			2 23 32	-12 30.9	7	4.88	.005	-0.03	.008	-0.04	.026	26	B9.5 Vn	3,15,1075,1425,2012*
	+58 00470	SZ Cas			2 23 33	+59 14.2	4	9.65	.074	1.35	.043	0.93	.039	4	F8 Ib	934,1399,1772,6011
15115	+5 00338				2 23 38	+06 04.1	1	6.76		0.39		-0.03		3	F4 IV	985
15144	-15 00426	HR 710, AB Cet	★	AB	2 23 38	-15 33.9	8	5.83	.016	0.15	.010	0.10	.009	31	A6 Vp Sr	15,1007,1013,1063*
15144	-15 00426	IDS02212S1547		C	2 23 38	-15 33.9	1	9.30		0.62		0.13		4		3024
		IC 1795 - 125			2 23 39	+61 50.0	1	12.39		0.82		-0.26		6		341
	+60 00478				2 23 40	+60 29.2	4	11.66	.361	3.19	.068	3.44	.690	7	M1 Ib	148,1083,8016,8023
		IC 1795 - 129			2 23 40	+61 42.1	2	12.34	.124	1.66	.040	1.22	.268	5		341,504
		LDS 981			2 23 42	-49 17.	1	12.00		0.52		-0.06		2		3032
15000	+58 00471				2 23 45	+58 53.0	1	7.42		0.40		0.03		1	F5 II	3016
15093	+34 00437				2 23 46	+35 23.4	1	7.04		0.22		0.06		2	F0	1625
	+61 00413	IC 1795 - 269			2 23 47	+62 06.6	1	9.47		0.60		0.33		1	F4 V	341
15023	+58 00472				2 23 50	+58 44.8	2	8.30	.013	0.29	.005	0.12		45	F2 III	3016,6011
		IC 1795 - 148			2 23 54	+61 48.0	1	12.46		1.59		1.12		4		341
15071	+52 00581				2 23 56	+53 16.6	1	8.63		0.09		-0.07		2	B8	1566
		VY Per			2 23 56	+58 41.5	2	11.16	.515	1.16	.290	0.23		2	F5	1399,6011
		IC 1795 - 149			2 23 59	+61 46.7	2	12.28	.030	1.04	.010	0.55	.119	5		341,504
		IC 1795 - 133			2 23 59	+62 04.9	1	13.29		0.72		0.38		2		504
		IC 1795 - 150			2 24 03	+61 43.8	1	13.50		1.83		1.12		1		504
15141	+25 00398				2 24 05	+25 49.0	1	6.98		0.58		0.09		2	F5	105
		IC 1795 - 145			2 24 05	+61 52.7	1	13.60		0.65		0.55		2		504
		IC 1795 - 136			2 24 05	+62 03.6	1	12.15		0.67		0.20		3		504
		IC 1795 - 168			2 24 07	+61 44.3	1	15.35		1.22		0.58		1		504
15142	+22 00347				2 24 08	+22 39.3	1	8.02		1.06		0.77		2	G5	1648
		IC 1795 - 166			2 24 08	+61 45.7	1	14.15		1.00		0.60		3		504
15152	+26 00409	HR 711			2 24 13	+26 47.4	2	6.12	.003	1.44	.017	1.78	.050	3	K5 III	3077,4001
		IC 1795 - 153			2 24 13	+61 58.6	1	12.93		0.80		0.32		2		341
	+61 00414	IC 1795 - 161			2 24 14	+61 53.4	3	9.50	.025	0.45	.026	-0.04	.026	7	F4 V	341,504,1689
	+61 00415				2 24 15	+61 34.7	1	9.51		0.36		0.21		3	A0 V	1689
		IC 1795 - 151			2 24 15	+62 03.4	1	14.61		1.07		0.50		2		504
15220	-20 00455	HR 713			2 24 16	-20 16.1	4	5.88	.006	1.25	.008	1.33		13	K2 III	15,1075,2007,3005
		VES 746			2 24 19	+60 20.6	1	12.51		1.17		0.07		1		1543
		VES 747			2 24 20	+60 30.1	1	12.41		1.40		0.39		1		1543
	+60 00481				2 24 20	+61 15.1	1	9.82		0.37		0.26		3	A0 V	1689
	+59 00490			V	2 24 22	+59 30.8	1	9.00		2.07				3	M2	369
15379	-67 00154	HR 722, TZ Hor			2 24 24	-66 43.1	3	6.38	.014	1.51	.016	1.45	.000	19	M5 III	15,1075,2038
	+30 00397			A	2 24 27	+30 45.4	2	10.11	.010	1.21	.015	1.04	.015	2	K0	801,3072
	+30 00397			B	2 24 27	+30 45.4	1	12.55		1.50		1.06		1	F8	801

Table 1

HD	DM	Other Id	N Rem	α_{1950}	δ_{1950}	S	V	σ_V	B–V	σ_{B-V}	U–B	σ_{U-B}	n	Spectrum	References
		IC 1795 - 163		2 24 27	+61 49.7	1	12.80		0.88		0.42		2		504
15176	+31 00427	HR 712		2 24 30	+31 34.7	1	5.54		1.12				2	gK1	71
15138	+49 00666	HR 709		2 24 30	+50 20.9	1	6.14		0.40				1	F4 V	254
15069	+61 00416	IC 1795 - 177		2 24 30	+61 59.7	2	7.78	.025	0.65	.000	0.21	.030	6	G1 V	504,1689
15137	+51 00579	LS I +52 003		2 24 34	+52 19.6	2	7.87	.010	0.03	.020	-0.89	.005	5	O9.5V	399,401
		IC 1795 - 179		2 24 34	+61 56.0	1	14.20		0.64		0.50		3		504
15124	+56 00630			2 24 35	+57 03.1	1	8.05		0.26		-0.34		2	B5 III	401
		Steph 271		2 24 43	-24 16.6	1	11.71		1.30		1.21		1	K7	1746
15287	-29 00884			2 24 43	-29 38.8	1	9.92		0.23		0.00		4	A2 III	152
15228	+9 00323			2 24 44	+09 58.6	2	6.50	.005	0.44	.014	-0.07	.009	4	F5 V	1003,3016
		LS I +63 198		2 24 44	+63 54.3	1	12.78		-0.08		-0.99		3		41
		G 159 - 65		2 24 45	-03 18.8	1	11.79		0.44		-0.21		1		3056
15227	+16 00293			2 24 46	+16 25.2	1	7.25		0.36		-0.03		2	F0	1648
15339	-46 00722			2 24 53	-46 13.4	2	7.13	.009	1.10	.000			8	K1 II/III	1075,2033
	+7 00385	G 73 - 58		2 24 54	+07 29.5	2	9.91	.015	0.95	.018	0.72	.023	2	K8	333,1620,3072
15089	+66 00213	ι Cas	★ AB	2 24 55	+67 10.8	6	4.50	.026	0.12	.013	0.05	.011	15	A5p Sr	15,1118,1263,1363*
15089	+66 00213	HR 707	★ C	2 24 55	+67 10.8	1	8.40		0.72		0.18		2		3016
		LP 885 - 60		2 25 06	-29 38.1	1	12.13		0.54		-0.08		2		1696
15395	-55 00432			2 25 07	-54 45.9	1	9.44		0.56				3	G0wF4 V	1594
	+60 00482	IC 1805 - 2		2 25 08	+61 16.6	1	9.52		1.52		1.74		1		1718
15285	+3 00339	G 73 - 60	★ AB	2 25 09	+04 12.3	4	8.71	.019	1.41	.022	1.11		9	K7 V +K7 V	1197,1381,1632,3072
15371	-48 00637	HR 721		2 25 09	-47 55.7	5	4.24	.007	-0.14	.006	-0.50	.009	21	B5 IV	15,1034,1075,2012,8015
15299	-7 00432			2 25 10	-07 09.1	1	7.04		0.98		0.69		3	G5	1657
15257	+29 00417	HR 717		2 25 14	+29 26.8	2	5.30	.005	0.29	.005	0.10	.000	4	F0 IV	1733,3026
	-5 00463	G 75 - 15		2 25 14	-05 15.0	1	9.29		0.74		0.29		1	G0	1620
		IC 1805 - 5		2 25 18	+61 16.9	2	13.07	.068	1.59	.024	1.08	.433	6		47,983
	+61 00417	IDS02215N6140	AB	2 25 18	+61 53.8	1	10.34		0.39		-0.26		1		8084
	+25 00404	G 94 - 70		2 25 19	+25 37.5	1	9.37		0.53		-0.10		2	G3	333,1620
		IC 1805 - 7		2 25 22	+61 25.1	1	14.44		0.84		0.16		2		983
15328	+1 00431	HR 719	★ AB	2 25 25	+01 44.3	2	6.44	.005	0.97	.000	0.80	.005	7	K0 III	15,1071
		IC 1805 - 8		2 25 26	+60 57.7	1	13.39		0.78		0.17		2		983
15318	+7 00388	HR 718		2 25 30	+08 14.2	20	4.28	.010	-0.05	.006	-0.11	.022	139	B9 III	1,3,15,244,1006,1020*
		LP 353 - 76		2 25 30	+22 39.0	1	11.40		1.05		1.04		2		1723
		IC 1805 - 15		2 25 34	+61 24.0	1	14.14		0.87		0.03		2		983
		IC 1805 - 18		2 25 41	+61 18.9	2	12.62	.005	0.84	.000	-0.13	.010	5		47,983
		IC 1805 - 21		2 25 43	+61 16.3	3	11.30	.007	0.98	.012	-0.11	.022	7	B0 III	47,983,1718
15402	-37 00934	IDS02237S3708	AB	2 25 44	-36 55.0	1	8.61		0.47				4	F5 V	2012
		St 7 - 3		2 25 46	+60 25.8	2	10.98	.045	0.49	.020	0.16	.030	2	B9.5V	841,8084
	+31 00434	G 74 - 27	★ AB	2 25 47	+32 02.1	1	9.58		1.36		1.23		1	M0p	3072
		IC 1805 - 22		2 25 47	+61 13.1	1	11.35		0.64		0.09		1		1718
	+60 00486	St 7 - 4	★ A	2 25 48	+60 26.0	2	9.87	.030	0.39	.000	0.04	.105	2		841,8084
15351	+13 00395			2 25 50	+13 39.3	1	8.61		0.31		-0.27		2	F5	1648
	+71 00145	G 245 - 55		2 25 51	+71 57.8	1	9.20		0.60		0.09		1	G0	1620
15335	+29 00423	HR 720		2 25 52	+29 42.5	2	5.88	.014	0.58	.000	0.01		3	G0 V	71,3077
		IC 1805 - 23		2 25 52	+60 59.2	2	11.46	.010	0.84	.000	-0.12	.010	3	B1 V	47,983
15253	+54 00557	HR 716	★ AB	2 25 53	+55 18.8	2	6.52	.009	0.09	.000	-0.01	.002	6	A2p Shell	401,595
15239	+60 00487	St 7 - 28	★ A V	2 25 53	+60 26.1	4	8.48	.044	0.32	.020	-0.39	.008	7	B2.5V+Shell	379,399,841,8084
	+60 00485	IC 1805 - 24		2 25 53	+60 54.4	1	9.38		0.40		0.05		3	A9 V	1689
		St 7 - 1	★ B	2 25 54	+60 26.	1	11.18		0.43		0.16		1		841
	+17 00372	G 94 - 71		2 25 55	+17 35.1	1	10.20		0.83		0.43		1	K0	333,1620
		St 7 - 27	★ B	2 25 55	+60 26.0	2	11.35	.160	0.34	.120	0.12	.055	2		841,8084
	-8 00456			2 25 55	-07 35.3	2	9.50	.071	1.56	.000	1.94		2	C3,0	864,1238
15427	-34 00905	HR 724		2 25 55	-34 02.1	2	5.13	.005	0.10	.000			7	A2 V	15,2027
15238	+60 00488	St 7 - 23	★ V	2 25 57	+60 26.8	3	8.54	.165	0.33	.049	-0.31	.138	6	B2.5V+Shell	379,399,841
		LS I +58 083		2 26 01	+58 00.2	1	9.86		0.48		-0.44		2	B1 III	1012
		IC 1805 - 26		2 26 02	+61 22.3	3	12.04	.017	0.67	.012	0.57	.020	8		47,983,1718
15426	-26 00882			2 26 02	-26 39.3	1	7.62		1.04		0.83		3	K0 III	1657
15251	+60 00492	St 7 - 18		2 26 03	+60 28.7	1	9.39		0.40		-0.02		1	B8 Vp	841
		IC 1805 - 29		2 26 03	+61 35.0	1	13.29		0.63		0.24		2		983
	+60 00491	St 7 - 29		2 26 04	+60 32.0	1	9.70		1.29		1.07		1	G8 IV	841
	+59 00497	LS I +59 135		2 26 05	+59 38.0	1	10.47		0.74		-0.32		2	B0 V:	1012
15250	+60 00490			2 26 05	+60 44.3	1	8.52		0.18		0.13		3	A3 V	1689
		RZ Cet		2 26 05	-08 35.0	2	11.23	.012	0.14	.019	0.10	.026	2	A3	597,699
		St 7 - 24		2 26 06	+60 26.5	1	11.64		0.46		0.22		1		841
		IC 1805 - 28		2 26 06	+60 55.7	1	13.49		0.76		0.11		2		983
	-9 00466			2 26 07	-09 13.0	1	11.19		0.44		-0.22		2		1696
15410	-10 00497			2 26 09	-09 56.9	1	8.58		0.13		0.10		4	A0	152
15386	+14 00408			2 26 10	+15 22.7	1	8.79		0.57		0.05		2	G5	1648
		Steph 276		2 26 12	+11 52.0	1	12.01		1.57		1.17		1	M1	1746
	+22 00353	G 94 - 72	★ AB	2 26 16	+22 38.9	1	9.23		0.88		0.57		1	K0	333,1620
		IC 1805 - 37		2 26 16	+61 00.6	1	13.86		0.85		0.10		2		983
15481	-42 00829			2 26 17	-42 38.8	2	8.21	.005	0.45	.009			8	F5/6 V	1075,2033
		IC 1805 - 39		2 26 20	+61 17.8	2	12.92	.107	0.87	.019	0.64	.019	3		983,1718
		LP 941 - 91		2 26 20	-32 55.8	1	13.96		-0.12		-0.94		2		3060
15316	+57 00576	LS I +57 061		2 26 21	+57 35.9	4	7.23	.012	0.77	.000	0.42	.010	7	A3 Iab	1012,1083,1110,8023
15385	+22 00354	HR 723		2 26 22	+23 14.8	3	6.19	.000	0.15	.004	0.14	.004	10	A5 m	1501,1648,8071
15303	+58 00476			2 26 22	+58 51.7	1	9.43		0.47				48	G0	6011
		CS 22175 # 22		2 26 25	-09 26.9	1	12.53		-0.02		-0.07		1		1736
15325	+56 00635	LS I +57 062	★ AB	2 26 26	+57 01.6	2	8.51	.000	0.42	.000	-0.47	.000	3	B1 IV	64,1012
15471	-31 00990	HR 727		2 26 26	-31 19.5	2	6.10	.006	1.10	.006	1.07		6	K2 III	58,2007

HD	DM	Other Id	N	Rem	α_{1950}	δ_{1950}	S	V	σ_V	B–V	σ_{B-V}	U–B	σ_{U-B}	n	Spectrum	References
	+81 00080				2 26 27	+82 07.4	1	9.26		0.59		0.02		2	G5	1375
15398	+20 00404				2 26 28	+21 22.3	1	8.02		0.52		0.06		2	F5	1648
		IC 1805 - 42			2 26 28	+61 10.5	1	13.45		0.90		-0.05		2		983
		IC 1805 - 43			2 26 28	+61 32.9	2	12.05	.175	1.23	.375	0.67	.380	4		47,983
		AS 61			2 26 30	+57 05.	1	12.00		0.58		-0.39		1		1543
15365	+45 00614				2 26 31	+45 48.6	2	6.70	.010	1.12	.005	1.02	.005	6	K1 III	985,1501
15346	+53 00532				2 26 32	+53 37.8	1	8.01		1.08		0.74		2	G5	1566
	+57 00579	LS V +57 039		⋆ A	2 26 37	+57 27.2	2	10.09	.000	0.49	.000	-0.40	.000	3	B1 V	64,1012
	+62 00409				2 26 37	+62 58.0	1	10.27		0.34		-0.26		2		1375
		IC 1805 - 49			2 26 39	+61 14.4	3	12.79	.023	1.10	.033	0.03	.020	6		47,983,1718
		IC 1805 - 51			2 26 41	+60 57.5	1	11.32		0.38		-0.35		2		983
		IC 1805 - 52			2 26 41	+61 19.6	2	13.26	.005	1.08	.000	0.02	.020	5		47,983
15468	-20 00465	IDS02243S2026	A		2 26 41	-20 12.3	2	8.86	.020	1.12	.005	1.04	.040	5	K4 V	1518,3072
15468	-20 00465	IDS02243S2026	AB		2 26 41	-20 12.3	2	8.78	.008	1.17	.014	1.05	.024	6	K4 V	22,1518
15468	-20 00465	IDS02243S2026	C		2 26 41	-20 12.3	1	12.85		1.61		1.23		3		1518
		IC 1805 - 50			2 26 42	+60 52.6	1	12.75		0.59		0.12		2		983
		IC 1805 - 53			2 26 42	+61 06.5	2	12.80	.024	0.79	.015	-0.09	.058	3		983,1718
		IC 1805 - 55			2 26 42	+61 27.0	1	13.82		0.87		0.49		2		983
		IC 1805 - 57			2 26 44	+61 16.9	1	13.90		0.86		0.31		2		983
		IC 1805 - 56			2 26 45	+60 52.4	2	13.33	.131	0.60	.122	0.24	.107	3		47,983
		CS 22175 # 17			2 26 47	-08 32.9	1	11.94		0.25		0.08		1		1736
15420	+38 00491				2 26 49	+38 54.9	1	6.76		1.10		1.04		2	K2 III	105
		IC 1805 - 60			2 26 50	+61 23.2	1	13.93		0.78		0.73		2		983
		IC 1805 - 62			2 26 52	+61 07.9	2	12.55	.010	0.85	.000	-0.14	.040	4		47,983
		IC 1805 - 63			2 26 52	+61 18.6	1	13.33		0.78		0.40		2		983
		CS 22175 # 24			2 26 52	-10 37.8	1	14.45		0.14		0.18		1		1736
15646	-64 00174	HR 734			2 26 54	-64 31.4	3	6.35	.007	-0.03	.009	-0.04	.005	18	A0 V	15,1075,2038
15453	+8 00385	HR 725		⋆ AB	2 26 55	+09 20.6	2	6.06	.005	1.02	.000	0.86	.005	7	K2 III	15,1256
		LP 993 - 47			2 26 56	-40 58.1	1	12.68		1.23		1.21		1		1696
15591	-51 00593				2 26 57	-51 31.7	1	9.03		0.45		-0.09		1	F5 V	1097
		IC 1805 - 68			2 26 58	+61 03.0	1	13.16		0.63		-0.11		2		983
	-26 00892	R For			2 27 02	-26 19.2	1	8.47		5.44				2	C6,3	864
	+60 00493	IC 1805 - 70			2 27 04	+60 57.4	6	8.39	.045	0.79	.012	-0.30	.025	10	B0.5Ia	47,64,983,1012,1083,1718
		IC 1805 - 69			2 27 04	+61 04.0	3	12.32	.017	0.67	.017	-0.25	.026	5		47,983,1718
		CS 22175 # 19			2 27 04	-08 46.0	1	12.01		0.22		0.16		1		1736
		IC 1805 - 71			2 27 05	+61 32.1	1	13.83		0.69		-0.03		2		983
		IC 1805 - 72			2 27 10	+61 03.3	3	12.62	.146	0.70	.063	-0.20	.184	5		47,983,1718
	+5 00345				2 27 14	+05 41.6	1	9.64		1.53				1	K7	1632
		IC 1805 - 74			2 27 14	+61 27.7	3	11.38	.027	1.96	.047	1.79	.099	6	K0 II	47,983,1718
15590	-42 00834				2 27 14	-42 17.8	2	7.98	.000	0.65	.000			8	G3 V	1075,2012
15464	+33 00445	HR 726			2 27 16	+33 36.8	1	6.25		1.07				2	K1 III	71
15382	+60 00495				2 27 17	+60 29.4	2	8.08	.045	0.41	.015	-0.04		5	F5 IV	379,1689
15555	-24 01106				2 27 17	-24 19.6	2	7.33	.009	1.05	.005	1.01		5	K1 III	2033,3008
		IC 1805 - 79			2 27 22	+61 30.6	1	14.32		0.75				2		983
15589	-36 00938				2 27 22	-36 22.3	3	8.93	.019	1.15	.005	0.88	.024	7	G8 II	565,1731,3048
232665	+53 00535				2 27 24	+53 43.7	1	9.79		1.74				4	M7	369
	+28 00426				2 27 28	+29 22.3	1	10.59		0.03		-0.18		7		272
		IC 1805 - 82			2 27 32	+61 16.9	2	12.48	.015	0.82	.005	-0.12	.015	5		47,983
		HIC 11716			2 27 34	+58 18.6	1	9.88		1.29		1.30		3		1723
		IC 1805 - 85			2 27 36	+61 32.6	1	13.83		0.69		-0.34		2		983
15588	-23 00942	HR 730		⋆ A	2 27 38	-22 54.3	3	6.76	.007	0.16	.034	0.15		12	A5/7 III	15,2012,8071
		Feige 22			2 27 39	+05 02.6	5	12.79	.012	-0.05	.004	-0.81	.005	103	DA	281,989,1298,1705,1764,3016
15524	+24 00358	HR 728		⋆ A	2 27 39	+25 00.9	1	5.94		0.40		0.04		2	F6 IV	254
		IC 1805 - 86			2 27 42	+61 21.9	1	12.79		0.54		0.37		2		983
		LDS 982			2 27 42	-49 17.	1	13.74		0.74		0.16		1		3032
15450	+56 00642	LS I +56 055			2 27 44	+56 40.6	1	8.84		0.34		-0.64		2	B1 IIIe	1012
15449	+57 00580				2 27 49	+57 48.2	2	6.58	.005	1.42	.000	1.60	.005	5	K2 III	1501,1502
15550	+19 00365	HR 729, UU Ari			2 27 50	+19 38.1	3	6.17	.028	0.24	.009	0.08	.013	8	A9 V	253,254,3058
		IC 1805 - 91			2 27 50	+61 11.3	1	13.02		0.55		0.16		2		983
15461	+55 00639				2 27 52	+55 45.9	1	9.35		0.29		0.14		2	A3	1502
	+62 00411	LS I +63 190			2 27 53	+63 11.8	1	8.46		0.32		-0.60		2	B1 Ib-II	1012
15598	-5 00471				2 27 56	-05 15.1	1	7.65		0.06		0.06		4	A0	244
15634	-25 00979	HR 733, TY For			2 27 59	-25 24.5	3	6.50	.007	0.29	.005	0.04		9	dA9 n	15,2012,3026
		IC 1805 - 94			2 28 00	+61 21.6	1	13.91		0.78		-0.10		2		983
		IC 1805 - 96			2 28 01	+61 19.3	1	13.50		0.56		-0.10		2		983
		IC 1805 - 99			2 28 04	+61 27.8	1	13.44		0.75		0.60		2		983
		LP 590 - 201			2 28 05	+00 23.5	1	11.57		0.69		0.24		2		1696
15596	+17 00380	HR 731			2 28 08	+17 29.0	6	6.22	.010	0.90	.005	0.54	.009	20	G5 III-IV	15,1003,1105,1355*
	+60 00496	IC 1805 - 103			2 28 08	+61 18.9	4	10.56	.017	0.50	.010	-0.41	.011	8	B0 V	47,49,983,1718
	+60 00497	IC 1805 - 104			2 28 08	+61 23.5	6	8.78	.015	0.57	.000	-0.50	.010	13	O7	47,983,1011,1012,1209,1718
16701	-86 00028				2 28 09	-85 56.5	1	7.86		0.37		-0.08		2	F0 IV/V	1628
		IC 1805 - 105			2 28 10	+61 08.3	2	11.48	.015	0.73	.000	0.50	.020	4	A1 II-III	47,983
15633	-0 00378	HR 732			2 28 12	+00 02.1	4	6.01	.009	0.17	.003	0.15	.009	14	A7 III-IVs	15,231,1061,1328
		IC 1805 - 108			2 28 14	+61 13.6	2	12.99	.161	0.78	.010	0.15	.034	3		49,1718
		IC 1805 - 107			2 28 14	+61 16.3	2	12.02	.000	1.37	.000	0.94	.000	3		49,1718
15497	+57 00582	V425 Per		⋆ ABC	2 28 15	+57 28.6	6	7.02	.025	0.78	.011	-0.07	.008	12	B6 Ia	1012,1077,1083,1110*
	+65 00267				2 28 16	+66 16.2	1	9.66		1.95				3		369
15652	-23 00947	HR 735			2 28 16	-22 46.0	3	6.13	.030	1.58	.008	2.00	.015	10	M0 III	1024,2007,3035
		Steph 279			2 28 17	+28 00.8	1	10.74		1.28		1.28		2	K7	1746
		IC 1805 - 109			2 28 18	+61 16.6	3	13.98	.018	0.80	.025	0.06	.109	5		49,983,1718

Table 1

HD	DM	Other Id	N	Rem	α_{1950}	δ_{1950}	S	V	σ_V	B–V	σ_{B-V}	U–B	σ_{U-B}	n	Spectrum	References
	−9 00471				2 28 20	−08 58.7	1	10.46		0.78		0.42		2		1696
		IC 1805 - 110			2 28 21	+61 13.3	1	13.26		0.78		0.36		2		983
		IC 1805 - 111			2 28 21	+61 25.0	3	11.53	.018	0.49	.004	-0.43	.011	6	B2 V	47,983,1718
		G 4 - 24			2 28 22	+08 09.6	1	10.90		1.34		1.25		1	K4	333,1620
	+60 00498	IC 1805 - 112			2 28 22	+61 19.8	8	9.92	.014	0.53	.011	-0.47	.012	17	O9.5V	47,49,64,983,1011,1012*
		MCC 394			2 28 23	+08 09.7	1	10.88		1.31				2	K4	1619
		IC 1805 - 113			2 28 23	+61 08.5	2	10.92	.000	0.88	.000	-0.18	.020	4	O9 Ve	47,983
		LS I +59 136			2 28 25	+59 48.3	1	10.52		1.08		0.06		2	B0.5 V	1012
	+60 00499	IC 1805 - 118			2 28 28	+61 19.9	4	10.28	.039	0.53	.017	-0.50	.008	7	O9 V	47,49,983,1718
		IC 1805 - 119			2 28 28	+61 28.3	1	11.51		1.76		1.47		1		1718
15548	+56 00647	LS I +56 056		★ A	2 28 30	+56 26.7	2	9.26	.005	0.28	.000	-0.58	.000	3	B1 V	64,1012
		IC 1805 - 121			2 28 30	+61 14.6	8	11.59	.009	0.63	.007	-0.30	.008	8	F3	47,49,983,1718
		IC 1805 - 122			2 28 31	+61 19.5	1	13.73		1.05		0.49		2		983
236966	+59 00504				2 28 33	+60 23.6	1	9.54		0.38		0.25		3	A2 V	1689
		IC 1805 - 123			2 28 33	+61 15.0	1	13.88		0.57		0.31		2		983
		IC 1805 - 124			2 28 33	+61 25.4	1	14.40		0.78				2		983
15522	+61 00422	IC 1805 - 125		★ AB	2 28 33	+61 35.2	1	8.30		0.18		0.11		3	A0	1689
	+60 00500	IC 1805 - 127			2 28 36	+61 08.1	1	9.74		1.29		1.05		3	G9	1689
		IC 1805 - 128			2 28 36	+61 19.8	1	13.40		0.70		-0.29		2		983
15579	+45 00620				2 28 37	+46 21.9	1	7.00		0.36		0.08		3	A8 V	1501
		IC 1805 - 129			2 28 39	+61 13.9	1	14.06		0.67		0.05		2		983
15036	+83 00056				2 28 39	+83 37.0	1	6.66		1.36		1.62		4	K0	985
		IC 1805 - 131			2 28 42	+60 53.2	2	12.53	.010	0.98	.010	0.45	.010	4		47,983
		IC 1805 - 130			2 28 42	+61 13.9	1	13.26		0.66		-0.14		2		983
		IC 1805 - 134			2 28 44	+61 25.3	1	14.07		0.63		0.06		2		983
15593	+49 00683				2 28 46	+49 59.8	1	7.72		0.06		-0.14		2	A0	401
		IC 1805 - 136			2 28 46	+61 19.0	3	11.02	.013	0.61	.004	-0.33	.016	6	B1 V	47,983,1718
15793	−58 00217	TV Hor			2 28 46	−58 01.9	2	6.76	.048	1.59	.031	1.58		13	M4/5 III	2012,3042
15571	+56 00648	LS I +57 066			2 28 47	+57 12.5	1	8.33		0.57		-0.43			B1 II	1012
	+60 00501	IC 1805 - 138			2 28 48	+61 15.2	8	9.58	.019	0.45	.007	-0.60	.015	16	O6.5V	47,49,64,983,1011,1012*
		G 173 - 57			2 28 49	+45 39.2	1	11.26		0.95		0.76		1	K5	1658
		IC 1805 - 139			2 28 49	+61 19.0	2	13.04	.078	0.54	.015	-0.07	.019	3		983,1718
15557	+61 00424	IC 1805 - 140			2 28 51	+61 30.8	1	7.39		0.36		-0.01		3	F0	1689
		IC 1805 - 144			2 28 52	+61 14.1	1	10.62		0.52		-0.45		3	O5	450
		IC 1805 - 143			2 28 52	+61 14.8	1	11.40		0.55		-0.51		2		983
		IC 1805 - 146			2 28 52	+61 17.6	1	13.53		0.56		-0.09		2		983
		IC 1805 - 142			2 28 52	+61 25.4	1	11.30		0.48		-0.36		2		983
		IC 1805 - 147			2 28 53	+61 14.4	1	13.34		0.64		0.04		2		983
15558	+60 00502	IC 1805 - 148		★ AB	2 28 54	+61 14.1	11	7.87	.030	0.51	.018	-0.56	.009	23		47,49,64,401,450,1011*
15558	+60 00502	IC 1805 - 148		★ D	2 28 54	+61 14.1	1	13.10		0.63		-0.23		1	O5 III(f)	8084
		IC 1805 - 149			2 28 54	+61 16.4	4	11.25	.036	0.47	.018	-0.39	.014	7	B3 V	47,49,983,1718
	−51 00599				2 28 54	−51 00.6	1	10.55		0.45		-0.23		2	F5	1696
15694	+1 00438	HR 737			2 28 55	+02 02.8	4	5.27	.020	1.25	.025	1.40	.014	12	K3 III	15,1071,1355,3016
		IC 1805 - 152			2 28 55	+61 13.4	1	12.96		0.73		-0.14		2		983
15472	+70 00182				2 28 55	+70 44.2	2	7.92	.044	0.06	.010	-0.60	.010	5	B3	379,1212
	+60 00503	LS I +60 239			2 28 56	+60 44.0	2	9.95	.000	0.66	.000	-0.30	.005	3	B1.5V	64,1012
		IC 1805 - 155			2 28 56	+61 11.4	1	14.39		1.32				2		983
		IC 1805 - 154			2 28 56	+61 12.1	1	14.07		0.77				2		983
	+56 00650	NGC 957 - 1			2 28 57	+57 16.0	1	9.49		1.16		0.94		1		49
		IC 1805 - 156			2 28 57	+61 14.6	3	12.00	.052	0.56	.032	-0.35	.015			49,983,1718
		IC 1805 - 157			2 28 58	+61 18.3	3	13.43	.034	0.61	.072	-0.21	.027	5		49,983,1718
		IC 1805 - 158			2 28 59	+61 13.9	1	12.73		0.98		-0.05		2		983
15570	+60 00504	IC 1805 - 160			2 29 01	+61 09.5	13	8.11	.010	0.68	.015	-0.41	.008	24	O4 I f+	47,49,64,401,983,1011*
		IC 1805 - 159			2 29 01	+61 13.6	1	13.74		0.71		-0.17		2		983
15801	−49 00699				2 29 01	−49 35.4	1	10.39		0.15		0.13		4	A0 III/IV	152
15656	+35 00497	HR 736			2 29 03	+35 55.6	2	5.16	.005	1.47	.000	1.77	.005	7	K5 III	1355,3002
15655	+38 00499				2 29 03	+38 53.9	1	8.89		0.13		0.11		3	A1 V	833
	−17 00484				2 29 03	−17 12.3	7	10.47	.011	0.44	.014	-0.19	.008	22	G2 VI	742,1003,1097,1594,1696*
15417	+76 00081				2 29 04	+76 30.0	1	6.77		1.77		1.97		4	M1	985
15900	−69 00127				2 29 04	−68 49.9	1	7.82		1.54		1.96		9	K5/M0 III	1704
		IC 1805 - 163			2 29 06	+61 09.8	2	12.23	.034	0.71	.049	-0.22	.044	3	A8	983,1718
		IC 1805 - 162			2 29 06	+61 16.5	3	12.47	.049	0.47	.010	-0.23	.039	5		49,983,1718
		IC 1805 - 161			2 29 06	+61 25.6	2	10.85	.005	0.46	.000	-0.45	.040	4	B1.5V	47,983
		LB 1628			2 29 06	−48 09.	1	14.53		-0.26		-1.04		1		132
		L 28 - 13			2 29 06	−75 45.	1	12.39		0.48		-0.26		2		1696
		IC 1805 - 164			2 29 07	+61 10.8	1	13.92		0.77		0.01		2		983
		LP 770 - 34			2 29 07	−15 49.6	1	13.26		0.99		0.46		2		3062
		IC 1805 - 166			2 29 09	+61 14.1	2	11.98	.010	0.53	.000	-0.23	.020	4	B3 V	47,983
		IC 1805 - 165			2 29 09	+61 16.4	3	12.75	.030	0.61	.036	-0.25	.035	5		49,983,1718
		HIC 11838			2 29 10	+57 59.3	1	10.36		1.36				2		1726
15619	+57 00583				2 29 10	+58 23.4	1	8.08		1.04				30	G9 III	6011
		IC 1805 - 169			2 29 11	+61 09.2	3	11.71	.012	0.64	.034	-0.19	.058	5	B2 IV	47,983,1718
		IC 1805 - 168			2 29 11	+61 12.6	1	13.72		0.78		0.26		2		983
		IC 1805 - 167			2 29 11	+61 23.3	1	12.25		0.48		-0.30		2	F6	983
	−20 00470				2 29 11	−20 15.0	1	10.26		1.27		1.20		1	K7 V	3072
	+60 00505	IC 1805 - 170			2 29 12	+61 13.2	4	10.07	.011	0.37	.018	0.30	.028	9	A2 II	47,49,983,1718
15671	+35 00498	IDS02262N3516		AB	2 29 13	+35 29.3	1	8.01		0.69		0.71		2	G5	1601
15670	+38 00500				2 29 14	+39 09.0	1	8.12		0.58		0.05		2	G2 V	105
15620	+57 00584	LS I +57 068			2 29 14	+57 42.6	3	8.35	.000	0.92	.000	0.17	.000	5	B8 Iab	64,1012,1083
15621	+56 00652	NGC 957 - 214			2 29 15	+57 18.5	1	8.53		0.46		0.02		1	F6 V	49

HD	DM	Other Id	N Rem	α_{1950}	δ_{1950}	S	V	σ_V	B–V	σ_{B-V}	U–B	σ_{U-B}	n	Spectrum	References
		IC 1805 - 172		2 29 15	+60 56.2	1	13.86		0.77		0.03		2		983
		IC 1805 - 171		2 29 15	+61 15.5	1	13.13		0.60		-0.06		2		983
		NGC 957 - 18		2 29 16	+57 19.6	1	12.28		0.58		-0.20		1		49
		IC 1805 - 175		2 29 16	+61 13.9	3	13.06	.047	0.60	.041	-0.13	.045	5		49,983,1718
		IC 1805 - 174		2 29 16	+61 15.1	4	11.57	.036	0.50	.016	-0.36	.006	7	B2 V	47,49,983,1718
		NGC 957 - 19		2 29 17	+57 20.3	1	13.20		0.65		0.04		1		49
	-47 00765			2 29 17	-46 54.2	1	11.24		1.03				1		1705
15767	-15 00447			2 29 20	-15 29.5	1	8.74		0.94		0.62		2	K2 V	3072
15705	+27 00394	IDS02264N2807	AB	2 29 21	+28 20.8	1	8.24		0.06		0.02		4	A0	1371
		IC 1805 - 180		2 29 21	+61 14.5	3	12.89	.026	0.64	.049	-0.25	.018	5		49,983,1718
		IC 1805 - 182		2 29 21	+61 18.7	1	13.38		0.62		-0.12		2		983
		NGC 957 - 23		2 29 22	+57 18.7	1	13.07		0.70		-0.01		1		49
	+60 00506	IC 1805 - 183		2 29 23	+61 13.9	4	11.15	.006	0.59	.027	-0.48	.011	7	B1 IV	47,49,983,1718
		IC 1805 - 185		2 29 23	+61 14.2	4	11.57	.048	0.53	.012	-0.37	.037	7	B2 V	47,49,983,1718
		IC 1805 - 184		2 29 23	+61 17.7	2	13.57	.000	0.77	.034	0.06	.024	3		49,1718
15642	+54 00569	LS I +55 037		2 29 24	+55 06.5	4	8.53	.011	0.07	.005	-0.84	.000	7	B0 III	1012,1083,1110,8023
		NGC 957 - 100		2 29 24	+57 20.6	1	14.84		1.36		1.04		1		49
		NGC 957 - 167		2 29 24	+57 22.9	1	12.45		0.59		-0.57		1		49
15533	+70 00183			2 29 24	+71 04.5	1	6.55		1.23		1.38		4	K0	985
		IC 1805 - 187		2 29 26	+61 10.9	2	12.37	.005	0.71	.010	0.44	.019	3		49,1718
	+8 00390			2 29 27	+08 49.1	1	9.79		0.95				1	K7	1632
15716	+29 00432			2 29 27	+29 39.3	1	8.56		1.13		0.94		2	K0	1371
		LP 830 - 4		2 29 27	-21 23.4	1	11.55		0.56		-0.07		2		1696
		NGC 957 - 103		2 29 29	+57 20.5	1	15.11		1.01		0.41		1		49
15640	+59 00505	IDS02258N5933	AB	2 29 29	+59 46.8	1	7.59		0.05		-0.06		2	A0 IV	401
		IC 1805 - 188		2 29 29	+61 18.9	2	12.66	.024	0.56	.010	-0.31	.024	3		983,1718
		IC 1805 - 189		2 29 30	+61 13.0	3	13.75	.034	0.71	.010	-0.10	.038	5		49,983,1718
15629	+60 00507	IC 1805 - 192		2 29 31	+61 18.1	8	8.42	.017	0.42	.008	-0.63	.013	16	O5 V((f))	47,49,64,983,1011,1012*
15733	+21 00349			2 29 32	+22 06.8	1	7.82		0.37		0.20		2	F0	1648
		IC 1805 - 191		2 29 32	+61 19.1	1	12.96		0.64		-0.26		2		983
		NGC 957 - 105		2 29 33	+57 20.6	1	11.98		0.54		-0.19		1		49
		NGC 957 - 168		2 29 33	+57 23.5	1	12.09		0.85		0.47		1		49
	+60 00508	IC 1805 - 194		2 29 36	+61 07.3	1	10.65		0.58		0.06		3	F6 V	1689
		IC 1805 - 195		2 29 37	+61 07.7	3	13.01	.016	1.46	.030	0.93	.055	5		47,983,1718
15779	-1 00353	HR 739		2 29 37	-01 15.3	3	5.36	.013	1.00	.012	0.84	.008	10	K0	15,1256,3016
		NGC 957 - 94		2 29 38	+57 14.7	1	11.13		0.66		-0.38		1	B3 V	49
		IC 1805 - 198		2 29 39	+60 59.3	2	13.16	.015	1.56	.000	0.80	.280	4		47,983
	+9 00332			2 29 40	+09 50.4	1	10.80		0.63		0.06		2		1375
		NGC 957 - 86		2 29 40	+57 18.4	1	11.92		0.54		-0.27		1		49
		G 246 - 6		2 29 40	+66 37.5	1	10.41		0.81		0.47		2		1658
		IC 1805 - 200		2 29 41	+61 15.6	1	13.90		0.68		-0.08		2		983
		IC 1805 - 199		2 29 41	+61 22.1	2	11.53	.005	1.63	.035	1.44	.000	5		47,983
		PHL 4144		2 29 42	+02 47.0	1	11.08		0.49		0.02		2	B6 V	1026
236970	+55 00643	LS I +56 057	⋆ AB	2 29 43	+56 05.9	1	8.96		0.94		0.20		2	A2 Iab	1110
		NGC 957 - 198		2 29 43	+57 21.0	1	13.67		1.03		0.87		1		49
		IC 1805 - 201		2 29 43	+61 09.2	2	11.94	.010	0.55	.034	-0.28	.010	3		49,1718
15798	-15 00449	HR 740		2 29 43	-15 27.8	8	4.75	.010	0.45	.006	-0.03	.011	25	F5 V	15,58,1008,1075,1425*
		NGC 957 - 195		2 29 46	+57 21.8	1	14.14		0.62		0.10		1		49
		IC 1805 - 207		2 29 46	+60 55.3	1	14.12		0.76		0.15		2		983
		IC 1805 - 204		2 29 46	+61 22.3	1	14.54		0.41		0.50		2		983
	+57 00586	LS I +58 086		2 29 47	+58 22.8	1	10.11		0.61		-0.33		2	B1 II	1012
		IC 1805 - 205		2 29 47	+61 22.4	1	13.88		0.65		-0.01		2		983
15848	-37 00953			2 29 49	-37 08.8	1	9.20		1.16		1.09		4	K0/1 III	1731
15755	+33 00454	HR 738		2 29 51	+34 19.4	1	5.83		1.08				2	K0 III	71
		IC 1805 - 208		2 29 51	+61 21.2	1	12.95		0.59		0.35		2		983
		NGC 957 - 194		2 29 52	+57 21.8	1	13.33		0.62		-0.05		1		49
		NGC 957 - 115		2 29 54	+57 19.8	1	13.79		0.60		-0.06		1		49
	+56 00655	NGC 957 - 192		2 29 54	+57 21.0	1	10.72		0.48		-0.37		1	B1 V	49
		IC 1805 - 211		2 29 54	+61 13.1	3	10.99	.064	0.57	.017	-0.34	.019	5	B1 V	47,983,1718
15690	+56 00656	NGC 957 - 213	⋆ A	2 29 55	+57 19.1	5	8.00	.012	0.66	.007	-0.26	.010	8	B1.5Ib	49,1012,1083,1110,8023
		NGC 957 - 156		2 29 55	+57 24.6	1	12.38		0.80		0.46		1		49
		IC 1805 - 212		2 29 55	+61 13.2	1	13.01		0.58		-0.21		2		983
	+56 00657	NGC 957 - 116	⋆ B	2 29 56	+57 19.4	1	9.86		0.58		-0.32		1	B1 V	49
		IC 1805 - 213		2 29 57	+61 17.3	1	13.64		0.56		-0.21		2		983
		CS 22189 # 3		2 29 57	-15 19.9	1	13.54		0.27		-0.88		1		1736
		NGC 957 - 189		2 29 59	+57 20.2	1	12.26		0.55		-0.24		1		49
		IC 1805 - 218		2 29 59	+61 19.0	1	13.61		0.54		-0.03		2		983
		IC 1805 - 215		2 29 59	+61 27.8	3	12.83	.020	2.10	.018	1.68	.061	5		47,983,1718
		IC 1805 - 214		2 29 59	+61 31.5	1	13.70		0.81		0.31		2		983
		NGC 957 - 117		2 30 00	+57 19.1	1	13.82		0.90		-0.04		1		49
	+60 00509	IC 1805 - 219		2 30 00	+61 21.4	1	9.07		0.61		0.12		3	F8 V	1689
		NGC 957 - 188		2 30 01	+57 20.7	1	11.99		0.54		-0.28		1		49
		IC 1805 - 220		2 30 01	+61 27.1	2	13.17	.010	0.70	.000	0.15	.020	4		47,983
		IC 1805 - 221		2 30 02	+61 00.6	3	11.58	.037	0.50	.061	0.22	.052	5	B8 II-III	47,983,1718
		NGC 957 - 118		2 30 03	+57 19.4	1	15.14		0.75		0.69		1		49
		IC 1805 - 225		2 30 04	+61 05.3	1	13.63		0.72		-0.05		2		983
		IC 1805 - 224		2 30 04	+61 07.7	1	13.56		0.74				2		983
		IC 1805 - 222		2 30 04	+61 14.7	1	14.09		0.70				2		983
		CS 22189 # 4		2 30 04	-14 51.9	1	13.43		0.39		-0.18		1		1736
		GD 29		2 30 05	+36 05.9	1	13.68		0.50		-0.10		1		3060

Table 1

143

HD	DM	Other Id	N Rem	α_{1950}	δ_{1950}	S	V	σ_V	B–V	σ_{B-V}	U–B	σ_{U-B}	n	Spectrum	References
		NGC 957 - 153		2 30 05	+57 22.9	1	11.50		0.52		-0.30		1		49
		CS 22189 # 5		2 30 06	−14 45.0	1	14.18		0.31		0.08		1		1736
15876	−37 00955			2 30 06	−37 22.8	1	9.40		0.53		0.07		4	G6/8 V	1731
		IC 1805 - 228		2 30 08	+61 05.6	1	14.02		0.83				2		983
		NGC 957 - 119		2 30 09	+57 18.1	1	14.32		0.67		0.19		1		49
15814	+14 00419	HR 741		2 30 10	+14 48.9	1	6.06		0.53		0.02		4	F8 V	254
	+56 00658	NGC 957 - 151		2 30 11	+57 22.0	1	10.52		0.64		0.14		1	G1 V	49
	+60 00512	IC 1805 - 230		2 30 11	+61 10.7	1	10.26		0.68		0.12		2	F8 V	1012
		IC 1805 - 229		2 30 11	+61 34.4	1	13.51		0.65		0.09		1		983
		CS 22184 # 3		2 30 11	−09 09.2	1	14.95		0.17		0.16		1		1736
15889	−36 00957	HR 742	3	2 30 11	−36 38.9	3	6.30	.005	1.02	.005	0.74		9	K0 III	15,2029,3051
		NGC 957 - 179		2 30 13	+57 20.2	1	15.23		0.62				1		49
	+60 00513	IC 1805 - 232	7	2 30 14	+61 10.1	7	9.40	.011	0.49	.005	-0.53	.017	12	O7.5:	47,64,983,1011,1012*
		LP 710 - 47		2 30 15	−14 25.1	1	15.76		0.75		-0.01		5		1663
15743	+58 00486			2 30 17	+58 57.8	1	8.34		0.38				27	F2	6011
15833	+7 00398			2 30 20	+07 40.1	1	8.89		1.14		0.76		2	G5	1375
		IC 1805 - 238		2 30 22	+61 25.0	1	14.13		0.47		0.54		2		983
		IC 1805 - 236		2 30 22	+61 29.6	1	13.67		0.54		-0.08		2		983
		NGC 957 - 145		2 30 24	+57 22.5	1	13.23		0.72		0.39		1		49
		IC 1805 - 241		2 30 24	+60 58.3	1	13.71		1.08		0.39		2		983
		IC 1805 - 240		2 30 25	+61 29.0	1	13.54		0.30		0.30		2		983
		LP 530 - 26		2 30 25	+07 36.6	1	12.02		1.50		1.33		1	M0	1696
	+59 00510	LS I +59 140		2 30 26	+59 48.2	1	10.28		0.56		-0.40		2	B1 III	1012
		LP 710 - 50		2 30 26	−09 39.0	1	14.07		1.44				1		1773
		IC 1805 - 248		2 30 27	+61 04.7	1	12.93		0.63		0.35		2		983
		IC 1805 - 244		2 30 27	+61 22.1	2	12.83	.015	0.68	.005	0.37	.088	3		983,1718
15871	−9 00478			2 30 28	−08 55.2	1	8.62		0.51		0.04		4	F5	1731
15752	+57 00589	V362 Per		2 30 31	+58 11.2	2	8.74	.000	0.49	.000	-0.52	.000	3	B0 III	64,1012
		IC 1805 - 252		2 30 31	+60 53.5	2	12.76	.000	0.88	.005	0.08	.020	4		47,983
		IC 1805 - 251		2 30 31	+61 08.1	1	14.04		0.67		0.40		2		983
		IC 1805 - 250		2 30 32	+61 35.4	1	12.66		0.56		-0.34		2		983,1718
	+60 00514			2 30 34	+60 30.7	1	10.34		0.59		0.17		3		1689
		IC 1805 - 254		2 30 34	+61 10.6	1	14.23		0.79		0.21		2		983
		IC 1805 - 255		2 30 34	+61 27.9	1	13.43		0.78		0.18		2		983
		IC 1805 - 253		2 30 34	+61 33.2	1	14.25		0.66				2		983
		LP 710 - 51		2 30 34	−09 38.5	1	12.72		1.00		0.81		1		1773
		IC 1805 - 256		2 30 35	+61 19.7	1	13.90		0.65		0.12		2		983
15910	−20 00477			2 30 36	−20 28.9	4	10.11	.005	-0.05	.005	-0.18	.016	20	B9 IV/V	152,272,1026,1775
		G 75 - 21		2 30 38	+01 28.3	1	11.11		1.14		1.13		1	K7	1620
		G 4 - 26		2 30 39	+07 58.1	1	11.35		0.96		0.67		1	K3	333,1620
	+61 00430			2 30 40	+61 39.6	1	9.91		0.52		0.05		3		1689
		IC 1805 - 260		2 30 41	+61 17.4	3	11.52	.005	0.46	.010	-0.34	.012	6	B2 V	47,983,1718
		IC 1805 - 258		2 30 42	+61 33.9	1	12.98		0.76		-0.17		2		983
		IC 1805 - 259		2 30 42	+61 34.0	1	11.01		0.60		-0.33		2		983
	+70 00188			2 30 42	+71 20.4	1	10.08		0.17		0.18		3		922
		G 4 - 27		2 30 44	+12 26.6	1	14.83		1.63		1.18		3		203
		IC 1805 - 267		2 30 45	+61 01.8	1	13.50		0.66		0.38		2		983
		IC 1805 - 262		2 30 45	+61 34.0	1	12.20		0.58		-0.28		2		983,1718
		IC 1805 - 265		2 30 46	+61 16.3	1	13.91		0.55		0.19		2		983
		IC 1805 - 264		2 30 46	+61 32.1	1	12.80		0.56		-0.09		2		983
		IC 1805 - 266		2 30 47	+61 33.8	1	13.47		0.61		-0.08		2		983
15869	+18 00325			2 30 48	+18 39.6	1	6.82		0.27		0.08		2	A5	1648
		IC 1805 - 269		2 30 48	+61 10.4	1	13.91		0.70				2		983
		IC 1805 - 270		2 30 48	+61 19.5	1	13.70		0.76		0.18		2		983
		UY Per		2 30 50	+58 36.7	2	10.87	.004	1.30	.031	1.10		2	F7	1399,6011
		PG0231+051E		2 30 51	+05 06.6	1	13.80		0.68		0.20		2		1764
		IC 1805 - 272		2 30 52	+61 34.1	1	14.17		1.00				2		983
		LP 770 - 42		2 30 54	−19 54.2	1	12.37		0.52		-0.17		2		1696
		PG0231+051D		2 30 56	+05 06.3	1	14.03		1.09		1.05		2		1764
15866	+30 00410			2 30 56	+31 21.7	1	8.00		0.66		0.26		8	G0 III	1371
15830	+42 00550	G 74 - 31		2 30 57	+42 34.1	1	7.61		0.66		0.18		1	G0	333,1620
		LP 710 - 88		2 30 57	−11 30.8	1	9.95		0.64		0.06		2		1696
		G 74 - 30		2 30 58	+40 04.8	1	11.63		0.60		-0.08		2		1620
	+58 00488	LS I +59 141		2 30 58	+59 17.3	2	9.85	.000	0.68	.000	-0.26	.000	3	B0.5V	64,1012
		IC 1805 - 276		2 30 58	+61 10.5	1	12.88		0.54		-0.28		2		983
		IC 1805 - 277		2 30 58	+61 13.2	2	12.68	.034	0.59	.015	-0.19	.010	3		983,1718
		IC 1805 - 279		2 30 59	+61 11.3	1	13.56		0.64		-0.06		2		983
		IC 1805 - 278		2 30 59	+61 22.6	3	11.56	.000	0.47	.008	0.32	.029	5	A0 II-III	47,983,1718
15976	−37 00960	IDS02290S3738	AB	2 31 00	−37 24.6	1	9.07		0.43		-0.02		4	F3/5 V	1731
		G 4 - 28		2 31 01	+14 47.1	1	13.76		1.63				1		906
15785	+59 00513	LS I +60 240		2 31 01	+60 20.0	2	8.34	.000	0.57	.000	-0.50	.000	3	B1 Iab	64,1012
		CS 22184 # 4		2 31 02	−08 46.9	1	14.22		0.41		-0.19		1		1736
15975	−35 00877	HR 744		2 31 02	−34 52.1	1	5.90		1.06				4	K0 III	2032
		PG0231+051A		2 31 03	+05 04.5	1	12.77		0.71		0.27		4		1764
		PG0231+051C		2 31 03	+05 07.2	1	13.70		0.67		0.11		1		1764
		PG0231+051		2 31 04	+05 05.5	1	16.10		-0.33		-1.19		4	sdO:	1764
		IC 1805 - 280		2 31 04	+61 12 8	1	13.76		0.63		0.47		2		983
		IC 1805 - 281		2 31 06	+61 17.8	1	14.21		0.60				2		983
		PG0231+051B		2 31 08	+05 04.4	1	14.74		1.45		1.34		4		1764
		IC 1805 - 283		2 31 09	+61 31.1	1	14.30		0.72				2		983

HD	DM	Other Id	N Rem	α_{1950}	δ_{1950}	S	V	σ_V	B–V	σ_{B-V}	U–B	σ_{U-B}	n	Spectrum	References
	+32 00467	G 74 - 32		2 31 12	+32 47.0	1	10.53		0.60		0.00		1	G0	333,1620
15894	+34 00462			2 31 15	+34 55.9	1	7.38		0.10		0.07		2	A2	1625
		IC 1805 - 288		2 31 15	+61 15.0	3	11.12	.007	0.47	.000	-0.42	.011	5	B1.5V	47,983,1718
	+62 00419	LS I +63 191		2 31 15	+63 21.9	2	9.72	.000	0.35	.004	-0.61	.008	8	B0:V:	1012,1775
		IC 1805 - 290		2 31 19	+61 22.4	3	11.74	.016	0.60	.023	0.20	.011	5	F1 V	47,983,1718
15971	-13 00479	U Cet		2 31 20	-13 22.0	2	8.12	.950	1.38	.030	0.70	.210	2	M4/5e	635,975
15996	-20 00480	HR 745		2 31 22	-20 13.2	3	6.20	.007	1.10	.000			11	K1 III	15,2013,2030
		G 4 - 29		2 31 23	+17 33.4	2	14.90	.005	1.69	.009	0.97	.018	5		316,3078
		LB 3264		2 31 24	-64 14.	1	11.92		0.45		-0.12		3		45
15784	+67 00215	LS I +68 014		2 31 26	+68 09.0	2	6.63	.010	0.43	.040	0.25		2	F4 II	592,6009
	+2 00397			2 31 30	+02 34.8	1	9.60		1.40				1	K7	1632
15984	-8 00480			2 31 30	-08 10.8	1	9.35		0.46		-0.02		4	G0	1731
16032	-33 00866			2 31 30	-33 02.7	1	7.79		1.03		0.81		5	K0 III	1673
		IC 1805 - 299		2 31 33	+60 54.4	1	13.84		1.26		0.71		2		983
15931	+28 00435			2 31 34	+28 25.3	1	8.52		1.53		1.73		7	K0	1371
		GD 31		2 31 37	-05 24.9	1	14.24		0.21		-0.68		1	DA	3060
		LP 770 - 72		2 31 37	-20 37.9	1	12.96		0.73		0.14		2		1696
		IC 1805 - 303		2 31 38	+60 53.9	3	11.89	.018	0.67	.011	0.51	.021	5	A0 II-III	47,983,1718
15994	-6 00502	IDS02291S0605	AB	2 31 38	-05 51.2	1	7.14		1.08		0.93		2	K1 III	404
	-48 00678			2 31 38	-47 43.9	1	9.48		0.76		0.06		2		3077
		IC 1805 - 301		2 31 39	+61 21.0	3	11.94	.018	0.66	.004	0.11	.019	5		47,983,1718
		IC 1805 - 300		2 31 39	+61 29.8	1	12.85		0.75		0.37		1		1718
16046	-28 00819	HR 749	★ A	2 31 39	-28 27.1	1	4.97		-0.05		-0.15		3	B9.5V	542
16046	-28 00819	HR 749	★ AB	2 31 39	-28 27.1	8	4.89	.016	-0.05	.015	-0.13	.016	20	B9.5V +A3 V	15,542,938,1075,1079*
16046	-28 00819	IDS02295S2840	B	2 31 39	-28 27.1	2	7.85	.019	0.23	.044	0.20	.058	3	A3 V	542,542
15983	+10 00340			2 31 43	+11 23.2	1	6.98		0.00		-0.03		2	A0	1733
15952	+24 00369			2 31 46	+24 40.5	1	8.38		0.29		-0.01		2	F0	1625
15862	+59 00515	G 220 - 13		2 31 46	+59 34.5	2	8.94	.010	0.65	.010	0.12	.029	3	G5 V	1502,1658
15581	+60 00515	IC 1805 - 311		2 31 46	+60 53.4	1	8.42		0.69		0.23		3	A2	1689
		IC 1805 - 310		2 31 46	+61 06.9	1	13.15		0.47		0.42		2		983
		IC 1805 - 308		2 31 46	+61 25.9	3	10.94	.017	1.53	.016	1.48	.029	5	G5 Ib-II	47,983,1718
16031	-13 00482			2 31 46	-12 36.0	6	9.78	.007	0.44	.003	-0.21	.013	15	F0 V	742,1003,1594,1658*
232677	+50 00587			2 31 47	+50 58.3	1	9.35		0.76		0.38		2	F8	1566
15966	+22 00368			2 31 48	+22 45.0	1	8.45		0.29		0.13		2	A5	1648
	+60 00516			2 31 53	+60 30.1	1	10.10		0.41		-0.14		3		1689
		IC 1805 - 315		2 31 56	+61 25.9	1	12.85		0.49		0.24		2		983
	+4 00415	G 73 - 67		2 31 58	+05 14.2	5	9.82	.016	0.91	.013	0.66	.028	19	K3 V	333,830,1003,1620,1783,3025
16054	-16 00462			2 31 58	-16 20.3	1	10.23		0.11		-0.20		4	A1 III/IV	152
236971	+56 00660	LS I +57 071		2 31 59	+57 16.1	2	9.55	.000	0.50	.000	-0.40	.000	3	B1 IV	64,1012
		G 134 - 37		2 32 00	+44 51.3	1	15.91		1.55				1		906
	+60 00519	IC 1805 - 321		2 32 02	+60 59.9	1	8.92		1.78		1.85		3	K2	1689
	+60 00518	IC 1805 - 320		2 32 03	+61 17.4	1	9.90		0.69		0.22		3	F8 IV	1689
		IC 1805 - 322		2 32 06	+61 32.3	1	13.51		0.75				2		983
		MtW 46 # 48		2 32 10	+30 27.2	1	14.70		1.09		0.72		5		397
		IC 1805 - 324		2 32 10	+61 26.4	1	14.08		0.86				2		983
		CS 22184 # 7		2 32 10	-09 00.4	1	14.17		0.44		-0.72		1		1736
		MtW 46 # 52		2 32 11	+30 28.0	1	15.72		0.60		-0.12		5		397
	-45 00815			2 32 11	-45 33.8	1	9.70		0.39				4		2012
	+57 00593	LS I +57 072		2 32 12	+57 34.3	1	9.92		0.61		-0.31		2	B1 III	1012
	+60 00520	IC 1805 - 331		2 32 12	+60 56.7	1	9.71		0.39		0.09		3	A3 V	1689
16170	-51 00611	HR 755		2 32 12	-51 18.7	2	6.25	.011	0.51	.008	0.07		6	F6 V	58,2007
		MtW 46 # 61		2 32 14	+30 28.7	1	13.81		0.70		0.20		5		397
		IC 1805 - 332		2 32 14	+61 12.1	1	12.98		0.52		-0.30		2		983
16074	-8 00484	HR 752	★ A	2 32 14	-08 04.6	4	5.74	.008	1.39	.008	1.65	.009	16	gK4	15,1071,1075,3016
		IC 1805 - 334		2 32 15	+61 08.6	1	14.24		0.56		-0.21		2		983
		IC 1805 - 330		2 32 15	+61 34.5	1	13.78		0.36				2		983
16043	+17 00395			2 32 17	+17 50.5	1	8.27		0.69		0.26		2	G5	1648
		CS 22184 # 13		2 32 17	-11 13.7	2	13.07	.000	0.25	.000	0.09	.009	2		966,1736
15992	+44 00534			2 32 19	+44 25.3	1	7.56		0.00		-0.17		3	B9	1601
		IC 1805 - 339		2 32 19	+61 00.2	2	13.42	.005	0.71	.029	0.41	.015	3		983,1718
		IC 1805 - 336		2 32 19	+61 20.8	1	14.37		0.58				2		983
16004	+39 00573	HR 746	★ A	2 32 20	+39 26.8	2			-0.10	.000	-0.35	.005	6	B9p HgMn	1049,1079
16226	-63 00169	HR 762		2 32 20	-62 48.3	2	6.76	.005	-0.06	.000			7	B9 V	15,2012
	+62 00424	LS I +62 219		2 32 23	+62 43.8	3	8.83	.000	0.45	.000	-0.57	.000	5	O8	64,1011,1012
16042	+30 00414			2 32 24	+30 27.8	1	8.24		1.07		0.81		6	K0 V	397
	+55 00659	Tr 2 - 10		2 32 24	+55 36.4	1	10.26		0.21		-0.13		1	B9.5V	49
		PHL 8374		2 32 24	-21 20.	1	12.50		0.01		0.14		2		3060
16060	+6 00392	HR 751		2 32 25	+07 15.3	5	6.18	.011	1.05	.013	0.89	.005	14	K0	15,253,1061,1355,3031
		IC 1805 - 341		2 32 25	+60 59.1	3	12.05	.007	0.60	.007	0.25	.015	5		47,983,1718
		CS 22184 # 6		2 32 28	-08 51.7	2	13.13	.000	0.27	.000	0.07	.009	2		966,1736
16157	-44 00775	CC Eri		2 32 28	-44 00.6	3	8.80	.074	1.36	.009	1.08	.027	24	M0 Vp	1705,1738,3072
		PHL 1376		2 32 30	+03 31.0	6	12.41	.010	-0.20	.006	-1.17	.008	118	DAwke	281,989,1264,1663,1764,3016
		IC 1805 - 345		2 32 31	+60 56.7	1	13.91		0.68		-0.01		2		983
15963	+57 00594	LS I +57 073		2 32 32	+57 51.5	1	8.02		0.05		-0.05		2	Bp	379
16028	+36 00519	HR 748	★ A	2 32 33	+37 05.7	1	5.71		1.39				2	K3 III	71
15964	+57 00595			2 32 34	+57 30.6	1	8.88		0.09		-0.12		1	B9 V	401
		IC 1805 - 346		2 32 37	+61 04.6	2	13.30	.005	0.63	.005	0.35	.029	3		983,1718
16348	-73 00178			2 32 37	-72 54.5	1	8.90		1.04		0.99		1	K4 V	3072
15990	+55 00660	Tr 2 - 3		2 32 39	+55 42.9	1	8.52		0.18		-0.21		1	B8 II:	49
		IC 1805 - 349		2 32 39	+61 08.8	1	14.60		0.89				2		983

Table 1   145

HD	DM	Other Id	N Rem	α₁₉₅₀	δ₁₉₅₀	S	V	σ_V	B–V	σ_B–V	U–B	σ_U–B	n	Spectrum	References
16115	−10 00513			2 32 40	−09 39.7	4	8.15	.017	1.23	.010	0.82	.049	7	C2,3i	109,742,864,1238
		IC 1805 - 350		2 32 41	+61 17.9	1	14.41		0.63				2		983
16522	−79 00066	HR 776		2 32 41	−79 19.7	3	5.27	.011	0.98	.000	0.74	.005	11	G8 III	15,1075,2038
		Tr 2 - 20		2 32 43	+55 45.2	1	12.55		0.78		0.29		1		49
		IC 1805 - 352		2 32 43	+61 19.5	1	13.01		0.45		0.29		2		983
16058	+34 00469	HR 750	⋆A	2 32 44	+34 28.2	3	5.38	.012	1.65	.009	1.91	.015	12	M3 IIIa	71,814,3055
	−10 00514			2 32 46	−09 38.1	2	10.14	.010	0.63		0.05		4	G5	742,1594
16141	−4 00426			2 32 49	−03 46.3	4	6.83	.015	0.67	.013	0.20	.009	11	G5 IV	196,1003,2033,3026
	+34 00470	G 74 - 33		2 32 50	+34 52.0	1	9.64		0.69		0.17		3	G5	333,1620
	+61 00435			2 32 51	+61 51.3	1	10.15		0.39		0.03		3		1689
		VES 750		2 32 53	+61 51.3	1	12.36		0.50		-0.22		1		1543
	+61 00436			2 32 53	+61 58.4	1	10.01		0.38		0.24		3		1689
16112	+15 00354			2 32 55	+16 23.7	1	8.35		1.15		1.14		2	K2	1648
	+70 00193	AB Cas		2 32 55	+71 05.6	1	10.18		0.31		0.11		10	A3	922
16012	+57 00599			2 32 56	+57 30.2	2	8.41	.059	0.08	.005	-0.12	.005	5	B9 III	379,401
16153	−14 00486			2 32 56	−14 11.0	1	10.03		1.00		0.85		1	K4	1696
	+61 00437			2 32 58	+61 30.6	1	10.53		0.62		0.07		3		1689
16152	−9 00484			2 32 58	−09 34.1	1	7.12		0.02		-0.04		1	A0	742
16040	+55 00663	Tr 2 - 4		2 33 01	+55 42.5	1	8.58		0.18		-0.23		1	B9 III	49
		G 4 - 30		2 33 03	+20 00.3	1	10.70		1.51		1.17		1	M3	680
16240	−42 00865			2 33 04	−42 19.8	1	7.06		1.54		1.62		4	M4 III	3007
		Tr 2 - 14		2 33 05	+55 43.6	1	11.25		0.60		0.34		1		49
16039	+55 00665			2 33 05	+55 58.8	1	7.36		1.59		1.68		2	K1 III	37
	+55 00664	Tr 2 - 30		2 33 06	+55 40.4	1	9.80		0.20		-0.18		5	B9p	256
16025	+57 00600			2 33 06	+57 54.2	1	8.97		0.17		-0.02		2	B9 V	401
	+5 00364	G 73 - 68		2 33 09	+06 23.1	1	10.49		1.07		0.96		1	K3	333,1620
		Tr 2 - 13		2 33 09	+55 45.0	1	10.79		0.26		0.06		1	A0 V	49
		Tr 2 - 22		2 33 09	+55 45.3	1	13.30		0.53		0.30		1		49
		Tr 2 - 31		2 33 14	+55 41.6	1	10.17		0.29		0.03		2		256
15920	+72 00140	HR 743		2 33 14	+72 36.1	2	5.16	.010	0.88	.010	0.58		5	G8 III	1118,1355
16161	+4 00418	HR 754	⋆A	2 33 15	+05 22.6	1	4.97		0.87		0.56		3	G8 III	542
16161	+4 00418	HR 754	⋆AB	2 33 15	+05 22.6	9	4.86	.010	0.87	.009	0.53	.007	28	G8 III	15,244,542,1007,1013*
16161	+4 00418	IDS02306N0509	B	2 33 15	+05 22.6	1	9.08		0.56		0.19		1		542
16068	+55 00667	Tr 2 - 1		2 33 17	+55 41.9	1	7.38		1.79		1.90		1	K3 II	49
		Tr 2 - 26		2 33 19	+55 45.0	1	13.86		0.69		0.46		1		49
16160	+6 00398	HR 753	⋆AP	2 33 20	+06 39.0	18	5.82	.010	0.97	.010	0.80	.015	99	K3 V	1,15,22,333,1003,1006*
16160	+6 00398	IDS02307N0626	B	2 33 20	+06 39.0	6	11.66	.019	1.61	.007	1.11	.016	10		1,694,1006,3078,8003,8006
		LP 197 - 27	A	2 33 21	+39 48.7	1	10.73		0.83		0.54		2		1723
16083	+51 00599			2 33 21	+51 44.7	1	7.26		0.95		0.38		2	A5	1566
		CS 22184 # 14		2 33 21	−11 15.9	2	13.48	.005	0.06	.002	0.14	.015	2		966,1736
		Tr 2 - 27		2 33 23	+55 44.4	1	14.19		0.43		-0.33		1		49
16080	+55 00668	Tr 2 - 8		2 33 25	+55 41.7	1	9.36		0.26		-0.06		1	B7 V	49
16200	−9 00486			2 33 25	−08 47.0	1	8.99		0.62		0.16		4	G5	1731
16081	+55 00669	Tr 2 - 6		2 33 26	+55 31.9	1	8.89		0.21		-0.24		1	B7 III	49
	+61 00440			2 33 29	+61 36.9	1	10.34		0.83		0.28		3		1689
16024	+65 00280	HR 747		2 33 29	+65 31.7	1	5.78		1.56		1.85		3	K5 III	1355
		G 73 - 71		2 33 32	+06 38.2	2	11.66	.000	1.62	.000	1.09	.000	11		1620,3078
16212	−8 00489	HR 759	⋆A	2 33 32	−08 02.9	4	5.52	.005	1.59	.005	1.93	.000	15	M0 III	15,1071,1075,3055
		Tr 2 - 11		2 33 35	+55 35.1	1	10.52		0.40				1		49
16222	−4 00430			2 33 35	−04 23.3	1	9.16		0.42		-0.03		3	F5	196
16223	−4 00431			2 33 35	−04 29.5	1	8.47		1.01		0.82		3	K0	196
		Steph 288		2 33 36	−19 57.3	1	11.99		1.52		1.17		1	M0	1746
		Tr 2 - 17		2 33 39	+55 46.2	1	12.10		0.52		0.31		1		49
214198	+46 03754	IDSS22316N464	AB	2 33 42	+46 57.6	1	9.15		0.43		-0.01		3	F5	1601
		Tr 2 - 23		2 33 43	+55 42.8	1	13.40		0.72		0.30		1		49
16107	+55 00671	Tr 2 - 2		2 33 43	+55 44.6	1	8.45		0.36		0.22		1	A8 III	49
16187	+30 00418	HR 757		2 33 44	+31 23.4	1	6.10		1.05		0.79		2	K0 III	252
		Tr 2 - 25		2 33 44	+55 45.8	1	13.50		0.90		0.22		1		49
		Tr 2 - 19		2 33 45	+55 40.0	1	12.49		0.55		0.31		1		49
16270	−24 01154			2 33 45	−23 44.3	1	8.43		1.08		0.95		1	K3 V	3072
		LP 410 - 64		2 33 46	+17 34.4	1	12.44		0.64		-0.02		2		1696
		Tr 2 - 24		2 33 46	+55 40.1	1	13.43		0.78		0.21		1		49
16280	−27 00904			2 33 48	−27 23.7	1	9.55		1.07		0.95		1	K2 V	3072
16176	+38 00515	HR 756		2 33 49	+38 31.1	3	5.91	.011	0.48	.005	0.05	.005	7	F5 V	71,1501,3053
16234	+11 00360	HR 763	⋆	2 33 54	+12 13.9	2	5.68	.020	0.48	.005	-0.08	.025	7	F8 V	254,3037
		Tr 2 - 28		2 33 54	+55 39.6	1	15.26		1.01		0.36		1		49
		VES 751		2 33 54	+60 03.0	1	12.94		0.88		0.02		1		1543
		G 76 - 13		2 33 55	+11 59.5	1	11.73		1.35		1.30		1	K7	333,1620
16247	+7 00402	HR 766		2 33 56	+07 30.8	3	5.81	.008	1.04	.000	0.86	.005	9	gK0	15,1256,1355
		Tr 2 - 21		2 33 59	+55 39.3	1	13.03		1.55				1		49
16210	+33 00470	HR 758, R Tri		2 34 00	+34 02.9	3	6.41	.290	1.50	.158	0.87	.025	3	M4 IIIe	765,814,3001
		Tr 2 - 16		2 34 00	+55 38.3	1	11.96		0.51		0.29		1		49
16307	−30 00958	HR 767		2 34 00	−30 15.7	3	5.73	.017	1.01	.005			11	G8/K0 III	15,1075,2035
		Tr 2 - 15		2 34 02	+55 40.8	1	11.59		0.48		0.28		1		49
	+62 00428	LS I +62 220		2 34 02	+62 42.6	1	10.19		0.69		-0.33		2		405
16232	+24 00375	HR 764	⋆B	2 34 04	+24 25.9	1	7.09		0.51		0.01		3	F4 V	1084,3024
16220	+32 00473	HR 761		2 34 05	+32 40.5	2	6.25	.015	0.47	.005	0.00	.005	5	F7 V	254,272
		Tr 2 - 18		2 34 06	+55 38.9	1	12.16		0.49		0.35		1		49
		L 297 - 72		2 34 06	−47 53.	3	12.00	.026	0.55	.009	0.02	.044	6		158,1696,3062
16246	+24 00376	HR 765	⋆A	2 34 07	+24 25.9	2	6.54	.040	0.41	.010	0.00	.030	4	F6 III	1084,3024

HD	DM	Other Id	N Rem	α_{1950}	δ_{1950}	S	V	σ_V	B–V	σ_{B-V}	U–B	σ_{U-B}	n	Spectrum	References
16159	+55 00674	Tr 2 - 5		2 34 09	+55 45.4	1	8.60		0.27		-0.13		1	B9 III	49
16287	-3 00410	G 75 - 28		2 34 09	-03 22.4	3	8.10	.004	0.95	.008	0.73	.000	6	K0	258,1620,2033
16219	+39 00582	HR 760		2 34 12	+39 40.8	2			-0.12	.000	-0.48	.005	4	B5 V	1049,1079
16371	-45 00826			2 34 15	-44 58.6	1	8.09		0.87		0.59		1	G8 III	1738
236978	+55 00676	Tr 2 - 9		2 34 17	+55 40.9	1	9.74		0.23		-0.13		1	B8 III	49
		Tr 2 - 12		2 34 19	+55 41.3	1	10.60		0.31		0.07		1	A0 V	49
	+43 00538	G 134 - 40		2 34 24	+43 47.5	2	10.17	.012	0.53	.005	-0.07	.039	5	F8	1620,7010
16315	-8 00493			2 34 29	-08 33.6	1	8.57		0.38		0.02		4	A5	1731
	+60 00526	LS I +60 245		2 34 31	+60 51.2	1	8.85		0.64		-0.43		2	B2	405
232684	+51 00602	IDS02312N5115	A	2 34 33	+51 28.5	1	9.98		0.21		0.01		2	B8	401
16183	+60 00527			2 34 33	+60 52.4	1	7.94		0.46		-0.03		3	F5 V	1689
		LP 354 - 122		2 34 42	+22 37.0	1	14.15		1.56		1.28		1		1773
16302	+19 00389			2 34 45	+20 21.5	2	6.91	.005	0.92	.005	0.57	.005	6	G9 III	1501,1648
		G 76 - 16		2 34 47	+14 14.0	1	13.98		0.96		0.70		2		1658
16284	+38 00523			2 34 47	+38 42.2	1	7.62		0.14		0.11		3	A2	833
236979	+56 00673	YZ Per		2 34 47	+56 49.8	5	7.94	.172	2.34	.017	2.61	.061	11	M1 Iab:V	369,1083,1110,8016,8023
16417	-35 00903	HR 772		2 34 54	-34 47.5	7	5.78	.006	0.65	.007	0.18	.028	25	G1 V	15,164,258,1075,2012*
		G 75 - 30		2 34 55	+00 08.5	2	15.15	.009	1.68	.014	1.23		4		419,906
16257	+55 00679	Tr 2 - 7		2 34 55	+55 40.8	1	9.00		0.40		-0.03		1	F0	49
16312	+22 00375			2 34 57	+22 54.8	1	8.40		0.51		0.04		2	F5	1648
16243	+57 00602	LS I +57 076		2 34 57	+57 36.1	1	8.26		0.56		-0.27		2	B2 II:	1012
16353	+11 00365			2 35 03	+12 03.2	1	7.58		1.01		0.70		2	G5	1375
16218	+61 00444	IDS02312N6209	A	2 35 03	+62 22.6	1	6.66		-0.02		-0.11		2	B9	401
16230	+61 00445	IDS02312N6209	B	2 35 08	+62 21.6	1	7.62		0.11		0.06		1	A0	401
16264	+56 00676	LS I +57 077		2 35 10	+57 22.5	1	9.25		0.44		-0.44		2	B1:V:	1012
		CS 22189 # 7		2 35 10	-13 09.3	1	13.21		0.58		-0.07		1		1736
16400	-4 00436	HR 771		2 35 11	-03 36.7	2	5.64	.005	1.02	.000	0.86	.005	7	K0	15,1256
16327	+37 00588	HR 768	⋆ A	2 35 12	+37 30.7	1	6.16		0.49		0.05		2	F6 IV	3053
16327	+37 00588	HR 768	⋆ AB	2 35 12	+37 30.7	1	6.20		0.46		0.06		2	F6 IV	254
16327	+37 00588	IDS02321N3718	C	2 35 12	+37 30.7	1	11.37		1.25		1.17		2		3032
		CS 22184 # 9		2 35 13	-09 13.0	2	15.16	.000	-0.06	.000	0.12	.013	2		966,1736
16399	+7 00405	HR 770		2 35 21	+07 28.8	3	6.39	.008	0.43	.010	-0.03	.014	12	F6 V	15,1061,3053
16350	+37 00591	HR 769		2 35 21	+37 52.4	1			-0.03		-0.03		4	B9.5V	1049
	+61 00446			2 35 23	+61 29.5	1	9.14		0.32		0.09		3	B9 V	1689
	+49 00727			2 35 24	+49 26.5	1	9.34		0.99		0.61		2		1375
		L 126 - 11		2 35 24	-61 08.	1	12.19		0.55		-0.16		2		1696
16446	-23 00998			2 35 29	-23 12.5	2	6.82	.021	1.04	.019	0.85	.011	7	K0 III	1628,3040
		AJ89,1229 # 155		2 35 31	+09 13.7	1	10.83		1.69				1		1532
16397	+30 00421	G 36 - 28		2 35 31	+30 36.4	6	7.36	.014	0.59	.010	0.00	.008	13	G0 V	22,908,1003,1620,1658,3026
16377	+33 00476			2 35 32	+34 09.1	1	7.85		0.54				1	F5	765
	+33 00478			2 35 37	+34 07.7	1	8.29		1.22				1	K0	765
16310	+58 00498	LS I +58 090		2 35 39	+58 51.1	4	8.07	.019	0.65	.008	-0.35	.004	7	B1 II:	1012,1083,1110,8023
		AO 0235+164 # 3	⋆	2 35 43	+16 23.8	2	12.92	.000	0.75	.005	0.27	.010	7		487,1595
		CS 22189 # 13		2 35 43	-15 42.2	1	12.40		0.06		0.06		1		1736
16386	+38 00527			2 35 44	+38 56.1	1	7.87		-0.06		-0.19		2	B9 V	105
		G 75 - 31		2 35 45	+02 13.8	3	10.53	.018	0.46	.004	-0.16	.004	5	F8	516,1620,1696
16555	-53 00457	HR 778		2 35 45	-52 45.5	3	5.31	.023	0.28	.004			8	A6 V	15,2016,2027
		AO 0235+164 # 2	⋆	2 35 46	+16 23.1	2	12.72	.005	0.83	.010	0.40	.010	8		487,1595
		AO 0235+164 # 1	⋆	2 35 50	+16 23.5	3	13.04	.009	0.58	.018	0.00	.015	15		487,495,1595
16420	+29 00448			2 35 50	+30 19.6	1	8.56		1.24		1.18		2	K0	1625
16565	-53 00458			2 35 50	-53 05.4	1	9.05		0.60		0.12		3	F8 V	1731
		AO 0235+164 # 8	⋆	2 35 53	+16 24.	2	16.62	.080	1.61	.063	0.74		7		495,1595
		AO 0235+164 # 7	⋆	2 35 54	+16 24.4	2	14.99	.010	1.22	.005	1.05		3		487,495
		CS 22189 # 14		2 35 56	-16 44.8	1	16.33		0.41		0.05		1		1736
16432	+21 00362	HR 773		2 35 58	+21 44.8	3	5.40	.055	0.16	.000	0.16	.024	7	A7 V	985,1049,3023
	-17 00508			2 35 58	-17 33.6	1	10.91		0.95		0.71		1	K4	1696
		AO 0235+164 # 6	⋆	2 35 59	+16 25.9	1	14.00		0.68		0.13		2		487
		Feige 25		2 36 00	+05 15.0	1	12.05		-0.04		-0.39		2		1026
16467	+2 00406	HR 775	⋆ AB	2 36 01	+03 13.7	2	6.20	.005	1.00	.000	0.77	.005	7	G9 III	15,1061
		AO 0235+164 # 5	⋆	2 36 03	+16 27.4	1	14.46		0.88		0.45		2		487
		CS 22184 # 23		2 36 04	-11 03.9	1	15.19		0.03		0.03		1		1736
	+58 00501	GP Cas		2 36 05	+59 23.0	3	9.37	.042	2.71	.007	2.59		10	M2 Iab	148,369,8032
16349	+61 00448			2 36 05	+62 23.3	1	7.44		0.05		-0.12		4	A0	1502
16422	+42 00565	NGC 1039 - 494		2 36 08	+42 35.6	1	8.56		1.27		1.26		2	K2	566
16538	-30 00973	HR 777		2 36 09	-30 24.5	2	5.82	.005	0.48	.000			7	F5 V	15,2012
16480	+14 00439	IDS02335N1425	AB	2 36 15	+14 38.7	1	7.05		1.12		0.43		2	F7 IV	8100
16591	-42 00875			2 36 16	-42 07.2	3	7.23	.011	0.94	.002	0.64		10	K1 IV/V	1075,2012,3008
		AO 0235+164 # 4	⋆	2 36 17	+16 23.4	1	10.35		0.68		0.14		3		487
		G 221 - 2		2 36 22	+74 30.0	1	15.93		0.20		-0.52		1	DAn	782
16589	-38 00875	HR 780		2 36 24	-38 12.3	2	6.48	.005	0.52	.000			7	F6 V	15,2029
17020	-82 00045			2 36 26	-81 42.9	2	9.18	.015	0.53	.010	0.06		3	F5 V	742,1594
16586	-25 01044			2 36 35	-25 34.8	1	8.38		0.33				4	A9 V	2012
16464	+42 00566	NGC 1039 - 25		2 36 36	+42 46.0	1	9.20		0.06		-0.09		4	A0	566
		NGC 1027 - 14		2 36 39	+61 26.4	1	13.10		0.40		0.28		1		49
16634	-45 00844			2 36 39	-44 55.4	1	7.86		1.00		0.68		5	G5	1673
16410	+60 00540	IDS02328N6103	AB	2 36 40	+61 16.2	1	7.72		1.31		1.16		3	G8 III	1689
	+53 00553			2 36 40	+53 34.3	1	10.67		1.86				3		369
		LS I +61 303	⋆ V	2 36 41	+61 00.8	2	10.76	.034	0.82	.015	-0.32	.010	3	B1Ibe	405,1543
16600	-26 00953			2 36 41	-25 57.6	1	7.14		1.20				4	K2 IV	2012
		CS 22184 # 27		2 36 43	-09 27.0	1	11.92		0.11		0.08		1		1736

Table 1 147

HD	DM	Other Id	N Rem	α_{1950}	δ_{1950}	S	V	σ_V	B−V	σ_{B-V}	U−B	σ_{U-B}	n	Spectrum	References
	+9 00345			2 36 44	+10 02.8	1	10.75		1.26		1.17		2	K2	1375
		CS 22189 # 12		2 36 44	−14 27.4	1	13.46		0.28		0.06		1		1736
		NGC 1027 - 20		2 36 49	+61 20.6	1	14.54		0.54		0.38		1		49
16509	+34 00487			2 36 50	+35 12.8	1	9.11		-0.02		-0.07		2	A0	1625
		CS 22184 # 21		2 36 50	−11 31.2	1	11.81		0.27		0.10		1		1736
		VES 755		2 36 51	+61 44.3	1	12.19		0.44		-0.24		1		1543
16461	+52 00609			2 36 52	+52 35.4	1	7.49		0.14		0.08		2	A0	1375
16429	+60 00541	V482 Cas	★ AB	2 36 54	+61 04.1	4	7.67	.000	0.62	.000	-0.38	.000	7	O9.5III	64,1011,1012,1083
16582	−0 00406	HR 779, δ Cet		2 36 55	+00 06.8	10	4.07	.007	-0.22	.008	-0.85	.014	28	B2 IV	15,244,1020,1034,1075*
16460	+55 00688	LS I +56 061		2 36 56	+56 03.4	1	7.49		0.47				3	F2 V	6009
16623	−26 00957			2 36 56	−26 31.8	3	8.76	.005	0.60	.004	0.01	.005	9	F7wG3 V	158,742,2033
	+61 00452			2 36 57	+62 21.1	1	10.36		0.35		-0.10		2		1375
		G 134 - 42		2 36 59	+39 41.2	1	15.00		1.50				3		419
16581	+0 00442			2 37 00	+01 09.2	6	8.19	.003	-0.06	.004	-0.28	.015	107	B9	147,989,1509,1729,1775,6005
16448	+56 00683			2 37 00	+57 05.7	1	7.06		1.15		1.18		3	K2 III	37
		CS 22189 # 8		2 37 01	−12 48.6	1	13.30		0.30		0.00		1		1736
		NGC 1027 - 10		2 37 04	+61 16.9	1	12.15		0.75		0.18		1	F0	49
16393	+69 00171			2 37 08	+69 29.8	1	7.60		0.03		-0.32		3	B8 V	555
16620	−12 00501	HR 781	★ AB	2 37 09	−12 05.0	5	4.83	.009	0.44	.006	-0.01	.007	16	F5 V+F6 V	3,15,1008,2012,3053
	+41 00508	NGC 1039 - 503	★ AB	2 37 10	+42 03.4	2	9.61	.019	0.83	.005	0.46		3	G5	566,1620
16534	+41 00507	NGC 1039 - 88		2 37 10	+42 16.3	1	9.16		0.03		-0.09		4	A0	566
16608	+1 00462			2 37 11	+01 55.0	3	8.33	.001	1.51	.004	1.80	.012	73	K5	147,1509,6005
		G 76 - 20		2 37 16	+06 34.0	1	12.95		0.74		0.14		2		1620
	−50 00776			2 37 16	−49 40.9	2	10.12	.010	0.78	.010	0.25		7		1594,3077
16619	−0 00407	IDS02347S0017	AB	2 37 18	−00 03.9	3	7.83	.004	0.65	.007	0.15	.008	4	G4 V +G4 V	258,292,3030
	+46 00603			2 37 23	+47 00.2	1	9.41		0.02		-0.38		2	B5	1601
		NGC 1027 - 6		2 37 26	+61 36.5	1	11.31		0.51		0.21		1	A1 V	49
		NGC 1027 - 24		2 37 27	+61 23.3	1	15.11		0.72		0.20		1		49
		NGC 1027 - 18		2 37 28	+61 22.8	1	14.04		0.96		0.24		1		49
16440	+67 00221	LS I +68 015		2 37 29	+68 15.4	1	7.88		0.71		0.07		3	B7 II	555
16717	−41 00754			2 37 29	−40 56.9	1	8.69		0.97				4	G8/K0 III	2012
		Feige 26		2 37 30	+03 43.0	1	13.99		-0.26		-1.27		1		308
		LP 830 - 22		2 37 30	−22 22.6	1	12.15		1.27		1.22		1		1696
		G 74 - 39	A	2 37 31	+34 07.1	3	11.86	.013	1.42	.013	1.19	.061	7		316,906,3016
		G 74 - 40	B	2 37 32	+34 07.1	2	13.05	.005	1.47	.009	1.25		4	K6	316,906,3016
		G 74 - 41	C	2 37 32	+34 07.9	2	14.32	.045	1.56	.015			2		906,3016
16523	+56 00686	V493 Per	★	2 37 33	+56 31.0	1	10.01		0.16		-0.14		2	C6	1359
	+61 00456	NGC 1027 - 2		2 37 34	+61 30.4	1	9.77		0.39		0.09		1	A0 V	49
		Feige 27		2 37 36	+08 48.0	1	11.79		0.08		-0.23		3		1026
	−58 00223			2 37 36	−58 24.1	1	9.65		1.39		1.10		1	M0 Ve	3072
16647	+5 00374	HR 783		2 37 37	+05 53.9	5	6.26	.017	0.39	.011	-0.05	.009	15	F5 V	15,253,254,1256,3016
	+57 00612			2 37 40	+58 24.0	1	10.61		0.44		-0.33		1		1543
		NGC 1027 - 16		2 37 41	+61 16.1	1	13.54		0.95		0.19		1		49
		NGC 1027 - 12		2 37 41	+61 34.9	1	12.52		0.53		0.33		1	A5	49
16714	−34 00968	IDS02356S3420		2 37 42	−34 06.5	1	8.29		0.71				4	G5 V	2012
16578	+47 00679			2 37 43	+47 28.6	2	9.01	.014	0.08	.002	0.03	.007	46	A0	588,1375
16646	+17 00414			2 37 45	+18 23.1	1	8.10		0.39		-0.04		2	F5	1648
16628	+26 00443	HR 782	★ A	2 37 45	+26 50.8	2	5.30		0.08	.005	0.11	.015	7	A3 V	1049,3024
16628	+26 00443	IDS02348N2638	B	2 37 45	+26 50.8	1	9.56		0.70		0.24		2	A2	3024
16605	+42 00572	NGC 1039 - 154		2 37 46	+42 39.4	2	9.66	.008	0.03	.002	-0.18	.000	8	A1p	566,1111
16673	−10 00525	HR 784		2 37 46	−09 40.0	2	5.77	.005	0.52	.000	-0.01	.010	7	F8 V	15,1415
16638	+25 00436	IDS02349N2612	AB	2 37 47	+26 24.5	1	7.55		0.54		0.02		1	F5 V +G3 V	292
		NGC 1027 - 13		2 37 47	+61 35.4	1	12.95		0.56		0.31		1		49
	+42 00573	NGC 1039 - 157		2 37 49	+42 41.5	2	9.47	.008	0.55	.004	0.07	.004	8	G0	566,1111
	+38 00536	IDS02347N3839	AB	2 37 50	+38 51.9	1	9.12		0.84		0.48		2	K2	1601
		NGC 1027 - 19		2 37 50	+61 32.3	1	14.09		0.63		0.33		1		49
	+42 00575	NGC 1039 - 159		2 37 51	+42 45.8	1	10.62		0.19		0.10		4	A2	566
		NGC 1039 - 161		2 37 52	+42 39.8	1	11.94		0.47		0.02		1		1111
		LP 941 - 146		2 37 52	−34 18.6	1	11.75		1.52		1.09		1		3078
	+42 00577	NGC 1039 - 162		2 37 53	+42 27.2	2	10.96	.009	0.23	.012	0.10	.000	10	A2	566,1111
16733	−31 01081	HR 786		2 37 54	−30 50.8	3	6.51	.005	1.04	.000			11	K0 III	15,1075,2032
16754	−43 00814	HR 789	★ A	2 37 54	−43 06.3	5	4.75	.005	0.06	.003	0.06	.003	21	A2 V	15,1075,2012,3023,8015
		NGC 1027 - 26		2 37 57	+61 19.1	1	15.57		0.66		0.34		1		49
16627	+42 00578	NGC 1039 - 166	★ AB	2 37 58	+42 27.9	1	9.23		0.08		-0.02		10	Am	566,1111
16593	+52 00612			2 37 58	+52 27.4	1	8.45		1.22		1.01		2	K0	1566
		NGC 1027 - 8		2 37 59	+61 33.2	1	11.95		0.41		0.21		1	A0	49
		CoD -58 00540		2 38 00	−57 48.6	1	11.23		0.76		0.27		2		1696
		NGC 1027 - 11		2 38 03	+61 27.2	1	12.27		0.46		0.24		1	A0	49
		NGC 1027 - 21		2 38 03	+61 31.3	1	14.87		1.57				1		49
16505	+67 00222			2 38 03	+67 51.1	2	7.05	.025	1.23	.005	1.40	.020	5	K3 III	1003,1355
	+42 00579	NGC 1039 - 174		2 38 04	+42 30.7	2	10.52	.004	0.20	.001	0.09	.012	7	A3	566,1111
19589	−88 00027			2 38 06	−88 22.1	2	8.30	.008	0.90	.043	0.54	.010	12	G8 IV	1117,1704
	+0 00444	G 75 - 34		2 38 07	+00 58.9	5	9.51	.016	1.19	.016	1.10	.010	5	K7	680,1017,1620,1705,3072
	+42 00580	NGC 1039 - 179		2 38 07	+42 39.9	2	10.31	.000	0.11	.009	0.09	.011	11	A0	566,1111
	+63 00351	LS I +64 092		2 38 09	+64 22.5	1	10.50		0.73		-0.23		2		405
		CS 22184 # 28		2 38 10	−09 20.9	2	13.71	.010	0.19	.005	0.10	.002	2		966,1736
16708	+2 00412			2 38 11	+02 41.1	1	7.09		1.01		0.71		1	G5 IV	3040
	+42 00581	NGC 1039 - 188	A	2 38 11	+42 36.2	1	10.16		0.11		0.07		1	A2	1111
	+42 00581	NGC 1039 - 188	AB	2 38 11	+42 36.2	1	9.93		0.19				6		566
	+42 00582	NGC 1039 - 191		2 38 12	+42 37.8	1	10.09		1.06		0.66		2	G5	566

HD	DM	Other Id	N	Rem	α_{1950}	δ_{1950}	S	V	σ_V	B–V	σ_{B-V}	U–B	σ_{U-B}	n	Spectrum	References
		NGC 1027 - 9			2 38 13	+61 28.3	1	12.02		0.27		0.26		1	A8	49
16683	+24 00381				2 38 16	+25 24.1	1	8.16		0.22		0.18		2	A0	105
	+41 00513	NGC 1039 - 197			2 38 16	+42 20.8	1	10.68		0.31		0.10		2	F0	566
		VES 757			2 38 17	+58 06.0	1	11.94		0.42		-0.39		1		1543
16694	+18 00337	IDS02355N1822	AB		2 38 18	+18 35.2	1	7.04		0.05		-0.02		2	A0 V	3032
16694	+18 00337	IDS02355N1822	AB		2 38 18	+18 35.2	1	9.47		0.25		0.11		2	A7 IV	3032
		CS 22184 # 19			2 38 18	-12 03.9	1	13.93		0.16		0.16		1		1736
16655	+41 00514	NGC 1039 - 200			2 38 19	+42 22.8	2	8.47	.004	0.06	.004	-0.04	.000	14	A0 V+Shell	566,1111
	+46 00610	G 78 - 1			2 38 25	+47 08.9	2	9.16	.000	0.52	.000	-0.10	.005	3	F8	1620,1658
	+42 00583	NGC 1039 - 221			2 38 26	+42 39.9	2	10.45	.009	0.16	.002	0.07	.018	5	A2	566,1111
		NGC 1027 - 25			2 38 26	+61 26.5	1	15.36		0.80		0.07		1		49
		NGC 1027 - 15			2 38 26	+61 29.5	1	13.18		0.59		0.27		1		49
		NGC 1039 - 222			2 38 27	+42 29.5	1	11.21		1.09		0.81		2	G5	1111
16670	+42 00584	NGC 1039 - 509			2 38 27	+42 55.8	1	8.92		-0.01		-0.28		5	B8	566
16891	-64 00192	HR 798			2 38 27	-64 29.8	3	6.55	.003	-0.08	.003	-0.23	.005	21	B8 V	15,1075,2038
	+42 00585	NGC 1039 - 226			2 38 28	+42 33.5	2	10.46	.000	0.18	.008	0.08	.008	13	A2	566,1111
16784	-30 00990				2 38 28	-30 21.0	4	8.02	.005	0.57	.005	-0.05	.010	11	F8 V	158,1696,2012,3077
16663	+44 00558				2 38 29	+45 16.9	1	8.28		0.47		0.03		2	F8	1601
	+9 00352	G 76 - 21			2 38 32	+09 33.4	3	10.17	.014	0.44	.004	-0.25	.000	5	F2	516,1620,1696
		NGC 1039 - 238			2 38 34	+42 30.1	1	11.21		0.35		0.07		4	A8	1111
		NGC 1039 - 241			2 38 35	+42 28.5	1	12.20		0.64		0.08		3		1111
16679	+42 00586	NGC 1039 - 244			2 38 35	+42 33.4	2	8.98	.000	0.00	.000	-0.21	.012	13	B9.5V	566,1111
		WLS 240 10 # 7			2 38 37	+07 50.6	1	10.79		0.61		0.08		2		1375
		NGC 1039 - 250			2 38 37	+42 38.8	1	11.47		0.36		0.05		1	A5	1111
		VES 758			2 38 38	+58 26.7	1	12.67		0.57		-0.17		1		1543
	+41 00515	NGC 1039 - 254			2 38 39	+42 14.0	1	10.87		0.25		0.07		5	A5	566
16765	-1 00377	HR 790	⋆AB		2 38 40	-00 54.4	5	5.72	.021	0.52	.008	-0.03	.012	11	F7 IV	15,254,938,1256,1477
		CS 22184 # 30			2 38 40	-11 28.2	2	12.99	.000	0.21	.000	0.15	.008	2		966,1736
		NGC 1039 - 258			2 38 41	+42 27.6	1	11.13		1.05		0.75		4	G5	1111
		NGC 1027 - 22			2 38 41	+61 29.1	1	15.00		1.09				1		49
16815	-40 00689	HR 794			2 38 42	-40 04.1	4	4.11	.003	1.02	.000	0.74	.003	26	K0 III	15,1075,2012,8015
16693	+42 00588	NGC 1039 - 263	⋆B		2 38 43	+42 34.6	1	8.52		0.00		-0.30		3	B8.5Vp	1111
16626	+60 00548	NGC 1027 - 1			2 38 43	+61 23.0	1	6.99		0.39		-0.07		1	A9 III-IV	49
		NGC 1039 - 267			2 38 45	+42 33.3	1	11.95		0.47		0.01		3		1111
16705	+42 00589	NGC 1039 - 266	⋆A		2 38 45	+42 34.7	1	8.46		0.01		-0.29		5	B9 IV	1111
16705	+42 00589	NGC 1039 - 266	⋆AB		2 38 46	+42 34.7	1	7.73		0.00		-0.31		4		566
	+5 00378				2 38 46	+05 46.5	1	10.57		1.25				1	M0	1632
		NGC 1039 - 269			2 38 46	+42 26.5	1	11.45		0.35		0.06		3	A5	1111
	+42 00590	NGC 1039 - 274			2 38 48	+42 32.1	2	9.71	.009	0.14	.024	0.07	.005	8	A2.5V	566,1111
	-0 00409				2 38 48	-00 18.4	1	9.56		0.57		0.02		3	G0	196
16905	-61 00218	IDS02376S6147	A		2 38 48	-61 33.9	1	9.50		1.03		0.88		1	K3 V	3072
16905	-61 00218	IDS02376S6147	B		2 38 48	-61 33.9	1	14.05		1.52		0.91		1		3072
		NGC 1039 - 276			2 38 49	+42 18.2	1	12.17		1.10		0.32		1	G0	936
16978	-68 00161	HR 806			2 38 49	-68 28.8	5	4.10	.006	-0.06	.003	-0.12	.012	20	B9 V	3,15,1075,2038,3023
		NGC 1039 - 278			2 38 50	+42 34.2	1	11.77		0.28		0.14		3	A8	1111
		VES 759			2 38 50	+57 40.7	1	11.99		0.66		-0.05		1		1543
	+41 00516	NGC 1039 - 280			2 38 51	+42 20.0	2	9.88	.004	0.10	.001	0.07	.009	9	A2.5V	566,1111
16879	-53 00467				2 38 51	-53 01.3	1	8.70		0.53		0.01		3	F8 V	1731
16719	+42 00591	NGC 1039 - 282	⋆A		2 38 52	+42 29.7	1	8.33		-0.01		-0.28		3	B8.5III	1111
16719	+42 00591	NGC 1039 - 282	⋆AB		2 38 52	+42 29.7	1	8.28		0.01		-0.30		10		566
		PHL 4260			2 38 52	-08 10.9	2	15.06	.000	0.08	.000	0.00	.010	2		966,1736
16878	-52 00328				2 38 52	-52 00.0	1	8.04		0.51		0.01		3	F5 V	1731
		NGC 1039 - 284			2 38 53	+42 33.2	1	10.74		0.28		0.12		4	A2	1111
16851	-41 00760				2 38 56	-41 30.8	1	7.66		1.62				4	K3/4 II/III	2012
	+42 00592	NGC 1039 - 294			2 38 57	+42 32.4	1	11.18		0.34		0.08		3	A5	1111
	+42 00594	NGC 1039 - 301			2 38 59	+42 34.4	1	10.01		0.12		0.05		4	A0	1111
		NGC 1039 - 306			2 39 00	+42 24.5	1	11.09		0.40		0.08		3	A5	1111
16728	+42 00598	NGC 1039 - 307	⋆AB		2 39 00	+42 29.2	2	7.92	.014	0.01	.011	-0.22	.005	20	A0 III-IV	566,1111
	+42 00595	NGC 1039 - 303			2 39 00	+42 32.0	1	9.93		0.12		0.06		4	A1	1111
	+42 00596	NGC 1039 - 308			2 39 00	+42 33.9	1	8.80		0.06		-0.07		4	A0 IV	1111
	+42 00597	NGC 1039 - 311			2 39 01	+42 39.2	1	9.30		0.05		-0.08		3	B9 V	1111
	+42 00599	NGC 1039 - 312			2 39 01	+42 40.1	1	10.45		0.20		0.18		1		1111
	-30 00995				2 39 02	-30 18.5	1	10.90		0.62		0.09		2		1730
		NGC 1039 - 315			2 39 03	+42 40.0	1	11.45		0.38		0.06		1	A5	1111
		NGC 1039 - 316			2 39 04	+42 23.3	1	11.97		0.68		0.13		2	G0	1111
16739	+39 00610	HR 788	⋆		2 39 05	+39 59.0	5	4.92	.015	0.58	.014	0.13	.021	12	F9 V	15,1118,1363,3026,8015
		NGC 1039 - 320			2 39 05	+42 34.0	1	12.55		0.65		0.15		1	F8	936
		CS 22184 # 29			2 39 06	-10 22.7	2	13.20	.000	0.03	.000	0.12	.010	2		966,1736
16920	-55 00446	HR 802			2 39 06	-54 45.8	2	5.20	.006	0.41	.009	-0.01		6	F4 IV	58,2032
	+42 00600	NGC 1039 - 515			2 39 08	+42 59.8	1	10.15		0.33		0.18		2	F0	566
		NGC 1039 - 331			2 39 09	+42 23.3	1	10.96		0.95		0.53		3	G5	1111
	+42 00601	NGC 1039 - 328	⋆AB		2 39 09	+42 32.8	2	8.87	.009	0.02	.003	-0.11	.004	12	B9 V	566,1111
		NGC 1039 - 334			2 39 11	+42 36.8	1	11.99		1.05		0.76		1		936
	+60 00552	NGC 1027 - 4	V		2 39 11	+61 22.8	2	10.15	.275	0.41	.046	0.30	.202	11	A0 V	49,943
16691	+56 00693	LS I +56 063			2 39 12	+56 41.5	5	8.70	.004	0.48	.004	-0.56	.015	10	O5 e	64,1011,1012,1083,1209
16825	-15 00478	HR 796			2 39 12	-14 45.8	2	5.97	.004	0.42	.004	-0.02		6	F5 V	58,2007
	+42 00602	NGC 1039 - 339			2 39 14	+42 32.7	2	9.31	.000	0.04	.009	-0.05	.005	8	B9.5Vn	566,1111
		G 4 - 33			2 39 15	+12 38.8	1	12.89		1.51		1.22		1		333,1620
	+41 00519	NGC 1039 - 343			2 39 16	+42 23.3	1	10.97		1.01		0.69		3	G5	1111
16824	-3 00421	HR 795			2 39 17	-03 25.6	3	6.04	.003	1.18	.009	1.12	.012	20	K0	15,361,1256

Table 1   149

HD	DM	Other Id	N Rem	α_{1950}	δ_{1950}	S	V	σ_V	B–V	σ_{B-V}	U–B	σ_{U-B}	n	Spectrum	References
	+42 00603	NGC 1039 - 346		2 39 18	+42 45.6	1	10.74		0.23		0.15		2	A3	566
		NGC 1027 - 5		2 39 18	+61 23.2	1	11.13		0.26		-0.17		1	B9 V	49
		CS 22189 # 9		2 39 19	−13 41.0	1	14.04		0.72		0.11		1		1736
		NGC 1039 - 349		2 39 20	+42 29.3	1	12.79		0.62		0.08		1		936
		NGC 1039 - 351		2 39 21	+42 32.1	1	11.50		0.36		0.07		2	F0	1111
16771	+41 00521	NGC 1039 - 355	AB	2 39 22	+42 22.7	2	7.32	.004	0.94	.000	0.59	.022	26	G5	566,1111
	+42 00604	NGC 1039 - 358		2 39 23	+42 42.1	1	10.42		0.24		0.16		2	A2	566
		LP 942 - 17		2 39 23	−36 31.0	1	12.58		1.15		1.15		1		1696
16718	+58 00508	IDS02357N5828	AB	2 39 24	+58 40.6	1	7.58		0.15		0.08		2	A1 III	401
		NGC 1039 - 362		2 39 25	+42 37.0	1	11.45		0.41		0.11		2	A8	1111
16931	−51 00634			2 39 25	−51 39.3	2	8.55	.005	0.64	.005	0.04		7	G3 V	1731,2012
16735	+52 00616	HR 787	★ A	2 39 27	+53 18.9	2	5.85	.014	1.13	.005	1.08		3	K0 II-III	71,4001
16727	+54 00598	HR 785		2 39 27	+54 53.6	2	5.74		-0.13	.000	-0.48	.004	3	B7 III(p)	985,1079
		VES 760		2 39 29	+61 34.3	1	10.61		0.09		-0.41		1		1543
16898	−30 00999			2 39 30	−30 31.6	1	9.49		0.52		0.03		2	F6/7 V	1730
	+42 00605	NGC 1039 - 370	★ B	2 39 31	+42 36.7	1	10.27		0.14		0.08		2	A2	1111
16811	+19 00403	HR 793	★ A	2 39 32	+19 48.0	2	5.74		-0.03	.005	-0.05	.020	6	A0 V	1049,1733
16782	+42 00607	NGC 1039 - 374	★ A	2 39 32	+42 36.5	1	8.26		0.01		-0.25		4	B8.5IV	1111
16782	+42 00607	NGC 1039 - 374	★ AB	2 39 32	+42 36.5	1	8.14		0.04		-0.24		2		566
	+42 00606	NGC 1039 - 517		2 39 32	+43 00.0	1	9.19		0.57		0.17		2	G0	566
		NGC 1039 - 376		2 39 33	+42 15.1	1	10.64		0.19		0.18		2	A0	566
	+48 00739	G 78 - 2		2 39 33	+48 45.4	4	9.85	.021	0.95	.008	0.65	.021	11	K8	1003,1620,1775,3072
	+42 00608	NGC 1039 - 380		2 39 34	+42 28.0	1	10.62		0.19		0.14		5	A2	566
		LS I +61 309		2 39 36	+61 50.9	1	11.49		0.36		-0.54		1		1543
		LS I +60 257		2 39 38	+60 51.1	1	10.43		0.75		0.02		2		405
	+41 00522	NGC 1039 - 386		2 39 39	+42 24.0	2	10.30	.012	0.12	.009	0.11	.020	7	A2	566,1111
16780	+47 00683	HR 792	★ AB	2 39 40	+48 03.2	2	6.29	.011	1.14	.005	0.88	.000	4	G5	1601,1733
	+42 00609	NGC 1039 - 391		2 39 43	+42 36.4	2	10.88	.031	0.14	.007	0.12		5	A2	566,1111
16916	−30 01000			2 39 44	−30 37.0	1	7.66		1.13		1.01		2	K0/1 III	1730
	+60 00565	NGC 1027 - 3		2 39 45	+61 18.3	1	10.72		0.48		0.36		1	A0	49
16861	+10 00360	HR 797		2 39 47	+10 31.8	5	6.30	.005	0.06	.005	0.06	.012	15	A2 V	15,1007,1013,1049,8015
		NGC 1027 - 7		2 39 50	+61 20.1	1	11.64		1.12		0.52		1	G5	49
	+12 00378			2 39 53	+12 31.4	1	9.22		0.50		-0.01		2	F8	1375
		NGC 1027 - 23		2 39 54	+61 17.2	1	15.02		0.82		0.27		1		49
	−31 01092			2 39 54	−31 20.3	1	10.52		1.13		1.14		3	K3 V	3072
16779	+57 00620	LS I +57 081		2 39 56	+57 37.0	3	8.85	.005	0.74	.000	-0.26	.000	5	B2 Ib	64,1012,1083
	+2 00418			2 39 58	+03 09.3	2	10.16	.010	1.35	.012	1.27		5	dK7	196,1619
	+60 00558	IDS02361N6057	AB	2 39 58	+61 10.2	1	10.19		0.52				1		592
		NGC 1039 - 423		2 40 02	+42 44.4	1	10.97		0.35		0.17		1	A0	566
		CS 22189 # 11		2 40 02	−14 18.3	1	13.99		0.52		-0.04		1		1736
		NGC 1039 - 428		2 40 03	+42 38.2	1	10.79		0.19		0.14		1	A3	566
		NGC 1068 sq1 4		2 40 03	−00 05.4	1	12.24		1.26		1.32		2		327
	+42 00611	NGC 1039 - 430		2 40 05	+42 34.7	2	10.60	.016	0.55	.008	0.07	.008	8	G5	566,1111
16834	+38 00544			2 40 06	+39 09.3	2	8.65	.010	0.06	.005	0.08	.000	5	A1 V	1501,1601
16778	+59 00535	LS I +59 147		2 40 06	+59 36.7	4	7.71	.019	0.90	.004	0.34	.007	7	A2 Ia	1012,1083,1110,8023
	+46 00616			2 40 07	+46 55.0	1	7.95		0.08		-0.23		2	F8	1601
16975	−38 00894	HR 805		2 40 07	−38 35.8	4	6.00	.009	0.92	.003	0.65		19	G8 III	15,978,1075,2032
	+42 00612	NGC 1039 - 443		2 40 12	+42 33.7	2	9.57	.003	0.05	.004	-0.04	.007	8	A0 V	566,1111
		WLS 240 65 # 10		2 40 12	+65 49.1	1	11.55		0.66		0.03		2		1375
	+66 00228			2 40 14	+67 24.2	1	10.29		1.56		1.28		2		1375
	−24 01209			2 40 14	−23 53.6	1	9.63		0.47		0.10		15	G0	1699
		NGC 1027 - 17		2 40 16	+61 19.0	1	13.57		0.93		0.38		1		49
16856	+42 00613	NGC 1039 - 520		2 40 18	+42 59.7	1	9.22		0.15		0.07		2	A0	566
16857	+42 00615	NGC 1039 - 456		2 40 19	+42 24.6	2	8.90	.003	-0.01	.003	-0.22	.000	13	B9 V	566,1111
16855	+42 00614			2 40 20	+43 19.6	1	6.76		0.25		0.10		2	A2	3016
17006	−47 00832	HR 807		2 40 20	−46 44.1	3	6.10	.005	0.88	.000			14	K1 III	15,1075,2012
		CS 22189 # 17		2 40 22	−15 47.1	1	12.76		0.24		0.08		1		1736
16808	+57 00622	LS I +58 091		2 40 23	+58 06.9	3	8.60	.000	0.56	.000	-0.47	.000	5	B0.5Ib	64,1012,1083
16458	+80 00086	HR 774		2 40 26	+81 14.4	1	5.78		1.30				2	K0	71
		NGC 1068 sq1 1		2 40 26	−00 11.8	1	12.25		0.74		0.43		4		327
16909	+18 00339	G 4 - 35		2 40 30	+19 13.1	2	8.28	.000	1.07	.009	0.96	.009	4	K5	333,1620,3072
16769	+67 00224	HR 791		2 40 30	+67 36.8	1	5.95		0.21		0.17		1	A5 III	592
16908	+27 00424	HR 801		2 40 31	+27 29.7	5	4.66	.012	-0.14	.005	-0.63	.005	15	B3 V	15,154,1223,1363,8015
		NGC 1068 sq1 3		2 40 32	−00 06.5	1	11.27		0.60		0.10		3		327
16832	+56 00703	LS I +56 065		2 40 33	+56 26.8	1	8.85		0.40		-0.58		2	B0p	1012
		G 75 - 37		2 40 34	+05 15.9	1	14.07		1.34		0.84		5		203
		G 4 - 36		2 40 37	+13 13.3	3	11.48	.013	0.48	.004	-0.22	.004	6		333,927,1620,1696
16970	+2 00422	HR 804	★ AB	2 40 42	+03 01.6	12	3.46	.009	0.09	.006	0.07	.008	44	A3 V	15,1007,1013,1020*
	−0 00414			2 40 43	−00 07.5	2	8.99	.015	1.55	.005	1.86	.025	5	M0	196,327
16944	+23 00362			2 40 45	+23 51.7	1	7.58		1.47		1.65		2	K0	1625
16895	+48 00746	HR 799	★ AB	2 40 46	+49 01.1	10	4.11	.017	0.49	.006	-0.01	.009	32	F7 V	1,15,254,1077,1197,1355*
16901	+43 00566	HR 800		2 40 49	+44 05.2	1	5.43		0.90		0.65		6	G0 Ib-II	1355
		DF Cas		2 40 51	+61 16.7	1	10.55		0.95				1	F8	592
17051	−51 00641	HR 810		2 40 51	−51 00.9	5	5.40	.011	0.56	.008	0.07	.008	16	G0 V	15,243,258,678,2012
17255	−74 00226			2 40 52	−74 32.2	1	8.09		1.13		1.10		7	K1 III	1704
	+27 00426			2 40 53	+27 59.5	1	9.00		1.00		0.72		2		3072
16955	+25 00441	HR 803	★ AB	2 40 57	+25 25.6	2	6.36		0.09	.015	0.10	.005	6	A3 V	105,1049
		CS 22184 # 33		2 40 57	−12 18.1	1	13.78		0.32		0.00		1		1736
17070	−50 00800			2 40 59	−49 56.5	1	7.62		0.43		0.00		4	F5 V	243
17215	−71 00156	IDS02406S7153	AB	2 41 02	−71 40.4	1	7.76		0.72				4	G6 V	2012

HD	DM	Other Id	N	Rem	α_{1950}	δ_{1950}	S	V	σ_V	B–V	σ_{B-V}	U–B	σ_{U-B}	n	Spectrum	References
17000	−0 00415				2 41 04	−00 33.3	1	8.97		0.51		0.00		3	G0	196
16900	+55 00701				2 41 09	+55 31.7	1	8.98		0.44		0.21		2	A2	1502
	−0 00418				2 41 11	−00 00.0	1	9.24		1.19		1.27		4	K5	327
	+57 00626	LS I +57 086			2 41 14	+57 26.5	1	9.95		0.65		−0.29		2	B1 Ib	1012
16831	+68 00190				2 41 16	+68 25.2	1	8.96		0.33		0.14		1	A5	634
		G 74 - 42			2 41 17	+36 40.7	1	11.13		1.36		1.32		3	k7	7010
		G 36 - 31		⋆ V	2 41 18	+25 19.1	2	10.60	.009	1.55	.000	1.11		3	m2	680,1017
236995	+57 00629	LS I +58 092			2 41 19	+58 20.4	1	8.64		0.52				1	A0 Ia	138
17126	−52 00330				2 41 21	−51 53.1	1	8.71		1.10				4	K1 III	2013
16942	+54 00602				2 41 23	+54 45.2	1	8.99		0.21		−0.09		2	A0	1733
17098	−41 00769	HR 814		⋆ AB	2 41 24	−40 44.4	2	6.35	.005	−0.02	.000			7	B9.5V	15,2012
		G 36 - 32			2 41 25	+27 26.8	1	13.71		0.86		0.22		2		1658
17084	−38 00899	UX For			2 41 26	−38 08.3	1	8.06		0.75		0.23		1	G5/8 V +(G)	1641
17017	+17 00426	HR 808			2 41 31	+17 33.2	1	6.40		1.04		1.10		3	K2 III	3077
		LP 197 - 44			2 41 36	+42 18.9	1	11.10		0.94		0.70		3		1726
17007	+28 00455	IDS02387N2902	AB		2 41 38	+29 15.0	1	7.13		0.37		−0.01		3	F0	938
		MCC 401			2 41 39	+10 45.0	2	11.08	.058	1.40	.000			3	K5	1017,1619
16893	+67 00226	IDS02374N6723	A		2 41 39	+67 35.9	1	8.53		0.39		0.37		8	A2	1524
16893	+67 00226	IDS02374N6723	B		2 41 39	+67 35.9	1	11.67		0.80		0.33		2		1524
	+62 00457	LS I +62 224			2 41 40	+62 34.4	1	9.69		0.55		−0.60		2		405
		VES 762			2 41 42	+57 49.6	1	12.66		0.62		0.31		1		1543
		LP 651 - 4			2 41 42	−05 39.5	1	12.04		0.39		−0.23		2		1696
		Wolf 1132			2 41 42	−05 39.5	1	11.92		1.46		1.07		1	M2	3079
17081	−14 00519	HR 811			2 41 44	−14 04.2	9	4.24	.006	−0.14	.005	−0.42	.025	33	B7 IV	3,15,1034,1068,1075*
17155	−46 00790				2 41 46	−46 39.5	3	9.04	.012	1.04	.009	0.90		10	K4 V	1075,2012,3072
17036	+14 00457	HR 809			2 41 47	+15 06.1	2			−0.01	.000	−0.20	.009	5	B9 Vn	1049,1079
		BPM 16907			2 41 48	−53 27.	1	14.38		0.47		−0.27		2		3065
		CS 22189 # 18			2 41 51	−16 00.1	1	15.33		0.36		−0.23		1		1736
		G 4 - 37			2 41 55	+08 16.3	4	11.42	.015	0.47	.004	−0.20	.016	4		516,1620,1658,3077
		G 37 - 3			2 41 55	+24 37.8	1	10.78		1.00		0.74		3	K0	7010
	−22 00926				2 41 57	−22 14.6	1	10.38		0.51		−0.02		2		742,1594
17134	−26 00996	IDS02398S2555	APB		2 42 01	−25 42.4	1	6.83		0.66				4	G3 V	2007
17144	−25 01083	UY For			2 42 08	−25 03.7	2	8.26	.007	1.19	.004	1.00	.019	10	K1 III	1641,1738
		CS 22189 # 27			2 42 11	−14 15.6	1	14.11		0.42		−0.22		1		1736
17094	+9 00359	HR 813			2 42 14	+09 54.3	15	4.26	.009	0.31	.005	0.08	.023	62	F1 III-IV	1,3,15,39,254,1020*
17093	+11 00377	HR 812, UV Ari			2 42 14	+12 14.2	3	5.19	.017	0.23	.012	0.10	.034	12	A7 III-IV	3,1363,3058
17022	+49 00761	LS V +49 001			2 42 15	+49 59.6	1	9.13		0.09		−0.66		2	F8	401
	−21 00492				2 42 15	−21 03.8	1	10.93		0.44		−0.03		2	F4	1026
17152	−24 01225				2 42 15	−24 37.4	2	8.39	.002	0.77	.001	0.39	.002	28	G8 V	978,1738
17168	−33 00943	HR 817			2 42 15	−32 44.1	2	6.21	.005	0.04	.000			7	A1 V	15,2027
17233	−55 00454				2 42 20	−55 00.2	2	9.04	.005	0.79	.000	0.32		3	G8wG0 IV/V	742,1594
		G 78 - 4			2 42 21	+44 44.4	1	10.81		1.40		1.27		2	K6	7010
		WLS 240 10 # 9			2 42 22	+10 03.7	1	10.98		0.74		0.24		2		1375
	−0 00423	IDS02399N0000	AB		2 42 28	+00 13.1	1	9.84		0.96		0.48		5	K0	3077
17254	−53 00475	HR 821			2 42 34	−52 46.9	2	6.14	.005	0.09	.000			7	A2 V	15,2012
17326	−67 00181	HR 823		⋆ AB	2 42 34	−66 55.5	4	6.25	.004	0.53	.000	−0.01	.009	21	F7 V	15,1075,2038,3077
17119	+29 00461				2 42 36	+30 06.6	1	8.50		0.40				4	F5 V	20
		CS 22184 # 42			2 42 37	−09 00.9	2	14.38	.000	0.21	.000	0.04	.008	2		966,1736
	+4 00436	G 76 - 28			2 42 38	+05 22.1	1	10.56		1.11				1	K5	1632
		CS 22189 # 19			2 42 42	−17 25.1	1	15.33		−0.08		−0.92		1		1736
17163	+4 00437	HR 816			2 42 43	+04 30.1	4	6.05	.032	0.31	.016	0.07	.010	10	F0 IV-V	15,39,254,1256
		GD 33			2 42 44	+39 24.1	1	16.77		0.19		−0.62		1		3060
17206	−19 00518	HR 818			2 42 46	−18 47.0	9	4.46	.007	0.48	.011	0.00	.015	30	F6 V	3,15,58,1075,1197*
17288	−60 00210				2 42 50	−60 16.0	4	9.83	.027	0.57	.008	−0.09	.007	13	F7 V	1097,1696,2013,3077
17176	+16 00342				2 42 56	+16 48.7	1	7.22		1.38		1.42		3	K0	1648
17088	+57 00632	LS I +57 089			2 43 08	+57 31.5	3	7.50	.000	0.82	.000	0.05	.000	5	B9 Ia	64,1012,1083
17242	−25 01093				2 43 09	−24 51.7	1	8.24		0.41		−0.02		39	F3 V	978
17277	−41 00777				2 43 09	−41 30.3	1	7.95		1.42				4	K4/5 III	2012
17086	+59 00541	LS I +60 262			2 43 18	+60 21.7	1	6.50		0.52				2	F0 V	6009
17190	+25 00449	G 37 - 4			2 43 20	+25 26.5	3	7.87	.032	0.84	.008	0.46	.023	6	K1 IV	680,1620,7008
17252	−17 00529				2 43 26	−16 40.1	1	9.66		0.79				4	K1 V	2033
	−41 00783				2 43 29	−41 09.7	1	11.07		0.93		0.80		2		1696
17325	−46 00797	HR 822			2 43 29	−46 29.8	3	6.84	.008	1.36	.000			11	K2 III	15,1075,2035
		CS 22189 # 33			2 43 30	−12 54.9	1	15.00		−0.15		−1.00		1		1736
	−42 00917				2 43 32	−41 46.2	2	10.10	.005	0.77	.002	0.34	.000	5	K0 V	1097,3008
17230	+11 00383	G 4 - 41			2 43 34	+11 34.1	3	8.60	.026	1.28	.004	1.24	.005	5	K6-7	196,1620,1705
17114	+58 00518	LS I +59 149			2 43 35	+59 04.7	3	9.16	.004	0.50	.000	−0.45	.000	16	B1 V	41,64,1012
		CS 22184 # 47			2 43 36	−08 03.6	1	14.58		0.28		0.07		1		1736
		LP 651 - 7			2 43 39	−05 10.3	1	15.86		1.50				3		940
17145	+57 00634	LS I +57 092			2 43 41	+57 28.1	4	8.15	.013	0.82	.008	−0.01	.004	7	B8 Ia	1012,1083,1110,8023
17322	−38 00910				2 43 41	−38 22.3	1	7.64		0.58		0.05		54	G0 V	978
		G 246 - 10			2 43 43	+64 04.7	1	10.76		1.00		0.83		2	K3	7010
	+64 00348			V	2 43 46	+64 32.9	1	10.60		2.28				3		369
		CoD -58 00562			2 43 51	−58 35.6	1	8.85		1.04		0.91		3		3062
17228	+35 00553	HR 819			2 43 52	+35 46.5	3	6.26	.026	0.94	.004	0.64	.020	7	G8 III	71,1733,3051
		L 127 - 50			2 43 54	−61 32.	1	15.20		0.95		0.27		2		3062
		L 53 - 75			2 43 54	−74 13.	1	10.49		1.28		1.31		1		1696
17320	−25 01099				2 43 55	−25 32.6	1	7.36		1.13		1.07		3	K1 III	1657
	+57 00635	LS I +57 093			2 43 57	+57 53.4	1	10.58		0.65		−0.25		3		41
17240	+34 00513	HR 820		⋆ AB	2 43 58	+35 20.8	1	6.25		0.34				3	A9 V	1733

Table 1   151

HD	DM	Other Id	N Rem	α_{1950}	δ_{1950}	S	V	σ_V	B–V	σ_{B-V}	U–B	σ_{U-B}	n	Spectrum	References
		G 75 - 39		2 43 58	−02 39.7	1	15.53		0.36		−0.51		3		538
		L 297 - 59		2 44 00	−47 19.	1	13.40		0.63		−0.07		2		1696
		vdB 9 # 13		2 44 02	+68 50.8	1	10.17		0.74				1	F2 IV	207
17218	+44 00576	LS V +45 001		2 44 03	+45 03.7	1	9.31		−0.07		−0.76		1		555
		CS 22189 # 29		2 44 06	−14 09.5	1	14.20		0.38		0.04		1		1736
		CS 22189 # 28		2 44 06	−14 18.1	1	14.00		−0.13		−0.54		1		1736
		LS I +57 094		2 44 08	+57 29.1	1	11.66		0.83		−0.13		2		41
	−9 00528			2 44 15	−08 42.9	1	11.62		0.27		0.03		1		1736
	+68 00193			2 44 16	+68 39.6	1	9.40		0.33				1	A0	207
17393	−35 00954			2 44 17	−35 07.1	1	8.32		0.16		0.08		2	A0/1 IV	1730
		CS 22189 # 24		2 44 22	−15 38.5	1	11.78		0.13		0.16		1		1736
17504	−64 00196	HR 833	★A	2 44 23	−63 54.8	1	5.74		0.93		0.63		2	G8 III/IV	58
		VES 763		2 44 24	+61 37.3	1	12.28		0.65		−0.52		1		1543
		G 246 - 11		2 44 25	+62 41.3	1	10.71		0.60		−0.01		1	G0	1658
		NGC 1097 sq1 3		2 44 27	−30 24.5	1	13.04		1.23		1.15		4		1687
		NGC 1097 sq1 5		2 44 28	−30 33.3	1	13.71		0.68		0.14		4		1687
		LS I +58 093		2 44 29	+58 49.0	1	10.55		0.42		−0.40		3		41
17390	−22 00479	HR 826		2 44 29	−21 50.9	2	6.48	.005	0.39	.005			7	F3 IV/V	15,2029
		CS 22189 # 21		2 44 30	−17 02.5	1	12.54		0.17		0.10		1		1736
	−10 00548			2 44 31	−10 20.3	2	10.41	.015	0.78	.024	0.15		3		742,1594
		NGC 1097 sq1 4		2 44 35	−30 35.5	1	13.27		0.64		0.09		3		1687
17332	+18 00347	IDS02418N1857	AB	2 44 38	+19 09.9	1	6.87		0.69		0.11		3	G0	938
17364	+8 00424			2 44 42	+09 06.0	1	7.68		1.34		1.60		3	K0	1733
		LS I +56 068		2 44 46	+56 25.3	1	12.54		0.68		−0.11		3		41
17566	−68 00169	HR 837		2 44 46	−67 49.6	4	4.83	.003	0.05	.005	0.08	.005	20	A2 IV/V	15,1075,2038,3023
	+68 00194			2 44 47	+68 37.7	1	10.03		0.40				1	A1	207
		NGC 1097 sq1 1		2 44 49	−30 27.8	1	11.45		0.65		0.17		4		1687
		NGC 1097 sq1 2		2 44 50	−30 29.3	1	11.51		0.73		0.30		3		1687
	+6 00422	G 76 - 30		2 44 52	+06 31.4	1	10.18		0.76		0.22		1	K0	1620
		TX Per		2 44 53	+36 45.5	1	10.33		1.54		1.48		1		793
17361	+28 00462	HR 824		2 44 55	+29 02.4	16	4.51	.007	1.11	.006	1.07	.009	127	K1 III	15,61,71,985,1355,1363*
17438	−23 01061	HR 827		2 44 56	−22 41.6	2	6.46	.005	0.40	.005			7	F2 V	15,2029
17477	−37 01043			2 44 57	−37 33.2	1	7.94		1.55		1.87		5	K5 III	1673
		LP 298 - 33		2 44 58	+28 30.2	1	11.07		1.46		1.19		4		1723
	+59 00546	LS I +59 150		2 44 59	+59 56.9	1	10.19		0.63		−0.44		3		41
		RL 65		2 45 00	+61 42.	1	12.68		0.39		−0.56		2		704
	+68 00195			2 45 00	+68 35.4	1	10.19		0.32				1	B9	207
	−35 00962			2 45 01	−34 57.2	1	11.36		0.97		0.80		2		1730
		LS I +59 151		2 45 02	+59 12.7	1	11.32		0.54		−0.42		3		41
		LS I +59 152		2 45 02	+59 17.8	1	11.97		0.64		0.00		3		41
17483	−38 00917			2 45 04	−37 56.9	1	8.33		0.43		−0.02		40	F5 V	978
		LS I +56 069		2 45 05	+56 15.5	1	12.43		0.62		−0.24		3		41
		LS I +57 095		2 45 08	+57 24.3	1	11.45		0.69		−0.24		3		41
		CS 22189 # 31		2 45 09	−13 51.0	1	13.67		0.26		0.06		1		1736
17306	+53 00574	LS V +53 003		2 45 10	+53 57.7	1	7.93		1.30				1	K3 Iab	138
17382	+26 00465	G 36 - 33	★A	2 45 12	+26 51.8	2	7.61	.010	0.82	.007	0.42	.019	3	K1 V	333,1620,3016
17414	+14 00469			2 45 13	+15 18.0	1	6.67		1.37		1.53		3	K0	1648
17653	−71 00165			2 45 13	−71 26.7	1	6.67		0.45		−0.04		6	F6 V	1628
		VES 764		2 45 14	+58 13.7	1	11.69		0.61		0.01		1		1543
17179	+72 00145	IDS02404N7230	A	2 45 14	+72 42.4	1	7.85		0.24		−0.45		2	B0.5V	555
17494	−35 00964			2 45 14	−34 59.8	1	9.06		1.14		1.13		2	K2 III	1730
17396	+29 00473			2 45 19	+30 08.2	1	8.50		0.60				3	G0 V	20
		CS 22184 # 40		2 45 20	−09 24.8	1	13.63		0.60		−0.06		1		1736
	+7 00429			2 45 24	+07 43.0	1	10.19		0.50		0.00		2	F5	1375
237006	+57 00641	LS I +57 096		2 45 24	+57 48.4	5	9.17	.015	1.62	.021	0.09	.035	12	M3	41,138,196,369,405
17359	+48 00762	IDS02420N4846	A	2 45 25	+48 58.6	2	7.56	.002	0.04	.012	0.06	.106	7	B9	401,595
		0245+18 # 1		2 45 29	+18 02.	1	15.61		0.81		0.31		2		551
		0245+18 # 2		2 45 29	+18 02.	1	15.91		1.56		1.57		2		551
		0245+18 # 4		2 45 29	+18 02.	1	15.32		0.68		0.02		2		551
		0245+18 # 5		2 45 29	+18 02.	1	13.24		0.69		0.16		2		551
		0245+18 # 6		2 45 29	+18 02.	1	14.37		1.21		0.79		2		551
17435	+19 00424			2 45 30	+19 48.2	2	8.38	.005	0.21	.000	0.10	.000	5	A3	1648,1733
17346	+56 00717			2 45 32	+56 49.6	1	6.82		1.26		1.06		1	G9 II	272
17491	−13 00530	HR 832, Z Eri		2 45 32	−12 40.1	2	6.81	.135	1.59	.041	1.52		5	M5 III	2035,3076
17528	−36 01050	HR 835		2 45 32	−35 45.5	1	6.51		0.96				4	K0-III	2032
	+17 00441			2 45 36	+17 57.4	1	9.04		1.15		0.83		2	K0	551
		TonS 280		2 45 36	−29 31.	1	13.95		0.00		−0.02		1		832
	−36 01052			2 45 36	−36 18.9	1	9.94		0.46				2	F2	1594
17433	+30 00448	VY Ari		2 45 42	+30 54.6	1	6.76		0.96		0.62		3	K0	196
		LS I +58 094		2 45 42	+58 52.6	1	11.61		0.45		−0.32		3		41
	−12 00525			2 45 42	−11 57.9	1	10.78		1.42		1.20		4	M0	158
17459	+17 00442	HR 828		2 45 44	+18 04.6	2	5.85	.023	1.13	.070	1.12		4	K1 III	71,3051
	+22 00396	G 5 - 1		2 45 46	+22 23.8	3	10.08	.017	0.58	.008	−0.06	.010	4	G4	1620,1658,3029
		LS I +57 097		2 45 46	+57 52.4	1	12.37		0.65		−0.09		3		41
17378	+56 00718	HR 825, V480 Per		2 45 48	+56 52.6	10	6.25	.009	0.89	.008	0.51	.016	63	A5 Ia	1,15,41,1012,1083*
		AJ79,864 # 1		2 45 48	+58 14.	1	13.02		0.71		−0.41		3	O5	373
		0246+18 # 4		2 45 51	+18 07.4	1	11.95		0.46		0.12		2		551
17471	+24 00396	HR 830, VZ Ari		2 45 51	+24 58.8	2	5.89		−0.05	.012	−0.13	.024	8	A0 V	1049,1733
17327	+64 00351	LS I +64 094	★A	2 45 51	+64 25.4	1	7.53		0.36		−0.03		7	B8 II	1524
17327	+64 00351	LS I +64 094	★AB	2 45 51	+64 25.4	1	7.43		0.34		−0.07		3	B8 II	24

HD	DM	Other Id	N	Rem	α_{1950}	δ_{1950}	S	V	σ_V	B–V	σ_{B-V}	U–B	σ_{U-B}	n	Spectrum	References
17327	+64 00351	IDS02418N6413	B		2 45 51	+64 25.4	1	10.33		0.51				3		1524
		TonS 281			2 45 54	−30 16.	1	13.68		0.09		0.08		1		832
		LP 711 - 8			2 45 56	−12 57.6	1	13.67		0.65		−0.07		2		1696
232715	+51 00633				2 45 57	+51 26.2	1	8.91		1.09		0.87		2	K2	1375
		0246+18 # 5			2 45 58	+18 04.5	1	10.76		0.32		0.33		2		551
		0246+18 # 7			2 45 59	+18 07.8	1	13.26		0.94		0.53		2		551
		0246+18 # 6			2 46 05	+18 03.3	1	11.60		0.91		0.47		2		551
232716	+53 00576	LS V +53 004		★ AB	2 46 07	+53 43.2	1	8.90		0.41		−0.22		2	B3	1566
17576	−37 01050	IDS02441S3724	A		2 46 07	−37 11.4	1	7.89		0.56		−0.20		18	G0 V	1732
		LS I +59 153			2 46 08	+59 29.6	2	11.07	.042	0.62	.000	−0.36	.009	4	O9.5V	41,342
		LS I +61 313			2 46 16	+61 53.1	1	9.87		0.68		−0.46		2		405
		LS I +58 095			2 46 18	+58 29.3	2	12.29	.029	0.48	.029	−0.31	.010	6		41,373
		0246+18 # 3			2 46 19	+18 11.7	1	13.72		0.74		0.16		2		551
		0246+18 # 2			2 46 19	+18 12.0	1	12.12		0.71		0.19		2		551
17484	+36 00566	HR 831			2 46 19	+37 07.2	1	6.47		0.42		0.12		2	F6 III-IV	254
		0246+18 # 1			2 46 24	+18 07.	1	15.63		0.88		0.44		2		551
17543	+16 00355	HR 836		★ AB	2 46 30	+17 15.5	3	5.25	.061	−0.07	.027	−0.49	.011	9	B6 V	154,1203,3024
		CS 22189 # 38			2 46 30	−14 15.8	1	13.74		0.24		0.02		1		1736
17559	−4 00476				2 46 32	−04 25.9	1	6.91		1.62		1.96		3	K5	1657
17597	−19 00533	IDS02443S1927	AB		2 46 35	−19 14.1	1	8.48		1.57		1.81		4	M3 III	3007
		LS I +56 071			2 46 36	+56 52.3	1	10.73		0.69		−0.26		3		41
		L 298 - 26			2 46 36	−46 42.	2	13.70	.035	0.91	.000	0.36	.074	5		1696,3062
		LP 994 - 2			2 46 37	−43 05.0	1	12.90		1.25		1.15		1		1696
237007	+59 00549	IC 1848 - 3			2 46 43	+60 11.7	3	9.43	.020	0.33	.005	−0.60	.000	4	B0 V	49,64,1012
	+68 00199				2 46 45	+68 37.5	2	8.07	.097	1.48	.015	1.59		3	K5	207,1524
		ApJ283,254 # 9			2 46 49	+68 39.7	1	13.48		0.82		0.28		2		1524
		IC 1848 - 10			2 46 50	+60 09.4	1	13.05		0.81		0.15		1		49
		CS 22189 # 39			2 46 51	−15 18.1	1	14.36		0.24		0.02		1		1736
		CS 22184 # 45			2 46 52	−08 36.5	2	14.64	.000	0.03	.000	0.09	.011	2		966,1736
		LS I +55 039			2 46 54	+55 41.3	1	11.31		0.60		−0.24		2		41
		VES 765			2 46 54	+60 14.7	1	12.43		0.62		0.06		1		1543
		LS I +55 040			2 46 55	+55 52.4	1	12.08		0.84		−0.08		2		41
237008	+56 00724	W Per			2 46 55	+56 46.6	2	9.62	.000	2.62	.000	2.56	.000	2	M7	8016,8027
		IC 1848 - 17			2 46 55	+60 10.6	1	14.50		0.55		0.12		1		49
		LS I +57 098		★ V	2 46 56	+57 07.8	1	11.43		1.18		−0.12		2	B0Iab:e	41
237009	+55 00712	IDS02434N5529	B		2 46 59	+55 41.6	2	8.50	.004	0.12	.007	−0.07	.004	15	A0	150,1084
17652	−32 01025	HR 841		★ AB	2 47 00	−32 36.9	4	4.46	.005	0.99	.005	0.69	.004	20	G8 IIIb	15,1075,2012,8015
		LS I +58 096			2 47 01	+58 35.5	1	12.22		0.51		0.11		2		41
17573	+26 00471	HR 838		★ A	2 47 02	+27 03.3	7	3.61	.016	−0.10	.008	−0.36	.019	17	B8 Vn	15,1049,1079,1363*
17573	+26 00471	IDS02441N2651	B		2 47 02	+27 03.3	2	10.25	.054	0.32	.029	−0.84	.024	3		8072,8084
17573	+26 00471	IDS02441N2651	C		2 47 02	+27 03.3	1	10.35		0.32		−0.73		1		8084
17573	+26 00471	IDS02441N2651	D		2 47 02	+27 03.3	1	8.60		1.10		1.26		1		8084
17506	+55 00714	HR 834		★ A	2 47 02	+55 41.4	10	3.77	.020	1.69	.007	1.88	.017	31	M3-Ib-IIa	15,150,1084,1119,1355*
		IC 1848 - 18			2 47 02	+60 10.7	1	14.60		0.78		0.22		1		49
17616	+0 00469				2 47 03	+00 42.9	1	6.79		1.02		0.87		3	K2 V	985
		IC 1848 - 7			2 47 03	+60 12.3	1	12.31		0.57		0.33		1		49
		LS I +57 099			2 47 04	+57 32.0	1	11.36		0.72		−0.03		2		41
		4HLF4 # 1			2 47 06	+00 52.	1	11.89		0.35		0.08		3	A3	208
		LS I +55 041			2 47 07	+55 25.4	1	11.11		0.67		−0.29		2		41
17443	+67 00230				2 47 09	+67 36.6	2	8.74	.000	0.30	.010	0.15	.019	12	B9 V	206,1524
17715	−45 00913				2 47 10	−45 17.8	1	6.78		1.01				4	K0 III	2033
		IC 1848 - 19			2 47 12	+60 11.1	1	15.35		0.81		0.21		1		49
		CS 22184 # 38			2 47 13	−10 30.4	1	12.91		0.41		−0.16		1		1736
17505	+59 00552	IC 1848 - 1		★ AB	2 47 15	+60 12.7	5	7.07	.014	0.40	.004	−0.64	.007	8	O6.5V((f))	49,64,1011,1012,1083
		G 36 - 35			2 47 17	+27 54.4	1	11.94		1.23		1.20		1	K7	1658
		WLS 240 65 # 9			2 47 17	+64 38.5	1	11.72		0.48		0.35		2		1375
237010	+57 00647	V648 Cas			2 47 19	+57 39.0	5	9.27	.082	2.74	.019	2.67		14	M2 Ia	138,148,196,369,8032
		IC 1848 - 15			2 47 19	+60 13.9	1	13.84		0.47		−0.05		1		49
		ApJ283,254 # 1			2 47 19	+68 40.6	1	11.03		1.56		1.37		2		1524
		BK Eri			2 47 20	−01 37.5	1	12.10		0.17		−0.04		1		699
		IC 1848 - 12			2 47 21	+60 12.5	1	13.22		0.48		−0.20		1		49
		IC 1848 - 8			2 47 21	+60 14.3	1	12.71		0.49		−0.15		1		49
		CS 22189 # 36			2 47 21	−13 20.0	1	14.38		0.43		−0.24		1		1736
		IC 1848 - 13			2 47 22	+60 10.1	1	13.38		0.49		0.23		1		49
17520	+59 00553	IC 1848 - 2		★ AB	2 47 22	+60 10.3	8	8.26	.021	0.32	.006	−0.68	.012	12	O9 V	49,64,1011,1012,1083*
		IC 1848 - 14			2 47 23	+60 10.3	1	13.60		0.47		−0.04		1		49
17606	+31 00490				2 47 24	+31 46.0	1	6.61		0.03		−0.01		2	A0	252
17584	+37 00646	HR 840		★ A	2 47 25	+38 06.8	10	4.22	.006	0.34	.007	0.08	.013	29	F2 III	1,15,39,254,1004,1077*
17639	+13 00456				2 47 28	+13 30.3	1	8.14		0.44		−0.01		2	F5	1648
17605	+36 00569				2 47 28	+36 44.6	1	6.44		0.61		0.14		2	G0	1601
		IC 1848 - 9			2 47 28	+60 10.1	1	13.02		0.35		−0.33		1		49
17463	+68 00200	HR 829, SU Cas			2 47 29	+68 41.0	3	5.78	.020	0.64	.018	0.48		4	F5:Ib-II	71,592,6007
		CS 22184 # 34			2 47 29	−12 01.6	1	12.80		0.21		0.13		1		1736
		ApJ283,254 # 2			2 47 32	+68 38.3	1	12.60		0.74		0.27		4		1524
		vdB 9 # 16			2 47 34	+68 36.1	2	11.02	.052	0.47	.000	0.37		4	A2 V	207,1524
		IC 1848 - 11			2 47 37	+60 13.5	1	13.17		0.43		−0.05		1		49
17713	−25 01120	HR 844		★ AB	2 47 38	−24 45.9	3	6.13	.004	1.08	.005			11	K0 III	15,1075,2032
	−31 01135				2 47 38	−31 07.6	2	10.29	.015	0.38	.005	−0.02		6	F2	1696,2034
		ApJ283,254 # 4			2 47 40	+68 34.9	1	12.33		1.60		1.10		2		1524
		IC 1848 - 16			2 47 42	+60 13.4	1	13.92		0.48		0.03		1		49

Table 1   153

HD	DM	Other Id	N	Rem	α_{1950}	δ_{1950}	S	V	σ_V	B–V	σ_{B-V}	U–B	σ_{U-B}	n	Spectrum	References
17663	+10 00374				2 47 44	+10 25.4	1	9.30		0.89		0.52		3	G5	3016
17729	−28 00903	HR 845			2 47 44	−28 08.9	2	5.38	.005	0.02	.000			7	A1 V	15,2027
	+56 00727	LS I +56 072			2 47 46	+56 44.6	1	10.72		0.77		-0.36		2		41
		LS I +57 100			2 47 48	+57 42.3	1	11.86		0.82		-0.06		3		41
17660	+15 00395	G 4 - 43			2 47 49	+15 30.6	4	8.89	.013	1.15	.043	1.20	.053	10	K5	1705,1746,3072,7008
		IC 1848 - 6			2 47 51	+60 12.6	1	12.03		0.39		-0.24		1		49
		WLS 248 50 # 10			2 47 52	+48 55.8	1	11.27		1.29		1.08		2		1375
17592	+51 00639				2 47 52	+52 13.0	1	8.50		1.19		0.85		2	K2	1566
		IC 1848 - 5			2 47 55	+60 10.2	1	11.48		0.42		-0.37		1	B5 V	49
17848	−63 00188	HR 852			2 47 55	−63 00.8	3	5.25	.005	0.10	.000	0.06		8	A2 V	15,2012,3050
		VES 766			2 47 56	+60 27.3	1	12.17		0.63		-0.25		2		1543
		Steph 303			2 47 57	+15 42.0	1	9.91		1.73		2.07		2	M0	1746
		vdB 9 # 19			2 47 57	+68 34.7	2	10.67	.045	0.74	.014	0.07		5	B5 III	207,1524
		ApJ283,254 # 7			2 47 58	+68 43.3	1	11.99		0.75		0.65		4	A5 V	1524
17581	+57 00651	HR 839		⋆ A	2 47 59	+58 06.6	3	6.46	.004	0.10	.006	0.09	.012	8	A1 m	70,3058,8071
17674	+29 00484		A		2 48 04	+30 04.9	2	7.54	.019	0.59	.005	0.05		3	G0 V	20,3016
17674	+29 00484		B		2 48 04	+30 04.9	1	9.79		1.28		1.16		1		3032
17603	+56 00728	LS I +56 073			2 48 05	+56 50.6	5	8.45	.005	0.64	.003	-0.42	.005	10	O8.5I(f)	41,64,1011,1012,1083
		AJ79,864 # 3			2 48 06	+58 17.	1	13.53		0.49		-0.26		2	B25	373
17673	+29 00485				2 48 07	+30 18.3	1	7.63		1.20				2	K1 III	20
	+59 00556	IC 1848 - 4			2 48 08	+60 13.1	1	10.06		0.36		-0.12		1	B9 V	49
		LS I +56 074		⋆ V	2 48 09	+56 42.6	1	11.47		0.58		-0.26		3		41
17793	−36 01067	HR 848		⋆ AB	2 48 13	−36 03.0	2	5.91	.000	0.90	.000	0.61		6	K0 III	404,2032
		LS I +56 075			2 48 15	+56 49.5	1	11.78		0.64		-0.17		3		41
		LS I +58 097			2 48 15	+58 46.2	1	10.11		0.47		-0.44		3		41
	+62 00480	LS I +62 229		⋆ V	2 48 15	+62 34.6	1	9.25		0.35		-0.37		2	B1 V	41
17647	+45 00669	IDS02449N4535	AB		2 48 17	+45 46.7	3	8.66	.019	0.70	.007	0.14	.030	7	K2 V	22,1003,3016
17656	+46 00648	HR 842			2 48 20	+46 38.2	1	5.88		0.89		0.58		2	G8 III	252
		LB 1642			2 48 24	−45 20.	1	12.67		0.11		0.09		2		45
17709	+34 00527	HR 843			2 48 26	+34 51.3	7	4.54	.012	1.56	.013	1.92	.008	18	K7 III	15,369,1355,1363,3001*
		LS I +56 076			2 48 27	+56 34.5	1	12.48		0.70		-0.10		3		41
17638	+56 00731	LS I +56 077			2 48 29	+56 43.8	2	10.39	.040	0.46	.031	0.13	.013	5	C6	41,1359
		ApJ283,254 # 6			2 48 37	+68 40.8	1	12.51		0.85		0.71		2		1524
17829	−36 01070	HR 851			2 48 39	−35 52.8	2	5.47	.005	1.25	.005	1.31		6	K2 III	2032,3035
		Steph 306			2 48 40	+13 11.0	1	11.17		1.38		1.13		2	K7	1746
17780	+1 00502	IDS02461N0117	AB		2 48 42	+01 29.8	1	7.93		0.25		0.07		2	A2 Ib	208
17769	+14 00480	HR 847			2 48 44	+14 52.6	2	5.48		-0.09	.002	-0.44	.012	4	B7 V	154,1203
		LS I +61 314			2 48 46	+61 25.2	1	11.42		0.39		-0.47		3		41
17824	−21 00509	HR 850		⋆ A	2 48 46	−21 12.5	5	4.76	.010	0.91	.009	0.62	.014	16	K0 III	15,1075,2012,3016,8015
		LP 830 - 51			2 48 47	−22 24.6	1	13.59		0.66		-0.04		2		1696
		LS I +56 078			2 48 50	+56 49.7	1	12.19		0.60		-0.32		3		41
237015	+59 00558	LS I +60 277			2 48 51	+60 11.2	1	9.44		0.24		-0.43		3	B6 V	41
17864	−40 00736	HR 853			2 48 52	−40 08.2	3	6.35	.007	0.05	.000			11	A0 V	15,2013,2029
17865	−44 00863				2 48 52	−44 17.0	2	8.18	.000	0.55	.000			8	F8 V	1075,2012
	+55 00716	LS I +56 079			2 48 54	+56 21.1	1	9.73		1.03		0.03		2		41
	+56 00733	LS I +57 101			2 48 55	+57 15.5	1	9.96		1.07		0.09		3		41
	+33 00529	G 37 - 8			2 48 56	+34 12.3	3	9.58	.027	1.33	.017	1.04	.023	5	K6:	196,979,1691
	+68 00201				2 48 56	+68 32.8	1	9.65		0.21				1	B8	207
	−2 00502				2 48 57	−02 35.5	1	11.02		1.37				2	K6	1619
	+10 00378	G 76 - 33			2 49 02	+10 27.5	1	10.01		1.19		1.13		1	M0	1620
17795	+15 00397				2 49 05	+16 17.6	1	8.13		1.67		1.71		3	K5	1648
		4HLF4 # 2			2 49 06	+01 03.	1	10.46		0.41		0.19		2	A7	208
		G 75 - 44			2 49 11	+06 01.5	1	13.19		1.58				3		333,1532
	+68 00202				2 49 13	+68 53.9	1	8.87		0.66				1	G5	207
17744	+48 00783				2 49 14	+48 36.7	1	7.05		0.02		0.07		5	A0	595
		LS I +60 278			2 49 14	+60 23.9	1	10.62		0.41		-0.50		3		41
17820	+10 00380	G 4 - 44			2 49 15	+11 10.3	5	8.38	.006	0.55	.012	-0.06	.009	8	G5	516,979,1620,3077,8112
		CS 22189 # 40			2 49 16	−15 29.4	1	11.87		-0.03		-0.16		1		1736
		LP 886 - 51			2 49 16	−31 01.9	1	11.53		0.48		-0.18		2		1696
17835	+2 00440		A		2 49 17	+02 42.5	1	8.94		0.24		0.16		2	A5 IV	208
17835	+2 00440		B		2 49 17	+02 42.5	1	10.13		0.36		0.15		1		221
17743	+52 00640	HR 846		⋆ AB	2 49 18	+52 47.6	1			0.07		-0.28		1	B8 III	1079
	−8 00535				2 49 18	−08 28.4	2	9.85	.051	1.22	.047	1.13		3	K7 V	1017,3072
		CS 22189 # 41			2 49 19	−16 00.8	1	15.31		0.01		0.00		1		1736
	+68 00203				2 49 21	+68 36.0	1	10.19		0.25				1	B9	207
	+60 00585				2 49 23	+60 39.0	2	9.16	.000	1.06	.019	1.03		4	K5 V	1017,3072
		CS 22189 # 42			2 49 25	−17 24.6	1	11.39		0.28		0.09		1		1736
17937	−50 00843				2 49 27	−50 30.7	1	8.97		0.35		0.02		4	F0 V	243
		G 36 - 38			2 49 28	+26 46.4	1	10.86		1.26		1.62		1	M0	1658
17706	+63 00364	IDS02454N6402	AB		2 49 32	+64 14.9	2	8.45	.005	0.37	.019	-0.19	.019	11	B8	24,1524
237019	+59 00562	LS I +60 279			2 49 35	+60 15.3	5	9.73	.000	0.47	.000	-0.53	.005	9	O8 V	41,64,1011,1012,1083
	+11 00403				2 49 37	+11 58.5	1	11.05		0.58		0.00		2	G0	1375
	+57 00655	LS I +58 098			2 49 39	+58 07.0	1	10.07		0.51		-0.41		3		41
17873	−1 00401				2 49 39	−00 51.3	1	7.81		0.14		0.13		2	A3 III	208
		LS I +61 315			2 49 41	+61 31.6	1	9.97		0.51		-0.27		3		41
14369	+88 00009				2 49 44	+88 55.3	2	8.08	.000	0.41	.012	-0.06	.022	20	F0	1332,1334
17895	−8 00536	RR Eri			2 49 47	−08 28.3	1	6.93		1.64		1.45		1	M5 III	3076
17926	−31 01148	HR 858			2 49 49	−31 01.2	2	6.39	.005	0.48	.000			7	F6 V	15,2012
	−48 00776	BY Eri			2 49 49	−47 59.9	1	12.16		0.23		0.10		1		700
		4HLF4 # 3			2 49 54	+02 00.	1	13.40		0.36		0.11		2	A7	208

HD	DM	Other Id	N Rem	α_{1950}	δ_{1950}	S	V	σ_V	B–V	σ_{B-V}	U–B	σ_{U-B}	n	Spectrum	References
17841	+35 00583	G 37 - 10		2 49 54	+35 26.2	1	8.38		0.79		0.38		1	K0	1620
	+68 00204			2 49 54	+68 39.9	1	10.08		0.34				1		207
17818	+47 00723	HR 849	⋆ AB	2 49 56	+48 22.0	2	6.24	.023	1.19	.017	0.96	.013	5	G5 I	595,1733
	+63 00365	LS I +63 194		2 50 00	+63 37.9	2	9.17	.020	0.95	.000	0.39	.016	4	A1 Ia	41,1215
	+5 00405	G 75 - 46		2 50 02	+05 31.0	2	10.32	.019	1.03	.015	0.90		3	K3	202,1620
		G 75 - 47		2 50 07	+01 43.9	2	14.74	.009	1.54	.005			4		419,3078
17925	−13 00544	HR 857		2 50 07	−12 58.3	11	6.04	.013	0.87	.008	0.56	.009	58	K1 V	15,678,770,1013,1075*
	−14 00551			2 50 09	−14 21.2	1	9.77		1.47		1.71		2	K5	1375
		Steph 311	AB	2 50 10	−09 26.1	1	11.28		1.46		1.22		1	M0	1746
		4HLF4 # 1002		2 50 12	−01 13.	1	11.43		0.37		0.08		1	F0	221
		LS I +62 230		2 50 14	+62 39.3	1	11.32		0.61		−0.39		3		41
17970	−33 00992			2 50 14	−33 39.3	2	8.09	.005	0.83	.012	0.41		6	K1 V	2012,3008
	+60 00586	LS I +60 280	⋆ AB	2 50 16	+60 26.8	5	8.47	.010	0.30	.000	−0.67	.005	10	O7	41,64,1011,1012,1083
		4HLF4 # 1003		2 50 18	+00 41.	1	12.04		1.05		0.86		2	K5	221
		LS I +60 281		2 50 18	+60 27.0	1	11.61		0.36		−0.50		3		41
18023	−50 00848			2 50 18	−50 04.4	1	7.05		1.04		0.80		4	G8/K0III/IV	243
		KUV 02503-0238		2 50 19	−02 37.5	1	14.73		0.11		−0.70		1	DA	1708
18035	−50 00849			2 50 22	−50 25.0	2	7.80	.005	1.28	.005	1.29		8	K0 III	243,2012
17943	−10 00569	HR 859		2 50 24	−09 38.7	3	6.31	.010	0.19	.005	0.11	.010	14	A7 IV	15,1415,2012
	−24 01292	TT Eri		2 50 24	−23 58.8	1	12.92		0.25		0.10		4		3064
		LB 3286		2 50 24	−66 19.	1	12.58		0.30		0.04		3		45
		LS I +57 102		2 50 32	+57 02.1	1	12.35		0.72		−0.13		4		41
17904	+37 00655	HR 855	⋆ AB	2 50 33	+38 08.1	2	5.38	.023	0.41	.009	0.03	.005	8	F2 V +F6 V	254,3030
17904	+37 00655	IDS02474N3756	C	2 50 33	+38 08.1	1	9.68		0.78		0.38		2		3030
17891	+46 00652	IDS02473N4645	AB	2 50 38	+46 57.5	1	6.78		0.05		−0.09		5	B9	595
		LS I +57 103		2 50 38	+57 49.8	1	11.37		0.78		−0.23		2		41
	+51 00648			2 50 39	+51 51.5	1	9.56		0.28		−0.06		10	B8	1655
17878	+52 00641	HR 854, τ Per	⋆ A	2 50 42	+52 33.6	6	3.95	.012	0.74	.016	0.46	.003	13	G4 III+A4 V	15,401,1118,1363*
		LS I +64 095		2 50 43	+64 17.3	1	11.92		0.82		0.08		3		41
18293	−75 00204	HR 872		2 50 46	−75 16.3	3	4.74	.004	1.33	.000	1.56	.000	12	K3 III	15,1075,2038
		WLS 300 30 # 6		2 50 50	+29 54.2	1	11.65		1.29		0.95		2		1375
	−0 00446			2 50 51	−00 11.3	1	9.54		1.08		0.85		4	G6	281
17902	+50 00654			2 50 53	+50 57.8	1	6.53		1.10		0.98		3	K2 III	1501
		LS I +58 099		2 50 54	+58 59.6	1	12.06		0.58		0.25		2		41
17940	+31 00499	IDS02479N3118	A	2 50 55	+31 29.6	1	7.96		0.25		0.15		1	A2	695
		HA 94 # 168		2 50 56	+00 09.5	1	11.82		1.03		0.76		8	K2	281
17922	+42 00646			2 50 58	+42 23.2	1	6.86		0.53		0.06		3	F5 V	3026
17963	+29 00495			2 51 02	+29 54.3	1	9.04		0.48				2	F6 V	20
18012	+1 00509			2 51 03	+01 46.1	1	6.74		0.94		0.54		3	G8 V	3077
		HA 94 # 171		2 51 05	+00 05.1	2	12.66	.004	0.82	.001	0.29	.016	13	G5	281,1764
	−0 00447			2 51 05	−00 08.9	2	9.12	.003	1.05	.002	0.77		11	G5	281,6004
17962	+31 00500			2 51 06	+32 08.0	1	8.13		0.21		0.08		2	A3	1375
	+56 00739	LS I +57 104		2 51 06	+57 14.4	3	9.96	.000	1.01	.005	−0.09	.015	7	O9.5Ib	41,1011,1012
		LS I +59 154		2 51 07	+59 03.8	2	10.67	.004	0.56	.004	−0.35	.004	5	B1 V	41,1012
17921	+47 00732	LS V +47 004		2 51 08	+47 56.5	1	7.75		0.02		−0.28		5	A0	595
17857	+63 00367	LS I +63 195		2 51 13	+63 57.3	3	7.69	.023	0.75	.000	−0.01	.005	8	B8 Ib	41,64,1012
18185	−63 00197	HR 866		2 51 14	−63 06.8	5	6.03	.005	1.25	.006			22	K1 III	15,1075,2013,2018,2030
		L 127 - 97		2 51 14	−63 53.5	2	11.38	.005	1.56	.015	1.20		2		1705,3078
		LP 830 - 61		2 51 16	−24 51.3	1	12.91		1.24		1.07		2		1696
		LP 886 - 60		2 51 17	−28 17.4	1	12.21		0.77		0.32		2		1696
		4HLF4 # 4		2 51 18	−01 21.	1	13.03		0.24		0.14		2	A3	208
19014	−85 00034			2 51 18	−85 22.0	1	8.20		1.71		1.92		1	K2 III	565
17911	+59 00567			2 51 20	+59 37.6	1	7.98		0.28		0.12		2	A5 V	1502
	−13 00551			2 51 20	−12 40.0	1	10.10		1.26		1.26		2		1375
18071	−22 00503	HR 862	⋆ AB	2 51 20	−22 34.7	2	5.94	.005	1.04	.000			7	K0 III	15,2027
		LS I +57 105		2 51 22	+57 06.9	1	10.93		0.84		−0.10		2		41
		CS 22963 # 13		2 51 26	−06 54.5	2	13.54	.010	0.30	.006	−0.12	.013	2		966,1736
18019	+19 00432			2 51 27	+20 21.8	1	7.12		1.26		1.21		2	K0	1648
17856	+68 00206			2 51 28	+68 46.4	2	8.72	.020	0.34	.026	0.21		2	A2	207,634
		4HLF4 # 5		2 51 30	+00 37.	1	14.00		0.27		0.02		3	A2	208
18100	−26 01057			2 51 30	−26 21.5	2	8.48	.006	−0.24	.001	−0.95	.009	19	B5 II/III	55,1732
		G 36 - 42		2 51 34	+22 58.4	1	13.68		1.44		1.08		1		1658
		4HLF4 # 6		2 51 36	−01 49.	1	13.08		0.31		0.10		2	A3	208
	+27 00453			2 51 37	+28 07.3	1	9.73		0.95		0.60		2	K0	1375
18149	−38 00948	HR 863		2 51 37	−38 38.4	3	5.92	.005	0.44	.005	0.07		9	F5 V	15,2027,3053
	−50 00857			2 51 40	−50 23.3	1	9.93		0.59		0.12		4		243
		4HLF4 # 1004		2 51 42	+01 08.	1	11.88		0.82		0.49		2	G5 III	221
		HA 94 # 188		2 51 43	+00 07.3	1	11.77		1.02		0.85		4	G5	281
18058	+9 00370	IDS02490N0922	A	2 51 43	+09 34.2	1	9.60		0.58		0.05		2	G5	3016
18058	+9 00370	IDS02490N0922	BC	2 51 43	+09 34.2	1	10.26		0.83		0.41		2		3016
		LS I +59 155		2 51 46	+59 12.6	1	10.50		0.56		−0.43		3		41
		BPM 16961		2 51 48	−58 06.	1	15.84		0.44		−0.30		1		3065
18169	−41 00832			2 51 50	−41 27.3	2	8.44	.005	0.42	.000			8	F3 V	1075,2013
	−0 00448			2 51 55	+01 17.7	1	10.87		0.50		0.02		1		97
17948	+60 00591	HR 860	⋆ A	2 51 58	+61 19.1	2	5.61	.005	0.44	.005	−0.09	.025	2	F4 V	254,3037
237027	+58 00527	LS I +59 156		2 51 59	+59 22.7	1	9.80		0.60		0.27		3	F0	41
	−0 00449			2 52 00	+01 14.5	1	10.03		0.95		0.55		1		97
17971	+59 00569	LS I +60 282		2 52 00	+60 11.5	2	7.74	.004	1.07	.004	0.86	.027	13	F5 Ia	41,1355
18168	−36 01091	IDS02500S3618	A	2 52 00	−36 06.3	3	8.23	.014	0.93	.006	0.65	.005	10	K1 V	158,2012,3062
		LP 942 - 54		2 52 00	−36 06.9	1	13.10		1.52		1.21		1		3062

Table 1   155

HD	DM	Other Id	N Rem	α₁₉₅₀	δ₁₉₅₀	S	V	σ_V	B−V	σ_B−V	U−B	σ_U−B	n	Spectrum	References
18091	+17 00454			2 52 07	+17 32.0	1	7.00		0.28		0.03		2	A3	1648
		HA 94 # 285		2 52 12	+00 17.1	1	13.24		0.70		0.23		1		97
18242	−50 00860	HR 868, R Hor		2 52 12	−50 05.5	2	10.62	.273	1.92	.151	0.43		3	M7e	975,8029
	−54 00487			2 52 12	−54 11.9	1	11.40		1.42				1	G3	1705
18040	+47 00737			2 52 13	+48 08.3	1	7.16		0.14		0.32		5	A2	595
18145	−0 00450			2 52 13	−00 15.1	4	6.53	.017	1.04	.009	0.88	.016	24	G8 II	97,281,3016,6004
		HA 94 # 286		2 52 15	+00 17.2	1	11.68		0.77		0.33		3		97
18065	+37 00660			2 52 15	+37 47.8	1	8.46		1.04		0.82		2	G5	1601
17958	+63 00369	HR 861	★ A	2 52 16	+64 07.9	2	6.26	.025	1.97	.079	2.41		3	K3 Ib	138,1355
		HA 94 # 288		2 52 17	+00 15.4	1	12.75		0.98		0.80		1		97
		G 4 - 45		2 52 17	+09 42.8	1	14.05		1.19		0.96		1		1658
17929	+68 00208			2 52 17	+68 38.1	4	7.86	.022	0.29	.014	-0.16	.009	10	B5 V	207,245,555,634
		LS I +59 157		2 52 18	+59 09.6	1	10.18		0.61		-0.37		3		41
		LS I +59 158		2 52 28	+59 03.9	1	12.27		0.58		-0.08		3		41
18265	−51 00683	HR 871		2 52 29	−51 04.5	3	6.21	.005	1.57	.009	1.88	.023	10	K4 III	58,243,2007
	−4 00495			2 52 31	−04 01.4	1	10.82		0.13		0.09		3	A2	1026
18183	−14 00557			2 52 32	−14 13.6	1	7.07		0.33		0.12		4	F2 III/IV	1628
18143	+26 00484	G 36 - 43	★ C	2 52 38	+26 40.3	1	13.86		1.58		1.17		4		3032
18175	−0 00451			2 52 40	+00 14.1	6	7.04	.043	1.14	.008	1.13	.010	33	K0 II	97,221,281,975,3040,6004
	+18 00375	G 4 - 46		2 52 40	+18 34.9	1	9.76		0.80		0.33		1	K0	1620
18112	+41 00570			2 52 40	+41 36.3	1	7.92		0.30		0.10		3	F0	1733
18143	+26 00484	G 36 - 44	★ A	2 52 41	+26 40.5	2	7.60	.007	0.93	.004	0.71	.000	5	G5	1355,3032
18143	+26 00484	G 36 - 44	★ AB	2 52 41	+26 40.5	1	7.42		0.98		0.67		3	G5	938
18143	+26 00484	IDS02497N2628	B	2 52 41	+26 40.5	1	9.80		1.40		1.50		2	M0 V	3032
		HA 94 # 36		2 52 41	−00 16.7	1	17.45		0.70				1		97
		G 36 - 45		2 52 42	+27 55.8	1	11.06		1.40		1.30		2	K7	7010
		LP 246 - 71		2 52 42	+34 16.4	1	10.23		1.06		0.94		1		1773
	+0 00480			2 52 44	+00 35.2	1	11.13		0.64		0.11		4		97
		WLS 248 50 # 9		2 52 44	+49 56.2	1	11.06		0.67		0.32		2		1375
		G 74 - 47		2 52 45	+34 15.1	1	13.51		1.20		1.03		2		1773
		HA 94 # 296		2 52 46	+00 16.1	3	12.26	.008	0.74	.012	0.24	.008	15	G3	97,281,1764
		HA 94 # 297		2 52 48	+00 15.9	2	12.07	.014	0.74	.009	0.28	.009	12	G5	97,281
		LS I +57 106		2 52 48	+57 04.1	1	11.02		0.95		0.06		2		41
	−0 00452			2 52 48	−00 10.6	1	10.66		0.64		0.20		1		97
		LB 1644		2 52 48	−47 31.	1	13.14		0.09		0.13		2		45
		HA 94 # 213		2 52 49	+00 09.7	1	14.51		0.57		0.01		2		97
		HA 94 # 212		2 52 49	+00 10.7	1	14.65		0.65		0.07		2		97
		G 174 - 19		2 52 50	+55 14.4	1	10.48		1.39		1.19		1		906
		G 174 - 20		2 52 52	+55 14.7	1	11.65		1.42		1.10		1		906
18217	−1 00414			2 52 54	−00 40.2	1	8.51		0.48		0.06		1	F3 V	97
18216	+1 00512			2 52 56	+01 49.0	1	6.60		-0.04		-0.11		2	B8.5V	208
	−0 00453			2 52 58	+01 16.0	1	10.52		0.63		0.21		1		97
		LS I +58 100		2 52 58	+58 31.8	1	11.77		0.70		0.05		3		41
		HA 94 # 42		2 52 59	−00 08.9	1	16.53		0.88		0.70		1		97
18191	+17 00457	HR 867, RZ Ari		2 53 00	+18 07.8	4	5.82	.048	1.45	.052	1.14	.020	23	M6-III	71,3042,6002,8032
		HA 94 # 300		2 53 01	+00 17.7	3	11.52	.008	1.10	.013	1.02	.015	18	K2	97,281,397
18076	+58 00534	LS I +58 101		2 53 02	+58 53.3	2	9.06	.000	0.59	.004	-0.45	.018	5	B0 II-III	41,1012
		HA 94 # 217		2 53 03	+00 11.8	2	13.53	.002	0.63	.000	0.06	.039	6		97,397
18125	+49 00801			2 53 03	+49 35.5	1	7.80		0.24		0.40		3	A2	595
		HA 94 # 301		2 53 04	+00 14.8	2	15.31	.020	0.59	.000	0.02	.073	6		97,397
		MtW 94 # 22		2 53 05	+00 11.0	1	15.59		0.88		0.28		3		397
17243	+85 00050			2 53 05	+85 40.0	1	8.71		0.06		0.04		6	A0	1219
		HA 94 # 382		2 53 06	+00 25.7	1	14.50		1.30		1.06		3		97
	+0 00482			2 53 06	+01 17.4	1	10.87		1.07		0.88		3		97
		Case *M # 31		2 53 06	+57 20.	2	9.92	.039	2.76	.034			3	M2 Ib	138,148
	+60 00594	LS I +61 316		2 53 06	+61 12.9	6	9.30	.019	0.36	.012	-0.63	.021	12	O9 V	41,64,1011,1012,1069,1083
		HA 94 # 383		2 53 07	+00 24.1	1	12.80		0.64		0.20		3		97
		BSD 9 # 1401		2 53 07	+61 12.7	1	10.76		0.38		0.25		2	B9	1069
		HA 94 # 43		2 53 07	−00 15.3	1	17.12		0.78		0.36		1		97
18155	+46 00658	HR 865	★ AB	2 53 10	+46 57.8	1	6.02		1.34				2	gK3	71
		LS I +57 107		2 53 10	+57 54.3	1	11.81		0.78		0.13		2		41
		WLS 248 50 # 7		2 53 11	+48 10.7	1	10.62		1.15		0.85		2		1375
		VES 767		2 53 11	+63 04.1	1	12.51		0.65		-0.23		3		1543
	+60 00596	LS I +60 283		2 53 12	+60 27.6	2	9.63	.003	0.36	.003	-0.53	.003	14	B1 V:n	41,1012
		HA 94 # 302		2 53 14	+00 16.9	1	15.04		0.81		0.35		2		97
		HA 94 # 303		2 53 15	+00 22.0	1	13.12		0.93		0.62		2		97
		HRC 8	V	2 53 17	+19 51.3	1	14.56		0.84		-0.40		1		776
		HRC 9	V	2 53 18	+19 51.5	1	14.64		1.39		0.19		1		776
18153	+50 00665	HR 864		2 53 19	+51 03.6	2	6.22	.004	1.57	.001	1.89	.013	5	K5 III	595,1733
		BSD 9 # 823		2 53 19	+61 10.6	1	10.47		0.63		0.44		2	B8	1069
		HA 94 # 223		2 53 20	+00 03.5	1	13.16		0.59		0.10		2		97
		HA 94 # 304		2 53 21	+00 12.6	1	16.25		0.62		0.04		2		97
17992	+72 00153			2 53 21	+72 28.4	1	7.72		1.12		0.97		2	G5	1502
		HA 94 # 388		2 53 23	+00 27.4	1	15.32		1.07		0.75		1		97
18423	−64 00206	HR 880		2 53 24	−64 38.3	3	6.55	.003	1.40	.005	1.59	.000	19	K2/3 III	15,1075,2038
		HA 94 # 389		2 53 25	+00 25.7	1	13.83		0.60		0.07		2		97
18257	+13 00476			2 53 27	+13 42.8	1	9.05		0.81		0.46		2	K0	1625
		HA 94 # 391		2 53 29	+00 23.9	1	14.79		0.94		0.46		2		97
	−0 00455			2 53 31	+00 22.9	3	11.03	.018	0.65	.009	0.13	.013	9	G1	97,196,281
	+2 00453			2 53 31	+03 00.2	1	9.90		0.39		0.16		2	F0 V	208

HD	DM	Other Id	N	Rem	α_{1950}	δ_{1950}	S	V	σ_V	B–V	σ_{B-V}	U–B	σ_{U-B}	n	Spectrum	References
18222	+31 00507				2 53 31	+31 45.0	1	8.63		0.38		0.12		1	F0	695
18056	+68 00209	IDS02490N6848		AB	2 53 31	+68 59.6	1	7.71		0.40				1	F2	207
18324	−36 01101				2 53 32	−35 45.8	1	8.60		1.08				4	K0 III	2033
		HA 94 # 306			2 53 33	+00 14.2	1	14.81		1.13		0.82		2		97
18262	+7 00450	HR 870			2 53 33	+08 10.9	4	6.00	.024	0.47	.011	0.04	.012	12	F5 IV	15,254,1061,3053
		4HLF4 # 7			2 53 36	+00 01.	1	13.94		0.31		0.08		2	A2	208
18256	+17 00458	HR 869			2 53 36	+17 49.5	3	5.62	.052	0.43	.008	-0.03	.010	6	F6 V	71,254,3053
18362	−50 00866				2 53 36	−49 51.7	1	9.29		1.04		0.82		4	G8/K0 (III)	243
18272	−0 00456				2 53 37	−00 24.1	1	8.59		1.25		1.20		1	G8 III	97
18179	+50 00670				2 53 38	+50 39.9	1	9.52		0.12		-0.06		2	A2	1566
18305	−17 00568				2 53 38	−17 10.5	1	8.88		1.21		1.18		2	G8/K0 IV	1375
		HA 94 # 394			2 53 39	+00 23.1	3	12.26	.004	0.54	.012	-0.04	.005	16		97,281,1764
		LS I +57 108			2 53 39	+57 34.5	1	11.02		0.86		-0.27		3		41
18286	−0 00457	IDS02511N0007		A	2 53 40	+00 19.2	8	8.74	.008	0.49	.008	0.00	.006	71	F3 V	97,196,281,989,1371,1509*
18152	+60 00597				2 53 40	+60 55.7	1	7.66		0.14		0.01		2	B9 III	1069
		HA 94 # 227			2 53 41	+00 08.3	1	13.26		0.69		0.11		3		97
		HA 94 # 309			2 53 41	+00 20.7	1	16.24		0.72		0.24		1		97
		HA 94 # 396			2 53 42	+00 24.4	1	15.87		0.78		0.00		3		97
		LB 1645			2 53 42	−44 58.	1	13.07		0.12		0.09		3		45
		LS I +59 159			2 53 45	+59 09.3	1	11.97		0.55		-0.26		3		41
		HA 94 # 310			2 53 46	+00 16.1	1	12.01		0.71		0.26		3		97
		HRC 10, WY Ari			2 53 47	+19 53.5	1	12.46		0.83		-0.46		1		776
		CS 22968 # 6			2 53 47	−54 03.9	2	14.94	.000	0.24	.006	-0.12	.012	2		966,1736
	−0 00459				2 53 48	+00 07.7	3	10.48	.004	0.40	.010	-0.01	.018	8	F2p	97,196,281
		HA 94 # 311			2 53 48	+00 13.6	1	16.06		0.74		0.35		1		97
	−0 00454				2 53 48	+00 18.9	10	8.89	.014	1.43	.014	1.58	.012	98	K5	97,196,281,397,975,989*
18200	+51 00657				2 53 48	+52 18.1	1	8.10		1.12		0.81		2	K0	1566
		L 90 - 44			2 53 48	−68 28.	1	11.91		0.50		-0.21		3		1696
		G 76 - 42			2 53 49	+03 40.8	2	13.15	.045	0.72	.010	0.05	.005	2		1620,1696
		HA 94 # 229			2 53 50	+00 04.6	1	12.05		0.44		0.00		3		97
		HA 94 # 230			2 53 53	+00 12.0	1	15.36		0.90		0.31		2		97
18246	+38 00594				2 53 53	+38 51.9	1	8.63		-0.03		-0.24		3	B9 V	833
		RL 66			2 53 54	+60 18.	1	15.50		-0.17		-1.08		2		704
		HA 94 # 313			2 53 55	+00 15.9	1	14.07		0.79		0.19		2		97
		BSD 9 # 838			2 53 55	+61 05.2	1	12.12		0.45		0.27		2	A0	1069
		HA 94 # 401			2 53 56	+00 28.1	2	14.31	.013	0.63	.009	0.12	.020	3		281,1764
		HA 94 # 402			2 53 56	+00 30.6	1	12.89		0.60		0.09		9	G0	281
	+60 00598	LS I +60 284		⋆ A	2 53 56	+60 24.1	2	9.94	.031	0.33	.004	-0.51	.040	5	B6	41,1069
		LP 411 - 26			2 53 58	+18 43.6	1	12.27		0.70		0.09		2		1696
18322	−9 00553	HR 874			2 53 59	−09 05.8	10	3.89	.010	1.10	.009	1.00	.008	29	K1-IIIb	3,15,58,1075,1425*
		HA 94 # 405			2 54 01	+00 25.0	1	15.65		0.57		0.03		1		97
	+1 00514				2 54 03	+01 33.1	1	9.79		0.13		0.15		2	B7 V	208
		HA 94 # 316			2 54 06	+00 20.5	1	13.73		0.80		0.45		2		97
		HA 94 # 406			2 54 06	+00 29.4	2	12.34	.004	0.55	.013	0.00	.004	11	G3	97,281
18331	−4 00502	HR 875			2 54 07	−03 54.8	31	5.17	.008	0.08	.006	0.07	.023	213	A3 Vn	1,3,15,116,418,985*
18696	−78 00077				2 54 07	−78 03.9	1	7.71		1.62		2.03		8	K5 III	1704
	+56 00749				2 54 11	+57 22.2	1	9.88		2.70		2.22		3	M1	8032
		LS I +57 109			2 54 12	+57 38.3	2	10.25	.000	0.82	.009	-0.17	.009	5	B2 III	41,1012
18296	+31 00509	HR 873, LT Per			2 54 15	+31 44.1	5	5.10	.004	-0.01	.011	-0.24	.009	15	B9p Si	695,1049,1202,1363,3023
		BSD 9 # 1426			2 54 15	+61 28.3	1	11.30		0.46		-0.04		2	B8	1069
		CS 22963 # 4			2 54 16	−05 03.5	1	14.98		0.53		-0.14		1		1736
237039	+59 00574				2 54 17	+59 47.0	1	9.46		0.28		0.23		2	B9 V	1069
		G 76 - 43			2 54 18	+10 07.2	1	12.61		1.33		1.04		1		333,1620
		HA 94 # 236			2 54 19	+00 07.9	2	10.47	.015	1.25	.020	1.20	.000	6	K3	97,281
		LS I +60 285			2 54 19	+60 10.2	1	11.86		0.76		0.25		2		41
		LS I +60 286		A	2 54 22	+60 29.7	1	12.25		0.45		-0.27		3		41
		LS I +60 286		AB	2 54 22	+60 29.7	1	11.56		0.46		-0.37		2		180
		CS 22181 # 2			2 54 25	−11 23.8	1	14.06		0.21		0.04		1		1736
18345	+3 00410	HR 877			2 54 27	+04 18.0	3	6.16	.057	1.65	.040	1.82	.020	10	M4 IIIab	15,1061,3016
		LS I +60 287			2 54 31	+60 11.7	2	11.02	.033	0.37	.014	-0.33	.009	4		41,1069
237040	+58 00538				2 54 32	+59 00.7	1	9.31		0.30		-0.09		2	B9p	1069
18369	−0 00460				2 54 36	+00 14.8	6	6.62	.011	0.32	.007	0.04	.012	32	A5 Ib	97,196,221,281,975,6004
		CS 22181 # 3			2 54 36	−11 48.4	1	11.21		0.06		0.03		1		1736
	−50 00869				2 54 37	−50 23.8	1	9.71		0.97		0.83		4		158
18368	+1 00515	IDS02521N0129		AB	2 54 39	+01 41.3	2	7.57	.014	0.58	.005	0.00		4	F9 V +G1 V	1381,3030
18358	+9 00381				2 54 40	+10 16.9	1	8.94		0.45				1	A5	1242
	−50 00870				2 54 40	−50 24.5	1	9.61		1.28		1.44		4		158
18357	+15 00414				2 54 41	+16 05.6	1	6.88		1.20		1.17		2	K0	1648
		LS I +59 160			2 54 41	+59 06.4	1	11.14		0.63		-0.33		3		41
		LP 711 - 27			2 54 41	−13 53.6	1	16.33		1.67				4		1663
		4HLF4 # 8			2 54 42	+00 31.	1	13.88		0.33		0.09		2	A2	208
		G 5 - 7			2 54 43	+10 35.6	1	13.06		1.57				1		1705
18317	+38 00598				2 54 43	+38 51.1	1	8.79		-0.02		-0.20		3	A0	833
	−36 01106	IDS02529S3633		A	2 54 43	−36 21.5	1	9.78		0.72		0.13		1	G5	3062
		BSD 9 # 247			2 54 45	+59 30.9	1	12.27		0.53		0.32		2	A0	1069
	+29 00503	Hyades vB 153			2 54 46	+29 27.7	3	8.92	.004	0.86	.008	0.56	.015	6	G5	1127,3016,8023
		HA 94 # 242			2 54 48	+00 06.6	5	11.73	.006	0.30	.004	0.10	.009	89	A5	208,281,989,1729,1764
	+0 00485				2 54 49	+00 46.3	1	10.72		0.77		0.38		2	K0 IV	281
		HA 94 # 597			2 54 50	+00 48.0	1	10.85		0.79		0.46		2	G5	281
		G 76 - 47			2 54 50	+09 10.8	1	11.31		1.03		0.90		1		333,1620

Table 1 157

HD	DM	Other Id	Rem	α_{1950}	δ_{1950}	S	V	σ_V	B–V	σ_{B-V}	U–B	σ_{U-B}	n	Spectrum	References
		LP 942 - 72		2 54 50	−36 20.8	1	14.18		1.51				1		3062
18339	+38 00599	HR 876		2 54 52	+38 24.9	2	6.04	.007	1.40	.003	1.61		5	K5 III	71,1501
		LS I +58 102		2 54 52	+58 35.5	1	11.02		0.84		-0.02		3		41
	+2 00458			2 54 54	+02 41.8	1	10.26		0.23		0.15		2	A3 V	208
		G 74 - 48		2 54 54	+38 17.7	1	12.59		0.69		0.05		3		1620
	−30 01121			2 54 54	−30 36.6	1	10.46		0.76				4	G0	955
		4HLF4 # 10		2 55 00	+00 10.	1	12.55		0.35		0.06		1	A7	208
		BSD 9 # 861		2 55 01	+60 56.9	1	11.09		0.57		0.03		2	B9	1069
	+61 00506			2 55 01	+61 28.1	1	10.18		0.17		0.01		2	B9 V	1069
18445	−25 01168	IDS02528S2522	C	2 55 01	−25 10.5	3	7.84	.006	0.96	.006	0.75	.004	8	K2 V	214,1414,3032
		LS I +57 110		2 55 02	+57 50.8	2	10.04	.000	0.81	.009	-0.22	.004	5	B1 II-III	41,1012
18455	−25 01169	IDS02528S2522	AB	2 55 02	−25 10.1	3	7.35	.012	0.86	.005	0.54	.007	6	K1/2 V	214,1414,3032
		LS I +58 103		2 55 03	+58 27.2	1	12.20		0.68		-0.22		3		41
	−0 00463			2 55 04	+00 21.8	2	10.06	.005	0.73	.005	0.28	.000	5	G3	97,281
		BSD 9 # 863		2 55 06	+60 33.9	1	12.53		0.63		0.34		2	A0	1069
18466	−30 01122	HR 884		2 55 06	−30 03.3	1	6.29		0.47				4	A2/3(V)+(G)	2007
		LS I +61 317		2 55 07	+61 59.5	1	12.19		0.41		-0.23		3		41
		VES 768		2 55 08	+60 42.1	1	13.28		0.77		0.20		1		1543
	−2 00524			2 55 08	−02 11.8	4	10.31	.011	-0.11	.006	-0.63	.016	34	B8	221,989,1026,1036,1729
		BSD 9 # 259		2 55 09	+59 45.8	1	9.84		0.32		-0.18		2	A0	1069
		BSD 9 # 1451		2 55 10	+61 42.7	1	10.27		0.40		0.05		2	B9	1069
18454	−24 01336	HR 883		2 55 10	−24 03.7	2	5.44	.005	0.24	.005			7	A5 V	15,2027
		LS I +56 080		2 55 12	+56 34.8	1	11.59		0.83		-0.21		3		41
	−68 00179			2 55 12	−68 29.1	1	10.73		0.78		0.32		2		1696
		HA 94 # 251		2 55 13	+00 04.1	4	11.21	.005	1.22	.005	1.28	.011	79	K2	281,989,1729,1764
18404	+20 00480	Hyades vB 154	★	2 55 13	+20 28.2	5	5.80	.018	0.41	.004	0.00	.016	13	F5 IV	15,254,1127,3077,8015
	+51 00659	LS V +51 001		2 55 14	+51 54.7	1	9.37		0.40		-0.52		2	B0 IIInn	1012
		LS I +60 288		2 55 14	+60 27.4	2	10.85	.005	0.64	.010	-0.27	.055	5		41,180
	+57 00668			2 55 15	+57 25.7	1	11.11		2.32				3		369
		BSD 9 # 862		2 55 17	+60 13.8	1	11.76		0.49		0.32		2	A0	1069
		LS I +60 289		2 55 22	+60 16.8	1	11.91		0.35		-0.46		2		41
		BSD 9 # 868		2 55 23	+60 15.7	1	9.98		0.29		0.16		2	B8	1069
	+60 00606	LS I +60 290		2 55 24	+60 23.9	3	9.45	.000	0.59	.016	-0.56	.009	5	B0	41,180,1543
		G 221 - 3		2 55 26	+70 36.4	1	12.26		0.89		0.55		2		1658
		HA 94 # 86		2 55 26	−00 16.7	1	11.52		0.77		0.33		4	G1	281
18326	+59 00578	LS I +60 291		2 55 27	+60 22.0	5	7.87	.036	0.36	.016	-0.63	.020	10	O7 V	41,180,1011,1012,1069
18463	−2 00526			2 55 27	−02 32.2	1	7.71		0.20		0.07		1	A2	1776
		BSD 9 # 874		2 55 29	+60 13.1	1	12.31		0.52		0.45		3	A0	1069
		LS I +60 292		2 55 29	+60 22.3	1	11.36		0.47		-0.49		2		41
		LS I +58 104		2 55 30	+58 32.8	1	11.33		0.91		-0.01		3		41
		HA 94 # 155		2 55 30	−00 01.3	1	11.67		0.50		0.02		4	G0	281
		HA 94 # 699		2 55 32	+00 55.3	1	11.44		0.53		0.04		2	G0	281
	+60 00607			2 55 32	+60 43.6	1	10.55		0.38		0.31		2	A1	1069
18411	+39 00681	HR 879		2 55 33	+39 27.8	5	4.70	.005	0.06	.009	0.12	.000	18	A2 Vn	15,1049,13/
18337	+58 00540			2 55 33	+59 23.6	1	7.68		0.26		0.14		3	A1 V	1069
	+34 00548	G 37 - 18		2 55 35	+35 21.8	2	9.48	.015	1.05	.010	1.02	.005	2	K2	1620,165?
		BSD 9 # 878		2 55 35	+60 15.7	1	10.82		0.35		-0.47		2	B4	1069
18402	+38 00601			2 55 36	+38 36.2	1	8.65		-0.07		-0.50		3	B9	833
		LS I +57 111		2 55 36	+57 30.9	1	11.42		0.85		-0.31		2		41
		BSD 9 # 879		2 55 36	+60 23.9	1	11.67		0.59		-0.22		2	B5	1069
18546	−38 00976	HR 893		2 55 36	−38 23.5	2	6.40	.005	-0.03	.000			7	A0 Vn	15,7
		HA 94 # 702		2 55 38	+00 58.9	4	11.60	.002	1.42	.005	1.62	.028	85	G5	28?
		LS I +60 293		2 55 38	+60 21.8	1	11.68		0.61		-0.24		3		4?
		LS I +60 294		2 55 40	+60 13.5	1	10.88		0.33		-0.51		3		
		HA 94 # 90		2 55 42	−00 13.5	1	11.89		0.66		0.09		4	G1	
		4HLF4 # 11		2 55 42	−01 35.	1	11.49		0.38		0.22		1	A3	
		TonS 294		2 55 42	−29 14.	1	13.44		0.04		0.35		1		
		CS 22963 # 12		2 55 43	−06 25.8	2	14.36	.000	0.18	.000	0.22	.008	2		
		BSD 9 # 11		2 55 44	+59 03.6	1	12.45		0.91		0.40		3	A0	
		LP 942 - 79		2 55 46	−36 49.6	1	13.00		1.01		0.66		2		
	−35 01023			2 55 47	−35 36.4	1	9.98		0.52		-0.10		2	F8	
18352	+60 00608	LS I +61 318		2 55 49	+61 05.5	3	6.83	.011	0.21	.009	-0.64	.026	7	B1 V	
		HA 94 # 163		2 55 49	−00 03.4	1	12.35		0.60		0.08		6	F9	/29,6004
		BSD 9 # 12		2 55 50	+59 05.8	1	11.76		0.64		0.52		2	A0	
18497	−2 00529	IDS02533S0159	AB	2 55 52	−01 46.8	1	7.55		0.35		0.02			F0	
18535	−24 01343	HR 889		2 55 52	−23 48.4	3	5.83	.007	1.33	.005			11	K2 III	,985,1355,1363*
		L 54 - 5		2 55 54	−70 34.	1	14.08		0.23		-0.59		3		
18496	−0 00465			2 55 55	+00 14.3	5	9.04	.009	1.00	.006	0.72	.008	41	G8	
		BSD 9 # 13		2 55 56	+59 05.0	1	11.26		0.63		0.49		5	A0	
		LS I +60 295		2 55 56	+60 21.7	1	12.26		0.47		-0.32		3		
18449	+34 00550	HR 882		2 55 57	+34 59.0	16	4.93	.005	1.24	.007	1.28	.007	98		
		LP 771 - 94		2 55 57	−20 05.9	1	11.52		0.58		-0.04		2		1648
		CS 22968 # 9		2 55 57	−53 37.5	1	14.82		0.39		-0.11		1		41,1355
		KUV 02560+0259		2 55 58	+02 58.9	1	16.02		-0.22		-1.15				41,1069
		CS 22968 # 3		2 56 00	−55 04.9	1	13.30		0.29		0.12				1069
18495	+13 00484			2 56 01	+13 24.4	1	7.27		1.08		0.84				1628
18391	+57 00672	LS I +57 112		2 56 01	+57 27.9	2	6.89	.000	1.94	.009	1.69	.075			
		LS I +59 162		2 56 03	+59 15.6	2	10.50	.039	0.64	.010	-0.25	.02?			
		BSD 9 # 893		2 56 07	+60 45.7	1	12.15		0.69		0.46				
18637	−55 00476			2 56 10	−55 12.8	1	6.76		0.36		0.05				

HD	DM	Other Id	N Rem	α_{1950}	δ_{1950}	S	V	σ_V	B–V	σ_{B-V}	U–B	σ_{U-B}	n	Spectrum	References
18543	−3 00470	HR 892	⋆ AB	2 56 11	−02 58.9	3	5.22	.005	0.00	.000	0.04	.000	8	A3 V	15,1061,3023
	+58 00542	LS I +58 105		2 56 12	+58 26.3	1	10.76		0.77		-0.14		3		41
		CS 22181 # 7		2 56 17	−12 14.7	1	13.36		0.09		0.15		1		1736
		LP 887 - 18		2 56 19	−29 16.0	1	13.96		1.22		0.90		2		3062
18508	+19 00440			2 56 20	+19 47.5	1	7.35		0.36		0.01		3	F0	1648
18520	+20 00484	HR 888	⋆ AB	2 56 21	+21 08.5	8	4.63	.006	0.04	.009	0.08	.010	20	A2 Vs	15,1007,1013,1049*
		LS I +57 113		2 56 21	+57 35.1	1	11.03		0.81		-0.12		2		41
		LS I +60 296		2 56 22	+60 09.5	1	10.85		0.41		-0.48		3		41
18557	−10 00585	HR 895		2 56 22	−09 58.5	4	6.14	.011	0.22	.017	0.12	.010	16	A2 m	15,355,2012,8071
18622	−40 00771	HR 897	⋆ AB	2 56 22	−40 30.3	4	2.91	.005	0.12	.003	0.12	.004	20	A4 III+A1 V	15,1075,2012,8015
		LS I +57 114		2 56 24	+57 14.8	1	11.31		1.00		0.42		3		41
18482	+40 00639	HR 886		2 56 25	+40 50.1	1	5.89		1.45				2	K2	71
		LS I +60 297		2 56 25	+60 22.5	2	10.92	.009	0.50	.014	-0.43	.023	4		41,180
18409	+62 00504	LS I +62 231		2 56 25	+62 31.4	3	8.37	.010	0.42	.000	-0.58	.005	7	O9 Ib	41,1011,1012
18474	+46 00669	HR 885		2 56 26	+47 01.3	1	5.47		0.89		0.61		7	G5:III	1355
		CS 22968 # 7		2 56 28	−53 55.4	2	14.93	.010	0.37	.006	-0.17	.010	3		1580,1736
	+60 00609			2 56 29	+61 12.2	3	10.54	.029	0.68	.016	-0.06	.057	7	B8	41,180,1069
		CS 22963 # 11		2 56 29	−06 21.3	2	13.38	.000	0.15		0.10	.008	2		966,1736
		CS 22968 # 1		2 56 29	−57 03.0	2	14.73	.010	0.39	.007	-0.24	.009	3		1580,1736
		LS I +57 115		2 56 30	+57 32.2	1	12.46		0.83		0.09		2		41
		LB 1646		2 56 30	−45 48.	1	12.72		0.20		0.06		2		45
18571	+0 00490			2 56 42	+01 02.8	2	8.64	.010	0.04	.000	-0.02	.005	9	A0 V	208,1371
	+59 00580			2 56 43	+59 47.6	2	9.92	.009	2.32	.000	2.68		4	M1 Ib	138,8032
		G 78 - 11		2 56 44	+48 49.1	1	14.70		1.53				4		538
		LS I +57 116		2 56 44	+57 42.4	1	11.94		0.89		0.01		2		41
		BSD 9 # 903		2 56 44	+60 51.8	1	12.56		0.64		0.17		2		1069
	+57 00676			2 56 50	+57 57.9	1	9.60		2.11				3	A2	369
		CS 22181 # 8		2 56 52	−12 37.2	1	12.37		0.28		0.09		1		1736
18605	−13 00565	IDS02545S1258	AB	2 56 52	−12 46.1	1	9.55		0.64				4	G6/8 V	2033
18458	+59 00581	IDS02531N5916	B	2 56 53	+59 27.3	1	8.05		0.06		-0.12		3	B8 II-III	1069
		4HLF4 # 12		2 56 54	+00 39.	1	12.41		0.32		0.18		2	A2	208
18541	+38 00606	ST Per	⋆ A	2 56 54	+38 59.5	1	9.54		0.25		0.15		1	A3 V	627
18541	+38 00606	IDS02537N3848	B	2 56 54	+38 59.5	1	11.76		0.62		0.02		3		627
		BSD 9 # 908		2 56 56	+60 57.2	1	10.69		0.40		0.34		2	B9	1069
18597	+2 00460	XY Cet		2 56 57	+03 19.2	1	8.78		0.26		0.14		8	F4 V	588
18481	+56 00759			2 56 57	+56 47.7	1	9.24		0.43				2	A5	70
18650	−29 01106	HR 900		2 56 59	−29 06.3	3	6.13	.008	1.05	.005			11	K1 III	15,1075,2009
18561	+26 00496			2 57 00	+26 41.5	1	8.47		1.41		1.50		2	K2	1733
		4HLF4 # 1008		2 57 00	−01 57.	1	12.90		0.27		0.06		2	A5	221
18552	+37 00675	HR 894		2 57 01	+37 56.0	2			-0.06	.005	-0.39	.015	6	B8 Vne	1049,1079
18604	+8 00455	HR 896	⋆	2 57 02	+08 42.6	8	4.70	.010	-0.11	.008	-0.44	.019	30	B6 III	15,154,1203,1256,1363*
		KUV 02571+0026		2 57 03	+00 25.7	1	16.61		0.26		-0.71		1		1708
		4HLF4 # 13		2 57 06	+01 00.	1	12.78		0.32		0.14		2	A5	208
		LS I +57 117		2 57 08	+57 19.1	1	10.82		0.87		-0.12		3		41
18633	−3 00475	HR 899		2 57 10	−02 39.8	4	5.55	.005	-0.08	.004	-0.18	.008	10	B9.5V	15,1079,1256,3023
18709	−44 00915			2 57 10	−43 56.7	1	7.40		0.59		0.00		6	F8/G0 V	1628
		CS 22968 # 10		2 57 13	−53 38.6	2	15.31	.005	0.17	.015	0.04	.013	2		966,1736
94	+24 00419			2 57 15	+25 02.9	1	8.48		0.38		0.23		2	A2	1625
		BSD 9 # 915		2 57 15	+61 09.0	1	10.89		0.39		-0.20		2	B8	1069
		CS 22181 # 9		2 57 15	−12 35.3	1	15.16		-0.21		-0.97		1		1736
	-51 00665	HR 890	⋆ A	2 57 18	+52 09.3	4	5.24	.049	-0.05	.003	-0.43	.022	11	B7 V	15,154,1009,1048
		BSD 9 # 911		2 57 18	+60 24.1	1	10.12		0.74		-0.23		2	B4	1069
		4HLF4 # 14		2 57 18	−01 54.	1	12.07		0.21		0.21		4	A0	208
	00665	HR 891	⋆ B	2 57 19	+52 09.3	2	6.74	.000	0.00	.000	-0.15	.000	7	B9 V	15,1009
	0546	IDS02535N5838	A	2 57 19	+58 50.3	1	8.58		0.23		0.23		3	A2	1069
		LS I +60 298		2 57 22	+60 22.5	2	10.32	.190	0.69	.034	-0.30	.029	3		41,180
		4HLF4 # 1009		2 57 24	−01 51.	1	10.73		1.12		0.92		2	G8	221
				2 57 24	−11 31.9	1	10.53		0.55		-0.14		2	G5:	1696
		HR 901	⋆	2 57 24	−25 28.4	3	5.70	.014	0.40	.008	0.01		12	F3 V	15,2012,3026
		CS 22963 # 16		2 57 31	−07 13.7	2	13.64	.000	0.15	.000	0.20	.009	2		966,1736
		I +58 106		2 57 32	+58 10.1	1	11.37		0.81		-0.15		3		41
		903		2 57 35	−32 42.3	2	6.30	.005	0.00	.000			7	A0 V	15,2012
				2 57 36	−03 52.2	1	7.72		0.44		-0.03		1	F5	1776
		59 163	⋆ A	2 57 41	+59 08.7	2	9.61	.018	0.44	.005	-0.30	.014	7	B3 II-III	41,1069
				2 57 42	+60 22.8	1	9.59		0.28		0.23		3	A0	1069
18866				2 57 43	+22 37.9	1	7.12		1.12		0.91		2	G5	1648
	981			2 57 43	+56 33.3	1	11.31		1.11		0.28		3		41
				2 57 43	+78 30.5	1	10.01		0.77		0.31		2		1723
				2 57 47	+00 16.8	1	9.94		-0.03		-0.40		2	B9	208
				2 57 47	+59 00.0	1	12.02		0.71		-0.19		3		41
18702 +5 00.. 237052 +57 006.. 237051 +58 00548				2 57 51	−64 16.2	4	4.98	.003	0.12	.005	0.14	.005	21	A3/5 IIIm	15,1075,2016,2038
				2 57 52	+02 01.0	1	10.25		0.37		0.02		2	F2 IV:	208
				2 57 54	+58 05.5	1	11.92		0.77		-0.14		3		41
18700 +10 00401 HR				2 57 54	−06 09.5	2	11.94	.000	0.45	.005	-0.19	.000	3		1620,1696
LS I +				2 57 55	+05 47.4	4	8.12	.007	0.84	.005	0.55	.010	8	K0 V	22,333,1003,1620,3079
				2 57 56	+58 07.8	1	8.45		0.37		-0.32		4	B3	41
			B	2 57 57	+58 39.2	1	9.56		0.54		-0.15		3	B8	41
				2 57 59	−13 35.3	1	13.05		1.52		1.39		1		1696
				2 58 01	+10 40.4	2	5.98	.025	1.56	.029	1.95		5	K5	71,3016
				2 58 01	+57 05.0	1	11.54		1.24		0.28		3		41

Table 1   159

HD	DM	Other Id	N Rem	α_{1950}	δ_{1950}	S	V	σ_V	B–V	σ_{B-V}	U–B	σ_{U-B}	n	Spectrum	References
	+31 00523			2 58 04	+31 55.3	1	9.27		0.43		0.19		2	A5	1375
		BSD 9 # 327		2 58 05	+60 03.1	1	12.33		0.58		0.14		2	B5	1069
18717	+14 00502			2 58 07	+14 50.1	1	7.30		0.53		0.07		2	F8	1648
	+27 00465			2 58 07	+28 02.1	1	10.40		0.15		0.14		2		1375
18665	+35 00607			2 58 07	+35 55.3	1	7.25		1.45		1.46		3	K2	1601
		BSD 9 # 933		2 58 09	+61 06.9	1	10.72		0.30		0.26		2	B9	1069
		CS 22968 # 2		2 58 09	−56 29.6	2	15.86	.000	0.08	.000	0.07	.010	2		966,1736
		CS 22181 # 10		2 58 10	−11 38.0	1	12.05		-0.09		-0.52		1		1736
		4HLF4 # 15		2 58 12	+00 37.	1	13.38		0.20		0.13		2	A0	208
18760	−3 00478	HR 904, CV Eri		2 58 20	−03 04.6	4	6.12	.011	1.74	.029	2.07	.036	11	M1 III	15,1061,1311,3055
		AJ89,1229 # 191		2 58 26	+03 21.9	1	10.31		1.65				1		1532
		LP 771 - 97		2 58 29	−17 28.4	1	12.73		0.47		-0.19		2		1696
		LS I +57 119		2 58 30	+57 10.6	1	11.03		1.00		0.01		3		41
		LS I +57 120		2 58 33	+57 37.1	1	12.18		0.81		-0.15		2		41
18791	−21 00535			2 58 33	−20 53.5	1	7.37		1.64		1.96		3	M0 III	1657
18819	−28 00982			2 58 35	−27 50.2	2	7.60	.000	0.60	.005			8	F8 V	1075,2034
		LS I +57 121		2 58 37	+57 26.9	1	11.39		0.98		0.04		2		41
		LS I +57 122		2 58 38	+57 05.1	1	11.62		0.96		-0.13		3		41
		Steph 328	AB	2 58 39	+11 12.1	1	10.88		1.27		1.16		2	K7	1746
18730	+38 00617			2 58 42	+38 40.9	2	6.65	.000	0.24	.004	0.19	.004	18	A7 V	627,833
18784	−8 00562	HR 907		2 58 42	−07 51.5	4	5.75	.004	1.05	.004	0.87	.007	16	K0 II	15,1075,1415,3008
		LP 887 - 23		2 58 42	−31 29.1	1	11.82		0.91		0.62		1		1696
		CS 22963 # 36		2 58 44	−02 52.1	2	13.84	.000	-0.18	.000	-0.88	.000	2		966,1736
		BSD 9 # 342		2 58 48	+60 02.0	1	10.97		0.49		0.40		2	B9	1069
		WLS 240 65 # 8		2 58 48	+64 27.3	1	11.95		0.73		0.31		2		1375
		CS 22181 # 12		2 58 49	−11 11.5	1	13.56		0.36		0.08		1		1736
		G 78 - 12	A	2 58 50	+42 32.7	1	10.15		0.72		0.16		2		7010
		G 78 - 12	AB	2 58 50	+42 32.7	1	10.07		0.74		0.19		1	G8	1620
		G 78 - 12	B	2 58 50	+42 32.7	1	13.66		1.19		1.09		4		7010
237056	+57 00681	LS I +57 123		2 58 50	+57 25.0	2	8.57	.166	0.70	.030	-0.46	.043	6	B0.5:V:pe	41,1012
18789	+0 00499			2 58 51	+00 57.6	1	7.37		0.21		0.17		2	A5 V	208
		BSD 9 # 341		2 58 52	+59 22.3	1	11.05		0.40		0.29		2	B9 p	1069
18769	+25 00477	HR 905		2 58 57	+26 15.9	5	5.91	.014	0.14	.008	0.13	.012	13	A3 m	985,1049,1199,3058,8071
		BSD 9 # 344		2 58 58	+59 24.0	1	12.04		0.48		0.29		2	A0	1069
		LS I +59 164		2 58 59	+59 16.3	1	11.76		0.60		-0.20		3		41
		G 75 - 59		2 59 01	+05 46.4	1	13.45		0.82		0.34		2		1620
18836	−17 00587			2 59 01	−17 25.7	1	8.32		0.57		0.07		2	G1 V	1375
		LS I +61 320		2 59 02	+61 29.9	2	11.34	.035	0.53	.030	-0.48	.084	5		41,180
18806	+1 00528			2 59 03	+01 58.3	1	9.15		0.06		-0.12		2	B9 III	208
		Case *M # 32		2 59 06	+59 38.	1	9.74		2.07				1	M3 Ia	138
		CS 22181 # 32		2 59 08	−09 07.8	1	14.14		0.17		0.19		1		1736
		BSD 9 # 943		2 59 09	+60 59.7	1	11.42		0.37		-0.09		2	B9	1069
		4HLF4 # 1010		2 59 12	+00 38.	1	12.89		0.47		-0.02		2	F0	221
18832	+4 00485	HR 908		2 59 14	+05 08.4	5	6.24	.052	1.04	.009	0.82	.014	13	K0	15,1061,1355,3016,8040
18826	+2 00465			2 59 15	+02 56.5	1	8.16		0.11		0.09		2	A2 IV	208
18949		T Hor		2 59 16	−50 50.3	1	8.69		1.45		1.05		1	M2/3e	975
		GD 38		2 59 20	+37 49.3	1	15.55		-0.03		-1.06		1	DA	3060
18438	+78 00103	HR 881	⋆ AB	2 59 21	+79 13.4	1	5.49		1.57				2	M1 III+F7IV	71
	−15 00531			2 59 23	−15 01.4	1	10.47		0.45		0.00		2	G0	1375
		CS 22968 # 4		2 59 23	−54 43.7	1	15.75		0.12		0.13		1		1736
		TonS 307		2 59 24	−30 41.	1	14.64		0.21		0.18		1		832
		LS I +63 196		2 59 26	+63 28.2	1	11.17		0.69		-0.17		3		41
18768	+46 00678			2 59 27	+46 54.9	1	6.74		0.59		0.13		5	F8	595
18803	+26 00503			2 59 28	+26 24.9	2	6.62	.022	0.72	.000	0.31		5	G8 V	1619,7008
18907	−28 00987	HR 914		2 59 28	−28 16.9	2	5.88	.005	0.79	.000			7	G5 IV	15,2012
		GD 39		2 59 29	−01 33.5	1	15.11		0.54		-0.08		2		3060
18873	−3 00482			2 59 29	−03 08.2	1	8.86		0.41		0.20		1	A5	1776
18885	−10 00594	HR 912		2 59 31	−10 09.5	3	5.82	.004	1.10	.010			11	G5	15,1075,2035
18736	+61 00512			2 59 33	+61 46.5	1	8.63		0.17		0.10		3	A1 V	1069
	−17 00588			2 59 34	−16 47.0	1	10.54		1.69				1		1746
		LS I +58 111		2 59 39	+58 04.0	1	10.54		0.66		-0.36		3		41
18884	+3 00419	HR 911, α Cet		2 59 40	+03 53.7	14	2.53	.011	1.64	.007	1.94	.013	67	M1.5 IIIa	3,15,814,1075,1194*
18894	−7 00537	HR 913		2 59 41	−06 41.3	3	6.19	.005	0.60	.005	0.15	.005	9	G0 IV-V	15,1061,3077
		4HLF4 # 1011		2 59 42	−02 08.	1	12.63		0.05		0.20		2	A1	221
		LS I +58 112		2 59 43	+58 01.3	1	11.08		0.81		-0.14		3		41
18883	+3 00420	HR 910		2 59 45	+04 09.4	5	5.62	.033	-0.11	.004	-0.41	.013	13	B7 V	15,154,252,253,1061
		WLS 300-15 # 5		2 59 45	−12 42.7	1	11.38		0.72		0.22		2		1375
18921	−18 00516			2 59 45	−18 24.2	1	7.53		0.33				4	F2 V	2009
		BSD 9 # 365		2 59 46	+59 21.0	1	11.26		0.55		0.28		2	B9 p	1069
		BSD 9 # 360		2 59 46	+59 27.8	1	11.68		0.42		0.28		2	B9	1069
		4HLF4 # 1012		2 59 48	+02 09.	1	10.85		0.64		0.18		1	A5	221
	+29 00514			2 59 48	+29 39.0	1	9.22		1.74		2.09		2	M0	1375
		BSD 9 # 1528		2 59 48	+61 47.1	1	10.96		0.48		-0.07		2	B6	1069
		BSD 9 # 955		2 59 51	+60 53.6	1	11.48		0.56		0.47		2	B8	1069
		CS 22963 # 35		2 59 53	−03 00.6	2	13.22	.000	0.13	.000	0.12	.009	2		966,1736
		G 246 - 21		2 59 55	+61 08.1	1	15.28		1.73		1.27		1		906
		LP 531 - 46		2 59 59	+04 38.9	1	10.97		0.82		0.42		2		1696
18757	+61 00513	IDS02559N6121	A	3 00 03	+61 31.2	2	6.64	.027	0.63	.009	0.13	.022	5	G4 V	22,1003
18903	+16 00380			3 00 05	+17 22.2	1	8.97		0.45		0.14		2	F5	1648
		TonS 308		3 00 06	−26 13.	1	12.64		-0.16		-0.70		1		832

HD	DM	Other Id	N Rem	α₁₉₅₀	δ₁₉₅₀	S	V	σ_V	B–V	σ_B–V	U–B	σ_U–B	n	Spectrum	References
		LP 771 - 100		3 00 11	−16 45.5	2	12.96	.024	0.60	.029	-0.04	.015	3		1696,3062
18978	−24 01387	HR 919		3 00 11	−23 49.2	8	4.08	.015	0.16	.005	0.08	.004	30	A4 V	15,1007,1013,1075*
		MCC 411		3 00 12	+22 10.0	1	10.78		1.14				1	M0	1017
18953	−8 00568	HR 917	★ AB	3 00 15	−07 52.9	2	5.31	.005	0.94	.000	0.73	.000	7	K0 II-III	15,1061
		LS I +57 124		3 00 16	+57 34.5	1	10.56		0.71		-0.33		3		41
		VES 770		3 00 17	+59 36.8	1	12.67		0.85		0.15		1		1543
		BSD 9 # 377		3 00 17	+59 45.6	1	10.89		0.50		0.32		2	B9	1069
		CS 22181 # 14		3 00 17	−11 32.1	1	12.38		0.13		0.13		1		1736
		GD 40		3 00 21	−01 20.2	1	15.56		-0.05		-0.92		3		3060
	+16 00382			3 00 25	+16 49.1	1	10.02		1.30				1	F5	1245
18842	+50 00689			3 00 25	+50 45.8	1	8.88		0.46		0.01		2	F8	1566
18975	−2 00538	IDS02580S0229	A	3 00 29	−02 16.9	1	7.51		0.52		0.02		2	F5	3026
18975	−2 00538	IDS02580S0229	B	3 00 29	−02 16.9	1	9.92		1.06		0.97		2		3024
18878	+47 00760	V509 Per		3 00 30	+47 39.2	1	6.50		0.30		0.29		5	F0	595
		CS 22181 # 15		3 00 30	−12 11.5	1	12.74		0.10		0.04		1		1736
		LP 771 - 72		3 00 30	−18 21.0	1	11.80		1.54				1		1705
		BSD 9 # 382		3 00 34	+59 15.9	1	11.37		0.44		0.21		2	B9	1069
237060	+58 00554			3 00 36	+59 15.1	1	9.15		0.33		-0.11		4	B8	1069
		4HLF4 # 17		3 00 36	−01 41.	1	12.10		0.32		0.08		2	A5	208
	−5 00566			3 00 36	−05 14.5	1	9.19		0.49		0.01		1	F5	1776
		LS I +56 082		3 00 43	+56 01.8	1	11.97		0.94		-0.09		3		41
		LS I +57 125		3 00 43	+57 33.0	1	10.81		0.67		-0.24		3		41
		BSD 9 # 37		3 00 45	+59 07.2	1	10.98		0.51		0.24		2	B9	1069
18972	+13 00494			3 00 47	+14 16.6	1	7.86		1.05		0.77		3	K0 IV	1648
		G 36 - 50		3 00 49	+29 24.1	4	12.02	.016	0.64	.010	-0.11	.007	6		316,927,1620,3060
18877	+59 00589			3 00 58	+59 49.9	1	8.33		0.18		-0.10		2	B7 II-III	1069
		4HLF4 # 18		3 01 00	+02 43.	1	12.92		0.36		0.09		2	A3	208
		4HLF4 # 19		3 01 06	+00 29.	1	12.76		0.16		0.18		2	A0	208
		4HLF4 # 1013		3 01 06	+01 54.	1	12.60		1.44		1.73		2	K0 III	221
18876	+62 00512			3 01 09	+62 50.1	1	7.47		0.04		-0.35		2	B8	1375
19034	−6 00594	G 75 - 62		3 01 09	−05 51.5	3	8.09	.012	0.67	.000	0.12	.009	5	G5	742,1620,3026
18925	+52 00654	HR 915	★ AP	3 01 10	+53 18.7	5	2.93	.010	0.70	.011	0.45	.004	17	G8 III+A2 V	15,1008,1118,3026,8015
19019	+5 00444			3 01 11	+05 56.3	1	6.76		0.52		-0.04		2	F8	8040
		4HLF4 # 20		3 01 12	+00 30.	1	12.48		0.36		0.05		2	A3	208
		BSD 9 # 1541		3 01 12	+61 38.8	1	11.12		0.41		0.29		2	A0	1069
19141	−47 00932	HR 929		3 01 13	−47 10.2	1	5.82		1.30				4	K2/3 III	2032
	+1 00532			3 01 14	+01 26.5	1	10.43		0.20		0.18		2		208
	+58 00555			3 01 17	+59 07.4	1	10.18		0.38		0.23		2	A0	1069
	+61 00516			3 01 23	+61 26.3	1	9.74		0.24		0.03		2		1069
18982	+45 00700			3 01 27	+46 08.6	1	8.73		1.04		0.74		2	G5	1733
		4HLF4 # 1014		3 01 30	−01 40.	1	11.64		1.07		0.96		2	G8 III	221
		BSD 9 # 1552		3 01 31	+61 51.8	1	13.24		0.55		0.32		3	B5	1069
18981	+49 00841			3 01 33	+49 31.5	1	8.36		0.24		0.17		3	A0	595
18992	+47 00763			3 01 34	+47 42.8	1	8.41		0.05		-0.58		2	B9	1375
		BSD 9 # 399		3 01 35	+59 40.3	1	11.35		0.55		0.37		2	B9	1069
		4HLF4 # 21		3 01 36	−01 16.	1	13.59		0.03		-0.01		2	A0	208
19082	−5 00568			3 01 37	−05 26.3	1	7.18		1.60		1.95		3	K5	1657
	+0 00508			3 01 39	+01 21.6	1	10.55		0.42		0.19		1		221
19115	−18 00527	CW Eri		3 01 41	−17 55.9	1	8.43		0.36		-0.02		11	F2 V	588
18970	+56 00767	HR 918		3 01 46	+56 30.7	3	4.76	.012	1.02	.006	0.85	.015	10	K0 II-III	1355,1363,3016
19108	−9 00583			3 01 46	−08 39.5	1	8.27		0.34		0.11		1	A3	1776
18963	+59 00590			3 01 47	+60 07.6	2	9.20	.014	0.27	.014	0.13		4	A0 V	70,1069
		KUV 03018+0011		3 01 48	+00 10.9	1	16.82		0.33		-0.54		1		1708
		CS 22968 # 11		3 01 48	−53 43.0	1	14.68		0.22		0.13		1		1736
18969	+57 00685	IDS02580N5721	AB	3 01 49	+57 32.4	1	9.18		0.38				2	A2	70
19107	−8 00572	HR 925		3 01 49	−07 47.7	4	5.26	.004	0.21	.005	0.08	.010	15	A8 V	15,1256,2027,3023
19059	+34 00567	G 95 - 4	★ AB	3 01 50	+34 34.8	1	8.66		0.61		0.06		1	G0	333,1620
		CS 22968 # 12		3 01 52	−54 12.0	1	15.27		0.32		-0.06		1		1736
19080	+15 00430	HR 924		3 01 53	+15 39.8	2	6.37	.000	1.39	.005	1.53	.000	5	K3 III	1648,1733
	−75 00215			3 01 54	−75 41.4	1	10.05		0.61		0.05		1		1696
18991	+55 00738	HR 920		3 01 55	+55 52.5	1	6.11		1.02				2	G9 III	70
		BSD 9 # 42		3 01 57	+58 47.8	1	10.64		0.56		0.41		2	B9	1069
19058	+38 00630	HR 921, ρ Per		3 01 58	+38 38.9	6	3.39	.009	1.65	.014	1.75	.058	15	M4 II	15,814,1363,3053,8003,8015
		GD 41		3 02 02	+02 45.3	2	14.83	.033	-0.23	.023	-1.13	.037	4		308,3060
19121	+1 00534	HR 926		3 02 03	+01 40.2	5	6.05	.014	1.04	.008	0.86	.009	13	K0 III	15,221,252,1415,3016
19057	+40 00663			3 02 04	+40 55.6	1	7.28		0.07		0.05		2	A0	401
	+59 00591			3 02 05	+59 28.4	1	10.48		0.28		0.20		3	B8 III	1069
19066	+40 00664	HR 923		3 02 06	+40 23.3	2	6.05	.002	1.01	.001	0.80		5	K0 III	71,1501
19079	+29 00518			3 02 07	+30 00.1	1	8.78		0.42		-0.01		2	F7 IV	1375
19400	−72 00219	HR 939	★ AB	3 02 08	−72 05.9	4	5.52	.007	-0.14	.010	-0.50	.012	16	B3 V+A0 IV	15,1034,1075,2038
18990	+61 00520	IDS02581N6123	AB	3 02 09	+61 34.7	1	8.75		0.18		0.07		2	A0 V	1069
	−0 00489			3 02 09	−00 08.2	1	10.60		0.74		0.34		2	G5 III	221
		WLS 248 50 # 8		3 02 10	+49 39.2	1	11.70		0.66		0.16		2		1375
19214	−39 00893			3 02 10	−39 21.9	1	8.28		1.04		0.85		2	K0 III	1730
		LS I +58 113		3 02 11	+58 19.2	2	11.78	.023	0.80	.005	0.15	.019	4		41,1543
		GD 426		3 02 11	−62 10.8	1	14.95		0.22		-0.57		1	DA	782
19031	+52 00656			3 02 12	+52 30.8	1	8.13		0.20		0.00		2	A0	1375
	+57 00686			3 02 14	+57 54.3	1	9.86		2.14				3		369
		LS I +58 114		3 02 16	+58 38.0	1	10.61		0.70		-0.35		3		41
19254	−51 00699			3 02 16	−51 36.3	1	8.21		0.52				4	F7 V	2012

Table 1   161

HD	DM	Other Id	N	Rem	α_{1950}	δ_{1950}	S	V	σ_V	B–V	σ_{B-V}	U–B	σ_{U-B}	n	Spectrum	References
		4HLF4 # 22			3 02 18	+01 02.	1	13.22		0.22		0.11		2	A2	208
		G 246 - 24			3 02 19	+63 34.4	1	11.18		1.08		1.05		1	K5	1658
19009	+59 00592				3 02 21	+59 23.5	1	8.89		0.27		0.23		2	A0 V	1069
18962	+66 00242				3 02 21	+67 13.1	2	7.91	.024	0.61	.004	0.07		3	G0	1375,6007
	+44 00619	G 78 - 14			3 02 22	+44 53.3	2	9.72	.004	0.78	.007	0.29	.014	3	G8	1620,7010
		4HLF4 # 23			3 02 24	+01 09.	1	12.29		-0.03		-0.13		2	A0	208
19178	-12 00590				3 02 26	-12 21.9	1	8.58		0.48		-0.05		1	F5 V	1776
19319	-60 00236	HR 934			3 02 26	-59 55.9	2	5.10	.005	0.35	.005			7	F0 III/IV	15,2012
	+57 00687	LS I +57 126			3 02 27	+57 39.1	2	10.01	.007	0.92	.011	-0.17	.004	12	B1 Ib	41,1012
19054	+56 00772				3 02 29	+56 41.7	1	8.92		0.27				2	A0	70
		CS 22968 # 15			3 02 29	-54 57.3	1	14.18		0.17		0.10		1		1736
19039	+56 00771	LS I +57 127			3 02 30	+57 19.0	2	7.93	.005	0.64	.019	0.50		4	F0 I	41,6009
19168	+6 00474	IDS02599N0649		A	3 02 33	+07 01.2	1	9.10		1.63		1.92		3	K5	3016
19168	+6 00474	IDS02599N0649		BC	3 02 33	+07 01.2	1	11.25		0.76		0.27		3	K5	3016
19134	+24 00431	HR 927		★ABC	3 02 33	+25 03.7	3	5.46		-0.03	.005	-0.38	.000	6	B7 Vn	1049,1079,3032
19134	+24 00431	IDS02596N2452		E	3 02 33	+25 03.7	1	13.27		0.72		0.21		2		3032
	+53 00611	LS V +53 005			3 02 34	+53 48.3	1	10.22		0.66		-0.30		2		405
		BSD 9 # 46			3 02 34	+59 03.6	1	11.52		0.64		0.50		3	B9	1069
19153	+20 00501				3 02 38	+20 42.6	1	7.73		0.30		0.08		2	A2	1648
		CS 22963 # 18			3 02 41	-06 43.9	2	13.64	.000	0.23	.000	0.16	.008	2		966,1736
		G 174 - 31			3 02 43	+50 52.0	1	12.96		1.52		1.16		1		906
		G 5 - 12			3 02 46	+11 38.2	1	12.86		1.50		1.26		1		1620
	-39 00896				3 02 46	-39 24.2	1	10.74		0.94		0.60		3		1730
		G 5 - 13			3 02 47	+11 38.6	1	12.51		1.45		1.13		1		1620
		CS 22963 # 22			3 02 47	-04 59.5	2	15.23	.000	-0.10	.000	-0.51	.002	2		966,1736
19493	-72 00221				3 02 48	-72 39.2	1	8.14		0.70		0.28		4	G3 IV/V	1731
19165	+27 00478				3 02 49	+27 29.7	1	8.45		0.49		-0.01		2	F6 V	1003
	-6 00602				3 02 49	-06 33.6	1	9.45		0.15		0.12		1	A0	1776
	-39 00898				3 02 57	-39 26.0	1	10.12		0.81		0.37		2	G5	1730
		4HLF4 # 24			3 03 00	+00 12.	1	12.43		0.40		0.07		2	A7	208
		4HLF4 # 25			3 03 00	+00 38.	1	13.27		0.27		0.15		2	A5	208
		Hyades J 201			3 03 00	+05 46.	1	12.13		0.95		0.81		6		1058
		LB 1650			3 03 00	-53 33.	1	12.67		0.24		0.06		2		45
		LS I +55 042			3 03 05	+55 32.8	1	11.20		1.23		0.26		3		41
		Case *M # 33			3 03 06	+55 33.	2	10.90	.083	3.15	.024			3	M3 Iab	138,148
19065	+63 00390	HR 922			3 03 06	+63 51.9	3	5.90	.015	-0.03	.005	-0.07	.047	9	B9 V	252,1079,1379
237065	+56 00774	LS I +57 128			3 03 08	+57 21.6	1	9.57		0.57		0.22		3	F4 II	41
	-0 00493				3 03 09	+00 19.9	1	10.95		0.45		0.04		2	F5	221
		LP 299 - 18			3 03 11	+28 41.7	1	11.60		1.02		0.86		5		1723
		L 298 - 69			3 03 12	-49 38.	1	14.02		0.73		0.08		2		1696
19529	-73 00199	IDS03033S7333		AB	3 03 14	-73 21.9	1	9.02		1.04		0.77		5	G8 III	1731
	+67 00244	RX Cas			3 03 15	+67 23.1	1	8.60		1.12		0.18		2	G3 III:	880
		G 245 - 65			3 03 15	+75 51.3	2	9.79	.016	1.37	.024	1.26		4	M0	196,1017,3072
19216	+33 00578	V383 Per			3 03 26	+33 26.0	1	7.83		0.00		-0.17		1	B9	695
19184	+42 00700				3 03 26	+42 23.8	1	9.30		0.46		0.19		7	F5	1655
		LS I +59 165			3 03 26	+59 43.5	1	11.62		0.67		-0.19		3		41
19206	+38 00635				3 03 30	+39 09.2	1	9.00		0.15		0.10		3	A3 V	1501
	+50 00703	Alpha Per 7			3 03 30	+51 06.1	1	10.00		0.56		0.30		3	F2 V	1048
		LS I +64 096			3 03 30	+64 01.3	1	10.77		1.02		-0.05		3		41
	+63 00392				3 03 32	+63 23.1	1	9.67		2.02				3		369
19258	+11 00434				3 03 33	+11 28.4	1	7.22		1.82				1	M1 III	3040
		CS 22963 # 33			3 03 33	-02 44.0	2	14.86	.000	0.26	.000	0.15	.009	2		966,1736
		KUV 03036-0043			3 03 34	-00 42.8	1	16.21		0.23		-0.70		1	DA + dM	1708
	-15 00539				3 03 37	-14 50.0	1	9.98		1.17		1.05		2	M0	1375
	+39 00710	G 78 - 16		★AB	3 03 38	+40 10.1	2	9.66	.019	1.19	.014	1.06		2	K6	1017,3072
19270	+12 00436	HR 931			3 03 39	+12 59.7	1	5.62		1.08				2	K3 III	71
	+59 00594				3 03 39	+60 17.9	3	9.02	.139	2.79	.017	2.62		8	M1 Iab	138,148,8032
19424	-57 00497				3 03 39	-57 36.0	1	8.41		0.50				4	F5 IV	2012
237069	+55 00739				3 03 40	+55 33.8	1	10.65		3.09		1.66		5	K0	8032
		G 76 - 57			3 03 42	+05 41.9	1	11.49		0.68		0.13		2	K2	1696
		CS 22181 # 20			3 03 44	-12 20.7	1	14.72		-0.11		-0.54		1		1736
18778	+80 00097	HR 906		★A	3 03 48	+81 16.8	4	5.95	.005	0.15	.002	0.09	.000	13	A7 III-IV	15,1007,1013,8015
		LB 1652			3 03 48	-47 13.	1	12.44		-0.01		-0.08		2		45
19305	+1 00543	G 76 - 58			3 03 50	+01 47.2	6	9.07	.018	1.37	.022	1.24	.000	14	M0 V	1003,1017,1197,1705*
		LP 831 - 70			3 03 52	-22 30.8	1	12.43		0.40		-0.27		2		1696
19205	+51 00678				3 03 53	+52 08.5	1	8.69		0.22		0.16		4	A0	1566
		AO 848			3 03 58	+40 44.6	1	10.01		1.20		1.05		1		1722
		BSD 9 # 52			3 03 59	+59 07.3	2	12.33	.040	0.57	.013	-0.11	.013	5	B5	41,1069
		4HLF4 # 26			3 04 00	+02 45.	1	11.49		0.09		-0.03		2	A0	208
		AO 849			3 04 02	+40 33.9	1	10.55		1.63		2.03		1		1722
	+58 00559	LS I +59 167			3 04 05	+59 06.6	2	10.68	.027	0.54	.013	-0.11	.018	5	B5 p	41,1069
19349	-6 00606	HR 935		★A	3 04 05	-06 16.8	3	5.26	.004	1.59	.005	1.78	.010	11	M3 III	15,1061,3005
		BSD 9 # 1606			3 04 08	+61 58.4	1	10.74		0.54		0.02		2	B5	1069
		LS I +60 299			3 04 09	+60 29.8	1	11.88		0.62		-0.16		3		41
	-9 00593				3 04 10	-08 44.1	1	11.06		0.90		0.67		2		1696
19346	+2 00478				3 04 11	+02 31.4	1	7.97		0.11		0.13		2	A2 III	208
	+58 00561				3 04 15	+58 59.2	1	10.50		0.45		-0.07		2	B6	1069
19301	+38 00640				3 04 19	+38 53.7	1	7.83		0.32		-0.04		2	F3 V	105
19279	+46 00692	HR 933			3 04 21	+47 07.0	1	6.41		0.12		0.11		2	A3 Vnn	252
	+48 00851	Alpha Per 12			3 04 21	+48 55.0	1	10.09		0.51		0.03		3	F6 V	1048

HD	DM	Other Id	N Rem	α_{1950}	δ_{1950}	S	V	σ_V	B–V	σ_{B-V}	U–B	σ_{U-B}	n	Spectrum	References
		CS 22963 # 21		3 04 22	−05 02.3	2	14.30	.000	0.00	.000	-0.10	.011	2		966,1736
	+60 00628	LS I +61 321		3 04 23	+61 12.6	3	10.17	.038	0.61	.018	-0.30	.035	6	B0	41,180,1069
		BSD 9 # 448		3 04 26	+60 09.2	1	11.40		0.72		0.32		2	B8 p	1069
19268	+51 00681	HR 930		3 04 28	+52 01.3	1	6.30		-0.01		-0.44		3	B5 V	154
19308	+36 00632	G 78 - 17	⋆ A	3 04 29	+36 25.8	1	7.37		0.66		0.18		1	G0	1620
	+49 00855	Alpha Per 29		3 04 29	+50 22.1	1	10.62		0.60		0.10		4	F8 IV-V	1048
19361	+14 00513			3 04 30	+15 00.7	1	8.03		1.54				4	K3 III	882
		BSD 9 # 451		3 04 31	+59 46.1	1	10.62		0.47		0.31		2	B9	1069
	+40 00671			3 04 33	+40 52.3	1	8.87		1.46		1.66		1	K0	1722
19436	−30 01185			3 04 34	−30 10.7	1	7.12		1.35		1.49		5	K3 III	1673
19374	+17 00493	HR 938, UW Ari		3 04 36	+17 41.3	1	6.10		-0.12		-0.80		2	B1.5V	154
		G 5 - 15		3 04 37	+12 33.9	1	13.50		1.24		0.99		1		1696
19256	+56 00778			3 04 37	+56 49.4	1	6.65		0.59				3	G0	70
19193	+66 00244			3 04 39	+66 43.6	1	8.07		1.24		1.23		19	G5	879
		BSD 9 # 453		3 04 42	+59 36.2	1	11.74		0.49		0.34		2	B9	1069
	+10 00415			3 04 43	+10 29.2	3	10.38	.012	0.57	.000	0.16	.009	17	F5 III	830,1242,1783
	−0 00494			3 04 45	−00 11.5	1	11.67		0.29		0.09		2	A3	208
		LS I +58 115		3 04 47	+58 59.4	2	11.51	.080	0.92	.014	-0.12	.019	6		41,1069
19243	+61 00525	LS I +62 232		3 04 48	+62 11.6	4	6.49	.100	0.29	.047	-0.66	.033	10	B1 V:e	41,180,1012,1212
19390	+14 00514			3 04 49	+15 18.7	1	9.72		1.12				2	G8 V	882
19488	−34 01122			3 04 49	−33 46.4	3	8.60	.011	1.15	.005	1.14	.010	16	K1 III	158,1775,2034
19322	+50 00706	Alpha Per 40		3 04 53	+50 58.6	1	7.02		0.20		0.36		5	A2	595
	+1 00548			3 04 54	+02 01.8	1	10.68		0.08		0.13		2		208
19356	+40 00673	HR 936, β Per	⋆ A	3 04 54	+40 45.9	3	2.12	.003	-0.05	.001	-0.37	.000	18	B8 V	15,1363,8015
19389	+23 00407			3 04 57	+23 29.8	1	7.98		0.29		0.04		2	A5	1625
		LS I +59 168		3 04 57	+59 05.1	2	11.71	.030	0.68	.000	-0.14	.015	6		41,1069
19467	−14 00604			3 04 57	−13 57.0	4	6.97	.012	0.65	.003	0.13	.006	18	G3 V	158,1775,2007,3031
		AO 851		3 04 58	+41 02.1	1	10.57		0.44		-0.02		1		1722
19288	+57 00696			3 04 58	+58 09.5	1	7.46		0.30				2	F0	70
19409	+15 00436			3 05 03	+15 52.9	1	9.41		1.18				2	K0	882
		CS 22968 # 14		3 05 04	−54 42.0	2	13.71	.010	0.75	.019	0.05	.005	3		1580,1736
		CS 22963 # 32		3 05 06	−02 55.3	2	15.32	.000	0.16	.000	0.25	.009	2		966,1736
		CS 22181 # 21		3 05 07	−11 56.0	1	13.43		0.10		0.17		1		1736
19431	+9 00397			3 05 08	+09 43.9	2	7.93	.009	0.29	.009	0.15		4	A0	1242,1320
		MCC 413		3 05 09	−04 10.	1	10.84		1.40				2	M0	1619
	−8 00586			3 05 09	−07 54.4	1	10.20		0.43		-0.05		1	F2	1776
		BSD 9 # 69		3 05 10	+58 42.3	1	10.32		0.59		0.05		2	B6	1069
		BSD 9 # 1970		3 05 16	+62 16.9	1	13.10		0.72		0.12		3	B5	1069
		CS 22963 # 31		3 05 19	−02 59.9	2	14.34	.000	0.13	.000	0.21	.009	2		966,1736
19461	+13 00507			3 05 22	+14 05.3	1	8.69		1.68				4	K2	882
		G 221 - 5		3 05 23	+73 35.5	1	14.66		1.79				5		538
	+59 00597			3 05 25	+60 15.0	2	9.36	.027	0.21	.013	-0.24	.040	5	B9p	196,1069
		TU Per		3 05 26	+53 00.2	1	11.94		0.68		0.46		1	A5	668
		AO 852		3 05 27	+40 37.1	1	10.88		1.16		1.05		1		1722
19373	+49 00857	Alpha Per 44	⋆ A	3 05 27	+49 25.4	16	4.05	.008	0.60	.007	0.12	.010	69	G0 V	1,15,22,1004,1008,1077*
19341	+58 00563			3 05 27	+59 10.0	1	8.13		0.12		-0.17		2	B8 III	1069
19445	+25 00495	G 37 - 26		3 05 29	+26 09.1	10	8.06	.017	0.46	.007	-0.24	.006	21	A4p	22,333,792,908,1003,1097*
		Alpha Per 28		3 05 30	+47 09.4	1	10.64		0.75		0.30		4	G0 V	1048
19460	+18 00414	HR 940		3 05 31	+18 36.3	3	6.26	.013	1.58	.017	1.98	.024	7	M0 III	1355,1733,3051
		BSD 9 # 1637		3 05 32	+61 49.7	1	11.09		0.30		0.02		2	B8	1069
19277	+67 00246			3 05 34	+68 10.4	1	8.92		0.34		0.33		4	A2	1733
	−0 00496			3 05 35	+00 07.0	2	9.66	.005	1.21	.000	1.21	.005	7	K0 III	221,1775
		4HLF4 # 1020		3 05 36	+00 06.	1	11.75		0.47		0.09		2	F2	221
19512	−2 00556			3 05 36	−02 00.7	2	9.45	.009	0.25	.000	0.20	.033	8	A7 III	208,1775
19495	+12 00443			3 05 37	+13 21.4	2	8.83	.045	1.49	.017	1.60		5	K0	882,3040
		Alpha Per 33		3 05 37	+47 41.6	1	11.15		0.58		0.04		3	F8	1048
	+40 00674			3 05 40	+40 27.7	1	10.06		0.57		0.06		1	F8	1722
		CS 22968 # 13		3 05 42	−54 38.3	2	13.72	.000	0.17	.000	0.13	.008	2		966,1736
19545	−28 01028	HR 943		3 05 43	−28 01.3	2	6.18	.005	0.16	.000			7	A3 V	15,2012
		LS I +55 043		3 05 44	+55 05.5	1	11.54		1.54		0.83		3		41
19510	+9 00398	X Ari		3 05 48	+10 15.4	2	9.01	.042	0.29	.013	0.18	.019	2	A0	668,699
		HA 9 # 104		3 05 48	+60 29.6	1	10.78		0.60		0.36		3		196
		LTT 1477		3 05 48	−28 23.	1	13.10		1.65		1.22		3		3062
		LB 1653		3 05 48	−51 59.	1	13.14		0.29		0.00		2		45
	−28 01030			3 05 49	−28 24.4	2	10.20	.002	1.37	.046	1.08		7	K7 V	1619,3062
19459	+40 00676			3 05 50	+40 39.2	1	8.52		0.36		0.06		1	F2	1722
19443	+40 00675			3 05 50	+41 06.3	1	7.89		1.05		0.80		1	G5	1722
		LS I +56 083		3 05 55	+56 36.4	1	11.66		1.08		0.08		3		41
19518	+14 00518			3 05 57	+15 08.8	1	7.84		0.62		0.13		2	G8 V	1648
19525	+7 00478	HR 942		3 05 58	+08 16.8	4	6.28	.007	1.05	.007	0.88	.006	15	G9 III	15,1061,1355,1509
		HA 9 # 16		3 05 58	+60 17.8	1	11.91		0.44		0.35		3		196
		4HLF4 # 27		3 06 06	+00 04.	1	13.51		0.15		0.20		2	A0	208
19476	+44 00631	HR 941	⋆ A	3 06 07	+44 40.2	6	3.80	.011	0.98	.002	0.83	.000	17	K0 III	15,1355,1363,3051*
		VES 776		3 06 08	+60 33.5	2	10.92	.020	0.38	.005	0.21	.015	3		1069,1543
		OE 110 # 1		3 06 08	+10 15.3	1	10.43		0.55		0.05		2		487
		AO 856		3 06 08	+40 32.5	1	10.46		1.24		1.12		1		1722
	+61 00529			3 06 08	+61 45.7	1	9.54		0.31		-0.32		3	B3 V	1069
		OE 110 # 2		3 06 09	+10 16.4	1	12.39		1.78		1.77		5		487
		G 78 - 19		3 06 09	+45 32.8	2	10.16	.004	1.47	.033	1.20	.024	6		1625,7010
	−24 01458			3 06 13	−24 20.9	1	10.14		1.19				2	K4	1619

Table 1   163

HD	DM	Other Id	N	Rem	α_{1950}	δ_{1950}	S	V	σ_V	B–V	σ_{B-V}	U–B	σ_{U-B}	n	Spectrum	References
		G 76 - 62			3 06 17	+09 50.4	1	14.86		1.71		1.30		6		316
		OE 110 # 3			3 06 18	+10 18.6	1	15.32		0.98		0.39		2		487
	+61 00530				3 06 20	+61 37.7	1	10.77		0.29		-0.30		3	B9	1069
19441	+58 00567	LS I +59 169			3 06 22	+59 20.1	2	7.91	.028	0.36	.005	-0.39	.038	6	B3 III	41,1069
		Alpha Per 56			3 06 23	+48 16.9	1	10.84		0.81		0.36		3	G3 V	1048
		OE 110 # 4			3 06 26	+10 15.2	1	13.76		0.95		0.63		2		487
19275	+73 00168	HR 932			3 06 28	+74 12.4	4	4.86	.030	0.03	.016	0.04	.010	10	A2 Vnn	15,1363,3023,8015
	+9 00402				3 06 29	+10 01.4	1	10.33		0.68		0.15		7	F8	1320
19482	+55 00745				3 06 29	+56 02.4	1	9.21		0.24				2	A2	70
		TV Eri			3 06 29	-19 39.1	1	12.90		0.53		0.00		4		3064
	+40 00680				3 06 33	+40 44.6	1	8.60		1.55		1.81		1	K5	1722
19548	+28 00499	HR 944			3 06 36	+28 53.3	3	5.72		0.13	.022	-0.15	.007	14	B8 III	1049,1079,1405
19743	-62 00257				3 06 36	-61 54.5	2	7.06	.000	0.93	.009	0.59		8	G6/8 III	158,1075
		BSD 9 # 1671			3 06 37	+61 19.3	1	12.25		0.57		-0.13		2	B5	1069
		CS 22963 # 30			3 06 37	-02 25.0	1	13.53		0.32		0.10		1		966
		CS 22963 # 30			3 06 37	-02 25.0	1	13.53		0.32		0.12		1		1736
19585	+2 00488				3 06 38	+03 03.2	1	8.10		0.05		0.01		2	A1 IV	208
		OE 110 # 5			3 06 38	+10 17.5	1	12.99		0.63		0.08		2		487
		OE 110 # 7			3 06 39	+10 14.8	1	14.35		1.09		1.20		2		487
		OE 110 # 6			3 06 40	+10 16.2	1	12.97		1.06		0.60		2		487
	+62 00524				3 06 40	+62 36.7	1	10.50		0.68		0.11		2	F2 II:	684
		4HLF4 # 28			3 06 42	+00 58.	1	11.66		0.33		0.15		2	A2	208
		4HLF4 # 29			3 06 42	+01 47.	1	12.80		0.29		0.17		2	A2	208
	+61 00532				3 06 46	+61 24.9	1	9.34		0.17		0.07		2	B9 V	1069
		HA 9 # 30			3 06 48	+60 15.1	1	10.74		0.66		0.20		3		196
19583	+16 00396				3 06 52	+17 04.3	3	7.74	.010	1.54	.009	1.81	.024	7	K2	882,1648,3040
		4HLF4 # 30			3 06 54	+00 57.	1	14.58		0.04		0.26		2	B9	208
		G 95 - 11			3 06 56	+34 39.6	1	11.95		0.58		-0.04		1		333,1620
	+40 00681				3 06 56	+40 36.9	1	10.03		0.24		0.16		1	A2	1722
		LP 943 - 16			3 06 57	-34 22.1	1	11.12		0.97		0.79		1		1696
19620	+4 00501	IDS03043N0449		AB	3 06 58	+05 00.8	1	7.99		0.58		0.06		4	G1 V +G2 V	3030
237079	+59 00602				3 06 59	+60 20.5	1	9.37		1.14		0.80		3	K0	196
19501	+60 00633				3 06 59	+60 46.0	2	8.30	.094	0.24	.005	0.09	.042	4	A5 III	196,8100
		BSD 9 # 494			3 07 01	+59 27.5	1	11.82		0.60		0.42		3	B9	1069
237080	+58 00568				3 07 04	+59 22.4	1	8.92		0.40		-0.02		2	B8	1069
19609	+15 00438				3 07 05	+15 29.3	1	8.55		1.64				4	K2	882
19780	-57 00502	NGC 1252 - 35			3 07 06	-57 12.0	1	8.98		1.23		1.38		2	K2 III	995
		LP 772 - 7			3 07 07	-15 51.1	1	11.78		0.70		0.13		2		1696
		BSD 9 # 1683			3 07 08	+61 34.4	1	10.24		0.38		0.27		2	A0	1069
19600	+27 00480	HR 945			3 07 09	+27 37.9	2	6.42		0.01	.018	0.00	.028	13	A0 V	1049,1405
	+49 00863	Alpha Per 93			3 07 09	+50 20.2	1	11.09		0.70		0.16		3	G4	1048
19619	+13 00515				3 07 10	+13 30.3	1	8.50		1.80		2.17		6	K5	3040
		Alpha Per 92			3 07 12	+50 09.4	1	11.06		0.65		0.24		3	F2	1048
	+1 00557	IDS03046N0150		AB	3 07 13	+02 01.0	1	10.97		0.10		0.04		2	A0	208
19618	+14 00524	G 5 - 17			3 07 13	+15 11.3	5	9.06	.014	0.83	.008	0.50	.003	7	K0 IV-V	516,1003,1620,1658,3077
19819	-60 00242				3 07 13	-60 21.9	1	9.35		1.22		1.14		2	K5/M0 V	3072
		BSD 9 # 502			3 07 14	+59 46.7	1	11.38		0.42		0.08		2	B9	1069
19649	+0 00531				3 07 16	+01 17.7	1	7.65		0.36		0.03		2	F0	221
		G 79 - 5			3 07 16	+14 50.2	1	15.28		1.54				1		906
19537	+55 00747				3 07 18	+55 32.7	1	8.68		0.57		0.06		2	G0	1502
		G 79 - 6			3 07 19	+12 48.6	2	11.11	.020	1.02	.015	0.90	.015	2		333,1620,1696
	+68 00224				3 07 20	+69 03.3	1	9.89		0.60		0.03		3		1723
19558	+52 00663				3 07 21	+52 57.3	1	7.37		0.30		0.02		2	A5	1566
		HA 9 # 39			3 07 21	+60 13.2	1	10.68		1.26		0.93		3		196
		G 78 - 20			3 07 22	+37 11.5	1	10.89		0.91		0.53		2		7010
19940	-69 00174	HR 959			3 07 25	-69 27.3	3	6.13	.007	1.01	.003	0.84	.005	20	K1 III	15,1075,2038
		AJ89,1229 # 202			3 07 29	+05 13.6	1	12.83		1.84				2		1532
19637	+26 00516	HR 946			3 07 29	+26 42.4	2	6.02	.000	1.28	.000	1.28	.000	4	K3 III	15,1003
	+11 00444	G 5 - 18			3 07 30	+11 51.8	1	9.43		1.15		1.05		1	K5	333,1620
19536	+60 00536				3 07 31	+60 26.8	2	7.36	.079	0.09	.045	0.26		5	A2 II	1379,8100
19688	-5 00589				3 07 33	-05 32.7	1	8.51		0.32		0.05		50	F0	978
19557	+57 00702	V623 Cas			3 07 34	+57 42.9	2	7.58	.115	2.08	.079			4	R5	70,1238
	+60 00635				3 07 36	+61 13.0	1	9.50		0.22		-0.07		2	B7 III	1069
20003	-72 00230				3 07 37	-72 30.8	1	8.38		0.75		0.35		4	G8 V	1731
	+54 00644	LS I +55 044			3 07 39	+55 12.0	1	9.94		0.61		-0.16		3	A2	41
	+0 00532				3 07 43	+01 16.4	1	10.49		0.54		0.04		1		221
		HA 9 # 46			3 07 46	+60 21.4	1	11.86		0.68		0.14		3		196
19712	-2 00563	EE Dra			3 07 46	-01 53.0	2	7.35	.000	-0.03	.000	-0.11	.032	30	A2 V	208,1775
19645	+31 00553				3 07 47	+32 02.3	1	7.26		1.06		0.85		2	K0	1625
		Alpha Per 94		V	3 07 48	+47 59.3	1	10.42		0.64		0.13		4		1048
	+60 00637				3 07 52	+61 19.7	1	10.14		1.36				1	K5 III	1017
		CS 22181 # 35			3 07 54	-08 16.2	1	13.29		0.29		0.08		1		1736
19740	-9 00603				3 07 54	-09 18.9	1	9.42		0.54		0.07		1	F8	1776
19698	+11 00445	HR 948			3 07 55	+11 41.1	2	5.98		-0.06	.009	-0.28	.005	4	B8 V	1049,1079
	+61 00533				3 07 55	+62 20.1	1	9.18		0.07		-0.34		4	B8	1069
	+59 00603				3 07 59	+60 06.4	1	10.36		0.29		0.06		2	A0 III	1069
19697	+13 00519				3 08 00	+14 09.3	1	8.89		1.68				3	K2	882
19739	-0 00498				3 08 02	+00 01.5	1	7.29		0.24		0.09		2	A8 IV	208
19656	+39 00724	HR 947		⋆ A	3 08 03	+39 25.4	6	4.61	.022	1.12	.006	1.04	.011	16	K1 III	15,1355,1363,3016*
		LS I +59 170			3 08 03	+59 03.2	1	12.42		0.79		-0.11		3		41

HD	DM	Other Id	N	Rem	α_{1950}	δ_{1950}	S	V	σ_V	B–V	σ_{B-V}	U–B	σ_{U-B}	n	Spectrum	References
19624	+51 00689	Alpha Per 145			3 08 06	+51 58.5	1	6.88		0.03		-0.24		3	B5	1048
	+61 00534				3 08 06	+61 23.8	1	9.47		0.17		-0.04		3	B8 V	1069
		LP 299 - 31			3 08 08	+29 56.5	1	11.41		0.67		0.38		3		1723
19754	−5 00592	EL Eri			3 08 09	−05 35.0	1	8.20		1.12		0.75		1	K0	1641
		HA 9 # 57			3 08 10	+60 17.6	1	12.02		0.63		0.36		3		196
	+59 00604				3 08 10	+60 21.8	1	9.92		0.32		0.16		3	B9 III	196
	+55 00748	LS I +55 045			3 08 11	+55 52.9	1	10.15		0.52		-0.19		3		41
	+60 00639	G 246 - 28		★ AB	3 08 11	+60 44.4	1	9.34		1.02		0.85		3	K0 V	196
19655	+47 00776	Alpha Per 104			3 08 13	+47 52.0	2	8.62	.005	0.34	.000	0.02	.011	44	F2 Vn	588,1048
19556	+66 00247				3 08 14	+67 12.9	1	8.40		0.64		0.10		31	G5	880
	+49 00868	Alpha Per 135			3 08 17	+50 11.5	1	9.71		0.49		0.01		3	F5 V	1048
	+9 00409	Hyades vB 133			3 08 18	+09 48.1	1	9.67		0.61		0.06		4	G0	1127
	−35 01099				3 08 19	−34 51.8	1	10.24		0.69		0.17		2		1696
19760	+1 00561				3 08 20	+02 07.7	1	6.78		1.04		0.91		1	K0	379
19826	−24 01480	HR 953			3 08 23	−23 55.6	4	6.37	.004	0.93	.005			18	K0 III	15,1075,2020,2029
	+2 00493				3 08 25	+02 43.3	1	11.22		0.36		0.12		2		208
		HA 9 # 250			3 08 25	+60 41.2	1	11.86		0.52		0.24		3		196
19938	−55 00495				3 08 25	−55 40.6	1	7.62		1.43		1.70		5	K3/4 III	1673
		LS I +56 084			3 08 27	+56 20.4	1	10.47		0.61		-0.25		3		41
	+8 00469	G 79 - 10			3 08 28	+09 00.5	1	11.31		1.25		1.25		1	K6	333,1620
	+49 00870	Alpha Per 143			3 08 30	+50 12.3	1	10.47		0.71		0.22		3	F8 IV-V	1048
		BSD 9 # 1102			3 08 30	+60 58.1	1	11.48		0.44		0.21		2	B9	1069
19916	−51 00737				3 08 30	−51 01.3	1	7.56		0.58		0.14		5	F8/G0 V	1628
		BPM 17088			3 08 30	−56 35.	1	14.07		-0.11		-1.06		3	DB	3065
19949	−57 00505	NGC 1252 - 34			3 08 30	−57 08.2	1	8.68		1.58		1.96		2	K5 III	995
		CS 22167 # 3			3 08 34	−04 34.0	1	13.15		0.26		0.13		1		1736
		CS 22968 # 20			3 08 34	−56 54.3	2	14.35	.015	0.22	.025	0.14	.017	2		966,1736
19825	−11 00604	IDS03061S1131	B		3 08 35	−11 17.3	1	8.66		0.30		-0.01		8	A3	1320
	+30 00500	G 37 - 27			3 08 36	+31 04.3	1	10.57		1.12		1.01		1		333,1620
19789	+12 00452	HR 952			3 08 37	+12 51.6	2	6.12	.035	1.04	.015	0.86	.015	5	K0 IIIp	1733,3016
19644	+59 00607	V638 Cas			3 08 37	+59 43.9	3	8.26	.029	0.26	.022	-0.41	.019	6	B3 II-III	41,180,1069
		HA 9 # 163			3 08 37	+60 29.2	1	11.43		0.40		0.25		3		196
		G 5 - 19			3 08 42	+12 26.3	2	11.15	.019	0.60	.005	-0.08	.015	3		333,1620,1696
		HA 9 # 252			3 08 42	+60 39.2	1	12.00		0.52		0.38		3		196
19653	+60 00640				3 08 45	+60 36.8	1	8.90		0.27		0.18		3	B9p	196
19787	+19 00477	HR 951			3 08 46	+19 32.3	8	4.35	.013	1.03	.005	0.90	.023	32	K2 III	3,15,1105,1355,1363*
19836	−4 00540	HR 955		★ A	3 08 48	−04 00.0	3	6.04	.004	1.66	.005	1.93	.005	9	M1 III	15,1415,3001
19948	−49 00884	HR 960			3 08 50	−48 55.3	3	6.11	.008	1.12	.000			11	K1 III	15,1075,2032
19736	+41 00631	HR 950			3 08 51	+42 11.3	1	6.15		-0.09		-0.57		3	B4 V	1079
		LS I +68 019			3 08 51	+68 36.3	1	11.30		0.63		-0.09		3		41
		LB 1656			3 08 54	−46 12.	1	13.80		0.46		-0.04		2		45
		G 5 - 20			3 08 56	+19 29.3	1	11.06		1.49		1.07		1		1696
19887	−16 00587	HR 957			3 08 57	−16 12.8	4	6.25	.007	1.19	.005	1.27		13	K2 III	15,252,1075,2035
19735	+47 00779	Alpha Per 138		★ AB	3 08 58	+47 32.4	5	6.34	.011	1.44	.007	1.64	.025	8	K2 V	15,1003,1355,3010,4001
		BSD 9 # 535			3 08 58	+60 09.3	1	11.10		0.29		0.14		2	B9	1069
	+60 00641				3 08 58	+60 30.5	1	10.66		0.47		0.33		3	F2 V	196
		NGC 1252 - 20			3 08 58	−57 54.0	1	12.41		0.93		0.76		1		995
		LS I +59 172			3 09 02	+59 34.5	2	11.74	.054	0.94	.000	-0.12	.000	5		41,1069
		RX For			3 09 03	−26 40.5	2	11.36	.240	0.16	.055	0.09	.030	2		597,700
19756	+45 00726				3 09 05	+45 55.8	1	8.49		0.31		0.32		5	A0	595
	+58 00574	LS I +58 116			3 09 06	+58 45.3	2	10.06	.028	0.63	.005	-0.34	.014	4	B1	41,1069
		L 54 - 20			3 09 06	−74 35.	1	11.65		0.82		0.41		2		1696
19833	+15 00446				3 09 08	+15 58.7	1	7.45		1.61				1	K5	882
19661	+63 00402	LS I +63 197			3 09 08	+63 25.3	1	8.78		0.17		-0.45		3	B9	41
		HA 9 # 359			3 09 10	+60 44.2	1	11.89		0.44		0.07		3		196
19991	−51 00738	SV Eri			3 09 10	−50 56.1	1	9.61		0.29		0.18		1	K0 (III)	700
20052	−62 00260				3 09 10	−62 32.7	3	8.32	.004	0.58	.000	0.09	.000	9	F8 V	164,258,1075
20060	−64 00229	IDS03083S6417	AP		3 09 10	−64 06.1	1	6.65		0.13				3	A3 III/IV	173
	+60 00642				3 09 11	+60 41.8	1	9.64		0.46		0.26		3	A4 III	196
		CS 22968 # 46			3 09 11	−52 29.5	2	15.81	.015	0.14	.000	0.15	.020	2		966,1736
20313	−79 00091	HR 981, BN Hyi		★ A	3 09 11	−79 10.8	1	5.70		0.32		0.19		3	F0 III	3013
20313	−79 00091	HR 981, BN Hyi		★ AB	3 09 11	−79 10.8	3	5.55	.012	0.30	.004	0.07	.015	11	F0 III	15,1075,2038
20313	−79 00091	IDS03109S7922	B		3 09 11	−79 10.8	1	8.06		0.42		0.04		2		3013
		NGC 1252 - 29			3 09 12	−57 46.7	1	12.42		0.56		0.02		1		995
19747	+51 00693				3 09 13	+52 21.0	1	9.13		0.44		0.19		3	F0	1566
19823	+29 00535				3 09 14	+29 38.0	1	8.46		0.63				4	G0 V	20
19767	+47 00780	Alpha Per 151			3 09 14	+47 39.1	1	8.97		0.32		0.11		3	F0 Vn	1048
19832	+26 00523	HR 954, SX Ari			3 09 15	+27 04.2	4	5.73	.090	-0.11	.018	-0.44	.018	9	B9p Si	401,1049,1263,3033
		CS 22963 # 19			3 09 15	−06 35.3	3	15.20	.021	0.08	.003	0.15	.008	3		966,1736,1736
20038	−59 00258				3 09 15	−59 01.0	2	8.90	.005	0.84	.010	0.42		3	F7wG6 III	742,1594
		HA 9 # 262			3 09 16	+60 31.8	1	11.38		0.31		0.23		3		196
		G 5 - 21			3 09 18	+20 23.9	1	12.11		0.95		0.65		2		7010
		NGC 1252 - 28			3 09 23	−57 58.6	1	11.95		0.47		0.01		1		995
		WLS 300-15 # 8			3 09 24	−14 53.3	1	11.68		0.80		0.62		2		1375
19935	−17 00614				3 09 24	−16 45.9	1	8.33		1.24		1.22		2	K1 III	1375
		CS 22963 # 25			3 09 27	−03 59.7	3	13.78	.164	0.28	.071	0.15	.045	3		966,1736,1736
		NGC 1252 - 19			3 09 27	−57 53.4	1	13.21		0.83		0.65		1		995
		LP 299 - 51			3 09 28	+32 43.1	1	11.51		0.66		0.01		4		1723
19931	−11 00607				3 09 28	−11 32.6	2	9.59	.036	0.30	.032	0.12	.032	2	A0	668,699
		NGC 1252 - 18			3 09 30	−57 54.6	1	12.49		0.73		0.26		1		995

Table 1   165

HD	DM	Other Id	N	Rem	α_{1950}	δ_{1950}	S	V	σ_V	B–V	σ_{B-V}	U–B	σ_{U-B}	n	Spectrum	References
	−46 00943				3 09 32	−46 42.9	2	11.52	.060	1.45	.015	1.21		2	K4	1705,3062
19910	−0 00504				3 09 33	−00 16.6	1	9.32		0.30		0.12		2	A9 V	208
19805	+48 00862	Alpha Per 167			3 09 35	+48 49.4	1	7.94		0.12		0.03		3	A0 Va	1048
	+49 00873	Alpha Per 185			3 09 35	+50 15.8	1	9.92		1.06		0.93		2	K5	1723
20059	−58 00266	NGC 1252 - 1			3 09 35	−57 53.3	1	8.68		1.03		0.96		2	K1 III/IV	995
		NGC 1252 - 12			3 09 35	−57 58.4	1	11.95		0.50		-0.03		1		995
19928	−0 00506				3 09 36	−00 26.0	1	7.68		1.08		1.01		21	K0	221
		NGC 1252 - 27			3 09 41	−57 59.1	1	13.17		0.49		0.06		1		995
		BSD 9 # 1134			3 09 43	+60 57.2	1	10.96		0.23		0.11		3	B9	1069
		NGC 1252 - 16			3 09 45	−57 56.0	1	13.21		0.78		0.23		1		995
19926	+6 00496	HR 958		⋆	3 09 47	+06 28.4	4	5.55	.010	1.06	.014	0.66	.039	12	K1 IIIep+	15,1061,1355,6009
		CS 22968 # 45			3 09 47	−52 50.5	2	14.22	.000	0.12	.000	0.12	.009	2		966,1736
		NGC 1252 - 17			3 09 47	−57 53.1	1	11.92		1.51		2.00		2		995
		BSD 9 # 1137			3 09 48	+60 52.6	1	12.09		0.38		0.30		2	A0	1069
		NGC 1252 - 14			3 09 49	−57 57.6	1	12.91		0.60		0.06		1		995
20093	−57 00508	NGC 1252 - 30			3 09 52	−57 22.8	1	9.66		0.51		0.05		2	F5	995
19882	+38 00662				3 09 54	+38 47.2	2	7.97	.005	0.31	.000	0.12	.005	6	F3 IV	833,1601
	−57 00509	NGC 1252 - 24			3 09 54	−57 27.5	1	11.39		0.51		-0.04		1		995
		NGC 1252 - 11			3 09 54	−57 58.9	1	10.50		1.05		0.95		2		995
19845	+47 00782	Alpha Per 175		⋆	3 09 55	+47 59.4	1	5.90		0.97		0.81		3	G8 III	1048
		NGC 1252 - 15			3 09 56	−57 55.9	1	11.47		1.21		1.28		2		995
20010	−29 01177	HR 963		⋆ AB	3 09 57	−29 11.0	7	3.85	.018	0.52	.013	0.02	.035	18	F8 IV	15,938,1075,1425,2012*
		NGC 1252 - 25			3 09 59	−57 26.9	1	11.96		0.69		0.20		1		995
	+59 00608	LS I +59 173			3 10 00	+59 39.3	3	9.94	.038	0.41	.027	-0.40	.022	7	B2 III	41,180,1069
		NGC 1252 - 31			3 10 00	−57 20.8	1	11.17		0.88		0.58		2		995
	+60 00644				3 10 01	+60 41.3	1	8.50		1.40				3	K0 V	196
	−69 00177			A	3 10 02	−68 47.2	4	11.38	.017	0.03	.009	-0.68	.032	15	DA	45,1698,1705,1765
	−69 00177			B	3 10 02	−68 47.2	1	14.73		0.62				1	DA3	1705
		Alpha Per 192			3 10 03	+48 35.4	1	10.76		0.66		0.24		3	F5 IV-V	1048
19820	+59 00609	CC Cas			3 10 07	+59 22.6	7	7.11	.010	0.51	.014	-0.50	.010	29	O9 IV	41,180,1011,1012,1069*
	+29 00537				3 10 08	+30 14.2	1	10.24		1.32		0.99		2	G5	1375
		LS I +58 117			3 10 08	+58 50.5	1	11.49		0.50		-0.20		3		41
19963	+10 00425				3 10 11	+10 56.8	1	7.98		0.33				1	A0	1242
19957	+16 00403				3 10 11	+16 56.7	1	8.52		0.46		0.00		2	F5	1648
19943	+24 00451				3 10 11	+25 19.7	1	8.59		0.23		0.14		3	A2	1733
19984	+2 00498				3 10 13	+02 30.5	2	8.02	.010	0.15	.005	0.07	.010	3	A5 V	208,379
19994	−1 00457	HR 962		⋆ AB	3 10 13	−01 22.9	5	5.07	.024	0.57	.012	0.11	.011	15	F8 V	15,254,1197,1256,3026
232766	+54 00651				3 10 14	+54 45.0	3	9.23	.011	2.53	.026	2.58		9	M1 Iab	138,148,8032
19893	+49 00876	Alpha Per 212			3 10 18	+49 23.0	1	7.15		0.04		-0.11		3	B9 V	1048
19923	+38 00667	IDS03071N3846		AB	3 10 21	+38 57.5	1	8.09		0.39		0.00		3	F3 V	833
20007	−5 00596				3 10 21	−05 19.1	2	8.45	.005	0.50	.001	0.06	.005	46	F8	978,1776
		NGC 1252 - 21			3 10 21	−57 51.6	1	13.17		0.78		0.58		1		995
	+60 00645				3 10 22	+61 19.8	1	9.57		0.30		-0.02		2	B8 III	1069
20047	−29 01180				3 10 23	−29 36.6	1	8.47		0.41				4	F3 V	2012
20145	−57 00510	NGC 1252 - 32			3 10 25	−57 09.9	1	9.27		1.05		0.78		2	G8 III	995
19844	+61 00540				3 10 28	+61 31.8	1	7.86		0.06		0.03		2	A0 V	1069
	−57 00511	NGC 1252 - 10			3 10 28	−57 42.1	1	9.16		1.08		0.99		2	G8	995
		4HLF4 # 31			3 10 30	+00 46.	1	13.15		0.16		0.20		2	A0	208
		G 5 - 22			3 10 30	+18 39.2	1	14.29		1.46		1.26		2		1705,3078
		Alpha Per 201			3 10 30	+47 08.3	1	9.85		0.61		0.11		2	F8 V	1048
19908	+49 00877	Alpha Per 225		⋆ AB	3 10 30	+50 11.6	1	8.93		0.58		0.36		3	F4 III	1048
	−38 01058				3 10 30	−38 17.2	2	11.47	.040	1.55	.041	1.22		2	M5	1705,3078
		NGC 1252 - 22			3 10 30	−57 49.0	1	12.50		0.82		0.53		1		995
	+60 00646				3 10 34	+61 16.3	1	9.77		0.27		-0.17		3	B6 III	1069
	+3 00443	G 76 - 68			3 10 36	+03 38.2	1	9.96		0.93		0.52		3	K0	333,1620
		NGC 1261 sq1 1			3 10 36	−55 27.8	1	11.62		0.47		-0.06		3		789
		BSD 9 # 1155			3 10 37	+60 50.9	1	10.77		0.31		0.26		2	B9	1069
		G 77 - 31			3 10 40	+04 35.2	2	13.79	.009	1.83	.005	1.15	.000	4		203,3078
20121	−44 01025	HR 968		⋆ AB	3 10 40	−44 36.4	2	5.92	.005	0.44	.000			7	F7 III+A0 V	15,2027
		G 77 - 32			3 10 41	+03 03.7	1	12.30		0.65		-0.09		2		1696
20177	−56 00504				3 10 43	−56 35.4	2	7.19	.005	1.47	.002	1.80		6	K5 III	2012,3040
		LS I +60 301			3 10 46	+60 32.1	3	10.41	.023	0.81	.028	-0.23	.035	9		41,180,1069
19954	+48 00865	Alpha Per 220		⋆ A	3 10 47	+48 23.5	1	9.14		0.33		0.15		3	A9 IV	1048
		NGC 1261 sq1 3			3 10 47	−55 28.6	1	13.60		0.56		-0.04		3		789
20045	−0 00511				3 10 48	−00 00.7	1	8.49		0.22		0.13		2	A5 IV-V	208
20066	−13 00609				3 10 48	−12 54.7	1	9.59		1.38		1.34		2	K3 III	1375
		NGC 1245 - 9			3 10 50	+47 03.3	1	13.77		0.68		-0.31		1		49
		NGC 1245 - 6			3 10 51	+47 01.5	1	12.31		1.19		0.75		1		49
		NGC 1261 sq1 2			3 10 51	−55 17.6	1	13.41		0.79		0.22		5		789
	−58 00267	NGC 1252 - 9			3 10 52	−57 45.5	1	11.03		0.75		0.31		2		995
		NGC 1245 - 23			3 10 54	+47 04.6	1	16.15		0.98				1		49
		NGC 1261 - 1			3 10 54	−55 25.	2	14.13	.005	0.70	.015	0.16	.005	8		142,789
		NGC 1261 - 2			3 10 54	−55 25.	1	14.55		1.24		1.01		3		142
		NGC 1261 - 3			3 10 54	−55 25.	1	13.85		1.56		1.51		3		142
		NGC 1261 - 4			3 10 54	−55 25.	1	15.18		1.17		0.71		3		142
		NGC 1261 - 5			3 10 54	−55 25.	2	15.07	.010	1.04	.010	0.86	.000	7		142,789
		NGC 1261 - 7			3 10 54	−55 25.	1	16.50		0.52				3		142
		NGC 1261 - 9			3 10 54	−55 25.	1	13.37		1.69		1.55		3		142
		NGC 1261 - 10			3 10 54	−55 25.	2	13.77	.020	1.58	.015	1.54	.020	8		142,789
		NGC 1261 - 11			3 10 54	−55 25.	1	14.30		1.36				3		142

HD	DM	Other Id	N Rem	α_{1950}	δ_{1950}	S	V	σ_V	B–V	σ_{B-V}	U–B	σ_{U-B}	n	Spectrum	References
		NGC 1261 - 12		3 10 54	−55 25.	1	14.67		0.99		0.54		3		142
		NGC 1261 - 13		3 10 54	−55 25.	1	15.55		0.74		0.26		3		142
		NGC 1261 - 14		3 10 54	−55 25.	1	15.16		1.06				3		142
		NGC 1261 - 15		3 10 54	−55 25.	1	15.46		0.78		0.32		3		142
		NGC 1261 - 16		3 10 54	−55 25.	1	14.91		1.05		0.69		3		142
		NGC 1261 - 17		3 10 54	−55 25.	2	14.73	.010	1.12	.010	0.73		8		142,789
		NGC 1261 - 18		3 10 54	−55 25.	2	15.88	.015	0.84	.005			7		142,789
		NGC 1261 - 19		3 10 54	−55 25.	2	15.48	.005	1.03	.000	0.56	.100	7		142,789
		NGC 1261 - 20		3 10 54	−55 25.	1	16.74		0.57		-0.06		2		142
20037	−57 00265	NGC 1252 - 13		3 10 54	−57 30.1	1	6.62		0.89		0.55		1	G8 III	995
		L 90 - 3		3 10 54	−65 12.	1	12.14		1.38		1.09		2		3056
		G 76 - 69		3 10 56	+07 31.6	1	11.90		1.44		1.12		1	K5	333,1620
		NGC 1245 - 24		3 10 56	+47 04.1	1	16.24		1.10				1		49
		NGC 1245 - 19		3 10 56	+47 04.3	1	14.97		0.56		0.29		1		49
		NGC 1245 - 7		3 10 57	+47 01.4	1	12.75		0.20		-0.05		1		49
	+58 00578	LS I +58 118		3 10 57	+58 55.1	2	9.74	.036	0.42	.045	-0.46	.049	5	B3 III	41,1069
20089	−10 00638			3 10 57	−10 11.0	1	8.44		0.27		0.05		1	A2	1776
		NGC 1245 - 22		3 10 59	+47 01.8	1	15.81		0.59				1		49
		Alpha Per 228		3 10 59	+48 10.0	1	9.95		0.47		0.16		4	F0 V	1048
	+51 00697	G 174 - 38		3 11 00	+52 10.9	2	10.22	.015	1.29	.025	1.31		4	K5:	1625,7010
		CS 22167 # 6		3 11 00	−05 57.0	1	12.66		0.22		0.15		1		1736
	−58 00268	NGC 1252 - 8		3 11 00	−57 47.9	1	9.07		1.64		1.96		2		995
	−58 00269	NGC 1252 - 2		3 11 00	−57 53.4	1	9.59		0.51		0.01		2	G0	995
		NGC 1245 - 20		3 11 02	+47 04.6	1	15.23		0.91		0.48		1		49
		NGC 1245 - 14		3 11 03	+47 04.2	2	14.22	.015	1.16	.016	0.84	.053	2		49,4002
20043	+18 00432			3 11 04	+18 47.2	1	6.50		1.50		1.76		2	K5	1648
		NGC 1245 - 162		3 11 04	+47 03.2	1	13.86		1.16		0.85		1		4002
		NGC 1245 - 16		3 11 04	+47 03.9	1	14.26		0.44		0.34		1		49
20144	−36 01208	HR 970		3 11 04	−36 07.8	2	6.26	.005	-0.08	.000			7	B8 V	15,2012
		NGC 1261 sq1 7		3 11 04	−55 16.9	1	14.27		1.37		1.42		3		789
	+23 00423			3 11 05	+24 04.7	1	8.19		1.51		1.74		2	K2	1625
		NGC 1245 - 15		3 11 06	+47 04.9	1	14.24		0.65		0.14		1		49
		CS 22968 # 17		3 11 07	−56 00.5	1	14.34		0.23		0.03		1		1736
		NGC 1245 - 12		3 11 09	+47 00.7	1	14.03		0.56		0.35		1		49
		NGC 1245 - 125		3 11 09	+47 02.7	1	13.19		1.28		1.08		2		4002
		NGC 1245 - 25		3 11 09	+47 02.9	1	16.46		0.70				1		49
		NGC 1245 - 5		3 11 09	+47 03.9	1	12.03		0.93		0.66		1	K0 V	49
		NGC 1245 - 143		3 11 09	+47 05.5	1	12.39		1.87		1.66		1		4002
		NGC 1245 - 11		3 11 11	+47 01.8	1	13.96		0.43		0.33		1		49
237090	+59 00611	LS I +59 175		3 11 11	+59 43.6	4	8.99	.062	0.54	.010	-0.37	.038	12	B0.5IV:nn	41,180,1012,1069
		G 95 - 16		3 11 12	+34 32.4	2	12.27	.002	0.84	.017	0.40	.012	3		1620,1658
		NGC 1245 - 10		3 11 12	+47 03.5	1	13.83		1.14		0.90		1		49
		NGC 1245 - 13		3 11 12	+47 05.2	1	14.18		0.52		0.39		1		49
	−58 00270	NGC 1252 - 3		3 11 12	−57 53.4	1	9.61		0.86		0.41		2	G5	995
	+60 00648	LS I +60 302		3 11 13	+60 38.1	3	10.13	.018	0.22	.012	-0.36	.032	8	B5 V	41,180,1069
		Steph 349		3 11 13	−07 03.1	1	12.34		1.46		1.19		1	K7	1746
		G 78 - 24		3 11 14	+48 20.3	1	11.43		1.49		1.13		1	M1	333,1620
19992	+48 00868	Alpha Per 241		3 11 14	+48 23.4	1	8.26		0.33		0.21		3	A0	379
	−10 00639	UY Eri		3 11 15	−10 37.7	2	11.29	.241	0.45	.066	0.07		2		688,1399
		NGC 1245 - 21		3 11 16	+47 02.6	1	15.66		0.67				1		49
		NGC 1245 - 17		3 11 17	+46 58.3	1	14.64		0.45		0.27		1		49
237091	+59 00612	LS I +59 176	★ AB	3 11 17	+59 43.7	4	8.70	.028	0.64	.020	-0.40	.033	17	B1:V:penn	41,180,1012,1069
20234	−57 00513	HR 977, TW Hor		3 11 17	−57 30.5	6	5.73	.020	2.43	.136	2.91	.040	17	C5 II	15,109,864,2007,8015,8022
	+1 00565			3 11 18	+01 41.9	1	9.96		0.37		0.15		2	A8 V	208
		BPM 17113		3 11 18	−54 18.	1	14.75		0.52		-0.42		3	DF	3065
		NGC 1245 - 8		3 11 20	+46 56.9	1	13.58		0.81		0.33		1		49
20086	+15 00450			3 11 21	+15 24.3	1	7.18		0.07		0.04		2	A0	1648
		NGC 1245 - 18		3 11 22	+47 04.5	1	14.85		0.50		0.42		1		49
20023	+46 00713	NGC 1245 - 1		3 11 23	+46 58.2	1	7.95		0.01		-0.17		1	B9 III	49
	+59 00613			3 11 23	+60 03.2	1	10.52		0.30		0.22		2	A1 III	1069
		Case *M # 34		3 11 24	+54 42.	2	9.60	.039	2.84	.015			3	M3.3Iab-Ib	138,148
		4HLF4 # 32		3 11 24	−01 55.	1	13.53		0.25		0.02		2	A5	208
20117	−5 00598			3 11 24	−04 42.1	1	8.97		0.67				4	G5	2033
		NGC 1245 - 3		3 11 25	+47 04.8	1	11.22		0.68		0.14		1	G0 V	49
20017	+48 00870	Alpha Per 247		3 11 25	+48 30.6	2	7.93	.020	0.31	.020	-0.18	.034	5	B5 Ve	379,1212
		BSD 9 # 1174		3 11 25	+60 53.3	1	11.26		0.31		0.25		2	B9	1069
		G 79 - 15		3 11 26	+08 22.2	1	12.29		1.43		1.28		1	K5	333,1620
		UZ Eri		3 11 26	−14 23.7	1	12.34		0.20		0.08		1		699
		G 79 - 16		3 11 27	+08 22.7	1	13.09		1.44		1.21		1	K5	333,1620
		NGC 1245 - 382		3 11 27	+47 06.1	1	13.44		1.36		1.10		1		4002
20115	+0 00542	IDS03089N0022	AB	3 11 28	+00 33.2	3	7.36	.020	0.53	.024	0.08	.098	8	F8	176,938,3026
		NGC 1245 - 44		3 11 28	+47 01.1	1	13.93		1.13		0.79		1		4002
	−44 01028			3 11 28	−44 06.2	1	10.81		0.07		-0.80		17		1732
	+27 00485			3 11 29	+27 41.8	1	10.58		0.47		0.03		2	F2	1375
	+54 00655	V411 Per		3 11 29	+54 42.2	2	9.96	.328	2.76	.055	2.38		7	M3 Iab	369,8032
		Alpha Per 239		3 11 30	+46 57.3	1	10.81		0.51		0.10		2	F6	1048
	−57 00514	NGC 1252 - 23		3 11 31	−57 32.2	1	10.84		0.49		-0.09		2		995
20126	−0 00514			3 11 33	+00 13.4	1	9.01		1.08		0.96		2	K0	221
20176	−30 01238	HR 974		3 11 33	−29 59.4	3	6.16	.004	1.04	.005			11	K1 II	15,1075,2031
19968	+60 00651			3 11 35	+60 56.6	1	7.55		0.08		-0.39		3	B5 III	1069

Table 1   167

HD	DM	Other Id	N Rem	α_{1950}	δ_{1950}	S	V	σ_V	B–V	σ_{B-V}	U–B	σ_{U-B}	n	Spectrum	References
		G 79 - 17		3 11 36	+08 56.8	1	11.16		1.16		1.10		1	K4-5	333,1620
20063	+41 00638	HR 966		3 11 37	+42 19.1	2	6.06	.000	1.09	.014	1.05		5	K2 III	71,1733
		NGC 1245 - 236		3 11 39	+47 05.5	1	14.25		1.20		0.89		1		4002
		NGC 1245 - 342		3 11 42	+47 03.4	1	14.19		1.12		0.66		1		4002
20099	+32 00588			3 11 43	+32 28.6	1	7.97		0.96		0.55		2	G5	1375
20287	−59 00261			3 11 43	−59 35.8	1	9.47		0.29		0.01		2	A8 III/IV	1730
		NGC 1252 - 7		3 11 45	−57 45.1	1	11.97		0.93		0.54		2		995
20151	+1 00566			3 11 46	+02 16.4	1	9.08		0.44		-0.08		35	F8	588
		NGC 1245 - 4		3 11 47	+47 04.1	1	11.57		0.44		0.28		1	A4 V	49
	+2 00501			3 11 48	+02 46.9	1	11.13		0.26		0.11		2		208
		4HLF4 # 33		3 11 48	−01 28.	1	11.95		-0.05		-0.31		2	A0	208
20286	−58 00271	NGC 1252 - 4		3 11 48	−58 01.2	1	9.17		-0.02		0.03		2	B9.5V	995
		CS 22968 # 42		3 11 49	−53 29.9	1	15.59		0.18		0.12		1		1736
		CS 22968 # 43		3 11 51	−53 24.1	2	14.63	.010	0.50	.011	-0.17	.005	3		1580,1736
		NGC 1252 - 6		3 11 51	−57 45.3	1	11.19		0.73		0.26		2		995
	+5 00460			3 11 52	+05 33.4	1	10.30		0.68				1	G5	920
	+48 00871	Alpha Per 270		3 11 52	+49 15.3	1	10.11		0.51		0.01		3	F7 V	1048
	−58 00272	NGC 1252 - 5		3 11 53	−57 48.5	1	10.38		0.47		-0.01		2	F8	995
		4HLF4 # 34		3 11 54	+00 23.	1	12.07		0.25		0.15		2	A2	208
20077	+46 00717	NGC 1245 - 2		3 11 55	+47 05.6	1	9.03		0.25		0.26		1	A4 V	49
237096	+59 00615			3 11 55	+59 50.5	1	9.72		0.28		0.03		3	A2	1069
		CS 22172 # 2		3 11 56	−10 46.3	1	12.73		0.78		-0.14		1		1736
	−15 00560			3 11 56	−15 01.7	2	10.61	.005	0.38	.005	-0.05		4	F0 VI	742,1594
20263	−46 00959			3 11 56	−46 02.3	1	9.11		0.31		0.04		2	A9 IV/V	1730
20041	+56 00798	HR 964	★	3 11 57	+56 57.4	3	5.79	.019	0.71	.020	0.09		7	A0 Ia	70,138,206
		BSD 9 # 600		3 11 57	+59 34.1	1	12.13		0.71		0.47		2	A0	1069
		L 176 - 18		3 12 00	−56 18.	1	14.71		0.16		-0.55		3		3060
20150	+20 00527	HR 972		3 12 01	+20 51.6	5	4.90	.020	0.00	.018	0.00	.005	12	A1 V	15,1049,1363,3023,8015
		NGC 1252 - 26		3 12 03	−57 54.5	1	12.02		0.56		0.02		1		995
20165	+8 00482	G 79 - 18		3 12 04	+08 48.1	5	7.84	.014	0.86	.003	0.58	.010	19	K0	22,333,1197,1355,1620,3008
		LS I +59 177	A	3 12 04	+59 50.7	1	10.96		0.87		0.06		3		41
		LS I +59 177	AB	3 12 04	+59 50.7	1	10.28		0.31		0.19		1		180
20196	−2 00581			3 12 05	−02 31.1	1	7.16		1.60		1.92		3	M1	1657
278403	+34 00608	G 37 - 29		3 12 06	+34 26.8	1	9.65		0.92		0.76		1	K2	333,1620
20205	−10 00642			3 12 07	−09 52.3	1	7.66		0.44		-0.02		1	F5	1776
237098	+56 00799	LS I +57 129		3 12 10	+57 03.9	1	9.53		0.48		0.17		3	A0 II	41
		Steph 350		3 12 12	−09 51.0	1	11.63		1.55		1.14		1	M0	1746
	−32 01209			3 12 12	−32 08.4	1	10.08		1.20		1.19		1		1696
20222	−15 00562			3 12 13	−15 01.4	1	9.15		0.50		-0.07		1	F5/6 V	742
		Alpha Per 271		3 12 14	+47 29.9	1	10.41		0.60		0.09		2	F7	1048
		BSD 9 # 614		3 12 15	+59 35.9	1	12.36		0.75		0.08		2	A0	1069
	+15 00452	IDS03095N1556	AB	3 12 17	+16 08.0	1	8.83		0.67		0.21		1	G0	3016
20149	+30 00512	HR 971		3 12 17	+30 22.3	2	5.61	.000	0.01	.000	0.03	.048	7	A1 Vs	1049,1733
	+47 00791	Alpha Per 276		3 12 18	+47 44.0	1	9.93		0.64		0.16		1	F8 V	1048
		G 174 - 41	AB	3 12 18	+57 59.3	2	10.30	.111	1.55	.059			4		1017,1625
	−46 00962	IDS03106S4605	AB	3 12 20	−45 53.9	1	10.21		0.49		-0.04		2	F2	1730
		HA 71 # 1		3 12 21	+14 42.7	1	11.01		-0.04		-0.73		2		97
		CS 22167 # 7		3 12 21	−06 53.1	1	15.07		0.08		0.13		1		1736
		CS 22172 # 3		3 12 22	−10 49.5	1	14.44		0.64		0.09		1		1736
		G 77 - 34		3 12 25	+00 52.0	2	14.61	.056	1.47	.037	1.04	.037	4		333,419,1620
		Alpha Per 299		3 12 25	+50 13.3	2	11.15	.035	0.64	.002	0.11		5	F7	1048,1694
	+0 00549	G 77 - 35	★ A	3 12 29	+00 51.1	1	10.22		0.68		0.07		3	G5	333,1620
20135	+47 00792	Alpha Per 285		3 12 32	+47 50.6	1	8.09		0.21		0.15		3	A0p	1048
		CS 22185 # 9		3 12 33	−14 54.9	1	13.79		0.34		-0.14		1		1736
		Alpha Per 290		3 12 34	+47 59.5	1	10.71		0.64		0.15		2	F9 V	1048
20280	−26 01207			3 12 35	−26 37.9	6	9.17	.040	1.23	.016	1.10		21	K5 V	955,1020,1075,1705,2012,3072
20123	+50 00729	Alpha Per 308	★	3 12 37	+50 45.2	2	5.03	.000	1.15	.000	0.83		4	G6 Ib-IIa	1119,1355
20122	+50 00728	Alpha Per 314		3 12 37	+51 14.7	1	9.25		0.43		0.16		3	F2 V	1048
20352	−57 00517	NGC 1252 - 33		3 12 38	−57 06.1	1	8.64		0.52		0.04		2	F7 V	995
20162	+44 00648	HR 973		3 12 40	+45 09.7	2	6.19	.025	1.68	.003	2.00		4	M2 III	71,3055
20193	+32 00591	HR 975	★ A	3 12 41	+32 40.3	2	6.32	.015	0.37	.015	-0.02	.019	3	F4 V	254,3037
		NGC 1252 - 37		3 12 41	−57 56.1	1	11.14		0.31		-0.02		1		995
	+51 00704	LS V +51 006		3 12 42	+51 49.7	1	10.34		0.43		-0.39		2	B1 V:nn	1012
20301	−36 01218	TZ For		3 12 42	−35 44.6	1	6.87		0.74				4	G2 V	2006
		LP 831 - 47		3 12 44	−21 34.9	1	11.43		0.78		0.30		2		1696
		S116 # 3		3 12 44	−59 24.2	1	11.20		0.20		0.10		2		1730
	−16 00601			3 12 48	−15 44.8	1	11.46		0.30		0.11		1		1736
20259	−1 00465		A	3 12 49	−01 21.8	1	8.55		0.53		0.03		1	F8	3077
20259	−1 00465		B	3 12 49	−01 21.8	1	9.94		0.45		0.00		1		3016
	+49 00889	Alpha Per 309		3 12 51	+49 26.5	1	9.96		0.49		0.01		2	F5 V	1048
	−15 00564			3 12 51	−14 39.3	1	10.39		1.08		0.66		1	K0 IIIba	565
20293	−26 01210	HR 980		3 12 51	−26 17.1	2	6.24	.005	0.04	.000			7	A1 V	15,2027
20399	−59 00262			3 12 51	−59 17.8	1	7.89		1.09		1.01		2	K1 III	1730
		HA 71 # 11		3 12 52	+14 50.4	1	10.32		1.22		1.08		2		97
		BSD 9 # 632		3 12 52	+59 54.1	1	9.81		0.25		0.17		2	A0	1069
20210	+34 00610	HR 976, V423 Per	★ A	3 12 53	+34 30.3	2	6.24		0.28	.024	0.12	.002	9	A1 m	1049,1202
	+60 00655			3 12 53	+61 06.7	1	9.98		0.27		0.23		2	B8 IV	1069
20267	−2 00583			3 12 53	−02 21.6	1	8.77		0.53		0.03		1	F5	1776
	+59 00624			3 12 54	+59 47.9	1	10.51		0.31		-0.06		2	B5 V	1069
		BSD 9 # 146		3 12 55	+59 00.1	1	12.05		0.74		0.23		2		1069

HD	DM	Other Id	N Rem	α_{1950}	δ_{1950}	S	V	σ_V	B–V	σ_{B-V}	U–B	σ_{U-B}	n	Spectrum	References
20134	+59 00625	LS I +59 179		3 12 59	+59 53.0	5	7.45	.036	0.13	.020	-0.49	.028	10	B2.5IV-V	41,180,1069,1212,1423
		4HLF4 # 1025		3 13 00	-00 22.	1	12.00		1.10		1.00		2	K0 III	221
20291	-20 00605			3 13 00	-20 12.2	1	7.01		0.02		0.03		4	A0/1 V	1628
20351	-46 00965			3 13 00	-45 45.6	1	9.97		0.14		0.12		2	A2 V	1730
		BSD 9 # 638		3 13 04	+59 44.7	1	10.81		0.37		0.31		2	B9	1069
		HA 71 # 309		3 13 05	+15 34.6	1	10.78		0.69		0.24		2		97
	+60 00659			3 13 06	+60 50.8	1	10.50		0.28		0.22		2	B9 V	1069
		HA 71 # 122		3 13 07	+15 02.6	1	13.08		0.72		0.26		1		97
20104	+65 00338	HR 967	⋆ AB	3 13 07	+65 28.5	2	6.41	.052	0.08	.001	0.10	.031	5	A2 V +A4 V	379,595
		HA 71 # 188		3 13 11	+15 11.4	1	13.40		0.53		0.06		3		97
		POSS 413 # 1		3 13 11	+16 33.3	1	15.77		1.52				2		1739
		G 78 - 26		3 13 11	+37 55.6	2	10.68	.009	1.21	.022	0.97	.004	5		1064,3078
		Alpha Per 326		3 13 11	+49 41.6	2	12.13	.020	0.59	.005	0.41		4	A3	1048,1575
		HA 71 # 123		3 13 12	+15 10.9	1	12.90		0.59		0.12		2		97
20191	+50 00731	Alpha Per 333		3 13 13	+51 02.1	1	7.19		0.03		-0.19		4	B9	1048
20278	+11 00456	Hyades vB 158		3 13 18	+11 26.7	2	8.03	.005	0.61	.000	0.13	.002	4	G5	1127,8023
		HA 71 # 126		3 13 21	+15 09.5	1	11.69		0.72		0.21		2		97
	+59 00626			3 13 22	+59 49.9	1	10.23		0.25		-0.03		2	A0	1069
20320	-9 00624	HR 984		3 13 24	-09 00.3	9	4.80	.005	0.23	.007	0.08	.012	35	A5 m	15,216,355,1075,1199*
20407	-46 00968			3 13 25	-46 05.4	2	6.76	.000	0.58	.000	-0.02		5	G0/2 V	258,2012
	+49 00892	Alpha Per 334		3 13 26	+49 44.6	2	10.35	.020	0.55	.004	0.04		4	F7 V	1048,1575
20039	+71 00190	G 221 - 7		3 13 26	+72 05.8	1	8.89		0.75		0.21		1	F8	1658
20340	-17 00631			3 13 27	-17 00.8	1	7.97		-0.13		-0.62		3	B3 V	399
20277	+31 00576	HR 978		3 13 30	+32 00.1	1	6.06		0.99		0.68		7	G8 IV	1355
20319	-6 00636	HR 983	⋆ AB	3 13 32	-06 06.1	4	6.16	.005	-0.02	.003	-0.24	.029	15	B9 V	15,1071,1079,2027
		POSS 413 # 5		3 13 34	+16 44.8	1	17.61		1.57				2		1739
		BSD 9 # 645		3 13 35	+60 10.5	1	11.87		0.46		0.29		4	A0	1069
	+47 00797	Alpha Per 330		3 13 37	+47 26.8	2	9.88	.015	1.31	.004	1.11		3	G9 III	1048,1575
		POSS 413 # 2		3 13 38	+16 23.9	1	15.80		1.50				2		1739
20305	+13 00530			3 13 39	+13 40.1	2	8.49	.010	1.64	.039	1.68		3	K2	882,3040
		NGC 1252 - 36		3 13 39	-58 01.3	1	10.62		0.79		0.27		2		995
		HA 71 # 195		3 13 44	+15 13.1	1	12.47		0.79		0.39		1		97
		G 174 - 44	⋆ V	3 13 44	+52 06.7	1	13.77		0.97		0.21		1		1658
	+15 00458			3 13 46	+15 27.3	1	9.94		0.45		0.11		2	F2	97
	+48 00876	Alpha Per 338		3 13 48	+49 19.1	1	9.93		0.56		0.02		2	F7 V	1048
		CS 22172 # 7		3 13 49	-08 44.3	2	14.47	.000	0.19	.000	0.15	.008	2		966,1736
		POSS 413 # 6		3 13 52	+17 17.4	1	18.02		1.71				2		1739
	+30 00516	G 37 - 31		3 13 52	+30 51.1	1	9.40		0.90		0.62		1	K2	333,1620
		Alpha Per 340		3 13 52	+49 07.3	1	11.45		0.69		0.17		3	F9	1048
		LS I +60 303		3 13 53	+60 40.3	2	11.75	.005	0.77	.020	0.02	.025	5		41,180
20329	+15 00459			3 13 54	+15 28.6	3	8.76	.012	0.67	.007	0.19	.023	5	G5	97,882,3077
20283	+39 00743	HR 979	⋆ AB	3 13 54	+40 18.0	1	6.46		-0.03		-0.16		3	B9 p Si+A7m	1733
19978	+77 00115	HR 961	⋆ AB	3 13 54	+77 33.2	3	5.48	.030	0.19	.000	0.08	.025	3	A6 V	15,254,1008
		CS 22172 # 8		3 13 54	-08 13.0	1	13.79		0.47		-0.14		1		1736
		BPM 17132		3 13 54	-53 41.	1	12.60		0.81		0.35		2		3065
		HA 71 # 132		3 13 56	+15 10.2	1	14.10		1.19		1.09		1		97
		G 245 - 71	A	3 13 57	+79 47.0	1	11.23		1.60		1.17		5		7010
		HA 71 # 200		3 14 01	+15 16.2	1	12.08		0.56		0.06		2		97
		CS 22167 # 8		3 14 01	-07 17.8	1	15.57		0.03		-0.17		1		1736
20327	+25 00521			3 14 04	+25 24.2	1	8.25		0.42		0.00		3	F5	833
		BSD 9 # 660		3 14 04	+59 52.2	1	10.72		0.39		0.32		3	A0	1069
		MtW 71 # 12		3 14 05	+15 18.5	1	15.58		1.32		0.81		1		97
		4HLF4 # 35		3 14 06	+02 07.	1	11.34		0.17		0.13		2	A0	208
		Alpha Per 350		3 14 06	+48 39.2	2	11.06	.060	0.69	.013	0.16		5		1048,1575
		CS 22167 # 36		3 14 06	-06 55.8	1	15.68		0.06		0.12		1		1736
20369	+4 00519	IDS03115N0440	AB	3 14 07	+04 50.4	1	9.22		0.62		0.03		1	G5	1620
		HA 71 # 202		3 14 07	+15 19.2	1	15.15		1.13				1		97
		Alpha Per 347		3 14 08	+47 55.5	1	10.51		0.68		0.18		2	G0 V	1048
		BSD 9 # 156		3 14 08	+59 01.5	1	11.00		0.55		-0.21		2	B6	1069
20423	-31 01303	HR 990		3 14 08	-31 00.7	2	6.64	.005	-0.07	.000			7	B9 V	15,2012
20385	-4 00560			3 14 09	-03 42.8	1	7.48		0.52		-0.01		1	F5	1776
20395	-9 00627	HR 988		3 14 10	-09 20.3	4	6.14	.028	0.41	.010	-0.04	.010	15	F1 V	15,254,1061,2027
	+14 00544			3 14 12	+15 04.0	1	10.64		0.32		0.30		2		97
		HA 71 # 136		3 14 12	+15 10.2	1	14.46		0.59		0.00		1		97
20296	+46 00722	Alpha Per 341		3 14 12	+46 46.0	1	7.53		0.06		-0.01		5	A0	595
		MtW 71 # 25		3 14 13	+15 14.6	1	16.75		0.21		-0.77		1		97
		GD 427		3 14 13	+64 49.2	1	16.54		0.01		-0.67		2	DA	782
20404	-3 00525			3 14 13	-03 32.8	1	7.66		0.00		-0.10		1	B8	1776
		HA 71 # 203		3 14 16	+15 14.8	1	11.77		0.57		0.10		std		97
20403	-3 00524			3 14 16	-02 46.0	1	9.21		0.58		0.09		1	F8	1776
		HA 71 # 262		3 14 17	+15 22.8	1	13.81		0.72		0.14		1		97
	+61 00548			3 14 17	+62 10.2	1	10.00		0.24		0.22		2	A0 V	1069
		CS 22185 # 12		3 14 17	-13 24.9	1	13.33		0.82		0.42		1		1736
20394	+1 00573			3 14 18	+02 09.4	1	8.72		1.14		0.66		2	K0	3048
		BSD 9 # 1804		3 14 19	+62 09.0	1	12.09		0.53		0.41		4	A0	1069
	+61 00549			3 14 21	+61 45.2	1	9.94		0.18		0.00		2	B5 V	1069
		HA 71 # 141		3 14 23	+15 09.3	1	14.93		0.64		0.01		1		97
20368	+17 00525			3 14 23	+17 23.4	2	7.42	.024	1.70	.019	2.01	.039	3	K5	1648,3040
		BSD 9 # 666		3 14 23	+60 04.6	1	10.72		0.29		-0.22		2	B6	1069
		4HLF4 # 1026		3 14 24	+01 02.	1	11.95		0.61		0.14		2	G0 III	221

Table 1 169

HD	DM	Other Id	N Rem	α_{1950}	δ_{1950}	S	V	σ_V	B–V	σ_{B-V}	U–B	σ_{U-B}	n	Spectrum	References
		4HLF4 # 1027		3 14 24	+02 05.	1	12.31		1.64		1.47		1	M5 III	221
	+14 00545			3 14 24	+15 13.4	1	8.47		1.14		0.94		std	K0	97
		HA 71 # 206		3 14 24	+15 17.9	1	14.40		0.86		0.48		1		97
	+25 00522	Hyades J 202	⋆ AB	3 14 24	+26 08.2	5	11.11	.006	1.41	.009	1.29	.072	18	M0	680,950,1058,1570,1674
20315	+43 00674	HR 982		3 14 25	+43 50.6	4	5.47	.005	-0.06	.001	-0.34	.011	14	B8 V	15,154,1009,1048
	+61 00550			3 14 26	+62 07.9	1	10.31		0.24		0.06		2	A0	1069
	+2 00512			3 14 27	+02 28.9	1	9.38		0.38		0.19		2	A5	208
	+49 00896	Alpha Per 361		3 14 29	+49 27.7	1	9.68		0.44		0.01		3	F4 V	1048
		4HLF4 # 37		3 14 30	+00 12.	1	11.66		0.44		0.08		2	A7	208
		4HLF4 # 36		3 14 30	+00 59.	1	14.39		0.05		0.15		1	A0	208
20347	+38 00689	IDS03113N3816	AB	3 14 30	+38 27.4	1	7.28		0.61		0.14		1	G0	292
		Hyades J 203		3 14 31	+31 49.	1	11.10		0.92		0.63		5		1058
20346	+38 00690	HR 986		3 14 31	+39 06.1	2	5.97		0.09	.025	0.08	.045	6	A2 IV	105,1049
237114	+55 00761			3 14 31	+55 37.0	1	9.00		0.58				2	A5	70
	+49 00897	Alpha Per 365		3 14 32	+49 43.4	1	9.90		0.50		-0.01		3	F6 V	1048
237115	+55 00762			3 14 35	+55 50.4	1	9.12		0.31				2	A3	70
20367	+30 00520			3 14 36	+30 56.7	1	6.39		0.56		0.06		3	G0 V	1501
20421	+1 00574			3 14 37	+01 22.5	1	7.78		0.24		0.16		2	A3 IV	208
	+8 00489			3 14 41	+08 55.4	1	9.28		1.06		0.78		2	K0	1733
		HA 71 # 210		3 14 42	+15 20.6	1	12.24		1.10		0.91		1		97
		BSD 9 # 1262		3 14 43	+60 38.3	1	11.45		0.31		0.25		2	A0	1069
		CS 22172 # 6		3 14 43	-09 00.7	1	12.76		0.40		-0.07		1		1736
		HA 71 # 146		3 14 44	+15 01.9	1	13.78		0.73		0.19		2		97
20430	+7 00493	Hyades vB 1	⋆ AB	3 14 45	+07 28.4	4	7.39	.005	0.57	.004	0.12	.008	11	F8	1127,1355,3077,8023
		HA 71 # 211		3 14 45	+15 13.0	1	12.17		0.73		0.19		3		97
		G 37 - 32		3 14 45	+25 04.4	1	11.84		1.46		1.10		1	M1	333,1620
21190	-83 00064	IDS03200S8354	A	3 14 45	-83 43.1	1	7.64		0.40		0.11		6	F2/3 Ib/II	1628
		MtW 71 # 70		3 14 47	+15 15.1	1	17.48		0.34		0.02		1		97
20295	+59 00630			3 14 47	+59 27.2	1	8.35		0.24		-0.39		3	B5 III	1069
		4HLF4 # 38		3 14 48	+02 03.	1	13.27		0.01		-0.19		2	B9	208
		MtW 71 # 71		3 14 48	+15 13.2	1	16.83		0.74		0.10		1		97
		CCS 139		3 14 48	-01 42.	1	11.63		1.70		1.84		2		628
	+61 00552			3 14 49	+61 23.3	1	10.32		0.36		-0.22		2	B6	1069
		MtW 71 # 73		3 14 50	+15 14.2	1	15.82		0.70		0.08		2		97
		LS I +58 119		3 14 50	+58 56.2	2	10.48	.036	0.52	.007	-0.29	.022	5		41,1069
		MtW 71 # 75		3 14 51	+15 12.9	1	17.91		0.58				1		97
20439	+7 00494	Hyades vB 2	⋆ C	3 14 52	+07 30.5	4	7.77	.005	0.62	.004	0.16	.004	10	G0	1127,1355,3077,8023
275122	+37 00748	G 78 - 31		3 14 52	+38 04.6	1	10.28		1.48		1.21		2	M1	3072
20504	-31 01305			3 14 52	-31 32.1	1	7.14		0.20				4	A1mA5-F2	2006
		BSD 9 # 1814		3 14 54	+61 57.0	1	10.51		0.29		0.21		2	B9	1069
		4HLF4 # 39		3 14 54	-01 32.	1	12.09		0.38		0.07		2	A7	208
	+14 00547			3 14 55	+14 54.8	1	9.74		0.92		0.05		2	F5	97
		Alpha Per 1589		3 14 55	+50 16.6	1	12.31		0.85				1		1694
		CS 22167 # 9		3 14 55	-06 34.0	1	14.87		0.00		-0.09		1		1736
		G 5 - 26		3 14 56	+23 26.8	1	13.44		1.00		0.64		1		1620
		Alpha Per 373		3 14 59	+47 10.3	2	11.52	.025	0.78	.012	0.29		3	G3	1048,1694
	+54 00663			3 14 59	+54 28.1	1	8.98		0.40				2		70
20492	-21 00594			3 14 59	-20 52.0	1	9.65		1.08		0.96		1	K3 V	3072
20365	+49 00899	Alpha Per 383	⋆	3 15 03	+50 02.4	4	5.15	.013	-0.06	.005	-0.56	.009	22	B3 V	15,154,1009,1048
20458	+13 00535			3 15 06	+13 39.9	1	7.53		0.11		0.05		3	A0	1648
		CS 22968 # 29		3 15 06	-55 16.0	1	14.29		0.41		-0.27		1		1736
		Alpha Per 387		3 15 09	+50 12.3	1	10.28		0.59		0.08		3	F6 IV-V	1048
284303	+22 00669	Hyades CDS 6		3 15 11	+23 09.8	1	9.48		0.98				3	K0	950
20391	+49 00900	Alpha Per 386		3 15 11	+49 35.3	2	7.93	.005	0.12	.002	0.09	.020	9	A1 Van	1009,1048
		CS 22172 # 9		3 15 13	-08 44.9	2	13.91	.000	0.24	.000	0.11	.008	2		966,1736
	+51 00710	LS V +51 007		3 15 14	+51 50.9	1	9.75		0.59		-0.18		2	B5 Ib	1012
		BSD 9 # 166		3 15 14	+58 58.0	1	11.18		0.53		0.38		2	A0	1069
		HA 71 # 213		3 15 15	+15 17.4	1	11.90		0.51		0.11		1		97
	+55 00763			3 15 16	+56 18.1	1	9.34		0.27				2	A2	70
20273	+69 00205	IDS03105N6922	A	3 15 16	+69 33.0	1	6.60		-0.01		-0.03		4	A0	985
		CS 22185 # 7		3 15 16	-15 28.3	1	13.34		0.71		0.05		1		1736
		Alpha Per 1590		3 15 17	+49 33.0	1	11.17		0.67				1		1694
		4HLF4 # 40		3 15 18	+00 28.	1	14.22		0.13		0.07		1	A0	208
		Alpha Per 1591		3 15 20	+48 05.1	1	12.59		0.93				1		1694
		HA 71 # 151		3 15 22	+15 03.1	1	11.26		1.13		0.90		2		97
		BSD 9 # 1280		3 15 25	+60 27.4	1	10.12		0.26		0.18		3	A0	1069
		BSD 9 # 1279		3 15 25	+60 30.9	1	11.70		0.46		0.32		4	A0	1069
		HA 71 # 93		3 15 27	+14 59.8	1	11.59		0.07		-0.04		1		97
20444	+35 00663			3 15 27	+35 22.0	1	8.67		0.14		-0.31		2	A0	1601
		LP 772 - 32		3 15 27	-19 48.2	1	11.69		0.60		-0.09		2		1696
20500	+12 00460			3 15 29	+12 38.5	1	7.76		0.22		0.16		2	B8	1733
		LP 299 - 53		3 15 29	+32 28.0	1	11.33		1.47		1.37		2		1723
		Alpha Per 1592		3 15 29	+49 22.8	1	15.68		1.50				1		1694
20477	+17 00527			3 15 30	+17 59.5	1	7.54		0.62		0.17		2	G0	1648
		TonS 335		3 15 30	-22 20.	1	13.46		0.64		0.11		1		832
		HA 71 # 153		3 15 32	+15 06.6	1	12.02		0.62		0.18		1		97
	+14 00549			3 15 32	+15 14.9	1	10.27		1.40		1.37		2		97
		Alpha Per 1593		3 15 32	+48 00.1	1	11.99		0.93				1		1694
20418	+49 00902	Alpha Per 401	⋆	3 15 33	+49 54.8	5	5.03	.009	-0.06	.007	-0.54	.012	17	B5 Vn	15,154,1009,1048,1363
20336	+65 00340	HR 985, BK Cam	⋆ A	3 15 34	+65 28.3	9	4.83	.028	-0.15	.014	-0.77	.013	26	B2.5Ven	15,41,154,379,1118*

HD	DM	Other Id	N Rem	α_{1950}	δ_{1950}	S	V	σ_V	B–V	σ_{B-V}	U–B	σ_{U-B}	n	Spectrum	References
		BSD 9 # 1829		3 15 35	+61 25.8	1	11.23		0.20		0.07		2	B9	1069
		HA 71 # 158		3 15 36	+15 05.4	1	12.18		1.08		0.70		1		97
20468	+33 00619	HR 991		3 15 36	+34 02.5	4	4.82	.005	1.49	.003	1.56	.007	15	K2 II	15,1355,1363,8015
20551	−11 00632	IDS03132S1056	AB	3 15 36	−10 45.1	1	8.92		0.43		−0.01		1	F5	1776
		POSS 413 # 3		3 15 37	+16 35.6	1	16.58		1.62				2		1739
275054	+40 00707			3 15 39	+41 00.3	1	9.09		1.10		0.93		3	K0	196
20512	+14 00550	G 5 - 27	★ A	3 15 40	+15 00.0	3	7.40	.017	0.79	.010	0.35	.009	7	G5	97,1620,3032
20512	+14 00550	G 5 - 28	★ B	3 15 40	+15 00.0	2	15.50	.054	1.35	.044	0.31	.083	6	DMp	419,3032
	+15 00463			3 15 40	+15 32.8	1	10.74		1.20		1.08		1		97
	+61 00554			3 15 40	+61 45.3	1	9.59		0.14		−0.08		3	B8	1069
		Alpha Per 1594		3 15 42	+47 48.9	1	13.85		1.12				1		1694
		BSD 9 # 1286		3 15 43	+60 23.7	1	11.49		0.62		0.48		3	B9	1069
		Alpha Per 394		3 15 44	+47 09.9	1	10.55		0.47		0.18		2	F1	1048
		Alpha Per 407		3 15 44	+50 34.1	2	11.14	.035	0.66	.016	0.43		4	A2	1048,1575
		G 77 - 43		3 15 44	−02 36.7	2	14.49	.000	0.89	.025	0.28	.030	7		538,3059
		POSS 413 # 7		3 15 45	+17 08.2	1	18.42		1.35				3		1739
		POSS 413 # 4		3 15 47	+17 00.2	1	17.36		1.42				2		1739
	+48 00884	G 174 - 46		3 15 49	+49 07.7	1	10.13		0.73		0.24		3	K0	1723
20559	−1 00469	HR 992	★ A	3 15 49	−01 06.7	1	5.62		1.05		0.82		1	G8 IV	3026
20559	−1 00469	HR 992	★ AB	3 15 49	−01 06.7	4	5.39	.014	1.03	.005	0.81	.000	10	G8 IV +G4 V	15,1003,1061,3077
20559	−1 00469	IDS03132S0118	B	3 15 49	−01 06.7	1	7.66		0.92		0.56		1	G4 V	3026
20559	−1 00469	IDS03132S0118	C	3 15 49	−01 06.7	1	10.03		0.61		0.06		1		3009
20640	−48 00900	HR 998		3 15 49	−47 56.0	3	5.84	.004	1.23	.010			11	K2 III	15,1075,2032
		BSD 9 # 1289		3 15 50	+60 42.0	1	11.38		0.43		−0.18		2	B7	1069
		CS 22185 # 6		3 15 51	−15 42.5	1	14.21		0.45		−0.23		1		1736
		Steph 354		3 15 52	+10 07.9	1	12.28		1.49		1.22		1	M0	1746
		HIC 15446	A	3 15 52	+41 05.4	1	10.55		1.10		0.76		2		1723
		BSD 9 # 1290		3 15 56	+60 40.9	1	12.41		0.86		0.36		2		1069
20606	−29 01216	HR 993		3 15 56	−28 58.7	2	5.90	.005	0.34	.005			7	F3 V +A8/9	15,2012
		G 79 - 25		3 15 58	+08 16.4	1	13.02		1.52		1.13		1		333,1620
237118	+59 00632			3 16 00	+59 41.0	1	9.34		0.23		−0.23		2	B8	1069
		Ross 570		3 16 00	−07 20.	1	11.22		0.46		−0.25		2	A6:	3077
	+61 00557			3 16 01	+62 17.5	1	9.29		0.10		−0.19		2	B8	1069
20499	+38 00693			3 16 04	+38 52.5	1	7.27		0.35		−0.01		3	F2 II	833
		BSD 9 # 692		3 16 06	+60 10.9	1	11.14		0.27		0.09		2	B9	1069
20610		HR 994	★ AB	3 16 09	−22 41.6	4	4.88	.005	0.90	.008			12	G6 III	15,1075,1414,2031
20475	+48 00885	Alpha Per 421		3 16 10	+48 44.0	2	9.23	.000	0.46	.003	0.09	.015	10	F2 V	1009,1048
		CS 22968 # 30		3 16 14	−54 55.3	2	15.48	.010	0.44	.009	−0.19	.008	2		1580,1736
		CS 22167 # 17		3 16 15	−03 49.7	1	13.65		0.05		0.10		1		1736
20526	+36 00676			3 16 16	+36 41.7	1	7.95		1.05		0.76		2	K0	1601
20487	+48 00886	Alpha Per 423		3 16 16	+48 26.8	1	7.64		0.07		0.01		3	A0 Vn	1048
	+15 00464			3 16 19	+15 37.5	1	10.68		1.21		1.11		1		97
20601	−1 00473	IDS03138S0124	AB	3 16 20	−01 12.1	2	9.53	.025	0.72	.005	0.21		6	G5	1414,3016
20622	−14 00646			3 16 20	−14 26.1	1	7.76		1.28		1.19		1	K1 III	3077
		BSD 9 # 1840		3 16 22	+61 17.9	1	10.71		0.30		0.28		2	B9	1069
		HA 71 # 302		3 16 23	+15 24.6	1	10.87		1.24		1.01		1		97
		Alpha Per 1595		3 16 24	+49 41.3	1	12.28		0.88				1		1694
20631	−19 00651	HR 997	★ AB	3 16 24	−18 44.4	2	5.70	.005	0.38	.005			7	F3 V	15,2012
278543	+33 00622	G 37 - 34		3 16 30	+33 25.8	2	9.68	.000	0.85	.000	0.47	.000	2	K2	516,1620
		NGC 1275 sq1 3		3 16 30	+41 22.0	1	12.50		0.59		0.08		1		327
	+48 00887	Alpha Per 435		3 16 30	+49 12.4	1	9.86		0.61		0.18		3	F6 IV	1048
20510	+50 00738	Alpha Per 441		3 16 30	+50 47.3	1	7.05		0.05		−0.14		1	B9 V	245
20619	−3 00534			3 16 30	−03 01.4	5	7.05	.022	0.66	.015	0.11	.004	15	G0	742,908,1067,2033,3026
	−1 00474			3 16 32	−01 08.7	1	10.02		0.61		0.06		2	G0 V	3077
		GD 45		3 16 34	+34 31.6	1	14.16		0.08		−0.68		1	DAs	3060
		BSD 9 # 1842		3 16 35	+61 27.4	1	11.16		0.31		0.22		2	A0	1069
20600	+19 00505			3 16 37	+19 32.4	1	8.48		0.54		0.02		4	G0	3026
20766	−63 00217	HR 1006	★ B	3 16 41	−62 46.0	6	5.53	.007	0.64	.007	0.06	.011	26	G3/5 V	15,678,2012,2015,3078,8015
20695	−35 01161			3 16 43	−35 20.9	1	8.10		1.29		1.35			K2 III	1673
20630	+2 00518	HR 996	★ A	3 16 44	+03 11.3	23	4.83	.011	0.68	.006	0.19	.008	100	G5 V	1,3,15,58,116,1004*
		NGC 1275 sq1 4		3 16 44	+41 20.2	1	14.06		0.56		0.04		2		327
		BSD 9 # 1309		3 16 44	+60 37.6	1	11.35		0.44		0.21		3	A0	1069
20537	+51 00716	Alpha Per 450		3 16 46	+51 26.3	1	7.30		0.03		0.05		5	B9	595
237121	+58 00587	LS I +58 120		3 16 47	+58 40.5	2	8.93	.005	0.47	.005	−0.52	.005	4	B0.5V	41,1012
20508	+59 00634	LS I +59 180		3 16 52	+59 51.0	2	8.27	.057	0.45	.024	−0.41	.009	6	B1.5IV	41,1069
20646	−1 00475	X Cet		3 16 53	−01 14.8	1	8.96		1.83		2.03		1	M2	975
		Alpha Per 444		3 16 55	+47 45.7	2	11.14	.020	0.83	.000	0.35		4	G3	1048,1575
20722	−41 00954			3 16 55	−41 20.5	1	9.09		1.34		1.69		1	K3/4 (III)	751
20618	+26 00540	HR 995		3 16 56	+26 53.5	1	5.90		0.86		0.51		3	G8 IV	1355
20589	+41 00656			3 16 58	+41 43.0	1	8.09		0.26		0.25		3	A2	196
		4HLF4 # 41		3 17 00	+01 20.	1	12.63		0.28		0.16		2	A2	208
20628	+19 00507			3 17 00	+20 05.8	1	8.36		0.27		0.12		3	A2	1648
		NGC 1275 sq1 5		3 17 00	+41 22.3	1	13.70		1.26		1.28		2		327
		LS I +68 020		3 17 03	+68 08.8	1	11.76		0.68		0.20		3		41
		Alpha Per 1596		3 17 05	+50 05.0	1	14.56		1.52				1		1694
20807	−62 00265	HR 1010	★ A	3 17 07	−62 41.8	6	5.24	.015	0.60	.009	0.00	.004	26	G2 V	15,678,2012,2015,3078,8015
20645	+16 00423			3 17 09	+17 19.2	1	7.78		0.30		0.06		3	F0	1648
		Alpha Per 453		3 17 10	+47 18.6	2	10.40	.010	0.62	.007	0.12		3	F8 V	1048,1575
21024	−77 00134	HR 1025		3 17 10	−77 34.3	6	5.51	.004	0.44	.002	−0.01	.009	30	F5 V	15,278,1075,2012,2038,3077
		Alpha Per 1597		3 17 11	+48 13.8	1	12.08		0.87				1		1694

Table 1 171

HD	DM	Other Id	N Rem	α₁₉₅₀	δ₁₉₅₀	S	V	σ_V	B–V	σ_B–V	U–B	σ_U–B	n	Spectrum	References
20547	+58 00588	LS I +58 121	⋆ AB	3 17 12	+58 59.3	2	8.17	.013	0.37	.004	-0.47	.013	5	B3 III	41,1069
	+9 00424			3 17 14	+09 27.1	1	11.36		0.66				2		920
20665	+8 00495			3 17 15	+08 31.2	1	7.27		0.04		0.04		2	B9	8040
		Alpha Per 457		3 17 16	+49 08.6	2	11.71	.010	0.73	.003	0.27		4		1048,1575
		BSD 9 # 1315		3 17 16	+61 07.3	1	9.96		0.19		0.18		3	A0	1069
20720	-22 00584	HR 1003, τ4 Eri	⋆ AB	3 17 17	-21 56.3	7	3.68	.026	1.61	.009	1.81	.018	29	M3.5 IIIa	3,15,542,1024,2029*
20644	+28 00516	HR 999		3 17 18	+28 52.1	6	4.47	.009	1.55	.010	1.81	.042	14	K2 II-III	15,1355,1363,1769,3016,8015
		4HLF4 # 43		3 17 18	-01 05.	1	11.70		0.31		0.13		2	A2	208
		CS 22172 # 10		3 17 18	-09 22.5	2	14.38	.000	0.08	.000	0.21	.010	2		966,1736
		4HLF4 # 1028		3 17 24	-01 32.	1	11.80		0.72		0.23		2	G2 III	221
20729	-24 01578	HR 1004		3 17 24	-24 18.2	4	5.61	.009	1.65	.013	2.00		13	M1 III	15,1075,2007,3055
20888	-67 00217	HR 1014		3 17 25	-67 06.5	3	6.04	.003	0.13	.000	0.09	.005	25	A3 V	15,1075,2038
20663	+25 00536	HR 1000		3 17 27	+25 29.0	2	6.13	.003	1.23	.014	1.35		5	gK3	71,833
	+49 00905	Alpha Per 462		3 17 28	+49 33.5	2	11.44	.045	0.86	.026	0.28		4	G0	1048,1575
		4HLF4 # 1029		3 17 30	+00 07.	1	12.29		1.06		1.00		2	K0 III	221
		CS 22172 # 12		3 17 32	-10 22.4	2	13.74	.000	0.28	.000	0.17	.009	2		966,1736
	-33 01173			3 17 33	-33 01.0	1	10.94		0.36		-0.24		2		1696
		Alpha Per 1598		3 17 35	+48 15.4	1	12.80		1.00				1		1694
	+21 00440			3 17 37	+22 16.0	1	10.24		1.44		1.42		2	G0	1726
		BSD 9 # 1320		3 17 38	+60 36.7	1	12.86		0.75		0.27		3	A0	1069
		BSD 9 # 1856		3 17 38	+62 05.1	1	12.19		0.72		0.11		2	B5	1069
20653	+38 00699			3 17 39	+38 42.4	1	8.68		0.12		-0.14		3	A0	833
	+59 00636			3 17 39	+59 59.6	1	9.68		1.83				3		369
		Alpha Per 1599		3 17 40	+48 49.2	1	15.63		1.56				1		1694
20855	-58 00275			3 17 42	-58 09.7	3	7.45	.012	1.06	.008	0.87	.013	10	K0 III	158,1075,3077
	+7 00499	G 79 - 28		3 17 44	+08 16.2	2	9.63	.030	1.13	.000	1.09		2	dK7	1620,1632
		Alpha Per 1600		3 17 44	+48 24.3	1	12.80		1.13				1		1694
		LS I +59 181		3 17 45	+59 23.9	1	11.03		0.69		-0.22		3		41
		CS 22968 # 27		3 17 45	-55 27.8	1	15.42		0.13		0.17		1		1736
		Alpha Per 1602		3 17 48	+48 34.7	1	11.96		0.80				1		1694
		Alpha Per 1601		3 17 48	+49 46.3	1	13.85		1.25				1		1694
20567	+64 00387	LS I +65 027		3 17 48	+65 08.2	1	8.52		0.81		0.30		3	B8 Ib	41
		Alpha Per 477		3 17 53	+47 31.3	2	10.58	.015	1.46	.010	1.54		4	G5:	1048,1575
20727	+8 00496	G 79 - 29	⋆ A	3 17 54	+08 51.3	3	8.46	.012	0.68	.000	0.16	.007	4	G0	333,908,1620,3026
20744	-2 00607			3 17 55	-01 56.1	1	9.22		0.23		0.10		2	A3 V	208
20781	-29 01229	IDS03159S2913	B	3 17 56	-28 57.8	1	8.48		0.82				4	K0 V	2013
20794	-43 01028	HR 1008		3 17 56	-43 15.6	6	4.26	.007	0.71	.004	0.22	.005	23	G8 III	15,1075,2012,3078*
20782	-29 01231	IDS03159S2913	A	3 17 57	-29 02.0	1	7.38		0.65				4	G3 V	2012
		LS I +59 182		3 17 58	+59 10.8	1	11.69		0.58		-0.17		3		41
	+47 00808	Alpha Per 481		3 17 59	+48 18.9	2	9.17	.010	0.38	.017	0.13		4	F1 IVn	1048,1575
20676	+44 00677	IDS03147N4502	A	3 18 04	+45 12.5	1	7.41		0.06		-0.13		5	B8	595
20677	+42 00750	HR 1002		3 18 05	+43 09.0	4	4.95	.004	0.05	.015	0.08	.009	9	A3 V	15,1363,3023,8015
		CS 22185 # 13		3 18 06	-13 15.2	1	13.77		0.14		0.07		1		1736
	+48 00892	Alpha Per 490	⋆ B	3 18 07	+48 56.5	2	9.58	.015	0.44	.008	0.04		5	F3 IV-V	1048,1575
20763	-4 00570			3 18 08	-04 32.0	1	8.91		0.57		0.10		1	G5	742
20772	-0 00530			3 18 11	-00 20.9	1	8.16		0.21		0.15		2	A0	208
		LS I +57 130		3 18 14	+57 04.1	1	11.18		0.86		-0.01		3		41
		Alpha Per 493		3 18 17	+47 32.2	2	11.06	.010	0.55	.016	0.07		4	F7	1048,1575
232781	+51 00722	G 174 - 47		3 18 17	+52 09.4	1	9.07		0.99		0.84		2	K5	1625
20675	+48 00893	Alpha Per 497	⋆ AB	3 18 19	+48 53.6	3	5.93	.002	0.44	.014	-0.02	.019	5	F6 V	71,254,3026
	-4 00571			3 18 19	-04 30.2	2	9.57	.000	0.45	.000	-0.02		5	G2	742,1694
20756	+20 00543	HR 1005	⋆ A	3 18 20	+20 58.1	3	5.27	.012	-0.07	.008	-0.53	.001	8	B5 IV	154,1223,1363
		Alpha Per 1603		3 18 21	+47 29.8	1	15.72		1.64				1		1694
	+49 00901	Alpha Per 403		3 18 22	+51 06.8	1	10.74		2.29				3	F5	369
20669	+55 00770	LS I +55 046		3 18 22	+55 24.4	1	8.19		0.52		-0.27		3	B8	41
	+50 00745	Alpha Per 505		3 18 23	+51 09.5	1	9.69		2.01				3	K8	369
20803	-10 00657			3 18 23	-10 31.4	1	9.73		0.40		0.06		1	F8	1776
	-8 00632			3 18 25	-07 39.7	1	9.70		0.47		-0.01		1	F5	1776
	+48 00894	Alpha Per 501	⋆ V	3 18 26	+49 02.2	2	9.16	.020	0.34	.010	0.13		4	F0 IV	1048,1575
		KUV 03184-0211		3 18 27	-02 10.8	1	16.01		0.16		-0.63		1	DA	1708
20507	+73 00180	G 245 - 73		3 18 28	+73 59.7	1	6.90		0.45		0.00		1	F5	1658
20791	+3 00461	HR 1007		3 18 30	+03 29.8	5	5.68	.014	0.96	.008	0.76	.020	14	G8.5 III	15,58,361,1415,3016
		BPM 48189		3 18 30	-23 29.	1	13.00		1.22				1		3061
20701	+47 00809	Alpha Per 507		3 18 33	+47 45.4	2	8.35	.005	0.12	.007	0.05	.025	9	A1.5V	1009,1048
		Alpha Per 1604		3 18 33	+48 38.9	1	12.06		0.78				1		1694
		Alpha Per 1605		3 18 33	+49 12.4	1	13.82		1.21				1		1694
		BSD 9 # 1336		3 18 37	+60 11.9	1	12.15		0.62		0.28		2	A0	1069
20853	-27 01183	HR 1013		3 18 37	-26 47.2	2	6.38	.005	0.54	.000			7	F7 V	15,2012
		LS I +58 122		3 18 39	+58 49.3	1	11.51		0.82		-0.08		3		41
		G 79 - 32		3 18 46	+14 08.7	2	15.24	.008	1.52	.037	1.01		8		316,906
21090	-73 00208			3 18 46	-73 14.7	1	8.45		1.10		1.01		5	K0 III	1731
20714	+51 00723	Alpha Per 522	⋆ A	3 18 48	+51 29.0	1	9.13		0.29		0.22		3	A7 Vn	1048
		Alpha Per 520	⋆ V	3 18 49	+48 57.8	2	11.69	.006	0.79	.001	0.21		28	G3	1048,1575
		CS 22172 # 14		3 18 59	-11 26.3	1	14.66		0.35		0.01		1		1736
20977	-60 00248			3 19 01	-60 40.5	1	7.89		0.91		0.58		7	G8 III/IV	1704
		Alpha Per 1606		3 19 07	+49 30.0	1	12.96		1.01				1		1694
		Alpha Per 1607		3 19 07	+49 30.0	1	12.88		1.03				1		1694
20825	+27 00500	HR 1012		3 19 11	+27 25.8	2	5.53	.010	1.10	.000	0.91	.000	3	G5 III	252,1769
		G 6 - 3		3 19 12	+17 03.7	1	10.90		0.99		0.88		1		333,1620
	+61 00566	LS I +62 233		3 19 12	+62 03.7	3	9.73	.014	0.39	.009	-0.40	.018	10	B5	41,1069,1775

HD	DM	Other Id	N	Rem	α_{1950}	δ_{1950}	S	V	σ_V	B–V	σ_{B-V}	U–B	σ_{U-B}	n	Spectrum	References
20894	−24 01600	HR 1016			3 19 13	−23 48.8	7	5.50	.025	0.88	.007	0.59	.002	30	G6.5 IIb	15,1007,1013,1075*
	+61 00568				3 19 14	+61 52.4	1	10.33		0.53		−0.13		3	B8	1069
		Alpha Per 538			3 19 16	+50 05.6	2	11.59	.020	0.61	.044	0.28		4	A7	1048,1575
	−21 00613	VW Eri			3 19 16	−21 38.2	1	9.20		1.45		1.40		18	K7	3049
		Alpha Per 537			3 19 19	+48 10.7	2	10.14	.010	1.15	.012	0.93		4	G5	1048,1575
		G 77 - 46			3 19 19	−06 51.2	1	11.37		1.50				1	M2	1746
		LS I +68 021			3 19 21	+68 35.6	1	10.81		0.70		−0.07		3		41
21166	−74 00250				3 19 23	−74 09.9	3	7.23	.003	0.41	.005	−0.06	.015	10	F3 V	278,1075,2012
		LP 355 - 64			3 19 26	+26 58.7	1	11.03		1.43		1.34		2		1723
		CS 22968 # 26			3 19 30	−55 49.1	2	14.23	.010	0.44	.009	−0.19	.008	2		1580,1736
237130	+58 00595				3 19 31	+58 49.6	1	9.15		0.34		−0.34		3	B8	1069
		Alpha Per 1501			3 19 34	+49 06.8	1	14.87		1.40				2		1575
		Alpha Per 1503			3 19 35	+48 31.9	1	15.54		1.05				1		1575
20808	+48 00897	Alpha Per 554			3 19 35	+49 13.2	1	8.68		0.14		0.13		3	A2 V	1048
20778	+55 00776				3 19 36	+55 57.0	1	8.08		0.16				2	A0	70
20909	−12 00640				3 19 36	−12 30.6	1	8.33		0.42		−0.03		1	F3 V	1776
		Alpha Per 1504			3 19 38	+48 43.3	1	16.09		1.26				1		1575
20761	+60 00674				3 19 38	+61 16.3	1	9.74		0.15		0.08		2	B9 V	1069
20809	+48 00899	Alpha Per 557		★	3 19 40	+49 02.2	4	5.29	.018	−0.07	.007	−0.53	.011	22	B5 V	15,154,1009,1048
	+59 00640				3 19 41	+59 50.8	1	10.10		0.29		0.26		3	B9.5V	1069
		LS I +55 047			3 19 42	+55 01.7	1	11.25		1.08		−0.06		3		41
		4HLF4 # 1030			3 19 42	−00 44.	1	13.42		0.82		0.41		1	K0	221
20835	+45 00749	G 95 - 31		★	3 19 44	+45 43.8	1	10.64		0.58		−0.01		1	G5	333,1620
		BSD 9 # 1891			3 19 45	+62 04.1	1	11.65		0.81		0.31		2	A0	1069
20893	+20 00551	HR 1015			3 19 52	+20 33.9	3	5.10	.005	1.23	.009	1.25	.020	9	K3 III	1355,1363,3016
		CS 22185 # 2			3 19 52	−17 06.4	1	13.77		0.41		−0.17		1		1736
		CS 22172 # 16			3 19 53	−11 45.2	1	11.07		0.26		0.08		1		1736
		LB 1662			3 19 54	−51 05.	1	13.26		−0.15		−0.55		1		45
		CS 22968 # 35			3 19 55	−54 04.5	1	15.51		0.11		0.19		1		1736
21011	−48 00930	HR 1021			3 19 56	−47 57.3	3	6.39	.008	0.99	.010			11	K0 III	15,1075,2032
232784	+50 00751	Alpha Per 574			3 19 59	+50 29.3	2	8.95	.010	1.26	.008	1.15		4	K0 III	1048,1575
		CS 22968 # 22			3 19 59	−57 06.4	1	15.58		−0.02		−0.17		1		1736
		CS 22968 # 28			3 20 03	−55 17.8	2	14.48	.000	0.11	.003	0.16	.018	2		966,1736
		Alpha Per 1608			3 20 04	+48 48.2	1	12.92		1.03				1		1694
20842	+51 00728	Alpha Per 575			3 20 04	+51 35.6	1	7.85		0.10		0.04		3	A0 Va	1048
20798	+61 00570	LS I +61 322		★ AB	3 20 05	+61 21.7	3	8.37	.000	0.25	.016	−0.57	.020	8	B2 III-IV	41,180,1069
20084	+84 00059	HR 965			3 20 06	+84 44.4	2	5.61	.000	0.92	.000	0.49	.000	4	G3IIp:+F0:V	15,1003
20980	−26 01257	HR 1018			3 20 07	−25 45.9	2	6.34	.005	0.01	.000			7	A1 V	15,2012
	+43 00699	G 78 - 33			3 20 09	+43 47.9	1	8.98		0.98		0.71		1	K2	1620
	+47 00815	Alpha Per 577			3 20 09	+47 46.9	1	9.79		0.68		0.20		3	F3 V	1048
		Alpha Per 1506			3 20 09	+48 59.9	1	15.39		1.56				2		1575
		BSD 9 # 778			3 20 10	+59 13.0	1	10.80		0.36		−0.25		2	B6	1069
		CS 22185 # 14			3 20 10	−13 25.4	1	14.79		0.01		−0.05		1		1736
20863	+48 00903	Alpha Per 581			3 20 15	+48 25.7	1	6.99		0.01		−0.15		3	B9 V	1048
		CS 22968 # 23			3 20 16	−56 31.8	1	15.46		0.24		0.19		1		1736
20934	+16 00432				3 20 17	+16 23.9	1	8.33		1.43				2	K0	882
20710	+71 00197				3 20 17	+71 30.5	1	7.61		0.10		−0.18		3	B8	555
	+49 00914	Alpha Per 588			3 20 19	+50 07.8	2	10.00	.010	0.53	.012	0.06		4	F5 V	1048,1575
20797	+64 00391	HR 1009			3 20 19	+64 24.6	3	5.18	.028	2.04	.027	2.14	.108	7	M0 II	1118,3001,8032
21022	−33 01191				3 20 21	−33 10.3	3	9.20	.012	0.95	.009	0.41	.063	8	G6/8 (V)w	742,1594,3077
20709	+72 00172				3 20 25	+73 02.0	2	7.06	.020	1.32	.002	1.44	.008	9	K3 III	1003,4001
		BSD 9 # 1904			3 20 27	+62 07.7	1	10.73		0.20		0.14		2	B9	1069
		BSD 9 # 1366			3 20 28	+60 47.0	1	12.01		0.47		0.35		3	A0	1069
	−7 00589				3 20 28	−07 14.9	1	9.60		0.45		−0.08		1	F8	1776
		CS 22172 # 19			3 20 28	−11 34.5	1	14.29		0.42		−0.17		1		1736
20955	+16 00433	IDS03177N1613		AB	3 20 30	+16 23.3	1	7.86		1.12				2	K0	882
		G 37 - 37			3 20 30	+33 48.1	2	12.28	.005	0.49	.010	−0.17	.010	4		1620,1658
	+60 00676				3 20 31	+60 30.0	1	10.17		0.28		0.27		3	A0 V	1069
21021	−30 01299	IDS03184S3033			3 20 31	−30 22.2	1	9.35		1.05		0.96		1	K4 V	3072
		Alpha Per 1609			3 20 33	+49 14.3	1	15.86		1.53				1		1694
		Alpha Per 1507			3 20 34	+48 29.1	1	16.06		0.85				1		1575
		Alpha Per 595			3 20 35	+51 32.2	2	10.36	.015	0.50	.005	0.37		4	F1 IVn	1048,1575
		Alpha Per 1509			3 20 36	+48 31.2	1	13.48		0.80				1		1575
		Alpha Per 1510			3 20 37	+48 13.4	1	12.10		0.36				1		1575
		G 5 - 32			3 20 39	+11 30.7	1	12.20		1.56		1.25		1		333,1620
20903	+45 00754				3 20 39	+46 06.7	1	8.67		0.31		0.14		301	A2	630
		Alpha Per 1512			3 20 40	+48 54.6	1	14.96		1.06				1		1575
		Alpha Per 1511			3 20 40	+48 59.5	1	13.66		0.74				1		1575
		Hyades L 1			3 20 41	+27 16.2	1	12.08		0.81		0.26		4		1058
		LS I +58 123			3 20 41	+58 57.6	1	11.37		0.63		−0.38		3		41
		CS 22172 # 17			3 20 41	−12 10.7	2	13.40	.015	0.33	.009	−0.02	.004	2		966,1736
		Alpha Per 600			3 20 42	+49 01.9	2	11.88	.020	0.66	.023	0.27		3		1048,1575
		Alpha Per 601			3 20 43	+49 28.4	2	11.41	.015	0.73	.003	0.20		4		1048,1575
	+49 00916	Alpha Per 602			3 20 43	+49 29.4	1	9.53		0.16		−0.43		2	B2p Shell	1012
		BSD 9 # 786			3 20 43	+59 56.3	1	11.54		0.39		−0.20		2	B8	1069
20902	+49 00917	Alpha Per 605		★ A	3 20 44	+49 41.1	11	1.79	.010	0.48	.004	0.39	.012	66	F5 Ib	1,15,1009,1048,1118*
		BSD 9 # 780			3 20 44	+60 09.2	1	12.04		0.45		0.13		2	A0	1069
20919	+48 00905	Alpha Per 606		★ V	3 20 46	+49 02.7	1	8.98		0.33		0.12		3	A8 V	1048
		Alpha Per 1514			3 20 47	+48 36.7	1	11.94		0.83				2		1575
		CS 22172 # 20			3 20 48	−10 08.5	2	15.19	.000	0.22	.000	0.02	.008	2		966,1736

Table 1 173

HD	DM	Other Id	N	Rem	α_{1950}	δ_{1950}	S	V	σ_V	B–V	σ_{B-V}	U–B	σ_{U-B}	n	Spectrum	References
		Alpha Per 608			3 20 49	+47 17.4	2	11.16	.030	0.61	.007	0.11		4	F5	1048,1575
	+49 00918	Alpha Per 609			3 20 49	+50 09.0	2	9.22	.005	0.41	.006	0.17		4	F1 Vn	1048,1575
21019	−8 00643	HR 1024		★ AB	3 20 51	−07 58.1	4	6.20	.008	0.70	.004	0.16	.005	16	G2 V	15,1075,1415,3077
		Alpha Per 1515		★ V	3 20 52	+48 37.8	1	14.12		1.29				21		1575
		BSD 9 # 209			3 20 52	+59 06.7	1	12.78		0.64		0.55		3		1069
		4HLF4 # 1031			3 20 54	−00 25.	1	12.05		0.57		0.10		2	G0	221
20931	+48 00906	Alpha Per 612			3 20 57	+48 57.8	1	7.87		0.09		0.06		3	A1 Va	1048
		Alpha Per 1517			3 20 58	+49 07.9	1	15.27		1.55				2		1575
	+61 00571	LS I +61 323		★ AB	3 20 58	+61 55.0	1	10.01		0.31		−0.37		3		41
237134	+59 00647				3 20 59	+60 04.6	1	9.46		0.38		−0.19		2	A7	1069
		LP 356 - 20	A		3 21 00	+23 37.	2	10.43	.026	1.50	.007	1.28		3		1017,1723
21066	−17 00654				3 21 00	−17 37.0	1	6.72		1.01		0.76		6	K0 III	1628
21018	+4 00532	HR 1023		★ AB	3 21 01	+04 42.3	4	6.38	.018	0.86	.004	0.51	.013	9	G5 III	15,1061,3077,8100
21032	+0 00581				3 21 02	+00 44.1	1	6.48		1.15		1.15		2	K0	985
	+55 00778				3 21 02	+56 20.0	2	9.96	.020	2.16	.005	2.65		7	M3	369,8032
	+87 00026				3 21 03	+87 43.9	1	8.84		0.48		0.02		3	G0	1733
21006	+19 00523				3 21 09	+19 43.8	1	6.83		1.63		1.96		3	K5	1648
20898	+59 00648	LS I +60 304		★ AB	3 21 11	+60 18.5	3	7.96	.021	0.43	.007	−0.42	.013	8	B2 III	41,1012,1069
		L 54 - 9			3 21 12	−71 51.	2	13.61	.035	0.50	.035	−0.20	.025	5		1696,3062
		BSD 9 # 796			3 21 13	+59 55.2	1	11.21		0.45		−0.20		2	B6	1069
		BSD 9 # 795			3 21 13	+60 02.2	1	10.18		0.44		0.05		2	A0	1069
		LS I +59 183			3 21 14	+59 52.6	1	11.28		0.37		−0.26		3		41
	+47 00816	Alpha Per 621			3 21 15	+48 14.1	1	9.86		0.49		0.02		3	F4 V	1048
	+78 00114				3 21 16	+78 32.9	1	10.37		0.85		0.59		2		1726
		Alpha Per 622			3 21 17	+48 41.8	2	11.64	.020	0.80	.013	0.28		5		1048,1575
	+59 00650	IDS03173N5911	AB		3 21 17	+59 22.2	1	10.12		0.37		−0.10		2	B9	1069
		LS I +56 086			3 21 18	+56 30.6	1	10.74		0.83		−0.02		3		41
		Alpha Per 1520			3 21 19	+48 53.7	1	15.66		1.55				2		1575
20929	+58 00598				3 21 20	+59 04.4	1	9.19		0.29		0.23		2	A0	1069
21017	+24 00481	HR 1022			3 21 21	+24 32.9	1	5.50		1.19				2	gK4	71
20961	+47 00817	Alpha Per 625			3 21 21	+47 44.4	2	7.63	.005	0.12	.008	0.02	.020	9	B9.5V	1009,1048
		Alpha Per 1610			3 21 21	+49 15.3	1	12.27		0.92				1		1694
20995	+33 00636	HR 1019		★ AB	3 21 22	+33 21.6	2	5.79		−0.02	.005	−0.15	.009	6	A0 V	1049,1733
		BSD 9 # 1377			3 21 23	+61 02.6	1	11.28		0.43		0.34		2	B9	1069
		4HLF4 # 44			3 21 24	+02 08.	1	12.70		0.30		0.22		2	A0	208
		BSD 9 # 800			3 21 24	+59 36.4	1	11.43		0.43		0.02		2	B8	1069
21051	+12 00473	HR 1028			3 21 25	+12 27.2	1	6.04		1.23		1.09		2	K0 III-IV	252
	+46 00745	Alpha Per 632			3 21 25	+47 14.3	2	9.72	.010	0.47	.008	0.02		4	F4 V	1048,1575
		CS 22968 # 25			3 21 27	−56 00.9	1	15.75		0.10		0.18		1		1736
		Alpha Per 1521			3 21 28	+48 51.6	1	15.56		1.60				2		1575
20969	+49 00921	Alpha Per 635			3 21 29	+49 37.2	1	9.05		0.34		0.11		3	A8 V	1048
		L 55 - 89			3 21 30	−73 49.	2	13.13	.005	0.86	.010	0.40	.005	6		1696,3062
	+55 00780				3 21 31	+55 24.0	1	9.77		2.11				1	K5 Ib	138
21050	+20 00556	HR 1027			3 21 33	+20 37.7	2			−0.04	.005	−0.09	.015	4	A1 V	1049,1079
20986	+48 00907	Alpha Per 639			3 21 36	+49 04.6	1	8.15		0.12		0.11		3	A3 Vn	1048
		4HLF4 # 45			3 21 36	−01 15.	1	13.31		0.24		0.17		2	A2	208
		BSD 9 # 1915			3 21 38	+61 50.2	1	12.68		0.60		0.12		2		1069
21049	+21 00447				3 21 40	+21 51.9	1	6.92		0.11		0.02		2	A0	1648
		CS 22968 # 33			3 21 40	−54 47.1	1	15.28		0.57		−0.05		1		1736
21094	+0 00586				3 21 41	+01 10.9	1	8.44		0.30		0.12		2	A8 IV	208
		Alpha Per 1525			3 21 44	+48 11.9	1	12.25		0.88				1		1575
21149	−33 01202	HR 1031			3 21 44	−32 53.0	4	6.51	.011	1.37	.009			18	K3 III	15,1075,2013,2029
21209	−50 01015	IDS03202S5021	A		3 21 44	−50 10.3	1	8.59		1.08		0.92		1	K3 V	3008
21209	−50 01015	IDS03202S5021	AB		3 21 44	−50 10.3	2	8.34	.010	1.13	.005	0.98		8	K3 V + K0	158,2012
21209	−50 01014	IDS03202S5021	B		3 21 44	−50 10.3	1	10.32		1.39		1.31		1	K0	3008
		Alpha Per 1526			3 21 46	+48 55.8	1	15.92		0.89				1		1575
	+50 00753	Alpha Per 647			3 21 46	+50 37.1	2	10.33	.020	0.43	.006	0.26		4	A7	1048,1575
		BSD 9 # 1916			3 21 46	+62 08.2	1	10.62		0.40		−0.05		2	A0	1069
		LP 472 - 66			3 21 47	+14 41.8	1	11.65		1.17		1.13		1	K5	1696
21005	+48 00909	Alpha Per 651			3 21 47	+49 08.5	1	8.42		0.19		0.13		3	A5 Vn	1048
20959	+58 00600	LS I +59 184			3 21 47	+59 15.4	4	8.03	.015	0.27	.004	−0.46	.006	13	B3 III	41,1012,1069,1775
	+6 00525				3 21 49	+06 48.6	1	10.28		1.28				1	K8	3072
		CS 22968 # 38			3 21 49	−53 09.2	1	16.19		0.07		0.16		1		1736
21038	+40 00736	HR 1026			3 21 50	+41 04.9	1	6.52		0.03				2	A0 V	1733
21102	−1 00484	WX Eri			3 21 50	−00 52.8	1	9.41		0.39		0.10		3	A7	208
		BSD 9 # 1389			3 21 51	+60 25.1	1	10.45		0.49		0.33		2	A2	1069
275241	+41 00678	G 95 - 35			3 21 52	+41 57.0	1	9.43		0.78		0.42		1	K0	333,1620
		Alpha Per 1611			3 21 54	+47 32.8	1	12.20		0.69				1		1694
		LP 943 - 94			3 21 56	−34 46.8	1	12.27		0.83		0.42		2		1696
21100	+14 00559				3 21 57	+14 47.9	1	8.18		0.22		0.14		3	A2	1648
		BSD 9 # 1390			3 21 57	+60 23.7	1	12.24		0.59		0.43		2	A0	1069
	+47 00819	Alpha Per 658			3 21 58	+47 47.5	1	9.25		0.41		0.29		3	A8 V	1048
		CS 22167 # 30			3 21 58	−03 37.5	1	14.93		0.03		0.06		1		1736
21475	−75 00231				3 21 58	−74 46.3	2	8.04	.006	0.26	.002	0.10		6	A8 III/IV	278,2012
		4HLF4 # 46			3 22 00	+01 03.	1	13.08		0.49		0.03		2	A7	208
		Alpha Per 1612			3 22 00	+48 19.6	1	13.72		1.13				1		1694
	+49 00925	Alpha Per 660			3 22 01	+50 08.8	2	10.08	.020	0.56	.007	0.11		4	F5 V	1048,1575
		Alpha Per 656			3 22 01	+50 53.1	2	11.76	.090	0.54	.023	0.38		4	A3	1048,1575
21004	+53 00657	HR 1020			3 22 03	+53 44.8	1	6.51		0.29		0.07		1	A9 III-IV	39
		CS 22968 # 48			3 22 03	−52 20.0	2	13.17	.000	0.06	.001	0.13	.021	2		966,1736

HD	DM	Other Id	N Rem	α_{1950}	δ_{1950}	S	V	σ_V	B–V	σ_{B-V}	U–B	σ_{U-B}	n	Spectrum	References
	+11 00468	G 5 - 35		3 22 04	+12 05.3	3	10.77	.005	0.54	.006	-0.18	.017	5	G0	1036,1620,3077
21161	−16 00630	IDS03198S1600	AB	3 22 05	−15 49.7	1	7.51		0.61				4	G1/2 V+G1 V	1414
		4HLF4 # 47		3 22 06	−02 20.	1	12.53		0.34		0.11		1	A0	208
21120	+8 00511	HR 1030		3 22 07	+08 51.3	15	3.60	.008	0.89	.005	0.61	.011	81	G8 III	1,3,15,1006,1020,1075*
21046	+46 00748	Alpha Per 665	⋆ A	3 22 08	+46 50.7	1	8.64		0.29		0.04		3	A7 V	1048
		CS 22167 # 22		3 22 08	−06 31.1	1	13.70		0.34		0.13		1		1736
21060	+44 00695	IDS03188N4510	AB	3 22 11	+45 20.4	1	7.51		0.08		-0.35		2	B8	1601
21110	+31 00597			3 22 18	+31 33.4	1	7.29		1.51		1.69		4	K4 III-IV	206
		Alpha Per 1527		3 22 19	+48 26.3	1	16.46		1.14				1		1575
	+10 00439			3 22 22	+11 01.4	1	10.70		1.15		1.12		1	K5	1696
		Alpha Per 1528		3 22 22	+48 20.7	1	13.09		1.05				2		1575
21071	+48 00913	Alpha Per 675	⋆	3 22 24	+48 56.8	4	6.07	.014	-0.08	.005	-0.48	.012	16	B7 V	15,154,1009,1048
	+46 00749	Alpha Per 680		3 22 25	+47 17.3	2	9.69	.000	1.46	.003	1.58		4	G8	1048,1575
		Alpha Per 1529		3 22 26	+48 27.4	1	15.93		1.27				1		1575
		Alpha Per 676		3 22 27	+50 30.3	2	11.43	.000	0.65	.000	0.16		4	G3	1048,1575
		Alpha Per 1530		3 22 28	+48 31.4	1	12.31		0.80				2		1575
	+49 00928	Alpha Per 679		3 22 28	+49 37.4	1	8.95		0.33		0.25		2	A9 Vn	1048
		Alpha Per 1531		3 22 29	+48 54.2	1	14.35		0.83				1		1575
		Alpha Per 684		3 22 31	+48 37.6	2	10.57	.024	0.57	.002	0.04		7	F9 V	1048,1575
21197	−5 00642	G 77 - 49		3 22 32	−05 31.7	9	7.87	.015	1.16	.008	1.15	.011	60	K5 V	22,989,1003,1509,1729,1775*
		LP 832 - 50		3 22 32	−26 19.6	1	12.18		0.78		0.30		2		1696
		fld II # 50		3 22 33	+45 20.1	1	12.77		1.12		0.80		2		1755
21157	+15 00481			3 22 35	+15 39.1	2	7.89	.024	1.25	.010	1.29		3	K2	882,3077
		4HLF4 # 48		3 22 36	−01 41.	1	13.37		0.09		0.11		1	A2	208
	+44 00697	G 78 - 37		3 22 37	+45 17.0	1	10.18		0.85		0.56		1	K1	333,1620
		fld II # 51		3 22 39	+45 17.6	1	12.95		0.63		0.51		2		1755
21092	+47 00822	Alpha Per 694		3 22 39	+48 02.8	1	8.48		0.20		0.15		3	A5 V	1048
21091	+47 00821	Alpha Per 692		3 22 39	+48 12.6	2	7.49	.005	0.04	.003	-0.04	.020	8	B9.5IVnn	1009,1048
		fld II # 49		3 22 40	+45 20.7	1	10.85		1.34		1.24		2		1755
	+16 00440			3 22 41	+16 58.1	1	9.43		0.70		0.16		1	G0	3077
		Alpha Per 1533		3 22 41	+47 58.7	1	12.94		1.04				1		1575
	+60 00682	LS I +60 305		3 22 41	+60 40.7	1	9.65		0.66		-0.21		3		41
		fld II # 52		3 22 42	+45 13.8	1	12.75		1.07		0.79		2		1755
		Alpha Per 1536		3 22 43	+48 14.8	1	14.23		1.16				2		1575
		Alpha Per 1537		3 22 43	+48 40.0	1	12.61		0.96				2		1575
	+42 00762			3 22 44	+42 25.1	1	9.06		0.28		0.03		3	A0	1733
		DO 571		3 22 45	+04 54.9	1	10.83		1.71				2		1532
	+30 00540			3 22 45	+30 45.4	1	9.21		0.34		0.01		4	B8 V	206
		Alpha Per 1538		3 22 45	+49 03.1	1	11.61		0.76				3		1575
		Alpha Per 696		3 22 45	+49 03.1	1	11.42		0.74		0.18		3		1048
		4HLF4 # 49		3 22 48	+00 32.	1	13.55		0.00		0.12		1	A0	208
		Alpha Per 699		3 22 48	+49 15.2	2	11.29	.020	0.71	.004	0.20		4	G3	1048,1575
		fld II # 6		3 22 51	+45 10.5	1	13.22		0.59		0.19		2		1755
		POSS 301 # 5		3 22 52	+31 31.3	1	17.66		1.67				2		1739
		fld II # 7		3 22 53	+45 09.9	1	11.16		1.43		1.43		2		1755
		Alpha Per 1541		3 22 53	+48 09.7	1	12.03		0.85				1		1575
		Alpha Per 1543	⋆ V	3 22 54	+48 51.7	1	12.84		0.97				24		1575
		fld II # 48		3 22 55	+45 19.1	1	12.66		0.63		0.13		2		1755
21252	−15 00595			3 22 56	−15 12.6	1	7.97		0.67				4	G3 V	2012
		Alpha Per 1545		3 22 58	+48 31.9	1	15.20		0.92				2		1575
		IRC +50 096	AB	3 22 59	+47 21.5	1	14.75		3.38				2		8019
		Alpha Per 1546		3 23 00	+48 59.3	1	12.41		0.78				1		1575
21156	+38 00722			3 23 01	+38 53.6	1	8.64		0.11		-0.04		3	A0 V	833
		fld II # 47		3 23 01	+45 19.6	1	13.35		0.77		0.23		2		1755
		fld II # 5		3 23 02	+45 10.5	1	12.63		0.68		0.26		2		1755
21117	+50 00757	Alpha Per 703		3 23 02	+50 40.3	1	7.63		0.09		-0.02		2	B8	1566
21251	−7 00597			3 23 03	−06 51.3	1	9.55		0.94		0.74		2	K2 V	3072
		Alpha Per 1547		3 23 04	+48 48.7	1	13.69		0.87				1		1575
		fld II # 4		3 23 06	+45 11.9	1	13.16		0.27		0.35		2		1755
		Alpha Per 1548		3 23 07	+49 06.6	1	14.83		1.04				1		1575
	+49 00933	Alpha Per 707		3 23 07	+49 36.9	2	10.03	.015	0.77	.001	0.32		4	G3	1048,1575
	+47 00825	Alpha Per 721		3 23 08	+47 42.5	2	9.66	.000	0.51	.000	0.21		5	F2 Vn	1048,1575
		Alpha Per 715		3 23 08	+48 36.2	2	9.74	.014	0.48	.004	0.02		6	F4 V	1048,1575
		Alpha Per 709		3 23 08	+49 44.1	2	10.94	.015	0.68	.001	0.14		5	G0 V	1048,1575
		Alpha Per 716		3 23 09	+48 50.1	2	10.43	.010	1.20	.002	0.90		4	G4	1048,1575
21182	+36 00689			3 23 10	+36 50.4	1	8.13		0.42		0.06		2	F2	1601
	+44 00702			3 23 14	+45 11.6	1	10.19		1.36		1.57		2		1755
		fld II # 2		3 23 16	+45 12.6	1	13.56		1.20		0.74		2		1755
		AX Per		3 23 16	+49 33.5	2	10.98	.500	1.05	.330	-0.33	.300	2	M3III:pe	1591,1753
		Alpha Per 727		3 23 17	+48 37.1	2	10.31	.010	0.56	.004	0.04		7	F7 V	1048,1575
21152	+47 00826	Alpha Per 729		3 23 19	+47 44.5	1	7.72		0.11		0.03		2	B9 V	1048
		Alpha Per 1552		3 23 19	+48 51.2	1	15.15		1.25				1		1575
		Alpha Per 1553		3 23 20	+48 33.4	1	15.48		1.11				1		1575
		CS 22167 # 29		3 23 20	−04 16.4	1	13.56		0.25		0.08		1		1736
		fld II # 46		3 23 22	+45 19.5	1	13.68		1.85		3.00		2		1755
		fld II # 44		3 23 22	+45 23.6	1	12.62		0.75		0.30		2		1755
21305	−20 00636			3 23 22	−20 30.8	1	10.35		-0.16		-0.72		1	B3/5 III	55
	−1 00491	SS Cet		3 23 23	−00 40.7	1	9.90		0.19		0.16		2	A2	1768
		Alpha Per 1554		3 23 25	+48 51.9	1	12.90		0.86				2		1575
		fld II # 1		3 23 28	+45 12.9	1	14.63		0.84		0.33		2		1755

Table 1   175

HD	DM	Other Id	N Rem	α_{1950}	δ_{1950}	S	V	σ_V	B–V	σ_{B-V}	U–B	σ_{U-B}	n	Spectrum	References
	+48 00916	Alpha Per 733		3 23 30	+48 36.8	1	9.94		0.50		0.04		3	F6 V	1048
21242	+28 00532	UX Ari		3 23 33	+28 32.5	2	6.49	.023	0.88	.018	0.44	.014	10	G5 IV	1501,7008
		fld II # 8		3 23 33	+45 05.6	1	12.69		1.07		0.70		2		1755
21181	+47 00828	Alpha Per 735		3 23 33	+48 01.9	2	6.83	.005	-0.01	.008	-0.21	.020	9	B8.5Vn	1009,1048
232793	+50 00759	Alpha Per 732		3 23 36	+50 42.3	2	10.19	.020	0.55	.002	0.03		4	F5 V	1048,1575
		4HLF4 # 51		3 23 36	−01 25.	1	11.80		0.36		0.01		2	A5	208
		fld II # 45		3 23 37	+45 21.0	1	12.95		0.75		0.34		2		1755
		fld II # 9		3 23 38	+45 08.6	1	12.19		0.36		0.33		2		1755
		Alpha Per 1555		3 23 39	+47 58.6	1	13.91		1.11				1		1575
		fld II # 43		3 23 41	+45 25.1	1	13.35		1.33		1.02		2		1755
21316	−1 00495	IDS03212S0123	A	3 23 44	−01 12.5	1	8.68		0.56		0.06		1	F8	1776
		LS I +57 131		3 23 46	+57 33.0	1	10.35		0.85		0.20		3		41
		fld II # 41		3 23 47	+45 17.5	1	13.72		0.91		0.38		2		1755
	+64 00394	LS I +65 028		3 23 48	+65 06.5	2	8.80	.005	1.00	.005	0.64	.020	7	F4 II	41,1775
		BPM 17232		3 23 48	−52 33.	1	14.54		0.45		-0.24		2		3065
		fld II # 53		3 23 49	+45 13.9	1	17.10		-0.13		-0.72		2		1755
		Alpha Per 1614		3 23 49	+47 49.0	1	13.30		1.08				1		1694
21563	−70 00230	HR 1053		3 23 49	−69 48.0	3	6.14	.003	0.48	.000	0.24	.010	17	A3/5 V+G0/5	15,1075,2038
		fld II # 10		3 23 51	+45 09.5	1	13.55		0.66		0.16		2		1755
		fld II # 40		3 23 51	+45 18.0	1	11.69		1.85		2.08		2		1755
		Alpha Per 1556	⋆ V	3 23 51	+48 12.0	1	13.00		1.00				27		1575
		G 77 - 51		3 23 53	−04 39.3	1	11.84		1.36		1.21		1	K4-5	1658
		fld II # 39		3 23 55	+45 17.7	1	13.69		0.90		0.42		2		1755
		fld II # 11		3 23 56	+45 11.5	1	13.16		0.72		0.30		2		1755
21338	−2 00627			3 23 56	−01 58.6	1	9.31		0.22		0.10		2	A5 IV	208
		fld II # 29		3 23 59	+45 15.0	1	12.40		1.20		0.85		2		1755
		fld II # 38		3 23 60	+45 17.5	1	13.68		0.60		0.28		2		1755
21423	−36 01290	HR 1042		3 24 00	−36 05.7	2	6.38	.005	0.08	.000			7	A1 IV	15,2012
		fld II # 28		3 24 01	+45 14.8	1	14.54		0.72		0.06		2		1755
		G 5 - 36		3 24 02	+23 36.5	3	10.82	.015	0.53	.004	-0.11	.004	9		1064,1620,3077
21434	−36 01291	IDS03221S3618	A	3 24 03	−36 07.9	1	7.30		0.29				4	A9 V	173
		Alpha Per 750		3 24 04	+48 49.1	2	10.57	.024	0.59	.002	0.09		4	F9 V	1048,1575
		fld II # 42		3 24 05	+45 21.8	1	12.80		1.13		0.86		2		1755
21239	+47 00830	Alpha Per 756		3 24 05	+48 06.0	1	7.95		0.10		0.11		3	A3 Vn	1048
		Alpha Per 1560		3 24 06	+48 14.6	1	15.82		1.70				3		1575
		Alpha Per 1615		3 24 06	+48 19.3	1	14.13		1.64				1		1694
		Alpha Per 1559		3 24 06	+49 06.6	1	14.85		0.94				1		1575
237143	+55 00787			3 24 07	+55 33.3	1	9.23		0.53				3	A0	70
21411	−31 01384	IDS03221S3058	A	3 24 07	−30 47.7	3	7.88	.004	0.72	.004	0.22	.007	17	G8 V	158,861,2033
		fld II # 37		3 24 10	+45 17.5	1	14.32		0.69		0.27		2		1755
21335	+18 00484	HR 1036		3 24 12	+18 35.0	2	6.57		0.14	.015	0.11	.005	5	A3 V	252,1049
		fld II # 12		3 24 12	+45 07.8	1	14.27		0.52		0.23		2		1755
		fld II # 30		3 24 12	+45 15.6	1	12.38		2.10		2.65		2		1755
		LP 300 - 3		3 24 13	+27 12.8	1	11.77		1.54				3		1723
		Alpha Per 1561		3 24 14	+48 49.9	1	12.77		1.09				2		1575
21225	+54 00682	IDS03204N5449	AB	3 24 14	+55 00.1	1	8.66		0.28				2	A2	70
		CS 22185 # 3		3 24 14	−16 15.8	1	15.10		-0.28		-1.02		1		1736
		fld II # 36		3 24 15	+45 18.5	1	13.14		0.81		0.26		2		1755
		Alpha Per 1562		3 24 15	+48 45.0	1	16.84		1.43				1		1575
		fld II # 13		3 24 16	+45 08.1	1	11.78		1.41		1.39		2		1755
		LS I +64 097		3 24 16	+64 11.3	1	11.28		0.53		-0.33		3		41
21430	−27 01228	HR 1045		3 24 16	−27 29.5	3	5.93	.011	0.94	.005			14	K0 III	15,1075,2029
		Alpha Per 1563		3 24 17	+49 01.8	1	12.29		0.92				1		1575
		4HLF4 # 52		3 24 18	+01 35.	1	12.84		0.23		0.14		2	A2	208
		4HLF4 # 53		3 24 18	+01 54.	1	11.64		0.30		0.15		2	A3	208
21280	+43 00726	Y Per		3 24 18	+44 00.2	1	9.75		2.93				1	C0 e	1238
		fld II # 27		3 24 18	+45 13.2	1	13.74		0.74		0.26		2		1755
21203	+59 00657	HR 1033	⋆ AB	3 24 18	+60 05.0	1			0.02		-0.24		2	B9 V + A1	1079
	+29 00561	Hyades J 205		3 24 19	+29 36.0	1	9.86		0.65		0.11		5	G0	1058
		Alpha Per 767		3 24 20	+49 35.3	2	10.68	.014	0.62	.007	0.08		2	F9 V	1048,1575
21410	−18 00610			3 24 20	−18 23.9	1	9.31		0.25		0.16		6	A2 III	355
	+61 00588		V	3 24 22	+61 23.6	1	9.82		2.01				3	M0	369
21473	−42 01115	HR 1049		3 24 24	−41 48.7	4	6.32	.006	0.06	.005	0.05		15	A0 V	15,2012,2012,3050
		fld II # 31		3 24 25	+45 17.3	1	14.91		0.63		0.18		2		1755
		fld II # 35		3 24 25	+45 19.8	1	11.76		0.30		0.22		2		1755
	+45 00769		A	3 24 25	+46 03.8	1	10.17		0.47		0.33		2		1723
21279	+47 00831	Alpha Per 775		3 24 25	+47 33.8	1	7.26		0.05		-0.16		3	B8.5V	1048
21224	+59 00658	LS I +59 185	⋆ AB	3 24 25	+59 44.0	2	7.54	.000	1.03	.030	0.63		6	F8	41,6009
21212	+61 00587	LS I +62 234		3 24 25	+62 19.2	2	8.24	.067	0.58	.004	-0.46	.022	5	B2 V:e	41,1012
21364	+9 00439	HR 1038		3 24 27	+09 33.6	6	3.74	.014	-0.09	.006	-0.33	.010	22	B9 Vn	15,1079,1256,1363*
		Alpha Per 1565		3 24 27	+48 49.6	1	13.00		1.05				2		1575
		fld II # 26		3 24 28	+45 15.4	1	9.61		1.66		2.02		2		1755
		fld II # 32		3 24 28	+45 17.6	1	14.27		1.34		1.25		2		1755
		fld II # 34		3 24 28	+45 19.4	1	14.02		1.27		1.00		2		1755
21278	+48 00920	Alpha Per 774	⋆	3 24 29	+48 53.4	6	4.98	.005	-0.09	.008	-0.55	.007	37	B5 V	15,154,1009,1048,1363,8015
		Alpha Per 771		3 24 30	+50 21.8	2	11.12	.020	0.68	.001	0.12		4	G0	1048,1575
21379	+12 00477	HR 1039		3 24 33	+12 33.7	1			-0.02		-0.07		3	A0 Vs	1049
		fld II # 24		3 24 33	+45 15.4	1	14.30		1.32		0.91		2		1755
		fld II # 25		3 24 33	+45 15.9	1	14.84		1.13		0.52		2		1755
		fld II # 33		3 24 35	+45 18.6	1	14.06		1.19		1.16		2		1755

HD	DM	Other Id	N	Rem	α_{1950}	δ_{1950}	S	V	σ_V	B–V	σ_{B-V}	U–B	σ_{U-B}	n	Spectrum	References
		LS I +58 124			3 24 35	+58 09.0	1	11.49		0.70		-0.04		3		41
		G 221 - 10			3 24 35	+73 51.7	1	16.49		0.48		-0.49		1	DC	782
		LB 3316			3 24 36	−65 13.	1	11.94		0.40		0.22		2		45
		fld II # 14			3 24 37	+45 07.8	1	11.69		0.65		0.08		2		1755
21405	+1 00597				3 24 38	+02 06.4	1	7.40		0.12		0.07		3	A0	221
		fld II # 15			3 24 38	+45 08.1	1	12.85		1.25		1.09		2		1755
		Alpha Per 1568			3 24 39	+49 02.9	1	13.38		1.10				2		1575
		Xa Ret			3 24 39	−65 13.8	1	11.16		0.14		0.00		1		700
		fld II # 23			3 24 40	+45 14.6	1	12.75		1.95		2.52		2		1755
		LP 31 - 139			3 24 40	+73 52.3	1	16.48		1.69				3		1663
	+22 00492				3 24 42	+22 43.5	1	10.97		0.73		0.24		2		1723
		Alpha Per 1569			3 24 43	+48 13.8	1	16.15		1.13				1		1575
21302	+49 00938	Alpha Per 780			3 24 43	+49 46.8	1	8.09		0.17		0.11		66	A1 Vn	1048
	−24 01656				3 24 43	−23 53.6	1	10.89		0.41		-0.21		2		1696
21503	−41 01006				3 24 44	−40 44.6	1	8.39		1.51		1.67		5	K4/5 (III)	1673
	+20 00571	G 5 - 40			3 24 46	+20 52.4	1	10.79		0.57		-0.03		1	G0	333,1620
		Alpha Per 1570		⋆ V	3 24 46	+48 29.4	1	12.83		1.00				40		1575
		Alpha Per 1571			3 24 46	+48 51.1	1	13.46		0.76				1		1575
21259	+61 00589				3 24 47	+61 24.6	1	8.78		1.66				4		369
		fld II # 16			3 24 48	+45 11.2	1	12.96		0.69		0.27		2		1755
		fld II # 22			3 24 48	+45 14.4	1	13.37		1.15		0.84		2		1755
		Alpha Per 1572			3 24 48	+49 04.1	1	12.78		0.99				2		1575
		fld II # 17			3 24 49	+45 11.6	1	10.96		0.85		0.57		2		1755
		fld II # 21			3 24 50	+45 13.8	1	14.18		0.66		0.22		2		1755
		Alpha Per 1573			3 24 51	+48 43.4	1	12.05		0.68				2		1575
21392	+23 00456				3 24 52	+24 00.6	1	7.85		1.09		0.85		2	G5	1625
	+9 00440	IDS03222N0935		AB	3 24 56	+09 43.4	1	10.45		1.39		1.25		2	K7	1746
	+48 00923	Alpha Per 799		⋆ AB	3 24 58	+48 46.1	1	9.66		0.45		0.04		2	F4 V	1048
	+49 00939	Alpha Per 794			3 24 58	+50 05.7	2	10.07	.015	0.54	.003	0.07		4	F3 IV-V	1048,1575
		LS I +59 186			3 24 58	+59 55.6	2	10.58	.020	1.02	.004	0.79	.000	4	A2 Iab	41,1215
21291	+59 00660	HR 1035		⋆ AB	3 25 00	+59 46.1	10	4.21	.019	0.41	.008	-0.23	.009	29	B9 Ia	1,15,41,1012,1079*
21276	+60 00685				3 25 00	+60 23.3	1	9.53		0.55		0.00		11	G5	1655
		fld II # 18			3 25 01	+45 11.3	1	12.74		0.44		0.34		2		1755
		Alpha Per 1574			3 25 01	+48 23.6	1	16.08		1.20				1		1575
		fld II # 19			3 25 02	+45 11.8	1	13.18		0.80		0.41		2		1755
21345	+48 00924	Alpha Per 802			3 25 03	+49 12.9	1	8.41		0.17		0.12		2	A5 Vn	1048
		fld II # 20			3 25 06	+45 12.5	1	12.75		0.56		0.13		2		1755
21179	+71 00201	HR 1032			3 25 06	+71 41.5	2	6.38	.044	1.81	.010	1.94	.044	6	M2 III	1733,3001
		G 5 - 41			3 25 07	+21 51.3	2	11.41	.028	1.41	.030	1.14	.024	3	M1	1620,5010
		CS 22968 # 37			3 25 07	−53 09.3	2	14.99	.010	-0.08	.015	-0.31	.003	2		966,1736
	−53 00566				3 25 08	−53 20.3	2	9.72	.010	0.87	.005	0.49		8	G8	158,2034
		G 174 - 54			3 25 11	+50 42.5	1	10.98		0.83		0.32		2		7010
21402	+33 00656	HR 1041			3 25 12	+33 38.2	2	5.73		0.03	.010	0.10	.034	5	A2 V	1049,1733
		Alpha Per 1575			3 25 13	+49 06.1	1	13.82		1.27				2		1575
		Alpha Per 1617			3 25 14	+49 01.6	1	13.05		0.95				1		1694
		CS 22172 # 40			3 25 15	−08 10.1	1	13.07		0.23		0.08		1		1736
	−8 00655				3 25 15	−08 36.0	1	9.74		1.47		1.78		2	K5 V	3072
21362	+49 00944	Alpha Per 810		⋆	3 25 16	+49 40.6	4	5.58	.006	-0.04	.002	-0.44	.011	23	B6 Vn	15,154,1009,1048
	+29 00565				3 25 17	+29 37.6	1	9.16		0.53		0.30		4	F0 V	206
		Alpha Per 1577			3 25 17	+49 08.2	1	16.27		1.04				1		1575
21375	+48 00927	Alpha Per 817			3 25 19	+48 53.9	2	7.46	.005	0.11	.001	0.05	.020	9	A0.5IVn	1009,1048
21722	−69 00192	HR 1064			3 25 22	−69 30.7	2	5.95	.006	0.41	.005	-0.01	.009	20	F3/5 IV	15,1075,2038,3077
	−36 01303				3 25 23	−36 26.0	1	9.74		0.97		0.82		2	K5 V	3072
		Alpha Per 815			3 25 25	+50 10.4	2	11.18	.005	1.43	.008	1.35		4	K2	1048,1575
21267	+67 00270				3 25 26	+68 15.7	2	7.99	.005	0.02	.005	-0.28	.005	6	B8 V	196,555
		Alpha Per 828			3 25 27	+48 03.8	1	11.59		0.71				1	F8	1694
		CS 22185 # 20			3 25 27	−14 17.1	1	12.66		0.20		0.19		1		1736
21467	+22 00495	HR 1048		⋆ AB	3 25 31	+22 38.0	1	6.03		0.95				2	K0 IV	71
21532	−21 00630				3 25 33	−21 34.2	1	9.94		-0.10		-0.48		2	B7 V	3060
21398	+47 00835	Alpha Per 831			3 25 35	+48 07.9	2	7.36	.005	0.01	.006	-0.14	.020	9	B9 V	1009,1048
		G 38 - 1			3 25 36	+37 13.4	3	11.11	.005	1.32	.005	1.06	.000	6	K4	1064,1620,3078
		Alpha Per 833			3 25 36	+48 00.5	1	10.03		0.49		0.03		3	F6 V	1048
21531	−20 00643				3 25 36	−19 58.9	4	8.39	.010	1.34	.017	1.20	.027	17	K5 V	22,1197,2012,3077
21450	+35 00697				3 25 37	+35 38.3	1	7.17		1.16		1.18		2	G5	1601
21574	−36 01306	HR 1054			3 25 37	−35 51.3	3	5.71	.008	1.29	.008			14	K2 III	15,1075,2029
21530	−11 00667	HR 1050			3 25 38	−11 27.5	2	5.73	.000	1.10	.000	1.04		7	K2 III	2006,3005
21483	+29 00566				3 25 42	+30 12.2	4	7.06	.024	0.35	.006	-0.33	.012	10	B3 III	1009,1012,1252,1733
21428	+49 00945	Alpha Per 835		⋆ AB	3 25 47	+49 20.3	6	4.67	.005	-0.09	.006	-0.57	.009	31	B3 V	15,154,1009,1048,1363,8015
21940	−77 00135				3 25 47	−76 55.2	3	6.81	.004	0.21	.005	0.10		10	A2mA5-A8	278,1075,2012
		HRC 11		V	3 25 48	+30 33.	1	13.43		1.28		0.92		1	K2e	776
21448	+44 00714	LS V +44 003		⋆ AB	3 25 48	+44 52.6	2	7.15	.002	0.28	.028	-0.63		7	G8 III	226,399
		LP 772 - 70			3 25 50	−15 55.1	1	12.23		0.50		-0.17		2		1696
	+48 00929	Alpha Per 841			3 25 51	+48 47.5	2	10.29	.005	0.54	.005	0.05		3	F7 V	1048,1575
		Steph 366			3 25 52	+39 49.8	1	11.32		1.46		1.43		1	M1	1746
		Alpha Per 1578			3 25 52	+49 10.3	1	13.06		1.02				2		1575
		Steph 370			3 25 52	−06 37.2	1	11.12		1.38		1.30		1	K7	1746
21543	−7 00603	G 77 - 54			3 25 52	−06 42.0	10	8.24	.018	0.62	.002	0.03	.013	22	G2 V-VI	516,742,1003,1064,1097,1620*
	+12 00479	Hyades L 2			3 25 54	+13 00.0	3	9.73	.004	0.91	.005	0.70		8	K2	582,950,1058
		Alpha Per 1579			3 25 54	+48 01.9	1	10.00		0.60				2		1575
	+47 00837	Alpha Per 848			3 25 54	+48 01.9	1	9.99		0.59		0.07		3	F9 V	1048

Table 1   177

HD	DM	Other Id	N	Rem	α_{1950}	δ_{1950}	S	V	σ_V	B–V	σ_{B-V}	U–B	σ_{U-B}	n	Spectrum	References
21389 +58 00607		HR 1040		★	3 25 54	+58 42.4	9	4.54	.011	0.56	.007	-0.11	.005	30	A0 Iae	15,41,369,1012,1119*
21455 +46 00760		Alpha Per 861		★ A	3 25 56	+46 46.0	5	6.24	.008	0.13	.004	-0.25	.011	25	B7 V	15,154,226,1009,1048
278695 +31 00607					3 25 58	+31 57.8	1	9.24		0.42		-0.20		2	G0	3060
21626 −44 01139					3 26 00	−44 01.7	2	6.72	.005	0.50	.000			8	F6/7 IV	1075,2012
		Alpha Per 1580			3 26 03	+49 11.7	1	15.00		0.87				2		1575
		CS 22172 # 33			3 26 06	−09 36.6	1	13.79		0.57		-0.15		1		1736
		Alpha Per 1581			3 26 07	+49 09.0	1	12.43		0.90				2		1575
21749 −63 00231				AB	3 26 08	−63 40.1	3	8.08	.013	1.12	.008	1.08		6	K4 V	1705,2033,3077
21427 +58 00608		HR 1043		★ AB	3 26 09	+59 11.7	1	6.13		0.09				2	A2 V	70
21414 +59 00662		LS I +59 188			3 26 10	+59 51.1	1	8.36		0.60		0.36		3	F2	41,8100
21447 +54 00684		HR 1046		★ A	3 26 11	+55 16.8	3	5.10	.012	0.04	.004	0.04	.003	23	A2.5V	1625,1648,3024
21447 +54 00684		HR 1046		★ AB	3 26 11	+55 16.8	15	5.10	.011	0.04	.007	0.04	.008	88	A2.5V	1,15,1006,1007,1013,1077*
21447 +54 00684		IDS03224N5506		B	3 26 11	+55 16.8	1	10.02		1.00		0.80		2		3024
		HRC 12		V	3 26 12	+31 12.4	1	14.25		1.59		1.16		1	K2e	776
		Alpha Per 1582			3 26 12	+48 05.1	1	12.73		0.88				1		1575
21480 +48 00930		Alpha Per 862			3 26 12	+48 59.0	2	8.52	.000	0.32	.004	0.07	.010	9	A7 V	1009,1048
	+48 00931	Alpha Per 863			3 26 13	+48 50.3	2	9.21	.000	0.51	.003	0.02		5	F6 V	1048,1575
21635 −36 01310		HR 1058		★ AB	3 26 16	−36 01.5	2	6.49	.005	0.13	.000			7	A1 V	15,2012
21479 +48 00933		Alpha Per 868			3 26 17	+49 02.5	2	7.28	.000	0.10	.004	0.01	.000	9	A1 IV-V	1009,1048
21481 +47 00840		Alpha Per 875			3 26 18	+47 48.4	2	7.67	.010	0.10	.001	0.04	.020	9	A0 Vn	1009,1048
	+30 00549				3 26 19	+31 15.4	1	10.47		0.48		-0.14		4	B8:p	206
		Alpha Per 879			3 26 21	+47 56.6	2	10.39	.015	0.48	.006	0.08		4	F1	1048,1575
21581 −0 00552					3 26 21	−00 35.3	2	8.71	.000	0.82	.005	0.21		4	G0	742,1594
21465 +54 00685					3 26 23	+55 12.2	1	7.10		2.00				3	K5 I	70
		Alpha Per 859			3 26 24	+51 20.4	2	10.50	.020	0.56	.001	0.02		4	G0	1048,1575
		Alpha Per 865			3 26 27	+50 51.8	1	10.73		0.69		0.52		3	A7	1048
21568 +15 00493					3 26 31	+16 01.6	1	8.34		0.41		0.00		3	F0	1648
		LP 412 - 48			3 26 31	+18 44.7	1	12.06		0.87		0.41		2		1696
	+66 00268	G 246 - 38			3 26 31	+66 34.5	6	9.91	.005	0.66	.004	-0.09	.007	9	sdG2	979,1064,1658,1774,8112,3078
21488 +52 00699		IDS03229N5233		A	3 26 34	+52 43.5	1	7.29		1.39		1.34		3	K0	3016
21488 +52 00699		IDS03229N5233		B	3 26 34	+52 43.5	1	10.42		0.60		0.11		3		3016
	+1 00604	G 77 - 56			3 26 36	+01 50.1	1	10.45		0.75		0.31		1		1620
	+54 00686	LS V +54 007			3 26 38	+54 44.5	1	10.02		0.52				3		70
275419 +37 00774					3 26 42	+37 23.8	1	11.04		0.30		0.18		2	A0	1041
		CS 22172 # 23			3 26 45	−11 52.7	1	13.34		-0.26		-1.10		1		1736
21527 +48 00934		Alpha Per 885			3 26 46	+48 19.7	1	8.79		0.28		0.15		3	A7 IV	1048
		G 6 - 13			3 26 47	+12 38.0	1	11.47		0.81		0.35		1	G8	333,1620
		Alpha Per 1585			3 26 48	+48 47.1	1	16.08		1.39				1		1575
		Alpha Per 1586		★ V	3 26 50	+48 14.5	1	14.31		1.32				31		1575
		CS 22172 # 27			3 26 52	−10 59.5	1	13.19		0.37		-0.03		1		1736
21590 +16 00450					3 26 53	+16 35.5	1	7.10		-0.06		-0.41		2	A0p	3016
	+10 00449	G 79 - 42			3 26 54	+10 23.8	2	10.76	.029	0.51	.010	-0.17	.015	3	sdF8	333,1620,1696
	−12 00662				3 26 56	−11 50.4	5	10.00	.016	1.40	.021	1.30	.005	14	M0 V	158,1619,1705,1775,3072
21540 +46 00762		Alpha Per 903			3 26 57	+46 53.5	1	7.03		0.12		-0.16		1	B8	245
275494 +37 00776					3 26 58	+37 49.7	1	10.88		0.28		0.18		2	A0	1041
21522 +55 00793		IDS03232N5537		AB	3 27 01	+55 47.5	1	9.34		0.31				2	A2	70
21553 +47 00842		Alpha Per 906		★ V	3 27 02	+47 27.5	1	8.78		0.28		0.19		3	A6 Vn	1048
21552 +47 00843		Alpha Per 900		★	3 27 02	+47 49.5	6	4.36	.013	1.37	.016	1.55	.007	25	K1 III	15,1355,1363,3016*
21551 +47 00844		Alpha Per 904		★	3 27 04	+47 56.0	4	5.82	.004	-0.04	.004	-0.31	.010	19	B8 V	15,1009,1048,1079
	+54 00687				3 27 04	+54 50.5	1	10.04		0.21				2		70
		Alpha Per 1587			3 27 06	+49 03.9	1	15.91		1.02				1		1575
21550 +50 00771		Alpha Per 887			3 27 07	+50 51.6	1	9.00		0.20		0.14		2	A0	1566
21826 −58 00285					3 27 08	−58 03.2	1	6.74		1.58		1.83		6	K5 III	1628
21665 −7 00606		HR 1060			3 27 12	−06 58.5	4	5.98	.007	1.02	.000	0.88	.007	12	G5	15,1256,2006,3077
21703 −24 01679					3 27 12	−24 16.4	1	9.36		1.10		0.84		2	K4 V	3072
21688 −13 00662		HR 1062			3 27 14	−12 50.8	3	5.58	.007	0.16	.035	0.11		9	A5 III/IV	15,252,2012
		Alpha Per 917			3 27 15	+47 43.1	2	11.02	.085	0.68	.020	0.19		4	F4	1048,1575
		Alpha Per 1588			3 27 15	+48 07.2	1	14.01		0.76				1		1575
		G 79 - 43			3 27 18	+09 16.1	2	11.58	.024	0.48	.000	-0.16	.005	3	sdF5	333,1620,1696
		CS 22185 # 22			3 27 18	−15 11.9	1	13.77		0.27		0.08		1		1736
		AK 9 - 47			3 27 20	+16 10.5	1	10.96		0.59				1		987
275458 +39 00799		G 95 - 40			3 27 20	+40 09.1	1	9.54		0.91		0.69		1	K0	333,1620
237148 +57 00727				V	3 27 21	+58 19.4	1	8.62		1.88				2	K7	369
21589 +41 00693					3 27 23	+42 01.9	1	6.93		0.38		0.01		3	F2	379
	+58 00611	LS I +59 189			3 27 25	+59 07.3	1	10.49		1.02		0.53		3	B8 II	41
		NGC 1342 - 23			3 27 29	+37 07.2	1	13.04		0.72		0.25		1		49
	−74 00252				3 27 29	−73 57.8	1	10.46		0.47				4	F8	2012
		G 37 - 40			3 27 34	+33 51.9	2	12.74	.003	1.46	.036	1.26		5		1691,7010
		Steph 373			3 27 37	+19 55.7	1	10.79		1.39				1	K6	1746
21663 +19 00547		Hyades vB 3		★ A	3 27 37	+19 56.0	5	8.32	.017	0.76	.011	0.33	.002	12	G5	1127,1355,1674,3026,8023
21663 +19 00547		Hyades vB 3		★ B	3 27 37	+19 56.0	6	10.78	.015	1.40	.010	1.23	.005	16		582,950,1058,1570,1674*
21600 +49 00953		Alpha Per 921			3 27 39	+49 32.2	1	8.59		0.20		0.15		3	A6 Vn	1048
21686 +10 00452		HR 1061			3 27 40	+11 10.0	4	5.14	.006	-0.04	.006	-0.04	.017	14	A0 Vn	3,1049,1363,3023
	+59 00668	LS I +59 190			3 27 40	+59 53.1	1	10.88		0.38		-0.41		3		41
		CS 22172 # 26			3 27 40	−11 23.4	1	14.74		0.34		0.10		1		1736
		CS 22172 # 28			3 27 42	−10 48.4	1	13.95		0.30		0.05		1		1736
	−53 00570			AB	3 27 46	−53 40.4	2	10.72	.000	1.51	.005	1.20		5	F8	158,1705
275500 +36 00700		NGC 1342 - 9			3 27 48	+37 19.8	1	10.54		0.52		0.01		1	F0 V	49
		GK Per			3 27 48	+43 44.1	1	12.92		0.84		-0.42		1	sd:Be	698
		NGC 1342 - 28			3 27 49	+37 01.7	1	14.00		0.82		0.20		1		49

HD	DM	Other Id	N	Rem	α_{1950}	δ_{1950}	S	V	σ_V	B–V	σ_{B-V}	U–B	σ_{U-B}	n	Spectrum	References
		NGC 1342 - 29			3 27 51	+37 14.0	1	14.20		0.85		0.32		1		49
		NGC 1342 - 32			3 27 51	+37 15.5	1	14.92		0.93		0.40		1		49
21651	+38 00737				3 27 52	+38 58.7	2	7.31	.010	0.35	.005	0.06	.005	4	F4 V	105,1733
275505	+36 00703	NGC 1342 - 15			3 27 54	+37 12.5	1	11.20		0.36		0.28		1		49
	+48 00937	Alpha Per 935			3 27 54	+48 49.3	2	10.05	.005	0.63	.009	0.10		3	F9.5V	1048,1575
21619	+49 00954	Alpha Per 931			3 27 54	+49 44.0	1	8.75		0.26		0.20		3	A6 V	1048
21650	+41 00696				3 27 55	+41 33.4	2	7.33	.015	0.03	.000	-0.42	.000	5	B5	379,1212
21620	+48 00938	Alpha Per 934		⋆	3 27 55	+49 02.4	2	6.28	.004	0.08	.009	0.08	.038	5	A0 Vn	595,1733
		NGC 1342 - 34			3 27 57	+37 10.5	1	10.16		1.87		1.64		1		4002
		NGC 1342 - 33			3 27 58	+37 16.0	1	14.97		0.80		0.34		1		49
21685	+27 00513	IDS03250N2723	B		3 27 59	+27 33.3	1	7.87		0.15		0.09		3	A0	1084
		NGC 1342 - 31			3 28 00	+37 16.3	1	14.40		0.77		0.24		1		49
		Alpha Per 953		⋆ B	3 28 00	+47 42.0	1	10.61		0.95				4		1009
275510	+36 00705	NGC 1342 - 8			3 28 01	+37 09.1	2	10.29	.000	0.29	.020	0.20	.035	3	A2 V	49,1041
21641	+47 00846	Alpha Per 955		⋆ A	3 28 01	+47 41.6	2	6.76	.010	-0.02	.000	-0.26	.015	9	B8.5V	1009,1048
21700	+27 00514	IDS03250N2723	A		3 28 02	+27 33.7	1	7.44		0.04		0.00		3	B9	1084
		CS 22172 # 34			3 28 02	-09 41.9	1	14.69		0.47		-0.16		1		1736
21755	+5 00502	HR 1067			3 28 06	+06 01.1	3	5.93	.008	0.96	.005	0.64	.011	10	G8 III	15,1061,1355
21754	+12 00486	HR 1066			3 28 07	+12 46.0	6	4.10	.006	1.11	.014	1.02	.019	21	K0 II-III	15,1105,1355,1363*
		NGC 1342 - 17			3 28 07	+37 11.4	1	11.29		0.49		0.30		1		49
		Alpha Per 936			3 28 07	+49 55.9	2	11.41	.005	0.60	.012	0.22		4	F1	1048,1575
21790	-5 00674	HR 1070			3 28 08	-05 14.7	8	4.73	.005	-0.09	.005	-0.26	.019	27	B9 Vs	3,15,1075,1079,1425*
	+49 00957	Alpha Per 944			3 28 09	+49 22.0	1	9.62		0.43				2	F3 V	1048
275515		NGC 1342 - 14			3 28 10	+37 07.4	1	11.17		0.39		0.30		1	A0	49
		CS 22172 # 30			3 28 10	-10 43.0	1	14.42		0.47		-0.22		1		1736
21882	-43 01085	HR 1075			3 28 10	-42 48.3	2	5.77	.005	0.22	.000			7	A5 V	15,2027
275509	+36 00707	NGC 1342 - 6			3 28 13	+37 11.4	2	9.64	.015	1.20	.018	0.87	.009	2	K0	49,4002
	-22 00617				3 28 13	-22 14.0	1	10.40		1.17				4	K4.5	1619
21989	-64 00247				3 28 13	-64 08.0	2	8.15	.005	1.28	.000	1.20	.012	3	K1p Ba	565,3048
21661	+48 00942	Alpha Per 954		⋆	3 28 14	+49 13.9	1			0.09		-0.14		2	B9 III	1079
275502	+36 00708	NGC 1342 - 10			3 28 16	+37 16.3	2	10.62	.005	0.41	.030	0.29	.025	3	A2 V	49,1041
		Steph 376			3 28 17	+14 45.6	1	11.53		1.37		1.27		1	K7	1746
21744	+24 00503	IDS03253N2455	AB		3 28 17	+25 05.2	1	8.13		0.23		0.09		2	A3	1625
21672	+48 00943	Alpha Per 965			3 28 20	+48 33.9	2	6.62	.005	-0.02	.004	-0.31	.010	9	B8 V	1009,1048
		Alpha Per 968			3 28 21	+48 21.5	1	10.41		0.57		0.03		3	F8 V	1048
21774	+20 00578	G 5 - 42			3 28 22	+20 36.1	1	8.14		0.68		0.25		1	G5	333,1620
	+48 00944	Alpha Per 970			3 28 22	+48 24.9	1	8.19		0.19		0.13		3	A5 V	1048
	+49 00958	Alpha Per 958			3 28 22	+49 42.1	1	9.20		0.39		0.15		3	F1 V	1048
		CS 22172 # 29			3 28 22	-10 47.4	1	14.36		0.61		-0.07		1		1736
		G 221 - 14			3 28 23	+74 28.9	1	13.55		1.58				1		906
		Alpha Per 979			3 28 25	+48 00.5	2	11.67	.005	0.56	.003	0.06		4		1048,1575
	-36 01326	SX For			3 28 26	-36 13.3	1	10.70		0.26				1	F7	700
		NGC 1342 - 25			3 28 27	+37 12.3	1	13.35		0.85		0.25		1		49
		NGC 1342 - 16			3 28 27	+37 13.2	1	11.22		0.51		0.43		1		49
21899	-41 01029	HR 1076			3 28 27	-41 32.3	2	6.11	.005	0.48	.000			7	F7 V	15,2012
21728	+36 00710	NGC 1342 - 2			3 28 28	+37 09.7	2	8.73	.020	0.22	.020	-0.11	.030	3	B8p	49,1041
22001	-63 00234	HR 1083		⋆ A	3 28 30	-63 06.8	5	4.71	.007	0.39	.003	-0.05	.010	17	F3 IV/V	15,146,1075,2012,3026
22001	-63 00234	IDS03276S6317	B		3 28 30	-63 06.8	1	10.68		1.46		1.10		4		146
	+21 00474				3 28 32	+21 37.9	1	8.88		0.40		-0.03		2	F2	1648
		Pleiades 3408			3 28 33	+26 05.8	1	10.61		0.61		0.15		2		1723
21861	-18 00622				3 28 33	-18 38.1	1	7.36		1.16		1.11		3	K0 III/IV	1657
21753	+35 00708				3 28 34	+36 18.9	1	7.82		0.20		0.11		2	A4 V	1041
		NGC 1342 - 20			3 28 34	+37 09.2	1	11.82		1.00		0.66		1		49
		NGC 1342 - 19			3 28 34	+37 13.5	1	11.77		0.65		0.49		1		49
232799	+53 00672				3 28 35	+54 19.3	1	9.06		0.44				2	B5	70
21699	+47 00847	Alpha Per 985		⋆ V	3 28 36	+47 51.3	5	5.47	.008	-0.11	.010	-0.57	.016	22	B8 III	15,379,1009,1048,1079
21727	+43 00744	G 78 - 40		⋆	3 28 39	+43 30.1	1	8.56		0.70		0.28		1	G5	1620
	+60 00695				3 28 39	+60 32.8	1	8.57		1.86				3	M0	369
21773	+36 00713	NGC 1342 - 1			3 28 41	+37 18.5	1	8.44		0.42		0.06		1	F4 V	49
		CS 22185 # 23			3 28 41	-15 35.2	1	12.12		0.33		0.11		1		1736
22167	-73 00221				3 28 42	-73 43.3	3	8.18	.005	0.20	.005	0.13	.003	7	A4 IV/V	278,861,2012
		NGC 1342 - 27			3 28 43	+37 12.1	1	13.92		0.95		0.44		1		49
		Alpha Per 988			3 28 44	+47 58.3	1	9.80		0.56		0.07		3	F5 IV-V	1048
21804	+23 00462				3 28 45	+23 28.8	1	8.08		0.42		-0.04		2	F5	1625
275508	+36 00714	NGC 1342 - 11			3 28 45	+37 10.6	2	10.64	.015	0.35	.035	0.24	.035	3	A2 V	49,1041
21711	+51 00744	Alpha Per 959			3 28 45	+51 53.8	1	8.63		0.25		-0.52		5	B8	595
21877	-17 00683				3 28 45	-17 20.3	1	8.60		0.20		0.11		5	A2 III	355
		NGC 1342 - 21			3 28 46	+37 12.3	1	12.40		0.70		0.34		1		49
		NGC 1342 - 30			3 28 46	+37 17.6	1	14.36		1.42		0.59		1		49
		NGC 1342 - 38			3 28 47	+37 17.5	1	10.81		1.57		1.18		1		4002
275507	+36 00715	NGC 1342 - 7			3 28 48	+37 11.2	2	10.12	.010	1.21	.007	0.78	.019	2	G5	49,4002
21785	+37 00781	NGC 1342 - 3			3 28 48	+37 20.7	2	9.21	.005	0.23	.015	0.14	.030	3	A3 V	49,1041
21697	+54 00689				3 28 48	+55 01.0	1	8.86		0.40				3	F5	70
		NGC 1342 - 26			3 28 49	+37 10.5	1	13.55		0.77		0.21		1		49
275491	+37 00782				3 28 49	+37 39.7	1	9.74		0.45		0.33		2	A2	1041
21772	+44 00732				3 28 51	+44 40.0	1	10.47		0.25				3		6009
21771	+44 00732				3 28 51	+44 40.1	1	7.24		1.40				3	K0	6009
		CS 22172 # 24			3 28 51	-11 44.3	1	13.70		0.30		0.05		1		1736
21867	-3 00565				3 28 55	-03 04.1	1	8.93		0.40				8	F0	1268
		Alpha Per 1618			3 28 56	+49 00.4	1	11.98		0.81				1		1694

Table 1   179

HD	DM	Other Id	Rem	α_{1950}	δ_{1950}	S	V	σ_V	B–V	σ_{B-V}	U–B	σ_{U-B}	n	Spectrum	References
	+55 00795	LS I +55 048		3 28 56	+55 40.8	2	10.17	.030	0.42	.005	-0.32		5		41,70
275501	+36 00716	NGC 1342 - 4		3 28 57	+37 12.8	2	9.32	.015	1.33	.008	1.02	.012	2	G8 II	49,4002
21770	+45 00778	HR 1069		3 28 58	+45 53.4	7	5.31	.010	0.40	.009	-0.02	.000	20	F4 III	15,226,254,1008,1119*
21726	+53 00675			3 28 58	+54 05.2	1	8.70		0.24				2	A0	70
21938	-37 01326			3 28 58	-37 32.5	1	8.38		0.70		0.13		2	G0 V	3056
		NGC 1342 - 24		3 28 59	+37 14.9	1	13.23		0.74		0.46		1		49
		G 5 - 43		3 29 00	+14 09.6	3	12.26	.013	1.55	.005	1.15	.054	9		940,1036,1705,3029
21981	-47 01071	HR 1081, TU Hor		3 29 01	-47 32.7	2	5.98	.005	0.11	.000			7	A1 V	15,2012
278713	+35 00712			3 29 03	+35 58.0	1	10.05		0.24		0.10		2	A0	1041
		AK 9 - 155		3 29 06	+18 09.7	1	10.60		0.68				1		1570
		LP 413 - 2		3 29 06	+18 09.7	1	11.57		0.68		0.07		2		1696
		LP 300 - 80		3 29 07	+30 07.7	1	13.86		1.51				2		1674
		NGC 1342 - 22		3 29 07	+37 15.8	1	12.70		0.64		0.40		1		49
		Alpha Per 1005		3 29 10	+47 58.2	1	9.61		0.45		0.03		3	F3 V	1048
		Alpha Per 992		3 29 10	+50 11.2	2	10.86	.025	0.72	.002	0.23		4	G0	1048,1694
		NGC 1342 - 37		3 29 12	+37 11.7	1	10.82		1.78		1.83		1		4002
		CS 22172 # 36		3 29 12	-09 20.1	1	12.11		0.19		0.14		1		1736
		CS 22176 # 1		3 29 12	-09 20.1	1	12.06		0.18		0.13		1		1736
		CS 22172 # 31		3 29 12	-10 34.6	1	14.02		0.26		0.07		1		1736
21803	+44 00734	HR 1072, KP Per		3 29 13	+44 41.2	4	6.40	.011	0.03	.004	-0.71	.013	10	B2 IV	15,154,1012,1223
	+54 00690			3 29 13	+54 41.6	1	9.80		1.31				3		70
	+55 00796			3 29 14	+55 44.1	1	9.70		0.48				3		70
		NGC 1342 - 18		3 29 15	+37 11.8	1	11.60		0.58		0.39		1		49
275516		NGC 1342 - 12		3 29 16	+37 05.4	1	10.84		0.33		0.32		1	A1 V	49
		LP 832 - 25		3 29 20	-26 29.8	1	13.88		1.56				1		3062
		CS 22172 # 35		3 29 21	-09 29.9	1	14.30		0.43		-0.17		1		1736
	+26 00568			3 29 27	+26 34.0	1	10.08		0.83		0.46		2		1723
		POSS 301 # 3		3 29 28	+31 53.3	1	15.96		1.59				1		1739
21847	+35 00714			3 29 28	+35 29.5	1	7.30		0.49		0.04		2	F8	1601
237150	+58 00618	IDS03255N5826	B	3 29 28	+58 35.7	2	7.90		0.40	.013	-0.01	.025	5	A7	1211,3024
		KUV 03295-0108		3 29 28	-01 08.0	1	17.17		0.07		-0.67		1	DA	1708
21856	+34 00674	HR 1074		3 29 29	+35 17.6	6	5.89	.011	-0.07	.009	-0.84	.017	22	B1 V	15,154,1009,1012,1203,1252
21769	+58 00619	HR 1068	★ A	3 29 31	+58 35.9	2	6.40		0.15	.015	0.14	.008	5	A4 III	1211,3058
21951	-9 00690			3 29 35	-09 33.2	1	9.67		0.32		0.08		1	F0	1722
		LP 713 - 2		3 29 37	-10 18.0	1	13.17		1.12		0.87		1		1696
		CS 22185 # 25		3 29 37	-16 31.9	1	14.18		0.42		-0.21		1		1736
21807	+52 00702			3 29 39	+52 47.1	1	8.90		0.29				3	A2	70
21802	+55 00797			3 29 39	+55 23.5	1	7.89		0.16				2	A0	70
21610	+72 00178	HR 1055		3 29 41	+73 10.8	2	6.57	.005	0.04	.010	0.02		16	A0 Vn	252,1258
		AO 860		3 29 41	-09 55.5	1	11.24		0.82		0.52		1		1722
		AO 861		3 29 42	-09 40.9	1	12.01		0.64		0.05		1		1722
275517	+36 00718	NGC 1342 - 5		3 29 43	+37 07.3	1	9.60		0.60		0.43		1	F1 III	49
275499	+36 00719	NGC 1342 - 13		3 29 43	+37 17.7	1	10.93		0.65		0.17		1	F2	49
21794	+57 00730	HR 1071		3 29 43	+57 42.0	3	6.36	.009	0.49	.024	0.00	.051	8	F7 V	595,1501,1733
		LP 888 - 19		3 29 44	-32 42.8	2	13.00	.000	0.50	.005	-0.17	.035	4		1696,3062
		Hyades CDS 52	★	3 29 45	+15 59.2	1	14.21		1.64				3		1674
21997	-26 01333	HR 1082		3 29 45	-25 47.0	2	6.37	.005	0.12	.000			7	A3 V	15,2012
		UKST 762 # 19		3 29 47	-05 58.2	1	13.58		0.68		0.18		3		1584
21965	-6 00695			3 29 49	-06 09.8	1	8.77		0.23		0.11		4	A2	1584
21819	+54 00693	HR 1073		3 29 50	+54 48.4	1	5.98		0.11				2	A3 V	70
	+23 00465	Hyades vB 4		3 29 52	+23 31.5	2	8.89	.010	0.84	.004	0.54	.001	4	G5	1127,8023
		UKST 762 # 7		3 29 52	-05 48.2	1	11.23		0.50		0.06		2		1584
		CS 22172 # 37		3 29 52	-09 18.8	1	12.54		0.19		0.11		1		1736
21996	-21 00644			3 29 52	-21 25.0	2	9.38	.008	-0.14	.000	-0.63	.008	9	B3/5 V	55,1775
21933	+8 00528	HR 1079		3 29 53	+09 12.4	4	5.76	.007	-0.08	.007	-0.27	.023	15	B9 IV	3,15,1079,1415
		Alpha Per 1036		3 29 53	+49 01.9	2	11.36	.015	0.83	.006	0.52		3		1048,1694
275490	+37 00785			3 29 55	+38 00.8	1	10.70		0.39		0.20		2	A0	1041
21985	-3 00570	AS Eri		3 29 55	-03 28.9	1	8.31		0.17		0.08		5	A1:V:	588
21995	-9 00693	G 160 - 1		3 29 58	-08 46.1	2	8.05	.020	0.56	.025	0.02	.010	2	F8	1658,3056
22120	-55 00537			3 30 00	-55 25.1	1	7.13		1.53		1.87		6	M0 III	1628
		G 77 - 61		3 30 02	+01 48.3	2	13.89	.004	1.74	.027	1.16	.013	21		316,1663
	-49 00988			3 30 02	-49 02.7	1	9.17		1.22		1.23		2		1730
21832	+55 00799			3 30 03	+55 55.8	1	9.00		0.30				2	A0	70
		KUV 03301-0100		3 30 04	-00 59.7	1	15.86		-0.20		-1.10		1	DA	1708
		UKST 762 # 22		3 30 04	-05 34.5	1	11.12						1		1584
278801	+34 00678	IDS03269N3502	AB	3 30 07	+35 12.1	1	9.44		0.30		0.24		2	A2	1041
278784	+36 00722			3 30 10	+36 31.0	1	11.06		0.55		0.27		2	A5	1041
		KUV 03302-0143		3 30 11	-01 42.9	1	17.12		-0.31		-1.00		1	DA	1708
		UKST 762 # 30		3 30 11	-06 16.6	1	15.35		0.55				2		1584
		LS I +59 191		3 30 12	+59 40.4	1	10.15		0.27		-0.44		3		41
21843	+59 00672	LS I +59 192		3 30 14	+59 34.1	2	7.81	.052	0.74	.009	0.50	.056	4	B3 III	41,8100
		AO 862		3 30 14	-09 26.1	1	10.70		0.80		0.43		1		1722
	+49 00967	Alpha Per 1050		3 30 16	+50 07.8	1	9.48		0.40		0.29		3	Am	1048
21806	+63 00426	LS I +63 199		3 30 16	+63 43.3	3	7.72	.068	0.31	.015	-0.55	.017	11	B1 V (n)	41,399,555
21912	+39 00811	HR 1078, IW Per		3 30 17	+39 44.0	3	5.82	.029	0.13	.012	0.10	.007	8	A5 m	985,1049,3058
		CS 22172 # 38		3 30 18	-09 05.6	1	13.60		0.81		-0.38		1		1736
		UKST 762 # 14		3 30 19	-05 39.1	1	12.53		0.70		0.21		2		1584
278795	+35 00716			3 30 20	+35 36.2	1	10.88		0.27		0.16		2	A2	1041
22252	-66 00195	HR 1092		3 30 20	-66 39.5	3	5.82	.004	-0.06	.000	-0.32	.005	13	B8 IV	15,1075,2038
		LP 653 - 3		3 30 23	-03 40.6	1	11.64		1.24		1.17		1	K4	1696

HD	DM	Other Id	N Rem	α_{1950}	δ_{1950}	S	V	σ_V	B–V	σ_{B-V}	U–B	σ_{U-B}	n	Spectrum	References
21943	+37 00786			3 30 24	+37 50.7	1	7.24		-0.03		-0.27		2	B8 V	1041
22067	-24 01715			3 30 25	-23 47.8	1	10.02		-0.09		-0.41		2	B8/9 V	3060
		UKST 762 # 29		3 30 31	-06 10.8	1	15.35		0.59		0.00		2		1584
	+6 00549			3 30 32	+07 17.1	1	11.20		1.26		1.19		1	K7	1746
22017	+13 00568			3 30 34	+13 36.9	1	6.90		1.06		0.99		2	G5	1648
22049	-9 00697	HR 1084	★	3 30 34	-09 37.6	22	3.73	.010	0.88	.007	0.58	.013	109	K2 V	3,15,22,1004,1006*
237153	+58 00622	LS I +58 126		3 30 36	+58 28.7	3	9.22	.030	1.05	.010	0.20	.019	5	B6 Iab	41,138,1215
		POSS 201 # 4		3 30 37	+31 49.4	1	17.23		1.73				2		1739
21931	+48 00949	Alpha Per 1082	★ AB	3 30 39	+48 27.1	2	7.34	.000	0.03	.006	-0.12	.020	9	B9 V	1009,1048
21894	+58 00623	LS I +58 127		3 30 41	+58 25.1	1	7.94		0.21		-0.64		3	B8	41
278813	+33 00677			3 30 43	+34 07.1	1	9.38		0.22		0.14		2	A2	1041
278778	+36 00724			3 30 43	+36 50.3	1	9.26		0.38		0.23		2	A0	1041
21942	+46 00770	Alpha Per 1090		3 30 44	+47 06.5	2	9.24	.010	0.21	.000	0.07	.024	6	A0 V	1009,1048
21974	+38 00743			3 30 45	+39 08.4	1	8.13		1.30		1.43		3	K4 III	833
		UKST 762 # 21		3 30 45	-06 16.8	1	13.82		0.63		0.06		2		1584
		UKST 762 # 26		3 30 47	-06 14.5	1	14.74		0.62		0.02		1		1584
		UKST 762 # 10		3 30 48	-06 16.7	1	11.59		1.61		1.65		1		1584
22449	-75 00236			3 30 48	-74 45.8	2	7.53	.003	0.16	.002	0.11		6	A3/4 IV/V	278,2012
		UKST 762 # 9		3 30 50	-06 14.1	1	11.45		0.40		-0.06		2		1584
	-10 00695			3 30 52	-09 51.7	1	9.88		0.81		0.41		1		1722
22004	+35 00717			3 30 55	+35 55.1	1	9.36		0.15		0.03		2	B9.5V	1041
21903	+59 00675	HR 1077	★ AB	3 30 55	+59 52.5	3	6.43	.012	0.43	.027	-0.06	.004	7	F3 V +F9 V	292,938,3024
21903	+59 00675	IDS03268N5942	C	3 30 55	+59 52.5	1	10.66		0.67		0.11		3		3024
		Alpha Per 1100		3 30 57	+46 54.4	1	11.20		0.81		0.35		3	G3	1048
		UKST 762 # 6		3 30 58	-05 59.6	1	11.08		0.65		0.09		2		1584
	-6 00700			3 30 58	-06 22.0	1	9.79		1.20		1.26		2		1584
278806	+34 00681			3 31 01	+34 39.9	1	11.08		0.51		0.23		2	B8	1041
		UKST 762 # 13		3 31 02	-06 09.3	1	12.36		0.82		0.40		4		1584
		UKST 762 # 12		3 31 04	-06 23.1	1	12.12		0.63				2		1584
		AO 864		3 31 04	-09 40.2	1	11.19		0.47		-0.05		1		1722
22231	-50 01071	HR 1090		3 31 05	-50 32.9	2	5.68	.000	1.10	.000	1.16		6	K2 III	2032,3009
	-26 01339			3 31 06	-26 02.2	1	11.31		-0.35		-1.23		12		1732
	+74 00159			3 31 07	+74 28.3	1	9.65		1.44		1.45		2		1726
	+23 00472	Pleiades 3301		3 31 09	+24 10.7	1	9.54		0.51		0.01		1	F5	8023
		Alpha Per 1102		3 31 10	+47 43.2	1	11.03		0.60		0.13		3	G1 V	1048
	-26 01341			3 31 10	-26 14.2	1	9.91		0.51		-0.02		5		1732
		UKST 762 # 25		3 31 11	-05 52.7	1	14.37		0.77		0.30		2		1584
22130	-9 00698			3 31 11	-09 31.0	1	9.57		0.38		0.07		1	F0	1722
		AO 865		3 31 11	-09 45.1	1	10.21		0.64		0.13		1		1722
22072	+17 00575	HR 1085		3 31 17	+17 40.2	2	6.15	.025	0.89	.005	0.52	.020	8	K1 IV	1355,3077
		AO 867		3 31 18	-09 23.9	1	10.73		1.12		1.09		1		1722
	-45 01184			3 31 18	-44 52.4	2	11.46	.005	1.58	.015			3	M3.5	912,1705
		L 228 - 89		3 31 18	-52 40.	1	14.98		0.72		0.22		3		3062
22071	+22 00506	G 5 - 44		3 31 21	+22 49.4	1	9.18		0.61		0.10		1	G0	333,1620
21930	+63 00430			3 31 25	+64 03.5	1	8.43		0.19		0.01		3	A2 V	555
		UKST 762 # 28		3 31 27	-06 21.9	1	14.92		0.90				1		1584
		AO 868		3 31 27	-09 48.4	1	9.95		0.98		0.71		1		1722
22044	+38 00747			3 31 28	+38 38.4	1	7.91		0.40		-0.04		3	F4 V	833
	-6 00701			3 31 28	-05 41.1	1	10.18		0.58		0.04		2		1584
22773	-81 00083			3 31 28	-80 52.7	1	7.78		1.17		1.23		9	K1 III	1704
22091	+23 00473	HR 1086	★ AB	3 31 29	+24 17.9	4	5.94	.010	0.12	.005	0.16	.007	12	A3 V +A3 V	603,1049,1381,3024
22091	+23 00473	IDS03285N2408	C	3 31 29	+24 17.9	1	9.92		0.50		0.00		4	A2	3024
		UKST 762 # 23		3 31 30	-06 17.9	1	14.11		0.80				1		1584
		TonS 359		3 31 30	-22 57.	1	13.83		-0.06		-0.34		1		1036
		Alpha Per 1101		3 31 31	+49 31.7	1	11.25		0.69				1	G4	1694
275573	+37 00792	G 78 - 41		3 31 32	+38 09.9	1	10.21		0.67		0.13		1	K2	333,1620
22203	-22 00628	HR 1088		3 31 35	-21 48.0	8	4.26	.010	-0.11	.010	-0.35	.004	21	B8 V +B8 V	15,369,1075,1079,1425*
22039	+44 00744			3 31 36	+45 17.0	1	7.58		1.66		2.00		2	K2	1601
22287	-49 00998			3 31 37	-49 04.9	1	8.21		0.95		0.65		2	G8 II	1730
	-26 01344			3 31 38	-26 08.0	1	10.16		1.12		1.02		5		1732
		G 78 - 41a	★	3 31 41	+38 08.5	1	10.01		0.82		0.48		2		1620
22359	-61 00265	IDS03307S6055	AB	3 31 44	-60 44.6	1	7.55		0.50				4	F6 V	2012
22676	-78 00101	HR 1109		3 31 45	-78 31.2	3	5.68	.012	0.93	.000	0.65	.000	11	G8 III	15,1075,2038
232811	+53 00680	IDS03280N5334	AB	3 31 47	+53 44.2	1	9.77		0.45				2	A0	70
		G 37 - 43		3 31 49	+34 26.8	2	10.87	.005	1.23	.012	1.45	.242	3	M0	3003,7010
22090	+38 00749			3 31 50	+38 25.2	1	7.04		0.31		-0.02		2	F0	1601
		AJ89,1229 # 229		3 31 52	+05 51.6	1	10.95		1.15				1	K7	1632
		EY Cep		3 31 52	+80 51.4	1	9.80		0.37		-0.04		4	A5	1768
22248	-31 01449			3 31 52	-30 47.7	1	7.15		1.08				4	K0 III	2012
		G 38 - 3		3 31 54	+28 07.0	1	16.73		1.86				1		906
22124	+31 00616	IX Per	★ AB	3 31 54	+31 51.1	1	6.68		0.41		0.03		3	F2 IV-V	985
22382	-61 00267	HR 1096		3 31 54	-61 11.1	2	6.42	.008	1.03	.006	0.91		6	K0 III	58,2035
22114	+37 00794	V496 Per		3 31 56	+37 51.0	1	7.58		0.07		-0.35		2	B8 Vp	1041
		UKST 762 # 24		3 31 56	-05 43.9	1	14.25		0.61		-0.13		1		1584
22262	-31 01450	HR 1093	★ AB	3 31 56	-31 14.9	2	6.19	.005	0.48	.000			7	F5 V	15,2012
		Hyades L 5	V	3 31 58	+22 41.5	1	12.30		0.77		0.33		4		1058
	+61 00431			3 31 59	+61 33.7	1	10.50		0.38		0.26		3		1689
22146	+23 00477	Pleiades 3302		3 32 02	+23 21.9	1	8.93		0.36		0.02		1	A5	8023
22285	-35 01279			3 32 03	-34 57.8	2	8.78	.000	1.06	.005	0.60	.029	3	G/Kp Ba	565,3048
		POSS 301 # 2		3 32 04	+32 01.6	1	15.43		0.75				1		1739

Table 1 181

HD	DM	Other Id	N Rem	α_{1950}	δ_{1950}	S	V	σ_V	B–V	σ_{B-V}	U–B	σ_{U-B}	n	Spectrum	References
		LP 593 - 16		3 32 07	+01 11.5	1	12.33		0.55		-0.10		2		1696
		G 37 - 44		3 32 09	+32 02.2	2	15.49	.005	0.17	.040	-0.53	.020	6		316,3060
	-5 00691			3 32 09	-05 39.6	1	10.86		0.52		-0.02		3		1584
		LP 943 - 130		3 32 09	-34 11.5	1	12.01		0.69		0.10		2		1696
22211	+5 00511	HR 1089		3 32 10	+06 15.1	3	6.49	.003	0.63	.002	0.15	.003	12	G0 IV	15,1256,1509
		UKST 762 # 8		3 32 11	-05 55.7	1	11.31		0.53		0.06		2		1584
22243	-10 00704	HR 1091		3 32 13	-10 02.1	2	6.24	.005	0.02	.000			7	A2 V	15,2027
		UKST 762 # 17		3 32 14	-05 55.6	1	13.42		0.50		-0.04		2		1584
		LP 888 - 25		3 32 20	-31 14.4	1	15.27		1.36		1.03		1		1773
22240	-3 00576			3 32 23	-03 34.7	1	7.58		0.14				9	A2	1268
22158	+38 00753	IDS03291N3859		3 32 24	+39 08.7	2	8.91	.015	0.06	.005	-0.13	.010	6	B9 V	833,1601
	+69 00219	LS I +69 006		3 32 25	+69 20.2	1	9.35		0.32		-0.62		3	B0	41
22225	+18 00507			3 32 26	+18 44.3	2	7.53	.055	1.72	.005	2.02	.050	5	K5	1648,3040
		G 78 - 42		3 32 27	+41 32.3	1	12.56		1.53				1		1691
22136	+46 00773	Alpha Per 1153		3 32 27	+46 55.6	2	6.89	.000	-0.02	.005	-0.30	.015	9	B8 IV: sn	1009,1048
		UKST 762 # 11		3 32 27	-06 09.1	1	11.95		0.69		0.19		2		1584
22105	+55 00801			3 32 29	+55 43.3	1	6.82		0.11				9	B8	70
22322	-32 01358	HR 1095	⋆ AB	3 32 33	-32 02.5	1	6.40		1.40				4	K3 III	2032
22254	+10 00461			3 32 34	+11 13.9	1	8.33		0.58		0.16		2	F8	3077
	-31 01454			3 32 37	-31 13.9	1	11.45		0.59		-0.11		1	K2:	1773
22043	+64 00398			3 32 38	+65 11.4	1	8.11		0.49		0.08		3	F8	1733
22135	+52 00703			3 32 42	+52 46.0	2	7.25	.015	1.85	.010	2.15		4	K5 II	70,1733
		UKST 762 # 15		3 32 43	-06 14.4	1	13.05		1.01		0.89		2		1584
22193	+41 00714	IDS03294N4201	AB	3 32 46	+42 10.6	1	8.34		0.68		0.16		3	G6 V +G8 V	3030
21910	+74 00161			3 32 46	+74 34.4	1	7.46		1.03		0.75		2	G8 III	1003
		CS 22185 # 28		3 32 49	-14 39.1	1	11.98		0.36		0.12		1		1736
278818	+33 00681			3 32 51	+33 58.7	1	10.22		0.35		0.22		2	A2	1041
	+11 00495	G 5 - 45		3 32 54	+11 52.7	1	10.53		0.93		0.72		1	K2	333,1620
22192	+47 00857	Alpha Per 1164	⋆ V	3 32 56	+48 01.7	8	4.23	.017	-0.06	.014	-0.57	.019	37	B5 Ve+Shell	15,154,379,1009,1048*
275579	+37 00796			3 32 57	+37 55.7	1	9.89		0.22		0.14		2	A0	1041
		UKST 762 # 20		3 32 59	-05 50.4	1	13.76		0.52		-0.05		2		1584
	+53 00682			3 33 04	+53 25.0	1	9.51		0.38				3	A0	70
		UKST 762 # 27		3 33 04	-06 09.8	1	14.86		0.75				1		1584
		UKST 762 # 18		3 33 05	-06 26.0	1	13.54		0.63		0.05		3		1584
	-47 01087			3 33 07	-47 26.3	2	10.24	.020	0.57	.000	-0.06	.000	5	G8 V-VI	1097,1696
		LP 653 - 13		3 33 08	-08 39.0	1	14.32		0.77				1		1691
	-77 00137			3 33 09	-77 16.0	1	9.88		1.09				4	K5	2012
	+54 00695			3 33 12	+54 34.4	1	10.35		0.40				3		70
		L 91 - 93		3 33 12	-67 52.0	1	11.13		1.20		1.16		2		1773
		L 91 - 94		3 33 12	-67 52.0	1	11.47		1.28		1.19		2		1773
		UKST 762 # 16		3 33 14	-06 16.8	1	13.26		0.51				2		1584
22309	+15 00507	G 5 - 46		3 33 15	+16 18.4	4	7.64	.012	0.58	.009	-0.02	.017	6	G0	927,1620,1648,3060
		POSS 301 # 6		3 33 19	+31 57.3	1	18.41		2.40				2		1739
		Alpha Per 1151		3 33 20	+50 46.5	1	11.52		0.64		0.18		3	F5	1048
22222	+53 00684			3 33 20	+53 49.2	1	8.49		0.24				2	A0	70
21970	+75 00143	HR 1080		3 33 20	+75 34.6	2	6.27	.005	0.97	.005	0.65		24	G9 III-IV	252,1258
22413	-28 01205			3 33 20	-28 30.0	5	8.82	.007	0.30	.012	-0.02	.024	17	A9 V	55,1097,2013,2017,3077
		LP 832 - 55		3 33 22	-21 13.3	1	10.78		0.62		0.09		2		1696
		Alpha Per 1185		3 33 23	+48 34.9	1	11.19		0.72		0.19		3	F7	1048
22380	-9 00709	G 160 - 3		3 33 24	-09 13.4	3	8.63	.038	0.67	.016	0.20	.015	6	G5	1658,2012,3056
	+29 00580	Hyades CDS 4		3 33 25	+29 43.5	1	9.19		0.76				2		1674
	+29 00580	Hyades L 4		3 33 25	+29 43.5	1	9.72		0.57		-0.04		4	F8	1058
22496	-48 01011			3 33 26	-48 35.3	2	8.59	.032	1.31	.009	1.24		5	K5/M0 V	2012,3072
		CS 22176 # 7		3 33 29	-08 56.8	1	14.09		-0.25		-1.02		1		1736
		POSS 301 # 1		3 33 30	+32 18.0	1	14.52		1.46				2		1739
232814	+54 00696			3 33 30	+54 22.9	1	9.44		0.38				2	A2	70
22328	+19 00562			3 33 31	+19 54.2	1	7.49		0.48		0.01		1	F5	3077
22267	+46 00779	Alpha Per 1210		3 33 33	+47 05.8	1	7.68		1.77		2.19		15	K5	1048
		G 6 - 18		3 33 35	+13 41.5	1	14.88		1.61				4		1663
	+49 00975	Alpha Per 1183		3 33 35	+49 31.3	1	10.20		0.35		0.26		3	A2 V	1048
22409	-11 00696	HR 1098		3 33 35	-11 21.6	2	5.57	.005	0.91	.000	0.55		6	gG7	2006,3077
22306	+33 00682			3 33 36	+34 11.0	1	6.89		1.33		1.47		2	K0	1625
	+49 00976	Alpha Per 1187		3 33 37	+49 23.6	1	10.06		0.55		0.09		3	F6 IV-V	1048
		LP 533 - 24		3 33 39	+05 42.4	1	11.96		1.06		0.97		1		1696
22327	+34 00693			3 33 43	+34 53.3	1	7.44		0.14		-0.18		2	B9 III	1041
	+46 00780	Alpha Per 1218		3 33 45	+47 11.0	1	9.17		0.43		0.17		3	F3 IV	1048
	+54 00697			3 33 45	+54 38.0	1	10.37		0.88				3		70
		LS I +59 193		3 33 47	+59 22.5	1	9.84		0.36		-0.42		3		41
22421	-6 00704			3 33 47	-05 45.1	1	9.40		0.36		0.01		2	F0	1584
22253	+56 00824	LS I +56 087		3 33 48	+56 34.5	4	6.53	.009	0.34	.004	-0.57	.005	17	B0.5III	41,70,1012,8031
		POSS 301 # 7		3 33 49	+32 10.1	1	18.42		1.72				2		1739
22794	-75 00238			3 33 49	-75 18.0	2	8.62	.000	0.42	.002	-0.05		6	F3 IV/V	278,2012
		Hyades J 210		3 33 50	+00 42.	1	11.86		0.83		0.33		3		1058
22634	-66 00199	HR 1104		3 33 51	-65 55.8	3	6.74	.003	0.16	.005	0.14	.010	16	A3 V	15,1075,2038
		G 77 - 66		3 33 53	+00 15.6	1	14.27		1.48		1.00		3		3024
		G 5 - 48		3 33 55	+20 56.0	1	13.08		1.51		1.17		1		1773
278835	+36 00730	IDS03307N3650	AB	3 33 55	+37 00.0	1	10.48		0.29		0.14		2	A0	1041
22610	-62 00281			3 33 55	-62 37.4	1	9.49		0.90		0.57		1	K2 V	3072
22374	+22 00518	V486 Tau		3 34 01	+23 02.8	3	6.72	.009	0.12	.008	0.12	.007	9	A2p	603,1202,1733
22470	-17 00699	HR 1100, EG Eri		3 34 01	-17 37.9	3	5.22	.004	-0.14	.004	-0.49		8	B8/9 III	15,2012,3023

HD	DM	Other Id	N Rem	α_{1950}	δ_{1950}	S	V	σ_V	B–V	σ_{B-V}	U–B	σ_{U-B}	n	Spectrum	References
		SS Tau		3 34 03	+05 11.8	2	12.09	.214	0.44	.085	0.28	.054	2		668,699
22353	+39 00829			3 34 04	+39 55.6	1	7.00		0.05		-0.06		3	A0 V	1501
275693	+36 00731			3 34 08	+37 04.5	1	9.62		0.23		0.01		2	A0	1041
22326	+47 00862	Alpha Per 1225	⋆ AB	3 34 09	+47 57.2	1	8.88		0.49		0.03		3	F7 IV-V	1048
22403	+25 00580	V837 Tau	⋆	3 34 10	+25 49.8	1	7.32		0.68		0.26		1	G2 V	1620
22298	+54 00698	LS I +55 049		3 34 10	+55 00.4	2	7.60	.042	0.53	.005	-0.33	.019	6	B2 Vne	41,1212
22468	+0 00616	HR 1099, V711 Tau	⋆ AB	3 34 13	+00 25.6	5	5.80	.052	0.92	.003	0.46	.020	13	G9 V	15,986,1415,1641,3016
22468	+0 00616	IDS03317N0016	B	3 34 13	+00 25.6	1	8.83		0.99		0.79		3		3016
22351	+43 00773			3 34 13	+43 23.7	1	8.43		0.09		0.00		3	B9	595
		CS 22176 # 6		3 34 15	-09 49.4	1	14.51		0.43		-0.23		1		1736
22586	-52 00421			3 34 15	-52 43.3	1	8.03		-0.19		-0.91		1	B1/2 III	55
		LP 995 - 46		3 34 16	-44 40.2	1	13.03		1.90		1.52		1		3073
22484	-0 00572		A	3 34 19	+00 14.7	14	4.29	.019	0.57	.009	0.06	.018	38	F9 V	15,22,986,1020,1075,1197*
	+46 00782	Alpha Per 1247		3 34 19	+46 55.9	1	8.94		0.53		0.04		2	F8	1601
		AK 9 - 396		3 34 22	+15 10.6	1	11.19		0.84				1		1570
22316	+56 00826	HR 1094		3 34 23	+56 46.2	2	6.30		-0.13	.003	-0.36		11	B9p	70,1079
22444	+21 00489	Pleiades 3303		3 34 28	+22 11.3	1	9.23		0.53		0.00		1	G0	8023
	+53 00689			3 34 28	+53 56.1	1	9.80		0.47				3		70
	+18 00514	Hyades vB 159		3 34 30	+18 26.0	2	8.76	.015	0.52	.002	-0.01	.002	3	F8	1127,8023
	+51 00755	Alpha Per 1202		3 34 34	+51 25.6	1	10.32		0.59		0.29		3	F4 IV	1048
232816	+52 00705			3 34 35	+52 26.2	1	9.00		0.53		0.01		2	G0	1566
22402	+42 00795	HR 1097		3 34 37	+42 25.2	1			-0.06		-0.38		1	B8 Vn	1079
22389	+48 00962	Alpha Per 1235		3 34 37	+48 55.2	1	7.19		-0.03		-0.37		3	B9	1601
	+20 00598	Hyades vB 5		3 34 40	+21 10.8	3	9.36	.008	0.92	.004	0.70	.000	8	G5	950,1127,8023
		Hyades CDS 53	⋆	3 34 41	+17 41.7	1	12.74		1.52				4		1674
		Hyades CDS 54	⋆	3 34 42	+17 41.5	1	13.29		1.54				3		1674
		LP 413 - 19		3 34 42	+17 41.5	1	13.01		1.41		0.90		1		1696
22401	+47 00865	Alpha Per 1259		3 34 43	+47 24.8	2	7.45	.005	0.01	.003	-0.10	.025	9	A0 Vp	1009,1048
22371	+53 00690			3 34 47	+53 42.4	1	8.53		0.41				3	F5 V	70
22417	+48 00963	Alpha Per 1245		3 34 51	+49 02.6	1	6.87		0.06		-0.09		3	B9	595
22582	-28 01214	IDS03328S2828	A	3 34 54	-28 17.5	1	8.86		0.73				4	G8 V	2012
22772	-68 00222			3 34 54	-68 01.6	1	8.68		1.11		0.77		2	G8p Ba	3048
22480	+34 00700			3 34 56	+35 01.1	1	8.94		0.11		0.06		2	A2 IV	1041
		G 6 - 22		3 34 57	+19 39.8	1	11.07		0.78		0.32		1	G8	333,1620
22522	+14 00586	HR 1102	AB	3 34 59	+15 16.1	2	6.40	.015	0.17	.000	0.08	.015	4	A5 IV	1733,3016
22466	+38 00766			3 34 59	+39 02.0	1	8.99		0.21		0.10		3	A5 V	1501
	+64 00400	LS I +64 098		3 35 04	+64 38.6	1	8.58		0.62		-0.41		3	B2	41
22688	-53 00587			3 35 07	-52 56.5	2	7.65	.006	0.94	.012	0.62		8	G8 III/IV	1673,2012
22155	+74 00164	IDS03293N7504	AB	3 35 11	+75 14.4	1	9.46		1.10		0.99		1	K0	3072
278842	+35 00728			3 35 14	+36 05.3	1	10.33		0.25		0.16		2	A0	1041
		Steph 392		3 35 14	-11 06.4	1	11.40		1.40		1.14		1	M0	1746
	+48 00965	Alpha Per 1262		3 35 15	+49 14.6	1	9.05		1.06		0.79		3	G6	1048
22663	-40 01008	HR 1106		3 35 18	-40 26.4	4	4.58	.009	1.04	.005	0.77	.004	20	K1 III	15,1075,2012,8015
22451	+52 00706			3 35 21	+52 39.4	1	7.78		0.52				2	F7 V	70
22584	-3 00587		A	3 35 22	-02 41.0	1	9.63		0.96		0.73		3	K2 V	3072
22584	-3 00587		B	3 35 22	-02 41.0	1	12.94		1.41		1.11		2		3072
22427	+58 00632			3 35 27	+59 17.0	1	6.89		1.40		1.57		3	K2 III-IV	37
	+41 00727	G 78 - 44		3 35 30	+42 14.1	1	8.96		0.81		0.50		1	K2	333,1620
22399	+63 00437	IDS03312N6333	A	3 35 34	+63 42.6	1	6.80		0.46		-0.08		3	F5	3024
22399	+63 00437	IDS03312N6333	BC	3 35 34	+63 42.6	1	8.20		0.83		0.40		3	F5	3024
	+54 00700			3 35 35	+54 29.1	1	9.98		2.17				3		369
22578	+22 00523	Pleiades 3304		3 35 44	+22 29.9	1	6.77		0.00		-0.01		3	A0	8023
		G 160 - 5		3 35 50	-11 37.4	2	13.03	.019	1.63	.037	1.38	.002	3		1774,3078
237163	+55 00804	LS I +55 050		3 35 59	+55 35.8	1	8.75		0.64		0.54		3	A3 II	41
22518	+55 00805			3 36 01	+55 43.1	1	8.41		0.14		0.08		2	A2	1502
		CS 22176 # 13		3 36 01	-10 52.5	1	14.88		0.12		0.10		1		1736
22675	-7 00647	HR 1108		3 36 03	-07 33.2	2	5.84	.005	0.98	.000	0.77	.005	7	G8 III	15,1256
		G 160 - 6		3 36 03	-09 57.3	1	14.12		0.82		0.07		2		3059
22220	+76 00128			3 36 04	+77 00.5	1	7.88		0.10		0.09		3	A0	1733
	+61 00623	LS I +61 324		3 36 05	+61 41.0	1	9.03		0.48		-0.49		3	B2	41
22615	+20 00602	Pleiades 3306	⋆	3 36 06	+20 45.3	4	6.50	.000	0.15	.005	0.18	.004	11	A3 IV	196,1049,8023,8052
22614	+24 00527	Pleiades 3305		3 36 08	+24 32.5	1	7.13		0.04		0.03		1	A0	8023
22728	-31 01479			3 36 09	-31 27.8	1	7.25		1.57				4	M2 III	2012
		Alpha Per 1300		3 36 11	+49 10.0	1	11.35		0.63		0.09		3	G2:	1048
		Alpha Per 1316		3 36 12	+47 35.4	1	11.17		0.62		0.10		3	F8	1048
		CS 22176 # 16		3 36 13	-12 26.9	1	13.04		0.29		0.12		1		1736
22736	-24 01779			3 36 13	-24 10.2	1	9.68		0.47		-0.03		1	G0 V	742
	+23 00486			3 36 14	+24 18.1	1	10.40		0.62		0.11		2		1723
		CS 22176 # 8		3 36 16	-07 58.8	1	11.93		0.09		0.12		1		1736
22637	+21 00492	Pleiades 3307		3 36 18	+21 40.9	1	7.28		0.08		0.02		1	A0	8023
		Alpha Per 1325		3 36 21	+47 13.3	1	10.91		0.51		0.10		3	F5	1048
275682	+37 00808			3 36 25	+37 36.9	1	9.80		0.44		0.47		2	A0	1041
22682	+13 00579			3 36 30	+13 43.9	1	6.67		0.90		0.61		2	G5	1648
		TonS 365		3 36 30	-31 25.	1	13.98		-0.11		-0.12		1		1036
	-24 01782			3 36 31	-24 12.8	2	9.92	.009	0.63	.005	-0.05		5		742,1594
		Pleiades 9308	⋆ V	3 36 32	+23 20.3	1	14.23		1.37				2		1508
22713	-6 00713	HR 1111		3 36 33	-05 47.1	4	5.96	.005	0.92	.009	0.66	.000	16	K1 V	15,1075,1415,3077
22695	+16 00484	HR 1110	⋆ AB	3 36 36	+16 22.5	1	6.16		1.00		0.69		2	K0 III	3016
22695	+16 00484	IDS03338N1613	B	3 36 36	+16 22.5	1	13.80		0.76		0.20		3	G5	3016
	+55 00807			3 36 38	+55 50.2	1	10.50		0.57				2		70

Table 1 183

HD	DM	Other Id	N Rem	α_{1950}	δ_{1950}	S	V	σ_V	B−V	σ_{B-V}	U−B	σ_{U-B}	n	Spectrum	References
278874	+32 00652	IDS03335N3309	AB	3 36 39	+33 18.7	1	9.03		1.09		1.00		3	K5	196
22694	+17 00601	G 6 - 26	★	3 36 41	+18 13.6	1	8.28		0.81		0.40		2	G5	1648
		LB 1681		3 36 42	−51 28.	1	11.24		0.49		0.01		2		45
22789	−28 01225	HR 1114		3 36 43	−28 06.3	2	6.00	.005	-0.02	.000			7	A0 V	15,2027
22826	−35 01311			3 36 47	−35 22.0	1	7.04		1.28		1.37		2	K2 III	937
		RL 68		3 36 48	+56 13.	1	11.89		0.37		0.17		2		704
23128	−74 00268			3 36 48	−74 08.3	2	7.62	.008	1.13	.004	1.01		6	K0 III	278,2012
		LP 773 - 20		3 36 49	−19 36.3	1	12.55		0.82		0.37		2		1696
		Alpha Per 1351		3 36 50	+47 24.0	1	11.81		0.43		0.37		3	A0	1048
22800	−24 01787			3 36 50	−24 15.0	1	9.05		1.08		-1.00		1	K0 III	742
22702	+24 00528	Pleiades 3308		3 36 52	+25 02.1	1	8.80		0.35		0.04		1	A2	8023
22824	−28 01227			3 36 59	−28 10.9	1	8.31		1.02		0.77		3	K1 III	1657
22799	−10 00717	HR 1117		3 37 02	−10 35.8	1	6.19		1.03				4	G5	2035
22960	−60 00261	IDS03361S6006	B	3 37 02	−59 56.3	1	8.32		0.51				4	F5 V	2033
	−35 01314			3 37 03	−35 29.4	1	10.02		1.05		0.87		2		937
		G 160 - 8		3 37 04	−11 08.9	2	13.02	.020	0.78	.015	0.10	.025	4		1696,3059
22862	−36 01380			3 37 04	−35 47.8	1	8.09		1.09		0.96		2	K0 III	937
22601	+58 00637			3 37 07	+58 41.1	1	6.84		0.28		0.07		2	F0	1733
	+26 00595			3 37 08	+26 48.0	3	8.33	.006	1.06	.005	0.79	.026	9	G8 III	1003,1775,3077
22798	−3 00591	HR 1116		3 37 08	−03 33.2	2	6.22	.005	1.04	.000	0.90	.005	7	G5	15,1061
22989	−60 00262	IDS03361S6006	A	3 37 10	−59 56.3	1	6.94		0.40				4	F3 V	2033
	−35 01318	V1230 Ori		3 37 14	−35 25.8	1	10.92		0.67		0.09		1		937
22796	+2 00581	HR 1115		3 37 15	+02 53.7	5	5.57	.016	0.93	.010	0.67	.004	12	G5	15,1256,1355,1769,3016
278908	+34 00706			3 37 16	+35 03.7	1	10.59		0.71		0.44		2	F0	1041
	+53 00692			3 37 17	+53 25.6	1	10.59		0.73				3		70
	+52 00707			3 37 18	+53 08.6	1	9.66		2.23				2		369
		NGC 1399 sq1 6		3 37 18	−35 26.0	1	11.98		0.92		0.49		1		937
	+49 00991	Alpha Per 1344		3 37 19	+49 20.7	1	9.77		0.16		-0.13		3	B8 V	1048
22973	−56 00559			3 37 19	−56 18.3	1	7.96		0.96				4	G6 III	2012
		NGC 1399 sq1 8		3 37 21	−35 47.6	1	12.34		0.65		0.04		1		937
22611	+62 00594	IDS03332N6219	B	3 37 24	+62 32.7	1	9.63		0.21		-0.05		1		414
		G 160 - 9		3 37 26	−16 01.7	2	11.82	.005	0.69	.002	0.09	.025	4		1696,3056
22819	−1 00519	HR 1119		3 37 27	−01 16.9	2	6.11	.005	0.99	.005	0.78	.005	7	K1 III-IV	15,1415
	+4 00568	Hyades vB 134		3 37 29	+04 28.2	1	10.04		0.65		0.16		3	G0	1127
22611	+62 00596	U Cam	★ AB	3 37 29	+62 29.3	1	7.55		4.29		4.49		1	C6.4 e	414
		LP 533 - 40		3 37 32	+05 19.7	1	13.23		0.91		0.51		1		1696
22733	+38 00782	IDS03342N3848	A	3 37 32	+38 57.4	1	7.85		0.20		0.15		2	A3	105
	−10 00719			3 37 32	−10 31.1	1	10.68		0.70		0.09		2		3056
	−35 01322			3 37 34	−35 26.8	1	11.70		0.57		0.02		1		937
		NGC 1399 sq1 9		3 37 34	−35 47.6	1	12.92		0.92		0.29		1		937
22946	−43 01126			3 37 34	−42 55.3	2	8.27	.000	0.53	.000			8	F7/8 V	1075,2012
22732	+40 00813			3 37 35	+41 01.6	1	7.18		0.39		-0.03		3	F2	1601
		MCC 424		3 37 41	+55 04.0	1	11.34		1.43				3	M0	1723
		PHL 8732		3 37 42	−25 08.	1	13.82		0.20		-0.57		1		1036
22932	−36 01388			3 37 42	−36 06.1	2	8.41	.009	0.32	.000	-0.03		5	A9 V	937,2012
22805	+24 00529	HR 1118	AB	3 37 47	+25 10.2	2	6.11		0.07	.015	0.14	.020	4	A2 IV	603,1049
22649	+62 00597	HR 1105, BD Cam		3 37 48	+63 03.4	3	5.09	.021	1.63	.000	1.81	.005	11	S3.5	1118,3016,8032
22879	−3 00592	G 80 - 15		3 37 49	−03 22.5	9	6.69	.009	0.54	.002	-0.09	.010	22	F9 V	22,742,908,1003,1594*
	+49 00994	Alpha Per 1354		3 37 50	+49 57.6	1	10.64		0.80		0.32		2	G5	1048
		Alpha Per 551		3 37 51	+49 03.0	2	11.19	.015	0.65	.002	0.14		3	F8	1048,1575
	+48 00974	Alpha Per 1370		3 37 51	+49 07.4	1	9.41		1.73				3		369
22780	+37 00811	HR 1113		3 37 52	+37 25.2	2	5.56	.015	-0.07	.000	-0.42	.000	5	B7 Vne	154,1212
22905	−15 00634	HR 1120		3 37 53	−15 23.2	1	6.33		0.88				4	G8/K0III+G	2006
		LP 473 - 76		3 38 04	+10 00.3	1	12.66		0.79				2		1674
22897	−2 00689			3 38 04	−02 22.7	1	9.58		1.02		0.86		1	K2 V	3072
		Alpha Per 1375		3 38 05	+49 23.8	1	11.74		0.40		0.31		3	A3	1048
22920	−5 00715	HR 1121		3 38 10	−05 22.3	6	5.53	.014	-0.15	.004	-0.55	.010	23	B9 IIIp	15,220,361,1079,1256,2006
232822	+54 00703			3 38 11	+54 45.7	1	9.25		0.36				2	A0	70
22918	−2 00690	G 80 - 16		3 38 15	−02 29.4	4	6.96	.009	0.96	.008	0.74	.017	11	G5	1311,1620,2012,3077
23400	−77 00143			3 38 15	−77 15.1	1	8.20		1.12		0.96		8	K1 III	1704
23474	−78 00105	HR 1154		3 38 21	−78 29.1	3	6.28	.004	1.15	.000	1.03	.000	13	K2 III	15,1075,2038
22887	+23 00489	Pleiades 3309		3 38 25	+23 19.7	1	9.18		0.45		0.01		1	F5	8023
		Wolf 212		3 38 25	+46 31.9	1	10.73		0.46		0.19		2		1723
22859	+37 00814			3 38 28	+37 25.7	1	7.77		0.08		-0.01		2	A1 V	1041
23358	−75 00244			3 38 30	−74 59.1	3	6.88	.000	0.33	.004	-0.02	.009	10	F0 III	278,1075,2012
	+3 00515	G 80 - 17		3 38 34	+03 27.3	6	9.61	.022	1.37	.013	1.26	.043	9	K4-5	196,680,1017,1620,1705,3072
	+25 00591			3 38 34	+25 27.7	1	10.10		0.59		0.09		2		1723
22764	+59 00699	HR 1112	★ A	3 38 34	+59 48.6	4	5.75	.018	1.74	.020	1.76	.025	11	K4 Ib	542,1379,8032,8100
22764	+59 00699	IDS03345N5939	B	3 38 34	+59 48.6	1	8.73		0.10		-0.22		2		542
		G 160 - 12		3 38 36	−07 59.4	1	13.46		0.75		0.06		2		3056
		Hyades CDS 55	★	3 38 39	+23 11.1	1	13.87		1.52				3		1674
		LP 653 - 25		3 38 39	−03 39.4	1	10.87		0.81		0.47		2		1696
23010	−12 00689	HR 1125	★ A	3 38 51	−11 57.8	5	6.47	.012	0.38	.006	0.02		17	F2 III	15,1075,2018,2032,3013
		G 79 - 56		3 38 59	+09 14.2	2	11.83	.015	0.73	.000	0.08	.029	3	G8	333,1620,1658
		Pleiades 9035	★ V	3 38 59	+25 10.2	1	14.24		1.46				2		1508
22872	+50 00802	G 174 - 58		3 38 59	+51 01.0	1	7.94		0.56		0.08		2	F9 V	1566
22830	+57 00739	LS I +57 133		3 39 00	+57 24.4	1	8.13		0.26		-0.52		3	B8	41
		LS I +64 099		3 39 01	+64 26.4	1	11.25		0.76		-0.03		2		41
22858	+55 00816			3 39 02	+55 21.7	1	9.02		0.25				2		70
		Hyades CDS 56	★	3 39 03	+18 36.1	1	13.34		1.53				3		1674

HD	DM	Other Id	N Rem	α₁₉₅₀	δ₁₉₅₀	S	V	σ_V	B–V	σ_B–V	U–B	σ_U–B	n	Spectrum	References
275772 +36 00735		IDS03358N3657	AB	3 39 03	+37 06.5	1	9.24		0.66		0.21		2	B7	1041
23008 +6 00571				3 39 08	+07 09.5	1	8.17		0.75		0.28		2	G0	1733
22977 +22 00531		Pleiades 3310		3 39 08	+22 42.0	1	9.11		0.47		0.02		1	F8	8023
23055 −20 00687		HR 1128		3 39 08	−19 44.6	2	6.58	.005	0.09	.000			7	A3 IV/V	15,2012
22951 +33 00698		HR 1123	⋆ A	3 39 12	+33 48.4	8	4.97	.008	−0.02	.012	−0.83	.012	22	B0.5V	15,154,1009,1203,1234*
22951 +33 00698		IDS03360N3338	B	3 39 12	+33 48.4	1	10.07		0.26		0.18		3	A2 Vn	3016
		Pleiades 9036	⋆ V	3 39 13	+23 55.6	1	15.95		1.50				2		1508
22648 +74 00168				3 39 15	+74 23.1	1	6.71		1.03		0.74		4	G5	985
22961 +35 00738		V497 Per		3 39 17	+35 28.8	1	9.56		0.18		0.05		2	A1p (Sr)	1041
		RL 70		3 39 18	+52 21.	1	15.77		0.07		−0.53		2		704
−25 01508				3 39 19	−25 07.2	1	10.74		0.48		0.03		2	F5	1696
22928 +47 00876		HR 1122, δ Per	⋆ A	3 39 21	+47 37.8	10	3.01	.011	−0.12	.011	−0.51	.011	45	B5 III	1,15,154,1009,1048*
		Pleiades 9256	⋆ V	3 39 23	+24 30.4	1	14.48		1.41				2		1508
+14 00598				3 39 24	+14 37.9	1	8.28		1.73				2	K5	1648
		GD 47		3 39 24	−03 32.3	1	15.20		0.15		−0.59		1		3060
−6 00727		G 160 - 13		3 39 24	−06 06.2	2	10.02	.005	0.77	.010	0.31	.015	2	K2	1658,3056
23016 +19 00578		HR 1126		3 39 26	+19 32.5	4	5.71	.023	−0.02	.008	−0.28	.006	13	B9 Vne	985,1049,1079,1501
23065 −11 00716		G 160 - 14		3 39 28	−10 51.2	2	8.26	.025	0.76	.017	0.14		7	G0	2033,3056
+21 00504		Pleiades 3311		3 39 29	+21 18.9	1	9.86		0.55		0.03		1	G0	8023
		Pleiades 9289	⋆ V	3 39 29	+22 44.2	1	14.45		1.41				2		1508
		LS I +57 134		3 39 41	+57 10.8	1	11.49		0.70		−0.17		3		41
23521 −76 00242				3 39 41	−75 55.5	3	7.68	.004	1.20	.005	1.25	.000	7	K2 III/IV	278,861,2012
		G 79 - 59		3 39 42	+12 22.9	2	12.91	.010	1.54	.005	1.24	.010	2	K6	333,1620,3078
		Pleiades 9003	⋆ V	3 39 42	+23 49.9	1	15.70		1.45				1		1508
232827 +50 00804				3 39 42	+51 11.2	1	8.94		1.87				4	K7	369
+59 00704		G 246 - 45		3 39 44	+60 15.2	1	10.36		1.09		0.95		1	K8	3072
23052 +16 00497				3 39 47	+17 08.1	1	7.08		0.66		0.18		2	G0	1648
23061 +24 00536		Pleiades 25		3 39 56	+24 20.1	2	9.47	.000	0.48	.000	0.01	.000	3	F6 V	1029,8023
278917 +34 00720				3 39 58	+34 39.7	1	10.62		0.25		0.08		2	G0	1041
		Pleiades 34	⋆ V	3 40 04	+24 30.8	2	11.98	.031	0.93	.009	0.62		5		1029,1508
		Pleiades 4307		3 40 06	+24 22.	1	17.23		1.82		1.80		11		5008
		Pleiades 4308		3 40 06	+24 22.	1	17.88		0.91		0.88		15		5008
		Pleiades 4311		3 40 06	+24 22.	1	16.26		1.70		2.30		5		5008
		Pleiades 4330		3 40 06	+24 22.	1	18.18		0.28		−0.20		10		5008
		Pleiades 4331		3 40 06	+24 22.	1	18.72		1.84				29		5008
		Pleiades 4334		3 40 06	+24 22.	1	17.08		0.72		0.04		21		5008
23075 +25 00593		IDS03371N2522	A	3 40 06	+25 31.4	1	7.17		0.40		−0.02		2	F5	1625
23060 +33 00704				3 40 13	+33 57.5	4	7.48	.022	0.10	.009	−0.56	.004	9	B2 Vp	1009,1012,1041,1252
23214 −34 01376				3 40 13	−34 34.8	1	8.94		1.18				4	K2/3 III	2012
23227 −32 01430		HR 1134		3 40 15	−32 05.8	4	5.00	.009	−0.17	.009	−0.60	.004	20	B5 III	15,1075,2012,8015
		IC 348 - 1		3 40 16	+32 08.2	2	12.78	.000	1.54	.010	1.08	.019	3		386,1718
		LS I +64 100		3 40 16	+64 18.3	1	11.40		0.56		−0.28		3		41
		G 79 - 60		3 40 17	+11 22.5	1	11.44		1.04		0.83		1	K3	333,1620
		G 160 - 15		3 40 18	−07 54.1	2	11.68	.029	0.84	.027	0.37	.005	6		1696,3056
23295 −51 00881				3 40 20	−51 35.1	1	9.07		1.10		0.93		1	K4 V	3072
		Pleiades 81		3 40 21	+23 42.2	3	13.56	.008	0.91	.018	0.36	.018	10	G8 V	257,1029,5008
		Pleiades 83		3 40 22	+23 41.5	2	14.87	.015	1.03	.029	0.73	.015	11	K0 V	257,5008
23050 +42 00812		G 78 - 47	⋆ A	3 40 22	+42 27.0	3	7.48	.007	0.58	.010	0.03	.004	7	G2 V	1003,1620,3026
23332 −56 00570				3 40 24	−55 50.0	2	8.90	.009	0.90	.005	0.54		4	G6wG0 III	742,1594
+22 00534		Pleiades 102		3 40 27	+23 04.1	2	10.51	.000	0.72	.010	0.15	.000	3		1029,8023
		Pleiades 97	⋆ V	3 40 27	+24 50.2	3	12.64	.038	1.08	.000	0.85	.036	6	K3 V	364,1029,1508
		Pleiades 105		3 40 28	+23 01.8	2	13.78	.008	0.96	.012	0.44	.020	4	G5 V	257,1029
23049 +48 00984		HR 1127	⋆ AB	3 40 30	+48 22.0	2	6.25	.163	1.56	.003	1.83		5	K4 III	71,595
23133 +22 00535		Pleiades 3312	⋆ AB	3 40 32	+22 34.7	1	8.59		0.47		0.00		1	F5	8023
23308 −46 01143				3 40 33	−46 07.0	3	6.49	.000	0.52	.000			10	F7 V	1020,1075,2012
		Pleiades 120		3 40 34	+23 31.0	2	10.79	.000	0.71	.010	0.19	.000	3		1029,8023
		Pleiades 129		3 40 36	+23 36.3	1	11.47		0.88		0.40		2		1029
−24 01826				3 40 36	−24 37.2	4	9.19	.024	1.13	.018	1.11		9	K5 V	912,1705,2013,3072
23082 +44 00782				3 40 37	+44 43.6	1	7.52		1.85		2.00		2	K3 IIa	1601
		Pleiades 134	⋆ V	3 40 38	+24 04.5	3	14.39	.007	1.53	.020	1.07		6	K7 Ve	257,1506,5008
		Pleiades 133	⋆ V	3 40 38	+24 14.2	4	14.32	.014	1.37	.015	1.16	.044	14	K5.5Ve	257,921,1508,5008
		Pleiades 146	⋆ V	3 40 39	+23 17.8	3	14.55	.027	1.41	.000			6	K7 Ve	257,921,1508
		Pleiades 152	⋆ V	3 40 40	+23 22.8	2	10.73	.020	0.69	.010	0.17	.000	3		1029,8023
		Pleiades 9041	⋆ V	3 40 43	+24 25.0	1	15.42		1.63				1		1508
		IC 348 - 2		3 40 43	+32 03.7	2	12.95	.010	1.18	.019	0.61	.034	3		386,1718
23157 +23 00496		Pleiades 157		3 40 44	+23 29.5	3	7.91	.009	0.34	.000	0.11	.000	5	A9 V	1029,8023,8030
23155 +24 00537		Pleiades 153		3 40 44	+24 55.4	2	7.52	.005	0.15	.000	0.10	.000	3	A2	1029,8023
23158 +23 00497		Pleiades 164		3 40 45	+23 26.3	2	9.53	.010	0.49	.010	0.01	.000	3	F6 V	1029,8023
23156 +23 00495		Pleiades 158	⋆ V	3 40 45	+24 13.1	5	8.23	.011	0.25	.009	0.14	.000	11	A7 V	1,369,1029,8023,8030
23141 +25 00599				3 40 45	+26 13.4	1	7.28		1.16				5	K1 III	20
278982 +33 00706				3 40 46	+34 01.3	1	10.18		0.21		−0.11		2	B8	1041
23154 +24 00538		Pleiades 3313		3 40 47	+25 15.4	1	9.84		0.58		0.04		1		8023
23130 +36 00741				3 40 48	+36 43.2	1	8.86		0.14		0.07		2	A5 V	1041
		Pleiades 174		3 40 49	+24 50.9	1	11.62		0.85		0.43		4		1029
		Pleiades 173		3 40 49	+25 02.0	3	10.88	.015	0.85	.016	0.44	.000	7		1029,1508,8023
		CS 22190 # 34		3 40 49	−13 15.4	1	13.64		−0.25		−1.05		1		1736
		3C 93 # 1		3 40 50	+04 46.8	1	12.11		0.87		0.37		2		327
		Pleiades 186		3 40 50	+23 03.3	2	10.50	.013	0.79	.004	0.28	.018	5		257,1029
		IC 348 - 3		3 40 50	+32 00.3	3	11.34	.007	0.75	.010	0.19	.028	3		386,1300,1718
		Pleiades 189		3 40 51	+23 23.0	5	14.01	.016	1.37	.004	1.20	.015	41	K5.5Ve	257,352,921,1508,5008

Table 1   185

HD	DM	Other Id	N Rem	α_{1950}	δ_{1950}	S	V	σ_V	B–V	σ_{B-V}	U–B	σ_{U-B}	n	Spectrum	References
		Pleiades 9477	⋆ V	3 40 51	+23 48.	1	17.93		1.06				1		1508
		CS 22176 # 22		3 40 51	−09 24.8	1	13.80		-0.30		-1.14		1		1736
23249	−10 00728	HR 1136, δ Eri		3 40 51	−09 55.9	16	3.53	.011	0.92	.007	0.69	.011	61	K0+IV	3,15,22,1013,1020*
		Pleiades 193		3 40 52	+24 05.5	1	11.29		0.81		0.36		4		1029
23183	+19 00582	HR 1132		3 40 53	+19 30.5	4	6.15	.004	0.99	.012	0.73	.014	8	G8 III	15,1003,3077,4001
		Pleiades 191	⋆ V	3 40 53	+24 41.1	5	14.50	.030	1.43	.018	1.21	.074	30	K7 Ve	257,352,921,1508,5008
		Pleiades 212	⋆ V	3 40 57	+24 16.2	2	14.34	.043	1.38	.015	1.13		2	K7 Ve	1508,5008
23319	−37 01415	HR 1143	⋆ AB	3 40 59	−37 28.2	3	4.58	.005	1.20	.004	1.31	.000	14	K2.5 III	15,1075,2012
281160	+31 00641	IC 348 - 4	⋆ A	3 41 00	+31 57.9	3	9.94	.009	0.47	.009	0.36	.033	4	A0 V	386,1300,1718
281160	+31 00641	IC 348 - 4	⋆ AB	3 41 00	+31 57.9	1	9.59		0.30		0.15		2		1041
23195	+23 00499	Pleiades 233		3 41 01	+23 43.6	2	9.66	.000	0.52	.005	0.81	.000	4	F5 V	1029,8023
23194	+24 00540	Pleiades 232		3 41 01	+24 24.0	4	8.06	.012	0.20	.004	0.15	.000	10	Am	1,369,1029,8023
		IC 348 - 5		3 41 01	+31 57.7	3	9.98	.048	0.50	.006	0.38	.030	4	A2 V	386,1300,1718
	+16 00502	G 6 - 29	⋆ A	3 41 02	+16 31.1	3	9.91	.033	1.48	.026	1.27		4	K6	333,1619,1620,1705
	+16 00502	G 6 - 28	⋆ B	3 41 02	+16 31.1	3	10.77	.029	1.51	.024	1.22		4	M0	1619,1620,1705
23205	+15 00525			3 41 03	+15 52.0	1	8.57		0.42		-0.02		4	F0	1648
		Pleiades 248		3 41 03	+23 23.3	2	11.01	.005	0.78	.010	0.25	.000	3		1029,8023
		Pleiades 9150	⋆ V	3 41 03	+24 54.5	1	15.62		1.51				1		1508
		Pleiades 257		3 41 05	+24 05.3	2	12.58	.040	0.80	.004	0.32	.013	5		257,1029
	+24 00541	Pleiades 253		3 41 05	+24 20.9	2	10.69	.025	0.68	.000	0.14	.000	3		1029,8023
		Pleiades 250		3 41 05	+24 50.0	2	10.70	.020	0.69	.010	0.14	.000	4		1029,8023
		Pleiades 263		3 41 06	+24 07.2	1	11.54		0.88		0.42		4		1029
		AO 1056		3 41 10	+23 53.5	1	10.23		0.56		0.26		1	K0 V	1748
23139	+45 00804	HR 1130		3 41 10	+45 56.6	2	6.14	.029	0.27	.005	0.20	.015	3	A7 IV	39,252
23281	−10 00729	HR 1139		3 41 10	−10 38.6	4	5.60	.012	0.21	.016	0.09	.008	19	A5 m	15,216,355,2012
		IC 348 - 6		3 41 11	+32 00.1	3	11.92	.006	0.76	.015	0.27	.019	4		386,1300,1718
23180	+31 00642	HR 1131, ø Per	⋆ AB	3 41 11	+32 07.9	14	3.83	.014	0.05	.011	-0.75	.008	49	B1 III	1,15,154,272,1009,1012*
		RL 71		3 41 12	+52 14.	1	11.33		0.41		-0.11		2		704
		IC 348 - 7		3 41 13	+31 58.2	2	11.62	.009	0.80	.009	0.19	.000	2		1300,1718
		Pleiades 296	⋆ V	3 41 14	+23 13.4	3	11.49	.062	0.84	.004	0.45		5		965,1029,1508
	+23 00501	Pleiades 298	⋆ AB	3 41 14	+23 52.5	2	10.88	.015	0.88	.000	0.47	.000	3		1029,8023
		Pleiades 9071	⋆ V	3 41 15	+22 56.	1	15.43		1.42				1		1508
		Pleiades 293		3 41 15	+24 37.4	3	10.80	.011	0.70	.011	0.17	.005	7		196,1029,8023
23005	+66 00284	HR 1124		3 41 15	+67 02.8	1	5.82		0.34		0.05		1	F0 IV	254
	+23 00502	Pleiades 303	⋆ AB	3 41 16	+23 56.8	2	10.49	.005	0.90	.010	0.35	.000	3	G1	1029,8023
		IC 348 - 9		3 41 16	+32 00.8	3	11.56	.007	0.98	.019	0.55	.019	4		386,1300,1718
		IC 348 - 8		3 41 16	+32 01.6	3	13.41	.023	1.22	.009	0.53	.055	4		386,1300,1718
23140	+45 00805	G 78 - 49		3 41 16	+45 52.8	2	7.71	.005	0.86	.005	0.54	.005	4	K2	196,1620
23261	+2 00595	Hyades vB 135		3 41 17	+03 17.3	2	8.98	.005	0.88	.008	0.62	.001	4	G5	1127,8023
23193	+36 00742	HR 1133		3 41 17	+36 18.2	3	5.61	.015	0.06	.000	0.11	.016	9	A2 m	985,1049,8071
278968	+34 00724	G 95 - 53		3 41 18	+34 49.1	1	10.63		1.38		1.21		1	M0	3003
		G 79 - 63		3 41 21	+09 47.0	2	11.58	.020	0.82	.005	0.42	.010	2	K0	333,1620,1658
		Pleiades 320		3 41 21	+24 37.0	3	11.06	.029	0.87	.015	0.46	.020	7		196,1029,8023
	+24 00542	Pleiades 314	⋆ V	3 41 21	+24 38.4	3	10.62	.038	0.66	.008	0.11	.005	7		196,1029,8023
23232	+24 00543	Pleiades 316		3 41 21	+24 43.1	2	8.84	.019	1.72	.009	2.02	.047	4	K2	196,1748
		Pleiades 9325	⋆ V	3 41 22	+24 27.	1	17.81		1.54				1		1508
23204	+35 00744	IDS03382N3532	AB	3 41 22	+35 41.9	1	8.37		0.16		0.11		2	A0	1041
		Pleiades 324	⋆ V	3 41 23	+24 36.7	4	13.01	.037	1.05	.018	0.78	.005	19		201,364,1508,5008
		IC 348 - 11		3 41 23	+31 57.0	3	10.62	.004	0.94	.006	0.72	.005	4	F0 m	386,1300,1718
		IC 348 - 10		3 41 24	+32 01.	3	11.81	.013	0.82	.015	0.73	.008	4	A2 V	386,1300,1718
		IC 348 - 28		3 41 24	+32 08.	2	9.14	.000	0.73	.000	-0.08	.000	3		386,1718
		IC 348 - 53		3 41 24	+32 08.	1	10.97		1.33		0.68		1		386,1718
		Pleiades 335	⋆ V	3 41 25	+23 54.7	3	13.77	.008	1.26	.013	1.08	.000	9	K5 Ve	257,1508,5008
23247	+23 00503			3 41 25	+23 58.6	1	8.81		1.73		2.09		1	F3 V	1748
23247	+23 00503	Pleiades 338		3 41 25	+23 58.6	4	9.06	.012	0.46	.008	0.10	.000	8	K4 I	1,369,1029,8023
		Pleiades 9010	⋆ V	3 41 26	+25 53.1	1	15.39		1.49				1		1508
281159	+31 00643	IC 348 - 20	⋆ AB	3 41 26	+32 00.4	7	8.52	.020	0.68	.012	0.01	.014	15	B5 V	386,1012,1041,1252,1300*
23246	+23 00504	Pleiades 344		3 41 27	+24 14.3	3	8.17	.005	0.27	.002	0.09	.000	7	A8 V	1,1029,8023
	+24 00544	Pleiades 345		3 41 27	+24 26.0	2	11.63	.034	0.85	.004	0.36	.009	6		1029,5008
		IC 348 - 12	⋆ C	3 41 27	+32 00.7	4	10.20	.004	0.87	.019	0.67	.010	8	A2 III	386,1300,1718,3024
		Pleiades 347	⋆ V	3 41 28	+24 41.3	4	14.04	.015	1.44	.008	1.11		48	K7 Ve	352,364,1508,5008
		IC 348 - 13		3 41 29	+31 57.4	3	12.99	.027	1.63	.021	0.92	.051	3		386,1300,1718
		Pleiades 357	⋆ V	3 41 30	+24 00.9	5	13.35	.073	1.22	.040	1.02	.207	15	K6 Ve	257,921,965,1029,1508
		TonS 374		3 41 30	−24 49.	1	15.00		-0.24		-1.08		1		832
23258	+20 00621	HR 1137		3 41 33	+20 46.3	2			0.01	.010	0.07	.045	6	A0 V	1049,1079
23245	+27 00556	IDS03385N2735	A	3 41 35	+27 44.5	1	6.77		0.37		-0.01		2	F0	3016
	−77 00144			3 41 36	−77 05.9	1	10.35		1.11				4	K0	2012
		Pleiades 380		3 41 38	+24 59.0	1	13.33		1.21				1	K4 Ve	1508
23456	−51 00887			3 41 38	−50 48.5	3	6.96	.011	0.52	.004	-0.08		10	F6 V	1075,1311,2012
23709	−73 00236			3 41 38	−73 04.4	1	9.50		0.17				4	A3 V	2012
23089	+62 00604	HR 1129	⋆	3 41 39	+63 11.4	6	4.79	.011	0.80	.019	0.25	.005	16	G0 III+A3 V	15,1008,1118,1363*
23269	+24 00545	Pleiades 405		3 41 40	+24 40.6	3	9.83	.008	0.54	.001	0.05	.000	7	F8 V	1,1029,8023
		IC 348 - 14		3 41 40	+32 06.1	1	14.00		1.35				1		386,1718
23257	+27 00558	IDS03385N2735	B	3 41 41	+27 46.0	2	6.81	.050	0.64	.002	0.14		3	G2 IV	20,3026
		IC 348 - 15		3 41 41	+32 07.4	2	13.28	.015	1.12	.015	0.77	.005	2		386,1718
		G 6 - 30		3 41 42	+18 17.5	3	15.19	.010	0.30	.005	-0.53	.014	9		538,1281,3006
		IC 348 - 16		3 41 42	+32 09.7	2	11.24	.019	1.04	.010	0.33	.044	3		386,1718
		Pleiades 3441		3 41 44	+25 20.5	1	15.43		1.63				2		921
22828	+77 00133			3 41 44	+77 57.9	1	7.12		-0.05		-0.22		3	B8	555
		Pleiades 430		3 41 45	+24 04.5	4	11.41	.026	0.81	.013	0.36	.024	6		965,1029,1508,5008

HD	DM	Other Id	N	Rem	α_{1950}	δ_{1950}	S	V	σ_V	B–V	σ_{B-V}	U–B	σ_{U-B}	n	Spectrum	References
23230 +42 00815		HR 1135	★	A	3 41 47	+42 25.3	9	3.78	.023	0.43	.005	0.27	.027	22	F5 II	15,1008,1118,1119*
		BPM 31594	★	V	3 41 48	−45 58.	1	15.03		0.21		-0.66		1		3065
23497 −54 00583					3 41 48	−54 37.8	1	8.56		1.12		1.09		2	K1 III	1730
+52 00710		LS V +52 010			3 41 49	+52 56.6	1	9.69		0.75				2		70
23288 +23 00505		Pleiades 447	★		3 41 50	+24 08.0	8	5.46	.005	-0.04	.003	-0.33	.007	36	B7 V	1,15,154,369,401,1029*
275831 +37 00825					3 41 50	+37 48.1	1	10.56		0.28		0.14		2	A0	1041
23760 −74 00270					3 41 50	−74 05.3	2	8.40	.009	0.30	.003	0.01		6	F0 IV	278,2012
		Pleiades 451	★	V	3 41 51	+24 45.3	4	13.44	.031	1.21	.016	1.08	.034	15	K5 Ve	257,921,1508,5008
23219 +47 00881					3 41 53	+47 30.3	1	7.18		-0.01		-0.16		2	B9	401
23289 +22 00537		Pleiades 470			3 41 54	+23 06.8	2	8.91	.045	0.39	.005	0.01	.000	4	F3 V	1029,8023
23302 +23 00507		Pleiades 468	★		3 41 54	+23 57.5	9	3.70	.009	-0.11	.006	-0.41	.006	30	B6 III	1,15,154,1029,1203*
		L 12 - 63			3 41 54	−79 47.	1	13.60		0.72		0.03		2		1696
23314 +9 00479		G 79 - 65			3 41 55	+10 01.5	1	8.65		0.66		0.16		1	G5	333,1620
		Pleiades 476			3 41 56	+23 46.0	3	10.80	.012	0.81	.018	0.27	.000	4	F9 V	965,1029,8023
		Pleiades 489			3 41 58	+24 16.6	3	10.39	.015	0.63	.001	0.13	.000	7	G0 V	1,1029,8023
+36 00746					3 41 58	+37 14.3	1	10.36		0.67		0.45		2	A8 IV	1041
23190 +54 00707					3 41 58	+54 54.5	1	6.84		0.20		0.07		2	A5	1566
23363 −1 00526		HR 1146			3 41 58	−01 19.2	4	5.25	.003	-0.09	.003	-0.37	.006	54	B7 V	15,1079,1256,3047
		IC 348 - 17			3 41 59	+32 01.1	2	12.63	.040	1.21	.035	0.49	.050	2		386,1718
		LS I +62 235			3 41 59	+62 34.4	1	11.73		0.52		0.05		3		41
−54 00584					3 41 59	−54 30.2	1	10.09		1.13		1.08		2		1730
		IC 348 - 18			3 42 00	+31 51.1	2	13.05	.015	1.27	.000	0.69	.040	2		386,1718
		L 177 - 24			3 42 00	−56 09.	1	12.93		0.45		-0.20		2		1696
		BPM 3116			3 42 00	−67 20.	1	15.74		-0.06		-0.77		1		3065
		Pleiades 513			3 42 01	+23 14.0	3	13.77	.015	1.32	.004	1.27	.120	44	K7 Ve	352,1029,5008
23312 +21 00514		Pleiades 3314			3 42 03	+21 52.6	1	9.49		0.49		0.02		1	F5	8023
−77 00145					3 42 03	−77 22.7	1	10.60		0.07				4	A0	2012
		Pleiades 514			3 42 04	+25 06.2	2	10.71	.020	0.70	.010	0.17	.000	4		1029,8023
+11 00514		G 79 - 66			3 42 05	+11 45.7	2	9.15	.005	1.17	.005	1.16	.005	2	K8 V:	333,1620,3072
		Pleiades 522			3 42 05	+23 41.1	3	11.97	.012	0.92	.011	0.63	.009	7	K2 V	1029,1508,5008
23326 +23 00509		Pleiades 530			3 42 07	+23 32.8	2	8.95	.005	0.39	.000	0.02	.000	4	F4 V	1029,8023
23325 +23 00508		Pleiades 531			3 42 08	+24 06.5	4	8.58	.005	0.34	.004	0.16	.000	8	Am	1,1029,8023,8030
23324 +24 00546		Pleiades 541	★		3 42 10	+24 41.0	6	5.65	.005	-0.07	.002	-0.36	.000	32	B8 V	1,15,263,1029,1079*
23327 +21 00515		Pleiades 3315			3 42 11	+22 08.3	1	9.13		0.50		0.13		1	F5	8023
232830 +51 00770					3 42 12	+52 14.5	1	9.29		0.52				2	F0 II	1119
23243 +52 00711					3 42 12	+52 45.7	1	9.07		0.49				2	B9 III	1119
23338 +24 00547		Pleiades 563	★	A	3 42 14	+24 18.7	11	4.30	.013	-0.11	.006	-0.46	.011	34	B6 IV	1,15,154,263,272,369*
		Pleiades 559	★	V	3 42 14	+24 56.0	1	13.33		1.11				1		1508
		Pleiades 571			3 42 15	+25 08.1	2	11.23	.030	0.78	.000	0.30		4		1029,1508
237172 +59 00711		LS I +60 306		A	3 42 16	+60 02.0	1	10.38		0.85		0.05		3	G5	41
		Pleiades 9127	★	V	3 42 18	+23 57.9	1	14.72		1.47				1		1508
23484 −38 01264					3 42 18	−38 26.5	5	6.99	.012	0.88	.000	0.55	.030	13	K1 V	258,1020,1075,2012,3008
		L 228 - 55			3 42 18	−51 33.	1	15.96		0.29		-0.50		2		3060
		Pleiades 590	★	V	3 42 19	+24 56.6	2	14.35	.015	1.37	.005			3		364,1508
23301 +38 00803					3 42 19	+38 31.2	1	6.44		1.20		1.22		4	K2 III	1501
23508 −41 01119		HR 1157	★	AB	3 42 20	−40 48.9	4	6.45	.014	1.06	.012			15	K1 III	15,1075,2029,3002
23351 +24 00548		Pleiades 605			3 42 21	+24 46.0	4	8.99	.008	0.44	.000	0.08	.000	8	F3 V	1,1029,8023,8030
		G 38 - 13			3 42 21	+28 53.8	1	11.47		0.74		0.12		2	G8	333,1620
		Pleiades 625	★	V	3 42 23	+23 34.4	2	12.67	.047	1.17	.036	0.76		2	K0 V	1029,1508
23352 +24 00549		Pleiades 627			3 42 23	+24 43.6	2	9.67	.015	0.51	.010	0.03	.000	4	F7 V	1029,8023
23413 −0 00593		HR 1150			3 42 23	−00 27.2	3	5.56	.005	1.41	.003	1.71	.003	73	K4 III	15,1415,3047
		Pleiades 636			3 42 24	+23 19.0	2	12.41	.024	1.02	.012	0.78		10		1029,1508
		Pleiades 624	★	V	3 42 24	+24 41.8	2	15.28	.005	1.53	.005			3	M2 Ve	921,1508
		Pleiades 649			3 42 27	+23 19.8	1	11.94		0.95		0.34		2		1506
		Pleiades 659			3 42 28	+23 16.5	2	12.07	.042	0.94	.005	0.39		4	G4 V	1029,1508
23361 +23 00510		Pleiades 652			3 42 28	+23 52.8	3	8.03	.012	0.21	.003	0.18	.000	7	A2.5Van	1,1029,8023
23300 +45 00811		HR 1141	★	AB	3 42 29	+45 31.6	1			-0.07		-0.46		2	B6 V	1079
		Pleiades 676	★	V	3 42 32	+23 36.3	3	13.68	.039	1.31	.005	1.24	.007	9	K3.5Ve	1029,1508,5008
23592 −52 00442					3 42 32	−52 12.2	1	9.47		0.86				2	G6/8	1594
		Pleiades 686	★	V	3 42 34	+24 08.9	1	13.38		1.27				1	dK7e	364
23242 +59 00713					3 42 34	+59 39.0	1	9.57		0.37		0.16		2		1502
23548 −42 01226					3 42 34	−42 03.2	1	7.37		1.73		2.05		4	M1 Ib/II	1673
23375 +24 00550		Pleiades 697			3 42 36	+24 18.5	3	8.59	.008	0.35	.001	0.11	.000	7	A9 V	1,1029,8023
278990 +33 00712					3 42 36	+33 27.5	1	10.31		0.55		0.34		2	A2	1041
23388 +20 00624		Pleiades 3316			3 42 37	+21 05.5	1	7.77		0.18		0.13		1	A3	8023
+23 00511		Pleiades 708			3 42 37	+23 55.7	3	10.12	.007	0.61	.009	0.13	.000	7	F7 V	1,1029,8023
23387 +23 00512		Pleiades 717	★	AB	3 42 39	+24 10.8	4	7.18	.022	0.16	.000	0.08	.000	12	A1 Vp	1,1029,8023,8030
+59 00712				B	3 42 39	+60 17.7	1	13.95		0.94		0.53		2		7010
281161 +31 00646					3 42 40	+31 51.7	1	9.75		0.46		0.24		2	A0	1041
		Pleiades 738	★	V	3 42 41	+23 36.0	2	12.29	.017	1.17	.004	0.72		3	G9 V	1029,1508
+24 00552		Pleiades 727	★	V	3 42 41	+24 28.3	4	9.70	.005	0.55	.005	0.05	.000	8	F7.5V	1,1029,8023,8030
23386 +24 00551		Pleiades 739	★	V	3 42 41	+24 45.6	4	9.53	.041	0.62	.005	0.06	.000	7	G1 V	1029,1508,8023,8030
		Pleiades 9078	★	V	3 42 41	+25 31.9	1	15.30		1.47				1		1508
+59 00712		G 246 - 48		A	3 42 41	+60 18.6	2	9.99	.016	0.75	.010	0.37	.035	5	K0	1726,7010
		CS 22190 # 3			3 42 42	−17 02.0	1	14.90		-0.27		-1.11		1		1736
23402 +22 00544		Pleiades 3317	★	A	3 42 43	+22 32.4	2	7.83	.005	0.18	.005	0.13	.019	3	A0	1648,8023
282269 +23 00513		Pleiades 745			3 42 43	+24 08.0	4	9.45	.012	0.52	.005	0.11	.000	8	F4 V	1,1029,8023,8030
+23 00514		Pleiades 746			3 42 43	+24 16.6	2	11.29	.015	0.84	.075	0.23		3		1029,1508
		Pleiades 740			3 42 43	+24 54.2	2	13.43	.023	1.07	.009	0.95		4	K3 Ve	257,1508
237176 +56 00838					3 42 43	+57 08.1	1	8.36		1.95				3	K5 III	369

Table 1   187

HD	DM	Other Id	N Rem	α_{1950}	δ_{1950}	S	V	σ_V	B–V	σ_{B-V}	U–B	σ_{U-B}	n	Spectrum	References
		Pleiades 762	★ V	3 42 46	+23 55.2	3	14.33	.015	1.41	.009	1.37		22		201,1508,5008
		Pleiades 761		3 42 46	+24 04.0	2	10.56	.005	0.69	.020	0.17	.000	3	G2 V	1029,8023
23410 +22 00545		Pleiades 801	★ AB	3 42 51	+22 59.5	3	6.85	.000	0.04	.000	0.01	.000	4	A0 Va	1029,8023,8030
		Pleiades 793	★ V	3 42 51	+23 41.9	4	14.30	.013	1.41	.039	1.35	.166	17	dM0	257,921,1508,5008
23408 +23 00516		Pleiades 785	★	3 42 51	+24 12.8	8	3.87	.009	-0.07	.002	-0.40	.007	31	B7 III sn	1,15,154,1029,1203*
+23 00515		Pleiades 784		3 42 51	+24 16.3	1	10.88		1.33		1.18		1	dG5	1722
		Pleiades 799	★ V	3 42 52	+23 43.2	4	13.67	.033	1.33	.010	1.15	.064	14	K5 V	257,921,1508,5008
		Pleiades 9197	★ V	3 42 52	+25 42.4	1	15.27		1.49				1		1508
		Pleiades 803		3 42 53	+23 28.0	1	11.97		1.42		1.13		2		1506
23409 +23 00517		Pleiades 804		3 42 53	+23 53.1	3	7.85	.014	0.20	.004	0.16	.000	7	A3 V	1,1029,8023
		PHL 8745		3 42 54	−25 26.	2	13.93	.033	-0.04	.014	0.06	.009	2		832,1036
23432 +24 00553		Pleiades 817	★	3 42 55	+24 24.0	8	5.76	.006	-0.04	.002	-0.23	.010	42	B8 V	1,15,272,1029,1079*
23254 +62 00608				3 42 57	+62 19.2	1	8.06		0.05		-0.45		5	B5	555
232833 +50 00816				3 42 58	+51 02.8	1	9.94		0.19				3	A1 V	1119
23431 +24 00555		Pleiades 839		3 42 59	+24 53.6	1	9.72		0.38		0.22		1	A2	1748
23430 +24 00554		Pleiades 3318		3 42 59	+25 14.7	2	8.03	.028	0.22	.000	0.13	.005	4	A0	833,8023
23466 +5 00539		HR 1153	★ A	3 43 01	+05 53.7	5	5.34	.020	-0.11	.004	-0.62	.008	16	B3 V	15,146,154,1061,1223
23466 +5 00539		IDS03404N0544	B	3 43 01	+05 53.7	1	12.66		0.89		0.32		6		146
23441 +24 00556		Pleiades 859	★	3 43 04	+24 22.4	7	6.43	.007	-0.02	.003	-0.15	.000	40	B9 Vn	1,15,1029,1079,6001*
23189 +68 00278		G 221 - 23	★ A	3 43 04	+68 30.9	3	9.16	.105	1.32	.031	1.21	.019	12	K0	906,1197,3016
		Pleiades 870		3 43 05	+23 35.0	1	12.61		1.25				1		1508
		Pleiades 865		3 43 05	+24 03.2	1	16.01		1.05				2		1029
23359 +48 00990				3 43 05	+48 38.8	1	7.86		1.00				3	F8 Ib-II	6009
23189 +68 00278		G 221 - 24	★ BC	3 43 05	+68 31.3	2	10.65	.050	1.53	.015	1.21	.015	2	K0	906,3016
		G 6 - 33		3 43 06	+14 33.7	1	11.92		1.43		1.16		1	M1	333,1620
		Pleiades 882		3 43 06	+23 15.1	2	12.88	.124	1.07	.000	0.73		3	K3 V	1029,1508
		Pleiades 874		3 43 06	+23 53.4	1	10.65		0.63		0.06		1	K5	1748
		Pleiades 883	★ V	3 43 08	+24 24.5	3	13.05	.010	1.13	.023	1.06		13		201,1508,5008
		Pleiades 879		3 43 08	+24 24.8	3	12.80	.021	1.07	.008	0.97		13		201,1508,5008
		Pleiades 885		3 43 08	+24 42.8	2	12.07	.015	1.02	.005	0.84		3		1029,1508
23465 +18 00537				3 43 09	+18 24.7	1	9.16		0.28		0.20		2	B8	1648
		AO 1060		3 43 10	+23 57.6	1	10.12		0.54		-0.01		1	K0 V	1748
		Pleiades 915	★ V	3 43 11	+23 11.6	2	13.73	.080	1.23	.000			2	K6 Ve	1029,1508
		Pleiades 906	★ V	3 43 11	+24 31.1	1	15.20		1.52				1	M0 Ve	1508
+22 00548		Pleiades 923		3 43 12	+23 11.2	3	10.14	.017	0.62	.005	0.09	.000	6	G0 V	1029,1508,8023
		Pleiades 916		3 43 13	+24 28.1	1	11.71		0.87		0.48		2		1029
		Pleiades 930		3 43 14	+23 54.0	2	14.23	.005	1.35	.014	0.85		10		201,5008
		G 78 - 50		3 43 14	+45 22.0	1	12.12		1.02		0.89		1		333,1620
23464 +22 00549		Pleiades 948		3 43 15	+22 58.5	2	8.66	.000	0.60	.000	0.10	.000	3	F8.5V	1029,8023
		Pleiades 945		3 43 15	+23 03.3	2	13.31	.014	1.03	.019	0.97		2	K2 Ve	1029,1508
23463 +23 00519		Pleiades 938		3 43 15	+24 02.5	1	7.69		1.23		1.34		1	K2	1722
23479 +23 00520		Pleiades 956	★ AB	3 43 17	+24 02.1	2	7.97	.005	0.32	.000	0.07	.000	3	A9 V	1029,8023
23462 +24 00559		Pleiades 995		3 43 17	+25 05.5	1	10.04		0.87		0.53		1	K1 V	1748
23453 +25 00613		G 6 - 34		3 43 17	+26 03.9	4	9.62	.018	1.45	.024	1.23	.015	6	K5	196,680,1620,3072
23697 −54 00589		HR 1168	★ AB	3 43 17	−54 25.8	4	6.30	.006	1.03	.005			18	K1 III	15,1075,2013,2029
23588 −28 01276		IDS03412S2811	AB	3 43 19	−28 01.2	3	8.20	.012	1.00	.004	0.78		9	K2 V	173,2012,3072
23670 −48 01069		HR 1167		3 43 19	−48 12.9	3	6.49	.011	1.01	.005			14	G8/K0 III	15,1075,2029
+23 00521		Pleiades 975		3 43 20	+23 20.0	3	10.58	.008	0.82	.014	0.29	.000	4		965,1029,8023
278987 +33 00714				3 43 20	+33 41.8	1	10.38		0.33		0.25		2	A2	1041
		Pleiades 992		3 43 21	+23 22.0	1	12.26		0.97		0.48		2		1506
23480 +23 00522		Pleiades 980	★ V	3 43 21	+23 47.6	8	4.18	.007	-0.06	.003	-0.42	.007	26	B6 V	1,15,154,1029,1212*
		Pleiades 974		3 43 21	+24 37.9	3	13.88	.041	1.33	.009	1.23		13	K7 Ve	201,1508,5008
278943 +36 00751				3 43 23	+36 48.0	1	10.99		0.52		0.31		2	A0	1041
282963		Pleiades 996	★ V	3 43 24	+24 25.0	3	10.42	.008	0.65	.012	0.15	.000	7	G0 V	1,1029,8023
		NGC 1466 sq1 13		3 43 25	−71 47.5	1	13.93		0.54		0.00		5		919
282953		Pleiades 1008		3 43 26	+24 51.8	1	10.04		1.81		2.17		1	M2 III	1748
23384 +51 00774				3 43 26	+51 33.2	1	6.85		0.37				3	F0	1119
		CS 22176 # 20		3 43 26	−10 32.5	1	13.50		0.23		0.11		1		1736
		Pleiades 1021		3 43 27	+23 19.1	1	10.36		1.28		1.03		2		1506
		Pleiades 1015		3 43 28	+24 58.9	3	10.53	.020	0.65	.000	0.11	.009	4	G5	1029,1748,8023
275828 +37 00827				3 43 28	+37 48.2	1	11.37		0.42		0.25		2	A0	1041
23526 +6 00583		HR 1159		3 43 29	+06 38.9	3	5.91	.008	0.99	.005	0.76	.007	10	G9 III	15,1256,1355
		G 6 - 35		3 43 29	+15 18.4	1	11.80		1.40		1.17		1	M1	333,1620
23489 +23 00523		Pleiades 1028		3 43 29	+24 06.1	5	7.36	.022	0.10	.006	0.12	.000	11	A2 V	1,369,1029,8023,8030
		Pleiades 1032		3 43 29	+24 16.8	2	11.21	.119	0.80	.055	0.38		3		965,1029
		Pleiades 1029	★ V	3 43 29	+24 36.3	2	14.27	.065	1.45	.065			2		364,1508
		Pleiades 1039	★ V	3 43 30	+23 26.3	2	13.02	.055	1.23	.005	0.79		4	K2 V	364,1508
23478 +31 00649				3 43 32	+32 08.1	4	6.66	.019	0.08	.014	-0.57	.005	9	B3 IV	1009,1012,1041,1252
		Pleiades 1061	★ V	3 43 33	+23 57.8	4	14.23	.030	1.38	.037	1.20	.040	13	dK5	257,921,1508,5008
+7 00542		Hyades HG7 - 5		3 43 34	+07 32.3	1	10.41		0.68				1	F8	950
		Pleiades 3461		3 43 34	+25 41.4	1	15.27		1.49				1		921
23488 +25 00615		Pleiades 3319		3 43 34	+25 41.4	1	8.73		0.37		0.12		1	A0	8023
+44 00794				3 43 34	+44 00.7	1	9.47		2.06				1		369
23817 −65 00263		HR 1175	★ A	3 43 34	−64 57.8	6	3.84	.006	1.13	.004	1.11	.005	34	K1/2 III	15,1075,2013,2030*
		Pleiades 1081		3 43 35	+23 09.1	1	14.61		1.42				1	K7 Ve	1508
23512 +23 00524		Pleiades 1084		3 43 36	+23 28.2	7	8.11	.016	0.36	.005	0.29	.010	33	A2 V	1,369,1029,8009,8023*
23383 +55 00824		HR 1147		3 43 36	+55 46.1	2	6.10		-0.02	.011	-0.16		5	B9 Vnn	70,1079
		G 79 - 69		3 43 37	+11 34.0	1	14.93		1.43				3		538
		Pleiades 1088		3 43 37	+23 38.8	1	13.07		0.84		0.23		2		1506
278959 +34 00728				3 43 37	+35 18.0	1	9.29		1.37		1.05		2	K2	1733

HD	DM	Other Id	N Rem	α_{1950}	δ_{1950}	S	V	σ_V	B–V	σ_{B-V}	U–B	σ_{U-B}	n	Spectrum	References
23439 +41 00750		G 95 - 57	⋆ A	3 43 37	+41 17.4	4	8.17	.016	0.76	.011	0.21	.009	9	K1 V	979,1658,3078,8112
23439 +41 00750		G 95 - 57	⋆ AB	3 43 37	+41 17.4	2	7.67	.013	0.80	.009	0.29	.000	5	K1 V	22,1003
23439 +41 00750		IDS03401N4110	B	3 43 37	+41 17.4	4	8.77	.009	0.89	.009	0.48	.027	9	K2 V	979,1658,3078,8112
		Pleiades 1103	⋆ V	3 43 38	+23 15.5	1	14.77		1.48				1	K7 Ve	1508
		Pleiades 1100	⋆ V	3 43 38	+24 11.4	3	12.19	.056	1.13	.032	0.93	.124	6	dK3	364,1029,1508
		Pleiades 1095		3 43 38	+24 35.7	1	11.92		0.88		0.71		2		1029
23538 +13 00594				3 43 39	+13 21.3	1	6.89		0.07		-0.01		2	A0	1648
282954 +24 00561		Pleiades 1101		3 43 39	+24 48.4	3	10.26	.018	0.61	.000	0.07	.009	5	F9.5V	1029,1748,8023
		LP 593 - 52		3 43 39	−00 52.8	1	13.04		0.55		-0.19		2		1696
282975 +23 00525		Pleiades 1117		3 43 40	+23 38.0	5	10.20	.008	0.72	.008	0.31	.020	10	G6 V	1,369,1029,1506,8023
		Pleiades 1110		3 43 40	+24 22.0	2	13.35	.060	1.21	.020	1.11		4	K6.5Ve	257,1508
23036 +77 00134				3 43 40	+78 10.6	1	8.09		-0.01		-0.35		3	B8	555
23514 +22 00550		Pleiades 1132		3 43 41	+22 46.0	2	9.42	.000	0.50	.005	0.03	.000	3	F5 V	1029,8023
		Pleiades 1124	⋆ V	3 43 41	+23 52.6	2	12.25	.094	0.96	.028	0.51		6	K2 V	1029,1508
23511 +23 00526		Pleiades 1122		3 43 41	+23 57.0	5	9.29	.008	0.46	.004	0.03	.014	9	F5 V	1,1029,1722,8023,8030
		Pleiades 1114	⋆ V	3 43 41	+24 46.6	3	14.09	.009	1.39	.013	1.21		13		201,1508,5008
23513 +22 00551		Pleiades 1139		3 43 42	+22 57.4	2	9.37	.005	0.48	.000	0.00	.000	3	F5 V	1029,8023
		Pleiades 1136		3 43 42	+23 20.6	2	12.11	.079	1.00	.005	0.54		3	G8 V	1029,1508
275826 +37 00828				3 43 43	+37 56.9	1	9.39		0.68		0.41		2	A5	1041
23719 −47 01147		HR 1169		3 43 43	−47 30.9	2	5.72	.005	0.96	.000			7	K1 III	15,2029
		AO 1067		3 43 45	+23 53.6	1	10.58		0.46		0.03		1		1748
		Pleiades 1163		3 43 47	+23 40.3	1	13.14		0.78		0.16		2		1506
23614 −12 00707		HR 1162, π Eri		3 43 47	−12 15.4	6	4.42	.012	1.63	.010	1.98	.021	25	M1 III	3,15,1075,2012,3055,8015
23477 +43 00809				3 43 48	+43 55.2	1	7.07		0.01		-0.05		3	B9	595
+22 00552		Pleiades 1182		3 43 50	+22 45.6	2	10.47	.010	0.64	.000	0.10	.000	3	F8	1029,8023
		Pleiades 1173	V	3 43 50	+24 26.8	5	15.15	.049	1.51	.018	0.91		23		201,921,965,1508,5008
282955		Pleiades 1184		3 43 52	+24 58.4	1	10.62		0.34		0.28		1	A5	1748
		KUV 03439-0048		3 43 52	−00 47.9	1	14.92		-0.30		-1.20		1	sdO	1708
+22 00553		Pleiades 1200		3 43 53	+23 05.1	2	9.92	.020	0.54	.000	0.05	.000	3	F6 V	1029,8023
		KUV 898 - 6		3 43 53	−00 47.8	1	14.91		-0.29		-1.20		1	sdO	974
		LP 773 - 37		3 43 55	−19 53.8	1	12.45		1.00		0.61		1		1696
		Pleiades 1220		3 43 56	+22 43.6	1	11.74		0.88		0.54		1		1029
+23 00527		Pleiades 1215		3 43 56	+23 25.8	4	10.53	.013	0.64	.012	0.16	.000	12	G0 V	1,1029,1508,8023
282962		Pleiades 1207	⋆ V	3 43 56	+24 38.6	2	10.47	.000	0.63	.010	0.09	.000	3	G5	1029,8023
23452 +51 00775				3 43 56	+51 22.5	2	7.30	.052	0.15	.038			4	A0 V	70,1119
237181 +56 00843		LS I +56 088		3 43 56	+56 59.2	1	9.93		0.88		0.70		2	A4 Ib	41
23458 +49 01024				3 43 57	+50 12.6	1	7.72		1.45		1.32		2	K2	1601
232836 +53 00700				3 43 57	+53 57.5	1	9.06		0.54				2	F6 V	1025
−10 00742		G 160 - 20		3 43 57	−10 12.4	2	10.88	.010	1.06	.015	1.01	.000	2		1658,3056
		G 248 - 6		3 43 59	+79 33.5	2	10.80	.008	0.67	.001	0.17	.042	3		1658,5010
		Pleiades 3646		3 44 00	+23 58.	1	16.01		1.59				2		921
		Pleiades 4008		3 44 00	+23 58.	1	15.83		1.64		1.31		14		5008
		Pleiades 4011		3 44 00	+23 58.	1	15.65		1.13		0.80		14		5008
		Pleiades 4072		3 44 00	+23 58.	1	17.82		0.99		0.32		16		5008
		Pleiades 4183		3 44 00	+23 58.	1	16.59		0.99		0.56		12		5008
		Pleiades 4185		3 44 00	+23 58.	1	17.64		0.89		0.80		4		5008
		Pleiades 4189		3 44 00	+23 58.	1	16.50		1.92		1.35		12		5008
		Pleiades 4210		3 44 00	+23 58.	1	16.13		1.50		0.32		4		5008
		Pleiades 4242		3 44 00	+23 58.	1	18.42		1.16		0.36		11		5008
		Pleiades 4260		3 44 00	+23 58.	1	16.96		0.67		0.12		5		5008
		Pleiades 5715		3 44 00	+23 58.	1	15.08		0.95		0.44		5		5008
		Pleiades 5728		3 44 00	+23 58.	1	16.10		1.52		1.47		7		5008
		Pleiades 5928		3 44 00	+23 58.	1	13.77		1.03		0.58		2		5008
		Pleiades 5929		3 44 00	+23 58.	1	16.26		1.70		0.99		5		5008
		Pleiades 6083		3 44 00	+23 58.	1	16.94		0.80		0.41		8		5008
		Pleiades 6381		3 44 00	+23 58.	1	15.34		1.54		1.32		6		5008
23568 +24 00562		Pleiades 1234		3 44 00	+24 22.0	6	6.81	.012	0.02	.003	-0.07	.001	29	B9.5Van	1,369,1029,1245,8023,8030
23277 +70 00257		HR 1138		3 44 00	+70 43.1	4	5.42	.016	0.10	.005	0.10	.014	11	A2 m	15,3058,8015,8052
23716 −38 01280				3 44 02	−38 44.3	1	8.26		1.04		0.82		5	G8 III	1673
		Pleiades 1275		3 44 03	+23 20.5	2	11.46	.009	0.83	.005	0.43		4	K0 V	965,1029
23567 +24 00563		Pleiades 1266	⋆ AB	3 44 04	+24 40.0	3	8.28	.013	0.37	.010	0.12	.000	3	F0 V	1029,1748,8023
232837 +52 00712				3 44 04	+53 16.4	1	9.34		0.58				2	F6 V	1119
22499 +84 00066				3 44 04	+84 40.2	1	8.35		1.59		1.61		3	M1	1733
		G 160 - 21		3 44 04	−13 20.8	1	12.56		1.50		1.11		1		3056
		Pleiades 1280	⋆ V	3 44 05	+24 00.4	2	14.56	.014	1.36	.009	1.11		9	dK7	201,5008
		Pleiades 1272		3 44 05	+24 29.0	1	10.35		1.35		1.25		1	gG8	1722
24063 −74 00271				3 44 05	−74 09.9	3	7.54	.006	0.25	.010	0.10	.035	7	A2mA3-A9	278,1034,2012
		Pleiades 1286	⋆ V	3 44 06	+23 27.8	2	15.35	.005	1.68	.000			6		921,1508
23585 +23 00528		Pleiades 1284		3 44 06	+23 50.5	5	8.37	.000	0.30	.003	0.08	.000	11	F0 V	1,369,1029,8023,8030
24035 −72 00260				3 44 07	−72 45.9	2	8.51	.000	1.24	.007	0.93	.027	3	Kp Ba	565,3048
		Pleiades 1298		3 44 09	+23 33.7	2	12.28	.071	1.01	.009	0.78		4		1029,1508
		Pleiades 1305	⋆ V	3 44 10	+23 04.4	4	13.54	.036	1.19	.005	1.10		10		921,965,1029,1508
		Pleiades 1306	⋆ V	3 44 10	+23 33.4	5	13.43	.044	1.33	.018	1.17	.012	15	dK5(e)	257,921,1029,1508,5008
		Pleiades 1321	⋆ V	3 44 11	+23 35.3	5	15.28	.040	1.53	.055	1.05		18		201,921,965,1508,5008
23584 +23 00529		Pleiades 1309		3 44 11	+24 07.4	3	9.45	.010	0.47	.001	0.02	.000	7	F6 V	1,1029,8023
23583 +24 00565		Pleiades 1331		3 44 11	+25 09.5	1	9.00		1.25		1.16		1	K1 III	1748
+55 00825				3 44 13	+55 29.7	1	9.94		0.58				3		70
		Pleiades 1332		3 44 15	+23 33.7	3	12.47	.034	1.03	.009	0.81	.031	7	K4 V	257,1029,1508
		Hyades HG7 - 7		3 44 17	+11 32.3	1	13.92		1.23				2		950
278994 +32 00666				3 44 17	+33 10.3	1	9.71		0.30		0.08		2	B8	1041

Table 1   189

HD	DM	Other Id	N Rem	α_{1950}	δ_{1950}	S	V	σ_V	B–V	σ_{B-V}	U–B	σ_{U-B}	n	Spectrum	References
23608	+23 00531	Pleiades 1338	★ D	3 44 18	+23 58.5	3	8.65	.019	0.46	.000	0.03	.000	4	F5 V	1029,8023,8030
23609	+23 00535	Pleiades 1357		3 44 19	+23 34.4	1	7.00		0.50		0.09		2	F8 IV	401
		Pleiades 1348		3 44 19	+24 14.3	4	12.67	.040	1.16	.012	1.02	.025	10	K5	257,921,1029,1508
		Pleiades 1355	★ V	3 44 20	+23 53.0	3	14.03	.028	1.40	.028			5	K5 Ve	921,1029,1508
23607	+23 00534	Pleiades 1362	★ C	3 44 21	+23 59.2	2	8.26	.010	0.25	.005	0.10	.000	3	F0 V	1029,8023
23629	+23 00536	Pleiades 1375	★ B	3 44 22	+23 57.8	4	6.31	.012	0.02	.004	-0.02	.011	7	A0 V	1029,1084,8023,8030
275812	+38 00807	G 95 - 58		3 44 22	+39 00.5	1	10.23		1.20		1.18		1		333,1620
23632	+23 00537	Pleiades 1380		3 44 23	+23 39.0	3	6.99	.018	0.03	.000	0.05	.000	9	A0 Va	1,1029,8023
275800	+39 00862			3 44 23	+39 41.0	1	8.85		1.21		1.06		2	K0	1601
		Pleiades 1391		3 44 25	+23 24.0	1	10.94		0.61		0.16		2		1506
23628	+24 00566	Pleiades 1384		3 44 25	+24 26.1	5	7.67	.023	0.21	.005	0.12	.007	14	A4 V	1,401,1029,8023,8030
23406	+64 00408	IDS03399N6426	AB	3 44 25	+64 35.8	2	7.67	.005	0.42	.015	0.02		3	F5	292,1381
23738	−29 01413	HR 1171	AB	3 44 25	−29 29.5	2	5.89	.005	0.12	.000			7	A2 V	15,2012
23610	+22 00556	Pleiades 1407		3 44 26	+22 46.1	2	8.13	.005	0.25	.005	0.13	.000	5	A0	1029,8023
23631	+23 00538	Pleiades 1397	★ AB	3 44 26	+23 45.7	4	7.25	.010	0.06	.014	0.04	.004	10	A0mA1Va	1,1029,8023,8052
		Pleiades 9337	★ V	3 44 27	+22 05.5	1	14.54		1.39				1		1508
23627	+24 00567	Pleiades 1404	★ AB	3 44 27	+24 30.3	1	8.73		0.18		-0.06		2	A0	401
		NGC 1466 sq1 8		3 44 27	−71 45.3	1	14.64		0.63		0.06		12		919
		Pleiades 1420		3 44 28	+23 23.3	1	10.44		1.38		1.12		2		1506
		Pleiades 9018	★ V	3 44 29	+22 11.8	1	14.98		1.55				1		1508
23643	+23 00539	Pleiades 1425	★ V	3 44 29	+23 31.5	5	7.77	.012	0.16	.005	0.12	.000	15	A3.5V	1,369,1029,8023,8030
		Hyades HG7 - 8		3 44 30	+12 45.	1	12.27		0.90				1		950
23630	+23 00541	Pleiades 1432	★ A	3 44 30	+23 57.1	12	2.87	.006	-0.09	.006	-0.34	.011	64	B7 III	1,3,15,369,1029,1084*
		Pleiades 1426		3 44 30	+24 24.2	1	10.52		1.85		2.27		1	K5	1722
		RL 72		3 44 30	+52 01.	1	11.99		0.50		-0.06		2		704
		NGC 1466 sq1 3		3 44 30	−71 46.2	1	11.93		0.33		0.06		8		919
23642	+23 00540	Pleiades 1431		3 44 31	+24 08.1	3	6.81	.010	0.06	.004	0.02	.000	11	A0 Vp +Am	1,1029,8023
		Pleiades 1440		3 44 33	+24 45.7	1	15.64		1.01		-0.06		1		1029
		Pleiades 1454	★ V	3 44 34	+24 31.9	3	12.86	.017	1.12	.018	0.90	.020	5	K3 V	257,1029,1508
		Pleiades 9105	★ V	3 44 35	+23 32.4	1	15.75		1.50				1		1508
23606	+34 00732	IDS03414N3443	A	3 44 35	+34 51.9	1	9.50		0.11		-0.09		2	A0 V	1041
23552	+50 00825	HR 1160	★ AB	3 44 36	+50 35.0	3	6.14	.009	0.06	.004	-0.33		6	B8 Vne	70,1025,1079
		L 228 - 72		3 44 36	−52 15.	1	12.89		0.73		-0.01		2		1696
		AO 1064		3 44 37	+23 56.0	1	10.32		1.51		1.74		1		1748
	+11 00520	G 79 - 71		3 44 38	+11 19.2	1	11.21		1.20		1.12		1	K5	333,1620
23524	+51 00777	IDS03409N5144	AB	3 44 38	+51 53.1	2	8.70	.010	0.77	.005	0.28		3	K0 V +K1 V	1025,3016
23654	+23 00542	Pleiades 1498		3 44 39	+23 27.4	3	7.71	.019	1.23	.006	1.11	.009	5	K0	1,369,1029
		Pleiades 1491	★ V	3 44 39	+24 34.8	1	16.42		0.60		0.25		1		1029
		Pleiades 1485	★ V	3 44 39	+24 44.3	3	14.23	.006	1.35	.014	1.13		16	K5 Ve	201,1508,5008
		Pleiades 1512		3 44 40	+23 18.9	2	13.51	.005	1.30	.052	1.06		2	K6	1029,1508
23626	+31 00650	HR 1164		3 44 40	+32 02.5	1	6.25		0.47				2	F7 IV-V	71
282967		Pleiades 1514		3 44 41	+24 12.7	3	10.50	.031	0.65	.008	0.12	.004	4	G5	1029,1722,8023
23510	+56 00845	LS I +56 089		3 44 41	+56 52.1	1	8.52		0.28		-0.44		2	B8	41
		Pleiades 1516	★ V	3 44 42	+24 09.0	3	13.90	.048	1.28	.025	1.04	.007	7		1029,1508,5008
23625	+33 00717	HR 1163	★ AB	3 44 42	+33 26.8	6	6.56	.013	0.07	.011	-0.60	.009	18	B2.5V	15,154,1009,1012,1041,1252
23754	−23 01565	HR 1173		3 44 42	−23 23.8	6	4.22	.007	0.43	.007	0.01	.016	22	F3 III	3,15,1075,2012,3026,8015
		AO 1066		3 44 43	+22 58.1	1	10.69		1.33		1.27		1		1748
		Pleiades 1532	★ V	3 44 43	+23 35.2	3	13.95	.034	1.28	.011	1.04	.080	3		1029,1508,5008
		Pleiades 1531	★ V	3 44 43	+23 49.1	2	13.39	.043	1.21	.028	1.05		4	dK7e	257,1029
23679	+16 00512			3 44 44	+16 33.1	1	8.44		1.20		1.01		2	K0	1648
		Pleiades 1553	★ V	3 44 44	+22 46.6	3	12.35	.048	1.10	.026	0.80	.044	6	K2.5V	364,1029,1508
23665	+23 00544	Pleiades 1549		3 44 44	+23 23.4	4	8.78	.009	1.15	.008	0.81	.006	7	G6 V	1,369,1029,1245
		NGC 1466 sq1 16		3 44 44	−71 53.0	1	14.55		0.68		0.22		5		919
23798	−31 01541			3 44 45	−31 00.5	2	8.30	.010	1.10	.010	0.76		6	G8/K1 (III)	742,1594
23664	+24 00568	Pleiades 3320		3 44 47	+25 14.0	1	8.33		0.27		0.14		1	A2	8023
282957		Pleiades 1630		3 44 48	+24 51.6	1	10.51		1.26		1.19		1	K1 III	1748
		Pleiades 1593		3 44 50	+23 03.9	2	11.11	.005	0.76	.005	0.30	.000	5	G6 V	1029,5008
		Pleiades 9083	★ V	3 44 51	+22 24.2	1	15.25		1.57				1		1508
23565	+51 00778			3 44 53	+51 40.3	1	7.60		0.66				2	G5 V	1025
282973	+23 00545	Pleiades 1613		3 44 54	+23 47.3	4	9.87	.007	0.54	.000	0.11	.000	8	F6 V	1,1029,8023,8030
		NGC 1466 sq1 6		3 44 54	−71 46.8	1	15.05		0.60		0.08		16		919
		Pleiades 1628		3 44 55	+22 47.4	1	13.45		0.81		0.29		1	G0 V	1029
23475	+65 00369	HR 1155, BE Cam		3 44 55	+65 22.4	5	4.46	.033	1.88	.016	2.13	.008	13	M2 IIab	15,1363,3016,8003,8015
23581	+50 00828			3 44 56	+51 13.4	1	7.21		1.22				2	K0 V	1119
	+23 00546	Pleiades 1645		3 44 58	+23 18.8	1	11.33		0.71		0.43		3		1029
		G 175 - 8		3 44 58	+55 51.9	1	10.94		0.79		0.38		2	K0	1723
282970	+23 00547	Pleiades 1666		3 45 00	+23 56.9	1	10.02		0.64		0.11		1	G0	1722
		Pleiades 1653	★ V	3 45 00	+24 34.7	5	13.60	.034	1.23	.015	1.01	.028	64	K4.5Ve	257,352,921,1508,5008
		G 6 - 36		3 45 03	+16 36.0	1	12.38		1.49		1.18		1	K5r	333,1620
232843	+51 00779			3 45 03	+51 48.5	1	8.52		1.20				2	K0 IV	1119
23401	+70 00259	HR 1148	★ A	3 45 03	+71 10.9	7	4.63	.011	0.03	.007	0.07	.007	99	A2 IVn	15,61,71,1118,1363*
24115	−72 00261			3 45 03	−71 51.5	1	8.95		0.49		0.05		5	F6wF3 V	919
23712	+24 00571	Pleiades 1705		3 45 07	+24 50.2	4	6.44	.014	1.70	.001	2.05	.031	33	K5	1,369,1029,1245
23713	+23 00548	Pleiades 1726	★	3 45 08	+23 59.4	5	9.27	.019	0.54	.010	0.10	.034	11	F7 V	1,196,369,1029,8023
		Pleiades 1712		3 45 08	+25 00.8	1	14.36		1.01		0.47		2		1029
		NGC 1466 sq1 4		3 45 08	−71 50.1	1	14.49		0.64		0.08		15		919
281230	+31 00651			3 45 11	+31 31.5	1	9.04		0.76		0.41		2	K5	1371
		NGC 1444 - 15		3 45 11	+52 30.3	1	14.06		1.08		0.59		1		49
		NGC 1466 sq1 5		3 45 11	−71 49.6	1	13.59		0.94		0.68		9		919
		Hyades HG7 - 10		3 45 12	+13 26.	1	12.83		1.32				1		950

HD	DM	Other Id	N	Rem	α_{1950}	δ_{1950}	S	V	σ_V	B–V	σ_{B-V}	U–B	σ_{U-B}	n	Spectrum	References
281226	+31 00651				3 45 12	+31 31.0	1	8.95		1.87		2.12		3	A0	1371
23623	+50 00831				3 45 12	+50 41.3	2	7.33	.014	0.41	.000			4	F6 Vp	70,1025
23523	+62 00612	HR 1158			3 45 12	+63 08.7	1	5.85		0.18				4	A5 Vn	1118
		Pleiades 1756			3 45 13	+23 21.3	3	14.12	.025	1.37	.000	1.23		14		201,1508,5008
237183	+58 00656				3 45 13	+58 46.3	1	9.63		1.23				47	K7	6011
		NGC 1444 - 23			3 45 14	+52 30.5	1	15.77		1.33				1		49
23733	+23 00549	Pleiades 1762			3 45 15	+24 10.0	4	8.26	.010	0.36	.001	0.11	.014	10	A9 V	1,1029,1722,8023
23732		Pleiades 1766			3 45 17	+25 03.8	5	9.13	.009	0.47	.000	0.07	.012	9	F5 V	1,1029,1748,8023,8030
		Pleiades 1785		⋆V	3 45 18	+24 21.1	2	14.33	.095	1.42	.053	1.31		11		201,5008
282958		Pleiades 1776			3 45 18	+24 53.8	4	10.93	.021	0.72	.004	0.22	.005	8	G5 V	1029,1508,1748,8023
		NGC 1444 - 20			3 45 18	+52 27.0	1	14.99		0.93		0.66		1		49
		CS 22176 # 32			3 45 18	−09 55.3	1	13.59		0.27		0.10		1		1736
23856	−30 01494	HR 1179			3 45 18	−30 03.3	3	6.54	.004	0.49	.005			11	F5 V	15,2020,2028
		NGC 1466 sq1 7			3 45 18	−71 48.7	1	14.02		0.91		0.57		12		919
	+23 00551	Pleiades 1797			3 45 19	+23 29.1	3	10.11	.016	0.56	.001	0.07	.000	5	G0 V	1,1029,8023
282972	+23 00550	Pleiades 1794		AB	3 45 19	+23 44.3	1	10.36		0.64		0.13		2	F8	1029
	+31 00653				3 45 19	+32 09.3	1	10.43		0.47		0.31		2	A3 V	1041
		NGC 1444 - 17			3 45 19	+52 29.2	1	14.25		1.57				1		49
23594	+56 00846	HR 1161		⋆A	3 45 20	+56 58.0	1			0.06		0.01		2	A0 Vn	1079
		LS I +60 307			3 45 20	+60 36.5	1	12.16		0.49		−0.38		3		41
		Pleiades 9023		⋆V	3 45 21	+24 45.8	1	17.02		1.79				3		921
		LS I +58 128			3 45 21	+58 20.1	1	10.11		0.36		−0.40		3		41
		G 80 - 22			3 45 23	+02 38.4	4	11.04	.011	1.51	.008			6	dM0	1619,1625,1632,1705
23753	+22 00563	Pleiades 1823		⋆	3 45 23	+23 16.1	7	5.45	.005	−0.07	.001	−0.32	.000	43	B8 Vn	1,15,1029,1079,1258*
		Pleiades 1827		⋆V	3 45 24	+23 49.2	1	14.87		1.47				2		1508
		RL 73			3 45 24	+49 55.	1	16.57		0.18		−1.04		2		704
23742	+24 00572	Pleiades 1829			3 45 26	+25 10.3	1	8.52		0.82		0.43		1	K0 V	1748
279019	+35 00747				3 45 26	+35 27.6	1	10.58		0.27		0.14		2	B9	1041
		NGC 1444 - 16			3 45 26	+52 26.7	1	14.08		1.00		0.47		1		49
		Pleiades 9086		⋆V	3 45 28	+22 15.3	1	15.47		1.59				1		1508
282971	+23 00552	Pleiades 1856			3 45 28	+23 53.8	3	10.02	.004	0.56	.002	0.09	.000	7	F7 V	1,1029,8023
		Pleiades 9025		⋆V	3 45 29	+22 03.6	1	14.84		1.43				1		1508
		NGC 1444 - 21			3 45 29	+52 25.0	1	14.99		1.11		0.70		1		49
		NGC 1444 - 18			3 45 29	+52 25.3	1	14.47		0.82		0.54		1		49
		NGC 1444 - 4			3 45 29	+52 27.0	1	11.84		0.79		0.32		1	F2 V	49
		NGC 1444 - 14			3 45 29	+52 27.5	1	14.02		1.64		1.25		1		49
		NGC 1444 - 6			3 45 29	+52 28.1	1	12.74		1.60		1.34		1		49
23650	+58 00657				3 45 29	+58 37.0	1	9.01		0.58				46	G2 V	6011
		Pleiades 1883		⋆V	3 45 30	+23 08.9	3	12.65	.012	1.04	.015	0.74	.018	24	K2 V	364,1029,1508
	−76 00246				3 45 30	−76 06.6	1	10.55		0.57		0.00		2	G5	1696
		Hyades HG7 - 15			3 45 31	+06 59.6	1	10.85		1.34				2	k7	920
23763	+23 00553	Pleiades 1876			3 45 31	+24 11.6	4	6.95	.005	0.12	.004	0.09	.000	14	A2 V	1,1029,8023,8030
		NGC 1444 - 8			3 45 31	+52 30.3	1	12.82		0.63		0.31		1		49
23878	−24 01877	HR 1181			3 45 31	−24 01.7	3	5.23	.003	0.07	.001	0.17		15	A1 V	3,15,2012
23793	+10 00486	HR 1174		⋆AB	3 45 32	+10 59.5	3	5.06	.000	−0.13	.008	−0.61	.001	7	B3 V +F5 V	154,1223,1363
		NGC 1444 - 5			3 45 32	+52 24.7	1	12.39		0.69		0.24		1	F0 V	49
279032	+34 00736				3 45 33	+34 43.9	1	10.87		0.32		0.15		2	A2	1041
		Pleiades 9156		⋆V	3 45 34	+24 07.	1	16.01		1.59				1		1508
		NGC 1444 - 9			3 45 34	+52 30.7	1	12.91		0.56		0.17		1		49
		NGC 1444 - 13			3 45 35	+52 28.8	1	13.77		0.96		0.48		1		49
23778	+23 00554	Pleiades 1912		⋆AB	3 45 36	+24 01.8	2	9.05	.000	0.48	.010	0.02	.000	3	F7 V	1029,8023
		NGC 1444 - 10			3 45 36	+52 29.6	1	13.29		0.74		0.34		1		49
		NGC 1444 - 11			3 45 36	+52 30.9	1	13.61		0.68		0.47		1		49
		Pleiades 1924			3 45 37	+23 17.0	3	10.34	.005	0.61	.004	0.13	.000	7	F9 V	1,1029,8023
		NGC 1444 - 22			3 45 39	+52 28.1	1	15.38		1.50				1		49
24188	−72 00262				3 45 39	−71 48.8	2	6.27	.005	−0.12	.015	−0.44	.010	12	Ap Siλ4200	919,1628
23805	+8 00571	Hyades vB 136			3 45 40	+08 25.7	2	7.43	.010	1.11	.001	0.95	.001	3	G5	1127,8023
		Pleiades 9087		⋆V	3 45 40	+22 37.0	1	14.57		1.44				1		1508
23777	+25 00620				3 45 41	+25 38.9	2	8.44	.005	0.11	.005	0.06	.005	4	B9	105,1733
23728	+43 00818	HR 1170, V376 Per			3 45 41	+43 48.6	2	5.99	.017	0.27	.016	0.07	.014	7	A9 IV	595,1733
23675	+52 00714	NGC 1444 - 1		⋆AB	3 45 41	+52 30.2	4	6.77	.057	0.45	.024	−0.54	.011	8	B0.5III	49,1012,1119,8100
	+51 00781	IDS03419N5156		B	3 45 42	+52 05.8	1	10.18		0.36				2		1119
23792	+21 00530	Pleiades 3321			3 45 43	+21 46.4	1	8.40		0.40		−0.01		1	F0	8023
		G 6 - 37			3 45 44	+20 54.4	1	10.56		1.11		1.14		1	K3	333,1620
	−65 00267				3 45 44	−65 08.1	1	10.86		0.92		0.67		1		1696
23791	+22 00565	Pleiades 1993			3 45 46	+23 06.5	3	8.37	.011	0.28	.011	0.13	.000	7	A9 V	1,1029,8023
		HRC 21		V	3 45 46	+38 47.2	1	14.05		1.24		0.29		1	K2e	776
		NGC 1444 - 12			3 45 46	+52 28.9	1	13.66		0.95		0.69		1		49
23841	+9 00494				3 45 47	+09 29.6	4	6.69	.004	1.22	.010	1.16	.007	13	K1 III	1003,1355,1509,3040
		NGC 1444 - 24			3 45 47	+52 28.1	1	16.41		1.36				1		49
		Pleiades 2016		⋆V	3 45 48	+23 11.3	3	13.57	.032	1.22	.000	1.15		7	K4 Ve	921,1029,1508
279010	+35 00748				3 45 48	+36 14.1	1	9.72		0.13		0.06		2	A0	1041
		NGC 1444 - 19			3 45 49	+52 26.9	1	14.93		1.10		0.47		1		49
		Pleiades 2027			3 45 50	+24 07.0	4	10.90	.021	0.86	.005	0.50	.000	7		965,1029,1508,8023
		G 246 - 49			3 45 50	+64 38.2	1	13.06		0.79		0.19		2		1658
		Pleiades 2034		⋆V	3 45 51	+23 49.5	3	12.59	.058	0.99	.033	0.42	.124	6	K2.5V	364,1029,1508
23674	+54 00713				3 45 51	+54 33.9	1	8.21		0.47				2	F2 III	70
23855	+1 00664	IDS03432N0117		A	3 45 52	+01 26.8	1	9.18		0.39		−0.04		1	F0	1776
		NGC 1444 - 7			3 45 52	+52 28.1	1	12.77		0.49		−0.01		1		49
24049	−56 00578				3 45 54	−56 11.7	3	9.48	.009	0.62	.007	0.01	.024	7	G3 V	742,2012,3077

Table 1

HD	DM	Other Id	N	Rem	α_{1950}	δ_{1950}	S	V	σ_V	B–V	σ_{B-V}	U–B	σ_{U-B}	n	Spectrum	References
23940	−30 01497	HR 1184			3 45 55	−30 19.1	4	5.54	.007	0.97	.004	0.68		18	G6 III	3,15,1075,2032
24023	−51 00914				3 45 55	−50 54.4	1	6.58		1.33		1.50		6	K3 III	1628
		HRC 22		V	3 45 57	+38 47.5	1	15.27		0.93		−0.44		1		776
23958	−36 01453	HR 1186			3 45 57	−36 15.5	2	6.20	.005	−0.10	.000			7	B8 V	15,2027
23823	+22 00566				3 45 58	+22 38.8	1	8.12		0.05		−0.34		2	B9	401
23822	+23 00556	Pleiades 2087			3 45 58	+23 42.4	1	6.47		0.42		0.04		2	F0	401
23802	+31 00655	IDS03428N3157		AB	3 45 58	+32 06.7	1	7.44		0.17		−0.34		2	B5 Vn	1041
		Pleiades 2082	⋆	V	3 45 59	+24 57.2	3	13.95	.030	1.35	.007	1.26		3	M0 V	921,1029,1508
		Pleiades 2106			3 46 01	+23 03.0	1	11.53		0.86		0.46		2		1029
279020	+35 00749				3 46 02	+35 25.0	1	10.23		0.17		0.06		2	B8	1041
23727	+50 00835				3 46 02	+51 02.7	1	8.96		0.37				3	A0 IV	1119
23887	−0 00602	HR 1182			3 46 05	+00 04.5	5	5.91	.008	1.23	.009	1.31	.007	16	K3 III	15,252,1061,1355,3016
		Pleiades 2126			3 46 05	+23 06.1	2	11.69	.033	0.85	.005	0.48		4		1029,1508
	+4 00590				3 46 06	+04 53.1	1	9.50		0.52		0.03		1	F8	1776
		G 246 - 50			3 46 07	+63 18.4	1	11.39		1.42		1.23		5	M3	1723
		Pleiades 2144	⋆	V	3 46 08	+23 35.3	1	15.31		1.45				1		1508
		Pleiades 2147			3 46 08	+23 37.8	2	10.85	.020	0.81	.010	0.34	.000	3	K0 V	1029,8023
23926	−14 00752			AB	3 46 09	−13 48.2	1	9.45		0.90		0.56		2	G6 V	3030
23850	+23 00557	Pleiades 2168	⋆	AB	3 46 11	+23 54.1	9	3.62	.008	−0.09	.005	−0.36	.007	29	B8 III	1,15,1029,1079,1203*
		NGC 1444 - 3			3 46 11	+52 30.5	1	10.30		0.40		0.26		1	A5 V	49
23862	+23 00558	Pleiades 2181	⋆	V	3 46 12	+23 59.1	12	5.09	.008	−0.08	.005	−0.28	.025	45	B8 Vn+Shell	1,15,272,879,942,1029*
282966	+24 00574	Pleiades 2172			3 46 12	+24 29.1	2	10.43	.005	0.63	.005	0.09	.000	3	F9 V	1029,8023
		Pleiades 2199			3 46 13	+22 58.7	2	14.36	.033	1.10	.024			2	K3 V	257,1029
		Pleiades 2193	⋆	V	3 46 13	+23 24.2	2	14.20	.025	1.42	.050			2	K6 Ve	364,1508
23852	+22 00569	Pleiades 3322			3 46 14	+22 27.5	1	7.75		0.18		0.12		1	A0	8023
23863	+23 00559	Pleiades 2195			3 46 14	+23 44.1	3	8.12	.007	0.22	.000	0.10	.011	10	A8 V	1,1029,8023
		Pleiades 2209			3 46 15	+23 04.6	2	14.45	.047	1.40	.047			2	K6.5Ve	1029,1508
		Pleiades 2208	⋆	V	3 46 16	+24 24.9	4	14.44	.030	1.37	.017	1.18		30	K6 Ve	201,921,1508,5008
23840	+34 00739				3 46 16	+34 46.0	1	8.45		0.07		−0.27		2	B8 V	1041
275877	+38 00811	XY Per	⋆	AB	3 46 17	+38 49.8	1	9.36		0.52		0.52		4	A2 II	206
		GD 50			3 46 17	−01 07.5	4	14.05	.013	−0.28	.014	−1.19	.005	58	DAwk	281,974,1764,3060
		KUV 03463-0108			3 46 17	−01 07.6	1	14.05		−0.27		−1.20		1	DA	1708
		Hyades HG7 - 18			3 46 18	+19 39.	1	12.93		1.47				1		950
23872	+23 00560	Pleiades 2220	⋆	B	3 46 18	+24 14.7	3	7.51	.018	0.10	.005	0.11	.000	9	A1 Van	1,1029,8023
		Pleiades 9339	⋆	V	3 46 20	+25 40.0	1	13.97		1.35				1		1508
		Pleiades 2244	⋆	V	3 46 21	+24 37.5	3	12.61	.029	0.99	.011	0.62		11	K2.5V	921,1029,1508
23937	−7 00685	BR Eri			3 46 21	−07 10.0	1	7.00		1.65		1.55		10	M2	3042
23873	+23 00561	Pleiades 2263	⋆	A	3 46 23	+24 13.8	5	6.59	.014	−0.03	.004	−0.12	.000	14	B9.5Va	1,1029,1355,8023,8030
23848	+32 00667	HR 1177, V467 Per			3 46 23	+32 56.4	4	5.10	.020	0.07	.013	0.14	.045	8	A3 V	1041,1049,1363,3023
23978	−21 00703	HR 1187			3 46 24	−21 03.3	3	5.81	.008	1.63	.010			11	K5 III	15,1075,2032
		Pleiades 2284			3 46 26	+23 41.3	1	11.35		0.78		0.38		3		1029
		Pleiades 2287			3 46 26	+23 49.9	1	15.72		1.15		0.98		1		1029
282960		Pleiades 2278			3 46 26	+24 47.2	4	10.89	.050	0.87	.008	0.46	.000	5	K0	965,1029,1508,8023
23886	+23 00562	Pleiades 2289			3 46 27	+24 05.8	3	7.97	.010	0.18	.002	0.13	.000	11	A4 V	1,1029,8023
		CS 22176 # 33			3 46 27	−10 02.6	1	14.78		0.63		−0.10		1		1736
		Pleiades 2311			3 46 30	+23 33.7	2	11.35	.005	0.82	.005	0.39		2		965,1029
		CS 22176 # 34			3 46 32	−10 22.1	1	14.90		0.28		0.12		1		1736
23912	+22 00570	Pleiades 2345			3 46 35	+23 13.8	3	9.10	.019	0.44	.003	0.02	.000	7	F3 V	1,1029,8023
		Pleiades 2341			3 46 35	+23 38.7	2	10.86	.010	0.72	.005	0.24	.000	3	G4 V	1029,8023
279021	+35 00751	V498 Per			3 46 35	+35 26.9	1	9.49		0.37		0.12		2	A0	1041
23838	+44 00801	HR 1176			3 46 35	+44 49.0	1	5.66		0.76				2	G2 III+F2:V	71
		LP 995 - 86			3 46 35	−39 17.6	2	13.24	.074	0.47	.007	−0.30	.010	5		1696,3060
		Pleiades 2368	⋆	V	3 46 37	+23 18.2	1	14.02		1.34				1		364
		Pleiades 2366			3 46 38	+24 08.7	1	11.53		0.82		0.36		2		1029
23800	+52 00715	NGC 1444 - 2			3 46 38	+52 19.8	3	6.92	.038	0.33	.029	−0.53	.005	5	B1 IV	49,1012,1119
24202	−64 00272				3 46 38	−64 29.1	1	8.79		0.61				4	G1wF5	2012
23913	+22 00572	Pleiades 3323			3 46 41	+22 23.0	1	7.03		0.03		−0.02		1	B9	8023
		Hyades HG7 - 19			3 46 42	+10 33.	1	11.28		0.69				1		920
		Pleiades 2406			3 46 42	+23 08.4	1	11.10		0.76		0.26		3		1029
		RL 74			3 46 42	+57 04.	1	10.56		0.30		−0.44		2		704
		Pleiades 2417			3 46 43	+22 59.7	1	12.28		0.98		0.68		3		257
23924	+22 00573	Pleiades 2415			3 46 43	+23 11.4	3	8.12	.020	0.21	.005	0.13	.000	8	A7 V	1029,8023,8030
		Pleiades 2407			3 46 43	+24 18.8	2	12.23	.050	1.02	.075	0.96		3	K3	1029,1508
	+51 00786				3 46 43	+51 51.5	1	9.60		0.50				2	A4 III	1119
		Pleiades 2411	⋆	V	3 46 44	+24 10.0	2	14.18	.005	1.61	.025	1.04		7	dM3e	257,921
23923	+23 00563	Pleiades 2425	⋆		3 46 45	+23 33.7	6	6.17	.004	−0.05	.001	−0.19	.000	16	B8.5Vn	1,15,1029,1079,8015,8030
		G 95 - 59			3 46 45	+43 17.6	2	13.89	.009	1.58	.009	1.37	.014	5		419,3078
24072	−38 01297	HR 1190	⋆	AB	3 46 45	−37 46.3	4	4.27	.005	−0.01	.000	−0.04	.000	20	B9 V	15,1075,2012,8015
23885	+37 00833				3 46 46	+37 43.4	1	6.61		0.27		0.08		3	A8 V	1501
23662	+68 00286	HR 1166			3 46 46	+68 21.4	1			−0.08		−0.24		2	B9 IVp	1079
24124	−54 00598				3 46 46	−53 54.3	1	8.58		1.31				7	K1 III	955
	+60 00753				3 46 47	+60 22.9	1	8.67		2.15				3	M0	369
23820	+54 00714				3 46 50	+54 38.0	1	8.36		0.45				3	F8 V	70
		Pleiades 2462			3 46 52	+23 33.3	2	11.51	.028	0.84	.005	0.46		2		965,1029
23935	+25 00621	Pleiades 3324			3 46 52	+25 29.8	1	9.57		0.53		0.06		1	F8	8023
		G 38 - 16			3 46 55	+27 17.4	1	12.28		1.22		1.12		1	K5	333,1620
23949	+23 00565	Pleiades 2484			3 46 56	+24 04.0	5	9.17	.012	0.16	.002	0.17	.035	19	A0	1,830,1029,1245,1783
		Pleiades 9092	⋆	V	3 46 56	+24 35.5	1	16.26		1.48				1		1508
23948	+23 00567	Pleiades 2488			3 46 57	+24 11.9	4	7.54	.008	0.08	.001	0.08	.000	12	A1 Va	1,1029,8023,8030
23950	+21 00535	Pleiades 3325	⋆		3 46 58	+22 05.6	4	6.07	.000	−0.01	.004	−0.32	.004	13	B9 IVp sn	15,1049,1079,8015

HD	DM	Other Id	N Rem	α_{1950}	δ_{1950}	S	V	σ_V	B–V	σ_{B-V}	U–B	σ_{U-B}	n	Spectrum	References
	+22 00574	Pleiades 2506		3 46 59	+23 04.1	2	10.25	.020	0.60	.000	0.08	.000	3	F8 V	1029,8023
24002	+0 00659	G 80 - 24		3 47 00	+01 12.3	2	8.60	.005	0.83	.009	0.42	.000	4	K1 V	1003,3077
23964	+23 00569	Pleiades 2507	⋆AB	3 47 00	+23 41.9	3	6.72	.008	0.06	.000	-0.06	.000	4	B9.5Vp	1029,8023,8030
		Pleiades 9108	⋆V	3 47 00	+25 14.9	1	13.29		1.31				2		1508
23966	+19 00600			3 47 01	+19 24.9	1	8.62		1.03		0.75		2	K0	1648
23990	+8 00574	IDS03444N0906	A	3 47 03	+09 15.4	1	6.76		-0.03		-0.15		2	B9	1733
23860	+51 00789			3 47 03	+51 48.2	1	9.32		0.34				2	A0 V	1119
		Pleiades 2548		3 47 06	+23 58.4	2	14.04	.008	1.35	.008			9	K5.5Ve	352,1508
24121		CpD -41 00396		3 47 06	-40 47.7	1	9.78		0.64		0.13		2	F8/G0 V	1700
		LS I +59 194		3 47 09	+59 52.3	1	12.02		0.66		0.18		3		41
		Hyades HG7 - 21		3 47 10	+09 16.3	1	13.32		0.70				1		1570
		Pleiades 2591	⋆V	3 47 13	+24 08.7	1	13.28		1.23		0.92		1		364
		Pleiades 2588	⋆V	3 47 13	+24 22.9	4	13.25	.045	1.17	.011	0.96	.022	19	K3 V	352,1029,1508,5008
		Hyades HG7 - 20		3 47 14	+08 09.9	1	14.82		1.25				1		950
		Pleiades 2602	⋆V	3 47 14	+23 50.7	2	15.49	.000	1.63	.000			4	M2.5Ve	921,1508
		Pleiades 2601	⋆V	3 47 14	+24 12.1	4	15.01	.033	1.54	.022			11	M3 Ve	257,921,965,1508
		Pleiades 9214	⋆V	3 47 16	+24 04.	2	17.54	.000	1.97	.000			2		921,1508
23819	+60 00754			3 47 16	+61 12.9	1	8.51		1.10		0.80		3	G5	1733
24308	-67 00258	RX Ret		3 47 16	-66 50.8	1	8.65		1.53		1.70		27	G8 II/IIIep	3049
23975	+24 00576	Pleiades 3326		3 47 17	+25 13.8	1	9.64		0.52		0.01		1	G0	8023
		Hyades HG7 - 22		3 47 18	+13 16.	1	14.98		0.83		-0.12		1		832
23985	+25 00624	HR 1188	⋆AB	3 47 18	+25 25.8	4	5.25	.006	0.22	.011	0.09	.008	19	A3 V +A5 V	379,879,1381,3030
23963	+31 00657			3 47 20	+32 16.8	1	9.54		0.16		-0.07		2	A0 V	1041
		Pleiades 2655		3 47 22	+23 25.3	1	15.46		1.29				1	K4	1508
		Pleiades 2644		3 47 22	+24 19.0	3	11.06	.014	0.74	.005	0.26	.010	7		1029,5008,8023
24161	-41 01154			3 47 22	-41 03.0	1	9.31		0.41		-0.06		3	F3 V	1700
24014	+15 00537			3 47 23	+15 51.0	1	8.54		0.41		-0.04		2	F5	1648
		Pleiades 2665		3 47 24	+22 56.8	1	11.36		0.83		0.44		1		1029
24293	-65 00272	IDS03429S6507	B	3 47 24	-64 59.4	3	7.85	.005	0.66	.007	0.14	.010	10	G3 V	158,2033,3077
23895	+54 00717			3 47 25	+54 33.3	1	8.38		0.28				2	A0	70
24450	-74 00275			3 47 25	-74 39.7	4	8.01	.003	0.39	.004	-0.04	.002	11	F3 V	278,861,1075,2012
		Pleiades 9412	⋆V	3 47 26	+22 02.3	1	13.80		1.31				1		1508
24000	+29 00633			3 47 26	+29 35.7	1	8.76		0.34		0.24		2	A0 V	1003
232848	+50 00841			3 47 28	+50 57.8	1	8.79		1.33				2	G8 II	1119
24013	+24 00578	Pleiades 2690		3 47 29	+24 20.7	2	7.41	.005	0.13	.002	0.12	.000	10	A3 Vn	1,1029
232849	+50 00840			3 47 29	+51 12.5	1	9.70		0.26				2	A1 V	1119
232847	+53 00705	IDS03437N5315	AB	3 47 29	+53 23.7	1	8.98		0.73				2	G8 IV	1025
24029	+18 00546			3 47 30	+19 06.2	1	8.54		0.43		0.18		2	F0	1648
24040	+17 00638	G 7 - 6		3 47 31	+17 19.8	2	7.51	.009	0.65	.009	0.24	.004	6	G0	333,1620,8111
279041	+33 00725			3 47 33	+34 17.3	1	10.65		0.31		0.18		2	A0	1041
	+16 00516	Hyades HG7 - 23	⋆V	3 47 34	+17 05.8	1	9.51		0.85				2	K0	950
		Pleiades 2741		3 47 35	+24 21.5	3	12.65	.019	1.01	.003	0.80		5		921,1029,1508
23933	+52 00716			3 47 35	+52 21.3	1	8.66		0.36				3	A5 III	1119
24160	-36 01467	HR 1195		3 47 35	-36 21.0	4	4.17	.005	0.95	.005	0.68	.005	20	G8 III	15,1075,2012,8015
283067		Pleiades 2786		3 47 41	+23 47.0	2	10.30	.010	0.61	.005	0.07	.000	3	G0	1029,8023
24200	-45 01280			3 47 41	-44 55.9	1	9.30		0.49				4	F6 V	2011
24039	+31 00659			3 47 50	+32 01.2	1	9.30		0.45		0.29		2	A3 IV	1041
		L 440 - 36		3 47 50	-39 11.3	2	13.15	.004	0.47	.009	-0.28	.005	14		1698,1765
24038	+32 00669	IDS03447N3216	AB	3 47 52	+32 24.5	1	9.11		0.41		0.27		2	A7 V	1041
24120	-5 00758			3 47 52	-05 13.6	1	7.72		1.23		1.21		3	K0	1657
		Pleiades 2870		3 47 54	+23 10.8	2	12.48	.030	1.04	.030	0.89		4		1029,1508
24076	+23 00570	Pleiades 2866		3 47 54	+23 48.7	3	6.93	.005	0.09	.002	0.03	.000	9	A2 V	1,1029,8023
23945	+53 00706			3 47 54	+53 58.6	1	8.56		0.28				2	A9 II	1119
		GD 51		3 47 54	-13 44.4	1	14.91		0.34		-0.56		1	DA	3060
		Pleiades 2881		3 47 56	+23 41.1	3	11.56	.026	0.96	.010	0.56		5		965,1029,1508
		Pleiades 2880		3 47 56	+24 02.9	1	11.75		0.86		0.37		2		1029
24333	-61 00282			3 47 59	-61 29.6	3	10.09	.019	0.53	.007	-0.16	.000	7	F3/5 IV/V	1696,2013,3077
24512	-74 00276	HR 1208		3 47 59	-74 23.6	5	3.23	.013	1.61	.011	1.97	.032	25	M1 III	15,1020,1034,1075,2038
237187	+55 00830	IDS03441N5543	AB	3 48 00	+55 52.2	1	9.05		0.65				2	A0	70
		Pleiades 2908	⋆V	3 48 02	+24 54.3	2	13.43	.020	1.14	.010	0.92		5	K3 Ve	257,1508
275907	+36 00763			3 48 02	+37 03.4	1	9.09		0.28		0.05		2	A5	1041
24234	-43 01194			3 48 02	-43 14.0	2	9.05	.013	1.14	.001	1.10		10	K2 (III)	863,2011
283066	+23 00571	Hyades vB 170		3 48 04	+23 45.2	4	10.23	.012	1.16	.011	1.10	.047	7	K7	582,950,1058,1127
24011	+46 00809			3 48 04	+46 57.0	1	8.03		0.25		0.19		2	A2	1601
24133	+1 00667			3 48 06	+01 24.8	1	6.57		0.42		-0.06		1	F0	1776
		Pleiades 2927	⋆V	3 48 06	+24 35.2	4	13.88	.032	1.26	.017	1.05		14	K4 Ve	352,364,1029,1508
24087	+24 00580	Pleiades 3327		3 48 07	+25 15.5	1	9.45		0.54		0.09		1	F8	8023
		Pleiades 2940	⋆V	3 48 08	+24 19.9	1	13.98		1.32				1	M0	1508
		LP 833 - 56		3 48 08	-25 29.9	1	12.01		0.72		0.21		2		1696
24249	-43 01196			3 48 08	-42 52.7	3	7.35	.001	0.17	.004	0.13	.012	10	A5/6 V	278,1075,2011
24156	-5 00762	G 160 - 27		3 48 12	-04 46.2	2	8.15	.005	0.69	.015	0.09		6	G0	2012,3056
		Pleiades 2966		3 48 13	+23 47.0	2	14.88	.020	1.50	.000			6	M1.5Ve	921,1508
		MCC 431		3 48 17	+44 48.0	1	11.51		1.19		1.09		5	M0	1723
		Pleiades 2984	⋆V	3 48 18	+23 40.7	2	12.40	.019	1.00	.000	0.85		2		1029,1508
		G 160 - 28		3 48 18	-06 13.9	2	12.79	.000	1.70	.009	1.40		4		538,3078
24248	-32 01487			3 48 18	-32 26.2	1	6.92		1.54				4	K5 III	2012
	+53 00707			3 48 19	+53 47.3	1	9.86		0.54				2		70
279112	+33 00726			3 48 20	+34 06.8	1	10.01		0.25		0.12		2	A2	1041
		LP 248 - 23		3 48 20	+36 06.7	1	11.71		1.36		1.21		4		1723
		LS I +56 091		3 48 21	+56 52.5	1	11.61		0.46		-0.28		3		41

Table 1   193

HD	DM	Other Id	N Rem	α_1950	δ_1950	S	V	σ_V	B–V	σ_B-V	U–B	σ_U-B	n	Spectrum	References
279096	+34 00748			3 48 22	+34 59.8	1	9.73		0.13		0.02		2	A0	1041
24118	+24 00583	Pleiades 3010		3 48 25	+25 00.8	1	6.79		0.16		0.14		2	A2	401
279101	+34 00749			3 48 25	+34 43.3	1	10.07		0.12		0.03		2	A0	1041
		Pleiades 3019	⋆ V	3 48 26	+23 56.3	2	13.52	.030	1.21	.007	0.97		5		1029,1508
24281	−43 01198			3 48 26	−43 41.1	1	8.29		1.69		1.80		3	C4,2	864
24291	−45 01286			3 48 26	−45 32.1	5	6.93	.002	0.95	.003	0.70	.003	30	K0 III	278,863,977,1075,2011
		Pleiades 3030	⋆ V	3 48 27	+23 44.4	3	14.03	.030	1.37	.017	1.18	.008	6	dK7	257,1029,1508
279095	+34 00750			3 48 27	+35 13.0	1	10.28		0.44		0.22		2	A0	1041
24132	+24 00584	Pleiades 3031		3 48 28	+24 22.2	4	8.84	.024	0.38	.002	0.05	.000	8	F2 V	1,1029,8023,8030
24155	+12 00516	HR 1194, V766 Tau		3 48 29	+12 53.8	3	6.31		-0.06	.012	-0.48	.013	10	B9p Si	1049,1079,8100
		Pleiades 3034		3 48 29	+24 24.9	1	11.76		0.46		0.09		3		1029
		Pleiades 3063	⋆ V	3 48 31	+23 45.0	4	13.53	.043	1.17	.005	0.92	.056	10		257,921,1029,1508
		Pleiades 3069		3 48 33	+23 15.0	1	13.85		1.12		0.96		1	K2 V	1029
		Pleiades 3065	⋆ V	3 48 33	+24 28.1	1	12.97		1.39				2		364
279111	+34 00753			3 48 39	+34 19.9	1	9.19		0.13		-0.09		2	A0	1041
24154	+21 00539	HR 1193		3 48 40	+21 53.0	2	6.70	.005	1.12	.005	0.95	.005	4	K0 III	1648,1733
		Pleiades 3104	⋆ V	3 48 40	+23 02.0	2	13.46	.024	1.27	.008	1.10		4		1029,1508
		Pleiades 3096	⋆ V	3 48 40	+24 24.0	2	12.12	.030	0.96	.035	0.66		3		1029,1508
		Pleiades 3097		3 48 41	+24 50.1	2	10.95	.015	0.74	.000	0.25	.000	3		1029,8023
279119	+33 00729			3 48 42	+33 45.9	1	10.35		0.34		0.19		2	A2	1041
24131	+33 00728	HR 1191	⋆ A	3 48 42	+34 12.6	6	5.77	.011	0.00	.012	-0.80	.015	17	B1 V	15,154,1009,1012,1203,1252
24502	−70 00259			3 48 42	−70 30.6	1	7.33		1.08				4	K0 III	2012
232850	+50 00848			3 48 44	+51 13.5	1	9.31		0.51				3	F0 V	1119
24305	−36 01476	HR 1200		3 48 45	−36 34.5	2	6.85	.005	-0.04	.000			7	B9.5V	15,2012
23982	+63 00458			3 48 46	+63 20.1	1	8.05		0.27		-0.32		3	B5	1212
		Pleiades 3122		3 48 47	+25 03.2	1	11.92		0.73		0.19		1		1029
		GD 52		3 48 48	+33 58.6	1	15.20		0.10		-0.63		1		3060
		CS 22176 # 36		3 48 51	−10 06.3	1	14.52		0.16		-0.69		1		1736
24331	−42 01269			3 48 53	−42 43.4	4	8.61	.002	0.92	.003	0.64	.004	18	K3 V	977,1075,2011,3077
		Pleiades 3163	V	3 48 54	+24 14.3	2	12.75	.032	1.00	.012	0.64		4		1029,1508
24178	+25 00631	Pleiades 3328		3 48 56	+25 51.0	1	7.64		0.17		0.13		1	A0	8023
24167	+30 00582	HR 1197		3 48 57	+31 01.2	1	6.25		0.20		0.14		2	A5 V	252
24194	+23 00573	Pleiades 3179		3 48 58	+23 46.2	2	10.05	.005	0.57	.005	0.06	.005	3	G0 V	1029,8023
		Pleiades 9201	⋆ V	3 48 58	+23 54.	2	16.92	.000					2		921,1508
		Pleiades 3187		3 48 59	+23 11.4	3	13.18	.008	1.20	.008	1.04		50	K4.5Ve	352,1029,1508
24094	+53 00708	LS V +53 011		3 48 59	+53 20.1	1	8.30		0.40				2	B1 III	1119
		Pleiades 3197	⋆ V	3 49 03	+24 30.9	3	12.10	.062	1.08	.049	0.79	.078	6	K3 V	364,1029,1508
24636	−75 00251			3 49 03	−74 50.7	3	7.12	.001	0.40	.004	-0.06	.012	10	F3 IV/V	278,1075,2012
279128	+32 00674			3 49 05	+33 15.5	1	8.91		0.49		0.33		2	B8	1041
		LP 357 - 27		3 49 06	+24 47.2	1	16.52		-0.20		-1.10		2	DAwk	3028
24190	+33 00730			3 49 07	+34 04.4	5	7.44	.012	0.04	.008	-0.62	.004	20	B2 V	1009,1012,1041,1252,1775
24206	+22 00583	G 6 - 38		3 49 08	+22 31.6	3	7.57	.005	0.68	.004	0.22	.004	8	G0	333,1355,1620,3026
279075	+35 00756			3 49 09	+36 03.4	1	10.54		0.22		0.13		2	A0	1041
		Hyades HG7 - 24		3 49 12	+14 31.	1	12.48		1.15				1		950
279110	+34 00755	V499 Per		3 49 12	+34 18.9	1	9.58		0.07		-0.22		2	B8	1041
24129	+50 00849			3 49 15	+50 54.1	1	7.83		0.13				3	A0 III	1119,8100
		LP 773 - 49		3 49 15	−17 22.3	1	11.62		0.86		0.57		2		1696
24142	+51 00793			3 49 16	+52 06.4	1	8.34		1.12				2	G8 III	1119
24339	−26 01453	IDS03472S2614	A	3 49 18	−26 04.8	4	9.44	.006	0.55	.009	-0.07	.009	12	F7 V	1311,1696,2033,3077
		G 160 - 30		3 49 19	−05 41.8	1	15.92		0.93		0.20		2		3062
24263	+6 00594	HR 1199	⋆ AB	3 49 20	+06 23.2	4	5.66	.007	0.06	.007	-0.44	.010	11	B5 V	15,1079,1256,3016
		Pleiades 3241		3 49 20	+24 35.2	1	13.78		0.57		0.06		3		1029
24392	−45 01291			3 49 20	−44 46.3	1	9.12		0.98				4	G8 III	2011
		Pleiades 9115	⋆ V	3 49 21	+24 25.0	1	15.44		1.49				2		1508
24289	−4 00680	G 80 - 28		3 49 22	−03 58.5	3	9.96	.000	0.52	.004	-0.15	.008	4	G0	927,1620,1696
24406	−44 01303			3 49 25	−44 30.9	3	8.00	.002	0.39	.000	0.06	.017	10	F2 III	278,1075,2011
279133	+32 00676			3 49 26	+32 58.7	1	9.69		0.19		0.12		2	A0	1041
279078	+35 00757			3 49 26	+35 39.0	1	10.13		0.32		0.21		2	A1	1041
24278	+18 00550			3 49 30	+18 27.0	1	7.94		0.21		0.15		2	A0	1648
24228	+31 00662			3 49 30	+32 15.6	1	6.74		0.91		0.58		5	K0 III	1501
24501	−60 00271			3 49 36	−60 34.7	1	7.75		1.65		1.94		7	M2 III	1704
24242	+34 00758			3 49 37	+34 19.6	1	9.49		0.12		0.08		2	A2 V	1041
279093	+34 00757			3 49 37	+35 08.3	1	10.08		0.15		0.07		2	B9	1041
279067	+36 00769			3 49 38	+36 42.4	1	9.78		0.24		0.14		2	A0	1041
283044		Hyades L 7		3 49 39	+25 39.4	1	11.15		1.30				2	G5	582
24141	+57 00752	HR 1192		3 49 39	+57 49.7	3	5.79	.008	0.18	.006	0.10	.012	13	A5 m	595,1501,8071
		Steph 420		3 49 40	+25 39.4	1	11.14		1.31		1.22		4	K7	1746
24416	−40 01109			3 49 40	−40 14.1	1	7.66		1.53		1.46		2	M4 III	1700
		Hyades HG7 - 25		3 49 42	+17 54.	1	14.24		1.23				1		950
		L 91 - 15		3 49 42	−65 26.	1	10.72		1.05		0.94		1		1696
283036	+25 00636			3 49 45	+26 12.0	1	10.47		0.63		0.14		3	A2	1726
24390	−29 01456			3 49 48	−28 59.1	1	8.38		0.58				4	F8 V	2012
		Hyades HG7 - 26		3 49 49	+11 06.8	1	13.73		1.54				2	M1	950
24189	+52 00720			3 49 49	+52 25.0	1	8.35		0.54				2	F6 V	1025
24116	+62 00619	LS I +63 200		3 49 49	+63 10.2	2	8.54	.000	0.22	.010	-0.23	.010	7	B7 II	41,379,8100
		POSS 249 # 5		3 49 50	+34 30.9	1	16.45		1.65				2		1739
24302	+24 00589	Pleiades 3329		3 49 54	+24 34.1	1	9.45		0.48		-0.02		1	F8	8023
24203	+52 00722	IDS03461N5259	AB	3 49 54	+53 07.6	1	8.13		0.61				2	G2 V +G5 V	1025
		G 80 - 30		3 49 54	−03 08.0	1	11.70		1.21		1.20		1	K5	1696
		POSS 249 # 6		3 49 57	+35 43.1	1	17.90		1.58				2		1739

HD	DM	Other Id	N Rem	α_{1950}	δ_{1950}	S	V	σ_V	B–V	σ_{B-V}	U–B	σ_{U-B}	n	Spectrum	References
24227	+50 00851			3 49 59	+50 47.9	2	8.48	.020	0.11	.000			5	B9 V	70,1119
	−0 00612			3 49 59	−00 11.2	2	10.37	.005	0.57	.005	0.04	.015	5	F9	271,281
		Hyades HG7 - 28		3 50 00	+07 39.	1	11.70		0.88				1		920
		G 7 - 10		3 50 00	+10 58.2	1	12.42		0.78		0.22		2	G8	333,1620
24240	+48 01015	HR 1198		3 50 01	+48 30.2	1	5.76		1.05		0.95		2	K0 III	252
232852	+52 00723			3 50 04	+52 24.1	1	9.42		0.73				2	F8 V	1025
279098	+34 00759			3 50 05	+34 45.7	1	10.16		0.36		0.10		2	F0	1041
		CS 22190 # 7		3 50 05	−16 33.4	1	14.20		0.46		−0.06		1		1736
	+0 00669			3 50 06	+00 21.9	1	11.68		0.84		0.48		6		281
		HA 95 # 301		3 50 06	+00 22.4	5	11.22	.003	1.29	.003	1.30	.002	73	G5	281,989,1729,1764,5006
		HA 95 # 15		3 50 06	−00 14.3	3	11.31	.007	0.70	.018	0.15	.012	9	G0	271,281,1764
		HA 95 # 16		3 50 07	−00 14.0	2	14.26	.045	1.34	.026	1.32		6		281,1764
		HA 95 # 302		3 50 08	+00 22.4	1	11.69		0.82		0.45		20		1764
24300	+35 00763			3 50 08	+35 24.3	1	8.68		0.06		−0.39		2	B6 V	1041
24500	−47 01178			3 50 09	−47 28.0	4	6.70	.002	0.28	.001	0.11	.009	18	A8 III/IV	278,977,1075,2011
		LS V +53 012		3 50 10	+53 04.1	1	11.09		0.75		−0.14		2	B1 V	1012
		G 175 - 10		3 50 10	+56 40.3	1	14.41		1.66		1.33		1		906
279079	+35 00764			3 50 11	+35 39.1	1	9.45		0.11		−0.23		2	A0	1041
	+54 00718	LS I +55 051		3 50 11	+55 16.2	1	11.03		0.74		−0.33		3	B5	41
		HA 95 # 19		3 50 11	−00 15.9	2	11.75	.025	0.63	.015	0.08	.005	10	G2	271,281
279107	+34 00760			3 50 12	+34 29.1	1	10.42		0.45		0.19		2	A8	1041
		L 301 - 33		3 50 12	−46 05.	1	14.36		1.22		0.86		1		3060
24388	−5 00769	HR 1202	★ AB	3 50 13	−05 30.6	4	5.47	.009	−0.11	.006	−0.41	.009	15	B8 V	15,1061,1079,2027
237190	+58 00663	RW Cam		3 50 15	+58 30.4	3	8.21	.020	1.21	.039	0.85	.079	3	K2	934,1399,6011
24499	−46 01206			3 50 16	−46 18.7	1	7.86		0.40				4	F5 III	2012
		G 38 - 18		3 50 17	+34 57.2	1	11.89		0.96		0.67		1		333,1620
		Hyades HG7 - 29		3 50 18	+16 51.	1	12.55		1.25				1		950
24357	+16 00523	Hyades vB 6	★	3 50 18	+17 10.8	5	5.97	.013	0.34	.008	0.00	.003	18	F2 V	15,39,254,1127,8015
		Pleiades 9369	★ V	3 50 18	+22 43.4	1	16.83		1.67				1		1508
283048				3 50 18	+25 36.	1	10.24		0.21				3		6009
24322	+32 00679			3 50 19	+32 57.8	1	8.25		0.28		0.19		2	A5 III	1041
24321	+34 00761	IDS03471N3446	AB	3 50 19	+34 54.7	1	8.79		0.09		−0.31		2	A0	1041
24401	−0 00613			3 50 20	−00 08.6	8	10.01	.006	0.15	.005	0.07	.006	129	A0	271,281,975,989,1729,1764*
24369	+22 00588			3 50 23	+22 58.6	1	8.29		1.54		1.61		2	K2	1648
		HA 95 # 97		3 50 23	−00 09.2	1	14.82		0.91		0.38		1		1764
		LS V +46 003		3 50 24	+46 45.0	1	10.17		0.39		−0.48		2	B2:IV:nne	1012
24566	−56 00587			3 50 24	−56 26.6	1	7.46		0.92				4	G8 III	2033
		G 80 - 31		3 50 26	+00 51.8	1	10.94		0.72		0.19		1	G0	333,1620
		HA 95 # 98		3 50 26	−00 06.0	1	14.45		1.18		1.09		1		1764
		HA 95 # 100		3 50 27	−00 08.6	1	15.63		0.79		0.05		1		1764
		G 160 - 32		3 50 28	−09 47.7	1	15.82		1.09		0.50		2		3062
		HA 95 # 249		3 50 30	+00 18.1	1	11.53		0.60		0.04		4	F7	281
		HA 95 # 101		3 50 30	−00 06.0	1	12.68		0.78		0.26		1		1764
24367	+25 00640			3 50 31	+25 35.0	2	7.85	.010	0.23	.010	0.18	.005	5	A2	833,1625
24275	+51 00796		AB	3 50 31	+51 24.1	2	8.51	.010	0.25	.005			5	A2 V	70,1119
24368	+25 00641	Pleiades 3330		3 50 33	+25 32.2	2	6.35	.010	0.12	.000	0.16	.005	3	A2 V	105,8023
24365	+27 00589			3 50 33	+28 00.1	1	7.84		0.84				3	G8 V	20
		LS I +56 092		3 50 33	+56 42.0	1	10.26		0.69		−0.09		2		41
		HA 95 # 102		3 50 33	−00 07.7	1	15.62		1.00		0.16		1		1764
		HA 95 # 252		3 50 37	+00 18.5	1	15.39		1.45		1.18		3		1764
24435	−2 00745	IDS03481S0208	A	3 50 37	−01 59.1	1	8.96		0.44		−0.02		2	F8	3016
24435	−2 00745	IDS03481S0208	BC	3 50 37	−01 59.1	1	10.03		0.85		0.41		2	F8	3016
		HA 95 # 190		3 50 39	+00 07.5	1	12.63		0.29		0.24		22		1764
		POSS 249 # 4		3 50 41	+35 32.1	1	15.96		1.75				1		1739
24446	−7 00695			3 50 41	−06 46.9	1	6.57		−0.10				4	B8 V:nn	2035
24238	+60 00762	G 246 - 53		3 50 44	+61 01.4	3	7.86	.011	0.83	.008	0.46	.014	9	K0 V	22,1355,3072
		HA 95 # 193		3 50 46	+00 07.7	1	14.34		1.21		1.24		10		1764
		IRC +10 050	V	3 50 46	+11 15.7	1	13.39		3.64		−0.04		1		8022
		HA 95 # 105		3 50 47	−00 09.2	1	13.57		0.98		0.63		1		1764
		Steph 425		3 50 48	−05 26.5	1	11.85		1.46		1.11		1	K7	1746
		HA 95 # 374		3 50 51	+00 35.0	1	12.37		1.06		0.73		2	K5	281
		HA 95 # 107		3 50 51	−00 06.5	1	16.27		1.32		1.12		1		1764
		HA 95 # 106		3 50 51	−00 07.5	1	15.14		1.25		0.37		1		1764
		HA 95 # 313		3 50 52	+00 27.7	1	11.44		1.47		1.43		6	K5	281
24456	+1 00673			3 50 54	+01 58.3	1	6.72		−0.04		−0.19		1	B9	1776
		Hyades HG7 - 32		3 50 54	+07 39.	1	12.70		1.49				1		950
		LS I +57 135		3 50 54	+57 48.2	1	12.10		0.48		−0.38		2		41
		G 160 - 33		3 50 54	−15 35.0	2	11.31	.015	0.61	.000	−0.04	.007	5		1696,3056
	+54 00720	LS I +55 052		3 50 58	+55 03.7	1	10.15		0.47		0.32		2	A0 II	41
24497	−18 00691	HR 1206		3 50 58	−18 34.9	2	6.21	.005	0.88	.000			7	K0/1III+A2V	15,2012
		Pleiades 9061	★ V	3 50 59	+23 11.9	1	14.26		1.34				1		1508
24398	+31 00666	HR 1203	★ A	3 50 59	+31 44.2	6	2.85	.020	0.11	.011	−0.77	.016	14	B1 Ib	150,542,8003,8007*
24398	+31 00666	HR 1203	★ AB	3 50 59	+31 44.2	11	2.85	.015	0.12	.011	−0.76	.015	25	B1 Ib+B8 IV	1,15,150,154,542,1004*
24398	+31 00666	IDS03478N3135	B	3 50 59	+31 44.2	1	9.16		0.23		−0.07		2	B8 IV	150
24398	+31 00666	IDS03478N3135	D	3 50 59	+31 44.2	1	10.35		0.71		0.20		3		450
24398	+31 00666	IDS03478N3135	E	3 50 59	+31 44.2	1	9.90		0.33		0.17		3	A2 V	450
24442	+14 00624			3 51 00	+15 02.2	1	8.52		1.19		1.04		2	G5	1648
237193	+57 00755			3 51 00	+58 03.8	1	9.19		0.96		0.77		1	G8 V	934
24351	+49 01058			3 51 02	+50 12.0	1	8.81		0.08				3	A0	70
24577	−45 01309			3 51 02	−45 43.1	3	8.69	.003	1.08	.002	0.92	.004	18	K2 III	863,977,2011

Table 1   195

HD	DM	Other Id	Rem	α_{1950}	δ_{1950}	S	V	σ_V	B–V	σ_{B-V}	U–B	σ_{U-B}	n	Spectrum	References
24341	+51 00798			3 51 03	+52 16.5	4	7.85	.034	0.68	.009	0.12	.010	7	G1 V	70,908,1025,3026
24576	−45 01306			3 51 03	−45 05.7	4	7.59	.004	1.30	.005	1.42	.007	27	K3 III	743,977,1075,2011
24164	+71 00222	HR 1196		3 51 04	+71 40.6	2	6.32	.020	0.30	.000	0.00		5	A5 m	985,1733
		HA 95 # 112		3 51 06	−00 10.0	1	15.50		0.66		0.08		1		1764
		HA 95 # 41		3 51 07	−00 11.3	1	14.06		0.90		0.30		1		1764
24411	+32 00683			3 51 09	+32 47.8	1	8.94		0.32		0.19		2	A5 IV	1041
		HA 95 # 317		3 51 10	+00 21.0	1	13.45		1.32		1.12		11		1764
		HA 95 # 42		3 51 10	−00 13.4	1	15.61		-0.22		-1.11		18		1764
		HA 95 # 263		3 51 13	+00 17.8	1	12.68		1.50		1.56		10		1764
		WLS 420 15 # 6		3 51 14	+15 20.3	1	10.53		1.32		1.08		2		1375
		HA 95 # 115		3 51 14	−00 09.6	1	14.68		0.84		0.10		1		1764
		HA 95 # 43		3 51 15	−00 11.8	1	10.80		0.51		-0.02		10		1764
	−0 00615			3 51 15	−00 11.8	1	10.80		0.54		0.06		2	F3	271
281295	+31 00667			3 51 16	+31 47.9	1	10.17		0.37		0.20		2	A0	1041
		HA 95 # 116		3 51 16	−00 06.1	1	11.78		0.60		0.02		6	F9	271
22701	+86 00051	HR 1107		3 51 17	+86 29.3	2	5.86	.005	0.37	.000	0.00		5	F5 IV	985,1733
		LS I +55 053		3 51 20	+55 21.5	1	11.83		0.86		-0.02		2		41
24376	+51 00799			3 51 22	+52 14.9	2	9.28	.019	0.15	.000			6	B9 V	70,1119
25887	−85 00044	HR 1271	⋆ AB	3 51 23	−85 24.9	3	6.40	.003	0.00	.004	-0.10	.000	21	B9.5IV	15,1075,2038
	−37 01501			3 51 30	−37 11.9	4	12.13	.016	1.45	.023	1.10	.095	12		158,912,1705,3078
		L 301 - 39		3 51 30	−46 29.	1	12.30		0.96		0.78		1		1696
24520	+1 00676			3 51 31	+02 02.2	1	8.57		0.14		0.05		1	B9	1776
24940	−76 00251			3 51 31	−76 02.6	2	7.32	.034	1.58	.018	1.92		6	M0 III	278,2012
		G 7 - 11		3 51 33	+08 24.7	1	11.94		1.46		1.22		1	M1	333,1620
24386	+53 00712			3 51 33	+53 24.7	1	8.45		0.50				2	F8 V	1025
	+0 00673			3 51 34	+00 29.9	3	10.41	.010	0.68	.005	0.17	.000	5	F8	271,281,1776
24587	−24 01945	HR 1213, τ Eri		3 51 35	−24 45.6	4	4.64	.010	-0.14	.007	-0.48	.008	14	B5 V	15,1075,2012,8015
24496	+16 00527	IDS03488N1620		3 51 37	+16 28.3	1	6.81		0.73		0.30		3	G0	1648
	+58 00667			3 51 39	+58 20.5	1	9.90		0.51		0.34		10		1655
24538	−0 00616			3 51 40	−00 17.2	5	9.57	.006	0.53	.005	0.07	.010	51	F8	271,281,989,1729,6004
24537	+0 00617			3 51 41	+00 08.6	6	8.74	.004	0.50	.003	0.02	.012	57	F8	271,281,989,1509,1729*
		HA 95 # 271		3 51 43	+00 10.0	1	13.67		1.29		0.92		7		1764
232857	+52 00724			3 51 44	+52 38.7	1	9.44		0.50				2	F5 V	1025
		HA 95 # 328		3 51 45	+00 27.7	1	13.52		1.53		1.30		11		1764
24519	+16 00528			3 51 45	+17 11.9	1	8.33		0.15		0.04		2	B9	1375
24395	+56 00857	LS I +56 093		3 51 45	+56 46.4	2	6.92	.000	0.28	.000	0.18	.010	15	A7 II	41,1733
24626	−35 01455	HR 1214		3 51 45	−34 52.8	2	5.10	.005	-0.14	.005			7	B6 V	15,2012
	−74 00277			3 51 45	−74 44.6	1	10.30		0.35				4	A5	2012
24432	+48 01019	LS V +48 008		3 51 46	+48 53.4	2	6.93	.075	0.58	.019	-0.18	.009	4	B3 II	1012,8100
24554	−3 00631	HR 1211	⋆ B	3 51 47	−03 06.0	2	6.15	.005	0.07	.015	0.06	.005	5	A2 V	150,542
24555	−3 00631	HR 1212	⋆ A	3 51 47	−03 06.1	2	4.76	.025	0.93	.010	0.66	.025	4	G8 III	150,542
24555	−3 00631	IDS03493S0315	AB	3 51 47	−03 06.1	7	4.46	.018	0.68	.006	0.41	.010	19	G8 III	15,150,542,1075,2030*
24550	+4 00601	V479 Tau	⋆ A	3 51 48	+05 01.7	2	7.42	.009	0.38	.005	0.15	.018	5	F3 II-III	258,3024
24550	+4 00600	IDS03492N0454	B	3 51 48	+05 01.7	1	8.92		0.32		0.08		4		3024
24616	−23 01619			3 51 49	−23 16.7	3	6.70	.008	0.82	.005	0.27		7	G8 IV/V	1075,2017,3077
		HA 95 # 329		3 51 50	+00 28.3	1	14.62		1.18		1.09		6		1764
24431	+52 00726	LS V +52 013		3 51 50	+52 29.7	6	6.74	.019	0.37	.014	-0.61	.000	12	O9 IV-V	1011,1012,1077,1118*
279183	+33 00739			3 51 51	+34 16.2	1	10.21		0.22		0.10		2	A3	1041
24421	+51 00803			3 51 51	+52 04.8	1	6.79		0.51				2	F5	70
24505	+27 00597			3 51 56	+28 02.6	1	7.97		0.69				3	G5 III	20
		HA 95 # 330		3 51 57	+00 20.3	1	12.17		2.00		2.23		23		1764
232859	+51 00804	IDS03482N5205	AB	3 51 58	+52 15.4	1	9.30		0.20				2	A0	1119
24706	−47 01187	HR 1216		3 52 00	−47 02.4	10	5.93	.005	1.24	.007	1.35	.008	54	K2 III	14,15,182,278,388,977*
		Steph 430		3 52 01	−09 18.2	1	11.22		1.53		1.14		2	M3	1746
		Hyades HG7 - 33		3 52 03	+16 10.2	1	14.25		1.58				2	M2	950
		LS I +61 325		3 52 05	+61 32.6	1	11.81		0.55		-0.23		3		41
279171	+34 00766	IDS03489N3450	AB	3 52 06	+34 59.1	1	9.54		0.24		0.07		2	A0	1041
		LP 889 - 50		3 52 06	−30 03.5	1	12.98		0.90		0.68		1		1696
		KUV 03521+0150		3 52 08	+01 50.0	1	15.63		-0.15		-0.94		1	DA	1708
24696	−43 01221			3 52 08	−43 18.6	4	7.92	.003	1.10	.002	1.00	.008	18	K0 III	861,977,1075,2011
24430	+57 00757			3 52 09	+57 46.3	1	8.10		1.18		0.87		19	G5 III	230
	−7 00699	G 160 - 35		3 52 09	−06 58.8	5	9.03	.017	1.37	.022	1.27	.020	15	M0 V	158,265,1197,1705,3072
		HA 95 # 275		3 52 10	+00 18.6	1	13.48		1.76		1.74		20		1764
24757	−55 00577			3 52 10	−54 59.8	1	7.75		-0.13		-0.49		1	B5 III/V	55
24572	+12 00523			3 52 11	+12 45.6	1	9.45		0.27		0.16		2	A5	1375
	+60 00767	LS I +60 308		3 52 11	+60 35.5	1	10.14		0.19		-0.50		3		41
		HA 95 # 276		3 52 12	+00 17.1	1	14.12		1.23		1.22		7		1764
		RL 75		3 52 12	+46 34.	1	13.74		0.15		-0.77		2		704
		RL 76		3 52 12	+52 07.	1	12.67		0.35		0.08		2		704
286363	+12 00524	Hyades L 8		3 52 15	+12 20.4	1	10.12		1.07				2	M0	582
285252	+16 00529	Hyades vB 7		3 52 15	+16 51.2	3	8.99	.004	0.90	.004	0.63	.005	6	K2	950,1127,8023
24534	+30 00591	HR 1209, X Per	⋆ A	3 52 15	+30 54.0	10	6.34	.194	0.25	.070	-0.81	.020	29	O9.5 ep	15,154,571,1009,1011,1012*
24534	+30 00591	IDS03491N3045	B	3 52 15	+30 54.0	1	11.67		1.36		-1.12		1		654
		HA 95 # 60		3 52 15	−00 25.8	1	13.43		0.78		0.20		10		1764
		HA 95 # 218		3 52 16	+00 01.4	3	12.09	.011	0.71	.001	0.21	.006	29	G0	271,281,1764
	+42 00849	IDS03488N4303	AB	3 52 16	+43 12.1	1	10.15		1.26		1.18		2		1726
		HA 95 # 132		3 52 17	−00 03.4	5	12.06	.009	0.45	.004	0.30	.004	65	G0	271,281,989,1729,1764
24622	−0 00618			3 52 20	−00 08.1	5	8.61	.003	0.45	.006	0.04	.006	27	F3 IV	271,281,1509,5006,6004
24504	+47 00912	HR 1207		3 52 21	+47 43.6	2	5.36	.024	-0.08	.005	-0.48		4	B6 V	154,1119
		Hyades HG7 - 37		3 52 24	+15 57.	1	14.92		1.56				1		950

HD	DM	Other Id	N	Rem	α_{1950}	δ_{1950}	S	V	σ_V	B–V	σ_{B-V}	U–B	σ_{U-B}	n	Spectrum	References
		Hyades HG7 - 38			3 52 24	+19 18.	1	12.08		0.78				1		950
		LP 357 - 271			3 52 24	+26 03.4	1	13.63		0.76				1		1674
24685	−25 01637				3 52 24	−25 02.2	1	8.21		0.64				4	G3 V	2012
		HA 95 # 62			3 52 26	−00 11.6	1	13.54		1.36		1.18		11		1764
	−76 00253				3 52 27	−76 32.1	1	9.67		1.33				4	K0	2012
		Hyades HG7 - 39			3 52 30	+15 44.	1	14.71		1.41				1		950
		HA 95 # 137			3 52 30	−00 05.2	1	14.44		1.46		1.14		1		1764
24503	+52 00728				3 52 31	+52 23.4	2	8.90	.004	0.28	.004			6	A0 V	70,1119
		HA 95 # 139			3 52 31	−00 05.5	1	12.20		0.92		0.68		2		1764
		HA 95 # 66			3 52 33	−00 18.3	2	12.87	.009	0.70	.006	0.11	.031	9	G2	281,1764
		HA 95 # 227			3 52 34	+00 05.8	1	15.78		0.77		0.03		7		1764
		HA 95 # 142			3 52 35	−00 07.4	1	12.93		0.59		0.10		11		1764
		Pleiades 9099		⋆ V	3 52 36	+22 39.8	1	14.63		1.50				1		1508
		LS I +55 054			3 52 36	+55 38.5	1	11.77		0.58		−0.36		3		41
		G 95 - 60			3 52 37	+41 50.8	1	13.86		0.77		0.06		2		1658
		Hyades HG7 - 41			3 52 38	+09 38.6	3	14.48	.009	0.14	.024	−0.67		6		1281,1570,1674
24744	−40 01128	HR 1219		⋆ AB	3 52 38	−40 30.2	3	5.70	.010	0.60	.005			14	K0 III+A3 V	15,1075,2027
		LS I +57 136			3 52 39	+57 06.9	1	10.88		0.36		−0.55		2		41
24583	+34 00767				3 52 40	+34 24.0	1	8.97		0.07		−0.26		2	B7 V	1041
24693	−15 00687				3 52 40	−15 03.1	1	7.06		1.71		2.00		3	M1 III	1657
		HA 95 # 68			3 52 42	−00 16.9	1	12.79		0.76		0.14		2	G3	281
279195	+33 00742				3 52 44	+33 34.4	1	10.34		0.40		0.07		2	F2	1041
279187	+33 00741				3 52 44	+34 08.0	1	9.94		0.16		0.02		2	A2	1041
237198	+58 00670				3 52 44	+58 19.8	1	9.13		1.22		1.00		1	G5 III	934
24601	+31 00669	IDS03496N3152		AB	3 52 45	+32 00.6	1	8.65		0.21		−0.16		2	A0	1041
24600	+31 00670	IDS03496N3152		C	3 52 49	+32 01.1	1	8.92		0.24		−0.05		2	B8 V	1041
24560	+44 00816	LS V +44 005			3 52 50	+44 47.6	1	8.14		0.22		−0.55		3	B1.5Vne	1212
24480	+60 00768	HR 1205		⋆ AB	3 52 51	+60 57.9	3	4.99	.008	1.44	.016	1.22	.040	6	K3 I-II	1355,1363,3016
24546	+50 00860	HR 1210		⋆ AP	3 52 53	+50 33.2	6	5.28	.009	0.40	.010	−0.01	.014	10	F5 IV	15,70,254,1008,1025,3026
24656	+9 00512				3 52 54	+09 47.0	1	8.54		0.46				1	F5	1674
279184	+33 00743	IDS03497N3406		AB	3 52 55	+34 14.7	1	9.54		0.43		0.14		2	A7	1041
		G 175 - 11			3 52 55	+53 25.2	1	10.86		1.43		1.10		2		3016
24712	−12 00752	HR 1217, DO Eri			3 52 55	−12 14.6	2	5.99	.005	0.33	.005			7	Ap SrEu(Cr)	15,2012
		HA 95 # 338			3 52 56	+00 25.4	1	12.33		0.65		0.08		2	G3	281
24755	−36 01508				3 52 56	−35 51.1	1	7.69		0.45				4	F6 V	2012
		HA 95 # 74			3 52 57	−00 17.9	4	11.53	.001	1.13	.002	0.70	.011	62	G5	281,989,1729,1764
		HA 95 # 73			3 52 57	−00 19.2	1	12.31		0.71		0.18		8	G5	281
24479	+62 00628	HR 1204			3 52 59	+62 55.7	5	4.97	.047	−0.08	.020	−0.17	.009	9	B9.5Ve	15,1079,1363,3023,8015
		Hyades HG7 - 42			3 53 01	+10 58.4	1	11.60		0.74				2		950
24703	−1 00559				3 53 01	−01 13.6	1	9.49		0.30		0.11		2	A5	1776
24665	+22 00599	Pleiades 3331			3 53 03	+22 58.9	1	9.09		0.55		0.12		1	F5	8023
		HA 95 # 146			3 53 04	−00 08.7	1	11.26		1.24		0.99		4	G8	281
24805	−46 01226				3 53 04	−46 33.7	3	6.90	.003	0.16	.002	0.10	.000	10	A3 V	278,1075,2011
		HA 95 # 231			3 53 05	+00 02.0	1	14.22		0.45		0.30		13		1764
		RL 77			3 53 06	+56 59.	1	13.83		0.38		−0.33		2		704
24754	−24 01960	T Eri			3 53 06	−24 10.7	2	8.82	.705	1.45	.105	0.74	.095	2	M5/6 IIIe	635,975
24655	+21 00556	Pleiades 3332			3 53 07	+22 05.0	1	9.07		0.45		−0.03		1	F8	8023
		HA 95 # 284			3 53 08	+00 17.9	1	13.67		1.40		1.07		9		1764
		HA 95 # 149			3 53 10	−00 01.6	3	10.93	.004	1.58	.011	1.56	.002	19	K4	281,1764,5006
		HA 95 # 285			3 53 12	+00 15.0	1	15.56		0.94		0.70		1		1764
283132	+25 00644	Pleiades 3333			3 53 12	+25 37.0	1	10.49		0.67		0.16		1	G0	8023
24640	+34 00768	HR 1215		⋆	3 53 15	+34 56.2	7	5.49	.014	−0.03	.010	−0.75	.012	19	B1.5V	15,154,1009,1012,1041*
24863	−53 00628	HR 1227		⋆ A	3 53 15	−52 50.1	3	6.45	.005	0.16	.004	0.11		8	A4 V	15,258,2012
		Hyades HG7 - 44			3 53 20	+18 16.3	1	14.40		1.59				2	M1	950
		HA 95 # 233			3 53 22	+00 03.7	1	11.00		0.68		0.17		3	G2 p	281
	−45 01325				3 53 25	−45 29.0	1	9.57		0.97				4	K0	1075
24825	−39 01245				3 53 27	−38 54.2	1	6.81		−0.04		−0.04		5	Ap CrEu(Sr)	1700
24711	+22 00601	Pleiades 3334			3 53 30	+23 00.4	1	8.34		0.28		0.12		1	A0	8023
		LS I +56 094			3 53 30	+56 53.3	1	11.77		0.29		−0.55		2		41
24702	+22 00602	G 6 - 40			3 53 31	+22 32.0	1	7.84		0.68		0.20		1	G0	333,1620
		POSS 249 # 3			3 53 31	+35 06.7	1	15.39		1.50				2		1739
		LS I +56 095			3 53 33	+56 59.5	1	12.18		0.54		−0.40		2		41
24749	+0 00678			⋆ AB	3 53 34	+00 34.0	4	9.46	.007	0.47	.006	0.18	.003	17	F0	271,281,1776,6004
232862	+50 00864	IDS03499N5034		AB	3 53 36	+50 42.7	2	9.63	.137	0.76	.088	0.94		5	G8 II	1119,8100
232861	+51 00810				3 53 38	+52 01.8	1	9.72		0.28				3	A2 V	1119
		HA 95 # 236			3 53 39	+00 00.1	4	11.49	.005	0.73	.002	0.17	.011	61	G2	281,989,1729,1764
		RL 78			3 53 42	+56 48.	1	14.44		0.32		−0.33		2		704
		Hyades HG7 - 45			3 53 43	+17 25.2	1	13.88		1.41				1		950
		LS I +56 090			3 53 44	+56 18.9	1	12.24		0.60		−0.32		2		41
		LS I +60 309			3 53 53	+60 45.3	1	11.05		0.18		−0.69		3		41
24740	+22 00605	HR 1218			3 53 55	+22 20.1	2	5.64	.005	0.34	.033	−0.01	.019	4	F2 IV	254,3016
	−75 00252				3 53 55	−75 33.1	1	9.94		0.70				4	G5	2012
24791	+3 00540				3 53 58	+03 56.3	1	9.01		0.55		0.06		1		1776
		Hyades HG7 - 47			3 53 58	+16 01.6	1	14.13		1.57				1		950
24761	+19 00625				3 53 58	+19 56.4	1	8.79		0.32		0.26		3		1648
		G 6 - 41			3 53 58	+24 30.6	1	11.69		0.79		0.35		1	G8	333,1620
	+55 00837	LS I +55 055			3 54 02	+55 46.6	2	9.58	.009	0.70	.009	−0.26	.004	5	B2 Ib	41,1012
	+52 00729	LS V +52 015			3 54 03	+52 32.8	1	10.13		1.06		−0.01		2	B2 Iab	1012
		CS 22169 # 1			3 54 05	−15 18.0	1	12.87		−0.27		−1.10		1		1736
		CS 22190 # 9			3 54 05	−15 18.0	1	12.88		−0.27		−1.09		1		1736

Table 1 197

HD	DM	Other Id	N	Rem	α_{1950}	δ_{1950}	S	V	σ_V	B–V	σ_{B-V}	U–B	σ_{U-B}	n	Spectrum	References
24769	+22 00607	HR 1221, V817 Tau	AB		3 54 06	+23 01.9	2			0.02	.025	-0.01	.020	7	B9.5IV	1049,1079
24834	−14 00783				3 54 08	−13 44.5	2	6.41	.005	1.66	.005	2.02	.130	4	M1 III	1024,3040
24736	+32 00691				3 54 09	+32 44.7	1	8.53		0.12		-0.04		2	A0 IV	1041
24819	−1 00561	IDS03517S0152	A		3 54 10	−01 42.9	1	8.09		0.30		0.21		5	A2	1776
24688	+51 00812				3 54 11	+51 59.1	1	7.96		1.02				2	K0 III	1025
		RL 79			3 54 12	+55 38.	1	16.72		0.19		-0.82		2		704
		UKST 117 # 23			3 54 12	−60 20.7	1	16.07		0.52		-0.11		1		1584
		POSS 249 # 7			3 54 14	+34 53.2	1	18.39		1.82				3		1739
24832	−10 00793	HR 1225, DL Eri			3 54 14	−09 53.7	3	6.18	.004	0.28	.008	0.11		8	F1 V	15,258,2027
24777	+20 00669				3 54 15	+21 10.8	1	6.84		1.03		0.79		2	K0	1648
24830	−1 00562				3 54 19	−01 01.8	1	8.38		0.46		0.09		1	F2	1776
		UKST 117 # 13			3 54 19	−60 20.6	1	13.15		0.95		0.51		4		1584
24817	+5 00564	HR 1224			3 54 22	+05 53.8	2	6.08	.005	0.06	.000	0.04	.005	7	A2 Vn	15,1415
24678	+58 00672				3 54 22	+58 28.6	1	8.31		0.64		0.20		1	F8 IV	851
24708	+51 00814				3 54 24	+51 35.4	1	8.13		0.34				2	F0 IV	1025
24802	+24 00599	HR 1222			3 54 27	+24 19.1	1	6.16		1.37				2	K0	71
24760	+39 00895	HR 1220, ε Per	★ A		3 54 29	+39 52.0	3	2.89	.041	-0.20	.011	-0.95	.023	6	B0.7III	150,542,1048
24760	+39 00895	HR 1220, ε Per	★ AB		3 54 29	+39 52.0	14	2.89	.012	-0.18	.012	-0.98	.011	47	B0.7III+A2V	1,15,154,401,542,1004*
24760	+39 00895	IDS03511N3943	B		3 54 29	+39 52.0	3	7.51	.079	-0.04	.010	-0.50	.051	7	A2 V	150,542,542
		G 80 - 37			3 54 31	+07 31.5	1	12.67		1.27		1.06		1		333,1620
		Hyades HG7 - 49			3 54 33	+14 49.7	1	13.52		1.58				2	M1	950
24967	−43 01235				3 54 36	−43 15.6	3	7.54	.007	1.24	.002	1.36	.002	16	K2 III	977,1075,2011
25269	−75 00253				3 54 38	−75 27.3	1	8.22		1.63				4	M0/1 III	2012
24976	−43 01236				3 54 39	−43 45.7	3	8.13	.000	0.50	.004	0.02	.016	16	F5/6 V	977,1075,2011
25053	−60 00274				3 54 39	−60 07.6	1	8.83		0.17		0.13		2	A4 V	1584
		G 160 - 38			3 54 40	−10 00.7	1	14.30		0.86		0.39		1		3056
		LS I +59 195			3 54 41	+59 29.2	1	10.62		0.90		0.75		3		41
24723	+53 00717				3 54 43	+53 56.0	1	8.30		0.28				3	A7 III	1119
		LS I +57 137			3 54 44	+57 02.2	1	11.23		0.33		-0.52		2		41
	+4 00611				3 54 46	+04 44.8	1	9.69		0.54		0.37		1	A5	1776
24733	+53 00718				3 54 46	+53 50.8	1	7.02		0.25				2	A7 V	1119
	−45 01335				3 54 46	−45 39.6	1	10.04		0.55				4		1075
		RL 80			3 54 48	+46 20.	1	15.52		0.28		-0.57		2		704
24809	+34 00773	HR 1223	V		3 54 49	+34 40.3	1	6.53		0.21		0.06		2	A8 V	1733
24451	+75 00154	G 248 - 10			3 54 49	+76 01.6	4	8.24	.020	1.14	.010	1.04		10	K4 V	22,1197,1758,3072
25004	−41 01208				3 54 53	−41 29.3	1	8.94		1.21		1.14		2	K4 V	3072
		Hyades HG7 - 50			3 54 56	+09 13.8	1	14.68		0.94				1		950
	+51 00815	LS V +52 016			3 54 57	+52 14.3	1	10.40		0.16				3	B6 V	1119
24916	−1 00565	IDS03524S0127	A		3 54 57	−01 18.0	3	8.04	.007	1.11	.005	1.01	.015	5	K4 V	680,1705,3072
24916	−1 00565	IDS03524S0127	AB		3 54 57	−01 18.0	2	8.00	.000	1.13	.010	1.05		14	K4 V+M3	680,2033
24916	−1 00565	IDS03524S0127	B		3 54 57	−01 18.0	2	11.55	.065	1.50	.025	1.10	.015	4	M3	680,3072
		UKST 117 # 16			3 54 57	−60 32.0	1	13.82		0.76		0.30		2		1584
		RL 81			3 55 00	+54 39.	1	15.45		0.38		-0.67		2		704
		LS I +56 096			3 55 01	+56 02.7	1	11.26		0.71		-0.07		4		41
		UKST 117 # 21			3 55 01	−60 03.4	1	15.08		0.71		0.01		1		1584
	+51 00816				3 55 04	+51 38.6	1	9.78		0.45				2	F5	1025
24775	+51 00817	IDS03514N5113	AB		3 55 06	+51 21.4	2	7.71	.108	1.69	.039	1.39		5	K2 Ib	1119,8100
25061	−53 00630				3 55 06	−53 04.7	1	9.26		0.85				4	K1 V	2012
	−44 01337				3 55 08	−44 45.5	2	9.67	.001	1.43	.003	1.76		10		863,2012
24843	+38 00827	HR 1226			3 55 09	+38 41.9	3	6.33	.021	1.08	.013	0.96	.015	7	K1 III	71,833,3016
25119	−60 00276				3 55 11	−60 44.2	1	8.34		1.18		1.04		2	K0 III	1584
	+56 00864	LS I +57 138	A		3 55 12	+57 05.7	3	9.68	.005	0.28	.000	-0.68	.014	5	O6 nn	41,1011,1012
	+56 00864	LS I +57 138	B		3 55 12	+57 05.7	3	10.08	.005	0.27	.005	-0.69	.005	5	O6 V	41,1011,1012
	−45 01339				3 55 12	−45 38.9	2	9.90	.000	0.96	.006	0.76		9	K0	863,1075
281371	+31 00674				3 55 16	+32 13.0	1	10.01		0.22		0.09		2	A0	1041
		LP 156 - 63	AB		3 55 16	+49 04.4	1	10.92		0.98		0.76				1723
24899	+23 00594	Pleiades 3335			3 55 21	+23 56.4	1	7.20		0.04		-0.01		1	B9	8023
		RL 82			3 55 24	+51 31.	1	11.40		0.23		-0.12		2		704
25170	−63 00275	HR 1236			3 55 25	−63 36.5	2	6.14	.002	1.09	.005	1.02		6	K1/2 III	58,2009
	+56 00866	LS I +56 097			3 55 27	+56 58.6	1	10.30	.025	0.35	.010	-0.58	.015	6	O9 V	41,1011,1012
25087	−50 01209				3 55 29	−49 54.1	1	7.80		0.97		0.69		5	G8 III	1673
		Hyades HG7 - 54			3 55 30	+07 34.	1	11.43		0.71				1		950
		RL 83			3 55 30	+52 28.	1	10.84		0.21		-0.42		2		704
		UKST 117 # 8			3 55 33	−59 53.1	1	11.76		1.29		1.43		2		1584
		UKST 117 # 15			3 55 39	−59 53.8	1	13.45		0.75		0.35		2		1584
	−60 00278				3 55 39	−60 00.7	1	10.45		0.60		0.09		2	G3	1584
286436					3 55 41	+11 45.8	1	10.26		1.58				1	R2	1238
25025	−13 00781	HR 1231, γ Eri	★ A		3 55 42	−13 39.0	11	2.95	.011	1.59	.006	1.95	.013	185	M1 IIIb	3,9,15,58,1024,1075*
24912	+35 00775	HR 1228, ξ Per			3 55 43	+35 38.9	18	4.04	.015	0.02	.013	-0.92	.018	58	O7.5I	15,154,985,1009,1011*
24913	+32 00697				3 55 45	+32 36.9	1	8.38		0.20		0.07		2	A0 V	1041
		LS I +55 056			3 55 45	+55 10.3	1	11.36		0.75		-0.22		3		41
232864	+51 00819				3 55 46	+52 10.3	1	9.45		0.17				2	B1 IV	1119
		LP 833 - 20			3 55 48	−25 42.0	1	12.89		0.69		-0.11		2		1696
		E2 - 62			3 55 48	−46 29.	1	11.59		0.91				4		1007
		AK 8 - 111			3 55 49	+10 54.1	1	11.40		1.05				1		1570
		Hyades CDS 57	★		3 55 53	+25 04.7	1	14.03		1.56				3		1674
		LP 248 - 28			3 55 54	+37 49.5	1	11.76		1.11		0.90		2		1723
		LS I +55 057			3 55 54	+55 17.7	1	12.19		0.58		-0.34		3		41
		Hyades HG7 - 60			3 55 56	+15 15.0	1	14.25		0.80				1		950
		LS I +57 140			3 55 56	+57 29.4	1	11.02		0.35		-0.59		2		41

HD	DM	Other Id	N Rem	α_{1950}	δ_{1950}	S	V	σ_V	B–V	σ_{B-V}	U–B	σ_{U-B}	n	Spectrum	References
		LP 474 - 233		3 56 00	+09 28.	1	13.01		1.31				2		1570
281367	+32 00698			3 56 01	+32 33.5	1	9.86		0.52		0.29		2	A7	1041
24970	+26 00655			3 56 02	+27 03.4	1	7.46		0.14		0.04		2	A0	1625
24545	+77 00138			3 56 03	+78 03.8	1	6.98		1.04		0.87		3	K0	985
		Hyades HG8 - 59		3 56 06	+21 53.	2	13.01	.014	1.51	.009	1.23		6		950,1058
		Hyades HG8 - 43		3 56 06	+25 09.	2	12.09	.009	1.40	.005	1.26		6	K5	950,1058
		Hyades HG8 - 41		3 56 06	+26 22.	1	11.58		1.46				3		1674
		Hyades J 212		3 56 06	+26 22.	1	14.07		0.67		0.09		3		1058
232866	+50 00871	IDS03524N5037	AB	3 56 06	+50 45.2	1	9.98		0.71				2	K2	1726
		RL 84		3 56 06	+51 25.	1	11.23		0.29		-0.32		2		704
	−9 00784			3 56 06	−09 17.8	3	11.42	.000	0.88	.018	0.58	.026	6		265,1696,3056
25138	−45 01347	IDS03545S4501	AB	3 56 06	−44 52.8	3	8.69	.003	0.84	.004	0.42	.008	16	G8 IV	977,1075,2011
		Steph 435		3 56 07	+26 20.1	1	11.58		1.46				3	K5	1746
24909	+47 00920	IQ Per	★ A	3 56 07	+48 00.6	1	7.72		0.08		-0.23		1	B9	627
24909	+47 00921	IDS03525N4752	B	3 56 07	+48 00.6	2	9.30	.001	0.21	.000	0.14	.005	5	A3	626,627
		NGC 1499 sq1 14		3 56 08	+37 21.2	1	14.71		2.72				1		510
		CS 22173 # 9		3 56 08	−18 08.6	1	13.39		0.34		-0.06		1		1736
		NGC 1499 sq1 13		3 56 09	+37 19.3	1	11.24		0.55		0.48		1		510
		Hyades CDS 58	★	3 56 14	+32 10.8	1	13.42		1.54				3		1674
25442	−75 00254			3 56 14	−75 18.5	2	9.04	.004	0.23	.007	0.18		6	A3 V	278,2012
		Hyades HG8 - 47	AB	3 56 16	+26 43.3	2	13.50	.005	1.35	.015			3		920,1570
279213	+37 00848			3 56 17	+37 19.2	1	9.67		0.47		0.45		1	A0	510
		Hyades CDS 59	★	3 56 18	+20 17.2	1	13.57		1.58				3		1674
25169	−46 01243			3 56 18	−46 31.4	4	8.01	.001	0.47	.007	-0.10	.009	26	F6/7 V	743,977,1075,2011
24982	+38 00829	HR 1229	★ AB	3 56 20	+38 40.8	2	6.51		0.10	.000	0.09	.034	5	A1 V(p)	1049,1733
		NGC 1499 sq1 12		3 56 21	+37 19.1	1	13.84		1.18		0.67		1		510
237204	+56 00868	LS I +56 098		3 56 21	+56 45.6	2	9.16	.014	0.30	.005	-0.57	.000	4	B0.5V	41,1012
25423	−74 00280			3 56 22	−74 05.6	1	9.20		0.42				4	A3mA7-F0	2012
285401	+14 00635			3 56 23	+14 18.2	1	9.75		0.63		0.14		2	F5	1375
24959	+45 00858			3 56 23	+45 33.4	1	8.66		0.45		0.31		2	A0	1601
	−75 00255			3 56 23	−75 16.4	1	10.11		0.52				4	G0	2012
		Hyades J 216		3 56 24	+26 43.	1	12.41		0.70				1		1058,1570
25069	−5 00789	HR 1232		3 56 24	−05 36.6	4	5.83	.013	1.00	.000	0.86	.004	13	G9 V	15,1256,2031,3077
		NGC 1499 sq1 15		3 56 30	+37 18.1	1	12.90		1.00		0.50		1		510
	−44 01351			3 56 32	−44 08.8	2	10.30	.002	0.47	.000	0.01		10	F5	863,1075
		Hyades L 9		3 56 37	+21 09.4	1	11.13		0.68				2		582
25023	+35 00779			3 56 37	+35 52.5	1	8.56		0.40		0.25		1	A2	510
25093	+0 00685			3 56 38	+00 49.0	1	9.24		0.45		0.27		1	A2	1776
		G 160 - 41		3 56 38	−07 04.5	1	14.58		1.00		0.77		2		3056
25450	−73 00246			3 56 41	−73 41.9	1	8.98		0.41				4	F0 IV	2012
		POSS 249 # 1		3 56 42	+34 54.3	1	14.20		0.86				1		1739
279235	+35 00780			3 56 42	+35 52.6	1	9.82		0.25		0.22		1	A2	510
		RL 85		3 56 42	+52 29.	1	10.19		0.20		-0.11		2		704
281377	+31 00680			3 56 43	+31 55.4	1	9.37		0.11		-0.28		2	B8	1041
24980	+47 00923			3 56 44	+47 25.9	2	8.36	.000	0.23	.000	0.12	.001	8	A2	626,627
	+55 00838	LS I +55 058		3 56 45	+55 20.6	2	9.29	.000	0.82	.000	-0.09	.007	10	B3 Ib	41,1012
		LF 6 # 59		3 56 46	+54 38.7	1	10.40		1.12		0.95		2		1723
25022	+39 00904			3 56 48	+39 52.4	1	6.89		0.12		0.11		3	A0 V	1501
279237	+35 00781			3 56 49	+35 39.1	2	9.93	.005	0.27	.015	0.19	.045	3	A0	510,1041
25245	−49 01165			3 56 49	−49 45.1	1	6.93		1.07		0.94		5	K0 III	1628
		UKST 117 # 10		3 56 50	−60 01.8	1	13.04		0.51		0.00		3		1584
		Hyades HG7 - 64		3 56 51	+16 48.0	1	13.67		1.56				2		950
25102	+9 00524	Hyades vB 8	★	3 56 56	+10 11.4	4	6.37	.011	0.41	.011	-0.01	.013	14	F5 III	15,254,1127,8015
24979	+52 00735			3 56 56	+52 50.7	1	9.07		0.25				2	B9 V	1119
	+52 00736	IDS03531N5306	A	3 56 58	+53 14.8	1	9.91		0.36				2		1119
	+52 00736	IDS03531N5306	B	3 56 58	+53 14.8	1	10.23		0.34				2		1119
	+54 00728	LS I +55 059		3 56 58	+55 05.2	1	10.32		0.94		-0.09		3		41
		G 8 - 3		3 56 59	+19 33.7	1	12.25		1.08		0.90		1		333,1620
	+3 00547			3 57 01	+03 18.3	1	9.81		0.55		0.11		1	F5	1776
		CS 22173 # 2		3 57 01	−20 50.9	1	13.47		0.40		-0.22		1		1736
	−60 00280			3 57 02	−60 02.5	1	10.61		0.52		0.03		2		1584
25137	+1 00689			3 57 06	+01 39.2	1	7.29		0.11		0.03		2	B9	1776
		POSS 249 # 2		3 57 06	+35 00.2	1	14.21		1.62				2		1739
25253	−44 01354			3 57 07	−44 26.0	1	9.02		0.90				4	G8 III	2011
25165	−12 00766	HR 1235		3 57 09	−12 42.9	2	5.60	.000	1.48	.000	1.76		6	K4 III	2006,3005
279262	+33 00753			3 57 12	+34 12.2	1	10.17		0.28		-0.04		2	A2	1041
25031	+47 00925			3 57 12	+48 06.3	1	8.73		0.38				2	F8	70
25189	−20 00755			3 57 13	−20 28.4	1	7.16		1.20		1.31		1	K1 III	8100
25154	−0 00627			3 57 14	−00 09.6	1	9.86		0.65		0.40		2	A5	1776
25030	+51 00827			3 57 15	+52 01.5	1	8.58		1.50				4	K1 Ib	1119
	−44 01355			3 57 15	−44 38.3	2	9.40	.001	1.04	.005	0.84		10	K0	863,2012
		UKST 117 # 18		3 57 17	−60 01.8	1	14.66		1.11		0.79		2		1584
		Hyades HG7 - 67		3 57 20	+12 29.7	1	14.79		0.62				1		950
	+23 00601		V	3 57 20	+23 23.6	1	10.30		1.50				1	R2	1238
279236	+35 00784			3 57 22	+35 44.5	1	10.50		0.56		0.01		1	F8	510
25153	+13 00625			3 57 28	+14 09.9	1	8.24		0.48		0.01		3	F5	3026
279216	+36 00801	LS V +37 001		3 57 29	+37 00.4	1	10.19		0.59		-0.13		1	B5	510
		NGC 1499 sq1 8		3 57 33	+35 42.2	1	11.18		1.86		2.16		1		510
		NGC 1499 sq1 7		3 57 36	+35 42.6	1	10.73		1.01		0.74		1		510
25346	−57 00606	HR 1245		3 57 38	−57 14.6	2	6.04	.005	0.44	.000			7	F5 IV/V	15,2012

Table 1   199

HD	DM	Other Id	N Rem	α_{1950}	δ_{1950}	S	V	σ_V	B−V	σ_{B-V}	U−B	σ_{U-B}	n	Spectrum	References
		NGC 1499 sq1 6		3 57 40	+35 44.6	1	11.51		1.53		1.17		1		510
		NGC 1499 sq1 5		3 57 41	+35 48.0	1	12.96		0.77		0.17		1		510
25029	+60 00773			3 57 42	+60 37.0	1	8.51		0.86		0.37		2	F8	3045
25301	−44 01358			3 57 42	−44 03.5	5	6.83	.007	1.10	.008	0.99	.001	37	K1 III	116,278,977,1075,2011
	+19 00641	Hyades vB 9		3 57 44	+20 14.5	2	8.67	.000	0.71	.001	0.24	.002	4	G4 V	1127,8023
25056	+53 00722			3 57 44	+53 43.6	5	7.03	.014	1.20	.010	0.92		10	G0 Ib	70,138,1119,1355,6009,8100
25175	+16 00544	HR 1237		3 57 45	+17 09.4	2			0.06	.000	0.01	.025	8	A0 V	1049,1079
		G 7 - 16		3 57 46	+08 06.0	1	15.90		0.69		0.04		2		1658
25143	+32 00703			3 57 46	+32 16.9	1	8.60		0.12		-0.20		2	B9 V	1041
25267	−24 02022	HR 1240, τ9 Eri		3 57 47	−24 09.4	6	4.64	.015	-0.13	.011	-0.41	.007	18	B6 V+B9.5 V	15,1063,1075,2012*
25186	+13 00627			3 57 49	+13 45.1	1	7.31		1.42		1.56		3	K0	1648
25204	+12 00539	HR 1239, λ Tau		3 57 54	+12 21.0	4	3.49	.088	-0.11	.006	-0.61	.024	11	B3 V +A4 IV	15,154,1363,8015
281457	+33 00758			3 57 54	+33 19.9	1	9.68		0.38		0.22		2	A2	1041
25203	+15 00565			3 57 55	+15 20.1	1	7.65		0.97		0.62		2	G5	1648
279276	+36 00804			3 57 55	+36 59.2	1	8.45		1.08		0.70		1	G5	510
25202	+17 00666	Hyades vB 137	★ A	3 57 56	+18 03.3	4	5.89	.008	0.32	.005	0.04	.007	7	F4 V	15,39,1127,8015
		Hyades HG7 - 71		3 57 57	+13 46.0	1	14.84		1.75				2	M4	920
25422	−61 00290	HR 1247		3 57 57	−61 32.5	4	4.57	.023	1.62	.008	2.00	.033	17	M2 IIIab	15,678,1075,2012
25152	+36 00805	HR 1234		3 57 58	+36 51.0	2			-0.01	.010	-0.06	.035	6	A0 V	1049,1079
		RL 86		3 58 00	+56 50.	1	13.20		0.37		-0.47		2		704
	−44 01360			3 58 00	−44 46.1	1	10.61		0.54				4		2012
25109	+50 00882			3 58 06	+50 30.0	1	8.03		0.14				3	A0	70
		LB 3333		3 58 06	−61 05.	1	12.24		0.52		0.15		1		45
25248	+4 00619	IDS03555N0448	AB	3 58 08	+04 56.6	1	8.55		0.34		0.39		4	A3	176
25230	+19 00643			3 58 12	+20 03.6	1	6.81		1.05		0.74		2	K1 III	3016
		RL 88		3 58 12	+51 51.	1	10.73		0.24		-0.33		2		704
		RL 87		3 58 12	+52 24.	1	10.66		0.14		0.25		2		704
25185	+31 00686			3 58 15	+32 07.8	1	8.90		0.30		0.19		2	A1 V	1041
	−60 00282			3 58 16	−60 06.8	1	10.40		0.45		-0.09		2	F9	1584
		LP 474 - 62		3 58 21	+13 31.2	1	11.26		0.79				2		1674
25228	+28 00609			3 58 21	+28 21.8	1	7.08		0.17		0.05		2	A0	23
25314	−27 01497			3 58 21	−26 58.6	1	8.43		0.51				4	F5 V	2012
232875	+51 00833			3 58 22	+51 31.6	1	9.12		1.17				3	K0 III	1119
232874	+53 00723	LS V +53 016		3 58 23	+53 36.9	3	8.82	.022	0.40	.007	-0.43		6	B0.5V	70,1012,1119
		Steph 439	AB	3 58 24	+11 57.4	1	11.48		1.43		1.23		3	M0	1746
25141	+52 00741			3 58 24	+52 44.3	1	8.92		0.12				3	B5 V	1119
		Hyades HG7 - 73	AB	3 58 25	+11 57.5	1	11.48		1.45				2		920
		E2 - 49		3 58 29	−44 29.	1	11.09		0.62		0.14		5		863
		Steph 442		3 58 30	−23 13.1	1	11.47		1.48		1.17		1	M0	1746
		E2 - Z		3 58 30	−44 29.	1	11.07		0.63				4		2011
25090	+62 00643			3 58 31	+62 17.0	3	7.30	.024	0.33	.017	-0.55	.018	13	B0.5III	399,555,1631
279299	+34 00790			3 58 33	+35 10.6	1	10.21		0.33		0.18		2	B9	1041
25132	+58 00685			3 58 34	+58 48.7	1	7.53		0.04		-0.53		3	B3 V	399
		Hyades HG7 - 76		3 58 36	+17 53.8	1	13.42		1.46				1		950
25239	+31 00687			3 58 36	+32 04.6	1	9.27		0.20		-0.06		2	B9.5V	1041
276021	+39 00912			3 58 37	+40 12.7	1	10.71		0.56		-0.05		3	F5	1064
		Hyades HG7 - 75		3 58 39	+18 12.8	1	14.59		1.44				2		1058
25150	+55 00843			3 58 39	+56 14.8	1	8.29		1.17		1.05		3	K0 III-IV	37
		POSS 655 # 5		3 58 39	−04 55.8	1	17.36		1.77				2		1739
		LS V +52 017		3 58 41	+52 43.0	1	10.64		0.87		-0.18		2	B1? p	1012
25371	−30 01597	HR 1246		3 58 41	−30 37.8	3	5.92	.005	0.04	.004			14	A1 V	15,1075,2027
		G 7 - 17		3 58 42	+18 36.1	2	15.45	.105	1.82	.017	1.14		4		1691,3028
25140	+58 00687			3 58 43	+58 47.2	1	8.14		1.36		0.95		1	G0 II	851
		LS I +55 060		3 58 45	+55 01.8	1	11.27		0.72		-0.19		3		41
281483	+31 00689			3 58 47	+31 50.7	1	9.85		0.45		0.23		2	A0	1041
	+52 00744			3 58 51	+52 36.9	1	9.21		0.86				2	G9 III	1119
		CS 22173 # 7		3 58 51	−18 42.2	1	13.77		0.52		-0.22		1		1736
25421	−44 01365			3 58 51	−44 38.0	2	9.03	.000	0.47	.000	-0.02		8	F5 V	1075,2011
25193	+51 00835	IDS03551N5152	AB	3 58 53	+52 00.5	1	9.04		0.89				2	G8 IV	1119
285290	+18 00571	Hyades HG8 - 72		3 58 55	+18 19.0	1	10.20		0.82		0.47		3	G5	1058
25470	−51 00975	HR 1250, XY Dor		3 58 55	−51 42.3	4	6.50	.004	1.64	.000	1.95		16	M1 III	15,1075,2030,3055
		UKST 117 # 11		3 58 55	−60 40.4	1	13.07		1.03		0.83		5		1584
		LS I +55 061		3 58 59	+55 11.1	1	11.64		0.72		-0.22		3		41
25340	−1 00572	HR 1244		3 59 00	−01 41.3	5	5.28	.005	-0.15	.005	-0.55	.005	13	B5 V	15,154,361,1223,1415
25330	+9 00528	HR 1243	★ AB	3 59 02	+09 51.6	3	5.66	.005	0.02	.008	-0.41	.008	9	B5 V	15,1061,1079
25339	−1 00571			3 59 02	−01 07.9	1	9.54		0.49		0.23		1	F0	1776
25215	+52 00745			3 59 03	+52 37.3	1	8.68		0.06				3	Ap Si	1119,8100
		AA45,405 S206 # 2		3 59 04	+51 12.1	1	12.28		0.87		0.41		2	F8 IV	797
		Hyades HG7 - 78		3 59 06	+13 02.3	1	14.66		1.43				1		950
279320	+33 00764			3 59 06	+33 54.5	1	10.81		0.27		0.18		2	A2	1041
		UKST 117 # 24		3 59 07	−60 40.1	1	17.02		0.80		0.20		1		1584
		Hyades HG8 - 44		3 59 09	+17 42.4	2	12.99	.010	1.45	.005	1.26		5		950,1058
		Steph 441		3 59 09	+17 42.4	1	13.00		1.45				2	K7	1746
25271	+38 00834			3 59 09	+38 57.6	1	7.71		0.26		0.09		2	A3	105
		POSS 655 # 4		3 59 09	−05 28.2	1	16.90		1.55				2		1739
25296	+27 00618	IDS03561N2751	AB	3 59 10	+27 59.3	1	7.31		0.95				3	G8 II	20
		LS I +58 129		3 59 10	+58 09.7	1	11.09		0.77		-0.32		3		41
237211	+56 00873	LS I +56 099		3 59 14	+56 24.1	3	8.99	.010	0.49	.000	-0.50	.005	7	O9.5I:p	41,1011,1012
25235	+52 00746			3 59 15	+52 29.5	1	8.79		0.17				2	B9 V	1119
	+21 00575	Hyades HG8 - 58		3 59 16	+21 56.6	2	9.81	.004	1.15	.004	0.80		5		950,1058

HD	DM	Other Id	N Rem	α_{1950}	δ_{1950}	S	V	σ_V	B–V	σ_{B-V}	U–B	σ_{U-B}	n	Spectrum	References
25392	−14 00803			3 59 18	−13 57.8	1	9.77		0.19		−0.51		12	A8/9 IV(w)	1732
		UKST 117 # 19		3 59 20	−60 32.6	1	14.70		0.53		0.02		2		1584
279308	+34 00793			3 59 22	+34 24.8	1	10.18		0.43		0.31		2	A6	1041
25772	−73 00247			3 59 22	−73 31.4	2	8.10	.004	1.24	.001	1.18		12	K2 III	1704,2012
286416	+12 00541			3 59 24	+12 40.5	1	9.65		1.16		0.90		2	K0	1375
		Hyades HG8 - 8		3 59 24	+24 33.	2	12.93	.024	1.54	.005	1.16		5		950,1058
25256	+50 00885			3 59 24	+50 28.0	1	8.91		0.21				2	A2	70
25309	+32 00708			3 59 25	+32 46.3	1	8.46		0.12		−0.31		2	B6 V	1041
	+51 00838			3 59 25	+51 19.8	1	9.92		0.15				2	A5 III	1119
237212	+57 00768			3 59 25	+57 56.0	1	9.47		1.47				36	K7	6011
	+51 00837			3 59 26	+51 41.0	1	10.06		0.38				2	A1	1119
		POSS 655 # 3		3 59 26	−04 28.7	1	16.86		1.74				2		1739
281465	+32 00709			3 59 27	+32 42.5	1	9.82		0.31		0.21		2	A0	1041
25308	+34 00794			3 59 27	+35 15.7	1	8.55		0.25		0.11		2	A0 IV	1041
		LP 474 - 854		3 59 30	+14 10.	1	10.79		0.73				2		1674
25503	−44 01369			3 59 31	−44 35.1	4	8.46	.001	0.49	.009	−0.06	.011	18	F6 V	395,977,1075,2011
	+50 00886	LS V +51 013		3 59 35	+51 10.6	3	11.24	.015	1.05	.009	−0.09	.015	6	O5	342,684,797
25513	−45 01368	IDS03580S4508	A	3 59 35	−44 59.3	1	9.22		0.92				4	G8 III	2011
25863	−75 00256	IDS04006S7502	A	3 59 35	−74 53.2	1	9.88		0.33				4	A2/3 V	2012
		L 301 - 28		3 59 36	−46 04.	1	12.99		0.56		−0.17		2		1696
		UKST 117 # 9		3 59 38	−60 28.1	1	11.83		0.92		0.61		1		1584
		UKST 117 # 12		3 59 38	−60 29.0	1	13.10		0.63		0.09		1		1584
25601	−60 00285			3 59 38	−60 29.6	1	9.53		1.53		1.78		1	K2/3 III	1584
25400	−0 00630			3 59 39	−00 03.5	1	8.33		0.33		0.25		1	A3	1776
		UKST 117 # 22		3 59 39	−60 11.5	1	15.18		0.93		0.40		1		1584
25537	−46 01268			3 59 44	−46 16.6	3	7.23	.010	1.02	.003	0.85	.006	16	K0 III	977,1075,2011
279307	+34 00797			3 59 45	+34 30.1	1	10.09		0.19		0.11		2	A0	1041
25413	−0 00631			3 59 45	−00 04.3	1	9.23		0.46		0.20		2	B8	1776
		UKST 117 # 14		3 59 45	−60 42.8	1	13.34		0.74		0.35		3		1584
25294	+47 00932			3 59 46	+48 05.3	1	8.98		0.05				2	A0	70
281486	+31 00694			3 59 47	+31 47.6	1	9.82		0.18		0.05		2	A0	1041
		LS I +57 141		3 59 48	+57 21.8	1	11.55		0.42		−0.35		2		41
		POSS 655 # 2		3 59 50	−04 58.0	1	14.81		1.54				2		1739
		UKST 117 # 20		3 59 50	−60 13.9	1	14.73		1.29		1.37		2		1584
232880	+51 00841			3 59 51	+51 59.3	2	9.62	.014	0.20	.005			4	B8	70,1025
25354	+37 00866	V380 Per		3 59 52	+37 55.0	1	8.36		0.12		0.10		3	A0p	1202
25319	+48 01030			3 59 52	+48 25.6	1	8.80		0.30				3	A5	70
25329	+34 00796			3 59 53	+35 09.3	9	8.50	.008	0.87	.008	0.35	.016	39	K1 V	22,792,908,1003,1197,1658*
		UKST 117 # 17		3 59 56	−60 14.0	1	14.14		0.46		−0.02		4		1584
25305	+51 00843			4 00 00	+51 45.7	1	8.88		0.27				2	Am	1119,8100
25292	+52 00750			4 00 00	+53 08.5	2	7.79	.009	0.58	.000			4	F8 V	70,1025
		Hyades HG7 - 81		4 00 01	+12 36.0	1	12.05		0.94				1		920
		G 8 - 6		4 00 03	+20 06.3	2	12.22	.005	0.74	.000	0.10	.005	3	G8	333,1620,1658
25457	−0 00632	HR 1249		4 00 03	−00 24.2	6	5.38	.011	0.52	.011	0.00	.008	21	F5 V	15,1013,1197,1415*
279382	+34 00799			4 00 04	+35 10.9	1	10.75		0.21		0.13		2	A0	1041
281452	+33 00766			4 00 07	+33 32.9	1	9.97		0.27		0.18		2	A0	1041
25535	−34 01491	IDS03582S3446	AB	4 00 08	−34 37.2	2	6.73	.000	0.63	.010	0.10		7	G1/2 V	1075,1311
		Hyades HG7 - 82		4 00 09	+13 19.5	1	11.02		0.83				1		950
25705	−62 00312	HR 1264, γ Ret		4 00 10	−62 17.9	5	4.50	.006	1.65	.017	1.80	.004	16	M4 III	15,1075,2012,3055,8029
		Hyades L 13		4 00 15	+24 52.5	1	10.84		0.93				2		582
		NGC 1496 - 52		4 00 15	+52 34.7	1	13.05		0.54		0.38		2		1668
25291	+58 00690	HR 1242	★	4 00 16	+59 01.1	7	5.04	.019	0.49	.007	0.48	.006	29	F0 II	15,41,1119,1363,3016*
25456	+14 00643			4 00 18	+14 20.8	1	8.48		0.59		0.13		2	F8	1648
279384	+34 00800			4 00 21	+35 00.7	1	9.94		0.15		0.03		2	B9	1041
		NGC 1496 - 16		4 00 22	+52 29.6	1	15.27		0.88		0.34		3		1668
		AJ79,864 # 4		4 00 24	+52 56.	1	13.86		0.63		−0.10		2	B3	373
279393	+34 00801			4 00 25	+34 41.6	1	9.74		0.15		0.01		2	B8	1041
		NGC 1496 - 51		4 00 25	+52 33.2	1	14.73		2.07				2		1668
237213	+55 00845	LS V +55 011		4 00 25	+55 52.1	2	8.80	.057	0.82	.038	−0.09	.024	4	B3 Ia	1012,8100
	−77 00156			4 00 25	−77 01.3	1	10.68		0.42				4	F5	2012
25363	+49 01092	IDS03568N4932	AB	4 00 26	+49 40.7	1	9.31		0.21				3	A0	70
		NGC 1496 - 15		4 00 28	+52 28.6	1	10.54		1.77		1.94		2		1668
25349	+52 00751			4 00 28	+52 37.7	2	9.53	.019	0.17	.019			4	B9.5V	70,1119
25290	+61 00665			4 00 29	+62 11.2	1	7.69		0.45		0.00		1	F5	758
25728	−61 00293	HR 1266		4 00 29	−61 13.1	3	4.96	.005	1.42	.000	1.70	.000	14	K4 III	15,1075,2012
25490	+5 00581	HR 1251		4 00 30	+05 51.1	8	3.90	.009	0.03	.004	0.09	.029	29	A1 V	3,15,1075,1363,1425*
25348	+52 00752	LS V +53 018		4 00 30	+53 11.5	3	8.38	.047	0.21	.019	−0.65	.005	7	B1 Vpenn	70,1012,1212
25337	+55 00846	IDS03565N5528	AB	4 00 30	+55 36.3	1	8.51		1.32		1.17		2	K0 III	1502
276029	+39 00916	G 81 - 2		4 00 32	+39 36.2	1	10.71		0.56		−0.04		1	F8	333,1620
		NGC 1496 - 10		4 00 32	+52 31.2	1	15.29		0.82		0.48		4		1668
		LP 474 - 92		4 00 38	+14 55.7	1	13.96		1.15				1		1570
		NGC 1496 - 5		4 00 38	+52 31.6	1	14.29		0.80		0.25		3		1668
25587	−27 01517			4 00 39	−27 37.3	1	7.39		0.54				4	F7 V	2012
25704	−57 00612			4 00 39	−57 21.0	1	8.10		0.55				4	F7 V	2012
		NGC 1496 - 6		4 00 40	+52 31.4	1	13.57		0.56		0.43		3		1668
		NGC 1496 - 1		4 00 40	+52 32.2	1	11.87		0.50		0.42		2		1668
25362	+54 00734	IDS03567N5447	AB	4 00 40	+54 55.8	1	6.54		0.46		−0.01		4	F2 IV	1501
25653	−45 01375			4 00 40	−44 48.1	10	8.20	.003	0.13	.003	0.15	.005	132	A3 IV/V	116,278,863,1075,1460,1628*
281451	+33 00769			4 00 41	+33 48.7	1	10.38		0.20		0.14		2	A0	1041
25521	+2 00640			4 00 42	+02 56.4	1	7.15		0.27		0.11		1	F0	1776

Table 1   201

HD	DM	Other Id	N Rem	α_{1950}	δ_{1950}	S	V	σ_V	B–V	σ_{B-V}	U–B	σ_{U-B}	n	Spectrum	References
		NGC 1496 - 7		4 00 42	+52 31.2	1	15.45		0.94		0.30		2		1668
		NGC 1496 - 36		4 00 44	+52 33.6	1	12.61		0.52		0.37		3		1668
265348	+19 00650	Hyades L 11		4 00 45	+19 19.1	3	10.15	.015	1.07	.008	1.03		6	F0	582,950,1058
25444	+39 00918	IDS03574N3914	AB	4 00 45	+39 22.5	1	7.13		0.68		0.24		1	G5	292
		NGC 1496 - 34		4 00 47	+52 32.1	1	11.01		1.29		1.16		2		1668
		NGC 1496 - 31		4 00 48	+52 31.4	1	13.52		0.56		0.43		2		1668
25361	+58 00694	RX Cam		4 00 49	+58 31.4	1	7.34		1.02				1	G0 Ia	6011
	+22 00625	Hyades HG8 - 55		4 00 50	+22 39.2	1	9.92		0.81		0.50		2		1058
		LS I +57 142		4 00 51	+57 19.3	1	10.87		0.74		0.45		2		41
		LP 833 - 29		4 00 52	−21 31.5	1	11.56		0.78		0.33		2		1696
		NGC 1496 - 20		4 00 53	+52 29.1	1	14.87		0.86		0.63		2		1668
		NGC 1496 - 22		4 00 53	+52 30.3	1	14.89		0.90		0.43		3		1668
		NGC 1496 - 42		4 00 53	+52 35.3	1	13.71		0.55		0.44		1		1668
		Hyades HG7 - 85		4 00 54	+14 51.	1	14.95		0.16				2		1570
25488	+27 00624			4 00 54	+27 34.4	1	8.80		0.38		0.06		12	A5	1603
		E2 - e		4 00 54	−45 01.	1	11.30		1.00				4		1075
		NGC 1496 - 23		4 00 56	+52 30.2	1	13.11		0.51		0.37		3		1668
		E2 - o		4 00 57	−44 55.1	1	14.09		0.57		0.04		5		1460
		Hyades HG8 - 26		4 00 58	+18 29.5	1	14.58		1.43				2		1058
		G 38 - 24		4 00 58	+30 34.6	1	12.77		1.58		1.21		1		333,1620
		NGC 1496 - 24		4 00 58	+52 30.4	1	12.94		1.01		0.50		3		1668
		NGC 1496 - 39		4 00 58	+52 34.0	1	15.35		0.78		0.29		2		1668
		UKST 117 # 7		4 00 58	−59 55.6	1	11.07		0.57		0.04		2		1584
		NGC 1496 - 25		4 01 00	+52 31.1	1	14.67		2.40				2		1668
25274	+68 00303	HR 1241		4 01 00	+68 32.7	1	5.87		1.54				2	K2 III	71
25938	−71 00234			4 01 00	−71 18.3	3	6.59	.009	0.08	.000	0.09		23	A1/2 V	1075,1408,2035
		E2 - s		4 01 01	−44 55.9	1	14.60		0.73		0.32		4		1460
	−74 00282			4 01 01	−74 35.0	1	10.16		0.90				4	K0	2012
		WLS 420 15 # 5		4 01 02	+17 05.2	1	10.95		1.13		0.69		2		1375
		NGC 1496 - 21		4 01 02	+52 28.9	1	13.67		0.78		0.23		2		1668
		NGC 1496 - 27		4 01 03	+52 30.5	1	14.35		0.60		0.48		2		1668
25558	+5 00584	HR 1253		4 01 05	+05 17.9	4	5.32	.006	-0.08	.005	-0.57	.013	11	B3 V	15,154,1061,1223
232882	+50 00889			4 01 05	+50 56.4	1	9.86		0.16				2	A3	70
		E2 - m		4 01 05	−44 53.1	1	13.10		0.81		0.49		5		1460
		NGC 1496 - 29		4 01 07	+52 30.7	1	14.65		0.64		0.43		2		1668
25173	+74 00184			4 01 08	+75 02.7	1	7.16		0.54		-0.08		2	F8 V	1003
		E2 - 50		4 01 08	−44 24.	4	10.95	.011	0.85	.005	0.44	.020	11		110,365,395,2011
		G 160 - 42		4 01 09	−05 15.7	1	14.85		1.45				3		3078
279370	+35 00794			4 01 10	+35 16.3	1	10.28		0.21		0.12		2	A0	1041
25532	+22 00626			4 01 12	+23 16.4	1	8.18		0.70		0.15		2	F6 IV-V	1003
283323		Hyades HG8 - 5		4 01 12	+26 26.	2	11.22	.000	1.42	.005	1.23		7	K6	950,1058
281563	+32 00711			4 01 12	+32 29.4	1	9.69		0.41		0.18		2	A5	1041
25631	−20 00769	HR 1258		4 01 12	−20 16.8	5	6.46	.006	-0.18	.012	-0.77	.019	18	B3 V	15,1034,1075,1732,2029
		L 373 - 7		4 01 12	−40 46.	1	14.70		0.30		-0.50		3		3062
25714	−44 01379			4 01 12	−44 26.1	6	8.81	.008	0.12	.008	0.13	.013	23	A1 V(m)	110,365,395,977,1075,2011
25570	+7 00592	Hyades vB 160	⋆	4 01 14	+08 03.6	6	5.46	.011	0.36	.012	0.00	.012	15	F4 V	15,254,1061,1127,3026,8015
25225	+73 00210			4 01 14	+73 52.0	1	6.49		0.99		0.83		2	K0	1502
		E2 - t		4 01 14	−44 53.3	1	15.17		0.63		-0.01		3		1460
		CS 22169 # 8		4 01 16	−12 57.9	1	14.99		0.43		-0.17		1		1736
		TonS 397		4 01 18	−29 30.	1	14.20		-0.03		-0.05		1		832,1036
		E2 - d		4 01 18	−44 46.	1	11.31		0.88				4		1075
25555	+23 00609	HR 1252	⋆ AB	4 01 22	+23 58.2	1	5.47		0.86				2	G0 III+A4 V	71
25675	−24 02062	DP Eri		4 01 22	−24 35.8	2	7.28	.015	1.54	.010	1.47		6	M3/4 III	2012,3042
25740	−44 01381			4 01 22	−44 31.7	5	8.12	.005	0.65	.004	0.12	.016	22	G3 V	110,395,977,1075,2011
25518	+38 00838			4 01 23	+38 45.9	1	8.16		0.51		0.22		2	F5 IV	105
232885	+50 00894			4 01 24	+50 56.9	1	9.50		0.28				2	A5	70
25661	−20 00770	HR 1259		4 01 24	−20 17.7	2	6.99	.019	1.23	.007	1.34		7	K2 III	1657,2032
	−44 01383			4 01 28	−44 19.4	2	9.94	.003	0.87	.006	0.50		10	K2	863,2012
25762	−44 01384			4 01 28	−44 43.7	5	8.76	.007	1.01	.005	0.85	.016	31	K1/2 III	863,977,1460,1628,2011
281518	+33 00772			4 01 29	+33 51.4	1	10.53		0.27		0.17		2	G5	1041
25408	+61 00667	UV Cam		4 01 31	+61 39.5	1	7.60		2.26				1	R8	1238
25539	+32 00714			4 01 32	+32 26.1	4	6.86	.016	0.04	.020	-0.60	.010	9	B3 V	229,1009,1041,1252
25621	+2 00645	HR 1257		4 01 33	+02 41.5	4	5.36	.015	0.50	.010	0.03	.008	10	F6 IV	15,254,1061,3077
25007	+80 00125	HR 1230	⋆ AB	4 01 34	+80 33.9	2	5.10	.000	0.57	.009	0.30		4	G8 III+A6 V	1363,3026
		Hyades HG8 - 7		4 01 36	+25 01.	5	13.80	.008	0.12	.010	-0.58	.008	17		203,1036,1058,1620,3060
25517	+43 00886	LS V +44 007		4 01 37	+44 08.6	1	9.27		0.25		-0.57		2	B1 V	1012
25874	−61 00295	IDS04008S6138	A	4 01 38	−61 29.7	2	6.73	.005	0.68	.005	0.17		6	G2 V	2033,3056
25874	−61 00295	IDS04008S6138	B	4 01 38	−61 29.7	1	11.79		1.50		1.17		1		3056
25538	+36 00813	IDS03584N3648	A	4 01 40	+36 57.1	1	8.00		0.17		0.13		3	A0	1601
25604	+21 00585	HR 1256	⋆ A	4 01 44	+21 56.8	7	4.37	.016	1.06	.009	0.97	.015	24	K0 III	15,1105,1355,1363*
25443	+61 00669			4 01 44	+61 58.0	4	6.74	.027	0.31	.015	-0.62	.016	11	B0.5III	758,1012,1631,8031
285391	+16 00554			4 01 46	+16 29.5	1	9.20		0.47		0.06		2	F2	1648
		E2 - l		4 01 46	−44 53.3	1	12.98		0.63		0.07		3		1460
25627	+17 00676			4 01 48	+17 22.9	1	6.65		1.10		1.00		2	K2 III	1648
		CCS 179		4 01 48	+23 05.	1	10.71		1.28				1	R0	1238
	+53 00728			4 01 48	+53 44.4	1	10.11		0.33				3		70
25515	+50 00896	LS V +50 005		4 01 52	+50 37.5	1	8.67		0.39		0.13		2	F3 III	1566
25700	−16 00770	HR 1263		4 01 52	−16 43.4	3	6.38	.004	1.26	.005			11	K2/3 III/IV	15,2013,2029
25795	−44 01386			4 01 52	−44 16.5	2	8.03	.001	1.40	.000	1.59		12	K3 III	977,2011
25682	−0 00636	G 80 - 41		4 01 57	+00 06.9	3	8.32	.015	0.77	.012	0.38	.010	9	G5	333,1355,1620,2012

HD	DM	Other Id	N	Rem	α_{1950}	δ_{1950}	S	V	σ_V	B–V	σ_{B-V}	U–B	σ_{U-B}	n	Spectrum	References
25425	+65 00391	HR 1248			4 01 57	+65 23.2	1	6.17		0.14		0.17		7	A3 m	8071
279390	+34 00805				4 01 58	+34 45.0	1	10.55		0.18		0.16		2	A3	1041
		TonS 398			4 02 00	−29 19.9	2	13.40	.009	−0.03	.042	−0.06	.009	2		832,1036
25723	−13 00806	HR 1265			4 02 02	−12 55.7	2	5.62	.010	1.07	.010	0.95		7	K1 III	196,2032
26169	−75 00258				4 02 04	−75 44.5	4	8.80	.022	0.74	.006	0.04	.010	11	G3w pec	119,742,1594,3040
25626	+27 00628				4 02 08	+27 28.5	1	7.91		0.09		0.02		std	A2	1390
232889	+51 00850				4 02 08	+51 52.1	1	9.19		1.89				3	M0	369
25842	−44 01388	IDS04006S4445	B		4 02 09	−44 35.7	4	9.50	.003	0.59	.004	0.09	.013	23		395,863,1460,2012
25842	−44 01389	IDS04006S4445	A		4 02 09	−44 37.0	12	8.48	.010	0.73	.010	0.22	.015	77	G5 V	23,110,365,395,977,1075*
25843	−45 01381				4 02 09	−44 52.2	10	7.64	.004	0.18	.005	0.12	.004	87	A1mA2-A7	278,395,863,1075,1460,1628*
		E2 - b			4 02 09	−44 54.9	2	11.57	.003	0.54	.010	0.02		7		1460,2012
25860	−48 01181				4 02 10	−47 59.9	4	6.63	.005	0.21	.003	0.12	.009	18	A4/5 IV	278,977,1075,2011
25873	−52 00491				4 02 10	−52 38.5	1	8.41		0.34		0.09		44	F0 III	978
284137	+22 00628	CF Tau			4 02 12	+22 21.7	1	10.11		0.93		0.33		5	G0	1768
25551	+53 00730				4 02 12	+53 27.6	1	8.35		0.20		0.17		2	A0	1566
		RL 89			4 02 12	+54 19.	1	15.39		−0.05		−0.74		2		704
		E2 - c			4 02 12	−45 02.	1	11.60		0.67				4		2011
		WLS 420 15 # 9			4 02 13	+15 25.7	1	10.37		0.55		0.04		2		1375
		LP 654 - 16			4 02 13	−03 11.2	1	11.78		1.27		1.10		1		1696
	+48 01040				4 02 14	+49 00.4	1	10.22		0.22				2	A0	70
25497		NGC 1502 - 60			4 02 14	+62 12.0	2	9.65	.027	0.32	.001	0.14	.008	5	A3 V	49,1631
25901	−52 00492				4 02 15	−52 20.0	1	8.85		0.03		0.03		43	A1 V	978
232890	+52 00761				4 02 16	+53 00.7	1	9.61		0.12				3	A0	70
	−45 01384		V		4 02 17	−44 47.6	1	10.41		0.75				4		2012
25498	+61 00670	NGC 1502 - 59			4 02 18	+62 12.6	2	7.99	.036	1.20	.005	0.79	.009	5	G0 V	49,1631
281549	+32 00717				4 02 21	+32 44.7	2	9.89	.004	0.17	.012	−0.07	.020	7	A0	1041,1775
25680	+21 00587	HR 1262	⋆ A		4 02 22	+21 52.5	5	5.90	.014	0.62	.008	0.12	.000	18	G5 V	15,1013,1197,1758,3077
25679	+27 00629				4 02 24	+27 44.7	1	8.42		0.42		0.19		13	A2	1603
		CS 22173 # 14			4 02 24	−17 29.3	1	13.73		0.37		−0.18		1		1736
232891	+52 00763				4 02 27	+52 57.3	1	9.18		0.17				2	B9	70
25803	−20 00774	HR 1267			4 02 29	−20 31.0	3	6.12	.008	1.16	.000			11	K1 II	15,1075,2035
		POSS 655 # 6			4 02 30	−05 07.8	1	17.88		2.16				2		1739
		Hyades L 12			4 02 31	+19 18.4	3	11.41	.004	1.35	.004	1.27	.020	7	K7	582,1058,1746
285433	+14 00648	G 6 - 44			4 02 35	+14 48.2	1	10.59		1.01		0.91		1	K0	333,1620
279397	+34 00809	IDS03594N3422	AB		4 02 38	+34 29.8	1	10.28		0.27		0.16		2	A2	1041
25720	+19 00658				4 02 42	+19 49.2	1	8.30		0.41		0.01		2	F5	1648
279394	+34 00811				4 02 42	+34 40.1	1	11.28		0.29		0.20		2	A0	1041
25602	+53 00732	HR 1255			4 02 42	+53 52.5	2	6.30	.005	0.99	.000	0.75		6	K0 III-IV	70,1355
25749	+13 00640				4 02 43	+14 09.1	1	7.30		1.22		1.28		1	G9 II-III	8100
	−44 01392				4 02 43	−44 19.6	1	10.92		0.44				4		2012
		G 7 - 20			4 02 47	+10 54.9	1	11.64		0.79		0.28		1	K0	333,1620
285367	+17 00679	Hyades L 10			4 02 47	+17 48.2	3	9.31	.013	0.89	.008	0.58		5	K2	582,950,1058
		LP 357 - 151	AB		4 02 48	+27 04.4	1	16.18		1.37				2		1570
281541	+32 00718				4 02 48	+32 52.7	1	10.14		0.22		0.14		2	A1	1041
		AJ79,864 # 5			4 02 48	+52 29.	1	12.85		0.83		−0.26		3	O5	373
279407	+33 00778				4 02 50	+34 03.4	1	10.21		0.16		0.14		2	A2	1041
25642	+49 01101	HR 1261			4 02 51	+50 13.1	8	4.28	.006	−0.01	.012	−0.04	.015	42	A0 IVn	15,61,71,369,1203*
283333	+25 00674	Hyades vB 161			4 02 57	+25 39.2	2	9.05	.000	0.84	.001	0.53		5	G5	950,1127
281540	+32 00719	G 38 - 25			4 02 57	+32 49.8	4	9.99	.010	0.97	.004	0.70	.009	10	K5	1003,1620,1775,3078
		NGC 1502 - 52			4 02 58	+62 13.1	1	12.25		0.70		0.45		1	B9	49
25594	+61 00673	NGC 1502 - 55			4 02 59	+62 15.4	3	7.95	.017	0.05	.017	0.04	.013	7	A2 Vn	49,305,1631
	−2 00801				4 02 60	−02 24.5	1	10.37		1.36		1.27		1	K6	1746
25899	−30 01638				4 03 00	−30 18.9	1	8.59		1.36		1.55		4	K3 III	1673
25768	+25 00675				4 03 02	+26 04.7	1	7.58		0.50		0.02		2	F8	1625
		NGC 1502 - 49			4 03 02	+62 13.7	2	10.73	.004	0.54	.022	−0.24	.007	6	B3 V	49,1631
25853	−13 00809				4 03 03	−13 07.7	1	9.53		0.42		−0.08		3	F3 V	196
283350	+24 00620	IDS04002N2427	A		4 03 04	+24 35.9	1	8.88		2.06		2.33		2	S:	1625
25852	−7 00738				4 03 05	−07 43.9	1	7.84		1.01		0.80		1	K0	1769
		NGC 1502 - 30			4 03 07	+62 11.2	1	9.68		0.59		−0.35		5	B1 V	1631
286513	+11 00570				4 03 08	+11 49.9	1	10.06		1.12				2	M1	1570
		NGC 1502 - 35			4 03 08	+62 11.9	1	10.50		0.52		−0.28		5	B3 V	1631
		Hyades HG7 - 89			4 03 09	+18 07.0	1	12.83		1.51				3		920
		Hyades HG7 - 90			4 03 10	+09 30.8	1	14.42		1.03				1		950
25966	−45 01388				4 03 12	−44 48.2	5	8.02	.006	1.60	.007	1.95	.011	61	M1 III	863,977,1460,1704,2011
279395	+34 00813				4 03 14	+34 31.8	1	10.91		0.36		0.16		2	A0	1041
		NGC 1502 - 27			4 03 16	+62 07.7	1	12.44		0.55		−0.08		1	B8	49
		NGC 1502 - 36	⋆ C1		4 03 16	+62 12.0	2	9.80	.031	0.56	.017	−0.32	.007	6	B8 V	305,1631
		NGC 1502 - 45			4 03 16	+62 13.9	3	11.44	.017	0.61	.013	−0.06	.008	7		49,305,1631
		NGC 1502 - 37	⋆ A1		4 03 18	+62 11.6	1	9.30		0.49		−0.28		1	B1.5V	1423
		NGC 1502 - 76	⋆ Q		4 03 18	+62 12.	1	10.39		0.58		−0.26		1	B8	542
	+61 00675	NGC 1502 - 26	⋆ G		4 03 19	+62 10.1	1	9.66		0.58		−0.35		8	B1.5V	1631
		Hyades CDS 11	⋆		4 03 21	+23 17.4	1	14.14		1.56				2		1570
	−31 01684				4 03 21	−31 19.9	1	8.30		0.62		−0.15		1		1774
		NGC 1502 - 44			4 03 22	+62 13.6	2	10.80	.010	0.60	.003	−0.19	.040	6	B8	305,1631
		Steph 451			4 03 24	+16 59.4	1	11.06		1.29		1.17		1	K7	1746
25639	+61 00676	NGC 1502 - 1	⋆ A		4 03 24	+62 12.0	4	6.95	.017	0.47	.016	−0.51	.009	11	B0 II-III	305,542,758,1631
		NGC 1502 - 43			4 03 24	+62 14.4	2	11.38	.005	0.55	.006	−0.11	.029	5		305,1631
25825	+15 00582	Hyades vB 10			4 03 26	+15 33.9	2	7.85	.000	0.59	.005	0.10	.000	4	G0	1127,8023
25638	+61 00676	NGC 1502 - 2	⋆ B		4 03 26	+62 11.8	4	7.00	.031	0.42	.019	−0.53	.008	12	B0 IIn	305,542,758,1631
		G 222 - 1			4 03 26	+82 47.4	1	10.80		1.42		1.25		2		7010

Table 1   203

HD	DM	Other Id	N	Rem	α_{1950}	δ_{1950}	S	V	σ_V	B–V	σ_{B-V}	U–B	σ_{U-B}	n	Spectrum	References
		NGC 1502 - 23			4 03 27	+62 08.5	1	10.75		0.62		-0.11		5	B3 V	1631
		NGC 1502 - 40			4 03 27	+62 13.0	2	11.23	.063	0.53	.020	-0.18	.008	7	B6	305,1631
25799	+31 00703	V490 Per			4 03 28	+32 15.1	3	7.03	.014	0.06	.015	-0.57	.000	10	B3 V	1009,1252,1775
		POSS 655 # 1			4 03 29	−05 13.1	1	14.65		1.52				1		1739
		LP 357 - 156			4 03 30	+23 34.	1	12.15		0.72				1		1570
		RL 90			4 03 30	+51 58.	1	10.51		0.16		-0.18		2		704
		NGC 1502 - 7			4 03 30	+62 12.3	2	11.53	.009	0.60	.022	-0.01	.044	6	B8	305,1631
25921	−10 00834	CY Eri			4 03 31	−10 25.8	1	7.10		1.61		1.79		4	M4 III	3042
25823	+27 00633	HR 1268, GS Tau			4 03 32	+27 28.0	4	5.19	.005	-0.14	.018	-0.49	.016	11	B9p Si	23,1049,1202,3023
25910	−9 00811	HR 1272			4 03 32	−08 59.4	5	6.26	.007	0.06	.003	0.09	.005	19	A3 V	15,688,1415,2012,2018
		CS 22173 # 15			4 03 32	−17 29.1	1	13.22		0.42		-0.25		1		1736
25945	−27 01540	HR 1275			4 03 34	−27 47.2	4	5.59	.014	0.32	.009	0.00	.020	13	F0 IV/V	15,158,2012,3013
25944	−20 00780	HR 1274			4 03 35	−20 38.8	3	6.34	.004	0.92	.000			11	G8/K0 III	15,1075,2031
25836	+21 00591				4 03 37	+21 49.4	1	7.92		1.71		1.93		2	K2	1648
		NGC 1502 - 42			4 03 37	+62 13.9	1	12.61		0.60		0.25		2	B9	305
		NGC 1502 - 10		⋆A2	4 03 39	+62 12.4	1	9.82		0.57		-0.22		4	B3	1631
25835	+25 00677				4 03 40	+25 35.3	1	7.97		0.18		0.20		2	A0	105
		NGC 1502 - 16			4 03 40	+62 10.9	2	11.62	.017	0.61	.014	-0.01	.006	7	B8	49,1631
25868	+18 00581				4 03 43	+19 01.1	1	7.49		0.36		-0.06		2	F2	1648
25833	+33 00785	AG Per		⋆AB	4 03 43	+33 18.8	3	6.70	.015	0.02	.013	-0.52	.005	8	B5 V:p	588,1009,1252
25834	+29 00672				4 03 44	+30 08.4	1	7.51		1.54				3	K1 II	20
		NGC 1502 - 11			4 03 44	+62 12.0	1	10.88		0.63		-0.03		5	B8	1631
25776	+48 01046				4 03 45	+48 47.7	1	9.02		0.38				4	A5	70
	+61 00678	NGC 1502 - 12		⋆D	4 03 45	+62 12.3	3	9.42	.005	0.48	.008	-0.43	.010	10	B2	305,1423,1631
26043	−47 01261				4 03 45	−47 21.4	3	8.77	.003	1.37	.002	1.60	.001	17	K2/3 III	863,977,2011
		Hyades HG7 - 92			4 03 46	+13 24.9	1	13.52		1.47				2	M1	950
		NGC 1502 - 13			4 03 46	+62 12.6	1	12.76		0.56		0.10		1		305
25821	+38 00841				4 03 47	+38 19.3	1	8.97		0.34		0.06		2	F0	1601
		RL 91			4 03 48	+52 24.	1	11.47		0.22		-0.06		2		704
		CS 22173 # 21			4 03 49	−18 44.2	1	14.00		0.42		-0.21		1		1736
26122	−58 00340				4 03 50	−58 18.8	1	10.16		1.00		0.80		1	F5IV/V	1696
		NGC 1502 - 14			4 03 50	+62 12.5	1	13.67		0.99		0.69		1		305
		Steph 455			4 03 52	−21 39.0	1	12.14		1.27		1.19		1	K7	1746
25867	+28 00619	HR 1269			4 03 55	+28 52.1	4	5.23	.026	0.34	.010	-0.01	.039	10	F1 V	23,229,254,3026
279467	+33 00786				4 03 57	+34 10.4	1	10.43		0.20		0.16		2	A2	1041
		Hyades HG8 - 34			4 03 58	+17 39.7	2	11.24	.005	0.78	.010	0.33		3		920,1058
25787	+51 00861	LS V +51 014			4 03 58	+51 19.2	1	7.65		0.02		-0.75		3	B2 V	399
26074	−46 01293				4 04 00	−45 48.1	3	7.56	.004	0.23	.002	0.14	.015	10	A7 II/III	278,1075,2012
	+62 00655	NGC 1502 - 61			4 04 02	+62 16.4	2	9.58	.013	0.44	.008	-0.34	.010	6	B2 Vn	49,1631
26004	−19 00820				4 04 03	−19 38.8	1	7.52		1.20		1.16		1	K0 III	8100
25933	+13 00644				4 04 04	+13 36.0	1	9.03		0.69				1	G5	882
269015		Sk -70 054			4 04 07	−70 07.1	1	14.29		-0.15		-0.95		4	WN4	980
26096	−47 01263				4 04 09	−46 56.1	3	8.50	.002	1.43	.001	1.69	.012	18	K3 III	863,977,2011
285507	+14 00653	Hyades L 15			4 04 11	+15 12.1	3	10.49	.004	1.18	.000	1.12		6	K5	582,1058,1746
		Hyades HG7 - 95			4 04 13	+18 54.3	1	11.49		0.87				1		920
		G 6 - 45			4 04 14	+15 51.3	1	9.09		0.69		0.20		2		333
25893	+37 00878	V491 Per		⋆AB	4 04 14	+37 56.7	4	7.13	.024	0.86	.024	0.47	.016	12	G5	938,1355,1620,3013
26414	−75 00259				4 04 18	−75 17.7	1	9.18		0.61				4	G0 V	2012
25665	+69 00238	G 247 - 13			4 04 23	+69 24.8	1	7.68		0.91		0.78		6	G5	7009
	−21 00784				4 04 23	−20 58.5	4	9.69	.014	1.21	.010	1.16	.016	15	K7	158,1705,1775,3073
25977	+13 00646				4 04 31	+13 24.3	1	7.86		0.31		0.10		2	A3	1648
		Hyades HG7 - 97			4 04 36	+13 17.6	1	12.16		1.09				1		920
		NGC 1502 - 78			4 04 39	+62 13.4	2	10.92	.009	0.50	.002	-0.20	.004	5	B3 V	49,1631
		G 221 - 29			4 04 40	+74 13.8	1	14.88		0.77		0.01		2		1658
		Tau sq 1 # 102			4 04 42	+25 25.3	1	14.09		1.12		0.63		6		23
25932	+42 00897	IDS04011N4255		B	4 04 44	+43 03.5	1	6.60		-0.03		-0.23		2	B8	985
25918	+44 00862	G 81 - 5			4 04 44	+44 31.9	1	8.29		1.02		0.83		1	G5	1620
284141	+20 00703				4 04 46	+21 07.8	1	9.18		0.63		0.13		1	F8	1261
25878	+52 00771	XX Cam			4 04 46	+53 13.7	2	7.29	.020	0.82	.005	0.30	.050	2	K0	759,842
		RL 92			4 04 48	+51 08.	1	14.03		0.15		-0.58		2	sdB	704
		Hyades HG7 - 100			4 04 52	+08 02.3	1	12.46		1.24				1		920
		LP 474 - 123			4 04 52	+14 05.1	1	10.81		1.42		1.28		1		1773
26015	+14 00657	Hyades vB 11		⋆AB	4 04 52	+15 01.8	4	6.03	.020	0.40	.005	0.01	.010	15	F3 V	15,254,1127,8015
285482	+16 00558	Hyades L 16			4 04 52	+16 23.1	3	9.94	.004	1.00	.012	0.89		7	K0	582,950,1058
26068	−10 00841				4 04 52	−09 53.4	1	6.74		1.18		1.14		3	K0	1657
	+54 00739			V	4 04 55	+55 02.4	2	10.11	.015	2.70	.000	3.30		6	M2	369,8032
	−63 00295				4 04 56	−62 56.9	1	9.14		1.24		1.18		2		3073
25975	+37 00881	HR 1277			4 04 57	+37 35.9	5	6.09	.004	0.94	.007	0.74	.014	21	K1 III	15,1003,1355,3016,4001
		Hyades HG7 - 103			4 05 01	+09 37.4	1	13.07		0.79				1		950
25940	+47 00939	HR 1273, MX Per			4 05 01	+47 34.9	10	4.04	.013	-0.03	.005	-0.55	.013	58	B3 Ve	15,61,71,154,1009*
26094	−13 00814				4 05 02	−12 47.5	1	8.36		0.96		0.64		3	G8 III	196
26039	+16 00559				4 05 04	+16 23.9	1	7.62		0.10		-0.08		2	B9 m	1648
		LP 414 - 103			4 05 06	+18 50.	1	13.26		1.18				1		1570
26038	+16 00560	HR 1280		⋆AB	4 05 07	+17 12.5	1	5.89		1.50				2	K5 IIIb	71
	+58 00707			V	4 05 08	+59 04.4	1	10.30		2.03				3	M3	369
	−31 01701				4 05 10	−31 26.4	1	10.52		-0.31		-1.17		5	O8p	222
25877	+59 00759	HR 1270			4 05 13	+59 46.6	2	6.32	.035	1.16	.020	0.91	.005	7	G8 IIa	1355,8100
25998	+37 00882	HR 1278			4 05 16	+37 54.6	2	5.52	.004	0.52	.038	-0.01	.021	7	F7 V	254,3013
26151	−27 01560				4 05 18	−27 33.3	3	8.49	.004	0.83	.004	0.42		10	K1 V	1075,2012,3008
26307	−60 00289				4 05 18	−60 00.5	1	7.86		0.94		0.67		7	K0 III	1704

HD	DM	Other Id	N	Rem	α_{1950}	δ_{1950}	S	V	σ_V	B–V	σ_{B-V}	U–B	σ_{U-B}	n	Spectrum	References
		Hyades HG7 - 104			4 05 19	+16 44.4	1	11.52		1.44				2		920
		Steph 458			4 05 19	+16 44.5	1	11.51		1.43		1.25		3	K7	1746
25914	+56 00884	LS V +56 056			4 05 19	+56 57.6	1	7.99		0.60		-0.28		2	B6 Ia	1012
	-41 01288				4 05 21	-40 53.4	1	10.99		1.13				1		1705
281621		G 38 - 26			4 05 23	+33 30.3	3	10.19	.028	1.53	.005	1.18	.014	5	K7	196,1620,1746
25774	+69 00240				4 05 23	+70 11.9	1	8.00		0.95		0.68		3	G5	1733
	+23 00621				4 05 24	+23 15.6	1	9.06		1.24		1.10		1	K0	1610
25948	+54 00740	HR 1276			4 05 24	+54 41.9	1	6.26		0.41				2	F5 V	1733
		Hyades L 21			4 05 26	+27 48.9	1	11.77		0.54				2		582
		CS 22169 # 13			4 05 27	-12 19.7	1	14.71		0.16		-0.62		1		1736
26091	+13 00647				4 05 30	+13 23.5	2	8.79	.020	0.90	.005			5	G5	882,987
26244	-45 01406				4 05 31	-45 04.7	3	8.35	.001	0.57	.000	0.05	.008	16	F8 IV	977,1075,2011
		CS 22177 # 9			4 05 34	-25 10.6	1	14.27		0.40		-0.23		1		1736
284155	+23 00622	Hyades L 18			4 05 36	+23 38.2	1	9.44		0.90				2	G5	582
26081	+25 00678				4 05 37	+25 44.8	1	7.18		1.42		1.26		3	G8 II	833
		NGC 1513 - 115			4 05 38	+49 22.7	1	11.05		0.62		0.47		2		1668
		NGC 1513 - 109			4 05 38	+49 26.2	1	15.28		0.85		0.65		1		1668
		Steph 459			4 05 40	+12 03.6	1	11.28		1.33				2	K5	1746
286554		Hyades L 14			4 05 40	+12 03.6	3	11.28	.004	1.33	.004	1.26		4	K2	582,950,1058
		Hyades HG8 - 53			4 05 41	+23 25.6	2	12.85	.018	1.51	.009	1.28		5		920,1058
		NGC 1513 - 113			4 05 41	+49 24.4	1	14.12		0.94		0.38		2		1668
26129	+11 00576	G 7 - 22			4 05 44	+12 12.7	1	8.61		0.90		0.67		1	K0	1620
	+52 00775	IDS04019N5219		AB	4 05 45	+52 27.4	1	9.92		0.32				3		70
		Hyades HG8 - 23			4 05 46	+19 23.2	2	12.27	.020	1.00	.000	0.76		3		950,1058
26262	-43 01304	HR 1285			4 05 46	-43 03.0	5	6.59	.009	0.93	.003	0.61	.007	21	K0 III	15,278,977,1075,2011
26090	+28 00624	IDS04027N2856		AB	4 05 47	+29 03.6	2	8.30	.023	0.64	.002	0.15		8	G1 V +G8 V	987,3016
		RL 93			4 05 48	+51 44.5	1	11.97		0.21		-0.32		2		704
		RL 94			4 05 48	+51 44.7	1	12.09		0.19		-0.29		2		704
26273	-44 01414				4 05 48	-44 07.9	2	9.66	.003	0.25	.005	0.07		10	A2/3 III	863,2011
26116	+23 00624				4 05 52	+23 44.4	1	6.97		1.20		1.02		2	K2	23
		NGC 1513 - 23			4 05 55	+49 20.2	1	15.17		0.94		0.65		3		1668
26141	+16 00561				4 05 57	+17 09.7	1	7.66		0.12		0.06		2	A0	1648
		NGC 1513 - 9			4 05 59	+49 22.4	1	13.38		0.68		0.45		4		1668
		NGC 1513 - 27			4 06 00	+49 20.9	1	14.46		1.08		0.41		3		1668
		NGC 1513 - 90			4 06 00	+49 25.0	1	13.27		0.66		0.36		2		1668
26020	+53 00738				4 06 00	+53 44.3	1	8.50		0.42		-0.02		2	F8	1566
		NGC 1513 - 92			4 06 01	+49 26.3	1	15.00		0.90		0.52		2		1668
		NGC 1513 - 26			4 06 02	+49 21.9	1	14.37		0.87		0.60		2		1668
26126	+28 00627	IDS04030N2823		AB	4 06 03	+28 30.8	1	8.21		0.50		0.11		1	F8 V	8082
		NGC 1513 - 2			4 06 03	+49 23.5	1	14.16		0.72		0.54		4		1668
		NGC 1513 - 1			4 06 03	+49 24.0	1	12.40		0.63		0.20		4		1668
		NGC 1513 - 94			4 06 06	+49 26.7	1	14.71		0.75		0.15		1		1668
		Hyades HG7 - 106			4 06 07	+16 00.2	1	12.54		0.90				1		950
		NGC 1513 - 3			4 06 07	+49 23.5	1	12.94		0.59		0.28		2		1668
		NGC 1513 - 88			4 06 08	+49 25.6	1	11.41		1.75		1.68		2		1668
		NGC 1513 - 38			4 06 10	+49 19.0	1	13.46		0.81		0.42		2		1668
		NGC 1513 - 4			4 06 10	+49 23.4	1	15.24		0.93		0.59		2		1668
		NGC 1513 - 62			4 06 11	+49 21.3	1	12.20		1.55		1.19		2		1668
26354	-52 00497	AG Dor			4 06 12	-52 42.0	1	8.69		0.95		0.64		1	K1 Vp	1641
26171	+13 00648	HR 1284			4 06 14	+13 16.0	3	5.95		0.05	.013	-0.07	.020	9	B9.5V	252,1049,1079
		NGC 1513 - 86			4 06 14	+49 25.2	1	13.68		0.60		0.34		3		1668
		Hyades HG7 - 107			4 06 15	+08 06.9	1	14.00		1.53				1		920
26162	+19 00672	HR 1283			4 06 15	+19 28.7	1	5.50		1.07				2	K2 III	71
		NGC 1513 - 65			4 06 15	+49 21.8	1	12.40		0.62		0.32		3		1668
232902	+52 00777				4 06 16	+52 22.0	1	9.70		0.26		0.14		2	A0	1566
		NGC 1513 - 61			4 06 17	+49 21.2	1	14.82		0.83		0.60		2		1668
		NGC 1513 - 101			4 06 18	+49 27.5	1	11.88		0.58		0.32		2		1668
		NGC 1513 - 56			4 06 21	+49 19.9	1	15.50		0.98		0.57		2		1668
		NGC 1513 - 42			4 06 22	+49 18.2	1	15.49		1.06		0.86		2		1668
		NGC 1513 - 55			4 06 22	+49 20.4	1	11.97		0.63		0.40		2		1668
	+57 00778				4 06 22	+57 43.7	1	9.55		2.12				3		369
26237	-9 00823				4 06 22	-08 48.1	1	7.16		0.03		0.01		30	B9	978
26301	-33 01569	IDS04044S3307		AB	4 06 22	-32 59.3	1	7.43		0.24				3	A7 IV/V	173
		NGC 1513 - 43			4 06 24	+49 18.0	1	15.60		1.07				4		1668
		NGC 1513 - 45			4 06 25	+49 19.1	1	13.01		0.76		0.51		2		1668
		NGC 1513 - 71			4 06 25	+49 22.7	1	11.55		0.63		0.40		2		1668
26140	+35 00809				4 06 27	+35 50.9	1	8.00		0.89		0.49		2	K0	1601
		NGC 1513 - 73			4 06 27	+49 21.7	1	15.16		1.02		0.38		1		1668
26594	-73 00249				4 06 27	-73 24.8	2	8.96	.003	0.25	.004	0.14		6	A4 V	278,2012
		Hyades HG7 - 110			4 06 28	+15 42.4	1	14.44		1.49				2	M1	950
		Hyades HG7 - 109			4 06 30	+07 05.	1	12.53		0.83				1		950
		NGC 1513 - 75			4 06 30	+49 21.0	1	15.36		0.97		0.43		2		1668
		NGC 1513 - 72			4 06 31	+49 22.1	1	12.17		0.61		0.35		2		1668
		NGC 1513 - 50			4 06 32	+49 20.4	1	12.01		0.60		0.33		2		1668
26352	-43 01309				4 06 33	-42 59.1	3	8.30	.006	0.32	.002	0.00	.016	16	F0 V	977,1075,2011
		LB 227			4 06 35	+17 00.2	1	15.35		0.09		-0.65		1	DA	3016
	+7 00604	Hyades HG7 - 113			4 06 38	+07 59.4	1	10.86		0.99				1		950
26114	+49 01112				4 06 38	+49 35.1	1	9.24		0.24				2	A0	70
26375	-45 01412				4 06 38	-45 20.9	3	8.94	.002	0.34	.001	0.08	.017	16	F2 III	977,1075,2011
284184	+22 00641	Hyades L 17			4 06 39	+22 33.1	2	10.79	.010	0.80	.005	0.44		3	G5	1058,1674

Table 1

HD	DM	Other Id	N	Rem	α_{1950}	δ_{1950}	S	V	σ_V	B–V	σ_{B-V}	U–B	σ_{U-B}	n	Spectrum	References
	−42 01385				4 06 39	−41 59.6	3	9.33	.005	1.41	.004	1.67	.007	17		863,1075,1460
26297	−16 00791				4 06 46	−16 01.3	1	7.47		1.11				3	G5/6 IVw	1594
		Hyades HG7 - 114			4 06 47	+09 57.9	1	14.94		1.13				1		1570
	+4 00643				4 06 48	+04 37.8	1	9.84		0.56		0.38		3	A2	1776
		TonS 401			4 06 48	−29 55.	1	12.28		−0.12		−0.50		1		832
26491	−64 00305 HR 1294				4 06 49	−64 21.5	7	6.37	.005	0.64	.005	0.11	.005	47	G3 V	15,1075,1075,2012*
26298	−16 00793				4 06 51	−16 32.0	6	8.15	.011	0.36	.007	−0.04	.010	20	F0/2 V	158,742,1003,1594,1775,3077
26294	−8 00798 IDS04045S0812		AB		4 06 55	−08 03.6	1	7.20		0.22		0.11		39	A2	978
26199	+41 00823		V		4 06 58	+42 02.7	1	7.75		1.96				3	M1	369
26200	+38 00848				4 06 59	+39 05.8	2	7.01	.010	0.39	.000	−0.03	.005	4	F4 V	105,1733
26413	−46 01314 HR 1291		★ AB		4 07 01	−45 59.8	8	6.58	.006	0.38	.005	−0.01	.006	45	F3 V	14,15,278,388,977*
	+8 00641				4 07 02	+09 07.2	1	10.36		1.43				2	K2	950
		Hyades L 22			4 07 02	+31 41.8	1	11.54		0.83		0.42		2		1058
26326	−16 00796 HR 1288				4 07 02	−16 31.0	2	5.43	.026	−0.15	.000	−0.53		5	B5 IV	1034,2007
26159	+49 01116				4 07 03	+49 49.2	1	9.01		0.14				3	A0	70
286572	+8 00642 Hyades HG7 - 115				4 07 05	+09 10.5	3	10.10	.011	1.20	.000			5	K5 V	920,1632,1746
		Hyades vA 6			4 07 05	+13 21.6	2	14.77	.005	1.44	.009	1.10		5		218,950
		G 7 - 23			4 07 10	+09 35.3	1	14.12		1.02		0.36		2		333,1620
26146	+55 00851				4 07 10	+55 42.8	1	9.05		0.28		0.19		3	A2	1502
	−15 00728				4 07 10	−14 50.1	2	10.65	.044	1.20	.044			3	M0	1017,1619
26337	−8 00801 EI Eri				4 07 15	−08 01.5	1	7.07		0.68		0.15		1	G2 IV-V	1641
		Met +28 033			4 07 16	+28 37.8	1	11.38		0.87		0.76		2	B8	8082
		Hyades vA 13			4 07 17	+17 54.7	1	14.14		0.17		−0.58		1	DA	3028
283366					4 07 18	+28 47.	1	10.60		0.77		0.33		3	K0 V	8082
285465	+16 00564 Hyades vA 14		★		4 07 20	+17 14.4	1	9.07		1.07		0.98		1	K5	333,1620
283367	+28 00628				4 07 21	+28 37.3	1	10.32		0.57		0.11		7	A0	8082
26389	−28 01423				4 07 21	−27 56.4	1	8.72		0.44				4	F3 V	2012
26323	+7 00607				4 07 25	+07 34.1	1	8.64		0.32		0.22		1	B9	1776
		MCC 435			4 07 26	+70 04.6	1	9.73		1.22		1.20		3	K7	1625
		Met +28 035			4 07 29	+28 49.6	1	12.02		0.77		0.62		1	F2	8082
		Hyades CDS 60		★		4 07 31	+32 48.4	1	12.10		1.44				3		1674
283420	+24 00629				4 07 34	+25 14.0	1	9.14		0.44		−0.04		2	F8	1625
		G 247 - 14			4 07 35	+62 26.9	1	12.06		1.42		1.20		2		7010
		Met +28 036			4 07 36	+28 28.0	1	12.95		1.53				1		8082
	−59 00314				4 07 43	−59 39.3	1	9.71		1.23				1	K4 V	1705
26196	+58 00715				4 07 44	+58 38.5	1	8.46		0.69				39	G0	6011
		Hyades vA 22			4 07 45	+14 35.3	1	14.99		1.45				2		1058
26311	+33 00807 HR 1286				4 07 46	+33 27.5	2	5.73	.010	1.45	.010	1.46	.010	8	K1 II-III	1355,8032
		Hyades HG7 - 117			4 07 47	+06 31.2	1	13.55		1.26				1		920
26322	+26 00686 HR 1287, IM Tau				4 07 47	+26 21.1	4	5.41	.029	0.34	.012	0.06	.016	8	F2 IV-V	23,39,254,3058
		Hyades HG7 - 116			4 07 48	+11 19.8	1	13.38		1.45				2		950
26101	+68 00310 HR 1282				4 07 48	+68 22.4	1	6.32		1.18				2	K0	71
		TonS 403			4 07 48	−28 29.9	1	12.92		−0.31		−1.08		1		832,1036
26345	+18 00594 Hyades vB 13				4 07 49	+18 17.6	2	6.62	.000	0.42	.000	−0.01	.000	4	F6 V	1127,8023
26667					4 07 54	−69 55.	1	10.00		1.04		0.90		1	C1,3	864
		CS 22177 # 10			4 07 55	−25 52.5	1	14.31		0.40		−0.20		1		1736
26409	−7 00758 HR 1290				4 07 56	−07 03.2	4	5.44	.004	0.94	.000	0.67	.000	12	G8 III	15,1071,1769,2032
		L 229 - 91			4 07 56	−53 30.7	1	11.80	.017	1.50	.022	1.18		5		912,1705,3073
		CS 22173 # 33			4 07 59	−19 56.7	1	13.61		−0.27		−1.13		1		1736
26310	+41 00827 IDS04046N4136		AB		4 08 00	+41 43.9	1	8.54		0.42		0.04		2	F8	379
	−43 01321		V		4 08 05	−43 29.8	1	10.10		0.87				4	K2	2012
26076	+71 00239 HR 1281				4 08 08	+72 00.0	1	6.03		1.01				3	K1 III	71
26441	−5 00841 IDS04058S0508		AB		4 08 16	−05 00.1	1	7.36		0.65		0.22		3	G2 IV +G2IV	3030
		RL 95			4 08 18	+53 10.	1	10.33		0.20		−0.45		2		704
26464	−9 00837 HR 1293				4 08 23	−08 56.9	4	5.70	.004	1.06	.007	0.93	.002	37	K0	15,978,1061,2032
	−18 00767 G 160 - 49				4 08 28	−17 54.6	1	9.62		0.91		0.57		1	G5	3056
26397	+29 00676 Hyades vB 138				4 08 30	+29 42.2	3	8.28	.018	0.87	.013	0.50	.000	4	K0	1127,1529,8023
		LP 301 - 85			4 08 31	+29 51.2	1	13.53		1.44				2		1674
26408	+26 00689				4 08 33	+26 48.7	1	8.88		0.46		−0.06		59	F8	23
		GD 56			4 08 33	−04 06.1	1	15.50		0.14		−0.54		1		3060
		Hyades HG7 - 122			4 08 37	+15 51.8	1	15.15		1.65				2	M1	950
26018	+75 00166				4 08 37	+76 09.9	1	8.19		0.84		0.52		4	G5	1625
26462	+5 00601 Hyades vB 14		★ A		4 08 40	+05 23.7	5	5.73	.011	0.36	.012	−0.01	.013	14	F1 IV-V	15,254,1127,1415,8015
26407	+34 00829				4 08 48	+35 13.9	1	7.59		0.32		−0.15		2	A0	1625
26370	+50 00926				4 08 50	+51 02.4	1	8.29		0.40		−0.05		2	F5	1566
283466	+26 00690				4 08 52	+27 01.1	1	9.55		1.13		0.79		5	G8 III	23
283467	+26 00691				4 08 53	+27 02.4	1	10.32		0.20		−0.14		5	A0	23
26575	−35 01588 HR 1299				4 08 54	−35 24.1	4	6.43	.009	1.07	.007	0.96		16	K2 III(p)	15,536,1075,2029
		WLS 420 15 # 8			4 08 55	+14 47.3	1	11.73		0.70		0.43		2		1375
284163	+23 00635 Hyades L 20				4 08 56	+23 30.5	3	9.36	.022	1.09	.003	0.80		5	K0	582,1610,1726
26823	−71 00249				4 08 58	−71 46.3	2	8.61	.005	0.48	.009			8	F6 V	1075,2012
26438	+31 00719				4 09 02	+32 03.5	1	8.59		0.36		0.29		2	A0	1625
276122	+41 00829 G 81 - 8				4 09 02	+41 51.9	2	10.53	.012	0.64	.000	0.03	.005	4	K0	1620,1658
283465					4 09 06	+27 02.	1	10.94		0.85		0.41		5	G2 III	23
26420	+41 00830 LS V +41 003				4 09 09	+41 59.5	2	7.83	.024	0.32		−0.44	.010	3	A5	379,1212
26612	−42 01400 HR 1302				4 09 09	−42 07.4	11	4.93	.006	0.33	.008	0.07	.009	50	A9 V	14,15,182,278,388,977*
26478	+21 00606 IDS04063N2117		AB		4 09 14	+21 24.5	1	8.75		0.66		0.31		2	F5	1648
		Hyades vA 43			4 09 15	+17 29.9	2	14.77	.005	1.65	.005			6	M1	950,1058
283444	+28 00630				4 09 15	+28 18.4	1	9.26		0.51		−0.04		7	G0 V	8082
	−74 00283				4 09 15	−74 05.4	1	9.40		1.20				4		2012

HD	DM	Other Id	N Rem	α_{1950}	δ_{1950}	S	V	σ_V	B–V	σ_{B-V}	U–B	σ_{U-B}	n	Spectrum	References
		L 230 - 188		4 09 17	−53 42.0	1	13.58		1.93		2.11		2		3078
26515	+14 00663			4 09 23	+14 48.3	1	8.61		0.44		0.01		1	G0	586
26591	−20 00801	HR 1300		4 09 24	−20 29.1	3	5.79	.005	0.17	.010	0.09		12	A2m	15,2027,8071
26514	+23 00642	IDS04064N2319	AB	4 09 25	+23 26.8	1	7.19		1.06		0.71		4	G6 III	206
26308	+66 00312			4 09 25	+66 25.3	1	8.47		2.08		2.20		3	K5	1733
26574	−7 00764	HR 1298, o1 Eri		4 09 25	−06 58.0	8	4.04	.006	0.33	.004	0.16	.013	148	F2 II-III	3,15,1075,1425,2012*
	+5 00603			4 09 28	+06 01.9	1	9.72		0.56		0.48		4	A2	1776
		Hyades vA 45		4 09 30	+16 07.4	1	13.99		1.53				2		920
		Hyades HG8 - 48		4 09 30	+26 28.	1	11.45		0.88		0.36		2		1058
		L 55 - 41		4 09 30	−71 57.	1	12.76		0.46		−0.20		2		1696
26546	+16 00569	HR 1295		4 09 39	+17 09.0	5	6.10	.013	1.08	.007	0.96	.013	10	K0 III	252,770,1375,3016,4001
		LP 714 - 40		4 09 44	−10 41.7	1	12.00		0.55		−0.06		2		1696
		Hyades vA 47		4 09 45	+11 52.5	1	11.59		0.88				1		1570
285554	+17 00694			4 09 45	+17 42.4	1	8.50		1.43		1.27		2	K2	1648
		WLS 420 35 # 6		4 09 47	+35 11.2	1	10.56		1.37		1.35		2		1375
284277	+19 00674	Hyades HG8 - 21	⋆ A	4 09 53	+19 52.5	1	10.38		0.94				2	K2	950
284277	+19 00674	Hyades HG8 - 21	⋆ AB	4 09 53	+19 52.5	2	10.02	.004	0.99	.004	0.84		5	K2	950,1058
284277	+19 00674	Hyades HG8 - 21	⋆ B	4 09 53	+19 52.5	1	11.40		1.17				2		950
26571	+22 00649	HR 1297		4 09 53	+22 17.2	2			0.19	.000	−0.27	.015	7	B9 III(p)	1049,1079
		CS 22169 # 35		4 09 53	−12 12.8	1	12.88		0.89		0.37		1		1736
	+50 00931			4 09 57	+50 26.1	1	9.91		0.06				3		70
		G 81 - 9		4 09 58	+42 52.5	1	12.29		1.39		1.20		1		333,1620
26482	+48 01059			4 09 59	+48 58.1	1	6.92		0.91		0.61		2	G5	1601
		WLS 420-20 # 6		4 10 00	−19 29.2	1	12.23		0.79		0.28		2		1375
26598	+10 00548			4 10 01	+11 02.7	1	8.08		0.37		0.13		2	F2	1733
26731	−44 01444	VX Hor		4 10 02	−44 45.7	1	9.4		0.50				4	F5/6 V	2012
		G 160 - 51		4 10 04	−11 25.5	1	15.46		0.31		−0.71		1		3056
		EG 31, HZ 2		4 10 06	+11 45.	1	13.86		−0.05		−0.88		1	DA:	3028
26544	+35 00823			4 10 07	+35 21.0	1	7.92		0.26		0.25		3	A0	1601
		AJ91,575 # 1	V	4 10 09	+28 11.6	1	13.73		1.49		1.04		1	M4 V	1606
26495	+50 00932			4 10 09	+50 21.1	1	7.79		1.28				2	K0	70
26744	−45 01439			4 10 10	−44 56.6	2	9.68	.006	0.57	.000	0.08		10	F7/8 V	863,2011
285597	+15 00600			4 10 12	+15 25.0	1	10.00		0.66		0.22		1	G5	586
		Hyades HG7 - 126		4 10 15	+18 52.0	1	16.29		0.12				1		1570
285571	+16 00570	Hyades vA 51		4 10 19	+16 38.4	2	9.18	.000	0.64	.000			2	G0	987,987
		Hyades vA 52		4 10 20	+14 36.9	2	11.73	.005	0.84	.005	0.43		4		582,1058
	−76 00264			4 10 20	−76 44.2	1	9.19		1.63				4	M0	2012
26570	+39 00956			4 10 21	+39 33.2	1	7.32		1.66		2.03		2	K5	1601
26779	−52 00502			4 10 22	−52 14.3	1	8.58		1.23		1.26		33	K1 III	978
	−30 01697			4 10 24	−30 20.0	1	10.12		1.14		1.06		2		1730
		G 160 - 52		4 10 28	−08 24.3	1	14.16		1.35		0.70		2		3056
		LP 774 - 32		4 10 30	−20 26.1	1	13.39		0.64		0.04		3		3062
	−50 01302			4 10 30	−50 30.4	1	10.47		1.26		1.18		4	K7	158
	+6 00648			4 10 33	+06 28.5	2	9.10	.000	1.29	.000	0.79		9	K0	1594,3077
		AK 8 - 744		4 10 33	+09 55.4	1	10.76		0.91				1		1570
	+19 00677			4 10 39	+19 43.9	1	10.38		2.04		2.36		3	M0	1746
26605	+37 00897	HR 1301		4 10 39	+37 50.4	1	6.46		1.05		0.85		2	G9 III	1733
26741	−21 00812			4 10 42	−21 29.8	1	8.81		1.17		1.11		2	K0 III	1375
26833	−53 00664			4 10 46	−53 32.3	1	7.07		1.16		1.18		5	K2 III	1628
26677	+8 00651	HR 1308	⋆ A	4 10 48	+08 45.9	3	6.52	.018	0.15	.005	0.10	.012	11	A2 m	15,146,1071
26677	+8 00651	IDS04081N0838	B	4 10 48	+08 45.9	1	12.80		0.78		0.27		6		146
26676	+9 00549	HR 1307		4 10 50	+10 05.2	3	6.22	.005	0.05	.004	−0.33	.014	9	B8 Vn	15,1079,1256
26690	+7 00617	HR 1309	⋆ AB	4 10 51	+07 35.4	6	5.29	.007	0.36	.010	0.00	.007	17	F2 V +F3 V	15,292,1256,1381,3030,8015
26771	−30 01701			4 10 52	−30 26.4	1	8.05		1.49		1.79		2	K3 III	1730
		NGC 1528 - 21		4 10 54	+51 03.3	1	13.34		0.75		0.33		1		49
		NGC 1528 - 28		4 10 54	+51 04.4	1	14.72		0.87		0.28		1		49
26553	+57 00785	HR 1296	⋆	4 10 55	+57 20.1	2	6.08	.000	0.61	.005	0.51	.014	7	A4 III	985,1733
		HA 3 # 222		4 10 56	+75 27.0	1	10.96		1.23		1.23		5		271
26820	−44 01450	HR 1316		4 10 56	−44 29.7	10	6.70	.007	1.48	.008	1.80	.006	59	K3/4 III	14,15,182,278,388,863*
	+50 00935	NGC 1528 - 6		4 10 58	+51 05.3	1	10.06		0.63		0.20		1		49
	+50 00936	NGC 1528 - 7		4 10 59	+51 00.5	1	10.08		0.37		0.27		1	F0 V	49
		NGC 1528 - 22		4 10 59	+51 04.5	1	13.50		0.74		0.27		1		49
26770	−28 01457	IDS04090S2848	AB	4 10 59	−28 40.0	2	7.45	.009	0.51	.005			8	F7 V	1075,2012
		KUV 04110+1434		4 11 00	+14 33.9	1	13.65		0.38		−0.12		1	sdB	1708
		KUV 684 - 1		4 11 00	+14 33.9	1	13.91		0.23		−0.18		1	sdB	974
		Hyades vA 54		4 11 02	+15 14.4	1	15.01		1.65				2		19,951
26703	+12 00564	HR 1310		4 11 03	+12 37.7	2	6.24	.005	1.15	.009	1.09	.021	4	K0	1375,3051
		AK 5 - 172		4 11 03	+22 57.3	1	11.65		1.22				1		1570
283447		V773 Tau		4 11 03	+28 04.	3	10.66	.026	1.39	.009	1.10	.009	7	K2e	649,5011,8082
26819	−44 01451			4 11 03	−44 00.1	1	8.25		1.37		1.63		2	K2/3 III	1673
26739	−1 00600	HR 1312		4 11 06	−01 16.5	5	6.43	.016	−0.12	.012	−0.55	.008	21	B5 IV	15,154,252,361,1061
	+50 00938	NGC 1528 - 9		4 11 07	+51 02.2	1	10.38		0.39		0.15		1	A3 V	49
		HRC 23, FM Tau		4 11 08	+28 05.3	1	13.98		0.78		−0.43		1		649
26854	−45 01445			4 11 08	−44 48.5	3	8.57	.002	0.31	.003	0.05	.009	16	A9 IV/V	977,1075,2011
		HRC 24, FN Tau		4 11 09	+28 20.4	1	14.95		1.69		0.84		1	K5:e	649
26603	+50 00940	NGC 1528 - 1		4 11 09	+51 06.4	1	8.75		0.22		−0.16		1	A0 V	49
		HRC 25, CW Tau		4 11 11	+28 03.4	2	13.47	.775	1.37	.015	0.38	.105	2	dK2e	649,5011
26722	+8 00652	HR 1311	⋆ AB	4 11 13	+09 08.3	6	4.84	.014	0.81	.009	0.49	.018	20	G5 III	15,1071,1355,1363*
26630	+48 01063	HR 1303	⋆ A	4 11 13	+48 17.1	6	4.15	.017	0.96	.016	0.64	.008	2	G0 Ib	150,542,8003,8007*
26630	+48 01063	HR 1303	⋆ AB	4 11 13	+48 17.1	6	4.14	.007	0.95	.004	0.65	.005	37	G0 Ib	15,61,71,1119,1355,1363

Table 1   207

HD	DM	Other Id	N Rem	α₁₉₅₀	δ₁₉₅₀	S	V	σ_V	B–V	σ_B-V	U–B	σ_U-B	n	Spectrum	References
26630	+48 01063	IDS04076N4809	C	4 11 13	+48 17.1	2	10.28	.030	0.46	.010	0.08	.015	4		150,542
		Hyades vA 57		4 11 15	+17 00.4	2	14.18	.039	1.53	.005			5		950,1058
26620	+50 00941			4 11 15	+50 33.7	1	7.45		1.59				2	K0	70
		CS 22186 # 5		4 11 18	−35 58.2	1	12.96		0.37		0.00		1		1736
		NGC 1528 - 23		4 11 20	+51 04.4	1	13.55		0.59		0.26		1		49
232921	+53 00748			4 11 20	+53 34.7	1	8.06		1.95				3	M0	369
26581	+58 00724	G 175 - 24		4 11 20	+58 24.0	2	8.67	.003	0.98	.031	0.85	.006	8	K0	196,7008
	+50 00943	NGC 1528 - 4		4 11 22	+51 04.1	1	9.97		0.92		0.53		1		49
232923	+50 00942	NGC 1528 - 2	⋆ AB	4 11 22	+51 07.3	1	9.61		0.24		0.13		1	A1 V	49
		NGC 1528 - 26		4 11 23	+51 05.1	1	14.28		0.82		0.22		1		49
		Hyades HG7 - 129		4 11 24	+09 16.9	2	11.10	.000	1.23	.000			4	K4	920,1746
		Hyades HG8 - 13		4 11 24	+21 19.4	1	13.80		1.31		0.99		3		1058
		AJ91,575 # 2	V	4 11 24	+28 44.0	1	11.64		1.34		1.25		1	K7 V	1606
26720	+19 00679			4 11 26	+19 27.3	1	8.18		0.50		0.04		2	F8	1648
26709	+27 00649			4 11 26	+27 50.1	2	8.42	.025	0.38	.005	-0.01	.010	5	F2	1625,8082
		NGC 1528 - 30		4 11 26	+51 04.8	1	15.34		0.90		0.29		1		49
		L 230 - 205		4 11 26	−54 00.1	1	13.87		1.56		1.13		1		3073
26719	+23 00648			4 11 27	+23 34.3	1	6.78		1.16		0.87		2	K0	1625
	+52 00790		V	4 11 27	+52 52.1	1	9.33		1.97				3	M2	369
26797	−17 00825			4 11 28	−17 36.0	1	8.52		0.26		0.04		2	A9 V	1375
26710	+25 00685			4 11 29	+26 07.9	4	7.22	.017	0.63	.006	0.11	.021	7	G2 V	23,979,3026,8112
26673	+40 00912	HR 1306		4 11 29	+40 21.5	4	4.70	.016	1.01	.008	0.64	.005	9	G5 Ib +A2 V	15,1363,6009,8015
26748	+14 00672			4 11 30	+14 25.4	1	7.53		1.80				3	K0	882
		NGC 1528 - 12		4 11 30	+51 02.5	1	11.28		0.27		0.14		1		49
26737	+22 00657	Hyades vB 16		4 11 32	+22 19.6	3	7.05	.008	0.42	.005	0.00	.010	6	F5 V	401,1127,8023
26736	+23 00649	Hyades vB 15		4 11 32	+23 27.0	3	8.08	.014	0.66	.001	0.20	.002	7	G3 V	1127,8023,8025
		NGC 1528 - 5		4 11 32	+51 06.9	1	10.05		1.17		0.86		1		49
26702	+37 00899			4 11 33	+37 25.1	1	6.31		0.97		0.73		2	G5	1375
		NGC 1528 - 19		4 11 33	+51 06.1	1	12.99		0.60		0.20		1		49
	+50 00948	NGC 1528 - 3		4 11 35	+51 03.9	1	9.91		0.26		0.12		1	A0 V	49
26756	+14 00673	Hyades vB 17		4 11 36	+14 30.0	2	8.46	.000	0.70	.002	0.24	.001	4	G5 V	1127,8023
		Hyades HG7 - 132		4 11 36	+18 36.3	1	14.49		1.64				2	M2	950
284248	+21 00607	G 8 - 16		4 11 36	+22 13.8	6	9.24	.018	0.42	.020	-0.20	.005	11	F2	516,979,1003,1620*
26767	+12 00566	Hyades vB 18		4 11 40	+12 18.6	2	8.06	.000	0.64	.001	0.17	.001	4	G0	1127,8023
		NGC 1528 - 20		4 11 40	+51 06.3	1	13.24		0.50		0.29		1		49
		AJ91,575 # 3	V	4 11 42	+27 45.1	1	12.08		1.52		1.26		1	M1 V	1606
		Hyades HG7 - 133		4 11 44	+18 44.2	1	11.84		0.73				1		950
		HRC 26, FP Tau		4 11 44	+26 39.0	3	13.88	.010	1.41	.064	0.60	.301	3	dM2.5e	640,649,1227
		HRC 27, CX Tau		4 11 44	+26 40.7	2	13.68	.035	1.50	.035	0.53	.045	2	dM1.5e	640,649
		CS 22169 # 19		4 11 44	−16 13.2	1	13.76		0.39		-0.22		1		1736
		NGC 1528 - 15		4 11 45	+51 03.6	1	12.45		0.47		0.26		1		49
		NGC 1528 - 27		4 11 47	+51 00.8	1	14.35		0.74		0.30		1		49
		Hyades HG8 - 1		4 11 48	+27 38.	2	12.69	.058	1.57	.004	1.19		5		950,1058
		G 8 - 17		4 11 48	+27 38.2	1	12.68		1.52		1.71		1		1620
26784	+10 00551	Hyades vB 19		4 11 49	+10 34.6	3	7.12	.015	0.51	.001	0.03	.018	9	F8 V	1127,1355,8023
		NGC 1528 - 17		4 11 49	+51 05.3	1	12.56		0.47		0.28		1		49
		NGC 1528 - 24		4 11 50	+51 00.3	1	13.99		0.93		0.43		1		49
		MCC 436		4 11 50	+66 46.9	1	10.28		1.22		1.17		2	M0	1726
26793	+9 00550	HR 1315		4 11 52	+09 53.2	5	5.22	.005	-0.10	.007	-0.34	.007	12	B9 Vn	15,1061,1079,1363,3023
		NGC 1528 - 25		4 11 52	+51 01.9	1	14.28		0.69		0.15		1		49
	+50 00951	NGC 1528 - 8		4 11 52	+51 13.1	1	10.38		0.31		0.21		1	A5 V	49
26794	+2 00665	G 80 - 42		4 11 53	+02 53.6	1	8.79		0.96		0.78		1	K3 V	333,1620
		G 160 - 55		4 11 53	−08 12.3	2	11.95	.015	0.83	.005	0.34	.028	4		1696,3056
26927	−40 01286	HR 1323		4 11 53	−40 29.0	4	6.36	.004	1.46	.005			18	K3 III	15,1075,2013,2029
		LP 414 - 119		4 11 54	+18 40.	2	14.65	.020	1.24	.025			2		950,1570
		NGC 1528 - 16		4 11 54	+51 03.4	1	12.52		0.40		0.25		1		49
		RL 97		4 11 54	+56 02.	1	11.09		0.44		-0.51		2		704
		NGC 1528 - 14		4 11 56	+51 07.3	1	11.94		0.37		0.25		1		49
26847	−13 00842			4 11 56	−12 51.5	1	10.55		0.24		-0.65		6		1732
		Hyades HG7 - 134		4 11 57	+19 00.9	2	12.35	.000	0.94	.000			2		987,987
		NGC 1528 - 10		4 11 58	+51 02.5	1	10.77		0.26		0.15		1	A0 V	49
	+47 00960			4 11 59	+48 02.8	1	9.25		2.35				3		369
		NGC 1528 - 11		4 11 59	+51 03.4	1	11.17		0.27		0.13		1		49
26846	−10 00867	HR 1318	⋆ AB	4 12 01	−10 22.8	6	4.86	.006	1.16	.016	1.14	.007	21	K3 III	15,1003,1075,2032*
26766	+29 00678			4 12 02	+29 46.8	2	7.11	.025	1.07	.000	1.02		4	K1 IV	20,23
281723	+32 00755			4 12 02	+32 55.9	1	10.35		0.71		0.19		2	K0	1375
		Steph 477		4 12 04	+12 55.8	1	10.72		1.21				1	K4	1746
286589		Hyades vA 68		4 12 04	+12 55.8	5	10.69	.044	1.20	.014	1.18		15	K5	19,582,582,951,3072
26920	−30 01710			4 12 04	−30 14.3	1	7.20		1.08		0.96		5	K0 III	1628
		NGC 1528 - 18		4 12 05	+51 04.0	1	12.64		0.41		0.28		1		49
		NGC 1528 - 29		4 12 06	+51 06.2	1	14.86		0.88		0.30		1		49
		CS 22186 # 2		4 12 08	−36 54.0	1	13.22		0.45		-0.18		1		1736
26746	+41 00837			4 12 09	+41 44.7	1	7.84		0.35		-0.02		1	F0	1462
		NGC 1528 - 13		4 12 09	+51 04.7	1	11.30		0.31		0.14		1		49
26934	−37 01664			4 12 09	−37 09.3	1	6.99		1.48		1.81		6	K3 (III)	1628
284234		Hyades L 27		4 12 11	+22 59.5	1	10.66		1.39				2	G5	582
		LP 474 - 201		4 12 12	+11 18.8	1	15.53		1.29				1		1570
		Hyades HG8 - 52		4 12 12	+23 29.6	2	12.57	.009	1.24	.009	1.19		2		920,1058
26802	+21 00610	IDS04093N2151	A	4 12 15	+21 58.4	1	8.80		0.22		-0.16		2		401
26981	−46 01345			4 12 15	−46 18.7	1	9.03		1.38		1.63		1	K2/3 (III)	1612

HD	DM	Other Id	N Rem	α_{1950}	δ_{1950}	S	V	σ_V	B−V	σ_{B-V}	U−B	σ_{U-B}	n	Spectrum	References
		G 175 - 25		4 12 16	+54 49.7	1	13.48		0.96		0.62		1		1658
		Steph 478		4 12 21	+14 16.4	1	11.54		1.38				1	K5	1746
285625		Hyades vA 72		4 12 21	+14 16.4	3	11.56	.021	1.37	.017	1.29		5	K2	582,950,1058
		Hyades vA 71		4 12 21	+14 57.2	2	15.62	.025	1.05	.000	0.32		2		1570,3028
26967	−42 01425	HR 1326		4 12 21	−42 25.0	9	3.85	.005	1.10	.005	1.00	.007	32	K2 III	14,15,182,278,388*
		Hyades HG8 - 66		4 12 24	+20 43.	1	14.99		1.30				3		1058
		RL 98		4 12 24	+52 08.	1	11.13		0.28		-0.21		2		704
26670	+61 00687	HR 1305		4 12 28	+61 43.6	1			-0.14		-0.52		2	B5 Vn +A2 V	1079
	−6 00855	G 82 - 5		4 12 28	−05 45.4	3	10.59	.009	0.70	.011	0.10	.024	7		1620,1658,3056
27019	−46 01347	IDS04111S4623	AB	4 12 37	−46 15.4	5	6.78	.006	0.58	.005	0.05	.008	27	F8/G0 V	278,863,1075,1612,2011
	−4 00782			4 12 40	−04 32.5	1	9.39		1.22		1.12		3	K5 V	3072
	−8 00813	G 160 - 59		4 12 42	−08 40.5	2	10.75	.078	0.46	.019	-0.15	.029	3		265,3056
285590		Hyades vA 75		4 12 43	+15 34.9	1	10.97		1.29				2	M2	19,920
		Steph 479		4 12 43	+15 35.0	1	10.97		1.29		1.24		3	K7	1746
		Hyades vA 76		4 12 43	+16 38.3	2	15.23	.015	1.56	.019	1.10		3		19,218,951
26842	+31 00737	IDS04096N3127	AB	4 12 45	+31 34.2	4	7.31	.015	0.59	.018	0.11	.015	10	F8	292,938,1381,3030
26775	+51 00893			4 12 45	+51 15.3	1	9.11		0.12				2		70
	−46 01350	SY Hor		4 12 45	−46 34.7	1	11.19		0.83		0.43		1		1612
26913	+5 00613	HR 1321, V891 Tau	⋆B	4 12 46	+06 04.6	7	6.93	.009	0.68	.011	0.20	.022	39	G0	15,770,1071,1084,1733*
26874	+20 00721	Hyades vB 162		4 12 46	+20 41.8	3	7.82	.013	0.70	.004	0.27	.006	6	G4 V	1127,1733,8023
		HA 3 # 172		4 12 47	+75 12.9	1	10.85		0.57		0.08		4	F5	271
26842	+31 00738	IDS04096N3127	C	4 12 48	+31 34.9	1	8.51		0.60		0.14		4	F8	3016
26764	+53 00750	HR 1314		4 12 48	+53 29.3	1	5.19		0.05				4	A2 Vn	1363
26923	+5 00614	HR 1322, V774 Tau	⋆A	4 12 49	+06 03.9	6	6.31	.012	0.57	.011	0.04	.013	31	G0 IV	15,770,1071,1084,1733,3077
26912	+8 00657	HR 1320		4 12 49	+08 46.1	9	4.28	.010	-0.06	.009	-0.52	.017	33	B3 IV	3,15,154,1071,1203*
26765	+52 00794			4 12 50	+52 34.5	1	9.27		0.22				2	A0	70
27054	−45 01458			4 12 55	−44 56.2	2	8.98	.000	0.38	.000	-0.02		8	F3 V	1075,2011
26911	+15 00603	Hyades vB 20	⋆A	4 12 56	+15 16.6	4	6.33	.021	0.40	.000	0.01	.016	8	F3 V	15,254,1127,8015
26965	−7 00780	HR 1325, DY Eri	⋆A	4 12 58	−07 43.8	17	4.42	.012	0.82	.005	0.45	.009	72	K1 V	3,15,22,58,369,1003*
26755	+57 00787	HR 1313		4 12 59	+57 44.3	1	5.71		1.09				2	K1 III	71
	+45 00901			4 13 03	+46 07.4	1	9.89		2.00				3		369
26976	−7 00781	IDS04108S0749	B	4 13 04	−07 44.1	4	9.51	.011	0.04	.034	-0.68	.008	14	DA	22,1281,1414,1732,3073
26976	−7 00781	IDS04108S0749	BC	4 13 04	−07 44.1	1	9.32		0.17		-0.66		4	DA +M4 Ve	158
26976	−7 00781	IDS04108S0749	C	4 13 04	−07 44.1	2	11.17	.005	1.58	.080	0.83		7	M4 Ve	1414,3078
		HA 3 # 176		4 13 06	+75 17.1	1	11.48		0.68		0.19		4	G0	271
		HA 3 # 97		4 13 08	+74 59.8	1	11.54		1.18		0.96		5	F9	271
		AJ89,1229 # 266		4 13 09	+03 08.3	1	10.90		1.70				1	M1	1632
		Hyades HG7 - 141		4 13 09	+18 44.1	1	14.07		1.52				2	M2	950
26994	−16 00820			4 13 11	−16 34.1	1	6.73		-0.07		-0.55		2	B7 III	252
	−72 00289			4 13 11	−72 31.5	1	9.99		0.25				4		2012
26858	+44 00896			4 13 12	+45 08.2	1	9.14		1.89				3	M1	369
		G 7 - 31		4 13 15	+07 46.6	1	11.50		0.76		0.28		1		333,1620
27050	−32 01683			4 13 16	−31 54.8	1	8.15		1.12		1.09		10	K1 III	1673
		Hyades L 25		4 13 18	+18 45.7	3	11.91	.005	1.45	.004	1.22		4	K5	582,950,1058
		AJ91,575 # 4	V	4 13 23	+28 00.2	1	12.49		1.47		1.20		1	K7 V	1606
		Met +28 065		4 13 25	+28 18.9	1	12.33		1.90				2		8082
26857	+50 00963			4 13 25	+50 44.6	1	6.60		1.71		1.43		2	K2	1566
		G 247 - 17		4 13 29	+70 00.5	1	10.96		1.08		1.06		4		7010
27346	−70 00287			4 13 29	−70 32.8	1	6.99		0.32		0.08		5	A9 IV	1628
26871	+46 00858			4 13 30	+47 04.6	1	9.73		0.89		0.35		1	F8	1306
284253	+21 00612	Hyades vB 21	⋆V	4 13 35	+21 47.1	3	9.15	.005	0.81	.004	0.47	.001	8	K0 V	950,1127,8023
	+8 00659	G 7 - 32		4 13 36	+08 36.0	1	9.85		1.15		1.15		1	K7 V	333,1620
		Hyades vA 88		4 13 36	+14 02.9	1	15.34		1.62				1		951
		Met +28 066		4 13 38	+28 26.7	1	12.52		1.89				2		8082
		SX Per		4 13 39	+41 36.6	3	10.71	.017	0.94	.025	0.69	.005	3	G0	592,1399,1462
	+47 00965	AR Per		4 13 39	+47 16.7	2	9.95	.007	0.46	.027	0.35	.004	2	A3	668,699
284229	+23 00661	Hyades L 29		4 13 40	+23 31.5	1	10.04		0.84				2	G0	582
26906	+45 00904	LS V +46 011	⋆B	4 13 40	+46 06.5	1	7.97		0.20		-0.69		2	B7	1733
26801	+61 00692	IDS04093N6140	A	4 13 45	+61 47.4	1	7.72		0.00		-0.13		2	B9	3032
26801	+61 00692	IDS04093N6140	B	4 13 45	+61 47.4	1	11.43		1.40		0.95		2		3032
26801	+61 00692	IDS04093N6140	C	4 13 45	+61 47.4	1	9.54		1.21		0.74		1		3016
27256	−62 00332	HR 1336	⋆A	4 13 47	−62 35.9	4	3.33	.016	0.92	.006	0.65	.014	18	G8 II/III	3,15,243,1075
		Hyades HG7 - 144		4 13 48	+07 17.4	1	12.33		1.11				1		920
		CS 22169 # 25		4 13 48	−14 41.4	1	14.15		0.32		0.08		1		1736
232932	+50 00966			4 13 50	+50 50.0	1	9.36		0.11				2	B7	70
232931	+52 00802			4 13 53	+52 15.1	1	9.55		0.51		0.02		1	G0	1566
26637	+74 00196			4 13 55	+75 05.9	2	8.63	.016	0.29	.012	0.06	.000	8	A5	271,1058
		Hyades HG7 - 145		4 13 58	+07 43.4	1	13.58		1.54				2		950
27029	+15 00607			4 13 58	+16 05.5	1	6.56		1.11		1.00		3	K1 III	1648
27039	+2 00670	IDS04114N0233	AB	4 13 59	+02 40.0	1	8.26		0.33		0.09		3	A5	3016
		Hyades vA 96		4 14 03	+16 14.1	2	14.40	.050	1.56	.010			4		1058,1674
27028	+19 00689	IDS04112N1926	AB	4 14 06	+19 33.3	1	7.10		0.41		-0.06		2	F4 V +F6 V	3030
276177	+41 00841			4 14 06	+41 35.2	1	10.73		0.73				1	K0	592
284310	+22 00666	Hyades L 28	⋆	4 14 08	+22 33.1	2	9.78	.015	1.23	.00	1.20		4	K5	582,3072
27304	−62 00334	HR 1340	⋆AB	4 14 09	−62 19.0	5	5.45	.011	1.11	.007	1.02	.015	16	K0 III	15,243,1075,2035,3051
27209	−46 01355			4 14 13	−46 05.5	4	8.92	.008	0.77	.002	0.34	.007	22	G8/K0 V	863,977,1075,2011
		LP 358 - 650		4 14 15	+22 32.6	1	11.12		0.68				1	K7	920
		IC 361 - 11		4 14 18	+58 06.4	1	13.17		1.11		0.80		2		735
		G 7 - 33		4 14 19	+11 30.9	1	13.32		1.42		1.10		1		333,1620
284321	+21 00614			4 14 19	+21 45.0	2	10.18	.003	0.40	.001	0.25	.008	2	A0	1722,1748

Table 1 209

HD DM	Other Id	N Rem	α_{1950}	δ_{1950}	S	V	σ_V	B–V	σ_{B-V}	U–B	σ_{U-B}	n	Spectrum	References
27045 +20 00724	HR 1329		4 14 20	+20 27.4	6	4.94	.020	0.25	.018	0.10	.006	20	A3 m	15,1199,1363,3023*
	IC 361 - 16		4 14 21	+58 05.7	1	14.78		1.13		0.69		2		735
	IC 361 - 10		4 14 22	+58 10.8	1	14.43		1.09		0.56		3		735
	G 160 - 61		4 14 22	−12 40.9	2	10.90	.016	1.51	.022	1.23	.073	6		3056,7009
	IC 361 - 9		4 14 24	+58 11.2	1	11.80		1.55		1.68		2		735
	G 160 - 62		4 14 25	−12 12.4	1	12.60		1.54		1.10		1		3056
	G 8 - 22		4 14 26	+23 21.0	2	11.36	.000	1.25	.015	1.01		2	K4	333,1620,1746
284350 +19 00691	Hyades HG7 - 148	A	4 14 28	+19 38.2	2	10.38	.015	0.80	.115			2	G0	920,950
284350 +19 00691	Hyades HG7 - 148	B	4 14 28	+19 38.2	1	11.72		0.84				1		950
	HRC 28, CY Tau		4 14 28	+28 13.5	1	13.50		1.28		-0.16		1	dM2e	640
26961 +49 01150	NGC 1545 - 1	⋆ V	4 14 28	+50 10.5	5	4.61	.009	0.04	.005	0.03	.005	23	A2 V	15,49,1363,3023,8015
	IC 361 - 20		4 14 28	+58 08.5	1	16.89		0.86		0.62		2		735
26974 +48 01076	LS V +48 015		4 14 29	+48 35.9	2	8.04	.010	0.96	.010	0.77		42	F5	1462,6011
	IC 361 - 7		4 14 29	+58 10.3	1	11.49		0.68		0.26		18		735
	CS 22186 # 17		4 14 29	−34 25.4	1	13.53		0.43		-0.21		1		1736
	Steph 480		4 14 30	+18 54.5	1	10.84		1.22				2	K5	1746
285630 +18 00614	Hyades L 26		4 14 30	+18 54.5	3	10.83	.012	1.22	.004	1.22		4	K5	582,950,1058
27004 +43 00935			4 14 30	+43 33.8	1	7.78		0.16		-0.20		2	B9	401
+48 01074			4 14 30	+48 59.0	1	10.51		1.12				1		592
	IC 361 - 8		4 14 30	+58 08.6	1	13.21		1.25		0.87		4		735
	L 56 - 67		4 14 30	−73 35.0	1	12.76		1.21		1.19		1		1773
+48 01075			4 14 31	+48 55.5	1	9.45		1.37				1		592
	Met +27 179		4 14 32	+27 13.7	1	11.36		1.06		0.79		2	G8 V	8082
	IC 361 - 17		4 14 32	+58 09.9	1	15.77		1.55		1.32		4		735
	AJ91,575 # 5	V	4 14 33	+28 25.7	1	13.56		1.50		1.37		1	M2 V	1606
	IC 361 - 21		4 14 35	+58 10.0	1	17.73		1.08				3		735
	L 56 - 66		4 14 36	−73 35.0	1	13.88		1.43		1.22		1		1773
	Hyades vA 105		4 14 37	+17 33.0	2	14.47	.035	1.51	.000			2		19,950
	Hyades vA 106		4 14 38	+14 46.7	2	14.47	.005	1.54	.005			4	M3	920,951
	IC 361 - 19		4 14 39	+58 11.6	1	15.29		1.51		1.29		1		735
27274 −53 00672			4 14 39	−53 26.3	4	7.64	.024	1.12	.005	1.02	.034	11	K5 V	1075,1311,2012,3072
27026 +41 00844	HR 1328		4 14 40	+42 01.2	2	6.21		-0.07	.007	-0.32	.011	2	B9 V	691,1079
	AA45,405 S208 # 7		4 14 40	+52 44.6	1	13.47		0.60		-0.31		3		797
	Met +27 180		4 14 43	+27 55.3	1	11.90		1.52		1.29		2	B5	8082
	AA45,405 S208 # 5		4 14 43	+52 44.6	1	10.94		0.60		-0.43		3	O9.5V	797
26684 +75 00173			4 14 43	+75 59.2	1	6.72		-0.04		-0.48		4	B5 III	555
27290 −51 01066	HR 1338, γ Dor		4 14 43	−51 36.7	5	4.24	.004	0.31	.005	0.02	.014	14	F0 V	15,278,1075,2011,3026
27108	BO Tau		4 14 45	+26 14.5	1	10.89		1.63				1	R3	1238
27025 +44 00901			4 14 46	+44 33.8	1	8.36		0.18		-0.26		2	B9	401
	AA45,405 S208 # 6		4 14 46	+52 43.6	1	12.65		0.63		-0.31		3	B0 V	797
27130 +16 00577	Hyades vB 22	⋆ V	4 14 47	+16 49.6	3	8.33	.029	0.76	.018	0.33	.001	4	G8 V	882,1127,8023
	RL 99		4 14 48	+56 13.	1	13.15		0.42		-0.41		2		704
	Hyades vA 112		4 14 52	+12 17.6	2	15.32	.005	1.75	.061			4	M2	19,950,951
	HA 3 # 105		4 14 52	+75 04.1	1	11.62		1.85				5	M4	271
27179 −6 00862	HR 1332		4 14 52	−06 35.6	5	5.94	.005	1.08	.004	0.95	.002	37	K0	15,978,1415,2031
27145 +13 00659			4 14 54	+13 42.9	3	6.93	.023	1.00	.012	0.59	.037	6	G5	882,1648,3040
	Hyades HG8 - 14	⋆	4 14 55	+21 25.0	5	12.38	.019	1.42	.011	1.20	.029	9	M0 V	333,950,1058,1620,1722,1748
	Hyades HG7 - 152		4 14 56	+18 21.2	2	13.88	.000	1.56	.005			6	M3	950,1058
	IC 361 - 12		4 14 58	+58 10.9	1	13.08		1.02		0.59		3		735
	Hyades vA 115		4 14 59	+13 32.4	5	12.56	.031	1.47	.016			8	M1	19,582,950,951,1570,1674
	IC 361 - 4		4 15 00	+58 08.7	1	13.34		1.77		1.46		3		735
	IC 361 - 15		4 15 02	+58 10.5	1	15.09		1.61		1.22		1		735
	IC 361 - 14		4 15 02	+58 11.6	1	15.37		1.59		1.29		3		735
	CS 22182 # 22		4 15 02	−31 52.0	1	13.02		0.40		-0.19		1		1736
	IC 361 - 5		4 15 03	+58 08.8	1	14.55		1.70		1.70		7		735
	LP 834 - 41		4 15 03	−26 10.6	3	11.82	.047	1.38	.010	1.06	.074	7		1696,1705,3073
	Hyades vA 119		4 15 04	+16 25.4	3	13.99	.025	1.53	.019			5	M2	19,950,951
	Hyades vA 122		4 15 06	+14 25.5	2	14.96	.054	1.63	.005			3		19,950
	LP 474 - 280		4 15 06	+14 25.5	1	13.16		1.69				1		950
	AA45,405 S208 # 1		4 15 06	+52 50.4	1	10.55		0.42		0.31		3	A0 V	797
	IC 361 - 2		4 15 07	+58 08.0	1	12.04		1.98		2.16		5		735
27149 +17 00703	Hyades vB 23		4 15 08	+18 08.1	3	7.53	.011	0.68	.008	0.23	.008	5	G5 V	1127,1355,8023
	IC 361 - 6		4 15 09	+58 08.5	1	15.00		1.05		0.61		3		735
	AO 871		4 15 11	+21 26.2	2	11.75	.013	1.24	.026	0.83	.002	2	G2 III	1722,1748
	IC 361 - 13		4 15 12	+58 10.8	1	14.57		1.70		1.63		2		735
285695	Hyades vA 125	AB	4 15 13	+15 24.4	1	11.25		1.29				3	F5	950
27126 +35 00840	G 39 - 5		4 15 13	+35 52.5	1	8.36		0.67		0.10		1	G5	333,1620
	IC 361 - 1		4 15 17	+58 06.7	1	11.13		0.77		0.31		14		735
27591 −73 00256			4 15 17	−73 46.1	1	8.90		1.32				4	K2 III	2012
	LP 774 - 48		4 15 19	−17 47.5	1	11.73		1.07		0.86		1		1696
	Hyades vA 129		4 15 20	+13 12.6	1	14.68		1.63				2	M2	920
−75 00264			4 15 21	−75 29.0	1	9.97		0.34				4	F0	2012
283520 +27 00652			4 15 22	+27 35.2	1	9.95		1.58		1.54		2	K3 III	8082
27159 +27 00651			4 15 22	+27 58.9	1	8.51		1.91		1.88		2	K1 III	8082
	HA 3 # 110		4 15 22	+75 06.7	2	12.22	.008	0.64	.004	0.23	.020	4	F8	271,1058
283518 +28 00637	V410 Tau		4 15 23	+28 20.7	2	10.87	.050	1.17	.010	0.89	.045	2	K5	5011,8082
	HA 3 # 189		4 15 24	+75 13.1	1	11.26		1.25		1.10		4	F9	271
27176 +21 00618	Hyades vB 24	⋆ A	4 15 25	+21 27.5	4	5.65	.004	0.28	.004	0.08	.004	10	A7 m	15,39,1127,8015
	HRC 30, DD Tau		4 15 25	+28 09.2	2	14.34	.151	0.96	.042	-0.52	.085	2	dK6e	649,1227
283550 +24 00643			4 15 26	+24 53.0	1	8.51		0.53		0.49		2	A2	1625

HD	DM	Other Id	N	Rem	α1950	δ1950	S	V	σV	B–V	σB–V	U–B	σU–B	n	Spectrum	References
		HRC 31, CZ Tau			4 15 26	+28 09.7	1	15.50		1.57		1.55		1	dM2e	649
26781	+74 00198				4 15 26	+75 02.0	1	9.65		1.28		1.23		5	K2	271
27084	+49 01155	NGC 1545 - 2		⋆	4 15 27	+49 55.7	3	5.46	.014	0.22	.001	0.12	.000	9	A7 V	15,49,70
285690	+15 00609	Hyades vB 25		⋆ V	4 15 28	+15 58.1	5	9.59	.013	0.98	.011	0.81	.004	8	K3 V	582,950,1127,8023,8025
285663	+17 00704	Hyades vA 135			4 15 29	+17 18.0	4	10.01	.012	1.10	.009	0.99	.000	5	K2	582,1058,1570,3072
		HA 3 # 113			4 15 33	+74 59.8	1	9.72		1.29		1.42		1		1058
		AO 872			4 15 34	+21 14.4	2	11.27	.010	1.58	.009	1.41	.001	2	K1 III	1722,1748
		AO 873			4 15 35	+21 32.9	2	11.66	.001	0.82	.008	0.26	.003	2	G0 III	1722,1748
27463	−61 00317	HR 1357		⋆ AB	4 15 35	−61 04.3	3	6.36	.007	0.07	.010	0.00		11	Ap EuCrSr	15,243,2012
27442	−59 00324	HR 1355		⋆ A	4 15 37	−59 25.3	6	4.44	.008	1.08	.005	1.07	.013	25	K2 IVa	15,243,1075,2012,8011,8015
27122	+48 01078				4 15 38	+49 07.0	2	8.45	.003	0.45	.006	0.42		41	A3	1462,6011
		AA45,405 S208 # 3			4 15 40	+52 51.5	2	14.00	.020	0.77	.010	-0.25	.010	8	B0 V:	797,7006
27236	+9 00558	HR 1334			4 15 41	+09 22.0	2	6.53	.005	0.16	.000	0.16	.005	7	A4 III	15,1415
		Hyades HG7 - 161			4 15 42	+11 42.0	1	11.31		0.74				1		950
		AA45,405 S207 # 4			4 15 44	+53 03.1	1	13.85		0.87		0.00		4		797
27157	+42 00938				4 15 45	+42 15.7	1	6.97		0.47		-0.02		1	F5	691
		AA45,405 S207 # 3			4 15 49	+53 03.1	1	14.20		0.99		0.42		3		797
		Hyades CDS 1		⋆	4 15 50	+27 10.5	1	14.47		0.15		-0.67		2		3006
283519	+27 00653				4 15 51	+27 50.3	1	9.74		0.66		0.08		2	G0	8082
		AA45,405 S207 # 2			4 15 53	+53 02.2	1	13.97		0.68		0.44		2		797
		E2 - 67			4 15 54	−46 03.	2	10.82	.002	1.01	.004	0.88		9		863,1075
		LB 3346			4 15 54	−61 33.	1	11.72		0.47		-0.06		1		45
		Hyades vA 141			4 15 55	+12 23.4	1	15.49		1.63				1	M1	951
27728	−76 00265				4 15 55	−75 55.8	3	7.24	.008	1.64	.004	1.98	.015	7	M1 III	278,2012,3040
27022	+64 00433	HR 1327			4 15 57	+65 01.3	6	5.27	.005	0.81	.005	0.47	.010	26	G5 IIb	15,1007,1013,1355*
		AA45,405 S207 # 1			4 15 58	+53 02.3	2	13.23	.000	0.75	.015	-0.34	.015	5	O9.5IV	797,7006
		Hyades vA 146			4 15 59	+13 14.7	3	11.99	.015	1.42	.015	1.15		5		19,582,950
284325	+21 00622				4 15 59	+21 35.9	1	10.49	.006	0.57	.008	0.46	.006	2	A2	1722,1748
27376	−34 01614	HR 1347		⋆ AB	4 16 00	−33 55.2	7	3.56	.004	-0.12	.003	-0.36	.008	24	B9 V	15,1020,1034,1075*
		E2 - 68			4 16 00	−43 31.	1	11.45		0.57				4		1075
279757	+35 00843				4 16 01	+35 32.5	1	9.73		0.50		0.33		2	A0	1375
		AA45,405 S207 # 9			4 16 01	+52 59.0	1	11.72		0.77		0.22		1		797
27250	+19 00694	Hyades vB 26		⋆ V	4 16 02	+19 47.2	2	8.62	.005	0.74	.001	0.33	.001	3	G9 V	1127,8023
27249	+20 00731	IDS04131N2104		AB	4 16 04	+21 10.8	3	8.18	.014	0.46	.005	0.01	.009	4	F5	401,1722,1748
	+48 01080	AS Per			4 16 04	+48 50.0	4	9.26	.038	1.15	.089	0.87	.029	4	F8	592,1399,1462,6011
		Met +27 185			4 16 05	+27 22.1	1	12.18		0.93		0.19		2	G0	8082
27362	−21 00831	HR 1345			4 16 05	−20 50.2	1	6.00		1.67				4	M4 III	2031
		G 7 - 35			4 16 06	+14 25.5	1	12.75		1.20		1.11		1		333,1620
		Hyades vA 152			4 16 06	+14 25.5	1	12.67		1.22				1		920
		RL 100			4 16 06	+56 32.	1	13.89		0.41		-0.42		2		704
		GD 429			4 16 06	+70 07.4	1	14.74		0.11		-0.49		1	DA	782
		LB 3347			4 16 06	−62 38.	1	12.09		0.25		0.07		1		45
281934		HRC 32, BP Tau			4 16 09	+28 59.3	3	12.03	.095	0.94	.058	-0.22	.053	3	dK5e	640,649,8013
		AA45,405 S207 # 8			4 16 11	+53 02.1	1	13.10		1.38		1.20		1		797
		ApJ110,424R259# 7			4 16 12	+29 00.	1	11.78		0.84		-0.26		1		1227
		LP 714 - 49			4 16 14	−13 19.6	1	11.52		0.47		-0.15		2		1696
27282	+17 00707	Hyades vB 27			4 16 15	+17 24.3	3	8.44	.028	0.73	.013	0.31	.002	4	G8 V	582,1127,8023
	+49 01156	NGC 1545 - 7			4 16 19	+50 09.5	2	10.56	.019	0.41	.003	0.20		3	A7 V	49,1655
		WK X Ray-1		V	4 16 23	+28 19.5	1	13.17		1.59		1.32		1		1529
		Hyades L 46			4 16 24	+22 31.4	1	14.01		0.67				3		582
		Hyades HG8 - 57			4 16 24	+22 32.	1	14.02		0.67		-0.05		2		1058
27192	+50 00973	HR 1333		⋆	4 16 24	+50 48.1	3	5.55	.004	-0.01	.004	-0.76	.001	8	B1.5IV	70,154,1223
26356	+83 00104	HR 1289			4 16 24	+83 41.6	1	5.56		-0.13		-0.52		3	B5 V	154
	+49 01157	NGC 1545 - 6			4 16 27	+50 09.2	2	10.32	.012	0.37	.001	0.03		15	A0 V	49,1655
27471	−45 01481				4 16 28	−45 46.3	4	7.53	.002	0.64	.006	0.16	.006	15	G2/3 V	278,861,1075,2012
27295	+20 00733	HR 1339, V1024 Tau			4 16 29	+21 01.4	3	5.35		-0.09	.019	-0.27	.015	6	B9 IV	1049,1079,3023
27411	−23 01856	HR 1353			4 16 29	−23 05.5	4	6.07	.009	0.31	.007	0.20	.000	17	A3mA5-F2	15,216,355,2012
		Hyades vA 162			4 16 31	+14 11.8	4	12.82	.022	1.48	.024			7	M1	19,582,950,951
		AO 876			4 16 31	+21 45.6	2	11.39	.015	1.76	.006	1.94	.048	2	K4 III	1722,1748
	−19 00877	IDS04142S1938		AB	4 16 31	−19 31.1	1	10.23		0.54				3	F8	176
		Hyades vA 165			4 16 32	+15 56.5	1	13.97		1.42				1		920
		Hyades L 44			4 16 32	+21 38.0	1	14.03		1.52				3		582
	+49 01159	NGC 1545 - 10			4 16 32	+50 03.5	1	11.02		0.36		0.09		1	A2 V	49
		NGC 1545 - 8			4 16 32	+50 17.4	1	10.88		0.21		0.02		1	A0 V	49
	−4 00797				4 16 35	−04 16.3	1	10.50		1.36				2	M0	1619
		AJ91,575 # 7		V	4 16 36	+27 42.6	1	12.32		1.39		1.18		1	K7 V	1606
		NGC 1545 - 24			4 16 36	+50 11.1	1	14.69		0.86		0.12		1		49
	−71 00259				4 16 36	−71 14.6	1	9.29		1.45		1.70		4	F5	119
		NGC 1545 - 22			4 16 38	+50 06.7	1	13.52		0.77		0.35		1		49
27309	+21 00623	HR 1341, V724 Tau			4 16 39	+21 39.3	2	5.38		-0.13	.005	-0.40	.005	5	A0p Si	1049,3033
	+57 00790	TW Cam			4 16 40	+57 19.3	1	9.51		1.43		1.18		1	F8 Ib	793
		G 38 - 28			4 16 41	+36 22.4	1	11.46		1.45		1.18		1		333,1620
		NGC 1545 - 25			4 16 42	+50 07.7	1	15.00		0.88		0.38		1		49
		Met +27 188			4 16 43	+27 12.0	1	12.15		0.86		0.80		2	A1	8082
		Be 11 - 1			4 16 46	+44 50.0	1	10.04		0.65		0.29		1		838
		G 8 - 26			4 16 47	+14 12.4	1	13.18		1.37		1.25		2		3016
27372	+13 00662	G 8 - 26		B	4 16 47	+14 12.4	1	13.18		1.35		1.34		3		3024
27278	+41 00852	HR 1337			4 16 47	+41 41.4	3	5.94	.007	0.96	.008	0.68	.023	7	K0 III	71,1462,1501
		Be 11 - 15			4 16 47	+44 49.5	1	12.95		2.41		2.37		1		838
276247	+41 00851	RW Per			4 16 48	+42 11.7	1	9.70		0.53		0.14		1	A5 V	691

Table 1

HD	DM	Other Id	N Rem	α_{1950}	δ_{1950}	S	V	σ_V	B–V	σ_{B-V}	U–B	σ_{U-B}	n	Spectrum	References
		Be 11 - 16		4 16 48	+44 48.9	1	14.83		0.89		0.37		1		838
286693	+12 00576	Hyades vA 171		4 16 49	+12 30.3	2	9.81	.000	0.92	.009			4	K2	950,1570
		Be 11 - 19		4 16 49	+44 47.9	1	13.85		0.77		0.02		1		838
		Be 11 - 17		4 16 49	+44 48.8	1	14.58		0.79		0.20		1		838
		Be 11 - 18		4 16 49	+44 48.9	1	15.18		0.84		0.16		1		838
		Be 11 - 22		4 16 51	+44 47.9	1	14.55		0.84		0.09		1		838
		NGC 1545 - 12		4 16 51	+50 06.1	1	11.82		0.40		0.27		1		49
		Be 11 - 20		4 16 52	+44 47.6	1	15.94		1.11		0.74		1		838
		HA 3 # 126		4 16 52	+75 03.5	2	10.89	.004	0.66	.011	0.12	.011	5	F6	271,1058
27554	−55 00626			4 16 52	−55 25.7	1	9.43		0.45		-0.06		4	F5 V	1731
27386	+9 00562	HR 1349		4 16 53	+10 00.1	3	6.31	.005	1.43	.005	1.60	.004	10	K0	15,1355,1415
		Hyades vA 173		4 16 53	+14 36.3	2	12.48	.005	1.08	.005			2		19,950
		Be 11 - 21		4 16 53	+44 47.7	1	14.71		0.88		0.24		1		838
		Be 11 - 23		4 16 53	+44 48.2	1	14.22		0.74		0.08		1		838
		NGC 1545 - 20		4 16 54	+50 06.3	1	13.47		0.77		0.38		1		49
27372	+13 00662	G 8 - 27	A	4 16 55	+14 09.4	4	7.53	.014	0.99	.019	0.84	.014	14	G7 III	333,882,1620,3077,4001
		Be 11 - 24		4 16 55	+44 48.0	1	12.42		0.70		0.16		1		838
27829	−76 00266			4 16 55	−76 06.8	1	8.66		1.75				4	M0/1 III	2012
27257	+49 01160	NGC 1545 - 5	⋆ C	4 16 56	+50 07.8	2	9.25	.021	0.22	.002	0.21		2	B9 V	49,1655
27247	+50 00975			4 16 56	+50 59.8	1	8.50		0.20				2	A0	70
27371	+15 00612	Hyades vB 28	⋆	4 16 57	+15 30.5	15	3.65	.012	0.99	.005	0.81	.011	62	K0 III	1,3,15,272,1006,1034*
		Hyades L 47		4 16 57	+21 53.8	1	13.69		0.99				3		582
		GD 60		4 16 57	+33 28.5	1	15.33		0.01		-0.73		1		3060
		Be 11 - 4		4 16 57	+44 49.7	1	15.09		0.78		0.25		1		838
		Be 11 - 5		4 16 58	+44 49.6	1	14.60		0.92		0.21		1		838
27224	+56 00904			4 16 58	+56 35.6	1	7.06		1.45		1.46		3	K1 III	37
27349	+31 00757	HR 1344		4 16 59	+31 50.1	1	6.16		1.71				2	K5	71
		Be 11 - 12		4 16 59	+44 47.0	1	13.92		0.74		0.04		1		838
		Be 11 - 9		4 16 59	+44 48.1	1	14.97		0.80		0.19		1		838
		Be 11 - 8		4 16 59	+44 49.1	1	15.13		0.84		0.28		1		838
27385	+12 00577			4 17 00	+12 58.0	1	7.47		1.52				3	K0	882
		G 38 - 29	⋆ V	4 17 00	+36 09.5	2	15.61	.020	0.21	.039	-0.54	.010	8		316,3060
		Be 11 - 11		4 17 00	+44 47.6	1	13.92		0.72		0.01		1		838
		Be 11 - 10		4 17 00	+44 48.0	1	13.60		0.80		0.09		1		838
		Be 11 - 3		4 17 00	+44 50.3	1	15.30		1.05		0.32		1		838
		RL 101		4 17 00	+55 22.	1	10.70		0.24		-0.02		2		704
27293	+42 00939			4 17 01	+43 07.2	1	7.30		1.80		2.03		4	K2	1355
		Be 11 - 2		4 17 02	+44 50.4	1	13.33		0.75		0.07		1		838
		NGC 1545 - 19		4 17 02	+50 08.3	1	13.13		0.60		0.33		1		49
27383	+16 00579	Hyades vB 29	⋆ AB	4 17 03	+16 24.2	2	6.89	.005	0.56	.000	0.09	.001	3	F8 V	1127,8023
27370	+23 00675			4 17 04	+23 28.8	2	7.19	.020	1.12	.035	0.79		2	G5	987,3077
		NGC 1545 - 18		4 17 04	+50 10.7	1	12.69		0.94		0.58		1		49
		Hyades L 43		4 17 05	+19 38.9	1	14.01		1.06				1		582
		NGC 1545 - 9		4 17 05	+50 13.7	1	10.96		0.24		0.07		1	A0 Vn	49
27657	−63 00316	HR 1372	⋆ AB	4 17 06	−63 22.6	1	5.87		-0.07				4	B9 III/IV	2007
27277	+49 01161	NGC 1545 - 4	⋆ B	4 17 07	+50 09.2	1	8.09		0.99		0.66		1	G6 III	49
		NGC 1545 - 16		4 17 07	+50 11.0	1	12.17		0.59		0.32		1		49
27397	+13 00663	Hyades vB 30	⋆ A	4 17 09	+13 55.0	5	5.59	.007	0.28	.008	0.08	.007	12	F0 IV	15,254,1127,8015,8025
		Hyades L 31		4 17 09	+20 39.4	1	11.67		0.75				2		582
27348	+34 00860	HR 1343	⋆ A	4 17 09	+34 26.9	4	4.93	.011	0.95	.009	0.69	.007	13	G8 III	37,1355,1363,3016
		Be 11 - 14		4 17 09	+44 48.0	1	13.97		0.76		0.21		1		838
27490	−34 01626	HR 1359	⋆ AB	4 17 10	−34 01.5	4	6.36	.004	0.13	.007	0.10		16	A2/3 V+F7V	15,404,1075,2027
27292	+49 01162	NGC 1545 - 3	⋆ A	4 17 11	+50 08.2	1	7.13		1.69		1.75		1	K5 III	49
		RL 102		4 17 12	+48 55.	1	16.70		0.00		-1.00		2		704
		Be 11 - 13		4 17 13	+44 48.7	1	11.75		0.43		0.31		1		838
27382	+27 00655	HR 1348	⋆ A	4 17 16	+27 14.0	2	4.95	.000	1.15	.000	1.08		5	K1 III	1355,1363
		NGC 1545 - 11		4 17 16	+50 09.8	1	11.67		0.39		0.26		1		49
		LS V +53 021		4 17 16	+53 03.2	1	10.93		0.64		-0.39		2		405
		NGC 1545 - 15		4 17 17	+50 04.8	1	12.01		0.43		0.30		1		49
27406	+18 00623	Hyades vB 31	⋆ V	4 17 18	+19 06.9	3	7.47	.008	0.57	.004	0.07	.013	6	G0 V	1127,1355,8023
27332	+45 00921			4 17 18	+45 20.8	1	7.66		0.11		0.09		2	A0	1601
	+45 00919	G 81 - 15		4 17 18	+45 43.6	2	9.90	.021	0.61	.010	0.02	.023	4	G5	1620,7010
		NGC 1545 - 13		4 17 21	+50 13.1	1	11.82		0.42		0.33		1		49
		NGC 1545 - 21		4 17 22	+50 14.8	1	13.49		0.68		0.22		1		49
		NGC 1545 - 17		4 17 24	+50 04.4	1	12.47		0.48		0.35		1		49
		LB 3350		4 17 24	−64 02.	1	11.57		0.49		0.03		1		45
		HA 3 # 56		4 17 25	+74 51.0	1	12.62		0.64		0.17		1		1058
		NGC 1545 - 23		4 17 26	+50 13.1	1	14.48		0.81		0.22		1		49
27245	+60 00800	HR 1335		4 17 26	+60 37.2	2	5.40	.007	1.50	.005	1.81		5	M0 III	71,3001
27604	−53 00679	HR 1365	⋆ AB	4 17 26	−52 58.9	5	6.08	.004	0.49	.006	0.02		19	F7 IV/V	15,1075,2018,2030,3053
285666		Hyades vA 191		4 17 27	+17 23.8	2	12.25	.015	1.05	.010			3	K1 V	582,1674
27405	+25 00703			4 17 27	+25 42.5	1	7.80		0.18		-0.27		2	B9	105
		Hyades J 280		4 17 31	+18 37.5	1	13.82		1.47		1.15		4		1058
27429	+18 00624	Hyades vB 32	⋆	4 17 31	+18 37.5	5	6.12	.024	0.37	.004	0.02	.022	17	F2 Vn	15,254,1127,8015,8025
		Hyades CDS 13	⋆	4 17 32	+21 15.6	1	15.50		1.73				2	M1	1570
27883	−75 00265			4 17 32	−75 28.0	3	8.46	.005	0.43	.001	0.17	.009	10	Fm δ Del	278,1117,2012
27518	−25 01842			4 17 37	−25 08.6	2	6.86	.009	1.44	.005	1.72		5	K3 III	2012,3040
		G 38 - 30		4 17 38	+37 21.9	1	12.72		1.47		1.15		4		333,1620
27507	−17 00853			4 17 39	−17 03.3	1	9.42		0.09		0.14		1	B9 V	732
27588	−44 01503	HR 1364	⋆ A	4 17 42	−44 23.2	10	5.33	.011	1.08	.008	0.95	.010	43	K2 III	14,15,182,278,388,977*

HD	DM	Other Id	N	Rem	α_{1950}	δ_{1950}	S	V	σ_V	B–V	σ_{B-V}	U–B	σ_{U-B}	n	Spectrum	References
27588	−44 01505	IDS04161S4430		B	4 17 42	−44 23.2	1	8.69		0.56		−0.03		4	F8 V	3077
		Hyades L 32			4 17 44	+20 48.9	1	11.73		0.80				2		582
27498	−2 00867	DQ Eri			4 17 44	−02 44.8	2	7.05	.009	1.54	.005	1.65	.017	10	M4 III	1311,3042
27485	−4 00801	G 82 - 12			4 17 45	−03 51.9	3	7.86	.016	0.65	.014	0.08	.005	7	G0	1620,2012,3056
27459	+14 00682	Hyades vB 33		⋆ V	4 17 46	+14 58.6	5	5.26	.007	0.22	.010	0.10	.004	12	F0 IV	15,254,1127,8015,8025
27322	+56 00905	HR 1342			4 17 47	+56 23.3	2	5.90	.015	0.11	.000	0.08	.005	5	A3 V	1501,1733
		L 178 - 47			4 17 47	−57 22.5	1	13.30		1.55		1.12		2		3073
		Hyades HG7 - 171			4 17 48	+19 31.5	2	11.84	.000	1.20	.000			4		950,1570
	−9 00872				4 17 50	−09 09.2	1	9.82		1.12		1.04		3	K5 V	3072
	−15 00767				4 17 52	−14 52.9	1	9.78		1.20		1.17		2	K5 V	3072
27528	−16 00838	IDS04156S1640		AB	4 17 53	−16 33.4	1	6.80		−0.03				4	B9 V	2006
	+47 00977				4 17 55	+48 13.1	1	9.63		1.17		1.09		1	K8	3072
27396	+46 00872	HR 1350, V469 Per			4 17 56	+46 22.9	4	4.85	.010	−0.03	.010	−0.52	.012	9	B4 IV	15,154,1363,8015
		Hyades vA 200			4 17 57	+15 07.1	2	13.85	.005	1.57	.009	1.18		6		920,1058
		CS 22186 # 23			4 17 57	−36 58.8	1	12.92		0.65		0.04		1		1736
27497	+5 00631	HR 1360			4 18 01	+06 00.8	6	5.77	.010	0.92	.002	0.68	.005	15	G8 III-IV	15,1355,1415,1488,1738,3016
		NGC 1545 - 14			4 18 02	+50 10.3	1	11.94		0.39		0.29		1		49
27483	+13 00665	Hyades vB 34		⋆	4 18 04	+13 44.8	6	6.17	.017	0.46	.004	0.01	.015	12	F6 V	15,254,1127,1355,8015,8025
		CS 22182 # 24			4 18 04	−31 29.0	1	12.88		0.38		−0.15		1		1736
27505	+8 00672	HR 1361			4 18 05	+09 06.4	2	6.52	.005	0.15	.000	0.10	.005	7	A4 V	15,1061
		Hyades vA 203			4 18 06	+14 44.5	1	16.70		1.62				2		951
27103	+75 00175				4 18 06	+75 20.9	1	8.91		0.18		0.07		3	A0	271
		BPM 17731			4 18 06	−53 59.	1	15.32		−0.10		−1.08		2		3065
27482	+27 00656				4 18 10	+27 13.9	2	7.32	.015	2.02	.030	2.39	.095	4	K5 III	23,8082
27536	−6 00875	HR 1362, EK Eri			4 18 11	−06 21.8	5	6.25	.011	0.90	.005	0.56	.005	14	G8 IV:	15,1415,1641,2031,3077
26836	+80 00133	HR 1317			4 18 14	+80 42.6	1	5.43		1.17				2	K0	71
27584	−21 00842				4 18 16	−21 27.2	1	7.98		1.02				4	G8/K0 III	2012
27563	−7 00798	HR 1363, EM Eri			4 18 17	−07 42.6	4	5.84	.005	−0.14	.006	−0.49	.011	15	B5 III	15,1079,1415,2027
27713	−55 00631				4 18 21	−55 00.6	1	8.13		1.03		0.81		4	K0 III	1731
27598	−17 00856	DG Eri			4 18 26	−16 56.9	2	7.02	.036	1.67	.014	1.85	.052	5	M3 III	3040,8100
		G 82 - 13			4 18 28	+05 59.1	1	11.86		1.18		1.07		1		333,1620
27616	−20 00831	HR 1367			4 18 28	−20 45.5	1	5.38		−0.02				4	A0 V	2031
		Hyades vA 208			4 18 33	+11 55.5	1	16.31		1.74				2		951
27524	+20 00740	Hyades vB 35			4 18 34	+20 55.4	3	6.80	.000	0.44	.004	−0.01	.008	6	F5 V	1127,8023,8043
27495	+39 00980	IDS04152N3942		A	4 18 36	+39 49.1	1	7.01		0.60		0.15		2	F8	1601
27534	+18 00629	Hyades vB 36			4 18 38	+18 18.0	2	6.81	.005	0.44	.000	0.00	.000	3	F6 V	1127,8023
27629	−19 00885	IDS04164S1934		AB	4 18 38	−19 27.2	1	7.08		0.86		0.49		2	G8 III	1375
285710	+14 00685	Hyades vB 171			4 18 39	+14 43.7	1	10.10		0.71		0.25		2	G5	1127
27402	+59 00793	HR 1352		⋆ AB	4 18 40	+59 30.0	1	6.18		0.09		0.11		2	A4 V	1733
		Hyades vA 211			4 18 41	+14 19.0	1	14.08		1.39				1		19
		LP 156 - 48			4 18 41	+45 30.0	1	10.32		0.67		0.20		3		1723
		BSD 3 # 402			4 18 43	+74 47.2	1	12.84		0.70		0.28		1	F8	1058
27561	+14 00687	Hyades vB 37			4 18 45	+14 17.5	2	6.61	.000	0.41	.002	0.00	.000	3	F4 V	1127,8023
		Hyades vA 216			4 18 45	+14 34.7	1	15.64		0.96		0.73		1	M2	3060
		Hyades vA 217			4 18 45	+15 00.6	2	12.40	.005	1.18	.015			2		950,1570
27560	+19 00704				4 18 48	+19 21.1	1	8.38		0.40		−0.05		2	G0	1648
26659	+82 00113	HR 1304			4 18 49	+83 13.6	2	5.46	.000	0.87	.000	0.46	.000	4	G8 III	15,1003
		HRC 33, DE Tau			4 18 50	+27 48.1	2	13.06	.110	1.36	.045	−0.22	.160	2	dM1e	640,649
283571	+28 00645	HRC 34, RY Tau			4 18 51	+28 19.6	6	10.58	.187	1.10	.082	0.57	.113	9	F8 V:e	206,640,649,8013*
		G 8 - 29			4 18 52	+21 12.9	1	13.03		1.56		1.08		1		333,1620
		Hyades HG8 - 15			4 18 54	+21 14.	2	12.84	.010	1.28	.005	1.15		5		920,1058
27611	−0 00687	HR 1366		⋆ A	4 18 54	−00 12.8	2	5.85	.005	1.32	.000	1.52	.000	7	K2	15,1415
		Hyades CDS 16		⋆	4 18 56	+23 18.1	1	13.44		1.53				2		1570
		LB 212			4 18 57	+15 19.5	1	16.62		0.14		−0.57		1		3016
		HA 3 # 275			4 18 59	+75 24.4	1	11.50		0.64		0.12		3		271
27504	+47 00981				4 19 00	+47 32.6	1	7.85		1.83				4	K5	369
27773	−54 00652				4 19 01	−54 48.6	1	8.44		1.16		1.16		4	K1 III	1731
		LP 415 - 625			4 19 03	+20 56.2	1	13.62		1.37				1		1570
		Met +24 208			4 19 03	+24 55.4	1	12.29		0.94		0.34		2	G5	8082
284419	+19 00706	HRC 35, T Tau			4 19 04	+19 25.1	6	10.32	.094	1.24	.018	0.74	.027	21	K0 IIIe	206,640,649,5011*
	+47 00982				4 19 08	+47 39.4	1	9.84		1.94				4	M0	369
		AJ89,1229 # 274			4 19 10	+02 54.8	1	11.36		1.77				1	M0	1632
27660	−6 00879				4 19 10	−06 24.1	1	6.45		0.06		0.09		35	A0	978
27570	+30 00659	Hyades vB 139			4 19 11	+30 18.5	2	9.09	.010	0.66	.010	0.14	.005	4	G5 IV	1127,1529
		CS 22186 # 22			4 19 13	−36 42.1	1	13.80		0.40		−0.19		1		1736
27628	+13 00668	Hyades vB 38		⋆ A	4 19 14	+13 57.6	4	5.72	.000	0.32	.008	0.11	.011	10	A2 m	15,1127,1199,8015
		Steph 484			4 19 18	+19 01.5	1	11.11		0.94		0.72		1	K4	1746
	+46 00873				4 19 22	+46 57.5	1	8.99		1.04		0.67		2	K0	1601
		RL 103			4 19 24	+50 01.	1	16.28		−0.02		−0.77		2		704
27639	+20 00744	HR 1370		⋆ A	4 19 26	+20 42.3	1	5.91		1.66				2	M0 IIIab	71
		Hyades L 45			4 19 26	+22 14.8	1	14.83		0.81				3		582
27710	−26 01642	HR 1374		⋆ AB	4 19 27	−25 50.7	5	6.03	.040	0.35	.014	−0.04	.011	18	F2 V	15,158,214,2007,3030
	−26 01643	IDS04174S2558		D	4 19 27	−25 50.7	4	8.25	.020	0.58	.005	0.03	.010	15	F9 V	158,214,1775,3024
		Hyades HG8 - 18		⋆	4 19 31	+20 26.3	2	11.69	.010	1.09	.010	0.98	.015	5	K3	333,1058,1620
27638	+25 00707	HR 1369		⋆ A	4 19 32	+25 30.8	5	5.39	.005	−0.04	.008	−0.11	.026	9	B9 V	23,1049,1079,1211,3032
27638	+25 00707	IDS04165N2524		B	4 19 32	+25 30.8	4	8.42	.010	0.59	.011	0.10	.001	29	G2 V	23,1211,1752,3032
		Hyades vA 237			4 19 33	+16 47.8	1	12.25		0.95				1		950
	−40 01349				4 19 33	−40 28.1	1	11.48		0.46		−0.16		2		1696
		L 302 - 89			4 19 36	−48 46.	1	14.40		0.52		−0.48		2		3073
		Hyades HG7 - 176			4 19 38	+11 46.4	1	14.21		1.48				1		950

Table 1 213

HD	DM	Other Id	N Rem	α_{1950}	δ_{1950}	S	V	σ_V	B–V	σ_{B-V}	U–B	σ_{U-B}	n	Spectrum	References
27723 −22 00809				4 19 38	−21 53.3	1	7.50		0.56		0.09		2	G0 V	1375
		Hyades HG7 - 177		4 19 39	+11 11.3	1	11.85		0.88				1		950
286770 +10 00568		Hyades L 33		4 19 39	+11 11.4	4	9.81	.007	1.18	.004	1.15	.005	7	K8	582,950,1058,3072
		WLS 420-20 # 5		4 19 40	−17 35.9	1	11.58		0.58		0.08		2		1375
		Hyades HG7 - 180		4 19 45	+10 19.1	2	12.64	.015	1.50	.025			4		920,1532
		Hyades vA 242		4 19 45	+18 09.2	2	13.01	.005	1.51	.005			4	K1 V	19,582,951
27917 −63 00320				4 19 45	−63 17.5	1	7.90		0.43		0.04		4	F3 IV	243
		Hyades L 41		4 19 46	+18 09.1	1	13.01		1.51				3		582
		Hyades HG7 - 181		4 19 46	+20 02.6	1	11.53		0.74				1		920
		LP 415 - 30		4 19 46	+20 27.2	1	14.97		1.64				2		950
27686 +14 00689				4 19 48	+14 47.6	1	8.13		1.54				2	G5	882
283603 +24 00647				4 19 49	+25 04.4	1	10.92		0.90		0.53		2	F2	8082
27685 +16 00585		Hyades vB 39		4 19 52	+16 40.5	4	7.85	.013	0.67	.005	0.22	.000	6	G4 V	1127,1355,1733,8023
		Met +27 217		4 19 53	+27 54.9	1	12.22		0.67		0.41		2	A7	8082
		WLS 420 35 # 5		4 19 53	+37 23.7	1	10.88		0.72		0.24		2		1375
27691 +14 00690		Hyades vB 40	⋆ AB	4 19 54	+14 56.4	3	6.99	.008	0.57	.005	0.09	.004	4	G0 V	1127,8023,8043
		Hyades CDS 49	⋆ V	4 19 54	+14 56.4	1	15.00		0.21				2		1570
26367 +85 00063				4 19 54	+85 25.1	1	6.56		0.51		0.04		3	F8	985
27894 −59 00335				4 19 55	−59 31.9	1	9.42		1.00		0.90		3	K2 V	3072
		S004 # 3		4 19 58	−84 15.4	1	11.02		1.18		1.29		1		1730
281989 +31 00769				4 19 59	+32 04.7	1	9.06		0.48		−0.01		2	F5	3032
27697 +17 00712		Hyades vB 41	⋆ A	4 20 03	+17 25.6	13	3.76	.010	0.98	.010	0.83	.009	65	K1 III	1,15,1006,1105,1127*
285762 +13 00671		Hyades vB 172		4 20 07	+14 08.0	2	10.13	.047	0.66	.016	0.13		4	K0 V	582,1127
27650 +42 00946		HR 1371	⋆ AB	4 20 07	+42 18.8	1			0.00		−0.16		2	A1p Si	1079
		Hyades vA 260		4 20 11	+15 06.7	1	16.71		1.66				1		951
283611 +24 00651				4 20 11	+24 40.8	1	9.28		1.31		0.98		1	G8 III	8082
		Hyades HG7 - 183		4 20 12	+13 12.0	1	13.65		1.65				1	M1	920
		Hyades vA 262		4 20 22	+15 35.8	1	15.85		1.77				1		951
		Met +28 141		4 20 22	+28 00.6	1	11.14		1.44		1.32		2	B5	8082
281962 +32 00781				4 20 22	+33 02.1	1	10.31		0.54		0.17		2	F0	1375
27764 +4 00683				4 20 24	+04 29.4	1	8.81		0.50		−0.09		1	F5	592
27732 +21 00635		Hyades vB 42		4 20 24	+21 15.8	2	8.85	.005	0.76	.000	0.35	.000	3	G9 V	1127,8023
284414 +19 00708		Hyades vB 43	⋆ V	4 20 27	+19 32.6	4	9.40	.000	0.91	.001	0.61	.000	9	K2 V	950,1127,8023,8025
		CS 22182 # 47		4 20 28	−27 41.8	1	13.24		0.48		−0.14		1		1736
27731 +24 00654		Hyades vB 44		4 20 29	+24 17.4	2	7.18	.005	0.45	.000	0.03	.001	3	F5	1127,8023
27772 +3 00596				4 20 30	+03 52.5	2	8.09	.015	1.00	.025	0.63	.075	2	G5	592,1462
27749 +16 00586		Hyades vB 45	⋆	4 20 33	+16 39.7	7	5.64	.006	0.30	.005	0.14	.014	19	A1m	15,39,1049,1127,1199*
		LP 156 - 52		4 20 33	+49 08.7	1	11.47		0.84		0.32		3		1723
		Hyades vA 275		4 20 34	+14 18.7	3	14.95	.007	1.55	.018	1.10		7	M2	218,950,951
		Hyades vA 276		4 20 34	+15 38.9	2	10.48	.028	1.23	.012	1.16		4	K5	1570,3072
285749 +15 00616		Hyades vB 173	⋆ V	4 20 34	+15 38.9	3	10.51	.020	1.23	.009	1.18	.000	6	K5 V	582,1127,8023
		Met +27 220		4 20 34	+27 19.8	1	11.61		0.90		0.68		2	A5	8082
27742 +20 00751		HR 1375		4 20 35	+20 52.0	2			0.03	.000	−0.26	.015	5	B8 IV-V	1049,1079
		Hyades HG7 - 186		4 20 36	+07 07.	2	14.37	.000	1.38	.000			4		950,950
		Met +27 221		4 20 36	+27 41.5	1	12.12		0.74		0.60		2	A0	8082
		G 82 - 16		4 20 37	+03 42.7	1	15.20		1.03		0.42		2		1658
27741 +27 00658				4 20 37	+28 04.4	2	8.27	.010	0.64	.010	0.11		5	G0 V	20,8082
283640 +26 00717				4 20 39	+27 05.8	1	10.00		0.73		0.29		5	G0	8082
27771 +14 00691		Hyades vB 46		4 20 42	+14 33.3	5	9.12	.018	0.86	.004	0.54	.000	7	K1 V	582,882,950,1127,8023
	+16 00588	Hyades vA 280		4 20 43	+16 17.8	2	12.10	.005	0.90	.005			2		19,950
27716 +34 00872				4 20 43	+35 06.1	1	7.91		0.36		0.18		2	F0	1625
27673 +50 00986		IDS04170N5037	AB	4 20 44	+50 43.9	1	6.86		0.45		−0.03		3	F7 V	1501
27810 −5 00895		IDS04183S0514	A	4 20 47	−05 07.2	1	7.59		1.27		1.41		3	K2	1657
		L 446 - 24		4 20 48	−37 17.	1	13.04		0.70		0.10		2		3062
283636				4 20 49	+27 42.5	1	10.93		1.21		1.14		2	K8	8082
		Hyades vA 282		4 20 52	+15 45.9	2	14.80	.025	1.57	.020	1.36		5	M2	19,218,951
27789 +13 00674				4 20 53	+13 42.8	1	7.60		1.09				2	G5	882
		Met +27 225		4 20 57	+27 02.9	1	10.74		1.23		0.77		2	G2	8082
27822 +3 00597				4 20 58	+03 52.1	1	9.95		0.18		0.15		1	A0	1462
27778 +23 00684		HR 1378	⋆ A	4 20 59	+24 11.2	6	6.36	.006	0.18	.014	−0.36	.020	13	B3 V	150,1049,1079,1084*
27778 +23 00683		IDS04180N2404	B	4 20 59	+24 11.2	4	8.18	.004	0.35	.011	0.24	.022	9	A1 Vp	150,1084,1211,3016
		Hyades vA 288		4 21 00	+14 48.4	4	13.33	.018	1.54	.009	0.98		10	M2	218,920,921,951
		LP 415 - 721		4 21 00	+20 06.1	1	11.53		0.90				1		1570
27729 +41 00861				4 21 00	+41 36.9	1	7.12		0.13		0.09		2	A2	1601
27881 −25 01862		HR 1384		4 21 00	−25 00.4	3	5.83	.008	1.51	.000			11	K5 III	15,1075,2035
		Hyades vA 290		4 21 01	+13 27.0	1	12.19		0.95				1		19,950
		Hyades vA 297		4 21 01	+13 27.0	4	12.55	.025	1.47	.008	1.28		16	M1.5Ve	352,582,1058,1570
		Hyades vA 292		4 21 04	+16 14.3	6	14.28	.016	−0.03	.020	−0.85	.027	13		951,974,1281,1570*
27821 +6 00676				4 21 05	+06 15.8	1	8.69		0.32		0.00		2	A7 V	1003
286734		Hyades vA 294		4 21 05	+15 36.2	6	10.89	.011	1.29	.009	1.28	.020	41	K5.5Ve	352,582,1058,1570*
	+44 00935		V	4 21 05	+44 46.0	1	10.29		2.04				3	M2	369
	−84 00047			4 21 06	−84 09.6	1	9.60		0.44		−0.04		1		1730
		Hyades HG7 - 192		4 21 07	+09 05.4	2	12.88	.005	1.52	.005			3	M1	920,950
27797 +14 00692				4 21 07	+14 17.9	2	8.75	.014	1.37	.000	1.24		4	K0	882,3077
27820 +9 00570		HR 1381	⋆ AB	4 21 08	+09 20.8	8	5.12	.008	0.07	.005	0.11	.007	25	A3 V +A4 V	15,1007,1013,1363*
27928 −37 01724				4 21 08	−37 22.7	3	9.56	.010	0.71	.005	0.00	.039	8	Gw	742,1594,3077
		Hyades L 38		4 21 09	+22 00.3	2	11.02	.000	1.25	.050			4	K5	582,1746
27861 −4 00818		HR 1383		4 21 11	−03 51.6	8	5.18	.008	0.07	.009	0.08	.004	20	A2 V	15,1007,1013,1415*
27819 +17 00714		Hyades vB 47	⋆ A	4 21 13	+17 19.8	5	4.80	.000	0.15	.007	0.12	.004	15	A7 IV	15,1127,1363,3023,8015
		CS 22182 # 33		4 21 13	−30 44.1	1	14.67		0.44		−0.18		1		1736

HD	DM	Other Id	N Rem	α_{1950}	δ_{1950}	S	V	σ_V	B–V	σ_{B-V}	U–B	σ_{U-B}	n	Spectrum	References
27777	+33 00853	HR 1377		4 21 14	+34 01.0	2			-0.07	.005	-0.34	.015	5	B8 V	1049,1079
27905	-25 01868			4 21 15	-25 30.4	1	7.81		0.63				4	G3 V	2012
27808	+21 00641	Hyades vB 48		4 21 16	+21 37.3	4	7.14	.005	0.52	.007	0.04	.009	11	F7 V	1127,1355,8023,8043
27837	+12 00584	Hyades vA 304		4 21 17	+12 51.5	1	7.70		1.00				3	K0	882
27941	-35 01687	HR 1386	⋆ A	4 21 18	-35 39.6	4	6.38	.004	1.24	.006			18	K1 III	15,1075,2013,2029
28525	-80 00116	HR 1426		4 21 18	-80 20.0	3	5.67	.006	0.84	.000	0.53	.000	22	K2/3 III+A	15,1075,2038
		Hyades vA 305		4 21 19	+14 58.4	2	15.10	.009	1.71	.038			6	M3	950,1058
27835	+16 00589	Hyades vB 49		4 21 21	+16 15.9	2	8.24	.000	0.59	.002	0.12	.002	3	G0 V	1127,8023
28093	-63 00324	HR 1395		4 21 21	-63 30.3	4	5.23	.007	0.95	.006	0.67	.024	10	G8 III	243,1409,2032,3000
27836	+14 00693	Hyades vB 50	⋆ V	4 21 22	+14 38.6	5	7.61	.024	0.60	.009	0.12	.004	8	G1 V	582,1127,8023,8025,8043
		Hyades vA 310		4 21 23	+17 53.3	1	10.00		1.04		1.00		3		3072
285720	+17 00715	Hyades vB 174	⋆ V	4 21 23	+17 53.3	6	9.99	.011	1.05	.011	0.95	.000	19	K4 V	582,582,1127,8023*
27786	+33 00854	HR 1379	⋆ AB	4 21 23	+33 50.8	2	5.78	.027	0.39	.012	-0.14		5	F4 V	71,254
28043	-55 00636			4 21 28	-55 12.7	1	8.47		0.97		0.70		4	G8 III	1731
27833	+19 00714			4 21 29	+19 52.0	1	9.40		0.91		0.61		1	G5	8027
232947	+53 00765	LS V +53 022		4 21 29	+53 18.1	1	9.32		0.64		-0.40		2	B0 Ia	1012
27848	+16 00591	Hyades vB 51		4 21 30	+16 57.9	3	6.97	.005	0.44	.002	0.00	.003	4	F6 V	1127,8023,8043
		Hyades HG8 - 17		4 21 30	+20 43.	2	13.58	.010	0.82	.045	0.06		5		920,1058
		Hyades HG7 - 197		4 21 35	+10 32.7	1	14.24		1.02				1		1570
	+51 00921	LS V +51 019		4 21 35	+51 55.8	1	9.63		0.73		-0.30		2	B0 II	1012
27859	+16 00592	Hyades vB 52	⋆ V	4 21 36	+16 46.3	3	7.80	.000	0.60	.001	0.13	.001	6	G1 V	1127,8023,8025
		G 39 - 11		4 21 36	+32 20.1	1	12.44		1.51		1.15		1		333,1620
286735	+13 00675	Hyades vA 320		4 21 37	+13 52.9	1	10.03		0.55				1	G0	1570
		Hyades vA 321		4 21 37	+15 46.2	1	14.93		1.58				1	M2	19
	+45 00930			4 21 38	+45 44.0	1	9.49		0.91		2.25		9		1655
276355	+40 00954			4 21 41	+40 49.8	1	9.28		1.85				3	M2	369
		Hyades vA 326		4 21 42	+13 48.9	1	16.62		1.87				2		951
29138	-84 00048			4 21 45	-84 36.3	2	7.19	.009	-0.07	.005			8	B1 Iab	1075,2012
		Met +26 384		4 21 46	+26 22.1	1	12.15		0.83		0.54		2		8082
		Hyades vA 331	⋆	4 21 48	+17 05.3	3	13.22	.023	1.54	.029	1.11	.059	8		333,950,1058,1620
27999	-39 01474			4 21 50	-38 58.8	1	9.52		0.91		0.58		4	G8 III	119
		Hyades vA 333		4 21 52	+13 24.4	1	11.94		0.86				1		1570
		AJ91,575 # 8	V	4 21 52	+27 05.1	1	13.00		1.47		1.07		1	M0 V	1606
27795	+45 00931	LS V +46 016		4 21 52	+46 01.8	2	7.41	.015	0.26	.015	-0.50	.025	5	B1 V	399,401
		CS 22182 # 25		4 21 52	-32 24.3	1	14.57		0.39		-0.19		1		1736
	+3 00601	SW Tau		4 21 55	+04 00.6	5	9.39	.102	0.49	.063	0.31	.140	5	A7	592,688,1462,1488,6011
27924	+9 00571			4 21 57	+09 56.0	1	7.22		0.21		0.08		2	A0	401
		Hyades vA 334		4 21 57	+15 45.6	3	11.66	.014	1.42	.007	1.18		5	M0.5Ve	582,1570,3072
27831	+38 00886			4 21 57	+38 56.4	2	8.23	.010	0.64	.005	0.28	.000	5	G2 III	833,1601
		CS 22186 # 34		4 21 57	-33 34.3	1	14.10		0.34		-0.04		1		1736
		Hyades HG7 - 199		4 21 59	+10 39.4	1	14.02		1.53				4	M1	950
		AK 5 - 748		4 22 00	+21 37.7	1	11.79		0.93				1		987
	+4 00685			4 22 02	+04 41.7	1	9.45		1.62				82	K2	6011
27901	+18 00633	Hyades vB 53	⋆	4 22 02	+18 55.7	4	5.99	.029	0.38	.010	0.03	.021	10	F4 Vn	15,254,1127,8015
27935	+4 00686	Hyades vB 140		4 22 04	+04 35.2	3	8.93	.001	0.76	.002	0.34	.001	86	G5	1127,6011,8023
		Hyades vA 342		4 22 07	+16 52.3	1	10.31		1.04				2		1570
	+25 00710			4 22 07	+25 38.2	2	7.85	.005	0.44	.005	-0.09	.044	3	F5	105,8082
285742	+16 00593	Hyades vB 175	⋆ V	4 22 08	+16 52.3	4	10.27	.012	1.03	.004	0.93	.001	33	K4 V	352,582,1127,8023
28028	-34 01664	HR 1393		4 22 09	-34 07.9	4	3.96	.008	1.49	.000	1.80	.000	20	K4 III	15,1075,2012,8015
		Met +27 227		4 22 11	+27 27.7	1	11.17		2.12				2	K5 III	8082
		OF 038 # 1		4 22 12	+00 29.	1	12.51		0.68		0.12		1		1482
		OF 038 # 2		4 22 12	+00 29.	1	13.96		0.62		0.13		1		1482
		OF 038 # 3		4 22 12	+00 29.	1	11.16		0.54		0.07		1		1482
		OF 038 # 5		4 22 12	+00 29.	1	15.05		0.65		0.09		1		1482
		OF 038 # 6		4 22 12	+00 29.	1	14.75		0.70		0.08		1		1482
		OF 038 # 7		4 22 12	+00 29.	1	11.84		0.62		0.12		1		1482
		OF 038 # 8		4 22 12	+00 29.	1	13.50		0.49		0.04		1		1482
27846	+45 00933	LS V +46 017		4 22 13	+46 07.2	2	8.07	.010	0.29	.015	-0.45	.024	6	B1.5V	399,401
		Hyades vA 347		4 22 14	+12 59.1	2	13.10	.020	0.76	.000			2		19,950
		Hyades vA 353		4 22 14	+17 34.9	1	11.11		0.79				1		1570
		Hyades CDS 61	⋆	4 22 18	+22 56.9	1	13.96		1.56				3		1674
		L 56 - 41		4 22 18	-72 21.	1	13.49		0.77		0.24		2		1696
		Hyades L 40		4 22 19	+18 51.6	1	12.82		1.48				3		582
		Hyades vA 351		4 22 20	+17 09.3	2	13.20	.009	1.54	.000			4	M2	19,582,951
		AJ91,575 # 9	V	4 22 23	+17 09.3	1	13.21		1.52		1.21		1	M4 V	1606
27934	+21 00642	Hyades vB 54	⋆ A	4 22 23	+22 10.9	7	4.22	.002	0.14	.005	0.12	.009	27	A5 IV-V	15,23,1127,1363,3023*
27946	+21 00643	Hyades vB 55	⋆ B	4 22 26	+22 05.2	6	5.27	.012	0.25	.016	0.09	.010	16	A7 V	15,23,254,1127,3023,8015
283648				4 22 27	+26 10.8	1	11.04		1.25		1.27		2	K8	8082
		G 8 - 33		4 22 28	+23 59.5	1	13.53		1.28		0.99		1		333,1620
		Hyades vA 354		4 22 31	+17 48.1	2	11.18	.000	1.25	.044			3	K5.5Ve	19,582,582
		Hyades HG7 - 206		4 22 36	+07 57.	1	12.61		1.58				1		950
27962	+17 00719	Hyades vB 56	⋆ AB	4 22 36	+17 48.9	6	4.29	.010	0.05	.008	0.08	.017	19	A3 V	15,1049,1127,1363*
282162	+30 00663			4 22 41	+30 15.5	1	10.28		0.50		0.36		10	A2 V	272
		Hyades vA 358		4 22 42	+13 34.2	1	12.72		0.96				1		1570
		Met +27 228		4 22 43	+27 24.7	1	12.24		1.09		0.98		2		8082
27856	+55 00881	IDS04187N5525	AB	4 22 43	+55 31.9	1	7.34		0.41		0.07		2	F0	1502
27991	+15 00621	Hyades vB 57	⋆ AP	4 22 46	+15 49.7	4	6.46	.004	0.49	.000	0.01	.018	7	F7 V	15,254,1127,8015
		AJ89,1229 # 279		4 22 48	+05 21.	1	11.35		1.71				1	M0	1632
27855	+57 00800	HR 1382	⋆ A	4 22 51	+57 28.4	2	6.31	.015	0.04	.005	0.02	.000	4	A0 III	985,1733
279846	+33 00858			4 22 52	+34 04.9	1	10.69		1.00				2	K2	1726

Table 1   215

HD	DM	Other Id	N Rem	α_{1950}	δ_{1950}	S	V	σ_V	B–V	σ_{B-V}	U–B	σ_{U-B}	n	Spectrum	References
		Hyades vA 362		4 22 53	+17 25.9	1	15.87		1.73				1		951
283637				4 22 53	+27 30.3	1	11.23		0.74		0.65		2	A0	8082
27990	+17 00721	Hyades vB 176		4 22 54	+17 54.2	4	9.00	.019	0.94	.005	0.66	.000	10	K2 V	582,1127,3072,8023
27971	+31 00776	HR 1390		4 22 55	+31 19.7	2	5.29	.009	0.98	.009	0.78	.014	6	K1 III	1355,3016
27989	+18 00636	Hyades vB 58	★ AB	4 22 57	+18 45.1	3	8.07	.334	0.68	.000	0.23	.005	8	G6 V	1127,3030,8023
		Hyades vA 366		4 22 58	+15 24.5	3	12.26	.072	1.44	.010	1.04		5	M0 Ve	19,582,1570
286777	+13 00677	Hyades vA 367		4 22 59	+14 03.4	1	10.30		0.58				1	G0	920
		Hyades vA 368		4 23 00	+14 53.4	2	16.26	.010	1.56	.015			4		950,951
		Hyades vA 371		4 23 03	+13 23.4	1	15.23		1.69				1		951
	+4 00689	G 82 - 18		4 23 04	+05 09.2	2	11.75	.015	0.83	.010	0.34	.000	3		1620,1658
		BSD 3 # 429		4 23 05	+74 57.1	1	9.84		0.49		0.07		1	F8	1058
28143	−35 01704	HR 1398	★ A	4 23 05	−34 52.2	3	6.56	.033	0.44	.005	0.09		9	F3/5 IV	15,2012,3053
		Hyades CDS 19	★	4 23 06	+21 31.2	1	13.02		1.48				2		1570
286806	+11 00608	Hyades vA 376		4 23 07	+11 49.3	2	9.57	.052	0.53	.028	-0.11		4	F8	218,582
		AJ89,1229 # 280		4 23 08	+03 21.1	1	13.16		1.41				2		1532
		Hyades vA 380		4 23 08	+17 10.1	1	13.66		0.91				1		19
		LP 655 - 15		4 23 10	−06 59.2	1	14.24		1.67				1		1691
		Hyades vA 383		4 23 14	+14 55.8	3	12.19	.035	1.45	.012	1.21		5		582,1570,3003
28255	−57 00659	HR 1405	★ AB	4 23 14	−57 11.0	5	6.28	.009	0.66	.006	0.17		22	G2/8 +G6 V	15,158,1075,2012,2020
28034	+15 00624	Hyades vB 59	★ V	4 23 15	+15 24.7	3	7.48	.014	0.54	.004	0.08	.004	4	F8 V	1127,1510,8023
28116	−17 00875			4 23 16	−17 17.8	1	8.25		1.52		1.84		1	K4 V	1746
28247	−56 00667			4 23 16	−55 56.9	1	7.92		1.48		1.87		6	K4 III	1673
28069	+4 00690	Hyades vB 61		4 23 18	+05 02.2	4	7.35	.018	0.51	.001	0.01	.006	13	F5	1127,1355,1509,8023
		KUV 04233+1502		4 23 18	+15 01.7	1	14.44		0.16		-0.86		1	sdO	1708
		KUV 684 - 4		4 23 18	+15 01.7	1	14.72		-0.15		-0.89		1	sdO	974
		Hyades HG8 - 54		4 23 18	+22 50.	1	13.43		0.94		0.37		4		1058
28024	+22 00696	Hyades vB 60	★ A	4 23 19	+22 42.1	7	4.28	.006	0.26	.006	0.14	.013	24	A9 IV-n	15,23,39,1127,1363*
28033	+21 00644	Hyades vB 62		4 23 20	+21 21.5	2	7.38	.000	0.54	.001	0.08	.002	4	F8 V	1127,8023
27682	+74 00203			4 23 21	+74 52.3	1	9.60		0.00		-0.07		1	A0	1058
	−64 00327			4 23 21	−64 16.8	1	10.39		0.73		0.29		2		1696
28331	−63 00327			4 23 22	−63 24.8	1	9.18		0.57		0.02		4	F5wG3	243
29245	−84 00049			4 23 24	−84 18.0	1	6.98		1.13		1.09		1	K0/1 III	1730
28052	+15 00625	Hyades vB 141	★ A	4 23 30	+15 30.4	6	4.49	.009	0.25	.004	0.14	.004	14	F0 IV-Vn	15,1127,1363,1510*
28068	+16 00598	Hyades vB 63	★ V	4 23 32	+16 44.5	3	8.04	.028	0.65	.019	0.17	.000	4	G5 V	582,1127,8023
	−6 00904			4 23 35	−06 17.2	1	9.73		0.87		0.56		3	K2 V	3072
		Hyades vA 391		4 23 37	+11 49.3	1	15.30		0.70		0.08		1		3060
28006	+39 00989			4 23 37	+40 11.3	1	8.35		1.13		0.88		2	K0	1601
28114	+8 00687	HR 1397		4 23 38	+08 28.7	3	6.06	.018	0.01	.010	-0.42	.000	10	B6 IV	15,154,1061
28085	+16 00600			4 23 39	+17 02.6	1	7.51		1.22				2	G8 II	882
		Hyades HG8 - 25		4 23 40	+19 05.1	1	15.37		1.13				1		1058
28196	−35 01713			4 23 40	−34 57.5	1	8.30		1.62		2.01		2	M1 III	1730
		KUV 04237+1649		4 23 42	+16 48.7	1	13.91		0.34		-0.62		1	sdOB	1708
		Hyades CDS 45	★	4 23 42	+16 48.8	3	13.91	.015	0.35	.031	-0.62	.005	4	sdOB	974,1570,1708
28398	−67 00316	IDS04235S6658	A	4 23 44	−66 51.1	6	6.99		0.32		0.10		6	A2mA9-F2	1628
28005	+46 00884	G 81 - 19		4 23 45	+46 44.7	1	6.73		0.68		0.26		1	G0	1620
28246	−44 01546	HR 1404		4 23 45	−44 16.5	2	6.38	.005	0.45	.005			7	F6 V	15,2012
28100	+14 00697	HR 1396		4 23 47	+14 36.1	5	4.69	.004	0.98	.004	0.72	.005	11	G7 IIIa	15,1355,1363,3016,8015
		Hyades HG7 - 213		4 23 48	+09 01.	1	13.14		0.80				1		950
28099	+16 00601	Hyades vB 64	★ V	4 23 48	+16 38.1	3	8.10	.031	0.67	.014	0.20	.001	4	G6 V	582,1127,8023
		Hyades L 58		4 23 50	+21 07.4	1	10.24		1.36				2	K5	582
		KUV 04239+1406		4 23 53	+14 05.6	1	17.56		0.39		-0.76		1	DA	1708
286798		Hyades vA 404		4 23 55	+12 34.5	6	10.50	.010	1.36	.006	1.24	.010	29	M0 Ve	19,352,582,950,1058,3072
		WLS 420-20 # 9		4 23 55	−20 06.3	1	11.92		0.66		0.11		2		1375
		MCC 445		4 23 59	+12 33.	1	10.44		1.42				2	M0	1619
28067	+36 00895			4 23 59	+36 24.6	1	6.75		1.08		0.91		3	K0 III	1501
		HRC 36, DF Tau		4 24 00	+25 35.7	5	11.51	.195	0.89	.084	-0.22	.101	7	dM0e	640,649,1529,5011,8013
		HRC 37, DG Tau		4 24 01	+25 59.6	3	11.79	.122	1.01	.015	-0.38	.037	5		640,649,8013
286820	+10 00576	Hyades L 59		4 24 02	+10 45.6	3	9.45	.008	1.03	.005	0.87		5	K5	582,950,1058
	−35 01716			4 24 03	−34 55.4	1	9.99		1.66		2.03		2		1730
232957	+54 00771			4 24 04	+54 28.0	1	8.57		1.98				2	M0	369
286789		Hyades vA 407		4 24 06	+13 01.6	3	10.47	.008	1.15	.013	1.02		5	K7	218,582,582
		L 230 - 119	A	4 24 06	−52 16.	2	13.12	.010	1.13	.030	0.95	.050	2		1773,3062
		L 230 - 125	B	4 24 06	−52 16.	2	13.09	.005	1.13	.015	0.99	.035	2		1773,3062
	+8 00690	Hyades HG7 - 219	★ A	4 24 07	+08 35.6	2	10.67	.005	0.76	.000			3		920,1674
	+8 00690	Hyades HG7 - 219	★ B	4 24 07	+08 35.6	1	12.21		1.19				2		1674
	−35 01717			4 24 07	−34 56.7	1	11.42		1.02		0.78		2		1730
		Hyades vA 409	★	4 24 08	+16 17.5	2	12.06	.015	0.81	.005	0.17	.005	4	idG	1058,3029
283667	+24 00657			4 24 10	+24 15.6	1	9.41		0.87		0.45		1	K0 III	8082
285815		Hyades vA 413		4 24 11	+15 15.7	1	11.05		1.02				2	K1	920
		WLS 420 35 # 9		4 24 12	+35 59.1	1	10.78		0.80		0.35		2		1375
		LP 415 - 911		4 24 13	+19 34.5	1	13.19		1.20				1		987
28149	+22 00699	HR 1399		4 24 18	+22 53.1	2	5.53	.010	-0.10	.000	-0.49	.010	5	B7 V	23,154
		G 39 - 12		4 24 19	+36 20.2	1	13.68		1.09		0.91		1		333,1620
28083	+43 00977			4 24 23	+43 13.2	1	7.74		0.03		-0.16		2	A0	401
28413	−61 00335	HR 1416		4 24 23	−61 21.1	5	5.94	.014	1.53	.006	1.86	.015	17	K4/5 III	15,243,1075,2007,3005
		Hyades vA 420		4 24 24	+17 07.8	5	13.05	.016	1.49	.013			8		19,582,950,951,1570
28191	+1 00753	HR 1400		4 24 25	+01 58.1	6	6.22	.012	1.09	.009	1.03	.012	18	K1 III	15,361,1355,1415,1488,3016
		Met +24 263		4 24 26	+24 28.9	1	11.80		1.64		1.26		2		8082
		L 56 - 35		4 24 30	−72 12.	1	12.59		0.59		-0.16		2		1696
		Hyades vA 432		4 24 35	+14 00.4	1	15.91		1.58				2		951

HD	DM	Other Id	N	Rem	α_{1950}	δ_{1950}	S	V	σ_V	B–V	σ_{B-V}	U–B	σ_{U-B}	n	Spectrum	References
		Hyades vA 434			4 24 35	+17 41.0	1	15.34		1.53				1		951
27932	+69 00258				4 24 35	+69 16.2	1	6.65		1.68		2.01		3	K0	985
		Hyades vA 435			4 24 36	+14 09.0	3	10.42	.088	1.10	.016	1.00		5		1570,1674,3072
285828	+13 00684	Hyades vB 177			4 24 36	+14 09.0	3	10.42	.045	1.09	.020	0.97	.020	7	K2	582,1058,1127
		Hyades HG8 - 3			4 24 36	+26 54.	1	13.60		1.94				3		1058
		Hyades HG8 - 60			4 24 37	+21 41.7	1	14.69		1.59				2		920
285816		Hyades vB 188			4 24 42	+15 15.3	2	11.05	.015	0.81	.002	0.39		5	K0 V	582,1127
28471	−64 00334	IDS04242S6419		A	4 24 42	−64 11.8	3	7.90	.005	0.65	.003	0.14	.006	10	G5 V	158,1730,2012
28217	+10 00577	HR 1402		⋆ AB	4 24 43	+11 06.1	3	5.88		0.05	.004	−0.34	.009	9	B8 IV	401,1049,1079
		G 8 - 35			4 24 44	+26 23.7	1	14.11		1.47		1.34		4		316
28205	+15 00627	Hyades vB 65		⋆ V	4 24 45	+15 28.7	4	7.42	.000	0.54	.004	0.06	.006	9	F8 V	1127,1355,8023,8025
		Met +24 265			4 24 46	+24 23.6	1	11.90		0.88		0.73		2	A2	8082
283668	+24 00659	G 8 - 36			4 24 50	+24 20.0	6	9.41	.018	0.89	.017	0.52	.017	19	K3 V	333,830,1003,1529,1620*
28312	−24 02343	HR 1410		⋆ AB	4 24 51	−24 11.5	3	6.11	.004	0.14	.005			14	A3 V	15,1075,2027
	+48 01098				4 24 56	+48 29.3	1	9.58		0.55		−0.39		11	B5	1655
285830	+14 00699	Hyades vB 179			4 24 57	+14 18.4	4	9.48	.021	0.93	.005	0.71	.003	9	K0	582,1058,1127,1570
28237	+11 00614	Hyades vB 66			4 24 59	+11 37.6	3	7.51	.005	0.55	.004	0.05	.004	5	F8	1127,1355,8023
285765	+18 00638	Hyades HG7 - 227			4 25 02	+18 57.0	2	11.29	.000	1.22	.000			4	K0	920,1746
28226	+21 00647	Hyades vB 67		⋆ A	4 25 02	+21 30.6	4	5.72	.000	0.27	.000	0.10	.011	12	Am	15,1127,1199,8015
		Hyades L 72			4 25 04	+18 39.0	1	14.22		1.58				3		582
285766	+18 00639	Hyades L 57			4 25 05	+18 23.4	3	10.16	.012	1.06	.008	0.96		4	K2	582,950,1058
28225	+27 00660				4 25 12	+27 40.2	1	7.78		0.51		0.41		2	A0	23
28258	+13 00685	Hyades vB 178			4 25 15	+13 45.4	5	9.02	.025	0.84	.007	0.51	.020	11	K0 V	582,582,1058,1127,8023
		LP 201 - 30			4 25 17	+42 15.2	1	12.29		1.22		1.18		2		1773
28283	+9 00584	IDS04226N0951		AB	4 25 18	+09 57.7	1	7.46		0.22		0.16		2	A0	401
		LP 201 - 31			4 25 18	+42 15.3	1	12.08		1.17		1.13		2		1773
	−29 01725				4 25 18	−29 18.1	1	9.74		0.50				7	F8	955
285804		Hyades vB 189			4 25 19	+16 21.6	4	11.09	.014	1.34	.010	1.24		24	K5 Ve	352,582,582,1127
29116	−81 00115	HR 1456			4 25 19	−81 41.8	3	5.77	.010	0.36	.003	0.03	.015	21	F0/2 III	15,1075,2038
		Hyades vA 478			4 25 22	+16 32.6	1	15.38		1.09				1		19
28388	−29 01728				4 25 27	−29 18.5	1	7.86		0.75				6	G6 V	955
28322	+1 00755	HR 1413			4 25 28	+01 44.9	3	6.14	.005	1.02	.004	0.81	.005	9	G9 III	15,1355,1415
28293	+15 00630				4 25 29	+16 11.5	1	8.17		1.47		1.41		1	K2	1748
28481	−53 00706				4 25 29	−53 48.9	1	8.28		0.50				4	F6 V	2012
		Met +24 270			4 25 32	+24 43.4	1	11.99		0.90		0.79		2	A5	8082
28294	+14 00702	Hyades vB 68		⋆	4 25 33	+14 37.9	4	5.90	.012	0.32	.007	0.06	.011	10	F0 IV	15,254,1127,8015
28292	+16 00605	HR 1407			4 25 35	+16 15.0	3	4.97	.007	1.14	.004	1.08	.033	9	K2 III	3,1355,3016
		Hyades vA 486			4 25 35	+17 35.2	5	12.15	.036	1.48	.009	0.91		7	M1	19,582,582,951,1674
		AJ91,575 # 10		V	4 25 35	+17 35.2	1	12.07		1.49		1.43		1	M0.5V	1606
		MCC 230			4 25 35	+26 07.0	1	10.92		1.39		1.24		4	M0	1723
283646					4 25 36	+26 06.	1	10.95		1.35		1.21		2	M0 V	8082
28454	−47 01383	HR 1418			4 25 38	−47 03.3	4	6.09	.005	0.47	.010	−0.07		16	F5/6 V	15,1075,2029,3077
28575	−63 00330				4 25 39	−62 51.9	1	8.54		0.94		0.60		4	G6 III	243
		MtW 48 # 231			4 25 40	+30 12.6	1	15.12		0.68		0.32		5		397
		AO 1076			4 25 41	+16 23.3	1	11.91		1.47		1.36		1	K1 III	1748
28291	+19 00727	Hyades vB 69		⋆ V	4 25 41	+19 37.9	3	8.63	.015	0.75	.002	0.33	.004	6	G8 V	1127,8023,8025
		MtW 48 # 233			4 25 41	+30 10.8	1	16.14		1.12		0.68		2		397
		MtW 48 # 232			4 25 41	+30 13.4	1	13.72		1.01		0.48		5		397
28305	+18 00640	Hyades vB 70		⋆ A	4 25 42	+19 04.3	14	3.54	.010	1.01	.006	0.88	.007	93	K1 III	1,15,1006,1034,1105*
28271	+30 00665	HR 1406		⋆ A	4 25 42	+30 15.1	2	6.39	.005	0.53	.014	0.09	.005	7	F8 IV	397,3053
28271	+30 00665	IDS04226N3008	B		4 25 42	+30 15.1	1	8.28		0.46		−0.04		2	G5	3024
28271	+30 00665	IDS04226N3008	C		4 25 42	+30 15.1	2	11.26	.044	0.47	.010	0.35	.010	7		397,3024
28307	+15 00631	Hyades vB 71		⋆ B	4 25 43	+15 51.2	11	3.84	.012	0.95	.004	0.72	.013	43	G9 III	3,15,1084,1127,1286*
28304	+20 00761	IDS04228N2027	A		4 25 45	+20 34.1	1	7.74		0.15		−0.32		2	B8	1648
		Hyades vA 490			4 25 47	+16 51.7	7	14.02	.017	−0.10	.012	−0.98	.014	16	DA	951,974,1281,1570,1674*
28319	+15 00632	Hyades vB 72		⋆ A V	4 25 48	+15 45.7	22	3.40	.008	0.18	.004	0.12	.009	127	A7 III	3,15,254,985,1008,1020*
28344	+16 00606	Hyades vB 73		⋆ V	4 25 55	+17 10.6	4	7.85	.004	0.61	.015	0.13	.005	6	G1 V	582,1127,1355,8023
28281	+34 00883				4 25 55	+35 09.4	1	7.41		0.44		0.38		2	A2	1625
	+48 01101			V	4 25 56	+48 18.8	1	9.96		1.60				1	K5	1772
28375	+1 00757	HR 1415			4 25 57	+01 16.3	3	5.54	.005	−0.10	.004	−0.54	.005	9	B3 V	15,1079,1415
285806		Hyades vB 190		⋆ V	4 25 59	+16 10.7	3	10.71	.005	1.30	.033	1.18		6	K7	582,582,1127
		Hyades vA 500			4 25 59	+16 10.8	2	10.69	.010	1.29	.017	1.15	.015	3	K4	1748,3072
28701	−70 00300				4 25 59	−70 06.3	1	7.85		0.65				4	G2 V	2012
28612	−64 00337				4 26 00	−64 08.3	1	9.73		0.32		0.08		2	A8 III/IV	1730
		Hyades vA 502			4 26 01	+15 52.3	3	12.00	.024	1.42	.015	1.25		5	K7 Ve	582,1570,3003
		Hyades vA 503			4 26 01	+16 25.0	2	13.83	.000	1.54	.000			4		950,1058
28355	+12 00598	Hyades vB 74		⋆	4 26 02	+12 56.3	6	5.03	.005	0.22	.012	0.12	.000	20	A7 m	15,1127,1363,3023*
28343	+21 00652				4 26 02	+21 48.7	2	8.30	.021	1.36	.011	1.22		3	K7 V	1758,3072
		Hyades L 71			4 26 06	+18 33.8	3	13.17	.054	1.49	.004	1.04		7		582,1058,1674
		AJ91,575 # 11		V	4 26 07	+18 35.9	1	13.15		1.52		1.06		1	M2 V	1606
		Hyades CDS 63		⋆	4 26 07	+26 08.1	1	12.93		1.49				3		1674
28363	+15 00633	Hyades vB 75		⋆ AB	4 26 08	+16 03.0	2	6.59	.000	0.53	.000	0.06	.000	3	F7 V +G0 V	1127,8023
		Hyades 512			4 26 08	+16 14.2	2	14.30	.015	1.52	.010	1.00		4	M1	951,3003
		Met +27 245			4 26 11	+27 20.7	1	10.15		0.96		0.88		2	A3	8082
284446		Hyades HG8 - 62			4 26 12	+21 32.	1	11.40		0.80		0.34		3	G5	1058
28354	+27 00661				4 26 14	+27 17.7	3	6.56	.027	−0.03	.024	−0.22	.028	6	B9 V	985,1501,8082
		AO 1078			4 26 15	+16 33.7	1	11.15		0.70		0.20		1	G1 V	1748
279853	+38 00899				4 26 18	+38 41.3	1	8.61		1.97				5	M0	369
		Hyades vA 529			4 26 22	+15 09.9	3	12.40	.027	1.47	.012	1.02		5	M0 Ve	19,582,1570
		Hyades vA 531			4 26 23	+15 19.0	1	12.57		1.26				1		19,950

Table 1 217

HD	DM	Other Id	N Rem	α_{1950}	δ_{1950}	S	V	σ_V	B–V	σ_{B-V}	U–B	σ_{U-B}	n	Spectrum	References
283704 +26 00722	Hyades vB 76			4 26 26	+26 33.8	5	9.19	.022	0.77	.005	0.39	.010	10	G5	950,1127,1570,8023,8082
28394 +17 00731	Hyades vB 77			4 26 27	+17 26.2	2	7.04	.010	0.50	.001	0.04	.000	3	F7 V	1127,8023
28479 −19 00931	HR 1421			4 26 27	−19 34.0	3	5.96	.004	1.21	.005			11	K2 III	15,1075,2035
		Hyades vA 537		4 26 29	+12 15.1	2	14.70	.050	1.62	.025			2	M1	19,951
28552 −42 01510	HR 1429			4 26 31	−42 04.2	5	6.47	.020	1.64	.000	1.95		21	M1 III	15,1075,2013,2029,3005
28257 +57 00806	RV Cam			4 26 32	+57 18.2	1	8.07		1.47		1.30		1	M4 II-III V	3001
28667 −63 00332				4 26 32	−62 54.6	1	6.86		-0.01		-0.04		4	A0 V	243
		Hyades vA 543		4 26 35	+12 41.1	1	13.56		1.26				1		19
		AO 1079		4 26 35	+16 27.3	1	10.26		1.50		1.44		1	K2 III	1748
285805 +15 00634	Hyades vB 181	⋆ V		4 26 36	+16 07.8	3	10.32	.011	1.15	.021	1.08	.000	7	K5	582,1127,8023
28406 +17 00732	Hyades vB 78	⋆ A		4 26 37	+17 45.3	5	6.91	.007	0.45	.002	0.00	.009	11	F6 V	1127,1355,6007,8023,8025
		HRC 38, DH Tau		4 26 37	+26 26.5	3	13.68	.188	1.41	.054	0.15	.184	3	dM0e	640,649,5011
285773 +17 00734	Hyades vB 79	⋆ B		4 26 38	+17 47.1	5	8.95	.017	0.83	.003	0.49	.000	10	K0 V	582,1127,8023,8025,8027
		Met +26 409		4 26 38	+26 02.2	1	11.53		0.94		0.42		1	G0	8082
		HRC 39, DI Tau		4 26 38	+26 26.3	3	12.84	.069	1.61	.005	1.36	.063	3	dM0.5e	640,649,5011
283697 +26 00724				4 26 38	+27 10.2	1	8.71		0.49		-0.07		2	G0 V	8082
		Hyades vA 548		4 26 39	+16 08.2	2	10.31	.010	1.16	.005	1.09	.015	3	K4	1748,3072
		Hyades L 73		4 26 39	+21 33.6	1	14.21		1.55				3		582
		LP 358 - 348		4 26 39	+22 46.5	1	11.52		1.30				1		987
		AO 1081		4 26 41	+16 20.9	1	9.95		1.47		1.56		1	K3 III	1748
28424 +13 00688				4 26 43	+13 47.2	3	7.71	.004	1.22	.005	1.09	.022	5	K1 III	1003,1733,3077
	+64 00454	WW Cam		4 26 43	+64 16.0	1	10.14		0.51		0.44		6		1768
		LP 715 - 20		4 26 44	−12 12.0	1	14.01		1.42		1.14		4		1773
		Steph 491	A	4 26 45	+11 06.9	1	12.32		1.44		1.19		1	M0	1746
		LP 715 - 21		4 26 46	−12 11.5	1	11.66		0.82		0.36		4		1773
		Hyades HG8 - 46		4 26 48	+27 32.	1	12.66		1.24				2		950
		RL 104		4 26 48	+40 18.	1	13.76		-0.02		-0.97		2		704
28497 −13 00893	HR 1423, DU Eri			4 26 48	−13 09.4	7	5.45	.060	-0.16	.027	-0.95	.046	49	B1 Vne	15,681,815,1212,2012*
28906 −75 00270				4 26 49	−74 58.9	1	9.52		0.60		0.15		4	G1 V	1117
		AK 5 - 1061		4 26 50	+19 44.6	1	11.41		1.69				2		987
		G 175 - 34	A	4 26 50	+58 53.3	2	11.09	.010	1.64	.015	1.18	.035	5	dM4	1366,3078
		G 175 - 34	B	4 26 50	+58 53.3	2	12.44	.005	0.32	.010	-0.51	.020	5	DC	1366,3078
		LP 358 - 347		4 26 52	+23 43.2	1	13.19		1.25				1		1570
		LB 3360		4 26 54	−67 08.	1	12.00		0.43		-0.04		1		45
28436 +17 00735	IDS04240N1728	AB		4 26 55	+17 34.2	1	7.73		0.24		0.02		2	B7 V	401
28475 +10 00583	HR 1420			4 26 58	+10 24.8	1			0.09		-0.38		3	B5 V	1079
		G 81 - 21		4 26 59	+39 44.7	1	14.30		1.62		1.16		3		419
28487 +4 00696				4 27 00	+05 03.4	1	6.80		1.74		1.94		1	M3 II	8100
		Met +26 412		4 27 01	+26 33.5	1	11.76		1.08		0.65		2		8082
		Hyades vA 559		4 27 03	+16 48.4	4	12.82	.041	1.48	.021			5		19,582,951,1674
28462 +16 00609	Hyades vB 180			4 27 05	+16 33.9	5	9.08	.034	0.86	.012	0.57	.016	9	K1 V	582,582,1127,1748,8023
		Met +26 413		4 27 05	+26 35.1	1	11.46		1.85				2		8082
		AO 1083		4 27 09	+16 16.6	1	11.25		0.58		0.44		1	G0 III	1748
285846 +17 00736	HRC 43, UX Tau	⋆ AB		4 27 11	−29 07.4	4	11.00	.278	1.09	.070	0.22	.644	5	dG5e	640,649,5011,8013
		CS 22182 # 37		4 27 11	−29 09.2	1	14.10		-0.31		-1.24		1		1736
28732 −62 00357	HR 1435	⋆ A		4 27 11	−62 37.8	2	5.73	.005	1.00	.005	0.82		8	K0 III	243,2006
		Hyades vA 564		4 27 12	+12 18.6	1	15.15		1.62				1		950
		Hyades HG8 - 6		4 27 12	+26 16.0	2	13.52	.019	1.33	.019	1.30		4		950,1058
28447 +27 00662				4 27 13	+28 01.4	1	6.53		0.70		0.22		2	G5	3016
28949 −74 00291				4 27 16	−74 48.8	2	9.36	.009	0.48	.005	0.14		8	F3 III	1117,2012
28485 +15 00636	Hyades vB 80	⋆ AB		4 27 17	+15 31.8	4	5.58	.010	0.32	.004	0.10	.011	10	F0 V+n	15,254,1127,8015
28731 −62 00358				4 27 17	−61 55.0	1	9.11		0.46		-0.04		4	F5 V	243
28505 +9 00590	HR 1425			4 27 18	+10 09.3	2	6.49	.030	1.03	.005	0.85	.025	4	G8 III	1733,3016
28460 +31 00784				4 27 18	+31 29.0	1	8.49		0.14		-0.28		2	A3	1371
283696				4 27 19	+27 19.4	1	11.22		0.91		0.69		5		8082
		HZ 1		4 27 20	+17 37.0	2	12.66	.028	-0.06	.000	-1.02	.019	4	sdOp	272,974
28483 +19 00731	Hyades vB 81			4 27 22	+19 44.0	4	7.10	.005	0.47	.004	0.02	.026	6	F6 V	592,1127,1355,8023
28482 +23 00701				4 27 22	+23 28.9	1	7.17		0.39		0.01		1	B8 III	8082
		LP 715 - 23		4 27 24	−09 55.2	1	12.22		0.85		0.42		2		1696
28459 +32 00806	HR 1419	⋆ A		4 27 25	+32 21.0	3	6.21		-0.04	.009	-0.14	.015	6	B9.5Vn	252,1049,1079
28758 −62 00360				4 27 29	−61 50.9	1	8.08		1.02		0.82		4	K0 III	243
284514 +20 00769				4 27 30	+20 26.4	1	9.74		0.67				1	F5	1570
283695				4 27 30	+27 23.	1	10.38		0.66		0.05		2		8082
		Hyades vA 575	⋆ V		4 27 31	+17 23.5	2	14.47	.015	1.53	.015			5	M2	950,951
285847 +17 00738	Hyades vA 578			4 27 31	+17 40.8	1	9.95		0.59				1	F2	950
28571 −3 00795	G 82 - 22	A		4 27 31	−03 10.1	3	8.98	.000	0.64	.007	0.10	.007	7	G0	1064,1620,3029
28383 +55 00889				4 27 37	+55 46.7	1	8.53		0.58		0.23		2	F8	1502
		AJ91,575 # 12	V		4 27 38	+22 48.1	1	12.96		1.52		1.06		1	M3 V	1606
	−46 01426			4 27 39	−46 38.5	2	8.77	.014	1.12	.005	0.95		5	K0 IV	2033,3077
		HRC 45, DK Tau		4 27 40	+25 55.0	4	11.78	.066	1.09	.117	-0.12	.140	4	dM0e	640,649,1529,5011
28204 +72 00227	HR 1401	⋆ A		4 27 41	+72 25.5	2	5.93	.010	0.29	.007	0.17	.010	6	A8 m	985,3058
28939 −72 00304	IDS04284S7251	AB		4 27 41	−72 44.6	1	7.72		0.88		0.55		7	G8 III	1704
		MCC 446		4 27 42	+00 52.4	1	10.45		1.36				2	K7	1619
28527 +15 00637	Hyades vB 82	⋆ A		4 27 42	+16 05.2	5	4.78	.000	0.17	.001	0.13	.002	16	A5 m	15,1127,1363,3023,8015
		AO 1084		4 27 42	+16 17.9	1	11.54		1.58		1.37		1	K0 III	1748
		ApJ110,424R259# 12	A		4 27 42	+25 57.	1	12.60		1.24		0.18		1		1227
		RL 105		4 27 42	+42 51.	1	14.60		0.04		-0.78		2		704
28571 −3 00795	G 82 - 23	B		4 27 42	−03 09.6	3	14.72	.016	0.70	.027	-0.15	.018	8	dFs	538,1620,3029
28545 +15 00638	Hyades vB 182	⋆		4 27 43	+15 37.6	4	8.94	.009	0.85	.002	0.49	.001	10	K0	582,582,1127,8023
		WLS 420-20 # 12		4 27 43	−21 31.3	1	11.38		0.50		-0.07		2		1375

HD	DM	Other Id	N Rem	α₁₉₅₀ δ₁₉₅₀	S	V	σ_V	B-V	σ_B-V	U-B	σ_U-B	n	Spectrum	References
		Hyades HG7 - 246		4 27 44 +14 38.4	1	14.68		1.56				2	M1	950
		AO 1085		4 27 44 +16 19.4	1	10.97		1.32		0.99		1		1748
283698	+26 00725			4 27 47 +27 05.5	1	9.55		2.05		1.94		2	K2 III	8082
28556	+13 00690	Hyades vB 84	⋆A	4 27 48 +13 37.0	6	5.40	.013	0.26	.007	0.10	.006	15	F0 IV	15,254,1008,1127,3023,8015
28546	+15 00639	Hyades vB 83	⋆A	4 27 48 +15 35.1	5	5.48	.006	0.26	.003	0.10	.001	11	A7 m	15,39,1127,3023,8015
28625	-13 00896	HR 1431		4 27 51 -13 42.0	4	6.22	.017	1.00	.004	0.73		13	K0 III	15,252,1075,2032
28700	-46 01427	HR 1433		4 27 51 -46 37.4	3	6.14	.017	1.06	.000			11	K1 III	15,1075,2032
28403	+57 00809			4 27 52 +57 44.1	1	8.00		1.38		1.50		3	K2 III-IV	37
		Met +27 258		4 27 53 +27 04.2	1	11.96		1.38		1.00		2		8082
28568	+15 00640	Hyades vB 85	⋆B	4 27 55 +16 02.5	2	6.51	.000	0.43	.002	0.01	.000	4	F5 V	1127,8023
		KUV 04280+1605		4 27 58 +16 04.7	1	16.24		0.32		-0.60		1	NHB	1708
270504	-72 00305			4 27 58 -72 30.9	1	9.70		1.07				4	K0	2012
28503	+39 01013	HR 1424	⋆AB	4 27 59 +39 54.2	1			0.04		-0.37		1	B5 V +B8 V	1079
	+44 00967	LS V +44 014	⋆AB	4 28 03 +44 49.6	1	9.34		0.18		-0.58		2	B2	401
28446	+53 00779	HR 1417	⋆A	4 28 04 +53 48.3	3	5.78	.009	0.17	.009	-0.73	.008	7	B0 III	154,405,542
28446	+53 00779	HR 1417	⋆AB	4 28 04 +53 48.3	1	5.47		0.17		-0.73		2	B0 III + B1	542
28446	+53 00779	IDS04241N5342	B	4 28 04 +53 48.3	2	6.95	.005	0.16	.000	-0.70	.000	5	B1	154,542
		Hyades HG8 - 29		4 28 05 +18 18.0	2	13.78	.030	1.46	.015			2		950,1674
		G 247 - 22		4 28 05 +68 40.8	1	15.95		1.92				1		906
		G 247 - 23		4 28 05 +68 40.9	1	16.30		1.90				1		906
		AJ89,1229 # 283		4 28 06 +01 48.0	1	12.38		1.51				3		1532
		LB 3361		4 28 06 -61 21.	1	12.41		0.14		0.06		1		45
		Met +27 259		4 28 09 +27 03.9	1	12.66		1.41				2	K5 III	8082
		Hyades vA 607		4 28 10 +12 11.8	3	14.00	.013	1.55	.035			5	M3	19,950,951
28608	+10 00588	Hyades vB 86		4 28 11 +10 38.7	3	7.04	.008	0.47	.003	0.00	.000	6	F5	1127,1355,8023
		Hyades L 67		4 28 11 +25 52.9	1	11.49		0.96		0.47		1		1058
28581	+23 00702			4 28 13 +23 14.4	2	7.07	.044	1.70	.034	1.87	.058	3	K2 III	23,8082
28580	+24 00662			4 28 14 +24 55.5	2	8.52	.023	0.80	.019	0.40	.014	4	G5	23,8082
285842		Hyades HG7 - 248		4 28 15 +18 42.9	1	10.66		0.89		0.64		3	K0	1058
28595	+14 00711			4 28 17 +14 59.9	2	6.32	.013	1.69	.004	2.04	.018	5	M3 III	1003,3009
		Hyades vA 610		4 28 19 +16 17.4	1	15.13		1.58				1		951
28593	+19 00733	Hyades vB 87		4 28 19 +20 01.6	2	8.59	.010	0.74	.001	0.35	.001	3	G8 V	1127,8023
		Tau sq 1 # 130		4 28 20 +25 12.4	1	14.80		0.76		0.27		7		23
28607	+14 00712			4 28 21 +14 35.0	1	8.18		0.50		-0.01		2	F5	401
279924	+37 00931	G 39 - 18		4 28 21 +38 12.0	1	10.98		0.65		0.05		1	G0	1620
28445	+59 00809			4 28 25 +59 45.9	1	9.06		0.14		-0.22		2		1502
		Tau sq 1 # 131		4 28 26 +25 12.2	1	14.26		0.69		0.41		6		23
		POSS 596 # 3		4 28 28 -00 36.7	1	16.18		1.39				2		1739
28622	+15 00643			4 28 30 +15 42.6	1	7.84		0.45		-0.05		2	F2	401
		RL 106		4 28 30 +41 42.	1	12.97		0.27		-0.28		2		704
		Tau sq 1 # 128		4 28 31 +25 09.2	1	14.82		1.23		0.78		4		23
29598	-83 00091	HR 1485	⋆A	4 28 33 -83 00.6	3	6.76	.004	0.21	.003	0.09	.010	20	A7 IV/V	15,1075,2038
285849	+17 00743	Hyades vA 622		4 28 35 +17 36.7	4	11.90	.027	1.43	.017	1.26		6	K7	582,582,951,3072
		Tau sq 1 # 132		4 28 35 +25 12.5	1	12.73		0.89		0.30		5		23
28621	+22 00707			4 28 36 +22 44.3	1	7.51		0.20		0.09		1	A0	8082
		L 56 - 75		4 28 36 -74 18.	1	11.41		0.46		-0.28		2		1696
28635	+13 00691	Hyades vB 88		4 28 40 +13 47.8	2	7.78	.005	0.54	.000	0.06	.000	4	F9 V	1127,8023
		Haro 6 # 14	V	4 28 42 +18 01.	2	13.77	.623	1.44	.015	0.25	.010	3		640,8013
		Hyades HG8 - 51	A	4 28 42 +23 23.	1	12.40		0.97				1		950
		Hyades HG8 - 51	B	4 28 42 +23 23.	1	12.73		0.92				1		950
28648	+13 00692	IDS04259N1308	AB	4 28 43 +13 14.6	1	8.63		0.42		0.25		1	A0	401
28634	+17 00744	Hyades vA 627		4 28 43 +17 36.2	4	9.53	.014	0.99	.023	0.72	.050	7	K2	218,582,1058,1570
		Tau sq 1 # 133		4 28 43 +25 12.9	1	10.96		1.31		1.05		22		23
		Tau sq 1 # 149		4 28 43 +25 16.2	1	13.33		0.91		0.46		5		23
		HRC 49, HL Tau		4 28 44 +18 07.6	2	14.57	.090	1.39	.070	0.09		2		649,5011
283703	+26 00726			4 28 45 +26 17.5	1	10.50		1.94				5	K0 II	8082
		HRC 50, XZ Tau		4 28 46 +18 07.6	3	14.19	.177	1.33	.295	0.13	.185	3		640,649,5011
		Tau sq 1 # 118		4 28 50 +25 11.0	1	13.39		1.57		1.52		14		23
		LP 715 - 26		4 28 51 -11 37.6	1	15.08		0.84		0.32		3		3062
28776	-35 01768	HR 1439	⋆A	4 28 51 -35 45.6	3	5.96	.008	1.01	.005			11	K0 II	15,1075,2035
		Tau sq 1 # 116		4 28 52 +25 09.6	1	13.97		0.74		0.47		7		23
		Hyades vA 637		4 28 53 +14 56.1	4	12.28	.031	1.47	.019			5	M0 Ve	582,950,1058,1674
		Hyades vA 638	A	4 28 54 +15 31.4	2	12.20	.035	1.44	.030	1.00		4	M0.5	19,582
		Hyades vA 638	AB	4 28 54 +15 31.4	1	11.93		1.46				1	M0.5	1570
		Tau sq 1 # 117		4 28 54 +25 09.5	1	13.16		0.86		0.32		5		23
		Tau sq 1 # 135		4 28 54 +25 17.0	1	13.16		0.68		0.46		5		23
		Tau sq 1 # 119		4 28 55 +25 11.0	1	13.88		0.83		0.28		6		23
28620	+36 00907			4 28 57 +36 56.2	1	6.81		0.46		-0.04		2	F5	1601
28596	+44 00971			4 28 57 +44 47.8	1	7.84		0.12		-0.02		2	A0	401
28677	+15 00645	Hyades vB 89	⋆	4 29 00 +15 44.8	4	6.02	.000	0.34	.009	0.03	.004	11	F2 V	15,254,1127,8015
		Tau sq 1 # 142		4 29 00 +25 06.0	1	15.92		1.22		1.56		2		23
285876		Hyades vB 191		4 29 01 +15 23.6	7	11.02	.024	1.29	.022	1.23	.037	33	K5 Ve	352,582,1058,1127*
		Met +24 274		4 29 03 +24 49.4	1	11.50		1.54		1.28		2		8082
		HRC 51, V710 Tau		4 29 04 +18 15.3	2	13.62	.015	1.64	.075	0.27		2	K7e	649,5011
		Tau sq 1 # 136		4 29 04 +25 16.7	1	14.43		0.93		0.40		3		23
285880	+14 00716	Hyades vA 647		4 29 05 +14 34.5	2	11.33	.015	0.83	.010			3	F8	19,950
		Hyades HG7 - 254	B	4 29 05 +14 34.5	1	13.25		0.80				2		950
		Tau sq 1 # 140		4 29 06 +25 17.9	1	13.36		1.00		0.41		5		23
		LP 358 - 397		4 29 07 +22 12.0	1	12.71		1.44				2		950
28763	-13 00904	HR 1438	⋆A	4 29 07 -13 45.0	2	6.20	.005	0.12	.000			7	A3 V	15,2027

Table 1 219

HD	DM	Other Id	N Rem	α_{1950}	δ_{1950}	S	V	σ_V	B–V	σ_{B-V}	U–B	σ_{U-B}	n	Spectrum	References
		Tau sq 1 # 139		4 29 08	+25 17.0	1	13.09		1.61		1.61		5		23
284474				4 29 09	+23 13.	1	11.28		0.65		0.45		6		8082
		Tau sq 1 # 144		4 29 09	+25 06.3	1	13.58		1.68		1.75		4		23
		Tau sq 1 # 121		4 29 11	+25 11.4	1	11.77		0.81		0.55		5		23
29032	−65 00349			4 29 13	−65 29.1	1	8.06		0.05		0.06		3	A1 V	1704
		Hyades vA 657		4 29 15	+17 33.5	1	15.25		1.57				1	M2	951
		Tau sq 1 # 114		4 29 17	+25 00.7	1	14.52		1.52		1.31		10		23
		Tau sq 1 # 122		4 29 18	+25 11.7	1	13.41		0.70		0.21		5		23
		Tau sq 1 # 123		4 29 18	+25 12.6	1	13.93		1.08		0.50		8		23
283699				4 29 18	+27 16.	1	10.94		1.09		0.56		2	G2 III	8082
28873	−45 01567	HR 1443		4 29 18	−45 03.6	4	5.06	.005	-0.20	.004	-0.79	.005	11	B2 IV/V	15,1034,1732,2012
		LB 3362		4 29 18	−62 26.	1	11.51		0.30		0.11		1		45
28697	+24 00663			4 29 19	+25 04.8	1	7.44		0.16		0.08		36	A2	23
28749	−0 00713	HR 1437		4 29 19	−00 09.0	11	4.91	.012	1.33	.007	1.42	.012	104	K3 II-III	15,58,1075,1363,1425*
		Tau sq 1 # 108		4 29 20	+24 51.1	1	11.86		1.18		0.63		5		23
		WLS 420-20 # 8		4 29 20	−20 03.0	1	11.94		0.96		0.57		2		1375
		Tau sq 1 # 109		4 29 22	+24 52.5	1	14.27		1.12		0.59		3		23
		Met +27 265		4 29 22	+27 13.5	1	11.30		1.89				2	K5 III	8082
28604	+52 00843	IDS04255N5236	A	4 29 22	+52 42.0	1	7.22		1.93				3	K5	369
		Tau sq 1 # 110		4 29 23	+24 56.5	1	12.31		1.00		0.49		5		23
		Tau sq 1 # 113		4 29 23	+25 00.2	1	12.86		1.04		0.46		6		23
		Tau sq 1 # 127		4 29 23	+25 08.9	1	15.60		1.00		0.44		6		23
		Tau sq 1 # 107		4 29 24	+24 50.0	1	13.74		1.19		0.60		5		23
28736	+5 00674	Hyades vB 90	⋆	4 29 25	+05 18.3	5	6.38	.023	0.41	.009	-0.01	.020	11	F5 V	15,254,1061,1127,8015
279984		G 39 - 19		4 29 25	+34 02.0	1	10.89		1.27		1.19		1	K5	333,1620
		Tau sq 1 # 115		4 29 27	+25 06.2	1	14.75		0.98		0.55		6		23
		Met +27 266		4 29 27	+27 20.8	1	12.61		1.35				1		8082
	−63 00334			4 29 27	−62 56.0	1	10.03		0.27		0.08		4	A5	243
		Hyades vA 673		4 29 30	+17 38.8	5	13.95	.040	0.31	.012	-0.69	.000	15	DA + dMe	582,951,1281,1674,3028
		Tau sq 1 # 105		4 29 30	+24 51.4	1	13.32		1.00		0.45		5		23
		G 39 - 21		4 29 30	+26 21.4	1	13.44		1.36		1.09		1		333,1620
		Tau sq 1 # 106		4 29 32	+24 50.5	1	13.64		1.17		0.65		5		23
		Met +26 419		4 29 34	+26 53.8	1	11.33		2.36				2	M0	8082
		G 39 - 22		4 29 34	+36 02.0	1	13.72		1.42		1.14		1		1658
28904	−45 01571			4 29 34	−45 28.0	1	8.26		0.64				4	G3 V	2012
283702	+26 00727	Hyades HG8 - 4	⋆ A	4 29 36	+26 22.	3	10.70	.005	0.86	.018	0.61	.010	4	K3 III	1058,1620,1773,8082
283702	+26 00727	Hyades HG8 - 4	⋆ B	4 29 36	+26 22.	1	13.42		1.38		1.12		2		1773
286839		Hyades vA 677		4 29 37	+13 00.5	4	11.01	.026	1.20	.009	1.03	.015	8	K0	582,950,1058,3072
		KUV 04296+1418		4 29 37	+14 18.2	1	11.00		0.27		0.07		1	NHB	1708
		HRC 54, GG Tau		4 29 37	+17 25.4	4	12.25	.067	1.39	.019	-0.03	.200	4	dK6e	640,649,1227,5011
283714	+24 00664			4 29 37	+25 01.0	1	9.98		0.52		0.14		26	A3	23
		Tau sq 1 # 151		4 29 39	+24 54.6	1	15.35		0.97		0.35		6		23
		HRC 52, UZ Tau	AB	4 29 39	+25 46.2	4	12.89	.058	1.22	.032	-0.21	.107	5		640,649,776,8013
		Tau sq 1 # 103		4 29 41	+24 55.0	1	14.55		1.00		0.43		6		23
	+39 01020		V	4 29 41	+39 50.1	1	10.72		2.26				4	M3	369
		Tau sq 1 # 124		4 29 42	+25 08.8	1	14.59		1.67		1.53		6		23
283694	+27 00663			4 29 43	+27 38.1	1	10.01		0.84		0.36		2	G5 V	8082
285836	+18 00647	Hyades L 64		4 29 45	+19 00.5	5	10.51	.007	1.10	.004	1.03		15	K5	582,950,1058,1570,1674
		Tau sq 1 # 125		4 29 45	+25 09.4	1	15.24		1.03		0.50		6		23
		POSS 596 # 1		4 29 45	−00 34.4	1	13.78		0.63				1		1739
		Tau sq 1 # 126		4 29 48	+25 09.0	1	14.86		1.08		0.48		6		23
		MCC 231		4 29 52	+04 02.2	3	11.00	.029	1.36	.032			4	K5	1017,1619,1632
28704	+42 00990	HR 1434	⋆ A	4 29 53	+42 57.5	2	6.11	.005	0.37	.005	-0.01	.005	5	F0 V	254,3037
28858	−17 00900			4 29 54	−17 39.3	1	9.23		0.30		0.11		2	A1mF0-F0	1375
28783	+15 00646	Hyades vB 91	⋆ V	4 29 58	+15 54.1	3	8.91	.035	0.88	.001	0.55	.002	5	K1 V	582,1127,8023
		LB 1735		4 30 00	−53 42.	1	13.71		-0.20		-0.53		1		45
		G 39 - 23		4 30 05	+33 57.0	1	10.96		0.89		0.62		1	K0	333,1620
28843	−3 00809	HR 1441, DZ Eri	⋆ A	4 30 07	−03 18.8	4	5.80	.020	-0.14	.011	-0.55	.016	10	B9 III	15,1061,1079,1469
28901	−29 01770			4 30 07	−28 54.6	1	7.42		1.09				4	K1 IV	2012
28805	+15 00647	Hyades vB 92	⋆ V	4 30 08	+15 42.9	3	8.66	.008	0.74	.004	0.34	.000	6	G8 V	950,1127,8023
29137	−68 00268			4 30 08	−67 59.4	1	7.68		0.71				4	G5 V	2012
		WK X Ray-2	V	4 30 11	+24 28.0	1	12.08		1.40		1.22		1		1529
	−8 00873			4 30 12	−08 15.4	1	9.47		0.58		0.07		2		1603
282195	+33 00878			4 30 18	+33 25.5	1	9.69		0.54		0.05		2	F8	1375
		CS 22186 # 50		4 30 21	−37 09.6	1	13.64		0.43		-0.24		1		1736
		G 82 - 28		4 30 22	+00 00.3	1	13.46		1.64		1.29		1		906
		KUV 04304+1339		4 30 22	+13 38.9	1	16.50		0.29		-0.86		1	DA + dM	1708
		Hyades L 74		4 30 22	+23 53.2	2	12.61	.019	1.52	.000	1.08		4		582,1606
		G 82 - 29		4 30 23	+00 00.0	1	11.54		1.48		1.23		1	M0	906
29029	−49 01366			4 30 25	−49 25.8	1	9.26		0.54				4	F7 V	2033
28819	+23 00705			4 30 27	+24 10.5	1	7.76		0.32		-0.01		2	F0	8082
28804	+31 00794			4 30 27	+31 33.3	1	8.26		0.21		0.05		5	A0	1371
28747	+45 00955	LS V +45 010		4 30 27	+45 31.9	1	7.70		0.38		0.12		2	B9	1601
	−74 00293			4 30 28	−74 49.4	1	10.48		0.51				4	F5	2012
29154	−66 00296			4 30 31	−66 05.6	1	8.89		1.12		1.26		2	C1,1	864
		HRC 56, GI Tau		4 30 32	+24 15.1	2	13.03	.055	1.29	.115	0.19	.230	2	K5e	649,5011
		HRC 57, GK Tau		4 30 33	+24 14.9	2	12.35	.020	1.30	.025	0.38	.005	2	K5(e)	649,5011
		HRC 59, IS Tau		4 30 33	+26 03.6	1	14.29		1.99				1		5011
		HRC 58, DL Tau		4 30 36	+25 14.4	3	12.99	.167	1.06	.040	-0.35	.069	3		640,649,776
		ApJ110,424R257# 2		4 30 36	+25 15.	1	13.55		1.19		-0.55		1		1227

HD	DM	Other Id	N Rem	α_{1950}	δ_{1950}	S	V	σ_V	B–V	σ_{B-V}	U–B	σ_{U-B}	n	Spectrum	References
29255 −72 00308		SL 4 sq # 1		4 30 38	−72 22.9	2	8.35	.020	0.49	.015	−0.06		7	F6 V	1646,2012
		Hyades vA 709		4 30 39	+12 56.5	3	13.17	.019	1.57	.031	0.76		5	M1	218,920,951
28867 +17 00750		HR 1442	⋆ AB	4 30 39	+17 54.8	2			0.07	.010	−0.09	.035	8	B9 IVn	1049,1079
284552 +20 00774				4 30 39	+21 02.8	1	10.57		1.26		1.20		2	K0	1726
284552 +20 00774		Hyades L 66		4 30 39	+21 02.8	3	10.69	.011	1.24	.005	1.13		4	K7	582,950,1058
29048 −46 01445				4 30 42	−45 54.3	1	8.05		1.20				4	K1 III	2012
28878 +16 00620		Hyades vB 93		4 30 45	+16 39.5	4	9.41	.013	0.89	.004	0.62	.000	10	K2 V	950,1127,1570,8023
		L 447 - 10		4 30 45	−39 06.6	1	11.75		1.30		1.08		1		3078
28879 +16 00621				4 30 46	+16 13.2	1	6.59		0.18		0.09		2	F0	401
		HRC 60, HN Tau		4 30 46	+17 45.6	2	13.41	.260	0.99	.050	−0.45		2		649,5011
28842 +31 00795				4 30 46	+32 02.3	1	8.58		0.51		0.04		2	F8	1371
285837		Hyades L 65		4 30 47	+18 54.6	3	10.74	.012	1.20	.004	1.17		4	K7	582,950,1058
279961 +35 00883				4 30 49	+35 12.4	1	10.95		0.56		0.09		2	A0	1375
28866 +24 00665				4 30 50	+24 23.9	1	9.07		0.54		−0.04		5	F8	8082
−75 00272				4 30 50	−75 43.9	1	9.25		1.13		0.97		2	K5	1730
		HRC 61, CI Tau		4 30 52	+22 44.3	4	13.12	.145	1.26	.187	−0.12	.095	4		640,649,776,1227
		Met +24 283		4 30 52	+24 07.9	1	12.40		0.71		0.41		2		8082
28734 +57 00817				4 30 54	+57 19.1	1	6.65		0.30		0.02		2	F0	1733
		HRC 62, DM Tau		4 30 55	+18 03.9	3	13.81	.049	0.86	.017	−0.48	.072	3	dK5e	640,649,1227
		Hyades vA 722		4 30 57	+12 36.4	4	14.20	.015	−0.04	.016	−0.89	.000	10	DA	1281,1570,1674,3006
283700				4 30 57	+27 00.	1	10.64		0.94		0.87		5	A3	8082
		LP 84 - 29		4 30 57	+60 38.6	1	11.66		0.98		0.68		2		1723
28911 +12 00608		Hyades vB 94		4 30 58	+13 08.9	2	6.62	.000	0.43	.000	−0.01	.002	4	F5 V	1127,8023
		G 39 - 24		4 30 59	+34 43.5	1	13.13		1.57		1.17		1		333,1620
28910 +14 00720		Hyades vB 95	⋆ V	4 31 00	+14 44.5	5	4.65	.006	0.25	.005	0.08	.004	18	A9 V	15,1127,1363,3023,8015
28970 −11 00900		HR 1447		4 31 00	−10 53.4	2	6.06	.002	1.38	.000	1.46		6	K0	58,2007
28922 +10 00595				4 31 01	+10 24.8	1	7.86		0.17		0.11		2	A0	401
		L 375 - 2		4 31 01	−39 52.0	1	12.57		1.52		1.14		2		3073
		L 663 - 3		4 31 02	−21 18.2	3	11.31	.016	1.01	.004	0.75		11	K4	2034,2034,3062
28930 +9 00600		HR 1446		4 31 04	+09 18.6	3	6.00	.005	1.06	.004	0.82	.010	9	G8 III	15,1061,1355
284531 +23 00708				4 31 05	+23 12.9	1	10.40		0.68		0.12		6	G5 V	8082
28995 −21 00910		IDS04290S2121	A	4 31 06	−21 14.1	5	9.80	.009	0.68	.011	0.14	.019	24	G5 V	158,1064,1775,2034,3062
		POSS 596 # 6		4 31 07	−01 22.5	1	18.16		1.29				3		1739
285931 +14 00721		Hyades vB 96		4 31 08	+15 03.6	4	8.49	.018	0.95	.304	0.49	.001	5	K0 IV-V	582,1127,1674,8023
29054 −35 01791				4 31 08	−35 48.3	1	8.06		0.94		0.62		10	K0 III	1673
28946 +5 00678		G 82 - 30		4 31 11	+05 17.1	2	7.92	.009	0.78	.014	0.35	.014	4	K0	196,333,1620
		Hyades L 78		4 31 11	+25 02.3	1	12.84		1.10		0.57		1		1058
		Hyades vA 731		4 31 16	+14 06.8	1	12.33		1.44				2		19,920
		SL 4 sq # 4		4 31 18	−72 25.5	1	10.71		0.71		0.21		3		1646
		Met +27 275		4 31 19	+27 25.6	1	11.80		1.09		0.85		2	A1	8082
284530 +23 00709				4 31 20	+23 20.5	1	10.14		0.65		0.39		6	A0	8082
29086 −37 01799				4 31 22	−37 03.0	1	8.79		0.99		0.84		1	K3/5 V	3072
		Hyades L 79		4 31 25	+11 27.2	3	11.73	.018	1.40	.008	1.25		4		582,950,1058
283692				4 31 25	+27 19.	1	11.04		0.88		0.71		4		8082
28944 +22 00709				4 31 28	+23 00.4	1	9.25		0.46		−0.05		6		8082
29009 −7 00838		HR 1449, EH Eri	⋆ AB	4 31 28	−06 50.5	5	5.71	.004	−0.13	.005	−0.46	.013	17	B9p Si	15,1061,1079,1469,2028
28978 +5 00679		HR 1448		4 31 29	+05 27.9	5	5.68	.008	0.05	.007	0.11	.013	17	A2 Vs	15,1007,1013,1061,8015
		Met +22 192		4 31 30	+22 27.2	1	11.72		0.76		0.64		2	A3 V	8082
28929 +28 00666		HR 1445	⋆ AB	4 31 30	+28 51.5	2			−0.06	.010	−0.35	.030	4	B9p Hg	1049,1079
		SL 4 sq # 92		4 31 32	−72 29.8	1	17.78		0.73				3		1646
28953 +22 00710				4 31 33	+22 36.5	1	8.84		0.45		0.01		3	G0	8082
29085 −30 01883		HR 1453		4 31 33	−29 52.0	8	4.51	.006	0.98	.005	0.72	.011	33	K0 III	15,58,1075,1311,1509*
		SL 4 sq # 2		4 31 35	−72 29.5	1	8.69		1.31		1.25		3		1646
283691 +27 00665				4 31 36	+27 25.9	1	9.12		0.51		−0.03		2	G2 V	8082
286891 +12 00610		G 83 - 18		4 31 38	+12 38.1	1	9.70		0.58		−0.04		2	K2	333,1620
285913 +16 00623		AO 877	A	4 31 38	+16 24.6	1	11.49		0.60		0.46		1	F0	1722
285913 +16 00623		AO 877	AB	4 31 38	+16 24.6	1	10.99		0.57		0.42		1	F0+G5	1722
28977 +15 00650		Hyades vB 183		4 31 40	+15 43.5	4	9.67	.025	0.92	.008	0.68	.000	4	K2 V	582,1127,1570,8023
+37 00939			V	4 31 40	+37 29.4	1	10.52		2.22				4	M3	369
		Hyades vA 747		4 31 41	+15 43.5	1	9.69		0.91		0.68		3		3072
28780 +63 00515		HR 1440		4 31 42	+64 09.6	2	5.92	.015	−0.02	.005	−0.03	.005	5	A1 V	985,1733
		Hyades HG8 - 24		4 31 43	+18 57.9	1	11.77		1.00		0.40		2		1058
		Met +24 292		4 31 43	+24 18.5	1	11.32		1.58		1.14		2	K0	8082
28992 +15 00651		Hyades vB 97	⋆ V	4 31 44	+15 24.1	3	7.89	.036	0.64	.011	0.15	.002	4	G1 V	582,1127,8023
28976 +22 00712		IDS04288N2229	AB	4 31 46	+22 35.4	1	6.78		0.48		0.09		6	F2	8082
283690 +27 00667		Hyades vB 98		4 31 47	+27 57.8	2	9.23	.000	0.45	.001	−0.01	.002	3	F5	1127,8023
29064 −8 00887		HR 1451, DV Eri		4 31 47	−08 20.1	3	5.11	.008	1.69	.005	2.00	.012	11	M3 III	15,1415,3055
28991 +17 00751		Hyades vA 751		4 31 48	+17 38.8	2	7.95	.000	0.40	.000			2	F2	987,987
28975 +23 00710				4 31 48	+24 08.5	1	8.99		0.70		0.63		6	A5 III	8082
29063 −7 00841		HR 1450		4 31 48	−06 56.4	3	6.08	.009	1.38	.005	1.64	.000	6	K2 III	15,1075,1415
29065 −9 00930		HR 1452		4 31 48	−09 04.3	5	5.22	.032	1.46	.014	1.75	.026	11	K4 III	15,1003,1415,3016,8100
		Hyades vA 750		4 31 49	+15 06.4	2	12.40	.005	1.45	.000			3	M0	582,1570
283701 +26 00728				4 31 49	+27 06.0	1	9.60		0.79		0.34		5	B8 III	8082
28927 +38 00915				4 31 49	+38 47.3	1	8.19		0.19		0.16		2	A2 V	105
284532 +22 00713				4 31 50	+22 47.2	1	9.70		0.55		−0.05		3	G0	8082
		TY Tau		4 31 52	+15 09.8	1	12.02		1.01		0.48		5	K0	1768
		Steph 494		4 31 53	+20 17.5	1	11.13		1.31		1.23		6	K7	1746
284561		Hyades L 77		4 31 53	+20 17.5	1	11.12		1.30				2	K7	582
		HRC 63, AA Tau		4 31 53	+24 22.7	3	12.57	.426	1.11	.170	−0.06	.115	3	dM0e	640,649,1227
+3 00620		Hyades J 296		4 31 54	+03 35.0	1	10.31		0.71		0.29		1		1058

Table 1   221

HD	DM	Other Id	N Rem	α_{1950}	δ_{1950}	S	V	σ_V	B–V	σ_{B-V}	U–B	σ_{U-B}	n	Spectrum	References
		LP 715 - 28		4 31 55	−14 01.9	1	14.23		1.26		0.92		2		3062
29023	+18 00652			4 31 58	+18 18.7	1	8.06		0.21		−0.20		2	A0	401
		POSS 596 # 4		4 31 58	−01 26.2	1	16.94		1.52				2		1739
237287	+55 00900	G 175 - 39	A	4 31 59	+55 18.9	4	8.34	.011	0.91	.006	0.68	.007	8	K2 V	22,979,1003,8112
		Met +22 199		4 32 00	+22 47.5	1	12.44		1.26		0.73		2		8082
237287	+55 00900	G 175 - 39	B	4 32 00	+55 18.8	1	11.78		−0.34		−1.20		2		316
		SL 4 sq # 6		4 32 01	−72 27.0	1	13.56		0.68		0.19		3		1646
		SL 4 sq # 7		4 32 01	−72 28.7	1	12.66		1.19		1.08		3		1646
29082	−2 00942			4 32 04	−02 11.1	1	8.20		0.00		−0.04		2	A0	401
29038	+16 00624			4 32 06	+16 53.6	2	7.21	.010	1.18	.000	1.31	.010	3	K3 III	1003,3040
		SL 4 sq # 3		4 32 10	−72 26.5	1	11.13		0.70		0.17		3		1646
		Met +25 349		4 32 13	+25 42.2	1	12.05		1.08		1.04		2	K5	8082
		G 83 - 20		4 32 15	+08 34.0	1	15.44		1.62				1		906
		AO 878		4 32 15	+16 22.4	1	10.84		0.67		0.17		1		1722
		AO 879		4 32 15	+16 37.5	1	12.45		1.21		0.65		1		1722
		SL 4 sq # 8		4 32 16	−72 28.8	1	14.71		0.79				3		1646
29050	+22 00715			4 32 17	+22 56.6	1	8.88		0.77		0.32		3	K1 V	8082
	+89 00005	G 265 - 26		4 32 17	+89 35.8	1	11.40		0.55		−0.05		2	G8	1658
		Hyades CDS 47	⋆	4 32 19	+08 33.4	2	11.76	.000	1.48	.000			4	K4	987,1746
28942	+52 00856			4 32 23	+52 24.3	1	8.67		0.17		−0.10		2	A0	1566
		HRC 65, DN Tau		4 32 26	+24 08.9	4	12.72	.596	1.34	.046	0.67	.046	4	dK6e	640,649,1227,5011
		BSD 24 # 1015		4 32 26	+46 25.5	1	11.84		0.56		0.39		3	B8	1069
		CS 22186 # 43		4 32 26	−33 59.5	1	14.22		0.28		0.04		1		1736
29220	−43 01472			4 32 26	−43 37.6	1	8.86		1.11		1.00		1	K3/4 V	3072
		G 8 - 43		4 32 33	+26 46.0	1	13.38		1.10		0.52		2		1620
		Hyades vA 763		4 32 37	+15 17.8	1	16.13		1.62				1		951
		AO 880		4 32 38	+16 23.3	1	11.55		0.52		0.44		1		1722
285908	+16 00626	Hyades vA 765		4 32 40	+16 51.7	1	10.29		0.63		0.07		2	F8	1058
		Hyades vA 764		4 32 41	+14 06.6	1	14.57		1.14				1		951
28628	+76 00167			4 32 41	+76 51.0	1	8.59		0.40				4	A2	2012
		POSS 596 # 5		4 32 42	+00 03.2	1	17.83		0.84				4		1739
286898	+11 00629	Hyades vA 771		4 32 47	+11 59.9	1	8.80		0.62				1	F8	582
29104	+19 00742	HR 1455	⋆ A	4 32 47	+19 46.8	2	6.35	.005	0.73	.009	0.38	.033	4	G5 II-III+A	592,1733
		G 247 - 27		4 32 47	+60 38.6	1	13.58		0.74		0.07		2		1658
	−75 00273			4 32 48	−75 30.6	1	9.96		1.05		0.92		2		1730
		G 39 - 26		4 32 49	+38 56.8	1	11.56		1.17		1.13		1		333,1620
29184	−20 00880	HR 1461		4 32 49	−20 01.4	4	6.12	.007	1.17	.006			18	K2 III	15,1075,2018,2029
286909	+9 00606	IDS04302N0957	B	4 32 50	+10 04.2	2	7.83	.010	0.54	.005	0.04	.005	6	G0	1084,3024
29173	−10 00959	HR 1460	⋆ AB	4 32 51	−09 50.3	2	6.36	.005	0.11	.000	0.09	.005	7	A1 m	15,1415
29140	+9 00607	HR 1458	⋆ A	4 32 54	+10 03.6	9	4.25	.008	0.18	.011	0.09	.020	24	A5 m	15,39,216,355,1061*
29305	−55 00663	HR 1465, α Dor	⋆ AB	4 32 55	−55 08.9	4	3.26	.004	-0.10	.005	−0.35	.005	16	A0IIIp+B9IV	15,1020,1075,2012
		AO 882		4 32 58	+16 19.0	1	10.60		0.58		0.03		1		1722
		BSD 24 # 415		4 32 58	+44 29.6	1	11.72		0.36		0.03		3	B8	1069
		Hyades vA 776		4 33 00	+13 11.2	1	14.91		1.63				4	M2	920
		SL 4 sq # 9		4 33 00	−72 24.3	1	12.67		0.69		0.11		3		1646
29383		R Ret		4 33 01	−63 07.9	1	10.63		1.54		−0.09		1	M5 e	975
29399	−63 00342	HR 1475	⋆ A	4 33 02	−62 55.6	5	5.79	.012	1.03	.008	0.91	.005	16	K1 III	58,146,158,243,2007
29399	−63 00343	IDS04325S6302	B	4 33 02	−62 55.6	1	9.40		0.94		0.77		4		146
29139	+16 00629	HR 1457, α Tau	⋆ A	4 33 03	+16 24.6	8	0.87	.017	1.54	.008	1.90	.022	53	K5 III	1,3,15,1363,3051,8003*
		LP 890 - 41		4 33 03	−29 58.7	1	11.15		0.70		0.17		2		1696
		LP 358 - 447		4 33 05	+23 46.5	1	12.24		1.01				1		1570
		Hyades HG8 - 71		4 33 07	+18 35.8	2	14.89	.024	1.52	.028			2		1058,1674
	+16 00630	Hyades vB 184		4 33 08	+16 26.4	1	10.83		1.26				2		582
280026	+33 00884	IDS04299N3400	AB	4 33 08	+34 06.5	1	9.20		0.67		0.13		4	G5	3016
		Hyades L 86		4 33 09	+18 47.3	1	13.51		1.51				3		582
		Met +24 304		4 33 09	+24 16.8	1	12.18		0.94		0.41		2	F0	8082
		BSD 24 # 776		4 33 12	+45 17.9	1	11.96		0.49		0.06		2	B8	1069
29094	+40 01000	HR 1454		4 33 13	+41 09.8	5	4.25	.025	1.22	.007	0.80	.030	12	K4 III+A3 V	15,1355,1363,8003,8015
29159	+15 00654	Hyades vB 99		4 33 14	+15 35.0	4	9.38	.012	0.87	.006	0.60	.001	9	K1 V	950,1127,1570,8023
29122	+36 00914			4 33 17	+36 51.2	1	6.62		1.23		1.37		2	K2 III-IV	37
		Met +26 423		4 33 18	+26 18.1	1	12.20		1.02		0.84		2	A2 V	8082
		BSD 24 # 419		4 33 22	+44 26.7	1	12.00		0.58		0.44		3	B9	1069
29658	−75 00274			4 33 26	−75 47.0	1	8.21		0.68		0.24		2	G3 IV	1730
29169	+23 00715	Hyades vB 100	⋆	4 33 28	+23 14.4	6	6.04	.044	0.39	.015	0.01	.025	12	F2 V	15,23,254,1127,8015,8082
		BSD 24 # 423		4 33 28	+44 41.4	1	12.06		0.50		0.38		2	B8	1069
29132	+37 00947			4 33 29	+37 20.3	1	7.62		0.07		0.00		2	A0	1375
29226	−2 00952			4 33 31	−02 19.9	1	7.69		−0.10		−0.40		2	B9	401
29227	−3 00830	HR 1462	⋆ AB	4 33 32	−03 42.8	3	6.32	.005	-0.10	.008	−0.42	.007	9	B7 III	15,1079,1415
29207	+11 00632			4 33 33	+11 18.7	1	6.71		0.31		0.04		2	A0	401
29181	+18 00658			4 33 33	+18 33.9	1	8.39		0.28		0.03		2	A2	401
283756	+24 00670			4 33 33	+25 10.1	1	10.80		0.78		0.43		2	G9 V	8082
		KUV 04336+1034		4 33 35	+10 33.7	1	11.12		0.04		−0.69		1	NHB	1708
29263	−21 00919			4 33 36	−21 26.4	1	8.16		0.66				4	G3 V	2012
29291	−30 01901	HR 1464		4 33 36	−30 39.8	6	3.81	.006	0.97	.005	0.73	.009	30	G8 III	3,15,58,1075,2012,8015
		AO 883		4 33 39	+16 14.0	1	9.32		1.49		1.34		2		1722
		G 39 - 27		4 33 39	+27 03.9	3	15.87	.070	0.65	.011	−0.06		5		1570,1674,3060
283749	+26 00729			4 33 39	+27 09.3	1	9.43		0.57		0.01		2	F8	8082
		BSD 24 # 12		4 33 39	+43 52.1	1	11.66		0.34		0.28		2	B9	1069
		AO 884		4 33 41	+16 41.2	1	9.27		0.65		0.33		1		1722
283750	+26 00730	Hyades HG8 - 2	⋆ V	4 33 42	+27 02.0	4	8.32	.105	1.11	.015	0.86	.018	9	K2	1058,1620,3072,8082

HD	DM	Other Id	N Rem	α_{1950}	δ_{1950}	S	V	σ_V	B–V	σ_{B-V}	U–B	σ_{U-B}	n	Spectrum	References
		AO 885		4 33 43	+16 28.0	1	12.47		0.78		0.49		1		1722
232979	+52 00857			4 33 43	+52 48.0	4	8.62	.017	1.42	.015	1.17	.027	16	K7	22,1197,1758,3072
		Hyades L 87		4 33 44	+18 31.0	1	13.26		1.51				3		582
283759				4 33 44	+24 07.	2	10.31	.033	0.77	.005	0.31	.010	7	F2 V	809,8082
	−3 00832			4 33 45	−03 38.2	1	8.65		0.11		0.12		2	B9	401
		Hyades L 117		4 33 47	+03 57.5	1	12.12		0.87				2		582
		BSD 24 # 784		4 33 47	+45 10.9	1	11.86		0.29		−0.18		2	B9	1069
		AO 886		4 33 48	+16 10.7	1	11.05		0.75		0.16		1		1722
29225	+15 00656	Hyades vB 101		4 33 49	+15 46.1	2	6.65	.000	0.44	.003	0.01	.001	3	F5 V	1127,8023
	+45 00968			4 33 49	+45 17.1	1	10.33		0.24		−0.31		2	B8	1069
29248	−3 00834	HR 1463, ν Eri	★ A	4 33 49	−03 27.2	5	3.93	.023	−0.21	.009	−0.88	.007	15	B2 III	15,154,1061,1278,8015
		G 83 - 22		4 33 52	+11 07.0	1	14.30		1.62		1.16		3		419
		Met +23 190		4 33 54	+23 06.5	1	12.01		1.22		0.58		2	K2	8082
29091	+56 00949			4 33 54	+57 00.7	1	8.28		0.14		0.07		2	A0	1375
		BSD 24 # 1026		4 33 56	+46 32.0	1	12.11		0.67		0.24		3	B8	1069
		AO 887		4 33 59	+16 24.8	1	11.45		0.75		0.22		1		1722
29370	−42 01564			4 34 02	−42 06.2	1	9.31		1.10		0.78		1	K0 II	565
		WLS 448 55 # 6		4 34 04	+55 09.5	1	10.32		0.43		0.22		2		1375
29273	+3 00626	G 82 - 36		4 34 05	+03 25.0	1	8.59		0.83		0.40		1	G5	333,1620
	−19 00967	IDS04319S1929	AB	4 34 07	−19 22.8	1	9.92		0.98		0.74		2		536
		BSD 24 # 23		4 34 08	+43 26.6	1	11.44		0.32		0.09		2	B8	1069
		Hyades L 89		4 34 10	+20 37.2	1	14.28		1.15				1		582
29180	+44 00997	LS V +44 016	★ AB	4 34 13	+44 35.8	1	8.25		0.22		−0.52		2	B2 V	401
	+50 01021	G 175 - 43		4 34 14	+51 04.4	1	10.36		0.77		0.39		2	G8	7010
		Met +23 191		4 34 17	+23 21.2	1	11.68		1.07		0.50		2	F5	8082
283755	+24 00671			4 34 19	+25 07.9	1	10.60		0.67		0.16		2	G1 V	8082
283751				4 34 19	+27 03.4	1	11.46		0.69		0.17		2	Fe	8082
29260	+18 00661	SZ Tau		4 34 20	+18 26.6	2	6.37	.021	0.77	.003	0.41		2	F5 Ib	592,6007
		BSD 24 # 25		4 34 20	+43 39.8	1	12.14		0.52		−0.10		2	B5	1069
29246	+25 00720			4 34 21	+25 37.7	2	7.47	.010	1.36	.010	1.30		4	K2 III	20,105
29286	+10 00598			4 34 23	+10 44.5	1	7.51		0.24		0.16		2	A0	401
	+46 00897			4 34 24	+46 43.3	1	9.08		0.63		0.12		2	A0	1069
		HIC 21574	A	4 34 26	+43 29.7	1	10.52		1.45		1.55		3		1723
		R43B		4 34 27	+11 03.0	1	12.86		1.50		1.20		1		1773
		BSD 24 # 28		4 34 27	+43 51.8	1	12.11		0.25		0.14		2	B9	1069
29203	+45 00969			4 34 27	+46 08.1	1	7.00		1.30		1.03		4	G8 V	7008
29446	−45 01604			4 34 33	−45 14.2	1	7.29		0.43				4	F3 V	2012
		POSS 56 # 2		4 34 35	+66 29.7	1	14.46		1.57				4		1739
29192	+52 00861			4 34 36	+52 39.3	1	8.30		0.32		0.00		2	A5	1375
29335	+0 00798	HR 1469		4 34 39	+00 53.9	3	5.31	.004	−0.13	.004	−0.46	.022	10	B7 V	15,154,1256
29310	+14 00728	Hyades vB 102	★ V	4 34 41	+15 02.8	3	7.54	.005	0.61	.009	0.11	.005	6	G1 V	1127,1355,8023
		BSD 24 # 794		4 34 42	+45 36.9	1	11.71		0.43		−0.17		4	B7	1069
		LP 358 - 460		4 34 43	+25 19.0	2	11.62	.025	0.95	.010	0.75		4	K2 III	1674,8082
	−15 00822			4 34 45	−15 06.9	1	10.47		0.84		0.43		1	F8	1776
		L 13 - 38		4 34 48	−82 44.	1	11.04		0.77		0.36		2		1696
		CS 22191 # 17		4 34 50	−40 20.8	1	13.86		0.39		−0.19		1		1736
29394	−14 00931			4 34 51	−14 13.0	1	8.92		0.45		−0.07		1	F5 V	1776
		Hyades HG8 - 20	★	4 34 53	+19 34.4	2	13.02	.005	1.21	.005	1.12	.070	4		333,1058,1620
284623		G 8 - 46		4 34 53	+19 58.2	1	11.03		0.88		0.54		1		333,1620
29435	−30 01911	HR 1476		4 34 54	−30 49.0	3	6.29	.010	−0.10	.004			14	Ap Si	15,1075,2027
		Reticulum sq # 1		4 35 02	−59 01.8	1	12.12		0.16				1		501
29309	+31 00803			4 35 03	+31 54.0	1	7.10		0.32		−0.31		3	B2 V	399
29392	−10 00968	IDS04327S1003	AB	4 35 03	−09 57.0	1	8.63		0.57				4	F8	1414
		G 39 - 29		4 35 04	+28 07.1	2	12.55	.040	1.65	.000	1.16	.130	2		1620,1658
		POSS 596 # 2		4 35 04	−00 47.7	1	15.85		1.10				1		1739
29391	−2 00963	HR 1474	★ A	4 35 05	−02 34.3	4	5.21	.009	0.28	.006	0.02	.011	14	F0 V	15,1415,1509,3023
		MCC 450	A	4 35 06	+48 31.0	1	11.89		0.80		0.05		4	M0	1723
		G 247 - 30		4 35 06	+67 29.6	2	10.77	.017	1.29	.027	1.27	.024	7		1723,7010
		Steph 497		4 35 06	−02 35.3	1	10.59		1.45		1.06		1	M1	1746
29334	+24 00672			4 35 08	+24 27.3	1	8.99		0.65		0.54		1	A5 V	8082
29751	−73 00269			4 35 09	−73 18.7	2	6.80	.004	0.96	.002	0.66		6	G8 III	278,2012
		Met +25 358		4 35 10	+25 04.8	1	11.54		1.87		1.41		2		8082
		Reticulum sq # 2		4 35 10	−59 01.0	1	14.53		0.39				1	A2 V	501
	−74 00295			4 35 10	−74 03.1	1	9.70		1.48				4	K5	2012
		Met +25 357		4 35 11	+25 54.6	1	11.18		1.04		0.56		2	F8	8082
29376	+7 00676			4 35 13	+07 13.1	1	7.00		−0.06		−0.59		3	B3 V	399
29147	+65 00422	T Cam		4 35 14	+66 02.9	1	9.49		2.35		1.07		1	S4.7 e:	635
29375	+15 00661	Hyades vB 103	★ A	4 35 17	+15 56.1	4	5.79	.008	0.31	.005	0.06	.013	7	F0 IV-V	15,254,1127,8015
29333	+29 00725			4 35 17	+29 17.3	1	8.79		0.67		0.50		2	A2	1625
		Hyades L 118		4 35 19	+04 34.2	3	11.51	.008	1.36	.008	1.33		4		582,950,1058
		Steph 498		4 35 19	+04 34.2	1	11.51		1.36				2	K4	1746
29365	+20 00785	HR 1471, HU Tau		4 35 19	+20 35.2	3	5.90		−0.04	.008	−0.33	.014	6	B8 V	985,1049,1079
	+46 00898			4 35 19	+46 38.8	1	10.00		0.68		−0.05		2	B4	1069
		IBVS2380 # 4		4 35 20	+22 43.8	1	12.35		0.87		0.35		1		1475
	+54 00790			4 35 21	+54 44.3	1	9.51		2.01				3		369
29388	+12 00618	Hyades vB 104	★ A	4 35 22	+12 24.7	5	4.27	.003	0.12	.005	0.12	.009	15	A5 IV-V	15,1127,1363,3023,8015
29769	−72 00311			4 35 22	−72 50.1	3	7.00	.002	0.22	.007	0.13	.011	10	A6 IV/V	278,1075,2012
29364	+26 00731	HR 1470	★ AB	4 35 23	+26 50.5	1	6.54		0.34		−0.03		2	F2 V	1733
		LP 655 - 42		4 35 23	−08 53.8	1	14.10		0.14		−0.65		2	C2	3078
		HRC 67, DO Tau		4 35 24	+26 04.9	3	13.81	.267	1.28	.237	−0.09	.724	3		640,649,1227

Table 1

223

HD	DM	Other Id	N	Rem	α_{1950}	δ_{1950}	S	V	σ_V	B–V	σ_{B-V}	U–B	σ_{U-B}	n	Spectrum	References
	+46 00899				4 35 24	+46 17.5	1	9.56		0.58		0.04		3	B8	1069
	−11 00916				4 35 24	−11 07.9	3	10.34	.000	1.51	.013	1.17		7	M1	158,1619,1705
		POSS 56 # 3			4 35 25	+66 53.6	1	15.86		1.41				6		1739
29387	+15 00662				4 35 26	+15 21.9	1	9.87		0.71				2	K0	987
		Reticulum sq # 28			4 35 26	−59 01.9	1	13.48		0.60				1		501
		G 83 - 26			4 35 28	+12 45.1	1	12.40		0.96		0.55		2		333,1620
		IBVS2380 # 3			4 35 28	+22 48.1	1	11.95		0.68		0.36		1		1475
		IBVS2380 # 5			4 35 30	+22 44.5	1	12.58		0.88		0.41		1		1475
285947		Hyades L 83			4 35 31	+17 26.6	3	10.16	.009	1.15	.000	1.11		5	K5	582,950,1058
	+44 01000				4 35 33	+44 30.1	1	9.91		0.41		−0.07		3	B7 V	1069
276575	+43 01027				4 35 34	+43 49.4	1	10.11		1.24		0.98		2	K7	1726
		IBVS2380 # 6			4 35 35	+22 46.0	1	13.12		0.86		0.23		1		1475
		POSS 56 # 4			4 35 37	+66 20.2	1	16.32		1.58				6		1739
		POSS 56 # 1			4 35 38	+66 34.7	1	14.37		1.47				5		1739
29482	−13 00937	IDS04334S1314		A	4 35 40	−13 07.7	1	7.40		0.04		−0.02		2	B9/A0 V	1776
		BSD 24 # 57			4 35 44	+43 18.2	1	12.06		0.48		0.08		4	B7	1069
		BSD 24 # 60			4 35 45	+43 43.0	1	10.90		0.21		−0.16		2	B8	1069
	−57 00680				4 35 45	−57 38.7	1	9.26		0.91				2	G5	1594
29597	−52 00539				4 35 46	−51 56.4	1	8.35		0.86		0.41		10	G5 V	1673
		Reticulum sq # 22			4 35 46	−58 59.0	1	12.51		0.96				1		501
	+44 01001				4 35 48	+44 24.5	1	10.28		0.36		0.07		4	B9	1069
	+44 01002				4 35 50	+44 45.9	1	10.49		0.55		0.33		3	B8	1069
	+45 00970				4 35 50	+46 04.8	1	9.85		1.17		0.76		4	B8	1069
29419	+22 00721	Hyades vB 105			4 35 51	+23 03.1	2	7.53	.000	0.58	.002	0.10	.002	3	F5	1127,8023
29502	−13 00939				4 35 52	−13 21.4	1	7.33		0.44		0.00		1	F3/5 V	742
283754					4 35 53	+25 07.	1	11.12		1.02		0.72		5	A3 V	8082
29503	−14 00933	HR 1481		⋆ AB	4 35 53	−14 24.0	7	3.86	.004	1.10	.005	0.99	.018	30	K2 IIIb	3,15,1075,1194,1425*
29441	+7 00678				4 35 54	+08 04.6	2	7.66	.032	−0.01	.009	−0.73		5	B2.5Vne	555,2033
		Hyades HG8 - 12			4 35 54	+21 42.	1	13.49		1.57				3		950
29317	+52 00865	HR 1467		⋆ AB	4 35 58	+52 58.9	3	5.06	.012	1.08	.008	0.89	.004	7	K0 III	1355,3016,4001
		Hyades CDS 33		⋆	4 35 59	+19 05.1	1	14.00		1.52				2		1570
		IBVS2380 # 27			4 36 00	+22 34.3	1	10.57		0.37				1		1475
		IBVS2380 # 7			4 36 00	+22 39.0	1	13.36		0.93		0.23		1		1475
29316	+53 00794	HR 1466		⋆ AB	4 36 00	+53 22.6	2	5.37	.005	0.32	.015	0.05	.000	5	A9 V +F8 V	254,3026
		IBVS2380 # 28			4 36 04	+22 35.1	1	11.24		0.62				1		1475
284583	+22 00724	IBVS2380 # 52			4 36 07	+22 36.9	1	10.04		0.26		−0.20		1	B5	1475
		IBVS2380 # 29			4 36 07	+22 38.7	1	11.53		2.09				1		1475
		G 247 - 31		AB	4 36 07	+60 56.9	1	12.25		1.39		1.25		4		7010
29461	+13 00702	Hyades vB 106			4 36 08	+14 00.5	3	7.96	.000	0.67	.010	0.21	.004	4	G5	1127,1355,8023
29712	−62 00372	HR 1492, R Dor		⋆ A	4 36 10	−62 10.5	5	5.44	.060	1.61	.032	0.65	.065	34	M8 III	58,975,2035,3042,8029
29557	−24 02465				4 36 11	−24 45.4	1	8.47		−0.15		−0.53		10	B5 Ib/IIp	1732
		IBVS2380 # 43			4 36 15	+22 43.8	1	14.32		0.92				1		1475
		VY Tau			4 36 17	+22 42.0	2	12.94	.455	1.00	.135	0.00	.095	2		1475,8013
		IBVS2380 # 42			4 36 17	+22 44.3	1	14.04		0.76				1		1475
29479	+15 00665	HR 1478		⋆ B	4 36 18	+15 42.2	4	5.07	.000	0.14	.010	0.19	.021	18	A4 m	1049,1199,1363,3058
		AJ91,575 # 15		V	4 36 18	+22 15.2	1	12.09		1.32		0.79		1	K5 V	1606
		BSD 24 # 804			4 36 19	+45 41.2	1	11.72		0.65		−0.01		5	B4	1069
29459	+24 00674	HR 1477			4 36 20	+25 07.3	4	6.21	.022	0.17	.017	0.11	.027	8	A5 Vn	985,1501,1733,8082
29497	+12 00620				4 36 21	+12 54.4	1	7.19		0.43		−0.05		2	F5	401
		BSD 24 # 73			4 36 21	+44 02.5	1	11.48		0.67		0.44		4	B9	1069
		Hyades L 84			4 36 22	+21 37.0	1	11.91		0.81		0.19		2		1058
29499	+7 00681	Hyades vB 107		⋆ A	4 36 23	+07 46.4	6	5.39	.010	0.26	.009	0.12	.011	12	A5m	15,39,254,1061,1127,8015
29488	+15 00666	Hyades vB 108		⋆ A	4 36 25	+15 49.2	5	4.69	.010	0.15	.008	0.13	.009	14	A5 IV-V	15,1127,1363,3023,8015
		IBVS2380 # 31			4 36 27	+22 39.2	1	13.70		0.85				1		1475
		IBVS2380 # 2			4 36 27	+22 52.8	1	11.21		0.75		0.15		1		1475
286941	+10 00605	Hyades J 307			4 36 31	+11 10.1	1	9.68		0.72		0.25		4	G0	1058
		IBVS2380 # 30			4 36 33	+22 38.0	1	13.31		0.99				1		1475
29573	−12 00955	HR 1483			4 36 34	−12 13.2	8	5.00	.008	0.07	.005	0.09	.015	27	A2 IV	15,196,216,258,355*
		AA45,405 S212 # 10			4 36 36	+50 22.5	1	12.65		1.19		1.25		1		797
29574	−13 00942				4 36 37	−13 26.6	2	8.34	.005	1.40	.010	1.19		5	G8/K0IIIwF7	742,1594
		AA45,405 S212 # 9			4 36 43	+50 22.1	1	13.07		0.92		0.44		1		797
29457	+40 01017				4 36 44	+40 41.5	1	6.52		0.40		0.02		3	F2	379
		AA45,405 S212 # 14			4 36 44	+50 23.8	1	14.34		0.74		0.36		1		797
		AA45,405 S212 # 8			4 36 46	+50 22.9	1	13.38		0.89		0.58		1		797
29528	+12 00622	G 83 - 27			4 36 47	+13 01.7	1	8.93		0.83		0.36		1	K0	333,1620
		AA45,405 S212 # 1			4 36 47	+50 21.9	1	14.37		0.74		−0.01		3		797
		AA45,405 S212 # 3			4 36 48	+50 17.	1	14.71		0.65		−0.11		3		797
		AA45,405 S212 # 5			4 36 48	+50 17.	1	15.70		0.71		0.05		1		797
		AA45,405 S212 # 7			4 36 48	+50 17.	1	14.80		0.66		−0.24		3		797
		AA45,405 S212 # 11			4 36 48	+50 17.	1	16.17		0.69		−0.03		2		797
		AA45,405 S212 # 12			4 36 48	+50 17.	1	14.94		0.86		0.20		1		797
		AA45,405 S212 # 13			4 36 48	+50 17.	1	13.57		1.73		1.86		1		797
		AA45,405 S212 # 2			4 36 48	+50 21.9	2	11.78	.005	0.56	.010	−0.42	.015	7	O5.5neb	797,7006
		AA45,405 S212 # 4			4 36 50	+50 21.5	2	13.14	.010	0.60	.020	−0.30	.020	7	B0 Vneb	797,7006
		AA45,405 S212 # 6			4 36 51	+50 22.1	1	12.48		2.14		2.16		2		797
29666	−40 01499				4 36 51	−40 17.1	4	9.19	.016	0.64	.008	0.10	.010	11	G2 V	742,1097,2012,3026
		BSD 24 # 808			4 36 55	+45 36.1	1	12.07		0.78		0.45		6	B7	1069
		LP 475 - 240			4 36 56	+09 24.6	1	16.54		0.74				1		1674
		IBVS2380 # 1			4 36 56	+22 50.8	1	10.29		0.74		0.37		1		1475
		Hyades L 85			4 36 56	+23 02.5	5	11.40	.011	1.38	.004	1.25		13		582,950,1058,1570,1674

HD	DM	Other Id	N Rem	α_{1950}	δ_{1950}	S	V	σ_V	B–V	σ_{B-V}	U–B	σ_{U-B}	n	Spectrum	References
286955	+9 00621	G 83 - 28	★ A	4 36 58	+09 46.8	3	9.19	.008	1.03	.010	0.80	.008	5	K3 V	196,333,1620,3072
28985	+79 00150			4 36 58	+79 33.8	1	6.70		-0.01		0.01		4	A0	985
286955	+9 00621	G 83 - 29	★ B	4 36 59	+09 46.1	1	14.17		1.68				1		3072
29550	+17 00762			4 36 59	+17 23.5	1	7.91		1.32		1.15		2	K0	1648
		Hyades HG8 - 61		4 37 00	+21 24.	1	13.64		1.48				3		950
		KUV 04370+1514		4 37 02	+15 13.9	1	15.83		-0.06		-0.75		1	DA	1708
29487	+43 01036			4 37 02	+44 00.0	1	7.43		0.21		-0.15		2	B8 V	401
29613	-14 00936	HR 1487		4 37 02	-14 27.3	3	5.45	.008	1.05	.005			11	K1 IVa	15,1075,2032
286929	+12 00623	Hyades L 80		4 37 03	+12 37.9	3	10.04	.012	1.07	.005	1.00		5	K5	582,950,1058
29921	-71 00275	IDS04377S7154	AB	4 37 03	-71 47.5	1	8.84		0.33		0.18		3	A7 V	204
284574	+23 00722	Hyades vB 109		4 37 05	+23 12.5	5	9.40	.005	0.81	.006	0.45	.001	11	K0 V	950,1127,1570,1674,8023
283810	+25 00722			4 37 05	+25 29.8	1	10.64		1.21		1.02		6	K5 V	8082
28941	+80 00147			4 37 11	+80 27.1	1	7.67		1.40		1.60		3	K0	1733
		Eri sq 1 # 18		4 37 13	-16 32.7	1	10.73		0.67		0.10		2		8098
29610	-1 00689	HR 1488		4 37 15	-01 09.0	3	6.10	.003	0.94	.002	0.66	.014	28	K0	15,1415,3047
29589	+11 00639	HR 1484		4 37 16	+12 06.1	2			-0.12	.010	-0.50	.000	6	B8 IV	1049,1079
		Hyades HG8 - 68		4 37 17	+19 11.5	2	13.63	.017	1.46	.004			3		950,1058
29685	-30 01932			4 37 18	-30 41.5	2	8.90	.005	1.01	.005	0.66	.029	3	G8 III	565,3048
		Met +26 428		4 37 20	+26 27.2	1	12.05		1.09		0.45		2	G2	8082
	-11 00930			4 37 21	-11 32.7	1	9.72		1.58		1.94		2	M2 V	3072
		BSD 24 # 491		4 37 22	+44 26.5	1	10.78		0.61		-0.36		3	B0 p	1069
29536	+36 00924			4 37 24	+36 45.3	1	8.41		0.58		0.44		2	A3	1601
		Eri sq 1 # 17		4 37 28	-16 33.7	1	13.15		0.79		0.39		2		8098
276667	+43 01038			4 37 29	+43 21.2	1	10.26		0.35		-0.13		2	B7	1069
		Hyades HG8 - 19		4 37 30	+19 37.8	1	13.96		0.85		0.16		4		1058
29634	+0 00815			4 37 31	+00 27.3	1	8.57		0.09				1	A0	1245
270592				4 37 32	-72 07.4	1	11.37		0.79		0.25		3	G2:V	149
29608	+16 00640	Hyades vB 185		4 37 33	+16 25.1	4	9.47	.004	1.10	.006	0.93	.009	7	K0	582,950,1058,1127
		Hyades CDS 46	★	4 37 34	+13 53.0	1	14.93		-0.07				2		1570
	+44 01005		V	4 37 37	+44 35.4	2	10.23	.010	2.39	.030	2.44		7	M2	369,8032
29632	+7 00688			4 37 40	+07 42.1	1	7.88		0.98				2	K0	987
29526	+48 01128	HR 1482		4 37 40	+48 12.4	1	5.67		-0.02		-0.03		2	A0 V	401
283811				4 37 45	+25 23.	1	11.20		0.95		0.86		5	K3 V	8082
270587				4 37 46	-68 44.5	1	11.90		0.65		0.08		4	F8	149
29621	+23 00723	Hyades vB 110		4 37 48	+23 42.9	3	8.85	.014	0.69	.008	0.25	.004	4	G5	1127,1529,8023
29400	+66 00343	G 247 - 32		4 37 48	+66 38.4	3	8.28	.007	0.73	.008	0.22	.011	8	G8 V	1003,1658,1775
29701	-16 00919	Eri sq 1 # 16		4 37 49	-16 40.6	1	9.98		0.13		0.19		2	A9	8098
		Hyades L 88		4 37 50	+18 39.4	2	13.56	.005	1.30	.009			2		582,1058
29805	-51 01207	HR 1498		4 37 50	-51 46.2	1	6.44		1.32				4	K2 III	2007
		Eri sq 1 # 3		4 37 53	-16 33.7	1	13.10		0.58		-0.01		2		8098
29709	-13 00947			4 37 54	-13 06.7	1	8.22		0.43		0.07		1	F0 IV/V	1776
		Eri sq 1 # 2		4 37 54	-16 33.1	1	11.86		1.04		0.91		6		8098
29907	-65 00361			4 38 00	-65 31.9	4	9.84	.023	0.64	.008	-0.12	.016	12	F8 V	1311,2012,3078,6006
		BSD 24 # 816		4 38 01	+45 30.7	1	10.76		0.67		-0.05		6	B8	1069
29803	-45 01627	IDS04365S4519	AB	4 38 01	-45 13.1	1	8.86		0.91				4	K1 V	1414
		AJ91,575 # 16	V	4 38 02	+24 45.4	1	12.51		1.53		1.19		1	K7 V	1606
		Eri sq 1 # 1		4 38 02	-16 40.3	1	10.23		1.52		1.92		3		8098
29737	-24 02488	HR 1495		4 38 02	-24 34.7	3	5.57	.004	0.93	.005			11	G6 III	15,1075,2031
		G 175 - 45		4 38 03	+57 37.0	1	11.37		1.41		1.32		4	M1	1723
29647	+25 00723			4 38 04	+25 53.8	1	8.31		0.91		0.47		5	B8 III	8082
29587	+41 00931	G 81 - 30		4 38 04	+42 01.7	6	7.29	.021	0.62	.013	0.04	.006	11	G2 V	22,908,1003,1620,1733,3026
		LP 835 - 3		4 38 04	-22 38.4	1	13.21		0.85		0.35		2		1696
		Steph 502		4 38 05	+02 08.2	1	11.24		1.47		1.26		1	M0	1746
	-9 00956	IDS04358S0924	AB	4 38 06	-09 17.1	2	10.27	.004	1.48	.015	1.16		6	M0 V	158,1619
		G 39 - 30		4 38 09	+32 36.8	1	13.62		1.51		1.02		1		1658
	+43 01041	LS V +43 012		4 38 10	+43 57.5	1	8.54		1.42		0.44		3		1069
29646	+28 00680	HR 1490	★ A	4 38 12	+28 31.2	3	5.75	.038	-0.02	.013	0.01	.033	6	A2 V	985,1049,3024
29646	+28 00680	IDS04351N2825	B	4 38 12	+28 31.2	1	11.26		1.11				1		3024
		Eri sq 1 # 5		4 38 13	-16 37.8	1	12.15		0.53		-0.07		2		8098
		Eri sq 1 # 7		4 38 13	-16 39.0	1	12.51		0.83		0.41		2		8098
31053	-85 00050			4 38 13	-84 54.5	1	8.66		1.62		1.48		1	M4/5 III	794
		Hyades CDS 3	★	4 38 14	+10 53.0	3	13.82	.009	-0.16	.009	-1.04		3		1570,1674,3028
		POSS 56 # 6		4 38 14	+66 10.1	1	17.59		1.46				5		1739
29755	-19 00988	HR 1496, DM Eri	★ AB	4 38 15	-19 46.0	9	4.32	.018	1.60	.008	1.82	.033	38	M4 III	3,15,1024,1088,1311*
		G 8 - 50		4 38 20	+22 49.1	1	12.63		1.01		0.79		1		333,1620
29697	+20 00802	V834 Tau	★	4 38 22	+20 48.6	5	8.00	.011	1.11	.045	0.92	.009	11	K3 V	196,516,1619,1620,3072
283809				4 38 22	+25 48.	1	10.72		1.42		0.49		5	F8	8082
		BSD 24 # 820		4 38 24	+45 24.9	1	10.40		0.59		-0.08		6	B7	1069
29746	-10 00982			4 38 24	-10 36.3	1	9.40		0.46		-0.09		25	F8	588
29645	+37 00954	HR 1489		4 38 26	+38 11.2	2	6.00	.010	0.56	.005	0.07	.045	5	G0 V	254,272
		BSD 24 # 94		4 38 27	+43 36.1	1	11.00		0.39		0.08		2	B9	1069
	-13 00952		B	4 38 28	-13 09.5	1	11.13		0.73		0.24		2	K2	742
284705	+19 00754	Hyades vB 186		4 38 29	+20 11.0	2	10.00	.014	1.41	.008	1.26		6	G5	950,1127
		Hyades HG8 - 9		4 38 30	+23 18.	2	12.02	.004	1.39	.004	1.33		5		950,1058
		BSD 24 # 96		4 38 33	+43 36.8	1	12.36		0.79		0.43		2	B5	1069
		Eri sq 1 # 8		4 38 38	-16 41.1	1	12.89		1.05		0.84		5		8098
29800	-23 02091			4 38 38	-23 08.9	1	8.75		1.16		1.09		3	K1 III	1657
		Hyades CDS 64	★	4 38 41	+11 54.9	1	12.88		1.50				3		1674
		Hyades L 81		4 38 41	+13 07.6	3	11.23	.012	1.45	.008	1.28		5		582,950,1058
29813	-28 01674			4 38 42	-28 17.6	1	7.74		0.62				4	G3 V	2012

Table 1   225

HD	DM	Other Id	N Rem	α_{1950}	δ_{1950}	S	V	σ_V	B–V	σ_{B-V}	U–B	σ_{U-B}	n	Spectrum	References
29844		R Cae		4 38 46	−38 19.9	1	9.72		1.47		0.40		1	M6e	975
29606	+59 00826	HR 1486	★ AB	4 38 57	+59 25.6	1	6.50		0.21		0.11		3	A7 IV	1733
29875	−42 01587	HR 1502	★ AB	4 38 57	−41 57.5	5	4.45	.009	0.34	.005	-0.01	.016	21	F1 V	15,1075,2012,3026,8015
270602				4 38 59	−68 14.5	1	11.55		0.90		0.30		2	K5	149
29329	+76 00174	HR 1468		4 39 02	+76 31.2	1	6.51		0.50				1	F7 V	254
29789	−1 00697	Hyades vB 163		4 39 02	−00 50.4	2	7.98	.000	0.38	.010	0.03	.001	4	F2	1127,8023
280042	+37 00958			4 39 05	+37 45.7	1	9.48		2.28				4	G0	369
		BSD 24 # 506		4 39 10	+44 12.3	1	11.45		0.67		0.47		4	B8	1069
		BSD 24 # 505		4 39 11	+44 45.3	1	11.96		0.74		0.41		4	B9	1069
284659	+22 00737			4 39 12	+22 50.9	2	7.16		0.06	.003	0.06	.016	6	A0	1084,1211
29930	−48 01454			4 39 12	−48 38.0	1	6.90		1.09		0.97		4	G8/K0 III	243
29763	+22 00739	HR 1497	★ AB	4 39 14	+22 51.8	9	4.27	.011	-0.13	.008	-0.58	.011	23	B3 V	15,23,154,1084,1203*
		Eri sq 1 # 19		4 39 15	−16 41.3	1	10.25		0.31		0.09		2		8098
		Eri sq 1 # 22		4 39 15	−16 41.3	1	12.38		0.58		0.10		1		8098
		Hyades CDS 65	★	4 39 17	+11 49.6	1	13.37		1.51				3		1674
29786	+15 00669			4 39 18	+15 52.6	1	8.09		0.38		0.00		2	F2	401
283808	+25 00724			4 39 18	+25 44.8	1	10.78		0.99		0.82		4	A3 III	8082
		Met +25 365		4 39 18	+25 45.5	1	11.34		1.05		0.94		2	A3	8082
		LP 358 - 511		4 39 20	+25 42.2	1	12.26		2.05				2		920
29722	+43 01043	HR 1494		4 39 21	+43 16.3	2	5.30	.005	0.01	.002	0.02		3	A1 Vn	1363,3023
29851	−12 00969			4 39 22	−12 34.2	1	6.67		0.13		0.16		2	A2 IV/V	1776
		Hyades HG8 - 76		4 39 24	+16 25.	1	13.16		1.33		1.40		2		1058
		BSD 24 # 507		4 39 24	+45 04.5	1	11.10		0.71		0.12		5	B8	1069
	+45 00973	LS V +45 013		4 39 24	+45 32.2	1	8.58		0.52		-0.35		6	B3 V	1069
30003	−59 00370	HR 1504	★ AB	4 39 29	−59 02.5	5	6.53	.005	0.68	.000	0.22	.008	13	G5 V	15,214,258,2012,3077
		ApJ110,424R259# 19		4 39 30	+25 10.	1	13.98		1.24		-0.43		1		1227
29819	+9 00628	IDS04368N0926	A	4 39 31	+09 32.4	2	6.74	.000	0.34	.010	0.03	.000	4	F2	401,1733
		POSS 85 # 5		4 39 31	+61 01.8	1	16.66		1.64				3		1739
		G 82 - 42		4 39 31	−04 22.2	2	12.49	.009	0.61	.000	-0.06	.000	4		1620,1696
		Hyades CDS 37	★	4 39 33	+20 21.6	1	13.28		1.54				2		1570
29721	+49 01230	HR 1493	★ A	4 39 33	+49 52.8	2	5.89		0.00	.013	-0.27	.004	5	B9 III	985,1079
		G 39 - 32		4 39 42	+29 23.8	2	16.01	.018	1.59	.007			3		538,1691
		BSD 24 # 108		4 39 44	+43 17.0	1	11.52		0.56		-0.04		2	B2	1069
		G 83 - 31		4 39 52	+16 25.4	1	11.30		1.27		1.16		1	K4	333,1620
		G 39 - 33		4 39 52	+37 41.5	1	14.57		1.18		0.60		2		1658
287023	+9 00630			4 39 53	+09 30.4	1	9.97		0.75				1	G5	1570
		ApJ110,424R259# 24		4 39 54	+25 15.	1	15.06		1.15		-0.25		1		1227
29836	+18 00684	IDS04370N1832	A	4 39 56	+18 37.7	1	7.12		0.67		0.27		2	G5	1648
		BSD 24 # 1060		4 39 57	+46 22.3	1	12.09		0.73		0.40		2	B5	1069
285968	+18 00683	G 8 - 55		4 39 59	+18 52.6	5	9.96	.013	1.52	.013	1.15	.009	18	M2	333,680,1017,1197*
29861	+17 00774			4 40 05	+17 13.1	1	8.23		1.70		1.99		2	K5	1648
		G 175 - 46		4 40 05	+51 00.8	1	15.96		0.21		-0.61		4	DA	940
285970	+18 00685	Hyades HG8 - 28		4 40 06	+18 37.	2	9.91	.000	1.13	.005	1.09		6	K8	950,1058
29869	+14 00738			4 40 09	+14 43.3	1	6.67		0.90		0.45		3	G5	1648
29859	+23 00733	HR 1499		4 40 12	+23 59.8	3	6.14	.008	0.54	.011	0.06	.008	6	F7 IV-V	23,71,3016
		KUV 685 - 9		4 40 13	+14 54.8	1	13.97		0.21		-0.81		1	sdO	974
	−24 02523			4 40 14	−24 02.7	1	10.07		0.19		0.05		2		126
		AK 6 - 290		4 40 15	+19 16.4	1	12.46		0.99				2		987
29985	−34 01822			4 40 15	−34 00.7	1	10.03		1.17		1.17		1	K5/M0 V	3072
	−13 00959		A	4 40 17	−13 19.9	2	10.99	.005	0.70	.000	0.17		6		742,1594
29992	−37 01867	HR 1503		4 40 17	−37 14.5	3	5.04	.004	0.38	.004	0.01		8	F3 V	15,258,2012
232997	+54 00803			4 40 18	+54 26.7	1	8.82		1.08		0.85		3		1733
		CS 22191 # 24		4 40 20	−39 42.7	1	13.57		0.69		0.14		1		1736
30042	−47 01472	IDS04389S4727	AB	4 40 20	−47 21.3	1	6.76		0.76				4	G8 III +G	2012
30554	−79 00152			4 40 20	−79 33.4	3	7.98	.007	1.09	.011	0.76	.034	7	G8/K0p Ba	565,2012,3048
29896	+16 00646	Hyades HG8 - 65		4 40 22	+16 58.6	1	9.87		0.99		0.80		1	K0	1058
		Hyades CDS 66	★	4 40 22	+24 12.3	1	14.07		1.47				3		1674
		BSD 24 # 113		4 40 22	+44 00.9	1	11.97		0.45		0.14		5	B7	1069
29896	+16 00646	Hyades L 90		4 40 23	+16 58.6	3	9.85	.005	1.00	.000	0.85		4	K0	582,950,1058
	+44 01011			4 40 23	+44 53.9	1	10.53		0.56		-0.09		4	B8	1069
		LP 359 - 14		4 40 25	+21 56.4	1	13.38		1.24				1		1570
29883	+27 00688	G 85 - 11		4 40 29	+27 35.9	4	8.00	.012	0.91	.009	0.63	.004	12	K5 III	23,196,333,1197,1620
29818	+46 00903			4 40 30	+46 52.5	1	8.77		0.70		0.22		2	G5	1601
		AK 6 - 305		4 40 32	+19 12.4	1	11.05		0.71				1		987
29833	+42 01033	LS V +42 008	★ AB	4 40 32	+42 19.6	1	7.51		0.31		-0.20		2	B9	401
29867	+32 00827	HR 1501		4 40 34	+32 46.4	1	6.48		0.29		0.02		2	A8 V	1733
30479	−77 00181	HR 1531		4 40 36	−77 45.1	5	6.04	.005	1.10	.004	0.95	.004	26	K2 III	15,278,1075,2012,2038
		G 82 - 44		4 40 39	−03 15.1	3	13.47	.019	0.59	.018	-0.17	.024	7		1620,1658,3056
286996	+10 00618	Hyades L 94		4 40 45	+11 04.5	3	10.27	.025	1.19	.025			5	K2	582,1017,1619
29866	+40 01032	HR 1500		4 40 45	+40 41.7	3	6.06	.023	0.06	.018	-0.29	.016	8	B8 IVne	154,379,1212
		LP 535 - 60		4 40 47	+06 00.0	1	13.41		1.46		1.14		1		1696
232999	+50 01043	LS V +50 009		4 40 51	+50 26.5	1	9.37		0.56		-0.31		2	B1 IV	1012
30229	−65 00369	IDS04407S6530	A	4 40 56	−65 24.4	5	9.40	.019	0.66	.010	-0.04	.018	19	G2-F0	149,742,1594,2034,3077
29882	+44 01013			4 40 59	+44 40.5	1	7.75		0.25		0.04		2	A7 V	1601
276758	+41 00944			4 41 01	+42 09.4	1	10.18		1.35		1.10		2	G5	1726
		LP 475 - 284		4 41 02	+09 15.3	1	14.58		1.11				1		1570
30070	−33 01886			4 41 07	−33 31.5	1	8.42		0.54				4	F6 V	2012
270639				4 41 07	−69 51.2	1	11.58		0.67		0.09		2	G5	149
29817	+59 00829			4 41 08	+59 46.8	1	7.89		0.40		0.09		2	F0	1502
29912	+38 00933			4 41 09	+38 39.5	1	8.72		0.31		0.19		3	A2	833

HD	DM	Other Id	N Rem	α_{1950}	δ_{1950}	S	V	σ_V	B–V	σ_{B-V}	U–B	σ_{U-B}	n	Spectrum	References
29900	+46 00906			4 41 10	+46 17.8	1	9.19		0.35		0.05		2	B9 V	1069
30020	−9 00969	HR 1505, DW Eri	⋆ A	4 41 11	−08 53.2	2	6.83	.005	0.38	.009	0.19	.019	2	F4 IIIp Sr	1279,3032
30021	−9 00970	HR 1506	⋆ A	4 41 11	−08 53.3	2	6.71	.009	0.90	.014	0.63	.014	2	G8 III	1279,3032
30021	−9 00970	IDS04388S0859	⋆ AB	4 41 11	−08 53.3	3	5.99	.008	0.64	.005	0.32	.005	11	G8 III +	15,1415,2006
		POSS 158 # 3		4 41 12	+49 00.7	1	15.41		1.37				2		1739
		POSS 158 # 4		4 41 13	+49 07.5	1	16.05		1.01				2		1739
30080	−30 01968	HR 1509		4 41 13	−30 51.4	4	5.67	.005	1.39	.014	1.60		13	K3 III	15,1075,2031,3005
		POSS 158 # 7		4 41 14	+49 05.9	1	18.01		1.21				2		1739
		POSS 158 # 1		4 41 14	+49 07.0	1	14.56		1.46				2		1739
29910	+42 01034	G 81 - 33		4 41 15	+43 06.7	1	8.74		0.87		0.58		1	G5	333,1620
		POSS 158 # 5		4 41 17	+49 03.7	1	17.26		1.13				2		1739
		POSS 158 # 6		4 41 18	+49 05.0	1	17.57		1.14				3		1739
283779		G 39 - 34		4 41 19	+27 46.5	1	11.26		1.53		1.21		1	M25	1620
29969	+23 00738			4 41 20	+23 59.5	1	8.18		0.62		0.13		1	G0	23
283812	+25 00727			4 41 21	+25 26.2	1	9.48		0.73		0.58		6	A0 III	8082
	−16 00931			4 41 21	−15 53.9	1	10.01		0.95		0.77		1	K2 V	3072
30050	−10 00993	RZ Eri		4 41 24	−10 46.5	3	8.29	.410	0.62	.029	0.03	.033	8	Am	2012,8071,8080
		G 39 - 35		4 41 28	+29 43.9	1	13.45		1.60		1.55		1		1620
30185	−50 01471	HR 1516		4 41 29	−50 34.5	4	5.30	.009	0.98	.000	0.74		15	K0/1 III	15,243,1075,2035
30184	−49 01422	IDS04402S4910	AB	4 41 31	−49 04.6	1	9.50		0.47		-0.04		4	F5 V	243
		BSD 24 # 123		4 41 32	+43 22.1	1	11.50		0.72		0.29		2	B5	1069
30057	−3 00869			4 41 36	−03 15.5	1	7.87		0.97		0.71		3	K0	1657
283807	+25 00728	G 85 - 13		4 41 37	+25 50.7	1	10.14		0.59		0.00		2		333,1620
		AK 2 - 1036		4 41 38	+06 46.1	1	12.33		1.33				1		987
30034	+10 00621	Hyades vB 111	⋆ A	4 41 39	+11 03.3	4	5.40	.000	0.25	.001	0.08	.000	7	A9 IV -	15,1127,3023,8015
		BSD 24 # 1075		4 41 39	+46 35.7	1	11.54		0.42		0.25		2	B9	1069
30076	−8 00929	HR 1508, DX Eri		4 41 41	−08 35.7	9	5.85	.028	-0.08	.016	-0.79	.013	67	B2 Ve	15,154,681,815,1212*
276757	+42 01037			4 41 45	+42 21.4	1	9.80		1.48		1.36		3	K0	1723
286013		NGC 1647 - 71		4 41 48	+18 43.5	1	13.94		0.88		0.27		1	A0	49
		BSD 24 # 536		4 41 48	+44 56.2	1	11.50		0.41		-0.09		6	B7	1069
30194	−41 01546			4 41 54	−41 50.7	1	7.75		1.20		1.11		10	K0 III	1673
		L 179 - 59		4 41 54	−59 07.	1	10.98		1.03		0.94		1		1696
30127	−18 00906	HR 1513		4 41 55	−18 45.5	3	5.53	.005	0.02	.005	0.04		9	A1 V	15,2027,3050
		BSD 24 # 537		4 41 58	+44 10.6	1	11.00		0.38		-0.05		2	B8	1069
		Met +25 379		4 42 04	+25 12.1	1	12.63		0.96		0.76		2		8082
29678	+75 00189	HR 1491	⋆ A	4 42 04	+75 51.2	1	6.08		0.27		-0.04		1	A9 IV	254
		LP 716 - 2		4 42 04	−14 25.8	1	11.88		0.75		0.16		2		1696
30202	−41 01549	HR 1518		4 42 05	−41 09.4	4	6.24	.004	1.46	.005	1.78		12	K4 III	15,2013,2029,3005
		KUV 685 - 10		4 42 07	+14 16.4	1	15.08		0.18		-0.75		1	sdO	974
30112	+0 00834			4 42 08	+00 28.6	2	7.20	.004	-0.18	.003	-0.73	.012	4	B2.5V	399,1732
		LP 359 - 34		4 42 08	+22 45.9	1	11.74		1.65				2		950
285992	+18 00699	NGC 1647 - 54		4 42 13	+19 02.7	2	10.10	.000	0.35	.010	0.00	.000	3	A0 V	49,1757
30277	−50 01474			4 42 14	−49 59.6	1	9.52		0.07		0.09		4	A0 V	243
286007	+18 00700	NGC 1647 - 10		4 42 15	+18 55.7	1	10.25		0.48		0.08		1	B7.5 IV	1757
		NGC 1647 - 11		4 42 16	+19 00.5	1	11.94		0.66		0.46		1		1757
276738	+43 01048			4 42 17	+43 54.2	1	9.52		0.28		-0.09		2	B7 V	1069
		NGC 1647 - 53		4 42 19	+19 02.5	1	11.48		0.49		0.42		2	A5	1757
285993	+18 00701	NGC 1647 - 51	A	4 42 20	+19 01.9	1	10.03		0.39		0.09		2	B9 IVnn	1757
285993	+18 00701	NGC 1647 - 51	B	4 42 20	+19 01.9	1	13.20		0.75		0.35		1	B9 IVnn	1757
		NGC 1647 - 52		4 42 22	+19 02.7	1	11.42		0.93		0.48		2		1757
284844	+18 00702	NGC 1647 - 55		4 42 22	+19 07.2	1	10.31		0.43		0.03		1	B7 V	1757
270668				4 42 23	−69 10.4	1	11.37		0.53		0.00		3	F2	149
30292	−48 01479			4 42 27	−48 25.4	1	6.90		0.50				4	F6/7 V	1075
		BSD 24 # 131		4 42 28	+43 53.1	1	11.40		0.35		0.06		2	B8	1069
		LP 157 - 25		4 42 28	+48 40.0	1	16.45		1.73				1		1691
		NGC 1647 - 9		4 42 31	+19 00.2	2	11.60	.010	0.60	.019	0.48	.005	3	A3	49,1757
284842	+19 00773	Hyades HG8 - 67		4 42 32	+19 15.6	3	9.33	.011	0.89	.004	0.57	.000	7	K0 V	49,950,1058
30317	−52 00566			4 42 32	−52 21.3	1	6.61		0.37		0.12		4	F3 IV/V	243
30015	+50 01051			4 42 34	+51 00.4	1	8.01		1.48		1.64		2	K2	1566
283877	+25 00729			4 42 38	+25 37.4	1	9.95		0.67		0.14		2		8082
30123	+19 00774	NGC 1647 - 15		4 42 39	+19 14.3	1	8.61		0.36		-0.05		1	B8 III	49
284843	+18 00707	NGC 1647 - 13		4 42 41	+19 09.0	1	10.82		0.54		0.44		1	A3 V	1757
30122	+23 00739	HR 1512		4 42 41	+23 32.3	2			0.06	.005	-0.45	.000	4	B5 III	1049,1079
		NGC 1647 - 358		4 42 42	+19 01.7	2	14.46	.010	0.98	.010	0.41	.095	2		49,1757
		G 84 - 9		4 42 43	+03 52.9	3	11.46	.009	0.56	.013	-0.06	.011	5	G3	333,1620,1658,1696
		NGC 1647 - 350		4 42 43	+19 10.9	1	12.54		0.67		0.50		1		1757
		NGC 1647 - 351		4 42 44	+19 10.9	1	12.16		0.60		0.47		1		1757
		HIC 22182	A	4 42 44	+44 48.0	1	11.55		0.96		0.44		3		1723
		BSD 24 # 135		4 42 45	+43 47.3	1	11.94		0.40		0.14		2	B9	1069
30240	−27 01875			4 42 45	−26 51.6	2	8.01	.010	0.96	.005	0.55	.032	3	K1 II	565,3048
	+43 01050	LS V +44 018		4 42 47	+44 08.3	1	10.13		0.54		-0.41		2	B2	1069
		NGC 1647 - 214		4 42 48	+18 00.0	2	13.63	.010	0.93	.005	0.28	.010	3		49,1757
284841	+19 00775	NGC 1647 - 102		4 42 53	+19 11.5	1	9.34		0.32		0.01		1	B9 III	49
276742	+43 01052			4 42 53	+43 25.1	1	9.38		1.16		0.85		3	G5	1723
30273	−33 01901			4 42 53	−32 57.9	1	9.86		0.44		-0.19		2	F0 V	1696
30238	−21 00966	HR 1521		4 42 55	−21 22.4	4	5.72	.006	1.47	.005			18	K4 III	15,1075,2013,2029
286015		NGC 1647 - 224		4 42 56	+18 52.0	1	11.22		0.40		0.34		1	A0 V	1757
	−37 01883			4 42 56	−37 12.8	1	9.61		0.95		0.96		1	K5 V	3072
		NGC 1647 - 426		4 42 57	+18 58.2	2	15.19	.020	1.49	.005	1.43		2		49,1757
	+18 00711	NGC 1647 - 4		4 42 58	+18 54.3	1	10.92		0.35		0.31		1	A0 Vn	1757

Table 1

227

HD	DM	Other Id	N Rem	α₁₉₅₀	δ₁₉₅₀	S	V	σ_V	B–V	σ_B–V	U–B	σ_U–B	n	Spectrum	References
	+44 01017	V422 Per		4 42 58	+44 37.3	1	10.09		2.49				3		369
286016	+18 00712	NGC 1647 - 3		4 42 59	+18 51.6	1	9.62		0.32		-0.06		1	B7.5 IV	1757
		NGC 1647 - 429		4 42 59	+18 58.4	1	16.11		0.88				1		49
30170	+18 00713	NGC 1647 - 45		4 43 00	+19 05.5	1	8.88		0.28		-0.11		3	B7.5 IVn	1757
30211	-3 00876	HR 1520		4 43 00	-03 20.7	10	4.01	.008	-0.15	.005	-0.57	.011	166	B5 IV	3,9,15,154,1075,1223*
30568	-71 00281			4 43 00	-71 35.3	1	8.24		0.45		0.02		3	F8 V:	204
		NGC 1647 - 46		4 43 01	+19 04.6	1	11.72		0.51		0.39		3	A2	1757
30169	+21 00694	IDS04400N2105	AB	4 43 01	+21 10.4	1	9.39		0.80				1	G5	987
285996	+18 00714	NGC 1647 - 44		4 43 02	+19 05.3	1	9.25		0.25		-0.04		3	B8 IVnn	1757
285995	+18 00715	NGC 1647 - 42		4 43 05	+19 01.9	1	9.70		0.33		0.22		1	B9.5 IV	1757
30168	+25 00731			4 43 07	+25 56.8	1	7.70		0.31		-0.01		3	A0	1625,8082
		NGC 1647 - 427		4 43 08	+18 54.0	2	15.40	.185	1.04	.155	0.99		2		49,1757
	+42 01046			4 43 08	+42 34.7	1	10.13		2.38				2	M0	369
286004	+18 00716	NGC 1647 - 37		4 43 10	+18 56.9	1	10.16		0.30		0.13		1	B9.5 IV	1757
286005		NGC 1647 - 41		4 43 12	+19 00.0	1	11.17		0.40		0.27		1	B9.5 V	1757
		NGC 1647 - 40		4 43 13	+18 59.2	2	11.12	.015	0.86	.010	0.40	.010	2	G3	49,1757
30210	+11 00646	Hyades vB 112	⋆ A	4 43 15	+11 36.9	6	5.37	.000	0.20	.010	0.15	.039	22	A5 m	15,1049,1127,1199*
		Met +25 386		4 43 15	+25 14.8	1	11.99		0.70		0.51		2	F2	8082
		NGC 1647 - 38		4 43 16	+19 07.4	1	12.45		0.89		0.43		1		49
		NGC 1647 - 428		4 43 17	+18 51.2	2	15.60	.049	1.05	.005	1.32		3		49,1757
		NGC 1647 - 97		4 43 17	+19 07.1	2	11.78	.005	0.52	.005	0.44	.000	2		49,1757
		NGC 1647 - 98		4 43 17	+19 07.4	1	12.42		0.88		0.44		1		1757
30138	+40 01045	HR 1514		4 43 17	+40 13.4	2	5.97	.000	0.93	.006	0.68		5	G9 III	71,1501
30361	-47 01497			4 43 17	-47 30.1	2	8.32	.005	0.61	.005			8	G0 V	1075,2033
286002	+18 00718	NGC 1647 - 39		4 43 18	+18 58.9	2	10.32	.005	0.91	.015	0.46	.015	2	G8 III	49,1757
		NGC 1647 - 96		4 43 20	+19 05.7	2	12.65	.005	0.67	.010	0.49	.000	2		49,1757
284840	+19 00776	NGC 1647 - 99		4 43 20	+19 11.7	1	10.09		0.41		0.12		1	B9 IV	49
30197	+18 00719	Hyades vB 164	⋆	4 43 21	+18 38.8	5	6.01	.003	1.21	.004	1.32	.001	22	K4 III	15,1127,4001,6007,8015
		NGC 1647 - 249		4 43 21	+18 52.2	1	14.01		0.79		0.30		2		1757
		NGC 1647 - 243		4 43 22	+18 50.2	1	13.15		0.93		0.40		2		1757
		NGC 1647 - 242		4 43 24	+18 49.5	2	14.10	.010	0.84	.015	0.31	.044	3		49,1757
282407	+32 00834			4 43 24	+32 33.1	1	10.15		0.81		0.50		3	K0	1723
		NGC 1647 - 95		4 43 25	+19 06.7	1	13.55		0.79		0.23		1		1757
30190	+27 00694			4 43 26	+27 48.7	1	8.41		0.45		0.25		2	A2	1625
284839	+18 00720	NGC 1647 - 94		4 43 27	+19 06.9	2	9.68	.005	0.23	.000	-0.16	.010	2	B9 III	49,1757
282485	+29 00734			4 43 29	+29 13.7	1	9.88		0.29		-0.08		2	B9 V	1003
		POSS 85 # 1		4 43 30	+60 38.5	1	14.75		0.93				2		1739
		NGC 1647 - 93		4 43 33	+19 06.1	2	11.71	.000	0.43	.005	0.34	.005	2	A4	49,1757
30152	+41 00956			4 43 33	+41 13.0	1	7.18		0.07		-0.25		2	B8 II	401
30612	-71 00282	HR 1541		4 43 33	-71 01.4	3	5.52	.006	-0.13	.003	-0.45	.005	31	B8 II/IIIp	15,1075,2038
30478	-59 00376	HR 1530		4 43 36	-59 49.4	1	5.27		0.20				4	A8/9 III/IV	2032
		NGC 1647 - 30		4 43 38	+18 59.9	2	11.00	.000	0.40	.000	0.34	.000	3	A2	49,1757
		NGC 1647 - 28		4 43 38	+19 01.9	2	11.37	.005	0.51	.010	0.35	.005	3	A3	49,1757
	+45 00983	G 96 - 1		4 43 38	+45 53.9	1	10.24		0.80		0.46		1	G8:V:	333,1620
30246	+15 00678	Hyades vB 142		4 43 39	+15 23.0	4	8.32	.015	0.67	.007	0.19	.013	6	G5	950,1058,1127,8023
30233	+19 00777	NGC 1647 - 105		4 43 40	+19 24.3	1	8.45		1.60		1.48		2	K2	1648
		LS V +46 022		4 43 40	+46 07.5	1	10.47		0.80		0.02		2		1069
	-59 00377	NGC 1672 sq1 3		4 43 40	-59 24.8	1	9.47		0.86		0.49		4	G4	1687
		Hyades L 119		4 43 41	+03 32.8	3	10.92	.009	1.28	.005	1.27		4		582,950,1058
		Steph 512		4 43 41	+03 32.8	1	10.92		1.28				2	K4	1746
285997	+18 00722	NGC 1647 - 29		4 43 41	+19 01.4	2	10.70	.000	0.31	.005	0.17	.000	3	A0 IV	49,1757
	-13 00969			4 43 47	-13 03.9	1	7.97		0.22		0.07		6	A3	1776
		Haro 6 # 37	V	4 43 48	+16 55.	1	14.54		1.38		-0.34		1		640
		NGC 1647 - 173		4 43 48	+18 52.2	2	14.11	.005	0.89	.024	0.31	.015	3		49,1757
276833	+43 01054			4 43 48	+43 25.6	1	10.62		0.16		-0.19		2	B5	1069
		LP 476 - 537		4 43 49	+10 54.3	1	12.42		0.88				2		1674
30121	+56 00973	HR 1511	⋆ AC	4 43 50	+56 40.3	2	5.31	.040	0.25	.000	0.15	.025	5	A3 m	3058,8071
30136	+53 00813	IDS04399N5307	AB	4 43 51	+53 12.8	1	6.74		0.43		0.05		3	F3 V	1501
29579	+81 00162	IDS04344N8119	A	4 43 51	+81 24.3	1	9.32		-0.07		-0.16		5		1219
	-9 00978			4 43 52	-09 34.0	2	9.31	.005	0.91	.012	0.55		4	K0	1594,6006
	+35 00906			4 43 53	+35 44.7	1	9.80		2.31				2	M3	369
		Hyades HG8 - 30		4 43 54	+18 13.	1	13.90		0.90				3		1058
		NGC 1647 - 179		4 43 54	+18 57.0	2	12.80	.010	0.66	.005	0.49	.000	3		49,1757
30321	-3 00884	HR 1522		4 43 54	-03 02.6	2	6.32	.005	0.04	.000	0.06	.000	7	A2 V	15,1415
30264	+17 00782	Hyades L 92		4 43 55	+17 39.6	3	9.59	.009	0.97	.005	0.77		5	K0	582,950,1058
30196	+43 01057	IDS04404N4313	AB	4 43 56	+43 18.7	1	8.47		0.43		0.12		std	F0	1337
30178	+45 00984			4 43 56	+45 54.6	2	7.82	.040	1.87	.010	2.08		7	M2 Ib	369,8100
		NGC 1647 - 178		4 43 58	+18 56.4	2	12.74	.000	0.77	.005	0.31	.000	3		49,1757
30397	-34 01859	HR 1524		4 43 58	-34 05.7	2	6.85	.005	0.00	.000			7	A0 V	15,2012
		NGC 1647 - 256		4 43 59	+18 59.1	1	13.37		0.75		0.24		1		1757
30263	+18 00725	NGC 1647 - 22		4 43 59	+18 59.5	2	9.08	.005	0.20	.000	-0.10	.010	3	B9 II	49,1757
30209	+42 01050	LS V +42 009		4 43 59	+42 13.9	1	8.32		0.11		-0.66		2	B1.5V	401
		HRC 72, DQ Tau		4 44 00	+16 54.7	2	13.64	.028	1.69	.028	0.94	.075	2	dM0e	640,1227
30144	+55 00928	HR 1515		4 44 00	+55 30.9	2	6.33	.009	0.31	.009	0.02	.005	5	F1 V	1733,3013
30311	+8 00759	Hyades vB 113		4 44 01	+08 55.7	3	7.26	.005	0.56	.005	0.07	.009	7	F5	1127,1355,8023
283868	+25 00732	RV Tau		4 44 02	+26 05.4	1	9.19		1.74		1.59		1	G2 Ia e	793
		NGC 1672 sq1 5		4 44 02	-59 25.3	1	12.95		0.60		0.10		4		1687
	-4 00938	G 82 - 47	⋆ AB	4 44 04	-04 41.8	2	10.04	.000	0.88	.005	0.58	.000	4		1620,1658
		HRC 73, V1001 Tau	AB	4 44 06	+16 57.3	1	13.42		1.32		0.20		1		776
		Hyades HG8 - 69		4 44 06	+18 58.	1	13.42		1.07				1		1058

HD	DM	Other Id	N Rem	α_{1950}	δ_{1950}	S	V	σ_V	B–V	σ_{B-V}	U–B	σ_{U-B}	n	Spectrum	References
		Hyades CDS 67	⋆	4 44 08	+23 55.9	1	12.28		1.48				3		1674
		LP 359 - 42		4 44 08	+23 55.9	1	12.31		0.65				1		920
		POSS 158 # 2		4 44 10	+49 35.2	1	14.94		1.45				2		1739
		NGC 1672 sq1 4		4 44 10	−59 24.6	1	11.50		0.97		0.64		4		1687
284785	+20 00820	Hyades HG8 - 64		4 44 11	+20 47.7	2	9.84	.009	1.06	.000	0.94		4	K2	920,1058
286037	+15 00679	Hyades HG8 - 78		4 44 12	+15 38.3	1	10.01		0.84		0.46		2	K0	1058
		ApJ110,424R257# 8		4 44 12	+16 52.	1	13.60		1.36		0.17		1		1227
		HRC 74, DR Tau		4 44 13	+16 53.4	3	12.27	.601	0.81	.125	−0.57	.154	3	dK5e	640,776,1536
	+72 00238			4 44 13	+72 49.8	1	8.61		1.66		1.97		3	K5	1733
30432	−39 01624	HR 1526		4 44 14	−39 26.8	4	6.04	.007	1.06	.007			18	K1 III	15,1075,2013,2029
		BSD 24 # 162		4 44 16	+44 00.0	1	12.16		0.26		0.08		2	B9	1069
270702				4 44 19	−69 44.2	1	12.64		0.60		0.04		1	F8	149
		NGC 1672 sq1 6		4 44 21	−59 20.3	1	13.97		0.65		0.15		4		1687
30342	+9 00651	IDS04416N0952	AB	4 44 22	+09 57.8	1	7.19		0.16		0.11		2	A2	401
30501	−50 01492			4 44 23	−50 09.6	4	7.58	.007	0.89	.004	0.57	.012	14	K1 V	158,243,2012,3008
		SK *G 6		4 44 23	−69 23.3	1	13.78		0.00		−0.42		4	B7 Ib	149
30282	+36 00937	AW Per		4 44 25	+36 38.1	3	7.07	.013	0.90	.013	0.64	.028	3	F0	851,1399,6007
	+49 01243		V	4 44 25	+49 23.0	1	9.61		2.05				3	M2	369
284653		Hyades L 93		4 44 26	+22 57.7	4	10.68	.005	1.11	.004	1.06		6	K2 III	582,950,1058,1570
30422	−28 01735	HR 1525		4 44 26	−28 10.6	2	6.18	.005	0.19	.000			7	A3 IV	15,2012
		BSD 96 # 370		4 44 28	+00 38.6	1	10.77		1.27		1.40		1		116
		BSD 24 # 595		4 44 30	+44 54.6	1	11.75		0.55		−0.16		4	B3	1069
		POSS 85 # 4		4 44 30	+61 09.6	1	16.14		1.33				2		1739
30610	−63 00365	HR 1540		4 44 30	−63 19.2	4	6.46	.011	1.08	.005	1.00		15	K0/1 III	15,243,1075,2035
		NGC 1672 sq1 8		4 44 34	−59 22.3	1	13.85		0.63		0.14		4		1687
		NGC 1672 sq1 2		4 44 35	−59 17.3	1	12.33		1.01		0.80		4		1687
		G 82 - 48		4 44 36	+02 04.3	1	11.34		1.48		1.20		1	M0	333,1620
		Hyades CDS 48	⋆	4 44 37	+06 21.9	2	11.33	.000	1.42	.000			4	K6	987,1746
283869	+25 00733			4 44 37	+26 03.8	2	10.61	.005	1.17	.005	1.10		5	K7 III	953,3016
		HRC 75 ,DS Tau	A	4 44 39	+29 19.9	1	12.62		0.94		−0.52		1	dK4e	776
		HRC 75, DS Tau	AB	4 44 39	+29 19.9	1	11.86		0.85		−0.13		1	dK4e + dF0	1536
		POSS 56 # 5		4 44 40	+66 36.7	1	16.98		1.85				3		1739
		BSD 24 # 599		4 44 41	+44 36.1	1	12.24		0.42		−0.15		4	B6	1069
		ApJ110,424R259# 2		4 44 42	+29 20.	1	12.85		1.17		0.15		1		1227
30542	−50 01494			4 44 42	−50 42.6	1	8.69		0.98		0.66		4	G8 III	243
30355	+17 00786	Hyades vB 114		4 44 43	+18 10.3	2	8.53	.005	0.72	.001	0.31	.001	3	G0	1127,8023
		SK *G 7		4 44 43	−69 45.9	1	13.51		0.06		−0.18		4	B8 Ib	149
		LP 416 - 549		4 44 44	+17 14.5	1	13.37		1.28				2		1674
		Hyades L 91		4 44 45	+14 48.0	3	11.53	.005	1.36	.012	1.29		5		582,950,1058
		WLS 448 55 # 7		4 44 45	+52 32.2	1	11.24		0.75		0.25		2		1375
30392	+9 00655			4 44 46	+10 04.9	1	8.20		0.34		0.08		2	A0	401
		BSD 24 # 178		4 44 46	+43 50.5	1	11.49		0.19		−0.13		2	B8	1069
	+43 01062			4 44 46	+43 50.5	1	11.65		0.24		−0.10		2	B7	1069
30439	−21 00977	IDS04427S2123	A	4 44 48	−21 18.0	1	7.12		1.02		0.83		6	K0 III	1628
30551	−49 01439	R Pic		4 44 49	−49 20.1	1	7.19		1.61		1.44		1	K2/3 (II)pe	975
		G 39 - 36		4 44 51	+33 04.4	1	12.36		1.06		0.70		1		1620
276930	+39 01070			4 44 51	+39 51.7	2	9.80	.015	2.01	.015	2.29		5	M2 III	369,8032
30390	+14 00757			4 44 52	+15 06.2	1	8.88		1.12		0.75		2	K0	1648
		BSD 24 # 1110		4 44 52	+46 40.7	1	11.58		0.31		−0.06		2	B8	1069
287066	+10 00630	NGC 1662 - 11		4 44 55	+10 46.2	1	10.48		0.53		0.38		1	A5	49
276914	+40 01054			4 44 59	+40 37.0	1	10.01		0.16		−0.24		2	A0	401
30085	+70 00322	HR 1510		4 44 59	+70 51.3	2	6.37		−0.09	.004	−0.24	.004	5	A0 IV	985,1079
		NGC 1672 sq1 7		4 44 59	−59 22.9	1	12.74		1.25		1.23		4		1687
		Hyades CDS 68	⋆	4 45 01	+18 56.2	1	13.09		1.50				3		1674
		NGC 1662 - 28		4 45 02	+10 47.6	1	12.70		0.90		0.28		1		49
270716				4 45 04	−68 43.6	1	11.32		1.61		1.36		2	M0	149
		NGC 1662 - 35		4 45 05	+10 43.8	1	15.12		0.87		0.29		1		49
30341	+42 01059			4 45 06	+42 30.4	1	8.42		1.22				23	K0	6011
276838	+42 01058			4 45 06	+43 04.5	2	9.51	.004	0.19	.048	0.21		12	A2	1119,1337
		NGC 1662 - 20		4 45 07	+10 47.7	1	11.50		0.57		0.36		1		49
		Hyades L 95		4 45 08	+16 58.1	4	11.17	.038	1.41	.004	1.23		6	K6	582,950,1058,1674
30436	−0 00771			4 45 08	−00 10.6	1	7.97		0.42		0.03		1	F4 III	116
		BSD 24 # 193		4 45 09	+43 24.1	1	11.78		0.33		0.31		3	B9	1069
		BSD 24 # 195		4 45 09	+43 35.2	1	11.74		0.27		−0.28		3	B3	1069
283879	+24 00688			4 45 11	+25 06.5	1	11.08		1.01		0.35		5	M0	8082
276812	+43 01067			4 45 11	+44 01.1	1	10.53		0.17		−0.25		2	B8	1069
30378	+29 00741			4 45 13	+29 41.2	2	7.41	.005	0.03	.005	−0.17	.015	6	B9.5V	206,1733
		NGC 1662 - 30		4 45 15	+10 47.2	1	13.70		0.81		0.22		1		49
287149	+10 00633	NGC 1662 - 10		4 45 15	+10 51.9	1	10.14		0.35		0.26		1	A1 V	49
		Hyades J 333		4 45 16	+13 59.	1	11.58		0.79		0.27		2		1058
		SK *G 8		4 45 17	−69 58.7	1	13.92		0.14		0.26		2	A0 II	149
		NGC 1662 - 25		4 45 19	+10 54.1	1	11.83		0.73		0.22		1		49
30353	+43 01069	KS Per		4 45 20	+43 11.3	2	7.80	.040	0.46	.000	−0.19	.020	3	A5 Iap	616,1012
287158		NGC 1662 - 18		4 45 21	+10 43.7	1	11.30		0.55		0.29		1	A7	49
	+44 01028			4 45 21	+44 13.7	1	10.14		1.70				3		369
30495	−17 00954	HR 1532		4 45 21	−17 01.5	9	5.50	.012	0.63	.012	0.14	.007	32	G3 V	15,196,258,1067,1075*
		NGC 1662 - 33		4 45 23	+10 52.1	1	14.53		0.88		0.21		1		49
30623	−50 01501			4 45 27	−50 29.0	1	8.76		0.50		0.03		4	F6 V	243
		NGC 1662 - 29		4 45 28	+10 51.3	1	13.20		0.93		0.36		1		49
		NGC 1662 - 19		4 45 29	+10 55.3	1	11.34		0.51		0.32		1		49

Table 1   229

HD	DM	Other Id	N Rem	α_{1950}	δ_{1950}	S	V	σ_V	B–V	σ_{B-V}	U–B	σ_{U-B}	n	Spectrum	References
	−0 00773			4 45 30	+00 08.8	1	10.14		1.74		1.96		1	M0	116
30457	+10 00634	NGC 1662 - 5		4 45 30	+10 55.8	1	9.12		0.34		0.24		1	A2 V	49
		HL Tau		4 45 30	+40 25.	1	11.74		0.20		-0.50		2	DA	704
		L 93 - 7		4 45 30	−65 24.	1	13.29		0.52		-0.20		2		1696
30486	+0 00855			4 45 31	+00 35.3	1	7.47		0.44		-0.07		1	F2	116
		NGC 1662 - 34		4 45 31	+10 51.5	1	14.86		1.06		0.57		1		49
276826	+43 01071	NGC 1664 - 733		4 45 32	+43 31.6	2	8.88	.017	1.79	.011	2.22	.047	28	K5	49,1655
30593	−36 01884	T Cae		4 45 32	−36 17.8	3	7.62	.029	2.44	.057	4.62	.974	7	C	109,864,3038
30418	+24 00689			4 45 33	+24 39.4	1	8.26		0.35		-0.03		2	F2	1625
		NGC 1662 - 32		4 45 36	+10 49.1	1	14.10		1.17		0.76		1		49
		NGC 1664 - 741		4 45 36	+43 36.8	1	12.44		0.28		-0.04		1		49
		NGC 1664 - 739		4 45 36	+43 38.8	1	11.67		0.47		0.27		1	A7	49
		Sk -67 001		4 45 36	−67 40.	1	13.60		0.02		-0.94		3		798
		NGC 1662 - 23		4 45 37	+10 48.0	1	11.72		0.58		0.26		1		49
		NGC 1664 - 738		4 45 38	+43 33.8	1	11.48		0.48		0.13		1	F0	49
287157		NGC 1662 - 14		4 45 39	+10 41.9	1	10.77		0.42		0.31		1	A2	49
30470	+10 00637	NGC 1662 - 8	⋆ E	4 45 39	+10 49.9	1	9.47		0.32		0.19		1	A1 V	49
		Hyades CDS 69	⋆	4 45 39	+16 18.1	1	12.42		1.47				3		1674
30523	−11 00965			4 45 39	−11 01.2	1	9.53		1.13		1.11		1	K7 V	3072
287147		NGC 1662 - 15		4 45 40	+10 55.5	1	10.96		0.42		0.27		1	A3	49
30456	+17 00789	IDS04428N1738	AB	4 45 41	+17 43.2	1	8.02		0.42		-0.05		2	A2	3016
30456	+17 00789	IDS04428N1738	B	4 45 41	+17 43.2	1	9.66		0.69		0.13		2	G5	3016
		NGC 1662 - 16		4 45 42	+10 44.6	1	11.11		0.44		0.31		1		49
		NGC 1662 - 21		4 45 43	+10 55.0	1	11.65		0.63		0.31		1		49
		NGC 1664 - 784		4 45 43	+43 26.7	1	10.78		0.22		-0.31		2	B5	1069
	+0 00856			4 45 44	+00 37.4	1	10.12		0.67		0.21		1		116
30485	+10 00639	NGC 1662 - 1	⋆ A	4 45 44	+10 50.6	1	8.34		1.24		0.93		1	G8 III	49
284787	+20 00823	Hyades vB 115		4 45 44	+21 00.9	3	9.06	.012	0.85	.004	0.53	.000	6	G5	950,1127,8023
	+63 00537	G 247 - 36	⋆ A	4 45 44	+63 14.9	1	9.92		1.27		1.26		6	K7	1723
	−59 00379	NGC 1672 sq1 1		4 45 44	−59 18.8	1	9.08		1.29		1.29		4	K0	1687
30468	+21 00707			4 45 45	+21 13.7	1	7.03		0.12		0.10		2	A0	1648
	−30 02009			4 45 45	−30 49.0	1	11.61		0.65		-0.02		2		1696
		BSD 96 # 951		4 45 46	+00 32.7	1	11.45		0.53		0.02		1	F2	116
287148	+10 00640	NGC 1662 - 2		4 45 46	+10 52.8	1	8.83		1.20		0.86		1	G7 III	49
30455	+18 00734	G 83 - 34		4 45 46	+18 37.7	3	6.96	.007	0.62	.010	0.08	.007	5	G2 V	333,1003,1620,3026
	+43 01074			4 45 46	+44 09.2	1	9.76		0.35		0.25		1	A5 III	401
		Hyades J 334		4 45 47	+29 28.	1	11.60		0.98		0.83		3		1058
		NGC 1664 - 742		4 45 47	+43 37.1	1	13.49		0.22		0.07		1		49
30818	−66 00321			4 45 48	−66 36.7	1	8.81		0.56		-0.03		3	G0 V	204
30492	+10 00641	NGC 1662 - 3		4 45 49	+10 51.5	2	8.96	.015	0.37	.005	0.24	.005	3	A2 V	49,401
31081	−76 00294			4 45 50	−76 23.8	2	7.74	.005	0.51	.009			8	F7 V	1075,2034
30409	+43 01075			4 45 51	+44 09.2	1	8.31		0.07		-0.05		2	B9 V	401
	+66 00355	G 247 - 37		4 45 51	+67 05.7	1	9.54		0.63		0.09		1	G0	1658
30410	+43 01076	NGC 1664 - 732		4 45 52	+43 27.3	1	7.50		0.97		0.64		1	G5 III	49
30608	−30 02011	HR 1539		4 45 52	−30 06.5	5	6.36	.003	1.07	.007			22	K1 IV	15,1075,2013,2020,2029
		Sk -69 001		4 45 54	−69 33.	1	13.40		-0.12		-0.94		3		798
287150	+10 00642	NGC 1662 - 6		4 45 56	+10 46.8	1	9.28		0.35		0.24		1	A3p	49
287146		NGC 1662 - 13		4 45 56	+10 54.3	1	10.71		0.38		0.25		1	A3 V	49
		NGC 1662 - 26		4 45 57	+10 45.3	1	12.30		0.70		0.18		1		49
30443	+34 00911			4 45 58	+34 54.9	1	8.89		2.08				1	R4	1238
30533	+0 00858			4 45 59	+00 43.7	1	8.51		1.42		1.46		1	K0	116
		Ross 379		4 45 59	+08 02.3	1	12.91		1.02		0.81		1	K8	1696
287151	+10 00643	NGC 1662 - 7		4 45 59	+10 48.1	1	9.37		0.30		0.18		1	A2 V	49
30454	+31 00816	HR 1529		4 46 00	+31 21.1	1	5.58		1.12		1.03		2	K2 III	252
		CS 22191 # 29		4 46 00	−39 12.7	1	14.05		0.41		0.00		1		1736
30544	+3 00679			4 46 02	+03 33.7	5	7.32	.005	-0.06	.004	-0.32	.006	108	B9	10,147,361,1509,6005
276828	+43 01077	NGC 1664 - 734		4 46 02	+43 33.6	2	9.99	.014	1.09	.006	0.77	.014	22	G5 V	49,1655
30684	−46 01553			4 46 02	−46 41.2	2	8.12	.000	0.82	.000	0.35		6	G8 IV	2012,3008
30453	+32 00840	HR 1528		4 46 05	+32 30.1	3	5.85	.008	0.25	.008	0.13	.014	11	A8 m	351,595,8071
30466	+29 00742	V473 Tau		4 46 06	+29 29.1	1	7.28		0.05		-0.16		5	A0p	1202
30545	+3 00681	HR 1534		4 46 07	+03 30.1	8	6.03	.008	1.20	.009	1.14	.013	119	K1 III	10,15,147,361,1355*
		NGC 1662 - 24		4 46 08	+10 37.5	1	11.83		0.62		0.32		1		49
30505	+18 00736	Hyades vB 116		4 46 08	+18 33.3	3	8.99	.012	0.83	.010	0.51	.000	6	F5	950,1127,8023
30452	+37 00968			4 46 08	+37 14.9	1	7.77		1.30		1.03		1	K0	851
	+43 01078			4 46 08	+44 08.6	1	9.82		0.19		0.08		2	B9	1069
30562	−5 01044	HR 1536, LSS 1969		4 46 08	−05 45.4	6	5.78	.007	0.63	.013	0.19	.008	27	F8 V	15,1075,1311,1415*
30694	−44 01714			4 46 08	−44 21.3	2	8.11	.009	0.32	.000			8	A2 IV/V	1075,2033
287156		NGC 1662 - 17		4 46 09	+10 39.7	1	11.29		0.53		0.29		1		49
		G 96 - 3		4 46 09	+40 43.0	1	13.94		1.37		1.16		1		1658
283882	+24 00692	Hyades vB 117	⋆ V	4 46 10	+24 43.0	7	9.55	.021	1.05	.008	0.90	.019	21	K3 V	830,920,1127,1674,1783*
		NGC 1664 - 737		4 46 10	+43 38.3	1	10.90		1.00		0.67		1	dG7	49
	+5 00735	G 84 - 12		4 46 11	+05 27.6	2	10.60	.035	0.96	.000	0.78	.015	2	K2	333,1620,1696
30670	−40 01583			4 46 11	−40 32.0	1	9.37		0.98		0.73		2	K3 V	3072
	+0 00862			4 46 15	+01 06.0	1	10.51		0.27		0.05		1		116
		NGC 1662 - 31		4 46 15	+10 39.4	1	13.85		0.98		0.39		1		49
		NGC 1662 - 22		4 46 15	+10 45.3	1	11.65		1.37		1.12		1		49
276887	+41 00968			4 46 15	+41 32.6	1	9.89		1.61				3	K0	369
		Hyades CDS 70	⋆	4 46 17	+17 37.8	1	14.00		1.52				3		1674
276861	+42 01064	SV Per		4 46 17	+42 12.2	3	8.49	.023	0.84	.036	0.68		3	F8	757,934,6011
30606	−16 00956	HR 1538		4 46 17	−16 25.0	6	5.76	.025	0.54	.015	0.06	.014	14	F8 V	15,254,258,1075,2035,3053

HD	DM	Other Id	N Rem	α_{1950}	δ_{1950}	S	V	σ_V	B–V	σ_{B-V}	U–B	σ_{U-B}	n	Spectrum	References
30573	+3 00682			4 46 18	+03 36.2	1	7.02		0.02		0.01		5	A0	1509
276862	+41 00969			4 46 21	+42 07.8	1	10.54		0.40		0.38		10	A2	1655
30703	−42 01628	IDS04448S4204	AB	4 46 23	−41 58.7	1	9.26		0.94		0.71		1	K3 V	3072
30956	−70 00316			4 46 24	−70 14.9	1	9.90		0.01		−0.18		4	A0 V	149
	+46 00924			4 46 25	+46 48.9	1	9.60		0.13		−0.28		2	A2	401
		NGC 1664 - 645		4 46 29	+43 34.3	1	11.55		0.70		0.18		1	dG0	49
		BPM 3523		4 46 30	−78 57.	1	13.47		−0.10		−1.02		2		3065
276829	+43 01080	NGC 1664 - 735		4 46 31	+43 25.9	2	10.15	.014	0.16	.001	−0.01	.026	11	A1 Vn	49,1655
		NGC 1662 - 12		4 46 32	+10 34.7	1	10.70		0.42		0.28		1		49
287155	+10 00646	NGC 1662 - 9		4 46 32	+10 38.5	1	9.68		0.39		0.26		1	A1 V	49
30504	+37 00969	HR 1533		4 46 32	+37 24.1	5	4.88	.008	1.44	.010	1.71	.024	16	K3.5 III	1355,1363,3053,4001,8100
276870	+41 00973			4 46 33	+41 54.2	1	9.84		1.03				1	G5	934
	−0 00778	IDS04440S0009	AB	4 46 35	−00 03.5	1	9.50		1.14		0.99		1	K0	116
		G 81 - 36		4 46 36	+45 53.9	1	11.79		1.47		1.21		1		333,1620
30589	+15 00686	Hyades vB 118		4 46 40	+15 48.2	3	7.74	.000	0.58	.005	0.10	.010	7	F8	1127,1355,8023
30598	+10 00649	NGC 1662 - 4		4 46 41	+10 54.1	1	9.12		0.31		0.23		1	A1p	49
276888	+41 00974	LS V +41 017		4 46 41	+41 35.5	1	9.33		0.17		−0.56		2	G5	1012
		BSD 24 # 652		4 46 41	+44 26.5	1	12.60		0.57		−0.05		5	B2	1069
	−0 00780			4 46 42	+00 06.9	1	10.54		1.17		1.04		1		116
276837	+42 01065			4 46 42	+43 02.4	1	10.06		1.77		2.18		3	K7	8032
		NGC 1662 - 27		4 46 43	+10 35.2	1	12.38		0.63		0.32		1		49
		LP 476 - 57		4 46 44	+10 09.6	1	17.47		0.97				1		1674
276871	+41 00975			4 46 46	+41 59.6	1	9.61		1.14				1	G5	934
		NGC 1664 - 397		4 46 46	+43 34.0	1	12.40		0.51		0.29		1	A3	49
30572	+23 00747	Hyades vB 165		4 46 47	+23 18.6	2	8.52	.000	0.62	.007	0.19	.000	3	G0	1127,8023
30465	+50 01070			4 46 49	+50 19.4	1	7.62		1.77		2.06		2	M1	1566
30605	+15 00687	HR 1537	⋆ A	4 46 52	+15 49.1	1	6.08		1.60		1.72		3	gK3	1355
284760		G 85 - 17		4 46 53	+22 24.6	1	10.46		0.66		0.06		2		333,1620
276830	+43 01086	NGC 1664 - 736		4 46 54	+43 22.7	2	10.28	.009	0.16	.001	−0.22	.014	9	A0 V	49,1655
		AJ91,575 # 18	V	4 46 55	+23 35.9	1	11.36		1.42		1.10		1	M0 V	1606
		NGC 1664 - 388		4 46 55	+43 41.0	1	11.98		0.35		0.22		1	A2	49
	−0 00781			4 46 57	−00 04.3	1	10.84		0.47		−0.01		1	F5	116
		LP 416 - 62		4 46 58	+20 42.5	1	11.20		0.76				1		1570
		NGC 1664 - 740		4 46 58	+43 23.1	1	11.87		0.76		0.24		1	F8	49
		NGC 1664 - 229		4 46 58	+43 33.2	2	12.09	.005	0.39	.000	0.22	.034	6	A0	49,1069
		NGC 1664 - 345		4 46 59	+43 29.1	1	12.14		0.32		0.22		1	A0	49
		NGC 1664 - 463		4 46 59	+43 42.2	1	12.11		0.26		0.19		1	A2	49
		HIC 23304		4 47 00	+84 29.7	1	10.88		0.44		0.01		3		1723
30788	−44 01720	HR 1548		4 47 01	−44 04.0	3	6.72	.004	0.95	.000			11	G8 III	15,1075,2035
270754	−67 00345	Sk -67 002		4 47 02	−67 12.2	5	11.28	.013	0.10	.009	−0.79	.023	20	B1.5Ia:	149,328,1277,1408,8033
		NGC 1664 - 171		4 47 03	+43 33.4	1	11.76		0.29		0.20		1	A0	49
30652	+6 00762	HR 1543	⋆ A	4 47 07	+06 52.5	28	3.19	.006	0.45	.009	−0.01	.014	152	F6 V	1,3,15,418,985,1004*
		BSD 24 # 918		4 47 07	+45 11.2	1	11.91		0.75		0.40		3	B8	1069
	−22 01833			4 47 07	−22 42.6	1	11.20		0.86		0.35		2		1696
286085	+16 00655	Hyades L 96		4 47 08	+16 19.6	2	10.61	.000	1.16	.000			6	K0	582,1746
		LP 416 - 65		4 47 09	+18 07.0	1	12.60		0.92				1		950
		NGC 1664 - 122		4 47 09	+43 33.6	1	13.61		0.92		0.46		1	dG0	49
		BSD 24 # 661		4 47 10	+44 07.6	1	11.39		0.26		−0.19		2	B7	1069
233021	+51 00980			4 47 10	+52 09.1	1	8.41		1.91				3	M0	369
		Hyades CDS 71	⋆	4 47 11	+06 01.4	1	14.73		1.63				3		1674
30503	+52 00885			4 47 12	+52 18.8	1	8.83		0.30		0.29		2	A0	1566
30850	−50 01512			4 47 13	−49 52.7	1	7.35		0.97		0.69		4	G8 III	243
		SK *G 10		4 47 13	−68 44.4	1	13.92		0.06		0.14		4	A1 II	149
276996	+42 01067			4 47 14	+42 14.5	1	9.22		0.16		−0.22		1	B8	757
30849	−49 01449	SY Pic		4 47 18	−49 15.3	1	8.86		0.30		0.18		4	Ap SrCrEu	243
		ApJS2,315 # 65		4 47 18	−68 22.	1	11.54		0.19				3		8033
		Sk -69 002		4 47 18	−69 35.	1	13.16		−0.13		−0.98		4		798
		KUV 04473+1737		4 47 19	+17 37.0	1	12.65		−0.02		−1.00		1	NHB	1708
30585	+43 01088	NGC 1664 - 578	⋆	4 47 19	+43 24.5	2	8.87	.005	0.15	.001	0.06	.016	28	A0 V	49,1655
		NGC 1664 - 55		4 47 19	+43 34.6	2	11.05	.009	1.23	.005	0.99	.009	5	gG8	49,305
29475	+84 00088			4 47 19	+84 47.6	1	7.74		0.12		0.16		3	A2	1733
30677	+8 00775			4 47 21	+08 19.4	3	6.85	.010	−0.05	.021	−0.84	.014	6	B1 II-IIIn:	399,401,555
	+8 00774			4 47 22	+08 26.1	1	8.29		1.67		2.08		1	G5	401
283873	+25 00737			4 47 22	+25 58.3	1	10.67		1.56		1.21		2	K0 III	8082
		NGC 1664 - 89		4 47 22	+43 34.8	1	10.78		0.59		0.09		1		305
30744	−16 00960			4 47 22	−15 53.6	1	8.28		0.50		0.02		3	F7 V	1320
30557	+48 01162	HR 1535		4 47 23	+48 39.4	1	5.66		0.99		0.80		2	G9 III	252
30687	+10 00651			4 47 24	+10 31.2	1	7.90		0.40		−0.01		2	F2	401
		NGC 1664 - 82		4 47 24	+43 37.7	1	13.06		0.38		0.28		1	A2	49
30442	+63 00543	HR 1527	⋆ A	4 47 24	+63 25.4	2	5.42	.016	1.56	.012	1.76		6	M3 IIIab	71,3009
30743	−14 00970	HR 1545		4 47 25	−13 51.2	5	6.26	.008	0.45	.008	−0.10	.015	19	F3/5 V	15,258,1075,2027,3053
30584	+44 01036	IDS04438N4448	A	4 47 28	+44 53.1	2	8.31	.035	0.26	.015	0.01		5	B9 Vsi+B9 V	1069,1119
	−15 00872	RX Eri		4 47 29	−15 49.6	3	9.22	.032	0.22	.016	0.17	.024	3	F2	668,688,699
30862	−50 01514	IDS04462S5004	AB	4 47 29	−49 58.3	1	8.13		0.47		−0.02		4	F5 V	243
31290	−76 00295			4 47 29	−76 23.6	1	8.64		0.94		0.59		4	G8 III/IV	119
30676	+16 00657	Hyades vB 119		4 47 30	+17 07.1	2	7.11	.005	0.56	.000	0.07	.002	3	F8	1127,8023
		NGC 1664 - 95		4 47 30	+43 31.4	1	14.28		0.62				1		49
		NGC 1664 - 4		4 47 30	+43 35.2	2	11.75	.000	0.29	.030	0.23	.015	2	A0	49,305
		NGC 1664 - 56		4 47 30	+43 35.6	1	12.66		0.35		0.26		2	A2	305
		NGC 1664 - 743		4 47 30	+43 37.	1	16.78		0.83		0.53		1		305

Table 1

HD	DM	Other Id	N Rem	α_{1950}	δ_{1950}	S	V	σ_V	B–V	σ_{B-V}	U–B	σ_{U-B}	n	Spectrum	References
30860	−46 01565			4 47 30	−46 37.8	1	8.92		0.52		−0.04		2	F5 V	8082
		NGC 1664 - 58		4 47 31	+43 34.0	1	15.31		0.87		0.37		3		305
		NGC 1664 - 16		4 47 31	+43 35.5	1	15.14		0.75		0.16		1		305
		NGC 1664 - 51		4 47 31	+43 37.8	1	12.74		0.44		0.31		2	A0	305
		NGC 1664 - 61		4 47 33	+43 32.7	1	15.08		0.64		0.17		1		49
		NGC 1664 - 62		4 47 34	+43 32.3	1	14.94		0.68		0.08		1		49
		NGC 1664 - 17		4 47 34	+43 34.7	1	11.40		1.09		0.77		2	gG8	305
		NGC 1664 - 37		4 47 35	+43 37.4	1	10.93		0.34		0.34		1	A0	305
30848	−42 01637			4 47 35	−42 28.2	1	7.26		0.17				4	A3 V(s)	1075
	+43 01093	NGC 1664 - 26		4 47 37	+43 33.4	2	10.58	.015	0.34	.005	0.30	.029	3	A3 V	49,305
		NGC 1664 - 21		4 47 38	+43 35.1	1	11.81		0.29		0.26		2	A0	305
		NGC 1664 - 35		4 47 38	+43 36.9	1	12.50		0.37		0.25		1	A0	49
	−0 00783			4 47 39	+01 07.1	1	10.59		0.76		0.19		1		116
		LP 476 - 66		4 47 40	+14 50.8	1	12.23		1.07				2		1674
		BSD 24 # 286		4 47 40	+43 55.4	1	11.84		0.25		0.13		6	B8	1069
		NGC 1664 - 77		4 47 41	+43 37.5	1	12.84		0.49		0.25		1	A2	49
		NGC 1664 - 13		4 47 41	+43 37.8	1	14.36		1.09		0.71		1		305
30712	+14 00770	Hyades vB 120		4 47 43	+14 59.9	3	7.66	.141	0.73	.008	0.31	.012	5	G5	1127,1355,8023
		NGC 1664 - 100		4 47 43	+43 32.1	1	11.44		0.37		0.27		1	A2	49
		G 39 - 37		4 47 44	+26 02.4	1	12.85		1.54		1.33		1		1620
30675	+28 00704			4 47 44	+28 13.8	1	7.53		0.34		−0.26		5	B3 V	399
	+44 01037			4 47 45	+44 29.9	1	9.76		0.59				2	F5	1119
		BSD 24 # 1153		4 47 45	+46 19.6	1	11.66		0.56		0.47		2	B9	1069
30877	−42 01639			4 47 45	−42 30.0	2	7.31	.000	0.20	.000			8	A2/3mA3-F2	1075,2033
268612				4 47 45	−69 19.7	1	12.47		0.23		0.20		3	A2:I	149
30710	+15 00691	V1060 Tau		4 47 46	+15 42.7	1	9.43		2.75		4.27		1	C4 II	109
		NGC 1664 - 5		4 47 47	+43 35.9	1	12.87		0.45		0.26		2		305
		G 85 - 19		4 47 49	+22 02.6	1	15.23		1.95		0.94		6		419
		NGC 1664 - 834		4 47 49	+43 19.0	1	12.69		0.34		0.34		6	B9	1069
		LP 891 - 24		4 47 49	−31 10.3	1	12.47		0.45		−0.19		2		1696
276810	+43 01095			4 47 50	+44 07.8	1	9.94		0.32				2	A3	1119
30763	+0 00867	IDS04453N0053	AB	4 47 52	+00 57.9	1	8.74		0.37		0.02		1	F0	116
		BSD 24 # 301		4 47 52	+43 51.0	1	12.32		0.32		0.26		4	B8	1069
30739	+8 00777	HR 1544		4 47 53	+08 49.0	8	4.36	.003	0.01	.003	−0.02	.018	86	A1 Vn	15,61,71,1118,1203*
		NGC 1664 - 27		4 47 55	+43 36.7	1	13.87		0.51		0.21		1	F5	305
30738	+15 00692	Hyades vB 121		4 47 56	+16 07.6	2	7.29	.000	0.50	.002	0.03	.001	3	F8	1127,8023
30814	−16 00964	HR 1549		4 47 56	−16 18.2	2	5.04	.010	0.99	.010	0.77		7	K0 III	2035,3016
30650	+43 01096	NGC 1664 - 293		4 47 58	+43 29.7	3	7.49	.020	−0.03	.009	−0.47	.055	5	B3 V	49,1025,1069
		HIC 22580		4 48 00	+42 33.7	1	10.13		1.76		2.01		2		1723
	−39 01648			4 48 00	−39 23.3	1	9.97		0.50		0.02		2	F5	1730
		Sk -70 001a		4 48 00	−70 39.	1	13.65		−0.05		−0.90		3		798
		Hyades L 99		4 48 02	+19 51.3	1	12.29		0.93		0.63		1		1058
282501	+32 00843	G 39 - 38		4 48 02	+32 48.5	2	9.26	.005	0.85	.005	0.52	.005	4	G5	1064,1620
		EW Aur		4 48 02	+38 06.	1	13.05		0.89				1		1772
277043	+39 01085			4 48 03	+39 54.2	1	10.39		2.08		2.13		3	A0	8032
30649	+45 00992	G 81 - 38	⋆A	4 48 03	+45 45.5	8	6.98	.016	0.59	.014	0.03	.030	17	G1 V-VI	22,333,792,1003,1243*
		L 376 - 1		4 48 05	−39 58.6	1	13.45		1.35				1		3062
		RL 109		4 48 06	+39 36.	1	12.83		0.28		−0.39		2		704
31009	−56 00734	XZ Dor		4 48 08	−56 45.1	2	6.59	.023	1.68	.000	2.00		5	M2 III	2012,3040
		Sk -69 003		4 48 08	−69 15.8	1	12.56		0.15		−0.07		2	A2 Ib	149
30796	+0 00871			4 48 10	+01 03.8	1	6.68		0.05		0.08		1	A2	116
30708	+35 00914			4 48 10	+35 44.0	1	6.79		0.68		0.25		2	G5	1601
30674	+41 00984	IDS04447N4150	AB	4 48 10	+41 55.3	1	9.37		0.23				2	B9.5V	1119
		BSD 24 # 681		4 48 11	+44 21.4	1	12.85		0.35		−0.07		5	B5	1069
30812	−0 00785			4 48 16	−00 10.7	2	7.26	.010	1.09	.014	0.95	.017	10	K1 III	116,1003,1509,3040
		HA 96 # 321		4 48 17	+00 00.7	1	10.59		0.43		−0.02		2	F8	281
		WLS 448 55 # 10		4 48 19	+54 30.4	1	10.40		1.48		1.54		2		1375
	−0 00786			4 48 19	−00 45.1	1	9.82		0.49		−0.06		1	F5	116
	−39 01653			4 48 19	−39 30.6	1	11.39		0.54		0.06		2		1730
30707	+41 00987			4 48 20	+41 57.8	1	7.60		0.98				2	G8 III	1119
	+0 00873			4 48 21	+00 29.6	1	10.12		0.53		0.01		3	F8	3077
		SK *G 15		4 48 21	−69 33.2	1	13.38		0.15		0.09		2		149
286053	+18 00742	IDS04455N1840	B	4 48 22	+18 46.1	1	10.40		0.69				1	F0	920
287116	+12 00664	Hyades J 336		4 48 23	+13 06.3	1	10.78		0.72		0.13		3	G0	1058
30755	+28 00707	TT Tau		4 48 23	+28 26.6	1	8.02		2.93				1	C5 II	109
30939	−39 01656	IDS04467S3921	A	4 48 23	−39 16.0	1	8.52		0.49		−0.05		2	F5/7 V	1730
		Sk -69 004		4 48 23	−69 43.3	1	12.52		0.06		−0.42		4	B1 Iab	415
		RL 110		4 48 24	+39 42.	1	12.66		0.24		−0.18		2		704
30754	+28 00706	Hyades J 337		4 48 25	+28 32.9	2	9.54	.009	1.13	.005	1.11		6		950,1058
30838	−0 00787			4 48 26	+00 01.3	3	9.22	.008	0.32	.004	0.09	.025	18	A0	281,1775,6004
30810	+10 00654	Hyades vB 122	⋆AB	4 48 26	+10 59.1	3	6.77	.008	0.54	.019	0.05	.000	5	F8 V +F8 V	1127,1381,8023
284953	+22 00800	NGC 1750 - 20		4 48 26	+23 42.5	1	8.59		1.27		0.94		2	K0	1648
30780	+18 00743	Hyades vB 123	⋆A	4 48 27	+18 45.4	7	5.10	.014	0.21	.007	0.12	.006	14	A9 V	15,254,1008,1127,1363*
276994	+42 01072			4 48 27	+42 11.3	1	9.60		0.29				2	A0	1119
		Case *M # 39		4 48 30	+39 58.	1	10.46		2.11				1	M1 III	148
31027	−54 00718	IDS04474S5404	AB	4 48 30	−53 58.2	1	7.81		0.83		0.45		1	K1 V	3072
30836	+5 00745	HR 1552		4 48 32	+05 31.3	20	3.68	.007	−0.17	.006	−0.81	.005	94	B2 III+B2IV	1,3,15,125,154,1004*
30809	+15 00695	Hyades vB 143		4 48 32	+15 21.0	2	7.89	.005	0.53	.002	0.06	.000	8	F8	1127,8023
284922	+19 00794	G 85 - 21		4 48 32	+19 16.8	1	10.85		0.59		−0.06		3	G0	333,1620
30698	+50 01076			4 48 34	+50 42.4	1	8.76		0.14		0.13		2	B9	1566

HD	DM	Other Id	N Rem	α₁₉₅₀ δ₁₉₅₀	S	V	σ_V	B-V	σ_B-V	U-B	σ_U-B	n	Spectrum	References
		G 247 - 38	AB	4 48 34 +69 09.5	1	9.78		0.71		0.21		3	G5	5010
		UY Aur		4 48 36 +30 42.2	1	11.91		1.06		-0.14		1	G5V:e	8013
	+44 01040			4 48 36 +44 53.4	1	9.18		1.12				2	K5	1119
276988	+42 01073	IDS04451N4234	AB	4 48 37 +42 39.2	1	9.95		0.24				2	M2	1119
		LS V +53 029		4 48 37 +53 35.3	1	11.10		0.48		-0.67		2		405
		HA 96 # 167		4 48 39 -00 12.8	1	10.90		0.62		0.17		4	F2	281
30985	-41 01593	HR 1557	★AB	4 48 39 -41 24.4	3	6.06	.005	0.38	.005			14	F2/3 V	15,1075,2029
268632				4 48 40 -69 12.5	1	12.38		0.12		-0.25		3	A0 I	149
		NGC 1664 - 145		4 48 42 +43 43.5	1	11.62		0.19		-0.20		3	B8	1069
30736	+45 00995			4 48 42 +45 51.5	1	6.69		0.55		0.13		1	F7 V	1355
		HA 96 # 21		4 48 42 -00 19.8	2	12.17	.005	0.49	.000	-0.02	.006	9	G2	281,1764
276973	+42 01075			4 48 44 +43 07.1	1	9.84		0.06				4	B9	1025
	+43 01098			4 48 44 +43 49.7	1	10.14		0.20		-0.20		3	B7	1069
		HA 96 # 517		4 48 46 +00 10.8	1	11.43		0.63		0.14		2	F8	281
		HA 96 # 25		4 48 47 -00 18.0	1	12.30		0.41		0.02		4	G0	281
		HA 96 # 171		4 48 49 -00 14.0	1	11.04		0.49		0.08		4	G0	281
	+43 01099		AB	4 48 50 +43 29.6	1	11.53		1.23		1.03		3		1723
		NGC 1664 - 454		4 48 51 +43 59.6	1	11.68		0.36		0.25		4		1069
		Ua Pic		4 48 52 -50 44.5	1	10.67		0.14		0.07		1	A9.5	700
		Sk -69 005		4 48 53 -69 47.0	1	12.86		0.07		-0.33		4	A0 Iab	415
		Hyades CDS 72	★	4 48 55 +17 11.4	1	11.38		1.28				3	K5	1674
30898	-0 00788			4 48 55 -00 37.8	1	7.85		0.27		0.15		1	A2	116
		HIC 22664		4 48 56 +43 12.6	1	10.98		1.28		1.10		3		1726
276978	+42 01078			4 48 57 +42 50.9	1	9.80		0.28				2	A2	1119
		NGC 1664 - 206		4 48 57 +43 51.1	1	12.17		0.17		-0.22		4	A0	1069
	+44 01041			4 48 57 +44 33.1	1	10.23		1.43		1.63		3		1723
		GD 63		4 48 58 +04 28.1	1	16.45		0.57		-0.01		1		3060
30870	+9 00668	HR 1553		4 48 58 +09 53.5	4	6.11	.004	0.08	.004	-0.44	.009	12	B5 V	15,154,401,1256
30794	+36 00948			4 48 59 +36 33.6	1	6.73		1.14		1.03		1	K0	851
276967	+43 01100			4 49 00 +43 22.4	1	10.88		0.22		-0.26		2	B8	1069
31249	-67 00352			4 49 00 -67 47.9	2	7.63	.015	0.19	.000	0.09	.005	11	A3 V	149,204
	+0 00875			4 49 01 +00 38.3	1	10.12		0.54		0.03		2	A2 V	1003
30869	+13 00728	Hyades vB 124	★AB	4 49 01 +13 34.3	4	6.27	.011	0.50	.008	0.06	.004	10	F5	401,1127,3030,8023
30793	+39 01090			4 49 02 +39 12.1	2	7.90	.005	1.33	.010	1.46	.005	6	K1 III-IV	37,833
	+43 01101			4 49 02 +43 29.0	1	11.00		0.17		-0.23		2	A0	1069
		MCC 454		4 49 03 +16 24.	1	11.23		1.35				2	M0	1619
		G 81 - 39		4 49 03 +40 38.2	1	13.40		1.61				2		940
30614	+66 00358	HR 1542		4 49 04 +66 15.6	6	4.29	.009	0.03	.016	-0.89	.012	16	O9.5Iae	15,154,1118,1119,8015,8100
		BSD 24 # 695		4 49 05 +44 18.1	1	11.44		0.30		0.28		2	B9	1069
	-0 00789			4 49 07 -00 10.2	5	8.93	.007	1.06	.011	0.85	.004	50	G8	281,989,1729,1775,6004
30685	+61 00739			4 49 09 +61 24.0	1	6.65		0.17		0.15		3	A2	1502
		HA 96 # 36		4 49 09 -00 15.2	4	10.59	.004	0.25	.001	0.11	.006	55	A3	281,989,1729,1764
280188	+38 00955	V346 Aur		4 49 10 +38 25.4	1	8.64		2.85		2.66		1	N	109
		POSS 85 # 2		4 49 11 +60 34.0	1	15.11		1.41				2		1739
30913	+9 00669			4 49 12 +09 47.4	1	6.79		0.42		-0.03		2	F2	401
287220	+10 00661	Hyades vB 144		4 49 12 +10 43.3	1	9.09		0.51		-0.03		3	F8	1127
276968	+43 01102			4 49 12 +43 13.7	1	9.24		0.12				2	B9 II:	1119
30752	+52 00891	HR 1546		4 49 12 +52 45.5	1	6.40		0.06		0.09		3	A2 V	1733
30842	+31 00821			4 49 14 +31 54.2	1	7.56		0.23		0.15		2	A3	1625
30823	+42 01081	HR 1550		4 49 16 +42 30.2	2	5.71	.005	0.11	.000	0.13		6	A3 III	401,1025
30834	+36 00952	HR 1551		4 49 17 +36 37.2	10	4.78	.012	1.41	.006	1.56	.014	44	K2.5 IIIb	15,224,660,1007,1013*
		WLS 448 55 # 5		4 49 17 +57 16.6	1	11.31		0.67		0.38		2		1375
		HA 96 # 541		4 49 18 +00 09.3	1	11.25		0.43		-0.03		2	G2	281
		Hyades L 97		4 49 18 +18 59.4	1	11.30		0.93		0.69		3		1058
		SK *G 18		4 49 19 -68 55.1	1	13.32		0.09		0.07		2	A0 Iab	149
		LP 476 - 85		4 49 20 +11 33.8	1	14.17		1.23				2		1674
	+0 00877			4 49 22 +00 48.7	1	10.45		1.15		0.92		1		116
		BSD 24 # 704		4 49 23 +44 21.0	1	10.90		0.50		-0.29		5	B2	1069
		G 82 - 52		4 49 24 -06 23.8	2	11.96	.020	1.59	.001	1.25		2	M2	333,1620,1705
286171		Hyades J 339		4 49 24 +15 54.	1	11.08		0.75		0.27		3	G0	1058
	+44 01042			4 49 24 +44 50.8	1	9.87		0.35				2	B6 V	1119
		HA 96 # 376		4 49 24 -00 03.6	1	11.76		0.69		0.18		2	G0	281
		HA 96 # 51		4 49 26 -00 15.4	1	10.64		0.52		0.06		4	G1	281
284930	+18 00746	Hyades L 98		4 49 28 +18 54.9	1	10.29		1.07				2	K0	582
		Sk -67 003		4 49 29 -67 47.8	1	13.20		-0.24		-1.02		3		149
268654	-69 00290	Sk -69 007		4 49 30 -69 32.4	4	10.50	.034	0.17	.010	-0.54	.014	19	B9 Ia	149,328,1277,8033
		BSD 24 # 703		4 49 31 +44 57.5	1	12.23		0.64		0.04		4	B7	1069
		HA 96 # 549		4 49 33 +00 09.4	2	10.96	.029	1.26	.049	1.20	.010	3	K2	116,281
	+48 01171		V	4 49 34 +48 31.0	1	9.57		2.04				3	M3	369
268657		Sk -69 008		4 49 34 -69 31.1	1	11.52		0.09		-0.62		4	B5 Ia	149
		G 83 - 39		4 49 35 +14 58.2	1	15.92		1.58				1		906
30979	-0 00790			4 49 35 -00 02.1	2	8.50	.002	1.04	.004	0.82		16	K0 III	281,6004
30978	+0 00878	IDS04470N0035	AB	4 49 36 +00 40.2	1	9.24		0.35		0.07		1	A2	116
276979	+42 01083			4 49 36 +42 51.5	1	8.69		1.52				2	K1 III	1119
30912	+27 00701	HR 1554		4 49 39 +27 48.9	2	5.97	.000	0.36	.007	0.16	.002	3	F2 IV	39,3026
	-0 00791			4 49 39 -00 11.3	2	10.87	.018	0.36	.009	0.00	.005	5	A8	116,281
31093	-35 01962	HR 1559	★AB	4 49 39 -34 59.3	3	5.85	.004	0.08	.000	0.09		9	A1 Vn	15,2012,3050
31131	-43 01583	IDS04481S4313	AB	4 49 39 -43 08.1	1	8.08		0.46				4	F3 V	2012
		LP 476 - 643		4 49 41 +13 04.0	1	13.08		0.96				1		1674
30604	+70 00327	G 247 - 40		4 49 41 +70 33.4	1	8.85		0.57		0.03		2	G0 V	1003

Table 1 233

HD	DM	Other Id	N Rem	α_{1950}	δ_{1950}	S	V	σ_V	B–V	σ_{B-V}	U–B	σ_{U-B}	n	Spectrum	References
30959	+14 00777	HR 1556, o1 Ori		4 49 42	+14 10.1	2	4.72	.020	1.79	.048	2.03		4	M3	1363,3055
30854	+43 01104			4 49 42	+43 43.8	1	8.51		0.67				5	G2 V	1025
		G 81 - 40		4 49 43	+48 36.2	1	12.02		1.26		1.21		1	K5	1620
284908	+19 00801	Hyades vB 145	A	4 49 45	+19 55.1	1	9.62		0.68				1	K0	950
284908	+19 00801	Hyades vB 145	AB	4 49 45	+19 55.1	2	9.31	.009	0.76	.008	0.36		6	K0	950,1127
284908	+19 00801	Hyades vB 145	B	4 49 45	+19 55.1	1	11.26		1.04				1		950
		HA 96 # 727		4 49 46	+00 15.5	1	10.66		1.11		0.95		2	G3	281
		LP 359 - 94		4 49 47	+22 33.9	1	11.54		1.30				4		950
30883	+41 00996			4 49 48	+42 01.8	1	8.82		0.12				2	B9 V	1119
31203	−53 00760	HR 1563	⋆ AB	4 49 48	−53 32.8	3	5.22	.017	0.34	.005			14	F0 IV	15,1075,2030
		Sk -68 001		4 49 48	−68 45.9	1	13.36		0.01		-0.48		3		798
276970	+42 01084			4 49 49	+43 07.4	1	8.55		1.45				2	K0 III	1025
31019	+0 00881			4 49 51	+00 53.0	1	8.68		1.76		1.94		1	K2	116
		BSD 96 # 1116		4 49 52	+01 03.3	1	11.41		0.46		-0.05		1	F2	116
30989	+12 00667			4 49 53	+12 18.2	1	7.28		0.16		-0.17		2	B8	401
30974	+14 00778			4 49 54	+14 32.5	1	7.52		0.59		0.08		2	G0	1648
		Sk -67 004		4 49 54	−67 47.	1	13.00		-0.16		-0.98		4		798
		LP 416 - 109		4 49 55	+18 18.0	1	12.35		0.97				1		1570
		G 39 - 40		4 49 55	+30 43.0	1	13.56		1.36		1.05		1		333,1620
		G 222 - 6		4 49 55	+84 02.8	1	13.21		1.45		1.24		1		906
		G 222 - 5		4 49 55	+84 03.8	1	16.09		1.60				1		906
		MtW 96 # 28		4 49 56	+00 04.1	1	16.22		0.48		0.01		2		397
	−0 00792			4 49 56	−00 02.8	5	9.66	.008	0.60	.008	0.04	.006	38	G1	116,281,989,1729,6004
284855	+23 00755			4 49 59	+23 24.4	1	10.56		1.08				2	K0	920
		POSS 85 # 3		4 49 59	+60 41.5	1	15.86		1.53				2		1739
		HA 96 # 736		4 50 00	+00 17.2	1	11.92		0.44		0.17		10		281
		HA 96 # 737		4 50 00	+00 17.5	4	11.72	.002	1.33	.003	1.15	.022	59	K0	281,989,1729,1764
		AAS21,109 # 1		4 50 00	−68 39.	1	13.98		-0.09		-0.80		3		417
		MtW 96 # 45		4 50 01	+00 05.0	1	14.55		0.82		0.49		2		397
		MtW 96 # 49		4 50 02	+00 02.2	1	13.95		0.99		0.98		4		397
276962	+43 01108			4 50 02	+43 46.8	1	9.07		1.77				4		1025
30973	+21 00717	IDS04471N2204	AB	4 50 04	+22 09.3	1	8.78		1.01		0.90		5	K5 V	7009
276964	+43 01109			4 50 07	+43 37.2	1	9.22		1.30				2	K2	1119
		Sk -69 009		4 50 07	−69 18.	1	12.56		-0.15		-0.96		3		415
31128	−27 01935			4 50 08	−27 08.8	2	9.13	.009	0.49	.000	-0.19		4	F3/5 Vw	742,1594
		LP 776 - 25		4 50 10	−16 54.1	1	11.69		1.47		0.94		1	M3,e:	3062
	+0 00884			4 50 11	+00 36.3	1	10.17		0.54		-0.01		1	F8	116
	+43 01110	IDS04466N4355	AB	4 50 13	+44 00.4	1	9.35		0.31		0.10		2	B8 V	1069
31261	−53 00763			4 50 13	−53 29.4	1	9.09		0.88		0.49		2	K2 V	3072
		HA 96 # 405		4 50 15	+00 02.6	2	10.66	.010	1.29	.010	1.49	.000	10	K0	281,397
31073	−0 00793			4 50 17	+00 02.2	6	9.30	.008	0.22	.008	0.14	.006	44	A2	116,281,397,989,1729,6004
		HA 96 # 748		4 50 17	+00 20.7	2	11.67	.015	0.76	.004	0.29	.011	10	G6	281,1375
		LP 359 - 100		4 50 17	+22 45.2	1	13.23		0.95				1		1570
		HA 96 # 219		4 50 17	−00 06.8	1	11.49		0.51		-0.01		4	G0	281
		RL 111		4 50 18	+43 32.	1	13.57		0.24		-0.40		2		704
		BSD 24 # 966		4 50 20	+45 34.1	1	10.91		0.38		0.12		2	B8	1069
		BSD 24 # 369		4 50 21	+43 20.2	1	12.75		0.13		0.02		3	B8	1069
268605	−67 00355	Sk -67 005		4 50 21	−67 44.6	4	11.34	.007	-0.11	.012	-0.96	.015	12	O9.7Ia	149,328,1277,8033
268625				4 50 22	−68 04.2	1	10.92		1.34		1.35		3	K0 III:	149
		HA 96 # 753		4 50 24	+00 19.5	1	11.42		0.99		0.60		1		281
		Hyades L 100		4 50 24	+19 22.4	1	12.21		0.97		0.60		3		1058
		Sk -70 002		4 50 24	−70 26.	1	13.53		0.01		-0.56		4		798
		HA 96 # 409		4 50 25	+00 04.1	2	13.79	.010	0.53	.006	0.00	.028	5		281,1764
		HA 96 # 83		4 50 25	−00 19.7	4	11.72	.004	0.18	.005	0.20	.009	60	A3	281,989,1729,1764
31109	−5 01068	HR 1560		4 50 26	−05 32.1	8	4.38	.011	0.25	.011	0.17	.029	24	F4III+A6III	15,1008,1075,1425*
		Sk -69 010		4 50 26	−69 24.	1	12.61		-0.23		-0.88		4		415
30971	+41 00998			4 50 31	+41 57.0	2	9.27	.017	0.31	.004			33	A5 V	1025,6011
30988	+41 00999			4 50 33	+42 07.0	1	7.94		1.55				2	K1 III	1119
		HA 96 # 90		4 50 36	−00 22.7	1	11.59		0.44		0.03		4	F8	281
		L 736 - 21		4 50 36	−16 56.	1	14.10		1.20		1.15		1		3062
31274	−47 01546			4 50 36	−46 56.1	2	7.12	.005	0.96	.000	0.78		6	G8/K0 III	536,2012
		Sk -69 012		4 50 36	−69 25.	1	12.77		0.11		-0.66		4		798
		Sk -69 011		4 50 36	−69 29.	1	12.85		0.08		-0.73		3		798
284858	+22 00769	Hyades vB 125		4 50 38	+22 56.4	2	9.31	.000	0.49	.000	-0.04	.002	4	G0	1127,8023
31123	+0 00887			4 50 39	+00 20.0	3	9.27	.005	0.43	.004	0.06	.005	16	G5	281,1775,6004
31072	+18 00747			4 50 40	+18 59.5	1	7.55		1.73		2.07		2	K5	1648
	+43 01113			4 50 40	+43 38.8	1	10.72		0.10		-0.28		4	B9	1069
		LP 359 - 104		4 50 42	+22 55.5	1	13.80		1.42				1		1570
31107	+9 00675			4 50 43	+09 24.6	1	8.61		0.33		0.02		2	F2 V	401
30987	+44 01044			4 50 43	+44 39.1	2	8.67	.033	0.38	.019	0.26		4	F0 V	1119,8100
	+73 00257	G 248 - 28		4 50 43	+73 46.1	1	9.45		0.69		0.24		2	G5	1658
		HA 96 # 767		4 50 44	+00 18.7	1	11.68		0.96		0.78		11	G8	281
		AK 7 - 6		4 50 44	+21 40.3	1	11.90		0.85				1		987
		HA 96 # 235		4 50 45	−00 10.0	4	11.14	.006	1.08	.005	0.90	.005	55	G8	281,989,1729,1764
31139	+2 00800	HR 1562		4 50 46	+02 25.6	4	5.33	.013	1.63	.009	1.93	.021	14	M1 III	15,1061,1355,3055
31140	+1 00843			4 50 47	+01 26.8	1	9.18		0.04		0.02		6	A2	1371
237328	+55 00940			4 50 47	+55 39.0	1	9.40		1.85		1.98		2	K5	1375
		G 84 - 16		4 50 47	−04 02.6	2	13.53	.024	0.71	.029	0.05	.073	3		1620,3056
		BSD 24 # 1191		4 50 52	+46 20.1	1	11.37		0.34		0.29		2	B9	1069
31155	+0 00888			4 50 53	+01 02.9	1	7.35		1.54		1.83		1	K0	116

HD	DM	Other Id	N Rem	α_{1950}	δ_{1950}	S	V	σ_V	B–V	σ_{B-V}	U–B	σ_{U-B}	n	Spectrum	References
		SK *G 26		4 50 56	−69 47.7	1	13.20		0.23		0.32		3	A2 II	149
30958	+55 00941	HR 1555	⋆ A	4 50 57	+55 10.7	2	5.52	.001	0.02	.014	−0.03	.036	6	B9.5V	595,1733
	+0 00889			4 51 01	+00 10.4	2	9.39	.002	1.09	.001	0.93		11	K0 III	281,6004
31311	−43 01590	Z Cae		4 51 01	−43 08.6	1	7.83		1.63				4	M2 III	2012
277073	+43 01115			4 51 02	+43 16.8	1	9.46		1.84				3	M2	369
		HA 96 # 249		4 51 02	−00 12.1	1	11.77		0.65		0.13		4		1729
31518	−70 00321			4 51 04	−69 54.8	1	7.20		1.68		1.92		9	M3 III	1704
31182	+2 00804			4 51 06	+02 28.4	1	8.92		0.47		0.02		2	F8	1375
31153	+16 00664	Hyades vB 146		4 51 10	+16 56.8	1	7.24	.000	0.53	.000	0.06	.001	3	F8	1127,8023
268687	−69 00296	Sk −69 013		4 51 12	−69 30.7	2	10.67	.020	0.47	.005	0.21		6	F6 Ia	149,1483
31085	+41 01002			4 51 13	+41 40.8	2	8.23	.010	0.42	.005	0.12		3	F2 V	1119,8100
		HA 96 # 449		4 51 13	−00 04.2	1	11.88		0.66		0.10		2	G3	281
	−22 01868			4 51 14	−22 14.9	1	10.40		0.22		0.13		2		126
31069	+43 01116	HR 1558		4 51 16	+43 58.9	3	6.09	.014	−0.02	.010	−0.08	.024	6	A0 V	252,1079,1119
31084	+43 01117			4 51 19	+43 18.4	1	8.56		0.58				4	F9 V	1025
		HA 96 # 452		4 51 19	−00 01.0	1	11.52		0.72		0.20		2	K0	281
31209	+1 00847	HR 1565		4 51 20	+01 29.3	4	6.61	.007	0.04	.001	0.03	.011	32	A1 Vn	15,252,978,1061
		LP 359 - 129		4 51 22	+23 52.0	1	14.26		1.47				2		1570
268733		Sk −70 003		4 51 23	−70 06.3	1	12.26		0.16		0.18		4	A2 I	415
		G 81 - 41		4 51 25	+45 39.3	3	11.08	.032	1.12	.005	1.05	.056	5	K5	1620,1658,1723
		SK *G 30		4 51 25	−69 02.8	1	13.33		0.12		0.15		2	A0 II	149
		KUV 04514+1603		4 51 26	+16 02.8	1	16.47		0.30		−0.31		1		1708
		BB Eri		4 51 26	−19 30.8	3	11.04	.042	0.19	.038	0.10	.024	3	A5	597,699,700
		SK *G 31		4 51 26	−69 16.5	1	12.61		0.15		0.11		2	A1 Ib	149
31181	+17 00807	Hyades vB 166		4 51 27	+17 31.7	1	9.59		0.51		0.07		2	F8	1127
	−22 01869			4 51 28	−22 15.3	1	9.90		0.41		0.00		2		126
31407	−55 00702			4 51 28	−55 46.7	1	7.70		−0.22		−0.85		3	B2/3 V	1732
	+1 00849			4 51 29	+01 44.1	1	9.03		1.34		1.46		6	K2	1371
31098	+42 01091			4 51 29	+42 51.1	2	7.57	.025	1.41	.005	2.17		3	K2 III	1119,8100
		LS V +45 017		4 51 29	+45 56.6	2	10.66	.085	0.62	.028	−0.50	.054	4		405,1069
31223	+1 00850			4 51 32	+01 14.1	1	8.40		0.52		0.00		6	F5	1371
31208	+7 00754	G 83 - 40	⋆ B	4 51 34	+07 17.5	2	8.33	.050	0.90	.002	0.62	.010	4	K0	1620,3008
31208	+7 00754	G 83 - 41	⋆ A	4 51 34	+07 17.7	3	8.15	.037	0.84	.010	0.54	.009	4	K0	1355,1620,3008
		LP 776 - 26		4 51 35	−17 50.7	1	10.89		1.56				4	M3	912
		LP 776 - 27		4 51 35	−17 50.7	2	10.91	.005	1.54	.011			2	M3	1705,1746
	−62 00394			4 51 36	−62 14.0	2	10.13	.005	0.84	.000	0.29		3		742,1594
		Sk −67 006		4 51 36	−67 33.	1	13.57		−0.21		−1.03		3		798
31136	+40 01085			4 51 38	+40 40.1	1	8.26		1.89				3	M1	369
31237	+2 00810	HR 1567, π5 Ori		4 51 39	+02 21.6	9	3.71	.015	−0.19	.007	−0.82	.006	25	B3 III+B0 V	15,154,1004,1075,1203*
277068	+43 01123			4 51 39	+43 33.8	2	9.62	.015	0.14	.005	−0.29		4	B4 III	1069,1119
		Sk −67 007		4 51 40	−67 38.7	1	12.91		0.01		−0.35		4	B9 Iab	415
31118	+43 01124			4 51 41	+43 20.2	1	7.01		1.81				2	K5 Ib	1119
		G 83 - 42		4 51 42	+06 59.0	2	13.33	.020	0.84	.005	0.29	.007	5		1620,1658
		BSD 24 # 738		4 51 42	+44 51.6	1	10.57		0.32		−0.27		3	B5	1069
	+0 00890			4 51 47	+00 37.7	1	10.15		1.01		0.73		1	G5	116
		LP 416 - 130		4 51 47	+19 36.1	1	13.89		1.54				1		950
	−20 00958			4 51 47	−20 38.1	1	10.12		1.25		1.22		1	M0 V	3072
		Sk −67 008		4 51 48	−67 49.4	1	13.62		−0.17		−0.95		4		149
		AAS21,109 # 2		4 51 48	−69 28.	1	13.42		0.04		−0.96		2		952
		Sk −70 004		4 51 48	−70 57.	1	13.74		0.08		−0.49		4		798
31135	+43 01126			4 51 50	+44 01.4	2	9.36	.005	0.28	.010	0.29		4	A0 V	1069,1119
		BSD 96 # 660		4 51 52	+00 40.7	1	11.56		0.43		−0.02		1	F5	116
	−0 00796			4 51 53	−00 16.6	1	9.92		0.48		−0.02		1	F2	116
	+0 00891			4 51 54	+00 46.6	1	10.12		0.47		0.03		1		116
31254	+11 00672			4 51 54	+11 56.8	1	7.45		0.14		−0.25		2	A0	401
277090	+42 01093			4 51 55	+42 35.7	1	8.79		1.00				2	G5	1119
		SK *G 39		4 51 55	−68 58.0	2	13.17	.045	0.55	.000	0.46		5	F8 I	149,1483
31269	+11 00674			4 51 58	+11 50.9	1	8.18		0.21		0.15		2	A0	401
268742	−70 00323			4 51 58	−70 19.6	1	10.97		0.61		0.03		3	F5	149
	+37 00988			4 51 59	+37 13.4	1	10.25		2.14				2	M3	369
31283	+11 00675	HR 1569	⋆	4 52 00	+11 20.8	2	5.19	.000	0.12	.003	0.10		4	A3 V	401,1363
		LP 359 - 145		4 52 00	+23 23.5	1	12.74		1.32				1		920
		HRC 77, GM Tau		4 52 00	+30 17.2	1	12.30		1.12		0.37		1	dK5e	776
		AAS21,109 # 3		4 52 00	−68 51.	2	13.89	.066	−0.12	.042	−0.82	.005	4		417,952
31236	+19 00811	Hyades vB 126	⋆	4 52 02	+19 24.4	4	6.37	.004	0.29	.000	0.06	.010	7	F0 IV-V	15,254,1127,8015
31178	+41 01003			4 52 02	+42 02.6	1	8.28		1.14				2	G8 III	1119
		Sk −68 002		4 52 02	−68 30.7	1	12.84		0.07		−0.18		4	B9 Iab	415
		Sk −69 014		4 52 03	−70 00.6	1	12.44		0.18		−0.17		4	A3 Ib	952
31296	+7 00755	HR 1571		4 52 05	+07 42.0	3	5.33	.005	1.22	.000	1.18	.000	9	gK1	15,1061,1355
		AAS21,109 # 4		4 52 06	−68 50.	1	14.56		−0.16		−0.99		3		417
		Sk −69 015		4 52 06	−69 51.	1	12.77		0.01		−0.70		3		798
277069	+43 01127	LS V +43 015		4 52 07	+43 34.6	2	9.42	.010	0.40	.025	0.24		4	B9 III	1069,1119
31295	+9 00683	HR 1570	⋆ A	4 52 08	+10 04.4	8	4.64	.011	0.08	.004	0.09	.008	85	A0 V	15,61,71,1049,1118*
286123	+18 00753	Hyades J 341		4 52 08	+18 35.	1	9.89		0.62		0.11		3	G0	1058
31134	+52 00898	HR 1561		4 52 09	+52 47.4	2	5.75	.006	0.10	.008	0.11	.003	6	A2 Vs	595,1733
31673				4 52 09	−69 28.5	1	12.63		0.09				3		8033
	−0 00797			4 52 10	−00 07.0	1	10.41		0.91		0.51		1		116
		BSD 24 # 753		4 52 12	+44 35.6	1	10.94		0.44		0.02		2	B9	1069
		AAS21,109 # 5		4 52 12	−68 54.	1	13.79		0.17		−0.86		2		417
31306	+8 00799	IDS04495N0826	AB	4 52 13	+08 31.2	1	6.94		0.15		0.08		2	A0	401

Table 1

HD	DM	Other Id	N Rem	α_{1950}	δ_{1950}	S	V	σ_V	B–V	σ_{B-V}	U–B	σ_{U-B}	n	Spectrum	References
31280	+20 00846			4 52 14	+20 14.1	1	8.50		0.34		0.26		2	A0	1648
	+44 01051	LS V +44 024		4 52 14	+44 11.0	1	9.94		0.29		-0.36		4	B5	1069
31345	-11 00995			4 52 14	-11 41.5	1	9.18		0.26		0.11		2	A5	1776
31331	+0 00893	HR 1574		4 52 16	+00 23.3	5	5.99	.009	-0.12	.005	-0.55	.008	22	B5 V	15,154,281,1415,6004
268718	-69 00299	Sk -69 016		4 52 16	-69 30.3	3	10.71	.013	0.24	.012	-0.62		15	B9 I e	415,1277,8033
31195	+44 01052			4 52 17	+44 57.3	3	7.86	.010	0.04	.013	-0.28	.049	7	B7 V	401,1025,1069
		G 81 - 42		4 52 18	+45 25.7	1	12.83		0.92		0.46		1		1620
268623	-66 00335	Sk -66 001		4 52 18	-66 48.	1	11.61		-0.06		-0.86		4	B2 Ia	952
		G 97 - 5		4 52 19	+05 18.1	3	12.39	.010	0.58	.005	-0.15	.020	7		203,333,1620,1658
31207	+42 01097			4 52 19	+43 04.6	1	8.27		0.40				4	F5 V	1025
31166	+51 00999	IDS04484N5156	AB	4 52 21	+52 01.2	1	8.18		0.45		-0.04		2	F8	3030
31166	+51 00999	IDS04484N5156	ABC	4 52 21	+52 01.2	1	9.62		0.60		0.09		2	F8	3030
	+0 00894			4 52 24	+00 15.0	1	9.95		0.75		0.17		1		116
	-0 00798			4 52 24	-00 16.5	1	10.91		0.26		0.15		1	A2	116
		Sk -68 003		4 52 24	-68 29.	1	13.13		-0.13		-1.02		4		798
280270	+34 00922			4 52 27	+34 17.0	1	9.32		2.22				4	M0	369
31206	+43 01130			4 52 27	+43 53.2	1	9.17		0.16				2	B9 V	1119
		LS V +47 022		4 52 27	+47 18.9	1	12.10		0.52		-0.42		2	B0 V	342
	+0 00896			4 52 28	+00 44.3	1	11.25		0.65		0.01		1		116
31220	+43 01131		V	4 52 29	+43 24.9	4	7.40	.020	1.79	.018	2.07	.071	9	M0 Ib	369,1119,8032,8100
		BPM 17964	A	4 52 30	-55 57.	1	11.13		1.57		1.08		1		3072
		BPM 17964	B	4 52 30	-55 57.	1	12.15		1.60				1		3072
		Sk -70 005		4 52 30	-70 51.	1	13.66		-0.19		-1.01		4		798
31205	+47 01074			4 52 31	+47 21.7	1	8.75		0.05		-0.36		2	A0	401
268708		Sk -69 018		4 52 31	-69 09.5	1	11.60		0.35		0.35		3	F0:I	149
31293	+30 00741	AB Aur		4 52 34	+30 28.4	3	7.08	.042	0.12	.016	0.04	.019	7	A0 ep	206,351,379
268763		Sk -70 006		4 52 34	-70 35.4	1	12.50		0.16		0.21		3	A2 Ib	149
		Sk -68 004		4 52 36	-68 06.	1	13.84		-0.12		-1.00		4		798
31305	+30 00742			4 52 37	+30 15.6	2	7.58	.016	0.12	.010	0.15	.047	12	A0	379,1655
	+44 01055			4 52 38	+44 39.9	1	9.87		0.25				2		1119
		SK *G 49		4 52 38	-68 39.8	1	13.30		0.14		0.04		3	B9 Ib	149
		G 83 - 45		4 52 40	+16 08.1	1	13.44		0.79		0.02		2		1658
		Sk -69 019		4 52 40	-69 35.6	1	12.82		-0.05		-0.92		3		149
287234	+9 00686	Hyades vB 147		4 52 41	+10 07.8	1	9.17		0.55		0.05		3	G0	1127
31722	-69 00300			4 52 41	-69 29.1	3	8.46	.011	0.23	.007	0.19	.016	17	A5 III	149,204,1408
		Sk -66 002		4 52 42	-66 35.	1	13.18		-0.06		-0.70		3		798
		AAS21,109 # 6		4 52 42	-68 54.	1	13.61		-0.11		-0.68		1		417
		Sk -69 017		4 52 42	-69 49.	1	13.28		0.00		-0.77		3		798
31447	-29 01944	LP 891 - 33		4 52 43	-29 07.4	1	10.28		0.40		0.02		2	G5 IV/V	1696
31338	+19 00815	G 85 - 28		4 52 44	+19 55.7	1	7.98		0.80		0.47		3	K0	1355
	+36 00964			4 52 46	+37 04.5	1	11.73		2.32				1	M3	369
		BPM 85821		4 52 48	+22 00.	1	12.08		0.85				1		987
282624	+30 00743	SU Aur		4 52 48	+30 29.3	3	9.16	.070	0.89	.018	0.45	.038	3	G2 III e	351,649,8013
31460	-31 02095			4 52 48	-31 30.6	1	8.04		0.58				4	G0 V	2012
		Sk -66 003		4 52 48	-66 45.	1	14.73		-0.18		-0.90		4	WN3	980
		AAS21,109 # 7		4 52 48	-70 10.	1	14.17		-0.14		-0.91		2		417
	-85 00055	UZ Oct		4 52 48	-84 53.8	1	9.30		0.54		0.00		1		794
31383	-0 00802			4 52 49	-00 25.9	2	8.16	.002	0.02	.002	0.00	.004	4	B9	116,1657
		Sk -67 009		4 52 51	-67 10.5	1	12.57		0.17		0.23		3	A3 II	952
277120	+41 01007	IDS04494N4120	AB	4 52 52	+41 25.1	1	9.43		0.34				3	F0	1025
31414	-16 00991	HR 1579		4 52 52	-16 49.2	3	5.70	.004	0.95	.005			11	G8/K0 III	15,1075,2031
268698				4 52 53	-68 47.6	1	11.64		0.67		0.05		5	G2 V	149
		AAS21,109 # 8		4 52 54	-68 05.	1	14.10		-0.21		-1.03		3		417
		Sk -70 007		4 52 54	-70 46.	1	13.52		-0.01		-0.57		4		798
277082	+42 01098	IDS04494N4301	AB	4 52 58	+43 06.6	1	9.41		0.30				2	K0	1119
31841	-73 00280			4 52 58	-73 04.6	1	7.90		0.64				4	G0 V	2012
31373	+14 00787	HR 1576		4 52 59	+14 57.7	3	5.78		-0.10	.013	-0.46	.004	9	B9 V	1049,1079,3039
31327	+35 00930	HR 1573	★	4 52 59	+36 05.4	3	6.07	.010	0.40	.010	-0.43	.010	8	B2 Ib	15,154,1012
31265	+47 01076			4 52 59	+47 15.7	1	8.80		0.06		-0.38		2	B9	401
277121	+41 01008			4 53 00	+41 25.9	1	8.63		1.22				3	K0	1025
		AAS21,109 # 12		4 53 00	-68 06.	1	13.48		-0.18		-1.04		4		417
		Sk -68 005		4 53 00	-68 08.	1	13.63		-0.34		-0.98		3		952
		AAS21,109 # 9		4 53 00	-68 52.	2	13.61	.019	-0.15	.015	-0.99	.015	3		417,952
		Sk -69 020		4 53 00	-69 25.	1	12.84		-0.10		-0.91		4		798
31374	+13 00737			4 53 03	+13 33.1	1	7.81		0.12		-0.31		2	B9	401
		G 39 - 41		4 53 03	+25 49.5	1	14.16		1.55		1.15		5		203
31314	+39 01109	LS V +40 018		4 53 04	+40 05.5	1	8.15		0.10		-0.61		2	B2 V	401
31444	-16 00992	HR 1581, R Eri		4 53 04	-16 29.8	3	5.72	.004	0.87	.005			11	G6/8 III	15,1075,2035
		Hyades L 102		4 53 06	+16 50.1	1	11.33		0.83		0.41		1		1058
		G 39 - 42		4 53 06	+25 48.8	1	13.67		1.65		1.22		3		203
		Sk -67 009a		4 53 06	-67 12.	1	13.72		-0.17		-0.98		3		798
		AAS21,109 # 11		4 53 06	-70 12.	1	14.63		-0.05		-0.51		3		952
31292	+49 01271			4 53 12	+49 50.8	1	6.95		0.39		0.00		2	F0	1601
		Sk -66 004		4 53 12	-66 59.	1	13.00		-0.15		-0.98		5		798
		AAS21,109 # 13		4 53 12	-68 08.	1	13.85		-0.19		-1.08		3		417
31362	+24 00709	HR 1575		4 53 13	+24 30.8	1	6.39		0.32		-0.10		3	F2 V	254
31439	-2 01069			4 53 14	-02 25.0	1	8.34		0.87		0.49		2	G5	1375
31529	-39 01691	HR 1584		4 53 14	-39 42.5	3	6.10	.004	1.42	.000			11	K3 III	15,1075,2032
		SK *G 54		4 53 14	-67 05.2	1	13.27		0.09		0.08		2	A1 II	149
		POSS 717 # 4		4 53 15	-13 06.2	1	16.82		1.63				3		1739

HD	DM	Other Id	N Rem	α_{1950}	δ_{1950}	S	V	σ_V	B–V	σ_{B-V}	U–B	σ_{U-B}	n	Spectrum	References
31278	+53 00829	HR 1568	⋆ AB	4 53 16	+53 40.5	6	4.46	.016	-0.02	.006	-0.02	.009	19	A1 V	15,1007,1013,1363*
31412	+4 00782	IDS04507N0431		4 53 17	+04 35.7	1	7.02		0.55		0.04		3	F8	196
31400	+11 00680			4 53 17	+11 09.9	1	8.56		0.29		0.21		2	A2	401
		G 84 - 22		4 53 17	-01 42.4	1	11.18		0.66		0.08		2		1620
272302	-43 01604			4 53 18	-43 17.7	2	10.70	.010	0.56	.005	-0.09		8	F5	158,2034
31411	+5 00769	HR 1578		4 53 19	+05 19.3	2	6.49	.005	0.02	.000	-0.01	.000	7	A0 V	15,1061
	+44 01060			4 53 20	+44 34.6	1	9.17		0.49				2	F6 V	1025
31326	+47 01077			4 53 20	+47 48.1	1	7.89		0.00		-0.14		2	A0	401
31853	-71 00295			4 53 22	-71 39.3	1	9.21		0.48		-0.01		2	F8 V:	204
277160	+40 01099			4 53 24	+40 15.1	1	9.77		1.78				3	M0	369
277086	+42 01100			4 53 24	+43 02.3	1	8.17		1.29				2	K0 III	1119
		Sk -70 007a		4 53 24	-70 44.	1	13.33		0.07		-0.50		4		798
31422	+8 00803			4 53 25	+09 03.5	1	8.93		0.06		-0.24		2	A2	401
31754	-66 00338	HR 1598		4 53 26	-66 45.3	4	6.40	.003	1.62	.005	1.96	.000	43	M0/1 III	15,1075,2012,2038
31517	-25 02115	HR 1583		4 53 27	-25 48.4	2	6.71	.005	0.28	.005			7	A8 V +F3:	15,2012
268653	-67 00358	IDS04534S6705	⋆ A	4 53 28	-67 00.1	4	10.74	.013	-0.02	.008	-0.80	.024	18	B3 Ia:	149,328,1277,8033
		NGC 1841 sq1 3		4 53 28	-84 02.9	1	14.91		0.96				6		1669
		Sk -67 010		4 53 30	-67 03.	1	12.98		-0.02		-0.71		3		798
		AAS21,109 # 14		4 53 30	-68 20.	1	14.63		-0.13		-0.80		3		417
		Sk -68 006		4 53 30	-68 20.	1	13.42		0.07		-0.79		4		798
		Sk -69 021		4 53 30	-69 49.	1	13.00		-0.12		-0.95		3		798
31455	-0 00806			4 53 31	-00 23.1	1	9.26		0.10		0.05		1	A2	116
		Sk -67 011		4 53 31	-67 09.3	1	12.08		0.10		-0.26		4	A0 Ia	149
		G 83 - 46		4 53 32	+14 50.7	1	12.46		0.90		0.31		1	K0	333,1620
31452	+2 00816	G 84 - 23		4 53 33	+02 51.6	2	8.43	.014	0.84	.005	0.56	.014	4	G5	196,333,1620
31421	+13 00740	HR 1580	⋆ A	4 53 33	+13 26.2	5	4.06	.008	1.16	.008	1.12	.017	14	K2 III	15,1355,1363,3052,8015
		Hyades J 343		4 53 35	+13 27.	1	12.11		0.89		0.41		1		1058
		Sk -68 007		4 53 35	-68 34.8	1	12.87		-0.04		-0.81		4		149
		Sk -70 008		4 53 36	-70 19.4	1	13.72		-0.02		-0.15		3	B8 Ib	415
		NGC 1841 sq1 1		4 53 38	-84 02.3	1	14.37		0.71		0.09		5		1669
		Sk -66 007		4 53 39	-66 45.0	1	12.95		0.04		-0.22		4	A0 Ib	415
31560	-28 01839			4 53 42	-28 38.3	2	8.13	.010	1.07	.010	0.95		6	K3/4 V	2012,3077
		Sk -66 005a		4 53 42	-66 59.	1	13.24		-0.04		-0.60		4		798
		Sk -66 006		4 53 42	-67 01.	2	12.73	.005	-0.13	.009	-0.93	.009	5		360,798
		Sk -70 009		4 53 42	-70 46.	1	13.61		-0.07		-0.77		4		798
31398	+32 00855	HR 1577		4 53 44	+33 05.3	9	2.68	.020	1.53	.017	1.78	.013	19	K3 II	15,263,1119,1194,1355*
268749		Sk -69 022		4 53 46	-69 29.3	1	12.10		0.04		-0.65		3	G0	415
		Sk -66 008		4 53 48	-67 01.	2	13.00	.005	-0.07	.023	-0.82	.005	5		360,798
		AAS21,109 # 16		4 53 48	-68 53.	1	14.54		0.02		-0.91		1		417
		AAS21,109 # 17		4 53 48	-68 53.	1	14.16		-0.11		-0.97		3		417
		AAS21,109 # 15		4 53 48	-69 42.	1	13.96		-0.12		-1.00		2		952
		GD 64		4 53 50	+41 51.5	2	13.93	.108	0.12	.075	-0.62	.009	4	DA	940,3060
		Sk -65 001		4 53 54	-65 40.	1	12.50		-0.09		-0.97		3		415
268729	-68 00288	Sk -68 008		4 53 54	-68 47.6	3	11.03	.039	0.09	.021	-0.66		8	B4 Ia	149,1277,8033
31387	+41 01015	IDS04504N4154		4 53 55	+41 59.2	1	9.14		0.20				2	A1 IV	1119
	+41 01014			4 53 55	+42 04.0	1	9.95		1.87				2		369
31975	-72 00332	HR 1606		4 53 55	-72 29.5	4	6.28	.006	0.52	.005	0.00	.010	26	F6 V	15,1075,1075,2038
31512	-5 01091	HR 1582	⋆ A	4 53 56	-05 14.9	3	5.50	.005	-0.13	.005	-0.55	.007	9	B6 V	15,1061,1079
		STO 491		4 53 56	-69 09.2	1	12.72		0.49		0.28		1	F0 Iab	726
		AK 7 - 108		4 53 58	+21 55.3	1	11.68		0.81				1		987
31571	-17 00994	IDS04518S1754	AB	4 53 59	-17 48.8	1	7.35		0.78		0.31		3	G1 IV	1657
31489	+8 00811			4 54 00	+08 44.7	1	7.49		0.22		0.11		2	A2	401
		Sk -67 012		4 54 00	-67 18.	1	13.29		-0.22		-1.04		3		952
		AAS21,109 # 19		4 54 00	-68 57.	1	14.55		-0.05		-0.91		3		417
		AAS21,109 # 18		4 54 00	-70 40.	1	13.80		-0.04		-0.90		2		417
		Sk -70 010		4 54 00	-70 40.	1	13.28		-0.15		-1.03		4		798
		BSD 96 # 765		4 54 03	+00 17.1	1	11.05		0.25		0.17		1	A3	116
31746	-58 00437	HR 1597		4 54 04	-58 37.6	2	6.11	.005	0.44	.000			7	F5 V	15,2012
		Hyades L 103		4 54 06	+26 39.3	1	11.57		1.02				2		582
277226	+41 01017			4 54 09	+41 50.8	1	9.13		1.85				3	K2	369
31599	-21 01019	U Lep		4 54 09	-21 17.6	2	10.02	.120	0.12	.040	0.09	.035	2	A8/F2 (I)	668,700
	+13 00741	Hyades L 101		4 54 11	+13 50.1	3	10.93	.009	1.17	.000	1.16		6		582,950,1058
		Sk -66 009		4 54 12	-66 46.	2	12.37	.025	0.06	.015	0.08	.005	7		415,952
		Sk -69 023		4 54 12	-69 31.	1	13.10		0.03		-0.67		3		798
268758				4 54 16	-69 27.7	1	12.52		0.10		-0.41		4	B9 Ia	149
31511	+20 00853	IDS04513N2021	B	4 54 17	+20 29.8	1	11.75		0.88				1		987
277089	+42 01107			4 54 17	+42 41.6	1	10.10		0.22				2	B9	1119
31511	+20 00853	IDS04513N2021	A	4 54 18	+20 25.6	1	9.59		0.55				1	G5	987
		AAS21,109 # 22		4 54 18	-67 04.	1	13.72		-0.18		-1.02		3		417
		AAS21,109 # 20		4 54 18	-69 40.	1	14.63		-0.11		-0.89		2		952
		AAS21,109 # 21		4 54 18	-69 43.	1	14.32		0.04		-0.94		3		952
31567	+3 00716			4 54 22	+03 12.5	1	7.32		0.06		0.08		20	A0	978
		SK *G 67		4 54 23	-69 50.6	1	13.41		0.09		0.04		4	A2 Ib	149
268774		Sk -69 027		4 54 24	-69 42.	1	12.27		0.85		-0.60		2	B7	415
		Sk -70 011		4 54 24	-70 10.	1	13.46		0.04		-0.45		4		798
		Sk -70 012		4 54 24	-70 13.	1	12.89		0.03		-0.39		4		415
287324	+9 00693			4 54 25	+09 58.6	1	10.09		0.66		0.16		2	G0	401
31554	+10 00676			4 54 27	+10 44.0	1	7.89		0.29		0.21		2	A0	401
31475	+35 00936			4 54 27	+35 39.7	1	8.35		1.39		1.68		2	G5	3049
268668				4 54 27	-67 10.4	1	12.52		0.02		-0.06		2	A2 Ib	149

Table 1 237

HD	DM	Other Id	N Rem	α_{1950}	δ_{1950}	S	V	σ_V	B–V	σ_{B-V}	U–B	σ_{U-B}	n	Spectrum	References
31539	+16 00672	HR 1585		4 54 29	+17 04.6	1	5.49		1.31				2	gK1	71
277084	+43 01139			4 54 29	+43 10.4	1	9.50		0.62				3	G1 V	1025
		Sk -70 013		4 54 29	−70 04.	1	12.29		-0.12		-0.91		4		415
277299	+39 01118			4 54 30	+39 21.8	1	9.14		1.92				3	M2	369
		Sk -67 013		4 54 30	−67 12.	1	13.03		-0.16		-0.97		3		798
		Sk -69 026		4 54 30	−69 10.	2	12.71	.154	0.72	.059	-0.78	.045	5		798,952
268757	−69 00303	Sk -69 030		4 54 30	−69 17.3	5	10.18	.093	1.53	.016	1.33	.048	24	G5 Ia	149,1277,1483,3074,8033
		Sk -69 027		4 54 30	−69 42.	1	12.41		-0.08		-0.88		3		952
31594	−0 00807			4 54 31	+00 08.7	2	7.83	.011	1.07	.017	0.85	.008	33	K0	116,978
268685	−67 00361	Sk -67 014		4 54 31	−67 20.1	2	11.53	.009	-0.11	.014	-0.92	.009	5	B1.5Ia	149,360
31520	+27 00712	Hyades vB 148		4 54 33	+27 18.1	2	8.95	.015	0.62	.012	0.08	.001	3	F5	1127,8023
		MCC 456		4 54 35	+51 44.9	1	11.09		1.41		1.38		3	M0	1723
		Sk -66 010		4 54 36	−66 27.	1	13.54		-0.08		-0.54		3		798
		Sk -69 029		4 54 36	−69 20.	2	12.87	.014	-0.07	.014	-0.97	.005	4		360,798
31875	−62 00398			4 54 38	−62 52.5	1	8.04		0.11		0.09		3	A3 V	1097
		Sk -69 028		4 54 38	−69 23.	1	12.78		-0.03		-0.83		3		415
31501	+34 00927	G 39 - 43		4 54 39	+34 11.7	3	8.20	.023	0.75	.011	0.30	.019	6	G8 V	22,333,1003,1620
		Sk -68 009		4 54 42	−68 27.	1	13.36		-0.19		-0.97		3		798
268690	−67 00362	Sk -67 015		4 54 43	−67 19.4	2	11.64	.005	0.01	.009	-0.58	.014	4	B7 Ia	149,360
31623	−1 00762	HR 1591		4 54 45	−01 08.6	2	6.22	.005	0.42	.000	0.05	.010	7	F2	15,1415
31553	+23 00777	HR 1586	★ AB	4 54 46	+23 52.3	1	5.79		1.11				2	K0	71
		AA45,405 S217 # 6	A	4 54 46	+47 53.8	2	14.83	.025	0.54	.020	-0.21	.110	4		797,7006
		AA45,405 S217 # 6	B	4 54 46	+47 53.8	2	14.98	.045	0.63	.015	-0.22	.065	4		797,7006
		AA45,405 S217 # 8		4 54 46	+47 55.1	1	14.33		0.48		0.12		2		797
31622	+0 00905	Hyades vB 149	★ AB	4 54 48	+00 54.9	3	8.49	.017	0.63	.022	0.08	.014	5	G5	116,1127,8023
		AA45,405 S217 # 10		4 54 48	+47 56.5	1	13.38		0.67		0.18		1		797
	+43 01140	IDS04512N4352	AB	4 54 49	+43 57.0	1	10.22		0.27				3		1119
		AA45,405 S217 # 9		4 54 50	+47 55.9	1	11.88		0.46		0.09		2	B9 V	797
		POSS 717 # 6		4 54 50	−13 37.8	1	18.04		1.55				3		1739
		AA45,405 S217 # 7		4 54 51	+47 53.5	1	14.41		0.46		0.15		1		797
		G 191 - 15		4 54 51	+50 52.3	1	10.93		1.46				1		1017
31488	+44 01065			4 54 52	+44 21.4	1	8.97		0.32				2	F0 V	1119
		POSS 717 # 5		4 54 52	−13 45.5	1	17.13		1.20				2		1739
		AA45,405 S217 # 5		4 54 53	+47 52.4	1	13.82		0.80		0.25		1		797
		Sk -66 011		4 54 53	−67 01.	1	12.56		0.11		0.20		3		415
268675	−66 00339	Sk -66 012		4 54 54	−66 49.9	4	10.81	.010	0.04	.008	-0.54	.018	16	A0 Ia	149,1277,1408,8033
		AA45,405 S217 # 4		4 54 56	+47 52.7	1	12.45		0.62		0.01		2		797
		AA45,405 S217 # 12		4 54 57	+47 56.1	1	14.37		0.94		0.28		1		797
		AA45,405 S217 # 13		4 54 57	+47 56.1	1	15.47		0.97		0.31		1		797
31730	−31 02116			4 54 57	−31 20.3	1	9.40		1.31				4	F7/G0	955
		SK *G 75		4 54 57	−67 05.9	1	13.46		0.06		0.10		4	A0 Iab	149
268760		Sk -68 010		4 54 59	−69 02.4	1	12.26		0.11		-0.41		4	B9 Ib	415
		Sk -69 031		4 54 59	−69 30.8	1	12.54		0.17		-0.43		3		798
31609	+13 00749	Hyades vB 127		4 55 00	+13 55.6	2	8.89	.000	0.74	.002	0.35	.002	3	G5	1127,8023
		LS V +47 024		4 55 00	+47 55.4	2	11.33	.010	0.41	.010	-0.56	.049	6	B0 V	342,5006
	+49 01280			4 55 00	+49 46.5	1	9.79		1.42		1.22		2	M2	3072
		AAS21,109 # 24		4 55 00	−66 36.	1	13.09		-0.15		-0.95		2		417
	+44 01066			4 55 02	+44 23.7	1	10.11		0.25				2	A1 V	1119
		Sk -66 013		4 55 03	−66 55.3	2	11.55	.005	0.17	.010	0.13		6	F0 Ia	149,1483
		SK *G 78		4 55 05	−70 26.1	1	12.71		0.22		-0.14		4	B9 Ib	149
31592	+24 00717	HR 1590	★ A	4 55 06	+24 58.5	3	5.81		0.00	.000	-0.03	.013	6	A0 V	985,1049,1079
31550	+37 01002			4 55 06	+37 15.2	1	6.72		0.33		-0.01		4	F4 V	1501
		Sk -67 016		4 55 06	−67 35.	1	13.52		-0.15		-0.95		3		798
		AAS21,109 # 23		4 55 06	−68 26.	1	14.12		-0.14		-0.96		3		417
		Sk -70 014	V	4 55 06	−70 59.	1	12.62		0.63		0.48		1		798
		AA45,405 S217 # 2		4 55 07	+47 54.8	1	13.19		0.67		0.12		1		797
277260	+40 01115			4 55 08	+40 49.0	1	9.35		1.60		1.86		8	K2	1655
31487	+51 01005			4 55 08	+51 52.0	1	8.11		1.34		1.11		1	K0	993
		NGC 1777 sq1 6		4 55 08	−74 12.8	1	11.28		1.46				3		1574
31534	+42 01113			4 55 09	+42 34.6	2	8.25	.010	1.56	.025	1.37		3	K1 II	1119,8100
32034	−67 00364	Sk -67 017		4 55 09	−67 14.8	4	9.69	.024	0.10	.007	-0.63	.020	13	B9 Iae	149,328,1277,8033,8100
31641	+8 00814			4 55 10	+08 39.7	1	8.45		0.19		0.12		2	A1 V	401
268788		Sk -69 032		4 55 10	−69 41.4	1	12.13		-0.01		-0.60		4	B7	149
		Sk -67 018		4 55 11	−67 15.9	2	11.99	.035	-0.21	.005	-1.00	.025	8	O9f	149,980
		Sk -66 015		4 55 18	−66 32.	1	12.81		-0.12		-0.95		4		798
		Sk -66 014		4 55 18	−66 53.	1	13.32		-0.19		-1.02		3		798
		Sk -69 033		4 55 18	−69 42.	1	13.22		-0.09		-0.83		4		798
		Sk -70 016		4 55 18	−70 07.	1	13.14		-0.09		-0.82		4		798
		Sk -70 015		4 55 18	−70 52.	1	13.18		-0.07		-0.62		4		798
31826	−44 01792			4 55 19	−44 16.2	1	7.81		1.15				4	K0 III	2012
		L 131 - 6		4 55 21	−61 14.5	3	12.06	.017	1.49	.017	1.11		6		912,1705,3078
		Sk -66 016		4 55 21	−66 27.	1	12.51		-0.09		-0.94		4		415
282622	+30 00748	LS V +30 002		4 55 22	+30 37.3	1	9.66		0.31		-0.41		2	B1.5V	1012
31564	+41 01022	LS V +41 020		4 55 23	+41 21.7	1	9.05		0.04		-0.70		2	B2 V	401
268719	−67 00367	Sk -67 019		4 55 23	−67 30.8	1	11.19		0.14		-0.10		4	A1 Ia	149
		AAS21,109 # 26		4 55 24	−66 30.	1	13.45		-0.05		-0.87		2		417
		Sk -69 034		4 55 24	−69 55.	1	13.32		0.05		-0.35		3		798
285101	+18 00765	IDS04525N1850	AB	4 55 26	+18 55.1	1	9.38		0.57		0.15		2	F2	1648
31726	−14 01003	HR 1595		4 55 27	−14 18.5	3	6.14	.004	-0.21	.006	-0.87		10	B2 V	15,1732,2028
31591	+42 01119			4 55 29	+42 28.3	1	9.10		0.35				2	A8 V	1119

HD	DM	Other Id	N	Rem	α₁₉₅₀	δ₁₉₅₀	S	V	σ_V	B–V	σ_B–V	U–B	σ_U–B	n	Spectrum	References
		SK *G 87			4 55 29	−67 37.3	1	13.39		0.05		0.16		3	A2 II	149
268787	−69 00306				4 55 29	−69 30.4	1	10.78		1.00		0.66		1	K0	360
		AAS21,109 # 25			4 55 30	−69 41.	1	13.73		−0.07		−0.68		3		952
		NGC 1777 sq1 5			4 55 31	−74 14.2	1	13.51		0.54				5		1574
32109		Sk −67 020			4 55 32	−67 34.6	1	13.86		−0.28		−0.78		4	WN3p	980
31710	+0 00906				4 55 33	+00 47.4	1	8.25		0.20		0.19		1	A5	116
		HV 5497			4 55 33	−66 30.	2	11.72	.000	1.01	.005	0.83	.020	2	G2 I	689,3075
31648	+29 00774				4 55 36	+29 46.1	1	7.65		0.16		0.08		2	A2	379
31739	−2 01080	HR 1596		⋆ A	4 55 40	−02 17.3	2	6.34	.005	0.10	.000	0.10	.005	7	A2 V	15,1415
31724	−0 00811				4 55 41	−00 37.9	1	8.88		0.31		0.06		1	A3	116
		Sk −69 036			4 55 41	−69 15.5	1	12.70		−0.11		−0.92		4		149
31738	+0 00908	V1198 Ori			4 55 43	+00 22.7	2	7.11	.016	0.70	.005	0.20	.001	2	G5	116,1641
30932	+80 00156				4 55 43	+80 42.6	1	9.14		0.08		0.05		6	A0	1219
		G 96 - 13			4 55 45	+48 42.6	2	10.88	.021	1.29	.024	1.24	.036	3		1620,5010
31616	+43 01146				4 55 46	+43 32.6	1	8.78		0.16				3	B7 V	1119
31617	+43 01147	LS V +43 017			4 55 47	+43 14.9	2	7.38	.035	0.00	.000	−0.77		5	B2 IV	1012,1119
31579	+52 00906	HR 1588			4 55 47	+53 04.9	1	6.08		1.46				2	K4 III	71
		Sk −66 018			4 55 48	−66 03.	1	13.50		−0.20		−1.07		3		798
32125		Sk −66 021			4 55 48	−66 21.	1	14.33		0.04		−0.76		4	WC5	980
		Sk −66 019			4 55 48	−66 29.	1	12.79		0.12		−0.78		4		798
32494	−78 00159				4 55 48	−78 13.5	1	8.46		0.45				4	F2 III	2012
		NGC 1777 sq1 2			4 55 49	−74 15.3	1	10.61		0.64				1		1574
31647	+37 01005	HR 1592		⋆ AB	4 55 51	+37 49.0	6	4.95	.022	0.04	.010	0.03	.015	17	A1 V	15,1049,1119,1363*
		Sk −66 017			4 55 51	−66 33.	1	12.87		−0.12		−0.97		4		415
272897	−48 01577				4 55 52	−48 45.8	1	9.46		0.00		−0.77		1	G5	1737
		LP 416 - 173			4 55 53	+18 05.7	1	11.21		0.94				2		1570
		Sk −66 020			4 55 54	−66 30.	1	13.35		−0.14		−1.01		3		798
		AAS21,109 # 29			4 55 54	−69 05.	1	14.77		−0.21		−1.07		3		417
		AAS21,109 # 27			4 55 54	−69 49.	1	13.86		−0.16		−1.02		3		952
		AAS21,109 # 28			4 55 54	−69 52.	1	14.37		−0.14		−0.79		2		952
268819	−70 00329	Sk −69 037			4 55 55	−70 02.4	2	10.08	.000	0.50	.005	0.39		7	F6 Ia	149,1483
31767	+1 00872	HR 1601			4 55 57	+01 38.3	8	4.47	.012	1.40	.005	1.53	.018	27	K2 II	15,1075,1355,1363*
		TOU27,47 T12# 32			4 55 57	+43 58.8	1	9.83		0.27				2		1119
31312	+74 00229	HR 1572			4 55 58	+74 11.7	3	6.05	.010	1.56	.010	1.85	.015	7	K5 III	15,1003,3001
31871	−35 02024				4 55 59	−35 15.9	2	8.95	.018	1.04	.005	0.88		5	K1 III	2012,3040
		Sk −69 039a			4 55 59	−69 31.2	1	12.35		0.10		−0.17		3	A3 Iab	415
277195	+43 01148				4 56 00	+43 40.1	1	9.59		0.24				3	B7 V	1119
		AAS21,109 # 30			4 56 00	−68 20.	1	14.53		0.10		0.23		3		952
		AAS21,109 # 31			4 56 00	−68 43.	1	13.96		−0.19		−0.99		2		417
		AAS21,109 # 32			4 56 00	−69 05.	1	14.52		−0.11		−0.89		3		417
		Sk −69 040			4 56 00	−69 35.	1	12.97		−0.05		−0.76		3		798
		Sk −69 038			4 56 00	−69 37.	1	12.85		−0.12		−0.92		4		798
		Sk −69 039			4 56 02	−69 56.8	1	12.69		0.07		−0.31		4	A0 Iab	415
		Sk −66 023			4 56 06	−66 22.	1	13.09		0.08		−0.64		4		798
		Sk −69 041			4 56 06	−69 20.	1	13.09		−0.12		−0.92		4		798
		AAS21,109 # 33			4 56 06	−69 50.	2	13.99	.000	−0.13	.023	−0.87	.023	4		417,952
31747	+14 00795	IDS04533N1423	B		4 56 07	+14 28.5	3	7.59	.014	0.06	.008	−0.29	.011	7	B6 V	146,150,1084
		NGC 1777 sq1 1			4 56 08	−74 19.0	1	11.24		0.90				2		1574
31766	+8 00820				4 56 09	+08 54.0	1	8.42		0.08		0.01		2	A0	401
31764	+14 00796	HR 1600		⋆ A	4 56 09	+14 28.1	6	6.10	.016	0.05	.023	−0.27	.012	13	B7 V	146,150,1049,1079*
		Sk −66 024			4 56 09	−66 32.	1	13.40		−0.19		−0.89		2		415
31054	+79 00159				4 56 10	+79 50.6	1	8.96		0.54		0.06		3	G5	1528
		SK *G 94			4 56 10	−69 46.9	1	12.69		0.09		0.03		3		149
31664	+41 01023				4 56 11	+41 47.9	3	6.66	.027	0.95	.005	0.71		8	K0 III	1025,1501,6007
		Sk −66 022			4 56 12	−66 36.	1	13.15		−0.17		−1.01		4		798
286295	+14 00797	IDS04533N1423	C		4 56 13	+14 28.1	1	9.80		0.13		0.09		1	A0	8084
	+40 01122	AN Aur			4 56 13	+40 45.7	3	10.20	.032	1.12	.032	0.61		3	F6	934,1399,1772
		Sk −66 026			4 56 15	−66 32.	1	12.91		−0.05		−0.81		4		415
		Sk −66 025			4 56 15	−66 35.	1	12.75		−0.14		−1.01		4		415
268873	−71 00300				4 56 15	−71 23.0	1	12.47		0.49		−0.17		2	K0	1696
31678	+43 01149	IDS04527N4310	AB		4 56 17	+43 14.7	1	8.95		0.24				2	B9 V	1119
31846	−12 01040				4 56 17	−12 18.5	1	8.70		0.47		−0.02		1	F5 V	1776
31798	+7 00768	R Ori			4 56 18	+08 03.4	2	9.50	.871	2.87	.573			11	CS	864,3039
		Hyades J 345			4 56 18	+11 60.	1	12.34		0.67		0.26		3		1058
		AAS21,109 # 35			4 56 18	−66 15.	1	14.22		0.00		−0.88		3		417
		SK *C 5			4 56 18	−66 18.	1	12.81		0.17		0.05		3		149
		NGC 1777 sq1 4			4 56 19	−74 15.8	1	11.68		0.55				1		1574
		G 247 - 44			4 56 21	+61 03.2	1	10.15		1.04		0.90		5	K5	7010
		Sk −66 027			4 56 21	−66 34.	1	11.82		0.02		−0.74		4	B2.5-3Ia	415
277239	+41 01024				4 56 23	+41 26.5	1	9.75		0.11				3	A0	1025
		Sk −67 021			4 56 24	−67 12.	1	12.84		−0.12		−0.76		3	A0 Iab	798
268807		Sk −69 042a			4 56 24	−69 19.	2	11.98	.278	−0.02	.034	−0.61		6	B7	415,8033
	+43 01150			V	4 56 25	+43 56.2	1	9.30		1.95				3	M3	369
		AK 7 - 232			4 56 26	+20 24.7	1	11.62		0.88				1		987
31691	+43 01151				4 56 26	+43 54.7	2	8.02	.020	1.20	.024	0.82		5	G8 III	1119,8100
268809		Sk −69 043			4 56 27	−69 20.1	4	11.93	.018	−0.06	.007	−0.88	.005	15	B1 Ia	149,1277,1408,8033
32228	−66 00343	IDS04564S6638		⋆ AB	4 56 29	−66 33.0	3	10.95	.173	−0.14	.010	−0.87		9	WC5	980,1277,8033
		SK *G 101			4 56 29	−67 24.2	1	13.62		0.06		0.10		4		149
32257		Sk −69 042			4 56 29	−69 31.3	1	14.27		−0.20		−0.64		4	WC5	980
		Sk −66 028a			4 56 30	−66 46.	1	13.70		−0.17		−0.99		3		798

Table 1   239

HD	DM	Other Id	N Rem	α_{1950}	δ_{1950}	S	V	σ_V	B–V	σ_{B-V}	U–B	σ_{U-B}	n	Spectrum	References
		SK *G 102		4 56 30	−68 00.5	1	12.50		0.27		0.26		4	A5 Ib	149
		Sk -69 044		4 56 30	−69 19.	1	13.03		-0.08		-0.89		4		798
		AAS21,109 # 34		4 56 30	−70 25.	1	14.15		0.04		-1.00		2		952
	+0 00909			4 56 31	+00 31.5	1	11.09		0.51		-0.01		1		116
	+46 00948			4 56 31	+46 10.3	1	9.27		1.95				3		369
31782	+25 00766			4 56 32	+25 51.8	4	7.27	.018	0.81	.005	0.42	.009	18	K0 III	20,1003,1355,1775
		SK *G 104		4 56 32	−66 57.2	1	13.13		0.29		0.30		3		149
268747		Sk -67 021a		4 56 32	−67 24.1	1	12.41		0.09		-0.29		4	A5	415
31781	+26 00771			4 56 33	+26 10.6	1	8.45		0.55				4	F8 V	20
		BSD 96 # 1366		4 56 34	+00 33.8	1	10.88		0.74		0.15		1	G4	116
	−27 01987			4 56 34	−27 08.4	1	9.72		0.91		0.67		2	K3 V	3072
268727		Sk -66 029		4 56 34	−66 49.1	1	11.59		0.08		-0.30		4	A0 Ia	149
		Sk -70 018		4 56 36	−70 04.	1	13.15		-0.14		-0.94		3		798
		Sk -70 017		4 56 36	−70 07.	1	13.26		-0.03		-0.59		4		415
31705	+44 01074			4 56 37	+44 48.0	1	8.12		0.37				3	F2 III	1025
32440	−75 00290	HR 1629		4 56 37	−75 00.9	4	5.46	.005	1.52	.005	1.86	.030	25	K4 III	15,978,1075,2038
		G 96 - 14		4 56 38	+39 17.4	1	12.21		1.00		0.83		1		333,1620
		POSS 717 # 1		4 56 38	−13 31.9	1	13.84		0.53				2		1739
	−32 02069			4 56 41	−32 11.2	1	10.37		1.04				5		955
277200	+43 01152	LS V +43 020		4 56 42	+43 23.4	1	9.62		0.12				2	B5	1119
		Sk -65 002		4 56 42	−65 36.	1	12.83		-0.15		-0.92		4		798
		Sk -66 030		4 56 42	−66 24.	1	13.00		-0.04		-0.76		3		798
		AAS21,109 # 36		4 56 42	−69 10.	1	14.94		-0.21		-1.01		2		417
31900	−11 01020			4 56 43	−11 28.3	1	9.36		0.12		0.08		1	A0	1776
		IRC +60 150	⋆ V	4 56 44	+56 06.9	2	14.92	.225	1.88	.385	0.06		2		8010,8022
		Sk -66 031		4 56 45	−66 33.	1	12.29		-0.11		-0.95		4		415
268820	−69 00309		V	4 56 45	−69 28.7	1	10.84		0.26				2	B5	8033
31614	+61 00746			4 56 46	+62 04.9	1	8.93		0.18		0.14		4	A0	1733
	+0 00910			4 56 47	+01 06.1	1	10.14		0.29		0.06		1		116
31806	+27 00716	IDS04537N2710	A	4 56 47	+27 15.1	1	6.96		0.15		-0.33		3	B7 V +A0 V	1625
31925	−16 01013	HR 1604	⋆ AB	4 56 47	−16 27.1	5	5.66	.013	0.43	.029	-0.08	.021	13	F3 V +F9 V	15,254,292,1311,2012
		POSS 717 # 3		4 56 48	−13 34.9	1	16.59		0.30				2		1739
		Sk -66 032		4 56 48	−66 33.	1	12.94		-0.16		-1.02		4		798
		AAS21,109 # 38		4 56 48	−67 41.	1	13.89		-0.11		-0.94		3		417
		SK *G 501		4 56 51	−67 09.7	1	12.68		0.49		0.23		1		726
	+0 00911			4 56 52	+00 57.6	1	9.79		1.20		1.03		1	K0	116
31845	+15 00713	Hyades vB 128	⋆ A	4 56 52	+15 50.6	3	6.75	.005	0.45	.000	-0.01	.006	6	F5 V	1127,1355,8023
285083	+19 00832			4 56 52	+19 53.6	1	9.82		0.78				2	G0	987
31761	+39 01133	HR 1599	⋆ AB	4 56 52	+39 19.3	2	5.96	.010	0.41	.010	-0.04	.010	6	F5 V	254,833
		NGC 1777 sq1 23		4 56 52	−74 20.1	1	13.06		0.71				4		1574
		Sk -70 019		4 56 54	−70 30.	1	13.59		-0.13		-0.85		3		798
286230		Hyades J 346		4 56 55	+16 48.	1	10.66		0.80		0.46		3	G5	1058
31759	+41 01027	IDS04534N4149		4 56 55	+41 53.3	1	8.90		0.45				2	F5 V	1119
		SK *G 112		4 56 55	−67 07.2	1	12.92		0.25		0.31		3	A7 Iab	149
		SK *G 111		4 56 55	−68 04.9	1	13.53		0.04		-0.12		3	A1 Ib	149
31780	+39 01134	HR 1602		4 56 56	+39 34.9	1	6.48		1.71		2.01		3	K4 I	1733
	−74 00309	IDS04582S7422	A	4 56 56	−74 17.8	1	10.53		0.38				4		1574
		MCC 111	⋆ V	4 56 59	+01 42.6	2	10.05	.000	1.39	.016	1.13		2	K6	680,1705
		Sk -66 034	V	4 57 00	−66 34.	2	12.54	.070	-0.07	.040	-0.90	.025	2		360,798
268732		Sk -66 035		4 57 00	−66 39.	2	11.58	.005	-0.08	.005	-0.90	.005	4	BC1 Ia	360,952
		AAS21,109 # 41		4 57 00	−69 17.	1	13.87		-0.18		-0.93		3		952
		Sk -67 021b		4 57 02	−67 51.0	1	13.20		0.03		-0.38		4	A1 Ib	149
31758	+43 01155			4 57 03	+43 29.6	1	9.05		0.07				4		1119
31952	−10 01063			4 57 03	−10 25.7	1	7.63		1.22		1.18		3	K0	1657
268726	−66 00346	Sk -66 036		4 57 03	−66 27.9	1	11.35		0.07		-0.76		4	B2 I+neb	415
		Sk -69 045		4 57 03	−69 23.	1	12.31		-0.05		-0.94		3		415
		Sk -70 020		4 57 04	−70 52.	1	12.87		0.00		-0.45		4		415
		LP 891 - 44		4 57 05	−27 59.4	1	14.24		1.32		1.06		1		3062
31662	+60 00853	HR 1593	⋆ AB	4 57 06	+61 00.5	1	6.03		0.41				2	F4 V	71
		KUV 04571+1620		4 57 07	+16 19.9	1	16.18		0.28		-0.60		1	sdB	1708
		POSS 717 # 2		4 57 08	−13 24.8	1	16.00		0.49				2		1739
268835	−70 00336	Sk -69 046		4 57 09	−69 55.2	3	10.64	.013	0.16	.010	-0.62		9	B7	149,1277,8033
		G 84 - 26		4 57 10	−00 27.0	1	15.12		0.14		-0.62		1		3062
277269	+40 01126			4 57 11	+40 42.2	1	11.11		0.45				1	A3	934
	+36 00974			4 57 12	+37 01.7	1	10.98		2.13				3	M2	369
32227	−58 00442			4 57 12	−58 17.0	1	6.86		1.50				4	K5 III	2012
		Sk -68 011		4 57 12	−68 29.	2	12.27	.000	-0.05	.000	-0.91	.014	4		360,798
		Sk -69 049		4 57 12	−69 20.	1	13.26		-0.16		-0.96		3		798
31867	+24 00722			4 57 13	+25 03.8	1	7.96		0.68				3	G2 V	20
31985	−11 01025			4 57 14	−11 03.6	1	9.84		1.17		1.09		2	K5 V	3072
		NGC 1777 sq1 10		4 57 14	−74 23.7	1	14.69		0.83				6		1574
31918	+10 00685	IDS04545N1014	A	4 57 16	+10 19.0	1	8.00		0.30		0.06		2	A3	401
31881	+20 00860			4 57 16	+20 51.3	1	8.28		0.12		0.08		2	A2	1648
277222	+42 01123	LS V +42 015		4 57 16	+42 21.0	1	9.84		0.31				2	B5	1119
31148	+80 00159			4 57 16	+80 33.4	1	8.62		0.34		0.00		2	F0	1528
		Sk -66 037		4 57 18	−66 29.	1	12.98		-0.09		-0.89		4		798
		AAS21,109 # 45		4 57 18	−66 30.	2	12.76	.019	-0.01	.009	-0.64	.019	4		417,952
268798	−68 00295	Sk -68 012		4 57 18	−68 29.6	1	11.49		0.00		-0.78		4	B7	415
		AAS21,109 # 44		4 57 18	−68 30.	1	13.79		-0.15		-0.97		3		952
		Sk -68 013		4 57 18	−68 51.	1	12.58		-0.10		-0.95		3		798

HD	DM	Other Id	N	Rem	α_{1950}	δ_{1950}	S	V	σ_V	B–V	σ_{B-V}	U–B	σ_{U-B}	n	Spectrum	References
31805	+43 01159				4 57 20	+43 30.4	1	8.66		0.28				3	F0 III:	1025
31996	−15 00915	HR 1607, R Lep			4 57 20	−14 52.8	5	8.08	.343	5.70	.449	1.40		7	C7,4e	109,864,975,3001,8029
		Sk −69 050			4 57 24	−69 24.	1	13.26		−0.13		−1.01		4		798
32402		Sk −68 015			4 57 26	−68 28.2	1	13.30		−0.19		−1.01		4	WC5	980
268804	−68 00296	Sk −68 014			4 57 26	−68 29.2	2	11.21	.000	0.08	.005	−0.81	.014	5	B7	360,415
32008	−10 01066	HR 1608			4 57 28	−10 20.1	3	5.38	.012	0.80	.006	0.34	.009	8	G4 V	58,2035,3016
		NGC 1777 sq1 9			4 57 28	−74 23.1	1	12.33		0.84				4		1574
		MCC 458			4 57 29	+30 50.0	1	10.84		1.24		1.27		5	M0	1723
		SK *G 119			4 57 29	−70 10.8	1	13.83		0.06		0.19		4		149
277273	+40 01132	IDS04540N4020		AB	4 57 30	+40 25.0	1	10.35		0.51				1	A7	934
		Sk −67 022			4 57 30	−67 43.	1	13.46		−0.18		−1.07		3		798
31993	+3 00733	V1192 Ori			4 57 31	+03 12.8	1	7.55		1.28		1.14		1	K2	1641
		SK *G 120			4 57 33	−67 44.2	1	13.27		0.09		0.04		3	A0 Ib	149
31676	+65 00449				4 57 34	+65 29.8	1	7.89		1.31		1.31		3	K2	1733
	+44 01076				4 57 36	+44 16.5	1	9.23		0.52				3	F6 V	1025
32045	−12 01047	HR 1611, S Eri			4 57 36	−12 36.6	7	4.78	.004	0.27	.006	0.15	.026	20	F0 V	15,1008,1075,1425*
		Sk −66 039			4 57 36	−66 20.	1	13.58		0.02		−0.89		4		798
		Sk −66 038			4 57 36	−66 56.	1	13.10		−0.14		−0.97		4		798
277197	+43 01161				4 57 37	+43 30.3	1	9.51		0.06				2	B1 IV	1119
31844	+45 01023				4 57 39	+45 22.4	1	7.93		0.04		−0.28		2	B9	401
268743	−66 00347	Sk −66 041			4 57 39	−66 32.2	1	11.69		−0.12		−0.97		2	B7	204
		Sk −67 023			4 57 39	−67 52.0	2	12.60	.056	0.07	.014	−0.72	.028	4		149,952
270915					4 57 40	−71 51.1	1	11.15		1.35		1.33		1	K5	149
31981	+12 00697				4 57 41	+12 30.1	1	8.49		0.08		−0.16		2	A0	401
		Sk −66 040			4 57 42	−66 37.	1	12.89		0.03		−0.82		3		798
31966	+14 00804				4 57 43	+14 18.6	2	6.87	.132	0.66	.019	0.24		6	G5 V	1619,7008
32253	−52 00619				4 57 45	−52 19.5	1	9.58		0.43				1	F3 V	174
32023	+0 00916				4 57 46	+00 56.0	5	9.11	.002	0.56	.004	0.05	.011	22	F8 V	158,830,1003,1783,3077
31675	+66 00370	HR 1594			4 57 46	+66 45.3	2	6.19	.005	0.49	.003	−0.05		4	F8	71,3026
	+43 01163				4 57 47	+43 49.8	1	8.74		1.45				2	K0 II	1119
		Sk −66 042			4 57 48	−66 24.	1	13.34		0.02		−0.85		3		798
		Sk −68 016			4 57 48	−68 29.	1	12.96		−0.15		−1.01		4		798
		Sk −69 048			4 57 48	−69 34.	1	12.94		−0.11		−0.97		3		798
268884		Sk −70 021			4 57 48	−70 52.	1	13.03		−0.07		−0.74		2	OB	798
32022	+4 00808	IDS04552N0457		AB	4 57 50	+05 01.6	1	8.02		0.57				2	G5	176
31895	+41 01033				4 57 51	+41 56.0	2	7.88	.028	1.37	.024	1.27		6	K3 Ib	1119,8100
31866	+44 01077				4 57 51	+44 19.6	1	7.31		0.22				2	A3 V	1119
		SK *G 123			4 57 51	−67 00.1	1	12.90		0.05		0.02		3		149
32039	+3 00736	HR 1609		⋆B	4 57 54	+03 32.5	1	7.02		−0.05		−0.24		2	B9 Vn	1733
32005	+12 00699				4 57 54	+12 31.9	2	8.28	.005	0.12	.005	0.06	.015	4	A0	401,1733
		AAS21,109 # 47			4 57 54	−70 04.	1	13.64		−0.05		−0.77		2		417
280276	+38 01001		V		4 57 55	+38 45.7	1	9.34		1.93				2	M0	369
31913	+39 01138	RX Aur			4 57 55	+39 53.3	2	7.34	.010	0.78	.002	0.56		2	G0 I	934,6007
32040	+3 00737	HR 1610		⋆A	4 57 56	+03 32.6	1	6.63		−0.08		−0.33		2	B9 Vn	1733
32040	+3 00737	HR 1610		⋆AB	4 57 56	+03 32.6	2	6.06	.005	−0.07	.000	−0.27	.005	7	B9 Vn	15,1415
		Steph 536			4 57 56	+07 07.0	1	11.38		1.37		1.18		1	K7	1746
32021	+10 00688				4 57 57	+10 50.5	1	6.81		0.04		0.01		2	B9	401
31563	+73 00264	HR 1587			4 57 58	+73 41.6	1	6.22		1.65		1.96		5	K0	1733
		SK *G 124			4 57 58	−68 29.2	1	13.06		0.15		0.14		3	A0 Ib	149
		G 84 - 27			4 57 59	+06 26.8	1	15.37		1.50				1		906
31965	+29 00784				4 58 00	+29 15.7	1	8.65		0.09		−0.05		2	A0	1625
31894	+43 01164	LS V +43 022			4 58 00	+43 21.8	2	8.41	.005	0.04	.005	−0.66		5	B2 IV-V	401,1119
		Sk −69 053			4 58 04	−69 35.2	1	12.57		0.09		−0.55		4		415
		G 84 - 28			4 58 05	−00 13.9	1	10.50		0.86		0.56		2		1696
32115	−2 01095	HR 1613			4 58 08	−02 08.3	2	6.31	.005	0.29	.005	0.02	.015	7	A8 V	15,1061
32427	−66 00349				4 58 10	−66 20.1	1	9.28		0.92				4	G6 IV	2033
		G 85 - 36			4 58 11	+24 48.4	1	11.84		0.56		0.25		1		333,1620
282747	+29 00788	Hyades J 347			4 58 11	+29 52.	1	10.18		0.91		0.63		3	K0	1058
		Sk −69 051			4 58 12	−69 45.	1	12.54		−0.13		−0.92		3		952
268867	−70 00338	Sk −69 052			4 58 12	−69 57.0	1	11.50		−0.03		−0.82		3	B2.5la	149
31949	+42 01128				4 58 13	+42 16.4	1	8.14		0.48				3	F8 V	1025
		LP 359 - 170			4 58 16	+25 27.2	1	10.14		1.91				2		920
31977	+38 01004				4 58 17	+38 43.0	1	7.56		1.12		1.01		3	K1 III	833
		Sk −67 024			4 58 18	−67 48.	1	12.97		−0.06		−0.68		4		798
		AAS21,109 # 48			4 58 18	−70 00.	1	13.60		−0.10		−0.88		2		417
31590	+73 00265	HR 1589		⋆AB	4 58 20	+73 59.8	1	6.12		0.00		−0.03		2	A1 V	1733
32147	−5 01123	HR 1614			4 58 20	−05 48.6	12	6.23	.015	1.05	.009	1.03	.017	61	K3 V	15,22,770,1013,1075,1197*
		LS V +37 004			4 58 21	+37 23.5	1	11.45		0.51		−0.21		2		405
277324	+42 01130				4 58 21	+43 05.2	1	10.24		0.17				2	A0	1119
31964	+43 01166	HR 1605, ε Aur		⋆AB	4 58 23	+43 45.1	8	2.98	.015	0.54	.004	0.32	.015	28	F0 Iae	15,542,1119,1119,1363*
31964	+43 01166	IDS04548N4341		⋆C	4 58 23	+43 45.1	1	11.26		1.83		1.31		1		542
		SK *G 131			4 58 23	−67 04.1	1	13.30		0.24		0.30		4	A5 Ib	149
		Sk −67 025			4 58 26	−67 33.2	1	12.47		−0.07		−0.76		3	A1 Iab	952
32145	+3 00739				4 58 28	+03 38.7	1	7.25		−0.14		−0.52		28	B8	978
		Sk −65 003			4 58 30	−65 08.	1	13.04		−0.19		−1.01		4		798
270918		Sk −65 004			4 58 30	−65 54.	1	12.25		−0.11		−0.95		3	B7	415
32036	+31 00845				4 58 32	+31 42.4	1	7.70		0.15		−0.10		2	B9	401
	+44 01078	G 96 - 16			4 58 33	+44 58.3	1	9.86		0.86		0.58		1		333,1620
32070	+24 00726	G 85 - 37			4 58 34	+24 34.1	1	8.07		0.69		0.17		1	G5	333,1620
277335	+42 01133	G 96 - 17			4 58 35	+42 26.3	2	10.19	.005	0.85	.005	0.46	.020	2	K0	1620,1658

Table 1 241

HD	DM	Other Id	N Rem	α₁₉₅₀	δ₁₉₅₀	S	V	σ_V	B-V	σ_B-V	U-B	σ_U-B	n	Spectrum	References
	+43 01167			4 58 35	+43 56.1	1	8.78		1.23				2	K0 IV	1119
31991	+44 01079			4 58 35	+44 57.3	1	8.26		0.33				3	F2 V	1025
		Sk -65 005		4 58 36	−65 43.	2	12.79	.009	-0.14	.005	-0.86	.000	5		360,798
		Sk -66 044		4 58 36	−66 16.	1	13.15		-0.15		-1.02		3		798
		Sk -67 026		4 58 36	−67 25.	1	13.07		-0.21		-1.02		3		798
		Sk -69 054		4 58 36	−69 47.	1	13.38		-0.09		-0.81		3		798
		Sk -69 055		4 58 36	−69 55.	1	13.47		-0.02		-0.92		3		798
		Sk -70 022		4 58 36	−70 41.	1	13.91		-0.16		-0.89		4		798
	+43 01168	LS V +43 024	⋆E	4 58 37	+43 47.4	2	9.46	.040	0.94	.013	0.14	.036	5	B9 Iab	1012,8100
		Sk -67 027		4 58 37	−67 29.2	1	13.01		0.04		-0.02		3	B9 Ia+	415
	+3 00740	G 84 - 29		4 58 38	+04 02.4	3	9.81	.004	0.36	.000	-0.20	.000	5	sdF0	516,1620,3077
32092	+26 00775	IDS04556N2631	AB	4 58 38	+26 35.9	2	6.78	.035	0.48	.000	-0.03		7	F5	20,3024
282707	+31 00846	Hyades L 105		4 58 38	+31 33.7	2	9.22	.005	0.61	.000	0.05		4	dG0	582,3016
		Sk -70 023		4 58 38	−70 29.	1	13.43		-0.07		-0.72		4		415
32192	−9 01050			4 58 39	−09 17.8	1	8.93		0.02		0.03		2	B9	1776
32093	+26 00776	IDS04556N2631	C	4 58 40	+26 34.7	1	8.22		0.70		0.17		4	G2 V	3024
		Sk -70 024		4 58 41	−70 32.8	1	13.39		0.04		-0.06		4	B9 Ib	415
		Sk -65 006		4 58 42	−65 44.	1	13.42		-0.22		-0.92		3		798
		Sk -67 028		4 58 42	−67 15.	1	12.28		-0.14		-0.97		3		952
		Sk -69 056		4 58 42	−69 55.	1	13.06		-0.06		-0.83		4		798
277351	+41 01039			4 58 43	+42 04.4	1	9.67		1.81				3	K7	369
	+44 01080			4 58 45	+44 11.9	1	9.12		0.85				2	B5 V	1119
		WLS 500 0 # 10		4 58 45	−00 08.1	1	11.14		0.85		0.49		2		1375
		Hyades L 104		4 58 46	+13 51.7	3	11.39	.005	1.37	.012	1.25		5		582,950,1058
		Sk -68 017		4 58 47	−68 50.2	1	13.00		-0.07		-0.85		3		798
	+44 01081			4 58 48	+44 35.2	1	9.46		0.22				2	B8 V	1119
32546	−67 00376			4 58 48	−67 10.5	1	9.37		1.06		0.70		3	G2 III	149
		Sk -67 029		4 58 48	−67 31.	1	13.64		-0.14		-0.99		3		798
		Sk -70 025		4 58 48	−70 17.	1	13.00		-0.13		-0.95		3		952
		SK *G 140		4 58 50	−67 07.6	1	13.23		0.04		0.13		4	A1 II	149
268845		Sk -68 018a		4 58 50	−68 54.9	1	12.40		0.10		-0.38		4	A1 Iab	415
32017	+43 01170			4 58 51	+43 41.5	1	8.53		0.15				2	A0	1119
32533	−66 00350			4 58 51	−66 16.6	1	9.81		0.54				4	F8/G0 V	2033
31911	+60 00855	IDS04545N6018	B	4 58 52	+60 21.1	2	7.42	.010	0.29	.010	0.15		5	A5	245,1118
		LP 891 - 48		4 58 52	−30 26.2	1	11.93		0.53		-0.12		2		1696
285000	+22 00803	Hyades L 106		4 58 53	+22 22.2	1	10.22		0.78				2	G5	582
277317	+43 01171			4 58 53	+43 20.2	1	9.51		0.19				3	A0	1119
268902				4 58 53	−70 28.	1	11.77		1.01		0.70		2	K5	3074
	−74 00311			4 58 53	−74 27.1	1	9.38		1.14				4	K0	1574
277352	+41 01042			4 58 54	+42 02.1	1	10.00		0.12				3	B7	1119
		LP 776 - 38		4 58 54	−20 57.6	1	13.50		0.88		0.50		2		1696
		AAS21,109 # 49		4 58 54	−69 57.	1	14.09		-0.01		-0.82		4		417
		Sk -70 026		4 58 54	−70 19.9	1	12.75		-0.01		-0.42		4	B8 Iab	415
32159	+19 00839			4 58 56	+19 53.7	1	9.18		0.36		0.06		2	F2	1648
31910	+60 00856	HR 1603	⋆A	4 58 58	+60 22.3	6	4.03	.011	0.92	.019	0.62	.009	14	G0 Ib	15,1119,1355,1363*
270920	−65 00398	Sk -65 008		4 58 58	−65 44.4	4	9.90	.032	0.92	.034	0.72	.022	16	G2 Ia	204,360,415,1483
		G 86 - 22		4 58 59	+39 06.4	2	11.40	.020	1.33	.005	1.20	.000	2	K7	1620,1658
32068	+40 01142	HR 1612, ζ Aur		4 58 59	+41 00.3	6	3.76	.008	1.22	.012	0.38	.006	35	K4 II +B8 V	15,617,660,1363,8003,8015
		Sk -65 007		4 59 00	−65 40.	1	13.30		-0.22		-1.02		3		798
		Sk -65 007a		4 59 00	−65 48.	1	13.36		-0.23		-0.98		4		798
		AAS21,109 # 50		4 59 00	−65 57.	1	13.71		-0.12		-0.88		2		417
		Sk -66 046		4 59 00	−66 20.	1	13.73		-0.10		-0.95		3		798
		Sk -66 045		4 59 00	−66 29.	1	13.62		-0.22		-1.00		3		798
32202	+11 00702	IDS04562N1114	A	4 59 01	+11 18.2	1	7.19		0.03		-0.20		2	B8	401
32249	−7 00948	HR 1617		4 59 01	−07 14.8	7	4.80	.006	-0.20	.007	-0.73	.014	19	B3 V	15,1075,1079,1425,1732*
	−85 00060	IDS05095S8520	AB	4 59 01	−85 15.5	1	9.48		0.52		-0.19		1		794
282719	+31 00849			4 59 02	+31 11.6	1	8.91		0.44		0.14		12	F2	1733
277346	+42 01139	LS V +42 016		4 59 05	+42 14.1	1	9.89		0.34				4	B9	1119
		WLS 500 0 # 5		4 59 06	+02 07.5	1	11.62		1.02		0.67		2		1375
		AJ79,864 # 6		4 59 06	+40 41.	1	14.24		0.38		-0.30		1	B3	373
32091	+42 01141	LS V +42 017		4 59 07	+42 26.0	2	9.11	.005	-0.02	.005	-0.67		6	B3 III	401,1119
32090	+42 01140			4 59 07	+42 59.4	1	9.31		0.20				2	A0 V	1119
		Sk -65 010		4 59 07	−65 45.	1	12.79		-0.12		-0.90		16		415
		Sk -65 009		4 59 07	−65 53.	1	12.72		-0.22		-1.07		2		415
		WLS 500 0 # 7		4 59 09	−02 13.3	1	11.02		0.58		0.00		2		1375
		Sk -70 027		4 59 09	−70 36.0	1	13.06		0.00		-0.37		3	B9 Iab	415
		SK *G 144		4 59 10	−69 30.0	1	12.77		0.30		0.30		3	A4 Ib	149
		Steph 539		4 59 12	+03 41.8	1	11.28		1.52		1.20		2	M1	1746
		Sk -68 020		4 59 12	−68 53.	1	13.00		-0.13		-0.92		4		798
		Steph 538		4 59 13	+09 54.9	1	11.48		1.52		0.98		2	M4	1746
32190	+23 00804	NGC 1750 - 228		4 59 13	+23 57.5	1	8.36		0.11		-0.75		6	B5	399
	+44 01082			4 59 13	+44 46.5	1	9.56		0.38				4	A8 V	1119
		LP 476 - 207		4 59 14	+09 54.7	1	11.45		1.54				2		1532
	−65 00399	Sk -65 011		4 59 14	−65 46.4	1	11.66		-0.04		-0.69		3	B8 Ia	415
32309	−20 00990	HR 1621		4 59 15	−20 07.4	7	4.90	.010	-0.05	.006	-0.13	.016	22	B9.5 Vn	3,15,1075,1079,2012*
32263	+0 00923	HR 1618		4 59 16	+00 39.1	4	5.91	.008	1.27	.004	1.39	.004	13	K0	15,252,1355,1415
	+52 00911	G 191 - 19		4 59 17	+53 04.8	1	9.93		1.41		1.16		1	M0	3078
		AK 7 - 399		4 59 18	+17 28.7	1	11.64		1.02				1		987
31509	+77 00179			4 59 18	+77 56.2	1	8.47		0.95		0.65		4	F5	1733
		Sk -65 012		4 59 18	−65 41.	1	13.70		-0.18		-0.94		3		798

HD	DM	Other Id	N Rem	α_{1950}	δ_{1950}	S	V	σ_V	B–V	σ_{B-V}	U–B	σ_{U-B}	n	Spectrum	References
		AAS21,109 # 52		4 59 18	−65 57.	2	13.61	.000	-0.25	.028	-0.98	.009	4		417,952
		Sk -67 030		4 59 18	−67 43.	1	13.09		-0.07		-0.79		3		798
		AAS21,109 # 51		4 59 18	−69 48.	1	13.73		-0.17		-0.92		3		952
32237	+13 00778	G 83 - 50		4 59 20	+14 01.0	2	8.23	.010	0.72	.000	0.24	.000	3	G5	333,1620,3026
		Sk -69 057		4 59 22	−69 42.	1	13.17		-0.09		-0.83		4		415
		VES 863		4 59 23	+40 25.3	1	13.87		0.32		0.17		1		1543
32273	+1 00886	HR 1619	⋆ AB	4 59 24	+01 32.3	3	6.23	.005	-0.04	.011	-0.40	.029	8	B8 V +A1 IV	15,1079,1256
32143	+42 01142	LS V +42 019		4 59 24	+42 29.9	2	7.45	.020	-0.07	.005	-0.83		6	B1 IV	401,1119
		Sk -65 014		4 59 24	−65 39.	1	12.63		-0.18		-0.96		4		798
		Sk -66 047		4 59 24	−66 18.	1	13.10		-0.13		-0.97		4		798
		G 83 - 51		4 59 25	+15 28.6	2	13.51	.005	0.84	.010	0.27	.044	3		333,1620,1658
		Sk -65 013		4 59 25	−65 50.	1	12.73		-0.12		-0.91		3		415
	+44 01083			4 59 27	+44 12.9	1	9.93		0.13				2	B8 V	1119
		AJ79,864 # 7		4 59 30	+40 44.	1	13.71		0.57		-0.11		2	B3	373
		Sk -67 031		4 59 30	−67 47.	1	13.09		-0.10		-0.74		4		798
		Sk -70 028		4 59 30	−70 50.	1	13.88		-0.06		-0.71		4		798
		Sk -65 015		4 59 31	−65 54.	2	12.16	.029	-0.11	.015	-0.91	.015	3	BC2 Iab	360,415
268892	−69 00317	XX Dor		4 59 32	−69 40.2	1	11.45		0.36		0.16		5	A5 V:	149
		AAS21,109 # 55		4 59 36	−65 59.	1	13.98		-0.14		-0.96		3		952
		AAS21,109 # 53		4 59 36	−69 47.	2	14.01	.173	-0.11	.019	-0.91	.033	4		417,952
		Sk -65 016		4 59 37	−65 49.	1	11.96		-0.04		-0.88		3	B1.5Ia	415
277337	+42 01144	IDS04561N4215	A	4 59 39	+42 19.1	1	9.01		0.32				3	F2 IV	1025
		Hyades J 349		4 59 42	+21 15.	1	12.37		1.12		1.21		3		1058
		L 232 - 21		4 59 42	−52 26.	1	11.45		0.46		-0.06		2		1696
		Sk -68 021		4 59 42	−68 40.	1	12.65		-0.12		-0.91		3		798
		AAS21,109 # 54		4 59 42	−69 59.	1	14.17		-0.18		-0.98		3		952
32425	−39 01743			4 59 45	−38 59.5	1	8.18		1.40		1.57		10	K3 III	1673
32188	+41 01044	HR 1615	⋆	4 59 48	+41 22.3	2	6.09	.026	0.20	.021	0.22		6	A2 IIIShell	1025,1501
277338	+42 01146			4 59 48	+42 21.2	1	10.26		0.03		-0.44		1	B8	401
32763	−70 00343	Sk -70 029		4 59 49	−70 16.0	4	11.54	.096	0.23	.019	-0.84	.033	8	OB(e)	149,952,1277,8033
32918	−75 00292	YY Men		4 59 50	−75 21.0	2	8.17	.011	1.07	.010	0.71		2	K1 IIIp	1641,1642
277339	+42 01147			4 59 51	+42 21.1	1	10.27		0.02		-0.53		2	B8	401
32198	+43 01174			4 59 51	+43 15.7	1	9.26		0.21				2	B8 V	1119
268836				4 59 51	−67 53.5	1	12.56		0.12		0.22		2	F7	149
268935	−70 00345			4 59 53	−70 47.5	1	11.01		0.36		0.18		1	A7 III	149
		AJ79,864 # 8		4 59 54	+40 26.	1	13.78		0.68		-0.02		1	B3	373
		Sk -65 017a		4 59 54	−65 53.	1	13.37		-0.24		-1.03		4		798
268865				4 59 54	−68 36.0	1	11.49		0.77		0.46		1	K0	149
32453	−39 01744	HR 1631		4 59 55	−39 47.4	3	6.03	.008	0.89	.010			11	G8 III	15,1075,2035
		Sk -70 030		4 59 55	−70 13.	1	12.08		0.03		-0.65		3		415
286340	+15 00726	GP Ori		4 59 56	+15 14.7	2	9.20	.109	2.94	.117	3.50		10	SC	864,3039
270933	−65 00404	Sk -65 018		4 59 57	−65 50.2	3	10.98	.011	0.01	.016	-0.59		10	B9 Ia	415,1277,8033
32234	+39 01152	LS V +40 026		4 59 59	+40 00.4	1	8.11		0.02		-0.60		2	B8	401
268847				5 00 00	−68 01.	1	14.61		-0.33		-0.88		4	WN3	980
		GD 65		5 00 02	+24 59.8	1	15.50		1.20		0.58		2		3060
		FK X Ray-3	V	5 00 02	+25 18.6	1	13.02		1.52		1.13		1		1529
32762	−68 00300			5 00 03	−68 39.3	4	8.02	.021	0.25	.008	0.16	.011	18	A5 III	149,204,1408,2033
		Sk -69 058		5 00 05	−69 55.	1	12.99		-0.08		-0.82		4		415
32301	+21 00751	Hyades vB 129	⋆	5 00 06	+21 31.2	5	4.64	.005	0.16	.007	0.14	.005	15	A7 IV	15,1127,1363,3023,8015
		AAS21,109 # 56		5 00 06	−66 00.	2	13.79	.056	-0.18	.014	-1.02	.009	4		417,952
		Hyades L 108		5 00 08	+19 56.8	1	11.87		0.89		0.62		3		1058
		LP 776 - 61		5 00 08	−19 35.8	1	11.70		1.10		0.80		2		3073
32436	−26 01975	HR 1628		5 00 08	−26 20.7	4	5.01	.013	1.07	.005	0.97	.011	16	K1 III	15,1075,2012,3016
		LP 776 - 62		5 00 09	−19 35.8	1	14.94		1.50		1.04		2		3073
		MN83,95 # 101		5 00 09	−68 41.4	1	12.36		0.65				1		591
		Steph 541		5 00 10	+18 57.0	1	11.86		1.41		1.32		1	K7	1746
		Sk -67 033		5 00 12	−67 12.	1	13.09		-0.09		-0.98		4		798
32348	+11 00704			5 00 14	+11 59.3	1	7.25		0.35		-0.06		2	F2	401
		G 84 - 30		5 00 16	+05 07.3	1	10.63		0.90		0.64		1	K3	333,1620
32393	−4 01019	HR 1625		5 00 16	−04 16.8	2	5.84	.005	1.21	.000	1.36	.000	7	K3	15,1415
		SK *G 152		5 00 17	−70 17.7	1	12.64		0.07		0.07		1		149
32366	+9 00713			5 00 18	+09 18.7	1	7.59		-0.02		-0.51		2	A0	401
32347	+13 00783	Hyades vB 187		5 00 18	+13 39.6	4	9.00	.013	0.76	.004	0.37	.007	7	K0	582,950,1058,1127
32233	+47 01089			5 00 18	+47 35.7	1	7.12		1.31		1.08		5	K0	1733
		Sk -70 030a		5 00 19	−70 12.	1	13.46		-0.07		-0.94		3		415
32450	−21 01051	IDS04582S2124	AB	5 00 20	−21 19.4	6	8.31	.019	1.44	.018	1.17	.016	26	M0 V	1197,1509,1619,2012*
32744	−66 00353			5 00 22	−66 05.1	1	9.72		1.07				4	G8/K1 III	2033
		HV 883		5 00 23	−68 31.	2	11.66	.080	0.93	.150	0.58	.125	2	G2:I	689,3075
		G 83 - 52		5 00 24	+07 23.1	1	14.43		1.08		0.76		1		1658
		VES 867		5 00 24	+44 14.0	1	13.06		0.94		-0.15		1		1543
32282	+40 01154			5 00 25	+40 36.8	1	8.04		0.01		-0.38		2	B8	401
32515	−31 02163	HR 1635		5 00 29	−31 50.6	3	5.93	.004	1.17	.000			11	K2 III	15,1075,2007
		SK *C 11		5 00 30	−68 06.	1	13.62		0.04		-0.50		1		149
32391	+9 00718			5 00 31	+09 57.6	1	9.04		0.07		-0.17		2	A2	401
		WLS 600 80 # 6		5 00 31	+80 27.5	1	11.97		0.64		0.08		2		1375
		Sk -70 031		5 00 34	−70 47.5	1	12.61		-0.21		-0.51		3	B8 Iab	415
268822	−66 00356	Sk -66 048		5 00 35	−66 32.7	2	10.73	.060	0.57	.010	0.40		9	F6 Ia	149,1483
277327	+42 01149			5 00 36	+42 49.7	1	9.64		1.76				3	K5	369
		Sk -67 033a		5 00 36	−67 36.8	1	13.26		0.02		0.03		3	A0 Ib	952
		SK *G 156		5 00 36	−68 00.6	1	13.36		0.13		0.11		1	A1 Ib	149

Table 1

HD	DM	Other Id	N Rem	α₁₉₅₀	δ₁₉₅₀	S	V	σ_V	B–V	σ_B–V	U–B	σ_U–B	n	Spectrum	References
		Sk -70 031a		5 00 36	−70 08.	1	13.39		0.02		−0.49		4		798
		Sk -70 032		5 00 36	−70 15.5	1	13.10		−0.21		−1.01		3		149
32503		HR 1634		5 00 38	−22 51.9	4	5.76	.008	1.19	.009	1.14		13	K1 IV	15,1075,2032,3005
32316	+38 01012			5 00 39	+38 48.0	1	8.13		0.25		−0.14		3	B8 V	833
		SK *G 158		5 00 41	−67 50.1	1	13.20		0.13		0.20		1	A5 II	149
277342	+42 01151			5 00 43	+42 15.1	1	9.34		0.09				3	B8 V	1119
32420	+8 00840			5 00 45	+08 57.5	1	9.57		0.07		0.01		2	A0	401
32858	−69 00320			5 00 46	−69 25.3	4	8.71	.012	0.05	.005	0.05	.017	16	A0 V	149,204,1075,1408
		LP 416 - 251		5 00 47	+18 15.2	1	10.73		0.96				1		987
32712	−58 00452			5 00 47	−58 35.5	2	8.51	.015	1.18	.017	0.94	.024	3	K0p Ba	565,3048
		G 84 - 33		5 00 48	+05 01.8	2	11.70	.010	1.45	.024	1.24		3	M1	1620,1746
32387	+24 00739			5 00 48	+24 54.3	1	7.43		0.81		0.33		2	G8 V	1625
32330	+41 01046	LS V +41 024		5 00 48	+41 32.0	2	8.87	.024	0.04	.000	−0.55		4	B2 IV	401,1119
		Sk -67 034		5 00 48	−67 11.	1	13.42		−0.15		−0.84		3		952
		G 86 - 23		5 00 50	+28 48.6	1	9.81		1.06		0.98		1		333,1620
61535	−22 01962			5 00 50	−22 11.9	1	10.24		0.14		0.11		1	A3/5 II	137
		ApJ144,259 # 7		5 00 54	−15 40.	1	15.45		−0.33		−1.35		4		1360
		Sk -68 023		5 00 54	−68 11.	1	12.81		0.22		−0.61		4		798
270948	−66 00359	Sk -65 019		5 00 56	−66 00.4	2	12.05	.000	−0.09	.019	−0.96	.000	4	O8	360,415
32328	+43 01177			5 00 57	+43 39.5	2	7.58	.014	−0.05	.019	−0.35		4	B8 V	401,1025
		Sk -68 024		5 00 58	−68 41.8	2	12.61	.014	0.07	.009	−0.25	.023	4	A1 Iab	149,952
		Sk -68 023a		5 01 00	−68 10.	1	13.09		−0.09		−0.91		3		798
		Sk -70 033		5 01 00	−70 29.	1	13.73		−0.15		−0.93		3		798
	+42 01153	IDS04575N4224	AB	5 01 01	+42 27.9	1	9.88		0.37				3		1119
		AK 7 - 523		5 01 02	+17 48.2	1	10.91		0.78				1		1570
		Sk -70 034		5 01 02	−70 34.1	1	12.61		0.00		−0.48		3	B8 Iab	415
32406	+30 00772	HR 1626		5 01 03	+30 25.6	1	6.14		1.21				2	K0 II-III	71
270949	−65 00410	Sk -65 020		5 01 05	−65 53.0	4	11.15	.015	−0.07	.029	−0.79	.024	13	B3 Ia	415,1277,1408,8033
268910				5 01 05	−68 56.0	1	11.66		0.50		−0.03		1	F5	149
32358	+44 01086	LS V +44 028	★ AB	5 01 06	+44 53.4	1	9.11		0.37				2	B6 V	1119
		Sk -67 035		5 01 06	−67 10.	1	13.08		−0.16		−0.91		4		798
32377	+39 01157			5 01 10	+39 54.1	2	8.33	.044	0.13	.019	−0.39	.034	3	B9	401,934
277326	+42 01154			5 01 10	+43 03.0	1	9.62		0.10				2	B8 III	1119
		SK *G 163		5 01 10	−69 42.8	1	13.13		0.07		0.13		1	A2 Ib	149
32462	+16 00688	Hyades vB 150		5 01 11	+17 03.4	2	8.78	.005	1.04	.002	0.62	.000	3	K0	1127,8023
	+40 01160		V	5 01 12	+40 18.6	1	9.71		1.89				3	M2	369
270952	−66 00361	Sk -65 022		5 01 12	−65 57.	1	12.07		−0.19		−1.03		4	O6f	798
268874				5 01 12	−67 49.8	1	11.76		0.61		0.06		1	K0	149
		Sk -68 025		5 01 12	−68 21.	1	13.05		−0.06		−0.77		4		798
		Sk -70 035		5 01 12	−70 24.	1	13.45		−0.11		−0.83		3		798
		Sk -65 021		5 01 13	−65 46.	1	12.02		−0.16		−1.06		3		415
32386	+39 01159			5 01 15	+39 44.9	1	8.13		1.15		1.02		1	K2	934
	−23 02363			5 01 15	−23 19.3	4	9.27	.016	1.28	.012	1.22	.007	11	K5 V	158,1705,2034,3072
		AAS21,109 # 57		5 01 18	−68 16.	1	13.97		−0.20		−1.00		3		952
277340	+42 01156			5 01 19	+42 22.2	1	10.32		0.07				2	B1	1119
32778	−56 00767	IDS05004S5614	A	5 01 20	−56 09.7	7	7.02	.025	0.63	.004	0.01	.025	25	G0 V	158,1075,1311,1518*
32778	−56 00768	IDS05004S5614	B	5 01 20	−56 09.7	4	10.53	.034	1.38	.013	1.14	.010	15		158,1518,2034,3062
32428	+32 00879	HR 1627		5 01 22	+32 15.2	2	6.61	.012	0.27	.000	0.13	.004	8	A4 m	351,8071
		KUV 05014+1225		5 01 23	+12 24.7	1	16.85		0.67		0.13		1		1708
32482	+21 00755	HR 1633		5 01 23	+21 12.6	2	6.18	.005	1.32	.006	1.47		4	K0	71,3051
32447	+30 00775		V	5 01 23	+30 18.7	2	8.25	.033	1.87	.009			8	M2	20,369
		Sk -67 036		5 01 25	−67 24.	1	12.01		−0.08		−0.81		4		415
33032	−72 00341			5 01 25	−72 31.6	2	9.02	.005	0.49	.000	0.03		9	F5 IV/V	1075,1408
32418	+41 01050			5 01 27	+41 48.7	1	7.23		0.16				3	A4 V	1119
277320	+43 01179	G 96 - 19		5 01 28	+43 14.8	1	9.54		0.61		0.03		1	G0	333,1620
		STO 834		5 01 29	−71 11.8	1	12.54		0.62		0.05		1	F8 Iab	726
32480	+27 00723	HR 1632		5 01 30	+27 37.7	1			0.24		0.12		2	F0 III	1049
	+52 00913	LS V +52 021		5 01 30	+52 45.8	1	10.03		1.05		0.91		2	O VI	1726
32743	−49 01541	HR 1649	★ A	5 01 30	−49 13.3	4	5.37	.010	0.42	.003	−0.01	.000	12	F2 V	15,243,258,2012
		G 83 - 53		5 01 35	+15 45.5	2	14.79	.023	1.11	.028	0.57	.070	4		538,3059
32612	−14 01027	HR 1640		5 01 35	−14 26.3	5	6.40	.004	−0.19	.005	−0.76	.003	17	B2 IV	15,688,1075,1732,2027
268921				5 01 35	−69 09.3	1	12.40		0.04		−0.26		4	F8	149
		VES 868		5 01 39	+40 36.6	1	10.77		0.22		−0.15		1		1543
240534	+9 00720			5 01 41	+10 07.2	1	10.32		0.59				1	F8	950
32460	+39 01163			5 01 41	+39 12.6	1	8.91		0.21		0.15		3	A1 V	1501
		Sk -65 023		5 01 42	−65 42.	1	13.65		−0.19		−0.95		3		798
32549	+15 00732	HR 1638, V1032 Ori		5 01 43	+15 20.2	6	4.68	.010	−0.06	.012	−0.10	.013	17	A0p Si	15,1049,1079,1363*
32520	+29 00802	IDS04586N2950	AB	5 01 47	+29 54.5	1	7.82		1.53				3	K0	1724
32446	+44 01088			5 01 47	+44 09.4	2	8.24	.000	0.19	.005	−0.45		3	B5 III	401,1119
32343	+58 00804	HR 1622, BV Cam	★ A	5 01 47	+58 54.3	2	5.09	.090	−0.08	.009	−0.70	.019	6	C2.5Ve	154,1212
268973		Sk -71 001		5 01 47	−71 03.0	1	13.47		−0.03		−0.31		3	A0 Ia	415
		Sk -68 026		5 01 48	−68 15.	1	11.67		0.13		−0.75		5	B2 Ia	798
		Sk -68 027		5 01 48	−68 18.	1	12.55		0.16		−0.65		4		798
		Sk -70 036		5 01 48	−70 38.	1	13.42		−0.04		−0.88		3		798
32667	−24 02795	HR 1645		5 01 49	−24 27.4	3	5.60	.007	0.09	.009	0.07		10	A2 IV	15,2012,8071
32562	+12 00719			5 01 50	+12 37.8	1	8.73		0.04		−0.15		2	A0	401
32357	+58 00805	HR 1623, BM Cam	★ B	5 01 51	+58 57.3	2	6.11	.023	1.13	.009	0.85	.000	5	K0	1733,4001
		AAS21,109 # 58		5 01 54	−66 11.	1	13.24		−0.06		−0.49		3		952
268885				5 01 54	−67 32.9	1	11.78		0.71		0.10		4	G5 V	149
33031	−69 00321			5 01 54	−69 33.9	2	8.11	.010	0.26	.000	0.12		8	A7 V	1075,1408

HD	DM	Other Id	N Rem	α_{1950}	δ_{1950}	S	V	σ_V	B–V	σ_{B-V}	U–B	σ_{U-B}	n	Spectrum	References
		Sk -70 038		5 01 54	−70 20.	1	12.95		−0.06		−0.53		3	B0 I	952
32356	+60 00857	HR 1624		5 01 58	+61 06.2	2	6.04	.005	1.37	.010			7	K5 II	1118,1379
286320	+17 00832			5 01 59	+17 42.9	1	9.21		0.42		0.07		2	F2	1648
277559	+40 01166	G 96 - 20		5 01 59	+40 11.4	2	9.66	.000	0.42	.000	−0.16	.000	4	F8	516,1620
		Sk -70 040		5 01 59	−70 11.6	1	12.87		0.03		−0.39		3	A1 Ib	415
32646	−10 01085			5 02 00	−10 12.6	2	9.70	.009	0.80	.000	0.40	.005	4	M4	1064,1696
		Sk -65 024		5 02 00	−65 53.	1	13.26		−0.19		−0.93		4		798
		Sk -67 037		5 02 00	−67 36.	1	13.59		−0.20		−1.04		2		798
		Sk -68 028		5 02 00	−68 33.	1	13.44		−0.24		−1.05		3		798
293782	−4 01029	UX Ori		5 02 01	−03 51.4	1	9.61		0.20		0.21		1	A3e	1588
		ApJS17,467 # 103		5 02 01	−68 05.4	1	12.63		0.33				1		642
		Sk -70 039		5 02 02	−70 20.	1	12.92		−0.05		−0.64		3		415
268957	−70 00353			5 02 09	−70 07.2	3	9.78	.005	0.48	.005	0.08		14	G0	149,1594,2034
33030	−69 00322			5 02 12	−69 03.5	1	9.25		1.66		2.00		1	K5 III	726
		LP 835 - 60		5 02 13	−24 22.5	2	11.55	.025	0.74	.005	0.25	.010	5		1696,3062
240571				5 02 14	+04 26.	1	9.38		0.14		−0.23		2	G0	401
268971		Sk -70 041		5 02 15	−70 40.0	1	11.64		0.02		−0.47		2	A0 Ia	149
		VES 870		5 02 16	+39 51.7	1	13.22		0.41		−0.21		1		1543
268964	−70 00354	Sk -70 037		5 02 16	−70 39.9	1	12.00		−0.01		−0.38		5	A0 Ib	149
32445	+54 00859	HR 1630	⋆ AB	5 02 18	+54 20.3	1	7.14		1.04		0.85		3	G5	1733
32820	−41 01690	HR 1651		5 02 18	−41 48.9	3	6.30	.004	0.53	.000	0.04		8	F6 V	15,612,2012
		Sk -65 025		5 02 18	−66 00.	1	13.61		−0.08		−0.70		3		798
		Sk -67 037a		5 02 18	−67 50.	1	13.85		−0.03		−1.09		4		798
277487	−41 01055			5 02 19	+41 52.1	1	9.93		0.38				2	F0	1119
	−43 01182			5 02 21	+43 55.2	1	9.96		0.23				2	B9	1119
32498	−52 00915			5 02 24	+52 36.4	1	8.71		0.08		0.06		2	A0	1375
32686	−3 00998	HR 1646		5 02 24	−03 06.4	4	6.05	.016	−0.12	.013	−0.53	.008	10	B5 IV	15,154,247,1061
268962		Sk -70 043		5 02 24	−70 05.6	2	12.21	.009	−0.04	.009	−0.76	.023	4	B7	149,952
240569				5 02 26	+14 40.	1	9.50		0.49		0.04		2	G5	401
		MN83,95 # 201		5 02 26	−71 14.4	1	12.17		0.58				1		591
32660	+8 00852			5 02 29	+08 52.5	1	7.46		−0.03		−0.42		2	B9	401
32548	+42 01164			5 02 29	+42 46.2	1	8.46		1.23				2	K0 III	1119
	−43 01183			5 02 29	+43 52.2	2	9.44	.010	2.19	.015	2.40		6	M2	369,8032
273011	−42 01743			5 02 29	−42 25.7	4	10.04	.007	0.74	.008	0.20		12	K0	158,912,1705,2012
		Sk -70 042		5 02 30	−70 44.	1	13.48		−0.18		−0.95		4		798
		Sk -71 002		5 02 30	−71 47.	1	13.12		−0.01		−0.48		4		798
32642	+19 00847	HR 1642	⋆ AB	5 02 35	+19 44.4	2	6.44		0.22	.020	0.23	.035	5	A5 m	355,1049
		G 84 - 35		5 02 35	−02 28.7	1	10.95		1.14		1.12		1	K4	1620
268934				5 02 35	−68 30.9	1	11.92		0.63		−0.04		5	K0	149
32831	−35 02089	HR 1652	⋆ AB	5 02 36	−35 33.0	4	4.55	.003	1.20	.004	1.19	.005	20	K2 III	15,1075,2012,8015
32641	+22 00818	IDS04596N2256	AB	5 02 37	+23 00.2	1	6.80		0.12		−0.46		3	B5	1648
32685	+7 00793			5 02 38	+07 29.7	1	7.59		−0.05		−0.52		13	B9	8040
32846	−35 02090	HR 1653, X Cae		5 02 39	−35 46.4	4	6.33	.011	0.31	.008	0.04	.005	9	F2 IV/V	15,258,612,2012
268984		Sk -70 044		5 02 39	−70 44.1	1	12.04		−0.04		−0.72		3	B2.5Ia	149
32608	+35 00973	HR 1639		5 02 40	+35 52.2	2	6.51	.010	0.16	.005	0.10	.005	4	A5 V	985,1733
	+8 00853			5 02 41	+08 13.6	1	10.38		1.55				1	K7	1632
32478	+59 00850	IDS04583N5912	AB	5 02 41	+59 16.0	1	8.22		0.17				4	A0	1118
		Sk -70 043a		5 02 42	−70 48.	1	14.03		−0.24		−1.12		3		798
32803	−22 00995	T Lep		5 02 43	−21 58.3	1	9.62		1.59		0.22		1	M6e-M9e	975
33029	−61 00403			5 02 43	−61 29.0	2	8.74	.010	0.52	.000	−0.04		8	F7 V	1075,1408
277474	+42 01168			5 02 44	+42 23.6	1	9.94		0.06				2	B5	1119
32537	+51 01024	HR 1637	⋆ AB	5 02 45	+51 32.0	4	4.98	.018	0.34	.010	−0.02	.009	15	F0 V	15,1118,3024,8015
32537	+51 01024	IDS04588N5128	C	5 02 45	+51 32.0	1	9.43		1.15		1.01		3		3072
269006	−71 00308	Sk -71 003		5 02 46	−71 24.5	5	10.14	.367	0.12	.057	−0.69	.045	17	B2.5lpe	204,328,415,1277,8033
33117	−68 00305			5 02 48	−67 54.7	2	8.26	.010	0.97	.000	0.62	.000	9	G8:III	149,204
32736	+0 00939	HR 1648, W Ori		5 02 49	+01 06.6	6	6.09	.130	3.49	.143	6.50		16	C6 II	15,109,864,3001,8015,8022
32717	+8 00854			5 02 51	+08 24.1	1	8.48		0.04		−0.03		2	A0	401
32633	+33 00953	HZ Aur		5 02 51	+33 51.1	1	7.08		−0.06		−0.40		5	B9p	1202
277495	+41 01057	LS V +41 026		5 02 51	+41 42.2	1	9.58		0.10				2	B8	1119
		Sk -70 045		5 02 51	−70 32.	1	12.53		0.01		−0.46		3	A0 Iab	415
		VES 871		5 02 52	+41 49.1	1	12.27		0.30		−0.37		3		1543
		SK *G 176		5 02 52	−70 05.9	1	13.15		0.08		0.13		5		149
33519	−78 00165	HR 1682	⋆ A	5 02 53	−78 22.2	3	6.27	.011	1.52	.005	1.87	.000	16	K5/M0 III	15,1075,2038
		Sk -70 045a		5 02 57	−70 39.	1	13.36		−0.20		−0.99		2		415
		Sk -66 049		5 02 58	−66 32.	1	12.33		0.03		−0.14		4	A0 Iab	415
240629	+6 00829	Hyades vB 151	A	5 02 59	+06 23.9	2	9.92	.004	0.95	.002	0.70		5	K2	950,1127
240629	+6 00829	Hyades vB 151	B	5 02 59	+06 23.9	1	12.40		0.69				1		950
32619	+44 01091			5 02 59	+44 40.0	1	7.48		0.16				4	A4 V	1119
268993		Sk -70 046		5 02 59	−70 45.2	4	11.98	.008	0.03	.012	−0.51	.000	10	A0 Ia	149,952,1277,8033
32630	+41 01058	HR 1641	⋆	5 03 00	+41 10.1	16	3.17	.008	−0.18	.009	−0.67	.005	80	B3 V	1,15,154,247,369,1006*
32684	+29 00806			5 03 02	+29 41.9	1	8.35		0.35				2	F0/2 V/IVn	1724
32753	+7 00796	IDS05004N0803	AB	5 03 03	+08 06.9	1	9.16		−0.06		−0.38		2	B9	401
		SK *G 518		5 03 04	−66 30.8	1	12.63		0.34		0.27		1	F0 Iab	726
233077	+51 01027			5 03 06	+51 38.4	1	8.00		1.59		1.85		2	K5	1566
		SK *C 13		5 03 06	−67 54.	1	13.53		0.21		0.43		1		149
33133		Sk -66 051		5 03 07	−66 45.0	2	12.69	.020	−0.19	.035	−0.93	.000	7	WN8	415,980
32672	+38 01020	LS V +38 007		5 03 09	+38 27.4	2	7.75	.000	0.16	.005	−0.55	.025	5	B2 IV	399,401
268907	−67 00383	Sk -66 050		5 03 09	−67 01.6	5	10.63	.023	0.00	.012	−0.67	.004	23	B5 Ia+	149,1277,1368,8033,8100
		Sk -65 026		5 03 12	−65 53.	1	13.53		0.00		−0.84		3		798
		VES 872		5 03 13	+46 36.9	1	12.42		0.51		0.00		3		1543

Table 1   245

HD	DM	Other Id	N Rem	α_{1950}	δ_{1950}	S	V	σ_V	B–V	σ_{B-V}	U–B	σ_{U-B}	n	Spectrum	References
32890	−26 02005	HR 1655		5 03 14	−26 13.1	4	5.72	.004	1.17	.005			18	K2 III	15,1075,2013,2029
32655	+42 01170	HR 1644		5 03 15	+43 06.5	4	6.20	.015	0.44	.010	0.32	.015	8	F2 IIp:	39,1025,6009,8100
32887	−22 01000	HR 1654		5 03 21	−22 26.2	7	3.17	.018	1.47	.008	1.78	.009	38	K4 III	3,15,1075,2012,3016*
33285	−71 00309	HR 1677		5 03 21	−71 23.0	3	5.30	.003	1.00	.003	0.78	.005	23	G8 III	15,1075,2038
240648	+17 00841			5 03 23	+17 45.0	1	8.82		0.73				2	K0	987
277470	+42 01171			5 03 23	+42 32.0	1	9.08		0.39				2	F2 V	1119
268949		Sk -68 029		5 03 23	−68 37.6	1	11.76		0.07		−0.36		4	A0 Ia	149
		G 84 - 36		5 03 24	+04 16.2	1	11.54		1.51		1.17		1	M0	333,1620
		Sk -66 052		5 03 24	−66 10.	1	13.28		−0.12		−0.84		3		798
32901	−21 01068			5 03 26	−21 29.2	2	8.39	.015	1.26	.005	1.11	.032	3	K0 III	565,3048
32902	−23 02396			5 03 28	−23 21.8	1	8.54		0.52				4	F5 V	2012
32653	+50 01122	IDS04596N5010	A	5 03 30	+50 14.4	1	7.38		0.09		−0.20		1	B9	8084
32653	+50 01122	IDS04596N5010	B	5 03 30	+50 14.4	1	11.53		0.28		0.24		2		8072
32653	+50 01122	IDS04596N5010	C	5 03 30	+50 14.4	1	12.07		1.22		0.92		1		8084
32653	+50 01122	IDS04596N5010	D	5 03 30	+50 14.4	1	13.81		1.02		0.78		1		8084
268960		Sk -69 059		5 03 30	−69 06.	1	12.13		−0.12		−0.93		4	O8	415
		Sk -67 038		5 03 36	−67 56.	1	13.73		−0.27		−1.07		4		798
32629	+55 00959			5 03 37	+55 41.6	1	6.93		1.34		1.46		3	K0	1502
		Sk -68 029a		5 03 38	−68 59.7	1	12.47		0.12		0.12		2	A2 Ib	149
33042	−49 01562	HR 1663		5 03 40	−49 38.7	6	5.03	.013	1.50	.018	1.88	.017	24	K5 III	15,243,1088,2035,3055,8015
		Sk -68 030		5 03 42	−68 17.	1	13.21		−0.23		−1.02		4		798
		Sk -69 060		5 03 43	−69 32.4	1	13.20		0.05		0.04		2	A0 Ib	149
		AQ Eri		5 03 44	−04 12.1	1	17.52		0.06		−1.03		1		1471
		VES 874		5 03 45	+43 55.0	1	10.86		0.29		−0.03		2		1543
32853	+10 00712			5 03 46	+10 34.7	1	8.24		0.05		−0.16		2	B9	401
32752	+41 01061			5 03 46	+42 01.8	1	8.37		−0.02				2	A0	1025
	+42 01173		V	5 03 46	+42 14.4	1	10.26		1.90				3	M2	369
		Sk -66 053		5 03 48	−66 29.	1	13.28		−0.14		−0.89		3		798
		VES 875		5 03 50	+40 10.3	1	11.81		0.32		−0.20		2		1543
		Sk -70 047		5 03 50	−70 25.	2	12.97	.009	−0.14	.000	−0.99	.000	4		360,415
32850	+14 00831	G 97 - 19		5 03 51	+14 23.0	1	7.74		0.80		0.41		1	K0 V	333,1620
		L 179 - 27		5 03 54	−57 09.	1	11.33		0.77		0.29		1		3060
		Sk -66 054		5 03 54	−66 08.	1	14.16		−0.18		−1.04		3		798
268980				5 03 56	−69 29.7	2	11.71	.005	0.43	.005	0.05	.000	6	F4 V	149,204
33563	−76 00303	IDS05060S7646	AB	5 03 56	−76 41.7	2	7.53	.004	0.48	.001	−0.04		45	F5 V	978,2012
32925	−6 01086			5 03 57	−06 22.8	1	9.21		0.33		0.05		2	A2	1776
		LP 716 - 26		5 03 57	−09 32.5	1	13.24		0.90		0.57		1		1696
32835	+26 00787			5 03 58	+26 55.8	1	7.65		0.87				3	F5 V	6009
33116	−54 00768	HR 1667		5 03 58	−54 28.5	2	6.26	.004	1.55	.005	1.90		6	M2 III	58,2035
		Sk -67 038a		5 04 00	−67 21.	1	13.51		−0.19		−1.04		3		798
269018	−70 00357	Sk -70 048		5 04 01	−70 46.0	1	11.67		−0.02		−0.72		4	B3 Ia	149
32800	+36 01004	NGC 1778 - 1		5 04 02	+36 48.5	2	9.04	.010	0.25	.000	0.19	.000	2	A5 V	49,436
		Sk -70 049		5 04 02	−70 27.	1	12.68		−0.11		−0.91		3		415
240709	+29 00810			5 04 05	+29 10.8	1	9.81		2.13				3	M0	369
33146	−55 00739			5 04 05	−55 31.0	1	7.56		1.01				4	K0 III	2012
280467	+36 01005	NGC 1778 - 4		5 04 06	+36 52.7	2	10.64	.025	0.38	.005	0.21	.000	2	A7	49,436
		Sk -68 031		5 04 06	−68 03.	1	13.56		−0.15		−1.03		4		798
		Sk -69 061		5 04 06	−69 11.	1	12.56		0.04		−0.67		4		798
273763	−48 01630			5 04 07	−47 57.1	2	11.03	.010	0.11	.000	0.07	.000	5	A0 V	1097,3077
32518	+69 00302	HR 1636		5 04 08	+69 34.6	2	6.41	.005	1.11	.000	1.02	.005	6	K1 III	985,1733
		Sk -67 040		5 04 09	−67 05.4	1	13.11		−0.13		−0.61		3	B8 Ia	798
32908	+11 00716			5 04 10	+11 45.9	1	9.10		0.11		−0.15		2	A2	401
240706	+32 00890			5 04 10	+32 50.9	1	9.20		1.83				5	M2	369
		Sk -66 055		5 04 10	−66 30.	1	12.53		0.10		0.12		4		415
33132	−51 01367			5 04 11	−51 30.1	1	8.67		0.64				4	G2 V	2012
32906	+15 00742			5 04 12	+15 52.3	1	7.59		1.28				1	K2	987
		LP 716 - 36		5 04 12	−13 19.8	1	13.06		1.35		1.05		1		1696
		Sk -67 039		5 04 12	−67 24.0	1	12.76		0.05		−0.46		4	A2 Ia	415
		Sk -68 032		5 04 12	−68 32.	1	13.71		−0.19		−0.99		3		798
269029		Sk -71 004		5 04 12	−71 09.	1	11.90		0.04		−0.59		3	B7	415
		KUV 05042+1632		5 04 13	+16 32.2	1	16.50		0.15		−0.35		1		1708
		NGC 1778 - 135		5 04 13	+37 00.4	1	13.96		1.70		1.53		1		436
	+45 01044			5 04 13	+45 40.8	1	9.89		0.08		−0.50		2		401
		Sk -68 033		5 04 13	−68 15.9	1	12.69		0.01		−0.30		4	B7 Ib	415
		NGC 1778 - 12		5 04 14	+36 59.2	2	12.77	.035	0.35	.015	0.01	.005	2		49,436
269009	−70 00358	Sk -70 050		5 04 14	−70 15.9	1	11.20		−0.04		−0.71		6	B3 Ia	149
268939	−67 00386	Sk -67 041		5 04 16	−67 19.1	2	10.98	.025	0.09	.010	−0.80		7	B1 e	415,8033
		NGC 1778 - 13		5 04 17	+37 01.0	2	12.88	.020	0.34	.025	0.07	.030	2		49,436
		NGC 1778 - 14		5 04 17	+37 03.0	3	13.45	.008	0.62	.015	0.35	.008	4		49,436,1552
32964	−4 01044	HR 1657	⋆ A	5 04 17	−04 44.3	5	5.10	.023	−0.06	.007	−0.17	.014	12	B9.5V	15,1061,1079,3024,8015
32964	−4 01044	IDS05018S0447	B	5 04 17	−04 43.2	1	10.71		1.25		1.25		1	K5 V	3024
		NGC 1778 - 126		5 04 18	+36 53.2	1	14.24		0.59		0.44		1		436
233081	+52 00923			5 04 18	+52 18.2	1	9.07		0.71		0.23		2	F5	1566
32996	−13 01063	HR 1661		5 04 18	−13 11.3	1	6.05		−0.06				4	B9.5/A0 IV	2006
		Sk -67 042		5 04 18	−67 11.5	1	12.61		−0.01		−0.44		3		952
		Sk -68 034		5 04 18	−68 15.	1	13.72		−0.17		−0.91		3		952
		AAS21,109 # 62		5 04 18	−70 45.	1	13.86		−0.13		−0.80		3		952
237354	+55 00960	G 191 - 23		5 04 19	+55 21.7	2	9.33	.000	0.63	.000	−0.01	.000	3	G2 V	1003,3025
		NGC 1778 - 124		5 04 20	+36 53.9	1	14.55		0.52		0.27		1		436
		NGC 1778 - 127		5 04 20	+37 01.4	1	13.99		0.74		0.30		1		436

HD	DM	Other Id	N Rem	α_{1950}	δ_{1950}	S	V	σ_V	B–V	σ_{B-V}	U–B	σ_{U-B}	n	Spectrum	References
269013		Sk -70 051		5 04 20	-70 17.9	2	11.95	.009	0.01	.009	-0.77	.009	4	B2 Ia	360,415
293815	-3 01013			5 04 22	-03 23.3	4	10.06	.014	0.20	.006	0.06	.008	18	B9 V	830,1004,1783,8055
32680	+61 00760			5 04 23	+61 36.0	1	8.74		0.34		0.18		4	A2	1371
		Sk -70 053		5 04 23	-70 37.3	1	12.68		-0.03		-0.63		3	A0 Iab	415
		NGC 1778 - 62		5 04 24	+36 57.3	1	14.32		0.89		0.24		1		436
	-53 00794			5 04 24	-52 56.0	1	10.55		0.48				3	G0	1594
		Sk -66 056		5 04 24	-66 12.	1	12.83		0.02		-0.06		3		798
		AAS21,109 # 63		5 04 24	-70 23.	1	13.84		-0.20		-0.90		2		952
33115	-44 01860			5 04 25	-44 50.7	1	7.53		0.94		0.67		9	G8 III	1673
		NGC 1778 - 88		5 04 26	+36 54.8	1	15.42		0.62		0.20		1		436
		NGC 1778 - 37		5 04 26	+36 59.9	1	11.58		0.72		0.17		1		436
		NGC 1778 - 11		5 04 27	+37 01.8	3	12.68	.019	0.37	.004	0.20	.000	4		49,436,1552
		NGC 1778 - 16		5 04 27	+37 02.3	2	13.68	.005	0.62	.005	0.36	.020	2		49,436
		Sk -70 052		5 04 27	-70 38.	1	13.00		0.04		-0.30		3		415
		NGC 1778 - 34		5 04 28	+36 57.3	1	11.38		1.62		1.28		1		436
		NGC 1778 - 90		5 04 28	+37 01.4	1	15.48		0.73		0.16		1		436
268943	-67 00388			5 04 28	-67 21.7	1	10.88		0.05		0.07		4	A0 V	149
34172	-82 00106	HR 1716		5 04 29	-82 32.4	4	5.84	.004	0.93	.005	0.65	.016	13	G8/K0 III	15,258,1075,2038
		Hyades J 352		5 04 30	+17 17.	1	11.84		0.72		0.04		3		1058
32923	+18 00779	HR 1656	⋆ AB	5 04 30	+18 34.8	6	4.92	.034	0.65	.005	0.14	.006	15	G4 V +G4 V	22,292,1080,1355,1363,3077
		NGC 1778 - 20		5 04 30	+36 55.5	1	14.59		1.78				1		49
		NGC 1778 - 50		5 04 30	+37 01.1	1	13.46		0.48		0.25		1		436
		HRC 82	V	5 04 30	-03 24.	1	14.18		1.49		0.81		1		825
		Sk -67 042a		5 04 30	-67 20.	2	13.39	.000	-0.17	.009	-0.96	.005	5	B9 Ib	360,798
		NGC 1778 - 48		5 04 31	+36 56.2	1	13.35		0.43		0.08		1		436
		NGC 1778 - 24		5 04 31	+36 57.2	2	15.36	.120	0.66	.120	0.20		2		49,436
280464		NGC 1778 - 33		5 04 31	+37 00.2	1	10.96		0.25		0.10		1	A0	436
32747	+57 00857			5 04 31	+57 20.3	1	7.81		0.11		-0.22		2	B9	1375
		SK *S 30		5 04 31	-67 52.9	1	12.85		-0.13		-0.76		4		149
		NGC 1778 - 79		5 04 33	+36 55.1	1	14.51		0.64		0.31		1		436
		NGC 1778 - 73		5 04 33	+37 01.6	1	14.78		0.88		0.16		1		436
33073	-28 01965			5 04 34	-28 37.5	3	9.57	.070	0.69	.007	0.16		8	F3 V	742,955,1594
269020				5 04 35	-70 19.8	1	12.60		-0.12		-0.94		5	B0.5I	149
		NGC 1778 - 60		5 04 37	+36 57.8	1	14.21		0.45		0.29		1		436
		NGC 1778 - 39		5 04 37	+36 58.3	1	11.73		0.24		-0.08		1	B7 IV-V	436
		Sk -65 028		5 04 37	-65 52.	1	12.62		0.00		-0.22		2		415
240764	+30 00792	HRC 800, RW Aur	⋆ AB	5 04 38	+30 20.2	4	10.37	.115	0.59	.026	-0.29	.048	4	dG5:e	649,1536,1591,8013
		NGC 1778 - 54		5 04 38	+36 57.7	1	13.59		0.39		0.21		1		436
		Sk -65 027		5 04 38	-65 55.	1	12.08		-0.02		-0.60		4		415
		NGC 1778 - 38		5 04 39	+36 55.4	1	11.70		0.28		-0.02		1	B7 IV-V	436
		NGC 1778 - 129		5 04 39	+37 01.5	1	14.16		0.42		0.38		1		436
33262	-57 00735	HR 1674		5 04 39	-57 32.4	6	4.71	.008	0.52	.004	-0.04	.007	22	F7 V	15,688,1075,2012,3026,8015
		Steph 544		5 04 40	+18 37.7	1	11.43		1.19				4	K4	1746
		NGC 1778 - 49		5 04 40	+36 54.2	1	13.64		0.67		0.33		1		436
		NGC 1778 - 52		5 04 40	+36 55.6	1	13.41		0.57		0.36		1		436
280463		NGC 1778 - 31	⋆ A	5 04 40	+36 59.2	1	10.19		0.28		-0.02		1	B9 III:e	436
		NGC 1778 - 78		5 04 40	+36 59.4	1	14.91		0.54		0.34		1		436
		NGC 1778 - 42	⋆ C	5 04 40	+36 59.4	1	12.99		0.30		0.10		1		436
269028	-70 00359			5 04 40	-70 44.8	1	9.57		1.23		1.08		1	K0	360
		NGC 1778 - 2		5 04 41	+36 53.7	2	10.01	.000	1.59	.005	1.33	.000	2	G8 II-III	49,436
280462	+36 01009	NGC 1778 - 32	⋆ B	5 04 41	+36 59.0	1	10.29		0.24		-0.11		1	B6 III-IV	436
	-21 01074	Steph 546	⋆ ABC	5 04 41	-21 38.9	1	9.91		1.52		1.15		3	M2+M3	1705,1746
	-21 01074	Steph 545	⋆ BC	5 04 41	-21 38.9	1	11.30		1.51				1	M3	1746
280461	+36 01010	NGC 1778 - 3	AB	5 04 42	+36 54.6	2	10.23	.005	0.32	.005	-0.12	.000	2	B6 IV-Ve	49,436
		NGC 1778 - 22		5 04 42	+36 57.3	3	14.88	.025	0.59	.005	0.30	.005	4		49,436,1552
		NGC 1778 - 57		5 04 42	+36 58.4	1	13.72		0.37		0.19		1		436
		NGC 1778 - 142		5 04 42	+36 59.	1	16.58		1.33		1.63		2		1552
		NGC 1778 - 47		5 04 42	+36 59.5	1	13.13		0.27		-0.04		1		436
	-21 01074	Steph 546	⋆ A	5 04 42	-21 39.0	1	10.29		1.52				1	M2	1746
		Sk -68 035		5 04 42	-68 31.	1	13.47		-0.22		-0.93		3		798
		Sk -69 062		5 04 42	-69 23.	1	13.39		-0.17		-0.82		3		798
		NGC 1778 - 53		5 04 43	+36 54.3	1	13.73		0.41		0.26		1		436
		NGC 1778 - 119		5 04 43	+36 54.6	1	14.91		0.62		0.35		1		436
		Sk -71 005		5 04 43	-71 19.	1	12.69		-0.07		-0.69		3		415
		NGC 1778 - 56		5 04 44	+36 53.8	1	13.66		0.71		0.35		1		436
		NGC 1778 - 118		5 04 44	+36 54.4	1	14.61		0.86		0.25		1		436
		NGC 1778 - 41		5 04 44	+36 57.5	1	12.00		0.20		-0.10		1		436
32731	+61 00761			5 04 44	+61 26.8	1	8.86		0.18		0.15		4	A0	1371
		NGC 1778 - 5	AB	5 04 45	+36 56.8	2	11.18	.035	0.23	.025	-0.08	.035	2	B8 IV-V	49,436
		NGC 1778 - 15		5 04 45	+37 00.2	2	13.58	.030	1.56	.010	1.22		2		49,436
		KUV 05048+1653		5 04 46	+16 52.7	1	16.64		-0.10		-1.04		1		1708
		NGC 1778 - 72		5 04 46	+36 53.2	1	14.68		0.77		0.36		1		436
		NGC 1778 - 59		5 04 46	+36 57.9	1	13.94		0.40		0.25		1		436
		NGC 1778 - 71		5 04 46	+36 58.4	1	14.61		0.95		0.23		1		436
	-29 02058			5 04 46	-29 10.0	1	9.64		0.03		0.11		6		1775
		Sk -67 043		5 04 46	-67 51.0	2	12.72	.009	-0.08	.061	-0.73	.000	4		149,952
240896	+5 00811			5 04 47	+05 42.4	1	10.90		0.70				1	G5	920
		Sk -66 057		5 04 48	-66 19.	1	12.83		-0.08		-0.71		4		798
		Sk -70 057		5 04 48	-70 42.	1	13.36		-0.21		-1.01		4		798
		NGC 1778 - 7		5 04 49	+36 53.2	2	11.59	.015	0.33	.015	-0.06	.015	2	B6 IV-Ve	49,436

Table 1 247

HD	DM	Other Id	N	Rem	α_{1950}	δ_{1950}	S	V	σ_V	B–V	σ_{B-V}	U–B	σ_{U-B}	n	Spectrum	References
		NGC 1778 - 58			5 04 49	+36 53.7	1	13.76		0.88		0.12		1		436
		NGC 1778 - 36			5 04 49	+36 57.0	1	11.63		0.23		-0.01		1		436
		Sk -68 036			5 04 49	-68 48.9	1	12.18		0.16		-0.33		4	B8 Iab	415
32963	+26 00789				5 04 50	+26 15.9	2	7.60	.014	0.67	.005	0.20	.005	4	G5 IV	1003,3026
		Sk -67 043a			5 04 50	-67 42.	1	12.96		0.02		-0.22		4		415
32977	+20 00885	HR 1658			5 04 51	+20 21.3	2	5.29	.010	0.09	.005	0.11	.005	6	A5 V	985,1733
280460	+36 01012	NGC 1778 - 140		⋆A	5 04 51	+36 55.7	1	9.64		0.20		-0.14		1	A0	721
		NGC 1778 - 141		⋆B	5 04 51	+36 55.8	1	11.42		0.32		-0.25		1		721
		NGC 1778 - 18			5 04 51	+36 59.3	3	14.34	.008	0.88	.021	0.28	.008	4		49,436,1552
		NGC 1778 - 66			5 04 52	+36 56.2	1	14.55		0.82		0.29		1		436
32715	+64 00500	HR 1647			5 04 52	+64 51.5	2	6.38	.025	0.41	.005	-0.02	.000	5	F6 V	985,1733
		Sk -70 056			5 04 52	-70 50.	1	12.60		-0.11		-0.94		2		415
		NGC 1778 - 21			5 04 53	+36 58.9	2	14.65	.040	0.75	.065	0.22	.040	2		49,436
33021	+9 00736	HR 1662		⋆A	5 04 54	+09 24.8	4	6.16	.005	0.62	.005	0.09	.014	10	G1 IV	15,247,1061,3077
		G 85 - 41			5 04 54	+17 55.2	2	11.76	.045	1.66	.000	1.32		2	M2	1620,1746
		NGC 1778 - 10			5 04 54	+37 00.7	2	12.70	.015	0.43	.030	0.35	.010	2		49,436
		NGC 1778 - 26			5 04 54	+37 02.4	2	15.46	.005	1.12	.019	0.97		3		49,1552
		AAS21,109 # 65			5 04 54	-66 32.	1	13.67		0.04		-1.05		3		952
		Sk -68 038			5 04 54	-68 09.	1	13.16		-0.15		-0.99		4		798
		Sk -68 037			5 04 54	-68 28.	1	13.75		-0.26		-1.04		3		952
		NGC 1778 - 82			5 04 55	+36 55.9	1	14.95		0.63		0.23		1		436
		NGC 1778 - 40			5 04 55	+37 00.1	1	11.99		0.23		-0.12		1		436
		WLS 500 0 # 9			5 04 56	+00 53.6	1	11.31		1.19		1.04		2		1375
32991	+21 00766	HR 1660		⋆	5 04 56	+21 38.4	3	5.89	.045	0.19	.010	-0.58	.013	9	B2 Ve	154,1212,1223
240819	+32 00894				5 04 56	+32 25.4	1	9.60		2.01				3	M0	369
	-57 00737				5 04 56	-57 37.5	1	9.02		1.40		1.20		2	K7 V	3072
		VES 876			5 04 57	+42 38.8	1	13.57		0.48		-0.30		3		1543
30881	+86 00066				5 04 57	+86 14.1	1	7.95		-0.01		-0.05		3	B9	1219
		NGC 1778 - 43			5 04 58	+36 53.9	1	12.65		0.33		0.02		1		436
		NGC 1778 - 6			5 04 58	+37 02.6	1	11.47		1.65		1.72		1		49
		NGC 1778 - 35			5 04 59	+37 00.1	1	12.03		0.33		-0.10		1		436
33095	-19 01102	IDS05028S1932		AB	5 04 59	-19 27.6	1	6.44		0.62		0.10		5	G2 V	265
		NGC 1778 - 27			5 05 00	+36 55.4	1	15.75		0.90				1		49
		NGC 1778 - 63			5 05 00	+37 01.8	1	14.22		0.86		0.16		1		436
		AAS21,109 # 66			5 05 00	-67 14.	1	13.95		-0.03		-0.92		3		952
		NGC 1778 - 45			5 05 01	+37 00.1	1	13.17		1.14		0.66		1		436
33144	-28 01971				5 05 01	-28 43.3	1	9.74		0.42		-0.06		1	F2/3 V	742
		NGC 1778 - 61			5 05 02	+36 54.4	1	14.33		0.56		0.26		1		436
32784	+62 00730				5 05 02	+62 25.0	1	6.60		0.25		0.11		4	A5	1501
		NGC 1778 - 8			5 05 03	+36 53.4	3	11.60	.017	0.28	.015	-0.07	.033	4	B8 IV-V	49,436,1552
		NGC 1778 - 28			5 05 03	+36 59.9	2	15.84	.058	0.66	.097	0.17		3		49,1552
33069	-8 01035				5 05 03	-08 43.1	1	7.02		-0.11		-0.44		2	B8	401
32990	+24 00755	HR 1659		⋆A	5 05 04	+24 12.1	2	5.50	.000	0.06	.000	-0.55	.008	6	B2 V	154,1223
		Sk -68 039			5 05 04	-68 12.	1	12.01		-0.02		-0.82		4	B2.5Ia	415
33093	-12 01076	HR 1665			5 05 05	-12 33.3	2	5.96	.005	0.60	.000			7	G2 V	15,2012
		AAS21,109 # 64			5 05 06	-70 58.	1	12.92		-0.29		-1.10		3		952
		NGC 1778 - 29			5 05 07	+36 51.8	1	16.13		1.00				1		49
32936	+42 01179				5 05 08	+42 53.1	1	8.89		0.40				21	F5	6011
33054	+8 00866	HR 1664		⋆AB	5 05 09	+08 26.1	7	5.33	.010	0.34	.007	0.09	.012	17	Am	15,216,292,355,1008*
		VES 877			5 05 09	+34 18.2	1	14.05		0.79		0.32		1		1543
		NGC 1778 - 17			5 05 11	+37 00.3	1	14.29		0.58		0.30		1		49
268946	-66 00372	Sk -66 058			5 05 11	-66 48.2	3	10.29	.028	0.09	.005	-0.48		8	A0 Ia	149,1277,8033
33055	+8 00867	IDS05025N0816		AB	5 05 12	+08 20.3	1	8.		0.84		0.51		3	K2	3030
		Sk -70 058			5 05 13	-70 10.	1	12.81		-0.05		-0.90		4		415
33045	+15 00749				5 05 14	+15 47.5	1	7.24		1.19		1.23		2	K2	1648
		NGC 1778 - 30			5 05 14	+36 53.2	2	16.30	.010	0.92	.102	0.38		3		49,1552
33486	-68 00311				5 05 15	-68 09.1	1	7.86		-0.03		-0.18		6	A0 IV	1408
		Sk -70 059			5 05 15	-70 37.	1	12.41		-0.10		-0.94		4		415
33004	+30 00796				5 05 16	+30 23.1	1	8.75		0.18		-0.22		2	A0	401
32903	+48 01226	IDS05015N4859		A	5 05 16	+49 03.5	1	6.67		0.19		0.12		2	A3	985
		NGC 1778 - 23			5 05 18	+36 52.5	1	15.04		0.82		0.20		1		49
		Sk -70 060			5 05 18	-70 19.	1	13.85		-0.19		-1.07		4		798
32961	+41 01075	LS V +41 028			5 05 19	+41 40.6	1	8.93		-0.02		-0.74		2	B9	401
		KUV 05053+1628			5 05 20	+16 28.4	1	16.12		0.16		-0.34		1	sdOB	1708
		LP 836 - 5			5 05 20	-23 15.4	1	13.71		0.63		-0.02		3		3060
		SK *G 193			5 05 20	-67 11.3	1	12.50		0.26		0.28		5	A5 Iab	149
		STO 842			5 05 20	-71 41.8	1	12.86		0.51		-0.02		1	F2 Ia	726
		NGC 1778 - 19			5 05 21	+36 57.2	1	14.52		0.75		0.28		1		49
33066	+7 00812				5 05 22	+07 50.2	1	6.66		0.98		0.78		13	K0	8040
		NGC 1778 - 25			5 05 22	+36 57.5	1	15.54		1.16				1		49
		Sk -70 061			5 05 22	-70 42.3	1	12.78		-0.09		-0.90		3		415
		Sk -70 059a			5 05 22	-70 46.	1	12.94		-0.16		-0.98		4		415
33111	-5 01162	HR 1666		⋆A	5 05 23	-05 09.0	19	2.78	.013	0.13	.006	0.12	.028	85	A3 III	1,3,15,418,1004,1006*
		NGC 1778 - 9			5 05 24	+36 54.6	2	12.35	.005	1.71	.005	1.82	.000	3		49,1552
		SK *C 15			5 05 24	-66 46.	1	13.35		0.14		0.13		4		149
		SK *C 16			5 05 24	-67 19.	1	13.26		0.35		0.12		4		149
	-68 00312				5 05 24	-68 06.2	1	11.71		-0.07		-0.79		3	B1.5Ia	149
		SK *S 33			5 05 24	-68 37.1	1	12.57		0.09		-0.25		1		149
240938	+13 00809	Hyades L 109			5 05 26	+13 19.0	1	10.84		0.89		0.53		1	K0	1058
32989	+39 01191	LS V +39 015		⋆AB	5 05 27	+39 25.5	1	8.04		0.06		-0.63		2	B2 IV	401

HD	DM	Other Id	N Rem	α_{1950}	δ_{1950}	S	V	σ_V	B–V	σ_{B-V}	U–B	σ_{U-B}	n	Spectrum	References
		Sk -70 062		5 05 27	-70 37.	1	12.99		-0.13		-0.97		2		415
		Sk -70 063		5 05 30	-70 33.	1	12.82		-0.08		-0.87		4		798
		BQ Eri		5 05 33	-05 54.5	1	10.74		0.44		0.09		4	A8	1768
		Sk -65 029		5 05 36	-65 47.	1	13.55		-0.25		-1.04		3		798
		AAS21,109 # 67		5 05 36	-68 19.	1	13.54		-0.11		-0.91		3		417
		SK *C 17		5 05 36	-68 34.	1	13.18		0.08		-0.17		1		149
	+54 00864			5 05 37	+54 47.6	1	9.90		0.72		0.25		2		1375
33016	+38 01035	TX Aur		5 05 38	+38 56.2	1	9.09		3.19				1	C5 II	109
		Sk -68 041		5 05 42	-68 14.	1	12.01		-0.14		-0.97		3	B0.5Ia	952
		Sk -70 064		5 05 42	-70 26.	2	14.72	.020	-0.17	.055	-1.00	.005	7	WN4	798,980
		Sk -70 065		5 05 42	-70 31.	1	13.55		-0.17		-0.91		4		798
		Sk -70 070		5 05 42	-70 32.	1	13.67		-0.13		-1.00		3		798
		Sk -70 067		5 05 42	-70 44.6	1	12.39		-0.08		-0.70		3		149
		G 85 - 43		5 05 44	+27 15.8	1	11.38		1.16		1.10		1	M1	333,1620
		Sk -70 066		5 05 45	-70 33.	1	12.82		-0.17		-1.00		3		415
33747	-75 00297			5 05 45	-75 05.3	1	8.74		0.99				1	G6/8 III	1642
		Sk -67 043b		5 05 46	-67 49.0	1	12.37		0.19		0.24		2	A4 Iab	149
33140	+2 00870			5 05 48	+02 23.9	1	9.25		0.41		-0.02		2	F2 V	1375
		SK *C 18		5 05 48	-67 10.	1	13.52		0.06		0.08		4		149
		Sk -70 068		5 05 48	-70 13.	1	13.17		-0.16		-0.91		4		798
		Sk -70 069		5 05 48	-70 30.	1	13.96		-0.25		-1.12		2		952
269062		Sk -70 072		5 05 48	-70 49.	1	11.40		0.04		-0.69		3	B7	415
33763	-75 00298			5 05 51	-74 59.2	1	8.36		0.88				1	G3/5 III	1642
	+49 01310			5 05 52	+49 25.4	1	9.43		0.40		0.08		1	F0	851
33121	+19 00853			5 05 53	+19 47.8	1	6.50		0.88		0.53		2	G5 III	1648
		SK *G 198		5 05 53	-70 09.7	1	13.53		0.02		0.09		4		149
		Sk -65 030		5 05 54	-65 44.	1	14.04		-0.16		-0.97		3		798
		Sk -66 059		5 05 54	-66 51.	1	13.23		-0.14		-0.95		4		798
33224	-8 01037	HR 1671	*A	5 05 56	-08 43.7	4	5.77	.008	-0.06	.010	-0.36	.021	12	B8 V	15,1078,1079,2006
33331	-44 01873	TU Pic		5 05 57	-44 53.2	1	6.90		-0.09				4	B5 III	2006
		Sk -70 071		5 05 58	-70 44.	1	12.59		-0.08		-0.89		3		415
33474	-59 00421			5 06 00	-59 07.0	2	9.39	.010	0.44	.010	-0.06	.039	5	F5 V	1097,3077
		AAS21,109 # 68		5 06 00	-69 04.	1	13.38		0.04		-0.79		4		417
		Sk -71 006		5 06 00	-71 16.	1	13.97		-0.03		-1.06		3		798
33061	+38 01040			5 06 02	+38 57.2	2	8.64	.005	0.13	.000	-0.14	.005	5	B9 V	401,833
33579	-68 00314	Sk -67 044		5 06 02	-67 57.1	5	9.13	.020	0.19	.006	-0.24	.011	34	A3 Ia0	52,149,1277,1368,8033,8100
32650	+73 00274	HR 1643, BN Cam		5 06 03	+73 53.2	2	5.45	.005	-0.13	.005	-0.35	.005	5	B9p Si	1733,3033
		AAS21,109 # 69		5 06 06	-68 13.	2	14.16	.037	-0.20	.019	-1.00	.037	4		417,952
269081	-71 00311			5 06 07	-71 23.7	1	9.43		1.21		1.24		2	K0	1730
33107	+29 00820			5 06 08	+29 32.0	1	9.03		0.72				2	G5	1724
		VES 879		5 06 08	+45 47.1	1	10.71		0.12		-0.42		2		1543
		Sk -68 042		5 06 10	-68 15.	1	11.95		-0.04		-0.88		3		415
		AAS21,109 # 71		5 06 12	-67 07.	1	13.98		-0.17		-1.02		3		417
		Sk -68 043		5 06 12	-68 11.	1	13.10		-0.17		-0.97		4		798
		Sk -68 045		5 06 12	-68 11.0	2	12.05	.014	-0.13	.014	-0.97	.019	4	B0 Ia	149,952
		AAS21,109 # 70		5 06 12	-70 05.	1	13.78		-0.07		-0.94		3		952
33077	+42 01184	IDS05027N4233	AB	5 06 13	+42 36.7	1	8.05		0.77				22	G5	6011
33088	+39 01192	TT Aur		5 06 15	+39 31.4	2	8.79	.321	0.06	.005	-0.63	.019	6	B1 Vn	401,588
33256	-4 01056	HR 1673		5 06 15	-04 31.2	5	5.12	.000	0.44	.011	-0.05	.017	10	F2 V	15,254,1008,1415,3026
		Sk -70 073		5 06 16	-70 45.	1	12.85		-0.07		-0.87		4		415
		HV 2294		5 06 17	-66 45.	2	12.15	.050	0.51	.050	0.37	.025	2	F7 I	689,3075
		Sk -68 044		5 06 17	-68 44.2	1	12.67		0.06		-0.24		3	B9 Ia	415
		G 85 - 44		5 06 18	+15 24.2	1	12.48		1.50		1.00		1		3059
240984				5 06 19	+06 50.6	1	10.75		1.10				1	A0	1632
269066	-70 00363	Sk -70 075		5 06 19	-70 35.9	1	11.96		-0.13		-0.92		4	A2 Ia	149
		LP 777 - 36		5 06 21	-18 12.9	3	10.30	.013	1.53	.009	1.04		8		158,912,1705
		Sk -67 045		5 06 22	-67 45.7	1	12.97		-0.09		-0.74		5		149
		Sk -70 074		5 06 22	-70 42.	1	12.49		-0.08		-0.88		3		415
		AAS21,109 # 72		5 06 24	-67 10.	1	14.55		-0.12		-0.99		2		952
		Sk -70 076		5 06 24	-70 48.	1	13.69		-0.12		-0.97		3		798
269087	-71 00312			5 06 24	-71 31.0	1	10.86		0.55		0.03		2		1730
33616	-65 00428			5 06 27	-65 26.5	1	7.81		1.22		1.37		6	K2 III/IVc	1704
		Sk -70 077		5 06 27	-70 20.0	1	12.81		0.00		-0.29		3	B9 Iab	415
33377	-35 02126	HR 1680		5 06 28	-35 46.9	4	6.51	.007	1.09	.009	0.94		12	K1 III	15,612,1075,2035
33118	+39 01194	IDS05031N4002	AB	5 06 31	+40 05.4	1	8.07		0.44		-0.13		1	B9	379
269086	-71 00314			5 06 31	-71 17.0	1	11.21		0.41				2		1730
33170	+28 00754			5 06 32	+29 04.9	1	8.78		0.79				2	G0	1724
33185	+29 00822	IDS05034N2940	A	5 06 32	+29 44.2	1	6.66		0.51				3	F8 V	1724
33254	+9 00743	Hyades vB 130	*A	5 06 34	+09 46.0	8	5.42	.007	0.25	.010	0.17	.021	26	A7 m	3,15,355,1127,1199*
33152	+36 01021	LS V +36 006		5 06 34	+36 56.5	1	8.10		0.31		-0.61		3	B1 Ve	1212
		G 96 - 21		5 06 34	+48 47.2	1	11.39		1.53		1.20		1	M2	333,1620
		Sk -68 047		5 06 36	-68 11.	1	13.34		-0.17		-1.01		4		798
		Sk -68 046		5 06 36	-68 37.	1	12.45		-0.04		-0.91		5		798
33204	+27 00732	Hyades vB 131	*A	5 06 37	+27 58.1	6	6.00	.012	0.27	.017	0.10	.041	17	A7 m	15,1127,3024,8015*
33204	+27 00732	Hyades vB 132	*BC	5 06 37	+27 58.1	2	8.62	.047	0.71	.023	0.26	.004	6	A3	1127,3024
33151	+38 01045			5 06 38	+38 18.7	1	8.27		1.82				2	M1	369
		Sk -68 048		5 06 38	-68 34.5	1	11.85		0.24		-0.26		4	B9 Ib	149
33253	+14 00840			5 06 39	+14 17.6	1	8.05		1.80		1.76		2	M1	1648
33599	-61 00411			5 06 41	-61 52.1	1	8.88		-0.18		-0.71		1	B5p Shell	55
33316	-6 01094			5 06 42	-06 30.1	1	8.19		-0.07		-0.38		2	B9	401

Table 1   249

HD	DM	Other Id	N Rem	α_{1950}	δ_{1950}	S	V	σ_V	B–V	σ_{B-V}	U–B	σ_{U-B}	n	Spectrum	References
33328	−8 01040	HR 1679, λ Eri		5 06 45	−08 49.0	8	4.27	.012	−0.19	.012	−0.88	.012	29	B2 IVne	3,15,154,247,1075*
241126	+21 00775			5 06 47	+21 51.3	1	9.87		0.66				1	K0	987
269074	−70 00367	Sk -70 078		5 06 47	−70 33.2	1	11.29		−0.08		−0.89		3	B1 Ia	149
277680	+40 01189	LS V +40 033	⋆ AB	5 06 48	+40 35.9	1	8.91		0.06		−0.67		2	B2:III:nn	1012
		AAS21,109 # 73		5 06 48	−65 00.	1	14.69		−0.26		−0.96		3		952
		Sk -66 060		5 06 48	−67 01.	1	14.06		−0.23		−1.04		4		798
		Sk -68 049		5 06 48	−68 26.	1	13.01		−0.03		−0.70		4		798
33276	+15 00752	HR 1676	⋆ A	5 06 50	+15 32.1	7	4.81	.010	0.31	.011	0.19	.021	21	F2 IV	3,15,254,1008,1363*
	+49 01316	BK Aur		5 06 50	+49 37.5	2	9.16	.042	0.92	.004	0.59	.008	2	F8	851,1399
241105	+27 00733	Hyades vB 152	⋆	5 06 52	+27 35.0	4	9.23	.015	0.93	.007	0.60	.003	7	G5	333,950,1127,1529,1620
33473	−41 01727	IDS05053S4121	AB	5 06 54	−41 16.8	2	6.72	.018	0.66	.005	0.14		5	G3 V	1075,1311
		SK *G 205		5 06 55	−68 24.6	1	13.11		0.24		0.26		2	A3 Ib	149
33203	+37 01067	HR 1669	⋆ AB	5 06 56	+37 14.4	2	6.06	.046	0.72	.002	−0.36		3	B2 II + K3	71,8100
		STO 349		5 06 56	−28 28.5	1	12.05		0.62		0.13		1	G2 Ia	726
33345	−2 01155			5 06 58	−02 11.8	1	6.52		1.02		0.83		2	G8 V	1375
33167	+46 00970	HR 1668		5 06 59	+46 54.2	2	5.69	.010	0.41	.005	0.01	.010	5	F5 V	254,3037
		AAS21,109 # 74		5 07 00	−68 30.	1	14.10		−0.22		−0.92		3		417
		AAS21,109 # 75		5 07 00	−68 42.	1	13.22		−0.11		−0.89		6		417
		L 232 - 29		5 07 02	−53 06.1	1	12.23		1.39		1.04		1	M2:	3078
		Sk -67 046		5 07 02	−67 41.	1	12.34		−0.06		−0.88		2		415
		Sk -70 079		5 07 03	−70 34.	1	12.73		−0.04		−0.90		3		415
33313	+6 00852			5 07 04	+07 07.1	1	7.41		0.46		−0.05		13	F8	8040
33202	+40 01194			5 07 04	+40 47.0	1	7.66		0.52		−0.01		1	F5	379
33876	−73 00288			5 07 05	−73 39.6	1	8.25		0.59				4	G0/2 V	2012
	+44 01113			5 07 06	+44 22.0	1	10.44		0.08		−0.38		2	B5 V	723
		Sk -67 047		5 07 06	−68 02.	1	13.74		−0.21		−1.08		4		798
		Sk -68 051	V	5 07 06	−68 25.	1	13.18		0.02		−0.80		1		798
		Sk -68 050		5 07 06	−68 34.	1	13.38		−0.12		−0.99		3		798
		AAS21,109 # 76		5 07 06	−68 57.	1	14.07		−0.12		−1.01		3		952
277667	+41 01094	LS V +41 031		5 07 07	+41 08.2	1	9.44		0.02		−0.65		2	B5	401
		Hyades J 354		5 07 07	−00 04.	1	12.72		0.95		0.75		1		1058
271018				5 07 07	−66 07.0	2	11.43	.000	0.58	.005	0.41		10	F6:Ia	149,1483
33875	−73 00286	HR 1700		5 07 09	−73 06.2	3	6.27	.004	0.00	.004	0.01	.000	20	A1 V	15,1075,2038
33684	−63 00420	HR 1695, WZ Dor		5 07 10	−63 27.7	1	5.20		1.65		1.85		1	M3 III	3055
33370	−5 01172			5 07 11	−05 36.9	1	9.06		0.09		0.04		2	A0	401
269076	−70 00368			5 07 12	−70 09.8	1	9.65		1.24		1.28		1	K0 III	360
		AAS21,109 # 77		5 07 12	−70 28.	1	14.99		−0.10		−0.78		3		417
		AAS21,109 # 78		5 07 12	−70 29.	1	13.73		−0.18		−0.88		3		952
33338	+9 00747			5 07 15	+09 24.9	1	9.13		0.10		−0.01		2	A0	401
		Sk -69 063		5 07 15	−69 59.3	1	13.14		−0.04		−0.57		2	B7 Iab	415
241253	+5 00824	G 84 - 37		5 07 16	+05 29.8	2	9.72	.000	0.52	.000	−0.13	.000	4	G0	516,1620
33232	+40 01196	LS V +40 034		5 07 18	+40 56.5	3	8.17	.005	0.04	.017	−0.43	.013	6	B2 Vne	379,1212,1419
		AAS21,109 # 79		5 07 18	−68 28.	1	13.41		−0.11		−0.97		4		417
63574	−22 02036			5 07 19	−22 12.1	1	10.06		0.09		0.09		1	A0 V	137
277620	+42 01190			5 07 21	+42 53.4	1	8.81		1.63		2.04		1	K5	934
33404	−5 01174	SY Eri		5 07 21	−05 34.6	3	8.41	.128	2.60	.080	2.12	.211	7	C6,3	109,864,3038
33299	+30 00804			5 07 22	+30 44.2	3	6.70	.017	1.60	.020	1.57	.025	9	K1 Ib	1080,1355,8100
33368	+9 00751			5 07 23	+09 54.0	1	7.70		−0.01		−0.38		2	A0	401
33419	−0 00867	HR 1681		5 07 30	−00 37.6	3	6.10	.005	1.10	.000	1.07	.005	8	K0 III	15,247,1415
		SK *C 21		5 07 30	−68 29.	1	12.79		0.28		0.35		1		149
32781	+76 00190	HR 1650	⋆ AB	5 07 31	+76 24.8	1			−0.02		−0.09		2	G0 V	1079
269050	−68 00318	Sk -68 052		5 07 33	−68 35.8	4	11.67	.053	−0.07	.009	−0.92	.030	13	B0 Ia	149,328,1277,8033
		Sk -70 079a		5 07 33	−70 34.	1	13.19		−0.03		−0.86		3		415
		Sk -68 053		5 07 34	−68 11.0	1	12.55		−0.03		−0.56		2	B7 Iab	149
277621	+42 01192			5 07 35	+42 50.2	1	9.17		0.33		0.07		1	A5	934
		Sk -66 061		5 07 36	−66 29.0	1	13.20		0.02		−0.13		6	A1 Iab	149
		AK 7 - 1182		5 07 36	+19 45.8	1	11.90		0.87				1		987
33449	−7 00989	IDS05052S0711	AB	5 07 37	−07 07.8	1	8.48		0.68				4	G0	2012
		Sk -71 007		5 07 37	−71 17.	1	12.92		−0.11		−0.92		3		415
		SK *G 211		5 07 38	−68 22.4	1	12.83		0.10		0.18		1	A4 II	149
33366	+24 00772	G 85 - 45	⋆ AB	5 07 40	+25 05.0	1	8.49		0.59		0.08		1	G5	333,1620
		SK *C 23		5 07 42	−66 26.	1	13.54		0.03		−0.19		5		149
		NGC 1807 - 16		5 07 43	+16 24.4	1	12.01		0.72		0.15		2		305
	+49 01319			5 07 43	+49 48.8	1	9.74		0.43		0.06		1		851
33923	−71 00315			5 07 45	−71 19.9	1	7.35		1.20		1.24		2	K2 III	1730
33402	+17 00862			5 07 46	+17 22.8	1	7.91		0.03		−0.53		2	B8	1648
		SK *G 212		5 07 46	−67 57.5	1	13.17		0.03		−0.08		4		149
		NGC 1807 - 17		5 07 47	+16 23.6	1	12.33		0.20		0.06		3		305
		RL 113		5 07 48	+40 49.	1	14.72		0.22		−0.07		2		704
		SK *C 24		5 07 48	−66 43.	1	13.11		0.15		0.24		4		149
		Sk -68 054		5 07 48	−68 16.	1	13.47		−0.16		−0.96		4		798
33400	+20 00897			5 07 49	+20 30.5	1	7.85		0.40		0.00		2	F0	1648
241260	+30 00807			5 07 50	+30 09.5	1	10.62		0.48				3		1724
33417	+16 00713	NGC 1807 - 12	⋆ AB	5 07 51	+16 25.9	1	8.89		1.10		0.76		2	G5	305
33416	+16 00711	NGC 1807 - 1		5 07 51	+16 32.1	1	9.38		0.53		0.11		1	G5	305
		LP 657 - 6		5 07 51	−07 55.1	1	12.79		0.58		−0.07		2		1696
277714	+38 01051	LS V +39 017		5 07 52	+39 04.5	1	9.65		0.14		−0.59		2	B3	401
33365	+30 00808			5 07 53	+30 14.5	1	8.51		0.82				4	K0	1724
33267	+53 00864			5 07 54	+53 23.5	1	7.06		0.99		0.77		2	K0	1566
		AAS21,109 # 80		5 07 54	−66 55.	1	13.88		0.04		−1.00		4		952

HD	DM	Other Id	N Rem	α_{1950}	δ_{1950}	S	V	σ_V	B–V	σ_{B-V}	U–B	σ_{U-B}	n	Spectrum	References
		NGC 1807 - 13		5 07 55	+16 25.0	1	12.16		1.15		0.84		2		305
		NGC 1807 - 8		5 07 55	+16 27.7	1	16.40		0.83		0.32		3		305
33324	+42 01193			5 07 55	+42 41.3	2	8.28	.002	0.01	.001	-0.32	.012	10	B9	783,1655
		NGC 1807 - 15		5 07 56	+16 24.0	1	13.04		0.94		0.49		2		305
33382	+29 00829			5 07 56	+29 43.4	1	8.49		0.44				2	F5	1724
		NGC 1807 - 67		5 07 57	+16 28.1	1	17.25		0.88		1.42		2		305
		NGC 1807 - 35		5 07 57	+16 29.0	1	15.00		1.21		1.31		1		305
33428	+16 00715	NGC 1807 - 18	★	5 07 58	+16 22.2	3	8.59	.022	1.23	.009	1.19	.020	7	K0	305,882,3040
		NGC 1807 - 27		5 07 58	+16 26.7	1	13.44		0.87		0.40		2		305
		NGC 1807 - 28		5 07 59	+16 27.3	1	15.18		0.72		0.17		3		305
33297	+46 00972			5 07 59	+46 52.9	1	7.82		0.06		-0.04		2	A0	401
		NGC 1807 - 29		5 08 00	+16 27.7	1	11.22		0.51		0.24		1		305
241303	+16 00714	NGC 1807 - 11		5 08 00	+16 27.7	1	10.30		1.22		1.03		4	K5	305
		AAS21,109 # 82		5 08 06	-68 10.	1	13.98		-0.14		-0.86		3		417
		AAS21,109 # 83		5 08 06	-69 16.	1	13.20		-0.05		-0.78		3		417
		AAS21,109 # 81		5 08 06	-70 35.	1	14.01		-0.07		-0.90		3		952
269107	-71 00317			5 08 06	-71 11.6	1	10.62		0.67		0.15		1	G0	360
		Sk -71 008		5 08 06	-71 16.	1	13.25		-0.14		-0.96		4		798
		Sk -71 008a		5 08 07	-71 13.	1	12.69		-0.13		-0.89		3		415
		STO 847		5 08 08	-71 16.3	1	11.55		0.44		-0.05		1	F0 Ib	726
241283	+29 00831			5 08 09	+29 33.4	1	10.29		0.63				2		1724
33364	+37 01076			5 08 10	+37 14.4	1	7.50		1.43		1.57		2	K0	1375
33357	+41 01101	SX Aur		5 08 10	+42 06.3	3	8.42	.024	0.00	.006	-0.73	.014	7	B1:V:e:n	588,783,1012
		Sk -66 062		5 08 12	-66 25.8	2	12.89	.005	0.04	.015	-0.13	.015	8		149,415
		Sk -71 010		5 08 12	-71 34.	1	14.16		-0.17		-0.90		4		798
33398	+35 01004			5 08 13	+35 53.8	1	7.08		1.52		1.76		2	K2	1733
	+44 01121			5 08 13	+44 16.6	1	10.60		0.11		-0.35		2		723
		Sk -71 009		5 08 13	-71 16.	1	12.63		-0.11		-0.95		3		415
		WLS 520 35 # 6		5 08 18	+35 22.2	1	11.06		0.43		0.32		2		1375
		RL 114		5 08 18	+41 36.	1	15.59		-0.07		-0.98		2	sdOB	704
		AAS21,109 # 84		5 08 18	-66 53.	2	14.42	.084	0.00	.005	-1.07	.014	4		417,952
		Sk -69 064		5 08 18	-69 07.	1	13.26		-0.02		-0.83		4		798
		Sk -71 011		5 08 18	-71 05.	1	12.74		-0.10		-0.96		3		415
33547	-5 01178			5 08 20	-05 13.8	1	8.41		-0.05		-0.26		2	B8 V	401
		VES 882		5 08 24	+40 09.6	1	12.56		0.40		-0.32		4		1543
269073	-68 00320			5 08 26	-68 47.8	2	10.10	.022	0.87	.005	0.37	.058	7	K0	2,1496
33463	+29 00833			5 08 27	+29 50.6	1	6.38		1.78				4	M2 III	369
33555	-2 01161	HR 1685		5 08 27	-02 18.8	4	6.24	.008	0.98	.007	0.82	.004	10	G8 III	15,247,1415,3077
		Sk -70 080a		5 08 27	-70 27.	1	13.25		0.03		0.04		3		149
		Aur sq 2 # 58		5 08 28	+44 12.1	1	11.15		0.14		-0.25		2		723
33266	+61 00766	HR 1675		5 08 28	+61 47.4	3	6.17	.014	0.03	.007	0.08	.000	10	A2 III	252,1379,3050
33412	+41 01105	LS V +41 034	★ AB	5 08 30	+41 50.3	1	8.53		0.09		-0.27		1	B8	783
33411	+42 01197	LS V +42 025		5 08 30	+42 20.4	1	8.18		0.06		-0.60		1	B8	783
273845	-49 01595			5 08 31	-49 07.5	1	9.58		1.13		0.82		1	K0	565
		Sk -70 080		5 08 31	-70 26.5	1	12.60		0.43		-0.63		3	A0 Ib	952
	-37 02057			5 08 36	-37 51.1	1	10.07		0.59		0.14		1	G2	742
		Sk -67 048		5 08 36	-67 45.	1	12.33		-0.03		-0.70		4		798
		Sk -68 055		5 08 36	-68 38.	1	13.41		-0.04		-0.65		4		798
		Sk -69 065		5 08 36	-69 19.	1	12.67		-0.06		-0.78		4		798
33590	-5 01179			5 08 37	-05 39.7	1	9.03		0.04		-0.04		2	B9	401
		NGC 1856 sq1 8		5 08 38	-69 05.0	1	12.51		0.96				3		1521
		Aur sq 2 # 61		5 08 39	+44 07.3	1	10.78		0.20		-0.12		2		723
33742	-45 01838			5 08 41	-44 55.2	1	8.96		-0.08		-0.19		2	B9 III/IV	1730
33610	-6 01104			5 08 42	-06 05.0	1	7.91		0.12		0.04		2	A0	401
	+44 01123			5 08 43	+44 14.5	1	11.09		0.17		-0.28		2		723
33667	-26 02045	HR 1694		5 08 43	-25 58.3	4	6.41	.008	1.25	.005			18	K1 III	15,1075,2013,2029
277707	+39 01204	LS V +39 018		5 08 44	+39 29.8	1	10.26		0.40		-0.55		2	Bpe	1012
33461	+41 01106	LS V +41 035		5 08 44	+41 09.3	2	7.79	.005	0.27	.000	-0.54	.005	5	B2:V:enn	1012,1212
		Sk -67 048a		5 08 44	-67 50.8	1	12.76		0.09		0.16		4	A1 Ib	149
33462	+39 01205			5 08 46	+40 02.5	1	6.98		0.06		-0.31		2	B7 III	401
277656	+41 01107	G 96 - 22		5 08 46	+41 23.6	1	8.93		1.01		0.88		1	K2	333,1620
33608	-2 01165	HR 1687		5 08 48	-02 33.1	3	5.89	.005	0.46	.000	0.03	.009	8	F5 V	15,247,1415
		AAS21,109 # 85		5 08 48	-68 50.	1	11.38		-0.10		-0.92		3		417
33554	+15 00759	HR 1684		5 08 49	+15 59.1	2	5.18	.000	1.50	.005	1.86	.000	5	K5 III	1355,3016
	-37 02059			5 08 49	-37 51.0	1	9.45		0.60		0.13		1	G2	742
241415	+21 00788	Hyades L 110		5 08 52	+21 43.3	1	9.70		0.66				2		582
33296	+62 00734	HR 1678		5 08 52	+62 37.9	1	6.50		0.21		0.12		1	A7 Vn	39
	-29 02099			5 08 52	-29 40.8	1	11.35		0.90		0.56		2		1696
	+44 01124			5 08 54	+44 30.6	1	10.39		0.04		-0.10		2	B9.5V	723
34297	-77 00192			5 08 55	-77 37.5	3	7.32	.006	0.65	.003	0.07		49	G5 V	978,1642,2012
		NGC 1817 - 240		5 08 58	+16 34.3	2	12.45	.015	0.74	.005	0.22	.000	3		305,557
273495	-45 01839			5 08 58	-44 59.0	1	10.25		1.38				2	K5	1730
		NGC 1817 - 246		5 09 00	+16 39.7	2	16.02	.050	0.72	.015	0.22	.040	5		305,557
33572	+16 00721	NGC 1817 - 247	★ A	5 09 00	+16 41.0	2	8.64	.030	0.50	.005	0.00	.005	4	G5	305,557
33459	+44 01126	LS V +44 032		5 09 00	+44 22.6	1	7.77		0.05		-0.48		2	B5 III	723
		NGC 1817 - 248	★ B	5 09 01	+16 41.1	2	11.17	.000	0.77	.005	0.40	.049	3		305,557
		AK 7 - 1312		5 09 02	+20 03.5	1	10.92		0.91				1		1570
		NGC 1817 - 239		5 09 03	+16 34.7	2	15.48	.005	0.72	.015	0.19	.015	2		305,557
33664	-12 01092	HR 1693, RX Lep		5 09 03	-11 54.6	4	5.64	.139	1.49	.076	1.24	.061	21	M6 III	897,1024,2035,3076
		Sk -67 049		5 09 04	-67 49.7	1	12.31		0.01		-0.36		4	A0 Ia	415

Table 1

HD	DM	Other Id	N	Rem	α_{1950}	δ_{1950}	S	V	σ_V	B–V	σ_{B-V}	U–B	σ_{U-B}	n	Spectrum	References
277622	+42 01201	SY Aur			5 09 05	+42 46.3	4	8.96	.140	0.98	.084	0.64	.003	4	G0.5	934,1399,1772,6011
290038	−2 01166				5 09 06	−01 54.9	1	9.52		0.93		0.69		1	K3 IV	3072
33771	−37 02066				5 09 06	−37 52.6	2	9.48	.029	0.86	.005	0.32		3	G6/8w(F)	742,1594
		NGC 1856 sq1 5			5 09 06	−69 12.5	1	12.61		0.06				3		1521
33647	+0 00974	HR 1690, V1085 Ori	⋆ AB		5 09 07	+00 27.3	6	6.67	.009	−0.07	.007	−0.34	.013	13	B9 Vn	15,1004,1079,1415*
33636	+4 00858	G 97 - 25			5 09 07	+04 20.7	1	7.04		0.57		0.08		1	G0	333,1620
		Aur sq 2 # 50			5 09 07	+44 21.0	1	11.22		0.10		−0.35		2		723
		STO 360			5 09 09	−68 08.9	1	13.24		0.55		−0.04		1	F5 Ia	726
33646	+0 00975	HR 1691	⋆ AB		5 09 10	+00 58.6	2	5.88	.005	0.66	.000	0.28	.005	7	F5	15,1415
277631	+42 01202				5 09 10	+42 16.5	1	9.12		0.00		0.00		1	F8	8084
		NGC 1817 - 167			5 09 12	+16 33.9	1	14.88		0.50		0.19		1		557
		NGC 1817 - 92			5 09 12	+16 37.9	2	13.80	.000	1.41	.015	1.40		2		305,557
		NGC 1817 - 284			5 09 12	+16 38.	2	16.29	.083	1.01	.058	0.56	.097	3		305,557
		NGC 1817 - 96			5 09 12	+16 39.3	1	13.31		1.33		1.20		1		557
		NGC 1817 - 97			5 09 12	+16 39.5	1	12.94		0.17		0.11		1		557
		NGC 1817 - 81			5 09 13	+16 34.7	2	12.18	.005	1.13	.005	0.78	.010	3		305,557
241486	+16 00722	NGC 1817 - 83			5 09 13	+16 35.5	2	9.40	.000	1.39	.010	1.28	.010	5	K0	305,557
		NGC 1817 - 177			5 09 13	+16 41.2	2	12.40	.049	1.12	.010	0.81	.005	3		305,557
33811	−44 01905				5 09 13	−44 38.0	1	8.71		0.77				4	G8 IV/V	2012
		Sk -67 050			5 09 13	−67 09.5	1	12.77		−0.06		−0.61		2	B6 Iab	149
		NGC 1817 - 82			5 09 14	+16 34.9	2	12.56	.000	0.47	.005	0.33	.005	3		305,557
277624	+42 01203		A		5 09 14	+42 11.9	1	7.31		−0.12		−0.36		1	B9	8084
277624	+42 01203		C		5 09 14	+42 11.9	1	12.32		0.41		0.03		1		8084
277624	+42 01203		D		5 09 14	+42 11.9	1	7.67		0.01		0.02		1		8084
277624	+42 01203		E		5 09 14	+42 11.9	1	11.99		1.73		2.08		1		8084
269096					5 09 14	−69 20.1	1	12.41		0.24		0.25		3	F7	149
33570	+30 00818				5 09 15	+30 09.5	1	8.62		0.53				3	F8	1724
		NGC 1817 - 84			5 09 16	+16 35.8	2	12.76	.019	0.51	.019	0.34	.005	3		305,557
33585	+26 00796				5 09 16	+26 23.8	1	6.78		1.02		0.64		2	G5 III	1625
33624	+16 00725	NGC 1817 - 48			5 09 18	+16 37.7	2	9.51	.005	0.23	.005	0.20	.000	6	A2	305,557
		Sk -66 063			5 09 18	−66 29.	1	13.95		0.01		−1.08		2		798
		NGC 1817 - 161			5 09 20	+16 31.9	2	11.51	.005	1.27	.005	1.14	.000	3		305,557
241495	+16 00724	NGC 1817 - 77			5 09 20	+16 32.8	2	9.57	.000	1.47	.010	1.63	.024	3	K7	305,557
		LP 597 - 5			5 09 21	+01 28.5	1	13.24		0.59		−0.18		2		1696
		WLS 520 20 # 6			5 09 22	+20 22.7	1	11.10		0.65		0.10		2		1375
		Sk -69 066			5 09 22	−70 01.0	1	13.17		−0.03		−0.46		4	B9 Iab	415
		Sk -70 081			5 09 22	−70 07.5	1	12.95		0.01		−0.49		3	B9 Iab	415
34171	−72 00350				5 09 22	−72 19.1	1	9.72		0.52		0.04		2	F8 III	204
		LP 717 - 8			5 09 24	−14 03.9	1	11.67		0.47		−0.16		2		3062
		AAS21,109 # 86			5 09 24	−66 28.	1	13.40		0.59		0.10		3		952
		NGC 1817 - 44			5 09 25	+16 35.2	2	11.36	.093	1.05	.015	0.69	.019	3		305,557
		NGC 1817 - 12			5 09 25	+16 38.2	2	12.61	.010	1.01	.015	0.57	.030	4		305,557
		NGC 1817 - 8			5 09 26	+16 37.2	2	12.12	.010	1.10	.030	0.81	.005	2		305,557
33605	+27 00737				5 09 26	+27 32.7	1	8.59		1.43				3	K0	1724
33569	+34 00963				5 09 26	+34 45.9	1	8.64		0.14		−0.47		2	B8	401
		NGC 1817 - 5			5 09 27	+16 36.7	2	13.86	.005	0.43	.005	0.33	.045	2		305,557
33542	+44 01128				5 09 30	+44 30.5	3	7.32	.021	0.06	.012	−0.49	.008	6	B5 III	401,723,1733
33661	+13 00828				5 09 31	+13 51.9	1	8.53		1.01				2	G5	882
33725	−9 01094	IDS05071S0913	A		5 09 31	−09 09.9	3	8.04	.008	0.80	.010	0.40	.017	9	K1 V	22,2012,3008
269110	−69 00333		V		5 09 32	−69 39.7	2	10.74	.075	0.90	.040	0.75		10	G0 Ia	149,1483
	+44 01130	LS V +44 034	⋆ AB		5 09 33	+44 24.9	1	9.42		0.08		−0.48		2	B8	723
33690	+8 00886				5 09 34	+09 03.9	1	8.68		0.13		0.10		2	B9	401
33807	−30 02232				5 09 34	−30 17.1	1	7.07		0.44		0.02		4	F3/5 V	1628
33659	+17 00869				5 09 36	+17 09.3	1	8.46		1.56				3	K2	882
	+44 01129				5 09 36	+44 23.2	1	9.94		−0.03		−0.26		2	B8 V	723
		Sk -71 012			5 09 36	−71 29.	1	13.31		−0.04		−0.97		4		798
241511	+23 00875	G 85 - 51			5 09 37	+23 50.4	1	10.32		0.88		0.67		1	K7	333,1620
		ER Aur			5 09 38	+41 57.	1	11.20		0.99				1		1772
		KUV 05097+1649			5 09 40	+16 48.6	1	14.35		0.65		0.20		1	DA	1708
33793	−45 01841				5 09 42	−44 59.9	4	8.84	.023	1.56	.009	1.21	.000	10	M0 V	1097,2012,3078,8105
		SK *C 26			5 09 42	−66 09.	1	13.39		0.04		−0.07		3		149
		NGC 1817 - 33			5 09 43	+16 34.7	1	12.18		0.27		0.19		1		557
		G 96 - 24			5 09 43	+45 57.7	1	10.74		0.97		0.67		1	K2	333,1620
269099					5 09 43	−69 01.	1	12.33		0.67		0.53		1	K7	360
		G 85 - 52			5 09 44	+19 36.3	1	10.82		1.55		1.21		1	M0	1620
33604	+40 01213	LS V +40 036			5 09 44	+40 08.1	3	7.40	.044	0.03	.011	−0.70	.015	7	B2 V:pe	401,1012,1212
34144	−69 00334				5 09 44	−69 13.2	5	9.36	.020	0.20	.015	0.11	.059	24	A3:V	2,149,204,1496,1521
		Sk -68 056			5 09 45	−68 28.4	1	12.54		0.02		−0.39		4	B8 Ib	415
33657	+27 00738				5 09 48	+27 42.9	1	9.24		0.49				2	F6/8 V/IV	1724
33603	+41 01114				5 09 48	+41 52.6	1	10.08		1.12		0.90		10	K2	1655
		AAS21,109 # 88			5 09 48	−66 34.	1	14.06		0.47		−0.14		3		952
		SK *P 2111			5 09 48	−66 48.3	1	12.06		0.52		−0.01		1	F2 Ia	726
		AAS21,109 # 87			5 09 48	−68 32.	1	13.72		−0.18		−0.97		3		952
		NGC 1817 - 144			5 09 50	+16 31.9	1	10.81		0.49		0.06		1		557
33872	−37 02071	HR 1699			5 09 52	−37 27.3	5	6.56	.004	1.62	.000	1.95		18	M1 III	15,1020,1075,2012,3005
34048	−62 00435				5 09 52	−62 44.9	2	9.97	.017	0.55	.000	0.05		3	G0wF2	1594,6006
		Sk -67 051			5 09 52	−67 58.	1	12.63		−0.16		−1.00		4		415
33908	−45 01845				5 09 53	−45 00.9	1	9.34		0.91		0.52		2	K0	1730
33708	+13 00829				5 09 54	+13 18.9	2	8.83	.010	1.20	.025	1.11		5	K0	882,3077
33643	+33 00977				5 09 54	+33 15.6	1	8.54		0.43		0.04		2	F2	1375

HD	DM	Other Id	N	Rem	α_{1950}	δ_{1950}	S	V	σ_V	B–V	σ_{B-V}	U–B	σ_{U-B}	n	Spectrum	References
33632	+37 01091	IDS05065N3713	A		5 09 54	+37 16.8	1	6.46		0.52		-0.02		3	F7 V	1501
		G 85 - 53			5 09 55	+16 15.1	1	10.59		1.23		1.22		2	K7	333,1620
241596	+19 00869	G 85 - 54			5 09 55	+19 40.3	1	9.87		1.02		0.86		1	K0	1620
33707	+17 00872				5 09 56	+18 05.8	1	8.52		0.88		0.60		2	G5	1648
33752	-0 00878				5 09 56	-00 34.3	1	8.89		0.04		0.02		2	A0 V	1375
	-11 01100				5 09 56	-11 51.7	1	8.23		0.45		-0.06		5	G0	897
		Sk -69 067			5 09 57	-69 22.3	1	12.29		0.11		0.05		2	A2 Iab	149
33802	-12 01095	HR 1696	⋆	A	5 09 58	-11 55.7	7	4.45	.009	-0.10	.011	-0.39	.011	21	B8 V	15,1075,1079,1425*
33641	+38 01063	HR 1689			5 10 00	+38 25.6	7	4.85	.024	0.19	.009	0.09	.005	19	A4 m	15,39,374,1363,6007*
		AAS21,109 # 89			5 10 00	-68 33.	1	13.89		-0.12		-0.95				952
34187		Sk -68 057			5 10 00	-68 57.	2	13.67	.040	-0.19	.035	-0.86	.010	7	WN3	798,980
		Sk -71 013			5 10 00	-71 24.	1	13.90		-0.16		-0.99		4		798
290050	+0 00981	G 99 - 16			5 10 01	+00 43.2	1	8.36		0.70		0.17		1	F8 V	1620
241659	+4 00865	G 84 - 39			5 10 02	+04 14.8	1	10.58		0.79		0.32		1	K2	333,1620
33734	+13 00831				5 10 03	+13 51.8	1	8.89		1.75				2	K0	882
273831	-48 01675				5 10 03	-48 27.5	1	9.93		0.58		0.00		4	F8	243
	-57 00752	IDS05085S5743	A		5 10 03	-57 39.0	1	9.82		0.87				3		1594
	-10 01124				5 10 04	-10 36.9	1	10.57		0.63		0.06		2		1696
269089					5 10 04	-67 42.2	2	12.08	.010	0.78	.000	0.28	.010	8	G8 V	149,204
		Sk -67 052			5 10 06	-67 18.	1	13.28		-0.10		-0.94		3		798
269101		Sk -68 058			5 10 06	-68 50.	3	12.03	.017	-0.03	.011	-0.71	.014	10	B3 Iab	415,1277,1408
269116		Sk -68 059			5 10 06	-68 59.	1	12.07		-0.05		-0.83		4	B0.7Ia+	415
		AAS21,109 # 90			5 10 06	-69 18.	1	14.18		-0.20		-0.94		2		417
		KUV 05101+1619			5 10 07	+16 19.2	1	16.36		0.50		0.19		1	DA	1708
33750	+13 00832				5 10 09	+13 54.4	1	8.84		1.17				2	K0	882
	+44 01134				5 10 10	+44 32.4	1	10.06		-0.02		-0.31		2	B8.5V	723
34359	-74 00318				5 10 10	-74 54.5	1	8.55		0.18		0.09		57	A3 IV/V	978
33671	+35 01012				5 10 11	+35 35.7	1	7.84		0.10		-0.05		2	A0	401
		Sk -68 060			5 10 11	-68 19.	1	13.42		-0.03		-0.40		4		415
		Sk -65 031			5 10 12	-65 26.0	1	12.00		0.05		-0.38		3		149
	+44 01135				5 10 13	+44 37.2	1	9.73		0.02		-0.10		2	Am	723
33688	+35 01014	IDS05069N3534	A		5 10 15	+35 38.6	1	8.32		0.10		-0.47		2		401
		Sk -69 068			5 10 15	-69 12.	1	12.52		0.01		-0.82		4		415
		NGC 1856 sq1 27			5 10 16	-69 10.6	1	12.54		-0.03				3		1521
		WLS 520 20 # 11			5 10 17	+21 45.0	1	11.51		0.70		0.35		2		1375
269172	-71 00319	Sk -71 014			5 10 19	-71 27.6	4	10.61	.011	0.08	.009	-0.46	.010	17	A0 Ia0:	149,415,1277,8033
33889	-26 02064	IDS05083S2640	AB		5 10 20	-26 36.6	2	8.33	.010	0.56	.001	0.06		42	F7 V	978,2012
33968	-43 01726	IDS05088S4300	A		5 10 20	-42 56.2	1	10.80		0.74		0.32		2		1696
33968	-43 01726	IDS05088S4300	AB		5 10 20	-42 56.2	1	9.81		0.72				4		2012
33833	-6 01109	HR 1697			5 10 21	-06 06.9	4	5.91	.014	0.96	.000	0.73	.015	15	G8 III:	15,247,1061,2030
269100		Sk -68 061			5 10 21	-68 18.2	1	12.41		0.28		0.35		3	A7 Ib	415
33774	+16 00732	IDS05074N1647	A		5 10 22	+16 49.6	1	8.94		1.43				2	K2	882
33952	-40 01781				5 10 23	-40 24.4	2	8.93	.010	1.11	.002	1.09	.013	14	K0 III	9,9
33704	+36 01047				5 10 26	+36 58.5	1	6.85		0.01		-0.02		2	A0 V	401
277787	+40 01214				5 10 28	+40 15.3	1	9.24		1.33				1	K0	934
		Sk -65 032			5 10 30	-65 26.	1	13.20		-0.01		-0.27		3		415
		AAS21,109 # 91			5 10 30	-66 38.	1	14.49		-0.18		-0.86		3		952
		AAS21,109 # 92			5 10 30	-66 49.	1	14.72		-0.13		-0.83		4		952
		Sk -67 053			5 10 30	-67 19.	1	13.44		-0.19		-0.92		4		798
269117		Sk -68 062			5 10 30	-68 29.	1	11.77		-0.01		-0.62		4	B7	415
	+44 01137				5 10 31	+44 19.8	1	11.10		0.18		0.01		2	K0	723
	-8 01057				5 10 34	-08 07.4	1	8.22		0.04		0.00		2	A0	401
34091	-57 00758				5 10 35	-57 40.3	1	9.37		1.17		1.13		1	K2 II/III	742
		Sk -68 064			5 10 36	-68 14.	1	12.57		0.29		0.15		3	A0 Ib	952
		Sk -70 082			5 10 36	-70 09.	1	13.39		-0.07		-0.93		4		798
		Hyades L 111			5 10 37	+15 12.4	1	11.61		0.82		0.41		3		1058
269128	-68 00326	Sk -68 063			5 10 37	-68 50.0	5	10.49	.053	0.00	.013	-0.77	.017	19	B2.5I e	149,328,360,1277,8033
241665	+27 00739				5 10 38	+27 11.0	1	9.62		0.53				3	G0	1724
33856	+2 00888	HR 1698	⋆	AB	5 10 40	+02 48.2	10	4.46	.013	1.18	.007	1.15	.018	31	K3 III	15,542,1075,1355,1363*
33749	+36 01049	IDS05073N3648			5 10 40	+36 51.5	1	7.65		0.08		-0.37		2	B8	401
33904	-16 01072	HR 1702, μ Lep			5 10 41	-16 15.8	9	3.29	.026	-0.11	.008	-0.38	.016	35	B9 IV:HgMn	3,15,1063,1075,1202*
	-40 01782				5 10 42	-40 32.0	2	9.69	.009	1.03	.008	0.84	.010	13	K0	9,9
		AAS21,109 # 94			5 10 42	-69 10.	1	13.65		-0.03		-1.00		2		417
269157					5 10 42	-70 14.3	1	10.01		1.59		1.76		3	K4 III	149
33654	+53 00872	HR 1692	⋆		5 10 43	+53 09.4	2	6.18	.020	0.08	.005	0.04	.005	5	A0 V	985,1733
269139		Sk -69 069			5 10 44	-69 12.9	1	11.42		0.22		-0.31		6	A0 Ia	149
34060	-49 01610				5 10 45	-49 07.2	1	7.82		-0.08		-0.18		4	B9 V(pSiCr)	243
		Aur sq 2 # 59			5 10 46	+44 25.9	1	10.53		0.12		-0.41		2		723
33618	+59 00857	HR 1688			5 10 47	+59 21.0	4	6.15	.009	1.18	.006	1.23	.040	15	K2 III-IV	37,252,1118,1379
33787	+28 00761				5 10 48	+28 10.0	1	9.24		0.59				3	G5	1724
		Sk -65 033			5 10 48	-65 21.	1	13.19		-0.13		-0.78		4		798
33902	-5 01191				5 10 52	-05 01.9	1	9.49		0.08		-0.04		2	B8 V	401
34073	-49 01611				5 10 52	-49 07.0	1	9.13		0.01		-0.03		4	A0 V	243
		KUV 05109+1739			5 10 55	+17 38.6	1	14.06		-0.06		-0.87		1	sdOB	1708
33949	-13 01092	HR 1705	⋆	AB	5 10 55	-13 00.0	7	4.36	.003	-0.10	.004	-0.36	.018	21	B7 V	15,1075,1079,1425*
33883	+1 00938	HR 1701	⋆	AB	5 10 56	+01 54.6	2	6.08	.005	0.42	.000	0.27	.010	7	A5 V	15,1415
33979	-27 02138				5 10 56	-27 13.0	1	7.34		1.38		1.51		3	K2 III	1657
33772	+39 01219				5 10 57	+39 18.4	2	8.02	.015	1.07	.000	0.85	.005	2	G7 II	1501,1601
241701	+32 00911				5 10 58	+32 22.9	1	9.64		0.22		-0.22		5		1726
33918	-4 01073	IDS05086S0447	A		5 11 00	-04 42.6	1	7.70		0.17		0.10		2	A0	401

Table 1 253

HD	DM	Other Id	N Rem	α_{1950}	δ_{1950}	S	V	σ_V	B–V	σ_{B-V}	U–B	σ_{U-B}	n	Spectrum	References
		Sk -67 055		5 11 00	−67 13.	1	13.21		−0.14		−1.00		4		798
233091	+50 01135	IDS05071N5047	A	5 11 01	+50 50.9	1	9.24		0.38		0.04		2	F0	1566
33829	+30 00827			5 11 02	+30 20.6	1	7.16		0.26				3	F0	1724
	+44 01139			5 11 03	+44 44.2	1	9.48		0.34		0.10		2	Am	723
		Sk -69 070		5 11 04	−69 25.	1	12.71		0.03		−0.52		4		415
33864	+21 00796			5 11 07	+21 10.0	1	7.19		0.21		0.07		2	A2	1648
33816	+38 01071	LS V +38 010		5 11 08	+38 10.1	1	8.78		0.00		−0.72		2	B5	180
33948	−8 01059	HR 1704		5 11 09	−08 12.3	2	6.36	.005	−0.14	.005	−0.53	.005	7	B5 V	15,1061
33653	+60 00864			5 11 12	+60 50.1	1	7.53		0.47				1	A0	1642
271067				5 11 12	−66 00.0	1	11.93		0.06		−0.22		4	G5	149
33946	+0 00988	HR 1703		5 11 13	+00 30.2	4	6.31	.010	1.46	.011	1.36	.021	10	M0 V	15,247,1355,1415
		Sk -65 035		5 11 13	−65 27.	1	12.64		0.03		−0.16		3		415
277792	+39 01221			5 11 14	+40 05.9	1	9.66		1.21				1	K0	934
	+44 01140			5 11 16	+44 25.9	1	10.04		0.23		0.12		2	Am	723
		SK *C 29		5 11 18	−65 54.	1	12.97		0.26		0.11		3		149
241814	+19 00872	G 85 - 55		5 11 19	+19 49.7	1	9.47		1.17		1.11		1	K5	333,1620
34089	−40 01788	IDS05097S4032	AB	5 11 19	−40 28.3	2	9.66	.007	0.47	.006	0.06	.014	13	F5 V	9,9
		VES 883		5 11 20	+32 42.4	1	11.94		0.34		−0.08		3		1543
33799	+44 01141			5 11 20	+44 26.2	1	9.57		0.00		−0.24		2	B9 V	723
	+44 01142	G 96 - 28	*AB	5 11 20	+44 29.2	1	10.33		1.06		0.97		1	K3	333,1620
		Steph 554		5 11 21	+05 19.5	1	12.17		1.44		1.20		1	M0	1746
		G 84 - 41		5 11 21	+07 57.1	1	15.89		0.41		−0.38		1	WD	782
		Hyades J 356		5 11 22	+03 15.	1	12.69		0.52		0.05		1		1058
241785	+29 00842			5 11 22	+30 03.5	1	8.46		1.84				4	M0	369
33994	−6 01112			5 11 22	−06 48.3	1	7.34		−0.06		−0.46		2	B8	401
33926	+16 00735			5 11 27	+16 37.3	1	8.27		1.32				3	K2	882
		SK *G 231		5 11 28	−68 27.1	1	13.82		0.21		0.36		1	F3 Ib	149
241801	+29 00843			5 11 32	+29 15.2	1	9.63		0.45				2		1724
241885	+14 00854			5 11 34	+14 25.6	1	9.28		1.48				1	K5	882
		Sk -67 056		5 11 36	−67 26.	1	13.18		−0.18		−0.99		3		952
		LS V +36 008		5 11 40	+36 38.6	1	11.13		0.64		−0.41		3		180
33861	+39 01225	UZ Aur		5 11 41	+40 04.6	1	8.14		1.80		1.97		8	M3 III	1655
		AAS21,109 # 96		5 11 42	−65 19.	1	13.65		−0.05		−0.41		3		952
		AAS21,109 # 95		5 11 42	−66 53.	2	14.59	.122	−0.15	.056	−1.02	.014	4		417,952
34045	−14 01074	HR 1710		5 11 43	−14 39.8	2	6.20	.005	0.38	.005			7	F2 III	15,2012
33853	+46 00985	IM Aur		5 11 47	+46 21.0	1	7.94		0.01		−0.48		2	B7 V	401
		Sk -65 037		5 11 48	−65 25.	1	12.57		0.02		−0.14		4	A0 Ia	415
		Sk -67 057		5 11 49	−67 13.	1	12.29		0.34		−0.52		4		415
		LS V +35 009		5 11 50	+35 40.6	1	10.97		0.73		−0.30		2	B1 V:p(e)	1012
		NGC 1851 sq1 7		5 11 50	−40 14.8	1	14.01		0.52		−0.07		2		473
		NGC 1851 sq1 13		5 11 50	−40 16.6	1	14.72		0.57		−0.03		2		473
34349	−65 00441			5 11 51	−65 14.1	6	7.03	.013	0.43	.013	0.01	.009	38	F5 V	14,361,1408,2001,2033,3021
		YZ Aur		5 11 53	+40 01.2	2	9.94	.006	1.18	.023			2	G5	934,1772
		NGC 1851 sq1 15		5 11 54	−40 15.1	1	14.90		0.67		0.09		2		473
		Sk -65 036		5 11 54	−65 40.0	1	13.31		0.02		−0.13		3		149
277784	+40 01222			5 11 56	+40 19.9	1	9.70		1.77				3	K5	369
241878	+29 00845			5 11 57	+29 10.4	1	9.30		1.90				3	K7	369
		NGC 1851 sq1 8		5 11 58	−40 16.5	1	14.05		0.55		−0.05		2		473
	+60 00866			5 11 59	+60 12.7	1	9.72		0.34				2	F2	1379
34167	−35 02176	HR 1715		5 12 00	−35 52.9	1	6.98		1.47				4	K4 III	2007
33541	+73 00280	HR 1683		5 12 01	+73 12.8	1	5.83		−0.04		−0.12		5	A0 V	1733
		IN Aur		5 12 02	+37 18.9	1	13.63		1.37				1		1772
34043	+4 00877	HR 1709		5 12 04	+05 06.0	4	5.49	.008	1.37	.009	1.55	.006	11	gK4	15,247,1061,1355
34101	−15 00978			5 12 04	−15 52.8	3	7.43	.007	0.72	.009	0.27		7	G8 V	258,2012,3026
34246	−48 01693			5 12 04	−48 33.0	1	9.18		1.04		0.80		4	K1 III	243
241894	+29 00846			5 12 06	+29 13.7	1	9.58		1.32				2		1724
		Sk -65 038		5 12 06	−65 30.	1	13.69		−0.23		−1.00		2		798
34085	−8 01063	HR 1713	*ABC	5 12 08	−08 15.5	10	0.14	.032	−0.03	.004	−0.67	.018	25	B8 Ia+B8III	1,15,1004,1034,1079*
273190	−41 01766			5 12 08	−40 57.1	1	10.78		0.47				4	K0	2034
33959	+32 00922	HR 1706, KW Aur	*AB	5 12 09	+32 37.9	5	4.99	.044	0.21	.015	0.16	.039	9	A9 IV	39,597,1049,1118,3058
33959	+32 00922	IDS05089N3234	C	5 12 09	+32 37.9	1	7.95		0.41		−0.06		3	F3 V	3024
		NGC 1851 sq1 12		5 12 09	−40 16.4	1	14.43		0.70		0.20		2		473
		NGC 1868 sq1 F2		5 12 10	−64 06.6	1	12.09		0.74				3		827
		SK *S 46		5 12 10	−69 24.3	1	13.30		0.15		0.28		2		149
34033	+12 00754			5 12 11	+12 57.5	2	8.66	.005	1.08	.010	0.88		6	G8 II-III	206,882
33814	+60 00867			5 12 12	+60 12.9	1	8.74		0.32				4	F0	1379
		SK *C 31		5 12 12	−67 15.	1	13.32		0.10		−0.05		2		149
		AAS21,109 # 97		5 12 12	−67 29.	1	13.89		0.11		−1.01		3		417
	−64 00425	NGC 1868 sq2 1		5 12 13	−64 05.7	1	11.09		0.95				3		827
34031	+19 00876			5 12 14	+20 00.0	1	7.76		0.67				1	G0	987
		NGC 1851 sq1 9		5 12 15	−40 18.3	1	14.09		0.65		−0.21		2		473
34328	−59 00444			5 12 17	−59 42.6	6	9.43	.016	0.49	.006	−0.24	.037	13	G0wA8 V	742,1097,1594,2012*
		Sk -69 071		5 12 18	−69 20.	1	13.15		−0.10		−0.83		4		798
34183	−26 02084			5 12 20	−26 51.0	1	7.41		1.36		1.50		34	K2 III	978
280662				5 12 21	+37 33.	1	10.12		0.10		−0.53		1	B8	401
269145	−67 00396	Sk -67 058		5 12 21	−67 22.8	1	11.34		0.00		−0.66		5	B5 Ia	149
		NGC 1851 sq1 4		5 12 22	−40 03.2	1	14.82		1.18		0.83		7		1638
		NGC 1851 - 3		5 12 24	−40 05.	1	13.68		1.30				3		9
		NGC 1851 - 20		5 12 24	−40 05.	2	16.14	.027	0.14	.014	0.16		10		926,1666
		NGC 1851 - 26		5 12 24	−40 05.	1	14.85		0.62				1		926

HD	DM	Other Id	N Rem	α_{1950}	δ_{1950}	S	V	σ_V	B–V	σ_{B-V}	U–B	σ_{U-B}	n	Spectrum	References
		NGC 1851 - 29		5 12 24	−40 05.	1	18.34		0.83				1		1666
		NGC 1851 - 37		5 12 24	−40 05.	2	14.92	.085	1.09	.075	0.69		4		9,473
		NGC 1851 - 41		5 12 24	−40 05.	2	14.96	.021	1.08	.016	0.75		10		926,1666
		NGC 1851 - 49		5 12 24	−40 05.	2	14.96	.007	0.71	.009	0.05		7		926,1666
		NGC 1851 - 50		5 12 24	−40 05.	1	14.91		0.83		0.58		2		1666
		NGC 1851 - 55		5 12 24	−40 05.	1	15.57		0.32		0.44		2		9
		NGC 1851 - 58		5 12 24	−40 05.	2	14.28	.034	1.29	.008	0.85		7		9,1666
		NGC 1851 - 79		5 12 24	−40 05.	1	16.87		0.00		-0.16		2		1666
		NGC 1851 - 90		5 12 24	−40 05.	3	13.18	.015	0.73	.006	0.30	.078	18		9,926,1666
		NGC 1851 - 94		5 12 24	−40 05.	1	16.91		0.76				2		926
		NGC 1851 - 95		5 12 24	−40 05.	4	13.60	.028	1.45	.026	1.51	.014	12		9,473,926,1666
		NGC 1851 - 102		5 12 24	−40 05.	1	15.87		0.23		0.14		1		1666
		NGC 1851 - 103		5 12 24	−40 05.	1	16.78		0.05		-0.07		1		1666
		NGC 1851 - 107		5 12 24	−40 05.	3	14.55	.064	1.21	.007	1.31		6		9,926,1666
		NGC 1851 - 108		5 12 24	−40 05.	1	16.17		0.49				8		926
		NGC 1851 - 109		5 12 24	−40 05.	3	14.88	.055	1.08	.034	0.83	.065	7		9,926,1666
		NGC 1851 - 112		5 12 24	−40 05.	3	13.81	.019	1.48	.001	1.47	.075	11		9,473,926,1666
		NGC 1851 - 113		5 12 24	−40 05.	1	17.24		0.74				2		1666
		NGC 1851 - 126		5 12 24	−40 05.	2	14.36	.005	1.27	.083			3		9,1666
		NGC 1851 - 136		5 12 24	−40 05.	1	15.71		0.88				1		9
		NGC 1851 - 151		5 12 24	−40 05.	1	13.75		1.33				3		9
		NGC 1851 - 164		5 12 24	−40 05.	2	14.37	.015	1.17	.021	0.89		7		9,926,1666
		NGC 1851 - 165		5 12 24	−40 05.	1	16.43		0.81				5		926
		NGC 1851 - 168		5 12 24	−40 05.	2	13.22	.004	1.70	.017	1.51		15		9,926,1666
		NGC 1851 - 183		5 12 24	−40 05.	1	15.45		0.83				4		926
		NGC 1851 - 187		5 12 24	−40 05.	2	16.43	.026	0.05	.004	0.14		6		926,1666
		NGC 1851 - 188		5 12 24	−40 05.	1	16.61		-0.06				7		926
		NGC 1851 - 189		5 12 24	−40 05.	1	15.78		0.96				7		926
		NGC 1851 - 191		5 12 24	−40 05.	3	14.75	.037	0.73	.038	0.02	.035	4		9,926,1666
		NGC 1851 - 195		5 12 24	−40 05.	1	15.88		0.87				1		1666
		NGC 1851 - 199		5 12 24	−40 05.	1	17.22		0.86				2		1666
		NGC 1851 - 200		5 12 24	−40 05.	1	16.57		0.02		-0.21		2		1666
		NGC 1851 - 209		5 12 24	−40 05.	1	14.39		1.28				2		9
		NGC 1851 - 210		5 12 24	−40 05.	1	14.52		1.02		0.45		2		9
		NGC 1851 - 216		5 12 24	−40 05.	1	14.85		0.83				1		9
		NGC 1851 - 236		5 12 24	−40 05.	2	14.58	.066	0.92	.028	0.42	.037	4		9,1666
		NGC 1851 - 239		5 12 24	−40 05.	1	17.25		0.67				1		1666
		NGC 1851 - 241		5 12 24	−40 05.	3	14.31	.031	0.55	.027	0.05		9		9,926,1666
		NGC 1851 - 249		5 12 24	−40 05.	3	15.33	.010	1.00	.019	0.65	.074	13		9,473,926,1666
		NGC 1851 - 251		5 12 24	−40 05.	2	15.77	.042	0.99	.075	0.38		4		9,1666
		NGC 1851 - 262		5 12 24	−40 05.	1	13.47		1.62				4		9
		NGC 1851 - 272		5 12 24	−40 05.	1	14.99		0.87		0.04		1		9
		NGC 1851 - 275		5 12 24	−40 05.	1	15.10		1.37				1		9
		NGC 1851 - 279		5 12 24	−40 05.	1	13.98		1.38				2		9
		NGC 1851 - 283		5 12 24	−40 05.	1	16.80		0.94				2		1666
		NGC 1851 - 285		5 12 24	−40 05.	1	16.10		0.72				8		926
		NGC 1851 - 293		5 12 24	−40 05.	2	15.57	.078	1.06	.034	0.93		3		9,1666
		NGC 1851 - 294		5 12 24	−40 05.	2	13.43	.030	1.63	.040	1.51		5		9,1666
		NGC 1851 - 295		5 12 24	−40 05.	1	16.22		0.61				2		1666
		NGC 1851 - 319		5 12 24	−40 05.	1	14.82		1.17		0.91		2		1666
		NGC 1851 - 320		5 12 24	−40 05.	3	14.94	.012	0.86	.004	0.39	.027	15		926,1638,1666
		NGC 1851 - 324		5 12 24	−40 05.	1	15.91		0.72				10		926
		NGC 1851 - 328		5 12 24	−40 05.	1	17.26		0.84				1		1666
		NGC 1851 - 329		5 12 24	−40 05.	4	13.45	.021	1.67	.017	1.98		27		9,926,1638,1666
		NGC 1851 - 333		5 12 24	−40 05.	4	14.05	.035	1.29	.012	1.29	.065	19		9,926,1638,1666
		NGC 1851 - 378		5 12 24	−40 05.	2	15.61	.013	0.57	.042	0.03		5		926,1666
		NGC 1851 - 379		5 12 24	−40 05.	1	15.16		0.86		0.24		2		1666
		NGC 1851 - 388		5 12 24	−40 05.	1	16.15		0.90				8		926
		NGC 1851 - 395		5 12 24	−40 05.	2	13.58	.049	1.59	.010	1.77		6		9,1666
		NGC 1851 - 397		5 12 24	−40 05.	1	15.12		0.72				1		926
		NGC 1851 - 398		5 12 24	−40 05.	1	16.17		0.16				6		926
		NGC 1851 - 399		5 12 24	−40 05.	1	15.16		1.01				1		926
		NGC 1851 - 400		5 12 24	−40 05.	1	17.01		0.76				2		1666
		NGC 1851 - 401		5 12 24	−40 05.	4	13.31	.026	0.70	.010	0.15	.017	22		9,473,926,1666
		NGC 1851 - 1125		5 12 24	−40 05.	1	15.19		1.10				3		9
		NGC 1851 - 1140		5 12 24	−40 05.	1	14.54		1.12		0.67		2		9
		NGC 1851 - 1504		5 12 24	−40 05.	1	14.06		0.97		0.39		3		9
		NGC 1851 - 1505		5 12 24	−40 05.	1	13.71		1.09		0.86		3		9
		NGC 1851 - 1512		5 12 24	−40 05.	1	13.37		1.10		1.20		1		9
		NGC 1851 - 1516		5 12 24	−40 05.	1	14.03		0.90		0.49		2		9
		NGC 1851 - 1517		5 12 24	−40 05.	1	14.91		0.66		0.44		1		9
		NGC 1851 - 1525		5 12 24	−40 05.	1	13.65		1.07		0.61		2		9
		NGC 1851 - 1526		5 12 24	−40 05.	1	13.12		1.08		0.59		2		9
		NGC 1851 - 1529		5 12 24	−40 05.	1	14.99		0.57		0.44		1		9
		NGC 1851 - 1536		5 12 24	−40 05.	1	14.37		1.14		0.43		2		9
		NGC 1851 - 1537		5 12 24	−40 05.	1	12.98		0.57		-0.02		3		9
		NGC 1851 - 1541		5 12 24	−40 05.	1	14.25		0.83		0.48		2		9
		NGC 1851 - 1542		5 12 24	−40 05.	1	13.11		0.92		0.33		1		9
		NGC 1851 - 1544		5 12 24	−40 05.	1	13.28		1.83				3		9
		NGC 1851 - 1545		5 12 24	−40 05.	1	13.37		1.31		1.02		3		9

Table 1

HD	DM	Other Id	N Rem	α_{1950}	δ_{1950}	S	V	σ_V	B–V	σ_{B-V}	U–B	σ_{U-B}	n	Spectrum	References
		NGC 1851 - 2001		5 12 24	−40 05.	1	9.56		1.05		0.91		8		1666
		NGC 1851 - 2002		5 12 24	−40 05.	1	13.96		1.04		0.68		1		1666
		NGC 1851 - 2004		5 12 24	−40 05.	1	11.44		0.56		0.05		8		1666
		NGC 1851 - 2006		5 12 24	−40 05.	1	11.73		0.59		0.01		8		1666
		NGC 1851 - 2022		5 12 24	−40 05.	1	13.16		1.62		1.10		1		1666
		NGC 1851 - 2033		5 12 24	−40 05.	1	18.33		0.76				1		1666
		NGC 1851 - 2039		5 12 24	−40 05.	1	14.34		0.95				1		1666
		NGC 1851 - 2046		5 12 24	−40 05.	1	17.95		0.68				1		1666
		NGC 1851 - 2047		5 12 24	−40 05.	1	13.42		1.54		1.73		1		1666
		NGC 1851 - 2052		5 12 24	−40 05.	1	18.49		0.43				1		1666
		NGC 1851 - 2053		5 12 24	−40 05.	1	13.83		1.79		2.15		1		1666
34053	+22 00864	HR 1711	⋆ A	5 12 27	+22 13.8	2	6.26		0.08	.000	0.13	.015	4	A2 V	351,1049
280661	+37 01110			5 12 29	+37 34.6	1	10.44		0.10		-0.52		2	B5	401
34198	−26 02085	UU Lep		5 12 29	−26 15.8	1	6.99		1.12		0.89		1	K2 III/IV	1641
271093				5 12 29	−72 05.2	1	12.97		0.26		-0.76		1	A3	149
34009	+36 01060	IDS05091N3637	A	5 12 30	+36 39.7	1	9.01		0.11		-0.40		2	B9	401
		SK *C 32		5 12 30	−65 53.	1	13.17		0.12		0.15		3		149
		AAS21,109 # 98		5 12 30	−67 31.	1	14.03		-0.17		-0.95		2		952
		AK 7 - 1578		5 12 31	+18 30.1	1	12.23		0.88				1		987
34063	+21 00801	G 85 - 58		5 12 36	+21 39.7	1	8.44		0.80		0.36		1	G5	333,1620
34052	+29 00847			5 12 36	+29 24.7	1	9.02		0.67				2	G2 V	1724
34052	+29 00847	IDS05094N2921	AB	5 12 36	+29 24.7	1	8.83		0.67		0.20		2	G2 V	1003
		UV0512-08		5 12 36	−08 52.0	1	11.32		-0.27		-1.11		1		1732
		Sk -67 059		5 12 36	−67 18.	2	12.84	.005	0.03	.050	-0.51	.125	8		415,798
34164	−5 01204			5 12 37	−05 18.5	1	9.48		0.12		0.01		2	A0	401
		NGC 1851 sq1 10		5 12 42	−40 18.8	1	14.21		0.71		0.21		2		473
		Sk -65 039		5 12 42	−65 09.	1	14.07		-0.19		-0.91		3		798
34802	−77 00196	YZ Men		5 12 42	−77 16.5	3	7.65	.041	1.09	.006	0.79	.008	7	K1 IIIp	1641,1642,1704
34266	−36 02127	HR 1721		5 12 43	−36 02.0	4	5.75	.007	1.01	.003			18	G8 III	15,1075,2013,2029
34347	−52 00677	HR 1727		5 12 43	−52 05.2	4	6.04	.017	1.39	.008	1.60		13	K3 III	15,1075,2007,3005
33988	+46 00989	LS V +46 026		5 12 45	+46 21.7	1	6.88		0.25		-0.74		3	B5	1212
34180	−1 00837	HR 1717		5 12 46	−01 27.9	3	6.15	.015	0.40	.005	-0.04	.009	8	F0 IV	15,247,1078
269154	−67 00398	IDS05128S6723	⋆ AB	5 12 47	−67 19.3	2	10.50	.005	0.50	.000	-0.06		6	F6 Ia	149,1483
34179	−0 00892			5 12 48	+00 03.3	3	8.04	.005	-0.05	.008	-0.43	.010	8	B8 V	1004,1731,8055
		AAS21,109 # 100		5 12 48	−67 05.	1	14.18		-0.20		-1.02		3		952
		AAS21,109 # 99		5 12 48	−68 46.	1	14.49		-0.19		-1.09		3		417
34062	+31 00905			5 12 53	+31 45.5	2	8.76	.010	0.06	.010	-0.19	.014	7	B9	401,1775
242029	+30 00840			5 12 56	+30 08.2	1	9.44		0.41				3		1724
271086		NGC 1866 sq1 6		5 12 57	−65 30.	3	12.04	.028	0.55	.018	0.04	.015	14	K0	14,2001,3021
		NGC 1868 sq1 3		5 12 58	−63 50.8	1	11.70		1.20				3		827
34029	+45 01077	HR 1708	⋆ AP	5 12 59	+45 57.0	7	0.07	.014	0.80	.012	0.44	.010	24	G5 IIIe +	15,1013,1118,1193*
34078	+34 00980	HR 1712, AE Aur	⋆ AB	5 13 00	+34 15.4	2	5.96	.015	0.23	.005	-0.70	.015	7	O9.5Ve:	154,206
33924	+60 00870			5 13 00	+60 07.8	2	6.97	.019	0.45	.010			6	F5	1118,1379
269211				5 13 00	−70 27.9	1	13.62		-0.36		-1.07		3		149
269221	−71 00321			5 13 04	−71 00.2	1	11.25		0.50		-0.18		3	F5	149
34116	+28 00766			5 13 05	+28 15.0	1	9.58		0.86				2	G5	1724
34114	+29 00849			5 13 05	+29 18.1	1	9.34		0.54				2	F8	1724
		AAS21,109 # 102		5 13 06	−67 14.	1	13.75		-0.18		-1.04		3		952
		Sk -69 074		5 13 06	−69 20.	1	12.90		0.00		-0.72		4		798
290136	−0 00895	AKN 120		5 13 07	−00 00.2	1	8.98		0.96		0.96		1	K2 V	899
	+11 00755	V431 Ori		5 13 09	+11 54.7	1	9.39		3.61				1	C6 II	109
233095	+53 00880	IDS05092N5328	B	5 13 11	+53 31.5	1	9.50		0.47		0.09		2	G0	542
		Hyades J 357		5 13 12	+19 07.7	4	10.97	.007	1.12	.000	0.99		12		950,1058,1570,1674
		AAS21,109 # 101		5 13 12	−69 27.	2	13.32	.023	-0.18	.023	-0.98	.061	4		417,952
		Sk -69 073		5 13 12	−69 56.	1	13.73		-0.16		-0.94		4		798
34093	+38 01087	IDS05098N3856	AB	5 13 13	+38 59.5	1	8.79		0.19		0.11		3	A0 V	833
34543	−65 00443			5 13 14	−65 17.6	6	8.36	.017	-0.05	.014	-0.28	.005	40	B8 V	14,361,1408,2001,2033,3021
34019	+53 00882	HR 1707, R Aur	⋆ A	5 13 15	+53 31.9	2	9.24	.750	1.74	.097	0.14	.161	3	M4	542,3001
242137		G 85 - 60		5 13 16	+16 53.4	1	10.54		1.09		1.04		1		333,1620
34147	+27 00743			5 13 16	+27 47.9	1	9.28		0.47				2	F6/8 V/IV	1724
		NGC 1851 sq1 1		5 13 16	−40 18.0	1	10.27		0.60		0.05		2		473
34203	+11 00756	HR 1718		5 13 17	+11 17.2	2	5.56		-0.03	.009	-0.02	.047	4	A0 V	985,1049
242106	+26 00800			5 13 18	+27 03.2	1	8.89		1.15				3		1724
		AAS21,109 # 103		5 13 18	−69 22.	1	14.19		-0.03		-0.72		3		952
242103	+29 00850			5 13 19	+29 19.3	1	10.11		0.71				2		1724
269171	−67 00400	Sk -67 061		5 13 20	−67 21.0	1	11.76		0.14		0.10		2	A3 Ib	149
34132	+37 01115			5 13 21	+37 34.9	1	8.86		1.79				4	M1	369
		KUV 05134+2605		5 13 22	+26 05.4	1	16.70		-0.05		-1.04		1	DB	1708
		AKN 120 # 1		5 13 22	−00 07.5	1	8.93		1.04		0.83		3		1687
		VES 884		5 13 23	+35 34.9	1	12.56		0.41		-0.13		3		1543
34160	+28 00769			5 13 24	+28 08.2	1	8.52		1.33				2	K0	1724
34310	−27 02161	HR 1723		5 13 24	−26 59.9	3	5.06	.004	-0.10	.000	-0.20		8	B9 V	15,2012,3023
		AAS21,109 # 106		5 13 24	−68 45.	1	14.33		-0.11		-1.04		3		417
		AAS21,109 # 104		5 13 24	−69 25.	1	14.00		-0.20		-1.05		1		417
		AAS21,109 # 105		5 13 24	−69 26.	1	13.85		-0.30		-1.00		1		417
		AKN 120 # 4		5 13 26	−00 13.6	2	12.32	.013	0.68	.006	0.13	.019	21		899,1687
		AKN 120 # 12		5 13 27	−00 10.3	1	14.95		0.66		0.11		4		1687
		NGC 1851 sq1 2		5 13 28	−40 12.7	1	12.52		1.09		0.80		2		473
		LB 1776		5 13 30	−47 32.	1	12.07		-0.08		-0.18		3		45
242119	+28 00770			5 13 31	+28 39.8	1	10.04		0.55				2		1724

HD	DM	Other Id	N Rem	α_{1950}	δ_{1950}	S	V	σ_V	B–V	σ_{B-V}	U–B	σ_{U-B}	n	Spectrum	References
		AKN 120 # 2		5 13 33	−00 11.0	2	14.68	.000	0.57	.013	0.02	.013	6		899,1687
273211	−41 01776	RY Col		5 13 33	−41 41.1	1	7.57		−0.03		−0.69		2	A7	401
		AKN 120 # 3		5 13 34	−00 12.2	2	13.92	.022	0.89	.004	0.57	.048	6		899,1687
	+61 00776			5 13 37	+61 09.8	1	8.55		1.52		1.57		4	K2	1371
242169	+21 00804			5 13 38	+21 10.7	1	10.32		0.67				2	G0	987
34264	−0 00900	IDS05111S0053	AB	5 13 38	−00 49.5	1	9.06		0.63		0.06		4	G5 V	1731
34851	−75 00300			5 13 39	−75 25.1	1	7.86		1.10		1.09		53	K2 III/IV	978
		AKN 120 # 5		5 13 40	−00 12.3	2	12.39	.017	1.07	.026	0.89	.044	6		899,1687
		AKN 120 # 6		5 13 40	−00 14.6	1	14.83		0.48		−0.10		5		1687
		AKN 120 # 7		5 13 42	−00 15.0	2	11.91	.023	0.66	.009	0.23	.009	4		899,1687
34632		Sk -67 063		5 13 42	−67 28.	1	13.06		−0.16		−1.00		4	WN4	980
		SK *P 2155		5 13 42	−68 18.5	1	12.15		0.45		0.03		1	F0 Ia	726
		G 96 - 29		5 13 43	+45 47.5	1	10.16		1.50		1.24		2	M5	3024
34280	−3 01051	IDS05113S0336	AB	5 13 45	−03 32.3	1	7.78		0.10		0.01		2	B8 V	401
34651	−68 00335			5 13 45	−68 04.6	1	8.37		0.34		−0.02		3	F0 V	1408
		AKN 120 # 14		5 13 46	−00 16.8	1	10.38		0.92		0.59		3		1687
34649	−67 00401	HR 1744		5 13 47	−67 14.5	4	4.82	.005	1.28	.002	1.38		17	K2.5 IIIa	15,1075,2012,2038
34251	+18 00812	IDS05109N1820	AB	5 13 48	+18 23.1	1	7.16		0.09		−0.45		6	B3 V +B6 V	399
32196	+85 00074	HR 1616		5 13 50	+85 53.7	2	6.52	.010	0.33	.005	0.11	.005	8	A5 m	985,1733
		AKN 120 # 8		5 13 53	−00 12.3	2	11.98	.316	0.94	.150	0.70	.337	4		899,1687
269216	−69 00341	Sk -69 075		5 13 53	−69 35.7	1	10.76		0.08		−0.77		4	B7:Iae	149
		AAS21,109 # 108		5 13 54	−69 22.	1	13.37		−0.13		−0.94		3		952
		AAS21,109 # 109		5 13 54	−69 23.	1	12.65		−0.08		−0.92		3		417
		Sk -70 083		5 13 54	−70 21.	1	13.86		0.01		−1.00		4		798
		HV 2369		5 13 55	−67 07.	2	12.21	.040	0.67	.050	0.44	.020	2		689,3075
		Sk -67 065		5 13 56	−67 22.	1	12.51		0.41		0.20		4		415
		SK *G 561		5 13 56	−67 26.0	1	11.60		0.35		0.05		1	F4 Iab	726
		AKN 120 # 9		5 13 57	−00 13.9	1	13.84		0.69		0.16		6		1687
		AKN 120 # 13		5 13 57	−00 19.0	1	8.46		1.19		1.18		3		1687
34664		Sk -67 064		5 13 57	−67 30.3	2	11.79	.068	0.22	.045	−0.70	.005	5	B0	149,360
269215	−69 00342	Sk -69 076		5 13 58	−69 21.2	1	11.94		−0.15		−1.00		4	B7	415
269217		Sk -69 077		5 13 58	−69 24.5	3	11.88	.013	0.06	.008	−0.66		7		149,1277,8033
34435	−35 02199	HR 1730		5 13 59	−34 58.9	6	6.65	.007	0.16	.010	0.15	.010	25	A2 III/IV	15,216,243,355,1075,2027
		AKN 120 # 11		5 14 00	−00 14.4	1	13.92		0.57		0.09		18		1687
269219	−69 00344	Sk -69 078		5 14 02	−69 23.2	1	10.87		−0.07		−0.94		4	B7	415
		Steph 555	AB	5 14 03	+64 01.2	1	11.03		1.78		1.95		3	M1	1746
34317	+1 00957	HR 1724		5 14 05	+01 53.6	6	6.42	.005	−0.02	.000	0.02	.008	13	A0 V	15,1004,1079,1415*
34250	+28 00772	LS V +28 002		5 14 06	+28 51.1	3	6.77	.012	0.69	.021	0.47		7	F0	1625,1724,6009
34342	−5 01208			5 14 06	−05 06.9	1	8.92		0.09		−0.08		2	B9 V	401
269189	−67 00404	Sk -67 066		5 14 06	−67 23.0	2	11.54	.014	−0.01	.005	−0.66	.028	4	B6 Ia	149,360
269195	−67 00406	Sk -67 067		5 14 06	−67 30.5	1	11.34		0.10		−0.45		5	A0 Ia	149
		Sk -71 015		5 14 06	−71 34.	1	13.66		−0.16		−0.92		3		798
34190	+45 01084			5 14 07	+46 04.5	1	7.25		1.48		1.61		3	K3 III	37
34159	+52 00940			5 14 07	+52 20.3	1	9.30		0.18		0.11		2	A0	1566
269187	−67 00405	Sk -67 068		5 14 07	−67 19.0	2	11.49	.009	0.25	.005	0.27	.005	5	A9 Ia	149,360
242265	+17 00885			5 14 08	+17 34.9	1	10.42		0.33				1	A0	987
34259	+27 00746	IDS05111N2739	A	5 14 12	+27 42.7	1	9.31		0.41				2	F5	1724
34304	+16 00742			5 14 13	+16 17.9	1	7.44		1.03				1	G8 III	882
34305	+15 00779			5 14 16	+15 12.6	2	8.27	.020	1.15	.005	0.89		4	K0 III	882,3040
242211	+32 00937	IDS05110N3301	AB	5 14 16	+33 04.3	1	8.65		0.05		−0.47		2	B3	401
269227		Sk -69 079		5 14 16	−69 35.1	4	11.83	.113	0.18	.045	−0.81	.062	11	WN9	149,980,1277,8033
33564	+79 00169	HR 1686	*AB	5 14 17	+79 10.7	2	5.04	.005	0.48	.005	−0.13		4	F6 V	1363,3026
34257	+33 00995			5 14 18	+33 32.4	1	8.05		0.10		−0.40		2	B8	401
		AAS21,109 # 110		5 14 18	−65 22.	1	14.22		−0.26		−0.98		3		952
242261	+25 00812	IDS05112N2558	A	5 14 19	+26 01.8	1	9.55		0.43				2	F5	1724
277913	+40 01236			5 14 22	+40 45.3	1	9.31		1.68				3	M0	369
277933	+40 01237			5 14 23	+40 19.5	1	9.94		0.06		−0.49		8	B3	1543
		G 84 - 44		5 14 23	−00 14.8	1	11.58		0.46		−0.23		2		1620
		Sk -67 069		5 14 24	−67 11.	1	13.09		−0.16		−1.05		4		798
34129	+59 00865			5 14 26	+59 42.6	1	9.25		0.18		0.12		2	A2	1502
34128	+60 00873	IDS05100N6008	AB	5 14 27	+60 11.8	1	9.08		0.39		0.24		4	A2	1371
		NGC 1868 sq1 4		5 14 27	−63 58.4	1	12.14		0.60				3		827
		Hyades L 112		5 14 28	+25 29.0	1	12.68		1.12				2		582
		Sk -72 001		5 14 28	−72 07.8	1	13.57		−0.05		−0.55		3	B9 Ia	415
34303	+23 00888			5 14 29	+23 57.5	1	6.86		1.07		1.02		2	K0	1625
34587	−52 00683	HR 1742		5 14 30	−52 14.2	3	6.48	.015	1.21	.007			17	K2 III	15,1075,2030
		POSS 418 # 5		5 14 31	+17 24.4	1	16.62		1.64				2		1739
34417	−7 01024			5 14 31	−06 59.4	1	8.49		−0.05		−0.25		2	A0	401
269209	−67 00407			5 14 32	−67 31.7	1	10.55		0.98		0.79		1	K2	360
34335	+19 00886	CD Tau	*A	5 14 33	+20 04.8	1	6.80		0.49		0.01		6	F7 V	3026
34335	+19 00886	IDS05115N2001	B	5 14 33	+20 04.8	1	9.88		0.81		0.44		6	G5 IV	3026
269236		Sk -69 080		5 14 33	−69 36.0	1	11.20		0.27		0.10		9	F2 Ia	149
34496	−33 02217			5 14 34	−33 35.5	2	6.95	.005	1.00	.000	0.74		8	K0 III	243,2012
34447	−17 01069	HR 1731		5 14 35	−17 11.7	1	6.56		−0.16				4	B2 V	2006
269250				5 14 35	−70 19.9	1	11.49		0.49		0.15		1	F2 Iab	149
		Sk -66 064		5 14 36	−66 31.	1	13.16		−0.15		−0.94		4		798
		AAS21,109 # 111		5 14 36	−69 22.	2	13.13	.005	−0.06	.037	−0.99	.047	4		417,952
34783		Sk -69 081		5 14 36	−69 22.7	1	14.54		−0.26		−0.78		4	WN3	980
34352	+17 00886			5 14 37	+17 36.6	1	9.09		0.29		0.20		2		1375
34248	+46 00993			5 14 40	+46 59.7	1	7.72		1.13		0.77		2	G5	1601

Table 1

HD DM Other Id	N Rem	α₁₉₅₀	δ₁₉₅₀	S	V	σ_V	B–V	σ_B-V	U–B	σ_U-B	n	Spectrum	References
34269 +42 01239 HR 1722, PU Aur		5 14 41	+42 44.4	3	5.61	.032	1.59	.017	1.64	.020	16	M5 III	71,1501,3001
AAS21,109 # 112		5 14 42	−68 44.	1	13.55		-0.11		-0.96		6		417
34247 +48 01248		5 14 43	+48 52.3	1	6.90		1.27		1.35		3	K2 III	1501
34516 −33 02218 IDS05129S3312	AB	5 14 43	−33 08.3	1	9.45		0.52		0.02		5	F6 V	1700
34302 +37 01127 IDS05114N3732	AB	5 14 46	+37 35.8	1	7.88		0.09		-0.20		2	B8	401
34004 +71 00299		5 14 46	+71 39.8	1	6.56		0.92		0.65		3	G5	985
269244 Sk -69 083		5 14 51	−69 33.1	1	11.67		-0.15		-0.99		4	B0.5Ia	149
34428 +5 00875		5 14 52	+05 25.8	1	7.13		1.00		0.84		2	G5	1733
34385 +19 00888		5 14 52	+19 55.4	1	8.66		0.44		-0.03		24	F5	588
269238 −69 00345 Sk -69 082		5 14 52	−69 17.2	1	10.92		0.04		-0.57		8	A0 Ia	149
34334 +33 01000 HR 1726	★ AB	5 14 54	+33 19.3	6	4.54	.011	1.27	.006	1.27	.007	34	K3 III	15,263,1118,1355,3016,8015
Sk -66 065		5 14 54	−66 31.	1	13.91		-0.21		-1.03		3		798
34333 +36 01073 EO Aur		5 14 59	+36 34.8	2	7.45	.255	0.07	.010	-0.63	.005	4	B3 V +B3V	401,588
34364 +33 01002 HR 1728, AR Aur		5 15 01	+33 42.9	3	6.14	.013	-0.06	.001	-0.14	.038	4	B9.5V	777,1079,1118
34900 −70 00377 IDS05157S7043	AB	5 15 05	−70 38.4	1	8.88		0.34		0.05		2	F0 III:	204
34350 +39 01245		5 15 06	+39 18.0	1	7.90		0.46		0.08		3	F5 II	1501
34554 −31 02328		5 15 06	−31 19.9	3	7.46	.005	0.47	.010	-0.09		10	F5 V	1075,2034,3037
34383 +30 00849		5 15 07	+30 52.9	1	8.95		0.52				2	F8	1724
LS V +37 009		5 15 07	+37 50.6	1	10.98		0.09		0.06		1		180
34481 −4 01090		5 15 08	−04 46.9	1	9.05		0.07		0.00		2	A0 V	401
34299 +47 01126		5 15 09	+47 11.7	1	8.10		0.03		-0.22		2	A0	401
34233 +57 00874 HR 1719		5 15 09	+58 04.0	2	6.12	.000	-0.02	.006	-0.48	.001	7	B5 V	154,1223
34332 +40 01240 HR 1725		5 15 11	+40 24.8	1	6.18		1.37				2	K0	71
34503 −7 01028 HR 1735	★ A	5 15 11	−06 53.8	9	3.59	.008	-0.12	.008	-0.47	.011	24	B5 III	15,154,1004,1034,1075*
280777 +37 01130 IDS05118N3739	AB	5 15 12	+37 42.2	1	9.47		0.02		-0.58		2	A0	401
Sk -66 066		5 15 12	−66 54.	1	13.07		-0.11		-0.91		4		798
Sk -67 070		5 15 12	−67 07.	1	13.28		-0.05		-0.73		4		798
269218 −67 00409		5 15 14	−67 19.7	1	10.44		0.25		0.14		3	A3 V	149
34454 +13 00852 V1057 Ori		5 15 15	+13 21.9	2	7.77	.080	1.75	.005	1.56	.040	7	M2 II-III	206,1733
34382 +33 01003		5 15 15	+33 06.8	1	8.16		-0.05		-0.37		2	B9	401
34414 +30 00851		5 15 16	+30 19.5	1	8.34		1.06				2	K0	1724
VES 886		5 15 17	+37 37.4	1	11.23		0.06		-0.33		3		8113
242386 +30 00850		5 15 18	+30 12.3	1	9.86		0.14		-0.34		4		1723
34485 −33 02216		5 15 19	−33 05.5	1	8.23		0.27		0.05		9	A7 V	1700
242404 +29 00858		5 15 22	+29 49.6	1	9.60		1.15				2		1724
34538 −13 01116 HR 1737		5 15 22	−13 34.3	5	5.49	.029	0.93	.011	0.61	.005	17	G8 IV	15,158,252,1075,2035
34598 −32 02246		5 15 22	−32 23.6	1	9.58		0.48		-0.01		4	F7 V	1700
34525 −10 01144 IDS05130S1048	A	5 15 23	−10 44.4	1	10.25		0.40		-0.06		1		55
34525 −10 01144 IDS05130S1048	B	5 15 23	−10 44.4	1	10.55		0.43		-0.05		1		55
34381 +40 01245		5 15 25	+41 02.6	1	6.66		1.52		1.82		5	K5 III	1501
34511 −0 00913		5 15 27	−00 05.4	3	7.39	.005	-0.12	.011	-0.67	.010	7	B5 V	1004,1731,8055
−64 00435 NGC 1868 sq1 1		5 15 27	−63 59.2	1	10.13		1.62				4	M2	827
LS V +43 027		5 15 31	+43 44.6	1	10.68		0.34		-0.76		2		405
34469 +21 00813		5 15 32	+21 44.4	1	7.31		0.26		-0.24		2	B8	1648
NGC 1868 sq1 2		5 15 34	−63 54.7	1	11.86		0.68				3		827
34425 +36 01078 IDS05122N3605	A	5 15 35	+36 08.7	1	8.25		0.09		-0.56		2	B5	401
+38 01111		5 15 35	+38 57.7	1	10.01		1.83				2	M2	369
34426 +35 01044		5 15 36	+35 48.8	2	8.98	.005	0.07	.000	-0.53	.014	4	B8	401,1375
34411 +39 01248 HR 1729	★	5 15 37	+40 03.4	13	4.71	.013	0.62	.015	0.12	.013	62	G2 IV-V	1,15,22,224,660,1118*
34642 −35 02214 HR 1743		5 15 41	−34 56.6	6	4.82	.009	1.00	.003	0.80	.008	29	K0 IV	15,243,1075,1311,2012,8015
34452 +33 01008 HR 1732, IQ Aur		5 15 42	+33 41.8	5	5.38	.004	-0.20	.005	-0.60	.005	188	A0p Si	777,1049,1118,1263,3033
34255 +62 00742 HR 1720		5 15 42	+62 36.2	1	5.60		1.75		2.00		5	K4 I:	1355
34479 +30 00852		5 15 44	+30 19.0	1	9.68		0.10		-0.35		2		401
VES 887		5 15 46	+44 37.8	1	13.87		0.83		0.49		2		8113
34467 +35 01046 V348 Aur	★ A	5 15 49	+35 44.5	2	9.20	.000	2.76	.015	4.28		2	C4 II	109,414
34467 +35 01046 IDS05125N3541	B	5 15 49	+35 44.5	1	12.90		0.53		0.30		1		414
269247		5 15 53	−67 53.4	2	11.19	.005	0.96	.005	0.40	.005	8	K0 III:	149,204
TonS 415		5 15 54	−30 52.	2	13.38	.009	-0.29	.028	-1.16	.009	2		832,1036
34616 −16 01096		5 15 55	−16 14.4	1	7.68		0.93		0.62		1	K0 IV	3040
34677 −33 02235		5 15 55	−33 29.1	3	7.87	.004	-0.07	.000	-0.22	.005	20	B9 Vp	243,1700,1775
34477 +34 00994 IDS05126N3447	AB	5 15 56	+34 50.4	1	7.01		0.18		-0.33		2	B8	401
280978 +34 00997		5 15 57	+34 11.5	1	9.30		0.07		-0.55		2	A0	401
34659 −27 02189 IDS05140S2736	AB	5 15 59	−27 32.7	1	9.14		1.01		0.75		1	K3 V	3072
34466 +40 01247 IDS05125N4056	AB	5 16 00	+40 59.6	1	8.56		-0.02		-0.33		2	A0	401
Sk -68 065		5 16 00	−68 32.	1	13.24		-0.14		-0.94		3		798
POSS 418 # 8		5 16 05	+18 32.7	1	18.09		1.45				3		1739
34493 +37 01141 LS V +38 011	★	5 16 05	+38 05.5	1	8.63		-0.06		-0.81		1	B5	180
34499 +33 01010 HR 1734	★ AB	5 16 06	+33 56.1	1	6.49		0.24		0.09		2	A7 V	401
Sk -67 070a		5 16 06	−67 11.8	2	12.87	.000	0.05	.010	-0.21	.005	6	A1 Iab	149,415
AAS21,109 # 113		5 16 06	−69 37.	1	13.30		-0.20		-0.83		3		417
34509 +30 00854		5 16 08	+30 57.3	1	9.20		0.11		-0.42		2	A0	401
NGC 1868 sq1 5		5 16 11	−64 06.6	1	12.37		0.50				3		827
SK *P 2182		5 16 12	−66 31.0	1	12.45		0.47		-0.03		1	F4 Ia	726
AAS21,109 # 116		5 16 12	−66 57.	2	14.13	.094	-0.03	.056	-0.77	.028	4		417,952
280868 +37 01142		5 16 15	+37 46.6	1	10.03		1.78				3	A0	369
34559 +21 00816 HR 1739		5 16 16	+22 02.8	4	4.92	.025	0.94	.005	0.69	.013	10	G8 III	1080,1355,1363,3016
34579 +19 00893 HR 1741	★ A	5 16 17	+20 05.0	1	6.12		1.02		0.84		2	G8 II-III	542
34579 +19 00893 HR 1741	★ AB	5 16 17	+20 05.0	2	6.09	.009	1.00	.005	0.81	.009	4	G8 II-III +	542,1355
34579 +19 00893 IDS05133N2002	B	5 16 17	+20 05.0	2	10.13	.614	0.58	.224	0.01	.229	3		542,542
34724 −33 02238		5 16 18	−33 29.6	1	8.56		0.17		0.12		4	A5 V	243

HD	DM	Other Id	N	Rem	α₁₉₅₀	δ₁₉₅₀	S	V	σ_V	B–V	σ_B–V	U–B	σ_U–B	n	Spectrum	References
		AAS21,109 # 114			5 16 18	−69 11.	1	13.92		−0.01		−0.71		1		417
		AAS21,109 # 115			5 16 18	−69 32.	1	13.62		−0.15		−0.86		2		417
		Sk −69 084			5 16 18	−69 39.9	1	12.13		0.08		−0.35		4	A0 Ia	415
34639	−9 01119				5 16 22	−09 05.7	1	9.61		0.00		−0.34		2	B9	1776
269251					5 16 22	−67 17.5	1	12.34		0.14		0.15		5	A0	149
34498	+44 01170	HR 1733			5 16 24	+44 22.5	1	6.48		1.35		1.46		2	K0	1733
		POSS 418 # 3			5 16 25	+18 08.6	1	15.61		0.95				3		1739
		NGC 1857 - 126			5 16 25	+39 16.1	1	12.46		0.36		0.30		1		305
35026	−68 00340				5 16 25	−68 11.2	1	8.86		0.26				4	A7 IV-V	8
		NGC 1857 - 111			5 16 28	+39 13.8	1	11.75		0.46		0.17		1		305
		NGC 1857 - 62			5 16 28	+39 16.6	1	13.55		0.39		0.08		2		305
		NGC 1857 - 63			5 16 28	+39 17.3	1	12.56		0.39		0.26		2		305,1693
34520	+39 01254	NGC 1857 - 203		★ E	5 16 28	+39 19.9	3	9.16	.024	0.01	.011	−0.41	.083	12		305,1501,1693
242534	+29 00864				5 16 29	+29 27.0	1	10.18		0.62				2		1724
		NGC 1857 - 202			5 16 29	+39 20.2	1	15.71		0.51		−0.07		1		305
		NGC 1868 sq1 F3			5 16 30	−64 03.0	1	11.39		0.46				3		827
269271		NGC 1901 - 2			5 16 30	−68 10.6	1	11.45		0.40				4	F2	8
		Sk −70 084			5 16 30	−70 39.	1	12.78		−0.09		−0.92		4		798
		NGC 1857 - 65			5 16 31	+39 17.6	1	12.94		0.36		0.01		1		305
		NGC 1857 - 140			5 16 31	+39 19.6	1	16.08		0.57		0.31		1		305
		NGC 1857 - 67			5 16 32	+39 18.1	1	14.23		0.42		0.21		1		305
		Sk −69 085			5 16 33	−69 12.	1	12.66		−0.14		−0.99		4		415
		LS V +38 012			5 16 34	+38 51.7	1	11.25		0.52		−0.54		3	O9 V	180,342
		NGC 1857 - 287			5 16 34	+39 21.2	1	13.44		0.68		0.17		2		305
34658	+2 00916	HR 1746			5 16 35	+02 32.7	5	5.34	.000	0.42	.006	0.08	.000	11	F5 II	15,1008,1061,3026,6009
		NGC 1857 - 58			5 16 35	+39 15.6	1	15.92		0.59				3		1693
		NGC 1857 - 28			5 16 35	+39 17.4	1	12.42		1.64		1.49		1		305
		NGC 1857 - 70			5 16 35	+39 18.2	1	12.22		1.34		1.10		1		305
		NGC 1857 - 141			5 16 35	+39 19.3	1	14.38		0.45		0.21		2		305,1693
		NGC 1857 - 23			5 16 37	+39 15.6	1	15.76		0.59				3		1693
34721	−18 01051	HR 1747		★ A	5 16 37	−18 10.9	5	5.96	.008	0.57	.006	0.03	.030	16	G0 V	15,158,1075,2035,3077
		NGC 1857 - 22			5 16 38	+39 15.8	1	15.40		0.49				3		1693
		NGC 1857 - 20			5 16 39	+39 16.1	1	15.29		0.51				3		1693
34545	+39 01257	NGC 1857 - 1		★ A	5 16 39	+39 17.6	2	7.38	.000	1.69	.015	2.06	.005	6	K5	305,833
		NGC 1857 - 142			5 16 39	+39 20.6	1	16.24		1.13		1.10		1		305
34686	−5 01219				5 16 39	−05 00.6	1	8.80		0.08		0.04		2	B9 V	401
34738	−22 01070	RZ Lep			5 16 39	−22 15.8	1	8.30		1.76		1.95		11	S	3039
		NGC 1857 - 76			5 16 40	+39 19.0	1	14.67		0.38				3		1693
34673	−3 01061	G 84 - 45		★ AB	5 16 40	−03 07.6	8	7.76	.016	1.04	.006	0.86	.025	21	K3 V	22,1003,1013,1197,1620,1758*
242582	+26 00805				5 16 40	+26 12.5	1	9.51		0.68				3		1724
		NGC 1857 - 316			5 16 41	+39 17.7	1	11.47		0.17				3		1693
		NGC 1857 - 10			5 16 42	+39 16.3	1	11.73		1.43				3		1693
		NGC 1857 - 315			5 16 42	+39 18.	1	11.87		1.65		1.35		1		305
		AAS21,109 # 117			5 16 42	−69 32.	1	13.81		−0.18		−1.08		2		417
34578	+33 01013	HR 1740		★	5 16 43	+33 54.5	6	5.03	.012	0.27	.008	0.42	.034	14	A5 II	1,15,1077,1355,1363,3023
		NGC 1857 - 146			5 16 44	+39 20.3	1	12.93		0.33		0.04		1		305
34557	+40 01253	HR 1738			5 16 44	+41 02.2	1	5.52		0.11		0.04		2	A3 V	401
34751	−21 01131	IDS05146S2130		AB	5 16 44	−21 26.7	4	9.36	.016	1.29	.017	1.24	.011	13	K3/5 V	158,680,1775,3072
		Sk −68 066			5 16 44	−68 25.2	2	12.68	.010	0.11	.000	−0.24	.005	4	A0 Iab	149,415
		POSS 418 # 4			5 16 45	+17 54.0	1	16.13		1.02				2		1739
34591	+28 00777				5 16 45	+28 08.3	1	9.23		0.56				3	F7/9 V/IV	1724
		NGC 1857 - 42			5 16 45	+39 15.3	1	10.39		0.30		0.20		2		305,1693
		NGC 1857 - 44			5 16 45	+39 15.8	1	14.45		0.42				3		1693
		NGC 1857 - 9			5 16 45	+39 16.3	1	15.77		0.64				3		1693
		NGC 1857 - 78			5 16 45	+39 19.2	1	12.60		0.50		0.27		1		305,1693
242581	+26 00806				5 16 46	+27 02.0	1	9.33		0.63				3	G5	1724
34577	+35 01054				5 16 47	+35 44.2	1	7.35		1.83				4	M1	369
		NGC 1857 - 80			5 16 47	+39 18.7	1	13.31		0.35		0.17		1		305
		NGC 1857 - 38			5 16 48	+39 16.5	1	10.61		1.20		1.05		2		305
		Sk −69 086			5 16 48	−69 21.	1	14.41		−0.04		−0.79		4	WN4	980
34576	+36 01086	LS V +36 010			5 16 49	+36 37.5	3	7.50	.023	−0.03	.013	−0.59	.029	7	B2 V	180,399,401
		NGC 1857 - 102			5 16 50	+39 14.0	2	12.10	.095	0.40	.200	−0.02		6		1693,8113
		NGC 1857 - 85			5 16 50	+39 17.8	1	12.87		0.38		0.08		1		305
		NGC 1857 - 94			5 16 51	+39 15.5	1	12.68		0.35		−0.41		2		305
278043	+41 01157				5 16 52	+41 14.1	1	9.42		1.77				4	K7	369
35094	−68 00341	NGC 1901 - 4			5 16 54	−68 14.4	1	9.08		0.20				4	A5 III-IV	8
		AAS21,109 # 118			5 16 54	−69 33.	1	12.60		−0.05		−0.80		4		417
		AAS21,109 # 119			5 16 54	−69 36.	1	13.00		−0.17		−0.95		3		952
34534	+46 00998	IDS05132N4652			5 16 55	+46 54.5	1	9.32		0.17		0.02		2		3016
34533	+46 00998	HR 1736		★ A	5 16 55	+46 54.9	1	6.54		0.60		0.37		2	A2 V+G III	3016
34736	−7 01036				5 16 56	−07 23.9	1	7.86		−0.08		−0.44		2	B9	401
273975	−48 01741				5 16 59	−48 54.9	5	10.68	.015	0.51	.007	−0.19	.000	13	F5	565,158,1594,1696,2034
		L 233 - 30			5 16 59	−53 43.0	2	13.19	.000	1.16	.000	0.73		3		1705,3060
277861	+42 01249				5 17 00	+42 15.1	1	9.72		0.10		0.10		1	A2	401
		Sk −67 071			5 17 00	−67 32.	1	13.53		−0.13		−0.87		4		798
269279	−68 00342	NGC 1901 - 3			5 17 00	−68 40.5	1	10.30		0.22				2	F0	8
		AAS21,109 # 121			5 17 00	−69 15.	1	13.96		−0.07		−0.99		2		417
277860	+42 01250				5 17 02	+42 15.8	1	9.42		−0.05		−0.39		2	A0	401
34734	−4 01102				5 17 02	−04 23.5	1	8.33		0.14		0.12		2	A1 V	401
34748	−1 00859	HR 1748			5 17 03	−01 27.7	7	6.33	.016	−0.11	.006	−0.75	.006	23	B1.5Vn	15,154,361,1004,1415*

Table 1

HD	DM	Other Id	N Rem	α_{1950}	δ_{1950}	S	V	σ_V	B–V	σ_{B-V}	U–B	σ_{U-B}	n	Spectrum	References
34683	+16 00755			5 17 05	+16 40.8	1	8.40		1.51		1.77		2	K2	1648
34798	−18 01055	HR 1753	⋆ A	5 17 06	−18 34.2	3	6.38	.015	-0.16	.016	-0.60	.008	7	B3 V	146,1079,1084
		Sk -67 072		5 17 06	−67 14.8	1	12.59		0.03		-0.49		2	A0 Iab	149
		AAS21,109 # 122		5 17 06	−69 12.	1	13.13		-0.17		-0.94		3		417
		AAS21,109 # 120		5 17 06	−70 34.	1	14.52		-0.22		-0.88		3		952
34797	−18 01056	HR 1754, TX Lep	⋆ B	5 17 07	−18 33.6	3	6.53	.015	-0.11	.011	-0.45	.015	7	Ap	146,1079,1084
269280	−68 00343	NGC 1901 - 5		5 17 07	−68 12.9	1	10.56		0.37				3	F0	8
269302		Sk -69 087		5 17 08	−69 49.8	1	11.86		0.04		-0.52		5	A1 Ia	149
34626	+36 01090	MZ Aur		5 17 10	+36 35.0	3	8.17	.016	0.01	.008	-0.72	.031	11	B1.5IVnp	180,399,401
34544	+54 00882			5 17 12	+54 12.2	1	6.63		1.16		1.12		2	G5	1566
		Sk -69 088		5 17 12	−69 40.	1	13.17		-0.23		-1.03		4		798
34682	+25 00818	IDS05142N2504	A	5 17 13	+25 07.3	1	8.37		0.23		0.11		2	A2	1625
34774	−5 01221			5 17 14	−04 55.6	1	7.36		0.16		0.16		2	A2 V	401
35798	−81 00134	HR 1815		5 17 14	−81 35.8	2	6.50	.005	1.11	.000	1.01	.005	7	K1 III	15,1075
34816	−13 01127	HR 1756		5 17 16	−13 13.6	6	4.29	.005	-0.27	.015	-1.01	.005	20	B0.5IV	15,1075,1203,1425*
		SK *S 57		5 17 16	−69 22.4	1	12.00		0.56		-0.03		2		149
		POSS 418 # 1		5 17 17	+18 32.8	1	13.88		1.42				2		1739
		AAS21,109 # 123		5 17 18	−69 49.	1	13.25		-0.15		-0.95		3		417
34656	+37 01146	LS V +37 010	⋆ A	5 17 19	+37 23.3	3	6.78	.017	0.01	.021	-0.90	.000	7	O7 I((f))	180,1011,1209
34613	+43 01250			5 17 20	+43 21.8	1	8.25		0.00		-0.27		2	B9	401
35140	−68 00344	NGC 1901 - 6		5 17 20	−68 11.2	1	9.73		0.19				4	A2 III	8
34719	+19 00898			5 17 21	+19 31.7	2	6.66	.020	-0.03	.000	-0.34	.025	4	A0p	1648,3033
		van Wijk 18		5 17 22	−68 09.8	1	12.00		1.12		0.68		4		1099
34625	+40 01255			5 17 24	+40 50.1	2	7.30	.000	-0.01	.005	-0.05	.020	4	B9	401,1733
34868	−27 02204	HR 1758		5 17 24	−27 25.1	2	5.98	.005	-0.04	.000			7	A0 V	15,2012
		GD 66, V361 Aur		5 17 25	+30 45.5	1	15.56		0.22		-0.59		1		3060
34813	−7 01041			5 17 26	−06 58.9	1	8.82		0.01		-0.08		2	A0	401
34814	−7 01042			5 17 26	−07 10.9	1	9.10		0.00		-0.10		2	A0	401
34897	−33 02251	T Col		5 17 27	−33 45.5	1	7.42		1.38		1.05		1	M5/6e	975
242736	+19 00899			5 17 30	+19 48.9	1	10.23		0.43		0.29		2	A2	1375
34635	+42 01253			5 17 30	+42 27.3	1	7.73		0.01		-0.17		2	B9	401
280897	+36 01094		A	5 17 32	+36 39.8	1	10.93		0.14		-0.44		2	K0	401
280897	+36 01094		B	5 17 32	+36 39.8	1	11.54		1.12		1.12		1		401
287727	+1 00976	EW Ori		5 17 33	+01 59.7	1	9.88		0.60		0.06		3	F8	1768
		VES 889		5 17 33	+39 46.8	1	11.18		0.38		-0.41		3		8113
34717	+28 00779			5 17 35	+28 50.6	1	8.55		0.83				2	G5	1724
34827	−5 01223			5 17 35	−05 15.5	1	7.21		-0.02		-0.22		2	B9	401
		G 84 - 46		5 17 36	−02 44.8	1	13.78		1.48		1.08		2		3056
35183	−68 00345	NGC 1901 - 7		5 17 36	−68 31.4	3	9.15	.010	0.14	.007	0.12		9	A5 IV	8,1408,2003
278091	+39 01264	LS V +39 024		5 17 37	+39 16.2	2	9.92	.035	0.32	.015	-0.43	.015	3	B2 V	180,1012
34793	+10 00758			5 17 38	+10 50.5	1	7.43		0.15		0.06		2	A0	1733
		G 99 - 2		5 17 38	−02 42.6	2	12.78	.029	0.82	.022	0.21	.056	3		1620,3056
34932	−33 02256			5 17 38	−33 28.2	1	9.56		0.21		0.12		4	A7 V	243
34863	−12 01132	HR 1757		5 17 40	−12 21.9	2	5.29	.000	-0.13	.010	-0.41		7	B7 IVnn	154,2032
		NGC 1901 - 8		5 17 40	−68 28.8	2	12.52	.040	0.61	.030	0.08		6	F7	8,1099
34835	−6 01141			5 17 41	−05 53.7	2	8.18	.007	-0.05	.001	-0.45	.009	10	B8	401,1732
269311	−69 00351	Sk -69 089		5 17 43	−69 49.8	1	11.39		-0.05		-0.82		5	B2.5Ia	415
34885	−15 01016			5 17 44	−15 53.5	1	8.79		1.00		0.73		2	K3 V	3072
35024	−51 01452			5 17 44	−51 37.8	1	7.57		0.89		0.52		9	G6 III	1673
		Sk -66 067		5 17 45	−66 05.	1	12.69		-0.13		-0.99		4		415
269301	−68 00346	NGC 1901 - 9		5 17 46	−68 24.7	3	10.89	.012	0.32	.004	0.03	.020	15	F2	8,317,1099
242767	+25 00820	IDS05147N2538	AB	5 17 47	+25 41.1	1	9.28		0.45				2		1724
34861	−7 01043			5 17 47	−07 09.2	1	8.97		0.02		-0.04		2	A0	401
		Sk -67 073		5 17 48	−67 24.	1	13.13		-0.20		-1.05		4		798
269314		Sk -70 085		5 17 48	−70 23.	1	12.30		-0.10		-0.95		3	G0	415
35230	−68 00347			5 17 50	−68 38.7	4	7.57	.005	0.88	.003	0.45	.010	15	G8 III	1408,1425,2012,2013
34762	+27 00758	HR 1750		5 17 51	+27 54.5	2			0.04	.000	-0.26	.020	5	B9 IV	1049,1079
34575	+59 00870	G 191 - 34		5 17 54	+59 14.0	1	7.07		0.75		0.37		5	G5	1355
		HRC 83, V534 Ori		5 17 54	−05 50.	2	13.38	.065	0.65	.055	-0.57	.000	2		776,825
		Hyades L 113		5 17 55	+16 29.1	1	11.15		0.82				2		582
34761	+32 00952	IDS05147N3229		5 17 57	+32 31.8	1	8.43		0.07		-0.34		2	B8	401
34588	+60 00878			5 17 57	+60 18.0	1	9.33		0.22		0.15		4	A2	1371
269316		Sk -69 090		5 17 58	−69 48.2	1	11.90		0.08		-0.12		2	A1 Ia	149
34810	+19 00902	HR 1755		5 17 59	+19 45.9	1	6.18		1.23				2	K0 III	71
34880	−5 01225	HR 1759	⋆ AB	5 17 59	−05 25.0	3	6.38	.005	-0.03	.004	-0.36	.008	9	B8 III	15,1078,1079
		AAS21,109 # 127		5 18 00	−69 12.	1	14.00		-0.20		-0.95		2		417
		AAS21,109 # 126		5 18 00	−69 13.	1	13.50		0.00		-0.89		2		417
		GD 67		5 18 02	+00 34.6	1	16.64		0.17		-0.63		1		3060
34790	+29 00869	HR 1752		5 18 02	+29 31.3	2	5.67		0.05	.010	0.09	.034	7	A1 Vs	1049,1733
34890	−5 01226			5 18 02	−05 51.7	1	8.95		0.09		-0.24		2	A0	401
34760	+33 01017	LS V +33 011	⋆ A	5 18 05	+33 20.2	1	8.33		0.04		-0.66		2	B8	401
34889	−5 01227			5 18 05	−05 20.2	1	8.73		-0.05		-0.35		2	B9	401
271151	−65 00452			5 18 06	−65 34.9	5	9.40	.013	1.35	.009	1.46	.021	35	M0	14,361,2001,2033,3021
		AAS21,109 # 128		5 18 06	−65 53.	1	13.82		-0.25		-1.02		3		952
35072	−50 01723	HR 1767		5 18 08	−50 39.5	3	5.44	.007	0.51	.004	0.01		12	F6 IV	15,1311,2012
242815	+18 00829	Hyades J 361		5 18 10	+18 24.	1	10.25		0.74		0.33		1	G5	1058
34970	−31 02366			5 18 10	−31 13.1	1	9.29		0.96		0.69		1	K2 V	3072
269327	−69 00353	Sk -69 091		5 18 13	−69 54.0	1	10.74		-0.03		-0.91		5	B7	415
		POSS 418 # 7		5 18 14	+17 50.3	1	17.96		1.08				3		1739
35293	−68 00348	NGC 1901 - 11		5 18 14	−68 24.3	1	9.25		0.20				2	A1 m	8

HD	DM	Other Id	N	Rem	α_{1950}	δ_{1950}	S	V	σ_V	B−V	σ_{B-V}	U−B	σ_{U-B}	n	Spectrum	References
269310	−68 00349	NGC 1901 - 10			5 18 14	−68 34.5	3	10.46	.011	0.25	.008	0.07	.044	14	A7	8,317,1099
34759	+41 01162	HR 1749			5 18 16	+41 45.4	4	5.22	.004	-0.15	.009	-0.58	.008	13	B5 V	154,1203,1223,1363
	+61 00781				5 18 16	+61 16.6	1	8.61		1.93		2.22		4	K5	1371
269321	−69 00352	Sk -69 092			5 18 16	−69 19.1	3	10.79	.076	0.09	.005	-0.65		13	B5 Ia e	415,1277,8033
		LS V +36 012			5 18 19	+36 35.9	1	10.45		0.10		-0.89		2		180
34968	−21 01135	HR 1762		★ AB	5 18 19	−21 17.3	5	4.70	.005	-0.05	.005	-0.10	.004	13	A0 V	15,1075,2012,3023,8015
34826	+32 00955				5 18 20	+32 15.9	1	8.74		1.04				2	K0	1724
		NGC 1866 sq1 7			5 18 21	−65 34.5	2	12.31	.025	0.73	.092	0.13		2		2001,3021
		G 248 - 38			5 18 24	+68 38.5	1	10.99		1.31		1.23		3	K7	1723
		Sk -67 074			5 18 24	−67 37.	1	13.22		-0.16		-0.93		5		798
269312	−68 00351	NGC 1901 - 12			5 18 24	−68 37.0	1	10.38		0.24				2	A7	8
		Sk -71 016			5 18 24	−71 18.	1	14.06		-0.16		-1.01		3		798
		G 86 -B1	A		5 18 26	+33 19.2	1	14.20		1.66		1.41		2		3016
		G 86 -B1	B		5 18 26	+33 19.2	1	16.20		0.31		-0.64		3	DAe	3060
269331	−69 00354	Sk -69 093			5 18 26	−69 36.6	2	10.31	.013	0.25	.000	0.02	.035	6	A5 Ia	149,415
34807	+39 01272	IDS05150N3928			5 18 27	+39 31.5	1	7.31		0.48				3	A2	6009
269315		NGC 1901 - 13			5 18 28	−68 36.6	1	11.37		0.45				2	F8	8
		G 99 - 4			5 18 30	+03 25.	1	13.61		1.26		1.18		2		3056
242810					5 18 30	+32 54.	1	10.70		0.28		0.23		2	B9	1375
		AAS21,109 # 130			5 18 30	−69 18.	1	12.55		-0.22		-1.05		3		417
242832	+27 00762				5 18 31	+27 23.0	1	9.08		2.07				3	M0	369
		G 84 - 48			5 18 32	+03 24.7	1	11.89		0.64		0.03		2		333
35046	−34 02198	HR 1766			5 18 32	−34 44.9	4	6.33	.004	0.33	.008	0.10	.008	15	F2 IV-V	15,243,1637,2012
34842	+32 00957	UV Aur		★ A	5 18 33	+32 27.9	2	8.95	.640	1.86	.440	0.09	.350	2	O9 II	414,1753
34842	+32 00957	IDS05153N3224	B		5 18 33	+32 27.9	1	10.96		0.21		-0.30		1		414
35342	−69 00355	IDS05189S6920	ABC		5 18 33	−69 16.0	1	10.10		-0.11				3		8033
35343	−69 00356	S Dor		★	5 18 34	−69 18.0	2	10.56	.349	0.05	.044	-0.98		18	A0 e	328,8033
269319	−68 00352	NGC 1901 - 14			5 18 36	−68 31.0	4	10.23	.007	0.23	.007	0.12	.031	52	A5	8,317,361,1099
		AAS21,109 # 133			5 18 36	−69 15.	1	13.05		-0.25		-1.02		2		417
		AAS21,109 # 131			5 18 36	−69 19.	1	13.50		-0.20		-1.05		2		417
		Sk -69 094			5 18 36	−69 19.	1	9.32		0.20		-0.66		3		952
34841	+33 01018	NGC 1893 - 364			5 18 39	+33 20.3	1	9.06		0.28		0.13		5	A2	263
		Sk -69 096			5 18 39	−69 55.5	1	12.96		0.13		0.09		4	A0 Iab	415
269333	−69 00357	Sk -69 095			5 18 40	−69 14.6	3	11.24	.022	-0.09	.000	-0.91		7	B2 I	980,1277,8033
34959	+3 00857	HR 1761			5 18 41	+03 57.8	7	6.59	.044	-0.09	.009	-0.48	.017	17	B5 Vp	15,154,401,1004,1256*
		NGC 1866 sq1 8			5 18 41	−65 34.2	1	12.67		0.74				1		2001
269324		NGC 1901 - 15			5 18 41	−68 30.3	4	11.56	.011	0.49	.013	0.01	.011	39	G5	8,317,361,1099
		Sk -69 097			5 18 42	−69 49.	1	12.94		-0.05		-0.86		4		798
34926	+18 00831	IDS05158N1848	A		5 18 44	+18 51.5	1	7.73		0.87		0.56		2	G0	1648
271163	−65 00453	Sk -65 040			5 18 45	−65 44.4	1	11.75		-0.07		-0.70		4	B3 Ia	149
269309		Sk -67 075			5 18 48	−67 08.0	1	11.99		0.04		-0.57		3	B7 Iab	149
		Sk -68 067			5 18 48	−68 34.	1	12.40		-0.04		-0.72		4		798
		Sk -69 098			5 18 48	−69 39.	1	12.29		0.19		-1.01		4		415
		NGC 1893 - 371			5 18 49	+33 27.8	1	13.67		0.38		0.12		2		263
		NGC 1893 - 372			5 18 50	+33 28.2	1	14.62		1.24		1.45		1		263
		NGC 1893 - 365			5 18 50	+33 28.8	1	10.65		0.06		-0.35		6		263
		G 86 - 38			5 18 50	+35 14.6	1	11.11		0.41		0.26		1		1620
		BSD 49 # 90			5 18 51	+28 55.2	1	11.48		0.31		0.25		4	B9	117
		NGC 1893 - 333			5 18 51	+33 24.1	1	12.93		0.52		0.31		1		263
		NGC 1893 - 373			5 18 51	+33 28.5	1	13.35		0.32		0.30		2		263
		G 96 - 34			5 18 52	+47 52.1	1	12.11		1.20		1.11		1		333,1620
		NGC 1893 - 366			5 18 53	+33 29.1	1	12.74		0.41		0.14		6		263
		POSS 418 # 2			5 18 54	+17 17.6	1	14.62		1.75				2		1739
271164		NGC 1866 sq1 5			5 18 54	−65 35.	2	11.36	.000	0.77	.041	0.39		3	K5	2001,3021
		Sk -68 068			5 18 54	−68 10.6	1	12.25		0.08		-0.42		2	B6 Ia	798
269334		NGC 1901 - 16			5 18 54	−68 30.7	1	11.64		0.41				1	F5	8
		Sk -69 099			5 18 54	−69 16.	1	11.83		0.04		-0.47		4		798
		AAS21,109 # 134			5 18 54	−69 30.	1	13.15		0.00		-0.25		4		417
34905	+31 00939				5 18 55	+31 19.5	1	8.04		0.48				3	F5	1724
		VES 891			5 18 55	+44 27.8	1	10.73		0.26		-0.18		3		8113
		Sk -65 041			5 18 56	−65 43.	1	12.82		-0.08		-0.81		4		415
35008	−1 00872				5 18 57	−01 35.6	2	7.12	.018	-0.10	.005	-0.33		5	B8 V	1004,1089
		BSD 49 # 809			5 18 58	+29 25.1	1	12.00		0.29		-0.06		2	B9	117
		NGC 1893 - 369			5 18 58	+33 26.3	1	12.56		0.90		0.50		2		263
		NGC 1893 - 370			5 18 58	+33 26.7	1	12.85		1.12		0.83		2		263
35007	−0 00929				5 18 59	−00 27.9	1	11.94		0.69		0.24		2	B3 V	1752
35007	−0 00929	HR 1764		★ A	5 18 59	−00 27.9	9	5.68	.013	-0.12	.014	-0.66	.010	27	B3 V	15,154,247,1004,1089*
34989	+8 00933	HR 1763		★	5 19 00	+08 22.8	7	5.79	.007	-0.13	.003	-0.88	.006	20	B1 V	15,154,1004,1061,1732*
34925	+33 01020	LS V +33 013		★ A	5 19 00	+33 45.1	2	8.69	.015	0.08	.015	-0.63	.024	3	B3	401,8084
34925	+33 01020	IDS05157N3342	B		5 19 00	+33 45.1	2	9.74	.065	0.06	.050	-0.52	.030	2		401,8084
	+47 01140	G 96 - 35		★ A	5 19 00	+47 52.1	1	9.23		0.59		-0.01		1	G0	333,1620
		AAS21,109 # 136			5 19 00	−69 13.	1	14.07		-0.19		-0.91		1		417
		AAS21,109 # 135			5 19 00	−69 17.	1	12.55		-0.22		-1.02		3		417
		Sk -70 086			5 19 00	−70 38.	1	13.54		-0.20		-0.95		4		798
		NGC 1893 - 256			5 19 01	+33 25.5	3	11.82	.031	0.22	.031	-0.43	.011	4	B2 V	49,263,306
269351		Sk -69 100			5 19 01	−69 35.4	1	11.74		-0.02		-0.62		3	B6:Ia	149
34925	+33 01021	IDS05157N3342	CD		5 19 02	+33 44.2	2	9.73	.045	0.11	.040	-0.44	.020	2	G5	401,8084
		VES 892			5 19 03	+40 18.9	1	12.74		0.56		-0.37		2		8113
269355		Sk -69 101			5 19 03	−69 48.8	2	11.23	.015	0.45	.000	0.39		10	F8 Ia	149,1483
35158	−43 01801				5 19 06	−43 35.0	1	6.80		1.66				4	M3/4 III	2012

Table 1
261

HD	DM	Other Id	N Rem	α_{1950}	δ_{1950}	S	V	σ_V	B–V	σ_{B-V}	U–B	σ_{U-B}	n	Spectrum	References
269338	−68 00354	NGC 1901 - 17		5 19 06	−68 37.2	1	10.24		0.23				2	A5	8
		AAS21,109 # 137		5 19 06	−69 14.	1	12.80		-0.15		-1.00		2		417
		AAS21,109 # 138		5 19 06	−69 16.	1	13.60		-0.15		-0.95		3		417
		NGC 1901 - 18		5 19 07	−68 35.6	2	12.57	.005	0.67	.019	0.20		6	G0	8,1099
		POSS 418 # 6		5 19 08	+18 14.3	1	17.78		1.20				3		1739
35113	−33 02269			5 19 09	−33 40.1	2	8.98	.005	0.86	.000	0.45	.005	13	K1 III +G	243,1775
		NGC 1893 - 158		5 19 10	+33 22.0	2	12.77	.045	0.40	.035	0.31	.015	2		49,263
34921	+37 01160	LS V +37 011		5 19 11	+37 37.7	3	7.48	.030	0.17	.022	-0.86	.015	6	B0 IVpe	180,1012,1212
34787	+57 00879	HR 1751	⋆ A	5 19 11	+57 29.9	2	5.26	.015	-0.02	.005	-0.08	.005	5	A0 Vn	985,1733
242908	+33 01023	NGC 1893 - 269		5 19 12	+33 28.0	6	9.07	.025	0.30	.012	-0.67	.044	19	B0	49,263,306,1011,1012,1209
35039	−0 00930	HR 1765		5 19 12	−00 25.8	11	4.73	.012	-0.17	.009	-0.79	.006	47	B2 IV-V	15,154,1004,1075,1089,1363*
		Sk -65 042		5 19 12	−65 54.	1	13.92		-0.16		-1.07		2		798
		AAS21,109 # 139		5 19 12	−69 33.	1	13.05		0.00		-0.80		3		417
	+54 00886	G 191 - 37		5 19 13	+54 45.5	1	9.46		1.04		0.91		1	K8 V:	3072
242911	+31 00942			5 19 14	+31 25.6	1	10.27		0.55		0.24		2	A0	117
278056	+40 01267	LS V +40 041		5 19 15	+40 54.6	1	8.34		0.05		-0.55		2	B5	401
		NGC 1866 sq1 9		5 19 16	−65 36.3	2	13.07	.029	0.63	.068	0.18		3		2001,3021
		NGC 1893 - 168		5 19 17	+33 26.5	2	12.39	.032	0.26	.023	-0.42	.018	5		49,263
		LP 777 - 21		5 19 17	−19 18.3	1	11.66		0.52		-0.11		2		1696
		NGC 1893 - 86		5 19 18	+33 22.0	1	14.16		0.57		0.10		1		49
34904	+40 01268	HR 1760		5 19 19	+40 58.9	1	5.54		0.12		0.12		2	A3 V	401
34887	+46 01006			5 19 19	+46 56.5	1	8.13		0.78		0.47		2	G5	1601
		NGC 1901 - 19		5 19 19	−68 33.7	2	12.30	.020	0.50	.020	0.02		10	F8	8,1099
269357		Sk -69 102		5 19 19	−69 15.9	1	12.10		-0.21		-1.03		7	O9.5I	149
		NGC 1893 - 82		5 19 20	+33 20.4	1	12.91		0.50		-0.40		1		49
		NGC 1893 - 85		5 19 20	+33 21.8	2	13.51	.025	0.45	.020	0.30	.020	2		49,263
		NGC 1893 - 88		5 19 20	+33 26.0	1	15.55		0.36		0.13		1		49
		NGC 1893 - 90		5 19 20	+33 26.8	2	12.36	.023	0.28	.009	0.21	.005	5		49,263
		NGC 1893 - 329		5 19 21	+33 14.9	1	11.97		0.19		-0.03		3		263
		NGC 1893 - 87		5 19 21	+33 25.4	1	16.03		0.60		-0.09		1		49
		NGC 1893 - 361		5 19 22	+33 14.3	1	14.86		0.85		0.16		2		263
		NGC 1893 - 38		5 19 22	+33 23.1	2	12.49	.019	0.49	.005	0.14	.009	4		49,263
		NGC 1893 - 39		5 19 22	+33 23.6	2	12.88	.209	0.55	.024	0.01	.000	3		49,263
		SK *G 261		5 19 22	−69 44.1	1	12.42		0.03		-0.45		3		149
242926	+33 01024	NGC 1893 - 330		5 19 23	+33 16.3	6	9.37	.017	0.34	.011	-0.63	.024	17	O6.5	49,263,306,1011,1012,1209
		NGC 1893 - 149		5 19 23	+33 19.5	3	10.19	.011	0.49	.014	-0.45	.010	12	O6	49,263,306
		NGC 1893 - 92		5 19 23	+33 26.6	1	11.86		0.37		0.22		4		263
		NGC 1893 - 307		5 19 24	+33 21.	1	11.79		0.12		1.31		1		263
		NGC 1893 - 328		5 19 24	+33 21.	1	11.95		0.19		0.00		1		263
		NGC 1893 - 374		5 19 24	+33 21.	1	11.23		0.25		-0.66		1	B1 V	306
		NGC 1893 - 375		5 19 24	+33 21.	1	9.44		0.24		-0.73		1	O6	306
		NGC 1893 - 376		5 19 24	+33 21.	1	11.40		0.31		-0.65		1	B1 V	306
		NGC 1893 - 377		5 19 24	+33 21.	1	11.30		0.21		-0.65		1	B1 V	306
		NGC 1893 - 378		5 19 24	+33 21.	1	10.92		0.31		-0.60		1	B1 V	306
		AAS21,109 # 140		5 19 24	−69 15.	1	13.05		-0.25		-1.05		2		417
		Sk -69 105		5 19 24	−69 47.	1	12.79		-0.18		-1.02		4		798
35462	−68 00357	NGC 1901 - 20		5 19 25	−68 37.8	1	9.88		0.26				2	A2 III	8
242954	+27 00765			5 19 26	+28 03.3	1	9.54		0.45				2		1724
		NGC 1893 - 327		5 19 26	+33 14.5	1	11.61		0.22		0.12		1		49
		NGC 1893 - 35		5 19 26	+33 22.3	2	13.53	.005	0.72	.010	-0.22	.005	2		49,263
269344	−67 00416			5 19 26	−67 54.7	1	9.77		1.69		2.01		6	K5 III	149
		NGC 1893 - 77		5 19 27	+33 20.2	1	13.55		0.38		-0.36		1		49
		NGC 1893 - 34		5 19 27	+33 21.9	2	13.77	.115	0.22	.135	-0.32	.015	2		49,263
	+33 01025	NGC 1893 - 1		5 19 27	+33 23.6	3	10.31	.014	0.25	.007	-0.66	.027	13	O7	49,263,306
35067	+3 00864			5 19 28	+03 31.5	1	7.42		1.62		2.02		2	K5	3040
		NGC 1893 - 73		5 19 28	+33 19.5	2	10.81	.007	0.27	.007	0.15	.004	12		49,263
		NGC 1893 - 33		5 19 28	+33 21.5	2	12.39	.042	0.25	.023	-0.51	.005	4		49,263
35079	−3 01075			5 19 28	−03 00.7	5	7.06	.005	-0.03	.005	-0.52	.008	19	B3 V	401,1004,1089,1775,8055
35165	−34 02207	HR 1772	⋆ AB	5 19 28	−34 23.6	7	6.08	.016	-0.19	.007	-0.68	.012	52	B5 IVnpe	15,243,681,815,1637*
242935	+33 01026	NGC 1893 - 14	⋆ AB	5 19 29	+33 22.4	3	9.43	.005	0.09	.115	-0.84	.105	6	O8	263,1011,1012
		NGC 1893 - 47		5 19 29	+33 25.9	1	14.30		0.77		0.26		7		263
34920	+45 01108			5 19 29	+45 35.7	1	9.10		0.03		-0.28		2	A0	401
		BSD 49 # 1633		5 19 30	+30 33.4	1	11.46		0.19		-0.18		2	B9	117
		Sk -66 068		5 19 30	−66 56.	1	13.87		-0.06		-0.91		4		798
		AAS21,109 # 141		5 19 30	−69 41.	1	12.20		-0.09		-1.01		3		417
		NGC 1893 - 72		5 19 31	+33 19.3	2	12.48	.019	0.79	.011	0.23	.053	11		49,263
		NGC 1893 - 48		5 19 31	+33 25.8	1	11.07		1.28		1.17		10		263
		NGC 1893 - 71		5 19 32	+33 19.9	2	14.45	.038	0.55	.019	0.37	.004	11		49,263
		NGC 1893 - 3		5 19 32	+33 24.0	1	11.25		0.23		-0.58		12		263
		NGC 1893 - 50		5 19 33	+33 25.4	1	14.39		0.72		0.12		2		263
35104	−13 01135	HR 1769		5 19 33	−13 48.2	3	6.59	.030	-0.09	.009	-0.47	.024	8	B8 II	252,1079,2035
		BSD 49 # 853		5 19 34	+29 12.7	1	11.22		0.18		-0.25		2	A0	117
		Sk -65 043		5 19 36	−65 44.	1	13.95		-0.09		-0.91		4		798
		NGC 1901 - 21		5 19 36	−68 30.3	1	12.75		0.58				3	F7	8
		L 31 - 84		5 19 38	−78 19.6	1	11.91		1.49		1.20		1		3078
34987	+28 00783	Hyades vB 167		5 19 39	+28 42.1	3	8.92	.013	0.48	.005	0.01	.003	6	F8	1127,1724,3026,8023
		NGC 1893 - 62		5 19 40	+33 23.2	2	12.67	.005	0.34	.019	-0.29	.029	3		49,263
		NGC 1893 - 51		5 19 40	+33 25.6	1	13.47		1.08		0.62		2		263
		van Wijk 13		5 19 40	−68 29.0	2	12.35	.008	1.06	.004	0.79	.020	14		361,1099
269369				5 19 40	−69 21.7	1	11.65		0.65		0.04		3	G5	149

HD	DM	Other Id	N Rem	α_{1950}	δ_{1950}	S	V	σ_V	B–V	σ_{B-V}	U–B	σ_{U-B}	n	Spectrum	References
		NGC 1893 - 228		5 19 41	+33 19.3	1	12.51		0.25		-0.51		1		49
		NGC 1893 - 59		5 19 41	+33 23.8	1	12.22		0.20		-0.50		2		263
35446	-65 00455			5 19 41	-65 38.4	7	9.44	.010	0.34	.012	0.01	.026	39	A9 V	14,204,361,1408,2001*
		NGC 1893 - 134		5 19 42	+33 21.3	2	13.18	.009	0.30	.014	-0.37	.009	5		49,263
		AAS21,109 # 142		5 19 42	-66 40.	1	14.09		-0.15		-0.90		3		952
		Sk -69 106		5 19 42	-69 41.	1	11.97		-0.19		-0.92		4	WC5	980
35517	-69 00361	IDS05201S6945	AB	5 19 42	-69 42.1	3	11.16	.164	-0.16	.022	-0.94		7	WC	415,1277,8033
242977	+28 00784			5 19 43	+29 03.0	1	10.64		0.21		-0.35		3	A7	117
		BSD 49 # 2442		5 19 43	+31 26.3	1	11.15		0.48		0.39		2	A0	117
		NGC 1893 - 106		5 19 43	+33 27.0	1	12.43		0.33		-0.44		5		263
35162	-24 03023	HR 1771	⋆AB	5 19 43	-24 49.2	2	5.06	.000	0.66	.015	0.16		6	G7 II-III +	2031,3016
269363		Sk -68 069		5 19 44	-68 54.7	1	11.98		0.08		-0.31		1	A0 Ia	149
		NGC 1893 - 290		5 19 45	+33 28.8	3	11.24	.016	0.23	.021	-0.62	.028	7	B1.5V	49,263,306
		G 85 - 67		5 19 47	+24 46.9	1	13.70		1.08		0.84		1		1658
		BSD 49 # 1652		5 19 48	+30 51.2	1	12.00		0.30		0.23		2	A0	117
		R 90		5 19 48	-69 41.	1	12.00		-0.21				1		1277
		Sk -70 087		5 19 48	-70 35.	1	13.20		-0.05		-0.56		4		798
35034	+29 00876			5 19 49	+29 41.0	1	8.02		0.03		-0.34		2	B9	401
34986	+35 01083	LS V +35 012		5 19 49	+35 39.7	1	8.66		0.09		-0.72		2	B8	401
35014	+28 00785			5 19 50	+29 02.7	1	7.94		1.37				1	K0	1724
		NGC 1893 - 123	AB	5 19 50	+33 25.2	2	13.02	.010	0.46	.010	-0.42	.015	2		263,263
242992	+31 00947	IDS05167N3104	B	5 19 51	+31 06.9	1	9.79		0.19		-0.15		2	B9	401
35035	+28 00787			5 19 52	+28 25.4	1	7.54		0.31		0.20		2	Am	3016
35033	+31 00948	IDS05167N3104	A	5 19 52	+31 06.7	2	9.05	.015	0.09	.005	-0.32	.040	4	A0	117,401
		G 86 - 39		5 19 53	+33 08.7	2	11.54	.000	0.84	.000	0.33	.000	3		333,927,1620
		NGC 1893 - 363		5 19 53	+33 17.3	1	13.14		1.31		1.21		8		263
35494	-66 00400			5 19 53	-66 39.3	1	9.65		0.42				1	F5/6 V	642
269373		NGC 1901 - 22		5 19 54	-68 31.6	3	11.28	.184	0.48	.110	0.27	.364	13	F5	8,317,1099
		AAS21,109 # 143		5 19 54	-69 27.	2	14.31	.047	-0.08	.037	-0.76	.080	4		417,952
35155	-8 01099			5 19 55	-08 42.8	1	6.77		1.80		1.71		6	S4.1	3039
35474	-66 00399			5 19 55	-66 06.4	1	7.51		0.38		0.02		4	F2/3 IV/V	1498
243007	+30 00867			5 19 56	+30 10.7	1	10.95		0.23		0.10		3	A3	117
		NGC 1893 - 359		5 19 56	+33 16.1	3	11.06	.008	0.41	.019	-0.43	.017	10	B1 V	49,263,306
		HV 2447		5 19 56	-68 44.	2	11.88	.040	1.18	.095	1.06	.280	3	G0 Ia	689,3075
243018		NGC 1893 - 340		5 19 57	+33 31.0	5	10.74	.051	0.29	.023	-0.62	.016	7	O9.5V	49,263,306,666,1012
243083	+17 00916			5 20 00	+17 35.2	1	9.67		0.32		0.24		2	A0	1375
		NGC 1893 - 362		5 20 00	+33 15.0	1	11.08		0.58		0.03		8		263
		NGC 1893 - 209		5 20 00	+33 23.4	1	12.75		0.18		0.06		2		263
		NGC 1893 - 302		5 20 01	+33 23.9	1	14.35		1.00		0.53		5		263
269349	-66 00402	ApJS17,467 # 202		5 20 01	-66 39.8	1	10.97		0.47				1		642
		NGC 1893 - 301		5 20 02	+33 24.4	3	10.97	.013	0.18	.016	-0.69	.033	9	B1 V	49,263,306
		NGC 1893 - 344		5 20 02	+33 29.4	3	11.23	.005	0.27	.013	-0.62	.032	3	B1.5V	49,263,306
		van Wijk 21		5 20 03	-68 07.0	1	12.26		2.39		1.70		1		1099
		NGC 1893 - 353		5 20 04	+33 27.0	2	10.70	.024	0.43	.029	0.15	.107	3		49,263
269384		NGC 1901 - 23		5 20 04	-68 30.8	2	11.86	.009	0.55	.005	-0.02		10	G5	8,317
	+33 01028	NGC 1893 - 299		5 20 05	+33 26.3	3	9.13	.011	1.19	.020	0.97	.012	10	K0	49,263,666
35179	-14 01113			5 20 05	-14 26.5	1	9.48		0.98				2	G8 IV	1594
35110	+14 00890	IDS05173N1416	AB	5 20 06	+14 18.9	1	7.66		1.13				4	K0 III	882
		Sk -65 044		5 20 06	-65 27.	1	13.65		-0.21		-0.98		3		798
		AAS21,109 # 145		5 20 06	-66 56.	1	14.42		-0.19		-0.96		3		952
		NGC 1893 - 367		5 20 07	+33 27.3	1	14.06		0.75		0.41		1		263
243035		NGC 1893 - 343		5 20 07	+33 31.4	2	10.92	.033	0.27	.024	-0.58		2	B3	666,1297
		Sk -67 076		5 20 08	-67 23.	1	12.42		-0.13		-0.81		3		415
269377				5 20 08	-68 34.	2	11.83	.020	1.00	.008	0.65	.020	11	K7	317,1099
35134	+2 00936			5 20 09	+02 45.0	1	6.74		0.07		-0.01		2	A0	401
35178	-7 01054			5 20 10	-07 38.4	1	9.02		0.04		-0.04		2	A0	401
35152	+0 01032			5 20 12	+00 09.8	1	8.77		1.06		0.96		7	K5 V	7009
35149	+3 00871	HR 1770	⋆A	5 20 12	+03 29.9	12	5.00	.005	-0.15	.008	-0.86	.011	42	B3 Vn	15,154,542,1075,1084,1211*
		AAS21,109 # 144		5 20 12	-69 24.	1	13.40		-0.20		-0.98		3		417
35148	+3 00872	IDS05176N0327	B	5 20 13	+03 30.3	3	7.17	.005	-0.12	.009	-0.62	.008	8	B1 V	542,1084,1211
35076	+28 00788	HR 1768		5 20 13	+28 53.5	2			-0.03	.025	-0.19	.020	5	B9 Vs	1049,1079
		Sk -67 077		5 20 13	-67 02.	1	12.25		-0.16		-0.97		4		415
269388				5 20 14	-69 33.0	1	11.51		0.16		-0.29		2	F7	149
		NGC 1893 - 368		5 20 16	+33 18.9	1	11.80		0.56		0.06		1		263
35301	-38 02001			5 20 17	-38 32.2	1	7.99		1.16		1.14		9	K1 III	1673
243070		NGC 1893 - 345		5 20 18	+33 31.5	1	10.84		0.31		-0.60		1	B7	666
		MCC 460		5 20 18	+65 07.0	1	10.42		1.21				1	K8	1017
269392		Sk -69 108		5 20 18	-69 56.0	1	12.10		0.27		-0.49		4	B7	415
35274	-32 02296			5 20 19	-32 35.2	1	7.89		1.06		0.93		5	K1 III	1700
		LS V +35 013		5 20 23	+35 24.5	1	11.37		0.64		0.12		4		8113
		Sk -66 069		5 20 23	-66 48.5	1	13.17		0.04		-0.47		3	A0 Iab	415
34917	+61 00783			5 20 24	+61 46.8	1	7.96		0.08		-0.10		4	B9	1733
269371	-67 00420	Sk -67 078		5 20 24	-67 20.9	1	11.26		-0.04		-0.73		4	B3 Ia	149
		Be 19 - 17		5 20 26	+29 31.7	1	10.44		0.35				2		839
	-24 03031			5 20 26	-24 25.0	1	10.95		1.09		1.03		1		1696
35225	-8 01103			5 20 28	-08 08.0	1	8.59		0.21		0.09		2	A0	401
		Be 19 - 18		5 20 30	+29 29.2	1	13.01		0.78				2		839
		Be 19 - 20		5 20 30	+29 32.1	1	14.03		0.83				1		839
		Sk -65 045		5 20 30	-65 31.	1	14.77		-0.28		-0.88		3	WN3	980
271178	-65 00456			5 20 30	-65 48.4	1	10.68		0.47		-0.01		1	F8	360

Table 1

263

HD	DM	Other Id	N Rem	α_{1950}	δ_{1950}	S	V	σ_V	B–V	σ_{B-V}	U–B	σ_{U-B}	n	Spectrum	References
35192	+0 01035	IDS05179N0058	AB	5 20 31	+01 00.6	1	7.10		-0.01		-0.15		2	A0	401
35223	-6 01158			5 20 32	-06 45.8	1	8.89		0.18		0.11		2	A2	401
35322	-35 02254			5 20 32	-35 20.6	1	8.39		0.43				4	F3 V	2012
35147	+16 00763			5 20 33	+16 56.3	1	8.12		0.57		0.02		2	F8	1648
35203	+0 01036			5 20 35	+01 05.6	2	7.98	.004	-0.09	.000	-0.47		4	B6 V	1004,8055
269376	-67 00422			5 20 35	-67 01.8	1	11.26		0.17		0.19		4	A0 V	149
		Be 19 - 19		5 20 36	+29 34.6	1	12.88		0.70				2		839
		Sk -65 046		5 20 36	-65 48.	1	13.84		-0.16		-1.00		3		798
		Sk -69 109		5 20 36	-69 14.	1	13.10		-0.19		-0.99		4		798
		Be 19 - 7		5 20 37	+29 30.1	1	13.71		0.58				1		839
269393	-69 00362			5 20 38	-69 05.3	1	10.22		0.44		0.00		1	F5	360
243129	+30 00871			5 20 39	+30 11.5	1	9.40		0.49				2		1724
35261	-8 01105	IDS05183S0812	AB	5 20 40	-08 09.1	1	7.57		0.00		-0.20		2	A0	401
35108	+38 01144			5 20 42	+38 31.4	1	8.63		0.00		-0.60		2	B5	401
35171	+17 00917	G 85 - 68		5 20 43	+17 16.7	3	7.95	.014	1.11	.023	1.00	.000	5	K2	333,882,1620,3072
		Be 19 - 8		5 20 43	+29 31.8	1	15.82		0.89				2		839
		Be 19 - 9		5 20 43	+29 31.9	1	15.78		0.83				2		839
		Sk -65 047		5 20 43	-65 30.	1	12.51		-0.18		-1.06		3		415
35189	+16 00765	HR 1774		5 20 44	+16 39.2	1			0.14		0.19		4	A2 IV	1049
		Be 19 - 6		5 20 44	+29 30.2	1	16.64		0.87				1		839
		Be 19 - 10		5 20 45	+29 32.6	1	17.31		1.12				1		839
243127	+30 00872			5 20 45	+30 59.5	1	11.79		0.29		0.29		3	A2	117
35132	+32 00966	IDS05175N3232	AB	5 20 45	+32 34.8	1	8.34		0.06		-0.51		2	B9	401
		G 85 - 69		5 20 47	+22 30.1	1	15.52		1.83		1.65		3		203
		BSD 49 # 946		5 20 47	+29 46.4	1	11.63		0.30		-0.01		2	B5	117
		BSD 49 # 943		5 20 47	+29 57.2	1	11.22		0.22		-0.04		5	B8	117
35339	-31 02403			5 20 47	-31 47.2	1	9.15		0.36		-0.05		1	F2/3 V	742
		Be 19 - 5		5 20 49	+29 31.0	1	15.58		0.74				1		839
		Be 19 - 12		5 20 49	+29 32.5	1	15.74		0.95				3		839
269401				5 20 49	-70 22.6	1	13.04		0.14		0.22		3	A5	149
243163	+29 00880			5 20 50	+29 40.7	1	9.47		0.54				2		1724
35242	+5 00905	HR 1777		5 20 51	+05 16.6	2	6.34	.005	0.12	.000	0.08	.005	7	A2 V	15,1078
		Be 19 - 14		5 20 51	+29 33.0	1	15.39		0.84				2		839
	-31 02404			5 20 51	-31 45.5	1	9.99		1.19		0.97		1	K0	742
		BSD 49 # 952		5 20 52	+29 18.7	1	11.18		0.23		-0.23		5	B6	117
		Be 19 - 13		5 20 52	+29 32.1	1	14.79		1.26				2		839
		Be 19 - 3		5 20 54	+29 29.7	1	16.28		1.29				3		839
281004	+37 01173			5 20 54	+37 31.0	1	10.05		0.27		-0.16		3	B8	8113
269399		Sk -69 110		5 20 54	-69 35.	1	11.65		-0.08		-0.79		4	B7	415
		Be 19 - 4		5 20 55	+29 28.6	1	14.92		1.17				2		839
		Be 19 - 15		5 20 55	+29 31.8	1	15.44		0.95				3		839
35281	-8 01107	HR 1778	⋆ AB	5 20 55	-08 27.7	3	5.99	.005	-0.03	.004	-0.36	.000	14	B8 III	15,1415,2012
271182	-65 00457	Sk -65 048		5 20 55	-65 50.8	7	9.70	.033	0.59	.037	0.45	.044	39	F8 Ia	149,360,1277,1368*
35187	+24 00826	IDS05179N2452	AB	5 20 57	+24 54.9	1	7.78		0.27		0.10		3	A2	974
		Be 19 - 2		5 20 57	+29 29.2	1	13.91		0.81				2		839
		Be 19 - 1		5 20 57	+29 30.6	1	12.36		0.76				2		839
35120	+41 01181			5 20 58	+41 46.9	1	7.80		-0.05		-0.35		2	B5	401
35086	+50 01166			5 21 00	+50 56.2	1	8.28		0.51		0.08		2	F5	1566
	+60 00880			5 21 00	+60 58.2	1	9.48		0.62		0.13		4	K2	1371
		Sk -65 050		5 21 00	-65 32.	1	13.23		-0.15		-0.93		4		798
		Sk -65 049		5 21 00	-66 00.	1	13.95		-0.09		-0.96		4		798
	-27 02233			5 21 02	-27 26.9	1	8.99		1.13		0.93		1	K2 III:	565
		WLS 520 35 # 5		5 21 03	+37 12.2	1	11.84		0.37		0.19		2		1375
		WLS 520 20 # 5		5 21 04	+22 04.8	1	11.37		0.53		0.40		2		1375
		BSD 49 # 1747		5 21 05	+30 06.8	1	11.00		0.20		0.10		3	B8	117
35388	-32 02303			5 21 06	-32 42.9	1	9.41		0.07		0.06		3	A0 V	1700
		Sk -66 071		5 21 06	-66 46.	1	13.18		-0.11		-0.88		3		798
		Sk -66 070		5 21 06	-66 57.	1	13.40		-0.04		-0.44		4		798
		Be 19 - 16		5 21 07	+29 30.3	1	11.52		0.33				2		839
		van Wijk 20		5 21 08	-68 21.0	1	12.36		0.65		0.16		4		1099
35299	-0 00936	HR 1781		5 21 09	-00 12.3	28	5.69	.010	-0.21	.007	-0.87	.010	186	B1.5V	1,3,10,15,125,154,361*
		G 97 - 39		5 21 10	+12 23.1	1	12.46		1.52		1.22		1	M2	333,1620
34886	+70 00351			5 21 10	+70 11.1	1	7.20		0.00		-0.02		3	B9	1502
35386	-26 02185	HR 1785		5 21 11	-26 45.1	2	6.48	.005	0.50	.000			7	F6 V	15,2012
		BSD 49 # 2483		5 21 13	+31 18.2	1	10.76		0.30		-0.08		4	B9	117
243202	+32 00970			5 21 13	+32 46.5	1	11.09		0.33		-0.12		4	A5	206
35337	-14 01119	HR 1783		5 21 13	-13 58.4	3	5.24	.006	-0.21	.006	-0.88		9	B2 IV	15,1732,2012
35298	+1 00996	V1156 Ori		5 21 14	+02 02.2	3	7.89	.010	-0.13	.017	-0.61	.015	4	B3 Vw He wk	160,1004,8055
35704	-68 00361			5 21 14	-68 24.0	2	9.34	.020	1.27	.005	1.31	.015	21	K2 III	317,361
35186	+37 01175	HR 1773	⋆ AB	5 21 15	+37 20.4	4	5.01	.020	1.42	.008	1.76	.015	11	K4 III	1080,1355,1363,3053
	-31 02409			5 21 15	-31 44.3	1	9.66		0.84		0.54		1	K0	742
		BSD 49 # 983		5 21 16	+29 10.8	1	11.05		0.23		0.14		2	B8	117
35215	+30 00873	LS V +30 004		5 21 16	+30 08.8	4	9.40	.011	0.08	.010	-0.65	.012	18	B1 V	830,1012,1724,1783
35305	+0 01041			5 21 17	+00 49.0	1	8.43		-0.09		-0.40		1	B6-7 IV-V	160
		BSD 49 # 1760		5 21 17	+30 10.6	1	12.03		0.15		-0.19		2	B9	117
		Sk -68 071		5 21 17	-68 05.	1	12.50		0.15		-0.39		2		415
		Sk -69 111		5 21 18	-69 35.	1	12.00		-0.06				2		415
35317	-1 00882	HR 1782	⋆ AB	5 21 19	-00 54.7	3	6.14	.033	0.48	.030	-0.02	.029	11	F7 V +F8 V	15,254,1415
		SK *G 267		5 21 19	-66 57.3	1	12.87		0.23		0.25		3	A4 Ib	149
35353	-8 01109			5 21 20	-08 20.1	1	7.69		0.22		0.01		2	A0	401

HD	DM	Other Id	N	Rem	α_{1950}	δ_{1950}	S	V	σ_V	B–V	σ_{B-V}	U–B	σ_{U-B}	n	Spectrum	References
35416	−31 02414	IDS05195S3150		A	5 21 20	−31 47.6	2	7.53	.005	0.39	.005	-0.11		4	F3 V	742,1594
35471	−42 01912				5 21 20	−42 22.1	1	8.68		0.72				4	G5 V	2012
243232	+31 00952				5 21 21	+31 28.5	1	10.57		0.33		0.27		4	A0	117
243233	+31 00953				5 21 22	+31 13.1	1	9.49		0.17		-0.03		5	B9	117
		BSD 49 # 2487			5 21 25	+31 03.0	1	10.66		0.18		0.11		4	B9	117
35239	+31 00955	HR 1776			5 21 25	+31 05.9	3	5.94		0.04	.008	-0.12	.035	7	B9 III	1049,1079,1118
35238	+31 00954	HR 1775			5 21 25	+31 10.8	2	6.28	.007	1.25	.002			4	K1 III	71,1724
		Sk -65 051			5 21 26	−65 47.	2	12.63	.009	-0.16	.009	-1.00	.000	19		361,415
		BSD 49 # 246			5 21 27	+28 47.4	1	11.19		0.37		0.11		4	B7	117
35580	−56 00840	HR 1801			5 21 27	−56 10.8	2	6.10	.005	-0.10	.000			7	B8/9 V	15,2012
271187	−66 00404				5 21 27	−65 56.4	2	10.63	.015	1.29	.000	1.28	.000	9	G8:III	149,204,360
243266	+29 00881				5 21 28	+29 10.4	1	10.36		0.16		-0.11		5	A2	117
		Sk -67 079			5 21 28	−67 14.6	1	12.40		-0.02		-0.66		5		149
35296	+17 00920	HR 1780		⋆A	5 21 30	+17 20.3	4	4.99	.015	0.53	.013	-0.04	.005	19	F8 V	1067,1197,1363,3072
		Sk -67 080			5 21 30	−67 03.	1	13.58		-0.25		-1.02		4		798
		AAS21,109 # 146			5 21 30	−67 34.	1	14.50		-0.20		-0.96		2		417
269409	−69 00364				5 21 30	−69 30.4	1	10.20		0.33		0.06		2	A3	149
35369	−7 01064	HR 1784			5 21 32	−07 51.2	6	4.12	.008	0.96	.000	0.70	.005	18	G8 III	15,1075,1425,2012*
		Sk -65 052			5 21 32	−65 48.	4	12.77	.014	-0.13	.006	-0.99	.008	26		361,415,1277,1408
35073	+61 00785	G 249 - 8			5 21 33	+61 12.1	1	8.33		0.70		0.28		5	G5	1371
		BSD 49 # 996			5 21 35	+29 51.6	1	11.67		0.20		-0.18		5	B9	117
271190	−65 00460				5 21 35	−65 41.3	1	10.10		0.84		0.26		5	G5 III	149
271191	−65 00461	IDS05215S6550		⋆AB	5 21 35	−65 47.7	5	10.09	.071	0.75	.007	-0.61	.064	11	B7	360,415,952,1277,8033
271192	−65 00462	Sk -65 053			5 21 35	−65 54.8	4	10.56	.034	0.12	.007	-0.45	.005	11	A0 Ia	149,360,1277,8033
		Sk -69 112			5 21 35	−69 29.	1	11.35		0.02		-0.51		4		415
		AAS21,109 # 147			5 21 36	−69 07.	1	13.46		-0.17		-0.96		2		417
35254	+32 00974				5 21 38	+32 47.4	1	9.03		1.05				3	G5	1724
		Sk -65 054a			5 21 38	−65 49.	4	12.37	.009	-0.15	.012	-0.97	.008	27		361,415,1277,1408
269400	−67 00424	Sk -67 081			5 21 40	−67 08.9	1	11.58		-0.01		-0.66		3	B5 Ia	149
		NGC 1904 sq2 13			5 21 41	−24 35.2	1	13.75		0.78		0.42		1		581
34653	+77 00195	HR 1745			5 21 42	+77 56.2	2	6.57	.005	0.14	.000	0.16	.000	6	A7 III	985,1733
		BSD 49 # 2500			5 21 43	+31 28.1	1	11.28		0.30		0.08		3	A0	117
		BSD 49 # 2499			5 21 43	+31 30.6	1	10.98		0.24		-0.28		4	B6	117
		STO 853			5 21 43	−71 30.1	1	11.47		0.55		-0.06		1	G0 Iab	726
		BSD 49 # 255			5 21 45	+28 28.6	1	11.09		0.26		0.20		5	B7	117
35515	−39 01940	HR 1793, SW Col			5 21 45	−39 43.4	3	5.76	.047	1.63	.005			11	M2 III	15,1075,2007
269414	−69 00367	Sk -69 113			5 21 45	−69 29.9	1	10.71		0.09		-0.42		3	A2 I	798
243314	+23 00909				5 21 46	+23 32.6	1	9.63		0.59				2	G0	1724
243308	+29 00882				5 21 46	+30 01.9	1	11.78		0.15		-0.18		2	A0	117
		VES 895			5 21 46	+36 26.9	1	13.32		0.70		0.45		3		8113
		BSD 49 # 1014			5 21 47	+29 48.3	1	11.48		0.50		0.21		4	A0	117
243312	+28 00792				5 21 49	+28 56.7	1	9.72		0.73				2		1724
		BSD 49 # 2502			5 21 49	+31 28.0	1	11.96		0.30		0.16		3	B7	117
		WLS 520 20 # 9			5 21 50	+20 45.5	1	11.54		0.59		-0.04		2		1375
243346	+23 00912	Hyades L 115			5 21 51	+24 03.5	1	10.48		0.67				2	G5	582
		G 100 - 5			5 21 51	+26 18.0	1	12.01		1.07		0.99		1	K3	333,1620
243326	+30 00879				5 21 53	+30 44.2	1	10.37		0.08		-0.48		3	B9	117
35295	+34 01031	HR 1779		⋆A	5 21 53	+34 48.7	3	6.53	.011	1.12	.005	1.15	.019	9	K1p III-IV+	37,1084,3024
35295	+34 01030	IDS05186N3446		B	5 21 53	+34 48.7	2	8.69	.217	0.48	.000	0.08	.005	6	F6 V	1084,3024
		Sk -65 055			5 21 54	−65 50.9	2	13.34	.015	-0.20	.015	-0.96	.020	10	WN7	149,980
		Sk -67 083a			5 21 54	−67 04.0	1	12.85		0.02		-0.30		4	A0 Iab	798
		Sk -67 082			5 21 54	−67 56.	1	13.15		-0.10		-0.96		4		798
35349	+17 00923				5 21 56	+17 09.2	1	7.60		0.00		-0.46		2	B5	401
35410	−1 00886	HR 1787			5 21 56	−00 05.6	4	5.07	.016	0.97	.008	0.69	.011	10	G9 III-IV	15,247,1363,1415
35528	−37 02176	HR 1797			5 21 56	−37 22.9	4	6.82	.006	1.04	.005			18	K1 III	15,1075,2013,2029
35906	−72 00369				5 21 56	−72 08.5	1	8.48		1.54		1.05		15	K4 III	430
35411	−2 01235	HR 1788, η Ori		⋆AP	5 21 58	−02 26.5	9	3.36	.026	-0.17	.007	-0.92	.009	34	B1 V +B2e	15,154,1004,1089,1203*
35407	+2 00947	HR 1786			5 22 00	+02 18.5	6	6.32	.008	-0.15	.002	-0.63	.008	18	B4 IVn	15,154,1004,1078,8015,8055
278126	+42 01280	LS V +42 030			5 22 00	+42 20.6	1	9.85		-0.01		-0.56		2	B8	401
		Sk -67 083			5 22 00	−67 55.	1	13.07		-0.10		-0.97		4		798
269412		Sk -67 084			5 22 00	−67 57.	1	11.95		0.00		-0.71		4	B7	415
271213		Sk -71 017			5 22 00	−71 59.	3	12.31	.006	-0.05	.008	-0.75	.009	21	B1 Ia	430,798,1277
243362	+29 00885				5 22 03	+29 57.6	1	10.94		0.27		-0.01		4	A0	117
		Sk -65 056			5 22 05	−65 54.9	1	12.53		-0.15		-0.98		3		149
		Sk -69 114			5 22 05	−69 31.	1	12.09		-0.02		-0.56		4		415
		BSD 49 # 1837			5 22 06	+30 39.7	1	12.09		0.23		0.21		2	A0	117
		AAS21,109 # 148			5 22 06	+69 42.	1	13.23		-0.15		-0.82		3		952
35347	+29 00886	LS V +29 002			5 22 07	+29 34.3	3	8.90	.012	0.14	.019	-0.72	.015	8	B1:V:e	117,1012,1212
35328	+30 00882	LS V +30 006		⋆AB	5 22 07	+30 07.4	1	8.76		0.09		-0.60		3	B8	117
		NGC 1904 sq2 12			5 22 08	−24 39.4	1	13.54		0.46		-0.01		2		581
35703	−59 00463				5 22 08	−59 40.8	1	8.24		0.89		0.49		10	G8 IV	1673
35439	+1 01005	HR 1789			5 22 09	+01 48.1	10	4.94	.020	-0.20	.009	-0.92	.017	34	B1 Vpe	15,154,1004,1075,1212,1363*
35456	−2 01237	IDS05196S0235		AB	5 22 09	−02 32.5	1	6.94		-0.05		-0.49		4	B8	1089
35511	−24 03059				5 22 11	−24 24.7	1	9.44		0.04		0.04		3	A0 V	581
243477		G 84 - 51			5 22 12	+03 12.6	1	10.90		1.29		1.17		1		333,1620
35395	+20 00948	LS V +20 003			5 22 12	+20 32.4	1	6.75		0.22		-0.67		3	B0.5III:	399
243395	+29 00887				5 22 12	+29 18.9	1	9.95		0.23		0.23		3	A2	117
		BSD 49 # 1830			5 22 12	+30 20.3	1	11.07		0.50		0.35		3	B9	117
243357	+32 00978	G 98 - 3			5 22 12	+32 22.2	2	9.76	.000	0.60	.000	-0.02	.000	4	G3	516,1620
35327	+34 01036	LS V +34 007			5 22 12	+34 08.6	1	6.67		0.63				3	F2	6009

Table 1   265

HD	DM	Other Id	N Rem	α_{1950}	δ_{1950}	S	V	σ_V	B–V	σ_{B-V}	U–B	σ_{U-B}	n	Spectrum	References
		NGC 1904 - 6		5 22 12	−24 34.	1	15.27		0.93		0.41		1		581
		NGC 1904 - 28		5 22 12	−24 34.	1	15.48		1.00				1		581
		NGC 1904 - 36		5 22 12	−24 34.	1	16.30		0.12		0.07		1		581
		NGC 1904 - 66		5 22 12	−24 34.	1	15.95		0.14		0.31		1		581
		NGC 1904 - 68		5 22 12	−24 34.	1	13.44		1.40		1.60		1		581
		NGC 1904 - 126		5 22 12	−24 34.	1	15.21		0.41		-0.04		1		581
		NGC 1904 - 127		5 22 12	−24 34.	1	16.32		0.04		-0.08		1		581
		NGC 1904 - 156		5 22 12	−24 34.	1	16.83		0.75		0.14		1		581
		NGC 1904 - 160		5 22 12	−24 34.	1	13.02		1.54		1.79		2		581
		NGC 1904 - 170		5 22 12	−24 34.	1	14.97		0.89		0.51		1		581
		NGC 1904 - 171		5 22 12	−24 34.	1	13.99		0.89		0.69		1		581
		NGC 1904 - 175		5 22 12	−24 34.	1	16.35		0.89				1		581
		NGC 1904 - 176	V	5 22 12	−24 34.	1	14.95		0.91		0.57		1		581
		NGC 1904 - 180	V	5 22 12	−24 34.	1	16.65		0.12		-0.27		1		581
		NGC 1904 - 202	V	5 22 12	−24 34.	2	13.15	.021	1.39	.000	1.37		8		581,1638
		NGC 1904 - 206		5 22 12	−24 34.	1	17.57		-0.12		-0.39		1		581
		NGC 1904 - 237		5 22 12	−24 34.	2	14.15	.020	1.10	.016	1.01		9		581,1638
		NGC 1904 - 240		5 22 12	−24 34.	1	13.99		1.13		0.87		1		581
		NGC 1904 - 274		5 22 12	−24 34.	1	14.17		1.19		0.78		1		581
		NGC 1904 - 286		5 22 12	−24 34.	1	17.06		-0.01		-0.49		1		581
35513	−24 03060			5 22 13	−24 44.0	1	8.75		0.38		0.10		4	F2 V	581
35348	+29 00888			5 22 14	+29 09.9	1	8.51		1.82				4	K5	369
		NGC 1904 sq3 3		5 22 14	−24 32.9	1	13.65		1.17		0.84		7		1638
35505	−17 01117 HR 1792			5 22 15	−17 01.2	2	5.64	.005	0.00	.000			7	A0 V	15,2012
		NGC 1904 sq3 2		5 22 15	−24 34.4	1	13.24		1.42				8		1638
35304	+43 01265			5 22 16	+43 11.3	2	7.90	.019	1.72	.009	1.81		4	M2	369,797
269423				5 22 17	−68 57.9	1	11.48		0.44		-0.01		1	F0	149
269408		Sk -66 072		5 22 18	−66 18.3	1	11.59		0.05		-0.46		4	A0 Ia	149
		Sk -67 087		5 22 18	−67 30.2	1	13.22		-0.03		-0.44		4	A2 Ia	798
		SK *G 274		5 22 19	−66 57.6	2	12.25	.010	0.50	.005	0.23		6	F6:Ia	149,1483
269433		Sk -69 115		5 22 19	−69 29.6	1	11.58		0.01		-0.56		4	A0 Ia	415
		G 96 - 38		5 22 21	+44 35.2	1	12.08		0.84		0.45		1		333,1620
278114	+42 01281			5 22 22	+42 56.6	1	9.37		0.68		0.22		1	G5	797
		NGC 1904 sq3 5		5 22 22	−24 34.5	1	14.32		1.10		0.71		9		1638
35378	+30 00885			5 22 23	+30 39.3	1	8.93		0.04		-0.27		5	A0	117
34506	+80 00168			5 22 23	+81 01.0	1	7.62		1.48		1.75		2	K0	1502
35345	+35 01095 LS V +35 015			5 22 24	+35 36.2	1	8.39		0.21		-0.74		3	B1 Vpe	1212
		Sk -67 086		5 22 24	−67 54.	1	12.57		-0.04		-0.63		4		798
		Sk -67 085		5 22 24	−67 56.	1	12.88		-0.17		-0.99		4		798
35379	+30 00887			5 22 25	+30 38.6	1	8.35		0.03		-0.45		4		117
35468	+6 00919 HR 1790		⋆A	5 22 27	+06 18.4	15	1.64	.007	-0.22	.007	-0.86	.012	62	B2 III	3,15,154,1006,1020*
		NGC 1904 sq1 2		5 22 27	−24 24.3	1	12.21		0.72		0.24		4		488
243410	+30 00886			5 22 28	+30 54.9	1	9.21		0.15		-0.30		5	A2	117
243433	+29 00891			5 22 29	+29 35.7	1	9.74		1.89				3	K5	369
271204	−65 00463			5 22 29	−65 46.8	1	10.88		0.45		0.02		1	F0:V	360
		Sk -66 073		5 22 30	−66 19.	1	13.07		-0.09		-0.87		4		798
		Sk -66 074		5 22 30	−66 40.	1	13.44		-0.08		-0.67		4		798
		SK *G 276		5 22 30	−69 08.5	2	12.65	.045	0.15	.035	0.17	.030	2	A1 Ib	149,360
		Sk -71 018		5 22 30	−71 24.	1	14.28		-0.11		-1.01		4		798
35502	−2 01241 IDS05200S0258		A	5 22 31	−02 51.6	3	7.35	.009	-0.04	.003	-0.54	.000	9	B5 V	1004,1089,8055
		Steph 564		5 22 32	+12 29.4	1	10.49		1.26		1.21		1	K7	1746
		NGC 1904 sq1 8		5 22 32	−24 26.1	1	13.16		0.54		0.02		5		488
	−24 03063			5 22 33	−24 32.6	1	11.50		0.54		0.06		4		581
35404	+30 00888			5 22 34	+30 18.3	1	9.12		0.10		-0.23		4	A0	117
35501	+1 01009 IDS05200N0150		AB	5 22 35	+01 52.8	2	7.42	.000	-0.05	.004	-0.39		3	B8 V	1004,8055
		NGC 1904 sq1 7		5 22 35	−24 22.6	1	13.06		1.07		0.93		4		488
		NGC 1904 sq1 11		5 22 36	−24 26.7	1	13.99		0.72		0.19		6		488
		Sk -66 075		5 22 36	−66 04.	1	13.52		-0.15		-0.81		4		798
		Sk -67 088		5 22 36	−67 20.1	1	12.44		0.02		-0.40		3	A0 Ia	415
		NGC 1904 sq1 9		5 22 37	−24 25.9	1	13.59		0.47		-0.10		4		488
		NGC 1904 sq2 16		5 22 37	−24 31.6	1	14.48		0.98		0.58		1		581
		Hyades L 114		5 22 38	+16 09.1	1	11.33		0.72				2		582
243446	+30 00889			5 22 38	+30 15.6	1	10.22		0.25		0.16		4	A0	117
		NGC 1904 sq1 4		5 22 38	−24 20.6	1	12.28		1.07		0.90		4		488
		NGC 1904 sq1 13		5 22 38	−24 23.6	1	14.59		0.44		-0.22		5		488
35536	−10 01178 HR 1799			5 22 40	−10 22.4	5	5.60	.009	1.54	.011	1.85	.034	14	K5 III	15,247,1075,2035,3005
		NGC 1904 sq1 12		5 22 40	−24 22.5	1	14.34		0.61		0.00		4		488
		NGC 1904 sq1 15		5 22 40	−24 25.0	1	15.10		0.54		-0.10		4		488
		NGC 1904 sq1 14		5 22 41	−24 19.7	1	15.08		0.65		-0.03		5		488
	−24 03065			5 22 42	−24 40.1	1	9.88		1.28		1.41		3	K0	581
		AAS21,109 # 149		5 22 42	−68 10.	1	13.91		-0.17		-0.97		3		952
290375	−1 00888			5 22 43	−01 15.5	1	9.86		0.35		0.12		3		808
		VES 896		5 22 44	+38 20.4	1	11.19		0.51		0.10		3		8113
		NGC 1904 sq1 5		5 22 45	−24 25.5	1	12.60		0.49		-0.04		4		488
		NGC 1904 sq2 17		5 22 45	−24 36.7	1	14.86		0.72		0.14		2		581
		NGC 1904 sq2 10		5 22 45	−24 37.8	1	12.66		0.61		0.03		3		581
		NGC 1904 sq1 6		5 22 46	−24 20.5	1	12.70		0.49		-0.03		4		488
	−24 03066			5 22 47	−24 35.4	1	11.95		0.82		0.31		2		581
	−62 00466			5 22 48	−62 21.9	1	8.36		4.04				1	N	864
35846	−61 00454 IDS05223S6144		AB	5 22 50	−61 41.0	1	9.43		0.26		0.06		2	A7 IV	849

HD	DM	Other Id	N Rem	α₁₉₅₀	δ₁₉₅₀	S	V	σ_V	B−V	σ_B−V	U−B	σ_U−B	n	Spectrum	References
		Sk −67 089		5 22 50	−67 14.8	1	11.89		0.10		0.05		7	A1 Ia	149
		NGC 1904 sq1 10		5 22 51	−24 20.0	1	13.83		0.78		0.28		5		488
		AAS21,109 # 150		5 22 54	−67 19.	1	14.43		−0.24		−0.80		2		417
		KPS 543 - 210		5 22 56	+25 12.6	1	10.14		0.79		0.58		3		974
35480	+25 00828			5 22 56	+25 43.0	1	8.40		0.48				2	F8 V	1724
290356	−0 00944			5 22 56	−00 22.6	1	10.07		0.23		0.05		3		808
243588	+6 00921			5 22 57	+06 32.7	1	9.08		1.15		0.80		4	K0	206
281113	+37 01191	VES 897		5 22 57	+37 09.2	1	10.09		0.15		−0.30		3	B5	8113
35548	−0 00945	HR 1800	⋆ AB	5 22 58	−00 35.2	4	6.57	.008	−0.05	.007	−0.19	.014	12	B9 p HgSi	15,1079,1089,1415
		BSD 49 # 1109		5 22 59	+30 00.9	1	11.51		0.56		0.14		2	A0	117
		AA45,405 S225 # 12		5 22 59	+40 19.2	1	10.22		0.44		0.07		1		797
		NGC 1904 sq1 1		5 22 59	−24 20.7	1	10.28		1.52		1.80		4		488
243504	+30 00891			5 22 60	+30 12.1	1	9.62		0.51				2	F2	1724
		G 85 - 70		5 23 00	+18 35.5	1	11.62		1.05		0.93		1		333,1620
		AAS21,109 # 151		5 23 00	−67 03.	1	13.77		−0.13		−0.75		5		417
		AAS21,109 # 152		5 23 00	−67 23.	1	14.00		−0.13		−0.88		3		952
269445		Sk −68 073		5 23 00	−68 04.	4	11.45	.010	0.27	.010	−0.87	.060	10	O7	317,415,1277,8033
269451		Sk −70 088		5 23 00	−70 39.	1	12.56		−0.05		−0.84		4	B7	415
		Sk −71 020		5 23 00	−71 19.	1	13.71		−0.20		−0.98		4		798
		Sk −71 019		5 23 00	−71 24.	1	14.20		−0.20		−1.05		4		798
290338				5 23 01	+00 18.	1	10.77		0.35		0.01		3	B9	808
		NGC 1904 sq2 7		5 23 01	−24 32.4	1	12.00		0.62		0.11		3		581
35560	+0 01054			5 23 02	+00 19.2	1	8.96		0.06		−0.03		3	A1.5V	808
		AA45,405 S225 # 6		5 23 02	+42 58.6	1	11.56		0.25		0.07		1		797
278115	+42 01286	LS V +42 031		5 23 02	+42 58.6	1	10.09		0.28		−0.64		2	B0.5V	1012
35479	+29 00893			5 23 03	+29 56.6	1	8.09		0.07		−0.25		2	B9p	117
		NGC 1904 sq1 3		5 23 03	−24 21.3	1	12.24		0.57		−0.02		3		488
243549	+22 00900			5 23 04	+22 47.0	1	10.13		0.47				1	F8	1724
269440	−67 00429	Sk −67 090		5 23 04	−67 14.1	1	11.29		−0.09		−0.90		3	B1 Ia	149
36063		Sk −71 021		5 23 04	−71 39.2	1	12.71		−0.25		−0.95		4	WN7	980
41301	−88 00058			5 23 04	−88 20.2	2	7.15	.000	1.70	.013	2.05		11	M2 III	1704,2012
		BD +22 00900a		5 23 05	+22 46.8	1	10.25		0.53				1	F8	1724
35575	−1 00889			5 23 05	−01 32.1	4	6.43	.007	−0.17	.000	−0.72	.007	11	B3 V	1004,1089,1732,8055
		Sk −68 072a		5 23 05	−68 06.	1	12.99		−0.05		−0.93		4		415
	−24 03071			5 23 06	−24 37.8	1	10.99		0.39		−0.02		3		581
		Sk −65 057		5 23 06	−65 59.	2	14.90	.000	−0.26	.005	−1.00	.005	8	WN3	798,980
		Sk −66 076		5 23 06	−66 38.	1	13.60		−0.07		−0.98		4		798
35497	+28 00795	HR 1791	⋆ A	5 23 08	+28 34.0	9	1.65	.008	−0.13	.002	−0.49	.004	43	B7 III	1,15,1006,1049,1079*
36062	−71 00335			5 23 09	−71 33.1	2	7.43	.005	0.20	.005	0.10		8	A7 III/IV	1408,2033
		van Wijk 17		5 23 10	−68 04.4	1	11.56		0.30		−0.80				1099
35532	+16 00775	HR 1798		5 23 12	+16 39.5	1	6.24		−0.08		−0.64		3	B2 Vn	154
		BSD 49 # 1893		5 23 12	+30 26.2	1	12.33		0.24		0.17		3	A0	117
		Sk −71 022		5 23 12	−71 23.	1	13.80		−0.07		−0.89		3		798
35588	+0 01056	HR 1803		5 23 13	+00 28.7	7	6.16	.008	−0.18	.005	−0.75	.009	18	B2.5V	15,154,1004,1415,1732*
		KPS 543 - 214		5 23 14	+25 02.9	1	10.71		0.73		0.12		3		974
243547	+28 00796			5 23 14	+28 17.7	1	9.76		0.29		0.07		4	A0	117
290389	−1 00890			5 23 16	−01 53.6	1	10.44		0.32		0.17		3		808
		G 86 - 40		5 23 17	+34 42.3	2	11.83	.015	0.58	.005	−0.16	.000	3		1620,1658
		BSD 49 # 1898		5 23 18	+30 19.5	1	11.54		0.18		0.03		4	B5 p	117
35477	+38 01159			5 23 18	+38 54.9	2	8.34	.015	0.04	.010	−0.06	.005	5	A0 V	833,1601
		Sk −66 077		5 23 18	−66 04.	1	12.93		−0.19		−0.99		4		798
		Sk −68 074		5 23 18	−68 16.7	2	12.43	.023	0.03	.014	−0.29	.023	4	A1 Iab	149,952
		Sk −67 091		5 23 20	−67 28.7	1	12.32		−0.01		−0.56		4		798
		AAS21,109 # 153		5 23 24	−67 12.	1	13.62		−0.19		−0.96		2		417
		LS V +40 046		5 23 25	+40 30.4	1	10.74		0.41		−0.58		3	O9 V	342
294046	−2 01246			5 23 25	−02 22.7	1	8.27		−0.09		−0.51		4	B8	808
35765	−44 02036	HR 1813		5 23 26	−44 16.1	1	6.08		1.20				4	K1 III	2006
		VES 898		5 23 27	+37 06.4	1	13.19		0.83		−0.31		3		8113
		Sk −66 078		5 23 29	−66 44.7	1	12.22		−0.06		−0.88		3	B1 Ia	149
		Sk −66 078a		5 23 29	−66 44.8	1	13.18		0.05		−0.07		3		149
35613	+0 01057			5 23 30	+00 41.0	1	9.18		0.07		0.00		3	A2 V	808
243610	+26 00817			5 23 30	+26 57.4	1	9.65		0.57				3	F8	1724
35520	+34 01040	HR 1795	⋆	5 23 30	+34 21.0	2	5.94		0.14	.000	0.21		6	A1p	1049,1118
35612	+0 01058			5 23 31	+00 47.5	1	8.30		−0.06		−0.34		3	B7 Vn	808
		WLS 520 35 # 9		5 23 33	+34 47.8	1	11.47		0.38		0.24		2		1375
35250	+67 00385	IDS05184N6750	AB	5 23 33	+67 53.0	1	8.24		0.20		−0.35		2	B8	555
35521	+33 01045	HR 1796		5 23 34	+33 13.2	1	6.16		1.15				2	K0	71
35519	+35 01102	NGC 1912 - 562	⋆	5 23 34	+35 24.9	3	6.15	.010	1.45	.004	1.68	.006	11	K2	15,49,49
35659	−7 01075			5 23 34	−07 00.3	1	8.33		0.26		0.05		2	A3	401
35859	−52 00717	IDS05225S5224	C	5 23 34	−52 21.4	2	6.79	.005	0.07	.005	0.07	.010	6	A2 V	146,158
		BSD 49 # 1915		5 23 35	+30 05.6	1	11.52		0.20		−0.36		4	B5	117
		AA45,405 S225 # 11		5 23 35	+40 18.4	1	11.14		1.14		0.90		1		797
35640	−5 01247	HR 1806		5 23 35	−05 33.7	7	6.23	.007	−0.05	.008	−0.23	.011	19	B9.5Vn	15,1004,1060,1079*
		Sk −70 089		5 23 36	−70 42.	1	13.50		−0.13		−0.93		3		798
243609	+31 00966			5 23 38	+31 07.4	1	9.65		0.14		−0.32		3	A2	117
294101	−2 01247	IDS05211S0222	AB	5 23 38	−02 18.9	1	9.32		0.32		0.13		3	F0	808
35860	−52 00718	HR 1818	⋆ AB	5 23 38	−52 21.6	2	6.26	.005	0.07	.000	0.06	.015	6	A0 V	146,158
35574	+24 00831			5 23 39	+24 58.2	1	8.39		0.31		0.15		3	A5	974
290372				5 23 39	−00 57.	1	11.30		0.22		0.15		5	A7	808
		Sk −71 024		5 23 39	−71 25.	1	13.17		−0.08		−0.80		2		415

Table 1   267

HD	DM	Other Id	N Rem	α_{1950}	δ_{1950}	S	V	σ_V	B–V	σ_{B-V}	U–B	σ_{U-B}	n	Spectrum	References
269475	−71 00336	Sk -71 023		5 23 39	−71 45.9	3	11.57	.015	-0.01	.004	-0.69		10	B5 Ia	149,1277,8033
269452	−66 00409	Sk -66 079		5 23 41	−66 41.7	1	11.53		0.00		-0.66		6	B7	149
		Sk -67 094		5 23 41	−68 00.	1	12.47		-0.07		-0.88		4		415
269463		Sk -68 075		5 23 41	−68 15.3	4	12.04	.018	-0.07	.016	-0.92	.011	38	B7	149,317,361,1099
278199	+42 01288	LS V +42 032		5 23 42	+42 15.3	2	9.48	.000	0.38	.000	-0.23		3	B8 Ib	592,1012
290412	+0 01061			5 23 43	+00 06.0	1	10.05		0.23		0.10		3	A3	808
35557	+30 00897	IDS05205N3012	AB	5 23 43	+30 14.9	1	9.08		0.13		-0.35		2	A2	401
35546	+34 01041			5 23 43	+34 43.6	1	8.36		0.05		-0.46		2	B8	401
287873	+0 01062			5 23 45	+00 45.4	1	9.85		0.13		0.07		3	B9	808
278235	+40 01288	IDS05202N4034	AB	5 23 45	+40 36.8	1	10.30		0.28		0.12		1	A0	797
35698	−13 01149			5 23 46	−12 56.9	1	7.12		1.31		1.36		3	K1/2 III	1657
		BSD 49 # 1156		5 23 47	+29 59.8	1	11.53		0.21		-0.02		4	B6	117
281151	+34 01043	LS V +34 010		5 23 47	+34 17.0	2	9.43	.000	0.16	.000	-0.69	.000	2	B0.5V	651,666
35657	+0 01063			5 23 48	+00 45.4	1	8.30		0.08		0.04		3	A1 IV	808
		Sk -71 025		5 23 48	−71 28.	1	13.70		-0.19		-0.98		4		798
35736	−19 01173	HR 1812	⋆ A	5 23 50	−19 44.3	1	5.91		0.39		-0.07		3	F5 V	254
35736	−19 01173	IDS05217S1947	AB	5 23 50	−19 44.3	2	5.63	.014	0.44	.000			7	F5 V	15,2035
36156		Sk -71 026		5 23 52	−71 23.6	1	12.79		-0.22		-0.97		4	WC5	980
287872	+0 01065			5 23 53	+00 47.8	1	9.82		0.20		0.08		3	A3	808
35586	+27 00771	IDS05208N2731	AB	5 23 53	+27 34.1	2	7.87	.005	0.49	.007	0.00		5	F8	1724,3016
35586	+27 00771	IDS05208N2731	C	5 23 53	+27 34.1	1	13.20		0.86		-0.19		1		4016
35673	+2 00961	IDS05213N0252	AB	5 23 54	+02 53.6	2	6.52	.012	0.00	.000	-0.20		4	B9 V	1004,8055
		Sk -70 089a		5 23 54	−70 28.	1	14.15		-0.23		-1.04		4		798
290424				5 23 55	−00 03.	1	11.73		0.35		0.11		3	A7	808
290466				5 23 55	−02 13.	1	11.89		0.42		0.18		3		808
243714		Hyades L 116		5 23 56	+18 22.7	1	10.81		0.74		0.31		3	K7	1058
35600	+30 00898	HR 1804	⋆	5 23 56	+30 10.0	4	5.72	.034	0.16	.040	-0.21	.028	10	B9 Ib	252,1049,1079,8100
35656	+6 00923	HR 1807		5 23 57	+06 49.6	3	6.41	.005	-0.02	.021	-0.04	.020	9	A0 Vn	15,1061,1079
281150	+34 01045	LS V +34 011	⋆ A	5 23 57	+34 23.2	2	9.74	.035	0.16	.005	-0.69	.000	2	B0.5V	651,666
		G 97 - 40		5 23 58	+09 46.7	1	14.24		0.73		0.04		2		1658
35601	+29 00897	V362 Aur		5 23 58	+29 52.8	1	7.35		2.20				1	M1 Ib	138
	+30 00899			5 23 58	+30 11.5	1	8.96		1.36				3	G5	1724
		St 8 - 1		5 23 58	+34 28.4	2	10.91	.000	0.30	.000	-0.49	.000	2		651,666
243681	+31 00970			5 24 00	+31 24.4	1	10.88		0.46		-0.10		4	A7	117
		Sk -66 080		5 24 00	−66 27.	1	12.96		0.02		-0.61		4		798
		AAS21,109 # 155		5 24 00	−67 56.	1	14.03		-0.20		-1.06		4		417
		AAS21,109 # 154		5 24 00	−69 24.	1	13.79		-0.14		-0.99		3		417
		Sk -71 027		5 24 00	−71 19.	1	13.03		-0.10		-0.94		4		798
35718	−5 01251			5 24 02	−05 41.3	2	8.70	.014	0.05	.014	-0.02	.024	4	A1 V	1060,8055
278234	+40 01290			5 24 03	+40 40.0	1	9.41		0.00		-0.42		1	B8	797
35544	+43 01272			5 24 03	+43 19.6	1	6.79		0.00		-0.16		2	B9 V	401
		Sk -71 028		5 24 03	−71 29.3	1	13.19		-0.05		-0.47		3	B9 Iab	952
		St 8 - 4		5 24 04	+34 21.2	2	10.06	.000	0.25	.000	-0.55	.000	2		651,666
290423	−0 00948			5 24 04	−00 00.	1	10.63		0.19		0.00		3		808
35555	+40 01291			5 24 05	+40 36.5	1	8.16		0.10		0.08		1	A0	797
		Sk -66 081		5 24 05	−66 27.	1	12.50		0.20		-0.05		4		415
		KPS 543 - 219		5 24 06	+24 54.3	1	14.47		1.09		0.66		3		974
		AAS21,109 # 157		5 24 06	−66 12.	1	13.51		-0.13		-0.84		3		952
		Sk -69 116		5 24 06	−69 39.	1	13.27		-0.20		-0.95		4		798
243713	+24 00833			5 24 07	+24 10.2	1	9.25		0.39				2		1724
294110				5 24 07	−02 36.1	1	10.20		0.27		0.08		4	A2	808
290465	−1 00893			5 24 09	−01 54.4	1	10.13		0.19		0.13		3	A0	808
35610	+30 00901			5 24 10	+30 56.4	1	9.34		0.44				3	K3 III	1724
		LS V +34 018		5 24 10	+34 22.6	2	10.01	.000	0.67	.000	-0.32	.000	2		651,666
290467	−2 01249			5 24 10	−02 11.5	1	10.40		0.20		0.10		2		1089
		Sk -66 082		5 24 11	−66 27.	1	12.80		-0.07		-0.90		4		415
		BSD 49 # 1949		5 24 12	+30 24.3	1	11.86		0.24		0.15		2	A0	117
35715	+2 00962	HR 1811, ψ Ori	⋆ AB	5 24 13	+03 03.2	9	4.59	.008	-0.21	.009	-0.93	.012	28	B2 IV	15,154,1004,1075,1363,1425*
294145	−5 01252	IDS05218S0514	A	5 24 13	−05 11.9	1	9.98		0.35		0.08		2	F1 Vn	1060
		LS V +34 021		5 24 14	+34 24.5	2	10.70	.000	0.94	.000	-0.11	.035	2		651,666
35671	+17 00928	HR 1808	⋆ AP	5 24 15	+17 55.3	2	5.42	.008	-0.10	.001	-0.54	.000	7	B5 V +B8 V	154,1223
35621	+31 00973	LS V +31 004		5 24 15	+31 21.5	2	9.03	.014	0.16	.018	-0.43	.005	7	B8	117,1212
35731	−0 00949			5 24 15	−00 41.3	1	9.30		-0.01		-0.10		2	A0	1089
35730	+3 00901			5 24 16	+03 34.4	1	7.20	.000	-0.14	.017	-0.69	.015	4	B5p	160,1004,8055
	+38 01162			5 24 16	+38 56.3	2	9.77	.005	2.43	.015	2.38		7	M2	369,8032
35311	+69 00325	AS Cam		5 24 16	+69 27.4	2	8.58	.006	0.00	.004	-0.26	.021	5	A0	588,1768
	+34 01047	St 8 - 6		5 24 17	+34 28.4	1	9.57		1.54				1		666
35619	+34 01046	LS V +34 022	⋆ AB	5 24 17	+34 42.9	4	8.57	.005	0.24	.003	-0.70	.004	136	O7	167,666,1011,1012
		St 8 - 2		5 24 18	+34 23.	1	11.46		0.21		-0.54		1		666
		St 8 - 3		5 24 18	+34 23.	1	12.16		0.24		-0.36		1		666
		St 8 - 5		5 24 18	+34 23.	1	10.94		0.54		-0.21		1		666
		St 8 - 7		5 24 18	+34 23.	1	11.85		0.28				1		666
		St 8 - 8		5 24 18	+34 23.	1	11.71		0.74				1		666
		St 8 - 9		5 24 18	+34 23.	1	11.24		0.40		-0.41		1		666
		St 8 - 10		5 24 18	+34 23.	1	11.16		0.65				1		666
		St 8 - 11		5 24 18	+34 23.	1	11.44		0.21				1		666
		St 8 - 12		5 24 18	+34 23.	1	11.74		0.32		-0.35		1		666
		St 8 - 13		5 24 18	+34 23.	1	11.66		0.33		-0.41		1		666
		St 8 - 15		5 24 18	+34 23.	1	12.57		0.27		-0.34		1		666
		St 8 - 16		5 24 18	+34 23.	1	11.89		0.28		-0.41		1		666

HD	DM	Other Id	N Rem	α_{1950}	δ_{1950}	S	V	σ_V	B–V	σ_{B-V}	U–B	σ_{U-B}	n	Spectrum	References
		St 8 - 18		5 24 18	+34 23.	1	11.89		0.25				1		666
		St 8 - 19		5 24 18	+34 23.	1	12.22		0.43		-0.27		1		666
		St 8 - 21		5 24 18	+34 23.	1	12.27		0.39		-0.41		1		666
		Sk -65 058		5 24 18	-65 52.	1	13.94		0.01		-0.95		4		798
		Sk -66 083		5 24 18	-66 24.	1	13.29		0.01		-0.87		4		798
		AAS21,109 # 158		5 24 18	-67 41.	1	12.87		1.47		0.87		3		952
		AAS21,109 # 159		5 24 18	-68 03.	1	13.82		0.05		-1.07		2		417
		Sk -69 117		5 24 18	-69 39.	1	12.49		-0.08		-0.89		3		798
35634	+30 00902			5 24 19	+30 50.8	1	8.72		1.08				3	K0	1724
290454	-1 00894			5 24 19	-01 06.7	1	10.26		0.29		0.08		3		808
35620	+34 01048	HR 1805	★ AD	5 24 20	+34 26.1	5	5.08	.008	1.40	.005	1.66	.010	17	K4 IIIp	37,1118,1355,3052,4001
	+40 01293			5 24 21	+40 18.2	1	10.96		0.35		0.14		1		797
290426	-0 00950			5 24 21	-00 11.8	1	9.60		0.21		0.13		3		808
35693	+15 00822	HR 1809		5 24 22	+15 13.0	2	6.16		0.08	.010	0.13	.025	6	A1 IV	252,1049
35854	-32 02337			5 24 22	-32 32.7	1	7.74		0.94		0.68		1	K2/3 V	3072
35633	+34 01049	IDS05210N3423		5 24 24	+34 29.5	2	8.05	.009	0.32	.000	-0.66	.000	4	B0.5IV	666,1012
		Sk -69 119		5 24 24	-69 55.	1	13.19		-0.09		-0.91		4		798
		Neckel # 20		5 24 27	+05 28.6	1	11.01		1.18		1.17		std		1783
290400	+0 01069			5 24 28	+00 41.7	1	9.90		0.27		0.02		3	A2	808
35653	+33 01049	LS V +33 033		5 24 28	+33 54.3	1	7.44		0.12		-0.75		2	B0.5V	1012
35777	-2 01250			5 24 28	-02 24.1	4	6.62	.005	-0.17	.013	-0.75	.009	11	B2 V	1004,1089,1732,8055
		NGC 1907 - 62		5 24 29	+35 17.7	1	12.52		1.32		0.91		1		4002
35762	+3 00903			5 24 30	+03 48.8	3	6.75	.004	-0.17	.006	-0.73	.008	6	B2 V	1004,1732,8055
290409	+0 01070			5 24 31	+00 23.2	1	10.00		0.13		0.06		3	B9	808
		NGC 1907 - 63		5 24 31	+35 17.2	1	14.16		0.77		0.65		1		49
	+40 01295			5 24 31	+40 17.4	1	10.26		0.15		-0.07		1		797
		G 191 - 40		5 24 32	+52 10.5	1	11.14		0.86		0.59		2	K1	7010
290401	+0 01071			5 24 33	+00 39.8	1	10.34		0.23		0.15		3	B8	808
35652	+34 01051	IU Aur		5 24 33	+34 44.5	3	8.39	.051	0.21	.015	-0.70	.013	9	B3 Vnne	399,401,666
		NGC 1907 - 71		5 24 33	+35 19.9	1	13.29		0.54		0.43		1		49
243814	+18 00854	G 85 - 71		5 24 34	+18 49.5	1	9.49		0.92		0.69		1	K0	333,1620
35618	+40 01296	IDS05211N4043	B	5 24 34	+40 19.6	1	9.15		0.06		-0.25		1		797
35793	-3 01102			5 24 34	-02 57.8	1	9.78		0.12		0.05		2		1089
290413	+0 01073			5 24 36	+00 06.1	1	10.32		0.16		0.10		3	A0	808
		NGC 1907 - 257		5 24 36	+35 12.5	1	12.44		0.46		0.07		1		49
		NGC 1907 - 73		5 24 36	+35 19.8	1	13.38		0.49		0.34		1		49
		NGC 1907 - 74		5 24 36	+35 19.8	1	14.74		0.56				1		49
278240	+40 01297	IDS05211N4043	A	5 24 36	+40 18.8	1	8.81		1.03		0.77		1	K2	797
		AAS21,109 # 161		5 24 36	-69 08.	2	14.33	.047	-0.11	.080	-0.94	.075	4		417,952
		AAS21,109 # 160		5 24 36	-69 24.	1	13.99		-0.14		-0.83		4		417
		NGC 1907 - 45		5 24 37	+35 13.8	1	13.30		0.54		0.41		2		305
		NGC 1907 - 102		5 24 37	+35 15.9	1	14.16		0.99		0.27		1		49
35792	-1 00897			5 24 37	-01 24.5	3	7.21	.007	-0.14	.005	-0.64	.009	9	B3 V	1004,1089,8055
290464	-1 00898			5 24 37	-01 52.4	1	9.88		0.29		0.03		3	A2	808
290399	+0 01074			5 24 38	+00 45.2	1	10.68		-0.03		-0.29		3	B5	808
35708	+21 00847	HR 1810	★ A	5 24 38	+21 53.8	5	4.88	.012	-0.15	.006	-0.77	.005	16	B2.5IV	15,154,1223,1363,8015
		NGC 1907 - 51		5 24 38	+35 16.2	1	12.27		0.37		0.28		1		305
		NGC 1907 - 75		5 24 38	+35 20.2	1	14.55		0.46		0.38		1		49
35807	-3 01103			5 24 38	-02 58.3	1	9.21		0.06		-0.09		2	A0	1089
243791	+29 00899			5 24 39	+29 45.3	1	11.01		0.27		0.11		2	A0	117
243793	+24 00836			5 24 40	+24 17.5	1	9.35		0.68				2		1724
		NGC 1907 - 42		5 24 40	+35 13.4	2	11.95	.015	0.30	.010	-0.28	.000	2	B9 V	49,305
		NGC 1907 - 85		5 24 40	+35 18.4	1	13.00		0.54		0.37		1		49
		NGC 1907 - 105		5 24 40	+35 16.2	1	14.78		0.66		0.33		1		49
		NGC 1907 - 76		5 24 40	+35 20.8	1	14.69		1.01				1		49
290421	-0 00953			5 24 40	-00 04.8	1	9.41		0.19		0.15		3		808
		NGC 1907 - 44		5 24 41	+35 13.1	1	14.14		1.00		0.05		1		49
		NGC 1907 - 124		5 24 41	+35 17.4	1	13.07		0.51		0.40		1		305
		NGC 1907 - 256		5 24 41	+35 18.8	3	11.32	.027	1.71	.011	1.70	.035	4		49,305,4002
278239	+40 01298			5 24 41	+40 19.6	1	9.65		0.06		-0.03		1	A0	797
		G 100 - 8		5 24 42	+16 54.2	2	13.78	.056	0.64	.005	-0.09	.007	3		1620,1658
		NGC 1907 - 121		5 24 42	+35 16.9	1	12.84		0.45		0.42		1		305
		NGC 1907 - 43		5 24 42	+35 13.6	1	14.52		0.60		0.33		1		49
		NGC 1907 - 77		5 24 42	+35 20.4	1	13.89		0.68		0.11		1		49
271234				5 24 42	-66 10.4	1	12.06		0.05		-0.55		4	B8:Ia	149
		Sk -67 096		5 24 42	-67 17.	1	13.19		-0.13		-0.74		3		798
269485		Sk -68 077		5 24 42	-68 34.	1	14.37		0.06		-0.84		3	WN4p	980
		Sk -69 120		5 24 42	-69 45.	1	12.73		-0.12		-0.98		3		798
		NGC 1907 - 123		5 24 43	+35 17.1	1	12.46		0.56		0.45		1		305
		NGC 1907 - 127		5 24 43	+35 17.5	1	13.99		0.56		0.43		1		305
35808	-5 01256			5 24 43	-05 01.3	1	9.01		0.21		0.09		2	A5 V	1060
		BSD 49 # 2571		5 24 44	+31 28.9	1	11.89		0.41		-0.18		3	B6	117
		NGC 1907 - 110		5 24 44	+35 16.6	1	12.70		0.46		0.38		1		305
		NGC 1907 - 113		5 24 44	+35 16.8	1	11.89		1.41		1.14		1		305
35788	+4 00933			5 24 45	+04 13.2	1	7.33		1.17		1.16		5	K0	897
		NGC 1907 - 37		5 24 45	+35 14.8	1	12.09		0.53		0.44		1		305
		NGC 1907 - 133		5 24 45	+35 17.2	2	12.82	.010	1.30	.002	0.94	.042	2		305,4002
		NGC 1907 - 131		5 24 45	+35 17.4	2	12.39	.010	1.31	.008	0.96	.024	2		305,4002
		NGC 1907 - 130		5 24 45	+35 17.7	1	12.85		0.50		0.37		1		305
35850	-12 01169	HR 1817		5 24 45	-11 56.5	2	6.33	.035	0.52	.020	-0.01		7	F8 V	254,2007

Table 1   269

HD	DM	Other Id	N Rem	α_{1950}	δ_{1950}	S	V	σ_V	B–V	σ_{B-V}	U–B	σ_{U-B}	n	Spectrum	References
		MN83,95 # 501		5 24 45	−66 57.8	1	11.73		0.56				1		591
		NGC 1907 - 116		5 24 46	+35 16.8	1	11.44		0.41		−0.01		1		305
	+34 01053	St 8 - 14		5 24 47	+34 25.2	2	10.92	.000	0.27	.000	−0.27	.000	2	B1.5V	651,666
		NGC 1912 - 139		5 24 47	+35 48.2	1	13.78		0.54		0.25		1		49
		NGC 1907 - 25		5 24 48	+35 14.0	1	16.62		0.82		0.27		1		305
		NGC 1907 - 34		5 24 48	+35 14.6	1	15.09		0.89		0.23		1		305
		VES 899		5 24 48	+35 43.9	1	10.33		0.47		−0.52		3		8113
		SK *C 40		5 24 48	−66 40.	1	13.34		0.02		−0.38		3		149
243829	+29 00900	LS V +29 004		5 24 49	+29 41.4	1	10.14		0.18		−0.43		3	B8	117
		St 8 - 22	⋆ AB	5 24 49	+34 22.8	3	10.92	.050	0.18	.008	−0.65	.042	3		651,666,8084
	+34 01054	St 8 - 23	⋆ AB	5 24 49	+34 23.0	4	8.89	.028	0.18	.005	−0.66	.012	5	B0 IV	342,651,666,8084
	+34 01054	St 8 - 24	⋆ F	5 24 49	+34 23.0	1	12.40		0.41		0.20		1		8084
	+35 01114	NGC 1907 - 31		5 24 49	+35 14.5	1	9.90		0.38		−0.58		2		305
281135	+35 01113	NGC 1912 - 608		5 24 49	+35 27.1	1	9.30		0.45		0.32		1	A2	49
		NGC 1912 - 782		5 24 49	+35 49.4	1	15.31		0.53		0.33		1		49
	+35 01112	NGC 1912 - 271		5 24 49	+35 53.0	1	10.50		0.23		0.07		1	A0 V	49
		NGC 1907 - 24		5 24 50	+35 13.9	1	16.62		1.07		0.83		1		305
		NGC 1907 - 27		5 24 50	+35 14.4	1	12.48		0.42		0.37		2		305
		NGC 1907 - 35		5 24 50	+35 14.8	1	13.85		0.55		0.49		1		305
294103	−2 01252			5 24 50	−02 15.8	1	9.76		0.11		0.04		2		1089
	+11 00809	CO Ori		5 24 51	+11 23.2	2	10.56	.000	1.07	.005	0.64	.005	2	G5 Vpe	8013,8027
243830	+29 00901			5 24 51	+29 33.4	1	9.89		0.22		−0.31		3	B5	117
	+34 01056	St 8 - 17		5 24 51	+34 21.3	3	9.83	.044	0.19	.005	−0.68	.008	4	B0.5V	342,651,666
		NGC 1907 - 36		5 24 51	+35 14.9	1	15.71		0.72		0.55		1		305
		NGC 1907 - 157		5 24 51	+35 15.6	1	12.75		1.39		1.30		2		305
		NGC 1912 - 141		5 24 51	+35 48.9	1	14.19		0.52		0.24		1		49
243827	+33 01056	LS V +33 036		5 24 52	+33 16.4	1	10.61		0.45		−0.50		2	B0 III	1012
	+35 01115	NGC 1907 - 254		5 24 52	+35 23.7	1	10.46		0.19		−0.52		1	B3 V	49
35770	+15 00826	HR 1814		5 24 53	+15 50.0	1			0.01		−0.05		4	B9.5Vn	1049
35744	+23 00916	IDS05219N2313	AB	5 24 53	+23 15.2	1	8.23		0.22		0.08		2	B9	1625
281143	+35 01116	NGC 1907 - 1		5 24 53	+35 14.3	1	9.97		0.39		0.04		5	F2	305
		NGC 1907 - 196		5 24 53	+35 19.1	1	12.54		1.45		1.47		1		4002
		Sk -66 086		5 24 54	−66 07.	1	12.89		−0.03		−0.91		4		798
		NGC 1907 - 259		5 24 55	+35 21.8	1	13.01		0.40		−0.23		1		49
35836	+0 01076			5 24 56	+00 13.0	1	8.91		−0.01		−0.11		1	A2	1089
		NGC 1907 - 204		5 24 56	+35 20.3	1	13.52		0.47		0.20		1		49
		NGC 1912 - 152		5 24 56	+35 51.4	1	14.19		0.71		0.10		1		49
		NGC 1907 - 5		5 24 57	+35 14.1	1	14.71		0.86		0.27		2		305
290417				5 24 58	+00 02.	1	10.76		0.24		0.18		3	A3	808
		BSD 49 # 1227		5 24 58	+29 28.0	1	12.38		0.38		0.33		2	A0	117
35743	+30 00906			5 24 58	+30 36.2	1	8.10		1.16				3	K0	1724
		NGC 1907 - 3		5 24 58	+35 14.3	1	15.41		0.77		0.27		1		305
		NGC 1907 - 4		5 24 58	+35 15.1	1	14.00		0.44		0.31		2		305
		NGC 1907 - 180		5 24 58	+35 16.9	2	12.14	.015	0.46	.000	0.32	.015	2		49,305
		NGC 1907 - 160		5 24 59	+35 15.6	1	13.69		0.48		0.36		1		49
35868	−6 01185			5 24 59	−06 39.9	1	9.64		0.32		0.06		2	A9 IV	1060
		Sk -65 060		5 25 00	−66 00.	1	13.10		−0.01		−0.59		4		798
		AAS21,109 # 162		5 25 00	−69 06.	1	14.05		−0.30		−1.15		2		417
		NGC 1907 - 215		5 25 01	+35 20.0	1	10.66		0.56		0.13		2		305
35834	+0 01078	IDS05224N0101	AB	5 25 02	+01 04.0	2	7.69	.012	−0.03	.008	−0.33		4	B8 V	1004,8055
243845	+29 00902	LS V +30 008		5 25 02	+30 00.7	1	9.80		0.14		−0.55		4	B5	117
		NGC 1907 - 258		5 25 02	+35 25.1	1	12.62		0.27		−0.28		1		49
		NGC 1907 - 8		5 25 03	+35 14.3	1	14.64		0.44		0.44		1		49
290420				5 25 03	−00 08.	1	10.75		1.16		1.05		3	A3	808
		NGC 1907 - 216		5 25 04	+35 19.1	1	13.39		0.43		0.24		1		49
35885	−7 01081			5 25 04	−06 58.1	1	9.68		0.12		0.04		2	A0	1060
35769	+22 00912			5 25 05	+22 53.5	1	8.67		0.68				2	G5	1724
		NGC 1912 - 26		5 25 05	+35 46.3	1	12.18		0.28		−0.34		1	B8	49
278195	+42 01295	Y Aur		5 25 05	+42 23.9	2	9.19	.035	0.72	.029	0.51		2	F6	592,1399
290461	−1 00900			5 25 05	−01 38.9	1	10.30		0.23		0.17		3		808
		NGC 1912 - 155		5 25 06	+35 54.5	1	13.17		0.53		−0.12		1		49
		Sk -66 087		5 25 06	−66 51.	1	13.76		−0.12		−0.97		4		798
35802	+17 00931	HR 1816		5 25 07	+17 12.0	1	5.77		1.63		2.01		2	M1 III	252,3001
290432	−0 00958			5 25 07	−00 18.6	1	10.53		0.35		0.07		4	A5	808
290469	−2 01253			5 25 07	−02 00.5	1	9.79		0.15		0.06		2	A2	1089
	+35 01119	NGC 1907 - 253		5 25 08	+35 25.2	1	9.44		0.27		0.16		1	A0	49
		SK *G 288		5 25 08	−69 40.4	1	12.22		0.24		0.14		1	A5 Ib	149
		Hyades J 359		5 25 09	+18 22.	1	12.67		0.68		0.03		2		1058
		G 100 - 9		5 25 09	+22 26.1	2	13.04	.004	0.92	.013	0.58	.017	6		203,333,1620
35867	−0 00958			5 25 10	−00 18.0	1	8.14		−0.07		−0.30		4	B9	1089
		NGC 1912 - 22		5 25 11	+35 45.1	1	11.87		0.29		0.10		1		49
		NGC 1912 - 780		5 25 11	+35 47.9	1	14.48		0.44		0.26		1		49
		NGC 1912 - 781		5 25 12	+35 47.9	1	15.12		0.69		0.29		1		49
278196	+42 01297			5 25 12	+42 23.9	1	10.16		0.25				1	A0	592
35901	−7 01083			5 25 12	−06 55.0	3	9.03	.005	−0.03	.005	−0.14	.005	5	A0	1004,1060,8055
		AAS21,109 # 163		5 25 12	−68 31.	1	13.64		−0.20		−0.99		2		952
		AAS21,109 # 164		5 25 12	−68 35.	1	13.95		−0.19		−1.06		2		417
		NGC 1912 - 49		5 25 13	+35 51.1	1	10.69		0.18		−0.02		1	A2 V	49
	+35 01121	NGC 1912 - 405		5 25 13	+35 58.9	1	9.76		1.71		1.88		1		4002
35882	−1 00901			5 25 13	−01 51.2	4	7.82	.012	−0.07	.003	−0.46	.014	15	B6 V	1004,1089,1732,8055

HD	DM	Other Id	N Rem	α_{1950}	δ_{1950}	S	V	σ_V	B–V	σ_{B-V}	U–B	σ_{U-B}	n	Spectrum	References
	+35 01120	NGC 1912 - 50		5 25 14	+35 50.5	1	10.24		0.19		-0.02		1	A0 V	49
		NGC 1912 - 164		5 25 15	+35 52.9	1	11.18		0.28				1	A2	49
		G 249 - 11		5 25 15	+68 51.9	1	15.22		1.68		1.33		1		906
35899	−2 01254			5 25 15	−02 11.1	5	7.52	.007	-0.14	.007	-0.62	.007	21	B5 V	124,1004,1089,1509,8055
		G 97 - 42		5 25 16	+09 36.7	3	12.46	.019	1.64	.002	1.24	.022	15		235,1705,1764
35833	+16 00782			5 25 16	+16 24.0	1	6.86		0.64		0.18		2	G0	1648
		NGC 1907 - 255		5 25 16	+35 16.0	1	11.10		0.73		0.26		1		49
		NGC 1912 - 478		5 25 16	+35 28.2	1	11.12		0.54		0.07		1		49
		BSD 49 # 2005		5 25 18	+30 20.5	1	11.22		0.77		0.39		2	B7	117
35881	+0 01082			5 25 19	+01 03.9	3	7.80	.011	-0.08	.017	-0.52	.035	4	B8 V	160,1004,8055
		BSD 49 # 2587		5 25 19	+31 21.9	1	11.57		0.33		0.18		3	B9	117
		Sk -67 098		5 25 19	−67 54.2	1	13.09		0.09		-0.11		2	A0 Ib	149
	+34 01058	LS V +34 036		5 25 20	+34 37.7	2	8.78	.000	0.26	.000	-0.73	.000	4	O8 nn	1011,1012
		BSD 49 # 428		5 25 21	+28 50.7	1	10.82		0.26		-0.23		5	B7	117
		Sk -67 097		5 25 21	−67 30.	1	12.97		-0.18		-0.98		3		415
	−5 01260			5 25 22	−05 34.6	1	10.27		0.29		0.19		2		1060
36257	−66 00412			5 25 22	−66 08.8	1	9.06		0.14		0.08		1	A2 Vn	360
	+34 01059	LS V +34 037		5 25 23	+34 58.4	1	9.22		0.19		-0.76		2	B0 IV-V	1012
	+35 01125	NGC 1912 - 3		5 25 23	+35 47.5	1	9.85		1.19		0.89		1	G0	4002
35832	+22 00914			5 25 24	+22 42.2	1	7.20		1.14				3	K0	1724
		NGC 1912 - 55		5 25 25	+35 51.3	1	10.90		0.19		0.01		1	A0	49
35913	−1 00904			5 25 25	−00 59.9	1	8.67		0.29		0.10		3	A8 IV	808
35993	−25 02445			5 25 25	−25 22.9	2	9.57	.015	1.13	.005	0.77	.015	3	Kp Ba	565,3048
35912	+1 01021	HR 1820		5 25 26	+01 15.5	7	6.41	.017	-0.18	.007	-0.74	.007	17	B2 V	15,154,1004,1078,1732*
		St 8 - 20		5 25 26	+34 12.4	2	10.68	.000	0.20	.000	-0.03	.000	2		651,666
35583	+62 00759	HR 1802		5 25 26	+63 01.7	2	5.43	.010	1.70	.004	2.02		4	M1 IIIa	71,3055
		G 99 - 8		5 25 27	+00 09.4	4	13.24	.031	0.96	.025	0.52	.034	6		419,1620,1773,3059
		NGC 1912 - 5		5 25 28	+35 47.2	1	11.50		1.21		0.96		1		4002
35910	+3 00910			5 25 29	+03 29.7	2	7.57	.004	-0.10	.000	-0.52		3	B6 V	1004,8055
	+35 01127	NGC 1912 - 97		5 25 29	+35 44.4	1	10.37		0.21		0.05		1	A0 Vn	49
35991	−21 01174	HR 1823		5 25 29	−21 25.0	4	6.08	.012	1.03	.007	0.87		13	K0 III	15,1075,2035,3051
36060	−41 01884	HR 1827	⋆A	5 25 29	−40 59.1	4	5.86	.005	0.24	.003	0.17	.004	15	A5mA5-F2	15,216,355,2027
		NGC 1912 - 66		5 25 30	+35 49.7	1	11.55		0.20		0.08		1	A0	49
290470	−2 01256			5 25 30	−02 03.7	1	9.76		0.11		0.03		2	A0	1089
		Sk -67 099		5 25 30	−67 39.	1	13.13		-0.16		-1.00		4		798
		G 99 - 9		5 25 31	+00 08.6	2	14.57	.045	1.27	.035	0.99	.005	2		1620,1773,3059
244042	+3 00912	FO Ori		5 25 31	+03 35.0	1	9.44		0.19		0.15		5		1768
290415	+0 01084			5 25 32	+00 10.4	1	10.96		0.28		0.21		3	A5	808
290435	−0 00959			5 25 33	−00 29.	1	10.68		0.40		0.09		5		808
35926	+0 01085	IDS05230N0042	AB	5 25 35	+00 44.8	2	8.34	.014	-0.08	.009	-0.41	.005	2	B7 IV	160,1089
36189	−59 00472	HR 1836		5 25 35	−58 57.3	3	5.13	.004	0.99	.004			11	G6 III	15,1075,2007
243940	+29 00905	LS V +29 005		5 25 36	+29 59.2	1	10.73		0.15		-0.52		4	B5	117
		Sk -66 088		5 25 36	−66 17.	1	12.70		0.20		-0.65		4		798
		AAS21,109 # 166		5 25 36	−68 40.	1	13.88		0.00		-1.01		2		417
		Sk -69 124		5 25 36	−69 05.	1	12.81		-0.18		-1.01		3		952
		NGC 1912 - 178		5 25 37	+35 53.6	1	13.55		0.47		0.23		1		49
		G 99 - 10		5 25 38	+02 56.5	3	12.82	.013	1.52	.062	0.90	.059	5	M3	203,1691,3078
		G 99 - 11		5 25 39	+02 48.7	1	12.89		0.83		0.45		1		1620
35948	−1 00906			5 25 39	−01 48.7	2	8.37	.007	0.03	.006	-0.05	.034	3	A0	1089,1732
35957	−2 01258	IDS05231S0201	A	5 25 41	−01 58.9	1	8.51		-0.07		-0.35		3		808
243952	+31 00981			5 25 42	+31 25.9	1	10.49		0.19		-0.28		4	B9	117
		NGC 1912 - 65		5 25 42	+35 50.9	1	10.93		0.52		0.17		1	A4	49
290471	−2 01258			5 25 42	−01 59.2	1	9.48		0.09		0.06		3		808
269504		Sk -67 100		5 25 42	−67 21.4	1	11.95		-0.09		-0.86		4	B1.5Ia	149
		NGC 1912 - 92		5 25 43	+35 43.2	1	11.66		0.21		0.05		1	A0	49
36037	−25 02452			5 25 43	−25 40.6	2	7.53	.005	0.14	.005	0.16	.024	6	A2 V	1628,1776
		NGC 1912 - 63		5 25 44	+35 51.3	1	12.56		1.11		0.70		1		49
35909	+13 00903	HR 1819		5 25 45	+13 38.4	1			0.15		0.14		2	A4 V	1049
35879	+23 00920			5 25 45	+23 55.1	1	8.87		0.57				2	F8	1724
290439				5 25 45	−00 46.	1	10.69		0.36		0.00		5	A5	808
269511				5 25 45	−68 53.	1	11.17		0.17		-0.28		4	F7	415
	+35 01131	NGC 1912 - 70		5 25 46	+35 49.1	1	9.79		1.20		0.89		1	F8	4002
290447				5 25 47	−01 12.	1	11.84		0.27		0.24		3	A3	808
36316	−65 00469			5 25 47	−65 47.0	3	7.94	.005	1.45	.002	1.71	.003	46	K3/4 III	978,1642,1659
		BSD 49 # 2034		5 25 48	+30 10.2	1	11.52		0.32		0.05		3	B8	117
		LP 119 - 48		5 25 48	+52 38.0	1	15.56		0.00		-0.97		3		940
35741	+53 01027			5 25 50	+53 23.6	1	7.14		0.38		-0.04		2	F0	1566
36137	−46 01860			5 25 51	−46 08.1	2	7.67	.019	0.41	.009			8	F2 V	1075,2033
		BSD 49 # 1271		5 25 52	+29 15.8	1	11.00		0.25		-0.02		6	B9	117
35971	−0 00960			5 25 52	−00 01.2	1	6.67		-0.06		-0.24		3	B9	1089
35972	−0 00961			5 25 52	−00 44.5	1	8.83		0.07		-0.27		2	A0	1089
290462	−1 00907			5 25 54	−01 41.0	1	10.32		0.27		0.08		3		808
		Sk -66 089		5 25 54	−66 52.	1	13.94		-0.23		-1.00		4		798
		Sk -71 029		5 25 54	−71 30.	1	14.05		-0.16		-0.98		4		798
		AAS21,109 # 165		5 25 54	−71 35.	1	14.30		-0.03		-0.92		3		952
269518		Sk -69 125		5 25 56	−69 16.1	1	12.14		0.20		0.11		6	A3 Iab	149
35945	+16 00786			5 25 57	+16 23.2	1	7.65		0.02		-0.12		2	B9	401
36003	−3 01110	G 99 - 12	⋆A	5 25 57	−03 31.7	6	7.64	.012	1.11	.009	1.07	.032	24	K5 V	22,680,1003,1758,2012,3072
		NGC 1912 - 212		5 25 58	+35 43.0	1	11.88		0.23		0.08		1		49
		KUV 05260+2711		5 25 59	+27 10.6	1	15.42		0.42		-0.15		1	DA	1708

Table 1 271

HD	DM	Other Id	N Rem	α_{1950}	δ_{1950}	S	V	σ_V	B−V	σ_{B-V}	U−B	σ_{U-B}	n	Spectrum	References
35944	+20 00961			5 26 00	+20 24.2	1	7.25		1.40		1.47		2	K0	1648
290404	+0 01087			5 26 01	+00 43.4	1	9.75		0.13		0.08		3	A2	808
35898	+32 01003			5 26 01	+32 09.6	1	7.05		0.48				3	F8	1724
290448				5 26 02	−01 15.9	1	11.31		0.33		0.26		4	A3	808
		Sk −66 090		5 26 02	−66 48.2	1	12.77		0.08		0.04		6	A2 Ib	149
35956	+12 00801	IDS05233N1227	A	5 26 03	+12 30.9	2	6.75	.000	0.58	.002	0.06	.012	2	G0 V	572,908
		Sk −67 101		5 26 03	−67 32.	1	12.63		-0.17		-0.98		4		415
244023	+30 00909			5 26 05	+30 29.7	1	10.19		0.14		-0.39		2	B9	117
		NGC 1912 - 193		5 26 05	+35 51.9	1	11.77		0.28		0.08		1		49
36017	−4 01141		AB	5 26 05	−04 44.2	1	7.53		0.26		0.12		2	A3	1060
36079	−20 01096	HR 1829	⋆ AB	5 26 06	−20 47.9	5	2.84	.005	0.82	.007	0.46	.013	18	G5 II	15,542,1075,2012,8015
		Sk −66 091		5 26 06	−66 57.	1	13.61		-0.14		-0.93		4		798
		Sk −67 102		5 26 06	−67 05.	1	13.58		-0.21		-0.92		3		798
		Sk −67 103		5 26 06	−67 27.	1	13.50		-0.22		-1.05		3		952
244140	+4 00941	G 97 - 43		5 26 08	+04 45.1	1	9.10		0.74		0.31		1	G5	333,1620
		G 102 - 4		5 26 08	+12 29.7	1	13.98		1.65		1.37		4		3016
244068	+23 00921			5 26 08	+23 39.3	1	10.06		0.72		0.23		4	F2	206
36014	−0 00963			5 26 08	−00 30.1	1	8.74		0.26		0.10		3	A8 V	808
36016	−4 01142			5 26 08	−04 14.5	1	9.01		0.30		0.03		2	A0	1060
	+39 01314			5 26 09	+39 22.0	1	9.60		1.89				3	M2	369
36013	+1 01026	IDS05235N0134	A	5 26 10	+01 36.3	3	6.91	.007	-0.15	.011	-0.63		8	B3 V:N	1004,2033,8055
36032	−2 01260			5 26 10	−02 46.4	1	9.16		-0.04		-0.22		4	B8	1089
36402	−67 00433	Sk −67 104		5 26 10	−67 32.3	1	11.44		-0.17		-1.01		4	WC5	980
		BSD 49 # 1290		5 26 11	+30 00.5	1	11.26		0.21		-0.43		4	B7	117
		NGC 1912 - 459		5 26 11	+35 37.8	1	11.12		0.18		-0.47		1		49
36012	+2 00974			5 26 12	+02 07.5	2	7.31	.038	-0.11	.004	-0.58		3	B5 Vne	1004,8055
35943	+25 00839	HR 1821	⋆ AB	5 26 12	+25 06.7	3	5.47	.005	-0.04	.004	-0.17	.018	6	B8.5V +A1 V	351,1049,1733
		AAS21,109 # 169		5 26 12	−67 31.	1	12.47		-0.21		-0.98		3		952
		Sk −66 092		5 26 14	−66 42.5	1	12.40		0.02		-0.58		4	B6 Iab	415
		Sk −67 105		5 26 14	−67 13.	1	12.42		-0.15		-0.96		3		415
		Sk −66 093		5 26 18	−66 33.	1	13.49		-0.15		-0.96		3		798
		AAS21,109 # 167		5 26 18	−69 13.	1	14.05		0.10		-0.90		3		417
		AAS21,109 # 168		5 26 18	−69 15.	2	13.71	.070	-0.06	.028	-1.02	.000	4		417,952
244066	+28 00808	IDS05232N2850	AB	5 26 19	+28 52.7	1	9.40		0.22		-0.45		5	B5	117
36046	−0 00964			5 26 20	−00 38.5	1	8.07		-0.10		-0.56		4	B9	1089
244138	+11 00819	GW Ori		5 26 21	+11 49.9	3	9.78	.126	1.02	.070	0.34	.086	3	K0	572,8013,8023
269525	−67 00435	Sk −67 106		5 26 21	−67 32.2	1	12.78		-0.17		-0.98		3	B7	415
		Hyades J 360		5 26 22	+18 48.	1	12.22		0.77		0.25		2		1058
35954	+27 00778			5 26 22	+27 48.3	1	9.18		0.39				2	A9/2 V/III	1724
35921	+35 01137	LY Aur	⋆ AB	5 26 22	+35 20.2	3	6.78	.050	0.19	.010	-0.78	.005	5	O9.5III	588,1011,1012
35940	+34 01064			5 26 23	+35 04.2	1	6.85		1.06		0.85		2	K0	1625
36057	−1 00909	IDS05239S0151	AB	5 26 23	−01 48.0	1	8.70		0.14		0.03		2	A0	1089
		Sk −66 094		5 26 23	−66 14.7	1	12.11		0.32		-0.03		7		149
35985	+18 00862	IDS05235N1817	AB	5 26 24	+18 19.6	1	6.74		0.12		0.05		2	A2	401
36056	+0 01088			5 26 27	+00 16.5	1	8.87		0.07		-0.01		1	A0	1089
36058	−3 01115	HR 1826	⋆ AB	5 26 27	−03 20.8	4	6.39	.005	-0.02	.008	-0.07	.016	12	A0 Vn	15,1079,1089,1415
		Sk −67 107		5 26 27	−67 32.	1	12.50		-0.12		-0.94		3		415
		NGC 1912 - 432		5 26 28	+35 51.1	1	13.27		0.54		0.09		1		49
		BSD 49 # 1308		5 26 29	+29 32.7	1	12.46		0.36		0.20		2	B9	117
35984	+29 00909	HR 1822		5 26 30	+29 08.9	3	6.24	.029	0.45	.008	0.00	.015	7	F6 III	254,1724,3037
290547	−2 01261			5 26 30	−02 02.3	1	9.40		0.35		0.08		3	B2	808
36089	−2 01263			5 26 31	−02 48.9	1	8.65		0.27		0.00		3	Am	808
244100	+28 00810			5 26 32	+28 30.4	1	10.46		0.23		-0.25		5	A3	117
36075	−2 01262			5 26 32	−02 12.9	1	8.68		-0.06		-0.23		4	B9	1089
36187	−37 02220	HR 1835		5 26 32	−37 16.3	4	5.56	.010	0.02	.005	0.03	.000	13	A0 V	15,1637,2012,3050
36090	−4 01146	S Ori	⋆ A	5 26 33	−04 43.9	3	10.88	.055	1.68	.050	-0.13	.142	3	M4	975,8022,8027
		Sk −67 108		5 26 33	−67 39.	1	12.56		-0.20		-1.05		4		415
35983	+30 00912			5 26 34	+30 33.6	1	8.66		1.16				3	K0	1724
		BPM 85872		5 26 35	+15 32.5	1	10.61		1.47		1.29		3		7009
		HRC 86, V649 Ori	AB	5 26 36	+11 49.6	1	12.03		1.19	.070	0.49	.025	2	dK3e	572,825
35952	+35 01139	NGC 1912 - 427		5 26 36	+35 55.0	3	8.86	.008	0.27	.000	-0.34	.017	4	B5 II-III	49,1012,8100
		Sk −67 109		5 26 36	−67 25.	1	13.15		-0.12		-0.94		3		798
		G 86 - 43		5 26 37	+32 55.0	1	12.88		0.86		0.45		1		333,1620
		Sk −66 094a		5 26 37	−67 00.2	1	13.44		-0.01		-0.19		4	B9 Ib	415
		G 86 - 44		5 26 38	+32 03.0	2	12.18	.035	1.49	.010	1.11	.040	2		1620,1658
	+35 01141	LS V +35 021		5 26 39	+35 09.9	1	9.59		0.10		-0.74		2	B0.5:V:n	1012
290488				5 26 40	+00 10.0	1	11.18		0.12		0.02		3	A2	808
	+3 00924	GS Ori		5 26 40	+03 26.7	1	10.26		2.58				1	N0	864
		BSD 49 # 483		5 26 40	+28 53.7	1	10.63		0.40		0.35		5	B9	117
290497	−0 00967			5 26 41	−00 12.5	1	9.48		0.15		0.02		1	B8	1089
34109	+85 00078	HR 1714		5 26 42	+85 38.3	2	6.73	.010	0.00	.010	0.01	.010	5	A2 V	1502,1733
36120	−5 01269			5 26 42	−05 49.8	3	7.97	.005	-0.04	.010	-0.35	.005	6	B8 Vn	1004,1060,8055
		Sk −66 095		5 26 46	−66 44.3	1	12.43		0.02		-0.52		4		415
35997	+34 01066			5 26 47	+34 39.0	1	8.77		0.35				2	F2	1724
		Orion P 106		5 26 47	−04 45.1	1	11.60		0.44		0.31		2		1060
36118	−2 01266			5 26 50	−02 02.8	1	8.88		-0.02		-0.19		2	A0	1089
		LP 777 - 27		5 26 50	−16 34.6	1	10.98		0.54		-0.16		2		1696
36521		Sk −68 080		5 26 50	−68 52.6	2	12.41	.010	-0.21	.020	-0.98	.020	6	WC5	415,980
	−27 02290			5 26 51	−27 09.0	1	10.64		0.35		0.39		3	A5	1776
		BSD 49 # 1326		5 26 52	+29 18.5	1	11.40		0.44		0.48		4	A0	117

HD	DM	Other Id	N Rem	α_{1950}	δ_{1950}	S	V	σ_V	B–V	σ_{B-V}	U–B	σ_{U-B}	n	Spectrum	References
		NGC 1978 sq1 22		5 26 52	−66 12.7	1	14.39		0.79				2		1525
	−29 02277			5 26 53	−29 55.7	7	11.62	.016	0.51	.014	−0.24	.009	18	F6	158,742,1696,1775,2017*
269540	−67 00437	Sk −67 110		5 26 53	−67 31.2	1	12.62		−0.07		−0.94		4	B7	415
269541	−68 00372	Sk −68 081		5 26 53	−68 54.4	1	10.40		0.21		−0.16		5	A8:Ia0	149
269542	−69 00376	Sk −69 129		5 26 53	−69 29.4	2	9.91	.015	0.38	.045	−0.10		7	F6 Ia	149,1483
		BSD 49 # 2087		5 26 54	+30 27.2	1	11.28		0.28		0.07		4	B9	117
36117	−0 00968			5 26 54	−00 04.8	1	7.97		0.10		0.02		1	A0	1089
36134	−3 01116	HR 1830		5 26 54	−03 29.1	4	5.78	.022	1.16	.028	1.00	.023	9	K1-III	15,247,1061,8015
269547		Sk −71 030		5 26 54	−71 36.	3	11.61	.005	0.05	.009	−0.67		9	B3 Ia:	415,1277,8033
36044	+29 00911	HR 1825	⋆ AB	5 26 55	+29 30.7	2	7.10	.015	0.85	.005	0.54		6	G9 III	1724,3024
36044	+29 00911	IDS05237N2928	C	5 26 55	+29 30.7	1	9.42		0.55		0.06		4	G5	3024
36028	+30 00913			5 26 55	+30 24.9	1	9.25		0.23		−0.02		4	A0	117
244156	+30 00914			5 26 55	+30 33.0	1	10.33		0.33		0.20		5	A0	117
36133	+3 00928	IDS05243N0304	ABC	5 26 56	+03 06.6	2	7.10	.061	−0.11	.007	−0.67		7	F8 III	1004,8055
244153	+33 01069			5 26 56	+33 21.0	1	9.54		1.78				3	K7	369
36115	+5 00934			5 26 57	+05 11.3	1	8.23		0.01		−0.35		2	B8	401
290502				5 26 57	−00 30.	1	11.37		0.38		0.26		3	A5	808
		Sk −67 111		5 26 57	−67 31.	1	12.57		−0.20		−1.03		3		415
	−6 01194			5 26 58	−06 01.4	1	10.83		0.36		0.10		2	A6	1060
36104	+12 00803			5 26 59	+12 13.9	1	7.00		−0.11		−0.50		1	B8	572
36054	+23 00922			5 27 00	+23 43.2	1	8.15		1.23				2	G5	1724
36151	−7 01092	IDS05246S0720	A	5 27 00	−07 18.0	3	6.69	.020	−0.12	.010	−0.58	.010	5	B5 V	1004,1060,8055
269546	−68 00373	Sk −68 082		5 27 02	−68 52.3	5	9.89	.014	−0.03	.013	−0.81	.010	19	B5 Ia	328,980,1277,1732,8033
290504				5 27 03	−00 12.	1	10.40		0.45		0.26		3	A5	808
36435	−60 00424			5 27 03	−60 27.3	4	6.99	.016	0.76	.011	0.30	.017	14	G6/8 V	158,1075,2033,3008
		NGC 1978 sq1 20		5 27 03	−66 12.7	1	13.24		0.52				2		1525
		Sk −69 132		5 27 03	−69 02.	1	12.28		0.00		−0.70		4		415
36139	−0 00969			5 27 04	−00 03.6	1	6.39		−0.01		0.03		1	A0	1089
		NGC 1978 sq1 21		5 27 04	−66 13.2	1	14.31		1.11				2		1525
269545	−67 00438	IDS05272S6744	⋆ AB	5 27 04	−67 41.9	1	11.90		−0.13		−0.98		3	B1 Ia	149
36042	+34 01069			5 27 05	+34 09.8	1	7.45		1.04				3	G7 III	1724
		Sk −65 061		5 27 06	−65 51.	1	13.92		−0.18		−0.98		4		798
		Sk −67 113		5 27 06	−67 22.	1	13.38		−0.23		−1.03		4		798
		AAS21,109 # 170		5 27 06	−69 08.	1	13.09		−0.17		−1.02		3		417
269549		Sk −69 133		5 27 06	−69 09.	1	14.66		−0.13		−0.86		3	WN4	980
36767	−75 00314			5 27 06	−75 43.7	3	7.17	.004	0.54	.010	−0.01	.000	9	F8 V	158,258,2012
36102	+21 00865			5 27 08	+21 20.6	1	8.73		0.51				2	F8	1724
		NGC 1978 sq1 35		5 27 08	−66 18.8	1	12.25		1.21				1		1525
36598	−70 00393			5 27 08	−70 06.2	1	8.05		1.30		1.08		1	Kp Ba	565
36150	−0 00970	IDS05246S0053	AB	5 27 09	−00 50.4	1	6.49		0.25		0.11		4	A8 IV	808
		Sk −67 114		5 27 09	−67 28.	1	12.13		0.48		−0.40		2		415
		Sk −66 096		5 27 10	−66 19.	2	13.00	.023	0.09	.014	−0.12		4		415,1525
269551	−68 00374	Sk −68 083	⋆ AB	5 27 10	−68 52.4	1	9.86		−0.03		−0.83		3	Of	952
244204	+27 00780			5 27 11	+27 40.8	1	9.36		2.03				4	M0	369
36255	−30 02421	HR 1838		5 27 11	−30 09.3	1	6.75		1.06				4	G8/K0 III	2031
271273				5 27 11	−66 08.4	1	11.93		0.08		−0.46		3	A0	149
		Haro 6 # 49		5 27 12	+12 12.	1	13.41		1.22		0.96		1		572
		BSD 49 # 2098		5 27 12	+30 27.3	1	11.71		0.44		0.08		3	B5	117
36167	−1 00913	HR 1834, CI Ori	⋆ A	5 27 12	−01 07.8	1	4.69		1.58		1.91		2	K5 III	542
36167	−1 00913	HR 1834, CI Ori	⋆ AB	5 27 12	−01 07.8	6	4.70	.005	1.56	.011	1.84	.005	24	K5 III	15,542,1075,1425,2029,3053
36167	−1 00913	IDS05240S0110	B	5 27 12	−01 07.8	2	9.79	.107	0.66	.063	0.12	.034	3		542,542
274903	−49 01734			5 27 13	−49 37.4	1	9.80		0.66				4	G5	2013
		HRC 89, GX Ori		5 27 14	+12 11.3	2	13.48	.205	1.11	.170	0.12	.150	2	dK3e	572,825
36071	+31 00991			5 27 15	+31 14.7	1	7.88		0.09		−0.12		3	A0	117
290500	−0 00971			5 27 15	−00 26.0	1	11.04		0.31		0.06		3	B8	808
36584	−68 00375	HR 1859	⋆ AB	5 27 15	−68 39.7	3	6.02	.003	0.34	.003	0.01	.015	25	F0 IV/V	15,1075,2038
36041	+39 01322	HR 1825	⋆ A	5 27 16	+39 47.3	1	6.37		0.97		0.76		2	G9 III	542
36041	+39 01321	IDS05238N3945	B	5 27 16	+39 47.3	1	7.62		0.94		0.69		2		542
36040	+41 01206	HR 1824		5 27 16	+41 25.5	2	5.99	.019	1.11	.002	1.12		5	K0 IIIp	37,71
36113	+20 00969			5 27 17	+20 30.8	1	7.08		−0.05		−0.38		2	B5	401
	+35 01146			5 27 17	+35 31.5	1	9.71		1.83				3	M2	369
36176	−1 00914			5 27 17	−01 41.8	1	8.67		−0.05		−0.29		2	A0	1089
		Sk −66 098		5 27 17	−66 24.	1	12.41		−0.07		−0.74		4		415
		Sk −66 097		5 27 18	−66 24.	1	12.50		0.02		−0.81		4		798
		AAS21,109 # 171		5 27 18	−66 38.	2	13.31	.005	−0.02	.014	−0.85	.019	4		417,952
		Sk −67 115		5 27 18	−67 13.	1	13.42		−0.10		−0.61		4		798
		Sk −70 090		5 27 18	−70 43.	1	13.81		−0.14		−1.03		3		798
36166	+1 01032	HR 1833		5 27 19	+01 45.1	7	5.78	.005	−0.20	.004	−0.84	.007	24	B2 V	15,154,1004,1061,1732*
36165	+1 01033			5 27 19	+02 06.2	1	8.13		−0.11		−0.47		1	B7 V	160
290545				5 27 19	−02 07.	1	10.57		0.39		−0.03		3	A2	808
244252	+25 00842			5 27 21	+25 59.7	1	9.18		0.42				2		1724
290521	−1 00915			5 27 21	−01 03.2	1	10.05		0.22		0.10		4		808
36085	+31 00992	IDS05241N3126	A	5 27 22	+31 28.4	1	7.54		1.77				3	K5	369
35961	+54 00902	G 191 - 41	⋆ A	5 27 23	+54 37.3	1	7.72		0.62		0.09		1	G1 V	3016
35961	+54 00902	G 191 - 41	⋆ AB	5 27 23	+54 37.3	1	7.53		0.64		0.13		2	G1 V + K3	1003
35961	+54 00902	IDS05233N5435	B	5 27 23	+54 37.3	1	9.71		0.99		0.80		1	K3	3032
		G 99 - 14		5 27 23	−03 28.4	1	12.03		1.54				1		1705
		NGC 1978 sq1 9		5 27 23	−66 07.3	1	13.51		0.27				3		1525
36185	−2 01268			5 27 24	−02 17.6	1	9.10		0.09		−0.03		4	A0 V	808
36203	−7 01096			5 27 24	−07 40.9	1	9.25		0.15		0.05		2	A3 Vs	1060

Table 1   273

HD	DM	Other Id	N Rem	α_{1950}	δ_{1950}	S	V	σ_V	B–V	σ_{B-V}	U–B	σ_{U-B}	n	Spectrum	References
269559 −67 00439		Sk -67 116		5 27 24	−67 15.3	1	11.29		-0.05		-0.73		3	B3 Ia	149
		AAS21,109 # 172		5 27 24	−67 31.	2	13.24	.009	-0.16	.000	-0.96	.033	4		417,952
		BSD 49 # 2633		5 27 25	+31 24.2	1	12.69		0.45		0.23		2	A0	117
290544				5 27 25	−02 03.	1	11.29		0.39		-0.05		3	A5	808
290518				5 27 26	−01 04.0	1	10.26		0.23		0.08		3	A2	808
290526 −1 00916				5 27 26	−01 29.2	1	10.66		0.38		0.33		4		808
		NGC 1978 sq1 15		5 27 26	−66 08.8	1	14.47		-0.03				2		1525
		Sk -67 117		5 27 30	−67 14.	1	12.92		-0.18		-1.03		4		798
		AAS21,109 # 173		5 27 30	−69 10.	1	13.00		-0.14		-1.01		3		417
		NGC 1978 sq1 10		5 27 31	−66 07.5	1	13.52		1.04				3		1525
36219 −1 00918		IDS05250S0150	AB	5 27 33	−01 47.2	3	7.65	.013	-0.06	.005	-0.34	.014	8	B9	1004,1089,8055
		Sk -67 117a		5 27 33	−67 30.	1	12.76		-0.16		-0.99		4		415
		Sk -67 117b		5 27 33	−67 30.	1	12.62		-0.15		-0.95		4		415
36162 +15 00837		HR 1832		5 27 34	+15 19.4	1			0.14		0.14		3	A3 Vn	1049
244249 +31 00993				5 27 34	+31 18.7	1	9.45		0.18		-0.07		2	A0	117
36274 −22 01129				5 27 34	−22 45.4	1	7.59		0.99		0.74		3	K0 III	1657
36111 +30 00917				5 27 35	+31 00.0	1	8.93		0.57				2	F8	1724
		NGC 1978 sq1 30		5 27 36	−66 13.0	1	13.81		1.57				2		1525
		AAS21,109 # 175		5 27 36	−66 34.	2	13.54	.009	-0.12	.023	-0.80	.009	4		417,952
		AAS21,109 # 174		5 27 36	−68 57.	1	12.90		-0.11		-0.87		5		417
290517 −1 00920				5 27 38	−01 04.0	1	9.00		-0.05		-0.29		2	A0	1089
271279 −65 00473		Sk -65 062		5 27 39	−65 43.2	3	11.26	.004	0.09	.005	-0.28		10	A0:Ia	149,1277,8033
		NGC 1978 sq2 100		5 27 39	−66 16.0	1	13.64		-0.15				5		1525
		Sk -67 117c		5 27 39	−67 30.	1	12.71		-0.16		-0.99		3		415
269563		Sk -67 118a		5 27 40	−67 21.4	1	11.94		0.12		0.12		6	A4 Ib	149
36217 +4 00949		HR 1837, CK Ori		5 27 41	+04 10.0	3	6.17	.027	1.27	.000	1.38	.000	12	K2 III	15,897,1415
36160 +22 00925		HR 1831		5 27 42	+22 25.5	2	6.28	.009	1.18	.000	1.15		4	K0	1375,1724
290535				5 27 42	−01 44.	1	10.61		0.33		0.06		3	A3	808
		Sk -67 119		5 27 42	−67 20.	1	13.33		-0.21		-1.04		4		798
		Sk -67 118		5 27 42	−67 21.4	1	12.99		-0.20		-1.03		4	A4 Ib	798
		Sk -70 090a		5 27 42	−70 32.	1	13.78		-0.08		-0.61		4		798
36234 −5 01274				5 27 43	−05 14.4	3	8.65	.005	-0.06	.005	-0.38	.010	6	B9 V	1004,1060,8055
		NGC 1978 sq1 33		5 27 43	−66 19.9	1	11.10		1.09				2		1525
290516 −1 00921				5 27 44	−00 55.7	1	9.47		0.03		-0.14		2	A0	1089
290527 −1 00922				5 27 44	−01 32.2	1	10.15		0.27		0.10		3		808
269566 −66 00415		Sk -66 099		5 27 44	−66 16.5	2	10.65	.005	0.40	.020	0.00		13	A8:Ia0	149,1525
35863 +67 00390				5 27 45	+67 58.8	1	6.83		0.36		-0.04		4	F0	1625
		Sk -68 084a		5 27 45	−68 56.	1	12.70		-0.11		-0.88		4		415
		Sk -69 139		5 27 45	−69 03.	1	11.85		-0.01		-0.74		4	B2.5-3Ia	415
		NGC 1978 sq2 7		5 27 47	−66 15.3	1	14.42		2.02				1		1525
244363 +12 00806				5 27 48	+12 23.3	1	10.68		0.53		0.11		1	G0	572
		HRC 90, V447 Ori		5 27 48	+12 34.6	1	14.47		1.17		0.27		1		776
290493 −0 00974				5 27 48	−00 27.2	1	9.99		0.12		0.05		3		808
290543 −2 01269				5 27 48	−02 04.2	1	10.42		0.32		-0.02		3		808
		Sk -66 100		5 27 48	−66 58.	1	13.26		-0.21		-1.06		4		798
		MN83,95 # 601		5 27 48	−71 11.9	1	11.73		1.06				1		591
		NGC 1978 sq1 31		5 27 49	−66 12.3	1	13.44		0.36				1		1525
		NGC 1931 - 2		5 27 51	+34 09.0	1	11.61		0.50		0.04		2		1605
		Sk -67 120		5 27 51	−67 30.	1	12.26		-0.19		-1.02		4		415
		Sk -68 085		5 27 51	−69 00.	1	12.43		-0.08		-0.87		3		415
290478 +0 01094				5 27 52	+00 34.4	1	10.10		0.30		0.04		3	A7	808
		NGC 1978 sq1 65		5 27 52	−66 06.1	1	11.36		0.89				2		1525
		NGC 1931 - 3		5 27 53	+34 10.5	1	13.46		0.36		0.31		2		1605
36650 −68 00377				5 27 53	−68 06.8	1	8.79		1.12		0.87		1	K0 III	565
244306 +29 00917				5 27 54	+30 03.4	1	9.96		0.37		0.36		2	A3	117
		POSS 57 # 1		5 27 54	+65 33.4	1	15.74		0.69				2		1739
36174 +26 00829		IDS05248N2631	AB	5 27 56	+26 32.9	1	8.90		0.83				2	F5	1724
290515 −0 00976				5 27 56	−00 53.1	1	9.28		0.04		-0.05		2	A0	1089
36285 −7 01099		HR 1840		5 27 56	−07 28.3	9	6.31	.008	-0.19	.006	-0.82	.009	91	B2 IV-V	15,154,1004,1060,1415,1732*
		POSS 57 # 5		5 27 57	+65 30.2	1	17.53		1.25				4		1739
		Sk -68 086		5 27 57	−68 53.	3	12.61	.004	-0.16	.007	-0.97	.010	10		415,1277,1408
		NGC 1931 - 5		5 27 58	+34 09.6	1	13.01		0.52		0.35		2		1605
36083 +48 01278				5 27 58	+48 53.9	1	8.32		0.06		-0.25		3	B8	555
		NGC 1978 sq1 3		5 27 58	−66 09.8	1	13.27		2.01				8		1525
		NGC 1931 - 7		5 28 00	+34 09.1	1	13.75		0.73		0.45		2		1605
	−7 01100			5 28 00	−07 02.4	1	10.86		0.26		0.05		2		1060
		Sk -67 123		5 28 00	−67 18.	1	13.33		-0.13		-0.86		3		798
		Sk -67 121		5 28 00	−67 50.	1	13.57		-0.24		-1.01		3		798
		AAS21,109 # 176		5 28 00	−71 36.	2	14.09	.112	-0.08	.054	-0.89	.000	3		417,952
36269 +0 01098				5 28 01	+00 19.7	3	7.49	.017	0.03	.034	0.04	.016	45	A7 IV-V	808,978,1659
36232 +12 00808				5 28 01	+12 19.2	1	8.74		0.33		0.05		1	F2	572
244328 +30 00918		LS V +30 009		5 28 01	+30 08.4	1	9.89		0.30		-0.32		4	B9	117
		Wolf 1452		5 28 01	−05 03.1	1	12.58		0.88		0.45		2		1696
		NGC 1978 sq1 54		5 28 02	−66 21.2	1	13.83		0.61				2		1525
269577 −67 00440		Sk -67 122		5 28 02	−67 23.0	1	10.88		0.00		-0.67		4	B6 Ia	149
		NGC 1931 - 11		5 28 03	+34 09.1	1	12.03		0.30		-0.30		2		1605
269578 −69 00379			V	5 28 03	−69 01.7	1	10.02		0.16				2	B7	8033
		HRC 91, V448 Ori		5 28 04	+12 06.4	1	14.28		1.14		0.52		1	dK4e	776
		NGC 1978 sq1 53		5 28 04	−66 20.6	1	13.18		1.01				2		1525
		BSD 49 # 2138		5 28 05	+30 05.0	1	11.47		0.54		-0.23		2	B3	117

HD	DM	Other Id	N	Rem	α_{1950}	δ_{1950}	S	V	σ_V	B–V	σ_{B-V}	U–B	σ_{U-B}	n	Spectrum	References
36267	+5 00939	HR 1839	★	AB	5 28 06	+05 54.7	7	4.20	.007	-0.14	.010	-0.55	.010	29	B5 IV +B7 V	15,154,1075,1223,1363*
36263	+10 00800	IDS05254N1011		AB	5 28 06	+10 13.1	1	7.47		-0.08		-0.57		2	B9	401
		NGC 1931 - 1			5 28 06	+34 13.	1	14.52		0.60		0.27		2		1605
		NGC 1931 - 4			5 28 06	+34 13.	1	14.96		0.57		0.48		2		1605
		NGC 1931 - 6			5 28 06	+34 13.	1	15.11		0.67		0.47		2		1605
		NGC 1931 - 8			5 28 06	+34 13.	1	12.47		0.37		-0.26		2		1605
		NGC 1931 - 9			5 28 06	+34 13.	1	14.50		0.47		0.42		2		1605
		NGC 1931 - 10			5 28 06	+34 13.	1	15.31		0.93		0.32		2		1605
		NGC 1931 - 12			5 28 06	+34 13.	1	15.33		0.97		0.61		2		1605
		NGC 1931 - 13			5 28 06	+34 13.	1	16.10		0.53		0.39		2		1605
		NGC 1931 - 14			5 28 06	+34 13.	1	15.10		0.41		0.32		2		1605
		NGC 1931 - 15			5 28 06	+34 13.	1	15.57		0.81		0.40		2		1605
		NGC 1931 - 16			5 28 06	+34 13.	1	13.58		1.21		0.84		2		1605
		NGC 1931 - 17			5 28 06	+34 13.	1	13.36		0.34		-0.17		2		1605
		NGC 1931 - 18			5 28 06	+34 13.	1	14.94		0.41		0.15		2		1605
		NGC 1931 - 20			5 28 06	+34 13.	1	13.91		0.77		0.24		2		1605
		NGC 1931 - 21			5 28 06	+34 13.	1	14.97		0.92		0.41		2		1605
		NGC 1931 - 22			5 28 06	+34 13.	1	15.62		0.75		0.04		2		1605
		NGC 1931 - 23			5 28 06	+34 13.	1	15.36		0.84		0.27		2		1605
		NGC 1931 - 24			5 28 06	+34 13.	1	12.13		0.53		-0.42		2		1605
		NGC 1931 - 25			5 28 06	+34 13.	1	12.99		0.52		-0.31		2		1605
		NGC 1931 - 27			5 28 06	+34 13.	1	13.98		0.52		-0.02		3		1605
	+34 01074	NGC 1931 - 19	★		5 28 06	+34 14.2	2	11.15	.019	0.31	.005	-0.43	.015	3		797,1605
		NGC 1931 - 26	★	A	5 28 06	+34 14.2	2	11.51	.019	0.39	.019	-0.52	.039	3		797,1605
		NGC 1931 - 26	★	ABC	5 28 06	+34 14.2	1	10.48		0.43		-0.46		1		8084
		Sk -66 101			5 28 06	-66 26.	1	12.93		-0.14		-0.95		4		798
		ApJS2,315 # 37			5 28 06	-67 33.	1	12.93		1.75		0.76		1	M6 III	35
		Sk -69 141			5 28 06	-69 12.	1	13.46		-0.20		-0.93		1	WN8	980
		Sk -69 140			5 28 06	-69 15.	1	12.76		-0.13		-0.90		3		952
244337	+30 00919				5 28 07	+30 45.8	1	10.87		0.49		-0.18		2	B8	117
36264	+10 00801				5 28 08	+10 07.6	1	7.09		-0.01		-0.13		2	A0	401
		Sk -67 123a			5 28 09	-67 30.	1	13.08		-0.12		-0.90		4		415
		Sk -68 087			5 28 09	-68 59.	1	11.83		-0.03		-0.73		4		415
	-69 00380	Sk -69 142			5 28 11	-69 03.3	1	10.70		0.26		0.17		4		415
36766	-71 00344				5 28 11	-71 15.4	1	8.74		0.39		0.02		3	F0 V:	204
36313	-0 00977	V1093 Ori	★	AB	5 28 12	-00 24.6	1	8.21		-0.07		-0.40		4	B9	1089
		Sk -70 091			5 28 12	-70 39.	1	12.78		-0.23		-1.04		4		798
		Sk -70 092			5 28 12	-70 39.	1	14.16		-0.27		-1.00		4	WN3	980
36215	+27 00783				5 28 13	+27 44.0	1	7.42		0.60				3	F9 V/IV	1724
244352	+30 00920				5 28 13	+30 39.9	1	10.87		0.31		0.08		2	A0	117
271289	-66 00417				5 28 13	-66 10.0	1	10.28		1.24				11	K5	1525
36751	-70 00396	IDS05288S7036		AB	5 28 13	-70 33.8	1	10.12		0.14		0.01		1	A0 V	204
36262	+11 00834				5 28 14	+12 03.6	1	7.61		-0.12		-0.66		6	B3 V	399
		NGC 1978 sq1 2			5 28 14	-66 10.9	1	13.59		1.50				6		1525
269587		Sk -66 102			5 28 14	-66 43.5	1	11.76		0.13		0.13		3	A5 I	149
	-69 00382	Sk -69 143			5 28 14	-69 04.4	1	10.81		0.11		-0.66		3	A4 Ia	415
36312	-0 00978				5 28 15	-00 00.5	1	8.14		-0.07		-0.36		4	B8	1089
290507					5 28 15	-00 23.	1	11.45		0.61		0.38		4	A5	808
		Sk -68 091			5 28 15	-68 58.	1	12.58		-0.16		-1.00		4		415
244397	+23 00927				5 28 16	+23 58.1	1	10.01		0.50				2		1724
		GD 68			5 28 16	+41 26.6	1	16.36		-0.01		-0.76		1		3060
36324	-5 01277				5 28 16	-05 31.7	3	9.02	.013	0.07	.009	0.07	.014	5	A1 V	1004,1060,8055
36066	+57 00889	HR 1828			5 28 17	+57 11.3	2	6.47	.010	0.57	.005	0.11	.005	4	F5 IV	985,1733
269588		Sk -66 103			5 28 17	-66 31.1	1	12.04		-0.03		-0.62		3	B6 Iab	149
244372	+31 00996				5 28 18	+31 35.8	1	10.76		0.23		0.21		2	A2	117
290490					5 28 18	-00 08.	2	9.91	.008	0.12	.012	-0.05	.206	4	A1:	808,1089
36341	-2 01274				5 28 18	-02 24.2	1	8.35		-0.04		-0.39		4	B9	1089
269585		Sk -70 093			5 28 18	-70 09.8	1	12.84		0.07		0.08		3	A1 Iab	415
269586					5 28 18	-70 29.7	1	11.18		0.45		-0.02		1	G0	726
36310	+4 00953	IDS05257N0436		AB	5 28 19	+04 37.9	1	7.95		-0.02		-0.48		1	B6 V	160
36214	+31 00998	IDS05251N3126		AB	5 28 19	+31 28.4	1	8.45		1.33				3	K0	1724
271291					5 28 19	-66 06.2	1	12.07		0.07		-0.38		6	B7	149
269584					5 28 19	-70 29.	1	10.83		0.52		0.01		1	A2	360
290506					5 28 21	-00 14.	1	11.06		0.40		0.31		3	A2	808
36283	+15 00838	G 97 - 45			5 28 22	+15 44.5	3	8.64	.005	0.67	.015	0.12	.008	5	G5 V	333,1003,1620,3077
269591		Sk -67 124			5 28 22	-67 04.6	1	12.12		0.12		0.10		6	A4 Iab	149
36342	-4 01152	IDS05259S0420		A	5 28 24	-04 17.5	1	7.52		0.14		0.10		2	A3 V	1060
		Sk -69 144			5 28 24	-69 16.	1	12.94		-0.19		-1.01		3		952
		Sk -71 031			5 28 24	-71 29.	1	13.91		-0.14		-0.93		4		798
36157	+47 01168				5 28 25	+47 09.4	1	7.79		0.04		-0.05		2	B9	401
36352	-2 01275				5 28 25	-02 44.1	1	9.20		0.02		-0.05		4	B9	1089
		NGC 1978 sq1 28			5 28 25	-66 22.1	1	11.74		0.49				2		1525
290482	+0 01101				5 28 26	+00 22.6	1	10.03		0.14		0.08		3	A2	808
244391	+31 00999				5 28 26	+31 35.1	1	10.32		0.31		0.06		2	B8	117
36213	+32 01016				5 28 26	+32 25.3	1	8.94		0.47				3	G0	1724
36212	+34 01077	LS V +34 048			5 28 26	+34 50.7	2	7.81	.040	0.21	.030	-0.46	.050	3	B3 II	1012,8100
		POSS 57 # 3			5 28 26	+65 19.5	1	16.10		0.91				2		1739
		NGC 1978 sq1 29			5 28 26	-66 21.7	1	13.88		-0.13				1		1525
269594	-67 00443	Sk -67 125			5 28 26	-67 26.7	2	10.60	.005	0.63	.010	0.43		6	F8 Ia	149,1483
36340	+3 00944				5 28 27	+03 19.0	3	7.97	.011	-0.15	.013	-0.77	.007	10	B2 V	247,399,1732

Table 1   275

HD	DM	Other Id	N	Rem	α_{1950}	δ_{1950}	S	V	σ_V	B–V	σ_{B-V}	U–B	σ_{U-B}	n	Spectrum	References
294180	−5 01278	Orion P 571			5 28 28	−05 00.0	1	10.84		0.47		0.35		1		8114
		POSS 57 # 2			5 28 29	+65 32.6	1	16.10		0.98				2		1739
294166	−3 01119				5 28 29	−03 37.7	3	10.32	.007	0.21	.008	0.12	.005	5	A2 Vn	1004,1060,8055
269593	−67 00444	Sk -67 126			5 28 29	−67 42.6	1	11.39		-0.01		-0.68		4	B5:Ia	149
		Orion P 575			5 28 30	−04 43.5	1	10.73		1.49		2.85		1		8114
294180	−5 01278				5 28 30	−04 59.6	1	10.82		0.52		0.35		2	A1	1060
271294					5 28 30	−65 42.	1	12.56		-0.16		-1.02		4	O8	415
		Sk -66 105			5 28 30	−66 39.	1	13.08		-0.16		-0.84		4		798
		SK *G 278			5 28 30	−68 54.5	1	12.94		0.07		0.16		1	A0 Ib	149
		Sk -71 032			5 28 30	−71 29.	1	13.74		-0.19		-1.02		4		798
35919	+69 00327	IDS05230N6935	A		5 28 32	+69 37.5	1	7.84		0.45		-0.01		3	F5	1502
		Orion P 588			5 28 32	−05 41.4	1	12.75		0.78		0.52		1		8114
36180	+46 01022				5 28 34	+46 06.3	1	9.19		0.22		0.17		1	A2	401
36366	−6 01204	IDS05261S0616	A		5 28 34	−06 13.7	3	8.15	.041	0.07	.032	-0.03	.010	6	A1 V +Am	1004,1060,8055
		HRC 93, HI Ori			5 28 36	+12 07.5	1	14.32		1.10		0.09		1	dK0e	776
		KUV 05286+2208			5 28 36	+22 07.8	1	14.75		0.24		0.02		1	NHB	1708
36475	−33 02368		AB		5 28 36	−33 34.3	1	9.14		0.85		0.49		4	K3 III	214
36705	−65 00475	AB Dor	⋆A		5 28 36	−65 29.3	5	6.87	.045	0.83	.019	0.37	.012	5	K1 IIIp	1498,1641,1642,1659,1675
		NGC 1978 sq1 32			5 28 36	−66 11.8	1	13.10		0.69				2		1525
		Sk -67 127			5 28 36	−67 26.7	1	12.41		-0.04		-0.63		4	B8 Iab	415
		Sk -68 092			5 28 36	−68 54.1	1	11.71		-0.07		-0.88		3	B0.5Ia	149
36351	+3 00948	HR 1842	⋆AB		5 28 37	+03 15.3	8	5.46	.008	-0.18	.010	-0.82	.007	22	B1.5V	15,154,1004,1203,1415,1732*
36281	+29 00921				5 28 37	+29 24.1	1	8.63		0.84				2	G5	1724
36519	−43 01885				5 28 37	−43 37.0	1	7.73		1.46				4	K4 III	2012
	−6 01205				5 28 38	−06 43.0	1	10.80		0.41		-0.01		2	A5	1060
36379	−10 01204				5 28 38	−10 06.8	1	6.91		0.56				4	G0	2012
	+59 00886	G 191 - 44			5 28 39	+59 07.3	2	9.89	.000	0.75	.000	0.33	.000	6	G5	516,1064
		Sk -69 145			5 28 40	−69 06.	1	11.37		0.29		0.10		4	A3 Ia+	415
		Sk -72 002			5 28 40	−72 01.7	1	13.39		0.05		-0.10		2	A0 Iab	415
		BSD 49 # 2157			5 28 42	+30 32.0	1	10.68		0.35		0.20		3	B9	117
		Sk -69 146			5 28 42	−69 17.	1	12.71		-0.20		-0.99		3		952
		HR Ori			5 28 43	+12 06.7	1	11.54		0.46		0.00		1		1588
		Sk -69 147			5 28 43	−69 02.2	2	11.44	.014	0.39	.005	0.37		7	F2 Ia	149,1483
269599	−69 00383	IDS05290S6913	⋆AB		5 28 43	−69 10.7	2	10.09	.079	0.19	.020	-0.61		5	B8:I	149,8033
	+39 01328	LS V +40 047			5 28 44	+40 01.8	2	9.85	.005	0.57	.000	-0.46	.000	4	O9 III:	1011,1012
290510					5 28 44	−01 12.1	1	11.69		0.27		0.17		3	A7	808
		Orion P 625			5 28 44	−04 41.9	1	12.57		0.68		0.23		1		8114
36282	+28 00813				5 28 45	+28 58.7	1	8.85		0.50				2	F8/9 V/IV	1724
290492	−0 00980	IDS05262S0034	AB		5 28 46	−00 31.6	1	9.27		0.12		-0.02		2	B8	1089
36518	−37 02242				5 28 46	−37 14.5	1	7.78		1.45		1.58		1	G3 V	3040
36553	−47 01884	HR 1856	⋆A		5 28 47	−47 06.8	5	5.45	.007	0.62	.002	0.23		19	G3 IV	15,1075,1279,2013,2029
36553	−47 01884	IDS05274S4709	BC		5 28 47	−47 06.8	1	11.08		0.75				1	F8/G2	1279
		BSD 49 # 2159			5 28 48	+30 25.7	1	10.96		0.39		0.31		3	B8	117
		Sk -67 128			5 28 48	−67 12.	1	12.80		-0.12		-0.83		3		952
36393	−2 01278				5 28 49	−02 01.0	1	8.48		-0.10		-0.49		4	B8	1089
		Orion B 61			5 28 49	−04 22.5	1	12.73		1.06		1.03		1		114
36552	−43 01889				5 28 49	−43 43.9	1	8.08		0.86		0.52		2	K2 III +(G)	536
290596					5 28 50	−01 32.	1	11.25		0.17		0.09		3	A2	808
36394	−2 01279	IDS05263S0210	AB		5 28 50	−02 08.1	1	8.08		0.08		0.03		3	A1 V+A3V	808
	−27 02312				5 28 50	−27 10.1	1	9.95		0.17		0.14		2	A0	1776
36689	−62 00479	HR 1867			5 28 50	−62 21.1	4	6.59	.009	1.53	.005	1.85		13	K4/5 III	15,1075,2035,3005
		SK *G 319			5 28 50	−66 03.8	1	12.51		0.21		0.25		3		149
269604	−68 00378	Sk -68 093			5 28 50	−68 56.2	1	10.74		0.13		-0.22		5	A1 Ia0	149
36280	+34 01079	LS V +34 049			5 28 51	+34 54.3	1	8.85		0.10		-0.77		2	B0.5IVn	1012
36412	−5 01281	EY Ori			5 28 51	−05 44.4	2	9.47	.004	0.73	.013	0.53	.009	5	A7 V	1060,8055
269606	−67 00446	Sk -67 130			5 28 52	−67 01.8	1	11.43		-0.02		-0.65		3	B7 Ia	149
		Sk -69 149			5 28 52	−69 10.	1	11.77		0.08		-0.57		4		415
		Sk -69 150			5 28 53	−69 14.3	1	12.27		0.10		-0.08		4	A0 Iab	415
36392	+1 01045				5 28 54	+01 39.2	2	7.54	.007	-0.12	.007	-0.63		5	B3 V	1004,8055
244463	+32 01020		V		5 28 54	+32 11.4	1	9.83		2.07				3	F8	369
290491	−0 00979				5 28 54	−00 14.	1	10.19		0.31		0.19		3	A0	808
36411	−4 01156				5 28 54	−03 58.1	3	9.71	.000	0.09	.004	0.06	.005	6	A3 V	1004,1060,8055
		Sk -69 148			5 28 54	−69 15.	2	10.82	.024	1.42	.005	1.36	.073	6		415,798
		Sk -71 033			5 28 54	−71 02.	1	13.80		0.02		-0.97		4		798
36395	−3 01123	G 99 - 15	⋆A		5 28 55	−03 41.1	20	7.97	.010	1.47	.009	1.21	.023	117	M1 V	1,22,116,124,125,680*
36430	−6 01207	HR 1848			5 28 55	−06 44.7	9	6.22	.008	-0.18	.009	-0.74	.010	21	B2 V	15,154,1004,1060,1078,1732*
244516	+24 00846	Hyades vB 155			5 28 56	+24 57.4	2	9.47	.005	0.54	.005	0.07		4	G0	1127,1724
36376	+9 00860				5 28 57	+09 11.4	1	7.79		0.09		-0.57		2	B8	401
36245	+44 01227				5 28 57	+44 19.5	1	7.95		0.07		-0.34		2	B8	401
		G 97 - 46			5 28 58	+16 11.3	2	12.24	.039	0.76	.000	0.23	.007	3	K5	1620,1658
290550	−2 01280				5 28 58	−02 10.1	1	9.53		0.29		0.09		3	A2	808
		Sk -69 152			5 28 58	−69 11.	1	11.80		-0.16		-0.97		4		415
36473	−20 01105	HR 1849			5 28 59	−20 54.0	3	5.54	.007	0.00	.005	0.03		9	A0 V	15,2027,3050
290540	−1 00927				5 29 00	−01 51.7	1	9.53		0.03		-0.11		3	A0	808
		AAS21,109 # 177			5 29 00	−66 33.	1	13.72		-0.22		-1.04		3		952
		Sk -66 106			5 29 00	−66 41.	1	12.72		-0.08		-0.91		4	B1.5Ia	415
		Sk -67 129			5 29 00	−67 36.	1	13.35		-0.17		-1.00		3		952
269609	−68 00380				5 29 00	−68 51.1	1	10.67		0.60		0.00		2	G0	10
		Sk -69 151			5 29 00	−69 34.	1	12.30		-0.03		-0.70		4		415
290474					5 29 01	+00 39.8	1	11.05		0.39		0.21		3	A7	808

HD	DM	Other Id	N	Rem	α_{1950}	δ_{1950}	S	V	σ_V	B–V	σ_{B-V}	U–B	σ_{U-B}	n	Spectrum	References
244514	+29 00922				5 29 01	+29 17.3	1	10.69		0.34		-0.40		4	G0	117
		BSD 49 # 2168			5 29 01	+30 55.7	1	12.09		0.40		-0.04		2	B8	117
290595					5 29 02	-01 27.	1	11.96		0.24		0.16		3	A0	808
36335	+29 00923				5 29 03	+29 09.6	2	7.95	.000	0.39	.000	-0.01		5	F5	1625,1724
36244	+45 01132				5 29 03	+45 27.4	1	8.00		0.91		0.58		2	G5	1601
36429	+2 00986	IDS05264N0245		A	5 29 04	+02 47.8	2	7.56	.000	-0.12	.004	-0.63		5	B5 V	1004,8055
		NGC 1978 sq1 40			5 29 04	-66 14.8	1	13.58		1.67				1		1525
		Sk -69 153			5 29 05	-69 16.0	1	12.54		0.03		-0.32		4	A1 Iab	415
290566					5 29 06	-00 07.	1	11.93		0.75		0.31		3		808
		Orion P 705			5 29 06	-05 03.2	1	13.38		1.51		0.71		1		8114
		Sk -68 095			5 29 06	-68 51.	2	12.71	.040	-0.07	.010	-0.89	.005	7		10,952
		NGC 1978 sq1 41			5 29 07	-66 13.2	1	13.61		1.77				1		1525
36444	-1 00928				5 29 08	-01 09.7	2	8.97	.004	-0.03	.000	-0.18	.034	9	B9	1089,1775
		Orion P 720			5 29 08	-06 13.4	1	12.35		1.86		2.25		1		8114
		NGC 1978 sq1 42			5 29 08	-66 12.3	1	14.86		0.07				1		1525
36443	-0 00981	G 99 - 16			5 29 10	+00 04.2	5	8.36	.007	0.70	.007	0.17	.008	16	G5 V	22,333,1003,1775,2033
		Orion P 725			5 29 10	-05 07.0	1	13.81		1.05		0.67		1		8114
		Orion P 732			5 29 10	-06 07.9	1	13.89		1.29		0.15		1		8114
		Orion P 730			5 29 11	-05 01.9	1	13.26		0.78		0.34		1		8114
		Sk -66 107			5 29 12	-66 37.	1	12.86		0.01		-0.55		4		415
		Sk -69 154			5 29 12	-69 31.	1	13.22		-0.14		-0.94		3		952
36291	+42 01323				5 29 13	+42 55.7	1	8.54		0.14		-0.27		2	B9	401
269612		Sk -69 155			5 29 14	-69 01.8	1	11.44		0.27		-0.02		3	A9 Ia	149
		Sk -67 131			5 29 15	-67 23.	1	12.41		-0.14		-0.95		4		415
36487	-7 01103				5 29 16	-07 05.1	3	7.79	.014	-0.11	.000	-0.53	.010	6	B5 V	1004,1060,8055
36389	+18 00875	HR 1845, CE Tau			5 29 17	+18 33.5	7	4.35	.025	2.07	.011	2.21	.010	17	M2 Iab-Ia	15,1363,3001,8003*
36242	+51 01083				5 29 17	+51 26.1	1	7.93		0.05		-0.23		2	B9	401
290615					5 29 17	-01 58.6	1	11.30		0.50		0.20		3	A5	808
		SK *G 323			5 29 17	-69 01.3	1	12.17		0.17		0.10		4		149
		L 234 - 26			5 29 18	-54 02.	1	10.54		1.30		1.00		1		3062
		Sk -66 108			5 29 18	-66 54.	1	12.87		-0.10		-0.84		4		798
		ApJS2,315 # 38			5 29 18	-66 57.	1	14.02		0.18		0.08		5	A7 Ib	35
		Sk -67 132			5 29 18	-67 08.	1	12.99		-0.15		-0.93		3		952
		AAS21,109 # 178			5 29 18	-67 14.	1	14.15		-0.07		-0.90		2		417
		AAS21,109 # 179			5 29 18	-67 14.	1	13.66		-0.17		-1.03		2		417
		Sk -69 156			5 29 18	-69 13.	1	12.40		-0.13		-0.88		3		952
		Sk -70 094			5 29 18	-70 21.	1	13.13		-0.14		-0.98		3		798
		WLS 520 35 # 13			5 29 19	+33 12.3	1	11.45		0.18		-0.32		2		1375
36408	+16 00794	HR 1847		★ AB	5 29 20	+17 01.4	1			-0.04		-0.24		1	B7 III+B7IV	1079
		NGC 1978 sq1 52			5 29 20	-66 06.6	1	14.03		0.04				1		1525
		NGC 1978 sq1 51			5 29 21	-66 06.2	1	14.10		0.65				1		1525
		POSS 57 # 4			5 29 22	+65 07.4	1	17.14		1.08				3		1739
		NGC 1978 sq1 25			5 29 22	-66 16.6	1	13.34		0.45				1		1525
36407	+17 00946				5 29 23	+17 42.5	1	7.99		0.46		0.05		2	F0	1375
36373	+29 00924				5 29 23	+29 22.1	1	9.06		0.65				2	G5	1724
36849	-66 00421				5 29 23	-66 28.6	1	8.00		0.16				4	A3 III	1525
		Sk -70 095			5 29 23	-70 49.0	1	13.23		0.04		-0.10		3	A1 Ia	415
		Orion P 784			5 29 24	-04 15.3	1	10.74		0.52		0.21		1		8114
		Orion P 790			5 29 24	-05 07.0	1	14.18		1.06		-0.02		1		8114
		Sk -67 133			5 29 24	-67 22.	2	12.61	.024	-0.07	.029	-0.77	.024	6		415,952
271318	-72 00378	R Vol			5 29 24	-72 27.3	1	11.05		5.24				2	G0	864
294219	-4 01158				5 29 26	-04 18.1	1	10.27		0.21		0.09		2	A8 IV	1060
36513	-7 01105				5 29 26	-07 45.1	3	9.49	.005	0.02	.005	-0.09	.020	6	A0	1004,1060,8055
36597	-35 02348	HR 1862			5 29 26	-35 30.4	5	3.87	.007	1.14	.010	1.08	.007	24	K1 IIIa	15,1075,1637,2012,8015
36471	+5 00951				5 29 27	+06 00.7	1	8.69		0.06		-0.23		4	B9	206
36485	-0 00982	HR 1851		★ C	5 29 27	-00 19.2	5	6.85	.022	-0.16	.012	-0.72	.014	12	B2 V	146,154,1084,1089,1211
36486	-0 00983	HR 1852, δ Ori		★ A	5 29 27	-00 20.1	15	2.23	.017	-0.22	.007	-1.05	.015	49	B0 III+O9 V	15,146,154,1004,1075*
		SK *G 324			5 29 27	-68 47.6	1	11.86		-0.15		-0.75		1	B1 Ia	149
36371	+32 01024	HR 1843		★	5 29 28	+32 09.4	8	4.77	.014	0.34	.013	-0.45	.015	29	B5 Iab	15,154,263,369,1118*
		G 97 - 47		★ V	5 29 30	+09 47.3	3	11.54	.023	1.62	.005	1.07		11	M3	235,694,1705
244599	+26 00838				5 29 30	+26 15.9	1	9.09		0.47				3		1724
269618		Sk -68 098			5 29 30	-68 47.	1	13.89		-0.21		-0.94		3	WN3	980
	-0 00984				5 29 31	-00 27.9	1	8.43		-0.10		-0.46		3	B8 III	808
		Orion P 812			5 29 31	-04 58.0	1	12.47		0.23		0.10		1		8114
		Orion B 20			5 29 31	-03 03.0	1	13.92		0.89				1		454
36512	-7 01106	HR 1855			5 29 31	-07 20.2	27	4.62	.013	-0.26	.007	-1.07	.008	128	B0 V	1,15,125,154,1004,1006*
		Sk -67 134			5 29 32	-67 11.	1	12.80		-0.08		-0.74		4		415
36502	-1 00931				5 29 33	-01 29.7	1	9.23		-0.02		-0.16		3	B9	1089
290598	-1 00932				5 29 33	-01 39.4	1	9.83		0.08		0.01		3	A0	808
290564					5 29 34	+00 06.	1	11.03		0.45		0.16		5	B5	808
		NGC 1978 sq1 43			5 29 35	-66 21.2	1	13.42		0.95				1		1525
		Orion P 825			5 29 36	-05 07.0	1	11.66		1.44		1.81		1		8114
		Orion P 835			5 29 37	-06 16.3	1	13.20		0.61		0.17		1		8114
	-7 01108	V1162 Ori			5 29 37	-07 17.2	1	9.95		0.32		0.06		2		1060
244650	+20 00982				5 29 38	+20 53.0	1	8.40		1.85				2	M2	369
		Sk -68 099			5 29 38	-68 48.	2	12.63	.014	-0.12	.005	-0.93	.009	5		10,415
		BSD 49 # 599			5 29 39	+28 08.8	1	11.13		0.59		0.25		4	B9	117
		G 96 - 45			5 29 39	+44 47.1	3	12.19	.007	1.60	.011	1.19	.041	9	M2	316,333,1620,1663
	-5 01283				5 29 39	-05 26.7	1	10.61		0.38		0.02		2		1060
		Orion P 845			5 29 39	-06 16.2	1	13.82		0.96		1.19		1		8114

Table 1   277

HD	DM	Other Id	N Rem	α_{1950}	δ_{1950}	S	V	σ_V	B–V	σ_{B-V}	U–B	σ_{U-B}	n	Spectrum	References
36542	−10 01210			5 29 39	−10 01.9	1	8.66		0.00		−0.12		2	B9 V	1003
36441	+26 00841	LS V +26 004		5 29 40	+26 42.8	1	8.25		0.06		−0.77		6	B1.5IV-Vn:	399
269619		Sk -68 100		5 29 40	−68 30.3	1	11.34		0.13		−0.48		5	B9 Ia	149
36526	−1 00933	V1099 Ori		5 29 41	−01 38.1	1	8.31		−0.11		−0.60		3	B9 p(SiSr)	1089
36541	−6 01209			5 29 41	−06 44.6	3	7.68	.010	−0.09	.010	−0.45	.005	5	B8 V	1004,1060,8055
36405	+32 01025			5 29 42	+32 37.5	1	8.54		0.49				2	A0	1025
		HRC 97	V	5 29 42	−03 08.	1	11.51		1.53		1.11		1	K7e	825
		Orion P 859		5 29 42	−06 12.3	1	10.49		0.47		0.05		1		8114
271308	−66 00422			5 29 42	−66 04.7	1	10.01		0.48				1	F8	1525
		Sk -67 135		5 29 42	−67 06.	1	12.90		−0.11		−0.88		4		798
290583			A	5 29 43	−00 54.	1	10.76		0.39		0.24		4		808
290583			AB	5 29 43	−00 54.	1	10.37		0.44		0.22		4		808
290583			B	5 29 43	−00 54.	1	11.70		0.55		0.08		4		808
290611				5 29 44	−01 50.	1	11.24		0.38		0.22		4	A7	808
36528	−4 01160	Orion P 857		5 29 44	−04 12.4	1	8.17		1.78		2.54		1	K5	8114
269628		Sk -67 136		5 29 44	−67 03.6	1	11.91		0.00		−0.55		6	B8 Iab	149
244666	+21 00885			5 29 45	+21 39.5	1	11.00		0.21		0.10		1	A2	1220
244607	+31 01004			5 29 45	+31 30.4	2	10.10	.024	0.25	.009	0.22		4	A0	117,1025
36425	+31 01003			5 29 45	+31 50.2	2	7.35	.024	0.07	.014	0.02		4	A2	401,1025
269620	−68 00381			5 29 45	−68 54.3	1	9.63		0.67		0.13		1	G5	360
36440	+28 00818			5 29 46	+28 18.1	1	8.75		0.45				2	F5 V	1724
36527	−3 01130			5 29 46	−03 36.0	3	9.49	.005	0.15	.005	0.10	.000	4	A1 Vn	1004,1060,8055
36540	−4 01162	V1101 Ori		5 29 46	−04 33.2	4	8.13	.009	0.06	.004	−0.48	.003	9	B7 III	1004,1060,8031,8055
		Orion P 869		5 29 46	−05 06.9	1	13.43		1.53		1.75		1		8114
269629		Sk -67 137		5 29 46	−67 23.0	1	11.93		0.04		−0.58		3	B9 Iab	149
37027	−70 00399			5 29 46	−70 51.7	1	8.07		−0.03		−0.07		3	B9.5IV/V	1408
	−6 01210	Orion P 874		5 29 48	−06 09.8	2	10.21	.005	0.55	.000	−0.07	.019	3	F9	997,1696
		Sk -66 109		5 29 48	−66 41.	1	13.12		−0.14		−0.97		4		798
	−5 01284	Orion P 880		5 29 50	−05 50.4	2	10.40	.010	0.35	.030	0.08	.005	3	A5	1060,8114
36550	−6 01211	IDS05275S0628	B	5 29 50	−06 26.6	1	9.35		0.02		−0.06		3	A0	8055
36469	+25 00852			5 29 52	+25 51.6	1	7.67		1.53				2	K0	1724
		G 98 - 8		5 29 53	+29 21.3	1	12.07		1.46		1.19		1		333,1620
		G 248 - 42		5 29 53	+73 15.3	1	15.87		1.32				1		906
		G 248 - 43		5 29 53	+73 15.4	1	16.41		1.52				1		906
36560	−6 01212	IDS05275S0628	A	5 29 53	−06 25.6	4	8.29	.063	−0.09	.014	−0.39	.011	6	B8 V	1004,1060,8055,8114
		AAS21,109 # 181		5 29 54	−66 28.	1	13.78		−0.02		−1.11		3		417
		Sk -66 110		5 29 54	−66 54.	1	12.97		−0.20		−1.01		4		798
		Sk -68 103		5 29 54	−68 56.	2	12.64	.014	−0.03	.009	−0.59	.014	4		360,952
269624		Sk -68 102		5 29 54	−68 56.	1	14.88		−0.28		−0.93		4	WN3	980
		Sk -69 157		5 29 54	−69 26.	1	12.53		−0.11		−0.93		3		952
		AAS21,109 # 180		5 29 54	−70 09.	1	14.79		−0.12		−0.98		3		952
		Sk -71 033a		5 29 54	−71 04.	1	11.75		0.04		−0.59		4	B5 Ia	952
36404	+41 01218	HR 1846		5 29 55	+42 04.5	2	6.55		−0.01	.009	−0.28	.004	3	B9 III(p)	252,1079
244641	+30 00927			5 29 56	+31 04.3	1	10.12		0.33				3	A3	1025
		BSD 49 # 2691		5 29 56	+31 29.7	1	11.38		0.38		0.23		2	A0	117
		Orion P 905		5 29 56	−05 06.8	1	13.57		1.40		2.87		1		8114
		Sk -67 138		5 29 56	−67 09.	1	12.13		−0.02		−0.68		4		415
244642	+30 00928		AB	5 29 57	+30 54.0	2	9.69	.010	0.36	.010	0.08		5	A2	117,1025
		Orion P 910		5 29 57	−05 20.3	1	13.08		1.56		1.52		1		8114
		NGC 1978 sq1 24		5 29 57	−66 12.0	1	12.33		1.09				1		1525
36876	−64 00452	HR 1882	★ AB	5 29 58	−63 57.9	2	6.18	.005	0.23	.005	0.15	.010	7	F0 IV	15,1075
		STO 859		5 29 58	−71 05.2	1	11.59		0.53		−0.01		1	F8 Ia	726
		Steph 570		5 29 59	+65 19.9	1	11.53		1.45		1.23		1	K5-M0	1746
36559	−4 01163			5 29 59	−04 36.6	3	8.81	.000	−0.04	.010	−0.22	.005	8	Am	1004,1060,8055
		Orion P 918		5 29 59	−04 55.8	1	13.59		1.12		0.71		1		8114
36453	+32 01027			5 30 00	+32 15.3	2	6.61	.033	−0.04	.028	−0.19		4	B9	401,1025
36423	+38 01204			5 30 00	+38 45.0	1	8.62		0.02		−0.39		3	B7 V	401,833
		Orion P 925		5 30 00	−05 05.2	1	13.00		1.12		0.76		1		8114
36619	−23 02824			5 30 00	−23 27.9	1	8.60		0.23		−0.69		2	A0mA5-A9	1011
		Sk -66 111		5 30 00	−66 42.	1	13.62		−0.17		−0.88		4		798
		Sk -67 139		5 30 00	−67 03.	1	12.89		−0.17		−1.00		4		798
36500	+24 00850			5 30 01	+24 19.9	1	8.87		0.47				2	F8	1724
244663	+28 00819			5 30 01	+28 23.2	1	9.71		0.75		0.31		5	A2	117
		Orion P 938		5 30 01	−06 14.7	1	12.49		1.20		1.95		1		8114
269634		Sk -67 140		5 30 01	−67 29.6	2	12.44	.032	0.08	.012	−0.23		4	A0 Ia	149,1277
36549	+1 01053			5 30 03	+02 03.4	1	8.56		−0.08		−0.40		1	B6 Vwp Hewk	160
		Orion P 941		5 30 03	−06 20.5	1	11.46		0.78		0.30		1		8114
35783	+78 00193	G 222 - 9		5 30 04	+78 19.8	2	7.69	.000	0.46	.000	−0.02	.000	3	F6 V	1003,3002
		Orion P 943	V	5 30 05	−04 58.5	1	12.64		1.21		1.42		1		8114
269638		Sk -67 141		5 30 05	−67 16.8	1	12.01		0.07		−0.03		3	A1 Ia	149
		Sk -67 142		5 30 06	−67 16.	1	13.17		−0.12		−0.84		4		798
		Sk -69 159		5 30 06	−69 06.	1	12.60		−0.14		−0.99		3		952
244694	+29 00927			5 30 08	+29 20.1	1	10.37		0.61		−0.20		4	B9	117
36452	+37 01226			5 30 08	+37 30.1	1	9.71		0.17		0.16		2	A2	1375
		Orion P 958		5 30 08	−05 25.1	1	12.16		1.17		1.00		1		8114
		Orion P 962		5 30 08	−06 03.7	1	12.54		0.54		0.40		1		8114
36524	+19 00953	G 100 - 17	★ AB	5 30 09	+20 00.3	1	8.81		0.90		0.58		1	G5	333,1620
290578				5 30 09	−00 41.	1	10.65		0.30		0.21		4	B9	808
36591	−1 00935	HR 1861	★ AB	5 30 09	−01 37.6	17	5.34	.008	−0.19	.008	−0.93	.011	78	B1 IV	1,15,154,247,401,1006
		Sk -65 064		5 30 09	−65 58.	1	13.17		−0.01		−0.57		4		415

HD	DM	Other Id	N	Rem	α_{1950}	δ_{1950}	S	V	σ_V	B–V	σ_{B-V}	U–B	σ_{U-B}	n	Spectrum	References
		Sk -66 112			5 30 09	−66 25.8	1	12.64		0.03		−0.15		3	B8 Iab	149
	−1 00934				5 30 10	−01 51.6	1	10.22		0.17		0.07		2	A3 Vn	1060
36592	−2 01285				5 30 10	−02 04.3	1	9.05		−0.01		−0.20		3	B9	1089
		Orion P 969			5 30 10	−06 22.9	1	13.23		0.77		0.29		1		8114
36484	+32 01028	HR 1850			5 30 11	+32 46.1	4	6.47	.005	0.09	.007	0.16	.034	11	A2 m	1049,1118,1199,8071
36702	−38 02077				5 30 11	−38 35.5	4	8.37	.014	1.22	.008	0.90	.049	8	F0wG/K	742,1594,3077,6006
36734	−46 01892	HR 1870			5 30 11	−45 57.7	4	5.85	.004	1.35	.000	1.54		13	K3 III	15,1075,2035,3005
		Orion P 976			5 30 12	−06 19.7	1	13.18		0.83		0.45		1		8114
		Orion P 977			5 30 12	−06 22.5	1	13.95		0.66		0.14		1		8114
269639	−67 00448	Sk -67 143			5 30 12	−67 17.9	1	11.46		0.05		−0.42		6	B9 Iab	149
		Sk -69 158			5 30 12	−69 12.	1	12.90		−0.05		−0.74		3		952
		Sk -71 034			5 30 12	−71 02.	1	13.42		−0.22		−1.01		3		798
		G 191 - 45			5 30 13	+51 11.0	1	11.07		1.50		1.20		3	M3	7010
		Orion P 972			5 30 13	−05 06.4	1	12.56		0.57		0.15		1		8114
		Sk -66 113			5 30 13	−66 58.	2	12.25	.018	−0.11	.005	−0.93	.005	5		360,415
36590	−1 00936				5 30 14	−01 04.0	1	9.36		−0.02		−0.18		3	B9	1089
36607	−6 01216				5 30 14	−06 54.9	3	9.21	.020	−0.03	.005	−0.16	.005	5	A0	1004,1060,8055
		Orion P 993			5 30 15	−06 24.4	1	13.75		0.61		0.05		1		8114
36548	+15 00851				5 30 16	+15 54.7	1	8.20		0.50		−0.01		2	F8	1648
244708	+29 00928				5 30 16	+29 12.4	3	9.85	.006	0.64	.013	0.52	.000	17	A5	830,1025,1783
36386	+52 00969				5 30 17	+52 11.4	1	8.68		0.43		−0.03		2	F2	1566
36605	−0 00988				5 30 17	−00 44.8	1	7.96		0.08		−0.20		3	B9	1089
		Sk -67 146			5 30 17	−67 01.0	1	12.20		0.03		−0.37		4	B9 Iab	415
		Sk -69 162			5 30 17	−69 02.3	1	11.80		−0.06		−0.80		3		149
36483	+36 01177	LS V +36 017		⋆AB	5 30 18	+36 25.5	2	8.18	.000	0.43	.000	−0.55	.000	4	O9.5III	1011,1012
37026		Sk -67 144			5 30 18	−67 28.	1	13.60		−0.31		−0.69		4	WC5	980
		Sk -69 161			5 30 18	−69 02.	1	12.83		−0.10		−0.88		3		952
		Sk -69 160			5 30 18	−69 09.	1	12.45		−0.02		−0.93		3		952
36499	+34 01083	HR 1854			5 30 19	+34 41.5	2	6.27		0.15	.020	0.17		5	A3 IV	1049,1118
290599	−1 00937				5 30 19	−01 36.6	1	9.72		0.03		−0.09		3		808
36617	−2 01286				5 30 19	−02 13.9	1	8.45		−0.01		−0.09		3	B9	1089
36606	−5 01285	Orion B 2			5 30 19	−04 59.7	2	8.77	.019	0.17	.000	0.14	.019	4	Am	114,1060
		Orion B 9			5 30 19	−06 10.7	1	13.02		0.82		0.24		1		114
		Orion B 6			5 30 20	−05 18.3	1	13.49		0.95		0.48		1		8114
		Orion B 8			5 30 21	−04 30.4	1	13.01		0.89		0.39		1		8114
		Orion B 15			5 30 21	−06 15.3	1	13.90		1.01		1.00		1		8114
244706	+32 01029				5 30 23	+32 10.7	1	10.22		0.23				2	B9	1025
		Orion P 1021			5 30 23	−04 07.7	1	12.34		0.52		0.31		1		8114
	−5 01286	Orion B 12			5 30 23	−05 17.4	1	10.64		0.50		0.04		4		454
269647	−66 00424	IDS05304S6657		ABC	5 30 23	−66 55.2	1	10.84		0.44		−0.43		3	A0 Ia	149
269644	−67 00450	Sk -67 145			5 30 23	−67 34.5	4	11.15	.026	−0.01	.007	−0.63	.014	10	B7 Ia	149,1277,1408,8033
		Orion B 18			5 30 24	−05 44.6	3	12.23	.012	0.71	.021	0.37	.032	6		114,454,8114
		SK *C 42			5 30 24	−65 57.	1	12.66		0.35		0.15		3		149
		Sk -66 114			5 30 24	−66 08.	1	13.58		−0.11		−0.85		2		798
		Sk -68 105			5 30 24	−68 51.	1	13.07		−0.15		−0.98		3		952
		Orion B 24			5 30 25	−06 15.4	1	13.18		1.44		0.74		1		114
	−6 01217	IDS05280S0634		A	5 30 25	−06 32.9	1	8.69		0.17		0.07		2	F8 IV-V	1060
		WLS 520 20 # 8			5 30 26	+20 15.2	1	11.88		0.61		0.15		2		1375
294225	−5 01287	Orion B 17			5 30 26	−05 11.5	3	10.03	.009	0.28	.009	0.08	.006	8	A8 V	454,1060,8055
		Sk -66 115			5 30 26	−66 31.0	1	12.74		0.02		0.00		3		149
36506	+32 01030				5 30 27	+32 42.9	1	6.49		1.58				3	K5	1025
		Orion B 19			5 30 27	−04 36.7	2	10.34	.005	0.71	.007	0.22	.009	8		454,1519
		Orion B 14			5 30 27	−05 01.0	1	13.26		2.12		1.78		1		8114
		Orion B 21			5 30 27	−05 20.1	2	11.16	.044	1.35	.000	1.08	.136	4	K1.5III	114,454
		Orion B 23			5 30 28	−04 53.2	1	13.92		0.91		0.58		1		8114
		Ross 796			5 30 28	−16 07.9	1	12.50		1.32		1.14		1		1696
290580					5 30 29	−00 46.	1	10.61		0.35		−0.01		5	A7	808
36629	−4 01164	Orion B 25			5 30 29	−04 36.0	5	7.66	.015	0.01	.005	−0.65	.006	16	B2.5IV	114,454,1004,1060,8055
		Orion B 30			5 30 29	−05 45.9	1	13.09		1.31		1.88		1		114
		STO 62			5 30 29	−66 39.5	1	12.46		0.34		0.29		1	F5 I	726
		Orion B 27			5 30 30	−05 09.6	1	14.09		1.00		0.90		1		8114
		Orion B 28			5 30 30	−05 16.5	1	11.88		0.66		0.17		4		454
		Orion B 34			5 30 30	−05 46.2	1	12.98		1.51		0.33		1		114
		AAS21,109 # 182			5 30 30	−66 29.	1	13.74		−0.02		−0.84		2		417
		Sk -67 147			5 30 30	−67 14.	1	12.84		−0.07		−0.63		4		798
		Sk -67 146a			5 30 30	−67 15.3	1	12.51		0.03		−0.30		2	B9 Ia	415
		Sk -67 148			5 30 30	−67 40.	1	13.25		−0.20		−1.07		3		952
36547	+23 00942	LS V +23 003			5 30 31	+23 18.5	1	8.81		0.38		−0.56		2	B1 III	1012
36628	−1 00938	Ring Ori # 27			5 30 31	−01 16.5	1	7.98		−0.04		−0.24		2	B9 V	1089
		Orion B 29			5 30 31	−04 32.8	2	11.83	.028	1.56	.028	1.61	.047	4	K2 IV	114,454
36673	−17 01166	HR 1865		⋆A	5 30 31	−17 51.4	13	2.58	.012	0.21	.008	0.25	.043	37	F0 Ib	1,3,15,254,1004,1020*
		Sk -66 116			5 30 31	−67 00.	1	12.26		0.13		0.19		4		415
36627	+3 00958				5 30 32	+03 05.8	2	7.58	.008	−0.10	.004	−0.54		4	B6 V	1004,8055
36602	+7 00929	RT Ori			5 30 32	+07 07.2	3	7.90	.028	2.99	.035	5.17	.042	7	C6,4	109,864,3038
244759	+26 00846				5 30 32	+26 30.1	1	9.07		1.16				3		1724
244756	+30 00931				5 30 32	+30 16.1	1	10.73		0.40		−0.02		3	B8	117
36468	+43 01301				5 30 32	+43 54.3	1	7.23		0.02		0.00		2	B9	401
		Orion B 43			5 30 33	−06 12.2	1	14.49		0.36		0.29		1		114
36733	−30 02464				5 30 33	−30 35.3	1	9.44		−0.16		−0.66		8	B5 V	1732
	−34 02306				5 30 33	−34 29.0	1	10.80		0.50		−0.11		2		1696

Table 1   279

HD	DM	Other Id	N Rem	α_{1950}	δ_{1950}	S	V	σ_V	B–V	σ_{B-V}	U–B	σ_{U-B}	n	Spectrum	References
290569				5 30 34	−00 02.3	1	11.26		0.29		0.15		3	B8	808
		Orion B 33		5 30 34	−04 21.4	1	14.34		1.16		0.60		1		454
		Orion B 42	⋆ V	5 30 35	−05 28.5	3	12.27	.028	0.93	.010	0.36	.032	6		454,825,8114
		Orion B 44		5 30 35	−06 10.2	1	13.59		1.05		0.69		1		114
		R43A		5 30 36	+12 19.2	1	12.57		1.52		1.25		1		1773
36576	+18 00877	HR 1858, V960 Tau		5 30 36	+18 30.4	2	5.67	.066	0.01	.033	-0.77	.028	6	B2 IV-Ve	154,1212
36646	−1 00939	HR 1863	⋆ AB	5 30 36	−01 45.1	8	6.53	.027	-0.10	.009	-0.64	.016	21	B4 VN	15,154,1060,1078,1089*
		Orion B 45		5 30 36	−05 21.8	1	13.89		1.09		0.79		1		8114
		R 108		5 30 36	−67 19.	1	12.90		1.19				1		1277
		Sk -67 149		5 30 36	−67 35.	1	13.63		-0.01		-0.97		4		798
		Sk -69 163		5 30 36	−69 17.	1	12.69		-0.06		-0.79		3		952
		BSD 49 # 2231		5 30 37	+31 01.7	1	10.89		0.38		0.30		3	B9	117
290617	−2 01287			5 30 37	−01 57.6	1	9.33		0.28		0.16		2	A2	1060
		Orion B 37		5 30 37	−04 19.5	1	12.69		1.07		1.09		2		454
		Orion B 49		5 30 37	−06 02.2	2	13.84	.005	1.35	.090	1.22		2		114,454
		Orion P 1091		5 30 37	−06 25.0	1	12.27		1.30		2.39		1		8114
		Sk -66 117		5 30 37	−66 51.	1	12.36		-0.06		-0.76		4		415
		Sk -67 150		5 30 37	−67 03.	1	12.24		-0.11		-0.92		4	B1.5Ia	415
		Sk -69 164		5 30 37	−69 23.1	1	12.52		0.06		-0.25		4	A1 Iab	415
290609				5 30 38	−01 45.3	1	8.64		0.06		-0.05		4	A0 V	808
294190	−2 01288			5 30 38	−02 42.6	1	10.34		0.17		0.09		3		808
	−5 01288	Orion B 46		5 30 38	−05 36.7	1	10.63		0.46		0.03		2		114
		Sk -67 150a		5 30 39	−67 05.2	1	12.72		0.04		-0.19		3	B9 Ib	415
36589	+20 00989	HR 1860		5 30 40	+20 26.4	3	6.21		-0.02	.015	-0.41	.006	29	B6 V	879,1049,1079
		Orion B 48		5 30 40	−04 32.3	1	14.03		1.01		0.03		1		114
		Orion P 1096		5 30 40	−05 18.4	1	14.31		1.57				1		8114
36655	−5 01289	Orion B 50		5 30 40	−05 22.5	4	8.62	.012	-0.04	.005	-0.23	.027	8	B9 V	114,1004,1060,8055
		Orion B 56		5 30 41	−05 59.3	1	13.69		1.20		1.34		1		114,454
244769	+30 00935			5 30 42	+30 22.2	1	9.98		0.34				3	B9	1025
		R 109		5 30 42	−67 18.	1	12.00		-0.01				1		1277
		Sk -71 035		5 30 42	−71 10.	1	13.02		-0.11		-0.93		4		798
290608	−1 00941	Ring Ori # 3		5 30 44	−01 45.2	1	9.11		0.01		-0.20		2		1060
		Orion B 57		5 30 45	−04 16.7	1	13.45		0.72		0.17		1		114
		Orion P 1123		5 30 45	−06 25.2	1	14.16		0.67		0.11		1		8114
269649	−69 00385	Sk -69 165		5 30 45	−69 21.7	1	10.69		0.09		-0.67		4	B7	415
	+24 00857			5 30 46	+24 59.7	1	10.43		2.10				2	M3	369
36671	−4 01165	Orion B 59		5 30 46	−04 40.1	2	8.71	.030	0.32	.009	0.11	.034	6	Am	114,1060
36575	+27 00791			5 30 47	+27 07.8	1	7.80		1.53				3	K0	1724
		Orion B 60		5 30 47	−04 30.2	1	13.03		0.72		0.11		1		114
		G 98 - 10		5 30 48	+29 19.8	1	12.57		1.46		1.12		1		333,1620
		Sk -70 097		5 30 48	−70 54.	1	13.33		-0.23		-1.00		3		798
36670	−4 01166	Orion B 62		5 30 49	−04 23.2	5	8.94	.011	0.01	.009	-0.05	.011	8	Am	114,454,1004,1060,8055
294185	−2 01290	IDS05283S0222	AB	5 30 50	−02 19.6	1	9.72		0.24		0.17		3		808
		Orion B 71		5 30 50	−05 58.0	1	13.19		1.42		0.84		1		114
		Orion B 68		5 30 50	−06 08.6	2	10.21	.000	0.45	.009	0.06	.009	5	F3 IV	114,454,461
	−6 01220	Orion P 1147		5 30 50	−06 26.6	1	10.45		0.59		0.06		2		1519
290574				5 30 51	−00 22.	1	10.87		0.33		0.28		3	B8	808
		Orion P 1152		5 30 51	−06 24.9	1	12.20		0.69		0.38		1		8114
36668	+0 01113	V1107 Ori		5 30 52	+00 35.3	2	8.07	.005	-0.12	.005	-0.46	.005	3	B6 Vwp Hewk	160,1089
36669	−0 00989			5 30 52	−00 49.1	1	8.94		-0.03		-0.27		2	B9 V	1060
36697	−7 01115			5 30 52	−07 37.3	3	8.64	.019	0.07	.005	-0.04	.000	6	Am	1004,1060,8055
	−61 00477			5 30 52	−61 29.4	1	9.29		1.47		1.76		2	K2	849
		SK *G 338		5 30 52	−67 11.4	1	12.20		0.09		0.09		3	A2 Ib	149
36684	−1 00942	IDS05283S0108	AB	5 30 53	−01 05.6	2	8.65	.010	0.00	.000	-0.13	.010	5	B9	1060,1089
		Orion B 70		5 30 53	−05 08.4	2	12.39	.086	2.19	.036	2.46	.162	5		454,8114
269655				5 30 53	−68 26.9	1	12.22		0.03		-0.82		4	B0:Ia	149
269651	−69 00386	Sk -69 166		5 30 53	−69 11.3	1	10.73		0.19		-0.18		5	A2 Ia0:	149
37181	−70 00403			5 30 53	−70 52.0	1	8.92		1.16		1.07		3	K0/1 III	204
		Orion B 67		5 30 54	−04 19.	1	13.80		1.18		0.22		1		114
		Orion B 78		5 30 54	−05 30.3	2	10.94	.015	0.60	.000	-0.02	.030	4		114,1519
		Orion B 79		5 30 54	−05 58.5	1	12.28		1.44		1.12		1		114
		Sk -66 119		5 30 54	−66 52.0	1	12.17		0.16		-0.34		4	A0 Iab	415
		Orion B 75		5 30 55	−04 36.2	1	14.17		1.12		1.44		1		114
		SK *G 341		5 30 55	−66 55.3	1	12.15		0.03		-0.38		3		149
		Sk -66 118		5 30 55	−66 57.	1	12.81		-0.05		-0.86		4	B2 Iab	415
		STO 228		5 30 55	−67 03.8	1	12.43		0.49		0.33		1	F Ib	726
36421	+59 00891			5 30 56	+59 43.9	1	8.14		0.20		0.17		2	A0	1502
		Orion B 86		5 30 56	−05 19.7	1	13.13		0.77		0.32		1		454
36745	−24 03191	IDS05289S2419	AB	5 30 56	−24 17.3	1	9.22		0.86		0.44		1	K2 V	3072
		Orion B 87		5 30 57	−05 23.3	1	11.65		0.58		0.08		3		454
		Sk -66 120		5 30 57	−66 43.3	1	12.50		0.03		-0.31		3		149
		Sk -66 119a		5 30 57	−66 49.0	1	12.16		0.07		0.00		3	B8 Ib	149
		Sk -67 151		5 30 58	−67 40.3	1	12.48		0.08		-0.11		4	B9 Ib	149
		Lambda Ori # 39		5 30 59	+09 39.2	1	10.47		0.27		0.18		1		572
36695	−1 00943	HR 1868, VV Ori		5 30 59	−01 11.4	7	5.34	.012	-0.18	.007	-0.91	.004	31	B1 V	15,154,1004,1060,1732*
		Orion P 1173		5 30 59	−03 59.5	1	12.84		1.55		1.14		1		3062
		Orion B 89		5 30 59	−04 32.4	2	11.47	.033	0.52	.009	0.01	.019	4		114,454
		Orion B 91		5 30 59	−05 10.3	2	12.05	.033	0.65	.009	0.18	.042	4		454,8114
36889	−47 01903			5 30 59	−47 30.6	1	7.37		0.67		0.21		43	G3/5 V	978
244908	+9 00871			5 31 00	+09 38.1	1	10.13		1.23		0.95		1	A2	572

HD	DM	Other Id	N Rem	α₁₉₅₀	δ₁₉₅₀	S	V	σ_V	B–V	σ_B-V	U–B	σ_U-B	n	Spectrum	References
	+35 01169	LS V +35 022	⋆ A	5 31 00	+35 46.8	2	9.37	.000	0.43	.020	-0.37	.030	5	B1:V:pe	1012,1212
290600				5 31 00	−01 32.	1	10.64		0.38		0.14		2		1060
37049	−61 00478			5 31 00	−61 29.3	1	8.84		0.15		0.16		2	A2 III	849
		AAS21,109 # 208		5 31 00	−67 26.	2	13.98	.028	-0.26	.023	-1.04	.014	4		417,952
		Sk -67 152		5 31 00	−67 36.	1	13.37		-0.14		-0.96		4		798
		Sk -69 168		5 31 00	−69 07.	1	12.84		-0.03		-0.79		3		952
		Sk -69 167		5 31 00	−69 17.	1	12.28		-0.10		-0.89		3		952
		AAS21,109 # 183		5 31 00	−70 18.	2	13.67	.005	-0.13	.033	-0.83	.000	4		417,952
		STO 861		5 31 00	−71 00.4	1	13.04		0.76		0.37		1	F Ia	726
		Sk -71 037		5 31 00	−71 22.	1	13.27		-0.12		-0.77		4		798
244827	+29 00930			5 31 01	+29 51.7	1	10.01		0.35				3	A2	1025
		Orion B 92		5 31 01	−04 35.8	1	14.12		1.01		-0.03		1		114
		Orion B 94		5 31 01	−04 53.7	1	12.99		1.14		1.40		1		8114
		Orion B 101		5 31 01	−06 11.4	1	11.83		1.15		0.87		1		114
		Orion B 95		5 31 02	−04 44.2	1	12.36		0.56		0.00		4	F8 V	114
36712	−6 01223	Orion B 100		5 31 02	−06 05.0	2	9.80	.009	0.15	.014	0.10	.019	4	A4 V	114,1060
		Sk -67 153		5 31 02	−67 19.	1	12.01		-0.02		-0.66		3		415
		Orion B 98		5 31 03	−04 59.7	1	12.46		0.70		0.27		1		114
	−5 01290	Orion B 102		5 31 03	−05 47.3	2	10.67	.000	0.65	.010	0.17	.015	5	G5 IV	114,1519
244907	+9 00873			5 31 04	+09 43.8	1	10.38		0.48		0.04		1	F8	572
36653	+14 00947	HR 1864		5 31 04	+14 16.3	1	5.63		-0.14		-0.63		3	B3 V	154
	−26 02288			5 31 04	−26 45.6	1	9.21		1.04		0.63		1	K2 V	3072
		LS V +37 013		5 31 05	+37 24.9	1	10.79		0.62		-0.20		1		8113
36694	−0 00991			5 31 05	−00 17.7	1	9.09		0.24		0.12		2	Am	1060
37066	−61 00479			5 31 05	−61 34.6	1	7.17		-0.11		-0.45		2	B8 V	849
244826	+30 00937			5 31 06	+30 21.7	2	9.82	.009	0.42	.009	-0.05		6	A2	117,1025
		Sk -67 154		5 31 06	−67 24.	1	12.61		-0.06		-0.87		4		798
		Sk -69 169		5 31 06	−69 14.	1	12.30		0.21		-0.77		3		952
	−1 00944			5 31 07	−01 37.9	1	10.20		0.11		0.09		2		1060
294224		Orion B 111		5 31 07	−05 07.0	4	11.39	.014	0.70	.012	0.42	.021	9	B8 V	454,461,1060,8055
		Sk -68 109a		5 31 07	−68 18.3	1	13.45		0.15		-0.13		3	A0 Iab	798
244847	+27 00793			5 31 08	+27 32.5	1	10.15		0.69				3		1724
		Orion B 107		5 31 08	−04 17.0	1	14.04		0.75		0.18		1		114
		Orion B 112	⋆ V	5 31 08	−05 03.6	1	15.22		0.79		-1.32		1		114
		Sk -68 109		5 31 08	−68 43.8	1	12.55		0.14		-0.03		4	A1 Iab	415
		G 99 - 17		5 31 09	+01 54.8	1	11.53		1.58				1		1705
244928	+9 00875			5 31 09	+09 26.6	1	10.84		0.48		0.06		1	F8	572
		Orion B 116		5 31 09	−04 23.9	1	13.66		0.97		0.51		1		8114
36969	−53 00888			5 31 09	−53 50.4	1	8.52		1.05				4	K0 III	2012
	−1 00945			5 31 10	−01 28.0	1	10.04		0.32		0.11		2		1060
		Orion B 119	⋆ V	5 31 10	−05 28.9	1	11.56		0.82		0.42		1		8114
36848	−38 02085	HR 1877		5 31 10	−38 32.8	4	5.47	.013	1.22	.005	1.34		13	K2/3 III	15,1075,2035,3005
36709	−0 00992			5 31 12	−00 03.7	1	8.34		-0.04		-0.19		2	B9 V	1060
269662	−69 00387	Sk -69 171		5 31 12	−69 05.1	2	10.34	.065	0.24	.000	-0.29		6	B9:I e	415,8033
		AAS21,109 # 184		5 31 12	−71 05.	1	13.81		-0.07		-0.92		2		952
36643	+27 00794			5 31 13	+27 17.8	1	8.09		0.38				3	F2 V	1724
294223	−5 01291	Orion B 127		5 31 14	−05 02.8	1	10.39		0.57		0.08		2	F6 IV	114
269661	−69 00388	Sk -69 170		5 31 14	−69 33.5	4	10.34	.023	0.13	.011	-0.54	.020	14	A0 Ia0	149,328,1277,8033
		G 100 - 20		5 31 15	+22 03.6	1	11.55		1.20		1.13		1	K3	333,1620
		Orion B 131		5 31 16	−04 42.0	1	12.82		0.59		0.06		2		114
269678		Sk -67 155		5 31 16	−67 17.2	1	11.60		0.09		-0.18		3	A0 Iab	149
269665	−68 00385	Sk -68 110		5 31 16	−68 47.0	1	11.18		-0.27		-1.12		4	B9	415
		Sk -69 172		5 31 16	−69 14.	1	12.08		0.04		-0.61		4		415
		G 97 - 51		5 31 17	+11 31.9	1	13.51		1.30		1.10		1		1658
		Orion B 132		5 31 17	−05 23.0	1	13.75		1.07		0.34		1		8114
36726	−0 00993	IDS05288S0009	A	5 31 18	−00 06.6	1	8.81		0.08		0.06		2	Am	1060
		HRC 105	V	5 31 18	−01 11.	1	13.43		0.92		-0.28		1		825
294188				5 31 18	−02 28.2	1	10.55		0.24		0.20		3	A2	808
		Orion B 136	⋆ V	5 31 18	−05 32.8	1	14.09		1.41		0.25		1		825
		Sk -66 122		5 31 18	−66 30.	1	13.35		-0.16		-0.97		4		798
		Sk -66 121		5 31 18	−66 38.	1	13.50		-0.19		-1.05		4		798
		AAS21,109 # 186		5 31 18	−67 40.	1	13.87		-0.15		-0.87		3		417
37248		Sk -71 038		5 31 18	−71 04.	1	13.10		-0.19		-0.86		4	WC5	980
		Sk -71 041		5 31 18	−71 08.	2	12.83	.000	-0.08	.009	-0.93	.005	5		360,798
		Sk -71 040		5 31 18	−71 59.0	1	13.22		0.05		-0.27		3	B9 Iab	149
288040	+0 01114			5 31 19	+01 00.3	1	9.73		0.13		0.06		3	A0	808
244876	+29 00932			5 31 19	+29 14.2	1	10.12		0.51		0.39		5	A0	117
36642	+30 00938			5 31 19	+30 35.0	2	9.04	.018	0.32	.014	-0.31		7		117,1025
	−1 00947			5 31 19	−01 34.9	1	10.44		0.16		0.13		2		1060
36742	−5 01292	Orion B 135		5 31 19	−05 11.6	1	9.71		0.41		0.03		2	F2 V	114
		Orion B 134	⋆ V	5 31 19	−06 06.4	1	12.77		1.80		2.33		1		114
269682		Sk -67 155a		5 31 19	−67 12.5	1	12.16		0.11		0.14		6	A2 Iab	149
		Orion B 141		5 31 20	−05 59.2	3	12.12	.013	0.90	.012	0.49	.092	3	K1 III:	114,461,825
36874	−35 02367	HR 1881		5 31 20	−35 10.4	2	5.77	.010	1.08	.010	1.00		6	K0 III	2035,3005
		Orion P 1285		5 31 21	−06 22.8	1	13.88		0.29		0.16		1		8114
269684		Sk -67 156		5 31 21	−67 11.5	1	11.51		0.01		-0.52		4	B8 Ia	149
269668		Sk -68 111		5 31 21	−68 56.1	1	12.01		-0.08		-0.89		3	B1 Ia	149
36741	+1 01058	HR 1871		5 31 22	+01 22.5	7	6.58	.010	-0.18	.017	-0.80	.010	23	B2 V	15,154,1004,1078,1732*
		BSD 49 # 663		5 31 22	+28 51.9	1	11.21		0.61		0.00		4	B7	117
	−5 01293	Orion B 145		5 31 22	−05 34.7	2	10.72	.022	0.42	.017	0.05	.004	6	F3 IV	454,461

Table 1   281

HD	DM	Other Id	N Rem	α_{1950}	δ_{1950}	S	V	σ_V	B–V	σ_{B-V}	U–B	σ_{U-B}	n	Spectrum	References
		Orion B 148		5 31 22	−05 47.7	2	10.74	.010	1.00	.005	0.79	.005	6	K3 IV-V	114,1519
	+34 01090	NGC 1960 - 466		5 31 24	+34 10.0	1	9.86		1.74				3		369
290570				5 31 24	−00 03.	1	11.13		0.32		0.19		2		1060
		Sk -66 123		5 31 24	−66 26.	1	13.35		-0.12		-0.75		4		798
		AAS21,109 # 185		5 31 24	−71 19.	1	14.27		-0.13		-0.97		3		952
	−5 01294	Orion B 146		5 31 25	−04 16.7	2	12.02	.014	0.71	.000	0.26	.063	4		454,8114
36783	−6 01226			5 31 26	−06 38.9	3	9.49	.005	0.01	.000	-0.10	.010	5	A0	1004,1060,8055
		Sk -68 112		5 31 26	−68 39.	1	12.76		-0.17		-1.02		4		415
244894	+27 00797	LS V +27 003		5 31 28	+27 33.6	1	9.86		0.45		-0.61		2	B1 III-IVpe	1012
		Orion B 156	⋆ V	5 31 28	−04 49.8	1	15.32		0.57		-0.41		1		825
36782	−5 01295	Orion B 161		5 31 28	−05 38.4	1	8.81		0.48		0.01		2	F7 IV	114
	−5 01296	Orion B 162		5 31 28	−05 40.8	1	10.14		0.42		0.23		4		114
		AE Tau		5 31 29	+26 09.9	1	11.46		1.01				1		1772
36760	−0 00996			5 31 29	−00 30.6	2	7.63	.000	-0.10	.000	-0.44	.015	4	B7 V	1060,1089
36781	−1 00948			5 31 29	−01 47.2	3	8.51	.005	0.00	.010	-0.41	.012	9	B6 V	1060,1089,8055
269660	−71 00348	Sk -71 042		5 31 29	−71 06.1	4	11.17	.013	-0.02	.010	-0.87	.019	12	B1.5Ia	952,1277,1408,8033
36665	+27 00798	LS V +28 005		5 31 30	+28 01.1	2	8.04	.024	0.33	.042	-0.59	.066	6	B8	117,1212
290605				5 31 30	−01 45.	1	10.13		0.17		0.09		2		1060
		Orion B 159		5 31 30	−04 27.0	1	12.59		0.87		0.59		2		454
		Orion P 1321		5 31 30	−06 20.8	1	13.14		1.25				1		8114
		Sk -66 124		5 31 30	−66 45.	1	12.46		0.04		-0.23		4		415
		Sk -69 173		5 31 30	−69 20.	1	12.61		-0.05		-0.80		3		952
		AAS21,109 # 187		5 31 30	−71 07.	1	13.36		-0.09		-0.92		3		952
36779	−1 00949	HR 1873	⋆ A	5 31 31	−01 04.1	9	6.22	.019	-0.17	.010	-0.81	.007	25	B2.5 V	15,154,1004,1060,1089,1415*
36778	−0 00997			5 31 32	−00 18.2	1	9.31		0.00		-0.08		2	B9.5V	1060
36780	−1 00950	HR 1874		5 31 32	−01 30.2	3	5.92	.005	1.55	.000	1.88	.012	8	K5 III	15,247,1078
		Orion B 164		5 31 32	−04 19.7	1	11.34		1.68		2.04		2		114
		Orion B 166		5 31 32	−05 02.0	2	12.54	.049	1.09	.019	0.67	.190	3	K0 IIInn	114,8114
		Orion B 173		5 31 32	−06 10.1	1	13.27		1.60		1.81		1		114
		Orion P 1329		5 31 32	−06 17.2	1	12.95		0.66		0.53		1		8114
269664		Sk -70 098		5 31 33	−70 43.7	2	11.61	.014	0.12	.028	-0.35	.014	4	A2 Ia+	149,952
269689		Sk -67 157		5 31 33	−67 26.8	1	11.95		0.00		-0.51		3	A0 Ia	149
		Orion B 168		5 31 34	−04 46.7	1	11.70		0.85		0.33		2		114
36664	+34 01092	NGC 1960 - 470		5 31 35	+34 15.0	1	8.86		1.50				3	K0	1724
		Orion B 172		5 31 35	−04 25.7	1	11.33		1.46		1.39		4		454
294186	−2 01293			5 31 36	−02 16.8	1	10.24		0.32		0.07		4		808
		Orion B 179		5 31 36	−05 56.9	1	14.42		0.79		0.23		1	K0 V(e)	114
		Orion B 180		5 31 36	−06 07.5	1	12.74		1.87		1.54		1		114
37004	−47 01909			5 31 36	−47 43.3	1	7.68		0.14		0.09		38	A3 V	978
		Sk -66 125		5 31 36	−66 23.0	1	12.74		-0.08		-0.63		3		149
		Sk -67 160		5 31 36	−67 19.	1	14.97		-0.24		-0.94		3	WN3	980
		Sk -67 158		5 31 36	−67 33.	1	13.38		-0.15		-0.87		4		798
		AAS21,109 # 189		5 31 36	−71 05.	1	13.45		-0.08		-0.91		3		952
290556	+0 01118			5 31 37	+00 23.1	1	9.54		0.12		-0.22		2		1089
		BSD 49 # 2262		5 31 37	+30 57.5	1	11.42		0.68		0.15		2	A0	117
		Sk -71 043		5 31 37	−71 37.5	1	13.48		0.01		-0.21		3	B9 Iab	415
288046	+0 01117			5 31 38	+00 49.9	1	10.35		0.28		0.05		4	A3	808
36681	+31 01013			5 31 38	+31 36.2	2	9.34	.024	0.30	.000			4	F0	1025,1724
36811	−1 00951	Ring Ori # 4		5 31 38	−01 56.1	3	7.08	.005	0.17	.004	0.12	.008	7	A1 M	1060,1089,8055
36777	+3 00964	HR 1872	⋆ AB	5 31 39	+03 44.1	6	5.36	.007	0.05	.004	0.07	.007	15	A2 V	15,1007,1013,1415*
		Orion B 181		5 31 39	−04 41.5	1	13.52		1.17		1.06		1		114
		Orion B 191	⋆ V	5 31 39	−05 28.0	1	14.40		1.20		0.30		3		8013
244913	+30 00941	IDS05285N3056	AB	5 31 40	+30 57.8	1	9.92		0.42				2	A2	1025
		Orion B 189	⋆ V	5 31 40	−05 15.8	1	15.90		1.20		0.30		1		8023
36813	−6 01227			5 31 40	−06 52.3	1	8.56		0.07		-0.23		2	A3 V	1060
		Sk -67 159		5 31 40	−67 40.	1	12.76		-0.16		-0.98		4		415
		Orion P 1355		5 31 41	−04 04.0	1	12.15		0.66		0.33		2		1060
		Orion B 182		5 31 41	−04 46.4	1	13.81		0.94		0.42		3	G8 V	114
		Orion B 183		5 31 41	−04 48.5	1	13.12		0.76		0.27		2	G8 IV-V	114
		Orion B 193		5 31 41	−05 12.3	2	11.64	.020	0.56	.005	0.03	.025	4		114,1519
244988	+22 00947		V	5 31 42	+22 08.6	1	9.67		1.84				2	K7	369
244937	+29 00934			5 31 42	+30 01.5	1	9.90		0.50				3	A3	1025
294202	−3 01140			5 31 42	−03 16.9	3	10.19	.019	0.27	.005	0.10	.010	4	A3 Vn	1004,1060,8055
		Orion B 188		5 31 42	−04 18.3	1	13.19		0.98		0.65		2		454
		Orion B 187		5 31 42	−04 21.7	1	11.69		0.62		0.08		2		114
		Orion B 195		5 31 42	−04 52.6	1	13.04		1.28		0.63		1		8114
		Sk -66 127		5 31 42	−66 19.	1	13.33		-0.14		-0.91		3		798
		Sk -71 044		5 31 42	−71 04.	2	13.05	.005	-0.14	.005	-0.95	.009	4		360,798
271345				5 31 42	−72 00.0	1	11.27		1.12		0.94		2	K2	149
		Orion B 192		5 31 43	−04 16.5	2	11.70	.023	0.60	.000	0.26	.036	5		454,8114
		Orion B 194		5 31 43	−04 19.4	1	11.88		1.09		0.71		3		454
	−5 01297	Orion B 203		5 31 43	−05 24.9	1	10.31		0.56		0.10		1	F5 IV	461
		Sk -66 126		5 31 43	−66 54.	1	12.17		-0.01		-0.64		3		415
		Sk -67 163		5 31 43	−67 04.	1	12.49		-0.09		-0.78		4		415
245042				5 31 44	+09 23.	1	10.51		0.60		0.06		1	G0	572
290695		Ring Ori # 5		5 31 44	−01 54.8	1	10.99		0.36		0.04		4		808
36827	−2 01296			5 31 44	−02 54.9	1	6.69		-0.17		-0.74		5	B5	1089
		Orion B 198	⋆ V	5 31 44	−04 37.7	3	12.08	.040	0.66	.017	0.10	.009	5		114,461,8114
244956	+29 00935			5 31 45	+29 16.6	1	9.34		0.75		0.33		3		1724
		Orion B 202		5 31 45	−04 40.4	1	12.82		0.76		0.33		3		454

HD	DM	Other Id	N Rem	α_{1950}	δ_{1950}	S	V	σ_V	B–V	σ_{B-V}	U–B	σ_{U-B}	n	Spectrum	References
269697	−67 00454	Sk −67 162		5 31 45	−67 30.2	1	10.28		0.41		0.30		2	F5 Ia	149
		Orion B 204		5 31 46	−04 31.2	1	14.11		0.84		−0.39		1		8114
		Orion B 205		5 31 46	−04 33.5	1	12.69		1.15		0.81		4		114
288099				5 31 47	+00 53.	1	11.44		0.53		0.04		4	A5	808
244936	+31 01015			5 31 47	+31 21.6	1	10.09		0.48				3	A5	1025
36826	−2 01297			5 31 47	−02 25.1	1	8.22		0.00		−0.51		4	B5 Vn	1089
		Orion B 213		5 31 47	−05 30.3	1	12.16		0.84		0.37		1		461
	−5 01299	Orion B 216		5 31 47	−05 38.9	2	10.16	.035	0.60	.025	0.09	.015	10	F6 IV-V	337,454
	−14 01171	IDS05295S1426	ABC	5 31 47	−14 23.6	1	10.13		0.81		0.59		1		8084
		Sk −69 174		5 31 47	−69 19.5	1	12.07		0.08		−0.24		4	A0 Ia	415
	−5 01298	Orion B 211		5 31 48	−05 13.8	2	10.62	.005	0.48	.005	0.04	.000	8	F7 IV	114,337
		HRC 115 ,SW Ori		5 31 48	−06 38.2	1	12.91		1.09		0.50		1		825
269692		Sk −67 161		5 31 48	−67 43.	2	14.57	.036	−0.16	.012	−0.91		4	WN3	980,1277
		AAS21,109 # 190		5 31 48	−68 30.	2	14.06	.066	−0.10	.047	−0.92	.000	4		417,952
		Sk −68 113		5 31 48	−68 55.	1	12.26		−0.03		−0.77		3		952
269687		Sk −69 175		5 31 48	−69 08.	1	11.90		−0.07		−0.97		4	B7	415
290694				5 31 49	−01 53.0	2	10.02	.002	0.45	.006	0.02	.004	65	F5	808,1060
290696				5 31 49	−02 03.	1	11.49		0.15		0.03		4	B8	808
		Orion B 209		5 31 49	−04 30.6	2	12.59	.091	1.12	.009	0.91	.220	4		454,8114
		Orion B 220	⋆ V	5 31 49	−05 38.7	3	11.48	.022	0.84	.000	0.34	.010	10	G5 IV-Ve	114,337,1519
		Sk −67 165		5 31 49	−67 03.	2	12.27	.019	−0.05	.009	−0.72	.014	4		360,415
36724	+26 00856			5 31 50	+26 56.5	1	7.56		0.53				2	F7 V	1724
		Orion B 215	⋆ V	5 31 50	−05 06.2	1	14.71		1.39				1		8054
269698		Sk −67 166		5 31 50	−67 40.2	4	12.26	.015	−0.22	.017	−1.03	.023	15	O6f	149,328,1277,1408
244957	+28 00824	G 98 - 13		5 31 51	+28 04.1	1	10.08		1.27		1.29		1		333,1620
		Orion B 217		5 31 51	−04 23.2	1	13.81		1.03		1.02		1		8114
		Orion B 224	⋆ V	5 31 51	−05 06.8	3	11.65	.053	0.85	.007	0.30	.025	8	F8 Vn(e)	1519,8013,8054
		Orion B 223	⋆ V	5 31 51	−05 09.0	1	14.74		1.40				1	M0 IV,Ve	8054
36825	−0 00999			5 31 52	−00 47.8	2	8.65	.005	0.03	.005	−0.12	.005	4	A2 V	1060,1089
		Orion B 231		5 31 52	−05 29.2	1	11.39		0.76		0.28		12	G5 IV-V	337
36723	+30 00942	IDS05287N3052	AB	5 31 53	+30 53.8	2	8.13	.024	0.43	.000			4	F5	1025,1724
		Orion B 225		5 31 53	−04 20.6	2	11.45	.030	0.62	.015	0.12	.015	4		114,1519
		Orion B 222		5 31 53	−04 23.5	1	11.82		0.71		0.26		4		454
		Orion B 228		5 31 53	−04 35.7	1	13.54		1.62				1		8114
		Orion B 230		5 31 53	−04 45.7	1	11.38		0.74		0.34		3		1014
		Orion B 240		5 31 53	−05 40.9	4	11.10	.026	1.06	.017	1.04	.018	13	K4 IV-V	114,337,1014,1519
		Orion B 242		5 31 53	−05 53.7	1	13.09		1.05		1.21		1		114
245010	+21 00894			5 31 54	+21 54.7	1	10.20		0.21		0.05		1	A2	1220
36680	+41 01226			5 31 54	+41 30.5	1	8.73		0.05		−0.16		2	A0	401
		Orion B 227		5 31 54	−04 22.5	1	13.17		0.77		0.41		4		454
		Orion B 238	⋆ V	5 31 54	−04 59.7	2	13.09	.098	0.95	.056	0.55	.056	4		451,454
37179	−62 00483			5 31 54	−62 43.4	1	9.70		0.34		0.09		2	F0 V	849
		AAS21,109 # 191		5 31 54	−67 07.	1	13.74		−0.26		−1.01		3		952
		Sk −67 164		5 31 54	−67 35.	1	13.08		−0.11		−0.89		4		798
245009	+22 00949			5 31 55	+22 32.2	1	8.51		1.11		0.87		2	G5	1648
294237				5 31 55	−02 43.5	1	11.21		0.20		0.09		3	A0	808
		Orion B 233		5 31 55	−04 32.2	3	12.03	.027	0.91	.004	0.37	.030	5		114,454,461
		Orion B 237		5 31 55	−04 44.8	2	12.59	.049	0.71	.034	0.24		3	G5 V	114,8114
244953	+32 01036	LS V +32 009		5 31 56	+32 27.0	1	9.52		0.19				4	B3	1025
		Orion B 245		5 31 56	−04 31.0	1	13.79		0.76		0.26		2		454,8114
		Orion B 244		5 31 56	−04 49.2	2	12.77	.009	0.88	.075	0.45	.033	4	K0 IV,V(e)	454,8114
36843	−4 01167	Orion B 243		5 31 56	−04 50.2	2	6.82	.013	0.21	.018	0.13	.013	5	Am	114,1060
36965	−29 02348	HR 1888		5 31 56	−29 52.9	1	6.53		−0.02				4	A0 IV	2007
		Sk −67 169		5 31 56	−67 05.	1	12.18		−0.12		−0.90		4	B1 Ia	415
		Sk −67 170		5 31 56	−67 14.3	1	12.48		0.05		−0.22		2	A0 Iab	149
269676	−71 00351	Sk −71 045		5 31 56	−71 06.2	1	11.47		−0.11				3	O6 e	8033
36722	+32 01037	IDS05287N3241		5 31 58	+32 43.0	1	7.97		0.11				1	A0	1025
36842	−4 01168	Orion B 246		5 31 58	−04 24.2	5	8.13	.017	−0.09	.009	−0.48	.021	10	B6 V	114,153,1004,1060,8055
		Orion B 252		5 31 58	−05 24.0	3	10.92	.010	0.64	.012	−0.16	.004	10	G0 IV-V	337,461,1014
274939	−48 01866			5 31 58	−47 58.1	4	9.43	.013	0.73	.015	0.09	.021	8	G0	742,1594,3072,6006
		Sk −67 167		5 31 58	−67 43.	1	12.54		−0.19		−1.04		4		415
245030	+24 00866			5 31 59	+24 27.2	1	10.12		0.56				2		1724
294261	−4 01169	Orion B 255		5 31 59	−04 35.3	3	9.62	.017	0.57	.005	0.14	.022	8	F9 V	114,153,8114
		Orion B 270		5 31 59	−06 06.5	1	13.47		1.62		0.12		1	G5 III	114
36867	−7 01122			5 31 59	−07 25.3	3	9.29	.010	−0.01	.010	−0.15	.010	5	A0	1004,1060,8055
36758	+24 00868			5 31 60	+24 15.5	1	6.81		1.05				1	K0	1724
36841	−0 01002			5 32 00	−00 25.1	3	8.59	.015	0.03	.008	−0.35	.013	6	O8	1060,1089,8055
		Orion B 276		5 32 00	−05 05.	1	18.04		0.73		−0.99		1		114
		Orion B 265		5 32 00	−05 26.3	2	12.07	.169	1.21	.025	1.03		3	G9 IV-V	1014,8054
		Orion B 272		5 32 00	−05 33.9	1	12.20		1.16		0.67		2		1014,8114
		Sk −66 128		5 32 00	−67 33.	1	12.71		−0.12		−1.01		4		415
		Sk −67 171		5 32 00	−67 23.0	1	12.04		−0.03		−0.57		2		149
		AAS21,109 # 192		5 32 00	−67 35.	1	13.78		−0.19		−1.04		2		417
269702		Sk −67 168		5 32 00	−67 37.	1	12.08		−0.17		−1.00		3	O8	415
		Sk −69 176		5 32 00	−69 33.	1	12.74		−0.04		−0.66		3		952
		Orion B 274		5 32 01	−05 26.9	1	13.78		0.99		0.46		1	K5 Ve	8114
		Orion B 278		5 32 01	−05 43.6	1	13.91		1.24		0.54		1		8114
		G 97 - 53		5 32 02	+13 55.3	1	11.51		1.27		1.15		1	K7	333,1620
	−4 01170	Orion B 257		5 32 02	−04 26.8	2	10.55	.000	0.60	.012	0.14	.010	7	G2	454,1519
		Orion B 283		5 32 02	−05 25.9	1	12.02		0.90		0.46		4		1014,8114

Table 1   283

HD	DM	Other Id	N Rem	α_{1950}	δ_{1950}	S	V	σ_V	B–V	σ_{B-V}	U–B	σ_{U-B}	n	Spectrum	References
		Orion B 284	⋆ V	5 32 02	−05 49.2	1	14.96		1.23		0.43		1		114
		Orion B 288		5 32 02	−06 12.4	1	13.73		1.20		-0.01		1	K5 Ve	114
36824	+5 00958			5 32 03	+05 37.7	3	6.71	.007	-0.15	.010	-0.70	.012	6	B3 V	1004,1732,8055
36662	+45 01144			5 32 03	+45 27.7	2	8.51	.035	0.12	.010	-0.33	.005	5	B8	401,555
36863	−1 00952			5 32 03	−01 46.6	2	8.28	.004	0.27	.004	0.19	.009	5	A0	1060,8055
		Orion B 266		5 32 03	−04 26.8	2	10.97	.010	0.63	.000	0.19	.005	6		114,1519
		Orion B 286		5 32 03	−05 29.4	1	10.89		0.54		0.08		6	G0 IV-V	337
36822	+9 00877	HR 1876	⋆	5 32 04	+09 27.4	10	4.41	.006	-0.16	.013	-0.97	.013	29	B0 IV	15,154,572,1061,1078*
36865	−4 01171	Orion B 281	⋆ AB	5 32 04	−04 31.2	5	7.43	.013	-0.07	.011	-0.43	.020	12	B7 V	114,153,1004,1060,8055
		SK *G 359		5 32 04	−66 41.5	1	12.51		0.24		0.23		2		149
290666	−0 01003	Ring Ori # 24		5 32 05	−00 55.6	1	9.81		0.09		0.01		2		1060
		Sk -69 177		5 32 05	−69 16.	1	12.11		0.06		-0.55		4		415
		Orion B 299		5 32 06	−05 25.0	1	13.08		1.03		1.19		3	G7 V	1014
		Orion B 302		5 32 06	−05 30.4	1	12.31		0.97		0.50		1		8114
36866	−5 01301	Orion B 304		5 32 06	−05 44.8	3	9.30	.010	0.14	.006	0.14	.018	6	A2 V	114,461,1060
	−5 01300	Orion B 300		5 32 06	−05 47.7	2	10.55	.013	0.24	.000	0.10	.000	6		114,1060
245024	+29 00936	Hyades vB 156		5 32 07	+29 14.4	2	10.36	.013	0.73	.003	0.32		5	G5	1127,1724
		Orion B 293		5 32 07	−04 29.3	2	12.66	.077	0.91	.023	0.57	.054	5		114,454
294265	−5 01302	Orion B 295		5 32 07	−05 05.1	4	10.26	.007	0.27	.013	0.14	.030	9	Am	114,1014,1060,8055
269700	−68 00386	Sk -68 114		5 32 07	−68 34.7	4	10.55	.011	0.02	.011	-0.87	.029	16	B1 Iae	149,328,1277,8033
		Sk -70 099		5 32 07	−70 25.3	1	13.07		0.03		-0.29		3	A2 Iab	415
36496	+66 00401	HR 1853	⋆ AB	5 32 08	+66 39.9	1	6.29		0.21		0.05		2	A8 Vn	1733
		Orion B 311		5 32 08	−05 34.1	1	13.37		1.15		0.75		3	K2 IV	1014,8114
36864	−5 01303	Orion B 312	⋆ V	5 32 08	−05 43.6	1	9.16		1.14		1.08		2	K2 III	114
269696	−69 00389	AA Dor		5 32 08	−69 55.1	2	11.13	.000	-0.28	.004	-1.11	.004	5	B7	158,1408
294234	−2 01298			5 32 09	−02 34.8	1	10.26		0.35		0.01		3		808
36774	+30 00944			5 32 10	+30 34.4	1	8.41		0.17				2	A3	1025
	−0 01004			5 32 10	−00 38.0	1	9.68		0.06		0.04		2		1060
		Orion B 319	⋆ V	5 32 10	−06 07.5	1	14.87		1.01		-0.38		1		114
36793	+25 00866			5 32 11	+25 47.8	2	7.89	.005	0.13	.010	-0.17	.020	5	A0	833,1625
		Orion B 308		5 32 11	−04 20.0	1	12.73		0.78		0.46		4		454
294232	−2 01299			5 32 12	−02 30.8	1	9.80		0.38		0.14		3		808
		Orion B 328		5 32 12	−05 12.1	4	10.73	.031	0.70	.016	0.29	.010	8	B8 V	114,1014,1060,8054
		Orion B 334		5 32 12	−05 26.4	1	11.35		1.27		1.11		10	K1 IV-V	337
		Orion B 336		5 32 12	−05 59.7	1	14.41		1.31		0.52		1		114
		Orion B 337	⋆ V	5 32 12	06 08.3	1	13.94		1.26		0.92		1		114
		Sk -66 129		5 32 12	−66 46.	1	12.16		0.06		-0.19		3		415
		Sk -67 172		5 32 12	−67 31.5	1	11.88		-0.07		-0.81		3	B2.5Ia	149
		AAS21,109 # 193		5 32 12	−67 35.	1	13.50		-0.20		-0.94		2		417
		ApJS2,315 # 44		5 32 12	−67 44.	1	13.97		0.04		-0.73		1	B2 Ib	35
		Sk -69 178		5 32 12	−69 36.	1	13.28		-0.20		-1.02		3		952
245025	+27 00803			5 32 13	+27 47.3	1	9.35		0.57				4		1724
		BSD 49 # 2276		5 32 13	+30 50.0	1	11.16		0.35		-0.09		3	B8	117
290693				5 32 13	−02 00.	1	10.57		0.12		-0.01		5	B8	808
		Orion B 340	⋆ V	5 32 13	−05 24.6	1	12.86		1.16		0.58		5		8013
		Orion B 335		5 32 13	−05 28.6	1	12.58		0.99		0.41		2	K3 III-IVe	114
		Orion B 339	⋆ V	5 32 13	−05 42.1	1	14.60		1.16		0.54		1		8023
		Orion B 341	⋆ V	5 32 13	−05 52.2	1	12.30		0.88		0.43		2	K2 IV(e)	114
		Orion B 329		5 32 14	−04 26.8	2	11.06	.015	0.46	.000	0.00	.005	6	F4 III-V	114,153
		HRC 118, IX Ori		5 32 14	−05 24.6	1	13.72		1.00		-0.65		1		825
245140				5 32 15	+09 53.	1	9.27		0.11		-0.25		1	B9	572
	+34 01096			5 32 15	+35 01.7	1	8.80		0.74		0.42		2	F8	1733
36883	−4 01172	Orion B 330	⋆ AB	5 32 15	−04 25.5	5	7.25	.010	-0.08	.011	-0.46	.012	11	B6 V	114,153,1004,1060,8055
36899	−5 01304	Orion B 342		5 32 15	−05 09.2	6	9.60	.015	0.04	.009	-0.01	.016	14	A0.5V	454,1004,1014,1060*
		Orion B 347	⋆ V	5 32 15	−05 14.2	1	13.61		1.73				2		1014
		Orion B 361		5 32 16	−06 07.8	1	15.28		1.40		0.63		1		114
		Orion B 346		5 32 16	−04 29.2	3	11.55	.000	0.79	.020	0.37	.010	6	G8 V	114,454,1519
		Orion B 352		5 32 16	−04 44.2	1	13.61		1.10		0.84		1		114
		Orion B 359		5 32 16	−05 20.4	2	11.11	.032	0.49	.004	-0.11	.016	8		337,1014
		Sk -67 173		5 32 16	−67 43.	1	12.04		-0.12		-0.96		4		415
269694		Sk -70 100		5 32 16	−70 45.0	1	12.25		-0.04		-0.60		4	A1 Ia	149
		Orion B 360		5 32 17	−04 58.0	1	12.85		0.76		0.46		1		8114
		Orion B 374		5 32 17	−05 27.0	1	12.92		1.19		1.36		1		8114
		Orion B 370		5 32 17	−06 00.3	2	14.68	.100	1.13	.005			2	K3 IV?	114,997
	−6 01230	Orion B 381	⋆ C	5 32 17	−06 01.4	4	10.26	.012	0.54	.003	-0.03	.009	26	F8 V	114,153,454,1519
269722	−66 00429	Sk -66 132		5 32 17	−66 26.0	1	12.06		0.28		-0.67		3	B7	415
245114	+19 00970			5 32 18	+19 40.8	1	9.50		0.43				2		1724
		Orion B 363		5 32 18	−04 30.9	1	13.34		1.03		0.75		2		454
		Orion B 368		5 32 18	−05 08.8	1	15.08		1.53				1		8054
		Sk -66 131		5 32 18	−66 10.	1	13.65		-0.22		-1.07		4		798
		AAS21,109 # 194		5 32 18	−67 55.	1	14.44		0.04		-1.06		2		417
245084	+24 00869			5 32 19	+24 59.4	1	8.55		1.17				2		1724
		Orion B 365		5 32 19	−04 19.4	1	12.45		0.80		0.34		1		8114
36917	−5 01305	Orion B 388	⋆ V	5 32 20	−05 36.2	6	7.98	.033	0.16	.017	0.02	.018	15	B9.5V +A0.5	337,379,1004,1014*
		Orion B 394		5 32 20	−06 00.8	1	13.07		1.12		1.39		1	K4 V(e)	8114
269714	−67 00456	IDS05325S6745	⋆ ABC	5 32 20	−67 43.1	1	11.67		-0.18		-0.98		3	B7	415
245168				5 32 21	+09 53.	1	9.65		0.12		-0.05		1	B9	572
		Lambda Ori # 28		5 32 21	+09 58.1	1	10.61		1.07		0.62		1		572
		Orion B 398		5 32 21	−05 45.8	1	13.00		1.05		0.45		1		8114
		Orion B 407		5 32 21	−05 53.9	4	11.24	.024	0.67	.011	0.17	.021	10	G2 IV	153,454,1519,8114

HD	DM	Other Id	N Rem	α_{1950}	δ_{1950}	S	V	σ_V	B−V	σ_{B-V}	U−B	σ_{U-B}	n	Spectrum	References
		Orion B 408	⋆ V	5 32 21	−05 59.9	2	13.38	.345	0.71	.010	-0.34	.060	2	K2 IVe	739,1536
36919	−6 01232	Orion B 438		5 32 21	−06 01.8	2	11.50	.010	0.50	.005	-0.02	.010	5	F5 V-IV	114,153
269721		Sk -67 177		5 32 21	−67 01.9	2	11.52	.000	0.04	.015	-0.53	.149	10	O8	149,415
290636	+0 01123			5 32 22	+00 04.2	1	9.55		0.21		0.02		2	A5	1089
	−0 01010			5 32 22	−00 24.0	1	10.30		0.35		0.06		2		1060
	−5 01306	Orion B 405		5 32 22	−05 20.7	3	10.20	.024	0.57	.017	0.59	.057	9	A2 VP	337,461,1014
		Orion B 414	⋆ B	5 32 22	−06 01.8	2	9.32	.020	0.00	.005	-0.06	.010	5	B9 V	114,153
36918	−6 01231	Orion B 417	⋆ A	5 32 22	−06 02.3	3	8.35	.022	-0.10	.004	-0.51	.015	8	B9 IV	114,153,8055
		Sk -66 130		5 32 22	−66 43.3	1	12.05		0.04		-0.37		3		149
		Sk -67 175		5 32 22	−67 01.9	1	11.57		-0.02		-0.38		1	A2 Ia	360
		Lambda Ori # 27		5 32 23	+09 41.0	1	12.77		0.87		0.22		1		572
36861	+9 00879	HR 1879	⋆ AB	5 32 23	+09 54.2	12	3.39	.015	-0.19	.008	-1.02	.008	36	O8 III((f))	15,572,1011,1020,1075*
36862	+9 00879	HR 1880	⋆ B	5 32 23	+09 54.2	1	5.61		0.04		-0.77		1	B0.5V	572
36898	−0 01005	IDS05298S0011	A	5 32 23	−00 09.3	5	7.13	.039	-0.08	.005	-0.44	.011	20	B7 V	1004,1060,1089,1499,8055
		Orion B 399		5 32 23	−04 45.6	1	13.60		1.41		1.75		1		8114
		Orion B 400		5 32 23	−04 49.9	1	13.58		1.27		1.39		1		454
269705		Sk -69 179		5 32 23	−69 21.2	1	11.71		-0.02		-0.76		3	B7	149
		Lambda Ori # 29		5 32 24	+09 58.7	1	12.16		0.67		0.09		1		572
245185	+9 00880			5 32 24	+10 00.5	1	9.93		0.14		-0.02		1	A5	572
		Lambda Ori # 34		5 32 24	+10 01.0	1	12.05		0.44		0.21		1		572
36819	+23 00954	HR 1875		5 32 24	+24 00.5	2	5.38	.008	-0.09	.001	-0.63	.004	10	B2.5IV	154,1223
		NGC 1960 - 152		5 32 24	+34 09.0	1	11.26		0.02		-0.39		2		1
36570	+64 00536	HR 1857	⋆ AB	5 32 24	+64 07.5	1			0.01		-0.08		5	A0 V	1079
		Orion B 424	⋆ V	5 32 24	−05 44.3	1	12.84		0.95		0.86		1	F7 IV	8114
269726		Sk -67 178		5 32 24	−67 35.	1	11.95		0.01		-0.53		4	O8	415
		AAS21,109 # 196		5 32 24	−67 52.	1	14.60		-0.20		-1.04		3		417
		Sk -68 115		5 32 24	−68 28.	1	14.99		0.30		-0.53		4	WN8	980
		Sk -71 046		5 32 24	−71 06.	1	13.25		0.02		-0.88		4		798
269693				5 32 24	−71 15.0	1	11.45		1.08		0.83		1	G5	149
36836	+22 00952			5 32 25	+22 24.4	1	8.34		0.54				2	F8	1724
	+35 01176		V	5 32 25	+35 13.8	1	10.09		1.87				3	M3	369
36916	−4 01173	V1045 Ori	⋆ V	5 32 25	−04 08.5	3	6.74	.009	-0.12	.031	-0.55	.023	5	B8 IIIp	1004,1060,8055
		Orion B 422	⋆ V	5 32 25	−05 05.4	1	14.35		0.73		-0.48		1		825
	−5 01307	Orion B 430		5 32 25	−05 05.3	4	10.24	.023	1.05	.047	0.96	.122	16	K1 III	114,337,454,1014
37144	−51 01540	IDS05312S5108	AB	5 32 25	−51 05.9	2	8.90	.020	0.99	.010	0.64		4	K2 V	173,3072
36853	+21 00896	IDS05294N2108	AB	5 32 26	+21 10.5	1	8.14		1.07				2	K0	1724
245060	+33 01094	NGC 1960 - 252		5 32 26	+33 57.9	1	9.12		0.00		-0.36		1	B8	1
		BSD 49 # 700		5 32 27	+28 09.8	1	11.06		0.33		0.10		3	A0	117
		VES 902		5 32 27	+37 14.9	1	12.81		0.57		-0.18		2		8113
36915	−0 01006	Ring Ori # 23		5 32 27	−00 50.8	2	8.02	.005	-0.01	.000	-0.34	.005	4	B9 V	1060,1089
294257	−4 01174	Orion B 425		5 32 27	−04 22.6	4	9.80	.009	0.47	.011	0.04	.022	9	F6 V	114,153,454,8114
		Sk -65 066		5 32 27	−65 54.	1	13.06		-0.14		-0.99		3		415
36895	+9 00881			5 32 28	+09 34.9	3	6.74	.005	-0.14	.022	-0.71	.005	8	B2 IV-V	572,1200,2033
		Lambda Ori # 31		5 32 28	+10 01.	1	12.45		0.75		0.17		1		572
36881	+10 00818	HR 1883	⋆ AB	5 32 28	+10 12.5	3	5.60		0.13	.025	-0.09	.021	7	B9 IIIp	572,1049,1079
245134	+21 00897			5 32 28	+21 15.8	1	9.22		0.61				2		1724
	+33 01095	NGC 1960 - 184		5 32 28	+34 01.9	1	9.98		0.97		0.60		std		1
36938	−4 01175	Orion B 437		5 32 28	−04 47.9	5	8.87	.013	0.07	.010	-0.18	.019	7	B8 V	114,454,1004,1060,8055
		Orion B 443	⋆ V	5 32 28	−05 25.1	2	11.61	.027	1.22	.019	0.77	.000	11		337,825
		Orion B 442		5 32 28	−05 32.3	6	9.00	.013	-0.02	.013	-0.27	.018	15	B8.5Vb	114,337,1004,1014*
		Orion B 444		5 32 28	−06 04.0	3	10.73	.049	0.66	.047	0.07	.000	7	F8 V	114,153,1519
269717	−67 00459	Sk -67 176		5 32 28	−67 43.1	1	11.82		-0.16		-1.01		4	B7	415
245203	+9 00882			5 32 29	+09 39.9	2	7.48		-0.09	.070	-0.64	.005	3	B8	572,1200
36719	+47 01178	HR 1869		5 32 29	+47 41.1	2	6.12	.000	0.26	.000	0.13	.023	4	F0 V	39,254
36678	+54 00914	HR 1866		5 32 29	+54 23.9	2	5.71	.023	1.63	.002	1.98		4	M0 III	71,3001
		Orion B 434		5 32 29	−04 21.2	1	13.18		0.84		0.42		1		8114
		Sk -66 133		5 32 29	−66 30.	1	12.38		-0.17		-0.97		4		415
36936	−4 01176	Orion B 439		5 32 30	−04 22.4	1	12.68		1.53		2.21		1	B8	8114
36939	−5 01308	Orion B 440		5 32 30	−04 23.2	6	7.57	.013	-0.12	.010	-0.56	.008	16	B5 V	114,153,454,1004,1060,8055
		Sk -66 134		5 32 30	−66 08.	1	12.83		-0.05		-0.62		4		798
		AAS21,109 # 197		5 32 30	−67 25.	1	13.75		-0.13		-0.86		2		417
		Sk -67 178		5 32 30	−67 44.	1	12.21		-0.14		-1.00		4		798
		Sk -69 180		5 32 30	−69 30.	1	12.90		-0.14		-0.93		3		952
		Sk -70 101		5 32 30	−70 31.	1	13.33		-0.23		-1.00		3		798
36937	−4 01177	Orion B 454		5 32 31	−04 33.5	4	9.64	.016	0.27	.010	0.11	.016	9	A7 V	114,153,1060,8114
		Orion P 1691		5 32 31	−06 20.4	1	11.22		0.69		0.18		2		1519
36894	+9 00883			5 32 32	+09 44.8	1	8.78		-0.05		-0.27		1	B5 IV	572,1200
		Orion B 467	⋆ V	5 32 32	−05 24.9	2	12.92	.110	1.19	.080	-0.54		2		776,8054
269723		Sk -67 178a		5 32 32	−67 44.1	5	9.91	.008	1.07	.042	0.60	.000	20	G2 Ia	149,1277,1483,3074,8033
		NGC 1960 - 123		5 32 33	+34 08.7	1	11.53		0.11		-0.25		2		1
	−5 01309	Orion B 464		5 32 33	−05 07.0	4	10.92	.008	0.43	.023	0.33	.036	8		114,1060,8054,8055
		Orion B 466	⋆ V	5 32 33	−05 20.2	4	10.16	.009	0.16	.018	0.00	.020	17	A1 Vn +Am	114,337,454,1014
		Lambda Ori # 30		5 32 34	+10 00.1	1	12.63		1.26		0.92		1		572
36959	−6 01233	Orion B 482	⋆ B	5 32 34	−06 02.5	9	5.67	.011	-0.23	.011	-0.91	.009	26	B1 V	15,114,153,1004,1060*
269728				5 32 34	−67 13.8	1	12.43		0.05		0.06		2	A7	149
		NGC 1960 - 145		5 32 35	+34 00.7	1	12.07		0.36		0.10		1		1
36957	−4 01178	Orion B 472		5 32 35	−04 25.0	5	8.85	.013	0.05	.012	-0.02	.016	11	A1 V	114,153,1004,1060,8055
		Sk -66 135		5 32 35	−66 28.	2	12.32	.014	-0.03	.018	-0.69	.014	5		360,415
36935	−0 01007			5 32 36	−00 18.1	2	7.52	.005	-0.12	.005	-0.56	.005	4	B7 V	1060,1089
		Orion B 474		5 32 36	−04 31.1	2	13.07	.025	0.81	.005	0.13	.075	2	G0 V	114,8114

Table 1 285

HD	DM	Other Id	N Rem	α_{1950}	δ_{1950}	S	V	σ_V	B–V	σ_{B-V}	U–B	σ_{U-B}	n	Spectrum	References
		Orion B 481		5 32 36	−04 48.6	1	13.07		1.10		1.14		1		8114
	−5 01310	Orion B 479		5 32 36	−05 07.6	4	10.46	.020	0.56	.011	0.25	.013	8		114,1060,8054,8055
36960	−6 01234	Orion B 493	⋆ A	5 32 36	−06 02.0	9	4.78	.013	-0.25	.006	-1.01	.009	35	B0.5V	15,114,153,1004,1060*
		Sk -66 136		5 32 36	−66 19.	1	13.28		-0.08		-0.57		4		798
		Sk -71 047		5 32 36	−71 45.	1	13.77		-0.20		-0.84		2		798
290700				5 32 37	−02 15.	1	10.96		0.65		0.13		3	B3	808
		Orion B 486		5 32 37	−04 45.2	1	11.06		0.67		0.08		1	G5 IV-V	451
36958	−4 01179	Orion B 480	⋆ V	5 32 37	−04 45.8	4	7.34	.016	-0.09	.011	-0.60	.013	8	B3 V	114,1004,1060,8055
		Orion B 490		5 32 37	−05 10.1	3	10.57	.031	1.28	.008	0.92	.035	15	K0 V	114,454,8054
		Orion B 499		5 32 37	−05 31.5	1	14.57		1.29				2		1014
		Orion B 508		5 32 37	−06 00.9	3	8.80	.004	-0.07	.017	-0.27	.013	7	B9 V	114,153,8114
		STO 241		5 32 37	−67 16.0	1	12.04		0.51		0.02		1	F2 Ib	726
		Lambda Ori # 32	V	5 32 38	+09 56.0	1	10.44		1.30		0.85		1		572
		Orion B 483		5 32 38	−04 20.7	1	12.65		0.65		-0.03		2	F9 V	114
		Orion B 492		5 32 38	−04 39.1	1	12.47		0.63		-0.03		2	G1 Vn	114
36981	−5 01311	Orion B 502		5 32 38	−05 14.2	7	7.84	.016	-0.12	.006	-0.56	.020	45	B4 V	114,337,454,1004,1014*
		Orion B 497		5 32 38	−05 16.7	5	11.11	.077	1.21	.033	0.70	.049	21		114,337,454,1014,1519
		Orion B 510	⋆ V	5 32 38	−05 27.2	2	11.70	.019	1.01	.034	0.00		11	K4 IV(e)	114,337
36913	+10 00819			5 32 39	+10 29.2	1	8.31		0.06		0.00		1	A3	572
	+33 01096	NGC 1960 - 113		5 32 39	+34 02.0	1	10.54		0.11		0.03		2		1
290665	−0 01008	Ring Ori # 22		5 32 39	−00 56.3	1	9.44		0.10		-0.04		2		1060
36880	+20 01009	IDS05297N2050	AB	5 32 40	+20 52.4	1	8.59		0.80				2		1724
36954	−0 01009	Ring Ori # 21		5 32 40	−00 46.0	4	6.96	.020	-0.10	.009	-0.65	.007	9	B3 V	1004,1060,1089,8055
		Orion B 509	⋆ V	5 32 40	−04 29.7	1	13.13		0.98		0.24		3		454
		Orion B 504		5 32 40	−04 41.0	2	12.22	.005	0.83	.119	0.33	.005	5	G8 IV	114,454
36879	+21 00899	LS V +21 009		5 32 41	+21 22.3	3	7.57	.008	0.20	.015	-0.80	.005	8	O7.5III	1011,1012,1209
		NGC 1960 - 87		5 32 41	+34 08.6	1	10.65		0.07		-0.40		2	B6 V	1
		Orion B 515	⋆ V	5 32 41	−04 29.5	2	13.20	.060	0.64	.210	-0.41	.315	2		776,825
		Orion B 517	⋆ V	5 32 41	−05 07.8	1	14.80		2.54				1		8054
36983	−5 01312	Orion B 520		5 32 41	−05 54.0	5	9.19	.023	-0.02	.020	-0.17	.022	11	B9.5V	114,153,1004,1060,8055
		Orion B 531		5 32 41	−06 11.9	1	13.64		0.76		0.78		1		8114
		Sk -69 181		5 32 41	−69 15.	1	11.93		-0.02		-0.72		4		415
		Lambda Ori # 33		5 32 42	+09 57.9	1	12.78		1.20		0.78		1		572
		Haro 6 # 72	V	5 32 42	+10 07.	1	13.88		1.26		0.58		1		572
		BSD 49 # 2307		5 32 42	+30 23.8	1	12.32		0.50		0.14		2	B9	117
36772	+45 01145			5 32 42	+45 38.4	1	8.38		0.11		-0.30		2	B9	401
		Orion B 518	⋆ V	5 32 42	−05 08.7	1	14.47		1.50				1		8054
36982	−5 01313	Orion B 530	⋆ V	5 32 42	−05 29.8	6	8.48	.059	0.11	.016	-0.57	.039	19	B2 V	114,337,1004,1014*
		Orion B 522	⋆ V	5 32 42	−05 31.9	1	13.10		1.10		0.09		1	K1 IV-Ve	8114
		Orion B 540		5 32 42	−06 03.9	3	10.50	.019	0.54	.017	0.04	.013	7	F6 IV	114,153,8114
		Orion B 538		5 32 42	−06 09.2	1	12.75		0.78		0.28		2	K3 V	114
	−6 01236	Orion P 1791	V	5 32 42	−06 16.2	1	9.86		0.58		0.14		1		8114
37001	−6 01238			5 32 42	−06 35.7	4	8.89	.005	-0.05	.008	-0.24	.027	14	B9 V	1004,1060,1655,8055
		AAS21,109 # 198		5 32 42	−67 15.	1	13.38		-0.15		-0.87		3		952
		Sk -68 116		5 32 42	−68 52.	2	12.58	.244	-0.11	.025	-0.95	.025	7		415,952
		NGC 1960 - 85		5 32 43	+34 07.4	1	11.30		0.14		-0.27		2		1
	+44 01248			5 32 43	+44 26.9	1	7.25		0.43		-0.05		3	K2	1733
36997	−2 01305			5 32 43	−02 24.8	1	8.36		-0.05		-0.38		3	B9.5IIIp	808
		Orion B 544		5 32 43	−05 40.3	1	12.88		1.16		0.66		1		8114
		Orion B 543	⋆ V	5 32 43	−05 48.4	2	12.80	.055	1.39	.035	1.16	.270	2		825,8114
37065	−23 02865			5 32 43	−23 29.7	6	8.80	.009	0.92	.011	0.65	.012	17	K2 V	158,912,1705,1775,2012,3072
37297	−64 00456	HR 1917		5 32 43	−64 15.6	6	5.34	.013	1.04	.001	0.86	.004	28	G8/K0 III	15,1075,1075,1498,1754,2038
290633	+0 01126			5 32 44	+00 13.6	1	10.43		0.28		0.06		4	A5	808
		NGC 1960 - 114		5 32 44	+34 02.8	1	11.82		0.30		-0.06		1		1
		NGC 1960 - 79		5 32 44	+34 04.2	1	11.93		0.12		-0.12		1		1
	+34 01097	NGC 1960 - 196		5 32 44	+34 14.2	1	9.91		1.26		1.04		2		1
	−2 01303	IDS05302S0158	AB	5 32 44	−01 56.0	1	9.01		0.04		-0.09		2	A0 IV	1060
290702	−2 01304			5 32 44	−02 09.9	1	10.19		0.16		0.09		4		808
	−3 01143			5 32 44	−03 14.3	3	9.87	.004	0.14	.016	0.08	.023	6	A4 Vn	1004,1060,8055
		Orion B 541	⋆ V	5 32 44	−05 18.9	3	12.76	.004	1.36	.009	0.62		6	K0 IV,V?	454,1014,8054
37000	−6 01237	Orion B 552		5 32 44	−05 57.5	5	7.45	.016	-0.14	.003	-0.66	.002	31	B5 V sn	114,153,1004,1060,8055
		Sk -67 180		5 32 44	−67 12.	1	12.62		-0.05		-0.66		4		415
269708		Sk -70 102		5 32 44	−70 45.3	1	12.36		0.00		-0.52		9	B8 Iab	149
	+34 01098	NGC 1960 - 81		5 32 45	+34 04.1	1	10.01		0.02		-0.57		2	B2 III	1
		NGC 1960 - 86		5 32 45	+34 08.3	1	10.65		0.06		-0.24		2	B8 III-IV	1
36998	−4 01180	Orion B 529		5 32 45	−04 37.6	5	8.99	.018	-0.01	.010	-0.19	.017	9	B9 Vn	114,153,1004,1060,8055
		Orion B 553	⋆ V	5 32 45	−05 45.2	1	14.09		1.20				1		8114
37226	−54 00854	HR 1912	⋆ A	5 32 45	−54 56.1	2	6.42	.005	0.55	.000			7	F8 V	15,2030
294264	−4 01181	Orion B 545		5 32 46	−04 53.7	5	9.52	.018	0.33	.022	-0.42	.013	12	B3 Vn	114,454,1004,1060,8055
		Orion B 561	⋆ V	5 32 46	−05 32.9	2	13.02	.040	1.14	.099	0.37	.005	3		825,1014
		Orion B 563	⋆ V	5 32 46	−05 41.4	2	12.36	.148	1.14	.008	0.63	.008	7		337,825
		Sk -66 137		5 32 46	−66 15.	1	13.09		-0.07		-0.69		4		415
		SK *G 370		5 32 46	−66 54.5	1	12.39		0.07		0.03		3	A0 Ib	149
245199	+20 01010			5 32 47	+20 24.4	1	9.41		1.42				2	F8	1724
36859	+27 00806	HR 1878		5 32 47	+27 37.9	2	6.25	.015	1.55	.005	1.65		6	K0	252,1724
		NGC 1960 - 92		5 32 47	+34 10.2	1	10.96		0.05		-0.44		2		1
		Orion B 550		5 32 47	−04 19.9	1	12.48		0.88		0.57		2		114
		Orion B 555		5 32 47	−04 46.6	5	11.11	.065	0.96	.023	0.50	.035	9		114,454,461,825,1519
		Orion B 574	⋆ V	5 32 47	−05 45.2	1	13.62		1.15		-0.16		1		825
36999	−5 01314	Orion B 581		5 32 47	−05 51.5	5	8.49	.028	-0.11	.015	-0.44	.009	11	B7 V	114,153,1004,1060,8055

HD	DM	Other Id	N	Rem	α_{1950}	δ_{1950}	S	V	σ_V	B–V	σ_{B-V}	U–B	σ_{U-B}	n	Spectrum	References
		Orion P 1879			5 32 47	−06 21.3	1	13.89		0.81		0.37		1		8114
290634					5 32 48	+00 13.	1	9.82		0.27		0.12		4	Am	808
		GD 69			5 32 48	+41 28.2	1	14.75		0.32		−0.54		1	DAs	3060
37020	−5 01315	Orion B 587		⋆A	5 32 48	−05 25.1	7	6.73	.011	0.02	.018	−0.90	.043	14	B0.5Vb	15,114,665,1004,1060*
		Orion B 588		⋆V	5 32 48	−05 41.0	1	13.45		1.36				1		8114
		Orion B 597		⋆V	5 32 48	−05 53.6	1	12.33		0.95		0.44		1		8114
		AAS21,109 # 199			5 32 48	−67 36.	1	13.70		−0.17		−0.87		3		952
		AAS21,109 # 200			5 32 48	−67 36.	2	13.61	.014	−0.15	.000	−0.88	.019	4		417,952
		Sk -68 117			5 32 48	−68 57.	1	13.48		−0.12		−0.86		3		952
		NGC 1960 - 47			5 32 49	+34 06.2	1	10.44		0.12		−0.25		2		1
	+34 01099	NGC 1960 - 48			5 32 49	+34 07.0	1	9.52		0.06		−0.46		2	B3 V	1
		NGC 1960 - 50			5 32 49	+34 07.6	1	11.61		0.20		−0.12		2		1
		NGC 1960 - 126			5 32 49	+34 12.5	1	11.00		1.00		0.66		2		1
36996	−1 00957				5 32 49	−01 03.9	1	9.57		0.04		−0.03		2	A2	1060
294263		Orion B 582		⋆V	5 32 49	−04 43.0	1	10.10		0.33		0.16		2		114
		Orion B 604		⋆V	5 32 49	−05 23.6	1	10.55		0.34		0.15		10	A2 V	337
37021	−5 01315	Orion B 595		⋆B V	5 32 49	−05 25.0	2	7.96	.000	0.24	.000	−0.49		3	B0 V	1004,1060
37022	−5 01315	Orion B 598		⋆C	5 32 49	−05 25.3	7	5.13	.005	0.00	.017	−0.95	.005	20	O6	15,114,1004,1060,1234*
		Orion B 609			5 32 49	−05 41.1	1	11.35		0.91		0.61		6	G8 IV-V	337
37025	−6 01240	Orion B 621			5 32 49	−06 03.9	5	7.16	.023	−0.13	.008	−0.62	.012	10	B3 V	114,153,1004,1060,8055
		Orion B 620			5 32 49	−06 10.1	2	11.56	.025	0.55	.010	0.00	.025	7	F6 V-IV	114,153
	+34 01100	NGC 1960 - 197			5 32 50	+34 15.0	1	9.62		0.05		−0.48		3		1
294262	−4 01182	Orion B 599			5 32 50	−04 42.8	1	9.81		0.21		−0.10		4		114
37019	−5 01316	Orion B 608			5 32 50	−05 05.8	7	9.36	.022	0.03	.013	−0.02	.035	11	B9.5Vb	114,454,1004,1014*
37023	−5 01315	Orion B 612		⋆D	5 32 50	−05 25.1	6	6.69	.009	0.08	.008	−0.81	.020	18	B0.5Vb	15,114,1004,1060,8015,8055
		Orion B 625			5 32 50	−06 12.6	1	12.11		0.90		0.47		3		114
		Sk -67 182			5 32 50	−67 15.	1	13.22		−0.16		−0.88		4		415
245252	+9 00885				5 32 51	+09 40.2	1	10.17		1.24		0.87		1	K0	572
		NGC 1960 - 46			5 32 51	+34 05.4	1	12.32		0.28		0.15		2		1
	+34 01101	NGC 1960 - 91			5 32 51	+34 09.7	1	10.37		0.04		−0.47		2		1
		Orion B 606			5 32 51	−04 38.9	1	14.09		0.94		0.73		2	K1 Vn?	114
		Orion B 626		⋆V	5 32 51	−05 18.5	1	13.05		0.93		0.74		1		8114
		Orion B 637			5 32 51	−06 00.3	1	13.62		1.17		−0.80		1		114
294236					5 32 52	−02 41.8	1	11.79		0.13		0.02		4	A2	808
		NGC 1960 - 44			5 32 53	+34 05.3	1	11.38		0.09		−0.29		2	B8 III-IV	1
37017	−4 01183	Orion B 632		⋆V	5 32 53	−04 31.5	9	6.56	.010	−0.14	.005	−0.77	.010	27	B1.5Vp	15,114,153,1004,1060*
		Orion B 646		⋆V	5 32 53	−04 52.6	1	12.00		0.85		−0.10		1		8114
		Orion B 643		⋆V	5 32 53	−05 22.8	1	13.22		1.00		0.77		1		114
	−5 01318	Orion B 655		⋆V	5 32 53	−05 23.6	4	9.67	.045	0.29	.020	−0.37	.028	7	B2 V + Shel	454,1004,8054,8055
269736		Sk -67 181			5 32 53	−67 22.7	2	12.31	.117	−0.12	.009	−0.81	.023	4	O8	149,952
290627	+0 01127				5 32 54	+00 30.0	1	9.64		0.24		0.10		3	A3	808
36910	+24 00873	CQ Tau			5 32 54	+24 43.1	2	10.39	.385	0.77	.010	0.51		4	F5	1625,1724
		NGC 1960 - 18			5 32 54	+34 07.1	1	10.78		0.08		−0.37		2		1
37016	−4 01184	Orion B 631		⋆AB	5 32 54	−04 27.4	8	6.24	.008	−0.15	.008	−0.71	.011	32	B2 V	15,114,153,1004,1060*
		Orion B 648		⋆V	5 32 54	−04 44.0	1	11.03		0.68		0.21		3		153
		Orion B 662		⋆V	5 32 54	−05 08.9	1	14.12		1.29		−0.20		1		8054
	−5 01317	Orion B 653		⋆V	5 32 54	−05 11.1	3	9.89	.015	0.72	.004	0.19	.021	5	F8 III-IV	114,1014,8054
		Orion B 656			5 32 54	−05 14.1	6	10.94	.028	1.08	.014	0.69	.098	19	G0 IV-III	114,337,454,461,1014,1519
		Sk -66 138			5 32 54	−66 27.	2	13.07	.005	−0.20	.014	−1.00	.014	5		360,798
		Sk -70 103			5 32 54	−70 48.	1	13.60		−0.07		−0.71		4		798
		Orion B 660			5 32 55	−04 45.3	1	13.23		1.13		1.31		1		8114
37018	−4 01185	Orion B 659		⋆AB	5 32 55	−04 52.2	10	4.59	.005	−0.19	.008	−0.93	.010	39	B1 V	15,114,1004,1060,1075*
		Orion B 671			5 32 55	−05 11.0	1	13.57		1.22		0.91		1		8054
37041	−5 01319	Orion B 682		⋆A	5 32 55	−05 26.8	12	5.07	.012	−0.10	.014	−0.93	.013	43	O9.5V	15,114,337,1004,1011*
		Orion B 684			5 32 55	−05 53.0	2	10.99	.005	0.66	.010	0.16	.078	3	G5 IVp	1519,8114
37015	+0 01128				5 32 56	+00 27.9	1	8.33		−0.05		−0.22		1	A0	1089
		NGC 1960 - 14			5 32 56	+34 05.3	1	10.68		0.09		−0.38		2	B8 III-IV	1
		Orion B 681			5 32 56	−05 12.7	1	14.45		1.62				1		8054
245275					5 32 57	+09 53.	1	10.32		0.27		0.10		1	A7	572
		NGC 1960 - 112			5 32 57	+34 01.6	1	12.30		0.21		0.04		2		1
		NGC 1960 - 13			5 32 57	+34 05.4	1	10.80		0.13		−0.28		2		1
		NGC 1960 - 127			5 32 57	+34 12.4	1	11.34		0.42		−0.40		2		1
		Orion B 690		⋆V	5 32 57	−05 32.7	2	12.41	.030	0.94	.045	0.41	.079	3	G8 V	1014,8114
		NGC 1960 - 77			5 32 58	+34 03.2	1	12.11		0.16		0.00		1		1
		Orion B 696		⋆V	5 32 58	−05 11.7	1	12.88		1.00		0.72		1	K0 IV-V	8054
		Orion B 693			5 32 58	−05 12.6	1	13.76		1.20		0.84		1		8054
		LP 658 - 124			5 32 58	−07 27.6	1	12.25		0.38		−0.25		5		3062
		Sk -66 140			5 32 58	−66 13.	1	12.75		−0.05		−0.65		4		415
		Orion B 688			5 32 59	−04 27.3	1	12.94		0.54		0.28		1		8114
		Orion B 691			5 32 59	−04 49.3	1	12.19		0.90		0.36		1		8114
		Orion B 698		⋆V	5 32 59	−05 10.6	3	12.09	.104	0.90	.022	0.44	.004	6	K1 III-IVe	114,454,8054
		Orion B 711			5 32 59	−05 15.1	2	11.87	.043	0.64	.022	0.18	.007	11	G3 IV-V	337,1014
		Orion B 703		⋆V	5 32 59	−05 17.0	1	11.71		0.84		0.45		10	G9 IV-V	337
37042	−5 01320	Orion B 714		⋆B	5 32 59	−05 26.9	8	6.40	.012	−0.08	.017	−0.92	.014	24	B0.5Vb	114,337,1004,1014*
		Orion B 713			5 32 59	−05 29.5	2	11.70	.066	0.90	.019	0.16		13	G3 IV-V	337,1014
	−5 01322	Orion B 720			5 32 59	−05 50.1	4	9.78	.019	0.17	.021	0.13	.012	7	B7 III	114,153,461,1060
	−5 01321	Orion B 718			5 32 59	−05 53.0	3	9.67	.072	0.50	.101	0.02	.031	8	F3 IV	114,153,1060
37043	−6 01241	Orion B 721		⋆AB	5 32 59	−05 56.5	22	2.76	.013	−0.24	.009	−1.07	.014	78	O9 III	1,3,15,114,153,542*
269741		Sk -66 139			5 32 59	−66 52.5	1	11.83		0.04		−0.39		3	A0 Ia	149
		Orion B 708		⋆V	5 33 00	−05 13.0	1	12.26		1.81		0.83		1	G3 IV-V	8054

Table 1   287

HD	DM	Other Id	N Rem	α_{1950}	δ_{1950}	S	V	σ_V	B–V	σ_{B-V}	U–B	σ_{U-B}	n	Spectrum	References
		Orion B 735		5 33 00	−05 50.7	1	12.20		0.84		0.84		1		8114
		Sk -68 119		5 33 00	−68 56.	1	13.66		-0.15		-0.99		3		952
37037	−0 01011			5 33 01	+00 04.1	1	8.50		0.01		-0.27		1	B9	1089
245299	+10 00821			5 33 01	+10 21.8	1	9.68		0.50		0.06		1	F8	572
		NGC 1960 - 41		5 33 01	+34 04.3	1	12.38		0.19		0.04		2		1
		NGC 1960 - 55		5 33 01	+34 10.4	1	11.62		0.10		-0.30		2		1
	+52 00976	G 191 - 47		5 33 01	+52 29.8	2	10.15	.004	1.12	.005	0.95	.008	7	K5	1723,7010
		Orion B 719		5 33 01	−04 56.9	2	12.62	.005	0.84	.002	0.29	.012	4		114,454
		Orion B 734	⋆ B	5 33 01	−05 28.2	2	9.58	.005	0.06	.023	0.01	.014	13	B9.5V	114,337
		BSD 49 # 75		5 33 02	+27 57.5	1	11.18		0.49		0.38		2	A0	117
		NGC 1960 - 21		5 33 02	+34 08.4	1	9.76		0.01		-0.60		2	B2.5V	1
	+34 01103	NGC 1960 - 16		5 33 02	+34 08.4	1	8.86		0.00		-0.66		2	B2 V	1
		NGC 1960 - 56		5 33 02	+34 10.4	1	12.42		0.44		0.24		1		1
		NGC 1960 - 57		5 33 02	+34 10.9	1	11.76		0.12		-0.12		2		1
36384	+74 00252	HR 1844		5 33 02	+75 00.9	2	6.16	.018	1.58	.016	1.95		4	M0 III	71,3001
	−0 01012			5 33 02	−00 31.4	1	10.25		0.19		0.12		2		1060
		Sk -68 120		5 33 02	−68 39.	1	13.24		-0.14		-1.00		4		415
		NGC 1960 - 110		5 33 03	+34 01.4	1	11.98		0.20		-0.07		2		1
	+34 01106	NGC 1960 - 8	⋆ B	5 33 03	+34 06.3	1	9.36		0.02		-0.58		2	B2 V	1
		NGC 1960 - 3		5 33 03	+34 07.8	1	10.72		0.10		-0.25		2		1
37040	−4 01186	Orion B 722	⋆ A	5 33 03	−04 23.7	1	6.56		-0.16		-0.79		1	B2.5 IV	114
		Orion B 722	⋆ AB	5 33 03	−04 23.7	9	6.31	.010	-0.14	.010	-0.71	.013	25	B2.5 IV	15,114,153,1004,1060*
		Orion B 722	⋆ B	5 33 03	−04 23.7	1	8.37		-0.04		-0.33		1		114
37059	−5 01323	Orion B 736		5 33 03	−04 56.1	4	9.08	.011	-0.02	.005	-0.23	.020	8		114,1004,1060,8055
		Orion B 737	⋆ V	5 33 03	−04 57.2	1	14.99		1.27		0.14		1	K7 Ve	114
		Orion B 740		5 33 03	−05 06.1	3	13.28	.318	1.11	.066	0.46		3	K4 IV	114,997,8054
		Orion B 744		5 33 03	−05 34.8	1	12.04		1.04		0.62		2	G8 V	1014,8114
		Orion B 755	⋆ V	5 33 03	−05 50.7	2	13.72	.027	1.24	.027	0.01	.171	5		8013,8114
		Orion B 749	⋆ C	5 33 03	−05 56.7	1	9.76		-0.01		-0.27		1		8114
	+34 01108	NGC 1960 - 38		5 33 04	+34 04.0	1	9.92		0.05		-0.49		2		1
		NGC 1960 - 11		5 33 04	+34 05.4	1	11.24		0.10		-0.30		2		1
	+34 01107	NGC 1960 - 9	⋆ A	5 33 04	+34 06.2	1	9.13		0.00		-0.66		2	B2 V	1
	+34 01109	NGC 1960 - 23		5 33 04	+34 08.7	1	8.96		0.01		-0.68		2	B3 V	1
		WLS 520 35 # 8		5 33 04	+34 57.0	1	10.22		0.34		-0.28		2		1375
290684	−1 00963	Ring Ori # 8		5 33 04	−01 48.7	1	9.53		0.15		-0.02		2		1060
		Orion B 757	⋆ V	5 33 04	−05 11.3	4	12.54	.022	1.38	.020	0.66	.063	7	M0 IVe	114,454,825,8054
37061	−5 01325	Orion B 747	⋆ V	5 33 04	−05 17.9	6	6.83	.014	0.24	.038	-0.63	.037	19	B0.5Vb	114,337,1004,1014*
37062	−5 01326	Orion B 760	⋆ C	5 33 04	−05 27.1	7	8.24	.037	0.04	.015	-0.49	.021	21	B5 Vb	114,337,454,1004,1014*
	−5 01324	Orion B 767	⋆ V	5 33 04	−05 35.1	4	9.85	.054	0.46	.008	0.20	.050	11	F2 III-IV	337,1014,1060,8055
		HRC 145, TW Ori		5 33 04	−06 47.3	1	15.01		0.68		-0.34		1		825
37379	−66 00431			5 33 04	−65 58.5	1	8.44		0.51				1	F6/7 V	1642
		Sk -67 183		5 33 04	−67 35.5	1	12.46		0.08		-0.40		2	B9 Ia	149
37058	−4 01187	Orion B 761	⋆ V	5 33 05	−04 52.1	3	7.33	.010	-0.15	.005	-0.76	.020	6	B2 V sn	1004,1060,8055
		Orion B 773		5 33 05	−05 33.1	1	11.77		1.20		0.84		5	G9 IV-Ve	337
37055	−3 01146	HR 1900	⋆ AB	5 33 06	−03 17.0	7	6.40	.009	-0.13	.007	-0.64	.010	21	B3 IV	154,1004,1060,1089*
37060	−5 01327	Orion B 776		5 33 06	−05 08.2	6	9.36	.018	0.01	.019	-0.06	.017	10	A0 Vb	114,1004,1014,1060*
		Orion B 780		5 33 06	−06 03.3	1	12.19		0.79		0.40		3		114
36892	+32 01048			5 33 07	+32 18.1	1	8.74		0.07				3	A2	1025
		Orion B 789	⋆ V	5 33 07	−05 23.4	1	14.08		0.88		-0.52		1		776
245207	+30 00947			5 33 08	+30 42.7	1	10.39		0.31				2	B8	1025
37078	−6 01242	Orion B 796		5 33 08	−06 06.9	4	9.42	.020	0.11	.011	0.14	.033	10	A2 V	114,153,1060,8055
	−6 01243	Orion P 2125		5 33 08	−06 23.8	1	10.51		0.56		0.06		2		1519
37054	+0 01129			5 33 09	+00 42.1	1	8.75		0.02		-0.18		3	B9.5IV	808
245312	+9 00886			5 33 09	+09 30.1	1	10.00		1.24		0.96		1	K5	572
		NGC 1960 - 30		5 33 09	+34 06.0	1	12.03		0.14		-0.14		2		1
	+34 01110	NGC 1960 - 27		5 33 09	+34 06.7	1	9.58		0.02		-0.58		2	B2 nn	1
37057	−3 01147			5 33 09	−03 36.5	3	9.29	.000	0.01	.005	-0.13	.009	5	A0	1004,1060,8055
		Orion B 785	⋆ V	5 33 09	−05 03.1	2	13.16	.240	1.65	.485	0.15	.600	2	K2 IV(e)	114,825
	−5 01328	Orion B 786	⋆ V	5 33 09	−05 14.3	6	9.89	.013	0.07	.011	0.08	.027	19	A0 V + Shel	114,337,1004,1014*
36909	+31 01020			5 33 10	+31 58.4	2	7.99	.005	0.52	.009			4	F5	1025,1724
		NGC 1960 - 33		5 33 10	+34 05.5	1	11.87		0.12		-0.12		2	B9 V	1
37056	−3 01148			5 33 10	−03 20.8	4	8.37	.007	-0.06	.012	-0.38	.013	12	B9 Vn	1004,1060,1089,8055
		Orion B 805	⋆ V	5 33 10	−05 28.5	1	13.64		1.39				4		1014
		NGC 1960 - 71		5 33 11	+34 02.5	1	12.16		0.16		-0.05		2		1
37077	−4 01188	Orion B 806	⋆ A	5 33 11	−04 53.2	9	5.26	.015	0.25	.019	0.16	.021	21	F0 III	15,114,247,254,1008*
		Orion B 808		5 33 11	−05 10.8	2	13.73	.225	1.25	.125	0.43	.350	2	M3 II,III?	114,8054
	−7 01129	Orion P 2158		5 33 11	−06 55.6	1	10.76		1.04		0.94		1		1519
37350	−62 00487	HR 1922, β Dor		5 33 11	−62 31.3	3	3.52	.085	0.66	.030	0.44	.029	4	F6 IVa	688,849,1754
		NGC 1960 - 69		5 33 12	+34 04.3	1	12.43		0.20		-0.38		1		1
37075	−1 00964			5 33 12	−00 57.1	1	9.37		0.03		-0.04		2	A0	1060
		Orion B 791		5 33 12	−04 27.1	1	12.80		0.98		0.65		2		454
		Sk -66 141		5 33 12	−66 51.	1	12.82		-0.13		-0.89		4		798
269748		Sk -67 184		5 33 12	−67 45.	1	13.21		-0.17		-0.96		4	WN4	980
37035	+9 00888			5 33 13	+09 30.1	1	8.64		-0.02		-0.48		1	B9	572
37034	+9 00887			5 33 13	+09 40.8	1	9.33		0.04		-0.14		1	A0	572
36931	+30 00949			5 33 13	+30 33.6	3	7.93	.032	1.06	.016	0.61		7	F8 II	1025,1724,8100
	+34 01111	NGC 1960 - 61		5 33 13	+34 09.0	1	9.14		0.01		-0.66		2	B2 V	1
36770	+56 01041			5 33 13	+56 27.5	1	7.45		0.98		0.69		3	G8 II-III	1502
		Orion B 816		5 33 13	−04 50.4	1	12.82		0.87		0.48		1		8114
		Orion B 824	⋆ V	5 33 13	−05 10.9	2	14.64	.080	1.91	.035	0.75	.755	2		114,8054

HD	DM	Other Id	N	Rem	α_{1950}	δ_{1950}	S	V	σ_V	B–V	σ_{B-V}	U–B	σ_{U-B}	n	Spectrum	References
		Orion B 812			5 33 14	−04 38.3	1	11.82		0.67		0.03		2		114
		Orion B 823			5 33 14	−05 22.1	2	11.30	.029	0.94	.011	0.76	.029	13	K5 V	337,1519
		Orion B 831		⋆V	5 33 14	−05 30.1	2	11.38	.027	1.02	.000	0.61	.008	11	K1 IV-Ve	337,825
		Orion B 829			5 33 14	−05 38.0	1	14.06		1.37				3		1014
		STO 804			5 33 14	−70 50.4	1	12.10		0.31		0.26		1	F2 Ia	726
37076	−1 00965	IDS05307S0103	A		5 33 15	−01 01.1	2	8.06	.035	−0.08	.010	−0.41	.000	4	B8 V	1060,1089
		Orion B 810		⋆V	5 33 15	−04 26.8	2	14.05	.039	1.32	.097	0.58	.151	3	K5 ne	114,825
		Orion B 819			5 33 15	−04 44.6	1	13.24		1.80		2.09		1		114,8114
		Orion B 832			5 33 15	−05 15.6	2	11.49	.027	0.71	.003	0.24	.020	14	F8 IV-V	337,1014
37091	−6 01244				5 33 15	−06 46.7	1	9.82		0.14		0.08		2	A3 V	1060
		Sk -67 185			5 33 15	−67 34.	1	12.46		−0.08		−0.90		4		415
		Sk -68 121			5 33 15	−68 51.	1	12.17		0.01		−0.76		4		415
		NGC 1960 - 109			5 33 16	+34 02.1	1	10.70		0.08		−0.41		2		1
290671	−1 00966	IDS05307S0103	B		5 33 16	−01 01.2	1	9.01		−0.02		−0.13		2	B5	1060
		Orion B 834		⋆V	5 33 16	−05 11.2	4	13.86	.094	1.29	.046	0.72	.078	5	K5 Ve	114,8013,8023,8054
		Orion B 833			5 33 16	−05 13.8	1	13.99		1.45				1		8054
		Orion B 850		⋆V	5 33 17	−05 09.1	2	12.77	.155	1.16	.035	0.63	.240	2	K2 Ve	114,8054
		Orion B 848			5 33 17	−05 34.0	1	12.77		0.94		0.89		5		1014
36754	+58 00845	IDS05289N5829	AB		5 33 18	+58 31.2	1	8.57		0.45		0.04		2	F5	1733
		Orion B 854		⋆V	5 33 18	−05 30.0	1	10.16		0.45		0.40		2		1014
		Sk -66 142			5 33 18	−66 44.	1	12.40		−0.02		−0.62		4		415
		Sk -67 186			5 33 18	−67 30.	1	12.91		−0.05		−0.67		3		798
		AAS21,109 # 201			5 33 18	−67 31.	2	13.22	.033	−0.04	.009	−0.89	.005	4		417,952
		Sk -67 189			5 33 18	−67 32.	1	11.82		0.02		−0.80		4		798
37051	+9 00889				5 33 19	+09 48.1	1	9.07		0.04		−0.16		1	B9	572
37102	−6 01245	Orion B 872			5 33 19	−06 02.6	2	8.99	.010	0.44	.005	−0.05	.005	6	F5 V	114,153
37502	−70 00407				5 33 19	−70 43.4	2	9.88	.000	0.40	.000	−0.05	.000	7	F0 V	149,204
		Orion B 870			5 33 20	−05 18.8	2	12.52	.030	0.95	.020	0.49	.045	3		1014,8114
	−6 01246	Orion P 2238			5 33 20	−06 43.6	1	9.82		0.65		0.25		1		997
245239	+33 01098	NGC 1960 - 138		⋆A	5 33 21	+34 02.0	1	9.04		0.02		−0.60		2	B3 V	1
290661		Ring Ori # 19			5 33 21	−00 44.5	1	10.74		0.34		0.07		2		1060
		Orion B 853		⋆V	5 33 21	−04 26.5	1	14.98		1.28		0.98		1		114
		Orion B 864			5 33 21	−05 12.4	1	12.08		1.04		0.45		1		8054
37336	−60 00440				5 33 21	−59 58.9	1	9.19		0.99		0.80		2	K0/1 III	1730
		Sk -67 187			5 33 21	−67 34.	1	12.37		−0.08		−0.90		3		415
245241	+31 01021				5 33 22	+32 02.6	1	10.21		0.16				3	B5	1025
288047					5 33 23	+00 58.	1	11.85		0.26		0.14		3	A3	808
		Lambda Ori # 40			5 33 23	+09 51.5	1	12.51		0.77		0.24		1		572
245310	+21 00901	LS V +21 010		⋆AB	5 33 23	+21 09.4	1	8.87		0.28		−0.67		2	B2:III:penn	1012
36995	+25 00878				5 33 23	+25 33.5	1	8.54		0.51				2	F7/8 V/IV	1724
	+34 01113	NGC 1960 - 101			5 33 23	+34 10.3	1	9.23		0.05		−0.69		std	B2 Ve	1
36891	+40 01346	HR 1884			5 33 23	+40 09.1	4	6.10	.016	1.03	.016	0.69	.005	13	G3 Ib	138,1355,6009,8100
294239	−3 01150				5 33 23	−02 56.0	1	10.54		0.17		0.09		3		808
		Orion B 875			5 33 23	−04 18.6	1	12.84		0.93		0.55		3		454
		Orion B 880			5 33 23	−04 53.0	2	13.39	.015	1.04	.005	0.38	.044	3		114,454
	−5 01329	Orion B 884		⋆V	5 33 23	−05 30.5	4	10.15	.134	0.43	.029	0.40	.046	8	A3 Vp	114,337,825,1588
		Orion B 885			5 33 23	−05 42.3	4	11.33	.050	0.63	.008	−0.11	.053	10	B4 V	114,337,461,1519
		S120 # 3			5 33 23	−60 02.7	1	11.24		1.60				2		1730
36994	+25 00879	HR 1889			5 33 24	+25 54.5	2	6.50	.009	0.43	.007			7	F5 III	71,1724
245287	+28 00829				5 33 24	+28 59.4	1	9.48		0.45				2	F5	1724
290654	−0 01014				5 33 24	−00 20.8	1	10.36		0.18		0.10		2	B8	1060
290691					5 33 24	−01 53.	1	10.30		0.20		0.14		1		1060
294256		Orion B 871			5 33 24	−04 17.0	2	11.04	.014	0.66	.009	0.14	.009	4		114,454
		Orion B 881			5 33 24	−04 32.0	1	13.04		1.15		0.86		1		454
		Orion B 883		⋆V	5 33 24	−05 09.0	1	14.76		1.68				1		8054
		Orion B 887			5 33 24	−05 10.0	2	12.49	.135	1.22	.030	0.66	.200	2	K1 V(e)	114,8054
		SK *C 50			5 33 24	−65 59.	1	12.43		0.79		0.37		3		149
		Sk -67 190			5 33 24	−67 24.	2	12.67	.035	−0.09	.020	−0.82	.040	7		415,952
		Sk -67 188			5 33 24	−67 48.	1	13.44		−0.10		−0.82		4		798
269719					5 33 24	−70 52.0	1	12.09		0.32		0.29		4	G0	149
	+39 01360			V	5 33 25	+39 55.2	1	10.05		1.84				3	M2	369
290662	−0 01015	Ring Ori # 20			5 33 25	−00 49.6	1	9.83		0.19		0.06		2		1060
		Orion B 892		⋆V	5 33 25	−05 07.0	4	11.57	.053	0.86	.014	0.39	.033	8	K1 IV	454,1519,8054,8114
37192	−33 02414	HR 1909			5 33 25	−33 06.8	4	5.76	.013	1.12	.000			15	K2 IIIa	15,2013,2020,2028
269735		Sk -69 182			5 33 25	−69 24.5	1	11.71		0.16		0.15		8	A5 Ia:	149
245259	+31 01022				5 33 26	+31 49.5	1	10.51		0.50		−0.24		4	B8	206
		Orion B 900			5 33 26	−04 49.9	1	12.21		0.87		0.32		2		114
		Sk -68 122			5 33 26	−68 43.	1	12.70		−0.06		−0.91		4		415
36949	+38 01226				5 33 27	+38 18.7	1	7.64		1.51		1.69		1	K0	1733
37113	−1 00968	Ring Ori # 9			5 33 27	−01 47.6	2	8.65	.000	0.04	.015	−0.14	.010	4	B9.5 V	1060,1089
		Orion P 2260			5 33 27	−04 15.1	1	11.27		0.54		0.04		1		1519
37115	−5 01330	Orion B 907		⋆AB	5 33 27	−05 39.5	8	7.08	.018	−0.08	.020	−0.54	.015	18	B7 Ve+Shell	114,337,379,1004,1014*
		Orion B 903			5 33 28	−04 38.4	2	11.38	.004	0.60	.009	0.08	.004	6		114,454
		Orion B 909			5 33 28	−04 58.8	1	14.16		1.24		0.82		1		114
		Sk -70 104			5 33 28	−70 59.1	1	13.08		−0.02		−0.46		3	B9 Ia	415
245370	+9 00890				5 33 29	+10 00.3	1	10.19		0.49		0.08		1	F8	572
		Lambda Ori # 38			5 33 30	+10 00.1	1	9.79		1.64		1.69		1		572
290678					5 33 30	−01 34.	1	10.69		0.21		0.05		2		1060
294229					5 33 30	−02 28.8	1	10.88		0.29		0.08		3	A7	808
		Orion B 910			5 33 30	−04 38.4	2	13.09	.015	1.26	.002	1.27	.105	2	K5 IV: n?	114,454

Table 1

289

HD	DM	Other Id	N Rem	α_{1950}	δ_{1950}	S	V	σ_V	B–V	σ_{B-V}	U–B	σ_{U-B}	n	Spectrum	References
		Orion B 916		5 33 30	−05 14.8	2	13.18	.045	1.99	.005	1.91		3		114,1014,8054
		Sk -67 192		5 33 30	−67 08.	1	13.14		−0.13		−0.81		4		798
		Sk -67 191		5 33 30	−67 33.	1	13.46		−0.21		−1.04		4		798
		Sk -69 183		5 33 30	−69 32.	1	14.42		−0.15		−0.97		3	WN3	980
		Lambda Ori # 37		5 33 31	+09 57.6	1	11.77		1.00		0.56		1		572
37112	−0 01017	Ring Ori # 18		5 33 31	−00 48.6	2	8.02	.010	−0.09	.005	−0.50	.015	4	B6 V	1060,1089
37114	−5 01331	Orion B 920		5 33 31	−05 24.4	5	8.98	.034	−0.02	.013	−0.12	.009	16	B9.5Vp	114,337,1004,1014,1060
245385				5 33 32	+09 57.	1	10.40		0.29		0.14		1	A0	572
		Orion B 928		5 33 32	−05 41.8	4	11.45	.064	1.05	.028	0.90	.034	12	G8 IV-V	114,337,454,1519
245386	+9 00891			5 33 33	+09 48.5	1	10.87		0.38		0.24		1	A2	572
37111	−0 01016			5 33 33	−00 21.5	1	8.81		0.04		0.00		2	Am	1060
294266	−5 01332	Orion B 923		5 33 33	−04 58.0	2	10.46	.000	0.42	.011	0.01	.007	5	F3 V-IV	454,461
37131	−6 01247			5 33 33	−06 18.3	2	8.19	.090	−0.05	.024	−0.37	.052	4	B9	1060,8055
		HRC 159, AV Ori		5 33 33	−06 44.4	1	13.98		1.28		0.31		1		825
		Orion B 926		5 33 34	−04 55.8	1	12.92		0.99		0.61		1		114
		Orion B 935		5 33 34	−05 19.5	2	13.48	.098	1.28	.058	1.19	.009	5		1014,8013
		STO 639		5 33 34	−69 13.7	1	11.28		0.56		−0.03		1	F8 Ia	726
		G 98 - 16		5 33 35	+37 04.4	1	12.06		1.08		1.00		1		333,1620
37130	−4 01189	Orion B 929		5 33 35	−04 46.9	6	9.97	.011	0.15	.020	−0.09	.015	13		114,454,461,1004,1060,8055
		Orion B 932		5 33 35	−04 52.1	2	13.70	.045	0.87	.015	0.63	.095	2	G8 V	114,454
269761		Sk -66 143		5 33 35	−66 53.2	1	11.75		−0.02		−0.64		3	B5:I	149
		Orion B 939		5 33 36	−04 46.5	2	12.62	.052	0.95	.023	0.46	.005	4		114,8114
		Sk -66 144		5 33 36	−66 59.	1	12.84		−0.06		−0.71		4		798
		Orion B 946		5 33 37	−05 01.6	1	13.44		0.85		0.48		1	K0 V	114
		Orion B 944		5 33 37	−05 09.1	2	12.54	.009	0.74	.009	0.23	.045	5	F6 IV	454,8054
37099	+9 00892			5 33 38	+09 43.4	1	8.56		1.01		0.73		1	K0	572
		Orion B 940		5 33 38	−04 27.4	6	7.14	.011	−0.16	.009	−0.71	.011	15	B2.5V sn	114,153,454,1004,1060,8055
36947	+43 01315			5 33 39	+44 02.5	1	7.27		1.02		0.45		2	F8	1733
		Orion B 950		5 33 39	−05 43.8	1	11.90		0.83		0.40		1		8114
37128	−1 00969	HR 1903, ε Ori	⋆ A	5 33 40	−01 13.9	17	1.69	.008	−0.18	.008	−1.03	.009	76	B0 Iae	1,3,15,154,1004,1006*
288101	+0 01130			5 33 41	+01 00.1	1	9.71		0.15		0.06		3	A2	808
		Orion B 963	AB	5 33 41	−06 02.5	2	11.37	.010	0.69	.005	0.14	.010	3	F9 V +K0 V	114,1519
37151	−7 01131	V1179 Ori		5 33 41	−07 25.6	2	7.40	.000	−0.08	.000	−0.40		3	B8 V	1004,1060
		AAS21,109 # 203		5 33 42	−66 30.	1	13.87		−0.20		−0.99		3		417
		AAS21,109 # 204		5 33 42	−66 51.	1	13.35		0.01		−0.78		2		417
		Sk -66 146		5 33 42	−66 54.	1	12.81		−0.14		−0.97		4		798
		Sk -66 145		5 33 42	−67 00.	1	12.57		−0.08		−0.81		4		798
		Sk -67 193		5 33 42	−67 29.	1	13.03		0.51		−0.07		4		798
245318	+31 01024			5 33 43	+31 17.0	1	10.11		0.60				2	F8	1025
	−1 00970			5 33 43	−01 32.2	1	10.04		0.20		0.08		2		1060
37141	−2 01309			5 33 43	−02 17.4	1	8.45		0.01		−0.21		3	B9.5V	808
37142	−5 01333	Orion B 958		5 33 43	−05 05.5	1	9.39		0.49		−0.01		2	F4 IV	114
		Orion B 965	⋆ V	5 33 43	−05 21.6	1	13.97		1.06		0.37		4		8013
245409	+11 00878			5 33 44	+11 17.9	2	8.81	.019	1.39	.015	1.18		4	K7	1705,3072
37110	+9 00893			5 33 45	+09 36.1	1	8.97		0.00		−0.28		1	A0	572
37140	−0 01018	V1130 Ori		5 33 45	−00 20.0	3	8.55	.008	0.10	.005	−0.41	.012	7	B8 p SiSr	1060,1089,8055
294258		Orion B 961		5 33 45	−04 23.4	4	10.94	.014	0.52	.018	0.06	.025	7	F7 IV	114,454,461,1519
37149	−1 00971	Ring Ori # 10		5 33 46	−01 39.9	2	8.04	.015	−0.10	.010	−0.50	.005	4	B8 V	1060,1089
		Orion B 969		5 33 46	−04 52.0	1	13.00		1.17		0.96		3	K4 IV-V	454
37212	−25 02539	SZ Lep		5 33 46	−25 46.2	3	7.57	.104	2.24	.045	3.22	.075	5	C7,3	109,864,1238
37072	+29 00941			5 33 47	+29 17.0	1	9.14		0.43				2	A3	1025
	−6 01252	Orion P 2370		5 33 47	−06 40.6	1	10.48		0.61		0.11		1		1519
36992	+38 01228			5 33 48	+38 58.4	1	8.79		0.29		0.23		3	A1 V	1501
290686				5 33 48	−01 52.	1	11.06		0.34		0.15		2		1060
37150	−5 01334	Orion B 980	⋆	5 33 48	−05 40.7	7	6.56	.017	−0.19	.016	−0.80	.014	17	B2.5V	15,114,154,1004,1014*
37402	−60 00441			5 33 48	−60 08.2	1	8.38		0.50		−0.02		2	F6 V	1730
		Sk -66 147		5 33 48	−66 22.	1	13.01		−0.16		−0.93		4		798
		AAS21,109 # 205		5 33 48	−66 30.	1	13.86		−0.16		−0.95		2		417
		Sk -67 198		5 33 48	−67 01.9	1	11.91		0.00		−0.58		3	A4 Iab	415
		Sk -67 195		5 33 48	−67 10.	1	12.82		0.01		−0.51		4		798
		Sk -67 194		5 33 48	−67 32.	2	13.75	.009	−0.15	.000	−1.03	.005	5		360,798
		Sk -67 196		5 33 48	−67 37.	1	13.08		−0.13		−0.97		3		952
		AAS21,109 # 202		5 33 48	−70 16.	1	13.49		−0.02		−0.57		3		952
245460	+9 00895			5 33 49	+09 34.2	1	9.64		1.08		0.85		1	K7	572
245458	+9 00894			5 33 49	+10 02.0	1	10.19		1.13		0.75		1	K2	572
245358	+23 00961			5 33 49	+23 24.3	1	8.83		0.73				2		1724
37032	+34 01118	LS V +34 059		5 33 49	+34 46.9	1	8.10		0.10		−0.78		2	B0.5V	1012
		Orion B 973		5 33 49	−04 24.3	4	10.97	.015	0.59	.010	0.09	.021	7		114,454,1519,8114
		Orion B 982	⋆ V	5 33 49	−05 13.5	1	13.56		1.54		1.50		1	K4 IV-Ve	114
		Orion P 2374		5 33 49	−06 25.7	2	10.91	.000	0.60	.010	0.35	.235	2		1519,8114
245459	+9 00896			5 33 50	+09 50.1	1	8.43		1.18		0.82		1	K0	572
		HRC 160, PQ Ori		5 33 50	−02 12.8	1	12.63		0.58		0.13		1	F0	825
36383	+77 00201			5 33 51	+77 30.6	1	8.05		0.11		0.11		2	A0	1375
37763	−76 00333	HR 1953	⋆ A	5 33 51	−76 22.7	5	5.18	.006	1.13	.000	1.18	.012	20	K2 III	15,1075,2017,2038,3077
245380	+23 00962			5 33 52	+23 08.6	1	9.64		0.08		−0.37		2		1726
		Orion B 983		5 33 52	−04 46.7	2	14.20	.060	1.27	.015	1.13		2		114,454
290659				5 33 54	−00 28.	1	11.02		0.38		0.17		2		1060
269763		Sk -66 148		5 33 54	−66 37.	1	12.15		−0.08		−0.86		4	B1.5Ia	415
		Sk -69 184		5 33 54	−69 21.	1	12.92		−0.10		−0.93		3		952
		G 98 - 18		5 33 56	+32 13.5	1	16.43		0.38		−0.59		3	DA	316

HD	DM	Other Id	N Rem	α_{1950}	δ_{1950}	S	V	σ_V	B–V	σ_{B-V}	U–B	σ_{U-B}	n	Spectrum	References
245335	+32 01050			5 33 56	+32 16.7	1	9.41		1.74				3	M2	369
		Orion B 989	★ V	5 33 56	−06 04.3	1	13.00		1.16		0.74		2		454
	−0 01020			5 33 57	−00 36.8	1	10.67		0.26		0.15		2		1060
		STO 864		5 33 57	−71 47.3	1	12.18		0.49		0.02		1	F6 Ia	726
37172	−1 00973			5 33 59	−01 16.0	1	8.30		0.12		0.04		2	A0	1060
37173	−2 01311			5 33 59	−02 00.8	1	7.86		-0.06		-0.55		2	B6 V	1089
		Orion B 993		5 33 59	−06 10.0	1	12.20		0.68		0.01		1	G4 V	114
274681	−45 02059			5 33 59	−45 08.4	2	10.05	.024	0.56	.005	0.03		3	F5	1594,6006
290656				5 34 00	−00 21.	1	10.60		0.25		0.09		2		1060
		Orion B 992		5 34 00	−05 26.3	3	9.20	.023	-0.02	.010	-0.12	.013	6	B9.5V	114,1004,1060
	−6 01253	V380 Ori	★ ABC	5 34 00	−06 44.8	4	10.26	.096	0.48	.043	-0.23	.012	4	A0	1588,8013,8023,8037
		AAS21,109 # 207		5 34 00	−66 27.	1	13.98		-0.14		-0.79		3		952
269766	−66 00432	Sk -66 149		5 34 00	−66 53.2	1	11.38		0.02		-0.59		3	B8 Ia	149
		Sk -69 185		5 34 00	−69 09.	1	12.62		0.18		-0.52		3		952
37148	+9 00897			5 34 01	+09 51.5	1	8.46		0.61		0.11		1	G0	572
37098	+26 00870	HR 1902	★ AB	5 34 01	+26 53.7	2			-0.06	.010	-0.37	.049	6	B8 IV+B8 IV	1049,1079
		Orion B 1000		5 34 01	−05 59.6	1	12.10		0.81		0.44		2		114
		Sk -66 150		5 34 01	−66 55.	1	12.20		-0.07		-0.88		4	B1.5Ia	415
	+20 01017			5 34 02	+20 43.9	1	10.05		1.93				2	M2	369
37188	−5 01336	Orion B 1001		5 34 02	−05 48.0	2	8.65	.018	0.29	.004	0.12	.018	5	F1 IV	114,1060
37276	−30 02513			5 34 02	−30 34.0	1	7.61		0.91		0.58		10	G8 III	1673
37124	+20 01018	G 100 - 27		5 34 04	+20 42.4	3	7.66	.015	0.66	.005	0.11	.009	6	G4 IV-V	196,333,1620,1724
		Orion B 1003		5 34 04	−05 44.8	1	12.30		0.82		0.37		1		114
37123	+22 00959			5 34 05	+22 05.3	1	8.90		0.10		-0.46		1	B8	1220
37187	−1 00974			5 34 05	−01 03.5	3	8.13	.007	-0.01	.000	-0.25	.015	7	B9 V	1004,1060,1089
290687	−2 01312	IDS05316S0161	AB	5 34 05	−01 59.1	1	9.68		0.20		0.08		3		808
		Orion B 1004		5 34 05	−05 25.6	3	11.28	.034	0.69	.010	0.19	.019	4	G8 V	114,461,1519
37210	−6 01254			5 34 05	−06 29.0	2	8.10	.000	-0.07	.000	-0.41		3	B9 IV-Vp	1004,1060
245480	+20 01019			5 34 06	+20 19.8	1	9.28		0.48				2		1724
		Orion B 1008		5 34 06	−05 55.7	1	12.94		0.81		0.60		1		8114
		HRC 165, BD Ori		5 34 06	−06 21.1	1	12.82		0.98		0.67		1		825
		ApJS2,315 # 50		5 34 06	−66 57.	1	13.43		1.26		0.30		2		35
		Sk -67 197		5 34 06	−67 34.2	1	12.30		0.01		-0.66		4	A0 Ia	798
		Sk -69 186		5 34 06	−69 29.	1	12.95		-0.09		-0.87		3		952
274680	−45 02061			5 34 07	−45 04.5	1	10.28		0.46		-0.05		1	G0	6006
37097	+31 01025			5 34 08	+31 18.2	3	8.27	.016	0.39	.022			8	F5	1025,1724,1733
		Sk -68 123		5 34 08	−68 43.	1	12.87		0.08		-0.28		4		415
37160	+9 00898	HR 1907		5 34 09	+09 15.9	10	4.08	.006	0.96	.008	0.63	.015	38	K0 IIIb	3,15,1003,1256,1355*
		Orion B 1012	★ AB	5 34 09	−06 05.7	1	8.90		-0.05		-0.49		1		114
37209	−6 01255	Orion B 1017	★ AB	5 34 09	−06 05.7	7	5.73	.020	-0.23	.007	-0.91	.009	16	B1 V	15,114,154,1060,1061*
37147	+16 00822	Hyades vB 168	★	5 34 10	+17 00.7	5	5.54	.023	0.22	.000	0.10	.008	9	F0 V	15,1127,3023,8015,8023
245389	+32 01053			5 34 10	+32 13.2	1	8.94		1.12				2	K0	1025
37208	−5 01338	Orion B 1016		5 34 11	−05 09.7	2	9.20	.018	0.28	.000	0.10	.004	5	A7 Vn	114,1060
	−5 01337	Orion B 1019		5 34 11	−05 21.6	2	10.72	.024	0.27	.009	0.08	.014	5	A9 IV	114,1060
		Orion B 1018		5 34 11	−05 30.4	1	10.67		0.71		-0.04		3	B5 V	114
245420	+30 00957			5 34 12	+30 37.9	1	9.86		0.30				3	B9	1025
290657	−0 01022			5 34 12	−00 25.5	1	9.41		0.06		0.03		2	B8	1060
		Orion B 1013		5 34 12	−04 26.1	1	14.09		0.92		0.97		1		8114
294259		Orion B 1015		5 34 12	−04 27.5	3	11.04	.010	0.62	.005	0.09	.012	6	F8 V	114,454,1519
		AAS21,109 # 211		5 34 12	−67 00.	1	13.31		-0.10		-0.90		3		417
	−3 01154			5 34 13	−03 32.1	3	9.68	.008	-0.01	.032	-0.10	.005	5	B8	1004,1060,8055
37286	−28 02298	HR 1915		5 34 13	−28 44.3	2	6.26	.000	0.15	.015	0.05		7	A2 III/IV	2031,8071
290688	−2 01313			5 34 15	−02 03.1	1	9.55		0.19		0.12		3		808
		Orion B 1026		5 34 15	−05 36.9	1	13.71		0.90		0.25		1		114
294241	−3 01155			5 34 16	−02 56.2	1	10.82		0.31		0.25		3	A0	808
		Orion B 1024		5 34 16	−04 27.7	1	14.18		1.57				1		8114
		Orion B 1025		5 34 16	−05 08.6	3	11.38	.005	0.52	.018	0.05	.016	7		114,454,1519
245416	+31 01027	IDS05310N3143	AB	5 34 17	+31 45.0	1	9.43		0.29				2	A2	1025
		Sk -66 151		5 34 17	−66 16.	1	12.50		-0.17		-1.03		4		415
37171	+10 00828	HR 1908		5 34 18	+11 00.3	3	5.91	.026	1.59	.017	1.92	.016	6	K4 II	15,1003,8015
245443	+30 00959			5 34 18	+31 03.1	1	10.49		0.24				4	A2	1025
37008	+51 01094	G 191 - 48		5 34 18	+51 24.9	3	7.74	.011	0.83	.011	0.45	.009	12	K2 V	22,1003,1355
290689	−2 01314			5 34 18	−02 05.1	1	9.97		0.10		-0.04		3		808
		Orion B 1027		5 34 18	−06 05.9	1	12.58		1.17		1.87		1		8114
		AAS21,109 # 210		5 34 18	−68 09.	1	14.65		-0.25		-1.00		2		417
		AAS21,109 # 212		5 34 18	−68 32.	1	14.70		-0.27		-0.92		2		952
		AAS21,109 # 213		5 34 18	−68 34.	2	14.24	.127	-0.14	.049	-0.76	.019	3		417,952
		Sk -69 187		5 34 18	−69 34.	1	12.52		-0.03		-0.69		3		415
37462	−58 00526			5 34 19	−58 54.1	2	6.71	.029	1.49	.034	1.80		6	K4 III	2007,3005
		Sk -67 200		5 34 20	−67 09.	1	12.59		-0.12		-0.94		4		415
	−1 00975			5 34 21	−01 26.8	1	9.80		0.34		0.19		2		1060
		Orion B 1032	★ AB	5 34 21	−05 30.8	3	11.13	.005	0.80	.007	0.44	.019	7		114,454,1519
275028	−49 01800			5 34 21	−49 33.1	1	11.50		0.47		-0.06		2	F8	1730
275031	−49 01801			5 34 21	−49 46.6	1	10.48		0.52		0.05		2	G0	1730
290640				5 34 22	−00 00.5	1	10.98		0.35		0.26		4	A2	808
294260	−4 01191	Orion B 1030		5 34 22	−04 27.4	5	10.76	.009	0.69	.009	0.18	.014	8	G0 IV	114,454,461,1519,8013
		Orion B 1031		5 34 22	−04 59.7	1	13.30		1.40		1.25		1		114
		Sk -66 152		5 34 22	−66 11.	1	12.49		-0.16		-1.04		4		415
245499	+26 00873			5 34 23	+26 37.4	1	9.32		1.83				2	K7	369
294267		Orion B 1034		5 34 23	−05 01.3	4	11.13	.002	0.62	.007	0.10	.012	7	F9 IV-V	114,454,461,1519

Table 1

HD	DM	Other Id	N Rem	α_{1950}	δ_{1950}	S	V	σ_V	B–V	σ_{B-V}	U–B	σ_{U-B}	n	Spectrum	References
		Orion B 1035		5 34 23	−05 51.3	1	13.67		1.55		0.84		1		8114
37235	−0 01023	Ring Ori # 17		5 34 24	−00 43.9	2	8.17	.000	−0.10	.000	−0.46	.020	4	B8 V	1060,1089
		Orion B 897	⋆ V	5 34 24	−05 53.6	1	14.05		1.32		0.29		1		8114
37392	−49 01803			5 34 24	−49 46.9	1	8.56		0.99		0.71		2	G8 III	1730
290725	+0 01133	IDS05328N0003	A	5 34 25	+00 04.2	1	9.42		0.20		0.03		3	B9	808
245531	+21 00906			5 34 25	+21 44.3	1	9.50		0.08		−0.39		1	B8	1220
		Orion B 1037		5 34 25	−04 41.7	2	12.11	.042	1.85	.052	1.91	.117	4		114,454
37501	−61 00488	HR 1936		5 34 25	−61 12.4	6	6.31	.005	0.85	.005	0.48		21	G5 IV	15,1020,1075,1754,2012,2020
269781	−67 00464	Sk -67 201		5 34 25	−67 03.2	3	9.89	.011	0.07		.000		19	A0 Iae	1277,1368,8033
269777	−67 00465	Sk -67 199		5 34 25	−67 20.2	2	11.06	.025	−0.01	.000	−0.73	.005	10	B3 Ia	149,415
37031	+50 01204	IDS05305N5102	AB	5 34 26	+51 03.7	1	8.92		0.24		0.21		2	A0	1566
	−1 00976			5 34 26	−01 20.8	1	10.18		0.15		0.09		2	A4 V	1060
		Orion B 1041		5 34 26	−04 58.9	1	14.07		0.94		1.45		1		114
		Orion B 1039		5 34 26	−05 01.7	1	14.49		1.69		1.32		3		8013
	−1 00977			5 34 27	−01 28.5	1	9.79		0.06		0.02		2		1060
		Orion B 902		5 34 27	−05 00.1	1	12.22		0.91		0.47		4		454
		Orion B 901	⋆ V	5 34 27	−05 06.2	2	13.88	.035	1.19	.065	0.56	.180	2		454,776
290719	+0 01134			5 34 28	+00 22.1	1	10.08		0.47		0.25		3	A3	808
37138	+33 01102	HR 1904		5 34 28	+33 31.8	2	6.33	.001	1.30	.008			5	K0	71,1724
290718	+0 01135			5 34 29	+00 28.0	1	10.70		0.36		−0.05		3	A3	808
271369		Sk -65 067		5 34 29	−65 40.8	1	11.44		0.05		−0.31		7	A0 Ia	149
	−5 01339	Orion B 1046		5 34 30	−05 54.1	2	9.60	.009	0.51	.009	0.02	.014	5		114,8114
		Sk -66 153		5 34 30	−66 21.	1	13.28		−0.13		−0.84		4		798
		Sk -67 203		5 34 30	−67 37.	2	13.03	.025	−0.12	.015	−0.93	.015	7		415,952
290770	−1 00978			5 34 31	−01 39.1	1	9.22		0.05		−0.18		2	B8	1060
		Orion B 1045		5 34 31	−04 54.1	1	13.21		0.78		0.03		1		114
269762		Sk -69 188		5 34 31	−69 01.0	1	11.37		0.08		−0.48		3	B9 Ia0:	149
245493	+33 01103	NGC 1960 - 505		5 34 32	+33 56.3	2	8.63	.000	0.02	.000	−0.73	.000	5	B2 Vpe	1012,1212
37258	−6 01257	Orion B 1051	⋆ V	5 34 33	−06 11.0	3	9.61	.010	0.13	.010	0.03	.009	8	A2 V	454,1060,8055
37234	+4 00989	IDS05319N0442	ABC	5 34 34	+04 44.4	1	7.75		0.04		−0.21		2	B9	401
245545	+23 00973			5 34 34	+23 06.6	1	10.22		0.10		−0.07		2		1726
		Orion B 1052		5 34 34	−05 43.4	2	13.49	.005	1.16	.002	0.84	.105	2		114,454,8114
		Orion B 1053		5 34 34	−05 51.1	2	12.84	.005	1.53	.075	1.14	.140	2		114,8114
269783	−67 00466	Sk -67 202		5 34 34	−67 09.4	1	11.53		0.22		−0.03		3	G0	415
		SK *G 387		5 34 34	−69 27.6	1	11.85		0.29		0.32		5	A9 Iab	149
37232	+8 01016	HR 1913	⋆	5 34 35	+08 55.4	5	6.11	.012	−0.17	.004	−0.83	.014	14	B2 IV-V	15,154,1004,1078,1732
		Orion B 1048	⋆ V	5 34 35	−04 38.8	2	12.58	.030	0.75	.030	0.04	.035	2	G8 V	114,997
294294		Orion B 1050		5 34 35	−04 40.2	3	11.36	.011	0.56	.013	−0.04	.002	5		114,454,1519
		Orion B 1054	⋆ V	5 34 35	−05 38.3	2	12.46	.075	0.80	.019	0.26	.052	4	G8 Ve	114,454,997
271374		Sk -65 069		5 34 35	−66 00.8	1	11.82		0.02		−0.56		3	B8 Ia	149
245546	+23 00974			5 34 36	+23 07.1	1	10.66		0.18		0.05		2		1726
290732				5 34 36	−00 24.	1	10.10		0.13		0.07		2		1060
		AAS21,109 # 214		5 34 36	−69 27.	2	13.20	.000	−0.20	.033	−1.03	.019	4		417,952
269758		Sk -70 105		5 34 36	−70 44.6	1	12.40		0.07		0.05		2	A2 Iab	149
37203	+19 00986			5 34 37	+19 45.0	1	8.87		0.12		0.08		2	A0	1648
		BSD 49 # 2389		5 34 37	+30 48.7	1	10.93		0.27		0.11		2	B9	117
37030	+55 01002			5 34 37	+55 43.2	1	7.91		0.45		−0.04		2	F5	1502
		Orion B 1059		5 34 38	−05 54.0	2	13.82	.005	2.05	.010	1.70		2		114,8114
269787	−67 00467	Sk -66 154		5 34 38	−67 00.5	2	10.81	.000	0.12	.005	−0.37		5	A0 Ia	149,8033
269764				5 34 38	−69 15.2	1	11.00		0.61		−0.05		2	G5	149
37202	+21 00908	HR 1910, ζ Tau	AB	5 34 39	+21 06.8	8	2.98	.037	−0.18	.011	−0.73	.048	23	B4 IIIpe	15,154,879,1212,1223*
37256	−0 01024			5 34 39	−00 13.5	1	8.93		0.06		−0.02		2	A0	1060
37257	−0 01025			5 34 39	−00 29.3	1	8.69		−0.03		−0.08		2	A0	1060
		Orion B 1057		5 34 39	−05 11.7	1	13.50		1.12		0.90		1		114
269779	−68 00389	MN83,95 # 701		5 34 40	−68 02.7	1	10.09		1.11				1		591
		Orion B 1060		5 34 41	−05 12.4	4	11.40	.016	0.59	.009	0.03	.022	7	G5 V	114,454,1519,8114
37434	−47 01940	HR 1927, TX Pic		5 34 41	−47 20.6	5	6.10	.004	1.16	.002	1.00		19	K2 III	15,1075,1641,2013,2029
		Sk -65 068		5 34 42	−65 52.	1	13.59		−0.23		−1.10		4		798
		AAS21,109 # 215		5 34 42	−66 23.	1	13.65		−0.25		−0.95		3		952
		Sk -66 155		5 34 42	−66 37.	1	12.70		−0.18		−0.98		4		798
		AAS21,109 # 216		5 34 42	−66 57.	1	14.59		−0.17		−0.89		3		952
37680		Sk -69 191		5 34 42	−69 47.	1	13.35		−0.20		−0.57		3	WC5	980
290731				5 34 43	−00 21.	1	10.96		0.38		0.14		3	A3	808
37272	−1 00979			5 34 43	−01 41.8	2	7.91	.010	−0.11	.010	−0.56	.020	4	B5 V	1060,1089
		Orion B 1065		5 34 43	−04 58.1	1	13.87		1.08		0.77		1		114
	−6 01258	Orion B 1069		5 34 43	−06 08.0	4	10.73	.028	0.89	.011	0.43	.042	8	K0 IV(e)	114,454,1519,8114
		MN83,95 # 703		5 34 44	−68 08.4	1	12.86		0.53				1		591
		Sk -69 190		5 34 44	−69 02.4	1	12.25		0.14		−0.29		4	A0 Ia	415
37273	−3 01158			5 34 45	−03 51.8	3	9.92	.008	0.15	.004	0.05	.005	6	A2	1004,1060,8055
269790				5 34 45	−67 32.1	1	12.68		0.23		0.26		1	A5	149
245581	+25 00888			5 34 46	+25 59.0	1	9.24		0.52				3		1724
		Orion B 1073		5 34 46	−05 42.2	3	11.30	.016	0.38	018	0.17	.014	6		114,454,8114
245580	+28 00834			5 34 47	+28 57.0	1	9.89		2.15		2.35		2	K5	8032
	−6 01259	BF Ori		5 34 47	−06 36.8	1	10.30		0.31		0.37		1	A5 II-IIIe	1588
		Orion B 1077	V	5 34 48	−05 47.4	2	13.34	.155	1.16	.070	0.38	.230	2		114,8114
275029	−49 01809			5 34 48	−49 29.4	1	10.72		0.64		0.16		2	G0	1696
		AAS21,109 # 218		5 34 48	−65 58.	1	13.57		−0.11		−0.90		2		417
		AAS21,109 # 219		5 34 48	−66 55.	1	13.38		−0.15		−0.95		3		952
		Sk -67 205		5 34 48	−67 19.	1	13.71		−0.20		−1.07		4		798
37285	−1 00980	IDS05323S0149	AB	5 34 49	−01 48.0	1	9.02		0.04		−0.04		2	A0	1060

HD	DM	Other Id	N	Rem	α₁₉₅₀	δ₁₉₅₀	S	V	σ_V	B−V	σ_B−V	U−B	σ_U−B	n	Spectrum	References
		Orion B 1076			5 34 49	−05 03.9	2	11.19	.005	0.47	.034	−0.08	.058	3		114,8114
		Orion B 1074			5 34 49	−05 11.5	1	13.35		2.00				1		8114
		Orion B 1080			5 34 49	−05 44.1	1	13.92		1.02		1.09		1		114
37306	−11 01238	HR 1919			5 34 49	−11 48.3	2	6.10	.005	0.05	.000			7	A2 V	15,2012
37284	−0 01027	Ring Ori # 16			5 34 50	−00 43.5	1	8.97		0.12		0.04		2		1060
	−4 01192				5 34 50	−03 59.4	1	10.96		0.30		0.16		2		1060
		Sk -67 206			5 34 50	−67 05.	1	12.00		−0.11		−0.95		4	B0.5Ia	415
	+17 00964				5 34 51	+17 56.4	1	9.76		1.86				3	M3	369
37169	+37 01262				5 34 51	+37 43.0	1	7.56		0.04		−0.22		2	B9	401
37070	+56 01044				5 34 52	+56 20.0	2	7.03	.015	0.41	.015	0.00	.010	6	F5	1502,3026
	−6 01260	Orion P 2527		V	5 34 52	−06 28.0	1	10.65		0.49		0.05		1		8114
294230	−2 01306				5 34 53	−02 06.2	1	9.63		0.28		0.09		3		808
	−4 01193	Orion B 1082			5 34 53	−04 42.1	3	11.29	.013	0.66	.025	0.25	.020	5	A0	114,454,8114
		Orion B 1083			5 34 53	−05 34.1	3	11.96	.026	1.09	.019	0.76	.014	6		114,454,8114
245600	+29 00943				5 34 54	+29 04.1	1	11.03		0.42		−0.20		3	A0	117
	+37 01263				5 34 54	+37 41.0	1	9.14		0.08		−0.09		1	A2	401
290748					5 34 54	−00 57.	1	10.20		0.17		0.11		2		1060
	−1 00981	IDS05323S0110		A	5 34 54	−01 04.7	1	10.69		0.47		0.25		2		1060
		Orion B 1084			5 34 54	−04 58.3	1	12.63		0.78		0.15		1		114
		Sk -66 157			5 34 54	−66 31.	1	13.50		−0.07		−0.89		4		798
		Brey 52			5 34 54	−67 23.	1	14.66		−0.27		−1.04		4	WN4	980
		AAS21,109 # 220			5 34 54	−68 38.	1	13.77		−0.03		−0.95		2		417
		Sk -69 192			5 34 54	−69 12.	1	12.78		−0.03		−0.88		3		952
		AAS21,109 # 217			5 34 54	−69 25.	1	13.40		−0.07		−0.75		3		952
37304	−7 01137				5 34 55	−07 13.0	1	10.24		0.17		0.14		2	A3 V	1060
		Orion B 1093			5 34 56	−05 50.3	3	11.39	.004	0.91	.013	0.72	.025	6	K2 V	114,454,1519
	−6 01261	Orion B 1090			5 34 56	−06 02.6	2	10.78	.014	0.39	.000	−0.02	.023	5		114,8114
269769	−69 00392	Sk -69 193			5 34 56	−69 48.7	1	12.10		−0.09		−0.93		4	O8	415
37294	−0 01028				5 34 57	−00 16.2	2	8.35	.000	−0.09	.003	−0.32	.004	8	A0	1060,1732
269797	−67 00470	Sk -67 204			5 34 57	−67 23.2	3	10.88	.010	0.04	.004	−0.53		12	B9 Ia0:	149,1277,8033
		MN83,95 # 702			5 34 57	−68 06.0	1	12.26		0.59				1		591
		STO 808			5 34 58	−70 37.2	1	12.68		0.54		0.10		1	F0 Ia	726
290733					5 35 00	−00 24.	1	10.29		0.43		0.01		2		1060
		AAS21,109 # 221			5 35 00	−66 04.	1	14.54		−0.21		−0.79		3		952
		AAS21,109 # 222			5 35 00	−66 57.	1	13.84		−0.11		−0.88		3		952
		AAS21,109 # 223			5 35 00	−67 30.	1	14.18		−0.15		−0.96		2		952
37303	−6 01262	Orion B 1098		★	5 35 01	−05 58.0	10	6.04	.013	−0.22	.014	−0.92	.013	28	B2 V	15,114,154,1004,1060*
		Sk -69 194			5 35 01	−69 48.	1	11.98		−0.08		−0.96		3		415
		Orion B 1100			5 35 02	−05 25.5	1	13.71		1.17		0.70		1		114
269801	−67 00472	Sk -67 207			5 35 02	−67 23.1	4	10.51	.009	0.07	.012	−0.53	.000	15	B9 Ia0	149,1277,1408,8033
37321	−1 00982	Ring Ori # 11		★ AB	5 35 03	−01 27.0	4	7.11	.020	−0.08	.004	−0.54	.019	8	B4 V	1004,1060,1089,8055
		Orion B 1099			5 35 03	−04 40.5	1	13.17		1.04		0.62		1		114
37302	−0 01029				5 35 04	−00 54.9	1	9.11		−0.02		−0.11		2		1060
		Orion B 1103			5 35 04	−05 19.9	4	11.68	.013	0.55	.022	0.02	.026	4	F7 V	114,454,1519,8114
269807	−67 00471	Sk -67 208			5 35 04	−67 03.3	1	10.85		0.16		0.10		3	A5 Ia	149
288156	+0 01137				5 35 06	+00 58.8	1	9.86		0.39		0.30		3	A5	808
	+88 00025	G 265 - 27			5 35 06	+88 08.0	1	10.58		0.75		0.25		1	G2	1658
		AAS21,109 # 224			5 35 06	−66 59.	1	14.00		−0.17		−0.93		2		417
269810		Sk -67 211			5 35 06	−67 34.0	3	12.27	.015	−0.19	.024	−1.04	.045	10	O7	149,328,1277
		Sk -67 209			5 35 06	−67 43.	1	13.27		−0.19		−0.97		4		798
245660	+18 00899				5 35 07	+18 30.7	1	9.28		1.83				3	M0	369
37322	−3 01159				5 35 08	−03 41.2	3	9.80	.000	0.07	.000	0.02	.005	4	A0	1004,1060,8055
		Orion B 1108			5 35 08	−05 48.4	1	13.94		1.52		0.27		1		114
	−5 01343	Orion B 1111			5 35 08	−05 54.8	2	11.00	.005	0.53	.005	−0.02	.024	3		114,1519
		Orion P 2570			5 35 08	−06 21.8	1	11.16		0.53		0.05		1		8114
		Orion P 2571			5 35 08	−06 25.0	1	11.33		0.51		0.02		1		1519
37722	−69 00396				5 35 08	−69 30.1	2	8.91	.024	0.06	.005	0.10	.005	9	A1 V	149,204
37334	−5 01342	Orion B 1109			5 35 09	−04 57.8	4	7.16	.016	−0.16	.004	−0.77	.011	16	B1.5V	114,1004,1060,1499
		Orion B 1118			5 35 09	−06 07.5	1	12.68		0.63		0.19		1		114
37241	+26 00879	IDS05320N2651		AB	5 35 10	+26 53.4	1	8.77		0.50				2	F7/8 V	1724
37333	−2 01319				5 35 10	−02 28.3	1	8.52		0.06		−0.06		2	A0	1089
		Orion B 1115			5 35 11	−06 04.2	1	12.00		0.75		0.34		2	K0 IV	114
269788	−68 00393				5 35 11	−68 48.5	1	9.90		1.53		1.58		5	K4 III	149
290745					5 35 12	−00 36.	1	10.52		0.25		0.11		2		1060
290765	−1 00983				5 35 12	−01 19.1	1	9.07		0.01		−0.07		2	B8	1060
		Orion B 1112			5 35 12	−04 52.9	1	12.49		0.65		0.26		1		114
294296	−5 01344	Orion B 1113			5 35 12	−05 07.1	2	9.94	.014	0.64	.009	0.11	.005	4		114,1519
		Sk -67 210			5 35 12	−67 44.	1	13.80		−0.23		−0.98		4		798
		AAS21,109 # 225			5 35 12	−68 29.	1	13.59		0.08		−1.00		3		952
		TOU28,33 T13# 29			5 35 13	+32 24.0	1	10.51		0.15				2		1025
37332	−0 01031	Ring Ori # 15			5 35 13	−00 48.4	2	7.61	.015	−0.13	.010	−0.60	.000	5	B5 V	1060,1089
		Orion B 1117			5 35 13	−04 52.8	2	12.51	.035	1.44	.055	1.04	.120	2		114,8114
		Orion B 1120			5 35 15	−05 49.6	1	12.70		0.73		0.40		1		114
	−6 01263				5 35 15	−06 35.9	1	10.53		0.36		0.04		2		1060
37331	+0 01139				5 35 16	+00 17.3	1	7.81		0.30		−0.01		3	A2	808
		Orion B 1125			5 35 16	−05 52.8	1	13.66		1.56		0.90		1		114
37344	−1 00984				5 35 17	−01 36.9	1	8.75		−0.03		−0.17		2	B8 V	1060
37430	−27 02389	HR 1926			5 35 18	−27 54.0	1	6.14		0.34				4	F0 IV	2007
		Sk -66 158			5 35 18	−66 26.	1	13.53		−0.15		−0.93		4		798
		Sk -69 195			5 35 18	−69 16.	1	13.29		−0.14		−1.01		3		952

Table 1   293

HD	DM	Other Id	N	Rem	α_{1950}	δ_{1950}	S	V	σ_V	B–V	σ_{B-V}	U–B	σ_{U-B}	n	Spectrum	References
37330	+0 01138	IDS05327N0054	A		5 35 19	+00 56.4	3	7.46	.029	-0.10	.014	-0.53	.019	5	B6 V	1004,1089,8055
37320	+7 00953	HR 1920			5 35 19	+07 30.8	3	5.87	.005	-0.07	.004	-0.37	.010	8	B8 III	15,1061,1079
245654	+30 00962				5 35 19	+30 55.8	1	9.24		0.20				2	B8	1025
		Sk -69 196			5 35 19	-69 40.	1	12.33		-0.13		-0.98		4		415
245669	+29 00944				5 35 20	+29 26.2	1	11.00		0.54		0.49		2	F0	117
		Orion B 1127			5 35 20	-05 52.6	1	12.57		1.25		1.10		1		114
37583	-55 00839				5 35 20	-55 26.4	1	9.56		0.49		-0.02		4	F5 V	1097
37656	-62 00494				5 35 20	-62 49.8	3	9.33	.014	1.12	.010	1.07	.024	10	K5 V	158,2012,3008
37342	+0 01140	IDS05327N0054	B		5 35 21	+00 57.5	2	8.01	.005	-0.13	.010	-0.57		2	B5 V	1004,1089
37357	-6 01264	IDS05329S0646	AB		5 35 21	-06 44.2	3	8.86	.010	0.10	.014	0.05	.035	6	A0	1004,1060,8055
37251	+32 01058				5 35 22	+32 38.9	2	7.94	.045	0.95	.009			5	G5	1025,1724
		Sk -66 160			5 35 22	-66 01.	1	12.59		-0.09		-0.95		4		415
	-6 01266	IDS05330S0630	AB		5 35 23	-06 28.1	1	10.87		0.30		0.25		3		8055
37372	-6 01265				5 35 23	-06 30.6	1	9.50		0.31		0.05		2	A8 V	1060
245710	+24 00897				5 35 24	+24 57.9	1	9.57		0.64				3	G0	1724
37343	-0 01033	IDS05330S0014	E		5 35 24	-00 13.1	1	8.69		0.10		-0.02		2	A0	1060
		L 132 - 46			5 35 24	-62 50.	2	14.48	.084	0.66	.019	-0.18	.084	4	K5	1658,3060
		Sk -66 159			5 35 24	-66 30.	1	13.02		-0.15		-0.95		3		798
		Sk -67 212			5 35 24	-67 44.	1	13.44		-0.14		-0.92		3		798
		AAS21,109 # 226			5 35 24	-69 23.	1	13.50		0.03		-0.83		2		417
269784		Sk -69 197			5 35 24	-69 46.	1	12.15		-0.03		-0.88		4	O7	415
37269	+30 00963	IDS05322N3026		⋆ AB	5 35 25	+30 27.9	3	5.41	.004	0.45	.015	0.25	.005	8	B9.5 V +	1049,3030,6009
37269	+30 00963	HR 1914	C		5 35 25	+30 27.9	1	8.57		0.28		0.09		2	A2	3030
37356	-4 01196	HR 1923			5 35 25	-04 50.5	5	6.19	.015	-0.04	.003	-0.72	.003	14	B2 IV-V	15,154,1060,1415,8031
		Orion B 1130			5 35 25	-05 51.6	3	11.66	.020	0.51	.010	0.08	.073	5		114,454,8114
37373	-6 01267				5 35 25	-06 45.0	3	8.32	.015	-0.10	.010	-0.39	.025	5	B9	1004,1060,8055
245711	+22 00968				5 35 26	+22 23.2	1	9.92		1.81				3	K7	369
37252	+32 01059	IDS05321N3225	A		5 35 26	+32 26.9	2	7.90	.045	0.48	.009			5	G0	1025,1724
294295		Orion B 1131			5 35 28	-04 48.9	1	10.53		0.26		0.13		1		8114
		Orion B 1132			5 35 28	-04 54.4	1	10.98		0.61		0.25		1		8114
		Sk -66 163			5 35 28	-66 05.	1	12.32		-0.12		-0.96		4		415
		Orion P 2616			5 35 29	-05 27.1	1	12.50		0.89		0.36		1		8114
		AAS21,109 # 229			5 35 30	-66 04.	1	12.95		-0.17		-1.05		3		417
		Sk -67 213			5 35 30	-67 09.	1	15.33		-0.16		-1.05		4	WN3	980
		Sk -69 199			5 35 30	-69 01.	2	12.85	.032	0.09	.014	-0.76	.014	5		360,798
		AAS21,109 # 227			5 35 30	-69 16.	1	13.70		0.20		-0.83		2		417
		Sk -69 198			5 35 30	-69 46.	1	14.17		-0.22		-0.92		3	WN4	980
269786	-69 00399	Sk -69 200			5 35 30	-69 46.8	1	11.18		0.01		-0.83		3	B7	415
245684	+30 00964				5 35 32	+30 11.8	2	9.31	.005	0.42	.005			4	F5	1025,1724
		Orion P 2620			5 35 32	-04 40.8	1	13.91		1.40				1		8114
37629	-57 00851				5 35 32	-57 29.3	1	7.59		1.58		1.92		12	M1 III	1673
		Sk -66 161			5 35 32	-66 47.2	1	11.90		0.33		0.31		6	F0 Ia:	149
37267	+37 01270				5 35 33	+37 42.7	1	8.27		0.06		-0.16		2	A0	401
37370	-0 01034	IDS05330S0014	AB		5 35 33	-00 12.8	2	7.46	.010	-0.04	.000	-0.44	.020	4	B6 V	1060,1089
290764	-1 00983	Ring Ori # 12			5 35 33	-01 17.0	1	9.88		0.32		0.09		2		1060
245709	+27 00822				5 35 34	+27 59.2	1	9.76		1.18				3		1724
37390	-6 01268				5 35 35	-06 16.2	3	9.44	.027	0.16	.010	-0.07	.020	6	A0	1004,1060,8055
37371	-0 01035	IDS05330S0014	C		5 35 36	-00 12.6	1	8.66		0.14		-0.05		2	B9	1060
37371	-0 01035	IDS05330S0014	CD		5 35 36	-00 12.6	1	7.95		0.10		-0.15		2	B9	1089
37371	-0 01035	IDS05330S0014	D		5 35 36	-00 12.6	1	8.83		0.05		-0.13		2		1060
37389	-1 00985	IDS05331S0148	AB		5 35 36	-01 46.8	1	8.38		-0.07		-0.37		2	A0	1060
		AAS21,109 # 230			5 35 36	-68 34.	2	14.53	.028	-0.21	.028	-0.94	.009	4		417,952
		Sk -68 124			5 35 36	-68 40.	1	13.68		-0.05		-0.87		3		952
		AAS21,109 # 228			5 35 36	-70 55.	1	14.23		-0.10		-0.92		2		417
36972	+68 00398	S Cam			5 35 37	+68 46.3	1	8.18		2.49				1	C0 e	1238
		Orion P 2629			5 35 37	-05 33.3	1	13.09		0.67		0.20		1		8114
290750	-1 00986	Ring Ori # 14			5 35 38	-01 04.6	1	9.87		0.20		0.10		2		1060
		G 97 - 56			5 35 39	+11 48.6	2	14.13	.030	0.61	.005	-0.15	.020	2		333,1620,3060
37266	+37 01271				5 35 39	+37 57.0	1	7.53		0.04		-0.23		2	B9	401
		Orion P 2632			5 35 39	-04 41.0	2	11.88	.040	0.72	.060	0.63	.065	3		1060,8114
		Orion P 2636			5 35 39	-05 09.8	1	12.30		0.63		0.08		1		8114
		AN295,47 # 32			5 35 40	+26 25.1	1	11.77		0.66		-0.01		1	F9 V	601
		Orion P 2640			5 35 40	-05 17.8	1	12.67		1.42		0.67		1		8114
245771	+20 01027				5 35 41	+20 14.2	1	9.23		1.82				3	K5	369
37397	-1 00987	Ring Ori # 13			5 35 41	-01 11.8	3	6.84	.010	-0.16	.005	-0.74	.005	8	B2 V	1004,1060,1089
269833		Sk -67 214			5 35 41	-67 22.1	1	11.85		-0.01		-0.61		3	O8	149
37412	-6 01270				5 35 42	-06 27.5	2	9.79	.018	0.09	.013	0.05	.027	5	A0	1060,8055
		Sk -67 216			5 35 42	-67 02.	1	12.84		-0.16		-0.98		4		798
		Sk -67 215			5 35 42	-67 47.	1	13.83		-0.18		-1.00		3		952
		Sk -68 125			5 35 42	-68 51.	1	13.39		-0.16		-0.94		2		952
		Brey 56	V		5 35 42	-69 15.	1	13.61		-0.01		-0.85		3		980
		AAS21,109 # 231			5 35 42	-69 22.	2	13.05	.042	0.01	.023	-0.74	.000	4		417,952
245729	+29 00946				5 35 43	+29 44.0	1	10.30		0.34				2	B8	1025
37413	-7 01139				5 35 43	-07 46.9	1	9.72		0.23		0.11		2	A6 Vn	1060
37836	-69 00401	Sk -69 201			5 35 43	-69 42.3	4	10.61	.043	0.11	.015	-0.92	.034	15	Be	149,328,1277,8033
		Orion P 2648			5 35 44	-05 09.4	1	13.78		1.17		-0.16		1		8114
269842	-66 00435				5 35 45	-66 49.5	1	10.96		0.42		0.02		1	F8	360
37410	-4 01198				5 35 46	-04 08.2	1	6.85		0.12		0.07		2	Am	1060
		Orion P 2652			5 35 47	-05 13.8	3	11.43	.016	0.89	.013	0.33	.005	5		1060,8055,8114
37411	-5 01346				5 35 47	-05 26.9	3	9.82	.041	0.13	.019	0.12	.005	6	B9 V	1004,1060,8055

HD	DM	Other Id	N Rem	α_{1950}	δ_{1950}	S	V	σ_V	B–V	σ_{B-V}	U–B	σ_{U-B}	n	Spectrum	References
37495	−28 02321	HR 1935		5 35 47	−28 43.1	3	5.28	.023	0.49	.019	0.10	.000	8	F5 V	15,1008,2012
		Sk −69 203		5 35 47	−69 16.	1	12.29		0.01		−0.85		4	B0.7Ia	415
		Sk −69 202		5 35 47	−69 19.	1	12.24		0.04		−0.65		4		415
245770	+26 00883	V725 Tau		5 35 48	+26 17.3	2	9.09	.263	0.49	.040	−0.61	.060	3	B0pe	601,1012
37318	+28 00836	LS V +28 007		5 35 48	+28 26.0	1	8.39		0.35		−0.66		3	B1 Vne	1212
37290	+38 01239	LS V +38 016		5 35 48	+38 31.3	1	8.52		0.03		−0.66		2	B8	401
290766				5 35 48	−01 23.	1	10.64		0.30		0.20		2	A0	1060
		Orion P 2659		5 35 48	−05 56.0	1	10.97		0.50		0.03		1		8114
		Sk −66 164		5 35 48	−66 23.	1	13.70		−0.22		−1.07		3		798
		Sk −66 165		5 35 48	−66 41.	1	13.15		−0.18		−0.94		4		798
269844		Sk −67 217		5 35 48	−67 06.	1	11.79		−0.14		−0.87		3	B1.5Iab	415
		Sk −69 204		5 35 48	−69 29.	1	12.00		−0.10		−0.94		4		415
245725	+32 01060	IDS05326N3235	AB	5 35 49	+32 37.0	1	10.78		0.16				2	A0	1025
37329	+26 00884	HR 1921		5 35 50	+26 35.4	4	6.43	.000	0.98	.007	0.77	.004	10	G9 III	1501,1625,1724,1733
		Orion P 2661		5 35 50	−04 56.4	1	12.56		0.84		0.31		1		8114
		AN295,47 # 33		5 35 51	+26 24.4	1	11.34		0.31		−0.60		1	A5	601
		TOU28,33 T13# 35		5 35 52	+32 11.9	1	10.67		0.26				2		1025
		Orion P 2665		5 35 53	−04 52.5	1	13.16		1.44		1.83		1		8114
37428	−6 01271		AB	5 35 53	−06 10.0	4	8.68	.038	0.12	.012	−0.16	.016	11	A0	1004,1060,1732,8055
		Sk −67 218	V	5 35 54	−67 05.	2	14.35	.135	−0.22	.068	−0.95	.018	5	WN3	798,980
		Sk −69 205		5 35 55	−69 44.	1	12.17		−0.10		−0.95		4		415
		VES 904		5 35 56	+38 05.7	1	12.22		0.44		−0.40		3		8113
269841	−67 00476	Sk −67 219		5 35 57	−67 28.2	3	11.81	.044	0.11	.005	−0.23		9	A0 Ia	149,1277,8033
37427	−0 01036	IDS05334S0012	AB	5 35 58	−00 10.5	2	8.62	.000	−0.03	.000	−0.33	.030	4	B8 V	1060,1089
294273	−2 01322			5 35 58	−02 45.5	1	10.66		0.26		0.07		3	A3	808
	−4 01199			5 35 58	−04 03.0	1	10.67		0.35		−0.03		2		1060
		Sk −68 126		5 35 58	−68 59.	1	12.61		0.01		−0.76		4		415
37316	+38 01241			5 35 59	+38 28.8	1	8.30		0.15		0.02		1	B9	401
37444	−5 01347			5 35 59	−05 04.3	1	7.64		0.33		0.17		2	Am	1060
37993	−73 00316	HR 1964, WX Men		5 35 59	−73 46.3	4	5.78	.010	1.71	.003	1.83	.005	30	M3 III	15,1075,2012,2038
		Sk −66 167		5 36 00	−66 02.	1	13.80		−0.23		−0.97		4		798
		AAS21,109 # 234		5 36 00	−67 03.	1	13.35		−0.25		−1.06		3		952
		Sk −67 221		5 36 00	−67 31.	1	12.87		−0.02		−0.84		4		798
		Sk −69 208		5 36 00	−68 31.	1	12.77		−0.09		−0.87		3		952
		Brey 60		5 36 00	−69 01.	1	14.70		0.21		−0.71		4	WN3	980
		Brey 61		5 36 00	−69 01.	1	15.66		−0.02		−0.96		4	WN3	980
		Sk −69 206		5 36 00	−69 09.	1	12.84		0.14		−0.76		3		952
		Brey 57		5 36 00	−69 13.	1	13.44		0.31		−0.56		2	WN6	980
		Brey 58		5 36 00	−69 14.	1	12.40		0.69		−0.38		3	WN5	980
		AAS21,109 # 232		5 36 00	−69 43.	1	12.10		−0.03		−0.90		3		417
294279				5 36 01	−02 56.5	1	10.72		0.39		0.03		4	A2	808
		HRC 172, TX Ori		5 36 02	−02 46.3	1	13.71		1.37		1.13		1	K2:e	825
269863				5 36 02	−66 15.5	1	11.71		−0.04		−0.79		6	B2 Ia	149
		NGC 2058 sq1 8		5 36 02	−70 16.5	1	13.06		0.56		0.09		3		970
37352	+30 00966			5 36 03	+30 07.4	2	7.70	.011	0.11	.006	−0.15		10	A0	1025,1499
37455	−5 01348			5 36 03	−05 09.1	1	9.56		0.16		0.07		2	A3 V	1060
269854	−67 00477	Sk −67 222		5 36 03	−67 05.8	1	11.19		−0.01		−0.66		3	B5 Ia	149
269845	−67 00478	Sk −67 220		5 36 03	−67 29.3	3	11.79	.007	−0.02	.013	−0.80		8	B2.5Ia	149,1277,8033
294271	−2 01324	IDS05335S0238	A	5 36 04	−02 35.9	1	7.91		−0.11		−0.56		2	B5 V	1089
294272	−2 01323	IDS05335S0238	B	5 36 04	−02 35.9	1	8.48		0.03		−0.05		4		808
294272	−2 01323	IDS05335S0238	C	5 36 04	−02 35.9	1	8.77		−0.04		−0.30		2		808
	−6 01273			5 36 05	−06 27.0	3	10.27	.056	0.30	.010	−0.02	.025	6	B8	1004,1060,8055
269818		Sk −69 207		5 36 05	−69 12.4	1	13.32		0.29		−0.57		3	WC5	980
290744				5 36 06	−00 42.	1	11.16		0.33		0.13		2		1060
294272	−2 01323	IDS05335S0238	BC	5 36 06	−02 34.9	1	7.85		−0.01		−0.18		3	B8 V	808
		Orion P 2694		5 36 06	−05 48.4	1	12.26		0.80		0.44		1		8114
		AAS21,109 # 233		5 36 06	−70 23.	1	13.88		0.06		−0.98		2		952
37388	+23 00981	IDS05331N2314	A	5 36 07	+23 15.7	1	8.90		0.53		0.06		3	F8	3026
37388	+23 00981	IDS05331N2314	B	5 36 07	+23 15.7	1	9.63		0.51		0.02		3	F8	3026
37388	+23 00981	IDS05331N2314	D	5 36 07	+23 15.7	1	11.12		1.49		1.51		3		3024
37367	+29 00947	HR 1924	⋆	5 36 07	+29 11.3	1	5.95		0.16		−0.51		2	B2 IV-V	1223
		Orion P 2697		5 36 07	−06 00.2	1	13.33		0.62		−0.02		1		8114
37530	−27 02395			5 36 07	−27 14.3	1	6.53		1.56				4	K4/5 III	2035
37470	−6 01274			5 36 08	−06 11.5	4	8.23	.007	0.04	.015	−0.24	.011	10	B8 p(Si)	1004,1060,1732,8055
		HRC 174	V	5 36 08	−07 52.0	1	13.15		1.17		0.59		1	K5e	825
		Steph 574		5 36 09	+61 10.8	1	11.27		1.91		2.21		1	M0	1746
269809	−69 00407			5 36 09	−69 53.0	2	11.82	.005	0.26	.000	0.29		5	F0 Ia	149,1483
37366	+30 00968	LS V +30 011		5 36 11	+30 51.8	2	7.62	.029	0.10	.005	−0.75		5	O9.5V	401,1025
37339	+37 01275			5 36 11	+37 57.7	1	6.96		0.00		−0.33		2	B9	401
37469	−4 01200			5 36 11	−04 42.3	3	9.62	.000	0.20	.013	0.03	.019	5	A0	1004,1060,8055
		Orion P 2705	V	5 36 11	−04 55.1	1	12.02		0.69		0.44		1		8114
37387	+23 00982	IDS05331N2314	C	5 36 12	+23 17.8	2	7.50	.045	1.92	.072	1.98		5	K1 Ib	138,206
37481	−6 01275	HR 1933		5 36 12	−06 36.1	6	5.95	.008	−0.22	.008	−0.90	.009	17	B1.5IV	15,1004,1060,1061,1732,2032
		Brey 63		5 36 12	−68 55.	1	14.59		−0.21		−1.12		3	WN4	980
37468	−2 01326	HR 1931	⋆ AB	5 36 14	−02 37.6	11	3.80	.021	−0.24	.008	−1.02	.013	37	O9.5V+B0.5V	15,154,490,1004,1075*
37468	−2 01326	IDS05337S0239	C	5 36 14	−02 37.6	1	8.79		−0.02		−0.25		1		1206
37468	−2 01326	IDS05337S0239	D	5 36 14	−02 37.6	2	6.81	.048	−0.18	.017	−0.81	.004	6	B2 V	490,808
		Orion P 2713		5 36 15	−04 30.3	1	14.16		0.93		−0.28		1		8114
37467	+2 01028			5 36 16	+02 50.1	1	7.89		−0.10		−0.41		1	B7 IV-V	160
37479	−2 01327	HR 1932, V1030 Ori	⋆ E	5 36 16	−02 37.3	7	6.66	.031	−0.18	.014	−0.87	.014	14	B2 Vp	154,490,808,1089,1206*

Table 1   295

HD	DM	Other Id	N	Rem	α_{1950}	δ_{1950}	S	V	σ_V	B–V	σ_{B-V}	U–B	σ_{U-B}	n	Spectrum	References
37386	+29 00949	HH Aur			5 36 18	+29 48.4	2	8.54	.019	1.28	.014			4	G5 IV	1025,1724
		Sk -67 224			5 36 18	−67 10.	2	13.02	.000	-0.19	.020	-0.99	.025	7		415,952
		AAS21,109 # 236			5 36 18	−67 30.	2	13.53	.005	-0.03	.000	-0.85	.019	4		417,952
		AAS21,109 # 237			5 36 18	−67 40.	1	13.98		-0.12		-1.02		2		417
		Sk -68 127			5 36 18	−68 31.	1	13.70		-0.16		-0.94		3		952
		Brey 64			5 36 18	−69 01.	1	13.26		0.04		-0.79		3	WN9	980
		Sk -69 210			5 36 18	−69 04.	2	12.59	.005	0.35	.014	-0.60	.014	4		360,952
		Sk -69 209			5 36 18	−69 33.	1	11.79		-0.06		-0.90		4		415
245869	+30 00969				5 36 19	+30 07.1	1	10.04		0.52				2	F8	1025
37480	−4 01201				5 36 21	−04 17.9	2	9.21	.000	0.01	.000	-0.06		2	A0	1004,1060
		Orion P 2731			5 36 21	−05 09.9	1	12.46		1.84		1.50		1		8114
269828	−69 00409	IDS05367S6915	⋆ AB		5 36 21	−69 13.6	1	11.25		0.01		-0.82		4	WN/0f?	980
		RR Tau			5 36 24	+26 20.8	3	10.62	.193	0.49	.016	0.19	.023	10	A2II-IIIe	351,601,1655
		LS V +35 024			5 36 24	+35 52.3	1	10.79		0.82		-0.30		5	O9 V	342
		AAS21,109 # 235			5 36 24	−70 53.	1	14.65		-0.02		-0.98		2		417
37439	+21 00918	HR 1929			5 36 27	+21 44.2	2	6.43		0.07	.005	0.06	.029	5	A1 Vn	1049,1733
		Orion P 2737			5 36 27	−04 01.1	1	10.96		0.76		0.34		1		1519
	−5 01350	Orion P 2740			5 36 27	−05 11.7	1	11.07		0.50		0.05		1		8114
	−42 02048				5 36 27	−42 21.2	1	9.29		1.41		1.44		1	K5	565
290722	+0 01144				5 36 28	+00 21.8	1	10.28		0.19		0.11		3	A5	808
37507	−7 01142	HR 1937			5 36 28	−07 14.4	6	4.80	.011	0.14	.012	0.11	.008	23	A4 V	15,1075,1425,2012*
269832	−69 00414	Sk -69 211			5 36 28	−69 25.8	1	10.36		0.09		-0.58		5	A0 Iae:	149
37384	+37 01277	IDS05330N3754	AB		5 36 29	+37 55.9	1	7.32		0.10		-0.18		1	B9	401
		PKS 0537-441 # 4			5 36 29	−44 06.1	1	12.79		1.04		0.91		4		1687
269840	−69 00412	Sk -68 128			5 36 29	−68 57.5	1	10.32		0.42		0.21		6	F3 Ia	149
		Sk -69 212			5 36 29	−69 14.	1	12.23		-0.06		-0.91		4		415
		Orion P 2752			5 36 30	−06 26.3	1	10.91		0.73		0.31		1		1519
37655	−43 01954				5 36 30	−42 59.7	4	7.43	.007	0.60	.006	0.07	.000	20	G0 V	258,1075,1499,2033
		PKS 0537-441 # 5			5 36 30	−44 05.5	1	12.81		0.62		0.13		4		1687
		AAS21,109 # 240			5 36 30	−66 28.	1	14.00		-0.28		-0.93		3		417
		AAS21,109 # 242			5 36 30	−66 28.	2	14.23	.070	-0.16	.019	-0.92	.066	4		417,952
		AAS21,109 # 238			5 36 30	−68 57.	1	13.37		-0.03		-0.82		2		417
37365	+39 01373				5 36 31	+39 51.4	1	8.40		0.07		-0.27		2	A2	401
37525	−2 01328				5 36 31	−02 40.6	2	8.08	.004	-0.09	.003	-0.58	.019	3	B5 V	1089,1732
37835	−63 00466				5 36 31	−63 11.0	1	10.38		0.16		0.10		2	A2 V	849
37490	+4 01002	HR 1934, ω Ori			5 36 33	+04 05.7	7	4.58	.023	-0.11	.007	-0.76	.004	23	B3 IIIe	15,154,1004,1212,1363*
37524	−2 01329				5 36 35	−02 01.7	1	8.74		-0.04		-0.23		2	B9	1089
37526	−5 01351				5 36 35	−05 13.3	3	7.60	.005	-0.13	.010	-0.54	.020	5	B3 V	1004,1060,8055
37527	−6 01277				5 36 35	−06 31.5	1	8.67		0.15		0.09		2	A0 V	1060
		Sk -69 213			5 36 35	−69 13.	1	11.97		0.10		-0.75		4	B1 Ia	415
37423	+31 01040				5 36 36	+31 11.9	2	8.49	.000	0.46	.005			4	F8	1025,1724
290768					5 36 36	−01 32.	1	10.41		0.56		0.35		2		1060
37547	−6 01278				5 36 36	−06 06.8	3	9.30	.023	-0.02	.014	-0.09	.010	6	A0	1004,1060,8055
		AAS21,109 # 243			5 36 36	−66 29.	2	14.27	.000	-0.15	.034	-0.96	.019	3		417,952
		AAS21,109 # 244			5 36 36	−66 29.	2	14.31	.000	-0.13	.005	-0.92	.009	4		417,952
		Sk -69 214			5 36 36	−69 34.	1	12.19		0.01		-0.84		4	B0.7Ia	415
37438	+25 00902	HR 1928			5 36 38	+25 52.3	5	5.17	.008	-0.15	.002	-0.69	.003	14	B3 IV	154,601,1203,1223,1363
37545	−3 01165				5 36 38	−02 58.2	1	9.31		-0.02		-0.15		2	B9	1089
	−5 01352				5 36 39	−04 57.6	2	10.44	.013	0.62	.022	0.43	.009	5		1060,8055
		LP 3 - 297			5 36 41	+86 03.4	1	10.91		0.69		0.00		2		1726
269868					5 36 41	−67 42.3	1	12.34		0.50		0.37		6	F5:I	149
		AN295,47 # 34			5 36 42	+26 12.4	1	11.44		0.78		0.13		1	G1 V	601
37328	+52 00983				5 36 42	+52 32.7	1	7.21		0.97		0.66		3	G5	1566
290761					5 36 42	−01 21.	1	11.38		0.50		0.32		2		1060
		AAS21,109 # 246			5 36 42	−66 28.	1	13.66		-0.08		-0.91		3		417
		AAS21,109 # 245			5 36 42	−66 30.	1	13.70		-0.26		-1.09		3		417
		Sk -68 130			5 36 42	−68 20.	1	13.37		-0.11		-0.89		2		952
		Sk -68 129			5 36 42	−69 00.	1	12.77		0.03		-0.84		3		798
269846	−69 00417	Sk -69 215			5 36 42	−69 08.5	1	11.63		0.07		-0.80		3	F5	415
		AAS21,109 # 241			5 36 42	−69 14.	1	12.16		-0.01		-0.87		3		417
		AAS21,109 # 239			5 36 42	−70 27.	1	13.89		-0.13		-1.04		3		952
		G 191 - 50			5 36 43	+55 42.0	1	14.21		1.48				3		940
246067	+3 00996	G 99 - 21			5 36 44	+03 56.4	3	10.36	.004	0.64	.004	0.04	.004	4	G0	333,927,1620,1696
37453	+30 00970	IDS05335N3002	AB		5 36 44	+30 04.2	2	8.09	.015	0.80	.029			5	F4 III	1025,1724
37564	−2 01330				5 36 44	−02 33.2	1	8.46		0.23		0.15		3	A8 V	808
	−0 01038				5 36 45	+01 00.6	1	10.54		0.25		0.11		2		1060
37706	−46 01936	IDS05354S4609	AB		5 36 45	−46 07.6	2	7.33	.000	0.78	.000			8	G6 V	1075,2012
		AN295,47 # 27			5 36 46	+26 18.0	1	11.68		0.78		0.26		1	G7 V	601
	+47 01184				5 36 46	+47 05.8	1	9.43		0.39		-0.04		3	F2	1601
294281					5 36 46	−03 09.1	1	10.42		0.35		0.25		4	A0	808
245975	+25 00904				5 36 47	+25 54.5	1	10.09		1.12				2		1724
37466	+24 00909				5 36 48	+24 11.9	2	7.11	.005	0.35	.005	-0.06		4	F3 V	1625,1724
37437	+31 01043				5 36 48	+31 52.4	2	8.05	.020	0.04	.000	-0.33		5	B9	401,1025
290743	−0 01039				5 36 48	−00 46.9	1	10.61		0.34		0.14		2		1060
290758					5 36 48	−01 12.	1	10.95		0.34		0.22		2		1060
269889		Sk -66 169			5 36 48	−66 40.	1	12.56		-0.13		-1.00		4	O9.7Ia+	415
269879		Sk -66 168			5 36 48	−66 47.5	2	10.89	.040	1.08	.000	0.85		10	G2 Ia	149,1483
		AAS21,109 # 247			5 36 48	−67 37.	1	13.63		0.08		-0.79		2		417
245906	+26 00887				5 36 50	+26 18.1	1	10.47		0.42		0.11		1	B8 V	601
37974	−69 00420	Sk -69 216			5 36 50	−69 24.6	4	10.92	.012	0.18	.016	-0.93	.010	11	Be	149,328,1277,8033

HD	DM	Other Id	N Rem	α_{1950}	δ_{1950}	S	V	σ_V	B−V	σ_{B-V}	U−B	σ_{U-B}	n	Spectrum	References
		AN295,47 # 30		5 36 51	+26 21.0	1	10.66		0.36		0.44		1	A3 V	601
269857	−68 00397	Sk −68 131		5 36 51	−68 55.8	1	10.29		0.35		0.11		2	A9 Ia	149
245972	+30 00971			5 36 52	+30 12.5	1	9.67		0.42				3	A5	1025
290740				5 36 54	−00 41.	1	11.66		0.23		0.18		2		1060
37577	−1 00994			5 36 54	−01 36.5	2	9.28	.020	0.06	.010	-0.05	.010	5	B9	1060,1089
		Sk −66 170		5 36 54	−66 23.	1	12.88		-0.09		-0.96		5		798
37935	−66 00439	HR 1960		5 36 55	−66 35.3	3	6.29	.010	-0.06	.004	-0.14	.005	22	B9.5Ve	15,1075,2038
245971	+31 01044			5 36 57	+31 20.7	1	9.42		0.21				3	B3	1025
		AN295,47 # 29		5 36 58	+26 19.5	1	11.07		0.61		0.12		1	G0 V	601
288151				5 36 59	+01 01.	1	10.96		0.37		0.22		3	A7	808
	−38 02136			5 36 59	−38 22.8	1	9.55		1.05		0.86		1	K5 V	3072
		Sk −65 070		5 37 00	−65 49.	1	13.92		-0.27		-1.08		4		798
		Sk −66 171		5 37 00	−66 41.	1	12.19		-0.15		-1.02		4		415
		Sk −67 225	V	5 37 00	−67 48.	1	13.60		-0.14		-0.92		1		798
		Sk −68 132		5 37 00	−68 25.	1	13.97		-0.22		-0.98		3		952
		Sk −69 217		5 37 00	−69 31.	1	11.97		-0.11		-0.99		4		415
		Sk −69 218		5 37 00	−69 32.	1	12.00		-0.11		-0.98		4		415
269839	−70 00417	NGC 2058 sq1 23		5 37 00	−70 19.5	1	10.43		0.54				2		970
	−0 01040			5 37 01	−00 25.0	1	10.19		0.17		0.10		2	A5 V	1060
37594	−3 01166	HR 1940		5 37 02	−03 35.5	2	5.99	.005	0.28	.005	-0.01	.015	7	A8 Vs	15,1061
		Sk −69 216a		5 37 02	−69 51.	1	13.62		-0.19		-1.02		3		952
290723				5 37 04	+00 04.	1	11.40		0.50		0.31		3		808
37643	−17 01199	HR 1944	★ AB	5 37 04	−17 52.6	2	6.38		-0.10	.009	-0.41		6	B7 V	1079,2031
269860	−69 00422	Sk −69 219		5 37 04	−69 10.4	2	11.61	.056	0.41	.009	0.31	.042	4	F0 Ia	149,952
37622	−11 01251			5 37 05	−11 14.2	2	7.98	.025	-0.09	.000	-0.73	.013	7	B3 Vn	55,399
290739				5 37 06	−00 41.	1	11.44		0.31		0.17		2		1060
37607	−1 00995			5 37 06	−01 23.5	1	9.29		0.03		-0.08		2	B9.5V	1060
		Sk −66 173		5 37 06	−66 39.	1	13.13		-0.23		-1.07		4		798
		Sk −67 226		5 37 06	−67 47.	1	13.43		-0.14		-1.02		3		798
		AAS21,109 # 248		5 37 06	−69 38.	1	13.20		-0.13		-0.98		4		417
245986	+32 01061			5 37 07	+32 32.6	1	9.00		0.41				3	F5	1025
37717	−40 01999	HR 1947		5 37 07	−40 44.1	1	5.81		-0.09				4	B8 V	2035
269858	−69 00427	IDS05375S6933	★ AB	5 37 07	−69 31.4	1	11.16		-0.07				3		8033
		NGC 2058 sq1 4		5 37 07	−70 08.9	1	12.39		0.67		0.25		4		970
37592	+0 01146			5 37 08	+00 52.9	1	8.35		-0.06		-0.30		1	A0	1089
37635	−9 01197	HR 1942		5 37 08	−09 44.0	4	6.50	.015	-0.11	.018	-0.49	.017	11	B7 V	15,154,247,1078
		PKS 0537-441 # 6		5 37 08	−44 05.0	1	14.48		0.63		0.13		4		1687
		HRC 177, V510 Ori		5 37 09	−02 32.9	1	13.21		0.48		-0.81		1		825
246080	+24 00912			5 37 11	+24 28.3	1	9.76		0.46				2		1724
37538	+24 00911			5 37 11	+24 56.4	1	9.26		0.45				2		1724
		NGC 2058 sq1 6		5 37 11	−70 11.1	1	12.52		0.55		0.05		3		970
		Sk −66 175		5 37 12	−66 20.	1	13.03		-0.18		-1.00		5		798
		Sk −66 174		5 37 12	−66 24.	1	13.33		-0.18		-0.99		4		798
38030		Sk −69 222		5 37 12	−69 28.	1	13.00		-0.22		-0.84		4	WC5	980
269859	−69 00428	Sk −69 221		5 37 12	−69 31.5	2	10.55	.075	-0.06	.000	-0.90		6	B1 Ia	149,8033
269869	−68 00401			5 37 13	−68 53.6	1	10.30		1.07		0.95		2	M0	149
37606	+1 01088			5 37 14	+01 27.9	1	6.90		-0.07				1	B8 V	1004
37558	+21 00923			5 37 14	+21 15.6	1	8.54		0.41				2		1724
37539	+24 00913			5 37 14	+24 30.7	1	7.12		1.06				2	K0	1724
		Orion P 2840		5 37 15	−06 46.4	1	11.09		0.43		0.06		2		1060
37653	−19 01231			5 37 15	−19 48.8	1	8.24		0.89		0.50		46	G8 IV/V	978
		Sk −68 133		5 37 15	−68 48.	1	13.13		-0.10		-0.98		4		415
37633	−2 01332	V1147 Ori		5 37 16	−02 42.2	1	9.04		0.03		-0.36		3	B9	1089
37634	−5 01355			5 37 16	−05 04.9	1	8.91		0.64		0.08		2	G5	1064
246132	+12 00853	G 102 - 20		5 37 17	+12 09.9	3	10.22	.008	0.65	.008	-0.02	.010	5	G5	516,1620,1696
37394	+53 00934	HR 1925	★ A	5 37 17	+53 27.8	8	6.22	.014	0.84	.006	0.51	.006	28	K1 V	15,22,1013,1197,1355,1758*
38029	−69 00429	Sk −69 223		5 37 17	−69 13.2	1	11.59		0.11		-0.73		4	WC5	980
		Sk −66 177		5 37 18	−66 43.	1	12.87		-0.10		-0.91		4		798
		Sk −66 176		5 37 18	−66 59.	1	13.59		-0.23		-1.03		4		798
		Sk −69 225		5 37 18	−69 18.	1	12.17		0.20		-0.15		3		952
		PKS 0537-441 # 2		5 37 20	−44 03.1	1	13.20		0.59		0.09		4		1687
37605	+5 00985	G 99 - 22		5 37 21	+06 02.3	1	8.67		0.82		0.56		1	K0	333,1620
		G 100 - 28	AB	5 37 21	+24 46.8	2	14.87	.043	1.88	.000			10		538,906
37519	+31 01048	HR 1938		5 37 22	+31 20.0	6	6.04	.009	0.04	.012	-0.21	.013	29	B9.5III-IV	15,1025,1049,1079*
		Sk −67 227		5 37 24	−67 51.8	1	12.87		0.03		-0.39		3	A2 Ia	149
		Sk −69 226		5 37 24	−69 32.	1	11.94		-0.11		-0.92		3		415
		NGC 2058 sq1 5		5 37 24	−70 09.4	1	11.93		0.93		0.63		5		970
246099	+25 00909			5 37 25	+25 58.1	1	9.38		0.33				2		1724
37557	+28 00846			5 37 25	+28 57.1	3	7.03	.009	1.15	.003	0.96		15	K0	1025,1499,1724
246047	+32 01062			5 37 25	+32 04.9	2	8.95	.034	0.15	.010	-0.50		5	B9	401,1025
37289	+65 00485	HR 1916		5 37 25	+65 40.4	1	5.60		1.25				2	gK5	71
37641	−1 00997			5 37 25	−01 57.2	3	7.56	.012	-0.06	.004	-0.38	.000	7	B7 V	1004,1060,1089
37620	+10 00841			5 37 26	+10 13.6	1	7.93		0.36		-0.06		2	F0	1733
246114	+21 00924			5 37 26	+21 42.3	1	9.20		1.02				2		1724
37642	−3 01167	V1148 Ori		5 37 26	−03 21.4	2	8.06	.000	-0.14	.004	-0.58	.022	5	B9	1089,8055
37536	+31 01049	HR 1939, NO Aur		5 37 27	+31 53.7	4	6.16	.037	2.10	.019	2.19	.055	9	M2 Iab	148,1025,3001,8032
233153	+53 00935	G 191 - 51	★ B	5 37 27	+53 28.3	3	9.78	.022	1.47	.011	1.16		4	M1	694,1017,3072
		Sk −69 228		5 37 29	−69 22.	1	12.12		0.07		-0.76		4	B1.5Ia	415
246070	+31 01050			5 37 30	+31 51.4	2	9.27	.035	0.19	.015	-0.53		6	B5	401,1025
37663	−6 01281			5 37 31	−06 48.1	2	9.19	.000	-0.02	.000	-0.09		3	A0	1004,1060

Table 1

HD	DM	Other Id	N Rem	α_{1950}	δ_{1950}	S	V	σ_V	B–V	σ_{B-V}	U–B	σ_{U-B}	n	Spectrum	References
		PKS 0537-441 # 3		5 37 31	−44 02.9	1	14.08		0.67		0.16		4		1687
		Sk -69 227		5 37 31	−69 40.	1	12.23		0.04		-0.81		4		415
37575	+27 00833			5 37 32	+27 45.6	1	8.15		0.38				2	F2/3 IV/V	1724
37660	−1 00999	LS VI -01 002		5 37 32	−01 27.1	2	6.99	.014	0.61	.000	0.55	.047	4	A2	490,808
37662	−1 00998			5 37 32	−01 44.3	1	8.56		0.13		0.13		2	A0	1060
		G 98 - 20		5 37 33	+29 10.7	1	12.08		0.92		0.60		1	K3	333,1620
290712				5 37 34	+00 37.	1	11.63		0.15		-0.03		4	A2	808
	−44 02164			5 37 35	−44 08.2	1	10.68		0.49		0.06		4		1687
		Sk -69 229		5 37 36	−69 32.	1	12.66		-0.09		-0.93		4		415
246108	+31 01051			5 37 37	+31 21.4	1	9.69		0.19				3	A0	1025
	+35 01201	LS V +35 025		5 37 38	+35 49.3	2	10.54	.000	0.88	.005	-0.19	.005	7	O9 V	342,684
246128	+26 00899			5 37 39	+26 58.4	1	9.02		0.60				3		1724
37574	+32 01064			5 37 41	+32 52.2	1	6.73		0.45				2	F8	1724
37659	+0 01150			5 37 42	+00 56.8	1	8.88		0.10		0.05		3	A1.5V	808
37602	+25 00910			5 37 42	+25 14.4	1	8.54		1.04				2	K0	1724
37674	−1 01001			5 37 42	−01 29.3	2	7.68	.005	-0.08	.000	-0.63	.050	4	B3 Vn	1060,1089
37686	−2 01335			5 37 42	−02 32.4	1	9.23		0.02		-0.09		2	B9	1089
		Sk -66 178		5 37 42	−66 43.	1	12.07		-0.11		-0.97		4		415
		AAS21,109 # 249		5 37 42	−69 01.	1	13.32		-0.01		-0.95		2		417
		AAS21,109 # 250		5 37 42	−69 12.	1	13.40		0.31		-0.71		2		417
246144	+29 00952	IDS05345N2902	AB	5 37 43	+29 03.9	1	9.15		0.56				2		1724
290798	−0 01046			5 37 44	−00 46.1	2	10.40	.013	0.39	.004	0.30	.013	5	A2	1060,8055
		Orion P 2909		5 37 47	−05 36.7	1	10.89		0.71		0.28		1		1519
	−5 01356			5 37 47	−05 37.2	1	10.44		0.71		0.50		2	A2	1060
		Sk -68 134		5 37 48	−68 53.	1	13.71		-0.01		-0.85		3		952
		AAS21,109 # 251		5 37 48	−68 59.	2	13.82	.005	-0.08	.023	-0.96	.014	4		417,952
		Sk -69 230		5 37 48	−69 28.	1	12.09		0.12		-0.32		3		952
37699	−2 01336			5 37 49	−02 27.7	1	7.62		-0.13		-0.69		2	B5	1089
269900	−67 00484	Sk -67 228		5 37 49	−67 44.8	1	11.49		-0.05		-0.82		3	B1 Ia	149
294301				5 37 50	−02 41.8	1	11.04		0.40		-0.02		4	G0	808
37687	−3 01168			5 37 50	−03 27.1	1	7.05		0.03		-0.44		2	B8	1089
37795	−34 02375	HR 1956	★ A	5 37 50	−34 06.0	9	2.65	.010	-0.12	.004	-0.41	.025	70	B7 IVe	15,200,681,815,1075*
38043	−64 00464			5 37 51	−64 56.2	2	9.42	.015	0.25	.010	0.00		3	A/Fw pec	1594,6006
290799	−0 01047			5 37 52	−00 47.6	1	10.63		0.21		0.06		2		1060
		NGC 2058 sq1 11		5 37 52	−70 19.7	1	12.94		0.97		0.72		4		970
246139	+32 01065			5 37 53	+32 42.1	1	9.45		0.06				4	B8	1025
	+35 01204			5 37 53	+35 57.3	1	10.11		1.87				3	M2	369
290813				5 37 54	−01 48.	1	11.00		0.51		0.20		2		1060
		Sk -69 231		5 37 54	−69 23.	1	13.80		0.05		-1.00		3	WC5	980
		G 99 - 23		5 37 57	+04 32.9	2	12.44	.350	1.48	.025	1.22	.065	2	M2	333,1620,3056
246178	+28 00848			5 37 57	+28 11.6	1	10.45		0.50				2	F2	1724
37617	+33 01120			5 37 57	+33 53.7	1	6.91		1.05				3	K0	1724
37587	+38 01249			5 37 57	+38 53.0	1	8.38		0.30		0.24		2	A3 V	105
37514	+48 01297	IDS05341N4825	A	5 37 57	+48 26.4	1	7.86		0.14		0.10		2	A2	1733
37700	−4 01210			5 37 57	−04 26.8	3	8.01	.039	-0.10	.010	-0.47	.010	5	B6 V	1004,1060,8055
246197	+26 00901			5 37 58	+26 06.2	1	8.81		0.38				2		1724
37811	−32 02479	HR 1958		5 37 59	−32 39.3	5	5.44	.003	0.91	.004			38	G6/8 III	15,1075,2013,2029,8033
		AAS21,109 # 253		5 38 00	−69 03.	1	13.88		-0.13		-1.02		3		417
		Brey 71		5 38 00	−69 10.	1	13.96		0.09		-0.66		4	WN7	980
269883		Sk -69 233		5 38 00	−69 10.	1	11.46		0.14		-0.74		3	B1 I	980
		Sk -69 232		5 38 00	−69 44.	1	13.33		-0.06		-0.86		4		798
		AAS21,109 # 252		5 38 00	−70 15.	1	13.19		-0.09		-0.71		3		952
37648	+24 00918			5 38 01	+24 19.5	1	7.60		0.47				2	F8	1724
294319	−4 01211			5 38 01	−04 23.0	3	10.26	.007	0.14	.004	0.10	.008	7	A2 V	1060,8055,8055
		HRC 179	V	5 38 03	−08 08.7	1	13.79		1.39		0.62		1		825
37614	+38 01250		AB	5 38 05	+38 09.8	2	8.19	.005	0.12	.005	-0.62	.009	4	B2 III	401,1012
37639	+31 01056			5 38 06	+31 09.7	2	7.45	.010	-0.02	.005	-0.08		5	A0	401,1025
290787	−0 01050			5 38 06	−00 19.0	2	10.14	.009	0.58	.009	0.12	.004	5		1060,8055
		Orion P 2942		5 38 06	−04 30.8	2	12.24	.004	0.71	.013	0.51	.063	5		1060,8055
269932		Sk -66 179		5 38 06	−66 20.	1	12.74		-0.19		-1.04		4	O8	415
		Brey 73		5 38 06	−69 11.	1	12.11		0.13		-0.70		4	WN4	980
269888		Sk -69 234		5 38 06	−69 16.	1	14.78		0.00		-0.48		3	WC5	980
37744	−2 01337	HR 1950	★	5 38 07	−02 51.0	7	6.21	.009	-0.21	.007	-0.90	.017	21	B1.5V	15,154,490,1004,1089*
37747	−7 01148			5 38 08	−07 41.1	1	7.42		1.16		1.13		3	K0	1657
269896	−68 00406	Sk -68 135		5 38 08	−68 56.7	3	11.34	.017	0.01	.012	-0.88	.029	12	O9.5I	149,328,8033
269891	−69 00442	Sk -69 235		5 38 09	−69 06.7	1	11.34		0.17				3	B0:	8033
37646	+29 00953	HR 1945	★ A	5 38 10	+29 27.8	2	6.43	.005	-0.10	.010	-0.39		5	B8 IV	1025,1084
37647	+29 00954	IDS05350N2926	B	5 38 10	+29 28.2	2	7.17	.010	-0.05	.010	-0.14		5	A0 V	1025,1084
		PASP87,769 T3# 1		5 38 10	−64 09.4	1	12.61		0.63		0.20		4		431
	+16 00840	DX Tau		5 38 11	+16 34.3	1	10.33		1.97				2	M3	369
37745	−4 01212			5 38 11	−03 57.5	2	9.20	.000	0.02	.000	-0.07		2	A0	1004,1060
		Sk -69 236		5 38 12	−69 26.	1	12.43		0.02		-0.81		4		415
37684	+19 01014			5 38 14	+20 02.3	1	8.12		1.73		1.97		3	K5	1648
37742	−2 01338	HR 1948	★ AB	5 38 14	−01 58.0	13	1.76	.012	-0.21	.004	-1.06	.002	49	O9.5Ibe	15,200,1004,1011,1034*
		G 96 - 48		5 38 15	+38 18.2	1	11.99		0.54		-0.09		2		1620
		PASP87,769 T3# 2		5 38 16	−64 10.3	1	12.92		0.54		0.12		2		431
246254	+27 00837			5 38 18		1	8.91		0.35				2		1724
37756	−1 01004	HR 1952		5 38 18	−01 09.2	8	4.94	.012	-0.22	.005	-0.83	.013	25	B2 IV-V	15,154,1004,1060,1075,1089*
		Sk -67 229		5 38 18	−67 39.	1	13.38		-0.06		-0.57		4		798
246323	+12 00857			5 38 19	+12 37.6	1	9.61		0.31		0.26		1		1722

Catalogue of mean UBV data

HD	DM	Other Id	N Rem	α_{1950}	δ_{1950}	S	V	σ_V	B–V	σ_{B-V}	U–B	σ_{U-B}	n	Spectrum	References
294326	−4 01213			5 38 20	−04 45.4	1	10.10		0.35		0.15		2	A7 V	1060
		PASP87,769 T3# 3		5 38 22	−64 09.9	1	13.70		0.69		0.26		1		431
37711	+16 00841	HR 1946	⋆AB	5 38 24	+16 30.6	5	4.86	.009	−0.13	.010	−0.63	.004	13	B3 V +B4 V	15,154,1223,1363,8015
37776	−1 01005	LS VI −01 003		5 38 24	−01 31.9	3	6.98	.010	−0.14	.005	−0.86	.000	7	B2 IV	1004,1060,1089
294304	−2 01342			5 38 24	−02 45.	1	10.03		0.29		−0.14		3	B8	808
		Sk −66 180		5 38 24	−66 37.	1	13.99		−0.24		−1.00		4		798
		Sk −69 237		5 38 24	−69 24.	1	12.08		−0.03		−0.86		4	B1 Ia	415
		Sk −69 238		5 38 24	−69 29.	1	12.21		−0.12		−0.97		4		415
37808	−10 01258	HR 1957, V1051 Ori		5 38 25	−10 26.0	3	6.50	.032	−0.15	.014	−0.48	.025	11	B9.5IIIp	220,1079,2035
37710	+17 00979			5 38 28	+17 29.9	2	6.95	.049	1.87	.015	2.02		6	M1	369,1733
37789	−1 01006			5 38 28	−01 11.0	1	8.80		0.10		0.09		2	A2	1060
246372	+12 00858			5 38 30	+12 27.5	1	10.03		0.79		0.41		1		1722
37788	+0 01152	HR 1955		5 38 31	+00 18.8	2	5.92	.005	0.31	.005	0.02	.015	7	F0 IV	15,1061
37847	−20 01149	TW Lep		5 38 31	−20 19.4	1	7.67		1.07		0.67		1	K0 IV	1641
269902	−69 00446	Sk −69 239		5 38 31	−69 07.9	2	10.23	.015	0.36	.010	−0.38		7	B9 I	415,8033
269927	−69 00468	Brey 77	⋆AB	5 38 31	−69 07.9	1	10.96		0.12		−0.76		3	WN+O?	980
246292	+29 00955			5 38 32	+29 41.7	1	9.62		0.31				3	B5	1025
37805	−2 01343			5 38 32	−02 19.7	1	7.51		0.29		0.04		3	F0 V	808
37806	−2 01344			5 38 32	−02 44.5	1	7.90		0.03		−0.26		2	A0	1089
38112	−64 00465			5 38 32	−64 07.1	3	9.40	.014	1.06	.028	0.98	.003	9	K0 II/III	357,358,431
37709	+25 00918			5 38 33	+25 29.4	1	8.45		0.03		−0.62		2	B3 V	105
37671	+34 01142			5 38 34	+34 52.9	1	8.27		1.01				3	K0	1724
37828	−11 01258			5 38 34	−11 13.4	2	6.87	.005	1.13	.005	0.86		4	K0	742,1594
37670	+35 01207			5 38 35	+35 36.4	1	6.88		0.02		−0.01		2	A0	1601
246316	+23 01005			5 38 36	+23 26.9	1	9.06		1.14				3		1724
37846	−8 01197			5 38 36	−08 04.7	1	8.05		0.08		0.08		2	A0	401
		Sk −69 239a		5 38 36	−69 25.	1	12.97		0.01		−0.79		4		415
246315	+24 00919			5 38 37	+24 16.3	1	9.37		0.29				2		1724
246294	+28 00850			5 38 37	+28 05.6	1	9.56		0.63				3		1724
		PASP87,769 T3# 5		5 38 37	−64 05.9	1	12.56		0.52		0.05		4		431
290775	+0 01153			5 38 38	+00 38.4	1	10.32		0.29		0.24		3	A2	808
246289	+32 01067			5 38 39	+32 27.5	1	9.33		0.31				3	A3	1025
37807	−3 01171			5 38 39	−03 39.4	4	7.90	.011	−0.11	.008	−0.63	.014	7	B8	1004,1060,1089,8055
38602	−78 00195	HR 1991, ι Men		5 38 39	−78 50.9	4	6.04	.003	−0.01	.006	−0.32	.016	14	B8 III	15,258,1075,2038
37754	+13 00954	IDS05358N1306	AB	5 38 40	+13 07.4	1	9.23		0.18		−0.12		2		1648
		GD 439		5 38 40	+68 11.7	1	13.82		0.63		−0.11		1	sdG	782
		PASP87,769 T3# 6		5 38 40	−64 09.4	1	13.06		0.73		0.33		4		431
37786	+9 00925			5 38 41	+09 10.5	1	7.48		−0.14		−0.60		2	B8	401
		AO 890		5 38 41	+12 42.6	1	10.11		1.86		2.12		1		1722
246312	+26 00904	IDS05356N2658	AB	5 38 41	+26 59.8	1	9.09		0.68				3		1724
		Sk −68 137		5 38 42	−68 54.	1	13.26		−0.07		−0.96		3		952
		Sk −68 136		5 38 42	−68 59.	1	13.60		−0.01		−0.85		3		952
269908		Sk −69 241		5 38 42	−69 07.	1	14.52		0.04		−0.51		4	WN4p	980
37725	+29 00958			5 38 43	+29 16.4	1	8.33		0.17				3	A3	1025
37739	+22 00989			5 38 44	+22 50.9	1	7.92		0.43				2	F5	1724
37657	+42 01376	LS V +43 029		5 38 44	+43 02.2	2	7.21	.088	0.06	.049	−0.64	.064	5	B3 Vne	401,1212
246369	+26 00905			5 38 46	+26 14.1	1	8.29		1.75				4	M0 III	20
246340	+29 00960			5 38 46	+29 14.6	5	8.48	.021	0.52	.009	−0.04	.038	10	F8 V	979,1025,1724,3026,8112
246370	+25 00920			5 38 47	+26 00.9	1	8.91		0.96				3		1724
246309	+30 00976			5 38 47	+31 00.1	1	10.21		0.16				3	A0	1025
37601	+56 01050	HR 1941		5 38 47	+56 33.5	1	6.05		0.95		0.73		3	K0 III	1355
37802	+9 00926			5 38 48	+09 08.4	1	9.49		0.18		0.14		1	A0	401
246338	+29 00961	LS V +29 007		5 38 48	+29 29.9	1	8.88		0.35				3	B3	1025
37845	−1 01009			5 38 48	−01 21.0	1	8.79		−0.01		−0.12		2	A0	1060
		AAS21,109 # 254		5 38 48	−65 55.	2	14.70	.037	−0.20	.061	−1.06	.000	4		417,952
		Sk −67 231		5 38 48	−67 20.	1	13.36		−0.12		−0.99		4		798
		Sk −67 230		5 38 48	−67 37.	1	13.57		−0.11		−0.89		4		798
		Brey 80		5 38 48	−69 07.	1	12.89		−0.30		−0.66		4	WN7	980
		Sk −69 240		5 38 48	−69 31.	1	12.96		0.26		−0.54		3		798
37824	+3 01007	V1149 Ori		5 38 49	+03 45.2	1	6.70		1.14		0.90		1	G5	1641
246342	+28 00853	IDS05357N2811	AB	5 38 50	+28 12.8	1	9.35		0.53				2	G0	1724
		PASP87,769 T3# 10		5 38 50	−64 00.3	1	11.55		0.33		0.00		2		431
		HRC 180, V614 Ori		5 38 51	+09 06.4	1	13.96		1.33		0.57		1	K5:e	825
290791	−0 01053			5 38 51	−00 19.7	1	10.46		0.41		0.22		3		808
290779				5 38 52	+00 14.1	1	10.82		0.29		0.24		4	A3	808
37752	+23 01007	HR 1951		5 38 52	+23 18.2	2	6.57		−0.06	.000	−0.53	.020	5	B8p	1049,1079
246363	+30 00979			5 38 52	+30 16.6	1	9.19		0.41				3	F2	1025
37738	+27 00839			5 38 53	+27 44.8	1	8.28		0.17		−0.42		3	B3 IV/V	1733
246366	+28 00854			5 38 54	+28 13.1	1	9.14		0.52				2	F5	1724
		Sk −66 181		5 38 54	−66 31.	1	13.31		−0.15		−0.84		4		798
		Brey 83	V	5 38 54	−69 08.	1	11.92		0.12		−0.77		4	WN6	980
269905	−69 00455	Sk −69 242		5 38 54	−69 47.3	1	10.96		0.18		−0.24		5	A1 Ia0	149
37586	+59 00905			5 38 55	+59 40.9	1	8.67		0.59		0.14		2	F5	1502
		3C 147 # 3		5 38 56	+49 49.9	1	13.22		1.42		1.38		1		327
	−4 01214			5 38 56	−04 24.4	1	10.66		0.36		0.18		2		1060
246361	+31 01058			5 38 57	+31 51.0	1	9.04		0.51				2	F8	1025
		3C 147 # 2		5 38 57	+49 50.8	1	11.23		0.56		−0.01		1		327
37889	−7 01151			5 38 57	−06 57.5	2	7.67	.000	−0.13	.005	−0.68	.014	4	B2 V	1060,8055
246455	+12 00865			5 38 58	+12 19.5	1	9.31		0.34		0.03		1		1722
37769	+22 00993			5 38 58	+22 31.7	1	7.72		1.01				2	K0	1724

Table 1 299

HD	DM	Other Id	N Rem	α_{1950}	δ_{1950}	S	V	σ_V	B–V	σ_{B-V}	U–B	σ_{U-B}	n	Spectrum	References
		V 335 Aur		5 38 58	+37 37.0	1	11.90		0.97				1		1772
	−63 00472			5 38 58	−63 45.2	1	11.45		0.54		0.08		2	A3	849
246454	+12 00863			5 38 59	+12 29.3	1	9.40		1.77		2.08		1		1722
246380	+29 00962			5 38 59	+29 13.6	1	9.42		0.35				3	B8	1025
37962	−31 02652			5 38 59	−31 22.2	1	7.85		0.65				4	G5 V	2012
		AAS21,109 # 255		5 39 00	−67 26.	2	14.29	.033	-0.20	.019	-0.94	.014	4		417,952
		Sk -68 139		5 39 00	−68 32.	1	13.96		-0.21		-1.00		3		952
		R 139	V	5 39 00	−69 07.	1	11.97		-0.09		-0.72		3	WN/Of?	980
		SK *C 55		5 39 00	−69 47.	1	11.59		0.52		-0.03		2	G5:V:	149
		3C 147 # 4		5 39 01	+49 47.6	1	13.19		0.66		0.15		1		327
		HRC 181, DL Ori		5 39 01	−08 07.1	1	13.07		0.77		-0.49		1		776
37784	+22 00996	HR 1954		5 39 02	+22 38.2	1	6.35		1.21		1.33		2	K2	1733
37886	−3 01172			5 39 02	−02 59.9	1	9.00		-0.04		-0.34		2	B9	1089
37874	−4 01216			5 39 02	−04 01.7	1	9.69		0.32		0.14		2	A2	1060
37888	−6 01286			5 39 02	−06 47.6	3	9.21	.000	0.03	.015	-0.04	.015	5	A0	1004,1060,8055
		HA 169 # 479		5 39 02	−44 58.3	1	13.40		0.51		-0.06		std		2
		Sk -69 138		5 39 03	−69 01.9	1	12.23		0.49		-0.02		4		415
38268	−69 00456	IDS05394S6909	⋆ ABC	5 39 03	−69 07.6	2	9.45	.121	0.14	.014	-0.75		49	OB +WN	980,8033
246450	+18 00915	DY Tau		5 39 04	+18 31.0	1	9.58		1.93				2	M2	369
246421	+22 00997		A	5 39 04	+22 59.4	1	10.19		0.10		-0.53		5		1723
	−10 01261		AB	5 39 04	−10 20.7	1	10.04		0.44		0.20		4		206
		HA 169 # 482		5 39 05	−44 57.1	1	13.06		0.59		-0.06		std		2
269920	−69 00459	Sk -69 244		5 39 06	−69 25.	1	11.48		0.00		-0.69		3	B3 Ia	952
37903	−2 01345	LS VI -02 005		5 39 07	−02 17.0	5	7.83	.009	0.10	.005	-0.63	.011	16	B1.5V	490,1004,1089,1732,8055
37887	−3 01173			5 39 07	−03 45.3	2	7.70	.000	-0.01	.000	-0.11		3	A0 V	1004,1060
246400	+30 00981			5 39 08	+30 35.6	1	10.03		0.25				2	A3	1025
246417	+29 00963			5 39 09	+29 42.2	1	9.08		0.22				3	B5	1025
37737	+36 01233	LS V +36 018		5 39 09	+36 10.6	3	8.07	.022	0.34	.020	-0.60	.080	7	B0 II:	342,1012,8100
		3C 147 # 1		5 39 09	+49 50.5	1	12.70		1.18		0.88		1		327
37904	−2 01346	HR 1959	⋆ AB	5 39 10	−02 55.2	2	6.41	.005	0.31	.005	0.05	.015	7	A9 IV-V	15,1415
		Sk -71 048		5 39 11	−71 40.	1			-0.01		-0.75		3		415
		HA 169 # 635		5 39 12	−45 00.1	1	12.45		0.65		0.00		std		2
		Sk -67 233		5 39 12	−67 31.	1	13.72		-0.16		-0.93		4		798
		G 100 - 31		5 39 14	+22 36.0	1	14.09		1.00		0.66		1		1658
		Sk -67 232		5 39 14	−67 54.2	1	13.08		0.03		-0.07		2	B9 Ib	149
38282	−69 00462	Sk -69 246		5 39 14	−69 03.7	4	11.13	.018	-0.13	.018	-0.90	.010	11	WN7	415,980,1277,8033
		G 102 - 22		5 39 16	+12 29.0	6	11.55	.043	1.64	.017	1.19	.033	24		1620,1705,1722,1764*
269923	−69 00466	Sk -69 247		5 39 16	−69 31.4	1	10.42		0.17		-0.54		4	A0 Ia+	415
		AO 894	A	5 39 17	+12 23.1	1	9.89		0.35		0.09		1		1722
		AO 894	B	5 39 17	+12 23.1	1	9.98		0.33		0.09		1		1722
		AO 894	C	5 39 17	+12 23.1	1	12.67		0.53		0.28		1		1722
37820	+26 00908			5 39 17	+26 25.0	1	8.15		1.06				2	K0	1724
37819	+28 00856	V356 Aur		5 39 17	+28 58.5	2	8.08	.015	0.57	.010			5	F8	1025,1724
37800	+29 00964	IDS05361N2948	AB	5 39 17	+29 49.6	2	7.26	.005	0.56	.010			5	F8 IV	1025,1724
	+23 01008			5 39 18	+23 08.6	1	10.21		1.95				2	M2	369
246440	+26 00907			5 39 18	+26 54.0	1	8.98		0.60				2		1724
		Sk -68 140		5 39 18	−68 58.	2	12.73	.010	0.09	.005	-0.80	.015	5		415,952
269928	−69 00464	Sk -69 248		5 39 18	−69 07.6	2	12.01	.005	0.10	.015	-0.77	.020	7	WN7	415,980
37767	+36 01236	LS V +36 019		5 39 19	+36 07.6	1	8.94		0.15		-0.48		2	B3 V	1012
		PASP87,769 T3# 7		5 39 21	−64 04.8	1	13.47		0.70		0.22		1		431
37927	−2 01348			5 39 22	−02 49.3	1	8.44		-0.08		-0.46		2	B9	1089
37914	−0 01056			5 39 23	−00 36.0	1	9.43		0.21		0.09		3	A7 V	808
		Sk -68 141		5 39 23	−68 32.6	1	12.84		-0.01		-0.54		4	A0 Ia	415
		Sk -69 250		5 39 23	−69 18.	1	12.14		0.39		-0.19		3		415
		NGC 2024 - 1		5 39 24	−01 52.	3	12.16	.023	1.41	.000	0.28	.016	5		8004,8007,8055
		Sk -68 142		5 39 24	−68 14.	1	13.48		-0.07		-0.60		3		952
		AAS21,109 # 256		5 39 24	−69 07.	1	13.00		0.20		-0.66		3		952
		Brey 91		5 39 24	−69 31.	1	10.86		-0.11		-0.92		3	WN9	980
	+37 01292	LS V +37 018		5 39 25	+37 57.7	1	9.08		0.20		-0.52		2	B3 V:pe:	1012
		HA 169 # 795		5 39 25	−45 13.0	1	12.31		0.50		-0.09		std		2
		AO 895		5 39 26	+12 42.9	1	11.15		1.58		1.51		1		1722
246437	+31 01060			5 39 26	+31 57.0	1	10.08		0.38				3	A0	1025
		HRC 182	V	5 39 26	−08 02.3	2	11.85	.015	1.03	.030	0.46	.010	2		776,825
		HA 169 # 796		5 39 26	−45 11.5	1	11.36		0.44		-0.08		std		2
		Sk -67 234		5 39 26	−67 08.2	1	12.74		0.04		-0.46		4	B9 Iab	149
246486	+30 00983			5 39 28	+30 44.1	1	10.31		0.54				2	F8	1025
37971	−16 01208	HR 1962		5 39 28	−16 45.0	2	6.20	.005	-0.14	.005			7	B3 IVp	15,2028
		Sk -68 143		5 39 28	−68 13.3	1	12.80		0.03		-0.42		2	A0 Iab	149
	−6 01287			5 39 29	−06 16.6	1	10.60		0.24		0.09		4		206
		AAS21,109 # 257		5 39 30	−67 37.	2	14.29	.033	-0.22	.033	-0.97	.014	4		417,952
246637		G 99 - 24		5 39 32	+07 22.9	1	10.85		1.02		0.82		1		333,1620
37881	+18 00920			5 39 32	+18 57.9	1	7.31		1.34		1.50		2	K0	1648
37638	+61 00816	HR 1943		5 39 32	+61 27.3	1	6.15		0.90				2	gG5	71
		PASP87,769 T3# 8		5 39 33	−64 06.9	1	13.23		0.57		0.06		2		431
		Sk -69 252		5 39 35	−69 19.	1	12.35		0.00		-0.83		4		415
290802				5 39 36	−00 14.	1	11.15		0.44		0.34		3	A7	808
38344		Sk -69 251		5 39 36	−69 04.	1	12.92		0.07		-0.71		4	WN6	980
38056	−33 02483	HR 1966		5 39 37	−33 25.4	1	6.34		-0.04				4	B9.5V	2007
37986	−15 01126			5 39 38	−15 39.1	2	7.35	.005	0.79	.000	0.44		5	G8/K0 IV	258,2012
38305	−67 00486			5 39 39	−67 04.0	1	8.33		0.48		0.00		2	F6 III	149

HD	DM	Other Id	N	Rem	α_{1950}	δ_{1950}	S	V	σ_V	B–V	σ_{B-V}	U–B	σ_{U-B}	n	Spectrum	References
246564	+26 00911				5 39 40	+26 17.8	1	9.41		1.74				2	K0	369
37859	+33 01127	IDS05364N3316	A		5 39 40	+33 17.5	1	6.92		0.44				2	F5	1724
	−5 01365				5 39 41	−05 20.1	1	10.12		0.26		0.11		2		1060
		PASP87,769 T3# 9			5 39 42	−64 07.9	1	14.03		0.58		0.16		1		431
269962	−67 00487	Sk −67 236			5 39 44	−67 23.3	1	12.28		−0.06		−0.87		4	B7	415
37841	+41 01257	IDS05362N4104	A		5 39 45	+41 06.0	1	7.47		0.01		−0.47		2	B8	401
		Sk −67 235			5 39 45	−67 21.	1	12.50		−0.05		−0.84		4		415
246655	+12 00870				5 39 47	+12 17.4	1	9.48		0.45		0.01		1		1722
	+34 01150	LS V +34 063			5 39 47	+34 19.6	1	9.58		0.20		−0.49		2	B2 V	1012
	−45 02113	HA 169 # 806			5 39 48	−45 18.3	1	10.82		0.45		−0.08		std		2
	−45 02113	HA 169 # 806			5 39 48	−45 18.3	1	13.44		0.54		−0.02		std		2
		Sk −67 237			5 39 48	−67 22.	1	13.12		0.07		−0.57		4		798
		AAS21,109 # 258			5 39 48	−69 01.	1	13.80		0.10		−0.96		2		417
246560	+30 00986				5 39 49	+30 52.8	1	8.07		1.35				3	K0	1025
		Sk −68 144			5 39 51	−68 40.6	1	13.23		0.00		−0.29		4	A4 Ia	415
39780	−84 00075	HR 2059, TZ Men			5 39 51	−84 49.0	3	6.20	.004	−0.01	.010	−0.11	.000	12	B9.5IV/V	15,1075,2038
37393	+74 00257	G 248 - 49			5 39 52	+74 35.5	1	7.37		0.70		0.28		3	G0	1733
37984	+1 01105	HR 1963			5 39 53	+01 27.1	8	4.91	.009	1.16	.005	1.06	.009	25	K1 III	15,1003,1078,1355*
269936	−69 00473	IDS05403S6930	★	AB	5 39 53	−69 28.6	1	11.23		−0.02		−0.81		6	09.5:I	149
		Sk −67 238			5 39 54	−67 19.	1	13.06		−0.10		−0.91		4		798
246627	+22 01002				5 39 55	+22 53.1	1	8.77		0.60				2		1724
246557	+32 01069				5 39 55	+32 15.7	1	9.28		0.36				2	A2	1025
		AO 897			5 39 56	+12 27.9	1	10.98		1.50		1.33		1		1722
38023	−8 01199				5 39 57	−08 09.4	2	8.87	.005	0.32	.005	−0.31	.005	6	B4 V	206,401
37924	+25 00934	IDS05369N2519	AB		5 39 58	+25 20.5	1	8.15		0.25		0.05		3	A0	833
269926		Sk −69 245			5 39 58	−69 02.	1	13.07		−0.13		−0.97		4	WR	980
38923	−80 00160	IDS05440S8025	A		5 39 59	−80 23.2	1	8.87		0.52		0.05		2	F7 V	1730
		AAS21,109 # 259			5 40 00	−69 01.	1	13.90		−0.05		−0.97		2		417
246604	+29 00969				5 40 01	+29 55.6	2	8.56	.029	0.73	.005			5	G0	1025,1724
38054	−17 01214	HR 1965			5 40 02	−17 33.2	1	6.15		1.38				4	K3 III	2035
		AO 898			5 40 04	+12 23.9	1	10.63		1.33		1.05		1		1722
246620	+31 01064				5 40 06	+31 13.4	1	9.35		0.20				2	B9	1025
		AAS21,109 # 260			5 40 06	−69 29.	1	13.58		−0.15		−0.93		3		952
	−69 00474	Sk −69 254			5 40 06	−69 46.0	6	12.04	.040	0.25	.018	−0.58	.015	29	B5 I	358,415,430,431,1277,8033
		PASP87,769 T1# 13			5 40 07	−69 47.8	1	13.19		0.53		−0.74		1		431
38014	+2 01041	G 99 - 26	★		5 40 08	+02 39.8	3	8.55	.010	0.89	.020	0.62	.023	6	K1 V	22,333,1620,3077
		G 99 - 27			5 40 08	+02 40.6	2	13.22	.010	1.46	.007	1.14	.000	3		333,1620,3056
37981	+14 00991				5 40 08	+14 09.4	2	6.73	.004	1.09	.006	1.01	.002	13	K1 IV	1355,1499
38090	−22 01194	HR 1968			5 40 08	−22 23.8	2	5.86	.005	0.11	.000			7	A2/3 V	15,2012
		PASP87,769 T1# 2			5 40 08	−69 44.3	1	12.55		0.59		0.08		3		431
246621	+30 00987				5 40 10	+30 43.6	1	9.43		0.09				3	B5	1025
	+18 00926				5 40 11	+18 05.1	1	9.77		1.90				2	M2	369
37956	+29 00970				5 40 12	+29 10.7	2	6.62	.019	1.08	.019			4	K1 III	1025,1724
		SK *G 421			5 40 13	−67 07.4	1	13.33		0.12		0.14		4	A2 Ib	149
38052	−6 01291				5 40 14	−06 50.3	1	9.56		0.24		0.12		2	A7 Vs	1060
37967	+23 01015	HR 1961, V731 Tau			5 40 17	+23 10.9	4	6.20	.040	−0.06	.004	−0.65	.015	10	B2.5Ve	154,879,1212,1223
38138	−30 02571	HR 1972			5 40 17	−30 33.5	2	6.18	.005	0.01	.000			7	A0 V	15,2012
		Sk −69 256			5 40 17	−69 18.	1	12.61		0.03		−0.83		4		415
37838	+52 00991				5 40 18	+52 40.4	1	8.20		0.09		−0.05		3	B8	555
38051	−4 01223	LS VI -04 001			5 40 18	−04 39.5	4	8.48	.020	0.35	.017	−0.30	.003	8	B8	490,1004,1060,8055
		AAS21,109 # 261			5 40 18	−69 09.	1	14.32		−0.19		−0.77		1		417
		AAS21,109 # 262			5 40 18	−69 17.	1	12.73		0.05		−0.74		2		417
		HA 169 # 663			5 40 21	−45 06.5	1	14.28		0.55		−0.12		std		2
38048	+0 01158				5 40 22	+00 30.8	1	9.29		0.13		0.00		1	B9	1089
37979	+24 00931				5 40 22	+24 03.8	1	7.75		1.54				2	K0	1724
246659	+32 01072				5 40 22	+32 14.6	1	10.27		0.22				3	B5	1025
246660	+31 01065				5 40 24	+31 44.3	1	9.43		0.06				3	A2	1025
37978	+26 00920				5 40 25	+26 17.0	1	8.33		0.39				2	F0/2 V/IV	1724
246706	+29 00971				5 40 25	+29 43.9	1	9.10		0.06				4	B8	1025
	−69 00476	Sk −69 257			5 40 26	−69 45.4	6	12.45	.055	−0.07	.018	−0.94	.012	45	O9 II	358,415,430,431,1277,8033
38170	−34 02401	HR 1973			5 40 27	−34 41.5	2	5.29	.014	−0.05	.002	−0.08		8	B9.5V	1637,2035
38089	−6 01293	HR 1967	★	AB	5 40 28	−06 49.2	7	5.97	.023	0.43	.016	−0.05	.016	19	F5 V +F6 V	15,247,254,292,1078*
38089	−6 01293	IDS05380S0651		CD	5 40 28	−06 49.2	1	10.35		0.46		0.00		1		3030
		G 191 - 52			5 40 30	+56 14.4	1	13.25		0.67		−0.09		2		1658
38087	−2 01350				5 40 30	−02 20.1	3	8.30	.004	0.12	.010	−0.45	.011	6	B5 V	1089,1732,8055
		AAS21,109 # 264			5 40 30	−69 01.	1	13.60		0.10		−0.98		2		417
		NGC 2081 - 1			5 40 30	−69 25.	1	14.65		−0.03		−0.93		1		10
		NGC 2081 - 4			5 40 30	−69 25.	1	13.61		−0.09		−0.94		1		10
		NGC 2081 - 6			5 40 30	−69 25.	1	14.07		−0.17		−1.00		1		10
		NGC 2081 - 7			5 40 30	−69 25.	1	12.90		−0.08		−0.79		1		10
		NGC 2081 - 9			5 40 30	−69 25.	1	14.72		−0.08		−0.96		1		10
		NGC 2081 - 10			5 40 30	−69 25.	1	14.72		0.22		−0.68		1		10
		NGC 2081 - 11			5 40 30	−69 25.	1	13.14		1.31		−0.20		1		10
		NGC 2081 - 12			5 40 30	−69 25.	1	13.94		−0.07		−0.90		1		10
		NGC 2081 - 13			5 40 30	−69 25.	1	14.56		−0.11		−0.76		1		10
		NGC 2081 - 14			5 40 30	−69 25.	1	14.71		0.14		0.17		1		10
		NGC 2081 - 15			5 40 30	−69 25.	1	13.45		−0.12		−0.90		1		10
		NGC 2081 - 16			5 40 30	−69 25.	1	13.20		−0.20		−0.91		1		10
		NGC 2081 - 17			5 40 30	−69 25.	1	15.23		0.72		−1.29		1		10
		NGC 2081 - 18			5 40 30	−69 25.	1	14.93		−0.04		−0.81		1		10

Table 1   301

HD	DM	Other Id	N Rem	α_{1950}	δ_{1950}	S	V	σ_V	B–V	σ_{B-V}	U–B	σ_{U-B}	n	Spectrum	References
		NGC 2081 - 19		5 40 30	−69 25.	1	13.13		-0.18		-1.00		1		10
		NGC 2081 - 20		5 40 30	−69 25.	1	13.36		0.80		0.36		1		10
		NGC 2081 - 21		5 40 30	−69 25.	1	13.76		-0.01		-0.75		1		10
		NGC 2081 - 22		5 40 30	−69 25.	1	13.13		1.84		0.74		1		10
		NGC 2081 - 25		5 40 30	−69 25.	1	13.39		-0.12		-0.82		1		10
		NGC 2081 - 26		5 40 30	−69 25.	1	11.89		0.33		-1.00		1		10
		AAS21,109 # 263		5 40 30	−69 41.	1	13.02		-0.10		-0.94		2		952
		G 96 - 50		5 40 31	+40 56.2	1	15.12		1.56		1.11		2		316
38088	−4 01224			5 40 31	−04 51.5	3	9.68	.005	0.16	.000	0.06	.009	5	A0 V	1004,1060,8055
		HA 169 # 669		5 40 31	−44 59.0	1	13.82		0.52		-0.12		std		2
		EF Tau		5 40 34	+19 24.0	1	12.74		0.77				1		1772
38010	+25 00941	LS V +25 003		5 40 34	+25 25.1	1	6.84		0.03		-0.72		3	B1 Vpe	1212
246770	+21 00947			5 40 35	+21 51.1	1	8.35		1.79				2	K7	369
		Sk -68 146		5 40 36	−68 26.	1	13.53		-0.14		-0.72		3		952
38472	−69 00479	Sk -69 258		5 40 36	−69 25.4	1	13.26		-0.17		-0.91		4	WN4	980
		AAS21,109 # 265		5 40 36	−69 42.	1	12.46		-0.11		-0.96		3		952
290842		GG Ori		5 40 37	+00 42.6	1	10.36		0.53		0.32		4	A2	1768
		HRC 183, V625 Ori		5 40 37	+09 04.8	1	13.12		1.01		0.03		1		825
38099	−1 01012	HR 1970, V1197 Ori		5 40 37	−01 38.1	3	6.30	.005	1.47	.005	1.76	.008	8	K4 III	15,247,1415
38448	−69 00478	Sk -69 255		5 40 37	−69 24.1	1	13.08		-0.10		-0.77		4	WC5	980
38489	−69 00479	Sk -69 259		5 40 37	−69 25.4	1	13.86		-0.09		-0.83		3	Be	149
38109	−5 01369			5 40 38	−05 49.8	1	9.45		0.34		0.14		2	F2 IVs	1060
38211	−37 02356			5 40 38	−37 28.8	1	7.81		1.13				1	K2 III	2012
269953	−69 00481	Sk -69 260		5 40 38	−69 41.4	6	9.95	.011	0.88	.026	0.62	.015	20	G0 Ia	149,431,1277,1483*
246742	+29 00972	IDS05375N2934	AB	5 40 42	+29 35.6	1	9.26		0.14				2	B9	1025
	−16 01217			5 40 42	−16 47.6	2	9.98	.020	1.08	.047	1.00		4	C2,0+	864,1238
		SK *C 56		5 40 42	−69 24.	1	12.06		0.27		-0.97		2		149
	−69 00483			5 40 42	−69 32.3	1	10.99		0.42		0.00		4	A5:Ia	149
38098	+5 01001			5 40 44	+05 20.2	2	6.74	.005	-0.04	.002	-0.22	.007	44	B9 III	401,978
37880	+55 01012	IDS05366N5540	AB	5 40 44	+55 41.7	1	8.87		0.12		0.04		2	A0	1502
38120	−5 01370			5 40 44	−05 01.1	4	9.08	.016	0.03	.004	-0.06	.013	9	A0	1004,1060,1732,8055
		HA 169 # 522		5 40 45	−44 56.0	1	14.93		0.62		-0.14		std		2
270003		Sk -67 239		5 40 47	−67 05.4	1	12.10		0.03		-0.53		4	A0 Ia	149
	−69 00485	Sk -69 262		5 40 48	−69 15.9	1	11.42		0.36		0.08		3	A2 Ia	149
		Sk -69 261		5 40 48	−69 38.	1	12.16		-0.05		-0.87		3		952
		Sk -70 106		5 40 48	−70 30.	1	12.83		0.09		-0.62		4		798
		Sk -71 049		5 40 48	−71 14.	1	13.10		0.08		-0.77		4		798
38065	+16 00852			5 40 50	+16 20.2	1	9.15		0.15		0.12		4	A2 V	206
		GV Aur		5 40 50	+37 34.0	1	11.58		0.96				1		1772
		Sk -67 240		5 40 51	−67 23.	1	12.98		0.04		-0.35		4		415
38017	+30 00992	LS V +30 016	⋆ AB	5 40 52	+30 54.8	4	8.11	.034	0.27	.010	-0.42	.047	9	B3 II	401,1012,1025,8100
38034	+29 00974			5 40 53	+29 28.8	1	8.29		0.08				2	B5	1025
269955	−69 00489			5 40 54	−69 53.1	1	9.76		1.01		0.73		14	G5	430
38108	+6 01005			5 40 55	+06 52.0	1	7.24		-0.06		-0.32		2	B8	401
38033	+31 01069			5 40 57	+31 59.0	1	7.82		1.05				2	G5	1025
	−69 00490			5 40 59	−69 34.9	1	11.67		0.18		-0.41		3	B6 Iab	149,3078
		Sk -67 241		5 41 00	−67 23.	1	13.16		-0.08		-0.84		4		798
		Sk -69 263		5 41 00	−69 21.	1	12.04		0.24		-0.50		3		952
246821	+26 00929	LS V +26 007		5 41 01	+26 44.9	1	8.62		0.79				2	A8 II	1724
246841	+24 00936			5 41 02	+24 37.8	1	9.28		0.46				2		1724
37783	+65 00489			5 41 02	+65 35.7	1	8.92		0.45		-0.01		3	F8	1733
	−22 02416			5 41 03	−22 03.2	1	10.22		0.04		-0.56		2		401
38145	+3 01018			5 41 04	+03 59.0	1	7.89		0.32		0.04		49	F0	978
		LP 778 - 16		5 41 04	−19 17.8	1	12.44		0.61		-0.12		2		1696
		G 99 - 30		5 41 06	+03 33.9	3	13.26	.036	0.67	.011	-0.03	.047	5		1620,1658,3056
246836	+26 00930			5 41 06	+26 59.3	1	9.44		0.50				2		1724
38062	+29 00975			5 41 06	+29 59.9	1	8.25		0.13				3	B9	1025
	+62 00780	G 249 - 18		5 41 06	+62 13.7	1	9.02		1.40		1.23		1	M0	3072
269964	−69 00492	AA Dor	⋆	5 41 06	−69 44.6	3	11.28	.271	-0.10	.304	-0.84	.474	13	G0	45,360,431
38184	−7 01158			5 41 08	−07 18.8	1	9.53		0.04		-0.12		2	A0	401
		Sk -67 242		5 41 08	−67 15.	1	12.61		-0.08		-0.76		4		415
39091	−80 00161	HR 2022		5 41 08	−80 30.5	5	5.65	.004	0.60	.002	0.10	.008	22	G1 V	15,1075,2012,2038,3078
246890	+20 01068			5 41 09	+20 49.8	1	9.04		1.77				2	K7	369
38206	−18 01172	HR 1975		5 41 10	−18 34.7	3	5.72	.005	-0.01	.004	-0.12		9	A0 Vs	15,1079,2012
40857	−86 00072	R Oct		5 41 10	−86 25.2	1	9.45		1.41		0.06		1	MD	975
38061	+32 01076			5 41 11	+32 13.2	2	8.15	.042	0.41	.014			4	F0	1025,1724
38165	−1 01013			5 41 11	−00 57.6	2	8.83	.014	0.25	.005	-0.22	.028	4	B9	1089,8055
		Sk -69 265		5 41 11	−69 19.	1	11.88		0.12		-0.63		4		415
246866	+23 01021			5 41 12	+23 03.7	1	9.23		0.52				3		1724
38207	−20 01162			5 41 12	−20 12.6	1	8.45		0.40		-0.03		49	F2 V	978
		Sk -69 266		5 41 12	−69 28.	1	14.96		-0.12		-0.70		4	WN3	980
		Sk -68 147		5 41 14	−68 33.	1	12.96		-0.10		-0.84		4		415
38185	−9 01213			5 41 16	−08 57.5	1	7.61		-0.11		-0.70		2	B8 III	401
38134	+12 00880			5 41 17	+12 41.8	1	8.51		1.07				54	K0	6011
38154	+6 01007			5 41 18	+06 53.7	1	8.15		-0.08		-0.32		2	B9	401
38094	+30 00993			5 41 18	+30 28.0	2	7.36	.005	0.88	.005			4	G5	1025,1724
		AAS21,109 # 266		5 41 18	−67 33.	1	13.22		0.22		-0.75		3		952
		Sk -68 148		5 41 18	−68 55.	1	13.40		-0.10		-1.01		3		952
246879	+26 00931			5 41 20	+26 52.8	1	9.13		0.41				2		1724
246878	+27 00850	LS V +27 008	⋆ AB	5 41 20	+27 12.6	1	9.38		0.16		-0.59		2	B0.5V:pe	1012

HD	DM	Other Id	N Rem	α₁₉₅₀	δ₁₉₅₀	S	V	σ_V	B–V	σ_B–V	U–B	σ_U–B	n	Spectrum	References
269982	−69 00495	Sk −69 267		5 41 20	−69 16.9	2	10.79	.039	0.36	.005	0.07		6	A7:Ia	149,8033
		Sk −67 243		5 41 22	−67 27.0	1	13.22		−0.01		−0.10		4	B9 Ib	798
	−69 00496			5 41 22	−69 21.9	2	11.30	.005	0.64	.010	0.54		8	F6 Ia	149,1483
		Sk −70 107		5 41 24	−70 02.	1	13.32		0.04		−0.83		4		798
		Sk −68 149		5 41 25	−68 44.2	1	13.25		0.12		−0.04		3	A2 Iab	798
		Sk −69 268		5 41 25	−69 40.	2	12.62	.014	−0.03	.014	−0.85	.090	5		360,415
38143	+18 00939			5 41 26	+18 30.4	1	8.41		0.35		0.09		1	A2	376
38116	+28 00868	IDS05383N2858	AB	5 41 27	+28 59.8	2	7.88	.009	0.20	.005	−0.47		4	B5	401,1025
38105	+31 01072	IDS05382N3116	AB	5 41 28	+31 17.1	2	8.36	.022	0.45	.018			5	G0	1025,1724
		Sk −69 269		5 41 30	−69 42.	1	13.58		−0.07		−0.85		4		952
		Sk −70 108		5 41 30	−70 41.	1	13.12		0.11		−0.71		5		798
		Sk −71 050		5 41 30	−71 29.	1	13.44		−0.12		−1.01		3		798
38115	+30 00994			5 41 31	+30 36.2	2	8.08	.005	1.22	.005			5	G5	1025,1724
38142	+24 00940			5 41 33	+24 53.8	2	7.81	.035	1.01	.010			4	G8 III	20,1724
38113	+33 01137			5 41 34	+33 42.2	1	7.22		0.98				3	G5	1724
38114	+32 01077			5 41 35	+32 22.3	1	8.15		0.95				3	G5	1724
246901	+33 01138	LS V +33 042		5 41 35	+33 30.5	2	8.05	.019	1.26	.014	−0.11	.066	4	B1:	1012,8100
38239	−6 01297			5 41 37	−06 45.2	3	9.22	.000	0.05	.010	0.01	.000	4	A0	1004,1060,8055
269997	−69 00498	Sk −69 270		5 41 41	−69 06.3	1	11.27		0.14		−0.66		3	B3 Ia	149
38181	+20 01070			5 41 43	+20 41.0	1	9.11		0.52				2	F5	1724
38238	+0 01170	V351 Ori		5 41 45	+00 07.4	1	8.87		0.35		0.22		1	A7 III	1588
38131	+35 01223	LS V +35 028		5 41 45	+35 08.6	1	8.19		0.21		−0.69		2	B0.5V	1012
246996	+23 01026			5 41 46	+23 07.5	1	9.17		0.67				2		1724
38192	+20 01073			5 41 48	+20 31.2	1	8.16		−0.11		−0.48		2	B9	401
38141	+33 01139			5 41 48	+33 36.4	1	6.78		0.40				2	F5	1724
		Sk −67 244		5 41 48	−67 31.	1	13.18		−0.23		−1.03		3		798
		Sk −70 109		5 41 48	−70 55.	1	13.85		−0.11		−0.84		3		798
		SK *G 431		5 41 49	−65 24.9	1	13.40		0.03		0.06		5	A0 I:	149
38616	−67 00492			5 41 49	−67 25.5	4	7.06	.013	−0.02	.010	−0.11	.046	25	A2 Ib/IIp	1075,1408,1425,1499
		Sk −69 271		5 41 49	−69 38.	1	12.01		0.00		−0.69		4		415
	−46 01969	W Pic		5 41 50	−46 28.5	1	7.77		4.75		5.60		3	N	864
38385	−39 02140	HR 1981		5 41 51	−39 25.7	1	6.25		0.37				4	F3 V	2035
	−5 01377			5 41 52	−05 16.3	2	9.78	.004	0.45	.009	0.34	.000	5	A4 II-III	1060,8055
38274	−6 01300			5 41 52	−06 38.3	1	10.25		0.31		0.05		2	A8 Vs	1060
	+34 01162	LS V +34 067		5 41 53	+34 04.6	1	8.93		0.12		−0.58		2	B2 V:nn	1012
269998	−69 00501	Sk −69 272		5 41 53	−69 20.8	1	11.81		0.27		0.28		6	A9 Ia:	149
		AAS21,109 # 267		5 41 54	−68 34.	2	14.27	.173	−0.17	.098	−1.00	.028	4		417,952
		Sk −69 273		5 41 54	−69 26.0	1	12.02		0.42		−0.27		4		415
269992	−69 00503	Sk −69 274		5 41 55	−69 49.4	7	11.22	.014	0.05	.012	−0.76	.009	34	B2 Ia	149,328,358,430,431*
38191	+21 00958	LS V +21 013		5 41 56	+21 26.4	1	8.73		0.14		−0.69		2	B1:V:e:n	1012
		HRC 185, V631 Ori		5 41 57	+08 55.8	1	13.91		1.25		1.01		1		825
38396	−37 02365			5 41 59	−37 43.8	1	8.47		0.47				4	F3 V	2012
38219	+16 00855			5 42 00	+16 04.0	1	7.02		0.03		0.00		2	A0	1648
246988	+32 01078	IDS05387N3221	A	5 42 00	+32 22.5	1	9.23		0.27				2	B9	1025
		Sk −70 110		5 42 00	−70 33.5	1	13.06		0.16		−0.32		3	A0 Iab	415
247068	+22 01019			5 42 01	+22 08.8	1	12.97		0.53				1	G0	3061
38104	+49 01398	HR 1971		5 42 01	+49 48.4	8	5.46	.010	0.03	.005	0.07	.014	22	A0 p Cr	15,401,1007,1013,1063*
		Sk −66 182		5 42 01	−66 46.	1	12.21		0.54		0.50		3		415
		Sk −69 276		5 42 01	−69 35.	1	12.41		−0.16		−0.01		4		415
		Sk −69 275		5 42 01	−69 38.	1	12.77		−0.05		−0.83		4		415
	−8 01208			5 42 03	−08 44.6	1	10.22		1.45		1.24		4		206
	+34 01164			5 42 05	+34 16.7	1	9.31		1.83				3	M2	369
270033	−67 00494	Sk −67 246		5 42 05	−67 22.1	1	11.70		0.10		−0.09		3	A1 Ib	149
38459	−47 01997			5 42 06	−47 50.8	1	8.50		0.87				4	K0 V	2012
		Sk −67 245		5 42 06	−67 40.	1	14.01		−0.03		−0.99		4		798
		Sk −69 278		5 42 06	−69 24.	1	12.70		0.09		−0.55		4		415
38312	−6 01302			5 42 07	−06 53.1	1	6.78		0.17		0.15		2	A2	401
269993	−70 00435			5 42 07	−70 02.1	1	11.85		−0.07		−0.87		3	B1.5Ia	149
		Sk −69 277		5 42 08	−69 51.	1	12.65		0.02		−0.72		3		415
38218	+24 00943	TU Tau		5 42 10	+24 24.0	3	8.41	.058	2.88	.106	1.36		3	C5 II	109,414,1238
247042	+29 00981			5 42 10	+29 08.3	1	9.52		0.16				3	A7	1025
		Sk −69 280		5 42 11	−69 20.	1	12.66		0.09		−0.74		4		415
		G 99 - 31	A	5 42 12	+09 13.7	3	11.81	.013	0.56	.000	−0.17	.005	7		1064,1620,3077
		G 99 - 31	B	5 42 12	+09 13.7	1	12.64		0.82		0.15		1		1064
247060	+31 01078			5 42 12	+31 37.8	1	10.15		0.20				2	A7	1025
247040				5 42 12	+32 12.	1	11.09		1.63		1.67		3	G0	1723
		Sk −66 183		5 42 12	−66 41.	1	13.57		−0.22		−0.98		4		798
38262	+13 00971	ST Tau		5 42 13	+13 33.4	3	7.86	.116	0.71	.068	0.54	.050	3	G0	592,1462,6011
247135	+18 00948			5 42 13	+18 23.1	1	9.44		1.77				2	K7	369
		Sk −69 279		5 42 13	−69 37.	1	12.79		0.05		−0.84		4		415
38247	+18 00950			5 42 15	+18 41.1	1	6.61		1.62		1.71		8	G8 Iab	1355
38200	+31 01080			5 42 15	+31 18.3	2	6.49	.024	1.13	.009			4	K0	1025,1724
38311	−0 01073			5 42 15	−00 04.9	1	8.71		0.05		−0.04		1	A0	1089
270037		Sk −67 247		5 42 15	−67 18.7	2	12.22	.009	−0.06	.005	−0.77	.005	5	B7	149,360
38458	−45 02131	HR 1984		5 42 16	−45 51.3	4	6.38	.009	0.29	.005	0.10		15	A9 V	15,258,2012,2012
38290	+6 01010			5 42 17	+06 22.0	1	9.01		0.09		0.09		2	A2	1732
38091	+56 01058	HR 1969		5 42 17	+56 05.9	1	5.94		0.17		0.12		1	A4 Vn	39
38468	−45 02132			5 42 17	−45 38.6	1	7.93		1.01		0.76		9	K0	1673
38291	+6 01012			5 42 19	+06 19.8	2	7.18	.001	−0.07	.004	−0.32	.018	5	B8	401,1732
		G 100 - 33		5 42 19	+23 00.1	1	13.13		1.22		1.16		1		333,1620

Table 1   303

HD	DM	Other Id	N Rem	α_{1950}	δ_{1950}	S	V	σ_V	B−V	σ_{B-V}	U−B	σ_{U-B}	n	Spectrum	References
38189 +40 01403		HR 1974		5 42 19	+40 29.3	2	6.57	.010	0.25	.000	0.02	.005	9	δ Del	1501,8071
38382 −20 01171		HR 1980		5 42 19	−20 08.8	3	6.34	.004	0.58	.000			14	F8/G0 V	15,1075,2029
38232 +29 00983		LS V +29 009		5 42 21	+29 16.7	4	7.42	.013	0.64	.020	0.35		6	F5 II	1025,1724,6009,8100
38392 −22 01210		HR 1982	⋆ B	5 42 21	−22 26.2	7	6.16	.012	0.95	.007	0.74	.000	25	K2 V	15,124,1013,1075,2013*
		Sk −69 281		5 42 22	−69 14.0	1	11.84		0.22		−0.01		4	A1 Iab	415
		Sk −71 051		5 42 22	−71 21.	1	12.71		−0.09		−1.00		3		415
38309 +3 01025		HR 1978	⋆ A	5 42 23	+03 59.3	3	6.09	.012	0.32	.008	0.07	.011	8	F1 V	15,247,1078
38199 +37 01308				5 42 23	+37 38.4	1	7.47		1.74				3	M1	369
38393 −22 01211		HR 1983	⋆ A	5 42 23	−22 27.8	9	3.59	.011	0.48	.011	−0.01	.012	45	F7 V	3,15,1007,1008,1013*
38231 +30 00997		LS V +30 017		5 42 24	+30 08.9	1	8.40		0.05				2	B5	1025
		Sk −66 184		5 42 24	−66 41.	1	13.66		−0.19		−1.01		3		798
		SK *G 439		5 42 24	−68 28.8	2	11.87	.005	0.56	.000	0.37		5	F6 Ia	149,1483
		Brey 97		5 42 24	−70 36.	1	14.68		−0.12		−0.89		4	WN4	980
		SK *G 440		5 42 25	−70 26.8	1	13.17		0.28		0.32		5		149
270004				5 42 26	−69 45.1	1	12.30		0.14		0.20		2	A5	149
38289 +13 00974				5 42 27	+13 41.3	2	8.80	.020	0.17	.020	−0.23	.025	2	B9	592,1462
38260 +25 00958				5 42 27	+25 42.7	1	8.47		0.41				2	F3 V/IV	1724
38261 +25 00961				5 42 28	+25 05.9	2	7.47	.015	1.14	.015			4	K2 III	20,1724
		Sk −68 151		5 42 28	−68 11.9	1	13.04		0.10		0.23		2	A0 Ib	798
		SK *C 57		5 42 30	−64 22.	1	12.02		1.25		1.22		3		149
		Sk −66 185		5 42 30	−66 20.	1	13.11		−0.19		−0.94		4		798
247125 +30 00999		LS V +30 018		5 42 31	+30 45.4	1	9.11		0.47				2	A5	1025
38352 −2 01363				5 42 31	−02 31.3	1	9.16		0.09		−0.01		2	A0	1089
38188 +44 01278				5 42 32	+44 45.9	1	7.93		0.14		−0.23		2	B5 V	401
247152 +27 00857				5 42 33	+27 50.2	1	8.81		1.20				2		1724
270019 −69 00507				5 42 33	−69 03.0	1	12.06		0.00		−0.76		4	F5	149
38308 +12 00889		IDS05397N1245	AB	5 42 34	+12 46.2	4	7.46	.008	0.77	.007	0.43	.009	59	G0	592,1308,1462,6011
38488 −41 02037				5 42 34	−41 07.6	2	8.58	.005	1.39	.007	1.46	.012	3	K2/3 III	565,3048
38230 +37 01312		G 96 - 51	⋆ A	5 42 35	+37 16.4	7	7.36	.010	0.83	.010	0.51	.003	17	K0 V	22,333,1013,1197,1355,1620*
270050 −67 00495		Sk −67 248		5 42 35	−67 21.1	3	10.84	.007	0.40	.005	0.32	.028	8	F6 Ia	149,360,1483
		Sk −69 283		5 42 36	−69 52.	1	13.56		−0.16		−0.89		2		952
247176 +26 00943		LS V +26 008		5 42 37	+26 32.5	1	9.59		0.18		−0.55		2	B2 V	1012
38179 +47 01193		IDS05388N4753	A	5 42 37	+47 53.1	1	7.98		0.10		0.04		2	B9	401
		HRC 186, FU Ori		5 42 38	+09 03.0	2	9.06	.119	1.39	.023	0.96	.038	2	F2:p I-II	1588,8037
		AV Tau		5 42 38	+27 03.1	1	11.85		1.22				1		1772
		SK *G 444		5 42 39	−68 23.6	1	12.95		0.24		0.15		2	A4 Ib	149
38307 +20 01083		HR 1977, Y Tau	A	5 42 40	+20 40.6	7	6.76	.175	3.01	.035	5.81		17	C5 II	15,109,1238,3001,6002*
38426 −21 01252		IDS05406S2142	A	5 42 43	−21 40.7	1	6.77		−0.18		−0.72		2	B3 V	1732
38350 +6 01014				5 42 44	+06 16.7	1	7.21		0.06				2	A2	401
38321 +18 00955		EU Tau		5 42 44	+18 38.3	2	7.93	.025	0.62	.005	0.43	.050	2	G5	592,1462
38467 −29 02447				5 42 44	−29 56.0	1	8.26		0.67		0.26		2	G3/5 V	1730
247187 +29 00985				5 42 45	+29 06.6	1	8.89		1.17				3		1724
38246 +39 01405				5 42 45	+39 03.4	1	7.92		1.00		0.71		3	G5 III	833
270038 −68 00423				5 42 46	−68 05.0	1	10.13		1.59		1.71		4	K4 III	149
		Sk −70 112		5 42 48	−71 00.	1	13.30		0.10		−0.53		4		798
247297 +14 01018		G 102 - 27		5 42 49	+14 40.3	1	9.05		0.65		0.11		1	G0	1620
38335 +21 00975				5 42 53	+21 42.7	1	8.04		0.94				3	K0	1724
		Sk −69 285		5 42 54	−69 30.8	1	11.92		0.15		−0.43		4	B9 Ia	415
38484 −29 02450				5 42 55	−29 51.3	1	9.44		0.44		0.02		2	F5 V	1730
		Sk −69 284		5 42 55	−69 43.	1	12.60		−0.08		−0.91		4		415
		Sk −68 152		5 42 57	−68 15.0	2	12.80	.034	0.16	.005	−0.10	.029	3	B9 Ib	149,952
247209 +31 01084				5 42 58	+31 22.1	1	9.32		0.03				2	A7	1025
		Sk −69 286		5 42 58	−69 18.0	1	12.94		0.07		−0.59		3	A2 Ia+	415
270011 −70 00437				5 42 59	−70 14.9	1	11.25		0.92		0.36		6	G5 III	149
38363 +21 00978		IDS05400N2117	AB	5 43 00	+21 17.9	1	7.70		0.35				2	F2	1724
247206				5 43 00	+32 51.	1	10.43		0.55		0.10		2	G0	1723
		Sk −69 287		5 43 00	−69 06.	1	12.41		0.05		−0.81		3		952
		Sk −70 112a		5 43 03	−70 37.5	1	12.98		0.14		−0.47		2		415
38258 +47 01194				5 43 05	+47 26.9	2	7.47	.005	0.03	.005	−0.14	.015	4	B8	401,555
270039 −68 00425		Sk −68 153		5 43 05	−68 34.9	2	12.23	.005	0.06	.005	−0.45	.010	3	A0 Ia	149,415
247289 +25 00966				5 43 06	+25 17.2	1	8.70		1.05				3		1724
		Sk −69 288		5 43 06	−69 05.	1	12.73		−0.06		−0.87		3		952
		Sk −68 154		5 43 07	−68 14.2	1	13.10		0.06		−0.48		4		149
		G 96 - 53		5 43 08	+43 37.9	1	17.09		0.14		−0.63		1	DA	782
38361 +22 01025				5 43 09	+22 45.3	1	8.19		1.23				3	K0	1724
247258 +31 01085				5 43 10	+31 53.1	1	9.77		0.13				2	A0	1025
38333 +32 01083				5 43 10	+32 37.9	1	7.41		1.32				3	K0	1724
38510 −27 02472				5 43 10	−27 00.9	2	8.24	.015	0.51	.015	−0.09		5	F5/6 V	265,1594
38360 +25 00967		IDS05401N2540	AB	5 43 11	+25 41.0	1	8.14		1.41				3	G5	1724
38377 +18 00959		IDS05403N1849	AB	5 43 12	+18 50.5	3	7.59	.062	0.38	.119	0.13	.020	3	F2	592,629,1462
	−69 00514	Sk −69 289		5 43 15	−69 34.2	1	11.40		−0.01		−0.82		2	B1.5Ia	149
		HIC 27252		5 43 16	+33 30.0	1	11.39		1.53		1.50		3		1723
		LP 718 - 20		5 43 16	−10 08.4	1	12.98		1.50		1.01		1		1696
	−69 00513			5 43 16	−69 01.2	1	12.21		−0.04		−0.82		3		149
		Sk −68 157		5 43 18	−58 58.	1	13.22		−0.12		−0.93		2		952
270060		Sk −67 249		5 43 18	−67 25.	2	12.43	.005	−0.06	.005	−0.85	.005	5	O8	360,415
		Sk −69 291		5 43 18	−69 29.	1	12.92		−0.10		−0.97		4		415
38359 +33 01146				5 43 19	+33 07.3	1	8.57		0.39				2	F2	1724
		Sk −68 156		5 43 22	−68 44.5	1	13.18		0.05		−0.25		4	A0 Iab	415
		Sk −68 155		5 43 22	−68 58.	1	12.72		0.03		−0.82		4		415

HD	DM	Other Id	N	Rem	α_{1950}	δ_{1950}	S	V	σ_V	B–V	σ_{B-V}	U–B	σ_{U-B}	n	Spectrum	References
38443	+5 01010				5 43 23	+05 04.9	1	8.60		0.98		0.74		2	K0	1733
		Sk -69 292			5 43 25	-69 43.5	1	13.11		0.03		-0.53		2	B9 Iab	415
		Sk -69 293			5 43 27	-69 58.9	1	12.81		0.06		-0.63		4	A1 Ia	415
		Sk -67 250			5 43 29	-67 53.	1	12.68		-0.17		-1.03		3		415
38495	-4 01235	HR 1986		⋆ A	5 43 34	-04 17.2	1	6.34		1.05		0.99		1	K1 III	542
38495	-4 01235	HR 1986		⋆ AB	5 43 34	-04 17.2	5	6.25	.004	1.02	.010	0.82	.008	17	K1 III+G0IV	15,247,542,1061,2029
38495	-4 01235	IDS05411S0418		B	5 43 34	-04 17.2	1	8.97		0.55		0.08		1	G0 IV	542
		Sk -66 186			5 43 36	-66 03.	1	13.62		-0.13		-0.96		4		798
		Sk -67 251			5 43 36	-67 18.	1	13.62		-0.19		-0.94		3		798
		Sk -70 113			5 43 36	-70 44.	1	13.46		-0.10		-0.97		3		798
38358	+42 01396	HR 1979			5 43 40	+42 30.6	1	6.29		1.35		1.59		2	K0	252
38572	+30 01014	FU Aur			5 43 41	+30 36.6	1	8.31		2.67		4.31		1	C4 II	109
		Sk -70 114			5 43 42	-70 45.	1	12.99		0.04		-0.68		4		798
38451	+21 00981				5 43 46	+21 11.0	2	8.85	.000	0.19	.000	0.14	.009	8	A2 IV-V V	1678,1733
		SK *G 453			5 43 46	-68 20.1	1	12.70		0.25		0.18		2	A4 Ib	149
38465	+20 01093				5 43 48	+20 09.4	1	7.63		0.37		0.01		2	F2	1648
270046	-69 00516	Sk -69 294		⋆	5 43 48	-69 16.1	2	10.33	.045	0.93	.015	0.65		11	G0 Ia	149,1483
38942	-68 00431				5 43 50	-68 06.1	1	7.49		0.97		0.70		5	G8 III	1704
271517	-65 00494	IDS05437S6532		AB	5 43 51	-65 30.5	1	9.60		0.26		0.17		2	A3 III	204
38401	+38 01289				5 43 52	+38 58.5	1	8.85		0.21		0.17		3	A3 V	1501
38478	+15 00926	HR 1985			5 43 53	+15 48.3	2			-0.06	.009	-0.44	.014	5	B8 IIIp Hg	1049,1079
		Steph 577			5 43 57	+49 57.3	1	12.04		1.35		1.26		1	K7-M0	1746
38441	+31 01091				5 43 59	+31 38.6	1	7.64		0.03		-0.02		2	A0	401
38893	-65 00495				5 43 59	-65 07.3	2	9.47	.015	0.72	.003	0.19		2	G3wF5	1594,6006
38529	+1 01126	HR 1988			5 44 00	+01 09.1	4	5.94	.007	0.77	.005	0.41	.007	13	G7 III	15,1061,1355,3077
		Sk -71 052			5 44 00	-71 17.	1	12.57		0.00		-0.87		4		798
	-29 02470				5 44 02	-29 51.4	1	11.14		0.64		0.12		2		1730
38400	+46 01051				5 44 05	+46 58.1	1	8.22		0.59		0.13		3	F8	1601
		AAS21,109 # 268			5 44 06	-67 16.	2	13.54	.005	-0.20	.000	-0.97	.019	4		417,952
38527	+9 00954	HR 1987		⋆ A	5 44 07	+09 30.3	3	5.78	.006	0.89	.016	0.58	.008	8	G8 III	15,247,1061
38666	-32 02538	HR 1996			5 44 08	-32 19.5	6	5.17	.006	-0.28	.008	-1.07	.015	34	O9.5 V	15,1034,1311,1732,2012,8033
247505	+23 01042			V	5 44 09	+23 09.3	1	8.51		1.73				3	M0	369
38563	+0 01177	LS VI +00 001		⋆ A	5 44 10	+00 03.6	3	10.42	.013	0.59	.017	-0.06	.014	7	B5	1004,1732,1732,8055
38563	+0 01177	IDS05416N0002		B	5 44 10	+00 03.6	2	10.58	.060	1.15	.032	0.39		6	B1	1004,8055
		SK *G 455			5 44 10	-67 39.1	1	12.81		0.29		0.23		3	A3 Iab	149
		AAS21,109 # 269			5 44 12	-67 16.	2	13.50	.173	-0.09	.042	-0.79	.169	4		417,952
		Sk -67 252			5 44 12	-67 32.	1	13.26		-0.10		-0.88		4		798
38491	+29 00997	IDS05410N2937		AB	5 44 13	+29 38.5	2	7.20	.010	0.52	.015	0.01		4	F8	1724,3016
38504	+22 01031				5 44 14	+22 30.7	2	7.65	.005	1.05	.005	0.82		4	K0	1648,1724
270083					5 44 17	-67 31.2	1	11.79		1.03		0.61		3	K0	149
38892	-61 00512				5 44 18	-61 48.5	1	9.56		0.22		0.16		2	A3 Vs	849
		AAS21,109 # 270			5 44 18	-67 16.	2	13.91	.098	-0.21	.005	-1.03	.052	4		417,952
38284	+62 00784	HR 1976		⋆ AB	5 44 21	+62 47.5	1	6.21		0.11		0.09		2	A4 V	1733
38545	+14 01025	HR 1989			5 44 22	+14 28.3	1			0.04		0.16		3	A3 Vn	1049
38410	+50 01229				5 44 24	+51 00.1	1	8.74		-0.07		-0.43		3	B5 III	555
		Sk -66 187			5 44 24	-66 25.	1	13.83		-0.22		-1.01		4		798
		AAS21,109 # 271			5 44 24	-67 16.	2	13.20	.033	-0.18	.047	-0.91	.009	4		417,952
		Sk -69 295			5 44 24	-69 33.	1	13.70		-0.20		-1.00		3		952
38525	+22 01032				5 44 27	+22 54.3	1	7.54		1.14				2	K0	1724
		Sk -67 253			5 44 27	-67 22.3	1	12.51		0.11		-0.59		6		149
38524	+25 00978				5 44 29	+25 33.1	3	6.44	.018	1.19	.013	1.05		7	K1 III	20,105,1724
		AO Aur			5 44 29	+31 59.7	1	10.57		0.98				1	F5	851
		Sk -66 188			5 44 29	-66 21.	1	12.49		-0.12		-0.93		3		415
270094	-67 00499	Sk -67 255			5 44 29	-67 13.4	1	11.42		-0.06		-0.71		6	B3 Ia	149
38543	+18 00966				5 44 30	+18 42.8	4	7.79	.038	1.08	.023	0.94	.033	4	K0	376,592,629,1462
		AAS21,109 # 272			5 44 30	-67 16.	2	13.25	.037	-0.20	.033	-1.01	.005	4		417,952
		Sk -67 254			5 44 30	-67 30.	1	13.58		-0.13		-0.90		4		798
38558	+17 01004	HR 1990			5 44 31	+17 42.7	2	5.53	.036	0.27	.023	0.25	.014	5	F0 III	39,254
247559	+26 00955				5 44 31	+26 54.4	1	8.52		1.10				3		1724
290861	+0 01181				5 44 34	+00 16.9	1	9.90		1.05		0.31		4	B5	206
38420	+52 01000				5 44 35	+52 34.6	1	8.12		0.21		0.11		2	A3	1566
		STO 673			5 44 35	-69 58.1	1	12.90		0.51		0.00		1	F0 Ia	726
38503	+35 01239	IDS05412N3507		AB	5 44 36	+35 08.6	2	6.58	.000	0.75	.020	0.35		7	F8 Ib-II	6009,8100
38623	+8 01072				5 44 39	+08 31.9	1	8.21		-0.09		-0.32		2	A0	401
		Sk -67 256			5 44 39	-67 16.	1	11.90		-0.08		-0.89		4	B1 Ia	415
		HRC 187		V	5 44 40	+00 08.1	1	13.79		1.11		0.26		1		825
38678	-14 01232	HR 1998			5 44 41	-14 50.4	17	3.54	.011	0.10	.008	0.08	.022	82	A2 Vann	3,15,125,418,1006*
39014	-65 00496	HR 2015			5 44 41	-65 45.3	5	4.33	.006	0.21	.005	0.11	.010	25	A7 V	15,688,1075,2038,3023
38439	+51 01112				5 44 42	+51 08.9	1	7.98		-0.02		-0.31		3	B8	555
		Sk -66 189			5 44 42	-66 28.	1	12.82		-0.17		-1.04		4		415
247593	+23 01046				5 44 43	+23 38.3	1	9.37		1.02				3		1724
		Steph 578			5 44 44	+00 01.7	1	10.99		1.45		1.15		1	M0	1746
		Sk -67 257			5 44 48	-67 13.	1	13.28		-0.19		-1.04		4		798
270100					5 44 48	-67 31.	2	11.83	.005	1.05	.015	0.84		6	G2 Ia	204,1483
		Sk -68 159			5 44 48	-68 29.	1	13.83		-0.03		-0.52		3		952
38650	+4 01038				5 44 51	+04 05.0	1	7.67		-0.06		-0.20		2	B9	401
38622	+13 00979	HR 1993			5 44 52	+13 53.0	1	11.97		0.55				2	B2 IV-V	1752
247638	+24 00964				5 44 52	+24 27.2	1	9.30		0.63				3	G0	1724
38584	+24 00963	IDS05417N2439		A	5 44 52	+24 40.2	2	6.77	.005	1.53	.005	1.78	.010	8	K7 III	1501,1625
38622	+13 00979	HR 1993		⋆ A	5 44 53	+13 53.0	4	5.28	.009	-0.17	.005	-0.67	.009	12	B2 IV-V	154,1203,1223,1363

Table 1   305

HD	DM	Other Id	N	Rem	α_{1950}	δ_{1950}	S	V	σ_V	B–V	σ_{B-V}	U–B	σ_{U-B}	n	Spectrum	References
38713	−16 01244	HR 2000			5 44 53	−16 15.3	1	6.17		0.89				4	G8 Ib-II	2035
		Sk -68 158			5 44 54	−68 24.	1	14.10		0.02		−0.90		3		952
		Sk -69 297			5 44 54	−69 22.	1	12.73		0.11		−0.78		4		415
		Sk -69 296			5 44 56	−69 49.3	1	13.25		0.17		−0.22		2	B9 Ia	149
38462	+53 00950				5 44 57	+54 01.0	1	8.49		0.03		−0.16		2	B9	1566
		Sk -68 160			5 44 57	−68 54.	1	13.34		0.05		−0.40		3		952
38583	+30 01015	IDS05418N3030	A		5 44 58	+30 31.1	1	7.00		1.50		1.70		2	F1 V	3016
38583	+30 01015	IDS05418N3030	B		5 44 58	+30 31.1	1	10.18		0.48		0.06		2		3016
	+22 01038				5 45 00	+22 29.7	1	9.65		0.96				2	F5	1724
270105	−67 00501	Sk -67 258			5 45 00	−67 21.0	2	12.12	.057	−0.03	.035	−0.61	.000	6	O8	149,360
38871	−46 01999	HR 2008		⋆ A	5 45 04	−46 36.9	3	5.30	.004	1.04	.000			11	K0/1 II	15,1075,2035
247682	+23 01052				5 45 05	+23 21.6	1	9.58		0.96				3		1724
38735	−10 01281	HR 2001, V1031 Ori		⋆	5 45 06	−10 33.0	2	6.02	.005	0.16	.000			7	A4 V	15,2027
		Sk -67 259			5 45 06	−67 12.	1	14.24		−0.24		−1.08		4	WN4	980
39194	−70 00447				5 45 06	−70 10.8	4	8.09	.014	0.76	.004	0.28	.010	16	G8 V	158,1097,2012,3078
	+36 01261	LS V +36 020			5 45 07	+36 11.7	1	9.09		0.28		−0.47		2	B2 III:p:	1012
38804	−28 02449	HR 2005			5 45 07	−28 39.4	1	6.22		−0.15				4	B5 III	2007
270111	−67 00502	Sk -67 260			5 45 07	−67 11.8	3	10.22	.022	0.68	.016	0.49	.017	10	F8 Ia	149,360,1483
38789	−26 02466				5 45 08	−26 16.5	1	8.42		0.62				4	F8 IV	2012
38672	+12 00902	IDS05424N1223	A		5 45 14	+12 24.1	1	6.68		−0.10		−0.39		2	B5	401
		Sk -69 298			5 45 14	−69 33.5	1	12.90		0.22		−0.24		3	A0 Ia	415
		G 99 - 33			5 45 15	+08 21.7	2	14.15	.039	1.44	.010	1.34		3		203,3078
38671	+17 01011				5 45 17	+17 28.8	1	8.59		0.06		−0.05		2	A0	401
38710	+6 01027	HR 1999		⋆ AB	5 45 19	+06 26.3	3	5.26	.005	0.24	.005	0.15	.009	8	A5 V	15,1061,3023
38755	−6 01313				5 45 21	−06 27.1	1	7.68		−0.11				1	B6 V	1004
270072					5 45 21	−69 32.7	2	12.19	.000	0.64	.000	0.06	.005	9	G5 III	149,204
38670	+20 01105	HR 1997		⋆ AB	5 45 23	+20 51.2	3	6.07		−0.09	.009	−0.39	.025	9	B9 Vn	1049,1079,3032
38604	+39 01416				5 45 23	+39 31.1	1	6.80		0.81		0.41		2	G0	1601
38771	−9 01235	HR 2004			5 45 23	−09 41.2	15	2.06	.010	−0.18	.007	−1.02	.010	75	B0.5Ia	1,3,15,154,198,1004*
38803	−20 01185				5 45 23	−20 15.4	1	10.14		−0.13		−0.50		5	B8 II	1732
247754	+25 00989	LS V +25 005			5 45 24	+25 05.9	1	9.65		0.26		−0.60		2	B1 V:nn	1012
		Sk -67 263			5 45 24	−67 08.	1	14.80		−0.04		−0.68		4	WN4	980
		Sk -67 264			5 45 24	−67 11.	1	13.68		−0.21		−1.05		3		798
38693	+20 01106	IDS05424N2050	C		5 45 25	+20 50.0	1	8.33		0.03		0.01		1		3016
		Sk -67 261			5 45 26	−67 12.	1	12.50		−0.13		−0.95		4		415
39062	−61 00517				5 45 29	−61 14.8	3	7.40	.009	1.28	.009	1.38	.013	19	K2 III	1075,1408,1499
		Sk -67 262			5 45 30	−67 22.	1	13.52		−0.23		−1.05		3		798
		G 100 - 37			5 45 31	+21 17.6	1	13.94		1.67		1.18		1		1773
38658	+28 00902	LS V +28 011			5 45 32	+28 18.5	1	8.35		0.20		−0.59		2	B3 II	1012
38940	−45 02160				5 45 32	−45 39.9	2	7.41	.008	0.50	.003	0.05		14	F5 V	1499,2012
		WLS 600 45 # 6			5 45 33	+45 11.4	1	11.26		0.33		0.23		2		1375
38885	−35 02509	HR 2009			5 45 33	−35 41.5	3	6.31	.007	1.19	.010			11	K1 III	15,2013,2029
38709	+17 01013				5 45 34	+17 25.1	1	7.29		0.06		0.07		2	B9	401
36905	+85 00080	HR 1885			5 45 34	+85 10.5	2	6.10	.005	1.56	.004	1.90		4	M0 III	71,3001
		Sk -67 265			5 45 34	−67 53.3	1	12.40		0.03		−0.29		2	A1 Iab	149
		Ki 8 - 18			5 45 37	+33 41.6	1	12.72		0.65				2		1623
		Sk -66 190			5 45 37	−66 54.0	1	12.58		−0.15		−0.96		4		149
270086	−69 00520	Sk -69 299			5 45 37	−69 00.9	4	10.28	.020	0.24	.007	−0.26		25	A1 Ia0	149,1277,1368,8033
		SK *G 465			5 45 38	−67 42.4	1	12.82		0.18		0.26		4	A3 Ib	149
38800	−6 01314				5 45 40	−06 08.7	1	8.46		0.02		−0.19		2	B8	401
38656	+39 01418	HR 1995		⋆ A	5 45 42	+39 10.0	20	4.51	.007	0.95	.006	0.69	.005	260	G8 III	15,61,71,985,1007,1013*
38708	+29 01005	LS V +29 010			5 45 43	+29 07.2	2	8.15	.066	−0.04	.033	−0.50	.009	4	B3:pe:Shell	379,1012
38688	+29 01004				5 45 43	+29 44.0	1	8.25		0.46				2	F2	1724
247770	+31 01104				5 45 43	+31 31.0	1	9.25		0.15		−0.46		2	A7	401
		Ki 8 - 19			5 45 45	+33 38.1	1	13.27		0.68		0.24		2		1623
38687	+30 01020				5 45 46	+30 41.8	2	8.36	.161	0.44	.015	−0.11		3	F0	379,1724
38824	−8 01219	IDS05434S0825	A		5 45 47	−08 24.0	1	7.29		−0.11		−0.54		2	B9	401
38798	+4 01046	IDS05432N0440	AB		5 45 49	+04 41.3	1	7.29		0.03		−0.07		2	A0	401
		Ki 8 - 8			5 45 50	+33 37.8	1	15.82		0.37				1		1623
	+36 01263				5 45 50	+36 22.1	1	9.73		1.94				3	M3	369
	−36 02458				5 45 53	−36 20.6	3	10.72	.012	1.48	.013	1.11		8	M2	158,912,1705
247795	+31 01106	LS V +31 007			5 45 54	+31 49.2	1	9.61		0.11		−0.79		2	B7	405
247818	+28 00906				5 45 55	+28 12.7	1	9.52		1.85				2	A0	369
		Ki 8 - 44			5 45 55	+33 37.3	1	16.18		0.59				1		893
38686	+38 01303				5 45 55	+38 43.4	1	7.20		1.17		1.12		3	K2 III	833
38989	−41 02066				5 45 55	−41 36.4	2	6.80	.000	1.58	.019	1.75		12	M3 III	2012,3040
38751	+24 00970	HR 2002		⋆	5 45 57	+24 33.2	6	4.88	.014	1.02	.008	0.81	.010	14	G8 III	1080,1355,1363,1724*
		Ki 8 - 11			5 45 57	+33 37.5	1	12.20		0.57				3		1623
		L 810 - 58			5 45 57	−11 09.3	1	11.00		1.42				1		1746
		Sk -67 266			5 45 57	−67 15.2	1	12.01		−0.13		−0.95		2		149
		Ki 8- 109			5 45 58	+33 35.9	1	15.42		1.06		0.59		1		1623
		Ki 8 - 89			5 45 58	+33 37.8	1	12.20		0.57				3		893
38797	+11 00945	IDS05432N1157	A		5 45 59	+11 58.9	1	6.93		−0.10		−0.38		2	B8	401
		Ki 8- 120			5 45 59	+33 34.6	1	15.82		0.37				1		893
		Ki 8- 107			5 45 59	+33 35.7	1	15.55		1.20		0.95		1		1623
38723	+34 01182				5 46 00	+34 11.7	1	7.50		0.94				3	G5	1724
	−48 01982				5 46 00	−48 31.9	1	9.74		1.38		1.22		2	M0 V	3072
38750	+25 00991				5 46 01	+25 38.1	2	7.22	.020	1.42	.020			4	K2 II	20,1724
247835	+30 01021				5 46 01	+30 05.1	1	8.54		1.36		1.46		1	K2	401
		Ki 8- 121			5 46 01	+33 35.1	1	13.27		0.68		0.24		2		893

HD	DM	Other Id	N Rem	α_{1950}	δ_{1950}	S	V	σ_V	B–V	σ_{B-V}	U–B	σ_{U-B}	n	Spectrum	References
		Ki 8 - 6		5 46 01	+33 36.0	1	11.27		0.56				2		1623
		Ki 8 - 23		5 46 02	+33 37.1	1	14.05		0.54				1		893
		Ki 8 - 79		5 46 03	+33 37.5	1	15.42		1.06		0.59		1		893
38707	+36 01264	IDS05427N3609	A	5 46 03	+36 10.1	1	8.77		0.48				2	F5	1724
		Ki 8 - 67		5 46 04	+33 36.2	1	11.27		0.56				2		893
		Ki 8 - 12		5 46 05	+33 38.9	1	14.05		0.54				1		1623
		Ki 8 - 71		5 46 06	+33 36.8	1	15.55		1.20		0.95		2		893
38858	−4 01244	HR 2007		5 46 06	−04 06.4	4	5.96	.014	0.64	.000	0.10	.008	10	G2 V	15,247,1061,3026
39060	−51 01620	HR 2020		5 46 06	−51 05.0	5	3.85	.005	0.17	.005	0.09	.010	16	A5 V	15,611,1075,2012,3023
		Sk -68 161		5 46 06	−68 06.	1	14.12		-0.15		-0.97		4		798
		Ki 8 - 3		5 46 07	+33 36.5	1	16.18		0.59				1		1623
		Ki 8 - 14		5 46 08	+33 39.7	1	14.00		0.48		0.29		3		893
		SU Tau		5 46 09	+19 03.1	1	10.12		0.99		0.62		1	G0	842
		Ki 8- 135		5 46 09	+33 36.8	1	12.72		0.65				2		893
		Ki 8 - 1		5 46 09	+33 37.5	2	10.21	.000	1.40	.000			4		893,1623
38868	−5 01406			5 46 09	−05 50.7	1	8.23		0.06		0.05		2	A0	401
270123	−67 00508	Sk -67 267		5 46 09	−67 51.5	1	11.47		0.04		-0.47		6	A0:Ia	149
		Ki 8- 134		5 46 10	+33 36.3	1	11.13		0.61		-0.03		2		893
		Ki 8 - 2		5 46 10	+33 37.2	2	13.93	.000	2.19	.000			3		893,1623
38856	+0 01184			5 46 11	+00 42.6	1	7.25		-0.14		-0.55		2	B8	401
		SK *G 468		5 46 11	−67 25.9	1	13.04		0.05		0.20		2	A0 Ib	149
		Ki 8- 133		5 46 12	+33 36.3	1	11.60		0.47		0.17		2		893
39110	−54 00892	HR 2023		5 46 12	−54 22.6	3	6.18	.008	1.41	.005			11	K3 III	15,1075,2007
38796	+21 01003			5 46 13	+21 07.3	1	7.32		-0.07		-0.36		2	B9	401
		G 99 - 36		5 46 13	−03 39.3	1	15.06		1.47		1.20		1		3056
		Ki 8 - 20		5 46 14	+33 37.7	1	14.00		0.48		0.29		2		1623
39002	−35 02518			5 46 16	−35 20.7	1	8.27		0.92		0.56		9	G8/K0 IV	1673
38808	+24 00973			5 46 17	+24 12.5	1	7.53		1.01				2	G3 Ib-II	1724
38618	+56 01065	HR 1992	⋆A	5 46 17	+56 54.3	1	6.59		0.09		0.12		2	A4 IV-V	1733
247920	+23 01069			5 46 18	+23 36.6	1	9.18		0.61				3	G0	1724
		GD 70		5 46 19	+26 30.5	1	17.18		0.14		-0.54		1		3060
248018	+9 00970			5 46 20	+09 52.7	1	8.95		1.20		1.25		2	K2	3077
247940	+23 01070			5 46 21	+23 54.9	1	9.41		1.13				3		1724
39040	−40 02085	HR 2017		5 46 22	−40 40.1	1	6.61		1.12				4	K1 III	2035
270132	−67 00510			5 46 22	−67 35.0	1	11.50		0.14		-0.11		5	A8:I	149
247967	+20 01116			5 46 23	+20 34.2	1	9.02		-0.03		-0.52		2	F7	401
247912	+30 01022	LS V +30 020		5 46 25	+30 08.6	1	9.67		0.21		-0.61		2	B3	401
38730	+45 01181	IDS05428N4504	AB	5 46 26	+45 04.7	1	8.15		0.10		-0.01		2	A0 V	401
38438	+71 00324			5 46 27	+71 16.6	1	7.14		0.15		0.13		2	A3	985
38900	+4 01048			5 46 29	+04 11.9	1	7.82		-0.06		-0.21		2	B9	401
247909	+32 01098	IDS05432N3256	A	5 46 29	+32 57.5	2	8.47	.005	0.74	.000	0.32		6	K0	1625,1724
247909	+32 01098	IDS05432N3256	B	5 46 29	+32 57.5	1	8.73		0.79				2		1724
38807	+27 00880			5 46 30	+27 32.2	1	6.99		1.48				2	K0	1724
39027	−33 02547			5 46 30	−33 26.8	1	6.85		1.38				4	K3 III	2012
		G 98 - 25		5 46 33	+36 50.0	1	12.19		1.45		1.24		1		333,1620
38777	+39 01421	IDS05431N3933	AB	5 46 34	+39 33.7	1	8.61		0.09		-0.28		2	A0	401
		Ki 8 - 17		5 46 35	+33 39.0	1	11.13		0.61		-0.03		1		1623
39039	−34 02459			5 46 35	−34 57.1	2	7.29	.008	0.94	.003	0.66		15	G8 III	1499,2012
		Sk -68 162		5 46 36	−68 13.	1	13.74		-0.02		-1.07		4		798
		SK *S 93		5 46 37	−66 51.3	1	13.75		0.04		0.08		2		149
		Ki 8 - 16		5 46 38	+33 39.1	1	11.60		0.47		0.17		1		1623
		S016 # 3		5 46 40	−80 31.3	1	11.14		1.27		1.17		2		1730
38899	+12 00912	HR 2010	⋆A	5 46 44	+12 38.2	11	4.90	.016	-0.07	.004	-0.16	.022	56	B9 IV	1,3,15,1006,1034,1049*
		Sk -68 163		5 46 48	−68 10.	1	13.70		-0.14		-0.89		4		798
270069		Sk -71 053		5 46 48	−71 15.0	2	12.64	.010	0.16	.005	0.16	.010	5	A2 Ib	149,415
248051	+25 00997			5 46 54	+25 22.9	1	9.73		1.01				3		1724
38775	+49 01411			5 46 54	+49 03.2	1	8.39		0.29		0.14		3	B8	555
270149		Sk -67 268		5 46 54	−67 11.	1	14.51		-0.18		-0.87		4	WN3	980
		SK *G 470		5 46 57	−68 47.3	1	13.00		0.18		0.20		3		149
38765	+51 01117	HR 2003		5 46 58	+51 30.1	2	6.30	.005	1.05	.002	0.90		5	K1 III	71,1501
38864	+32 01102			5 47 00	+32 08.5	1	8.78		0.85				3	K0	1724
38818	+42 01415			5 47 00	+42 58.2	1	8.48		0.52		0.04		2	F5	1375
39000	−0 01095			5 47 00	−00 41.8	1	7.62		-0.07		-0.38		2	A0	401
248048	+31 01114			5 47 06	+31 58.9	1	9.61		1.34		1.28		2		1726
38909	+31 01115	LS V +31 008		5 47 15	+31 03.1	3	8.18	.031	0.01	.005	-0.61	.023	13	B3 II-III	1012,1775,8100
39192	−42 02163			5 47 15	−42 20.4	1	8.32		0.56				4	F7 V	2012
		STO 871		5 47 15	−71 29.3	1	13.09		0.54		-0.01		1	F Ia	726
		Sk -67 269		5 47 16	−67 28.	1	13.24		-0.08		-0.83		4		415
39007	+9 00978	HR 2014	⋆AB	5 47 17	+09 51.5	6	5.79	.009	0.87	.007	0.62	.007	13	G8 III	15,247,1061,1355,3051,6009
		Sk -67 270		5 47 18	−67 59.	1	13.54		-0.20		-0.98		3		798
270133	−68 00439			5 47 19	−68 40.0	2	10.65	.010	0.45	.000	0.01	.000	8	F2 V:	149,204
39070	−14 01251	HR 2021	⋆AB	5 47 20	−14 29.8	3	5.49	.008	0.87	.010			11	G8 III	15,1075,2007
248104	+32 01105		V	5 47 23	+32 25.1	1	9.22		1.88				3	K7	369
		G 102 - 33		5 47 29	+17 18.8	1	15.64		1.40				1	sdK:	782
248175	+23 01077	G 100 - 43		5 47 29	+23 27.1	1	8.83		0.69		0.14		2	G5	333,1620
		NGC 2099 - 1861		5 47 29	+32 23.0	1	14.32		1.88		2.10		2		93
38980	+27 00886			5 47 31	+27 29.5	1	7.11		0.44				2	F2 V	1724
248149	+31 01117			5 47 31	+31 11.6	1	8.94		1.77				4	K7	369
		NGC 2099 - 1680		5 47 31	+32 23.0	1	14.50		1.24		0.90		2		93
38645	+68 00412	HR 1994		5 47 33	+68 27.6	1	6.20		0.95				2	G9 III	71

Table 1 307

HD	DM	Other Id	N	Rem	α_{1950}	δ_{1950}	S	V	σ_V	B–V	σ_{B-V}	U–B	σ_{U-B}	n	Spectrum	References
39051	+4 01052	HR 2019			5 47 34	+04 24.6	4	5.96	.010	1.36	.012	1.57	.009	11	gK2	15,247,1061,1355
39018	+17 01029				5 47 37	+18 00.8	1	7.72		-0.08		-0.31		2	B9	401
39019	+14 01041	HR 2016			5 47 38	+14 17.6	1	5.52		1.01				2	K0	71
38944	+37 01336	HR 2011			5 47 38	+37 17.6	4	4.74	.005	1.62	.006	1.91	.014	12	M0 III	15,1363,3016,8015
38998	+27 00887				5 47 39	+27 40.4	1	7.49		1.68				2	M1	369
		Sk -68 164			5 47 42	-68 02.	1	13.47		-0.16		-1.01		4		798
39082	+4 01054				5 47 44	+04 56.6	1	7.42		0.02		-0.06		2	B9	401
248209	+28 00912				5 47 46	+28 36.7	1	9.78		0.74				3		1724
248207		NGC 2099 - 1217			5 47 48	+32 35.4	1	11.27		0.41		0.22		4	F0	93
39190	-23 03135	HR 2026			5 47 48	-22 59.1	3	5.87	.005	0.06	.005			14	A1 V	15,1075,2029
39004	+27 00888	HR 2013		⋆ A	5 47 49	+27 57.3	3	5.57	.012	0.97	.008			11	K0	71,1724,6002
248210	+28 00914				5 47 51	+28 31.5	1	9.57		1.21				3	K0	1724
38831	+58 00863	HR 2006			5 47 53	+58 57.1	1			-0.04		-0.14		2	A0 Vs	1079
39118	+1 01148	HR 2024			5 47 54	+02 00.7	2	5.97	.005	0.91	.000	0.30	.000	7	G8 III+A0IV	15,1415
39280	-44 02271				5 47 55	-44 41.8	2	7.73	.009	0.93	.009			8	G8 III	1075,2033
248271	+24 00984				5 47 57	+24 55.9	1	9.15		1.26				3	K0	1724
39047	+25 01005				5 47 57	+25 13.0	1	9.08		0.48				2	F5	1724
39098	+14 01047	IDS05451N1424	AB		5 47 58	+14 25.8	1	6.75		-0.07		-0.34		2	B9	401
		NGC 2099 - 1075			5 47 59	+32 37.4	1	14.16		1.31		1.39		2		93
		Sk -69 300			5 48 00	-69 34.	1	14.14		-0.09		-0.86		3		952
39099	+14 01048				5 48 01	+14 02.1	2	6.57	.030	1.05	.005	0.86	.050	3	K1 III	1003,3040
39003	+39 01429	HR 2012		⋆ A	5 48 01	+39 08.2	16	3.96	.007	1.13	.007	1.08	.006	222	K0 III	15,61,71,985,1118,1355*
39003	+39 01429	IDS05446N3907	B		5 48 01	+39 08.2	1	11.40		1.83				1		3024
39161	-5 01419				5 48 01	-04 59.4	1	8.94		0.09		0.05		2	A0	401
		WLS 600 80 # 10			5 48 02	+80 40.4	1	12.10		0.55		0.05		2		1375
248296	+22 01061				5 48 03	+22 10.3	1	10.05		0.46				2	F2	1724
		GD 257			5 48 04	+00 05.2	2	14.90	.148	-0.29	.005	-1.20	.010	7	DAwk	1727,3060
39169	-1 01038	IDS05456S0127	AB		5 48 06	-01 26.5	1	7.85		1.06		0.87		4	G5	3024
		G 102 - 34			5 48 07	+10 56.4	1	14.53		1.31				1	sdK:	782
		NGC 2099 - 934			5 48 07	+32 36.5	1	15.52		1.45		1.04		2		93
39170	-1 01039	IDS05456S0127	C		5 48 07	-01 26.6	1	9.38		0.24		0.13		4	F0	3024
39312	-44 02274	HR 2032			5 48 07	-44 53.3	1	6.38		1.27				4	K1 III	2035
		Sk -69 301			5 48 08	-69 00.5	1	13.44		0.07		-0.21		4	A0 Iab	415
248291	+23 01082				5 48 09	+23 51.6	1	9.56		0.49				3	G0	1724
39045	+32 01109	HR 2018		⋆ A	5 48 10	+32 06.7	7	6.25	.019	1.75	.010	2.01	.019	22	M3 III	15,148,369,1003,3001*
		WLS 600 25 # 6			5 48 11	+25 38.6	1	11.80		0.40		0.17		2		1375
		NGC 2099 - 810			5 48 11	+32 30.3	1	12.83		0.29		0.21		3		93
		Sk -67 271			5 48 11	-67 40.4	1	13.30		0.07		-0.49		3	B8 Ib	415
		NGC 2099 - 1191			5 48 12	+32 20.1	1	14.06		1.53		1.62		2		93
		Sk -67 272			5 48 12	-67 47.	1	13.80		-0.20		-0.95		5		798
		Sk -68 165			5 48 13	-68 14.0	1	12.89		0.04		-0.16		5	B9 Ib	149
	+30 01031				5 48 14	+30 36.7	1	9.60		0.35		0.14		2		1726
		NGC 2099 - 685			5 48 21	+32 27.4	1	10.63		1.48		1.64		2	G8	93
		NGC 2099 - 687		⋆ A	5 48 22	+32 28.8	1	11.85		0.28		0.22		3	A0	93
		NGC 2099 - 706			5 48 23	+32 36.9	1	13.34		1.38		1.46		2		93
39096	+28 00918				5 48 24	+28 04.8	1	8.09		0.01		-0.19		2	B8 III	379
		NGC 2099 - 571			5 48 24	+32 32.5	1	14.51		0.79		0.37		1		49
270202	-66 00458	MN83,95 # 801			5 48 24	-66 34.4	1	10.56		0.32				1		591
		NGC 2099 - 572			5 48 25	+32 32.8	1	14.82		0.54		0.40		1		49
		MN83,95 # 802			5 48 25	-66 39.6	1	11.50		0.62				1		591
		Be 21- 509			5 48 28	+21 44.7	1	12.39		0.51		-0.01		3		775
39095	+33 01172				5 48 29	+33 36.6	1	8.65		0.53				3	G5	1724
		Be 21- 508			5 48 31	+21 44.7	1	12.97		0.58		-0.02		3		775
		Be 21- 335			5 48 31	+21 47.2	1	14.04		0.71		0.13		2		775
248356	+27 00891				5 48 31	+27 49.8	1	8.29		1.00				3	K0	1724
39116	+30 01034	IDS05453N3043	A		5 48 31	+30 44.3	1	8.17		0.44				2	F4 V	1724
		Sk -68 166			5 48 31	-68 16.	1	12.76		-0.13		-0.98		4		415
		Be 21- 250			5 48 32	+21 47.0	1	12.80		0.92		0.31		4		775
		NGC 2099 - 422			5 48 32	+32 33.3	1	14.36		1.28		1.02		2		93
	-78 00181				5 48 32	-78 23.9	1	9.16		0.81				4	K0	2034
		NGC 2099 - 843			5 48 33	+32 41.4	1	14.89		0.51		0.32		2		93
		Be 21- 254			5 48 34	+21 45.4	1	13.53		1.27		0.57		2		775
248412	+24 00989				5 48 35	+24 41.6	1	8.93		1.35				3	K0	1724
		NGC 2099 - 433			5 48 35	+32 36.2	1	13.22		0.40		0.35		1		49
		NGC 2099 - 1723			5 48 36	+32 51.3	1	14.20		1.25		1.08		2		93
		Be 21- 159			5 48 37	+21 46.0	1	18.46		1.04		0.20		1		775
		NGC 2099 - 285			5 48 38	+32 31.9	1	12.76		0.42		0.26		1		49
		Be 21- 260			5 48 39	+21 45.0	1	13.65		0.87		0.15		2		775
39136	+32 01111	LS V +32 012			5 48 39	+32 13.7	2	8.79	.009	0.10	.005	-0.46	.005	4	B3 III	401,1012
		NGC 2099 - 1654			5 48 39	+32 13.8	1	18.80		0.09		-0.49		4		93
		NGC 2099 - 401			5 48 39	+32 29.3	3	11.32	.127	1.24	.071	1.22	.187	5	gG5	49,4002,8092
		NGC 2099 - 440			5 48 39	+32 37.0	1	14.09		0.53		0.41		1		49
		NGC 2099 - 716			5 48 40	+32 39.6	1	11.62		1.14				3		8092
		MN83,95 # 805			5 48 40	-66 34.3	1	13.95		0.84				1		591
39210	+10 00908	IDS05459N1013	A		5 48 41	+10 14.4	1	8.81		0.01				2	A0	6009
		NGC 2099 - 286			5 48 41	+32 32.5	1	14.05		0.52		0.36		1		65
		Be 21- 521			5 48 42	+21 46.	1	11.13		0.53		0.02		1		775
		Be 21- 168			5 48 42	+21 46.0	1	16.88		1.42		0.98		1		775
		NGC 2099 - 291			5 48 42	+32 33.7	1	14.12		0.61		0.30		1		49
		NGC 2099 - 437		⋆ A	5 48 42	+32 36.3	1	11.42		0.33		0.31		1	A0	49

HD	DM	Other Id	Rem	α_{1950}	δ_{1950}	S	V	σ_V	B–V	σ_{B-V}	U–B	σ_{U-B}	n	Spectrum	References
		Be 21- 227		5 48 43	+21 50.6	1	15.46		0.88		-0.28		4		775
		NGC 2099 - 281		5 48 43	+32 31.7	1	13.54		0.49		0.33		1		49
		NGC 2099 - 290		5 48 43	+32 33.6	1	13.58		0.45		0.37		1		49
		G 102 - 35		5 48 44	+11 47.9	1	13.16		1.52		1.13		1		1620
		Be 21- 513		5 48 45	+21 51.0	1	13.48		1.31		0.42		3		775
	+26 00980	LS V +27 012		5 48 45	+27 00.9	2	10.15	.000	0.41	.005	-0.54	.000	6	B0 V	342,684
248411	+28 00920	LS V +28 013		5 48 45	+28 15.0	1	8.95		-0.02		-0.06		2	B5	379
		Be 21- 347		5 48 46	+21 44.8	1	12.46		1.36				1		775
		Be 21- 504		5 48 46	+21 49.3	1	11.02		0.34		0.02		4		775
	+32 01112	NGC 2099 - 277		5 48 46	+32 30.0	3	10.31	.022	1.41	.014	1.35	.092	5	gG8	49,4002,8092
		NGC 2099 - 167		5 48 46	+32 32.3	1	12.11		0.34		0.32		1	A3	65
		G 99 - 37		5 48 46	-00 11.2	3	14.56	.033	0.48	.022	-0.46	.025	16	DGp	308,940,3060
		Be 21- 312		5 48 47	+21 50.5	1	16.21		0.91		0.44		1		775
248429	+25 01008			5 48 48	+25 40.2	1	9.49		0.75				3	K0	1724
		NGC 2099 - 533		5 48 48	+32 26.1	1	11.28		0.32		0.28		3	A0	93
		NGC 2099 - 161		5 48 48	+32 31.4	1	14.23		0.50		0.30		1		65
		NGC 2099 - 168		5 48 48	+32 32.4	1	12.87		0.36		0.28		1		65
		Sk -67 274		5 48 48	-67 33.	1	13.82		0.00		-1.13		4		798
		ApJS2,315 # 63		5 48 48	-67 36.	1	11.61		0.38				3		8033
		Sk -67 273		5 48 48	-67 50.	1	13.68		-0.07		-1.10		4		798
		NGC 2099 - 391	⋆B	5 48 49	+32 27.8	1	10.92		1.47				3	gG8	8092
39810	-72 00418	HR 2062		5 48 49	-72 43.0	3	6.52	.003	1.07	.005	0.98	.005	18	K0 III	15,1075,2038
		NGC 2099 - 160		5 48 50	+32 31.3	1	13.41		0.29		0.21		1		65
		NGC 2099 - 447		5 48 50	+32 37.4	1	14.75		0.83		0.06		1		49
		Be 21 - 90		5 48 51	+21 47.7	1	15.32		0.93				2		775
		NGC 2099 - 180	⋆B	5 48 51	+32 33.6	2	10.73	.010	0.12	.007	-0.20	.020	2	A0	49,65
		NGC 2099 - 148		5 48 52	+32 29.9	2	11.07	.015	1.26	.006	1.01		5	gG5	4002,8092
		NGC 2099 - 46		5 48 52	+32 31.3	1	11.70		0.36		0.35		1	A0	65
		NGC 2099 - 178	⋆A	5 48 52	+32 33.3	1	10.65		0.25		0.26		1	A0	65
		NGC 2099 - 183		5 48 52	+32 34.4	2	12.87	.031	0.40	.012	0.26	.020	2		49,65
		Be 21- 200		5 48 53	+21 46.7	1	11.32		0.50		0.05		3		775
		NGC 2099 - 45		5 48 53	+32 31.3	1	13.20		0.45		0.27		1		65
		NGC 2099 - 55		5 48 53	+32 32.5	1	13.60		0.68		0.07		1		65
		NGC 2099 - 181		5 48 53	+32 33.5	1	14.17		0.54		0.12		1		65
248496	+19 01105			5 48 54	+19 05.3	1	9.47		1.67				2	K0	369
		Be 21- 514		5 48 54	+21 50.5	1	13.75		0.74		0.25		2		775
39227	+19 01106			5 48 55	+19 30.6	1	7.31		-0.03		-0.38		2	B9	401
		Be 21- 211		5 48 55	+21 47.8	1	15.10		1.43				2		775
		NGC 2099 - 44		5 48 55	+32 31.6	1	13.95		0.57		0.27		1		65
		NGC 2099 - 14		5 48 55	+32 33.7	1	14.42		0.70		0.26		1		65
39523	-56 00946	HR 2042		5 48 55	-56 10.7	4	4.50	.006	1.10	.000	0.98	.011	16	K1 III	15,1075,2012,3077
		NGC 2099 - 454		5 48 56	+32 38.0	1	11.16		1.24				3	gG8	8092
		NGC 2099 - 35		5 48 57	+32 31.5	1	13.48		0.46		0.24		1		65
		NGC 2099 - 64	⋆B	5 48 57	+32 33.6	2	11.15	.000	1.27	.061	0.95		4		65,8092
		NGC 2099 - 190		5 48 57	+32 35.0	1	15.52		0.83		0.36		1		65
		NGC 2099 - 455		5 48 57	+32 37.7	1	15.01		0.72				1		49
		NGC 2099 - 600		5 48 57	+32 38.8	1	11.53		0.31		0.26		3	A0	93
39291	-7 01187	HR 2031		5 48 57	-07 31.8	6	5.35	.007	-0.20	.008	-0.83	.010	20	B2 IV-V	15,154,1004,1415,1732,2012
39427	-43 02082			5 48 57	-43 42.9	1	8.70		0.69				4	G6 V	2012
		NGC 2099 - 319		5 48 58	+32 36.3	1	15.02		1.33		1.31		2		93
		NGC 2099 - 456		5 48 58	+32 37.2	1	14.75		0.59		0.21		1		49
271586				5 48 59	-64 47.8	1	11.16		0.92		0.46		3	K0	149
		Be 21- 501		5 49 00	+21 47.0	1	11.63		1.34		0.96		4		775
		NGC 2099 - 4		5 49 00	+32 33.2	2	11.54	.023	1.19	.007	0.84		4		4002,8092
		NGC 2099 - 67	⋆B	5 49 00	+32 34.1	3	11.08	.040	1.20	.012	0.94	.056	5		65,4002,8092
		NGC 2099 - 195	⋆A	5 49 00	+32 35.7	1	11.01		0.75		0.43		1	F2	65
		NGC 2099 - 196	⋆B	5 49 00	+32 36.1	1	12.35		0.29		0.20		1	A0	49
39739	-68 00440			5 49 00	-68 24.4	1	10.00		0.14				14	A5 III/IV	1368
		NGC 2099 - 16	⋆C	5 49 01	+32 32.3	2	11.24	.028	1.21	.000	0.94		4		4002,8092
		NGC 2099 - 69	⋆A	5 49 01	+32 35.4	1	11.99		0.34		0.34		1		65
		NGC 2099 - 120		5 49 02	+32 30.6	1	11.36		1.05				3		8092
	+32 01113	NGC 2099 - 9	⋆A	5 49 02	+32 32.5	2	9.18	.022	1.56	.045	1.24	.075	4	F8	65,8032
39608	-60 00487			5 49 02	-60 41.4	1	7.36		1.52		1.90		5	K5 III	1704
		NGC 2099 - 460		5 49 03	+32 40.1	1	11.40		0.25		0.20		3	A0	93
		NGC 2099 - 76		5 49 04	+32 34.4	1	10.69		0.42		0.43		1	A2	65
	+36 01276	IDS05457N3616	A	5 49 04	+36 17.0	1	9.47		0.62				2	G0	1724
	+36 01276	IDS05457N3616	B	5 49 04	+36 17.0	1	10.43		1.63				1		1724
		NGC 2099 - 12		5 49 05	+32 32.8	2	11.67	.009	1.17	.001	0.81		4		4002,8092
		NGC 2099 - 80		5 49 05	+32 33.6	1	15.20		0.76		0.39		1		65
248472	+30 01036			5 49 06	+30 02.7	1	9.75		1.03				4	K0	1724
		NGC 2099 - 2111		5 49 06	+32 32.	1	15.77		0.68		0.17		1		65
		NGC 2099 - 2112		5 49 06	+32 32.	1	16.37		0.92		0.44		1		65
		NGC 2099 - 2113		5 49 06	+32 32.	1	16.68		0.95		0.48		1		65
		NGC 2099 - 2114		5 49 06	+32 32.	1	17.70		0.89		0.30		1		65
		NGC 2099 - 2115		5 49 06	+32 32.	1	17.09		1.02		0.48		1		65
		NGC 2099 - 2116		5 49 06	+32 32.	1	14.15		1.86				2		93
		NGC 2099 - 2117		5 49 06	+32 32.	1	14.65		1.45		1.33		2		93
		NGC 2099 - 2118		5 49 06	+32 32.	1	15.59		1.27		1.37		2		93
		NGC 2099 - 2120		5 49 06	+32 32.	1	10.58		0.17		-0.37		3		93
		NGC 2099 - 2121		5 49 06	+32 32.	1	18.77		0.55		0.52		4		93

Table 1 309

HD	DM	Other Id	N	Rem	α_{1950}	δ_{1950}	S	V	σ_V	B–V	σ_{B-V}	U–B	σ_{U-B}	n	Spectrum	References
		NGC 2099 - 2122			5 49 06	+32 32.	1	15.53		1.38		0.95		2		93
		NGC 2099 - 2123			5 49 06	+32 32.	1	15.30		1.36		1.55		3		93
		NGC 2099 - 2124			5 49 06	+32 32.	1	18.38		0.76				4		93
		NGC 2099 - 2125			5 49 06	+32 32.	1	15.36		1.45		1.06		2		93
		NGC 2099 - 2126			5 49 06	+32 32.	1	16.87		1.00				4		93
		NGC 2099 - 2127			5 49 06	+32 32.	1	17.32		0.73				6		93
		NGC 2099 - 2128			5 49 06	+32 32.	1	16.02		0.82				2		93
		NGC 2099 - 2129			5 49 06	+32 32.	1	15.26		0.41		0.16		2		93
		NGC 2099 - 3118			5 49 06	+32 32.	1	14.94		0.74		0.24		1		49
		NGC 2099 - 725			5 49 07	+32 42.7	1	10.89		1.21		1.01		3	G0	93
		MN83,95 # 803			5 49 07	−66 34.5	1	12.62		0.60				1		591
		NGC 2099 - 255			5 49 08	+32 28.1	1	10.92		0.62		0.46		1	F2	49
		NGC 2099 - 93			5 49 08	+32 32.7	1	12.19		0.40		0.46		1		65
		NGC 2099 - 81	★	A	5 49 08	+32 34.3	1	11.73		0.33		0.21		1	A0	65
		NGC 2099 - 86			5 49 09	+32 33.8	1	15.03		0.78		0.28		1		65
		Be 21- 511			5 49 10	+21 47.3	1	9.97		1.12		0.68		3		775
39364	−20 01211	HR 2035			5 49 10	−20 52.9	8	3.78	.022	1.00	.010	0.68	.019	19	G8 III/IV	15,22,1075,2012,2017*
39182	+39 01435	HR 2025			5 49 11	+39 33.8	2	6.39		0.09	.000	0.12	.039	5	A2 V	1049,1733
39425	−35 02546	HR 2040			5 49 12	−35 47.2	8	3.11	.008	1.16	.005	1.20	.009	32	K1.5III	3,15,58,1075,2012*
		Sk -69 302			5 49 12	−69 29.	1	13.54		0.07		-0.46		3		952
		NGC 2099 - 107			5 49 13	+32 31.8	3	11.30	.011	1.24	.029	0.99	.000	5	G5	65,4002,8092
		NGC 2099 - 208			5 49 13	+32 34.8	1	13.81		0.50		0.32		1		49
		NGC 2099 - 239			5 49 14	+32 31.3	1	11.51		0.32		0.33		1	A0	65
		MN83,95 # 806			5 49 14	−66 34.0	1	14.67		0.67				1		591
		NGC 2099 - 897			5 49 16	+32 20.8	1	14.01		1.58		1.80		3		93
		NGC 2099 - 1574			5 49 16	+32 51.6	1	12.22		0.26		0.16		2		93
		NGC 2099 - 1176			5 49 17	+32 17.9	1	13.46		0.44		0.31		3		93
		NGC 2099 - 508			5 49 17	+32 27.1	1	10.98		1.28				3	gG8	8092
		MN83,95 # 804			5 49 18	−66 36.4	1	13.66		1.11				1		591
39224	+34 01202				5 49 19	+34 15.5	1	7.46		0.35				2	F0	1724
39348	−7 01190				5 49 19	−07 55.9	1	8.31		0.03		0.00		2	A0	401
		NGC 2099 - 656			5 49 20	+32 24.5	1	11.45		1.46				3		8092
39225	+33 01179	HR 2028			5 49 21	+33 54.4	3	5.99	.019	1.61	.011	1.96	.015	9	M1.5 II-III	369,1003,8100
270145	−70 00454	Sk -70 115			5 49 21	−70 04.9	1	12.24		-0.10		-0.98		3	O8	415
		NGC 2099 - 339			5 49 22	+32 35.3	1	13.49		0.53		0.33		1		49
39385	−22 01246	HR 2036			5 49 23	−22 56.3	2	6.16	.005	1.02	.000			7	K0 III	15,2012
39547	−52 00791	HR 2044	★	AB	5 49 23	−52 46.8	3	6.34	.004	0.74	.015			11	K0/1III+A4V	15,1075,2035
		NGC 2099 - 1172			5 49 24	+32 19.0	1	12.27		0.30		0.25		3		93
		NGC 2099 - 350			5 49 24	+32 31.6	2	10.34	.023	1.52	.028	1.59		4	gK0	65,8092
		NGC 2099 - 348			5 49 25	+32 32.3	2	11.66	.052	0.33	.008	0.26	.008	2	A0	49,65
248587	+19 01111	LSS 3			5 49 26	+19 08.3	1	7.94		0.65		0.12		2	A0 Iab	1012
39286	+19 01110	HR 2030			5 49 26	+19 51.4	2	6.06		0.54	.019	-0.02	.005	4	G2III+B8III	401,1049
		NGC 2099 - 342			5 49 26	+32 33.7	2	12.09	.062	0.33	.061	0.21	.008	2	A0	49,65
40410	−80 00164				5 49 26	−80 19.7	1	8.95		0.31		0.01		2	A8 III	1730
248562	+24 00995				5 49 27	+24 07.2	1	8.93		1.11				3	K0	1724
		Sk -67 276			5 49 27	−67 17.	1	12.49		-0.01		-0.47		4		415
		Sk -67 275			5 49 29	−67 16.5	1	12.79		-0.07		-0.65		4	A3 Ia	415
	+32 01115	NGC 2099 - 485			5 49 30	+32 33.5	3	10.22	.024	1.44	.022	1.32	.038	5	gG8	65,4002,8092
248521	+32 01116	NGC 2099 - 489	★	A	5 49 31	+32 32.3	2	9.95	.006	0.64	.009	0.41	.001	2	F5	49,65
		NGC 2099 - 488	★	B	5 49 31	+32 32.7	1	11.16		1.28		0.86		1		65
39317	+14 01060	HR 2033, V809 Tau			5 49 32	+14 09.6	3	5.55		-0.05	.012	-0.07	.013	5	B9 p SiCr	1049,1079,3033
39376	−7 01192	IDS05471S0720		A	5 49 32	−07 19.3	1	7.88		-0.08		-0.39		2	B9	401
248581	+25 01013				5 49 33	+25 55.5	1	9.06		1.03				3	K0	1724
		NGC 2099 - 621			5 49 33	+32 37.6	1	14.74		0.67		0.24		2		93
39251	+34 01203	IDS05462N3425		A	5 49 33	+34 26.1	1	7.75		0.57				2	F8	1724
		GD 71			5 49 34	+15 52.7	4	13.03	.006	-0.25	.005	-1.11	.006	119	DAwk	281,989,1764,3060
39274	+29 01027	IDS05464N2945		AB	5 49 34	+29 45.5	1	8.91		0.54				2	G0	1724
		NGC 2099 - 763			5 49 34	+32 26.0	1	13.11		0.38		0.29		1		49
		NGC 2099 - 626			5 49 34	+32 36.0	1	11.81		0.40		0.33		1		49
		NGC 2099 - 627			5 49 34	+32 36.0	1	13.30		0.40		0.32		1		49
39275	+29 01027	IDS05464N2945		C	5 49 35	+29 45.6	1	9.08		0.72				2	G0	1724
270218		Sk -67 277			5 49 35	−67 09.8	1	12.30		0.04		-0.25		2	A4 Ia	149
270151	−70 00455	Sk -70 116			5 49 35	−70 03.4	3	12.02	.025	0.10	.005	-0.72		7	B2 Ia	149,1277,8033
	+15 00964			V	5 49 37	+15 31.8	1	9.76		1.82				3	M3	369
		CN Ori			5 49 40	−05 25.7	1	12.63		0.04		-0.60		1		1471
248559	+32 01117	NGC 2099 - 1124			5 49 41	+32 45.2	1	10.16		1.06		0.76		3	K5	93
		SK *G 475			5 49 41	−64 33.2	1	13.42		0.15		0.16		6	A0 V	149
		NGC 2099 - 862			5 49 42	+32 40.4	1	12.42		0.36		0.30		3		93
		NGC 2099 - 748			5 49 44	+32 35.4	1	11.72		1.10				3		8092
		NGC 2099 - 746			5 49 44	+32 36.2	1	12.89		0.35		0.32		3		93
39421	−9 01255	HR 2039			5 49 45	−09 03.2	3	5.95	.012	0.10	.000	0.07	.005	11	A2 Vn	15,1061,2006
39640	−52 00794	HR 2049			5 49 45	−52 07.2	3	5.16	.005	0.99	.005	0.72	.005	14	G8 III	15,1075,2012
248602	+27 00895	IDS05466N2759		AB	5 49 46	+27 59.3	1	9.49		0.56				3	G0	1724
39303	+29 01028	IDS05466N2905		AB	5 49 46	+29 06.1	1	7.89		1.00				2	G5	1724
		NGC 2099 - 880			5 49 48	+32 27.7	1	12.90		1.34		1.15		3		93
39419	−6 01337				5 49 48	−06 51.5	1	9.09		0.08		-0.20		2	B9	401
		NGC 2099 - 865			5 49 49	+32 36.1	1	11.49		1.50		1.57		2		93
39400	+1 01151	HR 2037	★	A	5 49 51	+01 50.7	5	4.77	.016	1.38	.005	1.46	.005	19	K1.5 IIb	15,1075,1355,1363,1425
248640	+21 01021				5 49 51	+21 37.2	1	9.28		1.94				2	M0	369
39755	−62 00529				5 49 51	−62 05.5	1	7.82		0.40				4	F2 V	2012

HD	DM	Other Id	N Rem	α_{1950}	δ_{1950}	S	V	σ_V	B–V	σ_{B-V}	U–B	σ_{U-B}	n	Spectrum	References
		Sk -69 303		5 49 54	−69 20.	1	13.69		-0.15		-0.94		2		952
39439	−7 01194			5 49 56	−07 23.2	1	8.86		-0.01		-0.08		2	A0	401
39398	+5 01035			5 49 57	+05 09.2	1	8.44		1.06		0.79		4	G6 II-III	206
39844	−66 00463	HR 2064		5 49 57	−66 54.8	3	5.09	.007	-0.14	.004	-0.48	.010	19	B6 V	15,1075,2038
39340	+26 00985	V593 Tau		5 49 59	+26 26.1	1	8.13		0.10		-0.56		3	B3 V	1212
		Sk -67 278		5 50 00	−67 55.	1	13.87		-0.19		-1.01		4		798
39316	+32 01118	LS V +32 013		5 50 01	+32 10.2	1	8.75		0.53				2	F0 II	1724
248742	+16 00898			5 50 02	+16 12.7	1	10.14		0.05		-0.62		2	B8	401
248665	+27 00897			5 50 02	+27 36.9	1	9.14		0.43				2	F5	1724
		NGC 2099 - 992		5 50 03	+32 31.1	1	15.64		0.33		0.21		2		93
39602	−41 02104	IDS05485S4142	AB	5 50 03	−41 41.6	1	8.01		0.83				4	G8 III +(G)	2012
		HIC 27825	A	5 50 04	+30 32.3	1	10.86		1.50		1.54		2		1723
39543	−29 02556	HR 2043		5 50 04	−29 27.6	4	6.45	.013	1.47	.010	1.72		13	K3 III	15,1075,2007,3005
	−33 02585			5 50 04	−33 15.4	1	9.94		-0.19		-0.76		4	B5	1732
		NGC 2099 - 1591		5 50 06	+32 46.2	1	11.86		0.22		0.03		3		93
39533	−25 02734			5 50 06	−25 57.3	3	6.86	.008	0.93	.004			10	G8 IV	1020,1075,2012
		Sk -68 167		5 50 06	−68 02.	1	13.40		-0.15		-0.87		4		415
39339	+30 01041			5 50 07	+30 26.4	1	8.38		0.58				2	F8	1724
		Sk -68 168		5 50 08	−68 15.4	1	12.77		0.05		-0.17		3	A0 Ib	149
		NGC 2099 - 1147		5 50 09	+32 33.1	1	12.19		0.34		0.31		3		93
		Sk -70 117		5 50 10	−70 09.	1	12.84		-0.20		-1.01		3		415
39357	+27 00899	HR 2034	⋆	5 50 11	+27 36.1	7	4.57	.020	0.02	.060	0.05	.012	26	A0 V	15,369,1049,1363,1784*
248740	+18 00997			5 50 12	+18 56.5	1	8.47		1.74		2.17		1	K7	401
		Sk -68 169		5 50 12	−68 10.	1	13.80		-0.14		-1.03		4		798
248756	+20 01155			5 50 13	+20 14.9	1	9.87		0.31		0.24		1	A7	1722
39330	+36 01282			5 50 14	+36 07.2	1	7.38		1.12		0.82		2	K0	1601
39393	+21 01025			5 50 15	+21 31.9	1	7.69		0.96		0.68		3	G5	848
		Sk -68 170		5 50 15	−68 05.7	1	13.28		-0.05		-0.52		4	B9 Ib	415
		NGC 2099 - 1291		5 50 17	+32 33.6	1	11.95		1.66		1.84		3		93
39394	+18 00998			5 50 18	+18 57.7	1	8.51		0.02		-0.34		2	B9	401
39655	−44 02295			5 50 18	−44 01.6	2	8.54	.000	0.37	.000			8	F0/2 V	1075,2012
270167	−70 00456			5 50 18	−70 02.1	1	10.68		1.21		1.23		1	K0	360
		Sk -70 118		5 50 18	−70 09.	1	13.34		-0.18		-1.00		2		798
39417	+20 01156	HR 2038		5 50 21	+20 17.4	1			-0.07		-0.34		2	B9 V	1079
	+37 01350			5 50 22	+37 29.6	1	9.15		1.50		1.50		2	K0	1601
248796	+16 00900			5 50 23	+16 18.0	1	8.80		1.76				2	K7	369
		G 98 - 30		5 50 24	+26 08.3	1	12.05		1.10		1.03		1	K5	333,1620
		G 99 - 41		5 50 25	+04 42.6	1	12.19		1.43		1.26		1		333,1620
39416	+25 01020			5 50 25	+25 03.8	4	7.50	.010	1.05	.007	0.74	.010	13	G3 Ib-II	848,1625,1702,1724
248753	+25 01019	LS V +25 007		5 50 25	+25 44.0	1	8.44		0.23		-0.71		3	B1 Venn	1212
294443	−3 01220	G 99 - 40		5 50 25	−03 30.0	2	9.19	.035	0.56	.005	-0.01	.010	2	F8	1620,1658
		STO 684		5 50 26	−69 25.6	1	12.98		0.56		0.03		1	G Ia	726
		Sk -70 119		5 50 28	−70 10.	1	12.92		-0.10		-0.98		2		415
39220	+59 00920	HR 2027, TU Cam		5 50 29	+59 52.8	3	5.27	.202	0.01	.008	0.03	.005	8	A0 V	1501,1733,3016
39436	+24 01007			5 50 30	+24 16.6	1	8.08		-0.06		-0.40		3	B8	848
		G 192 - 7		5 50 30	+47 14.5	1	12.75		0.55		-0.14		1		1658
39902	−65 00507			5 50 30	−65 16.3	1	7.91		0.05		0.06		2	A0 V	204
39455	+18 01001			5 50 31	+18 09.6	1	7.59		0.46				3	F5 II	6009
39435	+25 01021			5 50 32	+25 35.0	2	8.52	.005	0.09	.015	-0.42	.015	6	A0	833,1625
248768	+27 00900	LS V +27 014		5 50 32	+27 31.2	1	9.07		1.05				2	B9 II	1724
	−6 01339			5 50 34	−06 00.0	2	9.72	.024	1.33	.012	1.25		3	M0 V	1705,3072
		AO 900		5 50 35	+20 31.7	1	10.24		0.35		0.31		1		1722
270220		Sk -68 171		5 50 36	−68 12.0	2	12.04	.042	-0.07	.033	-0.94	.075	4	B0.5I	149,360
39453	+22 01084			5 50 37	+22 25.1	1	9.23		0.10		-0.09		3		848
		Sk -66 191		5 50 37	−66 49.3	1	13.20		0.04		-0.10		3	B8 Ib	798
39283	+55 01027	HR 2029		5 50 39	+55 41.9	4	4.97	.022	0.05	.004	0.11	.009	10	A2 V	15,1363,3050,8015
39493	+14 01067			5 50 46	+14 25.4	1	9.02		0.02		-0.04		2	A0	401
		SK *C 60		5 50 48	−67 05.	1	13.27		0.52		-0.05		2		149
		AO 901		5 50 49	+20 19.6	1	10.61		0.64		0.05		1		1722
39595	−12 01302			5 50 51	−12 18.5	1	9.43		0.02		-0.25		2	B9 (IV)	1375
39720	−37 02457	HR 2053	⋆ A	5 50 51	−37 38.4	2	5.63	.000	1.05	.005	0.87		6	K1 III	58,2035
39478	+26 00992	LS V +26 013		5 50 53	+26 24.8	1	8.17		0.08		-0.74		3	B2 V	1212
39508	+16 00904			5 50 54	+16 08.7	1	8.48		-0.08		-0.75		1	B5	401
		MC +25 023		5 50 54	+25 07.8	1	12.74		2.00				1	K5 III	1702
		Sk -68 172		5 50 55	−68 11.	1	13.19		-0.05		-0.72		3		415
39507	+21 01030			5 50 58	+21 49.6	1	8.69		0.07		-0.43		2	A0	1648
39492	+25 01024			5 50 58	+25 32.1	1	8.61		0.23		0.08		3	A2	848
		WLS 600 25 # 12		5 50 58	+27 03.2	1	11.06		0.33		0.27		2		1375
248894	+20 01158	LS V +20 004		5 50 59	+20 52.0	2	9.29	.000	0.24	.000	-0.73	.000	4	O8:V:nn	1011,1012
248893	+22 01090	LS V +22 005		5 51 01	+22 05.9	1	9.69		0.44		-0.56		2	B0 II-III	1012
	+25 01025			5 51 01	+26 00.8	1	10.57		1.12				1	K0	1702
39614	−6 01343	IDS05486S0626	AB	5 51 02	−06 25.1	1	9.41		0.12		-0.24		2	B9	401
39477	+30 01045	IDS05478N3028	AB	5 51 03	+30 29.1	1	7.69		0.17		-0.19		2	B5	401
		HRC 188	V	5 51 05	+01 37.6	1	13.60		1.21		0.26		1	K0:e	825
		G 102 - 39		5 51 05	+12 23.8	3	15.87	.028	0.00	.064	-0.75	.009	6		316,1620,3060
39414	+45 01194			5 51 07	+45 20.1	2	8.84	.015	0.03	.005	-0.29	.005	5	A0	401,1733
39752	−38 02270	HR 2055	⋆ AB	5 51 07	−38 32.1	1	6.70		1.08				4	K1 III	2035
39963	−64 00486	HR 2073		5 51 07	−64 02.7	3	6.34	.006	0.87	.004	0.57	.000	21	G8 III	15,1075,2038
39570	+12 00937	IDS05484N1224	A	5 51 10	+12 24.8	1	7.76		0.58		0.08		2	F8	3016
248907	+23 01103	IDS05481N2357	AB	5 51 10	+23 57.9	1	9.51		1.08				1	K2	1702

Table 1 311

HD	DM	Other Id	N Rem	α_{1950}	δ_{1950}	S	V	σ_V	B–V	σ_{B-V}	U–B	σ_{U-B}	n	Spectrum	References
		MCG8-11-11 # 2		5 51 10	+46 25.	1	14.34		0.72		0.26		1		899
		MCG8-11-11 # 3		5 51 10	+46 25.	1	14.78		0.72		0.12		1		899
		MCG8-11-11 # 4		5 51 10	+46 25.	1	14.72		0.66		0.10		1		899
		MCG8-11-11 # 5		5 51 10	+46 25.	1	16.34		0.65		0.28		1		899
		MCG8-11-11 # 6		5 51 10	+46 25.	1	16.38		1.05		0.60		1		899
		SK *G 478		5 51 10	−68 54.7	1	13.62		0.05		0.17		5	A0 Ib	149
248906	+26 00995			5 51 12	+26 37.9	1	8.78		0.54				3	G0	1724
		LP 781 - 13		5 51 12	−16 10.6	1	13.54		0.50		-0.01		1		3060
39689	−17 01278			5 51 12	−17 16.5	1	8.65		1.07		0.86		2	K0 III	1375
248924	+24 01011			5 51 14	+24 19.2	1	9.66		0.18		-0.31		3	B8	1723
		WLS 600 5 # 6		5 51 15	+04 59.1	1	11.46		0.59		0.10		2		1375
39647	−5 01434			5 51 15	−05 42.8	1	7.09		-0.01		-0.06		2	B9	401
270234				5 51 17	−68 06.	1	10.96		1.01		0.81		1	G5	360
		NGC 2112 - 101		5 51 18	+00 23.	1	11.47		0.58		0.38		2		1573
		NGC 2112 - 102		5 51 18	+00 23.	1	13.53		1.04		0.43		1		1573
		NGC 2112 - 106		5 51 18	+00 23.	1	11.76		0.72		0.30		2		1573
		NGC 2112 - 110		5 51 18	+00 23.	1	13.44		1.12		0.61		2		1573
		NGC 2112 - 116		5 51 18	+00 23.	1	10.04		2.25		2.62		3		1573
		NGC 2112 - 118		5 51 18	+00 23.	1	14.92		0.99		0.52		3		1573
		NGC 2112 - 202		5 51 18	+00 23.	1	13.75		1.00		0.38		1		1573
		NGC 2112 - 216		5 51 18	+00 23.	1	13.39		1.66		1.46		2		1573
		NGC 2112 - 220		5 51 18	+00 23.	1	12.85		0.85		0.38		2		1573
		NGC 2112 - 303		5 51 18	+00 23.	1	12.90		0.95		0.44		2		1573
		NGC 2112 - 306		5 51 18	+00 23.	1	15.16		1.11		0.47		3		1573
		NGC 2112 - 308		5 51 18	+00 23.	1	14.26		1.74				2		1573
		NGC 2112 - 309		5 51 18	+00 23.	1	14.12		1.22		0.60		2		1573
		NGC 2112 - 310		5 51 18	+00 23.	1	13.52		1.68				2		1573
		NGC 2112 - 316		5 51 18	+00 23.	1	12.34		2.11		2.77		1		1573
		NGC 2112 - 317		5 51 18	+00 23.	1	11.79		2.15		2.64		1		1573
		NGC 2112 - 318		5 51 18	+00 23.	1	13.42		1.49		1.01		1		1573
		NGC 2112 - 401		5 51 18	+00 23.	1	12.30		1.51		1.17		2		1573
		NGC 2112 - 402		5 51 18	+00 23.	1	11.47		1.47		1.15		2		1573
39764	−33 02599	HR 2056, λ Col		5 51 18	−33 48.7	7	4.87	.010	-0.15	.006	-0.56	.010	37	B5 V	15,1034,1075,1409*
	+13 01021		V	5 51 21	+13 12.1	1	10.32		1.63				3	M2	369
39301	+63 00616			5 51 21	+63 16.4	1	7.78		0.21		0.10		2	A5	1733
		HRC 189	V	5 51 22	+01 43.5	1	13.77		1.00		-0.39		1	K2:e	825
		AO 902		5 51 22	+20 09.0	1	10.88		1.62		1.56		1		1722
		AO 903		5 51 24	+20 32.6	1	9.68		0.38		0.22		1		1722
39587	+20 01162	HR 2047		5 51 25	+20 16.1	14	4.40	.010	0.59	.009	0.06	.012	63	G0 V	15,254,814,1007,1008,1013*
39411	+52 01010			5 51 26	+52 19.9	1	9.59		0.08		0.00		2	A2	1566
39632	+10 00927	HR 2048		5 51 27	+10 34.7	2	6.11	.004	1.46	.004	1.36	.043	8	G9 II	1355,1509
288313	+1 01156	IDS05489N0139	AB	5 51 28	+01 39.7	1	9.93		1.07		0.69		4		206
	+46 01065	MCG8-11-11		5 51 29	+46 33.9	1	9.16		1.33		1.58		1	K2	899
39937	−57 00901	HR 2072		5 51 30	−57 09.9	2	5.93	.005	0.66	.000			7	F7 IV	15,2012
248983	+22 01094			5 51 32	+22 56.9	1	9.24		0.41		-0.03		2	F5	1375
		AO 904		5 51 34	+20 05.7	1	10.19		0.85		0.52		1		1722
270239				5 51 35	−68 13.3	1	11.87		0.22		0.15		6	A5 III:	149
39685	+3 01071	HR 2051		5 51 38	+03 13.1	3	6.31	.005	1.29	.008	1.47	.008	9	K0	15,1355,1415
39716	−6 01347			5 51 38	−06 45.7	1	8.51		-0.02		-0.59		2	B5	401
249066	+11 00963			5 51 40	+11 38.7	1	9.15		1.69				3	M0	369
		Ross 797		5 51 42	−14 23.0	3	11.47	.005	0.54	.012	-0.16	.015	7	F9	1696,2013,3077
		HRC 190	V	5 51 43	+01 42.4	1	14.39		1.39		0.58		1		825
270278	−67 00519			5 51 43	−67 04.7	1	9.77		1.04		0.82		4	G0	119
39586	+31 01139	HR 2046		5 51 44	+31 41.8	2	5.91	.015	0.13	.005	0.12	.005	3	A5 IV	1733,3052
39683	+8 01107			5 51 45	+08 02.8	2	7.15	.010	-0.03	.005	-0.14	.005	4	B9	401,1733
39662	+11 00964	HR 2050	★ A	5 51 45	+11 45.3	2			0.03	.010	0.03	.040	5	A2 V	1049,1079
39661	+15 00976	IDS05489N1529	AB	5 51 45	+15 30.0	1	8.02		-0.01		-0.01		2	B9	401
249063	+14 01072			5 51 46	+14 42.5	1	10.13		1.72				3	A7	369
39789	−19 01293	HR 2060		5 51 48	−19 38.8	1	6.69		0.10				4	A3 IV	2035
249041	+25 01027			5 51 49	+25 05.3	1	9.96		0.35				2	F0	1724
		G 98 - 31		5 51 51	+35 38.8	1	11.33		1.37		1.28		1	M0	333,1620
39901	−42 02205	HR 2069		5 51 51	−42 55.8	1	6.55		1.37				4	K3 III	2007
39715	+2 01085	G 99 - 43		5 51 52	+02 08.6	2	8.84	.005	1.02	.005	0.87	.019	4	K3 V	196,3072
39645	+22 01096			5 51 53	+22 30.7	3	7.20	.009	1.25	.013	1.23		11	G7 III	848,1702,1724
39680	+13 01026	LS VI +13 002	★ A	5 51 55	+13 50.8	3	7.92	.070	0.04	.020	-0.96	.005	6	O6:pe	401,1011,1012
39644	+27 00906			5 51 55	+27 20.0	1	7.87		0.48				2	F5 V	1724
270196	−70 00461	Sk -70 120		5 51 55	−70 17.8	3	11.60	.012	-0.07	.007	-0.88		9	B1 Ia	149,1277,8033
39700	+13 01027	IDS05491N1350	B	5 51 57	+13 50.3	1	8.26		0.02		-0.15		1	A0	401
39917	−43 02114	SZ Pic		5 51 57	−43 34.0	1	7.91		0.81				4	G8 V	2012
249040	+27 00907			5 51 58	+27 05.4	1	9.04		1.02				3	K0	1724
39698	+19 01126	HR 2052		5 51 59	+19 44.5	2	5.92	.009	-0.17	.001	-0.76	.007	9	B2 V	154,1223
249073	+24 01015			5 52 01	+24 52.5	1	9.51		0.97				3	K0	1724
		MC +26 036		5 52 01	+26 09.8	1	12.21		0.82				1	G2	1702
		LP 781 - 14		5 52 02	−19 50.2	1	13.46		0.61		-0.11		1		3060
		SK *G 482		5 52 02	−64 51.0	1	13.60		0.18		0.16		6	A2 V	149
39699	+17 01051			5 52 04	+17 23.6	1	7.30		1.56		1.85		2	K5 III	1648
249074	+23 01106			5 52 06	+23 24.6	1	9.74		0.45		0.20		3	F2	848
39777	−4 01281	HR 2058		5 52 06	−04 04.3	6	6.55	.016	-0.18	.010	-0.80	.010	19	B1.5V	15,154,361,1004,1061,1732
270262	−68 00444			5 52 06	−68 03.8	1	9.75		1.06		0.85		2	K0 III	149
	+47 01213			5 52 08	+47 18.3	2	9.42	.005	-0.02	.009	-0.66	.005	4	B5	401,1375

HD	DM	Other Id	N	Rem	α_{1950}	δ_{1950}	S	V	σ_V	B–V	σ_{B-V}	U–B	σ_{U-B}	n	Spectrum	References
39775	+0 01208	HR 2057			5 52 09	+00 57.6	4	5.99	.010	1.33	.015	1.45	.018	13	K0 III	15,252,1355,1415
249071	+27 00909	LS V +27 015			5 52 10	+27 54.4	1	9.47		0.39		-0.47		2	B2 III:	1012
39677	+29 01037	IDS05490N2956	A		5 52 10	+29 57.4	1	7.07		0.35				2	F0	1724
	−9 01261				5 52 10	−09 24.6	3	10.71	.015	1.18	.012	1.04		6	M0	158,1017,1705
39773	+5 01044	IDS05496N0550	AB		5 52 15	+05 51.2	1	6.80		0.00		-0.36		2	B9	401
39727	+20 01168				5 52 15	+20 26.7	1	8.37		0.48		0.01		1	F2	1722
39551	+51 01128	HR 2045			5 52 16	+51 47.8	2	6.64	.005	0.16	.010	0.14	.010	7	A5 V	1501,1733
249106	+25 01028				5 52 17	+25 54.9	1	9.02		1.23				1	K5	1702
249090	+28 00944				5 52 17	+28 18.5	1	10.54		0.42				2	F0	1724
249124	+20 01169				5 52 18	+20 09.9	1	9.52		1.01		0.68		1	G5	1722
39713	+29 01039				5 52 18	+29 10.0	2	7.64	.000	0.99	.020			5	G5 III	20,1724
39891	−29 02595	HR 2068		⋆ AB	5 52 18	−29 09.3	2	6.35	.005	0.38	.005			7	F3 V	15,2027
39962	−42 02215				5 52 18	−42 14.4	3	7.96	.004	0.41	.009	-0.04		10	F3 V	1075,2033,3037
		Sk -68 173			5 52 18	−68 07.	1	12.08		0.00		-0.62		4		415
		Steph 580			5 52 22	−06 17.8	1	10.92		1.20		1.14		1	K7	1746
		WLS 600-15 # 6			5 52 22	−15 09.4	1	11.25		1.20		0.87		2		1375
39712	+30 01055	LS V +30 024			5 52 23	+30 42.3	1	8.38		0.11		-0.56		2	B2 IV	1012
39660	+41 01304				5 52 24	+41 19.0	1	6.63		0.28		0.13		2	F0	1601
39853	−11 01321	HR 2065			5 52 24	−11 46.9	4	5.64	.025	1.53	.008	1.84	.000	10	K5 III	15,1003,2007,3005
		Sk -68 174			5 52 27	−68 04.5	1	13.51		-0.13		-0.87		3		798
		Sk -68 175			5 52 27	−68 04.6	1	12.86		-0.10		-0.86		3		149
39801	+7 01055	HR 2061, α Ori		⋆ AP	5 52 28	+07 24.0	1	0.48	.099	1.86	.017	2.07	.058	16	M1/2 Ia-Iab	15,1078,1363,8003*
39429	+66 00413	HR 2041			5 52 30	+66 05.4	2	6.25	.009	1.36	.005	1.65	.010	6	K0	985,1733
39746	+27 00914	LS V +27 016			5 52 31	+27 42.5	2	7.03	.005	0.22	.000	-0.67	.010	4	B1 II	1012,1207
	+24 01017				5 52 32	+24 41.9	1	10.24		1.38				1		1702
249141	+26 01003				5 52 32	+26 28.3	1	8.86		1.02				3	K0	1724
		Sk -68 176			5 52 36	−68 17.	1	13.23		-0.18		-1.01		3		798
270273	−68 00445				5 52 38	−68 04.6	1	11.47		0.55		0.00		2	G0 V	149
39785	+20 01171				5 52 39	+20 27.7	1	7.63		1.83		2.00		1	K5	1722
39725	+35 01283				5 52 40	+35 09.8	1	8.40		1.13				3	K0	1724
		G 99 - 44			5 52 40	−04 09.2	2	14.49	.035	1.05	.010	0.78	.025	10	DK	203,1705,3078
249181	+26 01004				5 52 42	+26 19.2	1	10.02		1.06				1	K2	1702
		LS V +20 005			5 52 43	+20 04.3	1	10.91		0.33		-0.56		2	B1 IV	1012
249202	+22 01099				5 52 44	+23 01.0	1	11.45		0.45				1	K2	1702
		AO 908			5 52 46	+20 19.4	1	10.20		1.86		2.18		1		1722
		G 99 - 45			5 52 47	+00 41.7	3	14.72	.069	1.22	.025	0.96	.072	8		940,1691,3056
39745	+35 01284				5 52 48	+35 13.8	1	7.78		1.05				2	K0	1724
39816	+20 01171	HR 2063, U Ori			5 52 51	+20 10.1	2	9.26	.540	1.53	.540	0.43		2	M4	814,3001
249220	+24 01022				5 52 51	+25 01.1	2	9.79	.000	1.18	.000			5	K0	1702,1724
		LS V +25 008			5 52 54	+25 23.7	1	10.89		0.68		-0.32		2	B1 III	1012
39628	+55 01028				5 52 55	+55 56.6	1	6.97		1.20		1.31		3	K2 IV	37
249222	+23 01111				5 52 56	+23 52.1	1	9.34		0.49				2	F5	1724
40105	−50 01977	HR 2083			5 52 57	−50 22.7	5	6.53	.023	0.90	.004	0.62	.015	21	G8 IV	15,1075,1311,2012,3077
39910	−4 01289	HR 2070			5 53 02	−04 37.4	3	5.86	.005	1.18	.005	1.21	.004	8	gK2	15,247,1061
	+58 00876	G 191 - 55			5 53 03	+58 40.9	2	10.48	.009	0.51	.008	-0.17	.009	4	F8	1658,7010
42556	−85 00077				5 53 03	−85 55.8	1	6.79		0.94		0.65		4	G8 III	1628
39815	+33 01193				5 53 06	+33 12.7	1	8.14		0.49				2	F8	1724
39867	+19 01136				5 53 07	+19 22.2	1	8.09		-0.10		-0.32		2	B9	401
39927	−4 01291	HR 2071			5 53 07	−04 47.7	2	6.27	.005	0.06	.000	0.06	.005	7	A2 III	15,1061
		Sk -68 177			5 53 07	−68 14.	1	12.93		-0.18		-0.96		4		415
249256	+27 00918				5 53 08	+27 09.9	1	9.09		0.29				2	A2 III	1724
	+24 01024				5 53 09	+24 28.4	1	10.10		0.59				2	F8	1724
	+24 01025				5 53 10	+24 30.3	1	9.79		1.63				1		1702
39882	+12 00951				5 53 11	+12 57.4	1	8.17		0.00				4	B5	2033
39881	+13 01036	HR 2067		⋆ A	5 53 11	+13 55.7	6	6.60	.013	0.65	.013	0.13	.009	23	G5 IV	15,1067,1197,1355*
39743	+49 01423	HR 2054			5 53 14	+49 04.6	1	6.47		0.99				2	G8 III	71
249328	+17 01058				5 53 15	+17 26.1	1	8.92		1.74				2	M0	369
40091	−39 02260	HR 2082			5 53 15	−39 57.9	2	5.56	.009	1.51	.007	1.85		6	M0 III	58,2035
249321	+25 01036				5 53 20	+25 30.1	2	8.64	.000	1.12	.000			6	K0	1702,1724
249300	+26 01006				5 53 20	+26 14.7	1	9.84		1.24				1	K5	1702
39866	+28 00952	HR 2066		⋆	5 53 23	+28 56.2	1			0.30		0.26		3	A2 II	1049
		MC +25 114			5 53 24	+25 11.1	1	11.03		1.19				1	G8 V:	1702
270296	−68 00450	Sk -68 178			5 53 25	−68 07.5	1	11.55		-0.01		-0.55		3	B9:I	149
	+23 01114				5 53 26	+23 49.6	1	10.20		1.05				1		1702
	−49 01945	HR 2089			5 53 26	−49 38.1	2	6.09	.005	-0.14	.005			7	B3 III	15,2012
39846	+35 01287				5 53 30	+35 43.6	1	8.60		0.60				3	G0	1724
		MC +25 125			5 53 31	+25 37.4	1	11.39		1.55				1	K3 III	1702
39951	+7 01062				5 53 32	+07 28.8	1	8.87		0.29		0.03		2	F0	1375
39864	+35 01288				5 53 33	+35 34.4	1	7.69		1.49		1.10		3	M2	1601
		CCS 438			5 53 34	−75 45.3	1	11.58		1.71				2	N	864
270261	−69 00541				5 53 36	−69 22.7	2	10.02	.034	0.54	.000	0.05	.005	6	F8:V	149,204
40011	−3 01238				5 53 38	−03 48.7	1	8.07		0.16		-0.01		2	B8	401
40953	−79 00202	HR 2125			5 53 39	−79 22.3	3	5.46	.004	-0.08	.008	-0.23	.005	13	B9 V	15,1075,2038
		G 99 - 46			5 53 42	+01 46.2	1	12.47		1.50		1.14		1		3056
288309	+1 01163				5 53 42	+01 51.5	1	9.17		1.19		0.80		4		206
		Sk -67 279			5 53 42	−67 31.8	1	12.49		0.21		-0.31		5		149
40008	+0 01221				5 53 43	+00 08.9	3	9.23	.004	0.58	.007	0.08	.023	15	G2 IV	271,281,6004
39985	+9 01016	HR 2075			5 53 43	+09 30.2	3	5.98	.005	-0.06	.011	-0.14	.005	9	A0 IV	15,1061,1079
40125	−29 02613				5 53 43	−29 08.6	1	8.03		0.42				4	F3 IV/V	2012
40409	−63 00498	HR 2102			5 53 43	−63 06.3	4	4.64	.005	1.05	.000	0.97	.005	16	K1 III/IV	15,1075,2012,3077

Table 1   313

HD	DM	Other Id	N Rem	α_{1950}	δ_{1950}	S	V	σ_V	B–V	σ_{B-V}	U–B	σ_{U-B}	n	Spectrum	References
290986	+0 01222			5 53 44	+00 15.6	2	10.44	.000	0.59	.005	0.07	.014	7	F8 V	271,281
40292	−52 00805	HR 2094		5 53 44	−52 38.7	2	5.28	.005	0.32	.005			7	A9/F0 IV/V	15,2012
249474	+11 00974		V	5 53 45	+11 02.2	1	8.26		1.78				3	K5	369
249384	+26 01008			5 53 45	+26 50.0	1	9.88		1.22				1	K7	915
		MN83,95 # 903		5 53 45	−69 05.2	1	11.55		0.88				1		591
		G 99 - 47	⋆ V	5 53 47	+05 22.0	3	14.08	.016	0.60	.012	-0.15	.028	9		419,1620,3078
		MC +25 136		5 53 47	+25 28.2	1	12.94		0.38				1	G2	1702
40176	−37 02487	HR 2087		5 53 47	−37 07.6	3	4.97	.008	1.10	.005			11	K1 III	15,1075,2035
249423	+21 01052			5 53 49	+21 25.3	1	9.70		1.97				2	M7	369
290987	+0 01223			5 53 50	+00 12.2	2	10.22	.010	0.59	.010	0.41	.019	10	A1 V	271,281
	+33 01194			5 53 50	+33 50.6	1	10.20		1.04				1	R2	1238
249421	+24 01032			5 53 51	+24 59.0	1	9.96		1.17				1	K7	1702
		MC +25 139		5 53 51	+25 24.8	1	12.17		0.59				1	G0	1702
249447	+18 01027			5 53 52	+18 56.1	1	8.72		0.46		-0.04		2	A0	401
39970	+24 01033	HR 2074	⋆	5 53 52	+24 14.7	5	6.02	.005	0.39	.003	-0.23	.027	31	A0 Ia	15,154,1012,1207,1784
249420	+27 00922			5 53 56	+27 35.7	1	8.75		1.03				1	K2	915
40306	−53 00965			5 53 56	−53 32.1	1	8.65		0.91				4	K1 III F/G	2033
40005	+16 00926			5 53 57	+16 21.0	2	7.22	.014	-0.14	.005	-0.73	.009	8	B2 V	399,401
39949	+27 00923			5 53 57	+27 18.7	3	7.23	.004	1.09	.007			4	G2 Ib	138,915,1724
249513	+10 00938	G 102 - 41		5 53 58	+10 29.5	1	9.84		1.05		0.90		1	K2	1620
40455	−64 00495	HR 2104		5 53 58	−64 29.4	3	6.61	.007	0.37	.007	0.09	.010	19	F0 III	15,1075,2038
271660				5 54 00	−65 46.0	1	11.70		0.66		0.10		2	F5	149
		MN83,95 # 904		5 54 00	−69 01.3	1	12.57		0.77				1		591
270282	−69 00542	MN83,95 # 901		5 54 00	−69 01.8	1	9.78		1.30				1		591
		HA 97 # 18		5 54 01	−00 13.2	1	11.92		0.55		0.01		4	F8	281
		HA 97 # 319		5 54 02	+00 15.1	1	11.71		0.95		0.77		4	G5	281
40020	+11 00975	HR 2076		5 54 02	+11 31.0	2	5.88	.003	1.11	.016	1.11		3	K2 III	71,3077
		HA 97 # 19		5 54 03	−00 10.6	1	12.26		0.61		0.06		4	F9	281
270305	−68 00451	Sk -68 179		5 54 05	−68 19.2	1	11.73		-0.04		-0.67		3	B5 Ia	149
39983	+22 01109		V	5 54 06	+22 50.0	1	7.99		1.62				1	M5 III	1702
39938	+35 01292			5 54 08	+35 46.3	1	7.87		0.83				2	G5	1724
40136	−14 01286	HR 2085		5 54 08	−14 10.5	10	3.71	.011	0.33	.007	-0.01	.013	35	F1 V	15,58,1008,1013,1075*
40151	−22 01269	HR 2086	⋆ A	5 54 08	−22 50.8	2	5.97	.009	1.11	.005	0.99		5	K0/1 III	2006,3077
40040	+15 00993	G 102 - 42		5 54 09	+15 44.4	2	8.21	.000	0.64	.000	0.08	.000	2	G0	1620,3002
249482	+23 01118			5 54 09	+23 37.9	1	9.29		1.40				1	K5	1702
249460	+27 00925	IDS05511N2721	A	5 54 10	+27 21.9	1	9.56		0.46				2	F5	1724
40039	+19 01145			5 54 11	+19 12.7	1	7.86		-0.07		-0.21		2	B9	401
		HA 97 # 142		5 54 11	−00 07.1	1	11.47		0.50		-0.04		2	G0	281
249483	+23 01120			5 54 12	+23 12.7	2	9.38	.004	1.11	.000			6	K0	1702,1724
40003	+23 01119	LS V +23 007		5 54 12	+23 24.9	3	8.59	.020	0.80	.011	-0.08	.000	6	B3 Ib	1012,1207,1724
249500	+22 01110			5 54 13	+22 21.5	2	8.89	.004	0.92	.004			6	G5	1702,1724
		HIC 28198		5 54 13	+33 33.9	1	10.32		1.23		1.13		3		1723
249499	+24 01036			5 54 14	+24 59.8	1	9.38		1.77				1	K5	1702
		MC +26 080		5 54 16	+26 10.9	1	13.04		1.26				1	K5	1702
40597	−69 00544			5 54 18	−69 14.7	1	8.92		1.50		1.58		2	K2/3 III	149
		HIC 28206		5 54 20	+32 00.7	1	10.45		1.68		1.93		3		1723
271664	−65 00519			5 54 20	−65 19.1	1	12.78		0.45		0.02		5	G5	149
		HA 97 # 327		5 54 21	+00 12.9	1	12.18		0.55		-0.01		4	F8	281
40001	+33 01199			5 54 22	+33 15.3	1	7.44		0.53				2	F8	1724
40285	−38 02301			5 54 25	−38 17.2	1	8.71		0.06		0.05		4	A0 V	1700
		STO 688		5 54 25	−69 02.9	2	12.59	.000	0.51	.005	0.02		2	G0 Ia	591,726
		LP 894 - 1		5 54 26	−27 51.7	3	12.81	.011	0.41	.007	-0.22	.011	17		1696,1698,1765
249548	+23 01122	IDS05514N2308	AB	5 54 27	+23 08.6	1	8.69		0.90				4	G0	1724
		MC +26 085		5 54 27	+26 07.6	1	11.95		1.25				1	K0 III:	1702
40248	−31 02848	HR 2092		5 54 28	−31 23.3	1	5.56		0.38		0.20		4	F2 III	1628
40235	−23 03263	HR 2090		5 54 29	−23 13.3	1	6.36		1.07				4	K0 III	2031
	+25 01046			5 54 32	+25 22.3	1	9.91		1.20				1		1702
40361	−45 02232			5 54 32	−45 57.1	2	9.02	.024	0.97	.007	0.59		3	G8 (III)	1594,6006
		HA 97 # 249		5 54 34	+00 00.9	4	11.74	.007	0.65	.003	0.10	.010	91	G2	271,281,989,1764
249569	+24 01040			5 54 34	+24 45.5	2	8.39	.000	1.10	.000			6	K0	1702,1724
270297	−69 00545	MN83,95 # 902		5 54 35	−69 06.4	1	10.61		0.32				1		591
		HA 97 # 42		5 54 35	+00 11.5	1	12.45		1.63		1.21		1		1764
		HA 97 # 252		5 54 36	+00 02.3	2	11.74	.034	0.60	.015	0.09	.020	8	G0	271,281
249544	+30 01066			5 54 37	+30 36.8	1	9.70		1.32		1.23		3	K5	1723
270284				5 54 38	−69 30.3	1	13.05		-0.02		-0.58		3	F7	149
249566	+29 01054			5 54 39	+29 48.5	1	9.27		0.63				3	G0	1724
249545	+30 01067			5 54 40	+30 26.6	1	9.46		1.96				2	M7	369
40267	−22 01271			5 54 40	−22 18.2	1	9.83		-0.17		-0.68		4	B3 II/III	1732
		HIC 28216		5 54 41	+22 31.7	1	10.77		1.02		0.66		2		1723
249590	+25 01049			5 54 41	+25 10.0	1	8.74		0.18		-0.38		3	B5	848
249613	+25 01050			5 54 47	+25 41.3	1	9.13		1.06				3	K0	1724
40165	+9 01020			5 54 48	+09 34.6	1	8.38		-0.03				2	A0	6009
249670	+14 01089			5 54 48	+14 55.8	1	9.13		0.08		0.12		2	A2	401
40037	+40 01469			5 54 48	+40 47.0	1	7.89		0.08		0.07		2	A2	401
249671	+14 01090			5 54 50	+14 55.0	1	9.39		0.01		-0.26		1	A2	401
		HA 97 # 345		5 54 51	+00 20.1	3	11.61	.001	1.65	.008	1.68	.005	27		281,1764,5006
40210	+0 01227	IDS05523N0001	ABC	5 54 52	+00 01.4	3	6.90	.003	0.00	.003	-0.14	.005	23	A0 V	271,281,6004
290985	+0 01228			5 54 53	+00 13.1	5	9.26	.008	0.60	.011	0.11	.006	52	G3 V	271,281,989,1729,6004
	+21 01060	LS V +21 019		5 54 53	+21 10.5	1	10.55		0.93		-0.13		2		405
40111	+25 01052	HR 2084	⋆	5 54 53	+25 57.0	7	4.82	.009	-0.06	.008	-0.92	.017	38	B1 Ib	15,154,1012,1207,1363*

HD	DM	Other Id	N Rem	α_{1950}	δ_{1950}	S	V	σ_V	B–V	σ_{B-V}	U–B	σ_{U-B}	n	Spectrum	References
39965	+52 01019			5 54 53	+52 29.4	1	8.63		1.09		0.85		2	G5	1566
40359	−31 02854	HR 2098		5 54 57	−31 58.9	4	6.44	.003	1.07	.006			18	G8 III	15,1075,2013,2029
40185	+14 01093			5 55 01	+14 57.7	1	9.10		0.05		−0.04		1	A0	401
290984	+0 01231			5 55 03	+00 13.4	7	9.78	.004	0.20	.003	0.09	.010	150	A0 V	271,281,989,1729,1764*
40144	+25 01053			5 55 03	+25 33.9	2	7.96	.015	1.00	.005			8	K0	1702,1724
40085	+38 01341			5 55 06	+38 53.4	1	7.20		1.17		1.14		2	K2 III	105
40244	+2 01095			5 55 09	+02 35.4	1	8.92		0.30		0.08		2	F0	1375
		MC +25 191		5 55 09	+25 25.8	1	12.03		0.73				1	G0	1702
40259	+2 01096			5 55 13	+02 03.6	2	7.86	.000	0.38	.000	−0.06		3	F0 V	742,1594
		MC +25 195		5 55 13	+25 06.6	1	11.28		1.21				1	G0	1702
		G 192 - 11		5 55 14	+58 35.5	1	10.27		1.52		1.25		2	M1	7010
249740	+12 00962			5 55 16	+12 07.9	1	9.61		1.78				3	M0	369
40282	+1 01168	HR 2093		5 55 19	+01 13.2	5	6.20	.012	1.48	.008	1.83	.013	16	M0 III	15,247,1355,1415,1509
249699	+23 01128			5 55 19	+23 33.7	1	9.41		0.52				3	G0	1724
290989	−0 01128			5 55 19	−00 04.0	2	10.46	.010	0.53	.015	0.35	.010	5	A1 V	271,281
40402	−27 02620	IDS05533S2721	A	5 55 20	−27 20.1	1	8.61		0.93		0.66		2	G8/K0 III	536
		HA 97 # 75		5 55 21	−00 09.7	3	11.48	.001	1.87	.006	2.10	.000	19	K0	281,1764,5006
40207	+23 01130			5 55 22	+23 09.1	2	7.60	.014	1.05	.005			5	K0	1702,1724
		HIC 28273		5 55 22	+23 57.8	1	10.63		1.01		0.58		3		1723
249695	+30 01071	LS V +30 027		5 55 23	+30 12.2	1	8.99		0.17		−0.65		2	B1:V:penn	1012
		HA 97 # 191		5 55 23	−00 08.1	1	14.11		0.78		0.51		2		281
		G 249 - 27		5 55 24	+68 08.8	1	13.37		1.56				1		1691
290991				5 55 24	−00 08.	1	11.19		0.52		−0.01		3	F6 V	271
40035	+54 00970	HR 2077	⋆A	5 55 25	+54 17.0	21	3.72	.006	1.01	.007	0.84	.011	389	K0 III	1,15,61,71,985,1004*
40483	−44 02343			5 55 25	−44 00.9	1	6.63		0.52				4	F6 V	2033
		G 249 - 28		5 55 27	+68 09.7	1	12.99		1.56				1		1691
		HIC 28279		5 55 28	+23 46.2	1	10.18		1.15		0.90		2		1723
249729	+25 01055			5 55 29	+25 17.5	1	8.84		1.32				1	K2	1702
249728	+26 01022			5 55 29	+26 21.3	1	10.00		0.85				1	K0	1702
40084	+49 01428	HR 2081		5 55 29	+49 55.3	1	5.89		1.24				2	G5 III	71
		HA 97 # 196		5 55 30	−00 08.0	1	12.65		0.62		0.07		7	G0	281
40376	−14 01292	IDS05532S1413	AB	5 55 30	−14 12.8	1	7.37		1.31		1.47		3	K2/3 III	1657
40241	+24 01045			5 55 32	+24 47.7	2	7.34	.000	1.03	.015	0.81		4	G5	1625,1724
	+25 01056			5 55 33	+25 23.4	1	10.47		0.47				2	F2	1724
290981	+0 01235			5 55 35	+00 18.2	2	10.17	.005	0.90	.000	0.54	.000	8	G5 III	271,281
40242	+23 01132		V	5 55 35	+23 39.7	2	7.84	.025	1.67	.010	1.96		6	M1	369,848
40131	+46 01074			5 55 35	+46 54.9	1	8.00		0.03		−0.26		2	A0	401
40062	+55 01036	HR 2079		5 55 35	+55 19.2	2	6.44	.005	0.31	.000	0.12	.005	8	A5 m	3058,8071
290990	−0 01129			5 55 36	−00 04.3	2	10.21	.010	0.54	.010	0.02	.035	5	F8 V	271,281
	−14 01294			5 55 38	−14 03.9	1	10.41		1.03		0.85		4		206
249750	+29 01061			5 55 39	+29 10.5	1	9.01		1.07				3	G5	1724
40347	−1 01078	HR 2097		5 55 39	−00 59.9	2	6.21	.005	1.14	.000	1.20	.000	7	K0	15,1415
40083	+54 00971	HR 2080		5 55 40	+54 32.7	2	6.14	.020	1.20	.008	1.33		5	K2 III	37,71
290982	+0 01236			5 55 41	+00 07.2	2	10.16	.014	0.86	.000	0.49	.024	7	K0 V	271,281
		HA 97 # 369		5 55 42	+00 12.9	2	11.03	.030	0.69	.010	0.25	.005	4	G4	271,281
		LP 838 - 4		5 55 42	−23 39.3	1	11.33		0.73		0.17		2		1696
		WLS 600 80 # 5		5 55 43	+82 24.1	1	11.17		0.66		0.08		2		1375
249788	+23 01135	LS V +23 009		5 55 44	+23 14.8	1	9.12		0.45		−0.45		2	B1 V	1207
		HA 97 # 210		5 55 44	−00 04.9	1	12.11		0.80		0.45		6	A5	281
40316	+16 00940			5 55 46	+16 35.7	1	7.39		0.00		0.01		2	B9	401
		G 98 - 32		5 55 46	+34 38.3	1	14.89		0.85		0.20		1		1658
40494	−35 02612	HR 2106	⋆A	5 55 46	−35 17.3	5	4.36	.008	−0.18	.004	−0.66	.004	21	B2.5IV	15,1034,1075,2012,8015
40280	+25 01058			5 55 47	+25 46.5	3	6.62	.006	0.96	.012	0.74		11	K0 III	105,1702,1724
40503	−37 02505			5 55 47	−37 51.2	1	9.28		0.96		0.69		2	K2/3 V	3072
40372	+1 01171	HR 2100, V1004 Ori	⋆A	5 55 49	+01 50.0	4	5.89	.009	0.21	.005	0.14	.017	10	A5m δ Del	15,39,146,1415
40372	+1 01171	IDS05532N0150	B	5 55 49	+01 50.0	1	10.57		0.55		0.32		4	F7 V	146
40160	+46 01075			5 55 49	+46 31.9	2	7.50	.015	−0.08	.005	−0.64	.005	6	B5	401,555
	+0 01237			5 55 51	+00 04.1	3	10.79	.005	1.37	.004	1.10	.011	22	G5 III	281,989,1729
		HA 97 # 284		5 55 51	+00 05.0	1	10.79		1.36		1.09		68	G8	1764
249786	+27 00936			5 55 51	+27 46.9	1	9.50		1.04				3	G0	1724
40183	+44 01328	HR 2088, β Aur	⋆A	5 55 52	+44 56.7	8	1.90	.006	0.03	.006	0.05	.008	21	A2 V	15,1007,1013,1363*
40329	+20 01199			5 55 56	+20 40.4	1	8.17		0.94		0.65		2	G5	1648
40297	+27 00938	LS V +27 020		5 55 57	+27 33.5	2	7.27	.005	0.28	.005	−0.08	.005	4	A0 Ib	1012,1207
290983	+0 01238			5 55 58	+00 06.2	1	10.87		0.60		0.13		2	G2 IV	281
40331	+18 01040			5 55 58	+18 49.0	1	7.06		1.29		1.33		2	K0	1648
		LP 894 - 3		5 55 58	−30 54.1	1	11.26		0.48		−0.20		2		1696
40430	−10 01331			5 56 00	−10 52.7	1	8.07		1.01		0.69		1	K0	565
40556	−40 02179			5 56 00	−40 07.9	2	8.56	.008	−0.17	.006	−0.69	.011	4	B2/3 V	55,1732
271689	−65 00525			5 56 01	−65 00.9	1	10.03		0.37		0.00		1	F2 V	204
		HA 97 # 219		5 56 02	−00 07.1	1	12.87		0.64		0.12		4	G7	281
40369	+12 00968	HR 2099	⋆AB	5 56 04	+12 48.3	1	5.70		0.89				2	K2 III+A5 V	71
		HA 97 # 222		5 56 04	−00 05.0	1	11.60		1.30		0.86		8	K0	281
		Sk -67 280		5 56 04	−67 38.	1	13.58		−0.10		−0.34		4		415
249826	+28 00961			5 56 05	+28 17.2	2	9.73	.063	1.92	.010			3	M2	369,915
		HA 97 # 111		5 56 06	−00 10.7	1	11.99		0.59		0.35		2	A5	281
249870	+25 01060			5 56 07	+25 03.6	1	9.35		0.42				2	F5	1724
	+25 01061			5 56 07	+25 20.5	1	10.09		1.62				1	K0	1702
233190	+51 01138			5 56 07	+51 26.2	1	8.68		1.18		0.97		3	K0	1733
40314	+31 01156	IDS05529N3104	A	5 56 09	+31 04.0	1	8.19		0.49				2	F5	1724
		HA 97 # 224		5 56 10	−00 05.4	2	14.11	.016	0.90	.009	0.34	.000	5		281,1764

Table 1

HD	DM	Other Id	N Rem	α_{1950}	δ_{1950}	S	V	σ_V	B–V	σ_{B-V}	U–B	σ_{U-B}	n	Spectrum	References
40665 −53 00978		HR 2114	*A	5 56 11	−53 25.8	2	6.44	.005	1.48	.000			7	K5/M0 III	15,1075
40367 +18 01043				5 56 12	+18 48.6	1	8.60		-0.02		-0.24		2	A0	401
291033 −0 01130				5 56 12	−00 04.9	3	10.22	.005	0.44	.004	-0.03		12	F6 V	281,5006,6004
40239 +45 01217		HR 2091, π Aur		5 56 13	+45 56.1	6	4.29	.055	1.70	.021	1.81	.016	18	M3 II	15,1363,3016,8003*
40446 +0 01239		HR 2103	*A	5 56 15	+00 33.0	4	5.21	.006	0.01	.003	0.01	.000	10	A1 Vs	15,1061,1363,3023
	+10 00956			5 56 15	+10 23.0	1	10.25		1.87				3	M2	369
249845 +32 01146		LS V +32 014		5 56 15	+32 53.1	2	8.76	.024	0.05	.019	-0.54	.009	4	B2:V:nn	379,1012
		Sk −67 281		5 56 18	−67 20.	1	13.24		0.48		-0.22		3		798
		MC +25 242		5 56 19	+25 30.6	1	13.04		0.60				1	G5	1702
40328 +32 01148				5 56 19	+32 47.6	1	8.25		0.33		0.08		1	A2	379
40312 +37 01380		HR 2095, θ Aur	*AB	5 56 19	+37 12.7	6	2.65	.029	-0.08	.005	-0.16	.021	17	B9.5IV	15,1049,1203,1363*
		Sk −68 180		5 56 19	−68 14.	1	13.01		-0.16		-0.99		4		415
		AS 113		5 56 21	+23 43.	1	11.34		0.80		-0.30		2		1207
		Sk −67 282		5 56 23	−67 24.5	1	13.74		0.03		0.02		5		149
	+23 01138			5 56 24	+23 43.4	1	10.26		2.52		2.38		3		8032
		G 99 - 48		5 56 27	+04 10.6	2	11.86	.034	0.71	.010	-0.04	.005	3		333,1620,1696
40424 +19 01160				5 56 32	+19 46.6	1	8.65		0.08		-0.36		2	B9	401
250034 +12 00974				5 56 33	+12 06.5	1	9.42		1.73				3	M0	369
249964 +23 01140				5 56 35	+23 12.2	1	9.90		1.13				1	K5	1702
249989 +24 01049				5 56 37	+24 15.4	1	9.65		1.47		1.73		2	M0	1375
40535 −9 01284		HR 2107, V474 Mon		5 56 38	−09 23.1	5	6.13	.020	0.30	.005	0.10	.014	10	F2 IV	15,39,247,1256,6009
40443 +21 01072				5 56 39	+21 36.1	1	6.91		-0.06		-0.24		3	A0	848
40325 +44 01332		HR 2096	*AB	5 56 39	+44 35.4	1	6.20		1.15		1.12		2	K2 III	1733
		Sk -67 283		5 56 39	−67 32.0	1	12.75		-0.10		-0.64		3		149
		MC +25 254		5 56 40	+25 44.2	1	13.15		0.82				1	G0	1702
40490 +9 01037			V	5 56 41	+09 15.3	1	8.76		0.03		-0.28		3	A0	1732
40423 +22 01130		IDS05537N2228	AB	5 56 41	+22 28.1	1	7.54		0.23		0.12		2	A0	1375
40536 −9 01285		HR 2108		5 56 42	−09 33.6	9	5.03	.015	0.19	.010	0.16	.012	26	A6 m	15,216,277,355,374*
		MC +25 259		5 56 44	+25 48.2	1	12.16		0.53				1	G0	1702
40534 −4 01316				5 56 46	−04 21.7	1	9.06		0.01		-0.27		2	A0	401
250028 +25 01065		LS V +25 012		5 56 49	+25 05.2	3	9.02	.060	0.25	.013	-0.65	.032	6	B2:V:penn	379,1012,1207
		ApJS39,135 # 8		5 56 49	+27 39.1	1	11.58		1.77				3		915
249959 +33 01208				5 56 50	+33 54.4	1	9.38		0.06		-0.23		2	B9	1723
		MC +25 262		5 56 51	+25 44.5	1	11.91		1.43				1	G8 III	1702
		ApJS39,135 # 9		5 56 52	+27 33.4	1	10.13		1.90				1		915
40440 +28 00965				5 56 52	+28 44.7	1	8.84		0.50				2	F5 V	1724
		WLS 600-15 # 10		5 56 52	−15 21.4	1	11.06		0.54		0.03		2		1375
270340				5 56 53	−69 22.2	1	13.00		0.07		0.15		4	F5	149
		MC +25 265		5 56 54	+25 55.8	1	11.37		1.54				1	K0 III	1702
40441 +28 00966				5 56 55	+28 07.4	1	6.72		1.62				1	K5 III	915
		WLS 600 80 # 7		5 56 56	+78 22.4	1	10.26		0.57		0.09		2		1375
40439 +33 01209				5 56 57	+33 08.1	1	6.89		0.16		0.14		2	A2	985
40460 +27 00943				5 56 58	+27 16.3	5	6.61	.007	1.02	.007	0.83	.015	8	K1 III	915,1003,1355,1724,3077
		RL 6		5 57 00	+25 28.5	2	14.58	.017	0.48	.000	-0.11	.011	5		704,1709
250142				5 57 04	+11 30.	1	10.07		1.40				1	A5	915
40459 +31 01158				5 57 04	+31 56.4	1	7.07		1.07				2	K0	1724
		ApJS39,135 # 10		5 57 06	+27 34.7	1	10.91		1.54				3		915
250047 +31 01159		G 98 - 34		5 57 06	+31 26.0	1	9.41		0.93		0.78		1	K2	333,1620
40140 +66 00420				5 57 06	+66 54.0	1	7.06		1.69		2.06		2	K5	1733
40574 −1 01083		HR 2109		5 57 06	−01 26.8	4	6.61	.012	-0.06	.017	-0.38	.024	11	B8 IIIn	15,252,1079,1415
40457 +35 01304		CO Aur		5 57 07	+35 18.6	2	7.66	.118	0.67	.043	0.36	.053	2	F5 Ib	793,1772
		ApJS39,135 # 7		5 57 08	+27 42.6	1	11.85		1.69				1		915
250096 +21 01073				5 57 09	+22 00.7	1	9.76		1.07				1	K0	1702
250094 +24 01052				5 57 09	+24 56.8	1	8.80		1.06				3	K0	1724
40733 −44 02363		HR 2117		5 57 09	−44 02.2	2	5.81	.000	1.06	.003	0.87		6	G8 II	58,2035
		ApJS39,135 # 11		5 57 11	+27 35.6	1	11.41		1.96				2		915
40394 +47 01227		HR 2101		5 57 11	+47 54.1	2	5.73	.024	-0.01	.010	-0.13	.010	7	B9.5p SiFe	220,401
40531 +19 01165				5 57 12	+19 59.4	1	7.90		-0.13		-0.59		2	B8	401
250092 +26 01027				5 57 13	+26 48.4	2	10.00	.009	0.56	.000	-0.07		4	G0	1375,1724
		ApJS39,135 # 6		5 57 13	+27 44.7	1	10.42		1.63				1		915
250163 +19 01166		LSS 10		5 57 15	+19 11.5	2	9.61	.010	0.60	.020	-0.35	.010	4	B1.5:V:pen	1012,1207
40510 +25 01069				5 57 16	+25 55.3	2	8.46	.005	0.33	.000	-0.03		4	F0 II	379,1724
		RL 9		5 57 17	+25 20.0	3	10.84	.006	0.13	.004	-0.60	.013	9		483,704,1709
250134 +22 01133				5 57 18	+22 02.9	1	9.40		1.01				1	K0	1702
		ApJS39,135 # 4		5 57 19	+28 19.0	1	11.12		1.56				2		915
		ApJS39,135 # 3		5 57 23	+28 19.7	1	12.18		1.48				3		915
		SK *G 490		5 57 23	−67 57.5	1	13.56		-0.10		-0.31		3		149
		RL 12		5 57 24	+25 32.0	1	14.68		0.58		-0.27		3		1709
250115 +30 01078		IDS05542N3005	AB	5 57 25	+30 05.0	1	9.94		0.66				4	G5	1724
		G 99 - 49		5 57 26	+02 42.3	1	11.33		1.68		1.10		1	M4	333,1620
40545 +22 01135				5 57 26	+22 53.9	1	7.06		0.10		0.10		2	A2	1648
		ApJS39,135 # 5		5 57 26	+27 53.8	1	11.45		1.46				2		915
40530 +28 00969				5 57 26	+29 00.3	1	7.51		0.36				3	F2 II	1724
		G 101 - 14		5 57 30	+46 27.1	1	11.34		0.69		0.09		1		1620
40602 +8 01138				5 57 31	+08 57.3	1	7.90		0.37		0.12		1	Am	355
40569 +16 00957				5 57 31	+16 17.8	1	6.54		1.54		1.87		4	K2	985
		RL 11		5 57 31	+23 35.6	1	12.53		0.41		-0.37		3		1709
250133 +29 01072				5 57 31	+29 16.3	1	9.01		1.25				3	K0	1724
	−26 02627	LP 838 - 8		5 57 32	−26 16.6	1	10.92		0.82		0.35		2		1696
		NGC 2129 - 113		5 57 33	+23 19.4	1	12.69		0.40		0.23		1		49

HD	DM	Other Id	N Rem	α₁₉₅₀	δ₁₉₅₀	S	V	σ_V	B–V	σ_{B-V}	U–B	σ_{U-B}	n	Spectrum	References
40657	−3 01256	HR 2113		5 57 33	−03 04.5	7	4.53	.021	1.22	.012	1.20	.013	22	K2 III	15,247,1075,1075,1425*
40781	−38 02338			5 57 33	−38 42.8	1	8.41		0.57		0.04		4	F7 V	1700
		RL 13		5 57 35	+25 28.8	1	13.00		0.48		−0.48		4		1709
40470	+45 01225			5 57 37	+45 09.6	1	7.69		0.08		0.08		2	A0	1601
40808	−42 02266	HR 2120		5 57 37	−42 49.0	4	3.96	.006	1.14	.003	1.08	.000	20	G8/K1 II	15,1075,2012,8015
		RL 14		5 57 40	+25 33.1	1	15.60		0.48		−0.06		2		704
270365				5 57 40	−69 11.4	1	11.44		0.97		0.48		3	G8 III	149
		MC +25 299		5 57 45	+25 21.2	1	12.01		1.66				1	K5 V	1702
250221	+22 01137			5 57 47	+22 19.4	1	9.89		0.51				1	G0	1702
39329	+82 00155			5 57 49	+82 27.7	1	8.27		1.39		1.61		2	K0	1733
270402		Sk −67 284		5 57 49	−67 58.9	1	12.17		0.06		−0.34		3	A2:I	149
		NGC 2129 - 116		5 57 50	+23 15.4	1	15.72		0.97				1		49
		NGC 2129 - 96		5 57 50	+23 16.1	1	11.38		1.01		0.80		1		49
294521	−3 01258	G 99 - 50		5 57 50	−03 38.6	1	10.24		0.60		0.05		1		1620
		NGC 2129 - 68		5 57 51	+23 18.0	1	15.06		0.93				1		49
250241	+23 01145	NGC 2129 - 70		5 57 51	+23 19.9	1	10.10		0.42		−0.47		1	B3 III	49
		NGC 2129 - 117		5 57 52	+23 13.3	1	16.54		0.84				1		49
40589	+27 00945	HR 2111	⋆ AB	5 57 52	+27 34.3	5	6.05	.011	0.25	.011	−0.26	.004	13	B9 Iab	15,1012,1049,1079,1207
40486	+48 01333	HR 2105		5 57 52	+48 57.6	1	5.96		1.45				3	K0	71
		NGC 2129 - 65		5 57 53	+23 16.7	1	15.34		0.72				1		49
		NGC 2129 - 67		5 57 53	+23 17.5	1	16.02		1.08				1		49
		NGC 2129 - 114		5 57 54	+23 12.8	1	13.82		1.23		0.73		1		49
250268	+21 01079	G 100 - 51		5 57 55	+21 01.4	2	10.06	.005	1.24	.005	1.18	.000	2	K7	333,1620,3072
		NGC 2129 - 64		5 57 56	+23 16.8	1	14.66		0.84		0.45		1		49
		NGC 2129 - 48		5 57 56	+23 19.5	1	12.55		0.44		−0.30		1		49
40588	+31 01164	HR 2110		5 57 56	+31 02.0	2	6.19		0.08	.000	0.10	.029	5	A2 V	1049,1733
		NGC 2129 - 63		5 57 57	+23 15.8	1	15.42		1.04				1		49
		NGC 2129 - 45		5 57 57	+23 17.9	1	11.26		0.44		0.05		1		49
		ApJS39,135 # 12		5 57 57	+27 28.7	1	12.27		1.85				1		915
		NGC 2129 - 74		5 57 58	+23 22.3	1	15.45		0.66				1		49
250263	+25 01073			5 57 58	+25 53.5	1	8.99		1.92				2	K7	369
40745	−12 01337	HR 2118		5 57 59	−12 54.0	3	6.22	.008	0.36	.004			14	F2 IV	15,1075,2027
40655	+17 01089			5 58 00	+17 53.4	1	8.18		0.98		0.54		2	G5	1648
40587	+31 01165	IDS05548N3132	A	5 58 00	+31 32.4	2	8.07	.000	1.08	.005	0.78		10	G5	588,1724
		ApJS39,135 # 20		5 58 01	+27 29.3	1	12.48		1.25				1		915
		NGC 2129 - 53		5 58 02	+23 21.7	1	12.62		0.56		−0.19		1		49
250289	+23 01148	NGC 2129 - 7	⋆ A	5 58 03	+23 20.3	3	8.26	.008	0.57	.009	−0.32	.004	5	B2 IIIe	49,1012,1207
		NGC 2129 - 52		5 58 03	+23 21.4	1	11.46		0.44		−0.40		1	B5 V	49
		NGC 2129 - 54		5 58 04	+23 22.6	1	14.98		0.80				1		49
		ApJS39,135 # 28		5 58 04	+27 34.6	1	13.43		1.19				1		915
250310	+20 01216	LS V +20 008	⋆ A	5 58 05	+20 14.1	1	9.16		0.20		−0.66		2	B3 V	1207
250290	+23 01149	NGC 2129 - 4		5 58 05	+23 18.3	5	7.37	.006	0.62	.005	−0.23	.007	24	B5 V	49,848,1012,1207,8031
250287	+25 01075			5 58 05	+25 23.2	1	9.18		0.99				4	K0	1724
		AAS42,335 T5# 2		5 58 06	+23 18.	1	13.50		0.59		−0.01		4		848
		NGC 2129 - 55		5 58 06	+23 22.0	1	15.95		0.78				1		49
40744	−7 01248			5 58 06	−07 28.2	1	7.34		1.26		1.29		3	K0	1657
		NGC 2129 - 8		5 58 07	+23 20.5	1	10.45		0.57		−0.32		1	B6 V	49
		NGC 2129 - 91		5 58 08	+23 15.5	1	12.83		0.50		0.11		1		49
		ApJS39,135 # 1		5 58 08	+28 08.7	1	10.56		1.15				5		915
250282	+28 00974			5 58 08	+28 48.3	2	8.79	.005	1.11	.005			5	K0	915,1724
40600	+32 01158			5 58 09	+32 38.4	1	7.34		0.89				2	G5	1724
250285	+27 00948			5 58 10	+27 17.0	1	9.10		0.54				4	G0	1724
40681	+16 00964			5 58 12	+16 59.5	1	7.69		−0.03		−0.13		2	B9	401
		NGC 2129 - 112		5 58 12	+23 12.5	1	12.36		1.34				1		49
		NGC 2129 - 42		5 58 12	+23 18.1	1	12.16		0.51		−0.24		1		49
		AAS42,335 T5# 3		5 58 12	+23 23.	1	12.83		0.55		0.00		3		848
		AAS42,335 T5# 4		5 58 12	+23 24.	1	13.76		0.70		0.20		3		848
40865	−37 02534	IDS05565S3704	A	5 58 12	−37 03.6	3	8.59	.016	0.63	.009	0.05		11	G5 V	158,912,2012
270416				5 58 12	−67 56.7	1	12.87		0.11		0.24		6	A2 I:	149
250332	+24 01062			5 58 14	+24 24.7	1	9.52		0.38				2	F5	1724
250308	+26 01034			5 58 15	+26 44.4	1	9.12		1.20				1	K2	915
		NGC 2129 - 81		5 58 16	+23 22.1	2	12.77	.009	1.35	.018	1.05	.045	5		49,848
40055	+75 00247	HR 2078		5 58 16	+75 35.3	1	6.29		1.53		1.83		2	K5	1733
250306	+27 00949			5 58 17	+27 32.8	1	9.89		1.23				1	K5	915
		NGC 2129 - 88		5 58 18	+23 17.4	1	15.92		0.99				1		49
		AAS42,335 T5# 6		5 58 18	+23 23.	1	13.44		0.54		0.00		4		848
		AAS42,335 T5# 7		5 58 18	+23 23.	1	14.26		0.62		0.35		4		848
		AAS42,335 T5# 8		5 58 18	+23 24.	1	14.08		0.64		0.12		4		848
		AAS42,335 T5# 9		5 58 18	+23 27.	1	13.10		0.53		0.10		3		848
		ApJS39,135 # 16		5 58 19	+27 40.7	1	11.16		1.20				3		915
		ApJS39,135 # 2		5 58 20	+28 04.0	1	11.72		1.86				1		915
		NGC 2129 - 115		5 58 21	+23 13.7	1	14.76		0.69		0.62		1		49
		NGC 2129 - 82		5 58 21	+23 21.7	1	14.55		0.64		0.22		1		49
		NGC 2129 - 83		5 58 23	+23 20.2	1	14.69		0.67		0.50		1		49
	+24 01063	LS V +24 007		5 58 25	+24 29.2	1	10.51		0.46		−0.09		2	B5	1207
250371	+23 01151		B	5 58 26	+23 08.5	1	11.05		0.21		0.02		4		627
250371	+23 01151		C	5 58 26	+23 08.5	1	11.83		0.73		0.26		3		627
40678	+23 01150			5 58 26	+23 42.2	1	7.35		0.08		−0.10		1	A0	1338
40696	+23 01152			5 58 28	+23 44.3	1	8.24		0.08		0.10		3	A0	848
		ApJS39,135 # 19		5 58 28	+27 29.8	1	13.18		1.21				3		915

Table 1   317

HD	DM	Other Id	N Rem	α_{1950}	δ_{1950}	S	V	σ_V	B–V	σ_{B-V}	U–B	σ_{U-B}	n	Spectrum	References
40695	+25 01079			5 58 29	+25 29.3	2	8.96	.010	0.53	.005	0.16		7	G0	848,1724
40887	−31 02902	IDS05566S3103	ABC	5 58 29	−31 03.0	3	7.85	.007	1.14	.008	0.94		7	K3/4 V+K5 V	258,1414,2012
		SK *S 108		5 58 29	−66 03.7	1	11.97		1.19		1.14		2		149
41004	−48 02083			5 58 31	−48 14.5	1	8.70		0.91		0.53		1	K1 V	3072
		WLS 600 5 # 10		5 58 33	+05 10.3	1	12.27		0.91		0.47		2		1375
		ApJS39,135 # 14		5 58 35	+28 00.5	1	12.18		1.39				5		915
250388				5 58 36	+23 09.	1	10.66		0.79		0.40		2	K0	1338
		ApJS39,135 # 18		5 58 36	+27 32.5	1	11.19		1.47				3		915
40676	+31 01168			5 58 36	+31 19.6	2	7.82	.007	1.06	.002	0.74		27	G5	750,1724
		G 192 - 15		5 58 36	+49 52.4	1	14.48		1.84		1.45		1		906
		MC +26 171		5 58 38	+26 14.0	1	12.94		0.70				1	G0	1702
		ApJS39,135 # 25		5 58 38	+27 29.2	1	13.94		1.31				1		915
250368	+27 00952			5 58 38	+27 42.4	1	9.59		1.08				1	K0	915
40669	+38 01357	IDS05552N3843	AB	5 58 38	+38 43.2	1	7.27		0.84		0.48		3	G5	833
40724	+22 01140	HR 2116		5 58 40	+22 24.1	3	6.34		-0.08	.004	-0.31	.019	10	B8 V	627,1049,1079
		ApJS39,135 # 21		5 58 40	+27 32.2	1	12.36		1.20				2		915
40917	−30 02764			5 58 40	−30 05.1	1	8.64		1.20				4	K1 III	2012
250424	+24 01064			5 58 42	+24 42.0	1	9.86		1.19				1	K7	1702
		ApJS39,135 # 22		5 58 42	+27 33.9	1	12.94		1.30				3		915
		ApJS39,135 # 17		5 58 42	+27 36.9	1	10.96		1.39				3		915
40951	−37 02539			5 58 42	−37 02.2	1	8.49		0.56				4	F7 V	2012
40740	+23 01154			5 58 44	+23 08.1	1	8.73		0.24		0.10		9	A5	1338
40568	+56 01086			5 58 48	+57 01.2	1	6.48		0.97		0.73		2	G5	1502
		ApJS39,135 # 29		5 58 50	+28 17.0	1	12.95		1.35				1		915
	+42 01467			5 58 50	+42 55.7	1	10.88		0.47		0.04		2		1375
40803	+7 01095		A	5 58 51	+07 36.6	1	8.92		0.14		-0.41		2	F5	401
40803	+7 01095		B	5 58 51	+07 36.6	1	9.15		0.51		0.05		1	F5	401
		MC +26 174		5 58 51	+26 15.2	1	12.92		0.56				1	G0	1702
250449	+27 00954			5 58 54	+27 06.2	1	10.02		1.22				1	K5	915
250448	+27 00955	IDS05558N2712	A	5 58 55	+27 12.5	1	9.00		0.66				3	G0	1724
		ApJS39,135 # 26		5 58 55	+27 35.5	1	12.74		1.56				2		915
40626	+49 01441	HR 2112		5 58 55	+49 54.4	1	6.05		-0.06		-0.11		2	B9.5IV	401
		LP 779 - 43		5 58 56	−19 33.7	1	13.02		0.65		-0.07		2		1696
		LS V +20 010		5 58 57	+20 11.4	1	11.50		0.61		-0.33		2	B1 V	1012
250473	+22 01141			5 58 57	+22 14.5	1	10.31		0.47				1	F0	851
250444	+29 01078			5 58 57	+29 59.9	1	9.59		0.96				3	K0	1724
38847	+84 00112			5 58 57	+84 59.6	1	8.86		0.58		0.12		2	G0 V	1003
		ApJS39,135 # 15		5 59 00	+27 52.0	1	11.21		1.28				3		915
		Sk -70 121		5 59 00	−70 35.	1	12.90		-0.01		-0.54		2		798
250447	+28 00982			5 59 01	+28 08.6	2	9.32	.024	1.20	.000			3	K0	915,1724
250497	+25 01083			5 59 02	+25 12.0	1	9.86		0.97				2	K0	1702
250500	+24 01070			5 59 03	+24 33.3	1	9.92		0.35				2	F0	1724
40694	+42 01469			5 59 03	+42 53.0	1	8.32		0.10		-0.30		3	B8	555
250549	+16 00971			5 59 04	+16 34.1	1	8.98		0.92		0.55		7	G5	1655
250496	+25 01084			5 59 04	+25 51.3	1	9.33		1.58				3	K7	1702
40649	+53 00981			5 59 05	+53 32.6	1	7.00		0.32		0.12		3	F5 II	1501
250550	+16 00974		V	5 59 06	+16 31.0	3	9.50	.018	0.01	.043	-0.27	.017	3	A0	351,825,1588
		AAS42,335 T5# 12		5 59 06	+23 55.	1	13.41		0.93		0.64		4		848
40788	+23 01156			5 59 06	+23 55.8	1	9.07		0.22		0.19		3	A2	848
		ApJS39,135 # 30		5 59 06	+28 10.9	1	12.06		1.48				2		915
	−34 02594			5 59 06	−34 13.7	1	9.55		0.59		0.00		2		1696
		MC +26 181		5 59 09	+26 08.1	1	11.85		1.05				1	K0 III	1702
		WLS 624 70 # 6		5 59 10	+69 26.3	1	12.41		0.54		0.03		2		1375
		ApJS39,135 # 31		5 59 11	+28 02.3	1	12.55		1.11				2		915
40972	−25 02865	HR 2129		5 59 11	−25 25.0	3	6.04	.007	0.02	.000			11	A0 V	15,2013,2029
		AAS42,335 T5# 13		5 59 12	+24 00.	1	12.98		0.72		0.26		4		848
	+11 01001			5 59 13	+11 41.3	1	9.69		1.83				3	M2	369
		NGC 2164 sq1 1		5 59 14	−68 29.5	1	14.77		0.29		0.26		7		347
250543	+22 01143			5 59 15	+22 10.2	1	9.41		0.46				1	F8	1702
		ApJS39,135 # 13		5 59 16	+28 06.9	1	11.66		1.17				3		915
40722	+43 01421	HR 2115		5 59 17	+43 22.8	1	6.35		1.21		1.32		2	K0	1733
		NGC 2164 sq1 5		5 59 18	−68 27.2	1	13.20		0.57		0.13		8		347
		MC +26 183		5 59 19	+26 09.8	1	11.91		2.09				1	K0 III	1702
40833	+24 01071			5 59 21	+24 46.5	1	9.28		-0.03		-0.27		3	A0	848
40813	+26 01039			5 59 21	+26 26.0	1	8.74		0.41				2	F2 V	1724
40787	+32 01165			5 59 21	+32 35.5	2	6.78	.007	0.41	.015			4	F5	71,1724
40862	+14 01115			5 59 22	+14 53.0	1	8.90		-0.06		-0.60		2	B8	401
250573	+25 01086			5 59 22	+25 08.5	1	10.27		1.27				2	K5	1702
		G 249 - 33		5 59 23	+64 45.1	1	10.93		0.55		0.00		3		7010
250542	+27 00959	G 100 - 52		5 59 25	+27 24.4	2	9.15	.025	0.69	.010	0.26		5	G5	333,1620
	+28 00987			5 59 25	+28 32.1	1	10.06		1.84				2	M2	369
250539	+28 00988			5 59 25	+28 33.2	1	10.06		1.89				1	K2	915
250540	+28 00989			5 59 26	+28 12.5	1	8.94		1.32				1	K2	915
40989	−17 01329			5 59 26	−17 18.0	1	9.76		0.16		0.15		2	A2/3 IV	1375
41047	−33 02681	HR 2131		5 59 27	−33 54.7	4	5.54	.004	1.58	.000	1.94		13	K5 III	15,1075,2035,3005
40967	−10 01349	HR 2128	⋆AB	5 59 29	−10 35.8	7	4.95	.010	-0.12	.006	-0.59	.010	23	B5 III	15,1075,1079,1425,1732*
		G 99 - 52		5 59 30	+00 18.3	3	12.75	.021	0.73	.019	0.08	.010	4		1620,1658,3056
		AAS42,335 T5# 14		5 59 30	+23 54.	1	12.50		0.72		0.41		2		848
		AAS42,335 T5# 15		5 59 30	+23 54.	1	15.20		0.81		0.08		2		848
		MC +25 376		5 59 30	+25 40.2	1	12.32		1.10				1	G0	1702

HD	DM	Other Id	N Rem	α_{1950}	δ_{1950}	S	V	σ_V	B–V	σ_{B-V}	U–B	σ_{U-B}	n	Spectrum	References
41298	−62 00558			5 59 31	−62 48.6	1	8.18		0.36				4	F0 III	2012
41355	−65 00531			5 59 32	−65 26.9	1	9.02		0.13		0.12		1	A3 IIIP	204
250642	+22 01146	RZ Gem		5 59 36	+22 14.1	2	9.49	.004	0.85	.012			2	G5	851,6011
		ApJS39,135 # 27		5 59 36	+27 30.2	1	12.28		1.22				1		915
40932	+9 01064	Hyades vB 169	★ AB	5 59 38	+09 38.9	10	4.12	.009	0.17	.019	0.11	.018	36	A2 m	15,216,355,1061,1127*
		MC +25 387		5 59 38	+25 45.7	1	11.15		1.14				1	G5 III	1702
		ApJS39,135 # 32		5 59 38	+28 02.3	1	13.51		1.18				2		915
40832	+32 01166	HR 2122		5 59 39	+32 38.4	3	6.25	.023	0.41	.014	-0.01	.015	10	F4 V	71,254,272
41214	−51 01713	HR 2138		5 59 39	−51 13.0	2	5.66	.005	0.21	.005			7	A1mA3-A7	15,2012
40897	+22 01147			5 59 40	+22 02.9	1	8.06		-0.05		-0.18		3	B9	848
40964	+1 01195	HR 2127		5 59 41	+01 41.7	2	6.58	.005	-0.05	.000	-0.26	.005	7	B8 V	15,1061
		AAS42,335 T5# 16		5 59 42	+23 10.	1	11.53		0.28		-0.14		2		848
		AAS42,335 T5# 17		5 59 42	+23 12.	1	11.76		0.62		0.26		2		848
		AAS42,335 T5# 18		5 59 42	+23 53.	1	11.68		1.27		1.03		4		848
40801	+42 01473	HR 2119		5 59 42	+42 54.9	5	6.09	.011	0.98	.011	0.78	.010	9	K0 III	15,1003,1355,3026,4001
		G 222 - 11		5 59 43	+82 07.5	4	10.49	.010	1.52	.010	1.18	.013	10	M3	196,906,940,1691
250663	+24 01072			5 59 44	+24 54.8	1	9.65		1.79		1.78		2	K7	8032
250664				5 59 45	+23 57.	1	10.83		0.25		-0.12		3	A7	848
		ApJS39,135 # 24		5 59 45	+28 16.2	1	11.20		1.20				2		915
40895	+25 01089	IDS05567N2553	AB	5 59 46	+25 53.2	2	7.90	.005	0.47	.005	0.03		4	F6 V	1625,1724
40983	+1 01196			5 59 48	+01 05.4	3	8.55	.005	0.00	.002	-0.05	.013	61	B9 V	147,1509,6005
		AAS42,335 T5# 19		5 59 48	+23 11.	1	11.71		0.26		0.05		2		848
40896	+23 01161			5 59 48	+23 19.9	1	8.62		0.17		0.11		1	A0	1338
		AAS42,335 T5# 21		5 59 48	+23 59.	1	13.94		0.79		0.32		3		848
250689	+22 01148	IDS05568N2217	ABC	5 59 49	+22 17.0	1	9.65		0.52				1	F5	851
40894	+28 00991	LS V +28 020		5 59 49	+28 40.7	1	7.56		-0.08		-0.74		2	B2 V	1012
40784	+50 01263			5 59 49	+50 14.2	1	8.34		0.02		-0.37		2	B5 III	555
237451	+59 00935			5 59 49	+59 52.5	1	8.88		0.68		0.26		2	G5	1502
41296	−58 00602			5 59 49	−58 06.1	1	6.96		0.11		0.10		4	A3 V	1628
40893	+31 01174	LS V +31 013		5 59 52	+31 03.2	1	8.90		0.16		-0.78		2	B0 IV:	1012
40963	+7 01099			5 59 53	+07 42.0	1	7.95		-0.01		-0.48		2	B8	401
250688	+22 01149			5 59 53	+22 58.4	1	9.60		0.12		0.10		1	A0	1338
		G 102 - 44		5 59 54	+13 04.8	1	10.84		0.71		0.17		1	G8	1620
		AAS42,335 T5# 22		5 59 54	+23 59.	1	14.19		0.67		0.24		3		848
40469	+72 00298	IDS05302N7356	AB	5 59 55	+72 24.2	1	8.51		0.16		0.15		2	A0	1375
250720	+19 01183			5 59 56	+19 28.9	1	9.08		1.71		1.68		2	K7	8032
40752	+55 01051			5 59 56	+55 31.9	1	9.37		0.12		-0.24		2	A3	1502
		MC +26 197		5 59 58	+26 01.7	1	10.82		1.64				1	K0 III	1702
		WLS 600 25 # 9		5 59 59	+23 58.1	2	11.96	.009	0.61	.013	0.03	.022	5		848,1375
250825	+4 01107			6 00 00	+04 04.6	2	10.07	.015	0.60	.000	0.08		3	F8	742,1594
		ApJS39,135 # 23		6 00 00	+28 17.3	1	11.14		1.11				2		915
41556	−71 00407			6 00 00	−71 11.5	1	8.50		1.14		1.10		1	K0 III	726
250770	+12 00995			6 00 02	+12 13.8	1	8.75		1.80				3	K0	369
40961	+17 01101			6 00 02	+17 40.1	1	7.91		0.96		0.70		2	K0	1648
		NGC 2141 - 5009		6 00 03	+10 31.5	1	14.62		1.28		1.17		1		307
40946	+22 01150			6 00 04	+22 22.1	1	8.20		0.56				34	F8	6011
250713	+25 01091			6 00 05	+25 42.6	1	9.41		1.85				2	K7	1702
270442	−68 00460	NGC 2164 sq1 9		6 00 05	−68 27.3	1	10.42		0.21		0.08		3	F0	347
41029	+1 01197			6 00 06	+01 06.9	3	8.18	.000	0.98	.002	0.76	.010	59	K0	147,1509,6005
		AAS42,335 T5# 24		6 00 06	+23 59.	1	11.08		1.26		1.02		3		848
		AAS42,335 T5# 25		6 00 06	+23 59.	1	12.70		0.73		0.26		3		848
		AAS42,335 T5# 26		6 00 06	+24 00.	1	13.69		0.61		0.03		4		848
		LB 1798		6 00 06	−45 48.	1	12.13		0.12		0.10		3		45
		NGC 2141 - 2135		6 00 07	+10 28.5	2	15.77	.005	0.20	.005	-0.13	.010	3		307,1552
40930	+31 01175			6 00 07	+31 23.5	2	7.08	.004	0.40	.002	-0.09		107	F8	750,1724
41172	−27 02680	IDS05581S2725	A	6 00 07	−27 25.5	2	7.16	.033	0.43	.019			8	F3 V	1075,2033
		MC +25 413		6 00 08	+25 29.3	1	10.19		1.38				2	K0 III	1702
41182	−34 02604			6 00 09	−34 30.7	1	8.68		-0.03		-0.06		2	A0 V	1730
		NGC 2141 - 4011		6 00 10	+10 30.7	1	12.91		0.63		0.12		std		307
250715	+24 01075			6 00 10	+24 13.2	1	9.16		0.49				2	F8	1724
250740	+22 01151			6 00 11	+22 20.7	1	8.18		1.49				35	K2	6011
		AAS42,335 T5# 27		6 00 12	+23 41.	1	10.89		2.13		2.36		3		848
		RL 115		6 00 12	+25 25.	1	12.83		0.13		-0.34		2		704
250738	+25 01092			6 00 12	+25 39.6	1	9.45		0.43				2	F5	1724
		LB 1799		6 00 12	−46 39.	1	11.08		0.01		-0.12		3		45
250759				6 00 13	+22 17.	1	11.60		0.40		-0.10		2	A7	1207
250792	+19 01185	G 100 - 54	★ A	6 00 15	+19 22.3	6	9.31	.012	0.62	.005	-0.04	.025	12	G0 V	22,979,1003,1620,3077,8112
250792	+19 01185	IDS05573N1922	B	6 00 15	+19 22.3	1	13.40		1.26		1.00		2		3024
		MC +26 206		6 00 16	+26 10.9	1	12.22		1.27				1	G8	1702
		Sk -70 122		6 00 16	−70 54.3	1	12.80		0.01		-0.13		3	A2 Ib	415
	+15 01038			6 00 17	+15 12.1	1	10.47		1.96				3	M2	369
41125	−14 01315	HR 2136		6 00 17	−14 29.7	3	6.20	.004	0.95	.010			11	K0 III	15,1075,2035
		NGC 2141 - 1405		6 00 18	+10 26.	1	14.71		1.58		1.70		1		307
		NGC 2141 - 1410		6 00 18	+10 26.	1	17.56		0.61		0.31		1		307
		NGC 2141 - 2242		6 00 18	+10 26.	1	16.26		0.96		0.54		2		307
		NGC 2141 - 2431		6 00 18	+10 26.	1	14.04		0.64		0.10		2		307
		NGC 2141 - 3138		6 00 18	+10 26.	1	14.60		0.69		0.14		2		307
		NGC 2141 - 3139		6 00 18	+10 26.	1	15.70		1.25		0.66		2		307
		NGC 2141 - 3143		6 00 18	+10 26.	1	16.44		0.88		0.25		1		307
		NGC 2141 - 3147		6 00 18	+10 26.	1	16.87		0.76		0.29		1		307

Table 1   319

HD	DM	Other Id	N	Rem	α_{1950}	δ_{1950}	S	V	σ_V	B–V	σ_{B-V}	U–B	σ_{U-B}	n	Spectrum	References
		NGC 2141 - 3154			6 00 18	+10 26.	1	16.27		0.79		0.23		1		307
		NGC 2141 - 3240			6 00 18	+10 26.	1	13.33		1.80				1		307
		NGC 2141 - 3245			6 00 18	+10 26.	1	14.95		1.34		0.96		1		307
		NGC 2141 - 3251			6 00 18	+10 26.	1	14.39		0.82		0.46		1		307
		NGC 2141 - 3252			6 00 18	+10 26.	2	14.46	.010	1.36	.010	0.97	.034	3		307,1552
		NGC 2141 - 4010			6 00 18	+10 26.	1	11.63		1.78		2.14		3		307
		NGC 2141 - 4017			6 00 18	+10 26.	1	10.95		1.10		0.82		std		307
		NGC 2141 - 4036			6 00 18	+10 26.	1	10.66		1.89		1.92		1		307
		NGC 2141 - 4041			6 00 18	+10 26.	1	11.67		1.00		0.71		1		307
		NGC 2141 - 4043			6 00 18	+10 26.	1	16.55		0.96		-0.05		1		307
		NGC 2141 - 4044			6 00 18	+10 26.	1	16.59		0.75				3		307
		NGC 2141 - 4045			6 00 18	+10 26.	1	16.50		0.83		0.25		std		307
		NGC 2141 - 4046			6 00 18	+10 26.	1	16.70		0.84		0.33		std		307
		NGC 2141 - 4047			6 00 18	+10 26.	1	18.40		0.73				2		307
		NGC 2141 - 4048			6 00 18	+10 26.	1	15.77		1.33				1		307
		NGC 2141 - 4049			6 00 18	+10 26.	1	11.05		0.49		0.01		2		307
		NGC 2141 - 5006			6 00 18	+10 26.	1	14.86		0.69		0.17		std		307
		NGC 2141 - 5007			6 00 18	+10 26.	1	12.67		0.40		0.23		std		307
		NGC 2141 - 5008			6 00 18	+10 26.	1	12.96		0.56		0.12		std		307
40959	+27 00963	IDS05572N2739		A	6 00 18	+27 38.7	2	8.76	.020	0.79	.020	0.41		8	sgG5	1724,3024
40959	+27 00963	IDS05572N2739		B	6 00 18	+27 38.7	2	9.67	.010	0.30	.000	0.16		6	dA7	1724,3024
40959	+27 00963	IDS05572N2739		C	6 00 18	+27 38.7	1	11.06		0.69		0.19		3		3024
		NGC 2164 sq1 10			6 00 18	-68 26.7	1	13.18		0.52		0.15		2		347
250754	+26 01044				6 00 19	+26 07.7	1	9.02		0.53				2	F8	1724
	+23 01164				6 00 20	+23 56.8	1	10.39		0.94		0.57		3		848
		MC +26 212			6 00 24	+26 03.8	1	10.56		0.82				4	G2 IV	1702
		NGC 2141 - 4023			6 00 25	+10 26.6	1	12.17		1.25		1.14		std		307
250784	+29 01088				6 00 25	+29 20.2	1	9.30		1.79				2	M0	369
		NGC 2141 - 4022			6 00 27	+10 27.1	1	11.05		1.13		0.89		std		307
		RL 20			6 00 27	+24 52.4	1	11.86		0.18		-0.12		2		1709
40996	+26 01046				6 00 27	+26 31.9	2	7.00	.010	0.03	.010	-0.02	.025	5	A0 V	1501,1625
		NGC 2141 - 5021			6 00 29	+10 28.0	1	13.86		0.60		0.12		1		307
41040	+19 01186	HR 2130		★	6 00 30	+19 41.6	4	5.14	.005	-0.11	.002	-0.42	.015	8	B8 V	1049,1079,1363,3023
		AAS42,335 T5# 29			6 00 30	+23 38.	1	11.17		1.34		1.38		3		848
		MC +25 449			6 00 31	+25 57.4	1	11.09		1.28				3	K0 III	1702
40873	+51 01146	HR 2123		★ A	6 00 31	+51 34.6	1	6.45		0.18		0.16		1	A7 III	39
	+26 01047				6 00 32	+26 11.5	1	10.32		1.72				1		1702
		G 101 - 17			6 00 32	+41 00.1	1	14.05		1.00		0.37		1		1658
250833	+24 01078				6 00 33	+24 50.0	1	9.96		0.49				3	G0	1724
41010	+29 01089				6 00 34	+29 22.8	1	8.54		1.03				3	G5	1724
	+44 01351				6 00 34	+44 17.0	1	9.10		1.01		0.87		2	K5	1375
		NGC 2164 sq1 13			6 00 36	-68 27.0	1	12.44		1.04		0.83		2		347
41076	+11 01009	HR 2133			6 00 37	+11 41.0	1			-0.04		-0.03		3	A0 Vs	1049
250830	+26 01049	LS V +26 019		★ A	6 00 38	+26 54.5	1	9.89		0.51				2	A8 II	1724
		LS V +30 031			6 00 38	+30 10.3	2	10.95	.000	0.31	.000	-0.61	.000	4	O9 V	342,797
250928	+10 00986				6 00 39	+10 39.2	1	9.64		0.52		0.02		std	G0	307
250831	+26 01050				6 00 39	+26 44.5	1	8.91		1.12				1	K2	915
250810	+31 01179	CQ Aur			6 00 39	+31 19.9	1	9.04		0.86		0.41		30	G0	750
40827	+59 00937	HR 2121			6 00 42	+59 23.8	1	6.34		1.10		1.10		3	K1 III-IV	37
	-9 01310				6 00 42	-09 42.5	1	10.12		0.19		-0.36		4		206
250956	+10 00987				6 00 43	+10 38.0	1	8.55		1.01		0.78		std	G5	307
41121	+8 01161				6 00 44	+08 34.9	1	8.40		-0.09		-0.44		2	B8	401
		MC +25 452			6 00 44	+25 25.5	1	11.46		1.14				1	K0 III	1702
		MC +26 220			6 00 44	+26 01.9	1	12.59		0.64				1	G0	1702
		NGC 2164 sq1 14			6 00 44	-68 27.2	1	11.68		0.35		0.04		1		347
250853	+26 01051				6 00 45	+26 21.3	1	8.67		1.80				2	K7	369
		G 99 - 53			6 00 45	-00 37.0	2	13.28	.025	1.20	.000	0.98	.005	2		1620,3056
		SK *S 109			6 00 45	-69 23.5	1	11.54		1.53		1.56		5		149
41323	-44 02395				6 00 46	-44 00.7	1	8.71		0.64				4	G3/5 V	2012
		WLS 600-15 # 5			6 00 47	-12 40.0	1	11.16		0.33		0.20		2		1375
41142	+8 01162				6 00 48	+08 35.6	1	7.70		0.32		0.01		1	F0	401
40957	+45 01235	IDS05572N4536		AB	6 00 52	+45 35.4	1	7.30		0.22		0.07		2	A2	1601
		AA45,405 S241 # 2			6 00 53	+30 15.6	1	12.74		0.69		0.20		1		797
		SK *G 495			6 00 53	-69 37.3	1	12.90		0.33		0.27		6	F2:I	149
251003	+10 00989				6 00 54	+10 08.0	1	9.06		0.42		0.02		2	K0	1733
40978	+46 01091				6 00 54	+46 35.3	2	7.22	.029	-0.05	.005	-0.72	.020	5	B3 Ve	401,1212
250915	+24 01079				6 00 56	+24 44.1	1	9.37		1.10				1	K0	1724
		WLS 600 5 # 7			6 00 57	+03 14.4	1	11.11		0.58		0.02		2		1375
41117	+20 01233	HR 2135		★	6 00 57	+20 08.5	14	4.63	.008	0.27	.011	-0.69	.015	43	B2 Ia	1,15,154,369,1004,1207*
250917		G 100 - 56			6 00 57	+23 00.8	2	11.03	.014	1.24	.024	1.25		2		333,1017,1620
		MC +26 225			6 00 59	+26 01.9	1	11.08		1.16				3	K0 III:	1702
41164	+14 01129				6 01 01	+14 58.0	1	8.28		-0.09		-0.39		2	B9	401
		V616 Mon			6 01 02	+00 03.0	1	12.11		0.27		-0.67		1		878
250941	+24 01080				6 01 02	+24 58.9	1	9.22		1.71				2	K7	369
41115	+25 01099				6 01 03	+25 05.8	3	8.63	.012	0.93	.009	0.55		6	K0	848,1702,1724
		RL 26			6 01 04	+23 31.9	1	13.69		0.51		-0.36		3		1709
250942	+23 01169				6 01 04	+23 49.4	1	9.38		0.97		0.67		3	K0	848
41116	+23 01170	HR 2134		★ AB	6 01 05	+23 16.1	8	4.16	.011	0.83	.017	0.52	.004	37	G7 III	15,1217,1363,1724,1784,3030*
	+26 01052				6 01 05	+26 03.5	1	10.44		1.05				2		1702
		MC +26 228			6 01 05	+26 13.9	1	10.98		0.92				2	K0 III	1702

HD	DM	Other Id	N Rem	α₁₉₅₀	δ₁₉₅₀	S	V	σ_V	B–V	σ_B–V	U–B	σ_U–B	n	Spectrum	References
		MC +26 229		6 01 05	+26 22.4	1	11.77		0.65				1	G0	1702
41140	+21 01099			6 01 06	+21 29.8	1	7.79		0.19		0.12		3	A3	848
41139	+25 01100			6 01 07	+25 26.9	3	6.94	.008	0.94	.004	0.71		12	K0	1625,1702,1724
41100	+28 00997			6 01 07	+28 18.3	2	8.11	.054	1.24	.000	1.10		3	K2	915,1625
		CM Ori		6 01 11	+08 14.6	1	12.13		0.37		0.29		1		699
		RL 28		6 01 11	+24 04.2	2	14.05	.055	0.45	.025	-0.29	.000	5		483,704
41338	-34 02615			6 01 13	-34 14.6	1	9.57		0.56		0.04		2	G0 V	1730
41312	-26 02675	HR 2140	★ AB	6 01 15	-26 17.0	6	5.02	.010	1.34	.007	1.46	.010	30	K3 II/III	3,15,418,2017,2029,3077
250989	+25 01101			6 01 16	+25 47.9	1	9.98		1.59				1	K5	1702
	+30 01096			6 01 20	+30 30.3	1	10.40		0.54		-0.35		4		206
41253	+2 01118	LS VI +02 001		6 01 21	+02 52.1	1	7.31		-0.01		-0.57		1	B5	247
		MC +25 494		6 01 27	+25 15.0	1	10.58		1.56				3	K3 III	1702
41074	+42 01477	HR 2132		6 01 28	+42 59.3	2	5.90	.005	0.33	.023	0.03	.009	5	F3 V	254,3026
		MC +25 496		6 01 30	+25 14.1	1	11.00		1.70				1	K0 III	1702
251072	+23 01172			6 01 32	+23 23.0	2	9.12	.015	0.43	.010	0.08		5	F5	848,1724
41586	-60 00537	HR 2151, SW Pic		6 01 32	-60 05.7	1	6.44		1.58				4	M4 III	2035
		MC +26 240		6 01 33	+26 22.4	1	11.74		1.04				1	K0 III	1702
251011	+28 01000			6 01 33	+28 17.7	1	10.05		1.23				1	K0	915
41162	+37 01405	HR 2137	★ AB	6 01 36	+37 58.1	1	6.32		0.82		0.55		2	K0 III+A2	1733
41221	+25 01105			6 01 37	+25 11.0	1	8.45		0.50				3	F6 V	1724
		MC +26 241		6 01 37	+26 20.8	1	11.86		1.76				1	K5 III	1702
41288	+0 01264			6 01 40	+00 37.0	2	8.81	.011	0.07	.004	-0.37	.005	11	A0	14,1732
		IC 2157 - 33		6 01 41	+24 05.1	1	13.57		0.36		0.08		2		704
41368	-25 02892			6 01 41	-25 35.4	1	9.73		-0.15		-0.60		4	B5 III/IV	1732
		LS V +23 020		6 01 42	+23 24.3	1	11.43		0.51		-0.45		4		848
251066	+27 00967			6 01 42	+27 49.0	1	10.17		1.19				1	K0	915
		IC 2157 - 3		6 01 43	+24 04.2	2	12.96	.012	0.30	.015	-0.08	.004	5		704,1709
251123				6 01 45	+23 27.	1	11.47		0.22		0.23		3	A0	848
		IC 2157 - 2		6 01 46	+24 03.5	3	11.04	.014	0.27	.012	-0.31	.004	7		483,704,1709
		IC 2157 - 5		6 01 46	+24 05.1	2	11.86	.010	0.28	.020	-0.29	.010	5		483,704
		RL 34		6 01 46	+24 05.1	1	11.85		0.26		-0.28		2		1709
42269	-78 00212			6 01 47	-78 16.8	1	9.99		0.88		0.57		2	K1/2 V	3072
		IC 2157 - 1		6 01 48	+24 03.8	2	11.59	.010	0.32	.015	-0.28	.010	4		483,704
		RL 116		6 01 48	+25 21.	1	14.45		0.26		-0.12		2		704
251117	+28 01001	LS V +28 021		6 01 48	+28 46.3	2	9.11	.000	0.16	.004	-0.77	.008	9	B0 IV	1012,1775
41335	-6 01391	HR 2142		6 01 48	-06 42.3	8	5.22	.008	-0.08	.010	-0.84	.011	49	B2 Ven	15,154,247,681,815*
		IC 2157 - 197		6 01 51	+24 10.3	2	13.59	.025	0.48	.050	-0.24	.015	4		483,704
		IC 2157 - 201		6 01 51	+24 11.5	1	14.63		0.44		-0.19		2		704
		MC +26 249		6 01 51	+26 21.0	1	11.72		0.92				1	G5 IV	1702
251149	+23 01175			6 01 52	+23 44.2	1	10.29		0.18		0.12		2	A2	848
40956	+63 00630	HR 2126		6 01 52	+63 27.5	1	6.46		1.01		0.82		2	k0	1733
251177	+22 01159			6 01 53	+22 19.0	2	9.56	.004	0.99	.017	0.74	.000	6	G5	848,1723
		IC 2157 - 21		6 01 53	+24 04.7	3	12.09	.010	0.32	.005	-0.25	.010	7		483,704,1709
41576	-52 00838			6 01 53	-52 06.7	1	8.45		0.40				4	F2/3 III	2012
		AAS42,335 T5# 35		6 01 54	+23 44.	1	11.67		0.47		0.02		3		848
		RL 117		6 01 54	+25 15.	1	13.40		0.08		-0.46		2		704
251175				6 01 56	+23 47.	1	10.49		1.93		2.23		3	K7	848
	+35 01332	LS V +35 030		6 01 56	+35 28.7	1	9.30		0.25		-0.68		2	B2	405
	+63 00631			6 01 56	+63 23.3	1	8.58		1.48		1.50		2		1726
41285	+16 00989			6 01 57	+16 39.8	1	7.80		-0.11		-0.66		2	B5	401
251176	+23 01177			6 01 57	+23 28.6	1	10.57		0.43		0.03		4	G0	848
41406	-20 01271			6 01 59	-20 37.0	2	8.69	.000	0.48	.002	-0.09		3	F7 +AII/III	1594,6006
41259	+25 01108			6 02 00	+25 48.3	1	8.04		-0.04		-0.31		3	A5	833
		RL 41		6 02 02	+23 40.8	2	11.39	.005	0.35	.027	-0.43	.024	4		483,1709
251204	+23 01179	LS V +23 022		6 02 03	+23 23.9	1	10.28		0.48		-0.55		2	B0 IV	1012
41161	+48 01339	IDS05582N4815	AB	6 02 04	+48 15.3	2	6.77	.005	-0.09	.015	-0.97	.025	6	O8 V	399,555
		V 530 Ori		6 02 04	-03 11.5	1	9.86		0.63		0.06		5	G0	1768
41333	+9 01080			6 02 05	+09 34.0	1	9.08		-0.02		-0.11		1		401
41282	+24 01086			6 02 05	+24 20.9	2	8.67	.015	0.04	.005	-0.29	.035	5	B9	848,1625
41304	+14 01136			6 02 07	+14 23.7	1	6.72		0.46		-0.01		3	F6 V	1648
251198	+29 01100			6 02 09	+29 03.2	1	9.32		0.46				3	G0	1724
41345	+9 01081			6 02 11	+09 35.4	1	8.62		-0.06		-0.26		2	A0	401
		ApJ144,259 # 13		6 02 12	+03 57.	1	19.87		0.19		-1.00		2		1360
41492	-35 02671			6 02 13	-35 06.8	2	8.82	.018	1.13	.009	1.12		4	K1 III	2012,3040
41269	+33 01236	HR 2139		6 02 16	+33 36.2	2			-0.08	.004	-0.25	.050	4	B9p Si	1049,1079
41361	+5 01085	HR 2144		6 02 18	+05 25.5	4	5.66	.007	1.05	.011	0.83	.014	10	K0	15,247,1355,1415
251288	+18 01080			6 02 18	+18 13.0	1	9.58		0.90		0.62		3	K0	848
41380	+4 01116	HR 2145		6 02 20	+04 09.8	4	5.62	.007	1.05	.008	0.77	.027	10	G4 III	15,247,1078,1355
	+26 01061			6 02 21	+26 12.3	1	10.93		1.12				2		1702
41378	+8 01176			6 02 22	+08 57.8	1	8.19		0.03		-0.34		2	A0	401
	+26 01059			6 02 22	+26 13.6	1	10.79		1.43				2		1702
41652	-53 01002			6 02 22	-53 53.2	1	7.46		1.15		1.09		27	K1 III	978
251315	+20 01244			6 02 23	+20 08.2	1	8.95		0.55		0.33		3	F2	848
41258	+40 01488			6 02 24	+40 41.7	1	8.40		0.12		-0.16		3	B8	555
251402	+6 01106	G 99 - 54		6 02 26	+06 27.4	2	10.52	.005	0.75	.005	0.30	.005	3	G5	333,1620,1696
		WLS 600 5 # 9		6 02 27	+04 27.9	1	10.98		0.67		0.21		2		1375
		RL 48		6 02 27	+24 49.4	1	12.01		0.18		-0.32		2		1709
251311	+23 01183	LS V +23 023		6 02 28	+23 00.6	3	8.81	.011	0.34	.003	-0.50	.005	14	B1.5:IV:n	848,1012,1207
41534	-32 02743	HR 2149	★ AB	6 02 29	-32 10.2	3	5.64	.006	-0.20	.007	-0.82		8	B2 V	15,1034,2012
41433	+0 01269	IDS06000N0052	A	6 02 34	+00 52.3	1	7.04		0.98		0.67		3	K0	3016

Table 1   321

HD	DM	Other Id	N Rem	α_{1950}	δ_{1950}	S	V	σ_V	B–V	σ_{B-V}	U–B	σ_{U-B}	n	Spectrum	References
41433 +0 01269	IDS06000N0052	B	6 02 34	+00 52.3	1	10.19		0.28		0.07		3		3016	
41317 +34 01255			6 02 37	+34 58.6	1	9.38		0.06				2		1724	
	MCC 234		6 02 37	+67 59.0	1	9.75		1.25		1.20		1	K8	3072	
251363 +24 01087			6 02 38	+24 09.0	1	10.89		1.24		0.82		3	G0	1723	
251333 +27 00972			6 02 44	+27 48.9	1	10.33		1.11				3	K0	1724	
	MC +26 264		6 02 45	+26 12.7	1	10.71		1.27				1	G8 III	1702	
+26 01064			6 02 45	+26 12.8	1	11.03		1.08				2		1702	
41511 −16 01349	HR 2148, SS Lep		6 02 45	−16 28.8	5	4.92	.012	0.23	.017	0.11	.061	12	A Shell 2e	15,272,379,2012,2016	
251359 +29 01106			6 02 47	+29 22.0	1	9.34		0.64				3	G5	1724	
41460 +0 01270			6 02 48	+00 37.1	1	7.02		1.60		1.98		7	K5	14	
41330 +35 01334	HR 2141	⋆A	6 02 48	+35 23.8	2	6.11	.009	0.60	.001	0.06		4	G0 V	71,3026	
251383 +26 01067	G 98 - 42		6 02 49	+26 33.9	5	9.42	.005	0.89	.006	0.61	.025	35	K2 V	333,1003,1620,1724,1775,6011	
251519 +6 01108			6 02 52	+06 55.9	1	9.71		0.51		0.03		2	G0	1375	
	MC +25 566		6 02 52	+25 09.8	1	10.95		1.16				2	K2 IV	1702	
41459 +7 01121			6 02 53	+07 32.5	1	8.25		1.71				4	K5	369	
251432 +25 01115			6 02 56	+25 17.3	2	9.38	.000	1.24	.005			5	K0	1702,1724	
41398 +28 01008	LS V +28 024		6 02 56	+28 56.4	1	7.46		0.32		−0.65		2	B2 Ib	1012	
	WLS 600-15 # 9		6 02 59	−14 47.0	1	12.27		0.67		0.12		2		1375	
251431 +25 01116			6 03 00	+25 22.1	1	10.62		0.38				2	F2	1724	
	MC +26 271		6 03 02	+26 21.3	1	10.50		1.19				1	K0 III	1702	
41700 −45 02300	HR 2157	⋆AB	6 03 02	−45 02.1	5	6.35	.005	0.52	.004	−0.02		20	F8/G0 V	15,158,1020,2012,2012	
251461 +26 01070			6 03 04	+26 01.7	1	9.21		0.98				2	K0	1724	
41547 −10 01368	HR 2150		6 03 05	−10 14.3	3	5.87	.009	0.38	.018	0.02		9	F2 V	15,254,2027	
41430 +29 01111			6 03 06	+29 06.2	1	7.51		1.24		1.39		2	K2 III-IV	37	
41357 +38 01377	HR 2143		6 03 08	+38 29.4	2	5.36	.005	0.24	.010	0.13	.020	5	A4 m	3058,8071	
41429 +29 01112	HR 2146, V394 Aur	⋆A	6 03 11	+29 31.1	1	6.08		1.73		1.94		3	M3 II	542	
41429 +29 01112	HR 2146	⋆AB	6 03 11	+29 31.1	4	6.03	.015	1.66	.008	1.83	.028	13	M3 II +F7 V	71,542,3001,8100	
41429 +29 01112	IDS06000N2931	B	6 03 11	+29 31.1	2	10.72		0.90		0.25		2	F7 V	542	
41501 +5 01089			6 03 12	+05 44.6	1	7.91		−0.12		−0.35		2	B9	401	
−54 00956			6 03 12	−54 27.9	1	9.54		1.03		0.81		2	G7	1730	
41667 −32 02753			6 03 13	−32 59.4	2	8.53	.005	1.01	.000	0.56	.000	5	G8 V	119,3077	
251487 +27 00978			6 03 14	+27 42.2	1	9.95		0.44				2	F8	1724	
41742 −45 02302	HR 2158	⋆AB	6 03 14	−45 04.7	4	5.92	.007	0.49	.005			18	F5 V	15,1075,2012,2018	
251611	G 102 - 47		6 03 15	+07 19.6	3	10.31	.016	0.64	.011	−0.05	.005	5		516,1620,1696	
	NGC 2203 sq1 19		6 03 15	−75 32.0	1	10.78		1.32		1.50		3		1646	
41456 +26 01074			6 03 17	+26 31.9	3	7.70	.001	0.92	.002	0.58		34	G8 III	1655,1724,6011	
	DY Ori		6 03 22	+13 53.5	1	11.59		1.52		1.12		1		793	
41469 +23 01188			6 03 22	+23 11.2	1	8.23		1.02		0.84		3	G5	848	
	MC +26 281		6 03 24	+26 16.9	1	13.25		0.35				1	G8 III	1702	
251595 +17 01124			6 03 26	+17 49.7	1	9.68		1.02		0.86		3	K0	848	
41499 +19 01203			6 03 27	+19 14.8	1	8.17		0.23		0.10		4	A5	848	
251549 +26 01075	AA Gem	A	6 03 28	+26 20.1	5	9.52	.138	1.03	.121			6	K0	934,1702,1724,1772,6011	
251549 +26 01075		B	6 03 28	+26 20.1	1	12.50		1.30				1		6011	
	L 181 - 1		6 03 28	−55 18.1	2	12.14	.005	1.46	.002	1.11		6		158,3073	
41824 −48 02124	HR 2162	⋆AB	6 03 29	−48 27.2	4	6.63	.031	0.73	.009	0.26		15	G6 V	15,214,1075,2035	
	MC +26 285		6 03 30	+26 12.4	1	11.54		1.15				1	K2 III	1702	
	G 98 - 43		6 03 31	+30 15.6	2	13.54	.004	0.84	.013	0.30	.017	6		203,1658	
41606 −4 01356			6 03 33	−04 33.5	2	9.01	.005	−0.02	.000	−0.33	.023	8	B9	401,1775	
251627 +21 01115			6 03 36	+21 13.8	1	10.20		1.86				2	K7	369	
41594 +3 01123			6 03 37	+03 20.5	1	8.00		1.46				1	K0	1254	
41759 −35 02684	HR 2160		6 03 41	−35 30.5	3	5.80	.004	0.02	.001	0.03		12	A0 V	15,1637,2012	
41580 +10 01004	IDS06009N1046	A	6 03 42	+10 45.4	1	7.19		−0.06		−0.22		2	A0	401	
251617 +25 01120			6 03 42	+25 33.6	3	9.94	.012	0.03	.003	−0.16	.002	16	B9 V	830,1003,1783	
41698 −24 03679	HR 2156, S Lep		6 03 42	−24 11.4	4	6.88	.182	1.59	.024	1.32	.015	14	M5 III	58,897,2035,8029	
251739 +8 01185			6 03 43	+08 51.0	1	8.89		1.73				4	M0	369	
−35 02685			6 03 43	−35 31.7	1	8.09		0.49				4	G2	2012	
251579 +28 01011			6 03 45	+28 12.8	1	9.45		0.66				2	G0	1724	
41543 +23 01192			6 03 46	+23 38.7	2	6.64	.010	1.36	.010	1.53		7	K0	1625,1724	
	MC +25 609		6 03 46	+25 43.7	1	11.80		1.05				1	G0	1702	
41479 +35 01339			6 03 46	+35 13.2	1	7.20		1.25		1.14		2	K0	1601	
41593 +15 01065			6 03 49	+15 33.0	1	6.76		0.81		0.42		3	K0	1355	
	G 106 - 26		6 03 51	+04 31.5	1	15.08		1.34		0.83		2		3073	
	G 106 - 25		6 03 51	+04 31.6	3	10.88	.019	1.01	.017	0.85	.023	7		158,680,3073	
	WLS 600 45 # 10		6 03 52	+45 35.3	1	10.87		0.77		0.28		2		1375	
41467 +41 01357	HR 2147		6 03 53	+41 51.7	1	6.12		1.22				2	K0 III	71	
41695 −14 01331	HR 2155		6 03 53	−14 55.8	7	4.67	.006	0.05	.009	0.03	.017	26	A0 V	15,379,1075,1356,2012*	
41603 +7 01131			6 03 54	+07 30.6	1	8.47		−0.06		−0.32		2	B9	401	
	RL 118		6 03 54	+23 52.	1	15.97		0.00		−0.88		2		704	
41563 +26 01079			6 03 55	+26 40.7	1	7.49		1.17				2	G6 III	1724	
	MC +26 297		6 03 56	+26 03.8	1	12.74		0.44				1	K5 III:	1702	
	MC +26 299		6 03 56	+26 03.8	1	12.15		1.03				1	G8 III:	1702	
251726 +19 01210	LSS 12		6 03 58	+19 02.5	2	9.39	.015	0.53	.010	−0.53	.005	4	B1 V:e	1012,1207	
	NGC 2203 sq1 24		6 03 58	−75 20.6	1	13.06		0.74		0.35		3		1646	
251693 +26 01080			6 03 59	+26 29.5	1	10.51		0.14				1	A0	934	
	NGC 2203 sq1 21		6 04 00	−75 17.9	1	11.19		0.34		0.27		3		1646	
+13 01110			6 04 01	+13 34.8	2	9.01	.000	1.80	.015	2.02		6	M0 III	148,8032	
+26 01081			6 04 10	+26 08.6	1	10.23		1.84				2		1702	
41692 −4 01362	HR 2154	⋆A	6 04 10	−04 11.2	5	5.38	.013	−0.14	.012	−0.53	.006	18	B5 IV	15,154,247,1078,2027	
41843 −29 02769	HR 2164		6 04 10	−29 45.1	2	5.80	.005	0.05	.000			7	A1 V	15,2027	
251755 +23 01200	LS V +23 027		6 04 11	+23 01.4	1	9.80		0.42		−0.46		3	B5	41	

HD	DM	Other Id	N	Rem	α_{1950}	δ_{1950}	S	V	σ_V	B–V	σ_{B-V}	U–B	σ_{U-B}	n	Spectrum	References
		MC +26 303			6 04 13	+26 20.2	1	11.57		1.17				1	G8 III	1702
		WLS 600 45 # 5			6 04 13	+46 51.3	1	11.10		0.53		-0.05		2		1375
		G 249 - 39			6 04 14	+61 04.5	1	15.87		1.52				1		906
41541	+42 01486	IDS06007N4241		AB	6 04 15	+42 40.4	1	7.04		0.04		-0.38		3	B5	555
42025	-54 00961				6 04 15	-54 24.2	1	7.17		0.94		0.69		31	G8 III	978
		NGC 2203 sq1 23			6 04 15	-75 19.4	1	13.37		0.70		0.12		3		1646
41614	+22 01175				6 04 16	+22 45.0	1	9.20		-0.03		-0.42		3	A0	848
251751	+26 01083				6 04 16	+26 17.5	2	9.87	.019	1.43	.029			3	K2	934,1702
		LS V +21 023			6 04 18	+21 13.9	1	11.69		0.36		-0.33		3		41
251785	+25 01121				6 04 19	+25 31.6	1	8.59		1.39				2	K2	1702
		NGC 2203 sq1 22			6 04 20	-75 21.4	1	11.64		0.77		0.05		3		1646
41658	+18 01095	IDS06015N1848		AB	6 04 24	+18 48.1	1	7.98		0.39		0.09		4	F3 III	848
		NGC 2158 - 3			6 04 24	+24 06.	1	13.14		1.71				2		915
		NGC 2158 - 5			6 04 24	+24 06.	1	16.66		0.95				1		639
		NGC 2158 - 37			6 04 24	+24 06.	1	10.76		0.10		-0.05		1		639
		NGC 2158 - 38			6 04 24	+24 06.	1	12.88		0.66		0.53		1		639
		NGC 2158 - 50			6 04 24	+24 06.	1	17.95		0.99		0.12		1		639
		NGC 2158 - 51			6 04 24	+24 06.	1	18.74		0.99		0.30		1		639
		NGC 2158 - 54			6 04 24	+24 06.	1	18.23		0.86		0.44		1		639
		NGC 2158 - 1520		AB	6 04 24	+24 06.	1	15.64		1.22				1		639
		NGC 2158 - 1523		AB	6 04 24	+24 06.	1	16.30		0.83				1		639
		NGC 2158 - 1528		AB	6 04 24	+24 06.	1	17.62		0.68		0.25		1		639
		NGC 2158 - 3268		AB	6 04 24	+24 06.	2	13.16	.130	0.66	.010	0.09	.025	2		639,915
		NGC 2158 - 3340			6 04 24	+24 06.	2	13.71	.130	0.66	.015	0.22	.035	2		639,915
		NGC 2158 - 3342		AB	6 04 24	+24 06.	1	12.90		0.44		0.17		1		639
		NGC 2158 - 3347			6 04 24	+24 06.	1	15.17		0.66		0.42		1		639
		NGC 2158 - 3401			6 04 24	+24 06.	1	14.40		1.55		1.44		1		639,915
		NGC 2158 - 4261		AB	6 04 24	+24 06.	1	14.91		1.32		0.77		1		639
		NGC 2158 - 4269		AB	6 04 24	+24 06.	1	13.11		1.86		1.75		1		639
		NGC 2158 - 4319			6 04 24	+24 06.	2	15.02	.020	1.32	.020	1.02	.070	2		639,915
		NGC 2158 - 4328			6 04 24	+24 06.	2	15.01	.015	1.33	.015	0.86	.088	3		639,915
		NGC 2158 - 4411		AB	6 04 24	+24 06.	1	16.10		0.79				1		639
		NGC 2158 - 4413			6 04 24	+24 06.	2	14.91	.005	1.34	.028	0.95	.126	4		639,915
		NGC 2158 - 4428			6 04 24	+24 06.	1	17.39		0.78		0.35		1		639
		NGC 2158 - 4433			6 04 24	+24 06.	1	15.00		1.35		1.12		1		639
		NGC 2158 - 4435			6 04 24	+24 06.	2	14.67	.060	0.55	.020	0.34	.020	2		639,915
		NGC 2158 - 4436			6 04 24	+24 06.	2	15.06	.088	1.34	.054			3		639,915
		NGC 2158 - 4448			6 04 24	+24 06.	1	17.45		0.74		0.35		1		639
		NGC 2158 - 4515			6 04 24	+24 06.	2	14.12	.110	0.50	.035	0.32	.080	2		639,915
		NGC 2158 - 4521			6 04 24	+24 06.	1	17.68		0.79		0.26		1		639
		NGC 2158 - 4531			6 04 24	+24 06.	1	18.97		1.10		0.67		1		639
		NGC 2158 - 4532			6 04 24	+24 06.	1	19.04		0.76		0.32		1		639
41883	-31 02995				6 04 25	-31 39.8	1	8.88		-0.06		-0.33		4	B8/9 V	1732
41756	-3 01297				6 04 26	-03 20.1	3	6.92	.012	-0.12	.018	-0.52	.010	7	B5	247,401,2033
251808	+27 00990				6 04 27	+27 25.0	1	9.54		1.83				2	K7	369
41841	-23 03431	HR 2163		⋆A	6 04 27	-23 06.2	3	5.47	.005	0.07	.014	0.09		10	A2 V	15,2027,8071
251807	+28 01015				6 04 30	+28 28.8	1	9.56		0.45				2	F2	1724
		V 643 Ori			6 04 30	-02 54.	1	9.38		1.21		0.79		2		1768
251887	+20 01269				6 04 31	+20 48.4	1	9.48		0.45		-0.01		3	F8	848
251847	+23 01203	LS V +23 028			6 04 31	+23 44.8	2	8.94	.004	0.09	.009	-0.78	.013	5	B1 IV	41,1207
41814	-11 01386	HR 2161			6 04 31	-11 10.0	2	6.65	.005	-0.16	.005			7	B3 V	15,2027
		MC +26 312			6 04 32	+26 04.7	1	11.73		1.17				1	K3 III	1702
		MC +26 319			6 04 32	+26 04.7	1	13.25		1.07				2	K0	1702
		Mon R2 # 1			6 04 32	-05 54.7	1	12.49		0.48		0.37		6	A0	502
	-57 00949				6 04 33	-57 52.3	1	9.27		0.90		0.55		1		6006
41879	-23 03432				6 04 34	-23 32.0	1	7.57		1.21		1.15		5	K1 III	897
251838	+30 01116				6 04 36	+30 06.9	1	10.10		0.51				4	G0	1724
		MC +25 651			6 04 37	+25 12.7	1	10.64		0.97				2	K0 IV	1702
		CS Ori			6 04 38	+11 09.2	1	10.85		0.72		0.57		1	F5	1462
41690	+21 01120	LS V +21 024		⋆AB	6 04 38	+21 52.8	4	7.73	.006	0.20	.006	-0.60	.008	27	B1 V	41,848,1012,1207
41754	+9 01094				6 04 39	+09 12.0	1	7.90		-0.05		-0.58		2	B8	401
41753	+14 01152	HR 2159			6 04 43	+14 46.6	6	4.42	.007	-0.16	.013	-0.66	.022	20	B3 IV	15,154,1223,1363,8015,8040
41791	+8 01193				6 04 44	+08 16.7	1	8.01		-0.04		-0.16		2	B9	401
		G 100 - 59			6 04 44	+27 46.8	1	12.70		1.48		1.21		1		333,1620
41793	+6 01125				6 04 45	+06 26.6	1	8.28		0.03		-0.20		2	A0	401
42120	-54 00966				6 04 47	-54 30.2	1	8.30		1.01		0.84		2	K0 III	1730
		LS V +21 025			6 04 48	+21 39.2	1	11.31		1.11		0.01		4		41
41789	+11 01032				6 04 50	+11 00.9	1	7.52		1.08		0.94		1	K0	1462
		LS V +22 013			6 04 50	+22 28.5	1	11.43		0.38		-0.47		3		41
41708	+27 00994				6 04 51	+27 26.2	2	8.04	.019	0.62	.000	0.14		3	G0 V	1724,3026
41636	+41 01365	HR 2153			6 04 51	+41 03.9	4	6.35	.005	1.04	.007	0.84	.017	8	G9 III	15,1003,1355,3016
41933	-21 01353	HR 2166			6 04 51	-21 48.3	2	5.74	.029	1.65	.005	1.79		6	M3 II/III	2007,3005
251971	+23 01206	LS V +23 029			6 04 54	+23 40.7	1	9.16		0.49		0.41		3	F0	41
41727	+25 01125				6 04 54	+25 27.8	3	8.74	.012	0.02	.012	-0.36	.011	8	B9	833,848,1625
41726	+25 01124				6 04 55	+25 55.9	1	8.60		0.40				2	F3/5 V	1724
		LS V +20 015			6 04 57	+20 51.4	1	12.39		0.46		-0.30		2		41
		LP 659 - 16			6 04 57	-07 09.9	1	11.74		0.48		-0.21		2		1696
		Mon R2 # 2			6 05 00	-06 37.6	1	12.33		0.67		0.41		8	A0	502
41786	+21 01125	IDS06020N2119		AB	6 05 03	+21 18.2	1	7.29		0.35		0.12		3	F0	848
41751	+30 01121				6 05 04	+30 15.6	1	8.19		1.10				2	G5	1724

Table 1

323

HD	DM	Other Id	N Rem	α_{1950}	δ_{1950}	S	V	σ_V	B–V	σ_{B-V}	U–B	σ_{U-B}	n	Spectrum	References
41590 +52 01035				6 05 04	+52 25.8	1	8.58		1.26		1.16		2	K2	1566
		Mon R2 # 3		6 05 04	−06 28.9	1	14.14		0.80		0.60		5		502
		NGC 2203 sq1 17		6 05 05	−75 31.3	1	11.94		0.77		0.16		3		1646
		NGC 2169 - 19		6 05 06	+14 02.8	2	12.04	.010	0.64	.005	0.14	.010	2		49,459
41787 +18 01101				6 05 06	+18 40.7	1	8.95		0.08		0.09		9	A1 V	1655
41766 +24 01109				6 05 06	+24 55.0	1	9.25		-0.04		-0.33		3	A0	848
252024 +26 01091				6 05 06	+26 12.6	1	10.30		1.09				2	K2	1702
−6 01415		Mon R2 # 4		6 05 06	−06 23.5	3	10.30	.023	0.65	.007	-0.20	.022	11	B2 V	206,490,502
+25 01126				6 05 07	+25 47.1	1	10.06		1.07				1		1702
		NGC 2168 - 627		6 05 08	+24 20.5	1	12.97		0.35		-0.31		1	A8	1784
42078 −42 02343 HR 2171				6 05 08	−42 17.5	4	6.15	.019	0.25	.004	0.12	.010	12	A2mA5-A9	216,355,2006,2018
41853 +8 01197 G 102 - 48				6 05 09	+08 13.9	1	8.99		0.89		0.55		1	K2 V	1620
252047		NGC 2168 - 617		6 05 10	+24 17.8	1	10.95		0.40		0.04		1	F2	1784
		NGC 2203 sq1 18		6 05 11	−75 29.1	1	12.94		0.73		0.22		3		1646
		HRC 194	V	6 05 12	+18 38.9	1	11.80		0.74		0.00		1	K0	776
252041 +28 01021				6 05 13	+28 59.1	1	8.82		1.34				3	K0	1724
		NGC 2168 - 625		6 05 14	+24 20.2	1	11.29		0.29		-0.22		1	A2	1784
		MC +25 683		6 05 14	+25 09.9	1	10.57		1.49				2	K0 III	1702
252044 +27 00999				6 05 15	+27 19.6	1	9.51		1.07				2	G0	1724
252044 +27 00999				6 05 15	+27 19.6	1	9.97		0.45				2	G0	1724
42054 −34 02655 HR 2170				6 05 15	−34 18.3	4	5.83	.013	-0.14	.012	-0.57	.018	45	B3/5 Vnn	681,815,1637,8035
		NGC 2168 - 451		6 05 18	+24 21.3	1	12.10		0.35		-0.14		1		1784
		NGC 2168 - 610		6 05 19	+24 14.9	1	12.03		0.42		0.13		1	A3	1784
41831 +22 01185				6 05 20	+22 13.0	1	9.16		0.16		-0.59		2	B3 V	1207
		Mon R2 # 5	A	6 05 20	−06 21.4	1	12.51		0.54		0.10		7	A5	502
		Mon R2 # 5	B	6 05 20	−06 22.6	1	14.22		0.83		0.80		4	A5	502
		NGC 2169 - 27		6 05 21	+13 56.2	1	13.19		1.81				1		459
		AA45,405 S247 # 8		6 05 21	+21 37.5	1	12.30		1.24		1.00		2		797
		NGC 2168 - 612		6 05 21	+24 16.5	1	10.63		0.88		0.25		1	F9	1784
41829 +25 01128				6 05 22	+25 58.5	2	8.12	.010	1.00	.005			4	K0	1702,1724
		Mon R2 # 6		6 05 22	−06 16.1	1	13.46		1.27		0.89		4		502
		Mon R2 # 7		6 05 22	−06 26.3	1	14.02		0.71		-0.10		6		502
		NGC 2169 - 25		6 05 24	+13 57.5	2	13.24	.015	1.04	.000	0.41	.020	2		49,459
		AA45,405 S247 # 6		6 05 24	+21 40.2	1	12.86		0.59		0.20		2		797
		NGC 2169 - 15		6 05 25	+13 58.1	2	11.07	.005	0.15	.005	0.01	.005	2	B9.5V	49,459
		AA45,405 S247 # 7		6 05 25	+21 40.7	1	12.13		1.43		1.41		2		797
252177 +21 01129 LS V +21 026				6 05 25	+21 53.7	1	12.10		0.42		-0.05		2	A3	41
		NGC 2168 - 607		6 05 25	+24 14.6	1	12.15		0.53		0.05		1		1784
		NGC 2168 - 454		6 05 25	+24 21.3	1	12.24		0.40		0.39		1	A6	1784
42286 −59 00584				6 05 25	−59 31.5	3	8.45	.005	0.84	.007	0.44	.002	10	K1 V	158,2012,3008
		G 100 - 60		6 05 26	+22 26.8	1	11.54		0.74		0.24		1	G8	333,1620
		NGC 2168 - 456		6 05 27	+24 22.4	1	10.82		0.31		0.03		1	A1	1784
252214 +13 01120 NGC 2169 - 2			⋆ V	6 05 28	+13 58.8	3	8.12	.012	-0.03	.037	-0.65	.058	3	B2.5V	49,459,8084
		NGC 2169 - 11		6 05 28	+13 59.8	2	10.58	.010	0.05	.000	-0.37	.000	2	B8 V	49,459
41909 +14 01160 NGC 2169 - 3			⋆ A	6 05 28	+14 00.3	2	8.41	.005	0.95	.005	0.69	.005	2	B5	49,459
+14 01161 NGC 2169 - 7			⋆ B	6 05 28	+14 00.7	3	9.35	.004	1.73	.007	1.89	.005	5	G5	49,369,459
252129 +22 01186 LS V +22 018			A	6 05 28	+22 59.0	1	9.74		0.56		-0.36		3	B3	41
252129 +22 01186 LS V +22 018			B	6 05 28	+22 59.0	1	9.93		0.63		-0.25		3		41
+26 01095			V	6 05 28	+26 41.5	1	10.00		1.71				3	M2	369
252215		NGC 2169 - 9		6 05 29	+13 53.7	2	9.99	.015	0.03	.010	-0.55	.005	2	B4 V	49,459
252153 +24 01116 NGC 2168 - 648				6 05 29	+24 27.5	1	9.77		0.20		-0.03		1	A0	1784
		NGC 2168 - 24		6 05 30	+14 02.0	2	13.12	.035	0.34	.005	0.17	.000	2		49,459
		HRC 195	V	6 05 30	+18 09.	1	14.22		0.73		-0.50		1		776
		AA45,405 S247 # 2		6 05 31	+21 40.2	1	11.01		1.02		0.97		2	K0 III	797
42042 −19 01361 HR 2168				6 05 31	−19 09.5	3	5.32	.013	1.66	.004	1.99	.010	11	M1 III	1024,2035,3005
		LS V +21 027		6 05 32	+21 37.1	1	11.07		0.62		-0.45		2	B0 III	41
		NGC 2168 - 601		6 05 32	+24 12.8	1	11.22		0.35		0.04		1		1784
		NGC 2168 - 266		6 05 32	+24 21.1	1	12.56		0.51		0.15		1	A0	1784
		NGC 2169 - 16		6 05 33	+14 00.0	2	11.18	.000	0.11	.000	-0.13	.005	2	B9 V	49,459
		NGC 2168 - 267		6 05 33	+24 22.2	1	11.89		0.51		-0.33		1	F8	1784
41765 +41 01369				6 05 33	+41 33.5	1	8.06		0.02		-0.32		3	B8	555
		NGC 2169 - 12		6 05 34	+13 57.3	2	10.80	.005	0.05	.015	-0.42	.015	2	B9 V	49,459
41870 +22 01187 SS Gem				6 05 34	+22 37.5	1	8.67		1.04		0.71		1	F8 Ib	793
41597 +58 00897 HR 2152				6 05 34	+58 56.7	2	5.36	.012	1.09	.003	0.92	.002	4	G8 III	1355,4001
42137 −37 02605				6 05 34	−37 01.7	1	7.64		1.53		1.93		12	K3/4 V	1673
252251 +11 01038				6 05 35	+11 24.3	1	8.70		0.94		0.60		1	G0	1462
		NGC 2169 - 32		6 05 35	+13 57.2	1	13.61		0.60		0.21		1		459
		NGC 2169 - 34		6 05 35	+13 58.4	1	14.61		0.44		0.28		1		459
		NGC 2168 - 279		6 05 35	+24 23.5	1	12.51		0.46		0.33		1		1784
42168 −45 02317 HR 2178				6 05 35	−45 05.1	4	6.50	.003	1.16	.005	1.13		12	K1 III	15,2013,2030,3005
		NGC 2169 - 35		6 05 36	+13 58.	1	13.69		0.21		0.24		1		459
		NGC 2169 - 36		6 05 36	+13 58.	1	14.24		0.89		0.59		1		459
		NGC 2168 - 277		6 05 36	+24 23.0	1	12.33		0.42		0.09		1		1784
252176 +24 01117 NGC 2168 - 285				6 05 36	+24 24.6	2	9.91	.005	0.23	.030	-0.17	.008	5	A0 V	49,1784
41850 +29 01123				6 05 36	+29 51.3	1	8.71		0.38				2	F2	1724
252248 +13 01123 NGC 2169 - 5			⋆ V	6 05 37	+13 56.4	2	8.77	.005	-0.05	.005	-0.68	.000	2	B3 Vn	49,459
		NGC 2169 - 13		6 05 37	+13 57.3	2	10.77	.015	0.04	.005	-0.34		2	B9 V	49,459
		NGC 2168 - 271		6 05 37	+24 22.0	1	12.06		0.46		0.34		2		1784
42004 −6 01417 Mon R2 # 9				6 05 37	−06 13.1	2	9.66	.010	0.31	.005	-0.43	.005	10	B2 V	206,502
		NGC 2169 - 26		6 05 39	+13 56.0	1	12.65		0.59		0.12		1		459

HD	DM	Other Id	Rem	α_{1950}	δ_{1950}	S	V	σ_V	B–V	σ_{B-V}	U–B	σ_{U-B}	n	Spectrum	References
		NGC 2169 - 31	⋆ C	6 05 39	+13 58.9	1	11.42		0.09		-0.27		1	B9 V	459
252208	+20 01276			6 05 39	+20 25.7	1	8.75		1.74				2	K7	369
		NGC 2168 - 412		6 05 39	+24 13.3	1	10.92		1.32		1.28		3		1784
		NGC 2168 - 284		6 05 39	+24 24.2	2	11.06	.009	0.26	.041	0.03	.014	5	B8	49,1784
		NGC 2168 - 286		6 05 39	+24 24.9	1	13.18		0.61		0.42		2		1784
252198		NGC 2168 - 464		6 05 39	+24 25.8	2	10.22	.001	0.20	.032	-0.23	.007	5	B9	49,1784
	-6 01418	Mon R2 # 8		6 05 39	-06 21.2	3	9.21	.004	0.31	.001	-0.56	.003	12	B1	206,502,1732
		NGC 2169 - 33		6 05 40	+13 56.1	1	14.25		0.24		0.33		1		459
		NGC 2169 - 14	⋆ V	6 05 40	+13 57.7	2	10.89	.015	0.13	.010	-0.08	.000	2	B9 V	49,459
252266		NGC 2169 - 6	⋆ E	6 05 40	+13 58.1	3	9.13	.030	0.02	.008	-0.53	.015	3	B3 V	49,459,8084
41943	+13 01124	NGC 2169 - 1	⋆ AB	6 05 40	+13 58.8	4	6.93	.013	-0.10	.008	-0.87	.014	6	B2 III	49,399,459,8084
252201	+24 01118	NGC 2168 - 410		6 05 40	+24 12.4	1	9.98		0.20		-0.17		1	A7	1784
252151	+26 01097			6 05 40	+26 31.5	1	9.94		0.36				2	F0	1724
		NGC 2169 - 28		6 05 41	+13 55.7	1	13.49		0.75		0.37		1		459
		NGC 2168 - 228		6 05 41	+24 15.0	1	10.91		0.36		0.21		1	F0	1784
		NGC 2168 - 229		6 05 41	+24 15.4	1	11.40		0.76		0.50		4		1784
		NGC 2169 - 18		6 05 42	+13 57.9	2	11.79	.010	0.12	.010	0.00	.000	2	B9.5V	49,459
		NGC 2169 - 4	⋆ D	6 05 42	+13 58.5	2	8.60	.005	-0.05	.005	-0.81	.005	2	B2.5IV	49,459
41890	+25 01131	OX Gem		6 05 42	+25 39.3	1	7.55		1.64		1.84		2	M1	8032
252202	+24 01119	NGC 2168 - 782		6 05 43	+24 06.4	1	9.88		0.16		-0.08		1	A0	1784
		NGC 2168 - 287		6 05 43	+24 24.3	2	11.89	.010	0.52	.010	0.15	.000	3	F5	49,1784
252200		NGC 2168 - 409		6 05 44	+24 13.8	1	10.12		0.27		-0.10		1	A7	1784
		NGC 2169 - 8		6 05 45	+13 58.2	2	9.90	.005	1.13	.005	0.94	.005	2	G8 III	49,459
		NGC 2168 - 290		6 05 45	+24 26.1	2	12.50	.023	0.43	.032	0.19	.018	5		49,1784
	+24 01122	NGC 2168 - 783		6 05 46	+24 05.8	1	10.01		0.12		-0.28		1		1784
252199	+24 01120	NGC 2168 - 293		6 05 46	+24 26.6	2	9.32	.006	0.18	.031	-0.20	.020	5	A0 V	49,1784
252197	+24 01121	NGC 2168 - 662		6 05 47	+24 30.6	1	8.51		1.31		1.50		3	K2	1784
		NGC 2168 - 119		6 05 48	+24 22.0	1	10.93		0.22		-0.04		1	A2	1784
42167	-37 02609	HR 2177		6 05 49	-37 14.7	2	5.01	.005	-0.12	.005			7	B9 IV	15,2012
252204	+23 01220			6 05 49	+23 30.0	1	10.02		1.99				2	A0	369
252236		NGC 2168 - 219		6 05 50	+24 13.9	1	10.41		0.23		-0.07		1	A7	1784
		NGC 2168 - 124		6 05 50	+24 22.7	1	10.88		0.23		-0.06		1	A5	1784
		NGC 2169 - 22		6 05 52	+13 58.9	1	12.78		0.64		0.07		1		49
		NGC 2169 - 23		6 05 52	+14 00.8	2	13.02	.005	0.82	.005	0.39	.015	2		49,459
		NGC 2168 - 129		6 05 52	+24 24.1	2	13.66	.033	0.56	.028	0.31		4		49,1784
		NGC 2168 - 130		6 05 52	+24 24.7	2	11.19	.015	0.28	.054	0.06	.015	3	A1	49,1784
41998	+9 01107			6 05 53	+09 38.6	1	7.78		0.00		-0.36		2	B8	401
41940	+24 01123	NGC 2168 - 123		6 05 53	+24 22.1	2	8.16	.008	0.11	.017	-0.54	.004	26	B2 V	49,1784
252230	+28 01024			6 05 53	+28 45.5	1	8.93		1.00				4	K0	1724
		NGC 2168 - 131		6 05 54	+24 24.3	2	10.46	.005	0.25	.044	-0.01	.007	4	A1	49,1784
		MC +26 335		6 05 54	+26 22.1	1	11.52		0.84				1	K0 III	1702
252234	+26 01099			6 05 54	+26 27.4	1	9.11		1.05				4	K0	1724
		NGC 2168 - 136		6 05 56	+24 22.5	1	10.64		0.21		-0.23		2	A2	1784
		NGC 2168 - 138		6 05 56	+24 24.8	1	11.65		0.31		0.13		1	A0	1784
252229	+29 01125			6 05 56	+29 34.0	1	9.36		0.74				3	K0	1724
252260	+24 01124	NGC 2168 - 81		6 05 57	+24 16.5	1	8.47		1.34		1.47		2	K5	1784
		NGC 2168 - 87		6 05 57	+24 17.7	2	12.06	.019	0.38	.049	0.30	.044	3		49,1784
		NGC 2168 - 42		6 05 57	+24 22.5	1	9.96		0.23		-0.21		2	B8	1784
41939	+25 01133			6 05 57	+25 31.3	2	8.60	.005	0.15	.005	0.16	.010	8	A0	833,848
42051	-6 01420			6 05 57	-06 32.4	1	8.90		0.40				4	B3	2013
		NGC 2203 sq1 1		6 05 57	-75 30.7	1	11.92		0.63		0.11		3		1646
	+14 01164			6 05 58	+14 07.3	1	9.64		2.01				3	M3	369
		NGC 2168 - 137	V	6 05 58	+24 23.0	2	10.54	.025	0.22	.052	-0.07	.015	5	A5	49,1784
		NGC 2168 - 44		6 05 59	+24 22.4	1	11.56		0.38		0.14		1	A0	1784
		MC +26 338		6 05 59	+26 19.1	1	10.30		1.17				2	G8 III	1702
41906	+31 01208			6 05 59	+31 44.6	1	8.84		0.38				2	F5	1724
42050	-5 01515	Mon R2 # 10		6 05 59	-05 19.9	2	8.10	.010	0.06	.005	-0.76	.005	6	B1 V	206,401
		NGC 2168 - 43		6 05 60	+24 22.1	1	12.35		0.41		0.07		1	A2	1784
		NGC 2168 - 139		6 06 00	+24 24.3	2	13.26	.035	0.68	.030	0.20	.020	2		1784,49
		NGC 2169 - 17		6 06 01	+13 59.9	2	11.63	.020	0.22	.020	0.17	.010	2		49,459
252325	+20 01278	LS V +20 016		6 06 01	+20 39.0	3	10.79	.005	0.57	.000	-0.43	.013	6	B1:V:	41,241,1012
252294	+22 01190	LS V +22 021		6 06 01	+22 26.0	1	10.74		0.30		-0.43		4	A3	41
		NGC 2168 - 41		6 06 01	+24 18.2	1	11.33		0.38		0.08		1	B9	1784
		NGC 2168 - 140		6 06 01	+24 23.8	1	12.31		0.41		0.36		8	A3	1784
42015	+7 01148			6 06 02	+07 13.8	1	8.61		-0.03		-0.27		2	A0	401
		NGC 2203 sq1 4		6 06 02	-75 30.1	1	14.14		0.74		0.30		3		1646
42035	+8 01202	HR 2167		6 06 03	+08 40.7	3	6.54	.005	-0.07	.015	-0.17	.013	9	B9 V	15,1061,1079
41997	+15 01079	LSS 13		6 06 03	+15 42.9	2	8.41	.010	0.38	.005	-0.62	.010	4	O7	1011,1012
		NGC 2168 - 75		6 06 03	+24 16.2	1	12.15		0.31		0.16		1		1784
		NGC 2168 - 74		6 06 03	+24 16.5	1	10.77		0.28		0.02		1	A3	1784
		NGC 2203 sq1 10		6 06 03	-75 25.1	1	12.42		0.67		0.10		3		1646
252320		NGC 2168 - 208		6 06 04	+24 13.9	1	10.49		0.30		0.03		1	A7	1784
		NGC 2168 - 204		6 06 04	+24 14.8	1	12.41		0.51		-0.06		1		1784
		NGC 2168 - 202		6 06 04	+24 15.4	1	11.12		0.21		-0.09		1	B8	1784
		NGC 2168 - 8		6 06 04	+24 18.9	1	11.67		0.39		0.15		1	A2	1784
		NGC 2168 - 144		6 06 04	+24 23.1	1	10.21		0.27		-0.11		1	A0	1784
		NGC 2168 - 142		6 06 04	+24 24.8	2	14.51	.148	0.55	.004	0.34		9		1784,49
		NGC 2168 - 149		6 06 05	+24 23.8	1	12.52		0.32		0.27		1	A2	1784
		NGC 2175 - 74		6 06 06	+20 36.4	1	12.96		0.67		0.17		1		423
		NGC 2168 - 207		6 06 06	+24 14.0	1	10.59		0.22		-0.02		1	A2	1784

Table 1

HD	DM	Other Id	N	Rem	α_{1950}	δ_{1950}	S	V	σ_V	B–V	σ_{B-V}	U–B	σ_{U-B}	n	Spectrum	References
		NGC 2168 - 150			6 06 06	+24 24.0	1	10.92		0.27		-0.10		1	A0	1784
		NGC 2168 - 151			6 06 06	+24 24.2	1	11.18		0.25		0.03		1	A0	1784
42525	−66 00493	HR 2194			6 06 06	−66 02.0	3	5.70	.004	-0.02	.005	-0.07	.000	15	A0 V	15,1075,2038
		NGC 2169 - 21			6 06 08	+13 59.1	1	12.50		0.25		0.17		1		49
		NGC 2168 - 6			6 06 08	+24 19.1	1	11.15		0.30		0.03		1	A0	1784
41888	+39 01533				6 06 08	+39 11.4	2	8.21	.005	0.19	.005	0.17	.010	5	A1 V	833,1601
252321	+23 01223	LS V +23 030			6 06 09	+23 54.0	2	9.26	.004	0.06	.009	-0.80	.013	5	B1 V	41,1207
		NGC 2168 - 307			6 06 09	+24 24.9	1	10.03		0.17		-0.26		1	B8	1784
		NGC 2175 - 78			6 06 10	+20 34.8	1	13.25		0.46		-0.22		1		423
		NGC 2168 - 197			6 06 10	+24 15.2	1	10.61		0.28		0.09		3	F0	1784
		NGC 2169 - 20			6 06 11	+13 56.8	1	12.36		0.45		0.12		1		49
		NGC 2175 - 73			6 06 11	+20 37.7	1	14.39		0.63		0.15		1		423
		NGC 2168 - 196			6 06 11	+24 15.6	2	11.97	.024	0.33	.063	0.15	.024	3	A2	49,1784
252414	+13 01128	NGC 2168 - 10			6 06 12	+13 55.0	1	10.10		-0.02		-0.30		1	B9 V	49
		NGC 2175 - 76			6 06 12	+20 33.6	1	12.63		0.46		0.35		1		423
		NGC 2168 - 380			6 06 12	+24 13.4	1	11.64		0.37		0.14		1		1784
		NGC 2168 - 195			6 06 12	+24 16.1	1	10.15		0.21		-0.30		2	B9	1784
		NGC 2168 - 66			6 06 12	+24 18.1	1	10.53		0.33		0.17		1	F0	1784
41996	+24 01126	NGC 2168 - 310			6 06 12	+24 25.7	2	7.41	.005	1.10	.015	0.88		8	G0	1724,1784
252353	+24 01125	NGC 2168 - 309			6 06 12	+24 25.7	1	9.01		0.18		-0.26		2	A7	1784
		NGC 2168 - 194			6 06 13	+24 16.1	1	11.65		0.58		-0.01		2		1784
252319		NGC 2168 - 164			6 06 13	+24 20.8	2	9.37	.010	0.21	.030	-0.11	.001	6	A0 V	49,1784
252318		NGC 2168 - 160			6 06 14	+24 22.7	1	9.79		0.22		-0.24		1	A7	1784
41995	+24 01127	NGC 2168 - 781			6 06 14	+24 37.7	1	8.81		0.15		-0.29		4	A9	1784
		MC +25 734			6 06 14	+25 21.5	1	11.54		1.14				1	K0 III	1702
252314	+29 01126				6 06 14	+29 07.0	1	8.91		0.91				2	K0	1724
41689	+62 00818				6 06 14	+62 19.4	2	8.43	.005	-0.07	.010	-0.86	.025	6	B1 V (n)	399,555
43013	+24 01128	NGC 2168 - 780			6 06 16	+24 35.4	1	8.55		0.16		0.04		2	A0	1784
		NGC 2203 sq1 11			6 06 16	−75 22.8	1	11.38		1.03		0.71		3		1646
		NGC 2168 - 52			6 06 17	+24 19.8	1	10.77		0.39		0.21		1	A5	1784
		NGC 2168 - 161			6 06 17	+24 22.7	1	10.65		0.26		-0.13		1	A0	1784
		NGC 2168 - 317			6 06 17	+24 24.2	1	11.10		0.35		0.14		1	A7	1784
		NGC 2175 - 80			6 06 19	+20 33.4	1	13.99		0.63		0.23		1		423
252381	+20 01280				6 06 19	+20 53.8	1	9.74		0.43		-0.01		1	F5	401
		NGC 2168 - 168			6 06 19	+24 20.4	1	11.42		0.34		0.11		2	A0	1784
41994	+27 01006				6 06 19	+27 12.2	1	7.83		1.07				2	G5 II	1724
42505	−62 00580				6 06 19	−62 40.6	1	9.42		1.10		1.04		2	K4 V	3072
		NGC 2168 - 374			6 06 20	+24 11.9	1	11.70		0.39		0.21		1	A3	1784
		NGC 2168 - 319			6 06 20	+24 24.8	1	12.49		0.51		0.56		1	F0	1784
42303	−42 02351	HR 2181		⋆ AB	6 06 20	−42 08.7	1	5.50		0.00				4	A0 V	2035
42111	+2 01139	HR 2174		⋆ A	6 06 21	+02 30.5	6	5.72	.011	0.07	.006	0.07	.035	16	A3 Vn	15,146,1084,1211,1415,3024
42111	+2 01139	IDS06038N0231		AB	6 06 21	+02 30.5	1	5.42		0.06		0.02		4	A3 Vn	1415
		NGC 2175 - 82			6 06 21	+20 31.8	1	14.00		0.64		0.08		1		423
		LS V +20 017			6 06 21	+20 38.9	1	11.74		0.33		-0.37		2		41
252373	+24 01129	NGC 2168 - 177			6 06 21	+24 19.1	1	9.20		0.13		-0.23		1	B7 III	49
		NGC 2168 - 169			6 06 21	+24 20.1	1	11.56		0.21		0.23		1		1784
		NGC 2168 - 332			6 06 21	+24 20.9	1	12.33		0.63		0.13		1		1784
		NGC 2203 sq1 5			6 06 21	−75 30.1	1	13.45		0.12		0.19		3		1646
		NGC 2168 - 563			6 06 22	+24 10.7	1	11.33		0.24		0.14		1	A0	1784
		NGC 2168 - 326			6 06 22	+24 22.1	1	12.46		0.37		0.51		1		1784
252404	+24 01130	NGC 2168 - 323			6 06 22	+24 23.4	1	9.64		0.25		-0.11		1	A7	1784
		NGC 2168 - 320			6 06 22	+24 24.9	1	11.42		0.27		0.18		1	B9	1784
42092	+2 01140	IDS06038N0231		B	6 06 23	+02 30.4	4	6.93	.008	0.03	.004	-0.05	.021	9	A0 V	146,1084,1211,3024
		NGC 2168 - 324			6 06 23	+24 22.6	1	11.61		0.32		0.12		1	F0	1784
252371	+26 01102				6 06 23	+26 11.3	1	10.05		1.38				2	K5	1702
		NGC 2175 - 61			6 06 24	+20 37.3	1	11.52		0.55		0.36		1		423
252409	+20 01281				6 06 24	+20 53.6	1	9.70		-0.03		-0.60		2	B8	401
252405	+24 01131	NGC 2168 - 364			6 06 24	+24 15.2	3	8.76	.039	0.13	.039	-0.32	.011	7	B6 III	49,1207,1784
		NGC 2175 - 56			6 06 26	+20 29.2	1	12.79		0.68		0.09		1		423
		LS V +21 028			6 06 28	+21 45.9	1	12.56		0.34		-0.22		2		41
		NGC 2168 - 334			6 06 28	+24 20.5	1	11.56		0.36		0.14		1	A5	1784
		NGC 2168 - 513			6 06 28	+24 24.0	1	11.09		0.34		0.14		1	F0	1784
252400	+26 01104	IDS06034N2631		AB	6 06 29	+26 30.4	1	9.93		0.55				3	G0	1724
252370	+28 01027				6 06 29	+28 22.2	1	9.05		1.17				2	K0	1724
42067	+19 01235				6 06 30	+19 14.5	1	8.47		0.44		-0.02		3	F2 IV	848
		NGC 2168 - 172			6 06 30	+24 20.2	1	12.78		0.50		0.31		2		1784
42049	+22 01198	HR 2169			6 06 31	+22 12.0	1	5.93		1.63				2	K4 I	71
		NGC 2168 - 559			6 06 31	+24 11.3	1	11.86		0.33		0.19		1	A5	1784
252427		NGC 2168 - 335			6 06 31	+24 21.0	1	10.51		0.34		0.07		2	A7	1784
252402	+26 01105				6 06 31	+26 21.4	1	9.73		1.53				1	K7	1702
42068	+17 01144				6 06 32	+17 50.6	1	8.42		0.53		0.02		3	F7 V	848
252464	+20 01282				6 06 32	+20 53.4	1	9.91		0.12		-0.20		1	A0	401
42033	+29 01128	IDS06034N2915		AB	6 06 32	+29 14.9	1	7.84		0.46				2	F5/6 V	1724
		NGC 2168 - 522			6 06 33	+24 23.3	1	11.28		0.26		0.04		1	A0	1784
		NGC 2168 - 785			6 06 33	+25 34.2	1	13.41		0.71		0.47		1		1784
		NGC 2168 - 549			6 06 34	+24 14.5	1	11.09		1.11		1.11		1		1784
252458		NGC 2168 - 336			6 06 34	+24 20.1	1	10.08		0.29		-0.15		1	A7	1784
		NGC 2168 - 784			6 06 35	+24 04.2	1	11.43		0.31		0.20		1		1784
		NGC 2168 - 760			6 06 35	+24 10.0	1	10.16		0.25		-0.10		1	B8	1784
		NGC 2168 - 554			6 06 35	+24 12.3	1	11.09		0.30		-0.01		1	A5	1784

HD	DM	Other Id	N Rem	α_{1950}	δ_{1950}	S	V	σ_V	B–V	σ_{B-V}	U–B	σ_{U-B}	n	Spectrum	References
42034	+28 01028			6 06 35	+28 24.5	1	8.46		0.36				2	F2/3 IV/III	1724
252399	+29 01129	IDS06034N2915		6 06 35	+29 14.3	1	10.35		0.50				2	G5	1724
42540	−62 00582	HR 2196		6 06 36	−62 08.7	1	5.05		1.25		1.35		2	K2/3 III	58
		NGC 2175 - 60		6 06 37	+20 38.2	1	13.04		0.83		0.42		1		423
		MC +26 349		6 06 37	+26 14.5	1	11.13		1.20				2	K0 III	1702
42012	+34 01279	G 98 - 47		6 06 37	+34 08.9	1	8.42		0.79		0.35		1	K0	333,1620
		LS V +23 031		6 06 38	+23 05.0	1	10.55		0.45		−0.61		3		41
		POSS 660 # 1		6 06 38	−06 30.2	1	14.42		0.69				2		1739
		LS V +22 022		6 06 39	+22 49.9	1	11.25		0.29		−0.43		3		41
		NGC 2168 - 764		6 06 40	+24 08.8	1	9.17		0.22		−0.26		3		1784
42066	+24 01133	NGC 2168 - 763		6 06 40	+24 09.3	1	8.64		0.28		0.18		1	A2	1784
42088	+20 01284	LS V +20 019		6 06 41	+20 29.9	10	7.55	.009	0.05	.005	−0.88	.013	48	O6	41,241,401,1011,1012,1207*
		NGC 2168 - 545		6 06 41	+24 15.0	2	12.12	.006	0.56	.001	−0.06	.041	11	F0	49,1784
42087	+23 01226	HR 2173	★ AB	6 06 42	+23 07.4	7	5.74	.024	0.21	.009	−0.63	.005	26	B2.5Ib	15,41,154,1012,1119*
252461		NGC 2168 - 753		6 06 42	+24 10.2	1	10.44		0.28		0.06		1	A7	1784
252459		NGC 2168 - 547		6 06 42	+24 13.5	2	10.42	.031	0.17	.043	−0.22	.025	6	A7	49,1784
252460		NGC 2168 - 546		6 06 42	+24 14.1	2	10.72	.012	0.25	.046	−0.08	.006	5	A7	49,1784
252537	+19 01236	G 105 - 13		6 06 44	+19 06.2	2	9.90	.010	0.94	.000	0.69	.015	2	K0	1620,1658
		NGC 2168 - 542		6 06 45	+24 17.4	1	11.94		0.37		0.12		1	A5	1784
42086	+24 01135	NGC 2168 - 744		6 06 46	+24 13.5	2	7.53	.002	0.22	.059	0.19	.034	11	A3 V	49,1784
252480	+24 01136	NGC 2168 - 724		6 06 46	+24 22.2	2	9.77	.026	0.16	.016	−0.30	.002	6	A0 V	49,1784
		NGC 2175 - 5		6 06 48	+20 20.	2	13.20	.050	0.44	.005	−0.28	.025	2		298,423
		NGC 2175 - 11		6 06 48	+20 20.	1	14.46		0.85		0.11		1		298
		NGC 2175 - 12		6 06 48	+20 20.	1	14.18		0.77		0.45		1		298
		NGC 2175 - 14		6 06 48	+20 20.	1	15.45		0.39		0.56		1		298
		NGC 2175 - 16		6 06 48	+20 20.	1	13.01		1.37		1.54		1		298
		NGC 2175 - 17		6 06 48	+20 20.	1	13.64		0.64		0.56		1		298
		NGC 2175 - 33		6 06 48	+20 20.	1	12.69		0.32		−0.23		1		423
		NGC 2175 - 100		6 06 48	+20 20.	2	11.03	.018	0.40	.014	−0.58		5		241,8064
		NGC 2175 - 101		6 06 48	+20 20.	1	14.23		0.51				1		8064
		Mon R2 # 11		6 06 49	−06 29.9	1	13.00		0.62		0.36		5		502
252503		NGC 2168 - 723		6 06 50	+24 23.0	1	10.58		0.23		−0.02		1	A7	1784
42204	−3 01308			6 06 50	−03 47.1	2	8.45	.000	−0.10	.010	−0.62		6	B5	401,2033
252535	+20 01287	LS V +20 020		6 06 51	+20 37.7	2	10.12	.030	0.19	.005	−0.75	.010	6	B2	41,241
		GD 72		6 06 51	+28 15.1	1	14.63		0.00		−0.82		1	DA	3060
252531	+22 01200	LS V +22 023		6 06 52	+22 40.0	1	10.38		0.18		−0.60		3	B8	41
42301	−22 01327	HR 2180		6 06 52	−22 25.0	2	5.49	.005	−0.01	.000			7	A0 IV	15,2027
		L 380 - 96		6 06 52	−44 34.8	1	14.09		0.72		0.38		1		3060
		NGC 2175 - 9		6 06 53	+20 30.8	2	14.33	.210	0.34	.415	−0.17	.015	2		298,423
		NGC 2175 - 49		6 06 53	+20 35.2	1	13.03		0.57		0.09		1		423
252505	+22 01199			6 06 53	+22 58.5	1	10.03		1.88				2	K7	369
252504		NGC 2168 - 731		6 06 53	+24 14.9	2	9.37	.000	0.31	.033	−0.18	.011	4	B9.5V	49,1784
		NGC 2175 - 44		6 06 54	+20 33.0	1	10.71		0.39		0.01		1		423
252477	+29 01131			6 06 54	+29 21.4	1	8.84		0.97				2	G5	1724
		NGC 2175 - 43		6 06 56	+20 33.0	1	10.58		0.22		0.09		1		423
252558	+20 01289	LS V +20 021		6 06 56	+20 39.1	2	10.59	.015	0.36	.000	−0.48	.010	6	B9	41,241
42085	+33 01260			6 06 57	+33 18.1	1	9.01		0.46				2	G0	1724
42504	−54 00973	TY Pic		6 06 57	−54 25.8	1	7.65		1.02		0.68		1	G8/K0 III	1641
		Mon R2 # 12		6 06 58	−06 18.0	1	12.83		0.82		0.38		6		502
42106	+30 01133			6 07 01	+30 33.6	1	7.76		1.05				2	G7 III	1724
42180	+14 01170			6 07 02	+14 51.8	1	7.30		−0.05		−0.37		2	B9	401
42159	+19 01237			6 07 02	+19 41.2	1	8.04		0.37		0.10		3	F2 III	848
42179	+15 01084			6 07 04	+15 20.4	1	8.55		0.18		−0.40		2	A0	401
42129	+30 01134			6 07 04	+30 26.2	1	8.31		0.55				2	G5	1724
42160	+17 01145	G 102 - 51		6 07 05	+17 56.9	1	8.49		0.67		0.12		1	G2 V	1620
		G 101 - 22		6 07 05	+41 55.4	2	12.59	.030	1.12	.015	1.01	.050	2	K5	1620,1658
42261	−6 01431	Mon R2 # 13	★ AB	6 07 05	−06 18.9	3	9.19	.022	0.18	.005	−0.49	.010	13	B3 V	206,502,2013
42259	−5 01521	LS VI -05 001		6 07 06	−05 03.4	2	8.49	.000	0.33	.050	−0.56	.020	4	B0 V	490,1212
42448	−44 02452	HR 2187	★ A	6 07 06	−44 20.8	1	6.27		−0.15				4	B8 II	2035
42158	+21 01137			6 07 07	+21 14.7	1	8.25		0.30		0.07		3	F0	848
42178	+16 01021			6 07 08	+16 46.0	1	9.38		−0.03		−0.23		2	A0	401
		MC +26 357		6 07 08	+26 16.5	1	11.53		1.03				1	G5 IV	1702
252554	+26 01111			6 07 08	+26 30.7	1	8.88		1.03				3	K0	1724
42327	−18 01316	HR 2182	★ A	6 07 08	−18 07.0	2	6.34	.005	−0.03	.000			7	B9 V(n)	15,2012
42278	−5 01523	HR 2179		6 07 09	−05 42.1	2	6.16	.005	0.35	.005	−0.01	.015	7	F3 IVw	15,1061
42322	−15 01283	IDS06049S1500	A	6 07 10	−15 00.3	1	9.48		0.18		−0.25		14		1732
252680	+14 01171			6 07 11	+14 05.1	1	9.10		0.04		−0.64		4	B3	206
	+25 01141			6 07 11	+25 02.4	1	10.16		1.19				2		1702
42323	−15 01285			6 07 11	−15 06.3	1	9.19		0.20		0.02		5	B(9) (V)	1732
42537	−52 00864			6 07 11	−52 31.7	2	8.92	.000	1.85	.015	1.92	.019	3	S	565,3048
42235	+9 01112			6 07 12	+09 28.7	1	7.29		−0.11		−0.38		2	A0	401
	+9 01111	CT Ori		6 07 12	+09 53.2	1	10.31		0.93		0.45		1	F9	793
		LS VI +13 004		6 07 12	+13 07.9	2	10.23	.000	0.46	.000	−0.56	.000	4	O9 V	1011,1012
		NGC 2175 - 20		6 07 12	+20 35.9	1	13.79		0.77		0.11		1		423
252583	+25 01142			6 07 13	+25 30.6	1	10.20		1.59				2	K5	1702
	+23 01229	LS V +23 033		6 07 18	+23 08.4	1	11.02		0.25		−0.46		3	A0	41
		Mon R2 # 14		6 07 18	−06 17.9	1	10.26		0.10		−0.49		6	B5 V	502
42341	−14 01348	HR 2183		6 07 18	−14 34.5	4	5.55	.004	1.15	.005	1.20		13	K2 III	15,1075,2035,3005
42252	+10 01027			6 07 21	+10 29.4	1	8.54		0.90		0.99		1	K5	401
42254	+7 01155			6 07 22	+07 38.3	1	8.07		−0.03		−0.42		2	B9	401

Table 1   327

HD	DM	Other Id	N Rem	α_{1950}	δ_{1950}	S	V	σ_V	B–V	σ_{B-V}	U–B	σ_{U-B}	n	Spectrum	References
		AA131,200 # 8		6 07 23	+12 49.3	1	15.79		0.86		-0.20		2		7005
252742	+15 01088			6 07 25	+15 48.3	1	8.76		0.12		-0.41		2	B8	401
252636	+26 01114			6 07 25	+26 59.0	1	10.21		0.42				1	F0	1724
42176	+30 01138	IDS06042N3059	A	6 07 26	+30 58.1	1	8.67		0.53				2	F7 V	1724
42936	-72 00451			6 07 28	-72 29.9	1	9.09		0.91				4	K0 IV/V	2012
252743	+15 01089			6 07 30	+15 45.5	1	8.76		1.05		0.73		1	G5	401
		MC +25 792		6 07 32	+25 35.6	1	12.39		1.47				1	K0	1702
252705	+22 01204			6 07 33	+22 35.4	1	9.75		1.30		1.23		2	F5	1726
42316	+9 01117			6 07 35	+09 32.8	2	8.89	.010	0.33	.012	0.21		43	A5	851,6011
		Mon R2 # 15		6 07 35	-06 18.0	1	11.40		0.19		-0.10		6	B8 V	502
		LS V +22 024		6 07 36	+22 18.1	1	11.82		0.23		-0.49		3		41
42443	-22 01330	HR 2186	★ AB	6 07 42	-22 45.9	2	5.70	.005	0.44	.000			7	F5 V	15,2027
	+18 01110			6 07 43	+18 55.4	1	10.20		2.04				2	M2	369
42083	+52 01041	HR 2172		6 07 45	+52 39.6	2	6.30	.000	0.14	.000	0.09	.009	8	A5 m	3058,8071
42294	+18 01111			6 07 46	+18 42.8	1	8.26		0.08		0.04		3	A2 IV	848
42272	+26 01117	TU Gem		6 07 47	+26 01.6	1	7.29		2.77				1	C6 II	109
42127	+48 01352	HR 2176	★ AB	6 07 47	+48 43.4	1	5.78		0.10		0.07		2	A3 V	401
42486	-26 02761	HR 2192		6 07 47	-26 41.4	2	6.28	.008	1.01	.001	0.78		6	G8/K0 III	58,2035
		Mon R2 # 16		6 07 48	-06 26.9	1	13.63		0.63		0.52		2		502
		MC +25 807		6 07 49	+25 59.9	1	11.62		1.16				1	G5 III	1702
41927	+65 00517	HR 2165		6 07 49	+65 43.9	3	5.38	.077	1.34	.005	1.47	.037	10	K2 II-III	1355,3016,8100
252846	+13 01145			6 07 50	+13 04.0	1	9.96		2.00				3	F5	369
		NGC 2175 - 3		6 07 50	+20 37.9	2	12.79	.010	0.41	.025	-0.18	.040	2		298,423
		G 101 - 24		6 07 50	+42 33.7	2	12.43	.010	1.10	.005	0.91	.015	2		1620,1658
		NGC 2175 - 2		6 07 52	+20 38.4	2	12.75	.050	0.41	.005	-0.31	.075	2		298,423
252812	+20 01293	NGC 2175 - 1	★ AB	6 07 54	+20 37.2	3	10.81	.242	0.41	.021	-0.13	.151	3	B5	41,298,423
		LS V +20 023		6 07 55	+20 37.4	1	11.11		0.47		-0.34		1		41
42499	-25 02978	IDS06059S2536	AB	6 07 55	-25 36.9	1	7.66		0.57				4	G0 V	2012
		NGC 2175 - 15		6 07 56	+20 36.9	2	12.87	.030	0.56	.015	-0.19	.080	2		298,423
		NGC 2175 - 40		6 07 57	+20 34.8	1	12.93		0.54		0.07		1		423
		NGC 2175 - 4		6 07 57	+20 37.2	2	10.95	.060	0.51	.020	-0.28	.020	2		298,423
		NGC 2175 - 8		6 07 57	+20 37.9	2	13.44	.095	0.62	.015	-0.09	.015	2		298,423
42352	+13 01147	LS VI +13 005		6 07 58	+13 40.3	1	6.93		-0.01		-0.80		6	B1 III	399
		NGC 2175 - 6		6 07 58	+20 39.0	1	13.29		1.36		0.78		1		298
		LS VI +13 006		6 07 59	+13 11.7	3	10.56	.017	0.45	.013	-0.55	.021	6	O9 V	1011,1012,1207
42333	+20 01296	NGC 2175 - 7		6 08 00	+20 35.0	3	9.23	.022	0.09	.012	-0.04	.084	4	A0	298,401,423
252804	+25 01148			6 08 01	+25 34.8	1	9.63		1.75				1	K5	1702
252962	+5 01124			6 08 03	+05 24.7	1	9.98		1.81				4	K7	369
252832	+25 01149			6 08 05	+25 03.7	1	9.39		2.09				1	K7	1702
42351	+18 01112	HR 2184		6 08 06	+18 08.5	2	6.33	.006	1.35	.000	1.44		4	K1 II	71,8100
		RL 119		6 08 06	+23 11.	1	15.63		0.55		0.14		2		704
42683	-49 02084	HR 2204		6 08 08	-49 33.2	2	6.48	.005	0.52	.000			7	F7 V	15,2012
	+10 01032	G 102 - 53	★ AB	6 08 09	+10 21.0	3	10.41	.013	1.45	.014	1.08		4	M3	694,1017,3045
42401	+12 01049		V	6 08 11	+12 00.4	2	7.44	.073	-0.04	.015	-0.62		6	B2 V	401,2033
252904	+18 01115			6 08 11	+18 11.7	1	9.15		-0.04		-0.12		2	B9 V	401
		Mon R2 # 17		6 08 11	-06 09.3	1	12.84		0.49		0.22		4		502
252924	+20 01301			6 08 16	+20 34.7	1	8.53		1.27		1.38		1	K1 IV	401
252956	+13 01148	LS VI +13 007		6 08 17	+13 09.4	1	10.18		0.42		-0.54		2	B0.5IV	1012
42379	+21 01143	LS V +21 029		6 08 18	+21 34.5	3	7.41	.014	0.35	.004	-0.56	.008	7	B1 II	41,1012,1207
42378	+25 01151			6 08 20	+25 34.0	1	9.23		0.39				4	G5	1724
		Mon R2 # 18		6 08 20	-06 12.0	1	15.12		0.94		0.29		1		502
42421	+15 01097			6 08 21	+15 34.4	1	7.59		1.08		0.99		2	G5	1648
		ApJ144,259 # 14		6 08 24	+11 47.	1	15.24		0.51		-0.01		5		1360
42400	+20 01302	LS V +20 024		6 08 24	+20 55.0	5	6.84	.008	0.18	.004	-0.45	.009	17	B5 II	41,848,1012,1207,1784
253030	+9 01121			6 08 25	+09 39.2	1	9.79		0.32		0.18		1	A3	851
42377	+27 01012			6 08 25	+27 05.1	1	9.54		0.05		-0.11		2	B8/9 V	1375
		LS V +23 034		6 08 26	+23 56.4	1	12.14		0.33		-0.28		3		41
		MC +25 841		6 08 26	+25 40.3	1	12.59		1.53				1	K0	1702
42398	+24 01151	HR 2185		6 08 28	+24 26.0	2	5.81	.016	1.11	.000			4	K0 III	71,1724
		Jenkins 1679		6 08 28	-14 16.5	1	9.93		1.35		1.24		4		158
42581	-21 01377			6 08 28	-21 50.6	6	8.15	.011	1.49	.018	1.23	.004	39	M1/2 V	22,1013,1197,1596*
42399	+22 01213			6 08 29	+22 31.7	1	8.82		0.00		-0.11		2	A0	1375
42397	+25 01153			6 08 30	+25 01.3	1	7.83		0.69				4	G0 IV	1724
		Mon R2 # 19		6 08 30	-06 13.7	1	13.37		0.90		0.13		4		502
		Mon R2 # 20		6 08 30	-06 15.7	1	13.39		0.80		0.55		5		502
42456	+14 01180			6 08 31	+14 29.2	1	7.56		1.36		1.25		2	G5 Ib	8100
42682	-40 02291	HR 2203		6 08 33	-40 20.6	1	5.58		1.68				4	M2 II/III	2007
		G 104 - 24		6 08 34	+22 14.7	1	12.76		1.41		1.12		6		1663
		MC +25 849		6 08 34	+25 53.0	1	11.30		1.39				3	K0 III	1702
		MC +26 377		6 08 34	+26 04.2	1	11.58		1.37				1	K0 III	1702
		Mon R2 # 21		6 08 34	-06 13.7	1	12.20		0.56		0.36		6		502
42536	-6 01439	HR 2195, V653 Mon		6 08 35	-06 44.5	3	6.15	.004	0.01	.008	0.03	.000	14	A0 p SrCr	15,1415,2012
42621	-27 02780	HR 2200		6 08 35	-27 08.5	2	5.72	.002	1.07	.001	0.98		6	K1 III	58,2035
		LS V +23 035		6 08 37	+23 49.9	1	11.96		0.29		-0.32		3		41
42366	+35 01356	IDS06053N3532	AB	6 08 37	+35 31.7	1	9.09		0.13		-0.08		3	A0	1601
42309	+45 01259			6 08 37	+45 03.6	1	8.32		1.07		0.87		2	K0	1601
42477	+13 01151	HR 2191		6 08 38	+13 39.1	1			0.00		0.04		3	A0 Vnn	1049
252966	+26 01125			6 08 38	+25 59.9	1	8.82		1.13				2	K2	1702
		G 102 - 56		6 08 40	+10 33.7	1	12.48		1.39		1.09		1		1658
		Mon R2 # 22		6 08 40	-06 11.9	1	12.23		0.63		0.12		6	B8	502

HD	DM	Other Id	N Rem	α_{1950}	δ_{1950}	S	V	σ_V	B–V	σ_{B-V}	U–B	σ_{U-B}	n	Spectrum	References
253021	+21 01145	LS V +21 030		6 08 42	+21 38.7	1	10.23		0.61		-0.23		2	B2	41
42476	+17 01154	IDS06058N1724	AB	6 08 44	+17 23.4	1	7.20		0.03		-0.03		2	A0 IV	401
253017	+23 01240			6 08 44	+23 02.8	1	9.12		1.81		2.01		2	K7	8032
253016	+24 01152			6 08 44	+24 00.5	1	10.77		0.91		0.60		2	K5	1726
253049	+20 01305	LS V +20 025		6 08 45	+20 08.7	2	9.55	.010	0.15	.010	-0.55	.010	4	B2 IV	1012,1207
42606	-13 01375			6 08 46	-13 47.3	1	8.82		1.00		0.82		2	K3 V	3072
		LS V +22 025		6 08 47	+22 50.3	1	11.51		0.28		-0.29		3		41
253042	+24 01153			6 08 48	+24 43.5	1	8.76		1.20				3	K0	1724
	+63 00639	G 249 - 41		6 08 48	+63 24.6	2	9.70	.033	1.21	.000	1.18		2		1017,3072
42563	-3 01330			6 08 48	-03 24.3	1	8.89		-0.01		-0.37		2	B9	401
		L 59 - 49		6 08 48	-73 22.	1	11.22		0.88		0.59		2		1696
		LS V +23 036		6 08 49	+23 44.5	1	12.03		0.30		-0.30		3		41
42350	+47 01265			6 08 49	+47 26.0	1	7.54		0.03		0.06		2	B9	1375
42532	+9 01124	GQ Ori		6 08 50	+09 38.6	3	8.48	.029	0.94	.024	0.61		47	K2	851,6011,6011
42475	+21 01146	HR 2190, TV Gem		6 08 51	+21 52.9	2	6.62	.069	2.25	.005	1.69	.099	8	M1:Ia	3034,8032
42062	+67 00417			6 08 51	+67 51.1	1	7.66		1.43		1.63		2	K2	1375
42474	+23 01243	WY Gem		6 08 54	+23 13.2	4	7.32	.051	1.81	.013	0.98	.182	5	M2 Iabpe	369,8032,8049,8049
42454	+29 01140	IDS06057N2931	AB	6 08 54	+29 30.3	4	7.35	.015	1.23	.022	0.76		9	G2 Ib	138,1724,6009,8100
		MC +25 869		6 08 55	+25 47.9	1	10.44		1.93				1	K5 III	1702
		MC +26 384		6 08 55	+26 08.7	1	11.41		1.25				1	G8 III	1702
253107	+20 01306	LS V +20 026		6 08 56	+20 19.1	1	10.28		0.15		-0.61		3	B9	41
253072	+24 01154	LS V +24 012		6 08 56	+24 53.0	1	9.76		0.51		0.50		3	A5	41
253136	+19 01251			6 08 58	+19 12.6	1	9.94		1.85				2	K5	369
		LS V +20 027		6 08 59	+20 41.1	1	10.89		0.39		-0.43		2		41
		LS V +20 028		6 09 00	+20 40.6	1	10.91		0.33		-0.42		2		41
43641	-78 00219			6 09 00	-78 56.9	1	9.92		0.55				2	F5/6 Vw	1594
253204	+6 01153			6 09 01	+06 06.9	1	10.10		0.17		0.10		1	K0	1732
42572	+8 01224			6 09 03	+08 47.9	1	8.86		-0.04		-0.57		2	B9	401
42531	+17 01158			6 09 04	+17 46.8	2	8.23	.000	-0.03	.010	-0.13	.015	5	B9.5V	401,848
42509	+19 01253	HR 2193	⋆ A	6 09 04	+19 48.2	3	5.77		-0.07	.008	-0.15	.022	5	B9.5V	985,1049,1079
42471	+32 01217	HR 2189		6 09 04	+32 42.4	1	5.78		1.66				2	M2 IIIa	71
43107	-68 00474	HR 2221		6 09 04	-68 50.0	5	5.05	.008	-0.07	.005	-0.20	.007	24	B8 V	15,258,1075,2016,2038
42560	+14 01187	HR 2199	⋆ A	6 09 06	+14 13.3	7	4.47	.007	-0.18	.010	-0.66	.008	19	B3 IV	15,154,369,1203,1223*
42544	+19 01254			6 09 06	+19 31.8	1	8.04		1.12		0.99		3	K2 III	848
253134	+22 01219	G 104 - 25		6 09 06	+22 11.9	1	10.32		0.97		0.76		1	K0	1620
253228	+5 01127	NGC 2186 - 18		6 09 07	+05 26.9	1	10.26		0.36		-0.02		2	F2	410
42545	+16 01035	HR 2198		6 09 10	+16 08.6	6	4.96	.022	-0.15	.016	-0.59	.008	19	B5 Vn	15,154,1203,1223,1363,8015
42559	+17 01161			6 09 12	+17 46.2	1	8.42		1.14		1.28		1	K2 III	401
		MC +26 390		6 09 12	+26 11.7	1	13.39		1.78				1	K0 III	1702
42729	-26 02784	HR 2206		6 09 13	-26 28.2	3	6.08	.007	-0.04	.000			11	B9.5IV/V	15,2013,2029
		MC +25 884		6 09 14	+25 35.7	1	10.43		1.82				3	K2 III	1702
42834	-45 02349	HR 2211		6 09 14	-45 16.2	2	6.30	.005	-0.03	.000			7	A0 V	15,2012
42657	-4 01393	HR 2202, V638 Mon	⋆ AB	6 09 15	-04 39.2	4	6.17	.009	-0.08	.006	-0.37	.011	16	B9 p HgMn	15,1078,1079,2027
		NGC 2186 - 19		6 09 16	+05 30.4	1	11.67		0.23		0.10		1		410
42597	+7 01178	LS VI +07 001		6 09 16	+07 24.3	2	7.04	.005	-0.10	.010	-0.76	.010	5	B1 V	399,401
		NGC 2186 - 12		6 09 17	+05 26.0	1	11.32		0.39		0.00		1		410
		NGC 2186 - 15		6 09 17	+05 27.9	1	13.07		0.25				1		410
253180	+21 01149	LS V +21 031		6 09 17	+21 56.3	2	9.77	.013	0.26	.000	-0.63	.009	5	B0.5V	41,1207
42543	+22 01220	HR 2197, BU Gem		6 09 17	+22 55.3	2	6.18	.304	2.04	.286	2.30	.258	4	M1 Ia-Iab	3001,8032
42618	+6 01155	G 102 - 57		6 09 18	+06 48.0	5	6.86	.024	0.63	.007	0.13	.004	7	G4 V	1003,1620,1774,3026,6007
253214	+20 01309	LS V +20 029		6 09 18	+20 05.9	3	9.47	.004	0.24	.007	-0.54	.004	7	B1.5:V:nn	41,1012,1207
		MC +25 885		6 09 18	+25 58.0	1	13.10		0.88				1	G0	1702
		NGC 2186 - 14		6 09 19	+05 26.9	1	12.65		0.22		-0.10		1		410
42933	-54 00980	HR 2212, δ Pic		6 09 19	-54 57.4	3	4.76	.032	-0.24	.007	-0.99	.009	12	B1/2 IIIn	611,1088,1732
		NGC 2186 - 20		6 09 20	+05 28.5	1	13.34		1.01		0.61		1		410
253124	+29 01144			6 09 20	+29 54.7	1	10.01		0.51				3	G0	1724
		NGC 2186 - 13		6 09 21	+05 26.3	1	12.57		0.28		-0.05		1		410
		NGC 2186 - 21		6 09 21	+05 28.5	1	14.14		0.18		0.10		1		410
		NGC 2186 - 23		6 09 22	+05 27.9	1	13.65		0.81		0.24		1		410
	-6 01444	Mon R2 # 23		6 09 22	-06 08.7	2	10.86	.035	0.44	.005	-0.20	.005	10	B4 V	206,502
42711	-12 01405			6 09 22	-12 35.2	1	7.95		0.35		-0.01		2	F2 V	1375
		NGC 2186 - 16		6 09 23	+05 27.8	1	12.79		0.20		-0.15		1		410
		NGC 2186 - 22		6 09 23	+05 28.6	1	13.43		0.22		0.05		1		410
		MC +25 891		6 09 24	+25 06.0	1	11.11		1.40				2	G8 IV	1702
		MC +26 393		6 09 24	+26 12.0	1	10.78		1.59				1	K0 III	1702
		LS V +24 013		6 09 25	+24 47.8	1	10.88		0.23		-0.58		3		41
		L 380 - 78		6 09 25	-43 24.5	1	12.39		1.60		1.29		1	M3.5	3073
		NGC 2186 - 2		6 09 26	+05 28.6	1	9.82		1.09		0.77		1		410
253209	+24 01159			6 09 26	+24 56.4	1	8.94		1.07				3	K0	1724
42690	-6 01446	HR 2205		6 09 26	-06 32.2	5	5.05	.005	-0.20	.007	-0.77	.005	19	B2 V	15,154,1061,1732,2027
253247	+18 01123	LSS 17		6 09 27	+18 01.8	4	9.80	.019	0.34	.005	-0.60	.010	10	B1 V	342,608,797,1012
		NGC 2186 - 3		6 09 28	+05 27.9	1	10.86		0.20		-0.09		1		410
		NGC 2186 - 4		6 09 28	+05 28.1	1	11.77		0.23		-0.09		1		410
		NGC 2186 - 5		6 09 29	+05 28.2	1	12.71		0.22		0.00		1		410
		NGC 2186 - 6		6 09 29	+05 28.5	1	12.47		0.20		-0.15		1		410
		NGC 2186 - 17		6 09 29	+05 29.2	1	13.71		0.25		-0.07		1		410
		NGC 2186 - 7		6 09 29	+05 29.3	1	13.21		0.62		0.21		1		410
		LS V +20 030	⋆ AB	6 09 29	+20 25.4	1	10.96		0.48		-0.35		1		41
		NGC 2186 - 10		6 09 30	+05 27.3	1	13.61		0.30		0.32		1		410
42655	+10 01044			6 09 30	+10 20.8	2	7.49	.000	-0.06	.020	-0.60	.005	5	B2 V	399,401

Table 1   329

HD	DM	Other Id	N Rem	α_{1950}	δ_{1950}	S	V	σ_V	B–V	σ_{B-V}	U–B	σ_{U-B}	n	Spectrum	References
		LS V +23 037	★ AB	6 09 30	+23 50.5	1	11.59		0.48		-0.16		3		41
		NGC 2186 - 11		6 09 31	+05 27.5	1	14.24		0.39		0.32		1		410
		NGC 2186 - 9		6 09 31	+05 28.1	1	12.46		0.16		-0.10		1		410
		NGC 2186 - 8		6 09 33	+05 28.7	1	12.93		0.22		0.02		1		410
		MC +26 396		6 09 34	+26 10.5	1	11.60		2.43				1	K3:	1702
		MC +25 896		6 09 35	+25 52.3	1	10.80		1.44				1	K0 III	1702
		MC +25 898		6 09 35	+25 52.3	1	11.37		1.47				1	G8 III	1702
		G 101 - 25		6 09 35	+38 55.6	1	10.79		0.82		0.56		1	K1	333,1620
		SS Aur		6 09 35	+47 45.2	1	14.55		0.23		-1.13		1		698
42637	+15 01107			6 09 36	+15 21.8	2	8.48	.030	0.89	.020	0.62		5	G5	882,3040
42745	-14 01359			6 09 36	-14 26.5	1	8.16		-0.06		-0.59		3	B5 III	399
42673	+5 01130	NGC 2186 - 1		6 09 37	+05 26.6	1	9.64		0.17		0.11		2	A0	410
253236				6 09 37	+22 53.4	1	9.89		0.19		-0.60		3	B1 V:	41
		AAS42,335 T6# 16		6 09 42	+19 42.	1	10.97		0.47		-0.01		3		848
		AAS42,335 T6# 17		6 09 42	+19 46.	1	10.91		1.34		1.17		3		848
		MC +25 905		6 09 44	+25 41.8	1	11.40		1.49				1	K0 III	1702
253327	+18 01124	LSS 18		6 09 49	+18 00.1	4	10.74	.022	0.57	.008	-0.39	.013	12	B0.5V:	342,608,797,1012
42466	+51 01163	HR 2188		6 09 49	+51 11.3	1	6.05		1.06				2	K1 III	71
		WLS 600 5 # 8		6 09 51	+04 27.0	1	11.01		0.51		-0.02		2		1375
		MC +26 397		6 09 51	+26 11.9	1	10.83		1.69				3	K0 III	1702
42689	+13 01161			6 09 52	+13 40.5	1	8.35		1.03				3	K0	882
253323				6 09 52	+23 05.8	1	10.84		0.22		-0.56		3	B8	41
42634	+27 01019			6 09 52	+27 26.8	1	8.39		0.35				2	F2 V	1724
42931	-41 02266			6 09 55	-41 33.7	3	9.30	.035	0.94	.006	0.66	.022	10	K2 V	158,2012,3072
253339	+24 01162	LS V +24 014		6 09 58	+24 03.4	2	10.24	.022	0.53	.004	-0.33	.004	5	B3	41,1207
42707	+20 01312			6 10 00	+20 05.1	1	8.44		0.38		0.10		3	A5	848
42708	+19 01256			6 10 03	+19 49.9	1	8.31		0.31		0.04		3	A9 III	848
253422	+13 01164			6 10 05	+13 09.5	1	9.13		0.63		0.14		2	G5	1648
253363	+23 01250	LS V +23 039	★ AB	6 10 05	+23 52.3	1	9.66		0.33		-0.36		4	B5	41
		AAS42,335 T6# 19		6 10 06	+19 46.	1	10.22		1.21		0.92		3		848
		LSS 19		6 10 09	+17 59.6	1	11.66		0.88		-0.21		4	B3 Ib	797
42759	+14 01195			6 10 10	+14 19.7	1	8.17		0.86				1	G5	882
42616	+41 01392	QR Aur		6 10 10	+41 42.7	3	7.14		0.08	.012	0.04	.006	11	A2p	1063,1202,1263
42771	+9 01134			6 10 11	+09 03.0	1	7.97		-0.08		-0.54		2	B9	401
42770	+10 01048			6 10 14	+10 18.1	1	6.59		-0.06		-0.30		2	B9	401
42758	+19 01259			6 10 14	+19 01.4	1	7.44		-0.10		-0.34		2	B8 III	401
253440	+19 01258			6 10 14	+19 27.1	1	9.28		0.18		0.07		3	F0 IV	848
		LS V +22 027		6 10 14	+22 21.0	1	12.01		0.36		-0.37		3		41
42824	-2 01512	HR 2210		6 10 14	-02 29.4	2	6.61	.005	0.05	.000	0.08	.000	7	A2 V	15,1078
42687	+31 01231			6 10 15	+31 25.6	1	9.17		0.68				4	G5	1724
42773	+7 01185			6 10 16	+07 17.5	1	7.00		1.56		1.94		2	K5	1375
42250	+70 00395	G 251 - 5		6 10 18	+70 48.2	1	7.43		0.77		0.37		3	G5	1355
42787	+6 01160			6 10 19	+06 01.8	1	6.46		1.63				4	M1	369
42760	+13 01165			6 10 19	+13 30.8	1	8.46		1.11				2	K0	882
		Mon R2 # 24		6 10 19	-06 11.6	1	13.36		0.47		0.38		5		502
		Mon R2 # 25		6 10 20	-06 10.0	1	12.23		0.76		0.29		5		502
42846	+2 01158	IDS06078N0252	AB	6 10 25	+02 51.7	1	8.48		0.07		-0.09		2	A0	1375
	-16 01396			6 10 25	-16 47.8	1	9.14		1.02		0.66		2	K0	1375
42845	+3 01164			6 10 26	+03 31.4	1	7.51		-0.13		-0.48		2	B8	401
42807	+10 01050	HR 2208		6 10 26	+10 38.7	4	6.44	.014	0.66	.015	0.16	.000	24	G8 V	1067,1197,1355,3077
42614	+51 01164	IDS06065N5112	AB	6 10 26	+51 10.9	1	8.04		0.02		-0.25		3	B8	555
253434	+28 01048			6 10 29	+28 22.2	1	9.73		0.47				2	F5	1724
253496				6 10 32	+22 53.7	1	10.28		0.45		-0.46		3	B0	41
	+33 1279a	G 98 - 53		6 10 32	+33 26.2	1	11.14		0.51		-0.09		1	K0	1620
		G 101 - 26		6 10 34	+38 51.5	1	10.37		0.99		0.85		1	K3	333,1620
42927	-17 01398	HR 2213		6 10 34	-17 44.9	2	6.52	.005	-0.17	.005			7	B5 II/III	15,2027
42988	-28 02792			6 10 34	-29 00.2	3	7.44	.004	0.50	.010	-0.05		23	F5 V	861,2012,2034
253516	+24 01165			6 10 36	+24 41.4	1	8.51		1.35				3	K0	1724
42784	+18 01129	HR 2207		6 10 37	+18 41.7	3	6.59		-0.09	.005	-0.45	.012	10	B8 Vnn	848,1049,1079
253520	+23 01255	LS V +23 040		6 10 43	+23 29.7	1	10.42		0.54		-0.27		3	F5	41
42861	+6 01161			6 10 44	+06 23.0	1	9.15		0.04		-0.32		4	A0	1732
		LS V +22 029		6 10 44	+22 37.6	1	11.03		0.42		-0.47		3		41
42860	+9 01141			6 10 45	+09 38.6	1	7.58		-0.13		-0.55		2	B8	401
42911	-4 01407			6 10 45	-04 59.7	1	7.34		1.02				4	G7 III	2033
43071	-36 02714			6 10 45	-36 33.0	1	6.88		-0.15		-0.70		4	B3 Vn	158
42841	+19 01262			6 10 47	+19 20.9	2	7.75	.010	1.23	.005	0.97	.005	5	G5 II	848,8100
		GD 73		6 10 47	+20 51.5	1	15.77		0.24		-0.66		1		3060
		LS V +23 041		6 10 48	+23 25.0	1	11.57		0.57		-0.16		3		41
		LS V +23 042		6 10 52	+23 29.9	1	11.18		0.58		-0.18		3		41
42806		SU Gem		6 10 52	+27 43.1	1	10.19		1.51		1.18		1	M3	793
42734	+42 01509			6 10 52	+42 49.0	1	9.11		0.01		-0.21		2	B9	1375
		AAS42,335 T7# 2		6 10 54	+19 21.	1	12.93		0.43		0.17		2		848
		AAS42,335 T7# 3		6 10 54	+19 26.	1	12.11		0.42		0.13		2		848
		WLS 600 25 # 8		6 10 56	+25 01.0	1	11.25		0.73		0.33		2		1375
42874	+14 01206			6 10 57	+14 08.3	1	8.32		1.38				2	K0	882
42908	+8 01238	LS VI +08 004		6 10 58	+08 43.7	1	8.18		0.04		-0.56		3	B2 Ve	1212
253591	+22 01233	LS V +22 030		6 10 59	+22 30.9	2	10.04	.013	0.31	.004	-0.54	.009	5	B1 V	41,1207
		L 812 - 9		6 11 00	-12 23.	1	12.28		0.53		0.03		2		3060
253659	+16 01046	LSS 21		6 11 01	+16 32.0	2	9.34	.000	0.37	.025	-0.51	.000	4	B0.5V:enn	405,1737
42872	+17 01177			6 11 02	+17 33.1	1	9.08		0.08		0.02		3	A2 V	848

HD	DM	Other Id	N Rem	α_{1950}	δ_{1950}	S	V	σ_V	B–V	σ_{B-V}	U–B	σ_{U-B}	n	Spectrum	References
253652				6 11 03	+19 20.	1	11.28		0.36		0.04		2	A3	848
253651				6 11 03	+19 24.	1	11.60		0.23		0.06		2	A2	848
42959	−2 01515			6 11 05	−02 14.3	1	7.79		0.00		−0.45		2	B8	401
253647	+21 01162	LS V +21 032		6 11 07	+21 50.4	1	10.52		0.31		−0.48		2	B8	41
253616	+28 01051			6 11 07	+28 43.2	1	9.64		0.74				2	G0	1724
		G 105 - 23		6 11 08	+15 11.6	2	14.66	.019	1.61	.029	1.27		3		203,3078
253682	+19 01263			6 11 08	+19 38.3	1	9.11		0.52		0.04		3	F6 IV	848
42896	+20 01322	LS V +20 031		6 11 08	+20 11.1	4	8.62	.019	−0.08	.003	−0.87	.006	14	B1:V:nn	41,848,1012,1207
43455	−65 00561	HR 2245		6 11 08	−65 34.7	4	5.00	.003	1.60	.008	1.85	.000	24	M2.5 III	15,1075,2038,3055
42907	+19 01264			6 11 10	+19 29.0	1	9.34		0.91		0.53		3	G5 IV	848
253683	+19 01265	LSS 24	★ AB	6 11 11	+18 58.8	3	9.29	.005	0.34	.000	−0.57	.004	7	B0.5IV	848,1012,1207
42633	+60 00938	HR 2201	★ A	6 11 11	+60 00.9	1	5.35		1.34				2	K3 III	71
		AAS42,335 T7# 7		6 11 12	+19 26.	1	12.66		0.53		0.03		2		848
42983	+2 01163			6 11 15	+02 49.7	1	7.38		0.96		0.80		2	K0	3008
42783	+46 01119			6 11 15	+46 24.7	1	7.52		−0.04		−0.41		3	B5 III	555
		AAS42,335 T7# 9		6 11 18	+19 19.	1	11.65		0.61		0.03		2		848
		RL 120		6 11 18	+25 55.	1	14.82		0.37		0.02		2		704
41804	+80 00202			6 11 18	+80 22.6	1	7.53		0.45		−0.01		2	F5	1375
42782	+48 01361			6 11 20	+48 51.0	2	7.18	.005	−0.16	.005	−0.63	.010	5	B5	401,555
253702	+22 01237	LS V +22 031		6 11 22	+22 25.3	1	9.71		0.32		−0.55		3	B5	41
		AAS42,335 T7# 10		6 11 24	+19 20.	1	13.56		0.61		0.11		1		848
43023	−3 01345	HR 2218		6 11 25	−03 43.6	3	5.83	.007	0.93	.007	0.63	.005	12	G8 III	15,1078,1509
		G 98 - 56		6 11 31	+37 53.5	2	11.36	.000	0.63	.000	−0.03	.000	2	G5	927,1620
		WLS 600-15 # 8		6 11 32	−15 13.4	1	11.18		0.53		0.12		2		1375
42954	+17 01182	HR 2214	★ AB	6 11 33	+17 55.3	2	5.88	.005	0.23	.014	0.15	.005	10	A6 m	3058,8071
253754	+22 01238	LS V +22 032		6 11 33	+22 55.5	1	10.29		0.60		−0.28		3	B5	41
43047	+3 01170			6 11 34	+03 55.1	1	7.31		−0.05		−0.35		2	B9	401
		AAS42,335 T7# 11		6 11 36	+19 35.	1	12.86		0.42		0.22		3		848
		AAS42,335 T7# 12		6 11 36	+19 36.	1	13.54		1.33		1.19		3		848
43179	−29 02883	HR 2226		6 11 37	−29 22.8	1	6.54		−0.10				4	B7 V	2035
42999	+11 01075	IDS06088N1151	B	6 11 39	+11 49.8	1	7.66		0.06		0.04		2	B9	401
42997	+17 01183	IDS06088N1727	AB	6 11 41	+17 26.0	1	8.36		−0.09		−0.49		2	B7 III	401
43162	−23 03577	HR 2225		6 11 41	−23 50.9	3	6.38	.007	0.72	.005			11	G5 V	15,2013,2028
		AAS42,335 T7# 13		6 11 42	+19 20.	1	13.50		0.72		0.23		2		848
		AAS42,335 T7# 14		6 11 42	+19 28.	1	13.29		0.55		0.14		2		848
		AAS42,335 T7# 15		6 11 42	+19 33.	1	14.20		0.60		0.11		3		848
		AAS42,335 T7# 16		6 11 42	+19 33.	1	13.14		0.71		0.14		3		848
42996	+17 01184			6 11 43	+17 27.4	1	9.35		−0.04		−0.35		1		401
253749	+27 01030			6 11 43	+27 22.8	1	9.12		1.10				3	K0	1724
43062	+5 01146	G 106 - 34		6 11 44	+05 11.3	1	8.42		0.88		0.52		1	K0	1620
43551	−65 00564			6 11 44	−65 32.1	1	8.22		1.49		1.93		5	K4 III	1704
43834	−74 00374	HR 2261		6 11 44	−74 44.2	7	5.08	.010	0.72	.000	0.32	.008	37	G6 V	15,1075,1409,2012*
		AA45,405 S269 # 4		6 11 45	+13 50.7	1	14.60		0.67		0.21		2		797
		AA45,405 S269 # 5		6 11 45	+13 50.7	1	15.36		0.80		−0.31		2		797
43019	+19 01269			6 11 46	+19 18.1	1	8.75		−0.05		−0.24		std	B9.5III	848
		AA45,405 S269 # 3		6 11 47	+13 50.1	1	14.96		0.74		0.33		2		797
		AA45,405 S269 # 1		6 11 47	+13 50.5	1	14.28		0.65		0.17		2		797
		AAS42,335 T7# 18		6 11 48	+19 21.	1	12.11		0.61		0.11		2		848
		AAS42,335 T7# 19		6 11 48	+19 27.	1	14.10		0.81		0.15		2		848
		AA45,405 S269 # 2		6 11 49	+13 50.7	1	13.82		1.07		−0.06		4		797
288532	+1 01262			6 11 51	+01 33.6	1	10.69		1.90				4	B8	369
43044	+14 01213	IDS06090N1437	AB	6 11 52	+14 36.2	1	7.03		0.01		−0.11		1	B8 V	401
42995	+22 01241	HR 2216, η Gem	★ AB	6 11 52	+22 31.4	4	3.28	.005	1.59	.023	1.64	.033	11	M3 III	15,369,3016,8015
		WLS 624 70 # 10		6 11 53	+70 47.2	1	10.77		0.49		0.03		2		1375
43042	+19 01270	HR 2220	★ A	6 11 54	+19 10.5	6	5.20	.008	0.45	.017	−0.01	.007	22	F6 V	15,254,1008,1363,1784,3026
		AAS42,335 T7# 20		6 11 54	+19 29.	1	12.45		0.66		0.10		2		848
253850	+23 01265			6 11 56	+23 15.6	1	10.80		0.29		−0.55		4	B8	41
42819	+56 01100	G 192 - 23		6 11 57	+56 57.2	1	7.47		0.58		0.10		1	G0	1658
		G 106 - 35		6 11 59	+02 31.7	1	13.20		1.53		1.26		4		316
253870	+25 01177	IDS06090N2530	AB	6 12 03	+25 28.9	1	8.83		1.01				2	F5	1724
43618	−65 00565	IDS06120S6530	A	6 12 04	−65 31.1	1	6.82		0.46		0.05		4	F6 V	158
		AA45,405 S271 # 1		6 12 05	+12 22.4	1	12.19		0.68		−0.37		2	O9 V	797
43097	+14 01216			6 12 06	+14 42.6	1	8.76		−0.05		−0.50		2	B5 II	8100
		AAS42,335 T7# 21		6 12 06	+19 33.	1	14.38		0.75		0.22		2		848
		AAS42,335 T7# 22		6 12 06	+19 34.	1	14.82		0.64		0.44		2		848
43639	−65 00566	IDS06120S6530	B	6 12 07	−65 31.3	1	8.23		0.54		0.08		4	F7 IV	158
43098	+11 01080			6 12 08	+11 40.1	1	8.17		0.00		−0.53		2	B8	401
		AA45,405 S271 # 2		6 12 08	+12 23.4	1	11.78		0.50		−0.04		2		797
43157	−4 01421	HR 2224	★ AB	6 12 08	−04 33.1	2	5.82	.005	−0.18	.005	−0.64	.005	7	B5 V	15,1415
253919				6 12 09	+19 33.	1	11.01		0.41		0.13		2	A5	848
43236	−19 01391			6 12 09	−19 30.6	1	7.52		1.64				4	M2 III	2034
43096	+16 01057			6 12 10	+16 39.0	2	7.94	.000	1.07	.020	0.94		4	K0	882,3040
43079	+21 01167			6 12 11	+21 13.5	1	8.93		0.58		0.06		2	G0	848
43039	+29 01154	HR 2219		6 12 11	+29 31.1	14	4.34	.019	1.02	.011	0.81	.007	41	G8.5 IIIb	15,245,263,592,1003,1355*
		AAS42,335 T7# 23		6 12 12	+19 23.	1	12.08		0.40		0.21		3		848
		AAS42,335 T7# 25		6 12 12	+19 34.	1	11.83		0.50		0.01		2		848
		AA45,405 S271 # 3		6 12 13	+12 20.9	1	13.28		0.66		−0.32		2	B1 V	797
253942	+19 01273			6 12 14	+19 27.5	1	9.55		0.60		0.17		3	G0	848
43078	+22 01243	LR Gem		6 12 14	+22 19.1	3	8.78	.011	0.35	.004	−0.56	.005	7	B0 IV	41,1012,1207
	+22 01244	LS V +22 034		6 12 16	+22 18.7	1	10.30		0.40		−0.42		3		41

Table 1   331

HD	DM	Other Id	N Rem	α_{1950}	δ_{1950}	S	V	σ_V	B–V	σ_{B-V}	U–B	σ_{U-B}	n	Spectrum	References
43017	+36 01388	HR 2217	★A	6 12 17	+36 09.9	1	6.92		0.45		-0.02		2	F0	3024
43017	+36 01388	IDS06089N3611	B	6 12 17	+36 09.9	1	7.54		0.42		-0.06		2	G5	3024
43017	+36 01388	IDS06098N3611	C	6 12 17	+36 09.9	1	12.32		1.60		1.74		2		3024
43112	+13 01173	HR 2222	★A	6 12 18	+13 52.1	7	5.91	.014	-0.23	.011	-0.95	.018	52	B1 V	1,3,10,15,154,361,1258
253911	+29 01158			6 12 19	+28 59.2	1	8.89		1.07				4	K0	1724
		Mon R2 # 26		6 12 20	−06 19.5	1	12.16		0.32		0.03		4		502
43190	+3 01177			6 12 21	+03 48.8	1	8.15		-0.09		-0.43		2	B9	401
		LS V +21 033		6 12 21	+21 03.7	1	12.03		0.34		-0.33		2		41
		G 104 - 27		6 12 24	+17 44.8	1	13.40		-0.15		-0.98		1		3028
		LS V +20 032		6 12 25	+20 54.5	1	12.05		0.41		-0.24		2		41
43232	−6 01469	HR 2227	★A	6 12 25	−06 15.5	10	3.97	.007	1.32	.006	1.41	.009	45	K3 III	3,15,58,418,1075,1425*
253954	+22 01246	LS V +22 035		6 12 26	+22 28.1	1	9.55		0.58		-0.31		3	B8	41
		Mon R2 # 27		6 12 27	−06 21.7	1	10.98		0.19		-0.23		5	B7 V	502
43213	+3 01178			6 12 28	+03 53.8	1	8.09		-0.09		-0.35		2	B9	401
43152	+16 01059			6 12 29	+16 26.9	1	6.97		1.49		1.82		2	K5 Ib	8100
253981	+21 01168	LS V +21 034		6 12 29	+21 15.9	1	9.55		0.38		-0.51		3	B0	41
		LS V +22 036		6 12 30	+22 31.8	1	11.48		0.73		-0.06		3		41
43251	−8 01361			6 12 31	−08 47.3	1	7.24		-0.10		-0.43		2	B8	401
43212	+6 01170	IDS06098N0633	ABC	6 12 32	+06 32.0	1	7.65		0.35		0.15		1	A0	1732
43153	+16 01060	HR 2223		6 12 32	+16 09.6	4	5.33	.019	-0.15	.012	-0.45	.008	5	B7 V	351,1049,1079,3023
43151	+17 01187			6 12 33	+17 46.0	1	8.09		1.66		1.56		3	M2 III:	848
253980				6 12 33	+22 03.5	1	11.12		0.44		-0.50		3	B7	41
	+45 01274			6 12 34	+45 04.5	1	10.38		0.07		0.01		2		1375
254111	+3 01179			6 12 35	+03 48.6	1	9.55		0.32		-0.05		1	A7	401
	+37 01458	G 98 - 58		6 12 36	+37 44.6	5	8.92	.000	0.60	.005	-0.11	.007	9	G0	516,979,1620,1658,8112
		NGC 2214 sq1 6	AB	6 12 36	−68 15.8	1	15.60		1.07				1		919
	−59 00613	V Pic		6 12 37	−59 53.9	1	8.68		1.70		1.69		1		3076
43248	−4 01426			6 12 39	−04 36.2	1	8.07		-0.10		-0.37		2	B9	401
43210	+12 01078			6 12 43	+12 06.4	1	8.03		0.46		0.04		3	F6 III	8100
254052	+16 01062			6 12 43	+16 28.1	1	8.96		0.46		-0.09		3	F3 V	261
44148	−77 00243			6 12 44	−77 05.5	1	6.83		0.04		0.04		3	A1 V	1628
254082	+15 01130			6 12 45	+15 24.0	1	9.36		1.05		0.80		3	K0	261
43185	+18 01141			6 12 46	+18 18.9	2	6.61	.000	1.55	.005	1.85	.019	6	K2 III	848,1648
	−33 02824			6 12 46	−33 36.3	1	10.70	.000	0.76	.005	0.30	.005	6		158,1696
254042	+24 01176	LS V +24 015		6 12 47	+24 04.8	3	8.93	.011	0.36	.004	-0.58	.000	7	B0.5:IV:nn	41,1012,1207
		NGC 2214 sq1 2	AB	6 12 48	−68 11.8	1	13.81		0.57		-0.03		4		919
43286	+3 01180			6 12 51	+03 58.5	1	12.39		0.62		0.13		3	B5	1752
43286	+3 01180	IDS06102N0359	A	6 12 52	+03 58.5	1	7.00		-0.13		-0.57		1	B5	401
43264	+7 01207			6 12 53	+07 40.2	1	7.54		-0.03		-0.28		2	B9	401
43208	+21 01172	LS V +21 035		6 12 53	+21 07.2	1	9.05		0.19		-0.64		3	B5	41
		Mon R2 # 28		6 12 55	−06 24.8	1	10.53		0.31		-0.25		5	B5 V	502
43247	+12 01081	HR 2229	★	6 12 56	+12 34.1	4	5.45	.077	0.00	.022	-0.18	.044	9	B9 II-III	1049,1079,3023,8100
	+47 01276	G 101 - 30	A	6 12 57	+47 05.1	3	9.20	.018	0.77	.007	0.30	.021	6	K0	333,979,1620,3026
	+47 01276		B	6 12 57	+47 05.1	1	9.46		1.15		1.03		2		979
43285	+6 01172	HR 2231		6 12 59	+06 05.0	4	6.06	.010	-0.13	.008	-0.52	.014	14	B6 Ve	15,154,1212,1415
43396	−20 01336	HR 2242		6 12 59	−20 15.3	2	5.91	.003	1.32	.006	1.37		6	K2 III	1596,2031
43548	−49 02129			6 12 59	−49 47.9	1	8.80		0.72				4	G6 V	2012
43230	+20 01337			6 13 00	+20 31.6	1	8.67		0.12		0.02		3	A3 V	848
43300	+8 01250			6 13 02	+08 28.5	1	8.05		0.00		-0.03		2	B9	401
		AA45,405 S267 # 2		6 13 02	+14 17.2	1	11.77		0.76		-0.31		3	O9 V	797
43318	−0 01234	HR 2233		6 13 02	−00 29.5	5	5.64	.028	0.50	.018	0.00	.010	14	F6 V	15,254,1003,1415,2012
43319	−4 01431	HR 2234	★AB	6 13 02	−04 53.8	3	5.98	.005	0.09	.008	0.10	.005	14	A5 IVs	15,1078,2027
43284	+6 01173			6 13 03	+06 34.2	2	8.34	.001	0.00	.005	-0.55	.016	6	A0	401,1732
43362	−8 01368	HR 2237	★AB	6 13 03	−09 01.1	3	6.09	.005	-0.08	.008	-0.31	.013	9	B9 III	15,1061,1079
254067	+28 01059			6 13 04	+28 13.1	1	9.38		0.51				2	F8	1724
		NGC 2204 - 4206		6 13 06	−18 27.5	1	10.01		0.36		0.03		2		464
43429	−18 01352	HR 2243		6 13 06	−18 27.5	3	5.99	.008	1.06	.005	0.94	.007	19	K1 III	58,464,2035
		NGC 2214 sq1 3		6 13 07	−68 11.5	1	13.76		0.95		0.67		4		919
43317	+4 01181	HR 2232		6 13 08	+04 18.1	3	6.64	.004	-0.17	.004	-0.65	.008	10	B3 IV	15,154,1061
		AA45,405 S267 # 1		6 13 10	+14 17.8	1	10.11		0.24		-0.38		1	B3 V:	797
254094	+29 01160			6 13 10	+29 48.9	1	9.71		0.37				2	F0	1724
		NGC 2214 sq1 7		6 13 10	−68 14.7	1	14.57		0.19		0.22		4		919
43299	+10 01071			6 13 11	+10 18.0	1	6.84		1.22		1.39		2	K3 III-IV	37
		AA45,405 S267 # 3		6 13 11	+14 15.4	1	12.13		0.43		0.91		1		797
		LS V +22 038		6 13 12	+22 46.9	1	10.76		1.01		-0.08		3		41
		RL 121		6 13 12	+25 37.	1	15.32		0.33		-0.24		2		704
43147	+44 01408	G 101 - 31		6 13 12	+44 43.7	2	9.09	.005	0.79	.005	0.31	.020	3	G9 V	333,1003,1620
43282	+19 01281			6 13 13	+19 04.4	1	7.72		1.32		1.13		2	G5 II	8100
		LS V +23 043		6 13 13	+23 37.6	1	11.62		0.43		-0.31		3		41
254119	+26 01151			6 13 13	+26 43.8	1	9.22		0.43				2	F2	1724
		NGC 2214 sq1 1		6 13 13	−68 10.6	1	12.24		0.57		0.06		4		919
43261	+24 01182	HR 2230		6 13 16	+23 59.3	6	6.06	.019	0.89	.006	0.62	.003	38	G8 III	15,1007,1013,1217,1784,8015
254116	+29 01161			6 13 16	+29 42.9	1	9.13		0.29				2	F0	1724
43389	−2 01530			6 13 16	−02 22.1	2	8.31	.005	1.48	.002	1.44	.012	3	K5	565,3048
43447	−18 01357	NGC 2204 - 1224		6 13 17	−18 35.2	1	8.81		0.96		0.67		10	G8 III	464
42973	+61 00869	HR 2215, UW Lyn		6 13 18	+61 32.1	1	4.95		1.83		1.96		6	M3 IIIab	3052
43358	+1 01275	HR 2236	★AB	6 13 19	+01 11.2	2	6.36	.005	0.46	.000	-0.02	.010	7	F5 IV:	15,1415
42818	+69 00371	HR 2209		6 13 20	+69 20.5	3	4.80	.007	0.03	.008	0.00	.000	10	A0 Vn	15,1363,8015
254147	+27 01040			6 13 21	+27 14.6	1	9.66		0.48				2	F5	1724
254152	+23 01274	LS V +23 044		6 13 22	+23 56.3	1	10.66		0.38		-0.50		3	B8	41

HD	DM	Other Id	N	Rem	α_{1950}	δ_{1950}	S	V	σ_V	B–V	σ_{B-V}	U–B	σ_{U-B}	n	Spectrum	References
43410	−3 01359				6 13 22	−03 42.1	1	9.80		0.06		−0.06		2	A0	401
44828	−82 00143				6 13 25	−82 01.9	1	7.20		0.31		0.06		4	F0 IV	1628
43637	−46 02278				6 13 26	−46 59.7	1	8.27		1.23		1.38		10	K2 III	1673
43423	−2 01533				6 13 27	−02 15.4	1	8.82		−0.02		−0.10		4	A0	1732
43445	−13 01411	HR 2244			6 13 27	−13 42.0	8	5.00	.004	−0.08	.005	−0.23	.033	28	B9 Vn	15,1075,1079,1356*
254203	+22 01251	LS V +22 039			6 13 28	+22 19.5	1	10.35		0.36		−0.40		3	B8	41
43335	+17 01191	HR 2235			6 13 29	+17 12.0	2	6.38	.010	1.55	.030	1.84	.059	5	K5 II	261,8100
		NGC 2204 - 1103			6 13 30	−18 38.	1	17.42		0.54				1		464
		NGC 2204 - 1128			6 13 30	−18 38.	1	16.63		0.44				2		464
		NGC 2204 - 1212			6 13 30	−18 38.	1	13.93		0.97				1		464
		NGC 2204 - 1227			6 13 30	−18 38.	1	12.20		0.47		−0.03		1		464
		NGC 2204 - 1309			6 13 30	−18 38.	1	13.65		1.02		0.73		2		464
		NGC 2204 - 1329			6 13 30	−18 38.	1	11.49		1.17		1.20		1		464
		NGC 2204 - 2107			6 13 30	−18 38.	1	17.13		0.42				3		464
		NGC 2204 - 2108			6 13 30	−18 38.	1	16.92		0.58				3		464
		NGC 2204 - 2120			6 13 30	−18 38.	1	11.66		1.29		1.19		2		464
		NGC 2204 - 2143			6 13 30	−18 38.	1	16.25		0.46		0.02		2		464
		NGC 2204 - 2207			6 13 30	−18 38.	1	16.43		0.48		0.02		3		464
		NGC 2204 - 2212			6 13 30	−18 38.	1	12.76		1.21				1		464
		NGC 2204 - 2222			6 13 30	−18 38.	1	13.82		1.29				1		464
		NGC 2204 - 2224			6 13 30	−18 38.	1	14.19		0.90		0.46		2		464
		NGC 2204 - 2228			6 13 30	−18 38.	1	15.84		0.50		0.03		1		464
		NGC 2204 - 2311			6 13 30	−18 38.	1	13.62		1.04		0.71		2		464
		NGC 2204 - 2318			6 13 30	−18 38.	1	12.50		0.53		0.08		1		464
		NGC 2204 - 3109			6 13 30	−18 38.	1	14.84		0.16		0.09		2		464
		NGC 2204 - 3112			6 13 30	−18 38.	1	17.25		0.43				1		464
		NGC 2204 - 3141			6 13 30	−18 38.	1	17.52		0.61				2		464
		NGC 2204 - 3213			6 13 30	−18 38.	1	13.28		0.38				1		464
		NGC 2204 - 3215			6 13 30	−18 38.	1	13.72		1.00		0.68		2		464
		NGC 2204 - 3222			6 13 30	−18 38.	1	15.62		0.37		−0.04		3		464
		NGC 2204 - 3223			6 13 30	−18 38.	1	15.44		0.44		0.03		2		464
		NGC 2204 - 3304			6 13 30	−18 38.	1	12.28		1.42		1.46		2		464
		NGC 2204 - 4111			6 13 30	−18 38.	1	15.79		0.37		0.11		3		464
		NGC 2204 - 4131			6 13 30	−18 38.	1	15.63		0.42		0.11		1		464
		NGC 2204 - 4210			6 13 30	−18 38.	1	13.78		0.98		0.70		2		464
		NGC 2204 - 4221			6 13 30	−18 38.	1	15.53		0.44		0.04		2		464
		NGC 2204 - 4223			6 13 30	−18 38.	1	13.86		1.02		0.58		2		464
43406	+5 01156				6 13 34	+05 07.9	2	7.16	.005	−0.06	.005	−0.28	.010	4	B9	401,1733
		NGC 2214 sq1 4			6 13 34	−68 13.3	1	15.74		−0.16		−0.56		3		919
254229	+25 01188	G 103 - 26			6 13 35	+25 14.1	1	9.37		1.07		1.08		2	K2	1625
44165	−75 00368				6 13 35	−75 04.1	1	7.34		0.96				2	G5 III/IV	1594
		G 103 - 27			6 13 36	+29 58.2	1	11.54		0.62		0.00		2		333,1620
43386	+12 01084	HR 2241		⋆ AC	6 13 38	+12 17.3	6	5.04	.007	0.41	.010	−0.03	.015	12	F5 IV-V	15,254,351,1008,1363,3026
	+49 01476	IDS06098N4942		AB	6 13 42	+49 41.1	1	10.23		0.67		0.27		1		8084
	+49 01476	IDS06098N4942		C	6 13 42	+49 41.1	1	11.64		0.15		0.11		1		8084
43334	+26 01156				6 13 44	+26 30.7	1	8.33		1.05				2	K0	1724
43461	+1 01278	HR 2246			6 13 46	+01 05.9	3	6.62	.005	−0.05	.004	−0.44	.007	9	B6 V	15,1078,1079
254318	+21 01177	LS V +21 036			6 13 47	+21 57.4	1	9.98		0.23		−0.61		3	B5	41
43383	+25 01191				6 13 50	+25 30.8	2	8.45	.015	0.54	.005	0.03		5	F8 V	1724,3026
43385	+22 01253				6 13 51	+22 11.4	1	8.56		0.25				2	F0	6009
43244	+46 01122	HR 2228			6 13 51	+46 26.6	5	6.52	.011	0.26	.016	0.08	.009	12	F0 V	15,254,1007,1013,8015
43544	−16 01415	HR 2249			6 13 54	−16 36.0	8	5.92	.025	−0.16	.007	−0.78	.016	53	B2.5 Vne	15,252,681,815,1732,2012*
254346	+22 01254	LS V +22 040			6 13 56	+22 12.8	1	9.67		0.46		−0.41		3	B2:III:	41
43384	+23 01275	HR 2240		⋆	6 13 56	+23 45.6	9	6.25	.009	0.45	.004	−0.38	.013	49	B3 Ia	1,15,41,154,245,1012*
254315	+27 01041				6 13 56	+27 52.2	1	9.36		0.64				3	G0	1724
43572	−23 03624				6 13 56	−23 08.4	1	8.55		0.59				4	F7 V	2012
254429	+12 01088				6 14 00	+12 05.7	2	8.49	.010	0.62	.005	0.20		5	F8 II	6009,8100
254392					6 14 02	+20 36.7	1	11.00		0.43		−0.44		2	F5	41
43636	−29 02936	HR 2252			6 14 02	−29 46.2	4	6.67	.008	1.54	.007			18	K4 III	15,1075,2013,2029
43227	+53 01003				6 14 03	+53 41.5	1	7.71		0.10		0.11		2	A0	1566
254428	+13 01182	LS VI +13 011			6 14 04	+13 31.5	1	9.19		0.28		−0.68		2	B0	401
43333	+39 01575				6 14 07	+39 52.7	1	7.26		1.24		1.17		2	K0	1601
43511	+6 01180	IDS06115N0602		A	6 14 11	+06 00.9	1	8.82		−0.01		−0.52		2	A0	401
43459	+18 01156				6 14 11	+18 23.9	1	8.21		0.94		0.63		3	G8 III	848
43592	−14 01385				6 14 11	−15 01.1	2	10.04	.005	0.46	.007	−0.03		3	F3 V	1594,6006
254362	+27 01043				6 14 12	+27 18.6	1	9.39		0.49				2	F8	1724
43458	+18 01157				6 14 13	+18 56.0	1	7.01		1.00		0.86		3	G5 III	848
43382	+35 01375				6 14 13	+35 10.1	1	6.70		0.41		−0.03		3	F6 V	1601
43496	+15 01139	IDS06113N1553		A	6 14 14	+15 52.2	2	7.31	.020	−0.09	.010	−0.45	.015	4	B8 II	401,8100
43526	+7 01216	HR 2248			6 14 16	+07 04.3	4	6.56	.007	−0.12	.022	−0.51	.007	10	B7 III	15,1078,1079,6007
43480	+18 01159				6 14 18	+18 42.1	1	8.04		0.97		0.23		2	G5 II	8100
43497	+13 01183				6 14 19	+13 44.0	1	7.98		1.55				2	K3 III	882
		Steph 589		AB	6 14 19	+47 55.1	1	10.90		1.29		1.24		2	K7-M0	1746
43527	+7 01217				6 14 21	+06 58.6	1	8.39		−0.06		−0.58		2	B9	401
43525	+9 01173	HR 2247		⋆ AP	6 14 21	+09 57.7	3	5.40	.013	0.11	.005	0.08	.004	12	A2 V +A2 V	15,1078,3030
43525	+9 01173	IDS06116N0959		B	6 14 21	+09 57.7	1	11.80		0.56		0.07		6	A2 V	3030
43402	+39 01579				6 14 24	+39 08.5	2	8.65	.000	0.14	.005	0.13	.010	5	A1 V	833,1601
254477	+22 01258	LS V +22 041		⋆ AB	6 14 25	+22 26.8	1	9.77		0.59		−0.31		3	B8	41
		G 101 - 32			6 14 26	+43 58.4	1	11.32		0.68		0.11		2		1658
		G 103 - 28			6 14 27	+28 40.1	1	13.24		0.98		0.43		1		1658

Table 1 333

HD	DM	Other Id	N Rem	α_{1950}	δ_{1950}	S	V	σ_V	B–V	σ_{B-V}	U–B	σ_{U-B}	n	Spectrum	References
254504 +18 01160				6 14 29	+18 00.6	1	10.26		1.36				1	K5	1017
		LS V +23 046		6 14 32	+23 17.0	1	11.06		0.61		-0.32		3		41
43380 +46 01124		HR 2239		6 14 33	+46 22.9	4	6.39	.015	1.11	.010	1.12	.007	10	K2 III	15,1003,1355,3009
43670 -15 01328				6 14 34	-15 06.4	1	7.89		1.21				4	K1 III	2012
254583 +14 01230				6 14 36	+14 54.8	1	10.05		0.95		0.63		3	K0	261
43587 +5 01168		HR 2251	⋆ A	6 14 37	+05 07.0	4	5.70	.004	0.60	.008	0.10	.007	11	G0.5Vab	15,1078,1733,3077
43587 +5 01168		IDS06120N0508	B	6 14 37	+05 07.0	2	13.35	.075	1.52	.115	1.16	.065	2		906,3003
254494 +26 01164				6 14 39	+26 23.3	1	9.19		0.66				2	G0	1724
43583 +14 01233		HR 2250		6 14 43	+14 04.7	3	6.59		-0.03	.016	-0.09	.018	7	A0 V	252,1049,1079
43561 +20 01352				6 14 44	+20 27.7	1	8.53		0.22		0.12		3	A2	848
43785 -35 02800		HR 2256		6 14 46	-35 07.4	4	4.37	.005	1.00	.003	0.83	.004	20	G8 II	15,1075,2012,8015
254647 +11 01100				6 14 48	+11 13.1	1	10.01		0.39		-0.68		2	Bpe	1012
		LS V +19 002		6 14 48	+19 50.8	1	12.06		0.58		-0.38		1		41
254577 +22 01263		LS V +22 042		6 14 53	+22 25.8	2	9.07	.009	0.81	.004	-0.21	.018	5	B0.5II-III	41,1207
43560 +23 01282				6 14 53	+23 54.1	1	8.35		1.71				2	M1	369
43607 +19 01291				6 14 54	+19 28.6	2	7.53	.000	-0.04	.015	-0.18	.005	4	A0 V	401,848
43848 -40 02356				6 14 54	-40 30.9	1	8.64		0.93				4	K1 IV	2012
		LS V +23 047		6 14 55	+23 06.5	1	10.77		1.05		-0.11		3		41
43847 -39 02491		HR 2262		6 14 56	-39 14.7	2	5.99	.005	0.16	.000			7	A2 V(m)	15,2012
43745 -22 01364		HR 2254	⋆ A	6 14 57	-22 41.5	5	6.05	.017	0.58	.013	0.08	.010	11	G0 V	15,146,1311,2012,3077
43745 -22 01364		IDS06129S2240	B	6 14 57	-22 41.5	1	10.44		1.02		0.88		4		146
43582 +22 01267		LS V +22 043		6 14 59	+22 40.7	2	8.79	.004	0.38	.009	-0.54	.013	6	B0 IIIn	41,1012,1207
43649 +13 01187				6 15 06	+13 03.1	1	8.55		0.05		-0.01		2	A3 V	401
254657 +21 01184		LS V +21 037	⋆ A	6 15 06	+21 51.8	1	10.56		0.25		0.08		1	A3	41
254657 +21 01184		IDS06121N2153	B	6 15 06	+21 51.8	1	10.67		0.26		0.07		1		41
291455 -0 01246				6 15 06	-00 56.6	1	9.76		2.95		5.40		2	C6,4	864
254681 +20 01356		LS V +20 034		6 15 10	+20 33.1	1	10.16		0.17		-0.68		3	B5	41
+47 01284				6 15 11	+47 03.2	1	9.06		0.07		0.01		2	A0	1601
43624 +27 01054				6 15 13	+27 13.9	3	6.64	.007	1.10	.007	1.14	.007	17	K1 III	1003,1724,1784
43378 +59 00959		HR 2238, UZ Lyn		6 15 13	+59 01.9	4	4.47	.018	0.01	.004	0.03	.000	13	A2 Vs	15,1363,3050,8015
43760 -10 01455		HR 2255		6 15 14	-10 42.3	2	6.74	.005	0.37	.015	0.20		7	δ Del	2006,8071
43683 +14 01235		HR 2253	⋆ AB	6 15 15	+14 24.2	4	6.15	.005	0.04	.019	0.09	.031	7	A0.5V	985,1049,1769,3024
43683 +14 01235		IDS06124N1425	C	6 15 15	+14 24.2	1	13.28		0.56		0.34		2		3024
254733 +16 01078				6 15 17	+16 47.1	1	9.72		0.12		-0.01		3	A2	261
43682 +16 01080		IDS06124N1559	ABC	6 15 18	+15 58.4	1	8.40		0.33		0.06		3	Am	261
		RL 122		6 15 18	+25 41.	1	15.24		0.35		-0.29		2		704
43899 -37 02707		HR 2263		6 15 19	-37 43.1	3	5.56	.015	1.13	.006	1.07	.021	7	K2 III	1311,2009,3009
43880 -34 02770				6 15 20	-34 43.0	1	9.07		1.26				4	K1/2 (III)	2012
254728 +20 01358		LS V +20 035		6 15 22	+20 35.2	1	10.66		0.19		-0.63		2	B8	41
254700 +23 01287		LS V +22 044		6 15 23	+22 58.8	1	9.80		0.40		-0.48		3	B5	41
254699 +23 01286		LT Gem		6 15 23	+23 35.5	3	8.99	.034	0.39	.005	-0.47	.004	7	B1 V	41,1012,1207
254695 +25 01204				6 15 24	+25 55.0	1	9.68		0.46				3	G0	1724
254696 +25 01205				6 15 25	+25 54.0	1	9.47		0.39				2	F2	1724
43940 -37 02708		HR 2265		6 15 27	-37 14.0	1	5.87		0.14				4	A2 V	2006
		LS V +23 050		6 15 28	+23 00.8	1	11.92		0.49		-0.31		3		41
43827 -16 01426		HR 2260		6 15 28	-16 47.7	4	5.15	.011	1.30	.008	1.23		13	K3 III	15,1075,2035,3016
43704 +16 01083				6 15 30	+16 43.8	1	8.41		0.83				1	G5	882
254755 +22 01273		LS V +22 045		6 15 30	+22 42.0	4	8.85	.006	0.60	.006	-0.44	.005	13	O9 Vp	41,1011,1012,1207
		L 182 - 61		6 15 30	-59 12.4	1	13.92		-0.06		-0.93		14	DB	1765
43644 +35 01380				6 15 33	+35 13.8	1	6.62		0.97		0.74		3	K0	1601
43420 +59 00960				6 15 33	+59 52.6	1	8.00		1.58		1.92		2	K5	1502
44105 -59 00618		IDS06149S5910	AB	6 15 33	-59 11.8	1	6.44		0.58		0.13		5	F3/5 V	3077
44105 -59 00619		IDS06149S5910	C	6 15 33	-59 11.8	1	14.09		-0.09		-0.95		3		3065
+38 01451		G 101 - 33		6 15 34	+38 33.3	3	10.22	.004	0.74	.006	0.23	.008	6	G6	516,1064,1620
43777 -0 01247				6 15 34	-00 21.0	1	7.77		0.16				4	B5	2033
43726 +19 01297				6 15 35	+19 35.6	1	8.77		0.11		0.03		2	A1 V	848
		LS V +23 051		6 15 35	+23 00.9	1	11.73		0.42		-0.30		3		41
43703 +23 01289		LS V +23 052		6 15 37	+23 01.7	3	8.62	.009	0.42	.004	-0.54	.004	9	B1 IV:p:	41,1012,1207
43522 +52 01058				6 15 37	+52 24.6	1	8.91		0.51		0.06		2	F5	1566
44120 -59 00619		HR 2274	⋆ A	6 15 37	-59 11.4	5	6.43	.011	0.59	.004	0.08	.012	11	G0 V	15,258,1311,2012,3077
44247 -66 00507				6 15 37	-66 16.4	1	7.34		-0.10		-0.40		2	B9 IV	401
254848 +15 01149				6 15 39	+15 39.6	1	9.19		0.10		0.06		3	A2 V	261
43702 +25 01207				6 15 39	+25 22.2	1	7.80		-0.05		-0.32		3	B8	833
43725 +21 01190				6 15 40	+21 12.1	1	7.83		1.25		1.34		2	K0	848
+23 01290		LS V +23 053		6 15 41	+23 10.4	1	10.47		0.55		-0.39		3		41
44408 -71 00425				6 15 42	-71 38.2	2	7.31	.004	1.65	.005	2.03	.006	13	M1/2 III	1628,1704
		RL 54		6 15 43	+21 36.8	1	14.69		-0.11		-0.97		2	sdB	704
		LS V +22 046		6 15 46	+22 54.7	1	11.41		0.46		-0.46		3		41
		LTT 11829		6 15 46	+67 45.0	1	11.32		1.00		0.82		3		1723
43740 +23 01293				6 15 52	+23 37.5	1	6.55		0.90		0.58		2	G5	1625
233230 +51 01176				6 15 53	+51 00.6	1	9.53		0.34		0.01		2	F0	1566
		HRC 199	V	6 15 54	+15 18.	1	12.05		1.18		-0.27		1		776
44447 -71 00426		HR 2283		6 15 54	-71 41.1	4	6.63	.009	0.56	.000	0.01	.008	25	F8 V	15,1075,1499,2038
+23 01294		LS V +23 054		6 15 55	+23 07.6	1	10.19		0.51		-0.44		3		41
43821 +9 01184		HR 2259		6 15 56	+09 04.1	3	6.24	.005	0.87	.000	0.52	.005	11	K0	15,1078,1355
		CCS 715		6 15 56	-36 11.0	1	11.06		1.78		2.15		2		864
43753 +23 01297		LS V +23 055		6 15 58	+23 01.3	4	7.89	.012	0.30	.006	-0.65	.005	10	B0.5III	41,1012,1207,8031
		LS VI +14 005		6 16 00	+14 32.0	1	10.89		0.30		-0.64		2	B0 V	1012
254992 +7 01234				6 16 01	+07 31.7	1	8.90		1.76				3	K7	369
254861 +29 01173				6 16 02	+29 24.5	1	8.57		1.10				3	K0	1724

HD	DM	Other Id	N Rem	α_{1950}	δ_{1950}	S	V	σ_V	B–V	σ_{B-V}	U–B	σ_{U-B}	n	Spectrum	References
254838	+29 01172			6 16 02	+29 56.8	2	8.82	.010	1.07	.000	0.91		6	G5	1724,1733
43841	+11 01110	IDS06133N1112	AB	6 16 03	+11 11.2	1	7.09		0.02		-0.06		2	A0	401
43820	+13 01194	IDS06132N1349	AB	6 16 04	+13 47.9	2	8.41	.307	0.11	.017	0.06	.042	5	A2 Ib	3016,8100
	−13 01434			6 16 04	−13 51.3	1	9.94		1.18		1.18		2	M0 IV	3072
43955	−19 01407	HR 2266		6 16 04	−19 56.8	2	5.51	.007	-0.18	.001	-0.71		7	B2/3 V	1732,2035
43856	+6 01191			6 16 07	+06 44.9	1	7.96		0.51				19	F6 V	6011
43819	+17 01203	HR 2258, V1155 Ori		6 16 07	+17 20.8	5	6.29	.031	-0.08	.015	-0.32	.017	10	B9 IIIp Si	261,401,1049,1079,3033
43773	+26 01177			6 16 07	+26 00.2	1	9.27		0.16				2	B9/A0 V	1724
	+26 01178			6 16 10	+26 55.9	1	9.10		1.09				4	G0	1724
43840	+12 01103	IDS06134N1244	AC	6 16 11	+12 43.4	1	7.44		0.02		-0.01		1	A0	401
255045	+10 01085			6 16 13	+10 20.8	3	8.94	.020	1.65	.009	1.80		7	M0	148,369,8032
254913	+29 01175			6 16 13	+29 27.9	1	10.06		0.37				2	F0	1724
43873	+12 01105			6 16 17	+12 46.1	1	6.74		-0.12		-0.42		2	B8	401
255017	+14 01242			6 16 17	+14 39.8	1	9.40		0.30		0.14		3	A5 Ib	8100
43818	+23 01300	LU Gem		6 16 17	+23 29.5	3	6.91	.005	0.29	.004	-0.69	.007	7	B0 II	41,1012,1207
43837	+20 01369	V963 Ori		6 16 18	+20 36.1	3	8.46	.034	0.63	.042	-0.29	.000	6	B2 Ib p:	41,138,1207
43839	+18 01171			6 16 20	+18 53.4	1	8.29		1.09		0.90		2	G5 II	8100
43836	+23 01301	LS V +23 057	⋆ A	6 16 20	+23 17.8	3	6.95	.007	0.48	.000	0.22	.013	10	A0 II	41,206,1207
43817	+25 01213			6 16 20	+25 42.4	1	8.89		0.47				2	F5/7	1724
43912	+6 01193	IDS06137N0608	A	6 16 23	+06 07.0	1	8.26		-0.06		-0.26		2	A0	401
	+23 01302	LS V +23 058		6 16 26	+23 57.1	1	10.29		0.26		-0.54		1		41
43993	−9 01411	HR 2267		6 16 28	−09 22.1	2	5.35	.005	1.24	.000	1.32	.000	7	K1.5III	15,1078
		G 106 - 38		6 16 32	+04 37.2	2	15.71	.019	1.61	.068			3		906,940
44007	−14 01399			6 16 32	−14 49.4	5	8.06	.011	0.84	.007	0.20	.020	16	G5(IV)wF3/5	1003,1594,1775,2012,3077
43910	+13 01196	LS VI +13 012		6 16 33	+13 00.7	1	7.45		0.64		0.18		3	A2 Ia	8100
44021	−14 01400	HR 2268		6 16 33	−15 00.2	5	6.05	.030	1.66	.017	1.94	.045	16	M0 III	15,252,1075,2035,3051
		LS V +23 059		6 16 35	+23 07.4	1	11.69		0.32		-0.41		3		41
43931	+13 01199	IDS06138N1328	A	6 16 36	+13 27.4	1	6.92		0.47		0.03		2	F7 V	3016
	+38 01456	G 101 - 34		6 16 37	+38 22.2	2	10.76	.000	0.74	.000	0.06	.000	3	G5	333,927,1620
43930	+13 01200	V1260 Ori	⋆ B	6 16 39	+13 28.2	1	7.67		1.13		0.92		2	K1 V	3016
43929	+16 01089			6 16 39	+16 12.0	1	8.03		0.40		0.13		3	F5 V	261
255055	+23 01304	LS V +23 060		6 16 39	+23 18.7	4	9.39	.009	0.25	.005	-0.71	.005	10	O9 V:p	41,1011,1012,1207
		LS V +23 061	⋆ AB	6 16 40	+23 56.4	1	11.77		0.34		-0.39		1		41
255103		LSS 34		6 16 44	+18 23.6	2	10.52	.000	0.55	.000	-0.57	.000	4	B7	405,1737
43907	+22 01280	IDS06137N2209	ABC	6 16 44	+22 08.0	2	8.69	.022	0.29	.009	-0.59	.004	5	B1 V:p:	41,1207
44037	−8 01386	HR 2270		6 16 44	−08 33.9	3	6.21	.005	-0.04	.004	-0.14	.010	8	B9 V	15,1075,1079
44267	−52 00902	HR 2278		6 16 45	−52 42.7	3	6.39	.012	1.46	.000			11	K2/3 III	15,1075,2007
255093	+22 01281	LS V +22 048	⋆ AB	6 16 46	+22 19.6	2	9.16	.004	0.32	.000	-0.49	.009	5	B1.5V	41,1207
		LS V +21 038		6 16 47	+21 10.4	1	11.07		0.28		-0.41		2		41
255091	+23 01306	LS V +23 062		6 16 48	+23 50.9	2	9.54	.000	0.21	.004	-0.57	.004	5	B2 V	41,1207
44019	−0 01255			6 16 48	−00 53.6	1	7.27		1.19		1.24		1	K2 IV	509
		LS V +22 049		6 16 49	+22 06.7	1	11.28		0.28		-0.49		3		41
43885	+28 01078	IDS06136N2828	AB	6 16 49	+28 27.0	1	7.27		0.24				5	A3	1381
255134	+23 01307	LS V +23 063		6 16 51	+23 17.3	2	9.17	.000	0.36	.000	-0.53	.009	5	B1 IVp	41,1207
44081	−20 01355	HR 2271		6 16 51	−20 54.2	2	5.80	.005	-0.16	.005			7	B3 II/III	15,2028
255133	+23 01308	LS V +23 064		6 16 54	+23 45.7	1	9.90		0.29		-0.53		3	B8	41
		LTT 2515		6 16 54	−06 37.6	1	13.06		1.70				1		1705
43984	+11 01118			6 16 55	+11 01.8	1	8.30		-0.04		-0.48		2	B8	401
255194	+19 01304			6 16 56	+19 14.6	1	9.89		0.21		-0.15		3	A0	848
255168	+23 01310	LS V +23 065		6 16 57	+23 51.5	2	9.69	.004	0.27	.004	-0.51	.000	5	B1 V	41,1207
255191	+24 01204	LS V +24 016		6 16 59	+24 15.1	1	10.32		0.68		-0.29		3	B0	41
		LS V +22 050		6 17 00	+22 11.1	1	11.47		0.22		-0.56		3		41
		IX Gem		6 17 00	+24 21.	1	10.56		0.89				1		6011
		LS V +24 017		6 17 07	+24 00.6	1	11.65		0.36		-0.39		3		41
44052	+4 01207			6 17 08	+04 39.7	1	8.40		-0.09		-0.52		2	B8	401
		LS V +20 037		6 17 10	+20 57.1	1	11.21		0.23		-0.49		2		41
255215				6 17 10	+23 53.1	1	11.01		0.23		-0.51		3	B8	41
44034	+14 01246			6 17 11	+14 15.5	1	9.23		0.42		0.07		3	A7 II	8100
44076	−1 01192			6 17 11	−01 11.1	1	8.42		0.31		-0.05		2	A0	509
43963	+28 01080			6 17 12	+28 06.0	1	8.52		1.06				2	K0	1724
44033	+14 01247	HR 2269		6 17 13	+14 40.4	5	5.67	.010	1.58	.022	1.89	.032	24	K3 Ib	252,261,3005,6007,8100
44018	+19 01305			6 17 14	+19 53.8	1	7.74		1.10		1.16		3	G5	848
43981	+27 01070			6 17 17	+27 39.0	1	8.73		0.84				4	K0	1724
44112	−7 01373	HR 2273		6 17 18	−07 48.0	4	5.26	.015	-0.20	.004	-0.75	.007	17	B2.5V	15,154,1415,2012
255282	+23 02313			6 17 21	+23 17.3	1	10.08		0.43		-0.46		3	B5	41
255236	+28 01081			6 17 22	+28 42.7	1	8.75		1.20				4	K0	1724
	+6 01201			6 17 23	+06 25.6	1	10.53		1.88				3	M2	369
255278	+26 01184			6 17 24	+26 52.7	1	9.56		0.68				4	G5	1724
255336	+21 01198			6 17 28	+21 33.6	1	9.60		-0.01				2	A2	6009
44182	−17 01446			6 17 28	−17 30.2	1	7.27		-0.08		-0.32		2	B8 II	401
44131	−2 01564	HR 2275		6 17 29	−02 55.3	3	4.90	.008	1.59	.005	1.95	.005	12	M1 III	15,1078,3016
255312	+23 01315	LS V +23 068	⋆ AB	6 17 30	+23 20.5	1	9.45		0.72		-0.29		3	B2	41
44030	+25 01223			6 17 30	+25 37.9	3	7.58	.007	1.43	.016	1.66	.026	14	K4 III	1003,1775,4001
44362	−50 02169	HR 2281		6 17 33	−50 20.2	3	7.03	.004	0.84	.005			11	F8 II	15,1075,2031
44577	−63 00541			6 17 33	−64 00.6	1	7.54		-0.08		-0.19		2	B9 V	401
44109	+7 01243	IDS06149N0746	AB	6 17 34	+07 44.5	1	6.76		-0.04		-0.22		2	B9	401
		RL 123		6 17 36	+25 39.	1	12.91		0.19		-0.48		2		704
43812	+59 00964	HR 2257	⋆ AB	6 17 37	+59 23.7	1	6.08		0.14		0.09		3	A3 V	1733
44179	−10 01476	IDS06153S1036	AB	6 17 37	−10 36.9	1	8.84		0.43		0.27		1	B8	1588
255330	+29 01188			6 17 38	+28 59.3	1	9.35		0.42				2	F2	1724

Table 1 335

HD	DM	Other Id	N Rem	α_{1950}	δ_{1950}	S	V	σ_V	B–V	σ_{B-V}	U–B	σ_{U-B}	n	Spectrum	References
44073 +18 01178		IDS06148N1806	A	6 17 39	+18 04.3	1	7.58		1.18		1.21		3	G7 III	848
		POSS 122 # 3		6 17 41	+53 47.9	1	16.57		1.50				3		1739
44063 +26 01187				6 17 42	+26 31.2	1	9.30		0.43				2	F5	1724
43905 +53 01008		HR 2264		6 17 42	+53 28.6	4	5.35	.010	0.43	.014	0.12	.003	12	F5 III	15,254,1008,3053
255408 +18 01180		IDS06148N1835	AB	6 17 44	+18 33.5	1	9.26		0.30		0.12		3	A3	848
44072 +25 01226				6 17 47	+25 43.6	1	8.98		0.44				3	F3/7 V/IV	1724
44175 −1 01199				6 17 51	−01 16.3	1	7.97		0.08		0.06		2	A0	509
44071 +29 01189				6 17 52	+29 23.7	1	7.01		0.37				2	F2 V/IV	1724
44323 −34 02795		HR 2279		6 17 53	−34 22.4	3	5.77	.004	-0.09	.007	-0.27		9	B9 V	15,401,2012
		AAS11,365 T12# 1		6 17 54	+16 39.	1	10.10		1.99		2.20		4		261
44153 +7 01246		LS VI +07 003		6 17 56	+07 35.8	1	8.46		0.37		0.37		3	F0	493
		LS V +23 069		6 17 56	+23 49.5	1	11.70		0.40		-0.40		3		41
255462 +20 01379				6 17 57	+20 41.0	1	10.13		1.74				3	K2	369
291526 +0 01388				6 18 00	+00 08.7	1	8.50		0.00		-0.20		1	A2	509
		AAS11,365 T12# 3		6 18 00	+16 04.	1	10.98		1.36		1.10		5		261
44092 +29 01190		HR 2272		6 18 00	+29 33.9	2	6.45		0.06	.005	-0.02	.030	5	A1 Vs	592,1049
255494 +16 01102				6 18 02	+16 50.8	1	10.24		0.15		0.08		3	A2	261
255534 +16 01103				6 18 04	+16 57.2	1	9.54		0.24		-0.29		3	B5	261
44029 +45 01289				6 18 04	+45 38.1	1	7.10		1.41		1.63		2	K0	1601
		G 105 - 30		6 18 05	+06 46.7	1	16.37		0.55		-0.17		1		940
44173 +11 01128		HR 2276	⋆	6 18 05	+11 46.8	2	6.53	.005	-0.10	.000	-0.57	.020	5	B5 III	154,401
255455 +23 01318				6 18 05	+23 45.2	1	10.58		0.08		-0.17		3	A0	41
		AAS11,365 T12# 4		6 18 06	+16 31.	1	10.78		0.59		0.04		3		261
		RL 124		6 18 06	+26 52.	1	14.15		0.21		-0.34		2		704
44139 +22 01291		LS V +22 051		6 18 07	+22 11.4	3	8.79	.004	0.29	.012	-0.58	.008	7	B0.5V	41,1012,1207
44213 +5 01198		V1056 Ori	⋆ A	6 18 08	+05 45.8	4	7.99	.016	1.93	.035	1.92	.035	13	M3 Ib-II	148,369,8032,8100
255451 +29 01192				6 18 08	+29 16.6	1	9.85		0.91				4	G5	1724
44172 +14 01254				6 18 10	+14 43.7	2	7.34	.000	-0.10	.000	-0.51	.010	5	B6 V	261,401
+16 01105				6 18 11	+16 12.1	1	10.57		1.06		0.79		4		261
		AAS11,365 T12# 5		6 18 12	+16 44.	1	11.85		1.49		1.32		5		261
+13 01212			V	6 18 15	+13 56.2	3	10.30	.059	2.57	.020	2.17		3	M4	148,369,8032
44171 +21 01203		IDS06153N2111		6 18 15	+21 09.4	2	7.32	.015	-0.08	.005	-0.38	.005	4	B8	401,848
255447 +34 01324		G 103 - 31		6 18 15	+33 59.4	1	10.27		0.58		0.02		1	G0	1620
		NGC 2215 - 30		6 18 15	−07 13.3	1	12.53		1.56				3		2042
		NGC 2215 - 19		6 18 15	−07 16.9	2	12.11	.070	0.29	.037	0.24		4		305,2042
44212 +9 01199		IDS06155N0946	A	6 18 16	+09 45.2	1	8.05		0.90		0.59		1	K2	401
		NGC 2215 - 13		6 18 16	−07 18.3	1	13.38		0.32				3		2042
44380 −29 03008		IDS06165S3001	B	6 18 18	−29 59.7	1	7.64		1.17		1.15		3	K1 III	1657
44235 +9 01200				6 18 19	+09 47.2	1	7.99		-0.16		-0.76		2	B8	401
		NGC 2215 - 34		6 18 21	−07 12.7	1	12.53		0.30				2		2042
44193 +21 01204				6 18 23	+21 13.4	1	7.59		0.06		-0.01		2	A0	401
44091 +44 01426				6 18 23	+44 05.0	1	6.88		1.02		0.86		2	G5	985
		NGC 2215 - 20		6 18 24	−07 17.3	1	11.62		0.41				3		2042
44402 −30 03038		HR 2282	⋆ A	6 18 24	−30 02.4	8	3.02	.003	-0.20	.008	-0.71	.017	25	B2.5V	15,200,1020,1034,1075*
		NGC 2215 - 33		6 18 25	−07 13.9	2	11.55	.039	0.27	.010	0.22		3		305,2042
		NGC 2215 - 52		6 18 25	−07 16.3	1	16.25		0.52		0.26		1		305
−7 01383		NGC 2215 - 26		6 18 26	−07 13.6	2	10.57	.074	1.15	.031	0.80		12		305,2042
		NGC 2215 - 50		6 18 26	−07 14.6	1	15.02		0.92		0.45		1		305
		NGC 2215 - 10		6 18 26	−07 19.0	1	11.95		0.25				3		2042
		NGC 2215 - 51		6 18 27	−07 15.7	1	15.33		0.99		-0.15		1		305
		NGC 2215 - 49		6 18 29	−07 15.0	1	14.37		0.57		0.16		1		305
		NGC 2215 - 1		6 18 29	−07 16.5	2	10.91	.044	0.34	.017	0.25		6		305,2042
		LS V +22 052		6 18 30	+22 58.1	1	11.05		0.53		-0.43		3		41
		NGC 2215 - 24		6 18 30	−07 17.1	1	11.72		0.52				5		2042
44234 +17 01214		HR 2277		6 18 31	+17 47.3	2	6.28	.000	1.10	.000	0.91	.005	4	K0	1648,1733
44321 −6 01507				6 18 32	−06 40.1	1	7.64		-0.05		-0.21		2	B9	401
255680 +16 01110				6 18 33	+16 18.7	1	9.74		1.08		0.83		3	G5	261
44283 +6 01207				6 18 35	+06 47.3	1	9.36		0.30				21	K0	6011
		NGC 2215 - 5		6 18 35	−07 16.2	1	12.01		0.32				7		2042
		RL 125		6 18 36	+26 55.	1	12.93		0.08		-0.65		2		704
		POSS 122 # 1		6 18 37	+54 57.6	1	14.24		0.68				3		1739
44253 +19 01313				6 18 38	+19 55.0	2	7.29	.005	1.02	.030	0.89	.000	5	G5	848,1648
255702 +16 01111				6 18 41	+16 29.3	1	9.89		0.26		0.16		3	A0	261
255812 +3 01207				6 18 42	+03 18.0	1	9.52		2.07				3	K8	369
−5 01600				6 18 42	−05 28.3	1	11.01		0.44		0.01		1		6006
		NGC 2215 - 43		6 18 42	−07 13.9	1	12.56		0.55				2		2042
		G 104 - 35		6 18 43	+16 20.1	1	12.60		1.57		1.24		1	K7	1773
44320 +6 01208		SV Mon		6 18 45	+06 29.7	5	7.74	.038	0.79	.036	0.48	.053	5	G5	934,1484,1587,1772,6011
255627 +27 01081		IDS06156N2734	A	6 18 46	+27 32.9	1	9.76		0.39				2	G0	1724
255627 +27 01081		IDS06156N2734	B	6 18 46	+27 32.9	1	10.28		1.37				1		1724
44232 +27 01082		IDS06156N2734	C	6 18 47	+27 32.0	1	9.55		0.41				2	F5	1724
44594 −48 02259		HR 2290		6 18 47	−48 42.8	6	6.61	.017	0.66	.005	0.19	.010	26	G3 V	15,621,1075,1311*
44506 −34 02806		HR 2288		6 18 48	−34 07.2	4	5.54	.023	-0.19	.007	-0.86	.017	13	B1.5 Ve	15,401,1637,2012
44333 +2 01197		HR 2280	⋆ AB	6 18 49	+02 17.6	4	6.30	.019	0.25	.007	0.08	.009	10	A3 V +A6 V	15,1078,1381,3030
255771 +15 01165				6 18 50	+15 23.9	1	9.15		1.08		0.90		3	G5	261
		AAS11,365 T12# 9		6 18 54	+16 41.	1	11.66		1.58		1.38		5		261
		FT Mon		6 18 57	+04 05.7	1	12.00		1.11				1		1772
		CCS 498, BN Mon		6 19 00	+07 22.	1	10.16		4.25		6.70		1	C4-5,4-5	864
−19 01422				6 19 00	−19 14.1	2	9.68	.015	0.80	.010	0.28		3		1594,6006
291593 −0 01265				6 19 01	−00 18.2	1	9.79		0.10		0.06		1	A0	509

HD	DM	Other Id	N Rem	α_{1950}	δ_{1950}	S	V	σ_V	B–V	σ_{B-V}	U–B	σ_{U-B}	n	Spectrum	References
44458	−11 01460	HR 2284, FR CMa	★ AB	6 19 05	−11 44.9	7	5.55	.014	0.01	.009	-0.84	.012	51	B1 Vpe	681,815,1212,2007*
44351	+14 01258	LS VI +14 006		6 19 08	+14 20.0	2	8.29	.020	0.09	.015	-0.49	.090	4		379,401
255866				6 19 10	+16 09.	1	11.09		0.29		0.10		4	A2	261
44420	−0 01266			6 19 10	−00 30.7	2	7.61	.010	0.69	.017	0.29		6	G4 V	2033,3016
44373	+11 01138	IDS06164N1101	AB	6 19 11	+10 59.8	1	8.50		-0.08		-0.55		2	B8	401
		RL 126		6 19 12	+26 22.	1	13.29		0.17		-0.48		2		704
	+16 01112		V	6 19 14	+16 11.4	1	10.07		1.91				3	M2	369
255887				6 19 17	+16 50.	1	10.71		0.20		0.10		3	A0	261
44123	+57 00964			6 19 17	+57 22.9	1	7.76		1.26		1.26		3	K1 III	37
255930	+10 01104	V1028 Ori		6 19 18	+10 55.2	1	9.81		0.11		-0.56		5	B5	493
44665	−44 02600	IDS06178S4445	AB	6 19 18	−44 46.0	1	8.37		0.74				4	G5 V	2012
		L 182 - 34		6 19 18	−57 17.	1	11.20		0.80		0.32		2		1696
255817	+26 01200			6 19 19	+26 46.4	1	9.15		0.45				2	F8	1724
44437	+8 01294			6 19 20	+08 54.8	1	9.11		0.08		-0.21		1	A0	401
44002	+66 00441			6 19 20	+66 32.5	1	8.30		0.35		-0.07		3	F0	1733
44482	−9 01431			6 19 21	−09 39.5	1	7.38		-0.08		-0.40		2	A0	401
44415	+14 01259	RS Ori		6 19 22	+14 42.2	3	8.12	.192	0.86	.136	0.69	.185	3	F4 Ib	379,851,6011
44210	+50 01297			6 19 26	+50 46.0	1	7.97		1.01		0.76		2	K0	1566
44573	−22 01389			6 19 26	−22 11.2	2	8.47	.023	0.92	.007	0.59		5	K2 V	2034,3072
44414	+14 01260			6 19 28	+14 53.4	3	8.31	.005	0.64	.008	0.22	.000	48	F5 Ib	851,6011,8100
		AAS11,365 T12# 12		6 19 30	+16 15.	1	12.04		0.58		0.05		4		261
63013	−89 00022			6 19 33	−89 18.2	1	8.86		0.26		0.17		4	A3mA5-A8	826
44547	−8 01404			6 19 34	−08 12.3	1	7.77		-0.02		-0.11		2	A0	401
44517	−5 01606			6 19 35	−05 25.8	1	7.95		0.54				2	G0	1594
	−31 03220			6 19 38	−31 51.3	1	10.15		0.39		-0.04		3		1732
44498	+8 01296			6 19 39	+08 21.2	1	8.83		-0.08		-0.59		2	B2.5V	1012
44391	+28 01097			6 19 39	+28 00.8	1	7.74		1.40				1	K0 Ib	138
291590	+0 01396			6 19 41	+00 02.1	1	9.20		0.62		0.10		1	G0	509
269352				6 19 41	−67 56.0	1	10.13		1.64		1.91		3	K5 III	149
		AAS11,365 T12# 13		6 19 42	+16 26.	1	11.88		1.76		1.73		6		261
291599				6 19 45	−00 29.	1	10.28		0.36		0.02		1	F2	509
44544	+3 01214	FU Mon		6 19 46	+03 26.7	2	8.95	.072	3.09	.135	4.50		5	SC	864,3039
44497	+12 01123	HR 2287		6 19 48	+12 35.8	1	6.00		0.30		0.05		2	F0 III	1733
		CoD -22 03005		6 19 51	−22 42.1	2	10.82	.060	1.43	.005	1.23		2	M2	1746,3062
256082	+15 01172			6 19 52	+15 50.3	1	9.83		1.53		1.49		4	K0	261
44390	+35 01392			6 19 52	+35 32.8	1	7.44		0.03		0.05		2	A0	1601
		AAS11,365 T13# 2		6 19 54	+15 52.	1	13.68		0.57		0.40		5		261
		AAS11,365 T13# 1		6 19 54	+15 53.	1	13.06		0.54		0.30		5		261
		AAS11,365 T12# 14		6 19 54	+16 43.	1	12.27		0.56		0.22		5		261
44496	+17 01224	IDS06170N1737	AB	6 19 55	+17 36.0	2	6.92	.015	0.03	.005	-0.03	.000	5	A0	261,401
44478	+22 01304	HR 2286, μ Gem	★ A	6 19 56	+22 32.5	11	2.87	.021	1.64	.012	1.88	.042	63	M3 IIIab	3,15,369,814,1118*
256035	+22 01303	LS V +22 053		6 19 56	+22 53.3	2	9.16	.000	0.54	.004	-0.46	.004	7	O9 V:p	41,1207
256176	+9 01212	LS VI +09 003		6 19 59	+09 11.6	1	9.95		0.12		-0.49		5	B3	493
256115				6 20 00	+16 35.	1	10.31		0.28		0.08		3	A2	261
		RL 127		6 20 00	+26 50.	1	14.22		0.12		-0.54		2		704
	+21 01211			6 20 01	+21 13.5	1	10.01		1.78				2	M3	369
44604	−0 01272			6 20 02	−00 28.3	1	8.79		0.73		0.37		1	K0	509
291595				6 20 03	−00 12.	1	9.08		1.90		1.95		1		506
44586	+5 01214			6 20 05	+05 09.1	1	8.37		0.08		-0.36		2	B9	401
256147		LSS 39		6 20 05	+15 58.2	2	10.93	.010	0.53	.005	-0.57	.030	4	B0 Ibe2	261,405
44603	+0 01397			6 20 06	+00 07.2	1	9.16		0.14		0.08		1	A2	509
		AAS11,365 T13# 5		6 20 06	+15 57.	1	11.90		0.49		0.33		4		261
44602	+4 01229			6 20 07	+04 14.2	1	6.85		0.10		0.10		2	A3	401
		SAL2,43 # 28		6 20 09	+00 23.7	1	12.08		0.46		0.17		2		664
44657	−7 01399			6 20 11	−07 36.0	1	8.17		-0.01		-0.15		2	A0	401
		AAS11,365 T13# 6		6 20 12	+15 57.	1	13.42		0.88		0.31		6		261
44639	−2 01581	V Mon		6 20 12	−02 10.1	1			1.39		0.77		1	M4	975
		SAL2,43 # 30		6 20 13	+00 20.5	1	12.88		0.65		-0.03		3		664
44600	+6 01218			6 20 13	+06 45.1	1	9.56		-0.04		-0.44		2	B9	3039
44512	+28 01101	IDS06170N2848	AB	6 20 13	+28 46.9	1	8.34		0.03		-0.29		3	B8	555
		SAL2,43 # 29		6 20 15	+00 22.1	1	12.22		1.37		0.85		2		664
		LS V +20 038		6 20 15	+20 13.1	1	11.46		0.55		-0.36		2		41
44584	+16 01118	IDS06174N1634	AB	6 20 16	+16 32.5	2	8.01	.005	0.01	.010	-0.23	.005	5	B9	261,401
		SAL2,43 # 31		6 20 17	+00 21.4	1	13.11		1.39		0.59		1		664
44585	+15 01173			6 20 17	+15 52.1	2	8.25	.010	-0.04	.015	-0.16	.000	5	B7 V	261,401
44654	−0 01273			6 20 17	−00 48.4	1	7.60		-0.04		-0.23		1	B9	509
44762	−33 02927	HR 2296		6 20 17	−33 24.6	5	3.84	.006	0.88	.008	0.52	.002	25	G7 II	15,1075,1637,2012,8015
		AAS11,365 T12# 17		6 20 18	+16 25.	1	11.26		0.35		0.13		4		261
44583	+19 01325			6 20 18	+19 41.7	1	8.29		0.06		-0.38		1	A0	401
		LP 720 - 6		6 20 18	−12 51.	2	13.08	.084	1.00	.000	0.59	.055	10		1663,3073
	−0 01274			6 20 23	−00 19.4	1	9.67		0.12		0.13		3		664
256333	+3 01219			6 20 24	+03 37.7	1	9.03		1.95				3	K2	369
291597				6 20 24	−00 19.	2	9.69	.009	0.14	.006	0.10	.009	4		506,509
		LS V +23 071		6 20 25	+23 13.0	1	11.47		0.50		-0.40		2		41
291596				6 20 27	−00 15.	2	9.99	.010	0.12	.004	0.08	.008	2		506,509
291598				6 20 27	−00 25.	1	10.32		0.29		0.03		1	F0	509
256259	+16 01119			6 20 30	+16 08.8	1	10.46		0.42		0.03		3	F8	261
44597	+20 01399	LS V +20 039		6 20 30	+20 25.1	3	9.03	.005	0.26	.000	-0.70	.005	7	O9 V	41,1011,1012
44743	−17 01467	HR 2294, β CMa	★ A	6 20 30	−17 55.8	8	1.98	.009	-0.24	.006	-0.97	.012	38	B1 II/III	15,200,1034,1075,1278*
44701	−3 01413	IM Mon		6 20 32	−03 15.0	1	6.57		-0.15		-0.65		2	B5 V	401

Table 1

337

HD	DM	Other Id	N Rem	α_1950	δ_1950	S	V	σ_V	B–V	σ_B–V	U–B	σ_U–B	n	Spectrum	References
44637	+15 01176	LSS 41		6 20 33	+15 07.7	3	8.03	.019	0.26	.020	-0.51	.029	6	B2 V:pe	851,1012,1212
256321	+14 01267	IDS06177N1435	AB	6 20 34	+14 33.4	1	9.01		0.56				43	F7 V	6011
256283	+16 01120			6 20 34	+16 19.4	1	10.29		0.28		0.12		3	A2	261
		AAS11,365 T13# 9		6 20 36	+15 47.	1	12.73		0.55		0.32		5		261
		AAS11,365 T13# 8		6 20 36	+15 51.	1	12.95		1.30		0.92		6		261
		AAS11,365 T12# 19		6 20 36	+16 32.	1	11.37		0.49		-0.08		4		261
44720	-3 01414			6 20 36	-03 29.4	1	7.22		-0.12		-0.49		2	B8	401
44718	+0 01399			6 20 38	+00 13.4	1	8.73		-0.01		-0.11		1	A0	509
256351	+10 01114	LS VI +10 005		6 20 39	+10 34.2	1	10.21		0.05		-0.59		5	A3	493
44700	+3 01221	HR 2292		6 20 40	+03 47.5	4	6.39	.010	-0.15	.005	-0.63	.007	14	B3 V	15,154,1223,1415
44615	+29 01210			6 20 42	+29 00.1	1	8.78		0.47		0.05		2	F6 V	3026
44676	+14 01270			6 20 43	+14 04.1	1	7.54		1.23				3	K1 III	882
256276	+22 01311	LS V +22 054		6 20 44	+22 25.3	2	9.25	.009	0.15	.004	-0.70	.009	5	B1.5:V:nn	41,1012
256315	+18 01201			6 20 45	+18 23.2	1	9.93		0.52		-0.01		3	F8	261
		POSS 122 # 4		6 20 48	+53 43.8	1	17.00		1.49				4		1739
44712	+11 01151			6 20 50	+10 57.6	1	7.76		0.00		-0.02		2	A0	401
	-7 01402			6 20 52	-07 26.2	2	9.95	.005	2.60	.058	3.62	.224	3	C4,4	109,864
	-15 01365	IDS06186S1550	AB	6 20 52	-15 51.1	1	9.60		0.79		0.45		1		8084
256378				6 20 53	+15 51.	1	11.62		0.56		0.07		4	F8	261
		AAS11,365 T13# 11		6 20 54	+15 52.	1	10.85		0.39		0.02		3		261
44756	-4 01484	HR 2295		6 20 54	-04 39.6	3	6.66	.005	0.06	.000	0.01	.018	9	A2 V	15,1061,1079
44821	-24 03969			6 20 54	-24 31.8	1	7.37		0.68		0.22			K0/1 V	1732
44674	+25 01251	LS V +25 029		6 20 59	+25 26.7	3	8.44	.004	0.04	.012	-0.84	.007	11	B1 Vne	399,405,833
		AAS11,365 T13# 12		6 21 00	+15 52.	1	12.30		0.50		0.00		4		261
44537	+49 01488	HR 2289, ψ1 Aur		6 21 03	+49 18.9	4	4.87	.051	1.96	.019	2.19	.053	10	K5 Iab	1355,1363,3001,8032
256413	+19 01331	LS V +19 003		6 21 04	+19 56.2	2	8.92	.005	0.35	.000	-0.38	.009	4	B5 III	41,1012
44896	-33 02935			6 21 04	-33 35.3	2	7.24	.025	1.61	.020	1.89	.000	2	Kp Ba	565,3048
44769	+4 01236	HR 2298	⋆ A	6 21 07	+04 37.2	2	4.44	.000	0.18	.005	0.13	.005	6	A5 IV	1355,3024
44769	+4 01236	IDS06185N0439	AB	6 21 07	+04 37.2	5	4.30	.007	0.20	.006	0.10	.023	14	A5 IV +F5 V	15,1075,1363,1425,8015
44770	+4 01237	HR 2299	⋆ B	6 21 07	+04 37.4	1	6.72		0.45		-0.05		3	F5 V	3024
44785	-0 01278			6 21 10	-00 19.4	1	9.36		0.17		0.09		3	A2	509
44784	+0 01402			6 21 11	+00 00.6	1	9.14		0.12		0.10		3	A2	509
		AAS11,365 T13# 14		6 21 12	+15 47.	1	14.41		1.50		0.92		2		261
		AAS11,365 T13# 13		6 21 12	+15 51.	1	14.21		0.92		0.37		2		261
44816	-9 01444	HR 2301		6 21 14	-09 50.8	2	6.18	.005	1.84	.000	2.04	.005	7	K5	15,1415
		LP 839 - 38		6 21 15	-22 33.5	1	12.52		1.23		1.05		3		1773
45057	-53 01070			6 21 15	-53 18.5	1	6.86		-0.14				4	B3 V	2012
256443	+26 01210			6 21 16	+26 13.2	1	9.84		0.51				2	F8	1724
		LP 839 - 39		6 21 16	-22 33.6	1	11.10		0.89		0.53		4		1773
256577	+8 01314	LS VI +08 005		6 21 17	+08 19.7	1	9.48		0.13		-0.75		2	B2 IV:pe:	1012
	+8 01313			6 21 17	+08 55.5	1	8.88		1.02		0.76		1		401
44782	+9 01223			6 21 17	+09 12.0	1	7.38		-0.07		-0.34		2	B9	401
44783	+8 01316	HR 2300		6 21 18	+08 54.8	4	6.26	.011	-0.09	.009	-0.30	.004	10	B8 Vn	15,258,401,1078
44694	+35 01397			6 21 18	+35 17.0	1	6.56		1.19		1.18		3	K0	985
44797	+1 01322			6 21 19	+01 43.0	1	9.36		0.21		0.08		2	B9	3039
44768	+15 01181			6 21 19	+15 52.7	1	7.87		-0.07		-0.17		2	A0 III	401
44956	-31 03245	HR 2307		6 21 22	-31 45.8	1	6.34		0.89				4	G8 III	2035
		AAS11,365 T13# 15		6 21 24	+15 49.	1	13.62		0.52		0.32		3		261
256575	+10 01119	LS VI +10 006		6 21 25	+10 37.9	1	10.13		0.19		-0.17		3	B3	493
44841	-4 01490			6 21 25	-04 42.1	1	7.26		1.01		0.77		3	K0	1657
44979	-34 02833			6 21 26	-34 58.2	1	6.49		-0.05		-0.30		2	Ap Si	401
44839	+0 01405			6 21 29	+00 20.7	1	8.89		0.04		0.01		2	A0	509
45077	-50 02190			6 21 29	-50 04.5	1	8.02		0.88				4	G5 III	2012
		AAS11,365 T13# 18		6 21 30	+15 47.	1	11.38		0.39		0.11		4		261
		AAS11,365 T13# 17		6 21 30	+15 50.	1	14.42		0.62		0.27		4		261
		AAS11,365 T13# 16		6 21 30	+15 51.	1	11.06		1.38		1.08		4		261
44891	-14 01428	HR 2303		6 21 30	-15 02.6	1	6.24		1.53				4	K2/3 III	2035
44813	+7 01266	IDS06188N0757	AB	6 21 31	+07 55.4	1	8.27		-0.06		-0.26		1	A3	509
44814	+7 01267			6 21 33	+07 06.0	1	8.69		0.04		0.09		9	A2	1655
44855	+0 01406	IDS06190S0000	AB	6 21 33	-00 01.6	1	9.34		0.13		0.05		2	A0	509
		G 104 - 37		6 21 37	+23 28.0	1	13.06		1.76				6		940
44953	-19 01435	HR 2306	⋆ AB	6 21 38	-19 45.5	2	6.60		-0.14	.004	-0.65		5	B8 III	1079,2006
44780	+25 01255			6 21 39	+25 04.6	1	6.33		1.21		1.15		2	K2 III	985
		G 192 - 26		6 21 39	+56 12.3	2	14.59	.005	1.77	.018	1.23		5		906,940
		G 101 - 38		6 21 40	+48 45.5	2	10.83	.034	1.06	.032	0.89	.011	4		1620,5010
	-26 02983		V	6 21 40	-27 02.5	2	8.81	.173	2.86	.276	1.35		4	C6 II	414,8100
256625	+16 01129			6 21 41	+16 31.8	1	9.86		0.53		0.04		3	F8	261
44811	+19 01335	LSS 44	⋆ AB	6 21 41	+19 44.0	3	8.42	.000	0.13	.004	-0.81	.015	6	O7.5V	410,1011,1012
44766	+29 01213	HR 2297		6 21 41	+29 44.1	1			-0.06		-0.40			B8 IIIn	1079
44908	-2 01591	LS VI -02 006		6 21 41	-02 56.2		9.35	.018	0.01	.011	-0.59	.007	8	A0	401,490,1732
44870	+5 01227			6 21 43	+05 21.1	1	8.83		-0.04		-0.18		2	A0	401
256699	+10 01124			6 21 44	+10 41.2	1	10.17		0.00		-0.55		2	G5	401
44028	+77 00239			6 21 44	+77 01.7	1	8.62		0.47		-0.02		2	F8	1733
44887	-0 01284			6 21 46	-00 43.0	1	8.36		-0.01		-0.08		1	A0	509
44796	+28 01109			6 21 50	+28 03.6	1	8.13		-0.01		-0.14		2	A0	1625
44951	-11 01478	HR 2305		6 21 50	-11 30.1	4	5.22	.007	1.23	.008	1.20		14	K3 III	15,1075,2035,3016
44535	+60 00962			6 21 51	+60 05.5	1	9.19		1.57		1.86		3	K2	1733
45270	-60 00604	IDS06213S6010	A	6 21 53	-60 11.6	1	6.50		0.60		0.05		5	G1 V	1628
44907	+4 01240			6 21 54	+04 12.9	1	7.30		-0.08		-0.38		2	B9	401
		RL 128		6 21 54	+26 53.	1	15.11		0.10		-0.90		2		704

HD	DM	Other Id	N Rem	α₁₉₅₀	δ₁₉₅₀	S	V	σ_V	B–V	σ_B–V	U–B	σ_U–B	n	Spectrum	References
45018	−25 03189	HR 2311		6 21 54	−25 33.0	3	5.62	.004	1.56	.000			11	K5 III	15,2013,2029
44648	+55 01076			6 21 57	+55 54.1	1	7.85		1.05		0.81		2	G8 III-IV	1502
		LSS 45		6 21 58	+19 47.6	1	12.59		0.24		-0.54		2	B3 Ib	410
		AAS11,365 T13# 19		6 22 00	+15 48.	1	12.21		0.49		0.33		4		261
44867	+16 01135	HR 2302		6 22 00	+16 05.2	2	6.33	.013	1.07	.017	.90		5	G9 III	71,261
44948	−1 01231	IDS06195S0121	AB	6 22 02	−01 23.4	1	6.73		-0.10		-0.44		2	B8	401
44996	−12 01470	HR 2309	⋆A	6 22 02	−12 56.0	3	6.11	.005	-0.09	.005	-0.64		9	B3/5 V	15,1079,2012
45098	−36 02869			6 22 02	−36 59.2	1	6.90		-0.10		-0.46		2	B5 V	401
45229	−56 01072	HR 2320		6 22 02	−56 20.5	4	5.60	.006	0.25	.012	0.13	.009	11	A1mA3-A9	15,216,355,2012
256725	+19 01339	LSS 46	⋆A	6 22 03	+19 52.6	2	9.91	.034	0.24	.005	-0.76	.019	3	B0	41,410
256725	+19 01339	LSS 47	B	6 22 04	+19 52.6	2	10.47	.044	0.22	.015	-0.68	.005	3		41,410
256761		LSS 48		6 22 07	+19 43.2	1	12.58		0.31		-0.51		2	G5	410
	−24 03988			6 22 07	−24 30.2	1	10.80		0.08		-0.90		4		1732
44904	+17 01235			6 22 12	+17 01.4	3	7.01	.011	-0.06	.013	-0.24	.012	7	B9	261,401,1733
44975	+0 01409			6 22 13	+00 10.4	1	9.12		0.20		0.09		2	A0	509
44691	+56 01125	HR 2291, RR Lyn		6 22 13	+56 18.9	4	5.64	.015	0.24	.007	0.12	.000	10	A3 m	15,1007,1013,8015
44647	+58 00925	IDS06181N5828	C	6 22 15	+58 26.9	1	7.90		1.06		0.81		2	G9 III	3024
256714	+29 01217			6 22 16	+29 36.0	1	8.81		0.68		0.28		5	G0	1371
		POSS 122 # 2		6 22 16	+54 21.1	1	15.02		1.38				2		1739
45145	−36 02873	HR 2316	⋆A	6 22 17	−36 40.8	3	5.61	.008	1.03	.008			11	K1 II/III	15,1075,2007
		LSS 50		6 22 25	+19 56.6	1	11.01		0.36		0.06		2	B3 Ib	410
44965	+11 01162	LS VI +11 009		6 22 26	+11 42.8	2	7.83	.004	0.30	.009	-0.44	.000	5	B3 II	1012,8100
43145	+83 00164			6 22 26	+83 47.9	1	9.16		0.38		0.13		2	F0	1733
45029	−7 01413			6 22 26	−07 34.4	1	8.55		-0.10		-0.44		2	B8	401
		LSS 51		6 22 27	+19 47.7	1	11.75		0.35		-0.63		2	B3 Ib	410
44708	+58 00927	HR 2293	⋆A	6 22 27	+58 26.8	3	5.21	.009	1.54	.004	1.88	.000	10	K4 III	1118,1355,3024
44708	+58 00927	IDS06181N5828	B	6 22 27	+58 26.8	1	11.90		0.54		0.10		2		3024
45011	−0 01286			6 22 28	−00 19.8	1	8.54		0.03		-0.10		2	A0	509
45010	+0 01410			6 22 29	+00 25.4	1	8.81		0.09		-0.01		2	A0	509
45291	−52 00913	HR 2322		6 22 30	−52 09.2	3	5.97	.004	1.04	.008			11	G8 III	15,1075,2035
44990	+7 01273	HR 2310, T Mon		6 22 31	+07 06.9	7	5.73	.074	0.94	.079	0.62	.060	7	F7 Iab	688,934,1355,1587,1754*
44927	+23 01347	HR 2304	⋆	6 22 31	+23 21.4	2	6.05		-0.01	.020	0.02	.044	5	A2 Vn	1049,1733
45228	−45 02481	IDS06211S4554	A	6 22 31	−45 55.3	2	7.86	.015	0.71	.010	0.14		8	G6 V	158,2012
44472	+70 00401	HR 2285	⋆AB	6 22 32	+70 33.9	1	6.04		0.10		0.13		4	A4 V	1733
295035	−4 01497	NGC 2232 - 36		6 22 33	−04 38.1	1	10.23		-0.02		-0.06		3	A0	163
291625	+0 01412	G 106 - 46		6 22 34	+00 40.9	1	9.09		0.57		-0.03		1	F5	1620
		LS V +20 040		6 22 34	+20 07.6	1	11.45		0.28		-0.63		1		41
		RL 62		6 22 34	+21 23.1	2	12.44	.002	0.21	.012	-0.34	.023	4		483,1709
291628	+0 01413			6 22 35	+00 24.7	1	9.68		1.12		0.83		2	K0	509
45095	−13 01483			6 22 35	−13 12.1	1	7.52		-0.03		-0.02		1	G2 V	427
44984	+14 01283	HR 2308, BL Ori		6 22 37	+14 45.1	8	6.21	.052	2.34	.019	3.36	.100	13	C5 II	15,109,261,864,3001*
45051	−4 01498	NGC 2232 - 16	⋆AB	6 22 37	−04 24.4	1	8.93		0.37		0.09		5	A3	163
295044	−4 01499	NGC 2232 - 30		6 22 40	−04 50.0	1	9.71		0.11		0.10		3	A0	163
44983	+17 01237			6 22 41	+17 49.9	1	9.16		0.14		0.00		3	A0	261
45461	−63 00561	HR 2337		6 22 42	−63 39.3	2	6.28	.005	1.61	.010	2.01		6	M1 III	2007,3055
45050	+1 01332	HR 2312	⋆AB	6 22 43	+01 31.8	3	6.66	.008	-0.04	.005	-0.22	.005	8	B9 V	15,258,1415
44974	+21 01232			6 22 43	+21 40.5	2	6.53	.000	0.93	.005	0.68	.005	5	K0 III	1501,1648
45067	−0 01287	HR 2313		6 22 43	−00 54.8	7	5.87	.007	0.56	.007	0.06	.008	27	F9 V	15,509,1311,1415,1509*
45184	−28 02981	HR 2318		6 22 47	−28 45.0	2	6.38	.005	0.62	.000			7	G2 V	15,2029
256940	+19 01346	LSS 53		6 22 48	+19 44.0	1	12.04		0.28		-0.62		2	F5	410
44901	+42 01552			6 22 48	+41 59.4	1	7.05		1.07		0.78		2	G5	985
295045	−4 01500	NGC 2232 - 32		6 22 49	−04 47.3	1	9.90		0.06		-0.23		3	B9	163
45348	−52 00914	HR 2326		6 22 50	−52 40.1	7	-0.72	.014	0.16	.005	0.07	.071	56	F0 II	15,198,200,1034,2016*
45289	−42 02503			6 22 53	−42 49.8	5	6.65	.009	0.67	.008	0.18	.019	15	G5 V	1075,1311,2012,2017,3077
45046	+13 01242			6 22 55	+13 52.2	2	8.13	.005	1.01	.010	0.88		5	K0 II	882,8100
		RL 63		6 22 59	+20 53.6	1	15.38		0.46		-0.51		2		704
	−32 03010			6 22 59	−32 05.9	1	9.49		0.55		-0.03		3	G4	803
45023	+25 01272			6 23 01	+25 32.5	1	8.59		-0.01		-0.23		2	A2	105
45450	−58 00692	HR 2336		6 23 02	−58 31.0	2	6.47	.005	0.12	.000			7	A3 V	15,2012
45306	−40 02440	HR 2323		6 23 07	−40 15.3	2	6.30	.005	-0.05	.000			7	B9.5V	15,2012
45669	−69 00607	HR 2352		6 23 07	−69 57.4	3	5.55	.007	1.51	.000			13	K5 III	15,1075,2038
45138	−0 01288			6 23 08	−00 25.5	1	8.40		0.39		-0.04		2	F0	509
45089	+15 01191			6 23 09	+15 10.8	2	7.17	.000	1.20	.000	1.22	.000	2	K0 III	1738,1769
45006	+29 01221			6 23 09	+29 39.6	1	8.90		0.03		-0.07		4	B9 V	1371
45137	+2 01227	HR 2315		6 23 10	+02 18.1	4	6.51	.008	-0.03	.004	-0.06	.013	11	A0 V	15,258,1078,1079
		LSS 54		6 23 11	+19 52.2	1	12.50		0.28		-0.44		2	B3 Ib	410
45153	−4 01501	NGC 2232 - 5		6 23 11	−04 48.2	1	7.30		-0.07		-0.29		5	B7 V	163
45088	+18 01214	OU Gem	⋆AB	6 23 14	+18 47.3	4	6.79	.019	0.95	.009	0.69	.023	24	K0	1080,1355,1738,1758
45136	+4 01251	IDS06206N0435	AB	6 23 15	+04 33.2	1	7.97		0.02		0.00		2	A0	401
257112	+20 01416		V	6 23 17	+20 13.6	1	9.72		1.78				2	M0	369
45168	−3 01425	HR 2317		6 23 18	−03 51.6	2	6.34	.005	1.02	.000	0.78	.005	7	K0 III	15,1061
45151	+4 01252			6 23 19	+04 27.1	1	7.89		-0.10		-0.50		2	B5/6V	401
45121	+20 01418			6 23 27	+20 05.1	1	9.19		0.16		0.12		2	A3	1648
291630	+0 01415			6 23 29	+00 03.5	1	9.74		0.21		0.05		1	B9	509
291634	−0 01289			6 23 30	−00 43.8	1	9.43		0.60		0.05		2	F8	509
45042	+39 01632			6 23 32	+39 00.3	1	8.32		0.40		0.02		2	F0 V	105
295095	−4 01503	NGC 2232 - 26		6 23 34	−04 23.1	1	9.52		0.93		0.69		3	K0	163
45239	−7 01422	HR 2321		6 23 34	−07 51.9	4	6.40	.004	0.15	.005	0.12	.005	21	A4 V	15,1075,1415,2027
45166	+8 01332	LS VI +08 008		6 23 36	+08 00.3	2	9.89	.006	-0.07	.003	-0.73	.017	6	Bpe	1012,1732
45557	−60 00608	HR 2345		6 23 37	−60 15.2	2	5.79	.005	0.00	.000			7	A0 V	15,2012

Table 1   339

HD	DM	Other Id	N Rem	α_{1950}	δ_{1950}	S	V	σ_V	B–V	σ_{B-V}	U–B	σ_{U-B}	n	Spectrum	References
		LS V +22 055		6 23 40	+22 33.6	1	11.50		0.28		-0.51		3		41
45238	-4 01504	NGC 2232 - 12		6 23 40	-04 35.9	2	8.41		-0.04		-0.16		4	B8.5V	163
45406	-36 02895			6 23 41	-36 16.1	1	10.03		1.61		1.92		4	F5 V	119
45509	-52 00919	HR 2341		6 23 42	-52 46.6	3	6.53	.023	1.77	.065			11	K5/M0 III	15,1075,2035
		POSS 122 # 6		6 23 43	+54 43.2	1	18.75		1.50				5		1739
		WLS 624 70 # 7		6 23 43	+68 05.2	1	13.15		1.06		0.94		2		1375
45383	-34 02864	HR 2329	⋆ A	6 23 43	-35 02.0	3	6.24	.007	1.35	.013	1.59		9	K3 III	15,2012,3005
45260	-9 01458			6 23 44	-09 22.0	1	9.01		0.07		-0.64		2	B8	401
45382	-29 03090			6 23 45	-29 40.4	1	6.89		-0.02		-0.04		2	A0 V	401
45180	+15 01197	IDS06209N1535	AB	6 23 46	+15 33.2	2	6.88	.000	-0.03	.005	-0.16	.020	5	B9 V	261,401
295094	-4 01505	NGC 2232 - 24		6 23 46	-04 23.8	1	9.51		0.40		0.13		3	F5	163
257310	+5 01238			6 23 48	+05 05.7	1	9.53		1.87				3	K7	369
295098	-4 01506	NGC 2232 - 35		6 23 48	-04 44.0	1	10.08		0.34		0.29		3	A3	163
45284	-7 01424			6 23 48	-07 19.9	1	7.37		-0.10		-0.54		2	B8	401
295096	-4 01507	NGC 2232 - 25		6 23 49	-04 30.8	1	9.52		0.02		-0.46		4	B5	163
45257	-0 01292			6 23 52	-00 44.8	1	6.78		-0.03		-0.06		2	B9	509
45193	+16 01148			6 23 53	+16 29.4	1	9.16		0.19		-0.19		3		261
45256	-0 01291			6 23 53	-00 29.6	1	9.35		0.15		0.13		2	A2	509
	-60 00611			6 23 53	-60 08.2	2	11.08	.000	0.42	.015	-0.24	.010	4		1696,3062
	-4 01508	NGC 2232 - 34		6 23 56	-04 29.3	1	10.02		1.32		0.87		4		163
45439	-35 02894			6 23 56	-35 40.0	1	7.88		-0.06		-0.37		2	Ap Si	401
45282	+3 01247			6 24 03	+03 27.4	3	8.03	.008	0.66	.005	-0.01	.030	7	G0	742,1594,3077
45105	+47 01299	HR 2314		6 24 05	+47 26.2	2	6.55		-0.04		-0.11	.014	7	A0 V	1079,1733
45701	-63 00568	HR 2354		6 24 05	-63 23.9	4	6.46	.007	0.66	.000			18	G3 III/IV	15,1075,1425,2027
45321	-4 01510	NGC 2232 - 2	⋆	6 24 06	-04 34.0	6	6.15	.009	-0.15	.006	-0.63	.008	26	B3 V	15,154,163,1078,1732,2029
		NGC 2232 - 31		6 24 06	-04 43.	1	9.72		1.56		1.10		4		163
45319	-0 01295			6 24 07	-01 02.6	1	7.76		-0.10		-0.44		1	A0	509
45320	-1 01242	HR 2324		6 24 08	-01 28.6	2	5.86	.005	0.08	.000	0.08	.005	7	A3 Vn	15,1256
257288	+26 01230			6 24 11	+26 42.0	1	9.84		0.46				2	F8	1724
257395	+8 01337			6 24 12	+08 08.0	1	9.44		0.00		-0.46		2	B8	401
	+15 01200		V	6 24 12	+15 56.0	1	9.39		2.01				3	M3	369
		BPM 4225		6 24 18	-75 40.	1	15.38		0.26		-0.73		1		3065
45192	+32 01300	HR 2319		6 24 19	+32 35.7	1	6.43		1.26				2	K0	71
45380	-7 01429	HR 2328	⋆ AB	6 24 20	-07 28.8	3	6.27	.004	-0.03	.010	-0.09	.005	11	A0 Vn	15,1415,2035
		SW Mon		6 24 21	+05 24.3	1	9.08		1.89		1.90		1	M4 III	3076
45357	+0 01421	HR 2327		6 24 24	+00 52.3	2	6.70	.005	0.04	.000	0.00	.000	7	A1 Vn	15,1061
45314	+14 01296	LS VI +14 007		6 24 24	+14 55.2	3	6.63	.010	0.15	.005	-0.88	.000	7	O9:pe	1011,1012,1212
45572	-48 02308	HR 2348	⋆ AB	6 24 24	-48 08.8	11	5.76	.008	-0.06	.004	-0.16	.009	59	B9 V	14,15,182,278,388,611*
45379	-0 01298			6 24 25	-01 01.3	3	8.09	.004	0.31	.012	0.04	.018	6	F3 III	401,509,2033
45237	+30 01232			6 24 27	+30 40.6	1	6.96		1.16		1.27		1	K0 IV	245
	-45 02505			6 24 29	-45 55.2	1	10.18		0.59		0.04		1		938
45398	-4 01512	NGC 2232 - 4		6 24 30	-04 25.7	1	6.84		1.38		1.42		5	K0	163
45399	-4 01513	NGC 2232 - 13		6 24 30	-04 35.6	1	8.47		-0.03		-0.21		5	B8.5V	163
45418	-4 01514	NGC 2232 - 3		6 24 32	-04 19.5	2	6.50	.000	-0.04	.005	-0.56		9	B4 V	163,2033
	+0 01424		V	6 24 35	+00 50.6	1	9.87		2.05				4	M2	369
45417	-2 01613			6 24 36	-03 01.5	1	8.37		1.68				4	K5	369
45796	-63 00572	HR 2360		6 24 37	-63 47.9	3	6.26	.016	-0.14	.007	-0.46		9	B6 V	15,401,2012
45378	+7 01293			6 24 39	+06 57.9	1	8.50		-0.03		-0.14		2	A0	401
45416	-0 01426	HR 2334		6 24 40	-00 19.8	6	5.19	.005	1.18	.018	1.13	.014	21	K1 II	15,678,1078,1355,1509,8100
45434	-4 01516	NGC 2232 - 28		6 24 40	-04 30.6	1	9.58		0.14		0.11		6	A1 V	163
257467	+24 01271	G 104 - 38		6 24 42	+24 21.3	2	9.35	.010	0.81	.002	0.42	.006	3	K0	1620,7010
45433	-0 01299	HR 2335		6 24 42	-00 14.7	2	5.54	.005	1.38	.000	1.61	.000	7	K5 III	15,1415
45435	-4 01517	NGC 2232 - 18		6 24 42	-04 44.9	1	9.18		0.05		0.05		6	B9.5V	163
45415	+2 01237	HR 2333		6 24 43	+02 56.4	3	5.54	.005	1.04	.004	0.87	.005	9	G9 III	15,1078,1355
257581	+6 01249			6 24 44	+06 55.6	1	9.20		0.26		0.22		1	F0	401
45352	+20 01427	IDS06218N2051	A	6 24 48	+20 49.3	1	6.52		1.26		1.28		3	K0	1648
	-4 01519	NGC 2232 - 27		6 24 54	-04 54.5	1	9.52		0.03		-0.44		3		163
45394	+20 01428	HR 2330		6 24 58	+20 31.7	1			0.01		0.18		3	A2 Vs	1049
257703	+5 01244			6 25 00	+05 34.5	1	9.72		0.00				1	A0	6009
		ApJ144,259 # 15		6 25 00	-25 21.	1	15.72		-0.31		-1.24		2		1360
		G 105 -B2	A	6 25 01	+10 02.5	1	13.25		1.48		1.23		3		3016
		G 105 -B2	B	6 25 01	+10 02.5	1	16.58		0.21		-0.72		2	DF	3060
45311	+39 01635			6 25 04	+39 08.6	1	7.82		0.34		0.08		5	F2 III	833
45351	+34 01355			6 25 10	+34 31.5	1	7.73		0.15		0.17		2	A3	1625
45588	-25 03237	HR 2349	⋆ A	6 25 10	-25 49.3	4	6.06	.003	0.52	.015	0.03	.000	14	F9 V	15,158,265,2012
45588	-25 03237	IDS06231S2547	B	6 25 10	-25 49.3	1	11.57		1.04		0.14		4	K7	265
257779	+7 01298			6 25 11	+07 09.7	1	8.83		0.09		-0.41		2	B8	401
257726	+15 01205			6 25 14	+15 20.8	1	9.42		1.94				3	M0	369
45610	-32 03046			6 25 14	-32 08.0	2	7.18	.016	0.80	.009	0.24		4	G3 IV/V(w)	1594,6006
45515	-2 01617			6 25 15	-02 37.7	1	7.88		-0.03		-0.15		2	B8	401
257684		SX Gem		6 25 17	+20 35.8	2	10.20	.010	1.75	.005			2	A0	8022,8027
45350	+39 01637			6 25 18	+38 59.8	1	7.87		0.74		0.36		2	K0 III	105
		NGC 2232 - 22		6 25 18	-04 40.3	1	9.28		-0.02		-0.06		5		163
45516	-4 01522	NGC 2232 - 8		6 25 20	-04 47.5	2	7.82	.010	-0.08	.000	-0.33	.029	7	B7 V	163,401
45517	-4 01523	NGC 2232 - 7		6 25 20	-04 54.0	1	7.56		0.15		0.14		4	A0	163
45412	+30 01238	HR 2332, RT Aur		6 25 21	+30 31.6	4	5.29	.276	0.55	.141	0.39	.115	6	F5.5Ib	245,592,1363,6007
45507	+6 01253			6 25 22	+05 57.3	1	7.74		1.03		0.49		1	G5	934
45609	-25 03240	IDS06233S2541	AB	6 25 22	-25 42.2	1	8.36		0.84				4	K0 V	1414
45532	-5 01644	NGC 2232 - 10		6 25 23	-05 07.4	2	8.06	.014	-0.06	.000	-0.27	.028	8	B7 V	163,401
45391	+36 01442	G 101 - 41		6 25 24	+36 31.0	4	7.15	.005	0.60	.005	0.02	.025	7	G0	908,1601,1620,3026

HD	DM	Other Id	N	Rem	α_{1950}	δ_{1950}	S	V	σ_V	B–V	σ_{B-V}	U–B	σ_{U-B}	n	Spectrum	References
	−4 01525	NGC 2232 - 29	V		6 25 25	−04 44.6	1	9.20		0.61		0.30		5	A2	163
45680	−37 02837	HR 2353		★A	6 25 25	−37 51.9	1	6.48		0.39				4	F3 V	2007
		NGC 2232 - 41			6 25 26	−04 47.9	1	10.98		0.41		-0.07		4		163
45547	−6 01560				6 25 26	−06 15.6	1	8.74		-0.01		-0.10		2	B9	401
45514	+6 01254				6 25 29	+05 59.5	1	7.37		1.57		1.52		1	K2	934
		NGC 2232 - 40			6 25 29	−04 40.8	1	10.72		0.43		0.00		3		163
45546	−4 01526	NGC 2232 - 1		★A	6 25 29	−04 43.8	8	5.05	.012	-0.18	.007	-0.75	.008	28	B2 V	15,154,163,1075,1425*
		GD 74			6 25 30	+41 32.8	1	14.99		-0.03		-0.81		1		3060
295103		NGC 2232 - 39			6 25 30	−04 51.9	1	10.34		0.16		0.13		3		163
295104	−4 01527	NGC 2232 - 17			6 25 30	−04 59.4	1	8.95		0.97		0.78		4	G5	163
45644	−26 03038				6 25 30	−26 02.7	1	8.84		-0.14		-0.64		5	B3 V	1732
295102	−4 01528	NGC 2232 - 14			6 25 31	−04 45.8	1	8.68		-0.05		-0.16		5	B9 V	163
45512	+10 01149	HR 2342			6 25 33	+10 20.2	2	6.15	.011	1.15	.004	1.18		5	K2 III-IV	37,71
45530	+5 01249	V648 Mon		★AB	6 25 34	+05 18.3	1	7.41		-0.04		-0.24		2	A1p	401
45565	−4 01529	NGC 2232 - 19			6 25 34	−04 42.6	1	9.24		0.06		0.05		5	B9.5V	163
		NGC 2232 - 43			6 25 34	−04 48.9	1	11.36		0.44		0.28		3		163
45506	+16 01159	HR 2340			6 25 35	+16 16.3	3	6.23	.013	0.89	.012	0.58	.025	7	G5	71,261,3077
45485	+25 01287				6 25 39	+25 39.2	1	8.55		0.00		-0.04		2	A0	105
45585	−6 01564				6 25 39	−06 53.6	2	8.79	.010	0.09	.010	-0.35		6	B5	401,2033
	−13 01503				6 25 39	−13 04.7	1	10.20		0.46		-0.01		2		272
45563	+2 01244	HR 2347			6 25 41	+01 56.7	3	6.47	.005	-0.06	.011	-0.16	.015	9	B9 V	15,1079,1415
45629	−13 01504				6 25 41	−13 07.7	1	7.10		-0.04		-0.32		1	B9 III	272
45583	−4 01530	NGC 2232 - 9		★V	6 25 43	−04 52.0	2	7.98	.014	-0.11	.033	-0.45	.028	8	B9p	163,401
295099	−4 01531	NGC 2232 - 33			6 25 44	−04 29.7	1	9.95		0.15		0.15		3	A3	163
		NGC 2232 - 42			6 25 45	−04 48.0	1	11.00		0.36		0.02		3		163
257787	+29 01236				6 25 46	+29 48.0	1	9.29		1.12		1.01		5	K2	1371
295100	−4 01532	NGC 2232 - 38			6 25 47	−04 34.3	1	10.30		0.29		0.09		4	A5	163
45504	+27 01122	HR 2339			6 25 49	+27 00.1	2	6.58	.005	0.51	.005	0.02	.000	5	F9 IV-V	1625,1733
257971	+11 01191	LS VI +11 010			6 25 52	+11 19.8	1	8.90		0.06		-0.79		2	B0.5III	1012
45626	−4 01534	NGC 2232 - 21		★A	6 25 54	−04 25.8	4	9.26	.055	0.05	.019	-0.46	.008	11	B7p e Shell	163,379,490,1012
46116	−69 00614	HR 2377			6 25 54	−69 39.7	5	5.37	.005	0.97	.006	0.68	.005	21	G8 III	15,1075,2012,2038,3051
45627	−4 01535	NGC 2232 - 15			6 25 55	−05 00.1	1	8.88		0.06		0.04		4	A0 V	163
257937	+20 01440	IDS06230N2017		BC	6 25 56	+20 16.4	2	8.00		0.11	.005	0.04	.040	6	A0	1084,1211
45660	−9 01475				6 25 58	−09 47.5	1	8.34		0.90				1	G5	207
257886	+27 01124	G 103 - 34			6 25 59	+27 02.9	4	8.60	.015	0.90	.014	0.60	.010	11	K2 V	22,333,1003,1620,1775
45677	−12 01500	FS CMa			6 25 59	−13 01.2	4	8.24	.059	0.04	.010	-0.68	.007	25	Bpec	815,1212,1588,8035
45542	+20 01441	HR 2343		★AP	6 26 00	+20 14.7	10	4.15	.008	-0.12	.025	-0.49	.015	31	B6 IIIe	15,41,154,1084,1203,1211*
45659	−6 01568				6 26 00	−06 30.9	1	8.04		-0.04		-0.19		2	B8	401
		G 106 - 48		AB	6 26 02	+02 18.0	1	11.12		1.69				1		1705
	−12 01502				6 26 03	−12 49.6	1	8.96		0.33		0.01		1	F0	272
45658	−4 01536	NGC 2232 - 11			6 26 05	−04 32.7	2	8.16	.109	0.35	.004	0.13	.013	6	F0	163,379
45892	−52 00930				6 26 05	−52 54.4	1	8.42		1.00		0.77		10	G8 III	1673
45676	−9 01476				6 26 08	−09 43.5	1	9.38		0.55				1	F5	207
45580	+17 01260	G 105 - 38			6 26 09	+17 46.9	1	7.62		0.60		0.05		1	G0	1620
258001	+18 01233				6 26 09	+18 24.5	1	9.26		0.02		-0.40		3	A0	261
45388	+55 01082				6 26 10	+55 51.1	1	8.25		1.15		1.22		2	K2 III	1502
257925	+30 01243				6 26 11	+30 47.4	1	10.67		0.46		0.06		11	F5	272
45638	+11 01193	HR 2351			6 26 13	+11 03.2	1	6.58		0.28		0.02		2	A9 IV	1733
45674	−0 01308	LS VI -00 002			6 26 14	−00 32.3	1	6.54		0.79				2	F0	6009
45597	+17 01261				6 26 15	+17 31.8	1	7.80		1.57		1.78		3	K0	261
45411	+55 01083				6 26 16	+55 43.9	1	7.75		1.52		1.82		2	K2	1502
45733	−12 01503				6 26 16	−13 01.0	1	7.52		0.32		0.08		2	Fm δ Del	272
45691	−4 01537	NGC 2232 - 20			6 26 17	−04 17.7	1	9.24		0.05		0.03		5	B9 V	163
45984	−57 01001	HR 2369			6 26 17	−57 58.2	3	5.82	.008	1.28	.000			11	K3 III	15,1075,2035
		G 101 - 42			6 26 18	+35 55.7	1	9.95		1.00		0.73		2		7010
45466	+46 01149	HR 2338			6 26 19	+46 43.2	1	5.90		1.44				2	gK4	71
	+46 01148				6 26 19	+46 56.0	1	9.87		0.05		0.00		2		1601
	−22 03113				6 26 19	−22 54.1	1	10.84		1.04		0.86		2	K3 V	3072
45813	−32 03066	HR 2361			6 26 19	−32 32.8	5	4.47	.010	-0.17	.009	-0.61	.004	25	B4 V	15,1075,1637,2012,8015
		AA131,200 # 9			6 26 20	−10 29.7	1	11.09		0.52		-0.13		2		7005
45725	−6 01574	HR 2356		★A	6 26 23	−07 00.0	1	4.60		-0.10		-0.63		3	B3 Ve	1212
45725	−6 01574	IDS06240S0658		ABC	6 26 23	−07 00.0	6	3.75	.012	-0.16	.007	-0.73	.005	19	B3 Ve + B2	15,154,1075,1425,2012,8015
45726	−6 01575	HR 2357		★BC	6 26 24	−07 00.0	1	5.40		-0.07		-0.52		3	B2	1212
45728	−8 01443				6 26 24	−08 26.8	1	8.30		-0.04		-0.39		2	B9	401
45765	−17 01506	HR 2359			6 26 24	−17 26.0	3	5.76	.004	1.12	.000			11	K0 III	15,1075,2035
45729	−9 01478				6 26 26	−09 22.2	1	9.70		0.14				1	A0	207
45410	+58 00932	HR 2331		★A	6 26 27	+58 12.1	3	5.87	.013	0.93	.009	0.66	.005	6	K0 IV	15,1003,1355
46291	−71 00441	AE Men			6 26 32	−72 00.7	1	8.30		1.13		0.96		1	K2 III F/G	1641
46043	−60 00623				6 26 33	−60 04.4	1	8.25		0.24		0.15		31	A6 V	978
45764	−11 01500				6 26 36	−11 13.8	1	7.79		-0.03		-0.49		2	B8	401
	−53 01089				6 26 37	−53 10.0	1	9.46		1.32		1.39		1		6006
45724	+2 01253	HR 2355		★A	6 26 38	+02 40.8	5	6.15	.010	1.54	.009	1.85	.013	16	M1	15,369,1355,1415,3001
45688	+17 01264	IDS06238N1703		AB	6 26 43	+17 00.7	1	7.88		1.10		0.90		3	G5	261
	−9 01480				6 26 43	−09 14.8	1	9.20		1.00				1	G5 IV:	207
		NGC 2236 - 167			6 26 44	+06 54.3	1	14.28		1.38		1.39		1		1741
45772	−3 01453				6 26 44	−03 07.4	1	8.52		1.40				2	K5	369
		NGC 2236 - 172			6 26 45	+06 53.3	1	15.34		0.61		0.14		1		1741
		NGC 2236 - 168			6 26 45	+06 54.0	1	14.49		0.58		0.01		1		1741
		NGC 2236 - 165			6 26 46	+06 55.0	1	14.59		0.67		0.68		1		1741
		NGC 2236 - 178			6 26 48	+06 51.8	1	15.07		0.55		0.05		1		1741

Table 1 341

HD	DM	Other Id	N Rem	α_{1950}	δ_{1950}	S	V	σ_V	B–V	σ_{B-V}	U–B	σ_{U-B}	n	Spectrum	References
		NGC 2236 - 164		6 26 48	+06 55.4	1	14.59		0.53		-0.05		1		1741
45871	-32 03072	HR 2364	⋆ AB	6 26 48	-32 20.3	3	5.74	.013	-0.16	.004	-0.68	.016	10	B4 Vnpe	401,1637,2035
		NGC 2236 - 203		6 26 49	+06 50.0	1	12.55		0.31		-0.42		1		1741
		NGC 2236 - 183		6 26 50	+06 51.1	1	14.53		0.71		0.40		1		1741
		G 106 - 49	⋆ V	6 26 51	-02 46.4	3	11.11	.021	1.61	.188	1.23	.058	3		680,801,3003
		NGC 2236 - 245		6 26 52	+06 48.6	1	12.61		0.64		-0.56		1		1741
258335		NGC 2244 - 371		6 26 55	+05 10.0	1	10.48		0.24		0.18		1	A3	892
		NGC 2236 - 80		6 26 55	+06 51.8	1	14.73		0.65		0.06		1		1741
	-4 01539	LS VI -04 003		6 26 55	-04 06.6	1	10.04		0.50		0.54		2		490
		NGC 2236 - 2		6 26 58	+06 51.4	1	14.02		0.32		-0.25		1		1741
45832	-9 01484			6 26 58	-09 52.7	1	9.18		1.66				1	M1	207
		NGC 2236 - 1		6 27 00	+06 51.7	1	10.72		0.32		-0.06		1		1741
45841	-9 01486			6 27 05	-09 30.4	1	8.90		0.34				1	F0	207
45757	+17 01268	IDS06242N1758	AB	6 27 06	+17 56.5	1	7.55		0.06		-0.08		3	A0	261
45983	-40 02482	HR 2368		6 27 06	-41 02.5	2	6.31	.005	0.41	.005			7	F2/3 V	15,2012
		Steph 598		6 27 07	+50 05.1	1	11.09		1.50		1.14		1	M0	1746
258394	+13 01279	BK Gem		6 27 09	+13 39.0	1	8.26		1.60		1.71		2	K5	1648
45856	-9 01489			6 27 10	-09 40.8	1	10.16		0.13				1	B9	207
		NGC 2244 - 369		6 27 11	+05 14.6	1	12.33		0.67		0.11		2		892
		NGC 2236 - 18		6 27 12	+06 51.6	1	12.76		0.66		-0.58		1		1741
45721	+28 01138			6 27 13	+28 14.8	1	6.86		0.14		0.13		3	A1 V	1501
		NGC 2244 - 370		6 27 14	+05 13.6	1	11.45		0.45		0.02		1		892
258429		NGC 2244 - 368		6 27 14	+05 15.6	1	10.59		0.08		0.13		3	B9	892
45789	+7 01314	IDS06245N0711	A	6 27 14	+07 08.8	1	7.11		-0.13		-0.67		3	B2.5IV-V	493
45789	+7 01314	IDS06245N0711	AB	6 27 14	+07 08.8	3	7.09	.007	-0.13	.000	-0.66	.035	8	B2.5IV-V	401,493,2033
258423	+10 01154			6 27 16	+10 21.3	1	7.89		1.42		1.76		8	G5	1655
		MtW 50 # 181		6 27 16	+29 52.0	1	13.69		0.56		0.02		4		397
45829	+8 01367			6 27 19	+07 57.4	4	6.63	.004	1.58	.012	1.62	.022	9	K0 Iab	1355,1754,6007,8100
45852	+2 01260	Cr 96 - 3		6 27 21	+02 48.8	1	9.32		0.00		-0.08		1	A0	410
45827	+9 01259	HR 2362	⋆	6 27 21	+09 03.9	3	6.55	.012	0.15	.010	-0.03	.042	10	A0 III	15,493,1415
		LS VI -01 004		6 27 21	-01 33.6	1	10.61		0.49		0.00		1		490
	-31 03344			6 27 21	-31 14.5	1	12.12		1.06		0.77		3		543
	-31 03345			6 27 21	-31 16.0	1	11.54		1.23		1.26		3		543
		MtW 50 # 243		6 27 22	+29 51.9	1	15.99		0.42		0.16		4		397
45883	-9 01490			6 27 23	-09 17.0	1	9.11		1.02				1	K0	207
		Cr 96 - 11		6 27 24	+02 47.7	1	12.01		0.55		0.00		1		410
		NGC 2244 - 373		6 27 24	+05 06.7	1	12.39		0.78		0.28		2		892
258491	+7 01316			6 27 25	+06 56.9	2	8.90	.014	1.73	.007			25	M2	369,6011
288744	+2 01261			6 27 27	+02 26.3	1	9.21		1.86				3	M3	369
45839	+6 01266			6 27 27	+06 55.7	1	8.72		0.00				22	A0	6011
		NGC 2244 - 372		6 27 28	+05 07.9	1	11.93		0.41		0.39		2		892
		NGC 2243 - 4301		6 27 29	-31 15.1	2	14.18	.000	0.84	.020	0.55	.064	5		427,543
		NGC 2243 - 4303		6 27 30	-31 15.4	2	13.72	.015	0.91	.030	0.51	.010	4		427,543
46039	-40 02486	IDS06259S4018	A	6 27 30	-40 20.3	1	7.88		1.25		1.13		2	K2 IV/V	1730
		Cr 96 - 8		6 27 31	+02 54.0	1	10.85		1.03		0.86		1		410
258452	+19 01375	G 104 - 43		6 27 31	+19 19.3	1	8.76		0.56		-0.01		1	G0	1620
		Cr 96 - 9		6 27 32	+02 49.4	1	11.25		1.34		1.17		1		410
		MtW 50 # 359		6 27 32	+29 51.6	1	11.74		0.24		0.20		4		397
		MtW 50 # 166		6 27 32	+29 51.7	1	15.72		0.47		0.27		4		397
295145	-4 01544	NGC 2232 - 37		6 27 32	-04 36.7	1	10.24		0.33		0.05		3	F8	163
	-31 03347			6 27 32	-31 12.1	2	11.20	.005	0.64	.010	0.18	.005	14		427,543
		Cr 96 - 7		6 27 33	+02 53.9	1	11.50		0.28		-0.16		1		410
	-45 02540		V	6 27 33	-45 19.4	2	10.75	.004	1.34	.014	1.24		9	K5	863,1075
		NGC 2244 - 374		6 27 34	+04 49.4	1	11.99		1.25		1.11		1		892
45784	+29 01248			6 27 34	+29 51.9	1	8.05		0.35		0.15		4	F2 V	397
		Cr 96 - 10		6 27 35	+02 51.4	1	12.74		0.58		0.05		1		410
		HRC 200	V	6 27 35	+10 33.9	1	10.89		0.89		-0.15		1		776
46040	-40 02487	IDS06259S4018	B	6 27 35	-40 20.8	1	7.88		1.25		1.14		1	K0p Ba	565
		NGC 2243 - 4209		6 27 36	-31 15.2	2	12.02	.015	1.41	.025	1.42	.025	4		427,543
	-31 03348			6 27 37	-31 05.2	1	10.59		1.02		0.74		4		543
	-31 03349	NGC 2243 - 2307		6 27 37	-31 14.1	1	12.18		0.79		0.49				427
		NGC 2243 - 4110	AB	6 27 37	-31 14.9	2	12.85	.005	1.11	.005	0.89	.099	5		427,543
		Cr 96 - 12		6 27 39	+02 50.4	1	13.47		0.32		0.33		1		410
288733		Cr 96 - 6		6 27 39	+02 50.7	1	11.01		0.10		0.12		1	A2	410
258590	+5 01266	NGC 2244 - 366		6 27 39	+05 26.7	1	10.60		0.12		0.16		2	A2	892
	-31 03350	NGC 2243 - 1219		6 27 39	-31 14.9	2	11.79	.015	1.07	.015	0.98	.019	4		427,543
45901	+2 01262	Cr 96 - 1		6 27 41	+02 53.0	3	8.79	.020	0.31	.020	-0.66	.013	10		399,410,493
258582	+9 01262			6 27 42	+09 03.6	1	9.02		1.81				3	K7	369
		vdB 80 # 24		6 27 42	-09 48.4	1	11.13		0.29				1	A5 V	207
		NGC 2243 - 4236		6 27 42	-31 17.2	2	15.14	.010	0.88	.010	0.43		3		427,543
		NGC 2244 - 367		6 27 43	+05 24.9	1	11.92		0.63		0.14		1		892
		Cr 96 - 5		6 27 45	+02 52.7	1	11.45		0.40		0.28		1		410
258546	+17 01272	LSS 70		6 27 45	+17 16.9	2	11.58	.000	0.15	.000	-0.57	.000	4	B3 III	405,1737
		Cr 96 - 14		6 27 48	+02 47.5	1	11.60		0.31		-0.34		1		410
45977	-11 01512			6 27 48	-11 46.5	1	9.14		1.12		1.07		2	K4 V	3072
		Cr 96 - 13		6 27 49	+02 49.9	1	12.63		0.53		0.25		1		410
258545	+19 01376			6 27 49	+19 33.2	1	9.12		1.76				3	M0	369
45976	-9 01493	HR 2367		6 27 49	-10 02.8	4	5.92	.026	1.38	.018			12	K0	15,207,1075,2035
45911	+4 01282	IDS06252N0424	AB	6 27 50	+04 22.1	2	7.36	.045	-0.14	.000	-0.71		9	B2 IV-V	493,2033
258660	+4 01281	NGC 2244 - 375		6 27 50	+04 47.5	1	9.52		1.89		2.38		2	G8 V:	892

HD	DM	Other Id	N Rem	α_{1950}	δ_{1950}	S	V	σ_V	B–V	σ_{B-V}	U–B	σ_{U-B}	n	Spectrum	References
45910	+5 01267	AX Mon		6 27 52	+05 54.1	5	6.80	.020	0.34	.006	-0.60	.022	37	B2:IIIpe	379,815,1012,1212,8035
45953	-6 01585			6 27 52	-06 23.7	1	7.97		-0.04		-0.22		2	B9	401
		Cr 96 - 4		6 27 53	+02 55.5	1	10.63		0.32		-0.11		1		410
258691	+4 01283	NGC 2244 - 376		6 27 54	+04 43.5	3	9.72	.022	0.55	.023	-0.44	.014	9	B0	493,892,1732
		NGC 2243 - 1121		6 27 54	-31 15.	1	14.30		0.02		0.00		2		543
		NGC 2243 - 1206		6 27 54	-31 15.	2	17.91	.070	0.69	.035			4		427,543
		NGC 2243 - 1229		6 27 54	-31 15.	1	13.69		0.89		0.64		1		543
		NGC 2243 - 1301		6 27 54	-31 15.	1	16.00		0.30				2		543
		NGC 2243 - 1315		6 27 54	-31 15.	1	16.73		0.63				3		543
		NGC 2243 - 1317		6 27 54	-31 15.	1	16.67		1.25				4		543
		NGC 2243 - 2227		6 27 54	-31 15.	1	15.69		0.57		-0.01		2		427
		NGC 2243 - 2230		6 27 54	-31 15.	2	16.61	.009	0.51	.023	-0.06		4		427,543
		NGC 2243 - 2233		6 27 54	-31 15.	1	17.23		0.61				3		427
		NGC 2243 - 2245		6 27 54	-31 15.	1	14.11		0.94				1		427
		NGC 2243 - 2308		6 27 54	-31 15.	2	13.62	.005	0.96	.019	0.54	.019	3		427,543
		NGC 2243 - 2314		6 27 54	-31 15.	2	13.63	.030	1.03	.045	0.89		2		427,543
		NGC 2243 - 3119		6 27 54	-31 15.	1	16.17		0.47		-0.02		2		427
		NGC 2243 - 3208		6 27 54	-31 15.	1	17.36		0.52				1		427
		NGC 2243 - 3216		6 27 54	-31 15.	1	15.86		0.52		0.02		2		427
		NGC 2243 - 3221		6 27 54	-31 15.	2	17.98	.034	0.71	.049			3		427,543
		NGC 2243 - 3222		6 27 54	-31 15.	1	16.92		0.59				2		427
		NGC 2243 - 3316		6 27 54	-31 15.	1	13.72		1.07		0.26		1		543
		Mon R1 # 1		6 27 57	+09 53.8	1	11.20		0.55		0.17		1	F5:	5012
		LP 660 - 11		6 27 57	-05 20.6	1	12.28		0.54		-0.15		2		1696
		NGC 2244 - 377		6 27 58	+04 42.5	1	12.28		0.62		0.37		2		892
45975	-4 01546	NGC 2232 - 6		6 27 59	-04 39.7	2	7.46	.014	-0.06	.005	-0.32	.005	7	B8 V	163,401
45930	+7 01321			6 28 01	+07 27.3	1	8.66		0.11		-0.19		2	A0	401
258686	+10 01158	IDS06253N1008	AB	6 28 02	+10 06.0	2	9.13	.009	-0.06	.000	-0.44	.000	5	B8	206,5012
	-40 02492			6 28 02	-40 19.1	1	11.14		1.26		1.28		2		1730
	-44 02691			6 28 02	-44 12.4	2	10.23	.040	0.44	.005	-0.27		6	G0:	1075,1696
		HRC 201	V	6 28 04	+10 35.3	1	13.36		1.21		0.08		1		776
	+12 01177	DH Gem		6 28 05	+12 31.0	1	9.82		3.89		2.35		2	C4,5	864
46007	-9 01495			6 28 05	-09 21.8	1	9.81		0.41				1	F5	207
		S121 # 3		6 28 05	-59 41.8	1	10.37		1.01		0.62		2		1730
258271	+5 01270	Cr 97 - 2		6 28 06	+05 54.4	1	8.71		0.33		0.01		1	F0	1640
258756	+4 01285	NGC 2244 - 380		6 28 07	+04 38.1	1	9.94		-0.01		-0.02		1	A0	892
		NGC 2244 - 378		6 28 07	+04 42.1	1	13.03		0.43		0.05		2		892
45998	-4 01547	NGC 2232 - 23		6 28 08	-04 43.3	1	9.37		0.27		0.08		3	A2	163
	-31 03358			6 28 08	-31 14.2	2	11.35	.005	0.67	.005	0.16	.000	8		427,543
	-40 02494			6 28 08	-40 24.4	1	9.73		0.63		0.13		2	G0	1730
		Cr 97 - 10		6 28 09	+05 50.9	1	12.87		0.85		0.35		1	K5	1640
	-31 03359			6 28 09	-31 24.1	1	11.20		0.71		0.27		1		543
258643	+19 01378			6 28 10	+18 58.7	1	10.03		1.71				3	M0	369
45972	+5 01271	Cr 97 - 1		6 28 12	+05 50.0	2	8.41	.010	-0.01	.005	-0.17	.175	3	A0	401,1640
258749	+9 01266			6 28 12	+09 50.8	2	9.81	.023	0.09	.009	-0.49	.023	5	B8	206,5012
45996	+4 01286	NGC 2244 - 379	⋆ AB	6 28 13	+04 41.1	1	9.06		-0.02		-0.32		2	B8 V	892
	+52 01088	IDS06243N5229	AB	6 28 15	+52 26.9	1	9.72		1.26		1.17		1	K4	3072
	-31 03362			6 28 15	-31 16.2	1	10.29		1.12		1.05		6		543
45951	+17 01275	HR 2366	⋆ A	6 28 16	+16 58.5	1	6.20		1.16		1.06		2	K2 III	3024
45951	+17 01275	IDS06254N1700	B	6 28 16	+16 58.5	1	10.40		0.76		0.33		3		3024
45951	+17 01275	IDS06254N1700	C	6 28 16	+16 58.5	1	11.41		1.24		0.90		3		3024
		LS VI +01 005		6 28 17	+01 24.8	1	10.04		0.84		-0.03		2		405
46064	-13 01519	HR 2373	⋆ A	6 28 17	-13 06.8	2	6.15	.005	-0.16	.005			7	B2 III	15,2012
	-31 03364			6 28 17	-31 16.3	1	11.53		0.44		-0.02		5		543
	-31 03363			6 28 17	-31 18.5	1	10.42		0.88		0.54		2		543
46332	-59 00658			6 28 17	-59 44.3	1	7.68		1.29		1.37		25	K1 III	978
	-31 03365			6 28 20	-31 05.0	1	9.42		1.48		1.78		2		543
		HRC 202, VY Mon		6 28 21	+10 28.3	2	14.18	.500	1.35	.425	0.29	.915	2		776,5012
		Mon R1 # 4		6 28 21	+10 29.7	1	11.08		0.52		-0.13		1	B2.5V	5012
46005	+10 01159	IDS06256N1001	AB	6 28 23	+09 58.5	1	7.87		0.22		-0.50		3	B8	1732
45995	+11 01204	HR 2370	⋆ A	6 28 23	+11 17.2	5	6.13	.012	-0.09	.014	-0.84	.019	13	B2 V:nne	15,154,542,1012,1212
45995	+11 01204	HR 2370	⋆ AB	6 28 23	+11 17.2	1	6.03		-0.11		-0.79		1	B2 V:e+B9IV	542
45995	+11 01205	IDS06256N1119	B	6 28 23	+11 17.2	2	9.10	.010	-0.05	.000	-0.29	.010	3	B9 IV	542,542
	-31 03367			6 28 23	-31 18.2	1	11.32		0.22		0.05		2		543
		NGC 2244 - 44		6 28 24	+04 49.7	1	12.94		0.47		-0.08		2		892
		Mon R1 # 6		6 28 24	+09 53.7	1	10.26		0.29		-0.37		1	B3 V	5012
		HRC 203, V687 Mon		6 28 24	+10 28.2	2	13.27	.030	1.16	.130	0.37	.275	2		776,5012
		vdB 80 # 23		6 28 26	-09 19.7	1	10.79		1.16				1	G8 III	207
	-9 01497			6 28 26	-09 36.7	1	10.76		0.48		-0.20		4	B8 IV	206
		NGC 2244 - 48		6 28 27	+04 48.7	1	12.44		0.47		0.14		2		892
46060	-9 01498	IDS06261S0935	A	6 28 27	-09 37.1	1	8.56		0.34		-0.50		2	B8	401
46033	+2 01266	Cr 96 - 2		6 28 28	+02 43.8	1	8.59		0.41		-0.04		1	A0	410
46114	-17 01515			6 28 28	-17 53.0	2	7.71	.010	0.82	.015	0.42		6	G8 V	2012,3016
		VZ Mon		6 28 29	+11 53.4	1	13.38		1.50				1		1772
46079	-9 01499			6 28 29	-09 13.0	1	9.86		0.15				1	A0	207
46630	-71 00446			6 28 29	-71 59.4	1	8.42		0.03		0.05		37	A0 V	978
258859		NGC 2244 - 45		6 28 30	+04 49.6	1	10.64		0.18		0.31		2	A0	892
258853	+9 01269	LS VI +09 009		6 28 30	+09 49.6	4	8.84	.010	0.01	.012	-0.58	.006	12	B3	206,493,1732,5012
		CCS 524		6 28 30	-05 29.	1	11.32		3.20				1	CS?	864
	-9 01500			6 28 30	-09 44.5	1	9.87		1.40				1		207

Table 1   343

HD	DM	Other Id	N	Rem	α_{1950}	δ_{1950}	S	V	σ_V	B–V	σ_{B-V}	U–B	σ_{U-B}	n	Spectrum	References
	−31 03369				6 28 30	−31 16.6	2	9.95	.015	0.33	.000	-0.02	.005	3	A8	427,543
258857	+5 01273	Cr 97 - 3			6 28 31	+05 54.8	1	8.69		0.99		0.66		1	K0	1640
		Mon R1 # 5			6 28 31	+10 24.8	1	13.34		0.71		0.51		1		5012
46131	−22 01434				6 28 31	−22 17.1	1	7.15		-0.18		-0.72		3	B3/5 V	1732
		NGC 2244 - 46			6 28 32	+04 49.6	1	12.26		0.50		0.26		2		892
	−31 03370				6 28 32	−31 04.6	1	10.97		0.52		0.02		2		543
		NGC 2244 - 47			6 28 33	+04 49.6	1	13.05		0.57		0.02		2		892
	+40 01623				6 28 33	+40 26.5	1	9.32		1.01		0.77		1	K8	3072
		V 484 Mon			6 28 34	−02 06.7	1	13.25		1.04				1		1772
		Cr 97 - 9			6 28 35	+05 51.0	1	12.26		0.39		0.04		1	F0	1640
46273	−50 02241	HR 2384		★ AB	6 28 35	−50 12.2	1	5.22		0.41				4	F2 V	2035
		Cr 97 - 14			6 28 36	+05 57.	1	15.24		0.85		0.78		1		1640
46355	−56 01095	HR 2389			6 28 36	−56 49.1	3	5.22	.004	1.10	.010			11	K0 III	15,1075,2035
		Cr 97 - 27			6 28 37	+05 51.3	1	13.91		0.74		0.34		1		1640
	+5 01274	Cr 97 - 4			6 28 37	+05 51.9	1	9.76		-0.05		-0.23		1	B5	1640
46057	+4 01290	NGC 2244 - 381			6 28 38	+04 39.6	2	8.93	.055	-0.06	.005	-0.39	.020	4	A0 III	401,892
		GD 75			6 28 38	+47 39.0	1	13.67		0.62		-0.09		1		3060
		NGC 2244 - 78			6 28 39	+04 53.2	1	12.29		0.33		0.13		4		450
		NGC 2244 - 81			6 28 40	+04 52.8	1	12.54		0.32		-0.17		2		450
		NGC 2244 - 83		★ B	6 28 41	+04 52.4	1	10.94		0.26		-0.46		2		450
		Cr 97 - 30			6 28 41	+05 47.7	1	13.81		0.80		0.57		1		1640
46056	+4 01291	NGC 2244 - 84		★ A	6 28 42	+04 52.2	2	8.24	.015	0.20	.009	-0.74	.021	5	O8 V(e)	450,989
46056	+4 01291	NGC 2244 - 84		★ AB	6 28 42	+04 52.2	6	8.17	.009	0.20	.011	-0.75	.014	18		593,1011,1012,1059,1651,1729
46031	+16 01178	HR 2371			6 28 44	+15 56.4	1	6.38		0.25		0.11		2	A8 V	1733
258728	+34 01378	G 103 - 36		★ A	6 28 44	+34 33.8	2	9.82	.009	1.24	.024	1.20		2	K5	333,1017,1620
	+5 01276	Cr 97 - 5			6 28 45	+05 49.9	1	9.57		0.64		0.27		1	F5	1640
		Cr 97 - 8			6 28 45	+05 51.3	1	12.72		0.38		-0.03		1	A0	1640
		Cr 97 - 25			6 28 45	+05 51.3	1	14.51		0.62		0.05		1		1640
258914	+7 01328				6 28 45	+07 00.9	1	9.16		1.74				3	M0	369
46189	−27 03051	HR 2380			6 28 47	−27 44.0	3	5.92	.011	-0.16	.004	-0.65		11	B3 IV/V	15,158,2012
258985	+4 01293	NGC 2244 - 88			6 28 49	+04 46.9	1	9.68		0.03		-0.26		2	B9.5III	493
258987	+4 01294	NGC 2244 - 382			6 28 50	+04 31.8	1	9.41		0.71		0.29		2	G0	892
		RZ Cam			6 28 50	+67 03.8	1	12.09		0.12		0.12		1		597
		NGC 2244 - 70			6 28 51	+04 59.5	1	12.37		1.24		0.99		2		892
		NGC 2244 - 69			6 28 51	+05 00.4	1	12.78		0.39		0.39		2		892
		Cr 97 - 24			6 28 51	+05 48.9	1	12.62		0.51		0.14		1		1640
		Cr 97 - 13			6 28 51	+05 52.4	1	12.73		0.49		0.05		1	F0	1640
		NGC 2244 - 79			6 28 52	+04 53.2	1	10.62		0.16		-0.56		4	B2 V	1059
		Cr 97 - 15			6 28 52	+05 50.9	1	14.22		0.73		0.27		1		1640
46075	+11 01207	HR 2374			6 28 52	+11 49.7	2			-0.13	.015	-0.42	.015	5	B6 III	1049,1079
291763	−1 01265	G 106 - 53			6 28 52	−01 31.7	3	10.07	.013	0.85	.004	0.54	.021	5	K0	803,1620,1658
		NGC 2244 - 383			6 28 53	+04 33.0	1	12.86		0.46		0.00		2		892
		Cr 97 - 23			6 28 53	+05 49.1	1	12.67		1.62				1		1640
259012	+4 01295	NGC 2244 - 80		★ AB	6 28 54	+04 52.8	3	9.30	.045	0.19	.012	-0.65	.005	14	B0.5V	593,1059,1651
		Cr 97 - 22			6 28 54	+05 49.5	1	14.10		0.67		0.21		1		1640
46076	+11 01208				6 28 54	+11 08.3	1	8.19		1.58				3	K5	369
46491	−64 00571				6 28 54	−64 12.1	1	8.85		0.02		0.02		1	A1 IV	919
		POSS 122 # 5			6 28 55	+55 31.4	1	18.39		1.02				3		1739
258982	+6 01275	LS VI +06 002			6 28 56	+06 12.6	1	9.21		0.20		-0.73		3	B7	493
46108	+4 01296	NGC 2244 - 384			6 28 57	+04 33.1	1	8.52		0.16		0.26		2	A6 V	892
46107	+5 01278	NGC 2244 - 127			6 28 57	+04 58.2	1	8.73		0.10		0.18		2	A2 V	892
258973	+10 01163				6 28 57	+10 22.6	2	10.03	.005	0.25	.005	0.17	.005	5	A2	206,5012
46106	+5 01279	NGC 2244 - 115			6 28 59	+05 03.8	6	7.93	.012	0.14	.006	-0.75	.011	18	B0 V	1,593,1059,2033,8007,8027
46030	+22 01364				6 28 59	+22 13.4	1	7.27		0.01		-0.05		2	A0	1648
		NGC 2244 - 106			6 29 00	+05 08.1	1	11.84		0.92		0.21		2		892
46105	+5 01280	IDS06263N0550	A		6 29 00	+05 48.4	1	7.13		-0.03		-0.19		2	A1p	401
46089	+11 01209	HR 2375			6 29 01	+11 34.9	3	5.23	.005	0.16	.015	0.14	.010	5	A3 V	1049,1363,3023
		NGC 2244 - 129			6 29 02	+04 57.7	1	11.76		1.40		1.49		2		892
		NGC 2244 - 109			6 29 02	+05 07.7	1	13.58		0.52		0.13		2		892
		NGC 2244 - 133			6 29 03	+04 54.7	2	11.70	.010	0.41	.000	0.08	.025	4	B9	892,1059
46288	−40 02508	IDS06274S4023	AB		6 29 03	−40 24.7	1	6.68		-0.12		-0.54		2	B3 IV	401
		NGC 2244 - 107			6 29 04	+05 08.0	1	12.30		1.09		0.65		2		892
46184	−12 01518	HR 2379		★ A	6 29 04	−12 21.3	2	5.16	.015	1.26	.010	1.38		6	K1 III	2007,3016
46185	−12 01520				6 29 05	−12 31.4	1	6.81		-0.16		-0.83		3	B2/3 (II)	1732
		NGC 2244 - 135			6 29 06	+04 54.3	1	11.64		0.57		0.12		2	F8	1059
		NGC 2244 - 130			6 29 08	+04 56.5	1	11.60		0.27		-0.38		2	B2.5V	1059
		NGC 2244 - 108			6 29 09	+05 07.8	1	11.36		0.23		-0.13		3		1059
46165	−5 01666				6 29 09	−05 19.9	1	7.46		-0.07		-0.45		2	B9	401
		NGC 2244 - 111			6 29 10	+05 06.9	1	12.02		1.02		0.84		2		892
		G 101 - 43			6 29 10	+37 56.4	1	11.49		1.18		1.09		1		1620
46167	−9 01503				6 29 10	−09 32.2	1	10.27		0.11				1	A2	207
46182	−9 01504				6 29 10	−09 42.9	1	10.10		0.11				1	A0	207
		NGC 2244 - 112			6 29 11	+05 06.6	1	13.48		1.97		1.16		2		892
46052	+32 01324	HR 2372, WW Aur			6 29 11	+32 29.5	2	5.84	.037	0.14	.000	0.15	.000	4	A3 m + A3 m	8071,8080
		NGC 2244 - 131			6 29 12	+04 56.5	2	13.58	.020	0.53	.012	0.35	.004	4		593,1059
		NGC 2244 - 124			6 29 12	+04 58.5	2	12.99	.004	0.36	.004	0.11	.004	6	B8	593,1059
		NGC 2244 - 110			6 29 12	+05 07.1	1	10.71		1.09		0.73		2		892
259105	+5 01281	NGC 2244 - 128		★ A	6 29 13	+04 58.2	3	9.37	.010	0.18	.017	-0.68	.019	13	B1 Vn	593,1059,1651
		NGC 2244 - 125		★ B	6 29 13	+04 58.5	1	11.94		0.30		-0.34		2	B7	1059
		NGC 2244 - 120			6 29 13	+04 59.3	1	12.32		1.95				2		1059

HD	DM	Other Id	N	Rem	α_{1950}	δ_{1950}	S	V	σ_V	B–V	σ_{B-V}	U–B	σ_{U-B}	n	Spectrum	References
46149	+5 01282	NGC 2244 - 114			6 29 13	+05 04.2	6	7.60	.021	0.17	.012	-0.80	.013	18	O8.5V	1,401,593,1011,1012,1059
		NGC 2244 - 113			6 29 14	+05 05.9	1	14.56		0.61		0.19		1		892
		NGC 2244 - 121		⋆D	6 29 15	+04 58.9	1	12.40		0.44				1		1059
		NGC 2244 - 116			6 29 15	+05 01.6	1	12.73		0.32		-0.07		1	A0	1059
		NGC 2244 - 123		⋆E	6 29 16	+04 58.6	1	11.67		0.32		-0.30		2	B5 V	1059
46150	+5 01283	NGC 2244 - 122		⋆AB	6 29 16	+04 58.8	11	6.74	.015	0.13	.007	-0.83	.010	38	O5 V((f))	1,116,593,1011,1012*
		NGC 2244 - 119			6 29 16	+04 59.5	1	12.30		0.34		-0.23		2		1059
46121	+15 01229				6 29 16	+15 09.3	1	8.36		-0.08		-0.12		2	A0	401
		NGC 2244 - 175			6 29 17	+05 02.1	2	13.41	.004	0.42	.019	0.12	.004	5		593,1059
291725	+0 01462	G 106 - 55			6 29 18	+00 41.3	1	9.40		0.68		0.10		1	G0	1620
		NGC 2244 - 181			6 29 18	+05 00.1	1	11.54		0.63		0.14		2	G0	1059
		NGC 2244 - 193			6 29 19	+04 57.9	1	10.31		0.23		-0.52		5	B2 V	1059
		NGC 2244 - 190			6 29 19	+04 58.5	1	11.21		0.26		-0.43		5	B2 Vn	1059
259135	+4 01299	NGC 2244 - 200		⋆V	6 29 21	+04 54.9	5	8.54	.007	0.18	.006	-0.72	.011	14	B0.5V	593,892,1012,1059,1651
46205	-5 01669				6 29 21	-05 18.6	1	8.17		1.69				3	K5	369
		NGC 2244 - 173			6 29 22	+05 03.8	1	10.24		1.07				1	gG2	1059
259172		NGC 2244 - 167			6 29 23	+05 07.4	2	10.68	.005	0.21	.010	-0.57	.010	6	B3 V	593,1059
46136	+17 01286	IDS06265N1751	A		6 29 23	+17 49.3	1	6.28		0.53		0.08		2	F6 V	3024
46136	+17 01286	IDS06265N1751	B		6 29 23	+17 49.3	1	6.95		0.38		0.04		2	G8	3024
		NGC 2244 - 197			6 29 24	+04 57.2	1	12.55		0.36		-0.10		2	A0	1059
		NGC 2244 - 183			6 29 24	+04 59.8	2	11.38	.004	0.49	.013	0.29	.013	5	A4	593,1059
46365	-40 02512	HR 2390			6 29 24	-40 52.8	3	6.19	.004	1.40	.000			11	K3 III	15,1075,2035
46072	+35 01437				6 29 25	+35 06.1	1	7.04		1.26		1.34		2	K0	1601
		NGC 2244 - 195			6 29 26	+04 57.5	2	12.14	.013	0.29	.017	-0.21	.013	6	B9	593,1059
		NGC 2244 - 188			6 29 26	+04 58.7	1	11.79		1.02				2		1059
46179	+6 01278				6 29 26	+06 04.4	1	6.69		-0.04		-0.20		2	B9 V	401
46229	-8 01462	HR 2381			6 29 26	-08 07.2	3	5.43	.005	1.38	.005	1.52	.000	14	gK2	15,1075,1078
46349	-35 02947	HR 2388			6 29 26	-35 13.3	2	5.83	.008	0.81	.007	0.53		6	G8 III	58,2035
		NGC 2244 - 201			6 29 27	+04 54.5	3	9.73	.012	0.18	.000	-0.71	.010	12	B1 V	593,892,1059
		NGC 2244 - 198			6 29 28	+04 57.2	2	12.66	.013	0.34	.000	-0.01	.004	6		593,1059
		CS Mon			6 29 28	+06 41.2	3	10.76	.030	1.06	.024			3		851,1772,6011
46148	+15 01230	IDS06266N1547	AB		6 29 28	+15 44.6	1	7.17		0.69		0.29		1	F5	851
46223	+4 01302	NGC 2244 - 203			6 29 30	+04 51.6	12	7.27	.015	0.22	.008	-0.77	.013	44	O4 V ((f))	1,116,593,892,1011*
		NGC 2244 - 192			6 29 30	+04 58.2	1	12.43		0.47		-0.07		2	A0	1059
		NGC 2244 - 172			6 29 30	+05 04.5	1	11.18		0.29		-0.40		1	B2.5V	1059
46202	+5 01286	NGC 2244 - 180			6 29 31	+05 00.2	7	8.19	.015	0.18	.011	-0.74	.013	19	O9 V	1,116,593,1011,1059*
259163	+9 01274				6 29 31	+09 11.3	1	9.40		0.04		-0.56		4	K7	1732
46244	-9 01507	IDS06272S0934	AB		6 29 33	-09 35.9	1	8.11		0.48				1	F5	207
		NGC 2257 sq1 9			6 29 33	-64 16.1	1	15.61		0.95				3		919
		NGC 2244 - 206			6 29 34	+04 49.9	1	11.97		0.56		-0.10		3		1651
46199	+8 01379				6 29 34	+08 51.3	1	6.98		0.28		0.12		11	A3	1603
46415	-43 02518				6 29 34	-43 40.9	5	6.68	.009	1.00	.004	0.73	.004	21	G8 III	278,861,977,1075,2011
		NGC 2257 sq1 5			6 29 34	-64 16.8	2	13.71	.014	0.74	.005	0.33	.014	13		919,1669
259210		NGC 2244 - 189			6 29 35	+04 58.7	1	11.14		0.19		0.11		2	A0	1059
		NGC 2257 sq1 8			6 29 35	-64 16.3	1	16.48		0.70		0.33		6		919
		NGC 2244 - 385			6 29 36	+04 32.7	1	11.44		0.37		0.08		1		892
		NGC 2244 - 194			6 29 36	+04 57.6	2	11.93	.011	0.35	.008	-0.26	.007	4	B6 Vne	1059,1651
46178	+11 01213	HR 2378		⋆A	6 29 36	+11 42.7	1	6.03		1.07				2	K0 III	71
46162	+17 01287				6 29 36	+17 41.1	1	8.48		-0.08		-0.31		2	A0	401
		NGC 2244 - 242			6 29 37	+05 04.5	1	12.84		1.34		1.05		2		892
		Hiltner 566			6 29 38	+03 36.	1	11.14		0.82		-0.10		2		1012
		NGC 2244 - 260			6 29 39	+04 55.0	3	12.40	.038	0.33	.008	-0.20	.009	9	B9	593,1059,1651
259238		NGC 2244 - 239			6 29 39	+05 05.6	1	11.06		0.28		-0.25		4	B7 V	1059
		NGC 2244 - 271		AB	6 29 40	+04 51.2	1	10.84		0.38		0.22		2		1059
46241	+4 01304	NGC 2244 - 266		⋆	6 29 40	+04 53.6	8	5.84	.012	1.00	.004	0.78	.006	49	K0 III	15,593,892,1059,1355,1415*
		NGC 2244 - 256			6 29 40	+04 57.1	1	12.77		0.41		-0.13		2		1059
		G 105 - 44			6 29 40	+07 06.1	1	11.91		0.72		0.17		1	K0	1620
259268		NGC 2244 - 241			6 29 43	+05 05.0	1	11.07		0.28		-0.23		4	B8 V	1059
	-9 01509				6 29 44	-09 33.7	1	10.39		0.98				1	G5	207
		NGC 2244 - 290			6 29 45	+04 40.1	1	11.83		0.32		0.28		1		892
259301		NGC 2244 - 386			6 29 46	+04 37.7	1	10.90		0.07		0.09		1	A0	892
46328	-23 03991	Cr 121 - 22		⋆A	6 29 46	-23 22.9	7	4.33	.006	-0.24	.007	-0.98	.002	22	B1 III	15,62,1075,1278,1732*
259299	+5 01289	NGC 2244 - 240			6 29 48	+05 04.8	1	10.05		0.32		0.06		2	A7 V	1059
259267	+5 01288	NGC 2244 - 221			6 29 48	+05 17.0	1	10.06		0.09		-0.41		3	B9	493
		NGC 2257 sq1 17			6 29 48	-64 15.1	2	11.95	.009	0.54	.005	0.04	.014	4		919,1669
259300		NGC 2244 - 253			6 29 50	+04 59.2	1	10.76		0.32		-0.30		4	B3 Vp	1059
46283	-7 01456				6 29 50	-07 22.2	1	7.27		0.25		0.15		2	Am	355
		NGC 2257 sq1 12			6 29 50	-64 20.0	1	16.07		0.75				3		919
259264	+9 01275				6 29 51	+09 31.7	1	8.93		0.51				2	A7 II	6009
46431	-36 02962	HR 2393, SX Col			6 29 51	-36 54.2	2	6.29	.010	1.62	.015	1.91		7	M2/3 III	2035,3005
46160	+27 01141				6 29 53	+27 51.8	1	7.42		1.51				2	K5 III	20
46730	-65 00610	HR 2408			6 29 55	-65 32.0	3	6.27	.006	0.33	.003	0.04	.015	25	F0 III	15,1075,2038
		LkHα 215		⋆	6 29 56	+10 11.4	1	10.70		0.58		0.13		1	B7 IIne	5012
46304	-5 01678	HR 2386		⋆AB	6 29 56	-05 49.8	3	5.60	.004	0.26	.005	0.06	.010	14	F0 Vnn	15,1415,2027
46462	-37 02882				6 30 02	-37 08.1	1	7.56		-0.10		-0.52		2	Ap Si	401
	-26 03096				6 30 04	-26 59.3	1	11.42		1.36				4		912
		LP 839 - 11			6 30 07	-26 59.5	1	11.41		1.39				1		1705
46342	-9 01510				6 30 08	-10 01.4	1	9.36		0.42				1	F5	207
46569	-51 01946	HR 2400			6 30 08	-51 47.4	2	5.59	.005	0.54	.000			7	F7/8 IV	15,2012
		NGC 2257 sq1 11			6 30 08	-64 13.9	1	15.43		0.59				2		919

Table 1 345

HD	DM	Other Id	N Rem	α_{1950}	δ_{1950}	S	V	σ_V	B–V	σ_{B-V}	U–B	σ_{U-B}	n	Spectrum	References
46921	−71 00449			6 30 08	−71 28.3	1	8.19		0.16		0.18		35	A3 V	978
		NGC 2257 sq1 37		6 30 09	−64 14.1	1	18.36		0.23				1		919
42855	+86 00079			6 30 10	+86 44.1	1	6.60		1.20		1.20		2	K3 III	1003
		NGC 2257 sq1 16		6 30 10	−64 16.8	2	11.11	.005	0.82	.000	0.38	.009	5		919,1669
		LP 87 - 70		6 30 11	+57 52.0	1	10.19		0.69		0.17		4		1723
46341	−6 01598			6 30 11	−06 27.0	4	8.62	.004	0.56	.000	-0.07	.015	7	G0	516,742,1064,3077
46300	+7 01337	HR 2385	★	6 30 12	+07 22.3	10	4.50	.008	0.01	.009	-0.22	.041	36	A0 Ib	15,369,401,1078,1119*
46264	+17 01291		V	6 30 12	+16 59.2	2	7.67	.044	-0.01	.010	-0.56		6	B5 Ven	401,2033
		NGC 2257 sq1 7		6 30 12	−64 17.9	2	11.99	.005	1.08	.005	0.87	.000	8		919,1669
		CU Mon		6 30 13	+00 04.9	1	13.23		1.24				1		1772
46339	−4 01558			6 30 13	−04 24.7	1	9.03		0.00		-0.77		2	B3	401
		NGC 2257 sq1 2		6 30 15	−64 19.3	2	14.75	.035	0.64	.005	0.07	.000	9		919,1669
		NGC 2257 sq1 14		6 30 16	−64 17.3	2	13.84	.010	1.03	.020	0.90	.030	7		919,1669
46380	−7 01462	LSS 76		6 30 18	−07 28.3	1	8.00	.005	0.40	.035	-0.57	.020	4	B2 Vne	490,1212
		NGC 2257 sq1 13		6 30 18	−64 18.6	1	16.17		0.88				3		919
259440	+5 01291	LS VI +05 011		6 30 19	+05 50.3	2	9.14	.027	0.55	.003	-0.56	.008	3	B0pe	1012,1588
259431	+10 01172	LS VI +10 009		6 30 19	+10 21.6	4	8.72	.022	0.27	.016	-0.54	.011	5	B6pe	351,1012,1588,5012
259481	+5 01293	NGC 2244 - 353	★ AB	6 30 20	+04 58.7	1	9.27		0.30		-0.50		3	B3	493
		NGC 2257 sq1 6		6 30 20	−64 15.3	1	15.68		0.82		0.19		3		919
46697	−58 00718	TZ Pic		6 30 22	−58 58.1	1	7.65		1.14		1.00		1	K1 III/IVp	1641
46101	+55 01093	HR 2376	V	6 30 23	+55 23.5	1	6.46		1.60		1.83		4	K0	1733
	−43 02523	IDS06289S4328	AB	6 30 23	−43 29.7	3	10.64	.085	1.47	.023	1.23		8		158,912,1705
		NGC 2257 sq1 1		6 30 23	−64 19.5	2	11.23	.005	1.04	.005	0.92	.000	12		919,1669
46251	+33 01356	HR 2383		6 30 26	+33 03.8	2	6.53		0.04	.010	0.10	.024	5	A3 V	1049,1733
46407	−11 01520	HR 2392, HR CMa		6 30 26	−11 07.7	7	6.26	.010	1.10	.014	0.75	.017	26	K0 III	15,565,993,2031,6001*
46375	+5 01295	NGC 2244 - 400		6 30 32	+05 30.2	1	7.84		0.86		0.63		1	K1 IV	892
259426				6 30 32	+18 20.	1	9.76		0.47		-0.07		1	A7	159
46277	+28 01160			6 30 32	+28 01.1	1	7.44		1.23				3	K0 II	20
45821	+72 00322			6 30 33	+72 03.2	1	7.79		0.65		0.24		3	G0	196
46426	−8 01469			6 30 36	−08 55.5	1	7.39		0.07		0.07		2	A0	401
	+18 01259		V	6 30 37	+18 31.7	1	9.77		2.02				3	M2	369
46568	−37 02889	HR 2399		6 30 38	−37 39.5	1	5.25		0.98				4	G8 III	2035
46388	+4 01314	NGC 2244 - 388		6 30 41	+04 41.3	2	9.17	.030	0.15	.005	-0.33	.015	4	B7 V	493,892
46792	−61 00669	HR 2410		6 30 41	−61 50.5	2	6.14	.005	-0.16	.005			7	B2 V	15,2012
291782	+0 01472	LS VI +00 007		6 30 44	+00 43.6	4	9.13	.014	0.71	.011	-0.22	.023	23	B2 Ib	830,1012,1783,8100
	−43 02527			6 30 44	−43 29.0	1	8.61		1.15				4	K1 III	2012
46374	+14 01339	HR 2391		6 30 46	+14 11.7	1	5.53		1.11				2	gK2	71
		NGC 2257 sq1 27		6 30 46	−64 19.7	1	10.65		1.67		1.69		1		919
46547	−31 03407	HR 2397	★ A	6 30 47	−31 59.5	3	5.74	.023	-0.18	.005	-0.81	.014	8	B2 V	150,542,1732
46547	−31 03407	HR 2397	★ AB	6 30 47	−31 59.5	3	5.68	.004	-0.18	.007	-0.80		9	B2 V +B7 V	15,401,2027
46547	−31 03407	IDS06289S3157	B	6 30 47	−31 59.5	2	8.70	.000	-0.07	.010	-0.43	.030	4	B7 V	150,542
259557	+14 01340			6 30 48	+14 17.5	1	9.09		0.97				28	G0	6011
259597	+8 01388	LS VI +08 009		6 30 49	+08 22.5	2	8.62	.033	0.12	.014	-0.73	.005	4	B0.5:V:enn	401,1012
46336	+27 01148			6 30 50	+27 05.0	1	7.76		1.05				2	K0 III	20
259637		NGC 2244 - 390		6 30 51	+04 48.3	1	11.64		0.15		0.26		1	A7	892
259635	+5 01296	NGC 2244 - 393		6 30 51	+05 04.4	1	9.69		0.76		0.29		2	G0	892
259634	+5 01297		A	6 30 52	+05 20.5	1	9.90		0.08		-0.07		1	B9 III	493
259634	+5 01297		AB	6 30 52	+05 20.5	1	10.94		0.17		-0.25		1	B9 III	493
259634	+5 01297		B	6 30 52	+05 20.5	1	10.95		0.16		-0.25		1		493
259634	+5 01297		C	6 30 52	+05 20.5	1	11.92		0.33		0.28		1		493
		NGC 2244 - 399		6 30 53	+05 19.7	1	11.06		0.42		0.33		1		892
		WW Mon		6 30 53	+09 15.0	1	12.10		1.00				1	F6	1772
46652	−45 02580			6 30 53	−45 16.3	9	7.16	.009	1.08	.010	0.87	.007	81	K0 III	116,278,402,863,977*
		NGC 2244 - 398		6 30 54	+05 18.9	1	11.88		0.31		0.35		1		892
46296	+38 01523	IDS06275N3809	A	6 30 54	+38 06.9	1	6.66		1.00		0.75		2	K0	1601
		NGC 2244 - 394		6 31 00	+05 05.7	1	11.13		0.40		0.03		2		892
259631	+8 01389	LS VI +08 010		6 31 00	+08 04.5	1	9.52		0.15		-0.69		3	B5	493
46423	+14 01343			6 31 02	+14 20.0	2	8.27	.003	0.48	.000	-0.05		33	F6 V	1462,6011
		HIC 31538		6 31 03	+66 23.3	1	11.27		1.06		1.05		3		1723
259696	+5 01299	NGC 2244 - 397		6 31 04	+05 15.3	1	8.85		2.19		2.39		2	K7	892
		DX Gem		6 31 04	+14 30.5	2	10.55	.005	0.87	.005	0.61		2		592,1462
46566	−22 01446			6 31 04	−22 34.2	1	8.08		1.05				4	K0 III	2012
259697	+4 01317	NGC 2244 - 391		6 31 05	+04 48.3	2	10.24	.014	0.27	.009	-0.35	.023	4	B8	493,892
		NGC 2244 - 396		6 31 05	+05 11.7	1	11.10		0.15		-0.09		1		892
		WD0631+107		6 31 05	+10 43.9	1	13.82		-0.17		-1.02		5	DA	1727
46487	−1 01274	HR 2395		6 31 06	−01 10.8	4	5.09	.004	-0.14	.008	-0.56	.006	12	B5 Vn	15,154,1078,1203
46497	−6 01603	LS VI -06 002		6 31 08	−06 44.0	1	9.54		0.56		0.58		1	A5	490
46469	+5 01300			6 31 09	+05 31.2	1	8.51		0.14		-0.57		2	B2 V	401
259616	+20 01485			6 31 10	+20 43.0	1	10.16		0.06		-0.09		3		1723
		NGC 2244 - 392		6 31 11	+05 04.0	1	11.12		0.43		-0.33		2		892
	−89 00023			6 31 11	−89 22.1	1	9.71		1.62		2.02		4		826
46485	+4 01318	NGC 2244 - 387		6 31 12	+04 33.9	3	8.25	.030	0.32	.008	-0.69	.009	5	O7 V:n(e)	892,1011,1012
46860	−58 00722	HR 2412	★ AB	6 31 14	−58 42.9	2	5.69	.000	-0.06	.000	-0.17		6	B9 Ve	404,2035
46484	+4 01319	NGC 2244 - 389		6 31 15	+04 42.1	2	7.68	.042	0.35	.005	-0.63	.014	4	B1 V	892,1012
259683	+13 01309			6 31 15	+13 06.1	1	9.72		0.23				1	F0	592
46602	−20 01437	HR 2403		6 31 18	−20 53.1	1	6.40		0.83				4	K0 III +F/G	2035
291819				6 31 21	−00 53.	1	9.76		1.90				2	K7	369
	+37 01537	G 103 - 42		6 31 22	+37 27.8	1	10.36		0.93		0.72		1	K5	333,1620
46742	−43 02534			6 31 25	−43 32.2	2	7.99	.035	1.65	.002	1.92		7	M1/2 (III)	2011,3040
46517	+5 01302	NGC 2244 - 395		6 31 28	+05 09.4	1	8.07		0.24		0.24		1	A5 V	892

HD	DM	Other Id	N	Rem	α_{1950}	δ_{1950}	S	V	σ_V	B–V	σ_{B-V}	U–B	σ_{U-B}	n	Spectrum	References
46483	+16 01193				6 31 28	+16 41.2	1	8.61		-0.08		-0.27		2	A0	401
46046	+69 00385				6 31 28	+69 23.9	1	8.30		0.24		0.13		2	A5	1733
46727	-38 02730	HR 2407			6 31 29	-38 35.1	1	6.44		1.00				4	G8 III	2035
259779	+13 01312				6 31 32	+13 12.1	1	8.86		0.36		-0.03		1	F2	1462
46562	-4 01566				6 31 32	-04 31.7	1	8.29		1.66				3	K5	369
46536	+9 01282				6 31 33	+09 07.4	1	8.70		-0.06		-0.20		11	A0	1603
259865	+4 01323	LS VI +04 021			6 31 34	+04 46.6	1	9.91		0.31		-0.50		3	B5	493
		WLS 624 70 # 9			6 31 34	+69 31.0	1	13.29		0.92		0.44		2		1375
45947	+73 00340	HR 2365			6 31 37	+73 44.3	1	6.26		0.37		-0.01		1	F2	254
259896	+5 01304				6 31 38	+05 06.8	1	10.37		0.41		-0.54		3	B3	493
46559	+2 01292	LS VI +02 008			6 31 39	+02 25.7	2	8.16	.004	0.59	.013	0.01	.013	6	B8 I	1012,8100
259892	+8 01396	NGC 2251 - 2			6 31 40	+08 27.4	1	9.65		0.56		0.06		1	F0 V	49
46557	+6 01288				6 31 41	+06 44.1	1	7.55		0.12		0.07		2	A2 V	401
		NGC 2251 - 12			6 31 42	+08 21.5	1	12.92		0.28		0.19		1		49
		LS VI +11 013			6 31 42	+11 12.8	1	10.53		0.58		-0.13		2	B3 III	1012
46516	+19 01391				6 31 42	+19 28.2	1	7.09		0.00		-0.08		2	A0	1648
	+44 01485	G 101 - 45		A	6 31 42	+44 41.6	2	10.27	.074	1.00	.005	0.88	.028	5	G5 V:	1726,7010
46319	+54 01050				6 31 42	+54 37.8	1	8.39		-0.03		-0.09		2	A0	1566
45560	+79 00208	HR 2346			6 31 42	+79 38.5	1	6.76		0.05		0.04		2	A1 V	1733
		NGC 2251 - 16			6 31 43	+08 24.8	1	13.38		1.33		1.01		1		49
	+13 01315				6 31 43	+13 07.6	1	10.49		1.28				1		592
	+44 01485	G 101 - 45		B	6 31 43	+44 41.5	1	12.64		1.09		0.87		2		7010
		NGC 2251 - 19			6 31 44	+08 23.4	1	13.82		0.79		0.18		1		49
		NGC 2251 - 24			6 31 44	+08 26.2	1	14.69		0.78				1		49
		NGC 2251 - 21			6 31 45	+08 27.9	1	14.24		0.50		0.13		1		49
		NGC 2251 - 20			6 31 46	+08 22.8	1	13.92		0.42		0.31		1		49
		BB Gem			6 31 46	+13 07.2	3	10.79	.029	0.58	.025	0.46	.016	3		592,1399,1462
		LP 720 - 24			6 31 46	-13 25.6	1	12.08		0.50		-0.19		2		1696
46573	+2 01295	LS VI +02 010			6 31 47	+02 34.4	3	7.93	.011	0.34	.008	-0.66	.000	8	O7	1011,1012,1209
		NGC 2251 - 18			6 31 47	+08 26.5	1	13.74		0.32		0.18		1		49
		AA131,200 # 10			6 31 47	+09 06.9	1	15.46		0.99		0.27		2		7005
46817	-45 02594				6 31 47	-45 20.3	5	8.04	.006	0.13	.003	0.08	.009	43	A1 IV	116,278,1066,1075,2011
		NGC 2251 - 23			6 31 48	+08 21.6	1	14.60		0.50		0.05		1		49
		NGC 2251 - 22			6 31 49	+08 22.9	1	14.53		0.78		0.28		1		49
46318	+56 01134				6 31 49	+56 25.7	1	6.56		0.38		0.10		3	F5 III	1501
46835	-46 02504				6 31 49	-46 07.4	1	9.50		0.15				4	A1 V	2011
46616	-4 01569				6 31 50	-04 51.6	1	7.26		-0.14		-0.61		2	B8	401
46663	-10 01583				6 31 51	-10 56.0	1	9.55		0.94		0.52		1	G5	1696
46335	+55 01096				6 31 52	+55 44.1	1	7.83		0.46		0.02		2	F8	1502
259954	+8 01399	NGC 2251 - 1			6 31 53	+08 23.5	2	9.15	.055	0.05	.005	-0.58	.050	3	B2 IV	49,1012
		NGC 2251 - 25			6 31 55	+08 21.0	1	14.84		0.61		0.07		1		49
		NGC 2251 - 31			6 31 55	+08 26.6	1	16.48		0.82				1		49
		NGC 2251 - 10			6 31 55	+08 27.5	1	12.44		0.28		0.20		1		49
291794	+0 01484				6 31 56	+00 14.7	1	9.50		1.63				1	M3	369
		NGC 2251 - 26			6 31 56	+08 22.5	1	15.49		0.67		0.26		1		49
		NGC 2251 - 17			6 31 56	+08 23.8	1	13.71		0.44		0.23		1		49
		NGC 2251 - 14			6 31 56	+08 24.2	1	13.08		0.53		0.01		1		49
	+1 01398				6 31 57	+01 26.7	1	9.80		1.79				3	M2	369
259988		NGC 2251 - 7			6 31 57	+08 27.1	1	11.33		0.20		0.06		1	A7	49
46294	+59 00996				6 31 57	+59 42.4	1	6.86		0.08		0.10		2	A0	1502
		NGC 2251 - 30			6 31 58	+08 21.8	1	16.23		1.07				1		49
46397	+49 01516				6 31 58	+49 55.4	1	8.24		0.16		0.11		2	B9	1601
		NGC 2251 - 13			6 32 00	+08 23.4	1	12.94		0.25		0.14		1		49
259911	+17 01302				6 32 00	+17 03.2	1	9.09		1.80				3	M0	369
46859	-44 02742				6 32 00	-45 01.4	2	8.67	.010	1.14	.001	1.09		12	K0 III	977,2011
259986	+8 01402	RW Mon			6 32 02	+08 52.0	1	9.36		0.01		-0.15		1	A0	159
		LS VI +02 011			6 32 03	+02 22.8	1	11.26		0.69		-0.18		2		405
		NGC 2251 - 6			6 32 03	+08 23.8	1	11.31		0.18		0.07		1		49
		NGC 2251 - 28			6 32 03	+08 24.8	1	15.93		0.80				1		49
46593	+16 01202				6 32 03	+16 48.3	1	7.22		-0.13		-0.51		2	B9	401
46553	+28 01168	HR 2398			6 32 03	+28 03.8	4	5.26	.052	-0.03	.004	-0.07	.010	7	A0 Vnn	300,1022,1363,3023
46045	+72 00324				6 32 03	+72 53.6	1	8.73		0.39		-0.01		2	F5	1733
46642	+7 01357	HR 2404		⋆ A	6 32 04	+07 36.8	2	6.44	.005	-0.02	.000	-0.03	.000	7	A0 Vs	15,1078
		NGC 2251 - 15			6 32 04	+08 24.6	1	13.38		0.42		0.23		1		49
46815	-36 02990	HR 2411			6 32 04	-36 11.6	4	5.42	.005	1.42	.012	1.72		13	K3 III	15,1075,2035,3005
		NGC 2251 - 9			6 32 06	+08 23.4	1	12.33		0.32		0.23		1		49
		NGC 2251 - 29			6 32 06	+08 25.3	1	16.16		0.70				1		49
46595	+15 01246	W Gem			6 32 06	+15 22.3	2	6.59	.002	0.73	.011	0.59		2	G5	851,6007
260021					6 32 07	+09 14.	1	10.32		0.12		-0.45		1	B8	159
259990	+8 01404	NGC 2251 - 3		⋆ AB	6 32 08	+08 22.0	1	10.35		1.25		0.98		1	A7	49
260022		NGC 2251 - 4			6 32 09	+08 21.0	1	10.98		0.13		-0.06		1	A0p	49
46858	-38 02740				6 32 10	-38 48.5	1	8.19		0.92				4	G8 III	2012
46716	-11 01536	IDS06298S1109		AB	6 32 11	-11 11.4	1	7.44		0.40				2	F0	173
		NGC 2251 - 8			6 32 12	+08 19.6	1	11.87		0.21		0.11		1		49
46660	+11 01232	LS VI +11 014		⋆ AB	6 32 14	+11 09.9	1	8.04		0.31		-0.61		2	B1 V	1012
47252	-68 00536				6 32 14	-68 39.9	1	8.27		0.79				4	G8/K0 V	2012
		NGC 2251 - 5			6 32 16	+08 18.6	1	11.18		0.57		0.02		1		49
		NGC 2251 - 32			6 32 16	+08 23.7	1	16.90		0.65				1		49
46738	-8 01480	IDS06299S0832		AB	6 32 17	-08 34.1	1	8.86		0.04		-0.54		2	B5	401
		NGC 2251 - 27			6 32 18	+08 23.4	1	15.60		0.91				1		49

Table 1   347

HD	DM	Other Id	N	Rem	α₁₉₅₀	δ₁₉₅₀	S	V	σ_V	B–V	σ_B–V	U–B	σ_U–B	n	Spectrum	References
47001	−52 00947	HR 2416			6 32 18	−52 17.4	3	6.18	.010	1.09	.000			14	G8 III	15,1075,2027
46714	+0 01486				6 32 19	+00 03.6	1	8.64		1.68				2	K0	369
		NGC 2251 - 11			6 32 19	+08 20.1	1	12.64		0.22		0.14		1		49
46711	+2 01299	LS VI +02 012			6 32 20	+02 48.2	2	9.09	.004	0.85	.013	−0.01	.009	6	B3 II	1012,8100
46552	+32 01338				6 32 22	+32 37.1	1	7.70		−0.06		−0.33		3	B5 III	555
	−46 02512				6 32 29	−47 00.9	1	10.61		0.80				7		955
46709	+10 01186	HR 2406			6 32 32	+10 01.8	3	5.88	.008	1.50	.004	1.82	.030	6	K4 III	37,71,3001
46688	+20 01496				6 32 38	+20 55.8	1	7.95		0.35		0.03		2	F2	1648
46769	+0 01491	HR 2409		⋆	6 32 41	+00 55.9	4	5.80	.009	−0.01	.008	−0.43	.019	13	B8 Ib	15,154,1415,8100
45866	+78 00227	HR 2363			6 32 44	+78 02.4	1	5.73		1.47				2	gK5	71
46936	−32 03168	HR 2415			6 32 44	−32 40.5	3	5.62	.005	−0.09	.004	−0.24		9	B9 V	15,401,2012
46784	+5 01312				6 32 49	+05 33.2	1	7.96		1.63				4	M0 III	369
46783	+9 01295	LS VI +09 010			6 32 55	+09 53.6	2	8.05	.004	0.30	.004	−0.12	.027	5	B9 Ib	1012,8100
46764	+17 01307				6 32 57	+17 14.5	1	7.44		0.01		−0.01		2	A0	401
46933	−22 01458	HR 2414			6 32 58	−22 55.4	6	4.54	.008	−0.04	.012	0.00	.032	22	A0 III	3,15,1075,2012,3023,8015
46747	+17 01306				6 32 59	+17 44.1	1	7.26		1.13		1.03		2	K0	1648
46782	+13 01327				6 33 04	+13 13.1	1	7.34		1.16		0.94		1	K0	1462
46997	−32 03171	IDS06312S3214	AB		6 33 04	−32 15.9	1	7.90		−0.12		−0.64		2	B5 IV	401
46996	−31 03438				6 33 05	−31 48.2	1	8.19		−0.01		−0.06		2	A0 V	401
46847	+2 01302	LS VI +02 013	AB	⋆	6 33 07	+02 45.2	1	8.96		0.67		−0.37		2	B0 III:p	1012
46687	+38 01539	HR 2405, UU Aur	A	⋆	6 33 07	+38 29.3	5	5.40	.135	2.69	.111			17	C5 II	15,109,8015,8022,8100
		L 182 - 44			6 33 07	−58 29.9	2	11.64	.020	1.52	.016	1.16		2		1705,3073
46480	+61 00893	HR 2394	A	⋆	6 33 08	+61 31.7	1	5.94		0.89		0.52		5	G8 IV-V	1355
	−6 01624			V	6 33 14	−06 07.8	1	10.36		1.73				3	M3	369
46867	+5 01315	LS VI +05 016	A	⋆	6 33 15	+05 21.1	1	8.30		0.22		−0.64		2	B0.5III-IV	1012
47119	−46 02524				6 33 16	−46 28.1	3	8.57	.008	0.29	.002	0.05	.017	16	F0 V	977,1075,2011
46904	−5 01698				6 33 17	−05 46.9	1	8.98		0.52		−0.05		2	F5	3061
46780	+27 01164	IDS06302N2722	AB		6 33 18	+27 19.3	2	6.90	.005	0.64	.005	0.15	.009	4	G0	292,3030
		G 105 - 46			6 33 20	+11 39.4	1	14.22		1.65				3		1663
	−17 01546				6 33 20	−17 44.7	1	10.24		0.64		0.05		2		1696
46885	+4 01335	HR 2413			6 33 21	+04 32.4	3	6.54	.005	−0.06	.004	−0.28	.010	9	B9 V	15,1078,1079
	+37 01545	G 103 - 44			6 33 22	+37 53.8	1	9.45		0.75		0.34		1	G5	333,1620
46590	+56 01136	HR 2402			6 33 23	+56 54.0	1	5.89		0.00		0.02		3	A2 V	1733
	−0 01356				6 33 23	−00 45.2	1	9.80		1.80				4	M3	369
46883	+10 01193	LS VI +10 011			6 33 24	+10 19.6	3	7.78	.010	0.40	.015	−0.47	.015	11	B0.5:V	399,405,493
260450	+4 01336	LS VI +04 022			6 33 25	+04 49.0	1	10.06		0.32		−0.42		3	B8	493
46463	+64 00593	IDS06285N6449	A		6 33 25	+64 46.7	1	7.58		0.47		0.01		2	F5	1733
47147	−45 02613	V338 Pup			6 33 28	−45 16.0	2	9.19	.005	0.40	.010	0.04		4	A2/5w	1594,6006
47061	−31 03451				6 33 34	−31 33.0	1	7.89		−0.13		−0.62		2	B5 V	401
288833	+2 01307				6 33 36	+02 00.0	1	9.40		1.91				4	M0	369
46900	+15 01255				6 33 36	+15 47.6	1	7.36		0.25		0.10		1	A0	851
47011	−15 01448	IDS06314S1603	A		6 33 36	−16 03.6	1	7.39		−0.04		−0.24		2	B8/9 III	401
47144	−36 03009	HR 2424	AB	⋆	6 33 40	−36 44.3	2	5.59	.000	−0.13	.010	−0.49		6	A0 V	401,2035
46947	+6 01302				6 33 42	+06 16.7	1	8.24		1.30		1.61		2	K2 II	8100
260472	+14 01353	G 105 - 47			6 33 43	+14 02.6	1	9.41		0.93		0.63		1	G5	1620
47188	−44 02764				6 33 43	−44 14.8	1	9.56		0.18				4	A5 V	2011
	−44 02765				6 33 43	−44 55.0	2	9.99	.013	0.31	.007	0.04		10	A3	863,2011
46966	+6 01303	LS VI +06 003			6 33 45	+06 07.5	3	6.88	.011	−0.05	.015	−0.91	.012	6	O8	116,1011,1012
		LS VI +10 012			6 33 45	+10 54.3	1	10.18		1.22		0.42		2		405
260537	+7 01369				6 33 50	+07 45.2	1	9.47		0.04		−0.59		2	B5	401
47306	−52 00953	HR 2435			6 33 52	−52 56.0	4	4.38	.004	−0.02	.000	−0.15	.011	15	A0 II	15,1075,2012,3023
260574	+7 01370	LS VI +07 007			6 33 58	+07 31.9	1	9.81		0.09		−0.61		1	B9	493
45618	+82 00177	HR 2350			6 33 59	+82 09.8	2	6.64	.010	0.17	.005	0.07	.010	5	A5 V	985,1733
260597	+9 01299				6 34 01	+09 34.9	1	9.52		1.95				3	K7	369
46985	+14 01356				6 34 01	+13 55.9	1	8.27		0.46		−0.07		1		1462
260643	+3 01317				6 34 02	+03 04.1	1	10.21		1.88				1	K7	851
260564	+19 01405				6 34 05	+19 47.4	1	10.18		1.20		1.14		2	K5	3072
47033	+0 01504	IDS06315N0037	AB		6 34 07	+00 35.0	1	8.38		0.40		−0.01		1	F0	3030
47032	+4 01341	LS VI +04 024			6 34 07	+04 44.2	2	8.83	.000	0.44	.007	−0.56	.008	5	B0 III	1012,1732
47267	−44 02771				6 34 07	−45 01.8	2	8.43	.001	1.17	.003	1.14		12	K1 IV	977,2011
47054	−5 01710	HR 2418			6 34 08	−05 11.0	4	5.52	.005	−0.09	.009	−0.39	.007	16	B8 V	15,1079,1415,2012
47230	−35 03005	HR 2431	AB	⋆	6 34 09	−36 02.7	3	6.35	.005	0.48	.005			14	F6 V	15,2012,2012
47138	−18 01480	HR 2423	A	⋆	6 34 11	−18 37.0	2	5.71	.010	0.86	.005	0.48	.005	5	G8 III	542,3077
47138	−18 01480	HR 2423	AB	⋆	6 34 11	−18 37.0	1	5.52		0.76				4	G8III+F3IV	2031
47138	−18 01478	IDS06320S1835	B		6 34 11	−18 37.0	2	7.66	.005	0.32	.000	0.01	.005	5	F2 V	542,3077
260706	+3 01319				6 34 12	+03 11.6	1	10.10		1.55				1	K7	851
47265	−40 02562				6 34 12	−41 00.8	1	7.87		1.03				4	K0 III	2012
46944	+28 01184				6 34 13	+28 01.2	2	8.90	.030	0.50	.005	0.04		4	F7 V	20,3026
		vdB 1 - 38			6 34 17	+03 07.9	1	17.27		3.14				1		673
		vdB 1 - 35			6 34 19	+03 08.1	1	16.91		1.41		1.11		1		673
		vdB 1 - 7			6 34 20	+03 06.1	1	13.33		1.01		0.62		1		673
260737	+4 01346				6 34 20	+04 44.2	1	9.19		0.09		0.12		2	A2	1732
260658	+16 01220				6 34 20	+15 59.3	1	9.32		1.95				3	K5	369
260655	+17 01320	G 104 - 49	AB	⋆	6 34 20	+17 36.1	8	9.63	.005	1.49	.021	1.18	.013	16	M1:	694,1006,1197,1355*
47072	+5 01326	IDS06317N0536	AB		6 34 23	+05 33.7	1	7.14		0.31		0.16		3	F0 II	8100
47020	+24 01328	HR 2417			6 34 23	+24 38.1	2	6.48		0.09	.009	0.06	.009	4	A3 V	1022,1733
		vdB 1 - 39			6 34 24	+03 06.6	1	17.70		1.30		0.70		1		673
		vdB 1 - 41			6 34 24	+03 07.	1	13.73		0.66		0.32		1		813
		vdB 1 - 42			6 34 24	+03 07.	1	13.95		0.62		0.26		1		813
		vdB 1 - 44			6 34 24	+03 07.	1	14.31		0.69		0.51		1		813

HD	DM	Other Id	N	Rem	α_{1950}	δ_{1950}	S	V	σ_V	B–V	σ_{B-V}	U–B	σ_{U-B}	n	Spectrum	References
		vdB 1 - 10			6 34 25	+03 04.9	1	13.51		0.54		0.01		1		673
		vdB 1 - 17			6 34 25	+03 07.1	1	14.26		0.74		0.23		1		673
		vdB 1 - 18			6 34 25	+03 07.1	1	14.30		0.76		0.25		1		673
		vdB 1 - 30			6 34 25	+03 08.6	1	16.07		0.89		0.31		1		673
		vdB 1 - 22			6 34 26	+03 05.7	1	15.22		0.78		0.11		1		673
		vdB 1 - 23			6 34 26	+03 05.9	1	15.37		1.56		0.97		1		673
		vdB 1 - 8			6 34 26	+03 08.0	1	13.39		0.82		0.44		1		673
		vdB 1 - 29			6 34 27	+03 05.2	1	16.05		1.09		0.68		1		673
		vdB 1 - 33			6 34 27	+03 06.2	1	16.37		1.06		0.51		1		673
		CV Mon			6 34 27	+03 06.3	1	13.60		0.70				1		6011
		vdB 1 - 14			6 34 27	+03 06.4	1	13.76		0.66		0.18		1		673
		vdB 1 - 45		⋆ V	6 34 27	+03 06.4	4	9.93	.013	1.17	.021	0.82		4		813,851,1772,6011
		vdB 1 - 27			6 34 27	+03 07.1	1	15.74		0.88		0.65		1		673
		vdB 1 - 28			6 34 27	+03 07.2	1	15.97		1.04		0.58		1		673
		vdB 1 - 13			6 34 28	+03 06.5	1	13.72		0.69		0.27		1		673
260799	+3 01321	IDS06319N0317	A		6 34 29	+03 14.8	1	9.01		0.11				22	B5	6011
47089	+4 01348				6 34 29	+04 42.0	1	9.37		-0.08		-0.50		3	B5 III	1732
47182	-13 01570	HR 2428			6 34 29	-13 16.6	3	5.96	.004	1.56	.000			11	K4/5 III	15,1075,2035
		vdB 1 - 37			6 34 30	+03 05.3	1	17.11		1.26		0.79		1		673
		vdB 1 - 6			6 34 30	+03 07.1	1	13.32		0.55		-0.01		1		673
		vdB 1 - 24			6 34 30	+03 07.2	1	15.37		0.76		0.56		1		673
47088	+6 01308	V649 Mon			6 34 30	+06 06.2	2	7.59	.014	-0.03	.000	-0.80	.010	7	B1 III	401,493
47205	-19 01502	HR 2429			6 34 30	-19 12.7	11	3.95	.012	1.06	.007	1.01	.010	42	K1 III (+M)	3,15,58,418,1020,1075*
		vdB 1 - 3			6 34 31	+03 08.1	1	12.19		0.41		0.20		1		673
		vdB 1 - 26			6 34 31	+03 08.9	1	15.53		0.96		0.79		1		673
260797	+5 01328				6 34 31	+05 08.5	1	10.12		0.27		-0.51		3	B8	493
		vdB 1 - 9			6 34 32	+03 07.4	1	13.47		0.52		0.03		1		673
	+48 01407	G 101 - 47			6 34 32	+48 50.3	2	10.12	.014	0.72	.009	0.18	.012	5	G5	1620,7010
		V 495 Mon			6 34 32	-02 46.8	1	12.14		1.03				1		1772
		vdB 1 - 34			6 34 33	+03 05.9	1	16.65		1.10		0.65		1		673
		vdB 1 - 11			6 34 33	+03 08.3	1	13.55		0.51		0.03		1		673
		vdB 1 - 21			6 34 34	+03 07.7	1	14.72		0.78		0.46		1		673
47107	+5 01329	LS VI +05 018			6 34 34	+05 51.0	3	8.01	.007	-0.03	.005	-0.76	.015	10	B1.5Ia	401,493,8100
		vdB 1 - 4			6 34 35	+03 05.0	1	12.99		0.58		0.37		1		673
		vdB 1 - 40			6 34 35	+03 08.4	1	18.03		2.14				1		673
47247	-22 01472	HR 2433		⋆ AB	6 34 35	-22 34.3	3	6.34	.005	-0.12	.005	-0.55		8	B3 V	15,1079,2027
		vdB 1 - 19			6 34 37	+03 07.7	1	14.43		0.64		0.13		1		673
		vdB 1 - 5			6 34 38	+03 05.7	1	13.18		0.56		0.34		1		673
		vdB 1 - 25			6 34 39	+03 05.8	1	15.48		0.73		0.54		1		673
		G 103 - 46			6 34 39	+34 33.8	3	14.79	.017	1.58	.008	0.84	.035	7		203,1691,3078
46509	+71 00359	HR 2396			6 34 39	+71 47.6	1	5.93		1.19				2	K0 III	71
47158	+3 01323				6 34 40	+03 02.2	1	8.59		0.89				22	K0	6011
		vdB 1 - 16			6 34 40	+03 06.3	1	13.89		0.61		0.09		1		673
		vdB 1 - 15			6 34 40	+03 07.4	1	13.81		0.58		0.25		1		673
		vdB 1 - 2			6 34 42	+03 04.9	1	11.76		1.30		1.37		2		673
		vdB 1 - 12			6 34 42	+03 08.5	1	13.65		0.58		0.01		1		673
		vdB 1 - 36			6 34 43	+03 06.8	1	17.07		1.02		0.49		1		673
47129	+6 01309	HR 2422, V640 Mon			6 34 43	+06 10.7	6	6.06	.014	0.05	.021	-0.88	.014	21	O8 V + O8 f	15,116,1011,1012,1075,1509
		CCS 542			6 34 43	-12 05.2	1	10.92		5.10				2	N	864
		vdB 1 - 1			6 34 44	+03 05.8	1	11.08		0.45		0.00		10		673
		vdB 1 - 20			6 34 45	+03 07.4	1	14.48		0.74		0.09		1		673
		vdB 1 - 32			6 34 46	+03 06.2	1	16.21		1.82		1.48		1		673
		vdB 1 - 31			6 34 47	+03 06.8	1	16.11		1.46		1.35		1		673
47246	-16 01542				6 34 47	-16 18.8	1	7.81		-0.02		-0.44		2	B6 IV/V	401
47105	+16 01223	HR 2421		⋆ A	6 34 49	+16 26.6	15	1.93	.011	0.00	.005	0.06	.035	62	A0 IV	1,3,15,1006,1022,1034*
47426	-45 02627				6 34 50	-45 45.1	1	9.46		0.23				4	A7 V	2011
47156	+10 01201	HR 2426			6 34 51	+10 53.8	1	6.37		1.36		1.52		2	K0	1733
		G 105 - 50			6 34 52	+12 07.1	3	11.15	.049	0.77	.037	0.18	.005	3	G8	927,1620,1696
	+39 01687	G 107 - 12			6 34 57	+39 47.3	1	10.44		0.98		0.81		3	K3	1723
47444	-44 02781				6 34 58	-44 29.9	3	8.50	.009	0.10	.001	-0.03	.009	16	A0 V	977,1075,2011
47445	-44 02782				6 35 00	-44 36.5	4	8.91	.006	-0.04	.005	-0.26	.002	22	B9 IV	863,977,1075,2011
260931					6 35 03	+10 39.	1	10.82		0.30		0.02		1	G0	159
47220	+2 01315	HR 2430			6 35 04	+02 44.9	4	6.16	.007	1.08	.000	0.94	.017	14	K1 III	15,361,1355,1415
47390	-28 03201				6 35 06	-28 39.7	1	7.49		0.97		0.67		3	K0 III	1657
		G 109 - 16			6 35 07	+22 22.9	2	15.45	.020	1.58	.050			2		906,940
		G 104 - 50			6 35 08	+22 22.8	3	13.45	.024	1.46	.003	1.12	.010	11		316,906,940
47278	-8 01496				6 35 08	-08 11.5	1	7.23		0.00		-0.21		2	B9	401
47391	-32 03202				6 35 10	-32 10.9	3	7.64	.007	0.71	.007	0.16	.000	6	G5 V	1311,2012,3002
47070	+39 01690	HR 2419			6 35 12	+39 26.2	2	5.68	.003	1.35	.006	1.56		4	K5 III	71,3001
47299	-8 01498				6 35 12	-08 36.8	2	8.90	.005	-0.08	.009	-0.63	.000	8	B2 V	399,401
261021	+3 01329	LS VI +03 007			6 35 13	+03 39.9	1	9.48		0.66		-0.33		2	B2 III	1012
47240	+5 01334	HR 2432		⋆	6 35 13	+05 00.1	9	6.15	.010	0.15	.006	-0.73	.014	33	B1 Ib	15,154,1012,1020,1075
47152	+29 01293	HR 2425		⋆	6 35 13	+29 01.7	4	5.76	.049	0.00	.013	-0.07	.007	7	B9 np Eu	252,1022,3033,6007
47300	-9 01549				6 35 14	-09 21.2	1	8.47		-0.01		-0.45		2	B5	401
47475	-41 02488	HR 2445			6 35 16	-41 30.8	1	6.34		1.15				4	K0 II	2006
260986	+10 01202	LS VI +09 011			6 35 17	+09 55.7	1	9.72		0.33		-0.38		3	B3	493
47100	+40 01665	HR 2420			6 35 20	+39 56.8	3	5.20		-0.07	.000	-0.40	.016	4	B8 III	1022,1079,3023
47463	-38 02782	HR 2444		⋆ A	6 35 20	-38 06.2	2	6.03	.008	1.03	.000	0.84		6	K0 III	58,2035
47366	-12 01566	HR 2437			6 35 22	-12 56.3	3	6.12	.004	0.99	.005			11	K1 (III)	15,1075,2035
		LS VI +02 014			6 35 25	+02 27.5	1	10.68		0.37		-0.40		2	B1 V	1012

Table 1   349

HD	DM	Other Id	N	Rem	α_{1950}	δ_{1950}	S	V	σ_V	B–V	σ_{B-V}	U–B	σ_{U-B}	n	Spectrum	References
47538	−45 02631				6 35 26	−45 58.9	3	8.63	.003	1.58	.006	1.94	.004	18	K4 III	863,977,2011
47292	+7 01382				6 35 30	+07 11.6	1	7.49		0.11		−0.42		2	B8	401
47500	−36 03031	HR 2446		⋆ AB	6 35 30	−36 56.8	2	5.71	.000	−0.11	.005	−0.52		6	B5/7 IV	401,2035
	−44 02789				6 35 30	−44 55.3	1	9.23		1.13				4	K2	2011
47403	−15 01459				6 35 32	−15 54.0	1	8.51		0.00		−0.23		4	B8 V	1732
47386	−9 01553				6 35 35	−09 13.5	1	8.14		−0.01		−0.16		2	B8	401
47314	+6 01316				6 35 38	+06 02.0	1	8.46		−0.04		−0.45		3	B8 Ib	8100
		AA45,405 S283 # 8			6 35 39	+00 46.8	1	12.30		0.51		−0.49		3	B0 V:	797
		AA45,405 S283 # 10			6 35 39	+00 46.8	1	14.16		0.53		−0.38		2		797
		AA45,405 S283 # 7			6 35 40	+00 47.0	1	13.65		0.53		−0.45		3		797
47442	−18 01492	HR 2443			6 35 41	−18 11.6	12	4.42	.012	1.15	.008	1.04	.011	38	K0 II/III	15,58,1007,1013,1020*
47582	−45 02635				6 35 42	−45 32.8	1	9.51		0.60				4	G0 V	2011
47360	+4 01360	LS VI +04 025			6 35 44	+04 40.1	1	8.19		0.13		−0.69		2	B0.5V	1012
47359	+5 01340	LS VI +04 026			6 35 46	+04 55.5	1	8.67		0.29		−0.69		3	B0.5Vpe:	493
47174	+42 01585	HR 2427		⋆ A	6 35 46	+42 32.1	10	4.80	.016	1.24	.006	1.30	.015	28	K3 III	15,1003,1080,1355*
261175					6 35 47	+06 11.	1	10.56		0.26		−0.52		3	G0	493
47337	+10 01205				6 35 47	+10 38.7	1	9.11		0.01		−0.12		1	A0 V	159
47601	−43 02570				6 35 49	−43 24.4	5	6.86	.002	−0.16	.011	−0.63	.004	29	B3 V	278,402,743,1075,2011
		AA45,405 S283 # 1			6 35 50	+00 47.4	1	13.26		0.49		0.01		3		797
47382	+4 01361	LS VI +04 027			6 35 50	+04 39.1	1	7.14		0.17		−0.76		2	B0 III	1012
47420	−2 01691	HR 2440			6 35 50	−02 29.9	2	6.13	.005	1.48	.000	1.77	.000	7	K2	15,1061
		LS VI +04 028			6 35 51	+04 41.2	1	10.56		0.18		−0.58		3		493
47255	+31 01369				6 35 51	+31 48.7	1	7.63		0.11		0.06		5	A0	627
		AA45,405 S283 # 2			6 35 52	+00 47.4	1	14.22		0.65		−0.40		5		797
		AA45,405 S283 # 3			6 35 52	+00 47.5	1	15.08		0.55		0.03		2		797
		LP 720 - 28			6 35 52	−12 14.2	1	11.29		0.45		−0.21		2		1696
		AA45,405 S283 # 4			6 35 53	+00 47.1	1	14.75		0.64		−0.10		3		797
	−49 02340	IDS06347S4957		AB	6 35 54	−49 59.6	3	9.62	.015	1.46	.013	1.20		8	K0	158,912,1705
47536	−32 03216	HR 2447			6 35 55	−32 17.8	5	5.25	.019	1.18	.010	1.12	.000	16	K1 III	15,1075,1311,2006,3005
	−40 02586				6 35 55	−41 01.1	2	11.14	.010	0.74	.010	0.13	.000	6	K0	158,1696
	−43 02572				6 35 56	−43 37.2	2	9.92	.008	0.79	.000	0.36		10	K0	863,1075
47645	−44 02797				6 35 57	−44 55.9	3	8.05	.003	−0.09	.004	−0.29	.015	16	B8 V	977,1075,2011
47398	+4 01363	LS VI +04 029			6 35 58	+04 40.3	2	8.67	.010	0.08	.025	−0.78	.000	6	B1 V	399,493
47600	−38 02793				6 35 59	−38 35.0	1	7.91		−0.12		−0.59		2	B4 III	401
261026	+31 01370	IDS06328N3141		A	6 36 01	+31 38.2	1	8.69		0.43		−0.02		3	F5	627
47418	+3 01335				6 36 02	+03 27.3	1	8.80		0.96		0.65		3	G5 II	8100
		LS VI -04 004		⋆ V	6 36 02	−04 19.1	1	11.09		0.38		−0.39		2		490
47432	+1 01443	HR 2442, V689 Mon		⋆	6 36 03	+01 39.5	4	6.22	.018	0.14	.009	−0.81	.015	14	O9.5II	15,154,1415,8100
261229	+10 01207				6 36 03	+10 45.9	1	10.47		0.28		−0.36		1	A0	159
261025		AK Aur			6 36 03	+31 39.5	1	10.36		0.19		0.12		1	A3	627
	−44 02800				6 36 03	−44 18.8	2	9.04	.010	1.12	.004	0.97		9		863,2011
47657	−45 02639	IDS06346S4530		AB	6 36 04	−45 33.0	3	8.27	.009	−0.07	.002	−0.29	.006	16	B8/9 V	977,1075,2011
47358	+22 01416	HR 2436			6 36 05	+22 04.6	1	6.04		1.03				2	G9 III	71
47417	+7 01386	LS VI +06 007		A	6 36 06	+06 56.8	2	6.97	.000	0.00	.005	−0.86		3	B0 IV	369,1012
47417	+7 01386	LS VI +06 007		B	6 36 06	+06 56.8	1	10.26		1.85				4		369
	+5 01342				6 36 08	+05 11.4	1	9.08		1.66				3	M2	369
47416	+8 01425				6 36 08	+08 01.0	1	7.79		−0.11		−0.53		2	B5 V	401
261307	+5 01343	LS VI +05 020			6 36 09	+05 07.1	1	9.31		0.32		−0.25		5	B8	493
47471	−9 01557				6 36 09	−09 16.6	1	7.48		1.64				3	M1	369
47431	+4 01365	HR 2441			6 36 10	+04 44.8	3	6.56	.005	−0.07	.004	−0.33	.013	9	B8 IIIn	15,1061,1079
261261					6 36 10	+11 04.	1	10.37		0.45		0.00		1	F2	159
261227	+16 01230				6 36 10	+16 20.9	1	9.38		1.74				3	A2	369
47670	−43 02576	HR 2451			6 36 14	−43 09.0	9	3.17	.009	−0.11	.003	−0.40	.009	24	B8 III	15,200,278,1020,1034*
47671	−44 02804				6 36 14	−44 27.1	3	7.69	.005	−0.02	.002	−0.02	.015	16	A0 V	977,1075,2011
		NGC 2264 - 1			6 36 15	+09 45.7	1	14.32		0.93		0.03		2		67
47485	+0 01520	G 108 - 43			6 36 20	+00 12.0	1	9.05		0.57		−0.02		2	A3	1620
47270	+44 01506	HR 2434			6 36 20	+44 03.6	1	6.34		1.24		1.19		2	K1 III	1733
261331	+9 01320	NGC 2264 - 2		⋆ V	6 36 21	+09 43.8	3	9.72	.016	0.26	.018	0.19	.034	6	A7 III-IV	67,1718,8078
		NGC 2264 - 3			6 36 22	+09 23.9	1	14.58		1.45		1.13		1		67
47561	−16 01554	HR 2448			6 36 22	−16 49.7	2	6.01	.015	0.03	.000	0.09		6	A0 IV	401,2009
261300	+15 01270				6 36 23	+15 23.4	1	9.26		1.84				3	K7	369
47395	+28 01196	HR 2438		⋆ AB	6 36 24	+28 18.5	3	6.03		−0.09	.008	−0.48	.008	5	B7 III	300,1022,1079
		HRC 207, R Mon			6 36 25	+08 47.0	1	12.18		0.62		−0.37		1		1588
		NGC 2264 - 4			6 36 25	+09 23.7	1	13.46		0.43		0.01		1		67
		NGC 2264 - 5			6 36 25	+09 33.1	1	14.11		0.54		0.09		3		67
261355	+9 01321	NGC 2264 - 6			6 36 25	+09 34.5	3	8.20	.021	0.36	.018	−0.06	.016	5	F5	67,1718,8078
	+8 01427	R Mon			6 36 26	+08 47.0	3	11.29	.730	0.53	.068	−0.11	.046	4		405,8013,8027
47469	+9 01322	NGC 2264 - 7		⋆ AB	6 36 26	+09 41.5	3	7.77	.026	−0.12	.013	−0.61	.025	5	B3 V	67,159,1718
	−17 01567				6 36 26	−17 55.4	1	9.42		0.27		0.13		1	A3	1732
		NGC 2264 - 8			6 36 27	+09 33.5	1	13.62		1.26		1.01		3		67
47415	+24 01343	HR 2439			6 36 27	+24 38.7	1	6.36		0.52				std	F8 IV	1217
	−17 01568				6 36 28	−17 53.2	1	9.12		0.07		−0.54		4	A0	1732
		NGC 2264 - 9			6 36 29	+10 12.1	1	11.98		2.04		2.16		2		67
		LP 87 - 145		A	6 36 29	+61 11.1	1	11.47		1.36				3		1726
		LP 87 - 145		B	6 36 29	+61 11.1	1	11.92		0.44		−0.01		3		1726
47720	−44 02807				6 36 29	−44 14.3	3	8.46	.007	0.47	.002	−0.02	.008	16	F5 V	977,1075,2011
		NGC 2264 - 12			6 36 30	+09 28.4	1	15.47		1.30		0.19		1		67
		NGC 2264 - 10			6 36 30	+09 35.1	2	11.24	.005	0.56	.034	0.18	.024	3		67,159
		CCS 546			6 36 30	+24 10.	1	10.55		1.22				1	R2	1238
		NGC 2264 - 11			6 36 31	+09 29.1	1	16.04		0.87		0.07		1		67

HD	DM	Other Id	N Rem	α_{1950}	δ_{1950}	S	V	σ_V	B–V	σ_{B-V}	U–B	σ_{U-B}	n	Spectrum	References
48584	−77 00270			6 36 31	−77 38.6	1	6.67		0.49		0.02		5	F6 V	1628
		LS VI -04 005		6 36 32	−04 20.2	1	10.90		0.27		-0.68		3		490
		NGC 2264 - 13		6 36 33	+09 39.8	1	12.04		1.68		1.46		2		67
		NGC 2264 - 14		6 36 34	+09 29.2	1	14.82		0.80		0.21		1		67
		NGC 2264 - 15		6 36 35	+09 29.6	1	14.64		1.22		0.68		2		67
261391	+7 01388			6 36 36	+07 35.3	1	9.75		0.41				1	F8	592
		NGC 2264 - 16		6 36 37	+09 28.6	1	16.10		0.96		0.36		3		67
		NGC 2264 - 17		6 36 37	+09 32.0	2	12.87	.005	0.38	.019	0.26	.010	3		67,8078
		NGC 2264 - 18		6 36 39	+09 27.0	1	15.20		0.85		0.24		2		67
261457	+3 01339	LS VI +03 008		6 36 40	+03 54.3	1	9.95		0.32		0.12		1	G5	493
		NGC 2264 - 19		6 36 42	+09 25.5	1	15.51		0.83		-0.02		1		67
261446	+9 01323	NGC 2264 - 20	★ V	6 36 43	+09 44.8	2	10.28	.010	0.42	.010	0.13	.024	7	F2 III	67,1718
261448	+7 01390			6 36 44	+07 41.0	1	10.63		0.13				1	B8	592
		NGC 2264 - 21		6 36 44	+09 29.4	1	14.06		0.78		0.26		2		67
47640	−16 01558			6 36 44	−16 45.8	1	8.62		0.00		-0.40		2	B8 V	401
		NGC 2264 - 23		6 36 45	+09 23.7	1	13.72		0.83		0.52		2		67
		NGC 2264 - 22		6 36 45	+09 32.5	1	17.06		1.08		0.20		3		67
		LS VI +03 009		6 36 48	+03 41.9	1	10.70		0.59		-0.33		2	B2 V	1012
47806	−45 02647			6 36 48	−45 14.0	4	8.27	.008	0.42	.006	-0.04	.014	22	F2 V	402,977,1075,2011
47553	+10 01211	NGC 2264 - 24		6 36 49	+10 17.2	2	8.56	.000	-0.06	.000	-0.35	.000	3	B9 III	67,1718
47719	−28 03230			6 36 49	−28 42.6	1	9.50		-0.12		-0.47		3	B7 IV	1732
47554	+9 01324	NGC 2264 - 25		6 36 50	+09 41.1	3	7.82	.013	0.37	.013	0.01	.016	6	F2 V	67,1718,8078
261490	+8 01432	LS VI +08 013		6 36 51	+08 23.8	1	8.88		0.24		-0.51		4	B2 IIIK	493
		NGC 2264 - 26		6 36 52	+09 25.9	1	11.78		0.48		-0.02		2		67
261340	+30 01292	G 103 - 48		6 36 52	+30 05.3	1	10.87		0.85		0.56		1	G5	333,1620
47668	−18 01498			6 36 52	−18 08.4	1	7.13		1.65		1.95		4	M1/2 III	1657
		NGC 2264 - 27		6 36 53	+09 30.0	1	12.04		0.53		0.21		2	A0 III	67
		HRC 209	★ AB	6 36 54	+09 02.	1	12.38		0.76		-0.32		1		776
		NGC 2264 - 301		6 36 55	+09 47.1	2	12.56	.015	0.63	.010	0.25	.025	4	F4 V	991,8114
		NGC 2264 - 302		6 36 56	+09 37.5	2	13.25	.085	0.73	.015	0.07	.070	4	F7 V	991,8114
		G 103 - 49		6 36 56	+28 38.2	1	11.93		1.50		1.19		1		333,1620
47805	−37 02970			6 36 57	−37 57.0	1	7.16		-0.10		-0.36		2	B7/8 V	401
47575	+13 01356	HR 2449	★ AB	6 36 59	+13 01.8	1			0.06		0.11		3	A2 V	1022
		G 103 - 50		6 36 59	+28 30.0	1	12.07		0.86		0.30		1		333,1620
47667	−14 01525	HR 2450		6 37 00	−14 06.0	7	4.82	.010	1.49	.007	1.67	.005	21	K2 III	15,1075,1425,2012*
47355	+53 01049			6 37 01	+53 26.7	1	7.50		0.51		0.15		4	F5	1733
		NGC 2264 - 305		6 37 02	+09 43.7	2	13.89	.005	0.84	.030	0.32	.045	4		991,8114
47872	−45 02650			6 37 02	−45 19.7	3	8.16	.003	0.19	.002	0.11	.001	16	A2 III	977,1075,2011
		NGC 2264 - 28		6 37 03	+09 28.9	2	12.28	.000	0.47	.000	-0.10	.000	3		67,1718
		NGC 2264 - 306		6 37 03	+09 51.7	1	13.44		0.62		0.51		1		8114
261583	+10 01212	NGC 2264 - 29		6 37 05	+10 18.6	2	10.16	.045	0.43	.020	0.02	.015	2	F5	67,1718
		NGC 2264 - 307		6 37 06	+09 31.9	2	13.95	.030	0.80	.040	0.43	.230	4		991,8114
47480	+37 01558			6 37 07	+37 18.5	1	8.72		0.38		0.16		2	F2	1733
261585	+9 01326	NGC 2264 - 30		6 37 08	+09 30.9	4	10.76	.014	0.04	.012	-0.01	.017	9	A0 V	67,991,8078,8114
261586	+9 01325	NGC 2264 - 31		6 37 09	+09 25.0	3	10.25	.008	0.33	.018	-0.01	.026	4	A7 IV	67,1718,8078
288936	+1 01449			6 37 10	+01 00.9	1	9.22		1.75				3	K7	369
		NGC 2264 - 32		6 37 10	+10 07.7	1	12.99		0.77		0.09		1		67
295320				6 37 10	−04 32.	1	9.29		1.80				3		369
		NGC 2264 - 312		6 37 11	+09 33.5	2	13.70	.034	0.59	.034	-0.04	.234	3		991,8114
		NGC 2264 - 33		6 37 11	+09 37.3	4	11.63	.029	2.42	.074	2.68	.126	8	K1 V	67,991,8078,8114
		NGC 2264 - 311		6 37 11	+09 46.4	2	14.22	.097	0.59	.019	0.23	.253	3		991,8114
261622		NGC 2264 - 34		6 37 12	+10 08.5	1	10.91		0.41		0.09		1	F0	67
47925	−46 02570			6 37 16	−46 07.8	3	8.38	.002	0.01	.002	0.01	.011	16	A0 V	977,1075,2011
		Aur sq 1 # 4		6 37 18	+44 12.7	1	12.50		0.71		0.14		6		578
47923	−43 02587			6 37 18	−43 46.7	3	7.56	.000	-0.04	.009	-0.06	.000	10	B9.5V	278,1075,2011
47973	−48 02417	HR 2462	★ A	6 37 18	−48 10.5	6	4.94	.025	0.88	.005	0.60	.012	19	G8 III	15,182,278,388,1075,3077
47973	−48 02417	IDS06360S4808	AB	6 37 18	−48 10.5	1	4.92		0.87				4	G8 III	2011
261658	+9 01327	NGC 2264 - 35		6 37 19	+09 25.5	2	10.32	.000	0.08	.024	0.03	.034	3	A0	67,8078
		NGC 2264 - 318		6 37 20	+09 35.7	3	13.04	.104	0.50	.036	-0.09	.071	4		991,8078,8114
261657		NGC 2264 - 36		6 37 20	+09 37.6	4	10.99	.030	0.02	.010	-0.02	.017	8	B9 V	67,991,8078,8114
		Aur sq 1 # 5		6 37 20	+44 13.6	1	12.75		0.56		0.24		4		578
		NGC 2264 - 320		6 37 21	+09 38.6	3	11.29	.023	0.60	.025	0.07	.022	4	F5 V	991,8078,8114
47851	−31 03520	IDS06355S3144	A	6 37 22	−31 47.4	2	7.68	.004	-0.18	.010	-0.76	.000	4	B2 V	55,1732
	+7 01394	BE Mon		6 37 23	+07 39.2	2	10.24	.003	0.95	.023			2	G5	592,1772
261683	+9 01328	NGC 2264 - 37		6 37 23	+09 20.8	3	8.08	.007	1.45	.018	1.70	.015	6	K5 III	67,1551,1718
		NGC 2264 - 321		6 37 23	+09 51.3	2	13.65	.063	0.53	.024	0.10	.243	3		991,8114
		Aur sq 1 # 9		6 37 23	+44 13.9	1	15.65		0.62		0.05		6		578
	+72 00328			6 37 23	+72 24.1	1	9.74		0.27		0.07		2	A2	1375
		NGC 2264 - 322		6 37 24	+09 44.6	2	13.15	.065	0.82	.040	0.48	.190	4	F5 V	991,8114
		Aur sq 1 # 6		6 37 24	+44 11.3	1	12.84		0.87		0.53		10		578
		Aur sq 1 # 3		6 37 25	+44 15.0	1	12.17		0.66		0.20		4		578
		GD 77		6 37 26	+47 43.3	1	14.80		0.13		-0.64		1	DA	3060
	−44 02820	IDS06360S4458	AB	6 37 28	−45 00.5	1	9.42		0.68				4		2011
48189	−61 00688	HR 2468	★ AB	6 37 28	−61 29.3	4	6.19	.008	0.63	.010	0.10		13	G1/2 V	15,404,1075,2035
		NGC 2264 - 325		6 37 28	+09 27.6	3	12.94	.049	0.92	.004	0.46	.072	4		991,8078,8114
		NGC 2264 - 38		6 37 29	+10 06.6	1	10.95		1.04		0.66		1		67
	+1 01451			6 37 30	+01 11.7	1	9.71		1.88				3	M3	369
		Aur sq 1 # 2		6 37 31	+44 11.7	1	11.42		0.59		0.11		7		578
47827	−23 04172	HR 2455	★ AB	6 37 31	−23 38.9	3	6.04	.004	-0.03	.000	-0.06	.000	11	A0 V	15,1075,2012
261711		NGC 2264 - 39		6 37 32	+09 20.0	2	11.34	.015	0.12	.010	0.10	.010	2	A2	67,8078

Table 1   351

HD	DM	Other Id	N	Rem	α_{1950}	δ_{1950}	S	V	σ_V	B–V	σ_{B-V}	U–B	σ_{U-B}	n	Spectrum	References
		NGC 2264 - 326			6 37 32	+09 32.0	2	13.33	.015	0.46	.015	0.04	.229	3		991,8114
47761	−4 01607	LS VI -04 006			6 37 33	−04 39.1	5	8.71	.023	0.16	.012	-0.60	.011	36	B2 V:pe	490,989,1012,1212,1729
		Aur sq 1 # 8			6 37 35	+44 13.3	1	13.77		0.81		0.26		8		578
		NGC 2264 - 42			6 37 36	+09 39.4	2	13.25	.010	0.77	.005	0.28	.034	3		991,8114
261737	+9 01329	NGC 2264 - 43			6 37 36	+09 44.7	5	10.56	.024	0.20	.018	0.15	.018	12	A7 III	67,991,1718,8078,8114
		Aur sq 1 # 10			6 37 37	+44 16.0	1	16.16		0.68		0.19		6		578
		NGC 2264 - 332			6 37 38	+10 02.8	2	13.85	.107	1.62	.039	1.18		3		991,8114
		Aur sq 1 # 11		V	6 37 39	+44 16.5	1	16.93		0.82		0.70		5		578
47758	+3 01348	IDS06350N0338		A	6 37 40	+03 35.3	1	7.65		1.62		1.77		3	K2 II	8100
		G 108 - 19			6 37 40	+04 31.5	1	13.05		0.68		-0.03		2		1658
261736	+9 01330	NGC 2264 - 46			6 37 40	+09 49.0	5	9.21	.022	0.21	.011	0.16	.019	16	A5 III	67,991,1718,8078,8114
		NGC 2264 - 335			6 37 40	+10 01.5	2	13.67	.068	0.91	.078	0.81	.253	3		991,8114
47681	+17 01346	IDS06348N1744		AB	6 37 40	+17 41.3	1	7.76		0.47		0.05		2	F8	1648
	−27 03173			A	6 37 41	−27 08.7	1	11.16		0.32		-0.05		2		129
	−27 03173			B	6 37 41	−27 08.7	1	12.31		0.35		0.20		2		129
		NGC 2264 - 336			6 37 42	+09 55.7	2	13.43	.070	0.69	.015	0.40	.115	4		991,8114
47572	+39 01701				6 37 42	+39 16.9	1	7.25		1.06		0.88		3	K1 III	833
		Aur sq 1 # 7			6 37 42	+44 13.2	1	13.35		0.48		0.06		1		578
47732	+9 01331	NGC 2264 - 50	⋆	A V	6 37 43	+09 51.9	7	8.15	.022	-0.14	.014	-0.71	.020	17	B2.5V	67,991,1651,1718,1732*
46588	+79 00212	HR 2401			6 37 44	+79 37.4	3	5.45	.011	0.50	.027	-0.01	.005	9	F8 V	22,254,1197
	−27 03174			A	6 37 44	−27 11.9	1	10.12		0.44		0.03		3		129
	−27 03174			B	6 37 44	−27 11.9	1	12.26		0.44		0.02		3		129
		NGC 2264 - 54			6 37 45	+09 53.0	2	14.27	.019	1.12	.107	0.81	.175	3		67,991
		G 108 - 20			6 37 45	−01 57.9	1	15.34		0.84		0.26		3		538
47999	−37 02979				6 37 45	−37 35.1	1	8.90		-0.11		-0.63		2	B3 V	401
		NGC 2264 - 58			6 37 46	+09 51.2	1	15.50		0.97		0.63		1		67
	−5 01737				6 37 46	−05 48.1	1	10.24		1.47				3	M2	369
		NGC 2264 - 59			6 37 47	+09 46.1	1	15.06		1.20		0.70		2		991
		NGC 2264 - 60			6 37 47	+10 04.7	1	12.46		1.03		0.67		1		67
47946	−30 03386	HR 2460			6 37 48	−30 25.3	4	5.71	.009	1.14	.005	1.10		12	K1 III	15,1075,2035,3005
		NGC 2264 - 62			6 37 49	+09 35.8	2	12.36	.025	0.50	.015	-0.18	.005	2		1718,8078
		NGC 2264 - 64			6 37 49	+10 00.6	1	15.34		0.94		0.42		1		67
292090	−0 01385	LS VI -0 003			6 37 49	−00 18.5	2	9.87	.000	0.31	.000	-0.67	.000	4	O8	1011,1012
		NGC 2264 - 65			6 37 50	+09 22.0	1	11.71		0.56		-0.08		1		67
47756	+6 01338	HR 2454			6 37 51	+06 25.1	4	6.51	.005	-0.14	.006	-0.46	.012	11	B8 III	15,401,1061,1079
		NGC 2264 - 70			6 37 52	+09 29.7	2	11.14	.080	0.58	.040	0.06	.015	4		67,8078
261783	+9 01333	NGC 2264 - 69			6 37 52	+09 38.6	5	8.27	.014	1.39	.023	1.54	.039	18	K3 II-III	67,1551,1718,8078,8100
		NGC 2264 - 66	⋆	C	6 37 52	+09 50.2	3	12.35	.046	0.73	.033	0.11	.099	5	A0 V	67,991,8114
		NGC 2264 - 67	⋆	B	6 37 52	+09 50.3	5	10.88	.034	0.61	.019	-0.40	.027	9	B2 V	67,545,991,1651,8114
		NGC 2264 - 68			6 37 52	+09 57.8	3	11.73	.027	0.63	.025	0.12	.021	6	G0 IV-V	67,991,8114
47821	−6 01664				6 37 52	−06 18.0	1	6.85		1.65				3	M3 III	369
47755	+9 01332	NGC 2264 - 74	⋆	A	6 37 53	+09 50.1	5	8.46	.024	-0.12	.011	-0.62	.018	12	B3 Vn	67,1651,1718,8078,8114
261782	+10 01216	NGC 2264 - 73			6 37 53	+10 00.6	4	9.34	.022	0.85	.004	0.50	.022	8	G5 IIIp	67,1551,1718,8078
		NGC 2264 - 76			6 37 54	+09 51.5	1	14.11		0.76		0.41		1		67
		NGC 2264 - 77	⋆	V	6 37 55	+09 37.9	3	14.51	.031	1.15	.041	0.59	.128	4		67,991,5011
48069	−43 02592				6 37 55	−43 22.8	7	7.17	.005	1.55	.005	1.84	.015	85	M2 III	460,743,863,977,1075*
		NGC 2264 - 79	⋆	V	6 37 56	+09 36.8	1	15.87		0.56		-1.21		1		67
		NGC 2264 - 78	⋆	V	6 37 56	+09 53.9	1	15.40		1.23		-0.42		1		67
		NGC 2264 - 355			6 37 56	+09 57.2	2	13.63	.058	0.79	.063	0.17	.083	3		991,8114
		NGC 2264 - 84			6 37 57	+09 36.5	4	11.96	.007	0.58	.022	0.10	.030	7	G0 V	991,1718,8078,8114
47777	+9 01334	NGC 2264 - 83			6 37 57	+09 42.2	8	7.94	.014	-0.16	.013	-0.83	.024	19	B2 IV	67,493,991,1651,1718,2033*
		NGC 2264 - 81			6 37 57	+09 52.7	1	16.27		1.26		0.08		1		67
		NGC 2264 - 80	⋆	V	6 37 57	+09 54.6	1	15.25		1.54		0.72		1		67
		NGC 2264 - 85			6 37 58	+09 35.2	1	14.98		1.09		0.78		1		67
261810	+9 01335	NGC 2264 - 88			6 37 58	+09 48.9	7	9.07	.016	-0.11	.009	-0.64	.020	26	B3 V	67,990,991,1651,2033*
		NGC 2264 - 89	⋆	V	6 37 58	+09 50.0	1	16.34		1.05		-0.59		1		67
261809	+10 01217	NGC 2264 - 87			6 37 58	+10 11.4	2	10.77	.030	0.22	.015	0.11	.040	2	A0 V	67,8078
		NGC 2264 - 90			6 37 59	+09 50.9	6	12.75	.051	0.15	.009	-0.05	.052	13	B4 V	67,351,991,1651,1718,8078
	−27 03176				6 37 59	−27 09.6	1	11.56		0.04		-0.05		3		129
		NGC 2264 - 366			6 38 00	+09 26.7	2	14.43	.054	1.01	.044	0.51	.224	3		991,8114
		NGC 2264 - 365			6 38 00	+09 31.6	2	14.43	.097	1.17	.161	0.46		3		991,8114
		NGC 2264 - 364			6 38 00	+09 51.3	2	14.62	.078	0.70	.073	0.34	.015	3		991,8114
		NGC 2264 - 92	⋆	V	6 38 01	+09 52.1	3	11.69	.023	0.85	.007	0.40	.017	34	K0 IVp	67,990,8078
		NGC 2264 - 91			6 38 01	+10 11.6	1	12.32		0.67		-0.05		2		67
		NGC 2264 - 370			6 38 02	+09 31.7	1	12.19		0.75		0.11		1		8078
		NGC 2264 - 95			6 38 02	+09 35.5	1	16.05		1.11		-0.01		1		67
		NGC 2264 - 93			6 38 02	+09 49.0	1	13.32		0.92		0.65		1		67
		NGC 2264 - 96	⋆	V	6 38 02	+09 52.3	3	14.00	.031	0.98	.048	0.46	.238	4		67,991,8114
261840	+10 01218	NGC 2264 - 94			6 38 02	+10 14.3	1	10.42		0.43		-0.08		1	F8	67
		NGC 2264 - 98			6 38 03	+10 12.7	1	11.75		0.62		-0.07		1		67
261842	+9 01337	NGC 2264 - 103	⋆	AB	6 38 04	+09 24.8	3	10.09	.023	0.01	.004	-0.10	.021	5	B5	67,991,8114
		NGC 2264 - 99			6 38 04	+09 27.1	2	10.84	.028	0.38	.066	-0.04	.005	4		67,8078
261841	+9 01336	NGC 2264 - 100	⋆	G	6 38 04	+09 54.6	5	10.03	.020	0.12	.013	0.06	.016	12	A2 IV	67,991,8078,8084,8114
		NGC 2264 - 104	⋆	S	6 38 04	+09 56.2	4	11.39	.017	0.23	.027	0.14	.015	9	A5 IV	67,991,8078,8114
		NGC 2264 - 105	⋆	V	6 38 05	+09 39.7	2	15.26	.122	1.01	.044	-0.21	.122	3		67,991
		NGC 2264 - 106			6 38 05	+09 40.3	1	13.28		0.72		0.41		1		67
	−27 03180				6 38 05	−27 21.1	2	9.24	.011	-0.10	.011	-0.56	.019	8	B5	129,1732
		NGC 2264 - 110			6 38 06	+09 46.3	2	14.45	.097	1.23	.205	0.67	.317	3		991,8114
		NGC 2264 - 108			6 38 06	+09 47.6	7	12.07	.044	0.59	.016	0.14	.029	17	G0 III-IV	67,351,990,991,1718*
261878	+9 01338	NGC 2264 - 109	⋆	F	6 38 06	+09 54.7	5	9.12	.032	-0.11	.020	-0.56	.021	8	B6 V	67,991,8078,8084,8114

HD	DM	Other Id	N	Rem	α_{1950}	δ_{1950}	S	V	σ_V	B–V	σ_{B-V}	U–B	σ_{U-B}	n	Spectrum	References
48012	+10 01219	NGC 2264 - 107			6 38 06	+10 04.6	3	8.85	.023	-0.06	.001	-0.42	.010	15	B7 V	67,1651,1718,8100
261879		NGC 2264 - 112			6 38 07	+09 42.0	4	10.81	.023	-0.02	.012	-0.15	.028	8	A0 V	67,991,8078,8114
		NGC 2264 - 114			6 38 08	+09 16.8	1	11.54		0.52		0.10		1		67
47627	+44 01513	IDS06345N4407		AB	6 38 08	+44 04.4	1	9.38		0.48		0.09		2	F5	578
47898	−8 01515				6 38 08	−08 53.4	1	9.11		0.07		0.08		2	B9	401
		NGC 2264 - 116			6 38 09	+09 33.5	5	11.65	.030	0.52	.031	0.06	.029	7	F5 III-IV	67,991,1718,8078,8114
		NGC 2264 - 115	⋆	V	6 38 09	+09 36.3	2	14.46	.029	0.94	.058	0.36	.029	3		67,991
		NGC 2264 - 117			6 38 09	+09 50.1	3	13.49	.020	0.72	.016	0.19	.134	4		67,991,8114
		NGC 2264 - 118			6 38 10	+09 22.9	1	11.78		0.56		-0.03		1		67
		NGC 2264 - 121	⋆	D	6 38 11	+09 57.0	2	11.76	.760	0.49	.205	-0.02	.078	3	F8 V	67,991,8114
261905		NGC 2264 - 128			6 38 12	+09 14.5	2	11.00	.010	0.23	.020	0.05	.055	2	A5	67,8078
		NGC 2264 - 127		V	6 38 12	+09 33.1	1	15.75		1.50		0.87		1		67
		NGC 2264 - 126	⋆	V	6 38 12	+09 40.7	1	15.34		1.06		0.32		2		991
		NGC 2264 - 125			6 38 12	+09 51.5	4	12.32	.045	0.60	.021	0.06	.015	5	F9 V	67,991,8078,8114
47752	+24 01357	G 109 - 20			6 38 12	+24 00.6	3	8.11	.026	1.03	.010	0.92		7	K2	1197,1758,3072
47731	+28 01207	HR 2453	⋆	A	6 38 12	+28 14.7	1	6.42		1.09		0.85		5	G5 Ib	1355
47839	+10 01220	NGC 2264 - 131	⋆	AB	6 38 13	+09 56.6	24	4.65	.013	-0.24	.011	-1.06	.013	117	O7 V((f))	15,67,154,991,1011,1020*
		NGC 2264 - 389			6 38 13	+09 59.5	2	13.76	.122	0.96	.034	0.81	.550	3		991,8114
		NGC 2264 - 134			6 38 14	+09 33.8	3	12.37	.042	0.86	.048	0.14	.165	5	G5 V	67,991,8078
261902	+9 01339	NGC 2264 - 132			6 38 14	+09 36.4	5	10.22	.017	-0.04	.011	-0.30	.018	10	B8 V	67,991,1718,8078,8114
		NGC 2264 - 390			6 38 14	+09 56.9	2	10.43	.093	0.02	.010	-0.47	.200	3	A3 V	991,8114
		NGC 2264 - 133	⋆	V	6 38 14	+09 58.3	4	13.82	.081	1.12	.047	0.48	.067	5		67,991,5011,8114
		NGC 2264 - 136	⋆	V	6 38 15	+09 38.1	1	15.16		1.58		1.07		1		67
		NGC 2264 - 396			6 38 15	+09 54.3	2	14.67	.010	1.06	.131	0.36		3		991,8114
	+10 01222	NGC 2264 - 137	⋆	C	6 38 15	+09 55.2	4	9.94	.023	-0.09	.013	-0.40	.016	8	B8 V	67,991,8078,8114
		NGC 2264 - 395			6 38 15	+10 01.8	2	13.55	.112	1.09	.102	0.62	.516	3		991,8114
48087	−38 02817	HR 2465			6 38 15	−38 06.7	1	6.58		1.18				4	K1 III	2006
48150	−43 02596			V	6 38 15	−43 21.3	4	7.39	.003	-0.14	.002	-0.66	.006	57	B3 V	278,863,1075,2011
261904		NGC 2264 - 138			6 38 16	+09 27.1	5	10.20	.020	0.06	.007	-0.02	.028	9	A0 V	67,991,1718,8078,8114
		NGC 2264 - 139	⋆	V	6 38 16	+09 35.6	3	13.40	.111	1.27	.039	0.75	.105	3	G8 V:	67,1718,5011
261938	+10 01223	NGC 2264 - 142	⋆	AB	6 38 16	+09 55.7	4	8.86	.038	-0.09	.016	-0.53	.024	7	B6 V	67,991,1651,8084
47703	+36 01482	HR 2452			6 38 16	+35 58.8	1	6.46		0.49				2	F8 V	71
48151	−45 02660				6 38 16	−45 30.8	3	8.58	.010	1.01	.003	0.78	.003	18	G8 III	863,977,2011
47892	+0 01542				6 38 17	+00 11.2	1	8.67		0.02		-0.41		4	A0	1732
		NGC 2264 - 147			6 38 17	+09 18.4	1	10.96		0.77		0.47		1		67
	+9 01340	NGC 2264 - 145			6 38 17	+09 31.1	4	10.66	.012	0.05	.007	-0.03	.032	10	A0 V	67,991,8078,8114
		NGC 2264 - 141			6 38 17	+09 37.0	1	14.70		1.15		0.87		1		67
		NGC 2264 - 146		V	6 38 17	+09 41.6	2	14.57	.088	1.18	.068	0.59	.141	3		67,991
		NGC 2264 - 144			6 38 17	+09 51.3	3	13.86	.034	1.33	.053	0.76	.140	4		67,991,8114
		NGC 2264 - 404			6 38 17	+09 54.8	2	13.00	.040	0.78	.005	0.16	.175	4	G8 V:	991,8114
		NGC 2264 - 143	⋆	B	6 38 17	+09 55.7	1	10.59		0.07		-0.14		1		67
261903	+9 01341	NGC 2264 - 152			6 38 18	+09 30.3	3	9.12	.019	-0.06	.008	-0.41	.043	8	B8 Vn	67,1718,8078
		NGC 2264 - 148			6 38 18	+09 37.2	1	13.60		0.69		0.09		1		67
		NGC 2264 - 149			6 38 18	+09 37.8	1	14.35		1.01		0.64		1		67
		NGC 2264 - 150			6 38 18	+09 38.1	1	15.44		1.28		1.22		1		67
		NGC 2264 - 151	⋆	C	6 38 18	+09 50.8	4	12.57	.030	0.49	.021	-0.03	.035	6	F5 V	67,991,8078,8114
		NGC 2264 - 373			6 38 18	+09 56.	1	10.06		0.00		-0.08		1		8078
		NGC 2264 - 416			6 38 19	+09 26.4	2	14.40	.060	0.70	.005	0.24	.200	4		991,8114
		NGC 2264 - 154			6 38 19	+09 34.2	4	12.71	.065	0.75	.041	0.13	.040	5	G2 III-IV	67,991,1718,8114
261940	+9 01342	NGC 2264 - 157			6 38 19	+09 35.9	4	10.08	.020	-0.06	.012	-0.34	.013	9	B8 V	67,991,1718,8078,8114
		NGC 2264 - 155			6 38 19	+09 38.2	1	16.48		1.43		0.98		1		67
		NGC 2264 - 153	⋆	V	6 38 19	+09 43.6	1	15.87		1.24		0.36		1		67
		NGC 2264 - 156			6 38 19	+09 51.2	1	14.66		1.25		0.69		1		67
		NGC 2264 - 412			6 38 19	+09 51.3	1	14.82		1.18		1.24		2		991
		NGC 2264 - 413			6 38 19	+09 54.4	1	14.92		1.11		0.65		2		991
		NGC 2264 - 411			6 38 19	+09 54.8	2	12.60	.102	0.90	.019	0.41	.093	3	G8 V	991,8114
261937	+10 01224	NGC 2264 - 158			6 38 19	+09 57.6	4	10.34	.017	0.36	.011	0.10	.021	12	F0 V	67,991,8078,8114
261939		NGC 2264 - 159			6 38 20	+09 39.3	4	10.97	.007	0.06	.011	-0.04	.024	7	A0 V	67,991,8078,8114
261941	+9 01343	NGC 2264 - 165			6 38 21	+09 25.8	5	10.98	.023	0.15	.009	0.10	.031	9	A2 V	67,991,1718,8078,8114
		NGC 2264 - 163			6 38 21	+09 38.7	1	14.05		0.89		0.52		2		991
		NGC 2264 - 164	⋆	V	6 38 21	+09 39.3	3	13.32	.053	0.87	.063	0.38	.067	4	G8 V	67,991,8114
		NGC 2264 - 161	⋆	V	6 38 21	+09 51.2	2	15.11	.117	0.84	.136	-0.58	.024	3		67,991
		HRC 230, SS Mon			6 38 21	+10 29.6	1	14.05		1.36		1.18		1		5011
		NGC 2264 - 424	⋆	V	6 38 22	+09 30.4	3	12.82	.078	0.96	.090	0.06	.037	5		776,991,8114
		NGC 2264 - 167			6 38 22	+09 38.6	1	14.09		0.89		0.14		2		991
		NGC 2264 - 168	⋆	V	6 38 22	+10 01.4	1	15.13		1.26		1.04		1		67
		NGC 2264 - 175		V	6 38 23	+09 33.6	1	16.12		1.41		0.48		1		67
		NGC 2264 - 169			6 38 23	+09 46.9	2	13.38	.010	0.75	.024	0.24	.015	3		67,991
261936	+10 01225	NGC 2264 - 172			6 38 23	+10 11.3	1	10.04		-0.05		-0.45		1	B8 V	67
		NGC 2264 - 174			6 38 24	+09 44.1	1	15.17		1.02				2		991
		NGC 2264 - 173			6 38 24	+09 45.7	1	16.41		1.29		0.77		1		67
		NGC 2264 - 428			6 38 24	+09 48.9	1	13.00		0.61		0.62		1		8114
261935	+10 01226	NGC 2264 - 177			6 38 24	+10 20.3	1	9.20		0.77		0.09		1	G5	67
47887	+9 01344	NGC 2264 - 178	⋆	A	6 38 25	+09 30.8	4	7.14	.022	-0.21	.016	-0.96	.014	7	B1.5V	67,1012,1651,1718
		NGC 2264 - 176			6 38 25	+09 38.3	1	15.79		1.59		0.85		1		67
261969	+10 01227	NGC 2264 - 179	⋆	A	6 38 25	+09 55.9	5	9.95	.014	-0.01	.016	-0.19	.008	11	B9.5V	67,991,1651,1718,8114
		NGC 2264 - 180			6 38 25	+10 03.6	2	12.86	.005	0.51	.000	0.05	.090	2		67,8114
		NGC 2264 - 181	⋆	B	6 38 26	+09 55.8	5	10.07	.025	-0.05	.017	-0.30	.015	9	B9 Vn	67,991,1651,1718,8114
262015		NGC 2264 - 182			6 38 27	+09 14.1	2	10.31	.005	0.06	.019	0.07	.015	3	A2 V	67,8078
		NGC 2264 - 184	⋆	V	6 38 27	+09 55.4	3	14.23	.149	0.93	.070	0.13	.068	4	K3 III	67,991,5011

Table 1   353

HD	DM	Other Id	N Rem	α_{1950}	δ_{1950}	S	V	σ_V	B–V	σ_{B-V}	U–B	σ_{U-B}	n	Spectrum	References
261971	+8 01448			6 38 28	+07 57.1	1	9.53		0.13		0.08		1	A3 V	159
		NGC 2264 - 439	⋆ V	6 38 28	+09 29.1	2	13.55	.000	0.87	.044	-0.05	.355	3		825,991
		NGC 2264 - 189		6 38 28	+09 30.4	3	11.33	.050	0.52	.028	0.05	.094	4	F2 V	67,991,1718
262013	+9 01346	NGC 2264 - 187		6 38 28	+09 38.7	5	9.24	.014	-0.08	.023	-0.35	.033	8	B8 V	67,991,1718,8078,8114
		NGC 2264 - 186		6 38 28	+09 58.0	1	15.62		1.56		0.94		1		67
261966	+13 01370			6 38 28	+13 40.2	1	8.64		1.81				3	M0	369
		NGC 2264 - 190		6 38 29	+09 58.6	3	12.34	.030	0.62	.041	0.13	.031	5	G0 V	67,991,8114
47863	+16 01242	HR 2457		6 38 29	+16 26.7	1			-0.01		-0.07		2	A1 V	1022
262014	+9 01347	NGC 2264 - 193		6 38 30	+09 34.2	3	9.80	.020	0.23	.020	0.10	.023	6	A7 IIIp	67,1718,8078
		NGC 2264 - 443		6 38 30	+09 49.6	2	14.12	.044	0.90	.063	0.90	.536	3		991,8114
47964	+0 01546	HR 2461		6 38 31	+00 32.6	3	5.78	.005	-0.10	.004	-0.36	.008	8	B8 III	15,1078,1079
		NGC 2264 - 447		6 38 31	+09 29.1	1	13.73		1.36				1		8114
		NGC 2264 - 194		6 38 31	+09 41.2	2	13.01	.039	0.60	.029	0.11	.000	3		991,8114
		NGC 2264 - 195		6 38 31	+09 55.1	2	12.64	.005	0.52	.014	0.09	.140	4		67,8114
47886	+11 01273	HR 2458		6 38 31	+11 03.1	2	6.16	.046	1.67	.005	2.03		5	M1 III	71,3001
		NGC 2264 - 451		6 38 32	+09 30.4	1	15.08		0.92		-0.24		2		991
		NGC 2264 - 192		6 38 32	+09 57.4	2	13.80	.080	0.50	.070	-0.11	.085	4		991,8114
		NGC 2264 - 197	⋆ V	6 38 32	+09 57.4	1	16.28		0.93		-0.13		1		67
		NGC 2264 - 196		6 38 32	+10 00.1	1	11.46		0.53		-0.05		2		67
		NGC 2264 - 199	⋆ V	6 38 33	+09 36.8	1	14.92		1.02		0.34		2		991
262042	+9 01348	NGC 2264 - 202		6 38 34	+09 15.7	4	8.99	.011	0.07	.015	-0.61	.025	8	B2 V	67,159,493,8078
		NGC 2264 - 203		6 38 35	+09 30.3	1	12.90		0.76		0.18		1		67
		NGC 2264 - 454		6 38 35	+09 33.3	2	13.97	.068	0.91	.175	0.78	.166	3		991,8114
		NGC 2264 - 455		6 38 35	+09 34.6	1	14.39		1.06		1.12		2		991
47836	+27 01194			6 38 35	+27 07.8	1	7.52		0.89				2	G8 III	20
262041		NGC 2264 - 205		6 38 36	+09 25.2	3	10.62	.032	0.32	.008	-0.02	.022	4	A8 V	67,1718,8078
		NGC 2264 - 457		6 38 36	+09 48.5	1	15.05		1.00		0.39		2		991
47934	+9 01349	NGC 2264 - 206	⋆ AB	6 38 37	+09 46.8	4	8.72	.013	-0.10	.013	-0.43	.017	8	B7 V	67,991,8078,8114
		NGC 2264 - 208		6 38 38	+09 30.3	3	12.60	.029	0.76	.024	0.20	.112	4		67,991,8114
48038	-12 01585	LSS 83		6 38 38	-12 08.1	1	6.91		0.03		-0.68		3	B2 III	399
		NGC 2264 - 209		6 38 39	+09 36.8	4	11.31	.012	0.35	.011	-0.01	.015	6	F2 V	67,991,8078,8114
		NGC 2264 - 466		6 38 40	+09 24.5	2	13.28	.078	0.60	.039	0.10	.010	3		991,8114
		NGC 2264 - 210		6 38 40	+09 29.3	3	13.37	.043	0.79	.021	0.23	.073	4		67,991,8114
48733	-69 00642			6 38 41	-70 02.9	1	9.32		1.36		1.30		1	R0	864
		NGC 2264 - 469		6 38 42	+09 26.8	3	12.69	.017	0.45	.019	0.13	.049	6		991,8078,8114
47961	+9 01350	NGC 2264 - 212	⋆ C	6 38 42	+09 54.1	6	7.49	.021	-0.15	.011	-0.75	.017	24	B2 IV	67,991,1651,1718,1732,8078
	-44 02834			6 38 43	-45 03.3	4	9.71	.009	0.95	.004	0.64	.012	17	K0	861,863,1075,1460
		NGC 2264 - 213		6 38 44	+09 41.6	2	13.86	.078	1.00	.005	0.52	.078	3		991,8114
		NGC 2264 - 214		6 38 44	+09 42.5	2	13.12	.005	0.81	.010	0.26	.054	3		991,8114
262066	+9 01351	NGC 2264 - 215	⋆ AB	6 38 45	+09 52.7	5	9.32	.026	0.06	.019	-0.15	.015	13	A0 IV-V	67,991,1718,8078,8114
		NGC 2264 - 216		6 38 46	+09 57.9	1	11.69		0.77		0.18		1		67
48060	-7 01524	IDS06364S0754	AB	6 38 46	-07 56.7	1	7.90		-0.09		-0.62		2	B9	401
	-55 01035			6 38 46	-55 33.8	1	9.94		1.31		1.17		2		3072
		NGC 2264 - 217	⋆ V	6 38 47	+09 29.8	4	13.73	.089	0.93	.125	0.02	.269	5	K5 II-III	67,776,991,5011
		NGC 2264 - 480		6 38 47	+10 00.9	2	14.45	.073	0.86	.005	0.58	.195	3		991,8114
262110	+9 01352	NGC 2264 - 220		6 38 48	+09 22.0	2	9.69	.005	0.46	.015	-0.05	.015	2	F8	67,1718
		NGC 2264 - 481		6 38 48	+09 41.0	1	14.57		1.27		0.83		2		991
48058	-6 01675			6 38 48	-06 35.4	1	8.62		1.64				3	K5	369
47984	+6 01346	LS VI +06 008		6 38 49	+06 09.6	1	6.97		0.27		0.07		2	B9 V	493
		NGC 2264 - 483		6 38 49	+09 54.3	2	14.95	.063	0.64	.122	0.46	.054	3		991,8114
262108	+9 01353	NGC 2264 - 222	⋆ D	6 38 49	+09 54.5	5	9.93	.033	0.13	.018	0.15	.026	9	A3 IV	67,991,1718,8078,8114
		NGC 2264 - 221		6 38 49	+10 05.3	3	12.14	.018	0.38	.021	-0.05	.049	3		67,1718,8078
48165	-28 03278			6 38 49	-28 48.3	1	8.10		-0.13		-0.58		2	B3 V	401
262109		NGC 2264 - 223		6 38 52	+09 46.6	2	10.91	.037	0.33	.028	0.03	.033	4	F5	67,8078
		G 251 - 16		6 38 52	+71 56.6	1	10.94		1.48		1.21		3	K7	7009
		NGC 2264 - 224		6 38 53	+09 23.9	4	11.51	.020	0.60	.049	0.17	.000	5	F5 V	67,991,1718,8114
		NGC 2264 - 225		6 38 54	+10 02.9	1	13.19		0.58		0.03		2		67
		NGC 2264 - 491	⋆ V	6 38 55	+09 43.4	3	13.91	.134	1.04	.054	0.27	.082	5		825,991,8114
		NGC 2264 - 490		6 38 55	+09 53.8	1	14.21		0.60		0.69		1		8114
47883	+31 01388	VW Gem		6 38 55	+31 30.2	1	8.14		2.42				1	C5 II	1238
		NGC 2264 - 492		6 38 56	+09 36.8	2	14.35	.102	1.97	.029	1.58		3		991,8114
		NGC 2264 - 493		6 38 57	+09 35.6	2	12.88	.010	0.55	.005	-0.03	.064	5	F3 V	991,8114
262138	+9 01355	NGC 2264 - 226		6 38 57	+09 50.5	3	9.61	.021	0.14	.010	0.14	.017	7	A3 III	67,8078,8114
		NGC 2264 - 227		6 38 57	+10 03.2	1	11.77		0.53		-0.11		1		67
262062	+19 01441			6 38 57	+19 47.6	1	9.35		0.46				1	F5	1245
	-4 01617	V 372 Mon		6 38 57	-04 33.1	1	9.98		3.11				1	SC	864
262141	+7 01405			6 38 58	+07 52.8	1	9.75		0.03		-0.23		1	A2	159
		NGC 2264 - 228		6 38 58	+09 46.5	5	11.12	.019	0.35	.010	0.02	.013	8	F0 V	67,991,1718,8078,8114
		NGC 2264 - 497		6 38 58	+09 47.2	1	14.85		1.01		0.53		2		991
262137	+10 01228	NGC 2264 - 229		6 38 58	+10 05.0	2	8.51	.015	1.19	.010	1.18	.005	5	K2 II-III	67,1551
		CCS 559		6 38 58	-20 06.2	1	11.02		3.24				1		864
		NGC 2264 - 500		6 38 59	+09 43.8	2	14.02	.039	1.01	.063	0.64	.151	3		991,8114
		NGC 2264 - 498		6 38 59	+09 49.5	1	14.68		0.72		0.43		2		991
		NGC 2264 - 502		6 39 02	+09 48.8	2	14.25	.078	0.61	.010	0.21	.029	3		991,8114
		NGC 2264 - 503		6 39 02	+09 55.0	2	13.39	.039	0.93	.010	0.54	.068	3		991,8114
		NGC 2264 - 504		6 39 03	+09 37.1	1	14.70		0.94		0.86		2		991
		NGC 2264 - 505		6 39 03	+09 45.7	2	13.19	.010	0.78	.044	0.28	.083	3		991,8114
48055	+9 01356	NGC 2264 - 231		6 39 05	+09 33.4	6	8.98	.010	-0.14	.008	-0.65	.019	14	B5 Vn	67,493,991,1732,8078,8114
262177	+10 01230	NGC 2264 - 232		6 39 06	+10 04.1	2	9.83	.058	0.00	.024	-0.05	.044	3	A0 V	67,8078
262210	+9 01357	NGC 2264 - 233	⋆ AB	6 39 07	+09 14.9	2	9.55	.010	0.57	.015	0.08	.029	3	G0 V	67,159

HD	DM	Other Id	N Rem	α_{1950}	δ_{1950}	S	V	σ_V	B–V	σ_{B-V}	U–B	σ_{U-B}	n	Spectrum	References
		NGC 2264 - 507		6 39 07	+09 56.7	2	13.81	.035	0.94	.025	0.62	.180	4		991,8114
47960	+25 01392	IDS06360N2534	A	6 39 07	+25 31.1	1	7.34		1.59		1.89		3	M0 III	833
		NGC 2264 - 509		6 39 08	+09 47.4	2	14.60	.019	1.10	.010	0.41	.390	3		991,8114
		NGC 2264 - 234		6 39 08	+10 01.6	2	12.41	.020	0.47	.020	-0.04	.005	2		67,1718
		NGC 2264 - 235		6 39 08	+10 02.4	3	13.47	.016	0.53	.066	0.09	.030	3		67,1718,8078
47882	+39 01713			6 39 08	+38 56.6	1	7.01		0.35		-0.04		2	F2 IV	105
		NGC 2264 - 511		6 39 09	+09 57.0	1	14.53		1.05		-0.33		1		8114
		NGC 2264 - 513		6 39 10	+09 44.4	2	13.11	.045	2.56	.020	1.17	.335	4		991,8114
		NGC 2264 - 514		6 39 10	+09 56.5	2	13.61	.039	1.07	.034	0.44	.365	3		991,8114
		NGC 2264 - 236		6 39 11	+09 53.0	2	11.37	.023	0.64	.019	0.15	.042	4		67,1718
		NGC 2264 - 516		6 39 13	+09 48.2	1	15.06		0.97		1.27		2		991
48008	+25 01393			6 39 13	+25 24.8	1	8.78		0.45		-0.04		2	F6 V	3026
48261	-30 03412			6 39 13	-30 35.8	1	7.23		-0.06		-0.17		2	B9 V	401
		NGC 2264 - 517		6 39 15	+09 33.5	2	14.06	.078	0.78	.024	0.60	.141	3		991,8114
48159	-7 01530			6 39 15	-07 55.6	1	8.27		0.03		-0.14		2	B9	401
48099	+6 01351	HR 2467	★	6 39 18	+06 23.7	12	6.36	.013	-0.04	.008	-0.94	.021	59	O6e	15,154,975,1011,1012*
		NGC 2264 - 519		6 39 18	+09 32.7	1	12.09		0.35		0.04		3		8078
48029	+18 01313			6 39 18	+18 50.1	1	8.60		-0.08		-0.84		2	B2 IV	555
		NGC 2264 - 520		6 39 19	+09 33.6	1	12.91		0.55		0.33		1		8114
		NGC 2264 - 521		6 39 19	+09 51.0	2	14.04	.010	0.63	.054	0.19	.399	3		991,8114
		NGC 2264 - 527		6 39 22	+09 31.6	2	15.09	.058	0.63	.024	0.29	.239	3		991,8114
		NGC 2264 - 526		6 39 22	+09 43.0	2	14.85	.083	0.58	.170	0.75	.224	3		991,8114
		NGC 2264 - 528		6 39 22	+09 52.8	2	13.57	.035	1.52	.145	0.99	.255	4		991,8114
48158	-6 01682			6 39 22	-06 06.2	1	8.16		1.58				2	M1	369
48342	-44 02841			6 39 22	-44 22.0	3	8.67	.006	0.63	.002	0.12	.009	16	G1 V	977,1075,2011
		NGC 2264 - 530		6 39 23	+09 39.4	1	12.37		0.76		0.65		1		8114
		Wolf 290		6 39 24	+17 24.7	1	13.16		0.74		0.14		2		1696
		NGC 2264 - 535		6 39 25	+09 43.0	2	13.23	.015	0.91	.025	0.66	.015	4		991,8114
		NGC 2264 - 536		6 39 25	+09 52.2	1	12.60		1.41		1.19		1		8114
	-45 02673			6 39 25	-45 25.4	1	10.54		0.39				4		2011
48402	-47 02520			6 39 26	-47 34.4	3	6.64	.004	-0.15	.005	-0.53	.013	10	B3/5 V	278,1075,2011
		NGC 2264 - 538		6 39 27	+09 47.2	2	11.30	.079	1.02	.005	1.06	.045	5	K5 V	991,8114
47914	+44 01518	HR 2459		6 39 27	+44 34.5	5	5.01	.016	1.49	.013	1.83	.014	12	K5 III	15,1003,1080,1363,3001
292072	-0 01399			6 39 27	-00 06.2	1	9.65		1.07		0.89		1	K2	1732
48384	-45 02674			6 39 28	-45 41.9	2	9.64	.009	0.46	.008	0.01		9	F3/5 V	863,2011
48403	-47 02521	HR 2476		6 39 28	-47 37.6	2	6.67	.019	1.59	.010	1.92		6	M2	2035,3055
48006	+32 01378			6 39 29	+32 36.5	1	6.74		1.47		1.75		3	G5	1625
		NGC 2264 - 542		6 39 30	+09 49.6	1	14.71		0.61		0.67		1		8114
48097	+17 01357	HR 2466		6 39 30	+17 41.8	3	5.17	.054	0.06	.003	0.01	.000	5	A2 V	1022,1363,3023
292060	+0 01550			6 39 31	+00 15.4	1	9.61		0.81		0.39		2	B8	1732
262320	+9 01358	NGC 2264 - 237		6 39 31	+09 35.6	2	9.44	.000	1.44	.000	1.27	.000	5	K2 II-III	67,1551
262318		NGC 2264 - 238		6 39 33	+10 12.5	2	9.96	.019	1.26	.015	1.56	.024	3	K0	67,1551
48217	-9 01601	HR 2469		6 39 33	-09 07.1	4	5.19	.008	1.52	.013	1.89	.000	12	M0 III	15,369,1078,3035
		G 103 - 52		6 39 34	+31 01.5	1	12.48		0.62		-0.12		2		333,1620
48144	+7 01409			6 39 35	+07 26.8	1	6.86		1.63				5	K5	369
262319	+10 01233	NGC 2264 - 239		6 39 35	+10 12.3	2	9.33	.000	1.16	.009	1.03	.014	4	K0	67,1551
		ApJ109,528 # 22		6 39 36	+44 43.	1	16.70		-0.05		-0.95		1	DA	3028
48383	-40 02625	HR 2475	★ AB	6 39 36	-40 18.1	5	6.11	.004	-0.14	.005	-0.60		18	B4 V	15,401,1020,1075,2012
48215	-5 01753			6 39 41	-06 03.9	1	7.01		-0.15		-0.55		3	B5 V	399
	-45 02675			6 39 41	-45 07.0	1	11.87		0.48				4		1075
48237	-9 01602			6 39 44	-09 49.8	1	9.31		-0.04		-0.33		2	A0	401
	-5 01754			6 39 45	-05 19.1	1	9.41		1.75				4	M2	369
48287	-15 01478	IDS06375S1555	AB	6 39 45	-15 57.5	1	6.95		-0.03		-0.41		2	B8/9 III	401
262375	+8 01456			6 39 46	+07 58.9	1	9.44		0.22		0.01		1	A3 V	159
48429	-44 02846			6 39 48	-45 03.8	3	8.65	.007	0.27	.003	0.10	.011	16	A7 V	977,1075,2011
48073	+37 01567	HR 2464		6 39 50	+37 11.9	1	6.20		1.03				2	K0	71
48282	-10 01651			6 39 51	-10 26.9	2	8.87	.007	0.29	.004	-0.38	.000	12	B3 III	1012,1775
48464	-45 02676			6 39 51	-45 13.6	8	8.99	.007	0.25	.006	0.06	.006	78	F0 V	977,1075,1460,1628,1657*
48381	-33 03156			6 39 53	-33 25.2	1	8.48		1.05				4	K5	2012
	-44 02849			6 39 54	-44 56.6	1	11.31		0.42				4		2011
	-45 02679			6 39 55	-45 10.2	1	10.90		0.47				4		2012
	-45 02678			6 39 55	-45 18.0	2	11.56	.010	0.13	.004	0.09		10		1460,2011
48486	-45 02681			6 39 57	-45 30.0	2	9.61	.009	1.15	.001	1.10		10	K1/2 III	863,1075
		NGC 2264 - 230		6 39 59	+09 29.9	3	12.45	.033	0.91	.116	0.72	.095	6	K0 V:	67,991,8114
		G 107 - 18		6 40 04	+47 20.3	1	12.71		1.58		1.07		2		538
48279	+1 01472	LS VI +01 010	★ AB	6 40 05	+01 46.0	3	7.89	.032	0.15	.024	-0.78	.017	7	O8 V	450,1011,1012
48279	+1 01472	IDS06375N0149	C	6 40 05	+01 46.0	1	9.00		0.35		0.00		3		450
48279	+1 01472	IDS06375N0149	D	6 40 05	+01 46.0	1	11.26		0.19		0.10		3		450
	-43 02617			6 40 06	-43 43.9	1	10.17		0.25				4	A2	2012
	-48 02445			6 40 06	-48 10.6	2	10.54	.010	0.39	.005	-0.22		6	F8	1075,1696
262433	+21 01349	AD Gem		6 40 09	+20 59.4	3	9.57	.043	0.53	.037	0.30		3	G0	592,1462,6011
47979	+53 01056	HR 2463		6 40 09	+53 21.0	1	6.27		1.08				2	K0	71
47954	+56 01149			6 40 10	+55 55.2	1	8.41		0.49		0.00		2	F8	1502
48311	-9 01606	IDS06378S0928		6 40 10	-09 31.0	1	8.52		0.06		0.04		2	A0	401
	-45 02683			6 40 10	-45 44.7	1	10.85		0.13				4		2011
48503	-39 02777			6 40 15	-39 44.3	1	7.75		-0.06		-0.21		2	B9 V	401
237536	+59 01010			6 40 16	+59 41.5	1	9.69		0.37		0.06		2	F0	1502
		G 108 - 23		6 40 18	+08 04.7	1	13.05		1.07		0.87		1		1658
48485	-31 03583			6 40 26	-31 37.0	1	7.80		-0.10		-0.34		2	B8/9 V	401
	-44 02854	IDS06390S4447	AB	6 40 26	-44 50.0	2	10.14	.014	0.12	.003	-0.02		10	A0	863,2011

Table 1 355

HD	DM	Other Id	N	Rem	α_{1950}	δ_{1950}	S	V	σ_V	B–V	σ_{B-V}	U–B	σ_{U-B}	n	Spectrum	References
292173	+0 01554				6 40 28	+00 35.3	1	10.20		1.61				4	M0	369
48348	+3 01371	HR 2474			6 40 29	+03 05.1	3	6.18	.005	1.37	.004	1.60	.020	9	K0	15,1078,1355
262629	+7 01418				6 40 31	+07 33.5	1	10.68		0.15		0.04		1	K0	159
		NGC 2269 - 10			6 40 32	+04 40.9	1	13.19		0.49		0.04		1		410
		LS VI -01 006			6 40 33	−01 20.0	1	11.62		0.16		0.15		1		490
		NGC 2269 - 5			6 40 35	+04 39.2	1	13.32		0.50		0.22		1		410
		NGC 2269 - 7			6 40 35	+04 41.2	1	12.64		0.37		0.15		1		410
48543	−38 02844	HR 2482		⋆ AB	6 40 35	−38 20.9	2	6.27	.005	0.35	.010	0.10		6	A3 V +F/G	404,2006
		NGC 2269 - 6			6 40 36	+04 40.5	1	13.91		0.32		0.20		1		410
	−45 02687				6 40 36	−45 08.9	1	10.40		0.85				4		2011
		NGC 2269 - 11			6 40 37	+04 40.5	1	13.74		0.42		0.09		1		410
		NGC 2269 - 1			6 40 38	+04 40.1	1	11.61		0.30		0.01		2		410
		NGC 2269 - 2			6 40 39	+04 39.8	1	12.51		0.37		0.20		1		410
	−9 01611				6 40 39	−09 51.9	1	9.70		1.80				4	M2	369
48501	−22 01505	HR 2481		⋆ A	6 40 39	−22 24.0	2	6.12	.005	0.33	.015	-0.01		6	F2 V	15,1279
48501	−22 01505	HR 2481		⋆ AB	6 40 39	−22 24.0	2	6.11	.010	0.37	.005			8	F2 V	2013,2029
48501	−22 01505	IDS06386S2221	B		6 40 39	−22 24.0	1	8.23		0.68		0.15		3		1279
		NGC 2269 - 3			6 40 40	+04 38.7	1	12.10		0.31		0.15		1		410
		NGC 2269 - 12			6 40 40	+04 39.5	1	14.03		0.46		0.47		1		410
		LS VI -01 007			6 40 41	−01 27.9	1	10.61		0.29		0.15		1		490
		NGC 2269 - 4			6 40 42	+04 39.4	1	13.19		0.43		0.19		1		410
262553	+25 01402	G 103 - 53			6 40 42	+25 35.4	1	10.19		0.68		0.09		1	G5	333,1620
262720	+4 01411	LS VI +04 032			6 40 43	+04 23.0	1	10.13		0.32		-0.36		3	B5	493
		NGC 2269 - 8			6 40 43	+04 41.3	1	12.53		0.32		0.04		1		410
262716	+5 01379				6 40 44	+05 41.1	1	10.83		1.82				3	A7	369
		NGC 2269 - 9			6 40 46	+04 40.8	1	12.11		0.94		0.48		1		410
		LS VI -04 007			6 40 46	−04 38.6	1	11.93		0.11		-0.72		2		490
48228	+40 01696	IDS06373N4044	A		6 40 47	+40 40.6	1	6.84		1.59		1.79		2	M4 III	1003
		CCS 568			6 40 48	−08 42.5	1	9.91		3.10		4.30		1	C4-5,4-5	864
	−33 03165				6 40 48	−33 35.6	1	9.94		0.99		0.72		1		6006
262645	+21 01355				6 40 51	+21 00.6	2	9.54	.010	0.36	.002	0.06		38	F0	592,1462,6011
48329	+25 01406	HR 2473		⋆ A	6 40 51	+25 10.9	9	2.99	.013	1.40	.010	1.47	.011	27	G8 Ib	15,542,1080,1194,1355*
48329	+25 01407	IDS06378N2514	B		6 40 51	+25 10.9	1	9.22		1.13		0.91		2	K0 III-IV	542
48272	+36 01494	HR 2471			6 40 51	+36 09.6	2	6.37		0.06		0.10	.009	5	A2 V	1022,1501
262741	+7 01424	LS VI +07 008			6 40 54	+07 17.7	2	10.48	.035	0.47	.015	-0.61	.010	8	B7	405,493
48676	−42 02683	IDS06393S4228	AB		6 40 54	−42 31.3	2	7.94	.000	0.60	.005			8	F7 V	1075,2012
48630	−36 03097				6 40 56	−36 34.8	1	7.86		-0.01		-0.05		2	A0 V	401
	−45 02690				6 40 58	−45 06.2	5	10.05	.007	1.16	.006	1.12	.008	23	G8	861,863,1075,1460,1628
48434	+4 01414	HR 2479		⋆	6 41 00	+03 59.0	4	5.88	.022	-0.02	.000	-0.89	.007	11	B0 III	1,15,154,1415
		L 523 - 10			6 41 00	−30 48.	1	13.18		1.00		0.53		4		3073
	−33 03169				6 41 00	−33 31.9	2	9.84	.005	0.45	.002	-0.03		3		1594,6006
262700	+21 01358				6 41 03	+20 58.4	2	8.90	.018	1.12	.006			39	K5	592,6011
48730	−45 02691				6 41 04	−45 18.3	6	9.53	.009	0.46	.006	0.10	.012	33	F3/5 IV	863,1460,1628,1673,1704,2011
	−42 02687				6 41 05	−42 59.9	1	10.91		0.47				8		955
		E3 - t			6 41 07	−45 02.4	1	15.51		0.73		0.17		4		1460
48629	−29 03388				6 41 09	−29 11.2	1	7.30		-0.01		-0.02		2	A0 V	401
		E3 - e			6 41 09	−45 06.1	1	12.87		0.58		0.02		4		1460
48433	+13 01390	HR 2478		⋆ A	6 41 10	+13 16.8	8	4.49	.007	1.17	.006	1.15	.007	33	K1 III	3,15,418,1355,1363*
		LS VI +06 010			6 41 11	+06 57.9	1	11.96		0.32		-0.52		2		405
		E3 - a			6 41 12	−45 12.	1	12.17		0.55				4		1075
48473	+7 01425	IDS06386N0735	AB		6 41 15	+07 32.2	1	8.23		0.04		-0.11		6	A0	1732
48555	−10 01667				6 41 18	−10 25.3	1	9.64		1.70				2	M1	369
		E3 - c			6 41 18	−45 03.	1	11.98		0.96				4		1075
48725	−37 03022				6 41 19	−37 19.1	1	8.00		-0.09		-0.42		2	B8 IV	401
		E3 - o			6 41 19	−45 07.8	1	14.80		0.55		-0.07		5		1460
		E3 - v			6 41 22	−45 07.1	1	16.25		0.51		-0.03		5		1460
262868	+7 01427				6 41 23	+07 19.1	1	9.80		1.72				3	K7	369
		E3 - k			6 41 27	−45 05.6	1	14.08		0.58		-0.01		5		1460
48326	+47 01335				6 41 33	+46 55.0	1	8.81		0.29		-0.02		2	A5	1601
48857	−50 02357	IDS06403S5021	A		6 41 34	−50 24.0	1	7.00		-0.15				4	B4 IV/V	2012
48450	+29 01327	HR 2480			6 41 35	+29 01.4	2	5.44	.005	1.44	.010	1.69	.045	4	K4 III	252,3016
		Dol 24 - 1			6 41 36	+01 38.	1	10.06		0.25		0.16		1		797
		Dol 24 - 2			6 41 36	+01 38.	1	13.78		0.58		-0.11		2		797
		Dol 24 - 3			6 41 36	+01 38.	1	13.38		0.54		-0.27		2		797
		Dol 24 - 4			6 41 36	+01 38.	1	13.93		0.85		-0.23		2		797
		Dol 24 - 5			6 41 36	+01 38.	1	10.23		0.16		-0.66		2		797
		Dol 24 - 6			6 41 36	+01 38.	1	13.58		0.40		-0.05		2		797
		Dol 24 - 7			6 41 36	+01 38.	1	13.06		0.57		-0.13		2		797
		Dol 24 - 8			6 41 36	+01 38.	1	14.49		0.86		0.04		2		797
		G 87 - 1			6 41 36	+32 52.0	1	12.15		1.14		0.99		1		333,1620
		V 447 Mon			6 41 38	+04 32.9	1	13.67		1.28				1		1772
48532	+20 01550				6 41 41	+20 02.4	1	8.68		-0.08		-0.56		3	B2 V:	555
48412	+39 01731				6 41 42	+39 25.3	2	6.93	.005	0.33	.010	-0.04	.005	5	F0 V	1501,1601
48797	−39 02798	HR 2488			6 41 43	−39 08.5	1	6.30		0.26				4	A8:m δ Del	2031
49268	−71 00476	HR 2505		⋆ AB	6 41 43	−71 43.5	3	6.48	.014	1.11	.000	1.07	.005	21	K1 III	15,1075,2038
		NGC 2287 - 194			6 41 45	−20 46.2	1	10.83		1.11		1.06		2		1626
	−45 02702				6 41 45	−45 07.0	1	11.53		0.95				4		1075
48855	−45 02703				6 41 47	−45 11.4	7	10.66	.006	0.05	.004	0.03	.010	47	B9 V	863,1460,1628,1657,1673*
48616	+3 01379	LS VI +03 012			6 41 48	+03 11.6	4	6.84	.009	0.79	.008	0.55	.025	15	F5 Ib	1355,1509,6009,8100
48250	+59 01015	HR 2470		⋆ AB	6 41 49	+59 29.7	3	4.87	.004	0.08	.007	0.08	.005	6	A1.5V +A2 V	15,1363,8015

HD	DM	Other Id	N Rem	α_{1950}	δ_{1950}	S	V	σ_V	B–V	σ_{B-V}	U–B	σ_{U-B}	n	Spectrum	References
		NGC 2287 - 195		6 41 50	−20 44.0	1	11.31		0.95		0.78		2		1626
263070	+6 01372			6 41 51	+06 08.4	1	10.47		0.37		0.16		1	G0	159
		AO 910		6 41 52	−16 29.4	1	10.38		0.55		0.06		1		1722
	−45 02704			6 41 53	−45 10.0	2	11.34	.016	0.42	.014	−0.09		10		1075,1460
	−45 02705			6 41 53	−45 34.4	1	10.63		0.56				4		2011
292179		Dol 25 - 11		6 41 54	+00 20.4	1	9.93		1.05		0.89		1	K2	410
48549	+22 01458	LS V +22 057		6 41 55	+22 05.2	1	8.68		−0.11		−0.70		3	B2 V	555
	−48 02470			6 41 55	−48 37.4	1	10.61		0.54		−0.17		2	G0	1696
48566	+19 01460			6 41 56	+19 34.8	1	8.15		0.18		0.12		2	F0	1648
48565	+20 01552			6 41 56	+20 54.8	1	7.21		0.53		−0.08		1	F8	1462
		AO 911		6 41 56	−16 36.7	1	10.83		1.03		0.61		1		1722
48906	−46 02633			6 41 57	−46 54.0	2	9.26	.000	0.14	.000	0.13		8	A2/3 V	1075,2011
48049	+69 00389	AW Cam		6 41 58	+69 41.0	1	8.24		0.02		−0.07		2	B9 V	1375
		NGC 2287 - 199		6 41 58	−20 28.8	1	10.91		1.21		1.29		2		1626
		NGC 2287 - 196		6 41 58	−20 43.9	1	11.44		1.19		1.14		2		1626
		NGC 2287 - 198		6 41 59	−20 35.6	1	10.66		1.23		1.28		2		1626
292180	+0 01573	Dol 25 - 10		6 42 01	+00 20.0	1	8.87		0.98		0.84		1	F0	410
	+35 01484	G 87 - 2		6 42 01	+35 35.4	1	8.85		0.67		0.15		1	G0	333,1620
48362	+52 01122			6 42 01	+52 17.0	1	8.70		0.00		−0.14		2	A0	1566
48691	+0 01574	LS VI +00 016		6 42 02	+00 38.5	2	7.83	.005	0.07	.025	−0.80	.000	8	B0.5IV	92,399
48905	−43 02644			6 42 03	−44 00.2	3	8.26	.003	0.17	.004	0.13	.010	16	A3 III	977,1075,2011
48664	+3 01381	CZ Mon		6 42 04	+03 22.1	1	9.45		3.19				2	C4,5	864
		Dol 25 - 12		6 42 06	+00 18.5	1	10.79		0.36		−0.65		2	B2	410
		NGC 2287 - 193		6 42 08	−20 57.9	1	10.62		0.16		0.08		2	A2	1626
	−45 02708			6 42 08	−45 28.1	1	10.99		0.59				4		1075
		NGC 2287 - 197		6 42 11	−20 42.8	1	9.65		0.04		−0.07		2	A0	1626
		Dol 25 - 9		6 42 12	+00 23.0	1	12.30		0.51		−0.46		2	A0	410
48717	+3 01382	LS VI +03 013	⋆ AB	6 42 12	+03 43.9	1	7.63		0.15		−0.52		3	B5	493
48689	+7 01433	IDS06395N0734	A	6 42 12	+07 31.0	1	8.32		−0.03		−0.12		1	A2	1732
		Dol 25 - 13		6 42 13	+00 16.4	1	11.95		0.36		−0.57		2	B8	410
		V 504 Mon		6 42 14	−03 54.1	1	11.61		0.99				1		1772
48663	+12 01267	IDS06394N1221	AB	6 42 15	+12 18.0	1	8.59		−0.01		−0.73		2	B1 V	1012
		AO 912		6 42 15	−16 39.8	1	10.06		2.00		2.29		1		1722
		NGC 2287 - 190		6 42 15	−20 30.8	1	9.44		1.36		1.55		2		1626
48591	+29 01332	IDS06391N2928	AB	6 42 16	+29 25.2	1	8.22		0.56		0.02		2	F8 V	3026
48754	−4 01644			6 42 17	−04 20.5	1	8.55		0.31		0.16		2	Am:	355
		Dol 25 - 14		6 42 19	+00 16.4	1	11.79		0.49		−0.01		1	B8	410
292167	+0 01576	LS VI +00 017		6 42 19	+00 40.4	2	9.24	.009	0.45	.009	−0.61	.009	6	O9 III:	92,1011
		NGC 2287 - 191		6 42 19	−20 33.6	1	11.00		0.08		0.05		2	A2	1626
292178	+0 01577	Dol 25 - 8		6 42 22	+00 24.5	1	10.46		0.12		−0.02		1	A2	410
	−16 01586			6 42 24	−16 40.6	1	8.62		0.09		0.03		1		1722
49835	−78 00240			6 42 25	−78 23.4	1	8.57		0.05		0.02		2	B9.5 V	1776
		Dol 25 - 16	A	6 42 28	+00 16.4	1	13.07		1.29		0.78		1		410
		Dol 25 - 16	B	6 42 28	+00 16.4	1	13.22		0.58		−0.33		1		410
		Dol 25 - 15		6 42 28	+00 16.6	1	11.65		0.66		−0.44		2	B8	410
48969	−42 02700			6 42 28	−42 39.9	3	8.53	.009	0.69	.003	0.23	.006	16	G5 V	977,1075,2011
		Dol 25 - 18		6 42 29	+00 17.5	1	13.44		0.56		−0.46		2	F5	410
263260	+7 01434			6 42 29	+07 24.1	1	8.34		0.33		0.14		3	K0	1732
48752	+9 01374	IDS06397N0947	AB	6 42 29	+09 44.4	1	8.30		−0.02		−0.26		1	B9 V	159
48737	+13 01396	HR 2484		6 42 29	+12 57.1	6	3.35	.014	0.44	.006	0.06	.011	25	F5 IV	3,15,1008,1363,3053,8015
		Dol 25 - 23		6 42 30	+00 13.9	1	12.85		0.51		0.38		1		410
		Dol 25 - 19		6 42 30	+00 16.1	1	13.63		0.50		−0.21		2	A5	410
		Dol 25 - 6		6 42 30	+00 24.3	1	11.99		0.48		0.41		1	A5	410
48851	−20 01520	NGC 2287 - 235	⋆ AB	6 42 30	−20 21.6	2	9.16	.012	−0.05	.008	−0.18	.003	3	B8 V	1626,1732
		Dol 25 - 17		6 42 31	+00 16.9	1	11.43		0.64		−0.42		2	B8	410
48589	+39 01736			6 42 32	+39 01.4	1	7.28		−0.04		−0.25		2	B9 V	105
292176		Dol 25 - 5		6 42 33	+00 24.6	1	10.74		0.06		0.27		1	A0	410
48432	+57 01004	HR 2477		6 42 34	+57 13.4	2	5.34	.005	0.96	.005	0.73		8	K0 III	1118,1355
48873	−20 01522	NGC 2287 - 234		6 42 34	−20 23.9	1	8.16		1.43		1.70		2	K5	1626
48917	−30 03484	Cr 121 - 23	⋆ A	6 42 34	−31 01.1	8	5.33	.040	−0.15	.018	−0.88	.019	48	B2 IIIe	62,401,681,815,1637*
		Dol 25 - 22		6 42 36	+00 14.4	1	12.09		0.43		−0.47		1	A5	410
		Dol 25 - 21		6 42 36	+00 15.6	1	12.73		0.53		0.07		1	F5	410
292175		Dol 25 - 4		6 42 36	+00 24.6	1	10.70		0.56		0.07		1	F5	410
292177	+0 01579	Dol 25 - 7		6 42 36	+00 25.4	1	10.53		0.10		0.02		1	A0	410
		Hiltner 600		6 42 37	+02 09.1	1	10.45		0.18		−0.57		2	B1 V	1012
		NGC 2287 - 188		6 42 37	−20 31.1	1	10.81		0.77		0.35		2		1626
48807	+0 01580	LS VI +00 018		6 42 38	+00 00.6	1	7.02		0.21		−0.40		5	B7 Iab	92
		Dol 25 - 20		6 42 38	+00 15.4	1	11.26		0.95		0.71		1	F5	410
48587	+42 01600			6 42 38	+42 19.2	1	6.88		1.60		1.70		2	K0	1733
		NGC 2287 - 189		6 42 38	−20 29.7	1	10.59		1.10		1.01		2		1626
		NGC 2287 - 192		6 42 39	−20 50.1	1	11.15		0.14		0.10		2		1626
48829	−1 01362			6 42 41	−02 02.6	1	9.22		0.07		−0.46		4	B9	1732
		NGC 2287 - 215		6 42 41	−20 44.5	1	11.15		0.13		0.08		2	A0	1626
		AO 914		6 42 44	−16 25.7	1	10.64		0.49		0.12		1		1722
		NGC 2287 - 216		6 42 44	−20 42.8	1	11.11		0.18		0.07		2	A0	1626
		NGC 2281 - 5		6 42 45	+41 13.5	1	11.99		1.39		1.41		1		741
	−16 01589			6 42 45	−16 44.6	1	8.45		1.22		1.55		1	K5	1722
		Dol 25 - 3		6 42 48	+00 25.1	1	11.60		0.21		0.12		1	A0	410
		NGC 2287 - 236		6 42 48	−20 22.6	1	10.87		0.05		0.02		2	A0	1626
263139		GX Gem		6 42 49	+34 29.0	1	11.31		0.58		0.06		3	G5	1768

Table 1

HD	DM	Other Id	N	Rem	α_{1950}	δ_{1950}	S	V	σ_V	B–V	σ_{B-V}	U–B	σ_{U-B}	n	Spectrum	References
		NGC 2287 - 219			6 42 49	−20 37.8	1	11.61		0.20		0.09		2		1626
49074	−46 02640				6 42 50	−47 01.4	4	8.16	.008	0.96	.002	0.72	.020	21	G8 III	861,977,1075,2011
263336	+10 01258				6 42 51	+09 55.5	1	10.31		0.13		−0.03		1	A2	159
263175	+32 01398	G 87 - 3		★	6 42 51	+32 36.5	2	8.78	.033	0.96	.009	0.72	.014	4	K2	22,333,1620
		NGC 2287 - 220			6 42 51	−20 37.0	1	11.72		0.20		0.08		2		1626
48938	−27 03248	HR 2493			6 42 52	−27 17.6	4	6.44	.007	0.54	.004	−0.04	.002	10	F7 V	15,258,2012,3037
		G 87 - 4			6 42 53	+32 36.4	1	12.17		1.53		1.23		1		333,1620
49285	−63 00628				6 42 53	−63 05.1	1	8.06		1.26		1.50		5	K2 III	1704
		Dol 25 - 1			6 42 56	+00 25.4	1	11.38		0.43		−0.61		5	B5	410
		NGC 2287 - 214			6 42 56	−20 46.8	1	11.06		0.13		0.07		2		1626
48915	−16 01591	HR 2491		★ A	6 42 57	−16 38.8	13	−1.45	.021	0.00	.007	−0.05	.015	35	A0mA1 Va	1,15,22,200,1007,1013*
48915	−16 01591	IDS06408S1635		B	6 42 57	−16 38.8	2	8.39	.066	−0.06	.042	−1.04	.005	8	DA	563,3078
	−20 01525	NGC 2287 - 218			6 42 57	−20 37.5	1	10.76		0.05		0.06		2	A2	1626
	−20 01524	NGC 2287 - 118			6 42 57	−20 44.3	1	10.57		0.06		0.04		2	A0	1626
		Dol 25 - 2			6 42 58	+00 25.3	1	11.87		1.17		0.91		1	A5	410
		NGC 2281 - 7			6 42 59	+41 05.5	1	11.74		0.41		0.09		1		741
48924	−20 01526				6 42 59	−20 25.3	1	9.24		0.02		−0.12		6	B9 V	1732
		NGC 2287 - 217			6 42 59	−20 38.5	1	11.40		0.44		0.00		2		1626
	−20 01528	NGC 2287 - 213			6 42 59	−20 50.2	1	10.90		0.09		0.06		2	B8	1626
	−20 01528	NGC 2287 - 212			6 42 59	−20 50.5	1	10.82		0.09		0.05		2	A0	1626
		NGC 2287 - 233			6 43 00	−20 22.3	1	10.35		1.50		1.80		2		1626
48924	−20 01526	NGC 2287 - 200			6 43 00	−20 25.3	1	9.27		0.01		−0.13		2	A2	1626
		NGC 2287 - 187			6 43 00	−20 36.4	1	11.67		1.81		1.96		2		1626
		NGC 2287 - 231			6 43 03	−20 40.1	1	10.89		0.46		0.01		2		1626
		NGC 2281 - 8			6 43 04	+41 29.9	1	11.12		0.29		0.16		1		741
	−2 01751	NGC 2286 - 266			6 43 04	−02 17.4	1	10.13		1.67				3		369
48843	+12 01275	HR 2489		★	6 43 06	+12 44.8	2	6.47	.000	0.34	.005	0.28		4	A9 III	1733,6009
48885	+4 01430				6 43 07	+04 04.8	1	8.30		1.11		0.86		1	G5	1462
		NGC 2281 - 220			6 43 07	+41 10.8	1	14.66		0.94		0.81		1		741
	−20 01529	NGC 2287 - 85			6 43 07	−20 54.2	2	10.28	.033	0.01	.002	−0.02	.028	3	B9.5V	701,1626
49028	−30 03495	Cr 121 - 36		★ AB	6 43 07	−30 32.0	5	6.53	.008	−0.12	.012	−0.48	.008	17	B8 IV	15,62,164,401,2029
48805	+23 01491				6 43 08	+23 25.5	2	6.52	.028	0.84	.008	0.39	.030	8	G8 V	1625,7009
48682	+43 01595	HR 2483		★ A	6 43 08	+43 37.8	5	5.24	.013	0.55	.007	0.06	.004	24	G0 V	150,1067,1197,1355,3026
48682	+43 01596	IDS06395N4341		B	6 43 08	+43 37.8	1	8.63		1.60		1.94		1		150
		NGC 2287 - 68			6 43 08	−20 36.4	1	11.10		0.18		0.10		2	A1	1626
	−20 01530	NGC 2287 - 80			6 43 08	−20 44.4	1	10.27		0.11		0.11		2	A3 V	1626
49192	−54 01096	IDS06422S5435		B	6 43 08	−54 40.7	1	6.69		−0.09		−0.25		3	B8/9 V	1732
	+36 01500	G 103 - 58			6 43 09	+35 55.8	1	10.00		0.61		0.02		1	G2	333,1620
48965	−20 01532	NGC 2287 - 93			6 43 09	−20 27.9	2	9.73	.025	−0.04	.005	0.08	.148	3	B9 V:	701,1626
	−20 01531	NGC 2287 - 94			6 43 09	−20 30.7	2	9.34	.001	1.17	.017	1.05	.008	3	K0	701,1626
49053	−34 03083				6 43 10	−35 04.1	1	8.05		−0.02		−0.13		2	B9 V	401
49219	−54 01097	IDS06422S5435		A	6 43 11	−54 38.5	1	6.47		−0.12		−0.46		3	B5/6 V	1732
49133	−46 02643				6 43 12	−46 43.1	2	7.68	.008	1.51	.002	1.79		12	K4 III	977,2011
48914	+2 01379	V505 Mon			6 43 13	+02 33.2	2	7.23	.025	−0.04	.005	−0.36		7	B5 Ib	493,2033
49029	−31 03634				6 43 14	−31 36.1	1	7.61		−0.10		−0.34		2	B8 V	401
		NGC 2281 - 14			6 43 15	+41 18.9	1	12.96		0.55		0.05		1		741
		NGC 2287 - 186			6 43 15	−20 56.3	1	11.74		0.55		0.15		2		1626
		AO 916			6 43 16	−16 30.7	1	10.49		1.23		1.38		1		1722
48982	−20 01533	NGC 2287 - 232			6 43 16	−20 22.5	2	9.41	.010	−0.07	.004	−0.28	.006	10	B9 IV	1626,1732
		NGC 2287 - 67			6 43 16	−20 37.2	1	10.77		0.03		0.00		2	B8	1626
48984	−20 01534	NGC 2287 - 108			6 43 16	−20 54.9	2	8.82	.006	−0.05	.001	−0.19	.007	4	B9 V:n	701,1626
		NGC 2287 - 96			6 43 17	−20 28.5	2	10.85	.227	0.08	.039	0.10	.019	3		701,1626
		AA45,405 S289 # 8			6 43 18	−07 14.9	1	13.62		0.72		0.34		1		797
		NGC 2287 - 95			6 43 18	−20 29.1	2	10.75	.068	0.06	.038	0.04	.056	3	A1 V	701,1626
49001	−23 04325	HR 2495			6 43 18	−23 24.5	2	6.05	.001	1.21	.004	1.29		6	K2 III	58,2035
		NGC 2281 - 16			6 43 19	+41 12.0	1	12.79		0.57		0.02		1		741
48983	−20 01535	NGC 2287 - 101			6 43 19	−20 47.2	3	8.29	.004	−0.02	.006	−0.06	.016	7	B9.5V:	701,1626,3019
		NGC 2287 - 90			6 43 19	−20 52.3	2	10.99	.015	0.55	.006	0.03	.005	3		701,1626
		AA45,405 S289 # 2			6 43 20	−07 15.2	1	12.66		0.21		−0.61		2	B1 V	797
		AA45,405 S289 # 3			6 43 20	−07 15.7	1	13.71		0.23		−0.56		2		797
	−20 01536	NGC 2287 - 89			6 43 20	−20 51.6	2	10.70	.000	0.91	.017	0.49	.020	4		701,1626
		NGC 2287 - 184			6 43 21	−20 47.2	1	11.68		1.03		0.68		2		1626
		NGC 2287 - 122			6 43 22	−20 43.1	1	10.89		0.07		0.00		2	A0	1626
		NGC 2287 - 88			6 43 22	−20 51.0	2	10.76	.003	0.72	.001	0.28	.002	4	G9	701,1626
		LSS 85			6 43 23	−07 15.4	1	12.17		0.14		−0.74		3	B3 Ib	797
	−20 01537	NGC 2287 - 201			6 43 23	−20 32.5	2	10.30	.018	0.02	.003	0.04	.040	6	B8	701,1626
		G 192 - 43			6 43 24	+58 42.2	2	10.32	.005	0.43	.000	−0.18	.005	5		1064,3025
		AA45,405 S288 # 1			6 43 24	−07 15.5	2	12.26	.055	0.63	.010	−0.31	.030	4		797,7006
		AA45,405 S289 # 5			6 43 24	−07 15.6	1	14.61		0.24		−0.41		2		797
		AA45,405 S289 # 6			6 43 24	−07 15.8	1	15.58		0.32		−0.41		2		797
		NGC 2287 - 203			6 43 24	−20 25.7	1	11.30		0.64		−0.21		1		701
		NGC 2287 - 73			6 43 24	−20 46.0	1	10.81		1.63		1.95		2		1626
	+41 01501	NGC 2281 - 18			6 43 25	+41 14.1	3	8.30	.026	1.14	.004	1.08	.025	4	K0 III	75,76,741
		AA45,405 S289 # 4			6 43 25	−07 16.2	1	13.58		0.21		−0.60		2	B0 V	797
	−16 01594				6 43 25	−16 52.1	1	8.81		0.34		−0.02		1	F5	1722
	−20 01540	NGC 2287 - 70			6 43 26	−20 34.0	3	9.13	.029	−0.04	.008	−0.13	.020	7	B9 V: +B9	701,1626,3019
	−20 01542	NGC 2287 - 61			6 43 26	−20 40.0	3	8.98	.008	−0.06	.006	−0.14	.022	7	B9 V: +B9	701,1626,3019
	−20 01539	NGC 2287 - 86			6 43 26	−20 56.0	2	10.43	.035	0.20	.006	0.21	.025	3	A6 V	701,1626
48981	−11 01623		V		6 43 27	−11 57.0	1	9.15		1.83				3	K5	369
48999	−17 01611				6 43 27	−17 07.4	1	6.94		−0.07		−0.44		2	B5 III	401

HD	DM	Other Id	N Rem	α_{1950}	δ_{1950}	S	V	σ_V	B–V	σ_{B-V}	U–B	σ_{U-B}	n	Spectrum	References
49023	−20 01543	NGC 2287 - 103		6 43 27	−20 37.6	4	8.39	.012	-0.05	.009	-0.23	.015	9	B9 IVp	49,701,1626,3019
		NGC 2287 - 62		6 43 28	−20 38.0	2	11.00	.043	0.06	.018	0.06	.018	3	A0	701,1626
49025	−20 01545	NGC 2287 - 105		6 43 28	−20 45.7	2	9.09	.000	-0.03	.002	-0.15	.033	4	B8.5p +B9V:	701,1626
		NGC 2287 - 113		6 43 28	−20 46.9	1	13.47		0.60		0.04		1		49
		AA45,405 S289 # 9		6 43 29	−07 15.1	1	13.31		0.35		0.51		1		797
49022	−20 01544	NGC 2287 - 106		6 43 29	−20 27.5	2	9.31	.024	-0.06	.009	-0.25	.029	4	B9 IVp	701,1626
		NGC 2287 - 76		6 43 31	−20 44.2	2	11.17	.013	0.08	.003	0.08	.018	4	A1	701,1626
49095	−31 03640	HR 2500		6 43 31	−31 44.1	6	5.92	.008	0.48	.005	-0.04	.005	22	F7 V	15,158,1020,1075,2012,3053
48979	−7 01551			6 43 32	−07 17.1	1	7.12		0.04		0.05		1	A1 V	797
49024	−20 01546	NGC 2287 - 116	★ A	6 43 32	−20 38.4	3	7.86	.010	-0.09	.012	-0.35	.037	6	B9 IIIp	701,1626,3019
		NGC 2281 - 19		6 43 33	+41 16.6	2	11.29	.049	0.32	.005	0.12	.000	3		75,741
		NGC 2281 - 77		6 43 33	−20 43.7	1	11.79		1.18		1.12		2		1626
		NGC 2281 - 20		6 43 34	+41 08.9	2	11.32	.039	0.33	.005	0.13	.005	3		75,741
		AO 918		6 43 34	−16 37.8	1	10.95		0.51		0.02		1		1722
49049	−20 01548	NGC 2287 - 117	★ B	6 43 34	−20 38.4	3	8.57	.022	-0.08	.019	-0.22	.016	6	B9 IV	701,1626,3019
49068	−20 01549	NGC 2287 - 75		6 43 34	−20 47.9	5	7.44	.013	1.24	.015	1.27	.030	13	K1 IIb	460,701,1415,1626,3034
		NGC 2281 - 22		6 43 35	+41 32.9	1	11.64		0.46		0.05		1		741
		NGC 2287 - 91		6 43 35	−20 27.6	2	11.04	.028	0.10	.014	0.09	.052	3	A0	701,1626
	−20 01550	NGC 2287 - 63		6 43 35	−20 35.5	2	9.60	.023	-0.06	.005	-0.19	.035	5	A0 Vp	701,1626
263584	+7 01444			6 43 36	+07 23.0	2	9.69	.015	0.78	.005	0.29		4	G5	742,1594
		NGC 2281 - 23		6 43 36	+41 17.5	2	11.69	.034	0.39	.010	0.07	.005	3		75,741
		NGC 2287 - 223		6 43 36	−20 55.0	1	11.63		0.20		0.09		2		1626
		NGC 2287 - 185		6 43 36	−20 55.6	1	11.97		1.24		1.04		2		1626
49050	−20 01551	NGC 2287 - 87		6 43 36	−21 00.5	3	7.87	.041	1.36	.033	1.49	.024	5	K2 IIIb	701,1626,3019
49217	−46 02652			6 43 36	−46 15.9	1	10.52		0.44		-0.33			F0 IV/V	1737
49131	−30 03505	HR 2501, HP CMa	★ AB	6 43 37	−30 53.7	4	5.80	.008	-0.19	.026	-0.90	.052	11	B2 Vn	15,401,404,2012
263650	+3 01396	LS VI +03 015		6 43 39	+03 41.2	1	9.72		0.12		-0.49		3	B8	493
		NGC 2287 - 78		6 43 39	−20 43.6	2	10.66	.003	0.22	.001	0.05	.029	3	A3 V	49,701,1626
		BSD 98 # 2663		6 43 40	+00 24.5	1	11.48		0.01		0.03		3	B9	821
	+41 01503	NGC 2281 - 25		6 43 40	+41 02.0	2	10.88	.058	1.09	.010	0.81	.015	3		75,741
		NGC 2287 - 64		6 43 40	−20 35.5	1	11.53		0.62		0.04		2	F6	1626
		NGC 2287 - 82		6 43 41	−20 47.0	2	11.05	.006	0.10	.028	0.08	.010	3	A5 m	49,701,1626
49069	−20 01552	NGC 2287 - 84		6 43 41	−20 48.5	3	8.36	.022	-0.06	.008	-0.17	.020	6	B9 V:	701,1626,3019
49234	−46 02656			6 43 42	−46 53.0	1	7.76		-0.14		-0.63		2	B4 III	401
48996	+0 01589			6 43 43	+00 09.3	1	8.79		-0.02		-0.02		1	A0	821
		AA45,405 S289 # 11		6 43 43	−07 03.5	1	14.47		0.27		-0.49		2		797
		AA45,405 S289 # 12		6 43 43	−07 03.5	1	13.22		0.21		-0.53		2		797
49048	−14 01573	HR 2498		6 43 43	−14 44.5	2	5.31	.005	0.06	.005	0.11		6	A1 IV/V	401,2035
	+40 01712	G 107 - 19		6 43 44	+40 04.8	1	10.79		0.80		0.32		2	K0	7010
	−20 01553	NGC 2287 - 119	★ A	6 43 44	−20 38.1	2	9.88	.030	0.03	.010	0.07	.069	3	A0 V: + B9	701,1626
	−20 01553	NGC 2287 - 121	★ AB	6 43 44	−20 38.4	2	10.27	.001	0.03	.019	0.00	.053	3	A0	701,1626
		NGC 2287 - 114		6 43 44	−20 45.0	1	13.96		0.71		0.46		1		49
	−44 02889			6 43 45	−44 24.1	1	9.73		0.47		-0.02		6		863
49067	−13 01641			6 43 46	−13 47.0	1	9.05		0.13				4	B3 II/III	2033
	−20 01554	NGC 2287 - 20		6 43 46	−20 39.9	3	9.43	.018	-0.02	.002	-0.09	.018	8	B9.5Vn	701,1626,3019
		PASP85,203 # 3		6 43 47	+08 03.1	1	11.22		1.42		1.39		4		266
	+40 01711	NGC 2281 - 27		6 43 47	+40 53.5	1	10.83		0.25		0.17		1		741
	−44 02890			6 43 47	−44 32.0	1	9.71		0.47				4		2012
49260	−47 02561			6 43 48	−47 10.1	6	7.21	.004	-0.16	.006	-0.74	.006	37	B2 III	278,743,863,1075,1732,2011
48977	+8 01486	HR 2494		6 43 49	+08 38.5	4	5.92	.008	-0.18	.004	-0.69	.010	12	B2.5V	15,154,1061,1732
49091	−20 01555	NGC 2287 - 21		6 43 49	−20 43.3	4	6.90	.004	1.51	.014	1.77	.018	11	K3 IIb	460,701,1626,3034
49259	−46 02657			6 43 49	−46 48.0	1	7.48		-0.14		-0.66		2	B3 IV/V	401
		PASP85,203 # 4		6 43 50	+08 03.1	1	10.03		1.57		1.89		4		266
		NGC 2287 - 19		6 43 50	−20 35.6	1	11.02		0.09		0.05		2	A1	1626
48781	+48 01436	HR 2487		6 43 51	+48 50.7	3	5.21	.006	1.13	.013	1.05	.005	6	K1 III	1080,1355,3016
		NGC 2287 - 22		6 43 51	−20 43.3	1	9.36		-0.01		-0.10		2		1626
		NGC 2281 - 28		6 43 52	+40 54.8	1	12.16		0.48		0.04		1		741
49089	−15 01504			6 43 52	−15 32.2	1	8.54		-0.01		-0.36		2	B8 V	401
	−16 01595			6 43 54	−16 25.7	1	7.99		1.29		1.34		1	K2	1722
49106	−20 01556	NGC 2287 - 2		6 43 54	−20 40.1	4	8.93	.018	-0.05	.015	-0.17	.018	9	B9 V: +B9	49,701,1626,3019
		NGC 2287 - 4		6 43 55	−20 41.8	2	10.73	.031	0.06	.013	0.07	.052	5	A2 V	701,1626
		NGC 2281 - 194		6 43 56	+41 08.3	1	13.16		0.27		0.17		1		741
		CCS 579		6 43 56	−12 49.6	1	10.13		3.35		4.00		1	N	864
49105	−20 01557	NGC 2287 - 97		6 43 56	−20 33.2	4	7.79	.012	1.14	.011	1.05	.006	8	K0 IIab-II	49,460,701,1626,3034
		NGC 2287 - 3		6 43 56	−20 41.2	3	10.28	.018	0.04	.020	-0.06	.050	7	A0 IV	701,1626,3019
		NGC 2287 - 9		6 43 56	−20 41.2	1	12.24		0.39		0.03		2		3019
	−20 01558	NGC 2287 - 23	★ AB	6 43 56	−20 44.1	1	9.33		0.02		0.18		1	B9.5V:	701
292293	+0 01592			6 43 57	+00 08.5	1	9.91		0.05		0.11		2	A0	821
		BSD 98 # 2682		6 43 57	+00 09.7	1	10.35		0.09		-0.38		2	B5	821
263716	+4 01437			6 43 57	+04 10.8	1	10.25		0.01				1	B9	592
		NGC 2287 - 24		6 43 57	−20 46.8	1	12.49		0.40		0.02		1		49
		NGC 2287 - 15		6 43 58	−20 36.7	2	10.65	.015	0.24	.049	0.10	.017	3	A3 m	49,701,1626
	−20 01559	NGC 2287 - 5		6 43 58	−20 39.2	3	9.57	.017	-0.02	.008	-0.11	.024	5	B9 V:+B9.5	701,1626,3019
		NGC 2287 - 8		6 43 58	−20 40.3	1	13.10		0.47		-0.05		2		3019
49126	−20 01560	NGC 2287 - 102	★ AB	6 43 58	−20 42.0	2	7.28	.002	0.58	.005	0.19	.021	12	F8 IV-V+B9	701,1626
49044	+3 01398			6 43 59	+03 41.8	1	8.56		0.49				1	F8	592
		PASP85,203 # 8		6 43 59	+08 06.7	1	12.97		0.66		0.20		8		266
		NGC 2287 - 17		6 43 59	−20 35.3	1	11.75		0.22		0.10		2		1626
		NGC 2287 - 125		6 43 59	−20 49.2	1	11.85		0.26		0.01		2		1626
		NGC 2287 - 124		6 43 59	−20 49.9	1	11.71		0.20		0.11		2		1626

Table 1   359

HD	DM	Other Id	N Rem	α₁₉₅₀ δ₁₉₅₀	S	V	σ_V	B−V	σ_B−V	U−B	σ_U−B	n	Spectrum	References
		NGC 2281 - 198		6 44 00 +41 13.5	1	14.00		0.60		0.16		1		741
		NGC 2287 - 6		6 44 00 −20 40.1	3	10.08	.015	-0.04	.012	-0.10	.028	6	A0 Vpn	701,1626,3019
		NGC 2287 - 11		6 44 00 −20 42.8	1	11.45		0.32		0.05		2	A5	1626
49188	−28 03397			6 44 00 −28 45.5	1	7.63		-0.12		-0.58		4	B5 III	158
	+41 01504	NGC 2281 - 30		6 44 01 +41 07.1	3	9.39	.026	0.17	.008	0.12	.013	4	A3 V	75,76,741
		LSS 86		6 44 01 −07 25.9	1	11.35		0.19		-0.74		1	B3 Ib	797
		NGC 2287 - 59		6 44 01 −20 32.2	3	11.02	.016	0.23	.021	0.11	.032	5	A7 m	49,701,1626
		PASP85,203 # 11		6 44 02 +08 05.5	1	11.70		0.12		0.01		1		266
		NGC 2287 - 10		6 44 02 −20 41.3	1	12.03		0.32		0.42		1		701
		NGC 2287 - 120		6 44 02 −20 45.5	2	9.93	.008	0.02	.005	-0.02	.042	4	A0 Vn	701,1626
		NGC 2281 - 31		6 44 03 +40 57.5	2	10.71	.039	0.22	.019	0.17	.000	3		75,741
48767	+55 01122	HR 2485	⋆ AB	6 44 03 +55 45.7	1	5.52		0.47				2	F5	71
		NGC 2287 - 16		6 44 03 −20 35.7	1	11.89		0.29		0.08		2		1626
		NGC 2287 - 110		6 44 04 −20 28.4	2	11.35	.006	0.13	.015	0.10	.009	3		49,701,1626
	−20 01561	NGC 2287 - 25		6 44 04 −20 51.0	3	9.99	.017	-0.03	.010	-0.12	.017	6	B9.5V	460,701,1626
49063	+4 01440			6 44 05 +04 24.9	1	7.92		1.14		1.02		1	G5	1462
		NGC 2287 - 12		6 44 05 −20 42.5	1	11.53		0.25		0.10		2	A5	1626
		NGC 2287 - 205		6 44 06 −20 25.2	1	11.00		0.37		0.14		1		701
49151	−20 01562	NGC 2287 - 44		6 44 06 −20 45.0	3	8.46	.015	-0.06	.009	-0.21	.021	6	B9.5V:	701,1626,3019
		NGC 2287 - 26		6 44 06 −20 49.8	1	11.47		1.43		1.64		2		1626
		PASP85,203 # 2		6 44 07 +08 09.5	1	12.23		0.52		-0.03		6		266
49150	−20 01563	NGC 2287 - 1	⋆ AB	6 44 07 −20 42.3	2	8.50	.003	-0.05	.018	-0.15	.001	4	B8.5V:+B9	701,1626
		NGC 2281 - 32		6 44 08 +41 03.0	2	12.02	.049	0.45	.010	0.02	.005	3		75,741
49123	−8 01549			6 44 08 −08 51.8	1	7.22		-0.01		-0.09		2	B9	401
		NGC 2287 - 111		6 44 08 −20 29.7	2	11.82	.000	1.23	.009	1.07	.046	3		49,1626
		NGC 2287 - 14		6 44 08 −20 39.1	3	10.41	.016	0.05	.011	0.03	.020	5	A1 V	701,1626,3019
289060	+2 01388			6 44 10 +02 45.1	1	10.42		0.36		0.06		2	F2 V	137
263775	+5 01403	LS VI +05 022		6 44 12 +05 39.4	1	10.52		0.19		-0.63		2	B2	493
		NGC 2281 - 163		6 44 12 +41 06.4	1	13.86		0.78		0.22		1		741
		NGC 2281 - 34		6 44 12 +41 10.5	1	12.56		0.50		0.01		1		741
		NGC 2287 - 32		6 44 12 −20 41.7	1	11.62		0.22		0.10		2		1626
		NGC 2287 - 31		6 44 12 −20 42.0	1	11.66		0.37		0.00		2		1626
		NGC 2287 - 27		6 44 12 −20 48.1	2	10.80	.002	0.09	.002	0.09	.028	4	A2 Vm	1626,3019
		PASP85,203 # 5		6 44 13 +08 03.3	1	11.90		1.17		0.88		6		266
		NGC 2287 - 226		6 44 13 −20 26.8	1	10.70		1.62		1.40		2		1626
		NGC 2287 - 46		6 44 13 −20 38.5	1	9.91		1.11		0.79		2	G8	1626
		NGC 2287 - 47		6 44 14 −20 37.8	2	11.30	.010	0.14	.005	0.11	.042	4		1626,3019
49185	−20 01564	NGC 2287 - 30		6 44 14 −20 45.8	4	9.30	.019	-0.02	.012	-0.14	.033	8	A0 V	49,701,1626,3019
		G 87 - 7		6 44 15 +37 34.9	4	12.07	.028	-0.08	.007	-0.92	.010	12		203,1281,1620,3060
		NGC 2281 - 36		6 44 15 +41 01.0	2	11.86	.049	0.41	.005	0.07	.000	3		75,741
		NGC 2281 - 35		6 44 15 +41 16.4	1	10.55		1.49		1.75		2		75
		NGC 2287 - 58		6 44 15 −20 33.7	2	11.31	.013	0.12	.010	0.11	.024	4		1626,3019
		NGC 2287 - 28		6 44 15 −20 47.5	1	10.96		1.03		0.82		2		1626
50506	−80 00196	HR 2559		6 44 15 −80 45.8	5	5.62	.014	0.20	.006	0.15	.010	20	A5 III	15,615,1075,2012,2038
289120	+1 01503	LS VI +01 011		6 44 16 +01 22.2	1	10.18		0.18		-0.71		4	B2 n(e)	206
		PASP85,203 # 1		6 44 16 +08 02.2	1	10.28		1.78		2.16		6		266
49184	−20 01565	NGC 2287 - 104		6 44 16 −20 37.0	2	9.19	.020	-0.01	.012	-0.09	.019	3	B9.5V:+B9.5	701,1626
		NGC 2287 - 43		6 44 16 −20 52.6	1	11.23		0.09		0.09		2		1626
49147	−9 01644	HR 2502		6 44 17 −10 03.2	4	5.66	.013	-0.05	.007	-0.11	.005	15	B9.5V	15,351,1078,2027
		NGC 2287 - 29		6 44 17 −20 46.7	2	11.39	.005	0.12	.011	0.11	.030	4	A2	1626,3019
49396	−52 00996	HR 2513		6 44 17 −52 08.8	1	6.57		1.08				4	G2 Ib	2007
	−20 01566	NGC 2287 - 49		6 44 18 −20 36.2	2	10.42	.004	0.08	.026	0.08	.039	3	A3 V	701,1626
	−20 01567	NGC 2287 - 202		6 44 18 −20 53.6	2	10.66	.009	0.05	.004	-0.25	.017	3		701,1626
		PASP85,203 # 6		6 44 19 +08 04.5	1	11.85		1.43		1.29		4		266
		NGC 2281 - 37		6 44 20 +40 52.6	1	12.10		0.52		0.01		1		741
		LSS 87		6 44 20 −07 19.1	1	10.74		0.65		0.54		2	B3 Ib	797
49233	−22 01527			6 44 20 −23 05.7	1	8.28		-0.15		-0.65		2	B3/5 V	401
	+41 01506	NGC 2281 - 38		6 44 21 +41 01.9	2	10.20	.058	0.15	.000	0.11	.000	3	A2 V	75,741
		NGC 2281 - 39		6 44 21 +41 05.4	2	11.96	.063	0.44	.010	0.01	.024	3		75,741
		NGC 2287 - 33		6 44 21 −20 45.0	1	11.56		0.25		0.08		2	A3	1626
		BSD 98 # 2712		6 44 22 +00 18.6	1	11.61		0.26		-0.10		2	B9	821
49355	−45 02732			6 44 22 −45 56.6	2	8.99	.008	0.57	.002	0.01		12	F7 V	977,2011
		NGC 2281 - 160		6 44 23 +41 04.8	1	13.60		0.86		0.37		1		741
	+41 01507	NGC 2281 - 40		6 44 23 +41 12.5	3	10.32	.035	0.14	.004	0.11	.004	4	A3 V	75,76,741
		NGC 2287 - 34		6 44 24 −20 42.9	1	12.90		0.46		0.10		1		49
49212	−20 01568	NGC 2287 - 107		6 44 24 −20 45.4	5	7.78	.008	1.15	.028	1.02	.014	11	K0 IIab	49,460,701,1626,3034
49319	−39 02831	HR 2507		6 44 24 −39 29.1	5	6.61	.012	-0.13	.008	-0.52	.003	18	B4 Vne	15,401,1637,2013,2029
	−20 01569	NGC 2287 - 208	⋆ B	6 44 25 −20 31.6	1	10.32		0.12		-0.09		2	A0	1626
		NGC 2281 - 159		6 44 26 +41 04.7	1	13.18		1.16		1.02		1		741
	−20 01569	NGC 2287 - 207	⋆ A	6 44 26 −20 31.4	1	8.81		1.55		1.86		2		1626
49211	−20 01570	NGC 2287 - 51		6 44 26 −20 39.5	3	9.38	.004	-0.05	.012	-0.17	.030	7	B9 V	49,701,1626
		NGC 2281 - 42		6 44 27 +41 02.0	1	12.74		0.55		0.01		1		741
49059	+18 01349	HR 2499	⋆ AB	6 44 28 +18 15.0	1			0.07		0.01		2	A1 V +A3 V	1022
		NGC 2281 - 43		6 44 28 +41 10.5	3	11.15	.037	0.29	.000	0.15	.008	4		75,76,741
		NGC 2281 - 44		6 44 28 +41 12.0	3	10.87	.034	0.26	.004	0.16	.010	4	A4 V	75,76,741
	+41 01508	NGC 2281 - 45		6 44 28 +41 13.9	3	10.25	.035	0.16	.015	0.11	.011	4	A1 V	75,76,741
263902	+4 01442	V508 Mon		6 44 29 +04 01.5	3	10.28	.023	0.79	.005	0.48		3	F8	592,1462,1772
49336	−37 03065	HR 2510, V339 Pup		6 44 29 −37 43.2	3	6.20	.018	-0.14	.004	-0.65		9	B3 Vne	15,401,2027
		NGC 2281 - 46		6 44 30 +41 00.3	2	11.75	.063	0.54	.015	0.02	.024	3		75,741
		NGC 2281 - 166		6 44 30 +41 09.7	1	14.20		0.58		-0.04		1		741

HD	DM	Other Id	N Rem	α₁₉₅₀	δ₁₉₅₀	S	V	σ_V	B–V	σ_B-V	U–B	σ_U-B	n	Spectrum	References
		NGC 2287 - 112		6 44 30	−20 41.4	2	12.53	.019	0.39	.024	0.00	.005	3		49,701
		NGC 2287 - 52		6 44 31	−20 38.5	2	10.69	.002	1.07	.010	0.99	.017	3		49,701,1626
		NGC 2287 - 57		6 44 32	−20 32.1	1	11.94		0.26		0.05		2		1626
		NGC 2287 - 53		6 44 32	−20 36.5	2	11.10	.036	0.54	.007	0.10	.019	3	F0	701,1626
		NGC 2281 - 48		6 44 33	+41 15.8	3	11.75	.032	0.41	.008	0.08	.017	4		75,76,741
		LS VI -06 003		6 44 33	−06 06.8	1	11.26		0.16		-0.71		1		490
		NGC 2287 - 206		6 44 33	−20 27.5	1	11.20		0.55		0.02		2		701,1626
		NGC 2287 - 54		6 44 33	−20 35.0	3	10.70	.041	0.03	.020	-0.01	.026	4	A1 V	49,1626,3019
49229	−14 01584	HR 2504		6 44 34	−14 22.3	4	5.28	.005	-0.04	.005	-0.26	.000	10	B8/9 III	15,1079,2012,3023
		NGC 2287 - 40		6 44 35	−20 51.5	1	11.11		0.21		0.12		2	A6	1626
49161	+8 01496	HR 2503		6 44 37	+08 05.6	7	4.76	.008	1.39	.008	1.65	.007	20	K4 III	15,1078,1080,1355*
	+41 01509	NGC 2281 - 49		6 44 37	+41 07.9	3	10.61	.037	0.24	.004	0.16	.008	4	A3 V	75,76,741
		NGC 2281 - 51		6 44 38	+41 15.9	3	11.60	.038	0.38	.012	0.10	.004	4		75,76,741
		NGC 2281 - 50		6 44 38	+41 17.4	3	11.19	.018	1.13	.025	0.85	.011	4		75,76,741
		NGC 2281 - 132		6 44 39	+41 05.5	1	13.49		0.39		0.07		1		741
	−20 01571	NGC 2287 - 55		6 44 39	−20 36.3	2	10.80	.005	0.16	.000	0.09	.006	3	A3 V	49,701,1626
49947	−72 00522	HR 2531		6 44 39	−73 03.8	3	6.36	.003	0.96	.000	0.67	.005	19	G8 III	15,1075,2038
		NGC 2287 - 52		6 44 40	+41 02.5	3	10.92	.037	0.39	.018	0.09	.007	4		75,76,741
		NGC 2287 - 123		6 44 41	−20 41.0	2	10.96	.034	0.10	.046	0.11	.093	3		701,1626
49277	−20 01572	NGC 2287 - 39		6 44 41	−20 44.5	4	10.42	.023	0.00	.005	-0.02	.017	8	A0 V	49,701,1626,3019
		NGC 2287 - 38		6 44 41	−20 45.9	1	12.91		0.44		-0.04		2		701,3019
		NGC 2287 - 229		6 44 41	−20 53.4	1	11.70		0.50		-0.03		2	A0	1626
		NGC 2287 - 41		6 44 43	−20 52.2	2	11.82	.007	0.21	.009	0.11	.013	4		1626,3019
	+41 01511	NGC 2281 - 55		6 44 44	+41 07.7	3	8.90	.034	0.98	.007	0.69	.013	4	G8 III	75,76,741
		NGC 2281 - 53		6 44 44	+41 10.1	3	11.09	.051	0.31	.018	0.13	.000	4		75,76,741
	+41 01510	NGC 2281 - 54		6 44 44	+41 14.0	3	10.87	.035	0.32	.005	0.14	.017	4		75,76,741
		NGC 2287 - 42		6 44 44	−20 53.2	1	10.85		0.19		0.13		2		1626
		NGC 2281 - 56		6 44 45	+41 09.1	3	11.45	.023	1.22	.017	1.19	.005	4		75,76,741
49517	−52 00998	HR 2515		6 44 45	−52 21.3	2	5.79	.005	1.55	.015	1.82		6	K3 III	58,2035
		NGC 2281 - 57		6 44 46	+41 11.1	3	10.63	.038	0.24	.033	0.14	.012	4	A3 V	75,76,741
		G 108 - 26		6 44 47	+02 34.4	1	15.68		0.32		-0.53		4		940
		NGC 2281 - 130		6 44 47	+41 04.6	1	13.86		0.64		0.05		1		741
	+41 01512	NGC 2281 - 58		6 44 47	+41 08.3	3	9.46	.037	0.12	.017	0.07	.008	4	A2 V	75,76,741
		NGC 2281 - 168		6 44 47	+41 12.9	1	14.60		1.17		0.76		1		741
48680	+67 00452			6 44 47	+67 34.2	1	7.67		1.56		1.87		2	K5	1375
		NGC 2281 - 59		6 44 48	+41 07.7	1	10.19		0.21		0.14		1		741
		NGC 2281 - 134		6 44 48	+41 10.0	1	13.82		0.64		0.03		1		741
49299	−20 01573	NGC 2287 - 56		6 44 48	−20 35.8	3	10.22	.016	-0.08	.016	-0.07	.052	5	A1p	49,701,1626
		NGC 2281 - 60		6 44 49	+41 05.8	1	12.71		0.56		0.07		1		741
		NGC 2281 - 131		6 44 49	+41 06.0	1	13.16		0.77		0.58		1		741
		NGC 2281 - 61		6 44 49	+41 07.6	1	10.65		0.18		0.15		1		741
49010	+41 01514	NGC 2281 - 62	★ AB	6 44 50	+41 08.4	1	9.04		0.07		-0.09		1	A0	741
49009	+41 01513	NGC 2281 - 63		6 44 50	+41 21.5	3	7.29	.029	1.34	.004	1.41	.016	4	K2 III	75,76,741
		NGC 2287 - 210		6 44 50	−20 39.9	1	11.18		0.56		0.00		2		1626
		V 510 Mon		6 44 51	+02 34.3	1	12.49		1.44				1		1772
		NGC 2281 - 64	★ C	6 44 51	+41 08.5	1	10.62		0.31		0.15		1		741
		BSD 98 # 2769		6 44 52	+00 21.9	1	11.67		0.28		0.09		4	B8	821
		NGC 2281 - 67		6 44 52	+40 54.0	1	12.16		0.53		0.03		1		741
49040	+41 01516	NGC 2281 - 65		6 44 52	+41 01.1	3	8.85	.033	0.11	.017	0.07	.014	4	B9p	75,76,741
		NGC 2281 - 66		6 44 52	+41 03.6	3	11.77	.047	0.42	.011	0.07	.012	4		75,76,741
	−20 01574	NGC 2287 - 115		6 44 52	−20 49.9	2	10.35	.010	0.07	.003	0.03	.019	4	B9	701,1626
292301	+0 01596			6 44 53	+00 07.6	1	10.42		0.28		0.06		1	A7 V	137
49333	−20 01576	NGC 2287 - 109	★ V	6 44 53	−20 57.6	7	6.06	.016	-0.18	.012	-0.63	.017	24	B7 II/III	15,49,62,701,1079*
49317	−20 01575	NGC 2287 - 92	★ A	6 44 54	−20 37.2	2	8.26	.050	1.17	.037	1.01	.010	5	K0 IIIa	460,3019
49317	−20 01575	NGC 2287 - 92	★ AB	6 44 54	−20 37.2	3	7.91	.113	0.99	.093	0.78	.200	3		49,701,1626
		NGC 2281 - 69		6 44 55	+41 03.7	1	12.50		0.53		0.06		1		741
		NGC 2281 - 70		6 44 55	+41 04.0	1	11.65		0.65		0.15		1		741
	+41 01515	NGC 2281 - 68		6 44 55	+41 27.0	2	10.63	.030	0.22	.010	0.15		2		76,741
49224	+0 01600			6 44 56	+00 44.9	1	8.27		0.06		0.08		5	A2	1371
		NGC 2281 - 71		6 44 56	+41 04.8	3	11.19	.031	0.52	.020	0.04	.010	4		75,76,741
49420	−36 03143			6 44 56	−36 32.4	1	7.95		1.09		0.93		11	K1 III	1673
		NGC 2281 - 73		6 44 57	+41 18.2	1	12.51		0.54		0.10		1		741
49334	−20 01577	NGC 2287 - 204		6 44 57	−21 02.8	1	7.78		1.23		1.11		1	K0 IIIa-II	701
		NGC 2281 - 75		6 44 58	+40 55.6	1	11.59		0.54		0.03		1		741
	+41 01517	NGC 2281 - 74		6 44 58	+41 08.2	3	9.52	.015	1.13	.022	1.04	.010	4	K0	75,76,741
49315	−15 01511			6 44 59	−16 00.7	2	7.83	.010	-0.08	.005	-0.62		6	B4 III	401,2033
		BONN82 T6# 9		6 45 00	+00 42.5	1	10.17		1.14		0.94		3	K4	322
		NGC 2287 - 174		6 45 00	−20 56.6	1	11.14		0.08		0.07		2	A0	1626
		NGC 2281 - 77		6 45 01	+41 00.9	3	11.76	.040	0.43	.000	0.11	.010	4		75,76,741
		NGC 2281 - 76		6 45 01	+41 19.7	1	12.70		0.61		0.09		1		741
		LTT 11921		6 45 01	+78 16.0	2	10.94	.007	0.87	.002	0.62	.020	4		1723,1773
49296	−9 01649			6 45 01	−09 54.4	4	7.94	.010	1.67	.025	2.03	.010	10	M2 II	148,369,8032,8100
	+41 01518	NGC 2281 - 78		6 45 02	+41 09.8	3	10.64	.027	0.21	.012	0.15	.010	4	A4 V	75,76,741
		NGC 2287 - 182		6 45 04	−20 26.0	1	10.43		0.27		0.12		2	A0	1626
		NGC 2287 - 177		6 45 04	−20 48.7	1	11.59		1.18		0.88		2		1626
49247		DF Mon		6 45 05	+00 43.8	1	10.60		4.20				2	C4-5,4-5	864
	+41 01519	NGC 2281 - 79		6 45 05	+41 08.8	3	10.35	.036	0.15	.018	0.12	.012	4	A3 V	75,76,741
		NGC 2287 - 183		6 45 05	−20 43.0	1	11.72		1.62		1.67		2		1626
49246	+6 01384			6 45 06	+06 21.2	1	7.96		-0.08		-0.20		3	B9	3039
		NGC 2281 - 81		6 45 06	+41 06.7	3	11.85	.043	0.50	.004	0.06	.015	4		75,76,741

Table 1

HD	DM	Other Id	N Rem	α₁₉₅₀	δ₁₉₅₀	S	V	σ_V	B–V	σ_B-V	U–B	σ_U-B	n	Spectrum	References
		NGC 2281 - 80		6 45 06	+41 19.3	1	12.06		0.46		0.05		1		741
	+41 01520	NGC 2281 - 82		6 45 07	+41 09.7	3	10.08	.028	0.14	.000	0.10	.004	4	A2 V	75,76,741
		NGC 2287 - 175		6 45 07	−20 56.6	1	10.30		0.00		-0.07		2	A0	1626
292298				6 45 09	+00 13.3	1	11.02		0.55		0.12		4	F8	821
		NGC 2281 - 84		6 45 10	+40 55.1	1	12.62		0.55		0.01		1		741
		NGC 2281 - 154		6 45 10	+41 01.9	1	13.62		0.95		0.53		1		741
49098	+41 01522	NGC 2281 - 86		6 45 10	+41 07.4	3	8.63	.044	0.22	.007	0.15	.021	4	A4 V	75,76,741
	+41 01521	NGC 2281 - 83		6 45 10	+41 10.7	1	11.05		0.29		0.15		1		741
		NGC 2281 - 85		6 45 10	+41 10.8	1	11.70		0.41		0.13		1		741
		LP 16 - 313		6 45 11	+78 16.2	1	14.52		1.56		1.17		1		1773
		NGC 2281 - 87		6 45 12	+41 16.6	1	12.87		0.58		0.05		1		741
48992	+52 01130			6 45 12	+52 32.3	1	8.51		1.07		0.91		2	K0	1566
49294	+0 01604	IDS06426N0027	AB	6 45 13	+00 23.8	1	7.00		0.16		0.10		3	A2	322
49292	+3 01406			6 45 13	+02 58.7	1	7.78		1.65				4	M1	369
		NGC 2281 - 88		6 45 13	+40 53.8	2	10.83	.029	1.20	.015	1.27	.044	3		75,741
		NGC 2281 - 89		6 45 13	+41 02.6	1	12.49		0.52		0.02		1		741
		NGC 2281 - 91		6 45 13	+41 08.5	3	11.41	.038	0.36	.008	0.12	.005	4		75,76,741
		NGC 2281 - 129		6 45 14	+41 06.6	2	12.67	.100	0.65	.055	0.06		2		76,741
49331	−8 01558	HR 2508		6 45 14	−08 56.5	6	5.05	.008	1.80	.016	1.88	.008	22	M1 II	15,369,1415,2035,3005,8100
		BSD 98 # 2806		6 45 15	+00 07.0	1	11.42		0.33		-0.17		2	B5	821
49293	+2 01397	HR 2506		6 45 15	+02 28.1	11	4.47	.010	1.11	.005	1.03	.011	41	K0 III	15,369,1075,1355*
264111	+4 01447	LS VI +04 035		6 45 15	+04 43.4	4	9.65	.015	0.03	.011	-0.72	.010	24	B2	272,308,493,843
49311	−0 01433			6 45 15	−00 47.6	1	8.71		0.21		-0.36		1	B8	821
	+41 01523	NGC 2281 - 92		6 45 16	+41 00.1	2	10.71	.070	0.22	.015	0.17		2	A4 V	76,741
		NGC 2287 - 181		6 45 16	−20 37.5	1	10.68		0.26		0.14		2		1626
		NGC 2287 - 178		6 45 16	−20 47.5	1	11.15		0.09		0.04		2		1626
		LP 895 - 34		6 45 16	−31 46.3	1	12.54		0.52		-0.12		2		1696
292297	+0 01605	IDS06427N0021	AB	6 45 17	+00 17.9	1	8.57		1.72		1.23		5	K2	1371
49559	−44 02908			6 45 18	−44 55.1	3	7.97	.009	1.49	.002	1.76	.002	18	K3 III	863,977,2011
		NGC 2281 - 145		6 45 19	+41 09.0	1	14.46		0.67		0.14		1		741
		BSD 98 # 1450		6 45 20	+00 39.1	1	11.18		0.29		-0.20		1	B5	821
49416	−20 01581	NGC 2287 - 180		6 45 21	−20 44.1	1	9.44		-0.06		-0.25		2	A0	1626
		NGC 2287 - 179		6 45 22	−20 48.0	1	10.85		0.32		0.06		2		1626
49330	+0 01607	LS VI +00 020		6 45 23	+00 49.9	1	8.88		0.26		-0.69		2	B0:penn	1012
49415	−17 01626			6 45 23	−17 27.1	1	7.07		-0.03		-0.13		2	B8/9 V	401
49614	−47 02574	IDS06440S4742	AB	6 45 23	−47 45.0	1	6.94		1.40				4	K3 III	2034
		NGC 2281 - 93		6 45 25	+41 07.2	3	11.37	.038	0.42	.023	0.08	.005	4		75,76,741
292296	+0 01608	LS VI +00 021		6 45 26	+00 22.7	2	7.90	.015	0.47	.005	0.40	.015	8	A2 III	322,1371
50002	−70 00560	HR 2536		6 45 26	−70 22.8	3	6.10	.003	1.34	.005	1.50	.000	19	K3 III	15,1075,2038
		NGC 2281 - 151		6 45 27	+41 02.7	1	13.02		0.73				1		76
		NGC 2281 - 149		6 45 27	+41 03.8	1	13.56		0.69		0.11		1		741
48948	+60 01003	G 250 - 22		6 45 27	+60 23.2	2	8.58	.000	1.22	.019	1.17	.019	3	M0p	801,3072
		NGC 2281 - 148		6 45 29	+41 04.1	1	14.04		0.69		0.02		1		741
49488	−28 03431			6 45 29	−28 30.1	1	7.83		-0.17		-0.75		3	B3 IV	1732
		LS VI +04 036		6 45 31	+04 52.7	1	10.69		0.33		-0.65		2		405
49329	+5 01411			6 45 31	+05 53.2	1	9.08		0.04		0.10		2	B9	3039
		NGC 2281 - 144		6 45 31	+41 10.2	1	14.61		0.87		0.44		1		741
		NGC 2281 - 94		6 45 31	+41 10.5	1	12.12		1.09		1.04		1		741
		NGC 2281 - 95		6 45 34	+41 03.5	3	10.62	.034	0.21	.012	0.14	.015	4	A3 V	75,76,741
289112	+1 01518			6 45 36	+01 26.7	1	9.56		1.14		0.97		1	K0 III	137
	−0 01439			6 45 37	−00 32.2	1	10.50		0.11		0.09		1	A0	821
292496	−1 01384			6 45 37	−01 56.1	1	9.29		0.78				34	F8	6011
49385	+0 01610	IDS06431N0025	AB	6 45 38	+00 21.7	1	7.93		0.56		0.07		3	G0	322
49591	−37 03080	HR 2518	⋆ A	6 45 39	−37 52.4	4	5.26	.007	-0.09	.008	-0.24	.009	14	B9 IV	15,401,1637,2012
		G 192 - 47		6 45 40	+55 56.5	2	10.38	.008	1.22	.078	1.12	.108	4	M0	3003,7010
49705	−54 01115	HR 2524	⋆ AB	6 45 40	−54 38.4	4	6.45	.009	0.86	.000			16	G6 III	15,1020,1075,2012
289116	+1 01521			6 45 41	+01 10.8	1	9.79		0.41		0.05		1	F5 V	137
49689	−51 02078	HR 2523		6 45 41	−51 12.5	4	5.40	.005	1.33	.007	1.24		13	K1 II/III	15,1075,2035,3005
49368	+5 01414	V613 Mon		6 45 42	+05 35.9	2	7.66	.023	1.78	.000	1.95		14	M2	369,3039
		G 87 - 8		6 45 42	+37 11.4	2	13.77	.008	1.66	.004	0.94	.004	8		316,3078
		NGC 2281 - 97		6 45 43	+40 55.8	1	10.58		0.64		0.20		2		75
		NGC 2281 - 98		6 45 43	+40 58.8	1	11.69		1.28		1.32		1		741
49458	−12 01632			6 45 44	−12 25.1	1	8.29		1.73				2	M2 III	369
48879	+67 00454	HR 2490		6 45 45	+67 37.8	2	5.14	.009	-0.17	.001	-0.64		4	B4 IV	154,1363
	−0 01436			6 45 45	−00 50.6	1	10.45		0.06		0.04		1		821
		NGC 2281 - 101		6 45 47	+41 16.4	2	12.05	.039	1.36	.005	1.56	.029	3		75,741
49434	−1 01386	HR 2514		6 45 47	−01 15.7	4	5.74	.008	0.29	.005	0.03	.009	69	F1 V	15,252,1078,3047
		NGC 2281 - 221		6 45 48	+41 07.	1	14.83		0.73		0.05		1		741
		NGC 2281 - 222		6 45 48	+41 07.	1	14.87		0.75		0.14		1		741
49433	−0 01438			6 45 49	−00 59.5	1	9.13		0.08		0.16		1	A0	821
49039	+61 00910	G 250 - 23		6 45 50	+60 59.4	1	8.61		0.79		0.41		2	G5	1658
49435	−1 01387			6 45 50	−01 45.7	4	7.59	.010	1.52	.010	1.76	.206	11	K5 III	1311,1509,2033,3040
		NGC 2281 - 102		6 45 51	+41 33.8	1	12.37		0.62		0.17		1		741
		NGC 2281 - 103		6 45 53	+41 04.1	2	11.39	.034	0.72	.005	0.20	.039	3		75,741
		LF11 # 5		6 45 54	+02 27.7	1	12.67		0.58		0.08		1		137
	−45 02746			6 45 54	−46 01.9	2	9.93	.005	1.48	.007	1.74		10		863,1075
49408	+8 01507			6 45 55	+07 59.9	1	8.57		0.06		-0.44		3	B8	3039
49409	+7 01457	G 108 - 29	⋆ AB	6 45 57	+07 41.1	2	7.92	.010	0.60	.000	0.10	.000	3	F8 V +G3 V	1003,3077
		NGC 2281 - 104		6 45 57	+41 29.7	1	12.43		0.62		0.19		1		741
49478	−1 01388			6 45 57	−01 36.3	2	7.65	.001	0.95	.005	-0.07		34	K0	1311,6011

HD	DM	Other Id	N Rem	α_{1950}	δ_{1950}	S	V	σ_V	B−V	σ_{B-V}	U−B	σ_{U-B}	n	Spectrum	References
292418	−0 06470			6 45 59	−00 06.5	1	9.64		1.41		1.45		1	K2 III	137
		NGC 2281 - 105		6 46 00	+41 08.5	2	11.97	.029	1.17	.010	1.03	.019	3		75,741
292397				6 46 04	+00 12.1	1	11.35		0.06		0.01		3	A2	821
49548	−15 01519	IDS06438S1513	A	6 46 04	−15 16.1	1	7.96		0.15		−0.30		2	B8III+G5/K0	401
		NGC 2281 - 109		6 46 07	+41 04.2	2	12.15	.054	0.48	.010	0.03	.015	3		75,741
		NGC 2281 - 108		6 46 07	+41 21.2	1	10.74		0.54		0.05		1		741
292376	+0 01614			6 46 08	+00 36.5	1	10.05		−0.10		−0.14		3	B8	821
289186	+1 01526			6 46 08	+01 40.0	1	9.60		−0.12		−0.42		1	B8 V	137
		NGC 2281 - 110		6 46 08	+41 09.6	1	12.71		0.57		0.01		1		741
292413				6 46 10	+00 05.0	1	10.27		0.98		0.70		1	K0 III	137
49365	+28 01247			6 46 10	+28 36.0	1	8.26		0.68		0.26		3	G0 IV	3016
49570	−11 01644			6 46 12	−11 47.7	1	8.54		−0.01		−0.34		4	A0	1732
49572	−12 01635	IDS06439S1220	AB	6 46 12	−12 22.9	1	8.83		0.73				2	G6 V	1414
		LF11 # 9		6 46 13	+02 33.9	1	12.69		0.37		0.23		1	A2 V	137
		Bo 2 - 4		6 46 14	+00 26.3	1	13.33		0.76		0.16		2		410
49754	−43 02696	IDS06447S4347	AB	6 46 14	−43 50.1	2	9.65	.017	0.15	.004	0.10		10	A1/2 V	863,2011
		Bo 2 - 1		6 46 15	+00 26.3	1	11.31		0.52		−0.55		3		410
49728	−44 02916			6 46 15	−44 39.8	3	7.77	.013	1.14	.004	1.12	.001	16	K1 III	977,1075,2011
	+0 01617	Bo 2 - 2		6 46 16	+00 26.0	1	10.86		0.57		−0.52		2		410
		G 87 - 9		6 46 16	+35 12.1	1	10.18		1.33		1.27		1	K5	333,1620
292396	+0 01618			6 46 17	+00 18.9	1	10.65		0.13		−0.03		3	A0	821
		Bo 2 - 3		6 46 17	+00 25.8	1	11.15		0.50		−0.53		2		410
		NGC 2281 - 113		6 46 17	+41 19.2	1	11.77		0.41		0.03		1		741
		Bo 2 - 5		6 46 18	+00 26.	1	13.77		0.62		−0.43		2		410
		Bo 2 - 6		6 46 18	+00 26.	1	14.52		0.62		−0.27		2		410
		Bo 2 - 7		6 46 18	+00 26.	1	13.63		0.71		0.41		2		410
		Bo 2 - 8		6 46 18	+00 26.	1	13.94		0.35		0.15		1		410
292452				6 46 18	−01 10.3	1	11.58		0.35		0.10		1	F8	821
49288	+46 01192	IDS06426N4618	A	6 46 19	+46 14.6	1	7.15		1.29		1.25		2	K0	1601
		G 192 - 50		6 46 19	+53 11.7	1	9.83		0.95		0.78		3		7010
49778	−47 02588			6 46 19	−47 14.6	1	7.11		1.04				4	G8 II	2033
49877	−55 01063	HR 2526		6 46 20	−55 29.0	3	5.65	.037	1.54	.020			11	K5 III	15,1075,2035
		BSD 98 # 2919		6 46 21	+00 07.8	1	11.51		0.38		−0.04		2	B8	821
		BSD 98 # 4250		6 46 22	+00 58.3	1	12.15		0.41		0.11		2	B7	821
292431	−0 06464			6 46 23	−00 22.8	1	10.16		0.99		0.80		1	G8 III	137
		LP 661 - 1		6 46 25	−04 55.4	1	12.93		0.64		−0.12		2		1696
49380	+32 01414	HR 2512, IS Gem		6 46 26	+32 39.9	2	5.72	.004	1.30	.003	1.45		5	K4 III	71,1501
		NGC 2281 - 116		6 46 27	+41 05.1	1	9.76		0.22				1		76
	+41 01531			6 46 27	+41 05.2	1	9.56		1.52		1.87		1	K5	1746
49799	−45 02750			6 46 28	−45 09.8	1	9.35		1.51				4	K2/3 (III)	1075
49567	+1 01531	HR 2517	★	6 46 29	+01 03.6	3	6.15	.005	−0.14	.004	−0.68	.008	9	B3 II-III	15,154,1078
49363	+41 01532	NGC 2281 - 117	★ AB	6 46 29	+41 03.3	1	8.88		0.05				1	A0p	76
		LF11 # 11		6 46 32	−00 45.8	1	11.25		1.53		1.72		1		137
49585	+0 01624	LS VI +00 027		6 46 34	+00 08.8	2	9.09	.030	0.09	.010	−0.60	.094	3	B0.5:V:nn	821,1012
292378	+0 01623			6 46 35	+00 37.1	1	9.93		−0.07		−0.15		3	B9	322
		LS VI -06 004		6 46 35	−06 34.0	1	11.20		0.09		−0.61		2		490
49647	−14 01597			6 46 35	−14 19.9	1	7.45		1.38		1.51		3	K3 III	1657
49798	−44 02920			6 46 35	−44 15.6	8	8.27	.004	−0.28	.007	−1.16	.011	87	sdO	390,1034,1460,1628,1657*
49500	+25 01446			6 46 38	+25 32.6	1	6.93		1.60		1.81		2	K0 III	105
		LF11 # 12		6 46 39	+00 02.2	1	12.70		0.44		0.29		1		137
		BSD 98 # 2962		6 46 40	+00 23.3	1	11.58		0.55		−0.39		4	B3	821
50035	−61 00716			6 46 40	−61 45.4	1	8.41		0.47				4	F5 IV	2012
49405	+38 01617			6 46 41	+38 53.1	1	8.06		0.23		0.10		2	A8 V	105
		LP 895 - 37		6 46 41	−28 00.3	1	12.56		0.66		−0.06		2		1696
292398	+0 01626	LS VI +00 029		6 46 42	+00 14.1	1	9.92		0.21		−0.29		3	B9	821
49662	−14 01599	HR 2522	★ AB	6 46 42	−15 05.2	3	5.39	.009	−0.10	.000	−0.52		9	B7 IV	15,401,2012
49643	−2 01776	HR 2521	★ AB	6 46 46	−02 12.8	2	5.74	.005	−0.10	.000	−0.46	.000	7	B8 IIIn	15,1256
		BSD 98 # 4269		6 46 46	+01 07.2	1	12.33		0.37		0.23		1	B9	821
		CCS 590		6 46 46	−12 18.3	1	10.87		3.08		3.20		1	C6,3	864
	+60 01004			6 46 47	+60 21.5	1	9.80		0.77		0.33		1		801
49850	−43 02703			6 46 47	−43 44.6	5	7.41	.012	0.16	.003	0.12	.012	23	A4 V	278,977,1075,1637,2011
		NGC 2281 - 121		6 46 49	+41 22.5	1	12.04		0.96		0.77		1		741
289229	+0 01627	LS VI +00 031		6 46 50	+00 47.5	2	9.51	.015	0.22	.010	−0.55	.055	6	B5	322,821
292379				6 46 51	+00 38.7	1	10.08		0.23		−0.18		4	A5	821
289149				6 46 52	+02 17.8	1	11.08		0.13		−0.04		1	A0 V	137
49642	−0 01448			6 46 52	−00 29.7	1	9.74		0.01		−0.04		1	B9	821
		BONN82 T6# 15		6 46 53	+00 38.7	1	13.58		0.53		−0.15		3	B3	322
		BONN82 T6# 12		6 46 54	+00 37.8	1	12.09		0.38		0.22		3	F0	322
		BONN82 T6# 13		6 46 55	+00 39.3	1	12.65		0.49		−0.42		3	B0	322
		BONN82 T6# 11		6 46 55	+00 42.2	1	11.84		1.37		1.20		3	K7	322
49521	+33 01415		B	6 46 55	+33 17.9	1	12.66		0.51		−0.01		2	A2	627
		BONN82 T6# 17		6 46 56	+00 37.7	1	14.62		0.80		0.16		3	G2	322
264600		LS VI +06 012		6 46 56	+06 17.0	2	10.85	.045	0.28	.000	−0.72	.010	5	B7	405,493
49606	+16 01298	HR 2519, OV Gem	★ A	6 46 57	+16 15.7	3	5.87		−0.14	.008	−0.51	.017	6	B7 III+A2 V	985,1022,1079
49699	−12 01642			6 46 57	−12 36.6	1	7.20		0.12		−0.62		1	B(5)(ne)	1588
		NGC 2298 sq1 15		6 46 57	−35 55.8	1	12.42		0.62		0.05		3		1665
292382	+0 01628			6 46 59	+00 42.4	1	10.75		0.39		0.17		3	F5	322
49833	−34 03120			6 47 01	−34 52.7	1	6.92		−0.04		−0.10		2	B9 V	401
		BONN82 T6# 22		6 47 02	+00 39.8	1	15.63		0.70		0.56		4	K0	322
289189				6 47 02	+01 29.1	1	11.45		0.11		0.04		1	A2 V	137

Table 1   363

HD	DM	Other Id	N Rem	α_{1950}	δ_{1950}	S	V	σ_V	B–V	σ_{B-V}	U–B	σ_{U-B}	n	Spectrum	References
49640	+7 01467	LS VI +07 009		6 47 02	+07 54.2	1	7.84		0.51		0.41		3	A5	493
289230	+0 01629			6 47 03	+00 47.7	1	9.03		1.04		0.92		4	K0	322
		NGC 2281 - 123		6 47 03	+41 21.7	1	11.43		0.59		0.21		1		741
292383	+0 01630			6 47 05	+00 40.5	1	9.85		0.44		0.02		3	F8	322
49745	−13 01670			6 47 06	−13 20.2	1	9.00		1.65				2	M1 III	369
49536	+35 01496			6 47 09	+34 56.0	1	8.95		0.49		0.06		3	G0	1601
		NGC 2298 sq1 18		6 47 09	−35 52.5	1	11.63		1.23		1.06		3		1665
49637	+17 01409			6 47 12	+17 38.9	1	7.95		1.10		1.07		3	K0	1648
		NGC 2298 - 2		6 47 12	−35 57.	1	15.50		0.80				6		1611
		NGC 2298 - 5		6 47 12	−35 57.	1	15.30		0.81				5		1611
		NGC 2298 - 7		6 47 12	−35 57.	1	13.54		1.28		0.86		2		1611
		NGC 2298 - 9		6 47 12	−35 57.	1	14.00		1.20		0.59		4		1611
		NGC 2298 - 16		6 47 12	−35 57.	1	15.30		0.98		0.21		4		1611
		NGC 2298 - 18		6 47 12	−35 57.	1	14.81		1.12		0.44		4		1611
		NGC 2298 - 20		6 47 12	−35 57.	1	12.48		0.62		0.08		5		1611
		NGC 2298 - 25		6 47 12	−35 57.	1	14.32		1.27		0.64		4		1611
49942	−43 02709			6 47 12	−43 44.6	5	7.33	.010	−0.11	.002	−0.39	.008	19	B8 V	278,743,977,1075,2011
49520	+41 01536	HR 2516	⋆ A	6 47 14	+41 50.5	4	5.00	.011	1.26	.010	1.35	.000	9	K3 III	15,1003,1363,3053
		BSD 98 # 4308		6 47 16	+01 04.3	1	12.06		0.18		−0.02		1	B8	821
		NGC 2298 sq1 2		6 47 18	−35 59.5	2	11.14	.018	1.31	.004	1.29	.028	6		365,1665
	−35 03152			6 47 18	−35 59.9	1	9.88		1.05		0.79		3		365
	+48 01446	G 107 - 25		6 47 20	+48 07.3	1	9.79		0.86		0.64		3	K0	7010
292462				6 47 21	−01 03.4	1	11.44		0.18		0.20		1		821
		NGC 2298 sq1 5		6 47 22	−36 04.4	2	13.56	.010	0.64	.007	0.06	.040	6		365,1665
292409				6 47 24	+00 07.0	1	11.62		0.11		0.08		3	A0 V	137
49812	−14 01601			6 47 24	−14 59.2	1	7.56		0.01		−0.06		2	B9 V	401
		NGC 2298 sq1 7		6 47 25	−36 01.5	2	14.21	.007	1.01	.014	0.57		6		365,1665
		NGC 2298 sq1 23		6 47 26	−35 50.5	2	12.08		0.61		0.02		3		1665
		NGC 2298 sq1 6		6 47 26	−36 03.5	2	13.61	.008	0.82	.005	0.38	.010	6		365,1665
49846	−21 01609	IDS06453S2104	A	6 47 27	−21 07.4	1	8.30		−0.07		−0.42		2	B7/8 II	401
49787	−5 01815	LS VI -05 002		6 47 28	−05 27.3	2	7.51	.030	−0.08	.030	−0.82	.020	4	B1 V:pe	490,1212
292390				6 47 29	+00 34.9	1	10.99		0.35		−0.15		2	A2	821
292466				6 47 29	−01 14.9	1	11.33		0.07		−0.01		1	F0	821
		NGC 2298 sq1 3		6 47 29	−36 00.9	2	12.78	.023	0.93	.004	0.38	.037	6		365,1665
49765	+0 01635			6 47 30	+00 49.0	2	8.70	.023	−0.10	.028	−0.42	.061	4	B9	322,821
		HS Aur		6 47 33	+47 44.0	1	10.11		0.74		0.27		3	G5	1768
		AA131,200 # 12		6 47 34	−07 35.3	1	13.45		0.61		0.29		2		7005
292391				6 47 35	+00 30.6	1	11.16		0.06		0.10		2	A0	821
49738	+13 01434	HR 2525		6 47 36	+13 28.3	1	5.65		1.34				2	gK3	71
		G 107 - 26		6 47 37	+40 23.8	1	13.40		0.49		−0.12		2		1658
49808	+0 01637			6 47 39	+00 11.8	1	7.98		0.38		0.04		8	F0 V	137
	−22 03557			6 47 39	−22 05.9	2	10.28	.020	0.82	.015	0.38	.037	7	K2	803,3008
49891	−23 04438	HR 2528	⋆ AB	6 47 39	−24 01.0	2	6.24	.068	0.05	.010	0.07		6	A0/1 V	401,2035
	−35 03157	NGC 2298 sq1 1		6 47 39	−36 01.6	1	9.92		1.05		0.79		3	K0	1665
233289	+50 01359			6 47 41	+50 43.5	1	10.04		0.20		0.12		2	A5 V	1003
50241	−61 00720	HR 2550		6 47 41	−61 53.2	7	3.26	.009	0.22	.006	0.11	.019	19	A5/7 III/V	15,1020,1034,1075*
264862	+6 01402	LS VI +06 013		6 47 42	+06 38.1	1	9.76		0.08		−0.63		3	B3	493
292392	+0 01638	LS VI +00 034		6 47 43	+00 30.4	1	10.01		0.33		−0.52		2	O8	821
49843	−5 01821			6 47 44	−06 05.8	1	8.86		−0.04				4	A0	2014
292393	+0 01639	IDS06452N0032	AB	6 47 46	+00 28.4	1	10.24		0.01		−0.02		2	M5 III:	821
49960	−31 03707	IDS06459S3109	A	6 47 46	−31 12.0	1	8.35		1.02		0.85		2	K2 IIIp+(G)	536
264895	+5 01437	LS VI +05 025		6 47 48	+05 36.6	1	9.63		0.37		−0.30		2	B8	493
49601	+47 01355	G 107 - 27		6 47 49	+47 26.2	3	8.98	.013	1.24	.007	1.16		8	K2	694,1758,3072
49888	−12 01652			6 47 49	−12 31.5	3	7.20	.015	−0.13	.009	−0.78		11	B5 Iab/b	1212,2014,2033
		NGC 2298 sq1 4		6 47 49	−36 00.7	2	13.10	.038	0.60	.000	0.07	.019	6		365,1665
50080	−46 02696			6 47 49	−46 39.0	1	7.36		0.01		−0.04		2	A1 V	401
49784	+20 01589			6 47 53	+20 03.0	1	7.99		1.55		1.87		2	K2	1648
		NGC 2298 sq1 24		6 47 53	−35 56.0	1	12.31		0.26		0.20		3		1665
49763	+19 01502			6 47 54	+19 09.6	1	9.38		−0.20		−0.66		3	B8	555
49886	−6 01764			6 47 54	−06 08.0	1	7.58		−0.10				4	B8	2014
292460	−0 01456			6 47 55	−00 59.0	1	11.02		0.14		0.26		1	A2 V	137
289199				6 47 57	+01 42.9	1	10.06		0.94		0.64		1	K0 III	137
		BSD 98 # 3081		6 47 58	+00 25.1	1	11.75		0.36		−0.16		2	B7	821
50013	−32 03404	HR 2538		6 47 58	−32 27.0	9	3.88	.039	−0.21	.021	−0.96	.015	50	B1.5IVne	15,681,815,1075,1637*
49955	−17 01645			6 48 03	−17 14.0	1	6.63		0.03		0.04		2	A1 V	401
50126	−45 02768	IDS06466S4527	AB	6 48 04	−45 30.5	4	7.11	.007	−0.03	.000	−0.09	.007	18	A0 V	278,977,1075,2011
49633	+46 01197			6 48 05	+46 33.6	1	7.50		1.05		0.82		3	G8 II	8100
50012	−27 03310	HR 2537		6 48 06	−27 16.5	3	7.04	.004	−0.19	.004	−0.82	.015	16	B2 III	164,1775,2035
49980	−16 01624	HR 2535		6 48 08	−17 01.5	3	5.78	.004	1.43	.005			11	K3 III	15,1075,2006
49977	−13 01682	LSS 89		6 48 09	−14 03.2	1	7.97		0.20		−0.59		3	B5/7ne	1212
49978	−15 01533	IDS06459S1555	AB	6 48 09	−15 58.3	1	8.16		−0.04		−0.45		2	B7 III/IV	401
	−0 01461			6 48 17	−00 12.1	1	9.19		1.15				5		6004
49933	−0 01462	HR 2530	⋆ AB	6 48 17	−00 28.7	4	5.76	.004	0.39	.003	−0.10	.007	72	F2 V	15,1415,2012,3047
49976	−7 01592	HR 2534, V592 Mon		6 48 18	−07 58.9	3	6.28	.030	0.01	.010	0.01	.010	12	A2p SrCrEu	15,1202,1256
49805	+33 01417			6 48 19	+33 20.6	1	8.47		−0.06		−0.31		8	B8	627
49340	+69 00394	HR 2511		6 48 19	+68 57.0	2	5.11	.005	−0.13	.002	−0.44		4	B7 III	154,1363
		TX Mon		6 48 20	−01 22.0	1	10.79		1.03				1	F6	1772
289236				6 48 21	+00 46.1	1	11.13		0.05		−0.19		3	B8	821
50096	−33 03278			6 48 21	−33 49.6	1	7.41		−0.05		−0.09		2	B9.5V	401
49910	+12 01319	LS VI +12 012		6 48 23	+12 31.1	1	8.01		0.79				2	G0	6009

HD	DM	Other Id	N	Rem	α_{1950}	δ_{1950}	S	V	σ_V	B–V	σ_{B-V}	U–B	σ_{U-B}	n	Spectrum	References
292570					6 48 23	−00 17.0	1	10.18		1.33		1.41		1	K0 III	137
292401					6 48 24	+00 17.6	1	11.12		0.06		0.07		3	A0	821
		EE Mon			6 48 24	−07 55.	1	12.56		0.82				1		1772
49992	−5 01827	LS VI -05 003			6 48 28	−05 16.3	2	9.01	.005	0.06	.038	-0.87	.005	4	B1:pen	490,1012
		LS VI -08 001			6 48 29	−08 03.8	1	11.45		0.23		-0.52		3		490
50123	−31 03717	HR 2545		⋆A	6 48 30	−31 38.8	2	5.74	.017	0.14	.000	-0.50	.015	6	B3 V	146,1637
50123	−31 03719	IDS06466S3135	B		6 48 30	−31 38.8	1	7.86		0.16		0.11		2	F3	146
50196	−45 02773	HR 2546		⋆AB	6 48 30	−45 23.4	7	6.54	.008	1.51	.004	1.83	.010	42	K5/M0 III	14,15,278,863,1075*
50223	−46 02703	HR 2548			6 48 30	−46 33.6	9	5.13	.005	0.45	.002	-0.03	.007	48	F5 V	14,15,278,977,977*
50028	−7 01594		A		6 48 31	−07 52.7	1	10.05		1.42				4	F2	2033
292544	+0 01646				6 48 33	+00 06.6	1	9.81		1.01		0.81		1	G8 III	137
49908	+21 01405	HR 2529		⋆A	6 48 33	+21 49.3	3	5.24	.034	-0.02	.005	0.01	.000	6	A2 V	1022,1363,3023
292543					6 48 35	+00 07.9	1	11.32		0.25		0.20		2		821
50070	−17 01649				6 48 35	−17 14.6	1	8.84		-0.06		-0.39		2	B8 IV	401
50093	−25 03691	HR 2544			6 48 35	−25 43.1	3	6.32	.002	-0.19	.005	-0.75	.004	11	B2 III/IV	164,1732,2035
50072	−20 01602				6 48 36	−20 46.6	1	7.60		-0.11		-0.42		2	B5 III	401
		LP 661 - 2			6 48 37	−09 06.8	1	13.35		1.46				1		3062
	−44 02948				6 48 37	−44 27.1	1	9.74		0.90				4	K2	1075
292593					6 48 39	−00 55.5	1	10.41		1.41		1.61		1	K4 III	137
		BSD 98 # 3166			6 48 40	+00 41.0	1	12.30		0.20		0.04		3	B8	821
		G 250 - 26			6 48 40	+64 07.6	2	16.64	.005	0.52	.000	-0.17	.005	5	DAswk	782,1663
49618	+59 01028	HR 2520		⋆AB	6 48 41	+59 30.6	2	5.35	.019	0.67	.010	0.46		6	G4 III	71,3026
50310	−50 02415	HR 2553			6 48 42	−50 33.3	5	2.93	.011	1.20	.007	1.20	.007	23	K1 III	15,200,1075,2012,8029
50118	−20 01603				6 48 43	−20 51.0	1	7.16		-0.18		-0.74		3	B2 III/IV	1732
50023	+1 01549				6 48 46	+00 57.3	1	9.51		-0.01		-0.03		2	B9	821
49695	+55 01135				6 48 46	+55 44.5	1	8.05		1.06		1.00		2	K0	1502
50337	−53 01168	HR 2554, V415 Car			6 48 46	−53 33.8	3	4.40	.004	0.92	.004	0.61	.000	14	G6 II	15,1075,2012
		LS VI -03 001			6 48 48	−03 15.8	1	11.46		0.27		-0.52		3		490
50067	−9 01680	IDS06465S0958	A		6 48 50	−10 01.5	1	7.13		1.52		1.80		4	K4 III	1657
289158	+1 01550				6 48 51	+01 50.9	1	9.80		0.15		-0.34		1	B8	821
		GD 78			6 48 51	+36 49.4	1	16.16		0.53		-0.04		1		3060
		LS VI -07 002			6 48 51	−07 37.5	1	11.47		0.13		-0.52		2		490
49951	+25 01469				6 48 53	+25 43.4	2	6.76	.005	0.15	.010	0.10	.005	4	A2	105,1733
		HA 98 # 961			6 48 53	−00 12.0	1	13.09		1.28		1.00		1		1764
50154	−24 04507				6 48 53	−24 20.3	1	8.94		-0.16		-0.65		4	B2/3 Vn	1732
292547	+0 01650	V519 Mon			6 48 55	+00 01.7	1	8.45		1.69				3	M5 III:	369
		HA 98 # 966			6 48 55	−00 12.8	1	14.00		0.47		0.36		1		1764
		HA 98 # 556			6 48 56	−00 21.2	1	14.14		0.34		0.13		3		1764
		HA 98 # 557			6 48 56	−00 21.5	1	14.78		1.40		1.07		1		1764
292630	−1 01412	SZ Mon		⋆A	6 48 56	−01 18.6	2	9.79	.030	0.75	.103	0.56		2	F8	657,6011
292630	−1 01412	IDS06464S0115	B		6 48 56	−01 18.6	1	12.10		0.30				1		6011
289144					6 48 57	+02 27.6	1	10.40		0.98		0.63		1	K0 III	137
		HA 98 # 975			6 48 57	−00 07.8	1	12.04		0.40		0.03		1	F4	281
		HA 98 # 562			6 48 57	−00 15.4	1	12.19		0.52		0.00		1		1764
49968	+23 01518	HR 2533			6 48 58	+23 39.8	2	5.67	.012	1.46	.008	1.96		6	K5 III	71,3001
		HA 98 # 563			6 48 58	−00 22.8	1	14.16		0.42		-0.19		5		1764
289321					6 48 59	+00 59.	1	10.98		0.07		0.14		2	A0	821
50003	+16 01313				6 48 59	+16 40.4	1	8.06		-0.09		-0.40		2	B9	555
		NGC 2301 - 26			6 48 59	+00 30.8	1	12.48		0.33		0.13		1		49
		NGC 2301 - 25			6 48 59	+00 31.0	1	11.28		0.19		0.08		1	A2	49
292561					6 48 59	−00 08.1	1	10.59		0.61		0.11		1	F8	281
50064	+0 01651	NGC 2301 - 143			6 49 00	+00 21.5	5	8.25	.037	0.78	.017	-0.25	.045	12	B6 Ia	49,821,1012,8031,8100
289285	+1 01556				6 49 00	+01 31.6	1	10.37		0.17		-0.29		2	A2	821
		HA 98 # 978			6 49 00	−00 07.8	3	10.58	.005	0.61	.005	0.10	.009	62	G3	989,1729,1764
		NGC 2301 - 24			6 49 02	+00 31.3	2	11.69	.005	0.12	.035	0.11	.005	?	F0	49,305
		NGC 2301 - 22			6 49 02	+00 32.8	1	14.05		0.74		0.26		1		49
50062	+3 01437	HR 2543			6 49 02	+03 06.2	3	6.37	.003	0.05	.005	0.09	.002	43	A2 Vs	15,1415,3047
50087	+0 01654	NGC 2301 - 145			6 49 03	+00 21.9	1	9.13		0.05		-0.16		1	B8.5V	49
50085	+0 01653	NGC 2301 - 28			6 49 03	+00 28.8	3	8.64	.018	-0.03	.008	-0.21	.123	4	B7 Vp	49,305,821
		NGC 2301 - 37			6 49 03	+00 30.6	1	11.90		0.69		0.18		1		49
		NGC 2301 - 23			6 49 03	+00 32.2	1	12.84		0.40		-0.04		1		49
50235	−34 03140	HR 2549			6 49 03	−34 18.4	4	4.98	.005	1.38	.000	1.56		14	K2/3 III	15,1075,2006,3052
		NGC 2301 - 20			6 49 04	+00 33.0	1	13.47		0.60		0.10		1		49
		BSD 98 # 3229			6 49 04	+00 38.7	1	13.18		0.53		0.29		1	A0	821
		NGC 2301 - 147			6 49 04	+00 38.7	1	9.82		0.20		0.00		1	A0	821
		BSD 98 # 3222			6 49 04	+00 44.1	1	13.29		0.48		0.20		2	A0	821
		NGC 2301 - 30			6 49 05	+00 29.1	2	12.02	.020	0.22	.000	0.14	.015	2	A0	49,821
		NGC 2301 - 19			6 49 05	+00 33.4	1	11.81		0.34		0.09		1	F3	49
		NGC 2301 - 38			6 49 06	+00 30.7	1	10.90		0.07		-0.02		1	A0 V(m)	49
		NGC 2301 - 21			6 49 06	+00 32.9	1	15.71		0.93		0.66		2		305
50083	+5 01448	LS VI +05 027			6 49 06	+05 08.7	1	6.92		0.06		-0.76		3	B2 Ve	1212
		HA 98 # 581			6 49 06	−00 22.1	1	14.56		0.24		0.16		3		1764
		98 L1			6 49 06	−00 23.0	1	15.67		1.24		0.78		2		1764
		HA 98 # 580			6 49 06	−00 23.1	1	14.73		0.37		0.30		2		1764
50086	+0 01655	NGC 2301 - 146		⋆AB	6 49 07	+00 25.8	2	8.26	.018	-0.04	.014	-0.23	.063	5	B8 V	49,821
		98 L2			6 49 07	−00 17.5	1	15.86		1.34		1.50		1		1764
		NGC 2301 - 7			6 49 08	+00 32.1	1	12.54		0.32		0.08		1		49
		NGC 2301 - 14			6 49 08	+00 33.9	1	16.99		0.89		0.37		1		305
292518		NGC 2301 - 98			6 49 08	+00 36.7	1	10.55		-0.02		-0.11		1	A0 V	49
50138	−6 01775				6 49 08	−06 54.4	2	6.64	.039	0.01	.005	-0.40	.004	4	B9	272,1588

Table 1 365

HD	DM	Other Id	N Rem	α_{1950}	δ_{1950}	S	V	σ_V	B–V	σ_{B-V}	U–B	σ_{U-B}	n	Spectrum	References
		NGC 2301 - 119		6 49 09	+00 26.8	1	10.91		0.06		0.06		2	A0	821
		NGC 2301 - 8		6 49 09	+00 32.3	1	12.88		0.49		0.04		1	A1	305
		NGC 2301 - 16		6 49 09	+00 33.2	1	13.64		0.55		0.07		1		305
		NGC 2301 - 15		6 49 09	+00 33.5	1	14.30		0.48		0.17		3		305
		98 L3		6 49 09	−00 11.6	1	14.61		1.94		1.84		2		1764
		HA 98 # 1002		6 49 09	−00 12.2	1	14.57		0.57		-0.03		2		1764
		98 L4		6 49 09	−00 12.9	1	16.33		1.34		1.09		2		1764
		HA 98 # 590		6 49 09	−00 18.7	1	14.64		1.35		0.85		2		1764
		BSD 98 # 3251		6 49 10	+00 32.5	1	10.23		0.04		-0.01		2	B8	821
		NGC 2301 - 95	★ A	6 49 10	+00 36.0	1	10.05		0.02		-0.07		7		821
292519		NGC 2301 - 97	★ B	6 49 10	+00 36.0	1	10.45		0.07		0.03		1	A2	821
289289				6 49 10	+01 13.4	1	10.33		0.04		0.01		2	A2 V	137
		NGC 2301 - 42		6 49 11	+00 29.5	1	13.49		0.53		0.01		1		49
	+0 01658	NGC 2301 - 9	★ A	6 49 11	+00 32.6	2	9.14	.004	-0.06	.004	-0.38	.004	6	B9 V	49,305
		NGC 2301 - 12		6 49 11	+00 33.8	1	12.10		0.26		0.11		1		305
		HA 98 # 600		6 49 11	−00 21.9	1	11.11		0.57		0.05		1	G1	281
	+0 01660	NGC 2301 - 1	★ A	6 49 12	+00 31.2	3	8.01	.010	1.23	.009	1.22	.012	4	G8 IV	49,305,1685
	+0 01659	NGC 2301 - 2	★ B	6 49 12	+00 31.6	2	9.10	.020	-0.10	.000	-0.44	.010	2	B8p	49,305
		NGC 2301 - 120		6 49 13	+00 27.7	1	13.37		0.59		0.16		1		49
		NGC 2301 - 10		6 49 13	+00 32.9	1	13.69		0.59		0.05		1		305
		NGC 2301 - 11		6 49 13	+00 33.9	1	14.12		1.24		0.89		1		305
	−41 02615			6 49 13	−41 50.3	1	10.34		2.54		3.06		1	R	864
		NGC 2301 - 44		6 49 14	+00 29.5	1	12.75		0.99		0.63		1		49
295689				6 49 14	−04 30.	1	9.44		1.79				2		369
		HA 98 # 614		6 49 15	−00 16.9	1	15.67		1.06		0.40		1		1764
292573				6 49 15	−00 22.7	1	11.15		0.16		-0.09		1	B9	281
		HA 98 # 613		6 49 15	−00 22.9	1	11.81		0.17		0.05		5	A2	281
		NGC 2301 - 56		6 49 16	+00 32.0	1	12.69		0.34		0.04		1		49
		BSD 98 # 3293		6 49 16	+00 42.4	1	11.39		0.18		0.19		2	A0	821
49837	+52 01141			6 49 16	+52 33.2	1	9.27		0.19		0.15		3	A3	1566
		HA 98 # 618		6 49 16	−00 17.6	1	12.72		2.19		2.14		7		1764
50264	−29 03557			6 49 16	−29 31.0	2	9.05	.005	0.64	.010	0.05	.012	2	Gp Ba	565,3077
		NGC 2301 - 45		6 49 17	+00 29.3	1	11.64		0.33		0.05		1	A5	49
		HA 98 # 624		6 49 18	−00 16.6	1	13.81		0.79		0.39		1		1764
		NGC 2301 - 63		6 49 19	+00 31.7	1	13.06		0.46		0.02		1		49
		NGC 2301 - 62		6 49 19	+00 32.2	1	12.42		0.33		0.08		1		49
292514	+0 01662			6 49 19	+00 43.9	1	10.17		0.01		-0.04		6	A2	821
		HA 98 # 626		6 49 19	−00 17.1	1	14.76		1.41		1.07		1		1764
		HA 98 # 627		6 49 19	−00 18.4	1	14.90		0.69		0.08		1		1764
292528		NGC 2301 - 64		6 49 20	+00 30.8	1	10.59		0.00		-0.05		1	B9 m	49
		NGC 2301 - 61		6 49 20	+00 32.0	1	11.79		0.18		0.10		1	A0	49
50170	−2 01801	LS VI -02 007		6 49 21	−02 06.9	4	6.86	.016	0.48	.011	0.30	.023	8	F2 II	490,1462,6009,8100
50571	−60 00712	HR 2562		6 49 21	−60 11.4	3	6.10	.007	0.46	.000			11	F5 V	15,2013,2030
		HA 98 # 634		6 49 22	−00 17.3	1	14.61		0.65		0.12		1		1764
292572				6 49 22	−00 22.6	1	11.54		0.15		-0.02		3		281
	−15 01543			6 49 22	−15 46.2	1	9.80		0.36		0.16		2	F0	401
292524				6 49 23	+00 22.3	1	11.37		0.24		0.01		2		821
289291	+1 01560	LS VI +01 015		6 49 23	+01 26.1	3	9.65	.010	0.27	.010	-0.64	.040	6	O8:	821,1011,1012
		HA 98 # 642		6 49 25	−00 17.9	1	15.29		0.57		0.32		1		1764
292625	−1 01416			6 49 25	−01 14.3	1	9.60		1.77				3	F8	369
		NGC 2302 - 4		6 49 25	−07 00.9	1	13.05		0.33		0.21		2		410
		NGC 2302 - 9		6 49 25	−07 02.4	1	12.75		0.27		0.13		2		410
		NGC 2302 - 14		6 49 26	−07 01.1	1	13.53		0.50		0.09		2		410
	−6 01779	NGC 2302 - 1		6 49 26	−07 01.1	1	10.79		0.12		-0.23		2		410
	−6 01780	NGC 2302 - 3		6 49 27	−07 00.5	1	10.24		0.09		-0.28		5		410
292530		NGC 2302 - 2		6 49 27	−07 01.0	1	10.97		0.08		-0.30		2		410
		NGC 2301 - 85		6 49 28	+00 33.7	1	10.59		0.09		0.03		2	A1 m	821
		NGC 2301 - 151		6 49 28	+00 39.1	1	11.10		0.09		0.10		1	A0	821
		HA 98 # 185		6 49 28	−00 23.7	1	10.54		0.20		0.11		37		1764
		NGC 2302 - 7		6 49 28	−07 02.4	1	11.85		0.21		0.00		2		410
292529		NGC 2301 - 70		6 49 29	+00 31.4	1	11.16		0.05		0.04		1	A5	49
50167	+1 01561	IDS06469N0122	AB	6 49 29	+01 18.8	6	7.86	.007	1.54	.008	1.73	.009	79	K5	10,147,989,1509,1729,6005
		Be 28- 173		6 49 29	+02 58.8	1	12.76		1.47				1		410
49949	+45 01359	HR 2532		6 49 29	+44 54.1	1	6.26		0.21		0.12		2	A8 Vn	252
		HA 98 # 646		6 49 29	−00 17.6	1	15.84		1.06		1.43		1		1764
292574	−0 01466			6 49 29	−00 23.7	3	10.54	.004	0.20	.004	0.11	.005	28	A1	281,989,1729
		NGC 2302 - 6		6 49 29	−07 01.8	1	12.61		0.25		0.16		2		410
		NGC 2302 - 10		6 49 29	−07 02.0	1	13.54		0.60		0.25		2		410
		Be 28- 203		6 49 30	+02 58.6	1	13.24		0.99		0.76		1		410
50019	+34 01481	HR 2540	★ A	6 49 30	+34 01.4	9	3.60	.010	0.10	.009	0.14	.006	46	A3 III	1,15,245,1022,1077*
		HA 98 # 648		6 49 30	−00 16.0	1	12.98		0.76		0.15		5		281
		HA 98 # 193		6 49 30	−00 23.6	4	10.03	.005	1.18	.003	1.15	.009	62	G8	989,1729,1764,6004
		NGC 2302 - 5		6 49 30	−07 01.0	1	11.62		0.17		-0.06		2		410
		NGC 2302 - 16		6 49 30	−07 02.7	1	14.95		1.28				1		410
		Be 28- 222		6 49 31	+02 58.3	1	10.69		0.07		0.01		2		410
		Be 28- 226		6 49 31	+02 58.9	1	13.84		1.27				1		410
		G 250 - 29		6 49 31	+60 56.5	2	11.02	.006	1.55	.005	1.21		3		1691,5010
		HA 98 # 650		6 49 31	−00 16.0	2	12.27	.005	0.16	.001	0.10	.010	29		281,1764
		HA 98 # 652		6 49 31	−00 18.3	1	14.82		0.61		0.13		1		1764
292575				6 49 31	−00 25.0	2	10.02	.006	1.19	.005	1.15	.019	11	K0	281,975

HD	DM	Other Id	N Rem	α_{1950}	δ_{1950}	S	V	σ_V	B−V	σ_{B-V}	U−B	σ_{U-B}	n	Spectrum	References
50188	−0 01467			6 49 32	−00 14.6	6	9.54	.004	0.00	.004	-0.10	.005	114	A0	281,975,989,1729,1764,6004
		NGC 2302 - 11		6 49 32	−07 00.3	1	14.47		1.38				1		410
		NGC 2302 - 8		6 49 32	−07 01.8	1	11.72		0.17		-0.01		2		410
		NGC 2302 - 15		6 49 32	−07 02.6	1	14.47		0.57		-0.02		2		410
		NGC 2302 - 12		6 49 33	−07 02.7	1	13.45		0.45		0.18		2		410
		NGC 2302 - 13		6 49 33	−07 02.9	1	13.65		0.78		0.27		2		410
50018	+39 01771	HR 2539, OX Aur	★ A	6 49 35	+38 55.9	2	6.13	.010	0.38	.005	0.13	.010	3	F2 V δ Del	39,3024
50018	+39 01771	IDS06462N3859	B	6 49 35	+38 55.9	1	10.20		0.68		0.19		2	G5 Vn	3024
		HA 98 # 666		6 49 36	−00 19.9	1	12.73		0.16		0.00		14		1764
		HA 98 # 211		6 49 36	−00 25.6	1	12.02		0.30		0.23		2	A6	281
	+0 01664	NGC 2301 - 72		6 49 37	+00 31.3	2	11.33	.019	0.10	.034	0.10	.044	3	A0	49,821
50209	+0 01468			6 49 37	−00 14.1	8	8.38	.009	0.03	.005	-0.34	.006	59	B9 Ve	281,397,975,989,1212,1509*
292599	−0 14690			6 49 37	−00 49.2	1	10.13		0.09		-0.25		1	B8 V	137
		HA 98 # 671		6 49 38	−00 14.7	2	13.39	.007	0.97	.006	0.70	.019	26		281,1764
		HA 98 # 670		6 49 38	−00 15.6	3	11.94	.020	1.36	.007	1.36	.042	34		281,397,1764
		MtW 98 # 498		6 49 38	−00 16.0	1	15.40		0.61		0.19		4		397
		NGC 2301 - 154		6 49 40	+00 21.4	1	10.20		0.06		-0.02		4		821
		HA 98 # 676		6 49 40	−00 15.7	3	13.06	.012	1.15	.008	0.68	.031	13		281,397,1764
		HA 98 # 675		6 49 40	−00 16.0	1	13.40		1.91		1.94		21		1764
		MtW 98 # 529		6 49 40	−00 16.0	1	13.34		1.87		1.89		4		397
		NGC 2301 - 155		6 49 42	+00 25.2	1	12.01		0.40		0.37		1		821
50056	+35 01511	HR 2542		6 49 43	+35 51.0	1	6.01		1.46				2	K3 III:	71
		98 L5		6 49 43	−00 15.9	1	17.80		1.90		-0.10		3		1764
		HA 98 # 682		6 49 43	−00 16.0	2	13.73	.023	0.65	.031	0.11	.019	9		281,1764
50303	−17 01654			6 49 43	−17 11.1	1	8.30		-0.05		-0.25		2	B8/9 V	401
50208	+0 01665	NGC 2301 - 132		6 49 44	+00 31.6	3	8.77	.018	0.94	.018	0.66	.003	7	G5 III	49,137,1685
50503	−47 02628			6 49 44	−47 15.9	2	7.30	.009	1.15	.000			8	K0 III	1075,2012
		HA 98 # 685		6 49 45	−00 16.6	2	11.96	.003	0.46	.001	0.10	.007	23	G3	281,1764
		HA 98 # 688		6 49 45	−00 19.9	1	12.75		0.29		0.25		11		1764
		HA 98 # 1092		6 49 46	−00 10.6	1	15.01		0.83		0.00		1		1764
		HA 98 # 1087		6 49 47	−00 12.1	1	14.44		1.60		1.28		5		1764
		LF11 # 26		6 49 48	+01 57.0	1	11.94		0.38		0.12		1	A0 V	137
50037	+38 01636	HR 2541	★	6 49 48	+38 30.1	1	6.30		0.49				2	F6 IV-V	71
		HA 98 # 696		6 49 48	−00 17.9	1	12.24		0.43		0.24		1	A5	281
50446	−40 02734			6 49 48	−40 29.3	1	7.45		-0.05		-0.16		2	B9 V	401
289244				6 49 49	+02 43.7	1	11.25		0.09		-0.02		1	A1 V	137
50132	+25 01479			6 49 49	+25 49.6	2	7.89	.005	0.31	.005	-0.01	.000	6	F0	833,1625
		NGC 2301 - 158		6 49 49	+00 28.2	1	11.75		0.22		-0.11		5	B8	821
50249	−1 01420			6 49 49	−01 21.2	1	8.60		1.20		1.43		1	K2 III	137
50248	+0 01667			6 49 52	+00 21.1	1	10.04		0.02		-0.09		2	A2	821
50281	−5 01844	IDS06474S0503	A	6 49 52	−05 06.7	7	6.59	.016	1.06	.012	0.95	.004	25	K3 V	657,1013,1075,1197,1758*
50281	−5 01844	IDS06474S0503	B	6 49 52	−05 06.7	2	10.07	.029	1.45	.037	1.22		6	M2	2013,3072
292550	+0 01668			6 49 53	+00 04.7	1	10.51		0.16		0.19		2	A0	821
50105	+35 01513			6 49 53	+35 01.4	1	7.87		1.34		1.47		2	K0	1601
50280	−2 01805			6 49 53	−02 17.3	1	9.34		-0.01		-0.08		1	B9	1462
		HA 98 # 1102		6 49 54	−00 10.0	1	12.11		0.31		0.09		8		1764
292584				6 49 55	−00 28.9	2	10.94	.005	0.22	.012	0.19		6	A3	281,6004
50282	−5 01845	HR 2552		6 49 55	−05 15.3	3	6.30	.008	0.98	.004	0.77	.005	11	K0	15,1078,2035
		CCS 602, W Mon		6 49 55	−07 05.1	1	9.95		3.39		3.60		1	C4,5	864
		LP 122 - 11		6 49 56	+54 41.6	1	15.22		1.10		0.43		2		1773
		HA 98 # 712		6 49 56	−00 16.6	1	11.60		0.16		-0.08		5	A0	281
50472	−38 02970			6 49 56	−38 09.2	1	7.88		-0.11		-0.47		2	B8 IV/V	401
50501	−44 02969			6 49 56	−44 23.7	2	8.16	.003	1.31	.004	1.35		12	K1 III	911,2011
50445	−36 03189	HR 2558		6 49 57	−36 10.1	1	5.96		0.18				4	A3 V	2031
50862	−66 00608			6 49 57	−66 14.0	1	6.99		0.88		0.59		5	G5 III	1628
50279	+1 01567			6 50 01	+01 09.1	4	8.17	.009	-0.06	.007	-0.29	.009	51	B8	10,147,1509,6005
		HA 98 # 1112		6 50 01	−00 11.7	1	13.98		0.81		0.29		2		1764
292526	+0 01670			6 50 02	+00 24.1	1	10.87		0.07		0.05		3	B9	821
50186	+25 01482			6 50 02	+25 22.4	1	7.38		0.28		0.13		3	Am	8071
		HA 98 # 1119		6 50 03	−00 10.8	1	11.88		0.55		0.07		4		1764
		HA 98 # 1124		6 50 04	−00 12.9	1	13.71		0.31		0.26		12		1764
		HA 98 # 1122		6 50 04	−00 13.4	1	14.09		0.60		-0.30		12		1764
		HA 98 # 724		6 50 04	−00 15.7	2	11.12	.001	1.10	.002	0.91	.019	8	G8	281,1764
292531	+0 01671			6 50 05	+00 35.8	1	10.17		-0.01		-0.04		9	A0	821
50277	+8 01543	HR 2551		6 50 06	+08 26.6	2	5.76	.005	0.27	.005	0.09	.010	7	F0 Vn	15,1061
		HA 98 # 733		6 50 06	−00 13.6	1	12.24		1.28		1.09		10		1764
		LF11 # 29		6 50 07	−00 56.0	1	12.38		0.44		-0.09		1	F0 V	137
		G 192 - 57		6 50 08	+54 41.9	1	12.27		0.69		0.06		2		1773
292552				6 50 09	+00 02.1	1	11.26		0.11		0.05		2	A0 V	137
		BSD 98 # 1902		6 50 09	+00 07.4	1	11.53		0.50		-0.40		3	B5p	821
50568	−44 02974	IDS06487S4411	A	6 50 10	−44 14.8	2	8.12	.010	0.96	.005	0.67		8	K0 IV	158,2012
50568	−44 02973	IDS06487S4411	B	6 50 10	−44 14.8	1	10.13		0.60		0.02		4	G0	158
		HA 98 # 313		6 50 12	−00 32.3	1	11.07		0.66		0.17		2	G0	281
		LS VI -01 009		6 50 12	−01 22.2	1	11.98		0.04		-0.11		1		490
	−4 01704	LS VI -05 004		6 50 12	−05 00.2	1	10.44		0.13		-0.57		3		490
50621	−48 02556	HR 2563		6 50 12	−48 13.9	2	6.41	.004	1.21	.000	1.23		7	K2/3 III	58,2035
292582	−0 01471			6 50 14	−00 32.8	4	9.18	.007	1.15	.010	1.14	.005	52	G8	281,989,1729,1775
		EK Mon		6 50 15	−02 23.7	2	10.80	.064	1.10	.007	0.63		2		1462,1772
292583				6 50 19	−00 29.6	2	10.27	.003	0.05	.005	0.02		9	A0	281,6004
		BSD 98 # 3422		6 50 22	+00 35.5	1	11.32		0.27		-0.09		4	B8	821

Table 1   367

HD	DM	Other Id	N Rem	α_{1950}	δ_{1950}	S	V	σ_V	B–V	σ_{B-V}	U–B	σ_{U-B}	n	Spectrum	References
		BSD 98 # 3426		6 50 22	+00 39.0	1	12.08		0.37		0.13		3	B9	821
50372	+2 01448			6 50 22	+02 48.3	1	7.09		1.14		1.06		3	G6 II	8100
50545	−38 02973			6 50 22	−38 34.3	1	8.59		-0.11		-0.54		2	B5 V	401
		LF11 # 31		6 50 25	−00 59.6	1	11.87		1.20		0.97		1		137
50320	+15 01359			6 50 26	+15 53.1	1	8.50		-0.10		-0.51		2	B9	555
		BSD 98 # 1940		6 50 27	+00 11.8	2	12.20	.014	0.11	.000	-0.13	.037	4	B5	137,821
		Steph 611		6 50 27	+32 22.1	1	10.55		1.25		1.20		1	K7	1746
		HA 98 # 365		6 50 27	−00 31.7	1	11.80		0.18		0.11		2	A5	281
		BSD 98 # 3441		6 50 28	+00 49.4	1	12.10		0.29		0.32		2	B8	821
48745	+82 00188			6 50 28	+82 19.5	1	8.33		1.40		1.60		2	K2	1375
50387	+0 01675			6 50 31	+00 15.9	1	9.43		0.17		0.01		2	B8	821
50204	+38 01638 HR 2547			6 50 31	+38 34.1	1			-0.05		-0.19		2	B9.5p Si	1022
50463	−16 01638			6 50 31	−16 09.0	2	7.13	.005	-0.10	.000	-0.68		6	B3 III	401,2033
292556	+0 01676			6 50 32	−00 05.7	1	9.81		0.19		-0.09		3	B8	821
50491	−21 01630			6 50 32	−21 07.3	1	7.47		-0.12		-0.58		2	B4 V	401
50462	−11 01673			6 50 35	−12 05.5	1	7.14		0.26		0.17		2	A2mA7-F0	355
50371	+11 01344 HR 2555			6 50 36	+11 03.7	1	6.24		0.97				2	K0 III	71
265779	+7 01496			6 50 40	+07 44.4	1	9.09		1.77				4	M0	369
292557				6 50 40	−00 08.9	1	11.51		0.45		0.17		2	A2	821
		HA 98 # 402		6 50 40	−00 29.5	1	11.66		0.12		0.11		2	A3	281
50535	−22 01574			6 50 42	−22 59.1	1	7.86		-0.10		-0.41		2	B8 V	401
292579				6 50 43	−00 26.	1	10.61		0.02		-0.05		4		281
50436	−4 01708 GY Mon			6 50 43	−04 30.8	3	8.12	.015	2.46	.102	3.80		4	C6,3	109,864,1238
		HA 98 # 428		6 50 46	−00 25.6	1	12.15		0.36		0.05		2	F5	281
295747				6 50 48	−03 12.	1	10.10		1.83				2		369
		L 183 - 38		6 50 48	−59 03.	1	12.42		1.01		0.72		1		1696
289261	+2 01451			6 50 51	+01 56.2	1	9.20		2.24		1.86		3	K2	8032
		GD 79		6 50 51	+39 42.1	1	15.96		0.09		-0.80		1		3060
50562	−21 01633 LSS 92			6 50 51	−21 46.4	1	8.03		0.18		0.15		2	B2/3 III	401
292615	−1 01434			6 50 53	−01 24.0	1	10.16		0.06		-0.20		1	K0	490
50590	−22 01576			6 50 53	−23 02.8	1	9.03		0.98		0.75		1	K2/3 V	3072
50592	−23 04523 Cr 121 - 20			6 50 53	−23 57.5	1	10.30		0.00		-0.17		3	B9 V	62
289310				6 50 54	+00 59.6	1	10.85		0.19		0.11		1	A2 V	137
292555	+0 01682 LS VI -0 008			6 50 56	−00 05.2	1	10.17		0.12		-0.46		2	B5	821
292580				6 50 56	−00 31.6	1	11.00		0.46		0.03		2	G0	281
		LF11 # 34		6 50 57	+02 35.5	1	10.95		0.99		0.66		1	K0 III	137
50787	−45 02794			6 50 57	−45 59.2	3	8.32	.003	0.64	.006	0.20	.009	16	G3 V	977,1075,2011
50591	−23 04525 Cr 121 - 18			6 50 58	−23 54.0	1	9.61		0.48		-0.02		1	F3 V	62
50786	−44 02982			6 50 59	−44 59.2	3	8.84	.009	0.56	.006	0.02	.016	16	F8 V	977,1075,2011
50648	−26 03529 HR 2567	⋆ A	6 51 00	−26 53.6	1	6.40		1.53				4	M6 III	2031
289307	+1 01575			6 51 01	+01 15.3	1	9.57		0.94		0.58		1	G8 III	137
50785	−42 02793 HR 2575			6 51 06	−42 26.5	9	6.52	.005	0.41	.004	0.14	.009	56	F5 II/III	14,15,278,863,977,977*
50644	−18 01591 HR 2565			6 51 08	−18 58.2	2	5.65	.011	0.28	.000	0.17		12	A9/F0 III	3,2035
50646	−24 04551 Cr 121 - 6			6 51 09	−24 06.3	3	7.72	.005	-0.20	.000	-0.84	.014	7	B2 II/III	62,401,401
	−24 04552 Cr 121 - 15			6 51 09	−24 11.9	1	9.46		0.56		0.05		2		62
	−31 03760			6 51 09	−31 42.9	1	9.41		1.55		1.90		1	R	864
292536	+0 01684			6 51 10	+00 32.1	1	10.33		-0.01		0.02		6	B8	821
50643	−18 01594 HR 2566			6 51 10	−18 52.2	2	6.13	.010	0.16	.005	0.13		7	A3 III	2031,8071
50680	−23 04532 Cr 121 - 8			6 51 11	−24 03.2	1	8.27		-0.20		-0.78		4	B3 II/III	62
	+71 00370			6 51 15	+71 07.5	1	9.74		0.75		0.33		2		1375
50860	−43 02756 HR 2579	⋆ AB	6 51 16	−43 54.8	8	6.46	.008	-0.10	.004	-0.38	.007	43	B8 V	14,15,278,977,977*
50883	−46 02733			6 51 16	−46 44.8	3	8.04	.007	0.96	.004	0.69	.015	16	G8 III	977,1075,2011
	−7 01619			6 51 17	−07 52.0	1	9.27		1.71				2	M2	369
50711	−24 04553 Cr 121 - 3			6 51 18	−24 37.3	1	6.46		0.12		0.17		4	A1 V	62
50559	+2 01454			6 51 20	+02 11.5	1	9.22		-0.02		-0.09		2	B8 V	137
292696				6 51 21	+00 08.7	1	11.32		0.22		-0.24		2	B8	821
		G 107 - 34		6 51 21	+43 35.1	2	15.44	.038	1.45	.007			3		538,1691
50583	−0 01479			6 51 21	−00 15.0	2	8.28	.000	-0.04	.000	-0.28	.020	9	B9	158,1775
50584	−0 01480 IDS06488S0036	A	6 51 23	−00 40.1	1	9.46		0.42		0.04		1	F5 V	137
50707	−20 01616 Cr 121 - 24	⋆ V	6 51 23	−20 09.7	7	4.82	.005	-0.22	.008	-0.95	.007	25	B1 III	3,15,62,1075,1732,2012,8015
50740	−23 04537 Cr 121 - 11			6 51 23	−24 02.9	2	8.79	.020	-0.05	.005	-0.13	.035	5	B9 V	62,1776
	−24 04556 Cr 121 - 19	⋆ AB	6 51 28	−24 20.8	1	10.16		0.81		0.36		1	B9	62
50806	−28 03554 HR 2576			6 51 34	−28 28.2	4	6.03	.011	0.71	.009	0.25	.005	13	G5 IV	15,158,2012,3077
265866		G 87 - 12		6 51 35	+33 20.3	4	9.94	.063	1.59	.013	1.22	.015	5	M4	333,694,1620,3003,8039
50384	+46 01202 HR 2556			6 51 35	+45 53.5	2	6.34	.005	0.94	.005	0.67	.023	4	K0 III-IV	252,3051
		G 87 - 13		6 51 36	+35 35.0	1	11.07		0.48		-0.12		1	sdF2	333,1620
		LS VI -03 002		6 51 36	−03 45.7	1	10.35		0.50		0.45		3		490
50781	−24 04559 Cr 121 - 7			6 51 36	−24 19.5	1	8.14		0.52		0.00		2	F6 V	62
50420	+44 01551 HR 2557, V352 Aur			6 51 38	+43 58.5	3	6.13	.013	0.30	.015	0.20	.013	6	A9 III	39,254,8071
292712				6 51 39	−00 22.7	1	10.81		0.00		-0.39		2	B3	821
		LF11 # 38		6 51 39	−00 52.8	1	12.42		0.20		0.10		1	A0 V	137
51269	−65 00663			6 51 39	−65 33.5	1	8.54		-0.04		-0.27		4	B9 IV	1700
50700	−5 01863 HR 2570	⋆ AB	6 51 41	−05 47.3	3	6.40	.005	0.17	.005	0.15	.010	14	A6 Vn	15,1061,2029
266098	+9 01429 LS VI +09 018			6 51 42	+09 31.2	1	8.77		-0.04		-0.74		3	B9	493
		GD 80		6 51 42	−02 04.6	1	14.82		-0.21		-1.19		1	DAwk	3060
	+55 01142			6 51 43	+55 42.2	1	10.23		1.35		1.29		3	K4	1723
51043	−53 01177 HR 2587			6 51 43	−54 01.6	1	6.57		1.08				4	G5 Ib/II	2035
50578	+17 01441			6 51 45	+17 44.9	1	7.75		1.65		1.89		3	K2	1648
295731	−2 01819			6 51 45	−02 37.5	1	9.19		1.00		0.76		2	K0	1732
50804	−24 04563 Cr 121 - 5			6 51 45	−24 10.0	1	7.47		0.19		0.29		5	A5 II/III	62

HD	DM	Other Id	N	Rem	α_{1950}	δ_{1950}	S	V	σ_V	B–V	σ_{B-V}	U–B	σ_{U-B}	n	Spectrum	References
		LP 205 - 37			6 51 46	+44 29.3	1	11.27		1.34		1.32		2		1723
	−7 01623	RU Mon			6 51 47	−07 31.8	1	10.50		0.08		−0.25		2	B9	1768
50696	−0 01691				6 51 48	+00 14.8	2	8.94	.019	0.02	.019	−0.70	.094	4	B1:V:enn	821,1012
		LS VI -01 011			6 51 48	−01 55.8	1	11.72		0.25		−0.54		1		490
	−1 01441	LS VI -02 008			6 51 48	−02 02.9	1	10.37		0.24		−0.56		3		490
50635	+13 01462	HR 2564	★	A	6 51 49	+13 14.6	1	4.74		0.31		0.03		2	F0 Vp	3032
50635	+13 01462	HR 2564	★	AB	6 51 49	+13 14.6	3	4.65	.004	0.30	.003	0.08	.005	8	F0 Vp	15,1363,8015
50635	+13 01462	IDS06490N1318	B		6 51 49	+13 14.6	1	7.68		0.72		0.20		2	G6 V	3016
50749	−2 01820	IDS06493S0236	A		6 51 50	−02 40.0	1	9.03		−0.05		−0.67		5	B8	1732
50801	−18 01596				6 51 50	−18 30.1	1	8.94		−0.06		−0.31		2	B9 IV	401
50900	−34 03177				6 51 50	−34 40.5	1	9.24		0.97				6	F0 V	955
50853	−24 04565	Cr 121 - 2	★		6 51 51	−24 28.5	3	6.22	.014	0.01	.005	0.10	.053	11	A0 V	62,401,2035
50747	−0 01487	HR 2572			6 51 52	−01 03.8	3	5.44	.002	0.17	.005	0.17	.006	48	A4 IV	15,1078,3047
50778	−11 01681	HR 2574			6 51 52	−11 58.5	7	4.07	.011	1.43	.006	1.69	.008	29	K4 III	15,1003,1075,1075*
50964	−42 02805				6 51 54	−42 09.5	1	8.21		1.16		0.93		3	K0 III/IV	388
50850	−18 01598				6 51 57	−18 13.4	1	9.29		0.03		−0.57		2	B(3) Vnne	401
51210	−59 00716	HR 2592			6 52 01	−59 16.7	2	6.40	.005	0.18	.000			7	A3mA3-A8	15,2012
51557	−70 00572	HR 2602			6 52 02	−70 54.1	3	5.39	.003	−0.11	.003	−0.37	.005	23	B7 III	15,1075,2038
289361	+1 01586				6 52 03	+01 49.7	1	9.44		1.02		0.76		1	K0 III	137
292700					6 52 03	−00 07.2	1	10.69		0.35		0.31		1	A2 V	137
50877	−24 04567	Cr 121 - 1	★	V	6 52 03	−24 07.2	8	3.84	.038	1.74	.016	1.98	.023	32	K3 Iab	15,62,1732,2013,2028*
292716					6 52 05	−00 32.7	1	10.65		−0.02		−0.55		1	B8 V	137
292693	+0 01694				6 52 07	+00 00.9	1	9.72		1.60				4	K7	369
	+12 01343	G 110 - 21			6 52 08	+12 14.0	1	10.53		1.26		1.23		4	M1	7008
50896	−23 04553	Cr 121 - 4	★	V	6 52 08	−23 51.9	7	6.87	.028	−0.27	.014	−0.88	.044	21	WN	62,1096,1359,1732,1776*
		AK Gem			6 52 10	+13 46.1	1	13.33		0.28		0.19		1	M6.5	699
50820	−1 01446	HR 2577	★		6 52 10	−01 41.5	4	6.19	.042	0.52	.028	−0.34	.008	17	B3IVe+K2 II	15,681,1415,8035
292670	+0 01695				6 52 12	+00 33.1	1	10.32		0.03		−0.06		6	B9	821
50692	+25 01496	HR 2569			6 52 14	+25 26.4	3	5.72	.017	0.57	.014	0.02	.009	20	G0 V	1067,1217,3077
50939	−22 01588				6 52 16	−22 59.0	2	8.88	.011	−0.13	.008	−0.59	.014	6	B2/3 V	401,1732
50844	−0 01489				6 52 18	−01 00.4	1	9.10		0.21		0.08		18	A2	588
50846	−1 01449	AU Mon			6 52 22	−01 18.7	2	8.26	.040	0.07	.005	−0.65		5	B5	1588,2033
50871	−6 01814	LS VI -06 005			6 52 22	−06 18.6	1	9.38		−0.04		−0.58		1	B8	490
50938	−17 01673				6 52 23	−17 51.2	1	7.66		−0.12		−0.69		2	B3 Ve	401
51042	−34 03182	IDS06506S3406	AB		6 52 25	−34 09.5	1	7.35		−0.03		−0.12		2	A1 IV	401
289329					6 52 28	+02 50.8	1	9.66		0.94		0.71		1	K0 III	137
50890	−2 01827	HR 2582			6 52 28	−02 44.3	2	6.03	.005	1.10	.000	0.86	.005	7	K0	15,1061
50891	−3 01643	LS VI -03 003			6 52 29	−03 37.6	2	8.84	.005	0.24	.015	−0.70	.000	5	B0:pe	405,490
		CCS 614, EM Mon			6 52 30	−07 57.5	1	10.98		3.20				2		864
	+35 01523				6 52 31	+35 43.1	1	9.75		0.16		0.11		2		1726
292691					6 52 33	+00 11.0	1	10.73		0.05		−0.41		3	A3	821
51040	−27 03393	IDS06506S2726	AB		6 52 33	−27 29.7	3	8.33	.010	0.68	.009	0.27	.005	14	K0 IV +F/G	214,1414,1775
50868	+5 01472	LS VI +05 029			6 52 34	+05 29.9	2	7.88	.043	−0.19	.016	−0.81		7	B2 Ven	1732,2033
50767	+24 01457				6 52 34	+24 04.4	1	7.70		−0.20		−0.84		2	B2 V	555
50936	−7 01628				6 52 34	−07 46.1	1	8.91		0.14		−0.08		2	B9	388
51013	−24 04578	Cr 121 - 12			6 52 37	−24 11.5	4	8.81	.026	−0.15	.014	−0.64	.029	10	B3 V	62,401,1732,1776
		MCC 483			6 52 39	+47 03.4	1	10.88		1.25				1	K7	1017
51038	−24 04580	Cr 121 - 13			6 52 39	−24 51.6	2	9.07	.045	−0.16	.005	−0.68	.005	5	B3 V	62,401
		G 103 - 71			6 52 40	+31 55.9	1	12.54		1.30		1.29		1		333,1620
51035		Cr 121 - 16			6 52 40	−24 06.6	1	9.48		−0.09		−0.32		3	B8 V	62
51036	−24 04579	Cr 121 - 10			6 52 40	−24 15.1	2	8.78	.010	−0.19	.000	−0.81	.010	6	B5	62,401
		LF11 # 43			6 52 41	+01 45.1	1	12.64		0.57		−0.02		1	F5	137
		AA45,405 S285 # 4			6 52 43	+00 27.6	1	15.70		0.56		−0.03		2		797
		LS VI -01 013			6 52 43	−01 21.1	1	11.83		0.24		−0.54		1		490
50981	−8 01617	IDS06503S0822	AB		6 52 43	−08 26.1	1	7.76		0.04		−0.18		2	B9 V	388
		AA45,405 S285 # 5			6 52 44	+00 27.6	1	16.00		0.41		−0.24		2		797
		AA45,405 S285 # 2			6 52 44	+00 27.7	1	14.42		0.33		−0.21		2		797
	−0 01491	LS VI -00 009	AB		6 52 44	−00 29.8	1	12.05		0.26		−0.65		4	B0 V	92
		LF11 # 44			6 52 44	−00 48.6	1	11.17		1.30		1.09		1	K0 III	137
298383	−51 02150				6 52 47	−51 09.5	2	9.67	.015	0.88	.015	0.23	.015	6	B2 V	125,362
49878	+77 00266	HR 2527			6 52 48	+77 02.7	4	4.54	.009	1.36	.005	1.64	.020	11	K4 III	15,1363,3035,8015
	−7 01629	HH Mon			6 52 48	−07 21.6	1	9.61		0.55		0.19		1	Am:	355
51266	−50 02458	HR 2594			6 52 48	−50 33.0	4	6.25	.007	0.99	.003			18	K0/1 III	15,1075,2013,2028
50816	+27 01270				6 52 49	+27 21.1	1	7.22		−0.07		−0.33		3	B9	1625
50982	−9 01729				6 52 49	−09 15.4	1	7.73		−0.02		−0.42		3	B8	388
		LF11 # 45			6 52 50	+01 08.5	1	11.95		0.33		0.24		1	A2 V	137
50658	+46 01203	HR 2568			6 52 50	+46 20.4	1			−0.06		−0.46		2	B8 IIIe	1079
51088	−24 04586	Cr 121 - 9			6 52 50	−24 39.2	2	8.27	.007	−0.16	.014	−0.44	.013	10	B9 V	62,1732
50931	+8 01562	HR 2584			6 52 51	+08 23.4	2	6.28	.005	0.04	.000	0.06	.000	7	A0 V	15,1061
	+80 00224				6 52 52	+79 54.1	1	9.98		0.45		0.08		2	F8	1375
51208	−42 02818	HR 2591, NP Pup			6 52 52	−42 18.1	6	6.33	.027	2.24	.031	2.82	.017	18	C3 II	15,109,864,8015,8022,8029
51008	−11 01691				6 52 53	−11 42.2	1	8.14		−0.06		−0.44		2	A0	388
51055	−20 01624	HR 2588	★	AB	6 52 53	−20 20.4	1	5.74		0.08				4	A2 V	2035
	−24 04587	Cr 121 - 14			6 52 53	−24 36.9	1	9.30		1.07		0.86		4		62
51087	−24 04588	Cr 121 - 17	★	AB	6 52 55	−24 11.6	1	9.51		0.43		0.07		3	F3 V	62
		LF11 # 46			6 52 56	+01 32.5	1	11.19		0.04		−0.12		1	B8 V	137
50551	+57 01017	HR 2561			6 52 57	+57 37.8	2	6.03	.015	1.50	.005	1.75		7	K3 III	252,1379
50522	+58 00982	HR 2560	★	AB	6 52 57	+58 29.4	4	4.35	.005	0.85	.002	0.51	.004	12	G5 III-IV	15,1363,3016,8015
50952	+9 01439				6 52 58	+09 03.6	1	8.33		0.14		0.12		3	A2	1732
51111	−24 04595	Cr 121 - 21			6 52 58	−24 06.6	1	10.40		−0.04		−0.18		2	B9 V	62

Table 1 369

HD	DM	Other Id	N Rem	α_{1950}	δ_{1950}	S	V	σ_V	B–V	σ_{B-V}	U–B	σ_{U-B}	n	Spectrum	References
	−33 03337			6 52 58	−33 40.8	5	9.07	.010	0.47	.007	-0.15	.005	13	F5	158,1594,2017,2034,6006
	+40 01758	G 107 - 37	⋆ A	6 52 59	+40 08.8	2	9.11	.019	1.13	.014	1.06	.019	4	K5 V	1003,3072
	+40 01758	G 107 - 37	⋆ AB	6 52 59	+40 08.8	1	8.60		1.12				1	K5 V	1017
	+40 01758	G 107 - 38	⋆ B	6 52 59	+40 09.4	1	11.10		1.43		1.20		2		3072
		LF11 # 47		6 53 00	+00 09.3	1	11.65		0.23		0.11		1		127
51085	−17 01679			6 53 02	−17 09.0	1	7.89		0.00		-0.34		2	B8 III	401
51113	−24 04594			6 53 02	−24 27.9	1	9.99		0.61		0.04		2	G3/G5V:	1696
51027	−9 01730			6 53 03	−09 35.0	2	8.53	.005	-0.04	.005	-0.26		7	B8 V	388,2014
50975	+9 01442			6 53 04	+08 56.6	1	7.54		0.68		0.02		3	F8 Ib	1732
51110	−21 01650			6 53 04	−21 48.4	1	8.71		-0.16		-0.65		2	B9 Ib/II	401
289348				6 53 06	+02 16.5	1	10.35		0.94		0.74		1	K0 III	137
47976	+85 00098			6 53 07	+85 38.6	1	8.13		1.65		1.67		2	M1	1733
51079	−8 01625			6 53 07	−08 21.1	2	7.94	.005	-0.05	.015	-0.30		6	B8	388,2014
292705	−0 01497			6 53 10	−00 11.6	1	9.89		0.80		0.31		3	G5 III	137
51082	−10 01758	IDS06508S1006	AB	6 53 12	−10 09.3	1	8.29		0.04		-0.05		2	A0 V	388
		UV0653-23		6 53 12	−23 28.8	1	9.52		-0.13		-0.55		3		1732
50763	+46 01205	HR 2573		6 53 14	+46 46.4	1	5.86		1.08				3	gK0	71
266594	+8 01566			6 53 15	+08 50.4	1	9.97		0.13		0.08		2	G0	1732
		CCS 620, IV Mon		6 53 15	+11 02.	1	10.63		2.89				2	C6,4-	864
		LF11 # 50		6 53 15	−00 05.6	1	11.58		0.13		0.06		1	A1 V	137
51081	−9 01733	IDS06509S0922	A	6 53 15	−09 25.4	1	8.72		0.56		0.03		2	G0	3016
51081	−9 01733	IDS06509S0922	B	6 53 15	−09 25.4	1	9.84		0.80		0.38		2		3032
51107	−8 01626			6 53 16	−08 53.5	1	8.89		1.66				2	M1	369
292704				6 53 17	−01 57.3	1	10.07		0.11		-0.16		1		821
51076	−3 01648	HI Mon		6 53 17	−03 58.6	1	9.31		0.11		-0.75		3	B8	490
		POSS 309 # 6		6 53 18	+30 06.8	1	18.43		1.52				3		1739
		LS VI -03 004		6 53 18	−03 56.8	1	11.38		0.22		-0.36		3		490
	−0 01498	V521 Mon		6 53 19	−00 10.5	1	10.05		0.13		-0.22		3	B7	1768
51289	−37 03163	IDS06516S3722	A	6 53 20	−37 26.1	1	8.11		-0.02		-0.36		2	B8 IV	401
51106	−1 01459			6 53 22	−01 31.2	1	7.36		0.23		0.18		1	A3 m	355
	−18 01604			6 53 23	−18 34.1	1	10.35		1.80				2	M3	369
51200	−21 01655	IDS06513S2155	A	6 53 25	−21 58.3	1	6.81		-0.20		-0.81		2	B2 III/IV	401
51199	−19 01610	HR 2590	⋆ AB	6 53 27	−20 04.3	5	4.66	.019	0.37	.009	0.04	.025	13	F2 IV/V	15,1075,2012,3026,8015
		G 87 - 15		6 53 29	+38 58.2	1	14.93		1.32		1.20		2		316
51150	−2 01835	LS VI -03 006		6 53 30	−03 06.5	1	7.96		-0.01		-0.42		3	B8	490
51255	−26 03582			6 53 30	−27 04.3	1	8.24		-0.16		-0.69		4	B3 V	1732
51151	−3 01650	LS VI -03 007		6 53 32	−03 08.2	1	7.91		0.20		0.26		3	A0	490
51772	−65 00670	IDS06534S6547	A	6 53 36	−65 51.0	1	8.25		1.63		1.69		4	M5 III	1700
51196	−11 01699			6 53 39	−11 28.4	1	8.95		0.04		-0.31		2	B8 V	388
51285	−24 04604			6 53 39	−24 36.8	3	8.13	.009	-0.15	.009	-0.84	.007	17	B2 V(n)	401,1732,1775
		CCS 621, BG Mon		6 53 40	+07 07.9	1	9.65		2.80		4.70		2	C4-5,4	64
51104	+10 01335	HR 2589		6 53 41	+10 01.4	3	5.91		-0.09	.012	-0.36	.006	5	B8 Vn	985,1049,1079
50812	+52 01149			6 53 41	+52 20.2	1	9.21		0.23		0.16		2	A3	1566
51283	−22 01602	Cr 121 - 25	⋆	6 53 41	−22 52.5	5	5.30	.021	-0.18	.008	-0.80	.015	14	B2 III	15,62,401,2012,8100
51000	+33 01433	HR 2586		6 53 44	+33 44.9	1	5.89		0.88				2	G5 III	71
	−4 01738	LS VI -04 008		6 53 46	−04 06.2	1	10.36		0.31		-0.58		2		490
		LS VI -02 009		6 53 47	−02 52.5	1	10.66		0.09		-0.28		3		490
51223	−8 01629			6 53 47	−08 32.4	1	8.93		-0.07		-0.45		2	B8 V	388
51344	−26 03592			6 53 48	−27 03.7	1	8.51		-0.07		-0.27		4	B8/9 V	1732
51250	−13 01741	HR 2593	⋆ AB	6 53 49	−13 58.6	3	4.97	.022	1.19	.005			11	K2/3 III	15,1075,2035
51193	−3 01651	LS VI -03 008		6 53 50	−03 44.4	2	8.23	.015	-0.01	.020	-0.74	.010	7	B1 V:nn	158,490
51608	−55 01095			6 53 51	−55 11.5	4	8.16	.018	0.78	.011	0.36	.007	11	G8 V	258,1075,2012,3008
266611	+30 1367a	G 87 - 16		6 53 52	+30 49.9	2	9.74	.005	1.37	.015	1.23	.005	3	A7	333,1620,3072
		LF11 # 51		6 53 54	−00 41.8	1	12.50		0.34		0.16		1	A0 V	127
51309	−16 01661	Cr 121 - 26	⋆ V	6 53 54	−16 59.3	12	4.38	.010	-0.06	.006	-0.70	.011	34	B3 Ib/II	1,15,62,1004,1075*
51340	−21 01658	IDS06518S2123	AB	6 53 56	−21 26.4	1	9.19		-0.11		-0.60		2	B5 V	401
50973	+45 01367	HR 2585		6 53 58	+45 04.3	6	4.90	.005	0.03	.004	0.05	.004	21	A2 Vn	15,1007,1013,1363*
51219	+1 01600	G 108 - 40		6 53 59	+01 14.2	5	7.40	.014	0.70	.012	0.22	.008	17	G8 V	22,1003,1355,1509,3026
51307	−12 01703			6 54 01	−12 47.3	2	9.12	.000	-0.06	.000	-0.44	.000	3	Ap Si	320,388
292847				6 54 04	−00 49.9	1	10.53		0.13		0.08		1	A2 V	137
		TY Mon		6 54 06	+00 15.4	1	11.49		1.11				1	K5	1772
		LP 840 - 12		6 54 07	−21 43.1	1	11.20		0.64		0.04		2		1696
		LF11 # 53		6 54 09	−00 38.4	1	12.11		0.30		0.18		1	A0 V	127
		BV Mon		6 54 10	+04 35.4	1	11.12		1.05				1		1772
51335	−8 01632			6 54 11	−08 16.0	1	6.93		0.24		0.15		2	A5 V	388
51361	−12 01705			6 54 12	−12 59.9	1	10.23		-0.04		-0.46		2	B9 Ib/II	388
51277	+0 01719			6 54 13	+00 52.1	1	7.45		1.65				3	K5	369
51360	−7 01640			6 54 15	−08 07.0	1	7.60		-0.07		-0.45		3	B7 III	388
51304	−3 01655	LS VI -03 009		6 54 16	−03 35.2	1	7.89		0.56		0.42		2	F0	490
292861	−1 01468			6 54 17	−01 10.1	1	10.42		0.08		0.01		1	A0 V	137
51302	+1 01604			6 54 18	+01 21.2	1	8.86		0.00		-0.10		1	B8	821
		LSS 107		6 54 19	−16 36.9	1	10.86		0.28		-0.61		5	B3 Ib	1737
51379	−8 01635	IDS06520S0854	A	6 54 21	−08 56.9	1	9.05		-0.11		-0.45		2	B7 V	388
51098	+41 01558			6 54 22	+41 46.4	1	6.84		1.02		0.82		4	K1 III	1501
	+77 00268			6 54 28	+77 40.9	1	8.68		0.94		0.66		2	K0	1375
		LS VI -04 009		6 54 28	−04 05.4	2	10.70	.035	0.31	.020	-0.53	.000	6		199,490
		POSS 309 # 4		6 54 29	+30 13.3	1	17.07		0.31				2		1739
	−1 01471	LS VI -01 015		6 54 29	−01 41.1	1	9.92		0.46		-0.49		2	B0.5III	1012
289434				6 54 32	+01 06.	1	11.00		0.19		0.26		1	A3	821
51403	−6 01838	IDS06520S0602	AB	6 54 32	−06 07.5	1	9.07		0.09		-0.08		7	B8	1732

HD	DM	Other Id	N Rem	α_{1950}	δ_{1950}	S	V	σ_V	B–V	σ_{B-V}	U–B	σ_{U-B}	n	Spectrum	References
		LP 781 - 21		6 54 32	−20 16.3	1	11.98		0.50		−0.16		2		1696
51426	−10 01771			6 54 34	−10 55.4	1	8.54		−0.11		−0.62		3	B6 III	388
51404	−6 01840			6 54 35	−06 09.1	1	9.36		0.00		−0.65		8	B9	1732
51424	−7 01642	HR 2599		6 54 36	−08 06.7	4	6.34	.005	0.64	.005	0.26	.008	13	K0II-III+A2	15,388,1061,2035
51607	−38 03028			6 54 36	−38 32.6	1	7.68		−0.01		−0.03		2	A0 Vn	401
51605	−37 03176			6 54 37	−37 27.3	1	8.28		0.04		−0.15		2	A0 V	401
51330	+12 01361	HR 2597		6 54 38	+11 58.5	3	6.28	.008	0.34	.005	0.26	.010	7	F2 Ib-II	1733,6009,8100
	−46 02774			6 54 39	−47 00.1	1	10.26		0.79				4	G5	2033
266913	+21 01443	AL Gem		6 54 41	+20 57.6	1	9.84		0.44		0.10		1	F6 V	1768
52449	−74 00421			6 54 41	−74 39.8	2	7.65	.005	0.53	.010	0.03		8	F8 V	158,2012
51481	−16 01669			6 54 42	−17 00.7	1	8.47		−0.06		−0.46		2	B8 II	401
289415				6 54 43	+02 14.4	1	10.76		1.02		0.85		1	K0 III	137
51454	−10 01772			6 54 43	−10 52.8	1	9.39		0.00		−0.40		3	B8 IV	388
51477	−8 01639			6 54 45	−08 28.5	3	8.08	.019	−0.09	.012	−0.73	.011	7	B5 Vn	388,399,401
51512	−14 01656			6 54 45	−14 21.0	1	9.22		0.05		−0.50		2	B3 III	1732
		BSD 98 # 3881		6 54 46	+00 27.1	1	10.68		0.14		−0.19		3	B5	821
		G 108 - 42		6 54 46	+02 45.1	2	16.20	.033	0.10	.019	−0.71	.014	4	DC	940,3060
266944	+21 01445	IDS06517N2108	A	6 54 46	+21 05.1	1	9.74		0.30		0.08		3	F0	1293
51479	−10 01773			6 54 46	−10 12.8	3	8.41	.005	−0.04	.016	−0.40	.005	11	B7 V	206,388,2014
51549	−20 01634			6 54 47	−21 02.0	1	7.81		−0.15		−0.70		2	B3 IV	401
51354	+18 01423			6 54 48	+17 58.2	1	7.12		−0.18		−0.65		3	B3 en	1212
51478	−8 01641	X Mon		6 54 48	−08 59.8	4	7.65	.147	1.56	.062	1.58	.028	11	M4	897,975,2033,3076
51480	−10 01774	V644 Mon		6 54 48	−10 45.4	5	6.92	.018	0.32	.026	−0.57	.029	15	Ape	388,399,405,1588,1737
51575	−24 04634	IDS06528S2423	AB	6 54 50	−24 26.8	1	8.95		−0.10		−0.50		2	B8 II/III	401
51576	−24 04635			6 54 50	−24 54.1	1	7.96		1.46		1.75		3	K3 III	1657
51452	−4 01745	LS VI -04 010		6 54 52	−04 07.6	3	8.09	.019	0.14	.014	−0.76	.016	8	B0:III:nn	199,399,490
	−14 01658		V	6 54 54	−14 34.2	1	9.27		1.77				2	M2	369
51572	−17 01699			6 54 56	−17 29.6	1	7.88		0.00		−0.13		2	B9 IV/V	401
51799	−48 02601	HR 2608		6 54 56	−48 39.3	5	4.95	.004	1.69	.006	1.92	.000	22	M1 III	15,1075,2012,3055,8015
51537	−7 01644			6 54 58	−07 12.8	1	8.88		0.04		−0.11		2	B9	388
51682	−35 03225	HR 2604		6 54 58	−35 16.5	3	6.27	.012	1.28	.010			11	K2 III	15,1075,2006
51539	−9 01747			6 54 59	−09 23.8	1	9.58		−0.04		−0.32		3	B9 V	388
51929	−56 01199			6 54 59	−56 53.1	3	7.41	.015	0.58	.004	−0.04	.009	8	F8/G0 V	1311,2033,3077
51509	−3 01664	LS VI -03 010		6 55 00	−03 37.8	1	8.71		0.41		0.37		3	A0	490
51541	−10 01779			6 55 00	−10 52.7	1	8.27		−0.06		−0.36		3	B7 V	388
51542	−11 01778			6 55 00	−11 03.0	1	9.51		0.00		−0.62		3	B3 V	388
292814	+0 01733			6 55 06	+00 30.5	1	10.27		0.85		0.54		1	K2 V	137
51507	+1 01610			6 55 06	+01 33.5	1	8.00		−0.11		−0.61		3	B3 V	399
		L 454 - 9		6 55 06	−39 06.	1	14.73		0.72		0.05		1		3060
51630	−22 01616	Cr 121 - 38	⋆ V	6 55 07	−22 08.1	7	6.62	.018	−0.20	.008	−0.80	.004	21	B2 IV-V	15,62,540,976,1075*
		LF11 # 57		6 55 09	+02 14.2	1	10.68		0.67		0.26		1	G5 V	137
51419	+22 01531			6 55 11	+22 32.6	1	6.94		0.62		0.07		3	G5 V	1648
51506	+6 01459			6 55 14	+06 37.0	1	7.47		−0.11		−0.70		3	B5	1732
		POSS 309 # 3		6 55 14	+29 22.5	1	16.50		1.71				1		1739
		LS VI -04 011		6 55 14	−04 07.6	2	11.30	.030	0.19	.025	−0.63	.015	6		199,490
51738	−30 03722			6 55 17	−30 25.5	2	9.03	.059	1.17	.009	1.09	.005	7	K1/2 (III)	158,1279
51624	−8 01647			6 55 18	−08 49.3	1	9.56		−0.02		−0.35		2	B8 V	388
51568	+0 01736			6 55 19	+00 06.5	1	8.85		1.23		1.20		2	K1 III	137
51739	−30 03723			6 55 19	−30 25.6	2	9.30	.050	1.12	.000	1.02	.005	7	G8/K0 III	158,1279
	−3 01668	LS VI -03 011		6 55 22	−03 41.7	1	9.63		0.18		−0.75		3		490
51625	−10 01783			6 55 22	−10 59.5	1	9.88		0.01		−0.25		5	B8 V	388
295821				6 55 24	−03 09.	1	9.08		1.71				2		369
51587	+1 01612			6 55 25	+01 16.7	1	8.57		1.58		−0.23		4	B8 V	137
51826	−36 03252			6 55 27	−36 49.1	2	7.58	.019	−0.05	.000	−0.53		6	B4 IV	401,2012
51733	−24 04648	HR 2607	λ AD	6 55 30	−24 33.8	4	5.45	.009	0.37	.010	0.03	.000	9	F3 V	15,1008,2035,3026
51025	−35 03233	HR 2612	⋆ AB	6 55 30	−35 26.4	7	6.23	.008	0.46	.011	−0.05	.012	23	F5 V	15,258,1637,2012,2012*
295879	−4 01748	LS VI -4 012		6 55 31	−04 25.5	2	10.58	.030	0.04	.015	−0.68	.010	6	B3	199,490
267123		G 88 - 1		6 55 32	+19 04.0	1	10.57		0.94		0.65		1	K3	333,1620
		POSS 309 # 2		6 55 35	+30 13.6	1	15.15		1.51				1		1739
51531	+19 01559			6 55 36	+19 17.4	1	7.32		1.30		1.46		2	K2	1648
		AA131,200 # 13		6 55 38	−07 52.0	1	13.72		1.48		1.11		2		7005
51440	+38 01656	HR 2600		6 55 39	+38 07.4	3	6.00	.021	1.22	.013	1.17	.021	6	K2 III	15,1003,1355
		BSD 98 # 3967		6 55 40	+00 32.8	1	12.12		0.10		−0.06		4	B8	821
267315	+3 01484			6 55 40	+03 34.5	1	8.52		1.55				3	M2	369
51620	+6 01462	RV Mon		6 55 41	+06 14.1	1	7.06		2.65				1	C5 II	109
50885	+70 00430	HR 2581	⋆ A	6 55 42	+70 52.7	2	5.69	.004	1.33	.007	1.52		4	gK4	71,1733
51530	+26 01405	HR 2601	⋆ A	6 55 43	+26 08.9	2	6.18	.060	0.47	.023	0.00	.000	4	F8 Vab vw	254,3037
51698	−7 01655			6 55 43	−07 46.6	1	9.05		1.70				2	M1	369
51724	−8 01649			6 55 43	−08 46.8	1	8.59		1.20		1.02		5	K2	897
51823	−27 03460	Cr 121 - 39	⋆	6 55 43	−27 28.2	5	6.22	.008	−0.15	.009	−0.66	.011	17	B2.5V	62,158,401,1732,2035
51725	−8 01650	V523 Mon		6 55 44	−08 57.5	2	7.11	.120	1.59	.019	1.56	.053	17	M2	897,3042
292827				6 55 45	−00 00.8	1	10.72		0.37		0.09		1	F0 V	137
51324	+52 01152			6 55 46	+52 38.5	1	6.63		1.39		1.53		2	K0	985
		LF11 # 61		6 55 47	+02 34.4	1	10.63		0.46		0.00		1	F2 V	137
289422	+2 01486			6 55 48	+01 55.6	1	9.90		0.55		0.11		1	G0 V	137
51789	−14 01665			6 55 49	−14 25.6	1	9.29		0.00		−0.04		2	B9/A0 IV	401
51697	−4 01751	LS VI -04 013	⋆ AB	6 55 50	−04 34.2	2	9.43	.234	0.22	.014	−0.61	.028	4	B8	490,797
		AA45,405 S287 # 11		6 55 50	−04 34.3	1	10.53		0.18		−0.71		1		797
51759	−7 01656			6 55 52	−07 52.3	1	9.57		0.02		−0.08		2	B9	388
51788	−12 01722			6 55 52	−12 56.7	2	9.64	.004	0.13	.004	0.17	.041	8	A1 IV	388,1775

Table 1   371

HD	DM	Other Id	N	Rem	α_{1950}	δ_{1950}	S	V	σ_V	B–V	σ_{B-V}	U–B	σ_{U-B}	n	Spectrum	References
		LS VI -04 014			6 55 53	−04 50.9	2	11.15	.030	0.17	.015	-0.65	.010	4		199,490
		LF11 # 63			6 55 54	+01 12.4	1	11.26		1.12		0.88		1	K2 III	137
51854	−22 01623				6 55 56	−22 48.5	2	8.87	.015	-0.17	.003	-0.85	.021	7	B2 V	401,1732
51693	+7 01539	HR 2606			6 55 57	+07 41.5	2	6.26	.005	0.11	.000	0.09	.005	7	A3 V	15,1078
51787	−10 01788				6 55 57	−10 53.4	1	9.43		0.01		-0.05		3	B9.5V	388
		POSS 309 # 5			6 55 58	+30 41.1	1	17.59		1.52				2		1739
51756	−2 01856	LS VI -02 010	⋆	AB	6 55 58	−02 57.3	2	7.20	.024	-0.09	.014	-0.92	.009	4	B0.5IV	490,1012
51786	−9 01761	IDS06537S1003		AB	6 55 58	−10 06.8	1	8.09		0.15		0.10		3	A0 V	388
51785	−9 01760				6 55 59	−09 12.8	1	9.51		-0.05		-0.46		3	B6 V	388
52024	−43 02824				6 56 00	−43 43.2	1	7.29		-0.05		-0.22		2	B9 V	401
267353	+6 01466				6 56 02	+06 29.7	1	9.82		1.74				3	K2	369
51560	+37 01628	IDS06527N3714		ABC	6 56 04	+37 10.1	1	7.92		-0.04		-0.41		2	B6 V +A1 Vn	555
51754	−0 01520				6 56 04	−00 24.1	6	9.02	.021	0.58	.015	-0.02	.013	13	G0	516,742,1064,2033*
267374	+9 01461	UY Mon	⋆		6 56 05	+09 41.6	1	9.19		0.45				1	F8 II	1772
51782	−0 01521				6 56 06	−00 55.8	1	8.94		-0.05		-0.33		1	B8 V	137
	−2 01857	LS VI -02 011			6 56 06	−02 53.4	1	9.78		-0.08		-0.79		2		490
		LF11 # 65			6 56 07	−00 36.1	1	11.73		1.98		2.12		1		137
51849	−12 01724				6 56 07	−12 55.3	5	9.15	.019	1.16	.012	1.06	.010	17	K4 V	158,1705,1775,2034,3072
51925	−26 03646	HR 2614	⋆	AB	6 56 07	−27 05.8	4	6.36	.004	-0.20	.005	-0.81	.013	13	B2 V	164,401,1732,2006
		LF11 # 66			6 56 08	+01 33.9	1	11.00		1.04		0.69		1	K0 III	137
		POSS 309 # 1			6 56 08	+30 21.9	1	14.66		1.55				2		1739
51876	−15 01581				6 56 09	−15 59.2	1	7.15		-0.08		-0.44		2	B9 IIw	401
	−2 01860	LS VI -03 012			6 56 10	−03 03.5	1	9.91		-0.04		-0.82		3		490
		LS VI -04 015			6 56 10	−04 33.1	2	10.88	.000	0.18	.000	-0.63	.010	6		199,490
		LS VI -04 016			6 56 11	−04 41.4	2	11.44	.005	0.23	.010	-0.53	.010	6		199,490
51845	−6 01859				6 56 11	−07 07.0	1	7.12		0.81		0.40		2	G0	388
51613	+33 01445				6 56 13	+33 37.4	1	8.72		0.12		0.11		2	A0	1733
51898	−20 01650				6 56 13	−20 27.6	1	8.58		-0.15		-0.66		2	B2/3 V	401
52251	−55 01111				6 56 14	−55 50.8	1	8.59		1.24		1.21		11	K1 III	1673
		BSD 98 # 4008			6 56 15	+00 02.0	1	12.10		0.12		-0.07		1	A3	821
	−44 03045	IDS06549S4409		AB	6 56 17	−44 13.6	4	10.83	.009	1.63	.022	1.22	.012	15	M4	158,863,1075,3078
51689	+25 01525		⋆	C	6 56 18	+25 18.3	1	7.63		0.52		0.06		2	F8 V	3026
51814	+3 01488	HR 2610	⋆	AB	6 56 19	+03 40.3	3	5.96	.010	1.06	.004	0.88	.005	9	G8 III	15,1078,1355
51688	+26 01411	HR 2605	⋆		6 56 23	+25 59.0	2			-0.12	.029	-0.47	.034	3	B8 III	1022,1079
52057	−36 03261				6 56 23	−36 12.4	2	9.33	.015	0.05	.005	-0.25	.005	4	B8 IV/V	55,1097
51913	−9 01765				6 56 24	−09 16.3	1	8.73		-0.05		-0.24		3	B8 V	388
51749	+20 01661				6 56 25	+20 30.9	1	8.09		1.04		0.86		3	G5	1648
289407	+2 01492				6 56 26	+02 22.0	1	9.75		0.07		0.02		1	A0 V	137
		LS VI -02 012			6 56 29	−02 22.2	1	12.08		0.26		-0.52		3		490
		G 109 - 35			6 56 30	+19 25.6	1	14.85		1.90				1		3078
51915	−9 01769				6 56 31	−10 01.9	1	9.17		-0.07		-0.45		3	B6 V	388
51986	−23 04678				6 56 31	−23 48.7	1	6.73		0.03		-0.01		2	A0 V	401
51984	−22 01627				6 56 32	−22 28.0	1	9.39		-0.12		-0.57		2	B5 III	401
51940	−10 01792				6 56 33	−10 30.7	1	9.43		0.16		0.01		3	A0 V	388
51982	−19 01635				6 56 33	−19 15.6	1	8.78		-0.07		-0.39		4	B8/9 II	1732
52018	−25 03864	Cr 121 - 40	⋆	A	6 56 33	−25 20.7	5	5.58	.005	-0.17	.004	-0.70	.007	17	B3 V	15,62,1075,1732,2012
52196	−45 02843				6 56 34	−46 01.6	5	6.83	.004	-0.06	.003	-0.22	.003	20	A0 IV	278,743,977,1075,2011
51939	−10 01793				6 56 35	−10 09.6	2	8.64	.020	-0.09	.010	-0.58	.010	5	B5 III	388,401
51981	−17 01709				6 56 35	−17 33.8	1	8.23		-0.02		-0.28		2	B8/9 V	401
52092	−33 03389	HR 2619			6 56 35	−34 02.6	3	5.06	.012	-0.17	.007	-0.66		9	B4 IV/V	15,401,2012
52298	−52 01042				6 56 36	−52 34.5	2	6.89	.026	0.48	.009	-0.11		6	F5/6 V	1311,2012
51892	+7 01544	HR 2613			6 56 38	+07 23.2	3	6.34	.005	-0.11	.008	-0.47	.007	9	B7 III	15,1079,1415
		LS VI -04 017			6 56 39	−04 30.2	2	10.48	.000	0.22	.010	-0.62	.005	6		199,490
51961	−10 01794				6 56 39	−10 48.8	1	9.57		0.16		-0.50		3	B3 V	388
52089	−28 03666	HR 2618	⋆	AB	6 56 40	−28 54.2	7	1.50	.003	-0.21	.009	-0.92	.007	33	B2 II	3,9,15,1075,2012,6002,8015
51960	−7 01665				6 56 41	−08 06.5	1	9.52		-0.02		-0.18		2	B8 V	388
	−45 02845			AB	6 56 41	−45 52.6	1	11.20		1.10				4	K2	1075
51959	−6 01863				6 56 44	−07 02.4	1	8.95		1.06		0.78		1	K0	565
51978	−10 01795	IDS06544S1039		AB	6 56 45	−10 42.9	1	8.33		-0.04		-0.33		3	B7 V	388
		MCC 484			6 56 47	+59 55.1	2	10.95	.019	1.47	.008	1.20	.015	5	K5	1723,1746
52140	−30 03757	Cr 121 - 41	⋆	A	6 56 49	−30 55.7	4	6.42	.015	-0.15	.013	-0.61	.005	13	B3 V	15,62,401,2027
52013	−10 01796				6 56 50	−10 38.5	1	9.43		-0.03		-0.21		3	B9 V	388
51819	−20 01645				6 56 51	−20 55.7	1	9.39		-0.11		-0.53		2	B5/7 V	401
296000	−4 01765	LS VI -4 018			6 56 53	−04 36.1	2	9.85	.005	0.21	.010	-0.62	.005	6	B8	199,490
		G 108 - 46			6 56 55	+09 09.8	1	11.02		0.58		-0.03		2		1696
51834	+29 01425				6 56 56	+29 50.6	1	7.13		1.53		1.79		3	K4 III	974
52012	−7 01668				6 56 56	−07 22.8	1	8.08		0.00		-0.15		2	B8	388
		AA45,405 S287 # 9		AB	6 56 58	−04 43.3	1	12.91		0.65		0.07		1		797
52115	−20 01657				6 56 58	−20 43.3	1	8.23		-0.04		-0.32		2	B8 II/III	401
289519	+1 01624				6 56 59	+00 54.6	1	9.27		0.09		0.04		2	A2 V	137
52138	−26 03670	IDS06550S2620		AB	6 56 59	−26 24.3	1	7.16		-0.18		-0.79		2	B2 V	401
		AA45,405 S287 # 10			6 57 00	−04 41.9	1	12.94		0.25		0.11		1		797
		LP 781 - 16			6 57 01	−15 48.0	1	11.72		0.72		0.15		2		1696
		LS VI -04 019			6 57 02	−04 44.5	4	10.80	.021	0.44	.016	-0.51	.014	10	O9.5V	199,342,490,797
292962	−0 01528	IDS06545S0039		AB	6 57 04	−00 42.8	1	9.92		0.33		0.10		2	F0 V	137
		AA45,405 S287 # 6			6 57 05	−04 44.8	1	11.46		0.07		-0.12		1		797
52081	−10 01799				6 57 05	−11 05.2	2	9.70	.000	0.02	.000	-0.08	.000	3	B8p	320,388
52165	−21 01685	IDS06550S2115		AB	6 57 06	−21 18.6	1	8.71		-0.10		-0.57		2	B3 V	401
52047	−4 01766				6 57 07	−04 43.8	1	8.85		0.40		0.04		1	F2	797
51610	+55 01154	R Lyn			6 57 11	+55 24.1	1	7.56		2.02		1.48		1	S3.9 e	3001

HD	DM	Other Id	N Rem	α_{1950}	δ_{1950}	S	V	σ_V	B–V	σ_{B-V}	U–B	σ_{U-B}	n	Spectrum	References
52113	−9 01774			6 57 11	−09 39.1	1	9.15		-0.02		-0.35		4	B8 V	388
52112	−9 01775			6 57 12	−09 25.9	2	8.79	.030	-0.10	.015	-0.68	.005	5	B3 V	388,401
51495	+63 00678			6 57 13	+63 44.9	1	6.62		1.50		1.80		2	K5	985
		AA45,405 S287 # 1		6 57 13	−04 47.7	1	14.48		0.64		-0.20		2		797
52362	−45 02850	HR 2626		6 57 14	−45 41.9	8	6.22	.005	0.00	.004	-0.06	.004	41	A0 V	14,15,278,977,977*
52162	−12 01729			6 57 17	−12 55.9	2	7.87	.010	-0.07	.010	-0.67		6	B3/5 II	388,2033
267729	+3 01496			6 57 20	+02 55.3	1	10.53		0.04		-0.01		1	B9 V	137
52159	−10 01802			6 57 22	−11 05.2	1	9.69		-0.03		-0.52		2	B5 Vne	388
52005	+16 01354	HR 2615		6 57 23	+16 09.0	4	5.68	.011	1.65	.013	1.79	.021	9	K3 Ib	1080,1355,8032,8041
51067	+75 00281	IDS06508N7522	A	6 57 26	+75 18.2	1	7.12		0.56		0.11		1	G0	3026
51067	+75 00281	IDS06508N7522	B	6 57 26	+75 18.2	1	8.24		0.75		0.34		1	G5	3032
		AA45,405 S287 # 7		6 57 26	−04 43.8	1	11.36		0.06		-0.26		1		797
289456				6 57 27	+02 40.5	1	11.77		0.21		0.14		1	A2 V	137
52244	−16 01694	LSS 119		6 57 31	−16 07.8	1	9.16		0.31		-0.58		2	B1/2e	1012
52273	−21 01689	HR 2623		6 57 31	−21 32.0	2	6.26		-0.17	.007	-0.79		5	B2 III	1079,2035
292940				6 57 32	−00 31.9	1	10.84		0.15		0.02		1	A2 V	137
289458				6 57 33	+02 23.7	1	10.83		0.16		0.12		1	B8 V	137
52212	−10 01804			6 57 33	−10 12.1	1	9.23		0.03		-0.26		2	B7 V	388
	−4 01771			6 57 35	−04 45.1	1	10.49		0.16		-0.36		1		797
		AA45,405 S287 # 4		6 57 35	−04 45.4	1	11.04		0.07		-0.01		1		797
52242	−14 01675			6 57 36	−14 57.1	1	7.41		0.40		-0.03		2	F2/3 V	401
52470	−46 02811			6 57 36	−46 57.2	1	7.27		-0.01		-0.07		2	A0 V	401
52035	+23 01566			6 57 39	+23 30.7	1	7.17		-0.09		-0.24		2	A0	1625
52622	−56 01211	HR 2638		6 57 40	−56 19.5	2	6.44	.005	0.40	.005			7	F2 II	15,2012
52603	−55 01116	HR 2634		6 57 41	−55 39.5	3	6.26	.004	1.16	.005			11	K2 III	15,2013,2028
	−13 01769		V	6 57 42	−13 38.2	1	8.80		1.79				2	M3	369
52271	−11 01728	IDS06554S1152	AB	6 57 44	−11 56.0	1	8.36		0.02		-0.16		2	B9 V	388
52467	−42 02870			6 57 45	−42 10.0	1	8.45		-0.09		-0.50		2	B5 V	401
52356	−28 03689			6 57 46	−28 19.7	2	6.98	.015	-0.18	.000	-0.76	.000	6	B3 V(n)	158,401
52270	−9 01780	IDS06555S0909	AB	6 57 47	−09 13.0	1	8.98		-0.01		-0.15		3	B9 Vnn	388
52395	−29 03740			6 57 47	−29 38.1	3	7.78	.004	0.53	.004	0.04		9	F7 V	1075,2012,3002
51866	+48 01469	G 107 - 42		6 57 49	+48 27.4	3	8.00	.006	0.99	.005	0.83	.010	10	K3 V	22,1003,1775
52071	+27 01296			6 57 51	+27 13.7	2	7.12	.000	1.25	.005	1.27		4	K2 III	1003,1355
52265	−5 01910	HR 2622		6 57 51	−05 17.8	4	6.29	.007	0.57	.000	0.09	.005	18	G0 III-IV	15,1415,2013,2029
	−45 02858	IDS08387S4521	AB	6 57 53	−45 19.2	1	10.96		0.54				4		1075
		KPS 476 - 203		6 57 54	+30 03.9	1	11.50		0.46		-0.05		3		974
52266	−5 01912	LS VI -05 005		6 57 54	−05 45.4	3	7.20	.025	-0.04	.025	-0.90	.005	6	O9 V	490,1011,1012
52238	+0 01752			6 57 56	−00 02.0	1	9.03		1.03		0.88		1	K0 III	137
52350	−22 01642	IDS06558S2235	B	6 57 56	−22 38.9	1	8.66		-0.09		-0.51		2	B8 III	401
	−2 01875	LS VI -02 013		6 57 57	−02 16.0	2	9.78	.018	-0.05	.011	-0.81	.014	7		490,1732
		To 1 - 1		6 57 57	−20 32.5	1	9.43		0.29		0.11		3		982
52349	−21 01693			6 57 57	−21 44.3	1	9.38		-0.07		-0.46		2	B5 IV	401
		KPS 476 - 204		6 57 58	+30 04.2	1	10.73		1.04		0.80		3		974
52348	−19 01644	HR 2625		6 57 58	−20 05.3	2	6.31		-0.13	.007	-0.63		5	B3 V	1079,2035
		To 1 - 2		6 57 59	−20 32.4	1	13.10		0.35		0.25		2		982
52391	−23 04717			6 57 59	−23 48.9	1	8.51		-0.04		-0.04		2	B9.5IV	401
52312	−8 01662	HR 2624	⋆ AB	6 58 00	−08 20.1	5	5.95	.008	-0.08	.004	-0.32	.013	22	B9 III	15,388,1256,2013,2029
52100	+32 01460	HR 2620	⋆ A	6 58 02	+32 29.2	1	6.59		0.27		0.14		1	A9 III	39
		To 1 - 4		6 58 04	−20 28.6	1	12.77		1.19		1.00		2		982
		To 1 - 3		6 58 04	−20 29.2	1	10.50		1.89		1.70		3		982
52329	−8 01664			6 58 05	−08 47.7	2	8.62	.005	-0.09	.005	-0.44	.005	8	B6 V	206,388
		LS VI -08 002		6 58 07	−08 48.0	1	12.31		0.69		0.20		2		490
52347	−15 01597			6 58 07	−15 11.3	1	7.91		-0.07		-0.39		2	B0 II/III	401
		To 1 - 5		6 58 10	−20 29.4	1	12.95		1.02		0.70		1		982
	+40 01776	IDS06547N3959	A	6 58 11	+39 54.1	1	10.56		0.12		0.20		1	A5	8084
	+40 01776	IDS06547N3959	B	6 58 11	+39 54.1	1	9.95		0.23		0.25		1		8084
	−8 01666	LSS 120	⋆ ABC	6 58 11	−08 47.8	2	8.54	.005	0.14	.010	-0.68	.000	6		206,388
	−37 03224			6 58 11	−37 18.2	1	10.55		0.49		0.00		2	F5	1696
		GD 81		6 58 12	+05 35.2	1	16.89		0.15		-0.73		1		3060
		To 1 - 9		6 58 12	−20 24.	1	14.59		0.66		0.24		2		982
		To 1 - 22		6 58 12	−20 24.	1	13.71		1.88		2.49		2		982
		To 1 - 6		6 58 12	−20 30.1	1	12.92		0.80		0.53		1		982
52437	−21 01695	HR 2628, FU CMa	⋆ AP	6 58 12	−22 02.9	11	6.52	.019	-0.17	.012	-0.80	.023	51	B3 Vnn	15,401,540,681,815,976*
292958	−0 01540			6 58 13	−00 45.4	1	9.68		0.95		0.59		1	K0 III	137
		To 1 - 25		6 58 13	−20 20.4	1	12.55		1.34		1.40		5		982
52463	−27 03504			6 58 13	−27 43.7	1	8.30		-0.16		-0.71		2	B3 V	401
		To 1 - 24		6 58 14	−20 22.6	1	10.38		0.41		0.15		5		982
		To 1 - 7		6 58 14	−20 29.0	1	13.62		0.49		0.29		1		982
		XZ CMa		6 58 15	−20 21.5	1	12.28		0.61		0.49		1		982
52382	−8 01667	HR 2627, LSS 122		6 58 16	−09 07.9	7	6.49	.013	0.19	.015	-0.74	.027	18	B1 Ib	15,154,388,490,1061*
		LF11 # 76		6 58 18	−00 30.9	1	11.52		1.26		1.02		1	K2 III	137
		To 1 - 23		6 58 18	−20 22.8	1	13.09		0.59		0.03		2		982
		To 1 - 12		6 58 18	−20 30.0	1	11.18		0.50		0.05		4		982
		To 1 - 13		6 58 18	−20 30.9	1	12.44		1.57		1.58		2		982
		To 1 - 8		6 58 19	−20 28.0	1	12.82		0.49		-0.02		2		982
		To 1 - 11		6 58 19	−20 29.5	1	13.80		0.69		0.20		1		982
		To 1 - 14		6 58 20	−20 31.1	1	13.48		1.20		0.85		1		982
		KPS 476 - 209		6 58 21	+29 55.3	1	13.52		1.18		0.96		3		974
52384	−9 01784			6 58 21	−09 33.9	2	8.99	.020	-0.08	.005	-0.47	.005	4	B6 V	388,401
52462	−22 01647			6 58 21	−22 34.4	1	7.27		1.55				4	K4 III	2012

Table 1   373

HD	DM	Other Id	N Rem	α_{1950}	δ_{1950}	S	V	σ_V	B–V	σ_{B-V}	U–B	σ_{U-B}	n	Spectrum	References
51970	+51 01270			6 58 22	+51 38.5	1	7.50		1.10		0.97		2	K0	1566
		To 1 - 10		6 58 22	−20 29.4	1	13.83		0.53		0.27		2		982
		To 1 - 15		6 58 24	−20 29.3	1	13.48		1.18		0.79		2		982
52484	−19 01648	IDS06563S1947	AB	6 58 26	−19 50.7	2	9.21	.012	−0.06	.011	−0.62	.015	6	B5 V	401,1732
52485	−20 01670	To 1 - 18	⋆ AB	6 58 26	−20 34.1	1	6.77		0.25		0.28		3	A3	982
292915				6 58 27	+00 05.1	1	11.21		0.15		0.02		1	A0 V	137
		To 1 - 16		6 58 27	−20 29.2	1	13.47		0.56		0.32		2		982
52620	−39 03002			6 58 27	−39 45.0	1	8.10		0.01		−0.21		2	B9/9.5III	401
52434	−8 01668			6 58 29	−08 31.0	1	9.12		0.35		0.00		2	A5 III	388
52511	−23 04732			6 58 31	−23 47.4	2	9.21	.016	−0.15	.008	−0.62	.009	6	B5 IV/V	401,1732
52432	−3 01685	V614 Mon		6 58 32	−03 10.8	5	7.27	.039	1.77	.047	2.36	.056	17	C4,5i	109,864,1238,2033,3039
52486	−20 01671	To 1 - 19		6 58 32	−20 36.2	1	7.28		1.25		1.26		3	K0	982
		To 1 - 17		6 58 33	−20 29.3	1	11.40		1.21		1.04		2		982
52433	−3 01686			6 58 36	−03 53.6	1	8.31		−0.09		−0.52		3	B9	3039
52305	+25 01540			6 58 37	+25 38.4	2	8.49	.000	0.11	.005	0.11	.005	5	A0	833,1625
292956				6 58 37	−00 52.7	1	11.36		0.11		−0.02		1	A0 V	137
		To 1 - 21		6 58 38	−20 25.2	1	10.98		0.54		0.11		2		982
52431	+0 01760			6 58 41	+00 04.9	1	7.58		0.41		−0.01		1	F5 V	137
52098	+50 01383	IDS06548N5048	AB	6 58 42	+50 43.5	1	8.17		0.13		0.12		2	A2	1566
52481	−8 01669			6 58 43	−08 25.1	1	9.52		−0.05		−0.33		2	B9 V	388
52506	−8 01670			6 58 44	−08 12.9	1	8.29		0.09		−0.19		3	B8 III	388
52619	−28 03711	HR 2637		6 58 44	−28 25.0	1	6.27		0.45				4	F5 V	2035
52304	+29 01434			6 58 45	+29 44.3	1	8.59		0.12		0.16		3	A2	974
52596	−25 03906	IDS06567S2530	AB	6 58 45	−25 34.3	3	7.35	.004	−0.14	.000	−0.71	.017	8	B2 IV	158,401,1732
52597	−25 03907			6 58 45	−26 01.3	1	7.87		−0.14		−0.70		2	B2/3 V	401
52480	−2 01883			6 58 48	−02 40.7	1	9.24		0.04		0.03		3	B9	3039
		GD 446		6 58 49	+62 27.3	1	15.55		0.06		−0.71		1	DA	782
52535	−7 01687			6 58 53	−07 39.2	1	8.60		0.00		−0.19		2	B9	388
52536	−8 01672			6 58 53	−08 07.9	2	8.04	.020	0.03	.010	−0.30		7	B8 III	388,2014
		G 108 - 48		6 58 56	+06 29.4	1	11.87		0.76		0.12		2		1696
	−16 01709			6 58 56	−17 03.2	1	10.24		1.68				2	M2	369
52533	−2 01885	LS VI -03 013	⋆ AB	6 58 57	−03 02.7	6	7.70	.009	−0.09	.007	−0.95	.015	44	O9 V	490,989,1011,1729,1732,2033
52593	−17 01729			6 58 58	−17 19.3	1	8.67		1.82				2	M2 III	369
		To 1 - 20	A	6 58 58	−20 29.5	1	9.29		0.01		−0.29		1		982
		To 1 - 20	B	6 58 58	−20 29.5	1	9.95		0.02		−0.34		1		982
292971	−0 01548			6 58 59	−01 03.0	1	10.24		0.37		0.08		1	F2 V	137
52703	−33 03415	HR 2641		6 58 59	−33 23.7	1	6.40		1.05				4	G8 II/III	2035
52479	+5 01513	HR 2629		6 59 02	+04 53.4	2	6.62	.005	0.06	.000	0.09	.005	7	A3 Vs	15,1415
52403	+25 01542			6 59 03	+25 29.9	1	7.05		0.11		0.14		2	A2	105
52670	−25 03911	HR 2640		6 59 03	−25 08.6	3	5.63	.009	−0.17	.005	−0.67		10	B2/3 III/IV	15,1637,2029
292948	−0 01551			6 59 04	−00 22.7	1	8.58		1.65				2	M0	369
52643	−20 01675	IDS06569S2026	A	6 59 06	−20 30.0	1	8.82		0.03		−0.32		2	B9 V	401
		KPS 476 - 211		6 59 08	+30 07.7	1	12.55		0.46		0.04		3		974
		G 250 - 33		6 59 08	+68 21.6	1	11.95		1.53		1.25		2		7010
52589	−6 01887			6 59 08	−06 14.4	1	8.06		−0.10				4	B8	2014
292904	+0 01765			6 59 10	+00 20.3	1	10.11		1.06		0.96		2	K0 III	137
52698	−25 03913			6 59 11	−25 52.6	5	6.71	.010	0.89	.011	0.60	.017	13	K1 V	258,1020,2012,2012,3008
		LF11 # 82		6 59 12	+01 21.2	1	11.90		1.19		1.83		1		137
52731	−27 03532			6 59 12	−27 36.9	2	8.10	.005	−0.17	.001	−0.75	.021	4	B3 V	401,1732
52587	+0 01766			6 59 13	−00 00.5	1	10.01		0.46		0.02		6	A0	1729
52642	−12 01740			6 59 13	−13 03.6	1	7.88		1.69				2	M1/2 III	369
	−18 01638	SV CMa		6 59 13	−18 58.0	1	10.03		1.70				2	M3	369
52559	+5 01514	HR 2633	⋆	6 59 15	+05 37.8	3	6.57	.012	−0.02	.000	−0.64	.005	9	B2 IV-V	15,154,1415
	−57 01101			6 59 19	−57 26.0	1	9.80		0.88		0.45		1		742
52610	−0 01556	V526 Mon		6 59 21	−01 03.5	2	8.50	.016	0.55	.016	0.45		2	G0	1462,1772
52611	−1 01509	HR 2636	⋆ A	6 59 21	−01 16.3	4	6.16	.020	1.29	.014	1.36	.024	10	K0	15,252,1415,1462
52497	+24 01502	HR 2630, ω Gem		6 59 22	+24 17.3	4	5.18	.013	0.94	.006	0.68		15	G5 IIa-Ib	1105,1119,1355,1363
268079	+31 01472	G 87 - 19		6 59 23	+31 38.6	1	10.17		0.87		0.49		1	G5	333,1620
	−22 01661			6 59 23	−22 29.4	1	9.91		−0.09		−0.77		2	B5	401
52724	−17 01731			6 59 24	−18 03.5	2	9.42	.007	−0.07	.000	−0.41	.015	6	B8 V	401,1732
53143	−61 00754			6 59 25	−61 16.1	2	6.81	.005	0.81	.005	0.43		5	K1 V	258,2012
52556	+15 01431	HR 2632		6 59 26	+15 24.6	1	5.75		1.16		1.09		2	gK1	1733
293049				6 59 28	+00 20.4	1	11.41		0.14		0.04		1	A2 V	137
52899	−40 02843			6 59 28	−40 43.4	1	7.35		−0.02		−0.10		2	A0 V	401
52666	−5 01926	HR 2639		6 59 29	−05 39.0	4	5.19	.016	1.67	.008	2.03	.022	14	M2 III	15,369,1415,3016
		LP 661 - 3		6 59 29	−06 22.3	1	15.27		0.88				1		1705
52694	−10 01818	IDS06571S1044	AB	6 59 29	−10 48.6	1	6.59		0.73		0.33		3	F8	388
52721	−11 01747	GU CMa	⋆ AB	6 59 29	−11 13.7	5	6.59	.020	0.06	.004	−0.77	.019	15	B2 Vne	206,388,740,1212,1588
		CMa R1 # 3		6 59 29	−11 16.3	1	11.75		2.15		2.19		1	M4 Ib-II	740
52554	+17 01479	HR 2631, NP Gem		6 59 31	+17 49.7	2	5.92	.026	1.62	.011	1.64		3	M1	71,3001
52812	−27 03540	Cr 121 - 27		6 59 33	−27 09.0	2	6.95	.029	−0.18	.010	−0.76	.044	3	B2 V	62,401
	+46 01219			6 59 34	+46 11.8	1	8.97		1.27		1.49		2	K5	1733
289464	+2 01519			6 59 36	+02 14.4	1	10.55		0.15		−0.35		1	B8 V	137
52690	−3 01694		AB	6 59 37	−03 40.9	1	6.55		1.57		1.18		4	M1 Ib	8100
289497	+1 01651			6 59 39	+01 24.8	1	9.57		1.42		1.56		4	K3 III	137
53047	−51 02224	HR 2652		6 59 39	−51 19.8	2	5.14	.000	1.64	.024	1.92		6	M1 III	2035,3055
52609	+16 01363	HR 2635		6 59 40	+16 44.9	2	5.90	.055	1.66	.012	1.92		7	M2 III	71,3001
	+38 01670	G 87 - 20		6 59 41	+38 13.2	4	9.45	.005	0.64	.004	0.03	.004	7	G0	516,979,1620,8112
52849	−23 04766			6 59 41	−23 23.2	1	7.70		−0.14		−0.67		2	B3 IV	1732
52718	−3 01695	LS VI -03 014		6 59 42	−03 12.1	1	9.38		−0.01		−0.71		3	B5	490

HD	DM	Other Id	N Rem	α₁₉₅₀	δ₁₉₅₀	S	V	σ_V	B–V	σ_B-V	U–B	σ_U-B	n	Spectrum	References
52745	−9 01804			6 59 42	−09 48.8	1	7.79		0.26		0.04		2	A5 III	388
52877	−27 03544	Cr 121 - 28	⋆ A	6 59 44	−27 51.7	13	3.46	.022	1.73	.014	1.88	.036	63	K4 III	3,15,62,369,418,1024*
54239	−79 00238	HR 2689		6 59 46	−79 21.0	3	5.45	.004	0.05	.000	-0.07	.000	15	B9.5 III/IV	15,1075,2038
54481	−80 00209			6 59 47	−80 46.1	1	7.75		1.35		1.59		4	K3 III	1704
52772	−9 01805			6 59 49	−09 08.2	1	8.23		0.01		-0.34		2	B8 III	388
268326				6 59 50	+09 14.	1	11.56		-0.04				2	K2	8033
293039	+0 01769			6 59 53	+00 34.4	1	9.35		0.14		-0.62		2	B3 III	1012
289501				6 59 55	+01 09.3	1	11.64		0.16		0.14		1	A0 V	137
		G 193 - 27		6 59 55	+52 47.2	2	13.32	.042	1.82	.041			8		940,1691
53501	−67 00686	HR 2662		6 59 56	−67 50.9	6	5.17	.006	1.40	.004	1.65	.000	34	K3 III	15,1075,1075,2012*
293050				6 59 57	+00 14.3	1	10.84		1.25		1.00		1	K0 III	137
52334	+56 01173	IDS06558N5635	AB	6 59 59	+56 30.9	1	9.39		0.48		0.02		5		3016
52334	+56 01173	IDS06558N5635	ABC	6 59 59	+56 30.9	1	10.00		0.49		0.03		5		3032
52993	−35 03282			7 00 00	−35 28.5	1	6.59		-0.18		-0.59		2	Ap Si	401
52892	−14 01690			7 00 01	−14 18.8	1	8.63		0.02		-0.08		2	B9 V	401
289538	+2 01523			7 00 04	+02 44.8	1	8.98		1.24		1.35		1	K2 III	137
		LP 781 - 18		7 00 04	−18 03.5	1	10.91		1.17		1.09		1		1696
	−8 01692			7 00 08	−08 09.8	1	9.28		1.85				2	M2	369
53071	−40 02856			7 00 09	−40 49.4	1	7.90		-0.14		-0.71		1	B2 IV	55
293051	+0 01770			7 00 10	+00 17.1	1	10.12		0.10		0.08		1	A2 V	137
52679	+25 01551			7 00 11	+25 25.2	1	8.33		0.23		0.07		3	A2	833
		NGC 2323 - 144		7 00 11	−08 19.0	1	12.10		1.39		1.34		1		49
		NGC 2323 - 145		7 00 11	−08 19.3	1	14.09		0.57		0.23		1		49
		LF11 # 90		7 00 12	+02 28.5	1	12.65		1.29		1.00		1		137
52030	+70 00432	HR 2617		7 00 13	+70 48.5	1	6.48		1.58		1.82		2	K0 III	1733
		NGC 2323 - 152		7 00 13	−08 20.3	1	14.62		0.43		0.31		1		49
		NGC 2323 - 85		7 00 14	−08 19.0	1	13.89		0.46		0.11		1		49
293074	−0 01561			7 00 15	−00 17.7	1	9.85		1.17		1.34		1	K1 III	137
		NGC 2323 - 132		7 00 15	−08 15.2	1	13.34		0.44		0.28		1		49
	−8 01695	NGC 2323 - 79		7 00 15	−08 17.5	1	9.93		0.10		-0.13		1	A0 V	49
52552	+47 01388			7 00 16	+47 07.5	1	7.00		0.41		-0.03		2	F5	1601
52971	−21 01711			7 00 16	−21 45.3	1	9.15		-0.10		-0.51		4	B5/7 (V)	1732
52919	−6 01902			7 00 17	−06 43.3	2	8.38	.035	1.08	.005	0.96	.015	5	K5 V	265,3072
	−8 01696	NGC 2323 - 4		7 00 17	−08 14.3	1	9.93		0.03		-0.32		1	B9.5V	49
52970	−21 01710			7 00 17	−21 39.7	2	8.25	.002	0.09	.001	0.02	.004	3	A0 IV/V	401,1732
53500	−65 00685	IDS07001S6546	AB	7 00 18	−65 50.4	1	8.27		0.61		0.12		6	G0 V	1700
52922	−8 01697	NGC 2323 - 157		7 00 19	−08 21.8	1	9.38		0.09		-0.22		1	B9 V	49
	−12 01748			7 00 19	−12 09.9	2	10.24	.009	0.24	.009	-0.20	.023	5	B8 III	206,740
53349	−58 00820	HR 2661		7 00 19	−58 52.1	2	6.01	.005	0.31	.005			7	A8 III	15,2012
52711	+29 01441	HR 2643		7 00 20	+29 25.4	4	5.93	.007	0.59	.017	0.06	.020	15	G4 V	22,254,1067,3026
		NGC 2323 - 127		7 00 20	−08 14.4	1	11.50		0.18		0.01		1		49
		NGC 2323 - 49		7 00 20	−08 18.4	1	12.89		0.64		0.15		1		49
		NGC 2323 - 89		7 00 20	−08 20.0	1	15.11		0.81		0.51		1		49
289537				7 00 21	+02 52.8	1	11.73		0.19		0.12		1	A2 V	137
		NGC 2323 - 68		7 00 21	−08 15.9	1	12.32		0.24		0.16		1		49
53040	−26 03755			7 00 21	−26 40.6	1	9.81		-0.08		-0.43		2	B8	1732
52938	−8 01699	NGC 2323 - 159		7 00 22	−08 22.7	3	7.82	.022	1.70	.004	2.01	.039	7	K3.5IIb	49,460,1522
52942	−11 01755	FZ CMa		7 00 22	−11 22.8	4	8.14	.032	0.15	.006	-0.60	.015	11	B3 n	206,388,432,740
52968	−12 01750			7 00 22	−13 06.9	2	7.85	.025	-0.03	.005	-0.26	.055	4	B8/9 IV	388,401
		NGC 2323 - 38		7 00 23	−08 18.5	1	12.52		0.38		0.16		2		305
	−8 01700	NGC 2323 - 39		7 00 24	−08 18.1	2	9.87	.024	0.09	.019	-0.27	.019	3	B8 V	49,305
52965	−8 01701	NGC 2323 - 3		7 00 25	−08 14.4	3	9.10	.041	0.14	.023	-0.31	.005	6	B9p	49,305,2042
		NGC 2323 - 44		7 00 25	−08 16.5	1	12.99		0.53		0.15		1		49
		NGC 2323 - 40		7 00 25	−08 17.7	1	13.63		0.77		0.26		1		305
52918	−4 01788	HR 2648, V637 Mon		7 00 26	−04 09.9	10	4.99	.012	-0.20	.006	-0.92	.013	35	B1 V	3,15,154,369,490,1203*
		NGC 2323 - 36		7 00 27	−08 17.4	1	11.93		0.39		0.00		1		305
52986	−17 01743			7 00 28	−17 30.5	1	7.45		-0.03		-0.10		2	B8/9 III/IV	401
53190	−45 02885			7 00 28	−45 26.8	1	9.01		0.98		0.67		2	G8/K0 III	1730
289559				7 00 29	+01 50.2	1	11.00		0.20		0.12		1	A0 V	137
		NGC 2323 - 63		7 00 29	−08 16.3	1	14.65		0.80		0.00		1		49
		NGC 2323 - 167		7 00 29	−08 21.5	1	12.45		0.31		0.17		1		49
52984	−12 01752			7 00 29	−13 01.3	2	7.79	.015	0.00	.000	-0.14	.050	4	B9/9.5V	388,401
		G 110 - 32		7 00 30	+12 30.6	1	13.36		1.42		1.13		1		1773
		G 110 - 31		7 00 30	+12 32.2	1	11.93		1.16		1.05		2		1773
		AJ84,127 # 2207		7 00 30	+17 32.	1	11.97		1.07		0.93		1		801
52764	+27 01307			7 00 30	+27 04.8	1	8.46		0.52		0.06		2	G0	1625
	−8 01703	NGC 2323 - 2		7 00 30	−08 13.5	3	9.27	.054	0.18	.031	-0.18	.000	6	B9 V	49,305,2042
289548	+2 01528			7 00 31	+02 16.2	1	9.85		1.04		0.87		1	G8 III	137
		NGC 2323 - 32		7 00 31	−08 18.0	1	12.40		0.34				4		2042
	−8 01704	NGC 2323 - 171		7 00 33	−08 20.6	1	10.72		0.15				4		2042
52913	+9 01496	HR 2647		7 00 34	+09 12.8	2	5.96	.005	0.12	.000	0.10	.005	7	A3 Vs	15,1078
52980	−8 01706	NGC 2323 - 1	⋆ AB	7 00 34	−08 16.4	3	8.32	.035	0.22	.012	0.00	.060	3	B8 III	49,305,2042
		LS VI -04 021		7 00 35	−04 46.0	1	11.15		0.30		-0.47		3		490
	−8 01708	NGC 2323 - 51		7 00 36	−08 18.4	2	9.94	.034	0.11	.000	-0.29		3	B9p	49,2042
		LF11 # 95		7 00 37	−00 45.9	1	11.52		0.61		0.37		1	A2 V:	137
		LF11 # 96		7 00 38	+01 28.0	1	11.28		1.81		2.06		1		127
		CMa R1 # 6		7 00 38	−11 23.0	2	11.18	.015	0.37	.000	-0.37	.005	3	B2 V	432,740
53063	−20 01689	IDS06585S2011	AB	7 00 40	−20 15.0	1	9.45		-0.06		-0.39		2	B9 III/IV	401
53010	−10 01833			7 00 41	−11 02.6	2	8.89	.004	0.06	.001	-0.56	.007	136	B2.5V	356,388
52737	+36 01555			7 00 42	+36 23.3	1	6.68		0.84		0.46		3	G5 III	1501

Table 1 375

HD	DM	Other Id	N	Rem	α₁₉₅₀	δ₁₉₅₀	S	V	σV	B–V	σB-V	U–B	σU-B	n	Spectrum	References
268472	+18 01461				7 00 43	+17 56.0	1	10.49		1.27		1.21		1	M0	801
		PASP94,905 # 7			7 00 43	−20 45.7	1	14.67		0.60				1		971
53253	−43 02882	HR 2658			7 00 43	−43 19.8	3	6.42	.004	−0.04	.000	−0.08	.005	11	A0 V	15,1075,2012
52978	+3 01523				7 00 44	+02 57.0	1	8.47		1.63				4	K5	369
		NGC 2323 - 29			7 00 45	−08 18.1	1	12.06		0.32				8		2042
53124	−26 03770				7 00 45	−26 10.1	1	9.07		−0.09		−0.48		4	B5 V(n)	1732
53252	−42 02915				7 00 46	−42 33.5	1	7.00		−0.14		−0.59		2	B5 V	401
53035	−10 01834				7 00 47	−11 07.5	3	7.87	.006	−0.06	.003	−0.53	.007	136	B5 III	356,388,401
		NGC 2323 - 22			7 00 49	−08 19.5	1	12.52		0.32				6		2042
		NGC 2323 - 241			7 00 50	−08 15.8	1	12.01		1.36				2		2042
		CMa R1 # 17			7 00 50	−11 34.0	1	11.83		0.28		−0.33		1	B8 II	740
		CMa R1 # 7			7 00 50	−11 34.0	1	10.60		0.24		−0.46		1	B3 III-IV	740
53123	−24 04761				7 00 50	−24 30.6	1	7.14		−0.07		−0.20		2	B9 V	401
	−45 02890				7 00 50	−45 25.6	1	10.53		0.58		0.12		2	G0	1730
		G 87 - 21			7 00 51	+30 07.2	1	14.22		0.86		0.40		1		1658
	−5 01935	VV Mon			7 00 51	−05 39.7	1	9.40		0.81		0.30		1	K2 IV	588
52960	+11 01428	HR 2649			7 00 52	+11 01.6	4	5.14	.009	1.39	.003	1.61	.017	8	K3 III	37,1080,1355,3016
		NGC 2323 - 239			7 00 52	−08 16.6	1	13.18		0.51				2		2042
53032	−2 01898				7 00 53	−02 40.2	1	9.32		0.06		−0.50		2	B6:IV:	1012
53060	−10 01836				7 00 53	−11 03.8	1	7.99		0.01		−0.08		6	B9 V	388
		Bo 3 - 1			7 00 54	−05 00.	1	11.19		0.06		−0.44		2		410
		Bo 3 - 3			7 00 54	−05 00.	1	11.66		0.12		−0.25		1		410
53056	−2 01899				7 00 55	−02 57.0	2	7.47	.009	0.14	.000	0.19		5	A3	657,2012
53122	−20 01691				7 00 55	−20 40.7	1	8.67		−0.05		−0.37		2	B8 V	401
53658	−65 00686				7 00 55	−65 38.0	1	7.33		0.46				4	F6 V	2012
		Bo 3 - 8			7 00 56	−04 56.1	1	13.03		0.24		0.03		1		410
53138	−23 04797	Cr 121 - 29		★	7 00 56	−23 45.5	9	3.03	.017	−0.09	.015	−0.80	.015	31	B3 Ia	3,15,62,1034,1075*
53033	−4 01793				7 00 58	−04 11.5	1	7.43		1.34		1.54		4	K0	1657
293101					7 00 59	−01 03.3	1	11.22		0.11		0.02		1	A0 V	137
		Bo 3 - 7			7 00 59	−04 55.1	1	13.55		0.17		−0.02		1		410
		Bo 3 - 5			7 00 59	−04 55.8	1	13.76		0.44		0.21		1		410
		Bo 3 - 6			7 00 59	−04 56.8	1	13.37		0.16		0.26		1		410
		Bo 3 - 2			7 01 00	−04 55.8	1	11.94		0.12		0.07		1		410
		Bo 3 - 4			7 01 01	−04 56.4	1	11.91		0.27		0.18		1		410
		PASP94,905 # 5			7 01 02	−20 39.3	1	13.33		1.15				1		971
52976	+12 01406	HR 2651			7 01 03	+12 40.2	1	5.98		1.58				2	K5	71
	−10 01839				7 01 04	−10 37.8	3	9.76	.007	0.36	.007	−0.32	.013	8	B2 Vn	206,388,740
268518	+20 01686	IDS06582N2043	C		7 01 07	+20 40.3	14	7.58	.004	0.61	.003	0.09	.006	194	G7	58,71,542,985,1313,1502*
293063					7 01 07	−00 05.8	1	10.92		0.11		0.02		2	B9 V	137
52973	+20 01687	HR 2650, ζ Gem		★ A	7 01 09	+20 38.7	6	3.72	.063	0.73	.035	0.52	.019	9	F7 Ib	245,542,934,1363,1754,6007
52973	+20 01687	IDS06582N2043	B		7 01 09	+20 38.7	2	11.48	.019	0.44	.010	0.04	.015	3	G1 V	245,3016
289568	+1 01659				7 01 10	+01 27.5	1	10.14		1.32		1.33		1	K2 III	137
	−20 01693	PASP94,905 # 1			7 01 12	−20 35.4	1	9.56		1.50		1.75		1	M0	971
53136	−7 01712				7 01 14	−07 46.3	1	8.93		0.11		−0.18		2	B8	388
		NGC 2324 - 271			7 01 15	+01 08.0	1	14.36		0.30		0.22		1		49
		NGC 2324 - 273			7 01 19	+01 07.3	1	15.36		0.44		0.27		1		49
53135	−3 01709	LS VI -03 015			7 01 20	−03 55.4	1	9.12		0.02		−0.67		3	B9	490
		PASP94,905 # 3			7 01 21	−20 45.8	1	12.03		1.10		0.82		1		971
		NGC 2324 - 166			7 01 22	+01 09.8	1	13.01		0.07		−0.12		1		49
53179	−11 01760	Z CMa			7 01 23	−11 28.6	5	9.30	.027	1.20	.019	0.58	.067	10	Bep	351,388,401,740,1588
52860	+47 01391	HR 2645			7 01 24	+47 51.1	1			−0.04		−0.24		1	B9 IIIn	1079
293055					7 01 25	+00 12.0	1	9.57		0.94		0.72		2	K0 III	137
	−8 01723	LS VI -09 002			7 01 25	−09 04.7	1	10.33		0.22		−0.50		2		490
		CMa R1 # 10			7 01 25	−10 59.3	1	12.88		0.41		0.11		1		740
		CMa R1 # 30			7 01 25	−10 59.3	1	10.53		0.22		−0.30		1	B5 V	740
53442	−45 02896	IDS07000S4536	AB		7 01 25	−45 40.6	2	8.14	.005	0.09	.002	−0.03	.011	4	A0 IV/V	401,1730
		NGC 2324 - 50			7 01 27	+01 06.9	2	12.86	.015	1.09	.010	0.80	.058	3		49,305
289580	+1 01661	NGC 2324 - 257			7 01 28	+01 08.6	1	10.48		0.17		0.03		4	B5	305
		NGC 2324 - 119			7 01 29	+01 04.5	1	14.07		0.19		0.05		1		49
		NGC 2324 - 118			7 01 29	+01 04.8	1	13.45		1.09		0.62		1		49
		NGC 2324 - 51			7 01 29	+01 06.9	2	12.71	.047	0.44	.023	0.34	.023	4		49,305
296134					7 01 29	−04 01.	1	9.14		1.63				2		369
		PASP94,905 # 2			7 01 29	−20 41.0	1	10.27		0.42		0.12		1		971
		NGC 2324 - 282			7 01 30	+01 05.4	1	12.82		0.54		−0.01		1		49
		NGC 2324 - 22			7 01 30	+01 08.2	1	14.20		0.32		0.08		1		49
53244	−15 01625	Cr 121 - 30		★	7 01 30	−15 33.5	11	4.11	.007	−0.12	.008	−0.47	.015	41	B8 III	3,15,62,369,1063,1075*
		BPM 18394			7 01 30	−58 46.	1	14.46		0.22		−0.72		2		3065
		NGC 2324 - 39			7 01 32	+01 08.5	2	13.43	.010	0.27	.005	0.21	.010	3		49,305
		PASP94,905 # 6			7 01 32	−20 50.1	1	10.47		1.08		0.98		1		971
	−11 01761				7 01 33	−11 30.1	3	9.27	.019	0.07	.005	−0.50	.013	8	B2 V	206,388,740
	−11 01762				7 01 34	−11 24.1	1	10.05		0.09		−0.54		1	B2 V	740
		NGC 2324 - 283			7 01 35	+01 02.7	1	14.01		0.23		0.03		1		49
		NGC 2324 - 258			7 01 35	+01 10.3	1	12.39		0.53		0.06		2		305
53240	−9 01818	HR 2656			7 01 35	−10 02.9	6	6.44	.012	−0.08	.005	−0.31	.004	23	B9 IIIn	15,351,388,1079,1415,2012
	−11 01763		AB		7 01 35	−11 29.8	3	8.96	.009	0.04	.007	−0.68	.016	8	B1.5V	206,388,740
53267	−12 01761				7 01 35	−13 01.3	1	7.02		1.41		1.41		1	K2/3 III	388
53403	−34 03298				7 01 35	−34 40.8	1	8.24		−0.06		−0.19		2	B8 V	401
		NGC 2324 - 130			7 01 36	+01 04.5	1	13.96		0.52		0.04		1		49
		PASP94,905 # 4			7 01 36	−20 47.7	1	12.62		1.00		0.57		1		971
		NGC 2324 - 28			7 01 37	+01 09.0	1	13.70		0.32		0.17		1		49

HD	DM	Other Id	N Rem	α_{1950}	δ_{1950}	S	V	σ_V	B–V	σ_{B-V}	U–B	σ_{U-B}	n	Spectrum	References
52708	+60 01026	HR 2642		7 01 37	+59 52.7	2	6.40	.005	1.19	.005	1.26	.000	4	G8 III:	1502,1733
53208	−5 01943	HR 2655		7 01 37	−05 14.9	5	5.61	.006	1.29	.008	1.24	.000	20	K3 III	15,1415,2013,2028,3077
		NGC 2324 - 131		7 01 39	+01 04.8	1	14.26		0.18		0.16		1		49
53344	−24 04785			7 01 39	−25 00.5	3	6.86	.003	−0.17	.009	−0.73	.017	8	B2/3 V	158,401,1732
		NGC 2324 - 87		7 01 40	+01 10.1	1	11.66		0.76		0.18		1		305
	−10 01845			7 01 40	−11 07.8	1	10.38		0.10		−0.49		1	B4 V	740
53342	−24 04786			7 01 40	−24 14.1	1	8.33		−0.07		−0.39		2	B8 V	401
		NGC 2324 - 45		7 01 41	+01 02.0	1	13.17		0.59		0.11		1		305
		NGC 2324 - 76		7 01 42	+01 07.2	1	13.42		0.51		0.08		1		49
		GI Gem		7 01 42	+13 30.	1	12.39		0.18		0.14		1		699
52859	+52 01165	HR 2644	⋆ AB	7 01 42	+52 50.1	2	6.15	.030	0.10	.010	0.08	.005	4	A3 Vs	1733,3016
53303	−12 01763			7 01 43	−12 12.8	1	7.42		−0.11		−0.57		2	B5 III	388
		NGC 2324 - 80		7 01 44	+01 08.9	1	15.46		0.50		0.04		1		49
53206	+1 01666	NGC 2324 - 256		7 01 44	+01 10.9	2	10.01	.019	0.00	.014	−0.11	.019	4	A0	49,305
		NGC 2324 - 148		7 01 45	+01 09.4	1	14.97		0.34		0.08		1		49
53205	+1 01665	HR 2654	⋆ A	7 01 45	+01 33.8	3	6.56	.005	0.01	.007	−0.07	.012	8	A0 V	15,1079,1256
53373	−25 03976			7 01 45	−25 34.7	2	7.97	.000	−0.17	.002	−0.74	.023	4	B2 III/IV	401,1732
		NGC 2324 - 299		7 01 46	+01 00.1	1	14.00		0.41		0.31		1		49
		G 88 - 4		7 01 46	+25 04.6	1	11.62		1.48		1.16		1	M0	333,1620
293095	−0 01576			7 01 47	−00 57.4	1	9.26		0.43		−0.59		2	G0	1012
		LF11 # 101		7 01 48	−01 01.6	1	12.18		0.31		0.11		1	A5 V	137
289543				7 01 51	+02 35.0	1	11.23		0.19		0.03		1	B9 V	137
53299	−2 01908	IDS06594S0254	AB	7 01 51	−02 58.5	3	8.17	.014	0.70	.016	0.20	.033	8	G5	292,2012,3030
		NGC 2324 - 278		7 01 52	+01 09.3	1	13.81		0.25		0.26		1		49
53236	+2 01532			7 01 52	+02 00.9	1	9.62		0.16		0.11		1	A2 V	137
53300	−5 01945	LS VI -05 006		7 01 52	−05 13.8	1	8.08		0.32		0.33		3	A0	490
		AJ79,1294 T4# 8		7 01 52	−08 39.0	1	11.99		0.51		0.07		3		396
	−10 01847			7 01 53	−10 29.8	1	10.41		0.00		0.01		10	A0	1655
53371	−21 01726			7 01 53	−21 41.9	1	8.42		−0.02		−0.38		2	B7/8 III	401
		NGC 2324 - 296		7 01 54	+01 02.4	1	12.86		0.12		−0.12		1		49
289581		NGC 2324 - 295		7 01 54	+01 03.4	1	11.11		0.00		−0.42		1	B8	49
53339	−11 01766	IDS06596S1115	AB	7 01 55	−11 19.6	1	9.33		0.11		−0.50		4	B3 V	388
53263	+1 01668	NGC 2324 - 292		7 01 56	+00 59.4	1	9.10		0.26		0.15		1	A3	49
		NGC 2324 - 298		7 01 56	+01 06.6	1	13.60		0.24		0.15		1		49
53340	−15 01629			7 01 56	−15 15.1	1	8.39		0.00		−0.69		2	B2/3 III	401
		Steph 623		7 01 57	+52 17.8	1	11.12		1.23		1.21		1	K7	1746
		AJ79,1294 T4# 7		7 01 57	−08 39.4	1	11.84		0.51		0.04		3		396
		Ross 54		7 01 57	−10 25.2	3	11.29	.010	1.52	.010	1.10		6	M5	158,1705,3073
		NGC 2324 - 294		7 01 58	+01 11.1	1	10.86		0.42		0.07		1		49
289579	+1 01669	NGC 2324 - 293		7 01 59	+01 12.5	1	10.34		0.09		−0.26		1	A0	49
53338	−8 01725			7 01 59	−09 01.7	1	8.18		0.12		0.11		2	A0 V	388
		NGC 2324 - 300		7 02 03	+01 03.4	1	14.65		0.30		0.32		1		49
		NGC 2324 - 297		7 02 04	+01 08.7	1	13.14		0.16		0.05		1		49
53367	−10 01848	LSS 140	⋆ AB	7 02 04	−10 22.7	5	6.96	.015	0.44	.004	−0.59	.012	11	B0 IV:e	388,740,1012,1212,1588
53433	−21 01727			7 02 05	−21 11.4	1	7.65		−0.01		−0.24		2	B9 V	401
		NGC 2324 - 259		7 02 06	+01 00.7	1	12.36		0.38		0.22		2		305
53366	−7 01719	IDS06597S0708	AB	7 02 08	−07 12.2	1	9.05		0.04		−0.06		2	A0	388
53395	−8 01726			7 02 09	−08 21.9	1	7.33		1.14		0.97		3	K0	388
		LSS 141		7 02 10	−08 36.0	1	11.50		0.54		−0.38		3	B3 Ib	396
		AJ79,1294 T4# 10		7 02 10	−08 38.4	1	12.57		0.47		0.01		3		396
53335	−0 01579			7 02 12	−00 24.1	1	9.09		0.38		−0.05		1	F0	1462
		LSS 142		7 02 13	−08 37.3	1	11.33		0.45		−0.22		4	B3 Ib	396
	−8 01727			7 02 14	−08 12.3	1	9.85		1.85				2	M3	369
		CMa R1 # 21		7 02 15	−11 00.2	1	14.58		1.25		0.57		1		740
53428	−8 01729	LSS 143		7 02 16	−08 46.2	5	8.35	.019	0.30	.013	−0.55	.010	15	B2 Ib	388,396,490,1012,8100
53288	+20 01695			7 02 18	+19 54.1	1	8.84		−0.02		−0.08		3	A0	1648
53257	+22 01566	HR 2659		7 02 18	+22 42.8	3	6.02	.008	−0.03	.000	−0.03	.048	10	B8 Vn	814,1022,1733
53456	−11 01770	LSS 144	⋆ AB	7 02 18	−11 26.9	3	7.22	.005	0.00	.005	−0.81	.005	12	B0 V	388,401,2033
53457	−11 01771			7 02 18	−11 54.6	1	9.28		0.02		−0.36		2	B9	388
53487	−17 01760			7 02 18	−17 19.9	1	8.72		−0.08		−0.40		2	B8 II	401
53680	−43 02904	IDS07009S4328	C	7 02 18	−43 29.5	3	8.67	.013	1.18	.012	1.08	.050	11	K5 V	1279,2012,3077
53360	+1 01674	NGC 2324 - 291		7 02 20	+01 10.2	1	8.26		−0.02		−0.09		1	A0	49
53486	−13 01806			7 02 20	−13 52.0	1	8.89		−0.10		−0.63		3	B3 II/III	1732
53269	−20 01695			7 02 25	−21 04.7	1	9.53		−0.08		−0.46		2	B9 III	401
53602	−30 03870			7 02 25	−30 41.9	1	7.92		−0.08		−0.27		2	B8/9 V	401
53705	−43 02906	HR 2667	⋆ A	7 02 25	−43 32.3	2	5.59	.039	0.62	.015	0.05	.010	8	G3 V	1279,3077
53705	−43 02906	IDS07009S4328	AB	7 02 25	−43 32.3	4	5.27	.008	0.66	.005	0.11		14	G3 V + K0 V	15,244,2012,8015
		CMa R1 # 14		7 02 26	−10 51.7	1	12.09		0.58		−0.10		1	B7 III	740
		CMa R1 # 24		7 02 26	−10 51.7	1	13.31		0.58		0.18		1	A0 V:	740
		CMa R1 # 25		7 02 26	−11 02.0	1	13.10		0.38		0.18		1	A1 IV-V:	740
53706	−43 02907	HR 2668	⋆ B	7 02 27	−43 32.5	2	6.83	.035	0.78	.020	0.35	.030	7	G8/K0 (IV)	1279,3077
53704	−42 02929	HR 2666		7 02 28	−42 15.7	6	5.20	.010	0.20	.014	0.11	.009	18	A3mA3-F0	15,401,2013,2028,8071,8080
53171	+40 01788			7 02 29	+39 56.4	1	8.98		0.21		0.17		2	A2	1601
53921	−58 00826	HR 2674	⋆ AB	7 02 29	−59 06.2	1	5.49	.013	−0.12	.007	−0.46		10	B9 IV	15,404,2012
53483	−7 01723			7 02 31	−07 33.3	1	9.68		0.05		−0.26		2	B9	388
53451	+0 01791	IDS07000N0029	AB	7 02 35	+00 24.4	1	7.07		0.98		0.77		1	F0:	8100
53514	−8 01732	LSS 145		7 02 35	−08 22.5	2	9.01	.025	0.59	.035	0.25	.055	6	B9 Ib	388,490
	+27 01311	IDS06595N2737	A	7 02 36	+27 33.2	1	10.22		1.32		1.20		1	M0 V	3016
	+27 01311	IDS06595N2737	B	7 02 36	+27 33.2	1	14.63		1.10		0.98		1		3016
53811	−49 02587	HR 2672		7 02 36	−49 30.6	5	4.92	.005	0.13	.008	0.12	.005	15	A4 IV	15,244,2016,2029,3050

Table 1   377

HD	DM	Other Id	N	Rem	α_{1950}	δ_{1950}	S	V	σ_V	B–V	σ_{B-V}	U–B	σ_{U-B}	n	Spectrum	References
	−7 01725				7 02 38	−07 34.8	1	9.31		1.02		0.74		3		388
53629	−21 01732	HR 2664			7 02 39	−21 57.3	2	6.08	.006	1.23	.003	1.35		6	K2 III	58,2007
53598	−20 01706				7 02 44	−20 45.2	1	6.92		1.68		1.96		3	K5/M0 III	8100
		G 250 - 34			7 02 45	+67 16.8	1	11.17		1.50		1.21		3		7010
53595	−10 01856				7 02 51	−11 01.7	1	9.88		-0.02		-0.52		4	B5 Vn	388
53567	−2 01917				7 02 52	−02 15.1	1	9.12		1.12		1.03		9		1655
53329	+34 01524	HR 2660			7 02 54	+34 33.1	1	5.55		0.91		0.55		3	G7 III	1355
53510	+9 01510	HR 2663			7 02 55	+09 15.8	4	5.78	.013	1.51	.009	1.88	.019	14	M0 III	15,1355,1415,3001
	+17 01491				7 02 57	+17 30.4	1	9.09		0.13		0.11		2	A2	1733
53623	−12 01771	LSS 148			7 02 57	−12 15.0	7	7.97	.016	-0.06	.007	-0.86	.018	24	B1 II/III	206,356,388,401,432*
53622	−11 01778				7 02 58	−11 12.1	1	9.28		-0.02		-0.46		4	B5 V	388
		G 110 - 34			7 03 00	+11 23.4	2	12.72	.012	0.50	.002	-0.24	.002	3		1658,1696
53668	−16 01749				7 03 01	−16 26.6	1	8.70		-0.10		-0.52		1	B6 III	401
53649	−8 01733	LSS 149			7 03 02	−08 56.0	4	9.09	.039	0.23	.018	-0.67	.007	11	B0.5III	388,396,490,1012
53728	−24 04820				7 03 03	−25 01.4	2	8.28	.002	-0.18	.005	-0.73	.019	4	B2 IV	401,1732
		L 455 - 111		AB	7 03 03	−38 30.6	1	12.31		1.49		1.05		2		3078
53695	−19 01681				7 03 05	−19 46.6	1	9.21		-0.12		-0.60		2	B8 II(p Si)	401
53667	−8 01734	LSS 151			7 03 11	−08 39.1	5	7.72	.015	0.24	.015	-0.71	.012	14	B0.5III	388,396,490,1012,1212
	+18 01479	G 88 - 5			7 03 12	+18 42.9	1	10.21		0.72		0.19		1	G8	333,1620
		G 87 - 23			7 03 15	+34 31.8	2	11.11	.015	1.28	.005	1.20	.005	2	K5	1620,1658
53693	−14 01710				7 03 16	−14 47.7	1	7.96		1.04		0.74		3	G8/K0 III	1657
53725	−20 01711				7 03 16	−20 19.9	1	9.22		-0.05		-0.29		2	B8 V	401
53532	+22 01569				7 03 17	+22 45.7	1	8.23		0.70		0.24		2	G0	3077
53691	−10 01859				7 03 17	−11 04.8	4	9.37	.013	-0.01	.004	-0.65	.010	10	B3 V	388,401,740,1732
		LSS 153			7 03 19	−13 41.9	2	11.70	.005	0.43	.009	-0.49	.005	5	B3 Ib	405,1737
		AJ79,1294 T4# 11			7 03 22	−08 32.0	1	12.70		0.74		0.26		5		396
54118	−56 01232	HR 2683, V386 Car			7 03 22	−56 40.4	1	5.17		-0.04				4	Ap Si	2006
53588	+17 01492				7 03 23	+17 49.3	1	7.20		-0.11		-0.40		2	B9	1648
53756	−12 01777	FM CMa			7 03 23	−12 44.1	2	7.35	.017	-0.08	.000	-0.77	.004	6	B2/3 II	388,1012
53754	−8 01737	LSS 154			7 03 27	−08 43.8	6	8.16	.012	0.21	.012	-0.66	.009	19	B1 II	388,396,490,1012,2034,8100
53755	−10 01862	HR 2670, V569 Mon		⋆ AB	7 03 28	−10 35.0	4	6.48	.003	-0.06	.009	-0.89	.007	16	B0.5V+F5III	15,388,1075,2012
53776	−10 01865				7 03 31	−11 01.0	1	9.13		0.05		0.01		2	A0	388
	−10 01867				7 03 34	−11 02.6	1	9.19		0.51		-0.01		3	G0	388
53775	−6 01925	IDS07012S0628		AB	7 03 41	−06 32.8	1	8.63		1.58				2	M1	369
53914	−27 03620				7 03 41	−27 13.1	1	8.76		-0.16		-0.67		2	B4 IV	401
53983	−38 03149				7 03 41	−38 09.4	1	7.79		0.03		0.01		2	A0 V	401
53751	+0 01798				7 03 42	+00 23.0	1	8.55		0.52		0.09		1	G0	1462
		NGC 2335 - 13			7 03 43	−09 59.1	1	11.07		0.09		-0.35		2		189
53952	−34 03327	HR 2677		⋆ AB	7 03 43	−34 42.0	3	6.13	.012	0.37	.012	0.00		9	F2 V	15,404,2027
	−9 01840	LS VI -09 003			7 03 44	−09 23.2	1	10.04		0.05		-0.72		3		490
		LSS 158			7 03 45	−08 29.8	1	12.45		0.46		-0.29		3	B3 Ib	396
53859	−16 01761				7 03 45	−16 25.7	1	9.09		-0.05		-0.62		2	B2/3 IV/V	401
53823	−9 01842	IDS07014S0908		AB	7 03 47	−09 12.2	3	8.61	.018	-0.04	.005	-0.29	.010	9	B8 V	388,401,2014
		NGC 2335 - 11			7 03 49	−09 51.7	1	10.89		1.13		0.70		2		189
53857	−12 01781	IDS07015S1248		AB	7 03 49	−12 52.4	2	8.24	.025	-0.05	.010	-0.62	.005	5	B3 V	388,401
	+24 01529	G 88 - 7			7 03 50	+24 02.9	1	10.16		1.22		1.19		1		333,1620
53824	−9 01844	NGC 2335 - 3			7 03 50	−10 05.3	1	8.12		-0.08		-0.46		4	B6 V	189
		NGC 2335 - 10			7 03 51	−09 49.6	1	10.71		0.54		0.45		2		189
53854	−7 01738				7 03 55	−07 13.4	1	9.79		0.06		-0.27		2	B9	388
		NGC 2335 - 19			7 03 56	−10 00.7	1	11.59		0.28		-0.13		3		189
	−8 01743	LSS 163			7 03 58	−09 05.3	1	9.87		0.33		-0.57		3		490
	−12 01782				7 03 58	−12 15.8	1	8.76		1.97				2	M2	369
		07H03M59S			7 03 59	+18 13.0	1	11.01		0.78		0.39		2		1696
	−9 01845	NGC 2335 - 6			7 03 59	−10 05.9	1	10.03		0.64		0.05		2		189
54179	−50 02561	HR 2687			7 04 00	−50 17.0	1	6.46		1.42				1	K3 III	2008
		NGC 2335 - 15			7 04 01	−09 48.5	1	11.43		0.56		-0.01		2		189
		NGC 2335 - 29			7 04 02	−09 51.9	1	12.14		0.55		0.03		2		189
		NGC 2335 - 41			7 04 02	−09 59.7	1	12.66		0.27		-0.39		2		189
		NGC 2335 - 14		AB	7 04 02	−10 00.0	1	11.33		0.29		-0.19		2		189
		NGC 2335 - 35			7 04 04	−09 50.7	1	12.37		1.71		1.42		3		189
53686	+34 01530	HR 2665			7 04 05	+34 05.3	2	5.92	.009	1.51	.001	1.78		5	gK4	71,1733
		G 108 - 55			7 04 05	−02 03.9	1	11.04		1.21		1.14		1		1696
		NGC 2335 - 9			7 04 05	−10 05.1	1	10.63		1.62		1.82		2		189
54031	−30 03907	HR 2680			7 04 05	−30 34.7	2	6.34	.000	-0.15	.010	-0.64		6	B3 V	401,2035
53907	−10 01871	NGC 2335 - 2			7 04 06	−10 08.4	1	7.85		1.15		1.04		4	K2 V	189
		NGC 2335 - 61			7 04 10	−09 58.2	1	13.50		0.64		0.33		3		248
		NGC 2335 - 18			7 04 10	−10 03.0	1	11.59		0.37		-0.05		3		189
53931	−11 01788				7 04 10	−11 27.6	2	8.73	.090	0.00	.010	-0.13	.025	4	B9	388,401
		ApJ144,259 # 25			7 04 12	−02 44.	1	18.94		-0.14		-1.15		2		1360
	−9 01847	NGC 2335 - 7			7 04 12	−09 49.2	1	10.10		0.94		0.63		3		189
		NGC 2335 - 59			7 04 12	−10 00.	1	13.99		0.55		0.37		2		189
		NGC 2335 - 62			7 04 12	−10 00.	1	14.03		0.59		0.34		3		248
		NGC 2335 - 63			7 04 12	−10 00.	1	14.26		0.62		0.29		3		248
293178	+0 01804				7 04 13	+00 28.1	2	9.60	.010	0.42	.000	-0.06		7	F2	742,1594
		NGC 2335 - 28			7 04 13	−09 55.4	1	12.08		0.33		0.10		2		189
		NGC 2335 - 44			7 04 13	−10 02.5	1	12.74		0.48		-0.01		2		189
53879	+8 01653				7 04 15	+08 40.6	1	9.76		-0.09		-0.61		2	B2 III	555
	−16 01767				7 04 15	−16 28.3	1	8.97		1.69				2	M2	369
		NGC 2335 - 38			7 04 16	−09 54.4	1	12.48		0.22		0.02		2		189
		NGC 2335 - 52			7 04 16	−10 02.6	1	13.37		0.46		0.14		3		189

HD	DM	Other Id	N Rem	α_{1950}	δ_{1950}	S	V	σ_V	B–V	σ_{B-V}	U–B	σ_{U-B}	n	Spectrum	References
53975	−12 01788	HR 2679, LSS 166		7 04 16	−12 18.9	10	6.47	.016	−0.10	.005	−0.96	.016	33	B7 Iab/b	15,252,361,388,1011*
53948	−12 01787			7 04 16	−12 37.3	2	9.71	.010	0.03	.000	−0.51	.010	6	B5	92,388
53744	+28 01314	HR 2669		7 04 17	+28 15.3	2			−0.10	.005	−0.26	.000	5	B9 V	1022,1079
		NGC 2335 - 43		7 04 17	−09 54.8	1	12.70		0.68		0.24		2		189
293217				7 04 18	−00 34.	1	9.24		1.81				2		369
		NGC 2335 - 20		7 04 18	−09 56.0	1	11.63		0.56		0.08		2		189
		NGC 2335 - 31		7 04 19	−09 57.8	1	12.21		0.40		0.19		2		189
		NGC 2335 - 25		7 04 19	−09 58.2	1	11.96		0.29		−0.03		2		189
54153	−38 03163	HR 2685		7 04 19	−38 18.3	2	6.10	.005	0.70	.000			7	G0 III	15,2029
		NGC 2335 - 32		7 04 20	−10 01.4	1	12.27		0.33		0.22		3		189
53974	−11 01790	HR 2678, FN CMa	⋆ AB	7 04 20	−11 12.9	5	5.38	.008	0.04	.009	−0.84	.007	17	B0.5IV	15,388,740,1075,2012
53878	+10 01416			7 04 21	+09 57.5	1	7.89		0.32		0.02		2	F0	1733
53791	+22 01577	HR 2671, R Gem	⋆ A	7 04 21	+22 46.9	4	10.97	.229	2.46	.187	1.69	.079	4	S3.9 e	814,3001,8022,8027
	−9 01848	NGC 2335 - 5		7 04 21	−10 04.7	1	10.03		−0.06		−0.25		3	B9 V	189
		NGC 2335 - 34	V	7 04 22	−09 59.1	2	12.21	.099	0.26	.005	−0.08	.025	5		189,248
54208	−42 02959	IDS07028S4210	A	7 04 22	−42 15.0	1	7.20		−0.02		−0.08		2	A0 V	401
		NGC 2335 - 37		7 04 23	−09 58.4	2	12.31	.115	0.32	.000	0.03	.000	6		189,248
		NGC 2335 - 42		7 04 24	−09 52.6	1	12.66		0.35		−0.07		2		189
		NGC 2335 - 51		7 04 24	−10 01.0	1	13.24		0.45		0.18		2		189
54063	−25 04038			7 04 24	−25 23.5	1	8.82		−0.12		−0.55		6	B5 V	1732
53973	−7 01741			7 04 25	−07 10.4	1	9.09		−0.04		−0.24		2	B9	388
		NGC 2335 - 22		7 04 25	−09 53.8	1	11.69		1.04		0.72		2		189
		NGC 2335 - 17		7 04 26	−09 53.7	1	11.50		0.28		−0.19		2		189
53929	+5 01543	HR 2676		7 04 27	+04 59.4	3	6.10	.005	−0.13	.005	−0.46	.011	9	B9.5III	15,1079,1256
		NGC 2335 - 8		7 04 27	−09 57.0	2	10.54	.025	0.27	.015	−0.04	.035	6		189,248
		NGC 2335 - 24	AB	7 04 27	−10 01.9	1	11.90		0.51		0.14		3		189
		LS VI -09 005		7 04 28	−09 05.6	1	10.63		0.24		−0.58		3		490
		NGC 2335 - 58		7 04 29	−10 00.4	1	13.85		0.37		−0.01		2		189
	−9 01851	NGC 2335 - 4		7 04 30	−09 58.2	2	9.48	.020	1.40	.010	1.09	.010	6	K0	189,248
53685	+44 01584			7 04 31	+44 07.2	1	6.79		1.07		0.90		3	G5	985
	−7 01742	RY Mon	A	7 04 31	−07 28.7	2	8.14	.180	4.61	.458			3	C6,5	414,864
	−7 01742		B	7 04 31	−07 28.7	1	12.29		0.49		0.01		1		414
		NGC 2335 - 21		7 04 31	−09 57.1	2	11.58	.070	0.25	.020	−0.10	.010	6		189,248
		NGC 2335 - 30		7 04 31	−09 57.1	1	12.15		0.29		−0.06		2		189
		NGC 2335 - 33		7 04 31	−09 57.1	1	12.30		0.30		−0.03		2		189
		G 108 - 56		7 04 32	+03 31.7	2	9.85	.023	1.29	.010			3		1619,1705
53972	−5 01967	LS VI -05 007		7 04 32	−05 23.6	1	7.30		0.41		0.35		2	F2	490
		NGC 2335 - 55		7 04 32	−10 01.4	1	13.52		0.45		0.22		2		189
		NGC 2335 - 23		7 04 32	−10 01.7	1	11.86		1.77		1.62		2		189
		NGC 2335 - 47		7 04 34	−09 54.4	1	12.87		0.41		0.22		2		189
		NGC 2335 - 57		7 04 34	−09 59.8	1	13.80		0.46		0.20		2		189
		NGC 2335 - 54		7 04 34	−10 02.7	1	13.47		0.64		0.05		3		189
		NGC 2335 - 12		7 04 35	−09 58.2	1	11.03		0.35		0.06		3		189
		NGC 2335 - 48	AB	7 04 35	−09 59.7	1	13.01		0.45		0.29		3		189
54025	−11 01793	IDS07022S1110	AB	7 04 35	−11 14.9	2	7.62	.000	−0.03	.010	−0.80	.010	8	B1 V	92,388
		RL 129		7 04 36	−03 05.	1	14.56		0.15		0.15		2		704
		RL 130		7 04 36	−03 06.	1	13.89		0.07		−0.02		2		704
54024	−7 01745	LS VI -07 003		7 04 36	−07 40.0	2	8.60	.018	0.66	.036	−0.04	.018	5	B3 I	388,490
		NGC 2335 - 39		7 04 36	−10 00.8	1	12.49		0.48		−0.02		2		189
54232	−37 03313			7 04 36	−37 37.9	1	7.50		−0.02		−0.25		2	B9.5V(n)	401
		NGC 2335 - 60		7 04 37	−10 01.9	1	14.03		0.65		0.10		2		189
		NGC 2335 - 45		7 04 38	−10 00.0	1	12.84		1.74		1.43		2		189
	−13 01825	LSS 172		7 04 40	−13 35.9	2	9.49	.005	0.98	.005	−0.06	.005	5		405,1737
54088	−16 01769			7 04 40	−16 40.1	1	9.54		−0.02		−0.26		2	B9 IV/V	401
		NGC 2335 - 49		7 04 41	−09 58.6	1	13.09		0.59		0.08		2		189
54084	−12 01790			7 04 41	−12 09.4	1	8.00		0.17		0.16		2	A3 III/IV	388
		NGC 2335 - 46		7 04 42	−09 54.8	1	12.86		0.37		0.16		2		189
53816	+36 01567	IDS07014N3601		7 04 43	+35 56.5	1	8.56		0.51		−0.02		2	F8	1601
		FG Mon		7 04 43	−07 43.0	1	12.70		1.07				1		1772
		KUV 07047+2826		7 04 44	+28 25.9	1	16.30		0.29		−0.88		1		1708
54057	−4 01827	LS VI -04 022		7 04 44	−04 56.1	1	8.56		0.36		0.37		3	F2	490
		NGC 2335 - 36		7 04 44	−09 51.9	1	12.37		0.57		0.42		2		189
		NGC 2335 - 53		7 04 44	−09 54.5	1	13.38		0.66		0.08		2		189
		LS VI -09 006		7 04 46	−09 06.5	1	10.19		0.34		−0.46		3		490
	−5 01971	LS VI -05 008		7 04 47	−05 08.7	1	10.04		0.08		−0.75		3		490
54083	−8 01748			7 04 48	−09 01.9	1	9.31		−0.05		−0.31		3	B8 V	388
54173	−24 04868	HR 2686		7 04 48	−24 52.9	2	6.07	.012	1.34	.002	1.51		6	K2 III	58,2007
	−47 02804			7 04 49	−47 30.3	1	10.53		1.17		1.17		4		158
		NGC 2335 - 26		7 04 50	−09 59.4	1	11.97		0.81		0.35		3		189
		NGC 2335 - 27		7 04 51	−09 51.6	1	12.03		0.47		0.01		2		189
		NGC 2335 - 16		7 04 52	−09 58.5	1	11.47		0.14		−0.33		3		189
		NGC 2335 - 40		7 04 53	−09 54.7	1	12.52		0.62		0.13		2		189
53927	+30 01423	G 87 - 24		7 04 54	+29 55.1	1	8.32		0.92		0.58		1	G5	333,1620
		NGC 2335 - 56		7 04 54	−09 53.5	1	13.74		0.56		0.25		2		189
		NGC 2335 - 50		7 04 55	−09 52.6	1	13.14		1.55		1.01		2		189
	−11 01797			7 04 56	−11 53.9	1	8.47		2.24				2	M3	369
54143	−12 01794			7 04 56	−13 00.3	1	9.41		0.06		−0.04		2	A0 V +B/A V	388
53899	+34 01533	HR 2673		7 04 57	+33 54.7	2	6.29	.010	1.35	.005	1.54	.000	5	K0	985,1733
		NGC 2343 - 13		7 04 57	−10 32.3	1	10.94		0.23		−0.49		3		182
54104	−9 01853			7 04 58	−09 30.2	2	8.33	.000	0.15	.010	−0.49	.000	5	B3 II-III	388,401

Table 1 379

HD	DM	Other Id	N	Rem	α_{1950}	δ_{1950}	S	V	σ_V	B–V	σ_{B-V}	U–B	σ_{U-B}	n	Spectrum	References
54224	−26 03880	HR 2688			7 04 58	−26 34.7	4	6.61	.004	−0.18	.007	−0.76	.014	13	B3 III	158,401,1732,2007
54141	−9 01854	NGC 2335 - 1			7 04 59	−09 54.4	1	7.00		0.08		0.13		4	A1 V	189
58805	−86 00105	HR 2848			7 04 59	−86 57.5	3	6.47	.008	0.42	.000	−0.08	.010	11	F3 V	15,1075,2038
		NGC 2343 - 22			7 05 04	−10 28.5	1	11.76		0.15		0.05		2		182
	−10 01875	NGC 2343 - 9			7 05 05	−10 29.4	1	10.59		0.05		0.01		3		182
54197	−13 01828	LSS 175			7 05 06	−13 58.4	2	8.00	.007	−0.07	.002	−0.79	.015	5	B2 II	401,1732
		L 455 - 79			7 05 06	−37 23.	1	13.13		0.65		−0.05		2		1696
54079	+7 01607	HR 2682			7 05 07	+07 33.1	3	5.74	.005	1.18	.005	1.08	.000	10	gK0	15,1078,1355
54046	+15 01473	G 109 - 44		⋆ A	7 05 09	+15 36.7	1	7.80		0.56		0.00		2	G0 V	3077
54195	−10 01877				7 05 09	−11 04.2	2	9.48	.000	−0.10	.000	−0.47	.000	4	B7p	320,388
		NGC 2343 - 44	V		7 05 10	−10 29.0	1	13.66		1.05		0.52		4		182
	−3 01748	LS VI -03 016			7 05 12	−03 37.2	1	9.94		−0.11		−0.73		3		490
53925	+37 01660	HR 2675			7 05 13	+37 31.5	1	6.16		1.21				2	K1 III	71
		NGC 2343 - 40			7 05 14	−10 32.7	1	13.12		0.56		0.05		3		182
		NGC 2343 - 35			7 05 15	−10 30.1	1	12.88		0.31		0.18		2		182
		NGC 2343 - 21			7 05 15	−10 35.6	1	11.56		1.65		1.56		2		182
54219	−3 01750				7 05 17	−03 17.5	1	7.90		1.76				2	M1	369
54309	−23 04908	HR 2690, FV CMa			7 05 17	−23 45.7	8	5.83	.024	−0.14	.009	−0.87	.010	45	B2 IVe	164,540,681,815,976*
		NGC 2343 - 45			7 05 19	−10 32.6	1	13.70		0.58		0.12		2		182
	−10 01878	NGC 2343 - 12			7 05 19	−10 35.3	1	10.79		0.10		−0.26		4		182
54100	+15 01476	G 109 - 45		⋆ B	7 05 21	+15 36.2	1	7.68		0.53		−0.06		2	F8	3077
		NGC 2343 - 47			7 05 22	−10 30.3	1	13.83		0.54		0.10		2		182
		CMa sq 1 # 12			7 05 22	−11 44.6	1	10.96		0.48		0.00		3		577
		NGC 2343 - 42			7 05 23	−10 34.4	1	13.54		0.52		0.39		2		182
54252	−8 01754				7 05 24	−08 54.5	1	9.80		0.06		−0.25		3	B8 V	388
	−4 01829				7 05 25	−05 07.9	1	9.17		1.69				2	M2	369
		NGC 2343 - 52			7 05 25	−10 27.4	1	14.69		0.75		0.27		2		182
		NGC 2343 - 49			7 05 25	−10 28.7	1	14.17		0.68		0.09		3		182
		NGC 2343 - 25			7 05 25	−10 37.4	1	12.00		0.19		0.11		4		182
	−11 01800				7 05 25	−11 43.5	1	9.12		0.84		0.56		3	K0	577
		NGC 2343 - 33			7 05 26	−10 37.4	1	12.72		0.56		0.04		2		182
		CMa sq 1 # 15			7 05 26	−11 51.0	1	12.00		1.18		1.41		3		577
54284	−12 01798				7 05 26	−12 12.5	2	9.35	.010	−0.04	.000	−0.39	.015	4	B8 II	388,401
293334	+0 01809				7 05 27	+00 05.4	1	9.40		0.84				1	G5	592
		Steph 626			7 05 27	+30 47.7	1	11.36		1.49		1.17		1	M0	1746
293337	+0 01810				7 05 27	−00 03.5	1	8.93		0.98		0.75		1	K0	1462
		NGC 2343 - 16			7 05 27	−10 35.4	1	11.39		0.42		0.29		2		182
54282	−10 01879	NGC 2343 - 8			7 05 27	−10 43.5	1	10.44		0.03		−0.11		3		182
54283	−10 01879	NGC 2343 - 7			7 05 27	−10 43.5	1	10.32		0.18		0.19		3		182
		NGC 2343 - 39			7 05 29	−10 26.4	1	13.07		0.63		0.08		2		182
	−28 03875				7 05 29	−28 11.5	1	9.05		−0.14		−0.66		1	A0	1732
54475	−40 02930	HR 2691			7 05 29	−40 48.8	2	5.80	.005	−0.18	.005	−0.68		6	B3 V	401,2031
54131	+16 01397	HR 2684		⋆ AB	7 05 30	+16 00.7	2	5.46	.018	1.02	.016	0.88		3	G8 III	71,3077
		CMa sq 1 # 14			7 05 32	−11 55.0	1	13.75		1.67		1.03		3		577
54503	−40 02932				7 05 32	−40 27.5	1	7.99		1.46		1.77		10	K3 III	1673
293335	+0 01811				7 05 33	+00 03.5	1	10.53		0.09				1	A0	592
		LS VI -02 014			7 05 33	−02 33.3	1	10.67		−0.01		−0.14		3		490
		NGC 2343 - 14			7 05 33	−10 28.6	1	11.03		0.35		0.11		3		182
54306	−11 01801				7 05 33	−11 50.0	5	8.78	.014	0.00	.004	−0.68	.010	16	B2 V	92,388,401,414,577
		CMa sq 1 # 23			7 05 33	−11 53.1	1	11.51		0.23		0.10		3		577
		NGC 2343 - 34			7 05 34	−10 30.9	1	12.74		0.32		0.23		3		182
54304	−10 01882	NGC 2343 - 4			7 05 34	−10 32.5	1	10.00		0.03		−0.45		4		182
54413	−28 03879				7 05 34	−28 11.0	2	8.48	.012	−0.14	.001	−0.63	.012	5	B3 III	401,1732
		G 87 - 25			7 05 35	+38 33.9	1	12.43		1.44		1.12		1		333,1620
		NGC 2343 - 43			7 05 35	−10 33.8	1	13.62		0.61		0.30		4		182
		V465 Mon			7 05 36	+00 00.8	3	10.21	.034	0.67	.053	0.41		3	G0	592,1462,1772
54044	+34 01536				7 05 36	+34 00.9	1	6.53		1.47		1.77		2	K0	1733
		RL 131			7 05 36	−02 48.	1	11.78		0.05		−0.75		2		704
		NGC 2343 - 46		AB	7 05 36	−10 33.4	1	13.79		0.62		0.12		3		182
	−57 01139				7 05 36	−57 23.3	4	9.54	.005	0.49	.016	−0.13	.017	10	F8	158,803,1696,3025,6006
		NGC 2343 - 15			7 05 37	−10 31.2	1	11.35		1.09		0.65		2		182
		NGC 2343 - 28			7 05 37	−10 32.6	1	12.49		0.27		0.20		5		182
		NGC 2343 - 37			7 05 37	−10 33.0	1	13.05		0.44		0.22		4		182
		NGC 2343 - 41			7 05 37	−10 35.4	1	13.27		0.44		0.18		2		182
		NGC 2345 - 64			7 05 38	−12 58.8	1	12.42		0.50		−0.18		2		381
		NGC 2343 - 27	V		7 05 40	−10 31.4	1	12.44		0.30		0.19		5		182
		CMa sq 1 # 27			7 05 40	−11 51.4	1	15.10		0.57		0.33		2		577
		CMa sq 1 # 20			7 05 42	−11 49.2	1	11.99		0.29		0.12		3		577
54362	−11 01804				7 05 42	−11 56.8	1	9.34		0.14		0.08		3	A0	577
	−10 01883	NGC 2343 - 5		AB	7 05 43	−10 31.5	1	10.07		0.12		−0.06		3		182
		NGC 2343 - 23			7 05 43	−10 34.6	1	11.89		0.63		0.13		2		182
54361	−11 01805	W CMa	A		7 05 43	−11 50.6	7	6.65	.102	2.42	.081	4.23	.119	11	C6,3	109,414,631,740,864*
54361	−11 01805		B		7 05 43	−11 50.6	1	8.89		0.04		−0.70		1		631
		NGC 2345 - 4			7 05 43	−13 08.9	1	12.17		0.52		−0.06		3		381
		NGC 2343 - 51			7 05 44	−10 32.2	1	14.41		0.64		0.12		4		182
		NGC 2343 - 50			7 05 44	−10 33.2	1	14.20		0.64		0.12		2		182
		NGC 2343 - 26			7 05 44	−10 34.5	1	12.29		0.37		0.23		3		182
		NGC 2343 - 29			7 05 44	−10 37.5	1	12.51		0.61		0.08		3		182
		NGC 2343 - 31			7 05 45	−10 27.9	1	12.55		0.33		0.25		2		182
		NGC 2343 - 36			7 05 45	−10 27.9	1	12.98		0.40		0.29		2		182

HD	DM	Other Id	N Rem	α_{1950}	δ_{1950}	S	V	σ_V	B–V	σ_{B-V}	U–B	σ_{U-B}	n	Spectrum	References
		NGC 2343 - 6	AB	7 05 45	−10 30.7	2	10.30	.010	0.05	.020	-0.27	.035	6		182,248
		NGC 2343 - 10		7 05 45	−10 32.9	1	10.69		0.10		-0.23		3		182
		NGC 2343 - 17		7 05 45	−10 35.7	1	11.71		0.19		0.12		2		182
		CMa sq 1 # 4		7 05 45	−11 47.3	1	10.28		0.94		0.59		3		577
		CMa sq 1 # 19		7 05 45	−11 49.5	1	11.73		0.16		-0.10		3		577
		CMa sq 1 # 18		7 05 45	−11 53.6	1	10.87		1.18		0.84		5		577
		NGC 2343 - 18		7 05 46	−10 33.4	2	11.37	.074	0.18	.045	0.14	.025	5		182,248
		CMa sq 1 # 24	V	7 05 46	−11 50.3	1	13.32		0.87		0.52		2		577
		NGC 2345 - 3		7 05 46	−13 10.2	1	13.84		0.41		0.13		3		381
293333	+0 01814			7 05 47	+00 08.4	1	9.95		1.71				2	M5	369
54359	−9 01858			7 05 47	−09 53.3	5	8.86	.019	0.95	.008	0.73	.017	18	K0	158,680,1775,2034,3072
		NGC 2343 - 55		7 05 47	−10 31.8	1	15.29		0.82		0.19		3		182
		NGC 2343 - 24		7 05 47	−10 32.3	2	11.85	.125	0.23	.005	0.15	.010	6		182,248
		NGC 2343 - 48		7 05 47	−10 32.3	1	13.87		0.59		0.13		2		182
		CMa sq 1 # 5		7 05 47	−11 46.4	1	10.25		1.07		0.96		3		577
		NGC 2345 - 22		7 05 47	−13 06.4	1	12.05		0.72		0.11		6		381
54501	−34 03355			7 05 47	−34 32.7	1	9.46		-0.10		-0.61		2	B4 III/IV	401
54073	+38 01693			7 05 48	+38 41.6	1	7.94		0.31		0.01		2	A3 V	105
		NGC 2343 - 38		7 05 48	−10 33.7	2	13.01	.055	0.42	.015	0.13	.085	6		182,248
		NGC 2345 - 23		7 05 48	−13 04.9	1	13.50		0.72		0.14		4		381
		CMa sq 1 # 25		7 05 49	−11 51.0	1	13.82		0.77		0.47		1		577
		CMa sq 1 # 26		7 05 49	−11 51.3	1	13.55		0.63		0.49		2		577
		NGC 2345 - 5		7 05 49	−13 08.5	1	13.42		0.54		-0.04		3		381
		NGC 2343 - 53		7 05 50	−10 28.9	1	14.69		0.70		0.26		3		182
54360	−10 01884	NGC 2343 - 3		7 05 50	−10 29.5	3	9.42	.017	-0.02	.011	-0.25	.014	9	A0	182,248,401
54627	−46 02918			7 05 50	−46 32.2	1	8.78		0.97		0.80		2	K1 (III)+G	536
		NGC 2343 - 19		7 05 51	−10 31.2	2	11.48	.050	0.92	.010	0.01	.015	6		182,248
		NGC 2345 - 21		7 05 51	−13 05.9	1	13.88		0.70		0.17		3		381
		NGC 2343 - 54		7 05 52	−10 33.2	1	14.79		0.81		0.19		2		182
		NGC 2345 - 2		7 05 52	−13 10.8	1	11.86		0.39		-0.06		4		381
		CMa sq 1 # 11		7 05 53	−11 48.2	1	11.89		1.61		2.66		2		577
		NGC 2345 - 24		7 05 53	−13 04.6	1	13.98		1.00		0.27		3		381
		NGC 2343 - 11		7 05 54	−10 32.5	1	10.72		0.16		-0.08		3		182
54387	−10 01885	NGC 2343 - 2	★ AB	7 05 54	−10 32.6	2	8.48	.015	1.25	.005	1.02	.060	6	G5	182,248
		NGC 2343 - 56		7 05 54	−10 34.	1	14.25		0.81		0.33		3		248
		NGC 2345 - 20		7 05 54	−13 05.8	1	13.90		0.69		0.07		3		381
		NGC 2345 - 18		7 05 54	−13 07.8	1	14.49		0.52		0.23		3		381
		NGC 2345 - 6		7 05 54	−13 09.0	1	13.44		0.46		-0.03		3		381
		NGC 2345 - 19		7 05 55	−13 05.8	1	12.84		0.76		0.06		3		381
		NGC 2345 - 16		7 05 55	−13 07.5	1	14.01		0.46		0.05		3		381
		NGC 2345 - 17		7 05 55	−13 08.2	1	14.33		0.57		0.31		3		381
54388	−10 01887	NGC 2343 - 1	AB	7 05 56	−10 42.1	1	8.39		0.30		0.32		3	A3	182
		CMa sq 1 # 13		7 05 56	−11 56.4	1	11.19		0.82		0.45		3		577
		NGC 2345 - 25		7 05 56	−13 05.4	1	13.55		0.68		0.14		3		381
		NGC 2345 - 15		7 05 56	−13 06.9	1	13.94		0.55		0.31		3		381
54411	−13 01838	NGC 2345 - 1		7 05 56	−13 11.3	1	9.16		0.51		0.50		49	A2 II	381
		NGC 2343 - 30		7 05 57	−10 35.2	1	12.53		0.33		0.25		3		182
		NGC 2343 - 20		7 05 57	−10 38.2	1	11.53		0.17		0.03		2		182
		NGC 2345 - 27		7 05 57	−13 05.2	1	13.66		0.63		0.06		3		381
	−13 01839	NGC 2345 - 7		7 05 57	−13 08.9	1	9.87		0.75		0.70		16	A3 II	381
		NGC 2345 - 8		7 05 57	−13 09.2	1	13.78		0.53		0.09		3		381
54300	+10 01428	R CMi		7 05 58	+10 06.3	2	7.80	.941	2.47	.122			6	CS	864,3039
		CMa sq 1 # 21		7 05 58	−11 47.2	1	11.26		0.25		-0.03		2		577
		NGC 2345 - 28		7 05 58	−13 04.9	1	12.41		0.55		-0.03		4		381
		NGC 2345 - 26		7 05 58	−13 05.4	1	13.28		0.67		0.14		3		381
	−11 01810			7 05 59	−11 48.8	1	10.13		0.06		-0.02		3	A0	577
	−11 01811			7 05 59	−11 52.8	1	9.62		0.26		0.12		3	A5	577
		NGC 2345 - 14		7 05 59	−13 06.7	1	10.73		2.06		2.30		16		381
52881	+78 00240			7 06 00	+78 50.1	1	6.97		0.25		0.06		3	A5	985
		NGC 2345 - 30		7 06 00	−13 02.2	1	13.74		0.60		0.09		3		381
		NGC 2345 - 29		7 06 00	−13 02.3	1	12.91		0.79		0.27		3		381
		NGC 2345 - 11		7 06 00	−13 07.7	1	13.88		0.53		0.17		3		381
		NGC 2343 - 32		7 06 01	−10 34.9	1	12.69		0.41		0.22		3		182
		CMa sq 1 # 17		7 06 01	−11 49.7	1	13.81		1.83		1.22		3		577
		NGC 2345 - 32		7 06 01	−13 04.9	1	13.86		0.61		0.22		3		381
54732	−51 02306	HR 2698		7 06 01	−51 53.3	2	5.96	.002	1.00	.002	0.82		6	K0 III	58,2035
		NGC 2345 - 31		7 06 02	−13 04.5	1	14.18		0.76		0.48		3		381
		NGC 2345 - 33		7 06 02	−13 05.7	1	13.91		0.53		0.19		3		381
		NGC 2345 - 13		7 06 02	−13 06.7	1	13.95		0.62		0.15		3		381
		NGC 2345 - 12		7 06 02	−13 06.9	1	14.00		0.65		0.38		2		381
		NGC 2345 - 10		7 06 02	−13 08.1	1	14.69		0.60		0.25		3		381
		NGC 2345 - 9		7 06 02	−13 09.1	1	13.84		0.43		0.11		3		381
54439	−11 01812	LSS 189		7 06 03	−11 46.3	4	7.69	.011	0.05	.005	-0.76	.006	14	B2 IIIn	388,577,1012,2014
54440	−15 01662			7 06 03	−15 35.8	1	8.57		1.65				2	M2/3 III	369
		NGC 2345 - 35		7 06 04	−13 05.5	1	10.93		0.51		-0.15		5		381
	−12 01805	NGC 2345 - 34	AB	7 06 04	−13 05.5	1	9.94		1.50		0.50		17	K3 II +B	381
		NGC 2345 - 57		7 06 04	−13 10.0	1	14.07		0.53		0.24		3		381
		NGC 2345 - 58		7 06 04	−13 12.1	1	12.30		0.75		0.37		3		381
54555	−28 03888			7 06 04	−28 29.8	1	8.54		-0.17		-0.66		2	B3 IV	401
		NGC 2345 - 52		7 06 05	−13 08.0	1	14.09		0.52		0.18		3		381

Table 1

HD	DM	Other Id	N Rem	α₁₉₅₀	δ₁₉₅₀	S	V	σ_V	B–V	σ_B-V	U–B	σ_U-B	n	Spectrum	References
		NGC 2345 - 54		7 06 06	−13 08.9	1	12.86		0.47		-0.07		3		381
		NGC 2345 - 55		7 06 06	−13 09.4	1	14.08		0.48		0.27		4		381
54352	+8 01665			7 06 07	+07 57.9	1	9.56		-0.35		-0.41		1	B5 III	555
54437	−7 01761			7 06 07	−07 27.5	1	9.60		-0.03		-0.39		2	B9	388
		NGC 2345 - 44		7 06 07	−13 07.2	1	13.32		0.40		-0.05		3		381
		NGC 2345 - 53		7 06 07	−13 08.7	1	13.64		0.50		0.22		3		381
		NGC 2345 - 56		7 06 07	−13 09.6	1	13.50		0.49		0.27		3		381
54242	+30 01431			7 06 08	+30 13.7	1	7.28		1.08		0.97		3	K0	1625
		NGC 2345 - 43		7 06 08	−13 06.4	1	10.70		1.81		1.90		16		381
		NGC 2345 - 50		7 06 08	−13 07.7	1	10.40		2.04		2.05		16	K3 II	381
		NGC 2345 - 51		7 06 08	−13 07.7	1	12.37		0.45		-0.21		5		381
54518	−18 01696		V	7 06 08	−18 56.6	1	8.42		1.74				2	M2/3 (III)	369
54520	−22 01712		V	7 06 08	−22 21.6	1	9.12		0.19		0.19		3	A3 IV	1768
	−11 01814			7 06 09	−11 55.7	1	10.04		0.60		0.08		3		577
		NGC 2345 - 42		7 06 09	−13 06.6	1	13.40		0.50		0.10		3		381
		NGC 2345 - 45		7 06 09	−13 07.2	1	13.56		0.46		0.09		3		381
		NGC 2345 - 49		7 06 09	−13 07.6	1	12.82		0.57		-0.04		2		381
		NGC 2345 - 48		7 06 09	−13 07.9	1	12.74		0.50		0.10		3		381
		NGC 2345 - 59		7 06 09	−13 10.8	1	13.05		0.39		-0.12		3		381
54551	−23 04932			7 06 09	−23 21.1	1	8.66		-0.07		-0.80		2	B1/2 II	401
		NGC 2345 - 46		7 06 10	−13 07.5	1	13.27		0.56		0.11		3		381
	−12 01807			7 06 11	−12 28.9	1	9.75		0.02		-0.61		2		401
		NGC 2345 - 47		7 06 11	−13 07.9	1	12.61		0.45		-0.09		5		381
		NGC 2345 - 61		7 06 11	−13 08.4	1	13.36		0.43		-0.06		3		381
		CMa sq 1 # 16		7 06 12	−11 52.0	1	13.98		0.53		0.26		3		577
		NGC 2345 - 38		7 06 12	−13 05.8	1	13.13		0.68		0.04		3		381
		NGC 2345 - 62		7 06 12	−13 08.5	1	13.85		1.46		1.07		3		381
		NGC 2345 - 60		7 06 12	−13 09.0	1	10.48		1.82		1.96		16	K3 II	381
54493	−12 01809			7 06 13	−12 48.3	3	7.19	.016	-0.01	.019	-0.76	.008	7	B2 IV	388,399,401
		NGC 2345 - 37		7 06 13	−13 05.3	1	12.69		0.33		-0.19		3		381
		NGC 2345 - 41		7 06 13	−13 06.9	1	13.10		0.46		-0.07		3		381
54519	−20 01740	IDS07041S2042	A	7 06 13	−20 46.8	1	6.82		1.51		1.71		4	K3 III	8100
54351	+15 01482	G 88 - 8		7 06 14	+15 30.4	1	8.00		0.62		0.12		1	G0	333,1620
		NGC 2345 - 39		7 06 15	−13 06.1	1	12.92		0.38		-0.18		4		381
		NGC 2345 - 40		7 06 15	−13 07.6	1	13.14		0.45		-0.07		3		381
		NGC 2345 - 36		7 06 17	−13 05.1	1	10.26		0.22		0.20		3		381
54464	−3 01762			7 06 19	−03 58.6	2	8.34	.012	-0.01	.000	-0.62	.004	7	B2:V:pe	1012,1775
		CMa sq 1 # 9		7 06 20	−11 49.3	1	10.44		0.21		0.02		3		577
54605	−26 03916	Cr 121 - 31	★	7 06 21	−26 18.8	11	1.83	.011	0.68	.010	0.54	.025	44	F8 Ia	3,15,62,1008,1020*
54542	−8 01761	IDS07040S0831	A	7 06 25	−08 35.8	1	8.34		1.25		1.18		1	K2 III	388
293396	−0 01618	V647 Mon		7 06 26	−00 56.1	3	9.33	.028	0.04	.011	-0.81	.005	22	B1 V:ne	830,1012,1783
54543	−8 01761	IDS07040S0831	B	7 06 26	−08 35.6	1	9.91		0.04		0.07		1	A0	388
55151	−68 00591	HR 2712		7 06 26	−68 45.5	3	6.46	.003	1.04	.000	0.88	.005	24	K0 III	15,1075,2038
54371	+25 01594	IDS07034N2554	A	7 06 31	+25 48.7	3	7.09	.000	0.70	.000	0.25	.015	9	G8 V	1080,1355,3026
54544	−10 01889			7 06 31	−11 06.1	1	9.13		-0.03		-0.13		2	B9 V	388
	−2 01948	LS VI -02 015		7 06 32	−02 19.9	1	10.31		-0.01		-0.53		1		490
54575	−15 01664			7 06 32	−15 51.2	1	8.47		0.28		-0.44		3	B5 III	1212
54602	−18 01701			7 06 32	−18 11.4	1	8.68		-0.07		-0.55		2	B5 V	401
54967	−59 00771			7 06 32	−59 38.2	1	6.46		-0.13		-0.54		1	B4 V	55
54574	−11 01816			7 06 33	−11 25.2	2	8.73	.005	-0.01	.005	-0.67	.005	5	B2 V	388,401
54601	−12 01813			7 06 37	−12 14.3	1	8.65		0.05		0.02		2	A0 V	388
		G 87 - 26	★	7 06 39	+38 37.4	3	11.49	.020	1.71	.005	1.19	.027	13	M3e	694,8039,3078
54403	+25 01595	IDS07034N2554	C	7 06 40	+25 49.0	1	7.77		0.30		0.04		3	F0	833
54600	−10 01890			7 06 43	−10 10.5	1	9.05		0.03		-0.05		2	B9.5V	388
54669	−23 04949	Cr 121 - 32	★	7 06 44	−23 57.8	6	6.64	.015	-0.19	.009	-0.80	.011	14	B2 V	62,401,540,976,1732,2007
	+37 01665	G 87 - 27		7 06 46	+37 21.7	1	9.84		0.81		0.39		1	G5	333,1620
		G 109 - 47		7 06 49	+18 10.5	1	14.32		0.76		0.14		2		1658
	−9 01870	LS VI -09 007		7 06 49	−09 16.2	1	10.49		0.34		-0.45		3		490
54941	−51 02316			7 06 50	−51 52.7	1	9.14		0.51		-0.05		3	F7 V(w)	1097
54538	+8 01668			7 06 51	+07 53.4	1	9.92		-0.08		-0.43		1	B8	555
		G 87 - 28		7 06 51	+37 45.4	2	14.58	.019	1.66	.015	1.15	.049	12	sdM6	203,3028
		G 87 - 29		7 06 52	+37 45.4	2	15.59	.050	0.32	.015	-0.56	.005	11		203,3060
		NGC 2345 - 63		7 06 52	−13 06.0	1	12.32		0.64		-0.06		2		381
		L 671 - 49		7 06 54	−22 44.	1	11.57		1.50		1.26		2		3060
		KUV 07069+2929		7 06 55	+29 29.0	1	15.45		0.09		-0.66		1	DA	1708
54660	−8 01762			7 06 56	−08 12.2	1	7.14		1.18				5	G5	897
54662	−10 01892	HR 2694, LSS 197		7 06 58	−10 15.9	10	6.21	.005	0.03	.005	-0.90	.013	29	O7 III	15,92,388,1011,1020*
54664	−10 01894			7 07 00	−10 34.5	2	8.93	.015	-0.05	.005	-0.41	.005	4	B7 V	388,401
54771	−27 03683	IDS07050S2800	AB	7 07 01	−28 04.7	1	7.75		-0.16		-0.72		2	B3 IV	401
		GD 82		7 07 03	+38 12.2	1	15.29		0.55		-0.18		1		3060
54727	−12 01814	LSS 199		7 07 06	−12 27.6	1	9.74		0.39		-0.51		3	B1 Ib/II	388
	−6 01954	AP Mon		7 07 08	−06 39.0	1	8.77		1.77				2	M3	369
54563	+21 01528	HR 2692	★	7 07 09	+21 20.1	4	6.43	.005	0.88	.011	0.51	.018	10	G9 V	22,333,1355,1620,3077
		KUV 07072+2901		7 07 10	+29 01.2	1	16.00		-0.14		-0.89		1		1708
		BONN82 T7# 5		7 07 10	−09 11.7	1	9.49		1.13		1.10		3	gK2	322
54893	−39 03105	HR 2702		7 07 10	−39 34.5	4	4.82	.005	-0.18	.000	-0.69	.004	14	B3 IV/V	15,1075,2012,8015
	−9 01873			7 07 11	−09 11.7	1	10.12		0.46		0.12		3	F5	322
54816	−27 03687			7 07 15	−27 16.9	1	8.53		-0.16		-0.60		2	B8/9 III	401
54630	+12 01444			7 07 16	+12 37.8	1	9.14		0.28		0.07		2		1733
54765	−16 01801			7 07 16	−16 51.3	1	9.10		0.03		-0.13		2	B9 IV/V	401

HD	DM	Other Id	Rem	α_{1950}	δ_{1950}	S	V	σ_V	B–V	σ_{B-V}	U–B	σ_{U-B}	n	Spectrum	References
54740	−10 01897			7 07 17	−10 57.4	2	9.06	.015	0.03	.000	−0.70	.005	4	B2 III	388,401
		RL 133		7 07 18	+02 49.	1	14.11		0.04		−0.48		2		704
		BONN82 T7# 13		7 07 18	−09 11.3	1	12.07		0.61		0.54		4		322
54764	−16 01802	HR 2699, LSS 202	★A	7 07 18	−16 09.2	5	6.04	.022	0.05	.008	−0.79	.002	16	B1 Ib/II	15,1569,2012,2014,8100
54762	−11 01819	IDS07050S1155	AB	7 07 20	−11 59.6	1	10.26		0.13		−0.33		3	B9	388
		G 88 - 10		7 07 21	+24 25.7	2	11.87	.000	0.44	.000	−0.24	.000	3		333,927,1620
	−11 01820			7 07 22	−11 59.5	1	9.44		0.25		−0.41		2	B3 III	388
54786	−15 01672	LSS 204		7 07 22	−16 00.9	1	8.98		0.23		−0.87		2	B1/2 I(b)	1012
54814	−18 01705			7 07 23	−19 02.0	1	8.00		−0.09		−0.61		2	B5 II/III	401
		Steph 629		7 07 25	+28 27.9	1	11.69		1.42		1.31		1	K7	1746
54784	−8 01768			7 07 25	−09 05.7	2	9.15	.005	0.00	.005	−0.31	.005	4	B8 V	388,401
55000	−45 02960			7 07 29	−45 15.0	1	7.00		−0.07		−0.31		2	B8 V	401
		07H07M30S		7 07 30	−01 12.9	1	11.82		0.51		−0.26		2		1696
		RL 134		7 07 30	−02 50.	1	14.39		0.16		−0.11		2		704
	−8 01770			7 07 32	−09 03.2	1	10.03		0.05		−0.02		3	B9.5	322
	−9 01875			7 07 32	−09 13.6	1	9.14		−0.02		−0.28		3	B8	322
54913	−28 03925			7 07 34	−28 21.8	1	8.27		−0.16		−0.67		2	B2 V	401
54834	−15 01676			7 07 37	−16 09.1	1	6.62		0.31				4	A9 V	2014
54690	+20 01724	G 88 - 11		7 07 39	+20 31.7	1	9.09		0.69		0.20		1	G5	333,1620
54912	−25 04120	HR 2704	★AB	7 07 39	−25 08.9	3	5.70	.011	−0.16	.003	−0.76	.020	9	B2 V	401,1732,2007
54759	+5 01566			7 07 40	+05 23.0	2	9.38	.007	0.47	.015	−0.05		3	F8	1594,6006
		BONN82 T7# 14		7 07 40	−09 12.5	1	12.84		0.57		0.08		4	G0	322
54832	−6 01958			7 07 41	−07 08.5	1	8.68		−0.07		−0.44		3	B9	388
		AA45,405 S301 # 4		7 07 41	−18 21.4	1	12.05		0.42		0.19		1		797
54979	−35 03374			7 07 41	−35 12.3	1	8.54		0.01		−0.31		2	B8/9 III	401
		LSS 208		7 07 43	−18 21.2	1	11.56		0.18		−0.70		2	B3 Ib	797
		LSS 207		7 07 43	−18 25.2	4	10.85	.014	0.44	.014	−0.57	.006	10	B3 Ib	342,432,797,1737
		G 107 - 55		7 07 44	+48 25.2	1	12.21		1.43		1.19		3		419
293471	−2 01955			7 07 44	−02 18.4	1	9.23		−0.11		−0.63		3	B2	1732
		AA45,405 S301 # 5		7 07 44	−18 20.8	1	11.88		0.17		0.08		1		797
54810	−4 01840	HR 2701	★A	7 07 45	−04 09.5	9	4.91	.008	1.02	.006	0.79	.009	28	K0 III	15,1003,1075,1425*
		BONN82 T7# 10		7 07 45	−09 11.5	1	10.68		0.09		0.17		3	A5	322
54884	−17 01811			7 07 45	−17 48.0	1	9.48		−0.02		−0.29		2	B8/9 IV/V	401
54858	−9 01880	LSS 206		7 07 48	−09 15.2	2	8.17	.005	−0.04	.005	−0.26	.015	5	A0 II	388,401
54879	−11 01822			7 07 48	−11 43.2	2	7.64	.013	−0.01	.004	−0.87	.004	5	O9.5V	388,1011
54876	−8 01773			7 07 49	−08 38.5	1	9.88		0.03		−0.28		3	B9	388
	+39 01881			7 07 51	+39 10.1	2	9.59	.000	0.06	.000	0.08	.010	6	A0	833,1501
54857	−9 01881			7 07 53	−09 13.8	2	7.78	.015	0.19	.005	0.15	.035	6	A1 V	322,388
54911	−15 01681	LSS 209		7 07 53	−15 36.1	3	7.35	.015	−0.08	.008	−0.82	.006	6	B1 III	1012,1569,8100
		LSS 212		7 07 55	−18 21.8	1	12.11		0.21		−0.54		2	B3 Ib	797
		BONN82 T7# 12		7 07 57	−09 10.2	1	11.81		0.55		0.02		4	F8	322
54958	−18 01711	HR 2705	★AB	7 07 57	−18 36.2	1	6.23		0.40				4	F3 V	2006
54719	+30 01439	HR 2697	★AB	7 07 58	+30 19.7	7	4.41	.012	1.26	.005	1.40	.006	21	K2 III	15,1080,1355,1363*
		BONN82 T7# 15		7 07 58	−09 10.8	1	12.97		0.77		0.19		4	G0	322
54932	−12 01821			7 07 58	−12 37.1	1	9.97		0.07		−0.43		4	B7/8 Ib	388
		BONN82 T7# 11		7 07 59	−09 17.0	1	10.97		1.04		0.71		3	K2	322
		BONN82 T7# 16		7 08 02	−09 10.0	1	13.62		0.67		0.20		4	G5	322
55019	−28 03939			7 08 02	−28 39.9	3	7.30	.007	−0.17	.010	−0.76	.016	12	B3 V	158,401,1775
		TZ Aur		7 08 07	+40 51.7	3	11.34	.394	0.21	.051	0.08	.076	3	A2	597,668,699
296555	−2 01962	LS VI -2 016	★AB	7 08 08	−02 45.1	1	9.03		0.55		0.42		3	G8	490
54929	−7 01783			7 08 08	−07 47.3	2	7.34	.010	0.00	.010	−0.05	.010	5	B9	388,401
54070	+72 00352	HR 2681		7 08 10	+71 54.1	1	6.35		1.12				2	K0	71
	−30 04020	Cr 132 - 26		7 08 10	−30 36.9	1	9.23		0.53		0.05		2	G0	594
54802	+24 01549			7 08 12	+24 44.8	1	7.98		1.61		1.73		3	M1	1625
54716	+39 01882	HR 2696		7 08 13	+39 24.2	4	4.93	.030	1.45	.003	1.75	.020	16	K4 II-III	1355,1363,3053,8100
	−5 02002			7 08 14	−06 06.4	1	10.04		1.80				2	M3	369
54976	−8 01778			7 08 14	−08 50.5	1	9.79		0.11		0.05		3	A0 V	388
		G 193 - 37	AB	7 08 16	+52 21.6	1	11.28		1.49		1.21		3	M0	7010
54801	+27 01327	HR 2700		7 08 17	+26 56.4	3	5.77	.014	0.11	.007	0.14	.005	6	A4 IV	985,1022,1733
	−31 04127	Cr 132 - 32		7 08 18	−31 07.6	1	9.86		0.62		0.13		2		594
55194	−46 02947			7 08 18	−46 16.8	1	8.19		0.97				4	K0 III	2012
55070	−27 03710	HR 2708		7 08 19	−27 24.5	1	5.46		1.00				4	G8 III	2031
		BONN82 T7# 9		7 08 20	−09 13.3	1	10.48		0.08		0.11		3	A3	322
55040	−24 04958			7 08 22	−24 15.8	1	8.56		−0.08		−0.40		2	B8 V	401
	−9 01888			7 08 24	−09 12.0	1	9.85		1.02		0.70		3	gK0	322
54995	−9 01887			7 08 24	−09 15.2	3	7.41	.005	−0.11	.005	−0.59	.016	12	B4 V	322,388,401
		RL 135		7 08 24	−09 59.	1	14.59		0.19		0.08		2		704
55014	−11 01826			7 08 25	−11 29.5	3	7.75	.014	−0.06	.004	−0.34	.005	8	B7 Vn	388,401,2014
54715	+44 01598			7 08 26	+43 55.6	1	6.87		0.02		0.02		3	A1 V	1501
55144	−30 04207	Cr 132 - 16	AB	7 08 26	−31 01.3	1	8.14		0.01		−0.03		3	A0 IV	594
55012	−7 01788			7 08 32	−07 53.8	1	9.76		0.02		−0.28		3	B9	388
55013	−9 01890			7 08 32	−09 22.1	1	8.75		−0.03		−0.27		2	A0	401
	−17 01823	LSS 218		7 08 32	−17 29.8	1	10.37		0.36		−0.53		4	B1 III	1737
55173	−30 04030	Cr 132 - 9	★V	7 08 35	−30 30.0	3	7.44	.003	−0.18	.007	−0.77	.012	9	B2 V + B2	540,594,976
54824	+35 01570	IDS07053N3509	AB	7 08 36	+35 04.4	1	8.62		0.10		0.09		2	A2	1601
	+41 01609			7 08 37	+40 51.8	1	10.83		0.41		0.05		1	F2	289
55066	−12 01824			7 08 38	−13 07.0	1	8.72		0.22		0.13		2	A1/2 III/IV	388
55036	−4 01845	LS VI -04 023		7 08 42	−04 37.2	3	7.00	.007	0.29	.007	0.12	.066	6	A3 Ib	490,6009,8100
55094	−11 01828			7 08 45	−11 27.1	1	10.37		0.05		−0.34		2	B7 V	388
55121	−12 01825			7 08 48	−12 26.3	2	8.68	.035	−0.03	.025	−0.67	.010	4	B3 II/III	388,401

Table 1 383

HD	DM	Other Id	N Rem	α_{1950}	δ_{1950}	S	V	σ_V	B–V	σ_{B-V}	U–B	σ_{U-B}	n	Spectrum	References
55091	−8 01779			7 08 49	−09 03.0	2	7.58	.009	0.36	.000	0.18	.023	8	F0 IV-V	322,388
55215	−30 04037	Cr 132 - 28	★ AB	7 08 49	−31 05.8	1	9.27		0.23		0.13		2	A7 V	594
55057	−0 01634	HR 2707, V571 Mon		7 08 50	−00 13.1	5	5.45	.010	0.30	.005	0.13	.013	14	F0 V	15,254,1008,1256,3023
		WLS 720 30 # 11		7 08 51	+32 01.2	1	10.30		0.98		0.67		2		1375
55118	−10 01906	IDS07065S1023	AB	7 08 53	−10 27.5	3	7.88	.007	-0.04	.004	-0.37	.000	9	B8 V	388,401,2014
55117	−9 01895			7 08 54	−09 13.0	1	8.84		-0.02		-0.28		3	B8 V	388
	−22 04134			7 08 56	−22 41.6	1	10.77		0.74				2		370
55214	−25 04156			7 08 56	−25 50.4	1	8.16		-0.14		-0.65		2	B7 III/IV	401
55134	−7 01792	LS VI -08 008		7 08 57	−08 05.9	3	9.36	.011	-0.02	.012	-0.88	.014	7	B0.5V	388,401,490
55135	−10 01908			7 08 59	−10 20.7	2	7.31	.009	-0.08	.019	-0.61	.009	6	B4 Vne	388,1212
55376	−45 02984			7 08 59	−45 36.1	1	8.66		0.40				4	Fm δ Del	2012
55113	−1 01586			7 09 02	−01 16.2	1	8.62		1.57				2	K5	369
		WLS 720-10 # 7		7 09 08	−10 33.7	1	12.46		0.32		0.22		2		1375
55211	−12 01832			7 09 09	−12 59.2	1	7.67		-0.01		-0.17		2	B9 III	401
55111	+5 01577	HR 2710		7 09 11	+05 44.3	2	6.08	.005	-0.02	.000	-0.05	.000	7	A1 V	15,1078
55213	−17 01828			7 09 11	−17 14.8	1	6.67		-0.04		-0.40		2	B7 III	401
55865	−70 00600	HR 2736	★ AB	7 09 11	−70 25.1	4	3.61	.007	0.92	.005	0.60	.000	30	K0 III	15,1075,1075,2038
55187	−3 01789	LS VI -04 024	★ AB	7 09 17	−04 03.2	1	8.08		-0.07		-0.56		3	B9	490
		RL 136		7 09 18	−01 18.	1	15.45		-0.01		-0.24		2		704
55185	−0 01636	HR 2714	★ A	7 09 19	−00 24.5	10	4.15	.008	-0.01	.006	0.05	.037	34	A2 V	3,15,975,1075,1363*
55052	+24 01558	HR 2706		7 09 24	+24 12.8	5	5.85	.021	0.36	.015	0.07		55	F5 IV	15,130,254,1217,1351
55266	−10 01911			7 09 26	−10 50.7	1	9.98		0.02		-0.50		2	B5 V	388
55397	−34 03400			7 09 27	−34 18.3	1	7.82		-0.12		-0.55		2	B7 II/III	401
55526	−48 02765	HR 2719		7 09 27	−48 51.1	6	5.13	.010	1.25	.011	1.28	.012	20	K2 III	15,244,678,1075,2035,3077
55184	+5 01580	HR 2713		7 09 28	+05 33.6	3	6.15	.005	1.14	.004	0.97	.004	9	K0 III	15,1078,1355
54944	+47 01411			7 09 29	+47 41.6	1	6.76		0.28		0.02		2	A5	1601
54895	+51 01295	HR 2703, UX Lyn		7 09 30	+51 30.8	3	5.49	.018	1.64	.015	1.93	.030	7	M3 III	71,1501,3001
		G 112 - 1		7 09 32	+06 11.9	2	14.30	.005	0.66	.014	-0.12	.056	4		1658,3062
55344	−20 01767	HR 2716		7 09 32	−20 47.9	2	5.83	.004	-0.04	.000	-0.06		6	A0 IV/V	401,2007
		G 107 - 56		7 09 40	+41 25.0	1	14.94		1.36				1	sdK	782
55130	+27 01337	HR 2711	★ AB	7 09 42	+27 18.7	4	6.45	.011	0.50	.020	-0.01	.033	9	F8 V +F8 V	254,292,1381,3030
55290	−5 02014			7 09 42	−05 21.6	1	7.24		1.45		1.65		3	K2	1657
55395	−19 01742			7 09 42	−19 48.6	1	8.36		-0.08		-0.41		2	Ap Si	401
55156	+25 01609			7 09 45	+25 50.0	2	7.07	.000	-0.01	.037	-0.05		5	A0	105,1351
55340	−9 01901	IDS07074S0946	AB	7 09 47	−09 51.0	1	9.05		0.08		-0.24		1	B9 V	388
55229	+15 01502			7 09 48	+15 21.8	1	9.18		1.33				2	K2	882
55369	−11 01837			7 09 53	−11 58.1	2	9.34	.010	-0.01	.005	-0.25	.000	4	B8 V	388,401
55367	−10 01914			7 09 58	−10 52.0	1	9.60		-0.02		-0.19		2	B9 V	388
55444	−19 01746			7 10 00	−19 11.1	1	8.51		0.01		-0.23		2	B9 IV/V	401
	−45 02997			7 10 00	−46 06.8	1	10.74		0.72		0.16		3		803
	−27 03748	LSS 241		7 10 02	−27 38.0	3	9.26	.012	-0.19	.004	-0.87	.005	6	B2 Vp	401,540,976
55417	−7 01802	IDS07076S0803	AB	7 10 03	−08 07.6	2	8.03	.009	-0.04	.000	-0.40	.000	5	B7 V	388,1375
55419	−9 01903			7 10 04	−10 04.2	2	8.88	.020	-0.06	.005	-0.44	.005	4	B7 V	388,401
55496	−22 01749			7 10 04	−22 53.9	2	8.36	.000	0.96	.015	0.48	.019	3	G(II)wp(Ba)	565,3077
55523	−27 03749	IDS07081S2710	A	7 10 05	−27 15.0	1	6.91		-0.14		-0.65		2	B4 V	401
55420	−10 01916	LSS 239		7 10 06	−10 54.2	1	8.63		0.81		-0.10		2	B3 Ia	388
55438	−8 01794	LSS 238		7 10 07	−08 37.4	1	8.36		0.55		0.38		3	F5 II	388
55568	−30 04081	Cr 132 - 2	★	7 10 08	−30 44.2	4	6.09	.013	0.27	.011	0.05	.014	13	A8 III/IV	15,594,2027,8071
55418	−9 01904			7 10 09	−09 23.4	1	9.96		-0.02		-0.21		3	B8 V	388
55441	−11 01841			7 10 09	−12 00.1	1	8.91		0.21		0.14		2	A1 V	388
55522	−25 04191	HR 2718		7 10 09	−25 51.5	2	5.91	.010	-0.17	.000	-0.72		6	B2 IV/V	401,2031
55442	−11 01842			7 10 10	−12 05.8	2	9.54	.010	0.00	.000	-0.64	.005	4	B5	388,401
		G 88 - 12		7 10 12	+24 11.9	1	10.99		1.21		0.98		1		333,1620
55439	−9 01905	IDS07078S0941	AB	7 10 12	−09 45.6	3	8.47	.005	0.09	.007	-0.57	.000	10	B2 Ve	388,1212,2014
55720	−49 02676			7 10 12	−49 21.1	3	7.49	.017	0.71	.009	0.11	.077	7	G8 V	1311,2012,3077
		G 193 - 39		7 10 13	+50 48.9	1	11.52		1.50				2	M3	1746
		G 89 - 4		7 10 14	+13 49.2	1	14.93		1.70		1.57		4		419
55493	−14 01768	LSS 240		7 10 15	−14 24.4	1	8.17		0.58		-0.15		3	A0/1 Ia	8100
		SS Cam		7 10 19	+73 25.2	1	10.11		0.79		0.33		3	G1 III:	588
55490	−8 01798			7 10 22	−08 40.4	1	10.03		0.00		-0.24		3	B9	388
	−21 01804			7 10 22	−21 58.7	1	9.71		-0.09		-0.70		4		1732
55385	+11 01506			7 10 23	+11 43.1	1	8.65		0.04		0.07		2	A0	1375
55595	−27 03761	HR 2724, HN CMa		7 10 23	−27 23.3	1	6.59		0.19				4	A5 IV/V	2035
	+17 01524	G 88 - 13		7 10 24	+17 31.3	1	10.27		0.80		0.36		1		333,1620
		GD 83		7 10 24	+21 39.2	1	15.29		0.19		-0.52		1	DAs	3060
		Steph 633		7 10 24	+40 44.9	1	12.46		1.46		1.27		1	K5	1746
55517	−11 01845			7 10 26	−11 59.5	2	8.40	.005	-0.03	.000	-0.38	.009	4	B7 V	388,1375
55515	−11 01846			7 10 29	−11 12.5	1	9.48		-0.02		-0.31		2	B8 V	388
55538	−15 01695	LSS 245		7 10 29	−15 25.0	2	7.82	.029	-0.04	.014	-0.71	.014	7	B2 Vn(e)	399,401
55383	+16 01417	HR 2717, BQ Gem	★ A	7 10 30	+16 14.7	2	5.07	.047	1.65	.005	1.77	.033	4	M4 IIIab	1003,3053
55663	−31 04174	IDS07086S3149	AB	7 10 30	−31 53.8	1	7.13		1.67				4	M1 III	2012
55514	−7 01808			7 10 32	−07 50.1	1	8.92		-0.03		-0.28		3	B9	388
55561	−12 01842			7 10 36	−12 17.6	3	8.10	.014	-0.12	.004	-0.61	.010	8	B8 II/III	388,401,2014
55719	−40 02987	HR 2727		7 10 36	−40 24.8	3	5.30	.004	0.06	.000	0.08	.005	11	A3p SrEuCr	15,1075,2012
55434	+15 01509			7 10 37	+15 18.4	1	8.87		1.00				3	K0	882
55562	−12 01843			7 10 39	−12 31.2	2	9.04	.020	-0.01	.000	-0.61	.010	4	B3 V	388,401
55718	−36 03421	HR 2726	★ AB	7 10 39	−36 27.5	6	5.94	.018	-0.15	.008	-0.64	.022	17	B4 V	15,401,404,852,1776,2012
55694	−30 04098	Cr 132 - 23	★ AB	7 10 40	−30 59.8	1	8.86		0.56		0.03		2	F5 V	594
55589	−11 01849	HR 2723		7 10 46	−11 09.9	2	5.76	.016	1.50	.004	1.76		6	K0	58,2009
		Cr 135 - 77		7 10 46	−36 15.4	1	11.75		1.26		1.14		1		852

HD	DM	Other Id	N Rem	α_{1950}	δ_{1950}	S	V	σ_V	B–V	σ_{B-V}	U–B	σ_{U-B}	n	Spectrum	References
		WLS 728 55 # 6		7 10 47	+54 48.8	1	10.47		0.32		0.17		2		1375
55742	−36 03426			7 10 47	−36 08.9	1	7.85		0.05		0.08		3	A1 V	852
	−36 03425			7 10 47	−36 09.2	1	8.86		0.24		0.12		3		852
	+28 01343			7 10 50	+28 38.7	1	8.52		0.58		0.20		3	G0	3026
55636	−16 01832			7 10 50	−16 33.9	1	8.72		−0.01		−0.24		2	B8 IV/V	401
55458	+25 01613	G 88 - 14	⋆ A	7 10 51	+25 05.9	3	8.40	.004	0.83	.016	0.38	.018	5	K1 V	333,1003,1620,3016
55128	+57 01052	IDS07067N5727	A	7 10 53	+57 21.9	1	8.37		0.59		0.04		2	F8	1375
		GD 448		7 10 53	+74 06.0	1	14.97		−0.06		−0.75		1	DA	782
55510	+15 01512			7 10 57	+15 27.1	1	8.78		1.63				4	K5	882
55692	−20 01782			7 11 00	−20 29.7	1	8.48		−0.15		−0.71		2	B3 V	401
		WLS 720 10 # 6		7 11 03	+09 43.8	1	12.12		0.54		0.02		2		1375
55606	−1 01603	LS VI -01 018		7 11 03	−01 59.5	2	9.09	.020	−0.04	.000	−0.79	.000	3	B1:V:penn	490,1012
	−25 04215	NGC 2354 - 293		7 11 06	−25 42.2	1	10.57		0.99		0.76		1		876
	−9 01910	LS VI -09 008		7 11 07	−09 12.7	1	9.93		0.22		−0.43		2		490
55892	−46 02977	HR 2740, QW Pup		7 11 08	−46 40.5	6	4.48	.007	0.32	.008	−0.02	.015	25	F0 IV	15,244,1075,2012,3026,8015
55178	+60 01038			7 11 10	+59 51.8	1	7.19		0.93				2	K0	1379
		NGC 2354 - 296		7 11 10	−25 38.0	1	14.35		0.44				1		876
55687	−10 01926			7 11 12	−10 24.2	2	9.35	.005	−0.08	.005	−0.72	.020	4	B2 V	388,401
	−25 04218	NGC 2354 - 291	⋆ AB	7 11 12	−25 39.8	1	10.14		0.17		0.12		1		876
		NGC 2354 - 295		7 11 14	−25 36.7	1	13.87		0.34				1		876
	−25 04222	NGC 2354 - 298		7 11 16	−25 48.9	1	12.06		0.94		0.73		1		876
		WLS 720 30 # 6		7 11 17	+29 59.7	1	11.03		0.55		0.12		2		1375
55762	−22 01756	HR 2730		7 11 17	−22 35.2	2	5.99	.024	1.48	.000	1.69		3	K4 III	2008,3005
55817	−30 04123	Cr 132 - 11	⋆ AB	7 11 17	−30 52.8	1	7.72		−0.08		−0.47		3	B5 V	594
55759	−17 01855			7 11 18	−17 40.7	1	8.65		0.00		−0.26		2	B9 V	401
55736	−8 01805	IDS07089S0845	AB	7 11 19	−08 50.3	1	8.12		0.05		−0.04		3	B9.5V	388
		LP 722 - 1		7 11 20	−13 21.8	2	14.49	.033	1.46	.009	1.20		4		1663,3078
	−25 04225	NGC 2354 - 292		7 11 22	−25 36.1	1	10.35		1.23		1.23		1		876
		NGC 2354 - 294		7 11 22	−25 38.5	1	12.65		2.10				1		876
55579	+24 01576	HR 2722	⋆ A	7 11 24	+24 47.9	1			−0.03		−0.08		1	A1 p SrCr	1079
		NGC 2354 - 297		7 11 26	−25 36.3	1	14.70		1.02				1		876
	−14 01782			7 11 27	−14 09.8	1	9.93		1.75				2	M2	369
293629	+0 01859		V	7 11 30	+00 08.9	1	9.70		1.68				3	K5	369
55578	+28 01346			7 11 31	+28 31.8	1	8.60		0.70		0.20		3	G8 V	3026
56023	−51 02360			7 11 31	−51 07.5	1	7.20		−0.14				4	B2 V	2012
55755	−9 01913	NGC 2353 - 4		7 11 32	−10 08.1	4	9.84	.023	−0.07	.014	−0.24	.030	6	A0p	49,320,388,1714
	−10 01927	NGC 2353 - 7		7 11 32	−10 09.3	2	10.99	.015	0.44	.005	−0.18	.070	6	B9 V	49,1714
55280	+59 01065	HR 2715		7 11 33	+59 43.7	3	5.20	.009	1.08	.010	1.01	.000	12	K2 III	1355,1379,3016
55753	−7 01819			7 11 33	−07 14.3	1	9.10		−0.02		−0.48		2	B5	388
55754	−9 01912			7 11 33	−09 22.1	2	7.83	.020	−0.05	.015	−0.34	.010	5	B9 V	388,401
		NGC 2353 - 12		7 11 34	−10 15.1	1	13.27		0.45		0.08		1		49
56239	−62 00789	HR 2754	⋆ AB	7 11 34	−63 06.3	2	6.01	.005	−0.02	.000			7	A0 IV/V	15,2012
55857	−27 03789	Cr 121 - 33	⋆ V	7 11 35	−27 16.2	4	6.11	.010	−0.24	.017	−0.99	.004	12	B0.5V	15,62,158,2012
	−30 04132	Cr 132 - 29	⋆ AB	7 11 35	−30 56.1	1	9.44		0.00		−0.29		2		594
55649	+20 01743			7 11 37	+20 36.3	1	7.19		1.60		1.97		2	K0	1648
	−30 04133	Cr 132 - 20		7 11 37	−30 53.8	2	8.58	.030	−0.09	.005	−0.51	.010	5		401,594
55776	−9 01914			7 11 38	−10 03.5	1	9.97		−0.01		−0.37		2	B8 V	388
		NGC 2353 - 11		7 11 38	−10 10.7	1	13.13		0.41		0.16		1		49
55777	−10 01928			7 11 38	−10 21.3	1	7.77		0.27		0.11		17	A7 V	388
55621	+25 01618	HR 2725	⋆ A	7 11 39	+24 58.4	3	5.80	.014	1.55	.005	1.81	.014	8	M1 III	15,1003,3077
		NGC 2353 - 141		7 11 39	−10 03.8	1	9.97		0.02		−0.39		3		1714
	−17 01859	RX CMa		7 11 40	−18 04.9	1	7.34		1.10		0.73		1		851
55856	−22 01761	HR 2733	⋆ A	7 11 41	−22 49.2	2	6.36	.002	−0.19	.018	−0.82	.009	6	B2 IV	542,1637
55856	−22 01761	HR 2733	⋆ AB	7 11 41	−22 49.2	3	6.27	.008	−0.14	.006	−0.82		8	B2 IV+G5 IV	15,542,2027
55856	−22 01760	IDS07096S2244	B	7 11 41	−22 49.2	1	9.07		0.77		0.35		2	G5 IV	542
55775	−3 01804	HR 2731	⋆ AB	7 11 42	−03 48.9	2	5.74	.005	1.58	.000	1.88	.000	7	K5 III	15,1256
56073	−48 02793			7 11 42	−48 42.0	1	8.37		0.14		0.16		1	A1 V	1776
55751	+3 01609	HR 2729		7 11 43	+03 11.9	3	5.35	.005	1.19	.010	1.18	.004	8	K2 II	15,1355,1415
56022	−44 03223	HR 2746, OU Pup		7 11 44	−45 05.7	5	4.86	.074	−0.03	.016	−0.07	.005	14	B9 V	15,244,1520,2027,3023
55730	+12 01469	HR 2728		7 11 45	+12 12.2	1	5.62		1.01		0.77		2	G6 III	252
55961	−36 03441			7 11 46	−36 14.8	1	9.96		1.15		1.15		2	K0/2 (III)	852
56145	−53 01280			7 11 46	−53 48.5	1	7.83		0.58				4	G0 V	2012
	+58 01015	IDS07075N5810	A	7 11 47	+58 04.6	1	10.02		0.73		0.30		2		1723
		NGC 2353 - 15		7 11 47	−10 13.3	1	14.71		0.55		0.02		1		49
		NGC 2354 - 299		7 11 47	−25 36.0	1	11.44		0.18		0.02		1		876
	−23 05109			7 11 49	−23 19.6	1	9.87		−0.07		−0.70		1	A0	1732
55811	−9 01918			7 11 50	−09 30.5	1	9.55		−0.05		−0.30		3	B8 V	388
55812	−10 01930			7 11 51	−10 21.8	1	10.01		−0.03		−0.62		2	B8	388
		NGC 2353 - 14		7 11 52	−10 08.0	1	13.95		0.55		0.28		1		49
55854	−12 01860			7 11 52	−12 43.4	2	7.74	.015	−0.03	.000	−0.22	.005	4	B9 III	388,401
55906	−27 03794			7 11 52	−27 12.4	1	8.85		−0.15		−0.73		5	B3 III	1732
55958	−30 04143	Cr 132 - 5	⋆ V	7 11 52	−30 59.8	8	6.57	.012	−0.18	.009	−0.76	.006	28	B2 IV	15,62,540,594,976,1075*
55831	−7 01821			7 11 53	−07 25.9	3	8.44	.008	−0.08	.000	−0.56	.010	8	B8	388,401,2014
55832	−9 01921	HR 2732		7 11 53	−09 51.6	4	5.89	.006	1.53	.004	1.74	.017	28	K3 III	15,388,1415,2035
55727	+18 01538			7 11 56	+18 39.0	1	7.72		1.10		1.09		2	K2	1648
55830	−7 01822			7 11 56	−07 17.3	2	8.80	.010	−0.01	.000	−0.09	.015	4	A0	388,401
55853	−9 01923	NGC 2353 - 17		7 11 59	−10 08.8	2	9.78	.033	−0.04	.009	−0.39	.005	4	B8 Vn	388,1714
		NGC 2353 - 9		7 11 59	−10 12.3	2	12.12	.015	0.04	.005	0.05	.000	2		49,1714
	+68 00467	G 250 - 40		7 12 00	+68 33.0	1	10.17		1.19		1.20		1	K8	3072
		NGC 2353 - 25		7 12 00	−10 08.2	1	11.46		0.06		−0.10		3		1714

Table 1   385

HD	DM	Other Id	N Rem	α_{1950}	δ_{1950}	S	V	σ_V	B–V	σ_{B-V}	U–B	σ_{U-B}	n	Spectrum	References
		NGC 2353 - 54		7 12 00	−10 13.5	1	12.86		0.28		0.15		1	A5 V	1714
		NGC 2353 - 18		7 12 00	−10 13.7	1	10.07		0.76		0.55		2		1714
55852	−7 01823			7 12 01	−07 29.4	1	9.00		0.27		0.11		2	A0	388
		NGC 2353 - 10		7 12 01	−10 03.9	1	12.93		0.83		0.29		1		49
		NGC 2353 - 38		7 12 01	−10 08.6	1	12.32		0.15		0.08		3	A5.5 V	1714
55985	−30 04146	Cr 132 - 4	⋆	7 12 01	−30 15.2	5	6.31	.014	-0.19	.010	-0.82	.014	16	B2 IV-V	62,401,594,2007,1732
56096	−44 03227	HR 2748, L_2 Pup	⋆ A	7 12 01	−44 33.4	6	4.73	.274	1.55	.027	1.25		36	M5 IIIe	15,1520,3076,8015*
		NGC 2353 - 30		7 12 02	−10 11.2	1	11.90		0.23		-0.18		1		1714
		NGC 2353 - 13		7 12 04	−10 12.1	1	13.61		0.37		0.34		1		49
55879	−10 01933	NGC 2353 - 1	⋆	7 12 06	−10 13.7	7	6.01	.010	-0.17	.007	-0.99	.013	40	O9.5II-III	15,49,388,401,1714*
56047	−35 03429			7 12 06	−35 59.8	1	8.41		0.21		0.15		3	A5/7 III/IV	852
55575	+47 01419	HR 2721		7 12 08	+47 19.9	3	5.64	.060	0.57	.005	0.03	.005	6	G0 V	15,1003,3026
		NGC 2353 - 89		7 12 08	−10 09.3	1	13.59		0.55		0.05		2		1714
55902	−11 01853			7 12 08	−12 08.4	1	8.66		0.01		-0.11		2	B9 III	388
56046	−31 04209	Cr 132 - 10		7 12 08	−31 33.4	1	7.59		-0.12		-0.51		3	B6 V	594
		NGC 2353 - 21		7 12 09	−10 02.0	1	11.07		0.29		-0.32		2	B5 V	1714
		NGC 2353 - 138		7 12 09	−10 09.1	1	13.07		0.48		-0.25		3		1714
		NGC 2353 - 41		7 12 09	−10 11.2	1	12.40		0.25		0.09		2		1714
	−9 01926	NGC 2353 - 5		7 12 10	−10 00.9	2	10.05	.014	0.62	.009	0.10	.023	5	B9 IIIp	49,1714
		NGC 2353 - 53		7 12 10	−10 10.6	1	12.68		0.24		0.14		2	A3 V	1714
56708	−72 00577			7 12 10	−72 40.2	1	8.22		1.00		0.63		11	G6 III	1704
		NGC 2353 - 139		7 12 11	−10 09.2	1	13.58		0.52		0.01		2		1714
		NGC 2353 - 26		7 12 11	−10 09.7	1	11.69		0.08		0.01		4		1714
55901	−10 01934	NGC 2353 - 135	⋆ A	7 12 11	−10 12.3	3	8.82	.045	-0.04	.009	-0.53	.021	11	B5 III	388,399,1714
56069	−36 03447			7 12 11	−36 12.5	1	10.84		0.13		0.13		2	B9/A0 (V)	852
		NGC 2353 - 136		7 12 12	−10 09.9	1	12.14		0.30		0.11		1		1714
		NGC 2353 - 137		7 12 12	−10 10.0	1	10.48		0.04		-0.16		1		1714
	−10 01935	NGC 2353 - 16	⋆ B	7 12 12	−10 12.0	2	9.38	.020	0.00	.005	-0.33	.010	4	B8 III-IV	388,1714
		NGC 2353 - 142		7 12 12	−10 13.	1	10.31		0.06		-0.07		4		1714
55619	+45 01409	G 107 - 59		7 12 13	+45 07.9	1	8.67		0.60		0.04		2		1658
		NGC 2353 - 74		7 12 13	−10 07.1	1	13.41		0.36		0.15		4		1714
		NGC 2353 - 49		7 12 13	−10 09.0	1	12.86		0.40		0.09		2		1714
56014	−26 04057	HR 2745, EW CMa	⋆ AB	7 12 13	−26 15.9	10	4.64	.027	-0.17	.010	-0.76	.031	72	B3 IIIe	15,401,681,815,1588*
		NGC 2353 - 65		7 12 14	−10 13.0	1	13.24		0.32		0.11		1		1714
296692				7 12 15	−03 10.	1	10.15		1.72				2		369
		NGC 2353 - 140		7 12 15	−10 09.8	1	15.01		0.59		0.05		1		1714
		NGC 2353 - 31		7 12 15	−10 12.4	1	11.90		0.16		0.10		2		1714
		G 88 - 16		7 12 16	+16 01.2	1	11.49		0.94		0.82		1		333,1620
55927	−8 01813			7 12 16	−08 49.0	1	9.32		0.44		-0.03		1	G0	388
		NGC 2353 - 72		7 12 17	−10 08.4	1	13.44		0.36		0.13		2		1714
		NGC 2353 - 22		7 12 17	−10 08.6	1	11.39		0.44		0.01		4	F2 IV	1714
		NGC 2353 - 86		7 12 17	−10 08.7	1	13.56		0.47		-0.26		1		1714
56044	−27 03806			7 12 17	−28 04.9	1	7.92		-0.16		-0.70		2	B3 V	401
	−31 04215	Cr 132 - 33		7 12 17	−31 27.7	1	10.09		0.40		0.03		4		594
55930	−9 01929	NGC 2353 - 2		7 12 18	−10 07.6	3	9.16	.016	-0.02	.007	-0.29	.004	8	B8 V	49,388,1714
		NGC 2353 - 112		7 12 18	−10 09.7	1	14.70		0.53		0.38		2		1714
		NGC 2353 - 60		7 12 18	−10 11.0	1	13.05		0.29		0.15		3	A4 V	1714
		NGC 2353 - 19		7 12 19	−10 11.8	1	10.55		0.04		-0.23		4	B8 V	1714
		NGC 2353 - 24		7 12 20	−10 09.1	1	11.43		0.03		-0.05		4	B9.5 V	1714
55928	−9 01931			7 12 21	−09 41.3	1	9.59		0.13		0.06		2	A0	388
		NGC 2353 - 77		7 12 21	−10 09.9	1	13.49		0.83		0.41		2		1714
		NGC 2353 - 27		7 12 21	−10 10.3	1	11.65		0.07		0.01		4	B9 V	1714
56013	−19 01767	IDS07102S1949	AB	7 12 21	−19 54.2	2	7.35	.009	-0.17	.005	-0.76	.005	8	B2 III	399,401
		NGC 2353 - 61		7 12 23	−10 11.0	1	13.10		0.28		0.17		3		1714
		NGC 2353 - 42		7 12 23	−10 12.8	1	12.42		0.18		0.14		1	A3 V	1714
	−17 01866			7 12 23	−17 18.1	1	10.33		3.00		5.00		1	C6-7,4	864
	−9 01932	NGC 2353 - 6		7 12 25	−10 01.3	2	10.86	.050	0.28	.005	0.15	.035	2		49,1714
56113	−36 03450			7 12 25	−36 13.5	1	9.92		0.45		-0.01		2	F8	852
55980	−8 01817			7 12 26	−08 46.2	1	9.41		0.53		0.02		1	G0	388
55848	+15 01515			7 12 27	+15 19.6	1	8.42		1.01		0.84		2	K0 III	3040
55981	−10 01937			7 12 27	−10 54.0	1	9.84		0.18		0.08		2	B9	388
56142	−36 03451			7 12 27	−36 20.4	1	7.48		0.64		0.28		3	F6/7 V	852
		NGC 2353 - 8		7 12 28	−10 04.8	2	11.78	.010	1.35	.015	0.85	.095	2		49,1714
		RL 137		7 12 30	+00 58.	1	12.22		-0.04		-0.67		2		704
		Cr 132 - 24		7 12 30	−31 05.	1	9.07		0.68		0.23		2		594
		Cr 132 - 35		7 12 30	−31 05.	1	10.63		0.05		0.01		2		594
56007	−8 01818			7 12 31	−08 58.6	1	9.42		0.02		-0.19		2	B8 V	388
		NGC 2353 - 20		7 12 31	−10 15.8	1	10.87		0.01		-0.25		1	B8 IV	1714
56012	−15 01720	IDS07103S1518	AB	7 12 32	−15 23.3	1	7.27		0.43		-0.04		3	F3 V	3024
56012	−15 01720	IDS07103S1518	C	7 12 32	−15 23.3	1	9.83		0.62		0.08		3		3024
55847	+22 01620	IDS07096N2208	A	7 12 34	+22 03.3	1	7.03		1.59		1.95		2	K5 III	1648
56006	−7 01829			7 12 34	−07 35.5	2	7.69	.010	-0.09	.000	-0.50	.005	4	B8	388,401
56008	−9 01936			7 12 35	−09 17.0	1	9.78		-0.02		-0.30		3	B8 V	388
56265	−48 02803			7 12 35	−48 12.8	1	7.45		0.33		0.15		2	F0 IV	1776
56009	−9 01935	NGC 2353 - 3		7 12 36	−10 04.9	2	9.44	.015	0.98	.005	0.69	.005	2	K0	49,1714
56094	−23 05137	LSS 267		7 12 36	−23 24.1	3	7.28	.010	-0.18	.003	-0.94	.007	9	B2 IV/V	540,976,1732
57336	−79 00243			7 12 36	−79 20.8	1	7.82		0.16		0.18		2	A0 IV	1776
56039	−11 01858			7 12 39	−11 46.9	3	8.28	.012	-0.08	.004	-0.45	.000	8	B5 V	388,401,2014
56162	−31 04224	Cr 132 - 12	⋆ AB	7 12 40	−31 24.3	1	7.80		0.07		0.06		3	A1 V	594
56038	−10 01939			7 12 41	−10 25.9	1	10.13		0.07		-0.18		3	B9 V	388

HD	DM	Other Id	N Rem	α_{1950}	δ_{1950}	S	V	σ_V	B–V	σ_{B-V}	U–B	σ_{U-B}	n	Spectrum	References
56161	−30 04164	Cr 132 - 6	V	7 12 41	−30 34.1	3	6.94	.024	0.76	.019	0.36		12	G5 IV	594,1075,2012
56187	−36 03456			7 12 41	−36 44.1	1	9.72		0.24		0.06		3	A8/9 V	852
55549	+60 01039			7 12 42	+60 00.4	1	7.83		0.66				3	F5	1379
56533	−63 00703			7 12 42	−63 16.0	4	9.08	.017	1.26	.017	1.25		12	K5 V	158,912,1705,2034
56213	−35 03442			7 12 44	−36 04.8	1	9.08		−0.06		−0.38		4	B8 V	852
56003	+0 01871	HR 2744	⋆ AB	7 12 46	−00 04.4	3	6.40	.005	0.90	.000	0.57	.000	14	G5 III	15,1415,2012
56139	−26 04073	HR 2749, ω CMa	9	7 12 47	−26 41.1	9	3.83	.053	−0.15	.015	−0.71	.021	87	B2 IV/Ve	3,15,401,681,815,1637*
65322	−88 00068			7 12 47	−88 41.0	1	7.38		0.01		−0.32		5	B8 IV	1628
56111	−14 01789			7 12 48	−14 44.8	1	9.75		0.03		−0.46		2	B5 V	401
	−20 01805	LSS 271		7 12 48	−20 31.9	1	10.43		0.23		−0.74		4	B2 IIIne2	1737
55870	+28 01350	HR 2738		7 12 50	+27 59.2	2	5.72	.005	1.60	.019	1.95	.033	4	M1 IIIa	1375,3001
56160	−26 04074	HR 2750		7 12 50	−26 57.0	2	5.58	.000	1.22	.000	1.29		5	K3 III	2035,3005
56238	−35 03443			7 12 51	−35 42.3	1	9.31		0.07		0.05		2	A0 V	852
55946	+14 01615			7 12 52	+13 57.2	1	8.15		1.16				2	K0	882
	−15 01724	LSS 270		7 12 52	−15 15.1	1	9.93		0.33		−0.70		2		1737
56260	−37 03408			7 12 52	−37 14.7	1	9.13		0.49		−0.03		3	F5 V	852
56085	−9 01941			7 12 54	−09 22.6	1	10.30		0.13		0.06		2	B9	388
56086	−10 01940			7 12 56	−10 58.6	1	10.13		0.03		−0.24		2	B9	388
56211	−28 04057			7 12 56	−28 52.6	2	7.40	.005	−0.18	.000	−0.73	.010	6	B3 Vn	158,401
56031	+8 01712	HR 2747		7 12 57	+08 04.0	4	5.82	.013	1.59	.009	1.73	.026	14	M4-IIIab	15,1061,1355,3001
56237	−35 03446			7 12 57	−35 09.6	1	8.68		−0.01		−0.08		2	A0 V	401
56284	−37 03411			7 13 01	−37 09.4	3	7.22	.000	−0.17	.006	−0.75	.007	7	B2.5V	401,852,1732
55999	+15 01520			7 13 02	+14 53.9	3	8.20	.016	1.47	.022	1.76	.005	5	M0 III	882,1648,3077
		G 87 - 32		7 13 03	+33 14.7	2	14.48	.000	1.76	.008	0.93	.135	7		203,906
56283	−36 03460			7 13 03	−36 16.3	1	8.77		0.01		−0.07		2	B9 V	852
56032	+7 01655			7 13 04	+07 41.1	1	8.99		−0.10		−0.34		2	F5 V	1375
		RL 138		7 13 06	+00 19.	1	12.36		−0.09		−0.42		2		704
56157	−13 01901			7 13 06	−13 42.7	1	8.41		1.69				2	M2 III	369
56132	−7 01835			7 13 07	−07 23.5	1	9.21		0.35		0.27		2	B9	388
56133	−10 01944			7 13 07	−10 19.1	1	9.81		0.05		−0.06		2	B9 V	388
56346	−37 03412			7 13 10	−37 09.9	1	8.78		0.01		−0.03		2	A0 V	852
56314	−36 03461			7 13 11	−36 54.8	1	8.84		0.04		0.06		3	A1 V	852
56438	−46 02999			7 13 11	−47 02.8	1	8.10		1.04		0.87		2	K1 III	536
56378	−40 03020			7 13 12	−41 04.8	1	7.55		−0.07		−0.28		2	B9 III	401
56183	−13 01902			7 13 13	−13 28.6	2	8.78	.000	−0.11	.010	−0.62	.005	4	B3 V	388,401
	+27 01348	G 109 - 55		7 13 14	+27 14.0	2	10.84	.003	1.60	.021	1.28	.009	5	M0	801,7008
56456	−48 02807	HR 2762	⋆ A	7 13 15	−48 11.0	7	4.75	.007	−0.10	.003	−0.29	.006	21	B8/9 V	15,244,1075,1776,2012*
56345	−36 03463			7 13 17	−36 45.0	1	11.20		0.69		0.21		2	A8/9 V	852
56410	−41 02906	HR 2759		7 13 19	−41 20.3	1	5.94		−0.16				4	B4 III/IV	2031
		LSS 274		7 13 20	−19 48.5	1	10.62		0.41		−0.63		2	B3 Ib	1737
56455	−46 03000	HR 2761, PR Pup		7 13 20	−46 45.7	3	5.71	.007	−0.11	.005	−0.46	.005	10	Ap Si	244,401,2006
56207	−10 01945	HR 2752		7 13 21	−10 29.7	2	5.95	.007	1.17	.003	1.18		6	K0	58,2007
		RL 139		7 13 24	+02 15.	1	13.83		0.00		−0.40		2		704
56279	−19 01780			7 13 24	−20 02.4	1	8.02		−0.08		−0.34		2	B8 V	401
56343	−31 04243	Cr 132 - 27		7 13 24	−31 50.1	1	9.24		−0.06		−0.32		3	B9 V	594
56376	−36 03464			7 13 24	−36 08.2	2	7.40	.020	−0.09	.005	−0.48	.000	5	B6 V	401,852
56342	−30 04184	Cr 132 - 1	⋆	7 13 25	−30 35.9	4	5.34	.019	−0.17	.009	−0.64	.009	13	B3 V	15,401,594,2027
56375	−30 04244	Cr 132 - 21		7 13 26	−31 51.0	1	8.66		1.15		1.09		2	K2 II/III	594
56374	−30 04186	Cr 132 - 30	AB	7 13 27	−30 24.5	1	9.53		−0.01		−0.11		3	B9 V	594
56232	−10 01947			7 13 32	−10 53.5	1	9.71		0.02		−0.12		2	B9 V	388
56254	−10 01949			7 13 33	−10 25.4	1	10.38		−0.03		−0.45		2	B9	388
56274	−12 01871			7 13 33	−12 57.8	3	7.75	.012	0.61	.009	−0.02	.004	9	G5 V	22,1003,2012
56257	−13 01904			7 13 33	−13 11.9	1	8.52		0.12		−0.25		2	B9	388
56256	−11 01863			7 13 34	−11 10.2	1	7.86		1.55				29	K0	6011
	−12 01870			7 13 34	−12 57.1	1	10.73		2.10				2	M3	369
		G 107 - 60		7 13 35	+44 09.6	1	13.12		0.80		0.27		1		1658
56373	−26 04090			7 13 37	−26 46.1	1	8.97		−0.08		−0.31		2	B9 V	401
56273	−12 01872			7 13 38	−12 27.2	2	7.92	.005	−0.08	.010	−0.41	.005	4	Ap Si	388,401
55866	+52 01188	HR 2737		7 13 39	+52 13.3	1	5.92		1.26				2	gK1	71
	−24 05089			7 13 39	−24 28.6	2	9.49	.006	−0.03	.011	−0.72	.004	4	B2 III	540,976
	−3 01821			7 13 40	−03 34.8	1	9.01		1.01		0.85		3	K2 V	3072
56310	−15 01732			7 13 41	−16 08.8	4	6.86	.007	−0.17	.014	−0.90	.008	9	B1/2 III	399,401,532,1732
56341	−23 05173	HR 2755		7 13 42	−23 39.1	1	6.33		0.00		−0.04		4	A0 V	1637
56453	−36 03469			7 13 42	−36 28.3	1	7.69		1.59		1.91		2	K5/M0 III	852
56306	−9 01947	IDS07113S0906	AB	7 13 43	−09 10.8	2	8.22	.005	−0.08	.005	−0.42		7	B7 V+B9p Si	388,2014
56737	−60 00803			7 13 44	−60 58.9	2	7.16	.015	0.43	.005	−0.08		6	F3 V	2012,3037
56334	−13 01905			7 13 45	−13 51.2	1	8.30		−0.03		−0.25		2	B8/9 III	388
56308	−11 01864			7 13 47	−11 40.0	1	9.35		0.04		−0.04		2	A0 V	388
56531	−41 02915	IDS07122S4155	AB	7 13 47	−42 00.0	1	6.99		−0.06		−0.30		2	B9 V	401
	−8 01824	LSS 277		7 13 48	−08 25.9	1	9.78		0.01		−0.78		1		490
56475	−35 03463			7 13 48	−35 49.8	1	9.71		0.31		0.12		2	A8 V	852
56474	−35 03464			7 13 50	−35 36.6	2	8.69	.020	−0.01	.000	−0.21	.010	4	B9 V	401,852
56200	+16 01433			7 13 52	+16 14.2	1	6.76		0.40		−0.08		2	F4 II	3016
		G 112 - 3		7 13 52	−02 00.0	1	10.71		0.99		0.76		1		3062
56366	−13 01908			7 13 53	−13 38.3	1	9.03		−0.10		−0.50		2	Ap Si	388
56124	+33 01497			7 13 55	+33 11.1	1	6.94		0.62		0.14		2	G0	1733
56176	+26 01508			7 13 58	+26 46.9	2	6.40	.005	0.88	.010	0.50	.005	4	G7 IV	252,3077
56405	−15 01734	HR 2758		7 13 59	−15 29.7	5	5.45	.009	0.08	.006	0.05	.010	14	A1 V	158,196,401,532,2007
56430	−17 01884			7 13 59	−17 12.4	1	7.83		−0.03		−0.36		2	B7/8 II	401
		AA45,405 S294 # 1		7 14 00	−09 19.4	1	10.75		0.00		−0.16		1		797

Table 1   387

HD	DM	Other Id	N	Rem	α_{1950}	δ_{1950}	S	V	σ_V	B–V	σ_{B-V}	U–B	σ_{U-B}	n	Spectrum	References
	+63 00700			AB	7 14 02	+63 36.9	1	9.55		0.52		0.04		2		1723
56401	−10 01954				7 14 02	−11 00.6	1	9.07		0.46				25	F5	6011
	−5 02046				7 14 04	−05 12.2	1	9.40		1.67				2	M2	369
		AA45,405 S294 # 13			7 14 04	−09 21.7	1	13.11		0.50		0.02		1		797
		fld III # 31			7 14 04	−10 09.3	1	13.28		0.31		0.19		2		1755
56557	−35 03466				7 14 04	−35 43.5	2	8.75	.020	−0.01	.000	−0.11	.015	4	B9 V	401,852
		AA45,405 S294 # 2			7 14 05	−09 18.2	1	13.25		0.22		0.15		1		797
	+73 00368				7 14 06	+73 38.8	1	9.48		1.02		0.81		2	K0	1733
		AA45,405 S294 # 9			7 14 06	−09 18.2	1	13.13		0.44		0.03		1		797
		fld III # 32			7 14 07	−10 08.6	1	13.05		0.58		0.10		2		1755
56662	−48 02814				7 14 07	−48 08.6	1	7.65		0.61				4	G2 V	2012
		AA45,405 S294 # 5			7 14 10	−09 19.6	1	14.39		0.63		0.23		1		797
		AA45,405 S294 # 4			7 14 10	−09 20.0	1	13.99		1.03		−0.03		2		797
56554	−30 04205	Cr 132 - 8			7 14 10	−30 20.7	3	7.14	.009	−0.18	.004	−0.81	.042	10	B2 III/IV	164,594,1732
56705	−52 01123	HR 2767			7 14 10	−52 24.7	2	5.96	.005	1.10	.000			7	K1 III	15,2029
56224	+26 01510				7 14 11	+26 27.3	3	7.26	.004	1.17	.008	1.16	.007	8	K1 III	1003,1775,3077
		AA45,405 S294 # 10			7 14 11	−09 15.2	1	11.41		0.52		0.02		1		797
		fld III # 48			7 14 11	−09 59.0	1	9.99		0.30		0.10		2		1755
		AA45,405 S294 # 11			7 14 12	−09 18.4	1	13.72		1.03		0.86		2		797
		AA45,405 S294 # 12			7 14 12	−09 19.1	1	14.00		0.47		0.21		2		797
		AA45,405 S294 # 3			7 14 12	−09 20.4	2	13.29	.019	0.52	.024	0.03	.005	3		432,797
56361	+1 01757	IDS07116N0109		AB	7 14 13	+01 04.2	1	8.58		0.81		0.46		2	G9 V +G9 V	3030
		fld III # 30			7 14 13	−10 10.4	1	11.91		0.55		0.13		2		1755
56555	−31 04262	Cr 132 - 34			7 14 13	−31 40.1	1	10.12		−0.02		−0.17		2	B9 V	594
		fld III # 43			7 14 16	−10 04.9	1	13.01		0.48		0.00		2		1755
56582	−31 04265	Cr 132 - 22			7 14 16	−31 11.0	1	8.81		−0.03		−0.14		3	B9 V	594
		fld III # 49			7 14 17	−09 56.3	1	14.51		0.45		0.18		2		1755
		fld III # 44			7 14 17	−10 04.3	1	13.60		1.11		1.14		2		1755
56450	−11 01867	RY CMa			7 14 17	−11 23.8	2	7.75	.040	0.68	.025			2	K0	688,6011
56171	+39 01908				7 14 18	+38 58.0	2	6.48	.000	1.00	.005	0.80	.005	6	K0 III	833,1501
56640	−40 03035				7 14 19	−40 56.6	3	9.08	.006	0.66	.004	0.09		9	G3/5 V	158,1705,2012
	−80 00217				7 14 19	−80 35.3	1	8.43		1.03		0.82		3	G0	1704
56624	−35 03469	IDS07126S3600		AB	7 14 20	−36 05.2	1	9.59		0.11		0.00		2	A0 V	852
56397	+1 01759				7 14 22	+01 46.0	1	8.96		−0.04		−0.18		1	B9	289
		fld III # 47			7 14 22	−10 00.8	1	13.57		0.69		0.36		2		1755
		fld III # 37			7 14 22	−10 04.1	1	13.87		0.65		0.29		2		1755
		GD 84			7 14 23	+45 53.4	1	15.19		0.08		−0.79		1	DC	3060
	−40 03037				7 14 23	−41 01.1	1	8.15		0.49		−0.02		2		633
		fld III # 45			7 14 24	−10 02.1	1	12.12		0.22		−0.36		2		1755
56501	−13 01913				7 14 24	−13 21.9	2	7.73	.010	−0.04	.000	−0.37	.010	4	B5 III	388,401
		fld III # 42			7 14 26	−10 04.2	1	13.38		0.49		0.03		2		1755
		fld III # 50			7 14 29	−09 54.9	1	11.62		0.43		0.02		2		1755
56525	−13 01914				7 14 29	−13 35.5	1	7.20		0.05		0.01		2	A0 V	388
		fld III # 29			7 14 30	−10 12.7	1	12.22		0.48		0.01		2		1755
56577	−23 05189	HR 2764		⋆ A	7 14 30	−23 13.5	6	4.80	.024	1.70	.012	1.87	.015	25	K3 Ib	15,369,1024,3052,8015,8100
56577	−23 05189	IDS07124S2308		AB	7 14 30	−23 13.5	2	4.49	.005	1.14	.005	0.54		8	K3 Ib	1075,2012
56579	−23 05190				7 14 30	−23 44.2	1	7.57		−0.18		−0.78		3	B3 V	1732
56494	−6 02030				7 14 31	−07 05.9	1	9.33		−0.04		−0.34		2	B9	388
		fld III # 46			7 14 31	−10 01.6	1	13.23		0.56		0.05		2		1755
56639	−36 03478	IDS07127S3617		AB	7 14 31	−36 22.0	1	9.29		0.26		0.08		2	A2/3 IV	852
56495	−7 01851				7 14 32	−07 26.2	2	7.59	.010	0.34	.000	0.06	.001	7	A3p	388,1202
		fld III # 51			7 14 32	−09 55.4	1	12.42		1.51		1.26		2		1755
56638	−31 04275	Cr 132 - 19			7 14 32	−31 11.6	1	8.50		−0.02		−0.21		3	A0 IV	594
		fld III # 41			7 14 33	−10 04.0	1	13.26		0.49		0.00		2		1755
56221	+41 01630	HR 2753			7 14 34	+40 58.5	2	5.87	.000	0.17	.000	0.12	.005	4	A5 Vn	985,1733
56658	−36 03480				7 14 34	−36 39.7	1	8.91		0.22		0.12		2	A5 V	852
56618	−27 03852	Cr 121 - 44		⋆	7 14 35	−27 47.5	9	4.65	.015	1.60	.005	1.88	.017	46	M2 III	3,15,62,418,1024,1075*
		fld III # 28			7 14 36	−10 13.2	1	12.38		0.68		0.14		2		1755
56446	+6 01594	HR 2760			7 14 37	+06 46.3	3	6.64	.005	−0.12	.005	−0.40	.005	8	B8 III	15,1079,1415
	−1 01633				7 14 37	−01 46.9	1	9.55		1.65				2	M2	369
		fld III # 40			7 14 38	−10 03.9	1	13.18		0.61		0.20		2		1755
56680	−36 03481				7 14 40	−36 52.2	1	8.98		0.47		0.01		3	F5 V	852
		fld III # 39			7 14 42	−10 04.0	1	13.89		0.54		0.11		2		1755
56574	−12 01882				7 14 42	−12 31.0	1	9.46		0.02		−0.11		2	B9 III	388
56679	−35 03476				7 14 42	−35 48.1	1	9.81		0.23		0.17		2	A1 III/IV	852
		G 107 - 61			7 14 44	+39 22.0	1	10.30		1.51		1.20		2	M1	7010
56169	+49 01612	HR 2751			7 14 44	+49 33.4	5	5.04	.012	0.08	.008	0.11	.049	23	A4 IIIn	15,1363,3023,6001,8015
		fld III # 27			7 14 44	−10 13.1	1	11.22		0.11		0.13		2		1755
56657	−30 04220	Cr 132 - 25		⋆ AB	7 14 44	−30 37.2	1	9.08		0.10		0.09		3	A1 V	594
56593	−11 01874	IDS07124S1151		AB	7 14 47	−11 56.6	1	6.72		0.43		−0.06		2	F5	3024
56593	−11 01874	IDS07124S1151	C		7 14 47	−11 56.6	1	9.65		0.88		0.48		3		3024
		G 112 - 5			7 14 48	−03 06.4	1	12.58		1.41		1.07		1		3062
56598	−15 01739	NGC 2360 - 69			7 14 48	−15 27.6	1	9.78		0.20				4	A1 III/IV	2042
56733	−38 03288	HR 2769			7 14 48	−38 13.7	4	5.80	.014	−0.13	.008	−0.59	.005	13	B4 V	15,158,401,2012
		fld III # 33			7 14 49	−10 09.4	1	9.68		1.29		1.39		2		1755
56597	−13 01919	IDS07125S1349		AB	7 14 49	−13 54.0	1	7.75		−0.02		−0.48		2	B3 III	388,401
56813	−46 03023	HR 2771			7 14 49	−46 41.1	6	5.65	.009	1.44	.011	1.67		23	K4 III	15,1075,1520,2013*
		fld III # 52			7 14 50	−09 56.6	1	13.30		0.34		0.13		2		1755
56386	+31 01529	HR 2757			7 14 52	+31 02.9	3	6.24		−0.03	.007	−0.10	.006	6	A0 Vn	252,1022,1079
	−26 04122				7 14 52	−27 04.0	1	9.86		−0.06		−0.69		4	B5	1732

HD	DM	Other Id	N Rem	α_{1950}	δ_{1950}	S	V	σ_V	B–V	σ_{B-V}	U–B	σ_{U-B}	n	Spectrum	References
	−10 01965			7 14 54	−10 12.6	1	10.92		0.27		0.20		2		1755
		fld III # 35		7 14 55	−10 07.2	1	12.16		0.42		0.02		2		1755
57697	−77 00302			7 14 55	−77 40.4	1	9.70		0.00		−0.65		1	B2 II/III	55
	−2 02030			7 14 56	−02 28.2	1	8.71		1.71				2	M2	369
		fld III # 34		7 14 56	−10 07.5	1	12.16		0.43		0.02		2		1755
56909	−54 01231			7 14 56	−54 52.3	1	7.37		1.22		1.36		7	K2 III	1673
		fld III # 36		7 14 57	−10 07.0	1	14.49		0.57		0.61		2		1755
		fld III # 3		7 15 00	−09 54.1	1	11.95		0.66		0.31		2		1755
		fld III # 53		7 15 00	−09 58.4	1	12.44		1.24		0.80		2		1755
		fld III # 2		7 15 01	−09 53.4	1	10.33		0.47		0.45		2		1755
		NGC 2360 - 79		7 15 01	−15 33.2	1	11.29		1.02		0.68		2		3018
56731	−30 04234	Cr 132 - 3	★ A	7 15 01	−30 48.4	4	6.31	.014	0.20	.000	0.17	.013	12	A9 II	146,164,355,594
	−30 04233	Cr 132 - 14	★ B	7 15 01	−30 49.0	2	7.99	.025	0.29	.005	0.07	.005	5		146,594
		fld III # 1		7 15 03	−09 51.6	1	10.14		1.16		0.94		2		1755
		NGC 2360 - 66		7 15 03	−15 29.0	2	11.19	.048	0.95	.013	0.54		6		2042,3018
56779	−36 03485	HR 2770		7 15 03	−36 30.1	4	5.02	.003	−0.18	.005	−0.69	.008	14	B3 V	15,852,1075,2012
		fld III # 57		7 15 04	−10 05.2	1	15.76		−0.08		−0.87		2		1755
56614	−6 02032	HR 2765		7 15 05	−06 35.4	4	6.29	.005	1.62	.003	1.96	.000	18	K2	15,1415,2013,2029
		NGC 2360 - 78		7 15 06	−15 32.6	1	11.16		0.05		0.06		1		3018
		fld III # 25		7 15 07	−10 12.3	1	12.91		0.41		0.06		2		1755
56654	−12 01886			7 15 07	−12 42.2	1	9.50		0.03		−0.35		3	B8 Ib/II	388
56676	−13 01920			7 15 07	−13 49.5	1	8.68		1.68				2	M2/3 III	369
	−20 01831			7 15 07	−20 28.3	2	10.46	.010	0.05	.000	−0.67	.006	6	B2 III	540,976
		NGC 2360 - 62		7 15 08	−15 28.0	2	11.23	.034	1.05	.059	0.58		7		2042,3018
		NGC 2360 - 77		7 15 08	−15 31.1	1	12.53		0.34		0.19		6		3018
56752	−32 03847			7 15 08	−32 57.3	1	7.80		−0.03		−0.16		2	B9 V	401
		fld III # 54		7 15 09	−09 58.8	1	12.95		0.58		0.04		2		1755
		NGC 2360 - 61		7 15 10	−15 26.9	1	13.10		0.34				2		2042
		LTT 2785		7 15 11	−13 54.7	2	12.10	.005	1.44	.007			5		158,1705
56778	−30 04237	Cr 132 - 18		7 15 11	−30 19.4	1	8.47		0.97		0.73		2	G8 III+(G)	594
		NGC 2360 - 84		7 15 12	−15 34.5	2	11.55	.024	0.46	.034	0.25		3		2042,3018
56537	+16 01443	HR 2763	★ AB	7 15 13	+16 37.9	14	3.58	.007	0.11	.007	0.10	.016	66	A3 V	1,3,15,125,934,1006*
56670	−9 01960			7 15 13	−09 19.5	1	9.56		0.15		−0.64		1	B0.5Ve	388
		NGC 2360 - 76		7 15 13	−15 30.8	1	12.70		0.38		0.07		18		3018
		NGC 2360 - 87		7 15 13	−15 31.9	1	13.40		0.41		0.12		2		3018
		fld III # 56		7 15 14	−10 02.5	1	13.31		0.43		0.14		2		1755
	−9 01961			7 15 15	−10 04.9	1	10.08		1.27		1.47		2		1755
		NGC 2360 - 170		7 15 15	−15 31.2	1	13.78		0.51		0.24		3		3018
		fld III # 24		7 15 16	−10 18.0	1	13.19		0.45		0.20		2		1755
		NGC 2360 - 99		7 15 16	−15 35.5	1	12.52		0.40		0.12		1		3018
		NGC 2360 - 88		7 15 17	−15 31.8	1	12.83		0.34		0.13		2		3018
		NGC 2360 - 85		7 15 17	−15 33.5	2	11.46	.020	0.99	.015	0.67	.002	4		304,3018
		RL 141		7 15 18	−02 32.	1	13.48		−0.05		−0.06		2		704
		RL 140		7 15 18	−02 34.	1	13.66		0.03		−0.05		2		704
		WLS 720-10 # 10		7 15 18	−10 17.2	2	11.58	.019	0.06	.005	0.09	.028	4		1375,1755
		NGC 2360 - 56		7 15 18	−15 29.1	1	12.16		0.29				5		2042
		NGC 2360 - 89		7 15 19	−15 32.1	2	11.06	.015	0.99	.010	0.69	.005	4		304,3018
		NGC 2360 - 86		7 15 19	−15 33.1	3	10.80	.024	1.02	.013	0.70	.013	7		304,305,3018
	−20 01834	LSS 290		7 15 19	−20 58.5	1	9.89		0.15		−0.57		3	B6 Iab	540
56856	−36 03487	IDS07136S3655	B	7 15 19	−37 01.4	1	8.07		0.02		−0.10		2	B9/A0 (V)	852
56652	−1 01644	LS VI -01 019		7 15 20	−01 53.3	1	8.27		0.36		0.37		1	F0	490
		NGC 2360 - 90		7 15 20	−15 32.0	1	11.46		0.30		0.20		3		3018
		NGC 2360 - 91		7 15 20	−15 32.4	1	12.90		0.38		0.10		3		3018
		NGC 2360 - 92		7 15 21	−15 32.4	2	11.32	.020	0.85	.010	0.51	.002	4		304,3018
		NGC 2360 - 93		7 15 21	−15 33.0	2	10.35	.010	1.16	.000	1.01	.005	4		304,3018
56747	−18 01768			7 15 21	−18 42.2	1	7.84		1.51				2	K3 III	369
56513	+27 01356	G 88 - 20		7 15 22	+27 20.8	2	8.04	.009	0.63	.005	0.05	.005	4	G2 V	333,1620,3026
		NGC 2360 - 94		7 15 22	−15 33.4	1	13.33		0.50		0.05		3		3018
		NGC 2360 - 96		7 15 22	−15 33.9	2	11.24	.070	0.77	.007	0.42	.003	4		304,3018
		NGC 2360 - 55		7 15 23	−15 28.4	1	12.90		0.00				8		2042
		NGC 2360 - 53		7 15 23	−15 30.4	1	13.57		0.43		0.05		2		3018
		NGC 2360 - 52		7 15 23	−15 31.1	3	11.27	.025	0.98	.004	0.68	.015	8		304,305,3018
		NGC 2360 - 98		7 15 23	−15 35.6	1	12.61		0.41		0.17		2		3018
	−29 04173			7 15 23	−29 57.9	1	10.97		0.50		0.08		2		1730
56855	−36 03489	HR 2773	★ A	7 15 23	−37 00.4	6	2.71	.014	1.62	.015	1.24	.006	29	K3 Ib	3,15,852,1075,2012,8015
		fld III # 4		7 15 24	−09 55.5	1	12.85		0.51		0.12		2		1755
		fld III # 22		7 15 25	−10 14.5	1	13.13		0.43		−0.02		2		1755
		NGC 2360 - 48		7 15 25	−15 33.6	1	12.76		0.33		0.12		1		3018
		NGC 2360 - 97		7 15 25	−15 35.5	1	13.46		0.42		0.14		1		3018
56854	−36 03488			7 15 25	−36 45.0	1	8.35		0.00		−0.11		2	A0 V	852
		fld III # 55		7 15 26	−09 59.1	1	10.79		1.48		1.61		2		1755
		NGC 2360 - 191		7 15 26	−15 30.5	1	14.88		0.75		0.09		1		305
		NGC 2360 - 169		7 15 26	−15 30.7	1	14.12		0.59		0.05		2		3018
		NGC 2360 - 51		7 15 27	−15 31.6	2	11.20	.020	1.10	.000	0.85	.015	5		304,3018
		NGC 2360 - 50		7 15 27	−15 32.1	2	11.07	.015	1.02	.010	0.72	.005	5		304,3018
		NGC 2360 - 47		7 15 27	−15 33.6	1	11.90		0.33		0.16		1		3018
		NGC 2360 - 116		7 15 27	−15 34.1	1	12.23		0.35		0.21		2		3018
56482	+38 01731			7 15 28	+38 45.9	1	6.91		0.96		0.61		2	G8 III	105
		NGC 2360 - 54		7 15 28	−15 29.1	1	12.84		0.96				4		2042
		NGC 2360 - 46		7 15 28	−15 33.6	1	13.02		1.24		1.12		1		3018

Table 1

389

HD	DM	Other Id	N	Rem	α_{1950}	δ_{1950}	S	V	σ_V	B–V	σ_{B-V}	U–B	σ_{U-B}	n	Spectrum	References
		NGC 2360 - 113			7 15 28	−15 35.8	1	11.74		0.36		0.20		3		3018
		NGC 2360 - 180			7 15 29	−15 31.5	1	17.24		1.08				1		3018
		NGC 2360 - 177			7 15 29	−15 32.1	1	15.96		0.66				1		3018
		NGC 2360 - 45			7 15 29	−15 33.4	1	13.30		0.40		0.09		2		3018
		NGC 2360 - 168			7 15 29	−15 33.4	1	14.37		0.57		0.10		2		3018
		NGC 2360 - 117			7 15 29	−15 34.0	1	13.12		0.38		0.12		2		3018
		NGC 2360 - 171			7 15 29	−15 34.5	1	13.91		0.59		0.06		1		3018
	−9 01963				7 15 30	−09 54.2	1	11.28		0.42		0.03		2		1755
		NGC 2360 - 173			7 15 30	−15 32.	1	14.30		0.61		0.06		2		3018
		NGC 2360 - 118			7 15 30	−15 34.1	1	12.62		0.35		0.17		1		3018
		NGC 2360 - 39			7 15 31	−15 30.9	2	12.22	.005	0.40	.015	0.15	.007	3		305,3018
		NGC 2360 - 43			7 15 31	−15 32.2	1	11.39		0.32		0.19		2		3018
		NGC 2360 - 44			7 15 31	−15 32.7	2	10.73	.005	1.01	.030	0.71	.015	5		304,3018
		NGC 2360 - 114			7 15 31	−15 35.4	1	11.64		0.43		0.22		3		3018
56808	−24 05139				7 15 31	−24 43.5	1	8.42		-0.16		-0.91		4	B1/2 III	1732
		fld III # 15			7 15 32	−10 04.5	1	12.21		0.27		0.10		2		1755
		fld III # 16			7 15 32	−10 05.0	1	13.83		0.54		0.40		2		1755
		NGC 2360 - 37			7 15 32	−15 30.3	2	11.33	.010	1.01	.005	0.69	.012	4		304,3018
		NGC 2360 - 38			7 15 32	−15 30.8	1	13.31		0.40		0.12		2		3018
56904	−36 03491				7 15 32	−36 49.4	1	10.13		0.26		0.15		2	A8/9 V	852
		fld III # 6			7 15 33	−09 53.9	1	14.35		0.32		0.14		2		1755
		NGC 2360 - 165			7 15 33	−15 30.6	1	13.80		0.67		0.13		1		3018
		NGC 2360 - 40			7 15 33	−15 31.2	2	12.04	.019	0.40	.015	0.13	.010	3		305,3018
		NGC 2360 - 41			7 15 33	−15 31.7	2	11.10	.029	0.23	.024	0.09	.039	3		305,3018
		NGC 2360 - 42			7 15 33	−15 32.4	1	12.00		0.43		0.16		3		3018
		NGC 2360 - 119			7 15 33	−15 34.2	2	11.04	.010	1.03	.005	0.75	.007	5		304,3018
		NGC 2360 - 120			7 15 33	−15 35.0	1	12.73		0.49		0.09		2		3018
		NGC 2360 - 166			7 15 34	−15 30.4	1	15.54		0.80		0.33		1		3018
		NGC 2360 - 178			7 15 34	−15 32.2	1	16.14		0.90				1		3018
56880	−36 03490				7 15 34	−36 11.8	1	9.89		0.43		-0.01		2	F3 V	852
56905	−37 03448				7 15 34	−37 12.4	1	9.68		0.46		0.00		2	F5/6 V	852
56744	−11 01880				7 15 35	−11 32.9	1	10.27		0.28		-0.14		2	B9	388
		NGC 2360 - 179			7 15 35	−15 32.1	1	15.53		0.72		0.18		1		3018
		NGC 2360 - 167			7 15 35	−15 32.7	1	13.68		0.52		0.12		2		3018
56903	−35 03488				7 15 35	−35 44.5	1	10.31		0.29		0.13		1	A3 II/III	852
		NGC 2360 - 175			7 15 36	−15 32.0	1	16.52		0.87				1		3018
		NGC 2360 - 174			7 15 36	−15 32.3	1	16.07		0.80		0.30		1		3018
	−36 03492				7 15 36	−36 26.5	1	11.05		0.43		-0.02		2		852
		NGC 2360 - 36			7 15 37	−15 31.2	2	12.02	.015	0.39	.019	0.19	.002	3		305,3018
		NGC 2360 - 176			7 15 37	−15 32.0	1	16.38		0.97		0.49		1		3018
		NGC 2360 - 33			7 15 37	−15 33.9	1	11.52		0.49		0.25		2		3018
		NGC 2360 - 122			7 15 37	−15 35.1	1	11.74		0.38		0.21		2		3018
		fld III # 21			7 15 38	−10 15.5	1	12.99		0.93		0.61		2		1755
		NGC 2360 - 35			7 15 38	−15 31.3	1	12.44		0.43		0.10		2		3018
		NGC 2360 - 31			7 15 38	−15 33.2	1	12.14		0.38		0.17		2		3018
		NGC 2360 - 32			7 15 38	−15 33.7	1	12.20		0.42		0.15		2		3018
56772	−13 01923				7 15 39	−13 36.8	1	9.16		0.34		0.12		2	A3	388
		NGC 2360 - 134			7 15 39	−15 36.4	1	11.52		1.02		0.68		1		304
	−31 04300				7 15 39	−31 16.2	1	10.53		-0.31		-1.17		8		976
56803	−13 01924				7 15 40	−13 47.8	1	7.53		0.43		-0.06		2	F3 V	388
		NGC 2360 - 172			7 15 40	−15 34.2	1	13.58		0.56		0.11		2		3018
		NGC 2360 - 123			7 15 40	−15 35.0	1	12.36		1.02		0.61		2		3018
		NGC 2360 - 34			7 15 41	−15 31.9	2	12.77	.044	1.12	.010	0.75	.039	3		304,3018
		NGC 2360 - 30			7 15 41	−15 33.5	1	12.18		0.34		0.20		4		3018
56932	−36 03494				7 15 41	−36 53.7	2	7.94	.005	-0.05	.005	-0.28	.000	4	B7/8 V	401,852
56563	+35 01593				7 15 42	+35 15.0	1	7.11		1.34		1.45		2	K0	1601
		NGC 2360 - 124			7 15 42	−15 34.9	1	12.86		0.43		0.11		2		3018
56978	−38 03301				7 15 43	−38 56.0	1	7.38		-0.11		-0.45		2	B7/8 V	401
		fld III # 18			7 15 44	−10 07.3	1	11.08		0.37		0.05		2		1755
		NGC 2360 - 29			7 15 44	−15 33.0	1	12.20		0.34		0.17		2		3018
		NGC 2360 - 164			7 15 44	−15 33.1	1	13.56		0.53		0.09		3		3018
		fld III # 20			7 15 45	−10 10.1	1	12.25		0.49		-0.01		2		1755
56830	−13 01926	IDS07134S1331	A		7 15 45	−13 36.1	1	8.64		0.03		-0.10		2		388
56831	−13 01925	IDS07134S1331	B		7 15 45	−13 36.5	1	8.53		-0.08		-0.50		2		388
		NGC 2360 - 28			7 15 45	−15 32.9	1	13.20		0.50		0.10		3		3018
		fld III # 17			7 15 46	−10 05.8	1	13.41		0.48		-0.02		2		1755
56800	−10 01971				7 15 46	−10 34.0	1	8.35		-0.05		-0.34		2	B8 V	388
56801	−10 01972				7 15 46	−11 09.0	1	10.24		0.03		-0.11		1	B9 V	388
		NGC 2360 - 22			7 15 46	−15 30.9	1	11.59		0.47		0.27		2		3018
56876	−26 04140	HR 2774			7 15 46	−26 42.4	4	6.43	.009	-0.14	.007	-0.60	.005	45	B5 Vn	158,401,1637,2031
		NGC 2360 - 23			7 15 47	−15 31.1	1	13.04		0.37		0.21		2		3018
56956	−36 03495				7 15 47	−37 07.1	2	8.33	.005	-0.03	.005	-0.27	.000	5	B9 V	401,852
		fld III # 9			7 15 48	−09 58.3	1	13.01		0.31		0.16		2		1755
		fld III # 10			7 15 48	−09 58.8	1	12.55		0.49		0.04		2		1755
		NGC 2360 - 162			7 15 48	−15 33.2	1	16.11		0.90		0.40		1		3018
56899	−27 03879				7 15 48	−27 30.3	1	9.38		-0.06		-0.35		2	B8/9 V	401
		NGC 2360 - 24			7 15 49	−15 31.0	1	11.83		0.34		0.14		2		3018
		NGC 2360 - 163			7 15 49	−15 32.0	1	16.09		0.87		0.40		1		3018
		NGC 2360 - 126			7 15 49	−15 34.1	1	12.30		0.46		0.05		2		3018
		NGC 2360 - 125			7 15 49	−15 34.4	1	12.74		0.38		0.12		2		3018

HD	DM	Other Id	N	Rem	α_{1950}	δ_{1950}	S	V	σ_V	B–V	σ_{B-V}	U–B	σ_{U-B}	n	Spectrum	References
		NGC 2360 - 161			7 15 50	−15 32.1	1	14.22		0.67		0.10		1		3018
		NGC 2360 - 127			7 15 50	−15 33.4	1	13.08		1.20		0.68		1		3018
		fld III # 11			7 15 51	−10 00.6	1	12.84		0.51		-0.02		2		1755
		NGC 2360 - 25			7 15 51	−15 30.6	1	13.54		0.53		0.08		2		3018
		fld III # 12			7 15 52	−10 01.5	1	13.46		0.50		0.26		2		1755
		NGC 2360 - 26			7 15 52	−15 31.2	1	12.96		0.40		0.17		2		3018
		NGC 2360 - 2			7 15 52	−15 32.3	1	12.94		0.35		0.12		3		3018
	−18 01772	CS CMa			7 15 52	−18 31.8	1	9.83		1.95				2	M3	369
		MCC 491			7 15 53	+50 32.0	1	11.22		1.13		1.08		2	K8	1723
		NGC 2360 - 11			7 15 53	−15 26.6	1	11.43		1.04		0.66		2		3018
56847	−15 01748	LSS 294			7 15 53	−15 32.2	4	8.96	.024	0.18	.008	-0.47	.005	16	B5 Ib	138,1012,3018,8100
		fld III # 7			7 15 54	−09 56.5	1	11.94		0.41		0.02		2		1755
56827	−11 01883	LSS 292			7 15 54	−11 52.0	1	9.68		0.19		-0.72		3	B2 Ib	388
		NGC 2360 - 27			7 15 54	−15 31.4	1	13.32		0.31		0.07		2		3018
		NGC 2360 - 8			7 15 55	−15 28.7	1	12.16		0.39		0.15		2		3018
		NGC 2360 - 4			7 15 55	−15 31.4	1	11.62		0.34		0.21		2		3018
		NGC 2360 - 3			7 15 55	−15 32.5	1	11.62		0.44		0.28		2		3018
56196	+65 00566	G 250 - 43		⋆A	7 15 57	+65 31.7	1	8.96		0.62		0.12		1	G5	3025
		fld III # 14			7 15 57	−10 03.5	1	13.24		0.36		0.23		2		1755
56715	+13 01626	IDS07132N1342		A	7 15 58	+13 37.3	1	7.78		1.19				3	K2	882
55075	+81 00242	HR 2709			7 15 58	+81 21.2	1			-0.04		-0.10		2	A0 III	1079
56896	−16 01881				7 16 02	−16 43.9	1	9.41		0.44				31	F2/3 V	6011
	−9 01967				7 16 03	−10 03.2	1	10.88		1.20		1.18		2		1755
		NGC 2360 - 7			7 16 03	−15 29.5	2	11.08	.020	0.99	.005	0.69	.005	4		304,3018
57095	−46 03046	IDS07146S4649		AB	7 16 03	−46 53.7	3	6.69	.031	0.98	.015	0.69	.025	12	K3 V	244,2012,3072
		G 112 - 6			7 16 04	+03 47.4	2	11.10	.010	0.67	.018	-0.01	.033	4	G3	1696,3062
57016	−36 03500				7 16 04	−36 23.0	1	10.91		0.09		0.07		2	A0 III/IV	852
57034	−36 03502			V	7 16 05	−37 04.1	2	8.13	.034	-0.05	.005	-0.24	.015	6	B9 V	401,852
56871	−7 01870				7 16 06	−07 44.7	1	8.87		-0.07		-0.32		2	B9	388
		fld III # 19			7 16 06	−10 09.7	1	11.92		0.76		0.21		2		1755
		fld III # 8			7 16 07	−09 56.1	1	14.78		1.46		0.89		2		1755
56873	−10 01975				7 16 07	−10 30.5	1	10.69		0.01		-0.33		2	B9	388
56998	−30 04260	Cr 132 - 13			7 16 07	−30 50.8	1	7.93		-0.08		-0.40		3	B7 III	594
56997	−30 04259	Cr 132 - 17			7 16 08	−30 21.7	1	8.25		-0.06		-0.26		3	B9 V	594
56925		LSS 299			7 16 10	−13 07.5	1	11.40		0.28		-0.47		5	WN	797
	−24 05162	NGC 2362 - 67			7 16 11	−24 50.2	1	9.31		-0.11		-0.76		5		282
		RL 142			7 16 12	+02 26.	1	13.47		-0.02		-0.44		2		704
57014	−29 04186				7 16 12	−29 59.6	1	9.75		0.06		0.08		2	A1 IV	1730
		BPM 18476			7 16 12	−59 02.	1	10.71		1.24		1.18		2		3072
	+33 01505	G 87 - 33			7 16 14	+32 55.8	2	9.16	.010	1.41	.015	0.76	.040	2	M2:	679,3072,7008
56893	−7 01872				7 16 14	−07 21.1	1	9.32		0.04		-0.23		1	B9	388
56951	−16 01884				7 16 14	−16 44.2	1	9.47		0.32				32	A7/8III+(F)	6011
56995	−24 05165	NGC 2362 - 68			7 16 17	−24 45.7	1	8.91		-0.13		-0.84		5	B2 IV/V	282
		NGC 2362 - 64			7 16 17	−24 53.5	2	11.80	.000	0.48	.005	0.02		6		1,282
57064	−36 03507				7 16 19	−36 25.4	1	11.48		0.08		0.06		2	A0/2 (V)	852
56167	+69 00417	RU Cam			7 16 20	+69 45.9	3	8.42	.039	1.22	.065	0.88		3	C0.1	755,1238,6011
56923	−9 01972				7 16 22	−09 29.5	1	9.87		0.03		-0.36		1	B7 V	388
	−24 05166	NGC 2362 - 69			7 16 22	−24 46.6	1	10.05		-0.08		-0.66		6		282
57094	−35 03498				7 16 22	−35 43.5	1	9.11		0.44		0.00		1	F5 V	852
		RL 143			7 16 24	+02 39.	1	11.74		0.01		-0.05		2		704
56790	+22 01642				7 16 25	+22 07.0	1	7.76		1.04		0.86		2	G5	1648
56761	+27 01362				7 16 25	+26 55.0	2	6.95	.000	0.97	.005	0.74	.005	5	G9 III	1501,1625
	−30 04269	Cr 132 - 31			7 16 25	−30 43.4	1	9.76		-0.02		-0.15		2	A0	594
		NGC 2362 - 47			7 16 28	−24 50.4	1	13.82		0.39				2		1
57010	−18 01777	IDS07143S1848		AB	7 16 29	−18 53.3	1	9.17		-0.04		-0.40		2	B6 V	401
		NGC 2362 - 6			7 16 29	−24 48.5	1	12.88		0.18				2		1
		NGC 2362 - 1			7 16 29	−24 55.4	1	11.39		0.00		-0.28		3		1
		NGC 2362 - 5			7 16 30	−24 49.8	1	10.78		-0.06		-0.52		3	B5 V	1
		NGC 2362 - 4			7 16 30	−24 50.7	1	12.44		0.08		0.02		3		1
		NGC 2362 - 2			7 16 30	−24 51.3	1	11.18		-0.02		-0.38		3		1
		NGC 2362 - 3			7 16 30	−24 51.9	1	11.05		0.09		-0.15		3		1
		NGC 2362 - 49			7 16 30	−24 52.5	1	12.30		0.39		0.02		1		1
		NGC 2362 - 66			7 16 30	−24 53.8	2	11.45	.010	-0.02	.005	-0.32	.005	6		1,282
57029	−19 01807				7 16 31	−20 02.3	1	8.64		-0.12		-0.58		2	B3 V	401
		NGC 2362 - 11			7 16 31	−24 49.9	2	11.93	.005	0.01	.019	-0.22		6		1,282
		NGC 2362 - 7			7 16 31	−24 52.7	1	13.44		0.20				2		1
		NGC 2362 - 8			7 16 31	−24 53.5	1	12.51		0.38				2		1
		NGC 2362 - 13			7 16 31	−24 56.3	2	10.49	.010	0.10	.005	0.08	.005	10		1,282
57197	−43 03093	HR 2789			7 16 31	−43 53.7	3	5.84	.012	-0.12	.014	-0.47	.005	10	B8 II/III	244,401,2006
		GD 85			7 16 32	+40 27.1	1	14.94		-0.11		-0.99		1	DBs	3060
	−24 05170	NGC 2362 - 9			7 16 32	−24 54.0	1	9.84		-0.08		-0.61		3	B3 V	1
		NGC 2362 - 10			7 16 32	−24 54.5	1	12.04		0.40				2		1
57150	−36 03512	HR 2787, NV Pup		⋆A	7 16 32	−36 38.5	11	4.66	.026	-0.10	.012	-0.79	.010	61	B2 V+B3IVne	15,681,815,852,1425,1637*
	−24 05171	NGC 2362 - 12			7 16 33	−24 50.6	1	10.03		-0.07		-0.57		3	B5 V	1
	−24 05172	NGC 2362 - 48			7 16 33	−24 52.2	1	9.54		-0.11		-0.69		3	B3 V	1
	−10 01977			V	7 16 34	−11 09.2	1	9.32		2.01				2	M2	369
		NGC 2362 - 17			7 16 34	−24 51.5	1	12.30		1.13				2		1
		NGC 2362 - 14			7 16 34	−24 52.8	1	9.60		-0.12		-0.73		3	B2 V	1
		NGC 2362 - 15			7 16 34	−24 53.5	1	11.76		0.03		-0.14		3	B9 V	1
57060	−24 05173	HR 2781, UW CMa			7 16 35	−24 28.0	3	4.96	.022	-0.15	.000	-1.00	.007	16	O7 f	15,1088,8015

Table 1   391

HD	DM	Other Id	N Rem	α_{1950}	δ_{1950}	S	V	σ_V	B–V	σ_{B-V}	U–B	σ_{U-B}	n	Spectrum	References
		NGC 2362 - 16		7 16 35	−24 50.8	1	10.57		0.07		−0.20		3	B9 V	1
		NGC 2362 - 65		7 16 35	−24 51.4	2	11.65	.019	0.24	.005	0.10		6		1,282
		RL 144		7 16 36	−03 04.	1	13.45		0.03		−0.29		2		704
		NGC 2362 - 21		7 16 36	−24 52.9	1	10.44		−0.07		−0.51		3	B6 V	1
		NGC 2362 - 70		7 16 36	−24 53.3	1	11.90		0.00		−0.20		3		282
		NGC 2362 - 43		7 16 37	−24 47.4	1	14.75		0.50				2		1
	−24 05175	NGC 2362 - 20		7 16 37	−24 54.6	3	8.77	.004	−0.16	.004	−0.87	.000	16	B2 V	1,14,182
57120	−30 04276	Cr 132 - 7	⋆ AB	7 16 37	−30 42.4	2	6.99	.011	−0.19	.002	−0.77	.007	6	B2 V	594,1732
		Steph 637		7 16 38	+30 01.4	1	11.71		1.32		1.25		1	K7	1746
		NGC 2362 - 24		7 16 38	−24 50.3	3	11.01	.024	−0.03	.023	−0.42	.005	8	B7 V	1,14,282
57061	−24 05176	NGC 2362 - 23	⋆ AB	7 16 38	−24 51.7	13	4.39	.013	−0.15	.008	−0.99	.008	69	O9 I	1,14,15,182,282,542*
		NGC 2362 - 39		7 16 38	−24 52.7	1	9.78		−0.09		−0.67		3	B2 V	1
		NGC 2362 - 22		7 16 38	−24 53.1	2	11.92	.010	0.01	.005	−0.19		6		1,282
57122	−32 03873			7 16 38	−32 15.3	1	6.98		−0.09		−0.27		2	Ap Si	401
		AA45,405 S298 # 2		7 16 39	−13 11.6	2	10.87	.015	0.17	.029	−0.53	.034	3	B2 IV	797,1721
		NGC 2362 - 40		7 16 39	−24 47.9	1	12.86		0.31				2		1
		NGC 2362 - 25		7 16 39	−24 48.4	2	10.77		0.00	.004	−0.42	.008	8	B6 V	1,282
		NGC 2362 - 53		7 16 39	−24 48.9	1	13.58		0.18				2		1
		NGC 2362 - 50		7 16 39	−24 52.8	1	10.20		0.00		−0.36		3	B6 V	1
		NGC 2362 - 51		7 16 39	−24 53.5	1	12.23		0.43				2		1
		NGC 2362 - 71		7 16 41	−24 43.4	1	10.60		0.15		0.09		4		282
		NGC 2362 - 59		7 16 41	−24 48.1	1	12.21		0.23				2		1
		NGC 2362 - 52		7 16 41	−24 53.2	2	11.92	.010	0.05	.015	−0.19		6		1,282
		NGC 2362 - 26		7 16 41	−24 54.0	1	10.43		−0.07		−0.52		3	B5 V	1
		WLS 720 10 # 10		7 16 42	+09 40.2	1	12.22		0.69		0.10		2		1375
		NGC 2362 - 41		7 16 42	−24 47.7	1	13.13		0.55				2		1
		NGC 2362 - 35		7 16 42	−24 50.1	1	12.98		0.30				2		1
		NGC 2362 - 27		7 16 42	−24 52.2	3	10.16	.008	−0.08	.005	−0.62	.021	10	B3 V	1,14,182
		NGC 2362 - 44		7 16 42	−24 53.6	1	14.81		0.22				2		1
	−36 03514			7 16 42	−37 01.0	1	10.63		1.41		1.55		1		852
		NGC 2362 - 32		7 16 44	−24 48.0	2	10.76	.004	0.28	.000	0.22	.004	6		1,282
	−24 05180	NGC 2362 - 30	⋆ D	7 16 44	−24 51.4	4	8.20	.010	−0.18	.006	−0.91	.005	19	B1 V	1,14,182,542
	−24 05182	NGC 2362 - 31		7 16 45	−24 52.3	3	9.32	.005	−0.13	.004	−0.77	.004	14	B2 V	1,14,182
		NGC 2362 - 28		7 16 45	−24 53.1	1	12.05		0.02				2		1
		NGC 2362 - 29		7 16 45	−24 53.8	2	11.33	.004	−0.04	.008	−0.33	.000	7		1,282
	−26 04162	LSS 311		7 16 45	−26 51.5	4	8.82	.011	−0.09	.007	−0.81	.009	10	B1 IVn	401,540,976,1732
	−63 00713			7 16 45	−64 05.9	1	11.07		0.29		0.10		2		1730
56989	+2 01640	HR 2778		7 16 46	+02 50.0	2	5.88	.005	1.07	.000	0.83	.005	7	G9 III	15,1078
		NGC 2362 - 73		7 16 46	−24 44.3	1	11.06		−0.05		−0.52		4		282
		NGC 2362 - 55		7 16 46	−24 49.9	1	14.49		0.49				2		1
		NGC 2362 - 54		7 16 46	−24 50.1	1	14.13		0.42				2		1
57194	−36 03516	IDS07147S3633	CD	7 16 46	−36 40.6	2	8.10		0.09		0.01		2	A0/1 V	852
57057	−12 01894			7 16 47	−13 10.1	2	8.64	.025	2.02	.095	1.42	.060	2	K5 (III)	797,1721
57089	−19 01812			7 16 47	−19 22.7	1	8.96		−0.02		−0.47		2	B5 III	401
		NGC 2362 - 37		7 16 47	−24 50.3	2	12.53	.030	0.37	.010	0.09		5		1,14
57195	−36 03517			7 16 47	−36 48.1	1	9.03		1.47		1.68		2	K0	852
56915	+14 01636			7 16 48	+14 02.2	1	8.19		1.03				3	K0 V	882
		NGC 2362 - 56		7 16 48	−24 49.7	1	12.01		0.13				2		1
		NGC 2362 - 74		7 16 48	−24 53.9	1	10.91		0.16		0.12		3		282
56965	+10 01495			7 16 49	+10 29.6	1	11.52		0.18		−0.40		1	M0 III	1721
		NGC 2362 - 34		7 16 49	−24 51.9	2	10.50	.005	−0.03	.009	−0.49	.000	5	B5 V	1,282
57146	−26 04164	HR 2786		7 16 49	−26 29.6	6	5.29	.013	0.97	.007	0.67	.029	18	G2 II	15,1415,1488,1637*
57170	−30 04283	Cr 132 - 15		7 16 49	−31 08.0	1	8.10		0.20		0.11		2	A1 V	594
	+68 00474			7 16 50	+68 22.0	1	10.08		1.14		1.07		1	K8	3072
		WLS 720-10 # 6		7 16 50	−07 50.3	1	12.61		0.09		−0.22		2		1375
57118	−19 01813	HR 2785, LSS 310		7 16 50	−19 11.3	6	6.09	.012	0.61	.015	0.37	.027	17	F2 Ia	15,258,1075,2012,6009,8100
		NGC 2362 - 57		7 16 50	−24 50.7	1	12.16		0.02		−0.12		2		1
		NGC 2362 - 38		7 16 50	−24 51.2	1	12.76		0.12				2		1
		NGC 2362 - 58		7 16 50	−24 51.9	1	12.23		0.07		−0.04		2	A0 V	1
		WLS 720 30 # 5		7 16 51	+31 45.1	1	12.08		0.41		−0.03		2		1375
57219	−36 03519	HR 2790, NW Pup	⋆ B	7 16 51	−36 39.0	7	5.11	.006	−0.17	.008	−0.67	.009	30	A0 V	15,852,1075,1637,1732*
57240	−38 03309	HR 2791		7 16 51	−39 07.1	3	5.25	.005	0.03	.011	0.06	.000	8	A0 V	401,2032,3050
57083	−12 01895			7 16 52	−13 09.4	1	9.85		0.98		0.69		1	G8/K0III/IV	797
57220	−36 03521			7 16 52	−37 08.0	2	8.32	.005	−0.09	.005	−0.55	.000	4	B8/9 III	401,852
57623	−67 00730	HR 2803		7 16 52	−67 51.9	4	3.96	.007	0.77	.007	0.45	.000	20	F8 Ib/II	15,1075,2038,3026
		NGC 2362 - 36		7 16 54	−24 52.2	1	10.76		0.02		−0.48		3	B3 V	1
	−26 04168			7 16 54	−26 50.9	1	9.07		1.10		1.08		1	K0	1732
57056	−11 01885			7 16 55	−11 22.1	1	8.52		1.67				2	M2	369
57078	−8 01839			7 16 58	−08 41.4	1	7.15		1.71				2	M1	369
56711	+52 01195			7 17 00	+52 21.7	1	7.93		1.27		1.45		3	K2 III	37
		NGC 2362 - 42		7 17 00	−24 50.7	2	11.34	.005	−0.02	.009	−0.38		9		1,282
		CCS 719		7 17 00	−42 50.5	1	10.49		1.60		1.82		1		864
		WLS 720 30 # 10		7 17 04	+30 02.9	1	12.89		0.45		0.10		2		1375
57006	+7 01684	HR 2779		7 17 06	+07 14.2	2	5.90	.005	0.53	.000	0.06	.010	7	F8 IV	15,1061
56987	+19 01685			7 17 06	+19 36.9	1	7.19		1.26		1.31		2	K0	1648
		NGC 2362 - 75		7 17 06	−24 44.9	1	10.39		0.44		−0.02		5		282
57193	−25 04354			7 17 06	−25 28.4	4	7.46	.011	−0.18	.007	−0.94	.005	11	B1 III	158,540,976,1732
	−63 00714			7 17 07	−64 04.7	1	10.13		0.73		0.24		2		1730
56986	+22 01645	HR 2777	⋆ AB	7 17 08	+22 04.6	9	3.53	.008	0.34	.005	0.04	.005	35	F1 IV-V	15,934,1008,1013,1197*
57192	−24 05188	NGC 2362 - 46	⋆ AB	7 17 08	−24 51.8	9	6.80	.012	−0.17	.009	−0.73	.007	42	B1.5V	1,14,164,182,282,540*

HD	DM	Other Id	N Rem	α_{1950}	δ_{1950}	S	V	σ_V	B–V	σ_{B-V}	U–B	σ_{U-B}	n	Spectrum	References
57282	−36 03525			7 17 11	−36 52.5	1	9.86		1.11		0.91		2	G8 (III)	852
57167	−16 01898	HR 2788, R CMa		7 17 12	−16 18.0	3	5.70	.004	0.36	.049	0.04		6	F2 III/IV	532,2031,3053
		NGC 2362 - 76		7 17 12	−24 47.9	1	9.79		−0.05		−0.55		5		282
57049	+15 01541	HR 2780		7 17 16	+15 14.2	1			−0.02		0.07		3	A2 Vn	1022
	−18 01782			7 17 16	−18 27.9	1	9.45		1.78				2	M3	369
	−24 05190			7 17 16	−24 35.6	2	9.80	.002	0.04	.022	−0.63	.003	4	B3 II-III	540,976
		NGC 2362 - 77		7 17 17	−24 49.1	1	10.75		0.27		0.13		5		282
57301	−34 03517			7 17 17	−35 05.5	1	7.82		−0.02		−0.18		2	A0 V	401
57189	−15 01766	IDS07150S1542	AB	7 17 18	−15 47.6	1	8.06		−0.05		−0.42		2	B8 III	401
57187	−11 01889			7 17 22	−12 07.5	1	9.34		−0.02		−0.40		2	B8 II	401
57236	−21 01874	LSS 315		7 17 22	−21 54.7	3	8.74	.004	0.19	.004	−0.72	.018	7	O9 V	540,976,1011
	−24 05192	NGC 2362 - 78		7 17 22	−24 50.9	1	10.70		0.01		−0.10		4		282
57299	−33 03696	HR 2794		7 17 22	−33 38.0	2	6.31	.006	1.31	.001	1.46		8	K2/3 III	1637,2007
58160	−76 00444			7 17 22	−76 56.3	1	8.81		0.94		0.69		1	K3 V	3072
57331	−35 03510			7 17 25	−36 04.4	2	8.16	.025	−0.09	.015	−0.33	.010	4	B9 IV	401,852
	−15 01767			7 17 26	−15 46.7	1	9.37		1.97				2	M2	369
57353	−37 03467			7 17 26	−37 13.6	1	8.48		0.05		0.01		3	A1 V	852
57415	−45 03088			7 17 27	−45 12.3	1	7.15		0.51				4	F6 V	2012
57234	−14 01834			7 17 29	−15 10.0	1	8.34		−0.07				4	B8 II/III	2014
56941	+42 01699	HR 2775		7 17 31	+42 45.0	2	6.33	.015	1.45	.010	1.59	.005	3	K0	1733,3051
57281	−23 05277		AB	7 17 32	−23 55.8	2	8.91	.094	−0.08	.034	−0.62	.055	5	B5 V	206,1732
		G 87 - 35		7 17 35	+29 26.5	1	11.10		0.76		0.26		1		333,1620
57280	−19 01818			7 17 35	−19 52.5	1	8.45		−0.01		−0.04		2	B9 V	401
	+70 00448			7 17 36	+70 05.8	1	9.06		1.09		0.97		10	K0	1308
57374	−35 03514			7 17 37	−35 59.6	1	9.84		0.00		−0.13		2	B9.5V	852
57326	−26 04186			7 17 38	−26 28.6	1	8.55		−0.11		−0.52		2	B6 V	401
56963	+45 01422	HR 2776		7 17 41	+45 19.4	1	5.79		0.31		−0.02		4	F1 V	254
57411	−37 03470			7 17 42	−37 24.8	2	7.84	.025	−0.05	.010	−0.29	.005	4	B9 V	401,852
		Cr 135 - 74		7 17 44	−35 57.1	1	11.33		0.52		0.02		1		852
57047	+39 01923			7 17 51	+39 11.2	1	8.48		0.01		−0.04		3	A0 V	833
51802	+87 00051	HR 2609, OV Cep		7 17 51	+87 07.6	2	5.02	.033	1.60	.019	1.92	.033	10	M2 IIIab	15,3055
	−35 03519			7 17 53	−35 56.6	1	11.06		0.38		0.01		1		852
56820	+60 01048	HR 2772	⋆AB	7 17 54	+59 59.8	1	6.35		0.27		0.15		2	A8m	1733
		NGC 2367 - 15		7 17 54	−21 47.9	1	15.87		1.28				2		183
57321	−8 01849	LSS 321		7 17 55	−08 27.3	1	7.15		0.36		0.40		1	F2 II	490
		CCS 720		7 17 55	−20 15.3	1	9.31		1.60		1.90		1	C4+,3	864
		NGC 2367 - 10		7 17 55	−21 47.5	1	13.12		0.28		−0.20		2		183
57276	−4 01912			7 17 56	−04 50.6	1	8.57		1.29		1.38		26	K5	588
		NGC 2367 - 2		7 17 56	−21 47.4	1	9.68		0.08		−0.68		2	B1 V	183
57370	−21 01880	NGC 2367 - 1	⋆A	7 17 56	−21 47.4	1	9.39		−0.03		−0.84		3	B0.5V	183
57370	−21 01880	NGC 2367 - 1	⋆AB	7 17 56	−21 47.4	2	8.75	.008	0.08	.003	−0.75	.019	7		92,540
		NGC 2367 - 11		7 17 56	−21 48.3	1	14.46		0.63		−0.03		4		183
57275	+0 01909	IDS07154N0036	AB	7 17 57	+00 29.9	1	6.90		−0.01		−0.05		2	B9	3032
57275	+0 01909	IDS07154N0036	C	7 17 57	+00 29.9	1	13.04		0.67		0.27		1		3032
57275	+0 01909	IDS07154N0036	D	7 17 57	+00 29.9	1	13.10		0.65		0.24		1		3032
57275	+0 01909	IDS07154N0036	E	7 17 57	+00 29.9	1	11.54		0.72		0.21		2		3032
		NGC 2367 - 7		7 17 57	−21 45.2	1	11.83		0.57		0.10		2		183
		NGC 2367 - 9		7 17 57	−21 47.1	1	13.18		0.36		0.28		2		183
		NGC 2367 - 12		7 17 58	−21 46.6	1	12.87		0.29		0.08		2		183
		NGC 2367 - 4		7 17 58	−21 48.0	1	10.49		0.23		0.16		3		183
		NGC 2367 - 13		7 17 59	−21 45.7	1	13.69		0.54		0.03		2		183
	−21 01881	NGC 2367 - 5		7 18 00	−21 49.0	3	9.81	.008	0.05	.009	−0.75	.015	8	B1 V	92,183,976
57393	−23 05296			7 18 00	−23 59.5	4	9.12	.019	−0.02	.004	−0.75	.010	9	B3 Vnne	540,976,1012,1732
57460	−35 03520			7 18 00	−36 02.2	1	9.64		0.08		0.09		2	A1 V	852
	−21 01882	NGC 2367 - 3		7 18 01	−21 47.8	1	10.32		0.05		−0.75		4	B2 IV	183
		NGC 2367 - 8		7 18 01	−21 48.2	1	12.08		0.18		−0.32		4		183
		Cr 135 - 76	AB	7 18 02	−36 26.1	1	11.53		0.31		0.27		2		852
57485	−36 03529			7 18 02	−36 53.7	1	9.68		0.29		0.18		3	A7 V	852
		NGC 2367 - 14		7 18 03	−21 45.7	1	13.32		0.24		−0.31		2		183
57291	+3 01649			7 18 06	+03 40.6	1	6.87		−0.16		−0.72		1	B5	555
		NGC 2367 - 6		7 18 06	−21 46.1	1	10.93		0.11		−0.63		2	B5 V:	183
57484	−36 03530			7 18 06	−36 23.2	1	10.21		0.34		0.06		2	F0 V	852
57809	−63 00718			7 18 07	−63 58.3	1	9.05		0.13		0.11		2	A0 Vs	1730
	−4 01914	G 112 - 7		7 18 11	−04 45.9	1	9.91		0.87		0.55		1		3062
		RL 145		7 18 12	+01 52.	1	15.84		0.28		0.08		2		704
57505	−36 03532			7 18 12	−36 31.6	1	9.55		0.21		0.16		3	A5 V	852
	−14 01841		V	7 18 14	−14 54.5	1	9.66		1.71				2	M2	369
57386	−8 01856	LSS 325		7 18 18	−08 19.6	2	8.01	.020	0.00	.015	−0.76	.000	5	B1.5:V:penn	1012,1212
57364	−4 01915	AR Mon		7 18 20	−05 09.9	1	8.63		1.08				4	K0 II	2012
57435	−14 01843			7 18 20	−14 46.6	3	8.59	.014	1.42	.045	1.49	.205	7	K3 III	1003,2017,3077
		RR Gem		7 18 23	+30 58.7	2	10.70	.042	0.12	.023	0.14	.034	2	A8	668,699
57407	−5 02076	LS VI -05 009		7 18 26	−05 14.5	1	8.95		0.60		0.49		2	F5	490
	−20 01866	LSS 330		7 18 26	−21 08.0	2	9.50	.007	0.21	.025	−0.66	.019	4	B0 III	540,976
57247	+31 01549			7 18 29	+30 58.1	1	9.52		0.12		0.20		1	A0	289
		LP 58 - 170		7 18 29	+66 53.0	1	10.96		0.71		0.12		3		1723
57578	−36 03534			7 18 29	−36 21.6	1	10.28		0.10		0.11		2	A1 V(n)	852
	−7 01900		V	7 18 30	−07 26.5	1	9.78		1.68				2	M3	369
57503	−24 05228			7 18 31	−25 04.7	1	8.88		−0.17		−0.86		3	B2 III/IV	1732
57574	−31 04378	Cr 140 - 24		7 18 32	−32 04.6	1	9.00		0.19		0.11		3	Ap	703
57339	+16 01456			7 18 35	+16 49.0	1	8.57		1.21				3	K2 V	882

Table 1 393

HD	DM	Other Id	N Rem	α1950	δ1950	S	V	σV	B–V	σB–V	U–B	σU–B	n	Spectrum	References
		G 250 - 44		7 18 38	+61 01.6	1	10.42		0.89		0.60		2		7010
		AC +66 2298		7 18 38	+66 16.4	1	10.81		0.88		0.67		3		1723
	-36 03537			7 18 39	-36 43.2	1	11.07		0.33		0.07		2		852
57478	-14 01846	HR 2796		7 18 41	-14 15.9	2	5.53	.063	0.97	.005	0.63		6	G8/K0 III	252,2031
57264	+37 01707	HR 2793	★ AB	7 18 42	+36 51.4	3	5.13	.010	1.09	.009	0.93	.010	8	K0 III	1355,1363,3016
57129	+53 01138			7 18 42	+52 56.1	1	7.72		1.26		1.31		2	K2	1375
57245	+39 01926			7 18 43	+39 14.4	1	8.13		0.20		0.15		3	A0	833
57573	-22 01823	HR 2799		7 18 46	-22 45.4	2	6.60	.005	-0.16	.005			7	B3 V	15,2012
57639	-36 03538			7 18 46	-36 49.4	1	9.51		0.79		0.33		2	G5 V	852
57263	+39 01927	HR 2792		7 18 49	+39 05.5	4	6.40	.004	1.22	.004	1.24	.007	11	K1 III	833,1501,1601,1733
57519	-12 01909	IDS07165S1206	AB	7 18 49	-12 11.5	2	8.50	.000	-0.03	.000	-0.57	.000	8	B3 V	401,1775
		LP 782 - 1		7 18 49	-15 13.3	2	9.78		0.74		0.20		2		1696
57618	-29 04260			7 18 52	-29 46.1	2	7.88	.010	-0.16	.010	-0.65	.010	6	B5 Vn	164,401
57517	-8 01862	HR 2798		7 18 53	-08 46.9	3	6.55	.005	0.54	.005	0.03	.008		F7 V	15,258,1415
57593	-26 04223	HR 2800, HQ CMa	★ AB	7 18 53	-26 52.1	2	6.00	.010	-0.17	.000	-0.71		6	B3 V	401,2009
57659	-31 04386	Cr 140 - 17		7 18 56	-31 56.8	1	8.70		0.70		0.50		2	G9 III+A2V	703
57637	-32 03910	Cr 140 - 7		7 18 56	-32 22.9	1	7.36		1.32		1.48		2	K0 III	703
57423	+20 01775	HR 2795	★ A	7 19 00	+20 32.4	2	5.09	.004	1.53	.004	1.87		5	M0 IIIab	1363,3052
57616	-25 04399			7 19 00	-26 01.2	1	8.70		-0.15		-0.78		2	B3 III	401
57568	-12 01914			7 19 01	-12 32.7	1	10.03		1.07		0.96		3	K3 V	803
57615	-25 04400	HR 2802		7 19 01	-25 47.8	4	5.86	.012	1.61	.013	1.91	.008	16	M3 III	1024,1637,2007,3005
57613	-20 01874	LSS 337		7 19 05	-20 40.5	2	9.30	.005	0.12	.002	-0.67	.015	7	B2 IV	540,976
	-32 03913	Cr 140 - 52		7 19 05	-32 09.8	1	10.12		0.21		0.12		2	A5 V	703
57566	-10 02005			7 19 06	-10 14.6	1	9.42		-0.03		-0.56		3	B8	1732
57690	-32 03912	Cr 140 - 21	★ AB	7 19 06	-32 12.5	1	8.94		0.53		0.02		3	F5 V	703
57539	-5 02080			7 19 07	-05 48.1	1	6.47		-0.10				4	B5 III	2012
57853	-52 01153	HR 2814	★ B	7 19 09	-52 13.0	1	6.60		0.59		0.07		4	G0 V:e	1279
57852	-52 01153	HR 2813	★ A	7 19 09	-52 13.1	1	6.05		0.42		-0.01		4	F0/2 IV/V	1279
57852	-52 01153	IDS07180S5207	★ AB	7 19 09	-52 13.1	3	5.52	.005	0.49	.005	0.01		9	F0/2 IV/V +	15,404,2012
	-46 03086	HH Pup		7 19 10	-46 36.8	1	10.46		0.16		0.11		1	A3 II-III	700
58140	-66 00680			7 19 10	-67 05.6	1	8.03		1.55		1.46		11	M4 III	1704
57496	+14 01644			7 19 11	+14 44.2	1	8.64		0.66				2	F8 V	882
	+36 01611		A	7 19 11	+35 59.8	2	10.87	.030	1.41	.000	1.26		2	K5	801,1746
	+36 01611		B	7 19 11	+35 59.8	2	11.44	.025	1.42	.020	1.32		2		801,1746
	-16 01918	RZ CMa		7 19 17	-16 35.0	1	9.34		0.86				1	F6	6011
		RL 146		7 19 18	+01 38.	1	14.29		0.20		-0.76		2		704
		MCC 492		7 19 18	+67 27.0	1	10.30		0.95		0.72		3	K5	1723
		Cr 140 - 57		7 19 19	-31 52.2	1	10.44		0.12		0.15		2	A2 V	703
57759	-31 04396	Cr 140 - 31		7 19 21	-31 55.9	1	9.32		0.03		-0.04		2	A0 V	703
		WLS 720-10 # 9		7 19 23	-09 42.1	1	12.40		0.20		0.05		2		1375
		WLS 720-10 # 8		7 19 24	-11 36.0	1	11.15		0.38		0.15		2		1375
		KS 21		7 19 24	-19 55.	1	10.60		0.05		-0.70		2		540
		G 193 - 44		7 19 25	+59 04.6	1	10.51		0.96		0.76		2		1658
57917	-51 02445	HR 2815		7 19 26	-51 59.4	1	5.39		-0.07				4	B9 II/III	2009
57608	+0 01915	HR 2801		7 19 29	+00 16.4	3	5.98	.005	-0.08	.011	-0.29	.012	9	B8 III	15,1079,1256
		G 107 - 64		7 19 30	+41 03.4	1	13.80		1.49		1.16		1		1658
	-12 01919			7 19 31	-13 09.2	1	10.48		2.08				2	M2	369
		G 87 - 36		7 19 33	+30 46.4	1	13.34		1.50		1.04		2		316
		RL 147		7 19 36	-01 20.	1	14.57		0.14		0.02		2		704
		G 107 - 65	A	7 19 37	+46 11.2	2	10.53	.000	1.50	.061			2	M1	680,1017
		G 107 - 65	B	7 19 37	+46 11.2	1	11.92		1.28				1		680
57682	-8 01872	HR 2806	★	7 19 38	-08 53.0	9	6.42	.008	-0.19	.006	-1.03	.005	94	O9 V	15,154,1011,1203,1256,1732*
57678	-1 01677	G 112 - 12		7 19 39	-01 46.2	3	8.72	.014	0.78	.009	0.33	.010	7	K0	1064,2012,3062
57309	+54 01137			7 19 41	+54 11.4	1	8.40		0.06		0.06		2	B9	1566
	+9 01627	G 89 - 13		7 19 42	+09 00.2	3	9.37	.017	0.91	.007	0.67	.010	6	K2 V	1003,1620,3077
	-14 01863	TW CMa		7 19 45	-14 12.7	1	9.27		0.81				1	F5	6011
57603	+14 01648			7 19 46	+14 02.1	2	8.00	.015	1.09	.017	0.97		5	K1 III	882,3040
57802	-23 05335	CW CMa		7 19 46	-23 41.9	1	8.58		0.14		0.07		1	A0 V	513
		G 89 - 14		7 19 48	+08 55.2	4	10.40	.000	0.47	.005	-0.18	.007	7		516,1620,1696,3077
57708	-2 02079	HR 2807		7 19 48	-02 53.0	2	6.22	.005	0.68	.000	0.34	.005	7	F5	15,1415
57868	-36 03549			7 19 48	-36 21.0	1	9.32		0.02		-0.10		3	B9	852
57892	-36 03550			7 19 49	-36 23.9	1	8.35		0.02		-0.03		3	A0 V	852
	-31 04409	Cr 140 - 70		7 19 50	-31 50.7	1	11.14		0.10		0.01		2	A2 V	703
57824	-23 05339			7 19 54	-23 40.7	1	9.56		-0.03		-0.33		1	B5/7 V	513
57867	-31 04410	Cr 140 - 42		7 19 55	-31 36.5	1	9.71		-0.07		-0.40		3	B8 III:	703
	-3 01866			7 19 56	-03 21.4	1	9.40		1.56				2	M3	369
	-19 01845	IDS07178S1953	AB	7 19 56	-19 58.4	2	8.54	.005	1.19	.019	0.10	.021	6	B2:V: +G:	540,976
57749	-5 02089	HR 2811		7 19 58	-05 53.2	4	5.82	.013	0.35	.005	0.19	.014	15	F3 IV	15,39,1078,2028
57821	-18 01806	HR 2812		7 20 01	-18 55.2	10	4.96	.007	-0.05	.005	-0.38	.009	37	B5 II/III	15,369,611,1075,1075*
57912	-31 04414	Cr 140 - 19		7 20 03	-32 05.7	1	8.83		-0.02		-0.21		2	B8 V	703
57846	-23 05343			7 20 04	-23 56.7	1	8.94		0.31		0.10		1	A1mA7-A7	513
57946	-36 03554	IDS07183S3626	AB	7 20 06	-36 31.9	1	8.41		-0.04		-0.30		3	Ap Si	852
57674	+20 01781			7 20 07	+20 39.1	1	7.74		0.32		0.05		2	F0	1648
	-31 04417	Cr 140 - 47		7 20 08	-32 02.6	1			0.11		0.11		2		703
		WLS 720 10 # 7		7 20 09	+08 22.3	1	10.86		0.46		-0.05		2		1375
57819	-15 01786	IDS07179S1523	A	7 20 09	-15 28.4	1	8.43		-0.03		-0.38		2	B7/8 II	401
57944	-31 04418	Cr 140 - 44		7 20 12	-31 55.5	3	9.77	.010	-0.04	.020	-0.22	.017	7	B9 V	74,703,716
57817	-9 02007			7 20 13	-10 08.5	1	8.95		1.65				2	M1	369
57890	-20 01889	GH CMa		7 20 13	-20 24.6	1	6.95		1.61		1.48		1	M4 III	3042
57943	-31 04416	Cr 140 - 29		7 20 13	-31 20.5	1	9.28		0.14		0.11		2	A3 V:	703

HD	DM	Other Id	N Rem	α_{1950}	δ_{1950}	S	V	σ_V	B–V	σ_{B-V}	U–B	σ_{U-B}	n	Spectrum	References
57945	−32 03934	Cr 140 - 6		7 20 15	−32 29.4	3	7.11	.005	-0.05	.010	-0.14	.004	7	B9 V	74,703,716
57994	−36 03557	IDS07185S3620	AB	7 20 15	−36 25.3	2	8.25	.020	-0.01	.000	-0.20	.178	5	B8 IV	401,852
57962	−31 04420	Cr 140 - 13		7 20 16	−31 56.1	3	8.37	.005	1.30	.000	1.53	.022	7	K3 III	74,703,716
		LS VI -05 010		7 20 18	−05 55.5	1	11.39		0.00		-0.63		2		490
57770	+13 01653	V Gem		7 20 21	+13 12.0	1	13.79		0.26		0.17		6	M4	3064
		G 88 - 23		7 20 22	+19 07.0	1	12.04		0.49		-0.23		2		333,1620
57963	−32 03937	Cr 140 - 20		7 20 23	−32 29.9	1	8.89		0.12		0.10		2	A2 V	703
		ApJ144,259 # 20		7 20 24	+01 52.	1	16.56		-0.27		-1.12		4		1360
57991	−32 03935	Cr 140 - 45		7 20 24	−32 09.6	2	9.84	.015	0.09	.010	0.07	.029	3	A1 V	703,716
57727	+25 01660	HR 2808		7 20 26	+25 08.9	5	5.04	.021	0.90	.007	0.58	.007	9	G8 III	1080,1355,1363,3016,4001
57201	+69 00420			7 20 26	+69 33.2	1	8.89		0.51				11	F8	6011
57888	−12 01924			7 20 26	−12 49.1	1	9.24		1.70				2	M1/2 III	369
	−31 04426	Cr 140 - 56		7 20 27	−31 33.7	1	10.26		0.22		0.14		4	A3 V	703
		Cr 140 - 73		7 20 27	−32 02.1	1	11.34		0.37		0.08		2	A9 V	703
57744	+23 01698	HR 2810		7 20 28	+23 02.6	2	6.17		0.00	.005	-0.01		4	A1 V	1022,1733
58038	−35 03551	IDS07187S3544	A	7 20 28	−35 49.2	1	7.09		0.51		0.04		2	F8	852
58038	−35 03551	IDS07187S3544	B	7 20 28	−35 49.2	1	8.09		0.57		0.09		1		852
57906	−12 01925			7 20 31	−12 41.2	1	9.41		0.01		-0.30		2	B9 II/III	401
	−31 04428	Cr 140 - 43		7 20 34	−32 01.1	2	9.77	.009	0.09	.005	0.05	.023	4	A2 V	703,716
57044	+73 00375	IDS07145N7316	AB	7 20 35	+73 10.9	1	7.36		0.32		0.04		3	F0	1084
58111	−39 03269			7 20 37	−39 09.3	1	8.85		0.88		0.52		1	K0/1 V	3072
57884	−3 01873			7 20 38	−04 07.4	4	9.09	.049	2.28	.031	3.81	.141	36	C5,5i	864,989,1238,1729
	−17 01943			7 20 40	−17 59.2	1	10.34		1.95				2	M2	369
58010	−24 05289			7 20 40	−25 04.5	1	8.72		-0.17		-0.87		4	B2 IV	1732
58011	−25 04439	LSS 352	V	7 20 40	−25 54.8	4	7.16	.043	-0.08	.029	-0.98	.017	15	B1Iab/IInne	540,976,1088,1737
57837	+15 01559			7 20 41	+15 24.3	1	8.38		0.73					F6 V	882
57669	+40 01852	HR 2805		7 20 41	+40 46.2	4	5.24	.033	1.25	.011	1.25	.019	16	K0 IIIa	37,1080,3016,8100
55966	+82 00201	HR 2742, VZ Cam		7 20 41	+82 30.8	2	4.88	.057	1.63	.018	1.76		4	M4 IIIa	1363,3051
58063	−31 04430	Cr 140 - 5		7 20 41	−31 57.0	5	6.82	.005	-0.17	.007	-0.68	.018	22	B4 V	74,164,401,703,716
		Haf 8 - 26		7 20 42	−12 13.5	2	12.26	.011	1.40	.009	1.10	.008	4		410,1685
57128	+73 00376	IDS07145N7316	C	7 20 43	+73 11.0	1	7.82		0.30		0.04		3	F0	1084
		Haf 8 - 48		7 20 43	−12 11.7	1	13.38		0.33		0.18		2		410
		Haf 8 - 99		7 20 43	−12 12.3	1	14.26		0.31		0.26		2		410
		Haf 8 - 25		7 20 43	−12 12.9	1	13.65		0.28		0.08		2		410
	−20 01896	LSS 349		7 20 43	−20 47.4	1	9.72		0.07		-0.75		2	B1 V	540
57700	+39 01932			7 20 44	+39 13.8	2	8.33	.005	0.43	.005	-0.02	.000	5	F5 V	833,1601
		Haf 8 - 96		7 20 44	−12 11.1	1	13.04		0.39		0.20		2		410
		Haf 8 - 97		7 20 44	−12 11.1	1	13.41		0.28		0.08		2		410
		Haf 8 - 95		7 20 44	−12 13.1	3	11.72	.007	0.48	.003	-0.16	.009	10		405,410,1737
		Haf 8 - 98		7 20 44	−12 13.2	1	14.17		0.43		0.29		2		410
	−17 01944	LSS 348		7 20 45	−17 51.2	1	10.68		0.25		-0.68		3		1737
		Haf 8 - 60		7 20 46	−12 10.9	1	10.82		0.58		0.09		2		410
58036	−26 04271			7 20 46	−26 16.0	1	8.36		-0.15		-0.65		2	B5 III	401
58109	−36 03565			7 20 46	−36 38.8	1	9.36		0.27		0.11		2	F0 V	852
	+20 01790	G 88 - 24		7 20 47	+20 31.0	1	9.93		1.15		1.02		1	K3	333,1620
		Haf 8 - 54		7 20 47	−12 11.6	1	13.54		0.34		0.35		2		410
		Haf 8 - 55		7 20 48	−12 11.3	2	12.53	.005	1.31	.004	1.07	.011	4		410,1685
		Haf 8 - 23		7 20 48	−12 13.1	1	12.60		0.79		0.24		2		410
		Haf 8 - 22		7 20 49	−12 12.9	1	12.30		0.33		0.15		2		410
	−31 04432	Cr 140 - 54		7 20 49	−31 54.3	3	10.23	.004	1.48	.000	1.69	.010	4	K4 III	74,703,716
57791	+28 01377			7 20 50	+27 53.6	1	7.67		1.32		1.47		2	K2	1375
58136	−36 03568			7 20 50	−36 57.1	1	8.80		0.99		0.86		2	K1 III	852
	−5 02093			7 20 55	−05 23.3	1	9.43		-0.06		-0.46		3		1732
58061	−25 04441	VY CMa	★ABC	7 20 55	−25 40.2	1	7.95		2.24		1.82		1	M3/4 (II)	611
	−32 03941	Cr 140 - 46		7 20 55	−32 09.3	2	9.89	.005	0.70	.019	0.23	.014	4	G2 V:p	703,716
	+12 01523			7 20 57	+12 10.7	1	9.69		0.77		0.37		2	K0	1375
	−14 01873	IDS07187S1440	AB	7 20 57	−14 45.8	1	8.65		0.61				21		6011
58032	−14 01872			7 20 57	−14 46.9	1	9.41		0.71				19	K0 III +F	6011
57668	+47 01442			7 20 58	+47 03.9	1	9.16		0.45		0.06		2	G5	1601
58156	−36 03571			7 20 58	−36 39.4	1	9.30		1.20		1.11		1	K1 III	852
57901	+13 01655	G 89 - 16		7 20 59	+13 04.1	2	8.20	.000	0.95	.007	0.80	.000	4	G5	333,1620,3040
	−25 04446			7 20 59	−25 25.4	2	9.18	.011	-0.16	.010	-0.77	.002	4	A2	540,976
	−31 04435	Cr 140 - 59		7 20 59	−31 54.3	3	10.62	.000	0.87	.005	0.50	.004	5	G9 III	74,703,716
57308	+69 00422	IDS07156N6941	A	7 21 01	+69 35.0	1	8.05		0.81				14	G0	6011
57646	+52 01205	HR 2804		7 21 04	+51 59.2	2	5.80	.021	1.61	.000	1.96		4	K5 III	71,3001
58180	−36 03572	IDS07193S3653	AB	7 21 05	−36 58.7	1	8.29		0.28		0.10		1	A8 V	852
58055	−12 01933	IDS07188S1228	AB	7 21 06	−12 34.1	1	9.23		0.95		0.68		2	B8	401
58155	−31 04437	Cr 140 - 3	★	7 21 06	−31 49.6	8	5.42	.007	-0.16	.009	-0.68	.017	47	B4 Vnp	15,401,681,703,815*
58238	−44 03369			7 21 07	−44 31.2	1	7.71		-0.10		-0.53		2	B3/5 III	401
	+45 01429			7 21 13	+45 17.6	1	9.37		1.19		1.05		2	K2	1375
58028	−5 02095	IDS07188S0516	A	7 21 13	−05 22.0	1	8.25		0.07		0.04		1	A0	1732
58131	−19 01854	LSS 357		7 21 14	−20 07.9	2	7.38	.005	0.33	.005	-0.35		8	B9 Iab	2033,8100
		LS VI -03 017		7 21 15	−03 06.4	1	12.03		-0.15		-0.82		3		490
	−31 04442	Cr 140 - 68		7 21 15	−32 05.8	2	11.05	.068	1.02	.175	0.74	.351	3	K1 III	703,716
	−31 04443	Cr 140 - 35		7 21 16	−31 49.6	3	9.45	.017	0.95	.004	0.65	.041	7	G7 III	74,703,716
		Cr 140 - 75		7 21 18	−32 03.9	1	11.48		0.64		0.08		2		703
		G 193 - 46		7 21 19	+50 38.1	2	15.08	.009	1.36	.009	1.15		4		538,782
	−31 04444	Cr 140 - 62		7 21 19	−32 07.6	1	10.89		1.23		1.24		2	K0 III	703
58127	−13 01981			7 21 20	−14 00.1	1	7.58		0.03		-0.41		3	B8 V	1212
58216	−31 04445	Cr 140 - 11		7 21 21	−31 56.9	4	8.25	.008	-0.10	.025	-0.40	.007	9	B7 V	74,401,703,716

Table 1

HD	DM	Other Id	N Rem	α_{1950}	δ_{1950}	S	V	σ_V	B–V	σ_{B-V}	U–B	σ_{U-B}	n	Spectrum	References
	+46 01264	G 107 - 68		7 21 22	+46 12.2	1	9.16		0.82		0.46		2	K0 V	1003
		NGC 2374 - 40		7 21 23	−13 09.9	1	11.75		1.34		0.91		3	K5	1568
58235	−36 03575		AB	7 21 24	−36 43.3	2	8.48	.025	-0.07	.010	-0.43	.005	5	B7 V	401,852
57953	+23 01704			7 21 25	+23 01.6	1	8.15		0.39		0.00		2	F5	1625
57927	+27 01374	HR 2816		7 21 27	+27 44.2	2	5.74	.029	0.32	.024	0.07	.010	3	F1 III	39,254
	−31 04448	Cr 140 - 60		7 21 27	−31 44.3	2	10.71	.010	0.10	.010	0.11	.010	3	F5 V	703,716
58261	−36 03576			7 21 27	−36 56.0	1	8.49		1.05		0.98		1	K1 III	852
		NGC 2374 - 45		7 21 28	−13 09.0	1	13.71		0.31		0.13		3	A5	1568
58215	−27 04020	HR 2822	AB	7 21 28	−27 44.2	2	5.39	.010	1.53	.005	1.83		6	K4 III	2031,3005
		NGC 2374 - 46		7 21 30	−13 09.6	1	13.72		0.25		0.02		3	A5	1568
		NGC 2374 - 47		7 21 31	−13 09.8	1	11.84		1.10		0.52		3	K0	1568
58195	−22 01850	BE CMa		7 21 31	−22 52.5	1	9.00		2.87		6.50		2	C5,5	864
58233	−31 04447	Cr 140 - 40		7 21 31	−31 23.6	1	9.61		0.13		0.09		2	A1 V	703
58260	−36 03578			7 21 32	−36 14.5	5	6.74	.015	-0.13	.006	-0.76	.006	15	B2/3 Vp	401,540,852,976,2012
		NGC 2374 - 48		7 21 33	−13 10.4	1	12.33		0.47		-0.16		3	F0	1568
	−31 04450	Cr 140 - 74		7 21 34	−31 50.4	2	11.40	.005	0.12	.000	0.13	.040	2	A0 V	703,716
		NGC 2374 - 49		7 21 35	−13 10.7	1	12.31		0.61		-0.07		3	F0	1568
		NGC 2374 - 81		7 21 35	−13 11.9	1	13.38		0.44		0.10		3	F5	1568
58049	+16 01468			7 21 36	+16 02.6	1	8.28		0.89				1	G5 V	882
		NGC 2374 - 77		7 21 36	−13 08.2	1	10.65		0.45		-0.12		3	B5	1568
		NGC 2374 - 79		7 21 36	−13 09.4	1	11.66		0.13		-0.07		3	B7	1568
58258	−31 04451	Cr 140 - 22		7 21 36	−31 40.2	3	8.96	.005	0.41	.005	-0.04	.007	6	F3 V	74,703,716
	−31 04452	Cr 140 - 66		7 21 36	−31 47.9	2	10.97	.005	0.32	.005	0.08	.039	3		703,716
58050	+15 01564	HR 2817, OT Gem		7 21 37	+15 37.0	2	6.40	.049	-0.17	.034	-0.86	.059	5	B2 Ve	154,1212
		Steph 645		7 21 37	+18 42.6	1	11.29		1.34		1.26		1	K7	1746
		NGC 2374 - 50		7 21 37	−13 10.9	1	11.87		0.18		0.04		3	A0	1568
		NGC 2374 - 51		7 21 37	−13 11.3	1	12.64		0.29		0.11		3	A0	1568
58286	−31 04454	Cr 140 - 2	*	7 21 37	−32 06.3	4	5.38	.006	-0.17	.012	-0.71	.003	27	B2 V	15,401,703,2012
		RU 149G		7 21 39	−00 25.8	1	12.83		0.54		0.03		12		1764
		NGC 2374 - 62		7 21 39	−13 10.1	1	12.03		0.18		-0.02		3	A0	1568
		NGC 2374 - 52		7 21 39	−13 11.2	1	13.76		0.24		0.09		3	A5	1568
58285	−31 04453			7 21 39	−31 12.1	1	7.98		-0.08		-0.35		2	B9 III	401
		NGC 2374 - 63		7 21 40	−13 09.7	1	11.54		0.08		-0.17		3	B7	1568
		RL 149		7 21 41	−00 26.6	1	13.83		-0.17		-0.85		2	sdB	704
		RU 149A		7 21 41	−00 26.8	1	14.50		0.30		0.12		12		1764
		NGC 2374 - 82		7 21 41	−13 09.7	1	13.23		0.40		0.18		3	F0	1568
		RU 149F		7 21 42	−00 25.0	1	13.47		1.12		1.03		11		1764
		RU 149		7 21 42	−00 28.0	1	13.87		-0.13		-0.78		30		1764
		NGC 2374 - 58		7 21 42	−13 10.1	1	13.26		0.31		0.20		3	A5	1568
	−13 01986			7 21 42	−13 35.3	1	9.81		1.89				2	M2	369
		RU 149D		7 21 43	−00 26.7	1	11.48		-0.04		-0.29		11		1764
		NGC 2374 - 65		7 21 43	−13 09.0	1	13.79		0.49		0.14		3		1568
		NGC 2374 - 59		7 21 43	−13 10.3	1	12.60		0.25		0.23		3	A0	1568
	+26 01547	IDS07186N2649	A	7 21 44	+26 43.6	1	9.16		1.32		1.45		2	K5	1733
		NGC 2374 - 69		7 21 44	−13 07.6	1	12.07		1.01		0.55		3	K0	1568
		NGC 2374 - 67		7 21 44	−13 08.2	1	13.88		0.38		0.21		3	F0	1568
		NGC 2374 - 53		7 21 44	−13 10.9	1	13.00		0.25		0.23		3	A5	1568
		RU 149C		7 21 45	−00 26.4	1	14.43		0.20		0.14		11		1764
		RU 149B		7 21 45	−00 27.0	1	12.64		0.66		0.15		10		1764
		NGC 2374 - 57		7 21 45	−13 09.7	1	12.53		0.22		0.21		3	A0	1568
		NGC 2374 - 26		7 21 45	−13 11.9	1	13.13		0.40		0.17		3	F0	1568
		RU 149E		7 21 46	−00 25.2	1	13.72		0.52		-0.01		8		1764
		NGC 2374 - 56		7 21 46	−13 09.0	1	13.04		0.35		0.12		3	A5	1568
		NGC 2374 - 54		7 21 46	−13 10.7	1	12.78		0.19		0.14		3	A0	1568
58256	−23 05395	LSS 360		7 21 48	−23 50.1	2	8.85	.006	0.36	.014	-0.45	.025	4	B2 Ib/II	540,976
58229	−14 01881		V	7 21 50	−14 38.4	1	8.87		1.67				2	M2/3	369
		Cr 140 - 51		7 21 52	−31 58.2	1	10.09		1.25		1.10		4	K2 III	703
	−31 04460	Cr 140 - 39		7 21 52	−32 03.6	2	9.60	.014	0.15	.005	0.12	.019	4	A4 V	703,716
	−15 01804			7 21 56	−15 15.6	1	10.27		1.69				2	M2	369
58325	−29 04322	Cr 121 - 45	*	7 21 57	−30 07.1	4	6.59	.013	-0.20	.010	-0.80	.005	14	B2 IV-V	15,62,401,2012
	−31 04461	Cr 140 - 53		7 21 58	−31 37.7	2	10.17	.024	0.15	.024	0.09	.005	3	A2 V	703,716
	−31 04462	Cr 140 - 36		7 21 58	−31 57.3	2	9.55	.005	0.29	.019	0.13	.037	4	A7 V	703,716
	−32 03961	Cr 140 - 64		7 22 02	−32 08.7	3	10.92	.000	0.55	.000	0.08	.000	5	F9 V	74,703,716
		MCC 495		7 22 05	+44 09.0	2	10.98	.034	1.41	.000	1.24	.004	5	K7-M0	1723,1746
	−25 04470			7 22 05	−25 34.9	1	9.16		1.05		0.89		1		1462
58350	−29 04328	Cr 121 - 34	⋆ A	7 22 07	−29 12.3	10	2.45	.013	-0.08	.010	-0.71	.011	42	B5 Ia	3,9,15,62,1034,1075*
58395	−31 04467	Cr 140 - 23		7 22 07	−31 48.7	3	9.00	.010	0.04	.010	-0.03	.007	7	A1 V	74,703,716
	−31 04466	Cr 140 - 34		7 22 07	−31 51.5	3	9.36	.020	0.87	.025	0.59	.018	6	G8 III:	74,703,716
	−22 04461		V	7 22 09	−22 44.4	1	9.70		2.06				2	M3	369
58377	−28 04286			7 22 09	−28 43.8	3	6.77	.024	-0.16	.012	-0.63	.004	6	B3 III	401,540,976
58394	−31 04468	Cr 140 - 10		7 22 09	−31 47.7	3	7.85	.026	1.47	.016	1.84	.014	5	K3 III	74,703,716
58398	−31 04469	Cr 140 - 12		7 22 09	−32 03.6	3	8.34	.010	0.00	.010	-0.03	.013	7	B8 V:	401,703,716
	−32 03964	Cr 140 - 58		7 22 09	−32 09.7	3	10.45	.005	1.03	.010	0.83	.004	5	G9 III	74,703,716
58346	−22 01855	Cr 121 - 46	*	7 22 10	−22 48.8	4	6.18	.022	-0.09	.012	-0.38	.030	12	B8 V	15,62,1079,2012
58420	−35 03569	HR 2829	⋆ A	7 22 10	−35 44.4	2	6.30	.005	-0.15	.000	-0.56		6	B5 V	401,2006
58187	+11 01578	HR 2820		7 22 12	+11 46.1	2	5.30		0.10	.000	0.13	.000	4	A5 IV	1022,3023
	−31 04470	Cr 140 - 41		7 22 12	−31 39.4	2	9.66	.010	0.51	.010	0.01	.010	3	G0 V	703,716
58396	−31 04472	Cr 140 - 32		7 22 12	−31 47.6	2	9.35	.005	-0.02	.029	-0.07	.007	6	A0 V	74,703,716
58397	−31 04471	Cr 140 - 18		7 22 12	−31 55.7	2	8.72	.000	0.01	.005	-0.16	.009	5	B8 V	703,716
	−4 01937	FS Mon		7 22 14	−05 03.3	1	9.60		0.39		0.00		3	F2	1768

HD	DM	Other Id	N Rem	α_{1950}	δ_{1950}	S	V	σ_V	B–V	σ_{B-V}	U–B	σ_{U-B}	n	Spectrum	References
58321	−20 01906			7 22 15	−21 00.6	2	9.29	.020	0.14	.022	-0.67	.004	4	B2/3 II	540,976
		AJ79,1294 T5# 10		7 22 19	−21 02.1	1	11.70		0.17		-0.07		4		396
	−32 03968	Cr 140 - 65		7 22 19	−32 10.2	3	10.95	.000	1.59	.005	1.94	.010	6	K4 III	74,703,716
	−32 03970	Cr 140 - 61		7 22 21	−32 08.7	3	10.74	.000	1.34	.005	1.37	.007	6	K1 III	74,703,716
		G 89 - 17		7 22 22	+06 45.0	1	12.21		1.44		1.05		1		333,1620
	−20 01909	NGC 2383 - 8		7 22 22	−20 51.1	2	9.76	.002	1.68	.006	1.97	.017	5	K3 III:	183,1685
58120	+35 01609			7 22 23	+35 25.3	1	8.74		0.43		0.02		2	F5	1601
58343	−15 01810	HR 2825, FW CMa		7 22 24	−16 06.1	4	5.36	.040	-0.06	.010	-0.58	.015	26	B2 Vne	681,815,1212,8035
	−19 01865		V	7 22 24	−19 27.6	1	9.72		0.75		0.25		1	K0	1696
58489	−39 03287			7 22 24	−39 56.7	1	9.76		1.07		1.00		2	K2/3 (V)	3072
		NGC 2383 - 7		7 22 25	−20 50.8	1	11.62		1.28		1.03		2		183
		NGC 2383 - 3		7 22 29	−20 49.5	1	12.48		0.18		0.06		2		183
		NGC 2383 - 6		7 22 29	−20 50.7	1	13.25		0.26		0.05		3		183
		NGC 2383 - 10		7 22 29	−20 51.4	1	12.75		0.09		-0.23		2		183
57742	+66 00502	HR 2809		7 22 30	+66 26.0	2	6.52		-0.09	.000	-0.19	.005	4	B9 III	985,1079
		NGC 2383 - 11		7 22 30	−20 50.9	1	13.56		0.23		0.17		2		183
		Ru 18 - 14		7 22 30	−26 06.0	1	13.22		0.49		0.06		2		410
	−25 04486	IDS07205S2551	AB	7 22 31	−25 56.8	1	9.35		0.65		0.27		1	G5	1462
		Ru 18 - 16		7 22 31	−26 06.8	1	13.05		0.52		0.11		2		410
	−31 04476	Cr 140 - 78		7 22 31	−31 56.6	1	10.82		1.25		1.21		1		716
58774	−62 00824			7 22 31	−62 10.4	1	7.08		0.98				4	K0 III	2012
		NGC 2383 - 5		7 22 32	−20 50.4	1	12.95		0.23		-0.16		2		183
58416	−20 01912	LSS 374		7 22 33	−20 36.7	3	9.34	.007	0.03	.017	-0.76	.011	7	B1/2 II/III	354,540,976
		NGC 2383 - 9		7 22 33	−20 48.8	1	11.96		1.22		0.85		2		183
		NGC 2383 - 4		7 22 33	−20 49.6	1	12.94		0.68		0.07		2		183
		Ru 18 - 15		7 22 33	−26 04.2	1	11.54		1.60		1.46		2		410
58657	−53 01334			7 22 33	−53 34.3	2	7.22	.000	-0.07	.000	-0.20	.005	5	B8/9 V	55,158
	−20 01913	NGC 2383 - 1		7 22 34	−20 50.3	1	9.92		0.02		-0.64		5	B0 III	183
58441	−26 04315			7 22 34	−26 34.8	1	8.04		-0.15		-0.71		2	B4 II/III	401
		WLS 720 10 # 9		7 22 35	+10 38.0	1	11.37		0.28		0.12		2		1375
		Ru 18 - 17		7 22 35	−26 07.4	1	12.74		0.57		0.28		2		410
58207	+28 01385	HR 2821		7 22 37	+27 54.0	8	3.79	.006	1.02	.014	0.85	.012	20	G9 IIIb	15,1119,1194,1355*
		Ru 18 - 19		7 22 37	−26 10.1	1	12.91		0.66		0.29		1		410
	−8 01900			7 22 38	−08 11.7	1	9.60		2.09		2.25		1	SC?	864
		Ru 18 - 20		7 22 38	−26 10.5	1	13.20		0.68		0.24		1		410
58439	−18 01825	HR 2831, LSS 375	⋆AB	7 22 39	−18 54.8	3	6.26	.017	0.29	.007	0.00		9	A2 Ib	2007,6009,8100
		Ru 18 - 10		7 22 39	−26 05.4	1	13.39		0.34		-0.04		2		410
		Ru 18 - 18		7 22 39	−26 07.9	1	13.83		0.71		0.47		2		410
	−32 03976	Cr 140 - 30		7 22 39	−32 17.0	1	9.31		0.90		0.56		3	G8 III	703
58440	−20 01914	NGC 2383 - 2		7 22 40	−20 50.3	1	9.77		0.19		0.19		4	A3 V	183
		Ru 18 - 9		7 22 40	−26 05.5	1	12.06		0.58		0.03		2		410
58513	−31 04478	Cr 140 - 38	AB	7 22 41	−31 58.2	3	9.59	.014	0.22	.000	0.12	.009	5	A5 V	74,703,716
	−25 04490	Ru 18 - 1		7 22 42	−26 06.4	1	10.98		-0.02		-0.28		40		410
	−25 04489	Ru 18 - 2		7 22 42	−26 06.9	1	11.90		0.62		0.08		2		410
		Ru 18 - 11		7 22 42	−26 07.	1	13.89		0.88		0.73		1		410
		Ru 18 - 12		7 22 42	−26 07.	1	14.23		0.46		0.23		2		410
		Ru 18 - 13		7 22 42	−26 07.	1	14.28		0.47		0.46		2		410
58385	−2 02101			7 22 43	−03 02.1	1	8.64		2.29		3.02		1	C6,3	864
		Ru 18 - 6		7 22 43	−26 06.3	1	13.13		0.74		0.39		2		410
		Ru 18 - 3		7 22 43	−26 07.4	1	13.54		0.65		0.23		2		410
		Ru 18 - 7		7 22 44	−26 05.2	1	12.92		0.57		0.30		2		410
		AJ79,1294 T5# 9		7 22 45	−20 59.8	1	11.42		0.12		-0.57		4		396
		Ru 18 - 8		7 22 45	−26 04.9	1	13.63		0.58		0.02		2		410
		Ru 18 - 4		7 22 45	−26 06.8	1	13.53		0.70		0.42		2		410
	−31 04479	Cr 140 - 79		7 22 45	−31 37.5	1	10.72		0.10		0.12		1		716
58297	+13 01663			7 22 46	+13 10.4	1	7.56		1.23				3	K2	882
	−19 01871	LSS 378		7 22 46	−20 0.9	1	10.62		0.18		-0.80		2		1737
		Ru 18 - 5		7 22 46	−26 06.1	1	13.35		0.85		0.69		2		410
58534	−31 04481	Cr 140 - 9	⋆ C	7 22 46	−31 41.0	4	7.59	.004	-0.10	.009	-0.48	.007	14	B7 V	74,401,703,716
	−31 04480	Cr 140 - 37		7 22 46	−31 45.0	5	9.57	.005	0.45	.005	0.08	.009	5	B7 V:	703,716
58185	+38 01757	IDS07194N3801	AB	7 22 47	+37 55.4	1	8.38		0.93		0.65		3	K0	1723
58535	−31 04482	Cr 140 - 1	⋆AB	7 22 49	−31 42.6	7	5.35	.011	1.07	.009	0.90	.011	33	G8 II	15,74,703,716,2013*
58461	−13 02001	HR 2832		7 22 50	−13 39.1	3	5.78	.005	0.42	.002	-0.04		9	F3 V	15,2027,3037
58462	−14 01887			7 22 50	−14 47.0	1	6.77		0.04		0.04		2	A1 V	401
		NGC 2384 - 2	⋆ B	7 22 51	−20 55.2	1	9.59		0.00		-0.79		2	B2 III	183
58465	−20 01915	NGC 2384 - 1	⋆AB	7 22 51	−20 55.3	1	9.05		0.02		-0.87		2	O8 V	183
	−20 01916	LSS 382		7 22 52	−21 03.6	3	8.31	.018	0.07	.008	-0.79	.015	10	B0.5III:N	396,540,976
		AS 150, BX Mon		7 22 53	−03 29.9	1	11.68		1.24		-0.22		1	Mpe	1753
		NGC 2384 - 3		7 22 53	−20 55.4	1	11.85		0.11		0.07		2	A5	183
		RL 150		7 22 54	+00 04.	1	13.05		-0.07		-0.46		2		704
		NGC 2384 - 4		7 22 54	−20 55.9	1	13.25		0.27		0.21		2	F0	183
		AJ79,1294 T5# 11		7 22 54	−21 00.4	1	11.73		0.06		-0.56		4		396
58367	+9 01643	HR 2828		7 22 55	+09 22.6	10	4.99	.010	1.01	.006	0.78	.006	28	G6.5 IIb	3,15,1007,1013,1080*
	−20 01918	NGC 2384 - 8		7 22 55	−20 54.8	1	10.21		0.06		-0.72		2	B2 V	183
		NGC 2384 - 7		7 22 55	−20 55.5	1	12.52		0.58		0.00		2		183
		NGC 2384 - 6		7 22 56	−20 55.5	1	12.98		0.06		-0.48		3		183
		NGC 2384 - 5		7 22 56	−20 55.9	1	11.97		0.11		-0.38		1	A0	183
58142	+49 01623	HR 2818		7 22 57	+49 18.8	4	4.62	.056	-0.02	.007	-0.01	.008	9	A1 V	15,1363,3023,8015
58510	−20 01919	IDS07208S2059	AB	7 22 58	−21 04.5	5	6.80	.005	0.11	.008	-0.74	.014	14	B1 Ia	540,976,1737,2033,8100
		NGC 2384 - 9		7 22 59	−20 55.5	1	13.96		0.17		-0.21		2	F5	183

Table 1 397

HD	DM	Other Id	N Rem	α_{1950}	δ_{1950}	S	V	σ_V	B–V	σ_{B-V}	U–B	σ_{U-B}	n	Spectrum	References
		L 184 - 52		7 23 00	−57 54.	1	12.41		0.95		0.62		1		1696
		AJ79,1294 T5# 13		7 23 01	−20 45.2	1	12.34		0.15		−0.38		3		396
58635	−37 03549	HR 2842/3	⋆ AB	7 23 01	−37 11.5	2	6.16	.005	0.26	.010	0.15		6	F0 V	404,2006
58337	+22 01679			7 23 03	+22 00.6	1	9.49		1.32				1	R4	1238
		AJ79,1294 T5# 14		7 23 03	−20 50.4	1	13.19		0.44		0.17		5		396
58509	−20 01920	NGC 2384 - 10		7 23 03	−20 55.4	6	8.58	.015	0.02	.017	−0.83	.021	16	B0.5III	183,396,540,540,976, 976
		NGC 2384 - 11		7 23 04	−20 55.9	1	10.10		0.01		−0.81		4	B3 IV	183
		NGC 2384 - 12		7 23 05	−20 55.0	1	11.73		0.06		−0.60		2	B8	183
58366	+14 01664	IDS07204N1418	A	7 23 06	+14 11.2	1	8.70		0.14		0.12		2	A2	3032
58366	+14 01664	IDS07204N1418	B	7 23 06	+14 11.2	1	11.74		0.64		0.15		2		3032
		NGC 2384 - 13		7 23 06	−20 54.7	2	10.79	.019	0.04	.019	−0.67	.015	6	B8	183,396
58631	−31 04488	Cr 140 - 26	⋆ AB	7 23 07	−32 07.6	3	9.14	.005	0.18	.020	0.09	.005	6	A4 V	74,703,716
58364	+22 01680			7 23 08	+21 59.5	1	9.20		1.52				1	R4	1238
		AJ79,1294 T5# 15		7 23 08	−20 43.1	1	13.55		0.30		0.15		4		396
	−20 01921	LSS 390		7 23 09	−20 33.7	1	10.81		0.02		−0.76		3		396
		NGC 2384 - 14		7 23 09	−20 54.6	1	10.35		1.50		1.77		2	A5	183
58585	−21 01925	HR 2839	⋆ AB	7 23 11	−21 53.0	3	6.04	.007	0.24	.004			11	A5/7 II	15,2013,2029
	−31 04491	Cr 140 - 71		7 23 11	−31 52.5	1	11.23		0.43		0.06		1	F2 V	703
		NGC 2384 - 15		7 23 12	−20 53.4	1	12.22		0.14		0.11		2	F0	183
58383	+14 01665	IDS07204N1418	A	7 23 13	+14 12.2	1	7.41		−0.09		−0.31		3	A0 V	3016
58383	+14 01665	IDS07204N1418	B	7 23 13	+14 12.2	1	9.52		0.04		0.02		2		3032
58383	+14 01665	IDS07204N1418	C	7 23 13	+14 12.2	1	8.70		0.14		0.12		2		3016
58383	+14 01665	IDS07204N1418	D	7 23 13	+14 12.2	1	11.74		0.64		0.15		2		3016
58676	−38 03365	IDS07215S3809	A	7 23 14	−38 15.8	2	9.48	.015	0.89	.010	0.60	.034	6	K1/2 (V)	158,3072
		AJ79,1294 T5# 12		7 23 17	−20 39.0	1	11.90		0.13		−0.37		3		396
		Cr 140 - 77		7 23 17	−31 51.5	1	11.89		0.10		0.09		2	A1 V	703
	−31 04493	Cr 140 - 67		7 23 18	−31 50.6	1	11.05		0.37		0.09		3	F1 V	703
58895	−58 00909	IDS07224S5818	A	7 23 18	−58 23.7	1	6.58		0.71				4	G3 IV	2012
58676	−38 03367	IDS07215S3809	B	7 23 19	−38 15.7	2	9.65	.010	0.94	.005	0.71	.024	6		158,3072
58612	−24 05366	Cr 121 - 47	⋆	7 23 21	−25 07.0	5	5.78	.005	−0.10	.009	−0.39	.008	13	B5 III	15,62,611,1079,2012
58526	−5 02112	HR 2833		7 23 24	−05 40.5	6	5.98	.010	0.92	.004	0.62	.007	22	G3 Ib	15,1415,1509,2027*
	−20 01926	LSS 394		7 23 25	−20 39.1	1	10.62		0.07		−0.64		3		396
58295	+43 01691			7 23 26	+43 21.6	1	6.67		1.37		1.53		2	K2	1601
		MWC 560		7 23 26	−07 38.1	1	9.54		0.31		−0.56		1		1753
		G 193 - 49	A	7 23 29	+52 32.3	1	11.36		0.90		0.46		2	K2	7010
58702	−31 04500	Cr 140 - 27		7 23 29	−32 08.0	2	9.17	.015	−0.12	.010	−0.55	.000	5	B8 V	401,703
	−51 02481			7 23 31	−51 14.3	1	10.18		0.92		0.75		3	K0	803
	+5 01660			7 23 33	+05 32.4	1	10.04		−0.06		−0.26		1		1748
58453	+18 01616	IDS07206N1842	A	7 23 33	+18 37.1	1	7.26		0.33		0.12		2	F0	3032
58605	−10 02037			7 23 34	−10 37.0	1	9.10		0.00		−0.14		2	B9	3039
58580	−4 01943	HR 2838		7 23 35	−04 26.2	4	6.75	.009	−0.01	.009	−0.11	.003	15	A0 IV	15,1079,1415,2012
58477	+18 01618	IDS07206N1842	B	7 23 37	+18 36.9	1	8.20		0.34		0.05		2	F2 IV	3032
58477	+18 01618	IDS07206N1842	C	7 23 37	+18 36.9	1	9.92		0.18		0.09		2		3032
58477	+18 01618	IDS07206N1842	D	7 23 37	+18 36.9	1	13.62		0.60		0.06		2		3032
58625	−11 01934			7 23 37	−11 54.2	1	8.42		0.03		−0.09		2	B9	3039
	−9 02038			7 23 38	−09 51.1	1	9.29		1.74				2	M2	369
58647	−13 02008			7 23 38	−14 04.7	1	6.81		0.04		−0.11		2	B9 IV	401
58671	−22 01864			7 23 38	−22 45.6	1	9.12		−0.02				4	B5 III	2014
		LS VI -05 011		7 23 40	−05 42.5	1	12.13		0.01		−0.20		3		490
		LOD 1- # 8		7 23 40	−23 01.4	1	10.58		1.28		1.19		2		126
	−31 04503	Cr 140 - 80		7 23 40	−31 34.2	1	10.28		0.71		0.25		1		716
58552	+10 01532	HR 2836		7 23 43	+10 42.5	1			0.09		0.06		3	A2 IV	1022
58767	−32 03999	Cr 140 - 15		7 23 43	−32 24.5	4	8.59	.005	−0.07	.013	−0.30	.004	8	B8 V	74,401,703,716
58644	−8 01909			7 23 44	−08 59.7	1	8.26		−0.03		−0.67		2	B2 IV	3039
		WLS 720 30 # 9		7 23 46	+29 27.0	1	13.45		0.49		−0.08		2		1375
		LOD 1- # 9		7 23 48	−23 02.9	1	10.47		0.80		0.36		2		126
58766	−31 04506	Cr 140 - 4	⋆	7 23 48	−31 38.3	6	6.30	.005	−0.18	.009	−0.70	.011	26	B2 V	15,74,401,703,716,2028
		Wat 7 - 9		7 23 49	−14 59.6	1	14.59		0.38		0.31		1		862
	−31 04507	Cr 140 - 50		7 23 49	−31 53.6	1	10.03		1.02		0.77		2	G4 III	703
		Wat 7 - 10		7 23 50	−14 58.9	1	12.80		0.49		0.09		1		862
	−14 01893	LSS 397		7 23 50	−14 59.1	1	11.96		0.23		−0.25		1		862
		LSS 399		7 23 50	−14 59.6	1	12.44		0.24		−0.25		1	B3 Ib	862
		Wat 7 - 8		7 23 50	−14 59.6	1	14.46		0.37		0.31		1		862
		LOD 1- # 10		7 23 50	−23 03.9	1	10.76		0.56		0.17		2		126
		Wat 7 - 7		7 23 51	−14 59.6	1	13.21		0.27		−0.16		1	B6 V	862
		LOD 1- # 7		7 23 51	−23 01.2	1	11.72		0.19		0.12		2		126
		Wat 7 - 5		7 23 52	−14 58.7	1	13.88		0.27		−0.21		1	B7 V	862
		Wat 7 - 2		7 23 52	−15 00.0	1	13.52		0.29		−0.16		1	B6 V	862
58551	+21 01596	HR 2835		7 23 53	+21 38.2	3	6.54	.007	0.46	.004			38	F6 V	130,1351,1758
58720	−14 01894	LSS 402		7 23 53	−14 59.2	3	9.24	.019	0.21	.011	−0.67	.005	4	B2 Ib/II	138,401,862
		Wat 7 - 4		7 23 53	−15 00.1	1	12.43		0.58		0.08		1		862
		LSS 403		7 23 54	−15 38.1	1	12.37		1.16		1.24		1	B3 Ib	862
58722	−17 01971			7 23 54	−18 01.2	1	7.32		−0.02		−0.41		2	B5 III	401
58599	+11 01588	HR 2840		7 23 55	+11 06.6	2			−0.13	.000	−0.47	.000	6	B7 II-III	1022,1079
58313	+52 01212			7 23 56	+52 28.2	1	8.89		0.46		0.02		2	G0	1566
		LOD 1- # 6		7 23 56	−23 00.2	1	10.80		0.34		0.00		2		126
58522	+24 01659			7 23 57	+24 01.7	1	8.84		−0.01		−0.06		2	A2	1625
		LSS 404		7 23 57	−15 52.6	1	11.79		0.56		−0.32		1	B3 Ib	862
58579	+20 01805	HR 2837		7 24 00	+20 21.5	2	5.93	.010	0.29	.019	0.08	.019	3	F0 III	39,254
58451	+39 01945			7 24 00	+39 25.0	2	7.00	.015	1.03	.005	0.93	.010	5	K1 III	1501,1601

HD	DM	Other Id	N Rem	α_{1950}	δ_{1950}	S	V	σ_V	B–V	σ_{B-V}	U–B	σ_{U-B}	n	Spectrum	References
58738	−21 01936			7 24 01	−22 07.0	1	8.64		−0.06				4	B3 V	2014
		LTT 12019		7 24 02	+24 15.7	1	10.95		1.19		1.10		2		1723
	−25 04516	SS CMa		7 24 02	−25 09.5	2	9.67	.293	1.16	.170	0.87	.113	2	F6	689,1587
		CCS 763		7 24 03	−19 39.3	1	10.86		5.50				1	C5,4	864
58797	−26 04355			7 24 03	−26 41.4	1	7.41		1.43		1.62		3	K3 III	1657
	−23 05459			7 24 04	−23 15.6	1	9.66		0.84		0.49		2	K0	126
		LOD 1- # 11		7 24 05	−23 04.1	1	11.15		1.66		1.72		2		126
		LOD 1- # 14		7 24 05	−23 06.5	1	12.87		0.56		0.09		2		126
	−32 04007	Cr 140 - 55		7 24 06	−32 10.1	1	10.25		1.24		1.10		2	G5 III	703
58890	−37 03569	IDS07224S3757	AB	7 24 06	−38 02.9	1	8.20		0.28		0.01		2	B8/9 V	401
	+19 01730	G 88 - 27		7 24 07	+19 12.2	3	10.74	.010	0.44	.004	−0.19	.000	7	sdF8	516,1620,1696
		WLS 728 55 # 10		7 24 07	+54 46.4	1	14.05		0.67		−0.02		2		1375
		AO 1087		7 24 08	+05 07.0	1	10.08		0.42		−0.04		1	F5 V	1748
	+5 01664	IDS07215N0522	AB	7 24 08	+05 16.5	1	9.66		0.47		0.01		1	F2	1748
	−20 01937	LSS 409		7 24 08	−20 47.3	1	10.20		0.02		−0.71		3		396
58823	−25 04518			7 24 08	−26 08.1	1	8.89		−0.14		−0.70		2	B2 V	401
	−32 04008	Cr 140 - 33	★AB	7 24 08	−32 14.7	1	9.37		1.01		0.80		2	K2 III	703
58760	−15 01822			7 24 10	−15 40.2	1	9.29		1.16		1.02		1	K4 V	3072
58497	+38 01760			7 24 11	+38 46.1	1	6.88		0.90		0.52		2	G8 III	105
		LOD 1- # 5		7 24 11	−22 56.2	1	10.62		0.50		0.04		2		126
		AO 1089		7 24 14	+05 28.3	1	10.29		0.53		0.10		1	F7 V	1748
		LOD 1- # 16		7 24 15	−23 16.8	1	10.83		1.02		0.86		2		126
58888	−31 04517	Cr 140 - 8		7 24 15	−31 32.6	2	7.49	.004	0.04	.034	0.03	.025	7	A2 IV	74,703,716
	−31 04519	Cr 140 - 76		7 24 15	−31 56.7	1	11.62		0.08		0.02		2	A0 V	703
58821	−18 01837			7 24 16	−18 39.4	1	8.28		−0.08				4	B8 V	2014
	−31 04520	Cr 140 - 69		7 24 16	−32 07.5	1	11.12		0.36		0.06		1	A8 V	703
	−32 04011	Cr 140 - 49		7 24 16	−32 18.5	1	10.00		0.01		−0.13		2	A0 V	703
58791	−12 01969			7 24 17	−13 05.3	1	6.93		0.00		−0.38		2	B8/9 II/III	401
		LOD 1- # 4		7 24 19	−22 55.8	1	10.56		0.25		0.14		2		126
		WLS 728 55 # 5		7 24 23	+57 07.8	1	11.47		0.27		0.16		2		1375
	−6 02111			7 24 24	−07 02.7	1	10.00		1.73				2	M2	369
	−32 04013	Cr 140 - 48		7 24 24	−32 17.8	1	9.95		1.06		0.91		2	G8 III	703
		LOD 1- # 17		7 24 25	−23 23.2	1	11.25		0.47		0.01		2		126
	−25 04523	VZ CMa		7 24 25	−25 49.5	2	9.14	.032	0.87	.000	0.46		2	F5 II(R)	1462,1488
	−32 04012	Cr 140 - 72		7 24 25	−32 11.0	1	11.30		0.09		0.06		3	A2 V	703
58715	+8 01774	HR 2845, β CMi	★A	7 24 26	+08 23.5	7	2.89	.011	−0.10	.008	−0.29	.012	17	B8 Ve	15,1034,1079,1363*
		LOD 1- # 20		7 24 29	−23 27.2	1	13.23		0.63		0.27		2		126
	+5 01666			7 24 30	+05 38.9	1	9.18		0.98		0.68		1	K0	1748
58886	−22 01871			7 24 31	−22 58.8	1	7.41		−0.01		−0.07		2	A0 V	126
		LOD 1- # 19		7 24 32	−23 26.9	1	11.32		0.04		0.04		2		126
58713	+15 01573	RY Gem	B	7 24 33	+15 45.7	2	11.91	.287	0.61	.090	0.11	.161	7	A2 V	627,3064
58906	−23 05476			7 24 33	−23 25.8	1	9.44		−0.11		−0.53		2	B5 III	126
58521	+46 01271	Y Lyn		7 24 34	+46 05.6	2	6.96	.024	1.74	.010	1.19	.019	3	M5 Ib-II V	3001,8032
58907	−23 05477	HR 2850		7 24 34	−23 36.7	1	6.56		0.01				4	A0 IV	2035
		Ru 20 - 2		7 24 34	−28 42.5	2	11.61	.020	1.04	.000	0.79	.015	5		183,1685
58752	+5 01667			7 24 35	+05 15.0	1	9.09		0.47		−0.02		1	G0	1748
58729	+15 01574	IDS07217N1531	AB	7 24 35	+15 25.1	1	7.22		−0.01		−0.02		8	A0 III+A0 V	627
		Ru 20 - 3		7 24 36	−28 41.8	2	11.66	.006	1.56	.006	1.75	.005	5		183,1685
		Ru 20 - 5		7 24 36	−28 44.6	1	11.79		1.28		1.19		2		183
58548	+48 01535			7 24 38	+48 09.8	1	7.99		1.04		0.82		2	G5	1733
		Ru 20 - 4		7 24 39	−28 42.8	2	11.69	.004	1.56	.014	1.99	.020	4		183,1685
58784	+2 01677			7 24 40	+02 34.0	2	8.58	.009	−0.07	.000	−0.61		5	B3 III	555,2034
		Ru 20 - 6		7 24 40	−28 41.2	1	12.58		0.21		0.20		2		183
58959	−28 04350	Ru 20 - 1		7 24 41	−28 43.8	1	8.81		0.30		0.11		5	F0	183
58712	+22 01687	IDS07217N2221	A	7 24 42	+22 14.6	1	6.71		1.68		2.05		1	K5	8085
58712	+22 01687	IDS07217N2221	C	7 24 42	+22 14.6	1	8.97		0.24		3.54		1	A5	8085
		Ru 20 - 11		7 24 42	−28 42.7	1	13.84		0.44		0.00		2		183
	+5 01668	G 89 - 19		7 24 43	+05 22.8	7	9.84	.021	1.55	.006	1.18	.045	41	M5	694,989,1006,1017,1705,1748*
58683	+27 01389			7 24 43	+27 23.8	1	7.33		0.92		0.63		1	B8 III	3077
58881	−11 01941	FX CMa		7 24 43	−11 37.1	1	8.60		2.15		2.70		11	S3.9	3039
58882	−11 01942			7 24 43	−12 07.3	1	7.92		0.00		−0.24		2	B8 V	3039
		LSS 415		7 24 44	−14 43.8	1	12.86		0.39		−0.58		1	B3 Ib	862
		Ru 20 - 10		7 24 45	−28 45.1	1	12.36		0.62		0.03		2		183
	+5 01669			7 24 46	+05 43.3	1	9.99		0.09		0.15		1		1748
58728	+21 01602	HR 2846	★A	7 24 46	+21 32.9	2	5.25	.010	0.42	.040	−0.04	.015	4	F5 IV-V	254,3026
58728	+21 01602	IDS07218N2139	B	7 24 46	+21 32.9	1	10.92		1.32		1.15		2		3024
		Ru 20 - 7		7 24 47	−28 41.1	1	11.69		0.19		0.19		2		183
	−32 04018	Cr 140 - 25		7 24 47	−32 09.7	1	9.05		0.95		0.61		2	G2 III	703
		Ru 20 - 8		7 24 48	−28 41.7	1	12.12		0.45		0.20		2		183
58902	−13 02018	LSS 416		7 24 49	−13 16.8	2	9.73	.010	0.16	.045	−0.85	.045	4	B1/2 Ia	401,3039
		Ru 20 - 9		7 24 49	−28 44.0	1	12.07		0.06		−0.11		2		183
	−20 01946			7 24 50	−20 48.1	2	9.80	.018	0.05	.011	−0.61	.014	5	B3 II-III	540,976
59027	−34 03607			7 24 50	−34 50.2	1	8.03		0.96				4	G8 III/IV	2012
59026	−33 03813	HR 2856		7 24 51	−34 02.3	3	5.90	.002	−0.17	.003	−0.68	.006	9	B2 IV-V	401,1732,2006
58978	−22 01874	Cr 121 - 35	★V	7 24 52	−22 59.0	9	5.60	.023	−0.11	.012	−0.99	.021	45	B0.5IVpe	62,126,379,399,681*
58955	−19 01890			7 24 53	−19 50.1	2	9.78	.001	0.09	.008	−0.65	.011	4	B9 III/IV	540,976
	−31 04532	Cr 140 - 63		7 24 53	−31 56.5	1	10.92		0.10		0.10		3	A2 V	703
58954	−17 01980	HR 2853	★AB	7 24 54	−17 45.8	1	5.63		0.32				4	F2 V	2031
		Mel 66 - 1101		7 24 54	−47 38.	2	17.30	.065	0.63	.055			2		468,790
		Mel 66 - 1104		7 24 54	−47 38.	1	16.98		0.64				2		468

Table 1   399

HD	DM	Other Id	N Rem	α_{1950}	δ_{1950}	S	V	σ_V	B–V	σ_{B-V}	U–B	σ_{U-B}	n	Spectrum	References
		Mel 66 - 1121		7 24 54	−47 38.	2	16.85	.010	0.61	.005	0.10		2		468,790
		Mel 66 - 1123		7 24 54	−47 38.	2	16.31	.020	0.67	.005			2		468,790
		Mel 66 - 1126		7 24 54	−47 38.	2	16.70	.020	0.61	.030	-0.02		2		468,790
		Mel 66 - 1230		7 24 54	−47 38.	2	15.90	.024	1.04	.019			3		468,790
		Mel 66 - 1321		7 24 54	−47 38.	2	13.11	.029	1.50	.011			4		706,1504
		Mel 66 - 2133		7 24 54	−47 38.	2	13.19	.012	1.50	.002	1.57		7		468,1504
		Mel 66 - 2141		7 24 54	−47 38.	1	16.35		0.66		0.04		2		468
		Mel 66 - 2142		7 24 54	−47 38.	1	16.50		0.62		0.01		2		468
		Mel 66 - 2159		7 24 54	−47 38.	1	17.57		0.66				1		790
		Mel 66 - 2202		7 24 54	−47 38.	1	14.16		1.02				1		706
		Mel 66 - 2206		7 24 54	−47 38.	1	12.63		1.46				2		706
		Mel 66 - 2217		7 24 54	−47 38.	4	14.09	.014	1.27	.012	1.17	.039	7		468,706,790,1504
		Mel 66 - 2238		7 24 54	−47 38.	2	16.32	.015	0.88	.054			3		468,790
		Mel 66 - 2239		7 24 54	−47 38.	3	11.84	.017	1.58	.000	1.84		10		468,706,1504
		Mel 66 - 2261		7 24 54	−47 38.	2	13.61	.008	1.25	.002	1.26		4		468,1504
		Mel 66 - 2277		7 24 54	−47 38.	1	13.80		1.84				1		468
		Mel 66 - 2309		7 24 54	−47 38.	1	13.46		1.26				1		706
		Mel 66 - 2329		7 24 54	−47 38.	1	14.34		1.06				1		706
		Mel 66 - 2338		7 24 54	−47 38.	1	14.20		1.08				2		706
		Mel 66 - 2342		7 24 54	−47 38.	1	13.42		1.30				2		706
		Mel 66 - 3148		7 24 54	−47 38.	1	14.78		1.15		0.69		2		468
		Mel 66 - 3213		7 24 54	−47 38.	1	14.42		1.15		0.91		2		468
		Mel 66 - 3227		7 24 54	−47 38.	2	13.05	.010	0.63	.010	0.08		3		468,790
		Mel 66 - 3229		7 24 54	−47 38.	1	13.97		1.25				2		706
		Mel 66 - 3252		7 24 54	−47 38.	2	15.73	.030	0.72	.030			2		468,790
		Mel 66 - 4108		7 24 54	−47 38.	1	17.44		0.58				1		790
		Mel 66 - 4110		7 24 54	−47 38.	1	14.55		1.12		0.78		2		468
		Mel 66 - 4121		7 24 54	−47 38.	2	17.61	.010	0.52	.030			2		468,790
		Mel 66 - 4151		7 24 54	−47 38.	1	12.69		1.57				1		468
		Mel 66 - 4215		7 24 54	−47 38.	1	12.96		1.07				1		468
		Mel 66 - 4264		7 24 54	−47 38.	2	14.75	.030	1.13	.010	0.76		2		468,790
		Mel 66 - 4326		7 24 54	−47 38.	4	12.76	.032	1.27	.009	1.31		7		468,706,790,1504
		Mel 66 - 4351		7 24 54	−47 38.	2	13.29	.041	1.36	.018			3		706,1504
58747	+26 01564			7 24 57	+26 19.8	1	7.36		0.12		0.07		2	A2	1625
59052	−31 04534	Cr 140 - 14		7 24 57	−32 06.4	1	8.42		1.32		1.45		2	K2 III	703
	−19 01891			7 25 03	−19 33.1	2	9.59	.079	0.06	.027	-0.53	.028	7		540,976
59219	−50 02761	HR 2862		7 25 05	−50 55.0	4	5.09	.004	1.05	.006			12	K0 III	15,1075,1488,2009
59100	−34 03610	IDS07233S3407	B	7 25 07	−34 12.9	3	8.17	.013	0.63	.006	0.10	.029	8	G5 V	158,1279,3062
58661	+48 01538	HR 2844	⋆ AB	7 25 08	+48 17.3	1			-0.10		-0.42		2	B9 p HgMn	1079
59099	−34 03611	IDS07233S3407	A	7 25 08	−34 12.7	4	7.02	.015	0.49	.014	-0.04	.013	13	F6/7 V	158,1279,1311,3062
	−23 05494	Tr 7 - 12		7 25 09	−23 53.3	1	10.99		0.14		-0.27		2	B7 V	183
59640	−71 00574			7 25 09	−71 22.2	1	6.49		0.12		0.08		5	A3 V	1628
	+5 01671			7 25 10	+05 39.5	1	10.18		0.42		0.02		1		1748
	−23 05495	Tr 7 - 1		7 25 11	−23 49.7	1	9.46		0.13		-0.31		6	B4 III	183
		Tr 7 - 16		7 25 11	−23 50.7	1	13.75		0.31		0.19		2		183
58973	−2 02123			7 25 12	−02 58.1	2	7.84	.005	-0.14	.009	-0.62		5	B5	555,2033
		Tr 7 - 3		7 25 12	−23 49.	1	12.27		0.21		-0.18		2		183
		Tr 7 - 13		7 25 12	−23 51.6	1	12.97		0.25		0.09		2		183
		Tr 7 - 15		7 25 13	−23 50.9	1	12.98		0.28		0.18		2		183
	−23 05497	Tr 7 - 2		7 25 14	−23 49.0	1	10.76		0.14		-0.25		3	B8 V	183
		Tr 7 - 4		7 25 14	−23 49.6	1	12.74		0.21		-0.06		2		183
		Tr 7 - 5		7 25 15	−23 49.9	1	11.33		0.29		0.17		2	A2 V	183
		Tr 7 - 14		7 25 15	−23 51.2	1	12.98		0.23		0.09		2		183
		LSS 419		7 25 16	−15 45.6	1	12.14		0.41		-0.28		1	B3 Ib	862
		AO 1095		7 25 17	+05 23.2	1	10.43		0.97		0.79		1		1748
	−23 05500	Tr 7 - 6	⋆ AB	7 25 17	−23 50.4	1	9.13		0.06		-0.73		3	B2 III	183
		Tr 7 - 7		7 25 18	−23 51.	1	11.58		0.09		-0.40		2		183
		Tr 7 - 9		7 25 18	−23 52.1	1	11.74		0.14		-0.29		2		183
	−23 05501	Tr 7 - 11		7 25 18	−23 53.3	1	10.80		0.12		-0.31		2		183
		Tr 7 - 8		7 25 20	−23 50.8	1	11.22		0.18		-0.30		2		183
		Tr 7 - 10		7 25 20	−23 52.0	1	13.87		0.34		0.37		2		183
58923	+7 01729	HR 2851	⋆ AB	7 25 21	+07 02.7	3	5.24	.011	0.22	.005	0.17	.012	8	F0 III	15,1415,3023
59074	−18 01846	IDS07231S1817	A	7 25 21	−18 23.4	1	6.95		-0.09		-0.46		2	B9 II	401
59197	−41 03077			7 25 21	−41 31.6	1	8.11		0.05		-0.36		2	B6 V	401
	+29 01536			7 25 22	+29 44.3	1	10.04		1.37		1.47		2		97
57724	+78 00254			7 25 22	+78 47.6	1	7.44		0.56		0.11		2	G0	1502
58830	+32 01559			7 25 23	+31 58.0	2	7.74	.002	0.94	.010	0.68		5	G0	1758,3072
59075	−18 01847	IDS07231S1817	B	7 25 24	−18 23.6	1	7.65		0.34		-0.27		2	B9 Ib	1012
58972		HR 2854	⋆ A	7 25 26	+09 01.7	6	4.31	.016	1.42	.009	1.54	.005	17	K3 III	15,369,1080,1363,1415,8015
	−19 01896	LSS 422		7 25 26	−19 21.4	1	9.70		0.04		-0.74		2		976
58899	+21 01606			7 25 27	+21 39.2	2	7.16	.000	0.92	.000			33	G5	130,1351
58971	+9 01661	G 89 - 20		7 25 28	+09 36.3	1	8.78		0.76		0.41		1	G5	333,1620
59166	−31 04543	Cr 140 - 16		7 25 29	−31 35.2	1	8.68		0.40		0.13		2	F0 III:	703
59067	−11 01951	HR 2859	⋆ AB	7 25 31	−11 27.2	3	5.80	.037	0.58	.000			9	G8 Ib-II +	15,2027,6009
59138	−26 04389			7 25 32	−26 55.3	1	8.53		-0.13		-0.55		2	B5 III	401
59094	−15 01837	LSS 423		7 25 35	−15 59.5	1	8.44		0.15		-0.70		2	B2 (V)ne	1012
59136	−22 01878	HR 2860		7 25 35	−22 45.4	4	5.96	.016	-0.09	.012	-0.40	.017	9	B5 III	126,401,1079,2007
59193	−31 04546	Cr 140 - 28		7 25 36	−31 50.2	1	9.19		0.00		-0.14		1	B9 V	703
	+19 01739	G 88 - 29		7 25 37	+19 45.6	1	9.26		0.77		0.33		1	K0	333,1620
57508	+81 00252	HR 2797		7 25 39	+81 00.2	1	6.51		0.98		0.75		2	gK7	1733

HD	DM	Other Id	N Rem	α_{1950}	δ_{1950}	S	V	σ_V	B–V	σ_{B-V}	U–B	σ_{U-B}	n	Spectrum	References
59114	−15 01839	LSS 424		7 25 39	−15 17.0	2	9.49	.014	0.48	.000	-0.54	.000	5	O7 Iaf	862,1737
58425	+68 00480	HR 2830		7 25 42	+68 34.2	1	5.64		1.11				2	gK2	71
		RL 151		7 25 42	−02 19.	1	11.98		-0.07		-0.30		2		704
58680	+57 01080			7 25 43	+57 04.6	1	8.48		1.05		0.81		2	G8 III	1733
59162	−21 01949			7 25 44	−21 46.6	1	8.80		-0.08				4	B8 IV	2014
		G 89 - 21		7 25 47	+08 14.6	2	13.26	.005	0.66	.000	0.01	.010	2		1658,3062
	+32 01561	G 90 - 1		7 25 49	+32 05.7	2	7.73	.005	0.93	.000	0.63	.015	3	K2 V	333,1375,1620
59215	−28 04378	IDS07238S2810	AB	7 25 49	−28 16.1	1	7.10		-0.12		-0.54		2	B5 V	401
		AO 1096		7 25 53	+05 12.9	1	9.56		0.43		0.01		1	F4 V	1748
58946	+32 01562	HR 2852	⋆AB	7 25 54	+31 53.1	15	4.18	.006	0.32	.004	-0.03	.009	225	F0 V	1,15,130,369,1006*
59059	+15 01579	HR 2858		7 25 57	+15 12.8	2	6.22		-0.05	.000	-0.11	.000	5	B9 IV	252,1022
59256	−28 04383	HR 2863		7 26 00	−29 03.2	2	5.54		-0.06	.007	-0.15		5	B9 V(p Si)	1079,2006
		G 251 - 36		7 26 01	+77 01.5	2	10.32	.013	0.72	.001	0.19	.012	4		1658,5010
		LSS 427		7 26 03	−15 15.1	1	11.66		0.37		-0.56		1	B3 Ib	862
59213	−15 01843			7 26 04	−15 52.6	1	10.19		0.05		-0.46		2	B2/3 III	401
59212	−12 01988	IDS07238S1245	AB	7 26 07	−12 51.2	1	7.95		0.05		-0.11		2	B9 II/III	401
59421	−46 03172			7 26 08	−46 59.6	1	8.64		0.19		-0.14		2	B8/9 IV/V	1776
58855	+49 01630	HR 2849	⋆A	7 26 09	+49 46.7	4	5.37	.015	0.45	.014	-0.07	.019	15	F6 V	15,254,1008,3037
233399	+50 01435	IDS07223N5011	A	7 26 10	+50 04.4	1	9.49		0.68		0.14		2	G2 V	1003
59281	−27 04108			7 26 11	−27 17.8	1	8.46		-0.10		-0.62		2	B3/5 Vne	401
59468	−51 02507			7 26 11	−51 18.0	3	6.72	.008	0.70	.000	0.25		9	G5 V	158,1075,1705
		ApJ144,259 # 21		7 26 12	+13 21.	1	15.99		-0.32		-1.26		4		1360
59211	−9 02069			7 26 12	−09 56.4	1	6.62		-0.12		-0.57		22	B3 V	1222
59579	−59 00829			7 26 12	−59 28.6	1	7.81		1.06		0.95		7	K1 III	1673
		LSS 428		7 26 13	−15 11.7	1	11.77		0.30		-0.60		1	B3 Ib	862
59037	+28 01396	HR 2857		7 26 14	+28 13.4	4	5.01	.022	0.11	.002	0.12	.001	9	A4 V	374,1022,1363,3052
		G 112 - 21		7 26 14	−03 11.3	2	11.48	.005	1.46	.061	1.18	.033	4		316,3062
59446	−47 03072			7 26 16	−47 18.7	1	7.58		0.08		-0.35		1	B6 II/III	55
		LSS 430		7 26 18	−15 05.1	1	11.47		0.38		-0.50		1	B3 Ib	862
59392	−37 03596			7 26 18	−37 53.3	2	9.71	.009	0.46	.009	-0.20		5	F0 V	1594,6006
	−6 02131			7 26 21	−07 10.7	1	10.79		1.79				2	M3	369
59321	−26 04413	IDS07244S2702	AB	7 26 22	−27 08.3	1	9.02		0.11		0.10		1	A1 V	401
		LSS 432		7 26 24	−15 35.9	1	12.28		0.48		0.04		1	B3 Ib	862
59346	−26 04415			7 26 25	−27 07.5	2	9.11	.001	-0.16	.005	-0.80	.017	4	B4 II/III	401,1732
	−26 04417			7 26 30	−27 08.5	1	10.03		-0.03		-0.18		1	A0	401
59364	−26 04419			7 26 34	−26 22.9	1	9.08		-0.12		-0.53		2	B5 V	401
59149	+19 01743			7 26 35	+19 44.3	1	6.72		1.28		1.37		2	K0	985
		G 90 - 2		7 26 37	+32 03.3	1	9.57		1.19		1.12		1		1620
		G 90 - 3		7 26 37	+32 58.5	2	10.43	.000	0.48	.000	-0.18	.000	4	sdF8	516,1620
59466	−37 03601	HR 2869		7 26 37	−37 42.4	2	6.57	.005	0.06	.000			7	A1 IV	15,2012
	+30 01515			7 26 39	+29 51.7	1	9.76		0.19		0.25	std			97
		HA 51 # 411		7 26 39	+29 57.4	1	12.62		0.42		0.06		4		97
59148	+28 01400	HR 2861	⋆A	7 26 42	+28 01.3	4	5.02	.018	1.12	.004	1.04	.020	7	K2 III	1080,1355,1363,3051
		HA 51 # 413		7 26 42	+30 02.5	1	11.55		0.83		0.55		3		97
59530	−43 03250			7 26 44	−43 51.6	1	7.57		-0.05		-0.26		2	B9 V	401
59311	−1 01738	HR 2865		7 26 47	−01 48.1	3	5.59	.008	1.49	.010	1.75	.005	11	K5 III	15,1415,3001
		HA 51 # 547		7 26 48	+30 12.7	1	11.77		1.12		0.90		1		97
	−13 02040			7 26 49	−13 39.9	1	8.72		0.05		-0.63		2	B3	401
		G 250 - 49		7 26 50	+68 43.6	1	10.79		1.59				1		1017
		HA 51 # 548		7 26 51	+30 05.5	1	12.28		0.58		0.07		2		97
56862	+84 00152			7 26 52	+84 18.4	1	7.56		1.62		1.94		3	K0	1733
		HA 51 # 416		7 26 54	+29 58.9	1	12.81		0.55		0.05		2		97
59409	−21 01957	IDS07247S2111	A	7 26 54	−21 17.1	1	8.84		-0.04				4	B8/9 V	2014
		HA 51 # 303		7 26 56	+29 44.2	1	10.96		0.58		0.01		3		97
59499	−31 04590	HR 2870	⋆AB	7 26 56	−31 44.7	2	5.94	.000	-0.17	.005	-0.67		6	B3 V	404,2007
		HA 51 # 417		7 26 57	+30 03.0	1	14.81		0.49		-0.03		2		97
		HA 51 # 419		7 26 59	+29 55.7	1	13.59		1.08		1.10		3		97
		HA 51 # 418		7 26 59	+30 02.6	1	14.62		0.38		0.04		2		97
59380	−7 01996	HR 2866		7 27 00	−07 26.9	3	5.86	.005	0.48	.000	-0.03	.010	14	F8 V	15,1061,2029
59381	−10 02067	HR 2867		7 27 00	−10 13.4	3	5.75	.005	1.61	.009	1.99		10	K5 III	15,2027,3005
		G 89 - 24		7 27 01	+08 32.9	1	12.58		0.99		0.71		2		3062
59294	+12 01567	HR 2864		7 27 01	+12 06.7	6	4.53	.028	1.27	.014	1.30	.046	21	K2 III	3,15,1355,1363,3016,8015
	−13 02043			7 27 01	−13 54.4	1	10.18		0.00		-0.46		2		401
		G 107 - 69		7 27 02	+48 19.1	3	13.53	.017	1.70	.007	1.18	.035	6	sdM5	203,1705,3078
59462	−21 01959			7 27 02	−21 43.7	1	8.40		-0.06				4	B8/9 IV	2014
		Steph 650	A	7 27 05	+62 07.7	1	11.94		1.44		1.26		1	K7	1746
59438	−14 01925	HR 2868	⋆AB	7 27 05	−14 53.5	4	6.04	.019	0.47	.017	-0.07	.015	20	F5 V	15,265,1211,2012
		G 107 - 70		7 27 06	+48 17.4	1	14.62		0.99		0.40		3	DC	3078
		L 816 - 110		7 27 06	−14 53.	1	11.78		0.67		-0.29		2		265
59527	−34 03634			7 27 06	−34 48.9	1	6.91		-0.12		-0.55		2	B5 V	401
59440	−16 01973			7 27 07	−16 52.3	2	8.50	.007	-0.02	.010	-0.37	.005	5	B7 III	401,1732
		Steph 652		7 27 08	+33 10.3	1	10.83		1.59		1.88		2	M1	1746
		Steph 650	B	7 27 08	+62 07.8	1	13.59		1.56				1	K7	1746
59550	−31 04593	HR 2873		7 27 09	−31 21.1	6	5.77	.008	-0.20	.005	-0.74	.011	20	B2 IV	15,540,976,1637,1732,2012
		HA 51 # 310		7 27 11	+29 51.8	1	14.62		0.76		0.42		1		97
		HA 51 # 428		7 27 11	+29 59.9	1	12.33		0.95		0.61		2		97
		WLS 720-10 # 5		7 27 12	−07 49.6	1	11.57		0.19		0.14		2		1375
59498	−21 01960			7 27 13	−21 45.7	3	7.79	.026	-0.09	.001	-0.77	.001	8	B5 IV/V	540,976,2014
	+29 01539	G 87 - 40		7 27 14	+29 29.3	2	9.32	.000	0.79	.005	0.41	.010	3	G5	97,333,1620
		HA 51 # 431		7 27 16	+29 54.1	1	15.20		0.91		0.51		4		97

Table 1   401

HD	DM	Other Id	N	Rem	α_{1950}	δ_{1950}	S	V	σ_V	B–V	σ_{B-V}	U–B	σ_{U-B}	n	Spectrum	References
		G 90 - 4			7 27 17	+27 22.6	1	11.89		0.64		0.04		2		1620
		MtW 51 # 72			7 27 17	+29 52.4	1	15.84		0.94		0.66		2		97
		HA 51 # 432			7 27 18	+29 57.7	1	13.77		0.52		-0.05		2		97
		LSS 442			7 27 18	−14 57.5	1	12.31		0.37		-0.58		1	B3 Ib	862
		RU 152F			7 27 19	−01 58.2	1	14.56		0.64		0.07		8		1764
		RU 152E			7 27 19	−01 58.8	1	12.36		0.04		-0.09		8		1764
		HA 51 # 433			7 27 20	+29 54.8	1	14.73		0.63		0.21		4		97
		HA 51 # 315			7 27 21	+29 51.1	1	13.50		0.26		0.27		2		97
		HA 51 # 316			7 27 22	+29 51.4	1	12.25		0.71		0.32		2		97
59635	−38 03400	HR 2875			7 27 22	−38 42.5	2	5.42	.005	-0.17	.005			7	B3 V	15,2012
		RU 152			7 27 23	−02 00.0	1	13.01		-0.19		-1.07		25		1764
		HA 51 # 319			7 27 24	+29 51.2	1	11.67		0.38		0.14		3		97
		RU 152B			7 27 24	−01 59.4	1	15.02		0.50		0.03		10		1764
		RL 152			7 27 24	−02 00.	1	13.00		-0.23		-1.01		2		704
		HA 51 # 437			7 27 25	+29 57.6	1	13.13		0.63		0.13		3		97
		RU 152A			7 27 25	−01 59.8	1	14.34		0.54		-0.09		8		1764
		MtW 51 # 100			7 27 26	+29 56.4	1	15.82		0.82		0.55		1		97
59360	+20 01822				7 27 27	+19 55.3	1	7.07		0.59		0.09		2	G0	1648
		RU 152C			7 27 27	−01 59.0	1	12.22		0.57		-0.01		9		1764
		HA 51 # 320			7 27 28	+29 52.4	1	14.81		0.90		0.47		5		97
		HA 51 # 321			7 27 29	+29 52.1	1	13.28		1.01		0.75		3		97
	+30 01517				7 27 29	+30 09.4	1	10.32		0.49		0.03		2		97
	+30 01518				7 27 31	+29 56.1	1	9.54		1.05		0.85		std	K0	97
		HA 51 # 439			7 27 31	+30 01.5	1	13.71		0.55		0.03		3		97
		RU 152D			7 27 31	−01 57.9	1	11.08		0.88		0.49		10		1764
59033	+62 00934				7 27 33	+61 52.0	1	6.70		0.99		0.73		2	G5	985
59374	+19 01749	G 88 - 31			7 27 34	+19 04.4	6	8.49	.015	0.52	.009	-0.11	.007	14	F8 V	516,1003,1620,1696,1775,3077
		HA 51 # 441			7 27 34	+29 53.9	1	14.08		0.65		0.14		5		97
		HA 51 # 442			7 27 35	+29 59.7	1	13.25		0.98		0.93		2		97
59546	−15 01858				7 27 35	−15 38.1	1	8.03		0.12		0.06		1	B8/9 II/III	862
		MtW 51 # 125			7 27 36	+29 51.5	1	17.80		0.85		0.35		2		97
		MtW 51 # 124			7 27 36	+29 55.1	1	16.52		0.59		0.05		3		97
59543	−13 02051				7 27 36	−13 53.0	2	7.18	.010	-0.08	.015	-0.67	.010	5	B3 IV/V	399,401
	−13 02052				7 27 36	−14 03.0	1	9.13		-0.05		-0.64		2	B3	401
		MtW 51 # 127			7 27 38	+29 59.0	1	17.12		0.70		0.20		2		97
	−10 02073	LSS 449			7 27 38	−11 09.4	2	10.32	.000	0.05	.000	-0.83	.000	4		405,1737
	+24 01676	G 88 - 32			7 27 39	+24 11.7	3	10.81	.009	0.36	.005	-0.22	.000	6	sdF	516,1620,3029
59717	−43 03260	HR 2878		⋆ A	7 27 39	−43 12.0	6	3.25	.008	1.51	.012	1.78	.015	28	K5 III	15,1075,1279,2012*
59717	−43 03260	IDS07261S4306		B	7 27 39	−43 12.0	2	8.82	.005	0.70	.005	0.19	.000	7	G5	1279,3077
		MtW 51 # 131			7 27 40	+29 56.6	1	15.34		0.83		0.34		2		97
		HA 51 # 323			7 27 41	+29 46.5	1	10.75		1.05		0.86		3		97
		MtW 51 # 135			7 27 41	+29 55.2	1	16.75		0.60		-0.04		3		97
		HA 51 # 443			7 27 41	+29 57.0	1	12.16		1.12		1.06		3		97
		HA 51 # 575			7 27 41	+30 07.0	1	11.86		0.22		0.30		2		97
		HA 51 # 574			7 27 41	+30 11.5	1	10.71		0.72		0.30		2		97
		MtW 51 # 140			7 27 44	+29 55.3	1	17.53		0.69		0.01		1		97
		LSS 453			7 27 44	−14 52.5	3	10.67	.004	0.51	.013	-0.53	.006	6	B3 Ib	405,862,1737
59612	−22 01897	HR 2874, LSS 457		⋆ AB	7 27 44	−22 55.2	9	4.85	.012	0.23	.009	0.15	.036	26	A5 Ib	15,258,379,1075,1075*
59611	−20 01971				7 27 46	−20 41.5	1	8.94		1.00				19	K0 III	6011
		HA 51 # 444			7 27 49	+29 55.6	1	14.63		0.79		0.15		4		97
		AA45,405 S305 # 2			7 27 49	−18 26.2	3	12.75	.012	0.87	.019	-0.22	.025	7		432,797,7006
		AA45,405 S305 # 3			7 27 49	−18 26.3	1	13.53		0.53		0.06		2		432
		LSS 458			7 27 49	−19 02.3	1	11.30		0.64		-0.44		2	B3 Ib	797
59589	−15 01862	LSS 455			7 27 51	−15 41.7	1	8.90		1.46		1.58		1	K1 V	862
59609	−15 01863	LSS 456			7 27 51	−15 45.7	1	9.56		1.09		0.81		1	K0 I	862
		AA45,405 S305 # 4			7 27 52	−18 26.7	3	12.91	.012	1.09	.016	-0.05	.035	6		432,797,7006
		AA45,405 S305 # 1			7 27 53	−18 24.8	1	12.08		1.64		1.84		3		432
		HA 51 # 447			7 27 57	+29 54.9	1	13.78		0.85		0.38		3		97
		HA 51 # 448			7 27 57	+29 58.2	1	14.60		0.45		0.00		2		97
		AA45,405 S305 # 6			7 27 57	−18 27.1	1	11.71		0.18		0.09		2		432
59652	−24 05487				7 27 57	−25 02.5	1	7.11		0.19		0.15		2	A4 III	126
		HA 51 # 449			7 27 59	+30 00.4	1	12.88		0.74		0.28		2		97
59011	+67 00493				7 27 60	+67 02.7	1	7.70		1.53		1.76		2	K5	1733
	−15 01864	LSS 460			7 28 00	−15 41.2	1	9.90		1.21		1.08		1	G5 Ib	862
59647	−14 01932				7 28 01	−14 53.5	1	8.11		-0.01		-0.51		2	B2/3 III	401
	−40 03183				7 28 04	−40 29.6	1	8.11		0.47		0.08		2		633
59674	−18 01867				7 28 05	−18 19.6	1	8.39		-0.04		-0.41		2	B8 V	401
		WLS 720 10 # 8			7 28 06	+10 01.2	1	12.94		0.56		0.05		2		1375
59371	+42 01725				7 28 07	+42 44.9	1	7.52		1.67		1.93		2	K5	1375
		HA 51 # 333			7 28 10	+29 51.7	1	13.61		0.41		0.01		2		97
	−15 01866				7 28 11	−15 41.2	1	10.05		1.33		1.18		1	K1 II	862
		LP 365 - 14			7 28 13	+23 21.8	1	11.22		1.17		1.02		3		1723
	+29 01545				7 28 14	+29 42.7	1	9.81		1.04		0.81		2		97
59791	−38 03411	IDS07265S3827		AB	7 28 14	−38 33.7	1	8.68		1.03		0.73		4	G8 IV	119
59868	−44 03467				7 28 16	−44 48.4	1	7.84		-0.04		-0.32		2	B8 IV/V	401
59583	+12 01574				7 28 17	+12 21.6	1	8.28		0.01		-0.06		2	A0	1375
59698	−15 01867	LSS 463			7 28 17	−15 45.6	1	9.73		1.12		0.93		1	G5	862
59869	−46 03196				7 28 17	−46 35.0	1	7.40		0.03		-0.13		2	A0 V	401
59582	+14 01684	G 89 - 26			7 28 18	+14 43.5	3	8.98	.021	1.11	.007	1.06	.000	5	K5	333,1620,1758,3072
59509	+27 01403	IDS07262N2749		A	7 28 19	+27 43.7	1	7.95		0.17		0.13		2	F8	1375

HD	DM	Other Id	N	Rem	α_{1950}	δ_{1950}	S	V	σ_V	B–V	σ_{B-V}	U–B	σ_{U-B}	n	Spectrum	References
59700	−17 02006				7 28 19	−17 13.0	1	8.50		0.05		−0.19		2	B9 V	401
	−14 01936	LSS 464			7 28 20	−14 52.5	3	10.40	.005	0.62	.008	−0.46	.004	6	O9 III	405,862,1737
		Bo 15 - 14			7 28 20	−33 18.6	1	9.73		−0.04		−0.45		1	B7 II	568
	−8 01950				7 28 21	−08 55.8	1	8.85		1.67				2	M2	369
		AA45,405 S299 # 2			7 28 22	−15 11.5	1	13.57		0.66		−0.25		3		797
		AA45,405 S299 # 3			7 28 22	−15 11.5	1	14.59		0.62		−0.22		3		797
		LSS 465			7 28 22	−15 11.5	1	12.54		0.58		−0.38		2	B3 Ib	797
59669	−4 01979	HR 2876			7 28 23	−05 07.2	4	6.24	.008	1.19	.009	1.06	.000	18	K0	15,1415,2013,2029
		LSS 467			7 28 23	−19 00.0	3	11.15	.029	0.61	.025	−0.43	.010	8	B3 Ib	342,432,797
59693	−9 02085	U Mon			7 28 24	−09 40.3	1	5.66		0.93		0.56		1	F8 Ibe	793
60816	−78 00265	HR 2919			7 28 26	−78 59.5	3	5.53	.004	1.29	.005	1.42	.000	11	K2/3 III	15,1075,2038
		SV CMi			7 28 27	+06 05.0	1	14.37		0.06		−0.59		1		1471
60150	−64 00721	HR 2888			7 28 27	−64 24.3	3	6.38	.003	1.55	.005	1.84	.000	17	K5 III	15,1075,2038
59507	+39 01958	HR 2872			7 28 32	+39 00.2	2	6.55		0.08	.010	0.05	.020	5	A2 V	105,1022
59752	−15 01869				7 28 32	−15 19.3	1	9.36		−0.04		−0.39		2	Ap Si	401
		LSS 473			7 28 35	−19 11.3	1	11.37		0.73		−0.32		2	B3 Ib	797
	+30 01521				7 28 36	+30 08.6	1	10.49		0.36		0.02		2		97
59730	−9 02086				7 28 36	−10 00.1	2	6.57	.020	1.62	.001	2.01	.025	24	K2	332,1222
59768	−14 01939	IDS07263S1501		AB	7 28 37	−15 07.1	1	9.64		0.03		−0.39		2	B5 II/III	401
59600	+24 01683				7 28 38	+24 06.4	1	7.64		1.05		0.88		3	K0	1625
		LSS 470			7 28 38	−15 06.0	1	11.09		0.62		−0.39		1	B3 Ib	862
	+36 01638	VV Lyn	★	A	7 28 40	+36 19.8	2	11.82	.055	1.57	.020	1.15		5	M2	694,3016
59814	−23 05603				7 28 40	−23 25.0	1	8.59		−0.14		−0.63		4	B3 IV/V	1732
59864	−33 03879	V350 Pup			7 28 42	−33 59.1	3	7.62	.008	−0.09	.007	−0.83	.008	8	B1/2 Ib/II	158,540,976
		G 89 - 27			7 28 46	+16 50.7	1	11.50		1.55				1		1705
	+20 01828	G 88 - 34			7 28 46	+20 04.5	1	9.67		0.89		0.64		1	K0	333,1620
59813	−18 01873	LSS 477			7 28 46	−18 49.4	3	8.97	.001	0.52	.012	−0.51	.012	6	B0 Ib	540,797,1737
59890	−30 04620	HR 2881			7 28 46	−30 51.4	6	4.64	.007	0.92	.004	0.64	.026	31	G2 Iab/b	3,15,1075,1637,2012,8015
		AA45,405 S306 # 3			7 28 47	−18 49.5	1	13.30		0.62		−0.32		1		797
60060	−52 01198	HR 2884			7 28 47	−52 32.8	2	5.87	.002	1.01	.002	0.83		6	K0 III	58,2009
		LSS 476			7 28 48	−14 18.7	2	11.64	.000	0.51	.000	−0.53	.000	4	B3 Ib	405,1737
59841	−24 05514	IDS07268S2427		A	7 28 51	−24 33.7	2	8.17	.010	−0.15	.005	−0.51		6	B8 II/III	126,2014
		AA45,405 S300 # 5			7 28 52	−15 18.4	1	15.68		0.48		−0.07		2		797
		LSS 478			7 28 52	−15 18.5	1	11.50		0.44		−0.52		2	B3 Ib	797
		AA45,405 S300 # 2			7 28 52	−15 18.6	1	14.59		0.57		−0.09		2		797
59913	−34 03655				7 28 52	−34 51.5	1	8.71		−0.02		−0.08		2	B9.5/A0 V	401
59643	+24 01686	NQ Gem			7 28 53	+24 36.6	1	7.80		2.24				1	C6,2	1238
		LSS 479			7 28 53	−15 50.2	1	11.96		0.57		−0.29		1	B3 Ib	862
59686	+17 01596	HR 2877			7 28 55	+17 11.6	1	5.42		1.13				2	gK2	71
		LSS 480			7 28 55	−15 23.0	2	11.30	.019	0.76	.005	−0.33	.034	3	B3 Ib	797,862
		LSS 481			7 28 55	−15 46.4	1	12.37		0.59		0.17		1	B3 Ib	862
	+47 01455				7 28 56	+47 16.0	1	9.00		0.41		−0.01		2	F2	1375
59967	−37 03637	HR 2882			7 28 56	−37 14.0	2	6.64	.005	0.63	.000			7	G3 V	15,2027
59857	−15 01875				7 29 01	−15 11.5	1	8.26		0.13		0.13		2	A2 III/IV	401
		HA 51 # 104			7 29 04	+29 32.0	1	11.28		−0.04		−0.29		2		97
		WLS 728 55 # 9			7 29 04	+56 01.4	1	12.31		0.50		0.00		2		1375
		G 112 - 26			7 29 08	−00 55.9	1	13.29		0.71		−0.03		2		3062
		G 88 - 36			7 29 11	+17 25.7	3	10.98	.051	1.40	.053	1.28		3	K5	333,1017,1510,1620
59989	−34 03660	RY Pup			7 29 11	−34 52.6	1	7.46		1.73		1.91		7	K4 III	1673
59764	+12 01582				7 29 12	+12 46.1	1	6.52		1.00		0.69		1	G5	3040
59911	−21 01985				7 29 12	−21 32.3	1	9.11		0.48				27	F3 V	6011
		LP 897 - 12			7 29 12	−27 30.0	1	12.80		1.57		1.03		2		3073
	+30 01522				7 29 13	+30 04.8	1	9.83		0.00		0.00		2	A0	97
		Bo 4 - 10			7 29 15	−17 05.2	1	12.04		0.20		0.14		2		410
		Bo 4 - 9			7 29 15	−17 08.0	1	9.92		0.07		−0.27		2		410
		Bo 4 - 11			7 29 16	−17 03.6	1	12.37		0.21		0.08		2		410
59852	−4 01986				7 29 17	−04 15.2	1	8.69		0.94		0.57		1	G5	565
		LSS 489			7 29 19	−15 12.6	1	11.30		0.58		−0.36		1	B3 Ib	862
59938	−21 01986				7 29 19	−21 38.2	1	8.53		0.36				28	F0 V	6011
59882	−5 02148	LS VI -05 012			7 29 20	−05 28.6	4	8.80	.006	−0.17	.021	−0.92	.014	11	B1 V	399,490,555,1732
		CCS 791			7 29 20	−09 21.6	1	10.99		2.55		3.90		1	C6,2	864
		Bo 4 - 6			7 29 20	−17 04.9	1	12.24		0.12		0.01		2		410
		Bo 4 - 4			7 29 21	−17 05.6	1	10.64		0.10		−0.21		2		410
59683	+39 01962				7 29 22	+39 09.4	2	8.74	.010	0.44	.005	−0.03	.000	5	F4 V	1501,1601
	+30 01524				7 29 23	+30 04.6	1	10.66		0.42		0.06		2		97
		Bo 4 - 5			7 29 23	−17 04.0	1	10.97		0.20		0.16		2		410
	−16 01994	Bo 4 - 3			7 29 23	−17 05.0	1	9.07		1.37		1.39		2		410
59934	−16 01995	Bo 4 - 1			7 29 23	−17 05.4	1	8.05		−0.03		−0.63		7	B2 II/IV	410
60228	−54 01294	HR 2892			7 29 23	−54 17.6	2	5.95	.004	1.57	.007	1.96		6	M1 III	58,2009
		Bo 4 - 7			7 29 25	−17 03.8	1	12.07		0.29		0.17		2		410
59961	−16 01997	Bo 4 - 2			7 29 26	−17 05.6	1	9.79		0.06		−0.27		2	B8	410
		WLS 728 55 # 7			7 29 27	+53 19.5	1	12.25		1.06		0.60		2		1375
59960	−16 01996	Bo 4 - 12			7 29 27	−16 57.2	2	9.80	.005	0.25	.010	0.17	.010	4	A1 V	410,432
	−16 01998	Bo 4 - 13			7 29 28	−16 56.5	2	9.74	.005	1.52	.015	1.83	.020	4		410,432
		Bo 4 - 8			7 29 29	−17 05.5	1	13.67		0.77		0.42		2		410
		LSS 494			7 29 29	−20 31.9	1	10.83		0.13		−0.60		3	B3 Ib	1737
59881	+2 01691	HR 2880			7 29 30	+02 01.3	3	5.24	.005	0.22	.000	0.18	.009	8	F0 III	15,1078,3023
		LSS 491			7 29 31	−15 48.2	1	12.50		0.52		−0.35		1	B3 Ib	862
59932	−8 01961	IDS07272S0830		B	7 29 32	−08 35.0	1	9.00		0.35		0.42		1	F0 IV	3016
59986	−16 01999	Bo 4 - 14			7 29 32	−16 53.3	2	9.47	.010	0.23	.005	−0.69	.045	4	B5 Ib/II	401,410

Table 1   403

HD	DM	Other Id	N Rem	α_{1950}	δ_{1950}	S	V	σ_V	B–V	σ_{B-V}	U–B	σ_{U-B}	n	Spectrum	References
59933	−8 01961	IDS07272S0830	A	7 29 33	−08 34.8	1	8.66		1.08		0.91		1	F0 III:	3016
		AA45,405 S309 # 1		7 29 33	−19 15.9	1	12.10		0.44		0.09		1		797
		AA45,405 S309 # 3		7 29 35	−19 15.5	1	14.16		0.60		-0.38		3		797
	−42 03304			7 29 35	−43 04.8	1	8.68		0.49		0.00		2		633
		AA45,405 S309 # 4		7 29 36	−19 15.3	1	13.90		0.87		-0.18		3		797
59828	+21 01630			7 29 37	+21 31.0	1	6.80		0.98		0.76		2	G5	1648
		Bo 4 - 15		7 29 37	−16 51.0	1	10.57		0.09		-0.27		2		410
		Bo 4 - 30		7 29 37	−16 51.0	1	11.45		0.08		-0.06		1		410
		AA45,405 S309 # 2		7 29 37	−19 21.1	1	12.90		0.71		-0.32		2		797
60098	−35 03650	HR 2885		7 29 37	−36 02.8	5	6.67	.010	-0.12	.006	-0.58		18	B3 V	15,401,1020,2012,2012
		G 193 - 52		7 29 38	+49 50.9	1	11.35		1.41		1.25		2	M3	7010
60057	−26 04492			7 29 38	−26 47.6	1	8.82		-0.01		-0.02		2	B9 III	401
60007	−15 01882			7 29 40	−15 18.0	1	9.87		-0.01		-0.48		2	B5/7 II	401
		Bo 4 - 16		7 29 40	−16 55.7	1	10.64		0.09		-0.20		2		410
		LSS 496		7 29 41	−19 20.3	1	12.12		0.31		-0.52		2	B3 Ib	410
59984	−8 01964	HR 2883	⋆ A	7 29 42	−08 46.3	4	5.90	.009	0.54	.000	-0.09	.007	13	F9 V	15,1078,2031,3077
		Bo 4 - 20		7 29 42	−16 45.8	1	12.48		0.38		-0.23		2		410
		G 87 - 45		7 29 48	+31 13.6	1	11.44		0.64		-0.05		2		333,1620
		Bo 4 - 17		7 29 48	−16 48.5	1	13.29		0.84		0.09		2		410
		LSS 498		7 29 49	−19 18.8	1	12.31		0.35		-0.55		2	B3 Ib	410
		Bo 4 - 18		7 29 50	−16 49.0	1	12.65		0.46		-0.27		2		410
59878	+23 01744	HR 2879	⋆ A	7 29 51	+22 59.8	1	6.54		1.01		0.77		3	K0 II-III	3016
59878	+23 01744	IDS07268N2306	B	7 29 51	+22 59.8	1	8.96		0.55		0.05		3	G5	3032
59201	+73 00380	G 251 - 38		7 29 51	+73 22.9	1	8.55		0.80		0.47		1	G5	1658
		LSS 499		7 29 51	−19 19.7	3	10.67	.032	0.49	.015	-0.52	.021	8	B3 Ib	342,410,432
		Bo 4 - 19		7 29 52	−16 49.5	1	13.55		0.58		-0.12		2		410
		WLS 720 30 # 8		7 29 53	+30 07.6	1	10.45		1.59		1.83		2		1375
60049	−13 02076	IDS07276S1355	AB	7 29 53	−14 01.5	1	9.29		-0.04		-0.53		2	B5 III	401
	−24 05538			7 29 54	−25 02.1	1	9.51		1.02		0.80		2	G5	126
60168	−35 03652	HR 2889, PS Pup	⋆ A	7 29 54	−35 46.8	2	6.62	.010	-0.08	.000	-0.37		6	B8 V	401,2031
		G 234 - 5		7 29 55	+63 03.0	1	10.42		1.60		1.26		3	M0	7008
		LSS 501		7 29 56	−19 22.3	1	11.72		0.46		-0.46		2	B3 Ib	410
		Bo 4 - 29		7 29 57	−16 52.0	1	11.87		0.32		0.27		1		410
		Bo 4 - 28		7 29 57	−16 52.3	1	12.10		0.38		0.19		1		410
60068	−16 02001	Bo 4 - 22		7 29 58	−16 51.5	2	7.80	.020	1.37	.005	1.41		4	K2 III	410,6009
59950	+8 01800	S CMi		7 30 00	+08 25.6	1	8.12		1.50		0.27		1	M4	635
		LSS 503		7 30 00	−19 24.9	1	12.24		0.45		-0.49		4	B3 Ib	410
60124	−24 05540			7 30 02	−25 03.0	1	8.62		0.40		0.01		2	F2 V	126
		Bo 4 - 27		7 30 03	−16 50.2	1	11.18		0.15		-0.60		1		410
56049	+86 00102			7 30 04	+86 27.1	1	8.67		0.04		0.05		2	A0	1219
		LOD 2- # 15		7 30 04	−24 53.5	1	10.77		0.44		-0.04		2		126
60002	+7 01756			7 30 05	+07 14.5	1	8.47		0.49		-0.03		2	F8	1375
		Bo 4 - 26		7 30 05	−16 46.5	1	12.05		0.75		0.17		1		410
60148	−24 05542			7 30 06	−24 54.8	1	9.59		-0.06		-0.25		2	B8 V	126
59621	+60 01062			7 30 07	+59 56.8	1	8.27		1.08		1.01		2	K1 III-IV	1502
60197	−29 04525			7 30 09	−29 31.6	2	7.74	.009	1.69	.000	1.92	.005	4	K3 III	565,3048
		LSS 505		7 30 10	−15 25.1	1	11.73		0.30		-0.63		1	B3 Ib	862
	−16 02003	Bo 4 - 21		7 30 10	−16 47.6	2	10.42	.005	0.14	.019	-0.56	.039	6		206,410
		LOD 2- # 17		7 30 10	−24 54.2	1	11.73		0.52		0.03		2		126
		LOD 2- # 16		7 30 11	−24 53.5	1	11.34		0.42		0.00		2		126
60196	−28 04485	LSS 507		7 30 12	−28 37.6	4	9.00	.015	0.00	.009	-0.83	.019	11	B1 Ib/II	540,710,976,1012
		G 88 - 38		7 30 13	+17 11.8	1	11.09		0.77		0.38		1	K0	333,1620
		Bo 4 - 24		7 30 16	−16 49.5	1	11.86		0.37		0.28		1		410
		Bo 4 - 25		7 30 16	−16 49.9	1	11.71		0.21		0.18		1		410
295866	+4 01738			7 30 18	+04 22.6	1	10.43		0.34		-0.56		2	B5	199
		Bo 4 - 23		7 30 18	−16 48.5	1	10.98		0.60		0.20		1		410
60194	−24 05550			7 30 18	−24 56.4	1	9.48		-0.06		-0.26		2	B9 II	126
60139	−9 02097			7 30 21	−09 18.2	1	7.36		1.42		1.63		2	K2	1375
	−27 04197	LSS 511		7 30 21	−28 01.6	3	9.40	.018	0.27	.013	-0.63	.026	9	B0.5IIInn	540,710,976
		LSS 506		7 30 25	−15 12.8	1	12.67		0.41		-0.42		1	B3 Ib	862
60219	−21 01997	VX Pup		7 30 28	−21 49.3	1	7.97		0.49				1	F3 IV	6011
		LSS 508		7 30 29	−15 12.9	2	10.79	.005	0.53	.024	-0.58	.024	3	B3 Ib	862,1737
		Bo 15 - 15		7 30 31	−33 33.7	1	12.45		0.13		-0.54		1		568
	−24 05557			7 30 32	−24 55.8	1	10.38		0.28		0.12		2		126
60156	−6 02166			7 30 33	−06 45.4	1	7.29		1.12		1.05		1	K0	1311
60312	−35 03659	HR 2895	⋆ AB	7 30 33	−35 51.2	2	6.34	.010	-0.10	.000	-0.37		6	B8 V	401,2006
60111	+3 01715	HR 2887		7 30 34	+03 23.9	2	5.58	.005	0.32	.005	0.08	.015	7	F2 V	15,1078
		LSS 510		7 30 36	−15 11.3	1	12.91		0.27		-0.49		1	B3 Ib	862
60265	−20 02009	TY Pup		7 30 36	−20 41.0	1	8.49		0.41		0.26		5	F3 V	1655
60266	−20 02011	X Pup		7 30 37	−20 48.1	3	7.88	.015	0.83	.018	0.67		3	K0	689,1484,6011
60284	−27 04205	LSS 514	⋆ AB	7 30 37	−27 45.1	5	9.11	.018	0.14	.008	-0.78	.019	12	B5/7 V	540,710,976,1012,8100
	−73 00453	S Vol		7 30 38	−73 16.4	1	12.97		1.58		0.51		1	M4e	975
60431	−46 03219	V343 Pup		7 30 40	−46 50.1	1	8.59		-0.11		-0.50		2	Ap SiMg	401
60136	+11 01607			7 30 42	+11 07.2	1	6.83		1.63		1.84		3	M1	1733
		GD 86		7 30 44	+48 47.9	1	14.96		-0.09		-0.74		1		3060
	−36 03700			7 30 44	−36 36.3	1	10.85		1.24		1.21		1		1696
60107	+16 01510	HR 2886	⋆	7 30 45	+15 56.1	3	5.17	.097	0.05	.001	0.06	.000	6	A1 Vn	1022,1363,3023
		NGC 2414 - 7		7 30 49	−15 19.4	1	13.37		0.33		-0.49		2	F0	183
60282	−21 02000			7 30 49	−22 10.7	1	8.34		-0.02		-0.34		2	B8 IV/V	126
		LP 365 - 22		7 30 50	+22 30.6	1	16.46		1.60				1		1691

HD	DM	Other Id	N Rem	α1950	δ1950	S	V	σV	B-V	σB-V	U-B	σU-B	n	Spectrum	References
60260	-11 01994	LSS 512		7 30 50	-11 30.2	1	9.04		-0.07		-0.75		2	B5	1375
		NGC 2414 - 6		7 30 50	-15 19.7	1	12.99		0.33		-0.47		2	A5	183
		NGC 2414 - 8		7 30 50	-15 20.0	1	14.02		0.37		-0.19		2		183
60429	-39 03388			7 30 52	-39 53.6	1	8.35		-0.06		-0.72		1	B1/2 V	55
		NGC 2414 - 4		7 30 53	-15 19.8	1	12.67		0.32		-0.40		2	B8	183
		NGC 2414 - 3		7 30 53	-15 20.4	1	12.19		0.30		-0.49		2	B8	183
		NGC 2414 - 9		7 30 54	-15 19.9	1	13.52		0.28		-0.50		2		183
		NGC 2414 - 2		7 30 54	-15 20.2	2	10.79	.005	0.30	.015	-0.62	.010	6	B1 IV	183,1737
60370	-32 04134	IDS07290S3247	AB	7 30 54	-32 53.0	2	9.19	.005	0.48	.005	-0.06		4	F6 V	1594,6006
		NGC 2414 - 5		7 30 55	-15 19.3	1	12.49		0.72		0.22		2	F0	183
		NGC 2414 - 10		7 30 56	-15 20.2	1	10.59		0.31		-0.32		2		183
60308	-15 01892	NGC 2414 - 1		7 30 56	-15 20.7	4	8.22	.011	0.40	.013	-0.49	.013	9	B1 Ib	183,1012,1569,8100
60344	-23 05673	LSS 521		7 30 56	-23 49.5	2	7.72	.015	-0.17	.003	-0.84	.009	12	B2 III	272,1732
		LOD 1 # 8		7 30 58	-22 10.6	1	11.05		0.26		0.15		2		126
		NGC 2414 - 11		7 30 59	-15 20.6	1	13.02		0.39		0.01		2	A5	183
60342	-20 02015			7 31 00	-20 35.7	1	8.64		-0.06				4	B8 IV/V	2014
60517	-49 02901			7 31 00	-49 44.6	1	8.87		0.70				4	G6 V	2012
60369	-28 04502	LSS 528	★ AB	7 31 01	-28 13.0	6	8.14	.014	0.01	.014	-0.87	.015	16	B7 Ib	164,401,540,710,976,1737
60325	-14 01966	NGC 2422 - 1	★	7 31 04	-14 13.8	2	6.22	.004	-0.05	.004	-0.71		5	B1 V	1211,2007
		NGC 2414 - 19		7 31 04	-15 20.6	1	11.80		0.39		-0.60		1		862
60345	-24 05566	HR 2900		7 31 04	-24 36.1	5	5.84	.005	0.16	.010	0.12	.020	14	A3 III	15,126,258,2012,2014
	-32 04140	LSS 532		7 31 06	-32 26.6	2	10.65	.023	0.20	.028	-0.73	.009	4		568,1737
60341	-19 01944	HR 2899		7 31 07	-19 18.1	2	5.64	.015	1.12	.000	1.06		8	K0 III	2031,3077
		LOD 1 # 12		7 31 07	-22 05.1	1	11.96		0.12		0.08		2		126
		LOD 1 # 10		7 31 07	-22 09.4	1	11.70		0.62		0.26		2		126
60368	-21 02002			7 31 08	-22 11.3	1	8.31		0.68		0.16		2	G3 V	126
		LSS 523		7 31 09	-15 04.3	1	12.73		0.45		-0.43		1	B3 Ib	862
		LOD 1 # 11		7 31 10	-22 08.3	1	12.01		0.54		0.20		2		126
		Bo 15 - 16		7 31 11	-33 39.6	1	12.06		0.25		-0.43		1		568
		LOD 1 # 13		7 31 12	-22 04.2	1	11.86		0.14		0.06		2		126
		LSS 527		7 31 14	-15 06.3	1	11.62		0.52		-0.49		1	B3 Ib	862
	-24 05571			7 31 15	-24 37.3	1	10.15		0.22		0.14		2		126
		LOD 1 # 14		7 31 16	-22 04.9	1	10.69		0.36		0.12		2		126
		G 193 - 53		7 31 17	+52 55.5	1	11.06		1.11		0.97		2	K5	7010
		LSS 535		7 31 17	-33 21.1	1	10.76		-0.02		-0.76		1	B3 Ib	568
60275	+10 01563	HR 2893		7 31 20	+10 40.7	1			-0.01		-0.06		3	A1 V	1022
	-24 05575			7 31 23	-24 37.3	1	9.96		0.51		0.06		2		126
60575	-48 03023	IDS07300S4900	AB	7 31 23	-49 06.2	1	8.42		-0.01		-0.48		4	B3 V	164
60204	+28 01416	IDS07283N2854	AB	7 31 24	+28 47.8	1	6.65		0.93		0.65		3	G7 III	1501
	-27 04224	LSS 534	★ A	7 31 24	-28 04.5	1	9.27		0.05		-0.77		2	B3	710
60179	+32 01581	HR 2891	★ A	7 31 25	+32 00.4	8	1.58	.005	0.03	.006	0.01	.004	65	A1 V	15,198,1013,1022,1119*
60179	+32 01582	IDS07282N3166	★ C	7 31 25	+32 00.0	4	9.07	.018	1.49	.005	1.03	.015	4	M1 Ve	3003,6002,8003,8006
	+29 01553			7 31 27	+28 50.6	1	8.48		0.59				2	G0 IV	20
		LSS 531		7 31 27	-15 42.9	1	11.05		0.32		-0.55		1	B3 Ib	862
60462	-24 05576			7 31 28	-24 12.5	1	8.85		-0.09				4	B7/8 III	2014
	-24 05577			7 31 29	-24 39.3	1	9.52		0.91		0.64		2	K0	126
60414	-14 01971	NGC 2422 - 9	★ V	7 31 30	-14 24.9	8	4.97	.039	1.43	.023	0.29	.027	24	M2 Iabpe+B	15,369,1024,1075,2007*
60559	-39 03398	HR 2907		7 31 31	-39 57.0	3	6.25	.004	-0.13	.004	-0.47	.005	11	B8 IV(p Si)	15,1075,2012
60498	-33 03915			7 31 32	-33 17.4	2	7.35	.003	-0.14	.002	-0.62	.003	6	B4 IV	164,1732
60235	+28 01417			7 31 33	+28 36.8	1	7.38		1.37				2	K3 III	20
60479	-27 04229	LSS 538		7 31 33	-27 52.1	6	8.42	.016	0.33	.011	-0.66	.021	12	O9.5Ib	540,710,976,1012,1737,8100
60574	-42 03325	HR 2908		7 31 37	-42 58.6	1	6.52		0.92				4	G8 III	2006
60357	+3 01719	HR 2901	★ A	7 31 38	+03 28.9	2	5.80	.005	-0.02	.000	-0.09	.000	7	A0 Vnn	15,1415
		LP 365 - 24		7 31 38	+22 26.9	1	11.53		1.50		1.32		2		1723
	-13 02098	NGC 2422 - 14		7 31 41	-13 21.0	1	9.12		1.63				2	K5	369
60496	-24 05585	IDS07297S2432	AB	7 31 41	-24 38.3	1	8.07		0.99		0.81		2	K0 III	126
	-23 05693			7 31 42	-23 47.1	1	8.31		0.50		0.04		4	F8	272
		G 112 - 27		7 31 43	+01 06.1	1	11.93		1.47		1.11		5		203
60319		G 88 - 40		7 31 43	+17 00.9	3	8.95	.004	0.51	.004	-0.11	.004	8	G0	516,1620,3077
		LOD 1 # 3		7 31 44	-22 09.5	1	12.16		0.12		0.08		2		126
		LOD 1 # 4		7 31 45	-22 12.2	1	11.76		0.19		0.07		2		126
	-14 01975	NGC 2422 - 13		7 31 46	-14 37.3	1	9.99		0.23		-0.64		1		862
	+56 01224	G 193 - 54		7 31 47	+56 16.4	1	10.84		1.14		1.04		2	K3	7010
	-30 04705	LSS 542		7 31 47	-30 11.1	1	11.27		0.06		-0.73		2		564
		LP 423 - 7		7 31 48	+18 17.3	1	11.09		0.80		0.38		2		1696
60298	+25 01709	G 88 - 41		7 31 48	+25 04.2	3	7.36	.024	0.65	.012	0.10	.004	4	G2 V	333,1003,1620,3077
		LOD 1 # 2		7 31 49	-22 10.8	1	9.92		1.03		0.78		2		126
		LF12 # 1		7 31 53	-22 47.8	1	10.83		0.96		0.72		1		137
		G 111 - 18		7 31 54	+46 17.5	1	13.08		1.47		1.11		3		419
60532	-21 02007	HR 2906		7 31 55	-22 11.2	12	4.44	.011	0.51	.009	0.06	.016	34	F6 V	15,126,258,369,611,1008*
60318	+31 01620	HR 2896	★ AB	7 31 58	+31 04.3	2	5.37	.029	1.01	.000	0.86	.010	5	K0 III	1080,3016
60513	-15 01904			7 31 58	-16 04.8	1	6.75		0.61		0.08		5	G2 V	1628
	-23 05701			7 31 59	-23 41.1	1	9.46		1.06		0.93		2		272
	-2 02182	LS VI -02 017		7 32 00	-02 13.2	1	9.76		-0.15		-0.79		3		490
60491	-6 02184			7 32 00	-06 47.2	1	8.16		0.90		0.50		2	K2 V	3072
60605	-35 03680			7 32 00	-35 51.1	1	8.59		0.89		0.39		4	G6 IV	119
60553	-19 01950			7 32 02	-20 01.7	5	6.91	.015	-0.18	.010	-0.81	.015	16	B2 II	158,399,401,540,976
60606	-36 03715	HR 2911, OW Pup		7 32 02	-36 13.7	7	5.45	.030	-0.08	.012	-0.73	.009	43	B2 Vne	15,401,681,815,1637*
		LSS 540		7 32 03	-14 46.8	1	11.73		0.43		-0.46		1	B3 Ib	862
		LSS 541		7 32 04	-15 48.7	1	12.39		0.30		-0.52		1	B3 Ib	862

Table 1   405

HD	DM	Other Id	N Rem	α_{1950}	δ_{1950}	S	V	σ_V	B–V	σ_{B-V}	U–B	σ_{U-B}	n	Spectrum	References
		LSS 546		7 32 04	−27 28.3	1	11.70		0.51		-0.42		3	B3 Ib	710
		L 384 - 24		7 32 06	−42 47.	1	14.16		0.11		-0.66		1	DAe	3073
60753	−50 02835			7 32 08	−50 28.5	1	6.69		-0.09		-0.58		4	B2 III	158
60489	+3 01723	HR 2904		7 32 09	+02 50.1	3	6.54	.011	0.23	.005	0.14	.009	8	A7 III	15,39,1415
60552	−13 02101	IDS07299S1339	AB	7 32 10	−13 45.6	2	6.71	.010	0.48	.005	-0.09		7	F7 II/III	158,1594
60584	−23 05709	HR 2909	★ AB	7 32 12	−23 21.8	4	5.08	.004	0.45	.011	0.03	.000	11	F5/7 V	15,1008,1075,2012
	+36 01650	G 87 - 47		7 32 15	+36 03.9	2	10.34	.000	0.66	.000	0.05	.000	3		333,927,1620
60668	−37 03680			7 32 15	−37 12.6	1	7.81		-0.04		-0.52		2	B2 III/IV	401
60686	−39 03407	HR 2917		7 32 16	−39 47.8	1	6.76		1.13				4	K2 III	2007
		Bo 15 - 12		7 32 18	−33 00.6	1	14.77		-0.23		-1.13		1		568
		Ross 390		7 32 20	−10 16.3	2	11.09	.005	0.66	.010	0.00	.044	6	G5	158,3077
60646	−33 03926	HR 2913		7 32 20	−33 21.2	2	6.10	.005	0.30	.005			7	A9 IV	15,2027
		LP 783 - 22		7 32 22	−20 18.6	1	10.29		0.69		0.17		2		1696
60335	+43 01711	HR 2898	★ AB	7 32 25	+43 08.6	1	6.41		0.33		0.10		2	F0	1733
60629	−25 04719	HR 2912		7 32 25	−26 00.4	3	6.65	.005	-0.01	.009	-0.06		9	A0 V	15,401,2012
60814	−49 02914			7 32 29	−49 38.5	1	8.48		1.12		0.95		7	K0 III	1673
	−25 04722			7 32 30	−25 39.5	1	11.13		0.01		-0.16		1	B9 V	137
60666	−26 04574	HR 2916		7 32 32	−26 54.2	3	5.77	.008	1.05	.010			11	K1 III	15,1075,2031
		LP 256 - 33		7 32 34	+33 00.4	1	10.55		1.02		0.92		2		1723
60640	−20 02039			7 32 34	−20 52.4	1	9.01		1.09				17	G8 IV	6011
60641	−21 02014			7 32 34	−22 08.6	1	9.28		1.15		1.00		2	G8/K0 III	126
60624	−13 02104	NGC 2422 - 21	★ AB	7 32 36	−14 03.1	1	7.56		-0.01		-0.39		2	B8 III	401
60794	−46 03242			7 32 38	−46 32.0	1	8.85		0.00		-0.59		2	B3/5 III	401
	−27 04257	LSS 555		7 32 41	−27 55.8	1	10.37		0.23		-0.66		3		710
	−15 01909	LSS 552		7 32 42	−16 7.7	1	10.57		0.25		-0.76		2	O7 III	1737
60294	+56 01227	HR 2894		7 32 43	+55 52.0	1	5.92		1.12				2	K2 III	71
		CpD -21 02364		7 32 50	−21 58.8	2	10.19	.012	0.52	.019	-0.42	.034	5		540,976
	−45 03283			7 32 50	−45 10.8	2	10.49	.024	0.61	.005	-0.02	.000	7	G8 V-VI	1097,3077
60522	+27 01424	HR 2905	★ A	7 32 51	+27 00.5	10	4.06	.010	1.54	.013	1.95	.021	25	K5 III	15,31,369,1118,1119*
	−21 02016			7 32 51	−22 06.4	1	8.86		2.05		2.20		2		126
		LSS 557		7 32 52	−20 52.6	1	12.33		0.19		-0.33		2	B3 Ib	844
	−24 05617			7 32 52	−24 15.5	1	11.24		0.00		-0.04		1	A0 V	137
60437	+46 01286	HR 2903		7 32 54	+46 17.6	2	5.66	.015	1.56	.020	1.74	.160	4	M0 III	252,3051
	−25 04732			7 32 56	−26 03.8	1	10.42		0.28		0.09		1	A7 V	137
		G 88 - 42		7 32 59	+19 19.0	1	14.38		0.64		-0.04		2		1658
	−23 05729			7 32 59	−23 15.9	1	11.28		0.23		0.14		2	A3 V	137
	−25 04734			7 33 00	−26 08.5	1	10.65		0.10		0.05		3	A2 V	137
		AA45,405 S307 # 1		7 33 01	−18 41.3	1	12.46		0.71		-0.37		2		432
	−18 01920	LSS 561		7 33 01	−18 41.3	3	10.41	.026	0.63	.010	-0.45	.005	10	O9 V	342,797,844
	−27 04264			7 33 01	−27 39.5	1	11.17		0.39		-0.55		1		211
		LSS 568		7 33 01	−32 26.5	1	11.75		0.17		-0.58		1	B3 Ib	568
60749	−21 02018			7 33 06	−22 06.0	1	9.88		0.28		0.12		2	A3/5 II	126
		LSS 564		7 33 08	−20 11.5	1	11.50		0.33		-0.47		4	B3 Ib	844
		G 112 - 28		7 33 09	+03 36.9	2	13.70	.015	1.12	.025	0.83	.010	4		316,3078
	−49 02919			7 33 09	−49 43.4	1	10.95		0.63		0.15		2		1730
60561	+35 01656			7 33 14	+35 32.1	1	8.22		0.03		0.01		2	B9	1601
		LSS 566		7 33 15	−18 42.5	2	11.62	.000	0.73	.010	-0.26	.000	6	B3 Ib	797,844
		AA45,405 S307 # 4		7 33 15	−18 42.8	1	13.42		0.83		0.27		2		797
	−21 02021	LSS 569		7 33 17	−21 28.5	2	10.12	.006	0.24	.009	-0.63	.014	4	B1 II-Ib	540,976
		Bo 15 - 11		7 33 17	−32 46.1	1	11.90		0.12		-0.52		1	B0	568
60929	−44 03549			7 33 17	−44 11.1	1	6.81		-0.03		-0.10		2	A0 V	401
	−27 04269			7 33 18	−27 27.0	1	10.72		0.33		-0.59		1		211
60782	−14 01987	NGC 2422 - 32		7 33 19	−14 33.5	1	8.81		0.00		-0.23		2	B9 (IV)	401
233422	+50 01450	IDS07295N5050	AB	7 33 20	+50 43.3	1	9.08		0.76		0.38		2	G5	1566
60833	−23 05739			7 33 20	−23 28.3	1	9.64		0.00		-0.08		4	B9 V	137
		AA45,405 S307 # 3		7 33 21	−18 38.9	1	12.47		0.79		-0.32		2		797
60863	−28 04566	HR 2922	★ A	7 33 22	−28 15.5	10	4.63	.014	-0.11	.013	-0.40	.025	26	B8 V	15,126,146,1075,1079*
	+47 01467			7 33 23	+47 29.9	1	8.65		1.03		0.92		2		1601
61031	−51 02571	HR 2925		7 33 23	−51 21.8	1	6.28		0.08				4	A1 III/IV	2007
		AO 1097		7 33 24	+05 34.1	1	9.85		1.61		1.93		1	M3 III	1748
60834	−25 04748			7 33 25	−25 17.0	1	8.09		0.31		0.31		1	A9 V	137
		LSS 570		7 33 26	−18 26.2	1	12.17		0.56		-0.08		4	B3 Ib	844
60863	−28 04567	IDS07314S2809	B	7 33 27	−28 16.4	2	9.15	.054	0.36	.029	0.07	.102	3	F1 V	146,8084
60863	−28 04566	IDS07314S2809	C	7 33 27	−28 16.4	1	10.49		1.21		1.22		1		8084
		AA45,405 S307 # 5		7 33 29	−18 41.5	1	11.75		0.48		0.19		1		797
		LSS 571		7 33 29	−20 35.1	1	11.98		0.25		-0.45		4	B3 Ib	844
60861	−25 04750			7 33 29	−26 02.5	1	7.95		0.92		0.67		2	B8/9 V	401
		ApJ144,259 # 22		7 33 30	+02 49.	1	19.59		-0.35		-1.17		1		1360
60779	−2 02197			7 33 32	−03 02.5	2	7.16	.010	0.56	.005	0.01		8	G5	158,2012
		LF12 # 10		7 33 32	−25 38.5	1	11.11		0.68		0.19		1	G2 V	137
60880	−27 04276			7 33 33	−27 12.3	1	8.35		-0.04		-0.11		2	A0 V	126
60806	−7 02061			7 33 34	−07 18.5	1	9.95		-0.08		-0.50		3	B9	1732
		NGC 2422 - 186		7 33 36	−14 25.6	1	13.85		0.63		0.13		1		49
	−14 01993	NGC 2422 - 41		7 33 36	−14 34.0	1	10.17		0.25		0.18		1	A7 V	49
60406	+61 00983			7 33 37	+61 38.7	1	7.36		0.46		-0.02		2	F5	1733
60859	−19 01959	LSS 574	★ AB	7 33 37	−19 40.3	1	8.71		0.53		-0.36		4	B5 Ib	844
		AO 1099		7 33 38	+05 26.3	1	10.20		1.05		0.97		1	K1 III	1748
60778	+0 02011			7 33 38	−00 01.5	2	9.11	.004	0.11	.004	0.22	.043	6	A1 V	1003,1775
60856	−14 01994	NGC 2422 - 42		7 33 39	−14 35.9	1	7.94		-0.04		-0.39		2	B6 Ve	401
		CoD -22 04761		7 33 39	−22 45.9	1	10.01		0.35		-0.64		3		976

HD	DM	Other Id	N Rem	α₁₉₅₀	δ₁₉₅₀	S	V	σ_V	B–V	σ_B–V	U–B	σ_U–B	n	Spectrum	References
	+39 01976			7 33 42	+39 00.8	1	9.32		0.34		0.08		4	F2	439
		LSS 576		7 33 44	−20 40.1	1	11.25		0.24		−0.54		4	B3 Ib	844
		Pup sq 2 # 3		7 33 44	−27 39.	1	11.47		0.44		−0.02		1		211
		POSS 544 # 4		7 33 45	+05 57.1	1	15.51		1.57				2		1739
		NGC 2422 - 46		7 33 45	−14 21.4	1	11.64		0.36		−0.25		2		1040
	−28 04576			7 33 45	−28 15.4	1	11.17		0.35		0.10		2		126
60855	−14 01999	NGC 2422 - 45	★ AB	7 33 46	−14 22.8	9	5.69	.025	−0.12	.012	−0.71	.010	32	B2 Vn	15,49,542,815,1040*
	−14 02000	NGC 2422 - 47	★ C	7 33 47	−14 22.9	1	9.71		0.03		−0.23		2	A0 V	542
	−28 04579			7 33 47	−28 10.9	1	11.37		0.20		0.12		2		126
		G 111 - 21		7 33 48	+45 13.3	1	13.85		1.50		1.20		1		906
		NGC 2422 - 49		7 33 48	−14 26.9	1	10.94		0.54		0.05		1		49
61008	−35 03710			7 33 48	−36 03.1	1	7.85		−0.13		−0.61		2	B8 II	401
61077	−46 03263			7 33 48	−46 28.0	1	7.61		0.06		0.00		2	A1 V	401
		LOD 1 # 20		7 33 49	−22 12.2	1	10.90		0.23		0.09		2		126
	−33 03950	LSS 581		7 33 49	−33 30.3	1	12.05		0.30		−0.27		1		568
60898	−14 02002	NGC 2422 - 50		7 33 51	−14 17.8	2	7.82	.015	1.22	.010	1.10	.005	3	K1 III	49,1040
60899	−14 02001	NGC 2422 - 53		7 33 51	−14 35.1	2	7.92	.010	1.04	.010	0.79	.020	4	K0 III	49,1040
60952	−23 05759			7 33 51	−23 29.2	1	9.63		1.89		2.30		1	C4+,3	864
60853	−7 02065	HR 2920		7 33 52	−08 12.0	3	6.27	.008	1.54	.000	1.88	.000	11	K2	15,1415,2032
	−14 02004	NGC 2422 - 55		7 33 52	−14 33.1	1	10.50		0.17		0.16		1	A3 Vm	49
61005	−31 04778			7 33 52	−32 05.6	1	8.23		0.75		0.28		8	G3/5 V	1732
60654	+40 01903	HR 2915		7 33 53	+40 08.3	2	6.35	.010	1.59	.005	1.94	.000	4	M1	1733,3001
		NGC 2422 - 185		7 33 53	−14 15.4	1	12.86		0.61		0.12		1		49
	−14 02006	NGC 2422 - 57		7 33 54	−14 31.2	1	10.95		0.32		0.17		1		49
		NGC 2421 - 12		7 33 54	−20 30.1	1	12.02		0.94		0.66		1	F5	410
60921	−21 02029	IDS07318S2204	AB	7 33 54	−22 10.9	1	7.76		−0.15				2	A0 V	126
60803	+6 01729	HR 2918		7 33 55	+05 58.4	3	5.90	.008	0.60	.000	0.10	.005	8	G0 V	15,1078,3077
60826	+2 01715	BE CMi		7 33 56	+02 10.9	4	8.94	.077	2.70	.078	5.11	.109	31	C5,5	864,989,1238,1729
		NGC 2419 sq2 1		7 33 56	+38 52.5	1	15.26		0.49				1		1661
		NGC 2419 sq2 2		7 33 56	+39 04.4	1	16.03		0.98				1		1661
		NGC 2421 - 1		7 33 56	−20 31.2	2	11.39	.100	0.27	.030	−0.38	.020	9	B8	410,844
		NGC 2421 - 24		7 33 57	−20 30.6	1	12.61		0.34		−0.30		2	F0	410
61006	−33 03953			7 33 57	−33 16.8	1	7.21		−0.13		−0.68		4	B2 III	158
		NGC 2422 - 61		7 33 58	−14 25.6	1	11.53		0.45		0.01		2	F0	1040
	−14 02008	NGC 2422 - 58		7 33 58	−14 28.7	2	10.72	.005	0.22	.000	0.14	.030	4	A2 m	49,1040
		NGC 2421 - 25		7 33 58	−20 30.7	1	13.04		0.28		−0.15		2	F5	410
		NGC 2422 - 141		7 33 59	−14 29.8	1	13.68		0.39				3		1040
60951	−21 02030	HR 2923		7 33 59	−22 02.9	2	6.38	.029	0.97	.015	0.79		6	K1 II	126,2031
	−20 02057	NGC 2421 - 3		7 34 00	−20 29.3	1	10.66		0.27		−0.40		4	B5	410
		NGC 2421 - 26		7 34 00	−20 29.5	1	13.08		0.36		0.01		2		410
		NGC 2422 - 162		7 34 01	−14 26.8	1	12.62		0.46		0.11		1		49
		NGC 2421 - 8		7 34 01	−20 30.8	1	11.84		0.26		−0.21		2	B8	410
		NGC 2421 - 11		7 34 01	−20 33.4	1	12.69		0.40		0.07		2	F0	410
	−28 04585			7 34 01	−28 12.4	1	11.06		0.22		−0.62		2		126
		NGC 2422 - 63		7 34 02	−14 23.8	1	10.45		0.16		0.13		1	A2 Vm	49
60945	−16 02041			7 34 02	−17 04.8	1	8.95		0.04		−0.30		2	B7/8 III	401
60694	+39 01978	IDS07307N3905	AB	7 34 03	+38 58.9	2	7.94	.000	0.15	.000	0.09	.014	13	A0	439,1775
		NGC 2421 - 27		7 34 03	−20 28.3	1	13.28		0.43		0.00		2	F0	410
		NGC 2421 - 13		7 34 04	−20 28.0	1	12.25		0.32		−0.19		2	A5	410
	−27 04289			7 34 04	−27 37.6	1	10.64		0.07		−0.06		2		126
60944	−15 01926			7 34 05	−16 10.1	1	7.72		2.17				2	M2 III	369
		NGC 2421 - 23		7 34 05	−20 33.6	1	12.95		1.85		2.80		1		410
60941	−14 02010	NGC 2422 - 62		7 34 06	−14 28.5	1	9.11		0.06		−0.08		3	A0 V	1040
		NGC 2421 - 14		7 34 06	−20 28.2	1	12.52		0.35		−0.27		2	F0	410
		NGC 2421 - 6		7 34 06	−20 29.5	1	11.54		0.21		0.12		2	A0	410
		NGC 2421 - 28		7 34 06	−20 30.	1	13.89		0.72		0.20		2		410
61026	−27 04291			7 34 06	−27 23.5	1	7.00		1.49		1.55		2	K2 III	126
61168	−49 02927			7 34 06	−49 43.1	1	8.35		1.18		1.14		2	K1 III	1730
		NGC 2421 - 18		7 34 07	−20 29.7	1	13.25		0.50		−0.01		2		410
		NGC 2421 - 19		7 34 07	−20 30.0	1	13.11		0.43		0.15		2		410
	−20 02059	NGC 2421 - 4		7 34 07	−20 30.1	1	10.91		0.31		−0.32		2	B8	410
		NGC 2421 - 20		7 34 07	−20 30.2	1	13.70		0.40		0.05		2	F0	410
		NGC 2421 - 22		7 34 07	−20 31.9	1	13.78		0.31		0.01		2	F5	410
		NGC 2421 - 10		7 34 07	−20 33.4	1	12.07		0.26		−0.24		2	A5	410
	−27 04292			7 34 07	−27 32.1	1	10.71		0.88		0.53		2		126
60653	+45 01459			7 34 08	+45 10.3	1	8.36		0.38		0.13		2	F2	1375
		NGC 2421 - 7		7 34 08	−20 31.2	1	11.98		0.25		0.12		2	A5	410
	−20 02060	NGC 2421 - 5		7 34 09	−20 28.3	1	10.45		0.95		0.61		2	G0	410
61087	−37 03709	NGC 2451 - 131		7 34 09	−37 20.0	2	7.26	.005	−0.08	.005	−0.20		4	B9 V	401,2037
		NGC 2422 - 140		7 34 10	−14 16.9	1	12.54		0.47		0.05		1		1040
60969	−14 02016	NGC 2422 - 71		7 34 10	−14 29.0	1	7.01		−0.08		−0.44		3	B5 IV	1040
		NGC 2419 sq1 3		7 34 11	+38 58.8	1	12.29		0.59		0.08		3		439
60652	+49 01653	HR 2914		7 34 11	+48 53.2	2	5.90	.015	0.21	.010	0.19	.025	5	A5 m	252,8071
		NGC 2421 - 17		7 34 11	−20 30.4	1	12.24		0.26		−0.36		2	B8	410
		NGC 2421 - 21		7 34 11	−20 30.7	1	13.04		0.26		−0.21		2		410
		NGC 2421 - 9		7 34 11	−20 31.4	1	12.17		0.27		−0.25		2	A5	410
		NGC 2422 - 142		7 34 12	−14 19.	1	14.03		−0.18		−0.52		1		1040
60848	+17 01623	BN Gem		7 34 13	+17 01.0	1	6.85		−0.21		−1.13		3	O8 V:pe	399
60999	−14 02018	NGC 2422 - 75		7 34 13	−14 33.2	1	8.79		0.01		−0.10		1	B9 V	49
61022	−19 01965	LSS 583		7 34 14	−20 05.9	1	9.90		0.29		−0.48		5	B5 Ib/II	844

Table 1

407

HD	DM	Other Id	N Rem	α_{1950}	δ_{1950}	S	V	σ_V	B–V	σ_{B-V}	U–B	σ_{U-B}	n	Spectrum	References
		NGC 2421 - 15		7 34 14	−20 27.7	1	11.84		0.54		0.10		1	A5	410
		NGC 2421 - 16		7 34 14	−20 28.2	1	12.35		0.31		−0.21		2	A5	410
		LF12 # 11		7 34 14	−24 43.9	1	9.31		1.16		1.17		1	K0 III	137
	−27 04295			7 34 16	−27 20.8	1	10.74		−0.03		−0.15		2		126
60996	−14 02019	NGC 2422 - 78	⋆ J	7 34 17	−14 19.1	2	8.79	.030	−0.02	.005	−0.20	.040	4	B8.5V	49,1040
	−20 02061	NGC 2421 - 2		7 34 17	−20 33.4	2	10.51	.015	0.23	.005	−0.65	.015	8	B2	410,844
	−14 02022	NGC 2422 - 79	⋆ B	7 34 18	−14 20.8	2	7.78	.025	−0.05	.005	−0.40	.040	4	B5 V	49,1040
		LSS 586		7 34 18	−19 47.5	1	10.80		0.46		−0.34		4	B3 Ib	844
	−27 04297			7 34 18	−27 33.2	1	10.90		0.55		0.15		2		126
		NGC 2422 - 149		7 34 23	−14 09.6	1	11.80		0.67		0.25		1		49
61017	−14 02025	NGC 2422 - 83	⋆ A	7 34 23	−14 19.8	2	6.70	.020	−0.05	.010	−0.33	.020	4	B7 Vn	49,1040
		NGC 2419 sq1 8		7 34 24	+39 04.3	1	13.57		1.22		1.14		4		439
60771	+39 01979			7 34 25	+38 59.5	2	7.19	.005	0.27	.005	0.07	.005	6	A7 V	105,439
61016	−12 02052			7 34 25	−12 44.4	1	8.47		−0.12		−0.63		3	B3/5 V	1732
61084	−27 04299			7 34 25	−27 27.9	1	8.85		0.37		0.02		2	F2 V	126
61248	−52 01231	HR 2934		7 34 25	−52 25.3	3	4.94	.008	1.40	.005	1.63	.000	14	K3 III	15,1075,2012
61071	−25 04775	IDS07324S2507	AB	7 34 27	−25 13.2	3	6.83	.008	−0.08	.009	−0.76	.004	8	B2 III	206,540,976
61045	−14 02029	NGC 2422 - 89		7 34 29	−14 26.9	1	8.02		−0.12		−0.44		2	B9p	1040
61068	−19 01967	HR 2928, PT Pup		7 34 29	−19 35.4	6	5.71	.013	−0.18	.008	−0.87	.012	15	B2 III	401,540,976,1732,2007,8100
		NGC 2419 sq2 3		7 34 30	+39 05.9	1	12.82		0.58				1		1661
	−13 02123	NGC 2423 - 43		7 34 32	−13 44.4	1	10.44		1.04		0.70		2		1685
	−26 04642			7 34 32	−27 09.3	1	10.48		0.00		−0.01		2		126
		NGC 2419 sq1 2		7 34 34	+39 05.5	2	11.95	.014	1.09	.005	0.95		5		439,1661
		NGC 2422 - 139		7 34 34	−14 16.1	1	12.40		0.52		0.13		2		1040
	−14 02032	NGC 2422 - 91		7 34 34	−14 17.6	1	9.76		0.04		−0.06		4	A0 V	1040
	−14 02033	NGC 2422 - 93		7 34 34	−14 24.4	2	10.49	.015	0.08	.010	0.00	.040	4	A1 V	49,1040
	−27 04302			7 34 34	−27 44.7	1	10.64		0.06		−0.10		2		126
		NGC 2419 sq2 6		7 34 35	+39 04.6	1	18.60		1.09				1		1661
		NGC 2419 sq2 5		7 34 35	+39 06.0	1	17.26		0.90				1		1661
		NGC 2422 - 184		7 34 35	−14 03.4	1	12.00		0.39		0.09		1		49
		LF12 # 12		7 34 35	−25 58.3	1	12.19		0.45		0.01		1	A2 V	137
	−31 04800	LSS 591		7 34 35	−32 06.2	2	10.52	.002	−0.30	.005	−1.18	.010	17	sdO	540,1732
60915	+19 01784			7 34 36	+19 01.9	1	6.85		0.33		−0.01		2	F0	1648
		LSS 587		7 34 36	−18 56.7	1	10.79		0.57		−0.15		4	B3 Ib	844
61193	−41 03191			7 34 36	−41 58.2	2	8.21	.009	−0.06	.008	−0.76	.004	5	B2 V(n)	401,1732
		NGC 2419 sq2 7		7 34 37	+39 06.7	1	18.40		0.47				1		1661
	−19 01968	LSS 588		7 34 37	−19 42.5	1	11.00		0.50		−0.26		4		844
		NGC 2419 sq2 8		7 34 39	+39 06.3	1	19.02		0.60				1		1661
		NGC 2419 sq2 9		7 34 39	+39 07.2	1	16.86		0.76				1		1661
		NGC 2419 sq1 6		7 34 40	+39 01.9	1	13.47		0.58		−0.01		3		439
	−13 02126	NGC 2423 - 15		7 34 40	−13 48.1	1	9.77		1.21		1.13		2		1685
60914	+22 01735			7 34 41	+22 27.3	1	6.96		1.05				11	G5	1351
		NGC 2419 sq2 10		7 34 41	+39 04.0	1	20.27		0.97				1		1661
		NGC 2419 sq2 11		7 34 41	+39 06.6	1	14.50		0.47				1		1661
61245	−44 03573	V344 Pup		7 34 41	−44 50.7	1	6.95		1.05		0.81		1	K2 III	1641
62039	−77 00319			7 34 41	−77 58.5	1	8.12		0.59				4	F8 V	2012
	+54 01165	G 193 - 57		7 34 42	+54 33.5	2	9.93	.007	0.88	.006	0.53	.013	5	K0	1723,7010
		AO 1102		7 34 43	+05 20.8	1	10.35		1.08		0.99		1	K1 III	1748
		NGC 2419 sq1 25		7 34 43	+39 04.0	2	16.89	.025	0.93	.045	0.25		2		439,1661
		NGC 2419 sq2 14		7 34 43	+39 07.0	1	16.31		0.60				1		1661
		LSS 589		7 34 43	−19 25.6	1	11.84		0.56		−0.46		4	B3 Ib	844
		NGC 2419 sq1 1		7 34 44	+38 53.7	1	11.12		0.87		0.47		4		439
		LF12 # 13		7 34 46	−26 00.9	1	12.26		0.11		0.08		2	A2 V	137
61136	−27 04304			7 34 46	−27 10.8	1	8.45		1.22		1.28		2	K1 III	126
		NGC 2419 sq1 4		7 34 47	+38 56.7	2	13.00	.005	0.52	.028	0.02		4		439,1661
		NGC 2419 sq2 15		7 34 47	+39 06.1	1	12.67		0.56				1		1661
		NGC 2419 sq2 17		7 34 47	+39 06.5	1	13.19		0.56				1		1661
60799	+44 01650			7 34 47	+44 17.9	1	8.78		1.11		1.00		3	K0	1502
61064	−3 01979	HR 2927	⋆ A	7 34 47	−03 59.9	5	5.13	.011	0.44	.004	0.10	.016	18	F6 III	15,254,1008,1078,3026
	−26 04650			7 34 47	−26 22.7	1	9.79		0.84		0.59		1	K0 III	137
		NGC 2419 sq1 10		7 34 48	+39 04.0	1	14.33		0.63		0.14		3		439
		LF12 # 15		7 34 48	−25 30.5	1	11.96		0.10		0.02		1	A0 V	137
		NGC 2419 sq2 18		7 34 49	+39 07.0	1	18.41		0.94				1		1661
61114	−14 02039	NGC 2422 - 100		7 34 50	−14 12.7	2	7.80	.010	1.68	.000	1.82		5	M2 III	369,1040
		LSS 590		7 34 50	−19 12.2	1	12.09		0.50		−0.28		4	B3 Ib	844
61163	−27 04306			7 34 50	−27 38.2	1	9.31		0.96		0.73		2	G8 II/III	126
	−27 04307			7 34 50	−27 42.5	1	9.45		1.56		1.83		2		126
	−34 03746	LSS 594		7 34 50	−34 18.5	2	10.16	.010	0.27	.004	−0.64	.021	3	B0.5II-III	540,568
		NGC 2419 sq2 19		7 34 51	+38 56.2	1	19.03		0.91				1		1661
	−14 02040	NGC 2422 - 103		7 34 51	−14 26.5	2	10.50	.025	0.15	.005	0.09	.020	3	A1 pec	49,1040
		NGC 2419 sq2 20		7 34 52	+39 05.7	1	18.67		0.68				1		1661
	−13 02130	NGC 2423 - 3		7 34 52	−13 47.5	1	10.08		1.22		1.16		2		1685
		Bo 15 - 18		7 34 52	−33 38.5	1	12.14		0.47		−0.03		1		568
		NGC 2419 sq1 5		7 34 53	+38 56.2	2	13.40	.000	0.53	.023	−0.04		4		439,1661
		NGC 2419 sq2 23		7 34 54	+38 54.9	1	17.06		0.55				1		1661
		NGC 2419 sq2 22		7 34 54	+39 04.0	1	18.09		1.67				1		1661
		NGC 2419 sq1 9		7 34 56	+39 02.1	1	13.75		0.67		0.17		3		439
		NGC 2419 sq2 24		7 34 57	+39 04.7	1	16.85		0.61				1		1661
		Mel 71 - 102		7 34 57	−11 56.1	1	15.48		0.81				1		1607
		LF12 # 16		7 34 57	−26 28.6	1	12.19		−0.06		0.12		1		137

HD	DM	Other Id	N Rem	α_{1950}	δ_{1950}	S	V	σ_V	B–V	σ_{B-V}	U–B	σ_{U-B}	n	Spectrum	References
		Mel 71 - 105		7 34 58	−11 56.9	1	14.86		0.48				1		1607
61394	−55 01282	HR 2941		7 34 58	−55 46.5	3	6.38	.004	1.18	.000			11	K1/2 III	15,1075,2006
	−11 02026	Mel 71 - 83		7 34 59	−11 59.0	1	9.43		1.74				1		1607
		LF12 # 17		7 34 59	−23 04.6	1	12.62		0.02		−0.03		1	B9 V	137
		NGC 2420 - 17		7 35 00	+21 42.3	2	11.61	.010	0.35	.030	0.15	.030	4		384,1137
61037	+14 01712			7 35 01	+14 27.3	1	6.76		−0.07		−0.25		3	B9	1648
		Mel 71 - 233		7 35 01	−11 53.2	1	14.25		1.05				1		1607
61191	−26 04655			7 35 01	−26 42.1	1	6.88		1.06		0.99		2	K1 III	126
		NGC 2419 sq2 25		7 35 02	+39 03.2	1	17.96		1.18				1		1661
		NGC 2420 - 27		7 35 03	+21 42.5	1	14.51		0.45		−0.03		1		1137
	+37 01748	G 87 - 48	⋆ AB	7 35 03	+36 49.5	1	10.16		0.86		0.53		1	K5	333,1620
		Mel 71 - 120		7 35 03	−11 55.4	1	15.56		0.62				1		1607
		Mel 71 - 62		7 35 03	−11 59.6	1	11.85		0.15				1		1607
61209	−27 04310	IDS07330S2712	AB	7 35 03	−27 18.5	1	7.13		0.32		−0.20		2	B8 IV/V	126
		POSS 544 # 1		7 35 04	+06 01.6	1	14.01		1.54				1		1739
		MCC 116		7 35 04	+34 33.8	1	11.26		1.33		1.19		1	K4	3003
		Mel 71 - 109		7 35 04	−11 58.2	1	13.85		0.40				1		1607
61333	−44 03579			7 35 05	−44 52.5	3	7.20	.003	−0.12	.007	−0.69	.002	42	B3 IV	401,978,2012
		NGC 2420 - 34		7 35 07	+21 43.7	1	13.06		1.00		0.76		1		528
60935	+33 01560			7 35 07	+33 18.1	1	6.85		1.25		1.30		3	G5	1625
		NGC 2419 sq2 26		7 35 07	+38 56.0	1	20.25		0.32				1		1661
		LF12 # 18		7 35 07	−23 13.5	1	12.80		1.04		0.71		1		137
61288	−38 03500	NGC 2451 - 140		7 35 07	−39 04.4	1	9.55		0.09				3	A0 V	2037
		NGC 2420 - 41		7 35 08	+21 43.7	1	12.60		1.04		0.78		1		528
		NGC 2419 sq1 7		7 35 08	+38 55.9	2	13.50	.014	0.65	.028	0.04		4		439,1661
		NGC 2419 sq1 23		7 35 08	+38 56.3	2	16.14	.015	1.18	.029	0.69		3		439,1661
		NGC 2419 sq2 27		7 35 08	+38 57.0	1	19.33		0.77				1		1661
	−13 02133	NGC 2423 - 63		7 35 08	−13 42.2	1	10.83		0.34		0.12		3		1040
		NGC 2419 sq2 30		7 35 09	+39 01.6	1	18.30		1.10				1		1661
		LSS 595		7 35 10	−20 14.1	1	11.61		0.53		−0.44		4	B3 Ib	844
61227	−23 05791	HR 2933, LSS 597		7 35 10	−23 39.7	3	6.37	.005	0.54	.000	0.40		10	F0 Ib	15,2029,8100
		NGC 2419 sq2 31		7 35 11	+39 01.4	1	19.82		2.04				1		1661
		G 111 - 25		7 35 11	+41 04.0	1	10.71		1.12		1.00		2	K7	7010
		LF12 # 19		7 35 11	−24 15.5	1	11.53		1.15		1.06		2	K1 III	137
61310	−38 03501	NGC 2451 - 141		7 35 11	−38 42.5	1	8.85		0.01				1	B9 III-IV	2037
	+6 01736	G 89 - 33		7 35 12	+05 50.2	2	10.32	.040	0.63	.010	0.12	.080	2	G0	1620,3062
	+13 01721			7 35 12	+12 55.1	1	8.98		0.96		0.74		3	G5	1648
61035	+24 01727	HR 2926		7 35 12	+24 28.4	1	6.37		0.32		−0.02		3	F2 V	1733
		NGC 2419 sq2 32		7 35 12	+38 53.3	1	16.68		1.51				1		1661
		PKS 0735+17 # A		7 35 13	+17 49.	2	13.41	.009	0.47	.005	0.01	.019	8		326,1595
		PKS 0735+17 # B		7 35 13	+17 49.	1	14.94		0.55		0.05		2		326
		PKS 0735+17 # C		7 35 13	+17 49.	2	14.43	.037	1.06	.028	0.82	.052	8		326,1595
		PKS 0735+17 # D		7 35 13	+17 49.	2	15.84	.052	0.69	.014	0.22	.070	8		326,1595
		NGC 2419 sq2 33		7 35 13	+38 58.1	1	17.58		0.95				1		1661
	−22 01954			7 35 13	−22 41.0	1	10.37		1.00		0.83		2	K0 III	137
61262	−32 04212			7 35 13	−32 19.3	1	9.21		−0.03		−0.14		3	B9 V	1732
		Bo 15 - 19		7 35 13	−33 09.4	1	10.98		0.25		−0.55		1	B1 III	568
		G 91 - 4		7 35 14	+16 50.7	1	14.66		1.54		1.20		1		906
		PKS 0735+17 # 2		7 35 14	+17 49.	1	15.04		0.60		0.12		4		487
		LSS 596		7 35 14	−20 13.7	1	11.96		0.66		0.01		4	B3 Ib	844
		PKS 0735+17 # 3		7 35 15	+17 47.3	1	13.33		1.02		0.69		3		487
		NGC 2420 - 66		7 35 15	+21 38.7	1	14.57		0.85		0.42		1		57
60818	+54 01167			7 35 15	+54 01.0	1	6.68		0.06		0.08		2	B9	985
		NGC 2420 - 68		7 35 16	+21 43.5	2	14.10	.028	0.61	.005	0.10	.005	2		384,1137
60986	+35 01662	HR 2924	⋆ A	7 35 16	+35 09.7	1	5.56		0.93				2	K0 III	71
		NGC 2419 sq2 34		7 35 16	+39 02.9	1	19.11		0.85				1		1661
61332	−39 03444			7 35 16	−40 04.7	2	7.06	.005	1.22	.000	1.10	.029	3	Kp Ba	565,3048
		NGC 2419 sq2 35		7 35 17	+39 02.5	1	19.62		0.62				1		1661
61207	−15 01941			7 35 17	−15 34.1	1	8.15		−0.10		−0.73		2	B3 III	401
	−23 05794			7 35 17	−23 54.5	1	10.22		0.19		0.16		1	A7 V	137
		PKS 0735+17 # 1		7 35 18	+17 46.9	1	14.21		0.60		0.14		4		487
		NGC 2420 - 75		7 35 18	+21 39.9	1	15.15		0.39		−0.01		1		57
		NGC 2420 - 76		7 35 18	+21 44.9	4	12.66	.007	0.99	.004	0.72	.011	5		57,384,528,1137
	−16 02051			7 35 18	−16 26.0	1	9.86		1.87				2	M2	369
		NGC 2420 - 79		7 35 19	+21 38.8	3	14.24	.011	0.66	.040	0.19	.050	3		57,384,1137
		NGC 2420 - 80		7 35 19	+21 44.2	3	13.10	.014	0.40	.008	−0.06	.018	4		57,384,1137
		NGC 2419 sq2 36		7 35 19	+38 57.3	1	18.24		1.16				1		1661
		NGC 2419 sq2 37		7 35 19	+38 58.7	1	19.42		0.80				1		1661
	−23 05795			7 35 19	−23 20.1	1	11.75		0.17		0.02		1	B8 V	137
		NGC 2420 - 91		7 35 20	+21 38.9	1	12.61		0.97		0.73		1		57
		NGC 2420 - 90		7 35 20	+21 44.7	1	15.26		0.73		0.25		1		57
		Mel 71 - 196		7 35 20	−11 54.9	1	14.01		0.46				1		1607
61391	−48 03069	HR 2940		7 35 20	−48 43.0	2	5.71	.005	−0.06	.000			7	B9.5V	15,2012
		AO 1104		7 35 21	+05 13.8	1	10.35		1.08		0.89		1	K0 III	1748
61224	−14 02053	NGC 2422 - 125	⋆ A	7 35 21	−14 19.6	5	6.52	.007	−0.02	.013	−0.31	.007	26	B7 Vn	15,1040,1075,1079,2012
60845	+52 01232			7 35 22	+52 40.9	1	7.41		1.19		1.26		2	K0	1566
		Mel 71 - 185		7 35 22	−11 53.9	1	11.71		1.12				1		1607
		Mel 71 - 23		7 35 22	−11 58.1	1	10.96		1.30				1		1607
		LF12 # 22		7 35 22	−26 23.2	1	11.59		0.97		0.63		1		137
61390	−44 03586			7 35 22	−44 54.9	1	7.92		1.01		0.71		36	G8 III	978

Table 1   409

HD	DM	Other Id	N	Rem	α_{1950}	δ_{1950}	S	V	σ_V	B–V	σ_{B-V}	U–B	σ_{U-B}	n	Spectrum	References
61127	+5 01731	G 89 - 34			7 35 23	+05 42.2	1	8.21		0.61		0.06		1	G0	333,1620
		NGC 2420 - 111			7 35 23	+21 38.5	2	12.61	.010	0.90	.015	0.56	.010	2		57,384
		NGC 2420 - 112			7 35 24	+21 36.2	2	13.49	.005	0.49	.000	0.01	.005	3		384,1137
		NGC 2420 - 115			7 35 24	+21 40.7	2	11.56	.010	1.12	.005	0.82	.020	2		57,384
		NGC 2420 - 119			7 35 24	+21 43.6	3	12.55	.008	1.00	.005	0.69	.011	3		57,384,528
		NGC 2420 - 122			7 35 25	+21 39.0	1	14.99		0.44		-0.02		1		57
		NGC 2420 - 129			7 35 25	+21 39.8	2	14.40	.005	0.86	.005	0.42	.010	2		57,384
		Mel 71 - 191			7 35 25	-11 55.6	1	16.07		0.68				1		1607
		NGC 2420 - 140			7 35 26	+21 39.7	3	11.51	.030	1.24	.008	1.23	.029	5		57,384,1137
	-10 02135				7 35 27	-10 45.9	1	10.52		1.66				2	M3	369
		NGC 2420 - 156			7 35 28	+21 39.8	1	14.86		0.40		-0.05		1		57
		NGC 2420 - 171			7 35 29	+21 40.0	2	13.94	.005	0.43	.000	0.02	.000	2		57,384
		NGC 2420 - 176			7 35 29	+21 40.1	2	13.52	.005	0.97	.010	0.66	.000	2		57,384
		NGC 2420 - 173			7 35 29	+21 40.4	3	11.76	.007	0.90	.010	0.54	.023	3		57,384,1137
		NGC 2420 - 172			7 35 29	+21 42.6	2	12.44	.005	0.20	.005	0.12	.005	2		57,384
		NGC 2420 - 174			7 35 29	+21 45.3	3	12.40	.012	1.07	.078	0.76	.078	3		57,384,528
	+22 01740				7 35 29	+21 50.5	1	9.31		0.88		0.58		1	G5	1137
		NGC 2420 - 325			7 35 30	+21 41.	1	15.79		0.70		0.10		1		57
		NGC 2420 - 326			7 35 30	+21 41.	1	17.26		0.69		0.10		1		57
		NGC 2420 - 327			7 35 30	+21 41.	1	17.51		0.72		0.13		1		57
		NGC 2420 - 328			7 35 30	+21 41.	1	15.94		0.47		-0.04		1		57
		NGC 2420 - 329			7 35 30	+21 41.	1	16.25		0.59		0.05		1		57
		NGC 2420 - 330			7 35 30	+21 41.	1	16.50		0.55		-0.01		2		384
		NGC 2420 - 331			7 35 30	+21 41.	1	16.83		0.59		0.05		2		384
		NGC 2420 - 190			7 35 31	+21 41.4	1	12.36		0.16		0.25		2		1137
61330	-34 03755	HR 2937		⋆ AB	7 35 31	-34 51.3	6	4.53	.004	-0.09	.007	-0.31	.006	25	B8 IV/V	15,1075,1637,2012*
		POSS 544 # 2			7 35 32	+05 10.7	1	14.67		1.49				1		1739
		NGC 2420 - 189			7 35 32	+21 35.6	2	11.11	.020	0.94	.005	0.58	.010	2		57,384
		NGC 2420 - 192			7 35 32	+21 41.7	4	12.94	.003	1.02	.023	0.82	.041	4		57,384,528,1137
		NGC 2420 - 195			7 35 32	+21 42.3	3	15.03	.016	0.42	.005	-0.05	.010	3		57,384,1137
61375	-38 03506	NGC 2451 - 145			7 35 32	-38 56.5	1	9.15		0.27				2	B9.5V	2037
61376	-38 03507	NGC 2451 - 146			7 35 32	-39 05.3	1	8.91		0.43				2	F5 V	2037
		NGC 2420 - 198			7 35 33	+21 41.4	1	15.98		0.37		-0.01		3		1137
		NGC 2420 - 196			7 35 33	+21 42.7	2	14.68	.010	0.41	.000	-0.03	.020	2		57,384
		NGC 2419 sq2 38			7 35 34	+39 06.5	1	15.84		0.68				1		1661
61355	-32 04219				7 35 34	-32 17.0	3	8.65	.013	-0.13	.006	-0.69	.008	13	B3/5 V	540,976,1732
		NGC 2420 - 211			7 35 35	+21 42.2	2	15.25	.005	0.47	.005	-0.07	.014	2		57,1137
		WLS 736 45 # 7			7 35 35	+43 37.1	1	12.52		0.47		-0.07		2		1375
61374	-37 03730	NGC 2451 - 148			7 35 36	-37 34.4	1	9.03		-0.03				2	B8 V	2037
	-13 02138	NGC 2423 - 60			7 35 37	-13 42.1	2	9.48	.041	1.41	.010	1.46	.033	4	K2	1040,1685
		LF12 # 23			7 35 37	-23 40.4	1	12.47		0.25		-0.22		2		137
61199	+5 01734				7 35 38	+05 03.4	1	7.96		0.23		0.13		1	A3	1748
		NGC 2420 - 235			7 35 39	+21 40.7	2	14.34	.005	0.47	.010	0.02	.010	2		57,384
		NGC 2420 - 233			7 35 39	+21 43.7	1	14.82		0.36		-0.03		1		1137
		NGC 2420 - 236			7 35 40	+21 41.1	2	12.59	.005	1.00	.010	0.71	.010	2		57,384
	-23 05805				7 35 40	-24 07.1	1	10.74		-0.05		-0.15		1	B8 V	137
		AO 1106			7 35 42	+05 24.4	1	9.79		0.09		0.09		1		1748
	+21 01652				7 35 42	+21 46.8	1	9.58		0.30		0.15		1	F2	1137
61351	-27 04324				7 35 43	-27 10.1	1	8.52		-0.08		-0.30		2	B8 II/III	126
		NGC 2420 - 250			7 35 44	+21 37.9	3	11.40	.012	1.38	.007	1.17	.022	3		57,384,1137
		NGC 2419 sq2 39			7 35 44	+39 02.2	1	14.43		1.08				1		1661
	-26 04676				7 35 46	-26 40.2	1	11.44		0.24		0.08		1	A2 V	137
		NGC 2420 - 257			7 35 47	+21 37.4	3	12.41	.014	0.47	.007	-0.01	.020	3		57,384,1137
	+5 01735				7 35 49	+05 11.3	1	8.81		1.68		2.11		1	K7	1748
61372	-26 04679				7 35 51	-27 07.3	1	8.66		0.47		0.01		2	F5 V	126
	-25 04814				7 35 52	-25 26.7	1	10.39		0.28		0.18		2	A7 V	137
	-18 01938	LSS 604			7 35 53	-18 16.0	1	10.85		0.28		-0.59		4		844
61110	+34 01649	HR 2930			7 35 54	+34 42.1	6	4.90	.012	0.41	.008	0.10	.012	22	F3 III	15,254,1008,1118,3026,8015
61409	-34 03760	HR 2942			7 35 54	-35 09.8	3	6.59	.004	1.13	.005			11	K1 III	15,1075,2031
61452	-37 03735	NGC 2451 - 153			7 35 56	-37 50.3	1	8.97		0.50				3	F7 V	2037
61347	-13 02143	LSS 603			7 35 57	-13 44.2	3	8.43	.005	0.17	.000	-0.76	.000	5	O9.5Iab	1011,1012,8100
	+12 01624				7 35 59	+12 32.9	2	10.21	.005	1.09	.005	1.03	.010	3	F5	1003,1696
61387	-25 04819				7 35 59	-25 53.9	1	9.93		-0.06		-0.20		1	B9 V	137
61453	-37 03736	NGC 2451 - 154		⋆	7 35 59	-37 53.8	2	6.40	.019	1.51	.039			6	K4 III	2032,2037
	+6 01742	G 89 - 35			7 36 00	+06 06.1	2	9.68	.000	1.00	.015	0.74	.010	3	K2	1620,3062
	-23 05814	LSS 606			7 36 00	-23 48.1	3	10.47	.032	0.33	.005	-0.64	.019	3	B0.5 V(n)e2	540,1737
	-26 04684				7 36 00	-26 36.6	1	8.72		1.06		0.91		2	K1 III	137
	-25 04821				7 36 01	-25 15.4	1	9.11		1.54		1.56		2	K0 III	137
		G 90 - 15			7 36 02	+30 01.5	1	10.92		1.28		1.29		5		1723
		Bo 15 - 7			7 36 05	-32 34.8	1	13.03		0.64		0.15		1		568
	-15 01953	GG Pup			7 36 07	-15 56.0	1	10.35		2.17				2	M2	369
	-25 04827	LF12 # 30			7 36 09	-26 10.0	1	9.39		1.23		1.24		4	K2 III	137
61219	+24 01730	HR 2931			7 36 11	+24 20.3	2	6.18		0.02	.005	0.04	.000	6	A2 V	1022,1733
61429	-25 04828	HR 2944, PU Pup		⋆ AB	7 36 13	-25 15.0	6	4.69	.013	-0.11	.005	-0.35	.010	16	B8 IV	15,1075,1079,2012*
61448	-28 04636				7 36 16	-28 19.2	2	9.22	.003	-0.06	.005	-0.15	.000	7	B9 III/IV	1480,1655
	-28 04637				7 36 17	-28 23.1	1	11.06		0.30		0.24		1		1480
		LF12 # 31			7 36 18	-22 44.9	1	11.23		0.00		-0.09		2	B9 V	137
		LF12 # 32			7 36 19	-23 37.5	1	9.84		0.26		0.10		1	F0 V	137
61540	-39 03455	NGC 2451 - 158			7 36 20	-39 21.4	1	8.65		1.12		0.86		3	B9 V	80
	+5 01738				7 36 21	+05 24.2	1	8.83		0.30		0.17		1	F0	1748

HD	DM	Other Id	N Rem	α_{1950}	δ_{1950}	S	V	σ_V	B−V	σ_{B-V}	U−B	σ_{U-B}	n	Spectrum	References
		G 90 - 16		7 36 22	+33 34.9	1	11.85		1.45		1.17		1		333,1620
61383	−5 02196			7 36 23	−05 20.9	2	7.57	.007	0.60	.012	0.03		4	G0	1594,6006
62153	−73 00457	HR 2979	★ AB	7 36 24	−74 09.7	3	6.33	.002	-0.03	.004	-0.24	.000	47	B9 IV	15,1075,2038
		LSS 607		7 36 26	−18 45.8	1	11.66		0.38		-0.26		4	B3 Ib	844
	−26 04697			7 36 26	−26 12.1	1	10.48		0.10		0.12		1	A3 V	137
		ApJS53,765 T4# 4		7 36 26	−28 24.0	1	12.44		0.29		0.18		1		1480
61442	−14 02064			7 36 28	−14 38.8	1	9.13		1.78				2	M1/2	369
61492	−26 04701			7 36 28	−26 28.9	1	8.85		-0.15		-0.64		2	B3 III	401
61338	+18 01701	HR 2938		7 36 35	+17 47.4	2	5.06	.005	1.56	.002	1.92		4	K5 III	1363,3001
		ApJS53,765 T4# 5		7 36 35	−28 21.7	1	13.11		0.31		0.19		1		1480
	−28 04645	VZ Pup		7 36 35	−28 23.0	2	8.98	.004	0.83	.006	0.61	.017	2	F7.5	689,1587
61512	−26 04703			7 36 36	−26 12.1	1	7.15		-0.11		-0.42		2	A(p Si)	401
		PKS 0736+01 # C		7 36 37	+01 40.7	1	13.22		0.79		0.38		1		829
		G 193 - 58		7 36 39	+53 59.9	1	11.82		1.18		1.05		4	K5	7010
		Bo 15 - 21		7 36 40	−33 16.7	1	13.36		0.23		-0.43		1		568
61421	+5 01739	HR 2943	★ AB	7 36 41	+05 21.3	15	0.37	.009	0.42	.008	0.02	.017	89	F5 IV-V	1,3,15,22,1034,1075*
		LSS 609		7 36 41	−21 54.9	1	11.62		0.26		-0.68		2	B3 Ib	1737
61621	−38 03520	NGC 2451 - 161		7 36 41	−38 35.4	1	8.09		0.05				4	A0 V	2037
61623	−39 03463	HR 2952		7 36 41	−39 52.6	3	6.58	.004	-0.05	.000	-0.16	.005	11	B9 V	15,1075,2012
		PKS 0736+01 # 1		7 36 42	+01 44.	1	12.41		1.22		1.08		4		1595
		PKS 0736+01 # 2		7 36 42	+01 44.	1	15.37		0.48		0.02		4		1595
		PKS 0736+01 # D		7 36 42	+01 44.	1	15.54		1.45		0.70		1		829
		POSS 544 # 5		7 36 42	+06 20.4	1	16.29		1.68				2		1739
61295	+32 01599	HR 2936		7 36 42	+32 07.6	1	6.17		0.35		0.13		1	F1 III	39
61106	+57 01093	HR 2929		7 36 42	+57 11.9	2	6.09	.019	1.45	.009	1.66	.005	4	K5 III	1375,3001
		PKS 0736+01 # A		7 36 44	+01 39.8	1	10.81		1.71		2.15		1		829
	−23 05847			7 36 44	−24 08.1	1	10.72		1.15		1.01		1	K1 III	137
61622	−39 03464	NGC 2451 - 162		7 36 44	−39 34.7	1	8.59		-0.02				2	Ap	2037
	−28 04648			7 36 45	−28 24.8	1	11.37		0.02		-0.06		1		1480
61555	−26 04707	HR 2948	★ AB	7 36 46	−26 41.2	6	3.82	.012	-0.17	.014	-0.56	.013	19	B6 V	15,404,1075,1079,2012,8015
	+2 01729	G 112 - 35		7 36 47	+02 18.1	3	9.66	.036	1.39	.034	1.30		4	M0	1017,1705,3072
		LP 365 - 32		7 36 47	+21 11.7	1	12.25		1.45				3		1723
61252	+41 01699			7 36 47	+41 16.8	1	6.84		-0.05		-0.15		2	A0	1601
61642	−38 03521	NGC 2451 - 163	★	7 36 48	−38 39.9	5	6.20	.018	1.02	.011	0.77		15	G8 III	15,460,1075,2006,2037
		ApJS53,765 T4# 6		7 36 49	−28 22.9	1	13.19		0.98		0.60		1		1480
		LSS 614		7 36 51	−33 17.6	1	12.23		0.12		-0.46		1	B3 Ib	467
61294	+38 01803	HR 2935		7 36 53	+38 27.6	2	5.75	.026	1.63	.006	1.95		4	M0 III	71,3001
		LSS 616		7 36 53	−33 17.5	1	12.44		0.26		-0.81		1	B3 Ib	467
61715	−48 03091	HR 2957, MY Pup		7 36 53	−48 29.2	3	5.68	.031	0.66	.018	0.45		7	F7 Ib/II	58,1488,2007
61554	−18 01946	HR 2947		7 36 54	−18 33.8	2	6.72		-0.07	.011	-0.51		5	B5/7 V	1079,2009
61641	−36 03773	HR 2954		7 36 55	−36 22.9	4	5.79	.006	-0.16	.004	-0.72	.008	12	B3 III	15,401,1732,2012
		CCS 844		7 36 56	−35 17.1	1	10.39		4.00				1	N	864
61639	−35 03751			7 36 60	−35 22.1	1	8.76		0.25		0.12		2	A5	1730
	−24 05721	LSS 612		7 37 00	−24 38.1	1	10.94		0.48		-0.58		3		1737
		PKS 0736+01 # B		7 37 02	+01 45.4	1	12.11		0.43		-0.06		1		829
		LSS 611		7 37 04	−20 41.5	1	12.34		0.37		-0.33		4	B3 Ib	844
		LSS 618		7 37 05	−27 50.8	1	11.33		0.89		0.09		3	B3 Ib	710
61502	+5 01741	IDS07345N0530	AB	7 37 07	+05 23.4	1	8.59		0.27		0.13		1	A5	1748
61590	−17 02078			7 37 08	−18 07.9	1	9.78		-0.01		-0.38		2	B7 II	401
		LF12 # 37		7 37 08	−23 58.8	1	12.46		0.17		0.05		1	A2 V	137
61712	−43 03425			7 37 08	−43 28.7	1	8.96		-0.02		-0.41		1	B7/8 V	401
59664	+83 00191			7 37 09	+83 11.4	1	7.87		1.38		1.53		2	K0	1502
	−18 01948	LSS 615		7 37 12	−18 41.6	2	10.59	.078	0.36	.049	-0.61	.010	6		844,1737
	−1 01792	G 112 - 36		7 37 17	−01 24.2	6	9.23	.009	0.82	.000	0.29	.010	14	G6 V	516,1003,1064,1620*
		Bo 15 - 6		7 37 17	−32 26.7	1	11.80		0.30		-0.54		1		568
	−27 04363	LSS 623	★ AB	7 37 18	−27 30.2	2	10.23	.000	0.42	.005	-0.48	.014	4		211,710
	+47 01474			7 37 19	+47 06.2	1	8.82		0.49		0.04		2	F8	1601
	−11 02043	LSS 617		7 37 22	−12 08.6	2	9.50	.000	0.09	.000	-0.89	.000	4		405,1737
61709	−31 04864	LSS 627	★ AB	7 37 22	−31 11.2	2	8.39	.006	0.20	.013	-0.59	.016	6	B1/2 Ib	540,976
61672	−26 04722	HR 2956		7 37 24	−26 44.8	2	6.50		-0.11	.022	-0.49		5	B7 V	1079,2007
61632	−12 02079			7 37 26	−12 23.7	2	8.29	.005	0.63	.010	0.05		3	G3 V	1594,6006
	−32 04257	LSS 628		7 37 26	−32 24.1	2	9.89	.044	0.23	.013	-0.65	.028	8	B1:Vnnp	540,568,976
61687	−26 04723	IDS07354S2641	AB	7 37 27	−26 48.2	1	6.79		-0.13		-0.53		4	B3/5 V	164
61563	+5 01742	HR 2950	★ AB	7 37 28	+05 20.9	2	6.01	.005	-0.04	.000	-0.17	.005	7	A0 III	15,1415
		Bo 15 - 26		7 37 28	−33 31.4	1	13.09		0.32		-0.52		1		568
	+5 01743			7 37 29	+05 18.0	1	8.97		0.47		-0.07		1		1748
61500	+19 01794			7 37 29	+19 28.5	1	8.66		0.49		0.02		2	F5	1648
61606	−3 02001	G 112 - 37	★ A	7 37 29	−03 28.7	1	7.18		0.96				1	K2 V	1705
61606	−3 02001	G 112 - 37	★ A	7 37 29	−03 28.7	3	7.18	.011	0.96	.006	0.74		13	K2 V	1197,2012,3072
61363	+48 01561	HR 2939		7 37 31	+48 15.0	1	5.56		1.01				3	K0 III	71
		CCS 846		7 37 31	−27 35.1	1	9.45		5.15				1	C5,3	864
	−30 04887	LSS 629		7 37 31	−30 40.2	1	10.18		0.25		-0.56		2		564
		G 251 - 44		7 37 32	+72 56.5	1	13.10		1.33				6		940
61606	−3 02002	G 112 - 37	★ B	7 37 33	−03 29.0	3	8.93	.004	1.33	.000	1.24		11	K2	2013,2033,3072
		LF12 # 38		7 37 33	−26 45.4	1	11.38		0.04		-0.20		1	A0 V	137
	+12 01631	G 89 - 38		7 37 35	+12 14.3	1	9.77		0.91		0.70		1		333,1620
		Bo 15 - 25		7 37 35	−33 27.0	1	11.85		0.21		-0.65		1	B0.5V	568
61902	−50 02899			7 37 35	−50 57.9	3	8.23	.010	0.48	.007	-0.10	.010	10	F5/6wF2	158,2012,6006
		LSS 634		7 37 37	−33 14.3	1	12.08		0.32		-0.48		1	B3 Ib	467
	−38 03530	NGC 2451 - 173		7 37 37	−38 19.7	1	8.28		1.48				3		2037

Table 1   411

HD	DM	Other Id	N	Rem	α_{1950}	δ_{1950}	S	V	σ_V	B–V	σ_{B-V}	U–B	σ_{U-B}	n	Spectrum	References
	+60 01072	G 234 - 7			7 37 40	+60 42.1	3	9.70	.003	0.70	.005	0.19	.011	9	G8	1658,1723,7010
	−26 04727				7 37 40	−26 20.5	1	10.37		1.13		1.05		2	K2 III	137
		LSS 630			7 37 40	−27 20.8	1	13.56		-0.28		-1.24		1	B3 Ib	494
61831	−38 03531	NGC 2451 - 175		★	7 37 42	−38 11.5	8	4.84	.008	-0.19	.008	-0.65	.009	47	B2.5V	15,80,1075,1637,1732,2012*
	−26 04728	LSS 633			7 37 44	−27 10.6	2	10.68		0.33		-0.52		3		710
61832	−38 03532	NGC 2451 - 178			7 37 45	−38 44.9	1	8.05		1.40				2	K2 III	2037
61948	−52 01241				7 37 45	−52 50.8	1	8.51		-0.04		-0.58		1	B4/5 V	55
61830	−37 03762	NGC 2451 - 179			7 37 46	−37 14.6	1	9.10		0.03				2	B9 V	2037
		GD 87			7 37 47	+42 59.7	2	12.67	.010	0.20	.020	0.07	.045	5		272,3060
61966	−52 01242	HR 2971, V390 Car		★ A	7 37 48	−53 09.5	3	6.05	.004	-0.12	.004	-0.40	.005	11	B9 IV/Vp Si	15,1075,2012
61499	+39 01988				7 37 50	+39 19.7	2	7.96	.005	0.05	.005	0.07	.000	5	A0 V	833,1601
		LF12 # 40			7 37 50	−26 25.7	1	11.55		1.26		1.17		1	K0 III	137
61947	−48 03102	IDS07365S4849		A	7 37 52	−48 56.1	1	6.92		-0.10		-0.71		3	B2/3 IV	1732
61699	−11 02052				7 37 53	−11 57.8	1	7.96		-0.10		-0.46		2	B9	3039
		LSS 632			7 37 53	−18 58.3	1	12.03		0.23		-0.35		4	B3 Ib	844
		LSS 636			7 37 53	−27 45.4	1	12.01		0.59		-0.20		3	B3 Ib	710
61827	−32 04266	LSS 640			7 37 54	−32 27.7	5	7.65	.011	0.54	.026	-0.29	.029	13	O8/9 (Ib)	467,540,976,1034,1737
		G 88 - 46			7 37 57	+19 42.5	1	9.95		1.14		1.14		1	K5	333,1620
		LF12 # 41			7 37 58	−22 45.5	1	11.63		0.12		-0.03		1		137
61878	−37 03767	NGC 2451 - 182		★ AB	7 37 58	−38 01.4	3	5.73	.000	-0.13	.007	-0.50		10	B7 V	401,2031,2037
61630	+14 01721	HR 2953			7 37 59	+13 53.3	1	6.21		1.40		1.60		2	K0	1733
61603	+23 01780	HR 2951			7 37 59	+23 08.2	2	5.89	.006	1.56	.016	1.86		4	K5	71,3051
		LF12 # 42			7 37 59	−23 31.6	1	11.97		1.25		1.25		1		137
		LSS 644			7 37 59	−33 41.0	1	12.26		0.23		-0.66		1	B3 Ib	467
		LP 783 - 2			7 38 01	−17 17.3	1	12.68		0.47				1		1691
61774	−19 02003	HR 2960		★ AB	7 38 01	−19 32.7	2	5.92	.003	1.16	.000	1.07		6	K0 III	58,2007
		L 745 - 46		A	7 38 02	−17 17.4	2	13.04	.005	0.25	.002	-0.62	.024	11	DF	1698,1765
		LP 783 - 3		★ A	7 38 02	−17 17.4	5	13.01	.023	0.28	.029	-0.62	.008	6	DF	940,1281,1411,1705,3060
		LSS 641			7 38 02	−27 05.2	1	12.96		0.42		-0.64		3	B3 Ib	710
61851	−31 04879	LSS 643			7 38 02	−31 23.8	2	8.98	.011	0.09	.000	-0.75	.013	7	B1 II	540,976
61899	−37 03768	NGC 2451 - 184		★	7 38 02	−38 08.7	4	5.75	.005	-0.07	.000	-0.55		12	B2.5V	15,401,2012,2037
	+49 01658				7 38 03	+49 20.5	1	9.73		1.19		1.06		1	K4	3072
		LSS 637			7 38 05	−18 57.1	1	11.63		0.22		-0.50		4	B3 Ib	844
61898	−36 03788	NGC 2451 - 185			7 38 05	−37 02.7	1	8.54		1.36				2	A0 (V)	2037
61946	−42 03396	IDS07365S4303		AB	7 38 05	−43 09.8	1	7.16		0.01		-0.45		2	B5 IV/V	401
61772	−14 02082	HR 2959			7 38 06	−15 08.8	3	4.95	.005	1.56	.004	1.78	.015	6	K3 II	2009,3005,8100
61850	−29 04737				7 38 06	−30 03.1	1	7.07		1.09		0.80		2	K2 (II)+(A)	126
	−23 05894				7 38 07	−24 03.4	1	11.35		0.15		-0.06		2		126
		LSS 642			7 38 07	−27 16.1	1	12.22		0.43		-0.17		3	B3 Ib	710
61926	−38 03538	NGC 2451 - 186			7 38 07	−38 25.9	2	7.96	.005	-0.04	.010	-0.50		6	B7 V	401,2037
62689	−77 00321	HR 3000			7 38 07	−77 31.2	3	6.70	.007	1.72	.009	2.02	.005	18	M0 III	15,1075,2038
		LSS 638			7 38 08	−21 05.4	1	11.23		0.39		-0.64		4	B3 Ib	844
		LSS 639			7 38 08	−21 14.6	1	11.73		0.34		-0.67		4	B3 Ib	844
		NGC 2439 - 59			7 38 08	−31 33.9	1	15.67		0.65		0.22		1		407
		NGC 2439 - 56			7 38 08	−31 34.7	1	14.30		0.74		0.23		1		407
		NGC 2439 - 58			7 38 09	−31 33.7	1	13.30		0.69		0.15		1		407
61749	−7 02118	HR 2958			7 38 10	−08 04.1	3	6.01	.005	0.15	.008	0.11	.005	14	A3 IV	15,1415,2027
		LF12 # 43			7 38 10	−26 29.0	1	11.45		0.08		0.06		2	A0 V	137
	−35 03770				7 38 10	−35 30.5	1	11.39		1.20				2		1730
		NGC 2439 - 55			7 38 11	−31 34.9	1	12.46		0.52		-0.02		1		407
61924	−37 03771	NGC 2451 - 188			7 38 11	−37 15.5	1	8.48		0.01				2	B9 V	2037
61925	−37 03770	NGC 2451 - 187		★	7 38 11	−37 27.8	9	6.01	.011	-0.04	.006	-0.44	.015	61	B6 IV	15,80,681,815,1075*
	−69 00757				7 38 11	−69 57.2	2	8.64	.005	0.34	.000			8	F0 IV	2012,2034
		NGC 2439 - 57			7 38 13	−31 34.3	1	15.23		0.50		0.34		1		407
		LSS 647			7 38 13	−33 25.5	1	12.19		0.32		-0.45		1	B3 Ib	467
61645	+26 01625				7 38 14	+26 00.4	1	8.13		1.10				2	K2 III	20
61660	+21 01661				7 38 15	+21 33.9	1	7.87		1.17		1.06		2	K2	1648
	−23 05901				7 38 15	−24 05.8	1	10.93		0.13		-0.40		2		126
		LSS 648			7 38 16	−33 27.2	1	12.15		0.31		-0.42		1	B3 Ib	467
	+40 01926				7 38 18	+40 13.6	1	10.14		1.12				1	K8	1017
	−23 05907				7 38 19	−24 07.1	1	11.40		0.59		0.10		2		126
61922	−29 04747				7 38 22	−29 57.5	1	9.68		-0.05		-0.13		2	A0	126
	−29 04748				7 38 23	−29 59.2	1	9.87		-0.02		-0.08		2		126
61866	−14 02086				7 38 24	−15 09.8	1	9.11		1.62				2	M1 III	369
61945	−35 03773				7 38 24	−35 25.1	1	9.57		0.19		0.00		2	B8/9 II/II	1730
62036	−45 03362				7 38 24	−45 51.2	1	7.35		1.71				4	M1 III	2012
61522	+50 01457				7 38 25	+50 41.1	1	9.05		0.30		0.09		2	F0	1566
	−29 04750				7 38 26	−30 00.8	1	9.97		-0.02		-0.04		2		126
	−31 04880	NGC 2439 - 74			7 38 27	−31 37.3	1	11.56		0.00		-0.19		1		407
61944	−29 04751				7 38 28	−29 57.7	2	6.94	.016	0.00	.004	-0.05	.014	4	B3/5 (V)	126,1732
		LSS 646			7 38 30	−19 13.3	1	11.57		0.48		-0.01		4	B3 Ib	844
		LF12 # 44			7 38 30	−26 46.7	1	11.67		1.18		1.22		2	K0 III	137
61943	−29 04752				7 38 30	−29 55.9	1	7.80		-0.16		-0.65		4	A0 V	1732
	−31 04890	NGC 2439 - 68			7 38 31	−31 32.2	2	10.73	.000	0.20	.039	-0.61		6		407,2042
		NGC 2439 - 67			7 38 31	−31 33.5	2	12.67	.068	0.22	.068	0.11		6		407,2042
233440	+52 01238				7 38 32	+52 25.7	1	8.33		0.89		0.56		2	G5	1566
61893	−21 02072				7 38 32	−21 49.6	1	8.69		-0.11				4	B8 II/III	2014
62009	−39 03480	NGC 2451 - 190			7 38 32	−39 29.3	1	9.55		0.14				2	A1 IV	2037
		NGC 2439 - 66			7 38 34	−31 34.8	1	12.61		0.16		-0.39		2		407
	−23 05919				7 38 35	−23 43.7	1	9.46		1.01		0.75		1	K0 III	137

HD	DM	Other Id	N	Rem	α_{1950}	δ_{1950}	S	V	σ_V	B–V	σ_{B-V}	U–B	σ_{U-B}	n	Spectrum	References
	−23 05921				7 38 35	−24 06.2	1	10.51		0.22		0.02		2		126
		NGC 2439 - 75			7 38 35	−31 38.2	1	12.16		0.14		-0.70		1		407
	−31 04916	NGC 2439 - 16			7 38 35	−31 38.2	1	10.45		0.98		0.74		2	M3 Ia0-Ia	407
	−27 04390	LSS 650			7 38 36	−27 32.2	2	10.21	.000	0.56	.005	-0.33	.019	4		211,710
		NGC 2439 - 65			7 38 36	−31 33.0	1	12.38		0.21		-0.46		2		407
		LSS 656			7 38 36	−33 23.7	1	10.08		0.46		-0.55		1	B3 Ib	467
62034	−39 03481	NGC 2451 - 191			7 38 36	−39 18.2	2	6.67	.030	1.03	.010	0.82		4	K0 III	460,2037
	−23 05924				7 38 37	−24 02.1	1	11.03		0.18		0.11		2		126
		LOD 2 # 14			7 38 37	−24 07.6	1	13.52		0.17		0.08		2		126
62088	−45 03366				7 38 38	−45 13.9	1	7.29		1.57		0.99		6	K4 III	1673
	−23 05926				7 38 39	−24 03.1	1	10.79		0.00		-0.21		2		126
		NGC 2439 - 64			7 38 39	−31 33.1	1	12.81		0.21		-0.46		2		407
		NGC 2439 - 76			7 38 40	−31 37.7	1	12.42		0.57		-0.04		1		407
61961	−23 05927				7 38 41	−23 58.2	1	8.09		0.19		0.08		2	A3 III	126
61988	−29 04757	IDS07367S2953		AB	7 38 41	−29 59.8	1	9.45		0.02		-0.09		2	A0 V	126
		NGC 2439 - 52			7 38 41	−31 30.5	1	13.08		0.41		0.03		2		407
	−31 04897	NGC 2439 - 69		⋆ AB	7 38 41	−31 34.6	1	11.29		0.21		-0.27		2		407
61987	−27 04393	HR 2972			7 38 42	−27 49.7	2	6.76		-0.16	.013	-0.65	.034	3	B5 II	401,1079
		NGC 2439 - 77			7 38 42	−31 38.4	1	12.81		0.18		-0.52		1		407
62007	−29 04760				7 38 43	−29 40.2	1	9.30		0.27		0.14		2	A8/9 V	126
	−31 04900	NGC 2439 - 81			7 38 43	−31 31.0	1	11.30		0.16		-0.58		2	B1 V:	407
	−31 04898	NGC 2439 - 72			7 38 43	−31 35.5	1	11.83		0.12		-0.59		1		407
	−11 02058			V	7 38 44	−11 24.3	1	8.81		1.71				2	M3	369
	−23 05931				7 38 44	−23 58.7	1	8.86		1.98		1.36		2		126
61984	−23 05930				7 38 44	−24 09.4	1	7.70		1.36		1.34		2	K1 (III)	126
		NGC 2439 - 71			7 38 44	−31 34.7	1	13.44		0.22		-0.40		2		407
		NGC 2439 - 73			7 38 44	−31 35.7	1	12.78		0.37		0.16		1		407
		NGC 2439 - 51			7 38 45	−31 31.2	1	12.10		0.17		-0.58		2		407
		NGC 2439 - 63			7 38 45	−31 34.7	1	12.49		0.46		-0.10		1		407
		NGC 2439 - 70			7 38 45	−31 34.9	1	12.11		0.22		-0.68		2		407
		NGC 2439 - 78			7 38 45	−31 38.7	1	11.98		1.14		0.82		1		407
		POSS 544 # 3			7 38 46	+04 45.6	1	15.23		1.43				3		1739
		LSS 652			7 38 46	−19 48.3	1	12.16		0.26		-0.21		4	B3 Ib	844
		NGC 2439 - 53			7 38 46	−31 30.6	1	13.86		0.56		0.02		2		407
61497	+59 01103	HR 2946		⋆ A	7 38 47	+58 49.8	4	4.98	.008	0.09	.015	0.08	.004	12	A3 IVn	15,1363,3024,8015
61497	+59 01103	IDS07346N5857		B	7 38 47	+58 49.8	1	11.15		0.60		0.06		4		3024
		LOD 2 # 12			7 38 47	−24 10.1	1	12.38		0.16		-0.26		2		126
		NGC 2439 - 54			7 38 47	−31 30.3	1	14.81		0.32		0.19		1		407
		LF12 # 47			7 38 48	−22 51.6	1	12.38		0.86		0.50		1		137
61985	−25 04895				7 38 48	−25 23.8	1	9.49		0.18		0.12		2	A0 III/IV	137
62008	−29 04764				7 38 48	−30 03.1	1	9.59		-0.03		-0.22		2	B9 IV/V	126
		NGC 2439 - 82			7 38 48	−31 34.7	1	13.94		0.20		-0.30		2		407
		L 33 - 56			7 38 48	−76 57.	1	11.70		0.61		-0.12		1		1696,3060
61986	−26 04750				7 38 49	−26 14.5	3	8.68	.011	0.62	.008	0.07		23	G5 V	1696,2012,6011
62006	−29 04763				7 38 49	−29 37.5	1	10.02		0.20		0.12		2	A9 V	126
	−29 04765				7 38 50	−30 09.0	1	10.20		0.16		0.09		2		126
		NGC 2439 - 24			7 38 50	−31 29.4	1	12.03		0.15		-0.54		2		407
		NGC 2439 - 5			7 38 50	−31 32.9	1	10.77		0.24		-0.61		2	B1 III-IV	407
	−31 04902	NGC 2439 - 6			7 38 50	−31 32.9	1	10.48		0.21		-0.61		2	B2 V:	407
	−31 04905	NGC 2439 - 20			7 38 50	−31 36.6	1	10.51		0.16		-0.57		2	B3 V	407
62087	−38 03547	NGC 2451 - 192			7 38 50	−38 46.3	2	8.09	.025	1.43	.005	1.68	.000	4	K2 III	80,460
61935	−9 02172	HR 2970			7 38 51	−09 26.0	10	3.93	.006	1.02	.006	0.88	.013	44	K0 III	3,15,58,369,418,1075*
	−25 04898				7 38 51	−25 59.1	1	10.94		0.00		-0.13		1	B9 V	137
62033	−31 04906				7 38 51	−31 24.1	1	8.39		-0.08				4	B9 IV/V	2042
		NGC 2439 - 371			7 38 51	−31 28.2	1	12.23		0.22				5		2042
		NGC 2439 - 41			7 38 51	−31 35.5	1	14.34		0.29		-0.15		1		407
		NGC 2439 - 26			7 38 52	−31 22.6	1	12.30		0.15		-0.55		2		407
		NGC 2439 - 62			7 38 52	−31 31.0	1	15.27		0.23		-0.03		1		407
		NGC 2439 - 34			7 38 52	−31 35.5	1	12.96		0.23		-0.35		2		407
		NGC 2439 - 21			7 38 52	−31 35.9	1	11.42		0.19		-0.61		2		407
62086	−38 03546	NGC 2451 - 193			7 38 52	−38 23.3	1	9.20		0.37				2	F0 V	2037
		LOD 2 # 11			7 38 53	−24 09.3	1	12.67		0.34		0.09		2		126
62032	−29 04767				7 38 53	−29 53.6	2	8.31	.002	-0.12	.003	-0.50	.004	3	B8 V	401,1732
62030	−24 05775				7 38 54	−24 28.8	2	8.81	.010	0.48	.005	0.05	.034	3	F5 V	137,1462
		NGC 2439 - 22			7 38 54	−31 22.2	1	12.21		0.18		-0.54		2		407
		NGC 2439 - 61			7 38 54	−31 31.3	1	14.92		1.30				1		407
		NGC 2439 - 303			7 38 54	−31 32.	1	11.95		0.25				12		2042
	−31 04907	NGC 2439 - 23			7 38 54	−31 36.4	1	11.25		0.20		-0.59		2	B2 IV	407
	−31 04908	LSS 668			7 38 54	−31 49.5	2	9.43	.016	0.19	.008	-0.60	.009	5	B2.5II-Ib	540,976
		G 111 - 28			7 38 55	+43 03.5	1	10.26		1.06		0.89		2		7010
61957	−16 02081				7 38 55	−17 01.6	1	8.01		-0.06		-0.53		2	B3 V	401
		LP 783 - 15			7 38 55	−17 25.4	1	13.06		0.57		-0.13		2		1696
		NGC 2439 - 35			7 38 55	−31 35.3	1	14.11		0.23		-0.29		1		407
		NGC 2439 - 27			7 38 55	−31 35.6	1	11.47		1.60		1.75		2		407
		NGC 2439 - 33			7 38 55	−31 36.3	1	12.84		0.22		-0.47		2		407
	−23 05937				7 38 56	−23 34.7	1	9.48		1.19		1.19		1	K1 III	137
		NGC 2439 - 31			7 38 56	−31 34.7	1	12.67		0.24		-0.35		2		407
62058	−31 04910	NGC 2439 - 1		⋆ V	7 38 56	−31 32.6	6	6.60	.025	1.17	.019	0.89	.049	31	G2 O-Ia	15,407,1088,2031,2042,8015
		NGC 2439 - 25			7 38 56	−31 35.0	1	11.58		0.13		-0.57		2		407
	−31 04911	NGC 2439 - 2		⋆ A	7 38 57	−31 34.3	1	8.90		0.16		-0.76		3	B1.5Ib	407

Table 1   413

HD	DM	Other Id	N Rem	α_{1950}	δ_{1950}	S	V	σ_V	B–V	σ_{B-V}	U–B	σ_{U-B}	n	Spectrum	References
61887	+3 01758	HR 2966		7 38 58	+03 44.6	2	5.94	.005	-0.04	.000	-0.08	.005	7	A0 V	15,1256
		LOD 2 # 4		7 38 58	−24 02.0	1	11.80		1.07		1.10		2		126
	−31 04909	NGC 2439 - 15		7 38 58	−31 39.5	2	10.92	.034	0.11	.044	-0.16		6		407,2042
		LOD 2 # 10		7 38 59	−24 07.9	1	12.98		0.55		0.14		2		126
62057	−29 04772			7 38 59	−29 37.5	1	9.73		-0.03		-0.08		2	A0	126
		NGC 2439 - 4		7 38 59	−31 34.3	1	10.68		0.18		-0.62		2	B2 V	407
		NGC 2439 - 36		7 38 59	−31 35.3	1	13.69		0.30		-0.25		1		407
		NGC 2439 - 13		7 39 00	−31 27.7	1	10.80		0.36				3		2042
		NGC 2439 - 9		7 39 00	−31 33.8	1	12.58		0.22		-0.52		2		407
		NGC 2439 - 39		7 39 00	−31 35.3	1	14.71		0.24		-0.21		1		407
		NGC 2439 - 49		7 39 00	−31 36.1	1	14.78		1.21		0.50		1		407
		NGC 2439 - 32		7 39 00	−31 37.4	1	12.62		0.17		-0.55		2		407
		POSS 544 # 6		7 39 00	+07 45.5	1	17.32		1.79				3		1739
62001	−18 01967			7 39 01	−18 52.6	2	8.19	.006	0.05	.002	0.05	.000	13	A0 V	677,3077
	−23 05939			7 39 01	−23 45.2	1	11.39		0.00		0.02		1	A0 V	137
	−23 05941			7 39 01	−24 01.6	1	11.60		0.02		0.00		2		126
	−14 02111	NGC 2437 - 242		7 39 02	−14 41.6	2	10.22	.024	1.10	.015	0.83	.034	5		460,1040
		LOD 2 # 9		7 39 02	−24 07.3	1	11.87		0.24		-0.34		2		126
61885	+13 01737	HR 2965		7 39 04	+13 35.9	3	5.79	.025	1.67	.006	1.99	.000	7	M2 IIIab	71,1733,3001
		NGC 2439 - 47		7 39 04	−31 35.8	1	14.92		0.63		0.25		2		407
62000	−14 02112	NGC 2437 - 2		7 39 05	−14 42.1	2	8.70	.015	0.15	.010	-0.04	.030	3	A1 II/III	1040,8100
		LSS 662		7 39 05	−17 46.8	1	11.94		0.04		-0.49		4	B3 Ib	844
	−31 04917	NGC 2439 - 3		7 39 06	−31 33.9	4	8.88	.163	2.24	.094	2.30		13	M2 Iab	138,407,1522,2042
62193	−45 03374			7 39 06	−45 38.4	1	8.68		-0.10		-0.76		1	B2 III/IV	55
62083	−23 05944			7 39 07	−24 02.9	1	8.65		0.55		-0.07		2	F6 V	126
		NGC 2439 - 373		7 39 07	−31 29.0	1	12.87		0.26				5		2042
		NGC 2439 - 46		7 39 07	−31 35.6	1	14.17		0.28		-0.28		2		407
		LSS 666		7 39 09	−17 22.3	1	12.30		0.33		-0.66		4	B3 Ib	844
		NGC 2439 - 45		7 39 09	−31 36.5	2	13.39	.010	0.24	.045	-0.54		5		407,2042
61805	+30 01549			7 39 10	+30 17.2	1	6.66		1.35		1.47		3	K2 III	1501
62108	−29 04775			7 39 10	−30 05.5	1	9.49		-0.10		-0.43		2	B8 V	126
		NGC 2439 - 18		7 39 12	−31 39.3	1	13.24		0.42		0.38		2		407
		NGC 2437 - 70		7 39 13	−14 42.8	1	12.90		1.36				3		1040
		NGC 2437 - 74		7 39 13	−14 43.5	1	13.23		0.24		0.20		2		1040
		LOD 2 # 8		7 39 13	−24 06.6	1	12.71		0.43		0.19		2		126
62152	−37 03785	NGC 2451 - 197		7 39 13	−37 50.1	1	8.83		0.41				3	F2 IV/V	2037
61913	+14 01729	HR 2967, NZ Gem		7 39 14	+14 19.6	5	5.55	.024	1.65	.008	1.89	.015	18	M3 II-III	15,1007,1013,3055,8015
	−23 05949			7 39 14	−24 05.8	1	11.43		0.20		0.04		2		126
		LOD 2 # 7		7 39 14	−24 06.0	1	12.72		0.43		-0.14		2		126
		NGC 2439 - 17		7 39 14	−31 38.9	2	12.13	.030	0.23	.035	-0.45		4		407,2042
62082	−21 02077	HR 2976		7 39 15	−22 13.1	2	6.17	.010	1.63	.010	1.95		6	M1 III	2007,3055
		NGC 2439 - 19		7 39 15	−31 34.7	1	12.61		0.23		-0.52		2		407
		NGC 2437 - 71		7 39 16	−14 43.1	1	11.96		0.16		0.01		3		1040
	−14 02120	NGC 2437 - 78	⋆ AB	7 39 16	−14 44.4	1	10.45		0.13		0.03		2		460
62106	−25 04908			7 39 16	−26 09.0	1	9.22		1.26				18	K1/2 III	6011
62176	−37 03786	NGC 2451 - 201		7 39 17	−37 30.8	1	8.93		0.98		0.64		2	G8 III	80
61912	+21 01668			7 39 18	+21 05.7	1	7.90		1.15		1.21		3	K1 III	37
62049	−9 02176			7 39 18	−09 21.8	1	9.26		0.04		-0.07		2	B9	3039
	−28 04707			7 39 18	−28 52.9	1	11.02		0.58		0.05		1		211
62150	−32 04287	LSS 675		7 39 18	−32 31.6	5	7.68	.013	0.48	.026	-0.32	.025	9	B3/5 Iab	467,540,568,976,1034
		NGC 2437 - 29		7 39 20	−14 36.1	1	10.86		1.19		0.90		2		460
62212	−42 03413			7 39 21	−43 04.0	1	7.33		-0.04		-0.27		2	B8/9 V	401
		WLS 736 45 # 5		7 39 23	+47 23.0	1	12.16		0.51		0.00		2		1375
	−27 04404			7 39 24	−27 25.7	1	11.36		0.37		-0.54		1		211
62210	−39 03494	NGC 2451 - 200		7 39 25	−39 11.6	1	9.39		0.15				2	A1 V	2037
		LF12 # 52		7 39 27	−23 06.2	1	10.06		1.01		0.77		1	K0 III	137
		NGC 2437 - 25		7 39 28	−14 39.3	2	11.07	.020	0.55	.005	0.04	.005	2		460,1040
61859	+34 01657	HR 2962		7 39 29	+34 07.1	1	6.06		0.44		0.03		2	F7 Va	1733
		Orsatti 4		7 39 29	−28 11.4	1	12.77		0.51		-0.28		1		1766
62226	−38 03556	NGC 2451 - 203	⋆	7 39 30	−38 24.9	3	5.42	.007	-0.15	.005	-0.57		9	B5 III	15,401,2029
62227	−38 03558	NGC 2451 - 202		7 39 30	−39 07.1	2	7.73	.000	-0.05	.000	-0.11		5	B9 V	401,2037
		NGC 2437 - 173		7 39 33	−14 47.1	1	12.91		0.16		0.14		2		460
	−14 02136	NGC 2437 - 174		7 39 34	−14 47.4	1	10.62		1.09		0.84		2		460
62129	−14 02134	NGC 2437 - 481		7 39 34	−14 52.4	1	9.39		0.03		-0.14		4	B8 IV/V	1040
		NGC 2437 - 483		7 39 37	−14 52.3	1	13.35		0.20		0.14		1		1040
62191	−26 04779			7 39 37	−27 06.9	1	7.92		-0.11		-0.44		2	B5 III/IV	401
		Orsatti 6		7 39 37	−27 57.5	1	13.46		0.82		-0.14		1		1766
		LSS 678		7 39 39	−33 43.9	1	12.29		0.42		-0.49		1	B3 Ib	467
		LF12 # 53		7 39 40	−26 08.2	1	10.86		1.04		0.89		2	K0 III	137
62251	−37 03795	NGC 2451 - 207		7 39 41	−38 03.3	1	8.95		-0.05				2	B9p	2037
62278	−40 03322			7 39 41	−41 02.8	1	7.40		-0.06		-0.34		2	B7 III/IV	401
62125	−3 02019	IDS07372S0317	A	7 39 42	−03 24.0	1	8.06		1.25		1.41		3	K5	1657
		LF12 # 54		7 39 42	−25 27.5	1	11.54		1.10		0.93		3		137
	−31 04929	NGC 2439 - 29		7 39 43	−31 38.5	1	11.09		0.10		0.06		2		407
61997	+22 01754			7 39 45	+22 20.6	2	7.13	.000	0.41	.000			35	F5	130,1351
	−24 05797	EK Pup		7 39 45	−24 55.5	1	10.52		0.84		0.57		1		1462
	−32 04296	LSS 679		7 39 45	−32 30.6	3	10.15	.014	0.35	.013	-0.60	.007	5	B0 V	540,564,568
		LF12 # 55		7 39 46	−24 14.1	1	10.69		1.53		1.59		1	K5 V	137
62207	−23 05966	IDS07377S2323	A	7 39 47	−23 30.1	1	7.84		-0.08		-0.36		4	B8 V	1732
	−33 04043	LSS 681		7 39 47	−33 42.8	3	9.84	.006	0.43	.008	-0.59	.010	5	O5	467,540,564

HD	DM	Other Id	N	Rem	α_{1950}	δ_{1950}	S	V	σ_V	B–V	σ_{B-V}	U–B	σ_{U-B}	n	Spectrum	References
62318	−44 03655	HR 2984			7 39 48	−44 30.8	2	6.40	.005	0.00	.000	−0.19		6	B7 IV/V	401,2007
62317	−43 03471				7 39 49	−43 31.0	1	7.04		0.01		−0.18		2	A(p Si)	401
62165	−11 02067				7 39 52	−11 13.8	1	8.80		0.05		−0.22		2	B9	3039
		LSS 677			7 39 52	−20 25.0	1	11.75		0.21		−0.52		4	B3 Ib	844
	−24 05803				7 39 52	−24 22.8	1	10.46		0.97		1.00		2	K0 III	137
	−32 04300				7 39 52	−32 58.1	2	9.85	.000	0.32	.010	−0.46	.013	6	B1.5V	540,976
	−34 03814	LSS 682			7 39 52	−34 40.3	1	10.35		0.30		−0.73		1		568
	−25 04923	WX Pup			7 39 54	−25 45.5	1	8.76		0.79				1	G0	6011
62164	−10 02171	SU Mon			7 39 55	−10 45.6	1	8.25		2.60				8	S3.6	3039
		LF12 # 57			7 39 55	−22 46.3	1	10.32		0.92		0.64		2	K0 III	137
62248	−29 04796				7 39 55	−30 05.0	1	9.68		0.12		0.07		2	A3 V	126
	−24 05804				7 39 56	−25 05.8	1	10.52		1.04		0.79		2	K0 III	137
62224	−26 04792	IDS07379S2615		AB	7 39 56	−26 21.8	1	9.12		0.52		0.30		1	A2/4 +G/K	137
62246	−26 04797				7 39 59	−26 39.0	1	9.15		1.40		1.46		1	K2/3 III	137
	−14 02152	NGC 2437 - 482			7 40 02	−14 49.3	2	10.96	.010	0.11	.005	0.05	.054	5		460,1040
	−29 04801				7 40 04	−30 06.0	1	11.37		0.46		−0.01		2		126
	−25 04928				7 40 07	−25 20.9	1	9.80		1.08		0.91		1	K0 III	137
62399	−45 03386				7 40 07	−45 12.1	1	11.48		0.86		−0.19		3	Ap CrEuSr	1747
		WLS 736 45 # 9			7 40 08	+45 19.4	1	12.90		0.81		0.32		2		1375
62220	−15 01989			V	7 40 08	−16 12.9	1	8.78		1.67				2	M3 III	369
62315	−29 04805				7 40 08	−30 04.1	1	6.95		−0.16		−0.67		4	B3/5 V	164
62044	+29 01590	HR 2973, σ Gem	⋆	A	7 40 11	+29 00.4	6	4.26	.032	1.12	.006	0.97	.006	15	K1 III	15,1080,1363,3077*
62376	−38 03564	NGC 2451 - 209	⋆		7 40 12	−38 24.6	3	6.54	.010	−0.09	.005	−0.40		9	B8 V	401,2006,2037
62375	−38 03566	NGC 2451 - 210			7 40 14	−38 13.8	2	7.76	.035	1.29	.030	1.20		5	F3 V	460,2037
61995	+44 01666				7 40 16	+43 54.9	1	7.18		0.90		0.58		2	G5	985
62312	−26 04806				7 40 16	−27 03.9	1	7.88		−0.12		−0.51		2	B8 III	401
62483	−52 01251				7 40 16	−53 04.6	2	8.25	.014	−0.05	.019	−0.81	.005	2	B2 II	55,1034
61931	+50 01460	HR 2969	⋆	AB	7 40 17	+50 33.2	1	5.27		0.00		0.00		1	A0 IIIn	3023
		LSS 683			7 40 17	−20 36.9	1	11.44		0.20		−0.45		4	B3 Ib	844
	+18 01719	G 88 - 48			7 40 18	+18 18.1	1	11.32		1.40		1.05		1		333,1620
	−26 04807				7 40 21	−26 54.4	1	9.84		2.96		3.70		1	C8,2	864
62359	−29 04812				7 40 21	−30 07.3	2	8.12	.005	−0.13	.000	−0.65	.005	7	B8 IV	401,1775
62141	+22 01756	HR 2978			7 40 24	+22 31.1	1	6.21		0.93				2	K0 III	71
62358	−27 04429				7 40 24	−27 36.6	1	7.40		−0.06		−0.29		2	B8/9 V	401
	−15 01994				7 40 25	−15 33.8	1	10.28		1.76				2	M3	369
62415	−38 03569	NGC 2451 - 211			7 40 26	−38 24.0	1	8.18		0.28				3	A7 V	2037
	−14 02165				7 40 28	−14 52.9	1	11.19		1.94				2	M2	369
	−34 03821	LSS 690			7 40 28	−34 38.9	2	10.51	.005	0.48	.005	−0.51	.014	5	B1 III	568,1737
	−24 05819				7 40 30	−24 20.6	1	11.25		0.08		−0.58		1		211
		L 528 - 64			7 40 30	−33 30.	1	13.66		1.34		1.04		2		3073
62264	+0 02054	HR 2982	⋆	AB	7 40 31	+00 18.5	2	6.18	.005	1.02	.000	0.75	.005	7	K0 III	15,1256
62373	−24 05823				7 40 32	−24 22.9	1	8.69		0.28		0.08		1	A9 V	137
62414	−35 03796				7 40 32	−36 01.8	1	8.11		−0.08		−0.43		2	B8/9 V	401
		LF12 # 63			7 40 33	−22 37.4	1	10.31		0.19		0.07		2	A2 V	137
	−24 05825				7 40 38	−24 25.2	1	9.25		1.02		0.88		1	K1 III	137
	−18 01983				7 40 39	−18 21.7	1	9.34		1.65				2	M2	369
		Orsatti 14			7 40 40	−27 50.4	1	11.33		0.42		−0.23		1		1766
		LF12 # 65			7 40 43	−25 56.3	1	12.40		0.05		0.15		1	A2 V	137
62479	−37 03806	NGC 2451 - 214			7 40 43	−37 59.3	1	9.14		0.27				2	A7 V	2037
62351	−16 02093	IDS07385S1650		AB	7 40 44	−16 56.6	1	6.53		0.81		0.40		4	G5/6 IV	214
62394	−22 01991				7 40 44	−22 37.3	1	8.65		1.24		1.25		1	K1/2 IV	137
62412	−26 04824	HR 2988			7 40 44	−26 13.9	2	5.63	.006	0.99	.002	0.80		6	K1 III	58,2007
		CCS 871			7 40 45	−20 25.5	1	10.62		2.27				1		864
62447	−36 03820				7 40 45	−36 15.3	1	7.61		−0.11		−0.51		2	B8 V	401
62391	−20 02136	LSS 689			7 40 46	−21 02.9	1	10.04		0.03		−0.79		2	B3/5 Ib	540
		Orsatti 16			7 40 46	−27 27.1	1	12.93		0.30		−0.41		1		1766
62480	−38 03574	NGC 2451 - 219			7 40 46	−38 37.2	2	8.33	.015	1.41	.000	1.59	.030	4	K2 III	80,460
62503	−38 03575	NGC 2451 - 218			7 40 46	−39 04.9	2	7.26	.000	−0.08	.005	−0.25		4	B9 V	401,2037
62326	−4 02060				7 40 47	−05 03.3	1	9.47		0.00				3	A0	803
62429	−29 04823				7 40 47	−29 23.9	1	9.42		0.07		−0.03		2	A0 IV/V	126
		Orsatti 17			7 40 47	−30 01.9	1	12.91		0.50		−0.18		1		1766
		BPM 18615			7 40 48	−57 02.	1	15.06		−0.16		−0.79		1	DF	3065
		LF12 # 67			7 40 49	−23 43.8	1	11.71		0.87		0.64		1		137
62445	−28 04750	IDS07389S2832		AB	7 40 53	−28 38.6	1	8.34		0.04		−0.25		2	B5 Ib	401
62230	+24 01755				7 40 57	+24 21.9	1	7.07		0.24		0.14		2	A5	1625
	−19 02026	LSS 691			7 40 57	−19 17.5	1	10.56		0.16		−0.60		4		844
	−37 03808	NGC 2451 - 2			7 40 57	−37 55.8	1	10.22		0.21		0.13		2		80
		Orsatti 18			7 40 58	−30 10.0	1	12.04		0.20		−0.54		1		1766
62542	−41 03304				7 40 58	−42 06.6	1	8.04		0.17				4	B3 V	2012
		LOD 1+ # 20			7 40 59	−21 07.4	1	13.06		0.54		0.04		2		126
		LOD 1+ # 19			7 41 00	−21 05.9	1	12.30		1.30		1.29		2		126
		LF12 # 68			7 41 01	−23 15.4	1	12.67		1.22		0.63		2		137
	−34 03831	LSS 695			7 41 02	−34 11.9	2	10.97	.009	0.97	.009	−0.17	.023	5	B0 III	568,1737
62367	−4 02062				7 41 03	−04 33.6	1	7.14		−0.08		−0.43		3	B9	1212
62285	+26 01633	HR 2983			7 41 04	+25 54.3	4	5.31	.008	1.53	.020	1.88	.017	11	K5 III	20,1080,1355,3016
62559	−38 03580	NGC 2451 - 221			7 41 05	−38 47.9	1	7.72		0.34				2	F2 IV	2037
	−37 03810	NGC 2451 - 3			7 41 06	−37 56.3	1	10.83		0.56		0.05		2		80
		CCS 874			7 41 06	−44 04.9	1	9.79		3.36				1		864
62443	−22 01993				7 41 07	−22 15.9	1	10.01		0.07		0.02		1	A1 IV	137
	−37 03809	NGC 2451 - 1			7 41 07	−37 28.8	1	11.30		0.12		0.10		2		80

Table 1   415

HD	DM	Other Id	N Rem	α_{1950}	δ_{1950}	S	V	σ_V	B–V	σ_{B-V}	U–B	σ_{U-B}	n	Spectrum	References
	−25 04952			7 41 08	−25 22.0	1	10.08		1.07		1.06		1	K0 III	137
62517	−29 04837			7 41 10	−29 26.2	1	8.58		0.42		0.07		2	F3 IV	126
	+0 02058	G 112 - 43	⋆ A	7 41 11	+00 03.3	6	10.20	.033	0.45	.013	-0.17	.007	7	sdF	516,1620,1658,1773,3062,6006
	+0 02058	G 112 - 44	⋆ B	7 41 11	+00 03.5	5	11.26	.018	0.50	.019	-0.19	.011	6	sdF	1620,1696,1773,3062,6006
	−30 04978			7 41 11	−30 44.0	1	10.43		0.13		-0.43		2		564
		Orsatti 20		7 41 12	−28 19.1	1	13.31		0.44		-0.28		1		1766
	−26 04841			7 41 13	−26 24.8	1	10.15		0.03		-0.03		2	B8 V	137
62538	−31 04966			7 41 14	−31 37.6	1	9.28		0.64		0.06		4	G1 V	1775
		G 193 - 61		7 41 15	+51 30.8	1	12.90		0.79		0.22		2		1658
62758	−58 00967	HR 3006		7 41 15	−58 30.7	1	6.43		-0.10				4	B3 V	2007
		LOD 1+ # 16		7 41 17	−21 03.3	1	10.73		0.06		-0.10		2		126
		Orsatti 23		7 41 17	−30 07.3	1	11.66		0.18		-0.50		1		1766
		LF12 # 70		7 41 18	−23 45.8	1	12.71		0.50		-0.01		1	F5 V:	137
	−37 03811	NGC 2451 - 4		7 41 18	−37 41.8	1	11.51		0.54		0.04		2		80
	−37 03812	NGC 2451 - 5		7 41 19	−38 00.8	1	10.39		0.08		-0.01		3	A1 V	80
62578	−35 03809	HR 2994		7 41 22	−35 55.8	4	5.59	.008	-0.14	.005	-0.55		13	B5 V	15,401,2013,2028
62595	−38 03583	NGC 2451 - 223	⋆	7 41 22	−38 44.6	3	6.89	.014	1.03	.019	0.77		8	G6 III	460,2007,2037
62644	−44 03675	HR 2998		7 41 24	−45 02.7	5	5.04	.009	0.77	.008	0.32	.000	14	G6 IV	15,678,1488,2012,3008
62535	−20 02145			7 41 25	−21 09.7	1	7.96		-0.04		-0.28		2	Ap Si	126
		LF12 # 71		7 41 25	−25 56.4	1	12.15		1.24		1.20		1		137
62558	−29 04846			7 41 25	−29 14.4	1	8.20		0.60		0.17		2	G2 V	126
62345	+24 01759	HR 2985	⋆ AB	7 41 26	+24 31.2	13	3.57	.010	0.93	.007	0.69	.009	94	G8 IIIa	1,15,247,369,1006*
62407	+13 01750	HR 2987		7 41 27	+12 58.8	1	6.40		1.40		1.64		2	K0	1733
	−29 04849	LSS 701		7 41 29	−29 12.3	5	10.01	.011	0.21	.017	-0.69	.032	9	O9.5II:	126,211,540,1011,1012
		NGC 2451 - 7		7 41 29	−37 57.5	1	11.75		1.30		1.11		3		80
		LF12 # 72		7 41 30	−23 09.0	1	13.09		0.97		0.46		1		137
62437	+2 01761	HR 2989, AZ CMi		7 41 31	+02 31.6	2	6.46	.005	0.21	.005	0.13	.010	7	F0 III:	15,1256
62532	−17 02120	LSS 698	AB	7 41 31	−17 49.5	2	8.40	.010	0.07	.005	-0.70	.015	3	B3 (V)nne	1012,1737
62576	−28 04767	HR 2993	⋆ A	7 41 31	−28 17.5	6	4.58	.005	1.63	.006	1.97	.025	22	K3 Ib	15,1075,2012,3051*
	−37 03814	NGC 2451 - 6		7 41 32	−37 54.6	1	11.40		0.28		0.15		2		80
62301	+39 01998			7 41 33	+39 41.2	3	6.77	.018	0.54	.013	-0.04	.009	7	F8 V	22,1003,1733
62554	−22 01998			7 41 33	−22 24.7	1	10.04		0.21		0.13		4	A3/5 II	137
62555	−25 04966	HR 2992	⋆ AB	7 41 34	−25 23.0	1	6.55		0.06				4	A3 III	2031
		LOD 1+ # 15		7 41 35	−20 58.4	1	11.75		0.36		0.04		2		126
		G 112 - 45		7 41 36	+00 55.1	1	11.81		1.20		1.11		1	K5	3062
	−23 06039			7 41 38	−24 05.2	1	10.98		0.04		-0.07		1	A0 V	137
62642	−37 03815	NGC 2451 - 228		7 41 42	−37 37.8	2	7.60	.018	0.00	.018	0.02		5	A0 Vn	1367,2037
62660	−37 03816	NGC 2451 - 230		7 41 44	−38 01.0	4	8.31	.011	1.00	.010	0.70	.020	10	G8 IIIab	80,460,1367,2037
62850	−59 00883			7 41 44	−59 10.8	1	7.19		0.64				4	G2/3 V	2012
62593	−22 01999			7 41 45	−22 47.5	1	10.18		0.49		-0.05		2	A2 III	137
62756	−49 03014	HR 3005		7 41 45	−49 52.3	1	6.57		0.07				4	A2 V	2007
		LF12 # 75		7 41 47	−23 23.7	1	13.54		0.45		-0.04		1		137
62623	−28 04774	HR 2996, LSS 704		7 41 48	−28 50.1	9	3.96	.010	0.18	.008	-0.09	.044	25	A2 Iab	15,126,1075,1470,1637*
62659	−36 03832			7 41 48	−36 58.1	1	8.55		-0.07		-0.37		2	B8 V	401
	−20 02148			7 41 49	−20 58.4	1	10.23		0.52		0.06		2		126
		Orsatti 25		7 41 50	−28 10.9	1	13.99		0.56		-0.10		1		1766
	−28 04775			7 41 50	−28 51.6	1	10.35		0.25		-0.74		1		211
62549	−4 02069			7 41 51	−04 56.2	2	7.69	.005	0.61	.010	0.07		6	G1 V	2012,3016
62589	−16 02101	IDS07396S1642	AB	7 41 51	−16 48.7	1	8.09		-0.08				4	B3 III	2034
62605	−20 02149			7 41 51	−20 54.8	1	8.84		0.46		0.07		2	F5 V	126
62658	−31 04981			7 41 52	−32 00.1	1	9.27		-0.05		-0.37		3	Ap Si	396
	−37 03817	NGC 2451 - 9		7 41 52	−37 18.1	1	11.58		0.63		0.12		2		80
	−37 03819	NGC 2451 - 10		7 41 52	−38 06.9	1	11.00		0.46		0.25		3		80
62066	+65 00593	HR 2975		7 41 54	+65 34.7	1	5.92		1.18				2	gK2	71
	−31 04983			7 41 55	−32 06.5	1	11.39		0.25		0.20		3		396
62619	−22 02001			7 41 56	−22 40.4	2	9.41	.005	0.29	.005	0.12	.028	4	F0 V	137,8100
	−28 04778	LSS 707		7 41 56	−28 50.9	1	10.38		0.23		-0.73		3		1737
62140	+63 00733	HR 2977, BC Cam		7 41 57	+62 57.2	2	6.49	.005	0.26	.000	0.13	.009	5	F0p SrEu	39,220
62640	−28 04777			7 41 57	−28 18.6	1	8.10		-0.12		-0.57		2	B9 III(pSi)	401
62712	−37 03820	NGC 2451 - 233	⋆	7 41 57	−38 04.9	2	6.41	.009	-0.15	.009	-0.48		6	B7 III	1367,2037
62897	−57 01305	HR 3012		7 41 59	−58 06.6	2	6.22	.005	1.05	.002	0.84		6	K0 III	58,2007
		NGC 2451 - 12		7 41 59	−37 27.1	1	11.70		0.31		0.24		2		80
		NGC 2451 - 18		7 41 59	−37 27.1	1	11.77		0.65		0.10		3		80
62257	+55 01218			7 42 00	+55 45.9	1	7.58		0.14		0.13		2	Am	1502
62637	−24 05870	IDS07400S2442	AB	7 42 00	−24 48.0	1	8.08		0.73		0.41		1	F3 V	1462
	−25 04977			7 42 00	−26 05.7	1	10.71		0.12		-0.03		1	B9 IV	137
62713	−40 03377	HR 3002		7 42 00	−40 48.6	6	5.13	.017	1.10	.009	1.01	.030	22	K1 III	15,678,1075,1637,2006,3077
	−37 03821	NGC 2451 - 13		7 42 01	−37 23.9	1	10.84		1.00		0.56		3		80
		WLS 728 55 # 8		7 42 02	+55 20.5	1	11.81		1.03		0.79		2		1375
	−37 03824	NGC 2451 - 14		7 42 02	−37 33.2	2	9.29	.010	0.17	.000	0.13		6	A2 V	80,2037
	−37 03823	NGC 2451 - 15	⋆ AB	7 42 02	−37 46.5	2	10.56	.005	0.51	.005	0.02		4		80,2037
62783	−45 03411			7 42 02	−45 44.6	1	7.23		1.72				4	K5 III	2012
		G 50 - 4	⋆ V	7 42 03	+03 40.7	3	11.13	.025	1.61	.007	1.09	.093	4	M4e	1620,1705,3062
	−37 03825	NGC 2451 - 16		7 42 03	−37 50.8	3	9.92	.000	0.13	.011	0.13	.018	10	A2 V	80,1421,2037
62752	−39 03531	IDS07403S3938	AB	7 42 03	−39 45.2	1	8.11		-0.02		-0.21		2	Ap Si	401
		Orsatti 27		7 42 05	−28 50.4	1	11.36		0.21		-0.42		1		1766
62737	−37 03826	NGC 2451 - 237		7 42 05	−37 36.0	2	8.07	.019	-0.03	.005	-0.52		6	B8 III	1367,2037
62753	−40 03379			7 42 05	−40 11.3	4	6.60	.012	-0.13	.009	-0.81	.005	12	B2 Vn	158,540,976,1034
		LOD 3 # 20		7 42 06	−26 37.0	1	12.91		0.24		-0.24		2		126
		LF12 # 78		7 42 07	−24 45.7	1	11.54		0.03		-0.01		1	A0 V	137

HD	DM	Other Id	N Rem	α_{1950}	δ_{1950}	S	V	σ_V	B–V	σ_{B-V}	U–B	σ_{U-B}	n	Spectrum	References
	−26 04867	LSS 710		7 42 10	−26 36.5	1	9.64		0.12		−0.63		2		126
61994	+70 00474			7 42 11	+70 19.9	2	7.09	.005	0.66	.009	0.27	.003	6	G0	985,7008
62679	−23 06064	NGC 2447 - 3		7 42 11	−23 46.1	1	8.19		1.46		1.69		3	K2 (III)	1657
	−23 06063	NGC 2447 - 1		7 42 11	−23 47.3	1	10.37		0.18				3		2042
62682	−26 04869			7 42 11	−26 34.2	2	8.63	.024	−0.07	.019	−0.34		6	B8 V	126,2014
62654	−21 02102			7 42 12	−21 37.5	2	9.86	.010	0.20	.012	−0.46	.013	5	B8	540,976
		NGC 2447 - 50		7 42 12	−23 42.8	1	11.67		0.94				3		2042
62510	+20 01893	HR 2991		7 42 13	+20 26.3	2	6.54		−0.01	.005	−0.01	.005	5	A1 V	1022,1733
		LOD 3 # 17		7 42 13	−26 31.7	1	12.35		0.20		0.01		2		126
		NGC 2447 - 2		7 42 15	−23 46.7	1	11.73		0.22				3		2042
		LSS 712		7 42 15	−32 13.8	1	12.16		0.52		−0.43		3	B3 Ib	396
62509	+28 01463	HR 2990	★ AC	7 42 16	+28 08.9	10	1.14	.008	1.00	.008	0.85	.008	47	K0 IIIb	1,15,22,1077,1119*
		NGC 2451 - 17		7 42 16	−37 30.6	1	12.25		0.72		0.24		3		80
62782	−37 03827	NGC 2451 - 238		7 42 16	−38 01.3	4	8.55	.009	1.01	.008	0.66	.020	8	G8 IIIab	80,460,1367,1421
	−23 06074	NGC 2447 - 42		7 42 17	−23 43.5	1	9.76		0.39				4		2042
62803	−38 03598	NGC 2451 - 239		7 42 17	−38 41.8	1	7.41		−0.08				3	B9 V	2037
	−23 06076	NGC 2447 - 7		7 42 18	−23 45.6	1	8.46		1.44				3		2042
		LSS 713		7 42 18	−31 47.2	2	11.80	.000	0.65	.000	−0.25	.000	4	B3 Ib	396,1766
	−26 04871			7 42 19	−26 28.4	1	10.43		0.35		−0.06		2		126
	−37 03828	NGC 2451 - 21		7 42 19	−37 46.3	3	10.60	.000	0.27	.005	0.10	.018	7		80,1421,2037
62678	−21 02104			7 42 20	−21 22.7	1	9.95		0.20		−0.53		2	B3/6 (III)	540
62781	−35 03825	HR 3007		7 42 20	−35 56.6	3	5.80	.007	0.32	.004	0.00		9	F0 IV	15,401,2027
	−37 03829	NGC 2451 - 20		7 42 20	−37 23.3	1	9.80		1.19		1.04		3		80
62802	−37 03830	NGC 2451 - 22		7 42 22	−37 51.8	3	9.35	.017	0.06	.004	0.00	.021	4	B9 V	80,1367,2036
	−37 03831	NGC 2451 - 23		7 42 22	−38 01.6	2	10.45	.004	0.09	.009	−0.11	.021	6	A0 IV	80,1421
62704	−20 02152			7 42 23	−20 52.8	1	9.21		1.47		1.60		2	K3 III	126
	+2 01766			7 42 24	+02 16.0	3	10.18	.019	1.39	.020	1.34	.025	10	M0	1705,1746,7009
		NGC 2447 - 103		7 42 24	−23 38.7	1	12.27		0.31				4		2042
		NGC 2451 - 125		7 42 24	−38 12.3	1	12.72		0.57		0.11		2		80
	−25 04991			7 42 27	−25 31.2	1	11.02		−0.03		−0.16		1	B8 V	137
	−37 03833	NGC 2451 - 24		7 42 27	−37 31.6	2	9.86	.000	0.11	.000	0.08		4	A1 V	80,2037
63295	−72 00627	HR 3024	★ AB	7 42 27	−72 29.2	3	3.93	.006	1.03	.004	0.84	.005	21	K0 III	15,1075,2038
		NGC 2447 - 104		7 42 28	−23 38.6	1	13.06		0.17				2		2042
62747	−24 05885	HR 3004		7 42 28	−24 33.2	2	5.61	.005	−0.20	.005			7	B2 II	15,2012
62567	+26 01638	IDS07394N2614	AB	7 42 29	+26 06.7	1	7.35		1.53				2	K5 III	20
		NGC 2447 - 17		7 42 29	−23 46.4	1	11.60		0.07						2042
62731	−22 02004			7 42 30	−22 43.9	1	8.91		1.23		1.14		1	G8 III	137
		Haf 14 - 25		7 42 31	−28 11.8	1	10.55		0.65		0.35				1667
	−37 03835	NGC 2451 - 25		7 42 31	−37 23.4	1	10.44		0.61		0.12		3		80
		NGC 2451 - 130		7 42 31	−38 14.0	1	12.09		1.68		1.94		2		80
62780	−26 04881	LSS 716		7 42 32	−26 49.4	5	9.17	.013	0.15	.012	−0.78	.016	9	O9/B0e	540,976,1295,1376,1737
		Orsatti 29		7 42 32	−28 41.5	1	13.46		0.82		−0.13				1766
	−31 04994			7 42 32	−32 04.1	1	9.92		0.63		0.16		3		396
	−32 04348	LSS 719	★ AB	7 42 32	−33 06.0	4	8.90	.021	1.09	.010	0.11	.042	7	B7 Iab	540,564,976,1737
		NGC 2447 - 94		7 42 33	−23 39.6	1	12.88		0.33						2042
		Haf 14 - 4		7 42 33	−28 18.3	1	14.16		1.93		4.13		2		1667
		AJ79,1294 T6# 5		7 42 33	−31 57.7	1	11.46		0.46		0.01		3		396
		LOD 3 # 15		7 42 34	−26 29.9	1	11.40		0.15		−0.33		2		126
62600	+14 01744			7 42 35	+14 43.3	1	8.59		0.02		0.00		3	A0	1648
		LF12 # 81		7 42 35	−22 29.4	1	10.46		0.91		0.45		1	G8 III	137
		Haf 14 - 5		7 42 35	−28 18.6	1	13.61		1.57		1.10		2		1667
	−37 03836	NGC 2451 - 26		7 42 35	−37 31.9	2	10.39	.034	1.06	.005	0.71		3		80,2036
62729	−15 02014			7 42 36	−16 00.0	1	8.31		−0.02		−0.63		3	B2 V	1212
		NGC 2451 - 27		7 42 36	−37 48.2	3	11.40	.011	0.43	.009	0.00	.026	8		80,1421,2037
		Haf 14 - 2		7 42 37	−28 16.7	1	13.44		1.57		2.33		2		1667
		NGC 2447 - 19		7 42 38	−23 46.3	1	10.78		0.14				6		2042
62778	−26 04884			7 42 38	−26 30.3	1	9.45		0.38		−0.01		2	F2 V	126
	−34 03855			7 42 38	−34 13.0	1	10.70		0.04		−0.04		2		396
		Haf 14 - 3		7 42 39	−28 17.6	1	13.23		0.40		0.20		2		1667
	−31 04996			7 42 39	−31 17.1	1	9.26		0.70		−0.03		1		354
62844	−31 04998	LSS 721		7 42 39	−32 07.1	3	8.05	.007	0.71	.016	−0.29	.029	11	B2 Ia	396,540,976
62745	−15 02015		V	7 42 40	−15 34.8	1	8.56		2.06				2	M1 (Ib)	369
		Haf 14 - 16		7 42 40	−28 14.7	1	13.43		0.53		0.36		2		1667
62875	−37 03837	NGC 2451 - 243		7 42 40	−37 32.5	3	8.30	.013	−0.03	.004	−0.27	.000	8	B8 V	80,1367,2037
		NGC 2447 - 12		7 42 41	−23 49.5	1	10.07		0.21				4		2042
62876	−37 03838	NGC 2451 - 244		7 42 41	−37 52.0	4	8.62	.015	0.12	.012	0.11	.022	11	A4 IV	80,1367,1421,2037
62726	−11 02088			7 42 42	−11 38.5	1	8.12		−0.06		−0.42		2	B8	3039
		Haf 14 - 15		7 42 42	−28 16.1	1	13.31		0.51		0.08		2		1667
		Haf 14 - 6		7 42 42	−28 19.0	1	14.50		2.12		0.86		2		1667
62794	−20 02156	IDS07405S2049	AB	7 42 43	−20 55.9	1	9.06		−0.02		−0.31		2	Ap Si	126
	−23 06102	NGC 2447 - 28		7 42 43	−23 45.1	1	9.96		0.82				3	A2	2042
		LF12 # 82		7 42 43	−23 00.1	1	11.23		0.99		0.64		1	K0 III	137
		Haf 14 - 1		7 42 43	−28 16.9	1	11.54		2.47		3.52		2		1667
	−38 03604	NGC 2451 - 245		7 42 43	−38 39.4	1	9.24		0.25				2	A5	2037
62773	−17 02133			7 42 45	−17 36.2	1	8.84		1.68				2	M2/3 III	369
		Haf 14 - 14		7 42 45	−28 16.4	1	15.49		1.74		1.05		2		1667
	−37 03840	NGC 2451 - 30		7 42 45	−37 38.0	3	9.56	.010	1.08	.010	0.79	.038	6		80,1367,2037
		NGC 2451 - 124		7 42 46	−37 53.2	2	11.87	.016	0.62	.004	0.12	.024	4		80,1421
		NGC 2451 - 31		7 42 46	−37 54.6	3	10.97	.042	0.63	.015	0.15	.018	6		80,1421,2036
		Haf 14 - 17		7 42 47	−28 15.2	1	14.79		0.52		0.35		2		1667

HD	DM	Other Id	N	Rem	α_{1950}	δ_{1950}	S	V	σ_V	B–V	σ_{B-V}	U–B	σ_{U-B}	n	Spectrum	References
		Haf 14 - 13			7 42 47	−28 16.0	1	14.47		0.53		0.18		2		1667
62893	−37 03841	NGC 2451 - 246		⋆	7 42 47	−37 49.3	4	5.88	.008	−0.12	.003	−0.36	.004	22	B7 V	15,80,1367,2037
		Haf 14 - 7			7 42 48	−28 20.4	1	12.02		0.41		−0.01		2		1667
	−25 04996				7 42 49	−26 03.1	1	9.91		0.78		0.41		1	G8 IV	137
		Haf 14 - 12			7 42 49	−28 16.2	1	14.00		0.55		0.34		2		1667
		Haf 14 - 8			7 42 49	−28 17.7	1	13.37		0.56		0.01		2		1667
	−23 06096	NGC 2447 - 25			7 42 50	−23 46.7	1	9.98		0.90				8	A3	2042
		Haf 14 - 24			7 42 51	−28 12.4	1	11.36		0.45		0.10		2		1667
		Haf 14 - 11			7 42 51	−28 16.9	1	14.02		0.53		0.38		2		1667
		Haf 14 - 9			7 42 52	−28 17.5	1	13.64		0.69		0.28		2		1667
		NGC 2451 - 34			7 42 52	−37 31.3	2	10.82	.000	0.34	.004	0.04	.016	4		80,1421
	−37 03844	NGC 2451 - 33			7 42 52	−37 34.0	1	11.41		0.13		−0.11		3		80
		NGC 2451 - 32			7 42 52	−37 50.3	1	12.03		1.14		0.68		2		80
	−37 03843	NGC 2451 - 35			7 42 52	−38 05.9	1	10.24		0.46		0.14		2		80
63008	−50 02947	IDS07415S5013	A		7 42 52	−50 20.2	2	6.64	.000	0.53	.005	−0.04		5	F8 V	330,2033
63008	−50 02948	IDS07415S5013	B		7 42 52	−50 20.2	2	7.57	.000	0.70	.000			8	G0	2013,2033
		Ru 32 - 14			7 42 53	−25 23.3	1	14.67		0.68		−0.02		2		410
		Haf 14 - 10			7 42 53	−28 16.7	1	12.71		0.47		0.02		2		1667
		Ru 32 - 16			7 42 54	−25 24.	1	15.41		0.49		−0.32		1		410
		LSS 720			7 42 55	−19 18.8	1	12.60		0.22		−0.61		5	B3 Ib	844
		Haf 14 - 23			7 42 55	−28 12.5	1	14.35		0.66		0.31		2		1667
		Haf 14 - 18			7 42 55	−28 15.3	1	12.19		0.57		0.15		2		1667
		AJ79,1294 T6# 17			7 42 55	−32 06.2	1	14.28		0.57		0.07		3		396
	−37 03845	NGC 2451 - 36			7 42 56	−37 58.6	3	8.64	.010	0.08	.004	−0.02	.024	8	A0 Vp	80,1367,2037
		Haf 14 - 20			7 42 57	−28 15.0	1	14.06		1.55		1.30		2		1667
		Ru 32 - 15			7 42 58	−25 24.1	1	13.13		0.55		0.06		2		410
		Haf 14 - 21			7 42 58	−28 15.0	1	13.08		0.34		0.34		2		1667
		AJ79,1294 T6# 13			7 42 58	−31 49.5	1	13.30		0.66		0.23		3		396
	−37 03846	NGC 2451 - 38			7 42 58	−37 43.7	3	10.67	.024	1.47	.070	1.82	.043	7		80,1421,2036
62938	−37 03847	NGC 2451 - 248			7 42 58	−37 55.9	3	7.58	.010	−0.02	.000	−0.01	.029	8	A0 V	80,1367,2037
62871	−25 04999	Ru 32 - 10			7 42 59	−25 29.1	1	9.55		0.00		−0.02		2	B9 III/IV	410
		Haf 14 - 22			7 42 59	−28 14.0	1	13.90		1.64		1.41		2		1667
62961	−38 03611	NGC 2451 - 249			7 42 59	−38 53.9	2	8.10	.000	0.04	.000	0.01		5	A0 V	401,2037
		KS 75			7 43 00	−24 01.	1	10.38		0.28		−0.72		2		540
		Ru 32 - 8			7 43 00	−25 23.6	1	13.54		0.41		−0.17		3		410
		Haf 14 - 19			7 43 00	−28 15.8	1	11.69		0.51		0.01		2		1667
		NGC 2451 - 39			7 43 00	−37 44.8	3	11.51	.015	0.71	.054	0.18	.018	6		80,1421,2036
	−38 03608	NGC 2451 - 40			7 43 00	−38 10.6	1	10.44		1.77		1.85		3		80
		NGC 2451 - 37			7 43 01	−38 01.3	1	12.62		0.70		0.21		4		80
		NGC 2451 - 8			7 43 01	−38 04.6	1	13.94		0.69		0.09		3		80
62866	−20 02159	LSS 723			7 43 02	−20 41.3	3	9.04	.013	0.09	.011	−0.79	.014	8	B0.5Ib/II	540,844,976
		Ru 32 - 13			7 43 02	−25 24.0	1	14.33		0.33		−0.27		2		410
		Ru 32 - 7			7 43 02	−25 25.2	1	11.59		0.18		0.13		2		410
62870	−25 05001	Ru 32 - 9			7 43 02	−25 26.3	1	9.76		−0.05		−0.23		2	B9 V	410
62910	−31 05004	LSS 731			7 43 02	−31 47.2	1	10.10		0.34		−0.36		2	WN	564
		NGC 2451 - 19			7 43 02	−38 07.3	1	12.76		0.49		0.11		2		80
	−37 03848	NGC 2451 - 42			7 43 02	−38 07.3	1	11.59		0.24		0.17		3		80
	−24 05909				7 43 03	−24 58.2	1	10.65		1.07		0.86		2	K0 III	137
	−26 04901				7 43 03	−26 27.7	1	10.61		1.22		1.06		2		126
	−37 03851	NGC 2451 - 44		⋆ AB	7 43 03	−38 00.5	3	11.16	.004	0.41	.008	0.12	.017	7		80,1421,2036
	−38 03609	NGC 2451 - 108			7 43 03	−38 16.2	1	10.59		0.28		0.12		2	A8 V	80
62888	−24 05910	LSS 726			7 43 04	−24 31.8	2	7.81	.010	0.32	.005	−0.11		7	B9.5Iab	2014,8100
		Ru 32 - 11			7 43 04	−25 25.0	1	14.92		0.30		−0.32		2		410
		Ru 32 - 12			7 43 04	−25 25.0	1	14.88		0.38		−0.17		2		410
		AJ79,1294 T6# 11			7 43 04	−31 50.1	1	12.24		0.51		0.03		3		396
	−37 03850	NGC 2451 - 41			7 43 04	−37 32.8	2	10.71	.008	0.30	.008	0.09	.016	4	B2 III	80,1421
		NGC 2451 - 121			7 43 04	−37 44.2	2	12.48	.000	0.80	.000	0.42	.013	6		80,1421
	−37 03849	NGC 2451 - 43			7 43 04	−37 45.3	2	10.90	.000	1.12	.005	0.89	.019	6		80,1421
		Orsatti 32			7 43 06	−28 55.2	1	11.94		0.63		−0.13		1		1766
		LP 723 - 18			7 43 07	−14 54.7	1	12.20		0.97		0.55		1		1696
		LSS 733			7 43 07	−32 01.3	1	11.92		0.42		−0.48		2	B3 Ib	396
		NGC 2451 - 122			7 43 07	−38 04.5	1	13.35		1.53		1.20		3		80
		Ru 32 - 5			7 43 08	−25 26.1	1	11.33		0.23		0.11		5		410
	−25 05003	Ru 32 - 4			7 43 08	−25 26.1	1	11.69		0.12		−0.67		4		410
		NGC 2451 - 109			7 43 08	−38 11.8	1	11.83		0.51		0.00		2		80
	−38 03613	NGC 2451 - 110			7 43 08	−38 20.3	1	11.22		0.69		0.28		2		80
	−37 03853	NGC 2451 - 46			7 43 09	−37 52.9	1	9.68		0.49		0.06		2		80
	−37 03852	NGC 2451 - 45			7 43 09	−37 57.5	3	11.10	.017	0.40	.000	0.13	.021	7		80,1421,2036
		NGC 2451 - 127			7 43 09	−37 59.2	1	13.14		0.61		0.06		4		80,1421
		NGC 2451 - 129			7 43 09	−37 59.2	1	12.11		0.80		0.34		2		80
		LP 423 - 21			7 43 11	+15 14.4	1	11.73		1.11		1.05		1		1696
62720	+21 01679	IDS07402N2122	AB		7 43 11	+21 14.7	2	7.41	.000	0.38	.001			32	F2	130,1351
62864	−14 02194	HR 3010		⋆ AB	7 43 11	−14 34.1	1	5.64		0.11				4	A2 V	2031
		Ru 32 - 2			7 43 11	−25 27.4	1	12.03		0.07		−0.61		2		410
		Ru 32 - 3			7 43 11	−25 27.4	1	12.02		0.13		−0.64		2		410
		NGC 2451 - 117			7 43 11	−38 10.6	1	12.18		0.58		0.04		3		80
		Ru 32 - 6			7 43 13	−25 27.2	1	13.15		0.36		0.10		2		410
62959	−30 05036	LSS 736			7 43 13	−30 36.0	1	9.28		0.09		−0.70		1	B1/2 Vne	1470
62974	−37 03855	NGC 2451 - 47		⋆ AB	7 43 13	−37 58.3	3	8.30	.005	0.14	.005	0.11	.028	15	A2 V	80,1367,2037
		NGC 2451 - 11			7 43 13	−38 00.4	1	13.17		2.10		2.00		4		80

HD	DM	Other Id	N	Rem	α_{1950}	δ_{1950}	S	V	σ_V	B–V	σ_{B-V}	U–B	σ_{U-B}	n	Spectrum	References
		NGC 2451 - 106			7 43 13	−38 16.2	1	11.64		0.70		0.22		2		80
62721	+18 01733	HR 3003			7 43 14	+18 38.0	8	4.88	.013	1.45	.010	1.76	.010	23	K5 III	15,369,1003,1355,1363*
		AJ79,1294 T6# 16			7 43 14	−32 03.6	1	14.27		0.52		0.02		3		396
		NGC 2451 - 48			7 43 14	−37 52.7	3	11.66	.038	0.51	.035	0.01	.024	5		80,1421,2036
	−37 03857	NGC 2451 - 49			7 43 14	−38 04.2	1	10.90		0.30		0.17		3		80
63006	−40 03410				7 43 14	−40 20.5	1	8.43		−0.02		−0.25		2	B8/9 V	401
	−37 03856	NGC 2451 - 52			7 43 15	−37 39.8	2	10.37	.008	1.08	.000	0.74	.004	4		80,1421
	−37 03859	NGC 2451 - 50			7 43 15	−37 55.0	2	10.03	.004	0.42	.020	−0.05	.004	4		80,1421
62992	−37 03860	NGC 2451 - 250			7 43 15	−38 02.2	3	7.87	.009	0.12	.000	0.18	.039	17	A7 IV	80,1367,2037
63052	−45 03427				7 43 15	−45 50.9	1	8.69		0.00		−0.29		2	B8 V	401
		AJ79,1294 T6# 6			7 43 16	−31 53.3	1	11.62		0.51		0.28		2		396
		NGC 2451 - 56			7 43 16	−37 30.5	1	11.27		1.23		1.09		3		80
		NGC 2451 - 51			7 43 16	−38 07.6	1	12.38		0.67		0.05		2		80
		NGC 2451 - 118			7 43 16	−38 18.5	1	12.18		0.60		0.06		2		80
	−25 05005	Ru 32 - 1			7 43 17	−25 26.5	1	11.11		0.04		−0.80		4	B0 V	410
	−26 04911				7 43 17	−26 23.9	1	10.95		0.03		−0.48		2		126
		NGC 2451 - 128			7 43 17	−37 58.9	1	12.83		0.68		0.14		4		80
	+1 01900	G 112 - 47			7 43 18	+00 56.8	1	9.06		0.68		0.09		1	G2	3062
62767	+16 01551				7 43 18	+16 33.7	1	7.55		1.44		1.67		2	K0	1648
62991	−37 03861	NGC 2451 - 251		⋆	7 43 18	−37 45.9	8	6.53	.010	−0.10	.006	−0.63	.014	40	B2 IV	15,80,540,976,1367*
63429	−71 00612				7 43 18	−71 25.3	1	7.29		1.55		1.79		8	K5 III	1704
62647	+37 01769	HR 2999			7 43 19	+37 38.4	1	5.18		1.58		1.94		4	M2 IIIb	3055
		LOD 3 # 10			7 43 20	−26 24.4	1	11.33		0.10		−0.01		2		126
	−38 03615	NGC 2451 - 111			7 43 20	−38 16.1	1	11.24		0.38		0.17		2		80
	−20 02162				7 43 21	−20 58.2	1	10.96		0.33		0.14		2		126
	−38 03616	NGC 2451 - 112			7 43 21	−38 14.3	1	11.40		0.56		0.01		3		80
		NGC 2451 - 54			7 43 23	−37 52.7	1	11.44		0.56		0.05		4		80
	+51 01354	G 193 - 66			7 43 24	+50 57.4	1	11.36		0.96		0.75		3	K5	7010
		AJ79,1294 T6# 12			7 43 24	−31 50.5	1	13.08		0.62		0.09		4		396
	−26 04916				7 43 25	−26 51.6	1	11.52		0.18		−0.43		1	B3	1376
		NGC 2451 - 53			7 43 26	−37 42.9	1	12.33		0.49		0.16		4		80
		NGC 2451 - 116			7 43 27	−38 05.0	1	12.38		0.85		0.58		3		80
		NGC 2451 - 28			7 43 27	−38 13.9	1	13.10		0.68		0.33		3		80
		LOD 1+ # 9			7 43 28	−21 01.2	1	11.28		0.37		0.29		2		126
		Haf 15 - 13			7 43 28	−32 42.6	1	14.27		1.01		0.15		2		183
	−34 03870				7 43 28	−34 14.0	1	9.26		0.53		0.03		5	F8	396
		NGC 2451 - 55			7 43 28	−37 42.0	1	11.99		0.48		0.07		3		80
63032	−37 03863	NGC 2451 - 254		⋆	7 43 28	−37 50.8	8	3.60	.005	1.73	.005	1.70	.010	41	K4 Ib	15,80,1075,1367,2012*
		NGC 2451 - 123			7 43 28	−38 10.4	1	12.79		0.56		−0.08		3		80
		NGC 2451 - 126			7 43 29	−37 47.3	2	12.10	.012	0.71	.024	0.26	.016	4		80,1421
	−37 03862	NGC 2451 - 57		⋆ AB	7 43 29	−38 04.5	1	9.95		0.54		0.01		2		80
		AJ79,1294 T7# 16			7 43 30	−34 09.1	1	13.47		0.97		−0.04		3		396
62832	+11 01670	HR 3008			7 43 31	+10 53.5	3	5.25	.026	0.01	.005	−0.03	.015	6	A1 Vnn	985,1022,3023
		NGC 2451 - 59			7 43 31	−37 55.5	2	10.51	.000	0.31	.004	0.20	.016	4		80,1421
		NGC 2451 - 120			7 43 31	−38 10.0	1	13.11		0.53		0.26		3		80
		NGC 2451 - 119			7 43 31	−38 10.8	1	12.76		0.41		0.25		3		80
		Haf 15 - 11			7 43 32	−32 42.2	1	14.14		1.01		0.12		2		183
	−32 04371	Haf 15 - 1			7 43 32	−32 42.9	1	10.45		2.81		3.40		2		183
		NGC 2451 - 58			7 43 32	−37 46.8	1	11.52		0.63		0.10		2		80
	−20 02164				7 43 33	−21 03.6	1	10.64		0.54		0.03		2		126
		Haf 15 - 5			7 43 34	−32 43.4	1	13.02		0.98		0.04		1		183
		Haf 15 - 4			7 43 34	−32 43.5	1	11.84		0.94		0.01		2		183
	−37 03866	NGC 2451 - 60			7 43 34	−37 29.9	1	10.69		1.12		0.89		2		80
		NGC 2451 - 61			7 43 34	−37 42.5	2	11.82	.004	0.63	.007	0.11	.082	5		80,1421
		NGC 2451 - 29			7 43 34	−37 57.4	1	13.08		0.77		0.04		3		80
62902	−6 02281	HR 3014			7 43 35	−06 38.9	4	5.49	.009	1.38	.008	1.67	.005	12	K5 III	15,1003,1075,1415
	−20 02165				7 43 35	−20 55.8	1	10.98		0.40		−0.09		2		126
		Haf 15 - 6			7 43 35	−32 43.7	1	13.70		0.95		0.02		2		183
		Haf 15 - 7			7 43 36	−32 44.4	1	13.91		0.81		−0.04		2		183
	−33 04105				7 43 36	−33 49.6	1	11.33		0.12		0.11		2		396
		NGC 2451 - 400			7 43 36	−37 51.	1	9.22		0.20		0.18		2		1367
		NGC 2451 - 401			7 43 36	−37 51.	1	9.87		0.16		0.20		1		1367
		Haf 15 - 3			7 43 37	−32 42.0	1	12.22		0.91		0.05		2		183
		Haf 15 - 9			7 43 37	−32 42.8	1	12.65		1.03		−0.07		1		183
	−32 04372	Haf 15 - 2			7 43 37	−32 43.6	4	10.96	.016	0.80	.027	−0.23	.010	8		183,564,1737,1766
		NGC 2451 - 62			7 43 37	−37 27.0	1	11.19		0.92		0.50		2		80
		NGC 2451 - 64			7 43 37	−37 58.9	1	11.58		1.06		0.83		3		80
63080	−39 03565	NGC 2451 - 255			7 43 37	−39 13.4	2	7.19	.010	−0.03	.000	−0.18		5	A0 IV	401,2037
		Haf 15 - 12			7 43 38	−32 44.8	1	14.41		0.85		−0.08		2		183
	−37 03867	NGC 2451 - 63		⋆ AB	7 43 38	−38 07.2	1	10.38		1.79		1.79		2		80
62522	+60 01082	IDS07393N6033		AB	7 43 39	+60 25.3	3	7.03	.024	0.58	.004	0.03	.009	8	F9 V +F9 V	766,1379,3030
62952	−14 02199	HR 3015			7 43 39	−14 26.5	4	5.04	.005	0.33	.005	0.09	.000	13	F2 V	15,1008,2007,3026
		Ru 34 - 15			7 43 39	−20 14.3	1	13.64		0.78		0.39		2		410
		Ru 34 - 16			7 43 39	−20 14.8	1	13.17		0.42		0.01		2		410
		Haf 15 - 8			7 43 39	−32 43.1	1	13.02		0.85		−0.08		2		183
63177	−52 01273				7 43 39	−52 50.0	1	8.31		−0.04		−0.63		2	B2 V	401
		Ru 34 - 10			7 43 40	−20 15.5	1	12.93		0.28		0.11		2		410
		Ru 34 - 15			7 43 40	−20 15.5	1	14.44		0.55		0.35		2		410
		Haf 15 - 10			7 43 40	−32 42.7	1	14.21		0.93		0.13		2		183
	−33 04108				7 43 40	−33 44.5	1	9.90		0.07		0.01		2		396

Table 1

419

HD	DM	Other Id	N Rem	α_{1950}	δ_{1950}	S	V	σ_V	B–V	σ_{B-V}	U–B	σ_{U-B}	n	Spectrum	References
63079	−37 03868	NGC 2451 - 256		7 43 40	−37 28.4	3	6.98	.021	−0.07	.000	−0.13	.030	9	B9.5p	80,1367,2037
	−37 03869	NGC 2451 - 66		7 43 40	−38 04.4	1	11.41		0.55		0.09		3		80
62901	−3 02053	LS VI -03 018		7 43 41	−03 50.3	1	7.63		0.00		0.02		1	A0	490
		Ru 34 - 14		7 43 41	−20 13.9	1	13.77		0.47		0.42		2		410
		Ru 34 - 12		7 43 41	−20 15.0	1	14.05		0.50		0.13		2		410
		LF12 # 86		7 43 41	−22 32.5	1	11.51		1.10		0.89		1		137
		NGC 2451 - 65		7 43 41	−37 50.8	1	11.92		0.38		0.23		3		80
		Ru 34 - 1		7 43 42	−20 13.8	1	11.48		0.23		0.09		4		410
		Ru 34 - 5		7 43 42	−20 16.	1	13.36		0.58		0.11		2		410
		Ru 34 - 13		7 43 42	−20 16.	1	14.49		0.70		0.29		2		410
		Ru 34 - 17		7 43 42	−20 16.	1	13.90		0.45		0.03		2		410
63001	−20 02167			7 43 42	−20 34.2	1	8.86		−0.07		−0.37		4	A0	1732
		AJ79,1294 T7# 14		7 43 42	−34 15.1	1	12.94		0.48		0.40		4		396
	−37 03865	NGC 2451 - 67		7 43 42	−37 24.5	1	11.26		0.41		0.00		3		80
63118	−43 03534	HR 3020		7 43 42	−43 37.8	4	6.03	.005	−0.07	.005			16	B5 V	15,1020,2012,2012
63000	−20 02166			7 43 43	−20 30.8	1	8.48		1.05		0.83		1	G8/K1	1732
63077	−33 04113	HR 3018	A	7 43 43	−34 04.4	7	5.36	.016	0.58	.015	−0.07	.007	18	G0 V	15,258,688,1201,2012*
63077	−33 04113		B	7 43 43	−34 04.4	1	16.59		1.20				9		1516
		Ru 34 - 3		7 43 44	−20 13.0	1	12.66		1.09		0.63		2		410
		Ru 34 - 9		7 43 44	−20 15.9	1	13.61		0.25		0.16		2		410
63002	−20 02168			7 43 44	−21 05.5	1	9.49		1.19		1.15		2	K1/2 III	126
	−25 05024			7 43 44	−25 15.8	1	10.81		0.26		0.17		1	A2 V	137
		Orsatti 36		7 43 44	−31 17.7	1	12.86		0.28		−0.47		1		1766
		Ru 34 - 8		7 43 45	−20 15.9	1	11.65		0.22		0.06		2		410
63028	−23 06155			7 43 45	−24 08.0	1	6.73		−0.15		−0.62		4	B3 IV/V	164
63005	−26 04925	LSS 742		7 43 45	−26 22.2	7	9.13	.009	−0.02	.019	−0.92	.026	12	O5/6	126,540,976,1011,1012*
63202	−52 01275			7 43 45	−52 54.8	1	8.52		0.01		−0.03		4	K2/3 III	3043
	−20 02169			7 43 46	−20 50.7	1	10.64		0.21		−0.23		2		126
		POSS 544 # 7		7 43 47	+05 44.2	1	18.37		1.58				6		1739
		Ru 34 - 6		7 43 47	−20 15.8	1	11.19		1.31		1.13		2		410
		AJ79,1294 T7# 15		7 43 47	−34 13.4	1	13.01		1.39		1.27		3		396
		Ru 34 - 7		7 43 48	−20 16.4	1	11.66		0.51		0.05		2		410
	−28 04824	LSS 745		7 43 48	−29 1.7	1	11.80		0.48		−0.38		1		1766
		LSS 739		7 43 49	−18 34.1	1	12.53		0.29		−0.63		4	B3 Ib	844
	−23 06156	IDS07417S2323	A	7 43 49	−23 30.5	1	10.08		0.01		−0.05		1	A0 V	137
		LOD 3 # 8		7 43 49	−26 22.1	1	13.03		0.29		−0.23		2		126
	−37 03870	NGC 2451 - 68		7 43 51	−37 39.3	1	11.60		0.69		0.13		2		80
63153	−44 03709			7 43 51	−44 45.6	1	7.57		0.09		0.04		2	A2 V	401
63116	−38 03629	NGC 2451 - 259		7 43 52	−38 30.6	1	9.37		0.40				3	F0 IV-V	2037
		LP 783 - 19		7 43 53	−17 00.4	1	11.59		1.00		0.84		1		1696
	−33 04116	LSS 750		7 43 53	−34 09.5	1	10.96		1.07		−0.03		4		396
	−37 03872	NGC 2451 - 71		7 43 53	−38 03.7	1	12.23		0.45		0.28		2		80
		GD 89		7 43 54	+44 16.3	2	14.87	.050	0.10	.030	−0.71	.025	2	DA	782,3060
	−37 03871	NGC 2451 - 69		7 43 55	−37 37.3	1	11.23		0.51		−0.01		3		80,1421
	−26 04931			7 43 56	−26 19.7	1	11.07		0.91		0.65		2		126
	−37 03873	NGC 2451 - 70		7 43 56	−37 41.8	2	10.42	.004	0.06	.004	0.01	.013	3		80,1421
		LSS 747		7 43 57	−28 18.3	1	11.97		0.52		−0.33		1	B3 Ib	1470
63044	−20 02173	IDS07418S2059	AB	7 43 58	−21 05.9	1	9.25		0.46		0.02		2	F5/6 V	126
		LF12 # 89		7 43 58	−22 22.7	1	11.29		0.09		0.04		1	B8 V	137
		LOD 3 # 6		7 43 58	−26 19.5	1	12.51		0.50		0.08		2		126
63099	−34 03879	LSS 753		7 43 58	−34 12.5	1	10.49		0.93		0.05		3	WC	396
	−26 04932			7 43 59	−26 17.0	1	10.91		0.11		0.02		2		126
		LOD 3 # 5		7 43 59	−26 19.0	1	12.57		0.29		0.18		2		126
	−29 04913	LSS 749		7 43 59	−29 25.9	2	11.50	.030	0.35	.000	−0.61	.050	2	09 Vpe	211,1766
	−31 05033	LSS 752		7 43 59	−31 58.3	1	11.64		0.42		−0.54		3		396
	−31 05034			7 43 59	−31 58.3	2	11.08	.015	0.52	.005	−0.51	.059	5	B3 IIIne	564,1766
	−37 03874	NGC 2451 - 72		7 44 00	−37 48.8	1	11.72		0.36		0.20		2		80
63152	−38 03632	NGC 2451 - 260	★ AB	7 44 00	−38 50.9	1	8.89		0.28				2	A8 V	2037
63133	−37 03875	NGC 2451 - 261		7 44 01	−37 25.4	1	9.14		0.02				3	A0 Vn	2037
63175	−44 03710	IDS07424S4408	AB	7 44 01	−44 15.1	1	8.77		0.02		−0.33		2	B8/9 III	401
	−15 02034			7 44 03	−15 53.3	1	9.74		1.67				2	M2	369
		LSS 751		7 44 03	−27 22.6	1	11.37		0.53		−0.35		1	B3 Ib	1470
63132	−35 03849			7 44 04	−35 13.9	1	8.23		−0.11		−0.63		3	B3 II/III	1732
		LSS 756		7 44 05	−32 43.9	1	11.75		0.91		−0.19		3	B3 Ib	1766
	−37 03876	NGC 2451 - 73		7 44 05	−37 27.0	1	9.91		0.42		−0.02		2		80
		LOD 3 # 4		7 44 06	−26 19.6	1	12.19		0.42		0.08		2		126
63150	−36 03865	LSS 759		7 44 06	−36 22.5	3	8.71	.025	0.12	.007	−0.69	.024	8	B0.5Vnn(e)	401,540,976
		Orsatti 41		7 44 09	−31 13.6	1	12.18		0.37		−0.44		3		1766
		UY Pup		7 44 10	−12 49.6	1	14.00		0.09		−0.79		1		1471
	−37 03877	NGC 2451 - 74		7 44 10	−38 08.6	1	12.04		0.28		0.19		2		80
		Orsatti 43		7 44 11	−31 20.8	1	11.92		0.28		−0.55		2		1766
		LP 898 - 12		7 44 12	−29 28.5	1	11.40		1.05		0.93		1		1696
	−34 03888	LSS 761		7 44 12	−34 16.7	1	11.28		0.77		−0.32		4		396
	−37 03879	NGC 2451 - 75		7 44 12	−37 37.6	1	9.55		0.29		0.12		2	F0 V	80
	−37 03884	NGC 2451 - 77		7 44 13	−37 32.9	1	11.64		0.48		0.02		3	F0 V:	80
	−20 02177			7 44 14	−20 39.9	1	9.86		1.06		0.96		2		126
		LOD 1+ # 2		7 44 15	−20 44.6	1	10.74		0.12		−0.42		2		126
	−33 04121			7 44 15	−34 07.9	1	9.59		1.06		0.89		3		396
		NGC 2451 - 76		7 44 15	−37 53.0	1	11.99		0.64		0.11		2		80
	−38 03635	NGC 2451 - 265	AB	7 44 16	−38 51.0	1	8.89		0.20				2		2037

HD	DM	Other Id	Rem	α₁₉₅₀	δ₁₉₅₀	S	V	σ_V	B–V	σ_B-V	U–B	σ_U-B	n	Spectrum	References
62898	+33 01585	HR 3013	★ A	7 44 17	+33 32.4	2	5.14	.000	1.59	.005	1.92	.024	7	M1 IIIa	1355,3001
		LP 423 - 22		7 44 18	+15 05.4	1	12.68		0.79		0.34		2		1696
63218	−41 03363	W Pup		7 44 18	−42 04.4	1	7.73		1.40		1.14		1	M2/3e	975
63148	−28 04835			7 44 19	−28 47.2	1	7.85		0.25		0.17		5	A8 V	124
	−33 04123	LSS 765		7 44 19	−33 54.1	1	11.59		0.99		−0.11		4		396
		Pup sq 2 # 9		7 44 20	−25 33.	1	10.83		0.19		−0.55		1		211
		LSS 755		7 44 21	−18 44.6	1	12.03		0.12		−0.66		4	B3 Ib	844
63147	−26 04945			7 44 21	−26 15.9	1	8.07		−0.06		−0.49		2	B6 III	126
		LSS 760		7 44 21	−27 22.1	1	14.00		0.48		−0.48		2	B3 Ib	1766
	−29 04923	LSS 763		7 44 21	−29 44.4	1	10.90		0.18		−0.66		2		409
63067	−4 02085			7 44 22	−04 52.0	1	9.20		0.97		0.72		1	K0	6006
	−22 02017	LF12 # 90		7 44 22	−22 55.6	1	10.12		0.27		0.12		1	A7 V	137
	−37 03883	NGC 2451 - 79		7 44 22	−37 21.1	1	11.39		1.43		1.45		3		80
		NGC 2451 - 80		7 44 23	−37 20.4	1	12.32		0.61		0.30		3		80
63215	−37 03886	NGC 2451 - 267	★	7 44 23	−37 48.6	5	5.87	.011	−0.12	.005	−0.47	.000	15	B6 Vnn	15,80,1367,2012,2037
63216	−37 03887	NGC 2451 - 266		7 44 23	−37 58.7	1	8.51		−0.01				3	A0 IV	2036
63584	−69 00770	HR 3038		7 44 23	−69 42.0	3	6.17	.003	−0.05	.005	−0.08	.005	19	A0 IV/V	15,1075,2038
	−37 03885	NGC 2451 - 78		7 44 24	−37 24.5	1	11.32		0.73		0.20		3		80
63112	−12 02135	HR 3019		7 44 25	−12 33.1	4	6.38	.004	−0.02	.003	−0.08	.006	12	B9 III	15,1075,1079,2012
	−26 04948			7 44 25	−26 18.3	1	10.97		0.13		0.03		2		126
		LSS 767		7 44 25	−33 11.4	2	11.22	.005	1.16	.010	−0.02	.015	7	B3 Ib	1737,1766
63513	−65 00806	HR 3036		7 44 25	−65 57.0	4	6.38	.009	0.95	.004	0.67	.010	27	G6/8 III	15,1075,1637,2038
63066	−4 02086			7 44 27	−04 49.8	2	9.10	.010	0.65	.012	0.09		3	G0	1594,6006
	−37 03888	NGC 2451 - 81		7 44 27	−37 09.3	1	11.50		0.50		0.00		2		80
	−38 03636	NGC 2451 - 113		7 44 28	−38 20.6	1	11.52		0.59		0.08		2		80
63065	+0 02079	IDS07419N0016	A	7 44 29	+00 08.5	1	8.32		−0.02		−0.12		1	B9	8084
63065	+0 02080	IDS07419N0016	B	7 44 29	+00 08.5	1	9.27		0.07		0.05		1	A2 V	8084
63065	+0 02079	IDS07419N0016	C	7 44 29	+00 08.5	1	12.33		0.34		−0.06		1	F5 IV	8084
	−25 05040			7 44 29	−25 54.3	1	11.42		0.21		0.15		1	A2 V	137
63169	−23 06180			7 44 30	−23 22.9	1	9.82		−0.13		−0.75		4	B2 III	1732
	−24 05950			7 44 30	−25 01.7	1	10.33		0.38		0.12		1	F5 V	137
		LSS 770		7 44 30	−34 08.1	2	11.44	.015	0.76	.010	−0.30	.005	6	B3 Ib	396,1766
		NGC 2451 - 82		7 44 30	−37 47.	1	10.36		0.18		0.12		3	A1 V	80
	−26 04955	LSS 766		7 44 32	−26 34.2	3	10.46	.016	0.35	.046	−0.79	.019	4		409,1376,1737
		LSS 772		7 44 32	−33 56.4	2	11.38	.000	0.87	.000	−0.23	.000	5	B3 Ib	396,1766
	−33 04129			7 44 32	−34 09.7	1	11.43		0.65		−0.44		3		396
63251	−37 03890	NGC 2451 - 268		7 44 32	−37 46.4	4	7.71	.014	0.02	.008	−0.33	.010	12	B8 V	80,1367,2037,3043
	−30 05070	LSS 768		7 44 33	−31 08.1	2	11.36	.134	0.50	.081	−0.32	.072	7		564,1766
63382	−56 01420	HR 3031	★ A	7 44 33	−56 36.0	2	6.11	.005	0.40	.005			7	F0 II	15,2012
		LSS 764		7 44 34	−20 32.4	1	12.50		0.31		−0.45		4	B3 Ib	844
	−37 03891	NGC 2451 - 84		7 44 34	−37 26.9	1	11.58		0.13		0.11		2		80
		Orsatti 50		7 44 35	−31 10.0	1	11.60		0.41		−0.41		2		1766
	−33 04131	LSS 773		7 44 36	−33 18.4	1	10.61		1.42		0.23		4		1766
63427	−58 00986	IDS07437S5843	AB	7 44 37	−58 49.9	1	7.84		0.90				4	G5 III	2012
		Orsatti 52		7 44 39	−30 16.6	1	13.04		0.57		−0.21		2		1766
	−37 03892	NGC 2451 - 85	★ AB	7 44 39	−37 36.8	1	10.00		0.86		0.56		2		80
	−26 04960	IDS07427S2701	AB	7 44 42	−27 08.2	2	10.00	.005	0.07	.013	−0.66	.002	4	B2 II-III	540,976
233453	+54 01175	G 193 - 70	★ AB	7 44 45	+53 46.9	1	8.89		1.10		0.94		3	K5	22
	−37 03893	NGC 2451 - 86		7 44 45	−37 41.7	1	10.95		1.23		1.11		2		80
	−26 04963			7 44 46	−26 45.1	2	9.93	.003	0.07	.005	−0.71	.013	4	B0.5V	540,976
		Orsatti 53		7 44 46	−31 28.5	1	12.35		0.49		−0.32		3		1766
63291	−37 03895	NGC 2451 - 269		7 44 46	−37 38.9	3	6.26	.033	1.35	.012	1.53	.038	7	K3 II-III	1367,2037,3034
	−37 03894	NGC 2451 - 88		7 44 47	−37 51.8	1	11.40		0.40		0.22		2		80
63307	−39 03584	NGC 2451 - 270		7 44 47	−39 29.2	1	9.73		0.06		0.04		2	A0 V	80
63343	−45 03444			7 44 47	−46 02.2	2	7.59	.010	−0.07	.005	−0.48	.005	6	B5 V	164,401
	−20 02183	LSS 771		7 44 48	−21 02.3	2	10.01	.004	0.29	.004	−0.75	.005	5		540,1737
63250	−29 04937			7 44 49	−29 46.3	1	8.66		1.23		1.46		2	K2/3 (III)	409
		Orsatti 54		7 44 49	−31 15.4	1	11.55		0.41		−0.42		3		1766
	−37 03897	NGC 2451 - 89		7 44 49	−37 59.3	1	9.82		0.52		0.01		2		80
	−37 03898	NGC 2451 - 90		7 44 49	−38 06.8	1	10.58		1.12		0.83		2		80
		Orsatti 55		7 44 50	−32 32.7	1	13.50		1.16		0.04		3		1766
63308	−39 03587	HR 3025		7 44 50	−39 56.2	2	6.56	.005	−0.13	.000	−0.68		6	B2 III	401,2006
63274	−29 04938			7 44 51	−29 49.5	1	8.00		−0.12		−0.60		2	B5 V	409
	−37 03896	NGC 2451 - 91		7 44 51	−37 40.8	1	9.73		0.04		−0.08		2		80
		Pup sq 2 # 10		7 44 52	−27 19.	1	11.86		0.12		−0.27		1		211
		Orsatti 56		7 44 52	−34 14.8	1	12.64		0.74		−0.16		2		1766
		NGC 2451 - 92		7 44 52	−37 42.7	1	11.69		0.21		0.19		2		80
		G 253 - 6		7 44 54	+83 31.1	1	13.10		1.64				1		906
		NGC 2451 - 93		7 44 54	−37 36.2	1	10.74		1.37		1.21		2		80
		LF12 # 93		7 44 55	−24 56.8	1	11.52		1.23		1.20		1	K2 III	137
		Orsatti 57		7 44 55	−30 25.8	1	11.79		0.18		−0.55		2		1766
	−23 06196			7 44 56	−24 09.5	1	11.24		0.12		0.09		1	A2 V	137
	−22 02025			7 44 58	−22 18.7	1	10.35		−0.09		−0.42		2	B8	126
		NGC 2451 - 87		7 44 58	−37 42.1	1	12.70		0.64		0.06		2		80
	−37 03901	NGC 2451 - 94		7 45 02	−38 01.5	1	9.51		0.42		0.00		2		80
63290	−27 04533	LSS 778		7 45 03	−27 47.1	6	9.26	.026	0.50	.028	−0.42	.022	11	B1II/III(p)	211,409,540,976,1012,8100
63271	−22 02027	HR 3023		7 45 04	−22 23.7	4	5.88	.009	−0.19	.006	−0.82	.025	12	B2 IV-V	15,126,1732,2012
		Orsatti 58		7 45 04	−32 24.5	1	11.55		0.95		−0.13		3		1766
		NGC 2451 - 95		7 45 04	−37 32.0	1	11.77		1.13		0.66		3		80
		Pup sq 3 # 44		7 45 05	−27 49.6	1	12.96		0.38		0.25		2	A5	409

Table 1

HD	DM	Other Id	N Rem	α_{1950}	δ_{1950}	S	V	σ_V	B–V	σ_{B-V}	U–B	σ_{U-B}	n	Spectrum	References
63341	−38 03641	NGC 2451 - 275	★ AB	7 45 05	−38 20.6	1	9.55		0.22				3	A5 III+A2V	2037
		LOD 2+ # 3		7 45 06	−22 16.6	1	11.97		0.55		0.01		2		126
63266	−14 02211			7 45 07	−14 29.2	1	8.30		1.66				2	M2 (III)	369
63270	−19 02062			7 45 07	−19 48.5	1	8.17		−0.04		−0.28		2	B8/9 III	401
	−25 05054			7 45 07	−25 37.4	1	9.44		1.40		1.64		1	K3 III	137
		NGC 2451 - 83		7 45 07	−37 42.0	1	13.03		0.21		−0.51		4		80
63304	−26 04975			7 45 09	−26 27.0	1	8.83		0.02		−0.54		2	Ap Si	409
	−37 03900	NGC 2451 - 96		7 45 09	−37 35.8	1	11.44		1.97		2.12		3		80
		LOD 2+ # 4		7 45 10	−22 17.8	1	12.55		0.20		0.12		2		126
	−37 03904	NGC 2451 - 97		7 45 10	−38 09.6	1	10.63		0.57		0.05		2		80
	−38 03642	NGC 2451 - 115		7 45 10	−38 13.9	1	10.09		0.42		−0.02		2		80
	−37 03916	NGC 2451 - 114		7 45 11	−37 30.9	1	11.84		0.46		0.06		2		80
62946	+52 01249			7 45 14	+52 38.5	1	8.28		0.55		0.06		2	F5	1375
		LF12 # 96		7 45 14	−26 24.3	1	11.80		0.24		0.00		1	A0 V	137
		Orsatti 62		7 45 14	−32 49.3	1	13.37		1.04		−0.03		3		1766
	−37 03905	NGC 2451 - 98		7 45 15	−37 22.7	1	10.57		1.14		0.92		2		80
		G 91 - 11		7 45 16	+20 30.2	2	11.46	.010	1.51	.010	1.08	.015	2		333,1620,3078
		CCS 897, GO Pup		7 45 16	−11 49.7	1	10.43		4.76				2		864
		Orsatti 63		7 45 16	−33 55.2	1	12.92		1.42		0.24		1		1766
		G 90 - 18		7 45 17	+37 20.0	3	10.42	.019	1.27	.007	1.27	.005	3	K6	333,801,1017,1620
		Orsatti 64		7 45 17	−31 05.5	1	11.79		0.44		−0.29		3		1766
63375	−38 03644	NGC 2451 - 276		7 45 17	−38 23.7	1	9.65		0.12				2	A1 V	2037
	−33 04141	LSS 779		7 45 19	−33 13.0	2	9.77	.010	1.31	.010	−0.29		4	B2 Ve	564,1766
		Orsatti 65		7 45 20	−33 35.0	1	13.33		0.85		−0.17		2		1766
63401	−39 03595	NGC 2451 - 277	★ V	7 45 21	−39 12.4	5	6.33	.015	−0.16	.007	−0.56		15	B8 III	15,401,1075,2007,2037
63302	−15 02049	HR 3026, QY Pup	★ C	7 45 22	−15 52.0	4	6.35	.033	1.75	.029	1.90	.025	10	K3 Iab/b	15,1415,3001,8100
		Orsatti 66		7 45 22	−31 03.1	1	12.15		0.45		−0.29		2		1766
63241	+5 01790	IDS07428N0539	AB	7 45 24	+05 32.1	1	6.60		1.58		1.91		2	K0	985
63400	−37 03906	NGC 2451 - 280		7 45 25	−37 50.8	2	7.44	.059	0.98	.000	0.77		5	K0 IIIa	460,2037
63424	−38 03647	NGC 2451 - 281		7 45 25	−38 58.7	1	7.13		0.36				2	F3 IV	2037
63360	−26 04980	NGC 2453 - 11		7 45 26	−27 03.1	2	9.45	.010	0.09	.025	0.12	.040	5	B9.5V	248,382
63425	−41 03384	IDS07438S4116	A	7 45 26	−41 22.7	5	6.93	.006	−0.18	.003	−0.94	.008	14	B1/2 Ib/II	401,540,976,1732,2012
		NGC 2453 - 10		7 45 27	−27 02.6	2	13.86	.000	0.41	.000	0.28	.000	5		248,382
		NGC 2453 - 16		7 45 27	−27 03.7	2	12.56	.059	0.38	.005	0.00	.030	5		248,382
		NGC 2453 - 25		7 45 27	−27 03.7	1	11.88		0.11		0.12		2	A2 V	382
63361	−26 04986	NGC 2453 - 79		7 45 27	−27 07.3	1	9.38		0.15		0.15		1	A0 V	409
		NGC 2451 - 99		7 45 27	−38 10.3	1	12.62		0.83		0.31		2		80
63359	−26 04982	LSS 780		7 45 28	−26 33.4	5	9.51	.035	0.03	.014	−0.85	.027	10	B0 V	409,540,976,1012,1376
		Orsatti 67		7 45 28	−32 24.9	1	13.06		1.35		0.45		3		1766
	+37 01776	G 90 - 19		7 45 29	+36 48.0	1	10.02		1.15		1.16		1	K4.5	333,1620
63323	−15 02052	HR 3027	★ A	7 45 29	−15 53.4	2	6.42	.010	1.75	.045	1.97		7	K5 II/III	2007,3005
		NGC 2453 - 12		7 45 29	−27 03.3	1	13.67		0.35		−0.12		2		382
		NGC 2453 - 9		7 45 30	−27 03.1	2	13.93	.000	0.54	.000	0.41	.000	5		248,382
		NGC 2451 - 100		7 45 30	−38 10.8	1	11.43		0.50		−0.01		2		80
63372	−28 04864			7 45 31	−28 26.4	1	10.39		0.20		0.14		2	A1 IV	409
		Pup sq 2 # 11		7 45 32	−25 16.	1	10.87		0.06		−0.68		1	B2 Vp	1376
	−26 04987	NGC 2453 - 50		7 45 32	−27 04.1	2	10.45	.015	1.98	.020	2.08	.159	5		248,382
		NGC 2453 - 52		7 45 32	−27 04.6	1	14.02		0.39		−0.17		2		382
	−37 03908	NGC 2451 - 101		7 45 32	−37 37.6	1	9.01		1.82		1.15		2		80
	−23 06221			7 45 33	−23 34.4	1	10.55		1.11		1.15		1	K2 III	137
		NGC 2453 - 7		7 45 33	−27 02.6	2	12.87	.109	0.26	.015	−0.38	.000	5		248,382
		NGC 2453 - 61		7 45 33	−27 04.1	1	12.92		0.31		−0.33		2		382
		NGC 2453 - 63		7 45 33	−27 04.2	1	12.50		0.25		−0.37		2		382
		NGC 2453 - 57		7 45 33	−27 04.5	1	13.85		0.24		−0.01		2		382
		NGC 2453 - 58		7 45 33	−27 04.5	1	12.76		0.22		−0.38		2		382
		NGC 2453 - 54		7 45 33	−27 04.8	1	11.38		0.34		−0.24		2	B5 V	382
		Orsatti 70		7 45 33	−31 43.2	1	11.53		0.92		−0.12		3		1766
63208	+23 01812	HR 3021	★ AB	7 45 34	+23 16.0	1	6.20		0.54		0.27		2	G2 III+A4V	1733
63358	−19 02066			7 45 34	−19 51.6	1	8.01		−0.10		−0.59		2	B3 III	401
		NGC 2453 - 8		7 45 34	−27 03.3	1	13.06		0.32		−0.32		2		382
		NGC 2453 - 45		7 45 34	−27 03.8	1	12.64		0.31		−0.34		2		382
		NGC 2453 - 66		7 45 34	−27 04.1	1	14.46		0.40		0.11		2		382
63467	−43 03569	IDS07439S4309	AB	7 45 34	−43 16.2	1	7.58		−0.05		−0.50		2	B9 III/IV	401
63185	+31 01668			7 45 35	+31 02.1	1	8.54		0.31		0.03		2	F0	1625
		NGC 2453 - 64		7 45 35	−27 04.5	1	14.09		0.50		0.24		2		382
63336	−11 02106	HR 3029	★ AB	7 45 36	−12 04.1	2	5.47	.006	0.49	.008	−0.02		6	dF5	58,2007
		NGC 2453 - 38		7 45 36	−27 04.3	1	13.14		0.24		−0.41		2		382
63531	−49 03062			7 45 36	−50 03.6	2	7.08	.010	0.04	.005	−0.60		6	B2 III	401,2012
		LF12 # 99		7 45 37	−24 03.4	1	13.24		0.23		−0.01		2	A0 V	137
		Pup sq 5 # 55		7 45 37	−26 05.0	1	13.02		0.09		−0.05		2	B9 V	947
		NGC 2453 - 44		7 45 37	−27 03.8	1	13.79		0.31		0.00		2		382
		NGC 2453 - 40		7 45 37	−27 04.1	1	12.89		0.36		−0.38		2		382
63423	−30 05098	LSS 783		7 45 37	−30 25.4	4	7.86	.008	0.30	.024	−0.60	.011	9	B1 Ib/II	211,540,976,1766
		Orsatti 72		7 45 38	−32 00.7	1	13.49		0.49		−0.21		3		1766
		Pup sq 5 # 54		7 45 39	−26 05.6	1	12.34		0.15		−0.39		2	B9 p	947
63465	−38 03650	NGC 2451 - 283	★ AB	7 45 39	−38 23.2	4	5.08	.007	−0.11	.005	−0.65	.000	11	B2 III	15,401,404,2012
63685	−61 00888			7 45 39	−61 18.6	2	7.38	.000	0.76	.000			8	G8 IV	1075,2012
63563	−52 01278			7 45 40	−52 13.3	1	6.94		−0.13		−0.59		2	B8/9 II	401
	−37 03911	NGC 2451 - 103		7 45 42	−37 45.4	1	10.79		0.50		−0.05		2		80
	−25 05070			7 45 43	−25 14.5	1	11.27		0.02		−0.01		1	B9 V	137

HD	DM	Other Id	N Rem	α_{1950}	δ_{1950}	S	V	σ_V	B–V	σ_{B-V}	U–B	σ_{U-B}	n	Spectrum	References
		Pup sq 5 # 52		7 45 43	−26 16.2	1	11.81		0.46		−0.09		2	F5 V	947
		Pup sq 2 # 12		7 45 45	−28 04.	1	12.61		0.47		0.27		1		211
		Orsatti 73		7 45 46	−34 10.8	1	13.09		0.87		−0.11		3		1766
		Pup sq 5 # 53		7 45 47	−26 11.7	1	11.65		0.26		0.12		2	A7 V	947
		Orsatti 74		7 45 48	−33 45.8	1	13.03		1.13		0.03		3		1766
	−37 03910	NGC 2451 - 102		7 45 48	−37 37.2	1	11.38		0.59		0.04		2		80
	−37 03913	NGC 2451 - 104		7 45 48	−37 37.2	1	11.10		0.09		0.08		4		80
	−37 03912	NGC 2451 - 105		7 45 48	−37 43.3	1	11.15		0.48		0.04		2		80
62976	+60 01084			7 45 49	+60 28.1	3	6.77	.014	0.09	.004	0.12	.013	8	A2	252,766,1379
	−25 05077			7 45 49	−26 05.4	1	11.08		1.83		2.18		2	M6 III	947
		LSS 786		7 45 49	−33 40.0	2	12.01	.030	1.06	.000	0.02	.015	5	B3 Ib	396,1766
63488	−37 03914	NGC 2451 - 284		7 45 49	−37 53.1	4	8.29	.017	0.08	.006	−0.28	.029	10	B9 Vn	80,401,2037,3043
	−29 04955			7 45 51	−29 23.5	1	11.54		0.19		0.12		2		124
63444	−23 06230			7 45 52	−24 08.1	1	9.05		−0.08				4	B8 V	2014
63463	−29 04956	LSS 784		7 45 52	−30 00.6	2	8.69	.020	0.45	.010	0.10	.005	4	A0 Ib	409,1766
		Orsatti 76		7 45 52	−33 53.1	1	12.46		1.21		0.09		3		1766
	−26 04997			7 45 53	−26 14.2	1	10.23		1.05		0.75		2	K0 III	947
63511	−38 03654	NGC 2451 - 285		7 45 53	−38 33.1	2	9.15	.482	0.16	.000	0.11		3	A3 V	80,2036
64033	−73 00469			7 45 53	−73 43.1	1	11.34		0.54		0.41		2	F2 V	1776
		AJ84,127 # 2214		7 45 54	+37 15.	1	11.87		1.12		0.87		1		801
63445	−24 05991			7 45 54	−24 57.1	1	9.14		0.97		0.85		1	K1 III	137
		Pup sq 5 # 116		7 45 54	−26 28.2	1	11.87		0.17		−0.64		2	B1 V:	947
		Orsatti 78		7 45 54	−33 57.4	1	12.93		1.24		0.15		3		1766
63416	−21 02125			7 45 56	−22 13.4	1	8.45		0.09		0.09		2	A1 IV/V	126
		Orsatti 80		7 45 56	−30 57.6	1	11.91		0.56		−0.15		3		1766
		Orsatti 79		7 45 56	−34 24.0	1	12.47		0.94		−0.19		2		1766
		LF12 # 103		7 45 57	−25 01.4	1	13.44		1.19		0.78		1		137
63446	−25 05079	AD Pup		7 45 58	−25 27.1	1	9.59		1.01		0.67		1	F5-G2	689
	−30 05103			7 45 58	−30 22.1	1	9.34		0.80		0.31		2	K0 V	3008
63529	−37 03915	NGC 2451 - 287		7 45 58	−37 20.2	1	9.92		0.07		0.03		2	A1 V	80
	−37 03917	NGC 2451 - 107		7 45 59	−37 45.4	1	10.44		0.48		0.18		2		80
63462	−25 05081	HR 3034	★ A	7 46 00	−25 48.7	10	4.48	.012	−0.07	.015	−1.01	.009	50	B1 IV:nne	15,681,815,1034,1075,1637*
		Pup sq 5 # 115		7 46 00	−26 27.9	1	11.64		1.01		0.73		2	K2 III:	947
	−29 04958			7 46 01	−29 48.6	1	10.84		0.15		0.05		2		124
63578	−46 03435	HR 3037		7 46 01	−46 29.0	3	5.21	.012	−0.15	.005	−0.85	.002	8	B2 III	1034,1732,2006
		Orsatti 81		7 46 02	−32 07.7	1	12.75		0.84		−0.23		3		1766
		LOD 2+ # 6		7 46 03	−22 11.7	1	12.21		0.04		−0.05		2		126
		Pup sq 5 # 50		7 46 03	−26 13.7	1	11.64		0.22		0.11		2	A2 V	947
63368	−0 01828			7 46 04	−00 37.1	3	8.43	.003	0.94	.000	0.57	.014	34	K0 III	147,1509,6005
		LF12 # 104		7 46 04	−26 20.6	1	12.28		0.15		−0.40		1	B8 V	137
63508	−29 04959			7 46 04	−29 15.2	1	8.94		−0.06		−0.26		2	B9 V	124
63579	−46 03437			7 46 05	−46 53.3	1	6.98		−0.11		−0.71		2	B3 V	401
63353	+5 01797	W CMi		7 46 06	+05 31.1	2	8.93	.151	2.50	.009	3.09		2	C7,2	864,1238
	−31 05083			7 46 07	−31 29.5	2	11.22	.030	0.56	.025	−0.25	.005	4		126,1766
	−31 05087	LSS 789		7 46 07	−31 29.5	1	11.71		0.42		0.09		2		126
63390	+0 02098			7 46 10	−00 08.0	3	8.73	.002	0.04	.007	0.05	.010	34	A1 V	147,1509,6005
63439	−11 02111			7 46 10	−12 05.6	1	7.78		1.59		1.73		3	K5	1657
		Pup sq 5 # 57		7 46 10	−25 58.3	1	12.37		0.43		0.00		2	F3	947
	−27 04556			7 46 10	−28 09.7	1	12.02		0.22		−0.10		2		126
63666	−53 01432	IDS07449S5305	A	7 46 10	−53 12.5	1	7.62		0.02		−0.25		2	B7 IV/V	401
		LOD 4 # 5		7 46 11	−28 06.5	1	12.09		0.41				2		126
63577	−39 03621			7 46 12	−39 45.4	1	8.48		−0.06		−0.29		2	B9 V	401
	−23 06243			7 46 13	−23 31.8	1	11.78		0.10		−0.02		1	A0 V	137
		Pup sq 5 # 58		7 46 13	−25 57.3	1	12.54		0.38		0.00		2	F3	947
63352	+13 01772	HR 3030		7 46 14	+13 29.8	1	6.04		1.38				2	K0	71
		LOD 2+ # 8		7 46 14	−22 06.9	1	12.65		0.71		0.17		2		126
	−26 05009			7 46 14	−26 28.2	1	10.95		0.46		−0.01		2	F3 V	947
		Ru 36 - 7		7 46 15	−26 09.1	1	13.18		0.08		−0.04		2		183
63525	−26 05011	Ru 36 - 1		7 46 16	−26 12.3	1	10.28		−0.04		−0.26		5	B9	183
		Pup sq 5 # 113		7 46 16	−26 28.2	1	11.61		0.15		−0.18		2		947
		LOD 4 # 6		7 46 16	−28 07.4	1	12.13		0.16		−0.30		2		126
	−27 04559			7 46 16	−28 10.6	1	9.98		0.99		0.55		2		126
	−33 04162	LSS 792		7 46 16	−33 35.6	1	11.48		1.09		−0.05		2		1766
		Pup sq 5 # 112		7 46 17	−26 27.0	1	11.78		0.42		0.04		2	F5 V	947
63547	−28 04885			7 46 17	−28 14.0	2	7.76	.010	0.99	.030	0.80	.005	4	G8 III	126,409
		Orsatti 87		7 46 17	−33 30.1	1	12.24		1.03		−0.05		3		1766
63333	+24 01777			7 46 18	+24 36.9	1	7.10		0.48		−0.03		2	F5	1625
		Ru 36 - 9		7 46 18	−26 10.7	1	12.80		0.52		0.07		2		183
	−26 05013			7 46 18	−26 26.8	1	11.78		0.18		−0.22		2	B7 V	947
		LOD 4 # 7		7 46 18	−28 04.6	1	12.64		0.58				2		126
		Pup sq 3 # 64		7 46 18	−28 09.0	1	12.70		0.30		0.05		2	B9	409
63604	−41 03405			7 46 18	−41 28.9	1	8.47		0.07		−0.58		2	B5 II	401
		LOD 2+ # 7		7 46 19	−22 10.0	1	12.01		0.16		−0.51		2		126
		Ru 36 - 6		7 46 19	−26 09.7	1	12.74		0.07		−0.09		2		183
	−25 05094	Ru 36 - 3		7 46 19	−26 09.8	1	12.08		0.12		−0.05		2		183
		Ru 36 - 5		7 46 19	−26 09.9	1	12.40		0.08		−0.15		2		183
		Pup sq 5 # 60		7 46 20	−25 57.9	1	11.46		0.07		−0.21		2	B9 V	947
63435	+4 01826	HR 3033		7 46 21	+04 27.6	3	6.53	.005	0.78	.004	0.38	.000	9	G2 III	15,1355,1415
		Ru 36 - 4		7 46 21	−26 08.7	1	12.23		0.10		0.00		2		183
		Orsatti 88		7 46 21	−31 25.3	1	13.70		0.55		−0.19		1		1766

Table 1   423

HD	DM	Other Id	N Rem	α_{1950}	δ_{1950}	S	V	σ_V	B–V	σ_{B-V}	U–B	σ_{U-B}	n	Spectrum	References
		Pup sq 5 # 59		7 46 22	−25 56.2	1	12.19		0.09		−0.29		2	B8 III	947
63546	−27 04561	Pup sq 3 # 45		7 46 22	−27 44.3	1	10.10		0.07		0.16		1	A0 Vp	409
		Ru 36 - 2		7 46 23	−26 10.5	1	12.89		0.12		0.04		2		183
		Orsatti 89		7 46 23	−31 35.0	1	12.04		0.58		−0.04		1		1766
		Orsatti 90		7 46 23	−33 58.9	1	11.74		1.01				3		1766
63602	−37 03922	NGC 2451 - 291	★ AB	7 46 23	−37 33.1	3	8.34	.012	0.01	.010	−0.27	.000	6	B8 V	401,2037,3043
		Pup sq 5 # 61		7 46 24	−25 57.9	1	12.28		1.05		0.90		2		947
		Ru 36 - 8		7 46 24	−26 10.	1	13.84		0.15		0.14		2		183
		Orsatti 91		7 46 24	−33 15.9	1	12.95		0.87		−0.19		3		1766
	−27 04562			7 46 25	−28 04.9	1	10.74		0.38		0.30		2		126
63543	−21 02130			7 46 26	−21 53.0	1	7.28		1.18		1.04		2	K0 II	126
	−24 06010			7 46 26	−25 03.5	2	10.30	.004	−0.05	.013	−0.74	.016	4	B2 IV	540,976
		Orsatti 92		7 46 26	−31 31.8	1	13.79		0.72		−0.15		1		1766
63640	−40 03490	HR 3041, T Pup		7 46 26	−40 31.6	2	6.14	.000	1.58	.000	1.90		6	M2 III	2031,3055
	−27 04568	Pup sq 3 # 62		7 46 31	−28 10.4	1	10.47		0.19		0.14		2	B8 V	409
63641	−42 03541			7 46 31	−42 22.9	1	7.77		−0.03		−0.18		2	B9 III/IV	401
	−23 06253			7 46 32	−23 25.2	1	10.11		1.06		0.85		1	K0 III	137
		Pup sq 3 # 63		7 46 34	−28 14.9	1	11.51		0.34		0.26		2	A2 V	409
	−26 05019			7 46 35	−26 31.3	2	10.20	.079	0.98	.005	0.73	.005	5	K0 III	137,322
		Orsatti 97		7 46 35	−30 32.0	1	11.70		0.24		−0.49		2		1766
	−33 04170	LSS 798		7 46 35	−33 16.6	2	11.69	.000	1.03	.000	−0.15	.000	6		1737,1766
		Pup sq 3 # 65		7 46 36	−28 16.1	1	13.09		0.31		0.07		1	A0	409
		Orsatti 98		7 46 36	−33 27.7	1	13.21		1.02		−0.14		1		1766
	−26 05020			7 46 37	−26 26.3	2	10.39	.035	0.42	.010	0.00	.005	5	F5	322,947
		LOD 4 # 9		7 46 37	−28 04.1	1	12.35		1.09				2		126
		LOD 2+ # 10		7 46 38	−21 58.6	1	11.04		0.25		0.11		2		126
63625	−30 05126			7 46 38	−30 37.8	1	7.20		−0.16		−0.62		4	B5 III	164
	−34 03951	LSS 800		7 46 38	−34 36.4	2	11.85	.010	0.62	.010	−0.44	.005	6		1737,1766
63745	−52 01284			7 46 38	−52 26.1	1	10.82		0.07		−0.04		3	Ap EuCrSr	1097
63765	−53 01433			7 46 38	−54 08.1	1	8.10		0.75				4	G6 IV/V	2012
		LOD 4 # 10		7 46 39	−28 04.5	1	11.99		0.15				2		126
63258	+50 01475			7 46 40	+50 43.0	1	8.53		0.91		0.52		2	G5	1566
		Pup sq 5 # 108		7 46 40	−26 25.3	2	10.86	.059	0.12	.020	0.15	.015	5	A2 V	322,947
		Orsatti 100		7 46 40	−30 36.1	1	12.20		0.28		−0.39		2		1766
	−25 05104			7 46 41	−25 15.1	1	9.76		1.01		0.98		1	K0 III	137
		LF12 # 109		7 46 41	−25 39.2	1	13.00		0.06		0.10		1		137
		Pup sq 5 # 10		7 46 41	−26 10.2	1	12.03		0.16		0.06		1	A0 V	947
63599	−26 05022			7 46 42	−26 29.6	2	9.04	.010	−0.02	.015	0.01	.020	5	A0 V	322,947
	−26 05023			7 46 44	−26 11.3	1	11.33		1.17		1.11		2		947
		BONN82 T8# 11		7 46 44	−26 33.3	1	12.11		1.57		2.03		3	gK5	322
63475	+13 01775			7 46 45	+12 56.4	1	6.41		0.99		0.79		2	K0	1648
63410	+26 01656			7 46 45	+26 23.4	3	6.81	.014	0.95	.000	0.62	.012	7	G8 III	1003,1080,1733
		Pup sq 5 # 11		7 46 45	−26 10.1	1	12.67		0.20		0.01		2		947
		BONN82 T8# 13		7 46 45	−26 31.3	1	12.24		1.07		0.69		3	K1	322
63637	−29 04973			7 46 45	−30 07.9	2	7.52	.013	0.65	.000	0.12		6	G2 V	1311,2012
	−31 05095			7 46 45	−31 28.1	1	10.71		0.05		−0.06		2		126
63638	−31 05096			7 46 45	−31 29.5	1	6.94		−0.14		−0.46		2	B7 V	126
		Orsatti 101		7 46 45	−33 43.4	1	11.28		1.12		0.01		3		1766
65001	−82 00216			7 46 45	−82 13.1	1	7.55		1.13		1.00		7	K0 III	1704
		LOD 2+ # 11		7 46 47	−22 04.6	1	10.45		0.17		0.13		2		126
	−24 06017			7 46 47	−24 24.9	1	10.50		0.26		−0.57		1		211
63598	−24 06019			7 46 47	−24 50.7	1	7.93		0.54				4	G2 V	2012
63623	−25 05108			7 46 47	−25 52.9	1	8.72		0.58		0.13		2	F7 V	947
	−26 05026			7 46 48	−26 23.3	1	10.08		1.58		1.98		5	gK5	322
		LOD 5- # 3		7 46 49	−31 28.1	1	10.72		0.56		0.10		2		126
		LF12 # 111		7 46 50	−22 42.5	1	13.04		0.50		0.05		1	F2 V	137
63432	+28 01482			7 46 51	+28 19.5	1	6.77		0.08		0.13		3	A2 V	1501
63744	−46 03451	HR 3046		7 46 51	−46 57.0	5	4.71	.006	1.06	.003	0.92	.007	21	K0 III	15,330,1075,2012,8015
		Pup sq 5 # 49		7 46 52	−26 14.7	1	10.91		0.83		0.48		3	G2	947
		LOD 4 # 13		7 46 52	−28 04.7	1	12.75		0.52				2		126
		Pup sq 5 # 46		7 46 54	−26 18.1	1	12.38		0.39		0.11		1		947
63724	−38 03674	NGC 2451 - 297		7 46 54	−38 42.1	1	9.47		0.38		0.00		2	F2 V	80
63660	−24 06022	HR 3043		7 46 56	−24 47.2	3	5.33	.005	0.75	.007	0.36		12	G0 III	15,1637,2012
63635	−26 05030			7 46 56	−26 25.1	2	8.82	.014	0.07	.005	0.07	.010	7	A0 V	322,947
		Orsatti 103		7 46 56	−32 41.4	1	13.50		1.13		0.26		1		1766
63706	−37 03930	NGC 2451 - 298		7 46 56	−37 20.0	1	9.60		0.16		0.17		2	A3 III	80
		BV Pup		7 46 57	−23 26.4	1	13.12		0.00		−0.84		1		1471
		Pup sq 5 # 45		7 46 57	−26 17.8	1	12.39		1.12		0.90		1		947
		Pup sq 5 # 106		7 46 57	−26 24.8	2	10.65	.020	0.66	.010	−0.23	.005	5	B2	322,947
		Pup sq 5 # 48		7 46 58	−26 14.0	1	11.95		0.14		−0.40		3	B5 V:	947
		LOD 4 # 14		7 46 58	−28 04.9	1	11.73		1.01		0.94		2		126
		Pup sq 5 # 47		7 47 00	−26 16.0	1	12.98		0.37		0.06		1	A8 V	947
63787	−45 03471			7 47 02	−45 20.3	1	8.15		−0.08		−0.41		2	B7 II/III	401
	−18 02035			7 47 03	−19 12.4	1	9.86		0.67		0.16		1		1696
		Pup sq 5 # 12		7 47 03	−26 10.0	1	11.92		0.60		0.20		1	F2 p	947
		Pup sq 5 # 103		7 47 03	−26 38.7	1	11.21		1.03		0.76		1	G2 I:r	947
		Orsatti 104		7 47 03	−33 51.9	1	13.95		0.53		−0.11		1		1766
63738	−37 03939	NGC 2451 - 300		7 47 03	−38 04.6	2	7.14	.000	0.49	.005	−0.06		7	F7 V	80,2037
63277	+56 01241			7 47 04	+56 30.2	1	8.51		1.66		1.99		2	K2	1733
	−22 02037			7 47 04	−22 18.0	2	10.05	.010	0.51	.122	0.07	.117	3	F0.V	126,137

HD	DM	Other Id	N	Rem	α_{1950}	δ_{1950}	S	V	σ_V	B–V	σ_{B-V}	U–B	σ_{U-B}	n	Spectrum	References
		LF12 # 113			7 47 04	−25 27.1	1	13.17		0.60		0.24		1		137
	−25 05113				7 47 04	−25 48.5	1	10.03		0.67		0.40		2	A5 k	947
		LOD 4 # 15			7 47 04	−28 03.6	1	13.11		0.58		0.22		2		126
		Pup sq 5 # 13			7 47 05	−26 08.7	1	12.44		0.63		0.16		2		947
		Orsatti 105			7 47 05	−33 29.2	1	13.12		0.99		-0.13		3		1766
		Pup sq 5 # 8			7 47 06	−26 11.6	1	11.87		1.49		1.36		1		947
		LOD 4 # 16			7 47 06	−28 05.0	1	13.42		0.28				2		126
63704	−28 04906				7 47 06	−28 35.7	1	8.95		0.92		0.64		2	G8 III/IV	126
		Pup sq 5 # 14			7 47 07	−26 08.2	1	13.29		0.20		0.09		1		947
		Pup sq 5 # 102			7 47 07	−26 39.5	1	11.16		0.18		-0.31		1	B8 V	947
63737	−37 03940	NGC 2451 - 301			7 47 07	−37 39.6	1	8.66		0.41				2	F0 III-IV	2037
	−27 04580				7 47 08	−28 06.7	1	9.87		1.20		1.24		2	K0	126
	−30 05135	PW Pup			7 47 08	−31 00.1	2	9.39	.140	0.89	.045	0.47	.025	4	F2 Iab e	564,8100
63655	−12 02164	HR 3042			7 47 09	−13 13.6	2	6.23		-0.08	.005	-0.48		7	B8 III	1079,2007
63332	+54 01177	HR 3028			7 47 10	+54 15.4	2	6.02	.005	0.46	.000	0.02		5	F7 IV-V	71,272
	−16 02144				7 47 10	−16 28.8	1	9.38		1.69				2	M2	369
		Pup sq 5 # 66			7 47 10	−25 49.7	1	12.88		0.33		0.32		2	A3 V	947
63700	−24 06030	HR 3045	⋆	AB	7 47 11	−24 44.0	8	3.33	.017	1.24	.009	1.15	.019	27	G6 Ia	15,58,369,1075,1637*
		Pup sq 5 # 7			7 47 11	−26 13.3	1	11.90		0.49		-0.01		2	F6	947
		Pup sq 5 # 44			7 47 11	−26 20.5	1	12.14		0.75		0.40		1	K0	947
		Orsatti 107			7 47 11	−33 28.4	1	11.94		1.01		-0.06		2		1766
	−25 05118				7 47 12	−25 45.5	1	10.17		0.09		-0.38		2	B9 II	947
		LSS 808			7 47 12	−33 30.0	1	11.94		1.02		-0.06		2	B3 Ib	1766
63679	−22 02039				7 47 13	−22 26.3	1	7.92		1.59		1.59		2	M2 III	126
		LOD 2+ # 12			7 47 14	−22 10.1	1	10.69		0.52		0.06		2		126
63806	−42 03557				7 47 14	−43 11.1	2	7.11	.005	-0.14	.005	-0.66		6	B2 V	401,2012
63702	−26 05038				7 47 15	−26 27.8	2	9.04	.015	0.48	.000	0.43	.029	2	F0 V	137,322,947
63926	−56 01437	HR 3056	⋆	AB	7 47 15	−56 20.7	2	6.33	.007	1.01	.002	0.38		6	K0 III	58,2007
		Pup sq 5 # 42			7 47 16	−26 18.1	1	11.76		1.02		0.72		2	G5 p	947
		Orsatti 108			7 47 16	−34 11.1	1	12.47		0.98		-0.14		3		1766
63722	−25 05121	IDS07452S2536	AB		7 47 17	−25 43.8	1	9.48		0.01		-0.17		2	B8 IV/V	947
		LP 423 - 40			7 47 18	+19 03.0	1	12.23		0.55		-0.16		2		1696
63721	−23 06285				7 47 18	−24 05.8	1	8.96		1.05		0.92		1	K0 III	137
		Pup sq 5 # 15			7 47 18	−26 09.7	1	10.90		1.02		0.73		2	K0 III	947
		Pup sq 5 # 3			7 47 18	−26 11.8	1	12.14		0.83		0.45		2		947
	−48 03219				7 47 19	−48 36.2	1	10.14		2.90				1	R	864
	−24 06031				7 47 20	−24 36.2	1	10.06		0.26		0.09		2	A5 V	137
		Pup sq 5 # 67			7 47 20	−25 49.1	1	12.23		0.19		-0.22		1	A5 V:	947
		Orsatti 111			7 47 20	−33 26.2	1	13.77		1.02		0.08		1		1766
63719	−20 02211				7 47 21	−21 09.5	1	7.97		-0.14		-0.66		3	B3/5 III	1732
	−25 05122	LSS 806			7 47 21	−26 07.6	3	10.83	.005	0.64	.010	-0.30	.018	4	B0 V:r	947,1295,1376
	−26 05039				7 47 21	−26 15.4	1	10.80		0.56		-0.01		2	F7 III	947
		LF12 # 117			7 47 22	−23 02.5	1	12.04		1.09		0.84		1	K0 III	137
63805	−38 03686				7 47 22	−38 39.5	1	10.00		0.04		-0.28		2	B8/9 V	3043
		Pup sq 5 # 2			7 47 23	−26 12.4	1	12.61		1.44		1.34		2		947
63786	−34 03970	HR 3049			7 47 23	−35 07.0	2	5.94	.005	-0.06	.010	-0.20		6	B9 V	401,2007
63735	−26 05040	LSS 809			7 47 24	−26 12.9	4	9.79	.040	0.08	.035	-0.78	.017	5	B3/5 Ib	409,947,1295,1376
	−26 05042				7 47 24	−26 21.3	1	11.02		0.23		0.19		1	A5 V	947
		Pup sq 1 # 7			7 47 24	−28 37.	1	13.26		0.53		0.13		3		124
	−32 04444	LSS 810			7 47 24	−32 29.7	1	10.54		1.08		0.32		4	B9 Iab	1737
		LF12 # 118			7 47 25	−23 32.3	1	11.20		1.35		1.21		1	K3 III	137
	−25 05124				7 47 25	−26 03.6	1	10.98		0.32		0.12		2	A9 V	947
		BONN82 T8# 12			7 47 25	−26 31.3	1	12.19		0.21		-0.10		3	B8	322
63760	−28 04916	IDS07454S2828	AB		7 47 25	−28 35.2	1	9.58		0.05		0.00		2	B9 IV	126
63804	−33 04186	LSS 811	⋆	A	7 47 25	−33 12.2	3	7.74	.018	1.18	.019	0.33	.070	8	A0 Ia(p)	211,540,976
63697	−16 02146	HR 3044			7 47 26	−17 06.0	2	5.18	.000	1.28	.005	1.49		7	K3 III	2031,3016
		Pup sq 5 # 21			7 47 26	−26 03.8	1	11.98		0.28		0.06		2	A5 V	947
		Pup sq 5 # 17			7 47 26	−26 08.2	1	12.16		0.36		0.35		2		947
		BONN82 T8# 10			7 47 26	−26 29.8	1	11.47		1.16		0.77		3	K2	322
		Pup sq 5 # 41			7 47 27	−26 16.9	1	12.64		0.65		-0.14		1		947
63736	−26 05044				7 47 27	−26 27.1	2	8.67	.010	1.76	.020	1.98	.035	5	M1/2 (III)	322,947
63827	−37 03943	NGC 2451 - 304			7 47 27	−37 17.7	1	8.55		0.35				3	A8 V	2037
		Pup sq 5 # 22			7 47 29	−26 03.2	1	12.23		0.44		-0.02		2	F3 V	947
		Pup sq 5 # 16			7 47 29	−26 09.5	1	12.75		0.15		0.13		1	A1 V	947
		Pup sq 5 # 39			7 47 29	−26 15.3	1	12.78		0.60		-0.36		2		947
	−25 05126				7 47 30	−25 57.6	1	11.36		0.29		0.10		1	A5 V	947
		Orsatti 113			7 47 31	−30 32.2	1	12.54		0.33		-0.33		2		1766
63733	−18 02040				7 47 32	−18 52.8	1	7.94		1.75		2.00		9	S	3039
		Orsatti 114			7 47 32	−32 43.2	1	13.57		0.99		0.00		2		1766
63868	−40 03512				7 47 32	−40 34.5	4	6.50	.005	-0.15	.006	-0.68	.000	13	B3 V	401,540,976,2012
63754	−19 02085	HR 3048	⋆	AB	7 47 33	−20 04.7	2	6.55	.005	0.58	.000			7	G0 V	15,2012
		Pup sq 5 # 33			7 47 33	−26 11.8	1	11.08		0.94		0.32		1	F2 III	947
		Pup sq 5 # 101			7 47 33	−26 40.2	1	11.26		0.11		-0.16		1	B9 V	947
		Pup sq 5 # 34			7 47 35	−26 12.9	1	12.28		0.42		0.12		2	F3 p	947
		Orsatti 115			7 47 35	−31 41.5	1	11.46		0.65		-0.17		2		1766
		Pup sq 5 # 38			7 47 36	−26 17.1	1	12.18		0.20		-0.16		2	B9 V	947
	−26 05047				7 47 36	−26 34.4	1	11.56		0.28		0.20		2	A3 V	947
		Pup sq 1 # 8			7 47 36	−28 38.	1	13.84		0.88		0.68		3		124
		Orsatti 116			7 47 36	−32 56.7	1	13.08		0.95		-0.13		3		1766
		Pup sq 5 # 35			7 47 37	−26 13.6	1	12.23		0.22		-0.02		2	A0 V	947

Table 1

HD	DM	Other Id	N	Rem	α_{1950}	δ_{1950}	S	V	σ_V	B−V	σ_{B-V}	U−B	σ_{U-B}	n	Spectrum	References
63867	−38 03690	NGC 2451 - 305			7 47 37	−38 19.7	1	9.19		0.32				2	F0 V	2037
		Pup sq 5 # 68			7 47 38	−25 48.0	1	12.44		0.36		0.09		1	A7	947
		Pup sq 5 # 24			7 47 38	−26 07.1	1	11.70		1.99		2.27		2	C3.4:	947
	−26 05050				7 47 38	−26 31.4	1	11.05		1.16		1.11		2	K3 III:	947
		Orsatti 117			7 47 38	−30 04.7	1	12.50		0.49		-0.30		2		1766
		Pup sq 5 # 74			7 47 40	−25 55.9	1	11.78		1.25		1.40		1		947
		Pup sq 5 # 36			7 47 40	−26 16.6	1	12.60		0.66		0.22		2		947
		LF12 # 119			7 47 41	−22 50.7	1	10.68		1.24		1.13		2	K2 III	137
		Pup sq 5 # 99			7 47 41	−26 33.1	1	11.33		0.33		0.08		2		947
	−26 05053				7 47 41	−26 33.4	1	10.77		0.95		0.61		2	K3 V	947
63852	−32 04451 HR 3052				7 47 41	−33 09.7	2	5.60	.000	1.61	.000	1.95		6	M0 III	2009,3005
63648	+15 01676				7 47 42	+14 58.2	1	7.30		0.39		-0.02		3	F2	1648
		Orsatti 118			7 47 42	−32 57.0	1	12.86		0.94		-0.11		3		1766
63949	−46 03460 HR 3058, QS Pup				7 47 42	−46 43.9	4	5.83	.014	-0.14	.004	-0.85	.008	12	B2 III	15,401,1732,2012
		Orsatti 119			7 47 43	−30 43.9	1	13.54		0.74		0.41		2		1766
	−31 05112	IDS07457S3120		AB	7 47 43	−31 27.7	1	10.30		0.20		0.10		2		126
63884	−42 03566				7 47 43	−42 33.2	1	8.43		-0.07		-0.35		2	B7/8 II	401
63922	−46 03458 HR 3055			⋆ A	7 47 43	−46 14.8	4	4.11	.005	-0.19	.007	-1.01	.000	20	B0 III	15,1075,2012,8015
		Pup sq 5 # 32			7 47 44	−26 15.7	1	13.37		1.12		0.62		1		947
63850	−31 05115				7 47 45	−31 26.7	1	9.62		0.14		0.10		2	A0 V	126
63882	−38 03694 NGC 2451 - 306				7 47 45	−38 52.4	1	7.29		1.16				2	G6 III	2037
63752	−8 02096 HR 3047				7 47 46	−09 03.4	4	5.60	.004	1.44	.005	1.56	.000	18	K3 III	15,1256,2013,2029
		Pup sq 5 # 69			7 47 46	−25 47.4	1	11.99		1.02		0.64		1		947
		Pup sq 5 # 70			7 47 49	−25 48.4	1	13.07		0.34		0.30		1	B9	947
63610	+31 01676 IDS07447N3152		A		7 47 50	+31 44.5	1	6.86		0.18		0.09		3	A5	1084
63589	+33 01601 HR 3040			⋆ A	7 47 50	+33 21.7	3	6.03		0.15	.000	0.14	.010	9	A2 m	1022,1199,8071
62692	+79 00258				7 47 51	+78 52.1	1	8.92		0.40		0.00		2	F5	1502
	−13 02253 GR Pup				7 47 52	−13 25.9	1	10.38		2.54		4.40		1		864
63822	−19 02089 HR 3051				7 47 53	−19 23.8	1	6.12		1.26				4	K1 III	2031
	−25 05131				7 47 53	−25 57.6	1	9.84		0.59		0.16		1	G0 V	947
		Pup sq 5 # 76			7 47 54	−25 55.3	1	11.95		0.51		0.01		1	F5	947
63948	−44 03762 HR 3057				7 47 54	−44 37.5	2	6.31	.006	0.96	.002	0.72		6	K0 III	58,2009
63646	+25 01773				7 47 56	+25 14.3	1	8.61		0.42		-0.03		2	F5	1625
63631	+31 01677 IDS07447N3152		B		7 47 56	+31 44.7	1	7.75		0.18		0.06		3	A5	1084
		LF12 # 120			7 47 56	−24 16.8	1	12.16		0.08		0.02		1	A0 V	137
		G 90 - 23			7 47 57	+37 37.3	1	12.12		1.41		1.21		1		333,1620
		LOD 2+ # 18			7 47 57	−22 16.8	1	10.87		0.11		0.08		2		126
	−26 05057				7 47 57	−26 28.9	1	10.61		0.11		0.16		2	A2 V	947
63136	+73 00387				7 47 59	+72 52.2	1	8.68		0.40		0.07		2	G0	1733
	−26 05059				7 48 00	−26 29.4	1	10.76		0.44		-0.02		2	F5 III	947
		Pup sq 1 # 9			7 48 00	−28 50.	1	11.62		0.18		0.14		2		124
64067	−56 01442 HR 3062			⋆ AB	7 48 02	−56 17.0	3	5.59	.009	1.12	.008	0.84	.015	9	G5 II	58,404,2009
	−25 05135				7 48 03	−25 55.7	1	9.65		0.94		0.65		1	G8 III	947
64014	−48 03226				7 48 03	−48 30.8	2	9.34	.001	-0.11	.003	-0.86	.006	6	B2 II	540,976
		Pup sq 5 # 30			7 48 04	−26 15.3	1	12.14		0.13		-0.33		2	B5 V	947
		Pup sq 5 # 72			7 48 05	−25 49.7	1	11.96		0.66		0.05		1	G0	947
		Pup sq 5 # 26			7 48 05	−26 09.3	1	12.66		0.19		-0.38		1	B3 V	947
63881	−26 05062				7 48 05	−26 23.9	1	10.09		-0.03		-0.28		1	B9 III/IV	1376
		LOD 2+ # 15			7 48 06	−22 21.6	1	10.94		0.12		-0.05		2		126
	−26 05064				7 48 06	−26 13.9	1	11.24		0.68		0.27		2	F7 V	947
63967	−39 03663				7 48 06	−39 37.7	1	8.26		0.20		0.06		2	B8 Ib/II	401
62613	+80 00238 HR 2997				7 48 07	+80 23.7	4	6.56	.009	0.72	.011	0.27	.010	17	G8 V	196,1067,1197,1758
		Haf 16 - 15			7 48 08	−25 20.9	1	12.66		0.50		0.04		2		183
	−26 05065				7 48 08	−26 28.9	1	10.85		0.19		-0.06		2	B9 V	947
		Haf 16 - 9			7 48 09	−25 18.5	1	12.32		-0.04		-0.58		2		183
		Pup sq 5 # 71			7 48 09	−25 48.7	1	12.75		0.31		0.06		1	A7 V	947
	−28 04945				7 48 09	−28 49.8	1	11.70		0.47		0.05		2		124
	−44 03764				7 48 09	−44 28.4	1	9.76		0.80				4	K0	2012
63776	+2 01804				7 48 10	+02 35.9	1	8.10		0.41		-0.06		1	F8	1272
63799	+3 01818 HR 3050			⋆ AB	7 48 10	+03 24.3	2	6.17	.005	1.12	.000	1.02	.005	7	K1 III	15,1078
63988	−40 03527				7 48 10	−40 47.2	1	7.07		-0.07		-0.29		2	B8 V	401
		Haf 16 - 11			7 48 11	−25 19.4	1	14.70		0.50				3		183
		Haf 16 - 14			7 48 11	−25 21.2	1	13.47		0.37		-0.04		3		183
		Haf 16 - 7			7 48 12	−25 19.	1	13.57		-0.01		-0.22		2		183
		Haf 16 - 8			7 48 12	−25 19.	1	14.43		0.42		-0.06		2		183
		Haf 16 - 10			7 48 12	−25 19.	1	14.13		0.49		0.09		2		183
		Haf 16 - 12			7 48 12	−25 19.	1	14.30		0.25		0.18		2		183
		Haf 16 - 13			7 48 12	−25 20.6	1	11.62		0.05		-0.60		2		183
	−25 05136				7 48 12	−26 09.7	2	9.64	.005	1.16	.014	1.07	.014	4	K0 III	137,947
		Haf 16 - 5			7 48 13	−25 18.9	1	13.76		0.09		-0.29		2		183
		Haf 16 - 6			7 48 14	−25 19.5	1	12.21		-0.01		-0.56		2		183
63896	−26 05067				7 48 14	−26 26.1	3	8.58	.009	-0.09	.013	-0.39	.013	7	B8 III	947,1376,2014
		LOD 2+ # 16			7 48 15	−22 24.7	1	10.83		0.29		0.15		2		126
	−23 06320				7 48 16	−24 06.9	1	10.72		0.20		0.09		2	A7 V	137
		Haf 16 - 2			7 48 16	−25 18.6	1	13.30		1.13		0.84		2		183
		Haf 16 - 3			7 48 17	−25 19.4	1	13.35		0.03		-0.37		2		183
		Haf 16 - 4			7 48 17	−25 20.0	1	13.36		0.03		-0.38		1		183
64013	−41 03437				7 48 17	−42 01.1	1	8.16		-0.01		-0.23		2	B9 V	401
	−23 06322				7 48 18	−23 43.3	1	9.55		1.15		1.17		1	K2 III	137
		Orsatti 128			7 48 19	−31 34.7	1	12.33		0.44		-0.24		2		1766

HD	DM	Other Id	N Rem	α_{1950}	δ_{1950}	S	V	σ_V	B−V	σ_{B-V}	U−B	σ_{U-B}	n	Spectrum	References
64011	−39 03672			7 48 20	−39 39.8	1	8.35		−0.04		−0.41		1	A5 III	401
		Haf 16 - 1		7 48 21	−25 19.1	1	12.69		0.10		−0.09		2		183
63772	+21 01698			7 48 23	+20 57.9	1	8.95		0.35				11		1351
		Pup sq 2 # 14		7 48 23	−27 16.	1	12.03		0.50		−0.02			B7 V:	1376
63964	−28 04950			7 48 23	−28 56.4	4	8.20	.007	0.82	.004	0.51	.015	12	F3/5 III	16,409,1012,8100
63965	−28 04951			7 48 23	−29 07.9	1	9.35		0.20		0.12		2	B9/A0 III	124
64185	−59 00908	NGC 2516 - 224	⋆A	7 48 23	−60 09.5	3	5.78	.005	0.41	.009	0.02		10	F3 IV/V	15,2012,1683
		LF12 # 124		7 48 25	−25 01.7	1	12.86		0.19		0.04		1	A2 V	137
	−26 05074			7 48 25	−26 28.0	1	10.49		0.22		0.28		2	B9 V	947
	−27 04612	ApJ113,309 # 85		7 48 25	−27 17.4	1	11.08		0.43		−0.42		1	B1 V	1376
63912	−22 02048			7 48 26	−22 23.5	1	9.27		0.48		0.05		2	F6 V	126
	−25 05140			7 48 26	−25 11.3	1	10.88		0.12		0.04		3	A2 V	137
	+29 01625			7 48 27	+29 02.2	1	10.59		0.55		−0.02		1	A0	333
		Pup sq 2 # 15		7 48 27	−24 44.	1	11.41		0.07		−0.60		1	B2.5V	1376
	−24 06051			7 48 28	−24 28.5	1	11.51		0.07		−0.02		2	B8 V	137
63985	−29 05012			7 48 31	−29 11.3	1	8.98		0.94		0.72		1	K0 III	1470
64184	−59 00907			7 48 31	−59 15.0	2	7.50	.010	0.67	.005	0.17		8	G5 V	158,2012
63894	−10 02253	HR 3054		7 48 33	−11 00.0	1	6.16		1.13				4	K0	2007
		Orsatti 134		7 48 34	−30 57.6	1	12.27		0.54		−0.50		2		1766
63962	−22 02049			7 48 35	−23 02.7	1	9.21		1.04		0.87		1	K0 III	137
	−23 06332			7 48 36	−23 25.6	1	10.69		0.00		−0.21		1	B8 V	137
64028	−35 03935			7 48 36	−36 06.8	2	7.27	.005	−0.07	.008	−0.27	.002	4	B8/9 V	401,3043
63586	+55 01228	HR 3039		7 48 37	+55 20.3	1	6.37		−0.01		−0.05		2	A0 Vn	1733
		LF12 # 129		7 48 41	−25 31.1	1	13.44		0.14		−0.13		2	B9 V	137
	−29 05015			7 48 41	−29 14.3	1	10.40		0.10		0.03		2		124
		Pup sq 1 # 14		7 48 42	−28 39.	1	13.21		0.20		−0.31		2		124
		Pup sq 5 # 89		7 48 43	−26 26.6	1	11.50		0.19		0.20		2	A1 V	947
64005	−25 05152			7 48 44	−25 57.7	1	8.66		1.51		1.82		2	K3/4	947
	−25 05153			7 48 45	−26 06.7	1	11.18		0.97		0.63		2	G8 III	947
	−25 05155			7 48 45	−26 09.4	1	9.58		1.37		1.38		2	K0 III	947
	−26 05083			7 48 46	−26 29.0	2	10.43	.010	0.00	.000	−0.09	.005	3	A0 V	137,947
64066	−35 03940			7 48 46	−36 07.8	1	8.15		−0.02		−0.60		3	B2 IV/V	3043
64026	−25 05156			7 48 47	−26 01.1	3	8.52	.012	−0.11	.008	−0.45	.007	7	B8 Ib/II	540,947,976
		Pup sq 5 # 90		7 48 48	−26 28.0	1	11.80		0.12		0.10		2	A3 Vp	947
	−38 03714			7 48 48	−38 39.9	1	10.37		2.97		3.60		1	N	864
		CCS 923		7 48 49	−47 45.0	1	11.56		5.02				1	N	864
	−25 05158			7 48 50	−26 00.8	1	9.31		0.89		0.55		2	G5 III	947
	−26 05086			7 48 51	−26 25.5	1	10.84		1.00		0.67		2		947
		Pup sq 5 # 81		7 48 52	−26 01.1	3	10.35	.019	−0.02	.010	−0.72	.018	4	B1 V	947,1295,1376
64086	−32 04472			7 48 52	−32 52.1	2	8.80	.010	−0.07	.010	−0.41	.005	4	B8 V	124,401
64042	−24 06060	HR 3060	⋆AB	7 48 53	−24 24.0	3	6.44	.007	−0.01	.000			11	B9.5V	15,2013,2029
		Pup sq 1 # 11		7 48 54	−28 51.	1	13.40		0.47		−0.14		2		124
		Pup sq 5 # 82		7 48 56	−26 02.6	1	11.42		0.12		−0.08		2	B9 V	947
64138	−39 03689			7 48 59	−39 35.8	2	9.11	.015	−0.02	.007	−0.55	.010	4	B3 III/IV	401,3043
		Pup sq 5 # 83		7 49 00	−26 02.4	1	11.77		0.24		0.10		2	A3 V	947
63889	+19 01854	HR 3053		7 49 02	+19 27.3	2	6.01	.013	1.13	.006	1.05		5	K1 III	71,1501
		LF12 # 131		7 49 02	−23 51.1	1	12.52		1.47		1.25		2		137
		Orsatti 136		7 49 02	−30 32.0	1	13.26		0.74		−0.16		1		1766
64225	−50 03004	HR 3071		7 49 02	−50 22.8	2	5.91	.000	1.09	.000	1.02		5	K2 III	2035,3005
		Pup sq 5 # 85		7 49 04	−26 05.0	1	11.66		0.99		0.73		2	K0 II	947
64102	−34 04003			7 49 04	−34 13.2	1	8.48		−0.12		−0.63		2	B3/5 V	401
64202	−47 03353			7 49 04	−48 09.6	2	8.33	.010	−0.07	.005	−0.16	.005	6	B5/7 II	164,401
64320	−59 00910	NGC 2516 - 225	⋆	7 49 04	−59 55.4	4	6.73	.008	1.24	.000	1.16		13	K0 II	15,1075,1683,2007
63975	+2 01808	HR 3059		7 49 06	+01 53.8	7	5.13	.013	−0.12	.008	−0.46	.021	20	B8 II	3,15,1078,1079,1363*
		ApJ144,259 # 24		7 49 06	+03 08.	1	17.18		−0.21		−1.15		1		1360
		LF12 # 132		7 49 06	−23 50.3	1	11.50		0.98		0.86		2	K0 III	137
		Orsatti 137		7 49 06	−31 44.3	1	13.47		0.66		−0.12		1		1766
64181	−44 03780	HR 3069		7 49 07	−44 27.1	1	6.45		0.90				4	G6 III	2006
	−25 05166			7 49 10	−25 56.2	1	10.00		0.15		−0.01		1	A2 V	137
64136	−31 05148			7 49 12	−31 29.6	1	9.66		0.04		−0.01		2	A0 IV	126
		WLS 736 45 # 8		7 49 13	+44 52.1	1	13.06		0.89		0.52		2		1375
	−23 06357			7 49 13	−23 47.4	1	9.97		1.54		1.71		1	K4 III	137
	−25 05169			7 49 14	−26 03.2	2	10.15	.002	0.07	.007	−0.75	.004	5		540,976
	−25 05168	LSS 821		7 49 14	−26 03.5	2	10.14	.015	0.09	.005	−0.81	.010	4	B1 III	409,947
64249	−48 03244			7 49 14	−48 24.9	1	7.92		−0.01		−0.23		2	B9 II	401
		LF12 # 135		7 49 15	−26 11.1	1	13.43		1.49		1.67		1		137
		LF12 # 136		7 49 16	−22 46.7	1	9.97		1.02		0.72		1	K0 III	137
		G 112 - 50		7 49 21	+00 08.0	2	13.32	.070	1.68	.005	1.12	.030	7		203,3073
64077	−12 02179	HR 3063		7 49 21	−12 41.4	1	6.37		0.39				4	F3 IV	2007
64135	−24 06068			7 49 21	−24 29.0	1	9.49		0.40		−0.03		1	F3 V	137
64157	−31 05154			7 49 21	−31 30.3	1	8.06		−0.03		−0.12		2	B9.5/A0 V	126
64484	−65 00827	HR 3081		7 49 21	−66 04.1	4	5.78	.007	−0.04	.000	−0.16	.004	24	B9 V	15,1075,1637,2038
	−38 03728	NGC 2477 - 6501		7 49 23	−38 29.3	1	10.84		0.68		0.42		3		259
		Pup sq 1 # 6		7 49 24	−29 02.	1	12.55		0.32		0.04		2		124
64096	−13 02267	HR 3064	⋆A	7 49 27	−13 45.8	8	5.17	.016	0.60	.009	0.06	.004	23	G2 V	15,292,1007,1013,1197*
		Pup sq 2 # 16		7 49 28	−24 36.	1	10.61		0.59		0.13		1		211
	−38 03730	NGC 2477 - 7504		7 49 28	−38 21.6	1	10.72		0.17		0.05		4		259
64052	+3 01824	HR 3061, BC CMi		7 49 30	+03 24.5	3	6.32	.025	1.58	.016	1.61	.000	8	M4 III	15,1415,3042
64152	−20 02235	HR 3068		7 49 32	−21 02.7	4	5.62	.006	0.96	.002	0.70		13	K0 III	15,2013,2028,3077
	−25 05180			7 49 34	−25 24.5	1	10.25		−0.06		−0.69		2	B5	540

Table 1   427

HD	DM	Other Id	N Rem	α_{1950}	δ_{1950}	S	V	σ_V	B−V	σ_{B-V}	U−B	σ_{U-B}	n	Spectrum	References
		LF12 # 138		7 49 39	−25 02.1	1	11.93		1.26		1.24		1		137
		NGC 2477 - 6502		7 49 39	−38 32.1	1	11.46		1.44		1.45		3		259
64199	−25 05183			7 49 41	−25 41.3	1	8.92		-0.08		-0.43		1	B8 IV	1376
		NGC 2477 - 7503		7 49 42	−38 19.9	1	12.47		1.24		0.99		2		259
64287	−42 03601	HR 3074		7 49 42	−42 58.0	3	6.31	.004	-0.18	.004	-0.75	.005	11	B2 IV-V	15,1075,2012
	−49 03115			7 49 42	−50 10.8	1	10.06		1.22				1	K2	1705
64267	−37 03981			7 49 43	−37 30.3	1	8.48		0.10		-0.28		2	B8 V	3043
	−25 05184			7 49 44	−25 40.3	3	9.79	.005	0.01	.005	-0.56	.004	6	B3 IV	540,976,1376
64221	−29 05040	IDS07477S2925	AB	7 49 44	−29 32.2	1	8.31		0.10		0.09		2	A2 V	124
64318	−46 03491			7 49 44	−47 05.2	3	6.53	.006	-0.12	.007	-0.69	.009	9	B3 IV/V	401,1732,2012
63991	+26 01665			7 49 45	+26 03.5	1	8.60		1.04		0.90		1	K0	1746
		NGC 2467 - 2		7 49 45	−26 09.1	2	12.00	.020	0.11	.016	-0.08	.040	4		1376,1681
	−28 04985	LSS 823		7 49 45	−28 14.7	3	10.18	.005	0.30	.003	-0.62	.032	4	O8	211,540,1766
64425	−55 01366			7 49 45	−55 27.9	1	9.20		1.22		1.03		1	K0p Ba	565
64200	−26 05106	NGC 2467 - 1		7 49 46	−26 21.9	2	9.59	.004	0.23	.026	0.33	.074	5	A2 III	1376,1681
	−31 05163			7 49 46	−31 26.4	1	10.35		0.49		0.00		2		126
		NGC 2477 - 7502		7 49 46	−38 20.9	2	13.79	.087	-0.15	.014	-1.00	.032	18	sdOB	259,3028
64266	−34 04019			7 49 47	−34 30.4	2	9.31	.020	0.04	.015	0.02	.000	4	A2 V	124,126
64219	−24 06078			7 49 48	−24 59.7	2	9.73	.007	-0.06	.004	-0.76	.019	4	B2 II	540,976
		G 91 - 17		7 49 50	+20 29.9	1	12.42		1.39		0.99		1		333,1620
	−24 06080			7 49 51	−24 23.0	1	9.52		1.15		0.97		1	K0 III	137
64072	+19 01860			7 49 52	+18 57.4	1	8.42		0.25		0.12		2	A3	1648
	−38 03736	NGC 2477 - 7501		7 49 52	−38 18.4	1	10.51		0.99		0.65		3		259
		NGC 2467 - 4		7 49 54	−26 17.4	2	12.87	.012	0.34	.020	0.28	.051	4		1376,1681
		NGC 2467 - 3		7 49 54	−26 18.4	2	12.26	.012	0.49	.012	0.01	.004	4		1376,1681
		NGC 2477 - 7336		7 49 54	−38 21.5	1	14.30		0.70		0.34		2		259
		NGC 2467 - 7		7 49 56	−26 08.9	1	12.99		0.25		0.00		1		1376
		NGC 2467 - 6		7 49 56	−26 14.9	2	12.78	.021	0.36	.013	0.22	.017	3		1376,1681
64316	−39 03706			7 49 56	−39 56.4	1	8.27		0.49				4	F5 V	2012
	−25 05187	NGC 2467 - 8		7 49 57	−26 08.5	1	10.78		1.82		1.70		1		1376
		NGC 2477 - 7338		7 49 57	−38 21.6	1	14.86		0.72		0.29		3		259
		NGC 2477 - 7328		7 49 58	−38 20.7	1	14.74		0.66		0.23		2		259
		NGC 2467 - 9		7 49 59	−26 07.4	2	11.01	.943	0.17	.137	-0.19	.170	2		1376,1681
	−26 05111	KY Pup		7 49 59	−26 37.4	1	11.56		0.01		-0.81		1		409
		NGC 2477 - 7324		7 49 59	−38 20.1	1	13.41		0.59		0.36		2		259
	−45 03511	CCS 931		7 49 59	−46 08.1	1	10.30		4.13				1	N	864
64285	−29 05045			7 50 00	−29 42.7	1	10.63		0.58		0.04		2	A0 IV	409
64301	−31 05165	IDS07480S3123	AB	7 50 00	−31 30.4	1	7.72		-0.10		-0.41		2	B7 V	126
		NGC 2477 - 6288		7 50 00	−38 26.1	1	11.36		1.06		0.71		3		259
64238	−14 02250	HR 3073, LSS 822		7 50 01	−14 43.0	3	5.70	.012	0.37	.004	0.21		7	F1 Ia	2006,6009,8100
		NGC 2477 - 8502		7 50 01	−38 17.7	1	13.38		0.58		0.41		2		259
64365	−42 03610	HR 3078, QU Pup		7 50 01	−42 45.5	3	6.04	.007	-0.19	.007	-0.85		9	B2 III	15,401,2027
		NGC 2467 - 10		7 50 02	−26 14.1	1	14.00		0.54		-0.23		1		1376
64283	−27 04648	LSS 825		7 50 02	−27 32.5	1	10.29		0.32		-0.50		2	A1 V	540
		NGC 2477 - 8379		7 50 02	−38 18.1	1	13.82		0.61		0.35		2		259
		NGC 2477 - 7266		7 50 02	−38 24.4	1	12.23		1.19		0.93		3		259
	+22 01802			7 50 03	+22 40.9	1	10.94		1.51				1	M0	1017
		NGC 2477 - 7311		7 50 03	−38 19.7	2	11.64	.020	1.19	.005	0.89	.005	5		259,259
		NGC 2477 - 7219		7 50 03	−38 21.8	1	14.09		0.66		0.33		2		259
		NGC 2477 - 7223		7 50 03	−38 22.3	1	13.34		0.73		0.21		2		259
		NGC 2467 - 11		7 50 04	−26 12.5	2	11.77	.009	0.91	.004	0.55	.017	3		1376,1681
		NGC 2477 - 7242		7 50 04	−38 22.8	1	12.76		1.22		0.96		2		259
		NGC 2477 - 6025		7 50 04	−38 25.0	1	13.38		0.59		0.29		2		259
64165	+3 01827	IDS07475N0339	AB	7 50 05	+03 30.9	1	4.99		0.12		0.11		3	A4 V	1022
		NGC 2477 - 6029		7 50 05	−38 25.8	1	12.84		0.55		0.32		2	A8	259
	−25 05191	NGC 2467 - 95		7 50 06	−25 22.2	2	11.10	.000	0.20	.004	-0.78	.000	2	B1 Vnne	1295,1376
		NGC 2467 - 13		7 50 06	−26 14.3	2	12.80	.013	0.41	.021	-0.01	.017	3		1376,1681
		NGC 2467 - 12		7 50 06	−26 20.8	2	12.47	.024	0.15	.004	-0.10	.032	4		1376,1681
		ApJ113,309 # 88		7 50 06	−27 31.	1	10.30		0.32		-0.49		1		211
	−31 05166			7 50 06	−31 21.2	1	10.36		1.26		0.94		2		126
	−23 06383			7 50 07	−23 51.4	1	11.49		0.00		-0.20		1	B8 V	137
	−26 05112	NGC 2467 - 14		7 50 07	−26 13.2	2	11.29	.017	0.77	.013	0.12	.128	3		1376,1681
		NGC 2477 - 8501		7 50 07	−38 16.2	1	13.63		0.69		0.20		2		259
		NGC 2477 - 6040		7 50 07	−38 26.6	1	11.88		1.00		0.76		3		259
64259	−13 02270	IDS07478S1336	AB	7 50 09	−13 43.9	3	6.61	.005	1.11	.007	0.99	.010	8	K0 III	1003,1075,3040
64263	−17 02198			7 50 09	−17 53.3	1	8.29		0.05		0.00		2	A0 V	3039
	−25 05192	NGC 2467 - 16		7 50 09	−26 06.4	2	10.99	.017	0.16	.034	-0.12	.047	3		1376,1681
		NGC 2467 - 15		7 50 09	−26 16.2	1	11.60		0.08		0.00		1		1376
		NGC 2477 - 6252		7 50 09	−38 28.5	1	13.15		0.55		0.34		3	A9	259
	−38 03745	NGC 2477 - 6254		7 50 09	−38 29.1	1	10.83		0.96		0.59		4		259
		NGC 2477 - 8355		7 50 10	−38 17.3	1	13.93		0.65		0.39		3		259
		LF12 # 141		7 50 11	−23 54.4	1	12.53		0.10		0.02		1	A0 V	137
		NGC 2477 - 5268		7 50 11	−38 30.9	1	13.38		0.80		0.37		3		259
64191	+1 01932	AD CMi		7 50 12	+01 43.7	1	9.22		0.25				1	F0 III	1497
		NGC 2467 - 17		7 50 12	−26 15.8	1	11.32		0.19		0.00		1		1376
		NGC 2477 - 6073		7 50 12	−38 26.7	1	13.59		0.48		0.26		2		259
		NGC 2477 - 5254		7 50 13	−38 30.2	1	12.62		0.57		0.32		1		259
		NGC 2477 - 8256		7 50 14	−38 20.5	2	12.05	.000	1.22	.000	0.89	.000	4		259,3043
		NGC 2477 - 8039		7 50 14	−38 21.0	2	12.33	.000	1.20	.000	0.95	.000	5		259,3043
		NGC 2477 - 6086		7 50 14	−38 27.5	1	14.63		0.65		0.18		2		259

HD	DM	Other Id	N Rem	α_{1950}	δ_{1950}	S	V	σ_V	B–V	σ_{B-V}	U–B	σ_{U-B}	n	Spectrum	References
		NGC 2477 - 6221		7 50 14	−38 29.2	1	14.01		0.61		0.29		3		259
		NGC 2467 - 26		7 50 15	−26 05.3	2	12.36	.000	0.34	.034	0.08	.013	3		1376,1681
		NGC 2467 - 18		7 50 15	−26 19.0	2	13.06	.028	0.47	.024	0.46	.292	2		1376,1681
64300	−25 05196			7 50 16	−25 48.4	1	9.30		0.28		0.12		2	A8/9 V	137
		NGC 2467 - 21		7 50 16	−26 01.9	2	11.14	.024	0.10	.000	-0.58	.036	4		1376,1681
64315	−26 05115	NGC 2467 - 19		7 50 16	−26 18.0	7	9.23	.027	0.22	.015	-0.74	.020	18	O6 Vnn((f))	540,976,1011,1012*
		NGC 2477 - 8345		7 50 16	−38 17.4	1	13.87		0.82		0.36		2		259
		NGC 2477 - 6069		7 50 16	−38 26.4	1	13.37		0.94		0.49		2		259
64299	−22 02058			7 50 17	−23 10.0	1	10.15		0.09		0.07		1	A1 V	137
	−25 05195	NGC 2467 - 25		7 50 17	−26 00.6	2	11.12	.016	0.44	.012	-0.05	.008	4		1376,1681
		NGC 2467 - 20		7 50 17	−26 12.8	1	13.54		0.25		-0.26		1		1376
		Pup sq 3 # 92		7 50 17	−29 36.3	1	12.50		0.57		-0.06		2	B5 V	409
64363	−31 05168			7 50 17	−31 21.4	1	9.20		0.21		0.12		2	A8 V	126
		NGC 2477 - 8069		7 50 17	−38 20.6	1	13.65		0.43		0.34		2		259
		NGC 2467 - 24		7 50 18	−26 09.4	1	12.27		1.44		1.25		1		1376
		NGC 2467 - 23		7 50 18	−26 11.0	1	12.86		0.90		0.43		1		1376
		NGC 2467 - 22		7 50 18	−26 16.0	1	11.98		0.28		0.12		1		1376
		NGC 2477 - 8326		7 50 18	−38 16.1	1	14.49		1.89				1		259
		NGC 2477 - 8245		7 50 18	−38 19.1	1	13.92		0.62		0.34		2		259
		NGC 2477 - 8067		7 50 18	−38 20.9	1	13.67		0.62		0.39		2		259
64625	−62 00902			7 50 18	−62 42.0	1	8.30		1.32		1.35		7	K1 III	1704
		NGC 2467 - 29		7 50 19	−26 13.1	1	12.08		1.16		0.93		1		1376
		NGC 2477 - 8244		7 50 19	−38 18.9	1	13.40		0.54		0.38		2		259
		NGC 2477 - 8046		7 50 19	−38 21.9	1	13.34		0.64		0.34		2		259
		CCS 936		7 50 19	−46 21.6	1	9.84		3.29		3.10		1	R?	864
		LP 543 - 39		7 50 20	+02 34.8	1	11.42		1.10		0.97		1		1696
64235	−5 02280	HR 3072	★ AB	7 50 20	−05 17.9	3	5.76	.008	0.41	.005	-0.01	.010	14	F5 IV	15,1256,2012
64298	−21 02155			7 50 20	−21 52.2	2	9.06	.001	-0.08	.003	-0.70	.005	4	B2/3 (V)nn	540,976
		NGC 2467 - 31		7 50 20	−25 59.9	1	11.37		0.37		0.03		1		1376
	−26 05116	NGC 2467 - 27		7 50 20	−26 21.3	2	10.96	.000	1.14	.020	0.93	.055	4		1376,1681
		Pup sq 3 # 91		7 50 20	−29 29.9	1	12.28		0.40		0.08		1	B8 V	409
64361	−29 05057			7 50 20	−29 30.1	1	9.79		0.10		-0.12		2	B9 V	409
		NGC 2477 - 8273		7 50 20	−38 20.1	1	13.57		0.63		0.37		2		259
		NGC 2477 - 5047		7 50 20	−38 27.6	1	11.61		1.25		1.07		7		259
		NGC 2467 - 30		7 50 21	−26 08.8	2	11.66	.012	0.13	.012	-0.37	.008	4		1376,1681
		NGC 2467 - 123		7 50 21	−26 18.0	1	12.54		0.28		0.00		2		1681
64090	+31 01684	G 90 - 25	★ A	7 50 22	+30 45.7	9	8.30	.021	0.62	.012	-0.13	.016	18	G2 VI	22,516,792,908,1003,1620*
	−26 05117	NGC 2467 - 28		7 50 22	−26 31.6	1	11.29		0.44		-0.50		1	B1 V	1376
		Orsatti 147		7 50 22	−30 43.5	1	12.02		0.71		0.27		3		1766
		NGC 2477 - 8073		7 50 22	−38 21.0	1	13.44		0.52		0.32		2		259
		NGC 2477 - 5237		7 50 22	−38 31.5	1	13.79		0.63		0.35		2		259
	−18 02065			7 50 23	−18 27.5	2	9.63	.016	0.95	.012	0.52		4		1594,6006
		NGC 2467 - 105		7 50 23	−26 13.2	1	12.10		0.05		0.10		2		1681
		NGC 2467 - 106		7 50 23	−26 13.5	1	12.50		0.55		0.05		2		1681
		NGC 2467 - 32		7 50 23	−26 19.4	1	13.10		0.36		0.18		1		1376
		NGC 2477 - 5019		7 50 23	−38 26.7	1	17.04		0.92		0.31		2		259
		NGC 2477 - 5238		7 50 23	−38 31.3	1	12.54		1.31		1.03		2		3043
		Haf 18 - 3048		7 50 24	−26 14.	1	14.55		0.16		0.18		1		398
		Haf 18 - 3051		7 50 24	−26 14.	1	14.56		0.38		0.23		2		398
		Haf 18 - 3056		7 50 24	−26 14.	1	14.95		0.58		0.01		2		398
		Haf 18 - 3057		7 50 24	−26 14.	1	14.67		0.56		-0.09		2		398
		Haf 18 - 3058		7 50 24	−26 14.	1	14.69		0.42		-0.03		2		398
		Haf 18 - 3059		7 50 24	−26 14.	1	14.40		0.12		0.08		2		398
		Haf 18 - 3060		7 50 24	−26 14.	1	14.43		0.20		0.07		2		398
		Haf 18 - 3060a		7 50 24	−26 14.	1	11.95		0.15		-0.71		5		398
		Haf 18 - 3062		7 50 24	−26 14.	1	14.10		0.51		0.30		2		398
		Haf 18 - 3063		7 50 24	−26 14.	1	13.56		1.33		1.36		2		398
		Haf 18 - 3064		7 50 24	−26 14.	1	13.80		0.46		0.31		2		398
		Haf 18 - 3065		7 50 24	−26 14.	1	14.64		0.54		0.12		2		398
		Haf 18 - 3066		7 50 24	−26 14.	1	14.52		0.95		-0.02		2		398
		Haf 18 - 3067		7 50 24	−26 14.	1	12.01		0.38		-0.55		4		398
		Haf 18 - 3070		7 50 24	−26 14.	1	13.37		0.44		-0.49		3		398
		Haf 18 - 3073		7 50 24	−26 14.	1	14.59		0.61		0.00		2		398
		Haf 18 - 3074		7 50 24	−26 14.	1	12.88		0.18		-0.12		2		398
		Haf 18 - 3075		7 50 24	−26 14.	1	14.55		0.53		-0.39		2		398
		Haf 18 - 3077		7 50 24	−26 14.	1	12.97		1.52		1.20		2		398
		Haf 18 - 3078		7 50 24	−26 14.	1	14.39		0.35		-0.34		2		398
		Haf 18 - 3081		7 50 24	−26 14.	1	11.18		0.30		-0.64		5		398
		Haf 18 - 3082		7 50 24	−26 14.	1	14.92		0.44		-0.34		2		398
64379	−34 04036	HR 3079	★ AB	7 50 24	−34 34.7	5	5.02	.005	0.46	.011	-0.05	.018	14	F5 V	15,404,1013,2012,3077
		NGC 2477 - 7007		7 50 24	−38 24.5	1	11.94		0.58		0.34		8		259
		NGC 2477 - 5025		7 50 24	−38 27.1	1	16.35		0.89		0.31		2		259
		NGC 2477 - 5035		7 50 24	−38 27.5	1	12.27		1.20		0.99		3		259
		NGC 2477 - 5042		7 50 24	−38 27.8	1	15.62		0.86		0.28		4		259
	−26 05119	NGC 2467 - 34		7 50 25	−26 12.9	2	9.91	.004	1.10	.000	0.90	.016	4		1376,1681
		NGC 2477 - 5020		7 50 25	−38 26.7	1	14.19		0.66		0.21		4		259
64145	+27 01499	HR 3067		7 50 26	+26 53.8	5	4.97	.014	0.09	.005	0.10	.015	15	A3 V	1,15,1363,3023,8015
64336	−19 02106			7 50 26	−19 39.4	1	8.26		-0.06		-0.51		2	B3 V	3039
		NGC 2477 - 8310		7 50 26	−38 15.4	1	13.43		0.61		0.38		2		259
		NGC 2477 - 8077		7 50 26	−38 20.8	2	12.18	.000	1.24	.007	1.03	.005	4		259,3043

Table 1 429

HD	DM	Other Id	N Rem	α_{1950}	δ_{1950}	S	V	σ_V	B–V	σ_{B-V}	U–B	σ_{U-B}	n	Spectrum	References
		NGC 2477 - 8061		7 50 26	−38 21.4	1	16.19		0.80		0.25		3		259
		NGC 2477 - 8022		7 50 26	−38 22.5	1	14.57		0.83		0.43		3		259
		NGC 2477 - 5066		7 50 26	−38 28.7	1	13.56		0.59		0.29		4		259
		LF12 # 144		7 50 27	−25 06.6	1	11.69		0.10		-0.18		2	B8 V	137
		NGC 2477 - 5068		7 50 27	−38 28.3	1	17.04		1.09		0.39		2		259
		NGC 2477 - 5277		7 50 27	−38 30.3	1	16.59		1.03		0.43		2		259
		NGC 2467 - 33		7 50 28	−26 08.0	1	12.50		0.24		-0.14		1		1376
64360	−26 05122			7 50 28	−26 59.1	1	9.54		0.41		-0.48		1	K1/2 (III)	409
		NGC 2477 - 8221		7 50 28	−38 18.8	1	14.46		0.63		0.30		2		259
		NGC 2477 - 5071		7 50 28	−38 29.0	1	14.25		0.58		0.26		1		259
	−38 03753	NGC 2477 - 5223		7 50 28	−38 31.0	2	11.36	.052	1.43	.005	1.44	.005	8		259,3043
	+42 01777			7 50 29	+42 32.3	1	9.14		1.05		0.85		2	K0	1375
		NGC 2477 - 8220		7 50 29	−38 18.8	1	15.23		0.81		0.24		2		259
		NGC 2477 - 8019		7 50 29	−38 22.6	3	12.10	.011	1.26	.004	1.03	.018	4		259,259,3043
		NGC 2477 - 8018		7 50 29	−38 23.0	1	12.06		1.22		0.94		3		3043
		NGC 2477 - 8017		7 50 29	−38 23.1	1	12.57		1.17		0.89		1		259
		NGC 2467 - 91		7 50 30	−26 15.	1	10.87		0.06		-0.68		1		1295
		NGC 2477 - 1064		7 50 30	−38 20.3	1	13.59		0.66		0.12		4		259
		NGC 2477 - 3481		7 50 30	−38 25.	1	11.96		1.22		1.00		2		259
		NGC 2477 - 5102		7 50 30	−38 25.	1	17.40		0.66		0.51		2		259
		NGC 2477 - 5103		7 50 30	−38 25.	1	17.81		1.09		0.79		2		259
		NGC 2477 - 8354		7 50 30	−38 25.	1	13.80		0.66		0.44		3	A8 pec	259
		NGC 2477 - 9001	⋆ V	7 50 30	−38 25.	1	10.50		3.05				1		923
		NGC 2477 - 5424		7 50 30	−38 30.6	1	14.40		0.60		0.23		3		259
		NGC 2477 - 5205		7 50 30	−38 31.9	1	13.44		0.66		0.27		2		259
64440	−40 03579	HR 3080		7 50 30	−40 26.8	4	3.72	.013	1.03	.015	0.76	.025	14	K1/2 II	15,1075,2012,8015
		NGC 2467 - 35		7 50 31	−26 10.7	2	12.66	.000	0.19	.000	0.15	.061	2		1376,1681
	−26 05123			7 50 31	−26 14.2	1	10.00		1.05		1.04		1	K0 III	137
		NGC 2477 - 8306		7 50 31	−38 17.0	1	13.99		0.63		0.35		2		259
	−38 03754	NGC 2477 - 4004		7 50 31	−38 25.6	2	10.80	.005	1.25	.005	0.98	.000	5		259,3043
		NGC 2477 - 4064		7 50 31	−38 27.9	2	12.55	.000	1.20	.000	0.98	.000	5		259,3043
		NGC 2467 - 122		7 50 32	−26 14.0	1	11.85		1.04		0.67		2		1681
		NGC 2467 - 124		7 50 32	−26 14.8	1	11.15		0.31		-0.69		4		1681
		Orsatti 150		7 50 32	−30 04.4	1	12.90		0.72		-0.32		2		1766
	−38 03756	NGC 2477 - 1069		7 50 32	−38 20.6	2	9.81	.005	1.88	.000	2.24	.062	18		259,3043
		NGC 2477 - 4067		7 50 32	−38 28.5	2	11.41	.000	1.42	.000	1.36	.000	4		259,3043
		Pup sq 3 # 68		7 50 33	−28 33.5	1	13.12		0.08		0.14		1	A0	409
		NGC 2477 - 4019		7 50 33	−38 26.6	1	14.03		0.57		0.24		2		259
		NGC 2477 - 5204		7 50 33	−38 31.6	1	13.04		0.52		0.35		3	A7	259
		NGC 2467 - 36		7 50 34	−26 03.6	2	11.48	.013	0.67	.004	0.19	.013	3		1376,1681
64401	−29 05064			7 50 34	−29 38.2	1	9.92		0.03		-0.35		2	B8/9 V	409
		NGC 2477 - 4027		7 50 34	−38 26.8	1	12.20		1.15		0.92		2		3043
		NGC 2467 - 125		7 50 35	−26 15.5	1	11.92		0.16		-0.73		4		1681
		NGC 2477 - 1071		7 50 35	−38 20.6	1	13.86		0.64		0.27		1		259
64334	−17 02203			7 50 36	−17 17.3	1	9.72		0.03		-0.05		2	B8/9 III	3039
		Haf 19 - 4020		7 50 36	−26 07.	1	14.40		1.11		0.98		1		398
		Haf 19 - 4023		7 50 36	−26 07.	1	14.56		0.19		-0.21		1		398
		Haf 19 - 4024		7 50 36	−26 07.	1	13.12		1.27		1.29		1		398
		Haf 19 - 4026		7 50 36	−26 07.	1	15.18		0.64		0.05		1		398
		Haf 19 - 4028		7 50 36	−26 07.	1	14.18		0.41		0.30		2		398
		Haf 19 - 4030		7 50 36	−26 07.	1	14.95		0.32		0.09		2		398
		Haf 19 - 4032		7 50 36	−26 07.	1	13.93		1.03		0.54		1		398
		Haf 19 - 4034		7 50 36	−26 07.	1	15.53		0.58		-0.19		1		398
		Haf 19 - 4035		7 50 36	−26 07.	1	15.13		1.02		0.28		2		398
		Haf 19 - 4036		7 50 36	−26 07.	1	14.60		0.52		-0.01		2		398
		Haf 19 - 4037		7 50 36	−26 07.	1	14.56		0.13		-0.21		2		398
		Haf 19 - 4038		7 50 36	−26 07.	1	15.16		0.15		-0.12		2		398
		Haf 19 - 4047		7 50 36	−26 07.	1	14.42		0.31		-0.46		2		398
		Haf 19 - 4048		7 50 36	−26 07.	1	15.34		0.18		-0.11		2		398
		Haf 19 - 4049		7 50 36	−26 07.	1	15.13		0.16		-0.14		2		398
		Haf 19 - 4050		7 50 36	−26 07.	1	12.28		0.19		-0.61		4		398
		Haf 19 - 4052		7 50 36	−26 07.	1	12.58		0.28		-0.53		2		398
		Haf 19 - 4053		7 50 36	−26 07.	1	13.05		0.24		-0.61		2		398
		Haf 19 - 4055		7 50 36	−26 07.	1	13.84		0.33		-0.15		2		398
		Haf 19 - 4099		7 50 36	−26 07.	1	15.06		1.03		0.74		1		398
		NGC 2477 - 1263		7 50 36	−38 19.2	1	13.68		0.64		0.38		2		259
	−25 05200			7 50 37	−25 25.9	1	9.34		1.43		1.64		2	K3 III	137
		NGC 2477 - 4032		7 50 37	−38 27.3	1	14.46		0.69		0.29		2		259
		NGC 2477 - 4447		7 50 37	−38 29.8	1	12.85		0.50		0.34		2		259
64526	−48 03264			7 50 37	−48 44.3	1	8.37		0.01		-0.26		2	B8 V	401
63383	+77 00303			7 50 38	+77 42.6	1	6.91		0.23		0.13		2	A3	985
		NGC 2467 - 37		7 50 38	−26 14.0	2	12.46	.021	0.37	.030	0.09	.030	3		1376,1681
		NGC 2477 - 1262		7 50 38	−38 19.2	1	14.04		0.67		0.40		3		259
		NGC 2477 - 4257		7 50 38	−38 30.5	1	13.51		0.60		0.36		3		259
		NGC 2467 - 38		7 50 39	−26 13.4	2	12.77	.024	0.27	.009	0.09	.024	2		1376,1681
		NGC 2467 - 119		7 50 39	−26 18.4	1	12.49		0.49		0.03		2		1681
		NGC 2477 - 1261		7 50 39	−38 19.6	1	14.55		0.71		0.27		3		259
		NGC 2477 - 1080		7 50 39	−38 20.7	1	14.57		0.62		0.24		2		259
		NGC 2477 - 3034		7 50 39	−38 26.5	1	13.27		0.48		0.32		2		259
		NGC 2477 - 4031		7 50 39	−38 26.9	1	14.22		0.58		0.25		2		259

HD	DM	Other Id	N Rem	α_{1950}	δ_{1950}	S	V	σ_V	B–V	σ_{B-V}	U–B	σ_{U-B}	n	Spectrum	References
		NGC 2477 - 4045		7 50 39	−38 28.7	1	14.08		0.68		0.21		3		259
		NGC 2477 - 4267		7 50 39	−38 31.5	1	12.86		0.74		0.41		2	F3	259
64399	−24 06101	NGC 2467 - 96		7 50 40	−24 58.3	4	8.30	.005	0.01	.011	-0.69	.009	8	B2 V	1012,1295,1376,8100
		NGC 2467 - 39		7 50 40	−26 09.6	3	12.32	.019	0.11	.012	-0.71	.022	6		398,1376,1681
		NGC 2477 - 1272		7 50 40	−38 18.7	2	12.28	.000	1.18	.000	0.81	.000	4		259,3043
		NGC 2477 - 1044	⋆	7 50 40	−38 22.0	1	11.81		1.28		1.07		2		259
		NGC 2477 - 4043		7 50 40	−38 28.3	1	14.60		0.70		0.25		2		259
		G 50 - 9		7 50 41	+14 58.5	1	11.94		1.10		0.99		1		333,1620
		LF12 # 147		7 50 41	−25 56.3	1	13.12		1.29		0.97		1		137
		NGC 2477 - 1259		7 50 41	−38 20.2	1	13.25		0.58		0.37		2		259
		NGC 2477 - 3023		7 50 41	−38 25.2	1	13.09		0.51		0.26		2		259
	−25 05202	NGC 2467 - 40	⋆ AB	7 50 42	−26 08.9	3	10.86	.065	0.22	.015	-0.67	.039	9	B1.5V	398,1295,1376
		NGC 2477 - 1284		7 50 42	−38 17.9	1	13.19		0.81		0.46		2		259
		NGC 2477 - 1078		7 50 42	−38 20.5	1	13.27		0.54		0.36		4		259
		NGC 2477 - 4042		7 50 42	−38 28.3	1	14.13		0.60		0.25		3		259
		NGC 2467 - 42		7 50 43	−26 14.0	2	11.02	.009	1.11	.013	1.05	.013	3		1376,1681
		NGC 2467 - 41		7 50 43	−26 17.7	2	12.80	.017	0.18	.030	-0.25	.009	3		1376,1681
		Pup sq 3 # 47		7 50 43	−27 08.4	1	12.07		0.16		0.21		2	A3 V	409
		NGC 2477 - 1258		7 50 43	−38 19.6	1	14.27		0.66		0.32		2		259
		NGC 2477 - 1051		7 50 43	−38 21.4	1	14.34		0.63		0.24		2		259
		NGC 2477 - 4034		7 50 43	−38 28.1	1	14.18		0.60		0.24		2		259
64332	−11 02141	NQ Pup		7 50 44	−11 29.7	1	7.55		1.73		1.88		10	S6.2	3039
		NGC 2477 - 1252		7 50 44	−38 20.2	2	11.49	.000	1.44	.000	1.30	.000	5		259,3043
		NGC 2477 - 2049		7 50 44	−38 22.3	1	14.16		1.14		1.21		1		259
		NGC 2477 - 3027		7 50 44	−38 25.8	1	13.58		0.51		0.34		3		259
		NGC 2477 - 3060		7 50 44	−38 26.9	1	14.80		0.68		0.26		3		259
		NGC 2477 - 4035		7 50 44	−38 28.4	1	11.88		1.95		2.31		3		3043
		NGC 2477 - 4037		7 50 44	−38 29.1	1	11.96		1.21		0.96		2		259
		NGC 2477 - 4445		7 50 44	−38 29.6	1	14.55		0.63		0.17		2		259
		LF12 # 148		7 50 45	−23 06.8	1	13.14		0.25		0.16		1		137
64420	−28 05008			7 50 45	−28 36.5	2	9.81	.005	-0.03	.005	-0.18	.020	4	B9 IV	124,409
	−30 05220	LSS 838		7 50 45	−30 37.1	2	11.31	.024	0.51	.019	-0.48	.029	3		211,1766
		NGC 2477 - 2054		7 50 45	−38 23.9	1	13.84		0.58		0.29		2		259
		NGC 2467 - 45		7 50 46	−26 14.8	2	12.15	.026	0.40	.004	0.23	.000	3		1376,1681
	−26 05126	NGC 2467 - 43		7 50 46	−26 20.4	3	11.37	.005	0.24	.025	-0.72	.013	5	B0.5V	1295,1376,1681
		Pup sq 2 # 21		7 50 46	−27 53.	1	11.18		0.45		-0.40		1		211
		NGC 2477 - 1388		7 50 46	−38 17.7	2	12.63	.000	1.22	.024	0.87	.034	6		259,3043
	−38 03765	NGC 2477 - 4431		7 50 46	−38 29.5	1	11.85		1.20		0.96		2	A8	259
		NGC 2477 - 4441		7 50 46	−38 29.7	1	12.89		0.50		0.29		2	A9	259
		NGC 2467 - 44		7 50 47	−26 21.0	2	13.15	.033	0.34	.042	0.27	.146	2		1376,1681
	−27 04670	Pup sq 3 # 48		7 50 47	−27 17.9	1	10.73		0.16		0.19		2	A3 V	409
		NGC 2467 - 48		7 50 48	−26 08.8	3	10.91	.009	-0.01	.012	-0.07	.025	6		398,1376,1681
	−26 05127	NGC 2467 - 47		7 50 48	−26 11.9	2	10.30	.022	0.12	.019	-0.35	.022	5		1376,1681
		NGC 2467 - 46		7 50 48	−26 16.7	1	12.94		0.41		0.14		1		1376
		NGC 2477 - 1247		7 50 48	−38 19.6	1	14.71		0.70		0.31		2		259
		NGC 2477 - 3059		7 50 48	−38 27.1	1	15.18		0.67		0.13		2		259
		NGC 2467 - 50		7 50 49	−26 09.8	3	12.35	.009	0.26	.014	0.11	.028	6		398,1376,1681
		Pup sq 3 # 69		7 50 49	−28 38.5	1	13.29		0.23		0.04		2	A2	409
	+19 01865	SZ Gem		7 50 50	+19 24.3	1	10.96		0.08		0.23		1	A0	668
	−25 05207	NGC 2467 - 52		7 50 50	−26 03.1	2	10.32	.000	1.15	.000	1.13	.024	4		1376,1681
		NGC 2467 - 51		7 50 50	−26 12.6	2	10.39	.000	1.24	.000	1.35	.030	5		1376,1681
		NGC 2477 - 1220	V	7 50 50	−38 20.4	1	13.00		3.45				1		923
		NGC 2477 - 1215		7 50 50	−38 21.3	1	13.13		0.55		0.36		2		259
		NGC 2477 - 2077		7 50 50	−38 22.3	1	14.63		0.68		0.26		2		259
		NGC 2477 - 2068		7 50 50	−38 23.3	1	13.79		0.56		0.29		2		259
64106	+47 01498	HR 3065		7 50 51	+47 31.0	1	6.25		1.15		1.07		2	gK5	1375
		NGC 2467 - 53		7 50 51	−26 05.8	1	12.59		0.95		0.75		1		1376
	−26 05129	NGC 2467 - 49		7 50 51	−26 20.9	3	10.72	.004	0.24	.022	-0.72	.012	6	O7	1295,1376,1681
	−34 04041			7 50 51	−34 22.9	1	9.33		0.31		0.14		2		126
64503	−38 03769	HR 3084, QZ Pup		7 50 52	−38 43.9	5	4.49	.005	-0.20	.005	-0.68	.005	30	B2 V	15,1075,2012,3043,8015
64578	−48 03270			7 50 52	−48 33.3	1	8.74		-0.06		-0.37		2	B8 IV	401
		NGC 2467 - 54		7 50 52	−26 05.1	1	12.77		0.14		-0.16		1		1376
64438	−28 05015			7 50 53	−28 39.4	1	9.56		0.30		0.07		2	B9 III/IV	409
		NGC 2477 - 2064		7 50 53	−38 24.0	1	12.21		1.20		1.00		4		259
		NGC 2467 - 56		7 50 55	−26 11.5	2	12.20	.013	0.43	.017	0.01	.004	3		1376,1681
	−26 05132	NGC 2467 - 57		7 50 55	−26 13.1	2	10.79	.011	0.94	.004	0.14	.030	5		1376,1681
		NGC 2467 - 55		7 50 55	−26 16.1	2	12.40	.013	0.49	.004	0.05	.030	3		1376,1681
		NGC 2477 - 2205		7 50 56	−38 21.3	3	11.41	.010	1.57	.015	1.65	.024	8		259,259,3043
	+19 01867			7 50 57	+19 31.2	1	10.30		0.14		0.11		1		289
		HA 99 # 107		7 50 57	−00 37.4	1	12.32		0.55		0.02		2	G2	281
64577	−47 03384			7 50 57	−47 15.7	1	8.39		-0.04		-0.43		2	B8/9 II	401
		NGC 2467 - 58		7 50 58	−26 02.3	2	11.54	.016	0.45	.004	0.13	.032	4		1376,1681
		NGC 2467 - 103		7 50 58	−26 19.7	1	10.96		0.76		0.39		3		1681
	−26 05136	LSS 840		7 50 58	−26 59.0	3	9.64	.022	0.37	.014	-0.61	.038	5	O7 f	211,540,976,1737
		NGC 2477 - 3468		7 50 58	−38 25.7	1	15.07		0.69		0.28		1		259
		LF12 # 149		7 50 59	−23 32.9	1	11.05		1.10		0.85		1		137
		NGC 2467 - 101		7 50 59	−26 19.6	1	11.10		0.03		0.02		4		1681
		NGC 2477 - 2203		7 50 59	−38 21.2	1	14.35		0.66		0.30		2		259
		HA 99 # 6		7 51 00	−00 41.7	3	11.06	.005	1.25	.007	1.29	.009	30	G7	281,989,1729
		G 113 - 3		7 51 00	−06 36.1	1	12.07		1.26		1.12		1		3062

Table 1   431

HD	DM	Other Id	N	Rem	α_{1950}	δ_{1950}	S	V	σ_V	B–V	σ_{B-V}	U–B	σ_{U-B}	n	Spectrum	References
64455	−26 05137	NGC 2467 - 60			7 51 00	−26 17.0	2	7.75	.000	-0.13	.011	-0.61	.007	5	B8 IV	1376,1681
		HA 99 # 8			7 51 01	−00 44.3	1	11.53		0.76		0.31		2	G6	281
		NGC 2467 - 61			7 51 01	−26 08.3	1	12.72		0.20		-0.26		1		1376
	−26 05134	NGC 2467 - 59			7 51 01	−26 20.9	3	11.11	.050	0.12	.021	-0.38	.022	5	B5	1295,1376,1681
		NGC 2467 - 62			7 51 04	−26 10.3	1	12.70		0.29		0.10		1		1376
		Orsatti 158			7 51 04	−31 48.3	1	11.96		0.88		0.08		2		1766
64144	+47 01499	HR 3066			7 51 05	+47 41.8	2	5.45	.000	1.46	.002	1.73		5	K3 III	71,1501
		HA 99 # 250			7 51 05	−00 23.9	1	12.05		0.92		0.55		9	G3	281
64980	−69 00789				7 51 06	−69 57.5	1	8.54		0.39				4	F2 V	2012
		HA 99 # 12			7 51 09	−00 41.3	1	12.37		0.46		-0.03		2		281
		NGC 2467 - 65			7 51 11	−26 06.4	4	11.82	.017	0.14	.022	-0.74	.014	7	B0.5V	409,1295,1376,1681
		HA 99 # 14			7 51 12	−00 40.9	1	12.54		0.88		0.46		2		281
		NGC 2467 - 64			7 51 12	−26 15.8	1	12.86		0.36		0.00		1		1376
		NGC 2467 - 63			7 51 12	−26 18.3	1	11.90		0.39		0.27		1		1376
	−34 04048				7 51 12	−34 15.4	1	11.68		0.24		0.17		2		126
64572	−36 03989	HR 3085			7 51 13	−36 14.0	3	5.42	.004	1.16	.005			11	K1 III	15,2013,2029
		NGC 2467 - 66			7 51 15	−26 14.5	1	11.80		1.02		0.84		1		1376
		Pup sq 2 # 24			7 51 15	−28 06.	1	11.68		0.24		-0.64		1		211
		G 193 - 77			7 51 16	+51 59.7	1	10.19		0.94		0.65		2		7010
64351	+21 01714				7 51 17	+21 14.2	1	6.84		1.63		1.93		2	M1	1648
64452	−8 02128				7 51 17	−09 05.4	1	8.80		-0.05		-0.24		2	B9	3039
		NGC 2467 - 67			7 51 17	−26 01.7	2	11.12	.009	1.90	.026	2.05	.137	3		1376,1681
		LOD 6- # 13			7 51 17	−34 12.7	1	12.03		0.33		0.09		2		126
	−25 05217	Pup sq 3 # 40			7 51 18	−26 05.3	1	10.78		0.13		0.29		1	B8 IV:	409
		Orsatti 162			7 51 18	−32 40.2	1	13.10		1.35		0.13		2		1766
		LOD 6- # 12			7 51 18	−34 14.4	1	12.22		0.44		0.03		2		126
		NGC 2477 - 2351			7 51 18	−38 22.4	1	12.04		1.28		1.16		3		259
64722	−54 01420	HR 3088, V372 Car			7 51 18	−54 14.2	6	5.69	.004	-0.15	.005	-0.87	.019	21	B2 IV	15,401,1020,1732,2012,2012
		NGC 2467 - 69			7 51 19	−26 05.4	2	10.54	.026	0.22	.013	0.26	.073	3		1376,1681
		NGC 2467 - 68			7 51 19	−26 13.9	1	12.02		1.16		0.90		1		1376
		HA 99 # 258			7 51 20	−00 25.3	1	10.99		0.85		0.58		1	G6	281
	−34 04051				7 51 20	−34 13.3	1	10.57		1.15		1.12		2		126
		NGC 2467 - 71			7 51 21	−26 10.7	2	11.45	.009	1.11	.009	0.89	.051	3		1376,1681
		NGC 2467 - 70			7 51 21	−26 11.8	2	12.26	.013	0.64	.013	0.08	.030	3		1376,1681
	−34 04052				7 51 21	−34 19.0	1	9.18		1.19		1.12		2		126
64571	−34 04053				7 51 21	−34 48.4	3	6.63	.013	0.82	.008	0.53	.024	8	F8/G0 Ib	119,126,564
	−0 01851				7 51 24	−00 14.3	6	9.60	.006	0.78	.007	0.51	.003	52	G1	271,281,989,1728,1729,6004
	−26 05149	NGC 2467 - 72			7 51 24	−26 18.7	2	10.91	.004	0.10	.019	0.03	.037	5		1376,1681
	−25 05222	NGC 2467 - 74			7 51 25	−26 03.4	2	10.37	.007	0.43	.007	-0.02	.000	5		1376,1681
	−34 04054				7 51 25	−34 15.6	1	11.09		0.31		0.05		2		126
		HA 99 # 266			7 51 27	−00 18.5	1	10.58		1.14		1.16		4	G8	271
		NGC 2467 - 73			7 51 27	−26 18.8	2	11.38	.030	0.89	.013	0.51	.017	3		1376,1681
		HA 99 # 362			7 51 28	−00 15.7	1	10.80		1.14		1.10		1	K2	281
		NGC 2467 - 75			7 51 28	−26 16.4	1	13.41		0.26		-0.30		1		1376
64717	−50 03022				7 51 28	−50 23.4	2	7.10	.000	0.03	.000	-0.63		6	B2 III	401,2012
		NGC 2477 - 3501			7 51 29	−38 29.3	1	11.61		1.23		0.98		2		259
		HA 99 # 267			7 51 30	−00 21.9	1	12.30		0.58		0.06		5	G6	281
		NGC 2467 - 116			7 51 30	−26 05.3	1	12.24		0.74		0.23		2		1681
		NGC 2467 - 110			7 51 30	−26 09.6	1	11.75		0.11		0.15		2		1681
	−25 05223	NGC 2467 - 76			7 51 30	−26 11.0	2	9.84	.008	1.00	.008	0.73	.020	4		1376,1681
	−28 05034	LSS 847		⋆ AB	7 51 30	−28 36.2	2	10.36	.010	0.45	.019	-0.64	.015	3	B0 II	211,409
	−34 04058				7 51 30	−34 14.6	1	10.53		0.13		0.13		2		126
64590	−23 06428				7 51 31	−24 07.6	1	10.67		0.29		0.18		1	B9.5 V	1770
64569	−26 05155				7 51 32	−27 09.0	1	7.99		-0.13		-0.54		1	B3 II/III	409
	−34 04059				7 51 32	−34 18.5	1	11.15		0.59		0.18		2		126
64568	−25 05228	NGC 2467 - 77			7 51 33	−26 06.2	7	9.39	.018	0.07	.019	-0.90	.026	17	O4 V((f))	409,540,976,1295,1376*
64324	+34 01705				7 51 34	+34 45.3	1	7.79		0.65		0.18		2	G0	1625
		HA 99 # 366			7 51 35	−00 16.9	2	11.51	.010	0.16	.005	0.06	.010	9	A4	271,281
64761	−52 01315				7 51 35	−52 50.4	1	7.33		0.06		-0.06		2	B9 II/III	401
64592	−26 05156	NGC 2467 - 78			7 51 36	−26 19.0	2	9.92	.014	0.13	.007	0.18	.042	6	A0	1376,1681
	−34 04062				7 51 36	−34 21.3	1	9.72		0.06		-0.05		2		126
65836	−81 00237				7 51 36	−81 28.1	1	6.79		0.16		0.10		5	Ap SrCrEu	1628
	−25 05229	NGC 2467 - 79			7 51 37	−26 05.1	3	10.29	.012	0.13	.014	0.18	.024	6	A2 V	137,1376,1681
64716	−44 03823				7 51 37	−45 05.3	1	8.42		-0.10		-0.40		2	B6 V	401
		HA 99 # 367			7 51 38	−00 17.7	6	11.15	.012	1.00	.006	0.83	.013	45	G8	1728,271,281,989,1729,5006
64740	−49 03137	HR 3089			7 51 39	−49 28.9	5	4.63	.006	-0.23	.007	-0.92	.005	22	B2 V	15,1075,1732,2012,8015
		LOD 6- # 20			7 51 40	−34 03.1	1	11.95		0.43		0.09		2		126
	−33 04271				7 51 40	−34 03.9	1	10.66		0.61		0.16		2		126
64619	−27 04693				7 51 41	−27 53.5	2	8.77	.000	1.18	.000	1.24	.000	4	K1/3 III	16,409
	−33 04272				7 51 41	−34 06.6	1	10.31		0.96		0.62		2		126
		HA 99 # 146			7 51 43	−00 33.0	1	12.74		0.52		0.05		3	F4	271
		NGC 2467 - 117			7 51 43	−26 06.3	1	13.13		0.40		0.07		1		1681
		NGC 2467 - 118			7 51 44	−26 06.3	1	12.26		0.48		0.09		2		1681
		NGC 2467 - 80			7 51 44	−26 17.3	2	12.05	.022	0.17	.011	0.15	.033	5		1376,1681
	−34 04067				7 51 44	−34 23.6	1	10.13		0.22		0.17		2		126
		HA 99 # 35			7 51 45	−00 43.7	2	10.76	.014	0.42	.005	-0.04	.019	4	F7	271,281
64639	−24 06131	LSS 848			7 51 45	−24 40.3	2	9.64	.002	0.18	.002	-0.76	.024	6	B(9)nne	540,976
64616	−25 05234	NGC 2467 - 81			7 51 45	−26 07.8	2	6.95	.011	0.92	.004	0.65	.011	6	G8 III	1376,1681
		HA 99 # 148			7 51 46	−00 32.2	1	11.54		1.64		1.40		6	M5	271
	−33 04273				7 51 47	−34 09.2	1	9.06		1.18		1.21		2		126

HD	DM	Other Id	N Rem	α_{1950}	δ_{1950}	S	V	σ_V	B–V	σ_{B-V}	U–B	σ_{U-B}	n	Spectrum	References
		NGC 2467 - 82		7 51 48	−26 18.2	2	11.38	.016	1.06	.004	0.89	.032	4		1376,1681
64810	−52 01319			7 51 49	−52 32.6	1	7.71		1.15		1.00		1	K0 III	6006
		HA 99 # 152		7 51 50	−00 29.8	1	11.26		1.64		1.74		4	M6	271
64760	−47 03396	HR 3090, LSS 850		7 51 50	−47 58.3	5	4.24	.003	-0.15	.005	-0.99	.000	24	B0.5 Ib	15,1075,1075,2012,8015
		LF12 # 151		7 51 52	−26 40.7	1	11.74		0.02		-0.12		1	B8 V	137
64657	−26 05161	NGC 2467 - 83	⋆ V	7 51 54	−26 13.0	3	7.03	.068	1.54	.036	1.27	.091	8	M5 III	1376,1681,2012
		G 193 - 78		7 51 55	+57 50.2	1	15.07		0.12		-0.76		1	DC	782
64493	+18 01778			7 51 56	+18 13.9	1	7.31		1.46		1.73		2	K4 III	1648
		HA 99 # 375		7 51 56	−00 16.5	1	12.35		0.25		0.07		5	A8	281
64468	+19 01869	G 91 - 19	⋆ A	7 51 59	+19 22.5	3	7.76	.012	0.96	.010	0.80	.000	7	K2	1355,1758,3077
		NGC 2467 - 102		7 51 59	−26 15.2	1	9.32		1.01		0.79		3		1681
	−32 04535			7 52 00	−33 06.6	1	10.27		0.02		0.10		2		124
64606	−1 01883	G 112 - 54		7 52 03	−01 16.8	6	7.44	.005	0.73	.005	0.16	.007	12	G8 V	1003,1311,1658,1774*
		HA 99 # 288		7 52 04	−00 20.3	2	11.46	.000	1.10	.004	0.97		10	K3	281,5006
	−24 06142	NGC 2482 - 4		7 52 04	−24 13.7	1	11.50		0.12		0.11		1		410
	−52 01324	IDS07508S5224	AB	7 52 04	−52 31.4	2	9.50	.018	0.49	.000	-0.17	.000	5	F8	158,3077
	−23 06445	NGC 2482 - 3		7 52 05	−24 09.5	1	11.77		0.14		-0.27		1		410
		Orsatti 169		7 52 05	−30 11.9	1	11.51		0.43		-0.29		1		1766
		NGC 2482 - 2		7 52 07	−24 09.7	1	11.88		0.98		0.57		1		410
		G 90 - 28		7 52 09	+36 30.1	1	16.07		0.26		-0.43		2	DAwk	316
64605	−0 01853			7 52 09	−00 29.5	4	8.81	.007	0.04	.006	0.04	.003	26	A1 V	271,281,1509,6004
		BX Pup		7 52 09	−24 11.7	1	13.76		0.02		-0.74		1		1471
		HA 99 # 294		7 52 10	−00 19.9	1	12.16		0.50		-0.01		9	G0	281
		Pup sq 3 # 16		7 52 10	−27 57.7	2	12.88	.015	0.21	.015	0.14	.005	6	A0 V	16,409
64535	+22 01808	IDS07492N2156	AB	7 52 11	+21 48.2	3	8.63	.010	0.33	.012	0.13	.022	17	A3	1301,1345,1775
64777	−33 04283			7 52 11	−33 53.4	1	7.46		-0.09		-0.42		2	B8/9 V	401
	−26 05166			7 52 12	−26 18.7	1	9.63		1.21		1.25		2	K2 III	137
64633	−0 01854			7 52 13	−00 21.7	7	8.45	.011	1.19	.004	1.26	.006	56	K2 IV	271,281,989,1509,1728*
64825	−40 03608			7 52 15	−40 51.1	1	8.62		-0.01		-0.13		2	A0 V	401
64632	+1 01944			7 52 17	+00 57.2	1	8.36		0.26		0.12		1	F2 V	1272
	−23 06450	NGC 2482 - 1		7 52 17	−24 10.4	1	8.93		1.68		1.76		3		410
	−24 06148	NGC 2482 - 5		7 52 17	−24 13.9	1	11.41		-0.02		-0.18		1		410
		Pup sq 3 # 17		7 52 18	−28 02.3	2	12.92	.009	1.08	.033	0.66	.037	4		16,409
64652	−0 01855			7 52 19	−00 37.5	6	8.35	.007	1.09	.006	0.98	.008	45	K0 III	271,281,989,1509,1729,6004
64882	−48 03292			7 52 19	−48 39.8	1	8.85		-0.04		-0.47		2	B7 III/IV	401
64802	−35 04002	HR 3091		7 52 20	−35 44.7	5	5.47	.014	-0.19	.007	-0.73	.008	17	B2 V	15,401,540,976,1732,2012
		Orsatti 172		7 52 21	−31 52.9	1	11.90		0.58		-0.32		2		1766
64827	−41 03520			7 52 21	−41 34.3	1	6.85		-0.11		-0.46		2	B8 II	401
64908	−50 03030	IDS07510S5039	A	7 52 21	−50 47.0	1	8.55		0.02		-0.37		2	B5/7 III	401
64908	−50 03030	IDS07510S5039	B	7 52 21	−50 47.0	1	9.90		0.35		0.20		1	B9	401
64908	−50 03030	IDS07510S5039	C	7 52 21	−50 47.0	1	10.82		0.15		0.06		1		401
64756	−27 04706			7 52 23	−27 58.4	2	6.61	.005	0.07	.005	0.12	.015	11	A2 V	16,409
64385	+50 01489	G 193 - 79		7 52 24	+50 40.6	1	8.83		0.52		-0.06		2	G5	1566
64347	+56 01253	HR 3077		7 52 24	+56 38.3	1	6.72		0.03		0.03		2	A2 IV	1733
64491	+35 01705	HR 3083		7 52 25	+35 32.8	3	6.23	.010	0.27	.011	-0.03	.010	7	A3p	985,1022,1501
		LSS 851		7 52 26	−27 11.8	1	11.42		0.68		-0.47		2	B3 Ib	409
		LSS 853		7 52 26	−28 32.0	1	12.43		0.16		0.01		1	B3 Ib	409
		HA 99 # 54		7 52 27	−00 39.8	2	11.34	.015	0.68	.000	0.30	.035	5	G2	271,281
	−34 04083			7 52 27	−34 41.2	1	11.34		0.15		0.12		2		126
		CCS 947		7 52 31	−32 13.0	1	11.00		3.49				1	C6,3	864
64488	+39 02031			7 52 33	+39 25.2	2	7.26	.000	0.04	.000	0.09	.015	5	A1 V	833,1601
		HA 99 # 59		7 52 34	−00 39.2	1	11.59		0.69		0.30		2	G5	281
64905	−44 03838			7 52 35	−45 09.7	1	8.23		-0.04		-0.32		2	B9 III/IV	401
64800	−24 06156			7 52 36	−24 58.5	1	9.32		0.37		0.03		2	F3 V	126
64901	−40 03618			7 52 39	−40 43.5	1	8.54		-0.12		-0.47		2	B8/9 II/III	401
	−0 01856			7 52 40	−00 17.6	6	9.81	.007	0.41	.005	0.04	.004	51	F6	271,281,989,1728,1729,6004
64820	−23 06463	NGC 2482 - 6		7 52 41	−24 10.4	1	8.02		1.23		1.12		1	G2 II	410
65094	−60 00916	NGC 2516 - 226		7 52 42	−60 55.3	1	7.85		0.01		-0.15		3	A0	1683
		HA 99 # 314		7 52 43	−00 19.1	1	11.50		0.78		0.37		3	G2	271
	−24 06160	NGC 2482 - 7		7 52 44	−24 13.2	1	10.28		1.06		0.85		1		410
	−24 06161			7 52 44	−25 01.3	1	10.87		0.27		0.18		2		126
		Pup sq 3 # 11		7 52 44	−28 06.1	2	11.20	.005	-0.01	.010	-0.07	.015	3	B9 V	16,409
64648	+20 01946	HR 3086		7 52 45	+20 01.1	3	5.39	.021	-0.04	.005	-0.07	.015	6	A0 Vs	985,1022,3023
		KUV 07528+4113		7 52 45	+41 12.9	1	16.97		-0.13		-1.02		1	DA?	1708
	−24 06164			7 52 45	−25 03.8	1	10.71		1.05		1.02		2		126
64841	−27 04713			7 52 45	−28 08.5	3	8.76	.005	1.07	.005	0.94	.013	5	K2 III	16,409,8100
	−24 06163	NGC 2482 - 8		7 52 46	−24 12.2	1	10.58		0.09		0.07		1		410
	−34 04089	IDS07509S3424	AB	7 52 46	−34 32.2	1	10.41		0.40		0.06		2		126
		Pup sq 3 # 14		7 52 47	−27 56.0	1	11.11		0.52		0.14		2	G0 III	409
64876	−34 04091	HR 3092		7 52 47	−34 42.8	3	6.14	.016	1.54	.006	1.86	.037	8	K4 III	58,126,2007
64685	+9 01815	HR 3087		7 52 48	+08 59.8	2	5.85	.005	0.36	.005	0.00	.015	7	F2 V	15,1078
		ApJ113,309 # 99		7 52 48	−27 39.	1	10.70		0.30		-0.46		1		211
	−27 04715			7 52 48	−27 55.1	1	11.06		0.47		0.18		1	G0:	16
		NGC 2482 - 28		7 52 49	−24 05.6	1	11.58		1.77		2.18		1		410
		LP 723 - 26		7 52 50	−14 37.6	2	13.53	.030	-0.04	.015	-0.80	.005	2	DA	3028,3060
64860	−27 04718			7 52 50	−28 09.8	3	8.62	.013	1.43	.017	1.67	.064	5	K0 III	16,409,8100
		NGC 2482 - 29		7 52 51	−24 05.7	1	11.66		0.14		0.09		1		410
	−23 06466	NGC 2482 - 27		7 52 51	−24 07.5	1	11.25		0.05		0.06		1		410
		Orsatti 179		7 52 52	−32 12.0	1	11.44		0.47		-0.35		2		1766
	−43 03714			7 52 52	−43 45.7	1	10.93		2.83				1	R	864

Table 1

HD	DM	Other Id	N	Rem	α_{1950}	δ_{1950}	S	V	σ_V	B–V	σ_{B-V}	U–B	σ_{U-B}	n	Spectrum	References
64747	+0 02127				7 52 53	−00 09.5	4	9.47	.003	-0.04	.005	-0.15	.002	27	B9 V	281,989,5006,6004
64898	−31 05226				7 52 53	−31 24.1	1	7.19		-0.11		-0.66		2	B2 III/IV	401
		L 97 - 12			7 52 54	−67 38.	1	14.08		0.66		-0.17		7		1705,3078
64794	−10 02289				7 52 55	−11 01.1	1	7.35		-0.09		-0.50		2	B8	3039
	−23 06469	NGC 2482 - 26			7 52 57	−24 08.7	1	10.77		0.09		0.09		1		410
		NGC 2482 - 25			7 52 58	−24 08.8	1	11.94		0.48		0.12		1		410
64682	+25 01794	IDS07500N2456	A		7 53 01	+24 47.9	1	7.70		0.99		0.72		2	K0	1625
		NGC 2482 - 10			7 53 01	−24 14.9	1	11.80		-0.03		-0.31		1		410
65038	−49 03165				7 53 01	−50 02.3	2	7.48	.005	-0.13	.005	-0.67		6	B3 IV	401,2012
	−23 06474	NGC 2482 - 20			7 53 02	−24 09.6	1	12.15		0.21		0.12		1		410
	−24 06174	NGC 2482 - 9			7 53 02	−24 14.9	1	10.27		1.11		0.94		1		410
		NGC 2482 - 32			7 53 03	−24 04.1	1	11.56		1.67		1.92		1		410
		NGC 2482 - 24			7 53 03	−24 08.4	1	11.11		0.11		0.13		1		410
64996	−40 03626				7 53 03	−40 53.7	1	9.80		0.01		-0.48		2	B8 II	401
64704	+24 01805	G 91 - 20		⋆ AB	7 53 04	+23 49.7	1	7.34		0.63		0.13		1	G0	333,1620
		NGC 2482 - 31			7 53 04	−24 04.9	1	11.43		1.73		2.17		1		410
	−23 06475	NGC 2482 - 30			7 53 04	−24 05.7	1	10.66		0.12		0.12		1		410
65058	−49 03167				7 53 04	−49 33.2	1	9.27		-0.08		-0.67		2	B2 V	401
	−23 06476	NGC 2482 - 23			7 53 05	−24 08.0	1	10.21		1.09		0.94		1		410
	−24 06176				7 53 05	−25 05.5	1	10.78		0.10		0.07		2		126
	−25 05265	Tr 9 - 60			7 53 05	−25 43.2	1	9.28		0.55		0.08		1		183
65152	−60 00919				7 53 05	−60 27.1	1	7.19		1.85		2.11		1	M0 III	3043
	−25 05266	Tr 9 - 59			7 53 06	−25 42.9	1	8.92		0.42		0.01		1	F8	183
		Orsatti 183			7 53 07	−32 25.2	1	14.04		0.79		-0.30		1		1766
64681	+30 01612				7 53 08	+29 59.1	1	6.80		1.42		1.62		2	K0	1625
		NGC 2482 - 21			7 53 09	−24 08.6	1	11.35		1.32		1.44		1		410
	−25 05267	Tr 9 - 61			7 53 09	−25 45.6	1	9.59		0.16		0.12		1		183
	−23 06479	NGC 2482 - 33			7 53 11	−24 04.9	1	11.27		0.10		0.07		1		410
		NGC 2482 - 22			7 53 11	−24 08.3	1	12.65		0.26		0.10		1		410
	−23 06481	NGC 2482 - 19			7 53 11	−24 09.5	1	11.43		0.06		0.08		1		410
		NGC 2482 - 34			7 53 12	−24 04.9	1	11.79		1.21		1.32		1		410
		NGC 2482 - 35			7 53 12	−24 06.8	1	11.73		0.29		0.24		1		410
		NGC 2482 - 18			7 53 12	−24 10.0	1	11.81		1.36		1.20		1		410
		NGC 2482 - 12			7 53 12	−24 12.2	1	11.66		1.75				1		410
64922	−25 05268	Tr 9 - 62			7 53 13	−25 48.0	2	7.38	.019	1.30	.005	1.45	.024	3	K2 III	126,183
		NGC 2482 - 17			7 53 14	−24 09.9	1	11.68		0.22		0.13		1		410
		NGC 2482 - 16			7 53 15	−24 10.3	1	12.59		0.48		-0.03		1		410
	−27 04726	LSS 859			7 53 15	−27 36.4	1	10.31		0.14		-0.73		2	B0.5V	540
65170	−58 01018	IDS07523S5818		AB	7 53 15	−58 26.4	1	7.81		0.52		0.01		4	F7 V	709
		NGC 2483 - 174			7 53 17	−27 47.7	1	12.87		0.48		-0.05		1		455
65074	−45 03558				7 53 17	−45 31.9	1	7.34		-0.10		-0.52		2	B6 V	401
	−23 06487	NGC 2482 - 36			7 53 18	−24 08.8	1	11.39		0.12		0.15		1		410
		NGC 2482 - 15			7 53 18	−24 11.1	1	12.30		0.48		0.05		1		410
		NGC 2482 - 11			7 53 18	−24 14.5	1	11.98		0.02		-0.57		1		410
	−23 06488	NGC 2482 - 13			7 53 20	−24 10.7	1	11.47		0.08		-0.38		1		410
64972	−27 04729	V341 Pup			7 53 20	−28 09.0	3	7.20	.011	-0.12	.004	-0.56	.010	8	B8/9 II	16,409,8100
64854	+0 02129				7 53 21	−00 08.8	7	9.40	.003	-0.15	.004	-0.72	.007	159	B2 III	281,989,1728,1729,1764*
		NGC 2482 - 14			7 53 21	−24 11.4	1	12.47		0.48		0.08		1		410
		Orsatti 185			7 53 21	−32 19.4	1	11.51		0.78		-0.34		2		1766
64645	+45 01509				7 53 22	+45 35.1	1	8.67		0.26		0.06		3	A3	1601
		PASP81,804 T1# 13			7 53 22	−28 06.3	2	13.60	.206	0.66	.009	0.22	.023	4	G0:	16,409
	−0 01857				7 53 23	−00 44.8	1	10.24		0.53		0.07		2	G2	281
		Tr 9 - 6			7 53 23	−25 44.4	1	13.64		0.13		-0.15		1		68
	−27 04731	Pup sq 3 # 10			7 53 23	−28 05.1	2	10.75		1.07	.010	0.86	.019	3	K0:	16,409
		HA 99 # 93			7 53 24	−00 44.6	1	12.24		0.47		-0.03		6	F7	281
	−24 06184				7 53 24	−25 04.0	1	10.41		0.56		0.05		2		126
		Tr 9 - 7			7 53 24	−25 43.9	1	13.86		0.16		-0.04		1		68
		LSS 862			7 53 24	−29 46.2	1	11.16		0.16		0.07		2	B3 Ib	409
65114	−49 03174				7 53 24	−49 42.4	1	9.36		0.84				4	G8/K1 III/V	2012
	−23 06493	NGC 2482 - 38			7 53 25	−24 08.1	1	11.99		0.14		0.08		1		410
	−25 05272	Tr 9 - 3			7 53 25	−25 43.3	2	10.80	.000	0.36	.024	0.12	.010	3		68,183
		PASP81,804 T1# 9			7 53 25	−28 05.3	2	11.63	.180	1.29	.000	1.21	.000	2	K2:	16,409
		NGC 2482 - 37			7 53 26	−24 07.8	1	12.31		1.19		1.12		1		410
65113	−49 03175				7 53 27	−49 33.1	1	8.77		-0.04		-0.47		2	B9 II	401
		Tr 9 - 28			7 53 28	−25 45.4	2	13.40	.005	0.34	.088	-0.09	.044	3		68,183
		NGC 2483 - 171			7 53 28	−27 50.0	1	11.92		0.22		-0.49		1	B2	455
64994	−28 05088				7 53 28	−28 17.6	2	8.96	.000	1.53	.000	1.60	.000	4	K3/4 (III)	16,409
		HA 99 # 99			7 53 29	−00 42.3	1	12.03		0.54		0.07		2	F8	281
64993	−23 06495	LSS 861			7 53 29	−23 55.5	2	7.67	.007	0.01	.009	-0.80	.021	4	B1 Ib/II	540,976
	−25 05277	Tr 9 - 56			7 53 30	−25 43.9	2	10.82	.018	0.17	.054	0.20	.059	5		68,183
		Tr 9 - 22			7 53 30	−25 44.5	1	12.95		0.14		-0.27		2		183
		HA 99 # 103			7 53 31	−00 45.0	1	11.68		0.30		0.07		2	F0	281
		Tr 9 - 23			7 53 31	−25 44.5	1	14.50		0.18		0.18		1		183
		Tr 9 - 57			7 53 31	−25 44.5	1	12.55		0.14		-0.27		2		183
		NGC 2483 - 128			7 53 31	−27 42.5	1	12.91		0.28		-0.25		1	B5	455
		Tr 9 - 16			7 53 32	−25 47.1	1	12.68		0.05		-0.24		2		183
	−27 04734	NGC 2483 - 100			7 53 32	−27 48.9	1	10.63		0.15		-0.74		1	O9 V	455
		Pup sq 3 # 8			7 53 32	−28 08.9	1	11.10		1.45		1.37		3		409
		PASP81,804 T1# 8			7 53 32	−28 09.0	1	12.53		0.34		0.31		2		16
	−27 04739	Pup sq 3 # 7			7 53 32	−28 10.9	2	10.69	.023	0.41	.005	-0.01		4	F4 V	16,409

HD	DM	Other Id	N Rem	α_{1950}	δ_{1950}	S	V	σ_V	B–V	σ_{B-V}	U–B	σ_{U-B}	n	Spectrum	References
		Orsatti 188		7 53 32	−31 38.5	1	11.49		0.40		−0.24		1		1766
64887	+0 02131			7 53 33	−00 12.7	6	9.42	.004	−0.07	.003	−0.22	.006	105	B9 V	281,989,1728,1729,1764,6004
		Tr 9 - 21	A	7 53 33	−25 44.6	1	11.77		−0.01		−0.31		1		183
		Tr 9 - 21	AB	7 53 33	−25 44.6	1	11.67		0.06		−0.33		1		183
		Tr 9 - 58		7 53 33	−25 45.3	1	14.25		0.10		0.06		2		183
		NGC 2483 - 41		7 53 33	−27 47.6	1	10.68		0.17		−0.73		1	B0 V	455
65010	−24 06188			7 53 34	−24 51.2	1	9.11		0.01		−0.08		2	B9.5V	126
		Tr 9 - 13		7 53 34	−25 46.8	1	12.54		0.14		0.05		2		183
		Tr 9 - 15		7 53 34	−25 47.5	1	13.18		2.12		2.01		2		183
		NGC 2483 - 56		7 53 34	−27 43.3	1	13.34		0.34		−0.23		1		455
		G 90 - 32		7 53 35	+30 32.4	1	12.45		0.70		0.20		2		1620
	−23 06501	NGC 2482 - 39		7 53 35	−24 08.3	1	11.52		0.09		0.08		1		410
	−24 06189			7 53 35	−24 55.2	1	9.63		0.89		0.69		2		126
	−25 05278	Tr 9 - 2		7 53 35	−25 43.2	2	10.04	.009	−0.07	.014	−0.23	.014	5		68,183
		Tr 9 - 20		7 53 35	−25 45.0	1	12.51		0.11		−0.26		2		183
	−25 05279	Tr 9 - 19		7 53 36	−25 45.1	1	10.28		0.07		−0.43		4		183
		Tr 9 - 17		7 53 36	−25 46.1	1	13.72		0.23		0.08		2		183
	−25 05280	Tr 9 - 4		7 53 36	−25 47.1	1	11.54		0.04		−0.45		2		183
		NGC 2483 - 129		7 53 36	−27 42.1	1	12.56		0.50		−0.03		1		455
		NGC 2483 - 57		7 53 36	−27 42.8	1	10.97		0.06		−0.09		1	B8 V	455
		NGC 2483 - 8		7 53 37	−27 45.9	1	11.05		0.24		−0.39		1	B5	455
65089	−41 03536	IDS07519S4134	AB	7 53 37	−41 42.0	1	7.54		0.69				4	G8 III +(F)	2012
		Tr 9 - 24		7 53 38	−25 45.7	1	12.13		0.10		−0.36		2		183
		Tr 9 - 12		7 53 38	−25 47.0	1	11.59		0.10		−0.38		2		183
		NGC 2483 - 14		7 53 40	−27 45.1	1	12.85		0.40		0.01		1	B5	455
65252	−58 01019			7 53 40	−58 36.0	1	8.35		0.61		0.14		4	G0 V	709
	−25 05282	Tr 9 - 18		7 53 41	−25 45.7	1	10.09		0.73		0.35		2	G3 II	183
		NGC 2483 - 2		7 53 41	−27 46.1	1	12.16		0.30		0.12		1	B3	455
	−24 06191			7 53 42	−24 57.8	1	10.56		0.66		0.30		2		126
		Pup sq 3 # 30		7 53 42	−28 03.9	1	14.88		1.14		0.60		2		409
	−29 05159	LSS 867		7 53 42	−29 25.9	1	10.78		0.41		−0.69		5	B0 V (n)e1	1737
65032	−25 05284	Tr 9 - 1		7 53 43	−25 42.4	2	8.32	.000	0.27	.005	0.33	.032	5	A2 III	68,183
		Tr 9 - 14		7 53 43	−25 44.7	1	12.74		0.15		−0.19		2		183
		Pup sq 3 # 28		7 53 43	−28 03.3	1	11.02		0.55		−0.02		2	F7 V	409
65189	−52 01333	HR 3100		7 53 43	−52 27.0	3	6.38	.009	−0.01	.000	−0.21		9	B8 III	15,401,2012
		Tr 9 - 11		7 53 44	−25 43.6	1	14.12		0.15		0.00		1		68
		Tr 9 - 8		7 53 44	−25 46.2	2	13.57	.054	0.48	.010	−0.06	.161	3		68,183
64833	+26 01684			7 53 45	+26 14.4	1	7.60		1.00		0.78		2	K1 III	1625
64938	+4 01860	HR 3093		7 53 46	+04 37.2	5	6.16	.008	0.98	.019	0.75	.007	17	G8 III	15,252,1061,1355,1509
		NGC 2483 - 22		7 53 46	−27 45.9	1	11.55		0.28		−0.30		1	B3	455
	−23 06506	NGC 2482 - 41		7 53 47	−24 07.9	1	11.26		0.10		0.11		1		410
		Tr 9 - 9		7 53 47	−25 47.4	1	13.31		0.45		0.04		1		68
		NGC 2483 - 142		7 53 47	−27 42.0	1	12.86		0.28		−0.33		1	B5	455
		Pup sq 3 # 29		7 53 47	−28 03.2	1	13.30		0.34		0.12		2		409
65053	−23 06508	NGC 2482 - 40		7 53 48	−24 07.5	1	10.04		0.12		0.15		1	A2	410
	−24 06196			7 53 48	−25 01.8	1	10.38		1.32		1.32		2		126
	−27 04742	LSS 868		7 53 48	−28 1.2	1	10.62		0.22		−0.80		3	F7 V	1737
		Tr 9 - 53		7 53 51	−25 42.7	1	13.00		0.62		0.36		1		68
		NGC 2483 - 78		7 53 51	−27 45.2	1	13.21		0.57		−0.01		1		455
		NGC 2483 - 169		7 53 51	−27 50.2	1	11.41		0.14		0.08		1	B6	455
	−28 05104	LSS 870		7 53 51	−28 29.8	3	9.97	.007	0.16	.013	−0.76	.006	7	O7 f	540,976,1737
65273	−56 01468	HR 3105		7 53 51	−57 10.2	5	5.62	.011	1.30	.010	1.47	.025	16	K3/4 III	15,709,2013,2028,3077
		AAS21,193 # 104		7 53 54	−28 29.	1	12.09		0.16		−0.69		2		525
65188	−46 03572			7 53 54	−47 10.8	1	7.11		1.62		1.99		7	M2 III	1673
65087	−28 05107	LSS 872		7 53 55	−28 24.3	2	9.52	.010	0.09	.016	−0.57	.010	4	B2/3 II/III	525,976,1011
		Tr 9 - 10		7 53 56	−25 41.4	1	13.27		0.44		0.43		1		68
		NGC 2516 - 227		7 53 57	−60 50.1	1	11.18		0.45		0.25		3	A0p:	1683
65086	−27 04746	NGC 2483 - 163		7 53 59	−27 47.5	1	9.26		0.07		0.11		1	A0 V(p)	455
		Pup sq 3 # 18		7 53 59	−28 04.6	2	11.69	.010	0.41	.005	0.07	.015	3	F4 III	16,409
	−27 04747			7 53 59	−28 05.2	2	9.93	.029	0.90	.005	0.57	.005	3	G8 III	16,409
	−30 05294			7 53 59	−30 40.8	1	10.53		0.13		−0.36		2		124
		Tr 9 - 25		7 54 00	−25 40.5	1	13.23		0.18		0.17		1		68
		Pup sq 3 # 32		7 54 00	−28 07.4	1	15.73		1.03				1		409
	−28 05111	LSS 873		7 54 01	−28 26.3	1	11.26		0.15		−0.71		2	B0 V	525
		Tr 9 - 5		7 54 02	−25 44.8	1	12.03		0.26		0.07		1		68
		Pup sq 3 # 31		7 54 02	−28 08.0	1	15.43		0.45		0.30		2		409
		G 113 - 6		7 54 04	−07 36.3	1	13.54		0.73		−0.06		2		3062
		NGC 2483 - 184		7 54 06	−27 45.7	1	10.26		0.03		0.07		1	B7	455
64960	+16 01590	HR 3095		7 54 09	+15 55.5	2	5.80	.016	1.28	.007			9	K3 III	71,6002
	−28 05114	LSS 874		7 54 09	−28 20.3	3	10.09	.010	0.21	.008	−0.68	.019	7	B0.5III	525,540,976
65211	−43 03737	HR 3101	★AB	7 54 09	−43 42.7	3	6.02	.007	−0.11	.005	−0.50		9	B5 IV/V	15,401,2027
		G 111 - 45		7 54 10	+40 10.5	1	11.31		1.13		0.97		2		7010
	−60 00922	NGC 2516 - 228		7 54 10	−60 41.1	1	10.93		0.47		0.13		3	A0p:	1683
65131	−27 04751			7 54 11	−28 02.9	3	7.41	.005	−0.07	.009	−0.20	.005	15	B9 III	16,124,409
		Pup sq 3 # 19		7 54 13	−28 00.2	2	12.03	.000	0.20	.000	0.19	.000	4	A2 V	16,409
64307	+74 00338	HR 3075		7 54 15	+74 03.3	2	5.40	.029	1.42	.005	1.64		12	K3 III	252,1258
	−22 05393			7 54 15	−22 16.1	2	10.13	.070	0.63	.014	0.03		4		742,1594
	−60 00923	NGC 2516 - 229		7 54 15	−60 30.3	1	10.47		0.21		0.15		3	A2(p):	1683
65270	−45 03574	V342 Pup		7 54 17	−45 50.8	2	6.75	.005	−0.12	.005	−0.58	.005	6	B5 V	158,401
65405	−60 00924	NGC 2516 - 113		7 54 17	−60 24.7	2	8.38	.000	0.16	.019	0.15	.009	4	A2 V	36,1683

Table 1   435

HD	DM	Other Id	Rem	α_{1950}	δ_{1950}	S	V	σ_V	B–V	σ_{B-V}	U–B	σ_{U-B}	n	Spectrum	References
		Pup sq 3 # 22		7 54 20	−28 02.8	1	10.76		0.08		0.02		2	B9 V	409
		OI 090 4 # 1	★	7 54 22	+10 04.	2	14.43	.004	0.49	.000	0.00	.032	9		1482,1595
		OI 090 4 # 2	★	7 54 22	+10 04.	2	12.98	.008	0.65	.008	0.11	.008	9		1482,1595
		OI 090 4 # 3		7 54 22	+10 04.	1	14.71		0.81		0.45		1		1482
		OI 090 4 # 4		7 54 22	+10 04.	1	15.87		0.49		0.08		1		1482
		OI 090 4 # 6		7 54 22	+10 04.	1	15.91		0.74		0.14		1		1482
		OI 090 4 # 7		7 54 22	+10 04.	1	13.21		0.67		0.32		1		1482
		OI 090 4 # 8	★	7 54 22	+10 04.	1	15.69		0.83		0.26		4		1595
	−27 04755			7 54 22	−28 01.8	1	10.76		0.08		0.02		2	B8 V	16
	−27 04757	Pup sq 3 # 20		7 54 23	−28 00.0	2	10.97	.000	0.02	.000	−0.32	.000	4	B8 III	16,409
65183	−29 05189	HR 3099, PX Pup		7 54 23	−30 09.0	3	6.34	.058	1.65	.013	1.50	.005	40	M6 III	16,409,2006
65182	−28 05121			7 54 26	−28 59.8	1	9.59		0.18		−0.29		2	B9 II	525
	−29 05191	LSS 876		7 54 26	−29 17.3	4	9.77	.004	0.21	.015	−0.71	.025	7	B1 Ia	409,540,976,1737
65079	+3 01848			7 54 27	+03 05.1	5	7.83	.007	−0.18	.013	−0.79	.005	36	B2 V(n)(e)	989,1212,1509,1729,2033
		Pup sq 3 # 21		7 54 28	−28 01.1	2	12.02	.000	0.21	.000	0.12	.000	4	B9 V	16,409
	−34 04120			7 54 31	−34 19.2	1	10.06		0.11		0.08		2		124
65467	−60 00927	NGC 2516 - 114		7 54 32	−60 42.6	2	9.44	.005	0.11	.005	0.04	.000	4	A0 IV/V	36,1683
65066	+9 01824	HR 3097		7 54 33	+08 46.6	3	6.04	.007	1.00	.000	0.86	.013	9	K0 III	15,1256,1355
65208	−27 04762			7 54 35	−28 03.7	2	9.22	.000	−0.12	.000	−0.40	.000	4	Ap Si	16,409
65362	−48 03318			7 54 38	−48 52.0	1	8.60		0.07		0.00		2	A0 V	401
65360	−46 03588			7 54 39	−47 10.9	1	8.35		−0.05		−0.16		2	B9 V	401
65315	−40 03655	HR 3107		7 54 41	−40 36.1	4	6.77	.009	−0.18	.008	−0.74	.006	11	B2 V	15,401,1732,2012
65123	+1 01959	HR 3098	★ AB	7 54 42	+01 15.7	5	6.36	.030	0.50	.008	−0.01	.008	14	F6 V	15,254,292,1078,1381
65228	−22 02087	HR 3102		7 54 42	−22 44.7	12	4.19	.008	0.72	.005	0.44	.023	57	F7/8 II	3,9,15,369,1008,1075*
65314	−40 03656			7 54 43	−40 26.9	1	8.80		0.95		0.60		1	G6 III	565
64958	+44 01693	HR 3094		7 54 46	+44 06.8	2	6.35	.010	1.05	.005	0.91	.005	4	K0	985,1733
		LTT 2995		7 54 46	−21 33.9	1	11.15		0.74		0.07		4		158
		G 111 - 47		7 54 47	+41 26.8	1	12.04		1.63		1.15		5		316
65227	−22 02088	V335 Pup		7 54 48	−22 41.4	1	8.72		0.80		0.45		5	F3 II	3072
65158	−0 01864			7 54 53	−00 30.0	1	7.17		0.01				4	A0 V	2012
65176	−1 01900	LS VI -01 020		7 54 53	−01 28.7	4	8.23	.113	−0.13	.037	−1.03	.023	7	B0 II	399,405,490,8100
65243	−14 02293			7 54 55	−14 27.5	1	7.99		0.63				4	G3/5 V	2012
65424	−49 03201			7 54 57	−49 50.8	2	7.84	.020	2.97	.080	4.73	.370	2	Nb	109,864
65312	−29 05209			7 54 59	−29 34.8	1	9.54		0.41		−0.07		2	F6 V	409
65223	−8 02148			7 54 60	−08 23.9	1	9.88		0.25		0.14		1	A3	1732
65313	−29 05210			7 55 00	−29 36.7	2	9.00	.000	−0.03	.000	−0.09	.000	2	A0 V	16,409
		LSS 882		7 55 01	−27 46.6	1	11.82		0.09		−0.66		1	B3 Ib	409
65554	−60 00930	NGC 2516 - 115		7 55 01	−60 19.6	2	9.06	.009	0.06	.000	−0.11	.009	4	B9 V	36,1683
65311	−28 05144			7 55 03	−29 06.6	1	9.14		0.84				15	F3 IV/V	6011
65578	−60 00932	NGC 2516 - 116		7 55 03	−61 02.1	2	8.20	.014	0.02	.009	−0.21	.000	4	B8 V	36,1683
		Pup sq 3 # 54		7 55 05	−27 43.4	1	10.77		1.08		1.61		1	K0	409
	−28 05145			7 55 05	−29 08.8	1	9.13		1.31				19	K5	6011
65041	+43 01754			7 55 06	+43 38.4	1	7.31		−0.19		−0.75		3	B2 V	399
		Pup sq 3 # 97		7 55 07	−29 39.4	1	12.72		0.29		0.07		2	A0	409
		Pup sq 3 # 53		7 55 10	−27 46.8	1	13.61		0.54		0.00		2		409
	−59 00936	NGC 2516 - 117		7 55 10	−60 07.4	2	9.62	.005	0.10	.005	0.05	.005	4	A2	36,1683
		G 50 - 11		7 55 13	+09 51.4	1	12.32		1.02		0.85		1		333,1620
65378	−32 04612			7 55 13	−33 06.9	1	7.30		−0.17		−0.65		2	B3 V	401
65599	−59 00937	NGC 2516 - 118		7 55 13	−60 06.5	3	8.87	.005	0.01	.009	−0.12	.005	6	B9 V	36,1683,3015
65377	−29 05216			7 55 16	−29 34.6	1	9.08		0.90		0.55		2	G8 III	409
		Pup sq 3 # 96		7 55 17	−29 32.0	1	13.33		0.31		0.23		1	B8	409
65440	−40 03673	IDS07536S4056	AB	7 55 17	−41 03.7	1	8.08		−0.10		−0.63		2	B5 II/III	401
65442	−42 03717	HR 3111		7 55 17	−42 16.3	2	6.09	.002	1.36	.001	1.61		6	K3 III	58,2035
65492	−47 03442			7 55 17	−47 20.0	1	8.10		−0.08		−0.45		2	B7/8 II	401
	−32 04617			7 55 18	−33 01.5	1	10.49		0.34		0.30		2		124
	−28 05151	LSS 884		7 55 19	−28 29.9	3	11.14	.004	0.29	.003	−0.66	.032	5	O8 V	353,469,525
65441	−41 03565			7 55 19	−41 30.4	1	7.82		0.00		−0.03		2	A0 V	401
65460	−43 03758	HR 3114		7 55 19	−43 21.9	3	5.36	.009	−0.18	.005	−0.74	.007	8	B3 IV/V	401,1732,2035
	−28 05154	LSS 885		7 55 23	−28 34.0	4	10.96	.048	0.36	.015	−0.63	.012	8	B0 V	353,396,469,525
65307	−8 02151			7 55 24	−08 42.1	1	9.76		−0.19		−0.92		9	B3	1732
	−28 05156	LSS 888		7 55 24	−29 01.9	3	10.20	.013	0.42	.012	−0.46	.015	6	B1 III	409,525,540
65241	+7 01879	HR 3103		7 55 25	+07 21.0	3	6.40	.005	−0.04	.016	−0.07	.021	8	A0 V	15,1079,1415
65277	−0 01866	IDS07529S0033	AB	7 55 25	−00 40.7	1	8.09		1.03		0.83		1	K5 V	3072
65662	−60 00935	NGC 2516 - 119	★	7 55 28	−60 23.5	8	5.75	.016	1.56	.007	1.80	.018	27	K4 III	15,36,188,460,1075*
65663	−60 00937	NGC 2516 - 120		7 55 28	−60 57.9	3	6.76	.010	0.01	.004	−0.26	.012	5	B8 IIIe	36,188,1683
65575	−52 01343	HR 3117, χ Car		7 55 30	−52 50.8	4	3.46	.005	−0.18	.005	−0.67	.010	34	B3 IVp	15,200,1075,2012
65691	−60 00939	NGC 2516 - 6		7 55 31	−60 40.0	4	8.93	.019	0.04	.002	−0.12	.013	11	B8.5V	36,188,1683,3015
65348	−8 02154			7 55 33	−08 36.8	1	9.66		0.09		0.12		7	A0	1732
		Pup sq 3 # 76		7 55 34	−28 41.6	1	12.02		0.14				2	A2	409
	−28 05165			7 55 35	−28 19.6	1	10.77		0.08		−0.62		3	B2 V	525
65712	−60 00941	NGC 2516 - 230		7 55 35	−61 08.9	1	9.38		0.02		−0.07		6	A0p	1683
		AAS21,193 # 118		7 55 36	−29 02.	1	11.52		0.24		−0.62		2		525
65437	−27 04797			7 55 39	−27 26.1	1	9.77		0.42		0.05		1	F7 V	409
65257	+16 01598	HR 3104		7 55 40	+16 39.3	1	5.99		1.47				2	K0	71
65456	−29 05236	HR 3113		7 55 40	−30 11.9	7	4.79	.005	0.15	.000	0.20	.024	53	A7 III	15,16,409,1075,2012*
65551	−43 03766	HR 3116	★ A	7 55 41	−43 58.5	2	5.08	.005	−0.18	.005			7	B2 III/IV	15,2027
	−31 05295			7 55 43	−31 29.5	1	10.57		0.16		0.10		2		124
	−31 05296			7 55 44	−31 23.7	1	10.20		0.18		0.18		2		124
		NGC 2516 - 103		7 55 44	−60 37.9	1	12.32		0.70		−0.08		1		188
		NGC 2516 - 102		7 55 44	−60 45.0	1	11.41		0.14		−0.16		2		188

HD	DM	Other Id	N Rem	α_{1950}	δ_{1950}	S	V	σ_V	B–V	σ_{B-V}	U–B	σ_{U-B}	n	Spectrum	References
65345	+2 01833	HR 3110	⋆A	7 55 45	+02 21.6	4	5.30	.023	0.93	.014	0.71	.010	10	K0 III	15,1003,1415,3016
65455	−28 05172			7 55 46	−28 41.5	1	9.84		0.28		0.14		2	A9 (IV)	409
	+45 01515	G 111 - 48		7 55 47	+45 35.0	2	10.05	.028	0.86	.006	0.50	.009	5	K3	1726,7010
	−60 00942	NGC 2516 - 7		7 55 49	−60 38.5	3	9.96	.011	0.16	.007	0.13	.020	7	A1 V	36,188,1683
		NGC 2516 - 104	⋆D	7 55 49	−60 40.7	2	11.17	.137	0.51	.024	0.10	.029	5		1279,1683
65598	−47 03457	HR 3118	⋆AB	7 55 51	−47 45.3	3	6.21	.004	-0.11	.009	-0.51		9	B5 V	15,401,2027
	−60 00943	NGC 2516 - 121	⋆A	7 55 51	−60 29.8	1	10.61		0.28		0.13		3		1683
	−60 00943	NGC 2516 - 121	⋆AB	7 55 51	−60 29.8	2	10.11	.000	0.32	.004	0.15	.004	6		36,1683
	−60 00943	NGC 2516 - 121	⋆B	7 55 51	−60 29.8	1	11.24		0.42		0.18		3		1683
		LSS 892		7 55 53	−32 00.5	2	11.53	.005	0.27	.010	-0.50	.010	5	B3 Ib	1712,1766
	−33 04354			7 55 53	−33 49.2	1	8.84		1.16		1.02		1	K5 V	3072
		NGC 2516 - 112	⋆C	7 55 53	−60 40.7	2	9.84	.013	0.27	.000	0.20	.004	8	A2 IV	1279,1683
65372	+3 01860			7 55 54	+03 04.3	1	6.67		0.14		0.20		2	A3	252
		NGC 2516 - 106		7 55 54	−60 46.5	1	11.22		0.39		0.06		2		460
65622	−45 03603			7 55 55	−46 11.5	2	7.07	.000	-0.12	.005	-0.53		6	B5 V	401,2012
65710	−57 01354			7 55 55	−57 17.7	1	7.44		1.04		0.88		4	K1 III	709
65750	−58 01028	HR 3126, V341 Car		7 55 55	−58 59.4	2	6.40	.195	1.93	.000	2.18		6	M0 III	709,2006
		KUV 343 - 16		7 55 56	+39 23.1	1	14.31		0.49		-0.06		1		974
65511	−27 04804	LSS 889		7 55 56	−27 42.0	3	8.07	.017	0.45	.012	0.01	.034	12	B9 Ib	1655,2014,8100
65486	−25 05342			7 55 57	−25 29.2	3	8.42	.025	1.04	.019	0.85	.037	10	K3 V	158,2012,3072
	−28 05180	Ru 44- 183		7 55 57	−28 27.3	3	10.92	.022	0.29	.005	-0.69	.010	5		353,396,469
		NGC 2516 - 209	⋆B	7 55 57	−60 40.7	3	8.80	.025	0.08	.009	-0.26	.023	9	B8 III	188,1279,1683
	60 00944	NGC 2516 - 208	⋆A	7 55 57	−60 41.0	3	8.35	.010	0.02	.007	-0.33	.010	9	B8p	188,1279,1683
65324	+21 01730			7 55 58	+21 17.4	1	8.28		0.24		0.07		2	A0	1648
	−60 00945	NGC 2516 - 5		7 55 58	−60 42.3	4	8.52	.009	0.06	.013	-0.23	.015	14	B8.5IV	36,188,1683,2012
		NGC 2516 - 4		7 55 59	−60 48.7	1	10.79		0.38		0.18		1	A7 III	188
65343	+12 01733			7 56 00	+12 15.4	1	8.41		1.05		0.92		3	G5	1733
		G 40 - 1		7 56 01	+18 43.6	1	13.18		0.84		0.30		2		333,1620
65592	−39 03831	AP Pup		7 56 01	−39 59.2	1	7.10		0.69				1	F8 II	1484
65371	+10 01677	G 50 - 13		7 56 02	+10 16.0	1	8.15		0.81		0.38		1	K0	333,1620
65619	−43 03774			7 56 02	−43 20.3	1	9.46		-0.13		-0.61		1	B2/5	401
	−28 05182			7 56 03	−28 26.4	1	11.14		0.16		-0.45		3	B2 V	525
	−28 05184			7 56 04	−28 48.3	1	9.82		0.50		-0.03		3	F8 V	409
65620	−43 03776			7 56 04	−43 21.8	1	8.17		-0.14		-0.58		2	B9 II/III	401
	−60 00947	NGC 2516 - 13		7 56 04	−60 27.8	5	8.06	.011	-0.03	.009	-0.32	.010	14	B8 V	36,188,1683,2012,3015
65658	−46 03616	IDS07546S4619	AB	7 56 06	−46 27.4	2	7.23	.000	-0.13	.000	-0.64	.005	6	B3 V	158,401
	−31 05309	LSS 893		7 56 08	−31 58.1	4	9.15	.012	0.26	.011	-0.63	.026	10	B1 II:	211,540,976,1712
	−60 00948	NGC 2516 - 8		7 56 10	−60 38.9	4	9.69	.012	0.13	.004	0.10	.016	8	A0 V	36,188,1683,3015
65275	+35 01722			7 56 11	+34 49.0	1	7.30		1.65		1.93		2	M1	1601
65656	−41 03583			7 56 13	−41 30.3	1	7.53		-0.11		-0.54		2	B7 IV/V	401
		NGC 2516 - 108		7 56 13	−60 28.5	1	11.15		0.40		0.10		1		188
	−35 04061			7 56 14	−35 46.1	3	10.44	.012	0.54	.007	-0.11	.017	7	G2p	158,1696,3060
65615	−37 04102			7 56 16	−37 26.6	1	7.47		-0.10		-0.55		2	B4 V	401
65685	−45 03611	HR 3121		7 56 17	−46 26.5	4	5.15	.013	1.26	.010	1.36		13	K2 III	15,1075,2031,3005
	−28 05193	LSS 894		7 56 19	−29 03.0	1	10.20		0.36				2		409
65589	−28 05196	AQ Pup		7 56 21	−28 59.6	4	8.21	.112	1.12	.116	0.77	.000	4	K5	689,1484,1587,6011
65835	−60 00949	NGC 2516 - 3		7 56 22	−60 45.2	2	9.44	.005	0.09	.012	-0.02	.002	3	A0 V	188,3015
	−26 05283	NGC 2467 - 97		7 56 25	−26 30.9	1	10.29		0.46		-0.44		1	B2 III	1376
	−26 05285	NGC 2467 - 98		7 56 26	−26 26.0	3	10.69	.003	0.32	.017	-0.62	.015	3	O6.5IIIf	540,1376
65683	−39 03839			7 56 27	−39 28.6	1	7.99		-0.03		-0.05		2	B9 V	401
		NGC 2516 - 100		7 56 27	−60 39.1	1	12.04		0.54		-0.03		2		188
	−60 00951	NGC 2516 - 101		7 56 28	−60 26.7	2	10.18	.020	0.19	.030	0.17	.032	2	A2 V	188,3015
		NGC 2516 - 158		7 56 28	−60 44.3	1	14.48		0.68		0.06		2		3015
		AJ79,1294 T8# 19		7 56 29	−28 26.4	1	12.65		0.20		-0.13		3		396
	−60 00952	NGC 2516 - 9		7 56 30	−60 40.6	4	9.29	.013	0.12	.015	0.01	.004	11	B9 V	36,188,1683,3015
65869	−60 00953	NGC 2516 - 10		7 56 31	−60 37.9	5	7.73	.020	0.03	.003	-0.20	.009	16	B8.5V	36,188,1683,2012,3015
		G 111 - 49		7 56 32	+43 43.4	1	16.28		0.11		-0.44		2	DC	782
65562	−12 02252			7 56 33	−13 10.6	1	8.82		0.87		0.56		3	K0 V	3072
		NGC 2516 - 159		7 56 33	−60 43.9	1	14.35		0.71		0.17		2		3015
65560	−10 02315			7 56 34	−11 05.8	1	9.07		-0.08		-0.40		4	A0	1732
		NGC 2516 - 107		7 56 34	−60 41.5	2	11.33	.015	0.52	.015	0.07	.019	6		188,460
		NGC 2516 - 200		7 56 35	−60 38.6	1	13.09		0.70		0.10		1		3015
65908	−62 00925	HR 3139		7 56 35	−63 09.7	2	6.13	.005	-0.11	.005			7	B8 V	15,2012
		Ru 44 - 84		7 56 36	−28 26.2	1	14.97		0.64		-0.10		2		354
65430	+21 01731	G 40 - 2		7 56 37	+20 59.3	3	7.72	.014	0.84	.012	0.49	.010	6	K0 V	22,333,1003,1620
		KUV 07566+4221		7 56 37	+42 21.1	1	16.88		0.27		-0.66		1		1708
		NGC 2516 - 197		7 56 37	−60 41.1	1	12.86		0.72		0.08		2		3015
		Ru 44 - 88		7 56 39	−28 24.5	1	14.80		0.49		-0.22		2		354
		NGC 2516 - 167		7 56 39	−60 42.2	1	14.57		0.94		0.55		1		3015
	−60 00954	NGC 2516 - 40		7 56 39	−60 42.2	3	9.92	.024	0.20	.012	0.15	.004	7	A1 V	36,188,1683
		NGC 2516 - 160		7 56 39	−60 43.3	1	13.46		0.79		0.18		2		3015
65852	−57 01358			7 56 40	−57 20.6	2	7.41	.000	0.04	.005	-0.24	.000	6	B8 V	401,709
	−28 05205	Ru 44- 184		7 56 41	−28 18.1	1	11.16		0.29		-0.70		1		353
		Ru 44 - 6		7 56 41	−28 26.8	1	14.80		0.52		-0.16		2		354
65896	−60 00955	NGC 2516 - 12		7 56 41	−60 29.7	4	9.17	.013	0.11	.011	0.11	.013	8	A0 V	36,188,1683,3015
		NGC 2516 - 198		7 56 41	−60 40.3	1	14.08		0.92		0.26		1		3015
		NGC 2516 - 196		7 56 42	−60 36.0	1	14.50		1.01		0.85		1		3015
	−28 05207	Ru 44 - 94		7 56 43	−28 24.6	3	11.29	.011	0.49	.015	-0.64	.018	5		353,396,469
		Ru 44 - 60		7 56 43	−28 28.5	1	13.72		0.40		-0.38		2		354
		Ru 44 - 59		7 56 43	−28 28.6	1	13.15		0.42		-0.46		2		354

Table 1 437

HD	DM	Other Id	N Rem	α₁₉₅₀	δ₁₉₅₀	S	V	σV	B-V	σB-V	U-B	σU-B	n	Spectrum	References
		NGC 2516 - 157		7 56 43	-60 45.7	1	14.07		1.05		0.65		1		3015
	-59 00943	NGC 2516 - 123		7 56 44	-60 09.1	2	9.70	.005	0.23	.009	0.23	.014	4	A0	36,1683
		Ru 34 - 4		7 56 45	-28 27.4	1	13.32		0.21		-0.02		2		410
65721	-34 04160			7 56 45	-34 48.7	2	7.94	.005	0.74	.000	0.24		8	G6 V	158,2012
		AJ79,1294 T8# 20		7 56 46	-28 32.3	1	12.75		0.26		0.13		3		396
65931	-60 00960	NGC 2516 - 39		7 56 46	-60 55.3	3	8.77	.015	0.08	.007	-0.13	.019	6	B9 V	36,188,1683
		Ru 44- 102		7 56 47	-28 23.3	2	13.56	.010	0.41	.005	-0.43	.005	5		354,883
		Ru 44 - 3		7 56 47	-28 27.6	1	12.66		0.42		-0.43		2		354
		NGC 2516 - 195		7 56 47	-60 34.1	1	13.35		0.80		0.30		1		3015
65522	+13 01811	HR 3115		7 56 48	+13 22.8	1	6.02		1.32				2	K5	71
65723	-36 04067			7 56 48	-36 52.9	3	7.01	.010	0.98	.000	0.69	.005	9	K0/1 V	1075,1311,3077
65818	-48 03349	HR 3129, V Pup	⋆ AB	7 56 48	-49 06.5	6	4.44	.120	-0.18	.002	-0.95	.013	20	B1 Vp+B3IV:	15,401,1075,1088,2012,3019
		NGC 2516 - 199		7 56 48	-60 40.0	1	13.78		0.83		0.32		1		3015
		NGC 2516 - 162		7 56 48	-60 42.2	1	11.66		1.50		1.68		2		3015
	-60 00962	NGC 2516 - 97		7 56 48	-60 45.7	2	10.87	.005	0.32	.002	0.24	.032	5		188,3015
65523	+13 01810	G 91 - 23		7 56 49	+13 07.3	1	8.33		0.77		0.41		1	G5	333,1620
65677	-23 06612			7 56 49	-23 17.0	1	9.15		0.95				19	G8 III/IV	6011
		Ru 44- 105		7 56 49	-28 24.1	1	15.08		0.44		-0.24		2		354
		Ru 44 - 14		7 56 49	-28 25.3	2	13.82	.010	0.45	.015	-0.32	.010	5		354,883
		NGC 2516 - 96		7 56 49	-60 45.3	2	11.19	.054	0.43	.044	0.11	.049	3		188,3015
	-60 00961	NGC 2516 - 2		7 56 49	-60 46.4	4	8.80	.022	0.07	.009	-0.07	.008	10	A0 IV-V	36,188,1683,3015
65607	-7 02291	UX Mon		7 56 50	-07 22.1	1	8.32		0.30		-0.19		1	G2 III	1588
		Ru 44 - 17		7 56 50	-28 26.1	1	14.55		0.47		-0.20		3		883
		Ru 44 - 19		7 56 50	-28 26.9	2	12.40	.184	0.34	.000	-0.51	.015	7		396,396
		Ru 44 - 2		7 56 50	-28 27.9	2	12.64	.015	0.54	.135	-0.16	.245	4		354,410
	-28 05216	Ru 44- 182		7 56 50	-28 36.8	1	11.27		0.27		-0.68		1		353
	-34 04162			7 56 51	-34 51.2	1	10.64		0.13		-0.17		2		540
	-60 00963	NGC 2516 - 57		7 56 51	-60 35.5	4	10.18	.013	0.15	.002	0.17	.013	9	A2 V	36,188,1683,3015
		Ru 44 - 24		7 56 52	-28 26.8	1	12.81		0.36		-0.44		2		354
		Ru 44 - 51		7 56 52	-28 28.0	2	14.33	.010	0.35	.010	-0.43	.015	5		354,883
		Ru 44 - 44		7 56 52	-28 29.9	1	14.49		0.56		-0.16		3		883
65907	-59 00944	NGC 2516 - 124	⋆ A	7 56 52	-60 10.1	7	5.60	.011	0.57	.004	0.01	.017	24	G0 V	15,36,1279,1683,2012*
		Ru 44- 114		7 56 53	-28 24.9	1	13.72		0.40		-0.40		2		354
		Ru 44 - 49		7 56 53	-28 28.3	1	15.03		0.63		-0.18		2		354
		CoD -59 01774		7 56 53	-60 16.6	1	9.93		1.35		1.13		2		3060
	-60 00964	NGC 2516 - 14		7 56 53	-60 29.7	4	9.43	.007	0.05	.009	0.02	.016	9	A0 V	36,188,1683,3015
		Ru 44 - 27		7 56 54	-28 27.2	1	11.93		0.35		-0.56		2		354
		NGC 2516 - 202		7 56 54	-60 38.9	1	14.27		0.77		0.15		1		3015
		NGC 2516 - 203		7 56 54	-60 39.3	1	13.83		0.76		0.20		2		3015
		NGC 2516 - 205		7 56 54	-60 40.5	1	12.23		0.43		0.19		1		3015
		Ru 44- 112		7 56 55	-28 24.6	2	13.58	.029	0.39	.005	-0.37	.049	6		354,883
		Ru 44 - 33		7 56 55	-28 25.3	1	11.60		0.37		-0.59		1	B0	353
		NGC 2516 - 95		7 56 55	-60 39.7	2	11.75	.010	0.49	.005	0.12	.049	6		188,3015
	-28 05218	Ru 44- 185		7 56 56	-28 15.2	1	10.58		0.31		-0.69		1		353
		Ru 44 - 41		7 56 56	-28 30.4	2	12.16	.005	0.36	.010	-0.52	.020	2		353,469
		NGC 2516 - 201		7 56 56	-60 38.4	1	14.53		0.80		0.26		1		3015
	-60 00965	NGC 2516 - 42	⋆ V	7 56 56	-60 38.9	2	10.59	.031	0.27	.000	0.21	.013	6	A3 V	188,3015
		NGC 2516 - 93		7 56 56	-60 40.5	2	11.11	.024	0.36	.041	0.14	.034	3		188,3015
	-60 00968	NGC 2516 - 41	⋆ V	7 56 56	-60 41.4	4	9.01	.022	0.04	.016	-0.10	.018	28	B9p Shell	36,188,1683,3015
65950	-60 00967	NGC 2516 - 126		7 56 56	-60 47.4	5	6.87	.008	0.00	.007	-0.31	.014	25	B8.5IIIp	36,188,1683,2012,3015
65699	-22 02104	HR 3123		7 56 57	-23 10.4	3	5.10	.012	1.12	.000	0.92		8	G8 II	15,565,2012
65949	-60 00966	NGC 2516 - 91		7 56 57	-60 28.4	4	8.38	.012	-0.02	.009	-0.36	.005	20	B9 IVp	36,188,1683,3015
	-28 05221	LS Pup		7 56 58	-29 10.3	1	10.12		1.11		0.78		1		1587
	-32 04657	LSS 905		7 56 58	-32 25.2	1	9.53		0.08		-0.55		3		1712
	-60 00970	NGC 2516 - 94		7 56 58	-60 40.8	4	9.59	.013	0.11	.007	0.05	.006	18	A0 IV	36,188,1683,3015
	-60 00969	NGC 2516 - 11		7 56 59	-60 32.8	5	8.55	.005	0.06	.009	-0.03	.012	15	B9.5V	36,188,1683,2012,3015
		NGC 2516 - 182		7 56 59	-60 34.3	1	15.40		0.81		0.20		1		3015
	-60 00971	NGC 2516 - 58		7 56 59	-60 35.1	4	9.66	.019	0.08	.012	0.04	.011	9	A0 V	36,188,1683,3015
		Pup sq 2 # 34		7 57 00	-26 54.	1	11.66		0.38		-0.52		1		211
		Ru 44 - 40		7 57 00	-28 29.6	2	13.81	.080	0.39	.005	-0.38	.025	6		396,883
65817	-47 03472			7 57 00	-47 15.3	1	8.23		-0.04		-0.30		2	B8 III/IV	401
64486	+79 00265	HR 3082	⋆ AB	7 57 01	+79 37.2	2	5.41	.015	-0.06	.005	-0.14	.005	5	A0p Si	985,1733
65867	-51 02784	HR 3133		7 57 01	-51 18.7	1	6.44		0.25				4	A8/9 IV	2035
		NGC 2516 - 90		7 57 01	-60 31.0	1	11.53		0.80		0.12		2		188
		NGC 2516 - 125		7 57 02	-60 09.9	3	9.86	.010	1.36	.021	1.08	.003	5		36,1279,1683
		NGC 2516 - 180		7 57 02	-60 33.9	1	15.35		0.87		0.61		1		3015
	-60 00972	NGC 2516 - 44		7 57 02	-60 36.9	2	10.87	.030	0.34	.005	0.19	.017	4	A7 III	188,3015
		NGC 2516 - 214		7 57 02	-60 38.3	1	13.21		1.43		1.58		3		1683
		NGC 2516 - 215		7 57 03	-60 38.8	1	13.14		0.76		0.20		3		1683
		Ru 44- 148		7 57 04	-28 28.6	2	12.12	.015	0.41	.025	-0.55	.010	2		353,469
		NGC 2516 - 181		7 57 05	-60 35.1	1	14.04		0.88		0.37		2		3015
		NGC 2516 - 216		7 57 05	-60 38.2	1	14.39		1.52		1.79		4		1683
		Ru 44- 186		7 57 06	-28 22.9	2	11.15	.010	0.37	.000	-0.63	.005	2		353,469
		NGC 2516 - 179		7 57 06	-60 34.2	1	13.10		1.03		0.92		2		3015
		NGC 2516 - 156		7 57 06	-60 44.8	1	12.74		1.15		0.97		2		3015
	-60 00974	NGC 2516 - 45		7 57 07	-60 36.1	4	9.87	.009	0.19	.003	0.18	.012	10	A2 V	36,188,1683,3015
	-60 00973	NGC 2516 - 43		7 57 07	-60 37.2	2	9.38	.020	0.10	.007	0.03	.020	4	A1 V	188,3015
		NGC 2516 - 161		7 57 07	-60 43.5	1	14.06		1.06		0.81		1		3015
		NGC 2516 - 152		7 57 07	-60 46.8	1	14.38		0.93		0.48		1		3015
		NGC 2516 - 163		7 57 08	-60 41.6	1	13.44		0.63		0.03		1		3015

HD	DM	Other Id	N	Rem	α_{1950}	δ_{1950}	S	V	σ_V	B–V	σ_{B-V}	U–B	σ_{U-B}	n	Spectrum	References
	−60 00975	NGC 2516 - 1			7 57 08	−60 42.6	2	8.99	.041	0.11	.018	0.09	.032	5	A0 V	188,3015
		NGC 2516 - 155			7 57 08	−60 45.4	1	14.55		0.92		0.46		3		3015
		KUV 343 - 14			7 57 09	+42 32.8	1	14.97		0.57		−0.09		1	sdB	974
65301	+59 01130	HR 3106			7 57 10	+59 11.1	1	5.79		0.38		−0.06		3	F4 V	254
	−26 05309				7 57 10	−26 31.8	1	10.97		0.08		0.01		2		126
		Ru 44- 128			7 57 10	−28 25.9	1	11.59		0.36		−0.66		1		353
		Ru 44- 251			7 57 12	−28 31.6	2	14.67	.049	0.50	.034	−0.32	.019	6		354,883
65987	−60 00976	NGC 2516 - 15		⋆ V	7 57 12	−60 28.7	5	7.63	.017	−0.07	.017	−0.39	.014	29	B9.5IVp	36,188,1683,2012,3015
		NGC 2516 - 88			7 57 12	−60 34.7	2	11.93	.045	0.54	.020	0.04	.005	2		188,3015
	−60 00977	NGC 2516 - 46			7 57 12	−60 35.5	2	10.37	.014	0.38	.005	0.11	.005	4	F0 V	188,3015
		NGC 2516 - 193			7 57 12	−60 38.6	1	12.05		0.61		0.05		1		3015
	+32 01660	G 90 - 35			7 57 13	+32 15.8	1	10.31		0.86		0.61		1		333,1620
		NGC 2516 - 85			7 57 13	−60 38.6	2	12.04	.035	0.61	.015	0.06	.000	5		188,1683
		NGC 2516 - 184			7 57 13	−60 38.6	1	12.25		0.54		0.01		2		3015
	−60 00978	NGC 2516 - 127		⋆ V	7 57 13	−60 40.6	4	8.95	.011	0.07	.009	−0.24	.009	27	A0p	36,188,1683,3015
		NGC 2516 - 151			7 57 13	−60 46.8	1	16.29		0.80		0.25		1		3015
65695	−3 02157	HR 3122			7 57 14	−03 32.5	4	4.93	.012	1.21	.005	1.20	.009	16	K2 III	15,1415,2012,3016
		NGC 2516 - 183			7 57 15	−60 36.9	1	12.74		0.62		0.05		2		3015
		NGC 2516 - 82			7 57 15	−60 39.5	3	11.97	.010	0.53	.029	0.05	.039	6		188,1683,3015
		NGC 2516 - 217			7 57 16	−60 40.4	1	13.32		1.24		1.11		5		1683
		NGC 2516 - 84		⋆ V	7 57 18	−60 38.3	2	10.68	.033	0.35	.005	0.17	.005	4	A7 V	188,1683
	−60 00979	NGC 2516 - 83			7 57 18	−60 38.6	3	8.39	.009	0.05	.006	−0.19	.018	6	B8.5V	188,1683,3015
		NGC 2516 - 86			7 57 19	−60 36.5	1	10.22		0.19		0.18		1		188
		NGC 2506 - 4402			7 57 20	−10 35.1	1	12.58		1.73		2.08		2		884
65905	−47 03480	IDS07558S4725	AB		7 57 20	−47 32.7	1	7.27		−0.09		−0.41		2	B7 V	401
		NGC 2516 - 80			7 57 20	−60 40.8	3	12.17	.012	1.20	.005	1.12	.039	7		36,1683,3015
	−60 00980	NGC 2516 - 128			7 57 20	−60 40.8	7	6.70	.017	1.28	.005	1.19	.013	28	K1 III	36,188,460,1683,1685*
	−28 05235	LSS 909			7 57 21	−28 46.1	3	10.07	.002	0.30	.018	−0.86	.043	4	O9 V	353,540,1737
65848	−39 03858				7 57 21	−39 17.0	1	7.24		−0.09		−0.39		2	B5 III/IV	401
		NGC 2516 - 218			7 57 21	−60 37.6	1	14.06		0.64		−0.01		3		1683
		NGC 2516 - 81			7 57 21	−60 41.4	3	12.00	.019	0.55	.009	0.04	.004	7		36,1683,3015
	−60 00981	NGC 2516 - 38		⋆ V	7 57 21	−60 43.7	5	9.51	.018	0.21	.007	0.15	.010	10	A2 Vp	36,188,460,1683,3015
65793	−29 05289				7 57 22	−29 12.8	1	8.41		0.52		0.01		5	F5 V	1655
	−33 04377	LSS 910			7 57 22	−33 32.1	1	9.92		1.04		0.75		3	F1 II	1712
		G 90 - 36			7 57 23	+29 09.0	1	12.69		0.58		−0.13		3		1620
		G 90 - 36			7 57 23	+29 09.0	1	12.70		0.58		−0.13		1		1774
	−26 05315				7 57 23	−26 46.3	1	9.46		0.43		−0.03		2	F8	126
	−60 00982	NGC 2516 - 129			7 57 23	−60 39.9	5	7.22	.011	0.05	.005	−0.09	.006	112	B9.5IV	36,188,1683,2012,3015
	−60 00985	NGC 2516 - 37			7 57 23	−60 44.2	5	8.21	.011	0.01	.012	−0.20	.007	14	B8.5V	36,188,1683,2012,3015
		NGC 2506 - 4331			7 57 24	−10 36.4	1	15.18		0.39				2		884
	−33 04378	LSS 911			7 57 24	−33 32.1	1	11.09		0.50		−0.13		3	B7 II	1712
	−60 00983	NGC 2516 - 59		⋆ V	7 57 24	−60 30.8	2	10.43	.018	0.29	.005	0.20	.000	5		36,1683
	−60 00984	NGC 2516 - 47			7 57 24	−60 35.3	4	9.67	.015	0.23	.005	0.15	.014	10	A2 V	36,188,1683,3015
		NGC 2506 - 3392			7 57 25	−10 39.0	1	13.12		0.91		0.50		1		884
65891	−41 03606				7 57 25	−41 27.2	1	8.50		0.56				4	G0/2 V	2012
	−60 00986	NGC 2516 - 50			7 57 25	−60 37.1	1	10.12		0.16		0.09		1		3015
65904	−44 03920	HR 3137			7 57 26	−45 04.7	2	5.98	.005	−0.15	.005			7	B3 V	15,2027
65583	+29 01664	G 90 - 36			7 57 27	+29 22.0	6	7.00	.016	0.71	.011	0.18	.009	19	G8 V	22,1003,1080,1197*
65339	+60 01105	HR 3109, AX Cam		⋆ AB	7 57 27	+60 27.8	4	6.01	.008	0.15	.013	0.05	.016	11	A2p SrCrEu	766,1063,1202,1263
	−28 05237				7 57 27	−28 43.0	1	10.76		0.07		0.06		2	B9.5V	396
		NGC 2516 - 164			7 57 27	−60 41.2	2	13.98	.005	0.74	.007	0.19	.028	4		1683,3015
		NGC 2516 - 165			7 57 27	−60 42.0	2	13.94	.033	1.46	.007	1.31		4		1683,3015
		NGC 2516 - 150			7 57 27	−60 46.7	1	14.02		0.87		0.41		1		3015
	+14 01802				7 57 28	+13 57.1	2	10.31	.006	1.28	.001	1.30		5	K5	1017,7008
		NGC 2506 - 3265			7 57 28	−10 38.4	1	13.19		1.04		0.58		1		884
65813	−28 05236				7 57 28	−28 17.7	1	7.70		0.32		0.08		4	A9 IV/V	396
		NGC 2516 - 111			7 57 28	−60 37.0	2	11.41	.005	0.55	.012	0.03	.039	3		188,3015
		NGC 2506 - 3258			7 57 29	−10 38.7	1	16.36		0.52				1		884
		NGC 2516 - 176			7 57 29	−60 33.8	1	12.39		0.60		−0.01		1		3015
		NGC 2506 - 4240			7 57 30	−10 36.6	1	13.13		0.99		0.69		2		884
		NGC 2506 - 3257			7 57 30	−10 38.6	1	15.90		0.59		0.19		1		305
		AJ79,1294 T8# 15			7 57 30	−28 44.7	1	11.97		0.44		0.03		2		396
		NGC 2516 - 177			7 57 30	−60 33.4	1	12.95		0.68		0.01		1		3015
	−60 00987	NGC 2516 - 55		⋆ V	7 57 30	−60 41.4	4	10.97	.004	0.33	.007	0.14	.011	8		36,188,1683,3015
		NGC 2516 - 154			7 57 30	−60 44.	1	14.55		0.47		0.30		1		3015
		NGC 2516 - 166			7 57 30	−60 44.	1	15.05		0.43		0.07		1		3015
		NGC 2516 - 194			7 57 30	−60 44.	1	12.92		0.75		0.08		1		3015
		NGC 2516 - 207			7 57 30	−60 44.	1	14.80		0.59		0.05		2		3015
		NGC 2516 - 149			7 57 30	−60 47.3	1	14.30		0.71		0.12		1		3015
		NGC 2506 - 3270			7 57 31	−10 38.4	2	16.69	.075	0.58	.065	0.28		2		305,884
		NGC 2506 - 3248			7 57 31	−10 39.2	1	17.18		0.52				2		884
		LOD 4+ # 15			7 57 31	−26 46.1	1	11.90		0.21		0.09		2		126
		NGC 2506 - 4228			7 57 32	−10 36.9	1	12.00		1.66		1.72		2		884
		NGC 2506 - 3255			7 57 32	−10 38.5	2	13.09	.025	0.84	.025	0.49	.005	2		305,884
		NGC 2506 - 3254			7 57 32	−10 38.8	2	11.13	.005	1.41	.005	1.47	.015	3		305,884
		NGC 2506 - 3241			7 57 32	−10 39.3	2	13.83	.010	0.12	.020	0.10	.010	4		305,884
65930	−48 03361	IDS07561S4806	AB		7 57 32	−48 14.3	2	6.84	.005	−0.13	.005	−0.74		6	B2 III/IV	401,2012
		NGC 2506 - 4254			7 57 33	−10 35.0	1	14.96		0.45		0.11		1		884
		NGC 2506 - 3231			7 57 33	−10 40.1	2	13.12	.000	0.97	.010	0.64	.054	3		305,884
		LSS 913			7 57 33	−31 56.4	1	11.82		1.01		−0.29		2	B3 Ib	1712

Table 1   439

HD	DM	Other Id	N	Rem	α_{1950}	δ_{1950}	S	V	σ_V	B–V	σ_{B-V}	U–B	σ_{U-B}	n	Spectrum	References
66066	−60 00988	NGC 2516 - 130	⋆ A		7 57 33	−60 43.2	1	7.18		0.01		−0.33		4	B8.5III	1683
66066	−60 00988	NGC 2516 - 130	⋆ AB		7 57 33	−60 43.2	4	6.77	.019	−0.02	.010	−0.31	.016	5		36,188,1683,3015
		NGC 2516 - 148			7 57 33	−60 46.8	1	14.21		0.94		0.42		2		3015
		NGC 2506 - 4401			7 57 34	−10 32.8	1	11.86		0.64		0.17		3		884
		NGC 2506 - 4241			7 57 34	−10 36.5	1	15.09		0.41		0.06		1		884
		NGC 2506 - 3239			7 57 34	−10 39.2	2	14.95	.025	0.44	.015	0.05	.005	2		305,884
	−60 00989	NGC 2516 - 36			7 57 34	−60 44.3	3	9.43	.023	0.08	.012	0.03	.014	6	A1 V	188,460,3015
		NGC 2506 - 4150			7 57 35	−10 36.5	1	16.71		0.59				3		884
		NGC 2506 - 4125			7 57 35	−10 37.1	1	15.63		0.36				2		884
	−26 05322				7 57 35	−26 42.8	1	10.42		1.25		1.42		2		126
		NGC 2516 - 147			7 57 35	−60 46.3	1	13.99		0.93		0.61		1		3015
		NGC 2506 - 4128			7 57 36	−10 37.1	1	13.02		0.88		0.50		2		884
	−48 03363				7 57 36	−49 02.5	1	8.75		−0.05		−0.37		2		401
		NGC 2516 - 49			7 57 36	−60 36.5	4	9.81	.027	1.35	.008	1.44	.013	10		36,188,1683,3015
		NGC 2506 - 3213			7 57 37	−10 40.4	1	14.33		0.59		0.09		2		884
	−26 05323				7 57 37	−26 46.3	1	9.94		0.07		0.05		2		126
	−60 00990	NGC 2516 - 132			7 57 37	−60 39.9	4	8.57	.013	0.05	.013	−0.16	.007	12	B9 V	36,188,1683,3015
		NGC 2506 - 4101			7 57 38	−10 37.9	1	15.92		0.42				1		884
65810	−18 02118	HR 3131			7 57 38	−18 15.7	10	4.61	.004	0.08	.000	0.08	.006	37	A2 Vn	15,1007,1013,1020*
	−59 00949	NGC 2516 - 131			7 57 38	−60 07.7	2	9.71	.014	0.42	.009	−0.01	.009	4	A0	36,1683
65666	+19 01896				7 57 39	+18 58.9	1	7.11		−0.02		−0.08		2	A0	1648
		NGC 2506 - 1305			7 57 39	−10 33.9	1	14.31		0.38				1		884
		NGC 2506 - 3206			7 57 39	−10 39.9	1	14.88		0.43		0.01		2		884
		NGC 2506 - 3204			7 57 39	−10 40.7	1	12.66		0.90		0.54		3		884
	−60 00991	NGC 2516 - 77			7 57 39	−60 33.6	4	9.71	.023	0.11	.002	0.09	.008	8	A0 V	36,188,1683,3015
		NGC 2506 - 2283			7 57 40	−10 40.2	1	15.21		0.38		0.07		2		884
	−60 00992	NGC 2516 - 60	⋆ V		7 57 40	−60 29.5	1	10.54		0.33		0.16		2	A8 V	188
	−60 00993	NGC 2516 - 48			7 57 40	−60 35.5	4	8.77	.028	0.01	.015	−0.24	.017	9	B8.5IV-V	36,188,1683,3015
		NGC 2506 - 1112			7 57 41	−10 37.5	1	12.95		0.98		0.60		2		884
65888	−36 04076				7 57 41	−37 09.9	1	7.43		−0.13		−0.60		2	B2/3 V	401
		NGC 2516 - 79			7 57 41	−60 38.3	2	11.65	.019	0.51	.007	0.08	.022	3		188,3015
		NGC 2516 - 190			7 57 41	−60 39.2	1	12.93		0.81		0.39		1		3015
	−60 00994	NGC 2516 - 34			7 57 41	−60 49.1	1	10.33		0.19		0.18		1	A3 V	188
		CCS 978			7 57 42	−04 37.	1	10.41		1.48				1	R2	1238
		NGC 2506 - 2402			7 57 42	−10 41.5	1	12.45		1.19		1.07		3		884
65925	−38 03908	HR 3140			7 57 42	−39 09.5	3	5.22	.014	0.40	.004	0.02		9	F3 V	15,2012,3053
		NGC 2516 - 191			7 57 42	−60 39.0	1	14.68		0.68		0.08		1		3015
		NGC 2516 - 146			7 57 42	−60 44.9	1	15.18		1.18		0.36		1		3015
		NGC 2506 - 2111			7 57 43	−10 38.4	1	11.59		0.57		0.11		2		884
		NGC 2506 - 2122			7 57 43	−10 38.9	1	11.73		1.13		0.98		2		884
		NGC 2506 - 2401			7 57 43	−10 41.9	1	11.09		1.59		1.94		3		884
		LSS 915			7 57 43	−28 52.6	1	11.33		0.43		−0.56		1	B3 Ib	469
		NGC 2516 - 78			7 57 43	−60 36.2	1	12.74		0.65				1		188
		NGC 2516 - 185			7 57 43	−60 36.2	1	12.87		0.66		0.09		1		3015
	−60 00997	NGC 2516 - 54	⋆ V		7 57 43	−60 41.2	4	10.55	.011	0.30	.013	0.15	.010	8	A3 V	36,188,1683,3015
	−60 00996	NGC 2516 - 35			7 57 43	−60 45.6	2	10.48	.029	0.25	.002	0.18	.015	3	A1 Vm	188,3015
65865	−28 05249	Ru 44- 187			7 57 44	−28 35.8	1	10.93		0.29		−0.60		1	O3	353
	−30 05382	LSS 918			7 57 44	−30 33.2	3	10.91	.033	0.44	.021	−0.65	.024	6	B1 IIInne2	211,1712,1737
	−60 00995	NGC 2516 - 16			7 57 44	−60 25.9	3	9.64	.021	0.12	.019	0.13	.022	6	A1 V	36,188,1683
	−60 00998	NGC 2516 - 52			7 57 44	−60 33.9	2	10.25	.029	0.41	.000	0.09	.027	3	A7 III	188,3015
		NGC 2506 - 1254			7 57 46	−10 37.2	1	16.89		0.54				2		884
		NGC 2506 - 1136			7 57 46	−10 37.5	1	14.70		0.40		−0.01		1		884
		NGC 2506 - 2212			7 57 46	−10 38.3	1	11.95		1.07		0.83		2		884
		Fld a # 4			7 57 46	−34 02.6	1	11.73		1.01		0.85		2		1712
		NGC 2516 - 186			7 57 46	−60 38.7	1	13.43		0.73		0.14		2		3015
		NGC 2506 - 1253			7 57 47	−10 37.0	1	17.77		0.76				2		884
	−26 05329				7 57 47	−26 34.5	1	11.66		0.09		−0.46		2		126
	−33 04389				7 57 47	−33 57.0	1	10.83		0.35		0.32		2		124
66108	−59 00950				7 57 47	−59 14.2	1	9.43		0.02		−0.54		4	B5/6 V	709
	−60 00999	NGC 2516 - 109			7 57 47	−60 25.4	1	10.81		0.29		0.13		1		188
	−60 01000	NGC 2516 - 18			7 57 47	−60 31.1	4	9.58	.024	0.10	.019	0.08	.014	9	A0 V	36,188,1683,3015
		NGC 2516 - 189			7 57 47	−60 36.9	1	15.29		0.69		0.38		1		3015
		NGC 2516 - 187			7 57 47	−60 37.8	1	12.51		0.62		0.01		1		3015
		NGC 2516 - 72			7 57 47	−60 42.1	3	11.72	.022	0.60	.034	0.05	.009	5		188,1683,3015
65736	+18 01816				7 57 48	+18 23.0	1	7.09		0.96		0.68		2	G5	1648
		NGC 2506 - 1258			7 57 48	−10 37.4	1	15.65		0.41		−0.04		1		884
		NGC 2516 - 168			7 57 48	−60 43.0	2	14.40	.027	0.48	.023	0.18	.005	5		1683,3015
		L 137 - 85			7 57 48	−63 48.	1	11.82		0.89		0.46		2		3060
65887	−30 05385				7 57 49	−30 16.5	1	7.69		1.26		1.25		2	K2 III	126
		NGC 2516 - 188			7 57 49	−60 37.2	1	14.33		0.94		0.44		1		3015
	−60 01001	NGC 2516 - 33			7 57 49	−60 52.6	1	9.40		0.19		0.17		1	A1 V	188
	−33 04390				7 57 50	−34 02.1	1	9.04		0.52		−0.01		2	F8	1712
66064	−54 01456				7 57 52	−54 24.5	1	7.82		1.08		0.75		5	G6 II/III	1673
65735	+20 01976	HR 3125			7 57 53	+19 57.3	1	6.25		1.11				2	K1 III	71
	−26 05332				7 57 53	−26 30.5	1	10.13		0.09		0.05		2		126
66137	−60 01003	NGC 2516 - 19			7 57 53	−60 33.8	5	7.83	.024	0.05	.010	−0.12	.013	17	B9 V	36,188,1683,2012,3015
65714	+25 01812	HR 3124			7 57 54	+25 31.9	2	5.84	.014	1.02	.001	0.88		4	K0	71,105
65886	−25 05396	LSS 919			7 57 54	−25 55.8	1	9.62		0.08		−0.71		1		1470
	−60 01004	NGC 2516 - 53			7 57 54	−60 38.4	3	10.53	.009	0.14	.004	0.10	.012	7	A0	36,1683,3015
		NGC 2516 - 145			7 57 54	−60 43.5	1	13.77		0.82		0.28		1		3015

HD	DM	Other Id	N	Rem	α_{1950}	δ_{1950}	S	V	σ_V	B–V	σ_{B-V}	U–B	σ_{U-B}	n	Spectrum	References
		LP 544 - 8			7 57 55	+05 19.4	1	12.62		1.14		1.11		1		1696
65759	+17 01731	HR 3128			7 57 55	+17 26.8	1	5.55		1.32				2	gK3	71
65923	−30 05390	IDS07560S3006		AB	7 57 57	−30 14.3	1	7.73		0.55		0.07		2	F7 V	126
65429	+61 01021				7 57 58	+61 08.0	1	6.86		0.38		−0.01		1	F2	766
		NGC 2516 - 56			7 57 58	−60 43.4	2	11.38	.005	0.39	.010	0.06	.010	3		188,3015
	−31 05352	LSS 921			7 57 59	−31 58.9	2	10.51	.005	0.53	.019	−0.46	.023	4		1712,1766
65982	−42 03764				7 57 59	−43 09.8	1	8.28		0.60				4	G2/6 +F	2012
		NGC 2516 - 206		⋆ B	7 57 59	−60 30.4	2	9.52	.010	0.09	.010	0.04	.015	6	A1 V	1683,3015
		NGC 2516 - 144			7 57 59	−60 46.0	1	14.34		0.94		0.38		1		3015
66167	−60 01005	NGC 2516 - 133		⋆ A	7 58 00	−60 30.4	2	9.23	.365	0.06	.012	0.04	.010	3	A0 V	1683,3015
66167	−60 01005	NGC 2516 - 133		⋆ AB	7 58 00	−60 30.4	3	9.26	.382	0.13	.034	0.04	.020	6		36,188,1683
	−28 05257	LSS 920			7 58 01	−28 42.1	2	11.35	.005	0.25	.005	−0.69	.005	2		353,469
		Fld a # 2			7 58 01	−34 05.6	1	12.55		0.70		0.22		2		1712
	−60 01007	NGC 2516 - 51		⋆ V	7 58 01	−60 31.8	2	10.76	.025	0.30	.015	0.15	.017	4	A8 V	188,3015
66194	−60 01006	NGC 2516 - 134		⋆ V	7 58 01	−60 41.2	11	5.81	.016	−0.09	.007	−0.77	.018	79	B2 Vne	15,36,188,681,815*
65757	+23 01866	HR 3127		⋆ A	7 58 02	+23 43.3	1	6.31		1.00		0.86		3	K1 III-I	1733
	−60 01008	NGC 2516 - 17			7 58 02	−60 27.9	4	9.77	.014	0.19	.008	0.10	.008	11	A1 V	36,188,1683,3015
		NGC 2516 - 71			7 58 03	−60 45.4	1	12.22		0.69		0.14		1		3015
		LOD 5 # 6			7 58 05	−30 21.6	1	12.16		0.52		−0.05		2		126
	−30 05396				7 58 06	−30 17.9	1	10.52		1.41		1.38		2		126
65664	+41 01771				7 58 07	+41 31.6	1	7.27		0.53		0.12		2	F5	1733
	−30 05397				7 58 07	−30 19.4	1	11.00		0.90		0.51		2		126
		LOD 5 # 5			7 58 07	−30 21.2	1	11.60		0.16		−0.02		2		126
		NGC 2516 - 143			7 58 07	−60 45.9	1	14.00		1.19		1.00		1		3015
	−26 05338				7 58 08	−26 29.4	1	9.22		1.05		0.88		2	K0	126
	−33 04398				7 58 10	−34 00.5	1	10.18		0.90		0.50		2		1712
		NGC 2516 - 142			7 58 10	−60 45.8	1	13.65		0.73		0.21		1		3015
	−26 05340				7 58 12	−26 30.2	1	9.76		0.40		−0.03		2		126
65875	−2 02379	HR 3135, V695 Mon	⋆		7 58 13	−02 44.6	9	6.51	.039	−0.10	.011	−0.80	.014	43	B2.5Ve	15,154,361,490,681,815*
		Fld a # 1			7 58 13	−34 08.6	1	12.00		0.52		0.07		2		1712
	−60 01010	NGC 2516 - 21			7 58 13	−60 39.6	2	10.79	.018	0.28	.005	0.13	.014	5	A8 V	188,3015
65944	−26 05341				7 58 14	−26 27.2	1	7.92		0.91		0.64		2	G8 III (+G)	126
	−31 05358	LSS 922			7 58 14	−31 48.1	1	11.54		0.74		0.45		3	A1 Iab	1712
	−23 06645	WZ Pup			7 58 15	−23 33.9	1	9.95		0.58				1	G0	6011
65943	−26 05342				7 58 15	−26 21.8	1	9.60		0.02		−0.01		2	A0 V	126
66020	−39 03869				7 58 15	−39 53.5	3	9.64	.018	1.33	.009	1.21		9	K(5) V	1705,2012,3072
66079	−46 03655	IDS07567S4702		AB	7 58 15	−47 09.9	1	6.71		−0.09		−0.37		2	B8 V	401
	−60 01011	NGC 2516 - 27			7 58 15	−60 41.8	4	10.41	.013	0.66	.022	0.09	.061	10		36,188,1683,3015
		NGC 2516 - 75			7 58 17	−60 36.1	2	12.12	.060	0.62	.015	0.05		2		188,3015
		NGC 2516 - 219			7 58 19	−60 42.5	1	15.34		0.58		0.03		4		1683
		LP 664 - 40			7 58 21	−03 56.8	1	14.67		1.19		0.71		2		1773
		G 113 - 9			7 58 21	−03 57.0	3	11.00	.015	0.46	.005	−0.21	.008	7		1696,1773,3062
66078	−45 03653				7 58 21	−46 01.7	1	7.99		0.78				4	G5 V	2012
		NGC 2516 - 169			7 58 21	−60 41.4	2	13.87	.070	0.66	.030	0.13	.075	4		1683,3015
		NGC 2516 - 220			7 58 22	−60 03.7	1	13.00		0.63		0.01		3		1683
		NGC 2516 - 170			7 58 22	−60 40.9	1	14.71		1.02		0.60		1		3015
	−63 00861	IDS07578S6322		A	7 58 22	−63 29.7	3	9.97	.030	0.53	.014	−0.11	.007	7	F8	158,1696,3060
66019	−35 04094				7 58 23	−35 51.2	1	7.11		0.46		0.12		2	A3 Vm	126
		NGC 2516 - 139			7 58 23	−60 43.2	1	15.67		0.67		0.16		1		3015
		NGC 2516 - 141			7 58 23	−60 45.9	1	14.86		0.86		0.52		1		3015
66259	−60 01012	NGC 2516 - 20			7 58 24	−60 26.9	5	8.38	.009	0.04	.009	−0.17	.012	19	B9.5IV	36,188,1683,2012,3015
		NGC 2516 - 140			7 58 24	−60 46.3	1	13.01		0.62		0.04		1		3015
		NGC 2516 - 221			7 58 26	−60 26.0	1	12.09		0.67		0.15		3		1683
	−60 01013	NGC 2516 - 28			7 58 26	−60 43.8	4	9.95	.005	0.15	.011	0.14	.007	17	A1 V	36,188,1683,3015
66106	−45 03655				7 58 27	−45 47.8	1	11.35		−0.30		−0.23		2	A3 III	1737
		NGC 2516 - 171			7 58 27	−60 41.1	1	13.57		0.65		0.15		1		3015
	+66 00532				7 58 28	+66 19.8	1	9.36		0.96		0.73		2	G5	1733
	−35 04095				7 58 28	−35 43.4	1	10.56		0.34		0.20		2		126
		NGC 2516 - 172			7 58 28	−60 40.7	1	13.79		0.70		0.21		1		3015
		NGC 2516 - 138			7 58 28	−60 44.3	1	16.27		0.83				1		3015
		NGC 2516 - 222			7 58 29	−60 26.4	1	11.96		0.51		0.03		3		1683
		NGC 2516 - 175			7 58 29	−60 36.8	1	14.48		0.63		0.10		2		3015
		LOD 5 # 7			7 58 30	−30 22.9	1	12.90		0.27		−0.03		2		126
	−59 00953	NGC 2516 - 135			7 58 31	−60 03.7	2	9.55	.000	0.07	.000	0.01	.004	7	B8	36,1683
		NGC 2516 - 73			7 58 31	−60 37.8	1	12.38		0.62		0.00		2		3015
		NGC 2516 - 174			7 58 31	−60 38.5	1	14.36		0.96		0.44		2		3015
66295	−60 01015	NGC 2516 - 26		⋆ V	7 58 31	−60 40.7	4	9.11	.007	0.03	.013	−0.17	.005	34	A0p	36,188,1683,3015
	−31 05365	LSS 923			7 58 32	−31 21.2	1	11.52		0.62		−0.03		3		1712
		NGC 2516 - 74			7 58 32	−60 37.2	2	11.94	.044	0.54	.054	0.04	.046	3		188,3015
	−60 01016	NGC 2516 - 25			7 58 33	−60 39.3	3	9.86	.005	0.54	.009	0.16	.011	14		188,1683,3015
66059	−33 04411				7 58 34	−33 59.1	1	7.48		0.05		0.05		2	A1 IV	401
65900	+5 01857	HR 3136		⋆ A	7 58 35	+05 01.1	2	5.64	.005	0.00	.000	0.01	.000	7	A1 V	15,1415
66017	−26 05354	IDS07565S2633		AB	7 58 35	−26 41.6	1	8.87		0.33		0.01		2	F0 V	126
		LP 366 - 52			7 58 36	+24 56.2	1	12.59		1.32		1.19		2		1723
	−35 04097				7 58 36	−35 42.2	1	9.87		0.13		−0.04		2		126
		NGC 2516 - 173			7 58 36	−60 40.3	2	14.90	.031	1.07	.007	0.82	.048	6		1683,3015
66318	−60 01017	NGC 2516 - 24			7 58 37	−60 39.5	3	9.65	.011	0.07	.008	0.09	.017	32	A0p	188,1683,3015
		NGC 2516 - 137			7 58 37	−60 43.8	1	14.55		0.76		0.14		1		3015
65873	+16 01612	HR 3134			7 58 39	+16 35.7	1			−0.02		−0.02		3	B9.5Vn	1022
		LOD 5 # 8			7 58 39	−30 24.0	1	12.00		0.32		−0.06		2		126

Table 1 441

HD	DM	Other Id	N	Rem	α₁₉₅₀	δ₁₉₅₀	S	V	σ_V	B−V	σ_B−V	U−B	σ_U−B	n	Spectrum	References
65801	+35 01731	HR 3130			7 58 40	+35 33.2	1	6.34		1.56				2	K0	71
65953	−0 01882	HR 3141, V645 Mon			7 58 41	−01 15.1	7	4.68	.005	1.49	.008	1.76	.011	30	K4 III	15,1075,1509,2013*
66015	−19 02192				7 58 41	−19 20.7	2	8.39	.005	0.44	.000	-0.06		3	F6wF2 V	1594,6006
		NGC 2516 - 223			7 58 42	−60 40.6	1	13.98		0.88		0.30		3		1683
65856	+25 01816	HR 3132		⋆ A	7 58 43	+25 13.7	1			0.01		0.03		3	A1 V	1022
66102	−33 04414				7 58 44	−34 02.3	1	8.65		0.98		0.69		2	G8/K0 IV	1712
66190	−45 03662	HR 3146			7 58 45	−45 19.1	1	6.61		1.27				4	K1 Ib/II	2007
		LOD 5 # 9			7 58 46	−30 26.9	1	12.05		0.57		0.03		2		126
		NGC 2516 - 210		⋆ B	7 58 46	−60 26.8	1	8.02		-0.02		-0.27		4	B9 V	1683
66342	−60 01018	NGC 2516 - 110		⋆	7 58 46	−60 26.9	8	5.19	.012	1.76	.017	1.97	.035	38	M1.5IIa	15,36,188,460,1683*
66341	−59 00954	NGC 2516 - 136		⋆	7 58 47	−60 04.2	7	6.32	.007	-0.06	.005	-0.40	.005	22	B8 III	15,36,188,1683,2013*
		NGC 2506 - 3401			7 58 48	−10 39.	1	11.64		0.93		0.66		2		884
		NGC 2506 - 4277			7 58 48	−10 39.	1	17.97		0.68				2		884
66210	−48 03384	HR 3148			7 58 48	−48 50.6	3	6.02	.007	0.04	.005	0.07		9	A2 III/IV	15,401,2012
	−30 05422	LSS 924			7 58 50	−30 52.4	1	11.49		0.11		-0.84		3		1712
66340	−58 01031				7 58 51	−59 04.4	2	9.23	.032	0.83	.002	0.50	.023	5	G8 IV	709,3072
	−30 05423	LSS 926			7 58 53	−30 51.0	1	11.49		0.10		-0.71		3		1712
	−60 01019	NGC 2516 - 22			7 58 53	−60 32.0	4	9.60	.018	0.22	.010	0.17	.013	7	A2 V	36,188,1683,3015
66073	−26 05361				7 58 55	−26 33.3	1	9.16		-0.06		-0.14		2	B9 IV	126
	−28 05286	LSS 925			7 58 55	−28 57.3	1	10.38		0.32		-0.63		1		1470
		LOD 5 # 10			7 58 55	−30 27.6	1	12.14		0.50		0.01		2		126
		LSS 927			7 58 56	−31 13.8	1	12.23		0.51		-0.31		3	B3 Ib	1712
	−32 04702	LSS 928			7 58 58	−32 28.7	1	11.80		0.39		-0.29		3		1712
	−32 04706	LSS 929			7 59 4	−32 52.8	1	10.92		0.34		-0.62		3		1712
	−35 04102				7 59 00	−35 59.4	1	9.98		0.18		0.11		2		126
66255	−48 03388	HR 3151, PY Pup			7 59 01	−48 44.0	1	6.12		-0.11				4	B9 Ib	2006
		LOD 5 # 11			7 59 05	−30 28.3	1	12.17		1.04				2		126
	−60 01021	NGC 2516 - 67			7 59 05	−60 33.7	3	10.77	.015	0.32	.000	0.18	.015	6	A7 V	36,188,1683
66409	−60 01022	NGC 2516 - 23			7 59 07	−60 36.5	4	8.39	.008	0.00	.007	-0.31	.005	11	B8.5IV	36,188,1683,2012
66011	+9 01843	HR 3144			7 59 08	+09 03.2	2	6.21	.005	0.57	.000	0.12	.005	7	G0 IV	15,1415
65970	+20 01982				7 59 10	+19 52.6	1	7.66		-0.04		-0.12		3	B9	1648
		KUV 343 - 11			7 59 10	+39 30.9	1	15.26		-0.28		-1.17		1	sdO	974
66148	−23 06667				7 59 10	−23 41.8	1	8.17		0.48				19	F5 V	6011
		vdB 1 # a			7 59 10	−45 18.8	1	12.45		0.30		-0.22		3		434
66150	−26 05370				7 59 13	−26 55.4	2	8.73	.039	-0.11	.005	-0.50		6	B8/9 II	126,2014
		NGC 2516 - 64		⋆ B	7 59 13	−60 38.0	3	10.04	.020	0.38	.013	0.04	.010	6		36,188,1683
66442	−60 01023	NGC 2516 - 63		⋆ A	7 59 13	−60 38.2	3	9.21	.005	0.11	.004	0.14	.011	7	A2 V	36,188,1683
		Orsatti 277			7 59 17	−31 31.1	1	13.73		0.40		-0.29		1		1766
66230	−34 04213	IDS07574S3423	AB		7 59 17	−34 31.1	1	8.15		-0.05		-0.44		2	B7 II	401
	−60 01025	NGC 2516 - 66		⋆ AB	7 59 17	−60 32.7	3	10.13	.015	0.21	.000	0.16	.008	5	A1 V	36,188,1683
66182	−26 05372				7 59 18	−26 27.7	1	9.53		0.02		-0.04		2	A0 V	126
	−60 01026	NGC 2516 - 31			7 59 18	−60 53.5	1	10.39		0.37		0.09		1		188
		LOD 5 # 12			7 59 20	−30 27.1	1	13.80		0.39				2		126
66365	−50 03104				7 59 25	−50 21.0	1	9.37		0.03		-0.26		2	B8/9 V	1311
		LOD 5 # 13			7 59 27	−30 25.5	1	10.80		1.61		2.01		2		126
		AJ79,1294 T9# 9			7 59 27	−31 03.0	2	12.60	.000	0.36	.023	0.17	.019	4		396,1712
66279	−35 04109				7 59 27	−35 45.9	1	7.39		1.62		1.60		2	M0/1 III	126
66311	−41 03648				7 59 27	−41 37.2	1	7.88		-0.10		-0.43		2	B8/9 V	401
		G 194 - 7			7 59 28	+52 59.5	1	11.35		1.49				2		940
		HA 148 # 1245			7 59 28	−30 25.5	1	12.63		0.52		0.17		9		1499
		AJ79,1294 T9# 8			7 59 28	−31 00.0	2	12.51	.014	0.17	.000	-0.36	.000	4		396,1712
66227	−26 05382				7 59 31	−26 47.0	1	9.52		-0.02		-0.09		2	B9.5IV/V	126
	−60 01027	NGC 2516 - 62			7 59 31	−60 40.9	1	11.00		0.35		0.12		1		188
		HA 148 # 1260			7 59 32	−30 25.7	2	12.26	.008	0.29	.008	0.10	.000	11		126,1499
65915	+46 01348				7 59 34	+45 45.6	1	7.90		1.05		0.97		1	K0	1776
		KUV 343 - 10			7 59 35	+41 23.1	1	15.37		-0.30		-1.21		1	sdO	974
66277	−30 05453				7 59 35	−30 27.1	1	7.63		1.52				2	K3/4	126
		G 90 - 38			7 59 36	+36 10.6	1	11.25		0.73		0.14		1		1620
	−30 05452				7 59 36	−30 23.8	1	9.56		1.15		0.94		2	F5	126
66441	−53 01505	HR 3156			7 59 36	−54 00.7	2	5.87	.000	-0.13	.005	-0.53		6	B5 Vn	401,2009
66293	−35 04112				7 59 37	−35 14.4	1	7.51		-0.08		-0.31		2	B8 IV/V	401
65914	+46 01349				7 59 39	+46 21.5	2	8.37	.005	0.42	.005	0.02	.005	3	F5	1733,1776
66141	+2 01854	HR 3145		⋆ A	7 59 40	+02 28.4	10	4.39	.010	1.25	.008	1.28	.015	34	K2 III	15,1003,1075,1075*
	+15 01733	UU Cnc			7 59 41	+15 19.1	1	9.20		1.65		1.93		31	K4 III	3049
	−3 02178				7 59 41	−03 53.4	1	9.46		1.14		1.13		1		929
66591	−63 00866	HR 3159			7 59 42	−63 25.7	5	4.81	.006	-0.17	.010	-0.63	.011	22	B4 V	15,1075,1637,1732,2012
	−30 05458				7 59 43	−31 01.2	2	11.11	.009	0.08	.000	-0.25	.005	4		396,1712
66309	−35 04114				7 59 44	−35 39.0	1	7.84		-0.04		-0.12		2	B8/9 V	126
	−60 01029	NGC 2516 - 65			7 59 44	−60 34.6	1	9.86		0.09		0.06		1	A0 V	188
	−3 02179	LS VI -03 019			7 59 45	−03 49.8	4	10.34	.010	-0.30	.010	-1.14	.011	9	sdO	405,490,929,1732
		AJ79,1294 T9# 6			7 59 45	−31 00.9	2	12.23	.009	0.21	.005	0.15	.014	4		396,1712
		NGC 2516 - 61			7 59 45	−60 42.4	1	11.38		1.09		0.76		1		188
		G 113 - 11			7 59 47	−07 25.6	1	13.98		0.84		0.23		2		3062
66291	−28 05316				7 59 47	−28 16.1	2	8.46	.000	1.31	.010	1.11	.029	3	K1 II/III	565,3048
65733	+66 00533				7 59 48	+66 39.6	1	8.75		0.15		0.10		2	A2	1733
	−27 04877				7 59 48	−27 47.3	1	10.11		0.04		-0.73		2		540
		Orsatti 280			7 59 48	−31 31.8	1	13.08		0.26		-0.48		1		1766
66358	−36 04116	HR 3154			7 59 48	−37 08.6	3	5.94	.018	0.14	.005	0.13		9	A3 IV	15,401,2012
	−35 04116				7 59 49	−35 53.5	1	10.54		0.57		0.17		2		126
64764	+82 00224				7 59 51	+81 51.3	1	8.24		1.11		1.05		2	K2	1733

HD	DM	Other Id	N	Rem	α_{1950}	δ_{1950}	S	V	σ_V	B–V	σ_{B-V}	U–B	σ_{U-B}	n	Spectrum	References
		G 90 - 39			7 59 53	+25 43.8	1	15.12		1.79				5		538
		Ru 46 - 4			7 59 53	−19 20.1	1	13.03		0.47		0.13		2		183
		LP 257 - 24			7 59 54	+35 06.6	1	16.09		1.72				1		1691
		Ru 46 - 9			7 59 54	−19 20.	1	13.71		0.50		0.02		2		183
	−35 04118				7 59 54	−35 50.0	1	10.42		0.19		0.15		2		126
66464	−49 03273				7 59 54	−49 28.8	1	7.24		−0.11				4	B2 II	2012
66920	−72 00654	HR 3171			7 59 54	−73 06.3	3	6.33	.003	0.13	.009	0.12	.005	17	A3 III	15,1075,2038
66290	−19 02204	Ru 46 - 1			7 59 56	−19 19.5	2	9.45	.021	1.03	.002	0.74	.009	6	G8 (III)	183,1685
		Ru 46 - 3			7 59 56	−19 20.5	1	12.36		0.55		0.01		2		183
66384	−35 04122				7 59 56	−36 02.4	1	7.72		0.25		0.02		2	F2 IV	126
		Ru 46 - 6			7 59 57	−19 20.0	1	14.15		0.64		0.24		2		183
		Ru 46 - 2			7 59 58	−19 19.0	1	11.19		1.07		0.88		2		183
		Ru 46 - 7			7 59 58	−19 19.7	1	13.98		0.54		0.06		2		183
		Ru 46 - 5			7 59 58	−19 19.9	1	13.26		0.65		0.13		2		183
66383	−32 04738				7 59 58	−33 09.6	1	8.26		−0.04		−0.38		2	B9	401
66242	−5 02339	HR 3150			7 59 59	−06 11.8	5	6.33	.005	0.62	.001	0.13	.006	21	G0 III	15,1078,1509,2029,3077
		Ru 46 - 8			7 59 59	−19 18.7	1	14.03		0.22		0.10		2		183
66463	−48 03403	IDS07585S4838		AB	7 59 59	−48 46.4	1	8.49		−0.03		−0.39		2	B9 IV	401
	−30 05467	LSS 935			8 00 1	−30 56.6	1	11.51		0.29		−0.30		3		1712
		MCC 249			8 00 00	+33 44.2	1	10.16		1.10				1	K7	1017
	−60 01032	NGC 2516 - 30			8 00 02	−60 52.8	1	10.29		0.19		0.17		1	A5	188
66381	−30 05470				8 00 03	−30 59.8	1	7.89		0.10		0.10		4	A2 IV/V	396
		Orsatti 284			8 00 04	−28 12.4	1	11.85		0.17		−0.53		1		1766
		BSD 4 # 546			8 00 05	+75 19.4	1	12.37		0.44		0.03		1	F8	1058
	−30 05474	Ru 47 - 8			8 00 08	−30 54.5	1	10.90		0.39		0.20		5		183
		Ru 47 - 9			8 00 08	−30 55.3	1	12.79		0.12		−0.35		2		183
66546	−54 01470	HR 3157		★A	8 00 10	−54 22.5	2	6.06	.049	−0.04	.000	−0.65		6	B2 IV	146,2007
66546	−54 01471	IDS07590S5414		B	8 00 10	−54 22.5	1	8.16		−0.01		−0.55		2	B9 IIIp	146
66522	−50 03111				8 00 12	−50 27.9	2	7.20	.010	0.05	.000	−0.62	.005	6	B2 III	164,401
		CCS 999			8 00 13	−01 59.1	1	9.14		2.73				1	C5,4	864
		Ru 47 - 7			8 00 13	−30 54.4	1	12.47		0.13		−0.33		2		183
66656	−60 01033	NGC 2516 - 29			8 00 13	−60 47.6	2	8.28	.019	0.04	.010	0.01		6	A0 V	188,2012
		Ru 47 - 10			8 00 14	−30 56.4	1	13.14		0.08		−0.39		2		183
66478	−44 03969				8 00 14	−44 27.6	1	6.55		1.33		1.40		3	K2 III	389
		Ru 47 - 6			8 00 15	−30 54.0	1	13.19		0.14		−0.24		2		183
66435	−36 04120	HR 3155			8 00 16	−36 54.6	2	6.33	.010	1.61	.015	1.88		6	M1 III	2031,3005
66396	−21 02233				8 00 17	−22 04.1	1	8.95		−0.13				4	B2 III	2033
		Ru 47 - 5			8 00 17	−30 55.2	1	13.16		0.15		−0.28		2		183
		Ru 47 - 4			8 00 17	−30 55.8	1	12.95		0.10		−0.32		2		183
	−30 05480	Ru 47 - 3			8 00 19	−30 56.3	2	11.46	.015	0.17	.020	−0.46	.000	8		183,1712
66417	−28 05336				8 00 20	−28 43.0	1	10.07		0.06		0.01		2	A0 III/IV	409
		Ru 47 - 2			8 00 21	−30 55.7	1	11.78		0.21		0.14		2		183
		Ru 47 - 1			8 00 22	−30 55.8	1	12.00		0.19		0.19		6		183
66607	−55 01419	HR 3161			8 00 22	−55 18.9	2	6.28	.000	−0.14	.005	−0.63		6	B4 V	401,2035
	−35 04125				8 00 24	−35 45.3	1	9.14		1.10		1.15		2		126
66216	+28 01532	HR 3149		★A	8 00 27	+27 56.2	4	4.96	.021	1.13	.010	1.09	.000	7	K2 III	1080,1355,1363,3016
		AJ79,1294 T9# 7			8 00 27	−30 57.6	1	12.33		0.37		0.24		3		396
	−35 04127				8 00 27	−36 11.4	1	9.51		1.03		0.59		2		126
		LOD 7- # 12			8 00 28	−35 59.6	1	11.37		0.09		−0.15		2		126
	+19 01908	UZ Cnc			8 00 30	+19 19.4	1	10.90		0.99		0.54		1	G0	1768
		LOD 7- # 16			8 00 31	−35 43.0	1	12.67		0.31		0.19		2		126
	−35 04129				8 00 31	−35 52.6	1	10.89		0.67		0.25		2		126
		LP 366 - 35			8 00 32	+21 26.8	1	11.91		1.41		1.27		3		1723
		LOD 7- # 13			8 00 32	−35 54.8	1	11.81		0.39		0.02		2		126
	−35 04131				8 00 32	−36 10.0	1	9.61		1.08		0.80		2		126
67516	−79 00257				8 00 32	−79 29.7	1	7.74		1.65		1.88		6	K3 III	1704
		LOD 7- # 17			8 00 34	−35 43.7	1	11.98		0.24		0.11		2		126
66500	−36 04125				8 00 34	−36 41.8	1	7.85		−0.04		−0.35		2	B8 II	401
66477	−27 04895	IDS07585S2716		C	8 00 38	−27 23.9	1	9.27		−0.05		−0.72		2	B3 (V)	409
	−6 02423	IDS07582S0608		AB	8 00 39	−06 16.4	2	9.27	.023	0.79	.005	0.38	.009	5	K0	214,3016
	−30 05491				8 00 39	−30 41.0	1	10.62		0.11		−0.47		2	B4 IV	540
		Pup sq 3 # 58			8 00 40	−27 28.8	1	13.09		0.31		0.08		1	A2	409
66197	+39 02048				8 00 41	+38 46.9	1	8.62		0.13		0.11		2	A2 V	105
		AJ79,1294 T9# 11			8 00 41	−31 06.1	1	12.87		0.10		−0.48		3		396
		BSD 4 # 287			8 00 42	+74 11.1	1	12.23		0.58		0.06		2	G0	1058
		Orsatti 292			8 00 42	−31 27.4	1	11.97		0.13		−0.51		2		1766
	−27 04896	Pup sq 3 # 59			8 00 43	−27 32.6	1	10.33		0.07		−0.04		2	A0 V	409
	−35 04133				8 00 43	−35 43.6	1	11.06		0.32		0.24		2		126
66069	+54 01195				8 00 44	+53 14.0	1	7.04		0.31		0.11		3	A2	1566
66540	−36 04128				8 00 46	−36 15.6	1	6.57		0.24		0.14		2	A7 V	126
65871	+68 00518	G 234 - 21		★A	8 00 47	+68 31.6	1	8.17		0.50		−0.03		1	G0	1658
		BPM 18764			8 00 48	−53 19.	1	15.76		−0.05		−0.89		1		3065
66539	−30 05500	IDS07589S3056		AB	8 00 49	−31 04.5	2	7.66	.005	−0.06	.023	−0.64	.016	6	B3 III	540,976
	−30 05499				8 00 49	−31 08.9	2	10.02	.014	0.30	.005	0.07	.009	4		396,1712
	+44 01710				8 00 50	+44 06.2	1	9.83		1.27		1.19		1	M0	3072
		G 50 - 17			8 00 52	+08 17.5	1	12.61		1.41		1.10		1		1620
66347	+22 01845				8 00 53	+22 12.7	1	6.84		1.19		1.14		2	K0	3016
		AJ79,1294 T9# 10			8 00 53	−31 04.4	1	12.78		0.19		−0.21		3		396
	−31 05430				8 00 53	−31 18.1	1	10.97		0.08		−0.56		2	B4 II-III	540
	−30 05501	LSS 938			8 00 54	−30 32.0	4	10.06	.015	0.14	.010	−0.59	.019	11	B3 II-Ib	540,976,1470,1712

Table 1

HD	DM	Other Id	N Rem	α_{1950}	δ_{1950}	S	V	σ_V	B–V	σ_{B-V}	U–B	σ_{U-B}	n	Spectrum	References
66653	−45 03697			8 00 54	−46 11.9	2	7.53	.014	0.65	.005	0.18	.000	4	G5 V	389,1311
66428	−0 01891			8 00 56	−01 01.1	1	8.26		0.75		0.32		1	G5	3062
66559	−35 04134			8 00 56	−36 06.5	1	8.32		0.92		0.67		2	K0/1III(+F)	126
66299	+33 01636	IDS07579N3319	A	8 00 57	+33 10.4	1	6.70		0.06		0.10		1	A0	695
66582	−32 04763			8 00 57	−33 00.7	3	7.34	.000	−0.13	.009	−0.67	.003	7	B3 IV	401,540,976
66749	−55 01424			8 00 57	−55 27.8	1	7.37		−0.03		−0.12		2	A0 V	401
	−30 05506			8 01 00	−31 06.3	2	10.98	.019	0.03	.009	−0.28	.005	4		396,1712
66624	−40 03776	HR 3162, V336 Pup	★ A	8 01 02	−41 10.1	1	5.54		−0.15		−0.47		2	Ap Si	146
66624	−40 03776	HR 3162, V336 Pup	★ AB	8 01 02	−41 10.1	1	5.46		−0.10				4	Ap Si	2035
66624	−40 03775	IDS07593S4102	B	8 01 02	−41 10.1	1	8.77		1.30		0.07		4	M6 III	146
	−51 02813			8 01 02	−51 18.0	1	7.27		1.09		0.94		1	K5	389
66581	−31 05438			8 01 03	−32 07.5	1	8.29		0.15				2	A2 V	126
		Ru 49 - 2		8 01 05	−26 37.1	1	12.80		0.20		0.04		2		183
		Ru 49 - 3		8 01 05	−26 37.8	1	12.61		0.28		0.19		2		183
66768	−55 01425			8 01 05	−55 24.0	1	6.69		−0.03		−0.63		2	B3 Vn	401
66138	+58 01093			8 01 06	+57 55.0	1	7.07		0.39		−0.06		2	F5	3016
		Ru 49 - 4		8 01 07	−26 38.6	1	12.94		0.19		0.17		2		183
66598	−32 04766	HR 3160	★ A	8 01 07	−32 19.3	3	5.81	.013	1.23	.009	1.28	.004	8	K2/3 III	58,146,2007
66598	−32 04768	IDS07592S3211	B	8 01 07	−32 19.3	1	8.92		1.76		2.01		4		146
		Ru 49 - 7		8 01 08	−26 38.0	1	13.67		0.24		0.17		2		183
	−26 05439	Ru 49 - 1		8 01 09	−26 37.5	1	10.11		0.16		−0.35		6		183
		Ru 49 - 6		8 01 10	−26 37.8	1	12.53		0.21		0.01		2		183
		Ru 49 - 5		8 01 10	−26 38.8	1	12.42		0.56		0.04		2		183
	−36 04137	AR Pup		8 01 10	−36 27.3	1	9.80		0.88		0.74		1	F0 Iab:	6006
66473	+4 01888			8 01 11	+04 09.4	1	9.39		−0.06		−0.21		3	B9	1732
		Orsatti 293		8 01 11	−31 39.6	1	13.29		0.23		−0.58		2		1766
		Ru 49 - 8		8 01 12	−26 36.9	1	13.41		0.25		0.14		2		183
		Orsatti 295		8 01 13	−31 36.2	1	12.68		0.20		−0.36		1		1766
		LSS 940		8 01 13	−33 20.3	2	12.38	.010	0.39	.020	−0.50	.015	5	B3 Ib	1712,1766
		Ru 49 - 9		8 01 14	−26 37.9	1	13.95		0.31		0.06		2		183
		LSS 941		8 01 16	−31 34.3	2	11.98	.005	0.14	.005	−0.52	.020	5	B3 Ib	1712,1766
66622	−29 05427			8 01 17	−29 49.5	1	10.20		0.17		0.00		2	A0 V	409
66319	+40 01989			8 01 19	+39 53.0	1	6.89		1.05		0.83		3	K0	974
		LOD 6 # 17		8 01 19	−32 09.7	1	13.18		0.55				2		126
		BSD 4 # 548		8 01 22	+74 55.7	1	12.57		0.46		−0.03		1	G0	1058
	−32 04772	LSS 943		8 01 23	−32 25.5	1	11.54		0.43		−0.28		3		1712
	−28 05365			8 01 24	−28 32.9	1	9.95		0.09		−0.95		1		211
66512	+4 01890			8 01 25	+04 10.1	1	8.56		0.50		0.07		1	F8	1732
66765	−47 03537			8 01 26	−48 11.0	3	6.62	.010	−0.16	.011	−0.76	.000	9	B1/2 V	540,976,2012
	−28 05369	LSS 942		8 01 28	−28 31.0	2	9.96	.016	0.10	.000	−0.82	.013	4	B3 II	409,540
66647	−28 05368			8 01 28	−28 40.8	1	10.01		0.10		−0.07		2	B9 IV/V	409
		BSD 4 # 549		8 01 30	+75 25.5	1	12.14		0.51		0.01		1	G0	1058
	−32 04773			8 01 30	−32 15.6	1	9.64		0.12		−0.18		2		126
66443	+21 01753			8 01 32	+21 09.0	1	7.53		0.15		0.12		3	A2	1648
66700	−31 05452			8 01 32	−31 32.8	1	8.28		−0.05		−0.45		2	B5 IV/V	401
		CCS 1013, IK Pup		8 01 33	−23 41.6	1	10.51		4.22				1	N	864
	−32 04774			8 01 33	−32 15.0	1	10.92		0.49		0.11		2		126
		LOD 6 # 15		8 01 36	−32 16.1	1	12.78		0.68		0.26		2		126
66509	+12 01759	IDS07588N1235	AB	8 01 37	+12 26.0	5	7.79	.037	0.85	.008	0.48	.000	14	G5	292,1197,1381,1758,3030
66509	+12 01759	IDS07588N1235	C	8 01 37	+12 26.0	1	11.40		1.30				1		3016
66695	−26 05453	LSS 944		8 01 38	−26 54.3	2	9.78	.003	0.00	.008	−0.80	.019	4	B2/3 Ib/II	540,976
66594	−4 02197			8 01 41	−04 41.1	2	7.57	.065	−0.13	.005			8	B3 V	2012,2034
		POSS 425 # 2		8 01 42	+16 08.1	1	15.01		1.53				4		1739
		Pup sq 3 # 102		8 01 43	−29 51.8	1	11.36		0.04		−0.18		2	B9 V	409
	−31 05459	LSS 946		8 01 44	−31 34.3	1	10.98		0.13		−0.52		3		1712
		LOD 6 # 13		8 01 45	−32 17.4	1	12.60		0.27		0.05		2		126
66553	+15 01742	G 40 - 5		8 01 46	+15 30.6	1	8.48		0.85		0.51		1	G5	333,1620
		Pup sq 3 # 103		8 01 46	−29 47.3	1	13.04		0.42		−0.01		1	B8	409
	−32 04779			8 01 47	−32 18.3	1	11.15		0.52		0.00		2		126
		LOD 6 # 11		8 01 48	−32 16.3	1	11.90		0.20		−0.09		2		126
66812	−42 03832	HR 3166		8 01 49	−42 48.4	3	6.28	.008	1.01	.002	0.76	.003	8	G8 II	58,389,2035
66811	−39 03939	HR 3165, LSS 949		8 01 50	−39 51.7	14	2.25	.012	−0.27	.009	−1.11	.012	70	O5 Iaf	3,9,15,200,1011,1020*
	−45 03708			8 01 51	−46 00.4	1	9.50		1.27		1.38		1		1770
66552	+19 01911	HR 3158		8 01 52	+18 59.1	3	6.21	.000	−0.05	.005	−0.10	.007	8	B9 V	1022,1501,1733
66286	+59 01136	IDS07577N5932	A	8 01 52	+59 23.5	1	6.69		0.09		0.16		1	A0	766
66892	−45 03710			8 01 53	−46 09.9	1	8.76		1.50		1.88		1	K3 III	389
		CCS 1019		8 01 55	−31 11.2	1	9.54		1.74				1		864
66810	−35 04153	IDS08001S3514	AB	8 01 58	−35 22.2	1	8.34		−0.05		−0.35		2	B9 V	401
		KPS 343 - 208		8 01 59	+40 23.7	1	13.45		0.96		0.12		3		974
		KPS 343 - 209		8 02 00	+40 23.3	1	11.24		0.66		0.09		3		974
		Pup sq 3 # 57		8 02 02	−27 20.6	1	13.43		0.11		−0.02		1		409
	−32 04787			8 02 02	−32 16.6	1	8.78		0.92		0.97		2		126
66789	−28 05383			8 02 03	−29 01.9	1	7.37		0.08		0.08		2	A2 V	126
66940	−49 03306			8 02 03	−49 21.4	1	6.64		1.02		0.81		1	K0 III	389
66788	−27 04933	LSS 948		8 02 04	−27 20.5	4	9.45	.010	−0.08	.014	−0.93	.019	7	O8/9 (Ib)	211,409,540,976
		Ton 296		8 02 06	+33 03.	1	15.94		0.15		0.24		1		1036
66665	+6 01867			8 02 08	+06 19.7	2	7.82	.016	−0.26	.003	−1.00	.027	6	B1 V	1345,1732
		KPS 343 - 211		8 02 08	+40 28.3	1	11.25		1.23		0.90		3		974
66738	−12 02315			8 02 08	−12 59.6	1	8.09		−0.15		−0.67		1	B5 III	55
		Orsatti 303		8 02 08	−31 48.9	1	12.90		0.20		−0.39		1		1766

HD	DM	Other Id	N Rem	α_{1950}	δ_{1950}	S	V	σ_V	B–V	σ_{B-V}	U–B	σ_{U-B}	n	Spectrum	References
	+74 00346			8 02 09	+74 19.3	1	9.25		0.30		0.08		1	F0	1058
		Orsatti 304		8 02 11	−31 30.7	1	12.11		0.14		-0.37		1		1766
	−31 05475	LSS 952		8 02 12	−31 15.8	3	9.74	.018	0.30	.011	-0.58	.022	8	B3 Vep	540,976,1712,1766
		BSD 4 # 290		8 02 13	+74 21.3	1	10.82		0.51		0.09		1		1058
		LOD 6 # 2		8 02 14	−32 31.0	1	12.60		0.74		0.53		2		126
		LOD 6 # 5		8 02 15	−32 23.5	1	12.51		0.41		0.16		2		126
66664	+13 01831	HR 3163		8 02 17	+13 15.7	3	5.11	.010	0.00	.007	-0.01	.005	6	A1 V	1022,1363,3023
		G 111 - 54		8 02 18	+38 41.0	2	15.56	.005	0.06	.000	-0.85	.019	4	DFp	316,3060
66888	−32 04796	HR 3170, MZ Pup		8 02 19	−32 31.9	3	5.38	.080	1.91	.013	2.01	.026	11	M1 Ib	3,126,3041
		Pup sq 3 # 101		8 02 21	−29 56.9	1	12.05		0.23		0.09		2	A5 V	409
		LOD 6 # 9		8 02 22	−32 19.0	1	13.51		0.39				2		126
		CCS 1025		8 02 24	−42 59.	1	10.49		1.68				1		864
66834	−19 02228	HR 3168		8 02 28	−19 35.1	3	6.13	.005	-0.16	.005			11	B3 III	15,2012,2014
	+34 01740	G 90 - 43		8 02 29	+34 13.3	1	10.17		1.32		1.30		1	M0 V:p	1620
		Pup sq 3 # 81		8 02 29	−28 27.4	1	12.77		0.47		0.62		1		409
	−30 05559	LSS 953		8 02 29	−30 49.7	3	10.44	.005	0.24	.003	-0.81	.016	6		540,1712,1737
66711	+18 01839			8 02 30	+17 45.9	1	7.89		1.59		1.75		2	K5	1648
		LOD 6 # 7		8 02 31	−32 20.2	1	12.51		0.35		0.01		2		126
	−32 04801			8 02 32	−32 20.7	1	10.58		1.19		1.38		2		126
66933	−32 04802			8 02 32	−32 25.6	1	8.35		0.26		0.15		2	A7 IV	126
66684	+27 01536	HR 3164	★ AB	8 02 33	+27 40.4	2			0.01	.037	-0.09	.028	4	A0 V	1022,1079
		Pup sq 3 # 82		8 02 33	−28 26.2	1	11.67		0.20		0.54		2	B5	409
		KUV 343 - 6		8 02 34	+41 18.2	1	15.28		-0.30		-1.27		1		974
		KPS 343 - 218		8 02 35	+39 55.8	1	13.71		0.61		-0.06		3		974
		LP 123 - 75		8 02 36	+56 02.4	2	11.49	.014	1.50	.009	1.28		3		1723,1746
66932	−29 05474			8 02 38	−29 49.4	2	7.40	.009	1.61	.014	1.80	.003	6	M4 III	409,1657
	−32 04808			8 02 39	−33 12.0	1	10.61		0.04		0.06		2		124
		Orsatti 308		8 02 41	−33 16.2	1	13.16		0.61		-0.29		1		1766
66171	+72 00395	G 251 - 52	★ A	8 02 44	+72 04.5	4	8.20	.016	0.62	.008	0.06	.010	8	G2 V	22,1003,1658,3002
	−27 04950	NGC 2527 - 10		8 02 44	−27 59.2	2	9.77	.009	0.98	.005	0.70	.000	4	G8 III	187,1551
67061	−43 03896			8 02 44	−44 05.7	1	7.41		1.01		0.71		1	G6 III	389
	−32 04809			8 02 45	−32 30.9	1	10.41		2.10		2.55		3		124
		G 251 - 53		8 02 46	+80 01.6	1	14.57		1.09		0.62		2		1773
66965	−28 05407			8 02 46	−28 24.3	1	9.46		0.01		-0.14		2	B9.5V	409
		LSS 954		8 02 47	−31 36.2	2	12.54	.023	0.17	.005	-0.52	.028	4	B3 Ib	1712,1766
66954	−27 04954	NGC 2527 - 43		8 02 48	−28 03.0	1	9.95		0.18		0.19		1	A3 IV	187
		LP 842 - 17		8 02 49	−25 50.2	1	11.64		0.65		-0.01		2		1696
	−27 04953	NGC 2527 - 14		8 02 49	−28 00.8	2	10.56	.009	0.27	.005	0.14	.008	4	A5 V	187,525
	−35 04168			8 02 49	−35 43.6	1	11.02		0.21		0.08		2		124
	−27 04956	NGC 2527 - 6		8 02 50	−27 57.6	2	11.22	.009	0.26	.010	0.10	.004	4	A4	187,525
		NGC 2527 - 9		8 02 51	−27 59.5	1	11.81		0.35		0.05		1	A8	187
		vdB 3 # d		8 02 51	−39 04.7	1	12.12		0.51		0.00		3		434
		BSD 4 # 558		8 02 52	+75 22.0	1	12.78		0.49		0.01		1	G0	1058
		NGC 2527 - 45		8 02 52	−28 02.6	1	13.08		0.84		0.30		1		187
	−32 04815			8 02 52	−32 22.7	1	8.94		2.02		2.05		3		124
		vdB 3 # a		8 02 54	−39 10.9	1	11.43		0.14		-0.17		3	B6 V	434
	−34 04298			8 02 55	−34 55.4	1	9.87		0.48		-0.28		2	B4 II-Ib	540
	+26 01715			8 02 57	+26 25.5	1	10.21		1.46		1.20		2	M0	1746
	−27 04962	NGC 2527 - 17		8 02 59	−27 59.1	2	11.26	.009	0.39	.001	0.02	.002	4	A5	187,525
67036	−29 05488			8 02 59	−29 48.2	1	8.11		1.39		1.41		1	K1p Ba	565
		NGC 2527 - 507		8 03 00	−27 57.3	1	13.30		0.22		0.17		3		525
		vdB 3 # b		8 03 00	−39 11.9	1	11.14		0.18		-0.10		4	B9 V	434
67127	−45 03728			8 03 00	−46 00.5	1	7.68		1.13		1.02		2	K0 III	389
66825	+18 01842			8 03 02	+18 08.2	1	9.41		0.21		0.03		1	A2	1776
		NGC 2527 - 101		8 03 04	−28 03.7	1	11.79		0.31		0.13		1		187
		vdB 3 # c		8 03 04	−39 03.4	1	12.27		0.27		0.16		3	A0 V	434
		NGC 2527 - 102		8 03 06	−28 03.5	1	12.87		1.09		0.76		1		187
		NGC 2527 - 100		8 03 06	−28 03.8	1	13.75		0.28		0.24		1		187
	+80 00245	G 251 - 54	★ A	8 03 07	+80 03.9	3	10.02	.023	0.54	.024	-0.17	.014	4	G0	516,1620,1773
	−27 04971	NGC 2527 - 124		8 03 09	−27 52.0	1	10.32		1.03		0.75		3	K0	1551
67053	−27 04972	NGC 2527 - 57		8 03 09	−27 59.1	2	9.62	.005	0.24	.006	0.18	.000	4	A8 V	187,525
	−34 04301			8 03 10	−34 39.5	1	10.07		0.00		-0.27		2		124
		Ton 299		8 03 12	+32 41.	1	15.50		-0.30		-0.99		1	sdB	1036
		NGC 2527 - 47		8 03 12	−28 01.	1	13.54		1.74		1.78		1		187
		NGC 2527 - 1001		8 03 12	−28 01.	1	16.02		0.85		0.46		1		187
		NGC 2527 - 508		8 03 12	−28 01.	1	13.74		0.50		-0.03		2		525
		NGC 2527 - 509		8 03 12	−28 01.	1	14.02		0.18		-0.16		3		525
		NGC 2527 - 510		8 03 12	−28 01.	1	14.81		0.81		0.37		3		525
		NGC 2527 - 511		8 03 12	−28 01.	1	15.42		0.91		0.27		3		525
		NGC 2527 - 512		8 03 12	−28 01.	1	15.94		0.84		0.36		3		525
		NGC 2527 - 513		8 03 12	−28 01.	1	16.62		0.70				3		525
		NGC 2527 - 191		8 03 13	−28 03.8	1	12.86		0.54		0.03		1		187
66950	−0 01903	HR 3172		8 03 16	−00 25.8	3	6.40	.008	1.06	.010	1.00	.018	11	K0	15,361,1256
67249	−50 03138	HR 3178		8 03 18	−50 26.8	3	5.95	.000	1.22	.004	1.02	.006	7	G5 II	58,389,2035
66708	+50 01508			8 03 20	+50 02.3	1	8.69		0.39		-0.02		2	F2	1733
66875	+23 01887	HR 3169, BL Cnc		8 03 21	+22 46.8	2	5.98	.014	1.65	.015	1.88		4	M3 III	71,3055
67097	−27 04978	NGC 2527 - 213		8 03 24	−27 54.7	2	8.60	.004	0.02	.014	0.02	.018	6	A0 V	187,525
		SS Cnc		8 03 28	+23 23.7	2	11.63	.137	0.12	.034	0.16	.029	2	A8.5	668,699
	−32 04828	LSS 955		8 03 28	−32 52.5	2	11.93	.144	0.31	.000	-0.42	.020	5		1712,1766
66660	+55 01247			8 03 29	+55 26.8	1	7.20		1.08		1.04		3	K1 III	37

Table 1

445

HD	DM	Other Id	N Rem	α_{1950}	δ_{1950}	S	V	σ_V	B–V	σ_{B-V}	U–B	σ_{U-B}	n	Spectrum	References
67072	−15 02250			8 03 29	−16 13.1	1	9.66		−0.19		−0.89		4	B2 III	1732
	−32 04830			8 03 29	−32 19.8	1	9.06		1.29		1.11		3		124
		NGC 2527 - 506		8 03 30	−28 06.9	1	12.35		0.12		−0.36		2		525
		Orsatti 318		8 03 30	−33 26.0	1	12.90		0.54		−0.08		2		1766
	−27 04983	NGC 2527 - 203		8 03 31	−28 00.3	1	9.54		0.98		0.71		1		187
67144	−28 05427			8 03 32	−28 56.1	1	9.50		0.54		0.11		2	F6 V(w)	126
67190	−38 04049	RT Pup		8 03 32	−38 38.0	1	8.52		2.38		3.28		1	C6,2	864
		LSS 956		8 03 33	−33 44.9	2	12.15	.005	0.65	.025	−0.06	.015	5	B3 Ib	1712,1766
66948	+22 01854			8 03 37	+22 36.1	1	7.22		0.75		0.26		2	G5 IV	1648
		Orsatti 320		8 03 38	−31 20.9	1	12.82		0.19		−0.48		1		1766
	−31 05517			8 03 38	−31 46.6	1	10.69		0.10		0.05		2		124
67028	+1 01994			8 03 39	+01 37.7	1	8.68		0.05		0.12		1	A2	1270
66730	+53 01218			8 03 39	+53 42.1	1	7.67		1.35		1.50		3	K5	1566
	−30 05607	LSS 958		8 03 41	−30 33.5	1	12.32		0.15		−0.47		3		1712
66824	+43 01770	HR 3167		8 03 42	+43 24.3	2	6.36	.005	−0.03	.005	−0.10	.025	4	A1 V	1733,3016
		BSD 4 # 561		8 03 43	+75 12.2	1	11.71		0.54		0.02		1	G0	1058
	−30 05609	LSS 959		8 03 43	−30 33.5	1	12.15		0.11		−0.51		3		1712
67186	−28 05432			8 03 44	−28 58.3	1	10.63		−0.04		−0.38		2	B7 III/IV	126
67364	−52 01376	HR 3180		8 03 46	−52 57.8	4	5.52	.007	1.35	.008	1.60		13	K3/4 III	15,1075,2035,3005
66773	+52 01279			8 03 47	+52 24.9	1	9.02		1.12		1.04		2		1566
	−28 05433			8 03 48	−28 44.3	1	10.37		1.55		1.63		2		126
		Pup sq 1 # 20		8 03 48	−33 21.	1	12.91		0.44		0.15		3		124
67341	−46 03764	HR 3179		8 03 48	−46 50.1	5	6.18	.020	−0.15	.016	−0.68	.004	19	B3 III	15,540,976,2012,2012
67624	−68 00725			8 03 48	−68 47.6	1	6.85		0.95		0.61		5	G6 III	1628
	+0 02186	G 113 - 15		8 03 49	+00 02.0	1	9.92		0.91		0.66		2		3062
67243	−33 04525	HR 3177	★ A	8 03 49	−33 25.5	3	6.10	.073	1.03	.091	0.90	.019	8	G1 Ib	58,542,2009
67243	−33 04526	IDS08019S3317	B	8 03 49	−33 25.5	1	8.66		−0.09		−0.52		2	B9 IV	542
	−28 05435			8 03 53	−28 59.6	1	9.79		1.13		0.81		2		126
		HA 4 # 369		8 03 54	+75 10.0	1	12.39		0.59		0.11		1		1058
	−32 04836			8 03 54	−32 57.7	1	8.83		0.88		0.58		8	G5	124
		Orsatti 324		8 03 54	−33 00.8	1	12.55		0.44		−0.20		2		1766
		Pup sq 1 # 26		8 03 54	−33 08.	1	12.37		0.39		0.18		3		124
	−33 04527			8 03 57	−33 35.0	1	11.01		0.12		0.11		2		124
	−31 05524	LSS 960		8 03 58	−31 39.8	2	11.39	.015	0.26	.005	−0.55	.035	6		1712,1766
67295	−35 04193			8 03 58	−35 50.2	1	8.16		−0.04		−0.61		2	B3 II	401
		POSS 425 # 5		8 03 60	+16 21.5	1	17.54		2.29				2		1739
	−32 04839	LSS 961		8 04 0	−32 51.7	2	11.62	.000	0.63	.020	−0.32	.025	5		1712,1766
	−28 05436			8 04 00	−28 43.3	1	11.09		0.27		0.06		2		126
67536	−62 00953	HR 3186, V375 Car	★ A	8 04 00	−62 41.6	2	6.29	.005	−0.11	.005			7	B2 Vn	15,2012
67263	−28 05440			8 04 01	−28 55.4	1	9.45		−0.02		−0.08		2	A0 IV/V	126
67158	−8 02221	IDS08016S0857	BC	8 04 02	−09 05.6	1	7.93		−0.03		−0.08		3	A0	1084
67159	−8 02222	HR 3174	★ A	8 04 03	−09 06.0	1	6.23		−0.06		−0.12		3	B9 V	1084
67159	−8 02222	HR 3174	★ ABC	8 04 03	−09 06.0	3	6.00	.004	−0.04	.000	−0.12	.005	11	B9 V	15,1075,2012
	−33 04533			8 04 04	−33 23.7	1	9.61		0.08		−0.20		2		124
		Orsatti 328		8 04 05	−33 13.3	1	12.99		0.62		−0.35		1		1766
67009	+33 01644			8 04 07	+33 23.6	1	8.21		0.40		−0.02		2	F5	1625
		HA 4 # 72		8 04 07	+74 23.6	1	11.73		0.54		0.06		1		1058
		BSD 4 # 563		8 04 07	+75 11.5	1	11.76		0.51		0.06		1	F5	1058
67385	−44 04032	IDS08025S4453	AB	8 04 07	−45 01.6	1	7.06		−0.14		−0.66		4	B3 V	158
	−33 04536			8 04 09	−33 19.7	1	11.13		1.02		0.78		2		124
67334	−34 04323			8 04 09	−34 36.6	1	8.06		−0.05		−0.42		2	B6 V	401
	−32 04845			8 04 12	−32 25.0	1	9.34		1.06		0.85		3		1712
	−32 04844			8 04 12	−32 32.0	1	10.26		1.07		0.92		3		124
		Orsatti 329		8 04 12	−32 52.2	1	11.71		0.32		−0.31		2		1766
	−30 05627			8 04 14	−30 31.4	1	11.58		0.11		0.14		2		124
	−32 04846			8 04 14	−32 22.3	2	10.54	.000	0.00	.005	−0.37	.005	5	B7 V	540,1712
	+40 01994			8 04 15	+39 49.2	1	9.28		0.27		0.12		2	A2	974
		LSS 963		8 04 15	−33 08.0	2	11.87	.005	0.66	.010	−0.30	.025	5	B3 Ib	1712,1766
		POSS 425 # 3		8 04 16	+16 35.8	1	16.31		1.60				4		1739
	−33 04538			8 04 16	−33 21.7	1	10.19		−0.04		−0.27		2		124
		Fld b # 3		8 04 17	−32 23.9	1	12.07		0.17		−0.04		3		1712
		HA 4 # 151		8 04 21	+74 33.2	1	12.74		0.59		0.14		1		1058
	−29 05535	BN Pup		8 04 21	−29 57.1	2	9.32	.091	0.91	.070	0.67	.033	2	G4	689,1587
		HA 4 # 15		8 04 23	+74 16.5	1	11.51		0.37		−0.02		1		1058
66425	+74 00347			8 04 24	+74 11.9	2	8.96	.010	0.42	.001	−0.02	.005	6	G0	269,1058
67358	−29 05538			8 04 26	−29 16.1	1	10.33		0.08		0.03		2	A0 III	126
		POSS 425 # 4		8 04 27	+16 06.8	1	16.65		1.58				2		1739
	−29 05540			8 04 27	−29 43.0	1	10.04		0.10		0.04		2	F5	124
		Fld b # 5		8 04 28	−32 19.4	1	13.14		0.50		0.12		3		1712
67065	+38 01861			8 04 29	+38 43.6	1	7.09		0.33		0.13		2	F0 V	105
		Fld b # 6		8 04 30	−32 18.6	1	11.86		0.52		−0.04		3		1712
		Pup sq 1 # 23		8 04 30	−33 14.	1	13.50		0.63		0.20		3		124
		Orsatti 334		8 04 30	−33 22.0	1	11.71		0.34		−0.28		2		1766
		Wray 157		8 04 32	−28 23.2	1	13.20		1.47		0.67		1		1753
	−32 04855			8 04 33	−32 17.2	2	9.93	.010	1.16	.015	1.09	.040	6		124,1712
		Fld b # 8		8 04 33	−32 19.5	1	12.68		0.33		0.24		2		1712
	−33 04543			8 04 33	−33 33.0	1	11.45		0.17		−0.16		2		124
		Pup sq 1 # 24		8 04 36	−33 14.	1	13.61		1.06		0.33		3		124
		HA 4 # 154		8 04 38	+74 32.3	1	11.51		0.93		0.67		1		1058
	−34 04336			8 04 39	−35 07.5	1	10.56		0.09		−0.18		2		124

HD	DM	Other Id	N	Rem	α₁₉₅₀	δ₁₉₅₀	S	V	σ_V	B–V	σ_B-V	U–B	σ_U-B	n	Spectrum	References
67006	+51 01391	HR 3173		★A	8 04 42	+51 39.2	4	4.82	.031	0.05	.006	0.01	.010	10	A2 V	15,1363,3023,8015
	+75 00325			V	8 04 43	+75 06.8	2	9.54	.008	-0.37	.008	-1.17		8	O5p	1058,1118
		NGC 2533 - 95			8 04 44	-29 44.7	1	12.43		2.02		1.68		2		1667
		HA 4 # 155			8 04 45	+74 33.6	1	10.85		0.43		-0.03		1		1058
		NGC 2533 - 100			8 04 45	-29 41.8	1	10.96		1.13		1.22		2	K0	1667
67409	-26 05530	IDS08027S2650		A	8 04 47	-26 58.1	1	7.08		-0.08		-0.38		5	B7 III	1628
		Pup sq 1 # 19			8 04 48	-33 30.	1	12.08		0.20		-0.15		3		124
67228	+22 01862	HR 3176			8 04 49	+21 43.7	3	5.29	.010	0.64	.008	0.20	.009	22	G2 IV	15,1351,3026
		GD 261			8 04 49	+35 57.9	1	15.35		0.75		-0.26		2		3060
	-39 03995				8 04 50	-39 46.7	1	9.45		0.41		-0.34		2	B3 II-III	540
		BSD 4 # 299			8 04 51	+74 10.4	1	12.38		0.65		0.20		1	G0	1058
	+75 00326				8 04 51	+74 56.1	1	9.55		0.58		0.11		1	G0	1058
67226	+24 01863				8 04 52	+24 09.9	1	8.05		1.16		1.18		2	K0	1625
67458	-29 05555				8 04 58	-29 15.2	5	6.80	.018	0.60	.002	0.04	.032	13	G2 V	126,1311,2012,2012,3026
	-29 05558	NGC 2533 - 52			8 04 58	-29 45.7	2	10.86	.029	0.05	.010	-0.30	.010	3	B3	452,1667
67457	-28 05464				8 04 59	-28 28.9	1	9.26		0.48		-0.04		2	F3/5 V	126
	-30 05659	LSS 964			8 05 7	-30 21.0	1	9.86		0.70		0.47		3	A2 II	1712
		NGC 2533 - 108			8 05 01	-29 39.5	1	12.76		1.22		1.22		2		1667
		NGC 2533 - 15			8 05 01	-29 43.9	2	12.18	.029	0.12	.005	0.01	.019	3		452,1667
67473	-29 05559	NGC 2533 - 11		★AB	8 05 01	-29 45.1	3	9.03	.019	0.63	.010	0.14	.030	4	G0 V	452,1667,3043
		NGC 2533 - 22			8 05 02	-29 43.0	1	12.49		1.00		0.97		2		1667
		NGC 2533 - 16			8 05 03	-29 44.1	1	12.71		0.13		0.03		1	B8	452
	-28 05470				8 05 04	-28 35.7	1	9.89		0.93		0.74		2		126
		NGC 2533 - 109			8 05 04	-29 39.0	2	11.34	.317	4.35	.394	2.03	.185	3		239,1667
		NGC 2533 - 115			8 05 04	-29 39.9	2	13.11	.010	0.29	.010	0.25	.058	3		1667,3043
		NGC 2533 - 17			8 05 04	-29 43.8	1	12.13		1.12		0.78		2		1667
		NGC 2533 - 47			8 05 04	-29 46.8	1	12.65		1.77		3.69		2		1667
67582	-44 04051	HR 3187			8 05 04	-45 07.3	2	5.04	.000	1.51	.005	1.72		12	K3 III	1088,2035
	+25 01853				8 05 05	+24 51.7	1	8.97		0.99		0.76		2	K0	1625
67644	-53 01533				8 05 05	-53 54.0	1	7.99		1.18		1.32		1	K3 III	389
67456	-20 02395	HR 3183			8 05 06	-20 24.5	4	5.36	.011	0.09	.005	0.17	.005	13	A3 Ib/II	15,2012,8071,8100
		NGC 2533 - 5			8 05 06	-29 45.6	1	13.73		2.14		2.23		2		1667
		G 234 - 24			8 05 07	+69 55.7	1	10.96		0.47		-0.19		2		1658
		NGC 2533 - 4			8 05 07	-29 45.1	2	12.22	.030	0.15	.020	-0.03	.010	2	B8	239,452
		BSD 4 # 568			8 05 09	+75 23.5	1	12.78		0.58		0.05		1	G2	1058
67581	-43 03948				8 05 09	-43 44.9	1	8.73		0.60		0.05		1	G5 V	389
66751	+70 00497				8 05 11	+69 52.2	1	6.60		0.60		0.01		7	F8	7008
67404	-2 02437				8 05 11	-03 15.6	1	6.83		1.60				4	M1	2012
		NGC 2533 - 118			8 05 12	-29 39.5	2	12.08	.049	0.39	.051	0.25	.066	3	B5	1667,3043
67621	-48 03480				8 05 12	-48 21.1	5	6.33	.009	-0.19	.014	-0.80	.004	14	B2 IV	401,540,976,1732,2012
65299	+84 00169	HR 3108			8 05 14	+84 12.5	2	6.47	.015	0.04	.010	0.04	.010	4	A3 IV	985,1733
		NGC 2533 - 3			8 05 14	-29 44.7	1	12.61		1.65		2.01		2		1667
67620	-45 03777				8 05 14	-45 41.4	1	7.24		1.07		0.86		2	G8 II/III	389
67346	+19 01934				8 05 15	+19 21.8	2	7.63	.005	0.63	.034	0.17	.010	3	F8	1648,1776
	-28 05472	LSS 965			8 05 15	-28 55.9	1	10.79		0.05		-0.76		2	B1.5II	540
		NGC 2533 - 31			8 05 15	-29 43.3	1	13.10		0.40		0.19		2		1667
67371	+16 01642				8 05 17	+16 33.8	1	8.80		0.43		-0.07		2	G0	1648
		HA 4 # 375			8 05 17	+75 07.3	1	12.11		0.59		0.11		1		1058
	-30 05667				8 05 17	-30 47.6	1	11.54		0.29		-0.30		3		396
	+21 01764	G 40 - 7		★A	8 05 19	+21 15.5	2	9.45	.025	1.35	.030	1.24	.000	2	K5	333,1620,6009
	+21 01764	IDS08024N2124		B	8 05 19	+21 15.5	1	11.20		1.45				1		6009
67321	+25 01854	IDS08023N2551		AB	8 05 19	+25 42.0	1	8.13		0.45		0.02		3	F5	833
67507	-22 02160	RU Pup			8 05 20	-22 46.0	1	8.45		3.62				1	C5,4	864
	-4 02226				8 05 23	-04 22.6	2	11.21	.005	0.39	.009	0.04		4	M3	1705,7008
	-29 05569	NGC 2533 - 121			8 05 23	-29 40.1	1	10.70		0.09		0.10		2		1667
		HA 4 # 79			8 05 24	+74 23.6	1	12.68		0.52		0.09		1		1058
66633	+74 00348	IDS07594N7439		AB	8 05 24	+74 30.7	3	8.09	.010	0.30	.008	0.03	.003	33	F0	269,1058,1118
		NGC 2533 - 39			8 05 24	-29 45.8	1	12.31		0.26		-0.33		1	B2	452
67523	-23 06828	HR 3185, ρ Pup		★A	8 05 25	-24 09.5	8	2.80	.020	0.42	.010	0.19	.011	26	F2mF5 IIp	15,597,1088,1497,2009*
		NGC 2533 - 82			8 05 25	-29 46.5	1	11.97		0.27		-0.35		1	B3	452
67554	-28 05479				8 05 26	-28 32.7	1	9.69		0.04		-0.56		2	B2 II/III	126
		NGC 2533 - 125			8 05 26	-29 41.6	2	12.05	.024	1.35	.005	1.39	.292	3		1667,3043
	-32 04872				8 05 26	-32 46.8	1	10.45		1.01		0.70		4		124
		HA 4 # 378			8 05 27	+75 02.1	1	12.55		0.63		0.14		1		1058
		vdB 5 # a			8 05 30	-35 48.	1	13.47		0.62		0.27		7		434
		NGC 2533 - 80			8 05 32	-29 46.6	1	12.72		0.45		0.03		1		452
		HA 4 # 379			8 05 33	+75 02.6	1	12.79		0.56		0.09		1		1058
		BSD 4 # 574			8 05 34	+75 25.9	1	12.56		0.63		0.13		1	G2	1058
67738	-51 02839				8 05 35	-51 39.8	1	8.46		0.65		0.20		1	G3 IV	389
		POSS 425 # 1			8 05 36	+16 37.1	1	14.65		0.94				2		1739
		HA 4 # 80			8 05 36	+74 25.5	1	13.06		0.51		-0.01		1		1058
		Pup sq 1 # 32			8 05 36	-33 06.	1	11.62		1.93		2.90		3		124
	-32 04876	LSS 966			8 05 37	-32 55.1	3	10.02	.010	0.45	.005	-0.45	.022	6		1066,1712,1766
	-47 03601				8 05 38	-47 47.1	1	8.69		-0.06		-0.31		1		389
67763	-51 02840				8 05 39	-51 47.6	1	7.13		1.26		1.31		1	K2 III	389
	-28 05482				8 05 40	-28 36.9	2	10.33	.005	0.29	.015	0.09	.010	3		126,211
		POSS 425 # 6			8 05 44	+16 11.2	1	17.56		1.33				4		1739
	+33 01646	IDS08026N3306		A	8 05 45	+32 58.0	1	9.88		1.36				2	M0.5V:e	1625
67402	+27 01544	IDS08027N2746		AB	8 05 46	+27 37.7	1	6.88		1.04		0.77		2	K0 III	3016
	-33 04574				8 05 47	-33 21.1	1	10.96		0.59		0.13		3		124

Table 1

HD	DM	Other Id	N	Rem	α_{1950}	δ_{1950}	S	V	σ_V	B–V	σ_{B-V}	U–B	σ_{U-B}	n	Spectrum	References
		Pup sq 1 # 31			8 05 48	−33 07.	1	10.48		1.29		1.28		3		124
67680	−37 04271				8 05 48	−37 17.4	1	7.84		1.25		1.26		5	K1 III	1673
		Orsatti 347			8 05 50	−32 15.3	1	13.37		0.88		−0.28		2		1766
67653	−32 04880				8 05 50	−32 56.7	1	8.82		−0.04		−0.32		2	B8/9 V	401
67762	−48 03495				8 05 50	−48 14.6	4	6.73	.012	0.96	.008	0.70	.005	10	K1 III	389,1075,1311,3077
67483	+14 01831	HR 3184			8 05 55	+13 47.3	2	6.27	.028	0.44	.026	0.00	.009	4	F6 IV	254,3053
	+25 01858	G 40 - 8			8 05 55	+24 46.6	2	9.64	.010	0.68	.005	0.12	.035	3	G3 V	333,1003,1620
67370	+42 01819	HR 3181			8 05 57	+42 34.7	1	6.27		1.27				2	gK3	71
67224	+58 01102	HR 3175			8 05 58	+58 23.8	2	5.91	.020	1.39	.000	1.52	.020	4	gK4	1733,3016
67678	−31 05565				8 05 58	−31 57.4	1	9.12		−0.02		−0.23		2	B9.5V	401
		HA 4 # 312			8 06 01	+74 53.2	1	12.88		0.49		0.01		1		1058
67758	−41 03784				8 06 01	−41 39.9	1	7.14		−0.12		−0.60		4	B2/3 III	164
		HA 4 # 23			8 06 03	+74 14.0	1	12.77		0.59		0.06		1		1058
		Orsatti 349			8 06 04	−31 27.0	1	12.83		0.63		−0.34		2		1766
67735	−33 04582				8 06 04	−33 52.4	1	7.92		0.04		−0.16		2	B9 III	401
67594	−2 02450	HR 3188		★ AB	8 06 05	−02 50.2	13	4.34	.011	0.97	.007	0.71	.016	58	G2 Ib	3,15,150,418,1075*
67594	−2 02449	IDS08036S0242		C	8 06 05	−02 50.2	2	8.81	.010	1.43	.010	1.69	.025	6	K5	150,1084
		Orsatti 351			8 06 06	−32 04.7	1	12.45		0.97		−0.20		2		1766
67848	−50 03167				8 06 08	−50 57.9	1	7.87		0.93		0.61		1	G6/8 III	389
67698	−23 06846	IDS08040S2320		AB	8 06 10	−23 28.3	1	6.62		−0.09		−0.51		4	B3 III/IV	164
		XX Pup			8 06 12	−16 23.5	2	10.51	.015	0.11	.035	0.12	.025	2	A2	668,700
		BSD 4 # 579			8 06 13	+75 15.9	1	11.26		0.50		0.02		1	G0	1058
	−32 04888	LSS 967			8 06 13	−33 4.0	1	10.34		0.66		0.59		3	A9 II	1712
67847	−48 03498	NGC 2547 - 3			8 06 13	−48 51.9	1	6.66		−0.08		−0.25		3	B8 V	935
	−33 04585				8 06 15	−33 26.2	1	10.76		1.02		0.76		3		124
67778	−32 04890				8 06 18	−33 11.6	2	7.47	.009	0.00	.000	−0.30	.009	10	B8 III	124,401
67588	+10 01738				8 06 20	+10 39.4	1	8.29		0.24		0.06		2	A2	1733
67777	−32 04892				8 06 20	−32 58.9	3	8.70	.014	−0.07	.010	−0.49	.007	10	B8 II	124,401,1775
67650	−10 02396	RT Mon			8 06 21	−10 38.6	2	8.19	.065	1.81	.028	2.09	.035	2	M3 III	975,3076
		HA 4 # 27			8 06 22	+74 16.8	1	10.99		0.44		−0.02		1		1058
		HA 4 # 26			8 06 23	+74 19.4	1	11.54		0.38		−0.21		1		1058
	+44 01719				8 06 24	+43 55.1	1	10.93		0.78		0.36		1		3022
67728	−19 02261	IDS08041S1934		BC	8 06 24	−19 41.6	1	7.54		1.12		1.00		2	G8/K0 III	536
67929	−53 01541				8 06 24	−53 55.1	1	7.60		1.01		0.78		1	G8 III	389
		L 97 - 3			8 06 24	−66 09.	2	13.83	.107	0.05	.005	−0.85	.063	3		832,3065
		LP 17 - 112			8 06 28	+79 53.1	1	10.75		0.71		0.17		3		1723
67954	−54 01517				8 06 28	−54 40.7	1	8.13		0.08		−0.56		2	B3 V	401
		HA 4 # 84			8 06 29	+74 23.2	1	10.27		0.55		0.07		1		1058
	−49 03360	NGC 2547 - 59			8 06 29	−49 19.7	2	10.92	.004	0.34	.000	0.07		7		935,2022
67542	+29 01696				8 06 30	+29 14.4	2	6.49	.024	0.80	.010	0.47		7	G0 II	6009,8100
		GD 262			8 06 30	+29 29.4	1	16.16		0.38		−0.45		1		3060
67751	−19 02262	HR 3190		★ A	8 06 31	−20 13.0	2	6.35	.005	0.16	.000			7	A3 III	15,2012
		LSS 969			8 06 31	−32 59.9	2	12.56	.015	0.38	.015	−0.26	.005	6	B3 Ib	1712,1766
	−32 04895				8 06 33	−32 35.3	2	8.84	.005	1.71	.005	2.04	.015	8		124,1775
67725	−10 02400	HR 3189			8 06 34	−11 11.5	3	6.31	.005	0.00	.012	−0.10		8	A0 Vn	15,1079,2012
		G 40 - 9			8 06 35	+22 03.2	1	11.84		1.53		1.22		1		1620
		HA 4 # 28			8 06 35	+74 20.4	1	13.24		0.47		−0.07		1		1058
67368	+59 01144				8 06 36	+59 20.5	1	8.03		1.19		1.23		2	K2 III	1733
67907	−44 04082				8 06 38	−44 45.8	1	8.61		0.70		0.23		1	G5 V	389
	−48 03502	NGC 2547 - 119			8 06 38	−49 00.7	1	10.73		0.25				1		2022
		HA 4 # 242			8 06 40	+74 42.9	1	11.12		0.53		0.12		1		1058
	−32 04900	LSS 970			8 06 43	−32 56.0	1	11.05		0.28		−0.63		3		1712
		HA 4 # 85			8 06 47	+74 21.6	1	11.12		0.51		0.05		1		1058
67797	−18 02190	HR 3192			8 06 48	−19 05.8	5	4.40	.004	−0.16	.010	−0.60	.005	20	B5 V	15,1075,1425,2012,8015
		Orsatti 357			8 06 48	−33 46.4	1	13.28		0.55		−0.27		2		1766
		AAS70,69 T4# 2			8 06 48	−35 55.3	1	12.76		0.51		0.06		1		1636
67888	−37 04288	HR 3195, PQ Pup			8 06 48	−37 32.1	8	6.34	.021	−0.04	.011	−0.55	.008	42	B3 V	15,164,401,681,815*
		AAS70,69 T4# 1			8 06 50	−35 53.1	1	12.64		0.31		0.23		2		1636
	−49 03366	NGC 2547 - 37			8 06 53	−49 19.5	3	10.11	.010	0.48	.010	0.10		10	F2	935,2012,2022
		BSD 4 # 583			8 06 55	+75 17.8	1	12.57		0.58		0.04		1	G0	1058
67587	+35 01767	IDS08037N3545		A	8 06 56	+35 36.4	1	6.61		0.53		0.04		2	G0 V	3026
		Orsatti 358			8 06 56	−31 39.3	1	12.83		0.68		−0.43		2		1766
67887	−28 05525				8 06 57	−29 05.6	2	9.40	.001	0.49	.026	0.03	.013	4	F3 V	540,976
		Orsatti 359			8 06 58	−32 03.1	1	13.87		0.52		−0.18		2		1766
	+45 01540				8 06 59	+44 53.2	1	9.88		0.48				std	F8	1334
67924	−32 04905				8 06 59	−33 05.9	1	7.76		−0.04		−0.60		2	B2 III/IV	401
	−30 05721	LSS 971			8 07 2	−30 16.3	1	10.22		0.33		0.08		3	A0 II	1712
		ApJ144,259 # 26			8 07 00	−32 31.	1	11.58		0.28		0.05		1		3074
67712	+17 01776				8 07 03	+17 24.0	2	8.17	.020	1.12	.015	1.12		4	K0	882,1648
		HA 4 # 31			8 07 03	+74 13.2	1	12.97		0.61		0.02		1		1058
		Orsatti 360			8 07 03	−33 40.9	1	13.35		0.45		−0.14		2		1766
		AAS70,69 T4# 3			8 07 04	−35 55.3	1	13.51		0.39		0.24		1		1636
67921	−29 05620	HR 3196		★ AB	8 07 06	−30 10.5	4	6.64	.004	1.40	.004	1.60		15	K3 III	15,1075,2029,3005
67768	+13 01851				8 07 10	+13 09.2	2	7.64	.025	1.04	.000	0.89		4	K0	882,3040
		Orsatti 361			8 07 10	−32 19.0	1	13.36		0.75		0.00		2		1766
67880	−15 02280	HR 3194		★ AB	8 07 11	−16 06.1	3	5.67	.009	−0.17	.008	−0.74		10	B2.5 V	15,1732,2012
		LSS 972			8 07 11	−33 45.6	2	12.46	.019	0.35	.005	−0.21	.014	4	B3 Ib	1712,1766
		LP 366 - 47			8 07 12	+22 27.6	1	11.65		1.05		0.70		2		1723
67690	+26 01728	IDS08042N2609		A	8 07 12	+25 59.6	1	6.41		1.42		1.60		2	K0	3016
		BSD 4 # 585			8 07 13	+75 21.6	1	12.86		0.51		0.01		1	G0	1058

HD	DM	Other Id	N Rem	α_{1950}	δ_{1950}	S	V	σ_V	B–V	σ_{B-V}	U–B	σ_{U-B}	n	Spectrum	References
67862	−11 02261	XY Pup		8 07 13	−11 50.3	2	9.29	.011	0.26	.022	0.20	.034	2	A3	588,1588
		Orsatti 365		8 07 13	−33 59.1	1	11.99		0.34		−0.30		1		1766
	−44 04095			8 07 13	−44 35.6	1	7.94		0.49		0.06		2		389
	−60 01066			8 07 13	−60 26.0	1	9.45		1.19		1.09		3		3072
67790	+13 01852			8 07 14	+12 59.4	2	8.08	.015	1.13	.002	1.12		4	K2	882,3040
68073	−48 03510	NGC 2547 - 17		8 07 14	−48 39.3	3	9.23	.009	0.12	.011	0.06		10	A2 V	935,2012,2022
68074	−49 03370	NGC 2547 - 8	★V	8 07 14	−49 21.0	3	8.26	.000	−0.11	.006	−0.49		13	B8 III	935,2012,2022
		HA 4 # 34		8 07 16	+74 18.8	1	11.51		0.37		0.00		1		1058
		NGC 2539 - 15		8 07 17	−12 50.6	1	11.30		1.63		2.07		2		75
67977	−35 04256	HR 3199		8 07 17	−35 18.4	1	6.20		0.89				4	G8 III	2006
68092	−46 03833			8 07 18	−47 01.6	2	7.18	.000	−0.15	.007	−0.66	.005	6	B3 V	540,976
	−49 03371	NGC 2547 - 50		8 07 18	−49 14.9	3	10.72	.007	0.30	.005	0.08		10		935,2012,2022
67901	−12 02360	NGC 2539 - 17		8 07 19	−12 45.0	1	9.13		0.20		0.16		2	A8 IV	75
67562	+53 01221			8 07 20	+53 23.7	1	6.79		0.51		0.08		2	F5	1566
68091	−46 03834			8 07 20	−46 39.6	1	8.40		1.02		0.80		1	K0 III	389
68030	−37 04297	IDS08055S3730	AB	8 07 21	−37 39.0	1	8.21		−0.01		−0.58		2	B3/5 III	401
67646	+45 01544	IDS08039N4457	AB	8 07 22	+44 48.2	1	9.43		0.46		0.03		std	F5	1334
		HA 4 # 246		8 07 22	+74 44.6	1	11.82		0.51		0.01		1		1058
68114	−48 03511	NGC 2547 - 21		8 07 22	−49 04.8	3	9.40	.004	0.04	.005	0.02		9	A1 V	935,2012,2022
		NGC 2547 - 71		8 07 22	−49 07.4	1	11.45		0.44		−0.01		2		935
68279	−60 01068	IDS08065S6047	AB	8 07 22	−60 55.8	1	6.66		0.86				4	G8 III	2012
		NGC 2539 - 19		8 07 23	−12 38.0	2	11.65	.000	0.50	.000	0.05	.015	4		75,1647
	−38 04120			8 07 23	−38 44.0	1	9.77		0.45		−0.39		3		434
		G 51 - 1		8 07 24	+33 03.8	1	12.40		0.84		0.39		1		1620
		NGC 2539 - 20		8 07 24	−12 37.5	1	11.34		0.57		0.09		2		75
68115	−48 03512	NGC 2547 - 27		8 07 25	−49 04.9	2	9.70	.000	0.03	.005	−0.02		5	A1 V	935,2022
67561	+55 01258			8 07 26	+55 38.3	1	9.25		0.41		0.00		2	F8	1502
		NGC 2539 - 18		8 07 26	−12 41.2	1	11.93		1.13		1.23		2		75
68089	−41 03810			8 07 26	−41 55.9	2	9.61	.005	0.60	.005	−0.01	.005	6	G3 V(w)	158,1696
67767	+25 01865	HR 3191	★A	8 07 27	+25 39.6	2	5.72	.009	0.82	.009	0.44	.004	8	G8 IV	1355,3077
		NGC 2539 - 1002		8 07 27	−12 45.3	1	10.47		0.97		0.17		2		1647
		HA 4 # 386		8 07 28	+75 10.0	1	13.28		0.63		0.04		1		1058
	−12 02362	NGC 2539 - 16		8 07 29	−12 48.4	1	10.56		1.49		1.88		2		75
	+75 00328			8 07 32	+75 20.2	1	10.40		0.65		0.22		1	G5	1058
	−37 04302			8 07 33	−38 04.6	2	9.66	.005	0.27	.002	−0.50	.009	5	B2 IV-V	540,976
68026	−27 05090	LSS 974		8 07 35	−27 17.5	2	9.08	.004	−0.02	.001	−0.81	.009	5	B1 II	540,976
	−48 03513	NGC 2547 - 74		8 07 35	−49 07.3	2	11.52	.005	0.20	.014	0.16		5		935,2022
		NGC 2539 - 21		8 07 36	−12 36.1	1	9.56		1.33		1.56		2	K5 I-II	75
67939	−6 02489			8 07 39	−06 35.8	1	7.79		1.62		1.94		4	M1	1657
		NGC 2539 - 1003		8 07 39	−12 37.7	1	13.20		0.98		0.96		2		1647
		HA 4 # 248		8 07 40	+74 43.4	1	13.19		0.31		0.06		1		1058
68161	−48 03516	NGC 2547 - 1	★	8 07 40	−48 32.2	2	5.65	.050	−0.12	.000	−0.42		7	B8 Ib/II	935,2035
	−38 04128			8 07 41	−38 42.2	1	9.72		0.49		−0.35		2	B2 II-III	540
	−12 02366	NGC 2539 - 14		8 07 42	−12 49.6	2	10.45	.030	0.34	.005	0.08	.015	4	F2p	75,1647
		NGC 2539 - 22		8 07 43	−12 33.4	1	10.98		1.20		1.34		2		75
		NGC 2539 - 1005		8 07 43	−12 38.7	1	13.01		0.42		0.01		2		1647
		NGC 2539 - 1006		8 07 43	−12 39.7	1	12.65		0.24		0.21		2		1647
		NGC 2539 - 1007		8 07 43	−12 40.0	1	13.62		0.52		−0.08		2		1647
		NGC 2539 - 1008		8 07 43	−12 46.7	1	13.28		0.60		0.09		2		1647
	−48 03517	NGC 2547 - 44		8 07 43	−48 37.4	1	10.34		0.16		0.13		1	A2	935
		NGC 2539 - 1009		8 07 44	−12 36.2	1	13.64		0.69				2		1647
		NGC 2539 - 1010		8 07 44	−12 36.4	1	13.40		0.73				2		1647
		MCC 514		8 07 45	+47 43.8	1	11.29		1.38				1	M0	1017
		NGC 2539 - 1011		8 07 45	−12 39.4	1	11.41		0.51		0.14		2		1647
	−48 03518	NGC 2547 - 72		8 07 45	−48 55.2	3	11.43	.014	0.37	.009	0.08		10		935,2012,2022
		NGC 2539 - 1012		8 07 46	−12 46.4	1	12.49		0.28		0.17		2		1647
		NGC 2539 - 1013		8 07 46	−12 46.7	1	12.71		0.70		0.38		2		1647
68423	−63 00896	HR 3217		8 07 46	−63 39.2	3	6.29	.024	−0.06	.003	−0.37		12	B6 V	15,1637,2012
		NGC 2539 - 1014		8 07 47	−12 43.5	1	12.17		0.19		0.25		2		1647
	−48 03519	NGC 2547 - 34		8 07 47	−48 46.4	1	9.98		0.52		0.96		2		935
68520	−68 00736	HR 3223	★AB	8 07 47	−68 28.2	3	4.33	.007	−0.12	.003	−0.45	.005	16	B5 III	15,1075,2038
		NGC 2539 - 59		8 07 48	−12 36.2	1	12.16		0.16		0.15		2		75
	−31 05607	LSS 976		8 07 48	−32 5.9	2	11.53	.000	0.28	.010	−0.56	.025	5		1712,1766
		NGC 2547 - 82		8 07 49	−48 45.9	1	11.97		0.42		0.05		1		935
		NGC 2539 - 23		8 07 50	−12 32.0	1	12.90		0.78		0.64		2		75
67447	+68 00524	HR 3182		8 07 52	+68 37.4	2	5.45	.086	1.05	.005	0.80	.009	7	G8 II	1355,8100
		NGC 2539 - 1016		8 07 52	−12 39.8	1	13.35		0.40		1.13		2		1647
		NGC 2539 - 58		8 07 53	−12 36.0	1	12.01		0.41		0.00		2		75
		NGC 2539 - 1017		8 07 53	−12 40.0	1	11.96		0.23		0.18		2		1647
		NGC 2539 - 55		8 07 54	−12 33.9	1	12.59		0.51		0.03		2		75
		NGC 2539 - 65		8 07 54	−12 41.8	1	10.68		1.03		0.78		2		1647
		NGC 2539 - 13		8 07 54	−12 51.2	1	11.44		0.17		0.13		2	A0	75
68276	−50 03184			8 07 54	−50 51.9	1	7.60		0.06		0.05		1	A0 V	389
		NGC 2539 - 54		8 07 55	−12 33.9	1	11.86		0.18		0.16		2		75
		NGC 2539 - 1023		8 07 55	−12 45.2	1	12.24		0.91				2		1647
		NGC 2539 - 1021		8 07 55	−12 47.1	1	13.38		1.07		0.74		2		1647
		NGC 2539 - 1020		8 07 55	−12 47.5	1	12.62		0.35		0.12		2		1647
		NGC 2539 - 1022		8 07 55	−12 48.1	1	13.46		1.10		1.19		2		1647
68244	−48 03521	NGC 2547 - 12		8 07 55	−49 01.9	3	8.62	.010	0.18	.003	0.17		11	A2 V	935,2012,2022
		NGC 2539 - 53		8 07 56	−12 33.7	1	13.79		1.30				2		75

Table 1

HD	DM	Other Id	N Rem	α_{1950}	δ_{1950}	S	V	σ_V	B–V	σ_{B-V}	U–B	σ_{U-B}	n	Spectrum	References
		NGC 2539 - 56		8 07 57	−12 34.9	1	12.43		0.19		0.15		2		75
		NGC 2539 - 1028		8 07 57	−12 38.7	1	11.72		0.93		0.73		2		1647
68217	−43 03998	HR 3204		8 07 57	−43 58.5	2	5.20	.005	-0.20	.005			7	B2 IV	15,2012
68243	−46 03846	HR 3206	⋆ A	8 07 57	−47 11.8	4	4.27	.008	-0.23	.012	-0.91	.010	13	B1 IV	15,200,1034,8015
68274	−48 03522	NGC 2547 - 16		8 07 58	−49 03.0	3	9.03	.009	-0.02	.005	-0.10		10	B9 IV-V	935,2012,2022
		NGC 2547 - 62		8 07 58	−49 05.7	2	10.95	.010	0.32	.005	0.08		3		935,2022
68273	−46 03847	HR 3207, γ2 Vel	⋆ A	8 07 59	−47 11.3	8	1.79	.046	-0.24	.013	-0.98	.017	53	WC8 +O7.5e	15,200,200,1034,1075*
		Orsatti 367		8 07 60	−31 28.7	1	11.20		0.26		-0.34		2		1766
		NGC 2539 - 52		8 08 00	−12 34.1	1	13.16		0.33		0.06		2		75
		NGC 2539 - 57		8 08 00	−12 34.9	1	12.15		0.17		0.15		2		75
		NGC 2539 - 1033		8 08 00	−12 36.9	1	13.28		0.60		0.09		2		1647
		NGC 2539 - 1036		8 08 00	−12 44.2	1	11.34		0.18		0.16		2		1647
		NGC 2539 - 1029		8 08 00	−12 48.4	1	11.99		0.92		0.55		2		1647
		Orsatti 368		8 08 00	−32 23.6	1	13.42		0.70		-0.31		1		1766
	−49 03379	NGC 2547 - 24		8 08 00	−49 17.1	4	9.52	.010	1.04	.007	0.61	.089	12	K2	657,935,2012,2022
		NGC 2539 - 1035		8 08 01	−12 37.2	1	12.78		0.60		0.19		2		1647
		NGC 2539 - 1034		8 08 01	−12 39.9	1	12.19		0.38		0.24		2		1647
		NGC 2539 - 1030		8 08 01	−12 45.3	1	14.24		0.79		0.67		2		1647
	−48 03524	NGC 2547 - 39		8 08 01	−48 57.4	2	10.21	.018	0.15	.005	0.10		5	A2 V	935,2022
67808	+43 01773			8 08 02	+43 33.4	1	8.56		0.49		-0.01		1	F8	1776
		NGC 2539 - 69		8 08 02	−12 41.8	1	10.91		0.97		0.65		2		1647
		NGC 2539 - 1038		8 08 02	−12 44.0	1	13.43		0.97		0.85		2		1647
		NGC 2539 - 1039		8 08 02	−12 45.0	1	12.96		0.37		-0.01		2		1647
67827	+39 02065	HR 3193		8 08 03	+38 52.9	1	6.59		0.59		0.09		2	G0 IV	105
		HA 4 # 249		8 08 03	+74 41.9	1	11.10		0.39		-0.02		1		1058
		NGC 2539 - 60		8 08 03	−12 35.7	1	10.96		0.69		0.29		2		75
		NGC 2539 - 1041		8 08 03	−12 42.8	1	11.48		0.14		0.15		2		1647
		NGC 2539 - 1042		8 08 03	−12 44.6	1	12.22		0.22		0.13		2		1647
	−48 03526	NGC 2547 - 65		8 08 03	−49 11.8	3	11.00	.033	0.38	.009	0.04		11		935,2012,2022
68275	−48 03525	NGC 2547 - 35		8 08 03	−49 13.0	3	10.00	.018	0.15	.003	0.15		10	A3 V	935,2012,2022
		NGC 2539 - 1043		8 08 04	−12 40.1	1	12.99		0.29		0.13		2		1647
		MCC 515		8 08 05	+51 28.3	1	10.89		1.23				1	K4-5	1017
		SU UMa		8 08 05	+62 45.5	1	14.18		0.04		-1.11		1		698
		NGC 2539 - 1044		8 08 05	−12 48.7	1	12.32		0.33		0.15		2		1647
	−34 04419			8 08 05	−34 21.5	1	9.16		0.54				14	G0	6011
67958	+16 01651			8 08 06	+16 30.4	2	8.15	.030	1.14	.012	1.22		4	K0	882,3040
		NGC 2539 - 1045		8 08 06	−12 47.7	1	13.00		0.55		0.18		2		1647
68242	−42 03944	HR 3205	⋆ AB	8 08 06	−42 29.5	2	6.27	.015	-0.04	.000	-0.32		7	B7 V	404,2006
		CCS 1071		8 08 07	−29 23.1	1	11.25		2.63				1		864
		NGC 2539 - 1046		8 08 08	−12 44.1	1	12.46		0.18		0.15		2		1647
68305	−48 03528	NGC 2547 - 14	⋆ AB	8 08 08	−49 02.8	3	8.85	.009	0.05	.000	0.02		10	A1 V	935,2012,2022
	−49 03381	NGC 2547 - 56		8 08 08	−49 18.1	3	10.82	.013	0.46	.009	0.01		11		935,2012,2022
68346	−51 02861			8 08 08	−51 41.7	1	6.53		0.60		0.15		1	G0 V	389
67959	+15 01775	HR 3198		8 08 10	+14 46.7	2	6.23		0.03	.005	0.05	.024	5	A1 V	252,1022
		NGC 2539 - 1047		8 08 10	−12 43.1	1	11.15		0.17		0.12		2		1647
	−48 03530	NGC 2547 - 45	⋆ AB	8 08 10	−49 09.8	3	10.40	.026	0.32	.009	0.06		10	A9 V	935,2012,2022
68371	−52 01396			8 08 11	−52 25.6	1	7.24		-0.02		-0.54		4	B3 V	164
68456	−60 01074	HR 3220		8 08 11	−61 09.0	4	4.75	.004	0.44	.005	-0.04	.010	15	F5 V	15,1075,2012,3026
		NGC 2539 - 1049		8 08 12	−12 44.0	1	11.69		0.22		0.17		2		1647
		NGC 2539 - 1048		8 08 12	−12 48.6	1	11.88		0.14		0.16		2		1647
		LSS 977		8 08 12	−31 40.8	2	12.50	.015	0.16	.015	-0.42	.010	5	B3 Ib	1712,1766
	−40 03913			8 08 12	−40 41.1	1	10.91		0.21		-0.41		2	B3 V	540
68270	−43 04000			8 08 12	−43 22.8	1	7.17		1.46		1.59		1	K2/3 III	389
68324	−47 03653	HR 3213, IS Vel		8 08 12	−47 47.3	3	5.23	.007	-0.21	.007	-0.89	.004	9	B1 IVe	401,1732,2035
		NGC 2539 - 46		8 08 13	−12 39.5	1	11.80		0.18		0.13		2	A3	75
		NGC 2539 - 44		8 08 13	−12 41.0	1	10.83		0.64		0.39		2	A2	75
		NGC 2539 - 1050		8 08 13	−12 42.9	1	11.55		0.97		0.67		2		1647
		HA 4 # 392		8 08 14	+75 03.1	1	11.39		0.43		-0.04		1		1058
		NGC 2539 - 1053		8 08 14	−12 43.0	1	11.60		0.17		0.14		2		1647
68325	−49 03383	NGC 2547 - 20		8 08 14	−49 27.9	2	9.38	.004	0.02	.008	-0.06		7	A0 V	935,2022
		NGC 2539 - 43		8 08 15	−12 40.1	1	11.22		0.17		0.15		2	A5	75
		NGC 2539 - 45		8 08 16	−12 39.7	1	12.43		0.22		0.17		2	A2	75
		NGC 2539 - 1056		8 08 17	−12 41.8	1	11.86		0.15		0.16		2		1647
	+16 01654			8 08 18	+15 56.0	1	9.34		0.38		-0.05		1	F2	1776
		NGC 2539 - 51		8 08 18	−12 35.8	1	11.28		0.93		0.71		2		75
		NGC 2539 - 50		8 08 18	−12 36.6	1	11.40		0.20		0.17		2		75
	−31 05621	LSS 978		8 08 18	−31 20.0	2	11.60	.010	0.15	.015	-0.49	.010	5		1712,1766
		Orsatti 371		8 08 18	−31 37.6	1	12.90		0.31		-0.03		2		1766
		NGC 2539 - 42		8 08 19	−12 40.0	1	11.14		0.93		0.70		2	K0	75
		NGC 2539 - 1061		8 08 19	−12 41.8	1	12.25		0.34				2		1647
		NGC 2539 - 1060		8 08 19	−12 47.7	1	14.48		0.43		0.03		2		1647
		NGC 2547 - 97		8 08 19	−49 04.2	1	12.39		0.63		0.08		3		935
	−48 03533	NGC 2547 - 51		8 08 19	−49 05.6	2	10.71	.026	0.31	.017	0.08		6		935,2022
		NGC 2547 - 73		8 08 19	−49 10.8	1	11.49		1.33		1.31		3		935
		NGC 2539 - 1062		8 08 20	−12 48.2	1	13.72		0.43		0.03		2		1647
	−33 04622			8 08 20	−33 17.1	1	10.77		0.28		-0.38		2		1066
	−48 03536	NGC 2547 - 60		8 08 20	−49 00.7	3	10.91	.013	0.40	.007	0.05		9		935,2012,2022
68146	−13 02420	HR 3202	⋆ A	8 08 21	−13 39.1	4	5.53	.005	0.49	.004	-0.02	.000	13	F7 V	15,164,2012,3077
		LSS 979		8 08 21	−30 48.4	1	11.57		0.52		0.24		3	B3 Ib	1712
	−48 03535	NGC 2547 - 23		8 08 21	−48 52.9	2	9.47	.005	1.02	.005	0.76		5	K0	935,2022

HD	DM	Other Id	N	Rem	α_{1950}	δ_{1950}	S	V	σ_V	B–V	σ_{B-V}	U–B	σ_{U-B}	n	Spectrum	References
	−49 03384	NGC 2547 - 25			8 08 21	−49 16.2	2	9.66	.017	1.72	.044	2.05		6		935,2022
	−12 02376	NGC 2539 - 26			8 08 22	−12 31.3	1	10.70		1.04		0.93		2	K0	75
		NGC 2539 - 1065			8 08 22	−12 41.9	1	12.34		0.41				2		1647
		NGC 2539 - 1063			8 08 22	−12 46.7	1	14.41		0.89		0.67		2		1647
		NGC 2539 - 1066			8 08 22	−12 49.7	1	13.07		0.58		0.11		2		1647
		NGC 2539 - 1064			8 08 22	−12 50.3	1	13.02		0.28		0.10		2		1647
	−48 03537	NGC 2547 - 41			8 08 22	−49 03.9	3	10.20	.014	0.16	.007	0.11		10	A3 IV	935,2012,2022
	−48 03539	NGC 2547 - 49			8 08 22	−49 04.4	3	10.57	.013	0.22	.021	0.10		9	A0	935,2012,2022
		NGC 2539 - 1068			8 08 23	+12 43.4	1	12.76		0.32		0.09		2		1647
		MCC 516			8 08 23	+52 05.0	1	10.20		1.32		1.22		2	M0	1723
		HA 4 # 250			8 08 23	+74 50.0	1	12.17		0.53		0.09		1		1058
68266	−32 04930				8 08 23	−32 53.8	1	8.33		-0.12		-0.51		2	B8 V	401
	−48 03538	NGC 2547 - 52			8 08 23	−48 51.4	3	10.74	.009	0.15	.005	0.16		10		935,2012,2022
	−48 03540	NGC 2547 - 19			8 08 23	−49 02.4	1	9.36		0.13		0.07		4		935
		NGC 2539 - 1069			8 08 24	−12 43.7	1	12.49		0.20		0.15		2		1647
	−48 03541	NGC 2547 - 26			8 08 24	−49 02.1	1	9.67		0.06		0.03		5		935
68434	−55 01467	HR 3218			8 08 24	−55 56.2	1	5.66		0.21				4	A3mA5-A9	2035
		NGC 2539 - 49			8 08 25	−12 37.1	1	12.48		1.05		0.96		2	K0	75
		NGC 2539 - 1071			8 08 25	−12 44.0	1	11.61		0.27		0.15		2		1647
	−12 02379	NGC 2539 - 48			8 08 28	−12 37.2	1	10.04		0.52		0.03		2	F5	75
		NGC 2539 - 47			8 08 28	−12 37.5	1	11.07		1.60		1.96		2		75
		LSS 981			8 08 28	−33 59.7	2	11.97	.014	0.22	.000	-0.32	.005	4	B3 Ib	1712,1766
68398	−48 03543	NGC 2547 - 13			8 08 28	−49 10.6	3	8.62	.015	0.00	.007	-0.18		11	B9 IV-V	935,2012,2022
	−12 02380	NGC 2539 - 38			8 08 29	−12 44.2	1	10.54		0.61		0.42		2	F0	75
68396	−48 03544	NGC 2547 - 15			8 08 29	−48 59.4	3	8.91	.005	-0.04	.019	-0.19		11	B9 IV-V	935,2012,2022
		NGC 2539 - 41			8 08 30	−12 41.8	1	12.43		0.25		0.20		2		75
		NGC 2539 - 40			8 08 30	−12 42.5	1	12.05		0.25		0.12		2	A3	75
		NGC 2539 - 39			8 08 30	−12 43.3	1	11.82		0.21		0.14		2		75
68300	−36 04279				8 08 30	−37 01.9	2	8.48	.015	-0.10	.019	-0.69		3	B2 V	401,2041
68395	−47 03657				8 08 30	−47 40.2	1	8.09		-0.11		-0.63		2	B4 V	401
68017	+32 01695	IDS08054N3246		A	8 08 31	+32 36.9	3	6.83	.014	0.68	.009	0.13	.005	7	G4 V	22,1003,3026
68397	−48 03545	NGC 2547 - 7			8 08 31	−49 07.3	3	8.15	.000	-0.01	.005	-0.18		11	B9 IV	935,2012,2022
		BSD 4 # 593			8 08 32	+75 26.0	1	12.92		0.48		0.03		1		1058
		NGC 2539 - 1078			8 08 32	−12 36.5	1	12.76		0.58		0.09		2		1647
		NGC 2539 - 1079			8 08 32	−12 43.0	1	13.59		0.91				2		1647
		Orsatti 376			8 08 32	−32 21.4	1	13.20		0.43		-0.48		2		1766
		NGC 2547 - 84			8 08 32	−48 49.5	1	12.03		0.23		0.17		3		935
		NGC 2547 - 68			8 08 32	−48 59.6	3	11.04	.029	0.46	.007	0.00		7		935,2012,2022
68099	+10 01746	HR 3201			8 08 33	+09 58.3	2			-0.10	.023	-0.42	.014	4	B6 III	1022,1079
67870	+49 01711				8 08 33	+49 06.5	1	6.88		1.32		1.38		2	K0	1601
		NGC 2539 - 11			8 08 33	−12 51.6	1	12.57		1.14		1.04		2		75
		NGC 2547 - 105			8 08 33	−49 13.2	1	12.84		1.17		0.74		2		935
		HA 4 # 251			8 08 34	+74 48.4	1	10.39		0.69		0.29		1		1058
		NGC 2539 - 27			8 08 35	−12 31.7	2	11.74	.015	0.25	.000	0.11	.010	4		75,1647
		NGC 2539 - 1082			8 08 35	−12 43.7	1	13.68		0.71				2		1647
		NGC 2539 - 1083			8 08 35	−12 48.3	1	11.90		0.48		-0.01		2		1647
		NGC 2539 - 12			8 08 35	−12 50.5	2	12.20	.045	0.50	.005	0.14	.005	4		75,1647
		NGC 2539 - 1081			8 08 36	−12 34.7	1	11.91		0.10		0.19		2		1647
	−48 03546	NGC 2547 - 46			8 08 37	−48 40.1	3	10.48	.010	0.38	.010	-0.01		9		935,2012,2022
	−48 03547	NGC 2547 - 31			8 08 38	−49 05.4	3	9.86	.009	0.08	.005	0.06		10	A1 V	935,2012,2022
68635	−64 00828				8 08 38	−64 37.7	1	7.44		1.07		0.92		4	K0 III	1704
68420	−48 03548	NGC 2547 - 30			8 08 39	−48 51.8	3	9.82	.023	0.10	.009	0.09		12	A3 V	935,2012,2022
68432	−48 03549	NGC 2547 - 9			8 08 39	−49 07.7	3	8.41	.005	-0.06	.005	-0.37		10	B8 IV	935,2012,2022
		NGC 2539 - 1085			8 08 40	−12 34.8	1	12.96		0.33		0.19		2		1647
		NGC 2539 - 1086			8 08 40	−12 45.8	1	13.16		0.27				2		1647
		MCC 517			8 08 41	+32 02.0	1	11.32		1.27		1.09		2	M0	1723
		NGC 2547 - 100			8 08 41	−49 12.0	1	12.52		0.61		0.07		3		935
	−12 02381	NGC 2539 - 32			8 08 42	−12 41.0	1	10.55		1.01		0.80		2		75
		L 242 - 66			8 08 42	−52 49.7	3	11.75	.021	1.52	.017	1.08		6		912,1705,3073
		NGC 2539 - 1088			8 08 44	−12 37.6	1	13.05		0.32		0.24		2		1647
		NGC 2539 - 1089			8 08 44	−12 41.9	1	12.61		0.22		0.20		2		1647
		NGC 2547 - 102			8 08 44	−49 04.2	1	12.57		0.62		0.05		2		935
		NGC 2547 - 76			8 08 44	−49 13.5	1	11.64		0.55		0.07		3		935
		NGC 2539 - 1091			8 08 45	−12 37.4	1	13.99		0.49		0.14		2		1647
		NGC 2539 - 33			8 08 45	−12 39.7	1	11.50		0.16		0.13		2		75
	−48 03551	NGC 2547 - 43			8 08 45	−49 10.2	3	10.27	.009	0.43	.005	-0.01		9	F3:V:	935,2012,2022
	−48 03550	NGC 2547 - 66			8 08 45	−49 11.8	3	11.02	.028	0.47	.031	0.04		10		935,2012,2022
68122	+16 01657				8 08 47	+16 22.0	1	7.30		1.20				2	K0	882
68451	−48 03552	NGC 2547 - 4			8 08 47	−48 53.2	4	7.33	.008	-0.14	.008	-0.64	.000	16	B3 V	164,935,2012,2022
		NGC 2547 - 110			8 08 47	−49 02.2	1	13.25		0.79		0.30		2		935
		NGC 2539 - 1094			8 08 48	−12 41.2	1	13.17		0.34		0.18		2		1647
		NGC 2539 - 31			8 08 48	−12 42.5	1	12.03		0.14		0.14		2		75
		NGC 2539 - 5			8 08 48	−12 46.2	1	11.94		0.20		0.15		2		75
		NGC 2547 - 64			8 08 48	−49 06.3	3	11.02	.007	0.34	.009	0.05		7		935,2012,2022
		NGC 2539 - 1092			8 08 49	−12 37.8	1	11.96		0.23		0.18		2		1647
		NGC 2547 - 77			8 08 49	−49 12.6	1	11.66		1.76		2.05		3		935
		NGC 2539 - 1100			8 08 50	−12 34.5	1	12.92		0.30		0.14		2		1647
		NGC 2539 - 1097			8 08 50	−12 36.5	1	12.79		0.23		0.24		2		1647
		NGC 2539 - 3			8 08 50	−12 47.8	2	12.01	.050	0.18	.005	0.12	.020	4	A3	75,1647
		NGC 2539 - 2			8 08 50	−12 48.1	2	12.89	.075	0.21	.030	0.15	.010	4	F0	75,1647

Table 1   451

HD	DM	Other Id	N Rem	α_{1950}	δ_{1950}	S	V	σ_V	B–V	σ_{B-V}	U–B	σ_{U-B}	n	Spectrum	References
		Orsatti 379		8 08 50	−32 13.5	1	13.08		0.58		-0.15		1		1766
68364	−32 04939	IDS08069S3218	AB	8 08 50	−32 27.3	1	8.18		0.08		0.02		2	A1 V +A1 V	401
		NGC 2547 - 80		8 08 50	−48 57.8	1	11.90		0.58		0.10		3		935
		NGC 2539 - 6		8 08 51	−12 45.0	2	10.77	.030	0.97	.030	0.77	.010	4	B5	75,1647
68260	−12 02383	NGC 2539 - 4	⋆ E	8 08 51	−12 46.9	2	9.29	.015	0.16	.010	0.10	.005	4	A5 V	75,1647
		NGC 2547 - 91		8 08 52	−49 00.7	1	12.25		1.33		1.40		2		935
		NGC 2539 - 1101		8 08 52	−12 36.9	1	13.15		0.25		0.17		2		1647
68452	−48 03554	NGC 2547 - 18		8 08 52	−48 54.7	2	9.30	.009	0.06	.009	-0.02		6	A1 IV	935,2022
68478	−48 03556	NGC 2547 - 2		8 08 52	−49 05.3	4	6.46	.005	-0.15	.004	-0.66	.010	16	B3 V	164,935,2012,2022
		NGC 2539 - 1103		8 08 53	−12 43.6	1	13.69		0.43		0.18		2		1647
		NGC 2539 - 1		8 08 53	−12 48.5	2	11.35	.005	0.13	.030	0.11	.005	4		75,1647
68363	−32 04941			8 08 53	−32 13.8	2	7.20	.000	1.34	.000	1.40		5	K2 III	2012,3040
68477	−47 03664			8 08 53	−47 55.0	1	8.57		0.00		-0.05		2	B9.5V	401
	−32 04942			8 08 54	−32 55.6	1	10.91		0.05		-0.13		2		124
	−48 03553	NGC 2547 - 92		8 08 54	−48 51.6	1	12.27		0.23		0.21		3		935
		NGC 2539 - 1104		8 08 55	−12 36.5	1	12.38		1.00		0.93		2		1647
		NGC 2539 - 34		8 08 55	−12 40.7	1	11.85		0.15		0.14		2		75
		NGC 2539 - 37		8 08 56	−12 39.2	1	13.29		0.35		0.08		2		75
68290	−12 02385	HR 3211	⋆ AB	8 08 56	−12 46.6	8	4.71	.007	0.95	.005	0.72	.019	28	K0 III	3,15,1075,1425,2012*
		NGC 2547 - 104		8 08 56	−48 52.9	1	12.83		0.64		0.10		2		935
		NGC 2547 - 108		8 08 56	−48 59.9	1	13.21		0.58		0.13		3		935
		NGC 2547 - 81		8 08 56	−49 14.3	1	11.96		0.24		0.18		3		935
	−49 03390	NGC 2547 - 61		8 08 56	−49 14.3	3	10.93	.004	0.20	.039	0.12		9		935,2012,2022
68415	−37 04330	NGC 2546 - 448		8 08 57	−37 18.1	1	8.95		0.04		-0.13		2	B9.5II/III	126
		NGC 2547 - 114		8 08 57	−48 59.2	1	13.37		0.94		0.54		2		935
		NGC 2547 - 85		8 08 57	−49 10.0	1	12.04		0.62		0.08		3		935
		NGC 2539 - 1109		8 08 58	−12 36.3	1	13.69		0.43		0.18		2		1647
		NGC 2539 - 36		8 08 58	−12 39.2	1	11.65		0.12		0.12		1		75
	−48 03558	NGC 2547 - 48		8 08 58	−48 57.9	2	10.55	.000	0.24	.009	0.10		5	A7 V	935,2022
		NGC 2547 - 113		8 08 58	−48 59.9	1	13.35		0.74		0.22		3		935
68168	+16 01659			8 08 59	+16 40.6	1	7.36		0.66		0.21		2	G0	3026
		NGC 2539 - 1110		8 08 59	−12 37.0	1	12.94		0.55		0.13		2		1647
		NGC 2539 - 1111		8 08 59	−12 40.5	1	13.49		0.54		0.19		2		1647
68388	−31 05641			8 08 59	−32 05.6	1	9.48		0.19		0.16		2	A4 V	124
68496	−48 03560	NGC 2547 - 6		8 08 59	−49 00.9	4	7.94	.014	-0.09	.013	-0.48		14	B6 V	935,2012,2022,2040
		NGC 2547 - 95		8 08 59	−49 02.9	1	12.32		1.08		0.77		2		935
	−48 03559	NGC 2547 - 63		8 08 59	−49 03.2	2	10.96	.061	0.50	.039	0.05		6		935,2022
		Orsatti 380		8 09 00	−32 36.2	1	14.47		0.38		-0.44		1		1766
		NGC 2547 - 112		8 09 01	−48 59.1	1	13.31		0.41		0.15		3		935
		NGC 2547 - 90		8 09 01	−49 00.0	1	12.17		1.41		1.39		2		935
		NGC 2539 - 1112		8 09 02	−12 36.7	1	12.78		0.46		0.16		2		1647
68515	−47 03665			8 09 02	−48 02.3	1	8.26		1.05		0.92		1	K0 III	389
68494	−48 03561	NGC 2547 - 10		8 09 02	−48 40.9	3	8.47	.005	0.26	.006	0.10		11	A9 IV	935,2012,2022
68495	−48 03562	NGC 2547 - 22		8 09 03	−48 57.5	3	9.45	.003	0.08	.010	0.04		10	A0 V	935,2012,2022
	−48 03563	NGC 2547 - 53	⋆ AB	8 09 03	−49 07.8	3	10.80	.025	0.44	.010	0.01		12		935,2012,2022
		NGC 2539 - 1115		8 09 04	−12 41.0	1	12.78		0.60		0.19		2		1647
		NGC 2539 - 10		8 09 04	−12 51.2	1	12.45		1.22		1.40		2		75
		NGC 2539 - 35		8 09 05	−12 38.9	2	10.97	.000	0.20	.000	0.16	.000	4		75,1647
		Orsatti 381		8 09 05	−32 37.2	1	13.45		0.55		-0.30		1		1766
		NGC 2547 - 89		8 09 05	−48 51.9	1	12.17		0.46		0.00		3		935
		NGC 2547 - 115		8 09 05	−48 58.7	1	13.42		1.38		1.30		3		935
		NGC 2547 - 106		8 09 05	−49 02.5	1	12.86		1.13		0.90		2		935
		NGC 2547 - 116		8 09 06	−48 59.3	1	13.51		0.49		0.25		3		935
		NGC 2547 - 99		8 09 06	−48 59.7	1	12.51		0.61		0.08		3		935
68516	−49 03393	NGC 2547 - 29		8 09 06	−49 23.7	3	9.76	.009	0.06	.009	0.00		9	A1 V	935,2012,2022
68312	−7 02378	HR 3212		8 09 07	−07 37.3	4	5.36	.008	0.89	.007	0.60	.000	16	G8 III	15,1415,2027,3016
		NGC 2539 - 28		8 09 07	−12 32.1	2	11.01	.005	0.98	.005	0.79	.000	4		75,1647
		NGC 2539 - 1117		8 09 08	−12 45.1	1	12.22		0.37		0.25		2		1647
	−48 03564	NGC 2547 - 36		8 09 08	−49 09.4	3	10.09	.005	0.28	.004	0.08		9	A5 IV	935,2012,2022
		NGC 2539 - 1118		8 09 09	−12 46.9	1	13.79		1.30				2		1647
68474	−38 04155	IDS08074S3846	AB	8 09 09	−38 55.7	1	7.37		-0.09		-0.35		2	B4 III	401
68224	+13 01860			8 09 10	+13 32.4	1	8.75		0.89				2	K2	882
		NGC 2539 - 1116		8 09 10	−12 37.2	1	12.64		0.62		0.11		2		1647
		NGC 2539 - 7		8 09 10	−12 51.2	1	11.54		1.07		1.08		2		75
	−48 03565	NGC 2547 - 42		8 09 10	−48 52.9	3	10.26	.018	-0.04	.005	-0.54		9	B9	935,2012,2022
		NGC 2539 - 9		8 09 11	−12 51.9	1	12.06		0.87		0.39		2		75
68450	−36 04291	NGC 2546 - 682	⋆ A	8 09 11	−37 08.6	7	6.44	.010	-0.02	.005	-0.88	.016	23	B0 II	15,126,540,976,2012*
68475	−42 03961			8 09 11	−42 39.4	2	8.78	.000	0.88	.025	0.62	.032	5	K2 V	803,3072
		NGC 2547 - 79		8 09 11	−48 55.7	1	11.77		0.49		0.05		3		935
		G 50 - 22		8 09 12	+08 58.9	3	12.83	.015	1.77	.011	1.30	.105	16		694,1620,3078
		Orsatti 383		8 09 12	−32 25.0	1	12.91		0.26		-0.54		2		1766
		NGC 2547 - 69		8 09 12	−49 07.	1	11.16		1.11		0.88		2		935
		NGC 2547 - 75		8 09 12	−49 08.8	1	11.63		1.18		1.06		2		935
68558	−48 03567	NGC 2547 - 28		8 09 13	−48 48.5	3	9.75	.005	0.10	.007	0.05		9	A0 V	935,2012,2022
		NGC 2547 - 118		8 09 13	−49 12.9	1	13.98		0.68		0.15		2		935
		NGC 2547 - 94		8 09 16	−48 52.1	1	12.31		0.54		0.01		3		935
68556	−47 03670	AX Vel		8 09 17	−47 32.9	2	8.17	.030	0.65	.015	0.43		2	F6 II	389,1484
	−48 03568	NGC 2547 - 47		8 09 17	−48 43.0	3	10.49	.036	0.45	.005	0.00		10	F5	935,2012,2022
		G 194 - 15		8 09 18	+51 04.8	1	13.46		0.62		-0.12		1		1658
	−12 02387	NGC 2539 - 61		8 09 18	−12 29.4	1	9.15		1.62		1.75		2	M4 III	75

HD	DM	Other Id	N Rem	α_{1950}	δ_{1950}	S	V	σ_V	B–V	σ_{B-V}	U–B	σ_{U-B}	n	Spectrum	References
	−48 03569	NGC 2547 - 55		8 09 18	−49 08.5	1	10.83		0.46		0.00		6		935
		NGC 2539 - 8		8 09 19	−12 52.1	2	10.71	.000	0.65	.000	0.26	.010	4		75,1647
68559	−48 03571	NGC 2547 - 38		8 09 20	−48 57.9	3	10.17	.009	0.39	.009	0.00		12	F0 V	935,2012,2022
		NGC 2547 - 109		8 09 20	−49 14.6	1	13.22		0.72		0.16		2		935
68257	+18 01867	HR 3208	⋆ AB	8 09 21	+17 48.0	1	5.05		0.54		0.06		1	F8 V + G5 V	3030
68257	+18 01867	HR 3208	⋆ ABC	8 09 21	+17 48.0	3	4.67	.001	0.53	.000			48	F8 V + G5 V	15,130,1351
68256	+18 01867	HR 3210	⋆ C	8 09 21	+17 48.0	1	6.20		0.60		0.13		1	F9 V	3030
		NGC 2539 - 29		8 09 21	−12 31.3	1	11.01		1.44		1.79		2		75
	−34 04452	LSS 984		8 09 21	−34 16.3	3	10.59	.018	0.14	.005	−0.60	.008	8		1066,1712,1766
	−48 03570	NGC 2547 - 54		8 09 21	−48 45.9	3	10.81	.009	−0.04	.005	−0.61		11	A0	935,2012,2022
		NGC 2547 - 83		8 09 21	−49 02.5	1	11.97		0.49		0.00		3		935
		Orsatti 384		8 09 22	−32 40.0	1	13.62		0.40		−0.33		1		1766
	−33 04649			8 09 25	−33 46.8	1	9.86		0.03		−0.02		2		124
	−37 04337	NGC 2546 - 99		8 09 25	−37 30.9	1	8.49		1.21		1.02		2	gG8	460
68578	−48 03572	NGC 2547 - 11		8 09 25	−48 42.5	2	8.59	.000	0.04	.004	0.03		7	A0 IV	935,2022
	−37 04340	NGC 2546 - 98		8 09 26	−37 31.6	1	9.64		0.42				1	A3	2041
68407	−12 02390	NGC 2539 - 30		8 09 27	−12 31.7	1	9.13		0.10		0.09		2	A2 IV	75
	−37 04338	NGC 2546 - 112		8 09 27	−37 27.7	2	9.62	.156	0.04	.029	−0.24		3	B6	126,2041
	−20 02429	IR Pup		8 09 28	−21 03.8	1	9.57		3.84		2.57		1	N	864
		Pup sq 4 # 4		8 09 28	−29 46.4	1	12.81		0.38		0.22		2		525
		NGC 2547 - 101		8 09 28	−49 00.0	1	12.55		0.43		0.04		3		935
	−48 03573			8 09 29	−48 21.1	2	7.53	.001	0.56	.005	0.12	.004	2	K0	389,1754
68633	−50 03193			8 09 29	−51 02.5	1	8.00		0.29		−0.35		2	B5 V	401
		NGC 2547 - 103		8 09 30	−48 59.4	1	12.78		1.05		0.68		2		935
		NGC 2547 - 96		8 09 30	−49 09.3	1	12.33		1.56		1.79		2		935
68254	+29 01712	IDS08064N2907	AB	8 09 31	+28 58.2	1	8.03		0.04		0.03		1	A0	695
		HA 4 # 400		8 09 31	+75 02.1	1	11.54		0.46		0.00		1		1058
		NGC 2547 - 111		8 09 31	−48 57.8	1	13.30		0.73		0.23		3		935
68608	−48 03574	NGC 2547 - 5		8 09 31	−49 08.1	4	7.90	.015	−0.11	.007	−0.53	.009	14	B7 III	401,935,2012,2022
		NGC 2547 - 98		8 09 32	−49 06.3	1	12.39		1.52		1.61		2		935
68332	+14 01850	HR 3214		8 09 34	+14 09.3	2	6.54	.000	0.15	.005	0.14	.010	4	A7 III	1648,1733
68444	−14 02406			8 09 34	−14 30.9	2	9.18	.004	−0.16	.004	−0.82	.008	7	B3 III	55,1775
	−37 04345	NGC 2546 - 266		8 09 34	−37 16.3	2	10.34	.180	0.09	.005	−0.52		3	B2	126,2041
	−37 04346	NGC 2546 - 265		8 09 34	−37 17.3	1	10.24		0.94		0.89		2	K0	126
68553	−39 04084	HR 3225, NS Pup		8 09 34	−39 28.1	6	4.44	.014	1.62	.009	1.83	.031	22	K4 III	15,1075,2012,2031*
	−48 03575	NGC 2547 - 58		8 09 34	−48 51.6	2	10.90	.034	0.33	.013	0.08		7		935,2022
		NGC 2546 - 116		8 09 35	−37 26.5	1	11.16		0.20				1	A0	2041
68604	−44 04136			8 09 36	−45 00.9	1	8.07		1.00		0.76		1	G8 III	389
		NGC 2547 - 78		8 09 37	−48 54.3	1	11.72		1.01		0.68		2		935
		NGC 2547 - 107		8 09 37	−49 04.7	1	13.00		0.46		0.21		3		935
		NGC 2547 - 117		8 09 37	−49 04.7	1	13.59		0.61		0.12		3		935
68331	+18 01869			8 09 39	+18 40.1	1	9.11		1.01				1	K2	1245
68253	+29 01713	IDS08066N2950	AB	8 09 39	+29 41.6	1	8.56		0.13		0.04		1	A2	695
		NGC 2546 - 474		8 09 39	−37 13.0	1	11.90		0.19		0.13		2	B9	126
68631	−48 03577	NGC 2547 - 33		8 09 40	−48 51.3	3	9.93	.009	0.10	.005	0.07		11	A2 V	935,2012,2022
	+75 00331			8 09 41	+75 03.2	1	10.00		0.93		0.66		1	G5	1058
68468	−13 02429			8 09 41	−14 01.1	1	8.76		−0.11		−0.39		6	B3np(Shell)	399
		NGC 2546 - 120		8 09 41	−37 22.4	1	11.95		0.38				1	A3	2041
68657	−48 03576	HR 3227		8 09 41	−48 18.7	3	5.83	.005	−0.16	.005	−0.64	.005	7	B3 V	258,401,2006
	−30 05818			8 09 42	−30 52.6	1	10.21		0.96		0.61		3		1712
		Orsatti 385		8 09 42	−32 34.1	1	12.77		0.19		−0.34		1		1766
	−48 03578	NGC 2547 - 67		8 09 42	−48 55.7	2	11.07	.030	0.36	.013	0.05		7		935,2022
68195	+43 01779			8 09 43	+43 09.1	1	8.31		0.41		0.02		1	F2	1776
68627	−45 03867			8 09 43	−45 40.1	2	6.95	.007	0.97	.004	0.76	.005	2	G8/K0 III	389,1754
		NGC 2547 - 93		8 09 43	−48 56.4	1	12.28		1.74		2.06		2		935
		Pup sq 4 # 3		8 09 44	−29 47.9	1	12.25		1.05		0.65		3		525
68552	−31 05661	LSS 985	⋆ AB	8 09 44	−32 00.2	5	9.13	.014	0.06	.007	−0.76	.013	12	B5 Ib/II	401,540,976,1712,1766
		Orsatti 386		8 09 44	−34 21.2	1	13.43		0.11		−0.51		1		1766
68601	−42 03979	HR 3226, LSS 986	⋆ AB	8 09 44	−42 50.2	6	4.75	.004	0.18	.000	0.07	.006	28	A5 Ib	15,1075,2012,2040*
68536	−29 05708			8 09 45	−29 46.2	1	9.47		0.26		0.16		8	A8 V	525
	−30 05820			8 09 46	−30 45.5	1	10.67		1.18		1.07		3		1712
	−31 05663			8 09 47	−32 02.1	1	10.10		0.62		0.34		1		401
		Orsatti 387		8 09 47	−32 23.8	1	12.89		0.56		−0.28		2		1766
68572	−37 04352	NGC 2546 - 475		8 09 47	−37 13.8	3	8.20	.007	0.01	.002	−0.78	.011	7	B1 V	401,540,976
	−37 04354	NGC 2546 - 272		8 09 48	−37 15.7	1	10.70		0.11		−0.38		2	B3	126
		NGC 2547 - 86		8 09 49	−48 57.3	1	12.06		1.02		0.62		3		935
		3C 196 # 2		8 09 51	+48 20.8	1	15.50		0.77		0.25		4		327
68194	+48 01621			8 09 51	+48 25.9	1	6.85		−0.03		−0.08		3	B9	985
68352	+22 01880			8 09 52	+22 31.9	1	9.33		0.18		0.12		2	A2	1375
68077	+56 01278	HR 3200		8 09 52	+56 36.3	2	5.88	.023	1.02	.003	0.79		4	G9 III	71,3016
	−30 05822			8 09 52	−31 00.0	1	11.42		0.35		0.10		3		1712
		Fld c # 8		8 09 52	−31 00.4	1	11.67		0.41		0.08		3		1712
		NGC 2547 - 87		8 09 52	−49 04.2	1	12.07		0.57		0.07		2		935
	−30 05823			8 09 53	−30 55.3	1	9.97		0.96		0.68		3		1712
	−29 05712	Pup sq 4 # 2		8 09 54	−29 41.0	1	11.02		0.09		0.06		3		525
		NGC 2547 - 70		8 09 54	−49 00.2	1	11.18		1.36		1.27		4		935
		Orsatti 388		8 09 55	−32 14.8	1	12.41		0.33		−0.35		2		1766
68698	−48 03583	NGC 2547 - 32		8 09 57	−49 03.4	2	9.87	.000	0.07	.004	0.04		6	A1 V	935,2022
		3C 196 # 1		8 09 58	+48 20.6	1	15.47		0.79		0.29		7		327
		NGC 2546 - 127		8 09 59	−37 19.8	1	10.18		0.12				1	B9	2041

Table 1 453

HD	DM	Other Id	N Rem	α_{1950}	δ_{1950}	S	V	σ_V	B–V	σ_{B-V}	U–B	σ_{U-B}	n	Spectrum	References
		NGC 2547 - 88		8 10 00	−49 01.9	1	12.15		0.46		0.22		3		935
68624	−37 04359	NGC 2546 - 30		8 10 01	−37 25.4	1	8.51		0.01		-0.33		2	Ap	126
		KUV 08100+3915		8 10 02	+39 14.7	1	16.85		-0.04		-0.97		1	DA	1708
68351	+30 01664	HR 3215, BM Cnc		8 10 03	+29 48.5	3	5.63	.004	-0.07	.002	-0.13	.006	9	B9p SiCr	695,1022,1202
68652	−37 04361	NGC 2546 - 282		8 10 03	−37 16.8	1	9.46		0.08				1	A0 V	2041
		LOD 7 # 20		8 10 04	−34 11.2	1	11.85		0.30		0.03		2		126
68678	−38 04164			8 10 04	−38 49.0	1	8.45		-0.05		-0.34		2	B9 III	401
68694	−43 04021			8 10 05	−43 53.1	1	8.79		0.43		0.02		1	G3 V	389
		Ru 55 - 18		8 10 07	−32 28.1	1	11.23		0.53		0.22		1		410
		3C 196 # 3		8 10 08	+48 22.3	1	13.10		0.68		0.19		16		327
		Orsatti 389		8 10 08	−34 42.1	1	13.46		0.51		-0.09		1		1766
68461	+16 01662	HR 3222		8 10 09	+16 39.9	1	6.01		0.89				3	G8 III	71
	−31 05671			8 10 09	−31 21.0	1	9.95		0.04		-0.25		2		124
68677	−37 04365	NGC 2546 - 132		8 10 09	−37 22.6	2	9.35	.005	0.06	.000	-0.22		3	B8 III	126,2041
		NGC 2546 - 487		8 10 10	−37 12.9	1	11.67		0.27				1	B9	2041
		NGC 2546 - 284		8 10 10	−37 17.3	1	11.76		0.32				1	A2	2041
	−49 03401	NGC 2547 - 57		8 10 10	−49 20.6	3	10.89	.010	0.45	.009	0.20		10		935,2012,2022
	−37 04367	NGC 2546 - 763		8 10 11	−37 54.0	1	10.33		0.30				1		2041
		3C 196 # 4		8 10 12	+48 22.9	1	12.94		0.42		0.00		14		327
	−32 04967	Ru 55 - 1		8 10 12	−32 25.0	6	8.55	.023	0.41	.011	-0.50	.020	21	B1.5Ib	211,410,525,540,976,1712
		LOD 7 # 19		8 10 13	−34 12.6	1	11.74		0.83		0.61		2		126
		NGC 2546 - 138		8 10 13	−37 18.7	1	11.40		0.36				1	A6	2041
	−38 04168	LSS 990		8 10 13	−38 59.6	2	9.28	.011	0.46	.006	-0.50	.026	5	O8	540,976
68765	−49 03403			8 10 13	−49 55.1	1	7.34		-0.01		-0.27		2	B8 II	401
		Ru 55 - 4		8 10 14	−32 30.3	2	10.87	.010	0.29	.000	-0.64	.005	4		410,1712
		Ru 55 - 5		8 10 15	−32 19.2	4	11.19	.097	0.36	.023	-0.64	.015	49		410,1712,1766,1770
68460	+18 01875			8 10 16	+18 06.8	1	8.87		0.26		0.01		2	A0	1776
		Ru 55 - 16		8 10 16	−32 26.1	1	12.00		1.04		0.61		1		410
69547	−78 00297			8 10 16	−78 32.7	1	7.04		0.98				4	G8 III	2012
		NGC 2546 - 285		8 10 17	−37 16.1	1	12.39		0.51				1	A4	2041
		Ru 55 - 22		8 10 18	−32 27.	1	13.96		0.59		0.04		3		525
		Ru 55 - 23		8 10 18	−32 27.	1	14.25		0.40		0.27		3		525
		Ru 55 - 24		8 10 18	−32 27.	1	14.47		0.34		-0.09		3		525
		Ru 55 - 25		8 10 18	−32 27.	1	15.60		0.36		-0.24		2		525
		Ru 55 - 26		8 10 18	−32 27.	1	15.77		0.44		0.29		3		525
		Ru 55 - 27		8 10 18	−32 27.	1	16.02		1.10		0.69		3		525
		Ru 55 - 28		8 10 18	−32 27.	1	16.81		0.79				2		525
		Ru 55 - 29		8 10 18	−32 27.	1	17.26		0.75				3		525
		WLS 820-5 # 6		8 10 19	−04 43.8	1	11.80		0.99		0.67		2		1375
	−34 04467			8 10 19	−34 13.5	1	10.24		0.98		0.58		2		126
	−5 02406	NGC 2548 - 848		8 10 20	−05 44.7	1	10.66		0.33		0.08		2	A8	75
67739	+74 00350			8 10 21	+74 38.8	3	8.24	.007	1.41	.006	1.68	.002	17	K5	269,1058,1118
	−32 04969	Ru 55 - 10		8 10 21	−32 23.3	1	10.16		1.64		1.95		1		410
		Ru 55 - 9		8 10 21	−32 25.0	1	11.01		1.56		1.76		1		410
	−5 02407	NGC 2548 - 870		8 10 22	−05 41.8	1	9.71		1.12		1.06		2	G5 III	75
	−30 05833			8 10 22	−30 55.1	1	9.14		1.68		1.97		3		1712
		Fld c # 3		8 10 22	−30 59.4	1	10.83		0.95		0.60		3		1712
68810	−49 03405	NGC 2547 - 40		8 10 22	−49 23.7	3	10.22	.005	0.08	.010	0.04		10	A0 V	935,2012,2022
	−32 04971	Ru 55 - 2		8 10 23	−32 26.0	2	10.88	.005	0.26	.005	-0.62	.000	6	B0 V	410,1712
		Ru 55 - 11		8 10 23	−32 26.1	1	12.67		0.31		-0.57		2		410
		Ru 55 - 12		8 10 23	−32 26.1	1	12.67		0.26		-0.62		2		410
		Ru 55 - 15		8 10 23	−32 26.8	1	12.49		1.11		0.78		1		410
		G 50 - 23		8 10 25	+09 16.0	1	12.58		0.86		0.46		1		333,1620
68586	−6 02514			8 10 25	−07 03.1	2	7.91	.005	0.90	.000	0.55	.015	9	K0	158,1775
		Ru 55 - 19		8 10 25	−32 21.3	1	12.17		0.14		-0.24		6		525
		Ru 55 - 13		8 10 25	−32 25.3	1	13.04		0.50		0.17		2		410
	−33 04675	LSS 993		8 10 25	−33 40.6	4	11.13	.026	0.50	.005	-0.64	.033	9		1066,1712,1737,1766
68714	−37 04370	NGC 2546 - 134		8 10 25	−37 22.7	1	8.69		0.06				1	B8 IV	2041
68763	−43 04028			8 10 25	−43 15.0	1	6.46		1.71		1.99		1	K3 (III)	389
		Ru 55 - 17		8 10 26	−32 27.0	1	12.74		0.57		0.11		1		410
		NGC 2546 - 491		8 10 26	−37 12.1	1	12.08		0.23				1	A1	2041
68808	−46 03902	HR 3232, AH Vel		8 10 26	−46 29.6	4	5.58	.093	0.54	.038	0.36	.019	5	F7 Ib/II	58,657,688,1754
		G 113 - 17		8 10 27	−07 25.0	1	13.75		1.44				1		3062
		Ru 55 - 8		8 10 27	−32 25.4	2	10.77	.005	0.47	.004	0.20	.029	5		410,525
68806	−45 03881			8 10 27	−45 50.6	1	7.39		1.05		0.86		1	K0 III	389
	−30 05837			8 10 28	−30 54.4	1	9.62		1.18		1.21		3		1712
		LSS 992		8 10 29	−31 05.3	1	12.42		0.41		-0.69		3	B3 Ib	1712
	−37 04372	NGC 2546 - 184		8 10 29	−37 41.0	2	10.98	.015	0.18	.005	0.10		3	A0	126,2041
68585	−5 02411	NGC 2548 - 920		8 10 30	−05 25.1	1	9.60		0.06		0.08		2	A0	75
		Ru 55 - 21		8 10 30	−32 26.7	1	13.49		0.32		0.22		3		525
	+17 01793			8 10 31	+17 33.3	1	9.09		1.26		1.28		1	K0	1776
		NGC 2548 - 924		8 10 31	−05 42.0	1	11.06		0.18		0.11		2	A8	75
		LP 843 - 1		8 10 31	−21 23.5	2	12.08	.080	1.57	.067			2		1705,1705
		Orsatti 391		8 10 31	−33 27.9	1	12.71		0.41		-0.22		1		1766
68734	−34 04471	IDS08085S3411	AB	8 10 31	−34 19.9	1	9.11		0.09		-0.19		2	A0 V	126
	−36 04319	AT Pup		8 10 31	−36 47.6	1	7.54		0.57				1	F5	1484
68785	−40 03959			8 10 31	−40 21.2	1	8.19		0.62				4	G3 V	2012
		HA 4 # 260		8 10 33	+74 47.9	1	14.93		1.04				1		269
		Ru 55 - 14		8 10 33	−32 24.3	1	12.09		1.37		1.13		2		410
		Ru 55 - 20		8 10 33	−32 26.5	1	12.71		0.56		0.05		2		525

HD	DM	Other Id	N Rem	α_{1950}	δ_{1950}	S	V	σ_V	B–V	σ_{B-V}	U–B	σ_{U-B}	n	Spectrum	References
68618	−5 02412	NGC 2548 - 975		8 10 34	−05 43.8	1	9.99		0.06		0.08		2	A0	75
	−32 04978	Ru 55 - 3		8 10 34	−32 25.5	2	10.84	.005	0.28	.010	-0.60	.005	5	B0 V	410,1712
		HA 4 # 177		8 10 36	+74 32.0	1	14.23		0.64		0.09		1		269
68761	−36 04322	LSS 996		8 10 36	−36 50.3	3	6.53	.008	-0.08	.005	-0.91	.012	11	B1/2 III	164,540,976
		G 113 - 18		8 10 37	−04 22.0	1	10.58		0.96		0.70		1		3062
	−37 04376	NGC 2546 - 181		8 10 38	−37 40.8	2	11.15	.010	0.09	.005	-0.07		3	B8	126,2041
68646	−5 02414	NGC 2548 - 1005		8 10 40	−05 29.5	1	9.26		0.04		0.04		2	A2	75
69218	−70 00732			8 10 42	−70 38.0	1	8.09		0.85				4	G2 III/IV	2012
		HA 4 # 179		8 10 43	+74 36.8	1	14.13		0.50		0.03		2		269
68784	−36 04325			8 10 43	−36 31.0	1	6.51		-0.05				1	K4 III	2041
68543	+23 01913	HR 3224		8 10 44	+23 17.4	1			0.11		0.12		3	A4 IVn	1022
	−5 02416	NGC 2548 - 1029		8 10 44	−05 37.2	1	10.38		0.10		0.10		2	A3	75
68758	−29 05738	HR 3230		8 10 45	−29 45.6	2	6.51		0.06	.000			7	A1 V	15,2012
	−37 04381	NGC 2546 - 178		8 10 45	−37 40.5	2	10.83	.029	0.09	.010	-0.09		3	B8	126,2041
	−37 04382	NGC 2546 - 578	★ AB	8 10 45	−37 49.4	1	10.98		0.19				1	B9	2041
	−37 04383	NGC 2546 - 581		8 10 45	−37 51.2	1	10.61		0.48				1	F4	2041
		LSS 998		8 10 46	−33 32.9	2	11.94	.050	0.24	.005	-0.38	.035	4	B3 Ib	1712,1766
	−36 04326	LSS 1000		8 10 46	−36 31.4	1	9.37		0.16		-0.58		2		1066
		NGC 2546 - 369		8 10 46	−37 43.1	1	12.13		0.25				1		2041
68582	+17 01797			8 10 47	+16 47.9	1	8.32		1.43				2	K2	882
	−37 04347	NGC 2546 - 94		8 10 47	−37 30.4	1	9.86		0.04				1	B6	2041
68898	−49 03414			8 10 47	−49 43.1	1	8.70		0.44		-0.04		1	F5 V	389
	−5 02418	NGC 2548 - 1070		8 10 48	−05 32.2	1	10.56		0.07		0.08		2	A3	75
	−37 04385	NGC 2546 - 500		8 10 48	−37 14.1	1	12.28		0.57				1	A0	2041
68667	−0 01938			8 10 49	−01 00.9	1	6.51		0.92		0.60		5	K0	1628
	−13 02439	IDS08085S1336	AB	8 10 49	−13 46.1	3	9.38	.016	1.40	.018	1.22	.016	9	M0 V	158,1775,3072
68669	−5 02419	NGC 2548 - 1073		8 10 50	−05 29.4	1	8.72		0.07		0.09		2	A0	75
		NGC 2546 - 298		8 10 50	−37 19.0	1	10.17		0.20				1	O8	2041
		Ru 55 - 6		8 10 52	−32 20.5	2	12.10	.020	0.19	.000	-0.47	.015	5		410,1712
	−37 04387	NGC 2546 - 142		8 10 52	−37 22.9	1	10.14		0.33				1	A3	2041
	−33 04681	LSS 1002		8 10 53	−33 38.5	2	12.13	.010	0.22	.010	-0.53	.015	5		1712,1766
		NGC 2546 - 365		8 10 53	−37 46.2	2	10.68	.005	0.05	.010	-0.13		3	B8	126,2041
68581	+22 01886			8 10 54	+22 25.9	1	8.78		0.46		0.06		2	F5	1648
	−5 02421	NGC 2548 - 1117		8 10 54	−05 36.2	1	9.84		0.03		0.03		2	A1	75
		NGC 2548 - 1124		8 10 56	−05 41.0	1	11.28		0.22		0.10		2	A3	75
68892	−44 04158			8 10 56	−44 54.5	1	7.97		0.85		0.49		1	G5 V	389
68895	−45 03892	HR 3234	★ AB	8 10 56	−46 06.8	3	6.02	.004	-0.12	.004	-0.56		8	B5 V	15,258,2027
68843	−35 04336			8 10 58	−36 04.6	2	7.62	.000	-0.13	.000	-0.59	.017	6	B6 Vnn	164,1066
	−5 02423	NGC 2548 - 1148		8 10 59	−05 40.8	1	10.04		0.02		0.06		2	A1	75
	−5 02424	NGC 2548 - 1169		8 11 00	−05 39.1	1	9.59		0.06		0.02		2	A1	75
		LP 664 - 26		8 11 00	−09 17.9	1	14.34		1.54				1		1691
		Ru 55 - 7		8 11 00	−32 25.7	2	11.78	.000	0.19	.010	-0.62	.015	5		410,1712
		LOD 7 # 15		8 11 00	−34 22.4	1	12.26		0.62		0.02		2		126
	−37 04390	NGC 2546 - 508		8 11 00	−37 14.5	1	12.36		0.17				1	gMO	2041
68845	−37 04393	NGC 2546 - 165		8 11 00	−37 34.0	1	9.07		0.02		-0.30		2	B9	126
68862	−37 04394	NGC 2546 - 575	★	8 11 01	−37 46.4	5	6.41	.013	0.11	.008	0.14	.018	14	A3 V	15,126,1637,2012,2041
68752	−15 02324	HR 3229		8 11 02	−15 38.2	5	4.99	.006	1.08	.005	0.82	.035	15	G5 Ib/II	15,2013,2029,3016,8100
		NGC 2548 - 1188		8 11 04	−05 40.4	1	12.38		0.47		-0.01		2		460
		LOD 7 # 16		8 11 04	−34 22.1	1	13.01		0.36		0.05		2		126
		WLS 820 25 # 6		8 11 05	+25 27.4	1	11.65		0.49		-0.02		2		1375
	−37 04396	NGC 2546 - 356		8 11 05	−37 39.8	3	8.34	.009	1.30	.008	1.25	.052	6	gK8	126,460,1551
		HA 4 # 182		8 11 06	+74 34.0	1	13.78		0.56		-0.12		2		269
		NGC 2548 - 1204		8 11 06	−05 44.5	1	11.51		0.24		0.13		2	A5	75
	−31 05703			8 11 06	−31 43.7	1	9.60		1.11		0.93		2		126
	−5 02426	NGC 2548 - 1218		8 11 07	−05 43.9	2	9.63	.005	0.93	.005	0.61	.025	5	G8	75,460
	−37 04397	NGC 2546 - 355		8 11 08	−37 38.4	2	9.90	.010	0.01	.005	-0.30		3	B5	126,2041
69026	−54 01545			8 11 08	−55 12.1	1	8.48		0.02		-0.69		2	B1/2 V	401
		NGC 2548 - 1227		8 11 09	−05 34.4	1	14.02		0.56		-0.05		1		460
		NGC 2548 - 1230		8 11 09	−05 42.3	1	11.34		1.20		1.38		2		75
68860	−34 04488	RS Pup		8 11 09	−34 25.6	2	6.58	.049	1.18	.045	0.87	.029	6	F8 Iab	126,688,689,3074,6011
68916	−40 03981			8 11 09	−40 23.0	2	9.39	.000	0.82	.000	0.42		8	G8 (IV)	158,2012
	+18 01880			8 11 10	+17 52.8	1	8.78		1.24		1.16		1	K0	1776
	−37 04399	NGC 2546 - 567		8 11 10	−37 45.8	2	10.36	.005	0.13	.005	-0.15		3	B6	126,2041
		NGC 2548 - 1250		8 11 11	−05 40.5	1	14.24		0.90		0.48		1		460
		NGC 2548 - 1256		8 11 12	−05 33.2	1	10.18		0.10		0.10		2	A2	75
	−5 02430	NGC 2548 - 1260		8 11 12	−05 37.7	2	9.06	.025	0.52	.005	0.25	.005	5	F2	75,460
	−5 02429	NGC 2548 - 1253		8 11 12	−05 38.1	1	9.85		0.12		0.12		2	A3	75
68887	−37 04400	NGC 2546 - 353		8 11 12	−37 37.2	1	8.96		0.03		-0.20		2	B9 IV	126
		HA 4 # 183		8 11 13	+74 36.9	1	14.81		0.74		0.04		1		269
		NGC 2548 - 1266		8 11 13	−05 37.8	2	11.01	.015	0.16	.005	0.13	.010	4	A3	75,460
		NGC 2548 - 1281		8 11 15	−05 32.4	1	10.05		0.02		0.03		2	A0	75
	+18 01881			8 11 16	+18 09.0	1	8.82		1.00		0.89		1	K0	1776
68779	−5 02432	NGC 2548 - 1289		8 11 16	−05 39.7	1	9.19		0.03		0.01		2	A1	75
68886	−31 05709			8 11 16	−32 00.2	1	7.62		-0.12		-0.60		2	B3 III	126
	−5 02433	NGC 2548 - 1296		8 11 17	−05 38.9	2	9.24	.025	0.76	.005	0.48	.015	4	gG0	75,460
	−5 02434	NGC 2548 - 1306		8 11 18	−05 37.2	1	10.36		0.15		0.06		2	A2	75
68778	−5 02436	NGC 2548 - 1320		8 11 19	−05 28.3	1	8.95		0.07		0.07		2	A2	75
68703	+18 01882	HR 3228	★ AB	8 11 20	+17 49.7	2	6.47	.008	0.30	.004	0.13	.013	7	A0 Vn δ Del	3058,8071
	−33 04692			8 11 20	−33 27.6	1	10.80		0.18		0.09		2		1712
	−5 02437	NGC 2548 - 1338		8 11 21	−05 35.4	1	9.99		0.04		-0.01		2	A1	75

Table 1

HD	DM	Other Id	N	Rem	α_{1950}	δ_{1950}	S	V	σ_V	B–V	σ_{B-V}	U–B	σ_{U-B}	n	Spectrum	References
68794	−5 02438	NGC 2548 - 1367			8 11 22	−05 32.9	1	9.52		0.05		0.04		2	A2	75
68913	−32 04990				8 11 22	−32 49.0	1	8.51		-0.01		-0.11		2	B9 IV	401
		Fld d # 4			8 11 22	−33 31.7	1	13.05		0.24		0.17		2		1712
68562	+43 01783	IDS08080N4320		A	8 11 23	+43 11.3	1	6.82		0.03		0.05		1	A0	1776
	−33 04694				8 11 23	−33 30.0	1	10.57		0.98		0.73		2		1712
68943	−35 04343				8 11 24	−35 53.2	1	8.63		-0.07		-0.25		2	B8 V	401
		NGC 2548 - 1366			8 11 25	−05 36.8	1	13.80		0.44		-0.13		1		460
68944	−35 04344				8 11 25	−36 11.4	1	7.33		-0.14		-0.58		2	B5 V	1066
	−5 02440	NGC 2548 - 1406			8 11 28	−05 38.7	1	10.68		0.09		0.10		2	A2	75
68962	−35 04346				8 11 30	−36 09.5	1	7.34		-0.15		-0.65		2	B2/3 V	1066
	−5 02441	NGC 2548 - 1424			8 11 31	−05 41.7	1	10.78		0.18		0.12		2	A1	75
	−30 05877	LSS 1004			8 11 31	−31 3.6	1	10.89		0.09		-0.71		3		1712
	−33 04700				8 11 31	−34 03.2	1	10.59		0.20		0.13		2		124
68725	+21 01792				8 11 32	+20 51.6	1	6.95		0.36		0.10		2	F2	1648
		NGC 2548 - 1434			8 11 32	−05 44.2	1	10.52		1.46		1.83		2	K0	75
	−33 04698				8 11 32	−33 33.1	1	9.96		1.57		1.93		2		1712
		LOD 7 # 13			8 11 33	−34 25.1	2	12.11	.190	0.64	.054	-0.07	.015	3		126,1766
68982	−38 04187				8 11 33	−38 17.2	3	7.54	.028	0.10	.005	-0.56	.005	9	B3 V	401,434,2012
68776	+13 01868	HR 3231		⋆A	8 11 34	+13 12.1	2	6.36	.005	1.03	.010	0.80	.005	6	G8 III	1648,1733
		NGC 2548 - 1444			8 11 34	−05 37.7	1	12.44		0.49		-0.01		2		460
	−5 02442	NGC 2548 - 1453			8 11 35	−05 32.2	1	10.22		0.02		0.00		2	A0	75
		NGC 2548 - 1454			8 11 35	−05 41.3	1	11.08		0.60		0.11		2	G0	75
		LP 843 - 13			8 11 36	−21 56.5	1	11.74		0.97		0.74		4		3062
	−34 04495				8 11 36	−34 18.1	2	11.06	.045	0.20	.025	-0.56	.015	5		1066,1766
68980	−35 04349	HR 3237, MX Pup			8 11 36	−35 44.9	10	4.74	.047	-0.12	.014	-0.97	.008	61	B1.5 IIIe	3,15,681,815,1066*
68978	−31 05719	IDS08096S3126		A	8 11 37	−31 35.1	2	6.70	.030	0.62	.004	0.07		6	G2 V	1311,2012
	−33 04702				8 11 38	−33 33.6	1	10.86		0.10		-0.09		2		1712
68457	+60 01119	HR 3221		⋆AB	8 11 39	+60 32.0	2	6.44	.015	0.20	.000	0.14	.038	3	A7 Vm	766,3016
68457	+60 01119	IDS08074N6041		C	8 11 39	+60 32.0	1	10.48		0.56		0.04		2		3032
68457	+60 01119	IDS08074N6041		D	8 11 39	+60 32.0	1	13.66		0.70		0.24		1		3032
		NGC 2548 - 1474			8 11 39	−05 36.9	1	12.24		0.31		0.05		2		460
		LP 843 - 14			8 11 39	−21 56.5	1	13.90		1.33		1.18		2		3062
68853	−5 02444	NGC 2548 - 1521			8 11 40	−05 24.9	1	9.69		0.41		0.00		2	F5	75
	−34 04496	LSS 1007			8 11 40	−34 19.6	6	9.28	.021	0.35	.020	-0.64	.018	13	O8	540,976,1066,1712,1737,1766
		LOD 7 # 11			8 11 40	−34 25.9	1	13.09		0.36		-0.24		2		126
	−37 04412	NGC 2546 - 333			8 11 41	−37 26.6	1	9.62		0.06				1	B3	2041
		LOD 7 # 12			8 11 42	−34 25.3	1	11.94		0.39		0.24		2		126
69003	−37 04413	NGC 2546 - 313			8 11 42	−37 19.9	1	8.83		0.02				1	A0	2041
		NGC 2548 - 1523			8 11 44	−05 45.2	1	12.27		0.99		0.60		2		460
69001	−31 05722				8 11 44	−31 23.8	1	8.80		0.07		0.04		2	A0 V	126
		LSS 1008			8 11 44	−31 44.6	1	11.94		0.41		-0.33		3	B3 Ib	1712
69002	−33 04705	HR 3238		⋆AB	8 11 44	−33 25.0	1	6.37		1.14				4	K2 III	2009
69004	−37 04414	NGC 2546 - 322			8 11 44	−37 22.5	1	9.27		-0.02				1	A(p)	2041
		HA 4 # 262			8 11 45	+74 47.9	1	14.84		0.82		0.60		1		269
		LSS 1009			8 11 45	−32 44.1	1	12.27		0.09		-0.70		3	B3 Ib	1712
	+26 01753	IDS08103N2635		A	8 11 47	+17 19.9	1	10.45		1.56		1.93		1	M0:	1746
68878	−5 02446	NGC 2548 - 1541			8 11 47	−05 34.1	1	9.16		0.10		0.12		2	A2	75
		LOD 7 # 7			8 11 47	−34 32.6	1	11.87		0.31		0.19		2		126
	−34 04498	IDS08099S3425		AB	8 11 47	−34 32.9	1	10.20		0.40		0.09		2		126
	−48 03606				8 11 47	−49 04.0	2	12.26	.005	-0.17	.000	-1.16	.000	2		1650,1737
68879	−5 02447	NGC 2548 - 1560			8 11 49	−05 44.9	2	8.21	.015	1.14	.000	1.05	.020	5	G8 III	75,460
	−31 05724				8 11 49	−31 22.7	1	10.98		0.47		0.01		2		126
68774	+27 01568				8 11 50	+27 15.3	1	8.49		1.20		1.14		2	K0	1375
69043	−37 04417	NGC 2546 - 762			8 11 50	−37 44.9	1	9.66		0.05				1	B9 III	2041
68834	+13 01870	G 50 - 27			8 11 51	+13 10.5	1	8.84		1.19		1.13		1	K5	333,1620
		NGC 2548 - 1572			8 11 51	−05 37.4	1	12.29		0.34		0.06		2		460
69042	−34 04501	IDS08099S3426		AB	8 11 51	−34 35.2	1	8.08		0.18		-0.19		2	B8 IV	126
		LOD 7 # 8			8 11 52	−34 30.8	1	12.61		0.80		0.26		2		126
69017	−31 05727				8 11 53	−31 24.7	1	9.27		1.03		0.84		2	G6 III	126
		LOD 7 # 9			8 11 53	−34 30.0	1	13.31		0.43				2		126
		HA 4 # 186			8 11 54	+74 35.7	1	14.62		0.52		0.08		1		269
		HA 4 # 188			8 11 55	+74 39.6	1	11.55		0.47		-0.06		3		269
	−34 04503				8 11 56	−34 37.5	1	9.13		0.42		-0.08		2		126
69066	−34 04504				8 11 56	−35 03.3	1	6.89		0.79				4	G8 III +(G)	2012
69067	−37 04420	NGC 2546 - 544		⋆A	8 11 56	−37 34.3	2	8.25	.039	-0.10	.010	-0.41		3	Ap	126,2041
	−5 02451	NGC 2548 - 1616			8 11 58	−05 35.1	1	9.96		1.15		1.18		2	G8	75
	−33 04713	LSS 1011			8 11 58	−33 54.6	2	11.36	.039	0.20	.010	-0.42	.029	3		1712,1766
		LOD 7 # 10			8 11 58	−34 28.7	1	11.96		0.60		0.18		2		126
	−27 05181	LSS 1010			8 12 0	−28 10.3	1	10.96		0.17		-0.74		3		1737
	+44 01727	IDS08085N4416		AB	8 12 00	+44 07.4	1	9.40		0.69		0.33		3		1723
	−5 02452	NGC 2548 - 1628			8 12 00	−05 33.1	2	9.47	.005	1.04	.000	0.89	.040	5	K0 III	75,460
		LOD 7 # 4			8 12 00	−34 36.4	1	12.64		0.33		-0.09		2		126
69130	−45 03904	IDS08104S4539		AB	8 12 01	−45 48.5	1	8.63		1.00		0.74		1	F2/3 V +F	389
69144	−46 03929	HR 3244		⋆A	8 12 02	−46 50.4	5	5.13	.007	-0.15	.008	-0.63	.003	13	B3 III	15,258,540,976,2029
		LOD 6+ # 21			8 12 06	−31 19.8	1	11.84		0.51		0.10		2		126
69081	−35 04358	HR 3240, OS Pup		⋆A	8 12 06	−36 10.2	3	5.08	.007	-0.20	.000	-0.85	.010	16	B1.5 IV	1066,1088,2031
69082	−35 04360	HR 3241		⋆B	8 12 06	−36 11.3	6	6.11	.007	-0.18	.011	-0.71	.005	25	B2 IV/V	15,1066,1075,1075*
		HA 4 # 189			8 12 07	+74 40.6	1	13.58		0.27		-0.05		2		269
69194	−49 03430	HR 3247			8 12 07	−50 02.6	3	5.55	.022	1.63	.013	2.00	.000	4	M1 III	389,2008,3005
68638	+57 01128	G 194 - 17		⋆A	8 12 08	+57 15.0	2	7.48	.015	0.75	.002	0.27	.002	5	G8 V	1355,3026

HD	DM	Other Id	N Rem	α_{1950}	δ_{1950}	S	V	σ_V	B–V	σ_{B-V}	U–B	σ_{U-B}	n	Spectrum	References
	−35 04359			8 12 08	−35 21.2	1	10.25		0.24		−0.63		4		1066
68702	+51 01399	IDS08083N5106	A	8 12 11	+50 56.4	1	8.70		0.35		0.05		2	F0	1566
69168	−46 03931			8 12 11	−46 25.6	3	6.48	.009	−0.16	.007	−0.71	.005	9	B2 V	164,540,976
69080	−31 05742	HR 3239	★A	8 12 12	−31 59.3	6	6.06	.004	−0.16	.010	−0.75	.008	16	B2 V	15,126,540,976,1732,2012
69106	−36 04359	LSS 1014		8 12 12	−36 48.0	3	7.13	.004	−0.10	.004	−0.92	.001	9	B1/2 II	540,976,1066
		HA 4 # 190		8 12 13	+74 40.3	1	14.80		0.59		0.14		1		269
69105	−36 04358			8 12 13	−36 16.4	1	7.67		−0.09		−0.41		2	B8 V	401
69122	−34 04508			8 12 16	−34 38.4	2	8.32	.020	0.90	.022	0.30		16	G6/8 IV	126,6011
69142	−39 04128	HR 3243	★A	8 12 16	−40 11.7	5	4.43	.004	1.17	.002	1.09	.004	18	K1 II/III	15,1075,1637,2012,8015
69104	−31 05743	IDS08103S3111	ABC	8 12 17	−31 20.0	1	10.02		0.16		0.01		2	A1/2 III/IV	126
	−31 05744			8 12 17	−31 48.0	1	8.72		1.28		1.32		2	K0	126
69123	−35 04365	HR 3242		8 12 19	−35 20.3	5	5.78	.004	1.02	.006	0.87	.000	13	K1 III	15,258,688,1075,2006
69191	−44 04190			8 12 20	−44 35.9	1	7.42		1.11		0.96		1	K0/1 III	389
	−34 04510			8 12 23	−34 39.9	1	10.13		0.07		−0.41		2		126
69190	−41 03911		V	8 12 28	−41 33.1	1	11.47		0.93		−0.49		1		1753
		HA 4 # 263		8 12 31	+74 41.9	1	14.00		0.85		0.53		1		269
69257	−46 03938			8 12 31	−46 46.0	1	7.31		0.47		0.02		1	F7/8 V	389
69283	−49 03437			8 12 31	−50 09.6	1	7.80		0.96		0.64		1	G8 III	389
69256	−46 03937			8 12 33	−46 29.5	1	8.20		1.10		0.99		1	K0 III	389
68874	+36 01775			8 12 34	+36 05.6	1	8.88		0.88		0.52		2	K0	1733
	−18 02238			8 12 36	−18 51.5	1	9.33		1.29				1	K7	1473
69238	−43 04070	IDS08110S4323	AB	8 12 38	−43 32.5	1	8.52		0.72		0.24		1	F7/8 V	389
69282	−48 03620			8 12 40	−49 04.9	1	8.21		−0.06		−0.38		2	B8 V	401
69235	−40 04012			8 12 41	−40 21.3	1	7.95		0.07		0.01		2	A1 V	123
		Pup sq 6 # 2		8 12 42	−18 50.9	1	12.27		0.42				1		1473
	−40 04013			8 12 43	−40 24.0	1	12.06		0.44		0.08		2		123
69253	−40 04014			8 12 44	−40 40.3	2	6.60	.024	−0.15	.015	−0.60	.024	6	B3 V	123,158
	−40 04015			8 12 46	−40 28.9	1	8.76		1.61		1.83		2	M0	123
		Pup sq 6 # 6		8 12 47	−18 53.1	1	13.34		0.43				1		1473
69302	−45 03914	HR 3250	★AB	8 12 47	−45 40.9	4	5.83	.019	−0.18	.024	−0.79	.010	11	B2 IV	540,976,1732,2006
69031	+18 01892			8 12 48	+17 43.6	2	7.68	.020	1.04	.020	0.90		4	K0	882,1648
69207	−30 05910			8 12 48	−30 45.9	1	9.02		−0.05		−0.40		2	B7 II	401
	−31 05752			8 12 50	−31 54.9	1	9.78		1.36		1.36		2		126
68771	+59 01154	IDS08088N5930	A	8 12 51	+59 20.6	1	6.58		1.31		1.48		4	K2 III	1501
	−32 05021	LSS 1015		8 12 53	−32 20.1	1	11.10		0.21		−0.71		3		1712
		Pup sq 6 # 4		8 12 54	−18 53.0	1	12.93		0.39				1		1473
	−31 05753			8 12 54	−31 51.4	1	10.04		1.00		0.71		2		126
69252	−37 04442			8 12 54	−37 45.9	2	8.13	.000	−0.15	.008	−0.70	.012	4	B3 IV/V	401,1732
	−33 04732	LSS 1016		8 13 0	−33 17.7	1	9.81		1.17		−0.18		3		1712
69249	−30 05917			8 13 00	−30 15.9	1	8.00		−0.02		−0.11		2	A0 III/IV	401
69094	+8 02015	IDS08103N0756	B	8 13 01	+07 46.5	2	10.41	.019	0.69	.014	0.17	.014	4		1003,1696
		Pup sq 6 # 8		8 13 01	−18 50.1	1	12.05		0.43				1		1473
	−35 04379			8 13 01	−35 53.6	1	9.68		−0.05		−0.53		3	B3 III/IV	1066
69072	+19 01963	IDS08102N1900	AB	8 13 06	+18 50.9	1	7.55		1.05		0.89		3	K0	1648
		LSS 1017		8 13 08	−34 44.8	2	11.63	.029	0.14	.015	−0.53	.029	5	B3 Ib	1066,1712
69384	−46 03948	IDS08116S4650	AB	8 13 08	−46 59.2	1	8.24		0.69		0.40		1	G8/K1 II+A2	389
69181	−6 02532			8 13 09	−07 07.0	1	10.35		0.07		0.06		2	A0	1375
	−33 04740			8 13 10	−33 53.1	1	11.11		0.09		−0.48		2		1066
69382	−45 03920			8 13 12	−45 40.1	1	8.41		0.92		0.60		1	G8 III	389
	+20 02030	G 40 - 14		8 13 13	+19 51.4	1	11.20		0.38		−0.24		2	sdF0	333,1620
	−32 05030	IDS08113S3237	AB	8 13 13	−32 46.0	1	10.37		0.17		0.13		2		124
		CCS 1106		8 13 13	−37 46.5	1	11.65		2.51				2	N?	864
	−31 05763			8 13 14	−31 26.2	1	11.39		0.18		0.10		2		126
69404	−46 03951			8 13 16	−46 19.9	3	6.43	.024	−0.15	.011	−0.77	.024	9	B2 Vnn	540,976,2012
68375	+76 00310	HR 3216		8 13 19	+75 54.8	1	5.54		0.90				2	G8 III	71
	−32 05032	LSS 1018		8 13 21	−32 38.2	1	9.45		0.51		0.48		2	F0 II	1712
		G 90 - 50		8 13 22	+31 07.5	1	11.16		1.11		0.98		1	K4:	1620
69247	−10 02443	IDS08110S1008	AB	8 13 22	−10 17.6	1	9.34		0.95		0.81		2	K5 V	3072
		LOD 6+ # 4		8 13 23	−31 47.5	1	10.77		0.36		0.22		2		126
	−35 04384	LSS 1020	★AB	8 13 23	−35 35.0	5	9.19	.013	0.43	.016	−0.57	.027	13	B0 III	432,540,797,976,1066
	−39 04147			8 13 23	−39 41.5	1	9.82		0.48		−0.18		2	B4 III	540
69379	−40 04029			8 13 24	−40 16.8	1	8.95		−0.05		−0.25		2	B9 V	123
	−30 05934	LSS 1019		8 13 27	−30 31.0	1	10.01		0.07		−0.53		3		1712
	−35 04386			8 13 29	−35 56.2	1	11.15		−0.01		−0.47		3		1066
	−34 04536			8 13 31	−34 44.3	1	11.42		0.15		−0.20		2		124
		G 113 - 20		8 13 34	+01 27.4	2	10.07	.029	1.52	.019	1.20		3	M2:	265,1746
69270	−4 02275	NGC 2548 - 2184	★	8 13 38	−05 05.3	1	9.36		0.59		−0.08		1	G0	3062
69425	−36 04390	LSS 1021		8 13 39	−36 57.8	3	9.15	.027	0.06	.027	−0.73	.010	7	B2N	540,976,1066
69493	−45 03929			8 13 39	−46 01.9	1	8.25		1.29		1.43		1	K2 III	389
69469	−44 04218			8 13 41	−44 38.8	1	8.79		0.31		−0.69		2	G8 III	797
68930	+60 01124	HR 3235		8 13 42	+59 43.6	3	5.64	.004	0.16	.000	0.13	.000	8	A7 IV	985,1501,1733
		G 252 - 8		8 13 42	+71 17.8	1	12.44		0.63		0.03		2		1658
	−31 05779			8 13 44	−31 33.5	1	11.47		0.20		0.18		2		126
	−37 04467			8 13 44	−37 44.7	1	10.87		0.57		−0.01		3	B6 II-III	540
69424	−30 05943			8 13 45	−30 39.4	1	8.59		−0.13		−0.69		2	B3 V	401
		G 40 - 15		8 13 46	+21 47.2	1	17.02		0.48		−0.29		1	WD	782
	−31 05780			8 13 47	−31 30.3	1	9.61		1.68		1.80		2		126
69267	+9 01917	HR 3249	★A	8 13 48	+09 20.5	40	3.53	.011	1.48	.006	1.77	.007	305	K4 III	1,3,15,30,124,369,418*
	−31 05781			8 13 48	−31 40.3	1	9.51		1.07		0.82		2		126
69243	+12 01803	HR 3248, R Cnc		8 13 49	+11 52.9	6	7.75	.079	1.52	.081	0.47	.125	6	M4	635,814,817,3001,8022,8027

Table 1

457

HD	DM	Other Id	N Rem	α_{1950}	δ_{1950}	S	V	σ_V	B–V	σ_{B-V}	U–B	σ_{U-B}	n	Spectrum	References
69423	−30 05945			8 13 49	−30 26.7	1	8.62		−0.01		−0.06		2	B9.5V	124
	−31 05782			8 13 49	−31 43.4	1	9.60		0.77		0.42		2		126
69221	+21 01803			8 13 50	+21 23.2	1	7.63		0.95				5	G5	1351
	−48 03636	LSS 1024	⋆ V	8 13 50	−49 04.0	3	9.54 .188		0.00 .021		−0.73 .017		7	O9: nne	389,540,976
69445	−30 05946	HR 3251	⋆ AB	8 13 52	−30 46.3	2	6.21 .010		0.78 .000		0.34		7	G5 III +(G)	404,2006
	−35 04394			8 13 53	−35 37.4	1	10.85		0.27		−0.62		4		1066
	−11 02296			8 13 54	−11 36.6	1	10.28		1.10		0.90		4		158
69371	−15 02351			8 13 54	−15 31.9	1	7.36		1.19				4	K0 III	2033
69464	−35 04396	LSS 1022		8 13 55	−35 28.6	4	8.80 .017		0.31 .010		−0.64 .028		10	O6/7	540,976,1066,1737
69596	−50 03227	HR 3256		8 13 57	−50 17.7	2	6.42 .014		1.52 .018		1.87		5	K3/4 III	389,2009
		LSS 1023		8 13 58	−34 04.7	1	12.21		0.02		−0.74		2	B3 Ib	1712
		vdB 9 # a		8 14 00	−35 59.	1	13.17		0.52		−0.33		3		434
69624	−48 03638			8 14 00	−48 53.8	1	8.39		0.07		−0.30		2	B7 V	401
69655	−51 02914			8 14 03	−51 54.3	1	6.63		0.58		0.05		2	G1 V	389
69511	−35 04401	HR 3253		8 14 05	−35 44.9	2	6.15 .005		1.55 .007		1.70		6	K2 III	58,2009
		vdB 9 # b		8 14 06	−36 00.	1	14.66		0.84		−0.17		3		434
69510	−31 05787			8 14 08	−31 36.3	1	9.15		0.46		0.00		2	F2 V	126
	−33 04756			8 14 08	−33 34.3	1	11.08		0.07		−0.52		2		1066
	−35 04402			8 14 09	−35 56.2	1	8.88		0.00		−0.10		2		432
69653	−49 03456			8 14 10	−49 50.3	1	7.60		0.00		−0.33		2	B7/8 III	401
	−31 05790			8 14 16	−31 33.6	1	10.97		0.68		0.23		2		126
69198	+43 01795			8 14 17	+43 39.3	1	8.73		0.43		0.00		1	F5	1776
69365	+11 01806			8 14 22	+11 30.4	1	7.71		1.02		0.81		5	K0	814
	+0 02245	G 113 - 22		8 14 23	+00 10.5	2	9.68 .015		0.59 .002		−0.07 .010		3	G0	1658,3062
69647	−43 04114			8 14 23	−44 03.4	1	8.79		0.47		0.03		1	F8	389
69648	−43 04113	LSS 1028		8 14 23	−44 10.1	3	8.05 .004		0.29 .014		−0.67 .031		6	O8.5I	540,976,1737
		LSS 1025		8 14 24	−33 51.8	1	13.04		0.09		−0.51		2	B3 Ib	1712
69149	+54 01215	HR 3246		8 14 25	+54 18.0	2	6.28 .005		1.56 .013		1.91		5	K5	71,3016
68744	+73 00406	G 252 - 9		8 14 25	+73 29.8	1	8.50		0.60		0.05		2	G0 V	1003
	−14 02453	IV Pup		8 14 25	−14 17.7	1	10.25		3.05				1		864
69242	+43 01796			8 14 27	+43 33.9	1	7.63		0.31		0.01		1	F0	1776
68788	+73 00407	G 252 - 10		8 14 29	+73 34.7	2	8.38 .000		0.85 .009		0.53 .000		5	K1 V	22,1003
	−35 04412	LSS 1026		8 14 31	−35 35.1	4	9.48 .010		0.42 .015		−0.57 .031		11	O9.5III	540,797,976,1066
		LSS 1027		8 14 31	−35 51.9	1	11.33		0.56		−0.49		2	B3 Ib	797
69863	−62 00985	HR 3260	⋆ AB	8 14 31	−62 45.7	3	5.16 .009		0.09 .014		0.09		10	A2 V	15,404,2012
69620	−35 04413			8 14 32	−36 02.8	1	7.17		−0.14		−0.63		2	B3 IV	401
69646	−39 04162			8 14 32	−39 41.6	1	8.60		0.03		−0.59		2	B3 III/V	1732
69730	−48 03647			8 14 32	−49 01.8	1	8.31		−0.02		−0.08		2	B9.5V	401
		G 111 -B9	A	8 14 33	+37 55.5	1	13.68		1.08		0.94		4	dK5	3029
		G 111 -B9	B	8 14 33	+37 55.5	1	15.43		0.65		−0.10		4	sdG	3029
69530	−14 02456	IDS08122S1451	A	8 14 33	−14 59.8	2	7.18 .009		1.49 .005		1.78		17	K3 III	1673,2033
69562	−21 02358			8 14 33	−21 43.7	1	6.96		−0.15				4	B3 III	2033
69311	+39 02082			8 14 37	+38 57.4	1	7.85		0.91		0.53		2	G7 III	105
		G 40 - 16		8 14 38	+21 09.3	1	12.54		1.49		1.19		1		1620
	−35 04415	LSS 1029	⋆ AB	8 14 38	−35 29.5	3	10.47 .026		0.64 .012		−0.37 .023		8	O7 IIInn	797,1066,1737
69479	+4 01945			8 14 41	+04 22.5	1	6.53		0.63				4	G0	2007
69589	−20 02467	HR 3255		8 14 42	−21 09.9	2	6.59 .005		0.02 .000				7	A0 V	15,2012
69501	−2 02511			8 14 43	−03 03.7	1	8.55		0.51		0.04		2	F5	1375
69670	−33 04771			8 14 43	−33 59.4	1	8.58		0.38		−0.14		2	B(5/7) +F/G	401
		G 90 - 52		8 14 45	+31 17.0	1	11.22		1.49		1.22		1	M1	1620
69526	−4 02288	IDS08123S0504	AB	8 14 47	−05 12.8	1	7.73		0.39		−0.01		3	F2	3030
69478	+9 01921	HR 3252		8 14 49	+09 01.3	3	6.28 .005		0.98 .000		0.75 .011		10	G8 III	15,1256,1355
	−33 04773	LSS 1030		8 14 51	−33 46.3	1	9.31		0.45		0.47		2	A8 II	1712
	−40 04055			8 14 53	−40 34.2	1	10.21		0.25		0.14		2	A2	123
70103	−70 00747			8 14 53	−70 40.2	1	8.89		1.10				4	K0/1 III	2033
	−40 04056			8 14 56	−40 32.6	1	10.42		0.51		0.06		2		123
69148	+62 00991	HR 3245		8 14 57	+62 39.8	1	5.71		0.89				2	G8 III	71
69891	−56 01569			8 14 58	−57 06.9	1	7.83		−0.02		−0.42		2	B8 Vn	401
69611	−3 02288	G 113 - 24		8 15 00	−03 49.7	3	7.72 .022		0.58 .000		−0.05 .026		4	F8	265,1658,3062
	−35 04422			8 15 01	−35 14.5	1	10.56		0.96		0.73		2		1696
69844	−53 01595			8 15 01	−53 19.9	1	8.90		0.48		0.04		1	F5 V	389
67934	+82 00235	HR 3197		8 15 03	+82 35.4	2	6.30 .020		0.02 .010		0.01 .005		4	A2 Vn	985,1733
	+31 01781	G 51 - 6		8 15 05	+30 46.1	5	8.85 .022		1.14 .007		1.11 .000		12	K4 V	22,1003,1197,1758,3072
69665	−15 02362	HR 3257		8 15 06	−16 07.7	2	6.26 .078		0.04 .024		0.08		6	A1 V	252,2031
	−33 04781	LSS 1032		8 15 08	−33 59.0	1	9.32		0.30		−0.14		2	A2 Ib-II	1066
	−40 04060			8 15 08	−40 24.2	1	10.96		0.45		−0.02		2		123
69841	−49 03470	IDS08137S4935	AB	8 15 09	−49 44.4	1	8.20		0.10		−0.32		2	B7 III/IV	401
69760	−34 04573	IDS08133S3411	AB	8 15 10	−34 20.5	1	7.60		0.04		0.01		2	A1 IV	401
69887	−52 01425			8 15 10	−52 36.3	1	7.14		−0.08		−0.47		2	B5 IV	401
69552	+14 01869			8 15 12	+13 47.8	1	8.71		1.48				2	K2	882
68951	+72 00409	HR 3236	⋆ A	8 15 12	+72 33.9	3	5.99 .021		1.54 .017		1.66 .205		4	M0 III	71,542,3001
68951	+72 00412	IDS08097N7243	B	8 15 12	+72 33.9	1	9.78		0.39		−0.02		1	F5 IV	542
69761	−35 04425			8 15 17	−35 45.7	1	7.83		1.01		0.80		1	B9 V	389
69913	−52 01427			8 15 17	−53 03.5	1	8.18		0.09		−0.34		2	Ap Si	401
	−40 04062			8 15 18	−40 27.0	1	11.57		0.48		0.00		2		123
69663	−1 02005			8 15 20	−01 38.7	1	7.45		0.86				4	G5	2033
	+0 02252			8 15 24	+00 31.3	1	9.38		0.55		0.04		31	G0	588
69856	−45 03959			8 15 24	−45 35.8	2	7.99 .007		1.06 .006		0.89 .017		5	G8/K0 III	389,1673
		vdB 10 # b		8 15 26	−42 04.7	1	10.83		0.66		0.08		1		434
		vdB 10 # a		8 15 27	−42 04.6	1	11.01		0.33		0.22		3	A0 V	434

HD	DM	Other Id	N Rem	α_{1950}	δ_{1950}	S	V	σ_V	B–V	σ_{B-V}	U–B	σ_{U-B}	n	Spectrum	References
69837	−39 04183			8 15 28	−40 05.1	1	7.80		0.03		0.00		2	A0 V	123
69975	−55 01494			8 15 29	−55 33.6	1	9.12		0.14		0.11		2	A3 V	1730
69818	−32 05087			8 15 30	−32 42.8	1	6.95		0.03		0.03		2	A1 IV	401
233511	+54 01216	G 194 - 22		8 15 33	+54 15.1	2	9.74	.031	0.48	.000	-0.21	.013	5	F8	22,1003
69882	−42 04090	LSS 1034	⋆ AB	8 15 34	−42 21.9	5	7.16	.016	0.31	.012	-0.57	.021	12	B1/2 Iab/b	540,976,1034,1737,2012
	−53 01602			8 15 34	−53 28.2	1	7.36		1.10		1.06		1		389
		G 91 - 45		8 15 36	+18 25.5	2	14.98	.005	1.66	.010			3		906,1663
69992	−53 01604			8 15 39	−53 32.8	1	7.76		0.05		0.03		2	A1 V	401
	−47 03767			8 15 42	−47 34.6	1	8.68		1.13		1.06		1		389
69931	−44 04255			8 15 44	−45 01.9	1	8.78		0.47		-0.01		1	F6 V	389
69736	+4 01948			8 15 45	+04 32.0	1	8.06		-0.06		-0.49		2	K0	126
	−35 04437	AAS70,69 # 1		8 15 46	−35 38.6	1	10.37		1.63		1.99		1		1636
		NGC 2567 - 175		8 15 47	−30 27.5	1	12.00		0.22		0.18		2		1602
69973	−47 03771			8 15 47	−47 45.8	2	6.65	.012	-0.08	.011	-0.63	.043	3	B4 V	401,1469,2012
	−39 04186			8 15 48	−40 11.8	1	9.78		1.14		1.07		2		123
		NGC 2567 - 180		8 15 51	−30 24.5	1	12.90		1.41		1.42		2		1602
		CCS 1116		8 15 51	−33 26.7	1	10.90		4.38				2		864
	−30 05999	NGC 2567 - 177		8 15 52	−30 25.9	1	9.93		0.99		0.80		3		1602
		NGC 2567 - 176		8 15 52	−30 26.2	1	12.79		0.30		0.16		3		1602
69881	−30 06001			8 15 53	−30 18.0	1	8.67		-0.07		-0.37		2	B8 III/IV	401
	−37 04513			8 15 53	−37 22.9	1	11.56		0.50		-0.02		2		124
69879	−29 05897	HR 3261		8 15 56	−29 50.8	4	6.44	.007	1.04	.000			18	K0 III	15,1075,2013,2029
		NGC 2567 - 178		8 15 56	−30 26.3	1	13.47		1.12		0.73		2		1602
		NGC 2567 - 179		8 15 57	−30 25.7	1	12.96		1.07		0.71		2		1602
		AAS70,69 T5# 2		8 15 58	−35 40.9	1	11.42		0.34		0.01		1		1636
69698	+21 01807			8 15 59	+21 39.1	1	7.88		1.23		1.39		3	K2 III	37
		NGC 2567 - 155		8 15 59	−30 28.7	1	13.14		1.26		1.01		3		1602
		AAS70,69 T5# 3		8 15 59	−35 42.7	1	12.92		0.11		0.15		1		1636
		G 113 - 25		8 16 01	−06 06.3	1	13.68		1.40				1		3062
69830	−12 02449	HR 3259		8 16 01	−12 27.7	8	5.97	.024	0.76	.006	0.34	.023	30	K0 V	15,22,1003,1067,1075*
69550	+50 01530			8 16 03	+50 26.1	1	8.01		1.20		1.25		2	K0	1566
		NGC 2567 - 154		8 16 04	−30 30.3	1	12.63		0.16		0.19		2		1602
		NGC 2567 - 156		8 16 05	−30 25.4	1	14.48		0.47				2		1602
	−37 04519			8 16 06	−37 27.5	1	9.83		0.08		0.02		2		124
	+44 01733			8 16 07	+44 18.2	1	9.11		1.52		1.69		1	K5	1776
		NGC 2567 - 160		8 16 07	−30 21.7	1	12.76		0.06		-0.20		2		1602
		NGC 2567 - 153		8 16 07	−30 32.1	1	14.05		1.23		0.86		2		1602
		vdB 11 # f		8 16 07	−41 59.2	1	11.81		0.02		0.02		2		434
		vdB 11 # g		8 16 08	−41 59.3	1	12.38		0.45		0.02		2		434
70201	−64 00843			8 16 08	−64 14.8	1	7.80		-0.02		-0.29		3	B8 V	1732
		NGC 2567 - 108		8 16 09	−30 28.0	1	14.44		0.48		0.06		2		1602
		NGC 2567 - 107		8 16 09	−30 28.8	1	13.04		0.20		0.17		4		1602
		NGC 2567 - 157		8 16 10	−30 23.8	1	14.14		0.34		0.19		2		1602
69970	−36 04441	IDS08143S3635	AB	8 16 10	−36 44.2	1	9.32		-0.01		-0.10		2	A0 V	401
69734	+24 01903			8 16 11	+24 19.9	1	7.22		0.93		0.63		2	G5	1625
		G 113 - 28		8 16 11	−02 05.9	1	14.78		0.77		0.10		2		3062
		NGC 2567 - 161		8 16 11	−30 20.5	1	12.36		1.17		1.05		2		1602
	−30 06006	NGC 2567 - 158		8 16 11	−30 23.7	1	11.94		0.12		0.08		2		1602
		NGC 2567 - 109		8 16 11	−30 27.6	1	14.32		0.32		0.27		1		1602
69054	+75 00334	IDS08101N7508	A	8 16 12	+74 58.7	1	6.35		1.06		0.85		1	K0 III	3032
69054	+75 00334	IDS08101N7508	AB	8 16 12	+74 58.7	2	6.39	.024	0.99	.024	0.85		7	K0 III	1058,1118
69054	+75 00334	IDS08101N7508	B	8 16 12	+74 58.7	1	10.79		0.68		0.18		1		3032
		NGC 2567 - 174		8 16 12	−30 36.4	1	12.87		0.30		0.13		3		1602
70004	−41 03962			8 16 12	−41 58.8	1	9.86		0.89		0.62		2	K1 (III)	434
		NGC 2567 - 159		8 16 13	−30 22.1	1	13.13		0.34		0.17		2		1602
	−30 06008	NGC 2567 - 152		8 16 13	−30 34.7	1	8.97		1.22		1.13		2		1602
		NGC 2567 - 111		8 16 14	−30 25.6	1	13.15		0.14		0.17		3		1602
		NGC 2567 - 151		8 16 14	−30 34.3	1	13.83		0.70		0.28		3		1602
		NGC 2567 - 162		8 16 15	−30 19.5	1	12.69		1.21		1.19		2		1602
		NGC 2567 - 103		8 16 15	−30 32.4	1	14.43		0.46		0.06		2		1602
		NGC 2567 - 105		8 16 16	−30 29.7	1	12.96		0.96		0.58		2		1602
		NGC 2567 - 150		8 16 16	−30 34.3	1	13.74		1.11		0.67		3		1602
		NGC 2567 - 104		8 16 17	−30 30.9	1	10.80		0.95		0.63		3	G2	1602
	−30 06011	NGC 2567 - 114		8 16 18	−30 23.5	1	10.87		1.11		0.87		3	G5	1602
	−30 06009	NGC 2567 - 61		8 16 18	−30 29.9	1	10.09		1.13		1.05		3	G8	1602
69788	+16 01687			8 16 19	+16 16.8	1	7.00		-0.04		-0.14		2	A0	1648
		NGC 2567 - 115		8 16 19	−30 23.6	1	12.47		1.16		0.96		2		1602
		NGC 2567 - 66	AB	8 16 19	−30 26.3	1	13.34		0.88		0.21		2		1602
		Pis 1 - 21		8 16 19	−36 56.5	1	12.90		0.58		-0.06		1		410
69986	−37 04523			8 16 19	−37 19.2	1	8.34		-0.05		-0.20		5	B9/A0 V	124
70024	−41 03965			8 16 19	−41 57.2	1	9.07		0.05		-0.36		3	B5/7 V	434
		NGC 2567 - 60		8 16 20	−30 30.5	1	14.04		0.36		0.18		4		1602
		NGC 2567 - 101		8 16 20	−30 32.6	1	13.70		0.63		0.14		2		1602
		NGC 2567 - 112		8 16 21	−30 24.3	1	13.43		0.24		0.18		2		1602
		NGC 2567 - 113		8 16 21	−30 24.3	1	14.33		0.54		0.10		2		1602
		NGC 2567 - 68		8 16 21	−30 26.0	1	13.72		0.57		0.19		2		1602
		NGC 2567 - 67		8 16 21	−30 26.2	1	13.90		0.33		0.18		2		1602
		NGC 2567 - 62		8 16 21	−30 28.8	1	12.46		1.37		1.33		3		1602
		NGC 2567 - 59		8 16 21	−30 30.3	1	11.14		0.41		0.29		3		1602
		Pis 1 - 22		8 16 21	−36 56.5	1	13.45		0.50		-0.12		1		410

Table 1 459

HD	DM	Other Id	N Rem	α_{1950}	δ_{1950}	S	V	σ_V	B–V	σ_{B-V}	U–B	σ_{U-B}	n	Spectrum	References
70003	−36 04443	HR 3267	⋆ AB	8 16 21	−37 13.1	4	6.70	.008	0.24	.017	-0.03	.005	14	A0 IV	15,124,404,2012
69090	+74 00356			8 16 22	+74 34.7	1	9.58		-0.01		0.00		4	A2	269
		NGC 2567 - 58	V	8 16 22	−30 30.8	1	12.98		0.26		0.11		4		1602
		NGC 2567 - 116		8 16 23	−30 23.2	1	11.99		1.16		0.94		2		1602
		NGC 2567 - 65		8 16 23	−30 26.9	1	12.57		0.17		0.13		3		1602
		NGC 2567 - 64		8 16 23	−30 27.5	1	13.43		0.32		0.19		3		1602
70002	−35 04452	HR 3266		8 16 23	−35 17.7	1	5.58		1.24				4	K2 III	2006
	−30 06012	NGC 2567 - 57		8 16 24	−30 30.5	1	11.18		0.18		0.14		3	B8	1602
69548	+58 01112	HR 3254		8 16 25	+57 54.1	1	5.91		0.38		-0.12		3	F4 V	254
	−29 05911	NGC 2571 - 113		8 16 25	−29 32.4	1	11.18		0.01		-0.10		2		457
		NGC 2567 - 117	AB	8 16 25	−30 22.9	1	13.67		0.68		0.14		3		1602
		NGC 2567 - 72		8 16 25	−30 25.6	1	13.95		0.37		0.12		3		1602
		NGC 2567 - 71		8 16 25	−30 25.9	1	12.85		0.17		0.16		4		1602
	−30 06015	NGC 2567 - 54	⋆ AB	8 16 25	−30 30.1	1	11.16		1.08		0.78		4		1602
		NGC 2567 - 56	AB	8 16 25	−30 30.4	1	11.43		0.24		0.20		3	A2	1602
	−35 04453	AAS70,69 # 13		8 16 25	−35 46.3	1	11.00		0.60		0.16		1		1636
	−14 02469			8 16 26	−15 02.8	2	9.84	.034	1.13	.015	1.08	.039	3	K3 V	801,3072
		NGC 2567 - 75		8 16 26	−30 25.2	1	13.30		0.49		-0.04		2		1602
		NGC 2567 - 8		8 16 26	−30 27.3	1	13.15		0.25		0.13		4		1602
	−30 06016	NGC 2567 - 11	AB	8 16 26	−30 29.4	1	10.80		0.27		0.20		3	B8	1602
		NGC 2567 - 148		8 16 26	−30 34.5	1	11.45		0.24		0.16		2		1602
		NGC 2567 - 149		8 16 26	−30 34.9	1	12.87		0.20		0.16		2		1602
		Pis 1 - 7		8 16 26	−36 56.5	1	13.88		0.42		-0.15		1		410
		Pis 1 - 6		8 16 26	−36 56.6	1	12.79		0.50		0.02		1		410
70043	−41 03967			8 16 26	−41 58.9	1	9.85		0.16		0.13		3	A0 V	434
		NGC 2567 - 118		8 16 27	−30 22.3	1	13.07		0.22		0.16		3		1602
		NGC 2567 - 70		8 16 27	−30 26.6	1	13.18		0.47		0.19		3		1602
		NGC 2567 - 9		8 16 27	−30 27.8	1	12.74		0.28		0.19		3		1602
		AAS70,69 T5# 12		8 16 27	−35 45.9	1	11.17		0.60		0.21		1		1636
		Pis 1 - 8		8 16 27	−36 56.4	1	14.31		0.49		-0.28		1		410
		Pis 1 - 5		8 16 27	−36 56.7	1	12.76		0.54		-0.05		1		410
		Pis 1 - 4		8 16 27	−36 57.0	1	13.46		0.43		-0.11		1		410
		Pis 1 - 13		8 16 27	−36 57.9	1	13.25		2.02		2.40		1		410
70084	−46 04025	IT Vel		8 16 27	−46 56.1	2	7.05	.016	-0.14	.007	-0.69	.005	5	B5 II/III	164,1469
		NGC 2571 - 114		8 16 28	−29 33.0	1	11.93		0.06		-0.06		2		457
		NGC 2567 - 76		8 16 28	−30 25.2	1	14.06		0.57		0.07		4		1602
		NGC 2567 - 10		8 16 28	−30 28.0	1	13.83		0.30		0.16		3		1602
		NGC 2567 - 52		8 16 28	−30 31.6	1	12.97		0.20		0.17		3		1602
		NGC 2567 - 99		8 16 28	−30 33.9	1	13.65		0.34		0.16		2		1602
		Pis 1 - 16		8 16 28	−36 55.7	1	11.29		0.49		0.05		2	F8	410
		Pis 1 - 9		8 16 28	−36 56.3	1	14.01		0.59		-0.11		1		410
69787	+21 01810			8 16 29	+21 24.9	1	8.03		1.18		1.32		3	K2 III	37
69433	+65 00626			8 16 29	+65 36.1	1	8.03		0.73		0.29		1	G0	1355
		NGC 2567 - 53		8 16 29	−30 32.0	1	14.35		0.49		0.05		2		1602
		vdB 11 # c		8 16 29	−41 58.4	1	11.72		0.57		0.39		2		434
	−25 05877			8 16 30	−25 44.8	1	10.13		2.51				2	N	864
		NGC 2567 - 78		8 16 30	−30 24.5	1	13.31		0.50		-0.04		3		1602
		NGC 2567 - 7		8 16 30	−30 27.6	1	13.73		0.50		0.15		1		1602
		NGC 2567 - 49		8 16 30	−30 30.6	1	13.97		0.40		0.32		2		1602
		NGC 2567 - 146		8 16 30	−30 35.6	1	13.28		0.54		0.27		2		1602
		LOD 8 # 2		8 16 30	−36 29.6	1	10.47		0.45		0.01		2		126
		Pis 1 - 18		8 16 30	−36 57.8	1	13.30		1.38		1.41		1		410
		GD 90		8 16 31	+37 40.9	1	15.74		0.22		-0.63		1	DAp	3060
		NGC 2567 - 123		8 16 31	−30 21.1	1	13.16		1.30		1.14		2		1602
		NGC 2567 - 119		8 16 31	−30 22.1	1	13.34		0.19		0.16		3		1602
		NGC 2567 - 77		8 16 31	−30 25.3	1	14.18		0.43		0.07		3		1602
		NGC 2567 - 5		8 16 31	−30 27.0	1	11.82		0.16		0.11		3	B8	1602
		NGC 2567 - 12		8 16 31	−30 29.4	1	12.93		0.14		0.13		3		1602
		NGC 2567 - 51		8 16 31	−30 31.3	1	13.77		0.42		0.12		2		1602
	−30 06018	NGC 2567 - 147		8 16 31	−30 34.7	1	11.47		0.24		0.20		2		1602
		NGC 2571 - 66		8 16 32	−29 35.4	1	12.37		0.11		0.08		2	B9	457
		NGC 2567 - 79		8 16 32	−30 24.5	1	12.54		0.17		0.11		2		1602
		NGC 2567 - 4	AB	8 16 32	−30 27.5	1	11.70		0.08		0.06		3	A0	1602
		NGC 2567 - 13		8 16 32	−30 30.3	2	12.50	.019	0.24	.021	0.12	.016	4		239,1602
		HA 124 # 970		8 16 33	−15 33.0	1	11.85		0.94		0.57		11		1499
		NGC 2567 - 122		8 16 33	−30 21.3	1	12.63		0.31		0.05		2		1602
		NGC 2567 - 6	AB	8 16 33	−30 26.8	1	14.07		0.24		0.19		4		1602
		NGC 2567 - 2		8 16 33	−30 27.8	1	11.89		0.13		0.12		4		1602
		NGC 2567 - 1		8 16 33	−30 28.1	2	13.68	.004	0.40	.005	0.14	.045	5		239,1602
		NGC 2567 - 18		8 16 33	−30 28.8	1	11.38		0.13		0.11		3		1602
		NGC 2567 - 97		8 16 33	−30 32.5	1	12.94		0.25		0.12		2		1602
		NGC 2567 - 98		8 16 33	−30 33.8	1	13.04		0.57		0.08		2		1602
		Pis 1 - 24		8 16 33	−36 56.4	1	13.15		0.37		-0.06		1		410
		Steph 693		8 16 34	+54 45.8	1	12.60		1.49		1.17		1	M0	1746
		HA 124 # 972		8 16 34	−15 32.5	1	12.14		0.51		0.03		11		1499
		NGC 2571 - 59		8 16 34	−29 38.5	1	12.30		0.11		0.06		3	B9	457
		NGC 2567 - 120	AB	8 16 34	−30 22.2	1	12.39		1.03		0.85		2		1602
		NGC 2567 - 3		8 16 34	−30 27.8	1	14.55		0.48		0.10		2		1602
		NGC 2567 - 19	V	8 16 34	−30 28.6	1	13.24		0.34		0.13		4		1602
		NGC 2567 - 17		8 16 34	−30 29.2	2	11.41	.009	0.13	.013	0.08	.005	4		239,1602

HD	DM	Other Id	N Rem	α_{1950}	δ_{1950}	S	V	σ_V	B–V	σ_{B-V}	U–B	σ_{U-B}	n	Spectrum	References
	−30 06019	NGC 2567 - 16	AB	8 16 34	−30 29.5	1	11.04		0.94		0.59		3		1602
		NGC 2567 - 15		8 16 34	−30 30.1	1	13.29		0.32		0.10		4		1602
		NGC 2567 - 14		8 16 34	−30 30.4	1	12.47		0.09		0.08		3		1602
		NGC 2567 - 145	AB	8 16 34	−30 35.8	1	12.90		0.42		0.17		3		1602
69767	+35 01803			8 16 35	+35 21.5	1	7.23		0.98		0.76		2	K0	1601
69713	+45 01563			8 16 35	+44 49.3	1	7.87		0.48		−0.01		1	F5	1776
		NGC 2571 - 67		8 16 35	−29 34.5	1	11.79		1.02		0.70		2		457
		NGC 2567 - 121		8 16 35	−30 22.0	1	12.46		0.67		0.31		2		1602
		NGC 2567 - 80		8 16 35	−30 24.8	1	14.46		0.66		0.19		2		1602
		NGC 2567 - 81		8 16 35	−30 25.2	1	12.86		0.12		0.06		2		1602
		NGC 2567 - 20		8 16 36	−30 28.8	1	13.02		0.25		0.13		4		1602
		NGC 2567 - 21		8 16 36	−30 29.1	1	13.97		0.33		0.11		3		1602
		NGC 2571 - 115		8 16 37	−29 31.9	1	11.93		0.54		0.05		3		457
	−29 05919	NGC 2571 - 68		8 16 37	−29 33.9	1	11.19		0.01		−0.26		2	B5	457
		NGC 2567 - 28		8 16 37	−30 26.6	1	14.14		0.50		0.06		1		1602
		NGC 2567 - 26		8 16 37	−30 27.7	1	13.28		0.28		0.15		4		1602
		NGC 2567 - 22		8 16 37	−30 29.2	1	13.94		0.21		0.08		3		1602
70059	−35 04460			8 16 37	−35 21.7	1	8.36		0.08		0.06		2	A0 IV/V	401
		NGC 2571 - 58		8 16 38	−29 38.4	1	12.73		0.53		0.27		3		457
		NGC 2567 - 47		8 16 38	−30 31.3	1	13.38		0.35		0.26		3		1602
69389	+69 00462			8 16 39	+69 11.4	1	8.24		0.43		−0.04		2	F2	1733
		NGC 2567 - 29		8 16 39	−30 26.2	1	11.94		0.14		0.09		3		1602
		NGC 2567 - 46		8 16 39	−30 31.4	1	13.35		0.32		0.25		3		1602
		NGC 2567 - 144		8 16 39	−30 35.8	1	13.94		0.25		0.27		3		1602
70081	−39 04203			8 16 39	−40 12.1	1	8.91		0.66		0.24		2	G3 IV/V	123
		NGC 2571 - 69		8 16 40	−29 34.0	1	12.55		0.18		0.08		3	A0	457
		NGC 2567 - 30		8 16 40	−30 26.5	1	13.64		0.48		0.01		2		1602
		NGC 2567 - 27		8 16 40	−30 27.1	1	14.16		0.42		0.10		1		1602
		NGC 2567 - 48		8 16 40	−30 30.9	1	13.18		0.31		0.14		3		1602
		NGC 2567 - 141		8 16 40	−30 34.6	1	14.00		0.23		0.29		2		1602
69682	+54 01217	HR 3258		8 16 41	+53 44.0	2	6.50	.005	0.29	.005	0.06	.005	5	F0 IV	1566,1733
		NGC 2567 - 23		8 16 41	−30 28.9	1	11.85		0.13		0.08		4	B9	1602
		NGC 2567 - 142	AB	8 16 41	−30 35.1	1	13.79		0.65		0.00		1		1602
70060	−36 04449	HR 3270		8 16 41	−36 30.2	8	4.44	.011	0.22	.010	0.11	.015	31	A4(m)A5-A7	3,15,126,1075,1489*
		NGC 2571 - 72		8 16 42	−29 32.3	1	11.77		0.06		−0.13		3	B6	457
		NGC 2571 - 22		8 16 42	−29 36.3	1	12.47		1.18		0.96		2		457
		NGC 2571 - 111		8 16 42	−29 41.1	1	11.47		0.02		−0.23		3	B5	457
		NGC 2567 - 82		8 16 42	−30 24.5	1	14.38		0.48				1		1602
		NGC 2567 - 31		8 16 42	−30 25.6	1	12.69		0.15		0.08		2	A1	1602
		NGC 2567 - 24		8 16 42	−30 28.4	1	12.40		0.10		0.08		3	A0	1602
		NGC 2567 - 42		8 16 42	−30 29.9	1	11.71		0.09		0.06		2	B9	1602
		NGC 2567 - 96		8 16 42	−30 32.5	1	14.26		0.23		0.23		1		1602
		NGC 2567 - 143		8 16 43	−30 35.7	1	13.02		1.32		1.32		2		1602
		G 111 - 71		8 16 44	+38 44.2	2	16.56	.015	0.30	.050	−0.57	.010	5	DA,F	538,3060
		NGC 2571 - 74		8 16 44	−29 32.8	1	12.73		0.24		0.14		2		457
	−29 05924	NGC 2571 - 24		8 16 44	−29 35.6	1	10.85		−0.03		−0.35		3	B5	457
		NGC 2571 - 54		8 16 44	−29 39.5	1	12.88		0.21		0.15		2		457
		LOD 8 # 3		8 16 44	−36 27.6	1	11.71		0.15		−0.14		2		126
	−29 05925	NGC 2571 - 75		8 16 45	−29 33.0	1	10.81		1.26		1.20		3	K0	457
	−30 06028	NGC 2567 - 173		8 16 45	−30 38.2	1	10.13		0.55		0.08		2		1602
		NGC 2571 - 26		8 16 46	−29 34.0	1	13.59		0.35		0.16		2		457
		NGC 2571 - 20		8 16 46	−29 36.8	1	12.40		0.09		0.06		2	A1	457
		NGC 2571 - 52		8 16 46	−29 39.3	1	13.28		0.73		0.29		3		457
		NGC 2567 - 32		8 16 46	−30 25.6	1	14.25		0.93		0.51		1		1602
		NGC 2567 - 33		8 16 46	−30 25.9	1	14.33		1.05		0.72		1		1602
		NGC 2567 - 37		8 16 46	−30 27.3	1	11.38		1.04		0.70		3	G8	1602
		NGC 2567 - 44		8 16 46	−30 31.1	1	13.52		0.29		0.15		3		1602
		NGC 2567 - 93		8 16 46	−30 31.6	1	14.19		0.42		0.12		3		1602
		NGC 2567 - 95		8 16 46	−30 33.5	1	12.14		0.17		0.16		2	A2	1602
70122	−43 04156	LSS 1039		8 16 46	−44 06.6	2	9.22	.005	0.52	.006	−0.36	.027	5	B2 Ib/II	540,976
		G 111 - 72		8 16 47	+38 44.3	2	13.14	.029	1.52	.000	1.34		6	sdM2	538,3028
		NGC 2571 - 25		8 16 47	−29 35.8	1	11.46		0.02		−0.19		2	B5	457
		NGC 2567 - 125		8 16 47	−30 22.1	1	13.37		0.24		0.18		3		1602
		NGC 2567 - 124		8 16 47	−30 22.5	1	12.82		0.15		0.08		2		1602
		NGC 2567 - 83		8 16 47	−30 24.5	1	13.99		0.37		0.13		3		1602
	−30 06024	NGC 2567 - 34		8 16 47	−30 26.1	1	11.23		0.09		0.07		3	B7	1602
		NGC 2567 - 35	V	8 16 47	−30 26.3	1	12.96		0.48		−0.02		3		1602
		NGC 2567 - 40	V	8 16 47	−30 28.8	1	13.79		0.29		0.16		3		1602
		AJ84,127 # 2216		8 16 48	−15 04.	1	9.75		1.71				1		801
		NGC 2571 - 19		8 16 48	−29 36.8	1	12.17		0.07		0.00		3		457
		NGC 2571 - 110		8 16 48	−29 40.2	1	11.84		0.05		−0.10		3	B8	457
		NGC 2567 - 126		8 16 48	−30 22.0	1	13.11		0.58		0.00		2		1602
		NGC 2567 - 36		8 16 48	−30 26.1	1	14.12		0.18		0.07		1		1602
		NGC 2567 - 38		8 16 48	−30 27.4	1	13.04		0.15		0.12		3		1602
		NGC 2567 - 140		8 16 48	−30 34.6	1	12.65		1.61		1.68		2		1602
		Pup sq 1 # 73		8 16 48	−36 38.	1	13.46		0.62		0.23		3		124
		Pup sq 1 # 72		8 16 48	−36 40.	1	14.08		0.86		0.17		3		124
		NGC 2571 - 18		8 16 49	−29 36.2	1	11.76		0.12		0.10		2	A0	457
		NGC 2567 - 39		8 16 49	−30 27.6	1	13.21		0.22		0.19		3		1602
		NGC 2567 - 43		8 16 49	−30 30.0	1	12.78		0.14		0.12		3		1602

Table 1   461

HD	DM	Other Id	N	Rem	α_{1950}	δ_{1950}	S	V	σ_V	B–V	σ_{B-V}	U–B	σ_{U-B}	n	Spectrum	References
		NGC 2567 - 94			8 16 49	−30 32.5	1	12.19		0.10		0.08		2		1602
		NGC 2571 - 76			8 16 50	−29 31.8	1	12.69		0.58		0.10		2	F5	457
		NGC 2567 - 163			8 16 50	−30 20.9	1	13.00		0.52		0.05		3		1602
		NGC 2567 - 86			8 16 50	−30 26.5	1	12.98		0.20		0.14		3		1602
70119	−39 04209				8 16 50	−40 10.3	1	7.94		1.03		0.87		2	K0/1 (III)	123
70120	−40 04084				8 16 50	−40 14.5	1	8.36		-0.09		-0.48		2	B3/5 II/III	123
69997	−9 02471	HR 3265, HQ Hya			8 16 51	−10 00.5	4	6.30	.017	0.33	.011	0.15	.014	15	F3 IIIp δ Del	15,39,1075,2012
		NGC 2567 - 91			8 16 51	−30 30.9	1	13.31		0.60		0.11		2		1602
		NGC 2571 - 15			8 16 52	−29 34.7	1	11.96		0.07		-0.07		2		457
	−30 06020	NGC 2567 - 164			8 16 52	−30 20.8	1	11.20		0.27		0.19		3		1602
		NGC 2567 - 87			8 16 52	−30 27.0	1	13.60		0.48		0.09		3		1602
		NGC 2567 - 172			8 16 52	−30 37.2	1	12.14		0.09		0.05		2		1602
		NGC 2571 - 77			8 16 53	−29 31.6	1	13.32		0.30		0.15		2		457
70058	−29 05928	NGC 2571 - 16			8 16 53	−29 35.3	2	8.95	.005	-0.09	.001	-0.65	.007	8	B8	457,1732
	−29 05929	NGC 2571 - 50			8 16 53	−29 39.5	1	11.23		0.06		-0.05		2	B6	457
		NGC 2567 - 129			8 16 53	−30 23.9	1	12.91		0.60		0.13		2		1602
		NGC 2567 - 90		AB	8 16 53	−30 28.7	1	13.17		0.40		0.09		2		1602
		S164 # 3			8 16 53	−55 35.1	1	11.17		1.13		0.78		3		1730
		NGC 2571 - 4			8 16 54	−29 38.0	1	14.13		0.48		0.15		2		457
		NGC 2567 - 128			8 16 54	−30 23.6	1	13.46		1.55		1.11		4		1602
		NGC 2567 - 92			8 16 54	−30 31.4	1	13.73		0.31		0.14		3		1602
70267	−58 01095	HR 3274		⋆ A	8 16 54	−59 00.6	2	6.41	.005	0.40	.005			7	F5 V	15,2012
		G 51 - 7			8 16 55	+34 52.0	2	12.99	.028	0.96	.005	0.58	.002	4		1620,1658
		NGC 2571 - 28			8 16 55	−29 33.9	1	13.03		1.04		0.71		2		457
		NGC 2571 - 14			8 16 55	−29 34.9	1	12.47		0.34		0.10		2		457
		NGC 2567 - 84			8 16 55	−30 24.7	1	13.05		0.20		0.10		4		1602
	−35 04465	AAS70,69 # 14			8 16 55	−35 39.6	1	10.48		1.21		1.15		1		1636
		NGC 2571 - 13			8 16 56	−29 34.5	1	12.81		0.20		0.12		3		457
70078	−29 05933	NGC 2571 - 10			8 16 56	−29 35.8	3	8.83	.008	-0.13	.010	-0.74	.020	9	B2 V	239,457,1732
		NGC 2571 - 6			8 16 56	−29 37.1	1	13.34		0.50		0.16		2		457
		NGC 2567 - 166			8 16 56	−30 19.9	1	13.53		1.07		0.79		2		1602
		NGC 2567 - 127			8 16 56	−30 22.1	1	13.23		1.09		0.95		3		1602
		NGC 2567 - 130			8 16 56	−30 24.1	1	12.57		0.36		0.15		2		1602
	−30 06031	NGC 2567 - 131			8 16 56	−30 24.6	1	11.61		0.09		0.07		3		1602
		LOD 8 # 4			8 16 56	−36 24.5	1	10.70		1.06		0.85		2		126
		NGC 2571 - 9			8 16 57	−29 35.9	2	11.88	.037	0.04	.019	-0.12	.037	4	B6	239,457
	−29 05934	NGC 2571 - 48			8 16 57	−29 39.8	1	10.36		-0.05		-0.49		4	B4	457
	−32 05124				8 16 57	−32 49.0	1	9.32		0.43		-0.01		2		124
	−33 04828				8 16 57	−33 15.8	1	10.40		-0.01		-0.39		2	B6 III	540
		NGC 2571 - 5			8 16 58	−29 37.5	1	13.77		0.49		0.17		2		457
		NGC 2571 - 47			8 16 58	−29 38.8	1	12.42		0.25		0.11		2		457
		NGC 2567 - 88			8 16 58	−30 27.2	1	14.19		0.66		0.09		2		1602
70252	−55 01501				8 16 58	−55 26.4	1	9.94		0.37		0.01		2	F0 V	1730
	−30 06032	NGC 2567 - 171			8 16 59	−30 35.5	1	11.28		0.39		0.15		2		1602
	−35 04469				8 16 59	−35 58.3	2	10.19	.001	0.29	.001	-0.67	.009	4	O9 III	540,1066
	−30 06034	NGC 2567 - 138			8 17 00	−30 31.5	1	11.26		0.08		0.07		2		1602
		NGC 2567 - 139			8 17 00	−30 33.4	1	12.14		1.73		1.57		2		1602
69897	+27 01589	HR 3262			8 17 02	+27 22.9	7	5.14	.005	0.46	.011	-0.06	.005	35	F6 V	15,254,1008,1013,1197*
	−29 05938	NGC 2571 - 12			8 17 02	−29 34.8	1	10.40		-0.02		-0.37		4	B4	457
		NGC 2571 - 11			8 17 02	−29 35.6	1	11.74		0.03		-0.14		2	B8	457
		NGC 2571 - 3			8 17 02	−29 38.0	2	12.93	.009	0.47	.017	0.26	.009	6		239,457
		NGC 2571 - 44			8 17 02	−29 39.9	1	12.74		0.57		0.07		2		457
		NGC 2571 - 99			8 17 02	−29 41.3	1	11.90		0.05		-0.08		2		457
	−30 06035	NGC 2567 - 132			8 17 02	−30 26.6	1	12.56		0.13		0.14		2		1602
	−35 04470	LSS 1040			8 17 02	−36 03.3	2	10.54	.020	0.43	.005	-0.54	.030	3	B0 IV	797,1066
		NGC 2571 - 46			8 17 03	−29 39.4	1	13.04		0.69		0.21		2		457
	−30 06036	NGC 2567 - 134			8 17 04	−30 28.4	1	11.21		0.07		0.07		2		1602
		NGC 2567 - 137			8 17 04	−30 29.9	1	13.60		0.51		0.16		3		1602
		Ru 59 - 17			8 17 05	−34 20.2	1	12.16		0.38		0.25		1		183
		NGC 2567 - 170			8 17 06	−30 33.0	1	12.80		1.10		0.84		2		1602
		NGC 2567 - 136			8 17 07	−30 29.7	1	13.24		1.00		0.74		2		1602
70158	−40 04091				8 17 07	−40 24.5	1	9.21		0.35		0.03		2	F0 V	123
	+19 01979				8 17 08	+19 36.6	1	8.82		1.65		1.98		2	M0	1648
70099	−26 05853				8 17 08	−26 47.2	1	9.02		-0.15		-0.71		3	B2/3 (V)	1732
		Ru 59 - 16			8 17 08	−34 21.0	1	12.07		0.71		0.20		1		183
70157	−38 04294				8 17 08	−38 55.3	1	7.10		-0.06		-0.25		2	B9 III	401
		NGC 2571 - 40			8 17 09	−29 37.5	1	11.40		1.11		0.87		2	K0	457
		NGC 2567 - 133			8 17 09	−30 27.1	1	13.29		0.20		0.16		3		1602
		NGC 2567 - 135			8 17 09	−30 29.7	1	13.56		1.39		1.20		2		1602
	−40 04093				8 17 09	−40 20.8	1	11.40		0.44		0.06		2		123
		NGC 2567 - 167			8 17 10	−30 24.9	1	13.20		0.16		0.18		2		1602
		Ru 59 - 21			8 17 10	−34 19.0	1	13.60		0.56		0.05		1		183
		Ru 59 - 15			8 17 10	−34 21.4	1	12.05		0.26		0.15		1		183
	−35 04471				8 17 10	−35 59.3	4	9.17	.003	0.28	.014	-0.68	.022	9	B1 V	540,797,976,1066
70195	−42 04113				8 17 10	−42 22.2	1	7.07		0.96		0.68		1	G8/K0 III	389
70156	−32 05133				8 17 11	−32 56.9	1	7.97		-0.07		-0.44		2	B8 III	401
70013	+4 01954	HR 3269		⋆ AB	8 17 12	+04 06.4	3	6.05	.005	0.97	.005	0.64	.007	11	G8 III	15,1078,1355
	−29 05941	NGC 2571 - 39			8 17 12	−29 36.7	1	9.36		1.41		1.42		2	K1.5IIb	457
	−30 06040	NGC 2567 - 169			8 17 12	−30 32.8	1	11.04		1.46		1.54		2		1602
		Ru 59 - 4			8 17 12	−34 18.	1	13.19		0.51		0.03		1		183

HD	DM	Other Id	N	Rem	α_{1950}	δ_{1950}	S	V	σ_V	B–V	σ_{B-V}	U–B	σ_{U-B}	n	Spectrum	References
		Ru 59 - 6			8 17 12	−34 18.	1	13.65		0.28		−0.32		1		183
		Ru 59 - 8			8 17 12	−34 18.	1	14.33		0.59		0.05		1		183
		Ru 59 - 9			8 17 12	−34 18.	1	14.03		0.27		−0.06		1		183
		Ru 59 - 10			8 17 12	−34 18.	1	13.61		0.40		−0.31		1		183
		Ru 59 - 11			8 17 12	−34 18.	1	13.18		0.28		−0.31		1		183
		Ru 59 - 12			8 17 12	−34 18.	1	13.21		0.54		0.07		1		183
		Ru 59 - 14			8 17 12	−34 18.	1	13.89		0.63		−0.01		1		183
		Ru 59 - 18			8 17 12	−34 18.	1	14.04		0.65		0.05		1		183
		Ru 59 - 19			8 17 12	−34 18.	1	14.53		0.34		0.25		1		183
		Ru 59 - 20			8 17 12	−34 18.	1	13.00		1.31		1.32		1		183
	−29 05943	NGC 2571 - 41			8 17 16	−29 38.4	1	11.35		0.48		0.04		3		457
		Ru 59 - 13			8 17 16	−34 19.5	1	12.92		0.27		−0.36		1		183
		NGC 2567 - 168			8 17 17	−30 30.6	1	12.82		0.32		0.23		3		1602
69962	+25 01901				8 17 18	+25 39.4	1	8.41		0.36		0.01		3	F5	833
		NGC 2571 - 91			8 17 18	−29 38.8	1	11.72		0.04		−0.13		2	B6	457
	−30 06041	NGC 2567 - 181			8 17 18	−30 25.2	1	11.18		0.16		0.17		3		1602
	−34 04620	Ru 59 - 1			8 17 18	−34 22.0	1	10.17		0.23		0.12		2		183
70192	−35 04474				8 17 20	−35 45.0	1	9.22		−0.03		−0.16		2	B9 V	401
	−42 04120	LSS 1043			8 17 21	−42 42.6	2	9.91	.012	0.47	.012	−0.58		6	O5	540,2021
		G 50 - 28			8 17 23	+06 25.7	1	10.73		1.05		1.00		1	K4	333,1620
70190	−32 05138				8 17 23	−32 52.0	1	8.73		−0.05		−0.28		2	B9 III/IV	401
		L 34 - 7			8 17 24	−75 18.	1	11.94		0.57		0.09		2		1696
	+3 01957				8 17 26	+03 14.9	1	9.29		1.50		1.79		2	K5	1746
69994	+21 01817	HR 3264			8 17 26	+20 54.4	3	5.82	.021	1.14	.009	1.06		14	K1 III	15,1351,3016
		Ru 59 - 7			8 17 26	−34 18.6	1	14.89		0.32		0.15		1		183
		Ru 59 - 5			8 17 26	−34 19.9	1	12.99		0.92		0.78		1		183
70236	−40 04096				8 17 26	−40 17.8	1	9.93		0.08		0.06		2	A1 V	123
70385	−59 01018				8 17 26	−60 09.2	1	7.80		1.21		1.00		4	G5 Ib/II	1704
69865	+47 01565	IDS08140N4743		D	8 17 27	+47 34.7	1	8.22		1.24		1.06		2	G5	3024
70138	−17 02442				8 17 27	−18 06.4	2	9.42	.113	1.67	.052			2	C5,5i	864,1238
		NGC 2571 - 116			8 17 27	−29 38.7	1	12.64		1.25		0.94		2		457
	−34 04625	Ru 59 - 3			8 17 28	−34 18.5	1	10.80		0.45		0.03		2		183
70191	−32 05139				8 17 29	−33 12.5	1	8.18		0.09		0.06		2	A2 IV	126
	−34 04624	Ru 59 - 2			8 17 30	−34 18.5	1	11.44		0.25		0.25		2		183
70072	+3 01958	RY Hya			8 17 31	+02 55.6	1	8.83		4.08				2	C5,4	864
70264	−45 03997	IDS08159S4553		AB	8 17 33	−46 02.8	1	7.16		1.36		1.53		1	K2 III	389
70309	−47 03799	IDS08160S4753		A	8 17 33	−48 02.4	3	6.44	.005	−0.15	.005	−0.64	.005	7	B3 III	401,1469,2012
70011	+24 01909	HR 3268			8 17 34	+24 10.9	2	5.93		−0.05	.015	−0.13	.005	5	B9.5V	1022,1733
69894	+47 01567	IDS08140N4743		AB	8 17 34	+47 34.4	1	8.55		0.57		0.02		2	F8	3024
69894	+47 01567	IDS08140N4743		CR	8 17 34	+47 34.4	1	9.84		0.82		0.36		2		3024
70235	−34 04627	HR 3273			8 17 34	−34 25.9	3	6.43	.012	−0.08	.005	−0.38		9	B9p HgMn	15,401,2012
69961	+36 01794				8 17 35	+35 56.3	1	7.16		0.15		0.12		2	A0	1601
	−35 04476	LSS 1044			8 17 35	−35 27.6	1	10.48		0.82		−0.06		2		1066
70234	−31 05883				8 17 38	−31 39.9	1	8.82		0.08		−0.29		2	B9 III(e)	401
70213	−29 05957				8 17 39	−30 01.5	2	8.62	.006	−0.14	.005	−0.77	.012	4	B2 III	401,1732
70110	−0 01966	HR 3271			8 17 40	−00 45.0	3	6.18	.005	0.60	.000	0.15	.005	8	F9 V	15,1078,3077
		S124 # 4			8 17 40	−59 56.3	1	11.24		0.99		0.74		2		1730
70030	+25 01903				8 17 41	+25 29.7	1	7.26		1.41		1.64		3	K3 III	833
		WLS 820 25 # 10			8 17 44	+25 54.8	1	11.42		0.60		0.00		2		1375
70291	−40 04103	IDS08160S4008		AB	8 17 44	−40 17.5	1	8.24		0.34		0.03		2	F0 V	123
		LSS 1046			8 17 45	−33 22.2	1	11.23		0.34		−0.60		3	B3 Ib	1737
	+44 01737				8 17 46	+44 34.8	1	9.64		1.26		1.33		2	K2	1776
70514	−65 00907	HR 3280			8 17 46	−65 27.3	4	5.06	.006	1.15	.005	1.19	.000	20	K1 III	15,1075,2006,2038
70148	−4 02303	HR 3272			8 17 48	−05 10.2	4	6.12	.007	1.33	.006	1.52	.000	18	K2 III	15,1078,2013,2029
70288	−34 04632				8 17 50	−35 09.5	1	8.13		0.02		−0.06		2	A0 III/IV	401
		LOD 8 # 5			8 17 53	−36 26.5	1	12.03		0.31		0.11		2		126
	−36 04475				8 17 54	−36 22.5	1	9.60		0.04		0.03		2		124
		LOD 8 # 6			8 17 54	−36 27.1	1	12.33		0.17		−0.12		2		126
	−36 04477				8 17 54	−36 38.6	1	10.02		1.03		0.81		2		126
	−39 04226				8 17 55	−40 06.5	1	10.01		0.39		0.15		2		123
70467	−59 01020				8 17 58	−60 00.4	1	9.41		0.63		0.10		2	G5	1730
70133	+17 01823				8 17 59	+17 06.6	1	8.82		1.05				2	K0	882
		CCS 1130			8 18 00	+05 22.	1	9.86		2.19				1	R8	1238
70468	−59 01021	IDS08170S5953		A	8 18 00	−60 02.2	1	8.76		0.80		0.48		2	G0	1730
70306	−36 04479				8 18 01	−37 02.7	1	8.47		0.88		−0.26		2	B9 III	401
70368	−46 04045				8 18 01	−46 16.9	3	8.54	.005	0.43	.005	0.02	.005	9	F3 V	285,657,2012
70367	−42 04131				8 18 04	−42 33.8	1	7.85		1.21		1.29		1	K1/2 III	389
70327	−36 04481				8 18 05	−36 27.2	2	7.57	.025	0.04	.040	0.00	.000	4	A0 III	124,126
	−36 04480				8 18 06	−36 39.2	1	11.06		0.11		0.04		2		126
	+14 01876	G 50 - 30			8 18 08	+14 14.0	2	9.81	.035	1.28	.005	1.21		2	M0	1620,6009
		WLS 820-5 # 10			8 18 08	−05 30.3	1	11.14		0.33		0.03		2		1375
	−32 05156				8 18 08	−33 08.1	1	10.29		0.15		0.07		2		126
	−32 05157				8 18 09	−33 10.4	1	11.40		0.13		0.00		2		126
		vdB 12 # b			8 18 09	−49 57.7	2	11.78	.035	0.75	.000	0.20	.025	4		434,599
	−33 04863				8 18 10	−33 18.4	1	9.91		0.07		0.03		2		126
		HA 76 # 280			8 18 13	+14 55.0	1	12.67		0.92		0.71		11		1499
		BSD 4 # 642			8 18 14	+75 02.4	1	12.18		0.54		0.10		1	G0	1058
70414	−49 03495				8 18 15	−49 22.7	1	7.14		1.19		1.21		1	K1 III	389
70302	−22 02233	HR 3276		⋆ AB	8 18 17	−22 45.9	3	6.11	.012	1.04	.000			11	G8/K0 III	15,1075,2006
		HA 76 # 281			8 18 18	+14 53.7	1	12.24		0.52		0.04		11		1499

Table 1   463

HD	DM	Other Id	N	Rem	α_{1950}	δ_{1950}	S	V	σ_V	B–V	σ_{B-V}	U–B	σ_{U-B}	n	Spectrum	References
		Pup sq 1 # 70			8 18 18	−37 15.	1	12.10		0.25		0.09		2		124
	−32 05161				8 18 21	−33 07.0	1	10.17		−0.05		−0.39		2		126
	−36 04487				8 18 21	−37 09.0	1	10.17		0.26		0.14		2		124
		LOD 8 # 8			8 18 26	−36 28.2	1	11.67		0.26		0.08		2		126
	−36 04491				8 18 28	−36 29.2	1	9.80		1.55		1.84		2		126
70178	+29 01739				8 18 30	+28 58.6	1	8.31		0.85		0.51		2	G5 IV	3016
70465	−47 03825				8 18 30	−48 13.7	1	8.83		−0.04		−0.42		2	B9 III/IV	401
	+35 01811	G 51 - 8			8 18 31	+35 40.1	2	10.69	.007	0.98	.000	0.74	.002	6		1723,7010
69976	+61 01043	HR 3263			8 18 33	+60 47.5	2	6.40	.010	0.98	.005	0.75	.010	4	K0 III	1502,1733
		OJ-131 # 1			8 18 36	−12 49.	1	15.43		0.59		0.31		1		1482
		OJ-131 # 3			8 18 36	−12 49.	1	12.12		0.68		0.30		1		1482
		OJ-131 # 4			8 18 36	−12 49.	1	14.31		0.38		−0.05		1		1482
		OJ-131 # 5			8 18 36	−12 49.	1	16.68		0.48				1		1482
		OJ-131 # 6			8 18 36	−12 49.	1	14.19		0.65		0.13		1		1482
		OJ-131 # 8			8 18 36	−12 49.	1	15.61		0.53		0.25		1		1482
		OJ-131 # 9			8 18 36	−12 49.	1	16.89		1.05				1		1482
	−34 04650				8 18 36	−34 20.9	1	9.05		0.00		−0.71		2		401
70509	−50 03271				8 18 36	−50 43.6	1	8.79		0.55		0.02		1	G0 V	389
	+75 00335				8 18 40	+74 58.7	1	10.00		0.44		0.02		1	F8	1058
		GD 264			8 18 41	+33 44.2	1	14.50		0.35		−0.02		1		3060
70277	+13 01899				8 18 44	+12 58.4	1	8.09		1.30				2	K2	882
70622	−59 01023				8 18 46	−59 56.7	3	7.53	.005	−0.05	.008	−0.23		9	B9.5IV	614,1075,1075
70340	−1 02017	HR 3278		⋆ AB	8 18 48	−01 26.5	2	6.49	.005	0.02	.000	−0.02	.000	7	A2p EuSrCr	15,1256
70621	−59 01024				8 18 50	−59 44.2	1	9.44		0.49		−0.28		2	K3 III	3073
70605	−56 01595				8 18 52	−56 41.9	1	8.02		−0.01		−0.25		2	B9 IV	401
70275	+23 01939				8 18 54	+23 06.8	1	8.23		0.27		0.03		2	A2	1625
	−44 04311				8 18 57	−44 59.4	2	9.75	.015	0.99	.000	0.65		3		1594,6006
	−35 04501		A		8 18 59	−35 42.5	1	10.02		1.40		1.70		1		92
	−35 04501		AB		8 18 59	−35 42.5	1	10.00		1.31		0.91		4		92
		vdB 13 # a			8 18 59	−36 03.3	1	10.18		1.39		1.46		3	K1 III	434
70319	+14 01878				8 19 01	+14 28.7	1	7.84		1.11				2	K0	882
		vdB 13 # b			8 19 01	−36 03.4	1	13.11		0.81		−0.22		1		434
	−36 04505				8 19 03	−36 35.2	1	10.39		1.00		0.78		2		126
70462	−30 06094				8 19 05	−30 25.4	1	8.83		1.14				4	K1 III	2033
70223	+45 01568				8 19 06	+45 30.8	1	8.00		1.09		1.01		2	K0	1601
70461	−25 05940	LSS 1051			8 19 06	−26 10.4	1	8.97		−0.13		−0.62		3	B5p	1732
70442	−19 02369	HR 3279			8 19 07	−19 55.1	6	5.58	.008	0.77	.008	0.50	.000	18	G2 III + A3	15,1007,1013,1075*
70584	−49 03502				8 19 07	−50 09.3	2	9.12	.045	0.29	.000	−0.14	.025	4	B8/9 V	434,599
70337	+15 01805				8 19 09	+14 55.7	3	7.26	.012	1.27	.023	1.28	.022	8	K0	882,1648,3040
		LOD 8 # 16			8 19 10	−36 44.9	1	12.44		0.38		0.05		2		126
70530	−34 04668				8 19 12	−34 44.1	1	7.90		−0.03		−0.09		2	B9 V	401
	−36 04508				8 19 12	−36 44.3	1	9.81		0.14		−0.22		2		126
70222	+51 01413				8 19 17	+51 06.9	1	9.17		0.56		0.08		2	G5	1566
	+77 00328	AY Cam			8 19 17	+77 22.8	2	9.69	.009	0.36	.009	0.13	.016	2	A5	588,1768
	−36 04509				8 19 18	−36 39.0	1	9.59		1.52		1.70		2	K5	126
		LOD 8 # 14			8 19 18	−36 40.6	1	11.71		0.16		0.02		2		126
70583	−43 04195	LSS 1053			8 19 18	−43 26.8	3	8.23	.008	1.02	.010	0.27	.043	10	A0 Ia/ab	362,540,976
70854	−68 00771				8 19 20	−68 50.2	1	8.65		0.61		0.05		4	G2 V	158
311757	−74 00498				8 19 21	−74 34.8	1	9.28		0.84				7	G5	955
70557	−36 04511				8 19 24	−36 40.9	2	9.12	.025	−0.05	.045	−0.38	.025	4	B8 IV	124,126
70272	+43 01815	HR 3275			8 19 25	+43 21.0	6	4.25	.008	1.55	.006	1.89	.026	16	K5 III	15,1080,1355,1363*
70555	−32 05185	HR 3282			8 19 25	−32 53.7	5	4.83	.009	1.45	.005	1.59	.032	22	K3 II	15,126,1075,2012,8015
70614	−42 04150				8 19 27	−42 15.1	1	9.27		0.47		−0.26		4	B5/7	362
70556	−36 04513	HR 3283		⋆ AB	8 19 28	−36 19.5	4	5.19	.011	−0.20	.003	−0.81	.014	11	B2 IV/V	15,401,1732,2012
		WLS 820 25 # 7			8 19 29	+23 11.7	1	10.68		1.04		0.84		2		1375
	−2 02544				8 19 29	−02 55.9	1	10.18		0.41		0.02		2		1375
	+75 00336				8 19 30	+75 00.1	1	10.46		0.39		−0.01		1	F5	1058
70662	−46 04077				8 19 33	−47 09.8	1	7.78		1.30		1.36		1	K1 III	389
		Pup sq 1 # 76			8 19 36	−36 37.	1	12.73		0.65		−0.12		2		124
70612	−39 04245	HR 3286			8 19 36	−39 27.6	4	6.15	.007	0.17	.005			12	A4 V	15,1489,2013,2029
70421	+15 01808	Z Cnc			8 19 37	+15 09.2	1	8.28		1.57		1.10		1	M6 III	3001
70523	−17 02464	HR 3281			8 19 38	−17 25.5	5	5.72	.020	1.05	.009	0.97	.024	16	K0 III	15,1003,1075,1311,2031
70642	−39 04247				8 19 41	−39 32.9	2	7.17	.000	0.71	.000	0.28		5	K1 (III)+F	258,2012
70433	+16 01704				8 19 42	+16 19.4	1	6.66		1.18		1.15		2	K0	1648
	−32 05194				8 19 49	−32 55.5	1	11.33		0.08		−0.05		2		126
71243	−76 00507	HR 3318			8 19 51	−76 45.7	4	4.06	.005	0.40	.004	−0.04	.009	12	F5 V	15,1075,2038,3026
70639	−31 05949				8 19 52	−31 53.0	1	8.59		−0.15		−0.77		2	B2 III/IV	401
	−32 05197				8 19 53	−32 58.1	1	10.62		1.17		1.00		2		126
	+34 01812				8 19 59	+34 31.1	1	10.10		−0.01		−0.05		1		8070
		vdB 15 # d			8 19 59	−41 56.1	1	13.57		0.56		−0.20		2		434
70717	−44 04330				8 19 59	−44 50.6	1	7.92		1.06		0.89		1	K1 III	389
70728	−45 04045				8 19 59	−46 10.5	1	8.15		1.61		1.95		1	M0 III	389
71046	−71 00677	HR 3301		⋆ A	8 19 59	−71 21.3	3	5.36	.005	−0.06	.000	−0.30	.005	8	B9 III/IV	15,1075,2016
70455	+27 01601				8 20 01	+27 15.3	1	8.16		1.40		1.60		2	K2	1625
70313	+53 01246	HR 3277			8 20 02	+53 23.0	2	5.52	.015	0.11	.005	0.09	.010	5	A3 V	985,1733
70574	−5 02512	HR 3285			8 20 02	−06 01.1	3	6.15	.008	0.22	.008	0.12	.010	14	A8 IV	15,1417,2012
		WLS 820-5 # 7			8 20 03	−07 05.4	1	11.86		0.38		−0.03		2		1375
	−32 05202				8 20 05	−33 13.4	1	9.65		0.08		0.06		2		126
	−36 04528				8 20 05	−36 38.4	1	10.97		0.13		0.07		2		124
70839	−57 01490	HR 3293			8 20 05	−57 48.8	2	5.96	.005	−0.08	.003	−0.82		7	B2 III	26,2035

HD	DM	Other Id	N	Rem	α_{1950}	δ_{1950}	S	V	σ_V	B−V	σ_{B-V}	U−B	σ_{U-B}	n	Spectrum	References
70696	−32 05204				8 20 06	−33 11.7	1	8.41		−0.09		−0.48		2	B8 V	126
	−41 04036				8 20 09	−42 07.0	1	10.11		0.27		−0.44		2	B2.5V	540
71066	−71 00678	HR 3302		⋆B	8 20 10	−71 20.8	2	5.64	.005	−0.10	.000	−0.30	.005	7	A0 IV	15,1075
	−22 06058				8 20 12	−22 17.9	1	10.75		0.58		0.11		1		796
	+34 01814				8 20 14	+34 00.7	1	10.55		0.44		0.06		1	G5	8070
		G 253 - 13			8 20 15	+84 34.2	1	13.40		1.62		1.26		1		906
	−40 04149				8 20 15	−40 19.1	1	9.63		0.34		0.33		3		434
		BSD 4 # 652			8 20 19	+75 04.6	1	12.90		0.51		0.14		1	G2	1058
70631	−6 02571				8 20 22	−06 32.0	1	7.63		0.02		0.01		24	A0	978
70743	−38 04361				8 20 22	−38 21.5	1	8.65		−0.10		−0.69		2	B3 III	401
		LP 35 - 347			8 20 24	+69 11.9	1	15.67		2.09				3		1663
70673	−12 02490	HR 3289			8 20 26	−12 53.6	3	6.09	.065	0.99	.009	0.89		9	K0 III	15,252,2029
	−15 02405	AC Pup			8 20 26	−15 45.3	1	9.15		3.47				2	C5,4	864
70982	−63 00940	HR 3298			8 20 26	−63 56.7	3	6.11	.008	0.93	.000			11	G6/8 III	15,1075,2006
70652	−7 02452	HR 3288			8 20 27	−07 22.9	5	5.95	.009	1.67	.005	1.85	.000	20	M1 III	15,1256,2013,2029,3005
70780	−43 04214				8 20 27	−43 30.1	1	8.03		1.57		1.94		1	K5 III	389
		G 9 - 1			8 20 29	+19 10.1	1	11.25		1.15		1.07		3		1696
70569	+18 01930	HR 3284			8 20 30	+18 29.7	3	5.95	.008	0.18	.005	0.14	.008	8	A5 V	985,1501,1733
70798	−40 04157				8 20 31	−40 16.6	1	10.36		1.24		1.10		4	A2 IV	434
		POSS 210 # 43			8 20 33	+43 28.4	1	17.06		1.38				1		1739
	+22 01921	G 9 - 2			8 20 34	+22 00.9	1	9.54		1.18		1.12		2	M0	3072
	+25 01910	G 40 - 20		A	8 20 34	+25 19.0	1	10.04		0.96		0.74		1	K5	333,1620
		POSS 210 # 53			8 20 37	+44 25.1	1	17.68		1.29				2		1739
70851	−47 03860				8 20 39	−47 37.3	1	7.58		−0.06		−0.18		2	B9 IV/V	401
70491	+45 01575	IDS08173N4517		B	8 20 41	+44 54.3	1	9.41		1.00		0.77		4	K2	1084
		Ton 914			8 20 43	+20 22.	2	15.03	.040	0.04	.080	−0.78	.005	7		286,313
70761	−25 05988	HR 3291		⋆AB	8 20 43	−26 11.2	4	5.90	.008	0.38	.005	0.27		13	F2 Iab	15,2012,6009,8100
70775	−31 05975				8 20 43	−31 32.3	1	8.83		−0.14		−0.76		2	B3 II/III	401
		vdB 15 # c			8 20 43	−41 56.9	1	12.23		0.46		−0.37		3		434
70566	+32 01725				8 20 44	+32 27.3	1	7.73		0.22		0.13		2	A3	1625
70796	−31 05976				8 20 45	−31 26.8	1	7.23		−0.11		−0.72		2	B2 II	401
70816	−37 04619				8 20 45	−37 38.6	1	8.99		0.01		−0.23		2	B9 V	401
70516	+45 01576	IDS08173N4517		A	8 20 48	+45 06.9	1	7.70		0.65		0.18		4	G2	1084
		POSS 210 # 55			8 20 50	+40 40.6	1	17.77		1.37				2		1739
	−42 04172				8 20 52	−42 16.9	1	12.70		0.44		−0.19		3		362
	−32 05221				8 20 53	−32 59.4	1	10.52		0.28		0.04		2		126
	−42 04174				8 20 56	−42 17.3	1	10.86		−0.02		−0.06		2		123
70837	−31 05979				8 20 58	−31 16.7	1	8.39		−0.03		−0.32		2	B9/A0 V	401
70813	−24 06871				8 20 59	−24 53.9	1	9.79		0.84		0.33		2	B9 V	540
70930	−48 03734	HR 3294, LSS 1058		⋆AB	8 20 59	−48 19.7	4	4.82	.004	−0.15	.005	−0.84	.000	20	B2 III	15,1075,2012,8015
		Ton 913			8 21 01	+20 48.	1	13.94		0.29		0.16		3		286
	−44 04349	IDS08194S4432		AB	8 21 01	−44 41.5	1	9.18		0.44		−0.02		1		389
	+2 01954	G 113 - 33			8 21 02	+02 20.6	1	9.31		0.69		0.27		1	G0	3062
70352	+66 00550	G 234 - 33			8 21 02	+66 37.9	2	8.89	.012	0.90	.012	0.66	.012	7	K2 V	1003,1775
	−41 04059				8 21 03	−42 01.7	3	9.87	.009	0.31	.011	−0.56	.049	10		123,362,2021
	−33 04934				8 21 04	−33 41.9	1	8.67		1.30		1.45		1		401
70868	−33 04932				8 21 05	−33 40.8	1	8.79		−0.06		−0.39		2	B9 III	401
	−42 04176				8 21 05	−42 18.9	1	10.81		0.20		0.15		2	B1 V	123
70951	−48 03736				8 21 05	−48 53.4	1	7.76		−0.06		−0.25		2	B8 V	401
		vdB 15 # b			8 21 06	−41 55.4	1	13.00		0.58		−0.25		3		434
	−41 04061				8 21 07	−42 02.6	1	10.29		0.24		0.11		2	F0	123
70867	−32 05226				8 21 08	−32 58.6	1	9.53		−0.07		−0.57		2	B6 II	126
70734	+11 01830	HR 3290		⋆AB	8 21 11	+10 47.7	2	6.11	.037	1.50	.014	1.54		3	M2 III	71,3001
	−47 03877				8 21 11	−47 59.4	1	9.14		0.00		−0.51		1		389
70688	+29 01747				8 21 14	+28 54.8	1	9.02		0.48		0.04		2	F6 V	3026
		POSS 210 # 49			8 21 17	+41 43.4	1	17.49		1.46				2		1739
	−32 05230				8 21 18	−32 55.6	1	9.73		0.60		0.13		2		126
70647	+42 01859	HR 3287		⋆A	8 21 20	+42 10.1	3	6.03	.009	1.59	.001	1.96	.015	8	K5 III	71,1501,3001
70706	+27 01606				8 21 23	+27 07.7	1	9.27		0.32		0.05		2	A2	1375
		vdB 15 # a			8 21 25	−41 56.2	1	11.83		0.78		−0.30		3	K0 III:	434
	+30 01705				8 21 26	+30 16.4	1	9.92		0.72		0.24		8		272
70946	−37 04638	HR 3296			8 21 26	−38 07.4	4	6.33	.017	1.63	.008	1.86		14	M2 III	15,1075,2006,3005
71129	−59 01032	HR 3307			8 21 29	−59 20.9	5	1.86	.011	1.27	.014	0.19	.028	20	K3 III	15,1034,1075,2012,8029
70845	−6 02579				8 21 31	−07 06.1	1	8.23		−0.05		−0.14		28	A0	978
71043	−51 02980	HR 3300			8 21 31	−51 57.7	1	5.90		0.01				4	A0 V	2009
71242	−66 00840				8 21 32	−67 10.4	1	7.52		0.96				4	K0 III	2012
	−44 04355				8 21 35	−44 47.6	1	9.22		0.16		0.19		1		389
		POSS 210 # 45			8 21 38	+42 17.5	1	17.16		1.51				1		1739
		WLS 820-5 # 9			8 21 41	−04 43.6	1	12.63		0.44		0.02		2		1375
	−32 05237				8 21 42	−32 59.7	1	9.95		0.02		−0.05		2		126
71018	−42 04186	IDS08200S4206		AB	8 21 42	−42 15.4	1	8.86		0.89		1.12		2	M3/4 III	123
70963	−28 05961	IDS08197S2839		A	8 21 46	−28 48.5	1	6.66		0.01		−0.01		5	A0 V	1628
70999	−38 04389				8 21 47	−38 17.3	1	8.06		−0.12		−0.61		2	B3 III	401
	−41 04076				8 21 47	−41 59.9	1	10.19		1.67		1.74		2		123
	−38 04394				8 21 51	−38 55.7	1	10.34		0.63		0.11		2		123
70771	+35 01819	HR 3292			8 21 53	+35 10.5	1	6.06		1.27				2	K0	71
70590	+60 01132				8 21 53	+59 52.2	1	8.82		0.20		0.11		2	F0	1502
70843	+17 01836				8 21 55	+17 20.8	1	7.09		0.50		0.06		2	F5	1648
70826	+20 02079	IDS08190N2030		A	8 21 55	+20 19.0	1	7.38		0.63		0.38		3	A5	1648
71127	−51 02985				8 21 57	−51 46.6	1	8.67		1.26				4	G8/K0 III	2012

Table 1   465

HD	DM	Other Id	N Rem	α_{1950}	δ_{1950}	S	V	σ_V	B–V	σ_{B-V}	U–B	σ_{U-B}	n	Spectrum	References
70923	−0 01987	IDS08195S0049	AB	8 22 00	−00 59.0	2	7.05	.000	0.60	.005	0.10		6	G0	2012,3026
71015	−32 05245			8 22 00	−32 44.1	2	7.14	.003	−0.15	.009	−0.72	.012	4	B2 IV	126,1732
	+33 01694	G 51 - 10		8 22 03	+32 46.9	3	10.10	.021	1.02	.019	0.85	.022	5	K5	333,1003,1620,3072
71077	−41 04079			8 22 03	−41 52.3	1	7.43		−0.06		−0.18		2	B8/9 V	123
70958	−3 02333	HR 3297		8 22 05	−03 35.3	7	5.60	.013	0.46	.008	−0.07	.010	23	F3 V	15,254,1013,1197,1415*
		Ton 318		8 22 06	+32 01.	1	14.75		0.20		0.08		1		1036
70937	−4 02328	HR 3295		8 22 07	−04 33.2	4	6.01	.008	0.46	.009	0.02	.019	17	F2 V	15,254,1415,2012
71012	−21 02417			8 22 07	−22 12.9	1	8.27		0.62				4	F6 V	2012
71057	−36 04578			8 22 07	−36 33.2	1	7.46		−0.14		−0.70		2	B5 III	401
	−42 04196			8 22 08	−42 22.5	1	11.40		0.57		0.17		2		123
		BB Pup		8 22 10	−19 23.7	2	11.60	.010	0.26	.040	0.20	.020	2	F0	668,700
71123	−42 04198			8 22 10	−42 23.5	1	8.29		−0.04		−0.28		2	B9 III	123
71701	−77 00383	HR 3340	⋆ A	8 22 11	−77 19.4	4	4.33	.007	1.16	.009	1.20	.026	23	K1 III	15,1075,1279,2038
71701	−77 00383	IDS08236S7710	B	8 22 11	−77 19.4	1	12.44		0.26		−0.18		3		1279
		POSS 210 # 38		8 22 14	+43 13.8	1	16.98		1.27				1		1739
70858	+30 01706			8 22 15	+29 48.9	1	8.18		1.56		1.92		4	K0	1371
	−38 04407			8 22 15	−38 56.4	1	11.04		0.48		0.01		2		123
70897	+24 01921	IDS08194N2440	A	8 22 20	+24 30.8	1	8.56		1.07		0.94		2	K0	1625
71181	−44 04376			8 22 28	−45 04.4	1	7.61		1.17		0.94		1	G6 Ib/II	389
71238	−49 03528			8 22 32	−49 16.0	1	8.93		−0.01		−0.43		2	B9 II/III	401
71576	−72 00694	HR 3334	⋆ A	8 22 33	−73 14.3	4	5.28	.003	0.00	.005	−0.01	.000	18	A0/1 IV/V	15,1075,2038,3023
		LOD 9 # 17		8 22 35	−38 54.2	1	12.59		0.73		0.46		2		123
		MCC 522		8 22 36	+04 26.0	1	10.39		1.03				1	K8	1017
	−41 04092			8 22 40	−41 59.5	1	9.63		0.62		0.14		2	G0	123
		AAS27,343 # 6		8 22 42	−39 43.	1	12.70		0.89		−0.31		4		362
71217	−41 04093			8 22 42	−42 03.9	1	9.22		0.18		0.15		2	A7 V	123
71236	−45 04089			8 22 44	−45 43.9	1	8.34		1.17		1.10		1	G8 III	389
70990	+15 01817			8 22 45	+15 18.0	1	8.47		1.36				2	K2	882
71141	−22 02262	HR 3308		8 22 45	−22 59.4	2	5.67	.005	0.06	.000			7	A1 III/IV	15,2012
71071	−6 02585	LU Hya		8 22 48	−07 00.4	1	7.37		0.96		0.68		1	G5	1641
		LOD 9 # 16		8 22 48	−38 53.5	1	12.02		0.83		0.63		2		123
71216	−40 04212			8 22 48	−40 35.0	2	7.08	.010	−0.16	.010	−0.60	.000	8	B5 Vnn	164,362
		POSS 210 # 40		8 22 51	+39 06.5	1	17.02		1.60				2		1739
	−24 06912			8 22 51	−25 12.1	1	11.29		2.52				1		864
71176	−23 07277	HR 3315	⋆ A	8 22 54	−23 53.0	2	5.28	.005	1.48	.005	1.83		6	K5 III	2006,3005
71793		R Cha		8 22 57	−76 11.6	1	10.55		1.39		0.50		1	MD	975
		LOD 9 # 15		8 22 59	−38 54.4	1	11.82		0.37		0.21		2		123
71095	+2 01965	HR 3305		8 23 00	+02 16.0	2	5.72	.005	1.53	.000	1.86	.000	7	K5 III	15,1078
71030	+17 01842	HR 3299		8 23 00	+17 12.7	2	6.13	.030	0.43	.025	−0.04	.015	6	F6 V	254,3053
		LOD 9 # 14		8 23 03	−38 52.7	1	11.40		0.22		−0.18		2		123
71053	+18 01939			8 23 04	+17 59.9	1	8.31		0.60		0.12		1	F9 V	3026
		POSS 210 # 21		8 23 04	+40 11.7	1	16.42		1.65				2		1739
71196	−20 02522	HR 3316		8 23 06	−20 53.0	1	6.01		0.40				4	F3 V	2031
	+52 01308	IDS08194N5230	A	8 23 07	+52 20.7	1	9.73		1.06		0.88		1	K4	3072
	+52 01308	IDS08194N5230	B	8 23 07	+52 20.7	1	10.14		1.15		1.05		1		3072
		G 113 - 34		8 23 08	−02 15.1	1	10.83		0.85		0.56		1		3062
71286	−41 04098			8 23 08	−41 52.4	1	8.78		0.17		0.12		2	A5 V	123
71317	−49 03534			8 23 09	−50 09.0	1	8.67		0.07		−0.07		2	B8/9 V	401
71155	−3 02339	HR 3314		8 23 10	−03 44.5	16	3.89	.006	−0.01	.008	−0.02	.012	70	A0 V	1,15,30,125,1006,1020*
70919	+54 01225			8 23 12	+53 47.7	1	7.90		0.24		0.06		2	F0	1566
		G 113 - 35		8 23 12	−06 43.6	1	12.18		0.95		0.79		1	K5	3062
	−38 04432			8 23 12	−38 51.9	1	10.98		0.20		0.04		2		123
	−38 04431			8 23 12	−38 52.9	1	11.47		1.15		1.03		2		123
71302	−42 04219	HR 3322	⋆ AB	8 23 13	−42 36.4	6	5.97	.017	−0.16	.015	−0.69	.019	30	B3 V	15,123,362,540,976,2012
71115	+8 02053	HR 3306	⋆ A	8 23 14	+07 43.7	7	5.13	.019	0.94	.009	0.64	.018	18	G8 II	15,1078,1080,1363*
71115	+8 02053	IDS08205N0753	B	8 23 14	+07 43.7	1	11.16		0.60		0.06		3		3032
71304	−43 04259	LSS 1061	⋆ A	8 23 15	−44 08.2	4	8.26	.023	0.53	.005	−0.48	.031	15	O9 (III)	362,540,976,1737,8100
		Wolf 306		8 23 20	+44 07.0	1	11.61		1.14		1.08		2		1723
71613	−69 00902			8 23 22	−70 05.1	1	7.04		1.46		1.54		2	K2 II/III	1730
71231	−16 02442	HR 3317		8 23 23	−17 16.5	1	6.43		1.24				1	K1 III	2008
71336	−42 04221			8 23 24	−43 12.1	2	7.97	.005	−0.11	.000	−0.62	.010	12	B3 III/IV	164,362
71093	+28 01602	HR 3304	⋆ A	8 23 25	+28 03.6	2	5.57	.002	1.42	.020	1.68		4	K5 III	71,3077
		POSS 210 # 28		8 23 27	+41 51.1	1	16.63		1.11				1		1739
	−42 04222			8 23 27	−42 53.1	1	10.89		0.42		−0.48		4		362
		He3 160		8 23 27	−51 18.8	1	15.30		0.71		−0.61		1		1753
71027	+40 02050			8 23 28	+40 03.4	1	7.07		1.16		1.09		2	K0	1601
		POSS 210 # 35		8 23 31	+39 25.9	1	16.91		1.48				2		1739
71386	−48 03774	IDS08220S4850	AB	8 23 32	−48 59.7	2	7.66	.014	0.78	.005	0.33	.005	5	G6/K2	214,389
		Ton 918		8 23 36	+22 49.	1	14.30		0.06		0.06		3		286
71267	−14 02517	HR 3320		8 23 36	−14 46.0	3	5.97	.007	0.17	.005	0.13		10	A3(m)A3-A7	15,2012,8071
71348	−38 04439			8 23 37	−38 51.5	2	9.36	.011	0.21	.025	0.06	.055	4	A3 V	123,1730
	−44 04392			8 23 38	−44 59.9	3	9.22	.021	0.19	.018	−0.61	.019	6	B2 IVe	389,540,976
	−38 04441			8 23 39	−38 53.5	2	10.95	.001	0.55	.011	0.03	.022	4		123,1730
71362	−39 04315			8 23 40	−39 57.9	1	9.23		0.00		−0.48		2	B3 (III)	401
71384	−42 04226			8 23 40	−42 30.0	1	9.33		0.03		−0.07		2	A0 V	123
71152	+25 01920	HR 3312	⋆ A	8 23 42	+24 42.0	1	7.04		0.30		0.06		3	F0 III	3032
71153	+25 01920	HR 3313	⋆ ABC	8 23 42	+24 42.1	1	7.79		0.53		−0.03		3	F7 V	3030
		LOD 9 # 11		8 23 42	−38 55.9	1	12.36		0.57		0.04		2		123
71150	+27 01612	HR 3310	⋆ AB	8 23 46	+27 06.0	1			0.18		0.09		3	A6 V	1022
71334	−29 06145			8 23 46	−29 45.7	4	7.81	.010	0.67	.005	0.16	.013	21	G5 V	158,1775,2012,3008

Catalogue of mean UBV data

HD	DM	Other Id	N Rem	α_{1950}	δ_{1950}	S	V	σ_V	B–V	σ_{B-V}	U–B	σ_{U-B}	n	Spectrum	References
	−38 04444			8 23 46	−38 47.3	1	11.13		0.39		0.11		2		123
		LOD 9 # 10		8 23 46	−38 56.1	1	12.97		0.66		0.15		2		123
		LP 242 - 13		8 23 48	−50 48.0	2	13.07	.025	0.62	.017	−0.12	.013	4		1696,3065
	−29 06146			8 23 49	−29 53.8	1	8.70		1.35		1.61		21	K2	1655
	−38 04447	LSS 1062		8 23 49	−38 36.7	1	9.95	.016	0.43	.005	−0.59	.039	6		362,540
		G 113 - 36		8 23 50	−05 48.4	1	15.45		1.14		0.64		1		3062
71280	−7 02482			8 23 52	−07 49.5	1	8.30		0.26				4	A5	2012
	−38 04450			8 23 54	−38 54.0	2	10.09	.003	0.22	.027	0.09	.082	4		123,1730
	−38 04451			8 23 55	−38 54.5	1	11.78		0.66		0.22		2		123
71297	−3 02345	HR 3321, LM Hya	⋆ A	8 23 57	−03 49.3	4	5.59	.011	0.22	.004	0.07	.007	9	A5 III-IV	15,39,1256,3023
71250	+13 01912	HR 3319, BP Cnc		8 23 58	+12 49.3	4	5.52	.022	1.61	.009	1.86	.068	22	M3 IIIab	15,71,3071,8015
		Ton 320	⋆	8 23 58	+31 40.0	2	15.75	.028	−0.35	.021	−1.21	.028	6	DA1	1036,1727
70470	+78 00287			8 24 00	+78 23.7	1	7.10		0.93		0.67		3	G5	985
		POSS 210 # 54		8 24 03	+41 31.9	1	17.69		1.37				2		1739
71510	−51 03004	HR 3330	⋆ A	8 24 06	−51 33.8	2	5.17	.004	−0.17	.005	−0.68		9	B2 Ve	1637,2007
71261	+15 01822			8 24 07	+14 59.3	1	8.60		1.17				2	K0	882
		WLS 820 25 # 9		8 24 07	+24 47.6	1	12.18		0.66		0.13		2		1375
71459	−41 04119	HR 3326		8 24 07	−41 59.3	4	5.45	.011	−0.15	.006	−0.63	.025	25	B3 V	15,123,362,2027
	+29 01754	G 51 - 12		8 24 08	+29 05.7	3	9.63	.024	1.11	.014	1.07	.000	3	K8	333,1017,1620,3072
71148	+46 01398	HR 3309		8 24 08	+45 49.4	2	6.34	.021	0.62	.006	0.12		5	G5 V	71,3026
71439	−39 04327			8 24 10	−39 43.6	1	8.21		−0.04		−0.55		2	B8 II	401
71260	+19 02012			8 24 12	+19 25.2	1	8.38		0.32		0.01		3	F0	1648
	−38 04457			8 24 13	−38 44.4	1	9.04		1.62		1.89		2	M1	123
71458	−32 05296			8 24 19	−32 46.9	1	7.72		1.20		1.17		1	K0/1 II	565
71377	−12 02524	HR 3324		8 24 20	−12 22.1	4	5.53	.010	1.18	.015	1.14	.000	10	K2 III	15,1003,2031,3005
71634	−57 01513			8 24 23	−57 58.1	2	6.66	.000	−0.02	.000	−0.39	.005	6	B5 III	158,401
	+0 02299	G 113 - 37		8 24 26	−00 09.9	1	9.98		1.01		0.97		2		265
71486	−38 04463			8 24 27	−38 46.3	1	8.39		−0.03		−0.55		2	B8	123
71487	−38 04462	HR 3327, NO Pup	⋆ ABC	8 24 27	−38 53.7	1	6.10		−0.02				4	B9 IV/V	2006
71309	+23 01960	IDS08215N2329	A	8 24 29	+23 19.1	1	8.06		0.07		0.08		2	A0	1625
	−10 02522	IDS08220S1013	A	8 24 29	−10 22.5	1	9.93		0.73				4		2033
	−10 02522	IDS08220S1013	B	8 24 29	−10 22.5	1	11.39		0.85				4		2033
71528	−41 04127	LSS 1066		8 24 31	−41 58.4	4	7.87	.007	0.03	.009	−0.71	.026	11	B0/1 III/IV	123,362,540,976
	−22 02273	IDS08224S2232	A	8 24 33	−22 42.2	1	9.84		0.64		0.14		4	G0	3062
		POSS 210 # 29		8 24 34	+44 14.4	1	16.65		1.65				1		1739
	−38 04465			8 24 35	−38 42.1	1	9.04		1.31		1.30		2		123
	−69 00909			8 24 35	−69 40.8	1	11.05		0.54		0.02		3		1730
	−22 02275	IDS08224S2232	B	8 24 42	−22 43.3	1	10.23		0.83		0.46		4	K5	3062
	−42 04247	LSS 1069		8 24 46	−43 03.1	1	11.21		1.30		0.20		5		362
	+11 01842			8 24 47	+11 25.2	1	9.10		0.91		0.60		2	G5	1733
71372	+13 01918			8 24 47	+12 56.6	1	8.39		1.02				2	K0	882
71523	−28 06048	HR 3331		8 24 47	−29 03.0	1	6.73		−0.05				4	B9.5V	2035
	−38 04469			8 24 48	−38 30.9	1	11.25		1.03		−0.27		4		362
71433	−5 02530	HR 3325		8 24 50	−06 14.6	3	6.58	.004	0.51	.000	0.05	.010	11	F4 III	15,1075,2012
71370	+16 01727			8 24 55	+16 03.2	1	8.82		0.91				2	K0	882
71609	−42 04251	LSS 1070		8 24 55	−43 14.6	3	7.82	.029	0.10	.005	−0.64	.028	9	B2/3 II	362,540,976
71607	−42 04250			8 24 56	−42 21.5	1	7.21		0.14		0.12		2	A4 V	123
71627	−42 04253			8 24 57	−42 42.4	1	9.34		0.08		−0.66		4	B3 II/III	362
71722	−52 01474	HR 3341		8 25 02	−52 38.5	1	6.05		0.06				4	A0 V	2007
71088	+67 00545	HR 3303		8 25 05	+67 27.9	1	5.88		0.97				2	G8 III	71
71606	−39 04343			8 25 05	−39 14.8	1	8.18		−0.02		−0.32		2	B8 III/IV	401
71695	−49 03553			8 25 05	−49 19.9	1	7.25		−0.06		−0.27		2	B8 V	401
		G 52 - 13		8 25 07	+04 28.4	1	15.30		1.35		1.00		2		3062
71518	−14 02526			8 25 07	−14 46.2	1	6.65		−0.18		−0.67		4	B2 V	399
71694	−47 03955			8 25 10	−47 16.4	1	9.13		0.00		−0.28		2	B8 III/IV	401
71499	−3 02356	IDS08227S0405	AB	8 25 11	−04 14.9	1	6.84		0.34		0.03		2	F0	3030
71863	−64 00878	HR 3346		8 25 11	−64 26.1	3	5.94	.017	0.97	.010	0.74	.005	12	G8/K0 III	15,1075,2038
71971	−69 00913			8 25 11	−69 41.2	1	8.86		−0.01		−0.11		2	B9.5 V	1730
71878	−65 00933	HR 3347		8 25 12	−65 58.2	3	3.76	.008	1.13	.004	1.14	.005	11	K1 III	15,1075,2038
71428	+21 01844			8 25 13	+21 19.0	1	6.86		1.02		0.83		2	G5	1648
		G 9 - 6		8 25 14	+15 33.4	1	14.09		1.31		1.24		1		1658
		G 51 - 13	AB	8 25 14	+35 11.2	2	10.76	.005	1.54	.005	1.22	.015	2	M0	333,1620,3078
		POSS 210 # 47		8 25 14	+44 05.7	1	17.24		1.11				2		1739
71622	−31 06079	HR 3336		8 25 16	−31 30.4	1	6.33		0.90				4	G8 III	2006
71693	−43 04296			8 25 16	−43 41.8	1	9.32		0.04		−0.56		4	B3 III/IV	362
71649	−36 04649	LSS 1071		8 25 18	−36 29.3	3	8.78	.008	0.16	.004	−0.71	.021	7	B2 Ib	401,540,976
		Ton 924		8 25 19	+19 06.	1	14.05		0.20		0.12		3		286
		G 9 - 7		8 25 20	+20 18.8	1	13.08		1.48		1.06		7		203
71581	−20 02538	HR 3335, VV Pyx	⋆ AB	8 25 20	−20 40.7	1	6.56		0.04				4	A1 V	2031
		Ton 322		8 25 24	+30 38.	1	13.75		0.02		−0.03		1		1036
		Ton 925		8 25 25	+21 13.	2	14.47	.050	0.03	.025	−0.07	.025	9		286,313
71805	−52 01480	IDS08242S5222	AB	8 25 35	−52 32.3	1	6.50		0.42				4	F5 V	2012
		G 194 - 31		8 25 36	+55 17.6	1	11.97		1.24		1.03		3		7010
71716	−34 04832			8 25 37	−34 51.7	1	8.65		0.03		−0.03		2	A0 V	126
71496	+24 01931	HR 3329, CX Cnc		8 25 39	+24 18.7	2	6.07	.024	0.23	.010	0.14	.005	3	F0 Vn	1733,3058
		MCC 524		8 25 40	+45 55.8	1	11.11		1.20		1.08		2	M0	1723
		G 234 - 35		8 25 40	+61 54.0	1	10.31		1.39		1.17		1	K7	801
71788	−47 03963			8 25 40	−47 38.2	1	8.57		−0.02		−0.28		2	A0 III/IV	401
71597	+0 02305			8 25 41	+00 24.6	3	7.30	.013	1.16	.000	1.13	.009	9	K2 III	1003,1355,3077
71717	−36 04654			8 25 42	−36 30.5	1	9.03		−0.04		−0.37		2	B8 III	401

Table 1   467

HD	DM	Other Id	N Rem	α_{1950}	δ_{1950}	S	V	σ_V	B–V	σ_{B-V}	U–B	σ_{U-B}	n	Spectrum	References
	−40 04269	LSS 1072		8 25 43	−41 14.7	2	10.31	.012	0.43	.001	-0.45	.028	6	B2.5II	362,540
72234	−74 00509			8 25 45	−74 44.9	1	7.21		0.67		0.17		4	G3 IV/V	1704
71688	−25 06109	HR 3339		8 25 46	−25 58.0	1	6.62		0.11				4	A1/2 V	2006
	−34 04836			8 25 49	−34 50.7	1	9.66		0.15		0.09		2		126
71802	−46 04184			8 25 49	−46 41.1	1	9.43		1.08				4	K1 (III)	2012
71555	+14 01899	HR 3333	AB	8 25 50	+14 22.7	3	5.96	.000	0.19	.000	0.13	.009	8	A5 V	985,1022,1501
		LOD 8+ # 6		8 25 53	−34 49.8	1	11.60		0.50		0.04		2		126
71823	−47 03967			8 25 53	−47 33.7	1	8.79		0.01		-0.59		2	B3 Vn	401
71665	−8 02374	HR 3338		8 25 54	−08 38.9	3	6.43	.008	1.19	.005	1.23	.000	11	K0	15,1417,2032
71494	+39 02106	IDS08227N3912		8 25 56	+39 02.5	1	7.94		0.20		0.14		2		105
71663	−2 02581	HR 3337, LO Hya	★ AB	8 25 58	−02 21.0	3	6.42	.026	0.33	.004	0.07	.008	13	A5m	15,1078,3030
71663	−2 02581	IDS08234S0211	C	8 25 58	−02 21.0	1	10.49		0.82		0.41		6		3032
71875	−49 03567			8 26 02	−49 35.3	1	9.25		0.07		-0.53		2	B3/5 (IV)	401
71801	−34 04842	HR 3343	★ AP	8 26 03	−34 56.8	4	5.75	.008	-0.15	.007	-0.68	.004	15	B2 V	15,126,1732,2012
71537	+33 01703			8 26 05	+32 51.6	1	6.92		-0.09		-0.26		2	A0	252
71771	−27 05512	LSS 1073		8 26 07	−27 23.9	4	7.51	.012	-0.11	.014	-0.79	.013	18	B2 II	158,540,976,1775
71369	+61 01054	HR 3323	★ AB	8 26 08	+60 53.2	8	3.37	.040	0.85	.009	0.52	.005	21	G4 II-III	15,1008,1118,1119*
71369	+61 01054	IDS08220N6103	C	8 26 08	+60 53.2	1	11.45		0.66		0.09		3		3024
71369	+61 01054	IDS08220N6103	D	8 26 08	+60 53.2	1	10.95		1.11		0.81		4		3024
	−26 06080			8 26 08	−27 05.6	1	9.44		2.97		4.20		2	N3	864
71919	−54 01647	HR 3349	★ AB	8 26 10	−54 50.7	1	6.53		-0.02				4	A0 V	2006
	−41 04161	LSS 1075		8 26 11	−41 54.2	1	11.20		0.99		-0.19		5		362
71594	+29 01759			8 26 12	+29 38.2	1	8.24		-0.02		-0.02		4	A0	1371
	−34 04844			8 26 12	−35 00.5	1	10.86		0.56		0.12		2		126
		vdB 16 # a		8 26 12	−50 59.	1	15.60		1.88		0.08		1		434
71935	−52 01484	HR 3350, GU Vel		8 26 14	−52 55.3	2	5.08	.005	0.26	.005			7	A9/F0III/IV	15,2027
71222	+71 00459			8 26 17	+71 11.6	1	8.21		0.50		0.03		2	F8	1733
	−41 04164			8 26 20	−41 17.6	1	11.60		0.61		-0.19		4		362
71934	−49 03572			8 26 21	−49 56.0	3	7.55	.039	0.08	.013	-0.66	.006	7	B2 (V)e	401,540,976
71856	−34 04848			8 26 23	−35 00.0	1	7.58		1.18		1.18		2	K1 III	126
	−61 01024			8 26 24	−61 55.3	1	9.28		1.15		0.85		1		565
71815	−22 02286	HR 3344		8 26 25	−22 54.3	3	6.50	.004	0.05	.000	0.08	.000	11	A1/2 V	15,1075,2012
		POSS 210 # 20		8 26 26	+44 10.2	1	16.39		1.87				1		1739
71766	−9 02532	HR 3342		8 26 26	−09 34.9	3	5.99	.010	0.43	.010	0.19	.010	14	F2 III	15,1417,2012
71915	−41 04169			8 26 29	−42 03.8	1	7.58		1.54		1.89		1	M1 III	389
		LOD 8+ # 7		8 26 31	−34 59.2	1	11.98		0.51		-0.04		2		126
		CCS 1187		8 26 31	−36 41.3	1	10.99		3.73				1	C6-,4	864
71637	+35 01831			8 26 33	+35 08.5	1	8.28		0.18		0.14		2	A2	1601
71811	−15 02454	IDS08243S1532	AB	8 26 37	−15 42.1	1	8.52		0.94		0.62		5	K1 V	3072
		LOD 8+ # 8		8 26 39	−35 00.1	1	11.51		0.51		0.04		2		126
	−41 04171	LSS 1080		8 26 39	−41 31.2	1	11.06		0.82		-0.29		3		362
71833	−20 02549	HR 3345	★ A	8 26 40	−20 47.0	4	6.66	.004	-0.06	.006	-0.41	.010	13	B8 II	15,1075,1079,2012
71730	+24 01934			8 26 42	+24 30.8	2	7.11	.066	0.99	.009	0.80		4	K0 III	20,1501
		GD 91		8 26 42	+45 30.6	1	15.08		0.20		-0.52		1	DAs	3060
		POSS 210 # 42		8 26 44	+44 38.5	1	17.05		1.60				2		1739
71913	−34 04858	LSS 1079		8 26 46	−34 33.9	1	7.68		-0.11		-0.82		2	B1/2 II	126
71554	+54 01231			8 26 47	+54 17.3	2	6.78	.000	-0.02	.005	-0.08	.010	6	A0 V	1501,1566
		LOD 8+ # 9		8 26 47	−35 00.3	1	12.47		0.68		0.25		2		126
71781	+9 01985			8 26 49	+09 36.2	1	8.39		0.18		0.10		3	A2	1776
		G 51 - 15	★ V	8 26 52	+26 57.1	3	14.84	.033	2.06	.009	2.11		6	DAs	538,906,1705
72017	−47 03990			8 26 56	−47 16.3	1	8.65		1.03		0.82		1	G8/K0 III	389
	−34 04867			8 26 58	−35 00.1	1	10.96		1.75		1.81		2		126
71928	−28 06102			8 27 00	−28 31.3	1	7.66		-0.17		-0.78		2	B2 II/III	540,976
71763	+22 01941			8 27 01	+22 11.9	1	8.63		1.69		1.85		2	M1	1648
		AJ84,127 # 2218		8 27 06	+62 01.	1	12.73		1.22		1.15		1		801
71947	−28 06106			8 27 06	−28 29.9	1	7.44		0.93		0.59		3	G6/8 III	1657
		POSS 210 # 11		8 27 07	+43 22.4	1	14.99		1.58				1		1739
71946	−27 05534			8 27 07	−27 36.4	1	9.27		-0.02				4	B9	2012
72014	−42 04290	LSS 1082	★ A	8 27 07	−42 25.2	6	6.26	.048	-0.06	.016	-0.97	.019	19	B1/2 (V)nne	362,540,976,1075,1737,2012
		G 113 - 40		8 27 08	−01 34.2	3	11.93	.030	0.93	.004	0.52	.010	6		316,1705,3073
	−27 05536	LSS 1081		8 27 08	−27 37.5	1	9.16		-0.09				4	B9	2012
72337	−69 00919	HR 3370		8 27 10	−69 55.6	3	5.51	.007	-0.02	.005	-0.04	.000	17	A0 V	15,1075,2038
71705	+47 01583			8 27 11	+46 55.1	1	7.46		0.41		0.01		2	F5	1601
71886	−2 02589			8 27 11	−02 59.0	1	8.80		0.96		0.59		2	G5	1375
		LOD 8+ # 11		8 27 11	−35 02.4	1	11.81		0.51		0.09		2		126
71887	−5 02550	RT Hya		8 27 13	−06 09.0	2	7.96	.020	1.53	.015	0.72	.090	2	M3	975,3076
	−55 01565			8 27 16	−56 04.1	1	12.45		0.50		-0.02		1		3062
		KUV 08273+4101		8 27 18	+41 01.3	1	15.92		0.11		-0.67		1	DA	1708
71997	−26 06103	HR 3353		8 27 21	−27 09.9	2	6.70		-0.12	.004	-0.55		5	B4 V	1079,2031
72922	−80 00258	HR 3393		8 27 22	−80 45.1	3	5.68	.004	1.02	.000	0.75	.005	11	G8 III	15,1075,2038
72066	−43 04336			8 27 23	−43 45.0	1	6.75		1.39		1.63		1	K3/4 (III)	389
		LOD 8+ # 12		8 27 25	−35 02.5	1	11.51		0.34		0.22		2		126
72067	−43 04337	HR 3356	★ AB	8 27 26	−43 59.6	9	5.82	.021	-0.16	.006	-0.73	.021	42	B2 Vne	123,362,540,681,815,976*
72048	−34 04881			8 27 29	−34 31.9	1	8.10		0.23		-0.15		2	A0 III/IV	126
72108	−47 04004	HR 3358	★ AP	8 27 30	−47 45.7	6	5.33	.010	-0.15	.005	-0.79	.007	20	B2 IV	15,401,404,1732,2012,2012
72107	−43 04338			8 27 31	−43 17.9	1	9.51		0.31		-0.53		2	B3 II/III	540
		G 51 - 16		8 27 32	+32 52.3	3	15.72	.015	0.31	.025	-0.55	.022	5		538,1620,3060
		POSS 210 # 22		8 27 32	+44 41.1	1	16.42		1.34				2		1739
72088	−44 04459		A	8 27 32	−44 43.0	1	9.07		0.03		-0.48		2	B3 III/IV	123
	−34 04884			8 27 34	−35 05.6	1	9.98		0.10		0.02		2		126

HD	DM	Other Id	N Rem	α_{1950}	δ_{1950}	S	V	σ_V	B–V	σ_{B-V}	U–B	σ_{U-B}	n	Spectrum	References
72126 −43 04341				8 27 38	−43 40.5	2	9.27	.002	0.13	.000	-0.60	.026	8	B2nne	362,540
+72 00421				8 27 39	+71 44.3	1	10.44		0.99		0.82		4		1723
72125 −43 04342				8 27 40	−43 28.1	1	8.32		1.23		1.04		1	K1 II	389
		LP 258 - 29		8 27 41	+33 28.7	1	11.43		0.94		0.65		2		1723
72063 −34 04887		LSS 1083		8 27 41	−34 45.8	2	9.34	.018	0.07	.011	-0.63	.031	4	B2 (V)ne	126,540
		HA 52 # 193		8 27 43	+29 30.5	1	12.41		0.45		0.00		13		1499
	−41 04186			8 27 43	−41 18.2	1	11.31		0.46		-0.37		3		362
72275 −59 01048		V Car		8 27 43	−59 57.3	1	7.10		0.72				1	F8 Ib/II	1484
71844 +40 02065				8 27 45	+40 13.5	1	7.08		0.16		0.20		1	A3	695
72127 −44 04462		HR 3359, LSS 1085	★ AB	8 27 46	−44 33.4	3	5.02	.010	-0.18	.007	-0.80	.007	7	B2 IV	389,1732,2009
		HA 52 # 56		8 27 48	+29 24.1	1	13.19		0.64		0.12		9		1499
72180 −47 04012				8 27 50	−47 38.1	1	9.28		0.05		0.01		2	A0 V	401
237724 +56 01305				8 27 51	+55 56.7	1	8.34		1.06		0.89		2	K0	1502
71866 +40 02066		TZ Lyn		8 27 52	+40 23.6	4	6.73	.010	0.09	.013	0.02	.010	9	A0 p,V	695,1063,1202,1263
	+2 01990			8 27 53	+01 54.8	1	9.91		0.49		0.01		31		588
71779 +54 01238				8 27 55	+53 43.9	1	8.13		0.61		0.14		3	G0	1566
72179 −43 04343				8 27 55	−43 55.9	1	8.14		-0.12		-0.54		2	B3/5 II/III	123
72160 −42 04306				8 27 57	−42 20.0	1	8.40		0.53		0.10		1	F6 V	389
	−43 04344			8 27 58	−44 10.7	1	9.52		1.50		1.39		2		123
	−44 04465			8 27 58	−44 39.6	1	10.26		0.18		-0.06		2	A0	123
71553 +69 00472		HR 3332		8 27 59	+69 29.4	1	6.31		1.35				2	K0	71
	−44 04466			8 27 59	−44 41.0	1	9.91		0.26		0.16		2	A0	123
		LOD 11- # 18		8 28 03	−44 10.4	1	11.56		0.52		0.10		2		123
72177 −41 04194			A	8 28 04	−41 44.9	1	9.35		-0.03		-0.69		1	B9	362
72177 −41 04194			B	8 28 04	−41 44.9	1	9.55		0.02		-0.63		2		362
71906 +37 01870		HR 3348		8 28 06	+37 26.1	2			-0.04	.033	-0.15	.023	4	A0 V	1022,1079
72055 −6 02620				8 28 08	−06 59.3	4	8.11	.006	-0.13	.004	-0.46	.009	30	B8	989,1375,1509,1729
	−41 04196	LSS 1086		8 28 08	−42 00.5	1	10.88		0.61		-0.29		6		362
72232 −45 04183		HR 3363		8 28 08	−46 09.8	1	5.99		-0.15				4	B5 III	2035
	−43 04345			8 28 10	−44 11.6	1	11.01		1.09		0.85		2		123
71987 +26 01794				8 28 17	+25 54.6	1	7.57		0.15		0.11		3	A2	833
71881 +51 01431				8 28 17	+50 47.5	1	7.43		0.63		0.14		2	G1 V	1003
	−34 04903			8 28 17	−34 42.3	1	10.65		0.18		0.09		2		126
72251 −45 04185				8 28 17	−45 38.6	1	8.47		1.19		1.17		1	K1 III	389
72250 −41 04201				8 28 19	−41 28.2	1	9.55		0.07		-0.60		4	B3/5 III	362
72322 −54 01667		HR 3368	★ AB	8 28 19	−55 01.3	2	6.35	.005	0.80	.000			7	G3 III	15,2012
	−53 01708			8 28 20	−54 06.5	3	9.51	.005	0.77	.000	0.34	.024	7	K0 V	1696,2033,3009
		CpD -40 02559		8 28 23	−40 55.0	1	11.40		0.58		-0.21		4		362
		NGC 2632 - 7626		8 28 27	+22 13.0	1	10.61		0.56		0.01		2		1723
72227 −31 06165		HR 3362	AB	8 28 28	−31 59.4	4	5.62	.014	1.50	.008	1.78		13	K3 III	15,1075,2009,3005
72389 −59 01049				8 28 28	−60 00.7	2	7.84	.000	0.98	.000	0.74		5	G8 III	614,1075
		G 113 - 41		8 28 30	−05 00.5	1	11.44		1.34				1		3062
72195 −27 05563				8 28 31	−27 39.9	1	7.83		0.38				4	F0 V	2012
72041 +24 01940		HR 3355		8 28 33	+24 15.1	1	5.77		0.27		0.03		4	F0 III	254
		MtW 52 # 26		8 28 33	+29 49.2	1	15.31		0.48		-0.12		6		397
72268 −36 04715		HR 3364		8 28 35	−36 33.1	2	6.67	.027	1.99	.064	2.06		7	M3/4 Ib	15,158
72404 −56 01674				8 28 39	−57 06.9	1	9.12		-0.03		-0.50		2	B4 V	401
		MtW 52 # 31		8 28 40	+29 50.0	1	15.95		0.61		0.15		6		397
72285 −36 04718				8 28 40	−36 24.3	1	8.10		-0.04		-0.47		2	B5/6 V	401
72095 +17 01864				8 28 41	+17 30.3	1	7.66		1.49		1.80		2	K2	1648
		MtW 52 # 37		8 28 43	+29 50.0	1	13.81		0.41		0.01		6		397
72094 +18 01963		HR 3357	★ A	8 28 45	+18 15.9	4	5.33	.013	1.57	.007	1.94	.011	13	K5 III	3,1118,1193,3001
	−34 04911			8 28 45	−34 44.6	1	9.18		0.70		0.32		2	G5	126
		MtW 52 # 41		8 28 46	+29 48.2	1	10.93		1.04		0.81		6		397
		POSS 210 # 19		8 28 46	+40 28.2	1	16.27		1.44				1		1739
72052 +30 01719				8 28 48	+29 52.9	1	8.19		0.38		-0.05		6	F3 V	397
		WLS 820-5 # 8		8 28 48	−05 01.5	1	10.68		1.52		1.73		2		1375
	−34 04913			8 28 48	−34 45.7	1	10.55		1.35		1.42		2		126
71952 +53 01259		HR 3351		8 28 49	+53 17.2	2	6.26	.015	1.01	.000	0.82	.010	5	K0 IV	1080,3016
		G 40 - 27		8 28 52	+25 55.4	1	12.38		1.37		1.22		1		1620
		G 113 - 42		8 28 55	−05 51.8	1	11.22		1.41				1		3062
72315 −32 05389				8 28 57	−33 06.6	1	8.48		-0.11		-0.75		2	B2/3 III	401
	−34 04914			8 28 57	−34 48.4	1	11.19		0.17		0.10		2		126
	−34 04915			8 28 58	−35 04.6	1	9.71		0.42		0.01		2		126
72350 −44 04477		HR 3371	★ AB	8 28 58	−44 34.1	6	6.30	.007	-0.03	.009	-0.50	.012	15	B3 IV	26,123,540,976,2006,8023
		G 113 - 43		8 29 00	−05 52.1	1	12.37		1.47				1		3062
72333 −34 04916				8 29 03	−35 04.4	2	8.57	.034	1.15	.005	1.13		6	K1 III	126,2012
		Ton 927		8 29 07	+20 42.	2	11.45	.025	-0.18	.005	-0.81	.005	7		286,313
72208 +10 01816		HR 3361		8 29 12	+09 59.1	1	6.83		-0.07		-0.17		3	B9 p (Hg)	220
	+29 01770			8 29 14	+29 38.2	1	8.73		0.50		-0.01		4	F8	1371
72402 −44 04485				8 29 14	−44 26.7	1	8.77		0.02		-0.02		2	A0 V	123
72310 −19 02438		HR 3367	★ AB	8 29 16	−19 24.4	2	5.41	.005	-0.06	.000			7	B9.5IV/V	15,2012
	−34 04923			8 29 22	−35 03.6	1	10.06		0.33		0.07		2		126
		G 113 - 44		8 29 24	−02 22.5	1	12.87		1.34				1		3062
72256 +13 01936				8 29 26	+12 58.4	1	7.26		1.15		1.07		2	K0	1375
72453 −44 04489				8 29 31	−44 19.7	1	9.23		-0.05		-0.37		2	B3/5	123
72454 −44 04490				8 29 32	−44 24.2	1	9.54		0.24		0.15		2	F0 (V)	123
72436 −38 04566		HR 3373	★ A	8 29 34	−38 53.7	1	6.52		-0.16		-0.57		2	B5 V	362
72436 −38 04566		IDS08277S3844	AB	8 29 34	−38 53.7	7	6.29	.015	-0.14	.005	-0.56	.012	23	B5 V	15,362,404,540,976*
	−75 00504			8 29 34	−76 05.0	1	11.09		0.12		0.07		3		45

Table 1 469

HD	DM	Other Id	N Rem	α_{1950}	δ_{1950}	S	V	σ_V	B–V	σ_{B-V}	U–B	σ_{U-B}	n	Spectrum	References
	−47 04047	AL Vel		8 29 35	−47 29.8	1	8.66		0.84		0.33		1	A3 IV:	389
72485	−47 04048	HR 3375		8 29 35	−47 41.8	6	6.38	.004	−0.15	.008	−0.66	.005	16	B2.5 V	15,401,540,976,1732,2029
		LP 725 - 33		8 29 36	−13 29.6	1	11.27		0.76		0.20		2		1696
	−39 04441			8 29 36	−40 03.4	1	10.26		0.37		−0.16		4		362
	−44 04491			8 29 36	−44 17.1	1	9.64		0.26		0.08		2		123
	+22 01950			8 29 37	+22 34.9	1	9.15		0.70		0.24		2	G0	1375
		AS 201		8 29 37	−27 35.3	1	11.80		0.73		0.18		1		1753
		POSS 210 # 26		8 29 38	+38 32.9	1	16.56		1.44				2		1739
72184	+38 01920	HR 3360		8 29 40	+38 11.4	3	5.90	.000	1.11	.005	1.16	.010	6	K2 III	15,1003,3009
72451	−34 04929			8 29 40	−35 01.6	1	8.83		−0.07		−0.41		2	B8 V	126
		POSS 210 # 30		8 29 41	+44 11.3	1	16.76		1.15				1		1739
72482	−44 04496			8 29 45	−44 20.7	1	9.91		0.08		0.03		2	A0 V	123
72516	−47 04050			8 29 45	−47 16.4	1	8.70		0.36				4	F2 V	2012
	−44 04497			8 29 48	−44 15.7	1	10.84		0.99		0.68		2		123
72579	−53 01721			8 29 48	−53 55.2	1	8.22		0.79		0.44		2	K0 V	389
72292	+20 02109	HR 3366		8 29 49	+20 36.7	4	5.34	.005	1.25	.006	1.39	.012	11	K3 III	37,1080,1118,3016
72539	−48 03865			8 29 49	−48 34.7	1	7.97		−0.07		−0.43		2	B5 V	401
72306	+16 01754			8 29 50	+15 54.6	1	7.99		1.00		0.84		2	K0	1648
	−44 04498			8 29 52	−44 16.0	1	10.35		0.04		−0.14		2	B9	123
72113	+59 01176			8 29 55	+58 46.7	1	6.90		−0.01		0.01		2	A0	985
72359	+10 01818	HR 3372		8 29 57	+10 14.2	1			−0.02		−0.04		3	A1 V	1022
72515	−43 04371			8 29 57	−44 14.8	1	8.74		0.13		−0.43		2	B3 III	123
	−20 02583			8 30 00	−20 34.2	2	10.01	.014	0.64	.014	0.10		5		1594,6006
72535	−41 04225	IDS08283S4142	AB	8 30 01	−41 51.8	1	8.20		0.07		−0.68		4	B2 III	362
		LP 899 - 13		8 30 02	−27 48.0	1	13.97		0.66		0.07		1		3062
72480	−31 06209			8 30 02	−31 32.4	1	7.71		−0.10		−0.53		2	B5 II	401
72324	+24 01946	HR 3369		8 30 03	+24 15.4	4	6.35	.007	1.03	.011	0.87	.009	8	G9 III	15,1003,1375,3077
72555	−46 04267		V	8 30 03	−47 04.2	2	6.94	.139	−0.15	.005	−0.71	.017	7	B2 V	26,164
72552	−43 04373			8 30 05	−43 41.3	1	8.23		1.17		1.10		1	K1 III	389
72554	−45 04219	LSS 1093		8 30 06	−45 57.4	2	8.17	.005	0.35	.016	−0.50	.026	5	B1/2 Ib	540,976
72576	−43 04374	LSS 1092		8 30 08	−43 39.1	3	8.38	.008	0.41	.012	−0.40	.032	8	B2 Ib/II	362,540,976
		TT Cnc		8 30 09	+13 21.7	2	11.01	.236	0.31	.068	0.14	.041	2	F0	668,699
72037	+65 00638	HR 3354		8 30 09	+65 19.1	3	5.46	.000	0.21	.014	0.09	.005	7	A2 m	1118,3058,8071
72650	−53 01729	HR 3382		8 30 09	−54 13.4	3	6.33	.008	1.30	.010			11	K3 III	15,1075,2006
72291	+36 01836	HR 3365		8 30 10	+36 36.4	2	6.26	.005	0.35	.000	−0.07	.034	6	F5 Vb	254,3037
72553	−43 04375			8 30 11	−44 14.0	1	8.77		0.96		0.66		2	G6 III	123
		POSS 210 # 32		8 30 12	+42 07.8	1	16.77		1.90				1		1739
72462	−14 02564	HR 3374	⋆ AB	8 30 14	−14 51.6	2	6.37	.005	0.28	.005			7	A7 IV/V	15,2012
72551	−35 04745	IDS08283S3546	AB	8 30 19	−35 56.0	1	9.35		0.02		−0.55		2	B3/5 V	401
72573	−39 04452			8 30 20	−39 47.0	1	7.37		−0.04		−0.17		2	B8/9 V	401
		WLS 840 15 # 6		8 30 22	+15 21.5	1	11.20		1.15		1.24		2		1375
		BPM 19061		8 30 24	−53 30.	1	14.47		−0.15		−1.15		2		3065
72631	−46 04274			8 30 25	−47 09.4	1	7.98		1.72		2.06		2	K4/5 (III)	389
	−49 03617			8 30 28	−50 01.4	2	10.79	.009	1.22	.008	1.10		5	K7	158,1705
72408	+29 01778			8 30 35	+29 22.7	1	8.68		1.02		0.82		4	K0	1371
72648	−43 04382	IDS08289S4335	AB	8 30 36	−43 45.6	3	7.61	.004	0.12	.002	−0.61	.023	9	B1/2 Ib	362,540,976
72409	+27 01627			8 30 37	+27 31.0	1	8.58		0.88		0.51		2	G5	1375
	+43 01827			8 30 39	+42 44.1	1	10.34		1.11		1.09		1	K5	3072
		G 9 - 10		8 30 40	+13 58.6	2	11.53	.000	1.37	.015	1.31	.015	2	K6-7	1620,1658
72676	−45 04227			8 30 41	−45 51.6	1	8.45		0.54		0.35		1	A5 V +(K)II	389
72737	−52 01517	HR 3386	⋆ AB	8 30 42	−53 02.5	2	5.68	.005	0.58	.000			7	G8 III	15,2029
		G 115 - 9		8 30 47	+37 09.8	2	16.01	.000	0.21	.000	−0.59	.000	2	DAwk	782,3060
		G 252 - 14		8 30 47	+77 33.4	1	15.89		1.59				1		906
72736	−50 03393	IDS08293S5038	AB	8 30 47	−50 48.0	1	8.12		0.82		0.55		1	K1/2 III+A5	389
72754	−49 03621	FY Vel		8 30 51	−49 25.8	3	6.88	.047	0.20	.004	−0.71	.024	10	B2 (Ia)pe	401,1637,1737
		HA 202 # 724		8 30 52	−75 06.6	1	12.49		1.27		1.18		9		1499
72673	−31 06229	HR 3384		8 30 54	−31 20.4	6	6.38	.010	0.78	.005	0.30	.017	18	K0 V	15,258,1705,2012,2012,3078
72626	−24 07089	HR 3381	⋆ AB	8 30 55	−24 26.1	2	6.18	.005	0.28	.005			7	F2 IV/V	15,2012
	−43 04388			8 30 56	−43 23.6	1	10.56		0.49		−0.52		4		362
72505	+13 01940	HR 3376		8 30 59	+13 25.8	1	6.28		1.17		1.23		2	K0 III	252
72392	+47 01592			8 30 59	+47 18.7	1	6.70		0.01		0.03		2	A0	252
72565	−2 02613			8 31 01	−02 48.3	1	7.74		1.44		1.77		3	K5	1657
72688	−34 04959	HR 3385, VX Pyx		8 31 01	−34 27.8	2	6.37	.011	0.95	.000	0.65		5	G8 III	1641,2006
		POSS 210 # 41		8 31 02	+41 45.0	1	17.04		1.63				1		1739
71973	+75 00342	HR 3352	⋆ AB	8 31 02	+74 53.8	1	6.27		0.30		0.11		2	A2m	1733
		HA 202 # 728		8 31 02	−75 06.6	1	13.15		0.45		0.17		10		1499
72752	−46 04287			8 31 03	−46 52.2	1	6.49		1.29		1.37		2	K2 III	389
72561	+5 01997	HR 3378		8 31 05	+04 55.7	3	5.86	.004	1.07	.000	0.88	.004	11	G5 III	15,1078,1355
	+63 00786			8 31 06	+63 28.1	1	10.02		0.67		0.24		3		1723
72503	+25 01950			8 31 08	+25 24.3	1	8.02		0.22		0.12		2	A3	105
		G 234 - 37		8 31 09	+68 14.7	1	11.64		1.57				1		1705
		G 234 - 37		8 31 09	+68 14.7	2	11.66	.015	1.57	.000			3		940,1705
	−43 04392			8 31 13	−43 21.3	1	10.96		0.58		−0.35		4		362
72772	−43 04393			8 31 14	−43 15.3	1	8.29		0.45		0.00		1	F5 IV/V	389
72858	−54 01683			8 31 17	−54 57.1	1	7.35		−0.02		−0.13		2	B9.5V	401
72800	−47 04072	LSS 1095		8 31 18	−47 26.1	2	6.62	.015	0.13	.010	−0.31		7	B9 Ib	2012,8100
		Ton 930		8 31 19	+20 03.	1	13.16		−0.23		−1.00		3		286
72798	−45 04236	IDS08297S4525	AB	8 31 22	−45 34.9	1	6.45		−0.14				4	B3 III	2012
72838	−48 03888			8 31 23	−48 59.0	1	7.30		1.81		1.97		1	K3 II +A/F	389
72600	−36 04763			8 31 26	−36 29.4	1	7.65		−0.04		−0.58		2	B3 III	123

HD	DM	Other Id	N	Rem	α₁₉₅₀	δ₁₉₅₀	S	V	σ_V	B−V	σ_B−V	U−B	σ_U−B	n	Spectrum	References
72816	−43 04401				8 31 26	−43 30.8	1	7.11		0.99		0.74		1	G8 III	389
72771	−34 04969				8 31 27	−34 28.6	2	7.85	.002	−0.10	.001	−0.72	.003	35	B3 III	164,978
72787	−37 04850	HR 3388			8 31 28	−38 12.0	4	6.48	.006	−0.17	.012	−0.66	.010	12	B2.5 V	15,540,976,2012
	−44 04518				8 31 28	−45 00.6	1	9.52		0.28				4	B3	2012
72660	−1 02074	HR 3383			8 31 30	−01 58.8	2	5.80	.005	0.00	.000	0.00	.000	7	A1 V	15,1256
72524	+36 01840	HR 3377			8 31 31	+36 35.6	1			0.04		0.02		3	A2 Vnn	1022
72617	+8 02077	HR 3380		⋆ A	8 31 32	+08 37.5	3	6.06	.033	0.33	.004	0.05	.032	11	F1 III-IV	15,254,1415
72616	+17 01871				8 31 32	+17 03.2	1	8.34		0.63		0.20		2	G5	1375
72458	+52 01322				8 31 32	+52 22.4	1	7.54		1.02		0.79		2	G5	1566
		POSS 210 # 46			8 31 36	+44 09.3	1	17.18		1.70				1		1739
72769	−22 02317				8 31 36	−23 11.1	2	7.19	.019	0.75	.010	0.38		6	K1 (IV)(+G)	2012,3026
72559	+28 01624				8 31 37	+28 37.0	1	8.57		0.50		0.05		3	F6 V	3026
		G 52 - 18			8 31 39	+10 19.2	1	11.63		1.31		1.16		1		333,1620
		KUV 08317+4117			8 31 43	+41 17.0	1	17.11		−0.15		−1.10		1	DA	1708
72543	+41 01853				8 31 44	+41 11.8	1	7.42		1.00		0.78		2	G8 III	1601
		POSS 210 # 33			8 31 46	+44 38.0	1	16.79		1.28				1		1739
72832	−38 04610	HR 3389			8 31 47	−38 40.6	2	5.95	.005	−0.15	.005			7	B5 III	15,2012
72876	−44 04525				8 31 51	−44 49.9	1	8.23		1.18		1.13		2	K0 III	389
72976	−55 01606				8 31 51	−56 13.5	1	7.51		−0.11		−0.50		2	Ap Siλ4200	401
		LP 665 - 50			8 31 52	−07 56.2	1	13.15		0.46		−0.03		2		1696
72899	−45 04243				8 31 52	−46 04.6	1	7.62		1.03		0.86		1	K0 III	389
		G 113 - 46			8 31 53	−00 58.1	1	12.73		1.62				1		3062
72900	−46 04300	HR 3390			8 31 53	−46 47.9	1	6.24		1.56				4	K3 III	2006
		G 113 - 47			8 31 54	−05 32.3	2	13.48	.030	1.20	.000	1.01		2		1658,3062
72680	+9 02004	G 52 - 19			8 31 56	+09 32.8	2	8.95	.005	0.80	.005	0.50	.005	2	K0	333,1620,1658
	+67 00552	G 234 - 38			8 31 56	+67 28.1	3	9.28	.017	1.43	.020	1.23		5	M2	1017,1197,3078
72939	−47 04087				8 31 57	−47 43.3	1	9.46		0.07		−0.64		2	B2 IV	401
	−44 04527	Pis 4 - 14			8 32 01	−44 19.4	2	8.75	.009	1.52	.014	1.82	.012	3	M0	410,1685
		NGC 2632 - 4346			8 32 06	+21 16.2	1	11.15		0.69		0.19		4		1723
72614	+42 01899	G 115 - 10			8 32 08	+41 55.3	3	8.78	.021	0.93	.009	0.68	.021	11	K2 V	22,1003,1775
	−36 04773				8 32 09	−36 47.3	1	10.15		0.04		−0.26		2		123
	−44 04534	Pis 4 - 15			8 32 11	−44 19.2	1	10.19		−0.04		−0.15		1		410
72854	−20 02593				8 32 12	−21 01.8	1	8.24		0.36				4	F2/3 V	2012
		LOD 9+ # 18			8 32 13	−37 21.7	1	11.94		0.33		0.15		2		123
	−37 04865				8 32 13	−37 23.5	1	11.69		0.17		0.00		2		123
	−44 04537	Pis 4 - 16			8 32 14	−44 24.3	2	9.31	.011	1.46	.020	1.78	.014	3	M0	410,1685
72338	+70 00523				8 32 15	+69 52.4	1	7.13		0.94		0.67		2	K0	985
		LOD 9+ # 3			8 32 17	−36 44.1	1	9.65		−0.03		−0.47		2		123
72936	−36 04774				8 32 17	−36 44.1	1	6.70		0.40		−0.01		2	F3/5 V	123
	−36 04775				8 32 25	−37 12.6	1	10.36		0.35		0.18		2		123
72997	−44 04539	Pis 4 - 13			8 32 26	−44 22.3	1	7.46		−0.18		−0.73		1	B5	410
72779	+20 02118	NGC 2632 - 1133		⋆	8 32 27	+19 45.8	7	6.58	.013	0.68	.004	0.26	.004	40	G0 III	15,1140,1351,1355*
		Steph 707			8 32 29	+24 26.1	1	11.56		1.30		0.97		1	K7	1746
72973	−36 04777				8 32 29	−36 31.3	1	8.22		−0.08		−0.34		2	B7 V	123
	−36 04778				8 32 29	−37 14.3	1	9.32		0.39		−0.05		2		123
		G 113 - 48			8 32 31	−04 03.5	1	14.78		1.07		0.65		4		3062
72954	−32 05465	HR 3397		⋆ AB	8 32 32	−32 25.4	4	6.43	.008	0.75	.004	0.23		16	G5 V	15,1075,2012,3077
	−40 04399				8 32 32	−40 32.2	1	10.42		0.21		−0.12		2	A0	123
73064	−51 03088	IC 2391 - 77			8 32 32	−51 58.9	1	10.39		0.59		0.04		1	A2	1030
	−37 04871				8 32 33	−37 19.6	1	10.10		0.23		0.13		2		123
		G 113 - 49			8 32 35	+03 38.4	2	12.53	.025	0.75	.005	0.22	.015	2		1658,3062
72993	−37 04873	HR 3399		⋆ AB	8 32 36	−37 26.3	2	6.28	.015	1.55	.005	1.74		6	M0 III	2006,3005
		LOD 9+ # 5			8 32 37	−36 58.0	1	12.12		0.26		0.01		2		123
		LOD 10 # 19			8 32 37	−40 35.4	1	12.69		0.56		−0.24		2		123
72928	−15 02506				8 32 40	−15 55.2	1	10.28		1.24		1.22		1	K5 V	3072
	−36 04780				8 32 42	−37 06.7	1	10.28		0.47		−0.03		2		123
		LOD 10 # 18			8 32 43	−40 34.3	1	11.80		1.19		0.53		2		123
	−40 04402				8 32 43	−40 37.1	1	11.03		0.41		−0.03		2		123
	−36 04781				8 32 44	−37 13.4	1	9.63		0.99		0.66		2		123
	−44 04543				8 32 45	−44 50.8	1	10.33		0.25				4	B1 V	2021
73105	−52 01532				8 32 46	−52 53.9	1	6.80		−0.11		−0.60		2	B3 V	401
		UV0832-01			8 32 48	−01 45.4	1	11.43		−0.29		−1.18		0		1732
		Pis 4 - 10			8 32 48	−44 06.	1	10.74		0.07		0.00		1		410
73059	−44 04544	Pis 4 - 11			8 32 48	−44 20.0	1	9.28		−0.12		−0.32		1	B9	410
72908	+3 02014	HR 3392			8 32 49	+02 55.0	3	6.33	.008	1.02	.000	0.87	.010	9	G9 III	15,1078,1355
	+44 01761				8 32 49	+43 46.6	1	10.74		0.48		0.01		4		1723
		Pis 4 - 7			8 32 49	−44 12.4	1	11.86		0.77		0.29		1		410
73077	−46 04322				8 32 49	−46 57.2	3	8.01	.011	1.12	.005	1.06	.000	6	K0/1 III	258,688,1075
		LOD 10 # 17			8 32 50	−40 34.7	1	12.52		0.56		−0.19		2		123
72846	+20 02123	NGC 2632 - 1166			8 32 53	+19 56.7	1	7.56		0.17		0.15		1	A5 V	8023
73041	−36 04783				8 32 53	−36 48.6	1	7.94		1.34		1.25		2	K3 III	123
	−40 04407				8 32 53	−40 41.1	1	10.85		0.17		0.11		2		123
73075	−43 04422	Pis 4 - 1			8 32 53	−44 05.8	3	7.32	.008	1.00	.004	0.84	.019	5	K0	389,410,1685
73076	−44 04547	Pis 4 - 9		⋆ AB	8 32 53	−44 15.9	1	9.12		−0.08		−0.38		1	B9	410
		G 113 - 50			8 32 54	+04 37.2	1	12.25		0.64		−0.04		1		3062
72778	+43 01834				8 32 54	+42 45.2	1	6.97		0.19		0.07		2	A2	985
		Pis 4 - 5			8 32 54	−44 11.0	1	10.81		1.01		0.76		1		410
		Pis 4 - 6			8 32 55	−44 11.7	1	11.50		0.39		0.06		1		410
	−43 04424	Pis 4 - 2			8 32 56	−44 07.9	1	9.30		0.25		0.15		1		410
73090	−44 04548	Pis 4 - 12			8 32 58	−44 21.3	1	7.92		−0.09		−0.45		1	B9	410

Table 1   471

HD	DM	Other Id	N Rem	α_{1950}	δ_{1950}	S	V	σ_V	B–V	σ_{B-V}	U–B	σ_{U-B}	n	Spectrum	References
	+29 01784			8 32 59	+29 12.7	1	10.48		0.56		-0.03		2		1723
		AAS27,343 # 38		8 33 00	-41 30.	1	13.10		0.51				2		362
	-43 04425	Pis 4 - 4		8 33 00	-44 10.3	1	9.58		0.41		-0.52		1		410
73127	-50 03407			8 33 00	-50 55.2	3	6.57	.009	-0.16	.004	-0.72	.001	8	B2 V	401,1732,2012
	-43 04427	Pis 4 - 8		8 33 01	-44 14.6	1	10.71		0.12		0.02		1		410
72968	-7 02540	HR 3398, HV Hya	6	8 33 02	-07 48.5	6	5.72	.005	-0.03	.006	-0.02	.009	27	A1p SrCrEu	15,1063,1202,1417*
	-43 04428	Pis 4 - 3		8 33 03	-44 10.8	1	10.34		0.15		0.06		1	A5	410
		POSS 210 # 57		8 33 04	+42 48.0	1	17.89		1.50				2		1739
73141	-49 03643			8 33 04	-49 28.6	1	8.39		-0.07		-0.50		2	B7 III/IV	401
73468	-72 00713	HR 3417		8 33 04	-73 11.1	3	6.11	.007	0.95	.000	0.66	.005	14	G8 III	15,1075,2038
72884	+24 01955			8 33 05	+24 13.5	2	6.99	.010	0.08	.000	0.10	.005	5	A2 V	985,1501
73054	-36 04787			8 33 05	-37 10.1	1	7.94		0.16		0.17		2	A5 V	123
72945	+7 01997	HR 3395	⋆ A	8 33 12	+06 47.7	1	5.99		0.52		0.04		3	F8 V	3026
72945	+7 01997	HR 3395	⋆ AB	8 33 12	+06 47.7	3	5.62	.008	0.56	.005	0.08	.010	13	F8 V	15,1197,1256
72946	+7 01997	HR 3396	⋆ B	8 33 12	+06 47.9	1	7.25		0.71		0.27		3	G0 m	3026
72944	+10 01831	IDS08305N1015	A	8 33 12	+10 05.0	1	8.68		0.67		0.17		3	G0	3016
72944	+10 01831	IDS08305N1015	B	8 33 12	+10 05.0	1	9.78		0.78		0.34		2		3032
		LP 425 - 182		8 33 12	+17 55.0	1	10.72		1.06		1.03		1		1696
	-44 04551			8 33 12	-45 04.9	1	10.45		0.03		-0.12		3		125,396
73155	-49 03646	HR 3407		8 33 12	-49 46.3	5	5.00	.009	1.34	.009	1.37	.012	25	K1/2 II	15,1075,1637,2012,8015
		POSS 210 # 58		8 33 13	+39 01.9	1	18.42		1.60				2		1739
73089	-33 05207			8 33 13	-33 51.1	1	6.69		1.37		1.56		30	K3 III	978
72943	+15 01851	HR 3394		8 33 19	+15 29.3	1	6.33		0.32		0.06		2	F0 IV	1733
73072	-26 06225	HR 3402		8 33 21	-26 40.2	2	5.95	.005	0.38	.005			7	A1 V +G III	15,2012
		NGC 2632 - 532		8 33 22	+19 47.8	1	12.57		0.98				3	dK2	902
		G 115 - 11		8 33 22	+39 13.0	1	11.80		1.29		1.07		3		7010
73121	-39 04519	HR 3404		8 33 22	-39 47.8	2	6.46	.005	0.58	.000			7	G1 V	15,2012
72942	+20 02125	NGC 2632 - 534		8 33 24	+20 30.9	2	7.50	.031	0.14	.014	0.11	.031	6	A4 V	29,8023
73119	-39 04520			8 33 24	-39 30.5	1	7.62		1.64		2.03		4	M0/1 (III)	402
		BPM 19090		8 33 24	-52 01.	1	12.97		0.54		-0.16		1		3065
		NGC 2632 - 533		8 33 25	+19 43.0	2	11.58	.005	0.90	.009	0.57		4	dK0	902,8023
73744	-76 00525			8 33 30	-76 45.5	3	7.61	.004	0.61	.004			10	G2 V	1020,1075,2012
	-40 04421			8 33 31	-40 38.2	1	11.52		0.57		0.15		2		123
73034	-1 02084			8 33 33	-01 32.0	1	8.44		0.13		0.15		5	A2	1732
		G 113 - 51		8 33 33	-03 23.7	1	11.54		1.23		1.01		1		1658
		vdB 17 # b		8 33 33	-40 26.2	1	12.13		0.28		-0.09		2		434
		NGC 2632 - 535		8 33 36	+20 24.2	1	12.65		0.95				3	dK0	902
73222	-49 03655			8 33 36	-49 41.6	1	9.56		0.00		-0.73		2	B2/5	401
	+19 02045	NGC 2632 - 536		8 33 39	+19 08.4	2	9.43	.037	0.45	.005	0.03	.000	5	F6 V	29,8023
73047	-1 02085			8 33 39	-01 23.4	1	9.45		0.17		0.10		1	A5	1732
73220	-47 04112			8 33 39	-47 47.2	2	8.19	.020	0.48	.010	0.08	.005	4	F5 IV/V	389,401
	-40 04425			8 33 41	-40 40.3	1	11.34		0.43		0.04		2		123
	-40 04427	LSS 1097		8 33 42	-40 29.6	3	9.90	.029	0.47	.005	-0.43	.024	11	B1 IV	92,362,540
73219	-47 04114			8 33 43	-47 41.1	1	9.97		0.20		0.17		2	A2/3 III/IV	401
73000	+22 01962	IDS08308N2231	AB	8 33 45	+22 20.9	1	8.33		0.44		0.07		2	G0	1648
	-40 04429			8 33 46	-40 33.9	1	11.59		0.36		0.30		2		123
		LOD 10 # 10		8 33 46	-40 38.8	1	12.46		0.55		0.41		2		123
73046	+12 01872			8 33 47	+11 52.7	1	8.10		-0.04		-0.14		3	B9	3010
		AJ79,1294 T10# 10		8 33 49	-45 07.9	1	12.98		0.63		0.12		3		396
73244	-47 04116			8 33 49	-47 23.9	3	7.05	.013	1.35	.004	1.49	.002	6	K2 III	258,688,1075
73287	-53 01760	IC 2391 - 1		8 33 49	-54 01.9	1	7.07		-0.11		-0.46		3	B7 V	1040
73218	-42 04418			8 33 51	-42 23.2	2	7.47	.005	0.44	.005	0.00	.005	3	F5 V	123,389
		NGC 2632 - 537		8 33 53	+20 17.9	2	11.65	.005	0.76	.001	0.35	.002	5	dG5	1140,8023
		NGC 2632 - 9		8 33 55	+19 56.3	1	11.39		0.80		0.39		1	dG5	8023
73217	-40 04433			8 33 55	-40 41.6	1	7.38		0.51		0.01		2	F5 IV/V	123
73273	-51 03099			8 33 55	-51 38.8	2	7.56	.004	1.13	.001	1.05	.010	6	K0 III	389,1673
73045	+19 02047	NGC 2632 - 538		8 33 57	+19 03.5	3	8.66	.032	0.31	.007	0.14	.013	9	Am	29,8023,8052
		vdB 17 # a		8 34 00	-40 30.0	1	10.02		0.46		-0.41		3	B1 V	434
		POSS 210 # 8		8 34 01	+41 12.3	1	14.49		1.54				1		1739
73216	-37 04906	IDS08321S3757	AB	8 34 01	-38 07.4	1	8.43		-0.10		-0.50		2	B9 IV	401
73390	-57 01590	HR 3415		8 34 06	-58 03.1	4	5.26	.005	-0.14	.007	-0.62	.002	11	B4 V	15,26,2012,8023
73389	-57 01591	HR 3414		8 34 09	-57 50.1	3	4.85	.006	1.00	.006	0.81	.005	14	K0 III	15,1075,2012
73081	+20 02128	NGC 2632 - 16		8 34 10	+19 46.8	2	9.20	.030	0.46	.007	0.01	.003	6	F6 V	29,8023
73258	-39 04539			8 34 10	-39 59.8	1	10.01		0.00		-0.16		7	B7/9 V	402
	-40 04437			8 34 10	-41 07.0	1	9.90		0.53		-0.36		2	B3	123
72582	+74 00370	HR 3379		8 34 12	+73 48.4	1	6.15		1.02				2	G7 III	71
	-39 04543			8 34 14	-39 52.7	1	10.76		1.21		1.22		5		402
	-78 00356			8 34 14	-78 28.3	1	10.86		0.22		0.14		3		45
73241	-30 06540			8 34 15	-31 09.8	1	7.59		-0.10		-0.47		2	B5 III	401
73305	-47 04119			8 34 15	-47 40.3	1	9.14		0.02		-0.32		2	B8 IV	401
	-36 04816			8 34 16	-36 42.8	1	10.60		0.40		0.21		2		123
		NGC 2632 - 23		8 34 19	+19 58.8	1	11.29		0.71		0.25		1	dG0	8023
		NGC 2632 - 540		8 34 19	+20 27.6	1	11.03		0.69		0.18		1	dG0	8023
73284	-36 04818			8 34 21	-36 44.1	1	9.40		0.19		0.10		2	A4 IV/V	123
73285	-39 04544			8 34 21	-40 03.1	2	8.95	.015	0.04	.000	-0.72	.015	11	B2/3 III	362,402
73340	-50 03417	HR 3413, HV Vel		8 34 22	-50 47.7	3	5.80	.009	-0.13	.005	-0.55		9	B8p Si	15,401,2012
73143	+10 01837	HR 3406		8 34 23	+09 49.8	3	5.90	.021	0.08	.009	0.14	.016	6	A3 V	695,1022,3050
73326	-46 04349			8 34 23	-46 19.6	4	7.27	.005	-0.02	.007	-0.76	.003	12	B2 III/IV	158,540,976,1732
		G 40 - 31		8 34 24	+26 23.2	3	14.18	.013	1.53	.007	1.05	.029	4		308,1620,3016
	-40 04440			8 34 24	-40 31.0	1	10.57		0.18		0.09		2	A0	123

HD	DM	Other Id	N Rem	α_{1950}	δ_{1950}	S	V	σ_V	B–V	σ_{B-V}	U–B	σ_{U-B}	n	Spectrum	References
	−43 04460			8 34 25	−43 33.8	1	10.44		0.39		−0.48		4		362
		NGC 2632 - 27		8 34 26	+19 52.5	1	11.44		0.73		0.32		1	dG5	8023
	−41 04300			8 34 27	−41 23.8	1	10.98		1.43		1.23		3		1696
		G 51 - 20		8 34 28	+31 43.7	1	11.84		0.76		0.29		1		333,1620
73095	+32 01764			8 34 29	+32 01.0	1	8.85		0.29		0.07		2	A3	1733
		NGC 2632 - 30		8 34 30	+20 21.1	1	11.40		0.73		0.22		1	dG5	8023
		LOD 10 # 5		8 34 31	−40 37.6	1	11.98		0.59		0.33		2		123
		NGC 2632 - 100		8 34 32	+20 17.0	2	10.55	.005	0.58	.001	0.09	.000	5	dG0	1140,8023
	+20 02130	NGC 2632 - 31		8 34 36	+19 44.3	1	9.79		0.56		0.07		1	F6 V	8023
		NGC 2632 - 32		8 34 36	+19 47.6	2	11.65	.000	0.78	.002	0.39	.000	3	dG8	1140
	+19 02050	NGC 2632 - 34		8 34 37	+19 20.3	1	9.50		0.43		0.01		1	F6 V	8023
		AJ79,1294 T10# 11		8 34 37	−44 47.7	1	13.32		0.44		0.26		3		396
73017	+53 01268	HR 3400		8 34 39	+53 34.6	2	5.67	.008	0.96	.002	0.65		5	G9 III	71,1501
73161	+20 02131	NGC 2632 - 38		8 34 41	+20 11.3	4	8.70	.017	0.30	.007	0.08	.006	9	A9 IVn	29,1722,1748,8023
		NGC 2632 - 541		8 34 42	+18 49.8	1	10.66		0.63		0.11		1	dG0	8023
		AAS19,45 T3# 116		8 34 42	−40 04.5	1	12.87		0.99		0.80		5		402
		NGC 2632 - 3012		8 34 44	+19 36.5	1	13.67		1.23				4		902
73174	+20 02132	NGC 2632 - 40		8 34 45	+19 54.5	4	7.78	.020	0.20	.011	0.15	.016	8	Am	29,1722,1748,8023
	+32 01766			8 34 45	+32 29.2	1	10.59		1.08		0.99		3	G0	1723
		NGC 2632 - 3014		8 34 46	+19 37.4	1	14.16		0.81				2		902
		LOD 10 # 4		8 34 46	−40 38.2	1	11.57		0.75		0.32		2		123
		AJ79,1294 T10# 9		8 34 46	−44 53.2	1	12.69		0.51		0.02		4		396
72905	+65 00643	HR 3391		8 34 47	+65 11.7	6	5.63	.009	0.62	.004	0.07	.000	17	G1.5Vb	15,1007,1013,1118*
73404	−46 04360	IDS08332S4657	AB	8 34 47	−47 07.7	1	9.40		0.07				4	B9 IV	1075
73175	+20 02133	NGC 2632 - 45	★ A	8 34 49	+19 41.6	3	8.26	.006	0.24	.003	0.10	.007	7	A9 Vn	29,1140,8023
73421	−47 04132			8 34 49	−47 25.4	2	9.20	.005	0.31	.005	0.16		6	A8/9 V	401,1075
73463	−51 03105	IC 2391 - 78		8 34 49	−52 09.6	1	11.31		0.64		0.15		1	F2	1030
	−36 04826			8 34 50	−36 57.4	1	9.78		1.25		1.30		2		123
		AAS19,45 T3# 117		8 34 50	−40 04.7	1	13.35		0.54		0.05		2		402
	+19 02052	NGC 2632 - 47		8 34 51	+19 18.5	1	9.87		0.48		0.02		1	F4 V	8023
	−39 04558			8 34 52	−39 56.9	1	11.37		0.20		0.17		4		402
		AAS19,45 T3# 114		8 34 54	−40 04.3	1	11.94		0.34		0.22		5		402
73420	−43 04467	LSS 1101		8 34 54	−43 54.3	3	8.84	.013	0.12	.010	−0.75	.022	10	B2 II/III	362,540,976
73210	+19 02053	NGC 2632 - 50		8 34 55	+19 26.6	4	6.75	.005	0.19	.003	0.12	.010	10	A5 III	29,1140,2001,8023
		NGC 2632 - 49		8 34 55	+19 36.8	1	10.66		0.58		0.10		1	dG0	8023
		NGC 2632 - 48		8 34 55	+19 46.5	3	12.32	.015	0.90	.005	0.64		7	dK0	802,957,1140
	−49 03675			8 34 55	−49 55.0	1	9.80		0.66				4	G5	2033
		NGC 2632 - 52		8 34 56	+19 17.0	2	12.28	.000	0.91	.003	0.65		5	K2 V	802,1140
		NGC 2632 - 55		8 34 58	+20 04.1	1	11.41		0.84		0.47		1	dG8	8023
		NGC 2632 - 54		8 34 58	+20 29.0	2	11.17	.008	0.45	.001	−0.08	.013	2	F5 V	1722,1748
73281	−4 02401	HR 3411		8 34 58	−04 45.5	4	6.19	.005	1.06	.007	0.89	.005	18	K0	15,1256,2013,2029
73322	−16 02525			8 34 58	−17 19.2	1	8.72		0.91		0.53		1	K1 V	3072
73503	−52 01545	IC 2391 - 2	★ AB	8 34 59	−52 51.7	1	8.35		0.04		−0.11		3	A0 V	1040
		NGC 2632 - 60		8 35 00	+19 35.4	1	12.92		0.98				3		802
		NGC 2632 - 58		8 35 00	+20 09.8	2	11.27	.010	0.68	.004	0.19	.001	4	dG5	1140,8023
	−39 04561			8 35 00	−40 06.0	1	11.02		0.22		−0.26		4		402
73262	+6 02001	HR 3410	★ A	8 35 01	+05 52.8	7	4.15	.014	0.00	.004	0.01	.009	24	A1 Vnn	15,1075,1363,1425*
	−45 04299			8 35 01	−46 05.9	1	10.47		0.23		0.12		2	A2	123
	−44 04591			8 35 02	−44 49.6	1	11.30		0.11		0.00		4	A0	125,396
		GD 92		8 35 03	−02 33.1	1	14.04		0.54		−0.15		2		3060
73461	−47 04135	IDS08334S4709	AB	8 35 04	−47 19.5	3	7.38	.015	0.32	.005	0.02	.005	7	A5/7 V	258,401,1075
		NGC 2632 - 70		8 35 06	+19 24.7	4	11.83	.031	0.79	.009	0.41	.001	7	dK0	802,957,1140,8023
73029	+60 01148	HR 3401		8 35 06	+60 07.0	1	6.46		0.03		0.02		2	A2 Vn	1733
		AJ77,733 T9# 9		8 35 06	−44 59.1	1	12.69		0.51		0.02		4		125
	−45 04300			8 35 06	−46 05.1	1	10.61		0.58		0.14		2		123
73478	−47 04136			8 35 07	−47 49.4	1	7.37		−0.10		−0.59		2	B3 IV	401
	−44 04593		AB	8 35 08	−45 00.2	1	10.28		0.13		0.03		3	A0	125,396
73226	+26 01816			8 35 10	+26 13.6	2	7.58	.015	0.63	.010	0.14	.019	7	G5	1355,3026
73132	+51 01440			8 35 10	+50 50.4	1	8.74		1.15		1.10		2		1566
73477	−45 04303			8 35 11	−46 08.9	2	8.50	.019	0.53	.000	0.00	.015	3	F7 V	123,389
73192	+33 01728	HR 3409		8 35 12	+32 58.7	1	5.94		1.12				2	gK2	71
		vdB 18 # a		8 35 12	−39 14.	1	15.59		1.22		0.75		1		434
		NGC 2632 - 3063		8 35 13	+19 46.6	1	13.61		0.84				2		902
73460	−40 04451			8 35 14	−40 36.0	1	8.70		0.02		−0.24		2	B9 V	123
		AJ79,1294 T10# 13		8 35 14	−44 47.8	1	13.72		0.59		0.18		3		396
73295	+10 01839			8 35 15	+10 13.4	1	9.41		0.10		0.13		13	A2	1603
		NGC 2632 - 3066		8 35 15	+19 46.0	1	14.19		1.09				2		902
		NGC 2632 - 79		8 35 15	+20 09.8	2	12.09	.015	0.91	.052	0.54		6	dK0	802,1140
		NGC 2632 - 542		8 35 15	+20 36.9	2	11.72	.005	0.78	.001	0.39	.002	4	dG8	1140,8023
73440	−37 04927			8 35 15	−37 44.1	1	8.26		1.19		1.02		5	K0 II/III	1673
	−45 04304			8 35 15	−45 20.8	2	9.10	.000	0.19	.005	−0.64		8	B2 III	1075,2021
73459	−36 04833			8 35 17	−36 57.4	1	9.39		0.03		−0.14		2	B9 IV	123
	−39 04570			8 35 17	−39 51.4	1	9.55		0.16		−0.58		4		362
73131	+53 01269	HR 3405		8 35 18	+53 06.1	1	6.32		1.35		1.47		2	K0	1733
73502	−43 04474	RZ Vel		8 35 18	−43 56.4	4	6.51	.160	0.83	.136	0.66	.129	4	G1 Ib	389,688,689,1587
	−44 04596	IDS08336S4456	AB	8 35 18	−45 06.9	1	9.30		0.19		−0.53		1		389
	−49 03682			8 35 19	−49 30.1	1	9.34		0.08		−0.55		1		389
	−39 04571			8 35 20	−39 56.8	1	10.11		0.00		−0.10		3		402
73501	−43 04475	IDS08336S4328	AB	8 35 20	−43 38.2	1	7.57		0.89		0.59		1	G6/8 (III)	389
	−36 04834			8 35 21	−36 55.5	1	10.43		1.24		1.16		2		123

Table 1 473

HD	DM	Other Id	N	Rem	α_{1950}	δ_{1950}	S	V	σ_V	B–V	σ_{B-V}	U–B	σ_{U-B}	n	Spectrum	References
		NGC 2632 - 543			8 35 22	+20 44.7	1	11.45		0.72		0.29		1	dG5	8023
73316	+10 01840	HR 3412			8 35 23	+09 45.0	3	6.54	.017	-0.01	.019	-0.05	.019	13	A1 V	15,1365,1415
		NGC 2632 - 90			8 35 23	+19 32.5	2	10.91	.010	0.70	.000	0.22	.000	3	dG0	1140,8023
73350	−6 02664	IDS08329S0628		A	8 35 24	−06 37.9	2	6.75	.010	0.65	.005	0.17		8	G0	158,2012
	−36 04835				8 35 25	−36 53.5	1	11.05		0.23		-0.25		2		123
		AAS19,45 T3# 115			8 35 25	−39 55.1	1	12.33		0.66		0.19		3		402
		NGC 2632 - 92			8 35 26	+19 23.4	1	13.32		0.64				2		902
		NGC 2632 - 93			8 35 26	+20 22.4	1	14.15		0.52				3		802
	+32 01769				8 35 26	+32 07.7	1	10.82		0.49		0.02		3		1723
73500	−39 04573				8 35 26	−39 30.8	1	9.04		0.26		-0.59		4	B2 III	362
73526	−40 04454				8 35 28	−41 08.7	1	8.96		0.74				4	G6 V	2012
	−44 04599				8 35 28	−44 44.7	1	11.99		0.42		0.08		4		125,396
73525	−40 04453				8 35 29	−40 31.0	1	7.29		0.60		0.09			K1 (III)+B9	123
73294	+20 02136	NGC 2632 - 94			8 35 30	+20 23.0	4	7.84	.033	0.47	.016	0.07	.010	4	F7 III	1140,1722,1748,2001
	+24 01966				8 35 30	+23 56.1	1	10.26		1.07		1.01		44	K0	588
73476	−33 05257	HR 3419	★	A	8 35 31	−33 34.2	3	6.50	.017	0.32	.011	0.02		24	F0 IV/V	15,978,2012
73524	−39 04574	HR 3421			8 35 31	−39 58.3	6	6.54	.011	0.60	.006	0.12	.013	14	G1 V	15,258,402,1705,2012,3077
		NGC 2632 - 99			8 35 32	+20 14.2	1	11.33		1.04		0.83		1		1722
		AJ77,733 T9# 10			8 35 32	−44 54.9	1	13.70		0.56		0.05		4		125
	−39 04575				8 35 33	−39 55.4	1	9.43		0.55		0.39		3		402
	−46 04372				8 35 33	−46 32.0	1	10.40		0.14		-0.12		2	B9	123
73278	+31 01848				8 35 34	+31 20.2	1	8.47		1.25		1.43		std	K2	1211
73171	+53 01272	HR 3408			8 35 36	+52 53.3	2	5.92	.012	1.17	.005	1.05		3	gK1	71,3051
		NGC 2632 - 559			8 35 37	+20 03.3	3	14.03	.009	1.25	.012	1.20		9		802,902,1140
73589	−46 04375				8 35 37	−47 01.4	2	8.89	.045	0.04	.000	-0.46		6	B3 V	434,1075
73568	−44 04602	LSS 1102			8 35 38	−45 01.9	5	8.34	.008	0.31	.011	-0.56	.017	20	B2 III/IV	125,540,976,1075,2021
73701	−59 01065				8 35 38	−60 09.1	1	7.85		-0.05				4	B5 V	1075
73346	+17 01896				8 35 39	+17 14.2	1	8.12		0.38		-0.01		2	F5	1648
		NGC 2632 - 107			8 35 39	+19 31.2	1	11.64		0.43		0.01		1		1748
	−45 04313				8 35 39	−45 56.7	1	10.53		0.11				4	B3	2021
	−46 04376				8 35 39	−46 36.6	1	10.52		0.65		0.18		2		123
	+19 02056	NGC 2632 - 108			8 35 40	+19 38.5	3	10.01	.008	1.02	.004	0.78	.004	13	G8 V	1140,1245,1748
	−45 04314				8 35 41	−45 26.4	1	9.00		0.37		0.18		1		389
	−46 04378				8 35 41	−46 37.4	1	10.58		0.09		-0.17		2		123
73347	+15 01862				8 35 42	+15 02.1	1	8.00		0.36		-0.02		2	F0	1648
73330	+20 02137	NGC 2632 - 109			8 35 42	+20 02.2	2	9.18	.016	1.02	.005	0.80	.014	2	K0 III	1722,1748
		Pis 5 - 9			8 35 42	−39 22.5	1	12.40		0.68		0.58		2		184
73635	−48 03952				8 35 42	−48 44.1	1	8.42		0.45		0.04		1	F3 V	389
73495	−25 06356	HR 3420	★	A	8 35 44	−26 04.7	2	5.26	.005	-0.04	.000			7	A0 V	15,2012
	−39 04582	Pis 5 - 1	★	AB	8 35 44	−39 23.8	1	10.45		0.38		-0.22		3		184
		Pis 5 - 2			8 35 44	−39 24.1	1	12.96		0.47		0.38		2		184
73345	+20 02138	NGC 2632 - 114	★	V	8 35 45	+20 10.0	5	8.14	.010	0.21	.004	0.11	.005	9	A8 V	29,1140,1722,1748,8023
		POSS 210 # 27			8 35 45	+40 16.9	1	16.59		1.46				1		1739
		NGC 2632 - 117			8 35 46	+19 49.4	1	10.56		0.51		-0.01		1		1748
		Pis 5 - 5			8 35 46	−39 23.9	1	14.00		0.90		0.30		2		184
73681	−52 01552	IC 2391 - 3			8 35 46	−53 05.0	3	7.85	.018	0.10	.017	0.03	.026	4	A1 V	1030,1040,3015
73344	+24 01968				8 35 49	+23 51.9	2	6.90	.000	0.54	.001	0.08		12	F8 V	1351,1733
73588	−40 04455	GO Vel			8 35 50	−40 15.6	2	6.86	.093	1.69	.060	1.80		17	M4 III	2012,3042
		NGC 2632 - 544			8 35 51	+20 45.9	1	12.44		0.91				3	dK2	902
73108	+64 00698	HR 3403			8 35 52	+64 30.3	7	4.60	.011	1.18	.009	1.15	.020	21	K1 IIIb	15,1118,1355,3016*
		Pis 5 - 6			8 35 52	−39 23.8	1	13.73		0.74		0.49		2		184
		Pis 5 - 7			8 35 52	−39 24.1	1	14.16		1.17		0.60		2		184
73451	−6 02669	HR 3416			8 35 53	−06 29.2	3	6.50	.004	0.45	.004	0.16	.010	11	A1 V +G	15,1075,2012
		Pis 5 - 8			8 35 53	−39 24.8	1	13.63		0.89		0.61		2		184
73606	−40 04460				8 35 53	−40 39.9	1	8.34		0.31		0.14		2	A5 III(m)	123
73634	−42 04451	HR 3426, LSS 1103			8 35 54	−42 48.8	8	4.14	.002	0.11	.006	0.16	.032	49	A7 II	15,200,1075,1637,2012*
		Pis 5 - 3			8 35 54	−39 24.	1	14.67		1.11		0.61		2		184
		Pis 5 - 4	★	C	8 35 54	−39 24.	1	12.48		0.29		0.24		2		184
73397	+20 02139	NGC 2632 - 124			8 35 55	+19 40.6	4	8.99	.007	0.32	.004	0.04	.007	10	F1 V	29,1140,1748,8023
73411	+9 02021				8 35 56	+09 34.6	1	8.81		0.33		0.08		20	F5	1603
		NGC 2632 - 127			8 35 58	+20 14.7	2	10.80	.005	0.60	.007	0.09	.001	5	dG0	1140,8023
		NGC 2632 - 129			8 35 59	+19 45.4	1	11.93		0.52		0.07		1		1748
	−44 04606				8 35 59	−44 25.6	1	9.89		0.23				4	B2 III	2021
		NGC 2632 - 130			8 36 00	+20 11.9	1	11.48		0.43		0.10		1		1748
73658	−45 04322	LSS 1105			8 36 00	−46 06.4	4	6.86	.004	0.04	.008	-0.77	.022	12	B1 Ib/II	123,158,540,976
73722	−52 01554	IC 2391 - 4			8 36 00	−52 28.2	3	8.90	.013	0.43	.005	-0.04	.016	5	F3 V	389,1040,3015
		NGC 2632 - 3164			8 36 01	+19 44.9	1	14.22		1.31				4		902
73657	−41 04328				8 36 01	−42 08.6	1	7.84		1.21		1.28		1	K1 III	389
73678	−46 04388	T Vel			8 36 03	−47 11.2	1	7.71		0.76		0.47		1	G0 II	688
73834	−63 01012				8 36 03	−63 27.1	1	8.10		-0.02		-0.71		1	B5ne	55
	+19 02058	NGC 2632 - 134			8 36 04	+19 22.5	2	10.18	.040	0.91	.013	0.53	.011	3	G8	1140,1748
		NGC 2632 - 560			8 36 04	+20 21.5	3	13.95	.017	1.31	.023	1.26		8		802,902,1140
	−43 04488	LSS 1106			8 36 05	−43 22.7	2	10.59	.005	0.63	.000	-0.39	.040	8	B0.5 Vn	362,1737
		NGC 2632 - 3172			8 36 06	+19 47.4	1	13.79		1.17				3		902
73740	−53 01780	IDS08347S5315		A	8 36 07	−53 25.0	1	8.18		0.78		0.43		1	G8/K0 IV	389
		NGC 2632 - 138			8 36 08	+19 36.1	2	11.18	.000	0.26	.012	0.09	.013	2		1140,1748
73561	−16 02534				8 36 08	−16 49.6	1	8.63		0.89		0.49		5	K1 III +(G)	897
73471	+3 02026	HR 3418			8 36 09	+03 31.1	8	4.45	.016	1.21	.007	1.27	.016	25	K2 III	15,58,1075,1355,1363*
		NGC 2632 - 139			8 36 09	+20 11.0	1	11.88		1.18		1.32		1	K5 V	1748
		NGC 2632 - 141			8 36 11	+19 30.2	1	12.39		0.93				3	dK0	802

HD	DM	Other Id	N Rem	α_{1950}	δ_{1950}	S	V	σ_V	B–V	σ_{B-V}	U–B	σ_{U-B}	n	Spectrum	References
	+20 02140	NGC 2632 - 142	★ C	8 36 11	+19 54.1	4	9.30	.009	0.49	.004	0.01	.006	8	F7 V	29,1140,1748,8023
73430	+20 02141	NGC 2632 - 143		8 36 11	+20 10.6	3	8.31	.002	0.22	.005	0.10	.009	6	A8 V	29,1140,8023
		NGC 2632 - 561	★ V	8 36 11	+20 13.2	3	14.43	.012	1.37	.008	1.24		9		802,902,1140
		NGC 2632 - 1751		8 36 12	+19 41.9	1	13.79		1.20				4		802
73653	-37 04949			8 36 12	-37 50.0	4	7.19	.008	-0.17	.001	-0.80	.011	11	B2 III	401,540,976,1732
73429	+20 02142	NGC 2632 - 146		8 36 13	+20 17.6	4	9.38	.011	0.40	.004	0.00	.005	7	F5 V	29,1140,1748,8023
		NGC 2632 - 149		8 36 14	+19 28.9	1	12.37		0.46		-0.05		3		1140
73449	+20 02143	NGC 2632 - 150	★ A	8 36 14	+19 51.2	3	7.45	.002	0.25	.002	0.14	.003	7	A7 Vnn	29,1140,8023
73778	-53 01784	IC 2391 - 5		8 36 16	-54 00.5	1	8.77		0.36		-0.07		3	F0 V	1040
73450	+20 02144	NGC 2632 - 154	★ V	8 36 17	+19 46.2	3	8.50	.005	0.25	.001	0.08	.006	9	A9 V	29,1140,8023
73395	+36 01850			8 36 17	+35 41.8	1	7.38		0.06		0.06		2	A2	1601
		G 9 - 12		8 36 18	+12 05.4	2	12.16	.000	1.10	.005	0.91	.015	2		1620,1658
		NGC 2632 - 155	★ B	8 36 18	+19 51.3	4	9.41	.005	0.41	.005	0.01	.004	7	F4 V	29,1140,1748,8023
		NGC 2632 - 563		8 36 18	+19 57.7	1	16.11		1.60				1		957
73699	-39 04603			8 36 19	-39 54.3	2	7.58	.010	0.04	.005	-0.70	.010	6	B3 V	362,401
	-44 04619			8 36 19	-44 46.7	1	10.69		0.70		0.29		4		125,396
73696	-39 04604			8 36 20	-39 21.6	1	9.10		1.12		0.91		3	M4 (III)	402
73698	-39 04605			8 36 20	-39 41.4	3	8.95	.009	0.16	.011	-0.57	.022	10	B3 III	362,540,976
		NGC 2632 - 162		8 36 21	+19 17.6	2	10.59	.010	0.57	.001	0.06	.002	4	dG0	1140,8023
		NGC 2632 - 164		8 36 22	+20 23.3	2	11.31	.000	0.71	.003	0.24	.000	5	dG5	1140,8023
73739	-46 04393		V	8 36 22	-46 43.7	2	7.48	.070	1.63	.050	0.68		6	M7 II/III	123,1075
		NGC 2632 - 566		8 36 23	+20 14.9	3	14.93	.020	1.46	.025	1.10		9		802,902,1140
		L 60 - 100		8 36 23	+67 50.1	1	14.68		1.68				2		1663
73583	-12 02618			8 36 24	-13 04.8	4	9.68	.021	1.11	.007	1.02	.022	13	K3/4 V	158,1705,1775,3072
73695	-38 04689			8 36 24	-39 14.5	1	8.83		0.33		-0.58		4	B2/3 II	362
73603	-19 02489	HR 3425	★ AB	8 36 25	-19 33.6	5	6.33	.013	1.59	.007	1.83		20	M1 III	15,1075,2013,2029,3005
		NGC 2635 - 1		8 36 25	-34 35.2	1	12.45		0.60		0.08		4		183
	-44 04624			8 36 25	-44 50.4	1	9.90		0.62		0.03		4	G0	125,396
73776	-49 03692			8 36 25	-49 54.3	1	7.68		1.09		0.95		1	K0 III	389
		NGC 2635 - 6		8 36 26	-34 35.1	1	14.19		0.31		0.32		2		183
73887	-62 01058	HR 3432	★ AB	8 36 26	-62 40.6	3	5.46	.007	1.03	.009	0.83		9	K0 III	15,404,2012
73487	+20 02145	NGC 2632 - 167		8 36 27	+20 13.8	2	9.23	.010	0.92	.004	0.59	.011	2	K3 V	1722,1748
		NGC 2635 - 2		8 36 27	-34 35.7	1	13.48		1.04		0.86		2		183
		NGC 2635 - 3		8 36 28	-34 36.0	1	12.96		1.22		0.95		2		183
		NGC 2632 - 172		8 36 30	+20 02.3	4	12.53	.015	0.94	.004	0.70		10	K2 V	802,902,957,1140
		NGC 2635 - 4		8 36 30	-34 36.0	1	12.99		1.08		0.60		2		183
		NGC 2635 - 5		8 36 30	-34 36.0	1	13.83		1.16		0.74		2		183
73757	-45 04332			8 36 32	-45 25.0	1	7.78		1.03		0.86		1	K0 III	389
	+19 02061	NGC 2632 - 181		8 36 33	+19 38.2	3	10.51	.020	0.58	.022	0.08	.000	4	G0 V	1140,1748,8023
		POSS 210 # 10		8 36 33	+42 06.1	1	14.70		1.48				1		1739
73448	+32 01774			8 36 34	+32 13.3	1	8.30		1.59		1.97		std	K5	1211
		NGC 2632 - 568		8 36 35	+19 42.6	2	14.12	.010	0.78	.006	0.37		3		902,1140
		NGC 2632 - 184		8 36 37	+19 39.1	2	11.66	.024	0.90	.005	0.64		3	dK0	802,8023
		NGC 2632 - 183		8 36 37	+19 57.8	4	12.62	.015	0.99	.010	0.76		10	K2.5V	802,902,957,1140
		WD0836+237		8 36 37	+23 44.7	1	16.64		-0.25		-1.21		4	DA2	1727
	+20 02146	NGC 2632 - 182		8 36 38	+20 14.8	2	10.34	.015	0.65	.030	0.16	.005	2	F8 V	1748,8023
		NGC 2632 - 192		8 36 41	+19 28.4	3	10.96	.007	0.25	.000	0.09	.005	13		1140,1245,1748
73533	+20 02147	NGC 2632 - 190		8 36 41	+20 20.8	6	9.06	.006	1.30	.009	1.43	.005	20	K2 V	830,1140,1722,1748,1748,1783
73599	+8 02099	HR 3424		8 36 44	+08 11.7	3	6.45	.005	1.08	.000	0.99	.008	11	K1 III	15,1355,1415
	+19 02063	NGC 2632 - 196		8 36 44	+19 03.3	1	10.74		0.59		0.09		1	dG0	8023
		POSS 210 # 56		8 36 44	+38 51.5	1	17.82		1.08				3		1739
		NGC 2632 - 3277		8 36 45	+19 26.3	1	14.38		1.35				3		902
		NGC 2632 - 569		8 36 45	+19 59.7	2	15.20	.000	1.51	.009			6		802,902
		NGC 2632 - 198		8 36 47	+19 37.1	3	12.61	.016	0.97	.006	0.77		9	dK2	802,957,1140
	-44 04634	LSS 1108		8 36 47	-44 52.0	3	9.81	.004	0.99	.012	0.04	.000	11	B5 Ia	125,396,1737
73509	+29 01797			8 36 48	+28 40.8	1	8.67		0.50				2	F8 V	20
73813	-46 04395	IDS08351S4625	A	8 36 48	-46 36.0	2	7.77	.000	-0.02	.009	-0.52	.018	3	B5 V	123,1732
73813	-46 04395	IDS08351S4625	AB	8 36 48	-46 36.0	1	7.58		-0.02				4	B5 V	1075
73849	-50 03447			8 36 48	-50 19.9	1	8.27		0.08		-0.60		2	B2/3 III	401
		NGC 2632 - 570		8 36 49	+20 10.3	1	15.35		1.49				2		902
		NGC 2632 - 571		8 36 49	+20 12.5	1	15.41		1.36				4		802
73575	+20 02149	NGC 2632 - 204	★ V	8 36 50	+19 57.4	3	6.66	.004	0.24	.003	0.16	.005	8	F0 IIIb	29,1140,8023
73574	+20 02148	NGC 2632 - 203	★ AB	8 36 50	+20 15.8	5	7.72	.013	0.23	.006	0.10	.007	8	Am	29,1140,1722,1748,8023
		Steph 717		8 36 51	+19 26.0	1	11.46		1.39		1.22		1	K7	1746
		KUV 08368+4026		8 36 51	+40 25.7	1	15.55		0.22		-0.59		1	DA	1708
		G 194 - 37		8 36 51	+49 09.9	1	15.31		0.84		0.22		4		1658
		NGC 2632 - 206		8 36 52	+19 36.1	1	11.41		1.40		1.40		1	K5 V	1748
73576	+19 02064	NGC 2632 - 207	★ V	8 36 53	+19 27.2	3	7.67	.003	0.19	.002	0.12	.010	9	A7 IV-Vn	29,1140,8023
		LB 390		8 36 53	+20 10.9	1	17.85		0.16		-0.68		3		3028
		NGC 2632 - 208		8 36 54	+19 32.7	3	10.67	.009	0.58	.008	0.10	.009	5	G1 V	1140,1748,8023
73808	-39 04624			8 36 54	-39 53.4	1	8.37		1.02		0.77		3	K0 III	402
73811	-42 04469			8 36 54	-42 16.2	1	8.24		0.02		-0.23		2	B8/9 III	123
		LB 1847		8 36 55	+19 56.9	1	18.23		0.05		-0.49		3		3028
		NGC 2632 - 209		8 36 55	+20 00.3	3	12.73	.007	0.98	.004	0.79		9		802,902,1140
73393	+56 01322	G 194 - 38		8 36 55	+55 51.0	2	8.02	.013	0.68	.000	0.22	.000	5	G3 V	1003,1733
73752	-22 02345	HR 3430	★ AB	8 36 57	-22 29.4	5	5.05	.005	0.73	.007	0.35		16	G6 IV	15,258,1075,1705,2029
73598	+20 02150	NGC 2632 - 212	★ D	8 36 59	+19 43.1	6	6.59	.009	0.96	.007	0.72	.009	16	K0 III	1140,1355,2001,3077*
		NGC 2632 - 213		8 36 59	+19 43.7	4	11.81	.017	0.78	.015	0.45	.040	7		802,902,1748,8023
73847	-46 04400	IDS08353S4626	AB	8 36 59	-46 36.3	3	8.21	.008	-0.04	.002	-0.47	.035	8	B7 III	123,1075,1732
		NGC 2632 - 215		8 37 00	+20 07.5	1	11.30		0.52		-0.09		1		1748

Table 1   475

HD	DM	Other Id	N Rem	α_{1950}	δ_{1950}	S	V	σ_V	B–V	σ_{B-V}	U–B	σ_{U-B}	n	Spectrum	References
73904	−53 01789	IC 2391 - 6		8 37 00	−53 32.7	1	7.67		0.08		−0.04		3	A1 Vn	1040
		NGC 2632 - 217		8 37 01	+19 29.4	3	10.24	.010	0.51	.005	0.02	.002	3	F8	1140,1748,8023
73597	+21 01882	NGC 2632 - 218		8 37 01	+20 44.3	2	9.36	.024	0.41	.008	0.03	.016	7	F6 V	29,8023
	+20 02151	NGC 2632 - 222		8 37 03	+20 14.6	3	10.11	.004	0.48	.004	0.00	.004	7	F4 V	1140,1748,8023
73668	+6 02007	IDS08344N0608	A	8 37 04	+05 56.7	3	7.26	.012	0.61	.010	0.08	.004	9	G1 V	1003,1084,3024
73668	+6 02008	IDS08344N0608	B	8 37 04	+05 56.7	2	8.42	.005	0.82	.005	0.45	.000	7		1084,3024
73618	+20 02152	NGC 2632 - 224	⋆ AB	8 37 04	+19 43.8	3	7.32	.000	0.19	.003	0.15	.009	8	Am	29,1140,8023
73447	+51 01443			8 37 05	+51 36.5	1	6.88		0.96		0.65		3	G5	1566
73641	+19 02066	NGC 2632 - 227		8 37 06	+19 22.6	5	9.49	.005	0.41	.004	0.00	.005	9	F6 V	29,345,1140,1748,8023
73619	+20 02153	NGC 2632 - 229	⋆ C	8 37 06	+19 43.2	3	7.54	.002	0.25	.002	0.13	.009	9	Am	29,1140,8023
73616	+20 02154	NGC 2632 - 226		8 37 06	+20 20.2	3	8.89	.002	0.32	.004	0.05	.003	8	F1 V	29,1140,8023
73884	−47 04171			8 37 06	−47 36.7	1	7.84		1.86				4	K2/3 Iab/b	1075
		G 52 - 23		8 37 07	+09 06.8	1	12.93		1.58		1.29		1		333,1620
73667	+12 01888	G 9 - 13	⋆ A	8 37 07	+11 42.4	7	7.62	.015	0.83	.008	0.45	.009	20	K1 V	22,333,1197,1355,1620*
		NGC 2632 - 230		8 37 07	+19 50.8	1	9.78		0.38		0.08		1		1140
73617	+20 02155	NGC 2632 - 232	⋆ P	8 37 07	+20 12.6	6	9.23	.008	0.38	.006	0.01	.006	17	F5 V	29,542,830,1140,1748,8023
		IC 2391 - 76	⋆ B	8 37 07	−52 41.7	1	9.69		1.17		0.84		1	A8 V	1030
		NGC 2632 - 236		8 37 08	+19 44.7	3	11.94	.007	1.00	.004	0.81		6		802,902,8023
		NGC 2632 - 237		8 37 08	+19 45.3	2	12.87	.010	1.00	.005			6		802,902
		NGC 2632 - 3342		8 37 09	+19 29.2	1	14.43		1.38				3		902
		NGC 2632 - 238		8 37 09	+19 59.1	3	10.30	.005	0.50	.008	0.00	.000	6	F8	1140,1748,8023
73640	+20 02156	NGC 2632 - 239		8 37 09	+20 18.8	4	9.66	.005	0.43	.005	−0.01	.007	11	F6 V	29,1140,2001,8023
73642	+19 02067	NGC 2632 - 234		8 37 10	+19 07.6	2	9.96	.000	0.25	.000	0.14	.035	2	A3	344,345
	+19 02068	NGC 2632 - 244	⋆ V	8 37 11	+19 10.6	3	10.05	.076	0.62	.012	0.11	.009	3	F8 V	344,345,8023
		LP 844 - 42		8 37 11	−25 13.6	1	11.65		1.09		1.05		1		1696
		NGC 2632 - 572		8 37 12	+19 56.	3	14.57	.024	0.97	.015	0.69		5	K2 V	802,957,1140
		NGC 2632 - 246		8 37 12	+19 57.7	3	12.05	.025	0.83	.009	0.50		7	K0.5V	802,957,1140
73596	+32 01776	HR 3423		8 37 12	+32 07.2	1	6.22		0.39		0.05		2	F4 III-IV	1733
	+20 02157	NGC 2632 - 250		8 37 13	+19 54.4	3	9.79	.000	0.47	.004	0.00	.005	3	F6 V	1140,1748,8023
		NGC 2632 - 250s	⋆	8 37 13	+19 54.4	1	15.10		0.85				1		902
73665	+20 02158	NGC 2632 - 253	⋆ A	8 37 14	+20 11.1	13	6.39	.006	0.98	.005	0.83	.005	84	K0 IIIa	15,29,130,542,1140*
		NGC 2632 - 256		8 37 15	+19 12.2	2	12.63	.010	1.01	.008	0.68		5	dK2	802,1140
		NGC 2632 - 257		8 37 15	+19 29.1	2	11.00	.005	0.77	.006	0.38	.001	2	dG5	1140,8023
		NGC 2632 - 258		8 37 15	+19 37.9	2	10.24	.005	0.57	.000	0.08	.005	2	F8	1748,8023
73554	+43 01844	G 115 - 16		8 37 15	+43 17.7	1	9.30		1.08		0.93		1	K2 V	3072
73921	−50 03453			8 37 15	−51 05.8	1	7.77		0.89		0.56		1	K0 III +A0	389
		Pis 6 - 8	⋆ K	8 37 16	−46 03.0	2	12.20	.015	0.55	.039	0.07	.005	3		184,778
		NGC 2632 - 263		8 37 18	+19 48.0	4	12.01	.011	0.82	.006	0.43	.000	10	K0 V	802,902,1140,8023
		Pis 6 - 29		8 37 18	−46 02.8	1	13.50		0.21		−0.12		1		778
		NGC 2632 - 266		8 37 19	+19 37.2	1	11.70		0.41		−0.03		2		1526
73666	+20 02159	NGC 2632 - 265	⋆ B	8 37 19	+20 08.9	7	6.61	.005	0.01	.005	0.02	.005	60	A0mA2IVs	29,542,1140,1245,1351*
73766	−9 02612	RV Hya		8 37 19	−09 24.5	1	8.10		1.62		1.14		10	M5 II	3042
73882	−39 04631	LSS 1110	⋆ AB	8 37 19	−40 14.5	5	7.22	.013	0.39	.012	−0.53	.030	24	O9 III	362,402,540,976,2012
73903	−45 04348	Pis 6 - 3	⋆ F	8 37 19	−46 03.0	3	8.97	.014	0.21	.017	−0.58	.015	7	B1.5IV	184,778,2021
		G 234 - 45		8 37 20	+59 41.2	2	15.06	.017	1.90	.046	1.54		7		906,940
73952	−52 01565	IC 2391 - 8	⋆ A	8 37 20	−52 54.8	4	6.46	.011	−0.10	.005	−0.34	.020	10	B8.5Vn	42,1030,1040,2006
	+20 02160	NGC 2632 - 268	⋆ R	8 37 21	+19 49.0	4	9.90	.009	0.47	.004	0.01	.004	9	F6 IV	29,1140,1748,8023
		Pis 6 - 51	⋆ L	8 37 21	−46 01.5	1	11.73		0.26		0.04		1		778
		Pis 6 - 7	⋆ D	8 37 21	−46 03.2	2	11.15	.010	0.26	.005	−0.43	.010	2		184,778
		NGC 2632 - 573		8 37 22	+19 55.7	1	16.06		1.50				2		957
		NGC 2632 - 267		8 37 22	+19 57.4	2	13.18	.010	1.10	.015			5		802,957
		Pis 6 - 16		8 37 22	−46 05.2	1	13.33		0.36		−0.07		2		778
		NGC 2632 - S 1391	⋆	8 37 23	+19 37.2	1	13.91		1.33				4		802
	+20 02161	NGC 2632 - 271	⋆ Q	8 37 23	+20 10.3	5	8.81	.001	0.32	.003	0.07	.010	13	F2 V	29,542,1140,1748,8023
		NGC 2632 - 272		8 37 24	+20 05.6	1	12.70		1.01				3	F1 V	802
73919	−45 04351	Pis 6 - 1	⋆ AB	8 37 24	−46 03.0	3	8.86	.005	0.18	.017	−0.58	.018	7	B2 IV	184,540,778,976
	−45 04352	Pis 6 - 2	⋆ C	8 37 24	−46 03.1	2	8.89	.024	0.16	.010	−0.63	.015	4	B2 IV	184,778
		Pis 6 - 6	⋆ E	8 37 24	−46 03.6	2	11.48	.005	0.25	.005	−0.33	.005	4		184,778
		Pis 6 - 15		8 37 24	−46 04.7	1	14.35		0.46		0.33		2		778
73951	−51 03132	IC 2391 - 7		8 37 24	−51 59.1	1	9.29		0.04		0.02		4	A2 Vn	42
		NGC 2632 - 273		8 37 25	+20 15.3	1	12.35		0.26		0.06		2		1140
		NGC 2632 - 545		8 37 26	+18 43.9	1	12.63		0.78				3	dG5	902
73711	+20 02163	NGC 2632 - 276	⋆ B	8 37 26	+19 42.6	4	7.54	.001	0.15	.005	0.13	.008	14	Am	29,1140,8023,8052
	+20 02162	NGC 2632 - 275		8 37 26	+19 57.9	5	9.96	.004	0.58	.005	0.09	.005	13	G1 V	29,1140,1748,2001,8023
		NGC 2632 - 3401		8 37 26	+20 22.1	1	12.97		1.21				3		902
		Pis 6 - 9		8 37 26	−46 03.7	2	13.86	.093	0.28	.068	0.17		3		184,778
	−45 04355	Pis 6 - 4	⋆ G	8 37 26	−46 04.1	3	9.46	.007	0.20	.030	−0.63	.016	7	B1 Vne	184,778,1737
		Pis 6 - 5	⋆ H	8 37 26	−46 04.1	2	10.61	.019	0.18	.009	−0.53	.014	4		184,778
		Pis 6 - 14		8 37 26	−46 04.9	1	14.18		0.50		0.28		2		778
73967	−52 01566	IC 2391 - 9		8 37 26	−52 25.5	1	9.96		0.32		0.14		4	F0 V	42
		IC 2391 - 70		8 37 26	−52 46.9	1	13.24		0.50		0.04		3		1030
73190	+73 00428			8 37 27	+73 20.9	1	7.12		0.00		0.00		1	A0	1776
73900	−36 04872	HR 3434	⋆ AB	8 37 27	−36 25.8	2	6.12	.005	0.43	.005			10	F1 IV	15,2012
73712	+19 02069	NGC 2632 - 284		8 37 28	+19 31.6	3	6.78	.008	0.26	.004	0.14	.007	6	A9 IVn	29,1140,8023
73901	−39 04633			8 37 28	−39 37.3	1	7.27		0.15		0.15		3	A2 V(m)	402
	−42 04480			8 37 28	−42 36.4	1	10.52		0.08				1	A2	177
		Pis 6 - 10	⋆ J	8 37 28	−46 04.1	2	12.28	.030	0.27	.005	−0.06	.010	5		184,778
73709	+20 02165	NGC 2632 - 279	⋆ C	8 37 29	+19 51.9	4	7.70	.000	0.20	.007	0.15	.007	8	A5 m	150,542,1140,8023
	−52 01567	IC 2391 - 56		8 37 29	−52 47.1	1	10.33		0.68		0.18		2		1040
73710	+20 02166	NGC 2632 - 283	⋆ A	8 37 30	+19 50.9	11	6.42	.015	1.02	.003	0.89	.015	36	K0.5IIIa	15,150,542,1140,1569*

HD	DM	Other Id	N	Rem	α_{1950}	δ_{1950}	S	V	σ_V	B–V	σ_{B-V}	U–B	σ_{U-B}	n	Spectrum	References
		NGC 2632 - 283s		★	8 37 30	+19 50.9	1	13.60		0.86				1		902
	+20 02164	NGC 2632 - 282			8 37 30	+20 17.1	3	10.08	.000	0.50	.003	0.02	.008	7	F2 III	1140,1748,8023
		NGC 2632 - 288			8 37 31	+19 38.6	3	10.70	.008	0.59	.005	0.09	.005	4	G0 V	1140,1748,8023
		NGC 2632 - 287		★ B	8 37 31	+19 51.1	1	10.37		0.59		0.11		2	dG0	542
73730	+20 02168	NGC 2632 - 286			8 37 31	+20 00.8	4	8.02	.007	0.19	.000	0.13	.006	13	Am	29,1140,8023,8052
		Pis 6 - 11			8 37 31	-46 04.1	2	13.15	.010	0.26	.020	0.09	.085	4		184,778
73966	-51 03133				8 37 31	-51 24.3	1	9.36		0.11		-0.24		2	B7 IV	401
72520	+82 00253				8 37 32	+82 25.2	2	6.89	.015	-0.03	.005	-0.08	.005	4	A0	985,1733
	-5 02603				8 37 32	-06 17.7	3	9.89	.040	1.40	.027	1.25	.039	7	M0	1705,3072,7008
	-43 04518			A	8 37 33	-43 53.6	1	11.72		0.33		-0.38		3		362
	-43 04518			AB	8 37 33	-43 53.6	1	11.43		0.40		-0.30		2		362
		NGC 2632 - 293			8 37 34	+19 39.3	2	9.87	.020	0.48	.006	0.01	.005	2		1748,8023
	+20 02170	NGC 2632 - 295		★ D	8 37 34	+19 51.9	5	9.37	.015	0.41	.004	0.02	.006	12	F5 V	29,542,1140,1748,8023
73729	+20 02169	NGC 2632 - 292		★ V	8 37 34	+20 21.6	3	8.20	.009	0.30	.004	0.10	.006	9	A9 Vn	29,1140,8023
73593	+46 01422	HR 3422			8 37 34	+46 00.7	3	5.38	.016	1.00	.005	0.75	.003	7	G0 IV	1080,1355,3016
		NGC 2632 - 3427			8 37 35	+19 23.8	1	12.90		1.04				3		902
73731	+20 02171	NGC 2632 - 300		★ A	8 37 35	+19 43.4	8	6.30	.007	0.17	.003	0.17	.008	32	A5 m	15,29,1022,1140,1569*
		Pis 6 - 39			8 37 35	-46 05.1	1	13.09		0.43		0.31		2		778
	-52 01568	IC 2391 - 52			8 37 35	-52 50.9	1	9.62		0.46		-0.08		2	F6 V	1040
		NGC 2632 - 301			8 37 36	+19 27.4	2	11.16	.005	0.66	.002	0.19	.000	3	dG5	1140,8023
		NGC 2632 - 299			8 37 36	+19 50.0	3	13.21	.020	1.08	.007			7	K4 V	802,902,957
		LB 393			8 37 36	+19 54.2	1	17.66		0.04		-0.60		3		3028
		NGC 2632 - 297			8 37 36	+20 29.5	1	11.64		0.86		0.48		1	dK0	8023
73844	-16 02541	AK Hya			8 37 36	-17 07.4	5	6.68	.074	1.55	.025	1.14	.024	40	M6 III	897,1075,2012,2017,3042
		Pis 6 - 41			8 37 37	-46 04.9	1	12.88		0.32		-0.03		2		778
74009	-52 01569	IC 2391 - 10		★ AB	8 37 37	-52 32.0	1	8.78		0.41		0.02		4	F2 Vnn	42
73898	-29 06544	HR 3433		★ A	8 37 38	-29 22.9	3	4.87	.011	0.90	.000			11	G5 III	15,2013,2029
	-45 04360				8 37 38	-46 14.9	1	10.18		0.15		0.11		2	A0	123
		WLS 840 15 # 7			8 37 39	+13 10.5	1	11.55		1.09		0.86		2		1375
		NGC 2632 - 304			8 37 39	+20 22.8	2	11.52	.000	0.74	.004	0.31	.000	6	dG8	1140,8023
73840	-11 02420	HR 3431			8 37 39	-12 17.9	5	4.98	.007	1.41	.009	1.62		21	K3 III	15,1075,2013,2029,3053
	-46 04409				8 37 39	-46 20.3	1	9.64		0.12		-0.53		2		123
		NGC 2632 - 309			8 37 40	+20 01.7	2	11.63	.000	0.76	.002	0.35	.002	5	G9 V	1140,8023
73746	+19 02072	NGC 2632 - 318		★ V	8 37 41	+19 22.4	3	8.66	.006	0.29	.001	0.08	.000	5	A9 V	29,1140,8023
		NGC 2632 - 313			8 37 42	+19 48.7	3	12.20	.009	0.88	.008	0.57		10	dK0	802,902,1140
		vdB 20 # a			8 37 42	-40 17.6	1	10.86		0.19		-0.23		4	B6 V	434
		NGC 2632 - 546			8 37 43	+18 51.2	1	11.62		0.77		0.34		1	dG8	8023
		NGC 2632 - 321			8 37 46	+20 04.9	2	11.70	.015	0.32	.006	-0.02	.007	3		1140,1748
		NGC 2632 - 547			8 37 47	+18 52.7	1	11.99		0.97				3	dG8	902
73763	+19 02073	NGC 2632 - 323		★ V	8 37 48	+19 24.4	2	7.81	.005	0.21	.007	0.09	.000	2	A8 V	1140,8023
		NGC 2632 - 322			8 37 48	+19 50.9	3	10.87	.000	0.68	.004	0.19	.005	8	G2 V	1140,1748,8023
		KUV 08378+3934			8 37 48	+39 33.8	1	16.54		-0.04		-0.91		1	DA	1708
	+62 01010	IDS08336N6223		A	8 37 48	+62 12.1	1	9.59		0.46		-0.03		3		1723
	+62 01010	IDS08336N6223		B	8 37 48	+62 12.1	1	10.43		0.57		0.02		3		1723
	+14 01947	G 9 - 14			8 37 49	+13 44.1	1	10.18		0.80		0.40		1		333,1620
74044	-51 03135	IC 2391 - 11			8 37 49	-52 08.2	2	8.51	.019	0.22	.002	0.09	.005	6	Am	42,3015
		NGC 2632 - 327			8 37 50	+20 11.6	1	11.63		0.67		0.23		1		1748
73986	-42 04487				8 37 50	-42 18.6	1	8.38		-0.05		-0.19		2	B8 V	123
		NGC 2632 - 325			8 37 51	+19 24.1	3	10.61	.005	0.61	.004	0.09	.005	5	dG0	1140,1748,8023
		NGC 2632 - 326			8 37 51	+19 44.7	3	11.34	.008	0.71	.008	0.27	.008	5	G7 V	1140,1748,8023
73785	+20 02172	NGC 2632 - 328			8 37 51	+19 53.9	5	6.84	.007	0.20	.004	0.15	.008	18	A9 IIIb-	29,1140,1569,1569,8023
	-52 01573	IC 2391 - 59			8 37 53	-53 06.2	1	11.57		0.35		-0.01		1		1040
74056	-52 01574	IC 2391 - 12			8 37 53	-53 06.4	1	9.38		0.12		0.03		2	A2 V	1040
	+19 02074	NGC 2632 - 332			8 37 55	+19 29.3	4	9.56	.003	0.42	.006	0.00	.001	6	F6 V	29,1140,1748,8023
		NGC 2632 - 333			8 37 55	+19 43.3	1	10.35		0.90		0.62		1	K3 V	1748
		NGC 2632 - 334			8 37 56	+19 50.3	3	11.01	.010	0.72	.005	0.27	.005	6	G7 V	1140,1748,8023
		NGC 2632 - 335			8 37 56	+20 06.0	3	11.02	.008	0.64	.010	0.17	.001	4	G3 V	1140,1748,8023
		NGC 2632 - 336			8 37 57	+19 04.9	1	11.46		0.71		0.25		1	dG5	8023
	-52 01575	IC 2391 - 53			8 37 57	-52 51.0	1	9.66		1.20		1.00		2	K5 V	1040
74071	-53 01796	IC 2391 - 13		★ V	8 37 59	-53 15.7	8	5.47	.015	-0.16	.009	-0.57	.016	23	B6 V	15,42,540,976,1030*
73798	+20 02173	NGC 2632 - 340		★ V	8 38 00	+20 26.7	3	8.48	.005	0.26	.004	0.08	.010	6	F0 Vn	29,1140,8023
		ApJ144,259 # 29			8 38 00	-20 43.	1	18.35		-0.23		-1.08		1		1360
	-42 04488			V	8 38 00	-42 48.8	1	10.41		0.34				1	F5	177
		NGC 2632 - 341			8 38 01	+19 39.7	4	10.31	.009	0.52	.003	0.03	.004	9	F8 V	29,1140,1748,8023
		NGC 2632 - 344			8 38 03	+20 06.8	3	12.11	.020	0.86	.005	0.52		9	K1 V	802,902,1140
74043	-45 04374				8 38 03	-46 12.4	1	7.81		1.65		2.02		1	M1 III	389
73819	+20 02175	NGC 2632 - 348		★ V	8 38 04	+19 45.5	5	6.78	.007	0.17	.003	0.15	.010	21	A6 IVn	29,1140,1569,1569,8023
		NGC 2632 - 347			8 38 04	+20 02.6	1	12.59		0.75				3		802
		NGC 2632 - 349			8 38 05	+19 54.8	1	12.23		0.86				2	dK0	802
73818	+20 02174	NGC 2632 - 350			8 38 05	+20 06.8	4	8.71	.006	0.32	.001	0.12	.006	7	Am	29,1140,1748,8023
		NGC 2632 - 350s		★	8 38 05	+20 06.8	1	14.60		0.63				2		902
		WLS 840 15 # 10			8 38 08	+14 36.0	1	13.55		0.69		0.07		2		1375
		POSS 210 # 24			8 38 08	+44 44.8	1	16.52		0.43				2		1739
		KUV 08381+3737			8 38 09	+37 37.1	1	16.99		0.03		-0.84		1	DA	1708
74006	-34 05128	HR 3438		★ A	8 38 09	-35 07.8	7	3.96	.013	0.94	.005	0.65	.007	236	G7 Ib-II	3,9,15,26,1075,2012,8015
		IC 2391 - 58			8 38 09	-53 08.7	1	11.48		1.82				1		1040
73857	+10 01848	VZ Cnc			8 38 10	+10 00.2	2	7.37	.126	0.23	.059	0.14		2	F2 III	668,1497
		NGC 2632 - 353			8 38 10	+20 38.2	1	12.35		0.89				3	dK0	902
	-52 01576	IC 2391 - 55			8 38 10	-53 08.4	1	10.16		0.19		0.03		3		1040
74148	-59 01075	HR 3443		★ AB	8 38 10	-60 08.4	2	6.35	.005	0.00	.000			7	A0 V	15,2027

Table 1

477

HD	DM	Other Id	N Rem	α_{1950}	δ_{1950}	S	V	σ_V	B–V	σ_{B-V}	U–B	σ_{U-B}	n	Spectrum	References
	−52 01577	IC 2391 - 54		8 38 12	−52 59.5	1	9.95		0.64		0.12		2		1040
		NGC 2632 - 356		8 38 13	+19 42.1	1	11.04		0.90		0.50		1		1748
	+19 02076	NGC 2632 - 365		8 38 16	+19 15.0	2	10.18	.000	0.64	.002	0.18	.001	2	dG0	1140,8023
		NGC 2632 - 363		8 38 16	+19 37.5	2	12.44	.010	0.93	.001	0.66		5	dK0	802,1140
		NGC 2632 - 361		8 38 17	+20 05.1	1	11.87		0.59		0.00		2		1140
		NGC 2632 - 574		8 38 17	+20 06.8	2	13.64	.004	1.22	.013	1.14		5		802,1140
74117	−52 01578	IC 2391 - 14		8 38 17	−52 47.2	4	9.10	.010	0.36	.006	-0.04	.026	7	F2 IV	42,1030,1040,3015
	+20 02176	NGC 2632 - 371		8 38 18	+19 41.3	3	10.10	.005	0.51	.008	0.03	.001	5	F8 V	1140,1748,8023
		NGC 2632 - 367		8 38 18	+20 02.0	2	10.70	.010	0.68	.000	0.19	.005	2		1748,8023
		NGC 2632 - 368		8 38 19	+19 59.9	2	11.51	.015	0.73	.010	0.30	.010	2	dG5	1748,8023
73854	+20 02177	NGC 2632 - 370		8 38 19	+20 00.5	4	9.04	.012	0.35	.007	0.05	.006	6	F2 V	29,1140,1748,8023
	−0 02041			8 38 19	−00 38.9	1	10.15		0.16		0.10		6	F5	1729
74086	−44 04665			8 38 20	−44 32.2	1	7.99		0.90		0.53		1	G6 II/III	389
		NGC 2632 - S 1556	⋆	8 38 21	+19 43.2	1	14.23		1.43				4	K7 V	802
73940	−11 02426	IDS08360S1134	A	8 38 21	−11 44.9	1	7.59		0.71		0.29		3	G5	1657
	−41 04364			8 38 21	−42 10.8	1	10.64		0.24		-0.30		2		123
	−57 01633			8 38 21	−58 04.6	1	9.55		0.49		-0.17		2		3077
73872	+20 02179	NGC 2632 - 375		8 38 22	+20 06.0	3	8.33	.003	0.19	.005	0.06	.012	6	A7 V	29,1140,8023
73871	+20 02178	NGC 2632 - 377	⋆ AB	8 38 22	+20 39.3	1	6.73		0.10		0.05		1	A1.5IV	1140
		NGC 2632 - 575		8 38 23	+20 13.0	2	14.30	.009	1.31	.028	1.22		5		802,1140
		NGC 2632 - 576		8 38 24	+19 55.6	2	13.71	.010	0.81	.006	0.40		4		902,1140
		NGC 2632 - 381		8 38 25	+19 39.3	1	13.50		0.75				2		902
73890	+19 02078	NGC 2632 - 385	⋆ AB	8 38 27	+19 26.4	3	7.92	.003	0.24	.002	0.11	.006	5	A8 V	29,1140,8023
		NGC 2632 - 2183		8 38 28	+19 48.8	1	13.56		1.17				4		802
		Ton 943		8 38 29	+33 44.	1	14.94		-0.23		-0.93		3		286
74067	−39 04653	HR 3439	⋆ AB	8 38 29	−40 05.1	5	5.19	.012	-0.02	.008	-0.12	.036	20	Ap (SiCr)	15,362,362,2027,2040
		NGC 2632 - 577		8 38 31	+20 25.7	2	14.67	.005	1.46	.005	1.24		7		802,1140
74000	−15 02546			8 38 31	−16 09.6	10	9.67	.021	0.42	.009	-0.25	.014	36	FW	22,742,1088,1097,1594,1658*
		NGC 2632 - 390		8 38 32	+19 06.8	1	12.96		1.04				3		802
		Wat 6 - 12		8 38 32	−45 56.6	1	14.10		1.57		1.50		2		778
74146	−52 01579	IC 2391 - 16	⋆ A	8 38 32	−52 52.6	7	5.18	.025	-0.15	.004	-0.57	.008	19	B5 V	15,42,540,976,1030*
73938	+11 01894			8 38 33	+11 13.7	2	7.55	.013	0.11	.004	0.08	.014	12	A0	1330,1383
		Wat 6 - 14		8 38 33	−45 57.4	1	13.62		0.73		0.24		1		778
	+20 02181	NGC 2632 - 392		8 38 34	+20 07.4	3	10.73	.015	0.60	.002	0.10	.008	8	G0 V	1140,1748,8023
74104	−41 04370	IDS08368S4202	B	8 38 34	−42 10.5	2	7.74	.010	0.74	.005	0.43	.005	3	(G)	123,389
74105	−41 04371	IDS08368S4202	A	8 38 34	−42 12.7	1	6.89		0.16		0.04		2	A3 V	123
		Wat 6 - 10		8 38 34	−45 56.5	1	11.86		0.73		0.21		1		778
	−51 03140	IC 2391 - 79		8 38 34	−51 44.8	1	11.80		0.29		-0.36		1		1030
74169	−52 01581	IC 2391 - 18	⋆ V	8 38 34	−53 05.0	4	7.24	.023	-0.03	.022	-0.10	.019	10	A2p	355,1030,1040,3015
	+20 02180	NGC 2632 - 396		8 38 35	+19 43.3	2	9.86	.015	0.47	.000	0.02	.015	2	F4 V	1748,8023
		Wat 6 - 11		8 38 35	−45 56.7	1	12.90		0.14		-0.14		1		778
74145	−52 01580	IC 2391 - 15		8 38 35	−52 31.5	3	8.55	.020	0.23	.014	0.05	.013	7	Am	42,1040,3015
73997	−8 02452	HR 3437		8 38 36	−08 52.4	3	6.62	.004	-0.02	.000	0.02	.000	11	A1 Vn	15,1075,2012
74130	−48 04008			8 38 36	−48 16.1	1	7.71		1.37		1.29		1	K3 III	389
		NGC 2632 - 399		8 38 37	+19 55.5	3	10.94	.017	0.62	.004	0.13	.004	5	G1.5V	1140,1748,8023
		Wat 6 - 9		8 38 37	−45 56.7	1	12.73		0.26		-0.12		1		778
		NGC 2632 - 401		8 38 39	+19 03.1	1	12.97		1.00				3		902
		Wat 6 - 1		8 38 39	−45 58.0	1	9.67		0.16		-0.54		2		778
74182	−53 01804	IC 2391 - 19	⋆ AB	8 38 39	−53 40.4	2	7.38	.020	0.28	.015	0.02	.050	4	A5 IV	42,1040
		KUV 08387+4026		8 38 40	+40 26.0	1	17.81		-0.43		-1.26		1		1708
74168	−51 03141	IC 2391 - 17		8 38 40	−51 45.8	1	7.51		-0.12		-0.47		4	B9p	42
		NGC 2632 - 403		8 38 42	+20 08.9	2	11.71	.000	0.78	.005	0.39	.005	7	dG8	1140,8023
		Wat 6 - 2		8 38 42	−45 56.7	1	9.18		0.07		-0.23		2		778
73936	+20 02182	NGC 2632 - 407		8 38 43	+19 50.5	1	8.50		1.05		0.92		1	K0 III	1140
		NGC 2632 - 3609		8 38 45	+19 17.2	1	14.17		1.29				3		902
73937	+19 02080	NGC 2632 - 411		8 38 45	+19 19.3	3	9.32	.006	0.39	.000	-0.02	.003	6	F4 V	29,1140,8023
		NGC 2632 - 2233		8 38 46	+19 42.0	1	13.58		1.24				4		802
	+21 01891	NGC 2632 - 2236		8 38 47	+20 51.1	1	9.68		0.45		0.02		1	F6 V	8023
		KUV 08388+4029		8 38 47	+40 29.3	1	16.46		0.27		-0.88		1		1708
		NGC 2632 - 415		8 38 48	+19 26.7	1	13.09		1.07				3		802
	−32 05591			8 38 48	−32 20.7	1	9.98		0.68		0.17		2		1696
		NGC 2632 - 3623		8 38 49	+19 16.8	1	14.33		1.04				2		902
		LP 844 - 8		8 38 49	−23 17.3	1	12.00		1.56				3		912
	+20 02183	NGC 2632 - 416		8 38 51	+19 50.4	2	9.58	.005	0.40	.006	0.00	.002	2	F5 V	1140,8023
		NGC 2632 - 417		8 38 52	+20 08.5	2	12.35	.005	0.85	.037	0.59		8	dK2	802,1140
	+20 02184	NGC 2632 - 418		8 38 52	+20 24.4	2	10.47	.000	0.57	.002	0.06	.000	4	dG0	1140,8023
		Wat 6 - 6		8 38 52	−45 56.6	1	13.14		0.42		0.27		1		778
	−46 04432			8 38 52	−47 01.8	1	9.81		0.29		-0.50		2	B2 III	540
74195	−52 01583	IC 2391 - 20	⋆ V	8 38 52	−52 44.6	8	3.62	.022	-0.18	.003	-0.65	.021	45	B2.5IV	15,42,1030,1034,1040*
74196	−52 01584	IC 2391 - 21	⋆	8 38 52	−52 50.2	4	5.59	.019	-0.14	.003	-0.50	.008	19	B6 Vn	42,1030,1040,2006
74167	−44 04679	HR 3444		8 38 53	−45 00.8	4	5.70	.006	1.65	.010	2.00		13	M0 III	15,2013,2029,3005
74212	−53 01808	IC 2391 - 22		8 38 53	−53 15.8	1	8.68		1.06		0.82		2	K0 IIIa	1040
	+19 02081	NGC 2632 - 421		8 38 54	+19 26.8	3	10.16	.007	0.52	.001	0.03	.007	4	F8	29,1140,8023
		LOD 11 # 8		8 38 54	−42 12.2	1	11.51		0.33		0.06		2		123
74166	−41 04380			8 38 55	−42 09.7	1	7.67		1.30		1.24		2	K2 III	123
		IC 2395 - 60		8 38 55	−48 03.0	1	11.89		1.03		0.56		2		1667
74405	−69 00946	HR 3460	⋆ A	8 38 55	−70 12.5	4	5.19	.007	0.01	.005	-0.03	.000	12	A0 V	15,1075,2038,3023
		NGC 2632 - 425		8 38 56	+19 35.5	2	11.41	.005	0.52	.003	0.03	.000	2		1140,8023
		NGC 2632 - 3635		8 38 57	+19 38.3	1	13.67		1.18				3		902
		HIC 42347		8 38 57	+71 00.1	1	10.98		0.54		0.09		2		1723

HD	DM	Other Id	N Rem	α_{1950}	δ_{1950}	S	V	σ_V	B–V	σ_{B-V}	U–B	σ_{U-B}	n	Spectrum	References
		NGC 2632 - 3637		8 38 58	+19 22.5	1	14.50		1.38				3		902
73974	+20 02185	NGC 2632 - 428		8 38 58	+20 03.2	6	6.90	.003	0.95	.010	0.76	.040	15	K0 III	1140,1355,2001,3077*
74180	−46 04438	HR 3445, LSS 1113	⋆ A	8 38 58	−46 28.2	11	3.82	.021	0.70	.015	0.33	.075	32	F3 Ia	12,15,524,1018,1034*
		IC 2395 - 58		8 38 58	−48 05.2	1	12.88		3.93		0.65		2		1667
		NGC 2632 - 430		8 39 00	+20 20.8	2	12.05	.010	0.84	.003	0.52		7	dK0	802,1140
73993	+20 02186	NGC 2632 - 429		8 39 01	+20 20.3	3	8.54	.008	0.29	.002	0.08	.008	6	A9 Vn	29,1140,8023
74180	−46 04438	LSS 1114	⋆ B	8 39 01	−46 27.8	2	10.85	.035	0.23	.005	-0.51	.005	4	F3 Ia	12,524
		IC 2395 - 59		8 39 01	−48 04.0	1	12.24		0.75		0.11		2		1667
74543	−73 00523			8 39 02	−73 53.9	1	6.82		1.04				4	K1 III	2012
		NGC 2632 - 434		8 39 03	+19 26.2	2	11.38	.025	0.72	.003	0.25	.001	3	dG8	1140,8023
	+20 02188	NGC 2632 - 433		8 39 04	+19 48.8	1	11.03		0.42		0.00		1		1140
		NGC 2632 - 432		8 39 04	+19 52.1	2	11.04	.005	0.64	.003	0.18	.000	2	dG5	1140,8023
		NGC 2632 - 3657	A	8 39 05	+19 45.1	1	15.26		0.67				1		902
		NGC 2632 - 3657	B	8 39 05	+19 45.1	1	15.27		1.23				2		902
73954	+29 01811			8 39 05	+28 50.5	1	8.99		0.24		0.11		2	A2	1733
		KUV 08391+3800		8 39 05	+37 59.8	1	16.32		-0.03		-0.81		1	DA	1708
	−41 04383			8 39 05	−42 11.1	1	9.85		0.07		0.04		2		123
74194	−44 04683	LM Vel, LSS 1116		8 39 05	−44 52.8	6	7.55	.015	0.23	.009	-0.73	.022	17	O8/9	390,540,976,1075,1737,2012
		IC 2395 - 64		8 39 05	−48 02.5	1	13.13		1.43		0.82		2		1667
74224	−50 03474			8 39 05	−50 50.1	1	7.54		1.54		1.85		1	K3 III	389
		CCS 1260		8 39 06	+07 36.	1	10.05		1.39				1	R5	1238
		NGC 2632 - 3658		8 39 06	+20 17.2	1	13.23		1.11				3		902
73994	+19 02082	NGC 2632 - 439		8 39 07	+19 05.5	2	9.44	.025	0.39	.005	-0.01	.001	4	F5 V	29,8023
	−45 04394	LSS 1117		8 39 10	−45 55.0	3	10.15	.011	0.21	.013	-0.64	.035	9	B2:Vne	778,1737,2021
	−47 04208	IC 2395 - 65		8 39 10	−48 02.0	2	9.60	.024	0.07	.005	-0.41	.039	3		389,1667
		IC 2395 - 63		8 39 10	−48 02.7	1	13.86		0.81		0.49		2		1667
		IC 2395 - 62		8 39 11	−48 03.5	1	11.74		0.22		0.17		2		1667
		IC 2391 - 69		8 39 11	−52 53.1	1	12.29		0.41		0.13		3		1030
	−40 04510			8 39 12	−40 42.2	1	10.21		0.18		-0.48		2	B2.5IV	540
74210	−45 04395			8 39 12	−45 35.0	1	7.96		1.24		1.12		1	G8/K0 (III)	389
74178	−38 04735			8 39 13	−39 03.4	1	9.13		-0.09		-0.42		2	B8 V	1732
		IC 2391 - 68		8 39 13	−52 54.2	1	12.09		0.47		0.03		3		1030
		NGC 2632 - 443		8 39 14	+19 46.8	1	10.63		0.92		0.66		1		1140
74028	+19 02083	NGC 2632 - 445	⋆ V	8 39 15	+19 35.5	3	7.96	.004	0.21	.004	0.11	.012	7	A7 Vn	29,1140,8023
74254	−50 03478			8 39 15	−50 51.5	1	8.16		0.65		0.19		1	F7/G0 IV/V	389
		IC 2395 - 61		8 39 16	−48 03.5	1	13.42		1.19		0.48		2		1667
74234	−47 04210	IC 2395 - 2		8 39 17	−48 02.8	5	6.94	.003	-0.16	.014	-0.80	.011	19	B2 IV	1040,1075,1637,1732,2012
		NGC 2632 - 446		8 39 19	+20 19.5	1	13.22		1.12		0.97		2		1140
74050	+19 02084	NGC 2632 - 449	⋆ V	8 39 20	+19 06.8	2	7.93	.019	0.20	.004	0.13	.001	5	A6 Vn	29,8023
		NGC 2632 - 448		8 39 20	+19 27.4	2	12.16	.005	0.87	.008	0.56		7	dK0	802,1140
		NGC 2632 - 3695		8 39 21	+19 23.6	1	13.70		1.20				3		902
		NGC 2632 - 3698		8 39 22	+19 26.8	1	13.50		1.18				3		902
74275	−52 01587	IC 2391 - 23		8 39 22	−52 37.4	3	7.29	.012	-0.01	.013	-0.03	.026	7	A0 Vb	1030,1040,3015
	+20 02190	NGC 2632 - 454		8 39 24	+19 52.1	2	9.88	.000	0.47	.003	0.01	.001	2	F5	1140,8023
		NGC 2632 - 3706		8 39 24	+20 16.3	1	14.26		1.30				3		902
74137	−15 02554	HR 3441	⋆ A	8 39 24	−15 45.8	8	4.87	.008	1.06	.007	0.92	.001	34	K0 III	15,58,1007,1013,1075*
		IC 2395 - 66		8 39 24	−48 01.8	1	12.00		0.40		-0.15		2		1667
74251	−47 04214	IC 2395 - 4		8 39 25	−47 53.3	1	7.76		-0.11		-0.57		3	B3 IV/V	1040
		NGC 2632 - 456		8 39 27	+20 35.4	1	12.42		0.93				3	dK0	902
		NGC 2632 - 456s	⋆	8 39 27	+20 35.4	1	13.70		0.56				2		902
		CCS 1268		8 39 27	−34 45.2	1	10.86		3.73				1	N	864
	+20 02189	NGC 2632 - 458	⋆ AB	8 39 28	+20 13.0	3	9.72	.007	0.55	.005	0.07	.007	6	F6 V	29,1140,8023
74058	+20 02191	NGC 2632 - 459		8 39 29	+20 21.7	3	9.23	.003	0.38	.002	0.02	.005	6	F3 Vn	29,1140,8023
		G 114 - 13		8 39 29	−01 23.6	1	13.64		1.60		1.05		5		316
		POSS 210 # 9		8 39 30	+39 08.1	1	14.66		1.42				2		1739
233565	+50 01571			8 39 30	+50 10.1	1	8.52		0.54		0.07		2		1733
		IC 2395 - 19		8 39 30	−48 01.	1	11.95		0.73		0.29		3		1040
		IC 2395 - 20		8 39 30	−48 01.	1	12.31		0.26		0.23		2		1040
74273	−48 04020	HR 3453		8 39 30	−48 44.6	8	5.90	.008	-0.21	.008	-0.90	.004	25	B2 III	15,401,540,976,1020,1732*
	−52 01588	IC 2391 - 57		8 39 30	−52 21.0	1	10.63		0.15		-0.36		3		1040
74341	−57 01644	HR 3455	⋆ AB	8 39 30	−57 22.0	8	6.33	.004	0.21	.004	0.13	.010	11	A3 V	15,1075,2012
		NGC 2632 - 460		8 39 31	+20 17.9	2	12.09	.005	0.89	.003	0.62		7	dG8	802,1140
	−44 04691	LSS 1119		8 39 31	−45 05.9	2	8.48	.000	0.40	.005	-0.50		8	B2.5II	1075,2021
		IC 2391 - 67		8 39 31	−52 54.0	1	11.69		0.82		0.46		3		1030
74375	−59 01080	HR 3457, V343 Car	⋆ A	8 39 31	−59 34.9	3	4.32	.005	-0.12	.005	-0.80	.000	14	B2 III	15,1075,2012
73971	+47 01606	HR 3436		8 39 32	+47 04.9	1	6.22		0.96				2	G8 III	71
		NGC 2632 - 462		8 39 33	+19 46.1	1	10.99		0.55		0.03		2		1140
		IC 2395 - 67		8 39 34	−48 00.0	1	11.81		0.23		-0.29		2		1667
	−45 04401			8 39 35	−45 34.5	2	8.84	.015	0.18	.005	-0.52	.029	3		389,401
74272	−46 04448	HR 3452		8 39 35	−47 08.3	7	4.77	.006	0.12	.006	0.12	.013	30	A5 II	15,283,1075,1637,2012*
	−32 05613			8 39 36	−32 46.9	4	11.86	.021	0.23	.012	-0.57	.016	16	DAs	1698,1705,1765,3073
74205	−25 06448			8 39 38	−25 29.8	2	9.21	.015	0.41	.005	-0.06		4	F3 V	1594,6006
74290	−47 04217			8 39 38	−47 23.7	1	8.51		0.20		-0.46		2	B2/3 IV(n)	401
		NGC 2632 - 466		8 39 41	+19 34.6	2	10.99	.000	0.65	.000	0.16	.000	3	dG0	1140,8023
		NGC 2632 - 3733		8 39 44	+20 19.3	1	13.99		1.27				3		902
74340	−52 01592	IC 2391 - 24		8 39 44	−52 42.6	3	9.84	.017	0.46	.015	-0.05	.007	6	F6 V	42,1040,3015
	−47 04221	IC 2395 - 68		8 39 45	−47 59.1	1	10.79		0.15		0.16		2		1667
		CpD -43 02871		8 39 46	−43 44.0	1	11.72		0.62		-0.38		5		362
74190	−11 02432	HR 3446		8 39 47	−11 47.2	3	6.44	.004	0.16	.000	0.12	.005	11	A5 m	15,1075,2012
		Ru 67 - 25		8 39 47	−43 11.0	1	13.49		0.77		0.59		2		410

Table 1 479

HD	DM	Other Id	N	Rem	α_{1950}	δ_{1950}	S	V	σ_V	B–V	σ_{B-V}	U–B	σ_{U-B}	n	Spectrum	References
74289	−42 04517				8 39 48	−42 24.8	1	8.71		0.58		0.15		2	G0 V	123
		Ru 67 - 27			8 39 48	−43 09.4	1	13.30		0.49		0.43		2		410
		Pis 8 - 26			8 39 48	−46 06.	1	9.67		0.16		-0.54		2		778
		NGC 2632 - 471			8 39 49	+19 18.8	2	12.14	.005	0.85	.001	0.53		6	dK0	802,1140
	+20 02192	NGC 2632 - 472			8 39 49	+19 43.4	2	9.77	.000	0.45	.001	-0.01	.001	5	F2 III	1140,8023
		Pis 8 - 3			8 39 49	−46 04.4	2	13.00	.025	0.54	.025	0.12	.135	2		184,778
		Ru 67 - 13			8 39 50	−43 12.2	1	13.67		0.69		0.12		2		410
		Pis 8 - 4			8 39 50	−46 04.9	2	13.22	.020	0.69	.005	0.21	.010	2		184,778
		IC 2391 - 66			8 39 50	−52 53.6	1	12.35		0.71		0.15		3		1030
		NGC 2632 - 476			8 39 51	+19 16.8	3	11.63	.005	0.76	.000	0.36	.002	7	dG8	802,1140,8023
		Ru 67 - 14			8 39 51	−43 11.3	1	12.72		0.31		0.10		1		410
		Pis 8 - 5			8 39 51	−46 04.3	1	14.54		0.65		0.50		1		184
		NGC 2632 - 474			8 39 52	+19 48.2	2	12.13	.010	0.99	.002	0.76		6	dK0	802,1140
74319	−44 04698	IDS08382S4438		AB	8 39 52	−44 48.7	1	6.69		-0.10		-0.56		2	B3 V	401
		Pis 8 - 9			8 39 52	−46 05.3	1	14.33		0.64		0.37		1		184
		Pis 8 - 8			8 39 52	−46 05.7	2	13.32	.080	0.58	.020	0.05	.015	2		184,778
	+20 02193	NGC 2632 - 478			8 39 53	+19 45.6	1	9.74		0.44		0.01		1	F6 V	8023
		Ru 67 - 26			8 39 53	−43 10.5	1	14.53		0.93		0.63		2		410
		Pis 8 - 16			8 39 53	−46 05.0	2	12.04	.000	0.94	.000	0.48	.000	3		778,778
		Pis 8 - 14			8 39 54	−46 06.8	2	12.75	.025	0.55	.030	0.05	.000	2		184,778
		Pis 8 - 13			8 39 54	−46 06.9	1	13.55		0.83		0.13		1		184
		Ru 67 - 23			8 39 55	−43 14.3	1	11.41		0.10		0.14		2		410
		Pis 8 - 7			8 39 55	−46 05.2	1	14.19		0.75		0.43		1		184
		Pis 8 - 6			8 39 55	−46 05.5	2	13.37	.000	0.43	.020	0.11	.040	2		184,778
74373	−52 01595	IC 2391 - 25			8 39 55	−53 06.6	2	10.17	.014	0.16	.014	0.11	.042	4	A2 V	42,1040
		KUV 345 - 30			8 39 57	+39 55.6	1	14.29		-0.27		-1.10		1	sdB	974
		Ru 67 - 15			8 39 57	−43 09.8	1	13.19		0.44		0.31		2		410
		Ru 67 - 11			8 39 58	−43 11.5	1	13.34		0.58		0.49		2		410
		Ru 67 - 5			8 39 58	−43 12.8	1	12.43		0.56		0.35		2		410
	−45 04411	Pis 8 - 1			8 39 58	−46 05.8	1	10.57		0.54		-0.24		2	B5 Vn	184
		IC 2391 - 65			8 39 58	−52 54.6	1	11.08		1.14		0.83		2		1030
		Ru 67 - 4			8 39 59	−43 10.3	1	11.56		0.37		0.10		2		410
		Ru 67 - 6			8 39 59	−43 13.3	1	12.04		0.61		-0.02		2		410
		Pis 8 - 2			8 39 59	−46 05.2	2	13.11	.024	0.53	.088	0.15	.093	3		184,778
		Pis 8 - 12			8 39 59	−46 05.4	1	14.32		0.78		0.12		1		184
	−52 01594	IC 2391 - 71			8 39 59	−52 48.2	1	9.74		1.21		1.14		2		1030
		Ru 67 - 10			8 40 00	−43 12.	1	14.29		0.60		0.82		2		410
		Ru 67 - 19			8 40 00	−43 12.	1	14.62		0.52		0.60		2		410
		Ru 67 - 24			8 40 00	−43 12.	1	14.45		0.85		0.38		2		410
		Ru 67 - 12			8 40 00	−43 12.3	1	14.15		0.83		0.55		2		410
		Pis 8 - 11			8 40 00	−46 05.5	1	14.32		0.77		0.30		1		184
		IC 2391 - 64			8 40 00	−52 54.6	1	12.26		0.38		0.08		2		1030
74541	−67 00979				8 40 00	−67 38.5	1	7.89		1.15				4	K0 III	2033
	+10 01857	G 46 - 1		AB	8 40 01	+09 45.0	6	9.64	.010	1.41	.017	1.28	.010	6	M0	333,679,680,1017,1620,3072
	+10 01857	IDS08373N0956		C	8 40 01	+09 45.0	4	11.83	.023	1.54	.010	1.26	.029	5		679,680,1620,3072
		Pis 8 - 10			8 40 01	−46 06.0	2	13.67	.065	0.70	.005	0.65	.065	2		184,778
	−47 04226				8 40 01	−47 16.1	1	10.16		0.13		0.07		1	A2	283
		Ru 67 - 8			8 40 02	−43 13.0	1	13.72		0.54		0.43		2		410
74387	−52 01597	IC 2391 - 26			8 40 02	−52 48.6	2	9.26	.055	1.05	.000	0.75	.005	6	K0 III	1030,1040
		Ru 67 - 20			8 40 03	−43 09.3	1	13.69		0.51		0.47		2		410
		WLS 840 15 # 5			8 40 04	+16 57.7	1	12.02		0.48		-0.01		2		1375
		Ru 67 - 7			8 40 04	−43 12.8	1	12.87		0.58		0.32		2		410
	−47 04231	IC 2395 - 53		V	8 40 04	−47 56.0	1	9.81		1.98		1.30		2	M6	1667
	−47 04230	IC 2395 - 52			8 40 04	−47 59.2	1	11.29		0.35		0.19		2		1667
		Ru 67 - 9			8 40 05	−43 11.5	1	13.39		0.49		0.27		2		410
	−45 04415				8 40 07	−45 22.3	1	9.43		0.38		0.08		2		401
		Ru 67 - 21			8 40 08	−43 11.5	1	13.74		0.81		0.24		2		410
	−47 04232	IC 2395 - 12			8 40 08	−47 56.1	2	9.79	.039	0.50	.005	0.04	.029	5	F8	1040,1667
		NGC 2632 - 488			8 40 09	+20 31.1	2	11.43	.010	0.73	.001	0.31	.001	4	dG5	1140,8023
	+24 01981				8 40 09	+24 20.8	1	10.10		1.33		1.26		3		1723
		Ru 67 - 18			8 40 09	−43 10.0	1	14.04		0.59		0.38		2		410
74357	−41 04403				8 40 10	−42 14.1	1	8.91		0.09		0.05		2	A0 V	123
		NGC 2632 - 492			8 40 11	+19 20.9	2	12.12	.009	0.85	.009	0.55		6	dK0	802,1140
		POSS 210 # 51			8 40 11	+40 49.2	1	17.61		1.59				2		1739
74300	−31 06433				8 40 11	−32 03.1	1	7.31		-0.02		-0.15		2	B9 V	401
		Ru 67 - 16			8 40 11	−43 10.8	1	12.82		0.51		0.19		2		410
		Ru 67 - 1			8 40 11	−43 10.9	1	10.67		0.32		0.02		6		410
		IC 2391 - 63			8 40 12	−52 53.7	1	10.93		0.12		-0.12		2		1030
		NGC 2632 - 3774			8 40 14	+19 58.4	1	13.67		1.18				3		902
		NGC 2632 - 3773			8 40 14	+20 08.4	1	13.75		0.79				2		902
		Ru 67 - 17			8 40 14	−43 11.8	1	14.33		0.53		0.46		2		410
	−46 04461				8 40 14	−47 06.5	1	10.20		0.09		-0.03		1	A1 V	283
	+19 02087	NGC 2632 - 495			8 40 15	+19 37.1	1	9.97		0.66		0.17		1	F8 V	8023
		Ru 67 - 3			8 40 15	−43 11.4	1	13.78		0.50		0.38		2		410
74371	−44 04704	HR 3456, LN Vel		★	8 40 15	−45 13.8	7	5.24	.013	0.21	.015	-0.52	.021	24	B6 Iab/b	15,1018,1034,1075,1737*
	−48 04029				8 40 15	−48 55.7	1	9.89		1.08		0.47		2		540
74186	+19 02088	NGC 2632 - 496			8 40 16	+19 14.9	2	9.59	.026	0.52	.014	0.03	.002	5	F8 V	29,8023
		NGC 2632 - 3777			8 40 16	+19 53.7	1	13.33		1.15				3		902
		Ru 67 - 2			8 40 16	−43 11.4	1	12.56		0.44		-0.01		2		410
		Ru 67 - 22			8 40 16	−43 11.6	1	13.54		0.46		0.43		2		410

HD	DM	Other Id	N Rem	α₁₉₅₀	δ₁₉₅₀	S	V	σ_V	B–V	σ_B–V	U–B	σ_U–B	n	Spectrum	References
	−47 04236	IC 2395 - 17		8 40 16	−48 05.3	1	11.21		0.27		−0.27		3		1040
74200	+18 02022			8 40 18	+18 19.9	1	8.03		1.26		1.36		2	K0	1648
74355	−40 04528			8 40 18	−40 31.5	1	8.40		−0.08		−0.43		2	B6/7 III	401
		NGC 2632 - 498		8 40 19	+19 42.4	2	11.78	.000	0.77	.006	0.34	.001	4	dG8	1140,8023
74438	−52 01598	IC 2391 - 27		8 40 20	−52 53.0	3	7.58	.024	0.25	.026	0.03	.028	13	A3 V	1030,1040,3015
74385	−42 04528	IDS08386S4234	A	8 40 22	−42 44.9	6	8.12	.012	0.90	.005	0.62	.020	16	K1 V	158,389,912,1705,2033,3072
74385	−42 04528	IDS08386S4234	B	8 40 22	−42 44.9	1	12.68		1.52		1.51		3		3072
74401	−47 04241	LSS 1126		8 40 22	−47 29.1	4	8.98	.024	0.27	.014	−0.75	.020	10	B1.5IIIne	401,540,976,1737
74198	+21 01895	HR 3449	★ A	8 40 24	+21 39.0	5	4.67	.014	0.01	.008	0.02	.010	13	A1 IV	15,1022,1118,3023,8015
	−43 04563			8 40 26	−43 57.7	1	10.37		0.54				4	B0	2012
74228	+13 01972	HR 3450		8 40 27	+12 51.7	2	5.64	.005	0.39	.000	0.26	.010	4	G0 III+A3 V	985,1733
		NGC 2632 - 508		8 40 29	+19 57.0	2	10.78	.010	0.59	.002	0.09	.000	4	dG0	1140,8023
74132	+45 01624			8 40 29	+45 38.1	1	8.25		0.34		−0.02		2	F5	1601
	+62 01013			8 40 29	+61 41.4	1	9.76		1.04		0.65		1	K2	565
74436	−47 04246	IC 2395 - 6	★ AB	8 40 31	−48 03.9	1	8.24		−0.08		−0.60		10	B3 V	1040
	+75 00348			8 40 34	+74 43.4	1	9.54		1.17				1	R0	1238
	−47 04248	IC 2395 - 16		8 40 35	−48 07.3	1	10.87		0.12		−0.08		3		1040
		NGC 2660 - 3101		8 40 36	−46 58.	1	12.45		1.26				1		283
		NGC 2660 - 9001		8 40 36	−46 58.	1	16.37		0.89		0.38		2		283
		NGC 2660 - 9002		8 40 36	−46 58.	1	14.86		0.75		0.41		2		283
		NGC 2660 - 9003		8 40 36	−46 58.	1	15.83		0.84				1		283
		NGC 2660 - 9006		8 40 36	−46 58.	1	14.90		0.89		1.00		2		283
		NGC 2660 - 9007		8 40 36	−46 58.	1	16.58		0.85		0.32		2		283
		NGC 2660 - 9008		8 40 36	−46 58.	1	14.64		0.73		0.48		2		283
		NGC 2660 - 9010		8 40 36	−46 58.	1	17.64		0.95		0.40		1		283
		NGC 2660 - 9011		8 40 36	−46 58.	1	17.52		0.96				1		283
		NGC 2660 - 9012		8 40 36	−46 58.	1	17.02		0.75				1		283
		NGC 2660 - 9015		8 40 36	−46 58.	1	15.12		0.54		0.45		1		283
		NGC 2660 - 9017		8 40 36	−46 58.	1	16.20		0.93				1		283
		NGC 2660 - 9018		8 40 36	−46 58.	1	16.23		0.90		0.26		1		283
		NGC 2660 - 9019		8 40 36	−46 58.	1	14.79		1.44		1.27		1		283
		NGC 2660 - 9020		8 40 36	−46 58.	1	16.98		0.98		−0.20		1		283
		NGC 2660 - 9021		8 40 36	−46 58.	1	14.84		0.67		0.44		2		283
		NGC 2660 - 9022		8 40 36	−46 58.	1	15.66		1.29		1.18		1		283
		NGC 2660 - 9026		8 40 36	−46 58.	1	16.23		1.47		0.91		2		283
		NGC 2660 - 9027		8 40 36	−46 58.	1	15.60		1.46		1.25		1		283
		NGC 2660 - 9031		8 40 36	−46 58.	1	18.48		1.61				2		283
		NGC 2660 - 9032		8 40 36	−46 58.	1	17.83		1.16		1.18		1		283
		NGC 2660 - 9033		8 40 36	−46 58.	1	17.10		1.06				1		283
		NGC 2660 - 9034		8 40 36	−46 58.	1	18.28		0.88				1		283
		NGC 2660 - 9035		8 40 36	−46 58.	1	17.78		1.56		0.97		1		283
		NGC 2660 - 9036		8 40 36	−46 58.	1	17.86		0.82				1		283
		NGC 2660 - 9037		8 40 36	−46 58.	1	17.76		1.02		0.64		1		283
		NGC 2660 - 9038		8 40 36	−46 58.	1	18.25		0.92		0.10		1		283
		NGC 2660 - 9039		8 40 36	−46 58.	1	17.36		0.94		0.41		1		283
74280	+3 02039	HR 3454, η Hya		8 40 37	+03 34.8	35	4.30	.008	−0.19	.007	−0.73	.009	220	B3 V	1,3,15,30,124,125,154*
	−45 04425			8 40 38	−45 32.8	1	9.46		0.19		−0.49		2		401
74435	−43 04566			8 40 39	−43 29.6	1	8.10		1.18		1.26		1	K2 III	389
	−47 04250	IC 2395 - 18		8 40 39	−47 53.3	1	11.51		0.09		−0.51		2		1040
74455	−47 04251	IC 2395 - 1	★ AB	8 40 39	−47 55.1	7	5.50	.015	−0.18	.011	−0.91	.012	26	B1.5Vn	15,26,540,976,1040*
74497	−52 01599	IC 2391 - 28	★ AB	8 40 40	−52 34.6	3	7.85	.027	0.66	.008	0.18	.015	13	G3 V	389,1030,1040
		NGC 2632 - 514		8 40 41	+19 55.5	2	12.34	.005	0.93	.004	0.66		6	dK0	802,1140
	−46 04472	NGC 2660 - 9005		8 40 41	−46 34.7	1	8.38		1.13		1.14		4		283
	−47 04253	IC 2395 - 15		8 40 41	−47 53.0	1	10.66		−0.02		−0.22		3		1040
	+46 01426			8 40 42	+46 30.5	1	8.81		1.10		0.95		1	G8 III	3040
74517	−52 01600	IC 2391 - 30		8 40 42	−52 58.8	2	8.62	.015	0.13	.020	0.04	.015	5	A3 Va	1030,1040
74454	−46 04474			8 40 43	−46 24.0	1	7.87		−0.02		−0.20		2	B9 V	401
	+20 02196	NGC 2632 - 515		8 40 44	+20 22.1	2	10.14	.006	0.50	.004	0.03	.003	7	F6 V	29,8023
		WLS 900 60 # 6		8 40 44	+60 08.2	1	11.56		0.73		0.13		2		1375
74516	−52 01601	IC 2391 - 29		8 40 44	−52 47.3	3	7.39	.011	0.02	.018	0.00	.034	8	A0 V	1030,1040,3015
		NGC 2660 - 9016		8 40 46	−46 58.9	1	13.98		1.36		1.01		1		283
	−52 01602	IC 2391 - 51	★ D	8 40 46	−52 55.3	2	9.59	.000	0.48	.028	−0.10	.033	6	F6 V	1040,3015
	−47 04257	IC 2395 - 13		8 40 47	−48 01.3	1	10.21		0.16		−0.02		3	A2	1040
74536	−52 01603	IC 2391 - 32		8 40 47	−53 11.4	2	7.97	.010	0.02	.000	−0.45	.050	4	A2 V	1030,1040
	−21 02560			8 40 49	−22 03.2	1	9.15		1.44		1.47		1	K7	1746
74537	−53 01825	IC 2391 - 33		8 40 49	−53 58.4	1	8.68		0.22		0.04		3	Am	1040
74496	−47 04258			8 40 50	−47 20.4	1	6.71		0.24		0.09		2	A0mA3-F0/2	283
		NGC 2660 - 9023		8 40 52	−46 58.9	1	13.49		1.35		1.10		2		283
74535	−52 01605	IC 2391 - 31	★ BC	8 40 53	−52 55.2	7	5.50	.018	−0.15	.009	−0.55	.010	20	B8.5IVp	15,540,976,1030,1040*
	+19 02089	NGC 2632 - 549		8 40 53	+18 58.9	1	10.13		0.48		0.02		1	F8	8023
		G 115 - 22		8 40 54	+36 25.9	1	10.96		0.58		−0.12		4		1064
	−43 04570	LSS 1127		8 40 54	−43 58.2	1	10.45		0.32		−0.65		4		362
		NGC 2660 - 9025		8 40 54	−46 56.9	1	12.98		1.48		1.46		2		283
74561	−53 01826	IC 2395 - 35		8 40 54	−53 43.3	1	9.38		0.44		−0.08		3	F3 V	1040
		NGC 2660 - 9009	★ V	8 40 55	−47 01.6	1	11.68		4.29		3.62		4		283
74214	+42 01919			8 40 56	+42 16.9	2	7.08	.010	1.43	.005	1.66	.005	5	K3 III	1501,1601
	−44 04717			8 40 56	−44 50.2	1	10.37		0.31		−0.52		2	B1 V	540
		NGC 2660 - 9024		8 40 56	−46 57.2	1	13.28		0.46		0.33		2		283
		NGC 2660 - 2121		8 40 56	−47 00.4	1	12.99		1.31				3		283
74308	+10 01859			8 40 58	+09 52.1	2	8.66	.009	0.24	.001	0.17	.027	3	A2	695,1330

Table 1

HD	DM	Other Id	N Rem	α_{1950}	δ_{1950}	S	V	σ_V	B–V	σ_{B-V}	U–B	σ_{U-B}	n	Spectrum	References
		NGC 2632 - 550		8 40 58	+20 20.5	1	12.37		0.67				2	dG5	802
74243	+37 01899	HR 3451		8 40 58	+37 06.0	1	6.55		0.43		0.02		2	F6 IV-V	1733
74531	−47 04261	IC 2395 - 3	⋆ A	8 40 58	−47 59.0	3	7.25	.008	-0.16	.005	-0.86	.018	8	B2 III	540,976,1040
74530	−47 04263	IC 2395 - 8	⋆ AB	8 40 59	−47 53.7	1	8.68		-0.12		-0.63		3	B3 IV/V	1040
74560	−52 01607	IC 2391 - 34	⋆ A	8 40 59	−52 56.0	10	4.83	.022	-0.17	.007	-0.66	.006	28	B3 V	15,26,42,540,976,1030*
		Ton 10		8 41 00	+26 14.	1	14.78		-0.11		-1.03		1		98
74150	+56 01328			8 41 00	+55 43.2	1	8.91		0.80		0.48		2	K0 III-IV	1502
74475	−35 04976	HR 3463		8 41 00	−35 45.8	3	6.42	.009	0.02	.000	0.00		9	A0 V	15,401,2012
	−46 04470	NGC 2660 - 9004		8 41 01	−47 03.6	1	11.72		0.16		-0.45		4		283
74528	−45 04435			8 41 02	−45 22.5	1	8.44		0.16				4	B1 V	2021
		NGC 2660 - 4104		8 41 02	−47 01.1	1	14.15		1.25				2		283
74622	−55 01688	HR 3471		8 41 02	−55 35.7	5	6.28	.010	1.18	.009	1.20		20	K2 III	15,1075,2013,2028,3005
		NGC 2660 - 1132		8 41 03	−47 00.6	1	14.12		1.31				1		283
		NGC 2660 - 1115		8 41 03	−47 00.8	1	14.68		1.33				1		283
		NGC 2632 - 551		8 41 04	+19 04.5	1	12.67		0.96				2		802
74529	−46 04482			8 41 04	−47 03.6	1	8.41		1.14		1.09		1	K0 III	389
74307	+19 02090	NGC 2632 - 552	⋆ V	8 41 05	+19 12.9	1	8.38		0.11		-0.05		1	A0	8023
		NGC 2660 - 9014		8 41 05	−47 04.5	1	12.64		0.63		0.47		2		283
74582	−52 01608	IC 2391 - 36		8 41 07	−52 20.7	2	9.63	.014	0.46	.000	0.16	.009	5	F3 III	42,3015
		GM Hya		8 41 08	+06 02.2	1	10.74		0.74		0.25		3	G2	1768
		POSS 210 # 59		8 41 08	+41 47.2	1	18.73		1.54				2		1739
74558	−46 04483			8 41 08	−46 37.7	3	6.92	.017	0.27	.009	0.00	.015	8	A9 V	401,2012,3037
		NGC 2660 - 4224		8 41 08	−47 02.5	1	11.99		1.82				1		283
74559	−47 04268	IC 2395 - 11		8 41 08	−47 54.3	1	9.24		-0.05		-0.44		3	B9 Ve	1040
	−52 01609	IC 2391 - 72		8 41 08	−52 50.8	1	9.66		0.46		0.02		2		1030
74292	+32 01782			8 41 10	+32 14.6	1	7.04		0.23		0.03		3	A3 V	1501
74579	−45 04437			8 41 12	−45 17.0	1	8.16		1.19		1.15		1	K1 III	389
		NGC 2660 - 4302		8 41 12	−47 02.2	1	14.27		1.30				1		283
74395	−6 02708	HR 3459	⋆ A	8 41 13	−07 03.2	8	4.63	.011	0.84	.003	0.49	.020	30	G2 Ib	15,1075,1425,1509*
		NGC 2632 - 530		8 41 14	+19 29.8	1	12.26		0.77				3	dG8	802
	+20 02198	NGC 2632 - 553		8 41 15	+20 15.5	2	10.15	.000	0.44	.003	-0.01	.000	4	F8	1140,8023
		NGC 2660 - 1302		8 41 15	−46 59.9	1	13.45		0.54				1		283
		NGC 2660 - 1301		8 41 16	−47 00.8	1	14.37		1.29				1		283
74580	−47 04271	IC 2395 - 7	AB	8 41 16	−47 56.8	2	8.63	.005	-0.04	.010	-0.54	.015	5	B3 V	976,1040
	−42 04546			8 41 20	−42 29.7	1	11.56		0.39		0.28		2		123
	−47 04273	IC 2395 - 10		8 41 20	−47 56.4	2	9.05	.008	0.29	.005	-0.61	.017	5	B8	540,1040
		IC 2391 - 75		8 41 20	−52 53.4	1	10.89		0.84		0.28		3		1030
	−52 01614	IC 2391 - 73		8 41 21	−52 50.0	1	10.36		0.68		0.17		2		1030
74393	+4 02029	HR 3458		8 41 22	+04 31.0	3	6.36	.005	-0.06	.004	-0.12	.005	9	B9.5III-IV	15,1079,1256
74681	−59 01088	IDS08402S5910	AB	8 41 22	−59 20.8	1	7.84		0.06		-0.04		2	A0 V	401
		LP 90 - 28		8 41 23	+60 11.8	1	10.64		0.81		0.33		2		1723
	−47 04274	IC 2395 - 14		8 41 23	−48 02.2	1	10.38		-0.01		-0.35		3	A0	1040
		IC 2391 - 60		8 41 23	−52 57.2	1	9.69		0.42		0.02		3		1030
74601	−45 04442			8 41 24	−46 02.7	1	7.98		1.06				1	G8 III	389
74599	−45 04441			8 41 25	−45 22.9	1	6.84		1.51		1.72		1	K3 III	389
74602	−46 04486			8 41 25	−46 50.1	1	8.95		-0.03		-0.42		2	B9 V	283
	−52 01616	IC 2391 - 74		8 41 25	−52 51.0	1	12.36		0.53		0.03		3		1030
74379	+21 01899			8 41 26	+21 03.5	1	8.06		1.09		1.06		2	K0 III	37
74576	−38 04789			8 41 26	−38 42.4	4	6.56	.013	0.93	.002	0.64	.014	9	K2 V	258,1705,2012,3072
	−46 04487			8 41 27	−46 53.7	1	10.82		0.28		-0.14		2		283
74619	−47 04275	IC 2395 - 5		8 41 27	−47 27.7	1	8.06		1.03		0.85		10	K0 III	1040
74621	−47 04277	IC 2395 - 9		8 41 27	−48 00.2	1	8.88		-0.05		-0.37		2	B8 V	1040
	+25 01981	G 9 - 16		8 41 28	+24 59.0	5	9.32	.013	0.31	.004	-0.15	.012	13	F0	516,1003,1620,3077,6006
74348	+28 01640	IDS08384N2849	A	8 41 28	+28 38.3	1	8.60		0.56				2	G0 IV	20
74348	+28 01640	IDS08384N2849	B	8 41 28	+28 38.3	1	9.84		0.58				2		20
74618	−46 04488			8 41 28	−46 54.6	1	8.64		1.01		0.78		2	G6 III	283
74620	−47 04276			8 41 29	−47 30.4	1	8.73		-0.07		-0.64		2	B3/5	401
74665	−53 01834	IC 2391 - 38		8 41 29	−53 34.2	3	8.16	.013	0.20	.009	0.07	.033	5	A7 Vnn	42,1040,3015
74492	−16 02566	IDS08392S1639	A	8 41 31	−16 50.2	1	8.36		0.56				4	F8/G0 V	2033
74492	−16 02566	IDS08392S1639	B	8 41 31	−16 50.2	1	10.63		0.90				4		2033
74664	−52 01619	IC 2391 - 37		8 41 32	−53 13.7	1	10.25		0.15		0.10		4	A3 V	42
		GD 93		8 41 33	+45 47.5	1	15.95		0.24		-0.53		1		3060
74575	−32 05651	HR 3468		8 41 35	−33 00.3	9	3.68	.006	-0.18	.003	-0.88	.007	230	B1.5III	3,9,15,26,1034,1075*
74678	−52 01622	IC 2391 - 39		8 41 37	−52 53.8	4	7.68	.014	0.08	.019	0.03	.028	9	A1 V	42,1030,1040,3015
	−46 04490			8 41 38	−46 16.0	2	10.00	.014	0.06	.009	-0.57		8	B2 V	1075,2021
74650	−47 04282			8 41 39	−47 37.5	1	7.35		-0.05		-0.16		2	B9 V	401
74327	+44 01783			8 41 40	+44 22.0	1	8.82		0.36		0.01		2	F8	1601
		CCS 1277		8 41 40	−25 24.9	1	12.40		4.77				1		864
74617	−39 04709			8 41 40	−39 25.1	1	7.76		1.55				4	K4 III	2012
		POSS 210 # 12		8 41 41	+40 18.9	1	15.65		0.82				1		1739
74662	−47 04283			8 41 42	−48 09.8	1	8.82		-0.07		-0.66		2	B3 V	401
	−45 04447	LSS 1131	⋆ AB	8 41 43	−45 57.5	2	10.77	.030	0.50	.005	-0.54	.010	4	O7 V:n	410,1737
74649	−45 04449			8 41 44	−46 08.2	1	8.89		-0.03		-0.28		2	B9 V	401
	−52 01624	IC 2391 - 61		8 41 45	−52 56.5	1	11.11		1.27		1.06		2		1030
	−42 04554			8 41 46	−42 31.7	1	11.60		0.63		0.16		2		123
74467	+15 01886			8 41 51	+15 04.0	1	8.73		0.33		0.03		2	F5	1375
74442	+18 02027	HR 3461	⋆ A	8 41 51	+18 20.4	6	3.94	.009	1.08	.005	0.99	.011	19	K0 III	3,15,1080,1118,3016,8015
	−45 04451	LSS 1132		8 41 51	−45 42.7	3	10.06	.017	0.34	.018	-0.42	.019	8	B5	396,401,410
	−45 04452			8 41 51	−45 44.4	1	10.96		0.51		-0.01		2		396
		Cr 197 - 10		8 41 52	−41 01.6	1	12.03		0.51		-0.04		2		184

HD	DM	Other Id	N Rem	α_{1950}	δ_{1950}	S	V	σ_V	B–V	σ_{B-V}	U–B	σ_{U-B}	n	Spectrum	References
74377 +42 01922		G 115 - 23	★ AB	8 41 53	+41 51.8	3	8.56	.031	0.94	.001	0.73		8	K3 V	22,1197,1758
74713 −51 03167				8 41 54	−52 01.7	1	7.80		1.08		0.96		1	K1/2 III	389
−52 01626		IC 2391 - 62		8 41 54	−52 57.2	1	9.42		0.97		0.69		3		1030
74677 −45 04455		LSS 1133		8 41 55	−45 55.2	5	8.62	.007	0.21	.008	-0.54	.013	13	B2 II	401,410,540,976,2021
74072 +72 00427				8 41 56	+72 34.1	1	7.50		0.42		-0.04		1	F2	1776
74360 +46 01431				8 41 57	+46 21.3	1	7.97		0.42		0.00		2	F4 V	1601
−45 04457		IDS08403S4511	AB	8 42 00	−45 21.8	2	9.27	.015	0.13	.005	-0.55	.015	3		389,401
74712 −46 04498		SW Vel		8 42 00	−47 13.3	5	7.69	.176	0.97	.192	0.83	.170	6	F8/G0 Ib	283,389,689,1484,1587
74521 +10 01864		HR 3465, BI Cnc		8 42 02	+10 15.8	4	5.64	.008	-0.10	.006	-0.22	.028	16	A1p EuCr	3,695,1022,1202
−40 04552		Cr 197 - 8		8 42 02	−41 01.1	1	10.59		0.36		0.16		2	F2	184
−45 04458				8 42 03	−45 46.0	1	9.38		-0.04		-0.51		2	A5	401
−42 04558				8 42 04	−42 27.5	1	10.50		0.48		0.04		2	F0	123
74693 −43 04581				8 42 04	−43 18.2	1	9.46		0.20		0.12		2	A3 III/IV	401
74734 −51 03173		IC 2391 - 80		8 42 04	−51 49.3	1	8.88		0.25		0.01		3	A2	1030
74676 −40 04555		Cr 197 - 9		8 42 05	−41 04.0	2	9.30	.005	-0.02	.000	-0.38	.010	5	B8 V	184,401
74753 −49 03761		HR 3476, LSS 1138		8 42 06	−49 38.5	6	5.15	.009	-0.20	.010	-1.01	.011	21	B1II/III(n)	15,540,976,1075,1732,2027
74711 −46 04504		LSS 1136		8 42 07	−46 37.0	4	7.11	.007	0.08	.005	-0.71	.016	13	B2 III	158,540,976,1075
74326 +57 01169				8 42 08	+57 37.1	1	8.01		0.14		0.15		2	A0	1375
		LSS 1135		8 42 09	−45 56.2	3	10.88	.000	0.39	.009	-0.65	.022	7	B3 Ib	396,410,1737
−45 04462		LSS 1137		8 42 10	−45 52.7	2	11.38	.000	0.50	.010	-0.53	.025	5		396,410
74762 −53 01842		IC 2391 - 40		8 42 10	−53 20.6	3	7.79	.005	0.19	.004	0.11	.032	7	A5 V	42,1040,3015
		GD 94		8 42 11	+38 12.9	1	16.02		0.29		-0.60		1		3060
−40 04556		Cr 197 - 7		8 42 13	−41 00.6	1	10.31		1.06		0.78		2		184
74389 +49 01766				8 42 15	+49 03.7	1	7.48		0.07		0.05		2	A0	1601
233568 +54 01253				8 42 15	+54 05.9	1	8.47		1.62		1.91		2	K5	1566
−44 04746				8 42 16	−44 56.0	1	9.59		0.35		-0.15		1	A2	401
−45 04463				8 42 17	−45 43.6	1	9.39		-0.04		-0.48		1		389
74485 +31 01876		HR 3464		8 42 18	+30 52.8	5	6.12	.013	0.93	.014	0.63	.000	12	G5 III	15,252,1007,1013,8015
−40 04560		Cr 197 - 11		8 42 18	−41 03.6	1	11.80		0.44		-0.08		2		184
74774 −50 03518				8 42 19	−50 28.3	1	7.25		0.37		-0.01		1	F3 V	389
		POSS 210 # 18		8 42 20	+44 25.4	1	16.26		1.50				1		1739
		Cr 197 - 14		8 42 21	−41 05.4	1	12.00		0.80		0.28		2		184
74591 +6 02030		HR 3469		8 42 22	+05 51.8	2	6.12	.005	0.20	.000	0.09	.010	7	A6 V	15,1078
		LSS 1140		8 42 23	−45 44.2	2	11.67	.015	0.75	.005	-0.32	.024	6	B3 Ib	396,410
74483 +37 01903				8 42 26	+37 28.5	1	6.96		0.28		0.03		2	F0	1601
−20 02665				8 42 27	−21 15.0	1	10.16		1.23		1.21		2	K7 V	3072
		Steph 724	BC	8 42 28	+31 18.6	1	12.23		1.56				1		1746
−43 04588		LSS 1141		8 42 28	−43 48.1	1	9.93		0.72		-0.34		4		362
		Steph 724	A	8 42 29	+31 18.4	1	10.94		1.30		1.18		1	K7	1746
		Cr 197 - 13		8 42 29	−41 08.0	1	12.42		0.41		-0.02		2		184
		HA 172 # 1500		8 42 30	−45 26.5	1	11.59		1.97		2.21		9		1499
74773 −46 04512				8 42 30	−46 56.0	3	7.23	.012	-0.11	.010	-0.60	.004	8	B3 IV	164,283,401
		HA 172 # 1502		8 42 31	−45 27.9	1	11.89		0.81		0.33		11		1499
		AJ79,1294 T11# 13		8 42 32	−45 53.6	1	13.59		0.54		0.12		3		396
+22 01985				8 42 34	+21 50.7	1	10.53		0.55		0.07		4		1723
		KUV 345 - 31		8 42 34	+39 45.0	1	15.14		0.36		-0.23		1		974
−40 04564		Cr 197 - 6		8 42 37	−41 00.3	1	11.39		0.36		-0.17		2		184
−40 04566		Cr 197 - 4		8 42 37	−41 04.0	1	11.42		0.58		0.00		2		184
74772 −42 04569		Tr 10 - 1	★ A	8 42 37	−42 28.0	9	4.06	.012	0.87	.006	0.52	.011	37	G8 IIIab	12,15,389,1040,1075,1637*
74772 −42 04569		Tr 10 - 31	★ B	8 42 37	−42 28.0	1	11.24		0.31		0.05		2		12
−26 06398				8 42 40	−26 23.4	1	10.94		0.23		-0.48		1	F8	1470
		Cr 197 - 3		8 42 40	−41 07.1	1	11.87		0.66		0.20		2		184
		CpD -42 02880		8 42 40	−42 59.1	1	11.10		0.70		-0.27		4		362
74706 −20 02667		HR 3473		8 42 41	−20 59.1	2	6.10	.005	0.22	.000			7	A5 V	15,2012
		POSS 210 # 50		8 42 42	+42 05.3	1	17.60		1.38				3		1739
74807 −43 04593				8 42 42	−43 33.4	1	8.57		0.96		0.70		1	G8 III	389
		CCS 1281		8 42 43	−35 31.8	1	11.46		2.97				1	N	864
−40 04567		Cr 197 - 5		8 42 43	−41 02.4	1	11.33		0.33		0.16		2		184
−45 04472		LSS 1144		8 42 43	−45 42.6	2	11.28	.005	0.64	.000	-0.43	.030	5		396,410
		G 52 - 30		8 42 44	+11 05.8	1	11.63		1.44		1.30		1		333,1620
74922 −59 01092				8 42 45	−59 24.3	1	8.18		0.03		-0.21		2	B9.5III/IV	401
74688 −2 02676		HR 3472	★ AB	8 42 49	−02 25.1	3	6.41	.008	0.52	.005	0.05	.010	14	F2 IV	15,1417,2012
		Cr 197 - 15		8 42 50	−41 00.1	1	12.07		0.47		-0.01		2		184
74804 −40 04569		Cr 197 - 1	★ AB	8 42 51	−41 05.7	8	7.34	.010	0.32	.014	-0.60	.017	23	B0 V	158,184,342,362,401*
		G 115 - 25		8 42 53	+38 24.9	1	14.91		1.65				1		906
		Cr 197 - 2	★ D	8 42 54	−41 11.	1	10.88		0.45		-0.35		2		184
74853 −47 04311				8 42 54	−47 48.2	2	7.72	.006	1.75	.020	1.70	.000	6	K2/3 III	389,1673
74824 −36 04980		HR 3479		8 42 56	−36 57.9	2	5.75	.005	-0.16	.005			7	B2 III	15,2012
74842 −42 04577				8 42 57	−42 26.9	6	7.21	.008	0.74	.006	0.28	.015	14	G8/K1 (V)	123,389,1705,2012,2033,3026
74624 +28 01644				8 42 59	+28 34.4	1	8.30		0.47				2	F5 III	20
74241 +72 00428				8 43 01	+72 06.4	1	9.07		0.49		-0.01		1	F5	1776
75416 −78 00372		HR 3502		8 43 05	−78 47.0	3	5.46	.004	-0.10	.000	-0.34	.005	11	B8 V	15,1075,2038
−45 04482		LSS 1145		8 43 06	−45 48.0	1	10.78		0.56		-0.39		2	WN8	410
74885 −48 04069				8 43 06	−48 29.2	1	8.20		0.76		0.32		1	G8 V	389
−37 05070				8 43 07	−38 11.6	1	11.40		0.60		1.00		2		362
−45 04484				8 43 07	−45 45.9	1	11.38		0.17		-0.51		4		396
74868 −44 04771				8 43 08	−44 21.7	3	6.57	.005	0.56	.010	0.11	.010	7	F8 IV/V	389,401,2012
		LSS 1146		8 43 10	−45 59.2	2	11.55	.020	0.51	.010	-0.50	.020	5	B3 Ib	396,410
74957 −57 01678				8 43 10	−57 35.3	1	8.12		0.59		0.04		4	G2 V	158
−40 04578		Cr 197 - 16		8 43 12	−41 00.2	1	10.87		0.16		0.12		2		184

Table 1

HD	DM	Other Id	N	Rem	α_{1950}	δ_{1950}	S	V	σ_V	B–V	σ_{B-V}	U–B	σ_{U-B}	n	Spectrum	References
74884	−45 04490	SX Vel			8 43 12	−46 09.6	2	8.02	.085	0.75	.000	0.55		2	F8 II	389,1484
74867	−41 04455				8 43 13	−41 20.0	1	8.73		−0.07		−0.43		2	B7 IV/V	401
		AJ79,1294 T11# 11			8 43 13	−45 52.7	1	12.94		0.25		0.12		4		396
74721	+13 01981				8 43 14	+13 26.9	3	8.72	.007	0.04	.004	0.17	.034	8	A0 V	1003,1775,3077
74669	+28 01645				8 43 14	+27 46.8	1	7.23		0.94				2	K1 IV	20
	−40 04579	Cr 197 - 22			8 43 14	−41 02.0	1	9.85		0.20		−0.46		1	B2 V	184
	−40 04580	Cr 197 - 21			8 43 15	−41 06.5	1	10.38		0.24		−0.36		2		184
74956	−54 01788	HR 3485		★ AP	8 43 19	−54 31.5	8	1.96	.008	0.04	.003	0.06	.016	25	A0 V	15,1020,1034,1075,1705,2012*
74883	−40 04582				8 43 21	−41 14.7	1	8.78		−0.05		−0.38		2	B9 V	401
	−40 04584	Cr 197 - 17			8 43 22	−41 00.8	1	11.33		0.30		0.10		2		184
74955	−52 01634	IC 2391 - 41			8 43 22	−52 53.8	3	7.58	.018	0.08	.002	0.05	.027	6	A2 V	42,1040,3015
74425	+66 00575				8 43 23	+66 23.7	1	7.83		0.46		−0.05		2	F8	1733
	−27 05879	R Pyx			8 43 24	−28 01.1	1	9.63		4.29				1	R	864
	−40 04583	Cr 197 - 19			8 43 24	−40 58.7	1	11.70		0.48		0.02		2		184
		Cr 197 - 18			8 43 24	−41 00.4	1	12.16		0.93		0.60		2		184
		Cr 197 - 20			8 43 25	−41 06.6	1	12.31		0.55		−0.08		2		184
		AJ79,1294 T11# 12			8 43 25	−45 52.5	1	13.20		0.63		0.14		2		396
	−45 04494	LSS 1147			8 43 25	−45 58.1	1	11.53		0.58		−0.54		2		410
75116	−67 00990	HR 3491			8 43 26	−68 01.8	3	6.31	.004	1.50	.000	1.78	.000	13	K3 III	15,1075,2038
74920	−45 04496	LSS 1148			8 43 29	−45 51.3	5	7.54	.008	0.03	.007	−0.90	.031	12	O7 IIIn	390,396,410,1075,1737
74794	−1 02125	HR 3478			8 43 30	−01 52.0	2	5.69	.005	1.10	.000	1.04	.005	7	gK0	15,1417
74936	−45 04498	LSS 1149			8 43 32	−45 43.4	1	8.26		0.22		−0.53		4	B3/5 II	1075
	−45 04499				8 43 32	−45 48.2	2	9.08	.000	0.10	.000	−0.62	.005	5		396,410
74740	+22 01988				8 43 35	+22 32.2	1	8.20		0.23		0.10		2	A2	1648
		GD 95			8 43 35	+35 49.5	1	14.81		0.22		−0.67		1		3060
	−45 04501				8 43 38	−45 46.6	1	9.07		0.09		−0.60		1		389
74999	−53 01857	IC 2391 - 42			8 43 38	−53 21.7	1	9.03		0.13		0.11		4	A2 V	42
74738	+29 01823	HR 3474		★ B	8 43 39	+28 57.0	3	6.58	.010	0.05	.008	0.03	.005	7	A3 V	150,1084,3050
74879	−24 07377	HR 3483			8 43 39	−25 12.3	2	6.09	.005	0.08	.000			7	A3 IV/V	15,2012
74952	−45 04502				8 43 39	−45 43.9	1	7.90		0.21		0.06		4	A2/3 V	1075
74781	+18 02037				8 43 40	+17 40.8	1	7.87		0.34		0.07		2	F5	1648
74739	+29 01824	HR 3475		★ A	8 43 41	+28 56.7	9	4.03	.014	1.01	.015	0.78	.008	23	G7.5IIIa	15,150,1080,1084,1355*
74860	−10 02634	HR 3480		★ AB	8 43 43	−10 49.4	4	6.24	.010	1.61	.010	1.90		13	K5 III	15,2013,2029,3005
74462	+67 00559				8 43 45	+67 38.1	2	8.74	.000	0.97	.000	0.53	.000	3	G5 IV	1003,3025
74981	−42 04591				8 43 48	−43 08.7	1	8.18		1.06		0.83		1	K0 III	389
74982	−44 04785				8 43 48	−44 48.2	1	8.50		0.95		0.68		1	G6 III	389
75171	−65 01013	HR 3495			8 43 49	−65 38.6	3	6.04	.007	0.21	.004	0.07	.010	13	A4 V	15,1075,2038
74950	−35 05026				8 43 50	−35 25.0	1	9.44		1.44		1.54		1	K2 (III)	565
75029	−53 01862	IC 2391 - 43			8 43 51	−53 40.5	1	9.56		0.29		0.11		3	A7 V	42
74966	−36 04998				8 43 52	−36 33.6	1	7.44		−0.14		−0.62		2	B4 IV	401
		POSS 210 # 52			8 43 53	+42 15.2	1	17.63		1.47				2		1739
	−48 04083	NGC 2670 - 3			8 43 53	−48 39.0	1	10.67		0.38		−0.02		3		1040
		POSS 210 # 14			8 43 54	+41 31.5	1	15.95		1.19				1		1739
75086	−58 01202	HR 3489		★ AB	8 43 54	−58 32.5	4	6.20	.009	−0.09	.010	−0.47	.000	11	B7 III	15,401,404,2012
74979	−40 04593				8 43 57	−40 25.9	4	7.24	.001	−0.04	.005	−0.82	.015	10	B2 III	362,401,540,976
		ApJ144,259 # 30			8 44 00	+18 04.	1	14.30		−0.07		−1.03		4		1360
	−48 04087	NGC 2670 - 4			8 44 00	−48 35.4	1	11.14		0.26		−0.21		3		1040
74918	−13 02673	HR 3484		★ A	8 44 01	−13 21.8	6	4.31	.007	0.90	.000	0.62	.023	24	G8 IIIb	15,1075,1425,2012*
		POSS 210 # 5			8 44 02	+41 36.0	1	14.26		0.43				1		1739
75028	−47 04329				8 44 02	−47 44.8	1	8.31		0.06		0.06		2	A0/1 V	401
75009	−43 04611				8 44 03	−44 03.9	1	6.70		−0.09		−0.37		2	B8 IV/V	401
75008	−41 04476	Tr 10 - 10			8 44 06	−42 09.3	1	8.20		0.03		0.03		3	A2 V	1040
75066	−51 03196	IC 2391 - 44			8 44 06	−52 14.1	1	9.34		0.39		0.14		4	F2 IV	42
75067	−52 01639	IC 2391 - 45			8 44 06	−52 45.1	2	9.43	.019	0.00	.005	−0.28	.010	6	B9 V	42,184
		NGC 2669 - 25			8 44 06	−52 47.3	1	11.88		0.20		0.10		2		184
74811	+28 01647				8 44 07	+28 20.8	2	6.55	.035	0.63	.000	0.22		4	G2 IV	20,3016
74874	+6 02036	HR 3482, ε Hya		★ ABC	8 44 08	+06 36.2	8	3.38	.014	0.68	.007	0.34	.021	27	G5 III+A8IV	15,254,1008,1034,1078*
		POSS 210 # 17			8 44 08	+42 41.0	1	16.11		1.49				2		1739
74873	+12 01904	HR 3481			8 44 12	+12 17.7	1			0.11		0.05		3	A1 V	1022
	+36 01870				8 44 13	+36 38.4	1	8.51		1.39		1.52		2	K2	1733
74225	+78 00293			V	8 44 14	+78 21.1	1	6.94		1.61		1.64		5	M5 III	985
75064	−47 04332				8 44 16	−47 23.6	1	8.84		0.06		−0.10		2	B9 IV	401
75105	−52 01640	IC 2391 - 46			8 44 16	−52 47.3	2	7.68	.029	−0.09	.000	−0.47	.005	6	B8 Vp	42,184
		vdB 21 # a			8 44 18	−43 28.7	1	12.22		0.43		−0.28		4		434
75062	−43 04615				8 44 18	−43 34.1	1	7.92		−0.06		−0.33		2	B8 III/IV	401
		NGC 2669 - 26			8 44 18	−52 45.9	1	12.76		0.46		0.23		2		184
74604	+67 00560	HR 3470			8 44 19	+66 53.6	1			−0.11		−0.41		1	B8 V	1079
	−50 03547				8 44 19	−50 47.1	1	9.87		0.43				4	B3 V	2021
75061	−43 04617				8 44 20	−43 16.7	1	9.49		0.13		0.10		2	A2 IV/V	401
75063	−45 04517	HR 3487			8 44 20	−45 51.5	6	3.90	.007	0.00	.004	−0.04	.011	26	A0 II	15,1075,1637,2012*
	−52 01642	NGC 2669 - 27			8 44 20	−52 39.3	1	10.33		0.04		−0.23		2	B8 III	184
	−12 02669				8 44 21	−13 10.2	7	10.24	.012	0.30	.005	−0.16	.004	14	A5	1064,1594,1696,2017*
75104	−48 04091	NGC 2670 - 1			8 44 22	−48 30.2	2	7.46	.010	1.16	.000	1.16	.005	4	G2 II	389,1040
		G 234 - 47			8 44 23	+61 20.4	1	15.17		1.82		1.51		1		906
75082	−41 04479				8 44 26	−41 51.1	1	8.41		−0.01		−0.36		2	B8/9 III	401
74926	+11 01913	IDS08417N1132		A	8 44 27	+11 20.6	1	7.69		0.10		0.09		1	A0	695
	−52 01644	NGC 2669 - 5			8 44 28	−52 42.8	1	11.00		0.25		0.28		2		184
75060	−39 04765				8 44 29	−40 12.8	1	7.98		−0.10		−0.42		2	B5 V	401
75022	−29 06734				8 44 30	−29 34.5	1	7.58		1.45		1.60		1	K2/3 III	414
75083	−42 04605	Tr 10 - 14			8 44 30	−42 34.5	1	8.69		−0.08		−0.31		3	B8 V	1040

HD	DM	Other Id	N Rem	α_{1950}	δ_{1950}	S	V	σ_V	B–V	σ_{B-V}	U–B	σ_{U-B}	n	Spectrum	References
75103	−46 04557	IDS08428S4702	AB	8 44 30	−47 12.7	1	8.11		0.43		−0.02		2	F3 V	401
		NGC 2669 - 28		8 44 30	−52 40.3	1	12.88		0.27		0.22				184
	+46 01436		AB	8 44 31	+45 42.8	1	9.21		0.79		0.29		4		1723
75021	−29 06735	UZ Pyx		8 44 31	−29 32.6	6	7.18	.063	1.96	.051	3.21	.178	14	C6,3	414,864,1238,2033*
		POSS 210 # 3		8 44 32	+43 51.2	1	13.93		0.68				1		1739
75081	−40 04602	HR 3488		8 44 34	−40 56.5	3	6.21	.005	−0.05	.000	−0.20		9	B9 Ve	15,401,2027
75057	−32 05710			8 44 36	−33 03.7	1	7.88		0.41				4	F2 V	2012
		G 47 - 5		8 44 38	+22 04.5	1	15.69		1.56				1		906
		NGC 2669 - 6		8 44 38	−52 43.7	1	12.73		0.31		0.29		2		184
75129	−47 04337			8 44 40	−47 22.0	1	6.87		0.26		−0.28		2	B5 Ib	401
74988	−1 02130	HR 3486		8 44 43	−01 42.8	3	5.28	.005	0.04	.000	0.08	.000	8	A3 V	15,1256,3023
		NGC 2669 - 22		8 44 43	−52 46.5	1	11.84		0.41		0.12		3		184
75167	−52 01647	NGC 2669 - 1		8 44 43	−52 50.6	2	8.60	.011	1.14	.002	1.14	.013	4	K2 I/III	184,1685
75150	−47 04339			8 44 44	−47 43.5	1	8.20		0.97		0.70		1	G8 III/IV	389
75126	−42 04610	Tr 10 - 3	⋆ AB	8 44 45	−42 22.9	1	7.08		−0.13		−0.51		3	B4 V	1040
		NGC 2669 - 23		8 44 45	−52 46.2	1	12.63		0.21		0.20		2		184
75185	−53 01878	IC 2391 - 48		8 44 45	−53 32.3	1	9.83		0.14		0.13		4	A3 V	42
75127	−42 04611	Tr 10 - 12		8 44 47	−42 36.4	1	8.67		−0.06		−0.34		3	B8 V	1040
	+14 01974			8 44 48	+14 21.0	1	9.08		0.39		0.02		2	F5	1648
	−45 04524			8 44 48	−45 20.3	1	10.38		0.31		0.08		2		540
75149	−45 04526	HR 3494, LSS 1151		8 44 48	−45 43.7	8	5.45	.015	0.28	.013	−0.54	.023	27	B5 Iab	15,26,410,540,976*
75184	−52 01648	IC 2391 - 47	⋆ AB	8 44 48	−52 56.7	1	9.11		0.08		0.02		4	A3 V	42
	−52 01649	NGC 2669 - 7		8 44 49	−52 43.2	1	11.15		0.10		−0.19		4		184
75112	−34 05243	HR 3490		8 44 50	−34 26.3	4	6.36	.003	−0.14	.013	−0.58	.018	15	B3 V	15,164,1075,2012
	−52 01650	NGC 2669 - 20		8 44 50	−52 47.3	1	10.64		0.17		−0.01		4	B8 V	184
	−53 01880			8 44 50	−53 34.2	1	10.28		0.62		0.01		2		540
		NGC 2669 - 21		8 44 51	−52 46.5	1	12.23		1.44		1.23		2		184
75125	−39 04773			8 44 52	−39 43.4	3	7.05	.004	−0.08	.005	−0.30	.008	6	B8 III/IV	401,540,976
		NGC 2669 - 8		8 44 52	−52 44.8	1	12.77		0.49		0.21		2		184
75148	−43 04624			8 44 53	−43 47.1	1	8.37		0.20		0.14		2	A3 V	401
74925	+28 01648			8 44 54	+28 09.9	1	8.70		0.52				2	G8 IV	20
	−0 02066			8 44 54	−00 02.3	2	10.04	.019	1.02	.019	0.81	.010	3		116,118
75202	−52 01652	IC 2391 - 49	⋆ AB	8 44 55	−52 39.6	3	7.72	.028	0.18	.013	0.12	.000	7	A5 V	42,184,3015
		NGC 2669 - 18		8 44 57	−52 46.0	1	12.26		0.21		0.14		2		184
	−52 01653	NGC 2669 - 19		8 44 57	−52 47.7	1	11.58		0.15		−0.03		2		184
75747	−78 00378	HR 3524, RS Cha		8 44 58	−78 53.3	3	6.04	.004	0.24	.004	0.07	.010	11	A7 V	15,1075,2038
	+0 02391			8 44 59	−00 19.8	2	9.57	.019	1.59	.010	1.92	.029	3	M0	116,118
75200	−47 04341			8 45 00	−48 09.3	1	8.84		0.45		0.10		2	F3 V	401
		NGC 2669 - 12		8 45 00	−52 45.2	1	11.92		0.14		0.03		2		184
75012	+0 02392	IDS08425N0027	AB	8 45 01	+00 15.7	5	7.82	.006	0.08	.004	0.08	.016	38	B9	116,118,147,1509,6005
	−52 01655	NGC 2669 - 9		8 45 01	−52 44.4	1	9.41		0.04		−0.48		3	B6 IV	184
		NGC 2669 - 10		8 45 02	−52 44.7	1	12.07		0.52		0.03		2		184
	−52 01656	NGC 2669 - 11		8 45 02	−52 45.2	1	11.20		0.10		−0.01		2		184
		NGC 2669 - 14		8 45 02	−52 45.7	1	11.85		0.88		0.48		2		184
	−52 01659	NGC 2669 - 13		8 45 03	−52 45.5	1	11.68		0.12		−0.06		2		184
75227	−52 01657	NGC 2669 - 4		8 45 04	−52 41.6	3	8.22	.011	1.21	.011	0.97	.008	5	K0 II	184,389,1685
		NGC 2669 - 17		8 45 05	−52 47.0	1	12.62		0.27		0.21		2		184
75212	−48 04097	NGC 2670 - 2		8 45 06	−48 34.9	1	8.42		−0.08		−0.29		3	B8 V	1040
		POSS 210 # 6		8 45 08	+41 37.4	1	14.34		1.54				1		1739
		NGC 2669 - 16		8 45 08	−52 47.4	1	12.70		0.17		0.16		2		184
75590	−74 00538			8 45 13	−74 36.3	1	6.66		1.22		1.33		5	K2 III	1628
	−52 01660	NGC 2669 - 15		8 45 14	−52 46.5	1	11.57		0.17		−0.10		2		184
75211	−43 04635	LSS 1154		8 45 16	−43 53.4	6	7.51	.008	0.40	.009	−0.58	.041	17	O8 II(f)	362,390,540,976,1075,1737
75035	+17 01941	IDS08524N1746	A	8 45 17	+17 34.8	1	7.22		1.38		1.63		2	K2	1648
		LDS 2351		8 45 18	−18 48.	1	11.63		1.00		1.00		2		3028
		LDS 2352		8 45 18	−18 48.	1	15.55		−0.06		−0.93		2	DB	3028
	−42 04617	Tr 10 - 13		8 45 21	−42 31.1	1	8.68		1.21		1.13		3	K2 IIIa	1040
75243	−47 04348			8 45 21	−47 38.7	1	9.24		−0.07		−0.59		2	B8 II/III	401
75241	−44 04818			8 45 22	−44 53.4	2	6.59	.005	−0.12	.005	−0.54		8	B4 IV	1075,2012
		NGC 2669 - 31		8 45 22	−52 47.9	1	13.02		0.29		0.28		2		184
75295	−53 01888			8 45 23	−54 00.8	1	8.25		0.02		−0.08		2	B9.5V	401
	−42 04619	Tr 10 - 23		8 45 24	−42 21.5	1	9.60		0.02		−0.07		3	A1 V	1040
		NGC 2669 - 30		8 45 24	−52 48.4	1	12.58		1.33		1.24		2		184
75311	−56 01865	HR 3498, V344 Car		8 45 25	−56 35.1	9	4.47	.020	−0.17	.014	−0.76	.020	45	B3 Vne	26,681,815,1088,1637,1732*
	−42 04620	Tr 10 - 18		8 45 26	−42 27.2	1	9.12		0.90		0.57		3	G5 III	1040
	−45 04539			8 45 26	−46 07.2	1	10.04		0.12		−0.59		2	B2 V	540
		POSS 210 # 31		8 45 27	+38 38.2	1	16.77		1.58				2		1739
75239	−41 04498	Tr 10 - 19		8 45 27	−42 12.0	1	9.16		−0.02		−0.20		3	B9p	1040
75222	−36 05030	LSS 1155		8 45 29	−36 34.0	4	7.41	.012	0.38	.003	−0.57	.025	10	B0 II/III	540,976,1737,2012
	−30 06812			8 45 32	−31 12.8	1	9.84		0.96		0.74		4	K5 V	3072
75294	−52 01665	NGC 2669 - 29		8 45 32	−52 48.1	1	9.89		0.07		−0.28		2	B6 V	184
	−42 04623	Tr 10 - 27		8 45 33	−42 32.3	1	10.16		0.14		0.06		3	A2	1040
75140	−6 02727	HR 3493		8 45 37	−06 22.4	2	6.08	.005	1.28	.000	1.42	.000	7	K0	15,1256
	−41 04504	Tr 10 - 16		8 45 37	−42 17.4	1	8.95		−0.05		−0.26			B9.5V	1040
75276	−45 04541	HR 3496, LSS 1156		8 45 37	−45 58.2	6	5.75	.008	0.55	.015	0.39	.033	23	F2 Iab	15,1018,1034,1415*
75258	−41 04502	Tr 10 - 5		8 45 38	−42 16.1	1	7.19		−0.12		−0.56		3	B4 V	1040
75275	−43 04643			8 45 39	−43 52.9	2	8.81	.010	−0.07	.005	−0.65	.005	6	B3 V	362,401
75293	−47 04358			8 45 39	−47 40.0	1	9.27		−0.01		−0.21		2	Ap Si(Cr)	401
75274	−42 04624	Tr 10 - 7		8 45 40	−42 35.0	2	8.04	.020	1.03	.010	0.89	.005	4	K2 IIIa	389,1040
75138	+1 02173	IDS08432N0055	A	8 45 44	+00 44.4	3	7.24	.013	1.47	.014	1.67	.144	33	K2	147,1509,6005

Table 1

HD	DM	Other Id	N Rem	α₁₉₅₀	δ₁₉₅₀	S	V	σ_V	B–V	σ_B–V	U–B	σ_U–B	n	Spectrum	References
	−42 04625	Tr 10 - 28		8 45 44	−42 19.5	1	10.26		0.17		0.02		3		1040
		AJ79,1294 T12# 9		8 45 45	−42 03.2	1	12.97		0.64		0.14		3		396
	−42 04627	Tr 10 - 26		8 45 45	−42 31.6	1	9.89		0.47		0.03		3	F8	1040
75351	−51 03221	IDS08442S5133	AB	8 45 45	−51 44.2	1	7.82		0.62		0.19		1	G2 V	389
	+7 02031	G 46 - 4	⋆ AB	8 45 46	+06 39.7	2	10.35	.013	1.28	.002	1.21		3	K4-5	1017,3072
	+0 02394			8 45 46	−00 05.7	2	10.91	.034	0.50	.015	−0.03	.054	3	F2	116,118
75290	−42 04628	Tr 10 - 8		8 45 46	−42 18.0	1	8.06		−0.06		−0.33		3	B9 V	1040
75368	−54 01816	IDS08444S5425	AB	8 45 46	−54 36.4	1	8.34		0.31		−0.57		6	G8 IV/V	396
75137	+6 02040	HR 3492	⋆ AB	8 45 47	+06 01.4	5	4.35	.016	−0.04	.007	−0.04	.004	13	A0 Vn	15,1078,1363,3023,8015
75272	−38 04879			8 45 47	−38 48.4	4	6.98	.006	−0.16	.007	−0.67	.007	9	B9/A0 V	362,401,540,976
75309	−45 04547	LSS 1157		8 45 47	−46 16.0	4	7.85	.010	0.01	.004	−0.81	.019	9	B2 Ib/II	540,976,1075,1737
75291	−42 04629	Tr 10 - 22		8 45 50	−42 22.4	1	9.55		0.09		−0.02		3	A3 V	1040
		vdB 21 # b		8 45 50	−43 31.7	1	13.13		0.52		−0.07		3		434
75289	−41 04507	HR 3497		8 45 51	−41 32.9	4	6.36	.006	0.58	.000	0.10	.000	18	G0 V	15,1311,2012,3077
66368	+89 00013	IDS07580N8856	AB	8 45 54	+88 46.2	1	7.14		0.15		0.09		3	A0	985
75156	+13 01994			8 45 55	+12 44.0	2	6.63	.008	1.64	.014	1.85	.004	8	M1	1375,3040
		MCC 117		8 45 56	+29 10.0	1	10.99		1.36				1	K7	1017
	−42 04632	Tr 10 - 29		8 45 56	−42 32.8	1	10.97		0.29		0.09		3		1040
75349	−46 04590			8 45 57	−47 01.3	1	9.14		−0.04		−0.25		2	B8/9 V	401
	+0 02397			8 45 58	+00 10.8	2	9.95	.023	0.28	.014	0.12	.009	4	F2	116,118
		BSD 100 # 421		8 45 58	+00 22.9	1	11.33		0.28		0.10		1	A1	116
75268	−30 06819			8 45 59	−30 44.0	1	8.79		1.35				4	K2/3 III	2033
75324	−41 04510	Tr 10 - 4		8 45 59	−42 05.3	2	7.13	.009	−0.12	.004	−0.52	.000	8	B5 V	125,396,1040
		Ton 11		8 46 00	+30 13.	1	16.92		0.02		−0.64		1		98
74940	+62 01021			8 46 01	+62 01.1	1	7.75		1.07		0.85		2	K0	1375
		GD 96		8 46 02	+34 40.9	1	15.71		0.28		−0.35		1		3060
	+37 01912	G 115 - 27		8 46 02	+36 42.8	1	11.17		1.20				1	M1	1746
75348	−43 04649			8 46 03	−44 02.8	1	9.09		0.27		0.11		2	A9 IV/V	401
75217	−0 02069			8 46 05	−00 51.6	2	6.42	.004	1.14	.004	1.01	.004	9	K0	116,118
75346	−42 04636	Tr 10 - 9		8 46 05	−42 21.8	1	8.07		0.41		−0.04		3	F0 III-IV	1040
		WD0846+249		8 46 07	+24 56.4	2	16.71		−0.31		−1.23		2	DA1	1727
75364	−42 04638	IDS08443S4259	A	8 46 07	−43 10.1	1	7.23		0.85		0.55		1	K1 III	389
	−43 04653	LSS 1159		8 46 07	−43 46.3	1	10.25		0.61		−0.45		5		362
		CCS 1307		8 46 07	−70 52.0	1	9.21		2.91		4.70		2		864
75347	−42 04637	Tr 10 - 17		8 46 08	−42 24.3	1	9.09		0.08		0.00		3	A2 V	1040
	+0 02398			8 46 13	+00 32.5	2	10.06	.029	0.88	.010	0.51	.005	3		116,118
	−42 04641	Tr 10 - 15		8 46 13	−42 22.9	2	8.91	.020	−0.05	.010	−0.21	.025	4	B9 V	389,1040
	−41 04514			8 46 14	−42 01.4	1	11.17		0.49		0.08		4		125
75386	−41 04515	Tr 10 - 21	A	8 46 17	−42 06.1	2	9.27	.017	0.10	.009	0.06	.009	6	A4 IV	125,396,1040
75386	−41 04515	Tr 10 - 21	B	8 46 17	−42 06.1	1	11.17		0.49		0.08		4		396
75387	−41 04516	Tr 10 - 2	⋆	8 46 21	−42 16.7	7	6.42	.002	−0.20	.005	−0.78	.010	49	B2 IV	15,362,540,976,1040*
75398	−43 04658			8 46 24	−43 45.0	1	8.82		0.03		−0.05		2	A0 IV	401
		AJ77,733 T10# 9		8 46 25	−41 59.8	1	14.29		0.64		0.08		4		125
	−44 04834	LSS 1160		8 46 29	−44 41.2	1	10.01		1.03		−0.01		2	B1 III:n	1737
75466	−52 01675	IC 2391 - 50	⋆	8 46 32	−52 39.9	4	6.29	.025	−0.10	.007	−0.31	.005	13	B8 Vnn	15,42,2029,3015
75482	−53 01901			8 46 33	−53 43.6	2	8.77	.005	0.43	.000	−0.06		5	F2 V	389,2012
		CpD -43 03006		8 46 35	−43 50.0	1	11.13		1.02		−0.06		5		362
		NGC 2682 - 2277		8 46 37	+12 15.6	1	12.15		0.53		0.10		3		1717
74904	+70 00536			8 46 37	+70 29.2	1	7.13		1.64				3	M1	70
75502	−53 01904			8 46 39	−53 24.4	1	8.60		0.05		−0.28		2	A1 V	401
75465	−46 04605	LSS 1162		8 46 40	−46 43.5	2	9.02	.019	0.21	.015	−0.52		6	B2 III/IVe	401,2021
75395	−29 06794			8 46 41	−30 07.8	1	9.51		0.18		0.16		2	A3 IV	1730
		AJ77,733 T10# 7		8 46 42	−41 59.1	1	12.42		0.58		0.11		3		125,396
75446	−41 04523	Tr 10 - 6		8 46 42	−42 12.9	1	7.38		−0.13		−0.58		3	B6 V	1040
75318	+4 02051	G 114 - 18	⋆ AB	8 46 45	+03 52.2	1	7.92		0.69		0.24		1	G5	1658
75442	−36 05052	IDS08449S3614	AB	8 46 49	−36 24.9	1	8.25		0.94				4	G8 IV	2012
75464	−42 04656	Tr 10 - 24		8 46 49	−42 26.5	1	9.70		0.20		0.06		3	A6 IV-V	1040
	−45 04571	IDS08451S4536	AB	8 46 50	−45 47.1	1	9.54		0.16		0.11		2	A0	401
75478	−47 04377			8 46 50	−47 21.1	1	8.56		0.29		0.03		2	A7/8 V	401
	+12 01913	NGC 2682 - S 258		8 46 51	+12 02.6	2	9.50	.005	1.35	.009	1.53	.028	4	K2	196,1553
75333	−2 02699	HR 3500, KX Hya		8 46 51	−03 15.4	6	5.30	.007	−0.09	.007	−0.32	.029	21	B9p HgMn	3,15,1063,1079,1256,3023
		AJ77,733 T10# 8		8 46 51	−41 59.5	1	13.53		0.84		0.48		4		125,396
75477	−42 04658	Tr 10 - 20		8 46 53	−42 18.8	1	9.21		−0.03		−0.18		3	A0 V	1040
75440	−29 06799			8 46 59	−30 02.0	1	8.39		0.66		0.23		2	G1 V	1730
75476	−39 04824			8 47 00	−40 06.3	1	7.96		−0.12		−0.58		2	B8 III	401
		AJ79,1294 T12# 7		8 47 00	−41 53.2	1	11.86		0.45		−0.02		3		396
75535	−48 04120			8 47 01	−49 02.5	1	7.68		1.09		0.97		1	K0 III	389
75501	−41 04533			8 47 02	−41 55.2	1	9.21		0.48		0.13		4	F3/5 IV/V	125,396
	−41 04532			8 47 03	−41 58.8	1	10.33		0.42		0.03		4		125
75534	−47 04381	LSS 1165		8 47 06	−47 34.6	4	7.82	.010	0.37	.013	−0.52	.028	10	B1/2 Ib	540,976,1075,1737
75611	−54 01834			8 47 07	−55 15.8	1	7.68		−0.05		−0.42		2	B7 IV	401
75355	+15 01912	IDS08444N1512	A	8 47 09	+15 01.2	2	7.66	.015	0.44	.018	0.05	.020	4	F0	1648,3016
75355	+15 01912	IDS08444N1512	B	8 47 09	+15 01.2	1	8.58		0.47		0.07		2		3032
	−46 04615			8 47 10	−47 06.4	2	9.62	.005	0.10	.010	−0.64	.011	8	B1.5V	401,540,976
		G 194 - 43		8 47 12	+53 05.0	1	15.40		1.78		1.31		1		906
75533	−41 04534	Tr 10 - 25		8 47 12	−42 04.7	1	9.80		0.06		−0.03		2	A2 V	1040
	+12 01917	NGC 2682 - S 364		8 47 13	+11 52.8	3	9.81	.010	1.32	.012	1.45	.025	6		196,1553,1717
		BSD 100 # 847		8 47 15	+00 01.6	1	11.06		0.28		0.04		1	A2	116
		POSS 210 # 4		8 47 17	+42 10.5	1	13.97		0.59				1		1739
75549	−43 04668			8 47 17	−43 34.5	3	7.31	.007	−0.14	.005	−0.58	.009	13	B3 V	362,1075,2012

HD	DM	Other Id	N Rem	α_{1950}	δ_{1950}	S	V	σ_V	B–V	σ_{B-V}	U–B	σ_{U-B}	n	Spectrum	References
		KUV 08473+3838		8 47 20	+38 38.5	1	17.67		-0.03		-0.86		1	DA	1708
	-31 06576	IDS08453S3200	A	8 47 21	-32 10.6	1	9.94		1.10		1.00		2	K7 V	3072
	-31 06576	IDS08453S3200	B	8 47 21	-32 10.6	1	13.00		1.49		1.05		2	K7 V	3072
	-43 04669			8 47 23	-44 06.4	1	10.90		0.76		-0.27		4		362
75587	-45 04581			8 47 24	-45 27.1	1	8.31		0.29		0.08		2	F0 IV/V	401
75332	+33 01765	HR 3499		8 47 26	+33 28.4	1	6.27		0.48		-0.01		4	F7 Vn	254
75570	-41 04541	Tr 10 - 11		8 47 26	-42 08.8	1	8.54		0.35		0.02		3	F2 IV	1040
75586	-43 04670			8 47 27	-43 43.5	1	8.25		1.12		0.99		1	K1 II	389
	+12 01919	NGC 2682 - 4202		8 47 28	+12 02.6	3	8.85	.021	1.57	.013	1.96	.017	9	K5	196,1553,1717
75661	-53 01914	LSS 1166		8 47 29	-54 05.2	1	8.88		0.12		-0.82		2	B2 Vne	401
	+0 02399			8 47 30	+00 34.5	2	9.08	.023	0.93	.019	0.59	.009	4	G5	116,118
75495	-20 02693	HR 3507		8 47 30	-20 51.7	2	6.46	.005	0.24	.005			7	A6 IV	15,2012
	+0 02400			8 47 31	-00 14.8	2	10.22	.009	0.38	.009	0.06	.005	5	F2	116,118
	-41 04542			8 47 31	-41 30.8	1	10.16		0.63		0.08		4	G0	158
		POSS 210 # 34		8 47 33	+43 55.6	1	16.80		1.26				1		1739
	-9 02670			8 47 33	-09 36.4	1	9.70		1.07		0.82		1		3072
	+12 01920	NGC 2682 - 4169		8 47 34	+12 06.6	3	9.97	.012	1.07	.014	0.92	.011	8		196,1553,1717
	+66 00582			8 47 34	+66 19.1	1	9.28		1.34		1.23		2	K5	3072
75608	-42 04676			8 47 34	-43 11.0	1	7.45		-0.09		-0.51		2	B6 V	401
75660	-52 01684			8 47 34	-52 55.0	1	8.18		-0.02		-0.12		2	B9/9.5V	401
75609	-46 04626			8 47 35	-46 21.9	1	8.61		0.39		-0.02		2	F2 V	401
		NGC 2682 - S 676		8 47 37	+12 32.4	2	10.51	.005	1.18	.005	1.33	.023	4		1553,1717
		G 51 - 27		8 47 37	+34 50.5	1	14.71		1.70		1.42		6		419
75631	-42 04677			8 47 38	-42 49.1	1	8.83		-0.05		-0.28		2	B8 V	401
		BSD 100 # 854		8 47 39	+00 06.2	1	11.32		0.07		0.02		1	A1	116
	+0 02401			8 47 39	-00 13.1	2	10.28	.019	0.14	.014	0.10	.009	4		116,118
		AAS27,343 # 67		8 47 42	-40 58.	1	12.70		0.50		-0.02		2		362
75607	-41 04543	IDS08459S4112	AB	8 47 42	-41 22.6	1	8.73		0.13		-0.55		2	B3 V	401
75658	-47 04393	LSS 1167		8 47 45	-47 18.1	4	8.13	.025	0.19	.012	-0.77	.035	10	B2/3ne	540,976,1075,1737
		G 46 - 5		8 47 46	+07 49.0	3	11.31	.020	0.74	.000	0.15	.004	7		1064,1620,3077
75630	-39 04838	HR 3514		8 47 47	-40 08.0	4	5.48	.005	0.07	.005	0.12		13	A2/3 IV	15,401,2013,2027
75677	-49 03856			8 47 48	-50 07.2	1	8.10		1.04		0.85		1	K0 III	389
76236	-79 00352	HR 3543		8 47 49	-79 19.2	3	5.78	.008	1.59	.005	1.96	.000	11	K5 III	15,1075,2038
75605	-32 05770	HR 3512		8 47 50	-32 35.6	3	5.20	.004	0.88	.005			11	G8 III	15,1075,2031
75470	+18 02057	IDS08450N1823	AB	8 47 51	+18 11.4	2	6.72	.000	0.86	.000			52	G5	130,1351
75657	-42 04684			8 47 51	-42 37.9	2	7.47	.005	-0.14	.010	-0.69	.010	5	B6 III	362,401
75676	-47 04399			8 47 51	-48 09.3	1	8.31		1.10		0.96		1	G8/K0 III	389
75863	-68 00845			8 47 52	-68 28.7	1	8.11		1.02		0.80		3	K0 III	1704
75530	-4 02468	G 114 - 19		8 47 53	-05 20.5	8	9.18	.016	0.74	.007	0.22	.011	29	G8 V	22,830,1003,1658,1783*
75469	+19 02110	HR 3504		8 47 55	+19 01.2	2	6.42		-0.01	.000	0.01	.005	5	A2 Vs	1022,1733
75629	-28 06600	HR 3513		8 47 56	-29 16.6	1	5.87		0.95				4	G8 III	2031
		WLS 840 15 # 8		8 47 58	+14 49.0	1	13.28		0.68		0.04		2		1375
75655	-41 04549			8 47 58	-41 33.4	4	7.90	.006	-0.03	.008	-0.79	.015	10	B2 III	362,401,540,976
75654	-38 04925	HR 3517, HZ Vel		8 47 59	-38 57.3	2	6.38	.005	0.24	.005			7	A5 III	15,2027
		BPM 5102		8 48 00	-73 02.	1	15.30		-0.09		-0.72		1		3065
	+8 02131	IDS08453N0814	AB	8 48 02	+08 03.1	3	9.06	.012	1.35	.011	1.20	.000	5		497,679,3072
		NGC 2682 - 15		8 48 04	+11 59.1	1	13.61		0.55		0.04		2		135
75710	-44 04861	HR 3520		8 48 04	-45 07.3	2	4.92	.005	0.04	.005	0.14		6	A2 III	401,2006
	-0 02073			8 48 05	-00 04.8	2	10.28	.024	1.16	.015	1.09	.029	3		116,118
		NGC 2682 - S 721		8 48 06	+11 46.4	1	11.24		1.06		0.95		1		1553
		NGC 2682 - 20		8 48 06	+12 00.5	1	12.95		0.87		0.47		2		135
		NGC 2682 - 514		8 48 08	+11 56.4	1	14.60		0.72		0.18		1		135
		NGC 2682 - 23		8 48 08	+11 59.4	1	12.66		0.54		-0.01		1		135
		NGC 2682 - 515		8 48 09	+11 55.0	1	14.41		0.74		0.17		1		135
		NGC 2682 - 27		8 48 09	+11 55.8	1	13.59		0.67		0.10		1		135
		NGC 2682 - 24		8 48 09	+11 58.8	1	13.31		0.58		0.07		1		135
		NGC 2682 - 1043		8 48 12	+11 56.4	1	16.33		1.19		0.70		1		135
		NGC 2682 - 1035		8 48 12	+11 57.7	1	16.25		1.09		0.92		1		135
		NGC 2682 - 30		8 48 12	+12 03.5	2	11.97	.054	0.56	.034	0.15	.054	3		135,1717
		NGC 2682 - 39		8 48 14	+11 53.5	1	13.57		0.49		-0.01		1		135
		NGC 2682 - 37		8 48 14	+12 03.6	1	12.89		0.98		0.65		2		135
75528	+15 01917	HR 3510, AE Lyn	AB	8 48 14	+15 32.2	3	6.38	.008	0.64	.002	0.22		21	G2 IV	15,1351,3077
	-29 06828			8 48 14	-29 58.0	1	9.79		1.09		0.92		2	K0	1730
		NGC 2682 - 1028		8 48 15	+11 57.0	1	16.64		1.00		0.70		1		135
		NGC 2682 - 43		8 48 15	+11 57.5	1	13.44		0.63		0.06		2		135
		NGC 2682 - 40		8 48 15	+11 59.6	1	13.79		0.57		0.05		1		135
		NGC 2682 - 42		8 48 15	+12 00.2	1	13.51		0.61		0.06		1		135
	+73 00436			8 48 15	+73 22.8	1	9.57		0.92		0.71		2	G5	1776
75649	-28 06610	HR 3516		8 48 15	-28 25.8	3	6.16	.004	-0.09	.004	-0.18	.005	11	B9 V	15,1075,2012
		NGC 2682 - 46		8 48 16	+12 05.8	1	12.79		0.81		0.38		2		135
75724	-41 04554			8 48 16	-41 55.2	5	7.69	.006	-0.05	.010	-0.78	.012	18	B2 III	92,164,362,540,976
		NGC 2682 - 49		8 48 17	+11 50.9	1	13.44		0.60		0.04		1		135
		NGC 2682 - 52		8 48 17	+11 59.1	1	13.52		0.60		0.07		2		135
		NGC 2682 - 51		8 48 17	+12 00.8	1	12.69		0.62		0.09		1		135
		NGC 2682 - 48		8 48 17	+12 01.4	1	12.77		0.75		0.26		2		135
75725	-42 04691			8 48 17	-42 22.4	1	7.81		0.92		0.58		1	G8 III	389
	-44 04865	LSS 1172		8 48 17	-44 23.4	3	9.43	.006	0.72	.008	-0.28	.012	7	B0.5III	540,1737,2021
		NGC 2682 - 518		8 48 18	+11 56.6	1	14.96		0.76		0.33		4		135
		L 187 - 77		8 48 18	-58 19.	1	14.60		0.60		-0.08		1		3062
		NGC 2682 - 55		8 48 19	+11 56.3	2	11.33	.010	0.29	.007	0.13	.002	6	Am	135,1717

Table 1

487

HD	DM	Other Id	N	Rem	α_{1950}	δ_{1950}	S	V	σ_V	B–V	σ_{B-V}	U–B	σ_{U-B}	n	Spectrum	References
		NGC 2682 - 54			8 48 19	+11 57.1	2	12.65	.005	0.63	.005	0.12	.019	2		135,303
		NGC 2682 - 1023			8 48 19	+11 57.5	1	16.19		0.80		0.30		2		135
		NGC 2682 - 61			8 48 21	+11 57.2	1	13.46		0.58		0.03		2		135
		NGC 2682 - 62			8 48 21	+12 00.8	1	13.79		0.58		0.01		1		135
		NGC 2682 - 64			8 48 22	+11 55.1	1	14.05		0.70		0.17		1		135
		NGC 2682 - 1016			8 48 22	+11 58.7	1	16.38		1.00		0.70		2		135
		NGC 2682 - 63			8 48 22	+12 02.4	1	13.38		0.55		0.04		2		135
75596	−0 02075				8 48 22	−00 28.3	3	8.63	.020	0.48	.014	−0.09	.009	8	F8	116,118,2012
		NGC 2682 - 1063			8 48 23	+11 52.3	1	15.70		0.82		0.63		1		135
		NGC 2682 - 1064			8 48 23	+11 52.8	1	16.43		1.35		0.00		1		135
		NGC 2682 - 1021			8 48 23	+11 56.9	1	15.87		0.78		0.29		2		135
		NGC 2682 - 65			8 48 23	+12 04.3	1	12.58		0.63		−0.02		1		1717
75558	+16 01833				8 48 23	+16 11.2	1	7.37		0.92		0.55		2	G5	1648
		NGC 2682 - 72			8 48 24	+11 58.5	2	12.41	.010	0.96	.022	0.77	.042	5		135,1717
		NGC 2682 - 70			8 48 24	+11 59.4	1	11.55		0.41		−0.04		2		135
		NGC 2682 - 659			8 48 24	+12 00.8	2	14.83	.000	0.69	.000	0.16	.000	2		134,135
		NGC 2682 - 71			8 48 24	+12 01.4	1	13.62		0.57		0.08		2		135
		NGC 2682 - 73			8 48 24	+12 02.2	1	13.84		0.56		0.06		2		135
75760	−44 04873		A		8 48 24	−45 08.2	1	8.93		0.68		0.49		1	A2/3 IV	401
75760	−44 04873		B		8 48 24	−45 08.2	1	9.88		0.06		0.00		2		401
		NGC 2682 - 79			8 48 25	+11 53.0	2	12.79	.010	0.76	.025	0.31	.035	2		135,8043
		NGC 2682 - 75			8 48 25	+11 57.5	1	13.41		0.58		0.02		3		135
		NGC 2682 - 77			8 48 25	+11 59.6	1	13.36		0.58		0.02		1		135
75691	−27 05986	HR 3518			8 48 25	−27 31.4	7	4.01	.008	1.28	.007	1.38	.009	231	K3 III	3,9,15,26,1075,2012,8015
	−32 05777				8 48 25	−32 43.4	2	9.69	.018	0.06	.000	−0.64	.011	6	B6 Ia	540,976
		NGC 2682 - 573			8 48 26	+11 57.6	1	14.55		0.70		0.22		2		135
	−42 04694	LSS 1174			8 48 26	−42 58.5	2	10.62	.005	0.67	.005	−0.40	.015	5	O9 V	362,1737
		NGC 2682 - 574			8 48 27	+11 57.2	1	15.08		0.92		0.54		2		135
		NGC 2682 - 299			8 48 27	+12 00.1	2	16.05	.000	1.07	.000	0.83	.000	4		134,135
		NGC 2682 - 81			8 48 28	+11 56.6	5	10.03	.010	−0.07	.004	−0.39	.005	28	B8 V	134,135,303,1717,8023
		NGC 2682 - 83			8 48 28	+11 57.6	2	13.24	.000	0.60	.002	0.07	.002	3		134,135
		NGC 2682 - 575			8 48 28	+11 58.5	1	15.20		0.79		0.27		2		135
		NGC 2682 - 663			8 48 28	+11 59.7	2	14.91	.029	0.71	.017	0.21	.002	3		134,135
		NGC 2682 - 82			8 48 28	+11 59.9	1	13.90		0.59		0.06		2		135
		NGC 2682 - 80			8 48 28	+12 00.1	1	13.86		0.59		0.06		2		135
		NGC 2682 - 88			8 48 29	+12 01.9	1	12.21		0.72		0.30		2		1717
		NGC 2682 - 87			8 48 29	+12 02.6	1	13.66		0.55		0.03		2		135
		NGC 2682 - 86			8 48 29	+12 02.9	1	13.70		1.02		0.68		2	G8 IV	135
		NGC 2682 - 84			8 48 29	+12 04.0	6	10.57	.025	1.11	.018	0.99	.048	8	G8 III	134,135,303,1553,1717,8023
75620	+1 02183				8 48 30	+00 39.4	2	8.01	.007	0.08	.007	0.07	.007	18	A0	116,118
		NGC 2682 - 488			8 48 30	+11 52.5	1	14.46		0.67		0.10		1		135
		NGC 2682 - 90			8 48 30	+11 56.3	1	10.89		0.41		0.15		3		1717
		NGC 2682 - 89			8 48 30	+12 02.0	1	13.41		0.57		0.05		2		135
		NGC 2682 - 95			8 48 31	+11 58.8	3	12.68	.009	0.51	.011	0.01	.000	6		134,135,8023
		NGC 2682 - 93			8 48 31	+12 00.6	1	14.18		0.61		0.09		1		135
		NGC 2682 - 94			8 48 31	+12 01.5	3	12.83	.008	0.56	.010	0.07	.005	5		134,135,8023
75758	−41 04559				8 48 31	−41 42.0	2	8.96	−0.02	.005	−0.70	.005	5	B3/5 V	362,401
	−45 04606	LSS 1177			8 48 31	−45 20.1	4	8.95	.026	0.37	.012	−0.53	.031	13	B1 II-Ib	540,976,1075,2021
		NGC 2682 - 97			8 48 32	+11 59.5	1	14.34		0.54		−0.05		4		135
		NGC 2682 - 98			8 48 32	+12 04.3	1	12.85		0.58		0.06		2		135
		NGC 2682 - 99			8 48 32	+12 05.5	1	13.60		0.61		0.16		1		135
75757	−41 04558				8 48 32	−41 32.0	1	8.37		0.06		−0.60		2	B1/2 (IV)	401
75759	−41 04560	HR 3525, LSS 1176			8 48 32	−41 54.1	7	5.99	.005	−0.10	.008	−0.95	.013	44	B1/2 III	125,158,362,396,401*
		NGC 2682 - 489			8 48 33	+11 53.9	1	14.54		0.73		0.16		1		135
		NGC 2682 - 108			8 48 33	+11 56.7	5	9.72	.005	1.38	.014	1.52	.039	13	K4 III	134,135,1717,8023,8092
		NGC 2682 - 102			8 48 33	+11 56.8	1	12.36		0.78		0.27		2		1717
		NGC 2682 - 107			8 48 33	+11 57.3	1	13.92		0.59		0.07		1		135
		NGC 2682 - 106			8 48 33	+11 58.3	1	13.09		0.56		0.03		2		135
		NGC 2682 - 105			8 48 33	+11 59.6	8	10.31	.008	1.26	.004	1.32	.011	32	K2 III	134,135,303,1553,1717,8023*
		NGC 2682 - 103			8 48 33	+12 01.4	1	13.20		0.55		0.06		2		135
		NGC 2682 - 101			8 48 33	+12 01.9	1	13.20		0.56		0.10		1		135
		NGC 2682 - 104			8 48 33	+12 02.1	2	11.17	.029	1.06	.005	0.94	.039	3	K1 III	135,1717
		NGC 2682 - 110			8 48 34	+11 50.9	1	13.61		0.52		0.07		1		135
		NGC 2682 - 112			8 48 34	+11 54.2	1	13.28		0.58		0.09		1		135
		NGC 2682 - 114			8 48 34	+11 54.7	1	13.44		0.57		0.06		1		135
		NGC 2682 - 111			8 48 34	+11 57.2	3	12.74	.000	0.56	.000	0.04	.002	4		134,135,8023
		NGC 2682 - 115			8 48 34	+12 00.6	3	12.65	.004	0.63	.000	0.11	.010	4		134,135,8023
		NGC 2682 - 109			8 48 34	+12 01.4	1	13.54		0.58		0.04		2		135
75790	−44 04875	IDS08468S4421	AB		8 48 34	−44 32.0	1	7.50		0.98		0.74		1	K0 III	389
		NGC 2682 - 122			8 48 35	+11 51.5	1	13.70		0.55		0.09		1		135
		NGC 2682 - 117			8 48 35	+11 58.3	4	12.61	.018	0.77	.005	0.27	.005	7		134,135,303,8023
		NGC 2682 - 120			8 48 35	+11 59.2	1	13.68		0.58		0.00		3		135
		NGC 2682 - 119			8 48 35	+12 01.4	3	12.57	.008	0.59	.004	0.07	.000	5		134,135,8023
		NGC 2682 - 757			8 48 35	+12 02.6	2	13.05	.024	0.86	.007	0.49	.019	3		135,8043
		NGC 2682 - 128			8 48 36	+11 57.2	1	13.16		0.58		0.05		2		135
		NGC 2682 - 129			8 48 36	+11 57.4	1	13.19		0.58		0.06		1		135
		NGC 2682 - 127			8 48 36	+11 58.0	3	12.77	.004	0.56	.004	0.06	.000	10		134,135,8023
		NGC 2682 - 124			8 48 36	+11 58.3	5	12.14	.009	0.45	.013	0.05	.036	10		134,135,303,1717,8023
		NGC 2682 - 297			8 48 36	+12 00.9	2	15.76	.015	0.85	.005	0.51	.000	3		134,135
		NGC 2682 - 126			8 48 36	+12 01.7	1	13.98		0.58		0.06		2		135

HD	DM	Other Id	N	Rem	α_{1950}	δ_{1950}	S	V	σ_V	B–V	σ_{B-V}	U–B	σ_{U-B}	n	Spectrum	References
		NGC 2682 - 123			8 48 36	+12 03.5	1	13.95		0.61		0.05		1		135
		NGC 2682 - 125			8 48 36	+12 04.1	2	13.86	.000	0.58	.000	0.04	.000	2		134,135
	+12 01924	NGC 2682 - S 1135			8 48 36	+12 29.5	2	9.36	.005	1.46	.015	1.74	.025	5	K2	196,1553
75506	+44 01794	HR 3508			8 48 36	+43 54.9	4	5.17	.030	0.97	.006	0.69	.010	7	K0 III	1080,1355,1363,3016
75824	−50 03602				8 48 36	−50 24.2	1	8.01		0.55		0.11		1	K1 III	389
		NGC 2682 - 132			8 48 37	+11 56.3	2	13.12	.005	0.61	.005	0.08	.005	6		134,135
		NGC 2682 - 300			8 48 37	+11 56.6	2	15.83	.000	1.08	.000	0.76	.000	2		134,135
		NGC 2682 - 134			8 48 37	+11 57.2	4	12.27	.000	0.57	.023	0.08	.051	8		134,135,1717,8023
		NGC 2682 - 135			8 48 37	+11 57.4	6	11.44	.007	1.06	.008	0.91	.008	12	K2 III	134,135,303,1717,8023,8092
		NGC 2682 - 130			8 48 37	+11 57.6	2	12.94	.080	0.49	.033	0.00	.002	8		134,135
		NGC 2682 - 682			8 48 37	+12 00.3	2	15.23	.000	0.84	.000	0.30	.000	2		134,135
		NGC 2682 - 131		★ V	8 48 37	+12 04.7	7	11.21	.022	0.42	.004	0.06	.006	8	F0 IV	39,134,135,303,1553*
		NGC 2682 - 140			8 48 38	+11 58.0	1	13.20		0.56		0.06		2		135
		NGC 2682 - 1004			8 48 38	+11 59.6	1	15.71		0.93		0.58		2		135
		NGC 2682 - 296			8 48 38	+12 02.8	2	16.19	.000	0.66	.000	0.17	.000	2		134,135
		NGC 2682 - 764			8 48 38	+12 02.8	1	13.34		0.57		0.06		2		135
		NGC 2682 - 137			8 48 38	+12 03.0	2	14.05	.000	0.64	.000	0.16	.000	2		134,135
		NGC 2682 - 136			8 48 38	+12 03.9	5	11.31	.012	0.62	.012	0.16	.021	8	G3 III-IV	134,135,303,1717,8023
		G 9 - 26			8 48 38	+12 49.0	1	13.76		1.06		0.88		1		1658
75524	+39 02161				8 48 38	+39 05.7	2	8.87	.010	0.59	.000	0.10	.010	5	G5 V	1501,1601
75488	+48 01693	IDS08452N4757		A	8 48 38	+47 45.3	1	8.18		0.51		-0.02		2	G2 V	1601
		CCS 1311			8 48 38	−35 52.9	1	11.77		2.23				1	N	864
		NGC 2682 - 141			8 48 39	+11 59.3	6	10.47	.008	1.11	.010	0.95	.020	13	K2 III	134,135,303,1717,8023,8092
		NGC 2682 - 683			8 48 39	+12 00.1	1	12.72		0.58		0.01		3		135
		NGC 2682 - 684			8 48 39	+12 00.5	1	13.30		0.47		-0.01		2		135
		NGC 2682 - 142			8 48 39	+12 05.4	2	14.17	.000	0.62	.000	0.11	.000	2		134,135
		NGC 2682 - 598			8 48 40	+11 58.8	2	14.91	.000	0.78	.000	0.26	.000	2		134,135
		NGC 2682 - 145			8 48 40	+11 59.7	3	12.82	.004	0.58	.010	0.05	.002	4		134,135,8023
		NGC 2682 - 143			8 48 40	+12 01.1	6	11.51	.017	0.87	.007	0.50	.006	11	K0 III	134,135,303,1717,8023,8092
		NGC 2682 - 149			8 48 41	+11 58.9	3	12.56	.000	0.60	.002	0.08	.002	4		134,135,8023
		NGC 2682 - 147			8 48 41	+12 00.3	2	13.28	.005	0.59	.005	0.06	.012	3		134,135
		NGC 2682 - 148			8 48 41	+12 03.9	1	13.32		0.57		0.09		2		135
		NGC 2682 - 298			8 48 41	+12 04.6	2	15.64	.000	0.89	.000	0.53	.000	2		134,135
		NGC 2682 - 153			8 48 42	+11 55.1	5	11.29	.018	0.13	.007	0.10	.010	11	Am	134,135,303,1717,8023
		NGC 2682 - 486			8 48 42	+12 00.	1	14.55		0.57		-0.06		1		135
		NGC 2682 - 150			8 48 42	+12 00.4	1	13.26		0.58		0.05		2		135
		NGC 2682 - 152			8 48 42	+12 00.6	1	13.51		0.57		0.06		2		135
		NGC 2682 - 151			8 48 42	+12 05.2	6	10.50	.006	1.10	.012	0.95	.037	13	K0 III	134,135,303,1717,8023,8092
		NGC 2682 - 602			8 48 43	+11 58.8	2	14.76	.000	0.48	.000	-0.02	.000	2		134,135
		NGC 2682 - 155			8 48 43	+12 00.0	3	10.51	.000	0.59	.004	0.06	.025	4	G2 V	135,303,1553
		NGC 2682 - 156			8 48 43	+12 03.2	5	10.97	.034	0.11	.002	0.09	.007	8	A2 V	134,135,303,1717,8023
		NGC 2682 - 157			8 48 43	+12 04.7	3	12.78	.000	0.58	.004	0.08	.002	4		134,135,8023
75789	−42 04701				8 48 43	−43 07.5	1	7.53		1.55		1.86		1	K4 III	389
75823	−48 04146	IDS08471S4812		AB	8 48 43	−48 23.5	2	8.77	.005	0.51	.005	0.03	.015	3	F6/8 V	389,401
		NGC 2682 - 161		★ V	8 48 44	+12 00.8	1	12.89		0.46		0.00		1		135
		NGC 2682 - 162			8 48 44	+12 03.6	3	12.84	.004	0.58	.004	0.07	.010	4		134,135,8023
		NGC 2682 - 159			8 48 44	+12 07.0	1	13.33		0.63		0.03		1		135
75820	−45 04613				8 48 44	−46 03.6	1	8.63		-0.02		-0.18		2	B9.5V	401
		NGC 2682 - 166			8 48 45	+11 56.8	3	12.93	.019	0.92	.009	0.60	.011	4		135,303,8043
		NGC 2682 - 604			8 48 45	+11 59.3	1	13.83		0.63		0.12		2		135
		NGC 2682 - 164			8 48 45	+12 01.8	5	10.56	.024	1.12	.012	0.97	.083	9	K1 III	134,135,1717,8023,8092
		NGC 2682 - 163			8 48 45	+12 03.3	2	12.67	.000	0.69	.005	0.19	.005	2		135,8043
		NGC 2682 - 165			8 48 45	+12 05.5	1	12.86		0.60		0.11		2		135
	−50 03605				8 48 45	−50 55.6	1	10.57		0.15		-0.56		2	B2 V	540
		NGC 2682 - 171			8 48 46	+11 55.1	1	13.15		0.59		0.06		2		135
		NGC 2682 - 170			8 48 46	+11 58.6	6	9.69	.012	1.35	.008	1.48	.008	12	K3 III	134,135,1553,1717,8023,8092
		NGC 2682 - 174			8 48 46	+12 00.5	1	12.68		0.60		0.09		1		135
		NGC 2682 - 168			8 48 46	+12 02.8	1	13.31		0.49		0.03		1		135
75822	−47 04421				8 48 46	−47 36.3	1	9.09		-0.07		-0.58		2	B8 II	401
		NGC 2682 - 176			8 48 47	+11 57.1	1	12.64		0.57		0.05		2		135
		NGC 2682 - 175			8 48 47	+12 04.6	2	13.71	.015	0.60	.002	0.08	.005	3		134,135
		NGC 2682 - 185			8 48 48	+12 00.2	1	11.07		0.23		0.16		1		1717
		NGC 2682 - 180			8 48 48	+12 01.4	1	12.64		0.59		0.11		1		135
		NGC 2682 - 184		★ V	8 48 48	+12 02.0	3	12.26	.008	0.26	.026	0.07	.036	4		135,303,1717
		NGC 2682 - 177			8 48 48	+12 04.5	1	13.08		0.58		0.08		2		135
		NGC 2682 - 179			8 48 48	+12 06.4	1	13.67		0.65		0.10		1		135
75523	+45 01649	HR 3509			8 48 48	+45 30.1	1	5.99		1.26		1.33		2	K0 III	252
		NGC 2682 - 189			8 48 50	+12 01.0	1	12.87		0.52		0.06		1		135
		NGC 2682 - 188			8 48 50	+12 02.2	1	13.70		0.55		0.06		1		135
		NGC 2682 - 190			8 48 50	+12 02.5	4	10.98	.021	0.24	.015	0.15	.004	5	A8 IV-V	134,135,1717,8023
		NGC 2682 - 187			8 48 50	+12 03.0	1	13.23		0.62		0.11		1		135
		G 9 - 27			8 48 50	+19 04.8	3	12.06	.021	0.58	.014	-0.11	.011	4		1620,1658,1696
		NGC 2682 - 192			8 48 51	+12 09.2	3	12.80	.008	0.75	.015	0.31	.017	3		135,303,8043
		NGC 2682 - 195			8 48 52	+11 57.9	2	12.70	.005	0.70	.002	0.21	.012	4		135,8023
		NGC 2682 - 194			8 48 52	+12 00.2	1	13.60		0.59		0.06		1		135
		NGC 2682 - 193			8 48 52	+12 04.9	4	12.26	.007	1.02	.010	0.77	.072	6	K0 IV	134,135,1717,8023
75556	+42 01935	HR 3511			8 48 52	+42 11.5	2	5.99	.000	1.25	.001	1.29		5	K1 III	71,1501
	−45 04615	LSS 1180		V	8 48 52	−45 21.6	4	9.19	.210	0.36	.031	-0.70	.079	9	B2 Vne	401,540,1737,2021
75821	−46 04661	HR 3527, KX Vel		★ AB	8 48 52	−46 20.5	7	5.09	.010	-0.21	.007	-0.98	.006	31	B1 Ib	15,1075,1637,1732,2012*
		NGC 2682 - 202			8 48 53	+12 01.4	1	12.60		0.80		0.10		1		135

Table 1

HD	DM	Other Id	N Rem	α_{1950}	δ_{1950}	S	V	σ_V	B–V	σ_{B-V}	U–B	σ_{U-B}	n	Spectrum	References
		NGC 2682 - 199		8 48 53	+12 06.3	1	13.22		0.60		0.06		1		135
		NGC 2682 - 200		8 48 53	+12 10.3	1	10.98		0.80		0.36		2		135
		NGC 2682 - 207		8 48 54	+11 48.4	1	12.32		0.36		0.01		1		1717
		NGC 2682 - 205	★ V	8 48 54	+12 02.3	1	13.31		0.53		0.01		1		805
		NGC 2682 - 209		8 48 55	+11 59.2	1	13.74		0.65		0.14		2		135
		NGC 2682 - 210		8 48 55	+12 01.4	3	12.26	.019	0.58	.015	0.07	.019	6		135,303,1717
		NGC 2682 - 785		8 48 55	+12 03.0	3	12.15	.020	1.00	.033	0.79	.029	4		135,303,805
		NGC 2682 - 3035		8 48 55	+12 03.1	1	12.12		1.00		0.78		2		1717
75874	−51 03267			8 48 55	−51 42.6	1	7.60		1.28		1.34		1	K2 III	389
		NGC 2682 - 213		8 48 56	+11 57.9	1	13.78		0.59		0.09		2		135
		NGC 2682 - 211		8 48 56	+12 04.0	1	13.80		0.61		0.06		1		135
		G 9 - 28		8 48 56	+18 18.8	1	11.54		1.48				6		940
		NGC 2682 - 214		8 48 57	+12 00.4	1	13.83		0.59		0.07		1		135
		NGC 2682 - 215		8 48 57	+12 05.8	1	12.79		0.60		0.09		1		135
		NGC 2682 - 216		8 48 58	+11 54.9	1	12.73		0.57		0.08		2		135
		NGC 2682 - 219		8 48 58	+12 01.2	1	13.54		0.59		0.11		1		135
		NGC 2682 - 218		8 48 58	+12 01.4	2	11.66	.025	1.06	.008	0.87	.015	4		135,1717
		NGC 2682 - 217		8 48 58	+12 02.7	3	11.26	.039	1.08	.035	0.90	.010	5		135,805,1717
		NGC 2682 - 221		8 48 59	+11 57.9	1	12.37		0.57		-0.04		2		135
		NGC 2682 - 224		8 49 00	+11 55.7	7	10.77	.010	1.12	.013	1.08	.010	13	K1 III	134,135,303,1553,1717*
		NGC 2682 - 225		8 49 00	+11 56.6	2	13.09	.010	0.55	.005	0.03	.010	3		134,135
		NGC 2682 - 226		8 49 00	+11 57.7	4	12.77	.007	0.75	.005	0.26	.015	5		134,135,303,8023
		NGC 2682 - 223		8 49 00	+12 08.0	5	10.56	.026	1.11	.011	0.97	.010	8		134,135,1717,8023,8092
	+26 01854	IDS08460N2606	BC	8 49 00	+25 54.9	2	8.45	.000	0.61	.010	0.17	.015	4	K2 III	1625,3016
		NGC 2682 - 228		8 49 01	+11 53.2	1	13.24		0.57		0.07		1		135
		NGC 2682 - 229		8 49 01	+11 56.3	1	13.40		0.54		0.06		1		135
		NGC 2682 - 227		8 49 01	+11 58.1	5	12.97	.011	0.91	.012	0.59	.018	6		134,135,303,8023,8043
		NGC 2682 - 231		8 49 01	+11 59.1	5	11.50	.010	1.05	.007	0.89	.010	11	K0 III-IV	134,135,1717,8023,8092
		NGC 2682 - 233		8 49 02	+11 55.5	1	13.42		0.55		0.06		1		135
		NGC 2682 - 235		8 49 02	+11 57.8	1	13.41		0.58		0.04		2		135
	+71 00480			8 49 02	+70 45.6	1	9.55		1.01				2	K0	70
		NGC 2682 - 547		8 49 03	+11 56.3	1	12.91		0.48		-0.05		1		135
	+71 00479			8 49 03	+70 51.4	1	10.18		1.07				2	K2	70
		NGC 2682 - 236		8 49 04	+12 02.5	3	12.54	.004	0.66	.012	0.17	.017	5		135,303,8043
	−0 02078			8 49 04	−00 09.7	2	11.13	.039	0.48	.015	-0.04	.010	3		116,118
		NGC 2682 - 240		8 49 05	+11 44.6	1	11.98		0.99		1.06		2		1717
		NGC 2682 - 2057		8 49 05	+11 56.4	1	12.94		0.99		0.68		1		135
		NGC 2682 - 238		8 49 05	+12 00.6	2	10.77	.140	0.22	.033	0.14	.023	4		135,1717
		NGC 2682 - 241		8 49 05	+12 01.0	3	12.68	.000	0.59	.004	0.09	.005	4		134,135,8023
75700	+12 01927	NGC 2682 - 242		8 49 05	+12 05.0	1	7.85		1.12		0.99		1	K0	135
		POSS 210 # 13		8 49 05	+41 01.4	1	15.77		0.50				2		1739
		NGC 2682 - 244		8 49 06	+11 57.4	7	10.76	.018	0.94	.004	0.66	.008	14	G8 III-IV	134,135,303,1553,1717*
		NGC 2682 - 243		8 49 06	+12 00.8	3	12.61	.004	0.61	.004	0.08	.007	4		134,135,8023
75860	−43 04691	LSS 1184	★ AB	8 49 06	−43 33.8	6	7.59	.013	0.73	.007	-0.28	.034	18	B1/2 Ia	362,540,976,1737,2012,8100
	−43 04690	LSS 1183		8 49 06	−43 39.1	3	9.54	.004	0.86	.013	-0.27	.030	12	O8:	92,362,390
75873	−45 04625			8 49 06	−46 07.3	1	8.90		0.48		0.40		2	A2/3 II/III	401
		vdB 22 # d		8 49 06	−48 54.5	1	12.83		0.49		-0.30		3		434
75737	−6 02743	HR 3523	★ AB	8 49 07	−06 59.3	3	5.53	.004	0.15	.000	0.13	.005	11	A4 m	15,1075,2012
	−41 04572	LSS 1182		8 49 07	−41 56.3	1	11.34		0.82		-0.16		4		362
		NGC 2682 - 255		8 49 09	+11 59.7	3	12.72	.000	0.55	.009	0.04	.000	3		134,135,8023
75487	+59 01198	HR 3506		8 49 10	+59 14.7	2	6.35	.000	0.39	.000	0.04		4	F4 IV	1733,6009
	+71 00481			8 49 10	+70 39.4	1	8.65		1.38				2	K2	70
75850	−41 04574			8 49 10	−41 45.5	1	8.67		-0.07		-0.30		2	B8/9 V	401
		NGC 2682 - 263		8 49 12	+11 59.5	1	14.00		0.61		0.11		2		135
		NGC 2682 - 259		8 49 12	+12 00.0	1	13.92		0.69		0.19		2		135
		NGC 2682 - 262		8 49 12	+12 01.6	1	12.94		1.01		0.73		1		135
		LSS 1188		8 49 13	−49 09.5	1	12.09		0.55		0.03		2	B3 Ib	1737
		NGC 2682 - 265		8 49 14	+11 58.2	1	12.50		0.60		0.04		2		135
75751	−0 02079	IDS08467S0039	AB	8 49 14	−00 50.3	3	9.45	.004	0.16	.009	0.12	.030	8	A2	116,118,1775
		NGC 2682 - 267		8 49 15	+12 04.3	1	12.83		0.60		0.09		2		135
		NGC 2682 - 266		8 49 15	+12 06.4	5	10.54	.023	1.10	.016	0.97	.077	8	K1 III	134,135,1717,8023,8092
		POSS 210 # 25		8 49 16	+38 36.8	1	16.53		1.56				2		1739
75486	+62 01027	HR 3505		8 49 17	+62 09.1	2	5.73	.005	0.28	.005	0.11	.005	4	F2 III	985,1733
		NGC 2682 - 268		8 49 18	+12 00.5	1	12.38		0.59		0.08		1		135
		NGC 2682 - S 1533		8 49 18	+12 37.2	2	10.01	.010	1.20	.005	1.23	.015	2		1553,1717
		G 9 - 29		8 49 19	+20 20.1	1	10.49		0.93		0.66		1	K3	333,1620
75869	−38 04945			8 49 19	−38 26.7	1	7.01		-0.20				4	B2 III	2012
75887	−47 04432			8 49 19	−47 40.7	1	9.39		0.12		0.03		2	A0 V	401
75871	−41 04576			8 49 21	−41 28.4	4	7.80	.002	0.01	.000	-0.70	.016	10	B3 V	362,401,540,976
75553	+55 01297	IDS08456N5520	AB	8 49 24	+55 08.5	1	7.42		0.61		0.08		4	G2 V +G2 V	3016
75870	−39 04866			8 49 26	−39 33.4	1	7.71		-0.16		-0.72		3	B3 V	362
		NGC 2682 - 2152		8 49 27	+11 43.2	2	10.91	.005	1.12	.010	1.15	.000	4		1553,1717
	+12 01930	NGC 2682 - 280		8 49 27	+11 55.4	5	10.69	.021	0.11	.006	0.10	.013	6	A3 V	134,135,303,1717,8023
		BSD 100 # 889		8 49 28	+00 43.1	1	10.79		0.21		0.10		1	B9	116
75716	+28 01659	HR 3521, BO Cnc	★ A	8 49 28	+28 26.9	2	6.23	.000	1.61	.012	1.52		4	M3 III	71,3001
75750	+19 02114			8 49 29	+19 32.4	2	8.49	.011	-0.13	.004	-0.55		5	B5	1345,2033
75616	+52 01343			8 49 29	+52 34.7	1	6.94		0.47		-0.03		2	F5	985
75927	−43 04700			8 49 29	−44 13.3	1	9.29		0.02		-0.05		2	A0 V	401
75698	+33 01770	HR 3519	★ AB	8 49 30	+32 39.8	2	5.67	.010	0.21	.010	0.08	.005	6	A8 V	1501,8071
		vdB 22 # e		8 49 30	−48 54.	1	14.15		0.72		0.03		4		434

HD	DM	Other Id	N Rem	α_{1950}	δ_{1950}	S	V	σ_V	B–V	σ_{B-V}	U–B	σ_{U-B}	n	Spectrum	References
		G 252 - 23		8 49 31	+72 43.1	1	13.93		1.60		1.28		1		906
	−43 04701			8 49 31	−44 08.5	1	11.02		0.95		0.59		3	K2	322
		NGC 2682 - S 1557		8 49 33	+11 31.0	2	10.09	.010	1.29	.030	1.45	.005	5		1553,1717
	−45 04635	LSS 1189		8 49 33	−45 29.3	3	8.98	.010	0.39	.011	-0.49	.017	10	B1 II	540,976,2021
75767	+8 02134			8 49 35	+08 15.3	1	6.57		0.63		0.06		2	G0 V	3077
		NGC 2682 - 286		8 49 35	+11 55.8	2	10.43	.019	1.09	.015	1.00	.015	3		1553,1717
	−14 02678			8 49 35	−14 27.8	1	9.81		0.99		0.60		1	K5	565
75732	+28 01660	HR 3522	⋆ A	8 49 37	+28 31.4	7	5.95	.020	0.86	.010	0.65	.010	31	G8 V	22,1067,1080,1197,1355*
		HA 100 # 233		8 49 37	−00 30.3	1	12.16		1.06		0.95		18	K0	281
75925	−41 04584			8 49 37	−41 40.2	1	8.95		0.02		-0.65		3	B3/5 Vnne	362
75784	+13 02007			8 49 39	+13 25.4	1	7.85		0.99		0.90		2	G5	1648
75926	−42 04723	HR 3530		8 49 39	−42 19.0	3	6.55	.005	0.04	.005	0.07		9	A1 Vn	15,401,2012
		WLS 900-20 # 6		8 49 40	−20 13.9	1	11.51		0.73		0.14		2		1375
	−43 04703			8 49 40	−44 10.3	1	9.57		1.11		0.93		3	K4	322
75732	+28 01660	IDS08467N2842	B	8 49 42	+28 30.5	3	13.15	.009	1.66	.004	1.20		15	M5	694,1663,3024
75955	−45 04641			8 49 42	−45 26.1	1	7.73		0.01		-0.02		2	B9 V	401
		vdB 22 # c		8 49 42	−48 55.9	1	11.58		0.30		-0.25		3	B5 V	434
75811	+5 02074	HR 3526	⋆ AB	8 49 46	+05 31.8	2	6.32	.005	0.12	.000	0.13	.005	7	A5 V	15,1078
		HA 100 # 236		8 49 46	−00 25.6	1	11.32		0.62		0.10		2	G4	281
		BONN82 T11# 13		8 49 46	−44 09.8	1	13.26		1.07		0.09		3	B0	322
	+12 01928	NGC 2682 - S 1553		8 49 47	+11 27.0	2	8.80	.034	1.63	.010	1.98	.044	6	K0	1553,1717
		NGC 2682 - S 1639		8 49 48	+12 17.8	1	12.61		0.59		0.10		1		1717
	−43 04706			8 49 48	−44 10.9	1	10.40		1.14		0.92		3	K4	322
		vdB 22 # a		8 49 49	−48 34.6	1	12.14		0.38		-0.24		3		434
	−0 02080			8 49 50	−00 37.8	2	9.77	.005	1.14	.000	1.04		10	K0	281,6004
		G 9 - 31		8 49 51	+22 45.0	2	10.83	.000	0.60	.000	-0.04	.000	3		333,927,1620
75968	−46 04683			8 49 51	−46 25.3	1	8.17		-0.11		-0.27		2	B9 III/IV	401
		BONN82 T11# 11		8 49 52	−44 03.4	1	12.52		0.54		0.27		3	A9	322
75991	−47 04441	LSS 1190		8 49 53	−47 23.0	3	8.98	.004	0.12	.013	-0.72	.025	10	B2 II	92,540,976
76270	−72 00747	HR 3544		8 49 53	−72 21.8	4	6.10	.015	0.21	.004	0.18	.013	20	A3mA6-A7	15,1075,1637,2038
76143	−66 00927	HR 3537	⋆ A	8 49 55	−66 36.4	4	5.34	.007	0.42	.005	0.05	.009	13	F5 IV	15,1075,2038,3026
76006	−49 03899	LSS 1191		8 49 56	−49 22.1	1	7.33		0.66		0.41		1	F5/6 III	389
	+0 02410			8 49 57	−00 16.5	2	9.64	.004	1.01	.006	0.86		8	K0 III	281,6004
75881	−12 02714			8 49 58	−12 37.1	1	6.85		0.55		0.05		5	F8 IV	1628
76005	−48 04169			8 49 58	−48 33.9	1	9.08		0.21		-0.60		4	B3 II	1075
75918	−18 02506			8 49 59	−18 25.6	1	7.62		0.46		-0.01		2	F6/7 V	1375
		BPM 5109		8 50 00	−61 43.	1	14.73		-0.02		-0.89		1	DA	3065
	−0 02081			8 50 01	−00 28.5	6	10.14	.005	0.16	.006	0.10	.004	74	A0	116,118,281,989,1728,1764
	−0 02082			8 50 01	−00 47.3	2	10.70	.024	0.55	.015	0.00	.015	3		116,118
75989	−40 04685			8 50 04	−40 47.9	1	6.49		-0.11		-0.35		2	Ap Si	401
76004	−43 04711			8 50 04	−43 57.7	4	6.37	.016	-0.15	.010	-0.62	.013	14	B3 V	92,164,322,362
		BONN82 T11# 12		8 50 04	−44 06.8	1	13.16		0.80		0.61		3	K1	322
76164	−64 00958			8 50 05	−64 52.1	1	8.55		-0.04		-0.32		2	Ap Si	401
75916	−12 02716	HR 3529	⋆ A	8 50 08	−13 02.6	2	6.12	.009	1.14	.002	1.14		6	K1 III	58,2007
		BONN82 T11# 21		8 50 13	−44 10.1	1	14.53		0.77		0.39		3	G9	322
		BONN82 T11# 10		8 50 15	−44 04.4	1	12.45		0.55		0.14		3	G0	322
		NGC 2682 - S 1808		8 50 16	+11 52.1	1	12.58		0.57		0.08		1		1717
		NGC 2682 - S 1863		8 50 17	+12 33.6	1	9.55		0.94		0.61		1		1553
76031	−43 04716	LSS 1192		8 50 18	−43 49.2	6	9.00	.014	0.48	.011	-0.44	.023	20	B1 Iab/b	92,342,362,540,976,8100
75880	+9 02076	G 41 - 4		8 50 19	+09 22.7	1	8.64		0.69		0.25		1	G0	333,1620
		POSS 210 # 15		8 50 20	+39 16.8	1	16.04		1.45				2		1739
76060	−45 04653			8 50 20	−46 06.0	1	7.87		-0.04		-0.43		2	B8 IV/V	401
76113	−57 01759	HR 3536		8 50 20	−57 26.7	3	5.59	.007	-0.11	.009	-0.42		9	B8 III	15,401,2012
	−48 04176	LSS 1195		8 50 21	−48 46.9	1	10.78		0.10				4	A2 II	2021
75864	+17 01966			8 50 23	+17 33.5	1	6.87		0.40		0.11		2	F2 IV	1375
		G 234 - 49		8 50 23	+63 57.0	1	11.06		1.42		1.25		2	K5	7010
75914	+0 02412			8 50 24	+00 01.9	9	8.65	.009	0.05	.007	0.10	.022	52	A0	116,118,281,989,1371,1509*
76001	−32 05814	HR 3533		8 50 24	−32 19.2	3	6.50	.004	1.45	.010			11	K2/3 III	15,1075,2031
76131	−55 01815			8 50 24	−55 37.2	1	6.69		-0.05		-0.41		2	B6 III	401
	+0 02413			8 50 25	−00 01.0	2	10.06	.003	0.49	.001	-0.01		8		281,6004
		BONN82 T11# 9		8 50 27	−43 55.4	1	11.46		1.84		1.83		3	gK7	322
		S260 # 18		8 50 27	−43 55.4	1	11.53		1.77				2		1730
		HA 100 # 153		8 50 29	−00 32.8	1	12.04		0.40		0.01		2	F7	281
76163	−57 01763	IDS08493S5759	ABC	8 50 31	−58 10.6	1	8.09		0.42				4	F0 III	2012
		G 41 - 5		8 50 33	+09 36.8	1	11.21		0.77		0.34		1		333,1620
		BONN82 T11# 19		8 50 33	−44 03.6	1	14.31		0.78		0.37		3	G9	322
		BONN82 T12# 8		8 50 33	−44 03.6	1	11.60		0.48		0.04		3	F7	322
		HA 100 # 157		8 50 34	−00 33.9	1	11.68		1.11		1.03		2	K3	281
76074	−43 04718	IDS08488S4347	AB	8 50 34	−43 58.4	1	9.06		0.10		-0.07		2	A0 V	401
		HA 100 # 39		8 50 37	−00 41.3	1	10.47		0.43		-0.05		2	F5	281
		vdB 22 # b		8 50 37	−48 35.4	1	11.79		0.70		-0.24		3		434
	−23 07843	LP 844 - 22		8 50 38	−23 32.8	1	10.81		0.90		0.57		2		1696
	−43 04719			8 50 38	−44 09.9	1	9.49		1.18		1.34		9		322
	−0 02083			8 50 39	−00 53.4	2	9.15	.019	1.41	.019	1.64	.044	3	K5	116,118
	+35 01890	G 51 - 29		8 50 40	+35 25.0	2	9.26	.019	1.14	.029	1.10	.024	3	M0 V:	333,1620,3072
		POSS 210 # 39		8 50 40	+39 17.6	1	17.00		1.50				2		1739
		NGC 2682 - S 1907		8 50 41	+11 42.9	1	10.90		1.43		1.89		1		1553
	−0 02084			8 50 41	−00 32.1	5	9.15	.004	1.28	.003	1.50	.002	106	K0	281,989,1729,1764,6004
76072	−36 05125	HR 3534	⋆ AB	8 50 41	−36 21.4	3	6.41	.004	0.55	.005			11	G8 III + A2	15,1075,2031
		HA 100 # 164		8 50 42	−00 38.2	1	12.12		0.86		0.50		2	G7	281

Table 1 491

HD	DM	Other Id	N	Rem	α_{1950}	δ_{1950}	S	V	σ_V	B–V	σ_{B-V}	U–B	σ_{U-B}	n	Spectrum	References
75614	+70 00538				8 50 43	+70 19.5	1	8.77		0.91				3	K0	70
	+47 01621				8 50 44	+47 03.0	1	8.86		0.46		-0.03		2	F8	1601
75632	+71 00482	G 252 - 24		⋆AB	8 50 44	+70 59.4	5	8.05	.023	1.39	.017	1.25	.021	15	K5 V	1003,1013,1197,1381,3078
		HA 100 # 267			8 50 44	-00 30.1	3	13.03	.001	0.48	.008	-0.04	.010	16		281,1499,1764
		HA 100 # 269			8 50 45	-00 29.8	3	12.35	.000	0.56	.004	-0.02	.007	38	G0	281,1499,1764
	-43 04721				8 50 46	-44 01.0	1	10.63		0.66		-0.36		3	B0	322
75896	+36 01883	HR 3528			8 50 47	+35 43.7	3	6.14	.005	0.05	.004	0.10	.007	8	A4 III	985,1022,1501
	-48 04184	II Vel		⋆AB	8 50 48	-49 02.7	2	9.53	.000	0.10	.003	-0.59	.004	6	B2 IV	540,976
75935	+27 01682				8 50 51	+27 06.2	1	8.52		0.76		0.33		2	G8 V	3026
75897	+35 01892				8 50 51	+35 03.6	1	8.18		0.42		-0.01		2	F5	1601
		CpD -43 03088			8 50 52	-44 00.9	1	10.92		0.67		-0.39		4		362
		AJ82,474 # 9			8 50 52	-48 03.9	1	12.13		0.80		-0.07		2	B0.5V	779
		G 47 - 10			8 50 53	+27 23.3	1	11.95		1.03		0.80		3		333,1620
		HA 100 # 272			8 50 53	-00 27.9	1	12.07		0.77		0.37		16	K0	281
76110	-38 04980	HR 3535			8 50 53	-38 32.1	2	5.81	.010	1.50	.000	1.89		6	M0 III	2031,3005
76187	-52 01734				8 50 55	-53 08.2	1	7.97		0.06		0.03		2	A1 V	401
75895	+39 02164				8 50 56	+39 26.7	1	7.58		0.37		-0.04		3	F2 V	833
76011	+3 02085	S Hya			8 50 57	+03 15.5	1	11.85		1.39		0.30		1	M4	975
		BONN82 T11# 15			8 50 57	-44 10.0	1	13.72		0.60		-0.05		3	B5	322
		HA 100 # 277			8 50 58	-00 24.1	1	13.86		0.54		-0.01		2		281
		POSS 210 # 23			8 50 59	+39 48.6	1	16.43		1.43				2		1739
	-47 04459	IDS08493S4712		AB	8 50 59	-47 22.9	1	9.66		0.22				4	B5 V	2021
76161	-47 04460	HR 3539			8 51 00	-48 10.2	5	5.90	.007	-0.15	.008	-0.59	.001	16	B3 Vn	15,26,2013,2029,8023
76233	-55 01820				8 51 01	-56 03.2	1	7.21		-0.08		-0.38		2	B6 V	401
		HA 100 # 280			8 51 02	-00 25.3	5	11.80	.004	0.50	.005	0.01	.007	96	F5	1728,281,989,1729,1764
76214	-54 01890	IDS08497S5458		AB	8 51 02	-55 09.1	1	6.78		0.01		-0.09		2	A0 V	401
		POSS 210 # 36			8 51 03	+41 44.7	1	16.95		1.16				1		1739
75974	+20 02232	IDS08482N2021		AB	8 51 04	+20 09.4	2	6.68	.001	0.70	.000			50	F8	130,1351
	-43 04725				8 51 05	-44 06.8	1	9.83		0.14		0.12		3	A5	322
76183	-43 04727				8 51 06	-44 05.8	3	8.26	.011	0.14	.008	0.23	.025	13	A1/2 (IV)	322,401,1730
76213	-50 03637				8 51 06	-51 14.0	1	8.24		0.02		-0.15		2	F3 IV/V	401
		HA 100 # 51			8 51 07	-00 41.3	1	11.13		0.16		0.04		2	A4	281
76160	-43 04726				8 51 07	-43 33.8	1	7.59		1.08		0.96		1	K0 III	389
76184	-44 04920				8 51 07	-45 14.8	1	9.36		-0.02		-0.20		2	B8 V	401
76186	-46 04707				8 51 07	-47 12.3	1	6.81		-0.03		-0.11		2	B9 IV/V	401
		LP 426 - 28			8 51 08	+18 13.8	1	13.20		0.58		-0.14		2		1696
		BSD 100 # 572			8 51 10	+00 18.7	1	11.12		0.18		0.02		1	A4	116
76230	-51 03303	HR 3542		⋆AB	8 51 10	-51 56.4	3	6.39	.005	0.00	.005	-0.14		9	A0 V	15,401,2012
		HA 100 # 285			8 51 11	-00 22.9	1	13.56		1.06		1.20		6		281
75959	+31 01907	HR 3532		⋆AB	8 51 12	+30 46.2	1	5.39		1.05				3	G7 III	71
76051	-0 02086	IDS08487S0012	A		8 51 13	-00 23.3	1	9.30		0.46		-0.04		1	A5	281
76051	-0 02086	IDS08487S0012	AB		8 51 13	-00 23.3	1	8.73		0.45				4	A5	6004
76051	-0 02086	IDS08487S0012	B		8 51 13	-00 23.3	1	9.60		0.46		-0.02		1		281
75933	+40 02119				8 51 15	+40 19.6	1	7.62		0.59		0.11		3	G5	1601
	+0 02414				8 51 15	-00 21.5	2	9.12	.010	0.76	.005	0.43	.015	13	G8 III	281,397
76200	-46 04712				8 51 15	-47 01.6	1	9.19		0.63		0.17		2	A1 V	401
		MtW 100 # 41			8 51 16	-00 22.1	1	13.97		0.50		0.00		5		397
		OJ 287 # 5			8 51 18	+20 13.7	1	11.44		0.55		0.03		2		331
233582	+53 01290				8 51 18	+53 08.8	1	7.75		0.98		0.77		3	K0	1566
76211	-44 04923				8 51 19	-45 11.0	1	9.47		0.08		0.04		2	A0 V	401
		MtW 100 # 50			8 51 20	-00 22.1	1	14.87		0.71		0.29		5		397
		HA 100 # 394			8 51 21	-00 20.9	3	11.37	.011	1.31	.007	1.46	.024	16		281,397,1764
75949	+44 01798	IDS08481N4358		AB	8 51 24	+43 46.6	1	7.70		0.62		0.11		3	G2 V	1723
		MtW 100 # 57			8 51 24	-00 23.3	1	15.95		1.05		0.73		5		397
76025	+22 02014				8 51 25	+22 24.3	1	7.83		0.61		0.17		2	G5	1648
76082	-0 02087				8 51 26	-00 25.3	9	8.41	.008	1.12	.008	1.09	.011	58	K0 III	281,397,989,1371,1509,1657*
		HA 100 # 61			8 51 26	-00 44.0	1	12.72		0.65		0.11		2	G5	281
76181	-34 05361	V Pyx			8 51 26	-34 37.7	1	8.30		1.55		1.70		17	K1 Ib/II	3049
		OJ 287 # 9			8 51 28	+20 17.7	1	14.21		0.84		-0.50		2		331
		MtW 100 # 67			8 51 28	-00 26.1	1	11.74		0.43		0.00		4		397
	+73 00437				8 51 29	+73 04.7	1	9.60		1.26		1.17		1	K0	1776
76283	-51 03308				8 51 29	-51 17.7	1	8.74		-0.06		-0.50		2	B9 III	401
		ApJ144,259 # 31			8 51 30	+09 06.	1	15.51		-0.31		-1.28		3		1360
	-43 04733				8 51 34	-44 09.2	1	10.25		0.76		-0.22		3		362
		LP 36 - 78		AB	8 51 35	+71 10.0	1	10.52		0.93		0.72		2		1723
76256	-43 04732				8 51 35	-43 33.8	1	8.72		-0.12		-0.48		2	Ap Si	401
		AC UMa			8 51 36	+65 09.7	1	10.30		0.23		0.22		1	A2	627
	+65 00671	IDS08472N6521	B		8 51 36	+65 09.7	1	9.97		0.50		0.03		4		627
		LP 606 - 23			8 51 36	-01 56.4	1	13.31		0.45		-0.22		2		1696
76268	-43 04735				8 51 36	-44 08.6	1	7.42		0.11		0.11		2	A1/2 V	401
		POSS 210 # 1			8 51 37	+39 14.8	1	13.71		0.62				2		1739
		OJ 287 # 1			8 51 38	+20 22.4	1	11.54		0.63		0.17		2		331
	-45 04676	LSS 1197		⋆AB	8 51 39	-45 50.7	3	8.92	.013	0.78	.010	-0.15	.025	11	B0.5III	540,976,2021
		LP 666 - 11			8 51 42	-07 52.9	1	17.41		1.75				1		1691
76346	-56 01918	HR 3549			8 51 44	-56 27.6	3	6.02	.004	-0.02	.005	-0.09		9	A0 V	15,401,2012
		AJ82,474 # 20			8 51 45	-47 31.7	1	12.27		0.54		-0.45		2	O9.5V	779
	-43 04739	IDS08500S4400		AB	8 51 47	-44 11.3	1	9.32		0.17		0.04		2		401
76224	-26 06554				8 51 48	-26 19.5	1	8.21		0.90				4	G5 IV	2012
76306	-44 04931	IDS08501S4455		AB	8 51 48	-45 06.2	1	9.06		0.00		-0.48		2	B5 V	401
76282	-43 04740				8 51 49	-43 20.5	1	7.57		0.04		0.05		2	A0 V	401

HD	DM	Other Id	N	Rem	α_{1950}	δ_{1950}	S	V	σ_V	B–V	σ_{B-V}	U–B	σ_{U-B}	n	Spectrum	References
		HA 100 # 199			8 51 50	−00 36.5	1	12.20		0.77		0.38		4	G5	281
76151	−4 02490	HR 3538			8 51 50	−05 14.6	9	6.00	.010	0.66	.012	0.22	.015	33	G3 V	15,621,1067,1417,1509,1705*
76325	−48 04208				8 51 56	−48 52.8	1	8.81		0.30		0.08		2	A3mA7-F0	401
76095	+26 01865	IDS08490N2636	AB		8 51 58	+26 23.7	1	6.72		0.67		0.13		2	G2 V +K0 V	3077
75544	+78 00297				8 51 58	+78 20.3	1	7.37		0.31		0.12		2	F0	1502
76304	−39 04924	HR 3548			8 51 58	−40 15.4	3	6.46	.007	0.96	.000			11	K0II/III+A5	15,2013,2029
311510	−74 00546				8 51 58	−74 45.7	1	9.50		0.12		0.04		2	A0	1730
76264	−31 06673				8 51 59	−31 50.3	1	9.61		0.11		0.14		4	A2 III/IV	158
		OJ 287 # 4			8 52 02	+20 16.2	2	14.17	.044	0.87	.004	0.49	.018	11		331,1595
		AJ82,474 # 22			8 52 02	−47 22.0	1	12.70		0.75		−0.19		3		779
76147	+15 01934				8 52 03	+14 57.5	1	8.55		0.09		0.07		1	A2	1776
		OJ 287 # 10			8 52 03	+20 17.7	1	14.56		0.45		−0.01		7		331,1595
		POSS 210 # 44			8 52 03	+44 29.9	1	17.06		1.12				1		1739
		OJ 287 # 11			8 52 04	+20 17.2	2	14.95	.009	0.56	.078	−0.06	.037	9		331,1595
76343	−49 03930				8 52 04	−49 23.3	1	8.49		0.10		−0.15		2	B8 II	401
		HA 100 # 207			8 52 05	−00 38.9	1	12.39		0.52		0.07		4	G7	281
76208	−0 02088				8 52 10	−00 31.2	5	8.78	.009	0.41	.010	−0.06	.013	21	F0	116,118,281,1371,6004
76358	−44 04933				8 52 10	−45 11.2	1	7.45		0.86		0.52		1	G5 III	389
76360	−47 04480	HR 3551		⋆AB	8 52 10	−47 19.8	3	5.31	.011	0.27	.007			11	A9 IV/V	15,2012,2031
76341	−42 04762	LSS 1198			8 52 11	−42 17.7	6	7.16	.013	0.30	.008	−0.64	.018	13	B1/2 Ib	164,362,401,540,976,1737
		POSS 210 # 7			8 52 12	+41 30.7	1	14.45		1.48				3		1739
		HA 100 # 212			8 52 15	−00 35.6	1	11.45		0.88		0.68		4	G5	281
76359	−46 04732				8 52 16	−46 59.6	2	8.70	.005	0.00	.005	−0.07	.000	5	A0 V	401,402
		S036 # 4			8 52 18	−74 47.4	1	10.92		1.36				2		1730
76357	−42 04766				8 52 19	−43 13.8	1	7.96		1.44		1.73		1	K3 III	389
76242	−0 02089				8 52 20	−00 46.6	5	8.92	.009	0.82	.008	0.39	.004	36	G9 V	281,989,1728,1729,6004
		POSS 210 # 2			8 52 22	+44 18.3	1	13.76		0.95				1		1739
75958	+65 00673	HR 3531			8 52 22	+64 47.8	2	5.60	.009	0.88	.009	0.57		6	G6 III	627,1118
	+0 02422				8 52 23	−00 01.3	2	9.77	.039	1.10	.019	0.92	.024	3	K0	116,118
		OJ 287 # 2			8 52 24	+20 21.2	2	12.82	.035	0.64	.009	0.05	.022	11		331,1595
	−45 04685				8 52 24	−45 58.0	1	10.17		0.02		−0.07		2	A0	401
76380	−43 04751				8 52 25	−43 18.5	1	6.97		1.64		1.98		1	K5 III	389
	−48 04221				8 52 26	−48 40.7	1	9.49		1.15		1.01		2	K0	401
76441	−51 03325				8 52 26	−52 16.9	1	7.57		−0.09		−0.55		2	B4 III	401
76614	−69 00976				8 52 27	−69 38.4	1	7.45		−0.01		−0.01		2	A0 III/IVp	401
		OJ 287 # 7			8 52 28	+20 12.4	1	13.36		0.79		0.37		2		331
76273	−3 02506				8 52 28	−03 21.9	1	7.02		0.20		0.19		2	A2	252
	+0 02423				8 52 29	+00 19.9	2	9.79	.015	0.96	.005	0.61	.015	3	F8	116,118
		OJ 287 # 8			8 52 29	+20 19.1	1	14.65		0.75		0.38		2		331
76440	−51 03323				8 52 29	−51 23.5	2	8.42	.014	0.66	.005	0.18		5	G5 V	389,2012
76442	−52 01751				8 52 30	−52 25.7	1	8.22		−0.10		−0.55		2	B3/5 V	401
		G 115 - 33			8 52 31	+46 42.3	1	12.42		1.31		1.10		1		1773
76425	−48 04222				8 52 31	−48 56.2	1	8.69		0.06		0.08		2	A1 V	401
	+2 02098	G 46 - 9			8 52 32	+01 45.6	3	9.97	.026	1.41	.024	1.25	.035	7	M1	1705,3078,7008
76424	−46 04735				8 52 32	−46 18.3	1	10.17		0.33		−0.50		3	Ap Si	402
76221	+17 01973	HR 3541, X Cnc			8 52 34	+17 25.4	5	6.65	.078	3.36	.083			15	C6 II	15,3001,6002,8015,8022
	−47 04490	LSS 1199			8 52 34	−47 18.0	2	9.63	.019	0.48	.010	−0.47		2	B1 III	779,2021
76458	−51 03326				8 52 35	−51 30.5	1	7.70		−0.03		−0.15		2	A0 V	401
76459	−52 01753				8 52 35	−52 42.7	1	9.53		0.03		−0.33		2	B8 IV	401
76219	+28 01666	HR 3540			8 52 40	+28 07.2	4	5.25	.029	1.00	.005	0.75	.007	8	G8 II-III	1080,1363,3016,8100
76439	−44 04939				8 52 40	−45 15.8	1	7.94		−0.14		−0.54		2	Ap Si	401
76538	−59 01174	HR 3560			8 52 40	−60 09.8	3	5.78	.005	−0.08	.005	−0.56		9	B4 V	15,401,2012
	+0 02425				8 52 41	−00 05.6	2	10.12	.025	1.43	.025	1.64	.030	2		116,118
	−48 04227	LSS 1200			8 52 42	−48 41.3	1	8.84		0.39		−0.42		1		389
		OJ 287 # 3			8 52 43	+20 19.9	1	13.62		0.50		−0.01		2		331
	+0 02426				8 52 44	+00 02.1	2	11.46	.010	0.42	.000	−0.08	.005	2		116,118
		OJ 287 # 6			8 52 44	+20 16.6	1	13.35		0.53		−0.01		2		331
76294	+6 02060	HR 3547			8 52 45	+06 08.2	8	3.10	.007	1.00	.006	0.79	.015	19	G8 III-IV	15,37,1078,1080,1363*
76378	−23 07884				8 52 47	−24 12.3	2	8.65	.005	1.00	.000			5	K2/3 III	1705,2012
	−0 02090				8 52 51	−00 41.4	2	9.75	.008	1.18	.004	1.19		10	G6	281,6004
		POSS 210 # 37			8 52 52	+41 16.2	1	16.96		1.79				4		1739
		KUV 08529+4004			8 52 53	+40 03.6	1	16.68		0.36		−0.43		1		1708
76376	−17 02691	HR 3554		⋆A	8 52 54	−18 03.0	2	5.74	.005	1.33	.005			7	K2/3 III	15,2029
		vdB 23 # a			8 52 54	−46 44.	1	10.89		0.32		0.28		3	A0 V	434
		POSS 210 # 48			8 53 00	+41 44.0	1	17.44		1.00				2		1739
76370	−7 02661	HR 3553		⋆AB	8 53 03	−07 46.7	3	6.08	.004	0.22	.000	0.12	.010	14	A2 m	15,1256,2029
		POSS 210 # 16			8 53 08	+43 46.5	1	16.05		1.45				1		1739
76260	+39 02174				8 53 10	+39 00.3	1	7.35		1.34		1.53		2	K5 III	105
	+71 00483				8 53 11	+71 36.2	1	9.16		0.47		−0.01		2	F5	1776
76351	+12 01941	HR 3550			8 53 12	+11 49.1	1	5.41		1.46				2	gK5	71
76318	+26 01869				8 53 13	+26 33.2	1	8.28		0.36		0.02		2	F2 V	3026
76400	−8 02525	T Hya			8 53 14	−08 57.0	2	7.93	.410	1.45	.110	0.89	.130	2	M4	635,975
76516	−43 04762				8 53 15	−43 38.4	1	7.98		0.03		0.03		2	A0 V	401
76292	+40 02125	HR 3546			8 53 16	+40 23.7	3	5.89	.008	0.36	.008	0.06	.008	6	F2 III	1733,3053,6009
	−33 05568	BPM 51820			8 53 16	−33 49.8	1	10.63		1.00		0.85		2	K3 V	3072
76517	−47 04502				8 53 16	−47 21.6	1	9.14		0.06		0.13		3	A1/2 V	402
76700	−66 00938				8 53 17	−66 36.7	2	8.09	.059	0.75	.005	0.41		5	G6 V	2033,3008
	−74 00548				8 53 17	−74 48.2	1	10.21		1.04		0.86		2		1730
76536	−47 04504	LSS 1202			8 53 18	−47 24.0	3	8.81	.019	0.33	.024	−0.09	.061	7	WC	16,402,779,1075
76145	+60 01159				8 53 19	+60 24.5	1	8.82		0.23				4	A3	1118

Table 1 493

HD	DM	Other Id	N Rem	α_{1950}	δ_{1950}	S	V	σ_V	B–V	σ_{B-V}	U–B	σ_{U-B}	n	Spectrum	References
76535	−46 04750	LSS 1203	⋆ AB	8 53 19	−47 13.4	5	8.62	.018	0.39	.012	-0.60	.020	17	O9/B0	92,125,396,540,779,976
76534	−42 04780	IDS08515S4305	AB	8 53 21	−43 16.5	4	8.02	.027	0.13	.012	-0.58	.008	12	B2 Vn	164,362,401,434,1598
76332	+29 01845			8 53 23	+28 51.7	1	8.64		0.62		0.12		2	G2 V	3026
76483	−27 06072	HR 3556	⋆ A	8 53 23	−27 29.3	6	4.88	.010	0.12	.011	0.17	.040	23	A3 IV	3,15,1075,2012,3023,8015
76023	+71 00484			8 53 26	+70 50.9	1	7.07		1.18				3	K2	70
76567	−45 04694	IDS08517S4559	AB	8 53 26	−46 10.1	1	7.56		0.02		0.01		2	A0 V	401
76556	−47 04505	LSS 1204		8 53 26	−47 25.0	9	8.19	.018	0.40	.013	-0.59	.022	26	O6	125,390,396,402,540,779*
76291	+46 01459	HR 3545		8 53 27	+45 49.5	3	5.73	.009	1.10	.009	1.06	.005	6	K1 IV	15,1003,1080
		AJ82,474 # 31		8 53 30	−47 43.7	1	12.69		0.74		-0.25		2		779
76216	+58 01159			8 53 33	+58 24.7	1	6.92		0.15		0.12		2	A2	1733
		AJ82,474 # 32		8 53 33	−47 44.4	1	13.69		0.87		-0.04		1		779
76479	−21 02639			8 53 34	−21 23.6	1	9.50		0.55		0.05		2	F6 V	1375
76554	−40 04736			8 53 34	−40 53.2	3	8.31	.008	0.12	.013	-0.69	.030	8	B2 Vn	362,540,976
76566	−44 04951	HR 3562, IY Vel	⋆ A	8 53 34	−44 51.0	7	6.25	.009	-0.17	.007	-0.64	.012	28	B3 IV	15,540,976,1020,1075*
76431	+2 02100			8 53 35	+01 52.2	1	9.22		-0.28		-1.03		4		1732
76565	−43 04764			8 53 36	−43 36.5	1	8.22		0.00		0.02		2	A0 IV	401
	−41 04637	LSS 1205		8 53 37	−41 23.8	3	9.84	.016	0.72	.025	-0.36	.017	6	O6 Ib(f)(n)	362,540,1737
	−47 04510			8 53 38	−47 19.8	1	9.92		0.57		0.08		3	F8	402
76640	−57 01790	HR 3568		8 53 38	−58 02.9	4	6.38	.012	-0.11	.007	-0.50		16	B5 V	15,401,2012,2012
	−46 04757	LSS 1206		8 53 39	−46 35.3	1	10.74		0.49		-0.37		3		1747
76588	−45 04698			8 53 40	−45 28.7	1	8.71		0.32		0.10		2	A1mF0-F0	401
76512	−23 07902	HR 3559		8 53 44	−23 37.6	3	6.38	.004	0.16	.000	0.11	.005	11	A4 V	15,1075,2012
76653	−54 01925	HR 3570		8 53 47	−54 46.3	5	5.71	.011	0.48	.004	-0.02	.008	18	F6 V	15,1075,1637,2012,3077
	+20 02243	T Cnc	⋆ A	8 53 49	+20 02.5	4	8.68	.366	4.89	.303	8.01	.014	4	C5,5	864,1238,8005,8022
76510	−13 02718		V	8 53 50	−13 42.7	1	8.11		-0.18				4	B1 Ib	2033
76398	+33 01785	HR 3555		8 53 52	+33 06.2	1	5.45		0.12		0.12		2	A7 IV	3052
76728	−60 01243	HR 3571	⋆ A	8 53 55	−60 27.2	7	3.84	.005	-0.10	.005	-0.44	.006	24	B8/9 II	15,1034,1075,1637*
76397	+35 01905			8 53 57	+35 13.0	1	7.80	0.9	0.70				2	K0	1601
76494	+4 02081	HR 3557		8 54 00	+04 25.8	4	6.14	.005	1.00	.000	0.78	.029	11	G8 II-III	15,1078,1355,8100
	−46 04767	LSS 1208		8 54 10	−46 45.8	1	11.70		0.63		-0.42		3		1747
76525	+0 02430			8 54 11	+00 35.0	2	8.10	.019	1.39	.010	1.45	.000	3	K2	116,118
76475	+22 02029			8 54 12	+22 03.2	1	7.04		0.91		0.61		2	G5	985
76545	+0 02431			8 54 14	−00 13.3	3	7.85	.015	1.35	.024	1.39	.019	3	K2	116,118
		AJ82,474 # 38		8 54 14	−47 43.6	1	13.10		0.96		0.18		1		779
76579	−16 02639	HR 3564		8 54 15	−16 31.0	3	5.95	.007	1.54	.000			11	K3 III	15,2013,2030
		GD 98		8 54 16	+40 27.9	2	14.88	.005	-0.07	.055	-0.89	.045	2	DA	974,3060
76508	+17 01979	HR 3558		8 54 20	+17 20.2	3	6.17	.001	1.00	.001			36	K1 III	15,130,1351
		MCC 260		8 54 21	+11 50.5	1	10.55		1.92		1.57		1	M1	679
	+12 01944	G 41 - 9		8 54 21	+11 50.6	3	10.60	.024	1.54	.011	1.21		3	M5	1017,1620,1705
	−62 01130			8 54 22	−63 00.6	2	10.59	.005	0.48	.005	-0.14	.005	6		158,1696
76668	−36 05192	IDS08524S3644	AB	8 54 23	−36 55.8	1	6.84		0.62				4	G1 V	1075
76396	+51 01462			8 54 24	+51 37.5	2	8.89	.000	1.20	.000			2	R5	1238,3025
77049	−74 00550			8 54 25	−74 41.6	1	7.44		1.32		1.40		2	K2 III	1730
76543	+15 01945	HR 3561		8 54 28	+15 30.9	1	5.20		0.15				3	A5 III	1363
	−42 04793	LSS 1209		8 54 28	−42 54.8	2	11.19	.030	0.72	.035	-0.24	.025	5		362,432
	+72 00436			8 54 29	+71 47.4	1	10.24		0.57		-0.02		1		1620
76725	−44 04963			8 54 30	−45 12.3	1	8.60		0.07		-0.45		2	B6 III	401
		AAS19,45 T4# 210		8 54 30	−47 18.3	1	12.34		1.51		1.20		5		402
	−47 04524			8 54 30	−47 19.0	1	11.99		0.47		-0.26		4		402
	−47 04523			8 54 30	−47 20.4	1	11.45		0.31		-0.04		5		402
		AAS27,343 # 89		8 54 36	−42 53.	1	11.69		0.94		-0.16		4		362
76601	−0 02094			8 54 38	−01 09.5	2	8.81	.014	0.56	.014	0.08	.019	4	G5	116,118
		vdB 25 # a		8 54 39	−42 54.4	1	11.88		0.92		-0.12		7	O9: I:pe	434
76600	+1 02210			8 54 40	+00 43.7	2	7.66	.015	0.94	.010	0.63	.010	3	G5	116,118
		KUV 08547+3830		8 54 43	+38 30.5	1	15.72		-0.23		-1.08		1	sdB	1708
		G 252 - 25		8 54 44	+73 09.3	1	14.34		1.71		1.25		1		906
		AJ82,474 # 41		8 54 46	−48 13.0	1	11.78		1.23		0.25		2		779
76777	−47 04531			8 54 47	−47 44.4	1	9.09		0.05		0.04		2	A0 V	401
76582	+16 01864	HR 3565		8 54 48	+15 46.5	1	5.67		0.20				2	F0 IV	1363
76824	−55 01853	IDS08534S5508	A	8 54 48	−55 20.1	1	7.69		1.70		2.01		8	K0	1673
76805	−52 01788	HR 3574	⋆ AB	8 54 49	−52 31.8	4	4.68	.005	-0.12	.007	-0.48	.009	19	B5 V	15,1075,1637,2012
		LP 36 - 98		8 54 50	+73 09.4	1	14.30		1.70				2		1663
76764	−42 04802			8 54 50	−43 00.5	2	9.26	.010	0.11	.010	-0.39	.000	5	B5 V	401,434
76776	−43 04782			8 54 51	−43 43.9	1	8.93		0.35		0.10		2	Fm δ Del	401
	−31 06724	BPM 51867		8 54 52	−32 10.7	1	10.30		1.07		0.93		1	K5 V	3072
		AJ82,474 # 42		8 54 54	−47 24.3	1	13.64		1.07		0.00		2		779
76115	+75 00355	IDS08491N7549	A	8 54 54	+75 37.4	1	8.69		1.08				1	R0	1238
76647	−0 02095			8 54 54	−00 55.4	2	8.26	.023	0.97	.019	0.74	.023	4	K2	116,118
76418	+60 01161			8 54 55	+60 20.9	1	8.88		0.52				4	F5	1118
	+0 02432			8 54 55	−00 01.0	2	8.30	.015	1.59	.010	1.93	.049	3	M0	116,118
	+11 01949			8 54 56	+10 43.0	1	8.76		1.14		0.99		5	K0	897
		WLS 900 60 # 10		8 54 56	+61 01.7	1	12.45		0.56		0.17		2		1375
76572	+30 01795	HR 3563	⋆ AB	8 54 57	+30 25.6	3	6.30	.012	0.42	.000	-0.03	.012	7	F6 V	254,272,3037
76646	−0 02096			8 54 58	−00 27.8	2	8.32	.014	0.37	.009	-0.06	.009	4	F0	116,118
76788	−43 04785			8 54 59	−43 27.8	1	7.51		1.35		1.60		2	K3 III	389
	−45 04719	LSS 1211		8 55 1	−45 42.2	2	10.02	.015	1.00	.020	-0.04	.025	4	B0.5 Ib	1737,1747
76629	+9 02093	HR 3567		8 55 00	+09 34.9	3	6.18	.004	0.98	.005	0.76	.005	11	G8 III	15,1078,1355
76803	−47 04532	IDS08533S4729	AB	8 55 01	−47 40.1	1	8.81		0.14		0.08		2	A0 IV	401
		G 9 - 36		8 55 03	+24 40.2	2	11.95	.000	0.57	.000	-0.14	.000	4		516,1620
76627	+19 02131			8 55 05	+19 28.7	1	8.67		0.34		0.01		2	F0	1648

HD	DM	Other Id	N Rem	α₁₉₅₀	δ₁₉₅₀	S	V	σ_V	B–V	σ_B-V	U–B	σ_U-B	n	Spectrum	References
		vdB 25 # k		8 55 12	−43 05.4	1	14.10		0.58		0.10		2		434
76347	+70 00540			8 55 14	+70 25.9	1	10.18		0.67				2	A2	70
	−46 04786	LSS 1212		8 55 14	−46 51.1	4	9.09	.019	0.81	.012	-0.24	.019	9	B1 Ib	540,2021,1737,1747
	−42 04806			8 55 15	−43 04.2	1	9.03		0.01		-0.63		2		434
		vdB 25 # j		8 55 17	−43 05.7	1	11.85		0.23		0.12		4		434
		G 46 - 12		8 55 18	+11 42.1	1	11.35		1.13		1.09		1		333,1620
		vdB 25 # f		8 55 18	−43 02.2	1	11.39		0.17		-0.04		2	B8 V	434
76852	−48 04265			8 55 18	−48 52.8	2	9.58	.002	0.32	.005	-0.49	.009	4	B3 II	540,976
	+0 02433			8 55 19	+00 25.4	2	8.83	.029	1.24	.024	1.27	.010	3	K2	116,118
76595	+36 01889	HR 3566	★	8 55 19	+35 59.8	3	6.65	.010	0.01	.010	0.01	.005	8	A2 V	1022,1501,1733
		vdB 25 # e		8 55 19	−43 03.1	1	10.16		0.14		-0.09		2	B8 Vp?	434
76838	−42 04808	IDS08535S4252	AB	8 55 19	−43 03.8	6	7.31	.004	0.00	.006	-0.71	.008	20	B2 IV	342,362,401,434,540*
		vdB 25 # l		8 55 21	−43 04.6	1	13.77		0.56		-0.04		1		434
76860	−49 03994			8 55 21	−49 29.3	1	7.13		1.71		1.82		1	K3 Ib	389
76875	−49 03995			8 55 21	−50 09.8	1	7.72		1.11		0.86		1	K2 III +A	389
		G 9 - 37		8 55 22	+19 55.3	1	14.81		1.66				4		538
		vdB 25 # h		8 55 24	−43 05.5	1	11.52		0.26		0.17		2	A1 V	434
		LP 786 - 23	V	8 55 25	−19 06.2	1	12.41		1.18		1.20		2		1696
		vdB 25 # g		8 55 25	−43 01.7	1	11.20		0.45		0.05		2		434
	+0 02434			8 55 26	−00 07.3	1	11.29		0.96		0.48		1		116
		G 9 - 38	★ A	8 55 27	+19 57.4	1	14.06		1.84				4		538
		G 9 - 38	★ AB	8 55 27	+19 57.4	1	13.72		1.87				1		906
		G 9 - 38	B	8 55 27	+19 57.4	1	14.92		1.93				4		538
76706	+18 02090			8 55 29	+18 30.2	2	6.55	.020	0.02	.010	0.00	.005	5	A1 V	985,1501
		vdB 25 # i		8 55 29	−43 04.0	1	11.30		0.39		0.03		2		434
76757	+2 02112	HR 3573	★ AB	8 55 33	+01 44.1	2	6.58	.005	0.06	.000	0.08	.005	7	A2 V	15,1256
76734	+11 01954	RT Cnc		8 55 33	+11 02.4	2	7.47	.078	1.47	.039	1.09	.031	6	M4 III	897,3076
76917	−50 03704	IDS08545S5041	AB	8 55 35	−50 52.6	1	7.85		0.45		0.03		1	F5 V	389
		DG Hya		8 55 36	−05 14.7	1	11.61		0.15		0.16		1		597
76617	+44 01809			8 55 37	+44 37.2	1	9.23		0.37		0.02		std	G2 V	1783
		AJ77,733 T11# 10		8 55 37	−47 28.0	2	13.25	.030	0.74	.005	0.53	.015	9		125,402
		G 252 - 26		8 55 40	+70 52.2	1	10.65		0.93		0.60		3		7010
76915	−47 04543			8 55 40	−48 00.7	1	8.96		0.07		-0.19		2	B9/A0 V	401
76898	−43 04794			8 55 41	−44 04.3	3	7.39	.000	-0.16	.010	-0.61	.000	10	B3/5 V	362,401,2012
76955	−52 01806			8 55 42	−52 30.5	1	8.11		-0.10		-0.57		2	B3 III	401
	−0 02098			8 55 43	−00 47.6	2	10.77	.015	0.52	.005	0.05	.015	3	F8	116,118
	+21 01949	G 9 - 39		8 55 44	+20 44.6	1	9.26		1.11		1.01		1	M0	3072
		G 253 - 18		8 55 44	+78 42.0	1	12.33		0.84		0.32		3		7010
76756	+12 01948	HR 3572	★ A	8 55 45	+12 03.2	7	4.25	.013	0.14	.017	0.18	.027	21	A5 m	3,15,1118,1363,3023*
77002	−58 01301	HR 3582, V376 Car	★ A	8 55 45	−59 02.1	6	4.91	.013	-0.18	.010	-0.74	.013	16	B2 IV-V	12,26,1732,2032,3019,8023
77002	−58 01301	IDS08545S5851	B	8 55 45	−59 02.1	2	6.84	.025	-0.12	.005	-0.48	.010	5	B9.5 V	12,321
76644	+48 01707	HR 3569	★ A	8 55 48	+48 14.4	11	3.14	.014	0.20	.013	0.07	.005	57	A7 IV	1,15,22,1006,1077*
		AJ82,474 # 53		8 55 48	−47 45.7	1	12.81		0.95		0.01		1		779
		AJ82,474 # 54		8 55 51	−47 39.2	1	11.65		0.97		-0.01		2	B0.5Vnn	779
	−42 04819	LSS 1213		8 55 52	−42 30.2	2	9.73	.005	0.41	.005	-0.42		7	B1 V	362,2021
76969	−51 03382			8 55 52	−51 28.8	1	9.36		0.80				4	K0 V	2012
76752	+25 02018			8 55 54	+25 36.1	1	7.50		0.65		0.12		1	G2 V	3026
76968	−50 03710	LSS 1214		8 55 54	−50 33.3	7	7.08	.013	0.13	.007	-0.81	.019	18	O9/9.5Ia/b	390,540,976,1034,1737*
		vdB 25 # o		8 55 56	−42 59.9	1	11.17		0.08		-0.12		1		434
		vdB 25 # p		8 55 56	−43 00.2	1	10.83		0.15		-0.10		1		434
		G 115 - 39		8 55 58	+36 38.4	1	11.85		1.45		1.20		3		7010
76766	+26 01873			8 55 59	+26 07.2	1	7.94		0.51		-0.02		3	F8 V	3026
77032	−58 01306			8 56 00	−58 23.5	2	8.69	.001	-0.05	.008	-0.68	.010	5	B5 Vne	540,976
		AJ77,733 T11# 11		8 56 01	−47 34.4	1	13.89		0.63		0.42		4		125
76954	−42 04822			8 56 03	−42 19.9	1	7.89		-0.12		-0.54		2	B8/9 III	401
76593	+61 01093			8 56 04	+61 17.8	1	8.92		0.17				4	A0	1118
		vdB 25 # m		8 56 04	−42 57.0	1	10.24		0.43		-0.02		1		434
76967	−42 04824			8 56 05	−42 57.6	3	9.15	.007	0.00	.009	-0.56	.018	6	B3/5 V	362,401,434
	−42 04827	vdB 27 # a		8 56 10	−42 38.1	1	10.11		0.09		-0.43		2	B3 V	434
		G 47 - 18		8 56 11	+33 08.7	4	15.22	.056	-0.01	.023	-0.91	.026	6		419,974,1620,3060
	−47 04550	LSS 1215		8 56 11	−47 34.0	5	10.27	.016	0.86	.010	-0.23	.016	14	O6 V	125,396,779,1737,1747
		vdB 27 # b		8 56 12	−42 29.9	1	13.00		1.07		0.37		3		434
	−47 04551	LSS 1216		8 56 13	−47 32.6	10	8.40	.021	0.89	.017	-0.18	.022	29	O6 f	125,389,390,396,401*
		G 41 - 14		8 56 14	+08 40.3	2	10.90	.005	1.67	.005	1.21		2	M3	1620,1746
76998	−44 04991			8 56 14	−44 40.5	1	9.19		-0.02		-0.12		2	B9 V	401
		AJ82,474 # 56		8 56 15	−47 32.3	1	11.98		0.90		-0.08		2	O9.5V	779
77020	−48 04282	HR 3583		8 56 16	−48 22.7	2	5.88	.005	1.06	.005	0.88		5	G8/K0 II	389,2006
	+0 02435			8 56 20	+00 01.6	2	10.45	.029	0.94	.019	0.63	.029	3		116,118
76982	−34 05444			8 56 20	−35 14.6	1	8.74		1.72				4	K4/5 (III)	2012
		AJ82,474 # 58		8 56 20	−47 27.6	1	13.31		1.30		0.30		2		779
76830	+18 02093	HR 3577		8 56 21	+18 19.8	3	6.38	.000	1.54	.005	1.65	.020	10	M4 III	15,1003,3001
76868	+4 02088			8 56 22	+03 51.1	2	7.99	.009	0.22	.014	-0.52		7	B5	1212,2033
		vdB 26 # a		8 56 22	−47 11.2	1	14.20		1.21		0.11		4		434
		AJ82,474 # 59		8 56 22	−47 29.6	1	12.90		1.24		0.27		2		779
	+0 02436			8 56 23	−00 02.8	2	10.68	.019	0.09	.014	0.05	.000	4		116,118
76932	−15 02656	HR 3578		8 56 23	−15 56.5	5	5.82	.023	0.52	.008	0.00	.075	11	F6 V	15,1003,2012,2017,3077
	−47 04555			8 56 27	−47 25.0	1	11.85		0.57		0.44		3		125,396
		AJ79,1294 T13# 10		8 56 27	−47 33.6	1	12.58		0.82		0.05		3		396
76813	+32 01821	HR 3575	★ A	8 56 29	+32 36.8	2	5.22	.017	0.92	.008	0.64		5	G8 III	71,1501
		KUV 08565+4129		8 56 29	+41 29.0	1	17.47		0.16		-0.89		1		1708

Table 1   495

HD	DM	Other Id	N Rem	α_{1950}	δ_{1950}	S	V	σ_V	B–V	σ_{B-V}	U–B	σ_{U-B}	n	Spectrum	References
		AJ79,1294 T13# 7		8 56 29	−47 20.7	1	11.55		1.68		0.64		3		125,396
	−47 04556			8 56 30	−47 20.7	1	11.94		0.55		0.33		3		125,396
76910	−0 02103			8 56 32	−00 25.7	2	8.48	.014	0.44	.014	-0.10	.019	4	F5	116,118
	−47 04557	LSS 1217		8 56 32	−47 26.3	2	11.23	.015	1.40	.019	0.34	.000	6		125,396,779
	−48 04287			8 56 32	−49 12.1	2	10.16	.029	0.49	.012	-0.44	.004	3	F0	540,976
		G 114 - 25		8 56 34	−06 11.7	1	11.93		0.59		-0.17		2		1658
	−5 02678			8 56 34	−06 11.7	2	10.66	.005	0.44	.018	-0.01		5	F7	2033,6006
77046	−45 04742			8 56 34	−46 00.2	1	8.58		0.81		0.42		1	G3 III	389
	+0 02437	G 163 - 78		8 56 36	+00 17.9	2	9.64	.094	1.70	.005	1.89	.080	4	M2	116,118
	+0 02438			8 56 36	−00 00.7	2	9.46	.023	0.45	.019	-0.04	.009	4	F8	116,118
		vdB 28 # a		8 56 36	−43 13.9	1	11.28		0.72		0.20		2	B5 Vp	434
	−3 02525	G 114 - 26		8 56 38	−03 49.2	7	9.67	.022	0.48	.007	-0.20	.008	14	F3	516,1064,1620,1658*
		AJ82,474 # 63		8 56 38	−47 28.3	1	13.19		1.14		0.16		2		779
	+72 00438			8 56 39	+71 56.3	1	9.83		1.75		2.21		2	K5	1776
		AJ82,474 # 64		8 56 40	−47 28.0	1	11.50		1.65		0.59		2		779
76749	+53 01300			8 56 41	+53 31.6	1	8.32		0.43		0.00		2	F5	1566
	−47 04560			8 56 41	−47 22.7	2	9.52	.010	0.49	.005	-0.05	.000	6	F5	125,396,402
76846	+34 01929			8 56 43	+33 58.1	1	9.35		1.39				1	R1	1238
	−43 04813			8 56 44	−43 34.5	1	9.82		0.16		0.13		2	A2	401
	−28 06788			8 56 45	−29 09.3	1	10.92		3.69				1	R	864
	−47 04564	LSS 1218		8 56 47	−48 05.5	4	9.80	.008	0.73	.011	-0.25	.029	11	B1.5II	540,779,1747,2021
76866	+25 02022	IDS08539N2501	AB	8 56 49	+24 49.1	1	8.98		0.41		0.04		2	F5 V	3026
77145	−56 01948			8 56 50	−57 59.5	1	8.52		0.03		-0.04		2	B9.5IV	401
	+30 01801	G 47 - 19		8 56 51	+30 26.2	1	9.71		0.91		0.62		1	K3:	333,1620
77119	−49 04017			8 56 54	−50 03.3	1	7.25		1.07		0.85		1	K0 III	389
		AJ82,474 # 66		8 56 55	−47 42.2	1	12.68		1.05		0.10		2		779
77120	−50 03727			8 56 56	−50 36.7	1	7.57		1.22		1.32		1	K2 III	389
		LP 901 - 1		8 56 58	−31 01.0	1	12.33		1.46		1.34		1	M4:	3073
77115	−46 04808			8 56 58	−46 33.9	1	9.36		-0.01		-0.10		2	A0 V	401
	−43 04816			8 56 59	−44 03.1	1	9.03		0.88		0.52		2		401
		L 532 - 21	A	8 57 00	−31 01.	1	13.80		1.64		1.20		2		3078
		L 532 - 21	B	8 57 00	−31 01.	1	16.28		0.80		0.37		2		3073
77185	−58 01319			8 57 04	−59 00.0	1	7.83		-0.06		-0.43		2	B8 III	401
77087	−28 06793	HR 3585		8 57 08	−28 36.7	2	6.24	.005	1.00	.000			7	G8 III	15,2027
77114	−43 04829			8 57 09	−43 25.7	1	9.23		0.06		0.06		2	A0 V	401
77140	−46 04810	HR 3588, FZ Vel	★ A	8 57 10	−47 02.4	7	5.17	.008	0.25	.007	0.18	.011	25	A6/7mA7-F0	15,16,258,355,1279*
77140	−46 04810	IDS08555S4651	B	8 57 10	−47 02.4	1	12.08		0.67		0.23		3		1279
77110	−34 05459			8 57 17	−35 06.9	1	8.85		0.62				4	G1 V	2012
76944	+38 01986	HR 3580		8 57 20	+37 48.0	1	6.20		1.59		1.76		2	K5	1733
		G 47 - 20		8 57 22	+25 07.7	1	11.15		1.12		1.19		1	K5	333,1620
77244	−59 01207			8 57 22	−59 32.6	1	7.64		-0.09		-0.52		2	B4/5 V	401
77084	−18 02536	HR 3584		8 57 23	−19 00.7	1	6.18		0.46				4	F5 IV/V	2031
76943	+42 01956	HR 3579	★ AB	8 57 24	+41 58.9	23	3.97	.006	0.43	.005	0.04	.006	498	F3 V +G5 V	1,15,22,61,71,130,292*
		G 194 - 50		8 57 26	+48 37.7	1	14.12		1.70				4		940,1705
77069	−1 02174			8 57 26	−02 21.1	1	7.59		1.38		1.58		3	K2	1657
76991	+25 02024			8 57 30	+24 41.6	1	8.27		1.61		1.86		3	M1	1625
77166	−42 04853			8 57 32	−42 24.4	1	9.29		-0.05		-0.17		2	B9 V	401
	−46 04812			8 57 32	−46 57.2	1	12.05		0.44				2	F0 V:	16
77233	−52 01829			8 57 32	−53 08.9	1	7.83		1.39		1.65		1	K3 III	389
77137	−27 06141	TY Pyx		8 57 34	−27 37.2	1	6.90		0.68		0.26		2	G5 V	588
77207	−48 04300	LSS 1219		8 57 35	−48 38.1	3	9.37	.015	0.57	.013	-0.42	.025	7	B7 Iab/b	540,962,976
		AJ82,474 # 69		8 57 39	−48 08.1	1	14.08		0.78		0.10		2		779
77126	−11 02520	IDS08553S1204	AB	8 57 42	−12 15.9	1	8.07		0.53		0.05		1	F7 V	3030
	−46 04813			8 57 42	−46 57.7	1	10.90		0.56		0.01		2	F5 V:	16
77224	−47 04571			8 57 42	−48 05.7	1	7.33		1.14		1.00		1	G8/K0 III	389
		AJ82,474 # 70		8 57 43	−47 05.2	1	11.82		1.53		0.34		2		779
		WLS 900-20 # 10		8 57 45	−19 43.7	1	11.77		1.52		2.01		2		1375
	−46 04816			8 57 49	−47 01.6	1	10.59		0.58		-0.02		2	F8 V:	16
	−41 04694			8 57 51	−42 14.7	3	9.85	.008	0.39	.008	-0.30	.000	16	B2.5V	362,1088,2021
		DH Hya		8 57 52	−09 35.3	1	11.35		0.17		0.10		1	A0	668
	−46 04817			8 57 53	−46 52.4	1	9.48		0.50		0.16		2	A0	401
77065	+22 02037	G 9 - 42		8 57 54	+21 39.3	1	8.80		0.82		0.51		1	G5	333,1620
76921	+59 01212			8 58 00	+59 36.8	1	7.86		1.06		0.96		3	G8 III	37
	−42 04864		A	8 58 01	−43 09.4	1	10.99		-0.03		-0.29		1		362
	−42 04864		B	8 58 01	−43 09.4	1	11.08		0.02		-0.24		1		362
77271	−46 04818			8 58 02	−47 06.6	1	8.76		0.91		0.54		2	G8 IV	16
76827	+68 00551	HR 3576		8 58 04	+67 49.6	5	4.75	.008	1.54	.014	1.86	.020	21	M3 III-IIIb	15,1118,3055,6001,8015
		AJ82,474 # 71		8 58 08	−47 12.8	1	11.85		0.95		-0.05		2	B1 Vn	779
		G 46 - 15		8 58 11	+05 26.6	3	12.38	.021	1.46	.005	1.07	.000	3		333,1620,1705,3016
77370	−58 01327	HR 3598	★ A	8 58 11	−58 53.5	3	5.16	.005	0.42	.005	0.00		9	F3 V	15,2012,3026
		G 46 - 16		8 58 13	+05 26.4	3	12.70	.015	1.49	.015	1.07	.010	3		333,1620,1705,3016
		SY Cnc		8 58 13	+18 06.1	1	12.97		0.25		-0.69		1	G8 V	1471
77258	−40 04810	HR 3591		8 58 13	−41 03.5	4	4.45	.004	0.65	.004	0.38	.004	20	G8/K1 III+A	15,1075,2012,8015
		AJ82,474 # 73		8 58 14	−47 21.3	1	11.23		1.05		-0.07		2	O5 V	779
	−56 01953			8 58 14	−57 13.2	1	11.08		0.78		0.30		2		1696
		G 114 -B8	A	8 58 17	−04 10.4	1	10.28		0.58		0.01		3	dG	3029
		G 114 -B8	B	8 58 17	−04 10.4	1	14.52		0.94		0.62		3	dK	3029
	−39 05049	LSS 1220	★ AB	8 58 18	−39 21.2	2	10.24	.031	0.25	.008	-0.60	.011	5	B0.5II-III	362,540
	−9 02717			8 58 19	−09 38.5	1	10.35		0.71		0.27		1		289
77104	+32 01829	HR 3587	★ AB	8 58 20	+32 26.9	2	5.87		0.04	.034	0.09	.000	5	A2 V	1022,1733

HD	DM	Other Id	N Rem	α₁₉₅₀	δ₁₉₅₀	S	V	σ_V	B–V	σ_B–V	U–B	σ_U–B	n	Spectrum	References
77493	−67 01018			8 58 20	−67 56.6	1	6.92		−0.06		−0.38		2	B7 V	401
77321	−49 04042	IDS08567S4910	AB	8 58 23	−49 21.7	1	6.88		1.32		1.63		1	K1/2 II	389
77093	+40 02138	HR 3586		8 58 28	+39 54.7	1	6.38		0.29		0.02		4	A9 Vn	254
	−46 04820			8 58 28	−47 08.3	1	10.82		0.36		0.15		2	A7 III	16
	−46 04821			8 58 30	−46 29.4	2	8.98	.010	0.32	.015	−0.49	.024	3		389,401
77175	+15 01957	G 9 - 43	⋆ AB	8 58 31	+15 28.0	2	8.69	.029	1.28	.019	1.22	.000	3	K5	1620,3072
77320	−42 04875	HR 3593, IU Vel		8 58 33	−42 58.6	11	6.06	.018	−0.16	.010	−0.80	.011	54	B2.5 Vne	15,26,164,362,540,681*
77103	+45 01666			8 58 34	+45 20.0	1	9.18		0.23		0.12		2	A3	1733
	−48 04314	LSS 1224	⋆ AB	8 58 35	−48 49.3	2	10.48	.040	0.92	.050	0.03	.035	4	B3 III	962,1737
		G 46 - 17		8 58 36	+02 08.4	1	11.82		1.50		1.19		1		333,1620
77122	+36 01897			8 58 36	+36 03.4	1	7.84		0.45		0.02		2	F5	1601
		AJ82,474 # 75		8 58 36	−47 10.2	1	12.57		1.27		0.20		2		779
		LP 312 - 57		8 58 38	+30 45.9	1	11.45		1.20		1.19		2		1723
77343	−43 04839			8 58 38	−43 43.2	1	9.03		0.22		0.13		2	A1mA4-A6	401
		GD 99, VW Lyn		8 58 41	+36 19.0	2	14.55	.000	0.28	.090	−0.75	.155	2	DA	974,3060
77236	−1 02181			8 58 43	−02 21.8	3	7.50	.011	1.16	.003	0.99	.015	11	K0 III	1003,1509,2012
77400	−46 04826			8 58 43	−47 03.6	2	9.05	.010	0.08	.005	0.07	.005	5	A1 V	16,401
77385	−48 04318			8 58 43	−48 31.1	1	8.06		1.23		1.34		1	K2 III	389
77401	−46 04827			8 58 44	−47 08.6	1	9.49		1.62		1.25		2	M7 III	16
77366	−37 05360			8 58 46	−38 13.4	1	7.27		−0.13		−0.57		4	B2/3 V	158
		Cr 205 - 15		8 58 47	−48 47.3	1	12.65		0.60		−0.02		3		184
77190	+28 01674	HR 3589	⋆ A	8 58 50	+28 06.0	1	6.07		0.22		0.08		2	A8 Vn	1733
		Cr 205 - 10		8 58 50	−48 48.0	1	12.41		0.62		−0.05		5		184
		Cr 205 - 7		8 58 51	−48 46.2	1	12.96		0.61		0.03		3		184
		Cr 205 - 9		8 58 51	−48 47.5	1	12.37		0.68		−0.09		3		184
	−48 04322	Cr 205 - 3		8 58 52	−48 46.5	1	10.07		0.43		−0.48		2	B1 III	184
77421	−48 04321	Cr 205 - 2	⋆ AB	8 58 52	−48 47.3	2	9.64	.001	0.43	.016	−0.41	.014	6	B1 V	184,540,976
		Cr 205 - 8		8 58 52	−48 47.5	1	12.18		0.58		−0.12		2		184
77250	+6 02087	HR 3590	⋆	8 58 53	+05 50.3	6	6.06	.010	1.11	.012	1.03	.015	19	K1 II-III +	15,1078,1084,1355*
	−48 04324	Cr 205 - 13		8 58 53	−48 49.1	1	10.10		0.31		−0.52		2	B0 V	184
		Cr 205 - 4		8 58 54	−48 47.	1	10.75		0.46		−0.46		2	B1 III	184
		Cr 205 - 11		8 58 54	−48 47.	1	13.63		0.59		0.01		2		184
		Cr 205 - 12		8 58 54	−48 47.	1	10.67		0.43		−0.28		2	B5 V	184
		Cr 205 - 16		8 58 54	−48 47.	1	13.86		0.65		0.12		3		184
		Cr 205 - 17		8 58 54	−48 47.	1	14.88		0.74		0.49		3		184
		Cr 205 - 18		8 58 54	−48 47.	1	15.77		1.25		0.61		2		184
77230	+17 01990			8 58 55	+17 16.7	1	7.38		−0.02		−0.06		2	B9	1648
		WLS 900-20 # 7		8 58 57	−21 51.4	1	11.98		0.54		0.16		2		1375
		AJ82,474 # 76		8 58 57	−47 28.0	1	11.44		1.32		0.22		2	O9 Vn	779
		Cr 205 - 6		8 58 57	−48 46.9	1	12.23		0.66		−0.22		2		184
		Cr 205 - 5		8 58 57	−48 47.2	1	12.52		0.60		−0.20		2		184
	+27 01706	WY Cnc		8 58 58	+26 52.7	2	9.52	.063	0.73	.005	0.17	.020	3	G5 V:	588,1768
77420	−43 04848			8 58 58	−43 39.6	1	8.46		0.26		0.16		2	A5 III	401
77434	−48 04325	Cr 205 - 1	⋆ AB	8 58 58	−48 47.6	2	9.32	.000	0.58	.000	0.04	.025	16	F8 V	184,540,976
77361	−26 06647	HR 3597		8 59 01	−26 28.0	1	6.20		1.13				4	K1 III	2006
77433	−45 04772			8 59 01	−45 41.9	1	8.33		0.22		0.06		2	A5 V	401
		Cr 205 - 14		8 59 01	−48 49.3	1	12.66		0.64		−0.06		3		184
	−14 02728			8 59 02	−14 29.0	1	10.04		0.53		−0.13		2		1696
77464	−51 03420	CV Vel		8 59 04	−51 21.5	3	6.71	.008	−0.17	.012	−0.67	.010	10	B2 V	401,1075,2012
76905	+71 00486			8 59 05	+71 30.2	1	7.07		0.31		0.03		1	F0	1776
	−46 04835			8 59 06	−47 06.0	1	11.63		0.65		0.10		1	G0:	16
299823	−53 02048			8 59 08	−54 15.6	1	9.76		0.11		−0.38		2	A2	401
77453	−45 04774			8 59 09	−45 26.1	1	8.20		0.21		0.10		2	A5 IV/V	401
77293	+6 02089			8 59 11	+05 51.0	2	7.18	.010	0.30	.005	0.04	.000	6	F1 IV	1084,3077
	+71 00487			8 59 11	+71 36.6	1	8.93		0.45		−0.06		2	F8	1776
77490	−53 02050			8 59 13	−54 10.0	1	7.69		0.02		−0.02		2	B9.5V	401
77525	−56 01968			8 59 14	−56 54.7	1	8.53		−0.04		−0.52		2	B5 Vn	401
77158	+52 01357			8 59 16	+52 25.2	1	8.69		0.51		0.05		2	F8	1566
77489	−49 04050			8 59 17	−50 14.4	1	7.95		1.01		0.79		1	K0 III	389
		PASP81,804 T2# 28		8 59 22	−47 13.9	1	12.41		0.74		0.43		2	K0:	16
77353	+0 02449	HR 3596		8 59 25	−00 17.2	5	5.66	.016	1.15	.010	1.03	.008	15	K0 III	15,252,1003,1415,1509
77511	−46 04836			8 59 26	−47 03.3	1	7.74		0.23		0.13		2	A5mA5-F0	16
77512	−50 03761	IDS08578S5007	AB	8 59 26	−50 19.0	1	8.57		0.54		0.03		1	F7 V	389
77475	−41 04720	HR 3600, IZ Vel		8 59 29	−41 40.0	5	5.54	.010	−0.14	.011	−0.57	.012	15	B5 III	15,362,540,976,2012
77005	+72 00439			8 59 31	+71 54.0	1	8.58		0.12		0.02		2	B9	1776
77758	−70 00850			8 59 31	−70 42.6	1	8.86		1.14				4	K1/2 III	2033
77450	−24 07646			8 59 32	−25 14.1	1	7.66		1.06		0.80		3	K0 III	1657
	−46 04838			8 59 32	−47 17.3	1	12.05		0.46		0.08		2	F5	16
		G 41 - 18		8 59 37	+08 40.0	2	11.76	.005	1.45	.010	1.11	.025	2		1620,1658
		WLS 900 60 # 5		8 59 37	+62 15.7	1	11.91		1.00		0.69		2		1375
77615	−60 01283	HR 3604		8 59 37	−60 46.0	4	5.79	.010	1.21	.009			18	G8 II	15,1075,2013,2028
77462	−26 06661			8 59 40	−26 22.2	2	7.24	.021	0.50	.004	−0.07		6	F7 V	1311,2012
77521	−43 04861	IDS08579S4351	AB	8 59 41	−44 03.2	1	8.33		0.86		0.58		1	G5 III/IV	389
77234	+50 01603	DH UMa		8 59 45	+50 17.5	1	9.43		1.44				1	R6	1238
77552	−47 04589			8 59 47	−47 23.2	2	9.28	.010	1.81	.035	2.22		5	M0 (III)	16,8100
77350	+25 02029	HR 3595		8 59 49	+24 39.1	7	5.44	.017	−0.04	.010	−0.11	.008	21	A0p Si	15,1007,1013,1022*
	+45 01668	TT Lyn		8 59 49	+44 47.1	1	9.52		0.25		0.13		1	A9	699
77377	+22 02039			8 59 53	+21 43.0	1	8.09		0.99		0.76		2	G5	1648
77566	−49 04057			8 59 53	−49 22.9	1	7.45		−0.12		−0.55		2	B5 V	401
		KUV 08599+4130		8 59 54	+41 29.7	1	14.15		−0.04		−0.84		1		1708

Table 1 497

HD	DM	Other Id	N Rem	α_{1950}	δ_{1950}	S	V	σ_V	B–V	σ_{B-V}	U–B	σ_{U-B}	n	Spectrum	References
		KUV 345 - 57		8 59 54	+41 29.7	1	14.47		0.14		-0.81		1		974
	-17 02722			9 00 04	-17 32.5	1	9.63		1.08		0.75		2		1375
77595	-49 04060			9 00 04	-50 03.1	1	8.47		-0.01		-0.33		2	B8 II	401
77445	+7 02066	HR 3599		9 00 05	+07 29.8	4	5.86	.020	1.10	.009	0.98	.005	14	gK3	15,1078,1355,3016
77655	-55 01900	IDS08587S5515	AB	9 00 05	-55 27.1	1	8.93		0.02		-0.31		2	B8 IV/V	401
77594	-48 04345			9 00 07	-48 33.1	1	8.72		0.55		0.12		1	G0 V	389
77580	-38 05154	HR 3602		9 00 11	-39 12.3	4	6.26	.008	1.00	.000			18	K1 III-IV	15,1075,2012,2020
77653	-51 03430	HR 3605	⋆ AB	9 00 11	-51 59.5	3	5.23	.005	-0.12	.009	-0.46		9	B9p Si	15,401,2012
		BPM 73811		9 00 12	-08 13.	1	10.94		1.09		0.91		2		3072
77408	+33 01800			9 00 13	+33 04.8	1	7.03		0.50		-0.04		2	F6 IV	1003
77327	+47 01633	HR 3594	⋆ AB	9 00 13	+47 21.3	5	3.58	.032	0.00	.005	0.01	.004	16	A0 IV-V+A0V	15,1118,1381,3016,8015
77581	-40 04838	GP Vel		9 00 13	-40 21.4	9	6.89	.018	0.47	.020	-0.51	.016	57	B0.5Ia	285,358,362,540,633,976*
	+38 01992	UV Lyn		9 00 14	+38 17.8	1	9.81		0.67		0.17		4	G0	1723
77309	+54 01272	HR 3592		9 00 21	+54 28.9	2	5.75	.005	0.02	.000	0.05	.005	4	A2 V	985,1733
77578	-31 06828			9 00 25	-32 05.5	1	8.90		0.41				4	F2 V	2012
77468	+23 02035			9 00 31	+22 48.6	1	8.08		0.90		0.56		2	G5	1625
		AJ82,474 # 81		9 00 31	-48 30.3	1	13.35		1.03		0.04		1		962
77887	-68 00879	HR 3610		9 00 35	-68 29.2	4	5.88	.008	1.62	.007	1.96	.002	21	M1 III	15,1075,1637,2038
77443	+39 02193	UX Lyn		9 00 36	+38 56.5	2	6.71		1.52	.005	1.37	.028	8	M3 III	105,1501
78086	-75 00554			9 00 36	-75 31.7	1	7.65		1.74		2.13		8	M2 III	1704
77669	-43 04873			9 00 38	-43 46.3	1	8.10		-0.05		-0.30		2	B9 III/IV	401
	-48 04352	LSS 1228		9 00 41	-48 30.0	3	10.35	.029	0.76	.011	-0.35	.015	8	O9 Vn	92,342,962
77684	-42 04906			9 00 45	-42 30.5	1	7.29		0.04		0.02		2	A0 V	401
77718	-46 04855	LSS 1230		9 00 45	-46 21.8	1	8.46	.020	0.68	.011	-0.21	.028	9	B2 Ib/II	540,976,8100
77645	-31 06835			9 00 48	-32 14.5	1	6.93		-0.02		-0.11		7	B9.5V	1628
		AJ82,474 # 84		9 00 54	-48 26.0	1	12.13		0.95		-0.15		2	O9.5V	962
		09H00M55S		9 00 55	+02 37.3	1	12.28		0.63		-0.09		2		1696
77665	-25 06829	HR 3607		9 00 57	-25 18.3	2	6.73	.005	-0.03	.000			7	B8 V	15,2012
77741	-45 04798			9 00 58	-45 58.5	1	9.78		0.08		0.01		2	A0 V	401
77795	-54 01982			9 00 58	-54 49.9	1	7.56		0.05		0.00		2	A0 V	401
77740	-44 05064			9 00 59	-44 26.7	1	8.54		0.06		0.00		2	A1 V	401
77739	-43 04877			9 01 01	-44 02.8	1	8.34		-0.09		-0.46		2	B8 II/III	401
77793	-50 03782			9 01 07	-50 37.6	1	8.14		-0.02		-0.21		2	A0 IV/V	401
77754	-42 04913			9 01 10	-42 41.1	1	8.66		-0.05		-0.17		2	Ap Si	401
77557	+28 01683	HR 3601		9 01 11	+28 05.8	1			0.00		0.05		4	A1 V	1022
77640	-4 02530			9 01 14	-04 58.3	1	6.80		-0.06				4	A0	2006
77677	-13 02751			9 01 15	-14 14.3	1	7.64		1.42		1.74		3	K3 III	1657
77812	-44 05072	IDS08596S4459	AB	9 01 21	-45 10.5	1	9.03		0.08		0.02		2	A0 IV/V	401
		AJ82,474 # 86		9 01 21	-48 33.9	1	14.12		0.97		0.12		1		962
77586	+29 01860			9 01 23	+29 28.1	1	7.60		1.61		1.76		2	M3 III	1003
77884	-54 01987			9 01 25	-55 13.8	1	8.77		0.02		-0.12		2	A0 V	401
77852	-48 04368	LSS 1231		9 01 32	-48 18.5	1	8.64		0.78		0.17		3	B8/9 Iab/b	962
77865	-49 04075			9 01 32	-49 29.9	1	7.04		1.26		1.28		1	K1 III	389
77661	+18 02114			9 01 33	+18 28.7	1	9.20		0.58		0.10		2		1648
77907	-53 02072	HR 3611		9 01 35	-53 21.1	3	6.40	.007	-0.09	.009	-0.36		9	B6 V	15,401,2027
77849	-43 04885			9 01 36	-43 51.5	1	9.33		-0.01		-0.10		2	B9.5V	401
78045	-65 01065	HR 3615		9 01 40	-66 11.8	8	4.00	.012	0.15	.004	0.14	.012	29	A2/3 IV/Vm	15,355,1075,1637,2006*
		WLS 900 60 # 9		9 01 46	+59 27.2	1	11.12		0.57		-0.02		2		1375
		AJ82,474 # 88		9 01 46	-48 20.8	1	12.04		0.96		-0.12		2		962
		LSS 1232	⋆	9 01 47	-44 06.8	2	11.51	.010	0.77	.010	-0.31	.010	8	B3 Ib	362,1737
77904	-46 04870			9 01 47	-47 12.3	2	8.10	.000	-0.08	.005	-0.43	.010	6	B6/7 IV/V	16,164
		LP 426 - 63		9 01 48	+16 38.0	1	10.71		1.07		1.09		1		1696
77570	+51 01478			9 01 48	+51 01.5	2	6.78	.000	0.41	.005	0.01	.000	5	F5 V	1501,1566
	-46 04872			9 01 49	-47 11.3	1	11.08		0.50		0.04		2	F6 V:	16
77660	+32 01837	HR 3606		9 01 52	+32 34.7	2	6.46	.005	0.23	.002	0.10	.008	4	A8 V	1733,3016
	-47 04620	IDS09002S4708	AB	9 01 55	-47 19.7	1	10.13		0.22		0.10		1	A7 III	16
77943	-47 04621			9 01 56	-47 21.0	1	11.07		0.49		0.02		1	F0 V	16
77601	+49 01801	HR 3603		9 01 58	+48 43.8	1	5.97		0.44		0.15		3	F6 II-III	254
	-44 05080			9 01 58	-44 42.0	1	11.08		0.30				4		2011
	+45 01673			9 01 59	+44 43.0	1	11.38		0.55		-0.10		4	G0 V	1723
77825	-15 02685			9 02 00	-15 42.9	1	8.78		0.96		0.71		2	K2 V	3072
	-47 04622			9 02 01	-47 19.0	1	11.46		0.30		0.10		1	A0 V:	16
		G 115 - 49		9 02 05	+39 00.2	2	11.60	.000	0.51	.010	-0.21	.035	4		1064,3062
77599	+56 01361	IDS08584N5556		9 02 06	+55 43.7	1	7.97		0.57		0.10		2	G0	1502
77730	+23 02040			9 02 07	+23 01.8	1	7.39		0.60		0.09		2	G2 V	1625
77729	+26 01895			9 02 08	+26 21.9	3	7.63	.005	1.40	.013	1.57	.005	15	K4 III	1003,1775,3077
	-46 04793			9 02 09	-47 11.5	1	10.30		0.51		0.02		2	F7 III	16
77959	-48 04382	LSS 1234	⋆ AB	9 02 10	-48 21.5	2	9.11	.010	0.88	.005	-0.21	.005	5	O8 Ib(f)	390,962
	-45 04826	LSS 1236		9 02 20	-45 28.6	3	10.10	.018	0.57	.020	-0.42	.013	7		362,1737,1747
		LP 666 - 38		9 02 22	-08 04.9	1	13.01		0.63		-0.11		2		1696
78025	-50 03798			9 02 22	-51 15.0	1	8.60		-0.05		-0.22		2	B9 V	401
78005	-46 04882			9 02 23	-47 14.5	5	6.44	.007	-0.15	.007	-0.64	.010	14	B3 IV/V	16,158,540,976,1732
		AJ82,474 # 91		9 02 23	-48 22.8	1	13.79		1.01		0.03		2		962
	+25 02037	G 47 - 22		9 02 25	+25 30.2	1	10.57		1.24		1.12		1		333,1620
77938	-31 06877			9 02 25	-32 14.9	2	7.82	.030	1.55	.015	1.67		13	M4/5 III	1024,2033
		LP 546 - 50		9 02 26	+02 38.3	1	13.20		1.62		1.21		1		1773
78004	-46 04883	HR 3614		9 02 26	-46 53.9	6	3.75	.003	1.20	.006	1.21	.014	37	K2 III	15,16,1075,1637,2012,8015
78040	-46 04885			9 02 33	-47 11.2	1	10.53		0.10		0.11		2	A0 VN	16
78023	-47 04628			9 02 34	-47 59.9	1	8.94		0.46		0.05		1	F5 IV	389
	-52 01897			9 02 36	-52 39.0	1	10.68		1.06		0.80		1		1696

HD	DM	Other Id	N Rem	α_{1950}	δ_{1950}	S	V	σ_V	B–V	σ_{B-V}	U–B	σ_{U-B}	n	Spectrum	References
		KUV 09026+4235		9 02 37	+42 35.3	1	16.35		-0.39		-1.19		1		1708
77836	+19 02153			9 02 39	+19 38.2	1	7.92		0.44		0.02		2	F5	1648
78039	-42 04936			9 02 47	-43 15.3	1	8.67		0.48				4	F5 V	2012
		G 115 - 50		9 02 49	+35 43.8	1	14.77		1.65		1.42		7		203
		PASP81,804 T2# 9		9 02 49	-47 21.3	1	12.98		0.65		0.28		1	F5:	16
77692	+59 01217 HR 3608			9 02 51	+59 32.7	2	6.44	.000	0.04	.015	0.13		8	A2 V	252,1379
77770	+50 01608			9 02 54	+49 48.7	1	7.51		-0.21		-0.76		3	B2 IV	399
78080	-42 04941			9 02 55	-43 04.0	1	9.48		0.03		-0.02		2	A0 V	401
		LP 786 - 61		9 02 56	-18 20.1	1	9.83		0.58		-0.01		2		1696
78135	-52 01904			9 02 56	-52 18.7	2	7.43	.005	0.57	.005	0.09		5	G6/K0 III+F	389,2012
		G 114 - 34		9 02 57	+02 28.0	1	14.36		1.50				1		906
77892	+17 02004			9 02 57	+17 35.5	1	8.10		0.14		0.11		2	A2	1648
78097	-44 05097			9 02 59	-44 36.6	1	7.62		1.67		1.44		1	K2 III +F/G	389
		LSS 1238		9 03 03	-48 09.9	2	11.20	.000	0.89	.035	-0.21	.040	5	B3 Ib	962,1747
78231	-60 01305			9 03 07	-60 22.0	1	7.82		1.12		1.09		3	K0 III	1704
	-19 02614			9 03 09	-19 54.3	1	10.22		0.26		0.21		2		1375
78190	-55 01924 IDS09017S5556		AB	9 03 09	-56 08.5	1	6.89		0.04		-0.46		2	B5 V	401
	-46 04894			9 03 10	-47 03.5	1	11.24		0.42		-0.04		1	F2 V:	16
78075	-28 06904			9 03 12	-28 49.7	1	9.55		-0.22		-0.32		3	A2/3 II	1732
78205	-51 03478			9 03 19	-51 46.5	1	8.45		-0.05		-0.29		2	B9 V	401
77996	+5 02116 HR 3613			9 03 20	+05 17.6	5	4.97	.025	1.22	.007	1.21	.011	13	K2 II-III	15,1080,1415,3016,8041
77912	+39 02200 HR 3612			9 03 21	+38 39.2	7	4.57	.012	1.04	.005	0.82	.005	23	G8 Ib-II	15,1080,1363,3016*
		G 194 - 56		9 03 21	+50 21.3	1	12.59		1.35		1.10		1		1658
	+30 01814 G 47 - 23			9 03 23	+30 29.3	1	10.69		1.20		1.18		1	K5	333,1620
78229	-52 01914			9 03 23	-52 53.5	1	8.60		-0.03		-0.24		2	B8/9 V(n)	401
78187	-47 04640			9 03 26	-48 15.6	2	10.15	.030	0.54	.011	-0.33	.017	3	B9.5/A0 V	540,976
		PASP81,804 T2# 4		9 03 27	-46 57.6	1	12.21		0.36		0.25		1	A0 V	16
78293	-57 01859 HR 3622		⋆AB	9 03 28	-57 39.1	3	6.43	.004	0.26	.004	0.18	.010	11	A8 III	15,1075,2012
77995	+15 01976			9 03 29	+14 42.0	1	9.56		0.70				2	G5 V	882
78011	+15 01977			9 03 35	+15 28.5	1	7.68		1.56		1.88		3	M1	1648
78146	-27 06240			9 03 35	-27 52.9	1	8.56		1.16		1.10		2	K1 III	536
77818	+59 01218			9 03 36	+59 04.7	2	7.65	.000	1.00	.005	0.82		7	K1 IV	1003,1379
78184	-42 04953			9 03 37	-42 49.9	1	8.89		0.45		-0.01		1	F3 V	389
78243	-47 04644			9 03 43	-47 27.0	1	9.09		0.03		0.00		1	M1 III	16
	-44 05113 LSS 1239			9 03 47	-44 46.5	1	10.85		0.50		-0.48		4		362
	-46 04901			9 03 47	-46 53.1	1	10.98		0.58		0.18		1	G0 V	16
78266	-46 04902			9 03 47	-47 13.6	1	10.02		0.28		0.09		1	A7 V	16
78050	+21 01969 G 9 - 47			9 03 52	+20 42.8	2	7.68	.005	0.80	.005	0.32	.019	3	G0	333,1620,1658
78265	-44 05115			9 03 52	-44 57.6	1	8.46		0.03		0.01		2	B9 V	401
78290	-49 04111			9 03 52	-50 05.2	1	7.27		-0.12		-0.59		2	B3 IV	401
78327	-55 01937			9 03 52	-55 50.1	1	8.73		-0.04		-0.49		2	B6 IV	401
77246	+81 00282			9 03 53	+81 01.9	1	6.35		0.36		0.03		2	F2	1733
		LP 666 - 30		9 03 54	-08 41.0	1	12.02		0.41		-0.23		2		1696
78241	-37 05454			9 03 54	-38 02.6	1	8.31		0.33				4	F0 V	2012
	-46 04904			9 03 57	-46 51.4	1	11.66		0.54		0.11		1	A6 IV/V	16
		G 115 - 53		9 04 02	+42 08.4	1	13.72		1.31				6		3062
77800	+67 00573 HR 3609			9 04 02	+67 04.6	3	5.14	.010	1.51	.007	1.82	.014	9	K5 III	1118,1355,3016
	-46 04908			9 04 04	-47 16.2	1	10.99		0.47		0.01		1	F5 V:	16
311169	-64 00997			9 04 04	-65 09.8	3	10.31	.061	0.65	.005	0.08	.015	10		158,1696,3060
	-47 04649 LSS 1241			9 04 07	-47 28.1	2	10.19	.010	0.53	.030	-0.39	.040	4	B1 V	962,1747
78344	-47 04650 LSS 1242			9 04 08	-47 34.0	7	8.99	.021	1.11	.016	0.00	.034	13	O9.5/B0(Ib)	16,540,962,976,1737,1747*
		G 47 - 24		9 04 12	+29 20.1	1	12.31		1.44		1.27		1		1620
	+84 00197			9 04 16	+83 52.0	1	10.10		-0.02		0.09		5	A2	1219
		AJ82,474 # 95		9 04 16	-47 06.9	1	12.12		1.71		0.45		2		962
78345	-47 04654 LSS 1243			9 04 16	-48 06.8	5	9.37	.015	0.91	.014	-0.05	.032	11	B1 Iab/b	540,962,976,1747,2021
78355	-48 04420			9 04 18	-49 02.0	1	7.94		1.50		1.53		1	K1 III	389
78406	-55 01945			9 04 19	-55 32.3	1	9.12		-0.05		-0.45		2	B8 V	401
	-8 02582 G 114 - 35			9 04 20	-08 36.5	3	9.49	.013	1.43	.014	1.26		7	M0	1705,2033,3072
79837	-85 00183 HR 3678			9 04 21	-85 28.0	5	5.41	.009	0.31	.005	0.05	.014	25	A8/9 IV	15,1075,2012,2038,3026
78405	-52 01924			9 04 24	-52 36.0	1	8.25		-0.10		-0.55		2	B5 IV	401
78196	+2 02145 HR 3618			9 04 25	+01 39.9	4	6.18	.017	1.64	.011	1.94	.018	13	M1 III	15,1355,1415,3055
237781	+59 01220			9 04 32	+59 17.7	2	9.20	.010	0.38	.010	0.08		9	F0	1371,1379
78507	-61 01174			9 04 32	-61 54.1	1	8.12		-0.04		-0.48		2	B6 V	401
78175	+23 02048 HR 3617		⋆AB	9 04 34	+23 10.8	1	6.41		0.42		-0.01		2	F5 V+F5 V	1733
78138	+37 01943			9 04 36	+37 16.6	1	8.19		0.44		-0.05		2	F5	1601
		UV0904-02		9 04 36	-02 54.0	1	11.96		-0.31		-1.19		7		1732
78444	-54 02019 IDS09032S5430		AB	9 04 38	-54 42.2	1	8.76		0.02		-0.23		2	B9.5V	401
	+51 01483 V UMa			9 04 42	+51 18.9	1	10.33		1.65		1.43		1	M5.5	635
78282	+1 02237			9 04 48	+00 48.1	2	7.16	.008	0.03	.004	0.05	.012	10	A0	14,379
78429	-42 04968			9 04 48	-43 17.6	5	7.31	.007	0.67	.005	0.20	.006	20	G5 V	389,395,977,1075,2011
78279	+13 02044			9 04 55	+12 55.7	1	9.46		0.97				3	K0	882
78506	-52 01932 IDS09034S5252		AB	9 04 56	-52 43.9	1	8.38		-0.03		-0.40		2	B7 IV	401
78251	+26 01901			9 04 57	+25 49.7	2	7.04	.005	0.19	.005	0.08	.000	5	A5	833,1733
		CCS 1386		9 04 57	-58 23.7	1	11.46		2.92				1	R	864
78333	-2 02794			9 04 59	-02 52.9	1	9.29		0.15		0.12		4	A2	1732
78235	+30 01817 HR 3621			9 05 00	+29 51.4	3	5.46	.013	0.89	.004	0.60	.157	6	G8 III	1080,3016,8041
78234	+33 01810 HR 3620			9 05 01	+32 44.6	2	6.48	.019	0.36	.000	0.01	.010	6	F3 V	1733,3053
		AJ82,474 # 97		9 05 01	-48 41.5	1	13.42		1.35		0.52		1		962
78530	-53 02122			9 05 01	-53 53.3	1	8.57		-0.09		-0.52		2	B5 IV	401
		G 46 - 21		9 05 02	+06 30.2	1	12.47		1.43		1.11		1		333,1620

Table 1   499

HD	DM	Other Id	N Rem	α_{1950}	δ_{1950}	S	V	σ_V	B–V	σ_{B-V}	U–B	σ_{U-B}	n	Spectrum	References
78316	+11 01984	HR 3623, khi Cnc		9 05 02	+10 52.2	5	5.23	.008	-0.11	.008	-0.44	.011	15	B8 IIIp Mn	1022,1079,1202,1363,3033
78791	−72 00779	HR 3643		9 05 02	−72 24.1	7	4.47	.007	0.61	.003	0.21	.017	33	F7 II	15,1075,2031,2038*
78278				9 05 05	+21 47.4	1	10.59		1.52				1	R5	1238
	−44 05138			9 05 05	−45 11.6	1	11.05		0.54		-0.15		4		362
78568	−58 01386			9 05 06	−59 00.3	1	7.82		-0.06		-0.41		2	Ap Si	401
78548	−55 01957	HR 3629		9 05 08	−55 36.1	6	6.10	.008	-0.15	.015	-0.68	.010	24	B2/3 V	15,362,401,1020,2012,2012
78529	−52 01937			9 05 09	−52 19.7	1	7.43		1.02		0.85		1	K0 II/III	389
78632	−63 01093	HR 3632		9 05 10	−64 17.9	3	6.35	.007	1.36	.009	1.58	.000	15	K3 III	15,1075,2038
78212	+41 01922			9 05 14	+41 20.9	1	8.48		1.56		1.91		2	K5	1601
78764	−70 00861	HR 3642, V345 Car		9 05 15	−70 20.2	8	4.70	.019	-0.15	.009	-0.81	.010	46	B2 IVe	15,681,815,1075,1637*
76990	+84 00196	HR 3581		9 05 16	+84 23.2	2	6.31	.025	0.31	.010	0.09	.010	4	F2 III	985,1733
78124	+61 01102	IDS09014N6057	AB	9 05 19	+60 44.8	1	9.04		0.40		0.13		4	B9	1371
	+73 00447	G 252 - 28		9 05 20	+73 36.8	1	10.17		1.26		1.28		3	K5	196
78209	+52 01365	HR 3619		9 05 21	+51 48.5	5	4.48	.008	0.27	.002	0.12	.005	20	Am	15,1008,1363,3023,8015
78467	−29 07143			9 05 22	−29 52.9	1	7.39		1.00		0.69		4	G8 III/IV	1657
78763	−69 01009			9 05 22	−69 22.4	1	8.30		-0.06		-0.36		2	B7 Vn	401
78391	−6 02825	IDS09029S0644	ABC	9 05 24	−06 56.1	1	8.24		0.67				4	G0	2012
78233	+51 01485	G 194 - 58		9 05 25	+50 59.6	1	8.14		0.79		0.47		2	G5	1566
	+28 01697	IDS09025N2756	A	9 05 26	+27 44.2	1	8.26		0.58		-0.02		2	G0 V	3026
	−45 04875			9 05 27	−45 51.4	1	10.44		0.53		-0.27		3		362
	+28 01698	IDS09025N2756	B	9 05 30	+27 44.7	1	8.25		0.58		-0.02		2	G0 V	3026
		AAS27,343 # 157		9 05 36	−55 21.	1	11.99		0.60		-0.22		2		362
78599	−50 03849	HR 3631		9 05 38	−51 00.6	4	6.72	.005	1.62	.006	1.91	.021	10	K3/5 III	15,389,2029,3005
78725	−62 01168			9 05 41	−62 38.9	1	7.53		-0.03		-0.31		2	B8 V	401
78522	−31 06939	IDS09037S3138	AB	9 05 46	−31 50.1	2	7.91	.010	1.39	.015	1.41	.017	3	K1/2 II	565,3048
78419	+17 02018			9 05 47	+16 54.3	1	7.58		1.05		0.97		2	K0	1648
78366	+34 01949	HR 3625		9 05 47	+34 05.2	2	5.96	.030	0.58	.013	0.04		5	F9 V	71,3026
78418	+27 01715	HR 3626	⋆ A	9 05 51	+26 50.2	2	5.97	.010	0.65	.005	0.20		6	G5 IV-V	1118,3077
78541	−25 06895	HR 3628	⋆ AB	9 05 51	−25 39.3	8	4.59	.017	1.58	.016	1.90	.016	32	K4 III	3,15,369,1024,1075*
78616	−44 05150	KK Vel	⋆ AB	9 05 54	−44 25.8	7	6.78	.009	-0.01	.007	-0.76	.011	46	B2 II/III	278,362,540,976,977*
78629	−47 04673			9 05 55	−47 21.6	1	8.38		0.48		0.03		2	F6 V	389
	−53 02143	LSS 1246		9 05 55	−53 51.9	1	10.96		0.15		-0.71		2		362
	+33 01813	MCC 539	AB	9 05 59	+32 40.0	1	10.66		1.16		0.99		2	K5	1723
78249	+59 01221			9 06 00	+59 21.5	4	7.10	.015	0.98	.018	0.82	.000	15	K1 IV	1003,1080,1118,1379
78154	+67 00577	HR 3616	⋆ ABC	9 06 01	+67 20.4	8	4.80	.013	0.49	.005	0.02	.007	29	F6 IV-V+K2V	1,15,292,1077,1118*
78685	−49 04139			9 06 01	−50 17.7	1	8.84		0.05		0.07		1	A0 V	6006
78451	+17 02019			9 06 02	+16 37.3	1	9.37		0.78				1	G5	882
		Steph 749		9 06 02	+32 40.4	1	10.70		1.14				1	K5	1746
	+59 01222			9 06 04	+59 26.3	1	10.36		1.23				2	G2	1379
		MCC 25		9 06 06	+27 38.0	1	10.26		1.24		1.22		2	K5	3072
78558	−14 02757			9 06 06	−14 56.3	5	7.29	.014	0.62	.005	0.05	.018	12	G3 V	22,742,1003,1311,2012
78684	−45 04883			9 06 07	−45 41.5	1	8.54		1.09		0.90		1	K0 III	389
78647	−42 04990	HR 3634, λ Vel	⋆ A	9 06 09	−43 13.8	6	2.20	.011	1.67	.013	1.81	.009	20	K4 III	15,1034,1075,1509*
		G 115 - 57		9 06 13	+40 18.6	1	14.53		1.82				3		538
78479	+18 02129			9 06 14	+17 40.4	4	7.19	.014	1.21	.008	1.31	.019	6	K3 III	1003,1080,3040,8041
78556	−8 02588	HR 3630	⋆ AB	9 06 15	−08 23.2	4	5.59	.009	-0.06	.000	-0.18	.005	15	B9.5III	15,1079,1256,2028
78913	−67 01038			9 06 17	−68 17.2	2	9.34	.080	0.09	.005	0.11	.030	7	A0 V	164,3077
78747	−49 04142			9 06 19	−50 16.7	3	7.72	.013	0.56	.005	-0.04	.020	4	F8/G2 V	389,1594,6006
		G 116 - 9		9 06 20	+39 06.0	1	14.34		0.86		0.35		2		1658
78612	−10 02754			9 06 26	−10 33.2	1	7.15		0.61				4	G0	2012
78643	−25 06905	IDS09043S2526	AB	9 06 26	−25 38.2	5	6.76	.010	0.58	.014	0.04	.013	16	G0/1 V	116,258,1020,1075,2012
78515	+22 02061	HR 3627	⋆ AB	9 06 29	+22 14.9	5	5.15	.015	0.96	.011	0.80	.013	8	K0 III	71,1080,1363,3016,8041
	+33 01814	G 47 - 26		9 06 29	+33 01.8	3	9.96	.029	1.39	.015	1.21	.022	3	M2	333,679,801,1620
77291	+84 00199			9 06 30	+83 50.6	1	8.58		0.01		-0.02		6	B9	1219
		3C 216 # 2		9 06 31	+43 06.6	1	14.02		0.91		0.60		4		327
		3C 216 # 1		9 06 31	+43 08.1	1	11.53		0.46		-0.11		5		327
78676	−26 06766	HR 3637		9 06 33	−26 33.9	2	6.14	.005	0.17	.000			7	A4 IV	15,2012
78364	+59 01223			9 06 35	+59 27.5	1	8.34		0.44				5	F5	1379
78785	−45 04889	LSS 1247		9 06 38	−46 03.0	4	8.61	.010	0.51	.004	-0.41	.022	10	B2 II	16,362,1075,1747
		G 253 - 21		9 06 39	+84 23.2	1	12.16		1.47		1.14		3		7010
78801	−50 03870	BG Vel		9 06 39	−51 14.0	2	7.55	.025	1.12	.005	0.86		2	F7/8 II	389,1484
78574	+16 01911			9 06 41	+16 01.1	1	8.88		1.10				2	K0	882
	−46 04945			9 06 44	−46 44.9	1	10.58		0.57		-0.28		3		362
		G 114 - 38		9 06 46	−07 11.3	1	11.96		0.57		-0.10		2		1696
78670	−16 02694			9 06 46	−16 33.3	1	7.98		0.55		0.11		3	G0/1 V	803
78702	−17 02765	HR 3638		9 06 46	−18 07.5	4	5.72	.004	0.00	.000	0.04	.021	18	A2 Vn	3,15,1075,2012
	−34 05622			9 06 46	−34 38.4	1	10.28		1.30		1.10		2		3072
		Ton 1020		9 06 47	+34 48.	1	14.70		-0.03		0.06		1		286
78668	−11 02565	HR 3636		9 06 48	−12 09.2	1	5.77		0.94				4	G6 III	2031
78362	+64 00723	HR 3624	⋆ A	9 06 49	+63 43.1	5	4.67	.005	0.35	.003	0.15	.005	21	Am	15,1008,1118,3026,8015
78753	−29 07177			9 06 52	−30 11.7	1	9.45		0.14		0.12		3	A2 IV/V	803
78737	−26 06774			9 06 53	−26 49.6	2	8.93	.002	0.41	.000	-0.09		5	F3 V	1594,6006
298298	−52 01968	LSS 1248		9 06 53	−52 22.5	4	9.05	.039	0.34	.022	-0.63	.033	11	A0 Ib	362,540,976,1737
78661	+12 01979	HR 3635		9 07 03	+11 46.2	2	6.49	.010	0.34	.007	-0.09	.002	4	F2 Vp	1733,3037
	−45 04896			9 07 03	−46 13.9	1	10.74		0.42		-0.24		3		362
79041	−68 00896			9 07 03	−68 33.1	2	7.54	.005	-0.08	.000	-0.49	.005	6	B6 V	164,401
	−37 05499	L 460 - 73		9 07 04	−37 49.5	1	11.08		1.09		0.88		2		3072
78909	−53 02158			9 07 05	−53 55.0	1	9.27		0.02		-0.50		2	B3/5 IV/V	401
78732	−8 02593	HR 3641		9 07 09	−08 35.0	3	5.45	.005	1.01	.005	0.78	.005	14	G8 II	15,1417,2012
78931	−53 02162			9 07 10	−54 10.0	1	7.42		-0.11		-0.59		2	B3 IV/V	401

HD	DM	Other Id	N Rem	α_{1950}	δ_{1950}	S	V	σ_V	B–V	σ_{B-V}	U–B	σ_{U-B}	n	Spectrum	References
−45 04899	LSS 1249		9 07 12	−46 05.9	1	10.65		0.80		−0.29		3		362
	LP 487 - 5		9 07 14	+10 26.8	1	12.58		0.59		−0.11		2		1696
78727 +5 02131			9 07 16	+05 24.5	1	8.38		1.01		0.91		1	K0	1746
	G 46 - 23		9 07 20	+09 13.9	1	15.37		1.74				1		906
	CCS 1394		9 07 22	−45 05.5	1	11.27		3.14				1		864
	G 115 - 58		9 07 27	+46 35.0	2	12.10	.020	0.45	.010	−0.17	.015	2		1658,3029
78715 +22 02063	HR 3640	⋆ AB	9 07 29	+22 12.0	1	6.01		0.90				2	G8 III	71
	XX Hya		9 07 29	−15 23.8	2	11.27	.010	0.13	.005	0.15	.005	2		597,700
78927 −42 05017	LSS 1250		9 07 35	−43 06.3	3	8.28	.007	0.42	.004	−0.52	.036	15	B1 II/III	362,540,976
78773 +14 02040			9 07 37	+13 51.4	1	8.89		1.11				1	K0	882
78712 +31 01946	HR 3639, RS Cnc		9 07 38	+31 10.1	2	5.90	.050	1.62	.045	1.05	.020	6	M6 IIIase	3016,8032
	GD 100		9 07 40	−02 42.0	1	16.14		−0.02		−0.73		1		3060
78986 −47 04696			9 07 45	−48 01.1	1	8.14		1.01		0.80		2	G8 III	389
−41 04851	IDS09059S4143	A	9 07 47	−41 54.7	1	10.52		1.10		1.03		2		3062
−41 04851	IDS09059S4143	B	9 07 47	−41 54.7	1	11.76		1.32		1.25		2		3062
78958 −43 04989	LSS 1251		9 07 47	−43 41.0	5	8.97	.014	0.68	.009	−0.30	.031	23	B0/1 Iab	362,540,976,2021,8100
78922 −29 07194	HR 3644	⋆ A	9 07 49	−30 09.6	3	5.59	.009	0.18	.014	0.16		10	A4 IV-V	15,2012,8071
78823 +14 02041			9 07 57	+14 27.8	1	9.02		0.93				2	G5	882
79024 −48 04469			9 07 58	−48 22.5	1	8.74		0.47		0.06		1	F5 V	389
−43 04994	LSS 1252		9 08 03	−43 30.5	3	10.00	.010	0.51	.007	−0.35	.045	8	B1.5V	362,540,976
78849 +19 02171			9 08 04	+19 05.6	1	8.08		1.46		1.73		2	K2	1648
79025 −48 04471	HR 3647		9 08 04	−49 13.2	1	6.47	.005	0.17	.000			7	A3/5 III(m)	15,2012
79039 −46 04968			9 08 08	−47 16.3	4	6.83	.007	−0.13	.009	−0.54	.011	21	B4 V	278,1075,1770,2011
78955 −22 02512	HR 3646		9 08 09	−22 58.3	2	6.52	.005	0.00	.000			7	A0 V	15,2030
+60 01175			9 08 13	+60 27.3	1	8.98		1.01		0.73		4	K0	1371
	G 114 - 42		9 08 14	−03 35.7	2	12.76	.005	0.56	.005	−0.11	.017	3		1658,1696
78985 −25 06930			9 08 17	−26 04.9	3	7.48	.007	−0.12	.007	−0.68	.010	11	B3/4 V	55,158,1775
	WLS 900-20 # 8		9 08 20	−19 40.2	1	11.46		1.26		1.14		2		1375
79071 −47 04706			9 08 22	−48 00.1	1	7.89		0.88		0.58		1	G6 III	389
79091 −51 03553			9 08 22	−51 52.8	2	6.22	.009	1.02	.005	0.82	.005	4	K1 III	16,389
79072 −48 04479			9 08 25	−49 05.1	1	7.03		−0.11		−0.52		2	B3 III	401
78803 +45 01688			9 08 28	+45 01.8	1	7.23		1.04		0.86		3	K0 III	1501
78792 +51 01488			9 08 28	+50 38.7	1	6.71		0.00		0.02		3	A0 V	1501
78998 −17 02772			9 08 29	−17 36.1	2	8.00	.004	0.18	.004	0.15	.002	14	A4 V	861,1375
79060 −36 05423			9 08 33	−36 38.4	1	7.50		0.92				4	G8/K0 III	2012
79141 −54 02071			9 08 33	−54 29.6	1	9.05		−0.05		−0.40		2	B6 IV	401
79140 −52 02008			9 08 37	−52 26.7	1	8.05		0.87		0.51		1	G6/8 III	16,389
	G 46 - 26		9 08 38	+01 39.8	1	14.09		1.49		1.10		1		1773
78952 +16 01920			9 08 39	+15 47.0	1	9.55		1.03				1	K0	882
79088 −41 04866	IDS09068S4157	AB	9 08 41	−42 09.1	1	9.04		0.45		0.04		1	F5 V	389
78968 +13 02051			9 08 42	+13 05.8	1	8.45		1.20				2	K2	882
−47 04710	LSS 1253		9 08 42	−47 28.4	3	10.44	.008	1.10	.015	0.06	.024	6	B0.5 Vnn	362,1737,1747
297392 −49 04182			9 08 54	−49 34.0	1	10.64		0.83		0.47		1		1696
79171 −52 02010			9 08 54	−52 25.0	1	9.26		0.48		−0.04		2	A5/7 IV	16
79011 +0 02477	IDS09064N0042	AB	9 08 55	+00 29.8	2	6.78	.008	1.32	.008	1.33	.012	10	K0	14,379
	LP 727 - 67		9 09 04	−13 57.5	1	10.86		0.59		−0.03		2		1696
79170 −44 05200	IDS09073S4454	A	9 09 04	−45 06.2	2	9.77	.010	0.79	.020	0.40		6	K0 V	912,1279
79170 −44 05200	IDS09073S4454	AB	9 09 04	−45 06.2	3	9.27	.011	0.81	.005	0.47		9	K0 V + G	158,1705,2012
79170 −44 05200	IDS09073S4454	B	9 09 04	−45 06.2	1	10.30		0.91		0.65		4		1279
	G 47 - 28		9 09 06	+28 07.0	2	12.33	.010	1.57	.004	1.20	.015	5		1620,1723
78865 +57 01208			9 09 07	+57 08.9	1	7.02		1.11		1.09		3	K1 III-IV	37
−44 05201	LSS 1254		9 09 07	−44 50.2	1	11.04		0.36		−0.55		3		362
79206 −52 02017			9 09 08	−52 50.4	1	7.73		−0.09		−0.60		2	B3/4 (V)ne	401
79009 +18 02138	IDS09063N1827	A	9 09 09	+18 15.0	1	6.89		0.06				9	A0	1351
	G 115 - 62		9 09 10	+37 02.8	2	13.65	.095	0.73	.010	0.17	.115	2		1658,3062
78633 +72 00444	HR 3633		9 09 14	+71 51.8	2	6.53	.024	0.96	.005	0.74	.010	6	G8 III-IV	985,1733
79528 −51 03565			9 09 14	−52 12.2	1	9.38		0.00		−0.35		3	B8 III	401
302105 −57 01908			9 09 14	−57 19.5	3	9.56	.021	−0.07	.017	−0.84	.009	6	B2 Ib	401,540,976
79186 −44 05206	HR 3654, GX Vel		9 09 15	−44 39.8	17	4.99	.014	0.22	.011	−0.57	.019	96	B5 Ia	14,15,182,278,362,388*
79066 +6 02120	HR 3649		9 09 18	+05 40.5	3	6.35	.005	0.33	.004	−0.02	.011	9	A9 IV δ Del	15,1078,3016
79197 −43 05015			9 09 20	−43 50.8	1	8.52		0.46		−0.03		1	F5 V	389
+15 02001			9 09 23	+15 09.2	1	10.61		0.47		0.90		1	F0	1748
	LP 607 - 14		9 09 23	−03 31.5	1	12.66		1.40		1.24		1		1696
79097 −6 02839			9 09 23	−06 46.4	4	7.60	.007	1.63	.002	1.92	.017	34	M1	989,1509,1729,2012
79243 −51 03570			9 09 23	−51 41.0	1	9.39		−0.02		−0.15		2	B9/A0 IV/V	16
	Steph 754		9 09 25	+13 31.5	1	10.74		1.51		1.83		1	K7	1746
	vdB 29 # c		9 09 30	−45 21.2	1	12.42		0.62		0.19		3		434
	vdB 29 # b		9 09 32	−45 23.4	1	11.97		0.45		−0.04		3		434
79278 −51 03575			9 09 33	−51 54.4	2	8.33	.020	−0.13	.010	−0.66	.005	4	B3 IV/V	16,401
79096 +15 02003	HR 3650	⋆ AP	9 09 34	+15 11.9	11	6.50	.013	0.73	.006	0.27	.013	84	G9 V	15,22,101,130,1067,1080*
79108 +4 02139	HR 3651		9 09 36	+04 04.4	2	6.13	.005	−0.01	.000	0.01	.000	7	A0 V	15,1415
79351 −58 01419	HR 3659, V357 Car		9 09 39	−58 45.7	8	3.43	.010	−0.20	.005	−0.70	.009	25	B2 IV/V	15,26,243,1020,1034*
79181 −19 02644	HR 3653		9 09 41	−19 32.5	3	5.71	.011	0.98	.000			11	G8 III	15,2013,2028
299896 −54 02079			9 09 42	−54 49.6	1	9.55		1.01		0.84		2	K3 V	3072
79241 −38 05358	HR 3656		9 09 44	−39 03.2	2	5.99	.005	−0.12	.005			7	B5 III	15,2012
−54 02080	LSS 1257		9 09 44	−54 20.3	1	10.92		0.40		−0.65		2		362
79275 −46 04987	HR 3658		9 09 47	−46 22.7	13	5.78	.009	−0.22	.007	−0.84	.014	74	B2 III/IV	14,15,16,278,362,388*
	PASP81,804 T3# 4		9 09 47	−51 54.5	1	13.20		2.05				1		16
	PASP81,804 T3# 3		9 09 47	−51 56.4	1	11.67		0.70		0.19		2	G8:	16
79368 −57 01912			9 09 47	−58 07.6	1	8.39		−0.09		−0.59		2	B4 V	401

Table 1 501

HD	DM	Other Id	N	Rem	α_{1950}	δ_{1950}	S	V	σ_V	B–V	σ_{B-V}	U–B	σ_{U-B}	n	Spectrum	References
		LP 427 - 13			9 09 50	+19 57.6	1	13.90		1.23		1.06		1		1696
		NGC 2808 sq1 32			9 09 50	−64 48.2	1	14.05		0.70		0.17		2		722
79126	+21 01988				9 09 52	+20 52.8	1	8.44		0.56		0.09		2	G0	1648
79179	−6 02844	IDS09075S0631		AB	9 09 57	−06 43.5	1	7.51		1.02		0.77		4	K0	1657
		G 235 - 16			9 09 58	+60 48.2	1	15.90		1.60				1		906
79193	−6 02845	HR 3655, KM Hya			9 09 58	−06 54.2	4	6.10	.004	0.22	.005	0.11	.007	12	A3 m	15,355,1415,8071
	−41 04886				9 09 58	−42 05.6	1	10.47		1.08				4		1075
79145	+17 02034				9 10 00	+16 50.2	1	8.17		0.92				2	K0	882
		G 114 - 46			9 10 00	−06 58.1	1	12.37		1.34				1		3062
	−45 04931	vdB 29 # a			9 10 00	−45 26.5	1	10.96		0.43		0.15		3	B5 V:	434
79367	−53 02203				9 10 01	−53 54.2	1	8.77		−0.04		−0.60		2	B3 III	401
		NGC 2808 sq1 20			9 10 02	−64 48.1	1	12.70		0.60		0.14		2		722
79332	−44 05228				9 10 04	−45 01.7	1	7.97		−0.09		−0.44		2	B5 V	401
79350	−50 03914				9 10 04	−50 42.7	1	7.76		1.08		0.94		1	K0 III	389
		PASP81,804 T3# 5			9 10 04	−51 55.4	1	12.73		0.20		0.12		1	G7 V	16
		NGC 2808 sq1 16			9 10 05	−64 44.5	1	12.02		0.33		0.28		1		722
79331	−43 05032				9 10 06	−43 59.0	1	8.93		0.44		−0.03		1	F3 V	389
79387	−53 02205				9 10 06	−53 49.9	1	6.72		−0.05		−0.19		2	B9 V	401
		WLS 920 35 # 6			9 10 08	+35 17.9	1	12.11		0.67		0.18		2		1375
79420	−57 01913				9 10 09	−57 24.8	1	7.55		−0.08		−0.53		2	B4 III	401
79447	−61 01201	HR 3663			9 10 09	−62 06.7	7	3.96	.014	−0.19	.006	−0.67	.013	23	B4 V	15,26,243,1034,1075*
79421	−57 01914	IDS09088S5733		AB	9 10 11	−57 45.7	2	6.59	.015	−0.14	.005	−0.64	.005	6	B3 V	164,401
		NGC 2808 sq1 12			9 10 13	−64 36.6	2	11.52	.010	1.16	.015	0.89		11		438,722
298285	−49 04197				9 10 14	−50 16.8	2	10.54	.029	0.28	.012	−0.41	.005	3	B2.5IV	540,976
		NGC 2808 sq1 17			9 10 18	−64 38.3	1	12.35		1.41				4		438
79491	−64 01009				9 10 19	−64 24.5	3	9.11	.005	1.21	.023	0.91	.000	12	G8/K0 III	89,438,722
79028	+62 01058	HR 3648		⋆ A	9 10 25	+61 37.9	3	5.17	.042	0.57	.017	0.08	.000	5	F9 V	254,1363,3026
79403	−45 04935				9 10 26	−45 38.5	4	6.66	.004	0.43	.002	−0.01	.013	11	F3/5 V	278,1075,1770,2011
79475	−58 01422				9 10 28	−58 28.9	1	8.62		0.02		−0.39		2	B5/7 III	401
	−51 03593				9 10 30	−51 55.8	1	11.60		0.63		0.08		2	G0	16
79446	−53 02209	IDS09090S5333		A	9 10 31	−53 45.4	1	7.73		−0.08		−0.54		2	B5 II/III	401
79158	+43 01893	HR 3652			9 10 32	+43 25.5	3	5.32	.000	−0.14	.000	−0.48	.019	5	B8 IIIp Mn	985,1079,3023
		AJ82,474 # 107			9 10 32	−49 54.6	1	12.03		1.00		−0.06		2		779
		DD Vel			9 10 32	−50 10.2	1	12.18		1.48		1.07		1		689
79476	−58 01423				9 10 32	−58 35.3	1	8.17		−0.05		−0.44		2	B7/8 V	401
		NGC 2808 sq1 30			9 10 32	−64 39.0	2	13.65	.065	0.95	.000			7		9,438
		NGC 2808 sq1 6			9 10 35	−64 38.7	3	10.45	.014	0.16	.006	0.05	.038	19		9,438,722
	−78 00417				9 10 35	−78 35.7	1	11.42		0.23		0.16		2		45
		AJ82,474 # 108			9 10 36	−49 39.4	1	12.37		0.81		−0.12		2		779
79416	−43 05041	HR 3661		⋆ AB	9 10 39	−43 24.4	13	5.57	.007	−0.12	.004	−0.48	.006	70	B8 V(p Si)	14,15,125,182,278,362*
78584	+79 00300				9 10 42	+79 29.4	1	8.18		−0.17		−0.63		3	B3	399
		LP 845 - 32			9 10 42	−24 51.7	1	13.24		1.54		1.05		4		3062
79629	−68 00904				9 10 45	−69 02.9	2	7.15	.005	0.38	.005	−0.01	.005	6	F3/5 V	158,401
80258	−81 00302				9 10 45	−82 07.2	1	7.67		0.53		−0.02		6	F7 V	1628
79248	+21 01991	HR 3657			9 10 46	+21 29.4	1			0.02		0.06		4	A2 V	1022
		NGC 2808 sq1 29			9 10 46	−64 48.9	3	13.58	.009	0.65	.012	0.15	.000	11		89,438,722
79235	+25 02062				9 10 47	+25 27.7	1	8.66		0.29		0.08		2	A5	105
		L 605 - 3			9 10 48	−24 53.	1	12.75		0.56		0.03		4	M0	3062
		NGC 2808 sq1 9			9 10 49	−64 45.7	3	11.07	.004	0.41	.000	0.15	.000	13		89,438,722
78935	+73 00452	HR 3645			9 10 54	+73 09.3	2	5.97	.005	0.17	.005	0.13		6	F0 III	252,1258
		NGC 2808 sq1 15			9 10 54	−64 55.2	3	11.90	.008	0.44	.011	0.18	.000	13		89,438,722
79606	−64 01011				9 10 54	−64 59.2	3	8.16	.004	0.00	.004	−0.39	.000	12	Ap Si	89,438,722
79513	−53 02217				9 10 56	−53 29.6	1	9.23		−0.01		−0.57		2	B6/7 V	401
79210	+53 01320	G 195 - 17		⋆ A	9 10 59	+52 54.1	4	7.64	.033	1.41	.017	1.21	.008	8	M0 V	22,497,1003,3078
79210	+53 01320	IDS09078N5307		AB	9 10 59	+52 54.1	3	6.95	.013	1.43	.014	1.27	.000	10	M0 V	1013,1197,3003
		NGC 2808 - 5			9 11 00	−64 39.	1	18.18		0.88		0.28		1		722
		NGC 2808 - 11			9 11 00	−64 39.	1	17.17		0.91		0.33		1		722
		NGC 2808 - 34			9 11 00	−64 39.	1	17.69		0.73		0.56		1		722
		NGC 2808 - 35			9 11 00	−64 39.	1	14.14		1.38		1.28		1		722
		NGC 2808 - 36			9 11 00	−64 39.	1	17.27		0.22		−0.30		2		722
		NGC 2808 - 63			9 11 00	−64 39.	2	14.05	.065	1.50	.095	1.25		2		9,722
		NGC 2808 - 65			9 11 00	−64 39.	1	16.11		0.35		0.23		3		722
		NGC 2808 - 88			9 11 00	−64 39.	1	17.84		0.97		0.38		1		722
		NGC 2808 - 106			9 11 00	−64 39.	1	14.60		1.17		0.82		1		9
		NGC 2808 - 111			9 11 00	−64 39.	1	14.55		1.39				1		9
		NGC 2808 - 118			9 11 00	−64 39.	1	17.06		0.17		−0.21		1		722
		NGC 2808 - 120			9 11 00	−64 39.	2	13.63	.080	1.63	.070	1.58		2		9,722
		NGC 2808 - 151			9 11 00	−64 39.	1	17.25		−0.01		0.03		1		722
		NGC 2808 - 153			9 11 00	−64 39.	1	17.03		0.20		−0.21		1		722
		NGC 2808 - 175			9 11 00	−64 39.	1	14.76		0.60		−0.03		1		9
		NGC 2808 - 190			9 11 00	−64 39.	1	13.44		0.43		0.15		1		9
		NGC 2808 - 222			9 11 00	−64 39.	1	16.82		0.16		−0.01		2		722
		NGC 2808 - 226			9 11 00	−64 39.	1	14.80		0.49		0.22		1		9
		NGC 2808 - 230			9 11 00	−64 39.	1	17.76		0.07		−0.34		1		722
		NGC 2808 - 237			9 11 00	−64 39.	1	17.02		0.08		−0.05		2		722
		NGC 2808 - 240			9 11 00	−64 39.	1	16.10		0.84		0.24		1		722
		NGC 2808 - 241			9 11 00	−64 39.	1	17.01		0.62		0.31		1		722
		NGC 2808 - 251			9 11 00	−64 39.	1	13.45		1.19		0.97		1		722
		NGC 2808 - 269			9 11 00	−64 39.	1	14.29		1.37		1.25		1		722
		NGC 2808 - 275			9 11 00	−64 39.	1	13.69		1.66		1.71		1		722

HD	DM	Other Id	N Rem	α_{1950}	δ_{1950}	S	V	σ_V	B–V	σ_{B-V}	U–B	σ_{U-B}	n	Spectrum	References
		NGC 2808 - 281		9 11 00	−64 39.	1	17.72		1.33		0.83		1		722
		NGC 2808 - 282		9 11 00	−64 39.	1	15.24		1.22		0.91		1		722
		NGC 2808 - 283		9 11 00	−64 39.	1	15.10		1.23		0.89		1		722
		NGC 2808 - 310		9 11 00	−64 39.	1	13.36		0.65		0.31		3		9
		NGC 2808 - 311		9 11 00	−64 39.	1	14.43		1.38		1.13		1		722
		NGC 2808 - 315		9 11 00	−64 39.	1	14.75		1.37				2		9
		NGC 2808 - 317		9 11 00	−64 39.	1	15.33		1.22				2		9
		NGC 2808 - 321		9 11 00	−64 39.	1	15.02		1.48				1		9
		NGC 2808 - 332		9 11 00	−64 39.	1	13.82		1.70				3		9
		NGC 2808 - 502		9 11 00	−64 39.	1	14.24		0.66		0.17		2		9
		NGC 2808 - 503		9 11 00	−64 39.	1	14.51		1.00				1		9
		NGC 2808 - 504		9 11 00	−64 39.	1	14.11		1.78				1		9
		NGC 2808 - 505		9 11 00	−64 39.	1	14.87		1.29				1		9
		NGC 2808 - 506		9 11 00	−64 39.	1	14.66		1.03				1		9
		NGC 2808 - 507		9 11 00	−64 39.	1	13.84		1.12		0.59		1		9
		NGC 2808 - 508		9 11 00	−64 39.	1	15.15		0.81				1		9
		NGC 2808 - 509		9 11 00	−64 39.	1	14.50		1.44				1		9
		NGC 2808 - 510		9 11 00	−64 39.	1	14.46		1.17		0.69		2		9
		NGC 2808 - 511		9 11 00	−64 39.	1	13.55		1.75				2		9
		NGC 2808 - 512		9 11 00	−64 39.	1	11.82		1.01		0.71		2		9
		NGC 2808 - 513		9 11 00	−64 39.	1	13.27		1.29		0.77		1		9
		NGC 2808 - 514		9 11 00	−64 39.	1	13.18		1.50		0.95		2		9
		NGC 2808 - 515		9 11 00	−64 39.	1	13.35		1.76				2		9
		NGC 2808 - 516		9 11 00	−64 39.	1	13.28		1.30		0.80		2		9
		NGC 2808 - 517		9 11 00	−64 39.	1	13.54		0.72		0.27		3		9
		NGC 2808 - 518		9 11 00	−64 39.	1	11.61		1.13		0.91		3		9
		NGC 2808 - 519		9 11 00	−64 39.	1	13.31		1.59				1		9
		NGC 2808 - 520		9 11 00	−64 39.	1	12.90		1.58		1.13		1		9
		NGC 2808 - 526		9 11 00	−64 39.	1	14.55		1.23				1		9
		NGC 2808 - 527		9 11 00	−64 39.	1	13.74		1.13		1.09		1		9
		NGC 2808 - 528		9 11 00	−64 39.	1	14.92		1.38				2		9
		NGC 2808 - 531		9 11 00	−64 39.	1	15.22		1.16		0.92		3		9
		NGC 2808 - 533		9 11 00	−64 39.	1	13.22		0.68		0.22		3		9
		NGC 2808 - 534		9 11 00	−64 39.	1	14.14		1.45				2		9
		NGC 2808 - 535		9 11 00	−64 39.	1	14.16		1.52				1		9
		NGC 2808 - 536		9 11 00	−64 39.	1	14.57		1.35		1.05		2		9
		NGC 2808 - 537		9 11 00	−64 39.	1	14.58		1.44				1		9
		NGC 2808 - 538		9 11 00	−64 39.	1	14.09		1.04		0.58		2		9
		NGC 2808 - 542		9 11 00	−64 39.	1	14.31		1.15		0.85		2		9
		NGC 2808 - 543		9 11 00	−64 39.	1	14.30		1.15		0.69		2		9
		NGC 2808 - 545		9 11 00	−64 39.	1	14.85		1.23				1		9
		NGC 2808 - 546		9 11 00	−64 39.	1	13.50		0.78		0.24		3		9
		NGC 2808 - 547		9 11 00	−64 39.	1	14.76		0.59		0.11		2		9
		NGC 2808 - 548		9 11 00	−64 39.	1	13.78		1.46		1.05		2		9
		NGC 2808 - 549		9 11 00	−64 39.	1	14.15		1.23				2		9
79211	+53 01321	G 195 - 18	⋆ B	9 11 01	+52 54.2	4	7.70 .030		1.42 .018		1.24 .041		8	K2	22,497,1003,3078
		NGC 2808 sq1 11		9 11 03	−64 43.8	3	11.42 .004		0.62 .000		0.15 .009		18		89,438,722
79319	+14 02048			9 11 05	+14 25.1	1	8.50		1.69				1	R4	1238
	−3 02616	G 114 - 48		9 11 11	−03 41.3	2	10.64 .010		0.66 .007		0.15 .005		3		1658,3062
79655	−62 01192			9 11 16	−62 42.6	1	8.25		-0.01		-0.37		2	B8 V	401
	+73 00453			9 11 19	+73 25.6	1	8.80		1.05		0.80		2	G5	1502
		G 161 - 4		9 11 19	−10 19.6	1	12.03		1.27				1		3062
	−29 07241	LP 901 - 39		9 11 19	−29 52.9	1	10.85		0.92		0.66		1		1696
		SZ Hya		9 11 24	−09 06.7	1	10.44		0.07		0.04		1	A7	668
79524	−41 04904	HR 3668		9 11 25	−42 04.0	4	6.29 .008		1.25 .005				18	K2 III	15,1075,2013,2029
		AJ82,474 # 110		9 11 26	−49 57.8	1	13.17		0.76		-0.21		2		779
79523	−38 05376	HR 3667		9 11 28	−38 24.5	3	6.30 .004		0.00 .000		0.01 .000		11	A0 V	15,1075,2012
		NGC 2808 sq1 19		9 11 28	−64 51.5	1	12.70		0.59		0.17		2		722
	−64 01013	NGC 2808 sq1 7		9 11 28	−64 54.9	2	10.46 .005		0.48 .000		0.02		5		438,722
79246	+56 01376			9 11 29	+55 39.0	1	8.91		0.53		0.03		2	G0	1502
79573	−49 04218	LSS 1259		9 11 31	−49 54.0	1	10.69		1.11		0.50		2	WC	779
		PASP81,804 T3# 8		9 11 32	−51 58.1	1	11.73		1.19		0.50		1		16
309635	−64 01014	NGC 2808 sq1 5		9 11 32	−64 21.9	3	10.30 .004		0.35 .007		0.14 .000		12		89,438,722
79373	+25 02065			9 11 33	+25 13.2	1	6.93		1.26		1.42		3	K3 III	1501
	+34 01965			9 11 33	+33 46.1	1	9.64		0.89		0.48		2	K0	1375
	−80 00328	IDS09136S8045	AB	9 11 33	−80 57.4	2	10.08 .010		0.59 .002		-0.16		6		2033,3073
79625	−52 02074			9 11 35	−53 17.9	1	7.25		-0.06		-0.15		2	B9 V	401
		G 114 - 50		9 11 36	+00 27.8	1	9.70		0.88		0.63		1	K3	3062
79604	−51 03611			9 11 36	−51 47.4	1	9.03		1.44		1.29		2	G8/K0 (III)	16
298332	−52 02073			9 11 37	−52 34.0	3	10.58 .026		0.03 .018		-0.73 .007		6	B0.5V	362,540,976
79698	−58 01432	HR 3673		9 11 37	−59 12.4	3	5.54 .004		0.85 .005				14	G6 II	15,1075,2029
79699	−60 01353	IDS09104S6030	AB	9 11 37	−60 42.6	1	6.36		-0.05		-0.22		2	B9 V	401
		NGC 2808 sq1 24		9 11 37	−64 49.8	1	13.06		0.70		0.21		2		722
79624	−51 03614			9 11 41	−51 58.5	2	8.20 .015		-0.10 .000		-0.47 .020		4	B8 III	16,401
		PASP81,804 T3# 7		9 11 43	−51 54.2	1	11.52		0.60		0.27		1	A9 V	16
	−51 03615			9 11 44	−51 52.6	1	9.51		1.85		2.13		2	M0: I	16
79469	+2 02167	HR 3665	⋆ A	9 11 46	+02 31.6	15	3.89 .009		-0.06 .009		-0.11 .017		70	B9.5V	3,15,30,1006,1020*
79670	−53 02227			9 11 46	−53 46.6	1	7.99		-0.10		-0.55		2	B5 IV	401
		PASP81,804 T3# 10		9 11 48	−52 00.6	1	12.12		1.08		0.74		2		16
79392	+39 02223			9 11 49	+38 49.0	2	6.75 .015		0.36 .005		-0.06 .005		5	F0 V	105,1501

Table 1 503

HD	DM	Other Id	N Rem	α_{1950}	δ_{1950}	S	V	σ_V	B–V	σ_{B-V}	U–B	σ_{U-B}	n	Spectrum	References
	−27 06378			9 11 49	−27 18.6	1	9.71		1.04		0.91		1	K5 V	3072
79621	−46 05010	HR 3670		9 11 49	−47 07.9	9	5.91	.004	−0.05	.007	−0.10	.006	55	B9 Ve	14,15,182,278,388,1075*
79622	−48 04525			9 11 49	−48 53.2	1	6.61		1.52		1.86		1	K5 III	389
		NGC 2808 sq1 18		9 11 49	−64 51.7	1	12.50		0.42		0.13		2		722
79371	+44 01847	G 115 - 64		9 11 51	+44 15.7	1	9.00		0.68		0.18		2	G0	3016
		NGC 2808 sq1 10		9 11 51	−64 44.2	2	11.23	.005	0.95	.005	0.78		7		438,722
79601	−41 04907			9 11 52	−42 06.1	4	8.02	.009	0.57	.000	−0.02	.010	17	G2 V	389,977,1075,2011
		NGC 2808 sq1 13		9 11 52	−64 40.0	2	11.74	.005	1.17	.005	0.99		7		438,722
		NGC 2808 sq1 31		9 11 52	−64 45.0	1	13.59		1.85				3		438
79535	−21 02733			9 11 53	−21 34.3	1	8.87		1.00		0.60		2	G8 IV	1375
	+77 00361	G 252 - 30	⋆ AB	9 11 57	+77 27.2	2	10.07	.026	1.37	.028	1.23		3	K5	1017,3078
		NGC 2808 sq1 14		9 12 00	−64 40.8	2	11.74	.005	1.44	.009	1.56		13		438,722
79424	+37 01956			9 12 02	+37 16.1	1	7.86		1.31		1.35		2	K0	1375
79697	−49 04224			9 12 04	−50 01.0	1	8.74		0.45		0.03		1	F3/5 IV/V	389
79354	+57 01211	HR 3660		9 12 08	+56 57.0	3	5.27	.005	1.56	.003	1.86	.011	7	K5 III	1080,1355,3016
79452	+35 01966	HR 3664	⋆ A	9 12 10	+34 50.5	4	6.00	.018	0.84	.012	0.36	.011	11	G6 III	15,1003,1080,4001
79499	+16 01930			9 12 11	+16 12.8	2	8.32	.003	0.45	.000			50	F5	130,1351
79668	−41 04915			9 12 15	−42 15.6	1	8.03		0.41				4	F3 V	2012
79555	+5 02143	IDS09097N0451	AB	9 12 17	+04 39.0	2	7.97	.000	1.00	.020	0.83	.025	5	K0	196,3072
		PASP81,804 T3# 11		9 12 17	−51 59.7	1	12.10		1.22				1		16
79694	−43 05068	HR 3672		9 12 18	−43 56.3	13	5.85	.010	−0.13	.009	−0.50	.008	66	B5 V	14,15,158,182,278,362*
		AAS27,343 # 162		9 12 18	−55 21.	1	12.11		0.53		−0.04		2		362
79566	−0 02158			9 12 19	−01 22.7	1	7.00		0.10		0.14		7	A2	1628
79777	−54 02108			9 12 19	−54 31.3	1	9.30		0.03		−0.18		2	B9 V	401
		NGC 2808 sq1 25		9 12 19	−64 49.2	1	13.15		1.22		1.10		2		722
	−51 03626			9 12 20	−51 59.6	1	10.98		0.66		0.18		2	G0 V:	16
		NGC 2808 sq1 27		9 12 20	−64 50.4	1	13.48		0.57		0.08		2		722
79736	−44 05269			9 12 22	−44 51.6	4	8.51	.004	0.48	.004	−0.02	.008	17	F6/7 V	389,977,1075,2011
79778	−55 02028	LSS 1260		9 12 22	−55 19.9	4	8.21	.023	0.13	.017	−0.75	.025	11	B2 Vne	362,401,540,976
79737	−48 04534			9 12 23	−49 16.4	1	7.51		1.30		1.02		1	G2 Ib	389
		G 195 - 19		9 12 27	+53 38.6	3	13.86	.051	0.33	.020	−0.56	.089	6	DC	308,1705,3078
79554	+15 02009	HR 3669		9 12 28	+15 09.0	2	5.34	.010	1.32	.000	1.37	.036	9	K1 III	3,252
	−51 03631			9 12 29	−51 49.6	1	11.76		0.36		0.04		2	A7:	16
		NGC 2808 sq1 4		9 12 30	−64 41.9	3	9.95	.000	1.69	.004	2.06	.000	12		89,438,722
		NGC 2808 sq1 21		9 12 31	−64 49.6	1	12.73		0.89		0.60		2		722
79735	−42 05086	HR 3674	⋆ AB	9 12 32	−43 01.2	10	5.24	.005	−0.14	.009	−0.56	.015	37	B4 V	14,15,278,362,362,388*
298322	−51 03634			9 12 32	−51 58.9	1	10.48		0.10		−0.15		2	B8	16
	+30 01835			9 12 33	+30 00.0	1	10.05		1.50		1.81		5		1723
		NGC 2808 sq1 8		9 12 33	−64 41.8	3	10.54	.005	1.55	.004	1.77	.000	12		89,438,722
79439	+54 01285	HR 3662, DD UMa		9 12 36	+54 13.8	6	4.83	.012	0.19	.008	0.09	.005	21	A5 V	15,1007,1013,1363*
79864	−60 01366			9 12 39	−61 04.3	1	6.99		−0.04		−0.40		2	B9 II/III	401
80007	−69 01023	HR 3685		9 12 40	−69 30.7	7	1.67	.013	0.00	.006	0.03	.007	20	A1 III	15,1020,1034,1075*
80194	−76 00574	HR 3695		9 12 45	−76 27.3	3	6.13	.003	1.09	.000	1.00	.005	16	K1 III	15,1075,2038
		NGC 2808 sq1 22		9 12 47	−64 31.9	2	12.77		1.27	.000			4		89,438
79846	−55 02035	HR 3679		9 12 49	−55 21.7	3	5.27	.005	0.99	.005			14	G8 II/III	15,1075,2029
79810	−42 05095			9 12 52	−43 15.4	1	6.82		1.14		1.08		1	K0 III	389
		PASP81,804 T3# 15		9 12 54	−51 54.6	1	11.66		0.68		0.15		2	G0 V	16
79807	−37 05578	HR 3677		9 12 57	−37 23.6	3	5.86	.008	0.83	.008			14	K0 III +A3	15,1075,2029
79950	−61 01212			9 13 00	−61 54.3	1	9.46		0.01		−0.29		2	B9 IV	401
79752	−14 02793	HR 3675		9 13 03	−14 49.0	3	6.35	.016	0.01	.009	0.00		9	A0 V	15,252,2027
		G 47 - 31		9 13 07	+29 32.7	1	12.37		1.43		1.11		1		1620
		NGC 2808 sq1 26		9 13 08	−64 43.0	1	13.23		0.96		0.61		2		722
79884	−49 04243			9 13 11	−50 06.6	1	10.12		0.14		0.16		3	A1 V	402
79680	+20 02293			9 13 14	+20 16.9	1	8.73		0.34		0.00		2		1648
297398	−49 04244			9 13 16	−49 45.1	1	9.92		0.36		0.10		3	F2	402
299927	−53 02246			9 13 16	−54 04.4	1	9.47		0.14		−0.46		2	B5	362
79902	−49 04246			9 13 18	−49 44.3	1	9.91		0.09		0.09		3	A2/3 V	402
79946	−55 02037			9 13 18	−55 25.9	2	7.16	.000	−0.11	.005	−0.57		6	B5 V	401,1075
79726	+14 02057			9 13 23	+14 20.3	1	8.36		0.62		0.14		2	G0	1648
79843	−33 05895			9 13 23	−34 08.9	1	8.32		1.01				4	G8 III	2012
		AJ82,474 # 114		9 13 24	−49 31.7	1	11.17		0.81		−0.30		2	B0 V	779
79983	−56 02060			9 13 24	−57 01.0	1	9.40		0.03		−0.35		2	B8 V	401
80060	−64 01017			9 13 24	−64 50.7	4	7.06	.008	1.07	.005	0.92	.004	18	K0 III	89,438,722,1704
		G 116 - 16		9 13 26	+44 12.5	2	15.35	.024	0.22	.015	−0.55	.019	6	DAs	940,3060
79900	−45 04982	HR 3680	⋆ AB	9 13 26	−45 20.8	12	6.25	.004	−0.07	.010	−0.26	.009	68	B8 V	14,15,182,278,388,863*
		LP 427 - 18		9 13 30	+20 34.7	1	12.11		1.46		1.20		3		1723
		G 114 - 51		9 13 31	−06 44.5	1	14.35		0.98		0.66		2		3062
79982	−52 02126			9 13 33	−53 06.7	1	7.80		−0.04		−0.16		2	B9.5IV	401
	−55 02040			9 13 34	−55 50.6	1	10.33		0.14		−0.65		3		362
	+32 01857			9 13 35	+32 23.2	1	8.82		0.49		0.00		2	F8	1733
		AAS27,343 # 165		9 13 36	−54 41.	1	13.24		1.75		0.44		1		362
79917	−38 05408	HR 3682		9 13 38	−38 21.6	4	4.94	.005	1.11	.005	1.06	.004	20	K1 III	15,1075,2012,8015
79765	+19 02187			9 13 39	+19 01.2	2	7.04	.010	0.29	.002	0.00	.005	4	A3	1733,3026
	−38 05410	LSS 1263		9 13 40	−39 03.0	1	10.10		0.01		−0.73		2	B2 II-III	540
		AJ82,474 # 117		9 13 40	−49 55.6	1	13.23		1.21		0.18		2		779
298310	−50 03959	LSS 1264		9 13 40	−51 10.6	2	10.08	.024	0.91	.016	−0.24	.055	5	O9.5IV	362,540
	−49 04255			9 13 43	−49 54.9	1	10.58		0.56		0.18		3		402
79940	−36 05505	HR 3684	⋆ AB	9 13 45	−37 12.2	6	4.62	.004	0.45	.005	0.10	.009	23	F5 III	15,1075,2012,2024*
		Pis 11 - 16		9 13 45	−49 47.7	1	11.89		0.66		0.12		1		524
79938	−35 05519			9 13 46	−36 12.9	1	9.62		0.03		0.03		3	B9 IV/V	1730

HD	DM	Other Id	Rem	α_{1950}	δ_{1950}	S	V	σ_V	B−V	σ_{B-V}	U−B	σ_{U-B}	n	Spectrum	References
	−36 05504	NGC 2818 - 3001		9 13 46	−36 29.4	3	11.43	.012	0.44	.000	−0.06		5		170,580,1730
298325	−51 03654			9 13 47	−52 17.7	2	10.29	.016	0.06	.010	−0.69	.009	3	B2 III	540,976
79723	+43 01901			9 13 49	+43 17.7	1	8.24		1.12		1.09		2	K0	1601
		NGC 2818 - 3031		9 13 49	−36 20.0	1	12.20		0.60		−0.01		1		170
		Pis 11 - 15		9 13 49	−49 48.0	1	12.02		0.60		0.04		1		524
		Pis 11 - 13		9 13 50	−49 49.8	2	12.09	.049	0.83	.019	−0.25	.005	3		524,779
		Pis 11 - 17		9 13 51	−49 46.0	1	10.08		1.11		1.02		1		524
		Pis 11 - 14		9 13 52	−49 48.1	1	12.14		0.51		0.05		1		524
		G 48 - 1		9 13 53	+07 44.1	1	11.32		0.98		0.74		1		333,1620
		NGC 2818 - 2087		9 13 53	−36 21.5	2	14.12	.005	0.53	.005	0.15		2		170,580
		NGC 2818 - 2068		9 13 53	−36 24.1	2	14.10	.005	0.34	.005	0.21		2		170,580
	−49 04259			9 13 53	−50 03.1	1	11.12		0.43		0.03		4		402
80094	−57 01949	HR 3691		9 13 55	−58 10.8	4	6.01	.008	−0.11	.004	−0.45	.012	11	B6 III	15,26,2012,8023
		NGC 2818 - 1073		9 13 56	−36 23.8	2	13.91	.005	0.40	.040	0.16		2		170,580
	+41 01944			9 13 58	+40 48.3	1	9.12		1.44		1.72		3		1723
80214	−67 01053			9 13 59	−68 04.5	1	8.97		−0.04		−0.33		2	B8 Vn	401
		G 195 - 22		9 14 00	+58 38.7	1	15.11		1.85				6		940
		NGC 2818 - 1061		9 14 01	−36 23.2	2	12.88	.000	0.80	.005	0.32		2		170,580
		L 749 - 34		9 14 02	−18 24.8	1	10.75		1.55				1		1746
		NGC 2818 - 1123		9 14 03	−36 26.8	2	16.89	.023	0.69	.028	0.00		4		170,580
80092	−54 02130			9 14 03	−54 45.4	1	8.69		0.04		−0.15		2	B9/A0 V	401
		Pis 11 - 12		9 14 04	−49 50.2	1	11.57		1.82		1.38		1		524
299937	−54 02131			9 14 04	−54 45.8	1	9.44		0.02		−0.10		1		401
79873	+1 02267	IDS09115N0109	AB	9 14 06	+00 56.3	1	6.71		0.41		0.05		2	F5	1625
79675	+58 01182			9 14 06	+57 41.4	2	7.62	.018	1.12	.022	1.10	.000	5	K0 III-IV	37,1375
		NGC 2818 - 2100		9 14 09	−36 21.2	2	13.86	.015	0.54	.030	0.16		2		170,580
		Pis 11 - 10		9 14 09	−49 47.9	1	13.48		1.09		0.10		2		524
79763	+47 01658	HR 3676		9 14 10	+47 01.6	1	5.97		0.04		0.05		1	A1 V	3050
	+62 01062			9 14 10	+62 18.0	1	9.81		0.30		0.04		2	F0	1375
		NGC 2818 - 1109		9 14 10	−36 24.7	1	15.72		0.72		−0.01		1		170
		Pis 11 - 9		9 14 10	−49 49.1	1	13.31		0.91		−0.13		2		524
80126	−57 01951	HR 3693		9 14 10	−57 22.1	4	6.32	.003	1.04	.005			18	G8 II	15,1075,2013,2029
		NGC 2818 - 1004		9 14 11	−36 22.2	2	12.07	.005	1.14	.015	0.76		2		170,580
		NGC 2818 - 1131		9 14 11	−36 27.6	2	15.59	.050	0.54	.040	0.04		2		170,580
	−49 04263	Pis 11 - 2		9 14 11	−49 48.2	2	11.05	.015	0.99	.010	−0.18	.010	4		524,779
		Pis 11 - 6		9 14 11	−49 50.1	1	12.91		0.95		−0.08		1		524
	−36 05514	NGC 2818 - 1001		9 14 12	−36 21.6	2	11.31	.010	0.70	.010	0.18		2		170,580
		Steph 761		9 14 13	+62 00.6	1	11.16		1.07		0.84		1	M0	1746
79910	−5 02762	HR 3681	⋆ AB	9 14 13	−06 08.6	6	5.23	.007	1.17	.007	1.22	.006	27	K2 III	15,1075,1415,2013*
80077	−49 04264	Pis 11 - 1	⋆	9 14 13	−49 45.9	6	7.58	.024	1.33	.022	0.22	.038	14	B2/3 Ia(e)	402,524,540,779,976,1737
		Pis 11 - 5		9 14 13	−49 47.5	1	12.85		1.00		−0.10		1		524
		Pis 11 - 4		9 14 13	−49 47.9	2	12.72	.080	0.98	.000	−0.08	.000	4		524,779
		Pis 11 - 3		9 14 13	−49 49.2	1	12.15		0.81		0.34		2		524
79931	−8 02623	HR 3683		9 14 14	−08 32.1	7	5.47	.009	−0.08	.012	−0.26	.033	28	B9 III	3,15,1079,1417,2013*
		NGC 2818 - 1148		9 14 14	−36 26.1	2	15.34	.065	0.40	.040	0.12		2		170,580
	−36 05518	NGC 2818 - 3120		9 14 14	−36 31.5	3	10.56	.011	0.03	.014	−0.01	.025	5		170,580,1730
80057	−44 05305	HR 3688, LSS 1265	⋆ AB	9 14 14	−44 41.4	12	6.04	.016	0.29	.005	−0.11	.024	50	A0 Iab	14,15,158,278,362,388*
		NGC 2818 - 1020		9 14 15	−36 24.1	2	15.16	.025	0.82	.015	0.36		2		170,580
	−49 04265	AJ82,474 # 124		9 14 15	−49 32.5	1	10.85		0.92		−0.19		2		779
		NGC 2818 - 1162		9 14 16	−36 25.1	2	16.20	.034	0.96	.005	1.01		3		170,580
		NGC 2818 - 1161		9 14 16	−36 25.3	2	12.71	.010	1.52	.000	1.71		2		170,580
		NGC 2818 - 1151		9 14 16	−36 26.4	2	16.51	.020	0.77	.015	0.19		2		170,580
		NGC 2818 - 1153		9 14 16	−36 27.4	2	17.01	.055	0.72	.020	0.04		2		170,580
80125	−54 02133			9 14 16	−54 18.9	1	8.81		−0.04		−0.50		2	B7 V	401
	−44 05308	LSS 1266		9 14 19	−44 42.6	2	11.17	.030	0.51	.000	−0.35	.005	7		362,1737
		NGC 2818 - 1015		9 14 20	−36 23.8	2	13.35	.040	0.48	.030			2		170,580
		AAS19,45 T5# 310		9 14 20	−50 01.5	1	11.68		0.36		0.16		4		402
		G 161 - 9		9 14 21	−05 11.0	2	13.72	.020	0.68	.000	−0.01	.005	4		1658,3062
		NGC 2818 - 1157		9 14 21	−36 25.6	2	15.92	.025	0.76	.020	0.23		2		170,580
		Pis 11 - 7		9 14 22	−49 46.2	1	12.63		0.61		0.03		2		524
80255		RU Car		9 14 22	−66 01.4	1	9.71		2.47		5.10		2	N3	864
		Pis 11 - 8		9 14 23	−49 46.8	1	13.39		2.09		1.93		1		524
		G 46 - 31		9 14 28	+03 14.2	3	10.85	.008	0.53	.014	−0.13	.004	5		333,927,1620,1696
		Pis 11 - 11		9 14 30	−49 45.8	1	12.68		0.90		0.80		1		524
79850	+39 02226			9 14 32	+39 24.5	1	7.97		0.46		0.03		3	F5 V	833
80108	−43 05103	HR 3692		9 14 32	−44 03.4	7	5.12	.011	1.67	.011	1.89	.028	27	K3 Ib	14,15,182,388,1075*
80157	−55 02049			9 14 32	−55 20.9	2	8.12	.005	0.41	.005			8	F2 V	1075,1075
79954	+13 02066			9 14 38	+13 17.5	1	8.45		1.16				2	K0	882
79929	+28 01729			9 14 39	+27 37.9	1	6.77		0.40		−0.03		3	F6 V	1501
80050	−13 02808	HR 3687		9 14 45	−14 21.8	3	5.83	.028	1.06	.005	0.90		9	K0 III	15,252,2028
80230	−57 01961	HR 3696		9 14 47	−57 19.9	3	4.34	.011	1.63	.005	1.98	.000	14	M1 III	15,1075,2012
79517	+74 00393	HR 3666		9 14 49	+74 13.7	2	6.42	.020	1.05	.005	0.88	.010	4	G8 III	1502,1733
79992	+17 02053			9 14 51	+16 54.9	1	8.38		0.08		0.08		3	A0	1648
79990	+23 02072			9 14 53	+23 17.4	1	6.95		1.14		1.14		2	K0	1625
		AAS19,45 T5# 315		9 14 53	−50 04.6	1	12.75		0.32		0.28		4		402
79969	+29 01883	IDS09120N2900	AB	9 14 56	+28 46.7	5	7.20	.014	1.00	.019	0.83	.023	11	K3 V +K3 V	22,1080,1381,1758,3077
80229	−55 02051			9 14 56	−55 26.2	1	8.22		1.50				4	K4 III	1075
79889	+46 01490	BE Lyn		9 14 58	+46 21.8	1	8.80		0.24		0.07		4	A3	1601
80210	−49 04273			9 14 58	−49 22.1	1	6.98		0.70		0.41		1	K0/1 III +F	389
80170	−38 05430	HR 3694		9 14 59	−39 11.5	4	5.32	.011	1.17	.005			18	K5 III-IV	15,1075,2013,2028

Table 1

HD	DM	Other Id	N Rem	α_{1950}	δ_{1950}	S	V	σ_V	B–V	σ_{B-V}	U–B	σ_{U-B}	n	Spectrum	References
80171	−39 05345			9 14 59	−39 54.4	1	7.09		0.98		0.75		6	K0 III	1673
298353	−51 03670	LSS 1271		9 14 59	−51 43.7	1	10.13		0.63		−0.44		4		362
80325	−62 01209			9 15 03	−62 40.2	2	9.54	.010	−0.11	.004	−0.78	.004	6	B2 III/IV	540,976
80105	−11 02601			9 15 05	−11 45.1	2	7.07	.001	1.03	.004	0.80		6	K0	861,2033
		AAS19,45 T5# 309		9 15 05	−50 01.8	1	11.56		0.78		0.37		4		402
80228	−50 03973			9 15 06	−50 20.3	1	9.96		0.13		0.12		3	A5 V	402
299942	−54 02147			9 15 06	−55 10.7	1	9.59		0.10		−0.63		3	B5	362
80205	−44 05319			9 15 08	−44 48.1	6	6.74	.004	−0.01	.003	−0.01	.003	56	A0 V	278,863,977,1075,1770,2011
80268	−54 02146			9 15 08	−54 26.8	2	7.93		0.01		−0.12		2	B9 V	401
80064	+12 02009	HR 3689	⋆ A	9 15 09	+11 42.7	1	6.41		0.07		0.11		3	A2 IV	1022
		AAS19,45 T5# 312		9 15 15	−50 06.3	1	12.01		0.66		0.22		3		402
		AAS19,45 T5# 313		9 15 19	−50 04.5	1	12.18		1.27		0.90		3		402
	−52 02174			9 15 20	−53 10.3	1	9.51		0.47		−0.22		1		6006
80024	+35 01971	HR 3686	⋆ AB	9 15 21	+35 34.5	2	5.95	.005	0.17	.000	0.14	.005	4	A8 V	1601,1733
80282	−50 03979			9 15 21	−51 13.7	1	7.62		−0.13		−0.48		2	Ap Si	401
	+27 01739			9 15 25	+27 31.2	1	9.55		1.34		1.26		1	K5	1746
80281	−47 04811			9 15 25	−47 31.5	1	7.75		1.29		1.30		1	K1 III	389
80279	−45 05009			9 15 27	−46 03.9	1	9.45		0.44		−0.42		3	B3 II/III	362
		G 235 - 20		9 15 28	+62 16.1	1	11.42		1.59		1.14		2		7010
		AAS19,45 T5# 314		9 15 33	−50 03.8	1	12.38		0.54		0.02		3		402
		AAS27,343 # 125		9 15 36	−45 39.	1	12.01		0.74		−0.08		3		362
		AAS19,45 T5# 311		9 15 42	−50 05.9	1	12.00		0.67		0.06		3		402
80081	+37 01965	HR 3690	⋆ AB	9 15 44	+37 00.9	5	3.81	.008	0.06	.002	0.05	.004	13	A1 V +A4 V	15,1022,1363,3023,8015
80404	−58 01465	HR 3699		9 15 45	−59 03.9	6	2.25	.009	0.18	.007	0.14	.017	18	A9 Ib	15,1020,1034,1075*
80347	−50 03990			9 15 46	−50 21.9	2	7.32	.014	0.98	.000	0.78	.023	4	K0 III	389,402
80436	−61 01224			9 15 46	−62 13.8	1	8.85		−0.07		−0.52		2	B6 III/IV	401
80459	−63 01114			9 15 50	−63 33.9	1	7.39		−0.01		−0.52		2	B6 Vne	401
		G 47 - 33		9 15 51	+26 58.1	1	11.77		1.50		1.23		1	M1	333,1620
		WLS 900 60 # 8		9 15 51	+60 02.6	1	13.99		0.65		0.03		2		1375
		AAS27,343 # 168		9 15 54	−55 11.	1	11.58		0.18		−0.05		2		362
80383	−52 02185	IL Vel	⋆ A	9 15 55	−52 37.7	2	9.14	.002	0.04	.000	−0.66	.008	6	B2 III	540,976
		AJ82,474 # 130		9 16 00	−49 34.6	1	13.38		0.93		−0.01		2		779
80419	−55 02059			9 16 03	−55 43.5	4	7.35	.004	1.09	.005	0.89	.002	10	K0 III	258,688,1075,1075
		RW Cnc		9 16 08	+29 16.7	2	11.42	.176	0.19	.055	0.06	.019	2	A5	668,699
79968	+65 00703	G 235 - 21		9 16 08	+65 13.7	1	7.74		0.74		0.38		4	G5	1355
80380	−45 05025			9 16 10	−45 25.5	2	7.21	.003	−0.15	.001	−0.67	.012	5	B3 IV	401,1732
80435	−53 02281	HR 3700		9 16 10	−54 17.1	5	6.33	.009	1.40	.007			25	K3 III	15,1075,1075,2013,2028
80218	+18 02165			9 16 12	+17 55.1	3	6.64	.006	0.47	.000	−0.06	.000	6	F5	979,3026,8112
		G 115 - 68		9 16 12	+38 43.9	3	14.59	.116	1.71	.053	1.03		3		419,906,3016
		G 115 - 69		9 16 12	+38 44.0	3	14.68	.044	1.68	.015	1.07		3		419,906,3016
		LP 18 - 237		9 16 15	+75 42.0	1	11.16		0.61		0.01		2		1726
		LP 487 - 27		9 16 24	+13 20.5	1	12.56		0.41		−0.25		2		1696
		L 99 - 8		9 16 24	−66 59.	1	13.50		0.41		0.08		2		3060
80433	−44 05348			9 16 25	−44 47.3	1	9.36		0.35				4	A9 V +F/G	2011
80456	−50 04001	HR 3703		9 16 26	−50 50.4	4	5.25	.011	−0.07	.010	−0.39	.004	10	B7/8 III	15,16,401,2012
		G 252 - 33		9 16 29	+73 19.7	1	15.02		1.90				1		906
80454	−45 05031			9 16 31	−45 38.0	1	8.77		−0.04		−0.30		2	B9 V	401
		AAS27,343 # 169		9 16 36	−54 24.	1	12.97		0.91		−0.57		1		362
80671	−68 00918	HR 3712	⋆ AB	9 16 36	−68 28.7	5	5.38	.003	0.41	.010	−0.05	.009	15	F5 V	15,1075,2024,2038,3026
	−51 03690	IDS09150S5112	AB	9 16 38	−51 24.4	1	10.18		0.77		0.39		1		16
	−55 02071		A	9 16 43	−55 48.9	1	11.10		0.10		−0.44		1		362
	−55 02071		AB	9 16 43	−55 48.9	1	10.75		0.30		−0.62		1		362
80484	−44 05355			9 16 44	−44 24.3	6	8.34	.004	0.09	.009	0.06	.017	39	A1 V	116,278,1066,1075*
	−44 05356			9 16 44	−45 03.9	3	9.79	.011	1.91	.002	2.14	.004	23		1628,1673,1704
296833	−45 05037			9 16 44	−45 19.4	1	9.30		0.98		0.76		6	K0	863
80545	−57 01976			9 16 44	−57 34.0	2	9.52	.019	1.19	.002	1.00		6	K/M V	2033,3072
80339	+10 01972			9 16 49	+10 00.1	1	7.17		1.00		0.78		2	K0	1733
		E4 - 106		9 16 52	−45 15.1	1	11.09		0.30		0.05		5		863
	−44 05358			9 16 52	−45 15.1	3	9.77	.004	1.90	.014	2.14	.005	23		863,1075,1460
		ApJ239,112 # 20		9 16 52	−77 00.6	2	14.04	.000	0.96	.000	0.63	.000	4		874,1533
80367	+1 02277			9 16 54	+01 06.6	1	8.17		0.87		0.61		3	K0	196
296832	−44 05359			9 16 54	−44 54.5	1	9.95		0.30		−0.32		4	B5	362
		vdB 30 # a		9 16 54	−48 11.	1	14.16		1.36		0.62		3		434
80327	+25 02083			9 16 56	+24 37.8	1	7.90		0.51		−0.02		2	G0	3026
80233	+53 01330			9 16 59	+52 39.8	2	6.56	.015	0.07	.000	0.06	.000	5	A1 V	1501,1733
80313	+38 02022			9 17 00	+37 58.1	1	7.52		1.05		0.84		2	K0	1601
80576	−56 02096	LSS 1273		9 17 00	−56 57.4	1	9.14		0.50		0.40		3	F0 V	8100
80710	−66 01002	HR 3713		9 17 00	−66 50.4	2	6.10	.005	1.26	.000	1.10	.000	7	K2 III	15,1075
80527	−44 05362			9 17 01	−44 47.8	12	7.19	.011	1.11	.011	0.94	.005	85	K0/1 II/III	116,125,278,395,402*
80515	−43 05152			9 17 02	−43 49.7	3	8.86	.001	0.21	.003	0.16	.002	16	A8 IV	977,1075,2011
80558	−51 03693	HR 3708, LR Vel	⋆	9 17 03	−51 20.9	10	5.87	.015	0.54	.014	−0.34	.041	44	B6 Iae	15,16,362,540,976*
	−51 03694		A	9 17 04	−51 22.6	1	11.44		−0.02		−0.15		1		362
	−51 03694		B	9 17 04	−51 22.6	1	10.15		0.49		−0.38		1		362
80447	−10 02804	HR 3702		9 17 09	−11 06.2	3	6.62	.005	0.08	.005			14	A2 Vs	15,2027,2033
80479	−15 02763	HR 3704	⋆ AB	9 17 11	−15 37.3	4	5.78	.013	1.29	.005	1.39		13	K4 III	15,1075,2031,3005
80425	+5 02158			9 17 14	+05 25.7	1	6.64		0.24		0.08		2	A5	985
		Steph 765		9 17 14	+58 08.4	1	11.03		1.23		1.18		1	K7	1746
80410	+16 01945			9 17 15	+16 09.0	1	8.64		1.00				2	G5	882
80572	−47 04831			9 17 15	−47 45.5	6	7.03	.006	1.01	.007	0.86	.010	28	K1 III	389,863,977,1075,2001,2011
	−76 00579			9 17 15	−76 57.0	2	9.73	.005	1.47	.005	1.63	.024	6		874,1533

HD	DM	Other Id	N	Rem	α_{1950}	δ_{1950}	S	V	σ_V	B–V	σ_{B-V}	U–B	σ_{U-B}	n	Spectrum	References
80290	+51 01495	HR 3697		★ AB	9 17 16	+51 28.6	2	6.14	.025	0.42	.000	-0.09	.025	5	F3 V	1733,3016
80571	−46 05084				9 17 20	−46 25.7	1	7.89		1.09		0.99		1	K0 III	389
		ApJ239,112 # 18			9 17 20	−76 59.8	2	13.84	.000	0.72	.000	0.30	.000	4		874,1533
		G 47 - 34			9 17 21	+32 34.7	1	13.55		1.53		1.15		1		333,1620
80499	−11 02609	HR 3706		★ AB	9 17 22	−11 45.8	7	4.77	.020	0.93	.003	0.65	.017	28	G8 III	3,15,1075,1425,2012*
	−76 00580				9 17 22	−76 58.4	2	10.61	.005	0.65	.010	0.25	.025	5		874,1533
		LP 313 - 42			9 17 25	+26 56.4	1	15.56		1.92				3		1663
80667	−55 02076				9 17 26	−55 50.8	1	7.86		1.18				4	K0 III	1075
80951	−74 00579	HR 3720		★ AB	9 17 32	−74 41.0	4	5.28	.003	0.02	.000	0.04	.000	14	A1 V	15,1075,2016,2038
		L 35 - 12			9 17 32	−77 37.	1	13.10		1.64		1.15		1		3078
80950	−74 00580	HR 3721			9 17 33	−74 31.4	3	5.86	.004	-0.03	.005	-0.04	.000	12	A0 V	15,1075,2038
81056	−76 00581				9 17 35	−76 59.1	2	9.56	.019	0.52	.015	0.14	.029	6	F6 IV	874,1533
	−49 04314				9 17 39	−50 12.1	1	11.44		0.85		-0.18		3		362
299961	−53 02300				9 17 40	−53 48.9	1	9.78		0.20		-0.49		2	K5	362
80726	−55 02080				9 17 41	−56 09.2	4	6.83	.013	0.16	.009	0.13	.000	10	A3 Vs	258,688,1075,1075
		G 161 - 14			9 17 43	−05 09.2	3	12.32	.098	0.58	.010	-0.09	.010	4		1658,1696,3062
80590	−33 05973	HR 3710			9 17 43	−33 53.5	3	6.38	.004	-0.10	.003	-0.35		12	B8 V	15,1637,2012
		Pis 12 - 12			9 17 47	−44 54.0	1	13.66		0.53		0.02		1		410
80692	−45 05053			V	9 17 48	−45 28.0	1	9.64		0.17		0.03		5	A0 III/IV	1732
80441	+38 02025	HR 3701		★ AB	9 17 52	+38 24.1	1	6.12		0.38		-0.06		2	F2 V +F4 V	272
		Pis 12 - 11			9 17 53	−44 53.5	1	13.13		0.60		-0.01		1		410
		Pis 12 - 10			9 17 54	−44 54.7	1	12.16		0.53		0.03		1		410
80550	−9 02801	IDS09156S0908		B	9 17 55	−09 23.8	2	6.99	.022	0.39	.009	-0.03	.009	10	F4 IV-V	1084,3024
		AJ82,474 # 136			9 17 56	−49 19.3	1	11.36		1.89		1.34		2		779
	+33 01843	G 47 - 35		★ AB	9 17 57	+32 36.3	1	10.48		0.80		0.33		1		333,1620
		Pis 12 - 14			9 17 57	−44 56.8	1	13.12		0.59		0.09		1		410
		Pis 12 - 15			9 18 00	−44 56.9	1	13.51		0.70		0.10		1		410
80493	+35 01979	HR 3705			9 18 01	+34 36.3	7	3.14	.013	1.55	.004	1.94	.011	20	K7 IIIab	15,1355,1363,3001*
80781	−54 02186	HR 3717			9 18 01	−54 58.5	6	6.27	.007	-0.10	.004	-0.57	.010	22	B3 V	15,26,362,1075,2012,8023
80586	−8 02643	HR 3709		★ A	9 18 02	−09 20.6	4	4.80	.016	0.93	.012	0.67	.009	28	G8 III-IV +	15,1075,1084,1425*
80586	−8 02643	IDS09156S0908		C	9 18 02	−09 20.6	1	11.25		1.15				1	K2 V	3024
		Pis 12 - 8			9 18 02	−44 52.3	1	11.67		0.53		0.40		1		410
		Pis 12 - 13			9 18 02	−44 56.2	1	12.98		1.75		0.23		1		410
80567	+0 02499	IN Hya			9 18 03	+00 23.7	1	6.50		1.70		1.84		8	M2	3046
80390	+57 01214	HR 3698, CG UMa			9 18 04	+56 54.8	2	5.59	.147	1.61	.006	1.59		3	M4 IIIa	71,3001
		Pis 12 - 9			9 18 05	−44 51.2	1	12.42		0.68		0.24		1		410
80492	+40 02194				9 18 06	+39 52.8	1	6.66		0.93		0.59		3	K0	1501
		Pis 12 - 4			9 18 06	−44 55.	1	14.23		1.31		0.96		1		410
		Pis 12 - 16			9 18 06	−44 55.	1	12.46		1.08		0.91		1		410
233622	+50 01631				9 18 08	+50 18.7	1	10.01		-0.21		-0.80		1	B7	1026
		Pis 12 - 7			9 18 08	−44 52.6	1	11.56		0.83		0.59		1		410
		Pis 12 - 6			9 18 08	−44 54.2	1	14.12		1.27		0.74		1		410
80864	−62 01219				9 18 11	−62 31.2	1	8.37		0.06		-0.01		2	B9 V	401
		Pis 12 - 1			9 18 12	−44 53.8	1	12.55		1.49		1.32		2		410
		Pis 12 - 2			9 18 13	−44 53.7	1	13.16		0.52		0.77		1		410
		Pis 12 - 3			9 18 13	−44 54.9	1	12.55		1.53		1.20		1		410
		GD 101			9 18 14	−07 32.7	1	15.57		0.80		0.20		2		3060
80761	−46 05095				9 18 14	−46 44.3	6	7.27	.008	-0.10	.005	-0.47	.008	33	B5 III	125,278,460,1075,1770,2011
		Pis 12 - 5			9 18 15	−44 54.3	1	13.91		0.86		0.70		1		410
80536	+25 02084				9 18 16	+25 22.7	1	7.29		0.63		0.13		1	G2 V	3026
80632	−5 02778	G 161 - 15			9 18 16	−05 32.4	2	9.09	.020	1.16	.007	1.12		2	K2	1705,3072
	+28 01741	IDS09153N2753		AB	9 18 18	+27 40.4	1	10.06		0.89		0.50		1	K0	3032
	+28 01741	IDS09153N2753		C	9 18 18	+27 40.4	1	10.64		0.48		0.03		1		3032
80759	−42 05183				9 18 18	−43 11.2	1	7.55		1.20				4	K1 III	2012
80583	+12 02022				9 18 20	+12 08.3	1	8.84		1.00				2	K0	882
	−45 05058	LSS 1275			9 18 20	−45 19.1	3	11.33	.017	-0.32	.003	-1.21	.018	8	sdO	362,401,540,976
		G 46 - 34			9 18 21	+03 35.6	2	13.33	.004	1.63	.052	1.38		6		1663,3078
80777	−44 05382	Pis 12 - 17		★ AB	9 18 25	−44 57.6	9	6.70	.007	1.23	.005	1.32	.006	90	K0	395,402,410,474,768*
		EX Vel			9 18 25	−52 38.5	1	11.24		1.36		1.00		1		689
80546	+33 01848	HR 3707			9 18 26	+33 06.9	1	6.16		1.09		1.04		2	K3 III	252
	−53 02313				9 18 26	−53 22.0	2	10.45	.024	0.37	.009	-0.40	.001	3	B3 II	540,976
80774	−37 05668	HR 3716			9 18 29	−37 22.1	3	6.04	.004	1.39	.005			11	K3/4 III	15,2013,2029
80613	+15 02027	HR 3711			9 18 30	+15 35.1	2	6.53		0.00	.015	0.03	.020	5	A1 V	252,1022
	−50 04029				9 18 30	−50 35.0	1	11.34		0.26		-0.57		3		362
80719	−14 02828	HR 3714			9 18 33	−15 24.2	5	6.33	.018	0.47	.010	-0.01	.016	14	F6 V	15,254,1075,2012,3077
80654	+13 02074				9 18 35	+13 19.5	2	6.69	.015	0.47	.000	0.02	.010	5	F8 V	1733,3016
	−53 02317				9 18 35	−53 36.4	1	11.05		0.42		-0.33		2		362
		AJ82,474 # 137			9 18 36	−49 37.7	1	11.63		1.30		0.13		1	O9.5Vn	779
80773	−31 07162	HR 3715		★ AB	9 18 37	−31 32.8	1	6.80	.011	0.01	.005	-0.02	.010	14	A0 V	15,404,1637,2029
80652	+17 02065				9 18 39	+16 48.7	1	7.02		0.25		0.09		2	A5	3016
80580	+32 01865				9 18 39	+32 28.7	1	6.80		0.13		0.10		2	A2 V	1375
	−51 03713				9 18 40	−51 25.0	2	10.74	.025	0.51	.004	-0.41	.019	3	B0.5II-III	540,976
80817	−43 05186				9 18 41	−44 04.2	4	8.02	.005	0.09	.004	0.08	.007	24	A0 V	460,977,1075,2011
80676	+12 02024				9 18 42	+12 34.2	1	8.27		1.35				2	K2	882
80836	−45 05068				9 18 44	−45 22.6	1	9.62		-0.03		-0.23		10	B9 V	1732
		PG0918+029D			9 18 46	+03 00.3	1	12.27		1.04		0.82		11		1764
		WLS 920 35 # 10			9 18 46	+34 30.1	1	12.83		0.62		0.12		5		1375
80834	−41 05006	LSS 1276			9 18 47	−41 57.9	4	9.73	.025	0.25	.009	-0.77	.013	8	Be	540,976,1737,2021
80899	−49 04326				9 18 50	−49 41.7	1	8.19		1.07		0.89		1	G8/K0 III	389
80880	−50 04037				9 18 51	−50 34.7	1	8.57		0.73		0.57		1	F3 II	389

Table 1   507

HD	DM	Other Id	N Rem	α_{1950}	δ_{1950}	S	V	σ_V	B–V	σ_{B-V}	U–B	σ_{U-B}	n	Spectrum	References
		PG0918+029		9 18 52	+27 15.3	1	13.33		-0.27		-1.08		7	sdB	1764
80802	-33 05985			9 18 52	-33 25.8	1	9.15		0.49		-0.01		2	F6 V	1696
	-16 02761			9 18 55	-16 48.6	1	9.62		0.96		0.62		2	K3 V	3072
80855	-42 05195			9 18 57	-42 36.9	1	8.05		0.44		0.06		2	F3 V	389
		PG0918+029B		9 18 58	+03 00.8	1	13.96		0.77		0.37		11		1764
80949	-60 01391			9 18 58	-60 20.3	1	8.98		1.13				4	G8/K1	2033
		PG0918+029A		9 18 59	+02 59.1	1	14.49		0.54		-0.03		14		1764
80699	+19 02201			9 18 59	+18 57.7	1	7.69		0.56		0.10		2	G0	1648
		PG0918+029C		9 19 06	+02 59.4	1	13.54		0.63		0.09		10		1764
		L 461 - 53		9 19 06	-36 43.	1	11.07		0.67		0.10		2		1696
		CCS 1435		9 19 06	-46 48.8	1	10.38		3.86				1	N	864
80936	-52 02248			9 19 08	-53 02.5	1	7.04		1.06		0.86		2	K0 III	389
80732	+22 02082			9 19 09	+21 42.7	1	7.71		1.32		1.54		2	K0	1648
80922	-44 05398			9 19 13	-44 49.7	4	7.63	.003	0.09	.001	0.12	.007	20	A3 V	402,977,1075,2011
		Steph 767		9 19 17	+05 53.8	1	11.12		1.39		1.26		2	K7	1746
		Ton 1054		9 19 17	+36 26.	1	16.07		-0.16		-1.00		3		286
80715	+40 02197	BF Lyn	⋆	9 19 17	+40 25.2	3	7.62	.010	0.99	.008	0.79	.025	10	K2 V	22,1355,7009
80874	-25 07114	HR 3718		9 19 17	-25 45.1	7	4.72	.005	1.65	.018	2.01	.031	32	M1 III	3,15,1024,1075,2012*
80923	-45 05076			9 19 17	-45 21.6	1	10.22		0.26				4	A2 III(m)	2011
81038	-61 01241			9 19 19	-61 46.8	2	6.91	.005	-0.10	.010	-0.60		6	B5 Vn	401,2012
80999	-55 02102	LSS 1278		9 19 23	-55 35.9	5	7.94	.016	0.38	.008	0.30	.032	16	A0 II	285,657,1075,1075,8100
80934	-42 05203			9 19 24	-42 36.7	1	7.29		1.00		0.82		1	K0 III	389
	+23 02086			9 19 29	+22 45.2	1	10.18		0.95		0.72		2		1726
298335	-49 04334			9 19 29	-50 17.9	3	10.28	.053	0.31	.021	-0.41	.026	7	B5 II-Ib	362,540,976
80811	+13 02079			9 19 30	+12 45.9	1	8.35		0.87				2	K0	882
80605	+63 00838			9 19 34	+63 23.3	1	8.63		1.39		1.60		2	K2	1733
		LSS 1279		9 19 37	-50 02.1	2	11.74	.005	0.57	.018	-0.34	.014	5	B3 Ib	362,779
80976	-44 05407	IDS09178S4503	AB	9 19 38	-45 15.6	1	9.83		0.12				4	A0 V	2011
80820	+15 02031			9 19 39	+15 06.1	1	8.86		1.04				2	K0	882
		Ton 13		9 19 42	+27 19.	1	12.57		-0.30		-1.10		1		98
81101	-61 01242	HR 3728		9 19 45	-62 11.5	5	4.80	.008	0.94	.001	0.62	.005	19	G6 III	15,285,688,1075,2012
81080	-60 01394			9 19 46	-60 48.4	1	8.80		-0.04		-0.27		2	B9/A0 V	401
	-44 05413			9 19 53	-45 07.8	1	10.73		0.51				4		1075
81078	-54 02209			9 19 53	-55 11.3	1	7.87		1.47				4	K1 III	1075
81035	-46 05121	IDS09181S4654	AB	9 19 54	-47 06.4	3	7.46	.000	0.13	.002	0.05	.009	16	A2 IV/V	977,1075,2011
81050	-49 04341			9 19 55	-49 30.9	1	8.13		1.05		0.86		1	K0/1 III	389
		WLS 920 35 # 5		9 19 56	+37 24.4	1	11.98		0.47		-0.12		2		1375
81034	-41 05023	HR 3726		9 19 56	-41 58.8	3	5.58	.005	1.63	.005	1.94		9	M1 III	15,2027,3005,8100
	-53 02331	LSS 1280		9 19 57	-53 43.8	2	11.26	.005	0.43	.010	-0.63	.015	5	O9 III	362,1737
81223	-67 01071			9 20 02	-67 54.5	2	8.41	.005	0.98	.005	0.56		3	K0wF7	1594,6006
	+11 02021	G 41 - 34		9 20 06	+11 29.1	2	9.63	.015	0.78	.010	0.40	.000	3		333,1620,1696
	-45 05093	IDS09183S4512	AB	9 20 08	-45 24.9	1	10.99		0.47				4		2011
80810	+47 01667			9 20 12	+46 59.2	1	9.44		0.18		0.09		3	A2	1601
81077	-46 05123	IDS09184S4620	AB	9 20 12	-46 32.7	5	7.67	.006	1.00	.007	0.83	.009	19	G8/K0 III	389,395,977,1075,2011
81138	-56 02115			9 20 12	-56 41.4	4	7.22	.004	0.99	.006	0.74	.005	10	K0 III	258,657,1075,1075
81076	-45 05094			9 20 14	-45 25.7	4	9.60	.005	0.20	.004	0.17	.004	26	Ap EuCr	863,1460,1770,2011
80991	-21 02774			9 20 15	-22 18.3	1	8.55		1.02				2	K0 III	1467
81044	-31 07195			9 20 16	-31 56.3	2	8.86	.019	0.81	.015	0.36		6	K1 V	2033,3077
		G 116 - 24		9 20 17	+36 56.9	1	11.35		1.24		1.21		2		7010
		Pis 13 - 6		9 20 17	-50 54.9	1	13.19		0.76		0.26		2		785
		Ton 14		9 20 18	+29 43.	1	14.66		-0.30		-1.06		1		98
		G 161 - 16		9 20 18	-09 52.0	1	15.06		1.58				1		3062
298369	-50 04066	LSS 1281		9 20 18	-50 57.7	5	9.96	.032	0.37	.008	-0.60	.031	12	B2:Ven	362,540,976,1737,2021
		Pis 13 - 7		9 20 19	-50 54.5	1	13.70		0.56		0.10		2		785
81157	-54 02213	HR 3732		9 20 19	-55 18.1	3	5.63	.010	0.19	.000	0.10		9	A3 IVs	15,401,2012
80902	+39 02237	IDS09172N3934	AB	9 20 20	+39 21.0	2	8.51	.005	0.45	.005	0.01	.005	5	F5 V	833,1601
		Pis 13 - 17		9 20 21	-50 53.2	1	14.54		0.74		0.26		2		785
		G 46 - 37		9 20 22	+01 16.2	1	16.64		0.54		-0.25		1		3060
80783	+61 01118			9 20 23	+60 39.7	1	7.38		1.41				4	K2	1118
		AJ82,474 # 141		9 20 23	-49 56.1	1	13.10		1.06		0.33		1		779
	-50 04070	Pis 13 - 1		9 20 24	-50 52.9	2	12.11	.005	0.52	.025	0.04	.015	4		184,785
		Pis 13 - 10		9 20 24	-50 53.1	1	13.87		0.60		0.32		3		785
		Pis 13 - 4		9 20 24	-50 53.2	2	12.83	.040	0.54	.015	0.05	.025	5		184,785
		Pis 13 - 19		9 20 24	-50 53.5	1	14.74		0.56		0.12		2		785
304636	-59 01362			9 20 24	-60 04.2	2	9.51	.015	1.48	.026	1.16		2	M0 V	1705,3072
81009	-9 02816	HR 3724, KU Hya	⋆ABC	9 20 25	-09 37.5	4	6.52	.004	0.22	.000	0.10	.009	12	A5p SrCrEu	15,402,1075,2012
		Pis 13 - 18		9 20 25	-50 53.4	1	14.72		0.61		0.28		2		785
		Pis 13 - 16		9 20 25	-50 53.5	1	14.47		0.69		0.18		2		785
		Pis 13 - 15		9 20 25	-50 53.8	1	14.47		0.72		0.18		2		785
		Pis 13 - 21		9 20 25	-50 53.8	1	15.32		0.62		0.41		2		785
		Pis 13 - 23		9 20 25	-50 54.1	1	16.05		0.82		0.30		2		785
		Pis 13 - 2	V	9 20 26	-50 53.6	2	12.67	.015	0.46	.024	-0.09	.019	3		184,785
		Pis 13 - 22		9 20 26	-50 54.2	1	15.68		0.64		0.42		2		785
		Pis 13 - 20		9 20 27	-50 52.6	1	15.03		0.62		0.37		2		785
		Pis 13 - 14		9 20 27	-50 53.0	1	14.35		0.55		0.09		2		785
		Pis 13 - 5		9 20 27	-50 53.1	2	12.84	.034	0.58	.010	0.16	.005	3		184,785
		Pis 13 - 9		9 20 27	-50 53.1	1	13.83		0.81		0.43		2		785
		Pis 13 - 12		9 20 27	-50 53.8	1	14.21		0.54		0.12		3		785
		Pis 13 - 13		9 20 27	-50 54.0	1	14.33		0.59		0.11		2		785
		Pis 13 - 8		9 20 28	-50 53.5	1	13.72		0.59		0.10		3		785

HD	DM	Other Id	N Rem	α_{1950}	δ_{1950}	S	V	σ_V	B–V	σ_{B-V}	U–B	σ_{U-B}	n	Spectrum	References
		Pis 13 - 3	AB	9 20 28	−50 54.7	1	12.79		0.44		0.27		2		785
		Pis 13 - 11		9 20 28	−50 55.1	1	13.94		0.45		0.28		2		785
	−44 05420			9 20 30	−44 55.1	1	11.45		0.31				4		2011
298387	−52 02266			9 20 33	−52 34.0	3	10.28	.017	0.15	.008	-0.52	.007	7	B7 Iab	125,540,976
81136	−45 05099	HR 3730		9 20 34	−45 50.0	12	5.74	.011	0.92	.004	0.65	.007	67	G6/8 III	14,15,125,182,278,388*
81188	−54 02219	HR 3734		9 20 34	−54 47.8	8	2.49	.012	-0.19	.009	-0.76	.012	28	B2 IV/V	15,26,362,1034,1075*
81135	−43 05215			9 20 35	−43 53.3	3	8.15	.007	0.02	.000	0.05	.002	16	A0 V	977,1075,2011
		G 161 - 17		9 20 36	−12 25.2	2	12.88	.000	0.73	.005	0.15	.010	3		1696,3062
	−53 02341	LSS 1283		9 20 36	−53 41.0	1	10.26		0.48		-0.68		2		362
80956	+25 02088	HR 3723		9 20 38	+25 23.9	2	6.43	.030	0.85	.030	0.51	.010	4	G5 III-IV	1733,3016
81134	−39 05446	HR 3729		9 20 39	−39 33.7	5	6.53	.006	1.12	.009			22	K2 III	15,1075,2013,2020,2028
81504	−75 00575			9 20 39	−75 30.0	1	7.40		0.05		-0.13		1	B9 V	55
81029	+4 02178	IDS09181N0356	A	9 20 40	+03 43.0	1	7.31		0.34		0.08		2	Am	3016
81029	+4 02178	IDS09181N0356	B	9 20 40	+03 43.0	1	8.32		0.41		-0.03		2	F0	3032
81172	−46 05134			9 20 41	−46 38.0	1	8.29		-0.11		-0.47		2	B7/8 III	401
	−53 02343	LSS 1284	A	9 20 41	−53 41.1	1	11.09		0.43		-0.19		3		362
	−53 02343	LSS 1284	AB	9 20 41	−53 41.1	1	10.99		0.43		-0.21		2		362
		ApJ239,112 # 19		9 20 42	−77 07.3	1	13.92		0.82		0.24		4		1533
	+40 02200	G 116 - 26		9 20 45	+40 23.3	1	10.22		0.69		0.17		2	G8	1658
81222	−55 02118	V Vel		9 20 45	−55 44.8	2	7.21	.000	0.60	.005	0.36	.010	2	F8 II	657,688
81202	−49 04352			9 20 46	−50 01.2	1	7.76		-0.09		-0.45		2	B5 II/III	401
81028	+8 02215	DF Leo		9 20 50	+07 55.8	1	6.85		1.63		1.78		21	M4 III	3042
80730	+71 00503			9 20 51	+70 39.6	1	8.57		1.10		1.05		1	G5	1776
81238	−53 02347	IDS09193S5319	AB	9 20 52	−53 31.3	1	7.95		0.49		0.09		2	F5 V	389
	−4 02606	G 161 - 18		9 20 53	−04 55.6	1	9.76		0.86		0.50		1		3062
		AJ82,474 # 143		9 20 54	−48 49.3	1	12.10		1.13		0.32		2		779
80808	+66 00612			9 20 57	+65 45.9	1	8.99		1.14		0.92		2	K0	1733
81293	−58 01501			9 20 57	−58 31.2	1	8.25		0.00		-0.42		2	B8 III	401
81040	+20 02314			9 20 58	+20 34.7	1	7.73		0.64		0.14		3	G0	1648
297433	−48 04643	LSS 1285		9 21 01	−49 17.4	5	9.35	.025	1.12	.019	0.02	.041	10	B1 Ib	362,540,779,976,1737
81169	−28 07196	HR 3733		9 21 02	−28 37.2	4	4.70	.017	0.91	.000	0.63	.000	16	G8 III	15,263,1075,2012,8015
		G 116 - 27	A	9 21 04	+40 17.4	1	10.54		1.15		1.09		3	K4:	7010
		G 116 - 27	B	9 21 04	+40 17.4	1	14.59		0.65		0.02		3		7010
	+81 00297	G 253 - 26		9 21 06	+80 48.2	1	9.30		1.23		1.20		2	K8	3072
	−45 05103			9 21 06	−45 19.4	1	11.40		0.47				4		2011
		ApJ239,112 # 13		9 21 07	−77 09.7	1	12.45		1.52		1.46		4		1533
81259	−47 04882			9 21 11	−47 27.5	1	8.30		0.43		0.04		1	F5 V	389
		AAS27,343 # 179		9 21 12	−53 02.	1	11.35		0.53		0.01		1		362
		G 117 -B15	A	9 21 13	+35 29.8	1	15.52		0.20		-0.56		3	DA	3060
		G 117 -B15	B	9 21 13	+35 29.8	1	16.10		1.60				1	dM2	3016
81257	−44 05429			9 21 15	−44 21.3	1	10.53		0.31		0.28		2	G8 (III)	395
296851	−44 05428			9 21 15	−44 42.9	1	10.53		0.29				4		2011
81039	+37 01978	HR 3727		9 21 17	+36 48.2	2	6.67	.005	0.20	.005	0.10	.000	4	F0 V	985,1733
81290	−48 04649			9 21 17	−48 50.6	2	8.88	.010	0.33	.000	-0.03		4	F3/5wa	1594,6006
	+37 01977			9 21 18	+36 55.8	1	10.17		-0.24		-1.26		2		1726
296854	−45 05109			9 21 18	−45 27.4	3	9.48	.001	1.27	.002	1.41	.001	22	K5	1460,1628,2011
		G 48 - 10		9 21 19	+00 21.3	1	11.54		1.51		1.23		1	M3	333,1620
81276	−44 05430			9 21 20	−44 45.0	6	7.94	.003	0.04	.007	0.06	.010	23	A1 V	365,395,768,977,1075,2011
	−44 05431			9 21 22	−45 07.4	1	11.51		0.96				4		2011
81371	−58 01507			9 21 22	−58 20.7	1	8.60		0.02		-0.42		2	B7 III	401
		ApJ239,112 # 16		9 21 22	−77 02.4	1	13.66	.010	0.13	.005	-0.37	.000	5		874,1533
		G 47 - 39		9 21 23	+23 50.8	1	13.18		1.48		1.25		1		1620
81354	−55 02126			9 21 23	−55 31.3	1	9.31		0.20		-0.57		3	B3/5 Ve	362
81166	−4 02608			9 21 25	−05 08.8	1	7.48		1.43		1.54		3	K2	1657
81314	−51 03753			9 21 25	−51 41.4	1	8.84		0.02		0.03		7	A0 V	402
80982	+54 01297	IDS09179N5349	AB	9 21 27	+53 36.4	1	8.22		1.01		0.77		2	K2	1566
81025	+52 01389	HR 3725	★ AB	9 21 28	+51 47.4	1	6.32		0.75				1	G2 III	71
81353	−53 02356	IDS09199S5323	A	9 21 30	−53 36.3	1	6.81		1.05		0.80		2	G8 II/III	389
		L 99 - 14		9 21 30	−67 17.	1	10.78		0.79		0.39		2		1696
81289	−43 05225			9 21 32	−43 23.2	3	8.42	.004	0.00	.004	-0.20	.010	16	Ap EuSrCr	977,1075,2011
	−51 03754			9 21 32	−51 38.4	1	10.39		1.30		1.14		11		402
81230	−23 08331	IDS09193S2314	AB	9 21 33	−23 25.7	1	7.93		0.47		0.00		42	F6 V	978
		E4 - n		9 21 34	−45 11.3	1	15.01		0.82		0.36		4		1460
81350	−51 03757			9 21 34	−51 35.3	2	8.19	.010	-0.03	.015	-0.28	.015	5	B8 II/III	362,402
80768	+76 00351	G 252 - 34		9 21 35	+76 09.1	1	9.03		1.19		1.18		1	K2	3072
	−48 04654	LSS 1288		9 21 35	−49 02.8	3	10.18	.011	1.06	.051	-0.02	.031	7	B2 Ib	362,779,1737
		E4 - d		9 21 36	−45 11.9	2	12.58	.017	0.67	.031	0.14		8		1460,2011
81348	−48 04656			9 21 38	−48 54.3	1	9.46		0.20		0.20		1	A6 V	6006
	−50 04091			9 21 38	−51 18.5	1	10.74		0.32		-0.51		3		362
		LSS 1289		9 21 38	−52 13.4	3	11.85	.026	0.99	.026	-0.21	.017	9	B3 Ib	125,362,846
		WLS 920 35 # 9		9 21 39	+35 46.0	1	10.83		0.46		-0.03		2		1375
80842	+70 00561			9 21 39	+70 33.9	1	7.33		1.04		0.91		1	G5	1776
81347	−47 04890			9 21 39	−48 04.3	6	6.27	.004	-0.15	.007	-0.60	.004	38	B3 IV	278,362,1075,1770,2001,2011
296853	−45 05116			9 21 40	−45 19.8	2	10.00	.006	0.55	.004	0.03		10	G0	863,2011
81333	−45 05117			9 21 40	−46 03.3	3	9.15	.002	0.18	.000	0.15	.006	14	A5 IV/V	863,1075,2011
81370	−52 02281	LSS 1290		9 21 40	−52 32.0	4	8.71	.009	0.10	.008	-0.80	.026	32	B1 (III)n	125,362,540,976
80953	+64 00733	HR 3722		9 21 44	+64 09.4	2	6.28	.005	1.46	.005	1.74		8	K2 III	252,1379
81309	−37 05721	HR 3735		9 21 44	−37 32.5	2	6.47	.005	0.18	.000			7	A1mA5-F0	15,2012
81146	+26 01939	HR 3731	★ AB	9 21 45	+26 23.9	5	4.46	.012	1.22	.017	1.30	.005	13	K2 III	15,1080,1363,3016,8015
		AAS19,45 T6# 422		9 21 48	−51 28.3	1	13.76		0.67		0.13		3		402

Table 1   509

HD	DM	Other Id	N	Rem	α_{1950}	δ_{1950}	S	V	σ_V	B–V	σ_{B-V}	U–B	σ_{U-B}	n	Spectrum	References
	−44 05439				9 21 50	−45 13.3	1	11.45		0.52				4		2011
		ApJ239,112 # 11			9 21 50	−77 10.3	1	12.09		0.94		0.38		4		1533
		G 48 - 13			9 21 51	+05 59.6	2	12.68	.015	0.71	.010	0.11	.000	3		333,1620,1658
81369	−46 05146				9 21 51	−46 41.6	6	6.21	.002	-0.10	.008	-0.38	.004	31	B7 III	278,402,977,1075,1770,2011
		ApJ239,112 # 17			9 21 52	−77 10.8	1	13.66		0.76		0.17		5		1533
		E4 - q			9 21 53	−45 13.1	1	16.12		0.97		0.61		5		1460
		AAS19,45 T6# 420			9 21 53	−51 28.1	1	13.39		0.65		0.12		3		402
		E4 - c			9 21 54	−45 09.	1	11.77		1.34				4		2011
		E4 - r			9 21 55	−45 13.1	1	16.73		1.09				4		1460
		AAS19,45 T6# 417			9 21 55	−51 25.2	1	12.12		0.66		0.25		3		402
81192	+20 02318				9 21 57	+20 00.2	3	6.53	.015	0.95	.004	0.57	.010	5	G7 III	1003,1080,3077
		E4 - h			9 21 59	−45 13.4	2	13.76	.000	0.61	.002	0.09		9		1075,1460
		E4 - 57			9 22 00	−45 09.5	2	11.34	.003	1.28	.017	0.84		22		1460,2011
299981	−52 02292				9 22 01	−53 03.9	3	10.26	.023	0.05	.014	-0.70	.034	5	B3 Ib	362,540,976
81401	−46 05151				9 22 02	−46 50.6	4	8.19	.009	0.00	.002	-0.06	.011	17	B9 IV/V	395,977,1075,2011
		AAS19,45 T6# 421			9 22 02	−51 27.9	1	13.55		0.53		0.37		3		402
	−44 05442				9 22 05	−44 45.6	1	11.17		0.34				4		2011
81451	−52 02296				9 22 07	−53 03.3	2	8.34	.005	0.48	.000	0.11	.010	4	F6 IV	389,401
81502	−59 01374	HR 3740			9 22 08	−60 05.2	4	6.30	.005	1.48	.005	1.68		12	K1/2 II/III	15,1075,2031,3005
		E4 - g			9 22 12	−45 13.	1	13.62		0.67				4		2011
81414	−44 05445				9 22 12	−45 15.5	10	9.41	.011	0.08	.003	0.08	.004	129	A1 V	365,395,1075,1460,1628,1657*
	−45 05125				9 22 15	−45 36.0	1	10.73		0.40				4		2011
298377	−51 03765				9 22 16	−51 44.8	3	10.25	.013	0.29	.017	-0.59	.024	10	B3:Vn	362,402,2021
81411	−38 05541	HR 3737		⋆ AB	9 22 17	−39 12.6	3	6.06	.009	0.19	.023	0.10		11	A6/7 III	15,2027,8071
		Ton 1059			9 22 18	+33 05.	1	14.83		-0.29		-0.97		1		286
81471	−51 03767	HR 3739, LSS 1292			9 22 19	−51 31.3	6	6.08	.016	0.57	.018	0.31	.058	18	A5 Ia/ab	15,402,1018,1034,2012,8100
	−51 03768				9 22 21	−51 23.4	1	11.51		0.27		-0.42		3		402
		AAS19,45 T6# 419			9 22 21	−51 27.8	1	13.11		0.47		0.32		3		402
81301	+15 02043	IDS09196N1515		AB	9 22 24	+15 01.6	1	9.05		0.75				3	K0	882
		AAS19,45 T6# 418			9 22 24	−51 27.2	1	12.42		1.72		1.83		3		402
298420	−52 02303	LSS 1293			9 22 25	−52 39.0	4	9.85	.020	0.31	.014	-0.64	.034	27	B0.5IV	125,362,540,976
		E4 - e			9 22 30	−45 10.	1	13.20		0.67				4		2011
		AAS19,45 T6# 423			9 22 30	−51 30.	1	13.78		1.33		0.96		3		402
		AAS19,45 T6# 424			9 22 30	−51 30.	1	13.85		0.68		0.52		3		402
		AAS19,45 T6# 425			9 22 30	−51 30.	1	13.98		0.56		0.35		2		402
		AAS19,45 T6# 426			9 22 30	−51 30.	1	14.52		0.90		0.42		3		402
		AAS19,45 T6# 427			9 22 30	−51 30.	1	14.62		0.56		0.42		2		402
		AAS19,45 T6# 428			9 22 30	−51 30.	1	15.21		0.80		0.30		3		402
		AAS19,45 T6# 429			9 22 30	−51 30.	1	15.52		0.76		0.17		3		402
		AAS19,45 T6# 430			9 22 30	−51 30.	1	15.66		0.71		0.54		3		402
		AAS19,45 T6# 431			9 22 30	−51 30.	1	15.70		0.59		0.46		2		402
		AAS19,45 T6# 432			9 22 30	−51 30.	1	15.92		0.77		0.16		3		402
		AAS19,45 T6# 433			9 22 30	−51 30.	1	16.07		0.96		0.98		3		402
81246	+38 02030				9 22 33	+38 06.4	1	8.22		0.48		0.03		2	F5	1733
	−45 05126				9 22 33	−45 19.7	1	10.42		0.36				4		2011
81410	−23 08347	IL Hya			9 22 34	−23 36.6	5	7.65	.051	1.05	.009	0.70	.008	35	K2 IV/Vp	1467,1641,1642,2012,3049
		Ru 76 - 2			9 22 35	−51 26.1	1	12.93		0.42		0.22		2		184
		E4 - b			9 22 36	−45 06.	1	12.35		0.44				4		2011
81340	+14 02092				9 22 37	+14 24.4	1	8.57		1.67				2	K5	882
		Ru 76 - 3			9 22 37	−51 25.5	1	12.55		0.36		0.25		2		184
81518	−51 03771	Ru 76 - 1			9 22 37	−51 27.0	1	9.31		0.00		-0.17		4	A0	184
		ApJ239,112 # 8			9 22 37	−77 13.5	1	11.32		0.86		0.25		4		1533
	−52 02306	LSS 1295			9 22 38	−52 37.0	3	10.32	.016	0.39	.009	-0.56	.021	6	B0 V	362,540,976
81408	−12 02889	G 161 - 24			9 22 39	−12 44.3	3	9.63	.018	0.96	.014	0.70		7	K1/2 V	1620,2012,3073
80930	+75 00377	HR 3719			9 22 40	+75 18.9	2	6.24	.045	0.09	.015	0.11	.000	4	A5 Vs	985,1733
81517	−48 04666				9 22 41	−48 26.3	2	8.97	.010	0.50	.000	0.04		4	F6 III/IV	1594,6006
		ApJ239,112 # 21			9 22 41	−77 03.8	1	14.11		0.73		0.13		4		1533
		AAS27,343 # 139			9 22 42	−50 06.	2	11.64	.013	0.68	.013	-0.27	.022	6		362,779
		AJ77,733 T12# 21			9 22 43	−52 31.0	1	11.92		0.35		-0.29		3		125
		Ru 76 - 4			9 22 44	−51 26.2	1	13.46		0.39		0.25		2		184
		Ru 76 - 7			9 22 44	−51 28.3	1	12.50		0.54		0.00		2		184
		Ru 76 - 6			9 22 45	−51 27.6	1	12.52		0.27		-0.07		2		184
81361	+17 02078	HR 3736			9 22 46	+16 48.1	2	6.29	.000	0.97	.002			16	K0	15,1351
		Ru 76 - 5			9 22 46	−51 26.8	1	12.79		0.36		0.16		2		184
		ApJ239,112 # 22			9 22 46	−77 05.3	1	14.36		0.80		0.22		3		1533
	−51 03773				9 22 48	−51 41.2	1	11.23		0.17		0.17		5		402
		AJ82,474 # 146			9 22 50	−49 58.0	1	13.14		1.05		0.08		1		779
81613	−61 01265	HR 3742			9 22 50	−61 25.9	2	5.98	.005	1.06	.000			7	K0 III	15,2012
298408	−51 03774				9 22 51	−51 44.4	1	10.67		0.08		0.06		10	A0	402
		Steph 772			9 22 52	+32 53.9	1	11.54		1.41		1.23		1	K7-M0	1746
	−52 02315	LSS 1296		⋆ AB	9 22 53	−52 54.1	2	10.45	.030	0.22	.020	-0.56	.005	5		125,362
		ApJ239,112 # 10			9 22 53	−77 09.1	1	12.00		0.74		0.19		6		1533
81420	−4 02616	HR 3738			9 22 54	−04 54.1	5	5.60	.008	1.52	.011	1.81	.009	20	K5 III	15,1256,2013,2028,3001
		G 161 - 25			9 22 54	−07 06.8	1	12.66		1.21		1.07		2		3073
	−52 02318				9 22 54	−53 05.7	1	10.90		0.32		-0.48		2		362
		G 46 - 40			9 22 58	+00 31.6	2	15.09	.026	1.48	.007			2		1691,3062
		ApJ239,112 # 15			9 22 58	−76 59.1	1	13.56		0.77				1		1533
298406	−51 03778				9 23 03	−51 33.6	1	10.75		0.19		-0.09		4	A2	402
81590	−52 02319				9 23 04	−52 56.7	1	8.44		1.76		2.01		1	K4/5 II/III	1480
	−53 02380	LSS 1297			9 23 04	−54 07.9	1	10.51		0.54		-0.51		2		362

HD	DM	Other Id	N	Rem	α_{1950}	δ_{1950}	S	V	σ_V	B–V	σ_{B-V}	U–B	σ_{U-B}	n	Spectrum	References
81387	+29 01900	IDS09202N2905		AB	9 23 06	+28 52.1	1	8.46		0.49		-0.04		4	F8	3030
81576	-44 05460	GI Vel			9 23 10	-45 09.6	1	7.90		1.66				4	M3 III	1075
	-52 02324				9 23 10	-52 48.3	1	11.12		0.30		-0.47		2		362
	-52 02325				9 23 11	-53 02.2	1	10.56		0.60		0.12		1		1480
81573	-41 05070				9 23 12	-42 02.2	1	8.41		0.44		0.05		1	F3 V	389
		G 161 - 29			9 23 13	-06 33.2	1	11.36		0.98		0.85		2		3062
	-44 05461				9 23 13	-45 04.5	3	10.59	.003	0.20	.005	0.17	.003	23		365,1460,2011
81624	-49 04384				9 23 13	-50 04.2	1	7.83		1.21		1.09		1	G8/K0 III	389
81700	-63 01122				9 23 13	-63 30.1	1	8.53		0.65		0.10		5	G2 V	897
	-52 02327	LSS 1298			9 23 14	-52 48.9	1	10.52		0.26		-0.57		2		362
		ApJ239,112 # 12			9 23 14	-77 03.0	1	12.32		0.72				1		1533
81575	-43 05262	GK Vel			9 23 15	-43 45.6	1	6.50		1.57		1.63		4	M3 III	1075
298405	-51 03785				9 23 15	-51 36.7	1	8.27		1.54		1.90		8	K7	402
81654	-58 01513				9 23 15	-58 28.4	3	7.93	.005	-0.01	.004	-0.69	.004	6	B2/3 Ve	401,540,976
81574	-43 05263				9 23 17	-43 26.1	1	8.30		1.09		1.00		1	K1 III	389
		ApJS53,765 T6# 6			9 23 17	-52 57.2	1	12.93		0.57		0.01		1		1480
	-52 02329				9 23 18	-53 14.1	1	10.42		0.07		-0.72		2		362
		ApJS53,765 T6# 4			9 23 19	-52 55.2	1	12.13		0.21		0.07		1		1480
		ApJS53,765 T6# 5			9 23 19	-52 56.5	1	12.36		0.35		0.30		1		1480
	-52 02331				9 23 19	-53 02.3	1	11.09		0.51		0.17		1		1480
81638	-52 02330				9 23 19	-53 11.7	1	9.41		0.06		-0.42		2	B7/8 V	401
		ApJ239,112 # 9			9 23 20	-77 19.3	1	11.48		0.51		0.22		1		1533
81610	-44 05464				9 23 22	-45 16.1	10	8.85	.008	0.98	.004	0.72	.007	134	G8/K0 III	365,395,977,1460,1628,1657*
		UV0923-28			9 23 24	-28 25.5	1	11.21		-0.22		-1.09		5		1732
81637	-51 03787				9 23 24	-51 37.5	1	9.18		0.56		0.13		3	F8/G0 V	402
		AJ77,733 T12# 25			9 23 24	-52 36.1	1	12.38		0.51		0.05		3		125
	-76 00583				9 23 25	-76 51.1	1	10.35		1.03		0.54		2		1533
	-28 07246				9 23 27	-28 25.2	1	9.23		1.48		1.82		6	K5	1732
81621	-46 05169				9 23 27	-46 43.6	1	7.12		1.36		1.47		1	K2 III	389
		EZ Vel			9 23 31	-52 59.5	1	12.12		2.32		2.16		1		689
	-52 02334				9 23 32	-52 27.8	2	11.14	.010	0.60	.019	-0.43	.005	6		125,362
81666	-51 03790				9 23 33	-51 45.5	1	8.96		0.57		0.12		6	F7 V	402
81667	-51 03791				9 23 33	-51 12.5	1	8.87		-0.02		-0.20		2	B8/9 IV/V	401
		LP 787 - 42			9 23 35	-15 51.8	1	12.16		0.60		-0.17		2		1696
81650	-46 05170				9 23 35	-46 45.7	1	8.91		0.98		0.71		1	G8 III	389
	-47 04923	LSS 1299			9 23 35	-47 37.6	1	10.74		1.16		-0.24		2		362
	-51 03792				9 23 37	-51 30.3	1	10.66		0.48		0.14		4		402
		AJ77,733 T12# 30			9 23 38	-52 25.6	1	13.36		0.40		-0.30		3		125
		AJ77,733 T12# 32			9 23 38	-52 25.6	1	13.53		0.49		0.35		5		125
		AJ77,733 T12# 24			9 23 38	-52 44.2	1	12.17		0.75		0.24		4		125
		AJ77,733 T12# 26			9 23 38	-52 45.1	1	12.52		0.65		0.16		4		125
81439	+41 01961				9 23 39	+41 14.5	1	7.44		1.40		1.64		3	K4 III	1501
		AJ77,733 T12# 31			9 23 41	-52 40.5	1	13.47		0.51		-0.22		3		125
	-32 06385	LP 902 - 62			9 23 42	-33 16.5	1	10.92		0.85		0.62		2		1696
296881	-45 05136				9 23 42	-45 26.4	3	10.85	.013	0.20	.000	0.15	.018	9	F0	365,395,2011,2011
		ApJS53,765 T6# 7			9 23 42	-52 56.8	1	13.36		0.70		0.34		1		1480
81770	-60 01422				9 23 42	-61 07.8	1	8.62		-0.01		-0.30		2	B8/9 V	401
81681	-51 03795				9 23 45	-51 25.1	1	9.58		0.16		0.05		4	B8/9 II/III	402
81665	-45 05139				9 23 48	-45 37.3	4	9.20	.007	0.21	.007	0.12	.014	13	A6 V	365,395,1075,2011
81632	-31 07262				9 23 49	-32 13.3	1	8.25		0.99				4	K0 III	2012
81680	-45 05140				9 23 49	-45 22.7	3	9.70	.003	-0.03	.008	-0.16	.007	11	B9 V	365,863,2011
		LSS 1301			9 23 49	-49 20.9	3	10.85	.075	1.59	.026	0.65	.053	9	B3 Ib	362,779,846
81567	-0 02195	HR 3741			9 23 50	-01 14.8	2	6.00	.005	1.32	.000	1.44	.000	7	K3 III	15,1078
		ApJ239,112 # 7			9 23 50	-77 11.7	1	11.17		0.68		0.23		4		1533
	-23 08367				9 23 51	-23 46.1	1	7.95		0.46				1	K0	1642
81540	+17 02084				9 23 52	+16 54.9	2	8.01	.035	1.56	.045	1.81	.010	4	M1	1648,3040
		AJ77,733 T12# 29			9 23 55	-52 49.5	1	13.20		0.37		0.25		4		125
		AJ77,733 T12# 34			9 23 56	-52 42.9	1	14.33		0.67		0.38		6		125
296882	-45 05142				9 23 57	-45 27.6	3	9.85	.044	1.22	.016	1.30	.061	8	K0	365,395,1075
		LSS 1302			9 23 58	-53 01.9	2	11.75	.020	0.90	.025	0.09	.005	10	B3 Ib	125,846
81734	-52 02346	IDS09224S5249		A	9 23 59	-53 02.1	1	7.05		0.51		0.00		2	F7 V	389
81769	-56 02154	V377 Car			9 23 59	-57 08.7	1	8.09		0.05		-0.47		2	B4 V	401
81910	-69 01057	IDS09234S6959		AB	9 24 02	-70 11.6	1	7.80		-0.04		-0.35		2	B9 III	401
		VZ Leo			9 24 03	+16 49.3	1	11.50		0.32		0.10		1	A5	588
81581	+15 02049	IDS09213N1456		A	9 24 04	+14 42.7	1	7.75		0.31				7	A3	1351
81563	+15 02048				9 24 04	+15 08.8	2	8.28	.000	0.48	.000			39	G0	130,1351
298415	-52 02348				9 24 05	-52 32.1	1	9.96		0.34		0.12		4	A3	125
82114	-77 00507	IDS09248S7728		A	9 24 10	-77 40.7	1	7.06		0.62				4	F8/G0 V	2012
81804	-54 02280				9 24 12	-55 04.2	1	8.93		0.06		-0.49		2	B7 V	401
81830	-61 01271	HR 3752		★ A	9 24 13	-61 44.0	2	5.77	.009	0.14	.004	0.13	.009	10	A4 V	1279,3052
81830	-61 01271	HR 3752		★ AB	9 24 13	-61 44.0	4	5.76	.007	0.15	.009	0.12	.005	13	A4 V	15,243,401,2012
81830	-61 01271	IDS09230S6131		B	9 24 13	-61 44.0	2	9.82	.025	0.72	.015	0.23		6		1279,3024
302177	-56 02156	IC 2488 - 1			9 24 14	-56 41.8	1	10.30		1.36		1.41		3	K7	1644
81695	-28 07264	IDS09221S2901		AB	9 24 15	-29 13.6	1	8.68		0.36		0.01		3	F0 V +F/G	404
	+40 02208				9 24 18	+39 43.7	2	9.86	.008	1.30	.012	1.17		4	K8	196,1017
81783	-47 04935				9 24 20	-47 32.7	2	7.71	.000	1.27	.000			8	K3 III	1075,2012
81803	-53 02401				9 24 21	-53 37.3	1	7.09		0.67		0.48		2	F5 Ib	389
		AJ77,733 T12# 20			9 24 23	-52 36.7	1	11.88		0.60		0.06		3		125
81694	-23 08373				9 24 25	-23 57.0	2	6.91	.001	0.00	.000	-0.16	.001	44	A0 V	978,1732
81850	-55 02163	IDS09229S5539		AB	9 24 27	-55 52.3	3	7.30	.012	-0.01	.010	-0.28		10	B8/9 IV	401,1075,1075

Table 1   511

HD	DM	Other Id	N Rem	α₁₉₅₀	δ₁₉₅₀	S	V	σ_V	B-V	σ_B-V	U-B	σ_U-B	n	Spectrum	References
81713	-20 02905			9 24 28	-20 25.9	2	8.91	.005	0.94	.007	0.59		3	G8 IVwG3/5	1594,6006
81780	-39 05507	HR 3746		9 24 30	-40 17.1	2	6.19	.005	0.26	.005			7	A7 III	15,2012
81753	-28 07271	HR 3745	★ AB	9 24 34	-28 34.2	5	6.10	.008	-0.10	.003	-0.55	.009	17	B6 Ve	15,1075,1079,1637,2012
		AJ77,733 T12# 28		9 24 34	-52 31.2	1	12.82		0.55		-0.01		4		125
81356	+70 00563			9 24 35	+69 36.1	1	9.24		0.42		0.34		2		1776
81802	-45 05150			9 24 35	-45 54.5	3	7.99	.002	0.18	.000	0.08	.004	16	A2/3 (III)	977,1075,2011
81827	-52 02354			9 24 35	-52 38.9	1	9.79		0.04		-0.03		6	B9 V	125
81812	-45 05151			9 24 38	-46 11.2	1	10.26		0.38		-1.12		1	F3 V	389
	-52 02356			9 24 38	-52 34.3	1	10.88		0.24		0.08		4		125
81848	-52 02360	HR 3753		9 24 40	-53 09.7	5	5.10	.022	-0.12	.005	-0.52	.007	15	B6 V	15,16,362,401,2012
298448	-52 02362	LSS 1303		9 24 43	-52 45.5	5	9.63	.029	0.33	.011	-0.60	.017	15	B0.5III	125,362,401,540,976
81847	-50 04147			9 24 45	-50 57.4	1	8.35		-0.07		-0.40		2	Ap Si	401
81825	-43 05284			9 24 47	-44 01.3	3	7.49	.012	0.02	.000	0.01	.011	16	A1 IV	977,1075,2011
298411	-51 03816		A	9 24 47	-51 56.5	1	10.64		0.22		0.50		1	G5	362
298411	-51 03816		B	9 24 47	-51 56.5	1	11.75		0.28		0.18		1		362
298411	-51 03816		C	9 24 47	-51 56.5	1	12.48		0.81		0.80		1		362
81728	-8 02678	HR 3744	★ AB	9 24 48	-09 00.3	4	6.53	.005	0.05	.000	0.02	.000	12	A2 V	15,402,1075,2012
81767	-16 02787			9 24 49	-16 26.6	1	9.49		0.99		0.76		1	K2 (III)	3072
81844	-45 05152			9 24 50	-45 54.8	3	8.58	.001	0.15	.006	0.11	.006	16	A3 V	977,1075,2011
81891	-54 02290			9 24 53	-54 39.4	3	7.17	.008	-0.15	.036	-0.60	.008	20	B2/3 IV	158,362,401
81656	+32 01884	IDS09219N3154	AB	9 24 55	+31 41.1	1	8.83		0.48		0.04		1	G0	3032
81656	+32 01884	IDS09219N3154	C	9 24 55	+31 41.1	1	12.27		0.65		0.15		1		3032
	-56 02167	IC 2488 - 10		9 24 57	-56 41.4	1	11.55		0.52		0.23		2		1644
81799	-21 02802	HR 3749		9 25 01	-22 07.4	5	4.69	.012	1.14	.007	1.15	.025	16	K1 III	15,1075,2012,3016,8015
		Steph 776		9 25 05	+10 21.4	1	11.05		1.13		1.10		1	K7	1746
302174	-56 02171	IC 2488 - 12		9 25 05	-56 41.4	1	9.88		1.31		1.18		2		1644
81797	-8 02680	HR 3748	★ A	9 25 08	-08 26.5	13	1.98	.007	1.44	.006	1.72	.008	62	K3 II-III	3,15,198,1020,1034*
81907	-51 03824			9 25 08	-51 33.1	2	8.04	.000	-0.01	.010	-0.28	.030	5	B8 II/III	401,402
	-52 02369	LSS 1306		9 25 08	-53 14.7	2	11.24	.020	0.54	.010	-0.45	.015	8		125,362
		LSS 1307		9 25 10	-52 48.5	1	11.99		0.46		-0.27		4	B3 Ib	125
81921	-48 04707			9 25 12	-48 28.6	1	6.83		-0.05		-0.17		2	B9.5II/III	401
		IC 2488 - 72		9 25 16	-56 42.1	1	12.23		1.32		1.24		1		1644
		IC 2488 - 74		9 25 16	-56 42.1	1	13.23		0.76		0.22		3		1644
298447	-52 02371			9 25 17	-52 43.3	1	10.15		0.13		0.09		2	A2	125
81809	-5 02802	HR 3750	★ AB	9 25 18	-05 51.1	6	5.37	.011	0.64	.006	0.13	.010	24	G2 V	15,101,1067,1417,2012,3077
	-56 02177	IC 2488 - 18		9 25 19	-56 44.6	1	11.33		0.12		-0.21		3		1644
	-56 02176	IC 2488 - 17		9 25 19	-56 46.6	1	11.98		0.18		-0.07		3		1644
81948	-51 03830			9 25 22	-52 14.9	2	8.63	.005	-0.02	.000	-0.13	.010	6	B8/9 V	125,401
	-56 02179	IC 2488 - 20		9 25 23	-56 44.2	1	10.37		1.52		1.55		3		1644
	-56 02180	IC 2488 - 21		9 25 23	-56 45.0	1	11.82		0.16		-0.03		3		1644
81688	+46 01509	HR 3743	★ A	9 25 24	+45 49.3	2	5.43	.016	0.98	.007	0.74		5	K0 III	71,3016
81990	-59 01394			9 25 24	-59 21.7	1	6.52		-0.09		-0.42		2	B7 V	401
81947	-51 03832	IDS09238S5121	AB	9 25 27	-51 34.1	1	8.74		0.10		0.08		3	A1 V	402
81946	-48 04712			9 25 28	-48 37.7	1	7.78		-0.11		-0.49		2	B5 III	401
	-56 02181	IC 2488 - 22		9 25 28	-56 45.6	1	12.10		0.16		-0.02		3		1644
82303	-76 00585			9 25 31	-77 02.1	2	7.73	.009	1.81	.014	2.23	.023	8	K5 III	874,1533
	-50 04163			9 25 33	-50 23.1	1	10.82		0.85		0.10		2		362
81919	-34 05895	HR 3756	★ AB	9 25 34	-34 47.3	2	6.64	.005	0.21	.005			7	A5 III/IV	15,2027
81703	+50 01644			9 25 36	+50 15.0	1	7.07		1.58		1.92		2	M1	1733
302218	-56 02184	IC 2488 - 24		9 25 38	-56 43.4	1	9.36		0.57		0.07		3	F5	1644
82126	-68 00945			9 25 38	-69 03.2	1	8.30		0.34				4	F0 IV	2012
	-53 02428			9 25 39	-53 44.7	1	10.67		0.35		-0.62		2		362
82068	-64 01037	HR 3761		9 25 39	-64 42.8	3	6.04	.004	0.14	.005	0.09	.005	11	A3 Vn	15,1075,2038
		IC 2488 - 70		9 25 41	-56 45.3	1	12.68		0.35		0.16		3		1644
		IC 2488 - 71		9 25 41	-56 48.1	1	12.89		1.33		0.96		1		1644
82259	-75 00582			9 25 41	-75 33.4	1	7.88		1.02		0.65		3	G5/8 III	1704
81904	-23 08402			9 25 42	-23 27.6	2	8.05	.012	0.97	.006	0.72		3	G8/K0 III	1467,1642
82084	-63 01126	IW Car		9 25 43	-63 24.7	2	7.90		0.84	.014	0.69	.018	24	F7/8	897,3049
		AJ82,474 # 153		9 25 44	-50 05.1	1	12.03		1.36		0.40		1		779
82003	-53 02433			9 25 44	-53 45.0	2	8.41	.012	-0.08	.010	-0.33	.007	5	B9 V	540,976
		IC 2488 - 76		9 25 44	-56 47.6	1	12.14		0.28		0.14		3		1644
		ApJ239,112 # 14		9 25 44	-77 00.5	2	12.60	.000	0.92	.000	0.48	.000	4		874,1533
302208	-56 02186	IC 2488 - 26		9 25 45	-56 40.2	1	9.46		-0.01		-0.26		3	B8	1644
81702	+56 01386			9 25 46	+56 27.7	1	6.89		0.38		-0.01		2	F2	1733
81858	+9 02188	HR 3754	★ AB	9 25 47	+09 16.5	4	5.41	.009	0.60	.004	0.13	.007	10	F9 V	15,292,1256,3026
		LSS 1309		9 25 47	-52 48.0	2	12.16	.075	0.66	.000	-0.39	.010	6	B3 Ib	125,846
	-52 02378	LSS 1310		9 25 47	-53 08.9	1	10.42		0.43		-0.52		2	B6	362
		IC 2488 - 73		9 25 47	-56 41.2	1	12.31		1.56		1.50		1		1644
302219	-56 02189	IC 2488 - 28		9 25 48	-56 44.3	1	11.01		0.12		0.09		3		1644
		IC 2488 - 75		9 25 48	-56 45.4	1	12.23		0.62		0.11		3		1644
81873	+8 02226	HR 3755	★ A	9 25 49	+08 24.4	4	5.74	.026	1.05	.016	0.91	.018	11	K0 III	12,15,1415,3024
81873	+8 02226	IDS09232N0837	B	9 25 49	+08 24.4	2	11.12	.085	0.85	.035	0.44	.040	4		12,3024
	-80 00349			9 25 50	-80 19.4	3	10.11	.021	0.58	.022	-0.14	.038	4	G0	1236,1705,3078
	+38 02037			9 25 51	+38 17.0	1	10.41		1.29		1.21		3	K5	1723
81971	-43 05302			9 25 51	-43 25.1	3	8.85	.000	0.23	.003	0.12	.003	16	A9 V	977,1075,2011
		G 41 - 39		9 25 52	+18 54.5	1	16.60		0.31		-0.65		2		3060
		HA 192 # 2289		9 25 53	-60 23.5	1	12.01		0.55		0.16		6		1499
82554	-80 00350	HR 3795		9 25 53	-80 34.3	5	5.34	.017	0.45	.006	0.04	.022	18	F3/5 III/V	15,158,1075,2038,3026
		HA 192 # 1174		9 25 54	-60 21.5	1	12.25		0.69		0.13		9		1499
	+21 02033	G 41 - 40		9 25 57	+20 55.6	1	9.44		1.06		1.05		1	K5	333,1620

HD	DM	Other Id	N	Rem	α_{1950}	δ_{1950}	S	V	σ_V	B-V	σ_{B-V}	U-B	σ_{U-B}	n	Spectrum	References
	-56 02191	IC 2488 - 30			9 25 58	-56 42.5	1	10.85		0.12		-0.26		3		1644
302209	-56 02192	IC 2488 - 31			9 25 59	-56 42.1	1	11.59		0.18		-0.07		3	F0	1644
		LP 547 - 52			9 26 02	+06 31.5	1	12.27		1.48		1.19		1		1696
302224	-56 02200	IC 2488 - 35			9 26 05	-56 46.6	1	11.09		0.16		-0.04		3		1644
302225	-56 02202	IC 2488 - 37			9 26 07	-56 47.2	1	8.87		0.13		-0.12		3	B8	1644
82102	-58 01530				9 26 07	-58 42.9	1	8.07		-0.08		-0.63		2	B8 II/III	401
302217	-56 02203	IC 2488 - 38			9 26 08	-56 44.9	1	10.90		0.16		-0.13		3		1644
302215	-56 02204	IC 2488 - 39			9 26 10	-56 46.2	1	9.40		1.40		1.30		3	K3	1644
81772	+58 01192	IDS09225N5834			9 26 12	+58 20.9	1	8.24		0.32		0.15		3	F0 IV	8071
81790	+56 01388	HR 3747			9 26 13	+55 57.9	1	6.51		0.36		-0.03		2	F3 Vs	1733
302211	-56 02208	IC 2488 - 42			9 26 13	-56 45.8	1	10.44		0.10		-0.32		3		1644
	-7 02815				9 26 15	-08 14.3	1	10.49		1.18		1.09		1	K5 V	3072
		G 161 - 32			9 26 17	-11 56.6	1	12.27		1.47				1		3062
	-44 05498				9 26 18	-44 22.0	2	9.45	.001	1.08	.000	0.91		10	K0	863,1075
		LP 370 - 26			9 26 21	+26 11.5	1	16.43		1.73				1		1691
		G 116 - 32			9 26 21	+34 17.9	1	15.34		1.48		1.38		1		3062
302214	-56 02215	IC 2488 - 48			9 26 21	-56 50.4	1	9.17		1.53		1.54		3	K5	1644
	-8 02689				9 26 23	-09 03.1	2	10.50	.014	1.42	.010			5	M0	1705,2033
82111	-54 02321				9 26 23	-55 07.0	3	7.79	.004	-0.11	.008	-0.54	.000	15	B5 IV/V	164,362,401
		G 161 - 33			9 26 25	-07 08.7	5	12.10	.025	1.52	.020	1.11		9		158,694,940,1705,3073
		G 161 - 34			9 26 28	-07 08.7	3	15.03	.008	1.85	.030	1.25		4		940,1705,3073
	-53 02454	LSS 1311			9 26 28	-53 29.3	1	10.94		0.60		-0.48		2		362
	-56 02214	IC 2488 - 47			9 26 29	-56 52.3	1	11.85		0.29		0.02		3		1644
81980	-0 02201	HR 3758		★ AB	9 26 30	-01 02.3	2	6.26	.005	0.28	.005	0.09	.015	7	F0 Vn	15,1415
298451	-53 02455				9 26 30	-53 23.3	1	9.76		0.93		0.63		6	K0	125
82080	-45 05177				9 26 32	-45 32.7	2	9.54	.004	0.31	.002	0.02		10	F0 V	863,2011
81954	+21 02036				9 26 33	+21 08.0	1	8.31		1.11		1.07		2	G5	1648
	+9 02190	G 41 - 41			9 26 35	+08 51.5	7	11.15	.011	0.39	.009	-0.20	.008	15	A0	158,333,516,927,1064*
296931	-46 05206				9 26 36	-47 05.3	1	11.23		0.59		-0.16		2	G5	1696
81997	-2 02901	HR 3759		★ A	9 26 37	-02 33.0	11	4.59	.015	0.46	.007	0.02	.017	64	F6 V	3,15,418,1008,1013*
81997	-2 02902	IDS09240S0220	B		9 26 37	-02 33.0	3	7.18	.004	0.88	.009			11		2013,2033,3024
82350	-71 00833	HR 3777			9 26 37	-71 23.1	4	5.47	.010	1.08	.000	0.98	.008	25	K2 III	15,1075,2038,3077
81940	+28 01765				9 26 40	+27 36.8	1	8.37		0.16		0.08		5	A0	8053
82109	-44 05503	IDS09248S4504	AB		9 26 42	-45 16.7	6	7.14	.008	-0.13	.012	-0.57	.007	27	B5 V	362,977,1075,1770,2001,2011
		G 161 - 35			9 26 45	-07 31.9	1	15.96		1.54		1.00		1		3062
81977	+18 02207				9 26 46	+17 52.3	1	7.27		1.49		1.82		2	K0	1648
82121	-44 05505	IDS09248S4504			9 26 48	-45 16.9	9	7.62	.008	-0.15	.007	-0.63	.008	58	B5 IV/V	395,402,474,768,977,1075*
81912	+47 01677				9 26 51	+47 03.3	1	9.27		0.62		0.14		2	G5	1601
		G 161 - 36			9 26 51	-03 57.0	2	14.77	.015	0.13	.015	-0.59	.005	5		1663,3060
		G 161 - 37			9 26 51	-03 57.3	1	15.22		1.61		1.25		2		3060
302212	-56 02225	IC 2488 - 56			9 26 51	-56 43.7	1	9.57		1.46		1.45		3	K7	1644
233640	+52 01397	G 195 - 30			9 26 52	+52 22.6	1	9.47		0.58		0.00		4	G2	1566
82043	-1 02268	HR 3760			9 26 52	-01 59.1	2	6.13	.005	0.22	.000	0.15	.010	7	F0 III	15,1078
82077	-20 02915	HR 3763			9 26 54	-20 31.8	2	5.66	.005	1.60	.000	1.85		6	M1 III	2031,3005
82152	-43 05319	IDS09250S4322	A		9 26 55	-43 35.4	7	8.11	.010	0.49	.009	0.00	.010	33	F7 V	389,460,861,977,1075*
81787	+68 00572	IDS09227N6759	A		9 26 56	+67 45.7	1	7.53		0.15		0.15		1	A2	1776
82282	-65 01110				9 26 56	-65 52.9	1	8.87		0.88		0.35		1	K1 V	3072
81963	+39 02252				9 26 59	+39 08.5	2	8.18	.005	1.13	.010	1.08	.005	5	K1 III	1501,1601
82074	-3 02693	HR 3762			9 27 02	-04 01.6	3	6.26	.005	0.84	.000	0.40	.000	8	G6 IV	15,402,1417
		AG +70 0369			9 27 03	+70 07.5	1	12.39		0.79		0.32		3		1723
	-36 05714				9 27 03	-37 18.6	2	9.91	.000	0.65	.015	0.08		5		1594,6006
	-50 04189	IDS09254S5028	AB		9 27 06	-50 41.0	1	9.48		1.95		2.26		1		864
	-4 02639	G 161 - 38			9 27 07	-05 09.2	5	9.72	.018	1.31	.014	1.29		5	M0:	158,6009
82150	-35 05724	HR 3765			9 27 11	-35 43.9	4	4.51	.005	1.44	.000	1.68	.000	20	K3 IIIa	15,1075,2012,8015
82568	-76 00587				9 27 11	-77 01.3	2	9.59	.000	1.23	.014	0.91	.019	8	G8/K0 III	874,1533
	-56 02232	IC 2488 - 62			9 27 12	-56 45.6	1	11.57		0.48		0.02		1		1644
82184	-44 05512				9 27 13	-45 14.2	2	9.07	.002	0.92	.003	0.64		10	G6 III	863,2011
	-56 02233	IC 2488 - 63			9 27 13	-56 46.6	1	11.19		0.39		0.01		1		1644
82183	-44 05511				9 27 14	-44 34.9	3	8.50	.006	1.13	.006	1.09		13	K0 III	977,2001,2011
82165	-37 05817	HR 3766			9 27 15	-38 11.0	2	6.18	.005	0.22	.00			7	A7 Vn	15,2027
82106	+6 02182				9 27 19	+05 52.4	7	7.20	.011	1.01	.014	0.86	.023	33	K3 V	22,1080,1088,1197*
	+48 01777				9 27 20	+48 28.9	2	10.74	.020	-0.31	.010	-1.22	.015	5	sdO	272,308
		LSS 1313			9 27 22	-53 10.6	1	11.81		0.61		-0.40		3	B3 Ib	125
82242	-52 02396	LSS 1314		★ AB	9 27 25	-52 50.2	2	8.76	.000	0.84	.015	0.52	.044	9	A3/4 II	125,8100
300023	-53 02474				9 27 25	-54 12.8	1	9.73		0.01		-0.71		2	B3	362
82207	-43 05326	IDS09255S4406	A		9 27 26	-44 19.2	6	7.00	.007	0.54	.007	0.04	.004	20	F8 V	389,977,1279,2011*
82224	-42 05328				9 27 27	-42 58.2	7	6.60	.006	0.48	.004	0.06	.008	33	F5 IV	278,389,863,977,1075*
297447	-48 04742				9 27 29	-48 33.1	2	10.30	.003	0.54	.016	-0.20	.013	4	B5 Ib	540,976
81953	+59 01238				9 27 32	+58 56.6	1	7.42		1.42		1.66		2	K2	1733
81882	+67 00594				9 27 32	+67 04.5	1	9.10		0.30		0.11		1	F0	1776
82181	-24 08145				9 27 32	-25 18.5	1	7.45		1.14		1.12		3	K1 III	1657
82347	-61 01277	HR 3776			9 27 32	-62 03.2	7	5.91	.034	1.10	.005	0.98	.009	25	K1 III	15,243,1075,1520,1754*
82406	-66 01018	HR 3783			9 27 32	-66 29.0	2	5.90	.005	0.01	.000	-0.02	.000	7	A0 V	15,1075
82105	+22 02100				9 27 34	+22 02.0	1	6.96		0.22		0.10		2	A3	1648
82180	-22 02623	HR 3767			9 27 34	-23 07.5	4	6.24	.013	1.56	.015	1.83		12	K2/3 III	15,1075,2031,3005
82241	-43 05332	IDS09255S4406	B		9 27 35	-44 18.8	6	6.98	.010	0.50	.006	0.00	.009	20	F7 V	389,977,1279,2011*
		G 48 - 19			9 27 36	-00 44.9	1	11.48		1.37		1.18		1		1620
82278	-51 03865				9 27 36	-52 12.3	2	7.25	.000	-0.06	.004	-0.44	.008	18	B8 II	125,401
81937	+63 00845	HR 3757		★ A	9 27 37	+63 16.9	7	3.66	.009	0.33	.005	0.10	.003	19	F0 IV	15,39,254,1008,1118*
81937	+63 00845	IDS09236N6330	B		9 27 37	+63 16.9	1	9.19		1.17		1.28		2		3024

Table 1

513

HD	DM	Other Id	N Rem	α_{1950}	δ_{1950}	S	V	σ_V	B–V	σ_{B-V}	U–B	σ_{U-B}	n	Spectrum	References
82221	−32 06441			9 27 37	−33 05.4	2	7.93	.010	1.32	.010	1.25	.019	3	K1p Ba	565,3048
297487	−49 04428			9 27 37	−49 43.9	1	10.25		0.82		-0.02		2	F8	362
	−35 05731			9 27 38	−36 13.5	1	10.98		0.59		-0.05		2		1696
	−56 02239	IC 2488 - 65		9 27 40	−56 43.5	1	11.44		0.59		0.00		1		1644
82205	−26 07117	HR 3770	⋆ AB	9 27 41	−26 22.2	5	5.48	.007	1.36	.006	1.49	.022	21	K2 III	15,1075,1637,2030,3005
82239	−39 05561			9 27 41	−39 45.2	2	8.03	.001	0.94		0.59		4	K0 III	861,2012
82087	+34 01999	HR 3764	⋆ A	9 27 42	+33 52.6	1	5.85		1.05				2	K0	71
82325	−54 02343			9 27 42	−54 34.6	1	8.14		-0.06		-0.35		2	B8 V	401
300158	−56 02240	IC 2488 - 66		9 27 43	−56 39.6	1	11.62		0.25		0.08		3		1644
82297	−48 04748			9 27 46	−48 29.7	1	7.49		0.89		0.57		1	G6 III	389
		WW Leo		9 27 47	+07 25.5	2	12.21	.134	0.27	.021	0.07	.045	2		597,699
82238	−35 05732			9 27 48	−36 15.8	1	9.55		0.42				4	F3 V	2012
	+33 01872	G 116 - 35		9 27 49	+32 57.2	1	10.19		0.77		0.24		3	G8	7010
82140	+15 02062			9 27 51	+15 28.8	1	8.38		0.63				14	F8	1351
82235	−25 07251			9 27 58	−25 47.7	1	8.21		0.53		0.01		2	G3 V	389
		LTT 12511		9 28 00	+27 11.	1	11.04		1.29				2	K5	1625
82232	−14 02867	HR 3772		9 28 00	−15 21.4	4	5.85	.013	1.19	.010	1.22		13	K2 III	15,1075,2031,3005
	+18 02213	G 41 - 42		9 28 03	+18 28.6	2	10.07	.015	1.04	.015	0.89	.017	3	K8	333,1620,3072
81936	+70 00564			9 28 09	+69 57.1	1	8.47		0.45		0.00		2	F5	1776
		G 48 - 20		9 28 12	+00 32.9	2	11.72	.005	1.60	.000	1.23	.058	3	M4	333,1620,3073
82174	+29 01908			9 28 12	+29 30.6	1	8.93		1.12		1.04		2	G5	1733
82700	−75 00584			9 28 15	−75 54.7	1	9.28		0.70		0.20		3	G5 V	1696
82468	−61 01278			9 28 18	−62 08.5	1	6.55		0.09		0.08		4	A2 V	243
82342	−31 07352		A	9 28 20	−31 53.2	1	8.38		0.99		0.69		2	K3/4 V	3072
82342	−31 07352		B	9 28 20	−31 53.2	1	13.08		1.46		0.97		1		3072
82214	+12 02049			9 28 23	+12 04.7	1	8.58		0.98		0.73		2	K0	1648
82191	+28 01768			9 28 23	+27 36.5	1	6.58		0.08		0.07		5	A0	8053
82419	−50 04204	HR 3784		9 28 23	−51 17.8	3	5.45	.009	-0.09	.013	-0.39		9	B8 III	15,127,2012
		LP 165 - 63		9 28 24	+47 44.9	1	11.21		1.29				3		1726
82420	−52 02415	NGC 2910 - 23		9 28 25	−52 34.9	1	9.24		0.28				7	B9 II/III	2042
82386	−44 05529			9 28 26	−44 38.4	4	7.57	.000	0.48	.004	0.02	.006	17	F6 IV/V	389,977,1075,2011
82198	+35 02015	HR 3769		9 28 30	+35 19.5	2	5.39	.016	1.52	.011	1.81		5	M1 IIIab	71,3016
298479	−52 02417	NGC 2910 - 20		9 28 30	−52 38.6	1	11.27		0.06				3	A2	2042
297486	−49 04443	LSS 1315		9 28 34	−49 40.7	1	10.23		0.72		-0.27		2	F5	362
82416	−44 05532			9 28 36	−44 39.5	4	8.48	.004	0.39	.002	0.02	.007	17	F2 IV/V	402,977,1075,2011
		L 64 - 40		9 28 36	−71 20.	1	15.44		0.17		-0.44		2		3065
82466	−54 02363			9 28 37	−54 43.1	1	8.49		-0.08		-0.33		2	B9 IV	401
82384	−31 07355	HR 3781	⋆ AB	9 28 38	−31 40.1	2	5.75	.010	0.06	.005	0.05		6	A1 V + A1 V	404,2031
82457	−51 03880			9 28 38	−52 05.1	1	7.99		-0.08		-0.54		2	B7 II/III	401
	−52 02424	NGC 2910 - 47		9 28 39	−52 40.6	1	11.82		0.26				2		2042
82436	−45 05202		V	9 28 41	−45 21.0	5	6.62	.005	1.20	.005	1.22	.003	30	K2/3 III	395,863,977,1075,2011
82434	−39 05580	HR 3786	⋆ AB	9 28 44	−40 14.8	5	3.58	.014	0.36	.004	-0.06	.025	15	F3 IV+F0 IV	15,1075,2012,2024*
298480	−52 02426	NGC 2910 - 46		9 28 44	−52 41.2	1	10.69		0.55				2	F5	2042
82455	−47 04996			9 28 45	−47 23.1	7	8.65	.008	0.65	.007	0.07	.013	29	G3 V	116,863,977,1020,1075*
		NGC 2910 - 43		9 28 45	−52 43.2	1	13.42		0.18				2		2042
82308	+23 02107	HR 3773		9 28 52	+23 11.4	6	4.31	.011	1.54	.006	1.88	.024	15	K5 III	15,1080,1363,3016*
82518	−54 02368			9 28 52	−55 00.7	3	7.29	.017	1.61	.020	1.88	.030	9	M2 III	16,2012,8100
82565	−62 01246			9 28 52	−62 29.1	1	9.38		1.12		0.89		4	K0 III	243
	−12 02918	IDS09265S1303	AB	9 28 53	−13 16.1	9	10.06	.013	1.53	.019	1.16	.019	30	M3	116,158,1006,1197,1764*
82332	+16 01984			9 28 54	+15 59.5	1	7.67		1.14		1.07		2	K0	1648
		G 161 - 43		9 28 54	−02 28.0	1	14.74		1.44				1		3062
	+36 01970	G 116 - 36		9 28 55	+36 33.7	1	10.16		1.59		1.26		3	M0	7008
298429	−51 03884	LSS 1316		9 28 55	−51 26.3	4	9.67	.015	0.56	.015	-0.49	.032	11	O9 III	362,540,976,2021
82309	+20 02332			9 28 56	+20 13.7	2	7.43	.030	1.28	.000	1.48	.005	3	K3 III	1003,3077
82536	−57 02090	HR 3793	⋆ AB	9 28 56	−58 08.5	4	5.87	.068	1.70	.011	1.68		19	M2 III	15,1075,2027,3005
298428	−51 03885			9 29 03	−51 24.6	2	9.73	.004	0.90	.028	0.51	.003	8	F5	540,976
82533	−52 02440			9 29 03	−53 18.1	1	8.34		0.67		0.16		2	G5 III	389
296923	−45 05210			9 29 05	−45 44.7	2	10.99	.090	0.60	.000	-0.08	.005	2	A7	78,3077
		NGC 2910 - 40		9 29 05	−52 44.3	1	13.60		0.54				2		2042
302285	−57 02093			9 29 07	−57 56.4	2	10.40	.011	0.11	.001	-0.30	.008	4	B7 III	540,976
82516	−46 05238			9 29 08	−47 09.4	3	8.50	.012	0.91	.009	0.65	.021	14	K2 V	977,2011,3008
82331	+27 01770			9 29 10	+27 03.0	1	7.91		1.03				2	G5	20
82515	−44 05539			9 29 11	−45 14.5	2	8.79	.003	0.91	.005	0.51		12	G5 III	977,2011
298425	−50 04215	LSS 1317	A	9 29 11	−51 12.0	4	9.71	.029	0.54	.012	-0.49	.047	9	O9 V	362,540,976,2021
298425	−50 04215		B	9 29 11	−51 12.0	1	10.03		0.49		-0.55		1		362
		Ton 16		9 29 12	+29 04.	1	16.51		0.10		-0.68		1		98
82428	−9 02856	HR 3785		9 29 13	−10 19.8	3	6.13	.004	0.25	.004	0.09	.010	11	F0 Vn	15,1075,2012
		Steph 780		9 29 14	+29 22.8	1	11.51		1.46		1.16		1	M0	1746
82395	+11 02053	HR 3782		9 29 15	+11 31.3	4	4.97	.005	1.04	.008	0.88	.008	15	K0 III	3,1080,1363,3016
82381	+10 02014	HR 3779	⋆ A	9 29 17	+09 56.2	4	5.08	.020	1.36	.008	1.52	.010	6	K3 III	12,542,1080,3077
82381	+10 02015	IDS09266N1009	B	9 29 17	+09 56.2	2	9.56	.015	0.44	.000	0.04	.005	2	F6 V	12,542
		NGC 2910 - 27		9 29 19	−52 40.4	1	12.92		0.15				2		2042
	−52 02452	NGC 2910 - 34		9 29 19	−52 42.6	1	11.97		0.21				2		2042
82353	+37 01995			9 29 21	+37 29.0	1	8.47		0.13		0.18		2	A2	1375
82552	−45 05215			9 29 21	−45 47.2	4	7.98	.003	0.29	.005	0.12	.013	17	A7 III	402,977,1075,2011
		NGC 2910 - 35		9 29 22	−52 44.7	1	11.10		1.85				2		2042
82394	+22 02102	IDS09266N2218	AB	9 29 23	+22 04.7	1	7.45		1.00		0.78		2	K0 III	1080
82513	−31 07369	HR 3789		9 29 24	−31 39.0	3	5.92	.007	0.26	.017	0.16		10	A9 IV	15,2027,8071
	−49 04459			9 29 25	−49 59.7	1	11.23		0.51		-0.01		2		362
82446	−0 02211	HR 3787		9 29 26	−00 57.8	7	4.56	.005	0.10	.004	0.09	.018	24	A3 V	15,1075,1363,1425*

HD	DM	Other Id	N	Rem	α_{1950}	δ_{1950}	S	V	σ_V	B–V	σ_{B-V}	U–B	σ_{U-B}	n	Spectrum	References
82551	−45 05217				9 29 26	−45 23.6	3	8.15	.003	0.07	.002	0.11	.003	16	A1 V	977,1075,2011
		G 47 - 50			9 29 27	+27 01.0	1	13.24		1.16		1.06		1		1620
82514	−35 05751	HR 3790		★ A	9 29 28	−35 29.5	6	5.87	.010	1.29	.007	1.47		24	K3 III	15,1075,2013,2020*
298476	−52 02456	NGC 2910 - 59			9 29 29	−52 37.1	1	9.81		2.27		3.64		1		864
82477	−9 02858	HR 3788			9 29 30	−10 09.0	6	6.13	.009	1.18	.005	1.17	.000	26	K0	15,402,1075,1417,2013,2029
82578	−47 05002				9 29 30	−47 43.9	6	6.53	.002	0.26	.002	0.04	.003	29	A9 IV/V	278,977,977,1075,1770,2011
82328	+52 01401	HR 3775		★ AB	9 29 31	+51 54.4	9	3.17	.010	0.46	.006	0.02	.008	31	F6 IV	15,22,1003,1007,1008*
	−29 07538				9 29 32	−30 09.1	1	11.04		0.50		−0.19		2		1696
82668	−56 02270	HR 3803, N Vel			9 29 42	−56 48.8	4	3.14	.014	1.55	.000	1.87	.005	20	K5 III	15,1075,2012,3055
82380	+50 01657	HR 3778			9 29 46	+49 39.6	1	6.78		0.10		0.10		2	A4 V	1733
82508	−10 02856	IDS09273S1024		A	9 29 46	−10 37.3	1	7.58		0.71		0.35		1	G1/2 III	1738
		LP 91 - 197			9 29 48	+56 21.0	1	11.00		0.67		0.03		3		1723
82443	+27 01775	IDS09269N2726			9 29 50	+27 12.8	3	7.00	.014	0.77	.005	0.34	.009	4	K0	1080,3026,8041
		Steph 782			9 29 51	+24 25.6	1	11.54		1.28		1.13		1	K7	1746
82664	−50 04230				9 29 52	−51 01.9	1	8.03		0.25		0.09		2	A4/5 IV/V	127
82595	−36 05745				9 29 54	−36 38.3	1	8.18		1.03		0.94		2	K2 II/III	536
	−49 04469	DR Vel			9 29 54	−49 26.0	2	9.16	.037	1.35	.026	0.97	.008	2		689,1587
82665	−51 03902				9 29 57	−51 43.0	1	8.88		1.12		1.02		2	K1 III+G5/8	127
82558	−10 02857	LQ Hya			9 30 01	−10 57.8	4	7.86	.038	0.92	.005	0.58	.032	18	K0	258,1738,2012,3062
82573	−18 02708	HR 3796			9 30 01	−19 10.7	2	5.74	.000	0.14	.005	0.12		7	A5 V	2031,8071
		CCS 1485			9 30 02	−63 10.2	1	11.13		4.32				1		864
296920	−44 05559				9 30 05	−45 12.8	1	11.79		0.61		−0.15		2	A0	362
82543	+2 02217	HR 3794		★ AB	9 30 06	+02 05.2	4	6.11	.009	0.61	.009	0.26	.008	12	F7 IV-V	15,1355,1415,3077
82210	+70 00565	HR 3771, DK UMa			9 30 06	+70 03.1	6	4.57	.015	0.77	.003	0.33	.012	21	G4 III-IV	15,1025,1118,1355*
81817	+81 00302	HR 3751			9 30 07	+81 33.0	4	4.28	.018	1.49	.009	1.70	.019	11	K3 III	15,1363,3016,8015
		LP 428 - 4			9 30 09	+15 22.8	1	12.63		0.61		−0.07		2		1696
82590	−16 02817				9 30 09	−17 12.7	4	9.42	.011	0.47	.006	0.05	.007	14	F5/7wA9	158,742,1594,1775
82189	+72 00462	HR 3768			9 30 13	+72 25.8	2	5.71	.064	0.47	.045	−0.03		3	F8 V	71,254
82695	−47 05007				9 30 13	−47 58.4	1	7.64		1.12		1.05		1	K0 III	389
82712	−48 04778				9 30 18	−49 10.8	1	7.72		1.01		0.75		1	G8 III/IV	389
82694	−40 05284	HR 3807		★ AB	9 30 20	−40 25.6	4	5.35	.008	0.90	.005			18	G8 III	15,1075,2013,2029
82764	−56 02282				9 30 22	−56 46.0	1	7.12		0.00		−0.36		2	B8 III	401
82523	+29 01913	HR 3792		★ A	9 30 23	+28 35.4	3	6.52	.005	0.13	.009	0.08	.010	8	A3 Vnn	985,1022,3016
82523	+29 01913	IDS09274N2849		B	9 30 23	+28 35.4	1	12.08		0.62		0.07		2		3032
82442	+53 01347				9 30 23	+52 40.0	1	7.74		0.20		0.07		2	A3	1566
		WLS 920 35 # 8			9 30 24	+34 53.7	1	12.57		0.71		0.26		2		1375
82737	−52 02475				9 30 25	−53 19.2	1	7.68		−0.08		−0.49		2	B8 V	401
82522	+37 01998	HR 3791			9 30 27	+36 42.6	2	6.18	.002	1.26	.004	1.37		5	K3 III	71,1501
	−48 04783				9 30 31	−48 48.2	1	11.06		0.77		−0.30		3		362
	+77 00374	G 252 - 39			9 30 32	+76 46.3	1	9.86		1.17				1	M0	1746
82660	−12 02926	HR 3802			9 30 32	−13 17.7	4	5.94	.009	1.50	.000	1.74		13	K4 III	15,1075,2006,3005
	+20 02337	AB Leo			9 30 33	+20 05.2	1	10.40		1.56		1.47		1		793
		G 235 - 30			9 30 33	+63 13.7	1	13.16		1.03		0.80		1		1658
82858	−66 01025	HR 3813			9 30 33	−66 29.9	3	6.26	.004	1.35	.000	1.38	.000	11	K1 III	15,1075,2038
82638	−7 02836	HR 3801			9 30 34	−08 17.0	3	6.12	.004	0.98	.005	0.71	.005	14	K0	15,1417,2029
298454	−50 04241				9 30 37	−51 09.3	1	10.44		0.58		0.11		2	B8	127
311573	−72 00831				9 30 40	−72 52.0	1	9.23		0.46		0.01		2	F5	6006
		Ton 17			9 30 42	+29 23.	3	15.90	.053	0.25	.044	−0.60	.012	6	DA	98,316,3060
82834	−60 01455	IDS09294S6048		AB	9 30 43	−61 00.8	1	7.14		0.06		−0.06		2	B9.5IV	401
	+33 01878				9 30 44	+32 45.4	1	9.77		0.85		0.60		2		1726
		POSS 92 # 6			9 30 46	+60 12.8	1	18.07		0.90				3		1739
81263	+85 00147				9 30 47	+85 19.1	1	8.24		1.50		1.89		2	K5	1733
83064	−74 00602				9 30 49	−74 32.9	1	8.20		0.54				4	F6/7 V	2012
82674	−6 02939	HR 3805			9 30 51	−06 58.0	4	6.25	.011	1.17	.006	1.25	.000	18	K0	15,1417,2013,2029
298460	−51 03915				9 30 53	−51 32.0	1	9.56		1.28		1.30		2	K5	127
82734	−20 02936	HR 3808			9 30 54	−20 53.6	1	5.00	.012	1.02	.004	0.87		16	K0 IV	15,1075,2030,3016
298471	−52 02486			A	9 30 55	−52 28.7	1	9.91		0.22		−0.58		2	A0 IVp	362
298471	−52 02486			B	9 30 55	−52 28.7	1	9.88		0.25		−0.55		2		362
298464	−51 03918				9 30 59	−51 51.8	1	10.28		0.04		0.01		2	A2	127
82812	−52 02489	IDS09293S5256		AB	9 30 59	−53 08.8	1	8.44		−0.05		−0.34		2	B8 IV/V	401
82901	−62 01253	HR 3816, R Car		★ AB	9 30 59	−62 34.0	2	5.10	.538	1.30	.092	1.11		4	M6/7 IIIepv	975,1520
		LP 668 - 44			9 31 00	−04 27.8	1	12.07		0.47		−0.19		2		1696
298455	−50 04249				9 31 00	−51 12.4	1	9.68		1.16		1.13		2	K7	127
304724	−61 12924				9 31 00	−62 19.	1	10.18		0.30		0.17		4	A3	243
		G 161 - 46			9 31 01	−10 57.6	1	10.30		0.89		0.58		1		3062
82811	−49 04485				9 31 02	−49 46.9	1	8.34		−0.07		−0.50		2	B7 V	401
298489	−52 02490	LSS 1321			9 31 02	−53 03.8	1	10.07		0.16		−0.65		2	B5	362
82582	+47 01683	HR 3797			9 31 03	+47 07.5	1	6.52		0.22		0.09		2	F0 V	252
82831	−50 04251				9 31 03	−51 11.2	1	9.77		0.09		0.04		2	A1 V	127
		LP 548 - 6			9 31 05	+05 11.7	1	11.92		0.44		−0.22		2		1696
82649	+27 01779				9 31 05	+27 34.5	1	7.58		0.45		−0.04		2	F5	8053
83019	−69 01079				9 31 05	−70 14.9	1	7.03		−0.04		−0.45		2	B5 IIIn	401
	−40 05297	CCS 1487			9 31 06	−41 05.2	1	10.44		4.38				1	N	864
298462	−51 03920				9 31 06	−51 34.7	1	10.61		0.23		0.22		2	A3	127
82785	−38 05676	HR 3812		★ A	9 31 07	−38 54.4	2	6.42	.005	0.34	.005			7	F2 IV/V	15,2012
82853	−51 03921				9 31 07	−51 39.6	1	9.39		0.08		0.00		2	A0 V	127
82670	+24 02104	HR 3804			9 31 08	+23 40.7	2	6.25	.000	1.45	.004	1.77		6	K7 III	71,1501
82747	−22 02645	HR 3810			9 31 09	−22 38.5	2	5.90	.005	0.02		0.00		2	B9.5V	15,2012
82635	+37 02004	HR 3800, SU LMi			9 31 10	+36 37.2	5	4.55	.008	0.92	.005	0.61	.015	12	G8.5 III	15,1080,1363,3016,8015
82327	+74 00402	HR 3774			9 31 10	+74 32.5	1			−0.10		−0.34		1	B9 V	1079

Table 1

HD	DM	Other Id	N Rem	α_{1950}	δ_{1950}	S	V	σ_V	B–V	σ_{B-V}	U–B	σ_{U-B}	n	Spectrum	References
		LP 728 - 65		9 31 11	−10 31.8	1	11.74		0.67		0.18		2		1696
82919	−56 02300	IDS09297S5639	A	9 31 15	−56 52.3	2	7.11	.000	−0.05	.005	−0.50		6	B6/7 II/III	401,2012
83095	−72 00835	HR 3821		9 31 15	−72 51.5	3	5.46	.004	1.56	.000	1.74	.000	15	K4 III	15,1075,2038
		WLS 940 0 # 6		9 31 17	−00 16.0	1	11.36		0.26		0.13		2		1375
82830	−46 05274	LSS 1322		9 31 18	−46 32.6	4	9.30	.021	0.25	.016	−0.75	.022	9	B0/2 Ie	362,540,976,1737
82829	−44 05573	S Vel		9 31 20	−44 59.2	1	7.77		0.25		0.15		1	A5(m)A5-A8	1588
298459	−51 03923	IDS09296S5119	AB	9 31 20	−51 31.8	1	10.06		0.33		0.21		2	A2	127
83093	−71 00847	IDS09309S7201	AB	9 31 23	−72 14.8	1	7.73		−0.03		−0.74		1	B2 V	55
82621	+52 01402	HR 3799		9 31 25	+52 16.5	5	4.50	.015	0.02	.022	0.03	.013	11	A2 V	15,1203,1363,3023,8015
		Steph 785		9 31 26	+61 28.1	1	11.95		1.54		1.21		2	M3	1746
		POSS 92 # 2		9 31 27	+60 21.0	1	16.38		0.83				2		1739
	−50 04256			9 31 32	−50 33.2	1	10.85		0.27		−0.50		2		362
		Steph 788		9 31 34	+26 12.8	1	12.14		1.28		1.25		1	K4	1746
82932	−52 02508			9 31 36	−52 36.4	1	8.00		0.01		−0.08		2	B9/9.5V	401
82795	−2 02924			9 31 43	−02 56.0	1	8.26		0.48		−0.05		2	F8	1064
82965	−49 04502	IDS09301S4919	A	9 31 52	−49 32.1	1	7.37		1.03		0.82		1	K0 III	389
82966	−51 03935			9 31 52	−51 52.1	1	9.44		0.51		0.07		2	F7 V	127
	−39 05624			9 31 56	−39 21.9	1	10.25		1.18		1.09		2	K5 V	3072
82741	+40 02224	HR 3809		9 31 57	+39 50.7	5	4.81	.005	0.99	.000	0.76	.005	14	K0 III	15,1080,1363,3016,8015
82984	−48 04802	HR 3817	★ AB	9 31 57	−48 46.9	5	5.12	.009	−0.12	.005	−0.58	.012	24	B5 III	15,362,401,404,2012
82986	−49 04504	IDS09301S4919	B	9 32 00	−49 30.1	1	7.56		0.95		0.59		1	G6 IV	389
82819	+8 02243			9 32 02	+08 24.6	1	7.34		1.67		1.97		3	M1	3040
83032	−57 02126	IDS09306S5756	AB	9 32 02	−58 09.8	1	7.90		0.02		−0.45		2	B7 III	401
82870	−5 02840	HR 3814		9 32 03	−05 41.4	3	5.56	.005	1.16	.000	1.13	.005	9	K1 III	15,1415,3077
304725	−62 01254			9 32 05	−62 21.4	1	10.07		1.02		0.69		4	K2	243
82767	+40 02225	IDS09291N4024	C	9 32 09	+40 12.8	1	8.36		0.44		−0.06		2	G0	3016
83060	−56 02319			9 32 10	−56 56.4	2	9.11	.003	0.09	.002	−0.66	.021	7	B2 Vnne	540,976
82817	+26 01974			9 32 15	+26 25.0	2	7.66	.000	0.05	.000	0.07	.018	24	B9	1382,1775
82780	+40 02226	HR 3811	★ A	9 32 15	+40 11.2	2	6.76	.000	0.35	.012	−0.03	.015	6	F2 V	1084,3016
82780	+40 02226	IDS09291N4024	B	9 32 15	+40 11.2	2	8.08	.005	0.43	.007	−0.05	.017	6		1084,3032
83043	−53 02571	LSS 1323		9 32 15	−53 24.7	4	8.44	.065	0.04	.018	−0.86	.014	10	B2/3 Vne	362,401,540,976
82865	+13 02117			9 32 17	+12 52.7	1	6.82		−0.11		−0.59		1	B9	16
		AAS27,343 # 211		9 32 18	−54 18.	1	11.70		1.02		0.12		2		362
82620	+70 00567		V	9 32 20	+69 51.7	1	7.23		0.32		0.11		1	F0	1776
83058	−50 04270	HR 3819		9 32 25	−51 01.9	5	5.00	.009	−0.19	.006	−0.92	.031	16	B2 IV	15,127,362,1732,2012
83059	−52 02535			9 32 25	−52 50.3	3	9.63		0.18		0.15		3	A2 IVs pec	402
82906	+14 02113	IDS09297N1432	A	9 32 26	+14 18.0	1	7.87		−0.10		−0.45		2	B9	3010
83139	−62 01257	IDS09312S6302	AB	9 32 28	−63 15.0	1	8.13		0.18		0.09		4	A3 V	243
83138	−62 01256			9 32 30	−63 00.5	1	8.03		1.60		1.88		4	K3 III	243
	+70 00566			9 32 33	+70 23.9	1	9.30		1.07		0.77		2	G5	1776
82885	+36 01979	HR 3815, SV LMi	★ AB	9 32 40	+36 02.2	18	5.41	.011	0.77	.006	0.44	.007	271	G8 IV-V	1,15,22,130,1006,1007*
	+33 01882			9 32 47	+32 49.9	1	9.69		0.26		0.05		2	A5	1375
83184	−61 01296			9 32 53	−62 13.8	1	8.18		1.40		1.38		4	K1 III	243
82939	+38 02052			9 33 00	+37 46.7	1	8.24		0.76		0.36		3	G5	1723
83183	−58 01576	HR 3825		9 33 00	−59 00.4	5	4.08	.010	0.01	.008	−0.54	.012	21	B5 II	15,243,1075,2012,8100
83123	−44 05601			9 33 01	−45 00.5	1	8.32		1.09		0.87		1	K0 III	389
83025	+0 02532	G 11 - 37		9 33 02	−00 08.3	2	8.16	.000	0.75	.005	0.27	.000	2	K2	1620,1658
83170	−51 03953			9 33 02	−51 48.3	1	7.33		0.41		0.02		2	F2 V	127
		AAS19,45 T7# 416		9 33 02	−52 55.5	1	13.38		0.36		0.03		2		402
83153	−50 04276			9 33 04	−50 20.3	2	7.99	.020	0.04	.010	−0.50	.020	4	B3/4 III	362,401
		AAS19,45 T7# 418		9 33 05	−52 53.9	1	13.91		0.33		0.30		2		402
83048	−14 02893	X Hya		9 33 07	−14 28.1	1	8.53		1.49		0.57		1	M7 e	975
83108	−35 05803	HR 3823		9 33 07	−35 36.0	2	6.48	.005	0.42	.000			7	F5 V	15,2027
82861	+57 01224			9 33 08	+57 11.7	1	7.05		0.14		0.11		3	Am	8071
	−52 02561			9 33 08	−52 50.4	1	11.99		0.38		0.41		3		402
83023	+15 02077	HR 3818	★ A	9 33 09	+14 36.2	2	6.36		0.02	.014	0.00	.038	8	A1 V	12,1022,3050
83023	+15 02078	IDS09304N1450	B	9 33 09	+14 36.2	1	9.51		0.50		0.03		4	F8	12,3024
237822	+58 01199	G 195 - 33	★ AB	9 33 11	+58 08.4	1	9.96		0.63		0.03		2	G3 V	1003
		AAS19,45 T7# 413		9 33 11	−52 54.8	1	12.32		0.46		0.01		3		402
		AAS19,45 T7# 412		9 33 11	−52 55.2	1	12.24		1.06		0.67		3		402
82685	+73 00470	HR 3806	★ AB	9 33 12	+73 18.3	1	6.38		0.32				2	F1 V+ F2 V	1733
83005	+20 02340			9 33 14	+20 16.2	1	6.99		1.03		0.85		3	K0	1648
83104	−18 02728	HR 3822	★ A	9 33 14	−19 21.5	3	6.31	.023	0.06	.004	0.03		8	A0 V	12,15,2027
83104	−18 02729	IDS09309S1908	B	9 33 14	−19 21.5	2	9.53	.022	0.58	.009	0.05	.004	6		12,1775
83201	−51 03956			9 33 15	−51 40.7	2	8.87	.003	0.53	.008	0.10		4	F5 IV	1594,6006
83220	−48 04818			9 33 28	−48 54.4	2	8.55	.000	0.40	.002	−0.09		6	F2/3 V	1594,6006
		AAS19,45 T7# 419		9 33 30	−52 55.	1	13.93		0.71		0.16		2		402
		AAS19,45 T7# 420		9 33 30	−52 55.	1	14.18		1.06		0.92		2		402
		AAS19,45 T7# 421		9 33 30	−52 55.	1	14.33		0.62		0.06		2		402
		AAS19,45 T7# 423		9 33 30	−52 55.	1	14.84		0.72		0.11		3		402
		AAS19,45 T7# 424		9 33 30	−52 55.	1	14.88		0.70		0.54		3		402
		AAS19,45 T7# 425		9 33 30	−52 55.	1	15.03		1.18		1.06		3		402
	+35 02028	G 116 - 41		9 33 36	+35 03.6	1	10.51		0.61		0.00		3		7010
		POSS 729 # 3		9 33 36	−13 07.0	1	15.52		1.58				2		1739
83216	−45 05270			9 33 37	−45 53.3	2	8.15	.009	1.11	.001	0.86	.016	6	K1 III	389,1673
		AAS19,45 T7# 411		9 33 37	−52 54.0	1	12.13		0.84		0.47		4		402
		POSS 729 # 6		9 33 38	−13 21.6	1	18.00		1.35				3		1739
83083	+19 02232	IDS09309N1923	AB	9 33 39	+19 09.2	1	8.76		0.77		0.35		2	G5	1648
83084	+12 02067			9 33 40	+11 39.4	1	8.12		1.16		1.16		2	G5	1648
		POSS 729 # 5		9 33 42	−13 14.7	1	17.51		1.63				3		1739

HD	DM	Other Id	N	Rem	α_{1950}	δ_{1950}	S	V	σ_V	B–V	σ_{B-V}	U–B	σ_{U-B}	n	Spectrum	References
		LP 788 - 6			9 33 43	−21 25.3	2	10.92	.000	1.59	.005	1.18		3		912,3078
		AAS19,45 T7# 415			9 33 43	−52 53.0	1	12.96		0.65		0.29		3		402
83162	−11 02677				9 33 44	−12 14.1	3	10.59	.003	0.10	.004	0.10	.003	21	A0 III/IV	830,1003,1783
	−52 02575				9 33 44	−52 22.6	1	10.91		0.38		-0.30		2		362
83069	+31 02011	HR 3820			9 33 45	+31 23.2	2	5.57	.009	1.57	.016	1.86		6	M2 IIIa	71,3001
		AAS19,45 T7# 414			9 33 45	−52 59.0	1	12.37		0.34		0.20		4		402
		G 161 - 52			9 33 50	−02 36.2	1	13.59		1.50		1.16		1		3073
		AAS19,45 T7# 422			9 33 50	−52 57.8	1	14.63		0.81		0.49		3		402
	−52 02582				9 33 51	−52 58.1	1	11.37		0.13		0.14		4		402
83337	−60 01466				9 33 51	−60 32.1	1	8.66		0.07		-0.07		2	B9 V	401
82839	+70 00568				9 33 53	+70 29.5	1	6.97		0.40		0.05		2	F2	1776
		UU Hya			9 33 54	+04 19.7	1	11.79		0.16		0.12		1	A5	597
		L 788 - 6			9 33 54	−21 25.	1	9.20		1.58		1.18		1		3073
83281	−49 04527				9 33 54	−49 59.7	3	9.59	.044	0.41	.005	-0.23	.021	10	B2/3 II	362,540,976
83161	−2 02934				9 33 56	−02 33.2	1	6.96		1.02		0.79		2	K0	1375
297517	−48 04826	LSS 1327			9 33 58	−49 01.5	3	9.57	.021	0.84	.016	-0.18	.028	5	B1 II-Ib	362,540,976
	−52 02585				9 33 58	−52 56.5	1	11.38		0.25		0.20		4		402
	+26 01981	Y Leo			9 33 59	+26 27.4	1	9.20		1.64		1.89		1	A3 V	3001
83212	−20 02955				9 34 01	−20 39.8	3	8.34	.024	1.07	.017	0.76	.020	9	G6/K0IIIwF7	742,1594,6006
83322	−52 02590				9 34 08	−52 53.0	2	8.90	.000	-0.07	.010	-0.29	.015	4	B8 V	401,402
83335	−53 02616				9 34 08	−53 56.6	1	7.96		-0.09		-0.50		2	B6 V	401
83391	−61 01303				9 34 10	−62 14.1	1	8.20		1.59		1.92		4	K5 III	243
		AAS19,45 T7# 417			9 34 11	−52 54.0	1	13.73		0.51		0.28		2		402
		PKS 0934+01 # 2			9 34 14	+01 21.1	1	14.33		0.80		0.44		2		487
83260	−23 08550				9 34 15	−24 00.2	2	8.83	.010	0.62	.005	0.10		4	G5/8 III(w)	1594,6006
83128	+38 02054				9 34 16	+38 08.2	1	8.85		0.48		0.03		2	F2	1601
		POSS 92 # 1			9 34 16	+61 30.0	1	15.25		1.28				2		1739
300110	−53 02617				9 34 16	−53 59.0	1	10.20		0.26		-0.26		2	B8	362
		PKS 0934+01 # 1			9 34 17	+01 20.9	1	15.56		1.12		0.45		1		487
83189	+17 02109	HR 3826			9 34 17	+16 39.8	1	5.69		1.25				2	gK1	71
83261	−24 08263	HR 3828		★ AB	9 34 18	−24 28.7	2	6.52	.005	0.39	.005			7	F3 IV/V	15,2027
83319	−41 05205				9 34 18	−41 59.3	1	9.48		0.43		-0.03		34	F3 V	978
		BPM 19652			9 34 18	−58 43.	1	15.46		0.00		-0.92		2		3065
83277	−27 06758				9 34 19	−27 44.3	2	8.31	.002	0.31	.002	-0.01		4	A5/7 (IIw)	1594,6006
	−52 02596				9 34 20	−52 45.8	1	10.45		0.58		0.00		8		402
		PKS 0934+01 # 4			9 34 21	+01 11.7	1	12.97		0.50		-0.06		9		487
	+23 02121				9 34 22	+22 55.3	2	9.48	.029	1.28	.015	1.27	.010	3	M0 V:	801,3072
83358	−51 03977	IDS09326S5118		AB	9 34 22	−51 31.2	1	8.05		-0.02		-0.37		2	B9 III	401
		POSS 729 # 2			9 34 23	−13 21.9	1	13.72		0.48				2		1739
		PKS 0934+01 # 3			9 34 26	+01 13.9	1	10.88		0.78		0.35		3		487
		G 48 - 21			9 34 28	+02 31.5	1	11.58		1.12		1.16		1	K5	333,1620
83369	−52 02599				9 34 30	−52 59.4	1	8.28		-0.08		-0.44		2	B8 II	402
83225	+15 02083				9 34 33	+15 28.6	1	8.05		0.38		-0.02		3	F3 V	1648
83240	+7 02160	HR 3827			9 34 34	+07 03.7	5	5.01	.020	1.05	.005	0.87	.010	13	K1 III	15,1078,1080,1363,3016
		PKS 0934+01 # 5			9 34 36	+01 16.5	1	14.53		0.85		0.35		2		487
		POSS 729 # 4			9 34 36	−12 22.6	1	16.75		1.62				3		1739
83368	−48 04831	HR 3831, IM Vel		★ AB	9 34 36	−48 31.6	3	6.17	.005	0.27	.005	0.12		9	Ap SrEuCr	15,404,2012
		KS 317			9 34 36	−59 48.	1	11.19		0.14		-0.75		2		540
		G 49 - 17			9 34 37	+30 27.5	1	12.00		1.01		0.75		1		333,1620
	+56 01395	VV UMa			9 34 37	+56 14.4	1	10.13		0.25				1	A2 V	585
	+41 01979				9 34 40	+41 01.0	1	9.03		0.47		0.01		2	F8	1733
298523	−52 02605	LSS 1328			9 34 40	−52 35.9	1	9.39		0.70		0.09		2	A2 II	362
		PKS 0934+01 # 6			9 34 42	+01 15.9	1	15.66		1.12		1.31		1		487
83224	+25 02126				9 34 43	+24 37.4	3	9.01	.005	0.53	.004	0.06	.006	18	F6 V	830,1783,3026
83332	−24 08272	HR 3830			9 34 45	−25 04.3	4	5.70	.006	1.12	.005			18	K0 III	15,1075,2013,2028
83386	−47 05063	IDS09329S4735		AB	9 34 45	−47 48.0	2	8.85	.004	0.57	.007	0.09		12	G/K +A/F	977,2011
83385	−45 05285				9 34 47	−46 06.9	4	8.40	.008	0.58	.003	0.07		17	G0 V	389,977,1075,2011
		GD 299			9 34 50	+55 19.3	3	12.13	.030	-0.22	.107	-1.17	.013	6	sdO	272,308,1726
	−54 02488				9 34 53	−54 49.3	1	11.39		0.67		-0.06		2		362
		G 195 - 34			9 34 54	+58 08.9	2	11.79	.007	0.47	.007	-0.18	.010	5	sdFp	308,1658
83523	−64 01049	HR 3841			9 34 56	−64 43.6	3	6.55	.004	0.09	.005	0.10	.005	14	A2 V	15,1075,2038
	−47 05066				9 34 57	−48 05.9	1	11.38		0.54		-0.05		1		362
	−54 02490	LSS 1330			9 34 57	−54 44.8	1	10.93		0.62		-0.46		2		362
83475	−59 01428	GW Car			9 34 57	−59 45.7	3	9.67	.089	-0.05	.004	-0.81	.014	10	B5/7 II/III	401,540,976
83488	−60 01470				9 34 57	−60 54.5	1	8.12		0.03		-0.45		2	B3 V	401
83273	+25 02127				9 34 58	+24 53.8	1	6.65		0.53		0.09		2	G0 III	3026
		G 195 - 35			9 35 00	+48 40.4	1	13.16		0.67		-0.10		2		1658
		G 161 - 53			9 35 00	−05 25.5	1	11.96		1.13		0.96		2		3062
		L 462 - 56		A	9 35 00	−37 08.2	1	14.30		0.13		-0.72		1		3060
83380	−31 07458	HR 3833			9 35 01	−31 57.2	2	5.62	.005	1.02	.000			7	K1 III	15,2029
	+56 01396				9 35 02	+56 05.5	1	10.15		0.46				1		585
83446	−48 04836	HR 3836		★ A	9 35 02	−49 07.8	5	4.34	.011	0.17	.005	0.12	.006	21	A5 IV/V	15,1075,2012,3023,8015
	−53 02631				9 35 02	−53 45.0	1	11.41		0.18		0.04		2	B8 V	16
83503	−62 01262				9 35 04	−62 26.9	1	9.57		0.32		0.09		4	A8/9 IV/V	243
83465	−52 02612	HR 3837			9 35 05	−52 43.2	6	6.18	.009	1.05	.007	0.89	.000	21	K0 III	15,389,402,1075,2018,2027
	+23 02124				9 35 08	+22 44.8	1	9.93		1.21		1.17		1	K8	801
		POSS 92 # 3			9 35 08	+61 22.6	1	16.93		1.38				3		1739
		PASP81,804 T4# 4			9 35 09	−53 31.1	1	12.00		1.62		2.14		1		16
	−52 02616	AE Vel			9 35 11	−52 48.4	1	10.62		1.52		1.18		7		402
83444	−46 05342				9 35 14	−46 45.9	3	8.56	.016	1.59	.008	1.72	.014	17	K4 III	863,977,2011

Table 1   517

HD	DM	Other Id	N Rem	α_{1950}	δ_{1950}	S	V	σ_V	B–V	σ_{B-V}	U–B	σ_{U-B}	n	Spectrum	References
83287	+40 02232	HR 3829		9 35 15	+40 27.9	2	5.24	.005	0.22	.001	0.12		4	F0 V	1363,3023
83442	−41 05219	IN Vel		9 35 15	−41 47.7	1	8.87		1.16		1.00		1	K2 IIIp	1641
83443	−42 05452			9 35 15	−43 02.7	5	8.23	.008	0.81	.003	0.53	.019	14	K0 V +G III	389,1020,1075,2012,3008
83463	−48 04841			9 35 15	−48 28.3	1	8.84		-0.07		-0.39		2	B9 II	401
83343	+15 02087			9 35 18	+14 34.4	2	6.65	.010	0.43	.004	-0.02		19	F2	1351,3016
		PASP81,804 T4# 2		9 35 18	−53 29.1	1	12.65		0.53		0.31		1	F0:	16
83126	+67 00602	HR 3824		9 35 21	+67 29.9	2	5.94	.005	1.53	.012	1.93		4	K5	71,3001
83441	−35 05833	HR 3835		9 35 23	−35 52.2	3	5.97	.007	1.12	.000			11	K2 III	15,2013,2029
83373	−8 02725	HR 3832		9 35 24	−09 11.9	3	6.40	.005	-0.04	.004	-0.09	.005	14	A1 V	15,1415,2029
83979	−80 00365	HR 3860, ζ Cha		9 35 26	−80 43.0	5	5.10	.028	-0.14	.008	-0.58	.010	22	B4 IV	15,26,1075,2038,8023
		G 48 - 24		9 35 31	+04 06.7	1	11.98		1.41		1.23		1	K4-5	333,1620
298604	−53 02645			9 35 31	−53 37.0	1	9.05		1.45		1.45		2	K5	16
83520	−53 02646	HR 3840	⋆ AB	9 35 32	−53 26.6	5	5.45	.005	0.14	.009	0.10	.024	16	A2/3 V	15,16,401,2027,8100
83340	+28 01785			9 35 33	+28 13.9	1	7.78		0.64				2	G0 IV	20
83362	+13 02128			9 35 34	+12 57.7	1	6.74		0.90		0.60		2	G8 III	1648
	−43 05457			9 35 34	−44 00.7	2	9.36	.001	1.45	.001	1.67		11	K5	863,1075
83408	+2 02230			9 35 42	+02 03.7	1	7.75		0.53		0.00		2	F8	1375
83529	−49 04546	IDS09339S4933	AB	9 35 42	−49 46.1	4	6.98	.005	0.59	.004	0.02	.016	9	G3 V	389,1311,2012,3037
83141	+70 00570			9 35 46	+70 31.5	1	9.01		0.61		0.17		1	G5	1776
		G 48 - 25		9 35 49	+02 54.7	2	11.90	.015	1.42	.015	1.11	.029	3	M1	333,1620,3073
	−54 02510	LOD 17- # 7		9 35 49	−54 30.4	2	10.98	.010	0.46	.020	0.05	.025	5		125,127
		PASP81,804 T4# 11		9 35 50	−53 29.2	1	12.21		1.29		1.22		1	K:	16
83425	+5 02207	HR 3834		9 35 51	+04 52.6	9	4.68	.007	1.32	.005	1.46	.012	37	K3 III	15,1003,1075,1080*
83558	−53 02650			9 35 53	−53 31.7	1	9.57		0.03		-0.05		1	A0 V	16
	+29 01925	G 49 - 19		9 35 56	+28 38.4	1	10.59		0.55		-0.02		1	G5	1620
83516	−34 06039			9 35 56	−34 51.0	1	8.63		0.97				4	K0 III	2012
83371	+43 01943			9 36 01	+43 22.3	1	6.64		0.94		0.66		2	K0	985
	−53 02653			9 36 03	−53 35.0	1	11.40		1.33		1.09		1	F7: V	16
83548	−42 05462	HR 3842		9 36 04	−42 57.9	11	5.50	.005	1.00	.005	0.69	.009	53	G8 II	14,15,182,278,388,389*
83186	+72 00465			9 36 05	+71 59.0	1	7.48		0.52		0.00		2	F9 V	1003
	−53 02654			9 36 05	−53 40.4	1	11.42		0.57		0.03		1	F6:	16
		AAS27,343 # 221		9 36 06	−51 56.	1	12.05		1.24		0.73		2		362
83597	−53 02656	LSS 1331		9 36 10	−53 27.3	4	9.22	.018	0.08	.012	-0.76	.021	9	B2 Ve	16,362,540,976
		LOD 17- # 6		9 36 11	−54 30.8	2	11.30	.015	0.28	.020	0.14	.000	4		125,127
83642	−59 01436			9 36 13	−60 09.4	1	9.83		-0.04		-0.54		2	B4 IV/V	401
83584	−46 05357			9 36 18	−47 09.7	1	8.28		1.08		1.00		1	K1 III	389
83469	+26 01989			9 36 20	+25 35.6	1	7.22		0.32		0.08		2	F0	105
300178	−54 02519	IDS09347S5415	A	9 36 20	−54 29.0	1	9.72		0.36		-0.32		2	B5	362
300178	−54 02519	IDS09347S5415	B	9 36 20	−54 29.0	1	9.94		0.47		-0.36		1		362
	+69 00527			9 36 21	+69 02.0	1	9.28		0.57		0.05		1	G5	1776
	−26 07297	IDS09342S2617	APB	9 36 23	−26 30.7	1	9.49		0.75		0.52		1		8084
83583	−45 05303			9 36 23	−45 21.3	3	8.90	.004	1.26	.008	1.40	.004	19	K1 III	863,977,2011
83625	−53 02664	IO Vel		9 36 23	−53 59.6	1			-0.12		-0.39		1	Ap Si	16
	−47 05080	LSS 1332		9 36 25	−47 26.8	3	10.45	.013	0.57	.016	-0.47	.017	2	O9.5 Ib	362,846,1737
	−44 05667			9 36 29	−44 35.2	2	9.73	.002	1.30	.003	1.30		11		863,1075
83509	+13 02131			9 36 34	+13 32.3	2	7.04	.002	0.47	.001			64	F7 V	130,1351
	−53 02669			9 36 34	−53 38.8	1	11.02		0.39		0.16		1	A6 V	16
		ApJ144,259 # 33		9 36 36	−02 34.	1	15.54		-0.16		-1.15		2		1360
83610	−39 05697	HR 3844	⋆ AB	9 36 39	−39 23.2	2	6.69	.005	0.49	.000			7	F6 V	15,2012
		LOD 17- # 9		9 36 40	−54 28.4	1	14.35		0.81		0.21		2		125
		LOD 17- # 8		9 36 40	−54 31.7	1	13.63		0.56		0.24		3		125
		LP 548 - 18		9 36 43	+04 57.1	1	12.88		0.48		-0.20		2		1696
298608	−53 02673			9 36 43	−53 37.7	1	10.99		0.16		0.02		1	A1 V	16
83742	−56 02393	IDS09353S5705	AB	9 36 54	−57 18.3	1	7.17		0.11		0.09		2	A1/2 V	401
		LTT 12555		9 37 04	+66 57.0	1	11.31		0.81		0.30		3		1723
83604	−3 02739			9 37 06	−03 59.1	1	8.21		-0.06		-0.25		4	A0	1732
83833	−61 01308	IDS09358S6206	AB	9 37 08	−62 19.8	1	7.58		-0.04		-0.40		4	B8 V	243
83719	−48 04872			9 37 09	−48 50.1	1	7.65		0.06		-0.01		2	A0 V	401
83834	−63 01144			9 37 09	−64 10.6	1	6.69		-0.01				4	B7 III	2012
83760	−50 04353			9 37 17	−50 37.1	1	8.17		1.04		0.86		2	G8/K0 III	389
83618	−0 02231	HR 3845		9 37 18	−00 54.9	7	3.90	.009	1.32	.004	1.46	.015	27	K3 III	15,1075,1363,1425*
83651	−10 02886			9 37 19	−10 29.1	1	8.14		1.63		1.99		3	K2	1657
83650	−9 02898	HR 3846		9 37 21	−10 20.6	4	6.30	.005	-0.03	.006	-0.14	.011	15	A0 Vn	15,1075,1079,2012
		G 161 - 56		9 37 22	−09 32.4	1	15.90		1.12		0.80		2		3062
		G 42 - 16		9 37 30	+10 37.5	1	13.51		1.24		1.11		1		1696
83866	−57 02208			9 37 34	−58 03.2	1	7.66		-0.09		-0.61		2	B8 II	401
	+40 02235			9 37 39	+40 15.0	1	10.40		1.14		1.09		3	K0	974
83853	−52 02663			9 37 41	−52 43.0	1	7.44		1.62		1.81		1	K2/3 III	389
83632	+26 01990			9 37 42	+26 13.9	4	8.05	.017	1.39	.020	1.49	.011	12	K2 III	20,1003,1775,3077
83490	+67 00608			9 37 43	+66 59.1	1	9.23		0.05		0.00		1	A2	1776
83564	+56 01397			9 37 46	+56 05.7	1	6.50		1.13		1.15		3	K1 III-IV	37
83864	−53 02693			9 37 47	−54 02.3	2	7.86		1.23	.009	1.24	.019	4	K1 III	16,8100
	−40 05404			9 37 48	−40 50.5	2	10.75	.025	1.53	.015	1.17		5	M3	912,3073
83865	−54 02554			9 37 48	−54 31.6	5	6.83	.011	-0.13	.013	-0.58	.013	16	B5 V	125,127,362,401,2012
		POSS 729 # 1		9 37 49	−14 14.5	1	13.59		0.81				2		1739
83852	−47 05105			9 37 50	−48 00.6	1	7.37		0.40		0.02		1	F3 IV/V	389
83945	−61 01313			9 37 51	−62 19.6	1	7.62		-0.01		-0.30		4	B9 V	243
83683	+13 02136			9 37 52	+13 17.0	1	6.97		0.47				21	F8	1351
83731	−10 02888	HR 3848		9 37 54	−10 32.5	2	6.36	.005	0.08	.000			7	A2 V	15,2029
83754	−13 02917	HR 3849		9 37 54	−14 06.3	8	5.05	.010	-0.15	.018	-0.55	.015	31	B5 V	3,15,263,1020,1075*

HD	DM	Other Id	N Rem	α_{1950}	δ_{1950}	S	V	σ_V	B–V	σ_{B-V}	U–B	σ_{U-B}	n	Spectrum	References
83882	−53 02696			9 37 54	−54 05.5	3	9.10	.009	0.33	.017	0.01	.013	14	A9/F0 V	16,125,8100
83849	−42 05497			9 37 56	−43 17.0	1	7.37		1.12		1.01		2	K0 III	389
300166	−53 02698			9 37 57	−53 59.1	2	9.93		0.04	.005	−0.39	.014	5	B8 V	16,125
83944	−60 01477	HR 3856		9 37 58	−61 06.1	7	4.50	.010	−0.07	.008	−0.21	.012	23	B9 IV/V	15,26,243,1075,2012*
83881	−52 02669			9 37 59	−53 01.8	3	7.55	.000	−0.11	.004	−0.45	.015	8	B8 III	401,402,2012
83489	+69 00531	HR 3838		9 38 00	+69 28.0	1	5.69		1.14				2	K0	71
83826	−30 07775			9 38 06	−30 41.7	1	7.28		1.59		1.80		8	K0 III	1673
	−51 04042	LSS 1333		9 38 07	−51 55.7	1	10.98		0.68		−0.40		2		362
298647	−53 02701			9 38 07	−53 51.7	1	10.05		0.29		0.06		4	A7	125
		G 48 - 29		9 38 08	+01 14.5	4	10.49	.013	0.38	.010	−0.22	.016	9		1064,1620,1658,3077
		G 161 - 59		9 38 09	−04 03.3	2	11.53	.008	0.70	.010	0.11	.030	4		1658,3062
		NGC 2972 - 9		9 38 10	−50 03.5	1	13.32		1.39		1.26		2		184
		NGC 2972 - 10		9 38 10	−50 03.6	1	14.06		0.42		0.35		1		184
83769	+1 02341	IDS09356N0129	A	9 38 11	+01 15.5	8	10.18	.018	0.40	.012	0.06	.014	8	F8	1064,2034,6006
83769	+1 02341	IDS09356N0129	B	9 38 11	+01 15.5	3	10.46	.016	0.37	.005	−0.21	.034	7		2034,3025,6006
		G 49 - 20		9 38 11	+22 15.3	1	14.20		1.77		1.10		2		316
83958	−56 02415			9 38 13	−57 04.6	1	8.84		−0.02		−0.34		2	B9 V	401
300167	−53 02702	LSS 1335		9 38 14	−53 56.8	2	8.82		1.00	.015	0.29	.019	3	B8 Iab	16,362
83929	−53 02704			9 38 14	−54 05.2	1	10.28		0.02		−0.03		4	A0 V	125
83698	+39 02271	IDS09352N3924	AB	9 38 17	+39 10.8	2	6.91	.020	0.73	.000	0.38	.015	5	G4 III	1501,1601
		NGC 2972 - 8		9 38 17	−50 04.4	1	13.99		0.42		0.35		2		184
		Ton 20		9 38 18	+28 37.	1	15.57		0.05		−0.70		1		98
	+23 02134			9 38 19	+22 43.9	1	9.87		1.20		1.14		2	G0	3072
		NGC 2972 - 7		9 38 19	−50 04.8	1	13.18		0.37		0.34		2		184
		G 235 - 35		9 38 20	+70 15.7	1	10.52		1.53				3		1625
83942	−54 02565	IDS09367S5420	AB	9 38 20	−54 33.4	2	9.10	.020	0.43	.005	−0.01	.010	5	F3 V	125,127
		NGC 2972 - 6		9 38 22	−50 05.2	1	12.67		0.39		0.34		2		184
		NGC 2972 - 5		9 38 22	−50 05.6	1	13.51		0.53		0.34		2		184
298646	−53 02707			9 38 22	−53 46.6	1	10.15		0.15		0.09		3	A2	125
83506	+72 00466	HR 3839		9 38 24	+72 28.9	1	5.16		1.04		0.90		2	K0 III	3016
		NGC 2972 - 4		9 38 24	−50 05.9	1	12.98		0.27		−0.30		2		184
		NGC 2972 - 3		9 38 25	−50 06.4	2	12.13	.006	1.27	.016	0.92	.016	5		184,1685
	−49 04583			9 38 26	−50 06.5	1	11.42		1.52		1.38		2		1730
		NGC 2972 - 2		9 38 26	−50 07.3	2	11.39	.010	1.49	.001	1.37	.013	4		184,1685
		NGC 2972 - 1		9 38 26	−50 08.3	1	12.28		0.42		0.44		2		184
84046	−62 01279			9 38 27	−62 43.0	1	6.42		−0.09		−0.40		4	B8/9III/IVw	243
83808	+10 02044	HR 3852	⋆A	9 38 29	+10 07.2	5	3.52	.013	0.49	.015	0.21	.005	15	A5 V+F6 II	15,1008,1118,3026,8015
83789	+12 02076	G 48 - 31		9 38 29	+11 47.3	1	8.77		0.66		0.31		1	G0	333,1620
		G 42 - 17		9 38 29	+13 26.3	1	10.40		1.50		1.25		4		3016
		NGC 2972 - 11		9 38 29	−50 05.8	2	12.09	.002	1.29	.012	0.91	.002	4		184,1685
		NGC 2972 - 12		9 38 30	−50 05.9	1	13.38		0.45		0.32		2		184
83998	−55 02401			9 38 30	−55 29.0	2	7.95	.005	0.15	.010	0.10	.015	5	A3 III	125,127
		NGC 2972 - 13		9 38 32	−50 06.0	1	13.00		0.42		0.36		2		184
	+40 02240			9 38 34	+40 27.5	1	9.08		0.43		−0.03		3	F5	974
	−7 02861	G 161 - 62		9 38 34	−08 13.6	1	10.96		0.89		0.50		2		3062
	−42 05514			9 38 36	−42 36.7	2	11.05	.020	0.23	.001	0.10	.014	3	A0 IV	540,976
	−49 04585	LSS 1336		9 38 36	−50 19.1	3	10.55	.021	0.33	.016	−0.52	.024	6	B1.5III	362,540,976
		G 235 - 36		9 38 37	+70 16.1	1	11.19		1.55				2		1625
83787	+31 02026	HR 3850, DR Leo	⋆A	9 38 38	+31 30.4	2	5.90	.006	1.58	.014	1.77		8	K6 III	71,1501
300168	−53 02713			9 38 40	−54 00.1	1	10.73		0.15		0.08		3	A2	125
	−49 04586	NGC 2972 - 14		9 38 41	−50 06.6	2	9.40	.009	1.76	.008	1.78	.067	4		184,1730
83807	+28 01797			9 38 42	+28 11.0	2	8.42	.005	0.56	.010	−0.01		5	F8 V	20,3026
		KPS 347 - 211		9 38 42	+40 08.6	1	13.66		0.62		0.03		3		974
		KPS 347 - 213		9 38 44	+40 09.9	1	11.92		1.05		0.75		3		974
		LP 548 - 22		9 38 46	+05 44.0	1	13.50		0.62		−0.13		2		1696
83821	+26 01991	HR 3853		9 38 46	+26 08.5	1	6.24		1.25				2	gK2	71
83805	+40 02241	HR 3851		9 38 55	+39 59.2	2	5.63	.014	0.94	.005	0.67	.009	4	G8 III	1080,3016
		AJ77,733 T13# 7		9 38 55	−53 48.1	1	12.10		0.22		0.09		5		125
		KPS 347 - 214		9 38 58	+40 09.8	1	14.82		1.54		1.62		3		974
298645	−53 02719	Ru 79 - 25		9 38 59	−53 38.6	1	11.20		0.10		−0.18		2		482
83953	−22 02684	HR 3858	⋆A	9 39 00	−23 21.8	11	4.76	.017	−0.12	.012	−0.57	.014	59	B6 Ve	3,15,681,815,1075*
83891	+16 02010			9 39 02	+15 59.0	1	7.97		1.03		0.85		2	K0	1648
	−26 07338			9 39 02	−26 58.7	1	10.54		0.96		0.69		1	K3 V	3072
		WLS 940 0 # 7		9 39 03	−02 03.9	1	12.69		0.58		0.00		2		1375
		Ru 79 - 29		9 39 05	−53 35.6	1	11.96		0.82		0.46		2		482
		Ru 79 - 32		9 39 06	−53 36.3	1	13.11		0.62		0.08		2		482
84044	−45 05336			9 39 07	−45 56.3	1	7.84		1.37		1.61		1	K3 III	389
		AJ77,733 T13# 8		9 39 07	−53 47.0	1	12.87		0.18		0.05		5		125
83661	+67 00611	IDS09350N6736	ABC	9 39 08	+67 22.7	1	8.83		0.35		0.01		1	A5	1776
84043	−42 05524			9 39 08	−42 41.8	1	8.52		0.73		0.47		1	G8/K0 III+F	389
		Ru 79 - 36		9 39 09	−53 46.7	1	14.14		0.52		0.38		2		482
	−53 02724	Ru 79 - 24		9 39 10	−53 37.2	1	11.20		1.16		1.04		2		482
		Ru 79 - 27		9 39 10	−53 37.5	1	11.42		0.18		0.13		2		482
		AJ77,733 T13# 9		9 39 10	−53 49.7	1	13.02		0.55		0.07		4		125
84121	−57 02228	HR 3863	⋆AB	9 39 10	−57 45.3	3	5.31	.005	0.20	.007	0.08		8	A3 IV	15,258,2012
83784	+56 01399			9 39 12	+55 53.7	1	8.91		0.85		0.52		2	K0	1502
		KPS 347 - 218		9 39 15	+40 04.4	1	11.74		0.51		−0.04		3		974
		Ru 79 - 15		9 39 15	−53 35.0	2	13.54	.015	0.39	.010	0.14	.040	4		410,482
		Ru 79 - 13		9 39 15	−53 35.9	1	12.57		0.50		0.33		2		410
		Ru 79 - 14		9 39 15	−53 35.9	2	12.89	.040	1.88	.025	0.40		2		410,482

Table 1   519

HD	DM	Other Id	N Rem	α_{1950}	δ_{1950}	S	V	σ_V	B–V	σ_{B-V}	U–B	σ_{U-B}	n	Spectrum	References
84101	−54 02579			9 39 17	−54 45.3	3	6.93	.005	−0.11	.007	−0.51	.011	9	B5 V	125,127,1732
		Ru 79 - 16		9 39 18	−53 34.7	2	13.36	.060	0.61	.015	−0.11	.030	4		410,482
		Ru 79 - 6		9 39 18	−53 35.7	2	13.87	.049	0.51	.010	0.03	.054	3		410,482
		Ru 79 - 11		9 39 18	−53 36.	1	13.98		0.56		0.27		2		410
		Ru 79 - 50		9 39 18	−53 36.	1	17.92		0.49				1		482
		Ru 79 - 41		9 39 19	−53 38.2	1	14.95		0.62				1		482
	−19 02783			9 39 20	−19 39.8	1	10.47		0.64		0.15		2		1696
84070	−45 05337			9 39 20	−46 09.2	1	7.91		1.06		0.98		1	K0 III	389
		Ru 79 - 5		9 39 20	−53 36.0	2	13.60	.058	0.63	.010	0.02	.054	3		410,482
		Ru 79 - 20		9 39 20	−53 36.5	2	13.66	.024	0.59	.015	0.11	.010	3		410,482
		Ru 79 - 21		9 39 21	−53 35.6	2	14.42	.050	0.68	.005	0.06	.035	4		410,482
		Ru 79 - 12		9 39 21	−53 37.0	2	12.43	.050	0.62	.005	−0.34	.025	4		410,482
		Ru 79 - 4		9 39 22	−53 36.3	2	11.47	.030	1.04	.020	0.59	.030	4		410,482
		Ru 79 - 37		9 39 22	−53 36.5	1	14.31		0.64				1		482
		Ru 79 - 30		9 39 22	−53 42.3	1	12.85		0.59		0.24		2		482
	−55 02426	LSS 1337		9 39 22	−55 31.5	1	11.56		0.36		−0.40		3		846
84068	−41 05275			9 39 23	−41 45.0	1	7.56		1.11		1.02		32	K2 III	978
		Ru 79 - 22		9 39 23	−53 35.6	1	13.99		0.47		0.24		2		410
		AAS27,343 # 228		9 39 24	−53 35.	1	11.40		1.10		0.33		2		362
		KM Vel		9 39 25	−49 09.0	1	14.98		1.17		0.65		1		1753
		Ru 79 - 46		9 39 25	−53 35.1	1	16.85		1.14				1		482
83869	+49 01868	HR 3854		9 39 26	+48 39.6	3	6.37	.008	0.01	.011	0.00	.000	8	A1 V	985,1501,1733
		Ru 79 - 10		9 39 26	−53 37.3	2	12.98	.020	0.40	.015	−0.15	.035	4		410,482
	+22 02118	G 49 - 21		9 39 27	+21 57.8	1	9.66		0.94		0.66		1	G5	333,1620
		Ru 79 - 3		9 39 27	−53 36.2	2	12.16	.005	0.35	.030	0.27	.010	4		410,482
		Ru 79 - 7		9 39 27	−53 36.6	1	13.79		0.37		0.34		1		410
84152	−56 02435	HR 3864		9 39 27	−57 01.8	3	5.80	.004	1.08	.012			14	K0/1 III	15,1075,2027
		Ru 79 - 47		9 39 28	−53 36.8	1	16.92		1.10				1		482
83935	+26 01993			9 39 29	+25 49.1	2	7.26	.000	1.09	.010	0.91		5	K1 III	20,833
		Ru 79 - 49		9 39 29	−53 35.3	1	17.23		1.04				2		482
		Ru 79 - 1		9 39 29	−53 35.7	2	12.23	.025	0.60	.009	0.16	.012	51		410,482
		Ru 79 - 42		9 39 30	−53 34.9	1	15.22		0.64		0.07		2		482
		Ru 79 - 17		9 39 30	−53 35.1	2	14.12	.015	0.64	.005	0.40	.005	5		410,482
		Ru 79 - 35		9 39 30	−53 35.9	1	14.04		1.08		0.44		1		482
		Ru 79 - 38		9 39 30	−53 37.5	1	14.39		0.83		0.43		1		482
		Ru 79 - 34		9 39 30	−53 40.2	1	13.80		0.58		−0.02		2		482
		Ru 79 - 33		9 39 31	−53 36.4	1	13.66		0.64		0.28		1		482
		Ru 79 - 9		9 39 31	−53 37.3	2	13.53	.039	0.60	.005	0.10	.015	3		410,482
		Ru 79 - 31		9 39 31	−53 38.7	1	13.10		0.68		0.26		2		482
	−49 04602			9 39 32	−49 46.4	1	10.97		0.16		−0.74		2		362
		Ru 79 - 2		9 39 32	−53 36.3	2	11.91	.010	1.55	.045	0.94	.005	4		410,482
		Ru 79 - 19		9 39 32	−53 37.1	2	14.48	.029	0.66	.034	0.02		3		410,482
		Ru 79 - 8		9 39 32	−53 37.4	2	13.12	.000	0.58	.020	−0.14	.025	4		410,482
83965	+17 02120	IDS09368N1732	A	9 39 33	+17 18.6	1	8.32		0.12		0.10		3	B9	1648
83696	+71 00509			9 39 33	+70 39.8	1	8.18		0.42		0.07		2	F2	1776
83887	+51 01537	IDS09362N5103	AB	9 39 34	+50 49.0	1	8.01		0.67		0.36		2	G5	1566
		Ru 79 - 18		9 39 35	−53 35.1	2	13.27	.070	0.56	.085	0.11	.020	4		410,482
	−53 02729	Ru 79 - 23		9 39 35	−53 39.6	1	11.11		0.87		−0.57		1		482
		Ru 79 - 28		9 39 36	−53 40.6	1	11.86		2.76				2		482
		Ru 79 - 44		9 39 37	−53 36.5	1	15.99		0.62		0.44		1		482
		Ru 79 - 43		9 39 37	−53 37.3	1	15.70		0.56		0.58		1		482
		Ru 79 - 40		9 39 38	−53 40.7	1	14.70						1		482
83886	+55 01345	HR 3855		9 39 40	+54 35.6	3	6.49	.014	0.12	.020	0.09	.018	7	A5 m	1501,3050,8071
		Ru 79 - 45		9 39 40	−53 36.4	1	16.41		0.68		0.15		1		482
84177	−58 01612			9 39 40	−58 54.6	1	8.35		1.26		1.26		6	K2 III	1673
83951	+35 02042	HR 3857		9 39 42	+35 19.4	1	6.14		0.36		0.00		3	F3 V	272
84136	−48 04902	LSS 1338		9 39 42	−48 22.7	3	8.78	.008	0.43	.015	−0.45	.029	8	B1/2 II	362,540,976
		Ru 79 - 48		9 39 42	−53 36.5	1	17.10		1.35				1		482
	−53 02732	Ru 79 - 26		9 39 43	−53 39.0	1	11.38		0.28		0.26		2		482
84038	−7 02867	G 161 - 65		9 39 44	−07 32.3	1	8.58		0.67		0.03		1	G0	3062
		Steph 791		9 39 45	+06 42.6	1	12.34		1.32		1.21		1	K7	1746
		Ton 447		9 39 48	+31 54.	1	15.47		0.14		0.08		1		1036
		GD 102		9 39 51	−08 42.4	1	14.43		0.76		0.26		1		3060
		WLS 940 0 # 10		9 39 52	+00 41.7	1	11.06		0.72		0.27		2		1375
83550	+78 00317	HR 3843		9 39 56	+78 21.9	1	6.23		1.35				2	K2 III	71
84117	−23 08646	HR 3862		9 39 59	−23 41.4	6	4.92	.011	0.53	.012	0.00	.016	26	G0 V	3,15,678,2012,3026,8017
84173	−45 05347			9 39 59	−45 46.1	1	8.41		0.96		0.68		1	G8 III/IV	389
84261	−65 01124			9 39 59	−65 51.2	1	6.85		0.86				4	G3 III	2012
		Ton 21		9 40 00	+26 14.	1	14.53		−0.35		−1.16		1		98
	−50 04400			9 40 02	−50 38.7	3	10.75	.018	0.23	.039	−0.47	.033	5	B2.5IV	362,540,976
	−77 00542			9 40 05	−77 52.7	1	11.16		0.61		−0.03		2		1696
84228	−54 02594	HR 3868		9 40 09	−54 59.1	6	5.99	.011	−0.13	.010	−0.57	.017	19	B4 IV	15,125,127,164,362,2012
		Ton 448		9 40 12	+31 25.	1	14.60		0.02		−0.01		2		1036
83950	+56 01400	W UMa	★A	9 40 15	+56 10.9	1	7.90		0.66				4	G2 Vn	3024
83950	+56 01400	IDS09367N5625	B	9 40 15	+56 10.9	1	12.35		1.70				2		3024
84035	+43 01953	G 116 - 42		9 40 17	+42 55.9	6	8.13	.019	1.13	.016	1.14	.020	16	K5 V	22,1013,1197,1733,1758,3072
		L 35 - 20		9 40 30	−79 29.	1	11.24		0.99		0.84		1		1696
84124	+20 02366			9 40 35	+20 25.4	1	6.68		1.19		1.29		2	K0	1648
84224	−34 06097	HR 3867		9 40 35	−35 16.4	3	6.40	.004	−0.06	.000	−0.36	.005	11	Ap Si	15,1075,2012
84107	+30 01901	HR 3861		9 40 38	+30 12.3	3	5.65	.019	0.11	.006	0.08	.013	8	A2 IV	252,1022,3050

HD	DM	Other Id	N	Rem	α₁₉₅₀	δ₁₉₅₀	S	V	σ_V	B–V	σ_B-V	U–B	σ_U-B	n	Spectrum	References
		G 117 - 36		AB	9 40 39	+36 38.8	1	10.78		1.32		1.25		3		7010
84375	−67 01111				9 40 40	−68 16.6	1	7.09		-0.05		-0.45		7	B3 IIIp Sh	1628
83962	+65 00731	HR 3859			9 40 42	+65 12.8	1	6.21		0.39		0.04		2	F3 Vn	1733
84308	−59 01453				9 40 44	−59 20.9	1	8.73		-0.04		-0.46		2	B7 IV	401
		WLS 940 0 # 9			9 40 49	−00 29.6	1	13.98		0.69		0.26		2		1375
84183	+11 02087				9 40 54	+10 45.0	1	7.00		0.54		0.11		3	F8 V	1648
	−45 05356				9 40 57	−46 12.5	1	9.79		1.92		2.14		8		1657
84123	+42 02041				9 40 59	+42 16.9	1	6.87		0.28		-0.04		2	F0p	3016
84386	−66 01038			V	9 40 59	−66 24.2	1	9.42		1.70		0.38		1	K1 III	8022
83339	+83 00263				9 41 00	+83 33.7	1	7.92		0.30		0.01		2	F0	1502
84195	+12 02082				9 41 01	+12 23.4	1	7.79		1.62		2.00		2	K5	1648
84194	+14 02136	HR 3866		⋆ A	9 41 01	+14 15.1	3	5.36	.013	1.60	.010	1.96	.005	10	M2 III	71,1501,3001
	−60 01489				9 41 01	−60 56.7	2	10.65	.011	0.03	.017	-0.69	.006	3	B2 III	540,976
		G 49 - 23			9 41 04	+27 12.1	2	12.08	.017	1.55	.044	1.23	.017	5		1620,5010
84326	−51 04094				9 41 05	−51 48.5	2	9.29	.012	0.54	.009	-0.04		4	G2 (III) +F	1594,2012,6006
	+22 02124				9 41 06	+22 17.8	1	8.96		1.03		0.86		2	K0	1733
		Ton 22			9 41 06	+28 03.	1	13.23		-0.28		-1.03		1		98
84305	−42 05552				9 41 07	−43 02.7	2	8.24	.014	0.57	.005	0.03		5	F8/G0 V	389,2012
		WLS 940 0 # 5			9 41 08	+02 06.3	1	11.64		1.22		1.08		2		1375
84361	−57 02254				9 41 10	−57 52.6	1	8.41		-0.01		-0.73		2	B2/3 V	401
		CCS 1533			9 41 10	−64 17.2	1	10.26		1.88		2.17		1		864
84257	−19 02795				9 41 11	−20 08.2	1	6.59		0.91		0.59		5	G8/K0 III	1628
	+45 01755	G 116 - 44			9 41 13	+44 42.6	1	10.36		1.06		0.93		3		1723
84359	−54 02618				9 41 16	−54 32.2	1	7.97		-0.06		-0.37		2	B8 III	401
309718	−62 01291				9 41 16	−63 13.5	2	10.41	.020	0.49	.023	0.01	.152	3	A5	540,976
		G 161 - 68			9 41 18	−06 49.9	1	16.41		0.57		-0.30		1		3060
	+40 02249				9 41 22	+40 09.3	1	9.35		1.60		1.91		3	M0	974
84254	−9 02911	G 161 - 69			9 41 22	−09 45.7	1	9.34		0.61		-0.07		1	G5	3062
83727	+79 00319	HR 3847			9 41 29	+79 22.1	2	6.14	.040	0.26	.010	0.12	.005	5	F0 V:	985,1733
84400	−50 04420	HR 3872, IP Vel		⋆ AB	9 41 41	−50 59.9	3	6.15	.009	-0.10	.009	-0.47		9	B6 V	15,401,2012
		CCS 1531, FK Vel			9 41 42	−46 31.1	1	8.59		4.32		5.50		1	N	864
84252	+19 02251	HR 3869		⋆ A	9 41 43	+19 05.7	2	6.46	.005	1.16	.000	1.16	.020	4	K0	1648,1733
		G 252 - 44			9 41 43	+76 17.0	2	10.61	.015	1.41	.015	1.09	.010	2		679,3078
84324	−13 02936				9 41 43	−13 48.2	1	8.61		0.48		-0.02		4	F6 V	1731
84238	+35 02048				9 41 45	+35 12.1	1	8.67		0.27		0.04		2	A3	1601
84380	−42 05562				9 41 45	−43 19.7	1	8.67		0.42		0.00		1	F3 V	389
84678	−75 00607				9 41 46	−75 26.1	3	8.98	.015	1.47	.008	1.38	.015	13	K0p Ba	158,565,993
84464	−59 01464	IDS09403S5934		A	9 41 48	−59 48.0	1	7.49		-0.06		-0.54		2	B5 V	401
84398	−42 05563				9 41 49	−42 37.3	1	7.61		0.01		0.03		1	F5 V	389
84433	−48 04947				9 41 56	−49 05.4	1	8.27		1.12		1.02		1	K0 III	389
84462	−56 02470				9 41 56	−56 43.4	1	7.73		-0.11		-0.58		2	B5/6 V	401
		L 100 - 115			9 41 57	−68 40.3	1	12.78		1.73		1.22		1		3078
		Steph 796			9 41 58	+01 00.2	1	12.33		1.68		2.08		1	M1	1746
84367	−27 06881	HR 3871		⋆ AB	9 41 58	−27 32.4	7	4.79	.005	0.51	.011	0.35	.021	23	G8 III+A7 V	15,176,1008,1075,2012*
84461	−53 02788	HR 3875			9 42 00	−53 39.7	4	5.55	.008	-0.04	.000	-0.10	.012	11	A0 IV	15,26,2012,8023
84192	+60 01206				9 42 03	+59 54.1	1	9.50		0.38		-0.04		2	F5	1502
84179	+64 00752	HR 3865		⋆ AB	9 42 06	+63 53.1	1	6.36		0.30		-0.04		1	F2 V	254
84447	−38 05850	HR 3874			9 42 13	−39 20.5	2	6.81	.005	0.31	.005			7	F2 IV/V	15,2012
84480	−47 05169	IDS09404S4727		AB	9 42 16	−47 41.1	1	8.74		0.48		0.07		1	F5 V	389
84493	−50 04432				9 42 16	−51 08.2	2	7.72	.010	-0.12	.020	-0.63	.000	5	B2 IV	362,401
		GD 103			9 42 17	+31 50.4	1	15.17		0.75		0.05		1		3060
298632	−52 02751				9 42 18	−53 05.6	1	9.60		0.31		-0.49		2	B3	362
84523	−53 02806				9 42 24	−53 30.9	3	7.95	.024	0.05	.010	-0.52	.012	6	B3/5 Ve	362,540,976
		G 161 - 72			9 42 32	−03 29.9	1	13.41		1.28				1		3062
		LP 788 - 30			9 42 32	−17 58.8	1	12.60		1.59		1.28		5		3078
84476	−30 07851	IDS09404S3043		AB	9 42 34	−30 57.1	1	9.03		0.41		0.07		3	F2/3 V	1700
84346	+35 02050	R LMi			9 42 35	+34 44.6	2	8.39	.035	1.51	.190	0.47	.085	2	M4	3001,8027
		Ton 23			9 42 36	+28 12.	1	16.37		-0.06		0.20		1		98
		PG0942-029D			9 42 38	−02 52.0	1	13.71		0.56		0.13		1		1764
		PG0942-029A			9 42 38	−02 56.4	1	14.73		0.78		0.34		1		1764
	−45 05378				9 42 39	−45 31.8	1	10.11		1.53		1.13		4	M1:	158
		PG0942-029B			9 42 40	−02 53.1	1	14.11		0.53		0.09		1		1764
		PG0942+029			9 42 40	−02 55.5	1	14.00		-0.29		-1.17		7	sdB	1764
84553	−50 04438				9 42 40	−51 01.5	1	8.93		0.46		0.08		1	F6 V	389
		PG0942-029C			9 42 43	−02 52.8	1	14.99		0.73		0.37		1		1764
84474		RR Hya			9 42 44	−23 46.9	1	10.48		1.53		0.49		1	M3e	635
84552	−47 05182				9 42 46	−48 19.1	1	6.69		-0.10		-0.46		2	B6/7 V	401
84571	−50 04440	LSS 1342			9 42 48	−50 35.7	1	8.71		0.35		0.27		3	A2 III	8100
	−44 05804				9 42 57	−44 45.6	1	10.58		0.60		0.11		1		1770
84335	+57 01231	HR 3870, CS UMa			9 43 00	+57 21.5	1	5.16		1.62		1.80		2	M3 IIIab	3016
84441	+24 02129	HR 3873, ε Leo			9 43 01	+24 00.3	11	2.98	.014	0.81	.009	0.45	.034	32	G0 II	15,1008,1080,1119*
84598	−49 04652	IDS09412S4933		AB	9 43 04	−49 46.8	1	7.47		0.99		0.74		1	K0 III	389
297602	−50 04447				9 43 04	−50 56.4	1	10.57		0.21		-0.44		2	F5	362
84455	+19 02254				9 43 05	+18 54.9	1	6.78		1.36		1.48		2	K0	985
	−3 02764	G 161 - 73			9 43 05	−04 25.8	2	10.85	.010	0.52	.017	-0.11	.024	3		1658,3062
		POSS 92 # 5			9 43 08	+61 30.3	1	18.06		1.20				4		1739
84567	−29 07758	HR 3878			9 43 10	−29 58.3	6	6.44	.010	-0.12	.014	-0.94	.004	24	B0.5 IIIn	15,158,1034,1075,1637,2012
		ApJ144,259 # 34			9 43 12	−12 56.	1	16.32		-0.26		-1.02				1360
84612	−48 04962	IDS09414S4902		A	9 43 12	−49 15.4	3	8.04	.004	0.52	.004	-0.04	.020	4	F8/G2 IV/V	389,1594,6006
84627	−48 04963	IDS09414S4902		B	9 43 13	−49 15.4	2	8.22	.005	0.53	.000	-0.02	.015	2	F8/G0 V	389,6006

Table 1

HD	DM	Other Id	N	Rem	α_{1950}	δ_{1950}	S	V	σ_V	B–V	σ_{B-V}	U–B	σ_{U-B}	n	Spectrum	References
84656	−55 02485				9 43 18	−55 47.2	1	8.16		−0.11		−0.58		2	Ap Si	401
84498	+14 02139				9 43 20	+13 40.9	1	8.93		0.31		0.14		2	F0	1648
		G 117 -B11	A		9 43 23	+33 05.3	1	15.17		1.60				2		3016
		G 117 -B11	B		9 43 23	+33 05.3	1	17.24		0.40		−0.51		2		3060
84610	−37 06041				9 43 24	−37 30.2	1	7.82		1.04		0.78		1	G8 II	565
84643	−47 05194				9 43 24	−48 17.1	1	8.37		0.43				4	F3 III/IV	2012
84853	−72 00860				9 43 24	−73 04.4	1	8.86		1.00		0.67		4	G8 III	158
	−50 04461	LSS 1343			9 43 29	−50 36.1	2	11.43	.005	0.25	.025	−0.54	.045	4		362,846
		G 116 - 52			9 43 30	+44 08.6	3	13.31	.005	0.07	.000	−0.55	.009	13		203,1281,3028
84453	+45 01762				9 43 31	+45 20.9	2	6.82	.000	0.95	.000	0.72	.020	4	K0 IV	1003,1080
84542	+7 02181	HR 3876			9 43 32	+06 56.4	5	5.80	.012	1.64	.004	1.96	.008	16	M1.5 IIIab	15,1078,1355,1509,3035
		LP 488 - 31			9 43 32	+11 30.5	1	12.23		1.17		1.14		1		1696
84674	−49 04657				9 43 32	−49 34.9	1	7.52		0.45		0.02		1	F5 V	389
84716	−52 02772				9 43 33	−53 12.2	1	8.26		0.98		0.75		1	K0 III	389
	−44 05820				9 43 34	−44 31.0	4	9.99	.004	0.35	.002	0.16	.007	29		1628,1657,1673,1770
84561	+12 02090	HR 3877			9 43 42	+12 02.5	1	5.63		1.49				2	gK4	71
	−53 02838				9 43 42	−53 35.4	2	11.12	.045	0.21	.010	−0.55	.025	5		362,846
84406	+63 00861				9 43 44	+63 28.9	2	6.94	.000	0.95	.005	0.68	.005	7	G5	1355,3026
84730	−53 02840	IDS09420S5350	AB		9 43 44	−54 04.1	1	8.78		−0.04		−0.32		2	B9/A0 V	401
300214	−53 02841	LSS 1344			9 43 44	−54 08.3	1	8.66		1.09		0.23		2	B7 Ia	362
84607	+2 02246	HR 3879			9 43 49	+02 01.1	6	5.65	.021	0.34	.009	0.13	.025	21	F4 IV	3,15,39,254,1415,3016
	+11 02098				9 43 49	+11 32.7	1	11.33		0.58		0.11		1	G5	1722
84636	−13 02946		A		9 43 50	−14 21.5	2	7.31	.005	1.33	.000	1.57		8	K4 III	1731,2034
84636	−13 02946		B		9 43 50	−14 21.5	1	9.45		0.72				4		2034
		LSS 1345			9 43 51	−51 35.9	1	12.61		0.42		−0.46		4	B3 Ib	846
84810	−61 01333	HR 3884, L Car			9 43 52	−62 16.6	3	3.36	.017	1.02	.011	0.76	.009	3	G5 Iab/Ib	688,689,1754
		LP 314 - 64			9 43 56	+28 59.2	1	11.41		1.15		1.11		3		1723
84729	−43 05600				9 43 57	−43 26.8	1	7.25		1.12		1.10		1	K1 III	389
		LSS 1346			9 43 59	−55 51.6	1	11.80		0.84		−0.23		4	B3 Ib	846
	+83 00264	G 253 - 32			9 44 01	+83 11.8	1	10.51		0.65		0.09		1	F8	1658
	−13 02948				9 44 03	−14 18.0	1	9.45		0.72				4	K0	2034
84743	−44 05832				9 44 03	−44 29.0	2	8.34	.000	0.00	.000	0.01	.009	2	A1/2 V	1311,3077
84791	−56 02498				9 44 03	−56 30.3	1	9.00		0.00		−0.48		2	B6 II	401
84809	−56 02499	HR 3883			9 44 04	−56 57.3	3	6.46	.010	−0.12	.004	−0.59		9	B8 III/IV	15,401,2012
84789	−50 04481				9 44 12	−50 38.3	1	8.02		1.07		0.92		1	K0 III	389
		AO 936			9 44 13	+11 46.1	1	11.96		1.33		1.23		1		1722
84774	−45 05411	IDS09423S4527	AB		9 44 13	−45 41.2	1	6.67		0.94		0.63		1	G8 II/III	389
		KUV 09443+4229			9 44 16	+42 28.6	1	16.43		−0.12		−1.02		1	DA	1708
		AO 937			9 44 17	+11 23.9	1	11.91		1.01		0.92		1		1722
84752	−30 07887				9 44 19	−31 02.6	1	7.74		1.06		0.92		2	K0 III	1700
84753	−30 07888				9 44 20	−31 07.7	1	8.89		0.55		−0.01		1	G0	1700
84850	−58 01640	HR 3887			9 44 24	−58 33.8	2	6.21	.005	0.46	.000			7	F5 III/IV	15,2012
		G 235 - 40			9 44 28	+60 29.5	1	12.73		1.64		1.39		2		7010
		AO 938			9 44 29	+11 47.6	1	12.11		0.53		0.06		1		1722
		Ton 24			9 44 30	+27 33.	1	16.11		0.01		−0.97		1		98
		G 49 - 24			9 44 31	+26 32.4	1	10.91		1.38		1.30		1	K7	333,1620
84816	−44 05846	HR 3886			9 44 33	−44 31.4	7	5.55	.007	−0.18	.007	−0.71	.008	21	B2 V	15,26,540,976,1732*
	+12 02093				9 44 34	+11 39.6	1	9.19		1.17		1.10		1	K2	1722
		GD 104			9 44 36	−09 05.9	1	15.89		−0.35		−1.23		1	sdOp	3060
84861	−53 02865	LSS 1347			9 44 41	−53 55.1	4	8.68	.018	0.52	.005	−0.04	.029	12	B9 Iab/b	362,540,976,8100
		G 48 - 37			9 44 42	+01 48.3	1	10.98		1.36		1.20		1	M1	333,1620
		G 116 - 53			9 44 43	+33 36.8	1	12.98		0.55		−0.09		2		1658
84722	+12 02095	HR 3880		★	9 44 45	+11 48.0	5	6.45	.017	0.26	.006	0.08	.014	34	A7 Vn	15,254,814,1351,3016
		G 116 - 54			9 44 46	+35 22.4	1	14.07		1.02		0.82		1		3062
85012	−69 01115	IDS09439S6939	AB		9 44 49	−69 53.2	1	7.14		−0.02		−0.33		2	B6 V	401
		POSS 92 # 4			9 44 50	+61 19.3	1	17.07		1.03				3		1739
84804	−24 08451				9 44 51	−25 01.0	1	9.29		1.08				4	K1 III	2012
84877	−52 02793	LSS 1348		★ AB	9 44 51	−52 43.7	3	8.96	.008	0.45	.011	−0.45	.031	8	B1/2 Ib/II	362,540,976
84748	+12 02096	HR 3882, R Leo			9 44 52	+11 39.7	7	7.39	.852	1.44	.159	0.25	.091	8	M4	635,814,975,3002,8019*
84764	+11 02104	IDS09423N1051	AB		9 45 01	+10 37.0	1	7.81		0.22		0.10		2	A5	1648
	+27 01811				9 45 05	+27 20.7	1	9.48		0.47		0.00		2	F6 V	3026
		LP 261 - 52			9 45 06	+37 37.4	1	10.02		0.63		0.11		3		1723
84889	−47 05231				9 45 06	−47 28.9	1	8.54		0.47		0.03		1	F5 V	389
		AO 941			9 45 07	+11 21.8	1	10.84		1.17		1.17		1		1722
	−50 04501				9 45 07	−51 19.4	2	10.89	.010	0.27	.010	−0.50	.020	5		362,846
		LSS 1349			9 45 08	−49 58.7	1	13.34		0.01		−0.95		2	B3 Ib	846
		LSS 1350			9 45 09	−57 13.9	1	11.78		0.63		−0.18		2	B3 Ib	846
		MCG-5-23-16 # 5			9 45 10	−30 43.8	1	12.65		1.20		1.16		4		1687
	−52 02798				9 45 12	−52 41.8	3	10.36	.005	0.44	.019	−0.44	.050	5	B1 IV	362,540,976
298658	−51 04170	IDS09435S5135	A		9 45 16	−51 48.4	2	9.70	.005	0.46	.005	−0.30	.031	7	B0 V	362,976
298658	−51 04170	IDS09435S5135	AB		9 45 16	−51 48.4	2	8.67	.002	1.18	.043	−0.03	.111	3	B0 V	362,540
		MCG-5-23-16 # 2			9 45 18	−30 42.2	1	12.14		0.55		−0.01		4		1687
84903	−40 05513				9 45 18	−41 13.1	3	8.01	.008	1.17	.015	0.64	.061	8	GwA/F (V)	1594,3077,6006
90105	−88 00095				9 45 19	−89 15.0	1	7.22		1.69		2.05		3	M1 III	1704
84737	+46 01551	HR 3881			9 45 22	+46 15.3	10	5.09	.017	0.61	.010	0.14	.022	28	G0.5 Va	15,101,245,374,1067*
84888	−29 07788				9 45 23	−29 27.4	1	7.89		1.01				4	K0 III	2012
		AO 942			9 45 24	+11 33.5	1	12.39		0.85		0.52		1		1722
		MCG-5-23-16 # 3			9 45 25	−30 44.5	1	12.32		0.66		0.09		4		1687
		MCG-5-23-16 # 1			9 45 27	−30 40.4	1	11.58		0.57		0.03		4		1687
		MCG-5-23-16 # 6			9 45 27	−30 41.8	1	13.87		0.82		0.44		4		1687

HD	DM	Other Id	N	Rem	α_{1950}	δ_{1950}	S	V	σ_V	B–V	σ_{B-V}	U–B	σ_{U-B}	n	Spectrum	References
		MCG-5-23-16 # 7			9 45 28	−30 42.	1	14.97		0.61		0.03		4		1687
		AO 943			9 45 30	+11 42.3	1	11.50		0.68		0.18		1		1722
84800	+44 01908				9 45 36	+43 53.9	1	7.79		0.10		0.11		3	A2 II	8100
		L 390 - 66			9 45 36	−44 18.	1	12.23		0.55		−0.21		2		1696
237842	+56 01408				9 45 37	+55 41.6	1	9.83		0.69		0.18		3	G0	3016
	+11 02105				9 45 40	+11 32.2	1	10.00		0.68		0.15		1	G5	1722
		MCG-5-23-16 # 4			9 45 40	−30 42.9	1	13.43		0.62		0.11		4		1687
300330	−56 02533				9 45 47	−56 31.8	1	7.69		−0.02		−0.38		2	B9	401
	−45 05428				9 45 50	−46 02.1	1	10.54		0.32		0.30		1		365
85289	−74 00639				9 45 50	−74 49.1	1	9.63		0.06		−0.47		1	B9 II	55
84916	−3 02782		3		9 45 51	−04 10.4		8.67	.004	1.16	.006	1.11	.007	52	K5	147,1509,6005
85123	−64 01084	HR 3890	4	⋆ AB	9 45 51	−64 50.4		2.96	.003	0.27	.007	0.11	.015	21	A8 Ib	15,1075,2031,2038
	+65 00737	G 235 - 45	4		9 45 53	+65 32.5		9.70	.010	0.67	.010	0.10	.000	8	G5	516,1064,1658,3026
84779	+57 01234				9 45 58	+57 27.3	1	7.67		1.13		1.14		3	K0 III	37
85083	−57 02322				9 45 59	−57 57.3	1	8.25		−0.10		−0.62		2	B5 III	401
302356	−57 02323				9 46 01	−57 56.8	1	9.51		−0.01		−0.68		1	B8	401
85082	−55 02536				9 46 04	−56 12.5	1	8.95		−0.04		−0.41		2	B8 IV	401
84937	+14 02151	G 43 - 3	9		9 46 12	+13 59.3		8.32	.022	0.39	.018	−0.21	.014	17	F5 VI	22,792,908,1003,1064*
84971	−2 02986		7		9 46 12	−02 28.8		8.64	.006	−0.15	.006	−0.76	.012	89	B2 V	55,147,989,1509,1728*
		G 161 - 77			9 46 12	−10 21.2	1	11.41		1.41				1		3062
84984	−9 02928				9 46 16	−09 41.1	1	6.99		0.01		0.00		7	A0	1628
85081	−46 05524	IDS09444S4700		AB	9 46 18	−47 14.2	1	8.31		0.90		0.53		1	G8 III/IV	389
84950	+18 02274				9 46 20	+18 17.5	1	7.76		1.25		1.25		2	K0	1648
	+44 01910				9 46 20	+44 31.5	1	10.93		0.42		−0.20		3	A5	1064
		LP 728 - 28			9 46 21	−12 16.2	1	15.72		1.75				1		1691
85396	−76 00598	HR 3902	3		9 46 21	−76 32.6		5.44	.007	0.90	.010	0.57	.000	13	G8 III	15,1075,2038
84949	+25 02157				9 46 22	+24 40.6	1	8.55		1.55		1.81		2	K2	1625
	+8 02269				9 46 25	+08 03.2	1	9.37		0.93		0.64		2	G5	3016
84747	+70 00582				9 46 25	+70 26.6	1	8.47		0.43		−0.02		1	F0	1776
84914	+37 02022				9 46 27	+36 58.9	1	6.61		1.57		1.94		6	K4 III	1501
84812	+66 00637	HR 3885			9 46 30	+65 49.6	1	6.31		0.28		0.12		1	A9 Vn	39
237843	+55 01351				9 46 37	+55 19.4	1	9.50		0.97		0.66		2	G5	3016
85100	−34 06170	IDS09445S3433		AB	9 46 37	−34 47.2	1	7.32		0.26		0.12		3	A9 IV/V	404
	−55 02545	NGC 3033 - 8			9 46 38	−56 13.3	1	10.59		0.41		0.06		2	F0 V	184
		G 235 - 46			9 46 39	+62 52.8	1	15.75		1.64				1		906
85167	−51 04196				9 46 40	−52 17.0	1	8.10		1.04		0.81		1	K0 III	389
		NGC 3033 - 18			9 46 43	−56 10.0	1	12.56		0.34		0.28		2		184
85286	−66 01085				9 46 44	−66 37.8	1	7.24		1.36		1.23		3	K2/3 III	1704
		NGC 3033 - 9			9 46 46	−56 12.0	1	12.75		0.40		0.37		2		184
		LSS 1352			9 46 46	−58 30.0	1	11.89		0.44		−0.58		3	B3 Ib	846
		NGC 3033 - 17			9 46 47	−56 10.5	1	13.96		0.72		0.27		2		184
	+57 01235				9 46 48	+57 21.	1	9.91		0.50		0.01		3		1723
		NGC 3033 - 16			9 46 48	−56 09.9	1	13.80		0.52		0.16		2		184
		NGC 3033 - 10			9 46 49	−56 11.7	1	12.59		0.45		0.17		2		184
		NGC 3033 - 7			9 46 49	−56 12.9	1	12.70		0.38		0.33		2		184
		NGC 3033 - 12	2		9 46 50	−56 10.9		12.01	.014	1.92	.002	1.89	.022	4		184,1685
298661	−51 04197			AB	9 46 51	−51 49.9	1	8.66		1.14				2	K0	176
	−55 02546	NGC 3033 - 11		⋆ AB	9 46 51	−56 11.5	1	11.19		0.59		0.43		2		184
		NGC 3033 - 15			9 46 52	−56 09.5	1	13.71		0.40		0.24		2		184
		NGC 3033 - 6			9 46 53	−56 13.5	1	14.16		0.67		0.03		2		184
		G 195 - 42			9 46 55	+53 29.2	1	15.20		0.13		−0.70		3	DC	940
		NGC 3033 - 13			9 46 56	−56 10.8	1	12.71		0.42		0.34		2		184
		NGC 3033 - 5			9 46 56	−56 13.4	1	12.80		0.38		0.29		2	B8 V	184
		NGC 3033 - 14			9 46 58	−56 10.3	1	13.30		0.39		0.30		2		184
		G 116 - 57	2		9 47 01	+41 01.2		12.37	.014	0.59	.005	−0.07	.009	4		1658,3060
85250	−55 02548	NGC 3033 - 1	4	⋆	9 47 01	−56 10.7		6.05	.007	0.94	.005	0.67		17	K0 III	15,184,1075,2029
		NGC 3033 - 2			9 47 01	−56 12.3	1	11.67		0.35		0.34		2		184
85040	+21 02113	HR 3889, DG Leo	2	⋆ AB	9 47 02	+21 24.8		6.09	.005	0.25	.005	0.18		2	F0 III	39,8076
85228	−52 02830	IDS09453S5209	3	AB	9 47 02	−52 23.1		7.92	.009	0.90	.008	0.64	.010	6	K1 V	258,389,2012
	−55 02550	NGC 3033 - 4			9 47 02	−56 13.7	1	11.46		0.13		−0.07		2		184
85030	+34 02038				9 47 03	+33 50.1	1	7.94		0.46		0.14		11	F5	1603
		NGC 3033 - 19	2		9 47 03	−56 08.9		11.34	.010	1.07	.005	0.86	.005	4		184,1685
		LSS 1351			9 47 05	−46 17.7	1	12.43		0.57		−0.16		2	B3 Ib	846
297624	−50 04539		4		9 47 05	−50 51.7		10.19	.012	0.23	.014	−0.49	.014	9	B2 V	362,540,976,2021
	−55 02551	NGC 3033 - 3			9 47 07	−56 13.1	1	10.92		0.09		−0.02		2	B9 V	184
298697	−52 02831	LSS 1353	3		9 47 08	−52 54.0		10.03	.057	0.39	.030	−0.65	.037	6		362,540,976
85249	−53 02912				9 47 08	−53 21.1	1	8.55		0.49		0.05		1	F7 V	389
85091	+11 02108	G 48 - 39	2		9 47 09	+11 20.5		7.63	.015	0.60	.005	0.03	.019	3	F8	1620,1658
85029	+40 02261				9 47 10	+39 51.9	1	6.65		1.63		1.97		3	K5	1601
85015	+44 01913	IDS09440N4428		A	9 47 10	+44 14.1	1	8.46		0.40		−0.03		3	F3 II	8100
85209	−42 05648	IDS09452S4301		AB	9 47 10	−43 15.0	1	6.54		1.28		1.29		1	K3 III	389
		LSS 1354		AB	9 47 10	−58 17.9	1	11.45		0.54		−0.09		2	B3 Ib	846
		LP 126 - 53			9 47 11	+50 57.3	1	11.83		1.25				3		1723
		WLS 940 0 # 8			9 47 13	+00 05.0	1	10.76		1.01		0.77		2		1375
		G 43 - 5	2		9 47 13	+06 50.8		12.49	.029	0.63	.005	−0.16	.005	3		333,1620,1696
85313	−59 01491				9 47 14	−59 54.0	1	7.52		0.96		0.70		3	G8 III	8100
297625	−50 04544		2		9 47 18	−50 55.4		10.52	.020	0.19	.030	−0.22		7	B5 V:e	362,2021
85206	−36 05955	HR 3892	2		9 47 22	−36 57.2		5.96	.005	1.24	.000			7	K2 III	15,2027
84999	+59 01268	HR 3888, υ UMa	8	⋆ A	9 47 27	+59 16.5		3.80	.019	0.29	.003	0.10	.005	28	F0 IV	15,39,254,1008,1025*
85297	−50 04549				9 47 30	−50 25.6	1	8.15		0.89		0.56		1	G5 III	389

Table 1 523

HD	DM	Other Id	N Rem	α_{1950}	δ_{1950}	S	V	σ_V	B–V	σ_{B-V}	U–B	σ_{U-B}	n	Spectrum	References
		G 43 - 7		9 47 36	+05 23.1	1	11.75		0.80		0.36		1		333,1620
85296	−35 05961	HR 3897		9 47 44	−36 02.1	4	6.36	.007	1.01	.003			18	K0 III	15,1075,2013,2030
85341	−51 04213			9 47 46	−51 53.6	1	7.59		-0.11		-0.55		2	B9 II/III	401
85066		CCS 1554		9 47 48	+52 53.	1	9.73		1.75				1	R3	1238
298682	−51 04214	LSS 1355		9 47 50	−51 38.3	3	9.06	.004	0.16	.008	-0.71	.023	8	B0.5II	362,540,976
	−1 02308			9 47 52	−02 24.8	1	9.56		1.46		1.62		2	K5	1375
85217	+5 02240	HR 3893		9 47 54	+04 34.7	4	6.25	.031	0.47	.005	-0.02	.011	12	F6 V	15,254,1415,3016
85218	+2 02255			9 47 56	+02 03.7	1	8.03		1.20		1.20		2	K0	1375
85355	−45 05470	HR 3898	⋆ A	9 48 00	−45 29.9	2	5.07	.005	-0.11	.005			7	B7 III	15,2012
85307	−30 07938			9 48 01	−31 07.9	1	8.32		1.39		1.55		5	K2 III	1700
	+71 00515			9 48 04	+70 34.5	1	9.77		0.66		0.18		2	G5	1776
85356	−47 05287	LSS 1356		9 48 05	−47 42.8	2	8.03	.005	-0.08	.007	-0.88	.009	6	B2 Ib	540,976
85390	−49 04727			9 48 11	−49 33.3	1	8.55		0.85		0.56		1	K1 V	389
85411	−50 04563			9 48 14	−50 23.5	1	7.67		0.48		0.04		1	F6/7 V	389
	+38 02075			9 48 18	+37 50.3	1	10.04		1.10		1.02		2	K8	3072
85409	−49 04730	IDS09465S4909	AB	9 48 19	−49 23.1	1	7.72		0.48		0.02		1	F5/6 V	389
85268	+13 02164	HR 3896		9 48 20	+13 18.1	3	6.47	.018	1.57	.009	1.93	.009	6	M0 III	1648,1733,3001
		G 161 - 79		9 48 22	−06 30.7	2	13.92	.052	0.86	.000	0.41	.032	5		1658,3062
85304	+1 02370			9 48 23	+01 02.4	1	9.14		0.17		0.13		3	A2	1776
85319	−1 02312	W Sex		9 48 26	−01 47.7	2	8.81	.020	2.53	.168	6.40		10	C4,4	864,3038
85453	−49 04734			9 48 32	−50 06.1	1	8.16		-0.07		-0.41		2	Ap (Si)	401
85496	−57 02367	IDS09470S5745	A	9 48 36	−57 58.9	1	7.98		-0.02		-0.36		2	B8/9 III	401
85238	+50 01689			9 48 39	+49 51.5	1	7.80		0.57		0.08		2	G0	1733
	−11 02741			9 48 40	−12 04.5	4	10.03	.030	1.44	.027	1.15		7	M0	1705,2033,3078,6009
		G 235 - 48		9 48 41	+60 31.1	1	12.62		0.94		0.63		2		1726
		G 161 - 81		9 48 41	−10 14.2	1	12.77		1.41				1		3062
		G 161 - 80		9 48 41	−12 04.9	1	10.10		1.48				1		3062
85235	+54 01331	HR 3894	⋆ AB	9 48 43	+54 17.9	4	4.59	.021	0.03	.004	0.08	.000	10	A3 IV+A3 IV	15,1363,1381,3023,8015
85364	−3 02794	HR 3899		9 48 43	−04 00.5	2	6.00	.005	0.17	.000	0.10	.010	7	A8 III	15,1415
85405	−22 02739	Y Hya		9 48 45	−22 46.9	2	6.54	.061	3.47	.243	7.30		4	C5,5	864,3038
85483	−46 05558	HR 3904		9 48 47	−46 42.0	3	5.72	.004	1.08	.007	1.02		8	K0 III	15,389,2029
85465	−42 05677			9 48 49	−43 08.0	1	8.09		1.31				4	K2 III	1075
		KUV 09488+3422		9 48 51	+34 21.5	1	14.51		-0.01		-0.71		1		1708
85380	−5 02923	HR 3901		9 48 52	−05 56.9	4	6.42	.006	0.58	.008	0.10	.005	16	F8 IV-V	15,1415,2029,3077
	−54 02768	LSS 1357		9 48 54	−55 01.9	1	12.04		0.62		-0.19		3		846
		G 161 - 82		9 48 56	−04 24.3	3	11.99	.004	0.60	.009	-0.11	.023	5		1658,1696,3062
85567	−60 01510			9 48 59	−60 44.0	2	8.57	.016	0.12	.013	-0.55	.016	4	B5 Vne	540,976
85541	−52 02872			9 49 00	−53 09.2	3	9.05	.010	0.37	.000	-0.60	.026	7	A2/3 IV/V	362,401,846
85431	−15 02918	IDS09466S1605	AB	9 49 02	−16 19.4	1	7.96		0.50				4	F5 V	2033
85376	+25 02169	HR 3900		9 49 03	+24 38.0	4	5.32	.011	0.22	.004	0.06	.013	14	A5 V	374,1022,3031,8076
85444	−14 02963	HR 3903		9 49 04	−14 36.7	9	4.11	.008	0.93	.006	0.65	.016	39	G6/8 III	3,15,1007,1013,1075*
85430	−2 02993	G 161 - 83		9 49 05	−02 38.5	1	9.46		0.82		0.50		1	K0	3062
85512	−42 05678			9 49 05	−43 15.7	4	7.64	.023	1.15	.023	1.09	.024	9	M0 V	258,389,1075,3072
298690	−51 04237			9 49 06	−52 14.3	1	10.34		0.24		-0.27		2	B5 IV	540
	−64 01106			9 49 07	−64 46.7	2	8.03	.031	1.13	.002	0.93	.006	10		9,9
		G 161 - 84		9 49 09	−03 35.9	2	12.21	.022	0.90	.005	0.43	.027	5		1658,3062
85461	−10 02940			9 49 14	−11 06.3	1	6.52		1.61		1.88		2	M1	3040
		Pis 16 - 5		9 49 18	−52 57.	1	13.39		0.39		-0.14		2		184
		Pis 16 - 3	⋆ B	9 49 18	−52 57.	1	10.50		0.42		-0.43		2	B2 V	184
		Pis 16 - 4	⋆ C	9 49 18	−52 57.	1	12.09		0.52		-0.24		2		184
85373	+38 02076			9 49 20	+38 09.0	2	6.73	.000	0.24	.005	0.12	.005	5	A8 V	985,1501
85563	−45 05499	HR 3910		9 49 23	−45 57.6	5	5.62	.007	1.17	.005	1.16		19	K2 III	15,389,1075,2013,2029
		G 116 - 61		9 49 24	+42 08.0	1	15.20		1.57				4		538
85594	−52 02883	Pis 16 - 1		9 49 25	−52 56.9	2	8.68	.010	0.35	.005	0.03	.005	12	F0	184,976
		Pis 16 - 8		9 49 26	−52 56.5	1	13.47		0.40		0.02		1		184
		Pis 16 - 6		9 49 27	−52 56.1	1	13.16		1.13		0.85		2		184
		Pis 16 - 7		9 49 27	−52 56.1	1	11.76		0.42		0.31		2		184
		Pis 16 - 9		9 49 27	−52 56.9	1	12.98		0.40		-0.21		1		184
		LP 488 - 47		9 49 28	+13 53.0	1	11.97		0.64		0.04		2		1696
85595	−52 02885	Pis 16 - 2	⋆ AP	9 49 30	−52 56.6	2	9.36	.025	0.34	.035	-0.58	.030	4	B0 IV	184,362
85656	−62 01335	HR 3914		9 49 31	−62 30.6	4	5.56	.013	1.32	.005	1.30		15	K1 III	15,243,1075,2006
85519	−15 02920	HR 3908		9 49 36	−16 17.9	1	6.08		1.04				4	K0 III	2032
85623	−52 02887			9 49 36	−52 39.3	1	8.80		0.56		0.10		1	G3/5wF7 V	389
85504	+3 02280	HR 3906		9 49 37	+02 41.3	5	6.02	.007	-0.04	.008	-0.07	.018	14	A0 Vs	15,252,1003,1320,1415
85488	+3 02279	G 48 - 45	⋆ A	9 49 37	+03 27.4	3	8.85	.020	1.23	.003	1.23	.023	7	K5	333,1620,1758,3072
85505	+0 02573	HR 3907		9 49 38	+00 18.7	3	6.36	.017	0.94	.005	0.71	.038	9	G9 III	15,252,1415
85655	−58 01673	HR 3913		9 49 38	−59 11.4	5	5.79	.004	1.36	.005	1.28		20	K2 III	15,1075,2013,2029,3005
	−56 02585	LSS 1359		9 49 44	−57 09.7	2	10.99	.030	0.35	.020	-0.61	.060	4		362,846
85622	−45 05508	HR 3912		9 49 45	−46 18.8	4	4.58	.005	1.20	.000	0.99	.000	20	G5 Ib	15,1075,2012,8015
		G 195 - 45		9 49 47	+51 45.4	1	12.84		0.80		0.28		2		1658
85701	−59 01515			9 49 49	−60 05.8	1	8.68		0.54				4	F7 V	2012
85503	+26 02019	HR 3905		9 49 55	+26 14.6	6	3.88	.005	1.22	.006	1.39	.007	17	K0 IIIb	15,37,1080,1363,3016,8015
	−64 01108			9 49 56	−64 48.9	2	8.83	.006	0.46	.000	0.00	.001	12		9,9
		Steph 805		9 49 57	+31 31.8	1	11.90		1.44		1.24		1	M0	1746
85439	+51 01553	IDS09467N5105	AB	9 50 00	+50 51.4	1	7.25		0.99		0.73		2	G5	1566
		BPM 19738		9 50 00	−57 12.	1	14.94		0.20		-0.58		2	DA	3065
	−56 02591			9 50 01	−57 09.4	1	11.07		0.27		-0.21		2		362
85558	−7 02909	HR 3909	⋆ AB	9 50 02	−07 52.1	4	5.05	.004	0.04	.000	0.06	.000	17	A1 V +A4 V	15,1417,2012,3024
85558	−7 02909	IDS09476S0738	C	9 50 02	−07 52.1	1	12.28		0.67		0.21		4		3024
85425	+58 01219			9 50 06	+58 15.9	1	8.64		1.11		1.06		3	K0 III-IV	37

HD	DM	Other Id	N Rem	α₁₉₅₀	δ₁₉₅₀	S	V	σ_V	B–V	σ_B–V	U–B	σ_U–B	n	Spectrum	References
85741	−60 01516			9 50 10	−60 21.5	1	7.26		0.95		−0.08		4	K0 III	2012
	−53 02960			9 50 11	−54 09.5	2	11.37	.123	0.35	.010	−0.08	.005	8	B7	322,402
	+71 00517			9 50 13	+70 55.5	1	11.32		0.55		0.08		2	G5	1776
85557	+22 02138			9 50 16	+22 21.5	1	8.01		1.72		2.06		2	K5	1375
85599	−7 02910			9 50 17	−07 33.0	1	9.10		−0.01		0.33		3		322
		LP 847 - 4		9 50 17	−25 52.8	1	12.35		0.95		0.78		1		1696
		G 195 - 47		9 50 18	+50 59.1	1	12.03		1.51		1.25		3		7010
		AAS19,45 T8# 507		9 50 22	−54 09.7	1	11.89		0.51		0.03		4		402
85740	−53 02964			9 50 23	−54 02.9	6	8.47	.016	0.31	.015	−0.61	.023	25	B1 Ib	322,362,401,402,540,976
	−50 04589	LSS 1360		9 50 25	−50 52.4	3	11.02	.030	0.31	.012	−0.61	.019	8		362,846,1737
		LSS 1361		9 50 25	−52 27.2	1	11.60		0.68		−0.27		2	B3 Ib	846
		LP 428 - 62		9 50 27	+18 59.8	1	16.96		1.54				1		1691
		AAS19,45 T8# 509		9 50 30	−54 05.	2	14.41	.047	0.69	.037	0.11	.080	4	G2	322,402
		AAS19,45 T8# 510		9 50 30	−54 05.	1	14.67		1.61		1.10		1	G8	322,402
		AAS19,45 T8# 511		9 50 30	−54 05.	2	14.59	.061	0.46	.033	0.28	.061	4	A8	322,402
		AAS19,45 T8# 512		9 50 30	−54 05.	1	15.69		0.92		0.40		1		402
		AAS19,45 T8# 513		9 50 30	−54 05.	1	15.90		0.87		0.07		1		402
		AAS19,45 T8# 514		9 50 30	−54 05.	1	16.11		0.96		0.90		1		402
85759	−54 02800			9 50 31	−54 21.7	2	9.05	.020	0.27	.040	0.15	.035	6	A9 V	322,402
85781	−59 01517			9 50 31	−59 42.4	1	9.00		−0.09		−0.59		2	B5 III	401
85758	−52 02903			9 50 33	−52 24.8	1	7.31		1.26		1.25		1	K1/2 III	389
85585	+35 02073			9 50 38	+35 13.2	1	6.70		1.32		1.47		3	K2	1601
85633	+11 02117			9 50 40	+11 24.4	1	8.75		0.94		0.72		2	G5	1648
	−2 03000	G 53 - 13		9 50 41	−03 26.8	1	10.54	.004	1.53	.000	1.21	.014	6	M0	158,1620,1705
85725	−26 07505	HR 3916	★ AB	9 50 44	−27 05.9	3	6.29	.004	0.62	.000			8	G1 V	15,2012,3026
297635	−48 05089	LSS 1363		9 50 45	−48 35.8	1	9.96		0.21		−0.45		2		846
		AAS19,45 T8# 508		9 50 45	−54 11.1	1	11.89		1.22		1.07		2		402
		WLS 1000 20 # 6		9 50 46	+19 52.3	1	11.35		0.49		−0.08		2		1375
85615	+26 02021			9 50 46	+25 54.3	1	7.01		1.19		1.25		3	K2 III	833
85777	−51 04265			9 50 47	−51 28.6	2	7.48	.010	−0.13	.010	−0.62	.000	5	B3 IV	362,401
		LSS 1362		9 50 48	−46 02.7	1	12.50		−0.23		−1.16		2	B3 Ib	846
85791	−52 02907	LSS 1364		9 50 51	−53 16.8	2	8.37	.009	0.79	.000	0.55	.042	4	F3 Ib	389,8100
85675	+2 02264	T Sex		9 50 53	+02 17.6	1	9.82		0.22		0.18		1	A9 II-III V	668
85809	−53 02976			9 50 54	−54 05.0	3	9.37	.046	0.19	.023	−0.58	.025	6	B2/3 III/V	362,401,846
		SU Leo		9 50 59	+08 12.9	1	12.81		0.02		0.05		1		699
309794	−64 01112			9 51 00	−65 02.0	2	8.14	.010	0.04	.000	−0.28	.000	13	F2	9,9
85709	+6 02224	HR 3915		9 51 05	+06 11.7	3	5.96	.013	1.66	.005	1.92	.015	10	M2 III	15,1256,3055
85806	−46 05587			9 51 08	−46 27.1	1	8.22		0.92		0.57		8	K0 III	1673
85838	−51 04273			9 51 12	−51 52.4	1	7.91		0.81		0.37		1	G3 IV	389
85774	−22 02753			9 51 16	−22 42.5	1	9.39		1.16				3	K1/2 III	1594
	−31 07745	IDS09491S3117	A	9 51 17	−31 30.8	1	10.21		1.38		1.15		2	M0 V	3072
	−31 07745	IDS09491S3117	B	9 51 17	−31 30.8	1	14.42		1.70				1	M0 V	3072
85821	−43 05706			9 51 17	−44 20.9	1	6.66		0.99		0.78		1	F5 IV	389
85837	−48 05100			9 51 17	−48 41.6	1	7.84		1.26		1.30		1	K2 III	389
85871	−54 02816	HR 3920, LSS 1365		9 51 17	−55 08.2	5	6.47	.007	−0.14	.007	−0.89	.020	19	B1 IIIn	15,362,401,2012,2012
		BONN82 T12# 9		9 51 18	−54 24.7	1	12.08		0.64		0.17		3	G2	322
		AA54,263 # 7		9 51 18	−54 47.3	1	13.08		0.53		−0.08		1		524
	+39 02287			9 51 19	+39 09.0	1	9.19		0.52		0.04		3	G0	1601
85773	−22 02754			9 51 20	−22 35.9	1	9.38		1.12		0.82		1	F/Gw	6006
85850	−50 04608			9 51 21	−50 41.8	1	8.80		0.51		0.01		1	F6/7 V	389
		AA54,263 # 8		9 51 22	−54 47.4	1	14.06		0.62		0.50		1		524
		AA54,263 # 6		9 51 23	−54 47.4	1	12.56		0.40		−0.27		2		524
		AA54,263 # 9		9 51 24	−54 47.	1	14.61		0.82		0.67		1		524
85891	−54 02820	IDS09497S5433		9 51 24	−54 47.0	1	8.66		2.26		2.68		2	K5 Ib	524
		LSS 1366		9 51 25	−54 46.9	1	10.73		0.47		−0.32		2	B3 Ib	524
		AA54,263 # 5		9 51 25	−54 47.5	1	12.55		0.54		0.30		2		524
85583	+61 01151	HR 3911	★ A	9 51 26	+61 21.2	1	6.27		1.05				2	K0	71
		AA54,263 # 3		9 51 26	−54 46.7	1	11.96		0.36		0.12		2		524
		AA54,263 # 4		9 51 26	−54 46.7	1	12.71		0.36		0.23		2		524
	−54 02821	LSS 1366	★ AB	9 51 26	−54 46.8	1	10.86		0.47		−0.28		1		846
		LSS 1368		9 51 26	−56 41.4	2	11.73	.035	0.73	.005	−0.33	.015	4	B3 Ib	362,846
	−56 02606	LSS 1369		9 51 27	−56 22.3	3	10.66	.111	0.68	.032	−0.37	.047	7		362,846,1737
		G 161 - 87		9 51 33	−12 00.9	1	12.50		1.37				1		3062
300349	−54 02825			9 51 33	−54 49.0	2	10.23	.005	0.33	.015	−0.47	.035	4	B3	362,846
	−56 02609	LSS 1370		9 51 33	−56 58.7	1	11.29		0.47		−0.33		2		846
		GD 105		9 51 36	+33 17.2	1	15.06		0.70		0.02		1		3060
	−48 05103	LSS 1367		9 51 36	−48 28.9	1	11.97		0.14		−0.59		3		846
	+2 02265			9 51 37	+02 01.5	1	10.02		0.64		0.19		4		1249
		CCS 1579, FP Vel		9 51 37	−52 16.4	1	10.73		2.65		3.77		1		864
		HA 101 # 308		9 51 38	−00 13.3	1	10.54		0.56		0.00		5	G0	281
85924	−53 02994			9 51 42	−54 03.2	4	8.10	.010	0.02	.010	−0.39	.010	19	B8 Ib/II	322,362,401,402
300292	−54 02830			9 51 42	−54 28.2	4	9.97	.099	0.45	.022	−0.36	.030	10	B8	322,362,402,846
	+0 02580			9 51 45	−00 00.9	4	9.96	.006	0.43	.006	0.01	.008	30	F2	281,989,1729,6004
86320	−79 00457			9 51 48	−79 49.5	1	6.51		0.06		−0.23		8	B8 IV	1628
	−81 00382			9 51 49	−81 43.3	1	10.46		0.79		0.45		2		1696
85940	−53 02997			9 51 51	−53 59.3	1	9.60		0.14		−0.10		5	B9 II	402
85817	+2 02266			9 51 52	+02 09.8	1	9.48		0.93		0.67		4	G5	1249
85859	−25 07585	HR 3919		9 51 56	−25 41.8	4	4.88	.004	1.23	.003	1.30	.003	21	K2 III	15,1075,1637,2012
	+41 02026	G 116 - 63		9 51 59	+40 37.3	1	11.39		1.37		1.25		2	K5	7010
86089	−70 00949			9 51 59	−71 13.1	1	7.24		1.15				4	K2 III	2012

Table 1 525

HD	DM	Other Id	N Rem	α_{1950}	δ_{1950}	S	V	σ_V	B–V	σ_{B-V}	U–B	σ_{U-B}	n	Spectrum	References
85953	−50 04622	HR 3924		9 52 00	−50 54.6	6	5.93	.005	-0.16	.012	-0.73	.011	18	B2 III	15,362,401,540,976,2012
85844	+0 02581			9 52 01	−00 10.1	6	8.23	.002	0.26	.004	0.01	.007	52	A2	281,989,1509,1729,1775,6004
		GD 300		9 52 02	+51 51.2	2	12.80	.095	-0.33	.010	-1.23	.005	6	sdO	272,308
		LTT 18089		9 52 07	+74 26.5	1	11.95		0.61		0.06		3		1723
		HA 101 # 88		9 52 08	−00 38.1	1	11.25		0.97		0.64		2		281
		LP 788 - 55		9 52 11	−19 06.8	1	11.80		1.01		0.87		1		1696
85905	−21 02935	HR 3921		9 52 13	−22 15.1	2	6.23	.005	0.04	.000			7	A2/3 III	15,2012
86000	−54 02842			9 52 14	−54 36.7	1	8.14		-0.02		-0.34		2	B8/9 IV	401
85879	+2 02267			9 52 15	+02 06.1	1	9.85		0.42		0.00		4	F5	1249
85894	−4 02753			9 52 15	−05 18.3	1	8.70		0.39		0.03		2	B8 III	540
	−55 02642	LSS 1372		9 52 16	−56 05.6	2	11.00	.040	0.65	.020	-0.36	.079	5		362,846
		HA 101 # 315		9 52 17	−00 13.2	3	11.25	.000	1.15	.003	1.06	.007	22	G7	281,1764,5006
85966	−42 05736			9 52 17	−42 40.8	1	7.31		0.97		0.73		1	G8 III	389
		HA 101 # 316		9 52 18	−00 04.3	2	11.55	.001	0.50	.003	0.04	.004	24	F7	281,1764
		HA 101 # 169		9 52 19	−00 32.9	1	11.74		0.97		0.74		6	G3	281
85980	−44 05987	HR 3925	⋆ AB	9 52 19	−45 02.8	5	5.70	.005	-0.12	.004	-0.57	.005	13	B5 V	15,26,404,2027,8023
		BONN82 T12# 10		9 52 23	−54 29.6	1	12.43		0.58		0.08		3	F9	322
85997	−46 05613			9 52 26	−46 24.0	1	6.67		0.98		0.78		1	G8 III	389
85904	+0 02582			9 52 27	+00 03.3	2	7.99	.020	1.58	.023	1.84		10	M7 V	281,6004
85795	+50 01698	HR 3917, SY UMa		9 52 28	+50 03.4	1	5.28		0.07		0.12		3	A3 III	3050
		HA 101 # 174		9 52 28	−00 26.8	1	11.67		0.62		0.12		2	G3	281
	+63 00869	G 235 - 49		9 52 29	+63 02.1	3	9.00	.005	1.45	.013	1.19		8	M2	1197,1758,3072
	−0 02267			9 52 30	−00 36.7	1	10.51		0.51		0.04		3	F7p	281
85951	−18 02810	HR 3923		9 52 31	−18 46.3	3	4.94	.000	1.56	.008	1.91	.035	15	M1 III	1024,2031,3053
86006	−45 05556			9 52 33	−45 29.6	3	8.17	.010	0.71	.004	0.32		9	G5 IV/V	389,1075,2012
86005	−42 05741			9 52 38	−43 05.1	3	7.20	.034	1.31	.011	1.03	.010	26	K2 IIIp	389,1621,1738
		LP 428 - 67		9 52 42	+16 47.1	1	10.81		0.50		-0.20		2		1696
85994	−25 07596	IDS09506S2604	AB	9 52 46	−26 18.1	1	8.47		0.99		0.40		1	K1 V	3072
86034	−42 05743			9 52 48	−43 18.6	3	7.89	.003	1.25	.005	1.32	.002	16	K1/2 III	389,1621,1738
		LSS 1373		9 52 48	−54 12.1	1	11.56		0.47		-0.45		3	B3 Ib	846
85932	+19 02283			9 52 49	+19 03.6	1	7.99		0.39		0.01		3	F0	1648
		LOD 18 # 15		9 52 54	−54 30.9	2	11.37	.055	0.26	.134	0.05	.030	5		125,127
		AAS27,343 # 264	A	9 52 54	−57 05.	1	13.79		0.53		0.41		1		362
		AAS27,343 # 264	B	9 52 54	−57 05.	1	14.44		0.98		-0.19		1		362
		AAS27,343 # 264	C	9 52 54	−57 05.	1	13.51		1.04		0.00		1		362
		101 L1		9 52 55	−00 07.4	1	16.50		0.76		-0.10		1		1764
300339	−54 02859			9 52 55	−54 32.2	1	9.98		1.06		1.01		3	A2	322
86118	−57 02418	QX Car		9 52 57	−58 11.0	3	6.63	.004	-0.18	.004	-0.85	.000	11	B2 V	158,243,1732
		G 43 - 14		9 52 58	+16 19.9	2	13.19	.022	0.59	.002	-0.10	.017	3		1620,1658
		G 49 - 29		9 52 59	+26 55.6	2	12.07	.017	1.14	.011	1.11	.008	5		1620,5010
		HA 101 # 320		9 52 59	−00 08.3	1	13.82		1.05		0.69		7		1764
86117	−56 02635			9 52 59	−56 56.0	3	9.21	.010	0.32	.018	-0.37	.029	7	B3/5	362,540,976
		HA 101 # 321		9 53 00	−00 04.5	1	12.85		0.65		0.14		10		281
86050	−40 05610			9 53 00	−40 31.4	1	8.16		0.52				4	F7 V	2012
86087	−49 04801	HR 3927		9 53 00	−50 00.4	4	5.71	.009	-0.01	.009	-0.01	.000	18	A0 V	15,1075,2012,2029
86140	−58 01695			9 53 00	−58 29.9	2	9.03	.015	0.96	.005	0.65		6	K3/4 V	2012,3072
		101 L2		9 53 01	−00 04.6	1	15.77		0.60		0.08		1		1764
298763	−53 03019			9 53 01	−53 27.3	2	10.24	.010	0.08	.010	-0.60	.025	4	B8	362,846
85990	−0 02270			9 53 02	−00 53.3	7	8.00	.006	1.12	.007	1.03	.007	50	K0 III	281,989,1509,1729,1775*
		He2 38		9 53 04	−57 04.7	1	12.97		1.23		-0.16		1		1753
		HA 101 # 404		9 53 07	−00 04.1	1	13.46		1.00		0.70		7		1764
86183	−62 01356			9 53 08	−63 12.8	1	8.73		-0.04		-0.49		2	B5 V	401
86112	−50 04640			9 53 09	−50 34.9	1	9.14		-0.06		-0.35		2	B9 III	401
85988	+18 02291			9 53 11	+17 46.8	1	8.41		0.42		0.00		2	F2	1375
	−0 02271			9 53 11	−00 37.9	1	10.77		0.57		0.09		3	F8	281
86085	−38 06018			9 53 13	−38 31.5	1	8.94		-0.09		-0.85		1	B3	55
86162	−58 01697			9 53 13	−59 01.8	3	9.17	.009	-0.06	.005	-0.83	.017	7	B0/1 IV	401,540,976
86161	−57 02420	V396 Car		9 53 14	−57 29.4	1	8.26		0.32		-0.64		4	WN	362
86013	+0 02585			9 53 18	+00 07.5	1	10.05		0.38		-0.04		3	F5	281
		AAS27,343 # 267		9 53 18	−54 12.	1	12.10		0.29		-0.15		2		362
	+0 02586			9 53 21	−00 08.9	6	9.74	.003	1.16	.006	1.15	.004	81	K0 III	281,989,1729,1764,5006,6004
86111	−40 05616	X Vel		9 53 23	−41 21.0	1	7.19		4.34				1	C4-5,4-5	864
	−61 01361			9 53 23	−62 17.4	2	10.97	.010	-0.03	.007	-0.68	.005	3	B2 V	540,976
86214	−59 01528			9 53 24	−59 35.5	2	9.21	.006	-0.02	.013	-0.79	.003	5	B2 Ib	540,976
86083	−20 03049			9 53 27	−20 30.5	1	7.71		0.59				4	G2 V	2012
		HA 101 # 109		9 53 28	−00 41.2	1	11.93		0.53		0.01		4	G0	281
86046	+0 02587			9 53 30	+00 11.6	2	8.99	.002	0.35	.003	0.04		7	A5	281,6004
		CCS 1584, EF Vel		9 53 30	−49 38.5	1	9.82		2.74		5.80		1	R	864
	+52 01432	G 195 - 49		9 53 33	+52 15.8	1	9.91		0.82		0.47		2		7010
300338	−54 02872			9 53 33	−54 30.6	2	10.91	.064	0.30	.010	0.14	.045	5	A2	125,127
		HA 101 # 408		9 53 34	+00 01.6	1	14.78		1.20		1.35		1		1764
		G 116 - 66		9 53 34	+41 19.5	1	10.33		1.00		0.92		1	K1	1658
		HA 101 # 326		9 53 34	−00 12.9	1	14.92		0.73		0.23		6		1764
		HA 101 # 262		9 53 34	−00 15.5	1	14.30		0.78		0.30		2		1764
		LOD 18 # 18		9 53 34	−54 29.5	1	11.72		0.31		0.10		2		127
		HA 101 # 410		9 53 35	+00 00.2	1	13.65		0.55		-0.06		1		1764
		HA 101 # 327		9 53 35	−00 11.6	1	13.44		1.15		1.14		15		1764
86012	+33 01920			9 53 36	+32 37.3	1	6.55		0.39		-0.05		2	F2	1625
86272	−62 01360			9 53 37	−63 18.2	2	9.46	.003	-0.04	.006	-0.67	.012	6	B5 Vne	540,976
		LSS 1375		9 53 38	−55 22.0	1	11.53		0.90		-0.01		2	B3 Ib	846

HD	DM	Other Id	N	Rem	α_{1950}	δ_{1950}	S	V	σ_V	B–V	σ_{B-V}	U–B	σ_{U-B}	n	Spectrum	References
86082	−6 03033				9 53 39	−07 24.4	2	6.73	.015	1.42	.015	1.60		7	K4 III	2006,3040
		HA 101 # 413			9 53 40	+00 02.3	1	12.58		0.98		0.72		6		1764
85945	+58 01224	HR 3922			9 53 45	+57 39.4	2	5.94	.016	0.89	.000	0.57		2	G8 III	71,3002
		HA 101 # 329			9 53 45	−00 12.1	1	11.99		0.64		0.13		12	G0	281
		HA 101 # 268			9 53 45	−00 17.6	1	14.38		1.53		1.38		5		1764
86080	+9 02262	HR 3926			9 53 47	+09 10.3	2	5.84	.005	1.13	.000	1.01	.005	7	gK2	15,1078
	+18 02293	G 43 - 15			9 53 47	+17 44.8	1	10.16		0.87		0.63		1	K1	333,1620
		HA 101 # 330			9 53 47	−00 13.0	1	13.72		0.58		-0.03		14		1764
		KUV 348 - 7			9 53 48	+41 29.6	1	15.21		-0.05		-0.79		1		974
85893	+67 00625				9 53 48	+67 00.5	1	8.46		1.19		1.20		2	K2	1733
300342	−54 02879				9 53 48	−54 35.8	2	10.16	.065	0.49	.010	0.07	.020	4	G0	125,127
		HA 101 # 415			9 53 49	−00 02.6	1	15.26		0.58		-0.01		1		1764
86269	−57 02434	IDS09522S5728		AB	9 53 50	−57 43.2	1	9.03		1.04		0.80		2	K3 V	3072
		HA 101 # 270			9 53 53	−00 21.4	1	13.71		0.55		0.05		1		1764
86289	−58 01707				9 53 57	−58 30.2	1	7.63		-0.08		-0.43		4	B8 II	243
85841	+73 00478	HR 3918			9 53 58	+73 07.1	1	5.83		1.14				2	gK3	71
		LOD 18 # 20			9 53 58	−54 22.4	2	10.88	.050	1.09	.125	0.62	.035	4		125,127
		OK 290 # 1			9 54 00	+25 29.	1	14.60		1.31		1.15		2		326
		OK 290 # 2			9 54 00	+25 29.	1	16.28		1.07		0.89		1		326
		OK 290 # 3			9 54 00	+25 29.	1	16.98		1.17		0.75		1		326
86211	−40 05626	HR 3930			9 54 02	−40 35.2	4	6.41	.012	1.61	.008	1.99		13	M1 III	15,1075,2031,3005
86288	−52 02955				9 54 04	−52 44.5	1	7.80		0.09		-0.39		2	B5 III	362
86135	+0 02588				9 54 06	−00 13.4	7	7.84	.006	1.48	.004	1.79	.011	52	M0 III	281,989,1509,1657,1729*
300341	−54 02886				9 54 06	−54 34.4	2	10.77	.079	0.35	.020	0.23	.000	5		125,127
86319	−58 01709				9 54 06	−58 52.1	2	7.23	.000	-0.07	.005	-0.39	.005	6	B7/8 III	243,401
	−53 03044	LSS 1377			9 54 08	−53 37.9	2	11.48	.010	0.32	.010	-0.47	.030	5		362,846
86249	−40 05628				9 54 09	−40 32.9	2	9.00	.010	0.96	.002	0.73		7	K2 V	2012,3072
		G 48 - 54			9 54 12	+04 02.7	1	13.48		1.51		1.14		1		1658
86318	−53 03045				9 54 12	−53 49.7	1	8.10		1.30		1.34		1	K2 III	389
298761	−52 02962				9 54 13	−53 17.9	2	9.47	.001	0.19	.006	0.13	.015	8	A0 V	540,976
86133	+20 02399	IDS09515N2014		A	9 54 17	+20 00.1	1	7.60		0.56		0.00		2	G0	3077
86133	+20 02400	IDS09515N2014		B	9 54 17	+20 00.1	1	8.40		0.70		0.17		2	G5	3077
300340	−54 02889				9 54 19	−54 33.1	2	10.80	.084	0.26	.035	0.13	.025	5	A3	125,127
	−11 02763				9 54 20	−12 01.9	1	9.95		0.97		0.62		1	K3 V	3072
86303	−44 06018	IDS09524S4447		AB	9 54 20	−45 01.1	1	7.72		1.02		0.81		1	K0 II	389
		HA 101 # 278			9 54 21	−00 15.3	1	15.49		1.04		0.74		1		1764
		101 L3			9 54 21	−00 16.1	1	15.95		0.64		-0.03		1		1764
86248	−30 08049				9 54 21	−31 12.2	2	9.64	.004	-0.17	.013	-0.91	.013	7	B2 II	55,1775
86317	−45 05588	IDS09524S4555		AB	9 54 23	−46 09.2	1	8.14		1.04		0.79		2	G8/K0 III	389
86245	−24 08603				9 54 24	−25 05.9	1	8.28		0.55				4	F8 V	2012
86267	−32 06895	HR 3932			9 54 24	−33 10.8	3	5.84	.004	1.20	.000			11	K1 III	15,1075,2031
86332	−47 05381				9 54 26	−47 31.0	1	7.08		0.93		0.62		1	G8 III	389
86353	−53 03052				9 54 27	−53 22.9	1	6.81		-0.09		-0.45		2	B7 V	401
86266	−25 07622	HR 3931		★ AB	9 54 30	−26 18.7	4	6.27	.005	0.21	.009	0.11	.007	16	A4 V	15,1075,1637,2012
		HA 101 # 281			9 54 31	−00 17.4	5	11.58	.004	0.81	.006	0.44	.009	63		1728,281,989,1729,1764
86352	−50 04662	HR 3935			9 54 31	−51 05.9	6	6.37	.008	-0.17	.014	-0.76	.010	17	B2 IV-V	15,362,401,540,976,2012
	+0 02589				9 54 32	−00 15.7	4	10.00	.002	0.43	.007	0.01	.001	32	F5	281,989,1729,6004
		101 L4			9 54 35	−00 17.2	1	16.26		0.79		0.36		1		1764
	−30 08053				9 54 36	−31 00.2	1	10.08		0.86		0.47		1		3072
	−54 02899				9 54 36	−54 51.2	1	10.10		1.56				3		279
		BPM 6082			9 54 36	−71 02.	1	13.48		0.12		-0.65		2		3065
86188	+19 02284				9 54 38	+19 31.7	1	7.95		1.15		1.18		3	K0	1648
86146	+41 02033	HR 3928			9 54 38	+41 17.7	4	5.14	.012	0.46	.004	0.00	.000	11	F6 Va	15,1008,1363,3026
		101 L5			9 54 38	−00 17.3	1	15.93		0.62		0.12		1		1764
86301	−26 07551	HR 3933			9 54 38	−27 14.2	2	6.31	.005	0.17	.000			7	A4 V	15,2012
		HA 101 # 121			9 54 39	−00 36.9	1	12.13		0.64		0.14		2	G3	281
		HA 101 # 421			9 54 42	−00 03.0	1	13.18		0.51		-0.03		1		1764
		HA 101 # 338			9 54 44	−00 06.7	1	13.79		0.63		0.02		1		1764
		HA 101 # 339			9 54 44	−00 10.7	1	14.45		0.85		0.50		1		1764
		HA 101 # 424			9 54 46	−00 02.1	1	15.06		0.76		0.27		1		1764
		LTT 12657			9 54 47	+67 18.0	1	10.38		1.32		1.21		3		1723
		G 116 - 69			9 54 48	+32 51.2	2	12.78	.000	0.71	.000	0.05	.025	2		1658,3062
86166	+46 01566	HR 3929			9 54 48	+45 39.2	1	6.30		1.11		0.95		2	K0 III	252
86385	−51 04334				9 54 49	−51 35.6	1	7.92		-0.10		-0.55		2	B5 IV	401
86238	+17 02156				9 54 50	+16 41.9	1	7.30		1.28		1.56		3	K0	1648
86279	+0 02590				9 54 51	+00 00.8	3	8.32	.007	1.41	.002	1.65	.005	13	M0 III	281,1509,6004
	−53 03067				9 54 51	−53 39.6	1	11.27		0.21		-0.40		5		362
86403	−53 03068				9 54 51	−54 10.8	1	7.96		0.01		-0.01		2	A1 V	127
		HA 101 # 427			9 54 52	−00 03.0	1	14.96		0.81		0.32		1		1764
		HA 101 # 341			9 54 56	−00 07.6	1	14.34		0.57		0.06		1		1764
		HA 101 # 43			9 54 56	−00 50.4	1	10.17		1.16		1.08		2		281
86384	−47 05387				9 54 56	−47 26.3	1	7.41		1.16		1.03		1	K0 III	389
300343	−54 02908				9 54 56	−54 34.8	3	9.82	.048	0.50	.017	-0.01	.016	10	G0	125,127,279
		HA 101 # 342			9 54 57	−00 07.5	1	15.56		0.53		-0.06		1		1764
		HA 101 # 343			9 54 57	−00 08.6	1	15.50		0.61		0.09		1		1764
		HA 101 # 429			9 54 58	−00 04.0	1	13.50		0.98		0.78		13		1764
		G 253 - 34			9 54 59	+81 23.9	1	15.26		1.68				1		906
86427	−54 02909	IDS09533S5436		AB	9 54 59	−54 50.7	2	7.87	.038	0.17	.010	0.16		7	A4 IV	127,279
		G 49 - 33			9 55 00	+24 47.4	3	15.08	.010	0.24	.006	-0.60	.042	9		419,1620,3060
	−54 02911				9 55 00	−54 55.5	1	10.19		1.02		0.77		3		279

Table 1 527

HD	DM	Other Id	N	Rem	α_{1950}	δ_{1950}	S	V	σ_V	B–V	σ_{B-V}	U–B	σ_{U-B}	n	Spectrum	References
300345	−54 02912				9 55 01	−54 46.9	1	9.26		0.99		0.77		3	G0	279
		HA 101 # 431			9 55 03	−00 03.5	1	13.68		1.25		1.14		13		1764
		101 L6			9 55 05	−00 03.5	1	16.50		0.71		0.18		1		1764
	+0 02591				9 55 06	−00 10.5	1	10.05		0.46		-0.05		3	F8	281
86440	−53 03075	HR 3940		★A	9 55 06	−54 19.7	10	3.53	.017	-0.09	.006	-0.62	.015	31	B5 Ib	15,26,125,127,362*
86237	+39 02295	IDS09521N3925			9 55 07	+39 19.7	2	8.24	.015	0.47	.005	0.02	.005	6	F5 III	833,1601
86454	−58 01714			V	9 55 09	−58 34.2	1	9.15		0.40		-0.03		4	F2 IV/V	243
	+14 02170				9 55 11	+13 50.1	1	8.22		1.02		0.88		2	K0	1648
		G 161 - 90			9 55 13	−02 35.1	1	14.49		1.55				1		3062
86467	−54 02915				9 55 15	−54 57.4	1	9.15		1.01		0.80		3	K0 III	279
		G 42 - 28			9 55 16	+10 13.8	2	13.51	.003	0.75	.007	0.19	.017	4		333,1620,1658
86606	−70 00953	HR 3944			9 55 18	−71 09.0	4	6.33	.024	-0.06	.029	-0.93	.014	11	B0.5III/IV	15,1034,1075,8100
		HA 101 # 207			9 55 19	−00 33.2	2	12.41	.004	0.52	.002	-0.07	.004	28	F7	281,1764
86439	−49 04831	IDS09534S4924		AB	9 55 20	−49 37.9	1	7.24		-0.01		-0.21		2	B9 V	401
86438	−45 05606	LSS 1378			9 55 21	−45 23.6	2	8.75	.001	-0.15	.000	-0.85	.000		B2 II	846,1732
86490	−55 02704	LSS 1379			9 55 21	−55 48.5	3	8.26	.018	0.46	.011	-0.30	.025	6	B6 Ib	362,540,976
86466	−52 02980	HR 3941, IV Vel			9 55 22	−52 24.0	6	6.11	.005	-0.13	.009	-0.61	.006	18	B3 III	15,26,362,540,976,2012
		HA 101 # 56			9 55 28	−00 52.2	2	11.89	.000	0.43	.004	-0.04		15		281,5006
86453	−48 05168				9 55 28	−49 19.0	1	7.18		0.79		0.42		1	G5 II	389
86369	+9 02269	HR 3938			9 55 29	+08 33.2	2	6.03	.005	1.36	.000	1.55	.000	7	gK3	15,1078
86359	+15 02141				9 55 29	+15 27.6	1	7.45		0.89		0.62		3	G5	1648
		Ton 1137			9 55 30	+36 05.	1	15.34		-0.29		-1.25		7		313
		HA 101 # 57			9 55 31	−00 47.0	1	11.58		0.78		0.42		12		281
		HA 101 # 58			9 55 31	−00 53.9	1	12.48		0.91		0.44		2		281
86360	+13 02183	HR 3937			9 55 32	+12 41.0	3	5.22	.052	-0.04	.003	-0.13	.000	8	B9 IV	1022,1363,3023
86774	−76 00606				9 55 32	−77 03.0	1	7.68		1.89				4	M1 III	2033
86358	+28 01824	HR 3936			9 55 34	+27 59.9	1	6.46		0.34		-0.01		3	F3 V	1733
		Vel sq 1 # 101			9 55 36	−55 11.	1	14.64		1.40		0.30		3		846
	−54 02924				9 55 42	−54 42.3	2	10.35	.029	0.26	.001	-0.46	.014	3	B2 V	540,976
		G 116 - 70			9 55 43	+43 24.1	2	14.80	.000	1.64	.000	1.09	.000	4		203,203
		G 116 - 71		A	9 55 43	+43 24.4	1	13.31		1.49		1.24		2		203
		HA 101 # 216			9 55 43	−00 29.2	1	12.13		0.69		0.12		4	G4	281
86525	−51 04342				9 55 43	−51 40.7	1	8.15		1.08		0.87		1	G8 III	389
86408	+0 02593				9 55 45	−00 11.2	7	9.87	.005	0.26	.006	0.13	.005	83	A5	281,989,1728,1729,1764*
	−29 07960	IDS09535S2955		AB	9 55 46	−30 09.5	2	9.41	.010	1.04	.000	0.81	.020	10	F5	119,1775
86523	−47 05399	HR 3943		★AB	9 55 47	−48 10.5	4	6.05	.005	-0.14	.005			16	B2 V	15,1020,2012
300344	−54 02925				9 55 47	−54 30.1	5	10.38	.028	0.25	.011	-0.47	.015	19	B5	125,127,279,362,846
		Vel sq 1 # 102			9 55 48	−54 35.	1	14.28		0.89		0.00		3		846
		Vel sq 1 # 9			9 55 52	−54 26.5	1	10.00		0.01		-0.29		3		279
300336	−54 02929				9 55 52	−54 26.5	1	9.61		0.74		0.57		3	A7	279
86571	−54 02931				9 55 54	−55 01.5	1	9.76		0.15		0.18		3	A3 IV/V	279
86634	−63 01233	HR 3948			9 55 54	−64 15.1	4	6.57	.007	1.12	.005	1.07	.008	20	K0/1 III	15,1075,1754,2038
	−54 02930				9 55 56	−54 26.5	1	11.22		0.05		-0.63		4		279
86659	−68 01011	HR 3949			9 55 56	−68 51.8	4	6.19	.006	-0.10	.006	-0.61	.015	19	B3 V	15,26,1075,2038
298812	−53 03099				9 55 58	−54 09.9	2	9.27	.015	0.50	.010	0.02		6	F8	1696,2012
	−15 02945				9 55 59	−15 41.2	1	10.65		0.59		-0.02		4		1696
86585	−54 02937				9 56 02	−54 40.8	1	9.57		0.01		-0.36		3	B9 II/III	279
86435	+11 02136				9 56 06	+10 41.9	1	7.12		1.43		1.67		3	K0	1648
297714	−50 04690	LSS 1380			9 56 09	−50 35.0	1	9.76		0.41		-0.21		2	B3	362
		Vel sq 1 # 13			9 56 09	−54 26.4	1	12.72		0.26		0.25		3		279
86476	+5 02263				9 56 12	+05 02.9	1	7.16		1.64		1.85		2	M2 III	3040
86601	−52 02996				9 56 12	−52 39.2	1	8.05		0.03		-0.12		2	B9.5III/IV	401
298743	−51 04355				9 56 16	−51 55.3	1	10.18		0.10		-0.44		2	B5	362
		Vel sq 1 # 14			9 56 17	−55 12.0	1	13.29		0.50		0.08		5		279
86633	−55 02725	IDS09546S5542		AB	9 56 17	−55 55.9	1	8.93		-0.04		-0.46		2	B8 IV/V	401
		Vel sq 1 # 15			9 56 19	−55 09.8	1	10.67		1.63		2.02		5		279
		LSS 1382			9 56 20	−54 14.6	1	11.50		0.82		0.28		2	B3 Ib	846
86632	−54 02943				9 56 20	−54 50.6	1	10.26		0.08		0.03		3	B9 V	279
86675	−61 01379				9 56 20	−61 41.6	1	7.06		-0.02		-0.10		2	A0 V	401
86646	−58 01737	LSS 1383			9 56 21	−59 00.3	1	9.24		0.54		0.48		3	F0 III/IV	8100
86631	−54 02945				9 56 22	−54 39.9	3	10.14	.047	0.07	.004	-0.06	.007	10	B9 V	125,127,279
86656	−58 01738				9 56 22	−54 50.8	1	8.89		-0.02		-0.26		4	B8 IV	243
86658	−61 01380				9 56 24	−61 26.5	1	9.22		-0.03		-0.38		2	B8/9 IV	401
298742	−51 04357				9 56 25	−51 52.5	1	11.01		0.22				4	B9 V	2021
86643	−54 02946				9 56 25	−55 03.7	1	10.20		0.05		-0.08		3	B9/A0 IV	279
86460	+01826	G 54 - 7			9 56 26	+27 45.8	1	7.75		0.57		0.57		1	G0 V	333,1620
86378	+57 01242	HR 3939			9 56 26	+57 03.1	2	5.49	.011	1.47	.008	1.82		3	K5 III	71,3001
86655	−58 01739	RR Car			9 56 27	−58 37.3	2	7.52	.039	1.62	.070	0.96	.083	6	M6 II/III	897,975
		G 42 - 30			9 56 28	+15 01.6	1	12.10		0.89		0.58		1		333,1620
	−45 05627	LU Vel			9 56 33	−46 10.5	3	11.27	.009	1.55	.009	0.90			M5	158,912,1705
86654	−54 02951				9 56 33	−54 44.1	3	9.04	.039	0.49	.008	0.04	.004	8	F5 V	125,127,279
86689	−59 01559	LSS 1384			9 56 33	−59 25.9	2	9.99	.024	0.02	.011	-0.77	.016	4	B0/2ne	540,976
		Vel sq 1 # 20			9 56 38	−54 18.9	1	11.23		0.57		0.06		3		279
		LP 18 - 384			9 56 40	+79 56.3	1	11.31		0.03		0.01		2		1726
300390	−54 02952				9 56 40	−54 40.0	3	10.07	.048	0.29	.009	0.15	.008	8	F0	125,127,279
86516	+22 02148				9 56 41	+21 33.7	1	6.74		0.20		0.11		2	Am	1648
300395	−54 02953				9 56 41	−55 01.3	1	9.30		1.65				5	K7	279
86513	+30 01946	HR 3942			9 56 43	+29 53.1	1	5.73		1.06				2	K0	71
86629	−35 06050	HR 3947		★A	9 56 43	−35 39.1	5	5.22	.010	0.32	.011	0.07	.012	16	A8 IV	3,9,15,2027,3037
	−8 02813				9 56 44	−08 56.3	1	9.88		1.09		0.92		1	K4 V	3072

HD	DM	Other Id	N Rem	α_{1950}	δ_{1950}	S	V	σ_V	B–V	σ_{B-V}	U–B	σ_{U-B}	n	Spectrum	References
	+14 02177	XX Leo		9 56 47	+14 01.5	1	10.70		0.43		0.09		1	A8	588
86612	−23 08898	HR 3946		9 56 48	−23 42.6	7	6.21	.015	-0.10	.006	-0.66	.010	41	B4 Ve	15,681,815,1075,1637*
300392	−54 02956			9 56 50	−54 45.1	3	9.75	.041	1.06	.012	0.89	.009	8	F8	125,127,279
	+33 01931			9 56 53	+33 20.3	1	10.85		1.36				1	M0	1017
300386	−54 02960			9 56 55	−54 21.9	1	10.11		1.03		0.86		2	K7	279
	−53 03124	LSS 1385		9 56 56	−53 23.0	3	10.84	.028	0.42	.013	-0.62	.028	7		362,846,1737
		LSS 1387		9 56 56	−59 33.7	1	11.82		0.17		-0.61		3	B3 Ib	846
86626	−19 02884			9 56 58	−20 07.0	1	6.70		1.46		1.63		8	K3/4 III	1628
86511	+43 01980			9 56 59	+42 33.4	1	7.42		1.08		1.08		2	K2	1601
86699	−46 05690			9 56 59	−47 15.0	1	7.55		1.45		1.73		1	K3 III	389
300388	−54 02962			9 56 59	−54 32.9	2	10.28	.075	0.18	.010	-0.15	.005	4	B8	125,127
86715	−47 05414			9 57 02	−48 15.0	1	8.01		1.10		1.00		1	K0 III	389
		Wool 9308		9 57 02	−49 45.8	2	11.69	.025	0.96	.010	0.70		7		158,912,1705
		Sextans B sq # 11		9 57 04	+05 30.5	1	13.02		1.11				2		1579
		Sextans B sq # 13		9 57 04	+05 30.5	1	11.17		0.47		-0.02		1		1579
86611	+4 02276	HR 3945		9 57 07	+03 37.5	3	6.69	.005	0.28	.005	0.08	.012	8	F0 V	15,39,1415
86716	−49 04851			9 57 07	−49 39.3	1	8.13		0.98		0.72		1	K0 III	389
86771	−58 01750			9 57 07	−58 31.2	1	9.47		-0.08		-0.64		4	B4 IV	243
304883	−60 01558			9 57 08	−61 17.0	2	10.64	.024	0.00	.008	-0.38	.003	3		540,976
298787	−51 04371	LSS 1386		9 57 09	−51 58.7	1	10.98		0.20		-0.43		2	F8	846
86758	−54 02966			9 57 09	−55 04.8	1	9.42		0.62		0.23		3	G0 IV/V	279
86714	−44 06062			9 57 11	−44 46.0	1	8.44		1.12		1.08		2	K0 III	389
86757	−54 02968			9 57 11	−54 23.0	1	8.34		1.54		1.74		3	K2 III	279
86590	+25 02191	DH Leo		9 57 13	+24 47.6	2	8.01	.339	0.93	.068	0.55	.160	7	K0 V	1355,3026
		Sextans B sq # 12		9 57 15	+05 31.4	1	14.06		0.63		0.15		2		1579
86739	−46 05693			9 57 15	−46 51.1	1	7.26		1.50		1.79		1	K3 III	389
	−24 08642			9 57 17	−24 37.0	2	9.66	.000	0.83	.007	0.34		5	K0	1594,6006
86786	−54 02975			9 57 17	−54 44.8	1	9.39		0.57		0.09		3	A1 V F8 IV	279
		Vel sq 1 # 28	AB	9 57 21	−54 48.1	1	14.93		0.65		0.46		6		279
86322	+75 00399			9 57 22	+75 00.0	1	6.90		1.05		0.94		2	K1 III	1003
		Vel sq 1 # 29		9 57 23	−54 44.8	1	15.89		1.29		0.72		6		279
		Vel sq 1 # 30		9 57 25	−54 43.5	1	16.57		1.11		1.03		2		279
86754	−44 06064	IDS09554S4429	A	9 57 26	−44 42.9	1	6.72		-0.07		-0.33		9	B8/9 V	1628
		Vel sq 1 # 31		9 57 28	−54 45.4	1	15.08		0.49		0.38		2		279
		AAS27,343 # 280		9 57 30	−50 40.	2	12.57	.020	0.28	.005	-0.46	.050	5		362,846
		Vel sq 1 # 32		9 57 31	−54 47.9	1	13.58		0.39		0.27		6		279
86663	+8 02301	HR 3950		9 57 34	+08 17.1	10	4.69	.018	1.60	.009	1.93	.017	34	M2 IIIab	3,15,31,1078,1119*
		Vel sq 1 # 33		9 57 34	−54 37.9	1	12.58		0.51		0.22		2		279
86823	−57 02510			9 57 34	−57 53.3	1	7.10		-0.03		-0.15		3	B9 Vn	761
		Sextans B sq # 50		9 57 35	+05 40.2	1	14.56		0.71				1		1579
86683	+6 02240	IDS09550N0644	A	9 57 36	+06 29.4	1	7.69		0.47		-0.03		1	F5	3016
86683	+6 02240	IDS09550N0644	B	9 57 36	+06 29.4	1	10.53		0.97		0.79		1	F5	3032
		MCC 561		9 57 36	+27 30.5	1	11.37		1.42				1	K7-M0	1619
		Sextans B sq # 24		9 57 40	+05 31.2	1	12.96		0.96				1		1579
		Sextans B sq # 22		9 57 40	+05 38.9	1	15.10		0.97				2		1579
		KUV 09577+3700		9 57 42	+36 59.6	1	17.01		-0.20		-0.87		1	NHB	1708
		Vel sq 1 # 34		9 57 42	−54 47.5	1	14.26		1.94				6		279
		Vel sq 1 # 35		9 57 42	−54 49.2	1	12.42		0.47		0.35		3		279
		Vel sq 1 # 36		9 57 43	−54 49.9	1	14.31		0.52		-0.01		3		279
		Vel sq 1 # 37		9 57 44	−54 48.5	1	13.57		0.51		0.08		3		279
86680	+28 01827			9 57 48	+28 25.0	1	7.99		0.61		0.12		4	G0 V	3026
		Vel sq 1 # 38		9 57 48	−54 50.7	1	13.90		0.53		-0.02		4		279
		BPM 6114		9 57 48	−66 39.	1	14.60		-0.20		-0.98		1	DF	3065
		Vel sq 1 # 39		9 57 50	−54 52.1	1	13.49		1.32		1.33		5		279
		G 117 - 61		9 57 52	+32 33.4	1	11.93		1.42		1.30		1		3078
		Vel sq 1 # 40		9 57 52	−54 47.9	1	17.02		2.00				4		279
		L 319 - 78		9 57 54	−47 48.	1	12.11		1.21		1.19		1		1696
86833	−49 04863			9 57 54	−49 36.8	1	8.88		0.51		0.10		1	F5/6 IV	389
300391	−54 02983			9 57 54	−54 45.8	1	10.36		0.36		0.28		15	A5	279
	−54 02984			9 57 55	−54 46.8	1	11.61		0.27		0.14		3		279
86820	−42 05829			9 57 57	−42 46.5	1	8.29		0.95		0.67		1	G6/8 III	389
86867	−55 02986			9 57 57	−55 07.4	1	8.89		0.42		0.28		3	A1 V F6/7 V	279
86731	+18 02303			9 57 58	+17 48.5	1	8.07		0.47		-0.02		3	F5	1648
300389	−54 02987			9 57 58	−54 40.1	3	10.24	.005	0.32	.022	-0.61	.011	13	B1	279,362,846
		Vel sq 1 # 45		9 57 58	−54 53.9	1	17.38		0.96				3		279
		WLS 1000 20 # 10		9 57 59	+21 00.5	1	12.57		0.61		0.08		2		1375
86847	−48 05202			9 57 59	−49 20.3	2	9.01	.021	0.44	.009	-0.04		4	F2 V	1594,6006
		Vel sq 1 # 46		9 57 59	−54 47.0	1	15.81		0.87		0.27		3		279
	−54 02989			9 58 01	−54 39.4	2	10.21	.023	0.34	.016	-0.56	.020	3	B0.5II-III	540,976
		Vel sq 1 # 47		9 58 01	−54 47.4	1	12.20		1.60		1.72		4		279
		Vel sq 1 # 48		9 58 01	−54 55.5	1	15.79		1.29		1.05		4		279
86865	−48 05203			9 58 02	−48 42.7	1	8.20		1.07		0.92		1	K0 III	389
		Vel sq 1 # 49		9 58 03	−54 43.6	1	16.38		1.86		0.92		5		279
		Vel sq 1 # 50		9 58 05	−54 45.4	1	14.84		0.71		0.19		4		279
	−54 02990	Vel sq 1 # 51		9 58 05	−54 53.6	1	10.70		0.43		-0.01		3		279
		Vel sq 1 # 52		9 58 06	−54 52.6	1	12.12		0.06		-0.03		3		279
		Vel sq 1 # 53		9 58 06	−54 53.2	1	12.28		0.60		0.13		3		279
	−54 02993			9 58 07	−54 54.2	1	11.12		0.55		0.03		5		279
		G 117 - 62		9 58 08	+29 13.9	1	14.78		1.54				1		906
86728	+32 01964	HR 3951	⋆ A	9 58 08	+32 10.2	6	5.37	.010	0.65	.006	0.26	.009	18	G2 Va	15,22,1003,1067,1243,3026

Table 1   529

HD	DM	Other Id	N Rem	α₁₉₅₀	δ₁₉₅₀	S	V	σ_V	B–V	σ_B–V	U–B	σ_U–B	n	Spectrum	References
300393	−54 02994			9 58 12	−54 57.6	1	10.86		-0.01		-0.47		14	A2	279
		Vel sq 1 # 55		9 58 13	−54 52.5	1	15.15		0.77		0.54		2		279
86661	+56 01421	G 195 - 52	⋆ AB	9 58 16	+55 49.9	1	7.97		0.73		0.31		3	G8 IV-V	22
	−54 02995			9 58 17	−54 53.4	1	10.14		1.14		0.98		3		279
86936	−59 01591	SZ Car		9 58 17	−59 58.7	1	7.51		2.90		5.80		1	N3	864
		GD 108		9 58 18	−07 19.1	3	13.56	.003	-0.22	.007	-0.93	.016	19	sdB	281,1764,3060
	−8 02823	G 162 - 12		9 58 21	−09 16.6	1	9.86		1.08		0.93		4	K3 V	3072
300394	−54 02998			9 58 23	−55 00.0	1	10.09		0.25		0.15		3	A0	279
		G 117 - 63		9 58 24	+33 06.4	1	13.09		0.83		0.45		2		1658
		LP 429 - 17		9 58 26	+14 33.2	1	11.38		1.04		0.94		1		1696
86981	−59 01600	NGC 3114 - 1		9 58 30	−60 05.8	1	9.64		-0.03				5	B9 IV	2025
		G 117 - 64		9 58 31	+29 00.6	1	13.02		0.51		-0.22		2		1658
		Vel sq 1 # 103		9 58 36	−54 32.	1	14.04		1.00		-0.25		3		846
		G 42 - 32		9 58 38	+08 32.3	1	11.82		1.41		1.23		1		333,1620
		NGC 3105 - 98		9 58 38	−54 33.2	1	11.52		0.32		0.24		2		535
		NGC 3105 - 99		9 58 39	−54 33.3	1	12.83		0.20		0.08		2		535
	+18 02306			9 58 40	+17 50.8	1	9.60		0.67				1	G0	1269
86997	−58 01779			9 58 41	−58 46.2	2	8.00	.035	0.51	.005	-0.05		9	F7 V	897,2012
86801	+29 01987			9 58 42	+28 48.4	1	8.78		0.58		0.04		3	G0 V	3026
86957	−50 04729			9 58 42	−50 51.0	1	8.60		0.84		0.49		1	G8 III	389
		AAS27,343 # 282	A	9 58 42	−55 08.	1	13.26		0.63		-0.07		1		362
		AAS27,343 # 282	B	9 58 42	−55 08.	1	12.85		0.72		0.51		1		362
		NGC 3105 - 75		9 58 43	−54 31.1	2	12.72	.030	0.52	.015	0.35	.005	4	A2	279,535
86760	+53 01380			9 58 45	+52 37.4	1	8.68		0.76		0.40		2	K0	1566
		KUV 348 - 10		9 58 46	+38 12.0	1	11.86		-0.06		-0.16		1		974
		NGC 3105 - 74		9 58 46	−54 31.2	1	13.32		0.74		0.36		2		535
		NGC 3105 - 76		9 58 46	−54 32.2	1	15.44		0.79		0.08		1		535
87031	−59 01609	NGC 3114 - 2		9 58 46	−59 44.7	2	8.72	.000	0.00	.014	-0.24		7	B9.5IV	1040,2025
		NGC 3105 - 77		9 58 47	−54 32.3	1	14.78		0.95		0.06		1		535
86995	−54 03007			9 58 47	−54 54.0	1	9.89		0.08		0.01		3	Ap Si	279
		WLS 1000 20 # 7		9 58 48	+17 54.7	1	11.95		0.87		0.33		2		1375
		WLS 1000 20 # 5		9 58 48	+22 23.3	1	12.20		0.45		-0.02		2		1375
		NGC 3105 - 72		9 58 48	−54 31.8	1	14.05		1.00		-0.24		4		535
87013	−58 01781			9 58 48	−58 32.2	1	7.84		1.57		1.70		4	K3 III	243
		NGC 3105 - 70		9 58 50	−54 32.1	1	15.57		0.84		0.33		2		535
		NGC 3105 - 32		9 58 50	−54 33.0	1	15.18		0.86		0.14		2		535
		NGC 3105 - 71		9 58 51	−54 31.8	1	16.11		0.82		-0.25		1		535
		NGC 3105 - 25		9 58 51	−54 32.2	1	15.19		0.85		0.01		2		535
		NGC 3105 - 28		9 58 51	−54 32.7	1	15.23		0.63		0.22		1		535
		NGC 3105 - 33		9 58 51	−54 33.2	3	12.56	.007	2.56	.013	2.60	.140	12		279,380,535
		NGC 3105 - 12		9 58 52	−54 33.0	2	14.99	.005	0.81	.005	-0.08	.049	6		380,535
		NGC 3105 - 34		9 58 52	−54 33.2	2	15.38	.037	0.84	.014	-0.10	.056	4		380,535
		NGC 3105 - 35		9 58 52	−54 33.5	1	16.51		1.01		0.11		1		535
87030	−56 02746	HR 3953	⋆ AB	9 58 52	−56 42.3	4	6.51	.007	0.98	.000			18	K0 III	15,1075,2013,2029
233697	+53 01381			9 58 53	+53 13.4	1	10.07		0.72		0.20		2	G5	3016
87042	−59 01614	NGC 3114 - 3		9 58 53	−59 45.1	2	8.35	.000	0.03	.005	-0.07		7	B9 III-IV	1040,2025
		NGC 3105 - 22		9 58 54	−54 32.2	2	14.79	.010	0.88	.025	0.51	.015	7		279,535
		NGC 3105 - 16		9 58 54	−54 32.5	2	14.71	.010	0.80	.005	-0.08	.024	6		380,535
		NGC 3105 - 14	A	9 58 54	−54 32.8	2	13.60	.054	0.86	.005	0.09	.000	6	B2	380,535
		NGC 3105 - 14	AB	9 58 54	−54 32.8	1	13.39		0.82		0.12		2		279
		NGC 3105 - 9		9 58 55	−54 33.2	3	13.30	.020	2.32	.047	2.34	.234	10		279,380,535
		NGC 3105 - 37		9 58 55	−54 33.4	3	13.42	.021	0.26	.005	-0.01	.015	7		279,380,535
87028	−54 03013			9 58 55	−54 47.6	1	9.18		1.01		0.87		std	K0 II/III	279
304848	−59 01615	NGC 3114 - 4		9 58 55	−59 43.5	2	9.18	.005	-0.04	.005	-0.35		7	B9 IV	1040,2025
		NGC 3105 - 24		9 58 56	−54 31.9	3	13.03	.011	1.76	.014	0.57	.055	10		279,380,535
		NGC 3105 - 20		9 58 56	−54 32.5	1	15.76		1.01		0.38		2		535
		NGC 3105 - 7		9 58 56	−54 32.9	3	13.23	.032	0.99	.011	-0.42	.023	10	B2	279,380,535
		NGC 3105 - 6		9 58 56	−54 33.0	1	16.25		0.82		0.12		2		535
		NGC 3105 - 8		9 58 56	−54 33.1	1	16.13		1.00		0.10		2		535
		NGC 3105 - 45		9 58 56	−54 35.3	1	12.02		0.40		0.22		4	A2	535
		NGC 3105 - 5	V	9 58 57	−54 33.0	2	15.12	.122	0.83	.033	-0.04	.084	4		380,535
		NGC 3105 - 4		9 58 58	−54 32.8	2	14.06	.005	0.85	.000	0.18	.020	4		380,535
		NGC 3105 - 1		9 58 59	−54 32.5	3	12.36	.007	0.92	.000	0.47	.014	14	K0	279,380,535
	+18 02308			9 59 00	+17 59.4	1	9.55		0.30				1	A2	1269
		NGC 3105 - 66		9 59 00	−54 30.9	2	13.35	.009	2.71	.028	2.61	.281	4		380,535
	−29 08019	L 535 - 3		9 59 01	−30 09.5	2	11.43	.000	1.49	.000	1.09	.009	5	M4	158,3078
		NGC 3105 - 55		9 59 04	−54 32.7	3	12.71	.019	1.81	.012	1.41	.075	7	K2	279,380,535
87071	−59 01627	NGC 3114 - 5		9 59 04	−59 56.0	2	8.95	.007	1.17	.011	1.01		6	K1 III	1685,2025
87026	−49 04876			9 59 05	−50 14.2	2	6.88	.000	-0.15	.000	-0.68		6	B3 IV	362,2012
		Ton 26		9 59 06	+28 15.	1	16.79		0.17		-0.80		1		98
		NGC 3105 - 56		9 59 06	−54 32.1	3	13.03	.013	1.07	.007	0.62	.060	7	G0	279,380,535
87058	−54 03015			9 59 07	−54 35.5	1	9.51		0.19		0.20		3	A2 IVs	279
		G 42 - 33		9 59 08	+14 55.9	1	15.37		0.36		-0.49		1	DC	3060
87025	−46 05723			9 59 08	−47 06.7	1	8.34		1.05		0.82		1	G8/K0 III	389
		LSS 1389		9 59 09	−54 48.9	2	12.19	.000	0.77	.000	-0.40		6	B3 Ib	279,846
	+48 01829	G 146 - 6		9 59 13	+48 20.8	1	10.07		1.37		1.13		3	M2	3078
86972	−14 03003			9 59 14	−15 11.0	1	8.64		1.02		0.75		1	K3 V	3072
87070	−57 02540			9 59 14	−57 52.4	1	9.10		-0.03		-0.13		5	B9 IV/V	761
86898	+12 02132			9 59 15	+12 29.2	1	7.80		1.04				7	K0	1351
87109	−59 01632	NGC 3114 - 6		9 59 19	−59 59.1	2	7.64	.008	1.26	.002	1.11		6	G8 II-III	1685,2025

HD	DM	Other Id	N	Rem	α_{1950}	δ_{1950}	S	V	σ_V	B–V	σ_{B-V}	U–B	σ_{U-B}	n	Spectrum	References
87125	−59 01633	NGC 3114 - 7			9 59 22	−59 43.9	1	8.56		0.05				4	B9 IV	2025
86856	+45 01791	G 116 - 76			9 59 23	+44 49.3	1	9.04		1.07		0.98		1	K0	3072
302469	−57 02541	IDS09577S5742		AB	9 59 23	−57 56.1	1	10.26		0.23		0.23		2	A0	761
		Ton 27			9 59 24	+29 42.	1	16.43		-0.33		-1.25		1		98
		Vel sq 1 # 104			9 59 24	−54 50.	1	14.48		1.02		0.00		3		846
86989	−12 03055				9 59 26	−12 39.4	1	7.93		0.89		0.51		4	G8 III/IV	1657
87056	−42 05850				9 59 27	−43 19.7	1	8.29		1.14		0.98		1	K0 III	389
		NGC 3105 - 101			9 59 28	−54 35.3	1	13.78		0.79		-0.08		2		535
87122	−55 02794	IDS09577S5537		A	9 59 28	−55 51.3	1	6.41		0.01		-0.27		2	B8	401
304864	−59 01642	NGC 3114 - 283			9 59 28	−60 03.1	1	7.69		1.24		1.03		2	G8 II-III	1685
87103	−54 03020				9 59 29	−54 28.6	1	8.58		1.19		1.19		3	K1 III	279
87137	−59 01643	NGC 3114 - 8			9 59 29	−60 05.1	3	8.29	.033	-0.05	.019	-0.25	.033	19	B9 IV	1040,2025,3021
	−59 01648	NGC 3114 - 10			9 59 35	−59 56.8	1	11.62		0.13				3		2025
	−56 02763	LSS 1390		V	9 59 39	−57 16.9	3	10.68	.031	0.54	.012	-0.42	.019	6	B0.5V	362,540,846
87154	−54 03025				9 59 43	−54 33.9	1	9.72		0.13		0.18		3	A1 III/IV	279
87169	−59 01653	NGC 3114 - 12			9 59 43	−60 05.6	1	9.34		-0.03				4	B9.5V:	2025
87170	−59 01654	NGC 3114 - 13			9 59 43	−60 12.6	2	9.07	.020	1.05	.001	0.82		5	G8 III	1685,2025
	−59 01655	NGC 3114 - 14			9 59 44	−59 56.7	1	11.15		0.07				3		2025
300422	−54 03026				9 59 46	−54 31.4	1	9.52		0.98		0.69		5	K0	279
	−54 03028				9 59 46	−55 00.6	1	10.21		1.65				3		279
86986	+15 02156				9 59 47	+14 48.1	2	7.99	.000	0.12	.000	0.15	.005	3	A1 V	1003,3077
87080	−33 06602				9 59 49	−33 26.7	2	9.39	.019	0.78	.012	0.12	.039	3	G8 V(w)	565,3048
	−54 03030				9 59 50	−55 14.3	1	10.36		1.11		0.95		6		279
87152	−52 03087	HR 3955			9 59 51	−53 07.4	9	6.19	.008	-0.14	.006	-0.69	.016	33	B2.5 V	15,26,362,540,976*
		GD 109			9 59 52	−00 25.8	1	15.40		0.55		-0.10		1		3060
		vdB 32 # b			9 59 52	−57 33.5	1	13.09		0.41		-0.21		4		434
	−57 02548			A	9 59 52	−57 38.0	3	10.68	.036	0.23	.016	-0.54	.015	5	B1.5V	362,540,976
	−57 02548			B	9 59 52	−57 38.0	1	11.14		0.24		-0.67		1		362
307467	−62 01399				9 59 52	−62 25.5	3	10.49	.045	0.15	.015	-0.88	.017	4		540,976,1737
86945	+39 02303				9 59 53	+39 30.2	1	8.47		0.39		-0.02		4	F3 V	1501
87166	−54 03033				9 59 53	−54 56.0	6	9.60	.012	0.30	.011	-0.43	.015	24	B3 III/IV	279,362,540,846,846,976
87205	−59 01667	NGC 3114 - 20			9 59 53	−59 49.6	1	9.43		0.00				4	B9 V	2025
87005	+15 02157				9 59 54	+14 34.9	1	8.49		0.28		0.03		3	A3	1648
87190	−54 03034				9 59 54	−55 14.6	1	8.73		0.56		0.13		3	G0 V	279
87222	−59 01671	NGC 3114 - 23			9 59 58	−60 06.5	3	8.52	.033	-0.11	.005	-0.57	.022	12	B2 IVp	1040,2025,3021
		G 43 - 23			9 59 59	+15 13.9	1	14.23		1.70				1		906
86965	+39 02304				9 59 59	+38 51.3	1	8.36		0.49		-0.04		2	F7 V	105
87165	−50 04750				10 00 00	−50 31.2	1	8.36		1.20		1.16		1	K2 III	389
87015	+22 02164	HR 3952			10 00 02	+22 11.5	1	5.66	.003	-0.19	.006	-0.72	.017	21	B2.5IV	3,154,1203
300428	−54 03036				10 00 02	−55 03.8	1	9.74		1.01		0.83		3	K0	279
		vdB 32 # a			10 00 03	−57 31.8	1	13.49		0.54		-0.29		3		434
87203	−53 03193				10 00 04	−53 57.8	3	8.56	.016	0.12	.008	-0.64	.021	6	B5 Ve	362,401,846
300426	−54 03040				10 00 06	−54 48.8	1	8.87		1.47		1.92		3	K7	279
87241	−59 01672	NGC 3114 - 24			10 00 06	−59 52.3	2	7.84	.014	-0.01	.009	-0.25		7	B8.5IVp	1040,2025
87202	−52 03097				10 00 08	−52 45.8	1	8.31		0.20		-0.17		13	B9 Ib/II	362
87240	−59 01673	NGC 3114 - 25			10 00 08	−59 36.5	2	9.64	.009	-0.05	.005	-0.44		8	B8 IVp	1040,2025
87221	−54 03041	IDS09585S5430		B	10 00 09	−54 45.7	2	7.90	.005	-0.03	.005	-0.28	.000	5	B8/9 V	279,401
	−64 01154	LSS 1394			10 00 11	−65 17.0	2	10.48	.001	0.18	.019	-0.87	.006	3		540,1737
		L 463 - 50			10 00 12	−36 58.	1	16.20		-0.36		-0.84		3		3060
		G 236 - 12			10 00 15	+64 19.7	1	10.86		0.50		-0.09		2		7010
86839	+71 00525	G 236 - 11			10 00 15	+71 06.7	1	8.45		0.66		0.14		1	F8	1776
	−54 03042				10 00 16	−54 24.4	1	10.24		1.57				3		279
87238	−56 02770	HR 3957			10 00 16	−57 06.5	3	6.20	.008	1.11	.010			11	K1 II	15,1075,2031
87266	−59 01683	NGC 3114 - 33			10 00 16	−59 44.4	2	8.21	.009	-0.11	.009	-0.63		7	B2 IV-Vp	1040,2025
87254	−54 03043	IDS09585S5430		A	10 00 19	−54 44.4	1	7.77		1.17		1.18		3	K2 III	279
304837	−59 01685	NGC 3114 - 34			10 00 20	−59 46.0	2	10.19	.005	0.01	.005			6	B9	2025,2025
304835	−59 01686	NGC 3114 - 42	★	AB	10 00 21	−59 55.4	1	9.07		0.02				4	A0	2025
87283	−59 01695	NGC 3114 - 48	★		10 00 25	−60 10.7	6	5.94	.011	0.26	.008	0.23	.030	27	F2 III-IV	15,243,1040,2012,2025,8100
	+20 02421	G 42 - 34			10 00 27	+20 05.2	1	10.70		0.85		0.41		1	K2	333,1620
		LSS 1395			10 00 28	−59 19.2	1	12.34		0.28		-0.43		2	B3 Ib	846
		Vel sq 1 # 105			10 00 30	−54 24.	1	14.46		0.92		-0.15		3		846
87280	−54 03045				10 00 32	−54 34.9	1	9.32		0.22		0.19		3	A5 IV	279
87281	−57 02558				10 00 32	−57 27.7	1	8.27		-0.04		-0.23		3	B9 V	761
		NGC 3114 - 62			10 00 34	−59 49.7	1	13.84		0.18				4		2025
		LP 489 - 18			10 00 35	+09 19.9	1	12.47		1.53		1.19		1		1773
87199	−29 08034	HR 3956			10 00 35	−30 20.1	1	6.54		1.19				4	K1 III	2006
		Vel sq 1 # 88			10 00 35	−54 51.4	1	12.74		0.34		-0.26		3		279
	−57 02560				10 00 35	−58 00.1	1	11.10		0.38		0.14		1	A8	890
	−57 02561				10 00 36	−57 59.4	1	11.82		0.40		-0.45		1	B2	890
87161	−8 02833				10 00 37	−08 38.7	1	8.04		0.67				4	G2 IV	2034
		NGC 3114 - 69			10 00 37	−59 50.6	1	13.06		0.32				4		2025
87324	−61 01416				10 00 38	−61 31.8	1	9.48		0.22		0.25		3	A5 V	8100
	+34 02073				10 00 39	+34 16.0	1	9.06		0.38		-0.01		2	F2	1733
87309	−59 01705	NGC 3114 - 67			10 00 39	−59 55.2	1	9.31		-0.02				4	B8 V	2025
	−59 01704	NGC 3114 - 66			10 00 39	−60 05.7	2	10.83	.024	0.02	.005	-0.14		5	A5	1040,2025
	−38 06123				10 00 41	−39 05.4	1	11.34		0.98		0.69		2		1730
	−59 01712	NGC 3114 - 76			10 00 43	−59 49.4	2	11.73	.010	0.08	.015			6		2025,2025
87295	−52 03120				10 00 44	−52 54.4	2	7.68	.004	-0.06	.004	-0.60	.008	16	B3/4 IV	362,401
		Vel sq 1 # 89			10 00 44	−55 07.8	1	13.24		0.86		0.49		5		279
	−54 03046	LSS 1396			10 00 44	−55 09.0	2	10.15	.000	1.05	.010	0.31	.079	8	B7 II	279,362

Table 1 531

HD	DM	Other Id	N Rem	α_{1950}	δ_{1950}	S	V	σ_V	B–V	σ_{B-V}	U–B	σ_{U-B}	n	Spectrum	References
		Lod 28 # 29		10 00 45	−57 57.5	1	12.00		0.61		0.11		1	F8	890
	−54 03048			10 00 48	−54 54.0	1	9.65		1.85				3		279
298828	−52 03122			10 00 49	−52 30.1	1	10.71		0.26		−0.47		2		362
		NGC 3114 - 84		10 00 50	−59 52.4	1	12.88		0.35				4		2025
		POSS 550 # 1		10 00 51	+06 14.6	1	14.05		1.37				2		1739
		KUV 348 - 13		10 00 52	+40 48.8	1	13.28		−0.30		−1.17		1	sdO	974
86321	+84 00225	HR 3934		10 00 52	+84 09.7	2	6.35	.029	1.53	.010	1.63	.014	7	K0	985,1733
		Lod 28 # 31		10 00 52	−57 57.6	1	12.27		0.35		0.11		1	A8	890
87127	+38 02096	IDS09579N3830	A	10 00 54	+38 15.9	1	6.80		0.52		0.03		4	F5	3016
87127	+38 02096	IDS09579N3830	B	10 00 54	+38 15.9	1	13.49		0.82		0.30		4		3032
86940	+70 00593			10 00 54	+70 28.3	1	8.56		0.13		0.05		3	B8	1776
		Vel sq 1 # 92		10 00 54	−54 40.7	1	12.48		0.62		0.04		3		279
	−59 01723	NGC 3114 - 85		10 00 54	−59 46.7	1	11.30		0.05				2		2025
	−57 02568			10 00 55	−57 58.0	1	11.62		0.56		0.09		1	F8	890
87349	−59 01724	NGC 3114 - 86		10 00 55	−59 38.6	2	8.62	.005	−0.04	.005	−0.34		7	B8 III	1040,2025
87176	+14 02193			10 00 56	+13 42.6	2	8.09	.001	0.54	.001			44	F5	130,1351
87293	−38 06129			10 00 59	−39 10.9	2	7.34	.003	1.31	.000	1.57	.016	8	K3 III	1673,1730
	−57 02570			10 00 59	−57 57.8	1	10.47		1.06		0.96		1	K3	890
302499	−57 02569			10 00 59	−58 03.6	1	10.14		0.22		0.17		1	A2	890
87366	−59 01727	NGC 3114 - 94		10 00 59	−59 54.3	1	9.42		−0.01				4	B9 V	2025
	−59 01726	NGC 3114 - 93		10 00 59	−60 04.3	2	10.15	.005	1.27	.005	1.21		5		1040,2025
	−59 01732	NGC 3114 - 101		10 01 04	−59 45.7	1	10.95		0.07				3		2025
		NGC 3114 - 97		10 01 04	−59 49.7	1	12.17		0.19				4		2025
	−59 01736	NGC 3114 - 273		10 01 04	−60 14.5	1	9.78		0.95		0.61		3		1685
		Lod 28 # 15		10 01 05	−58 03.8	1	10.98		1.28		1.15		1		890
	−38 06131			10 01 06	−39 07.5	1	10.45		0.20		0.11		2		1730
87406	−60 01602	LSS 1399		10 01 07	−61 03.8	2	8.19	.001	−0.08	.020	−0.83	.007	4	B1 II	540,976
87076	+60 01234	G 196 - 10	⋆ AB	10 01 08	+59 55.2	1	10.09		1.08		1.01		2	A3	7010
		GD 110		10 01 08	−03 23.4	1	15.43		0.25		−0.59		1		3060
	−54 03054			10 01 08	−54 33.7	1	10.27		0.99		0.82		4		279
302503	−57 02575			10 01 08	−58 03.6	1	10.70		1.72		1.82		1	A5	890
	−57 02578			10 01 09	−57 51.1	1	11.17		0.42		0.04		1	F0	890
	−53 03216	LSS 1397		10 01 10	−53 34.5	3	11.50	.018	0.61	.009	−0.40	.016	7	B1 III:	362,846,1737
	−57 02579			10 01 10	−57 51.8	1	11.74		0.22		0.18		1	A0	890
		NGC 3114 - 110		10 01 11	−59 52.8	1	13.75		0.53				4		2025
		POSS 550 # 3		10 01 12	+06 28.0	1	15.19		1.49				2		1739
		Ton 28		10 01 12	+29 13.	1	16.05		0.12		−0.90		1		98
302498	−57 02581			10 01 12	−57 57.8	1	10.96		0.05		−0.23		1		890
		vdB 33 # d		10 01 12	−59 02.	1	14.34		0.46		0.48		4		434
	−58 01830			10 01 12	−59 02.7	1	11.33		0.05		−0.06		5		434
87405	−59 01740	NGC 3114 - 108		10 01 12	−59 43.0	2	8.50	.004	−0.09	.000	−0.43		5	B8.5IIIp	1040,2025
		NGC 3114 - 111		10 01 12	−59 46.3	1	11.75		0.12				2		2025
87262	−8 02836	HR 3959		10 01 13	−09 19.9	4	6.11	.008	1.67	.005	2.01	.000	18	K0	15,1417,2013,2029
87384	−57 02580	LSS 1398		10 01 13	−57 31.2	3	9.07	.014	0.32	.018	−0.65	.031	6	B0 III	362,540,976
302497	−57 02582			10 01 13	−57 52.9	2	10.83	.054	0.04	.005	−0.13	.005	3	B9	761,890
87403	−58 01831			10 01 13	−59 02.4	1	9.30		0.04		0.00		4	A1 III	434
297725	−48 05251	BPM 34630		10 01 14	−48 24.1	1	10.44		1.12		0.96		1	K0	3072
87380	−53 03218			10 01 14	−54 19.0	5	9.11	.008	−0.04	.006	−0.66	.012	20	B1 IVn	279,362,540,846,976
87334	−43 05872			10 01 15	−43 35.9	1	7.49		1.18		1.19		1	K1 III	389
298817	−51 04435	IDS09595S5134	B	10 01 17	−51 49.0	1	7.37		0.19		0.08		1	B9	401
		WD1001+203		10 01 18	+20 23.7	1	15.35		0.57		−0.72		1	DA	1727
87141	+54 01348	HR 3954		10 01 18	+54 08.1	2	5.75	.014	0.48	.009	0.04	.027	5	F5 V	254,272
87292	−16 02953			10 01 18	−17 13.9	1	10.04		−0.01		−0.31		3	A5/7 V	761
	−59 01745	NGC 3114 - 119		10 01 18	−59 48.8	1	7.65		1.34		1.08		5		3021
		NGC 3114 - 113		10 01 18	−59 50.4	1	12.67		0.53				4		2025
87438	−61 01431	HR 3967		10 01 18	−61 54.9	3	6.40	.018	1.70	.014	1.91	.044	10	K3 Ib	243,2031,3005
		POSS 550 # 2		10 01 19	+05 39.5	1	14.41		1.43				2		1739
87364	−51 04436	IDS09595S5134	A	10 01 19	−51 48.3	1	7.36		0.19		0.08		2	A4/5 IV/V	401
	−59 01748	NGC 3114 - 122		10 01 21	−59 47.7	1	10.54		0.05				2	B9 IV	2025
		NGC 3114 - 120		10 01 21	−59 52.2	1	12.80		0.36				4		2025
304859	−59 01749	NGC 3114 - 238		10 01 21	−59 59.8	1	8.46		1.27		1.27		3	G9 III	1685
		G 42 - 35		10 01 22	+19 02.5	1	11.36		1.43		1.28		1		333,1620
87363	−46 05759	HR 3964		10 01 22	−46 23.6	2	6.11	.005	0.02	.000			7	A0 V	15,2012
	−59 01747	NGC 3114 - 121		10 01 22	−59 47.0	2	11.54	.010	0.08	.015			6		2025,2025
87140	+55 01362			10 01 23	+54 35.3	2	9.00	.020	0.70	.000	0.02	.005	6	K0	516,3016
87318	−24 08711	HR 3962	⋆ A	10 01 23	−25 04.4	2	6.70	.000	0.00	.002	0.00		9	A0 V	1637,2031
87320	−28 07862			10 01 23	−28 48.1	2	8.12	.004	0.68	.000	0.18	.008	7	G8 V	78,1775
302517	−58 01838			10 01 23	−59 02.0	1	10.40		−0.01		−0.37		4	B9	434
87436	−59 01752	NGC 3114 - 125	⋆	10 01 24	−59 56.2	2	6.18	.004	0.17	.000			13	A7 III	15,2025
		NGC 3114 - 127		10 01 25	−60 05.2	2	11.17		1.01	.000	0.72		4		1040,2025
	−57 02583	LSS 1401	A	10 01 26	−57 53.9	4	10.41	.042	0.53	.015	−0.31	.020	6	B3	362,761,846,890
	−57 02583	LSS 1402	B	10 01 26	−57 53.9	3	10.91	.070	0.25	.040	−0.58	.054	4	B2	362,846,890
	−59 01753	NGC 3114 - 130		10 01 26	−59 50.6	2	11.39	.005	0.34	.000			6		2025,2025
87271	+12 02138			10 01 28	+11 52.3	1	7.13		0.18		0.02		3	A0	1648
		Lod 28 # 22		10 01 28	−57 54.3	1	11.61		1.03		0.68		1	K0	890
		NGC 3114 - 134		10 01 28	−59 51.3	1	13.65		0.42				4		2025
87289	+1 02396	G 162 - 16		10 01 29	+00 50.0	3	9.82	.016	0.60	.019	0.07	.014	3	F8	1620,1658,1696
		Vel sq 1 # 95		10 01 29	−55 01.4	1	13.41		0.61		0.58		3		279
87435	−57 02584	LSS 1403		10 01 29	−57 51.4	1	8.81		1.02		0.82		1	F3 Ib	890
	−54 03058	LSS 1400	V	10 01 30	−55 07.3	1	11.2		0.74		−0.26		4		846

HD	DM	Other Id	N	Rem	α_{1950}	δ_{1950}	S	V	σ_V	B–V	σ_{B-V}	U–B	σ_{U-B}	n	Spectrum	References
87458	−59 01756	NGC 3114 - 136			10 01 31	−59 35.9	2	7.74	.014	-0.03	.000	-0.38		7	B8 III	1040,2025
87301	+3 02311	HR 3961			10 01 33	+03 26.7	4	6.46	.011	0.40	.010	-0.02	.007	11	F4 V	15,254,1415,3077
		Lod 28 # 32			10 01 33	−57 47.8	1	12.34		1.65		1.03		1		890
	−57 02588				10 01 34	−57 48.0	1	11.41		0.58		0.05		1	F8	890
	−57 02587				10 01 34	−57 51.9	1	10.88		0.41		0.15		1	A8	890
	−54 03061				10 01 35	−54 47.8	1	9.85		1.59				3		279
	−57 02589				10 01 35	−58 00.1	1	10.32		1.40		1.61		1	M0	890
		NGC 3114 - 139			10 01 36	−59 52.4	1	12.01		0.10				4		2025
87457	−54 03063				10 01 38	−54 41.6	1	9.14		0.01		-0.07		3	B9.5V	279
87471	−59 01768	NGC 3114 - 148			10 01 38	−60 06.1	1	9.35		-0.01				4	B9 V	2025
	−57 02591				10 01 39	−57 52.7	1	11.51		0.42		0.17		1	A8	890
87344	−17 03047	HR 3963		★ AB	10 01 40	−17 51.5	3	5.85	.004	-0.06	.000			11	B8 V	15,2013,2027
297126	−45 05713	IDS09597S4547		AB	10 01 40	−46 01.1	1	9.25		0.56		0.06		1	F8	389
	−57 02593				10 01 41	−57 49.9	1	10.26		1.08		0.88		1	K3	890
304832	−59 01770	NGC 3114 - 149			10 01 42	−59 52.2	1	9.82		-0.07				4	A0	2025
87479	−59 01771	NGC 3114 - 150			10 01 44	−59 42.3	2	7.93	.023	1.17	.001	0.94		8	K0 IIIa	1685,2025
		Lod 28 # 19			10 01 46	−57 51.0	1	11.32		1.04		0.83		1	K3	890
		NGC 3114 - 152			10 01 47	−59 52.0	1	13.19		1.13				4		2025
		Lod 28 # 25			10 01 49	−58 03.2	1	11.81		0.40		0.08		1	A9	890
304831	−59 01777	NGC 3114 - 156			10 01 49	−59 50.6	1	9.74		-0.03				2	A0	2025
297737	−49 04911	LSS 1404			10 01 51	−49 52.7	2	9.42	.030	0.37	.005	0.05	.084	5	A0 II	362,8100
		Lod 28 # 30			10 01 51	−57 45.4	1	12.04		0.13		0.02		1	A0	890
87243	+53 01384	HR 3958			10 01 54	+52 36.8	1	6.35		0.12		0.10		2	A5 IV	1733
87492	−55 02853				10 01 54	−55 28.6	1	8.61		0.24		0.25		2	A3 III/IV	837
		Steph 823			10 01 55	+05 48.6	1	12.66		1.50		1.22		2	M0	1746
	−57 02595				10 01 55	−58 00.1	1	11.18		0.37		0.07		1	A9	890
87494	−57 02596	LSS 1405			10 01 56	−57 51.3	5	9.61	.024	0.42	.017	-0.59	.021	10	O9.5Iab/b	362,540,846,890,976
87507	−59 01784	NGC 3114 - 164			10 01 56	−59 58.1	1	8.48		-0.05				3	B9 III	2025
	−55 02855				10 01 59	−55 37.8	1	11.86		0.38		0.12		2	A6	837
		Lod 28 # 27			10 01 59	−58 04.0	1	11.87		1.59		1.99		1		890
87427	−23 08973	HR 3965			10 02 02	−24 02.6	1	5.70		0.30				4	F0 V	2031
87416	−27 07171	IDS09598S2754		AB	10 02 02	−28 08.1	1	7.28		0.47				3	F6 V	173
	−55 02856				10 02 02	−55 39.7	1	11.25		0.81		0.38		2	K0	837
		Lod 28 # 28			10 02 02	−57 50.5	1	11.90		1.26		1.45		1	K9	890
87526	−59 01791	NGC 3114 - 170			10 02 02	−59 57.4	4	7.31	.005	0.89	.007	0.51	.004	20	G3 II-III	1040,1685,2025,3021
87543	−61 01441	HR 3971		★ AB	10 02 02	−61 38.5	6	6.14	.005	-0.04	.012	-0.45	.013	19	B7 IVne	243,401,540,976,1637,2031
		POSS 550 # 10			10 02 03	+06 37.2	1	18.01		0.83				5		1739
87478	−48 05257				10 02 03	−48 23.1	1	8.16		1.16		1.04		1	K1 III	389
87467	−42 05893	IDS10001S4211		AB	10 02 06	−42 26.0	1	9.00		0.95		0.58		1	K1 V	3072
	−55 02858				10 02 06	−55 40.6	1	11.40		0.34		0.29		2	A2	837
304862	−59 01797	NGC 3114 - 280			10 02 07	−60 06.2	1	9.01		1.36		1.40		2	K5	1685
	−57 02600				10 02 09	−57 54.6	1	9.81		1.67		1.98		1	M0	890
87971	−81 00399	HR 3983			10 02 09	−81 58.4	4	5.53	.008	0.04	.005	0.05	.005	26	A0 IV	15,1075,1088,2038
		KUV 348 - 14			10 02 10	+39 00.8	1	14.26		0.02		-0.14		1		974
87565	−59 01800	NGC 3114 - 190			10 02 13	−59 36.8	1	9.66		-0.09				4	B2 Vnne	2025
	−59 01804	NGC 3114 - 182			10 02 15	−59 40.7	1	11.62		0.04		-0.04		3		1040
87566	−59 01802	NGC 3114 - 181			10 02 15	−59 48.6	2	8.29	.008	1.28	.000	1.26		6	K2 II-III	1685,2025
87541	−55 02861				10 02 16	−55 39.9	2	8.02	.005	-0.03	.005	-0.27	.010	4	B9 III/IV	401,837
87567	−59 01806	NGC 3114 - 183			10 02 16	−59 52.9	1	9.65		0.00				4	B8	2025
	−59 01807	NGC 3114 - 184			10 02 16	−59 53.7	1	7.67		1.26		1.03		7		3021
87477	−39 06100	HR 3968			10 02 17	−39 44.0	3	6.42	.004	1.31	.010			11	K1 III	15,2013,2029
	−55 02863				10 02 17	−55 35.3	1	11.09		0.99		0.68		2	K2	837
		LSS 1407			10 02 19	−55 20.8	1	11.69		0.84		-0.09		3	B3 Ib	846
		POSS 550 # 8			10 02 21	+06 23.6	1	17.73		0.47				4		1739
87312	+52 01450				10 02 21	+51 38.1	1	7.87		0.18		0.08		2	A2	1566
	−60 01619	LSS 1408			10 02 21	−60 27.7	2	11.61	.000	0.01	.015	-0.81	.010	5	B0.5 V	846,1737
		LP 903 - 51			10 02 22	−29 57.9	1	14.66		1.52		1.08		1		1696
87593	−59 01813	NGC 3114 - 192			10 02 23	−60 01.3	1	9.62		0.08				3	A2 V	2025
87540	−47 05500				10 02 24	−48 12.2	1	6.84		1.02		0.81		1	K0 III	389
87412	+20 02430				10 02 26	+19 41.0	1	8.30		1.09		0.84		2	G5	1648
87521	−38 06148				10 02 28	−39 04.5	1	9.14		1.03		0.84		1	K3 V	3072
	−57 02604				10 02 28	−57 52.0	2	10.54	.005	0.54	.000	0.35	.009	5	Ap	761,890
		POSS 550 # 6			10 02 29	+05 36.4	1	17.61		1.08				3		1739
		Ton 29			10 02 30	+28 12.	1	16.71		-0.24		-1.17		1		98
87612	−59 01816	NGC 3114 - 196			10 02 31	−59 38.3	2	9.56	.000	-0.08	.000	-0.51		6	B4 Vn	1040,2025
		LSS 1410			10 02 32	−57 42.5	1	11.93		0.45		-0.55		3	B3 Ib	846
	−59 01815	NGC 3114 - 195			10 02 32	−59 36.1	1	10.16		0.45		-0.07		3	B4 V	1040
87613	−59 01818	NGC 3114 - 197			10 02 32	−59 58.1	4	8.95	.014	1.37	.013	1.34	.019	18	G9 III	1040,1685,2025,3021
	+17 02169				10 02 33	+16 42.7	1	9.07		1.21		1.23		2	K0	1648
300469	−55 02868				10 02 35	−55 39.4	1	10.50		0.48		0.23		2	F2	837
		POSS 550 # 7			10 02 36	+06 41.5	1	17.65		1.30				3		1739
	−54 03084	LSS 1409			10 02 37	−55 08.6	2	11.33	.010	0.45	.020	-0.39	.015	4		362,846
87504	−12 03073	HR 3970			10 02 41	−12 49.3	7	4.59	.004	-0.09	.008	-0.27	.006	23	B8 V	15,1075,1079,1425*
300468	−55 02869				10 02 42	−55 36.4	1	10.26		1.07		0.91		2	K0	837
		G 43 - 26			10 02 43	+12 15.4	2	12.90	.019	0.54	.015	-0.14	.010	3		333,1620,1658
		Lod 1 # 26			10 02 44	−55 30.2	1	12.70		0.33		0.13		2	A6	837
		GD 111			10 02 46	+43 02.6	1	16.16		-0.08		-0.87		1		3060
		GD 112			10 02 47	+45 52.1	1	13.85		0.40		-0.21		1		3060
		Lod 1 # 24			10 02 50	−55 27.8	1	12.43		0.29		0.12		2	A3	837
87643	−58 01865	LSS 1411			10 02 50	−58 25.3	3	8.70	.043	0.70	.024	-0.40	.026	9	B2 e	7,92,362

Table 1 533

HD	DM	Other Id	N	Rem	α_{1950}	δ_{1950}	S	V	σ_V	B–V	σ_{B-V}	U–B	σ_{U-B}	n	Spectrum	References
87663	−59 01825	NGC 3114 - 205			10 02 53	−59 53.7	2	9.74	.018	-0.02	.014	-0.34		7	B8 Vn	1040,2025
	−55 02870				10 02 54	−55 39.9	1	10.70		1.29		1.40		2		837
		LSS 1412			10 02 54	−60 24.9	1	11.00		0.52		0.03		2	B3 Ib	846
87442	+40 02286	IDS09599N4004	A		10 02 56	+39 49.6	1	7.30		0.25		0.09		2	A5	3032
87442	+40 02286	IDS09599N4004	B		10 02 56	+39 49.6	1	11.76		0.95		0.58		4		3032
		Lod 1 # 19			10 02 56	−55 32.8	1	12.05		1.16		0.97		2	K5	837
	+13 02203				10 02 58	+13 13.2	1	9.21		1.10		0.96		3	K0	1648
87500	+16 02077	HR 3969			10 02 58	+16 00.1	1	6.37		0.37		0.15		1	F1 III	39
87641	−46 05785				10 03 00	−46 41.6	1	7.88		0.91		0.54		1	G8 III	389
302501	−57 02613				10 03 00	−58 17.3	2	9.69	.016	0.64	.005	-0.43	.049	4	O9 III	362,540
	−55 02873				10 03 04	−55 35.4	1	10.10		1.15		1.01		2	K3	837
87606	−35 06130	HR 3972			10 03 05	−36 08.4	1	6.27		1.11		1.07		2	K1 III	58
87640	−45 05735	IDS10011S4525	AB		10 03 05	−45 39.2	1	7.31		1.08		1.01		1	K0 III	389
		POSS 550 # 9			10 03 06	+06 23.4	1	17.73		1.00				4		1739
		X LMi			10 03 06	+39 36.2	1	11.74		0.14		0.08		1	A5	668
		LSS 1414			10 03 09	−58 17.4	1	12.17		0.47		-0.37		2	B3 Ib	846
	−55 02875				10 03 11	−55 41.0	1	11.87		0.34		0.19		2	A5	837
87677	−55 02874				10 03 12	−55 33.4	1	8.81		0.05		0.06		2	A0 IV	837
	−53 03270				10 03 15	−54 12.8	1	10.64		1.07		0.77		1	K0 III	137
		Lod 1 # 18			10 03 15	−55 29.5	1	12.03		1.06		0.72		2	K2	837
87691	−57 02616				10 03 15	−57 59.6	1	9.73		0.63		-0.39		3	F0 IV	846
	−58 01876		A		10 03 15	−58 55.8	2	11.09	.011	0.11	.003	-0.22	.005	3	B7 IV	92,540
	−58 01876		AB		10 03 15	−58 55.8	1	10.86		0.12		-0.14		4		92
	−59 01837	NGC 3114 - 209			10 03 15	−59 39.9	1	11.08		0.00		-0.16		2		1040
		POSS 550 # 5			10 03 17	+05 46.4	1	15.90		0.85				3		1739
	+20 02431				10 03 17	+19 41.9	1	9.56		0.36		-0.01		2	A7	1375
87676	−47 05513				10 03 17	−47 32.8	1	8.29		0.52		0.22		1	F3 III	389
		Lod 1 # 22			10 03 17	−55 36.2	1	12.32		0.40		0.23		2	A6	837
87699	−55 02876				10 03 18	−55 31.4	1	7.99		1.58		1.86		2	K4 III	837
		Lod 1 # 29			10 03 18	−55 39.4	1	12.99		0.60		0.05		2	G0	837
		POSS 550 # 4			10 03 19	+06 43.3	1	15.84		0.87				3		1739
87638	−32 07040				10 03 21	−33 08.9	1	6.98		0.31				4	F0 V	2012
87698	−55 02877				10 03 24	−55 26.4	1	9.70		0.48		0.07		2	F0/2 V	837
87482	+56 01431				10 03 30	+55 54.1	1	8.07		0.35		0.05		2	F2	1502
87751	−59 01842	NGC 3114 - 210			10 03 30	−59 48.5	2	9.27	.005	-0.03	.000	-0.37		6	B9 III	1040,2025
87752	−59 01843	NGC 3114 - 211			10 03 31	−59 59.4	2	9.82	.000	-0.02	.010	-0.38		9	B9 III-IVp	1040,2025
		GD 113			10 03 33	+01 08.1	1	11.63		-0.09		-0.49		2	Bp	3060
	−55 02879				10 03 34	−55 32.3	1	10.65		1.09		0.87		2	K4	837
	−51 04459				10 03 35	−52 13.0	2	11.68	.050	0.23	.020	-0.54	.030	4		362,846
298884	−53 03281				10 03 38	−53 38.0	1	10.12		0.42		0.10		3	F2 V	137
302505	−58 01886	LSS 1415			10 03 40	−58 29.7	4	9.55	.018	0.42	.015	-0.59	.030	12	O9.5III	92,540,846,976
87686	−30 08175				10 03 42	−30 57.3	1	9.10		1.24				4	K2 III	2012
		G 116 - 77			10 03 43	+41 57.7	1	11.31		1.49		1.22		3		7010
297782	−50 04799				10 03 45	−51 08.1	1	9.87		0.38		0.04		1	F2 V	137
	−55 02881				10 03 45	−55 31.9	1	10.32		1.68		1.66		2	M5	837
		LP 549 - 26			10 03 47	+04 30.0	1	12.11		0.77		0.34		2		1696
	−55 02882				10 03 47	−55 25.4	1	11.13		0.04		-0.20		2	B8	837
87583	+42 02086				10 03 48	+42 32.0	1	8.26		1.16		1.22		3	K1 III	37
87801	−59 01850	NGC 3114 - 213			10 03 55	−59 39.8	1	8.98		-0.07				3	B4 V	2025
		NGC 3114 - 212			10 03 55	−59 59.4	1	11.67		0.35		0.17		5		1040
	−59 01851	NGC 3114 - 223			10 03 56	−59 48.3	1	10.03		1.06		0.79		3		1685
		LP 489 - 33			10 04 02	+13 58.7	1	11.12		0.58		-0.08		2		1696
87600	+45 01798				10 04 02	+45 18.3	2	7.63	.020	0.41	.010	0.00	.005	8	F2	1601,1648
		G 235 - 55			10 04 06	+69 29.3	2	14.34	.016	1.75	.000			8		538,906
300425	−54 03032				10 04 07	−51 09.2	4	10.05	.008	0.26	.026	-0.63	.021	9	B1 II-Ib	279,362,540,976
87833	−59 01854	NGC 3114 - 262			10 04 07	−60 04.4	1	8.60		1.15		0.91		4	K0 II	1685
87800	−53 03290				10 04 08	−54 20.9	2	7.85	.005	-0.07	.000	-0.56	.015	4	B6/7 II	362,401
		Steph 828			10 04 09	+05 15.3	1	12.79		1.54		1.31		2	M1	1746
87846	−60 01646				10 04 09	−60 55.7	1	7.05		1.44		1.62		4	K4 III	243
87832	−59 01855				10 04 10	−59 52.4	1	8.98		0.00		-0.15		2	B8/9 IV/V	401
87682	+6 02259	HR 3973			10 04 11	+05 51.4	3	6.20	.005	0.94	.000	0.71	.014	9	K1 III	15,1256,1355
87783	−46 05806	HR 3976	★ AB		10 04 12	−47 07.5	3	5.07	.004	0.88	.000	0.55		8	K1 III+G5 V	15,389,2012
		W -53 027			10 04 12	−53 42.	1	10.39		0.25		-0.36		3		846
87816	−51 04471	HR 3978, R Vel			10 04 15	−51 56.6	2	6.50	.014	0.99	.009	0.82		5	K1 III	137,2006
	−59 01857				10 04 15	−59 34.1	1	11.23		0.26		0.15		2	A0 V	540
	−59 01858	NGC 3114 - 214			10 04 15	−59 58.1	1	11.18		0.05		-0.11		5		1040
88147	−79 00475	IDS10047S7938	AB		10 04 16	−79 53.3	1	8.80		0.62		0.09		2	F0 IV	3016
		G 42 - 36			10 04 18	+12 55.5	1	12.07		1.51		1.19		1		333,1620
	+14 02202				10 04 19	+14 24.6	1	8.99		1.59		2.01		2	M0	1648
87860	−57 02643				10 04 21	−57 39.4	1	8.54		0.08		0.11		3	A0 IV	761
	+3 02316				10 04 22	+03 12.6	2	9.94	.005	1.39	.013	1.28		2	M0	1705,3072
87782	−34 06425	IDS10022S3442	AB		10 04 22	−34 56.2	1	8.17		-0.06		-0.38		1	B5/7 V	55
87680	+29 02000				10 04 23	+29 28.9	1	7.96		0.67		0.11		3	G2 V	3026
		BPM 19844			10 04 24	−50 56.	1	11.90		0.48		-0.18		3		3065
87813	−43 05921				10 04 25	−43 42.0	1	8.64		0.45		0.01		1	F3 V	389
		LP 62 - 35			10 04 27	+66 33.6	1	14.71		0.67		-0.10		1		1773
87645	+53 01387				10 04 28	+53 26.5	1	7.79		0.11		0.12		2	A0	1566
87696	+35 02110	HR 3974			10 04 29	+35 29.4	11	4.48	.009	0.18	.009	0.07	.006	55	A7 V	1,15,39,1006,1007*
298878	−53 03299				10 04 33	−53 29.3	1	10.45		0.14		0.05		2	A1 V	137
87843	−44 06191				10 04 34	−45 08.0	1	7.67		0.97		0.65		1	G8 III/IV	389

HD	DM	Other Id	N	Rem	α_{1950}	δ_{1950}	S	V	σ_V	B–V	σ_{B-V}	U–B	σ_{U-B}	n	Spectrum	References
87794	−29 08092				10 04 35	−29 43.2	1	9.21		0.62		0.08		1	G3 V	78
87737	+17 02171	HR 3975		⋆ AB	10 04 37	+17 00.4	8	3.49	.053	−0.03	.015	−0.21	.028	31	A0 Ib	3,15,1022,1119,1363*
87679	+45 01802				10 04 39	+45 22.9	1	8.29		0.28		0.02		5	F0	1648
	+14 02203				10 04 41	+14 11.2	1	9.22		1.35		1.63		3	M9	1648
87695	+43 01990				10 04 41	+43 15.8	1	8.87		0.96		0.73		2	G5	1733
87598	+69 00558	G 235 - 57		⋆ A	10 04 41	+68 41.1	1	8.80		0.78		0.38		1	G5	1658
		G 53 - 24			10 04 42	+00 32.7	1	11.74		0.94		0.74		1	K2	333,1620
	−50 04808				10 04 42	−50 53.7	1	11.72		0.25		−0.04		1	A0 V	137
87808	−16 02974	HR 3977			10 04 46	−16 53.8	3	5.59	.010	1.49	.016	1.68		9	K4 III	15,252,2028
87894	−53 03304				10 04 46	−53 27.1	1	8.09		0.98		0.78		2	G8/K0 III	137
	−13 03031				10 04 49	−14 04.0	2	10.17	.009	1.33	.014	1.20	.050	5	K7 V	158,3072
87828	−28 07907				10 04 50	−28 37.4	1	7.61		1.54				4	K4/5 III	2012
87777	+13 02208				10 04 54	+13 16.3	1	8.40		0.58				9	F8	1351
	+24 02183	RR Leo			10 04 56	+24 14.2	2	9.94	.005	0.09	.021	0.14	.010	2	F0	668,699
	−13 03035				10 04 56	−14 27.5	1	11.20		0.73		0.31		1		1696
	−56 02853	LSS 1418			10 04 56	−57 10.9	2	10.61	.030	0.75	.025	−0.38	.015	4	O8	362,846
87776	+15 02167				10 04 57	+15 24.2	1	7.14		0.63		0.15		2	G0 V	1648
87919	−52 03223				10 04 57	−52 24.1	1	9.21		0.08		−0.37		1	B9 III/IV	137
87872	−34 06438				10 05 01	−34 38.1	1	9.05		1.11		1.05		1	K2 III	78
87838	−5 02991	G 162 - 21			10 05 05	−06 11.8	2	7.73	.005	0.51	.010	−0.06	.020	2	G0	1620,1658
	−52 03229				10 05 09	−53 14.4	1	10.21		1.44		1.35		3	K4 III	137
88351	−80 00423	HR 3997			10 05 09	−81 19.3	3	6.59	.003	0.92	.000	0.53	.000	17	G6/8 III	15,1075,2038
87870	−21 02987	IO Hya			10 05 11	−22 14.7	2	6.98	.034	1.54	.015	1.53		12	M4 III	2012,3042
		LF13 # 10			10 05 14	−52 45.6	1	11.75		1.04		0.82		2	K0 III	137
87837	+10 02112	HR 3980		⋆ AB	10 05 15	+10 14.6	8	4.37	.012	1.44	.009	1.75	.013	26	K3.5 IIIb	3,15,1080,1363,3016*
87855	−6 03078				10 05 16	−07 23.2	2	6.64	.019	1.61	.000	1.92		6	M1	2033,3040
		G 54 - 14			10 05 18	+20 25.1	1	12.01		1.40		1.23		1		333,1620
87986	−53 03318	IDS10035S5405		AB	10 05 19	−54 19.1	1	7.60		0.08		0.02		1	A0 V	137
	+65 00751	IDS10017N6456		AB	10 05 22	+64 41.4	1	9.63		0.47		−0.02		3	K7	1723
87822	+32 01982	HR 3979		⋆ AB	10 05 23	+31 51.0	1	6.24		0.44		0.01		4	F4 V	3032
87822	+32 01982	IDS10025N3206		C	10 05 23	+31 51.0	1	13.70		1.51		1.20		3		3032
87887	+0 02615	HR 3981			10 05 23	−00 07.6	8	4.49	.007	−0.04	.003	−0.05	.033	30	A0 III	3,15,1075,1363,1425*
87947	−42 05942				10 05 23	−42 52.6	1	8.42		0.47		0.03		1	F5 V	389
87946	−42 05943				10 05 24	−42 42.9	1	7.77		1.50		1.82		1	K4 III	389
	−49 04959				10 05 24	−50 06.8	1	10.60		1.46		1.05		1	K2 III	137
87734	+64 00770				10 05 29	+64 11.9	1	6.57		1.64		2.02		3	K5	985
		LF13 # 13			10 05 29	−53 45.3	1	12.04		1.30		1.25		1		137
	−57 02676	LSS 1419			10 05 29	−57 46.2	1	11.11		0.18		−0.67		3	O9 V	846
		Ton 487			10 05 30	+24 17.	1	15.10		0.01		−0.02		1		1036
88003	−47 05560				10 05 32	−48 16.8	2	8.19	.007	0.96	.004	0.57	.021	6	G8 III/IV	389,1673
87884	+12 02147	IDS10030N1227		BC	10 05 33	+12 14.5	4	8.14	.006	0.87	.005	0.53	.018	9	K2 V	542,1084,1355,3016
	−56 02871	LSS 1420			10 05 35	−57 20.0	1	9.53		0.37		−0.46		3	B0 V	846
88015	−47 05561				10 05 37	−48 00.9	1	6.38		−0.16				4	B3 V	2012
88042	−54 03158	LSS 1421			10 05 40	−55 16.5	3	11.11	.012	0.45	.020	−0.43	.004	7		362,846,1737
302532					10 05 40	−56 50.	2	9.54	.015	0.38	.040	−0.47	.025	4		362,432
87901	+12 02149	HR 3982		⋆ A	10 05 42	+12 12.7	1	8.13		0.88		0.54		4	B7 Vn	1752
87901	+12 02149	HR 3982		⋆ A	10 05 43	+12 12.7	19	1.36	.013	−0.11	.005	−0.37	.011	153	B7 Vn	1,3,15,26,198,1006*
88013	−36 06156	HR 3984			10 05 52	−37 05.3	3	6.35	.007	0.98	.000			11	K0 III	15,2013,2029
		LF13 # 14			10 05 53	−51 45.3	1	12.57		0.45		0.04		1	A5 V:	137
87852	+51 01577				10 05 56	+51 04.0	1	7.84		0.18		0.09		2	A2	1566
304898	−59 01890	Tr 12 - 2			10 05 57	−60 04.9	1	10.60		−0.02		−0.19		3		1040
87998	−19 02926	IDS10036S1915		AB	10 05 59	−19 30.3	5	7.27	.012	0.62	.007	0.05	.010	22	G2/3 V	1003,1311,1775,2012,3077
88115	−62 01438	LSS 1423			10 05 59	−62 24.5	3	8.32	.012	−0.05	.011	−0.84	.009	6	B1 Ib/II	401,540,976
88092	−57 02693				10 06 02	−58 07.7	1	7.61		0.82				4	G0 Ib	2012
	−57 02692	LSS 1422			10 06 03	−57 22.7	1	11.77		0.27		−0.56		2		846
		LF13 # 15			10 06 05	−50 19.7	1	11.94		0.37		0.20		1	A1 V	137
	−32 07082				10 06 07	−32 48.8	1	11.78		0.74		0.18		2		1696
		G 53 - 25			10 06 10	+02 58.9	1	12.32		1.45		1.15		1		333,1620
88025	−14 03036	HR 3986, μ Hya			10 06 11	−15 22.0	3	6.27	.012	−0.01	.012	−0.02	.025	10	A0 V	252,1075,2012
		LF13 # 16			10 06 12	−51 56.8	1	11.59		1.02		0.60		2		137
304896	−59 01893	Tr 12 - 3			10 06 14	−60 03.0	1	10.70		0.31		0.17		3		1040
302551	−57 02696				10 06 17	−58 08.7	4	9.94	.009	0.19	.019	−0.66	.008	8	B1 IV	362,540,846,976
88024	−10 03000	HR 3985			10 06 18	−10 38.3	2	6.52	.005	0.02	.000			7	A2 V	15,2029
302554	−57 02699				10 06 19	−58 14.3	1	8.23		0.87		0.60		3	G5	803
		LP 261 - 222			10 06 20	+38 07.0	1	10.47		0.47		−0.15		3		1723
87955	+39 02318				10 06 20	+38 40.6	1	7.76		1.62		1.91		2	M3 III	105
88035	−19 02928				10 06 20	−20 04.1	1	9.13		1.10		0.80		1	G8p Ba	565
88158	−61 01479				10 06 23	−61 58.6	3	6.45	.008	−0.10	.002	−0.47	.006	8	Ap Si	243,401,3036
88009	+19 02307				10 06 24	+18 46.6	1	6.91		1.07		0.91		2	G8 II	1648
	+75 00403	G 252 - 49		⋆ AB	10 06 24	+75 23.0	1	9.48		1.40		1.19		1	K6	3072
	−58 01930	LSS 1426			10 06 26	−59 16.2	2	10.48	.005	0.31	.000	−0.61	.006	3	B0.5II-Ib	540,1737
		CS 15621 # 45			10 06 28	+26 16.4	1	14.00		0.62		0.07		1		1744
87912	+56 01435	G 196 - 6			10 06 28	+55 47.4	1	8.40		0.66		0.31		2	G5	1502
		LSS 1425			10 06 28	−57 33.3	1	12.25		0.41		−0.35		2	B3 Ib	846
304901	−59 01901				10 06 30	−60 13.6	1	10.53		−0.07		−0.83		2		540
88008	+25 02206				10 06 33	+24 47.5	1	8.48		0.74		0.26		3	G5 V	3026
88021	+21 02156	IDS10038N2049		AB	10 06 34	+20 34.7	3	6.71	.043	0.54	.011	0.32	.002	9	F5	1375,1381,3030
88175	−56 02892				10 06 35	−57 02.7	1	8.51		−0.02		−0.41		2	B8 III	401
		CS 15621 # 43			10 06 39	+25 04.8	1	14.41		0.05		0.07		1		1744
88210	−61 01480				10 06 39	−61 30.9	1	7.95		1.04		0.83		4	K1 III	243

Table 1 535

HD	DM	Other Id	N Rem	α_{1950}	δ_{1950}	S	V	σ_V	B–V	σ_{B-V}	U–B	σ_{U-B}	n	Spectrum	References
		LF13 # 18		10 06 44	−52 46.1	1	12.54		0.28		0.19		2	A2 V	137
38209	−59 01905	Tr 12 - 1		10 06 44	−60 00.5	1	9.98		−0.01		−0.24		3	A0	1040
		CS 15621 # 47		10 06 47	+26 14.3	1	13.63		0.87		0.37		1		1744
87925	+63 00886			10 06 47	+63 09.8	1	7.11		0.98		0.75		2	K0	1733
	−25 07792			10 06 47	−25 58.3	1	10.77		0.75		0.30		1		78
297817	−50 04834			10 06 48	−50 45.8	1	11.31		0.42		−0.05		1	A2	137
		LSS 1428		10 06 50	−61 10.8	2	12.00	.040	0.09	.020	−0.64	.045	5	B3 Ib	846,1737
88071	+10 02116			10 06 52	+09 50.3	1	7.08		1.64		1.88		2	M1	3040
88186	−50 04836			10 06 52	−50 33.1	1	9.31		−0.04		−0.18		2	B9 V(n)	137
88108	−12 03098			10 06 56	−13 07.0	2	7.08	.005	1.65	.002	1.93		6	M1 III	2034,3040
298906	−51 04505			10 06 58	−52 18.5	1	10.82		0.18		0.11		1	A2 V	137
	−32 07093	LP 903 - 62		10 06 59	−32 31.5	1	11.08		0.95		0.74		1		1696
88082	+16 02088			10 07 01	+16 10.8	1	8.63		0.94				1	G5	882
88206	−51 04507	HR 3990		10 07 02	−51 33.9	5	4.85	.006	−0.12	.008	−0.71	.024	16	B3 IV	15,26,362,2012,8023
88185	−43 05970			10 07 03	−43 27.0	1	7.27		1.09		0.96		1	K0 III	389
		CS 15621 # 42		10 07 04	+24 13.0	1	13.70		0.56		0.01		1		1744
	−57 02716	LSS 1429		10 07 04	−57 25.4	1	11.76		0.63		−0.31		1		846
88046	+50 01721			10 07 07	+49 45.2	2	7.15	.019	0.39	.016	0.03	.006	3	F2	1566,1722
		LF13 # 21		10 07 07	−50 50.1	1	12.87		0.24		0.23		1		137
298907	−52 03276			10 07 10	−52 27.3	1	9.66		1.14		1.04		2	K1 III	137
88242	−51 04511			10 07 12	−51 52.5	1	9.46		−0.02		−0.15		2	B9.5V	137
88294	−61 01484		A	10 07 17	−61 38.4	1	8.89		−0.02		−0.26		4	B8/9 V	243
		L 464 - 77		10 07 18	−30 32.	1	13.83		1.02		0.79		2		3062
88293	−60 01695			10 07 18	−60 56.8	1	7.49		1.08				4	G8 III	2012
88201	−32 07101			10 07 19	−32 36.1	1	7.44		0.56				4	G0 V	2012
88323	−65 01248	HR 3995	★ A	10 07 19	−65 34.2	4	5.27	.005	0.97	.004	0.75	.003	23	K0 III	15,1075,1637,2038
88218	−35 06194	HR 3992	★ AB	10 07 22	−35 36.7	4	6.14	.013	0.61	.012	0.15	.005	9	G1 V	15,1008,2012,3077
88263	−50 04847			10 07 22	−50 23.6	1	7.88		0.09		−0.61		2	B2 IIIe	362
88292	−57 02724			10 07 23	−57 47.7	1	7.50		−0.06		−0.41		2	B8 II	401
88133	+18 02326	G 43 - 30		10 07 24	+18 26.2	1	8.04		0.81		0.51		1	G5	333,1620
		Ton 30		10 07 24	+30 37.	1	15.96		0.06		−0.40		1		98
88275	−50 04850			10 07 26	−51 02.6	1	9.67		0.04		−0.07		1	A0 V	137
	−31 08004			10 07 27	−32 21.5	4	9.80	.020	0.61	.011	0.09	.003	8	G0	78,158,1696,3025
88182	−11 02818	HR 3988		10 07 29	−11 50.9	5	6.24	.011	0.19	.005	0.15	.010	19	A5 m	15,355,1075,2012,8071
300491	−54 03201			10 07 30	−54 39.0	2	9.98	.065	0.60	.015	−0.43	.010	8	A0	362,846
		LSS 1431		10 07 30	−57 11.9	1	12.50		0.40		−0.39		2	B3 Ib	846
304935	−59 01921			10 07 31	−59 46.0	1	10.04		−0.03		−0.52		1	A2	401
88239	−24 08819			10 07 34	−25 08.8	1	10.31		0.48		0.06		1	F2/3 V	6006
302566	−58 01948	LSS 1432		10 07 36	−58 59.7	1	9.96		0.29		−0.38		2		540
88343	−60 01697			10 07 37	−60 50.4	2	9.51	.015	−0.04	.005	−0.39	.017	4	B8/9 III/IV	401,3036
		G 43 - 31		10 07 38	+17 36.2	1	13.54		1.08		0.86		1		1658
88195	−7 02972	HR 3989		10 07 38	−08 09.7	3	5.90	.005	0.02	.000	−0.05	.000	14	A1 V	15,1415,2012
304936	−59 01925			10 07 38	−59 46.4	1	10.36		−0.06		−0.75		2	B5	401
304960	−60 01698			10 07 38	−60 58.9	1	10.56		−0.02		−0.11		1	A0	401
	−25 07802			10 07 39	−25 43.0	1	10.80		1.35		1.20		1		78
88215	−12 03101	HR 3991		10 07 40	−12 34.1	6	5.30	.014	0.37	.005	−0.01	.013	16	F5 V	15,254,1008,2013,2029,3026
88261	−36 06180			10 07 40	−36 30.8	5	8.08	.004	0.60	.006	−0.02	.000	12	G3 V	158,2012,2039,3077,6006
304959	−60 01699			10 07 40	−60 58.6	1	10.22		0.00		−0.31		2	A0	401
	−36 06181			10 07 43	−36 27.4	1	9.35		1.03		0.87		1	K0	6006
	−56 02914			10 07 43	−56 24.9	2	10.85	.000	0.30	.000	0.04	.000	3	A7	127,549
		AO 946		10 07 44	+49 57.5	1	10.30		0.58		0.01		1		1722
		G 49 - 38		10 07 46	+21 56.3	1	11.54		0.81		0.41		1		333,1620
		GD 114		10 07 46	+35 50.5	1	13.26		0.56		−0.08		2		3060
88366	−60 01701	HR 3999, S Car		10 07 46	−61 18.2	2	5.56	.117	1.80	.096	1.75		3	M2/3 IIIe	817,1520
298915	−53 03383			10 07 47	−53 30.6	2	10.26	.005	0.54	.010	−0.21	.030	4	B3	362,846
88303	−44 06252			10 07 49	−44 56.9	1	7.36		1.43		1.64		1	K3 III	389
233713	+50 01723	IDS10053N4958		10 07 51	+49 38.6	1	10.30		0.48		0.01		1		1722
88350	−56 02917			10 07 53	−56 24.5	2	8.28	.000	1.63	.000	1.45	.000	3	M5 III	127,549
		AO 948		10 07 55	+49 31.0	1	10.91		0.42		−0.03		1		1722
88232	+16 02090			10 07 58	+16 17.1	1	8.42		0.76				3	K2	882
88161	+41 02063	HR 3987		10 07 59	+40 54.5	1	6.32		1.25				2	gK3	71
	−53 03389			10 07 59	−54 16.1	1	10.68		1.12		0.74		1	K0 III	137
304958	−60 01705	LSS 1434		10 08 00	−60 55.6	1	9.90		−0.03		−0.81		2		846
304971	−60 01706			10 08 02	−61 09.9	1	9.60		0.24		0.10		4	A3	243
		CS 15621 # 76		10 08 03	+26 28.4	1	12.92		0.44		−0.02		1		1744
88384	−56 02918			10 08 03	−56 25.4	2	7.91	.000	0.97	.000	0.68	.000	3	G5	127,549
88385	−56 02919			10 08 03	−56 30.1	2	8.09	.000	0.02	.000	−0.09	.000	3	Ap CrEuSr	127,549
	−60 01707	LSS 1435		10 08 04	−60 48.1	1	11.23		−0.06		−0.79		2		846
		KUV 10081+3817		10 08 05	+38 16.8	1	16.29		0.21		−0.74		1	DA	1708
		AO 949		10 08 06	+49 47.4	1	11.53		0.49		−0.05		1		1722
88284	−11 02820	HR 3994	★ A	10 08 09	−12 06.4	8	3.61	.009	1.01	.008	0.91	.008	33	K0 III	3,15,37,1075,1425*
88412	−60 01708	LSS 1436		10 08 10	−60 45.0	3	9.40	.078	−0.12	.021	−0.95	.005	6	B0/2 Iab/II	540,976,3036
88257	+15 02171			10 08 15	+14 43.8	2	8.57	.020	1.01	.015	0.83		5	K0	882,1648
88231	+38 02110	HR 3993		10 08 15	+37 38.9	1	5.85		1.29				2	gK3	71
88473	−68 01034	HR 4002	★ AB	10 08 16	−68 26.2	4	5.81	.003	0.01	.005	−0.02	.002	18	A0 IV	15,1075,1637,2038
300520	−55 02980			10 08 18	−56 11.0	1	9.99		0.05		−0.26		1	B9	549
88230	+50 01725	G 196 - 9	★ A	10 08 19	+49 42.5	11	6.60	.014	1.37	.013	1.28	.011	84	K9 V	1,22,1077,1118,1197,1355*
		CS 15621 # 49		10 08 20	+24 49.6	1	13.22		0.25		0.13		1		1744
88318	−18 02856	Ring Hydra # 3		10 08 20	−18 28.3	1	6.27		−0.02		−0.05		3	K1 III	15
88270	+21 02159			10 08 23	+20 56.7	1	6.64		0.34		−0.06		3	F2 VI	1501

Catalogue of mean UBV data

HD	DM	Other Id	N Rem	α_{1950}	δ_{1950}	S	V	σ_V	B–V	σ_{B-V}	U–B	σ_{U-B}	n	Spectrum	References
	+3 02327	Z Sex		10 08 24	+02 48.3	2	9.28	.150	1.43	.045	1.40	.045	2	M4 II	693,793
88333	−7 02977	HR 3996		10 08 27	−08 10.3	4	5.64	.007	1.31	.009	1.42	.000	18	K2 III	15,1256,2013,2029
	−55 02984			10 08 27	−56 08.7	1	10.08		1.06		0.77		1	K2	549
298899	−51 04530			10 08 29	−51 51.0	1	10.05		1.13		1.06		1	K1 III	137
304946	−59 01937	LSS 1437		10 08 30	−60 06.9	1	9.12		0.25		-0.59		2	B2	846
88399	−41 05658	HR 4001		10 08 31	−41 28.0	3	5.98	.009	1.24	.000	1.40		9	K2/3 III	15,2027,3005
	−55 02988			10 08 34	−56 14.1	1	11.13		1.11		0.99		1	K5	549
		LF13 # 27		10 08 35	−54 04.1	1	11.58		0.53		0.11		1	G0 V	137
		LTT 12733		10 08 36	−71 49.1	1	10.74		0.86		0.57		2		1726
88457	−56 02930			10 08 37	−56 30.3	2	9.59	.000	0.07	.000	0.05	.000	3	A1 V	127,549
		CS 15621 # 41		10 08 40	+23 13.9	1	12.60		0.22		0.05		1		1744
307497	−61 01490			10 08 40	−61 47.4	1	10.57		-0.02		-0.66		2	B8	540
300523	−55 02991			10 08 41	−56 21.3	1	10.28		0.15				1	A0	549
		AO 951		10 08 42	+49 39.9	1	11.01		0.95		0.60		1		1722
		CS 15621 # 75		10 08 43	+26 22.7	1	13.69		0.46		-0.15		1		1744
	−56 02933			10 08 43	−56 27.1	2	10.80	.005	0.85	.015	0.55	.000	3	K3	127,549
	−57 02755	LSS 1438		10 08 44	−57 32.4	1	9.91		0.31		-0.47		2	B1.5III	846
88372	−6 03096	HR 4000		10 08 48	−07 04.2	3	6.24	.004	0.01	.000	0.01	.000	11	A2 Vn	15,1075,2012
		W -53 290		10 08 48	−53 33.	1	11.94		0.42		-0.30		2		846
		Steph 834		10 08 49	+27 07.4	1	11.99		1.45		1.19		1	K5-M0	1746
		AO 952		10 08 49	+49 51.9	1	11.30		0.58		0.05		1		1722
	−55 02994			10 08 49	−56 18.7	1	11.75		0.07				1	A2	549
		AO 953		10 08 50	+49 45.3	1	12.21		1.05		1.01		1		1722
88468	−51 04536			10 08 52	−51 28.3	1	9.82		0.10		0.17		1	A0 IV/V	137
88483	−56 02939			10 08 54	−56 41.4	2	9.38	.000	-0.07	.005	-0.44	.010	3	B5 III	127,549
88356	+12 02162			10 08 55	+12 17.0	1	7.78		1.61				3	K5	882
88528	−63 01319			10 08 55	−64 08.3	1	6.91		-0.02		-0.28		2	B8 V	401
		AO 954		10 08 56	+49 33.8	1	12.18		0.77		0.34		1		1722
88419	−18 02861	Ring Hydra # 2		10 08 56	−18 42.5	1	6.80		1.54		1.83		2	M3 III	3012
88355	+14 02217	HR 3998	★ AB	10 08 57	+13 36.2	4	6.44	.014	0.46	.003	0.02	.000	10	F7 V	15,1007,1013,8015
	−56 02941			10 08 57	−56 23.2	2	11.18	.006	0.16	.000	0.13	.000	3	A3	127,549
88499	−56 02942			10 08 58	−56 34.3	2	8.56	.000	1.32	.000	1.42	.000	3	K2/3 III	127,549
88371	+24 02193	G 54 - 17		10 09 02	+24 00.1	4	8.42	.012	0.63	.004	0.11	.012	6	G2 V	333,1003,1620,1658,3026
	−0 02304			10 09 02	−00 56.4	2	9.78	.008	0.36	.012	0.06	.012	7	F4 V	1003,1775
		LSS 1441		10 09 04	−59 14.0	2	11.19	.025	0.61	.000	-0.40	.010	5	B3 Ib	846,1737
88389	+17 02180			10 09 05	+17 31.4	1	8.78		0.47		0.02		2	F5	1648
88497	−50 04878			10 09 07	−50 35.0	1	8.14		-0.08		-0.25		2	B9 III	137
	−56 02948			10 09 08	−56 33.8	1	10.56		0.52		0.10		1	F6	549
88625	−70 00980			10 09 08	−70 49.3	1	8.74		1.58				1	N	864
88437	−14 03049	Ring Hydra # 14	★ AB	10 09 09	−14 48.5	1	6.13		0.03				3	F3 V	15
		LF13 # 30		10 09 09	−50 45.9	1	13.19		0.62		0.31		1	G0 V	137
300521	−55 03000			10 09 11	−56 09.5	1	10.28		-0.03		-0.37		1	A0	549
		AAS27,343 # 317		10 09 12	−57 25.	2	11.65	.140	0.80	.045	-0.37	.010	4		362,846
88327	+54 01355			10 09 13	+53 44.0	1	8.38		1.12		1.05		2	K0	1566
300584	−56 02949	LSS 1443		10 09 13	−56 54.6	4	10.14	.069	0.53	.016	-0.57	.033	7	B1 Ve	362,540,976,1737
		AO 955		10 09 14	+49 38.9	1	12.39		0.69		0.27		1		1722
88526	−53 03420			10 09 14	−54 10.0	1	10.11		0.05		-0.22		2	B8/9 III	137
88542	−58 01967			10 09 14	−58 54.4	1	7.44		0.06		-0.02		2	B9 V	401
	−55 03001			10 09 15	−56 10.8	1	11.11		1.10		0.96		1	K5	549
	−56 02952			10 09 19	−56 23.8	1	11.20		0.50				1	F6	549
	−49 05031	LSS 1442		10 09 21	−49 49.4	1	11.88		0.07		-0.53		2		846
300577	−55 03005			10 09 23	−56 21.7	3	10.71	.015	0.02	.012	-0.19	.011	4	A0	127,287,549
	−55 03007			10 09 25	−56 11.1	1	10.46		1.18		1.01		1	K5	549
88493	−31 08032			10 09 26	−31 56.2	1	9.41		-0.03		-0.10		1	A0 V	55
		LOD 19 # 20		10 09 26	−56 24.5	1	12.92		0.37		0.29		2		127
88556	−49 05032			10 09 27	−50 14.6	1	7.87		-0.11		-0.55		2	B5/6 III	362
300522	−55 03008			10 09 27	−56 08.8	1	9.95		0.12		0.02		1	A2	549
300583				10 09 27	−56 39.1	1	9.74		0.39		0.06		1		549
300576	−55 03009			10 09 28	−56 20.5	3	10.95	.022	0.07	.012	0.00	.015	4	A0	127,287,549
302584	−57 02773			10 09 28	−57 33.4	1	9.86		0.32		-0.51		1	B3	362
	−55 03010			10 09 30	−56 13.2	1	11.41		0.17		0.18		1	A2	549
88416	+27 01860			10 09 31	+27 21.1	1	8.67		0.83		0.44		4	K0 IV	8088
		G 53 - 28		10 09 31	−02 25.9	4	10.64	.009	1.58	.008	1.20	.040	8	M2-3	158,1620,1705,3073
88799	−78 00519	IDS10095S7817	AB	10 09 31	−78 31.8	1	9.26		0.02		-0.43		1	B8 II/III	55
	−55 03011			10 09 33	−56 12.3	1	10.46		0.91		0.57		1	K0	549
	+22 02191			10 09 34	+22 27.5	1	8.81		0.95		0.59		2	G5	1375
88554	−48 05384			10 09 34	−48 41.1	2	9.33	.019	0.32	.009	0.08		4	F0 V	1594,6006
	−55 03012			10 09 34	−56 11.5	1	10.69		0.39		0.06		1	F0	549
		LOD 19 # 19		10 09 34	−56 23.1	1	12.69		0.50				2		127
		LP 429 - 53		10 09 35	+16 59.2	1	12.06		0.96		0.75		1		1696
302583	−57 02777			10 09 35	−57 30.5	2	9.73	.030	0.25	.015	-0.62	.020	4	B8	362,846
88446	+18 02330	G 43 - 33		10 09 36	+17 33.0	4	7.89	.014	0.54	.003	-0.06	.005	8	Gp	516,1375,1620,3077
300578	−55 03013			10 09 37	−56 21.0	3	10.85	.022	0.04	.019	-0.21	.004	4	A0	127,287,549
88606	−57 02780	LSS 1445		10 09 38	−57 47.9	3	8.26	.022	0.90	.005	0.55	.058	8	A3 Ia/b	191,730,8100
88624	−61 01496			10 09 38	−61 31.3	1	7.46		1.07		0.88		4	K0 III	243
		KUV 10097+3552		10 09 40	+35 51.7	1	14.96		0.53		-0.06		1	NHB	1708
		CS 15621 # 40		10 09 42	+27 32.5	1	13.81		0.20		-0.03		1		1744
88539	−34 06528	AB Ant		10 09 42	−35 04.6	2	6.73	.095	2.32	.050	3.27	.210	2	C6,3	109,864
88634	−58 01974			10 09 43	−58 49.2	1	9.69		0.03		-0.07		2	B9 V	1730
		BD +15 02302a		10 09 44	+14 34.1	1	11.74		0.08		0.01		2		272

Table 1 537

HD	DM	Other Id	N Rem	α_{1950}	δ_{1950}	S	V	σ_V	B–V	σ_{B-V}	U–B	σ_{U-B}	n	Spectrum	References
	−3 02870	G 53 - 29		10 09 46	−03 29.7	5	9.27	.015	1.49	.016	1.19	.013	14	M3	158,1620,1705,1775,3072
88522	−27 07266	HR 4003	⋆ AB	10 09 46	−28 21.5	3	6.28	.008	0.01	.000	-0.01		8	A0 V	15,78,2012
300575	−55 03015			10 09 46	−56 12.7	1	10.47		0.30		0.12		1	A3	549
		Lod 27 # 42		10 09 46	−56 21.6	1	12.78		0.02		0.05		1		549
	−17 03088			10 09 47	−18 22.2	4	9.95	.014	1.47	.035	1.22	.002	8	K7 V	912,1705,3072,7009
	−58 01977			10 09 48	−58 39.1	1	10.08		0.96		0.71		2		1730
		KUV 10098+4138		10 09 49	+41 38.0	1	16.33		-0.05		-1.06		1	DB	1708
	−53 03423			10 09 50	−53 33.4	1	10.66		1.12		0.95		2	K0 III	137
88603	−49 05039			10 09 51	−49 54.6	2	7.92	.010	-0.11	.010	-0.51		6	Ap Si	362,2012
300574	−55 03019			10 09 51	−56 13.1	1	11.47		0.07		-0.04		1		549
		Lod 27 # 41		10 09 51	−56 23.2	2	12.75	.000	0.04	.000	0.04	.000	3		127,549
88476	+28 01852			10 09 52	+28 29.4	1	6.86		0.93		0.62		3	G8 III	1625
88590	−44 06285			10 09 52	−44 28.9	1	7.10		0.91		0.51		1	G6 III	389
88647	−58 01979	HR 4007, V368 Car		10 09 52	−58 34.8	4	6.21	.078	1.67	.009			15	M3 III	15,1075,2018,2031
	+0 02627	G 53 - 30		10 09 53	+00 23.0	2	10.24	.015	0.67	.000	0.17	.000	2		1620,1658
	−50 04891			10 09 53	−51 18.1	1	11.41		0.95		0.68		1	K0 III	137
	−56 02968			10 09 57	−56 28.9	1	11.28		0.46		0.07		1	F0	549
	−56 02969			10 09 58	−56 36.3	1	9.85		1.56		1.82		1	M0	549
88646	−55 03021			10 09 59	−55 44.9	1	8.51		0.01		-0.14		2	B9.5IV	401
		Bo 8 - 8		10 10 00	−57 45.	1	14.10		0.56		0.31		1		411
		Bo 8 - 6	AB	10 10 00	−57 45.	1	13.34		0.51		0.02		1		411
	−55 03023			10 10 02	−56 14.9	1	10.30		1.13				1	K0	549
88661	−57 02781	HR 4009, QY Car		10 10 02	−57 48.8	11	5.70	.022	-0.08	.011	-0.90	.013	58	B2 Vne	15,164,362,401,681*
		CS 15621 # 50		10 10 04	+23 38.0	1	13.92		0.58		0.01		1		1744
88597	−36 06202			10 10 04	−37 04.5	1	8.96		-0.09		-0.39		1	B8 V	55
88562	−15 03003			10 10 05	−15 38.5	1	8.54		1.42		1.48		1	K1 III	565
		CS 15621 # 74		10 10 06	+27 27.5	1	13.69		0.62		0.21		1		1744
88424	+58 01244			10 10 06	+58 14.5	1	7.05		0.10		0.08		3	A2	1733
297845	−49 05042			10 10 06	−49 28.6	1	10.21		2.76		4.70		1	F5	864
300547	−54 03265			10 10 11	−55 01.0	2	10.03	.245	0.29	.010	-0.47	.020	4	B3	362,846
88547	+5 02301	HR 4004		10 10 12	+04 51.8	2	5.76	.005	1.18	.000	1.11	.000	7	gK0	15,1415
		CS 15621 # 77		10 10 12	+25 04.7	1	13.87		0.43		-0.17		1		1744
88695	−60 01742			10 10 12	−61 17.7	1	7.16		-0.08		-0.54		4	B5 Vn	243
88595	−18 02870	HR 4005		10 10 15	−18 54.2	2	6.46	.019	0.49	.010	0.00		6	F7 V	2031,3077
88694	−58 01985	LSS 1447		10 10 15	−58 42.5	4	8.33	.007	0.66	.024	-0.13	.015	10	B6 Iab/b	191,362,1730,8100
		LF13 # 33		10 10 16	−52 09.1	1	12.60		0.29		0.20		1	A2 V	137
	−55 03031			10 10 16	−56 13.2	2	10.78	.000	0.41	.000	-0.04	.000	3	F2	127,549
88687	−55 03030			10 10 16	−56 15.7	2	8.96	.015	1.14	.029	1.24		3	K1 III	127,549
88533	+28 01855	IDS10075N2755	A	10 10 19	+27 40.2	2	8.32	.006	0.71	.019	0.26	.012	6	G2 V	3026,8088
	−52 03337			10 10 20	−53 09.6	1	11.74		0.43		-0.09		2	F2 V	137
		Bo 8 - 5		10 10 24	−57 49.3	1	13.18		0.43		0.17		1		411
		Bo 8 - 3		10 10 24	−57 49.7	1	12.81		0.38		-0.11		1		411
		Bo 8 - 1		10 10 25	−57 49.3	1	12.10		0.39		-0.03		2		411
		Bo 8 - 2		10 10 25	−57 49.5	1	12.48		0.40		0.00		1		411
298933	−51 04559	W -51 266		10 10 27	−51 45.6	1	10.42		-0.15		-0.65		2	B5	846
		Bo 8 - 4		10 10 27	−57 48.9	1	12.09		0.41		0.01		1		411
88693	−51 04560	HR 4010		10 10 29	−51 55.0	4	6.16	.003	1.17	.006			18	K2 III	15,1075,2013,2029
		CS 15621 # 73		10 10 30	+24 20.0	1	13.53		0.42		-0.21		1		1744
	+10 02122	G 43 - 34		10 10 31	+09 51.3	2	9.90	.014	0.89	.005	0.57	.009	4	K5	333,1620,3077
	−56 02978			10 10 31	−56 28.8	1	11.28		0.56		0.08		1	F8	549
	−55 03036	LSS 1448		10 10 33	−56 05.7	2	11.14	.010	1.34	.000	0.00	.099	5		362,846
		Bo 8 - 9		10 10 34	−57 49.4	1	12.04		1.01		0.84		1		411
		LP 127 - 183		10 10 36	+55 23.0	1	10.33		0.63		0.05		3		1723
88656	−27 07278			10 10 36	−28 15.9	1	9.10		0.88		0.44		2	K2 V	3072
	−84 00262			10 10 36	−84 52.	1	10.28		0.34				1		1705
88703	−49 05048			10 10 37	−50 01.6	1	9.11		0.05		0.03		1	A0 III/IV	137
300573	−55 03039			10 10 37	−56 11.2	1	10.22		0.31		0.12		1	A2	549
88715	−53 03437			10 10 41	−53 55.5	1	9.40		0.34		0.14		1	F0 III/IV	137
88692	−43 06043			10 10 42	−43 45.7	1	7.82		1.38		1.65		1	K3 III	389
233719	+53 01395	G 196 - 13		10 10 44	+52 45.9	2	9.53	.010	1.10	.000	0.94	.000	3	K5	1003,3072
	−52 03344			10 10 44	−53 07.9	1	11.49		0.23		0.12		2	A0 V	137
		G 55 - 4		10 10 47	−05 46.2	1	11.84		1.40		1.24		1		1620
88654	−7 02985			10 10 47	−08 11.8	1	7.68		0.87		0.53		3	G5	1657
88732	−50 04912			10 10 47	−51 10.1	1	8.85		0.15		0.12		1	A5 IV	137
		LF13 # 40		10 10 52	−51 19.8	1	12.10		0.38		0.16		1	A0 V	137
88757	−55 03048			10 10 54	−55 32.2	1	8.94		-0.01		-0.41		2	Ap Si	401
	−38 06282			10 10 55	−38 50.9	2	11.06	.000	0.85	.000	0.31	.000	4	K0	78,1696
88746	−46 05923	IDS10089S4659	AB	10 10 56	−47 13.8	3	8.18	.007	0.80	.000	0.33		9	G8 V	258,1075,2012
297202	−45 05858			10 10 57	−45 56.6	1	8.83		1.17		1.11		2	G0	401
	−12 03114	WZ Hya		10 10 59	−12 53.6	2	10.27	.000	0.19	.010	0.14	.025	2	A2	668,700
		Ton 33		10 11 00	+29 25.	1	14.01		-0.14		-0.43		1		98
88773	−55 03053			10 11 00	−55 54.2	1	7.96		0.01		-0.07		2	B9 V	401
		L 189 - 45		10 11 00	−57 35.	1	13.12		0.64		0.11		2		3060
88639	+27 01862	HR 4006		10 11 01	+27 23.0	2	6.05	.009	0.84	.006	0.46		4	G5 III-IV	71,8088
88699	−26 07752	HR 4011		10 11 01	−26 46.9	2	6.24	.005	0.32	.005			7	A9mA8-F3	15,2027
88756	−53 03444			10 11 01	−54 19.5	1	7.90		1.33		1.26		1	K0 Ib/II	137
		CS 15621 # 39		10 11 03	+25 33.3	1	13.60		0.06		0.13		1		1744
	+34 02100	G 118 - 38		10 11 03	+34 24.2	1	10.67		0.97		0.75		2	K3	7010
	−58 01997	LSS 1450		10 11 07	−58 39.2	2	10.86	.025	0.61	.025	-0.41	.000	6		362,846
88295	+79 00328			10 11 08	+79 11.7	1	6.94		0.06		0.09		2	A0	985

HD	DM	Other Id	N Rem	α_{1950}	δ_{1950}	S	V	σ_V	B–V	σ_{B-V}	U–B	σ_{U-B}	n	Spectrum	References
	−52 03354	LSS 1449		10 11 08	−52 45.6	2	11.82	.000	0.12	.005	-0.67	.025	4	B2 III	846,1737
298964	−53 03447			10 11 08	−54 07.2	1	10.34		0.12		0.02		1	B9 V	137
89499	−84 00263			10 11 08	−84 50.0	5	8.65	.011	0.72	.009	-0.09	.036	17	F8/G0	1088,1696,2012,2019,3077
88630	+45 01811			10 11 09	+45 20.1	1	7.90		1.16		1.25		3	K1 III	37
88782	−50 04919			10 11 09	−50 51.0	1	8.83		0.02		-0.43		2	B5 II	362
	+81 00330			10 11 10	+80 46.2	1	9.71		0.57		0.05		2	G5	1733
307569	−63 01333			10 11 10	−63 53.8	1	10.69		0.01		-0.39		4	A2	1730
88680	+11 02190			10 11 11	+11 05.4	1	8.02		1.10		1.01		2	G5	1648
88742	−32 07158	HR 4013		10 11 12	−32 47.1	5	6.37	.007	0.59	.007	0.08	.009	13	G0 V	15,78,158,2012,3026
88845	−63 01334	IDS10097S6343	ABC	10 11 12	−63 58.1	1	8.36		1.07		0.79		2	G8/K1 III	1730
88609	+54 01359			10 11 14	+53 48.6	3	8.59	.020	0.93	.010	0.42	.017	5	G5 III	979,1566,8112
88864	−64 01335			10 11 14	−64 08.8	2	7.89	.002	0.64	.006	0.16		6	F8 V	1730,2012
	+23 02195	IDS10085N2257	AB	10 11 15	+22 42.4	1	10.06		1.03		0.98		2	K5	3072
88697	−6 03109			10 11 15	−07 08.2	2	7.21	.000	0.50	.005	0.03		8	F8	158,2034
	−56 02988			10 11 16	−56 28.6	1	11.37		0.06		0.01		1	B9	549
297864	−50 04921			10 11 18	−50 50.9	1	11.44		0.17		0.17		1	A2 V	137
		LF13 # 44		10 11 18	−52 42.3	1	11.89		0.49		0.17		1	F0 V	137
88793	−49 05060	IDS10094S5007	A	10 11 20	−50 21.8	2	7.90	.014	1.19	.000	1.33		8	K1/2 III	1279,2012
88793	−49 05060	IDS10094S5007	B	10 11 20	−50 21.8	1	13.08		0.52		0.06		3		1279
88825	−59 01974	HR 4018		10 11 20	−59 40.2	7	6.10	.011	-0.08	.008	-0.50	.015	41	B5 IIIe	15,681,815,1637,2012*
	−52 03364			10 11 21	−53 01.7	1	9.87		1.17		0.87		2	K0 III	137
	−56 02992			10 11 22	−56 25.2	1	11.06		0.22		0.15		1	A4	549
88844	−60 01765	LSS 1453		10 11 24	−60 55.9	2	8.53	.000	-0.16	.013	-0.92	.003	4	B0.5III	540,976
	−52 03366	LSS 1451		10 11 25	−52 34.9	1	11.69		0.36		-0.24		2		846
88824	−50 04924	HR 4017, LW Vel		10 11 27	−50 59.1	2	5.27	.005	0.26	.005			7	F0 IV	15,2012
		CS 15621 # 38		10 11 31	+23 49.2	1	13.39		0.49		-0.09		1		1744
88863	−56 02997			10 11 31	−57 13.7	1	8.94		0.00		-0.37		2	B8 II/III	401
88725	+3 02338	G 44 - 6		10 11 32	+03 24.3	6	7.75	.016	0.61	.009	-0.01	.006	9	G1 V	908,1003,1620,1658*
		G 146 - 15		10 11 32	+44 09.7	1	13.60		1.64		1.33		5		203
88842	−51 04578	HR 4020	*AB	10 11 33	−51 30.5	2	5.77	.005	0.14	.000			7	A3 IV/V	15,2012
		MCC 124		10 11 34	+21 19.5	1	10.20		1.37		1.15		4	K5	3072
88841	−49 05062			10 11 35	−50 12.9	1	7.94		0.92		0.67		1	G8wG5II/III	137
297918	−50 04925			10 11 35	−51 02.2	1	11.02		0.44		-0.06		1	F2 V	137
	−56 02999			10 11 36	−56 27.8	1	10.74		0.50		0.03		1	F8	549
88879	−57 02810	LSS 1454		10 11 36	−58 06.9	4	8.43	.015	0.41	.000	-0.54	.030	9	B1 Ia/ab	191,362,730,846
88809	−39 06222	HR 4015	*AB	10 11 38	−40 05.9	5	5.91	.010	1.21	.008	1.25		20	K1 III	15,1075,2013,2028,3005
88764	−7 02989			10 11 39	−07 44.7	1	6.97		1.03				4	K0	2034
88651	+60 01246	HR 4008, U UMa		10 11 42	+60 14.0	1	6.22	.024	1.61	.009	1.90		4	M0 III	1025,3001
88907	−61 01517	HR 4022		10 11 43	−61 24.6	5	6.41	.014	-0.11	.006	-0.65	.008	18	B3 V	15,26,164,243,2012
88737	+21 02165	HR 4012		10 11 45	+21 25.1	2	6.03	.016	0.55	.008	0.11		4	F9 V	71,3053
88920	−61 01519			10 11 46	−62 15.5	1	8.57		-0.05		-0.45		2	B5/6 III	401
88723	+38 02117			10 11 48	+38 11.8	1	8.59		0.45		-0.03		2	F8	1601
88836	−39 06225	HR 4019	*A	10 11 48	−40 03.7	3	6.34	.004	0.94	.000			11	G8 III	15,2013,2029
88748	+16 02098			10 11 49	+16 23.2	1	7.09		1.39		1.54		2	K0	1648
88806	−23 09119			10 11 49	−23 33.9	1	6.34		1.63				4	M1 III	2033
		G 118 - 41		10 11 51	+39 12.8	1	11.64		1.13		1.08		2		1726
88877	−52 03372			10 11 52	−52 59.5	1	7.71		-0.02		-0.02		1	A1 IV	137
	−57 02821			10 11 58	−57 28.3	1	8.39		0.00		-0.11		2		401
		LSS 1455		10 11 59	−59 05.8	1	12.14		0.36		-0.33		2	B3 Ib	846
	−62 01480	LSS 1457		10 12 00	−62 45.2	2	10.92	.146	0.11	.057	-0.88	.027	4		540,1737
298956	−53 03374			10 12 01	−53 20.4	1	10.53		0.25		0.12		2	A5 V	137
87787	+86 00146			10 12 04	+85 41.1	1	9.21		0.10		0.11		7	A0	1219
88981	−65 01273	HR 4025		10 12 06	−66 07.5	5	5.15	.005	0.22	.008	0.18	.008	19	A4mA6-F0	15,355,1075,2016,2038
	+40 02309	IDS10091N4001	A	10 12 10	+39 45.3	1	10.16		1.25		1.16		3	K4	3072
	+40 02309	IDS10091N4001	B	10 12 10	+39 45.3	1	10.60		1.36		1.20		3		3072
	−52 03376	LSS 1456	A	10 12 11	−53 12.9	2	11.56	.135	0.25	.015	-0.50	.020	4		362,846
	−52 03376		B	10 12 11	−53 12.9	1	12.00		0.47		0.18		1		362
88786	+32 02005	HR 4014		10 12 14	+31 43.0	2	6.46	.000	0.86	.005	0.48		4	G8 III	1733,6009
	−53 03486			10 12 14	−54 09.5	1	10.70		1.12		0.80		2	K0 III	137
302596	−58 02018			10 12 14	−58 41.2	1	10.24		0.22		-0.33		2	B5 III	540
88945	−54 03325			10 12 15	−54 44.7	2	7.48	.010	-0.09	.015	-0.60	.005	4	B4 V	362,401
		LF13 # 52		10 12 20	−53 17.8	1	12.56		0.66		0.23		1		137
	−50 04939			10 12 23	−50 27.9	1	10.55		0.95		0.91		1	K0 III	137
		LP 847 - 28		10 12 24	−23 59.7	1	11.11		0.97		0.77		2		1696
		G 43 - 38		10 12 26	+08 21.3	1	16.16		0.41		-0.40		1	dFs	3060
		WLS 1024 40 # 6		10 12 26	+39 49.6	1	12.15		0.56		0.09		2		1375
88785	+42 02104			10 12 26	+42 07.5	1	8.24		0.32		0.08		2	F3 III	1375
88970	−54 03332			10 12 29	−54 53.5	1	8.18		-0.04		-0.45		2	B5/7 Vn	362
305047	−61 01529			10 12 29	−61 26.4	1	10.43		-0.09		-0.73		2	B5	540
	−64 01220			10 12 29	−65 05.6	1	10.83		-0.06		-0.71		2	B2.5II-III	540
88853	+12 02177			10 12 30	+11 55.4	1	7.21		1.03		0.89		2	G5	1648
307549	−62 01484			10 12 30	−62 29.8	1	9.92		0.03		-0.87		2	B3	540
88978	−51 04592			10 12 32	−51 33.9	2	7.39	.000	-0.15	.005	-0.64	.005	18	B5 III	158,362
89080	−69 01178	HR 4037		10 12 33	−69 47.4	9	3.31	.006	-0.09	.007	-0.31	.016	28	B8 IIIe	15,26,1020,1034,1075*
	−32 07173			10 12 35	−32 38.7	1	10.91		3.79				1		864
88979	−53 03492			10 12 35	−53 25.6	1	10.04		0.13		0.08		2	A1/2 V	137
88955	−41 05713	HR 4023		10 12 38	−41 52.4	6	3.84	.005	0.05	.005	0.07	.013	26	A1 V	15,26,1075,2012,3023,8015
88954	−34 06563			10 12 42	−34 41.7	1	8.95		0.29		0.13		1	A9 V	78
299931	−53 02240			10 12 42	−51 44.8	1	10.52		1.08		0.93		1	K1 III	137
88976	−41 05716			10 12 43	−41 28.7	2	6.56	.009	0.16	.005	0.06		8	A2 IV/V	1097,2012

Table 1

HD	DM	Other Id	N	Rem	α_{1950}	δ_{1950}	S	V	σ_V	B–V	σ_{B-V}	U–B	σ_{U-B}	n	Spectrum	References
		CS 15621 # 71			10 12 44	+27 15.3	1	13.03		0.41		-0.03		1		1744
298965	-53 03498				10 12 44	-54 07.4	1	10.26		0.48		-0.04		2	F5 V	137
		CCS 1643, AM Car			10 12 48	-60 28.5	1	9.36		2.18		2.07		1	N	864
89048	-61 01536				10 12 50	-62 12.4	1	9.38		-0.03		-0.51		2	B3/5 II/III	401
		CS 15621 # 36			10 12 52	+25 20.9	1	11.67		1.31		1.26		1		1744
		L 320 - 124			10 12 54	-46 55.	1	13.46		1.71		1.34		1	M4	3078
302625	-57 02847	LSS 1461			10 12 54	-58 03.9	2	9.35	.020	0.39	.015	-0.47	.025	4	B0.5III	362,730
88995	-43 06082				10 12 56	-44 13.1	1	6.59		1.32		1.49		1	K3 III	389
		L 17 - 47			10 12 57	-82 37.8	1	11.07		1.42		1.15		1		3062
		Y LMi			10 12 59	+33 06.5	1	12.49		0.25		0.14		1		699
88627		CCS 1632			10 13 00	+77 21.	1	10.83		1.52				1	R6	1238
		LF13 # 57			10 13 01	-50 27.2	1	12.57		0.36		0.07		1	F0 V	137
		LF13 # 58			10 13 04	-53 26.4	1	10.95		1.56		0.95		1		137
		W -52 441			10 13 06	-52 56.	1	12.41		0.18		-0.57		3		846
89015	-35 06260	HR 4029			10 13 09	-36 16.1	4	6.18	.006	1.06	.005	0.93		15	K0 III	15,78,1075,2029
		BPM 19929			10 13 12	-55 55.	1	15.10		0.68		-0.12		2	DF	3065
302617	-57 02861	LSS 1462			10 13 12	-57 50.0	1	9.84		0.27		-0.65		2	B0.5V	362
		CS 15621 # 37			10 13 18	+24 59.2	1	11.64		0.23		0.02		1		1744
		GD 115			10 13 18	-07 07.2	1	16.67		0.19		-0.54		1		3060
	-58 02028	LSS 1465			10 13 18	-58 54.6	1	11.10		0.42		-0.49		3		1737
	+35 02123				10 13 19	+34 39.0	1	9.35		1.07		0.89		2	G5	1601
		CS 15621 # 33			10 13 20	+26 10.5	1	13.76		0.60		-0.04		1		1744
89341	-78 00527				10 13 20	-79 06.7	1	8.56		0.10		-0.01		1	A0 V	55
		CS 15621 # 34			10 13 22	+26 04.5	1	13.70		0.87		0.41		1		1744
305019	-60 01793	LSS 1466			10 13 22	-60 40.4	4	10.47	.055	0.02	.028	-0.94	.034	7	B0 ep	191,540,976,1737
88960	+30 01981	HR 4024			10 13 24	+29 33.6	3	5.51	.005	0.01	.000	0.01	.014	8	A0 Vn	985,1022,3050
89062	-42 06074	HR 4036			10 13 25	-42 51.7	3	5.59	.007	1.52	.000			11	K4 III	15,2013,2029
89104	-54 03356	HR 4038			10 13 25	-54 43.5	5	6.16	.007	-0.17	.008	-0.79	.027	15	B2 IV	15,26,362,401,2012
89105	-55 03123				10 13 28	-55 26.4	1	7.90		-0.02		-0.22		2	B9 III	401
		LF13 # 59			10 13 30	-51 24.1	1	12.30		0.22		0.03		1	A0 V	137
89140	-57 02873	IDS10118S5721		A	10 13 31	-57 36.2	1	10.20		1.66		0.41		3	A1 IV	1737
88987	+18 02338	HR 4028		★ AB	10 13 32	+17 59.3	1	6.59		0.30		0.03		2	A9 IV	1733
		LF13 # 60			10 13 34	-54 13.6	1	12.33		0.07		0.05		1	A0 V	137
		G 53 - 38			10 13 35	-01 04.4	2	15.36	.019	0.28	.028	-0.49	.009	4	DA	538,3060
88986	+29 02021	HR 4027			10 13 38	+28 56.0	1	6.46		0.60		0.16		2	G2 V	3077
297907	-49 05094				10 13 38	-50 16.8	1	9.22		1.15		1.08		2	K0 III	137
		LP 903 - 54			10 13 39	-28 45.3	1	11.81		0.63		0.09		4		1696
89033	-10 03029	HR 4034		AB	10 13 41	-10 57.2	1	6.08		1.10				4	K0	2031
89121	-47 05704				10 13 41	-48 18.9	1	8.33		0.93				4	G8 III	2012
89075	-25 07888				10 13 42	-26 20.7	2	8.54	.000	-0.02	.008	-0.13	.021	7	A(p Eu)	55,1775
89137	-50 04966	LSS 1467			10 13 43	-51 00.5	4	7.98	.008	-0.04	.015	-0.93	.020	7	B0.5 III	362,540,976,1737
89010	+24 02207	HR 4030		★ B	10 13 46	+23 45.1	3	5.97	.005	0.65	.008	0.21	.005	7	G2 IV	71,1501,3077
89000	+26 02063				10 13 47	+25 32.8	2	8.34	.005	0.38	.000	-0.04	.005	5	F8	833,1625
89090	-27 07328				10 13 47	-28 21.9	2	7.21	.005	0.54	.005	0.03		5	F8/G0 V	2012,3037
	-29 08222	WY Ant			10 13 48	-29 28.2	1	10.34		0.17		0.12		1	A8	668
88849	+71 00534	HR 4021		★ A	10 13 51	+71 18.7	2	6.68	.019	0.32	.012	0.18	.019	3	A7 m	1776,3058
88815	+73 00489	HR 4016			10 13 51	+73 19.5	1	6.42		0.22		-0.04		1	F2 V	254
		vdB 34 # c			10 13 51	-60 03.5	1	12.26		0.17		-0.23		6	B8 V	434
88850	+71 00534	IDS10098N7134		B	10 13 52	+71 18.4	1	7.36		0.29		0.19		2	A8 V	3024
89175	-52 03420				10 13 52	-52 23.7	3	7.71	.014	1.13	.009	0.69	.003	10	G8 Ib/II	137,565,1673
89024	+26 02064	HR 4032			10 13 54	+25 37.2	2	5.84	.008	1.21	.004	1.22		4	gK2	71,105
89173	-50 04971				10 13 54	-51 04.6	1	9.82		0.00		-0.19		1	B8 V	137
		vdB 34 # d			10 13 54	-60 03.3	1	11.93		0.05		-0.48		3		434
		vdB 34 # e			10 13 54	-60 03.6	1	12.32		0.09		-0.37		3		434
89025	+24 02209	HR 4031		★ A	10 13 55	+23 40.0	12	3.44	.019	0.31	.014	0.19	.014	38	F0 III	15,39,254,1007,1008*
89174	-51 04610	LSS 1470			10 13 55	-51 57.3	3	7.95	.001	0.11	.017	-0.75	.028	7	B1 Ib/II	362,540,976
89203	-61 01553				10 13 55	-62 17.3	1	7.06		-0.04		-0.32		2	B7 III	401
89403	-78 00529				10 13 55	-78 45.3	1	7.70		-0.03		-0.62		1	B2/3 IV	55
		vdB 34 # f			10 13 57	-60 03.6	1	12.46		0.31		0.18		3		434
89056	+14 02228	HR 4035			10 14 00	+13 58.7	3	5.41	.015	1.60	.002	1.88	.149	12	M1.5 IIIab	3,71,3001
		vdB 34 # a			10 14 00	-60 02.9	1	12.65		0.15		-0.41		4		434
		vdB 34 # g			10 14 00	-60 04.1	1	11.59		0.10		-0.37		2		434
89202	-59 02003				10 14 01	-60 03.2	1	9.49		-0.02		-0.30		3	B9 III	434
89117	-28 08041	IDS10117S2829		AB	10 14 02	-28 43.7	1	7.72		0.40				3	F3 V	173
89021	+43 02005	HR 4033			10 14 05	+43 09.5	5	3.45	.004	0.03	.008	0.06	.000	27	A2 IV	1,15,1006,3023,8015
89201	-56 03055	LSS 1473			10 14 05	-57 07.5	3	7.84	.007	0.70	.012	-0.37	.017	13	B1 Ia/ab	6,362,730
89192	-54 03382				10 14 06	-55 16.2	1	6.83		0.06		-0.02		2	Ap CrEuSr	401
89055	+26 02065	G 54 - 21		★ A	10 14 08	+26 06.9	2	7.57	.019	0.62	.023	0.06	.009	4	G0 V	333,1620,3026
89132	-18 02885				10 14 09	-19 03.5	1	6.71		0.14		0.10		9	A3 V	1628
89191	-45 05913	IDS10122S4554		AB	10 14 13	-46 09.2	2	8.26	.014	0.03	.000	0.00	.005	18	A0 IV/V	156,547
	-45 05915				10 14 16	-46 03.1	1	11.69		0.33		0.23		3		156
298983	-51 04614				10 14 16	-52 07.8	1	10.40		0.34		0.13		1	F2 V	137
	-45 05918				10 14 17	-46 10.1	1	10.81		1.26		1.27		11		547
		CS 15621 # 48			10 14 18	+24 30.8	1	12.20		0.26		0.08		1		1744
		G 118 - 48			10 14 19	+30 43.6	1	14.70		1.45		1.12		4		316
		G 162 - 38			10 14 19	-11 42.4	1	10.96		1.47		1.10		2		1705,3073
89263	-59 02008	HR 4043		★ AB	10 14 20	-59 39.2	2	6.21	.005	0.21	.005			7	A5 V +A5 V	15,2012
		LSS 1476			10 14 22	-60 18.3	1	11.69		0.30		-0.75		2	B3 Ib	1737
89275	-63 01348				10 14 22	-63 30.9	1	8.30		0.01		-0.38		2	B7/8 IV/V	401
89169	-19 02964	HR 4040		★ AB	10 14 23	-20 25.2	3	6.57	.005	0.48	.000			11	F5 IV	15,2018,2028

HD	DM	Other Id	N Rem	α_{1950}	δ_{1950}	S	V	σ_V	B–V	σ_{B-V}	U–B	σ_{U-B}	n	Spectrum	References
		CS 15621 # 35		10 14 25	+25 40.4	1	14.59		0.40		-0.09		1		1744
		NGC 3201 sq2 22		10 14 25	−46 10.4	2	11.97	.014	1.63	.014	1.97	.024	17		484,547
		NGC 3201 sq2 21		10 14 25	−46 11.9	2	11.12	.014	0.37	.005	0.02	.024	18		484,547
88983	+65 00767	HR 4026		10 14 26	+65 21.5	3	5.78	.023	0.14	.012	0.11	.009	6	A8 III	985,1733,3052
89070	+45 01819			10 14 27	+45 16.2	1	8.21		0.89		0.56		2	G5	1601
	−57 02899	LSS 1477		10 14 28	−58 09.0	1	10.80		0.58		-0.61		2		362
		CS 15621 # 72		10 14 29	+26 05.2	1	12.83		0.29		0.07		1		1744
	−57 02900			10 14 29	−58 12.7	1	10.32		1.69		0.80		2		362
89125	+23 02207	HR 4039	* AB	10 14 30	+23 21.5	6	5.82	.003	0.50	.006	-0.05	.000	32	F8 Vb	15,1013,1197,1774,3037,8015
89249	−54 03389	LSS 1475		10 14 30	−55 20.9	3	8.78	.063	0.70	.022	-0.52	.025	6	Bp Shell	7,362,1737
298982	−51 04618			10 14 31	−52 01.8	1	10.55		0.09		0.03		1	A2 V	137
		CS 15621 # 51		10 14 33	+23 18.3	1	13.97		0.45		0.09		1		1744
		NGC 3201 sq2 24		10 14 38	−46 12.7	2	12.79	.015	0.57	.010	0.04	.005	13		484,547
		NGC 3201 sq2 23		10 14 39	−46 11.5	1	11.95		1.22		1.16		8		547
		NGC 3201 sq2 25		10 14 39	−46 14.6	2	12.82	.054	0.77	.025	0.23	.015	8		484,547
		NGC 3201 sq2 26		10 14 41	−46 14.7	1	14.10		1.02		0.42		3		547
		LP 609 - 60		10 14 42	−00 19.5	1	11.51		0.86		0.58		2		1696
89273	−50 04990	HR 4045, GY Vel		10 14 43	−50 57.3	3	6.24	.047	1.53	.010			11	M4/5 III	15,1075,2006
		G 118 - 49		10 14 44	+32 46.2	1	10.80		1.01		0.83		1	K3	3062
	−55 03151	LSS 1478		10 14 45	−55 34.0	1	10.87		0.39		-0.61		2	B2 Ve	362
		CS 15621 # 52		10 14 47	+23 13.6	1	13.42		0.44		-0.11		1		1744
		G 43 - 39		10 14 48	+14 17.2	1	11.22		1.01		0.86		1		333,1620
	+56 01448			10 14 54	+56 22.	1	10.17		0.47		-0.04		3		1723
		W -53 714		10 14 54	−53 06.	1	12.55		0.28		-0.51		3		846
89109	+52 01470			10 14 56	+52 03.5	1	8.36		1.14		1.00		3	K0	1733
89480	−76 00619			10 14 59	−76 50.3	1	8.73		0.25		-0.07		1	B8 Vn	55
89196	+20 02460	IDS10123N2030	AB	10 15 01	+20 15.2	1	8.07		1.56		1.79		2	K2	8088
		NGC 3201 sq2 5		10 15 02	−46 02.9	1	12.22		0.51		0.09		1		547
89208	+13 02233			10 15 05	+13 21.6	1	8.47		1.05				2	K0	882
		NGC 3201 sq2 1		10 15 06	−46 02.0	2	10.92	.017	1.40	.017	1.53	.004	6		484,547
	−51 04626			10 15 06	−52 14.8	1	11.54		0.33		0.05		2	A5 V	137
89254	−7 03001	HR 4042		10 15 09	−07 49.1	6	5.24	.009	0.32	.013	0.10	.014	18	F2 III	15,254,1008,1417,2028,3023
		NGC 3201 sq2 3		10 15 10	−46 00.3	2	11.60	.025	1.47	.015	1.52	.005	8		484,547
		CS 15621 # 29		10 15 12	+25 15.8	1	13.98		0.63		0.02		1		1744
298986	−51 04628			10 15 15	−52 14.3	4	10.06	.010	0.43	.004	-0.19	.027	7	sdF5	1236,1696,2012,3077
89358	−57 02913	LSS 1481		10 15 15	−57 39.8	1	10.75		0.51		-0.16		2	WC	362
	−45 05933	NGC 3201 sq2 2		10 15 18	−45 57.4	2	11.01	.029	0.43	.010	-0.03	.000	9		156,547
89239	+28 01867	HR 4041		10 15 22	+27 40.0	4	6.53	.015	-0.02	.016	-0.10	.016	9	A0 V	186,252,1022,1079
302648	−58 02056	LSS 1482		10 15 23	−59 15.8	1	10.25		0.42		-0.63		2		362
89222	+38 02125	IDS10125N3801	AB	10 15 25	+37 45.5	1	8.08		0.25		0.06		2	A2	1375
89388	−60 01817	HR 4050, V337 Car	* A	10 15 25	−61 04.9	7	3.36	.027	1.54	.015	1.67	.014	29	K3 IIa	678,1088,1520,2006*
89357	−49 05126			10 15 28	−50 20.0	1	9.35		0.05		0.02		2	A1/2 V	137
	−49 05125	XZ Vel		10 15 28	−49 50.8	1	7.98		4.92				1	N	864
		NGC 3201 - 1104		10 15 30	−46 09.	1	13.76		0.92		0.42		2		547
		NGC 3201 - 1117		10 15 30	−46 09.	1	11.93		1.60		1.62		10		547
		NGC 3201 - 1118		10 15 30	−46 09.	1	13.74		0.61		0.11		1		547
		NGC 3201 - 1120		10 15 30	−46 09.	1	13.48		1.14		0.65		4		547
		NGC 3201 - 1121		10 15 30	−46 09.	1	12.70		1.66		1.87		2		547
		NGC 3201 - 1222		10 15 30	−46 09.	1	13.37		1.05		0.57		4		547
		NGC 3201 - 1309		10 15 30	−46 09.	1	13.26		1.03		0.53		3		547
		NGC 3201 - 1312		10 15 30	−46 09.	1	11.77		1.74		1.73		6		547
		NGC 3201 - 1314		10 15 30	−46 09.	1	11.77		1.53		1.45		5		547
		NGC 3201 - 1315		10 15 30	−46 09.	1	13.32		1.11		0.62		4		547
		NGC 3201 - 1323		10 15 30	−46 09.	1	13.83		0.72		0.22		1		547
		NGC 3201 - 1410		10 15 30	−46 09.	1	12.87		1.33		1.08		2		547
		NGC 3201 - 1411		10 15 30	−46 09.	1	13.54		1.18		0.77		2		547
		NGC 3201 - 1501		10 15 30	−46 09.	1	12.85		1.27		0.94		2		547
		NGC 3201 - 2109		10 15 30	−46 09.	1	15.14		0.93		0.35		3		547
		NGC 3201 - 2110		10 15 30	−46 09.	1	13.31		1.31		1.28		3		547
		NGC 3201 - 2114		10 15 30	−46 09.	1	13.81		1.20		0.70		1		547
		NGC 3201 - 2201		10 15 30	−46 09.	1	13.82		1.11		0.58		3		547
		NGC 3201 - 2207		10 15 30	−46 09.	1	14.30		1.07		0.55		4		547
		NGC 3201 - 2211		10 15 30	−46 09.	1	14.70		0.25		0.23		4		547
		NGC 3201 - 2212		10 15 30	−46 09.	1	15.72		0.20		-0.06		2		547
		NGC 3201 - 2214		10 15 30	−46 09.	1	13.84		1.03				1		547
		NGC 3201 - 2310		10 15 30	−46 09.	1	14.56		0.30		0.24		2		547
		NGC 3201 - 2321		10 15 30	−46 09.	1	14.14		1.11		0.55		5		547
		NGC 3201 - 2405		10 15 30	−46 09.	1	13.67		1.16		0.74		3		547
		NGC 3201 - 2526		10 15 30	−46 09.	1	12.41		1.42				1		547
		NGC 3201 - 3101		10 15 30	−46 09.	1	13.66		1.16		0.65		4		547
		NGC 3201 - 3102		10 15 30	−46 09.	1	14.31		1.00		0.49		1		547
		NGC 3201 - 3105		10 15 30	−46 09.	1	14.97		0.64		0.16		4		547
		NGC 3201 - 3107		10 15 30	−46 09.	1	13.38		1.14		0.61		3		547
		NGC 3201 - 3112		10 15 30	−46 09.	1	15.72		0.92		0.21		1		547
		NGC 3201 - 3119		10 15 30	−46 09.	1	13.91		0.88		0.33		3		547
		NGC 3201 - 3204		10 15 30	−46 09.	1	12.29		1.44		1.24		6		547
		NGC 3201 - 3205		10 15 30	−46 09.	1	14.95		0.19		0.18		3		547
		NGC 3201 - 3209		10 15 30	−46 09.	1	16.03		0.76		0.30		1		547
		NGC 3201 - 3217		10 15 30	−46 09.	1	13.35		1.71		1.56		4		547
		NGC 3201 - 3218		10 15 30	−46 09.	1	12.37		1.57		1.47		5		547

Table 1 541

HD	DM	Other Id	N	Rem	α_{1950}	δ_{1950}	S	V	σ_V	B–V	σ_{B-V}	U–B	σ_{U-B}	n	Spectrum	References
		NGC 3201 - 3227			10 15 30	−46 09.	1	15.25		0.21		0.21		2		547
		NGC 3201 - 3304			10 15 30	−46 09.	1	14.10		1.07		0.45		7		547
		NGC 3201 - 3306			10 15 30	−46 09.	1	15.10		0.10		0.05		1		547
		NGC 3201 - 3313			10 15 30	−46 09.	1	14.77		0.47		0.32		1		547
		NGC 3201 - 3401			10 15 30	−46 09.	1	12.71		1.37		1.15		4		547
		NGC 3201 - 3504			10 15 30	−46 09.	1	13.26		1.22		0.82		1		547
		NGC 3201 - 3522			10 15 30	−46 09.	1	12.73		1.34		0.95		3		547
		NGC 3201 - 3525			10 15 30	−46 09.	1	13.93		1.10		0.60		2		547
		NGC 3201 - 4215			10 15 30	−46 09.	1	13.56		1.26		0.85		7		547
		NGC 3201 - 4301			10 15 30	−46 09.	1	14.40		0.95		0.29		3		547
		NGC 3201 - 4307			10 15 30	−46 09.	1	14.33		1.11		0.60		2		547
		NGC 3201 - 4318			10 15 30	−46 09.	1	12.54		1.45		1.21		4		547
		NGC 3201 - 4319			10 15 30	−46 09.	1	12.51		1.52		1.38		5		547
		NGC 3201 - 4403			10 15 30	−46 09.	1	13.82		1.13		0.64		3		547
		NGC 3201 - 4507			10 15 30	−46 09.	1	12.72		1.42		1.17		3		547
		NGC 3201 - 4521			10 15 30	−46 09.	1	13.45		1.19		0.81		2		547
		NGC 3201 - 4523			10 15 30	−46 09.	1	14.07		1.13		0.64		2		547
		NGC 3201 - 4524			10 15 30	−46 09.	1	11.97		1.57		1.49		2		547
89312	−20 03145				10 15 31	−20 46.5	1	7.33		1.40		1.70		5	K4 III	1657
307633	−63 01353				10 15 31	−63 58.9	1	10.43		0.05		-0.62		2	B5	540
89221	+43 02007				10 15 32	+43 18.0	1	6.53		0.90		0.68		2	G5	1601
		NGC 3201 sq2 9			10 15 36	−46 02.8	2	13.34	.059	1.17	.015	1.02	.000	8		484,547
89252	+39 02337	IDS10127N3037	AB		10 15 37	+39 21.7	1	8.84		0.67		0.18		3	G2 V	833
		CS 15621 # 28			10 15 39	+24 37.1	1	13.91		0.59		-0.01		1		1744
297264	−45 05938				10 15 39	−45 52.4	1	10.22		0.44		0.01		5	F8	156
89384	−53 03580				10 15 39	−53 33.0	1	9.67		0.06		-0.01		2	A0 V	137
89402	−57 02929				10 15 41	−58 16.9	1	8.77		-0.03		-0.22		2	B9 IV/Vn	401
89307	+13 02237				10 15 42	+12 54.4	1	7.03		0.56		0.06		2	G0 V	1648
		NGC 3201 sq2 4			10 15 43	−46 01.3	2	11.86	.015	0.65	.005	0.16	.005	7		484,547
	−53 03583				10 15 46	−53 29.2	1	10.47		1.31		1.24		2	K2 III	137
89269	+44 01973	IDS10127N4433	A		10 15 50	+44 18.2	2	6.63	.013	0.65	.004	0.16	.004	5	G5	1080,3031
89353	−28 08070	HR 4049, AG Ant			10 15 50	−28 44.5	1	5.34		0.24				4	B9.5Ib/II	2032
		LF13 # 70			10 15 50	−50 33.6	1	13.08		0.50		0.04		1		137
		CS 15621 # 30			10 15 51	+26 00.8	1	13.41		0.69		0.21		1		1744
	−58 02064	LSS 1484			10 15 52	−58 24.5	1	11.39		0.75		-0.23		2	B1 III	1737
89268	+47 01761	HR 4044		★ A	10 15 54	+47 00.7	2	6.42	.025	1.20	.010	1.17	.005	5	K1 III	1601,1733
298994	−52 03464				10 15 54	−53 14.0	1	9.57		0.97		0.70		2	G8 IV	137
	+70 00607				10 15 59	+70 07.9	1	8.60		1.38		1.61		3	K2	1733
89429	−53 03594				10 16 00	−54 22.3	1	7.89		0.01		-0.06		2	A0	137
		LF13 # 73			10 16 01	−53 03.7	1	12.76		0.22		-0.07		1	A0 V	137
89369	−22 02889				10 16 02	−22 30.3	1	8.29		0.26		0.19		14	Fm δ Del	1603
89428	−51 04638				10 16 02	−51 41.7	1	8.22		1.23		1.26		1	K1 III	137
89412	−45 05943				10 16 03	−46 00.4	2	9.38	.015	0.17	.010	0.09	.005	12	A3 IV	156,547
89391	−25 07916				10 16 06	−26 14.9	1	7.94		0.94				4	K0 V	2012
	−56 03094	LSS 1485			10 16 07	−56 52.7	1	10.82		0.72		-0.39		2		362
		NGC 3201 sq2 8			10 16 08	−46 02.1	1	13.26		1.39		1.39		2		547
		NGC 3201 sq2 10			10 16 11	−46 06.2	1	14.11		0.46		0.28		1		547
305080	−59 02029				10 16 11	−60 01.3	1	9.76		-0.08		-0.77		1	A0	8100
		Ton 498			10 16 12	+31 46.	1	15.08		0.32		-0.08		1		1036
	−60 01829	LSS 1488			10 16 12	−61 20.3	1	11.07		0.16		-0.85		3	B2 Ve	1737
		NGC 3201 sq2 6			10 16 13	−46 06.6	1	12.27		0.53		0.05		1		547
89497	−58 02069	LSS 1487		★ AB	10 16 13	−59 15.2	1	10.32		0.38		-0.54		2	B2 III	401
89344	+25 02231	HR 4048			10 16 14	+24 57.8	1	6.40		1.39		1.56		2	K0	1733
89069	+79 00329	DI Dra			10 16 17	+79 00.9	1	8.40		0.00		-0.01		3	A0	220
89461	−41 05765	HR 4056			10 16 20	−41 25.1	2	5.95	.005	-0.06	.000			7	B9 V	15,2012
	−45 05947				10 16 20	−45 57.2	1	11.38		0.55		0.30		5		156
89319	+49 01940	HR 4046			10 16 21	+48 39.0	1	6.00		1.02				1	K0	71
89305	+54 01365				10 16 22	+53 47.2	1	8.61		0.51		0.06		2	F8	1566
		LSS 1486			10 16 22	−53 53.8	1	12.56		0.23		-0.21		1	B3 Ib	846
		LP 904 - 3			10 16 24	−30 53.1	1	14.84		0.05		-0.66		3		3060
302686	−58 02074	LSS 1489			10 16 24	−58 24.6	2	9.96	.021	0.09	.019	-0.77	.042	4	B2 II	362,401,540
89362	+22 02204				10 16 25	+22 17.7	1	8.98		1.16		1.16		3	K0	1648
		Steph 844			10 16 26	+00 48.1	1	11.76		1.48		1.21		1	M0	1746
89361	+25 02232				10 16 26	+24 36.9	1	7.60		1.17		1.17		3	K2 III	1625
89442	−36 06281	HR 4053			10 16 26	−36 33.2	1	6.30		1.28				4	K2/3 III	2006
	−52 03477				10 16 26	−52 57.3	1	12.17		0.39		0.29		1	A0 V	137
		CS 15621 # 32			10 16 28	+26 42.5	1	14.86		0.04		0.13		1		1744
89493	−45 05949	IDS10145S4547	AB		10 16 34	−46 01.8	1	8.77		0.46		0.01		6	F5 V	156
89396	+15 02188				10 16 35	+14 55.2	2	8.45	.019	1.38	.000	1.59		3	K2	882,3077
		vdB 35 # a			10 16 36	−60 39.	1	14.92		0.63		0.04		3		434
89532	−51 04652				10 16 37	−51 29.7	1	7.65		0.95		0.66		1	G8 III	137
89539	−55 03211				10 16 39	−55 57.6	1	9.01		-0.08		-0.57		2	B3 V	401
89473	−22 02892	IDS10144S2240	AB		10 16 44	−22 54.8	1	8.53		0.44		0.01		21	F3/5 V	1603
297961	−49 05141				10 16 46	−50 15.9	1	11.15		0.12		0.07		1	A1 V	137
89554	−54 03443				10 16 46	−54 35.9	1	9.28		0.09		-0.47		2	B5 III	401
89569	−55 03220	HR 4061			10 16 48	−55 51.6	3	5.80	.004	0.48	.001			8	F7 V	15,1488,2012
89455	−11 02851	HR 4055		★ AB	10 16 49	−12 16.6	1	6.00		0.26				4	A8 III	2006
	−30 08345				10 16 50	−30 41.7	1	11.09		-0.02		-0.94		5		1732
		CS 15621 # 31			10 16 51	+26 38.2	1	14.49		0.27		0.11		1		1744
	+20 02465	AD Leo		★	10 16 54	+20 07.3	6	9.42	.017	1.54	.009	1.07	.019	15	M4.5:Ve	1,694,1197,3072,8006,8088

HD	DM	Other Id	N	Rem	α_{1950}	δ_{1950}	S	V	σ_V	B–V	σ_{B-V}	U–B	σ_{U-B}	n	Spectrum	References
89415	+30 01994				10 16 55	+29 37.4	2	9.24	.015	0.45	.005	−0.05	.000	6	F5 V	186,3026
	−54 03450				10 17 00	−54 22.7	1	11.29		1.04		0.52		1	K0 III	137
89450	+15 02189				10 17 01	+14 55.7	2	8.63	.005	1.28	.016	1.41		4	K0	882,3040
89449	+20 02466	HR 4054			10 17 01	+19 43.5	13	4.79	.008	0.45	.005	0.01	.010	231	F6 IV	1,3,15,130,1008,1034*
89490	−4 02840	HR 4059			10 17 01	−04 51.2	4	6.37	.005	0.89	.005	0.60	.000	18	K0	15,1256,2013,2030
89389	+54 01366	HR 4051			10 17 02	+54 01.8	1	6.45		0.54		0.08		2	F9 V	272
89587	−50 05046				10 17 04	−50 27.9	1	6.88		−0.15				4	B3 III	2012
		Ton 34			10 17 06	+28 01.	1	15.69		0.37		−0.86		1		98
89484	+20 02467	HR 4057		⋆AB	10 17 13	+20 05.7	7	1.98	.018	1.14	.013	1.00	.008	25	K0 III	15,1003,1080,1363*
89625	−59 02044	LSS 1492			10 17 14	−59 31.0	1	8.89		0.11		−0.83		2	B3/4 V	401
		CS 15621 # 26			10 17 15	+24 11.8	1	13.67		0.62		0.03		1		1744
		CS 15621 # 53			10 17 16	+23 09.1	1	13.59		0.25		0.04		1		1744
305095	−60 01838				10 17 16	−60 22.5	1	10.09		−0.12		−0.71		2	B2 III	191
89343	+69 00568	HR 4047, EN UMa			10 17 17	+69 00.0	2	5.93	.030	0.22	.020	0.12	.000	4	A7 Vn	985,1733
89414	+54 01367	HR 4052			10 17 18	+54 28.1	1	6.01		1.13				1	gK3	71
89585	−30 08356				10 17 23	−30 44.7	2	7.44	.005	0.89	.000	0.59		5	G8 III/IV	78,2012
	−7 03007	RW Sex			10 17 27	−08 26.9	1	10.39		−0.02		−0.83		1	DAe	801
89565	−8 02897	HR 4060			10 17 30	−08 48.4	3	6.31	.008	0.33	.004	0.01	.015	11	F1 V	15,1417,2031
302724	−58 02090	LSS 1493			10 17 32	−59 02.7	2	10.03	.029	0.42	.029	−0.57	.110	4	B2 Ve	540,1737
89715	−64 01248	HR 4065		⋆AB	10 17 32	−64 25.5	3	5.66	.004	0.04	.005	0.05	.000	15	A1 V	15,1075,2038
		LF13 # 79			10 17 33	−52 55.8	1	12.42		0.33		0.21		1	A2 V	137
299070	−53 03632				10 17 38	−54 14.9	1	9.44		0.08		−0.51		2	B8	401
89682	−54 03474	HR 4063, GZ Vel			10 17 44	−54 46.7	3	4.56	.004	1.62	.000	1.81	.000	11	K3 II	15,1075,2012
89574	+12 02193	IDS10151N1151	A		10 17 47	+11 36.3	1	7.72		1.22		1.27		2	K0	1648
89714	−56 03131	HP Car		⋆AP	10 17 47	−57 09.2	1	8.80		0.18		−0.75		2	B1/2 IIIn	362
		Ton 500			10 17 48	+31 47.	1	15.09		0.41		−0.09		1		1036
89501	+56 01450				10 17 50	+55 49.2	1	8.68		0.16		0.08		2	A3/4IV/III	1502
89619	+7 02282	IDS10153N0656	AB		10 17 55	+06 40.9	1	7.36		0.50		0.06		2	gF7	3016
89619	+7 02282	IDS10153N0656	C		10 17 55	+06 40.9	1	9.61		0.58		0.06		2		3032
89639	−22 02897	VY Hya			10 17 55	−22 54.0	1	10.37		0.24		0.12		1	A8/9 V	668
89677	−39 06302				10 17 55	−39 27.0	1	7.31		0.01		−0.14		2	B9 IV/V	401
89713	−50 05055	NGC 3228 - 1		⋆AB	10 17 55	−51 19.0	1	6.94		0.45		0.05		4	F5 III	1082
89740	−58 02095				10 17 55	−58 53.7	4	6.92	.010	−0.12	.010	−0.71	.009	13	B3 V	6,362,657,2012
305053	−59 02049	LSS 1499			10 17 58	−59 29.0	1	9.69		0.39		−0.54		2	B3	540
		GD 116			10 17 59	+36 41.9	1	15.96		0.00		−0.81		2		3060
297975	−50 05058				10 18 02	−50 40.1	1	9.88		1.25		1.13		1	K1 III	137
89723	−53 03639				10 18 03	−53 38.1	1	9.64		0.16		0.09		1	A0 IV/V	137
	−57 02993				10 18 03	−58 14.0	1	10.94		1.19		−0.03		2		362
89572	+42 02114				10 18 04	+42 06.1	1	6.79		0.02		0.04		4	A1 V	1501
89756	−56 03141	LSS 1498			10 18 04	−57 15.3	1	7.63		0.56		0.39		3	F0 II	8100
89668	−0 02326	G 53 - 40			10 18 13	−01 12.9	7	9.41	.025	1.08	.007	1.06	.026	32	K3 V	830,1003,1620,1775,1783*
89736	−47 05790	HR 4066			10 18 14	−47 26.8	4	5.64	.007	1.67	.003			18	K5/M0 III	15,1075,2013,2029
89767	−51 04673	LSS 1500			10 18 15	−52 21.0	3	7.21	.009	0.15	.012	−0.76	.012	8	B0.5/1 Iab	158,540,976
		CS 15621 # 27			10 18 20	+23 14.6	1	12.70		−0.27		−1.06		1		1744
		G 55 - 14			10 18 20	−06 25.8	2	13.14	.010	0.62	.000	−0.10	.025	4		1658,1696
		LF13 # 82			10 18 20	−54 03.3	1	11.66		1.12		0.88		2	K0 III	137
89631	+27 01879				10 18 22	+26 57.0	1	8.51		0.46		−0.04		1	F5 V	3077
89630	+27 01878				10 18 22	+27 08.3	1	8.66		0.56		0.05		2	F8 V	3026
297977	−50 03363				10 18 22	−50 52.0	1	11.68		0.24		0.05		1	A2 V	137
		CS 15621 # 46			10 18 24	+23 39.4	1	12.78		0.66		0.16		1		1744
89707	−14 03093	Ring Hydra # 19			10 18 25	−15 13.9	5	7.18	.007	0.55	.007	−0.04	.004	26	G1 V	158,1075,1311,2034,3037
	−43 06163				10 18 25	−43 39.7	1	10.57		0.85		0.58		4		158
89688	+3 02352	HR 4064, RS Sex			10 18 27	+02 32.5	3	6.66	.018	−0.08	.005	−0.66	.010	10	B2.5IV	15,154,1256
		LF13 # 84			10 18 27	−52 59.8	1	13.30		0.31		0.14		2	A0 V	137
89820	−58 02107				10 18 29	−59 17.8	1	8.87		0.00		−0.38		2	B8 II/III	401
89802	−57 03008				10 18 30	−57 24.4	2	8.86	.010	0.02	.005	−0.23	.024	3	B9 V	401,761
		CS 15621 # 25			10 18 31	+24 06.9	1	15.17		0.04		0.06		1		1744
89593	+60 01255				10 18 34	+59 48.3	1	9.06		0.70		0.23		3	G5	1502
89787	−50 05066				10 18 34	−50 54.0	1	9.22		0.36		−0.01		1	F0 V	137
89789	−51 04678				10 18 35	−51 47.5	1	9.91		0.01		−0.06		1	A0 V	137
89747	−17 03129	HR 4068			10 18 43	−17 43.9	2	6.50	.005	0.41	.005			7	F3 IV	15,2027
89819	−51 04680	NGC 3228 - 2			10 18 44	−51 25.1	1	8.14		0.96		0.46		4	G6 III/IV	1082
	−59 02060	LSS 1502		⋆A	10 18 44	−59 24.7	1	10.39		0.43		−0.51		2	O7 IIIn	1737
89876	−62 01531	IDS10172S6250	A		10 18 46	−63 05.1	1	7.96		0.00		−0.48		2	B5 IV	401
89844	−56 03159				10 18 47	−56 55.1	1	7.74		−0.10		−0.71		2	B2 IIN	401
		NGC 3228 - 23			10 18 48	−51 27.7	1	13.18		0.50		0.04		2		1082
89841	−54 03497	RY Vel			10 18 48	−55 04.1	4	7.91	.068	1.14	.054	0.82	.019	4	F5 Ib/II	657,689,1484,1587
		HQ Car			10 18 50	−60 59.8	1	11.84		0.63		0.39		1		689
		LP 430 - 8			10 18 51	+15 37.1	1	12.31		0.55		−0.13		2		1696
		NGC 3228 - 22			10 18 52	−51 28.4	1	13.30		0.60		0.30		1		1082
89777	−16 03030				10 18 53	−16 47.9	3	9.35	.019	0.79	.014	0.33	.009	8	G8 V	1003,2012,3077
		Ton 502			10 18 54	+32 12.	1	14.02		0.38		−0.10		1		1036
88044	+87 00085				10 18 55	+87 31.6	1	8.27		1.00		0.79		2	K0	1502
89856	−51 04685	NGC 3228 - 3			10 18 57	−51 36.6	1	9.09		−0.10		−0.58		4	B9 III	1082
89873	−57 03022				10 18 57	−57 30.6	1	9.50		0.01		−0.16		1	B9 V	761
		NGC 3228 - 21			10 18 58	−51 28.9	1	13.09		1.58		1.49		2		1082
298051	−51 04686	NGC 3228 - 20			10 18 59	−51 34.5	1	10.17		0.11		0.01		3	A1 V	1082
		NGC 3228 - 24			10 19 02	−51 29.5	1	13.59		1.24				3		1082
89890	−55 03286	HR 4074		⋆AB	10 19 03	−55 47.4	7	4.50	.013	−0.12	.023	−0.58	.015	25	B3 IIIe	15,362,362,815,1075*
		LP 790 - 47			10 19 04	−16 06.3	1	11.69		1.34		1.20		1		1773

Table 1

543

HD	DM	Other Id	N Rem	α_{1950}	δ_{1950}	S	V	σ_V	B–V	σ_{B-V}	U–B	σ_{U-B}	n	Spectrum	References
+16 02114				10 19 06	+15 47.0	3	9.98	.004	-0.15	.004	-0.68	.004	17	B1	830,1026,1783
		CS 15621 # 69		10 19 06	+26 22.1	1	13.58		0.64		0.11		1		1744
		W -52 753		10 19 06	-52 37.	1	12.05		0.35		-0.49		2		846
		CS 15621 # 22		10 19 07	+24 06.1	1	12.88		0.29		0.00		1		1744
89816	-22 02904	HR 4071		10 19 07	-23 27.5	3	6.49	.004	0.20	.007	0.13		8	A4 IV/V	15,78,2012
		CS 15621 # 23		10 19 08	+24 31.3	1	13.29		0.48		-0.05		1		1744
		CS 15621 # 70		10 19 08	+27 26.5	1	12.61		0.36		-0.07		1		1744
89774	+15 02192	HR 4070		10 19 09	+15 13.7	3	6.12		0.02	.009	0.00	.039	7	A1 V	252,1022,1079
		G 118 - 53		10 19 10	+37 10.7	1	10.57		1.40		1.22		2	K4	1726
89772	+20 02475			10 19 11	+20 18.4	1	10.63		0.95		0.61		2		8088
89744	+41 02076	HR 4067		10 19 13	+41 29.0	1	5.78		0.53		0.06		2	F7 V	254
89901	-51 04690	NGC 3228 - 4		10 19 13	-51 29.8	1	8.43		-0.09		-0.42		5	B9 IV	1082
89782	+11 02212			10 19 14	+10 57.2	1	7.77		0.45				37	F8	130
89828	-21 03045	HR 4073	★ AB	10 19 14	-22 16.6	1	6.51		0.07				4	A1 V	2031
89829	-29 08292			10 19 15	-29 41.1	1	7.81		0.89		0.51		3	K0 III +G	1657
89900	-51 04691	NGC 3228 - 5	★ AB	10 19 16	-51 27.7	1	8.19		-0.02		-0.24		4	A0 IV	1082
89916	-52 03519			10 19 18	-52 40.2	1	9.55		0.12		0.08		1	A0 IV	137
		G 118 - 54	AB	10 19 20	+37 04.4	1	10.91		0.89		0.60		2		7010
		NGC 3228 - 26		10 19 20	-51 30.3	1	14.53		0.97				1		1082
89925	-57 03035			10 19 20	-58 13.9	1	7.29		0.87		0.62		2	G0 Ib/II	191
		CS 15621 # 24		10 19 21	+25 42.1	1	14.64		0.42		-0.21		1		1744
89758	+42 02115	HR 4069		10 19 21	+41 45.1	6	3.05	.011	1.58	.010	1.89	.011	23	M0 III	15,1080,1363,3016*
89915	-51 04693	NGC 3228 - 6		10 19 22	-51 29.3	1	7.90		-0.03		-0.11		4	A0 V	1082
298052	-51 04707	NGC 3228 - 7		10 19 25	-51 30.6	1	10.55		0.42		0.01		3	F2	1082
298047	-51 04695	NGC 3228 - 8		10 19 27	-51 29.0	1	9.37		-0.05		-0.10		3	B9 V	1082
89922	-51 04696	NGC 3228 - 9		10 19 27	-51 34.7	1	9.34		0.14		0.03		3	A2 III-IV	1082
		G 43 - 43		10 19 29	+12 19.8	1	12.73		1.50		1.23		1		333,1620
89813	+12 02200	G 43 - 44		10 19 30	+11 34.0	1	7.84		0.75		0.49		1	G5	333,1620
298048	-51 04697	NGC 3228 - 11		10 19 30	-51 27.0	1	9.97		0.05		-0.02		3	B9 IV-V	1082
		NGC 3228 - 14		10 19 32	-51 25.6	1	11.03		0.26		-0.03		3		1082
		GD 117		10 19 33	+46 16.1	1	16.70		0.33		-0.63		2	DAwk	3060
89938	-51 04698	NGC 3228 - 12		10 19 33	-51 30.1	1	9.31		-0.06		-0.21		3	B8 V	1082
89937	-51 04699	NGC 3228 - 13		10 19 34	-51 26.3	1	9.03		-0.06		-0.31		3	B9 V	1082
89884	-17 03133			10 19 35	-17 46.9	2	7.13	.000	-0.11	.010	-0.56	.000	4	B5 III	55,1212
299053	-52 03523			10 19 35	-53 04.6	1	10.23		0.24		0.06		1	A2 V	137
		L 190 - 19		10 19 36	-55 49.	1	13.30		1.00		0.68		2		3062
89885	-19 02986			10 19 38	-20 19.4	1	7.11		1.17				4	K0 III	2033
89956	-51 04701	NGC 3228 - 10		10 19 38	-51 46.6	1	8.21		-0.12		-0.44		4	B8 V	1082
89955	-50 05081			10 19 40	-50 50.9	1	8.33		1.11		1.00		1	K0 III	137
89920	-43 06190	IDS10176S4359		10 19 41	-44 15.2	1	9.78		1.00				4	G8/K0	2033
89991	-60 01856	AQ Car		10 19 41	-60 49.2	1	8.56		0.76		0.54		1	F8/G0 Ib	1593
	+12 02201	G 43 - 45	★ AB	10 19 42	+12 23.9	2	10.37	.014	1.18	.009	1.15		2	K3-4	333,1017,1620
	+39 02340			10 19 43	+39 22.7	1	9.98		0.49		-0.01		2	F8	1375
89847	+18 02352			10 19 46	+17 46.0	1	7.76		0.35		-0.02		3	F0	1648
		NGC 3228 - 15		10 19 46	-51 35.5	1	11.26		0.33		0.03		4		1082
		WD1019+129		10 19 48	+12 57.2	1	15.60		0.01		-0.80		2	DA3	1727
		BPM 35019		10 19 48	-47 02.	1	14.48		0.72		0.00		2		3065
		NGC 3228 - 25		10 19 48	-51 28.	1	14.19		1.48				1		1082
89911	-19 02987	HR 4076		10 19 49	-19 36.8	1	6.12		0.03				4	A0 V	2027
298053	-51 04702	NGC 3228 - 16		10 19 50	-51 35.0	1	10.69		0.13		0.02		4	Am	1082
298045	-51 04704	NGC 3228 - 17		10 19 52	-51 31.6	1	9.54		1.99		1.98		3	M3	1082
298022	-50 05085			10 19 54	-50 32.2	1	10.43		0.02		-0.10		1	B9 V	137
	-53 03678	LSS 1509		10 19 54	-53 48.4	1	11.20		0.10		-0.72		2		846
299107	-54 03519			10 19 54	-54 31.3	1	9.59		1.65		1.93		1	K0	768
	-54 03521			10 19 58	-54 30.5	1	9.45		1.02		0.80		1	K2	768
	-54 03522			10 20 01	-54 29.7	1	10.25		1.24		1.20		1	K5	768
89883	+31 02131			10 20 02	+30 35.1	2	8.76	.010	0.27	.010	0.10	.005	6	A3	186,1371
298054		NGC 3228 - 18		10 20 02	-51 37.9	1	9.79		1.56		1.07		3	G0	1082
90019	-56 03191			10 20 02	-57 14.8	2	9.95	.000	-0.03	.000	-0.40	.010	4	B7 II/III	401,761
89906	+16 02116	IDS10174N1551	AB	10 20 03	+15 35.9	1	7.28		0.65		0.15		2	G5	1625
89965	-28 08125			10 20 05	-28 35.7	1	9.47		0.95		0.63		1	K3 V	3072
307650	-61 01603	LSS 1513		10 20 07	-61 58.1	1	10.37		-0.02		-0.72		2	B0	401
89905	+31 02133			10 20 11	+31 05.4	1	7.52		0.95		0.73		2	G8 III	1733
89998	-41 05809	HR 4080		10 20 11	-41 23.9	5	4.83	.005	1.12	.005	1.08	.008	23	K1 III	15,1075,2012,3005,8015
90036	-53 03683			10 20 12	-53 30.8	1	8.53		0.24		0.09		2	A2 V	137
89904	+34 02120	HR 4075		10 20 14	+34 09.7	2	5.90	.000	0.14	.000	0.10	.000	5	A6 V	985,1733
90050	-57 03065		AB	10 20 16	-57 36.4	1	9.08		0.12		0.11		1	A0 V	761
89862	+57 01266			10 20 18	+56 46.5	2	8.26	.040	0.94	.020	0.68	.020	4	K0 III-IV	1,3016
90049	-53 03688			10 20 18	-54 20.0	1	7.78		1.42		1.61		1	K2 III	768
	-53 03689			10 20 19	-54 10.5	1	10.16		1.18		1.11		1	K2 III	137
90075	-59 02079	LSS 1517		10 20 21	-59 56.0	2	8.88	.001	0.07	.023	-0.76	.002	4	B1 II	540,976
	-56 03205	LSS 1516		10 20 22	-57 15.8	1	10.41		-0.01		-0.45		2		730
89962	+7 02289	HR 4077		10 20 23	+06 47.8	3	6.05	.012	1.12	.000	1.14	.004	9	K3 III	15,1415,3077
	-54 03527			10 20 23	-54 28.6	1	10.08		1.00		0.71		1	K2	768
300637	-54 03528			10 20 23	-54 35.6	1	11.27		0.18		0.13		1	B9	768
		Ton 504		10 20 24	+31 26.	1	14.97		0.55		-0.09		1		1036
		SOS 1064		10 20 24	-54 10.	1	11.78		0.04		-0.32		2	B8	761
89742	+74 00433			10 20 25	+73 44.5	1	8.92		0.16		0.12		1	A2	1776
		Lod 46 # 49		10 20 25	-54 28.6	1	11.99		0.44		0.30		1	A0	768
90065	-54 03529			10 20 25	-54 48.0	1	7.90		0.17		0.10		1	A4 V	768

HD	DM	Other Id	N	Rem	α_{1950}	δ_{1950}	S	V	σ_V	B–V	σ_{B-V}	U–B	σ_{U-B}	n	Spectrum	References
90073	−53 03696				10 20 26	−53 30.7	1	8.23		0.97		0.72		2	K0 IV	137
	−54 03530				10 20 28	−54 27.9	1	11.58		0.31		-0.44		1	B3	768
89822	+66 00664	HR 4072			10 20 33	+65 49.2	6	4.95	.028	-0.06	.007	-0.12	.016	20	A0p Si	15,1063,1202,1363*
	−54 03533	SOS 1059			10 20 33	−54 25.6	2	11.69	.000	0.05	.005	-0.21	.010	2	B7	761,768
299104	−53 03699				10 20 35	−54 06.5	1	10.75		0.03		-0.20		2	A0	761
	−54 03534				10 20 35	−54 26.0	1	11.62		0.64		0.21		1		768
90087	−59 02080	LSS 1518			10 20 35	−59 30.1	3	7.78	.015	0.00	.013	-0.89	.017	6	B2/3 III	401,540,976
89882	+58 01254				10 20 36	+58 03.6	1	9.53		0.56		-0.01		3	G5	3016
		L 190 - 266	AB		10 20 37	−59 55.1	2	10.58	.090	1.44	.005	1.12		5		158,1705
	−54 03537				10 20 38	−54 35.6	1	11.46		0.35		0.19		1	A0	768
	−54 03538				10 20 38	−54 48.6	1	9.99		0.16		-0.60		1	B3	768
89995	+6 02301	HR 4079	★ A		10 20 39	+05 56.9	3	6.54	.004	0.45	.010	-0.06	.007	12	F7 IV-V	15,1415,3053
	−44 06435				10 20 41	−45 13.1	2	9.32	.000	0.44	.002	-0.07		4		1594,6006
90086	−50 05097	NGC 3228 - 19			10 20 42	−51 21.6	1	7.97		-0.09		-0.41		3	B6 V	1082
299106	−54 03541				10 20 42	−54 25.9	2	10.84	.005	0.02	.000	-0.10	.000	3	A0	761,768
	−54 03542				10 20 44	−54 28.6	1	11.15		1.12		0.92		1	K3	768
89994	+14 02237				10 20 45	+14 09.5	1	8.75		0.28		0.13		3	F0	1648
299108	−54 03543				10 20 46	−54 28.6	1	11.03		0.25		0.19		1		768
90102	−57 03088	IDS10189S5710	AB		10 20 47	−57 25.1	2	8.68	.015	0.16	.005	-0.70	.030	4	B2/3 II	362,401
		LF13 # 94			10 20 48	−50 18.9	1	13.16		0.44		0.19		2		137
	−54 03544				10 20 48	−54 49.5	1	11.07		0.44		0.02		1	F5	768
	−54 03547				10 20 49	−54 48.8	1	11.07		0.48		0.01		2		540
90139	−61 01611				10 20 49	−62 06.0	1	8.61		-0.02		-0.28		2	B9 IV/V	401
90438	−81 00432				10 20 51	−81 23.8	1	7.25		0.24		0.09		6	A6/7 IV	1628
89993	+30 02005	HR 4078			10 20 52	+29 52.2	2	6.39	.010	1.09	.010	0.94	.030	4	G8 III	186,252
	−57 03091	LSS 1520			10 20 53	−58 18.8	1	10.25		0.74		-0.31		2	B1 Ia	1737
		G 236 - 28			10 20 55	+66 12.4	1	15.89		1.70						906
90044	−3 02911	HR 4082, SS Sex			10 20 55	−03 49.2	3	5.96	.005	-0.10	.004	-0.16	.005	8	B9p Si	15,1079,1415
90071	−29 08306	HR 4083			10 20 55	−29 54.5	4	6.26	.004	0.32	.005	0.04	.014	12	A9 IV	15,78,1075,2012
	−53 03707				10 20 55	−54 11.8	1	11.44		0.20		-0.11		2	B7	761
	−53 03705				10 20 55	−54 17.0	1	10.12		1.14		1.10		1	K5	768
90151	−61 01612				10 20 56	−61 49.2	1	7.66		-0.02		-0.16		2	B9 III/IV	401
90135	−53 03709				10 20 57	−54 20.1	4	9.36	.010	0.08	.013	-0.71	.006	7	B1/2 Ib	540,761,768,976
90009	+26 02081				10 21 03	+25 49.3	2	6.77	.010	1.18	.000	1.22	.005	7	K2 III	833,1625
	−53 03713				10 21 03	−54 14.3	1	9.92		1.62		1.68		1	K9	768
		CS 15621 # 54			10 21 04	+24 10.7	1	13.80		0.36		-0.04		1		1744
90007	+41 02080	IDS10182N4104	AB		10 21 07	+40 49.1	1	8.90		0.69		0.25		2	G5	1723
90177	−58 02145	HR Car			10 21 07	−59 22.3	3	8.44	.112	0.90	.022	-0.25	.034	8	Be	6,1588,1737
90008	+39 02344				10 21 08	+39 17.4	2	7.83	.005	0.38	.005	0.00	.005	6	F3 V	833,1601
300638	−54 03553				10 21 10	−54 51.0	1	9.70		1.30		1.24		1	K5	768
	−57 03103				10 21 12	−57 28.7	1	11.32		0.15		-0.41		2	B4	761
90160	−54 03556				10 21 14	−54 36.8	1	9.92		0.08		0.09		1	A0 V	768
		RV Leo			10 21 15	+10 00.2	1	13.26		0.14		0.10		1		668
	−57 03104				10 21 15	−57 28.4	1	11.65		0.02		-0.06		2	B9	761
90040	+34 02123	HR 4081			10 21 16	+33 58.3	1	5.50		1.18				2	gK1	71
90176	−53 03717				10 21 17	−54 07.5	1	8.71		1.09		0.97		1	G8 III	137
90132	−37 06509	HR 4086			10 21 18	−37 45.3	2	5.32	.005	0.26	.005			7	A8 V	15,2028
		SOS 1063			10 21 18	−54 10.	1	11.83		0.13		-0.35		4	B3	761
	−54 03559				10 21 18	−54 31.6	1	11.76		0.17		0.16		1	A0	768
90187	−57 03105	LSS 1524			10 21 18	−57 44.6	5	8.81	.053	0.26	.021	-0.76	.022	10	B0/3ep	362,540,730,976,1737
	−54 03560				10 21 20	−54 24.3	1	11.71		0.26		0.21		1	A0	768
	+38 02134				10 21 21	+38 22.8	1	9.12		0.85		0.50		2	G5	1601
90219	−60 01874				10 21 22	−61 00.7	1	7.48		1.25		1.21		60	K0 II/III	1593
	−53 03722				10 21 24	−54 19.4	1	11.94		0.12		0.16		1	A0	768
		CS 15621 # 68			10 21 25	+23 05.2	1	10.31		0.39		-0.02		1		1744
90175	−49 05210	IDS10194S4951	AB		10 21 25	−50 05.7	2	7.92	.025	0.49	.010	-0.06		4	F6 V	1594,6006
	−54 03562				10 21 26	−54 50.2	1	9.96		1.15		1.11		1	K5	768
	−54 03564				10 21 29	−54 25.5	1	11.61		0.27		0.19		1	A0	768
90264	−66 01243	HR 4089			10 21 29	−66 38.9	5	4.97	.010	-0.13	.005	-0.52	.014	17	B8 V	15,1075,1637,2016,2038
	−53 03726				10 21 30	−54 19.6	1	11.30		0.32		0.08		1	A6	768
	−54 03565				10 21 30	−54 22.8	1	11.46		0.23		0.18		1		768
	−53 03727				10 21 31	−54 21.8	1	11.75		0.37		0.17		1		768
		CS 15621 # 20			10 21 32	+24 23.1	1	10.97		-0.05		-0.08		1		1744
		CS 15621 # 18			10 21 32	+26 37.8	1	14.10		0.67		0.17		1		1744
		WLS 1100 75 # 6			10 21 32	+75 13.6	1	13.41		0.85		0.74		2		1375
90170	−41 05833	HR 4087			10 21 32	−41 42.0	3	6.27	.016	0.88	.000	0.60		9	G8 IV	15,2027,3077
90156	−29 08316				10 21 37	−29 23.6	1	6.94		0.65		0.37		1	G5 V	78
90202	−48 05601	LSS 1526			10 21 37	−49 12.8	2	9.12	.009	0.16	.011	-0.72	.012	4	B0.5/1Iab/b	540,976
	−54 03569				10 21 37	−54 24.8	1	11.64		0.31		0.20		1	A0	768
90125	+3 02358	HR 4085	★ AB		10 21 38	+02 37.3	2	6.31	.005	1.00	.000	0.75	.005	7	G9 V	15,1415
		Steph 852			10 21 40	+12 12.6	1	12.46		1.57		1.28		1	M0	1746
90246	−61 01622				10 21 40	−61 27.8	1	8.26		1.06		0.84		3	K0 III	1704
90123	+11 02217				10 21 41	+10 50.5	1	6.80		1.11		1.02		2	K0	1648
302772	−57 03120				10 21 41	−57 50.6	1	10.58		0.04		-0.72		2	B5	362
299097	−53 03730				10 21 43	−53 39.6	2	9.82	.013	-0.06	.009	-0.84	.007	5	B3	540,976
	−9 03063	G 162 - 50			10 21 44	−10 08.4	4	9.98	.014	1.24	.012	1.16	.009	10	K4.5	158,1705,2033,3072
299109	−54 03570				10 21 44	−54 29.2	1	11.07		0.36		0.22		1	A0	768
299114	−54 03573				10 21 46	−54 38.2	1	11.67		0.23		0.22		1	A0	768
		LP 790 - 49			10 21 47	−19 56.9	1	11.04		0.69		0.18		2		1696
		LF13 # 96			10 21 47	−53 19.9	1	13.02		0.76		0.34		1		137

Table 1

HD	DM	Other Id	N Rem	α_{1950}	δ_{1950}	S	V	σ_V	B–V	σ_{B-V}	U–B	σ_{U-B}	n	Spectrum	References
90263	−58 02157			10 21 47	−58 24.2	1	9.01		−0.03		−0.43		2	B7 III	401
	−54 03577			10 21 50	−54 31.7	1	11.03		0.38		0.22		1	A5	768
90144	+15 02196			10 21 53	+15 23.9	1	9.44		1.04				3	K0	882
90197	−23 09258			10 21 54	−24 21.2	4	7.11	.017	0.66	.012	0.09	.020	11	G3/5 V	78,861,1075,1311
90273	−57 03125	LSS 1529		10 21 54	−57 23.3	4	9.07	.014	0.16	.015	−0.83	.016	10	O7	6,362,730,761
	−53 03734			10 21 55	−53 55.2	1	10.92		0.19		0.14		1	B9	761
299113	−54 03581			10 21 57	−54 37.6	1	10.95		0.14		0.10		1	A0	768
299090	−52 03547			10 21 58	−53 10.3	1	10.07		0.16		−0.01		1	B9 V	137
		CS 15621 # 66		10 22 01	+25 07.0	1	14.25		0.51		−0.07		1		1744
90289	−57 03127	HR 4091		10 22 01	−57 42.0	6	6.35	.009	1.52	.004	1.78	.008	22	K4 III	15,657,1075,2013,2029,3005
		Wes 2 - 8		10 22 02	−57 30.8	1	13.84		1.05		1.16		1		411
	−56 03249	SOS 1102		10 22 04	−57 20.7	1	11.08		0.12		−0.48		2	B0	761
	−4 02862	G 162 - 51		10 22 06	−05 15.7	2	9.80	.000	0.57	.000	0.00	.020	2		1620,1658
90288	−56 03250	IDS10203S5657	AB	10 22 06	−57 12.6	4	8.15	.016	−0.15	.006	−0.90	.018	12	B2 III/IV	6,362,761,1732
		Wes 2 - 4		10 22 06	−57 30.	1	13.47		1.53		0.26		2		411
		Wes 2 - 2	AB	10 22 06	−57 30.	1	13.27		1.26		−0.02		2		411
90315	−61 01626			10 22 06	−61 44.9	1	9.10		0.02		−0.54		2	B3 IV/V	401
90165	+14 02240			10 22 07	+13 50.6	1	8.92		1.11				2		882
	+42 02118			10 22 07	+42 13.4	1	10.18		0.49		−0.02		2		1375
	−53 03742			10 22 07	−54 19.4	2	10.99	.010	0.39	.005	0.17	.010	3	A1 I	761,768
90313	−58 02163	LSS 1534		10 22 08	−59 15.6	3	8.38	.030	0.34	.020	−0.54	.022	7	B1 Ib	191,540,976
		Wes 2 - 1		10 22 09	−57 30.2	1	12.96		1.18		0.97		3		411
		CS 15621 # 17		10 22 10	+27 55.1	1	11.09		0.09		0.09		1		1744
299084	−51 04736			10 22 10	−52 19.9	2	10.49	.007	0.10	.009	−0.65	.009	3	B8	540,976
90122	+52 01478			10 22 11	+52 25.6	1	9.54		0.51		0.02		2	F8	1566
		Wes 2 - 6		10 22 11	−57 28.9	1	13.12		0.35		0.22		1		411
		Wes 2 - 9	AB	10 22 11	−57 30.3	1	11.44		1.35		0.06		3		411
90164	+31 02136			10 22 12	+30 37.6	1	7.90		0.55		0.03		4	F8 V	1371
90242	−18 02927	HS Hya		10 22 12	−18 50.3	1	8.10		0.43		−0.07		6	F5 V	588
302748	−57 03133			10 22 12	−57 22.9	1	10.23		0.00		−0.44		3	B8	761
		Wes 2 - 5		10 22 12	−57 29.4	1	12.80		1.20		−0.03		1		411
		Wes 2 - 3		10 22 12	−57 29.8	1	13.79		0.69		0.14		2		411
	−54 03583			10 22 13	−54 37.0	1	11.70		0.06		0.01		1	B9	768
	+11 02218	G 43 - 46		10 22 14	+10 58.6	1	11.05		0.85		0.50		1	K0	333,1620
		GD 118		10 22 14	+47 07.9	1	14.90		0.46		−0.24		2		3060
	−53 03745	SOS 1122		10 22 14	−54 20.9	2	11.68	.023	0.14	.014	0.03	.009	4	A1	761,768
90183	+25 02247	IDS10195N2508	AB	10 22 15	+24 52.1	1	8.42		0.64		0.09		2	G0 V	3026
		LF13 # 98		10 22 17	−51 03.4	1	12.47		0.24		0.24		1	A0 V	137
		Wes 2 - 7		10 22 17	−57 30.5	1	14.01		0.39		0.21		1		411
90208	+13 02252			10 22 19	+12 59.1	1	8.25		0.36		0.14		2	A2	1648
90312	−53 03748			10 22 19	−53 28.4	1	7.84		0.98		0.72		1	K0 III	137
	−53 03749			10 22 19	−54 22.6	1	11.51		0.35		0.14		1	A4	768
90255	−17 03140			10 22 21	−18 23.4	1	8.77		0.44		−0.33		4		1732
302771	−57 03137	LSS 1536		10 22 21	−57 48.6	2	10.05	.015	0.15	.020	−0.74	.015	4	B3	362,730
		CS 15621 # 21		10 22 22	+23 56.1	1	12.49		0.19		0.14		1		1744
90331	−54 03588			10 22 22	−54 26.9	1	8.45		1.63		1.80		1	K3 III	768
		SOS 1124		10 22 24	−54 07.	1	11.63		0.13		0.10		2	B9	761
90268	−26 07896	IDS10201S2646	AB	10 22 25	−27 01.4	1	8.24		0.39				4	F0 V	2012
90329	−51 04739			10 22 25	−51 46.7	1	10.10		0.06		−0.02		1	A0 V	137
		LF13 # 101		10 22 26	−53 10.2	1	12.78		0.54		0.27		1		137
302742	−56 03260			10 22 27	−57 18.8	2	10.12	.045	0.02	.015	−0.71	.010	6	B1 V	434,761
		CS 15621 # 56		10 22 29	+24 30.2	1	14.05		0.50		−0.05		1		1744
90283	−24 09040			10 22 29	−24 43.1	1	7.97		1.57		1.90		1	K4/5 III	1732
		LP 489 - 73		10 22 31	+09 10.7	1	12.82		1.49		1.15		1		1696
90254	+9 02351	HR 4088, DE Leo		10 22 37	+09 02.4	3	5.60	.004	1.61	.010	1.96	.005	10	M3 IIIab	15,1415,3001
		V LMi		10 22 37	+29 02.3	2	11.31	.208	0.16	.050	0.04	.030	2	A8	668,699
		G 146 - 32		10 22 39	+38 59.2	2	15.42	.047	1.72	.028	0.80		4		316,906
		LF13 # 102		10 22 39	−52 21.1	1	12.32		0.18		0.02		1	A0 V	137
90398	−60 01890	IDS10210S6103	AB	10 22 39	−61 17.9	1	7.61		−0.05		−0.29		2	B9 III/IV	401
90252	+17 02219			10 22 41	+17 10.5	1	9.14		1.21				4	K0	882
89571	+84 00234	HR 4062		10 22 41	+84 30.4	1	5.51		0.23		0.06		2	F0 IV	1733
	−53 03757			10 22 41	−54 16.5	1	11.84		0.14		0.13		1	A2	768
302741	−56 03269			10 22 41	−57 18.7	1	10.26		0.12		−0.63		2	B9	761
		LF13 # 103		10 22 43	−50 15.0	1	12.64		0.40		0.35		1	A0 V	137
	−53 03759			10 22 43	−54 13.2	1	10.28		1.42		1.48		1	K8	768
		G 54 - 26		10 22 44	+26 38.9	1	13.19		1.57		1.20		1	M6-7	3062
90163	+59 01309			10 22 44	+59 12.8	1	8.54		1.24		1.17		2	K0	1733
	−9 03070	G 162 - 52		10 22 44	−09 58.6	3	10.18	.009	1.55	.031	1.19	.031	6	M0	1705,3073,7009
	−53 03760			10 22 44	−54 19.8	1	11.41		0.23		0.12		1	A0	768
90370	−51 04744			10 22 45	−51 32.9	1	9.06		0.19		0.21		2	A0 III	137
299111	−54 03595			10 22 48	−54 33.9	1	11.78		0.05		0.04		1		768
	−53 03762	LSS 1538		10 22 49	−54 10.4	1	11.56		0.11		−0.77		3		846
		G 162 - 53		10 22 50	−02 20.8	2	11.69	.015	0.88	.015	0.69	.010	2		1658,1696
	−54 03596			10 22 50	−54 33.8	1	11.71		0.42		0.01		1	F2	768
90250	+36 02065			10 22 52	+35 40.8	4	6.47	.027	1.10	.012	0.97	.012	7	K1 III	1003,1080,3009,8041
299110	−54 03597			10 22 52	−54 31.2	1	11.55		0.16		0.13		1		768
	−54 03600			10 22 54	−54 34.2	1	12.22		0.16		−0.35		1	B6	768
299100	−53 03765			10 22 55	−53 46.2	1	10.31		0.11		0.04		2	B9	761
90395	−49 05234			10 22 59	−50 09.2	1	9.62		1.77		2.03		1	R0	864
		LF13 # 105		10 23 00	−51 25.3	1	12.61		0.33		0.16		1	A2 V	137

HD	DM	Other Id	N	Rem	α_{1950}	δ_{1950}	S	V	σ_V	B–V	σ_{B-V}	U–B	σ_{U-B}	n	Spectrum	References
	−54 03603				10 23 00	−54 26.6	2	11.42	.010	0.08	.000	−0.26	.000	2	B8	761,768
305154	−59 02110	LSS 1542			10 23 01	−59 57.8	3	10.20	.026	0.10	.012	−0.80	.047	6	O7 V	191,540,1737
90277	+34 02128	HR 4090			10 23 03	+34 03.1	7	4.74	.008	0.25	.011	0.17	.008	22	F0 V	15,39,254,1008,1118*
		CS 15621 # 55			10 23 04	+24 11.0	1	13.28		0.42		−0.11		1		1744
	−54 03607				10 23 04	−54 25.0	1	11.46		0.38		0.19		1	A1	768
90276	+40 02327				10 23 07	+39 31.3	1	7.20		1.37		1.63		3	K4 III	1501
298074	−49 05236				10 23 07	−50 11.6	1	10.99		0.07		−0.09		1	A0 V	137
		CS 15621 # 19			10 23 08	+26 16.2	1	11.93		0.16		0.11		1		1744
90393	−41 05850	HR 4093			10 23 08	−42 12.8	1	6.18		1.00				4	G8 III	2032
299112	−54 03610				10 23 08	−54 34.2	1	11.17		0.13		0.01		1		768
		LF13 # 107			10 23 09	−54 02.3	1	12.74		0.46		0.09		1		137
90434	−56 03286				10 23 09	−56 49.9	1	8.21		−0.01		−0.31		2	B9 IV/V	401
		WLS 1024 40 # 7			10 23 10	+38 03.7	1	12.41		0.50		0.04		2		1375
90454	−57 03164	HR 4095			10 23 10	−58 19.3	3	5.94	.008	0.33	.004			8	F2 II/III	15,1488,2012
90362	−6 03146	HR 4092			10 23 14	−06 48.4	7	5.57	.030	1.53	.017	1.86	.008	31	M0.5III	15,1003,1417,2028*
298083	−50 05122				10 23 17	−50 46.2	1	10.86		0.22		0.13		1	A2 V	137
305137	−59 02116	LSS 1547			10 23 17	−59 58.4	1	10.47		0.01		−0.76		2		540
	−44 06469				10 23 20	−44 38.1	1	11.18		0.94		0.76		4		158
		CS 15621 # 67			10 23 21	+24 37.7	1	14.45		0.68		0.08		1		1744
90589	−73 00733	HR 4102			10 23 24	−73 46.6	6	3.99	.010	0.36	.005	−0.03	.009	21	F2 IV	15,1075,2006,2024*
90490	−58 02182				10 23 27	−58 37.1	1	6.98		−0.08		−0.55		2	B5 Vne	401
90385	+15 02197	DW Leo			10 23 30	+15 09.3	1	8.51		0.93				3	G0	882
		CS 15621 # 16			10 23 30	+28 02.7	1	13.36		0.64		0.04		1		1744
90489	−56 03300	LSS 1546			10 23 31	−57 18.7	3	8.64	.016	0.67	.005	0.47	.034	5	F5 Ib/II	127,730,8100
		CS 15621 # 15			10 23 34	+23 45.9	1	12.27		0.14		0.15		1		1744
	+26 02085				10 23 35	+25 59.9	1	8.88		0.58		0.11		2	G0	1733
	−53 03775				10 23 35	−53 41.9	1	10.13		1.15		1.19		2	K2 III	137
307718	−61 01644	LSS 1550			10 23 38	−62 11.3	1	10.47		0.13		−0.75		2	B2 Ve	540
90432	−16 03052	HR 4094			10 23 40	−16 34.8	5	3.82	.017	1.48	.005	1.82	.007	16	K4 III	15,1075,2012,3016,8015
		CS 15621 # 14			10 23 44	+24 06.1	1	13.74		0.67		0.17		1		1744
		L 753 - 28			10 23 45	−17 43.6	1	10.95		1.23				1		1746
		LP 790 - 19			10 23 45	−17 43.6	1	10.91		1.25		1.13		1	K4-5	3062
90630	−73 00735	HR 4105			10 23 45	−73 43.0	3	6.18	.004	0.07	.000	0.10	.005	12	A2/3 V	15,1075,2038
90420	+11 02225				10 23 48	+10 45.5	1	7.20		1.27		1.47		2	K0	1648
90564	−63 01403				10 23 49	−63 51.7	1	8.80		0.04		−0.38		2	B8 V	401
		CS 15621 # 11			10 23 53	+25 59.6	1	14.01		0.66		0.11		1		1744
90402	+29 02046				10 23 58	+28 37.8	1	8.81		0.34		0.11		3	F5	186
	+29 02045	IDS10211N2935		A	10 23 58	+29 19.3	1	11.04		0.97		0.65		2	G5	186
90475	−29 08340				10 23 58	−29 36.5	1	8.84		0.94		0.62		3	G8 III	1730
	−54 03628				10 23 58	−54 39.5	1	11.01		0.98		0.74		2		1730
90551	−55 03419	IDS10221S5607		AB	10 23 58	−56 22.0	1	8.55		−0.05		−0.73		1	B3 Vn	401
302797	−58 02191	LSS 1552			10 23 58	−58 40.5	2	9.98	.012	0.57	.003	−0.45	.017	4	O9 Ib	191,540
90428	+13 02256				10 24 00	+12 44.6	1	8.14		1.10				1	G5	882
		LOD 20 # 17			10 24 00	−57 19.8	1	12.35		0.35		−0.25		2		127
90518	−42 06222	HR 4099			10 24 01	−42 29.0	4	6.13	.008	1.13	.004	1.06	.013	10	K1 III	15,78,2012,3005
	−58 02194	LSS 1556			10 24 03	−58 31.8	1	10.29		0.34		−0.65		1	O8	768
90563	−58 02193	LSS 1554			10 24 03	−58 32.5	2	9.87	.010	0.43	.002	−0.55	.010	3	B2 Ve	540,768
90473	−0 02341				10 24 04	−00 44.0	1	6.31		1.46				4	K0	2033
		S436 # 4			10 24 05	−29 40.5	1	11.43		1.01		0.98		2		1730
90519	−44 06486				10 24 05	−45 16.6	1	7.82		1.36				4	K1 III	2012
90520	−44 06487				10 24 05	−45 18.5	2	7.51	.000	0.64	.005	0.19		5	G0 IV/V	78,2012
90502	−29 08341				10 24 07	−29 45.5	1	8.38		0.39				4	F2 III	2012
	−56 03309				10 24 09	−57 19.7	1	11.30		0.06		−0.47		2		127
90578	−57 03197	LSS 1557			10 24 09	−57 34.3	3	9.30	.031	0.10	.019	−0.78	.010	17	B1 Ib/II	6,362,730
		CS 15621 # 13			10 24 10	+25 34.6	1	13.46		0.49		−0.06		1		1744
90441	+30 02014	IDS10214N3011		A	10 24 11	+29 55.8	2	7.79	.005	0.35	.025	0.01	.035	5	F2	97,186
		G 55 - 17			10 24 11	−02 06.2	1	13.19		0.67		0.01		2		1658
	−5 03063	G 55 - 16			10 24 11	−06 15.7	2	9.75	.015	0.94	.010	0.66	.029	3	K0 V	803,1620,3072
		S436 # 3			10 24 11	−29 48.1	1	11.37		0.62		0.14		2		1730
	−56 03310				10 24 11	−57 18.8	1	10.17		1.44		1.76		2		127
90442	+27 01893				10 24 12	+26 53.9	2	8.23	.004	0.84	.013	0.47	.009	13	K1 V	186,3026
90416	+47 01775				10 24 12	+46 58.2	1	8.99		0.28		0.06		2	F2	1601
90485	−3 02921				10 24 15	−04 08.0	2	6.55	.007	0.96	.003	0.68		9	K0	1509,2033
302891	−58 02198				10 24 16	−58 29.8	1	10.20		0.11				1	B3	768
90472	+20 02487	HR 4097			10 24 17	+19 37.2	1	6.15		1.14				2	K0	71
90559	−42 06225	IDS10221S4254		AB	10 24 17	−43 08.9	2	8.17	.003	1.22	.003	1.25		9	K2 III	1673,2012
90599	−57 03207				10 24 17	−58 21.8	2	9.08	.107	0.09	.049	−0.56	.029	3	B3/5 III	401,768
90600	−59 02126	LSS 1562			10 24 17	−60 07.4	4	9.42	.016	0.17	.016	−0.76	.024	7	O7/8	401,540,976,1737
		G 43 - 48			10 24 21	+10 27.0	1	12.14		1.14		0.98		1	K3	333,1620
302890	−58 02201				10 24 23	−58 28.4	1	10.03		0.26				1	B5	768
90617	−57 03210				10 24 24	−58 21.7	2	9.49	.015	0.25	.010	0.14	.015	2	A2/3 III/IV	401,768
90400	+60 01263				10 24 27	+59 51.2	1	6.85		1.04		0.83		2	K0	1502
	+28 01879				10 24 29	+28 07.6	1	9.07		0.84		0.39		3	G5	186
		HA 54 # 182			10 24 29	+29 47.7	1	10.91		0.40		−0.05		3		97
90615	−56 03322	LSS 1563			10 24 29	−57 03.7	6	8.21	.018	0.25	.014	−0.65	.015	16	B1 II	6,362,540,730,976,8100
		CS 15621 # 12			10 24 30	+25 37.5	1	12.27		0.22		0.09		1		1744
90470	+42 02123	HR 4096		★	10 24 31	+41 51.4	1	6.02		0.17		0.08		2	A2 V	252
90612	−52 03577				10 24 32	−53 10.9	1	9.54		−0.05		−0.37		1	Ap Si	137
302870	−57 03217				10 24 32	−58 22.0	1	10.92		0.00		−0.27		1	B9	768
90512	+12 02211				10 24 33	+11 34.3	2	6.66	.017	0.86	.010	0.56		38	G5 III	130,252

Table 1

HD	DM	Other Id	N	Rem	α_{1950}	δ_{1950}	S	V	σ_V	B–V	σ_{B-V}	U–B	σ_{U-B}	n	Spectrum	References
	−57 03216	LSS 1565			10 24 33	−57 39.1	1	11.04		0.89		−0.18		2	B0.5 Ia	1737
90661	−62 01577				10 24 33	−63 02.6	1	7.08		−0.01		−0.07		2	B9.5Vn	401
		HA 54 # 183			10 24 34	+29 54.5	1	11.87		0.86		0.66		3		97
	+28 01880				10 24 36	+27 47.0	1	10.21		1.01		0.72		3	G5 V	186
90524	+13 02258				10 24 37	+12 42.0	1	7.78		1.55				2	K0	882
302893	−58 02203				10 24 37	−58 35.5	1	10.71		0.00		−0.30		1	B9	768
	+28 01882				10 24 38	+27 38.7	1	9.47		0.92		0.53		3	G5	186
	−35 06431				10 24 38	−35 24.4	2	9.97	.010	0.51	.010	−0.03	.015	3	F8	78,1696
302920	−58 02204	LSS 1567			10 24 38	−59 12.1	1	9.89		0.17		−0.70		2	B0.5IV	540
90660	−60 01904				10 24 38	−60 51.8	1	9.21		0.03		−0.06		2	B9 Vn	761
	+28 01881				10 24 39	+28 22.5	1	9.75		1.30		1.41		3	K0 III	186
	−29 08346				10 24 39	−29 57.2	1	10.57		0.52		0.06		2		1730
90657	−58 02205	V398 Car			10 24 41	−58 23.1	2	9.66	.009	0.41	.014	−0.54	.005	2	WR	768,1096
		NGC 3255 - 1			10 24 42	−60 25.	1	12.67		0.82		0.60		1		411
		NGC 3255 - 2			10 24 42	−60 25.	1	12.44		0.19		0.05		2		411
		NGC 3255 - 3			10 24 42	−60 25.	1	13.74		0.90		0.73		1		411
		NGC 3255 - 4			10 24 42	−60 25.	1	13.68		1.03		0.91		1		411
		NGC 3255 - 5			10 24 42	−60 25.	1	13.54		0.39		0.49		1		411
		NGC 3255 - 6			10 24 42	−60 25.	1	13.77		0.31		0.38		1		411
		NGC 3255 - 7			10 24 42	−60 25.	1	14.09		0.40		0.35		1		411
		NGC 3255 - 8			10 24 42	−60 25.	1	13.80		1.38		0.77		1		411
	+29 02047				10 24 45	+29 19.3	1	10.74		0.53		−0.02		2	G0	186
90538	+16 02122				10 24 47	+16 01.9	2	7.95	.014	1.42	.014	1.68		4	K2	882,3077
90643	−50 05148				10 24 47	−51 09.2	1	8.53		0.08		−0.03		1	B9.5V	137
300688	−54 03649				10 24 47	−54 43.8	1	10.58		−0.02		−0.21		2	B9	1730
305170					10 24 47	−60 21.9	1	11.43		0.16		0.08		2		761
		G 53 - 41			10 24 50	+01 39.5	2	11.04	.000	0.48	.000	−0.14	.000	3	sdF8	333,927,1620
		CS 15621 # 8			10 24 50	+28 03.4	1	11.24		−0.02		−0.06		1		1744
	+29 02048				10 24 51	+28 40.7	1	10.11		0.49		0.01		3	F2	186
90610	−30 08465	HR 4104			10 24 52	−30 48.8	5	4.24	.008	1.45	.008	1.63	.003	27	K4 III	3,15,1075,2012,8015
		IC 2581 - 123			10 24 52	−57 24.5	1	15.48		0.59		0.46		2		46
305174	−60 01905				10 24 52	−60 39.1	1	10.45		−0.04		−0.29		2	A0	761
		G 43 - 50			10 24 53	+05 59.7	3	13.03	.013	0.67	.012	−0.04	.015	5	sdG	1620,1658,3029
90676	−53 03808				10 24 53	−53 25.1	2	9.24	.050	0.49	.005	0.00		6	F6/7 V	1594,6006
90677	−54 03651	HR 4107			10 24 53	−54 37.3	4	5.58	.006	1.56	.000			18	K3 II/III	15,1075,2013,2029
		POSS 431 # 5			10 24 54	+17 31.8	1	17.99		1.44				2		1739
		IC 2581 - 14			10 24 54	−57 20.5	2	11.24	.028	1.11	.005	1.11		4		46,2041
302892	−58 02208				10 24 54	−58 34.3	1	10.23		−0.02		−0.26		1	B9	768
90572	+4 02333	IDS10223N0404		AB	10 24 55	+03 49.2	1	7.23		0.97		0.69		3	K0	1355
302864	−57 03229	LSS 1569			10 24 55	−58 02.8	1	9.77		0.34		−0.49		2		730
302895	−58 02210				10 24 55	−58 37.9	1	10.23		0.03		0.00		1	A0	768
		G 43 - 51			10 24 56	+13 14.2	1	10.69		1.20		1.18		1	K4	333,1620
		IC 2581 - 121			10 24 56	−57 23.2	1	13.98		0.35		0.07		2		46
302861	−57 03231	IC 2581 - 74			10 24 56	−57 53.7	2	12.34	.009	0.04	.023	−0.05		4	B9	46,2041
	+41 02094				10 24 57	+40 32.1	1	9.50		1.00		0.71		2	G5	1375
		GD 119			10 24 58	+33 25.7	1	16.16		0.51		−0.16		2		3060
	−24 09072	CZ Hya			10 24 58	−25 17.5	1	8.88		3.56				2	Ne	864
		IC 2581 - 120			10 24 58	−57 23.8	1	14.91		1.60		1.71		2		46
90508	+49 01961	HR 4098		★AB	10 24 59	+49 03.2	6	6.44	.014	0.60	.007	0.05	.015	22	G1 V	15,22,1003,1008,1197,3071
		IC 2581 - 131			10 24 59	−57 28.4	1	11.41		1.49		1.49		3		46
90537	+37 02080	HR 4100		★AB	10 25 00	+36 57.8	5	4.21	.003	0.91	.005	0.64	.010	11	G9 IIIab	15,1080,1363,3016,8015
	−54 03656				10 25 00	−54 48.9	1	11.00		0.45		0.01		2		1730
	−55 03444				10 25 00	−56 10.2	1	11.21		0.52		0.07		2	F0	807
		IC 2581 - 67			10 25 00	−57 17.6	2	12.92	.009	0.23	.009	−0.25		4		46,2041
90569	+10 02152	HR 4101, CX Leo		★A	10 25 01	+10 01.1	3	6.03		−0.06	.002	−0.10	.004	9	A0 p Si(Cr)	1022,1202,1263
	+28 01885				10 25 01	+27 59.5	1	9.77		1.02		0.78		3	G5 III	186
		LF13 # 112			10 25 01	−51 45.2	1	11.57		1.15		0.90		1	K0 III	137
		IC 2581 - 118			10 25 01	−57 23.5	1	15.35		0.77		0.34		2		46
		CS 15621 # 64			10 25 02	+23 31.0	1	13.55		0.53		−0.16		1		1744
90594	−0 02344				10 25 02	−00 42.5	1	7.49		1.10		0.99		5	K0	1657
90706	−56 03343	IC 2581 - 2			10 25 04	−57 21.1	10	7.07	.013	0.47	.014	−0.42	.020	38	B2.5Ia	6,46,127,285,362,657*
		IC 2581 - 13			10 25 05	−57 19.1	2	12.29	.005	0.17	.005	0.10		4		46,2041
90622	−22 02923				10 25 06	−22 56.7	1	9.00		0.62		0.09		1	G5 V	78
	−55 03445				10 25 06	−56 11.2	1	11.10		0.36		0.05		2	F0	807
		IC 2581 - 68			10 25 06	−57 18.5	2	12.92	.056	0.86	.000	0.43		4		46,2041
90707	−57 03237	IC 2581 - 3		★V	10 25 07	−57 25.2	5	8.70	.039	0.28	.019	−0.61	.035	16	B1 III:	127,540,730,976,2041
90567	+27 01895				10 25 08	+27 28.0	3	8.70	.018	0.49	.012	0.01	.000	6	F8 V	20,186,3026
		HA 54 # 187			10 25 09	+29 47.8	1	12.77		0.53		0.06		2		97
90705	−50 05156				10 25 09	−51 16.9	1	8.37		1.22		1.29		1	K2 III	137
	+28 01886				10 25 10	+27 56.5	1	9.48		0.63		0.06		3	F5	186
		CS 15621 # 7			10 25 10	+28 03.4	1	12.14		0.72		0.25		1		1744
90089	+83 00297	HR 4084			10 25 10	+82 48.9	3	5.26	.005	0.37	.000	−0.05	.005	7	F2 V	15,1008,3037
		IC 2581 - 129			10 25 10	−57 26.5	1	13.06		1.18		0.93		2		46
		CS 15621 # 65			10 25 11	+26 31.7	1	14.17		0.35		−0.10		2		1744
305245	−60 01909	IDS10234S6009		ABC	10 25 11	−60 24.6	1	9.69		0.54		−0.17		2	B5	761
90704	−50 05157	IDS10232S5009		AB	10 25 12	−50 23.9	1	9.43		0.13		0.08		1	A1 IV	137
		IC 2581 - 130			10 25 12	−57 29.7	1	13.93		1.43		1.16		2		46
		CS 15621 # 9			10 25 13	+27 48.0	1	13.93		0.28		0.02		1		1744
		HA 54 # 188			10 25 13	+29 49.4	1	12.66		0.65		0.15		2		97
	−56 03347	LSS 1572			10 25 13	−57 14.5	1	10.55		0.19		−0.37		2		730

HD	DM	Other Id	N	Rem	α_{1950}	δ_{1950}	S	V	σ_V	B–V	σ_{B-V}	U–B	σ_{U-B}	n	Spectrum	References
		POSS 431 # 1			10 25 14	+17 30.8	1	14.37		1.12				1		1739
90727	−53 03816				10 25 15	−54 19.7	1	10.33		0.04		−0.07		2	B9/A0 V	137
	−57 03241	IC 2581 - 73			10 25 15	−57 22.9	2	11.20	.033	1.06	.019	0.89		4		46,2041
		IC 2581 - 12			10 25 16	−57 19.0	1	10.94		1.02				1		2041
		IC 2581 - 111			10 25 16	−57 21.8	1	14.45		1.62		1.23		2		46
		IC 2581 - 59			10 25 18	−57 23.8	1	13.56		0.37		0.14		2		46
305238	−59 02138				10 25 18	−60 19.3	1	11.23		0.18		0.15		2		761
		CS 15621 # 10			10 25 19	+26 48.2	1	13.77		0.64		0.18		1		1744
	+29 02049				10 25 19	+29 28.8	2	10.18	.010	1.32	.005	1.55	.025	4	G8 III	97,186
		G 146 - 35			10 25 19	+48 29.6	1	13.25		1.67				5		1663
		IC 2581 - 109			10 25 20	−57 20.3	1	14.05		0.41		0.23		2		46
		IC 2581 - 57			10 25 20	−57 26.1	1	12.16		0.11				1		2041
302889	−58 02214	LSS 1575			10 25 20	−58 32.0	1	10.26		0.18		−0.62		1	B5	768
		IC 2581 - 61			10 25 21	−57 22.0	1	12.78		0.36				1		2041
		Lod 172 # 6			10 25 21	−60 15.6	1	11.56		0.18		0.04		2		807
307693	−61 01657				10 25 21	−61 45.0	1	8.95		0.77				4	G5	2033
		G 236 - 30			10 25 22	+63 15.2	1	12.85		0.63		−0.11		2		1658
		IC 2581 - 36			10 25 22	−57 19.3	2	12.75	.093	0.23	.010	−0.31		3		46,2041
		Ton 517			10 25 24	+29 35.	1	14.17		−0.03		−0.04		1		1036
		IC 2581 - 60			10 25 24	−57 22.9	1	12.74		0.24				1		2041
		IC 2581 - 117			10 25 24	−57 24.4	1	14.33		0.45		0.06		2		46
300777	−56 03354	LSS 1576		⋆ AB	10 25 25	−56 51.0	1	9.20		0.33		−0.54		1	B5	730
	+29 02050				10 25 26	+28 38.6	1	9.33		0.45		0.03		3	F5	186
		IC 2581 - 103			10 25 26	−57 27.1	1	14.18		0.89		0.45		3		46
		Lod 172 # 15			10 25 26	−60 25.7	1	12.42		0.11		0.08		2		807
300764	−55 03460				10 25 27	−56 11.1	1	10.46		0.15		0.11		2	A2	807
		IC 2581 - 102			10 25 27	−57 26.0	1	15.18		0.87		0.31		2		46
		IC 2581 - 128			10 25 27	−57 27.6	1	14.85		2.21		1.62		2		46
90774	−61 01659				10 25 27	−61 45.7	1	7.56		0.56				4	F8 V	2033
		LF13 # 116			10 25 28	−50 33.9	1	11.95		1.40		1.09		1		137
		IC 2581 - 33			10 25 28	−57 17.5	2	13.64	.112	0.11	.131	0.03		3		46,2041
		IC 2581 - 37			10 25 28	−57 19.3	1	13.57		0.29		−0.14		2		46
	−55 03461				10 25 31	−56 15.7	1	10.82		0.44		−0.03		2	F0	807
		IC 2581 - 42			10 25 31	−57 20.9	1	13.22		0.23		−0.34		3		46
	−57 03254	IC 2581 - 8			10 25 31	−57 24.4	2	10.64	.004	0.23	.026	−0.57		6	B1 V	46,2041
90772	−57 03256	IC 2581 - 1		⋆ V	10 25 32	−57 23.0	13	4.68	.024	0.50	.023	0.10	.067	36	A9 Ia	15,127,540,730,976*
305243	−60 01910				10 25 32	−60 23.6	2	10.04	.025	0.03	.005	0.00	.015	4	A2	761,807
90663	+7 02308	G 43 - 52			10 25 33	+06 59.2	1	8.53		0.92		0.70		1	K0	333,1620
90702	−20 03182				10 25 33	−21 26.8	1	8.50		0.69		0.28		1	G2 V	78
90740	−43 06280				10 25 33	−44 05.2	1	7.11		0.90				4	G8 III	2012
		IC 2581 - 116			10 25 33	−57 25.2	1	15.25		0.52		0.19		2		46
		IC 2581 - 105			10 25 33	−57 25.8	1	13.67		0.97		0.80		3		46
		IC 2581 - 101			10 25 34	−57 27.6	1	13.09		0.72		0.20		4		46
		IC 2581 - 17			10 25 35	−57 17.0	1	13.09		0.28				1		2041
90602	+45 01832	HR 4103			10 25 36	+45 28.1	1	6.35		1.32		1.40		2	K0	252
302883	−58 02220				10 25 36	−58 24.6	1	10.80		0.08		0.05		1	A0	768
		IC 2581 - 100			10 25 37	−57 26.6	1	13.39		0.81		0.27		3		46
90786	−59 02144	LSS 1581			10 25 37	−59 41.4	1	8.80		−0.06		−0.75		5	B2 III	1732
		Lod 172 # 7			10 25 38	−60 19.6	1	11.62		0.51		0.12		2		807
	−51 04777	LF13 # 117			10 25 39	−51 43.5	1	11.08		1.34		1.18		1	K2 III	137
		HA 54 # 190			10 25 40	+29 50.3	1	13.87		0.67		0.19		2		97
		IC 2581 - 119			10 25 40	−57 29.8	1	13.44		0.53		0.02		2		46
90785	−53 03834				10 25 41	−53 38.9	1	8.01		1.08		0.92		2	K0 III	137
	−56 03363	IC 2581 - 10			10 25 41	−57 20.6	1	11.28		0.03				1		2041
302842	−57 03261	IC 2581 - 7			10 25 41	−57 24.1	2	9.89	.170	0.22	.044	−0.69		3	B1 V:e	1737,2041
90802	−60 01911				10 25 41	−60 27.0	1	8.94		−0.08		0.00		2	B5 III	761
		IC 2581 - 114			10 25 42	−57 16.8	1	14.91		0.69		0.13		2		46
		IC 2581 - 115			10 25 42	−57 29.5	1	14.22		0.68		0.09		2		46
90711	−5 03071	G 55 - 22			10 25 43	−06 20.5	4	7.88	.017	0.80	.008	0.46	.020	6	K0 V	1003,1620,1658,3008
302841	−57 03263	IC 2581 - 78		⋆	10 25 43	−57 22.8	2	9.76	.100	0.17	.010	−0.64		2	B0.5V	46,540,2041
		IC 2581 - 25			10 25 43	−57 26.4	2	12.27	.009	0.22	.009	−0.41		4		46,2041
	−56 03365	IC 2581 - 29			10 25 44	−57 20.3	1	11.55		−0.04				1		2041
90684	+15 02205				10 25 45	+15 00.6	2	7.55	.000	1.04	.015	0.86		6	K0	882,1648
	−57 03265	IC 2581 - 79			10 25 45	−57 23.1	4	9.73	.149	0.16	.054	−0.71	.022	6	B0.5Vn	46,976,1737,2041
		IC 2581 - 80			10 25 45	−57 23.2	2	10.15	.170	0.22	.070	−0.67		2	B0.5V	46,2041
90801	−57 03266	LSS 1584			10 25 45	−58 08.1	2	9.48	.015	0.21	.000	−0.59	.000	7	B2 III	6,730
90683	+16 02123				10 25 46	+16 00.7	1	7.32		0.41		−0.01		2	F5 III	1648
	+30 02017				10 25 46	+30 15.7	1	10.90		1.14		1.14		2	K0 III	97
		GD 120			10 25 46	+36 20.6	1	14.05		0.45		−0.13		1		3060
302896	−58 02222				10 25 46	−58 38.8	1	10.96		0.03		−0.33		1	A0	768
90967	−74 00702				10 25 46	−74 53.7	1	7.82		0.30		0.13		4	Fm δ Del	1311
	+29 02051				10 25 47	+29 15.3	1	11.29		0.63		0.16		2	G0	97
		IC 2581 - 26			10 25 47	−57 25.7	2	13.18	.145	0.30	.103	−0.18		4		46,2041
90800	−56 03368				10 25 48	−57 11.7	1	9.22		0.54		0.12		2	F7 IV/V	127
		POSS 431 # 3			10 25 50	+17 44.8	1	16.38		1.77				1		1739
	−56 03370	IC 2581 - 11			10 25 50	−57 18.0	2	11.87	.117	0.12	.089	−0.44		4		46,2041
	+29 02052				10 25 51	+28 40.1	1	10.25		0.97		0.72		2		186
	+30 02018				10 25 51	+29 45.6	2	10.19	.010	1.03	.000	0.72	.069	5	G5 III	97,186
	−56 03371	IC 2581 - 18			10 25 51	−57 16.9	1	11.61		0.29				1		2041
90874	−65 01354	HR 4115			10 25 51	−65 27.0	3	6.01	.004	0.09	.000	0.10	.005	12	A2 V	15,1075,2038

Table 1   549

HD	DM	Other Id	N Rem	α_{1950}	δ_{1950}	S	V	σ_V	B–V	σ_{B-V}	U–B	σ_{U-B}	n	Spectrum	References
		G 43 - 53		10 25 52	+06 04.4	3	11.49	.017	1.42	.010	1.23	.005	10	sdM	940,1620,3029
305237	−59 02147			10 25 52	−60 15.9	1	11.04		0.07		−0.08		2		807
305242	−60 01914			10 25 53	−60 24.4	2	10.17	.005	0.22	.005	0.14	.000	4	A0	761,807
90836	−59 02148			10 25 54	−60 16.5	2	8.95	.015	−0.02	.000	−0.17	.015	4	B9 V	761,807
		HA 54 # 284		10 25 55	+30 14.2	1	11.62		1.02		0.79		2		97
	−23 09305			10 25 56	−24 00.6	1	9.69		1.17		0.90		2	K5 V	3072
		IC 2581 - 22		10 25 57	−57 21.7	1	11.95		0.62				1		2041
		IC 2581 - 107		10 25 57	−57 22.6	1	13.63		0.32		−0.18		2		46
		Lod 172 # 11		10 25 57	−60 18.5	1	11.98		0.27		0.16		2		807
	−53 03839			10 25 58	−53 25.3	1	10.92		1.12		0.91		1	K0 III	137
	−56 03377	IC 2581 - 20		10 25 58	−57 18.4	2	11.74	.040	0.22	.020	−0.38		2		46,2041
	+30 02020			10 25 59	+30 21.9	2	10.10	.000	1.14	.009	1.15	.049	11	K0 III	97,272
90798	−48 05655	HR 4111		10 25 59	−49 08.9	6	6.11	.013	1.51	.006	1.68	.038	21	K4 III	15,78,1075,2013,2029,3005
90831	−56 03376	LSS 1587		10 25 59	−57 04.8	1	9.40		0.35		−0.54		2	B2 III	6
		HA 54 # 11		10 26 00	+29 13.5	1	12.41		0.57		0.03		3		97
		IC 2581 - 21		10 26 00	−57 18.4	2	12.81	.200	0.14	.161	0.17		3		46,2041
	−56 03378	IC 2581 - 6		10 26 00	−57 20.0	2	11.01	.058	0.20	.010	−0.58		3		46,2041
		Lod 172 # 8		10 26 00	−60 13.4	1	11.72		0.29		0.16		2		807
		Lod 172 # 13		10 26 00	−60 15.7	1	12.33		0.54		0.09		2		807
		Lod 172 # 14		10 26 00	−60 15.7	1	12.35		0.13		0.00		2		807
90833	−56 03379	IC 2581 - 5		10 26 01	−57 17.7	3	9.79	.029	−0.03	.007	−0.22	.005	5	B9 V	46,127,2041
		IC 2581 - 108		10 26 01	−57 22.8	1	14.09		0.34		0.00		2		46
		Lod 172 # 16		10 26 01	−60 19.7	1	12.88		0.00		−0.52		2		807
90832	−56 03380			10 26 02	−57 08.3	3	9.12	.012	0.31	.000	−0.57	.013	5	B1 III	6,127,730
302840	−56 03382	IC 2581 - 4		10 26 02	−57 21.4	3	9.69	.051	0.17	.019	−0.57	.050	5	B0.5V	46,127,2041
90853	−58 02227	HR 4114		10 26 02	−58 29.0	6	3.82	.006	0.31	.007	0.25	.024	19	F0 Ib	15,688,1075,2012,3026,8100
305256	−60 01918			10 26 02	−60 34.1	1	10.49		0.04		−0.07		2	A0	761
	−56 03383	IC 2581 - 16		10 26 03	−57 22.1	2	10.90	.014	1.14	.009	0.84		4		46,2041
305241	−59 02150			10 26 03	−60 21.5	1	11.26		−0.05		−0.32		1	A7	761
		IC 2581 - 132		10 26 04	−57 29.4	1	13.07		0.55		0.06		2		46
	−55 03471			10 26 05	−56 11.7	1	11.43		0.23		0.17		2	A6	807
	−57 03277	IC 2581 - 9		10 26 05	−57 26.5	1	11.33		0.08				1		2041
		Lod 172 # 9		10 26 05	−60 20.0	1	11.76		0.30		0.11		2		807
		IC 2581 - 15		10 26 06	−57 19.7	1	12.27		0.19				1		2041
90873	−58 02229	LSS 1593		10 26 06	−59 05.8	1	8.93		1.31		1.03		1	A3 Iab/b	8100
		IC 2581 - 125		10 26 07	−57 25.8	1	13.45		0.19		0.05		2		46
90852	−56 03384			10 26 08	−56 31.0	1	9.93		−0.05		−0.26		2	B8 IV	807
		LP 213 - 12	A	10 26 10	+42 24.3	1	11.01		0.64		0.06		4		1723
90872	−56 03386			10 26 10	−56 56.5	1	6.85		−0.15		−0.53		2	B8 III	127
302897	−58 02230	LSS 1595		10 26 10	−58 42.5	4	10.09	.014	0.48	.011	−0.54	.026	6	O9 Ib	191,540,768,1737
	+29 02053			10 26 11	+29 13.1	1	11.58		0.73		0.28		3	G0	97
		IC 2581 - 127		10 26 11	−57 23.3	1	13.97		0.48		−0.03		4		46
		CS 15621 # 6		10 26 12	+23 17.9	1	13.36		0.46		−0.21		1		1744
90763	−3 02929	HR 4109		10 26 12	−03 29.2	2	6.04	.005	0.05	.000	0.04	.000	7	A1 p(Sr)	15,1417
		Lod 172 # 10		10 26 12	−60 18.4	1	11.80		1.03		0.80		2		807
		SOS 1365		10 26 12	−60 39.	1	11.42		0.12		−0.24		1	B7	761
90948	−66 01278			10 26 12	−67 06.8	2	8.36	.019	0.55	.007	0.00		3	F8 V	1594,6006
90717	+30 02021			10 26 13	+29 59.0	2	6.61	.005	1.18	.010	1.19	.025	6	K0 III	97,186
302882	−58 02231			10 26 13	−58 24.5	1	11.29		0.03		0.00		1		768
	−60 01919	SOS 1366		10 26 13	−60 29.6	1	10.97		0.05		−0.50		2	B1	761
90851	−51 04782			10 26 14	−51 48.1	1	9.66		0.09		0.03		2	A0 IV	137
90871	−55 03474			10 26 14	−56 17.9	1	9.35		0.01		−0.05		2	B9 V	807
		CS 15621 # 57		10 26 15	+23 37.1	1	13.16		0.41		−0.09		1		1744
		MtW 54 # 4		10 26 15	+29 39.3	1	15.92		0.77		0.02		1		97
	−57 03283			10 26 15	−58 22.1	1	10.52		0.16		0.13		1	A2	768
		Lod 172 # 17		10 26 15	−60 20.0	1	13.04		0.36		0.10		2		807
		HA 54 # 150		10 26 17	+29 43.8	1	15.66		0.60		0.05		2		97
300765	−55 03475			10 26 17	−56 21.4	1	10.05		0.06		0.09		2	A0	807
		HA 54 # 193		10 26 18	+29 45.4	1	14.13		0.66		0.16		2		97
		HA 54 # 240		10 26 18	+29 55.3	1	13.11		0.76		0.33		2		97
305236	−59 02152			10 26 18	−60 13.4	1	11.08		0.14		0.13		2		807
	−57 03285			10 26 19	−58 22.8	1	11.05		0.04		−0.09		1	B9	768
305273	−60 01920			10 26 19	−61 15.6	1	11.02		0.08		−0.09		2	A2	761
90812	−26 07942			10 26 21	−27 06.4	2	8.86	.009	0.82	.005	0.34		5	K1 V	2012,3072
		HA 54 # 151		10 26 22	+29 35.1	1	15.05		0.62		0.03		2		97
	−50 05175	LF13 # 121		10 26 22	−51 17.7	1	10.98		1.76		1.73		2		137
	+1 02447	G 55 - 24		10 26 23	+01 06.5	10	9.66	.007	1.51	.008	1.22	.013	67	dM2	694,830,989,1006,1494,1729*
90633	+66 00671	HR 4106		10 26 23	+65 53.0	1	6.32		1.14				2	gK2	71
302839	−57 03288	IC 2581 - 133		10 26 23	−57 26.3	3	9.73	.036	0.15	.012	−0.58	.033	5	B5	46,127,401
90901	−57 03287			10 26 23	−57 49.9	1	9.98		−0.02		−0.57		2	B4/5 V	401
		CS 15621 # 2		10 26 24	+26 14.3	1	11.84		0.14		0.11		1		1744
90748	+28 01887			10 26 26	+27 45.3	1	8.68		0.50		−0.03		2	F8	186
		HA 54 # 152		10 26 26	+29 36.5	1	12.41		0.48		0.03		3		97
90912	−58 02234	YZ Car		10 26 27	−59 05.6	2	8.27	.005	0.88	.006	0.68	.005	2	G5	689,1587
		HA 54 # 153		10 26 28	+29 39.1	1	14.43		0.82		0.33		2		97
		HA 54 # 194		10 26 28	+29 45.9	1	15.73		0.70		0.06		2		97
302888	−58 02235			10 26 28	−58 31.5	1	10.68		0.31		0.16		1		768
		HA 54 # 154		10 26 29	+29 41.2	1	16.27		0.86		0.47		1		97
	+29 02054			10 26 31	+29 19.2	1	11.84		0.63		0.11		2		97
90899	−50 05177			10 26 31	−50 49.0	1	8.10		0.02		−0.17		1	B9 III	137

HD	DM	Other Id	N Rem	α_{1950}	δ_{1950}	S	V	σ_V	B–V	σ_{B-V}	U–B	σ_{U-B}	n	Spectrum	References
		Lod 143 # 32		10 26 32	−58 28.9	1	11.04		1.18		0.92		1	K3	768
		HA 54 # 72		10 26 33	+29 17.6	1	12.21		0.93		0.77		2		97
		MtW 54 # 23		10 26 33	+29 48.7	1	16.85		0.94		0.41		1		97
		HA 54 # 242		10 26 33	+29 55.3	1	14.39		0.50		-0.02		2		97
	+28 01888			10 26 34	+28 07.3	1	9.55		1.12		1.07		4	G8 III	186
		MtW 54 # 24		10 26 34	+29 49.5	1	17.01		1.04		0.64		1		97
	−55 03481			10 26 34	−56 21.9	1	11.94		0.23		0.14		2	A5	807
	+30 02022			10 26 36	+29 56.3	3	8.78	.008	1.12	.010	0.92	.021	10	K1 III	97,186,1003
300774	−56 03398	LSS 1598		10 26 36	−56 56.3	1	9.64		0.61		-0.34		2	B1 II-Ib	540
90966	−62 01595			10 26 36	−62 54.5	1	6.45		-0.06		-0.72		2	B2/3 IIIne	401
		HA 54 # 118		10 26 37	+29 28.4	1	12.09		1.19		1.12		2		97
		G 43 - 54		10 26 39	+11 42.9	1	16.48		0.72		0.26		1		3060
	−22 02926			10 26 39	−23 03.8	1	9.51		0.74		0.30		1	G0	78
302838	−57 03301	LSS 1600		10 26 39	−57 26.9	2	9.91	.059	0.28	.010	-0.62	.044	4	B8	540,1737
		G 43 - 55		10 26 40	+15 56.3	1	12.62		1.03		0.78		1		333,1620
		CS 15621 # 5		10 26 41	+23 19.8	1	12.86		0.23		0.11		1		1744
	+62 01132	ZZ UMa		10 26 41	+62 04.1	1	9.83		0.59		0.05		1	F8	1768
90945	−58 02236			10 26 41	−58 35.8	1	7.55		1.25		1.07		1	G6 III	768
	+29 02055			10 26 42	+29 00.1	1	10.77		0.72		-0.07		3	G0	186
	+30 02023			10 26 42	+30 02.1	2	10.10	.005	1.06	.000	0.93	.010	5	G8 III	97,186
299128	−53 03851			10 26 46	−53 33.3	1	10.97		0.24		0.20		2	A2 V	137
302898	−58 02238			10 26 48	−58 40.8	1	10.30		0.13		-0.07		1	A2	768
90944	−51 04790			10 26 50	−52 08.0	1	8.60		1.00		0.78		1	K0 III	137
		CS 15621 # 58		10 26 51	+23 40.0	1	14.39		0.31		-0.04		1		1744
		HA 54 # 197		10 26 51	+29 52.1	2	14.22	.015	0.65	.005	0.06	.020	5		97,397
		HA 54 # 122		10 26 52	+29 34.3	1	13.18		0.95		0.76		2		97
		Lod 89 # 24		10 26 52	−56 25.0	1	12.06		0.42		0.20		2	A6	807
302887	−58 02239			10 26 54	−58 32.0	1	9.84		1.23		1.17		1	K0	768
90989	−58 02240			10 26 54	−58 43.7	1	9.84		0.04		-0.46		1	B3/5 III	768
		SOS 1437		10 26 54	−60 35.	1	11.84		0.02		-0.13		1	A0	761
		LF13 # 125		10 26 55	−51 47.8	1	12.41		0.33		0.07		1	A2 V	137
90980	−56 03413			10 26 55	−56 58.6	1	6.74		1.02		0.90		2	K0 III	127
	−58 02241			10 26 55	−58 45.9	1	9.98		1.33		1.48		1	K8	768
90882	−2 03155	HR 4116		10 26 56	−02 29.0	4	5.21	.013	-0.05	.010	-0.12	.005	9	B9.5V	15,1079,1417,3023
	−56 03415			10 26 56	−56 40.1	1	9.36		0.13		0.10		2	A1	807
	+30 02024			10 26 57	+29 46.1	3	9.37	.011	0.47	.013	-0.04	.007	18	F8 V	97,186,397
		HA 54 # 199		10 26 57	+29 48.8	1	13.78		0.68		0.15		2		97
		HA 54 # 198		10 26 57	+29 51.8	1	14.14		0.66		0.15		2		97
90979	−56 03414			10 26 57	−56 29.0	1	9.39		0.08		0.04		2	A0 V	807
90987	−57 03308	LSS 1601		10 26 57	−57 30.6	2	9.64	.035	0.12	.015	-0.69	.015	4	B1/2III/IVp	401,730
305255	−60 01924			10 26 57	−60 37.1	1	11.18		0.17		0.20		1	A0	761
90841	+29 02056	IDS10243N2906	B	10 26 58	+28 48.7	1	8.10		1.25		1.25		4	K2 III	186
90745	+64 00789	HR 4108		10 26 59	+64 30.9	2	6.11	.015	0.16	.000	0.12	.000	5	A7 III	985,1733
		HA 54 # 201		10 27 00	+29 45.0	2	9.98	.009	1.16	.005	1.07	.055	13		97,186
		HA 54 # 202		10 27 00	+29 50.0	2	13.58	.003	0.68	.010	0.15	.024	7		97,397
		Lod 89 # 26		10 27 00	−56 25.6	1	12.40		0.24		0.14		2	A6	807
91024	−60 01925	LSS 1602		10 27 02	−60 38.3	3	7.63	.028	0.26	.007	-0.33	.051	6	B7 Iab/b	191,761,8100
90862	+17 02233			10 27 03	+17 14.2	1	8.71		1.47				3	K2	882
		HA 54 # 203		10 27 03	+29 47.2	2	14.84	.001	1.05	.005	0.95	.067	7		97,397
91026	−61 01676	IDS10254S6204	B	10 27 03	−62 19.9	1	8.78		-0.04		-0.45		1	B5/7	401
	+28 01889			10 27 04	+27 59.7	1	10.02		0.63		0.12		3	F8	186
	−56 03418			10 27 04	−56 42.2	1	10.87		1.04		0.70		2	K2	807
91027	−61 01677	IDS10254S6204	A	10 27 05	−62 19.7	1	8.34		-0.03		-0.45		2	B5/7 V	401
90861	+29 02057	IDS10243N2906	A	10 27 06	+28 50.3	2	6.90	.023	1.13	.014	1.04		5	K2 III	20,186
90806	+51 01604	IDS10240N5130	A	10 27 06	+51 15.0	1	7.56		0.21		0.11		4	A7 V	1501
		MtW 54 # 67		10 27 07	+29 51.4	2	16.04	.005	1.02	.010	0.70	.131	6		97,397
91023	−58 02243	IDS10253S5817	A	10 27 08	−58 32.3	1	9.69		0.91		0.39		1	G8 V	287
91023	−58 02243	IDS10253S5817	AB	10 27 08	−58 32.3	2	9.40	.000	0.75	.000	0.40	.000	2	G8 V	287,768
91023	−58 02243	IDS10253S5817	B	10 27 08	−58 32.3	2	9.65	.000	0.83	.000	0.41	.000	2		287,768
	+30 02025			10 27 09	+30 21.2	1	10.35		0.50		-0.06		2		97
		HA 54 # 204		10 27 10	+29 46.8	1	15.11		0.81		0.32		2		97
90807	+51 01605			10 27 10	+50 49.6	1	6.77		0.35		0.06		2	F2	985
90957	−29 08381	HR 4117		10 27 10	−29 24.4	2	5.59	.010	1.42	.000	1.72		6	K3 III	2032,3005
91002	−55 03497			10 27 10	−56 20.4	1	9.70		0.14		0.13		2	A1 V	807
	−56 03419			10 27 10	−56 42.3	1	12.06		0.55		-0.32		2	OB	807
		CS 15621 # 59		10 27 11	+23 57.8	1	13.24		0.22		0.11		1		1744
90878	+28 01890			10 27 11	+28 05.0	1	7.80		0.48		-0.03		3	F8	186
		MtW 54 # 73		10 27 11	+29 50.5	1	17.64		0.34		-0.06		1		97
90840	+39 02357	HR 4113		10 27 12	+39 10.9	2	5.78		0.09	.015	0.15	.010	6	A4 V	105,1022
	−56 03422			10 27 12	−56 41.0	1	10.19		0.12		0.15		2	A3	807
91041	−59 02160	Cr 223 - 4		10 27 12	−59 50.6	2	8.12	.022	0.00	.064	-0.42	.008	5	A0	1732,1734
		HA 54 # 76		10 27 13	+29 18.1	1	11.89		0.42		-0.04		2		97
305253	−60 01931			10 27 13	−60 33.3	1	10.22		0.08		-0.34		1	B8	761
91055	−60 01930			10 27 13	−60 43.1	1	9.71		-0.01		-0.09		1	A0 V	761
91056	−63 01440	HR 4120		10 27 13	−63 55.0	4	5.28	.012	1.86	.000	2.10		13	M0 III	15,2013,2028,3005,8100
237903	+56 01458	IDS10242N5629	A	10 27 14	+56 15.4	3	8.69	.009	1.35	.016	1.25	.015	12	K5	679,694,3072
	−56 03423			10 27 14	−56 30.7	1	11.30		0.35		-0.22		2	B7	807
		HA 54 # 125		10 27 16	+29 31.4	1	12.51		0.93		0.73		2		97
		HA 54 # 125		10 27 16	+29 31.4	1	13.44		0.78		0.27		2		97
		HA 54 # 155		10 27 16	+29 36.5	1	12.81		0.79		0.35		2		97

Table 1

HD	DM	Other Id	N Rem	α_{1950}	δ_{1950}	S	V	σ_V	B–V	σ_{B-V}	U–B	σ_{U-B}	n	Spectrum	References
		HA 54 # 156		10 27 17	+29 36.1	1	12.73		0.81		0.34		2		97
		HA 54 # 249		10 27 17	+29 54.9	1	12.19		0.62		0.04		3		97
90972	−29 08383	HR 4118	⋆ AB	10 27 17	−30 21.0	3	5.55	.004	−0.04	.000	−0.18		9	B9.5V	15,404,2012
91054	−58 02245	LSS 1603		10 27 17	−59 21.3	1	7.67		0.63		0.39		2	A2 II	191,8100
		POSS 431 # 2		10 27 18	+17 45.1	1	15.04		1.46						1739
91039	−56 03425	UX Car		10 27 18	−57 21.4	3	8.16	.361	0.57	.130	0.42	.044	4	F2 II	127,657,1488
	−57 03323			10 27 19	−57 26.9	1	11.18		0.01		−0.19		2		127
		HA 54 # 126		10 27 22	+29 29.5	1	11.27		1.00		0.77		4		97
		Cr 223 - 5		10 27 22	−59 47.9	1	11.55		0.06		−0.49		2		1734
		HA 54 # 157		10 27 24	+29 35.9	1	13.77		0.49		−0.06		2		97
91051	−55 03502			10 27 24	−56 16.1	1	9.16		−0.02		−0.09		2	B9 IV	807
		Cr 223 - 39		10 27 25	−59 50.8	1	13.20		0.11		−0.20		1		1734
91094	−64 01310			10 27 25	−64 55.2	1	6.41		1.67				4	M1 III	2012
90839	+56 01459	HR 4112	⋆ A	10 27 26	+56 14.3	13	4.83	.007	0.52	.004	−0.01	.005	246	F8 V	1,15,130,1028,1077,1118*
90839	+56 01459	IDS10243N5589	B	10 27 26	+56 14.3	2	8.69	.000	1.36	.000	1.26	.000	4		1,1028
		Lod 89 # 21		10 27 26	−56 31.7	1	11.76		0.39		−0.20		2	B7	807
91052	−56 03428			10 27 26	−56 33.1	1	8.86		0.00		−0.02		2	A0 V	807
	+27 01898			10 27 27	+26 37.3	1	11.06		1.00		0.82		1	K5	1744
		Cr 223 - 6		10 27 27	−59 48.2	1	13.05		0.33		0.27		1		1734
305214	−59 02165	Cr 223 - 7	AB	10 27 28	−59 42.3	1	10.64		0.25		0.23		1	A2	1734
91049	−52 03611	IDS10255S5215	AB	10 27 29	−52 30.6	1	8.36		1.10		0.99		1	K1 III	137
		CS 15621 # 1		10 27 30	+26 16.1	1	12.36		0.66		0.21		1		1744
302835	−56 03429	LSS 1604		10 27 30	−57 21.6	1	9.55		0.44		−0.43		2	B1 V	540
90875	+60 01266	G 196 - 28		10 27 31	+60 00.7	1	8.77		1.15				1	K5	1746
		Cr 223 - 8	AB	10 27 31	−59 43.4	1	12.30		0.56		0.07		1		1734
		CS 15621 # 61		10 27 33	+27 10.5	1	11.85		0.59		0.10		1		1744
90932	+28 01891			10 27 34	+27 35.6	2	8.26	.037	1.21	.009	1.12	.009	8	K1 III	186,1775
	+72 00492			10 27 34	+71 47.3	1	6.97		1.03		0.84		2		1733
		G 118 - 66		10 27 35	+33 05.9	2	12.71	.010	1.43	.015	1.13		2	M6:e:	1746,3062
91071	−55 03506			10 27 36	−56 15.0	1	9.44		−0.02		−0.23		2	B8 III	807
		Cr 223 - 41		10 27 36	−59 49.9	1	12.89		0.15		−0.11		2		1734
		Cr 223 - 3		10 27 36	−59 55.9	1	11.51		0.11		−0.24		1		1734
		CS 15621 # 60		10 27 37	+24 13.0	1	13.31		0.65		0.18		1		1744
		HA 54 # 158		10 27 38	+29 36.3	1	15.09		0.75		0.45		2		97
298124	−50 05191			10 27 38	−51 05.9	1	10.45		0.40		−0.08		1	F2 IV	137
	−54 03736			10 27 38	−54 52.7	1	10.75		0.05		−0.48		2	B9 Iab	540
		CS 15621 # 3		10 27 40	+24 45.9	1	12.76		0.45		−0.08		1		1744
		Cr 223 - 40		10 27 41	−59 50.6	1	12.03		1.15		1.12		2		1734
		LF13 # 129		10 27 42	−51 00.0	1	11.90		0.71		0.15		1	A5 V	137
		LF13 # 128		10 27 42	−54 34.6	1	12.54		0.30		0.22		1		137
90931	+37 02088			10 27 43	+36 31.9	1	6.86		0.32		0.12		4	F2 II	1501
91091	−53 03878			10 27 43	−54 04.6	1	10.19		0.03		−0.10		2	B9 V	137
91092	−56 03433			10 27 43	−57 14.3	1	8.70		0.06		0.02		2	A1 V	127
90994	+0 02663	HR 4119, β Sex		10 27 44	−00 22.8	9	5.07	.018	−0.14	.003	−0.50	.017	40	B6 V	3,15,154,1075,1119*
		LF13 # 131		10 27 45	−53 17.1	1	13.15		0.25		0.18		2	A0 V	137
91113	−59 02169	Cr 223 - 2	⋆ ABC	10 27 45	−59 57.2	1	9.54		0.05		−0.16		1	A0	1734
		POSS 431 # 6		10 27 46	+17 38.0	1	19.12		1.48				5		1739
		HA 54 # 159		10 27 46	+29 36.6	1	15.39		0.88		0.61		2		97
90759	+74 00437			10 27 46	+74 02.2	2	8.28	.029	0.16	.005	0.09	.005	3	A0	1733,1776
305215	−59 02170	Cr 223 - 9		10 27 47	−59 43.6	1	11.21		0.07		0.07		1	A0	1734
	+30 02026			10 27 48	+30 04.4	2	9.99	.005	1.00	.000	0.68	.035	5	G8 III	97,186
		HA 54 # 205		10 27 49	+29 45.0	1	13.03		0.68		0.18		2		97
305234	−59 02171	LSS 1609		10 27 49	−60 04.2	1	10.44		0.01		−0.60		2	B2 V	191
		Cr 223 - 1		10 27 50	−59 57.7	1	12.44		0.06		−0.46		1		1734
		HA 54 # 160		10 27 51	+29 37.5	1	13.58		0.65		0.07		2		97
299153	−52 03617			10 27 51	−52 37.8	1	10.65		0.31		−0.03		2	A5 V	137
91128	−61 01686	LSS 1610		10 27 51	−61 51.2	1	7.54		0.57		0.32		2	A2 Ib	191,8100
	−55 03508	LSS 1608		10 27 53	−56 21.4	2	10.13	.005	0.60	.000	−0.40	.020	5	Be	807,1737
302899	−58 02255			10 27 54	−58 42.8	1	10.83		0.37		0.08		1	A3	768
		Cr 223 - 68		10 27 55	−59 51.2	1	12.63		0.12		−0.25		1		1734
91011	+2 02325			10 27 56	+02 24.4	1	6.98		1.03		0.82		4	K0	1355
		LF13 # 134		10 27 56	−50 28.6	1	13.56		0.15		0.16		1	A0 V	137
	−52 03618			10 27 57	−52 43.2	1	10.22		1.52		1.61		1	K4 III	137
305248	−60 01936			10 28 00	−60 26.2	1	10.71		0.01		−0.16		1	A0	761
		HA 54 # 161		10 28 01	+28 42.7	1	14.71		0.79		0.20		3		97
		Cr 223 - 74		10 28 01	−59 48.6	1	13.31		0.11		−0.10		2		1734
	−20 03194			10 28 02	−20 58.8	1	9.66		1.18		1.09		1	K5 V	3072
	−55 03511			10 28 03	−55 48.3	1	11.27		0.05		−0.02		2	A0	807
91126	−56 03441			10 28 03	−56 25.1	1	9.41		−0.04		−0.21		2	B9 V	807
		Cr 223 - 12		10 28 04	−59 39.2	1	12.82		0.10		−0.15		3		1734
		Cr 223 - 10		10 28 04	−59 39.8	1	12.68		0.82		0.30		3		1734
	+29 02058			10 28 05	+28 53.9	1	9.61		0.06		0.10		3	A0	186
		GD 121		10 28 06	−09 46.8	1	13.52		0.48		−0.18		1		3060
		BPM 20110		10 28 06	−53 17.	1	13.79		0.66		0.08		1		3065
		Cr 223 - 11		10 28 06	−59 39.3	1	12.25		0.06		−0.44		3		1734
		Cr 223 - 42		10 28 06	−59 45.1	1	12.59		1.29		0.66		2		1734
		Cr 223 - 75		10 28 07	−59 47.9	1	9.96		1.54		1.70		3		1734
		Cr 223 - 72		10 28 07	−59 49.7	1	13.55		0.17		0.07		2		1734
90990	+39 02359			10 28 08	+39 16.7	1	7.39		1.14		1.06		3	K2 III	833
	−50 05197			10 28 08	−50 23.8	1	11.20		0.38		0.02		2	F2 V	137

HD	DM	Other Id	N Rem	α_{1950}	δ_{1950}	S	V	σ_V	B–V	σ_{B-V}	U–B	σ_{U-B}	n	Spectrum	References
91156	−59 02192	Cr 223 - 73		10 28 08	−59 49.1	1	9.38		1.09		0.93		3	K0	1734
302885	−58 02257			10 28 09	−58 29.0	1	11.10		0.01		−0.13		1	A2	768
		Cr 223 - 38	AB	10 28 09	−59 55.0	1	10.50		1.01		0.69		1		1734
	+30 02027			10 28 10	+30 09.2	2	10.40	.005	0.81	.010	0.29	.020	5	G5	97,186
91155	−59 02173	Cr 223 - 14		10 28 10	−59 40.1	1	8.60		0.47		−0.03		3	F8	1734
		Cr 223 - 13		10 28 10	−59 40.2	1	12.43		1.05		0.67		1		1734
302886	−58 02259			10 28 11	−58 31.7	1	10.58		0.02		−0.13		1	A0	768
		Cr 223 - 67		10 28 12	−59 51.3	1	13.27		0.46		0.08		1		1734
		Cr 223 - 43		10 28 13	−59 44.8	1	12.70		0.14		0.08		2		1734
90343	+85 00161			10 28 15	+84 39.4	1	7.30		0.82		0.51		3	K0	196
		LF13 # 136		10 28 15	−53 55.4	1	13.22		0.29		0.19		1		137
	+30 02028			10 28 17	+29 42.1	2	9.85	.000	0.62	.010	0.11	.010	6	F8	97,186
		Lod 59 # 2		10 28 17	−53 51.9	1	12.60		0.13		0.04		1		287
		Lod 59 # 3		10 28 17	−53 51.9	1	12.72		0.19		0.18		1		287
		Cr 223 - 69		10 28 17	−59 50.7	1	11.89		0.09		−0.23		3		1734
		Cr 223 - 110		10 28 18	−59 48.6	1	13.37		0.31		0.09		1		1734
	−59 02175	Cr 223 - 66	AB	10 28 18	−59 51.4	1	10.51		0.90		0.08		3		1734
		Lod 59 # 1		10 28 19	−53 51.5	1	12.69		0.48		0.20		1		287
302987	−58 02261			10 28 19	−58 38.9	1	10.12		−0.02		−0.29		1	B9	768
		Lod 143 # 37		10 28 19	−58 43.2	1	11.24		0.27		−0.03		1	A8	768
		Cr 223 - 45	AB	10 28 19	−59 44.4	1	12.81		0.12		−0.17		1		1734
		G 119 - 11		10 28 20	+30 06.0	2	15.61	.014	1.11	.000	0.68		5	DK	538,3062
	+30 02029			10 28 20	+30 20.8	1	10.10		1.13		0.97		2	G8 III	97
		Cr 223 - 44		10 28 20	−59 44.5	1	11.70		0.51		0.03		1		1734
		Cr 223 - 70		10 28 21	−59 50.8	1	12.44		0.04		−0.44		3		1734
		Cr 223 - 15	AB	10 28 22	−59 39.5	1	11.99		0.63		0.21		2		1734
	+22 02231			10 28 24	+21 31.1	1	10.12		1.22		1.19		2	G0	3077
305219	−59 02178	Cr 223 - 71	AB	10 28 24	−59 51.1	1	11.07		0.07		−0.41		3	A2	1734
302989	−58 02262			10 28 25	−58 44.9	1	11.16		0.27		0.14		1		768
		Cr 223 - 49		10 28 26	−59 44.7	1	11.88		0.14		0.06		3		1734
		Cr 223 - 46		10 28 27	−59 43.3	1	13.15		0.15		0.15		2		1734
91218	−61 01694			10 28 27	−62 14.0	1	8.64		0.04		−0.06		2	B9/9.5V	401
91106	−6 03173	HR 4122	★ AB	10 28 28	−07 22.8	2	6.19	.005	1.38	.000	1.18	.000	7	K5 III+F6 V	15,1417
91121	−20 03196			10 28 28	−20 57.8	5	8.73	.022	0.55	.008	−0.09	.016	10	G1wF2/5 V	78,1696,2033,2039,3060
91188	−56 03448	LX Vel		10 28 28	−56 49.2	3	6.62	.037	−0.10	.017	−0.58	.012	8	B3 III	401,540,976
91198	−58 02263			10 28 28	−58 45.8	2	8.88	.000	−0.06	.005	−0.50	.005	3	B8 II	401,768
		Cr 223 - 108		10 28 28	−59 45.9	1	13.31		0.12		−0.03		2		1734
		Cr 223 - 77		10 28 28	−59 48.6	1	13.27		0.21		0.03		2		1734
		Cr 223 - 76		10 28 28	−59 49.2	1	11.58		0.08		−0.39		2		1734
	+46 01635	G 146 - 39		10 28 29	+45 47.7	2	8.88	.009	1.33	.002	1.28	.004	6	K7 V	1003,3072
302988	−58 02264	IDS10266S5826	AB	10 28 29	−58 40.9	1	9.86		0.37		0.07		1	A5	768
		Cr 223 - 107		10 28 29	−59 46.0	1	13.54		0.18		−0.05		2		1734
		Cr 223 - 105		10 28 29	−59 46.4	1	11.77		0.06		−0.33		3		1734
	+57 01274	IDS10253N5737	A	10 28 30	+57 22.2	2	9.73	.030	1.44	.037	1.10		12	M0	6009,7009
		Cr 223 - 50		10 28 30	−59 44.6	1	12.09		0.56		0.11		3		1734
		Cr 223 - 78		10 28 30	−59 49.1	1	11.95		0.50		0.16		3		1734
91029	+47 01782			10 28 31	+47 07.9	1	7.31		1.47		1.81		2	K2	1601
91135	−25 08084	HR 4125		10 28 31	−26 13.6	3	6.50	.004	0.54	.000			11	F7 V	15,2013,2028
	−59 02181	LSS 1614		10 28 31	−59 33.8	1	10.03		0.82		−0.30		2	O9.5 Ib	1737
305217	−59 02180	Cr 223 - 106	AB	10 28 31	−59 46.4	1	11.00		0.08		−0.32		3	F2	1734
	−59 02179	Cr 223 - 79	★ AB	10 28 31	−59 49.9	1	10.64		1.17		1.16		5		1734
91120	−12 03181	HR 4123		10 28 32	−13 19.9	2	5.58		−0.02	.007	−0.16		5	B9 Vne	1079,2006
		Cr 223 - 109	AB	10 28 32	−59 44.8	1	13.57		0.43		0.25		2		1734
		LF13 # 137		10 28 33	−51 57.5	1	12.70		0.63		0.23		1	F5 V	137
302990	−58 02265			10 28 34	−58 43.4	1	11.22		0.19		0.15		1		768
		Cr 223 - 82		10 28 35	−59 48.0	1	12.33		0.11		−0.29		5		1734
305218	−59 02182	Cr 223 - 80		10 28 35	−59 49.3	1	10.36		0.04		−0.79		3	B5	1734
		Cr 223 - 64		10 28 35	−59 52.7	1	12.50		1.83		1.80		1		1734
91272	−66 01291	HR 4129		10 28 35	−66 43.7	3	6.20	.012	0.00	.005	−0.53	.000	16	B4 V	15,1075,2038
305216		Cr 223 - 48	AB	10 28 37	−59 43.4	1	11.28		0.11		−0.05		3	A0	1734
		Cr 223 - 51		10 28 37	−59 44.4	1	13.15		0.14		−0.09		2		1734
		Cr 223 - 85		10 28 38	−59 50.5	1	10.18		1.66		1.88		3		1734
		Cr 223 - 63		10 28 38	−59 53.2	1	12.47		0.60		0.06		1		1734
	−20 03198	IDS10262S2107	ABC	10 28 39	−21 22.6	2	10.10	.085	1.39	.017	1.36	.005	2	K7 V	78,3072
	−20 03198	IDS10262S2107	C	10 28 39	−21 22.6	1	13.13		1.57				1		3072
		Cr 223 - 83		10 28 39	−59 47.3	1	12.70		0.05		−0.30		4		1734
		Cr 223 - 81		10 28 39	−59 49.0	1	12.92		0.21		0.10		2		1734
		Cr 223 - 47		10 28 40	−59 42.0	1	12.79		0.24		0.10		2		1734
		Cr 223 - 84		10 28 40	−59 47.9	1	11.29		0.28		0.23		5		1734
		Cr 223 - 86		10 28 41	−59 50.5	1	12.04		0.10		−0.30		3		1734
		POSS 431 # 4		10 28 42	+17 37.0	1	17.84		1.24				3		1739
		Cr 223 - 52		10 28 42	−59 44.5	1	11.61		1.00		0.71		2		1734
		Cr 223 - 88	AB	10 28 42	−59 49.4	1	13.30		0.19		0.06		2		1734
91509	−80 00462			10 28 42	−80 48.0	1	7.58		1.11		1.01		3	K0 III	1704
		Cr 223 - 87		10 28 43	−59 51.0	1	12.81		0.10		−0.23		3		1734
		Cr 223 - 65		10 28 43	−59 52.3	1	11.30		0.08		−0.36		3		1734
300815				10 28 44	−56 30.0	1	9.85		0.85		0.48		1		549
		Cr 223 - 104		10 28 45	−59 46.2	1	13.71		0.17		0.06		2		1734
91242	−53 03899	IDS10268S5358	A	10 28 46	−54 13.5	1	7.95		1.52		0.88		5	K2/4 III	1673
		Cr 223 - 92		10 28 46	−59 49.4	1	13.63		0.25		0.14		2		1734

Table 1   553

HD	DM	Other Id	N	Rem	α₁₉₅₀	δ₁₉₅₀	S	V	σ_V	B−V	σ_B−V	U−B	σ_U−B	n	Spectrum	References
	+28 01893				10 28 47	+28 17.9	1	9.81		0.41		-0.04		2	F0	186
91269	−60 01944	IDS10271S6051		B	10 28 47	−61 04.8	1	7.54		-0.08		-0.58		2	B3/5 Vne	401
		Cr 223 - 90			10 28 48	−59 50.1	1	13.76		0.20		0.01		2		1734
		Cr 223 - 89			10 28 48	−59 50.3	1	12.89		0.27		0.18		1		1734
		Cr 223 - 103			10 28 49	−59 45.6	1	12.20		1.16		1.00		1		1734
		Cr 223 - 93			10 28 49	−59 47.4	1	11.89		0.12		-0.27		2		1734
91345	−70 01111				10 28 49	−71 18.3	2	9.04	.010	0.56	.002	-0.10		5	G0wF3	1594,6006
		Cr 223 - 53			10 28 50	−59 41.6	1	12.04		1.05		0.75		3		1734
		Cr 223 - 91			10 28 50	−59 49.4	1	13.33		0.21		0.05		2		1734
	−50 05206				10 28 51	−51 17.7	1	11.00		1.00		0.82		2	K0 III	137
		Cr 223 - 54			10 28 51	−59 42.1	1	11.95		0.51		0.06		3		1734
		Cr 223 - 94			10 28 52	−59 50.2	1	12.93		0.25		0.14		1		1734
91270	−60 01945	HR 4128		⋆ A	10 28 52	−61 06.0	1	6.44		1.64		1.84		2	M2 IIIe	3005
91241	−53 03902				10 28 53	−53 45.4	1	8.07		-0.03		-0.04		1	A0 IV	137
		Cr 223 - 55			10 28 53	−59 42.6	1	13.47		0.31		0.09		1		1734
91150	+11 02239				10 28 54	+11 25.5	1	8.28		0.17		0.09		3	A2	1648
		L 249 - 17			10 28 54	−51 03.5	1	12.35		1.43		1.17		1		3062
		Cr 223 - 97			10 28 54	−59 49.6	1	13.20		0.11		-0.07		2		1734
		Cr 223 - 95			10 28 54	−59 50.2	1	13.23		0.23		0.07		1		1734
		Cr 223 - 96			10 28 54	−59 50.5	1	12.56		1.53		1.62		1		1734
	+29 02059				10 28 55	+29 26.6	1	11.75		1.19				1		97
		Cr 223 - 62		AB	10 28 55	−59 53.2	1	12.06		0.43		0.13		2		1734
	+30 02030				10 28 56	+29 49.2	2	10.57	.005	1.07	.005	0.91	.000	5	K0	97,186
91223	−41 05921	IDS10269S4143		B	10 28 56	−41 57.8	1	8.04		0.10		0.06		1	A1 V	401
		Cr 223 - 37			10 28 58	−59 54.2	1	12.98		0.40		0.19		1		1734
91268	−55 03521				10 28 59	−55 27.0	1	9.01		0.11		-0.25		1	B8 V	761
		Cr 223 - 102			10 28 59	−59 44.7	1	13.35		0.22		0.12		1		1734
		Cr 223 - 98			10 28 59	−59 49.2	1	12.58		0.11		-0.18		2		1734
		Cr 223 - 36			10 28 59	−59 54.4	1	13.00		0.37		0.18		1		1734
91148	+24 02238				10 29 00	+24 20.4	2	7.93	.023	0.71	.014	0.22		5	G8 V	20,3026
91130	+33 01999	HR 4124		⋆ A	10 29 01	+32 38.2	2			0.10	.025	0.05	.020	7	A0 IV	1022,1079
		Cr 223 - 99			10 29 01	−59 50.0	1	11.78		0.08		-0.35		1		1734
		Cr 223 - 34			10 29 01	−59 55.3	1	11.82		0.50		0.03		3		1734
		Cr 223 - 101			10 29 02	−59 48.0	1	13.44		0.19		-0.08		1		1734
		LP 550 - 127			10 29 03	+07 56.5	1	14.90		1.46		1.19		1		1696
91239	−41 05923	IDS10269S4143		A	10 29 03	−41 58.3	1	7.37		-0.08		-0.15		2	Ap EuCr(Sr)	401
305295	−59 02186	Cr 223 - 35		⋆	10 29 03	−59 55.1	2	9.88	.015	0.10	.010	-0.54	.015	5	B5	401,1734
305282	−59 02187	Cr 223 - 56		AB	10 29 04	−59 42.6	1	11.12		0.09		-0.25		4	A0	1734
		Cr 223 - 61			10 29 04	−59 51.6	1	12.86		0.13		-0.30		1		1734
		Cr 223 - 33		AB	10 29 04	−59 53.7	1	11.76		0.69		0.14		3		1734
91375	−71 01034	HR 4138			10 29 04	−71 44.1	5	4.73	.008	0.04	.000	0.07	.004	17	A1 V	15,1075,2038,3023,8052
		Cr 223 - 100			10 29 06	−59 49.6	1	12.88		0.47		0.01		1		1734
		Cr 223 - 31			10 29 06	−59 55.1	1	11.89		0.19		-0.16		2		1734
91208	−16 03075				10 29 07	−16 49.2	1	8.02		0.95		0.60		1	Kp Ba	565
		Cr 223 - 58			10 29 07	−59 45.5	1	12.00		0.19		0.14		3		1734
		Cr 223 - 32			10 29 07	−59 54.5	1	11.88		0.21		-0.17		1		1734
91267	−52 03639				10 29 08	−52 59.8	1	9.78		0.95		0.80		4	K1/2 (III)	158
305283		Cr 223 - 57			10 29 08	−59 44.1	1	11.34		0.20		0.16		4	A0	1734
	+29 02060				10 29 10	+28 54.2	1	9.98		0.43		0.00		3	A7	186
91295	−58 02268				10 29 11	−58 27.0	2	8.45	.010	-0.06	.015	-0.42	.015	3	B7 V	401,768
91163	+30 02031				10 29 13	+29 59.4	2	7.86	.005	0.61	.010	0.14	.005	4	G0	97,186
		Cr 223 - 30			10 29 13	−59 52.5	1	12.84		0.20		-0.01		2		1734
		G 54 - 30			10 29 15	+19 47.9	1	15.49		1.65				3		538
91192	+22 02232	IDS10265N2233		A	10 29 15	+22 17.8	1	8.12		0.27		0.05		2	F0	3016
91192	+22 02232	IDS10265N2233		B	10 29 15	+22 17.8	1	8.42		0.31		0.05		2		3032
		LF13 # 140			10 29 15	−52 37.8	1	11.32		1.15		0.87		1	K0 III	137
		GD 122			10 29 16	+32 55.6	1	16.07		0.00		-0.88		1		3060
305294	−59 02190	Cr 223 - 29			10 29 17	−59 53.1	1	10.37		0.16		0.12		3	A0	1734
		Cr 223 - 16			10 29 18	−59 41.6	1	13.24		0.21		0.05		1		1734
		Cr 223 - 18			10 29 18	−59 43.3	1	12.37		0.55		0.04		2		1734
		Cr 223 - 17			10 29 19	−59 42.1	1	12.44		0.62		0.15		2		1734
		Cr 223 - 27			10 29 19	−59 51.0	1	12.40		0.23		0.10		2		1734
		Cr 223 - 28			10 29 19	−59 51.2	1	12.33		1.16		0.91		1		1734
91342	−60 01949	IDS10276S6104		AB	10 29 20	−61 19.1	1	8.25		0.00		-0.30		2	B9 III	401
		Cr 223 - 19			10 29 21	−59 43.6	1	12.86		0.14		-0.19		2		1734
		Cr 223 - 59			10 29 22	−59 46.0	1	12.94		0.25		0.16		2		1734
299188	−53 03907				10 29 23	−54 19.9	1	9.89		1.10		0.99		2	K2 III	137
		Cr 223 - 60			10 29 23	−59 46.1	1	11.29		0.62		0.14		2		1734
91324	−53 03909	HR 4134		⋆ A	10 29 25	−53 27.7	5	4.89	.009	0.50	.000	-0.06		20	F5 V	15,1075,2013,2029,3037
	−51 04818				10 29 26	−52 02.8	1	11.58		0.35		0.28		1	A5 V	137
	−38 06535				10 29 27	−38 25.0	1	11.02		1.36		1.27		1	K7 V	3072
91306	−50 05214				10 29 27	−51 20.6	1	8.74		0.92		0.63		2	G8 III	137
305284		Cr 223 - 20			10 29 27	−59 43.8	1	11.47		0.09		-0.26		1		1734
91280	−27 07503	HR 4130			10 29 29	−27 58.8	3	6.03	.014	0.51	.000	0.01		9	F7 V	15,2027,3053
91181	+44 01995				10 29 30	+44 26.4	1	7.39		0.20		0.08		2	A7 IV	1601
	−55 03527				10 29 30	−56 22.3	1	10.53		0.90		0.60		1	G3	549
		Cr 223 - 22			10 29 31	−59 46.1	1	12.87		0.13		-0.28		2		1734
91232	+14 02255	HR 4127			10 29 32	+14 23.7	3	5.46	.005	1.69	.008	2.04		13	M1.5 IIIb	71,3001,6002
91220	+25 02260				10 29 32	+24 42.0	1	7.20		0.22		0.05		2	F0	985
91323	−43 06335	IDS10274S4358		AB	10 29 35	−44 13.6	2	7.20	.009	-0.14	.009	-0.63		5	B3/5 IV/V	78,2012

HD	DM	Other Id	N	Rem	α_{1950}	δ_{1950}	S	V	σ_V	B–V	σ_{B-V}	U–B	σ_{U-B}	n	Spectrum	References
		Cr 223 - 21			10 29 35	−59 42.7	1	11.82		0.15		0.15		2		1734
		Cr 223 - 25			10 29 36	−59 50.3	1	12.56		0.65		0.16		1		1734
	+30 02032				10 29 37	+30 05.8	2	9.06	.005	1.21	.010	1.21	.034	6	K0 III	97,186
	+28 01897				10 29 38	+28 07.9	1	10.26		0.63		0.08		2		186
298136	−49 03530				10 29 38	−50 17.8	1	10.77		0.18		0.05		2	A0 V	137
		Cr 223 - 23			10 29 38	−59 47.2	1	12.51		0.26		−0.07		2		1734
		Cr 223 - 26			10 29 38	−59 52.9	1	10.79		1.03		0.78		2		1734
		Cr 223 - 24			10 29 40	−59 46.6	1	12.64		0.33		0.15		2		1734
300805	−55 03533				10 29 41	−56 16.9	1	10.54		0.33		0.13		1	A3	549
		L 36 - 61			10 29 42	−77 06.	1	13.08		1.12		0.79		1		3062
91129	+68 00610				10 29 45	+68 09.1	1	8.90		0.97		0.71		3	K0	1733
	+30 02033	IDS10270N3014		AB	10 29 48	+29 58.8	2	10.49	.005	0.60	.005	0.03	.005	6	G0	97,186
91355	−44 06583	HR 4135		★ AB	10 29 49	−44 48.5	2	5.15	.005	-0.16	.005			7	B9	15,2012
299142	−51 04824				10 29 50	−51 44.8	1	9.74		1.12		0.84		1	K0 III	137
305296	−59 02193	LSS 1622			10 29 50	−60 01.7	1	9.32		0.13		−0.81		3	B2 Ve	1737
91496	−72 00981	HR 4142		★ A	10 29 52	−72 57.9	5	4.94	.016	1.68	.010	1.84	.028	21	K4/5 III	15,1075,2031,2038,3005
307781	−62 01622	LSS 1624			10 29 56	−62 59.4	2	9.51	.004	0.02	.009	-0.64	.010	5	B4 II-Ib	540,976
	−51 04826				10 29 58	−51 44.8	1	11.71		0.24		−0.01		1	F0 V	137
		L 465 - 10			10 30 00	−35 22.	3	12.17	.002	0.50	.010	-0.18	.017	13		1696,1698,1765
	+28 01898				10 30 05	+28 05.7	1	10.15		0.46		0.03		3		186
91276	+36 02082	IDS10272N3554		AB	10 30 06	+35 38.4	1	8.86		1.06		0.83		2	G5	1601
		Lod 112 # 24			10 30 08	−56 24.2	1	11.94		0.71		0.19		1	G2	549
	−55 03537				10 30 08	−56 15.9	1	11.61		0.32		0.22		1	A9	549
91421	−57 03387	LSS 1625		A	10 30 09	−57 58.3	3	8.95	.003	0.11	.013	-0.71	.011	5	B0/1 Ib	540,976,1096
91421	−57 03387			B	10 30 09	−57 58.3	1	10.00		0.10		−0.37		1		836
91452	−63 01465	LSS 1628			10 30 09	−63 41.0	4	7.50	.021	0.22	.013	-0.70	.028	8	B0 III	191,540,976,8100
91316	+10 02166	HR 4133, ρ Leo		★ AB	10 30 11	+09 33.9	20	3.85	.014	-0.14	.007	-0.95	.009	115	B1 Ib	3,15,26,30,125,154*
		SOS 1673			10 30 12	−55 12.	1	11.64		0.03		0.04		1	B8	761
91114	+74 00438	IDS10261N7421		A	10 30 13	+74 05.5	1	7.54		0.17		0.08		1	A3	1776
91465	−61 01704	HR 4140, PP Car			10 30 14	−61 25.7	5	3.31	.011	-0.10	.008	-0.71	.004	35	B4 Vne	681,815,1088,1637,2006
91286	+39 02363				10 30 15	+39 28.8	2	7.19	.005	1.03	.010	0.86	.005	5	K0 III	833,1601
	−54 03777				10 30 15	−55 09.4	1	11.36		0.00		−0.52		1	B5	761
300811	−55 03542				10 30 15	−56 22.1	2	9.82	.005	0.60	.000	-0.34	.005	2		287,549
300813	−56 03480	LSS 1627			10 30 17	−56 27.6	2	9.54	.000	0.47	.000	-0.49	.010	2	B0	287,549
		LP 550 - 140			10 30 18	+07 44.8	1	12.51		1.41		1.17		1		1696
91312	+41 02101	HR 4132		★ A	10 30 19	+40 41.0	4	4.75	.014	0.22	.013	0.07	.013	9	A7 IV	15,1363,3023,8015
		Lod 112 # 10			10 30 19	−56 13.1	1	10.86		1.31		1.36		1	K0	549
		G 146 - 41			10 30 21	+47 40.1	1	11.20		1.15		0.71		2	K5	7010
		LF13 # 147			10 30 21	−51 59.2	1	13.65		0.30		0.23		1	A0 V	137
		Lod 112 # 30			10 30 21	−56 23.6	1	12.73		0.47		0.02		1	F5	549
91449	−52 03655				10 30 22	−52 27.1	1	8.89		0.08		0.05		1	A0 V	137
91540	−70 01118				10 30 22	−71 16.7	2	9.38	.035	0.57	.007	0.12		4	F7 V	1594,6006
		Lod 112 # 21			10 30 23	−56 17.8	1	11.70		0.11		0.04		1	B9	549
91313	+38 02152	IDS10275N3811		AB	10 30 24	+37 55.6	1	9.25		0.58		0.00		4	G0	1723
91437	−43 06347	HR 4139			10 30 25	−44 21.6	4	5.90	.007	0.92	.000			18	G6/8 III	15,1075,2013,2029
300812	−56 03483				10 30 25	−56 24.4	2	9.76	.005	0.34	.005	0.16	.005	2	A7	287,549
	+30 02034				10 30 26	+29 52.1	1	10.34		1.28		0.80		2	K3 III	186
		Lod 112 # 27			10 30 28	−56 26.5	1	12.48		0.48		0.04		1	F4	549
91477	−57 03397				10 30 28	−58 17.3	1	7.35		−0.11		−0.59		2	B5 III	401
91103	+77 00404				10 30 29	+76 59.4	1	7.54		1.16		1.11		2	K0	1375
		LF13 # 149			10 30 29	−52 09.1	1	12.75		0.45		0.04		1	A2 V	137
300814	−56 03485	LSS 1630		★ AB	10 30 29	−56 28.9	3	9.27	.004	0.39	.020	-0.47	.004	4	B3	287,549,730
91348	+28 01899				10 30 30	+28 02.1	2	8.62	.010	0.75	.015	0.40		4	G8 III	20,186
300816					10 30 32	−56 35.1	1	10.88		0.14		0.15		1		549
	−40 06102				10 30 35	−41 11.8	1	11.11		0.90		0.57		4		158
91311	+54 01381	HR 4131			10 30 36	+53 45.4	2	6.53	.005	-0.03	.000	-0.03	.000	4	A1 V	1566,1733
		Lod 112 # 20			10 30 37	−56 16.3	1	11.61		0.57		0.04		1	F8	549
		Lod 112 # 22			10 30 38	−56 36.0	1	11.77		0.38		−0.34		1	B6	549
91365	+35 02154	HR 4137			10 30 40	+35 14.8	2	5.58		0.03	.005	0.03	.005	6	A2 Vn	1022,1733
91491	−52 03661				10 30 42	−52 52.9	1	7.78		1.12		0.93		2	K0/1 III	137
91505	−54 03783	UZ Vel			10 30 42	−54 28.4	1	10.17		3.75				1	A8 V	864
		Lod 112 # 25			10 30 42	−56 26.8	1	12.06		0.74		0.24		1	G5	549
	+29 02061				10 30 43	+29 08.0	1	9.43		0.51		0.03		3	G0 V	186
300809	−55 03548				10 30 44	−56 23.0	1	10.55		0.22		0.18		1	A0	549
		Lod 112 # 26			10 30 45	−56 24.0	1	12.10		0.54		0.14		1		549
91523	−55 03549				10 30 46	−55 28.9	1	9.53		0.02		0.02		1	A0 V	761
		Lod 112 # 29			10 30 47	−56 27.3	1	12.63		0.45		0.01		1	F5	549
	−56 03492				10 30 47	−56 29.9	1	10.86		0.46		−0.42		1	B4	549
		Lod 112 # 23			10 30 49	−56 31.4	1	11.82		0.35		−0.14		1	B7	549
233745	+51 01606				10 30 50	+51 00.5	1	9.52		0.42		−0.04		2	F5	1566
91504	−46 06205	HR 4143		★ A	10 30 50	−46 44.7	4	5.02	.026	1.05	.013	0.59		16	K1/2 III	15,1075,2029,3005
	−50 05225				10 30 50	−50 27.7	1	11.17		1.31		1.37		1	K2 III	137
	−55 03550				10 30 50	−56 21.4	1	11.39		0.26		0.18		1	A8	549
91190	+76 00393	HR 4126			10 30 54	+75 58.3	1	4.84		0.96				2	K0 III	1363
	−54 03787				10 30 55	−55 15.9	1	11.09		0.21		−0.42		1	B5	761
91533	−58 02285	HR 4144, LSS 1633			10 30 55	−58 24.5	7	6.00	.019	0.32	.011	-0.08	.034	17	A2 Iab	15,191,401,434,836*
91075	+81 00343	HR 4121			10 30 59	+80 45.2	2	6.49	.030	0.96	.010	0.73	.010	5	G5	985,1733
91530	−50 05230				10 31 00	−51 16.5	1	8.33		1.13		1.08		1	K0 III	137
305301	−59 02206	LSS 1634			10 31 00	−60 11.6	1	10.80		0.23		−0.78		2	B2 Ve	1737
91472	−21 03095				10 31 02	−22 19.6	1	8.30		1.42		1.72		1	K3 III	1746

Table 1   555

HD	DM	Other Id	N Rem	α_{1950}	δ_{1950}	S	V	σ_V	B–V	σ_{B-V}	U–B	σ_{U-B}	n	Spectrum	References
300806	−55 03553			10 31 02	−56 13.1	1	11.58		0.20		0.10		1	F0	549
		CCS 1697		10 31 06	−53 48.0	1	11.37		3.02				1		864
300810				10 31 06	−56 26.1	1	11.06		0.20		0.12		1		549
		LP 730 - 61		10 31 08	−14 13.8	1	11.50		1.08		1.02		1		1696
300793	−54 03791			10 31 08	−55 17.3	1	10.42		0.30		0.05		3	A2	761
300807	−55 03554			10 31 08	−56 13.9	2	11.33	.105	0.32	.085	0.07	.030	2		549,549
	−56 03496			10 31 08	−56 31.3	1	10.66		0.38		-0.48		1	B4	549
91599	−61 01707			10 31 08	−61 38.7	1	9.61		-0.02		-0.52		2	B6 III	401
91558	−53 03934			10 31 09	−53 43.4	1	9.09		0.03		-0.01		1	B9/A1 V	137
300790	−54 03790			10 31 09	−55 07.4	1	10.31		0.03		-0.10		1	B9	761
91598	−60 01967	LSS 1638		10 31 09	−61 16.4	1	9.06		-0.06		-0.84		2	B2/3 IV	401
91569	−55 03555			10 31 11	−55 38.5	1	9.04		0.15		0.10		1	A1 V	761
91557	−52 03671			10 31 12	−53 01.1	1	8.94		0.12		0.07		1	A0/2 IV/V	137
91597	−60 01968	LSS 1639		10 31 12	−60 35.2	4	9.79	.096	0.07	.050	-0.81	.026	8	B7/8 IV/V	191,540,976,1737
91610	−61 01708			10 31 14	−61 43.1	1	9.98		0.03		-0.44		2	B5/7 V	401
		G 162 - 66		10 31 15	−11 26.2	4	13.01	.008	-0.17	.006	-0.99	.007	116	DAn	281,989,1764,3060
	−55 03557		AB	10 31 16	−55 27.7	1	11.08		0.40		0.04		2	B2	761
300791	−54 03795	IDS10294S5452	A	10 31 17	−55 07.7	1	6.68		0.99		0.73		2	G5	401
91538	−39 06488			10 31 18	−39 58.8	1	6.68		0.92				4	G6 III	2012
91595	−57 03424	Y Car		10 31 18	−58 14.4	1	8.20		0.67		0.36		1	F0/3	401
91593	−54 03797	IDS10294S5452	BC	10 31 19	−55 07.3	1	7.33		0.04		0.02		1	A1 V	401
91572	−57 03423	LSS 1640		10 31 19	−57 54.7	5	8.22	.018	0.04	.011	-0.87	.027	10	O7	6,191,540,836,976
91571	−57 03425	IDS10294S5727	AB	10 31 20	−57 42.8	1	9.55		0.00		-0.85		2	B0/3	836
	−55 03559			10 31 25	−56 17.9	1	11.04		0.47		0.00		1	F3	549
	+28 01901			10 31 26	+28 12.5	1	10.67		1.18		1.12		3		186
305330	−60 01974	LSS 1645		10 31 26	−61 10.2	1	10.16		0.36		-0.77		2	B0 Ve	1737
		LP 550 - 152		10 31 27	+05 05.5	1	10.55		0.90		0.68		3		1696
		AAS9,163 T3# 45		10 31 27	+28 11.6	1	10.35		0.46		-0.01		3		186
91619	−57 03431	HR 4147, V369 Car		10 31 32	−57 55.9	10	6.14	.022	0.36	.019	-0.42	.019	24	B6 Iab	6,15,540,836,914,976*
		vdB 39 # a		10 31 36	−59 25.	1	14.48		0.60		-0.14		3		434
91550	−22 02946	HR 4145	★ A	10 31 38	−23 29.2	4	5.07	.004	1.60	.011	1.82		16	K5 III	15,1075,2027,3052
		LP 610 - 36		10 31 39	−02 33.3	1	13.18		0.54		-0.13		2		1696
91585	−28 08270			10 31 39	−29 16.7	1	9.42		0.91				4	K2 V	2012
91651	−59 02214	LSS 1647		10 31 40	−59 52.1	3	8.85	.008	-0.01	.009	-0.90	.017	7	O8/9 V	6,540,976
		SOS 1749		10 31 42	−55 30.	1	11.98		0.29		-0.28		2	B7	761
303008	−58 02296	LSS 1648		10 31 43	−58 57.0	1	10.13		0.43		-0.51		3	B0.5II-III	540
303004	−58 02298	LSS 1650		10 31 46	−58 45.8	1	9.97		0.67		-0.25		3	B1.5 II	434
	−55 03567			10 31 47	−55 30.6	1	11.37		0.90		0.51		3	OB	761
91698	−64 01339	IC 2602 - 1		10 31 49	−65 06.6	2	8.02	.000	0.01	.005	-0.49		6	B6 V	401,2005
		Steph 860		10 31 53	+73 09.6	1	10.96		1.74		1.99		1	M1	1746
		WLS 1040-10 # 6		10 31 53	−10 06.9	1	12.44		0.55		0.02		2		1375
		vdB 38 # a		10 31 54	−61 27.	1	13.04		0.57		-0.18		3		434
	+30 02036			10 31 55	+29 41.3	1	10.82		0.64		0.07		2	G0	186
		LF13 # 155		10 31 55	−51 47.3	1	11.15		1.16		1.06		1	K2 III	137
91480	+57 01277	HR 4141		10 31 57	+57 20.5	8	5.16	.008	0.34	.005	-0.02	.005	25	F1 V	15,254,1007,1008,1013*
		G 236 - 34		10 31 58	+69 42.8	1	11.95		1.55				1		1705
298141	−50 05237			10 31 59	−50 50.4	1	10.82		0.30		0.19		1	A2 V	137
91645	−36 06485	IDS10298S3652	AB	10 32 02	−37 07.7	1	6.90		0.03		-0.01		2	A0 IV	401
91545	+28 01902			10 32 03	+28 13.3	2	6.84	.010	1.07	.000	1.00		4	K2 III	20,186
91677	−51 04850			10 32 04	−51 55.4	1	9.23		0.99		0.82		2	G8/K0 III	137
91696	−53 03946			10 32 04	−54 05.2	1	9.86		0.32		0.05		1	A9/F0 V	137
	−55 03573			10 32 04	−55 33.8	1	11.49		0.27		0.02		2	B8	761
	−54 03809			10 32 05	−55 17.3	1	11.49		0.23		-0.04		1	B8	761
298143	−50 05242			10 32 10	−50 56.8	1	10.85		0.13		0.05		1	A0 V	137
91873	−75 00671			10 32 10	−75 46.8	1	8.15		1.26		1.07		2	K0 III	1704
91612	+7 02330	HR 4146		10 32 12	+07 12.7	6	5.08	.007	0.94	.005	0.64	.004	17	G8 II-III	15,1003,1080,1256*
	−54 03812			10 32 12	−55 13.8	1	11.45		0.28		-0.13		1	B8	761
		SOS 1819		10 32 12	−55 19.	1	12.12		0.09		-0.36		1	B7	761
92029	−81 00449	HR 4161	★ A	10 32 12	−81 39.8	3	7.01	.033	-0.06	.014	-0.50	.005	15	B5 III/IV	15,1075,2038
91564	+42 02131			10 32 13	+42 10.1	1	6.94		1.50		1.94		2	K5	1601
	−9 03102	G 162 - 68		10 32 15	−09 51.4	4	10.50	.030	0.68	.007	0.06	.014	5		927,1620,1658,1696
91611	+14 02261			10 32 17	+13 52.4	1	9.11		0.42		-0.03		2	G5	1648
	−58 02312	LSS 1653		10 32 17	−59 13.4	1	10.27		0.51		-0.55		4	O8	1737
91767	−60 01983	HR 4151		10 32 23	−60 43.7	6	6.22	.005	1.40	.005	1.56	.017	21	K2/3 III	15,657,1075,2013,2030,3005
91721	−51 04855			10 32 24	−51 26.7	1	9.38		0.08		0.03		1	A0 V	137
91636	+9 02374	HR 4148, TX Leo	★ AB	10 32 25	+08 54.6	2	5.66	.005	0.05	.000	0.05	.000	7	A2 V	15,1415
91765	−57 03451	LSS 1654		10 32 27	−58 21.2	2	8.97	.002	0.24	.019	-0.62	.009	4	B2 II	540,836
	+30 02038			10 32 28	+29 45.1	1	10.42		0.52		-0.05		2		186
91668	−11 02913			10 32 28	−11 31.3	1	9.18		1.17		1.14		1	K2 III	1738
91707	−30 08546			10 32 30	−31 05.2	1	7.37		0.44				4	F5/6 V	2033
		Lod 153 # 1		10 32 30	−57 50.7	1	11.79		0.23		0.15		1		287
	+28 01904			10 32 31	+28 01.1	1	10.01		1.00		0.82		3	G8 III	186
91603	+46 01642	IDS10295N4610	AB	10 32 32	+45 54.8	1	7.22		1.06		0.92		4	K2 III	1501
91764	−56 03521	LSS 1655		10 32 34	−57 05.3	1	8.76		0.23		-0.66		2	B1/2 Ib	730
		G 118 - 73		10 32 35	+39 16.6	1	13.42		1.43		1.16		1		3062
91621	+42 02133	IDS10296N4226		10 32 35	+42 10.1	1	9.05		0.32		0.08		2	A3	1375
91706	−22 02952	HR 4149		10 32 35	−22 55.1	6	6.10	.010	0.50	.007	0.01	.002	19	F6 V	15,116,1020,2012,2012,3053
91778	−60 01989			10 32 35	−60 40.4	1	9.50		-0.04		-0.41		2	B8 IV/V	401
303072	−57 03456	NGC 3293 - 40		10 32 41	−57 53.3	1	9.61		1.21				2	G5	781
	+28 01905			10 32 42	+28 13.6	1	9.75		0.54		0.04		3	F8	186

HD	DM	Other Id	N Rem	α_{1950}	δ_{1950}	S	V	σ_V	B–V	σ_{B-V}	U–B	σ_{U-B}	n	Spectrum	References
303073	−57 03457	NGC 3293 - 41		10 32 43	−57 56.8	1	10.72		0.66				4	B8	781
		G 44 - 25		10 32 45	+07 47.1	1	12.40		0.92		0.55		2		333,1620
	−35 06549			10 32 45	−36 16.2	1	10.00		1.25		1.18		1	K7 V	78
	+28 01906			10 32 46	+28 21.3	1	9.75		1.07		0.91		3	G8 III	186
303071	−57 03459	NGC 3293 - 39		10 32 46	−57 52.0	1	11.39		0.15		0.19		1		287
303070		NGC 3293 - 38		10 32 47	−57 51.9	1	11.20		0.16		0.23		1		287
299203	−52 03702	IDS10308S5239	AB	10 32 49	−52 54.2	1	10.26		0.18		0.14		1	A1 V	137
91839	−64 01350	IC 2602 - 5		10 32 49	−65 04.2	1	8.39		0.19				4	A1 V	2005
307771	−61 01720	LSS 1659		10 32 50	−62 16.8	2	10.05	.016	0.03	.009	-0.76	.050	4	B2 V	191,540
		LP 848 - 34		10 32 51	−23 45.7	1	12.38		0.73		0.10		2		1696
91826	−59 02235		A	10 32 51	−60 00.8	1	8.10		-0.10		-0.75		2	B2 III	401
91826	−59 02235		B	10 32 51	−60 00.8	1	9.60		0.18		-0.74		1		401
91824	−57 03463	NGC 3293 - 1		10 32 53	−57 53.8	4	8.18	.026	-0.05	.007	-0.96	.029	15	O7 V	540,781,836,976
	+30 02040			10 32 54	+29 53.0	1	11.52		0.89		0.27		2		186
91728	−12 03205			10 32 54	−12 35.9	1	6.78		0.34		0.02		2	F0/2 V	1375
		BPM 20160		10 32 54	−51 33.	1	13.44		0.60		-0.15		1		3065
303074	−57 03461	NGC 3293 - 42		10 32 54	−57 57.5	1	9.74		-0.02		-0.33		5	A0	781
303068	−57 03464	NGC 3293 - 11	⋆ V	10 32 55	−57 53.5	2	9.78	.015	0.00	.015	-0.81	.040	9	B1 V	781,836
	−57 03465	NGC 3293 - 129		10 32 55	−57 54.6	2	10.00	.010	0.03	.015	-0.77	.035	5		781,836
91685	+29 02062			10 32 56	+29 21.8	2	8.68	.005	0.50	.000	-0.04	.005	13	F7 V	186,3026
		AAS9,163 T3# 50		10 32 56	+29 53.4	1	10.93		0.74		0.13		2		186
	−59 02238	Bo 9 - 6	⋆ AB	10 32 57	−59 57.5	1	9.58		0.18		-0.78		2		411
91684	+39 02366			10 32 58	+38 47.4	1	8.21		0.95		0.55		2	G7 III	105
303059	−57 03467	LSS 1660		10 32 58	−57 43.9	1	10.11		0.37		-0.54		2	B0.5II-III	540
91837	−59 02239	Bo 9 - 7	AB	10 32 58	−59 55.7	3	8.52	.012	-0.10	.010	-0.97	.015	9	B1 (I)	401,411,1732
91793	−38 06579	HR 4153, U Ant		10 32 59	−39 18.2	5	5.54	.213	2.95	.178	7.10		15	C4-5,4-5	15,864,2006,8015,8022
91805	−43 06395	HR 4154	⋆ A	10 33 00	−43 24.4	4	6.09	.010	0.94	.000	0.71		13	G8 II-III	15,536,1075,2032
91836	−54 03826			10 33 00	−55 08.9	1	8.74		0.42				4	F2 V	2012
91850	−57 03468	NGC 3293 - 2		10 33 02	−57 55.7	4	9.09	.008	0.15	.009	-0.70	.025	15	B1 III	540,781,836,976
	+29 02063			10 33 04	+29 17.2	1	10.24		0.54		-0.03		3		186
299214	−52 03709			10 33 04	−53 21.8	1	9.94		1.06		0.90		1	K0 III	137
91848	−54 03829			10 33 07	−55 09.0	1	7.20		0.81				4	G2 III	2012
296797	−45 06211			10 33 09	−45 39.7	1	9.59		0.43		-0.01		2	F0	401
91773	−12 03208			10 33 11	−12 44.5	2	7.57	.007	1.03	.005	0.82	.018	5	K0 III	1546,1657
91754	+3 02392	IDS10306N0304	A	10 33 13	+02 48.8	1	9.58		0.37		0.00		3	G5	3016
91754	+3 02392	IDS10306N0304	B	10 33 13	+02 48.8	1	9.67		0.51		0.01		3		3032
		AAS9,163 T3# 56		10 33 13	+29 11.4	1	10.42		0.59		-0.02		3		186
91790	−17 03187	HR 4152		10 33 13	−18 18.6	2	6.48	.005	0.20	.000			7	A5 IV/V	15,2029
91906	−63 01501	IC 2602 - 8	⋆ A	10 33 13	−63 52.5	3	7.39	.067	0.05	.007	0.02	.005	7	A0 V	17,1053,2005
		IC 2602 - 196		10 33 14	−63 57.2	1	10.99		1.10		0.79		3		891
91868	−53 03969			10 33 15	−53 58.8	1	10.03		0.31		0.00		1	F0/2 V	137
307860	−63 01503	IC 2602 - 78	⋆ BC	10 33 15	−63 52.2	2	8.27	.015	0.21	.004	0.02	.004	5	F0	17,1053
307861	−63 01502	IC 2602 - 79		10 33 15	−63 57.1	1	10.01		0.50		0.22		2	A5	1053
		Ton 1253		10 33 19	+32 59.	2	14.53	.005	0.00	.005	0.15	.015	8		286,313
305367	−59 02255	Bo 9 - 16		10 33 19	−59 54.8	1	10.60		0.01		-0.47		2	B9	411
305373	−59 02256	Bo 9 - 18		10 33 19	−59 57.2	1	10.14		0.01		-0.10		2		411
303058	−57 03478	LSS 1666		10 33 23	−57 40.4	1	9.98		0.38		-0.52		2	B0.5IV	540
	−59 02259	Bo 9 - 20		10 33 24	−59 51.1	1	10.38		1.11		1.01		1		411
305372	−59 02260	Bo 9 - 17		10 33 25	−59 55.9	1	10.38		0.15		0.12		2	A0	411
		GD 123		10 33 26	+46 24.1	1	13.15		0.55		-0.04		2		3060
		G 44 - 27	⋆ V	10 33 27	+05 22.7	3	12.62	.015	1.59	.013	1.10	.014	6		1705,1764,3016
		NGC 3293 - 127		10 33 29	−58 00.7	1	15.79		0.64				1		836
	−59 02267	Bo 9 - 19		10 33 29	−59 54.2	1	10.08		0.99		0.80		1		411
91784	+16 02138	G 44 - 28		10 33 30	+16 08.0	2	9.08	.010	0.60	.000	0.03	.020	2	G0	333,1620,1658
91752	+37 02100	HR 4150		10 33 30	+36 35.2	3	6.29	.013	0.39	.004	-0.05	.005	6	F4 V	70,254,3037
		IC 2602 - 195		10 33 30	−64 41.5	1	11.70		1.26		0.94		2		891
91944	−63 01507	IC 2602 - 9		10 33 31	−63 57.7	1	8.68		0.05		-0.38		2	B7 V	1053
		IC 2602 - 263		10 33 31	−64 18.0	1	10.96		1.73		1.95		2		891
91768	+35 02161			10 33 32	+35 23.0	1	8.84		0.67		0.28		2	G5	1733
91769	+35 02162			10 33 33	+35 09.4	1	8.54		1.01		0.73		2	K0	1601
91816	−11 02916	LR Hya		10 33 33	−11 39.0	5	8.03	.007	0.86	.005	0.51	.028	20	K0	158,1546,1738,2033,3008
		NGC 3293 - 126		10 33 33	−58 00.8	1	10.95		1.23		1.18		3		836
	−58 02341	LSS 1668		10 33 33	−58 59.4	1	10.39		0.50		-0.34		3	B2 II	434
305431	−61 01731			10 33 33	−61 38.5	1	9.87		0.17		0.18		2	A0	837
		IC 2602 - 261		10 33 33	−64 13.9	1	11.63		0.34		0.10		7		891
	−57 03485	NGC 3293 - 92		10 33 34	−57 57.3	1	11.32		0.08		-0.52		3		836
		NGC 3293 - 125		10 33 34	−58 00.6	1	15.19		0.73		0.16		2		836
		IC 2602 - 221		10 33 35	−63 57.0	1	14.37		1.68				5		891
303067	−57 03486	NGC 3293 - 10	⋆ V	10 33 36	−57 56.6	2	9.54	.004	0.04	.004	-0.75	.004	11	B1 V	781,836
		NGC 3293 - 110		10 33 36	−57 57.2	1	12.98		0.15		-0.21		3		836
		NGC 3293 - 111		10 33 37	−57 57.4	1	13.45		0.61		0.10		3		836
		NGC 3293 - 128		10 33 37	−57 57.7	1	15.77		0.67				2		836
91959	−63 01509	IC 2602 - 10		10 33 37	−63 43.4	1	9.24		0.20				4	B8 II	2005
		IC 2602 - 262		10 33 37	−64 16.1	1	12.12		1.33		1.10		7		891
	+19 02356			10 33 38	+19 10.7	1	9.70		0.38				1	F2	1245
		NGC 3293 - 81		10 33 38	−57 58.5	1	12.70		0.17		-0.27		3		836
303075	−57 03490	NGC 3293 - 12	⋆ V	10 33 38	−57 59.8	3	9.92	.036	0.12	.013	-0.77	.030	14	B1 V	781,836,1737
91942	−56 03544	HR 4159		10 33 40	−57 17.9	4	4.45	.006	1.62	.000	1.79	.005	16	K3/4 II	15,1075,2012,3005
		NGC 3293 - 82		10 33 40	−57 58.2	1	13.26		0.19		-0.10		3		836
		NGC 3311 sq1 14		10 33 41	−27 07.1	1	13.37		0.74				1		498

Table 1   557

HD	DM	Other Id	N Rem	α_{1950}	δ_{1950}	S	V	σ_V	B–V	σ_{B-V}	U–B	σ_{U-B}	n	Spectrum	References
		NGC 3293 - 71		10 33 42	−57 57.6	1	13.15		0.11		−0.25		3		836
		NGC 3293 - 109		10 33 42	−57 59.9	1	13.19		0.19		−0.07		3		836
305366	−59 02271	Bo 9 - 21		10 33 42	−59 52.1	1	9.87		0.85		0.48		1	G0	411
91881	−26 08022	HR 4157	★ AB	10 33 43	−26 24.9	4	6.29	.005	0.48	.000			18	F5 V	15,1075,2013,2028
		NGC 3293 - 74		10 33 43	−57 56.2	1	13.72		0.24		−0.09		2		836
		NGC 3293 - 72		10 33 43	−57 56.6	1	12.78		0.11		−0.32		3		836
300854	−55 03594			10 33 44	−56 12.2	1	11.88		0.23		0.12		2	B9	807
	−57 03494	NGC 3293 - 73		10 33 44	−57 56.8	1	11.88		0.11		−0.34		3		836
303076	−57 03491	NGC 3293 - 43		10 33 44	−58 01.4	1	9.73		1.65				4	K5	781
		NGC 3293 - 78		10 33 45	−57 58.2	1	12.22		0.98		0.41		4		836
	−57 03500	NGC 3293 - 16	★ V	10 33 46	−57 57.2	2	8.76	.004	−0.01	.004	−0.81	.009	11	B0.5V	781,836
	−57 03497	NGC 3293 - 79		10 33 46	−57 58.4	1	10.22		0.07		−0.68		3		836
		NGC 3293 - 80		10 33 46	−57 58.6	1	12.25		0.17		−0.38		3		836
91940	−53 03977			10 33 47	−54 02.0	1	8.65		−0.02		−0.01		1	A0 IV	137
		IC 2602 - 203		10 33 47	−64 01.3	1	15.01		1.29				3		891
91858	−9 03108	HR 4155		10 33 48	−10 19.4	3	6.56	.007	0.30	.004			11	F0 V	15,2013,2028
	−34 06837			10 33 48	−34 38.9	1	10.33		0.75		0.28		1		78
91943	−57 03499	NGC 3293 - 3		10 33 48	−57 56.0	3	6.73	.009	0.06	.005	−0.87	.017	16	B0.5Ib	781,836,2012
		NGC 3293 - 75		10 33 48	−57 56.4	1	11.63		0.05		−0.51		3		781,836
		NGC 3293 - 69		10 33 48	−57 57.4	1	12.84		0.10		−0.28		3		836
	−57 03503	NGC 3293 - 65	★ V	10 33 49	−57 58.0	2	9.85	.000	0.03	.005	−0.75		5		781,836
	−57 03502	NGC 3293 - 21	★ V	10 33 49	−57 59.1	4	7.32	.047	2.05	.020	2.19	.048	26	M1.5Iab-Ib	781,836,1522,3043
	−57 03501	NGC 3293 - 83		10 33 49	−57 59.7	1	11.33		0.15		−0.46		3		836
		NGC 3293 - 105		10 33 49	−58 00.2	1	13.26		1.39		1.28		1		836
91880	−15 03087	HR 4156	★ AB	10 33 50	−16 05.1	4	6.02	.014	1.64	.005	1.89	.028	16	M1 III	15,1024,2029,3005
		NGC 3293 - 63		10 33 50	−57 58.2	1	11.24		0.05		−0.58		3		836
		NGC 3293 - 87		10 33 50	−57 58.9	1	11.15		0.06		−0.60		2		781,836
		IC 2602 - 202		10 33 50	−64 01.5	1	14.72		0.45		0.11		5		891
	−57 03504	NGC 3293 - 15	★ AB	10 33 51	−57 56.8	1	10.01		0.00		−0.66		5	B1 V	781,836
		NGC 3293 - 84		10 33 51	−57 59.9	1	11.87		0.23		−0.36		3		836
		NGC 3293 - 104		10 33 51	−58 00.5	1	12.67		0.30		−0.12		2		836
91997	−63 01510	IC 2602 - 12		10 33 51	−64 01.9	2	8.96	.009	0.07	.019	−0.33	.000	8	B8 V	891,1053
		NGC 3293 - 66		10 33 52	−57 57.8	1	12.77		0.17		−0.34		3		836
	−57 03506	NGC 3293 - 22	★ AP	10 33 52	−57 58.6	2	7.60	.005	0.09	.014	−0.75	.000	10	B1 II	781,836
		NGC 3293 - 67		10 33 53	−57 57.2	1	13.21		0.08		−0.25		3		836
		NGC 3293 - 31		10 33 53	−57 58.0	1	11.49		0.03		−0.63		3	B2 V	836
305364	−59 02275	Bo 9 - 5		10 33 53	−59 51.5	1	9.41		−0.04		−0.51		2	B8	411
	−59 02276	Bo 9 - 15		10 33 53	−59 53.5	1	11.95		0.21		0.17		2		411
		IC 2602 - 205		10 33 53	−63 59.0	1	14.12		0.86				4		891
	−57 03507	NGC 3293 - 14	★ V	10 33 54	−57 57.0	2	9.27	.005	−0.04	.005	−0.86	.025	10	B0.5V	781,836
	−57 03506	NGC 3293 - 23	★ BC	10 33 54	−57 58.7	2	9.16	.039	0.02	.005	−0.79	.005	9	B1 III	781,836
91969	−57 03508	NGC 3293 - 4	★ AB	10 33 55	−57 57.9	3	6.52	.007	0.00	.004	−0.93	.030	30	B0 Ib	781,836,2012
	−57 03509	NGC 3293 - 86		10 33 55	−57 59.3	2	10.71	.019	0.12	.005	−0.74	.033	7		781,836
	−57 03510	NGC 3293 - 85		10 33 55	−58 00.1	1	11.50		0.08		−0.52		3		836
		NGC 3293 - 103		10 33 55	−58 00.4	1	13.45		0.37		0.11		1		836
305383	−59 02279	Bo 9 - 8		10 33 55	−59 53.3	1	10.83		−0.01		−0.26		2		411
	−57 03512	NGC 3293 - 130		10 33 56	−57 56.1	1	11.88		0.04		−0.49		4		836
		NGC 3293 - 34		10 33 56	−57 56.6	1	12.87		0.03		−0.29		1	B8 V	781
91970	−58 02349	vdBH 99 - 1		10 33 56	−58 45.6	1	9.24		−0.03		−0.25		4	B9 V	1711
		NGC 3293 - 30		10 33 57	−57 56.8	1	12.69		0.11		−0.22		2	B5 V	836
		NGC 3293 - 58		10 33 57	−57 58.1	1	12.76		0.12		−0.28		4		836
		NGC 3293 - 32		10 33 58	−57 57.1	1	12.87		0.20		−0.16		2	B8 Ve	836
		NGC 3293 - 55		10 33 58	−57 57.6	1	12.75		0.12		−0.22		3		836
303090	−58 02350			10 33 58	−58 29.4	1	11.35		0.18		0.18		1	A2	287
91856	+22 02243	G 54 - 33		10 33 59	+21 51.8	2	8.75	.040	0.77	.020	0.36	.025	2	G5	333,1620,1658
	−57 03514	NGC 3293 - 17		10 33 59	−57 56.7	2	10.50	.005	−0.03	.010	−0.71	.015	9	B2	781,836
		NGC 3293 - 56		10 33 59	−57 57.8	1	13.53		0.29		0.22		3		836
		NGC 3293 - 60		10 33 59	−57 58.4	1	13.34		0.65		0.14		3		836
	−57 03517	NGC 3293 - 24	★ V	10 33 59	−57 59.2	2	9.20	.020	0.05	.010	−0.79	.010	8	B1 III	781,836
91842	+28 01908			10 34 00	+28 02.2	1	8.53		1.05				2	K1 III	20
	−57 03515	NGC 3293 - 133	V	10 34 00	−57 57.4	2	9.09	.020	−0.02	.005	−0.82	.010	8		781,836
	−57 03518	NGC 3293 - 29		10 34 00	−57 58.2	2	10.64	.031	0.02	.004	−0.67	.000	6	B2 V	781,836
91983	−57 03516	NGC 3293 - 7		10 34 00	−57 59.9	2	8.58	.005	0.04	.005	−0.81	.029	7	B1 III	781,836
305363	−59 02282	Bo 9 - 22		10 34 00	−59 46.9	1	8.30		1.82		1.88		1	M3 Ib	411
305368	−59 02281	Bo 9 - 4		10 34 00	−59 52.7	2	9.78	.010	0.00	.005	−0.16	.005	4	A0	411,1737
91919	−32 07472			10 34 01	−32 34.3	1	8.61		0.24		0.15		2	A6 V	401
	−57 03519	NGC 3293 - 131		10 34 01	−57 56.8	2	9.99	.025	−0.05	.005	−0.79	.015	12		781,836
		NGC 3293 - 52		10 34 01	−57 57.1	1	12.08		0.04		−0.52		2		836
		NGC 3293 - 35		10 34 01	−57 58.5	2	12.86	.023	0.13	.027	−0.28	.005	5		781,836
		NGC 3293 - 91		10 34 01	−58 00.1	1	12.99		0.16		−0.21		3		836
	−57 03522	NGC 3293 - 13		10 34 02	−57 55.9	2	10.14	.015	−0.02	.000	−0.79	.010	12	B2 V	781,836
	−57 03520	NGC 3293 - 132		10 34 02	−57 57.1	2	10.25	.015	0.01	.005	−0.68	.020	8		781,836
		NGC 3293 - 88		10 34 02	−57 58.6	1	13.00		0.18		−0.24		3		836
	−57 03521	NGC 3293 - 25		10 34 02	−57 59.0	2	8.13	.010	0.07	.005	−0.79	.020	11	B1 III	781,836
303088	−58 02355			10 34 02	−58 28.5	1	11.20		0.13		0.15		1	A2	287
91889	−11 02918	HR 4158	★ A	10 34 03	−11 57.7	6	5.69	.019	0.53	.007	0.00	.004	21	F7 V	15,22,1003,1311,2012,3013
91901	−13 03170			10 34 03	−13 35.0	1	8.71		0.92		0.60		1	K2 V	3072
296803	−43 06415		AB	10 34 03	−43 56.3	1	10.16		0.17				4	K7	2011
	−57 03524	NGC 3293 - 18	★ B	10 34 03	−57 56.8	2	9.25	.010	−0.04	.000	−0.80	.005	10	B1 V	781,836
	−57 03523	NGC 3293 - 20		10 34 03	−57 57.8	2	8.02	.005	0.02	.000	−0.84	.015	13	B1 III	781,836

HD	DM	Other Id	N Rem	α₁₉₅₀	δ₁₉₅₀	S	V	σ_V	B–V	σ_B–V	U–B	σ_U–B	n	Spectrum	References
		NGC 3293 - 90		10 34 03	−57 59.8	1	12.04		0.18		−0.28		3		836
305371	−59 02283	Bo 9 - 9		10 34 03	−59 53.6	2	10.29	.030	0.30	.040	−0.69	.085	4	O9 V	191,411
91877	+11 02252			10 34 04	+11 17.0	1	8.69		0.72		0.32		2	K0	1648
	−57 03524	NGC 3293 - 19	⋆ A	10 34 04	−57 57.0	2	9.07	.025	0.19	.005	−0.70	.005	11	B1 III	781,836
	−57 03526	NGC 3293 - 26	⋆ V	10 34 04	−57 58.7	2	8.53	.320	0.05	.010	−0.81	.005	10	B1 III	781,836
92007	−57 03526	NGC 3293 - 6		10 34 04	−57 58.9	3	8.24	.017	0.08	.035	−0.79	.031	8	B1 II	540,781,836
		NGC 3293 - 100		10 34 04	−57 59.1	1	13.00		0.13		−0.19		3		836
		NGC 3293 - 96		10 34 06	−57 57.6	1	13.70		0.73		0.13		3		836
		AJ86,209 # 1		10 34 06	−59 34.	1	11.74		0.54		−0.32		1		924
	−59 02285	Bo 9 - 3		10 34 06	−59 50.7	1	11.39		0.32		−0.76		3		411
92007	−57 03527	NGC 3293 - 27	⋆ V	10 34 07	−57 59.6	2	8.94	.005	0.06	.000	−0.74	.005	7	B0.5III	781,836
		NGC 3293 - 93		10 34 08	−57 57.4	1	12.52		0.11		−0.36		3		836
		NGC 3293 - 94		10 34 08	−57 57.6	1	13.08		0.12		−0.18		3		836
		NGC 3293 - 95		10 34 08	−57 57.8	1	13.60		0.21		0.01		3		836
		G 253 - 37		10 34 09	+76 26.1	2	10.73	.000	1.23	.007	1.20		3	K4	1746,7010
	−57 03528	NGC 3293 - 33		10 34 09	−57 59.1	2	10.67	.015	0.05	.005	−0.63	.000	8	B2 V	781,836
	−59 02288	Bo 9 - 2		10 34 09	−59 52.9	1	11.36		0.32		−0.68		2		411
92683	−85 00245			10 34 09	−85 49.9	1	6.67		0.12		0.09		6	A1/2 IV/V	1628
303065	−57 03530	NGC 3293 - 9		10 34 10	−58 00.4	2	9.99	.015	0.05	.005	−0.77	.030	12	B2 V	781,836
303089	−58 02362			10 34 10	−58 28.1	1	11.39		0.13		0.11		1	A0	287
92027	−61 01734	LSS 1686		10 34 10	−61 36.3	4	8.47	.023	0.05	.015	−0.74	.018	8	B2/3 Vne	401,540,837,976
91840	+47 01794			10 34 11	+46 48.5	1	7.24		1.09		0.97		4	K0	8100
91810	+57 01278			10 34 11	+56 41.5	1	6.55		1.17		1.26		3	K1 III	37
91981	−47 06028	IDS10321S4720	AB	10 34 11	−47 35.4	1	7.29		0.58				4	F7/8 V	2012
	−57 03531	NGC 3293 - 28	⋆	10 34 11	−57 58.9	3	10.32	.010	0.00	.012	−0.69	.015	7	B1 V	781,836,1737
	−57 03532	NGC 3293 - 89		10 34 13	−57 59.8	1	10.97		0.05		−0.67		3		836
		NGC 3293 - 113		10 34 14	−57 57.0	1	12.65		0.14		−0.27		3		836
92024	−57 03533	NGC 3293 - 5	⋆ V	10 34 14	−57 57.5	2	9.02	.005	−0.03	.010	−0.88	.033	7	B1 III	781,836
		NGC 3293 - 114		10 34 15	−57 55.4	1	13.23		0.09		−0.27		3		836
		NGC 3293 - 112		10 34 15	−57 57.0	1	13.33		0.12		−0.13		3		836
305370	−59 02290	Bo 9 - 10		10 34 15	−59 51.7	1	11.06		0.10		0.06		2	B9	411
92066	−63 01513	IC 2602 - 13		10 34 15	−63 59.0	1	8.40		0.00		−0.43		4	B7 Vne	1053
92157	−74 00714	IDS10333S7433	AB	10 34 15	−74 48.6	1	8.66		0.27		0.07		2	B9 V	1776
	−61 01735			10 34 16	−61 52.2	1	10.71		0.13		0.08		2	A0	837
		NGC 3293 - 118		10 34 17	−57 57.1	1	15.72		0.65		0.46		2		836
91965	−32 07479			10 34 18	−33 00.8	1	7.13		0.08		0.07		2	A1 V	401
		NGC 3293 - 120		10 34 18	−57 57.8	1	12.75		0.13		−0.32		2		836
92025	−59 02292	Bo 9 - 1		10 34 19	−59 55.3	2	7.87	.005	−0.10	.005	−0.88	.005	6	B1 III	401,411
		Lod 213 # 13		10 34 20	−61 37.9	1	11.57		0.60		0.09		2	F8	837
		G 119 - 18		10 34 21	+31 41.8	1	13.70		0.94		0.57		1		3062
303044	−57 03539	LSS 1687		10 34 21	−57 31.4	1	9.52		0.42		−0.58		2	O7	540
92044	−57 03540	NGC 3293 - 8		10 34 21	−58 01.1	4	8.25	.012	0.17	.015	−0.69	.026	15	B0.5III	781,781,836,976
	−59 02293	Bo 9 - 14		10 34 21	−59 50.0	1	11.53		0.29		0.17		2		411
305432	−61 01737			10 34 21	−61 39.0	1	10.50		0.50		0.23		2	A7	837
91964	−27 07567			10 34 22	−27 23.8	1	6.65		1.45				1	K4/5 III	498
		NGC 3293 - 124		10 34 22	−57 59.4	1	12.59		0.19		−0.26		2		836
91961	−7 03070			10 34 23	−07 31.4	1	8.57		0.39		−0.03		2	F2	1375
		NGC 3293 - 117		10 34 23	−57 55.0	1	12.17		0.26		−0.34		3		836
	−59 02294	Bo 9 - 13		10 34 23	−59 51.5	1	11.41		0.02		−0.38		2		411
		G 146 - 50		10 34 24	+42 21.7	1	15.42		1.62				1		906
		NGC 3311 sq1 13		10 34 24	−27 12.6	1	13.34		0.92				1		498
91979	−31 08376			10 34 25	−31 24.5	1	8.48		1.03		0.74		1	Kp Ba	565
		NGC 3293 - 116		10 34 26	−57 55.3	1	11.49		1.00		0.72		3		836
		NGC 3293 - 115		10 34 27	−57 55.3	1	12.11		0.04		−0.48		3		836
92060	−57 03545	LSS 1690		10 34 28	−57 41.0	4	8.59	.049	0.36	.015	−0.55	.020	9	B2 Ib/II	401,540,836,976
92063	−58 02371	HR 4164		10 34 28	−59 18.3	5	5.08	.005	1.17	.005	1.20		19	K1 III	15,1075,1075,2006,3005
92061	−58 02372	IDS10326S5826	AB	10 34 29	−58 41.5	1	8.97	.014	0.08	.013	−0.73	.017	14	B5	401,434,540,976,1711
92062	−58 02370			10 34 29	−58 46.5	1	8.16		0.71		0.49		4	F5/6 Iab/b	1711
		PG1034+001		10 34 30	+00 07.2	1	13.23		−0.36		−1.27		6	DO	1764
		WLS 1024 40 # 8		10 34 30	+40 04.1	1	13.01		0.78		0.20		2		1375
91962	−8 02963	IDS10320S0819	AB	10 34 30	−08 34.8	1	7.03		0.62				4	G0	2012
305369	−59 02298	Bo 9 - 12		10 34 30	−59 51.5	1	10.43		0.50		0.02		2	F5	411
		Bo 9 - 11		10 34 30	−59 51.6	1	10.93		0.56		0.12		2		411
92209	−75 00678	HR 4170		10 34 30	−76 03.0	3	6.29	.004	1.21	.005	1.27	.000	12	K2 III	15,1075,2038
		NGC 3293 - 119		10 34 31	−57 56.3	1	13.32		0.13		−0.14		2		836
		NGC 3293 - 123		10 34 31	−57 59.0	1	12.19		0.18		0.15		2		836
305433	−61 01739			10 34 31	−61 38.9	1	10.28		0.10		0.03		2	A0	837
		NGC 3293 - 121		10 34 34	−57 58.0	1	12.70		0.23		−0.31		2		836
91950	+25 02271			10 34 36	+25 20.5	1	8.58		0.68		0.17		5	G2 V	3026
		Lod 213 # 15		10 34 36	−61 36.8	1	11.78		0.63		0.01		2	F8	837
	−61 01740	IDS10328S6125	AB	10 34 36	−61 40.3	1	11.19		0.62		0.06		2	F6	837
	−58 02381			10 34 39	−58 55.4	1	7.54		−0.12		−0.59		1		401
92072	−58 02380	vdBH 99 - 2		10 34 39	−58 55.8	3	7.03	.007	−0.11	.004	−0.55	.019	10	B6 IV	158,401,1711
		NGC 3293 - 122		10 34 41	−57 58.5	1	11.89		0.07		−0.07		2		836
92086	−58 02382	vdBH 99 - 3	⋆ AB	10 34 41	−58 34.5	1	9.60		0.07		−0.09		2	B9 V	1711
91992	−10 03094	HR 4160		10 34 43	−11 29.3	2	6.51	.005	0.30	.005			7	F0 V	15,2029
303056	−57 03553			10 34 43	−57 44.6	3	9.69	.008	0.08	.003	−0.71	.015	7	B0 V	434,540,976
	+29 02067		A	10 34 44	+28 31.9	1	10.59		1.28		1.25		3	K6	1723
91874	+65 00791			10 34 44	+65 13.8	1	8.04		1.37		1.63		1	K0	1746
92088	−58 02386	vdBH 99 - 4		10 34 45	−58 05.7	1	8.34		−0.01		−0.31		2	B8 II/III	1711

Table 1   559

HD	DM	Other Id	N Rem	α_{1950}	δ_{1950}	S	V	σ_V	B–V	σ_{B-V}	U–B	σ_{U-B}	n	Spectrum	References
92087	−58 02387	vdBH 99 - 5		10 34 46	−58 54.5	1	7.55		-0.13		-0.63		4	B4 (V)	1711
92105	−58 02388			10 34 47	−59 02.9	1	8.56		0.27		0.13		4	A2/3 IV/V	1711
		Lod 213 # 18		10 34 51	−61 36.1	1	12.17		1.26		1.19		2		837
92121	−57 03563	NGC 3293 - 48		10 34 52	−58 02.3	1	8.64		0.23				4	A3	781
92036	−26 08033	HR 4162		10 34 53	−27 09.2	5	4.89	.005	1.62	.004	1.95		19	M1 III	15,1020,2012,2012,3053
92305	−77 00622	HR 4174		10 34 54	−78 20.9	7	4.11	.023	1.58	.007	1.95	.044	24	K5 III	15,1034,1075,1409*
		NGC 3311 sq1 2		10 34 55	−27 19.1	1	10.96		0.99				1		498
92144	−57 03566			10 34 58	−57 35.0	3	9.05	.010	0.35	.015	-0.49	.026	6	B0.5/1 III	540,836,976
92145	−57 03567	NGC 3324 - 6		10 35 00	−58 22.1	1	9.28		0.12		0.07		4	A1 IV	520
		ApJS53,765 T2# 5		10 35 00	−61 19.6	1	12.59		1.06		0.75		1		1480
92000	+34 02145			10 35 02	+34 20.3	1	6.41		1.36		1.52		3	K0	1625
	−60 02048			10 35 02	−61 18.2	1	11.27		0.49		0.07		1		1480
	−61 01743			10 35 03	−61 46.3	1	11.41		0.06		-0.34		2	B7	837
307817	−61 01742			10 35 03	−61 46.4	2	10.72	.052	-0.01	.001	-0.54	.007	4	B3	540,837
92055	−12 03218	HR 4163, U Hya		10 35 05	−13 07.4	8	4.95	.207	2.66	.086	5.54	.391	18	C5 II	15,864,897,2006,8005*
303079	−57 03570	NGC 3324 - 24		10 35 05	−58 21.8	1	11.97		0.01		-0.38		2		520
92012	+20 02507			10 35 06	+20 14.2	1	8.66		0.39		-0.06		2	F5	1733
		NGC 3311 sq1 3		10 35 06	−27 15.7	1	11.31		0.60				1		498
		NGC 3324 - 34	AB	10 35 06	−58 22.7	1	12.57		0.50		-0.10		1		520
		CT Car		10 35 06	−61 19.2	1	11.80		1.26		0.91		1		689
92175	−64 01363	IC 2602 - 14		10 35 06	−64 49.5	2	8.77	.005	1.20	.000			8	G8 III	2005,2012
	−57 03571			10 35 07	−58 06.7	1	11.86		0.33		-0.47		3		434
		NGC 3324 - 28		10 35 07	−58 19.9	1	12.24		0.24		-0.47		4		520
	−61 01745			10 35 07	−61 50.3	1	11.49		0.07		-0.75		2	B2	837
	−57 03573	NGC 3324 - 8		10 35 08	−58 23.0	1	10.68		1.03		0.58		2		520
305429	−61 01744			10 35 08	−61 34.7	1	10.11		-0.01		-0.52		2	B5	837
		NGC 3324 - 35		10 35 10	−58 19.1	1	12.66		0.37		-0.41		2		520
303102	−58 02400	vdBH 99 - 6		10 35 10	−58 42.3	1	8.81		0.00		-0.28		4	B9	1711
307859	−63 01518	IC 2602 - 77		10 35 11	−63 51.5	2	10.36	.009	0.19	.005	0.06	.005	2	A0	891,1053
92139	−47 06042	HR 4167	★ AB	10 35 12	−47 57.9	6	3.84	.008	0.30	.005	0.06	.017	23	A3mF0-F2	15,355,1075,2012,3026,8015
	−52 03737			10 35 12	−53 08.5	1	11.42		0.03		-0.40		2	B6 III	540
		vdB 41 # b		10 35 12	−58 24.	1	13.78		0.57		-0.38		3		434
		NGC 3324 - 37		10 35 12	−58 24.0	1	13.06		0.53		-0.13		3		520
		NGC 3324 - 38		10 35 13	−58 24.4	1	13.10		0.31		-0.35		3		520
		ApJS53,765 T2# 1		10 35 13	−61 22.5	1	11.39		0.72		0.30		1		1480
92155	−53 04002			10 35 15	−53 35.7	1	6.40		-0.15				4	B3 V	2012
		CCS 1717		10 35 17	−47 32.5	1	10.21		2.52		2.49		2	R	864
	−60 02052			10 35 17	−61 21.2	1	11.31		0.41		0.07		1		1480
		L 321 - 15		10 35 18	−45 46.	1	13.42		0.76		0.14		2		1696
		NGC 3324 - 7		10 35 18	−58 20.2	1	10.51		1.84		1.03		2		520
303080	−58 02402	NGC 3324 - 13		10 35 19	−58 25.2	1	11.38		0.09		-0.38		3		520
303079	−57 03578	NGC 3324 - 9		10 35 20	−58 20.7	2	10.84	.005	0.19	.010	-0.68	.020	5		411,520
		NGC 3324 - 31		10 35 20	−58 20.8	2	12.38	.024	0.23	.014	-0.41	.019	7		411,520
		NGC 3324 - 30	AB	10 35 20	−58 21.1	2	12.32	.034	0.35	.010	-0.25	.054	6		411,520
		NGC 3324 - 32		10 35 21	−58 22.4	3	12.39	.050	0.20	.018	-0.48	.017	6		411,432,520
92174	−58 02404		V	10 35 21	−58 58.7	1	9.15		-0.05		-0.38		22	B8 II/III	1711
	−61 01748			10 35 22	−61 37.2	1	11.19		0.57		0.02		2	F6	837
		LP 430 - 51		10 35 23	+14 24.4	1	12.83		0.53		-0.19		2		1696
		NGC 3324 - 45		10 35 23	−58 22.5	2	13.98	.015	0.30	.030	-0.06	.010	4		411,520
	−58 02405	NGC 3324 - 21		10 35 23	−58 26.8	1	11.77		0.91		0.39		2		520
		NGC 3324 - 41		10 35 24	−58 19.4	1	13.20		0.22		-0.37		2		520
	−57 03580	NGC 3324 - 4		10 35 24	−58 22.1	5	9.05	.013	0.14	.011	-0.78	.014	14	O8.5Vp	411,432,476,520,836
		NGC 3324 - 44		10 35 24	−58 22.3	2	13.44	.025	0.29	.010	-0.32	.050	4		411,520
		AJ86,209 # 2		10 35 24	−59 18.	1	11.38		0.74		-0.31		1		924
92136	−35 06590			10 35 25	−35 27.6	1	7.01		0.01		-0.02		2	B9 IV	401
		NGC 3324 - 27	AB	10 35 25	−58 22.1	3	12.23	.027	0.26	.025	-0.24	.007	6		411,432,520
		NGC 3324 - 19		10 35 27	−58 21.6	2	11.71	.010	0.21	.005	-0.62	.015	6		411,520
92206	−57 03584	NGC 3324 - 2	★ A	10 35 28	−58 21.8	3	8.22	.005	0.13	.008	-0.85	.012	8	O6.5V(n)	411,432,520
92206	−57 03584	NGC 3324 - 2	★ AB	10 35 28	−58 21.8	5	7.82	.033	0.15	.008	-0.81	.024	12		401,476,836,1711,1732
		NGC 3324 - 5	★ B	10 35 28	−58 21.8	3	9.15	.012	0.16	.015	-0.82	.023	6		411,432,520
92190	−58 02406	vdBH 99 - 7		10 35 28	−58 55.2	2	8.51	.017	-0.07	.004	-0.43	.007	8	B8 IV	401,1711
	−57 03585	NGC 3324 - 10		10 35 29	−58 19.9	1	10.98		0.20		-0.62		3		520
		Lod 213 # 14		10 35 29	−61 27.8	1	11.58		0.05		-0.25		2	B8	837
		AJ86,209 # 3		10 35 30	−59 27.	1	12.43		0.65		-0.44		1		924
		NGC 3324 - 11		10 35 31	−58 17.6	1	11.45		0.68		0.10		3		520
		NGC 3324 - 36		10 35 33	−58 18.0	1	12.79		0.34		-0.25		2		520
		NGC 3324 - 43		10 35 33	−58 21.7	2	13.32	.015	0.71	.005	0.08	.010	3		411,520
		NGC 3324 - 40		10 35 33	−58 23.7	1	13.12		0.26		-0.41		3		520
92207	−58 02411	NGC 3324 - 1	★ V	10 35 33	−58 28.4	11	5.47	.028	0.50	.014	-0.23	.031	32	A1 Ia+	15,434,476,520,1018,1034*
		NGC 3324 - 22		10 35 34	−58 26.0	1	11.87		0.41		-0.01		2		520
		vdBH 99 - 109		10 35 34	−58 54.3	1	16.41		1.16		0.70		5		1711
307842	−63 01520	IC 2602 - 69		10 35 34	−63 23.4	2	9.40	.038	0.75	.013	0.33		3	G5	1053,2005
		NGC 3324 - 33		10 35 35	−58 24.2	1	12.43		0.16		-0.32		3		520
	−58 02413	IDS10337S5835	AB	10 35 35	−58 51.1	1	9.29		0.41		0.18		3		1711
		AJ86,209 # 4		10 35 36	−59 27.	1	12.84		0.66		-0.18		1		924
		NGC 3324 - 39		10 35 37	−58 23.9	1	13.10		0.41		-0.04		3		520
		Lod 213 # 16		10 35 37	−61 33.0	1	12.08		0.45		0.21		2	A8	837
	−57 03589	NGC 3324 - 18	★ AB	10 35 38	−58 22.9	1	11.69		0.28		-0.48		2		520
		NGC 3324 - 23		10 35 39	−58 25.5	1	11.90		0.40		-0.01		4		520
		vdBH 99 - 108		10 35 39	−58 54.5	1	16.29		1.57		0.77		4		1711

HD	DM	Other Id	N	Rem	α_{1950}	δ_{1950}	S	V	σ_V	B–V	σ_{B-V}	U–B	σ_{U-B}	n	Spectrum	References
92234	−58 02418	vdBH 99 - 8			10 35 41	−58 52.5	1	9.43		-0.01		-0.19		7	B9.5 V	1711
92235	−58 02417				10 35 41	−59 02.6	1	10.05		0.00		-0.16		2	B9/A0 V	1711
303078	−57 03592	NGC 3324 - 12			10 35 43	−58 19.1	1	11.02		0.20		0.13		3		520
		NGC 3324 - 25			10 35 44	−58 28.0	1	12.11		0.19		0.06		1		520
	−61 01752				10 35 44	−61 33.4	1	10.96		0.60		0.07		2	F5	837
		WD1035+532			10 35 45	+53 14.6	1	16.34		-0.35		-1.01		1	DA1	1727
305422	−60 02066				10 35 45	−61 17.8	1	9.66		0.55		0.09		1	G0	1480
92127	+14 02275				10 35 46	+14 12.7	1	8.31		0.42		-0.07		1	F5	1400
		vdBH 99 - 107			10 35 46	−58 53.7	1	16.63		0.76		0.21		5		1711
92250	−58 02426	vdBH 99 - 12			10 35 47	−59 11.5	1	9.83		0.07		0.05		4	A1 m	1711
		LP 848 - 38			10 35 48	−25 20.	1	10.24		1.03		0.91		1		78
		NGC 3324 - 20			10 35 48	−58 25.6	1	11.73		0.49		0.01		3		520
92251	−58 02423	vdBH 99 - 10			10 35 50	−58 14.2	1	9.81		0.03		-0.08		4	A0 m	1711
303105	−58 02421	vdBH 99 - 9			10 35 50	−58 50.5	1	10.05		0.07		0.05		3	A0	1711
	−58 02422	NGC 3324 - 11		⋆ V	10 35 51	−58 24.6	1	11.00		0.41		-0.51		4		520
303103	−58 02424	vdBH 99 - 11			10 35 51	−58 42.1	2	9.94	.023	0.03	.016	-0.04	.018	4	A0	401,1711
		NGC 3324 - 16			10 35 52	−58 21.0	1	11.61		0.53		-0.03		3		520
		NGC 3324 - 26			10 35 52	−58 26.9	1	12.20		1.26		0.80		2		520
		NGC 3324 - 42			10 35 53	−58 18.2	1	13.22		0.41		0.02		1		520
92125	+32 02061	HR 4166			10 35 55	+32 14.2	4	4.71	.021	0.81	.005	0.55	.014	14	G0 IIa	15,1355,1363,8015
92272	−57 03599	NGC 3324 - 3			10 35 56	−58 18.9	1	8.93		-0.01		-0.25		4	B9 IV	520
		NGC 3324 - 14			10 35 56	−58 21.3	1	11.43		0.50		0.04		3		520
303108	−58 02427				10 35 56	−58 51.5	1	10.13		-0.04		-0.36		5	B9	1711
		vdBH 99 - 106			10 35 56	−58 54.6	1	15.09		0.52		0.20		4		1711
		vdBH 99 - 111			10 35 56	−58 56.6	1	13.05		0.49		-0.26		2		1711
92074	+56 01468				10 35 58	+55 46.4	1	8.64		0.64		0.19		2	G0	1502
		vdBH 99 - 105			10 35 59	−58 55.1	1	14.49		0.44		0.12		4		1711
		LP 213 - 48	A		10 36 00	+39 58.4	1	11.51		1.16		1.12		4		1723
92095	+54 01387	HR 4165			10 36 00	+53 55.8	3	5.52	.005	1.27	.005	1.34	.005	6	K3 III	15,1003,1080
		Lod 213 # 17			10 36 00	−61 38.8	1	12.12		0.15		-0.20		2	B8	837
300895	−55 03648				10 36 01	−55 45.1	1	11.01		0.00		-0.11		2		761
92149	+38 02165				10 36 02	+37 31.8	1	8.68		0.50		0.03		2	F5	1375
		NGC 3324 - 17			10 36 03	−58 23.2	1	11.67		0.49		0.03		2		520
92274	−59 02345	LSS 1701			10 36 03	−59 46.3	1	8.94		0.21		-0.14		1	A5/8 IV/V	8100
305382	−59 02347	LSS 1700			10 36 03	−60 15.5	1	10.42		0.22		-0.70		2	B2 Ve	540
		NGC 3324 - 29			10 36 04	−58 18.2	1	12.27		0.45		-0.02		2		520
92273	−58 02431	vdBH 99 - 13			10 36 04	−59 08.6	1	9.24		-0.03		-0.17		6	B9 V	1711
92287	−56 03588	HR 4173			10 36 06	−56 59.8	4	5.90	.006	-0.13	.015	-0.62	.005	18	B3 IV	6,15,1637,2012
92291	−60 02077	LSS 1702			10 36 06	−60 00.9	1	9.07		-0.07		-0.86		2	B2 Ib/II	401
92289	−58 02432				10 36 08	−59 09.5	1	9.60		-0.07		-0.79		6	B0/2 I/III	1711
92214	−16 03100	HR 4171			10 36 09	−16 37.0	3	4.91	.016	0.92	.000	0.68		9	G8 III	15,2027,3016
92288	−58 02434	vdBH 99 - 14			10 36 10	−58 48.8	1	7.89		-0.07		-0.50		5	B7.5 Vn	1711
		Steph 873			10 36 12	+25 21.4	1	12.34		1.53		1.20		2	M2	1746
		SSII 1			10 36 12	+50 05.0	1	11.18		0.33		0.07		1	F0	1298
		vdBH 99 - 101			10 36 13	−58 58.5	1	14.34		0.54		0.17		5		1711
92196	+16 02144				10 36 14	+16 23.3	1	6.62		0.40		-0.03		2	F5 V	985
305356	−59 02353				10 36 15	−59 33.3	1	11.33		0.39		-0.58		2	A0	1737
92168	+38 02166	HR 4168			10 36 16	+38 10.3	2	5.85	.000	0.57	.000	0.17		5	F9 V	1118,1355
		LP 848 - 41			10 36 16	−26 34.8	1	12.01		0.81		0.38		2		1696
	+10 02179	DN Leo			10 36 17	+10 19.4	4	9.97	.011	-0.18	.004	-0.87	.015	36		842,843,1026,1241
		vdBH 99 - 102			10 36 17	−58 59.5	1	15.21		0.64		0.10		4		1711
		AJ86,209 # 6			10 36 18	−59 04.	1	11.18		0.82		-0.25		1		924
92335	−62 01679	LSS 1708			10 36 19	−62 47.7	1	7.62		0.38		0.24		2	A2/3 II	8100
303110	−58 02440	vdBH 99 - 15			10 36 20	−58 53.9	1	10.91		0.20		-0.59		2	A0	476
303111	−58 02442	vdBH 99 - 16			10 36 20	−58 54.5	1	11.08		0.33		0.07		7		1711
92245	−12 02925	HR 4172			10 36 22	−12 11.0	2	6.03	.005	0.00	.000			7	A0 Vn	15,2028
303197	−58 02443	LSS 1707			10 36 23	−58 38.1	1	9.28		0.37		-0.60		2	B5	476
92334	−56 03597	LSS 1706			10 36 24	−57 09.8	1	10.47		0.45		-0.61		2	O7 III	1737
307849	−63 01525	IC 2602 - 74			10 36 26	−63 30.5	1	10.48		0.48				2	F1	2005
310105	−64 01373	IC 2602 - 145			10 36 26	−64 54.8	1	11.14		1.12				2	G5	2005
92438	−72 00997	IDS10353S7303	A		10 36 27	−73 19.1	1	9.18		0.47		0.00		4	F2/3 V	1776
		vdBH 99 - 103			10 36 28	−58 58.4	1	16.03		1.12		0.86		5		1711
		vdBH 99 - 104			10 36 29	−58 58.7	1	16.05		0.77		0.31		5		1711
92223	+15 02232				10 36 30	+14 59.6	1	8.36		1.01		0.76		3	K0	1648
	−58 02447				10 36 30	−58 55.1	1	10.78		0.65		0.17		5		1711
		VV Car			10 36 30	−58 56.6	1	11.77		1.78		1.36		24		1711
		NGC 3330 - 42			10 36 32	−53 48.8	1	13.42		0.50				3		2042
303199	−58 02451	vdBH 99 - 17			10 36 34	−58 54.0	1	10.08		0.11		0.09		6	A0	1711
92385	−64 01374	IC 2602 - 17		⋆ V	10 36 34	−64 46.9	5	6.75	.015	-0.08	.006	-0.28	.013	18	B9 Vn	17,1053,2005,2012,3015
	−58 02452				10 36 35	−58 54.7	1	10.10		1.07		0.80		9		1711
		vdBH 99 - 110			10 36 35	−58 58.2	1	13.46		1.53		1.45		2		1711
		Steph 872			10 36 36	+62 21.1	1	12.16		1.30		1.20		1	K6	1746
		NGC 3330 - 41			10 36 37	−53 49.4	1	12.72		0.29				3		2042
		NGC 3330 - 5			10 36 37	−53 54.3	1	12.06		0.06				2		2042
92328	−42 06390	HR 4175		⋆ AB	10 36 38	−42 29.6	5	6.12	.012	0.66	.000	0.41		19	G8 III+A7 V	15,78,1075,2018,2028
92348	−53 04021	NGC 3330 - 66		⋆ AB	10 36 38	−53 52.2	1	8.82		-0.02				3	B6 II	2042
92368	−55 03665				10 36 40	−55 28.2	2	9.28	.030	0.47	.005	-0.04		4	F5 IV/V	1594,6006
92383	−57 03621				10 36 40	−57 46.9	2	9.35	.010	0.07	.005	-0.80	.005	5	B3 V	6,434
		Feige 34			10 36 41	+43 21.8	3	11.22	.013	-0.30	.000	-1.33	.052	5	sdO	1026,1298,3028
92384	−58 02457	vdBH 99 - 18			10 36 43	−59 51.9	1	8.54		-0.05		-0.44		6	B9 III/IV	1711

Table 1   561

HD	DM	Other Id	N	Rem	α_{1950}	δ_{1950}	S	V	σ_V	B–V	σ_{B-V}	U–B	σ_{U-B}	n	Spectrum	References
299254	−53 04025	NGC 3330 - 37			10 36 44	−53 50.5	1	10.38		0.07				2		2042
	−58 02458	FS Car			10 36 45	−58 53.6	1	11.90		0.29		0.21		21		1711
92380	−53 04027	NGC 3330 - 2			10 36 47	−53 54.4	1	10.03		−0.02				6	B9	2042
299261	−53 04028	NGC 3330 - 1			10 36 48	−53 55.7	1	10.52		0.07				2		2042
		AJ86,209 # 7			10 36 48	−58 33.	1	12.39		0.81		−0.17		1		924
92398	−58 02460	vdBH 99 - 20		⋆ B	10 36 51	−58 55.1	2	7.91	.053	0.05	.024	−0.37	.016	10	B9 II/III	401,1711
92397	−58 02460	vdBH 99 - 19		⋆ A	10 36 51	−58 55.3	2	4.74	.086	1.63	.005	1.82	.030	9	K4/5 III	3041,1711
92397	−58 02460	vdBH 99 - 19		⋆ AB	10 36 51	−58 55.3	4	4.67	.013	1.53	.070	1.48	.406	21	K4/5 III	15,1075,1711,2012
92399	−58 02462	vdBH 99 - 21			10 36 51	−59 00.1	2	6.49	.006	−0.13	.004	−0.62	.015	6	B4/5 III/IV	401,1711
		CCS 1728, TV Car			10 36 51	−60 33.8	1	11.10		3.62				2		864
92406	−58 02461	LSS 1712			10 36 52	−58 27.9	3	9.09	.034	−0.03	.004	−0.44	.014	4	Bp Shell	401,476,8100
307844	−62 01688	IC 2602 - 70			10 36 54	−63 18.8	1	9.30		1.41		1.47		1	K6	1053
		G 146 - 56			10 36 55	+42 24.9	1	13.24		0.63		−0.04		2		1658
	−58 02469	LSS 1714			10 36 59	−58 26.7	1	10.89		0.39		−0.49		2		476
303143	−56 03611	LSS 1715			10 37 01	−57 19.8	3	8.62	.019	1.18	.005	0.12	.018	9	A1 Iae	533,540,1737
		NGC 3330 - 30			10 37 03	−53 51.1	1	12.08		0.53				3		2042
92420	−57 03627	LSS 1716			10 37 03	−57 31.0	1	8.77		0.28		−0.64		2	B0.5/1Ib/II	533
92421	−58 02471	vdBH 99 - 22			10 37 03	−59 00.5	2	7.72	.003	−0.08	.002	−0.52	.014	5	B8 III	401,1711
		He3 461			10 37 04	−51 08.6	1	12.24		1.47		0.38		1		1753
303175	−57 03628	LSS 1717			10 37 04	−58 21.4	2	9.41	.010	0.09	.010	−0.79	.020	4	B8	476,836
92323	+9 02382	IDS10345N0922		AB	10 37 05	+09 06.2	3	7.50	.005	0.45	.002	0.04	.019	28	F5	130,292,3016
	−32 07520	LP 904 - 43			10 37 05	−33 20.0	1	10.09		0.72		0.35		1		1696
92436	−58 02474	vdBH 99 - 23		⋆ A	10 37 05	−58 33.4	5	5.95	.026	1.65	.023	2.00	.017	10	M1 III	401,1279,1637,1711,3055
92436	−58 02474	vdBH 99 - 23		⋆ AB	10 37 05	−58 33.4		5.88	.011	1.42	.015			17	M1 III	15,1075,2012
92467	−63 01534	IC 2602 - 18		⋆ AB	10 37 07	−64 14.2	2	6.98	.010	0.03	.000	−0.16		12	B9.5III	1053,2003
303182	−58 02475	vdBH 99 - 24		⋆ B	10 37 08	−58 33.3	4	8.34	.031	0.25	.029	−0.71	.033	7	B..	401,836,1279,1711
92278	+47 01797				10 37 09	+47 06.1	3	7.28	.016	0.17	.009	0.07	.003	70	A3 IV	275,1287,8097
		G 55 - 35			10 37 12	−06 39.7	2	11.31	.005	1.52	.002	1.24		2		1620,1705
303181	−58 02478	vdBH 99 - 25			10 37 13	−58 46.5	1	9.01		0.05		−0.04		4	A2	1711
		LP 430 - 58			10 37 14	+15 24.4	1	12.48		0.47		−0.20		2		1696
		NGC 3227 sq1 2			10 37 14	+24 21.1	1	11.11		0.49		−0.10		3		327
92403	−37 06699				10 37 14	−37 33.0	1	8.07		0.90		0.56		9	G6 III	1673
92451	−59 02378	LSS 1720			10 37 14	−60 04.8	3	8.99	.008	−0.07	.013	−0.88	.004	6	B1/2 II/III	476,540,976
92307	+36 02100				10 37 15	+35 45.2	1	8.59		1.28		1.37		2	K0	1601
303176	−57 03635	LSS 1719		⋆ ABC	10 37 15	−58 21.5	1	10.07		0.12		−0.56		2	B9	476
92478	−64 01383	IC 2602 - 19			10 37 18	−64 42.8	4	7.57	.012	0.05	.003	0.01	.007	16	A0 V	17,1053,2003,3015
92449	−54 03915	HR 4180		⋆ A	10 37 19	−55 20.5	8	4.27	.007	1.04	.006	0.76	.009	36	G2/3 Ib	15,1075,1279,1637*
92449	−54 03915	IDS10353S5505		C	10 37 19	−55 20.5	1	12.11		0.52		0.07		4		1279
92449	−54 03915	IDS10353S5505		E	10 37 19	−55 20.5	1	11.50		1.30		1.40		3		1279
307818	−61 01770	LSS 1722			10 37 19	−61 53.4	3	10.25	.022	0.41	.002	−0.57	.029	2	B0 Ia	191,540,1737
303173	−57 03640				10 37 20	−58 17.0	1	10.37		0.01		−0.74		2	B9	476
92492	−63 01539	IC 2602 - 20			10 37 21	−63 24.1	1	9.27		0.12		0.04		11	B9.5V	1053
92464	−55 03682				10 37 24	−55 43.7	3	7.10	.000	−0.09	.009	−0.57		10	B4 V	401,1075,2012
92463	−54 03916	IDS10353S5505		B	10 37 25	−55 20.8	2	6.25	.006	−0.08	.000	−0.31	.005	8	B8 V	1279,3077
92415	−15 03101	IDS10352S1550		B	10 37 32	−16 04.0	2	8.99	.000	0.68	.000			8	G3 V	2013,2033
92505	−60 02122				10 37 37	−60 43.5	2	7.01	.010	−0.11	.005	−0.65		6	B3/4 IV	6,2012
92536	−63 01542	IC 2602 - 22			10 37 37	−63 51.1	3	6.32	.011	−0.07	.005	−0.30	.010	14	B8 V	1053,2003,3015
92535	−63 01544	IC 2602 - 21			10 37 38	−63 31.0	3	8.25	.011	0.23	.013	0.10	.002	15	A5 Vn	1053,2005,3015
92504	−56 03622	LSS 1724		⋆ AB	10 37 39	−57 12.0	8	8.42	.009	−0.05	.008	−0.93	.012	10	O8	6,533,540,976
92440	−15 03103	IDS10352S1550		A	10 37 42	−16 06.6	2	7.75	.000	0.61	.000			8	G3 V	2013,2033
		G 44 - 30			10 37 43	+11 27.4	3	11.30	.010	0.64	.012	0.01	.006	7	G0	516,1620,1696
307856	−63 01547	IC 2602 - 76			10 37 44	−63 49.7	1	10.58		1.22		1.15		2	K0	1053
92410	+15 02236				10 37 45	+15 13.5	1	8.16		0.21		0.09		3	A2	1648
	−64 01393				10 37 45	−64 59.3	1	10.92		0.21		−0.52		2		540
92570	−64 01394	IC 2602 - 25		⋆ AB	10 37 47	−64 49.3	2	8.89	.024	0.46	.009	−0.01	.019	2	F6 IV/V	17,1053
92568	−63 01552	IC 2602 - 23			10 37 48	−63 25.9	1	8.58		0.42		0.14		9	Am	1053
92569	−64 01395	IC 2602 - 24			10 37 48	−64 25.6	4	9.46	.005	0.26	.019	0.16	.005	18	A0 V	891,1053,2003,2005
92555	−60 02125	LSS 1728			10 37 49	−61 01.9	2	9.14	.002	−0.05	.011	−0.92	.010	4	O9 V	540,976
303167	−57 03652	LSS 1726			10 37 50	−58 07.8	4	9.88	.019	0.60	.019	−0.36	.030	8	B0.5Iab	191,476,836,1737
305479	−59 02395	LSS 1727			10 37 50	−60 21.6	1	10.84		0.16		−0.58		2	B1 V	191
92501	−48 05822				10 37 51	−49 01.1	1	6.95		1.58		1.97		6	K5 III	1628
92554	−60 02126	LSS 1731			10 37 54	−60 39.1	3	9.48	.019	0.09	.016	−0.81	.017	6	O9/B0	6,540,976
		LP 790 - 30			10 37 56	−19 07.0	1	13.11		1.45		1.04		1		1773
92552	−55 03695				10 37 57	−56 01.0	1	8.00		−0.06		−0.54		1	B3/5 V	401
92645	−72 01004				10 37 57	−73 17.9	1	8.60		0.06		−0.09		2	B9.5 Vn	1776
92457	+12 02242	IDS10354N1236		A	10 37 59	+12 20.4	1	8.18		1.19		1.26		2	K2	1648
92566	−60 02129	LSS 1735			10 37 59	−60 44.1	1	8.72		0.20		0.08		1	B9.5II	8100
92387	+54 01388				10 38 02	+53 53.3	1	7.89		1.06		0.87		2	K0	1566
303142	−56 03632	LSS 1734			10 38 03	−57 17.3	1	9.73		0.20		−0.76		2	B0	533
92456	+26 02116				10 38 05	+25 57.7	2	7.62	.005	1.30	.000	1.25	.005	6	K1 III	833,1625
92550	−46 06325				10 38 06	−46 53.9	1	6.99		−0.07		−0.27		3	B9 V	1770
		AJ86,209 # 8			10 38 06	−59 45.	1	12.03		0.37		−0.46		1		924
92682	−73 00758	HR 4186			10 38 06	−74 13.9	3	6.06	.004	1.72	.005	1.90	.005	11	K3 II	15,1075,2038
92585	−56 03635				10 38 07	−56 38.8	1	9.03		−0.02		−0.48		2	B6 V	401
		LP 550 - 225			10 38 08	+06 05.8	1	11.23		0.68		0.06		4		1696
92584	−55 03701	LSS 1736			10 38 09	−56 18.8	1	8.72		0.05		−0.87		2	B3 Ib/II	401
92354	+69 00583	HR 4176			10 38 17	+68 42.3	1	5.75		1.30				2	gK3	71
303209	−58 02515				10 38 17	−59 08.8	1	10.37		0.00		−0.26		2	A0	1711
	−39 06571				10 38 18	−39 42.0	1	10.80		0.93		0.59		3		1696
92607	−59 02404	LSS 1738			10 38 18	−59 32.5	1	8.23		0.00		−0.87		2	O9 II/III	476

HD	DM	Other Id	N Rem	α₁₉₅₀	δ₁₉₅₀	S	V	σ_V	B–V	σ_B-V	U–B	σ_U-B	n	Spectrum	References
		AJ86,209 # 9		10 38 18	−60 07.	1	11.29		1.30		0.16		1		924
303179	−58 02517	LSS 1737		10 38 20	−58 29.0	2	9.71	.025	0.45	.020	−0.48	.015	4	B0	476,836
92608	−59 02409			10 38 20	−60 12.7	1	9.31		−0.05		−0.42		2		401
	−38 06655			10 38 21	−38 44.9	1	10.43		0.65		0.18		1		78
	−57 03658	LSS 1740		10 38 24	−57 25.7	1	10.92		0.76		−0.16		2	B2.5 Ib	1737
92563	−26 08070			10 38 25	−26 31.4	1	7.99		1.08		0.89		4	K0 III	1657
92664	−64 01403	IC 2602 - 27	⋆ V	10 38 27	−64 50.3	7	5.51	.011	−0.16	.005	−0.58	.016	35	B8 IIIp	15,17,1053,1075,2003*
92630	−58 02519			10 38 28	−58 41.2	1	9.44		0.06		−0.22		4	B9 III/IV	1711
92424	+66 00678	HR 4178		10 38 33	+65 58.7	3	5.13	.015	1.21	.010	1.28	.005	7	K2 III	1355,1363,3016
92589	−35 06646	HR 4183	⋆ AB	10 38 34	−35 28.8	3	6.36	.005	0.92	.000			14	G8 III +F	15,1075,2029
92663	−63 01558	IC 2602 - 26		10 38 35	−63 44.8	2	7.81	.005	1.52	.015	1.68		21	K2 III	1053,2005
305443				10 38 36	−59 41.2	1	10.59		0.11		−0.66		2	B5	476
92644	−59 02416			10 38 37	−59 31.0	1	8.84		−0.02		−0.84		2	O9.5/B0 III	476
307926	−63 01559	IC 2602 - 95		10 38 37	−63 45.6	1	9.70		1.34		1.49		6	K0	1053
92662	−58 02524	IDS10368S5901	A	10 38 40	−59 16.9	1	8.15		0.61		0.52		2	F2 Ib	1711
92626	−47 06099			10 38 42	−47 45.8	4	7.08	.004	1.34	.013	1.14	.018	15	Rp	565,864,993,2033
307910	−62 01706	LSS 1746	⋆ AB	10 38 42	−62 43.5	2	9.62	.089	0.11	.073	−0.77	.027	4	B3	540,976
	−59 02420	LSS 1745		10 38 45	−59 49.9	1	9.84		−0.08		−0.82		2		476
	−30 08626			10 38 49	−30 31.3	2	9.70	.015	0.76	.012	0.25		5		1594,6006
		G 146 - 58		10 38 51	+37 52.5	2	12.96	.005	1.68	.000	1.17	.010	3		203,3078
	−36 06589			10 38 51	−36 38.3	2	9.97	.030	1.48	.012	1.17	.000	2	M0 V	78,3072
92588	−0 02364	HR 4182		10 38 52	−01 28.7	3	6.26	.013	0.89	.010	0.59	.005	9	K1 IV	15,1256,3077
92693	−57 03663	LSS 1749		10 38 56	−57 40.4	2	6.97	.015	1.06	.010	0.45	.005	5	A3 Ia	191,533
92704	−60 02150	LSS 1751		10 38 56	−61 20.3	3	9.06	.018	−0.04	.006	−0.87	.014	6	B1 III/Vnn	401,540,976
92715	−64 01408	IC 2602 - 28		10 38 57	−64 23.5	3	6.82	.000	−0.02	.005	−0.14	.000	15	B9 Vnn	1053,2003,3015
92587	+14 02281			10 38 58	+14 14.4	1	7.63		1.65		2.00		2	M1	1648
92681	−51 04956	HP Vel		10 38 58	−51 49.4	1	8.70		1.73		1.35		1	M4/5 (Ip)	3044
92539	+47 01799	IDS10360N4710	AB	10 39 01	+46 54.7	1	8.56		0.12		0.09		std	A5 V	1287
92703	−58 02533			10 39 01	−58 37.6	1	8.76		1.12		1.02		4	K0 III	1711
92702	−56 03653	LSS 1750		10 39 02	−57 20.4	1	8.14		0.19		−0.71		2	B1 Iab	533
92712	−56 03657	LSS 1752		10 39 07	−56 59.7	2	7.89	.020	0.03	.015	−0.82	.020	4	B2/3 II/III	401,533
92714	−57 03671	LSS 1753		10 39 07	−58 18.5	6	9.41	.009	0.23	.018	−0.78	.044	10	Oe	6,476,540,836,976,1737
92655	−26 08077	IDS10368S2633	AB	10 39 09	−26 49.0	1	8.99		0.71		0.23		1	G6 V	78
		G 45 - 2		10 39 10	+12 22.1	1	11.61		1.42		1.13		1	K7	333,1620
92572	+50 01758			10 39 10	+49 50.3	1	9.51		0.29		0.11		1	Am	275
	−59 02438	LSS 1754		10 39 10	−60 01.9	1	10.66		0.20		−0.65		2	B2 III	476
92678	−35 06656			10 39 13	−35 28.2	1	6.82		0.04		0.02		2	A0 V	401
		G 44 - 32		10 39 14	+14 31.6	1	16.55		0.29		−0.56		3	DC	3060
92725	−58 02538	Bo 10 - 12		10 39 14	−58 59.2	4	8.25	.012	0.10	.009	−0.77	.017	11	B0/2 (III)	401,411,476,1711
307919	−63 01571	IC 2602 - 91		10 39 15	−63 30.3	1	10.06		1.14		0.90		1	G5	1053
92610	+33 02019			10 39 16	+33 24.7	1	9.39		0.34		0.04		2	F0	1733
305469	−59 02444	LSS 1757		10 39 16	−60 11.5	1	9.52		−0.11		−0.93		2	B3	401
92739	−58 02540	Bo 10 - 10	⋆ AB	10 39 17	−58 53.4	5	8.59	.009	0.06	.008	−0.80	.017	11	B1 II/III	401,411,476,1491,1711
92741	−59 02447	LSS 1758	⋆ AB	10 39 18	−59 42.7	4	7.25	.009	−0.01	.008	−0.81	.047	12	B1/2 II	6,620,1075,2012
92743	−60 02159			10 39 20	−60 24.5	1	9.33		0.00		−0.49		2	B3 III	401
92783	−63 01573	IC 2602 - 29		10 39 20	−64 12.8	3	6.73	.009	−0.05	.000	−0.21	.014	19	B8.5Vnn	1053,2003,3015
		Bo 10 - 26		10 39 21	−58 52.7	1	13.82		0.52		0.16		1		1491
	−59 02451	LSS 1760		10 39 21	−59 42.0	1	10.61		0.42		−0.47		2		476
92759	−57 03674			10 39 23	−57 24.1	2	9.43	.020	0.34	.005	−0.70	.005	5	Be	533,1737
92740	−59 02450	HR 4188, V429 Car	⋆	10 39 23	−59 24.9	4	6.42	.013	0.08	.004	−0.83	.000	9	WN7 + OB	15,476,1096,2012
92620	+32 02066	HR 4184, RX LMi	⋆ A	10 39 24	+31 57.5	3	6.00	.019	1.62	.007	1.80	.019	10	M2 III	31,70,3035
92538	+66 00679			10 39 24	+66 16.7	1	8.30		0.58		0.11		2	G0 V	1003
	−64 01411	Mel 101 - 1		10 39 24	−64 53.1	1	11.80		0.18				1		2004
92757	−55 03729			10 39 26	−55 48.4	1	6.80		−0.01		−0.08		2	B9.5V	401
	−59 02454	LSS 1763		10 39 26	−59 50.9	1	11.18		0.15		−0.61		2		476
303190	−58 02542	Bo 10 - 11		10 39 27	−58 53.6	3	10.23	.023	0.04	.011	−0.70	.024	6	B5	411,476,1491
310122	−64 01413	Mel 101 - 2		10 39 27	−64 53.3	1	11.35		0.26				1	A2	2004
303206	−58 02543			10 39 29	−59 10.1	1	9.37		0.05		−0.45		2	B8	1711
305483	−60 02160	LSS 1765		10 39 29	−60 30.8	1	9.74		0.12		−0.73		2	B7 Ia shell	540
92523	+69 00586	HR 4181		10 39 31	+69 20.3	2	5.00	.000	1.40	.012	1.54		4	K3 III	1363,3053
92756	−53 04064			10 39 31	−54 00.7	1	9.33		−0.04		−0.13		2	B9/A0 V	3044
303188	−58 02545	Bo 10 - 9		10 39 34	−58 48.7	3	9.14	.011	0.79	.009	0.07	.005	7	B3	411,476,1491
	−35 06662		A	10 39 35	−36 22.3	1	10.19		1.46		1.19		1	K7 V	3072
	−35 06662		B	10 39 35	−36 22.3	1	11.67		1.51		1.11		1	K7 V	3072
		AJ86,209 # 10		10 39 36	−59 03.	1	11.64		0.60		−0.32		1		924
307974	−64 01416	Mel 101 - 3		10 39 38	−64 50.4	1	9.73		1.85				1	K7	2004
92686	+11 02269			10 39 39	+10 37.1	1	7.08		1.74		1.99		2	K5	1648
92821	−63 01578	IC 2602 - 30		10 39 39	−63 31.1	1	8.87		1.28		1.31		1	K2	1053
		Bo 10 - 41		10 39 40	−58 49.5	1	12.26		0.39		0.05		1		1491
303225	−59 03225	LSS 1767		10 39 40	−59 24.0	1	9.94		0.03		−0.67		2		476
	−7 03095			10 39 42	−07 55.2	1	9.53		1.57		1.97		2		1375
92809	−58 02546	LSS 1768		10 39 42	−58 30.6	2	9.01	.047	0.26	.028	−0.33	.005	4	WC	836,1096
		G 253 - 38		10 39 44	+82 49.2	1	14.42		1.60				1		906
92719	−13 03196			10 39 44	−13 31.4	2	6.79	.009	0.62	.004			6	G2/3 V	897,6009
92808	−55 03737			10 39 44	−56 04.8	1	9.02		0.22		−0.22		2	F6/8 III/IV	476
92900	−73 00760			10 39 44	−73 45.5	1	8.80		0.08		−0.13		2	B9 V	1776
		Bo 10 - 40		10 39 48	−58 50.9	1	12.53		0.39		0.15		1		1491
92837	−63 01583	IC 2602 - 31		10 39 48	−63 50.7	3	7.17	.008	0.00	.000	−0.08		21	A0 IVnn	1053,2003,2005
307975	−64 01418	Mel 101 - 4		10 39 48	−64 50.6	1	11.14		0.26				1	B9	2004
92835	−58 02548	Bo 10 - 14		10 39 53	−58 52.1	2	8.87	.010	1.41	.005	1.47	.000	2	K1/3 (III)	411,1491

Table 1
563

HD	DM	Other Id	N Rem	α_{1950}	δ_{1950}	S	V	σ_V	B–V	σ_{B-V}	U–B	σ_{U-B}	n	Spectrum	References
		WLS 1040-10 # 10		10 39 54	−09 21.8	1	12.29		0.69		0.06		2		1375
		Bo 10 - 32		10 39 55	−58 52.9	1	12.64		0.32		-0.44		2		1491
92684	+47 01803	IDS10370N4713	AB	10 39 56	+46 57.2	1	8.90		0.78		0.34		5	G5	1723
300978	−56 03674	LSS 1769		10 39 56	−56 27.1	1	9.52		0.03		-0.81		2	B3	401
		POSS 169 # 5		10 39 59	+46 51.4	1	18.18		1.69				8		1739
303202	−58 02553	LSS 1771		10 40 00	−59 00.5	2	9.77	.001	0.12	.007	-0.72	.018	5		476,1711
92852	−58 02552	Bo 10 - 15		10 40 01	−58 50.6	2	9.68	.005	-0.01	.015	-0.70	.080	2	B3/5	411,1491
		Bo 10 - 24		10 40 01	−58 54.7	1	11.22		1.11		0.86		3		1491
		3C 245 # 2		10 40 02	+12 19.2	1	12.73		0.62		0.05		3		327
92850	−56 03676	LSS 1770		10 40 02	−56 44.8	6	8.06	.015	0.04	.007	-0.85	.016	12	B0 Ia	6,138,533,540,976,8100
307973	−64 01419	Mel 101 - 5		10 40 02	−64 49.0	1	11.51		0.29				1	A2	2004
310124	−64 01420	Mel 101 - 6		10 40 02	−64 56.0	1	11.06		0.33				1	A0	2004
92770	−13 03197	HR 4190		10 40 03	−13 42.8	1	6.24		1.54				4	K3/4 (III)	2006
303189	−58 02555	Bo 10 - 8		10 40 03	−58 52.1	3	10.28	.020	0.05	.023	-0.76	.030	7	B8	411,476,1491
92851	−56 03678	LSS 1772		10 40 04	−57 11.8	1	10.04		0.01		-0.71		2	B2/3 Ib	533
303187	−58 02557	Bo 10 - 25		10 40 04	−58 49.9	1	11.63		0.10		0.08		2		1491
		Bo 10 - 37		10 40 05	−58 56.0	1	12.06		0.20		-0.23		1		1491
307921	−63 01585	IC 2602 - 92		10 40 05	−63 38.1	1	9.04		1.58		1.60		2	K5	1053
92834	−53 04068	HH Vel		10 40 06	−53 27.6	1	8.70		1.61		1.10		1	M5/6 III	3044
92849	−52 03833			10 40 07	−53 14.8	1	8.86		-0.07		-0.32		2	Ap Si	3044
92804	−32 07567			10 40 08	−33 23.6	1	7.70		0.36		0.04		2	F3 V	401
305452	−59 02476	LSS 1776		10 40 08	−59 52.9	2	9.48	.010	0.04	.005	-0.67	.010	4	B5	476,850
310123	−64 01423	Mel 101 - 7		10 40 09	−64 54.5	1	11.34		0.96				1	A0	2004
		Steph 877		10 40 11	+03 32.9	1	12.52		1.46		1.21		1	M0	1746
		POSS 169 # 1		10 40 11	+46 41.4	1	14.74		0.95				2		1739
92896	−62 01720	IC 2602 - 32	★ AB	10 40 11	−63 14.7	4	7.31	.008	0.22	.002	0.06	.000	17	A3 IV	1053,2005,2012,3015
92847	−46 06361	IDS10381S4642	A	10 40 12	−46 57.5	1	6.95		1.33		1.47		6	K2 III	1628
		AJ86,209 # 11		10 40 12	−58 42.	1	12.48		0.55		-0.36				924
		Bo 10 - 16		10 40 12	−58 52.9	2	12.02	.014	0.47	.014	0.08	.014	4		411,1491
92875	−58 02561	LSS 1777		10 40 12	−59 04.1	2	9.22	.064	0.33	.030	-0.56	.054	5	B9	476,1711
	−64 01424	Mel 101 - 8		10 40 12	−64 51.2	1	12.02		0.42				1		2004
		LP 610 - 56		10 40 13	−01 25.6	1	12.26		1.12				1		1696
92877	−59 02478	IDS10383S5923	AB	10 40 13	−59 38.7	1	8.50		-0.02		-0.70		2	B2 III	401
305439	−59 02479	IDS10384S5927	A	10 40 13	−59 42.3	2	9.57	.019	0.51	.015	-0.52	.010	3		850,1737
305439	−59 02479	IDS10384S5927	AB	10 40 13	−59 42.3	2	9.14	.010	0.50	.015	-0.53	.025	4		476,1737
305439	−59 02479	IDS10384S5927	B	10 40 13	−59 42.3	1	10.88		0.51		-0.73		1		1737
		3C 245 # 1		10 40 14	+12 17.7	1	15.38		1.08		0.86		2		327
	−58 02563	Bo 10 - 7		10 40 15	−58 52.6	4	10.58	.015	0.11	.015	-0.68	.022	11		411,476,1491,1737
305455	−59 02480	LSS 1782		10 40 15	−59 58.9	2	10.18	.045	0.30	.000	-0.67	.000	4		476,850
	−64 01425	Mel 101 - 9		10 40 16	−64 49.5	1	11.87		0.28				1		2004
92769	+27 01927	HR 4189	★ A	10 40 17	+26 35.3	2	5.51	.017	0.17	.004	0.09		5	A4 Vn	70,1022
303297	−58 02565	Bo 10 - 4		10 40 17	−58 54.0	2	9.56	.015	0.04	.000	-0.81	.005	3	A2	411,1491
		POSS 169 # 4		10 40 19	+47 30.6	1	18.06		1.47				3		1739
	−64 01426	Mel 101 - 10		10 40 19	−64 51.5	1	11.66		0.36				1		2004
		G 163 - 6		10 40 20	+03 03.1	2	14.72	.016	1.57	.023	1.23		4		906,1764
92894	−58 02567	Bo 10 - 5		10 40 20	−58 52.5	3	9.40	.005	0.11	.005	-0.76	.024	7	A0 V	411,476,1491
307922	−63 01587	IC 2602 - 93		10 40 20	−63 42.0	1	9.99		0.60		0.18		2	F5	1053
92845	−32 07572	HR 4194	★ A	10 40 24	−32 27.2	3	5.63	.004	0.00	.000	-0.01		9	A0 V	15,401,2012
	+21 02227	G 58 - 12		10 40 25	+20 36.9	1	10.58		0.61		0.04		2	G0	333,1620
		Bo 10 - 17		10 40 25	−58 53.4	3	11.80	.026	0.56	.023	0.01	.024	3		411,1491,1491
92909	−58 02570	vdBH 99 - 26		10 40 25	−58 54.6	3	9.33	.028	0.03	.014	-0.02	.016	4	B9 V	411,1491,1711
307972	−64 01430	Mel 101 - 11		10 40 25	−64 47.6	1	11.69		0.23				1		2004
		Bo 10 - 22		10 40 27	−58 52.4	1	12.29		0.77		0.32		1		1491
92938	−63 01589	IC 2602 - 33	★	10 40 27	−64 12.3	6	4.81	.017	-0.14	.008	-0.58	.009	42	B3 Vn	15,17,1053,1075,2003,2038
	−64 01432	Mel 101 - 12		10 40 27	−64 47.8	1	11.57		0.26				1		2004
		Bo 10 - 21		10 40 28	−58 52.6	1	12.31		-0.14		-0.44		1		1491
92764	+45 01857			10 40 29	+45 12.8	1	9.04		0.19		0.16		3	A7 V	8100
303296	−58 02573	Bo 10 - 3		10 40 29	−58 53.7	2	9.61	.010	0.04	.005	-0.78	.020	4	Be	411,1491
92860	−21 03137			10 40 30	−22 14.7	1	7.92		0.43				4	F3 V	2012
	−64 01433	Mel 101 - 13		10 40 30	−64 51.1	1	11.90		0.48				1		2004
		Bo 10 - 20		10 40 33	−58 50.0	1	11.35		0.60		0.06		1		1491
92728	+57 01286	HR 4187		10 40 35	+57 27.7	1			-0.02		-0.10		1	A0 Vs	1079
92787	+46 01657	HR 4191	★ A	10 40 37	+46 28.0	3	5.18	.000	0.32	.005	0.01	.000	4	F1 IV	15,1008,3026
92786	+48 01889	G 146 - 60		10 40 37	+48 28.4	1	8.02		0.74		0.27		1	G5	1658
		Tr 14 - 63		10 40 38	−59 11.1	1	12.88		0.29		0.28		2		998
92966	−63 01592	IC 2602 - 34		10 40 38	−64 08.2	3	7.28	.011	-0.01	.004	-0.10	.005	20	B9.5Vnn	1053,2003,3015
307924	−63 01593	IC 2602 - 94		10 40 39	−63 43.9	1	10.70		0.62		0.02		1	F8	1053
307971	−64 01434	Mel 101 - 14		10 40 39	−64 47.8	1	11.39		0.24				1		2004
	−57 03698			10 40 40	−57 44.6	1	11.62		0.22		0.16		1	B9	761
	−64 01435	Mel 101 - 15		10 40 40	−64 48.1	1	11.29		0.25				1		2004
92824	+26 02123			10 40 41	+26 01.7	1	8.84		0.54		-0.04		2	F8 V	3026
	−59 02495	Cr 228 - 30		10 40 41	−59 43.7	1	10.80		0.05		-0.69		2	B1.5V	472
	−64 01436	Mel 101 - 16		10 40 41	−64 49.5	1	12.17		0.23				1		2004
92825	+23 02253	HR 4192		10 40 42	+23 27.0	3	5.09	.010	0.04	.003	0.06	.012	9	A3 Vn	1022,1363,3050
92936	−56 03696	LSS 1788		10 40 42	−56 36.9	5	7.04	.009	-0.05	.009	-0.83	.007	14	B1 II	158,401,533,540,976
305451	−59 02496	Cr 228 - 29		10 40 42	−59 46.8	1	10.21		0.07		-0.36		2	B9.5Vp?	472
		Bo 10 - 27		10 40 43	−58 50.9	1	12.72		0.30		-0.32		1		1491
		Bo 10 - 31		10 40 44	−58 54.9	1	11.74		0.13		-0.33		2		1491
92964	−58 02581	HR 4198		10 40 44	−58 57.2	10	5.38	.010	0.26	.014	-0.66	.015	40	B2.5 Iae	15,285,401,620,681*
92841	+5 02384	HR 4193	★ AB	10 40 45	+05 00.6	2	5.78	.005	1.17	.000	1.09	.000	7	K3 III +	15,1417

HD	DM	Other Id	N	Rem	α_{1950}	δ_{1950}	S	V	σ_V	B–V	σ_{B-V}	U–B	σ_{U-B}	n	Spectrum	References
303295	−58 02582	Bo 10 - 2			10 40 48	−58 51.3	3	10.78	.000	0.06	.005	-0.71	.012	7	B8	411,476,1491
305438	−59 02502	Cr 228 - 24			10 40 48	−59 38.5	4	8.80	.007	-0.04	.013	-0.94	.017	10	O7.5V	390,401,472,850
		G 55 - 38			10 40 50	+06 03.8	1	12.85		1.30		1.33		1		333,1620
303290	−58 02584	Bo 10 - 38			10 40 50	−58 47.1	1	11.12		0.09		-0.28		1		1491
305437	−59 02503	Cr 228 - 23			10 40 50	−59 36.6	3	9.08	.008	-0.01	.014	-0.85	.021	8	B0.5V	390,472,850
307931	−63 01594	IC 2602 - 98			10 40 50	−63 55.2	1	10.11		1.21		1.07		14	K5	1053
307970	−64 01438	Mel 101 - 17			10 40 50	−64 44.8	1	11.46		0.52				1		2004
92906	−34 06931				10 40 51	−34 44.5	1	7.50		0.98				4	K0 III	2012
	−58 02585	Bo 10 - 19			10 40 51	−58 49.2	1	11.05		0.52		-0.40		3		1491
		Tr 14 - 69			10 40 51	−59 19.2	1	11.88		0.19		-0.43		2		998
	−59 02504	Cr 228 - 49			10 40 51	−59 45.2	1	11.20		0.01		-0.36		1		472
305453	−59 02505	LSS 1794		★ ABC	10 40 51	−59 56.3	1	9.73		-0.04		-0.80		2	K1 III	850
		G 254 - 6			10 40 52	+72 48.2	1	10.99		1.37		1.30		2		7010
92985	−60 02193	LSS 1796			10 40 52	−61 05.9	1	9.42		0.03		-0.82		3	B2/3 II	466
92989	−64 01439	IC 2602 - 35			10 40 52	−64 24.9	4	7.59	.008	0.04	.000	-0.01	.002	18	A0.5Va	1053,2003,2005,3015
92982	−57 03700				10 40 53	−58 02.7	4	8.76	.038	0.35	.012	-0.62	.035	8	B1 II	6,476,533,8100
303313	−58 02586				10 40 54	−59 09.8	1	10.31		0.15		-0.58		3	B8	476
92811	+50 01760				10 40 56	+50 03.8	2	7.02	.015	1.17	.010	1.19	.010	5	K1 III	37,1733
92948	−47 06136				10 40 56	−48 21.4	1	8.24		-0.05		-0.44		1	B7 III	1770
303295		Bo 10 - 23			10 40 57	−58 51.4	1	9.98		0.12		-0.42		1	B8	1491
305544	−59 02507	Cr 228 - 15			10 40 57	−59 47.3	2	8.59	.000	0.66	.000	0.07	.010	6	G5	472,1696
		Tr 14 - 64			10 40 59	−59 17.4	1	12.53		0.22		0.19		2		998
		WLS 1040-10 # 7			10 41 00	−12 21.4	1	12.13		0.40		0.01		2		1375
305535	−59 02509	Cr 228 - 25			10 41 00	−59 42.5	1	9.39		0.04		-0.44		3	B2.5Vn	472
	+47 01806				10 41 02	+46 32.6	1	9.12		0.62		0.09		2	G1 IV	1003
	−64 01442	Mel 101 - 19			10 41 02	−64 56.8	1	12.60		0.29				1		2004
92946	−31 08469				10 41 03	−31 52.5	1	7.16		0.00		-0.03		2	B9.5V	401
307977	−64 01441	Mel 101 - 18			10 41 03	−64 48.6	1	11.63		0.28				1		2004
93012	−64 01443	IC 2602 - 36			10 41 03	−64 56.0	3	9.25	.007	0.33	.002	0.03		9	A1 IV	2005,2012,3015
303291	−58 02590	Bo 10 - 18			10 41 04	−58 49.3	1	10.59		0.08		-0.55		6		1491
92855	+46 01658	IDS10377N4643	B		10 41 05	+46 28.2	1	7.32		0.55		0.00		2	F9 V	3026
		Cr 228 - 42			10 41 06	−59 45.	1	10.48		0.66		0.20		2		472
93002	−58 02592	Bo 10 - 1			10 41 07	−58 49.2	3	9.50	.011	0.03	.000	-0.76	.015	9	B2 III	411,476,1491
92763	+69 00587	R UMa			10 41 08	+69 02.4	1	6.97		1.43		1.02		1	M4	3001
92945	−28 08394				10 41 08	−28 48.1	1	7.75		0.89		0.56		1	K1 V	78
	−64 01444	Mel 101 - 20			10 41 09	−64 50.6	1	12.18		0.33				1		2004
		Bo 10 - 36			10 41 09	−58 56.3	1	12.08		0.25		0.17		1		1491
305515	−59 02513	Cr 228 - 44			10 41 09	−59 35.9	1	10.35		0.09		-0.59		1	B1.5V sn:	472
	−77 00638				10 41 09	−77 55.3	1	9.80		1.02		0.76		2	K5 V	3072
		Cr 228- 101			10 41 10	−59 36.9	1	11.08		0.00		-0.60		2		850
93003	−60 02199				10 41 10	−60 43.5	1	7.12		-0.02		-0.55		2	B3 IV	6
93030	−63 01599	IC 2602 - 37		★	10 41 10	−64 07.9	5	2.76	.010	-0.23	.005	-1.01	.012	22	B0 Vp	15,1053,1075,2003,8015
		LP 213 - 61			10 41 11	+43 18.5	1	12.17		1.22		1.19		3		1723
	−64 01445	Mel 101 - 21			10 41 11	−64 54.9	1	12.02		0.34				1		2004
	−57 03713				10 41 15	−58 04.6	1	10.99		0.15		-0.50		2	B6	761
305543	−59 02517	Cr 228 - 28		★ V	10 41 15	−59 46.4	2	9.84	.109	0.03	.020	-0.79	.020	5	B1 V +B1 V	472,850
	+45 01859				10 41 16	+44 56.5	1	9.81		0.32		0.07		4	A7 Ib	8100
		Bo 10 - 30			10 41 16	−58 50.2	1	11.82		0.26		0.18		2		1491
	−59 02518	LSS 1800		★ A	10 41 16	−59 28.6	3	9.64	.010	0.27	.010	-0.75	.011	7	O6 V	850,924,1737
93010	−60 02203	IDS10394S6039	A		10 41 16	−60 54.4	1	6.63		0.00		-0.56		2	B3 III	6
93010	−60 02203	IDS10394S6039	AB		10 41 16	−60 54.4	1	6.39		0.00				4	B3 III	2012
		Tr 14 - 58			10 41 19	−59 16.8	1	13.23		1.57		1.81		1		998
92987	−38 06693				10 41 20	−38 47.8	1	7.02		0.64		0.15		7	G2/3 V	1628
		Bo 10 - 13			10 41 20	−58 54.6	1	11.29		0.10		-0.60		2		411
93026	−58 02594	Bo 10 - 35		★	10 41 20	−58 54.7	1	9.66		0.03		-0.80		2	B2 III	1491
305516	−59 02519	Cr 228 - 31			10 41 20	−59 35.3	2	9.87	.000	0.04	.015	-0.80	.025	4	B0.5V:B	472,850
	−59 02520	Cr 228 - 26			10 41 20	−59 52.0	1	10.63		0.21		-0.03		2	A0 Vn	472
93028	−59 02521	Cr 228 - 27			10 41 20	−59 56.3	6	8.37	.016	-0.07	.011	-0.94	.034	14	O9 V	6,138,390,472,620,850
		LP 490 - 61			10 41 21	+13 03.9	1	12.22		0.48		-0.18		2		1696
		Tr 14 - 62			10 41 21	−59 13.3	1	12.72		0.38		-0.28		1		998
93237	−79 00548	HR 4206			10 41 21	−79 31.3	5	5.96	.013	-0.06	.006	-0.51	.007	20	B5 III	15,26,1075,1637,2038
		Tr 14 - 52			10 41 22	−59 20.3	1	13.52		0.55		0.07		1		998
93027	−59 02522	Cr 228 - 14			10 41 22	−59 52.3	2	8.72	.000	-0.02	.019	-0.88	.019	6	O9.5V	472,850
310125	−64 01449	Mel 101 - 22			10 41 22	−64 52.0	1	11.43		0.54				1	A5	2004
		Tr 14 - 51			10 41 23	−59 19.7	1	12.38		0.37		0.11		1		998
93025	−56 03704				10 41 24	−57 14.8	1	9.53		0.09		-0.71		2	B1/2 III/IV	401
		AJ86,209 # 12			10 41 24	−59 09.	1	12.42		0.33		-0.46		1		924
		Tr 14 - 59			10 41 24	−59 17.5	1	13.18		0.34		0.28		2		998
		IC 2602 - 198			10 41 24	−64 08.	1	12.42		0.54		0.40		9		891
307937	−63 01605	IC 2602 - 101			10 41 26	−64 01.1	1	9.76		0.11		-0.26		4	A0	1053
		IC 2602 - 311			10 41 27	−63 48.6	1	12.38		0.46		0.22		1		891
307933	−63 01607	IC 2602 - 99		★ AB	10 41 28	−63 55.3	1	8.90		0.43		0.25		1	A3	1053
310138	−65 01471				10 41 28	−65 31.7	1	10.75		1.90		2.07		1	F5	864
		IC 2602 - 330			10 41 29	−63 49.5	1	15.61		0.73		0.17		3	B9	891
	−59 02564	Cr 228 - 116			10 41 30	−59 37.0	1	10.51		0.67		0.13		2		850
		Bo 10 - 39			10 41 31	−58 51.9	1	11.30		0.10		-0.66		3		1491
		Tr 14 - 57			10 41 31	−59 16.9	1	12.63		1.19		0.88		1		998
		Tr 14 - 50			10 41 31	−59 20.0	1	12.69		1.35		1.30		1		998
		Cr 228 - 72			10 41 31	−59 49.5	1	13.27		0.18		-0.21		1		472
93055	−58 02602	Bo 10 - 33			10 41 32	−58 50.3	1	8.76		-0.02		-0.12		2	B8/9 II	1491

Table 1   565

HD	DM	Other Id	N Rem	α_{1950}	δ_{1950}	S	V	σ_V	B–V	σ_{B-V}	U–B	σ_{U-B}	n	Spectrum	References
93056	−59 02527	Cr 228 - 13	⋆ AB	10 41 32	−59 50.2	2	8.97	.005	-0.05	.014	-0.79	.010	7	B1 Vb:	472,850
92941	+20 02514	HR 4197		10 41 33	+20 01.3	1	6.27		0.17				2	A5 V	70
		Klemola 47		10 41 34	+39 28.9	1	11.28		-0.12		-0.48		2		1026
93057	−59 02528			10 41 34	−60 03.1	1	8.25		0.39		0.07		2	F0 III	401
	−58 02604	LSS 1807		10 41 35	−59 13.7	2	9.99	.015	0.34	.005	-0.61	.030	5	B0 V(n)	476,1737
		Tr 14 - 56		10 41 36	−59 17.9	1	13.32		0.30		0.25		1		998
	−59 02530	Cr 228 - 78		10 41 36	−59 47.5	1	11.44		0.18		-0.36		1		472
		POSS 169 # 6		10 41 37	+46 40.0	1	19.10		1.15				3		1739
92839	+68 00617	HR 4195, VY UMa		10 41 37	+67 40.5	7	5.99	.046	2.41	.024	4.72	.104	11	C5 II	15,765,1238,3001,8015*
	−58 02605	Tr 14 - 30		10 41 37	−59 19.4	4	10.07	.005	0.51	.006	-0.48	.041	9	B0 III-IV:	315,479,998,1737
		Cr 228 - 73		10 41 37	−59 49.3	1	12.92		1.14		0.80		1		472
307934	−63 01612	IC 2602 - 100		10 41 37	−63 57.4	1	10.08		1.63		1.90		10	K7	1053
93070	−59 02532	HR 4200		10 41 38	−60 18.2	4	4.56	.004	1.70	.012	1.84	.030	13	K4 III	15,1075,2012,8100
		IC 2602 - 308		10 41 38	−63 50.9	1	12.63		0.63		0.23		1		891
		Tr 14 - 53		10 41 40	−59 19.1	1	13.48		0.56		-0.32		2		998
		Car sq 2 # 2		10 41 41	−60 28.7	1	13.03		0.30		-0.52		3		466
93098	−63 01613	IC 2602 - 38		10 41 41	−63 48.4	3	7.61	.005	0.04	.004	0.01	.010	15	A1 Vs	1053,2003,3015
		Tr 14 - 54		10 41 42	−59 19.0	1	12.11		0.20		-0.21		1		998
	−58 02608	LSS 1809	⋆ AB	10 41 44	−59 20.4	1	10.98		0.55		-0.52		3	O7 V	1737
	−59 02533	Cr 228 - 55		10 41 44	−59 39.5	1	11.71		0.31		0.14		1		472
		Cr 228 - 71		10 41 44	−59 49.5	1	11.40		1.00		0.44		1		472
		Bo 10 - 28		10 41 45	−58 50.5	1	11.03		0.38		0.16		3		1491
		Tr 14 - 55		10 41 45	−59 18.1	1	13.15		0.58		0.31		1		998
		Cr 228- 108		10 41 46	−59 54.2	1	13.10		0.12		-0.26		2		850
		Tr 14 - 28		10 41 47	−59 18.3	1	12.50		0.39		-0.53		2	O7	315
		Tr 14 - 61		10 41 47	−59 20.0	1	13.32		0.62		-0.29		1		998
92992	+11 02273			10 41 48	+10 38.3	1	7.52		1.29		1.47		2	K2	1648
		Tr 14 - 27		10 41 48	−59 18.0	2	11.30	.015	0.31	.005	-0.58	.015	6	O9 V	315,998
		Tr 14 - 60		10 41 48	−59 19.6	1	13.63		0.49		-0.33		1		998
305518	−59 02536	Cr 228 - 22		10 41 48	−59 32.5	3	9.73	.027	0.35	.018	-0.63	.020	6	O9.5IV	6,472,850
305556	−59 02538	LSS 1811		10 41 48	−60 04.7	2	8.78	.005	0.08	.015	-0.80	.049	3		620,850
93115	−62 01731	IC 2602 - 39		10 41 48	−63 23.0	1	7.85		1.52		1.78		6	K2 III	1053
		Tr 14 - 10		10 41 50	−59 17.1	1	12.48		0.27		-0.50		1	B2 V	315
	−58 02611	Tr 14 - 20		10 41 50	−59 17.1	4	9.61	.007	0.30	.023	-0.73	.005	12	O6 V((f))	238,315,1737,2042
	−59 02537	Cr 228 - 43		10 41 50	−59 37.6	2	10.41	.005	0.13	.068	-0.68	.019	3	B2 Vb	472,850
93114	−58 02612	Bo 10 - 34		10 41 51	−58 50.4	1	8.59		0.22		0.21		1	A7 II/III	1491
93097	−59 02540	Cr 228 - 69	⋆ A	10 41 51	−59 50.0	2	9.78	.015	-0.02	.000	-0.85	.029	3	B0 Vn	472,850
		Cr 228 - 70	⋆ B	10 41 51	−59 50.1	2	10.83	.070	0.92	.055	0.47	.235	2		472,850
303248	−57 03729	LSS 1812		10 41 52	−57 45.2	1	10.07		0.41		-0.59		1	B0	761
		Cr 228 - 54		10 41 52	−59 40.5	1	12.72		0.64		0.17		1		472
	−59 02539	Cr 228 - 74		10 41 52	−59 52.6	1	11.66		0.18		-0.43		1		472
	−58 02614	Tr 14 - 21		10 41 53	−59 17.6	1	10.88		0.34		-0.63		3	O9 V	315
		Tr 14 - 22		10 41 53	−59 17.8	1	12.34		0.32		-0.48		2	B2 V	315
		Cr 228 - 76		10 41 53	−59 45.5	1	12.44		0.04		-0.45		1		472
		Tr 14 - 66		10 41 54	−59 12.4	1	12.18		0.22		0.16		2		998
305521	−59 02542	Cr 228 - 16		10 41 54	−59 41.6	2	9.84	.035	0.06	.005	-0.73	.045	5	B0.5Vn:	472,850
	−59 02541	Cr 228 - 77		10 41 54	−59 44.8	1	11.57		0.20		-0.58		1		472
	−59 02543	Cr 228 - 48		10 41 54	−59 53.2	1	11.00		-0.03		-0.62		2	B1.5Vb:	472
		KS 458		10 41 54	−64 16.	1	10.72		1.88		2.05		2		540
		Tr 14 - 24		10 41 55	−59 18.1	1	12.12		0.37		-0.49		3	B1 V	315
	−59 02546	Cr 228 - 75		10 41 55	−59 46.1	1	11.15		0.32		0.19		1		472
	−59 02545	Cr 228 - 9		10 41 55	−59 47.8	2	9.58	.020	0.06	.010	-0.01	.010	4		472,850
	−59 02547	Cr 228 - 53		10 41 56	−59 41.5	2	10.96	.019	0.09	.010	-0.69	.034	3		472,850
		IC 2602 - 309		10 41 56	−64 00.1	1	14.18		0.40		0.09		1		891
		Tr 14 - 65		10 41 57	−59 12.6	1	12.05		0.61		-0.45		2		998
93131	−59 02548	Cr 228 - 3		10 41 57	−59 51.3	4	6.48	.004	-0.03	.009	-0.89	.012	13	WN	158,472,850,1096
		Tr 14 - 3		10 41 58	−59 17.2	2	10.74	.055	0.29	.030	-0.63	.020	2	B0.5IV-V	315,479
93128	−58 02617	Tr 14 - 2	⋆ AB	10 41 58	−59 17.3	3	8.79	.042	0.24	.009	-0.75	.015	6	O4 V	238,315,1737
		Tr 14 - 5		10 41 58	−59 17.7	1	11.41		0.33		-0.59		4	O9 V	315
		Tr 14 - 67		10 41 59	−59 11.2	1	12.62		0.30		-0.30		2		998
		Tr 14 - 14		10 41 59	−59 16.2	1	12.62		0.23		0.37		2		315
		Tr 14 - 9		10 41 59	−59 17.0	2	9.89	.034	0.21	.000	-0.71	.015	3	O8 V	315,479
		Tr 14 - 4		10 41 59	−59 17.5	2	11.05	.041	0.26	.027	-0.68	.005	3	B0 V	315,479
		Tr 14 - 11	V	10 42 00	−59 16.7	1	12.78		0.39		0.05		6		315
		Tr 14 - 26	V	10 42 00	−59 19.0	1	11.93		0.37		-0.57		7	B2 V	315
93113	−57 03731	LSS 1818		10 42 01	−57 25.3	1	8.67		0.03		-0.76		3	B1/2 II/III	533
		Tr 14 - 13		10 42 01	−59 16.3	1	12.63		0.24		-0.40		2		315
93129	−58 02618	Tr 14 - 1	⋆ AB	10 42 01	−59 17.1	3	7.02	.019	0.22	.023	-0.78	.012	5	O3 (f)	315,390,1737
		Tr 14 - 25	V	10 42 01	−59 18.6	1	12.88		0.32		-0.20		5		315
		Cr 228- 107		10 42 01	−59 56.2	1	12.80		1.05		0.52		1		850
93083	−32 07598			10 42 02	−33 18.7	2	8.30	.000	0.95	.005	0.72		7	K3 V	2033,3072
		Tr 14 - 12		10 42 02	−59 16.8	1	12.64		0.24		-0.48		1	B2 V	315
		Tr 14 - 19	V	10 42 02	−59 17.2	1	11.58		0.34		-0.42		2	B1 V	315
		Tr 14 - 6		10 42 02	−59 17.8	1	11.23		0.19		-0.68		4	B1 V	315
		Tr 14 - 18		10 42 02	−59 18.1	1	12.15		0.27		-0.59		4	B0 V	315
	−59 02551	Cr 228 - 21		10 42 02	−59 49.7	3	9.32	.008	0.01	.010	-0.89	.022	8	O7.5Vn	472,850,1737
		Cr 228 - 66	⋆ B	10 42 02	−59 49.7	2	9.86	.058	0.03	.029	-0.83	.029	3	O9.5V	472,850
		Tr 14 - 36		10 42 03	−59 17.5	1	11.23		1.22				2		2042
		Cr 228 - 56		10 42 03	−59 40.4	1	13.42		0.25		-0.18		1		472
93163	−63 01619	IC 2602 - 40	⋆	10 42 03	−63 59.2	6	5.77	.015	0.00	.011	-0.55	.008	23	B3 IV	15,26,1053,2003,3015,8023

HD	DM	Other Id	N	Rem	α_{1950}	δ_{1950}	S	V	σ_V	B–V	σ_{B-V}	U–B	σ_{U-B}	n	Spectrum	References
93064	−23 09500				10 42 04	−23 43.3	1	6.62		1.28		1.31		5	K2 II	1628
	−58 02623	LSS 1822			10 42 04	−59 09.9	1	10.07		0.17		−0.70		2		476
		Tr 14 - 15			10 42 04	−59 16.0	1	12.01		0.30		−0.15		2	B7 V	315
		Tr 14 - 16			10 42 04	−59 16.2	1	13.60		0.27		−0.27		1		315
	−58 02620	Tr 14 - 8		⋆ AB	10 42 04	−59 16.6	2	9.35	.063	0.17	.005	−0.75		3	O7 V	238,315
		Tr 14 - 7	V		10 42 04	−59 16.7	1	12.12		0.32		−0.45		2	B1 V	315
		Tr 14 - 35			10 42 04	−59 17.4	1	11.61		1.00				2		2042
	−58 02621	Tr 16 - 126			10 42 04	−59 19.6	1	10.97		0.41		−0.47		3	B1 V	315
93130	−59 02556	Cr 228 - 1		⋆ A	10 42 04	−59 36.7	5	8.06	.024	0.23	.020	−0.74	.024	14	O6 III(f)	6,472,620,850,1737
		Cr 228- 110			10 42 04	−59 47.8	1	12.82		0.38		0.22		2		850
		G 236 - 38			10 42 05	+66 50.3	1	12.82		0.78		0.36		1		1658
	−58 02625	Tr 16 - 127			10 42 05	−59 20.0	1	10.70		0.35		−0.55		2	O7	315
		Cr 228 - 60			10 42 05	−59 36.7	1	12.71		0.37		−0.26		1		472
		Cr 228 - 61			10 42 05	−59 36.8	1	11.70		0.22		−0.59		1		472
93146	−59 02555	Cr 228 - 65		⋆ AB	10 42 05	−59 49.4	3	8.44	.016	0.02	.013	−0.91	.000	6	O6.5V	390,472,850
	−59 02554	Cr 228 - 67		⋆ AB	10 42 05	−59 50.2	2	8.79	.015	−0.01	.010	−0.86	.029	3	O9 V	472,850
		Cr 228 - 68		⋆ C	10 42 05	−59 50.3	2	10.31	.122	0.02	.024	−0.77	.034	3	B1 Vn	472,850
93109	−52 03863	HI Vel			10 42 06	−52 49.8	1	8.50		1.64		1.40		1	M3/5 III	3044
303255	−57 03735	LSS 1824			10 42 06	−57 48.9	1	9.67		0.39		−0.53		2	B3	533
		AJ86,209 # 13			10 42 06	−58 34.	1	12.47		0.37		−0.43		1		924
	−58 02627	Tr 14 - 39			10 42 06	−59 13.9	2	10.15	.005	0.26	.020	−0.71		5		476,2042
		Tr 14 - 17			10 42 06	−59 16.7	1	12.65		0.29		−0.32		2		315
		AJ82,163 # 1			10 42 09	−57 21.3	1	11.15		1.63		1.80		4		533
		AJ82,163 # 2			10 42 09	−57 21.5	1	12.94		0.26		0.19		5		533
		Tr 14 - 29	V		10 42 09	−59 17.9	1	11.97		0.50		−0.33		3	B1.5V	315
		Car sq 2 # 3			10 42 09	−61 01.6	1	12.60		0.29		−0.64		3		466
92880	+73 00504				10 42 10	+72 33.1	1	6.92		1.63		1.95		2	K0	1375
93124	−49 05492				10 42 10	−49 34.7	1	8.50		0.34		0.06		1	F0 V	78
	−58 02629	Tr 16 - 124			10 42 10	−59 19.4	1	11.13		0.24		−0.65		2	B1 V	315
		Cr 228 - 59			10 42 10	−59 37.8	1	12.89		0.60		0.02		1		472
305520	−59 02560	Cr 228 - 4			10 42 10	−59 43.9	3	8.70	.013	0.18	.007	−0.71	.013	14	B1 Ib	472,479,850
		IC 2602 - 306			10 42 10	−64 00.9	1	13.12		0.74		0.12		2		891
		AJ82,163 # 4			10 42 11	−57 18.0	1	11.21		0.44		0.41		2		533
93160	−58 02631	Tr 16 - 175		⋆ AB	10 42 11	−59 18.7	2	7.82	.005	0.17	.000	−0.77		3	O6 III(f)	238,315
	−57 03739	LSS 1829			10 42 12	−57 51.2	1	10.55		0.49		−0.48		2		761
	−58 02630	Tr 14 - 40			10 42 12	−59 14.3	1	11.15		0.37				2		2042
	−59 02559	Cr 228 - 45			10 42 12	−59 18.7	3	10.19	.019	0.99	.008	0.70	.007	5		472,479,850
93161	−58 02631	Tr 16 - 176		⋆ AC	10 42 12	−59 18.8	2	7.83	.010	0.21	.005	−0.70		3	O6.5V((f))	238,315
		Tr 16 - 143			10 42 13	−59 25.5	1	11.48		0.56				1		2042
		Cr 228 - 52			10 42 13	−59 42.6	1	12.50		0.40		0.25		2		472
93162	−59 02561	Tr 16 - 177			10 42 14	−59 27.4	3	8.10	.004	0.41	.006	−0.65	.020	9	WN6 -A	6,945,1096
		Cr 228 - 50			10 42 14	−59 44.9	1	12.04		0.26		−0.34		1		472
93144	−54 03991				10 42 15	−55 17.3	1	8.18		1.30				4	K0/1 III	2012
305536	−59 02563	Cr 228 - 5			10 42 15	−59 47.6	2	8.94	.005	0.04	.010	−0.83	.010	6	O8.5V	472,850
		G 58 - 14			10 42 16	+23 50.6	1	13.15		1.43		1.04		1		333,1620
		Tr 16 - 130			10 42 16	−59 18.1	1	12.72		0.30		−0.34		4		315
	−58 02633	Tr 16 - 129			10 42 16	−59 18.4	1	11.57		0.38		−0.28		2		315
	−59 02565	Tr 16 - 39			10 42 16	−59 27.4	1	12.82		0.45		−0.21		1		315
	−59 02588	Cr 228- 115			10 42 16	−59 36.9	1	11.40		0.54		0.06		2		850
305519	−59 02562	Cr 228 - 57			10 42 16	−59 39.7	2	9.88	.020	0.09	.005	0.08	.000	2	A2 V	472,850
303246	−57 03743				10 42 17	−57 43.2	1	10.39		0.38		−0.49		1	B9	761
		Tr 16 - 245			10 42 17	−59 27.2	1	12.08		0.51		−0.52		3	B0 V	945
		Cr 228 - 58			10 42 17	−59 39.9	1	12.71		0.35		0.10		1		472
		Tr 16 - 246			10 42 18	−59 27.1	1	11.94		0.44		−0.63		3	B0.5V	945
		Cr 228 - 64			10 42 18	−59 47.2	1	12.85		0.62		0.47		1		472
93194	−63 01623	IC 2602 - 41		⋆	10 42 18	−63 41.9	7	4.82	.015	−0.14	.006	−0.62	.006	24	B3 Vnn	15,26,1053,2003,2005*
		AJ82,163 # 5			10 42 19	−57 17.7	1	11.58		1.85		1.96		1		533
	−57 03744	LSS 1835			10 42 19	−57 54.8	1	10.84		0.34		−0.57		2	B0	761
	−59 02566	Tr 16 - 57			10 42 19	−59 27.1	1	12.90		0.37		−0.08		1		315
	−59 02567	Tr 16 - 36			10 42 19	−59 27.9	2	9.44	.000	1.64	.000	2.07	.025	7		6,315
305522	−59 02568	Cr 228 - 8			10 42 19	−59 44.3	2	9.69	.005	0.06	.005	−0.78	.019	6	B0.5V:	472,850
		Cr 228 - 51			10 42 19	−59 45.6	1	11.88		0.10		−0.61		1		472
		AJ82,163 # 3			10 42 20	−57 21.3	1	11.99		0.58		0.07		5		533
	−59 02569	Cr 228 - 20		⋆ AB	10 42 20	−59 52.0	1	10.41		0.67		−0.22		3		472
	−59 02570	Cr 228 - 19			10 42 20	−59 53.3	2	10.54	.020	0.06	.035	−0.71	.010	5	B1 V:	472,850
		Cr 228- 109			10 42 21	−59 48.8	1	12.89		0.20		−0.52		2		850
93209	−64 01456	IC 2602 - 42			10 42 21	−64 24.8	2	9.39	.015	0.35	.025	0.25		14	A3 V	1053,2003
	−58 02660	Tr 16 - 123			10 42 22	−59 16.9	1	12.59		0.22		−0.41		3		315
		Tr 16 - 59			10 42 22	−59 26.7	2	12.35	.010	0.21	.010	−0.54	.000	3		315,945
93190	−58 02637	LSS 1838		⋆ AB	10 42 23	−59 01.2	3	8.58	.012	0.33	.005	−0.78	.021	7	O9/B0e	6,834,1737
	−30 08687				10 42 24	−30 54.3	1	9.81		0.72		0.19		1	K0	78
		Tr 16 - 247			10 42 24	−59 27.3	1	13.56		0.24		−0.28		3		945
		Cr 228 - 62			10 42 25	−59 48.5	1	12.69		1.16		0.59		1		472
	+39 02376	IDS10396N3902		AB	10 42 26	+38 45.1	1	9.23		1.41		1.21		2	M2	3072
	−59 02571	Tr 16 - 11			10 42 26	−59 23.6	2	11.15	.127	0.23	.093	−0.66		3	B1.5V	6,2042
		AJ82,163 # 7			10 42 27	−57 16.2	1	12.19		1.35		1.54		1		533
93206	−59 02572	Cr 228 - 33		⋆ AB	10 42 27	−59 43.8	6	6.30	.039	0.14	.007	−0.81	.017	22	O9.7Ib(n)	6,472,479,620,850,1588
	−58 02640	LSS 1840			10 42 28	−59 14.8	1	10.24		0.13		0.09		5		1711
93208	−61 01842	LSS 1843			10 42 28	−61 48.4	3	8.75	.015	−0.01	.009	−0.77	.012	6	O9/B0	401,540,976
	−58 02642	Tr 16 - 122			10 42 29	−59 17.3	2	11.33	.010	0.16	.040	−0.57		4	B1.5V	315,2042

Table 1

567

HD	DM	Other Id	N Rem	α_{1950}	δ_{1950}	S	V	σ_V	B–V	σ_{B-V}	U–B	σ_{U-B}	n	Spectrum	References
		Tr 16 - 56		10 42 29	−59 25.0	1	13.43		0.33		0.03		1		315
	−59 02574	Tr 16 - 94		10 42 30	−59 25.2	1	9.86		0.14		-0.62		4	B1 Vn	6
		Cr 228- 100	⋆ B	10 42 30	−59 56.9	1	10.41		0.45		-0.63		2		850
		Tr 16 - 209		10 42 31	−59 22.7	1	13.46		0.47		0.71		3		945
93191	−59 02577	Cr 228 - 2		10 42 31	−59 37.3	2	8.48	.005	-0.02	.005	-0.17	.010	6	B9.5Vb	472,850
305602	−59 02745	LSS 1841		10 42 32	−57 34.8	1	10.20		0.26		-0.56		2	WN	476
	−58 02644			10 42 32	−59 04.3	2	10.69	.014	0.18	.000	-0.68	.041	9	B1 V	476,834
		Tr 16 - 58		10 42 32	−59 24.5	1	12.42		0.23		0.21		1		315
		Tr 16 - 61		10 42 32	−59 28.0	1	12.40		0.21		-0.41		1		315
94009	−86 00210			10 42 32	−86 38.1	1	7.51		0.77				4	G5/6 IV	2012
93203	−56 03737	VY Car		10 42 33	−57 18.1	3	7.19	.083	0.99	.126	0.66	.038	3	F7 Iab/b	689,1484,1587
	−58 02648	Tr 16 - 12		10 42 33	−59 22.8	2	11.43	.010	0.30	.000	-0.63	.010	5	B1 V	6,315
		Tr 16 - 208		10 42 33	−59 23.7	1	13.42		0.42		0.12		2		945
		Tr 16 - 18		10 42 33	−59 26.8	3	12.04	.026	0.23	.016	-0.63	.023	8	B2 V	6,315,945
		Tr 16 - 60		10 42 33	−59 28.0	1	12.47		0.19		-0.40		1		315
		Tr 16 - 248		10 42 33	−59 28.2	1	13.38		0.34		0.48		1		945
		LSS 1846		10 42 33	−60 50.5	1	11.45		0.28		-0.63		3	B3 Ib	466
93173	−43 06522			10 42 34	−43 41.8	2	9.02	.022	0.67	.015	0.14		6	G5 V	2012,6006
	−58 02647	Tr 15 - 23		10 42 34	−59 05.7	1	11.12		0.26		-0.50		6	B0 V:	834
	−58 02645	Tr 16 - 35		10 42 34	−59 16.5	1	9.85		1.14		1.02		1		6
	−58 02649	Tr 16 - 10		10 42 34	−59 21.6	2	9.81	.019	0.30	.005	-0.69		3	B0 Vn	6,2042
		Tr 16 - 211		10 42 34	−59 21.9	1	14.00		0.66		0.56		2		945
		Tr 16 - 207		10 42 34	−59 23.6	1	13.56		0.59		0.83		3		945
	−59 02581	Tr 16 - 17		10 42 34	−59 25.9	1	11.01		0.24		-0.74		2	B1 V	6
	−59 02579	Cr 228 - 41		10 42 34	−59 36.4	2	11.04	.015	0.19	.015	-0.64	.010	3		472,850
305523	−59 02580	Cr 228 - 32		10 42 34	−59 41.5	4	8.49	.007	0.16	.018	-0.78	.030	11	O8.5II-III	6,472,620,850
93102	+3 02408	HR 4201		10 42 35	+02 45.1	3	6.27	.005	1.21	.000	1.27	.009	12	K4 III	15,1417,1509
		Tr 15 - 25		10 42 35	−59 05.0	1	12.73		0.23		-0.34		3	B5 V	834
		Tr 15 - 37		10 42 35	−59 06.3	1	13.88		0.31		0.02		1		834
		Tr 14 - 42		10 42 35	−59 14.9	1	12.41		0.34				2		2042
		Feige 35		10 42 36	+25 08.0	1	12.16		0.05		-0.01		1		3060
	−57 03753			10 42 36	−57 54.3	1	11.09		0.33		-0.45		2	B0	761
		Tr 15 - 36		10 42 36	−59 07.0	1	13.76		0.41		-0.03		1		834
		Tr 16 - 80		10 42 36	−59 18.2	1	12.14		0.40		-0.31		3		315
		SOS 2621		10 42 36	−59 25.	1	8.29		0.07		-0.01		1		761
		Tr 16 - 206		10 42 36	−59 25.4	1	13.68		0.45		0.31		3		945
93204	−59 02584	Tr 16 - 178	⋆ B	10 42 36	−59 28.7	4	8.44	.016	0.10	.010	-0.88	.007	11	O5 V	6,390,850,1737
		Cr 228- 106		10 42 36	−59 49.9	1	13.26		0.17		-0.03		2		850
	−59 02585	Cr 228- 113		10 42 36	−59 49.9	1	11.44		0.06		-0.71		2		850
		L 101 - 80		10 42 36	−69 02.	1			-0.04		-0.84		2		3065
		Tr 16 - 79		10 42 37	−59 22.5	2	13.61	.058	0.46	.005	0.47	.107	3		315,945
		Tr 16 - 210		10 42 37	−59 22.7	1	13.28		0.48		-0.14		4		945
	−59 02583	Tr 16 - 13		10 42 37	−59 24.6	3	10.76	.031	0.22	.018	-0.63	.018	6	B2 Vnn	6,315,945
		Tr 16 - 231		10 42 37	−59 28.3	1	12.07		0.10		-0.60		1		945
93205	−59 02587	Tr 16 - 179	⋆ A	10 42 37	−59 28.5	3	7.75	.004	0.06	.015	-0.91	.004	10	O3 V	6,390,850
	−59 02582	Cr 228 - 40		10 42 37	−59 37.1	2	10.62	.005	1.12	.015	0.95	.090	2		472,850
		POSS 169 # 3		10 42 38	+46 10.2	1	16.22		1.05				2		1739
		AJ82,163 # 8		10 42 38	−57 20.2	1	12.88		0.46		0.03		3		533
		Tr 16 - 62		10 42 38	−59 27.3	2	14.00	.078	0.26	.005	-0.13	.024	3		315,945
		Cr 228 - 63		10 42 38	−59 47.0	1	13.63		0.23		-0.13		1		472
		Tr 15 - 24		10 42 39	−59 05.6	1	12.14		0.26		-0.36		3		834
		Tr 15 - 20		10 42 39	−59 07.6	1	12.70		0.29		-0.35		3		834
		Tr 16 - 78		10 42 39	−59 22.3	2	12.17	.019	1.16	.054	1.12	.151	3		315,945
		Tr 16 - 205		10 42 39	−59 25.2	1	13.45		0.89		0.80		2		945
		Tr 15 - 19		10 42 40	−59 07.7	1	12.71		0.30		-0.39		4	O9 V:	834
		Tr 15 - 18		10 42 40	−59 08.5	1	11.28		1.11		-0.09		15	O9 I-II:	834
		Tr 16 - 230		10 42 40	−59 26.2	1	13.94		0.68		-0.41		1		945
		Cr 228- 102		10 42 40	−59 44.1	1	11.33		0.09		-0.51		2		850
93222	−59 02590	Cr 228 - 6		10 42 40	−59 49.7	4	8.10	.010	0.05	.014	-0.89	.026	26	O7 III((f))	6,472,620,850
		Tr 15 - 22		10 42 41	−59 06.3	1	13.17		0.36		-0.08		3	B9	834
		Tr 15 - 21		10 42 41	−59 07.3	1	13.13		0.25		-0.17		2	B0 III:	834
303311	−58 02652	Tr 16 - 98		10 42 41	−59 17.1	2	9.03	.019	0.14	.010	-0.86		6	O5 V	6,2042
		Tr 16 - 14		10 42 41	−59 24.2	2	11.47	.040	0.44	.020	-0.56	.005	5	B0.5V	6,315
	−59 02591	Tr 16 - 21		10 42 41	−59 31.7	2	10.93	.005	0.46	.010	-0.54	.000	6		6,479,2042
	−59 02592	Cr 228 - 12		10 42 41	−59 38.6	3	9.48	.020	0.79	.015	-0.26	.014	8	B2.5Ia	472,850,1737
	−59 02593	Cr 228 - 36	⋆ B	10 42 42	−59 45.4	2	10.26	.040	0.10	.005	-0.66	.045	5	B0.5V+B05V:	472,850
		Tr 16 - 213		10 42 42	−59 22.5	1	16.09		0.06		-0.24		2		945
		Tr 16 - 63		10 42 42	−59 27.3	1	12.82		0.27		-0.32		1		315
	−59 02594	Cr 228 - 35	⋆ A	10 42 42	−59 45.2	2	10.19	.015	-0.01	.005	-0.16	.000	5	B9.5Vb:	472,850
		Tr 15 - 8		10 42 43	−59 06.0	1	12.92		0.36				4		834
	−59 02595	Tr 16 - 20		10 42 43	−59 32.4	2	10.21	.009	0.05	.042	-0.59		4	B1 V	6,2042
		AJ82,163 # 18		10 42 44	−57 15.2	1	11.87		0.21		-0.40		5		533
		AJ82,163 # 10		10 42 44	−57 18.9	1	12.36		0.57		0.14		4		533
		Tr 15 - 4		10 42 44	−59 05.4	1	11.00		0.19		-0.61		8	B1 Vn	834
	−58 02653	Tr 15 - 14		10 42 44	−59 06.7	1	10.57		0.27		-0.48		4	B2.5IV-V	834
	−59 02598	Tr 16 - 16		10 42 44	−59 26.0	2	10.79	.080	0.25	.010	-0.57	.005	4	B2 Vb	6,1737
		Tr 16 - 235		10 42 44	−59 27.8	1	12.96		0.32		0.01		3		945
	−59 02597	Cr 228 - 34		10 42 44	−59 43.7	1	9.82		1.66		2.01		2		472
		Tr 15 - 5		10 42 45	−59 05.8	1	11.49		0.26		-0.52		4	B5 V	834
		Tr 15 - 9		10 42 45	−59 06.3	1	12.59		0.35		-0.31		4	B1 V	834

HD	DM	Other Id	N	Rem	α_{1950}	δ_{1950}	S	V	σ_V	B–V	σ_{B-V}	U–B	σ_{U-B}	n	Spectrum	References
	−58 02655	Tr 15 - 13			10 42 45	−59 06.7	1	10.78		0.29		−0.62		5	B1 V	834
	−58 02656	Tr 15 - 15			10 42 45	−59 07.3	1	10.08		0.19		−0.67		9	B0.5IV-V	834
	−59 02596	Tr 16 - 15			10 42 45	−59 24.4	2	11.26	.020	0.42	.010	−0.54	.005	5	B0 V	6,315
		Tr 16 - 233			10 42 45	−59 27.5	1	12.36		0.17		−0.52		2		945
		Tr 16 - 49			10 42 45	−59 27.5	1	13.48		0.32		0.18		2		315
		Tr 16 - 234			10 42 45	−59 27.8	1	13.52		2.42				1		945
	−59 02600	Tr 16 - 100			10 42 45	−59 31.2	4	8.61	.012	0.21	.012	−0.78	.020	11	O5.5V	6,620,850,2042
	−59 02599	Cr 228 - 10			10 42 45	−59 47.7	1	9.40		1.87		1.36		3		472
		Tr 15 - 28			10 42 46	−59 05.0	1	12.16		1.54		0.75		1		834
		Tr 15 - 6		V	10 42 46	−59 05.9	1	12.49		0.33		−0.25		4		834
	−58 02657	Tr 15 - 7			10 42 46	−59 06.0	1	10.58		0.30		−0.51		6	B2.5Vn	834
		Tr 16 - 218			10 42 46	−59 21.7	1	14.71		1.02		−0.18		1		945
		Tr 16 - 212			10 42 46	−59 22.7	1	13.86		0.61		−0.52		3		945
		Tr 16 - 204			10 42 46	−59 25.9	1	13.79		0.43		0.22		3		945
		Tr 16 - 228			10 42 46	−59 27.0	1	12.90		0.33		−0.31		2		945
		WLS 1040-10 # 9			10 42 47	−10 29.6	1	12.59		0.85		0.31		2		1375
		Tr 15 - 2		⋆ B	10 42 47	−59 05.5	1	9.47		0.20		−0.73		5	O9 III	834
93249	−58 02659	Tr 15 - 1		⋆ AB	10 42 47	−59 05.6	2	8.36	.005	0.14	.005	−0.77	.025	8	O9 III	834,1737
		Cr 228- 114			10 42 47	−59 38.0	1	12.54		0.23		−0.46		2		850
93198	−43 06524				10 42 48	−43 36.2	1	8.91		0.18		0.16		1	A4 V	6006
		Tr 15 - 3		⋆ C	10 42 48	−59 05.7	1	10.57		0.18		−0.54		4	B2 Vn	834
93250	−58 02661	Tr 16 - 180			10 42 48	−59 18.1	7	7.37	.007	0.17	.011	−0.84	.019	28	O3 V((f))	6,164,315,390,620,850*
		Tr 16 - 50			10 42 48	−59 26.7	1	12.92		0.30		−0.27		1		315
		Cr 228 - 79			10 42 48	−59 46.3	1	11.13		1.22		1.21		1		472
		Tr 16 - 48			10 42 49	−59 27.6	2	11.91	.025	0.19	.005	0.05	.015	4		315,945
		Tr 16 - 243			10 42 49	−59 28.4	1	13.17		0.24		−0.25		3		945
305524	−59 02602	Cr 228 - 7			10 42 49	−59 38.9	3	9.30	.017	0.26	.030	−0.74	.012	7	O6 Vn	472,850,1737
93344	−70 01183	HR 4211		⋆ A	10 42 49	−70 35.8	3	6.25	.008	0.21	.004	0.16	.010	10	A5 IV/V	15,1075,2038
		Ton 1288			10 42 50	+22 18.	1	13.43		0.32		−0.07		2		313
303251	−57 03756	LSS 1858			10 42 50	−57 48.8	4	8.89	.016	0.46	.012	−0.50	.013	9	B2 I-II	533,540,761,976
	−58 02662	Tr 15 - 10			10 42 50	−59 06.1	1	11.55		0.28		−0.38		5	B2 V	834
303310	−58 02663	Tr 15 - 16		⋆ V	10 42 50	−59 09.0	1	8.82		2.39		2.49		24	M2 I	834
		Tr 16 - 43			10 42 50	−59 24.7	2	12.24	.009	0.57	.019	0.12	.028	4		315,479
		Tr 15 - 39			10 42 51	−59 06.8	1	12.72		1.50		1.19		1		834
		Tr 15 - 12			10 42 51	−59 06.9	1	13.37		0.32		−0.34		1		834
		Tr 16 - 44			10 42 51	−59 23.5	2	13.24	.295	0.59	.042	−0.19	.037	4		315,479
	−59 02603	Tr 16 - 104			10 42 51	−59 28.1	2	8.77	.005	0.14	.000	−0.79	.005	5	O7 V:	6,850
305534	−59 02604	Cr 228 - 11			10 42 51	−59 42.2	2	9.68	.015	0.11	.025	−0.77	.025	5	B0.5V:+B1V	472,850
93269	−63 01629	IC 2602 - 43			10 42 51	−63 32.8	3	8.16	.004	1.15	.007	1.04		14	G8 III	1053,2003,2005
		Tr 15 - 38			10 42 52	−59 06.5	1	13.68		0.43		0.00		1		834
		Tr 16 - 32			10 42 52	−59 17.0	1	11.05		0.23		0.14		3		6
93075	+57 01290				10 42 53	+57 11.0	1	7.11		0.31		0.07		3	A8 V	8071
		Tr 15 - 29			10 42 53	−59 04.8	1	12.92		0.29		−0.10		1		834
		Tr 15 - 17			10 42 53	−59 09.0	1	13.26		0.32		−0.22		4	B5 V	834
		Tr 16 - 203			10 42 53	−59 26.3	1	12.90		0.39		0.13		3		945
93248	−56 03744				10 42 54	−56 50.2	1	9.45		−0.02		−0.17		2	B9 V	401
		AJ86,209 # 14			10 42 54	−58 45.	1	12.59		0.62		−0.22		1		924
		Tr 15 - 30			10 42 54	−59 04.4	1	12.92		0.20		−0.27		1		834
		Tr 16 - 227			10 42 54	−59 27.1	1	14.60		0.99		0.94		2		945
	−59 02605	Cr 228 - 18			10 42 54	−59 40.0	3	11.08	.019	0.21	.023	−0.73	.019	6		472,850,1737
		Tr 16 - 225			10 42 55	−59 26.1	1	13.74		0.48		0.79		1		945
		Tr 16 - 47			10 42 55	−59 27.4	2	12.90	.010	0.31	.025	−0.31	.005	2		315,945
		Cr 228- 111			10 42 55	−59 41.6	1	13.06		0.17		−0.34		2		850
		IC 2602 - 304			10 42 55	−63 56.4	1	11.23		0.43		0.28		4		891
93247	−55 03800	SV Vel			10 42 56	−56 01.6	2	7.93	.005	0.79	.025	0.72		2	F7 II	689,1484
93268	−58 02665	Tr 16 - 181			10 42 56	−59 17.7	1	9.32		0.10		0.05		2	A0 V sn	6
		Tr 16 - 220			10 42 58	−59 24.2	1	13.30		0.37		−0.34		1		945
		Tr 16 - 202			10 42 58	−59 25.6	1	13.16		0.24		0.68		2		945
	−59 02606	Tr 16 - 5			10 42 58	−59 25.7	2	10.83	.000	0.24	.000	−0.67	.000	4	B2 Vn	6,1349
	−59 02610	Cr 228 - 39			10 42 58	−59 40.3	2	10.01	.090	0.24	.080	−0.78	.025	4	O8 V	472,850
		Tr 15 - 32			10 42 59	−59 02.9	1	12.71		0.14		−0.36		1		834
		Tr 16 - 224			10 42 59	−59 26.6	1	13.26		0.42		−0.24		2		945
	+13 02299				10 43 00	+12 43.6	1	9.15		1.06		1.01		2	K0	1648
	−58 02668	Tr 16 - 31			10 43 00	−59 17.2	1	10.44		0.27		−0.67		3	B0 Vn	6
		Tr 16 - 30			10 43 00	−59 19.4	1	11.17		1.06		1.20		2		6
		Tr 16 - 40			10 43 00	−59 24.2	1	12.79		0.40		−0.29		1		315
	−59 02576	Tr 16 - 41			10 43 00	−59 24.6	1	12.30		0.26		−0.40		1		315
		Tr 15 - 31			10 43 01	−59 04.8	1	13.02		0.22		−0.31		1		834
	−58 02667	Tr 16 - 29			10 43 01	−59 20.8	2	11.35	.010	0.42	.000	−0.48	.015	5	B2 V	6,315
		Tr 16 - 216			10 43 01	−59 22.6	1	13.88		1.21		1.24		2		945
93281	−59 02611	Cr 228 - 17			10 43 01	−59 40.3	2	7.76	.024	1.75	.029	0.88	.019	3	M1 Iab+B	472,850
	−59 02614	Cr 228 - 37			10 43 01	−59 45.0	2	10.84	.030	0.20	.005	−0.65	.020	4	B2 Vb	472,850
	−59 02612	Cr 228 - 46			10 43 01	−59 52.1	1	10.74		0.16		0.15		3	A1 Vb	472
93359	−70 01185	HR 4212		⋆ B	10 43 01	−70 35.5	3	6.43	.022	0.23	.008	0.13	.010	11	A6 IV	15,1075,2038
		Tr 15 - 33			10 43 02	−59 06.6	1	12.27		0.19		−0.32		3	A0 V	834
	−59 02645	Tr 16 - 42			10 43 02	−59 23.9	1	12.74		0.34		−0.33		1		315
		Tr 16 - 6			10 43 02	−59 25.2	3	11.83	.161	0.17	.014	−0.52	.026	6		6,945,1349
		Tr 16 - 200			10 43 02	−59 25.4	1	13.59		0.29		0.07		3		945
	−59 02613	Tr 16 - 53			10 43 02	−59 26.5	2	13.19	.070	1.32	.015	2.14		2		315,945
		Cr 228- 105			10 43 02	−59 49.0	1	12.06		0.07		−0.61		2		850

Table 1

HD	DM	Other Id	N	Rem	α_{1950}	δ_{1950}	S	V	σ_V	B–V	σ_{B-V}	U–B	σ_{U-B}	n	Spectrum	References
		Tr 16 - 46			10 43 03	−59 26.6	2	11.68	.034	0.33	.010	-0.37	.005	3		315,945
	−59 02621	LSS 1867			10 43 03	−60 24.4	1	11.23		0.48		-0.57		2	B0.5 V	1737
93372	−71 01118	HR 4213			10 43 03	−72 10.9	3	6.26	.004	0.48	.005	0.00	.010	11	F6 V	15,1075,2038
		Tr 15 - 35			10 43 04	−59 08.0	1	13.25		0.49		0.16		2		834
		Tr 16 - 217			10 43 04	−59 22.8	1	14.38		0.68		0.75		1		945
		Tr 16 - 38			10 43 04	−59 24.3	1	13.37		0.24		-0.26		1		315
	−59 02618	Tr 16 - 26			10 43 04	−59 27.4	2	11.66	.000	0.22	.024	-0.40	.005	6	B1	6,315
	−59 02616	Tr 16 - 25			10 43 04	−59 27.8	2	11.88	.010	0.26	.005	-0.65	.024	6	B3	6,315
93152	+31 02180	HR 4203		⋆ AB	10 43 05	+30 56.8	4	5.26	.043	-0.06	.009	-0.13	.017	10	A1 Vn	70,1022,1079,3023
		Tr 15 - 34			10 43 05	−59 06.3	1	13.18		0.37		-0.10		1		834
		G 196 - 42			10 43 06	+56 44.1	2	10.38	.044	0.78	.010	0.38	.004	4	K2	1658,1723
		Tr 16 - 201			10 43 06	−59 25.6	1	12.39		0.26		-0.47		1		945
	−59 02619	Tr 16 - 4		AB	10 43 06	−59 26.2	4	10.98	.038	0.25	.009	-0.64	.014	12	B1 V	6,315,479,1349
	−55 03802				10 43 07	−55 58.7	1	11.00		0.26		0.20		1		1480
	−59 02622	Tr 16 - 8			10 43 07	−59 24.4	4	10.87	.042	0.14	.005	-0.77	.029	13	B1.5Vb	6,479,945,1349
93308	−59 02620	Tr 16 - 183		⋆ AP	10 43 07	−59 25.3	2	6.20	.008	0.61	.004	-0.45	.004	15		6,1349
		Tr 16 - 64		⋆ F	10 43 08	−59 25.1	1	10.72		0.10		-0.74		2	B1.5V:b	315
303308	−59 02623	Tr 16 - 7		⋆ H	10 43 09	−59 24.3	8	8.16	.011	0.13	.012	-0.84	.024	28	O3 V((f))	6,315,390,620,850,945*
		Tr 16 - 66			10 43 09	−59 25.1	1	11.98		0.16		-0.57		2		315
		Tr 16 - 65		⋆ E	10 43 09	−59 25.1	1	11.09		0.14		-0.65		2		315
		Tr 16 - 68			10 43 09	−59 25.6	1	12.48		0.21		-0.41		2		315
		Tr 16 - 72			10 43 09	−59 25.9	2	12.11	.010	0.23	.010	-0.54	.015	2	B1 V	315,945
		Tr 16 - 75			10 43 09	−59 26.3	1	14.26		0.64		0.41		1		315
	−59 02624	Tr 16 - 9			10 43 09	−59 27.3	4	9.79	.044	0.22	.010	-0.73	.017	14	O9.5V	6,315,479,1349
	−59 02626	Tr 16 - 23			10 43 09	−59 29.0	3	9.95	.014	0.38	.005	-0.62	.015	8	O7 Vn	6,479,1737
		Tr 16 - 67			10 43 10	−59 25.4	1	13.70		0.81		0.31		2		315
	−59 02627	Tr 16 - 3			10 43 10	−59 26.1	3	10.17	.000	0.20	.005	-0.77	.015	11	O9 V:	6,315,1349
		Tr 16 - 73			10 43 10	−59 26.8	1	11.90		0.42		-0.46		2		315
	−59 02625	Tr 16 - 24			10 43 10	−59 28.5	1	11.58		0.16		-0.63		3	B2 V	6
		Cr 228- 104			10 43 10	−59 47.6	1	11.68		0.19		-0.47		2		850
		ApJS53,765 T5# 3			10 43 11	−56 03.0	1	13.45		0.42		0.36		1		1480
		ApJS53,765 T5# 4			10 43 11	−56 03.3	1	12.50		0.34		0.25		1		1480
	−59 02628	Tr 16 - 1		⋆ G	10 43 11	−59 25.0	4	9.53	.018	0.11	.013	-0.81	.023	11	O9.5Vn	6,315,1349,1737
		Tr 16 - 77			10 43 11	−59 25.2	2	12.05	.037	0.27	.000	-0.45	.056	4		315,945
		Tr 16 - 69			10 43 11	−59 25.8	2	13.03		0.36		0.28		2		315
	−55 03803				10 43 12	−56 00.7	1	10.15		1.33		1.48		1		1480
		Tr 16 - 52			10 43 12	−59 23.0	2	12.59	.010	0.35	.020	-0.34	.030	2		315,945
		Tr 16 - 51			10 43 12	−59 23.2	2	12.83	.065	0.33	.000	-0.44	.020	2		315,945
		Tr 16 - 125			10 43 12	−59 27.	1	12.15		0.65		-0.04		3		315
		Tr 16 - 131			10 43 12	−59 27.	1	13.48		0.32		-0.30		3		315
		Tr 16 - 215			10 43 12	−59 27.	1	12.54		0.64		-0.74		1		945
		Tr 16 - 236			10 43 12	−59 27.	1	14.60		1.02		1.00		1		945
		Tr 16 - 244			10 43 12	−59 27.	1	10.70		0.75		-0.51		3	O3-O4If	945
	−59 02629	Tr 16 - 22			10 43 12	−59 30.3	2	10.97	.040	0.49	.005	-0.56	.040	6	O8.5V	6,479
93321	−60 02251				10 43 12	−60 23.7	1	9.70		-0.06		-0.56		2	B3/5 III	6
		Tr 16 - 33			10 43 13	−59 24.3	1	11.83		0.27		-0.51		2	B2 V	6
		Tr 16 - 37			10 43 13	−59 24.9	1	12.95		0.24		-0.30		2		315
		Tr 16 - 70			10 43 13	−59 25.7	1	13.41		0.31		-0.31		2		315
		Tr 16 - 74			10 43 13	−59 27.1	1	11.70		0.27		-0.60		2	B1 V	315
	−59 02630	Tr 16 - 45			10 43 13	−59 27.1	1	13.03		0.70		0.18		1		315
		Cr 228- 112			10 43 14	−59 39.5	1	11.50		1.21		1.14		1		850
93132	+58 01281	HR 4202			10 43 15	+57 37.8	3	6.35	.010	1.57	.005	1.90	.000	6	M1 III	15,1003,3016
	−59 02632	Tr 16 - 2			10 43 15	−59 25.4	3	10.79	.011	0.15	.005	-0.74	.013	9	B1.5V: +A?	6,315,1349
		Tr 16 - 71			10 43 15	−59 25.5	1	13.65		0.38		-0.21		1		315
		Tr 16 - 54			10 43 16	−59 23.3	1	12.85		0.38		-0.33		1		315
		Tr 16 - 76			10 43 16	−59 27.0	2	11.16	.030	0.44	.020	-0.51	.060	4	B2:V	315,479
	−59 02636	Tr 16 - 110			10 43 16	−59 28.5	2	9.28	.025	0.34	.025	-0.67	.015	7	O8 V	6,390
93343	−59 02633	Tr 16 - 182		⋆ B	10 43 16	−59 28.6	2	9.52	.050	0.27	.015	-0.76	.010	7	O7 V(n)	6,390
	−59 02635	Tr 16 - 34		⋆ A	10 43 16	−59 28.7	3	9.26	.024	0.25	.012	-0.75	.008	7	O7 Vnn	6,390,1737
305533	−59 02639	Cr 228 - 47			10 43 17	−59 42.1	2	10.47	.117	0.18	.039	-0.64	.102	3	B0.5Vnn+Shl	472,850
		Tr 16 - 237			10 43 18	−59 25.9	1	13.11		0.34		-0.12		1		945
		Cr 228- 103			10 43 18	−59 45.1	1	10.96		0.51		0.02		2		850
		G 119 - 29			10 43 19	+32 33.4	1	13.83		1.57		0.90		3		316
	−18 03019				10 43 19	−18 50.5	5	11.03	.007	1.45	.018	0.97	.010	10	M0.5	158,912,1705,1774,3078
93342	−58 02674	LSS 1875			10 43 20	−59 07.8	4	9.10	.017	0.61	.013	-0.41	.023	9	B0 III	6,191,834,1737
	−59 02640	Tr 16 - 27			10 43 20	−59 24.1	2	11.06	.005	0.13	.010	-0.68	.005	5	B1	6,315
		Tr 16 - 238			10 43 20	−59 25.9	1	13.63		0.38		-0.11		1		945
	−59 02641	Tr 16 - 112			10 43 20	−59 27.8	2	9.27	.024	0.32	.005	-0.71	.014	7	O5 V	6,1737
	+50 01766				10 43 21	+49 34.9	1	10.37		0.08		0.06		2	A3 V	1240
		Tr 16 - 240			10 43 21	−59 25.3	1	14.02		0.36		-0.28		2		945
		Tr 16 - 239			10 43 21	−59 25.5	1	13.78		0.32		-0.19		2		945
305528	−59 02642	Cr 228 - 80			10 43 21	−59 38.9	1	10.28		0.13		-0.49		2	B2 V	472
303300	−58 02675	LSS 1877			10 43 23	−59 01.3	1	9.81		0.32		-0.64		2	B3	834
		Tr 16 - 55			10 43 23	−59 23.8	1	12.22		0.23		-0.53		1	B1.5V	315
	−59 02644	Tr 16 - 115			10 43 23	−59 27.1	3	10.07	.061	0.17	.010	-0.80	.020	8	O9 V	6,479,1737
93341	−56 03759	LSS 1876			10 43 24	−56 43.6	2	9.62	.010	0.01	.010	-0.73	.005	4	B1/2 II/III	401,533
93215	+26 02126				10 43 25	+26 01.5	1	8.05		0.67		0.19		3	G5 V	3026
93318	−53 04112	IDS10414S5318		AB	10 43 25	−53 33.7	1	8.64		-0.02		-0.09		2	B9 IV	3044
		Tr 16 - 242			10 43 25	−59 25.9	1	12.87		0.44		0.22		2		945
93214	+36 02119				10 43 26	+35 54.2	1	8.96		0.41		-0.07		2	F8	1601

HD	DM	Other Id	N	Rem	α_{1950}	δ_{1950}	S	V	σ_V	B–V	σ_{B-V}	U–B	σ_{U-B}	n	Spectrum	References
	−58 02677	Tr 16 - 28	V		10 43 26	−59 21.8	2	11.75	.055	0.27	.005	-0.58	.005	5	B2 V	6,315
237929	+55 01409	SV UMa			10 43 28	+55 18.0	1	9.34		1.12		0.73		1	K3 Ia:p:	793
93244	+7 02356	HR 4207			10 43 29	+06 38.2	2	6.36	.005	1.12	.000	1.12	.005	7	gK1	15,1256
		L 465 - 1			10 43 29	−35 05.6	1	12.55		1.40		1.02		2		3073
305541	−59 02649	Cr 228 - 87			10 43 36	−59 50.4	1	10.55		0.16		-0.12		3	B9 V	472
305532	−59 02650	Cr 228 - 38			10 43 37	−59 41.6	3	10.20	.004	0.35	.008	-0.71	.015	9	O5 V	472,850,1737
	−58 02678	LSS 1879			10 43 38	−58 48.1	1	10.54		0.75		-0.29		2		476
		G 163 -B9	A		10 43 39	−03 24.1	1	11.41		0.62		-0.01		2		3060
		G 163 -B9	B		10 43 39	−03 24.1	1	15.32		0.43		-0.21		2		3060
93357	−52 03887	HQ Vel			10 43 39	−53 21.9	1	7.40		1.98		2.18		1	M1/2 III	3044
	−59 02651	Cr 228 - 85			10 43 39	−59 48.7	1	11.20		0.78		0.50		1		472
		Car sq 2 # 5			10 43 41	−60 54.2	1	12.56		0.35		-0.24		2		466
		AJ86,209 # 15			10 43 42	−59 33.	1	12.05		0.86		-0.20		1		924
		IC 2602 - 310			10 43 42	−63 58.7	1	13.67		0.29		0.13		5		891
93257	+19 02371	HR 4208			10 43 43	+19 09.3	2	5.50	.010	1.13	.003	1.15		5	K2 III	70,1501
	−59 02652	Cr 228 - 84			10 43 43	−59 48.6	1	11.05		0.25		-0.44		2		472
	−35 06718				10 43 44	−35 48.6	1	9.98		1.20		1.23		2	K1	1730
93291	+14 02294	HR 4209			10 43 46	+14 27.5	2	5.50	.017	0.91	.007	0.59		4	K0 III	71,3016
93424	−64 01469	IC 2602 - 45			10 43 46	−64 26.4	3	8.13	.011	0.16	.014	0.09	.009	19	A3 Va	1053,2005,3015
93403	−58 02680	LSS 1881			10 43 47	−59 08.7	5	7.27	.019	0.22	.008	-0.76	.027	13	O6 f	6,620,834,1075,2012
		FS Vel			10 43 48	−43 43.3	1	11.54		0.38		0.22		1		700
93368	−54 04020	HK Vel			10 43 48	−54 27.8	1	9.30		1.57		1.37		1	M3/4	3044
		AJ86,209 # 16			10 43 48	−59 35.	1	12.34		0.92		-0.12		1		924
	−59 02653	Cr 228 - 83			10 43 48	−59 48.5	1	10.74		0.09		-0.60		3	B9.5V	472
	−59 02654	Cr 228 - 86			10 43 49	−59 50.7	1	11.06		0.98		1.28		4		472
305538	−59 02655	Cr 228 - 82			10 43 50	−59 49.4	2	10.50	.030	0.25	.005	-0.57	.040	4	B0 Vb	472,850
		IC 2602 - 312			10 43 52	−63 57.6	1	14.67		1.31				4		891
307963	−64 01470	IC 2602 - 117			10 43 52	−64 30.6	1	10.06		1.15		0.89		3	K0	1053
	−28 08426				10 43 53	−29 04.4	1	9.57		0.64		0.03		1	G5	78
93420	−58 02683	BO Car		⋆ A	10 43 53	−59 13.5	1	7.18		1.94		1.86		5	M4 Ib	834
93421	−59 02659	Cr 228 - 88			10 43 56	−59 55.7	1	8.79		0.14		0.17		3	B9 V	472
93445	−60 02265	LSS 1884			10 43 56	−61 20.5	4	8.10	.015	-0.01	.008	-0.85	.014	8	B1 II	401,466,540,976
		G 58 - 15			10 43 57	+22 59.3	2	13.08	.034	0.60	.005	-0.14	.002	3		1620,1658
	−59 02661	LSS 1883			10 43 57	−59 41.2	2	11.22	.078	0.32	.010	-0.65	.010	3		476,850
	−59 02660	Cr 228 - 81			10 43 57	−59 49.8	1	10.89		0.20		-0.62		2	B0.5Vb	472
	+47 01812				10 43 58	+46 44.6	1	9.59		0.33		0.17		5	A7 II	8100
93329	+11 02277				10 43 59	+11 27.0	1	8.80		0.08		0.10		2	A0	1648
303243	−57 03778				10 44 01	−57 37.4	1	10.83		0.18		-0.01		1	A0	761
93328	+17 02273				10 44 02	+16 48.4	1	7.59		1.29		1.40		2	K0	1648
	+35 02182				10 44 04	+35 11.1	1	10.63		0.76		0.30		2		1726
93365	−18 03023				10 44 06	−18 36.6	1	7.44		1.15		1.13		4	K0 III	1657
93439	−48 05906				10 44 08	−49 07.6	2	9.21	.005	0.91	.005	0.62		6	K0 V	2012,3077
305525	−59 02665	Cr 228 - 98		⋆ AB	10 44 08	−59 35.1	3	10.03	.019	0.70	.015	-0.35	.035	6	O6 V	472,850,1737
	−60 02266	LSS 1888			10 44 09	−60 44.5	2	11.08	.000	0.23	.010	-0.49	.020	5		466,1737
93416	−51 05020				10 44 10	−51 57.6	1	9.00		0.04		-0.02		1	B9.5 V	1770
	−57 03781	LSS 1887			10 44 12	−58 23.3	1	10.70		0.56		-0.48		2	O7 V(n)	1737
93380	−23 09524				10 44 13	−24 19.2	1	9.39		1.16		1.00		1	K5 V	3072
93253	+60 01288				10 44 14	+59 35.8	1	8.79		1.51		1.84		2	K0	1502
93469	−58 02692	IDS10423S5848	AB		10 44 15	−59 03.3	2	7.86	.019	0.27	.000	0.15	.034	3	A6/7 IV/V	127,834
93483	−60 02268	LSS 1890			10 44 15	−61 00.6	2	9.57	.010	0.11	.000	-0.68	.010	5	B2 II	401,466
93398	−35 06722				10 44 16	−36 00.8	1	8.43		0.21		0.18		2	A5 V	1730
93506	−64 01475	TZ Car			10 44 16	−65 21.0	1	8.93		2.17		2.66		2	N	864
93503	−60 02271				10 44 18	−60 33.7	1	8.83		0.45		0.26		2	A1/2 II	8100
93505	−63 01641	IC 2602 - 46			10 44 18	−63 46.6	2	8.90	.010	1.62	.000			8	K5	2003,2005
93484	−61 01865				10 44 19	−61 40.7	2	7.33	.010	-0.06	.005	-0.60	.000	6	B3 V	164,401
93410	−25 08237				10 44 20	−25 47.1	1	6.70		1.03				4	K0/1 III	1075
93502	−59 02671	HR 4217		⋆ AB	10 44 21	−60 20.4	4	6.25	.007	0.04	.003	0.04		19	A0 IV	15,1075,1637,2012
93517	−63 01642	IC 2602 - 47			10 44 21	−64 19.9	3	7.85	.010	0.10	.009	0.07	.004	17	A1 V	1053,2003,3015
93363	+13 02304	IDS10417N1334	A		10 44 22	+13 17.7	1	8.40		1.02		0.82		2	K0	1648
93363	+13 02304	IDS10417N1334	B		10 44 22	+13 17.7	1	9.39		1.09		1.00		2		1648
93270	+66 00682	IDS10410N6559	A		10 44 24	+65 43.4	1	7.68		0.55		0.07		2	F8	3016
93270	+66 00682	IDS10410N6559	B		10 44 24	+65 43.4	1	10.68		1.00		0.75		2		3032
		vdB 43 # a			10 44 24	−59 45.	1	11.30		0.29		-0.64		3	B1 Ve	434
93397	−16 03124	HR 4214			10 44 25	−17 02.0	3	5.43	.011	0.11	.005	0.14	.027	14	A3 V	3,2006,8071
305559	−59 02673	Cr 228 - 97			10 44 25	−59 37.5	3	10.38	.020	0.54	.018	-0.56	.026	5	O5 V	472,850,1737
93501	−59 02672	Cr 228 - 96			10 44 26	−59 45.5	1	9.08		0.10		-0.67		2	B1.5III:	472
93500	−58 02695	IDS10425S5855	AB		10 44 27	−59 10.9	1	8.79		0.07		0.07		1	Ap CrEuSr	834
308006	−63 01645	IC 2602 - 126			10 44 27	−63 34.4	1	9.78		1.53		1.56		5	R0	1053
	+46 01664				10 44 28	+45 42.9	1	10.39		0.55		0.05		3		1723
93540	−63 01646	IC 2602 - 48		⋆	10 44 28	−64 15.1	6	5.34	.010	-0.10	.003	-0.47	.018	32	B6 Vnn	15,1053,1075,2003*
	−59 02674	Cr 228 - 95			10 44 29	−59 52.9	2	10.98	.005	0.09	.065	-0.69	.055	4	B0 Vb	472,850
		AJ86,209 # 17			10 44 30	−58 23.	1	11.62		1.09		0.04		1		924
305560	−59 02677	LSS 1894			10 44 32	−60 18.0	3	9.75	.007	0.35	.008	-0.63	.040	5	O9.5III	191,540,1737
	−43 06552				10 44 36	−44 12.2	1	10.64		0.53		0.06		2		1730
	+29 02091	G 119 - 32			10 44 37	+28 40.7	3	10.24	.008	0.50	.008	-0.22	.011	4	F5	979,1064,3077
93497	−48 05913	HR 4216		⋆ AB	10 44 37	−49 09.3	5	2.69	.005	0.90	.003	0.57	.000	22	G5 IIIa	15,1020,1075,2012,8015
	+22 02271	G 58 - 17			10 44 38	+21 45.7	2	10.14	.010	1.21	.015	1.19	.005	3	K8	333,1620,3072
93391	+27 01936				10 44 38	+27 10.5	1	7.19		1.57		1.95		4	K5 III	1501
303304	−58 02697	LSS 1897			10 44 38	−59 21.3	3	9.66	.005	0.41	.009	-0.64	.024	6	O5	476,834,1737
308012	−63 01651	IC 2602 - 131			10 44 38	−63 52.1	1	10.06		0.49		0.03		2	F8	1053

Table 1 571

HD	DM	Other Id	N Rem	α_{1950}	δ_{1950}	S	V	σ_V	B–V	σ_{B-V}	U–B	σ_{U-B}	n	Spectrum	References
93405	−63 01637	IC 2602 - 44		10 44 39	−64 04.1	3	9.16	.008	0.44	.000	0.01		19	F3 V	1053,2003,2005
308015	−63 01650	IC 2602 - 133		10 44 39	−64 04.1	2	8.96	.009	1.18	.014	0.82	.038	9	G5	891,1053
301024	−56 03795	LSS 1895		10 44 40	−56 24.0	1	10.39		0.08		−0.71		2	B3	401
93549	−63 01649	IC 2602 - 49	⋆ AB	10 44 40	−64 00.0	8	5.23	.013	−0.08	.008	−0.48	.009	40	B6 V	15,1053,1075,2003*
303302	−58 02699	LSS 1899		10 44 41	−59 07.9	2	10.85	.055	0.19	.010	−0.50	.010	4		476,834
93474	−23 09532			10 44 44	−24 17.1	1	8.11		0.33		0.01		1	A8 V	1776
93779	−79 00554	HR 4231	⋆ AB	10 44 49	−80 12.3	4	5.46	.011	0.95	.004	0.76	.011	18	K0 III	15,1075,1637,2038
		IC 2602 - 197		10 44 51	−64 06.1	1	13.80		0.78		0.26		7		891
93600	−63 01654	IC 2602 - 50		10 44 52	−64 21.1	2	8.43	.010	0.52	.025	0.02		19	F7 V	1053,2003
		AJ86,209 # 18		10 44 54	−59 04.	1	11.83		0.79		−0.19		1		924
93561	−54 04029	LSS 1900		10 44 56	−54 34.0	4	9.09	.058	0.05	.024	−0.89	.033	7	B1/3e	540,976,1737,3044
93563	−56 03800	HR 4221		10 44 56	−56 29.6	4	5.24	.026	−0.07	.011	−0.34	.002	17	B8/9 IIIe	15,401,1637,2012
93599	−60 02275	LSS 1902		10 44 56	−60 55.7	1	9.13		0.45		0.00		2	B9.5Iab/b	8100
93576	−59 02687	Cr 228 - 93		10 44 57	−59 48.8	2	9.58	.015	0.25	.000	−0.70	.010	5	O9 Vn	472,850
305539	−59 02680	Cr 228 - 94		10 44 57	−59 48.8	2	9.90	.000	0.30	.030	−0.71	.030	4	O7 V	472,850
		CCS 1751		10 44 58	−43 38.4	1	9.80		1.79		2.18		2		864
	−35 06733			10 44 59	−35 57.2	1	10.14		0.30		0.09		2		1730
93546	−43 06556			10 44 59	−44 21.7	1	9.02		0.19		0.13		2	A3 III	1730
	−21 03153			10 45 00	−22 01.4	1	9.94		1.15		1.02		1	K5 V	3072
93607	−63 01655	IC 2602 - 51	⋆	10 45 02	−64 07.2	12	4.85	.009	−0.15	.005	−0.65	.010	56	B2.5Vn	15,26,1053,1075,1637*
308024	−63 01656	IC 2602 - 139		10 45 02	−64 15.1	1	10.10		0.15		−0.61		2	B9	540
	+53 01434		A	10 45 03	+53 02.8	1	10.49		0.55		0.01		3		1723
93596	−56 03805			10 45 03	−57 16.9	1	8.73		−0.05		−0.47		2	B5 III	401
93529	−24 09320			10 45 05	−25 10.4	2	9.30	.000	0.88	.014	0.37		5	G6/8w	1594,6006
93595	−56 03808			10 45 08	−57 08.1	1	9.89		−0.07		−0.44		3	B8	3044
	−60 02278	LSS 1904		10 45 08	−60 39.3	1	11.21		0.22		−0.62		3		466
93458	+40 02371			10 45 09	+40 00.4	2	6.78	.005	1.28	.010	1.37	.005	5	K3 III	1501,1601
93527	−14 03184	IDS10427S1505	A	10 45 09	−15 21.7	1	7.66		0.50				4	F7 II/III	2033
93527	−14 03184	IDS10427S1505	B	10 45 09	−15 21.7	1	8.88		0.56				4		2033
93545	−36 06685			10 45 09	−37 15.6	1	7.90		0.01		−0.05		2	B9	401
93526	−14 03186	HR 4218	⋆ A	10 45 10	−14 59.9	2	6.66	.005	−0.01	.000			7	A0 III	15,2029
93649	−68 01302			10 45 10	−68 56.7	2	6.36	.000	0.18	.000	0.20	.019	9	A2 IV	1628,1771
	−34 06982			10 45 11	−34 26.7	1	11.32		0.87		0.56		1		78
93620	−59 02692	LSS 1905		10 45 11	−59 31.6	1	9.15		0.12		−0.57		2	B2 II/III	476
93471	+45 01866			10 45 13	+45 04.0	1	7.81		1.19		1.29		3	K2 II	8100
93593	−54 04037			10 45 13	−54 37.5	1	10.09		0.00		−0.67		2	B3/5 Vn	540
		Bo 11 - 8		10 45 14	−59 49.6	1	12.57		0.15		−0.31		2		411
93632	−59 02696	Cr 228 - 92	⋆ A	10 45 15	−59 50.0	6	8.36	.031	0.30	.022	−0.73	.017	15	O4 III(f)	390,401,411,472,476,1737
305540	−59 02697	Cr 228 - 91		10 45 15	−59 55.9	1	10.05		0.09		−0.05		3	A0 Vn	472
		S264 # 3		10 45 16	−44 12.9	1	11.36		1.47		1.50		2		1730
305612	−59 02700	Bo 11 - 3		10 45 17	−59 50.2	1	10.48		0.16		−0.71		2		411
	−59 02698	Cr 228 - 89		10 45 17	−59 57.7	1	10.43		−0.03		−0.69		2	B2 V	472
93619	−56 03810	LSS 1906		10 45 18	−57 03.7	2	6.95	.000	0.14	.005	−0.72	.050	5	B1 Iab	191,533
		Bo 11 - 5		10 45 18	−59 50.	1	10.80		0.40		−0.60		2		411
		Bo 11 - 6		10 45 18	−59 50.	1	11.85		0.42		−0.57		2		411
		Bo 11 - 7		10 45 18	−59 50.	1	13.07		0.49		0.00		2		411
303387	−58 02705			10 45 19	−58 55.9	1	10.35		0.72		0.44		2	A0	127
305612	−59 02702	Bo 11 - 2	⋆	10 45 19	−59 49.8	1	10.25		0.26		−0.69		2		411
93648	−63 01660	IC 2602 - 52		10 45 19	−64 00.0	3	7.85	.011	0.12	.009	0.08	.000	16	A0 Vn	1053,2003,3015
		Car sq 2 # 10		10 45 20	−60 36.1	1	13.00		0.37		−0.44		3		466
93845	−79 00556	HR 4234		10 45 20	−80 16.6	6	4.43	.009	−0.18	.009	−0.70	.014	24	B3 V	15,26,1075,1637,2038,8023
		G 146 - 64		10 45 21	+43 01.7	1	14.76		1.66				1		906
93618	−56 03812	LSS 1910		10 45 21	−56 54.2	1	9.15		0.13		−0.66		2	B2 Ve	533
303402	−58 02706	LSS 1912		10 45 21	−59 11.6	1	10.69		0.32		−0.54		2	B1 V	1737
93647	−59 02703	Cr 228 - 90		10 45 24	−59 57.1	1	9.44		0.11		0.15		2	A2 Vn	472
	−59 02704	Bo 11 - 4		10 45 25	−59 50.1	2	11.26	.005	0.16	.000	−0.66	.015	4	B2 Ve	411,476
93631	−55 03847	LSS 1913		10 45 27	−55 31.3	3	9.55	.013	0.10	.019	−0.73	.020	6	B2/3 II	540,976,3044
303388	−58 02708	LSS 1915		10 45 27	−58 56.8	2	10.01	.001	0.39	.008	−0.51	.051	4	B0 II-III	127,540
93629	−52 03915	HR Vel		10 45 29	−53 05.5	1	9.60		1.61		1.55		1	M2/3	3044
	+26 02134			10 45 31	+26 15.8	1	9.65		0.62		0.06		2	F8	3016
93427	+65 00803	HR 4215		10 45 31	+65 23.8	1			−0.02		−0.05		1	A1 V	1079
303389	−58 02712			10 45 32	−58 55.8	1	10.26		0.20		−0.04		2	A2	127
93521	+38 02179			10 45 34	+37 50.1	1	7.04		−0.28		−1.09		3	O9 Vp	399
93684	−65 01505			10 45 35	−65 39.1	2	7.61	.024	0.12	.005	−0.44	.000	6	B5 V	164,401
93662	−56 03821	HR 4226		10 45 38	−57 12.2	4	6.32	.029	1.66	.031	1.90		18	M1 III	15,1075,2029,3044
303344	−57 03805			10 45 40	−57 39.7	2	8.81	.015	1.56	.025	−0.04	.005	4	F8	533,1737
93683	−59 02712	LSS 1919		10 45 42	−60 21.2	3	7.89	.012	0.11	.012	−0.82	.015	7	B0/1 Vne	401,540,976
93626	−25 08249			10 45 43	−26 08.0	1	7.90		0.53		0.02		1	G0 V	78
303371	−58 02714			10 45 44	−58 30.0	1	10.29		0.30		0.11		2	A3	127
	−60 02288	LSS 1920		10 45 44	−60 28.9	1	11.22		0.16		−0.72		3		466
93695	−59 02713			10 45 47	−59 36.7	3	6.47	.010	−0.13	.009	−0.62		10	B3 V	6,1075,2012
93714	−63 01670	IC 2602 - 53		10 45 50	−64 17.0	2	6.55	.020	0.02	.015	−0.60		11	B2 IV-Vn	1053,2003
93541	+51 01625			10 45 51	+50 52.4	1	8.87		0.48		0.00		2	F2	1566
93679	−55 03855	HL Vel		10 45 52	−55 23.7	1	9.80		1.57		1.35		1	M2	3044
93739	−68 01305			10 45 52	−69 10.4	1	6.35		−0.10		−0.70		4	B2 IV	1732
93657	−31 08536	HR 4225		10 45 53	−31 25.4	4	5.87	.004	0.03	.000	0.04	.000	12	A1 V	15,78,1075,2012
		AJ86,209 # 19		10 45 54	−58 51.	1	11.65		0.35		−0.51		1		924
93614	+3 02419			10 45 56	+02 39.1	1	8.87		0.29		−0.21		8	F2	1732
93712	−58 02717			10 45 56	−58 26.1	1	9.01		0.02		−0.04		2	A0 V	127
93677	−51 05058			10 45 57	−52 04.1	1	8.85		0.23		0.09		1	A6 V	1770

HD	DM	Other Id	N Rem	α_{1950}	δ_{1950}	S	V	σ_V	B–V	σ_{B-V}	U–B	σ_{U-B}	n	Spectrum	References
		LOD 21 # 14		10 45 57	−58 40.7	1	11.24		0.38		−0.20		2		127
93713	−58 02720			10 45 57	−58 49.6	1	9.74		0.02		−0.10		2	B9.5III	127
93711	−57 03811			10 45 58	−57 55.8	1	9.97		−0.07		−0.43		1	B8/9 II	761
	−58 02721	LSS 1922	⋆ V	10 45 58	−58 52.8	2	10.46	.020	0.72	.010	−0.16	.005	8	He Star	1514,1737
		GD 124		10 46 00	−01 45.3	1	15.81		−0.03		−0.88		1		3060
		LP 791 - 7	AB	10 46 01	−19 56.2	3	13.06	.047	0.85	.014	0.28	.046	8		1696,1773,3060
		LP 731 - 59	⋆	10 46 03	−19 56.4	2	11.47	.045	0.54	.002	−0.19	.005	6		1696,3060
		LP 791 - 8		10 46 03	−19 56.4	1	11.43		0.54		−0.18		3		1773
93738	−63 01672	IC 2602 - 54		10 46 03	−63 59.9	6	6.46	.014	0.01	.014	−0.16	.013	23	A0 Vnn	540,976,1053,2003*
93582	+44 02020			10 46 04	+44 11.6	1	7.86		0.50		0.06		2	F8 V	1601
93723	−58 02722			10 46 04	−59 23.2	1	8.55		0.03		−0.56		2	B3 III	6
93692	−51 05059			10 46 05	−52 02.4	1	8.40		−0.02		−0.21		2	B9 III	1770
93655	−1 02446	HR 4224		10 46 08	−01 41.7	3	5.93	.008	1.59	.005	1.92	.005	11	M2 III	15,1417,3005
93737	−59 02720	HR 4228		10 46 08	−59 39.3	6	6.01	.018	0.27	.009	−0.31	.031	20	A0 Ia/ab	15,191,401,1075,2012,8100
308023	−63 01673	IC 2602 - 138		10 46 10	−64 15.9	1	9.95		0.27		−0.54		2	B8	540
93636	+30 02072	HR 4223		10 46 12	+29 40.9	1	6.15		1.15		1.06		3	K0	1355
		LOD 21 # 15		10 46 12	−58 42.0	1	12.79		0.39				2		127
		AJ86,209 # 20		10 46 12	−59 29.	1	12.36		0.57		−0.21		1		924
305635	−60 02295	LSS 1926		10 46 14	−60 25.5	1	10.15		0.11		−0.82		3	B2 Ve	1737
93721	−43 06565			10 46 15	−44 06.8	2	7.87	.001	−0.02	.000	−0.07	.019	6	B9/A0 III	1770,1776
		G 146 - 66		10 46 16	+46 24.2	1	15.42		1.59				1		906
303384	−58 02730			10 46 16	−58 43.5	1	10.98		0.44		0.01		2	F8	127
		LOD 21 # 7		10 46 16	−58 53.1	1	10.44		1.04		0.83		2		127
		Car sq 2 # 12		10 46 16	−61 07.1	1	12.58		0.13		−0.54		3		466
93777	−63 01675	IC 2602 - 55		10 46 16	−63 37.0	3	8.50	.007	1.21	.013	1.04		11	G8 III	1053,2003,2005
305619	−59 02727	LSS 1928		10 46 18	−60 00.0	6	9.42	.010	0.44	.021	−0.56	.033	10	O9 Ib	6,191,476,540,620,1737
93796	−62 01759	IC 2602 - 56		10 46 21	−63 21.5	3	10.31	.016	0.63	.008	0.09	.018	6	G2 IV	17,1053,3015
93796	−62 01760	IC 2602 - 57		10 46 21	−63 21.8	3	9.98	.021	0.58	.004	0.03	.016	6	F8 V	17,1053,3015
	+38 02182			10 46 23	+38 15.6	1	11.25		−0.20		−0.75		2	B2	1026
303390	−58 02732			10 46 24	−58 52.8	1	10.13		0.39		0.19		2	F0	127
93751	−47 06217			10 46 25	−48 05.2	3	7.04	.004	0.43	.004	0.02		11	F5 IV/V	1075,1311,2033
93794	−58 02733			10 46 25	−58 57.0	1	8.77		1.24		1.29		2	K1/2 III	127
93795	−58 02734	LSS 1929		10 46 25	−59 16.6	2	8.54	.025	0.78	.015	−0.03	.010	4	A2/3p Shell	191,476
	+44 02022			10 46 26	+44 07.3	1	10.19		0.45		−0.01		2	F5	401
		LOD 21 # 16		10 46 34	−58 41.9	1	10.52		1.56				2		127
93827	−60 02298	LSS 1931		10 46 35	−60 40.3	2	9.32	.010	0.05	.012	−0.74	.011	4	B2 Ib/II	540,976
303383				10 46 36	−58 44.	1	11.61		0.12		0.04		2		127
93702	+11 02283	HR 4227		10 46 38	+10 48.6	3	5.32	.038	0.02	.010	0.04	.015	6	A2 V	985,1022,3023
93745	−30 08738			10 46 38	−30 47.6	1	7.49		0.60		0.12		1	G2 V	78
93807	−57 03825			10 46 38	−57 24.5	1	7.09		1.33				4	G8 Ib	2012
93843	−59 02732	LSS 1932		10 46 40	−59 57.5	7	7.32	.018	−0.04	.010	−0.94	.025	12	O5 f	6,476,540,620,976*
93844	−60 02300			10 46 43	−60 52.7	1	8.29		0.09		−0.31		2	B8 V	401
93742	−3 02999	HR 4229	⋆ AB	10 46 45	−03 45.6	2	6.60	.005	0.22	.000	0.07	.010	7	A2 IV	15,1256
93821	−54 04064			10 46 46	−55 05.7	1	7.94		−0.08		−0.28		3	Ap Si	3044
		Ton 547		10 46 48	+28 10.	2	15.41	.005	0.15	.005	−0.56	.015	4	DA	1036,3028
		G 44 - 38		10 46 50	+04 56.1	2	13.51	.035	0.97	.030	0.57	.020	2		1620,1658
93858	−59 02735	LSS 1934		10 46 50	−60 19.8	2	9.08	.004	0.17	.008	−0.68	.011	4	B3 II/III	540,976
93874	−63 01682	IC 2602 - 58		10 46 50	−63 34.1	2	8.18	.009	0.18	.004	0.09		6	A3 IV	1053,2005
93873	−58 02747	LSS 1935	⋆ AB	10 46 56	−59 10.9	4	7.73	.036	0.45	.017	−0.51	.030	7	B1 Ia/ab	191,540,976,1737
93892	−63 01684	IC 2602 - 59		10 46 56	−63 47.2	1	8.99		0.50		0.06		1	F5 IV	1053
93840	−46 06478			10 46 57	−46 30.8	3	7.76	.013	−0.05	.008	−0.79	.064	7	B1/2 Iab/b	191,1034,2012
93856	−55 03876	HM Vel		10 46 58	−56 20.3	1	8.85		1.59		1.14		1	M6 III	3044
303382	−58 02748			10 46 58	−58 49.1	1	10.54		0.00		−0.11		2	A0	127
93913	−66 01423			10 46 58	−67 02.5	1	7.81		0.15		−0.47		2	B3 V	401
		G 119 - 36		10 46 59	+35 49.4	3	13.01	.037	1.65	.018	1.15	.000	4		203,1705,3078
	−59 02741	LSS 1936		10 47 00	−59 33.1	1	10.68		0.23		−0.46		2		476
		AJ86,209 # 21		10 47 00	−59 45.	1	11.34		0.84		−0.02		1		924
		AJ86,209 # 22		10 47 00	−60 12.	1	12.37		0.64		−0.30		1		924
305633	−60 02305	LSS 1938		10 47 00	−60 27.4	2	10.45	.020	0.27	.000	−0.69	.050	5	O9 Ib	191,1737
93852	−48 05954			10 47 01	−49 20.0	1	9.14		0.01		−0.21		1	B9 IV	1770
		LP 491 - 21		10 47 02	+08 52.6	1	12.47		1.33		1.24		1		1696
93890	−58 02750	LSS 1939		10 47 04	−58 37.8	5	9.07	.032	0.76	.017	−0.25	.022	10	B1 Ia/ab	127,191,540,976,1737
	−59 02743	LSS 1940	AB	10 47 06	−59 32.1	1	11.00		0.40		−0.31		2		476
93765	+28 01931	HR 4230		10 47 09	+28 14.3	3	6.06	.012	0.37	.011	−0.03	.005	5	F5 V	70,254,3037
93813	−15 03138	HR 4232		10 47 09	−15 55.9	11	3.12	.009	1.24	.011	1.30	.012	49	K2 II	3,15,30,418,1075,1425*
93898	−57 03838			10 47 11	−57 42.3	1	8.02		−0.05		−0.35		2	B8/9 III	401
	+21 02244	G 58 - 23		10 47 12	+20 45.5	2	9.96	.000	0.60	.000	−0.03	.000	4		516,1620
93727	+52 01513			10 47 12	+51 31.4	1	8.62		1.12		1.11		3	K0 III	37
93925	−63 01688	IC 2602 - 60		10 47 12	−63 59.4	1	9.31		0.17		−0.14		9	B9 III	891
93833	−9 03147	HR 4233		10 47 13	−09 35.2	4	5.85	.009	1.08	.007	0.91	.005	18	K0	15,1417,2013,2028
93868	−43 06576			10 47 13	−44 01.2	1	7.96		0.05		0.14		2	A0 V	1776
93911	−59 02746	LSS 1944		10 47 15	−59 55.1	2	8.36	.040	0.23	.000	−0.49	.030	4	B2/3 Iab/b	191,476
	−59 02749	LSS 1945		10 47 16	−60 00.0	1	10.51		0.19		−0.50		2		476
93763	+53 01436			10 47 24	+52 32.0	1	8.50		0.33		0.06		2	F0	1566
		G 196 - 47		10 47 24	+56 42.5	2	12.59	.007	0.60	.002	−0.12	.016	4		1064,1658
93924	−61 01898	LSS 1948		10 47 24	−61 34.0	1	8.82		−0.01		−0.75		2	B2 II/III	401
93943	−58 02755	HR 4239	⋆ AB	10 47 25	−59 03.5	5	5.85	.014	0.00	.005	−0.06	.001	29	B9.5IV/V	15,127,1075,1637,2012
305599	−59 02758			10 47 26	−59 33.8	1	10.00		0.34		−0.62		2		476
93942	−58 02754			10 47 27	−58 31.1	1	10.19		0.14		0.05		1	A1 V	401
93635	+76 00402			10 47 28	+76 15.6	2	7.20	.014	0.21	.005	0.07	.000	4	A3	985,1776

Table 1 573

HD	DM	Other Id	N	Rem	α_{1950}	δ_{1950}	S	V	σ_V	B–V	σ_{B-V}	U–B	σ_{U-B}	n	Spectrum	References
305601	−59 02760				10 47 28	−59 36.9	1	10.23		0.32		−0.43		2		476
305606	−59 02759				10 47 28	−59 45.6	1	10.48		0.00		−0.70		2	B3	476
		PG1047+003			10 47 29	+00 15.4	1	13.47		−0.29		−1.12		8	sdB	1764
		PG1047+003A			10 47 29	+00 15.4	1	13.51		0.69		0.17		7		1764
303410	−58 02759				10 47 29	−59 16.5	1	10.59		0.24		−0.61		1		924
		PG1047+003B			10 47 35	+00 13.9	1	14.75		0.78		0.17		5		1764
93906	−35 06771				10 47 36	−35 32.4	1	7.34		0.09		0.10		2	A1/2 V	401
93905	−33 07288	HR 4238			10 47 37	−33 47.6	3	5.60	.004	0.04	.000	0.05	.000	11	A1 V	15,1075,2012
	−55 03888				10 47 39	−56 19.3	1	10.92		0.32		0.15		2	A6	807
94008	−68 01323	IDS10460S6813	AB		10 47 39	−68 28.9	1	9.66		0.37		0.15		1	A5/8 V	1771
93904	−8 03017				10 47 43	−08 43.4	1	7.50		1.02		0.73		5	K0	1657
	−55 03890				10 47 43	−56 14.2	1	11.07		0.53		0.06		2	F7	807
		PG1047+003C			10 47 44	+00 15.5	1	12.45		0.61		−0.02		6		1764
93876	+26 02137				10 47 46	+25 54.4	1	7.56		0.53		0.06		3	F5	833
93936	−40 06331	IDS10455S4025	A		10 47 46	−40 41.3	1	8.22		0.54				4	F8/G0 V	2012
93903	−8 03018	HR 4237	⋆ A		10 47 47	−08 37.9	5	5.79	.013	0.16	.004	0.12	.005	19	A3 m	13,15,1417,2030,8071
93903	−8 03018	IDS10453S0822	B		10 47 47	−08 37.9	1	11.65		0.30		0.18		6		13
93975	−54 04080				10 47 49	−54 35.6	1	9.16		1.70		2.04		3	M1/2	3044
		Car sq 2 # 13			10 47 50	−61 21.4	1	11.94		0.28		−0.73		3		466
	−55 03892				10 47 52	−56 22.7	1	10.44		0.24		−0.30		2	B7	807
93932	−14 03199				10 47 58	−14 50.3	1	7.53		0.62				4	G3 V	2033
303353	−57 03857				10 47 60	−58 02.0	1	10.76		0.61		−0.22		3	B8	1737
	−55 03896				10 48 00	−56 16.1	1	10.82		0.39		0.05		2	A9	807
94024	−57 03856	LSS 1952			10 48 00	−57 36.5	3	8.72	.008	0.13	.015	−0.82	.018	6	O8	401,540,976
93971	−36 06725				10 48 02	−36 40.1	1	9.73		0.21		0.23		1	B9.5II	55
	−58 02766	LSS 1955			10 48 03	−59 21.7	1	11.20		0.48		−0.40		1		924
	−55 03897				10 48 04	−56 19.5	1	11.14		0.39		0.14		2	A8	807
94022	−55 03898				10 48 06	−56 11.8	1	9.40		−0.02		−0.51		2	B5/7 III/V	807
93859	+57 01294	HR 4235			10 48 07	+56 50.8	1	5.67		1.12				1	K2 III	71
94066	−63 01699	IC 2602 - 61			10 48 09	−64 16.5	2	7.87	.019	0.10	.014	−0.38		9	B5 Vn	1053,2005
	−55 03899				10 48 10	−56 15.1	1	11.13		0.23		0.15		2	A4	807
	−59 02770	LSS 1956			10 48 12	−59 52.3	1	10.08		0.94		0.15		2		476
94065	−61 01905				10 48 12	−61 32.4	1	8.95		−0.05		−0.83		2	B0/1 III/IV	401
		POSS 169 # 2			10 48 14	+46 09.6	1	15.67		1.35				2		1739
93875	+60 01296	HR 4236			10 48 16	+59 35.2	2	5.58	.005	1.15	.005	1.15		7	K2 III	252,1379
	−55 03900				10 48 16	−56 12.6	1	11.12		0.22		0.20		2	A3	807
		G 44 - 40			10 48 18	+07 05.0	3	11.67	.010	1.65	.013	1.20	.022	8		1705,1764,8039
94017	−33 07296				10 48 18	−33 44.4	1	9.17		0.67		0.20		1	G6 V	78
301085	−55 03902	LSS 1957			10 48 18	−56 21.7	3	10.01	.010	0.10	.015	−0.66	.001	6	B2 II-III	540,807,976
		AJ86,209 # 23			10 48 18	−60 39.	2	11.78	.020	0.77	.005	−0.33	.037	4		466,924
94097	−61 01908				10 48 18	−62 22.2	1	7.20		0.08		−0.44		2	B3 V	401
93915	+47 01822				10 48 20	+47 03.7	1	8.11		0.70		0.26		2	G5	1601
	−55 03903				10 48 20	−56 05.7	1	9.90		0.32		0.11		2	A7	807
	−55 03904	LSS 1958			10 48 21	−56 18.1	1	10.24		0.13		−0.67		2	B2	807
		WLS 1040-10 # 8			10 48 22	−10 02.4	1	11.87		0.39		−0.03		2		1375
94063	−55 03905				10 48 23	−56 04.6	1	8.53		0.18		−0.26		2	B5/7 V	807
		LSS 1961			10 48 23	−59 24.6	1	11.54		0.32		−0.44		1	B3 Ib	924
	−55 03906				10 48 26	−56 09.9	1	11.15		0.22		0.20		2	A3	807
94115	−63 01702	IC 2602 - 62	⋆ AB		10 48 27	−63 42.8	1	8.84		1.32				4	K0 III	2005
305632	−60 02315				10 48 28	−60 33.5	2	11.89	.005	0.20	.015	−0.56	.005	6	M0 Ib	466,1737
94033	−24 09357	KZ Hya			10 48 29	−25 05.8	2	10.08	.145	0.29	.030	−0.01	.045	2	B9 III/IV	1696,1776
	−55 03908				10 48 32	−56 04.5	1	10.43		0.94		0.68		2	K1	807
94014	−2 03236	HR 4240			10 48 33	−02 49.6	2	5.94	.005	1.48	.000	1.83	.000	7	K2	15,1417
	−55 03909				10 48 34	−56 19.3	1	10.33		0.38		0.09		2	A7	807
94012	+10 02212				10 48 37	+09 29.6	1	7.85		0.49				18	F8	130
		Lod 189 # 23			10 48 38	−56 10.1	1	11.46		0.44		0.04		2		807
		LP 791 - 11			10 48 42	−18 24.1	1	12.07		0.77		0.27		2		1696
		AJ86,209 # 24			10 48 42	−59 02.	1	12.67		0.51		−0.34		1		924
		Car sq 2 # 16			10 48 43	−60 57.6	1	11.85		0.40		−0.51		3		466
94144	−60 02317				10 48 43	−61 00.2	3	6.84	.013	−0.01	.000	−0.54	.005	10	B4 III	158,401,1075
94028	+21 02247	G 58 - 25			10 48 48	+20 33.0	6	8.23	.011	0.47	.006	−0.18	.011	12	F4 V	22,1003,1064,1620*
	−55 03912				10 48 50	−56 12.1	1	10.03		1.23		1.27		2		807
	−61 01914	LSS 1965			10 48 50	−62 10.5	1	10.88		0.47		−0.47		2	B0 I	191
		LP 551 - 23			10 48 51	+03 25.2	1	13.39		0.52		−0.14		2		1696
94140	−50 05463				10 48 51	−51 09.6	1	8.19		1.21		1.26		6	K1 III	1673
		Lod 189 # 16			10 48 51	−56 09.7	1	11.06		1.02		0.83		2		807
94072	−12 03280				10 48 52	−12 33.3	1	8.36		1.12		1.05		2	K0 III	1375
	−36 06742				10 48 53	−37 12.5	1	10.18		0.90		0.52		1	G5	78
94174	−63 01708	IC 2602 - 63			10 48 54	−64 12.8	4	7.75	.009	0.11	.003	0.06	.002	12	A0 V	1053,2005,2012,3015
93991	+56 01482				10 48 58	+55 37.2	1	8.11		0.29		0.14		2	F0	1502
	+9 02421	G 45 - 9			10 48 59	+09 21.6	2	10.15	.019	0.75	.005	0.39	.010	3	G8	333,1620,1696
94055	+28 01937				10 48 59	+27 33.8	1	8.91		1.16		1.02		2	K0	1375
94156	−53 04189	IDS10469S5349	AB		10 48 59	−54 04.8	1	10.10		0.24		−0.61		2	B3/5 III	540
94011	+59 01332	G 196 - 49			10 49 00	+58 38.2	1	9.08		0.66		0.20		1		1658
94173	−59 02784	IDS10470S5925	AB		10 49 00	−59 41.5	2	6.79	.005	−0.04	.005	−0.26		6	B8/9 IV	401,1075
	−55 03916				10 49 06	−56 12.7	1	10.82		0.22		−0.52		2	B3	807
94123	−7 03129				10 49 10	−07 48.3	1	8.73		0.48		0.01		2	F5	1375
	−20 03283	V Hya	⋆ A		10 49 11	−20 50.1	1	9.70		4.53				2	C6,3(e)	864
94040	+51 01632				10 49 12	+50 31.9	1	6.63		1.43		1.71		3	K4 III	1501
301073	−55 03917	AA Vel			10 49 13	−56 08.7	1	9.43		0.25		−0.67		2	B5	807

HD	DM	Other Id	N	Rem	α_{1950}	δ_{1950}	S	V	σ_V	B–V	σ_{B-V}	U–B	σ_{U-B}	n	Spectrum	References
94201	−58 02789				10 49 13	−58 51.9	2	9.42	.030	0.08	.005	−0.68	.060	4	B1/2 II	127,432
94151	−21 03168				10 49 15	−21 48.4	1	7.82		0.72				4	G6 V	2012
	−55 03918				10 49 15	−56 02.9	1	10.34		0.78		0.35		2	G9	807
		G 146 - 69			10 49 16	+47 21.1	1	13.60		0.58		−0.09		2		1658
305644	−60 02325	LSS 1972			10 49 17	−60 47.2	2	9.93	.010	0.12	.005	−0.78	.055	4	O8 IIIn	191,1737
305642	−60 02326				10 49 19	−60 40.6	1	9.87		0.03		−0.64		2	B8	401
		Lod 189 # 15			10 49 20	−56 09.2	1	10.98		1.13		0.92		2		807
94230	−59 02791	LSS 1973			10 49 21	−59 28.7	4	7.78	.010	0.37	.014	−0.56	.036	7	B1 Iab	191,540,976,1737
	−59 02793	LSS 1974			10 49 22	−59 37.5	2	10.37	.040	0.39	.085	−0.55	.030	2		924,924
		G 119 - 38			10 49 23	+26 15.0	2	14.15	.015	0.73	.015	0.13	.010	2		1658,3062
		Car sq 1 # 201			10 49 26	−59 24.2	1	14.16		1.07		−0.02		4		466
94260	−61 01922				10 49 28	−61 52.6	2	8.52	.045	0.56	.005	0.12	.035	4	F6 IV/V	125,127
94147	+14 02308				10 49 29	+14 06.5	1	9.66		−0.02		−0.01		1	A0	1026
		G 44 - 42			10 49 29	+14 15.6	1	12.66		1.68		1.30		1		3078
	+0 02709	G 163 - 14			10 49 30	+00 07.1	1	10.18		0.86		0.67		5	K8	7009
		AJ86,209 # 25			10 49 30	−59 32.	1	12.32		0.52		−0.06		1		924
94083	+53 01439	HR 4241			10 49 31	+52 49.9	1	6.73		1.09		1.03		2	gG8	1733
	−55 03920				10 49 31	−56 06.6	1	11.30		0.20		0.18		2	A4	807
	−58 02794				10 49 32	−58 53.7	1	10.42		0.14		−0.67		2		540
305630	−60 02331				10 49 32	−60 32.2	1	10.61		0.02		−0.66		3	B9	466
94084	+53 01440	HR 4242			10 49 33	+52 46.2	4	6.44	.005	1.12	.005	1.03	.012	10	K2 III	15,1003,1080,1733
94164	+8 02422	IDS10470N0800	A		10 49 34	+07 43.6	1	8.21		0.87		0.44		4	G5	3016
94164	+8 02422	IDS10470N0800	B		10 49 34	+07 43.6	1	9.27		0.52		−0.04		4	G5	3032
94290	−66 01434				10 49 34	−66 33.0	1	7.47		0.10		−0.43		2	B4 V	401
	−59 02794	LSS 1976			10 49 36	−59 35.0	2	10.46	.024	0.40	.000	−0.59	.019	3	B0.5 Ia	924,1737
94258	−57 03887				10 49 38	−57 59.0	1	8.93		−0.11		−0.61		2	B3 V	401
94180	+1 02495	HR 4244		⋆ A	10 49 40	+01 17.5	2	6.37	.005	0.08	.000	0.12	.005	7	A3 V	15,1417
94117	+52 01519				10 49 40	+52 20.3	1	7.03		0.37		0.16		3	F5 II	1501
		Car sq 1 # 202			10 49 40	−59 18.5	1	14.74		0.98		−0.10		4		466
		Car sq 2 # 18			10 49 40	−60 40.6	1	12.75		1.04		−0.22		3		466
305645	−60 02336				10 49 43	−60 47.5	1	10.47		0.29		−0.56		3		466
94275	−56 03927				10 49 44	−57 00.4	1	6.29		0.23		0.17		2	A5/7 III	401
94162	+39 02386				10 49 46	+38 36.1	1	7.67		0.46		−0.06		2	F5 V	105
94289	−62 01787				10 49 46	−62 38.4	1	7.81		0.04		−0.54		2	B3 V	401
	+22 02283				10 49 51	+22 06.4	1	10.82		0.66		0.09		4		1723
303492	−58 02801	LSS 1979			10 49 52	−58 42.6	4	8.85	.012	0.56	.018	−0.48	.036	8	O9 Ia	191,540,976,1737
94303	−57 03893				10 49 54	−57 44.4	1	9.58		−0.07		−0.47		2	B5 V	401
94304	−57 03895	LSS 1980			10 49 55	−58 09.0	3	6.85	.014	0.49	.009	−0.24	.032	6	B8 Ia	191,540,976
	−57 03896	LSS 1982			10 49 58	−58 17.5	1	10.97		0.52		−0.47		2	B1 Ib	1737
94333	−61 01929	LSS 1983			10 49 58	−61 55.2	2	9.56	.055	0.05	.005	−0.65	.060	4	B3/5 II/III	125,127
94220	+5 02412	G 44 - 44		⋆ AB	10 50 00	+05 16.2	1	8.17		0.64		0.12		1	K0	1620
94204	+26 02144				10 50 01	+25 49.6	2	8.79	.005	0.13	.000	0.10	.025	5	A3	1625,1776
94237	+0 02710	HR 4245			10 50 02	+00 03.9	3	6.30	.005	1.50	.000	1.82	.000	8	K5 III	15,1415,3035
	+28 01943				10 50 02	+28 27.2	1	10.04		1.08		1.00		4	K4	1723
		vdB 44 # f			10 50 02	−55 56.9	1	13.19		0.48		0.01		2		434
		Car sq 2 # 20			10 50 02	−61 19.1	1	15.15		0.82		−0.18		3		466
94413	−73 00776				10 50 03	−74 20.0	1	9.73		0.42		0.41		2	A2 IV/Vs	1776
		WLS 1100 5 # 6			10 50 04	+05 06.2	1	12.95		0.49		0.01		2		1375
	−1 02457	G 55 - 44			10 50 04	−01 50.1	4	9.79	.029	1.06	.007	0.88	.012	9	K8	158,803,1620,3073
	−55 03930				10 50 05	−55 56.8	1	9.46		1.61		1.96		4	M0 II:	434
94346	−60 02338				10 50 05	−61 09.8	1	7.47		0.05		−0.31		2	B6/7 III	401
		MCC 275			10 50 06	+00 45.0	2	10.66	.009	1.29	.005	1.23		2	K4	801,1017
		vdB 44 # b			10 50 06	−55 56.5	1	12.03		0.21		−0.47		5	B2 Vpe?	434
94315	−56 03935				10 50 06	−56 39.5	1	9.25		1.71		1.66		1	M4/6	3044
		Car sq 1 # 203			10 50 06	−59 34.3	1	12.70		0.80		−0.24		3		466
		AJ86,209 # 26			10 50 06	−60 23.	1	11.80		0.64		−0.32		1		924
94132	+70 00634	HR 4243		⋆ A	10 50 07	+70 07.3	1	5.92		0.99		0.86		1	K1 III:	1355
94330	−56 03936				10 50 07	−56 37.0	1	8.71		−0.04		−0.16		3	B9 V	3044
		Car sq 1 # 204			10 50 07	−59 33.2	1	13.84		0.86		−0.25		3		466
94283	−32 07710				10 50 08	−33 23.4	1	7.99		1.39		1.57		34	K3 III	978
94327	−51 05111				10 50 09	−51 32.0	1	7.53		−0.06		−0.22		3	B9 V	1770
94345	−59 02802				10 50 10	−59 41.3	2	9.00	.009	0.06	.009	−0.76	.003	4	B0/1 IV	540,976
94434	−73 00777				10 50 15	−74 10.7	1	8.95		0.51		0.45		2	A2 V	1776
	+25 02309				10 50 18	+25 14.7	1	10.80		0.77		0.30		2		1375
94280	−6 03252				10 50 18	−06 33.2	1	7.18		0.56		0.56		6	F8	1628
303508	−58 02808	LSS 1988			10 50 18	−59 20.9	4	10.48	.026	0.54	.011	−0.49	.024	9	B0.5 Ib	191,279,662,1737
94370	−58 02809	LSS 1990		⋆ AB	10 50 22	−58 28.8	3	7.94	.011	0.08	.008	−0.82	.014	6	O8 f	401,540,976
94369	−57 03909	LSS 1991			10 50 23	−57 59.2	2	7.35	.019	0.26	.010	−0.37		3	B0.5Ia	6,138
94382	−60 02345	LSS 1993			10 50 25	−60 44.6	2	9.45	.020	0.02	.015	−0.75	.015	5	B2 III	401,466
94367	−56 03947	HR 4250			10 50 28	−56 58.5	5	5.25	.013	0.16	.020	−0.51	.009	13	B9 Ia	15,1018,1034,2001,2012
94381	−59 02807				10 50 30	−59 41.3	1	9.81		0.14		−0.39		3	B3 II/III	279
94264	+34 02172	HR 4247			10 50 31	+34 29.1	9	3.82	.018	1.04	.009	0.92	.011	25	K0 III-IV	1,15,1007,1080,1355*
		Car sq 1 # 3			10 50 32	−59 22.9	1	10.40		1.97				3		279
94247	+55 01418	HR 4246			10 50 33	+54 51.1	1	5.09		1.36		1.52		3	K3 III	1080
		HA 102 # 974			10 50 33	−00 00.8	1	12.32		0.87		0.51		2	K0	281
	−0 02381				10 50 38	−00 46.8	4	9.91	.002	0.49	.005	−0.01	.010	35	F7	281,989,1729,6004
94340	−19 03122				10 50 39	−20 21.5	3	7.07	.048	0.65	.005	0.17		11	G3/5 V	158,2033,3026
	+76 00404	G 253 - 40			10 50 40	+76 19.8	2	9.59	.090	1.10	.010	1.02		2	K8 V:	694,3072
94366	−34 07042				10 50 41	−35 13.4	1	6.77		−0.06		−0.35		2	B6 III	401
94422	−63 01724	IC 2602 - 64			10 50 41	−64 08.9	1	8.24		0.10		−0.39		2	B6 V	1053

Table 1   575

HD	DM	Other Id	N Rem	α_{1950}	δ_{1950}	S	V	σ_V	B–V	σ_{B-V}	U–B	σ_{U-B}	n	Spectrum	References
		L 466 - 50		10 50 42	−39 49.	1	11.27		1.00		0.78		2		1696
303505	−58 02819			10 50 42	−59 11.9	1	10.16		0.13		0.01		1	A3	662
308037	−61 01933			10 50 42	−61 54.1	2	10.14	.015	0.11	.015	-0.34	.040	4		125,127
94408	−57 03915			10 50 43	−57 54.7	1	10.10		1.55		1.40		1	M2	3044
		Car sq 2 # 22		10 50 43	−61 19.1	1	12.01		0.12		-0.53		3		466
94406		SS Vel		10 50 45	−53 09.4	1	9.59		3.00		4.65		2	N3	864
	+10 02216	G 44 - 45		10 50 49	+10 00.3	1	10.16		0.84		0.53		1	K2	1658
94336	+26 02145			10 50 51	+26 28.5	1	7.06		1.57		1.79		3	Ma III	1733,8100
	−49 05632	AF Vel		10 50 51	−49 38.3	1	10.78		0.22		0.16		1	F2.5	700
303479	−58 02821	LSS 1997		10 50 51	−58 25.2	1	9.76		0.07		-0.81		2	B0 V	540
94363	−1 02459	HR 4249	⋆ A	10 50 52	−01 59.3	5	6.13	.021	0.91	.016	0.60	.009	17	K0 III+G0 V	13,15,1084,1417,3024
94363	−1 02458	IDS10483S0143	B	10 50 52	−01 59.3	3	8.73	.013	0.56	.014	0.01	.004	12		13,1084,3037
94389	−32 07720	AH Ant		10 50 54	−32 43.4	1	8.52		1.23		1.02		1	K2 III	1641
		AJ86,209 # 27		10 50 54	−60 09.	1	12.38		0.35		-0.44		1		924
		AJ86,209 # 28		10 51 00	−59 33.	1	11.09		1.02		0.14		1		924
94387	−15 03155			10 51 01	−15 33.3	1	8.18		0.79				4	G8 IV	2033
94451	−58 02824			10 51 01	−59 15.9	2	9.77	.005	0.02	.005	-0.12	.047	4	B9/A0 V	279,662
	−35 06811			10 51 02	−36 21.9	2	10.14	.007	0.50	.005	-0.04		4		1594,6006
94388	−19 03125	HR 4251	⋆ A	10 51 03	−19 52.1	6	5.23	.018	0.48	.014	0.06	.012	23	F6 V	3,15,1008,2013,2028,3053
		HA 102 # 982		10 51 04	−00 00.9	2	13.20	.000	0.54	.000	0.00	.000	15	G0	281,1473
94386	−14 03213	HR 4252		10 51 04	−15 10.7	4	6.36	.007	1.18	.012			15	K2 III	15,2013,2020,2028
94465	−58 02826	V427 Car	⋆	10 51 04	−59 14.4	1	8.81		0.22		-0.71		4	B8	279
303511	−59 02810			10 51 04	−59 24.4	1	10.35		0.34		0.19		2	A5	279
305682	−59 02811			10 51 06	−59 45.9	1	10.32		0.21		-0.60		2	B8	279
94334	+43 02058	HR 4248		10 51 07	+43 27.4	4	4.69	.028	-0.05	.004	-0.07	.022	9	A1 Vs	15,1118,3023,8015
94316	+55 01422			10 51 07	+55 04.1	1	7.73		1.13		1.10		2	K0	1733
94372	−0 02382			10 51 07	−01 05.3	2	9.08	.003	0.54	.003	0.08		11	F8	281,6004
94402	−1 02460	HR 4253	⋆ A	10 51 11	−01 51.8	4	5.45	.019	0.97	.005	0.77	.025	11	G8 III	15,252,1256,3077
94463	−51 05126			10 51 11	−51 31.6	1	8.26		0.09		0.11		1	A1 V	1770
	−61 01940			10 51 11	−61 44.9	1	10.96		0.08		-0.62		2		127
94444	−43 06619			10 51 12	−44 08.7	3	8.08	.005	0.53	.010	-0.09		9	F7/8 V	258,1075,2033
94464	−57 03929			10 51 12	−57 35.0	1	8.87		0.48		-0.04		2	F6/8	401
	−61 01941			10 51 13	−61 44.0	1	11.07		0.13		-0.04		2		127
94190	+77 00412			10 51 15	+77 21.2	1	6.87		1.59		1.94		1	M3 III	3025
	+40 02381	G 146 - 71		10 51 16	+39 47.1	1	10.32		0.53		-0.05		2	G0	1658
94493	−60 02359	LSS 2002		10 51 16	−60 32.9	5	7.25	.023	0.01	.012	-0.85	.015	14	B0.5Iab/b	6,540,976,1075,2012
94401	+3 02429			10 51 17	+02 55.5	1	8.09		0.30		0.00		2	F0	1375
94492	−58 02831	LSS 2003		10 51 19	−59 20.6	1	9.81		0.49		0.29		5	B8/9 II	279
94491	−58 02830	IDS10494S5819	B	10 51 20	−58 37.6	2	6.25	.015	-0.10	.005	-0.59		6	B3 IV/V	6,2012
94509	−57 03939	LSS 2004		10 51 25	−58 09.4	1	9.08		0.02		-0.09		2	Bp Shell	3044
	−58 02835	LSS 2005		10 51 26	−59 12.2	2	11.35	.004	0.10	.004	-0.62	.000	6		279,662
94510	−58 02834	HR 4257	⋆ A	10 51 27	−58 35.2	4	3.78	.009	0.94	.005	0.65	.001	15	K1 III	15,1075,1754,2012
	−59 02815	LSS 2007		10 51 28	−59 35.6	2	10.32	.005	0.47	.005	-0.56	.015	4	B0 Ib	279,1737
	−61 01945			10 51 28	−61 43.3	2	11.34	.065	0.37	.025	0.18	.020	4		125,127
		G 236 - 44		10 51 32	+64 54.2	1	11.39		0.91		0.53		2		7010
		LP 611 - 27		10 51 33	+01 40.2	1	14.12		1.40		1.11		1		1773
94485	−42 06572			10 51 33	−42 51.1	1	7.48		1.06		0.88		6	K0 III	1673
94473	−26 08207	IDS10492S2613	A	10 51 34	−26 28.8	1	7.51		-0.06		-0.52		2	B5 III	1776
303503	−58 02840	HH Car		10 51 35	−59 11.1	3	10.70	.367	0.44	.024	-0.28	.302	11		279,279,662
		HA 102 # 997		10 51 36	−00 03.1	1	12.97		0.35		0.09		2	G0	281
		HA 102 # 739		10 51 38	−00 16.1	1	10.77		0.57		0.11		2	G2	281
		Car sq 2 # 23		10 51 38	−60 38.3	1	10.63		1.16		-0.09		3		466
94533	−58 02844			10 51 39	−58 30.8	1	8.78		-0.07		-0.43		2	B7 III	401
94717	−78 00589	HR 4268		10 51 39	−79 17.6	3	6.33	.004	1.46	.000	1.57	.000	18	K2 II/III	15,1075,2038
94460	+13 02322			10 51 40	+12 38.3	1	7.53		0.94		0.69		2	K0	1648
94482	−22 03031			10 51 41	−22 55.0	1	8.14		0.56		0.06		1	G1/2 V	78
		LP 551 - 25		10 51 42	+08 10.6	1	11.69		0.97		0.75		1		1696
94398	+53 01443			10 51 42	+52 38.6	1	9.15		0.35		0.12		2		1566
94547	−59 02818	IDS10497S5919	AB	10 51 42	−59 34.9	1	10.26		0.22		0.09		3	B1 II	279
94546	−58 02845	V428 Car	⋆	10 51 43	−59 14.8	2	10.57	.021	0.39	.004	-0.57	.004	3	WR	279,1096
		HA 102 # 743		10 51 44	−00 20.0	2	12.08	.003	0.75	.022	0.28		7	G2	281,6004
94559	−61 01951	LSS 2011		10 51 44	−61 23.9	5	7.72	.025	0.02	.006	-0.75	.040	12	O9/B0II/III	125,127,191,540,976
94558	−58 02846			10 51 46	−59 07.0	1	8.85		-0.02		-0.15		6	B9 V	279
		Car sq 1 # 16		10 51 46	−59 44.9	1	12.06		1.18		0.76		3		279
		Car sq 1 # 17		10 51 48	−59 06.1	1	13.17		1.38		1.33		4		279
94481	−12 03293	HR 4255		10 51 49	−13 29.5	2	5.65	.005	0.83	.000			7	K0 III +(G)	15,2027
		Car sq 1 # 18		10 51 49	−59 44.8	1	14.03		0.52		-0.25		2		279
94518	−30 08807			10 51 50	−30 53.3	4	8.35	.007	0.60	.005	-0.02	.019	11	G2 V	158,2012,2039,3077
	−59 02819	LSS 2012		10 51 50	−59 48.1	1	11.12		0.40		-0.28		3		279
94396	+63 00930			10 51 51	+63 13.0	1	8.44		1.25		1.28		2	K2	1733
	+80 00341			10 51 52	+80 03.9	1	10.96		-0.01		-0.09		8		1219
		HA 102 # 747		10 51 52	−00 22.0	3	11.77	.007	0.60	.022	0.03	.000	9	G2	281,1473,6004
94570	−58 02848			10 51 53	−59 19.0	1	8.76		0.00		-0.13		3	B9 III	279
		Car sq 1 # 205		10 51 54	−59 02.7	1	14.34		0.70		-0.28		4		466
	−57 03956	LSS 2013		10 51 57	−58 1.6	1	10.72		0.56		-0.52		2		1737
		Car sq 1 # 21		10 51 58	−60 00.0	1	11.91		0.38		-0.48		2		279
94480	+26 02147	HR 4254		10 51 59	+25 45.4	4	6.19	.021	0.29	.007	0.15	.019	10	A8 V	70,254,833,8071
		AJ86,209 # 29		10 52 00	−59 21.	1	12.07		0.63		-0.21		1		924
		GD 125		10 52 01	+27 22.9	2	14.12	.005	-0.09	.015	-0.96	.020	4	DAn	1727,3060
94500	−0 02384			10 52 01	−01 15.1	4	8.03	.008	0.39	.007	-0.01	.010	18	F4 IV	281,1509,1770,6004

HD	DM	Other Id	N	Rem	α_{1950}	δ_{1950}	S	V	σ_V	B–V	σ_{B-V}	U–B	σ_{U-B}	n	Spectrum	References
		HA 102 # 465			10 52 02	−00 40.6	2	12.28	.006	0.76	.008	0.38		23	G0	281,6004
94650	−70 01246	HR 4262		⋆ AB	10 52 03	−70 27.2	3	5.99	.008	−0.02	.005	−0.46	.008	8	B6 V	15,1075,1771
94516	−0 02385				10 52 04	−00 39.5	4	9.25	.004	1.06	.008	0.92	.004	37	K0 III	281,989,1729,6004
94598	−59 02822				10 52 06	−60 05.2	2	7.85	.005	0.51	.005	0.05		5	F5 IV/Vp	614,1075
		AJ86,209 # 30			10 52 06	−60 21.	1	12.05		0.93		−0.17		1		924
94498	+24 02285	IDS10494N2417	A		10 52 07	+24 01.3	1	8.77		0.32		0.14		2	A5	1776
301172	−56 03980	LSS 2016			10 52 07	−56 54.3	1	10.30		0.08		−0.25		2	A2	540
		Car sq 1 # 206			10 52 07	−59 21.9	1	11.96		0.60		−0.45		3		466
	−59 02821				10 52 07	−59 46.1	1	11.21		0.23		−0.59		4		279
		Car sq 1 # 207			10 52 08	−59 06.8	1	14.38		0.80		−0.24		4		466
		Car sq 1 # 208			10 52 09	−59 21.1	1	12.46		0.73		−0.16		3		466
94613	−61 01955	BZ Car			10 52 09	−61 46.6	1	7.71		2.18		2.01		2	M1/2 Ia/b	127
305720	−60 02373				10 52 10	−60 40.9	1	10.04		0.00		−0.70		2	B8	401
308075	−61 01956				10 52 11	−61 47.7	1	13.82		0.26		0.23		2		127
94497	+34 02178	HR 4256			10 52 12	+34 18.1	1	5.72		1.01				3	K0	70
305680	−59 02823				10 52 12	−59 45.3	1	10.48		0.26		0.21		3	A2	279
		AJ86,209 # 31			10 52 12	−60 01.	1	12.06		0.86		−0.27		1		924
		G 146 - 72			10 52 14	+47 30.6	2	12.73	.005	1.57	.000	1.21	.000	7		419,1663
		HA 102 # 616			10 52 14	−00 35.9	2	12.27	.008	0.47	.003	−0.07		23	F8	281,6004
		Steph 895			10 52 15	+33 26.7	1	12.30		1.36		1.23		1	K5	1746
	+0 02713				10 52 15	−00 19.4	3	10.28	.001	0.52	.008	−0.04	.000	7	G0	281,1473,6004
	−18 03055				10 52 15	−19 07.5	1	10.45		1.32				1	R2	1238
		MtW 102 # 12			10 52 17	−00 36.9	1	14.79		0.44		0.03		3		397
94538	−0 02386				10 52 17	−00 39.1	5	8.75	.005	1.02	.006	0.82	.009	41	G9 III	281,989,1371,1729,6004
94565	−38 06818	IDS10500S3813	A		10 52 17	−38 29.3	1	7.02		−0.04		−0.31		2	B8 III	401
94478	+51 01636				10 52 18	+51 11.0	1	8.50		0.95		0.72		2	G5	1566
94609	−55 03960				10 52 20	−55 59.5	1	9.95		1.68		1.62		1	M2	3044
		MtW 102 # 21			10 52 22	−00 37.2	1	16.70		0.27		−0.09		3		397
94649	−60 02378				10 52 22	−60 29.8	1	8.40		0.01		−0.21		2	B9 V	401
94648	−59 02827	LSS 2019			10 52 25	−59 35.7	2	9.72	.041	0.25	.017	−0.61	.027	6	B2 Ib/II	279,540
		MtW 102 # 36			10 52 29	−00 36.6	1	12.74		1.00		0.80		3		397
		BPM 20372			10 52 30	−55 46.	1	13.42		0.60		−0.10		2		3065
	−0 02387				10 52 31	−00 32.3	8	10.07	.005	1.09	.008	1.02	.013	95	K5	158,281,397,989,1728,1729*
94683	−61 01960	HR 4266		⋆ AB	10 52 31	−61 33.6	9	5.92	.010	1.76	.016	1.99	.040	33	K4 III	15,125,127,404,1075*
94643	−58 02856	LSS 2021			10 52 32	−58 32.0	1	9.35		0.13		−0.78		2	O9.5III	6
94643	−50 05503				10 52 33	−50 58.0	1	9.51		0.15		0.11		1	A1 V	1770
		Car sq 1 # 209			10 52 35	−59 31.1	1	11.62		0.42		−0.40		3		466
		Car sq 1 # 25			10 52 36	−59 28.6	1	10.25		1.41				3		279
305689	−59 02830				10 52 37	−59 51.5	1	9.62		1.37				2	K0	279
94603	−0 02388				10 52 44	−01 09.5	10	9.38	.005	0.06	.005	0.03	.009	89	A0	281,830,989,1371,1728,1729*
	−59 02832				10 52 44	−59 41.4	1	10.52		0.21		0.11		3		279
94619	−19 03134	HR 4261			10 52 45	−20 23.9	1	6.44		1.10				4	K1 III	2006
94660	−41 06220	HR 4263, KQ Vel			10 52 45	−41 59.0	3	6.10	.004	−0.08	.004	−0.24		8	A0p (SiCr)	15,78,2012
		Car sq 2 # 24			10 52 46	−61 02.2	1	13.26		1.03		−0.20		4		466
94715	−59 02833				10 52 48	−59 36.7	1	9.63		0.12		0.06		6	A0/1 V	279
		G 196 - 51			10 52 51	+56 18.1	1	12.73		1.54		1.09		3		7010
303501	−58 02858				10 52 51	−59 06.9	1	9.83		0.74		0.29		5	B7 II	279
94616	−0 02389				10 52 52	−00 32.8	7	8.89	.005	0.55	.002	0.04	.006	46	G2 V	281,397,989,1473,1728*
303558	−58 02859	LSS 2025			10 52 54	−58 27.3	3	9.82	.007	0.26	.010	−0.69	.039	5	O9 III	191,540,1737
305692					10 52 54	−59 57.5	1	9.73		1.08		0.84		2	K0	279
94734	−61 01962				10 52 54	−61 24.1	2	9.65	.045	0.18	.005	0.16	.030	4	A2 V	125,127
94601	+25 02314	HR 4259		⋆ AB	10 52 55	+25 01.0	3	4.34	.051	0.02	.005	0.01	.005	16	A1 V +A2 Vn	15,1022,1363,3023,8015
94733	−60 02382				10 52 55	−61 11.9	2	9.34	.000	0.07	.010	−0.33	.044	6	B9 II/III	125,127
		Car sq 1 # 31			10 52 58	−59 48.1	1	11.75		0.20		−0.42		2		279
		Car sq 2 # 25			10 52 58	−60 29.1	1	14.44		1.07		−0.03		4		466
		G 45 - 17			10 52 59	+08 12.2	1	12.16		1.48		1.34		1	M1	333,1620
94600	+34 02181	HR 4258			10 52 59	+33 46.5	3	5.02	.011	1.10	.008	1.01	.010	7	K1 III	1080,1363,3016
		L 250 - 52			10 53 00	−55 04.	1	14.32		0.09		−0.62		4		3065
		Car sq 1 # 210			10 53 00	−59 20.7	1	13.09		0.55		−0.36		3		466
		AJ86,209 # 32			10 53 00	−60 00.	1	11.42		0.93		−0.14		1		924
		WLS 1100 75 # 10			10 53 02	+75 31.1	1	11.21		0.58		0.05		2		1375
	+21 02263	G 58 - 28			10 53 06	+20 41.1	1	10.30		0.52		−0.05		1	G0	333,1620
94672	+1 02501	HR 4265		⋆ A	10 53 08	+01 00.2	2	5.90	.005	0.43	.005	−0.03	.010	7	F2 III	15,1256
		Car sq 1 # 211			10 53 08	−59 48.6	1	14.23		0.88		−0.14		3		466
		Car sq 1 # 212			10 53 09	−59 53.7	1	14.32		1.15		−0.06		4		466
94724	−42 06595				10 53 11	−42 45.2	1	6.42		0.06		−0.02		2	A0 V	1776
94709	−25 08325				10 53 13	−26 05.1	1	8.50		0.56		0.04		1	G1 V	78
	−59 02838	IDS10513S5912	AB		10 53 17	−59 28.4	1	10.53		0.97		0.68		3		279
94776	−59 02840	HR 4271, T Car			10 53 17	−60 15.1		5.92	.009	1.07	.009	0.88		12	K0 III	15,614,1075,2007
94775	−59 02839				10 53 18	−59 37.9	1	9.64		0.13		−0.22		3	B8 II/III	279
94777	−60 02386	WZ Car			10 53 19	−60 40.4	2	8.94	.280	1.05	.275	0.86	.263	2	F8	689,1587
		Car sq 1 # 213			10 53 20	−59 33.1	1	13.18		0.55		−0.30		3		466
94774	−59 02844				10 53 22	−59 29.1	1	10.01		0.18		0.09		3	A0	279
305679	−59 02846	LSS 2027			10 53 23	−59 34.7	1	10.03		0.20		−0.67		13	B8	279
94669	+42 02162	HR 4264			10 53 24	+42 16.6	4	6.03	.008	1.13	.004	1.08	.010	7	K2 III	15,1003,1080,3035
94706	−0 02390				10 53 24	−00 44.1	3	8.81	.012	0.49	.007	−0.03	.047	15	G0	281,1371,6004
94705	+6 02369	HR 4267, VY Leo			10 53 26	+06 27.2	4	5.79	.019	1.45	.005	1.12	.108	10	M5 III	15,1415,3001,8032
305690	−59 02849				10 53 29	−59 50.5	1	10.35		0.59		−0.02		2	B7 II	279
308096	−63 01748	LSS 2030			10 53 29	−63 57.8	3	9.53	.007	0.50	.005	−0.32	.024	8	B1 III	191,540,976
94786	−54 04165				10 53 32	−55 23.0	1	9.38		−0.07		−0.54		2	B3/5 V	3044

Table 1

577

HD	DM	Other Id	N Rem	α_{1950}	δ_{1950}	S	V	σ_V	B–V	σ_{B-V}	U–B	σ_{U-B}	n	Spectrum	References
305678 −59 02848				10 53 32	−59 28.5	1	10.27		0.31		0.17		3	A3	279
94804 −58 02868				10 53 33	−58 59.0	1	9.21		0.58		0.52		3	A2 II	279
94718 +28 01952		G 119 - 42		10 53 34	+28 00.5	1	8.40		0.73		0.24		3	K0	3026
94753 −32 07751				10 53 34	−32 47.7	1	8.76		0.46		0.00		40	F3 IV/V	978
		Car sq 1 # 39		10 53 35	−59 26.0	1	11.32		0.34		−0.44		5		279
94720 +23 02279		HR 4269		10 53 36	+22 37.1	2	6.14	.027	1.55	.016	1.94		5	K2	70,1375
+2 02370		G 45 - 18		10 53 38	+02 19.8	1	10.44		0.88		0.62		1	K0	333,1620
94812 −58 02870				10 53 38	−59 23.2	1	10.09		−0.01		−0.25		3	B9 V	279
94719 +24 02287				10 53 40	+23 31.1	1	7.33		0.92		0.59		3	G5	1625
305749 −59 02851		IDS10517S5911	AB	10 53 43	−59 27.0	1	10.63		0.12		−0.31		5	A2	279
94827 −56 04010				10 53 46	−56 38.0	1	8.38		1.24		1.36		2	K1/2 III	3044
		Car sq 1 # 214		10 53 48	−59 05.6	1	13.06		0.78		−0.06		3		466
		Car sq 1 # 215		10 53 51	−59 25.7	1	14.73		0.79		−0.12		4		466
94747 +26 02152		HR 4270		10 53 52	+25 46.1	2	6.36	.005	1.04	.014	0.89		4	K1 III	70,105
−59 02852				10 53 52	−59 56.6	1	11.08		0.31		0.03		3		279
		HA 102 # 363		10 53 53	−00 51.7	2	11.02	.001	0.47	.011	−0.04		6	A8	281,6004
305691 −59 02853				10 53 53	−59 56.8	1	10.62		0.24		−0.54		4		279
+57 01297				10 53 54	+56 54.	1	9.78		0.52		−0.06		2		1723
		HA 102 # 364		10 53 54	−00 52.2	2	12.07	.003	0.57	.011	0.09		22	G0	281,6004
−59 02854		LSS 2032		10 53 54	−59 58.3	2	10.48	.019	0.34	.010	−0.47	.024	10	B3 IIIe	279,1737
94765 +8 02434				10 53 55	+07 39.4	1	7.34		0.92		0.66		2	K0	1375
		G 56 - 1		10 53 55	+14 49.5	1	13.80		0.76		0.23		2		1658
		LTT 4015		10 53 58	−49 17.1	1	10.96		1.51		1.64		1		78
94841 −48 06046				10 53 58	−49 17.1	1	8.81		1.05		0.79		1	G8 III/IV	78
94878 −59 02855		GG Car		10 53 58	−60 07.5	2	8.78	.045	0.54	.016	−0.75	.010	9	B e	540,976
94924 −68 01363				10 53 59	−68 31.4	1	8.06		0.11		0.08		2	A1 V	1771
		HA 102 # 370		10 54 00	−00 54.5	2	11.23	.004	0.62	.000	0.05		6	G0	281,6004
		L 537 - 11		10 54 00	−49 17.	1	11.32		0.47		0.03		1		78
94877 −58 02873				10 54 00	−59 15.4	1	8.54		0.64		0.17		std	G3 V	279
		Car sq 1 # 46		10 54 00	−59 17.2	1	16.55		1.08		0.35		2		279
		Car sq 1 # 47		10 54 01	−59 12.7	1	16.27		0.93		0.64		4		279
		G 45 - 20	★ V	10 54 04	+07 18.8	5	13.52	.021	2.02	.018	1.20	.036	28		694,1705,1764,3078,8006
305742 −61 01968				10 54 05	−61 26.2	2	10.11	.050	0.07	.000	−0.08	.015	4	B8	125,127
94899 −59 02858				10 54 06	−59 32.2	1	10.02		0.14		0.06		2	B9 V	279
94808 −0 02392				10 54 07	−00 54.1	7	7.92	.006	0.31	.004	0.09	.009	49	A3	281,989,1371,1473,1509*
		Car sq 1 # 49		10 54 08	−59 12.8	1	15.76		0.91		0.49		4		279
		LP 791 - 50		10 54 09	−16 59.4	1	11.82		0.67		0.03		2		1696
94900 −59 02859		LSS 2034		10 54 09	−59 39.8	1	9.24		0.17		−0.66		2	B1/2 III	279
		Car sq 1 # 50		10 54 09	−59 59.0	1	10.23		1.60				2		279
		Car sq 1 # 216		10 54 10	−59 53.9	1	12.79		0.22		−0.42		3		466
94807 +0 02716				10 54 11	−00 04.7	2	8.28	.003	0.40	.001	−0.05		13	F2	281,6004
94910 −59 02860		AG Car		10 54 11	−60 11.2	3	6.85	.166	0.63	.014	−0.59	.025	8	B2pe	401,1588,1637
94898 −57 04002				10 54 12	−57 47.1	2	7.14	.009	1.59	.014	1.78		5	M3 III	2012,3044
		Tr 17 - 157		10 54 12	−58 57.	1	11.41		1.04		0.79		5		56
		Tr 17 - 160		10 54 12	−58 57.	1	11.56		0.33		−0.20		4		56
		Car sq 1 # 217		10 54 12	−59 15.7	1	13.96		0.64		−0.36		5		466
+42 02163		G 146 - 75		10 54 13	+42 09.3	1	9.72		0.82		0.39		2	K1 V	1003
		Car sq 1 # 52		10 54 13	−59 17.7	1	10.92		1.30		1.00		26		279
94923 −61 01969				10 54 14	−61 26.4	2	8.64	.045	−0.05	.005	−0.66	.059	5	B5	125,127
		Car sq 1 # 53		10 54 17	−59 36.3	1	10.74		1.88				2		279
305830 −61 01972				10 54 17	−61 31.4	2	8.86	.040	0.42	.010	−0.02	.040	5	F2	125,127
+16 02188				10 54 19	+15 32.8	1	10.04		1.28				1	R0	1238
		Car sq 1 # 54		10 54 19	−59 17.1	1	14.27		1.60		1.22		5		279
94909 −56 04016		LSS 2036	★ A	10 54 20	−57 17.0	4	7.35	.042	0.50	.036	−0.44	.034	7	B0/0.5 1la	6,138,540,976
+70 00639		G 236 - 46		10 54 21	+69 51.8	2	10.26	.020	1.42	.035	1.35		5	M0	1017,7008
−44 06945				10 54 21	−45 19.7	1	10.72		0.33		0.08		5		1770
		Car sq 1 # 218		10 54 21	−59 41.4	1	13.88		0.96		−0.19		3		466
94936 −60 02395		LSS 2037		10 54 21	−61 06.0	2	9.42	.002	0.18	.013	−0.54	.015	4	B3/5 Ib/II	540,976
94890 −36 06808		HR 4273		10 54 23	−36 52.1	4	4.60	.006	1.03	.005	0.84	.004	20	K0 III	15,1075,2012,8015
303575 −58 02877				10 54 23	−58 57.4	1	10.37		0.42		−0.13		6	B9	279
94935 −59 02862		LSS 2038		10 54 24	−59 45.8	1	8.66		0.15		−0.60		5	B3 II	279
		vdB 45 # b		10 54 24	−62 45.9	1	11.36		0.44		−0.52		3	B1 Vp?	434
94891 −39 06797				10 54 27	−40 18.3	1	8.41		0.57				4	F8/G0 V	2012
+0 02717				10 54 28	+00 04.8	5	9.90	.002	0.66	.005	0.26	.008	86	G5 IV	281,989,1729,1764,6004
94833 +26 02153				10 54 28	+25 32.4	1	8.84		0.51		0.03		2	F8 V	3026
−60 02398		LSS 2040		10 54 28	−60 49.8	1	10.66		0.04		−0.57		3		466
94835 +22 02290		G 58 - 30		10 54 29	+22 04.5	1	7.94		0.62		0.12		1	G0	333,1620
		HA 102 # 816		10 54 30	−00 18.8	2	11.83	.018	0.65	.031	0.07		8	G0	281,6004
94906 −30 08839				10 54 30	−30 56.2	2	7.42	.010	0.35	.000	−0.08		12	F0 V	1088,2012
		AJ86,209 # 33		10 54 30	−59 11.	1	11.61		0.66		−0.29		1		924
+28 01954		G 119 - 44		10 54 32	+27 29.7	1	11.13		0.67		0.05		2		7010
94864 +0 02718				10 54 35	−00 02.6	2	6.88	.001	0.42	.002	−0.01		14	F5	281,6004
		HA 102 # 821		10 54 35	−00 17.9	2	11.12	.011	0.61	.024	0.11		9	G0	281,6004
		Car sq 1 # 219		10 54 35	−59 33.6	1	15.00		0.87		−0.20		4		466
		vdB 45 # a		10 54 35	−62 43.9	1	12.23		0.41		−0.54		3		434
94963 −61 01975		LSS 2043		10 54 36	−61 26.5	6	7.16	.025	−0.09	.006	−0.95	.023	20	O6/7f	125,127,164,540,976,1732
		HA 102 # 825		10 54 39	−00 21.6	2	12.36	.000	0.99	.073	0.61		20	G0	281,6004
305765 −59 02865				10 54 39	−59 58.4	1	9.86		1.21		1.18		2		279
94989 −64 01557				10 54 39	−65 02.1	1	9.17		0.57		0.11		60	F7/8 IV/V	1593
+38 02197				10 54 40	+37 55.0	1	9.16		0.50		0.04		2	F8	1601

HD	DM	Other Id	N	Rem	α_{1950}	δ_{1950}	S	V	σ_V	B–V	σ_{B-V}	U–B	σ_{U-B}	n	Spectrum	References
		vdB 45 # c			10 54 41	−62 45.0	1	11.57		0.40		−0.56		3	B0 V	434
	+26 02155				10 54 43	+25 44.8	1	10.62		0.51		0.00		2		1375
303594	−58 02881				10 54 44	−59 17.2	1	10.86		0.28		0.21		3	A5	279
		vdB 45 # d			10 54 45	−62 45.2	1	10.75		0.43		−0.60		4	O6	434
	−59 02870	LSS 2049			10 54 47	−59 39.8	2	10.73	.005	0.29	.000	−0.69	.015	4	B0.5 Ib:	279,1737
94988	−59 02869				10 54 48	−59 52.0	1	9.35		0.07		−0.58		2	B2/4N	279
		Car sq 1 # 61			10 54 50	−59 58.5	1	14.02		0.46		−0.42		3		279
		Cha T # 3		★ V	10 54 50	−77 08.6	1	14.65		1.23		0.34		1	M0.5	437
		G 56 - 2			10 54 51	+13 55.5	1	11.76		1.32		1.21		1	K7	333,1620
94987	−56 04025				10 54 51	−57 18.6	1	9.40		−0.06		−0.46		2	B8 II	3044
	+0 02721				10 54 52	+00 01.2	2	9.64	.003	0.46	.002	0.03		10	F5	281,6004
		Car sq 1 # 220			10 54 54	−59 05.2	1	14.77		0.84		−0.07		4		466
		Car sq 1 # 221			10 54 54	−59 23.4	1	11.75		0.34		−0.44		3		466
		Car sq 1 # 62			10 54 55	−59 36.5	1	16.58		0.75		0.09		3		279
303565	−58 02883	Tr 17 - 159			10 54 56	−58 46.3	1	9.80		0.06		−0.04		6	B8	56
94862	+56 01490				10 54 57	+55 41.9	1	8.06		0.97		0.73		3	G8 II!	1502
	−32 07773				10 54 57	−33 14.3	2	10.13	.010	0.55	.005	−0.02	.015	3	F8	78,1696
94985	−50 05534	HR 4274, IW Vel			10 54 57	−50 29.9	2	5.90	.005	0.18	.000			7	A4 V	15,2012
		Car sq 1 # 63			10 54 57	−59 36.7	1	16.84		0.90				3		279
		G 254 - 10			10 55 00	+74 55.9	1	11.34		1.15		1.11		2		7010
94956	−28 08551				10 55 00	−29 00.8	1	8.45		0.96		0.73		2	K0 III	536
	−32 07774				10 55 01	−33 12.9	1	9.33		0.88		0.66		1	K0	78
		Car sq 1 # 64			10 55 01	−59 37.4	1	16.06		0.63		0.33		4		279
	+41 02143				10 55 04	+40 31.0	2	10.64	.019	0.15	.005	0.11	.010	3	Am	1026,1240
95012	−52 04044				10 55 04	−52 47.3	1	9.05		−0.03		−0.21		1	B9 IV/V	1770
95018	−59 02873				10 55 04	−60 11.3	1	9.32		0.03		−0.35		2	B7 II	401
		Car sq 2 # 27			10 55 04	−60 33.3	1	12.60		0.20		−0.52		3		466
		Car sq 1 # 65			10 55 05	−59 27.2	1	14.34		0.64		0.49		5		279
95017	−59 02874				10 55 05	−59 32.0	1	9.03		1.10		0.91		2	G8/K0 III	279
		Car sq 1 # 67			10 55 05	−59 37.5	1	15.70		0.47		0.33		3		279
94686	+80 00347	IDS10507N8013	AB		10 55 06	+79 56.6	4	7.32	.008	0.55	.004	0.04	.005	15	F8 V	1007,1013,1334,8023
		G 163 - 27			10 55 06	−07 15.4	3	14.33	.011	0.30	.008	−0.52	.016	50	DA	281,1764,3060
		Tr 17 - 5			10 55 06	−59 00.0	1	11.20		0.47		−0.15		7		56
	−44 06958				10 55 07	−44 57.6	1	10.54		0.26		0.12		1		1770
		Car sq 1 # 68			10 55 07	−59 20.0	1	14.53		1.45		1.27		8		279
		Car sq 1 # 69			10 55 07	−59 28.4	1	14.44		0.65		−0.04		3		279
		Car sq 1 # 222			10 55 08	−59 57.4	1	13.34		0.60		−0.51		3		466
		Car sq 1 # 70			10 55 09	−59 36.4	1	13.29		1.27		0.87		4		279
303576	−58 02885	Tr 17 - 1			10 55 11	−58 58.3	1	9.83		1.56				2	M0	279
		Car sq 1 # 72			10 55 12	−59 26.0	1	14.18		0.61		0.04		5		279
		Tr 17 - 4			10 55 13	−59 00.6	1	12.38		0.41		−0.28		3		56
		Car sq 1 # 73			10 55 13	−59 33.6	1	11.94		0.44		0.09		13		279
		Bo 12 - 12			10 55 13	−61 27.8	1	12.50		1.31		1.31		1		411
		G 163 - 28			10 55 14	−03 55.4	1	16.53		0.40		−0.46		1		3060
		Car sq 1 # 74			10 55 14	−59 32.1	1	13.86		0.58		0.27		4		279
		Car sq 1 # 75			10 55 15	−59 33.3	1	14.53		0.81		0.30		4		279
		Car sq 1 # 76			10 55 15	−59 35.8	1	14.11		0.40		0.24		6		279
	−59 02877	LSS 2052			10 55 15	−59 41.6	2	11.05	.020	0.36	.010	−0.65	.020	5	B3 Ve	279,1737
310331	−64 01564	XX Car			10 55 15	−64 52.0	2	8.67	.049	0.71	.012	0.56	.039	2	G0	689,1593
		Car sq 1 # 77			10 55 16	−59 35.4	1	16.68		1.77				4		279
		Car sq 1 # 78			10 55 16	−59 37.7	1	15.18		0.50		0.44		2		279
95086	−68 01373				10 55 16	−68 23.9	1	7.41		0.22		0.04		1	A8 III	1771
		Car sq 1 # 80			10 55 18	−59 22.8	1	14.29		0.63		−0.35		6		279
		Bo 12 - 4			10 55 18	−61 27.7	1	11.64		1.19		1.03		1		411
		Cha T # 4		★ V	10 55 18	−76 55.6	2	13.01	.020	1.19	.100	0.14	.220	2	M0:	437,825
		Ton 1312			10 55 19	+34 41.	2	15.55	.035	−0.03	.040	−0.88	.000	7		286,313
		Car sq 1 # 223			10 55 19	−59 55.1	1	14.36		0.68		−0.23		4		466
305780	−59 02879				10 55 19	−60 19.3	1	10.04		0.05		−0.57		2	B5	401
		Bo 12 - 11			10 55 19	−61 28.5	1	12.77		1.04		0.89		1		411
		Bo 12 - 8			10 55 20	−61 27.6	1	13.95		0.18		−0.08		1		411
		Bo 12 - 2			10 55 21	−61 27.4	1	12.20		0.40		0.24		1		411
	−16 03156				10 55 22	−16 30.6	1	10.71		0.72		0.14		1	K3 V	3072
		Car sq 1 # 81			10 55 22	−59 27.1	2	11.95	.101	0.15	.000	0.04	.000	7		279,1577
		Bo 12 - 9	AB		10 55 22	−61 27.2	1	13.35		0.32		0.18		1		411
		Car sq 1 # 82			10 55 23	−59 23.1	1	14.02		1.07		0.82		6		279
		Car sq 1 # 83			10 55 23	−59 37.1	1	12.77		0.47		0.04		3		279
		Bo 12 - 3			10 55 23	−61 27.8	1	12.88		0.17		0.04		1		411
		Car sq 1 # 84			10 55 24	−59 25.7	1	15.52		0.76		0.59		1		279
		Car sq 1 # 85			10 55 24	−59 34.3	1	15.02		0.48		0.42		4		279
		Car sq 1 # 86	AB		10 55 24	−59 50.7	1	10.57		0.19		−0.61		2		279
		Bo 12 - 1			10 55 24	−61 27.5	1	11.75		0.07		−0.45		2		411
		Bo 12 - 5			10 55 24	−61 28.	1	11.43		0.04		−0.38		1		411
		Bo 12 - 6			10 55 24	−61 28.	1	12.95		0.43		0.20		1		411
		Bo 12 - 7			10 55 24	−61 28.	1	13.28		0.18		−0.04		1		411
		Bo 12 - 10			10 55 24	−61 28.	1	13.32		1.50		1.40		1		411
		Car sq 1 # 87	AB		10 55 25	−59 48.3	1	12.01		0.52				4		279
		Car sq 1 # 224			10 55 25	−59 49.6	1	11.52		0.23		−0.60		3		466
		ApJS57,743 T1# 4			10 55 26	−59 27.4	1	13.32		0.48		0.23		1		1577
		Car sq 1 # 88			10 55 27	−59 35.7	1	12.89		1.51		1.24		5		279
	+49 02004				10 55 28	+48 33.7	2	10.52	.009	1.37	.005	1.28		2	M0	801,1017

Table 1

HD	DM	Other Id	N	Rem	α_{1950}	δ_{1950}	S	V	σ_V	B–V	σ_{B-V}	U–B	σ_{U-B}	n	Spectrum	References
303592	−58 02890				10 55 29	−59 18.2	1	10.05		0.97		0.67		2	G5	279
95111	−67 01601				10 55 29	−67 33.0	1	9.66		0.37		0.14		1	A9 V	1771
	+41 02144				10 55 31	+40 55.5	2	10.60	.019	0.04	.000	0.06	.010	3	A2p	1026,1240
		Car sq 1 # 225			10 55 32	−59 55.9	1	13.77		0.74		-0.36		3		466
		Car sq 1 # 90			10 55 33	−59 31.3	1	13.03		0.44		0.08		3		279
305767	−59 02882				10 55 33	−59 56.0	1	9.93		0.83		0.55		5		279
95122	−68 01377	IDS10538S6830	A		10 55 33	−68 46.2	1	6.63		-0.02		-0.43		2	B7 V	1771
		Car sq 1 # 92			10 55 36	−59 35.0	1	11.76		0.54		0.12		7		279
95095	−59 02883	LSS 2055			10 55 36	−59 48.7	3	9.53	.011	0.30	.008	-0.49	.018	20	B0/1 III	279,540,976
		LP 431 - 38			10 55 37	+19 17.6	1	11.46		0.87		0.66		2		1696
95094	−58 02892				10 55 37	−59 20.8	1	10.11		0.06		-0.03		5	B9 III/IV	279
94902	+70 00641	VW UMa			10 55 38	+70 15.4	1	6.91		1.70		1.86		1	M1	3001
94979	+37 02139	IDS10528N3734	A		10 55 39	+37 17.9	1	8.78		0.78		0.29		2	G5	3032
94979	+37 02139	IDS10528N3734	B		10 55 39	+37 17.9	1	10.60		1.15		0.96		2		3032
		Car sq 1 # 95			10 55 40	−59 34.4	1	11.51		0.98		0.56		1		279
		Car sq 1 # 226			10 55 41	−59 58.7	1	13.42		0.78		-0.43		3		466
95109	−59 02888	HR 4276, U Car			10 55 46	−59 27.8	6	5.97	.195	0.98	.101	0.82	.122	8	G3 Iab	279,689,1484,1587,1754,3074
95107	−57 04026				10 55 47	−57 40.8	1	8.96		-0.03		-0.26		2	B8	3044
		Car sq 1 # 97			10 55 48	−59 30.9	1	13.79		0.53		-0.07		4		279
		Car sq 1 # 98			10 55 49	−59 26.5	2	11.09	.066	1.96	.009	1.81	.023	4		279,1577
95208	−74 00755	HR 4279			10 55 50	−74 49.9	3	6.12	.004	1.53	.000	1.63	.000	14	K1 II	15,1075,2038
		Steph 897			10 55 51	+60 31.3	1	12.39		1.24		1.20		1	K6	1746
		Car sq 2 # 28			10 55 52	−61 06.3	1	13.61		0.88		-0.25		3		466
		Car sq 1 # 227			10 55 53	−59 48.5	1	14.36		0.62		-0.27		4		466
95047	+20 02538				10 55 57	+19 53.4	1	6.79		1.34		1.50		2	K0	1648
		LP 731 - 76			10 55 58	−10 30.4	1	14.50		1.70		1.08		1		1773
		ApJS57,743 T1# 2			10 56 00	−59 28.3	1	13.03		0.63		0.18		1		1577
94860	+78 00367	HR 4272			10 56 01	+78 02.3	1	6.20		0.97				2	G9 III	71
	−10 03166				10 56 01	−10 30.5	1	10.08		0.84		0.58		1		1773
		Car sq 1 # 228			10 56 01	−59 49.4	1	13.65		0.76		-0.26		3		466
		G 119 - 46			10 56 02	+35 19.6	1	11.13		1.12		0.99		5	K5	7010
		Car sq 1 # 229			10 56 04	−59 35.3	1	14.61		0.84		-0.10		4		466
95045	+45 01879				10 56 09	+45 28.0	1	6.92		1.23		1.26		2	K0	1601
		GD 126			10 56 09	−01 22.1	1	12.92		0.48		-0.18		2		3060
	+33 02059	IDS10534N3329	AB		10 56 12	+33 12.6	1	10.76		0.60		0.10		3	G0	1723
305754	−59 02894				10 56 12	−59 40.0	1	9.89		1.02		0.76		2	G5	279
308106	−61 02000	LSS 2060			10 56 12	−61 53.5	1	10.68		0.10		-0.69		2	B3	540
		Car sq 2 # 29			10 56 14	−60 57.6	1	14.34		0.81		-0.26		4		466
		Car sq 1 # 100			10 56 16	−59 50.1	1	11.87		0.59		-0.41		4		279
		Car sq 2 # 30			10 56 16	−60 55.8	1	12.89		0.80		-0.35		3		466
		Ho 09 - 5			10 56 18	−58 47.	1	14.41		0.37		0.33		1		411
		Ho 09 - 6			10 56 18	−58 47.	1	13.53		1.19		0.96		1		411
		Ho 09 - 7			10 56 18	−58 47.	1	14.14		1.25		1.14		1		411
		Ho 09 - 8			10 56 18	−58 47.	1	13.84		0.73		0.43		1		411
		Ho 09 - 9			10 56 18	−58 47.	1	13.98		0.34		0.35		1		411
95169	−51 05197				10 56 19	−51 41.1	1	9.31		-0.02		-0.13		1	A0 V	1770
		Ho 09 - 1			10 56 20	−58 47.4	1	11.54		0.05		-0.14		2		411
95057	+52 01528	HR 4275			10 56 21	+52 09.0	1	6.17		1.38		1.59		2	K2 III	1375
		Ho 09 - 3			10 56 21	−58 47.2	1	13.03		0.61		0.10		1		411
303590	−58 02900				10 56 21	−59 16.7	1	9.86		1.06		0.76		6	G5	279
		Ho 09 - 4			10 56 22	−58 47.2	1	13.40		1.21		0.94		1		411
		Ho 09 - 2			10 56 22	−58 47.4	1	12.54		0.28		0.16		1		411
		Car sq 2 # 31			10 56 22	−60 57.6	1	12.76		0.75		-0.41		3		466
		Car sq 1 # 230			10 56 27	−59 36.2	1	14.32		0.73		-0.19		4		466
	−46 06608				10 56 34	−46 57.1	1	10.83		0.86		0.45		1	K0	78
		Car sq 1 # 231			10 56 34	−59 10.0	1	14.02		0.52		-0.52		4		466
		Car sq 2 # 32			10 56 35	−60 51.8	1	12.83		1.21		-0.02		3		466
305771	−59 02899				10 56 36	−60 05.0	2	9.62	.027	0.19	.002	-0.54	.020	4	B2 III	540,976
		Car sq 2 # 33			10 56 36	−61 02.4	1	14.85		1.05		-0.14		4		466
		Car sq 2 # 34			10 56 37	−60 28.8	1	12.81		1.03		-0.08		3		466
		LP 431 - 40			10 56 40	+14 35.4	1	13.56		0.66		-0.03		2		1696
		G 119 - 47			10 56 40	+34 31.0	1	15.58		-0.11		-0.70		1	DC:	3060
95128	+41 02147	HR 4277			10 56 40	+40 41.9	9	5.05	.015	0.61	.010	0.12	.007	25	G0 V	15,101,245,1008,1013*
		L 17 - 26			10 56 42	−81 38.	1	13.41		0.68		-0.05		3		1696
95254	−59 02900				10 56 43	−59 27.6	1	9.84		0.01		-0.48		2	B5/7 II	279
305757	−59 02901	LSS 2062			10 56 43	−59 50.6	2	10.56	.005	0.23	.010	-0.63	.060	4	B1 III	191,279,540
95129	+36 02139	HR 4278			10 56 46	+36 21.7	4	6.00	.000	1.59	.005	1.91	.009	8	M2 III	15,1003,3055,8015
95098	+59 01338	IDS10539N5927	A		10 56 46	+59 10.5	1	7.12		1.20		1.25		4	K2 II-III	1084
		CCS 1791			10 56 47	−62 22.8	1	10.28		4.94				2	N?	864
237960	+59 01339	IDS10539N5927	B		10 56 49	+59 10.9	1	8.95		1.14		1.12		4	K0	1084
		Car sq 1 # 232			10 56 50	−59 09.5	1	14.10		0.63		-0.30		4		466
		Car sq 1 # 233			10 56 51	−59 53.4	1	11.16		0.21		-0.68		3		466
95124	+53 01453				10 56 52	+52 31.4	1	8.76		0.44		0.00		2	F5	1566
95221	−33 07401	HR 4282		★ AB	10 56 52	−33 28.1	2	5.70	.005	0.38	.005			7	F2 V	15,2028
		Car sq 1 # 234			10 56 52	−59 52.7	1	11.37		0.17		-0.60		3		466
		Car sq 1 # 235			10 56 54	−59 54.2	1	14.54		0.89		-0.27		2		466
		Car sq 2 # 35			10 56 54	−60 56.0	1	11.67		0.91		-0.23		3		466
95190	+10 02230	IDS10543N1028	AB		10 56 56	+10 12.0	1	7.24		0.25		0.17		2	A2	3016
95175	+18 02430	G 56 - 4			10 56 56	+18 16.3	1	8.85		0.98		0.79		1	K0	333,1620
		WLS 1100 5 # 10			10 56 57	+04 28.6	1	13.66		0.69		0.06		2		1375

HD	DM	Other Id	N Rem	α_{1950}	δ_{1950}	S	V	σ_V	B–V	σ_{B-V}	U–B	σ_{U-B}	n	Spectrum	References
305778	−59 02906	LSS 2066		10 56 58	−60 12.9	1	9.59		0.16		−0.62		2	B5	401
303589	−58 02906		AB	10 56 59	−59 10.6	1	10.38		0.02		−0.13		2	A0	279
308104	−61 02006	LSS 2067		10 56 59	−61 39.5	1	11.13		0.15		−0.80		3	B2 Ve	1737
		LP 214 - 17		10 57 01	+43 26.3	1	10.64		1.18		1.07		2		1726
95290	−58 02907			10 57 02	−58 54.0	3	7.67	.000	−0.08	.010	−0.45	.000	7	B7 III	401,614,1075
95292	−59 02907			10 57 02	−59 42.3	1	8.72		1.57		2.00		3	K4/5	279
95234	−15 03174	HR 4284		10 57 03	−16 05.1	4	5.88	.006	1.63	.004	1.91		14	M1 III	15,2013,2029,3005
95216	+12 02284	HR 4281		10 57 04	+11 58.4	2	6.53	.015	0.42	.010	−0.03	.010	4	F5 V	985,1733
	−58 02908			10 57 04	−59 09.3	1	10.89		0.20		−0.77		2	O9 V	279
		Car sq 1 # 107		10 57 05	−59 45.7	1	10.43		1.52				2		279
95289				10 57 06	−56 51.	1	10.80		1.55		1.22		1	M2	3044
		HBC 566, SZ Cha		10 57 06	−77 01.3	3	12.00	.111	1.21	.117	0.83	.226	3	K0e	437,1633,1763
95188	+26 02163			10 57 07	+25 33.5	1	8.49		0.76		0.28		2	G8 V	3026
95291	−58 02909			10 57 08	−59 07.8	1	8.28		0.05		−0.03		3	B9.5Vp Si	279
	+42 02173			10 57 10	+41 51.2	1	10.44		1.32				1	R2	1238
95324	−60 02433	HR 4290	★ AB	10 57 13	−61 03.1	5	6.15	.017	−0.06	.007	−0.30	.005	18	B8 IV	15,401,404,1075,2012
		Car sq 1 # 236		10 57 15	−59 38.0	1	12.45		1.19		−0.04		2		466
		Car sq 1 # 237		10 57 16	−59 21.3	1	14.92		0.77		−0.13		4		466
		Car sq 2 # 36		10 57 16	−61 15.6	1	13.70		0.86		−0.27		3		466
95322	−59 02909			10 57 17	−59 38.6	1	9.06		−0.01		−0.21		3	B9 III	279
		Car sq 1 # 238		10 57 17	−59 51.8	1	14.05		0.82		−0.36		3		466
95272	−17 03273	HR 4287		10 57 20	−18 01.9	10	4.08	.012	1.09	.010	1.00	.015	39	K1 III	15,968,1003,1075,1311*
95320	−56 04069			10 57 20	−56 29.4	1	8.94		1.70		2.00		3	M1	3044
		NGC 3496 - 98		10 57 20	−60 00.9	1	13.30		0.60		0.35		1		56
95338	−55 04033			10 57 22	−56 21.2	1	8.62		0.88				4	K1/2 III/V	2033
		GT Car		10 57 22	−59 12.9	1	12.69		1.34		1.09		1		689
95212	+46 01680	HR 4280		10 57 23	+45 47.7	1	5.47		1.47				2	K5 III	70
		G 58 - 32		10 57 25	+23 06.3	3	10.03	.009	1.54	.011	1.22		4	M4	694,1017,3016
		NGC 3496 - 99		10 57 25	−60 01.1	1	13.17		0.64		0.51		1		56
		Car sq 2 # 37		10 57 26	−60 29.7	1	12.12		0.09		−0.57		3		466
95357	−58 02916	LSS 2070		10 57 28	−59 21.6	4	8.89	.015	0.04	.015	−0.82	.015	12		279,401,540,976
		Car sq 2 # 38		10 57 28	−61 11.1	1	12.58		0.44		−0.58		3		466
95233	+52 01529	HR 4283		10 57 29	+51 46.2	1	6.39		1.02		0.87		3	G9 III	1733
		NGC 3496 - 6		10 57 29	−60 08.1	1	13.25		0.73		0.45		4		56
95241	+43 02068	HR 4285	★ A	10 57 31	+43 10.9	3	6.02	.011	0.57	.008	0.01	.009	10	F9 V	70,254,3026
301233	−56 04076	LSS 2069		10 57 31	−56 41.7	1	10.63		0.04		−0.68		2	B2.5III	540
		Car sq 1 # 112		10 57 31	−59 47.3	1	11.76		1.04		0.79		4		279
95377	−59 02911			10 57 32	−59 24.2	1	9.27		0.07		0.04		3	A0 III/IV	279
		NGC 3496 - 29		10 57 32	−60 05.4	1	12.67		0.58		0.44		1		56
		NGC 3496 - 18		10 57 32	−60 06.0	1	12.96		0.59		0.36		4		56
	−59 02912			10 57 33	−59 26.2	1	11.08		−0.01		−0.20		4		279
	−59 02913	NGC 3496 - 5		10 57 33	−60 07.8	1	10.01		1.35		1.44		7		56
95155	+70 00643			10 57 34	+69 51.2	1	8.30		1.11		0.99		3	G5	1733
		NGC 3496 - 30		10 57 34	−60 05.3	1	14.56		1.06		0.73		1		56
		WLS 1112 50 # 6		10 57 35	+49 54.8	1	12.94		0.55		−0.01		2		1375
		NGC 3496 - 17		10 57 35	−60 06.1	1	14.14		0.63		0.34		3		56
		AJ77,733 T17# 8		10 57 35	−60 17.9	1	12.75		0.68		−0.39		3		125
		Car sq 1 # 239		10 57 37	−59 39.6	1	14.57		0.76		−0.12		1		466
95393	−61 02011			10 57 37	−62 12.9	1	7.33		1.41		1.24		3	G3 Ib	1704
		NGC 3496 - 4		10 57 38	−60 08.0	1	13.14		1.76		1.84		4		56
		Car sq 1 # 240		10 57 39	−59 37.6	1	12.54		0.57		−0.30		3		466
		NGC 3496 - 3		10 57 39	−60 08.5	1	14.32		0.59		−0.41		2		56
		Car sq 1 # 241		10 57 40	−59 34.6	1	14.77		1.20		0.20		4		466
95314	−13 03271	HR 4289		10 57 42	−13 48.9	4	5.87	.006	1.50	.000	1.82		12	K5 III	15,2013,2029,3005
95347	−43 06692	HR 4292		10 57 43	−43 32.3	4	5.81	.007	−0.08	.004	−0.22	.003	12	B8/9 V	15,1770,1776,2012
		NGC 3496 - 43		10 57 43	−60 04.4	1	12.01		0.45		−0.32		1		56
308113	−61 02012	LSS 2073		10 57 43	−61 57.8	1	10.64		0.05		−0.70		2	B2 III	191
		NGC 3496 - 2		10 57 44	−60 08.1	1	12.33		0.58		−0.28		4		56
		Car sq 1 # 114		10 57 45	−59 46.8	1	13.28		0.92		0.46		2		279
		NGC 3496 - 9		10 57 45	−60 07.2	1	14.55		0.79				1		56
95414	−59 02914	NGC 3496 - 1	★ AB	10 57 45	−60 08.6	2	9.13	.000	0.20	.004	−0.65	.004	11	B1 III	56,401
	−34 07135			10 57 46	−34 49.4	1	11.22		1.53		1.15		1		78
		NGC 3496 - 10		10 57 47	−60 07.1	1	13.30		1.44		1.02		2		56
		Cha T # 7	★ V	10 57 47	−77 06.6	2	13.26	.175	0.95	.105	−0.17	.015	2	M0:	437,825
303604	−56 04084	LSS 2072		10 57 48	−57 14.8	1	10.17		0.32		−0.61		2	B3	540
95413	−58 02918			10 57 48	−59 18.5	1	8.72		−0.03		−0.20		3	Ap SiCr	279
	−76 00647	VR j		10 57 49	−76 49.8	1	9.55		1.66		1.53		30	K0 III	1763
95370	−41 06276	HR 4293		10 57 51	−41 57.4	5	4.38	.008	0.11	.005	0.12	.003	21	A3 V	15,1075,2012,3023,8015
		HRC 244	★ V	10 57 51	−76 45.5	3	11.33	.053	1.20	.012	0.66	.058	3	K2	437,1763,5005
95296	+43 02069	IDS10550N4316	A	10 57 52	+43 00.0	1	6.69		1.12		1.03		4	K0	1601
305841	−59 02915	LSS 2075		10 57 52	−59 28.0	1	10.09		0.26		−0.53		5	B8	279
305854	−59 02916			10 57 53	−59 35.2	1	10.86		0.09		−0.69		4	B8	279
		AJ77,733 T17# 6		10 57 56	−60 09.4	1	11.82		0.31		0.20		4		125
95256	+64 00824	HR 4286		10 57 57	+63 41.4	2	6.39	.005	0.16	.002	0.14	.002	7	A2 m	3058,8071
95428	−51 05219			10 57 57	−51 25.1	1	9.95		0.11		0.12		1	A2	1770
95429	−51 05220	HR 4296	★ A	10 57 57	−51 32.9	2	6.14	.005	0.18	.000			7	A3 III/IV	15,2012
95430	−51 05218			10 57 58	−51 58.4	1	9.86		0.04		0.05		1	A1 V	1770
		Car sq 2 # 39		10 57 58	−60 50.1	1	13.19		0.80		−0.26		3		466
95345	+4 02407	HR 4291		10 57 59	+03 53.2	8	4.84	.011	1.16	.007	1.11	.006	33	K1 III	3,15,1080,1105,1363*
	+41 02150			10 58 00	+40 58.3	1	10.41		1.32				1	R0	1238

Table 1   581

HD	DM	Other Id	N Rem	α_{1950}	δ_{1950}	S	V	σ_V	B–V	σ_{B-V}	U–B	σ_{U-B}	n	Spectrum	References
		Sh 1 - 2		10 58 00	−60 06.	1	11.08		1.16		0.04		2		411
		Sh 1 - 3		10 58 00	−60 06.	1	12.49		1.11		-0.06		2		411
		Sh 1 - 4		10 58 00	−60 06.	1	10.69		0.94		0.52		2		411
		Sh 1 - 5		10 58 00	−60 06.	1	12.63		0.37		0.31		3		411
		Sh 1 - 6		10 58 00	−60 06.	1	13.37		1.02		-0.21		2		411
		Sh 1 - 7		10 58 00	−60 06.	1	13.64		0.47		0.24		2		411
		Sh 1 - 8		10 58 00	−60 06.	1	14.19		1.04		0.29		2		411
		Sh 1 - 9		10 58 00	−60 06.	1	14.23		1.27		0.23		2		411
		Sh 1 - 10		10 58 00	−60 06.	1	14.34		0.40		0.38		2		411
		Sh 1 - 11		10 58 00	−60 06.	1	14.08		1.24		0.97		2		411
		AJ77,733 T17# 4		10 58 01	−60 10.6	1	11.47		0.07		-0.17		5		125
95461	−58 02922	LSS 2078		10 58 02	−59 05.9	6	8.83	.018	0.41	.015	-0.59	.033	11	B0 II/III	6,138,279,540,976,1737
		Car sq 1 # 242		10 58 02	−59 40.8	1	11.05		0.08		-0.61		2		466
95310	+39 02400	HR 4288		10 58 03	+39 28.9	5	5.08	.008	0.24	.005	0.17	.000	12	Am	15,1008,1363,3020,8071
95366	+16 02202	G 56 - 8		10 58 05	+15 43.3	2	9.41	.015	0.85	.010	0.58	.005	2	K2	333,1620,1658
95445	−48 06083			10 58 05	−49 22.1	1	9.03		0.03		0.02		1	A0 V	1770
	−24 09457	LTT 4045		10 58 07	−24 52.4	1	10.17		0.79		0.39		1		78
		Car sq 1 # 243		10 58 07	−59 44.2	1	14.46		0.96		-0.28		2		466
95382	+6 02384	HR 4294	★ A	10 58 09	+06 22.2	4	4.99	.007	0.17	.009	0.11	.006	11	A5 III	15,1363,1417,3023
95405	−25 08383			10 58 09	−25 35.2	2	8.30	.005	1.23	.090	1.04		2	K0 III	1238,3040
95459	−56 04094			10 58 10	−57 17.5	1	7.08		1.62		1.86		4	M2/3 III	3044
95330	+48 01909			10 58 11	+47 57.1	1	7.58		1.31		1.52		2	K2	1375
		Car sq 1 # 244		10 58 11	−59 38.3	1	13.74		0.85		-0.31		2		466
95496	−59 02923			10 58 12	−60 04.3	1	9.41		0.28		-0.51		7	B2 II/III	125
		AJ77,733 T17# 5		10 58 13	−60 15.6	1	11.73		0.26		-0.26		5		125
95363	+27 01968			10 58 15	+27 24.1	1	7.98		0.49		-0.06		1	F7 V	3026
95424	−17 03282			10 58 16	−18 05.9	1	9.07		0.50				3	F5 V	968
305840	−59 02924			10 58 16	−59 25.5	1	10.43		0.11		-0.60		4	B5	279
95364	+24 02300			10 58 17	+24 20.2	1	8.57		0.69		0.14		4	G2 V	3077
95456	−31 08696	HR 4298		10 58 18	−31 34.3	4	6.06	.006	0.52	.008	0.03	.021	10	F8 V	15,78,2012,3053
95471	−48 06087			10 58 19	−48 44.2	1	7.58		0.64		0.12		1	B9.5 V	1770
		Car sq 1 # 245		10 58 20	−59 44.2	1	13.79		1.04		-0.17		3		466
		Car sq 1 # 120		10 58 21	−59 06.6	1	12.68		0.27		0.30		3		279
		Car sq 2 # 40		10 58 21	−60 58.0	1	11.50		1.02		-0.15		3		466
95492	−51 05225			10 58 22	−51 40.7	1	7.66		-0.01		-0.14		3	B9.5/A0 V	1770
95491	−50 05585			10 58 24	−51 22.5	1	9.21		0.26		0.16		1	Ap SrCrEu	1770
		G 253 - 41		10 58 25	+79 30.2	2	11.76	.015	0.48	.017	-0.18	.017	4		1658,5010
		G 45 - 26		10 58 26	+01 16.3	1	12.88		1.48		1.27		1		333,1620
		Steph 902		10 58 27	+57 08.4	1	11.89		1.46		1.22		1	M0	1746
95470	−34 07143			10 58 27	−34 33.9	2	8.68	.011	1.30	.013	1.38	.020	8	K2/3 III	1664,1763
95441	−15 03178	HR 4297		10 58 28	−15 31.4	1	6.34		1.18				4	K0 III	2006
95453	+0 02725	G 62 - 64	★ AB	10 58 30	+00 18.9	1	9.23		1.21		1.22		1	A2	1620
305869	−59 02926			10 58 31	−60 06.0	1	8.76		1.27		0.57		8	K0	125
95508	−49 05749			10 58 32	−50 09.6	1	9.49		0.03		-0.03		1	Ap Cr(Eu)	1770
	−33 07473	LTT 4074		10 58 33	−33 08.0	1	10.08		1.22		1.21		1		78
95506	−39 06845	IDS10563S3957	AB	10 58 33	−40 13.8	1	6.78		0.53		0.03		1	F7 V	78
95540	−59 02927	LSS 2082		10 58 34	−60 02.1	1	7.91		0.38		0.03		12	B9 Ib	125
	−34 07146			10 58 35	−34 31.7	2	9.11	.013	1.17	.010	1.07	.004	6	K0	1664,1763
	−59 02928	Sh 1 - 1		10 58 37	−59 30.8	1	9.45		0.03		-0.50		3	B4 IV	411
95534	−51 05230			10 58 38	−51 48.3	1	6.81		0.00		-0.02		2	A0 V	1770
		LP 491 - 47		10 58 39	+08 22.3	1	12.57		1.12		1.02		1		1696
95788	−80 00509	HR 4304	★ AB	10 58 39	−81 17.3	4	6.70	.004	0.53	.008	0.03	.006	21	F7 V	15,1075,1637,2038
	−59 02929	LSS 2083		10 58 40	−60 11.1	4	9.50	.019	0.34	.011	-0.59	.033	9	B0 II	191,401,540,976
		Car sq 2 # 41		10 58 41	−60 35.9	1	13.25		0.83		-0.36		3		466
		G 45 - 27		10 58 42	+03 16.7	2	14.07	.034	1.73	.000	1.27		7		940,3078
	−43 06702			10 58 43	−44 23.7	1	11.00		0.63		0.02		1		78
95486	+15 02276			10 58 47	+15 17.0	1	7.66		0.94		0.48		2	K0	3016
95418	+57 01302	HR 4295		10 58 50	+56 39.1	7	2.37	.009	-0.02	.006	0.00	.009	23	A1 V	15,1007,1013,1203*
		LP 731 - 81		10 58 50	−10 19.6	1	11.31		0.99		0.84		1		1696
95589	−61 02022	LSS 2086		10 58 50	−62 08.4	4	9.65	.012	0.26	.013	-0.70	.029	10	O7	191,390,540,976
	+76 00410			10 58 51	+76 10.3	1	8.70		1.08		0.88		2	K0	1733
		LP 611 - 42		10 58 51	−00 11.9	1			1.02		0.76				1696
		LSS 2085		10 58 51	−58 48.1	1	10.62		0.85		-0.07		2	B3 Ib	1737
		Cha T # 9		10 58 55	−76 28.1	1	15.92		1.21				1		437
		Ros 4 - 4		10 58 56	−59 58.4	1	11.84		0.68		-0.15		3		18
		Ros 4 - 3		10 58 58	−59 57.7	1	11.58		0.65		-0.24		3		18
		AJ77,733 T17# 7		10 58 58	−60 15.2	1	12.28		0.48		0.36		5		125
		Ros 4 - 2		10 58 59	−59 57.8	1	14.51		0.94		0.11		3		18
		SZ Leo		10 59 00	+08 26.0	1	11.91		0.20		0.12		1		597
		Pis 17 - 3		10 59 00	−59 33.	1	11.27		0.20		-0.59		2		411
		Pis 17 - 4		10 59 00	−59 33.	1	11.32		0.18		-0.60		2		411
		Pis 17 - 5		10 59 00	−59 33.	1	12.90		0.36		-0.24		2		411
		Pis 17 - 6		10 59 00	−59 33.	1	14.18		0.35		-0.32		2		411
		Pis 17 - 8		10 59 00	−59 33.	1	14.25		0.70		0.32		2		411
	+60 01302			10 59 01	+59 48.0	1	9.68		0.44		-0.04		2	F5	1502
95616	−59 02936			10 59 02	−59 57.6	1	9.40		0.06		-0.52		2	B3 V	401
		Ros 4 - 1	AB	10 59 02	−59 57.7	1	13.24		0.84		0.18		3		18
		Ros 4 - 7		10 59 02	−59 57.9	1	15.75		1.48		0.45		3		18
		Car sq 2 # 42		10 59 03	−60 31.4	1	13.42		0.81		-0.42		3		466
95635	−60 02455	LSS 2088		10 59 08	−60 57.2	1	7.95		0.09		-0.51		2	B2/3 II	401

HD	DM	Other Id	N	Rem	α_{1950}	δ_{1950}	S	V	σ_V	B–V	σ_{B-V}	U–B	σ_{U-B}	n	Spectrum	References
95615	−58 02937				10 59 09	−58 52.7	1	9.20		0.19		−0.48		2	B3 II/III	191
		LSS 2089			10 59 10	−60 02.4	1	11.28		0.43		−0.66		2	B3 Ib	1737
		BPM 35623			10 59 12	−46 42.	1	13.37		0.48		−0.22		1		3065
	−59 02944	Pis 17 - 1	⋆	APQ	10 59 14	−59 34.4	2	10.45	.010	0.20	.010	−0.69	.015	7	B0 Ve	411,434
		Pis 17 - 2			10 59 15	−59 34.8	1	11.28		0.18		−0.62		4		411
	−59 02945				10 59 15	−59 34.9	1	11.33		0.21		−0.61		3	B2 Ve	434
		Pis 17 - 7			10 59 15	−59 35.0	1	13.75		0.39		−0.36		2		411
95499	+59 01345				10 59 16	+58 56.1	1	6.61		1.58		1.90		2	K2	1733
95578	−1 02471	HR 4299			10 59 17	−02 12.9	8	4.74	.013	1.61	.008	1.93	.018	34	K5 III	15,30,369,1075,1425*
		Pis 17 - 9			10 59 17	−59 35.1	1	12.91		1.40		0.83		2		411
	−64 01598	LSS 2092			10 59 20	−64 57.3	1	10.96		0.17		−0.61		2	B2 II-III	540
95610	−33 07435				10 59 22	−33 38.3	1	8.30		0.55		0.02		1	G1/2 V	78
95577	+15 02277	IDS10568N1509	A		10 59 23	+14 53.4	1	9.35		0.52		0.00		2	G0	3016
95577	+15 02277	IDS10568N1509	B		10 59 23	+14 53.4	1	10.24		0.62		0.06		2	G0	3032
	+3 02446				10 59 24	+03 01.3	1	10.46		0.47		−0.05		2		1375
		WLS 1100 25 # 5			10 59 24	+27 21.5	1	11.81		1.02		0.84		2		1375
		AO 956			10 59 24	+36 20.5	1	11.87		0.56		0.02		1		1722
		AO 957			10 59 28	+35 58.6	1	11.25		0.41		0.00		1		1722
		GL 409			10 59 29	−17 41.2	1	11.88		0.62				1		1705
		SSII 4			10 59 30	+35 57.9	1	11.33		0.42		−0.04		1	F2	1298
		Car sq 2 # 43			10 59 33	−60 48.2	1	12.30		0.61		−0.49		3		466
	+10 02234	IDS10570N1026	B		10 59 34	+10 09.9	1	10.64		0.66		0.14		2	F8 Vn	627
95628	−20 03324				10 59 34	−21 13.6	1	8.94		0.86		0.49		1	K0 V	78
		HBC 568, TW Hya			10 59 36	−34 27.5	2	10.91	.040	0.97	.082	−0.37	.040	2	K7 V	1664,1763
95707	−60 02460	LSS 2096			10 59 37	−61 17.6	2	7.56	.005	0.24	.015	−0.34	.030	4	B9 Iab	191,401
305921	−61 02030	LSS 2097			10 59 37	−61 29.5	1	10.47		0.11		−0.70		2	B0 V	191
	+37 02149				10 59 38	+36 32.0	1	10.84		0.64		0.13		1		1722
		CCS 1801			10 59 38	−67 55.4	1	10.81		3.22				2	N	864
95608	+20 02547	HR 4300			10 59 40	+20 26.9	5	4.42	.004	0.06	.010	0.05	.000	14	A1 m	15,1022,1363,3023,8015
95667	−37 06993				10 59 40	−37 33.9	1	7.53		1.69		2.07		9	K5 III	1673
96124	−83 00386	HR 4312			10 59 42	−84 19.5	4	6.18	.013	0.10	.005	0.12	.003	19	A1 V	15,1075,1637,2038
		Cha F # 10			10 59 45	−77 22.6	1	13.74		2.05		2.10		2	K3 III	1633
95498	+73 00514				10 59 48	+73 22.9	1	9.05		0.09		0.08		1	A2	1776
303675	−58 02944				10 59 48	−58 57.3	1	10.20		0.19		−0.86		2	B9	432
305850	−59 02955	LSS 2099			10 59 48	−59 44.7	1	8.79		0.13		−0.72		2	B3	401
95733	−61 02032	LSS 2100			10 59 48	−61 36.9	1	9.31		0.17		−0.65		2	B1 Iab	191
308162	−61 02033	LSS 2103			10 59 51	−61 47.1	1	9.53		0.10		−0.69		2	B1 III	191
95752	−63 01794	LSS 2104			10 59 51	−64 02.1	1	6.95		0.47		0.38		60	A7 Ib/II	1593
95731	−58 02945	LSS 2102			10 59 53	−59 06.5	2	9.07	.010	0.29	.105	−0.68	.060	4	B1/2 Ib	6,191
95651	+9 02441				10 59 56	+09 26.5	2	7.24	.000	0.11	.005	0.06	.017	2	A0	627,3016
		AO 959			10 59 56	+36 31.8	1	12.05		0.65		0.07		1		1722
95650	+22 02302	DS Leo			10 59 57	+22 14.2	4	9.57	.021	1.49	.044	1.25	.070	7		679,1017,3072,7008
95718	−48 06110				10 59 57	−48 31.7	1	7.74		0.09		0.11		1	A1/2 IV/V	1770
95698	−26 08302	HR 4302	⋆	AB	10 59 59	−26 33.6	3	6.22	.006	0.32	.005	0.06		8	A9 III/IV	15,78,2027
95786	−65 01593				10 59 59	−65 44.6	2	7.52	.005	0.04	.025	0.05	.000	4	A0/1 V	401,1771
		AO 960			11 00 00	+36 19.7	1	11.63		0.49		−0.04		1		1722
		Car sq 2 # 44			11 00 00	−60 53.6	1	12.77		0.16		−0.55		3		466
	−35 06917				11 00 03	−35 24.7	1	9.77		0.69		0.22		1	K0	78
95695	−2 03270	IDS10575S0258	A		11 00 04	−03 14.6	1	6.80		0.95				4	G8 III	2006
95916	−78 00650	Cha F # 13			11 00 05	−77 20.1	1	9.23		0.55		0.05		2	F8 V	1633
		Car sq 2 # 45			11 00 07	−60 32.3	1	11.41		0.20		−0.67		3		466
95658	+36 02146	IDS10574N3613	A		11 00 10	+35 56.8	2	8.72	.017	0.25	.003	0.15	.004	3	Am	1601,1722
305845	−59 02961	LSS 2109			11 00 15	−59 32.3	2	9.63	.006	0.17	.007	−0.62	.012	5	B2 II	540,976
95545	+75 00434				11 00 16	+74 39.2	1	8.76		0.15		0.11		1	A2	1776
308149	−63 01798	XY Car			11 00 17	−63 59.6	2	8.86	.007	0.97	.018	0.66	.023	2	G5	689,1593
95638	+62 01160				11 00 18	+61 55.5	3	7.15	.000	0.51	.007	−0.03	.008	9	F7 V	1084,1733,3016
		WLS 1100 25 # 7			11 00 19	+22 58.2	1	12.28		0.63		0.13		2		1375
95779	−50 05619				11 00 19	−51 19.9	1	8.15		0.11		0.12		1	A2 V	1770
95799	−58 02953	NGC 3532 - 649			11 00 19	−58 29.5	1	8.01		1.00		0.78		3	G8 III	1671
95743	−8 03068	G 163 - 38			11 00 20	−09 03.6	1	9.04		0.99		0.69		2	K3 V	3072
95826	−59 02964	LSS 2110	⋆	AB	11 00 20	−60 14.9	1	8.53		0.11		−0.75		2	B5	401
95741	−2 03272	G 163 - 39	⋆	A	11 00 25	−03 06.6	2	8.89	.005	0.88	.002	0.67	.005	3	G5	1658,3016
95741	−2 03272	G 163 - 40	⋆	B	11 00 25	−03 06.8	2	12.86	.044	1.42	.015	1.17	.080	3		1658,3016
95820	−52 04144				11 00 25	−52 26.8	1	9.42		0.08		0.11		1	A2 V	1770
95725	+29 02116				11 00 28	+29 12.0	2	7.08	.005	1.05	.000	1.00		4	K1 II	20,8100
		SSII 5			11 00 30	+38 46.9	1	11.34		0.28		0.06		1	F0	1298
		B2 11O1+38 # 2			11 00 31	+38 24.6	1	10.25		0.98		0.80		2		487
95690	+55 01439	IDS10576N5504	AB		11 00 31	+54 47.7	1	8.45		0.89		0.58		1	K2 V	292
		G 58 - 36			11 00 32	+22 16.6	1	11.92		1.03		1.01		1	K3	333,1620
	−61 02040	LSS 2115			11 00 32	−61 28.5	1	10.48		0.77		−0.33		2	B0.5 Ia	1737
		B2 11O1+38 # 1			11 00 35	+38 40.0	1	10.39		0.09		0.08		2		487
95735	+36 02147	G 119 - 52			11 00 37	+36 18.3	11	7.50	.013	1.51	.014	1.12	.015	35	M2 V	1,22,1077,1118,1197*
	+39 02407	IDS10578N3855	AB		11 00 37	+38 38.9	1	10.44		0.10		0.08		1	A2	1026
		B2 11O1+38 # 5			11 00 38	+38 26.5	1	13.43		0.95		0.62		2		487
95862	−58 02963	NGC 3532 - 601			11 00 38	−58 33.0	3	9.11	.012	0.37	.006	−0.56	.023	8		191,540,976
95689	+62 01161	HR 4301	⋆	AB	11 00 40	+62 01.3	13	1.80	.007	1.07	.007	0.90	.012	122	K0 II-III	1,15,61,71,667,1077*
95771	+0 02728	HR 4303			11 00 41	−00 28.3	3	6.14	.013	0.26	.005	0.06	.011	9	F0 V	15,252,1415
95880	−59 02972	LSS 2117	⋆	AB	11 00 41	−59 28.3	4	6.95	.025	0.32	.011	−0.39	.020	13	B3 Ib/II	191,540,976,1075
		Car sq 2 # 46			11 00 42	−61 15.9	1	12.11		0.10		−0.61		3		466
		G 58 - 37			11 00 43	+16 12.6	1	10.99		1.18		1.22		1	K5	333,1620

Table 1 583

HD	DM	Other Id	N	Rem	α₁₉₅₀	δ₁₉₅₀	S	V	σ_V	B–V	σ_B–V	U–B	σ_U–B	n	Spectrum	References
	−35 06934				11 00 44	−35 55.0	2	10.28	.007	0.59	.009	0.02		6		1594,6006
		Car sq 2 # 47			11 00 44	−60 50.3	1	11.50		0.86		0.04		3		466
95808	−10 03184	HR 4305	AB	⋆	11 00 45	−11 02.0	1	5.50		0.94				4	G7-III	2006
		LP 791 - 20			11 00 45	−16 04.3	1	11.79		1.30		1.31		1	K5	1696
95858	−40 06468	IDS10584S4030	A		11 00 45	−40 46.6	1	9.58		0.34		0.04		1	F0 V	6006
95858	−40 06468	IDS10584S4030	B		11 00 45	−40 46.6	1	10.54		0.68		0.12		1		6006
95879	−58 02968	NGC 3532 - 596			11 00 45	−58 26.0	1	7.93		0.99		0.69		3	G8 III	1671
303763	−58 02969	LSS 2118			11 00 45	−59 02.9	2	10.19	.027	0.22	.006	−0.80	.006	3		540,1737
305906	−60 02476				11 00 45	−60 57.7	1	9.87		0.05		−0.50		2	B5	401
		LB 1938			11 00 46	+59 06.6	1	13.32		−0.10		−0.50		2		272
	+18 02441	G 58 - 39			11 00 48	+18 24.6	1	11.03		0.66		0.09		1	G2	333,1620
95768	+45 01887	IDS10580N4452	A		11 00 50	+44 36.0	1	7.34		0.87		0.55		2	Am	1240
95768	+45 01887	IDS10580N4452	B		11 00 50	+44 36.0	1	9.60		0.27		0.13		2		1240
		NGC 3532 - 539			11 00 52	−58 41.2	1	11.36		0.30		0.01		2		491
	+33 02070				11 00 53	+33 25.9	3	11.56	.020	0.10	.011	0.10	.010	4	A1 Vp:	1026,1240,1298
95857	−31 08726	HR 4307			11 00 53	−31 41.5	2	6.46	.000	1.62	.005	1.98		6	M1 III	2006,3005
	+33 02071	IDS10582N3326	AC		11 00 56	+33 09.2	1	10.55		1.09		1.15		4		1723
		Car sq 2 # 48			11 00 59	−61 01.1	1	13.76		0.85		−0.27		3		466
95933	−60 02478	LSS 2121			11 01 01	−60 33.3	6	7.54	.008	0.57	.004	0.32	.007	68	F5 Ib/II	258,614,657,688,1075,1593
95849	+0 02729	HR 4306			11 01 03	+00 16.1	2	5.94	.005	1.22	.000	1.32	.000	7	K3 III	15,1417
		LP 431 - 53			11 01 03	+18 13.9	1	11.30		0.62		0.02		2		1696
95804	+26 02171				11 01 03	+26 02.7	2	6.99	.005	0.26	.000	0.04	.000	6	A7 V	833,1501
		L 538 - 34			11 01 03	−21 50.2	1	10.74		0.53		0.01		1	K0V	78
95544	+81 00359	G 253 - 42	A	⋆	11 01 04	+81 18.6	1	8.38		0.68		0.27		1	G0	1658
95872	−21 03221				11 01 04	−21 49.3	1	9.89		0.89		0.66		1	K0 V	78
95907	−45 06629	IDS10588S4555	A		11 01 05	−46 10.9	1	8.35		0.00		−0.15		3	B9.5 IV	1770
95803	+39 02410				11 01 06	+39 13.9	1	7.87		1.15		1.02		3	K2 III	833
95870	−12 03333	HR 4308			11 01 07	−13 09.9	4	6.34	.016	0.88	.000	0.60		13	G8 III	15,252,2013,2030
95972	−61 02046				11 01 07	−61 30.0	1	8.76		0.07		−0.85		2	B2 Vnn	401
		HBC 569, CS Cha			11 01 08	−77 17.6	2	11.65	.045	1.17	.015	0.56	.085	2	M0	437,1763
303752	−58 02980	NGC 3532 - 540			11 01 10	−58 43.7	1	10.35		0.23		−0.29		3	A0	491
		NGC 3532 - 726			11 01 15	−58 43.	1	12.15		1.76		1.63		4		491
		NGC 3532 - 727			11 01 15	−58 43.	1	17.46		0.58				1		491
		NGC 3532 - 728			11 01 15	−58 43.	1	13.36		0.34		0.18		2		491
		NGC 3532 - 729			11 01 15	−58 43.	1	15.54		0.84		0.03		2		491
	−29 08822		V		11 01 18	−29 25.3	1	9.82		1.00		0.60		1	K0	565
		B2 11O1+38 # 6			11 01 21	+38 37.0	1	13.31		1.19		1.16		2		487
		B2 11O1+38 # 4			11 01 22	+38 38.6	1	14.99		0.57		−0.06		1		487
95990	−58 02986	NGC 3532 - 473			11 01 23	−58 31.0	2	9.24	.009	0.38	.019	0.09		4		822,2026
		AAS25,287 # 4			11 01 25	+38 16.4	1	11.84		0.49		−0.07		2		511
96012	−58 02987	NGC 3532 - 471			11 01 25	−58 29.1	2	9.51	.014	−0.01	.000	−0.22	.023	5	B9 III	822,1493
95897	+17 02309				11 01 26	+16 31.6	1	8.83		0.41		0.06		2	F2	1648
96011	−57 04156	NGC 3532 - 467			11 01 27	−58 17.4	3	9.06	.018	0.02	.012	0.00	.005	11	A0 V	491,1493,2026
		AAS25,287 # 3			11 01 28	+38 21.4	1	13.52		0.81		0.56		2		511
		LP 214 - 26			11 01 28	+40 16.5	2	10.75	.007	1.40	.005	1.23		5		1723,1746
		AO 963			11 01 29	+36 10.8	1	11.93		0.76		0.18		1		1722
305947	−59 02984				11 01 29	−59 41.4	2	9.28	.004	0.09	.004	−0.76	.014	5	B0.5II-III	540,976
		B2 11O1+38 # 3			11 01 30	+38 41.3	1	14.91		0.44		−0.08		2		487
95884	+39 02413				11 01 31	+39 08.3	1	7.15		0.21		0.09		3	A7 V	833
96008	−50 05641	LL Vel			11 01 31	−51 05.0	4	6.75	.032	0.35	.013	0.04	.012	11	F0 V	78,1311,2033,3062
		G 147 - 18			11 01 32	+37 47.1	1	12.02		0.90		0.62		1	K3	1658
96042	−58 02992	LSS 2123			11 01 35	−59 09.8	1	8.23		0.18		−0.78		2	B1 Vne	6
95936	+15 02282	IDS10590N1447	AB		11 01 36	+14 31.1	1	8.33		1.26		1.27		2	K3 III	1648
96072	−68 01413				11 01 36	−68 29.6	1	9.78		0.28		0.14		1	A8/9 V	1771
	+8 02453	G 45 - 29			11 01 38	+08 26.2	3	9.67	.015	1.02	.012	0.86	.009	5	K5	333,1017,1620,3072
		WLS 1100 25 # 9			11 01 38	+24 44.5	1	13.05		0.69		0.23		2		1375
		AO 964			11 01 38	+36 22.2	1	12.12		0.57		0.09		1		1722
96041	−58 02993	NGC 3532 - 421	AB	⋆	11 01 38	−58 31.1	3	9.14	.018	0.08	.013	0.04	.005	7	A1 V	822,1493,2026
		AO 965			11 01 41	+36 02.7	1	10.87		0.81		0.45		1		1722
	−57 04167	NGC 3532 - 412			11 01 41	−58 15.9	1	10.81		0.16				2	A7 V	2026
96060	−59 02988	LSS 2125			11 01 41	−60 18.6	1	8.56		0.14		−0.69		2	B2 II	401
96040	−57 04168	NGC 3532 - 413			11 01 42	−58 18.4	2	10.01	.040	−0.02	.025	−0.17		7		491,2026
95980	+6 02398	G 45 - 30			11 01 45	+06 03.9	2	8.24	.005	0.63	.000	0.14	.025	2	G5	333,1620,1658
95934	+39 02414	HR 4309	AB	⋆	11 01 45	+38 30.7	2	6.00		0.15	.008	0.10		7	A3 III-IV	70,1022
96058	−57 04170	NGC 3532 - 586			11 01 47	−58 07.5	2	8.38	.020	0.02	.005	0.00	.000		A1 Va:	822,1493
96059	−58 02995	NGC 3532 - 420			11 01 47	−58 29.7	3	8.05	.025	−0.03	.000	−0.10	.020	10	A0 III	822,1493,2026
		Feige 36			11 01 48	+24 55.9	1			−0.25		−1.02		1	sdB	1264
		NGC 3532 - 414			11 01 48	−58 19.8	1	11.72		0.21				2		2026
303742	−58 02997	NGC 3532 - 426			11 01 49	−58 39.8	1	9.52		0.00		−0.38		1	A0	822
96068	−53 04317				11 01 52	−53 55.7	1	6.54		1.11		0.96		5	G8 III	1628
95978	+29 02120				11 01 53	+29 26.5	1	7.86		1.23				1	K2 III	20
96088	−57 04174				11 01 53	−57 41.1	3	6.15	.016	−0.17	.004	−0.79	.005	10	B2 III	6,285,2012
		NGC 3532 - 23			11 01 53	−58 36.1	1	9.59		0.06		0.02		1		822
		Cha F # 16			11 01 53	−77 04.9	1	11.49		1.48		1.00		3	G2 IV	1633
		NGC 3532 - 419			11 01 54	−58 27.0	3	11.53	.034	0.37	.029	−0.01	.042	6		491,822,2026
95977	+32 02102				11 01 55	+31 42.4	1	7.20		0.87		0.50		3	G5	1625
		NGC 3516 sq1 4			11 01 56	+72 47.5	1	13.10		0.47		0.02		2		327
		NGC 3516 sq1 5			11 01 56	+72 47.5	1	12.09		0.54		0.02		2		327
96103	−58 03002	NGC 3532 - 290	AB	⋆	11 02 01	−58 34.6	3	9.91	.020	0.12	.017	0.07	.010	5	A2 V	822,885,2026
306004	−60 02495				11 02 02	−60 57.3	1	9.95		0.03		−0.56		2	B3 III	837

HD	DM	Other Id	N	Rem	α_{1950}	δ_{1950}	S	V	σ_V	B–V	σ_{B-V}	U–B	σ_{U-B}	n	Spectrum	References
		AAS25,287 # 1			11 02 04	+38 33.4	1	14.35		0.57		0.04		3		511
		ApJS57,743 T2# 4			11 02 04	−60 39.5	1	10.79		1.17		0.85		1		1577
		Cha F # 18			11 02 04	−77 25.2	1	15.22		2.22		2.20		2	G8 III	1633
96122	−58 03003	NGC 3532 - 273			11 02 06	−58 27.5	4	7.94	.010	0.60	.021	0.47	.019	23	F2 Ib	491,822,1671,2026
96118	−57 04181	NGC 3532 - 522		★ AP	11 02 07	−58 11.6	1	7.69		1.23		0.67		3	K3 III	1671
	+36 02150				11 02 08	+36 05.1	1	9.51		1.12		1.04		1	K0	1722
		ApJS57,743 T2# 7			11 02 08	−60 41.1	1	10.83		1.90		2.21		1		1577
96064	−3 03040	IDS10596S0341	A		11 02 09	−03 57.0	1	7.64		0.77		0.33		3	M3 V	3032
96064	−3 03040	IDS10596S0341	BC		11 02 09	−03 57.0	1	10.25		1.38		1.15		3	M3 V	3032
	−58 03004	NGC 3532 - 285			11 02 09	−58 32.8	2	10.62	.019	0.20	.039	0.13		3	A5 III	822,2026
305996	−60 02497	XZ Car			11 02 10	−60 42.6	4	8.21	.148	1.08	.119	0.87	.197	4	K5	689,1484,1577,1593
		Cha F # 19			11 02 10	−77 24.6	1	14.55		2.70				2		1633
96016	+39 02419	IDS10594N3857	AB		11 02 11	+38 40.9	1	7.84		0.48		0.03		2	F7 IV	105
96139	−60 02498	LSS 2127			11 02 11	−60 39.6	3	8.72	.039	0.29	.032	−0.47	.028	9	B3 Ib	191,466,8100
96216	−70 01292				11 02 11	−70 43.9	1	7.88		0.07		−0.09		2	B9 V	1771
96137	−58 03005	NGC 3532 - 278			11 02 13	−58 29.8	3	8.22	.017	−0.01	.004	−0.08	.015	8	A0 III	822,1493,2026
		ApJS57,743 T2# 3			11 02 13	−60 40.1	1	10.57		1.58		1.77		1		1577
	−57 04188	NGC 3532 - 269			11 02 14	−58 24.0	2	11.15	.000	0.12	.034	0.03		3		822,2026
		ApJS57,743 T2# 8			11 02 14	−60 41.6	1	12.62		0.29		0.11		1		1577
		AAS25,287 # 7			11 02 16	+38 45.2	1	12.72		0.72		0.46		2		511
96113	−47 06466	HR 4311			11 02 16	−47 24.6	2	5.66	.005	0.25	.005			7	A8 III/IV	15,2012
		NGC 3532 - 270			11 02 16	−58 25.1	3	11.41	.015	0.35	.023	−0.06	.054	7		491,822,2026
96138	−58 03006	NGC 3532 - 296			11 02 16	−58 38.9	2	10.07	.015	0.12	.030	0.10		2	A2 V	885,2026
	−58 03007	NGC 3532 - 272			11 02 17	−58 27.4	1	10.28		0.15		0.11		1	A0	822
		G 45 - 31			11 02 18	+06 50.0	1	13.31		1.39		1.20		1		333,1620
96158	−59 02995	LSS 2130			11 02 19	−59 32.5	1	7.63		0.12		−0.72		6	B2 II/III	1087
		G 45 - 32			11 02 20	+06 50.0	1	13.34		1.39		1.15		1		333,1620
	−58 03009	NGC 3532 - 279			11 02 20	−58 30.3	2	10.77	.029	0.25	.005	0.10		3		822,2026
305938	−59 02998	LSS 2129			11 02 20	−59 29.3	2	9.40	.005	0.13	.005	−0.76	.065	6	B1:V	127,1087
96159	−59 02996				11 02 20	−59 35.9	2	7.83	.015	0.17	.055	−0.67	.040	7	B1 II/III	127,1087
96157	−58 03011	NGC 3532 - 178			11 02 24	−58 35.6	4	9.86	.027	0.11	.020	0.11	.012	7	A2 V	822,885,1493,2026
305941	−59 03001				11 02 24	−59 31.2	2	10.05	.009	0.13	.019	−0.59	.047	4	B2 IV/V	127,1087
		LOD 22 # 12			11 02 25	−59 26.2	1	11.03		1.15		1.23		2		127
96097	+8 02455	HR 4310		★ AB	11 02 26	+07 36.4	7	4.63	.006	0.33	.006	0.07	.010	27	F2 III-IV V	15,39,1008,1256,1363*
		AAS25,287 # 6			11 02 26	+38 45.1	1	12.05		0.78		0.57		2		511
		NGC 3532 - 156			11 02 26	−58 26.3	1	10.74		1.14				2		2026
96176	−59 03003				11 02 26	−59 40.2	2	8.99	.010	0.24	.015	0.17	.015	6	A2 V	127,1087
	−57 04191	NGC 3532 - 411		★ AB	11 02 27	−58 15.6	1	10.09		0.78				2		2026
		NGC 3532 - 151			11 02 27	−58 24.9	1	11.36		0.41		0.02		3		822
96174	−58 03014	NGC 3532 - 152			11 02 27	−58 25.5	4	7.79	.016	0.91	.000	0.59	.009	15	G8 III	460,822,1493,1671
		NGC 3532 - 148			11 02 28	−58 23.3	1	11.77		0.40		0.02		1		822
96175	−58 03016	NGC 3532 - 160			11 02 29	−58 29.1	5	7.67	.013	1.00	.006	0.76	.010	14	G8 III	460,822,1493,1671,2026
	−58 03015	NGC 3532 - 170			11 02 29	−58 32.4	2	10.19	.045	0.16	.015	0.09	.015	2	A2 V	822,885
	−58 03017	NGC 3532 - 159			11 02 30	−58 27.6	1	9.77		0.09		0.04		1	A0	822
305937	−59 03004	IDS11004S5913	ABC		11 02 30	−59 28.9	1	10.53		0.23		−0.37		2	B2.5 V	127
96146	−35 06954	HR 4313			11 02 32	−35 32.1	3	5.42	.007	0.03	.000			11	A0 V	15,2013,2029
96191	−58 03019	NGC 3532 - 150			11 02 32	−58 24.9	1	9.15		0.03		−0.03		1	A0	822
96192	−58 03020	NGC 3532 - 175			11 02 32	−58 33.3	2	9.74	.023	0.11	.014	0.05	.005	4	A1 V	822,1493
	−57 04196	NGC 3532 - 147			11 02 33	−58 23.5	1	11.15		0.32		0.10		1		822
	−58 03022	NGC 3532 - 162			11 02 34	−58 30.5	1	10.12		0.73		0.40		1		822
	−35 06955				11 02 35	−36 17.1	1	10.35		0.80		0.44		1	G8	78
	−58 03025	NGC 3532 - 163			11 02 35	−58 30.8	1	10.99		0.11		0.07		1		822
305945	−59 03008				11 02 35	−59 37.4	1	10.63		0.15		−0.56		2	B5	127
96094	+25 02335	G 58 - 41			11 02 36	+25 28.4	2	7.63	.005	0.56	.000	−0.03	.005	3	G0	333,1620,3026
96212	−58 03027	NGC 3532 - 154			11 02 37	−58 25.5	1	8.66		0.06		0.00		2	A0	822
96213	−58 03028	NGC 3532 - 155		★ AB	11 02 38	−58 26.0	2	8.28	.010	−0.01	.000	−0.19	.010	6	A0 IV	822,1493
	−57 04201	NGC 3532 - 261			11 02 39	−58 20.2	1	10.50		0.18		0.10		1		822
		NGC 3532 - 262			11 02 40	−58 20.4	1	10.51		0.16		0.11		1		822
	−57 04203	NGC 3532 - 143			11 02 41	−58 23.3	2	10.61	.015	0.18	.040	0.13		4	A2 m	822,2026
		GD 127			11 02 42	+00 32.6	1	15.25		0.14		−0.63		1		3060
96205	−47 06473	IDS11005S4739	A		11 02 42	−47 55.6	1	7.29		0.02		0.03		2	A0 IV	1770
		NGC 3532 - 179			11 02 42	−58 35.6	1	10.97		0.53		0.07		1		822
		G 176 - 8			11 02 43	+45 16.9	1	11.10		1.48		1.29		3	M2	7010
	−58 03031	NGC 3532 - 157			11 02 43	−58 26.9	3	7.97	.028	1.59	.027	1.88	.036	7	K5 III?	491,822,1671
	−58 03034	NGC 3532 - 164			11 02 43	−58 30.8	1	10.50		0.16		0.10		1		822
		LOD 22 # 14			11 02 43	−59 24.6	1	11.65		0.99		0.74		2		127
	−59 03010				11 02 43	−59 26.1	1	10.50		0.12		−0.58		2		127
	−57 04205	NGC 3532 - 141			11 02 44	−58 21.8	2	10.68	.019	0.11	.010	−0.16		3		822,2026
305946	−59 03011				11 02 44	−59 39.2	1	10.21		0.35		0.13		2	A3	127
96266	−66 01524				11 02 44	−66 37.1	1	8.16		0.29		0.23		2	A2mA4-A5	1771
		Cha T # 14	V		11 02 44	−76 11.1	1	12.51		1.23		0.32		1	K7:	825
		Cha F # 21			11 02 45	−76 36.0	1	11.41		1.90		2.00		2	K3 III	1633
		AAS25,287 # 5			11 02 46	+38 34.1	1	12.68		0.79		0.34		2		511
96106	+39 02421	G 147 - 19			11 02 47	+38 32.7	1	8.72		0.69		0.17		1		1658
96226	−57 04206	NGC 3532 - 409			11 02 47	−58 16.9	2	8.03	.005	−0.11	.010	−0.49		6	B8	822,2026
	−58 03036	NGC 3532 - 39			11 02 47	−58 26.9	1	9.45		0.14		0.06		1	A0	822
96187	−29 08840				11 02 48	−30 09.9	1	7.50		1.10		1.00		8	K1 III	1673
96227	−58 03037	NGC 3532 - 40			11 02 48	−58 28.8	3	8.20	.015	−0.01	.008	−0.06	.000	8	A0mA1III	822,1493,2026
96224	−48 06157	HR 4316			11 02 49	−49 07.3	4	6.12	.005	−0.02	.003	−0.06	.002	11	B9.5V	15,78,1770,2012
96245	−57 04208	NGC 3532 - 139			11 02 49	−58 23.8	2	8.36	.010	−0.03	.005	−0.09		6	B9.5III	1493,2026

Table 1   585

HD	DM	Other Id	N	Rem	α_{1950}	δ_{1950}	S	V	σ_V	B–V	σ_{B-V}	U–B	σ_{U-B}	n	Spectrum	References
		Lod 306 # 13			11 02 49	−60 47.2	1	10.17		−0.03		−0.71		2	B2	837
96287	−63 01819				11 02 50	−64 20.7	1	7.24		0.00		−0.08		2	B9.5V	401
96246	−58 03038	NGC 3532 - 50			11 02 51	−58 31.5	2	8.34	.010	0.02	.010	−0.04	.010	4	A0.5IV	822,1493
96264	−60 02505	LSS 2136		⋆ A	11 02 51	−60 46.9	6	7.61	.017	−0.06	.012	−0.94	.015	15	O8/9	164,401,466,540,837,976
96248	−59 03017	V414 Car		⋆	11 02 52	−59 35.3	3	6.55	.005	0.18	.033	−0.73	.019	12	BC1.5 Iab	6,158,1087
	+44 02051	WX UMa	A		11 02 53	+43 47.2	6	8.75	.018	1.53	.020	1.18	.008	22	M2 V	694,1003,1197,1733,3078,8006
96105	+50 01793				11 02 53	+50 26.6	1	7.05		0.50		0.01		2	F5	985
	−58 03040	NGC 3532 - 51			11 02 53	−58 32.0	1	9.34		0.11		0.03		1	B8	822
	−58 03041	NGC 3532 - 54			11 02 53	−58 33.7	4	9.89	.019	0.07	.017	0.08	.000	8	A2 V	822,885,1493,2026
96243	−52 04187				11 02 54	−52 56.3	1	9.52		0.04		0.03		1	A0 III	1770
96263	−59 03018	IDS11008S5958	AB		11 02 54	−60 13.8	1	8.60		0.08		−0.75		2	B1 III	401
96202	−26 08338	HR 4314		⋆ AB	11 02 55	−27 01.4	3	4.93	.005	0.37	.011	0.04	.000	7	F3 IV	15,1008,2012
96241	−51 05303	IDS11007S5142	AB		11 02 55	−51 58.4	1	9.27		0.12		0.12		1	A1 V	1770
96260	−58 03043	NGC 3532 - 37			11 02 55	−58 26.0	3	9.32	.023	0.04	.005	0.02	.010	10	A0 V	822,1493,2026
		Cha F # 22			11 02 55	−77 01.3	1	14.10		1.98				2	G8 III	1633
	+44 02051	G 176 - 12		⋆ B	11 02 56	+43 46.9	3	14.41	.024	1.99	.010	1.18		9	M5.5V:e	906,1663,3078
	−57 04210	NGC 3532 - 259			11 02 56	−58 19.6	1	10.22		0.30		0.11		1	A0	822
	−58 03044	NGC 3532 - 38			11 02 56	−58 24.9	2	9.57	.014	0.06	.005	0.06	.005	5	A1 V	822,1493
306016	−60 02506				11 02 56	−61 07.3	1	9.93		0.02		−0.64		2	B2 Ve	837
305932	−58 03046	LSS 2140			11 02 57	−59 23.5	2	9.24	.038	0.05	.033	−0.75	.052	4		127,1087
96261	−59 03019	LSS 2139		⋆ A	11 02 57	−59 26.6	2	7.78	.000	0.14	.000	−0.72	.000	8	B1 Ib/II	6,1087
96261	−59 03019	IDS11008S5910	B		11 02 57	−59 26.6	1	9.43		0.31		−0.79		2		1737
96262	−59 03020				11 02 57	−59 48.3	1	9.55		−0.03		−0.36		5	B9/A0II/III	1087
		AAS25,287 # 2			11 02 58	+38 25.3	1	12.86		0.66		0.22		2		511
233803	+51 01648	IDS11001N5122	AB		11 02 58	+51 05.4	1	8.70		0.90		0.60		2	K0	1566
96126	+53 01463				11 02 59	+52 35.8	1	7.26		1.22		1.31		3	K3 III	37
96285	−58 03048	NGC 3532 - 483			11 03 00	−58 44.8	2	9.04	.023	0.02	.009	0.00	.009	5	A0 V	885,1493
96284	−58 03049	NGC 3532 - 46			11 03 01	−58 30.5	1	9.37		0.04		0.04		1	A3	822
	−58 03050	NGC 3532 - 181		⋆ AB	11 03 01	−58 35.4	2	9.60	.023	0.03	.005	−0.05	.000	5	A0 V	822,1493
96286	−59 03024	LSS 2141			11 03 01	−59 41.2	4	8.38	.007	0.10	.020	−0.77	.035	14	B1 II/III	127,540,976,1087
		WLS 1100 75 # 7			11 03 03	+72 52.1	1	13.40		0.61		−0.20		2		1375
96220	−10 03190	HR 4315			11 03 03	−10 49.1	2	6.08	.005	0.31	.005			7	F0 Vn	15,2027
305944	−59 03026				11 03 03	−59 33.1	1	9.82		0.16		−0.65		2	B3 IV	127
	+12 02300				11 03 04	+12 21.8	1	8.88		1.01		0.90		2	K0	1648
		Lod 306 # 15			11 03 04	−60 50.0	1	10.46		1.22		1.05		2		837
		Car sq 2 # 51			11 03 04	−61 15.1	1	11.89		0.11		−0.56		3		466
		Cha T # 15			11 03 04	−77 09.6	1	16.11		1.62		1.07		1		437
		1103 385			11 03 05	+38 29.2	1	17.20		−0.28		−1.08		1		1727
	−58 03051	NGC 3532 - 182			11 03 05	−58 35.7	1	9.49		1.12		0.86		1		822
	−57 04220	NGC 3532 - 256			11 03 06	−58 20.0	1	11.26		0.34		0.06		2		822
96306	−58 03053	NGC 3532 - 180			11 03 06	−58 34.3	3	9.26	.022	0.01	.004	−0.02	.000	8	A0 V	822,1493,2026
		Lod 306 # 25			11 03 06	−60 50.2	1	11.31		−0.01		−0.53		2	B5	837
96308	−60 02508	LSS 2143			11 03 06	−60 54.4	2	8.81	.010	−0.01	.010	−0.80	.010	4	B1 III	401,837
96237	−24 09514				11 03 07	−24 44.9	1	9.51		0.39		0.19		1	Ap SrEuCr	1776
	−57 04221	NGC 3532 - 255			11 03 07	−58 19.6	1	10.65		0.35		0.09		1	F2 V	822
		G 253 - 44			11 03 08	+76 25.5	1	11.99		0.82		0.33		1		1658
96307	−58 03055	NGC 3532 - 612		⋆ AB	11 03 08	−58 48.8	1	9.00		0.02		−0.05		1	B9 IV	885
	−57 04223	NGC 3532 - 137			11 03 09	−58 22.5	1	10.49		0.20		0.12		1	A4 V	822
96305	−58 03056	NGC 3532 - 49			11 03 10	−58 30.8	4	8.57	.017	−0.01	.011	−0.08	.010	10	A0 III	822,822,1493,2026
		NGC 3532 - 7			11 03 11	−58 29.8	1	9.91		1.58		1.52		2		822
	−58 03057	NGC 3532 - 8			11 03 11	−58 30.0	2	9.69	.005	0.07	.009	0.03	.009	5	A2 V	822,1493
		Lod 306 # 26			11 03 11	−60 56.2	1	11.54		0.06		−0.12		2	B9	837
96304	−58 03058	NGC 3532 - 134			11 03 12	−58 24.4	2	9.58	.000	0.02	.000	0.01	.009	5	A0 V	822,1493
	−58 03059	NGC 3532 - 56			11 03 12	−58 32.9	2	11.15	.010	0.08	.000	0.04		2		822,2026
	−57 04228	NGC 3532 - 131			11 03 13	−58 22.4	1	10.73		0.27		0.09		1	A5 V	822
		WLS 1100 5 # 9			11 03 14	+05 27.8	1	12.30		0.68		0.09		2		1375
305933	−58 03062				11 03 14	−59 22.4	1	10.67		0.09		−0.52		2	A2	127
		NGC 3532 - 132			11 03 15	−58 23.0	1	11.30		0.53		0.05		1		822
		NGC 3532 - 6			11 03 15	−58 28.4	2	11.51	.068	0.60	.029	0.13		3		822,2026
	−58 03063	NGC 3532 - 183			11 03 16	−58 34.3	1	10.42		0.20		0.12		1	A2 V	822
96302	−47 06487				11 03 17	−48 21.1	1	8.10		0.18		0.15		1	A2/3 IV/V	1770
	−57 04233	NGC 3532 - 248		⋆ AB	11 03 18	−58 19.7	2	10.63	.000	0.21	.010	0.09		3	A5 V	822,2026
96341	−57 04232	NGC 3532 - 403			11 03 19	−58 16.5	2	9.53	.014	0.03	.014	0.00		4	A0 V	822,2026
	−58 03065	NGC 3532 - 310			11 03 19	−58 39.1	2	10.14	.024	0.08	.044	0.12		3	A0	885,2026
		NGC 3532 - 186			11 03 20	−58 35.4	1	11.39		0.36		0.05		1		822
96300	−42 06719				11 03 22	−42 50.9	1	7.52		0.27		0.12		1	F0 V	78
96320	−49 05824				11 03 22	−50 16.9	1	8.45		−0.07		−0.26		1	B8/9 III	1770
96355	−60 02511	LSS 2144			11 03 22	−61 09.8	3	9.72	.022	0.13	.017	−0.69	.013	6	B0/1 III	540,837,976
	−57 04237	NGC 3532 - 253		⋆ B	11 03 24	−58 21.0	1	9.46		0.08		0.02		2		822
96354	−57 04235	NGC 3532 - 252		⋆ A	11 03 24	−58 21.1	1	8.79		−0.01		−0.07		2	B9	822
96357	−61 02064	LSS 2145		⋆ AB	11 03 24	−61 34.5	2	9.05	.000	0.10	.015	−0.63	.008	5	B3/5 Ve	540,976
96368	−58 03066	NGC 3532 - 27		⋆ V	11 03 26	−58 27.6	2	8.94	.015	0.02	.010	−0.04		5	A0mA1IV	1493,2026
	−58 03067	NGC 3532 - 311			11 03 26	−58 38.4	1	10.99		0.23				1		2026
306005	−60 02513				11 03 26	−61 01.2	1	9.29		1.05		0.80		2	K0 V	837
		Cha F # 23			11 03 26	−77 00.1	1	12.93		2.55		1.90		3	M5 III	1633
96386	−57 04239	NGC 3532 - 249			11 03 28	−58 20.8	1	10.39		0.27		0.10		1	A0	822
	−58 03068	NGC 3532 - 129			11 03 28	−58 24.4	2	10.55	.045	0.24	.000	0.14		4		822,2026
96338	−46 06734				11 03 29	−47 10.3	1	6.82		0.04		0.02		2	A0 V	1770
	−58 03069	NGC 3532 - 4			11 03 29	−58 30.0	2	9.02	.000	0.07	.000	0.00	.009	5	A1 V	822,1493
		NGC 3532 - 313			11 03 29	−58 36.8	1	11.51		0.39				1		2026

HD	DM	Other Id	N	Rem	α_{1950}	δ_{1950}	S	V	σ_V	B–V	σ_{B-V}	U–B	σ_{U-B}	n	Spectrum	References
	−58 03071	NGC 3532 - 2			11 03 30	−58 29.2	2	9.92	.005	0.07	.000	0.07	.000	4	A1 V	822,1493
	−58 03070	NGC 3532 - 3			11 03 31	−58 29.6	1	9.68		0.08		0.05		1		822
96314	−26 08342	HR 4317, khi2 Hya			11 03 32	−27 01.0	4	5.69	.012	-0.07	.010	-0.15	.119	14	B8 V	15,588,1075,1079
	−57 04238	NGC 3532 - 125			11 03 32	−58 23.3	1	10.50		0.14		-0.55		1		822
	−58 03072	NGC 3532 - 62			11 03 32	−58 31.2	1	10.33		0.67		0.25		1		822
96349	−49 05826				11 03 33	−49 41.5	1	9.91		0.12		0.12		1	A1 V	1770
96386	−57 04239	NGC 3532 - 246			11 03 33	−58 19.6	3	9.85	.013	0.07	.010	0.07	.019	6	A2 V	822,1493,2026
96387	−58 03073	NGC 3532 - 1			11 03 33	−58 27.7	2	8.72	.005	0.06	.005	0.03	.005	5	A1 V	822,1493
96272	+7 02412				11 03 34	+07 24.6	1	7.26		0.43		-0.05		2	F5	1375
96388	−58 03075	NGC 3532 - 192			11 03 35	−58 33.9	3	8.84	.017	0.05	.017	0.03	.023	7	A1 IV	822,885,2026
96389	−58 03076	NGC 3532 - 434			11 03 35	−58 41.6	1	10.07		0.14		0.10		1	A2 V	885
305936	−59 03033				11 03 35	−59 26.0	1	10.10		0.10		-0.64		2	B2 Vn	191
		NGC 3516 sq1 2			11 03 36	+72 49.8	1	14.38		0.49		-0.07		2		327
	−22 03078				11 03 36	−22 50.7	1	10.41		0.72		0.20		1		78
95603	+86 00159				11 03 37	+85 48.9	1	8.33		0.18		0.10		7	A3	1219
	−58 03077	NGC 3532 - 122			11 03 38	−58 24.4	2	8.19	.002	0.95	.013	0.63	.033	5		822,1671
96413	−57 04247	NGC 3532 - 401			11 03 41	−58 16.3	3	9.22	.023	0.00	.007	-0.05	.005	8	A0 V	822,1493,2026
	−57 04246	NGC 3532 - 244			11 03 41	−58 19.0	1	10.23		0.16		0.08		1		822
	−57 04245	NGC 3532 - 245			11 03 41	−58 19.5	1	11.36		0.32		0.07		1		822
96414	−58 03079	NGC 3532 - 65			11 03 41	−58 33.6	2	9.05	.020	0.11	.000	0.08	.000	2	A1 V	822,885
	−58 03080	NGC 3532 - 21			11 03 42	−58 26.0	2	9.60	.009	0.08	.000	0.05	.005	5	A1 V	822,1493
96448	−64 01628	IDS11017S6459	AB		11 03 43	−65 15.1	1	9.04		0.08		-0.27		1	B5/7 III	1732
	−58 03081	NGC 3532 - 17			11 03 44	−58 27.7	3	9.26	.032	0.07	.013	0.08	.009	7	A1 V	822,1493,2026
96384	−50 05683				11 03 45	−51 21.5	1	9.96		0.14		0.12		1	A1 IV	1770
	−57 04251	NGC 3532 - 116			11 03 45	−58 22.8	1	10.35		0.16		0.10		1	A2 V	822
	−58 03084	NGC 3532 - 194			11 03 45	−58 34.6	1	11.03		0.26		0.10		1		822
	−60 02516				11 03 45	−60 52.3	1	9.83		0.07		-0.38		2	B7	837
	−57 04254	NGC 3532 - 243			11 03 46	−58 19.4	1	10.98		0.45		0.04		1		822
96430	−57 04253	NGC 3532 - 115			11 03 46	−58 23.5	2	8.49	.025	-0.11	.005	-0.46		4	B8	822,2026
	−58 03085	NGC 3532 - 114			11 03 47	−58 24.3	1	9.13		0.03		0.06		1	B8	822
96415	−59 03036				11 03 47	−59 42.2	2	9.02	.010	0.12	.029	-0.62	.049	5	B2 III	127,1087
	−58 03086	NGC 3532 - 67			11 03 48	−58 32.8	2	9.79	.033	0.17	.019	0.10	.037	4	A2 V	822,1493
	+73 00516				11 03 49	+72 50.9	1	10.53		1.11		1.08		5		327
	−58 03087	NGC 3532 - 22			11 03 49	−58 25.3	2	9.29	.024	0.06	.015	0.03		3	A3	822,2026
	−58 03086	NGC 3532 - 66			11 03 49	−58 32.9	2	9.13	.023	1.57	.037	2.01	.000	4	A0	822,1493
310376	−67 01645	QU Car		★	11 03 49	−68 21.8	1	11.55		0.05		-0.86		2		1737
		NGC 3516 sq1 3			11 03 50	+72 51.3	1	14.42		0.93		0.75		1		327
	−58 03089	NGC 3532 - 18			11 03 50	−58 27.9	4	10.00	.031	0.15	.009	0.10	.020	9	A2 V	460,822,1493,2026
		LSS 2149			11 03 50	−62 41.4	1	11.16		0.54		-0.36		2	B3 Ib	1737
96445	−58 03090	NGC 3532 - 19			11 03 51	−58 27.3	5	7.73	.015	0.96	.006	0.68	.006	15	G8 III	460,822,1493,1671,2026
		QU Car			11 03 51	−68 21.7	1	11.33		0.01		-0.84		1		1471
96407	−50 05686	HR 4318			11 03 52	−50 56.5	5	6.29	.004	0.94	.005	0.59		19	G8 III	15,78,1075,2020,2029
	−58 03091	NGC 3532 - 316			11 03 52	−58 37.0	1	11.00		0.27		0.17		1		822
		Lod 306 # 18			11 03 52	−61 04.4	1	10.62		0.42		0.14		2	F0	837
96344	+15 02289				11 03 55	+15 18.5	1	8.71		1.40		1.12		1	M0 II:	8100
96380	−29 08855				11 03 55	−29 41.8	1	8.38		0.61		0.11		1	G2 V	78
96525	−70 01298				11 03 55	−70 48.9	1	8.32		0.02		-0.24		2	B8/9 V	1771
	−58 03092	NGC 3532 - 100			11 03 56	−58 25.0	2	7.50	.001	1.09	.004	0.93	.004	6		822,1671
96447	−60 02517				11 03 56	−60 54.3	2	9.13	.015	0.00	.000	-0.78	.040	4	B2/3 V	401,837
306018	−60 02518				11 03 56	−61 11.5	1	10.37		0.14		0.09		2	B7 III	837
	−58 03094	NGC 3532 - 101			11 03 58	−58 24.7	1	9.54		0.08		0.03		1		822
	−57 04256	NGC 3532 - 106			11 03 59	−58 23.6	1	11.21		0.31		0.06		2		822
	−58 03095	NGC 3532 - 197			11 03 59	−58 35.9	3	10.73	.020	0.24	.009	0.11	.000	6		460,822,2026
96474	−58 03096				11 03 59	−58 58.3	1	7.99		-0.09		-0.52		2	B4 III	401
96446	−59 03038	V430 Car			11 03 59	−59 40.8	3	6.69	.009	-0.14	.014	-0.82	.030	9	B2 IIIp	127,1087,1732
96377	−12 03346	IDS11015S1253	A		11 04 00	−13 08.8	1	7.32		1.04		0.84		5	K0 III	1657
96426	−48 06174				11 04 00	−49 13.5	1	9.64		0.06		0.06		1	A0 Vn	1770
96472	−57 04257	NGC 3532 - 113			11 04 00	−58 22.1	3	8.59	.017	0.00	.008	-0.06	.005	8	A0 IVn	822,1493,2026
96473	−58 03097	NGC 3532 - 317			11 04 00	−58 37.3	3	8.44	.014	-0.04	.007	-0.12	.010	18	B9.5III	822,1493,2026
	−60 02519				11 04 00	−60 58.1	1	10.68		0.39		0.18		2	A5	837
	−57 04258	NGC 3532 - 105			11 04 01	−58 24.0	1	9.87		0.04		0.04		1	A0	822
		NGC 3532 - 98		★ B	11 04 02	−58 25.5	1	9.85		0.09		0.03		1		822
	−58 03099	NGC 3532 - 97		★ AB	11 04 03	−58 25.6	1	8.91		0.03		0.03		1	A0	822
		NGC 3532 - 80			11 04 03	−58 31.0	1	11.59		0.39		0.02		1		822
96492	−60 02522				11 04 03	−60 47.0	2	8.47	.005	-0.02	.010	-0.34	.015	4	B8 III/IV	401,837
306001	−60 02521				11 04 03	−60 52.9	1	9.50		-0.03		-0.77		2	B5	837
96493	−60 02520				11 04 03	−60 58.3	1	8.50		0.15		0.14		2	A2 IV/V	837
	+27 01981				11 04 04	+26 34.9	1	11.00		0.19		0.13		2	A3	1026
	−57 04261	NGC 3532 - 107			11 04 04	−58 23.4	1	11.14		0.33		0.08		2		822
	−58 03100	NGC 3532 - 86			11 04 04	−58 28.8	1	10.40		0.19		0.10		1	A4 V	822
	−58 03101	NGC 3532 - 83			11 04 04	−58 29.8	1	10.48		0.18		0.14		1	A2 m	822
96372	+18 02452				11 04 05	+18 00.5	2	6.40	.005	1.57	.015	1.94	.015	5	K5 III	985,1501
	−58 03102	NGC 3532 - 89			11 04 05	−58 28.0	3	8.50	.015	0.06	.007	-0.01	.028	6	A2	822,1493,2026
		Lod 306 # 23			11 04 05	−60 55.0	1	11.16		0.19		0.14		2	A2	837
		G 147 - 23			11 04 06	+38 54.0	1	15.38		1.62				1		906
96470	−52 04211				11 04 06	−53 20.0	1	8.65		0.07		0.09		1	A1 V	1770
	−58 03104	NGC 3532 - 73			11 04 06	−58 32.4	1	8.83		0.04		-0.04		3	B8	822
96489	−58 03103	NGC 3532 - 199			11 04 06	−58 34.4	2	8.03	.015	0.07	.005	0.06		7	A1 V	822,2026
	−60 02523				11 04 06	−61 02.4	1	10.49		0.07		-0.38		2	B5 III	837
	−58 03105	NGC 3532 - 81			11 04 08	−58 30.8	1	11.04		0.30		0.11		1		822

Table 1   587

HD	DM	Other Id	N Rem	α_{1950}	δ_{1950}	S	V	σ_V	B–V	σ_{B-V}	U–B	σ_{U-B}	n	Spectrum	References
96509	−57 04264	NGC 3532 - 579		11 04 09	−58 09.3	1	10.02		0.07		0.07		3	A2	1493
	+29 02123			11 04 12	+28 40.0	1	9.30		0.44		0.00		2	F2	1733
96484	−50 05693	HR 4321		11 04 13	−50 41.2	4	6.31	.005	1.16	.005	1.22		14	K2 III	15,1075,2018,8100
	−58 03106	NGC 3532 - 87		11 04 13	−58 28.7	3	9.80	.023	0.10	.007	0.06	.005	7	A1 V	822,1493,2026
96441	−28 08657	HR 4320		11 04 14	−28 27.4	2	6.76	.005	0.04	.000			7	A1 V	15,2029
	−30 08958	LTT 4089		11 04 14	−31 03.7	1	9.87		1.64		1.75		1		78
96508	−56 04205			11 04 14	−56 56.7	1	10.02		0.01		−0.61		2	B1/2 III	191
96507	−52 04217	IDS11021S5244	A	11 04 16	−53 00.1	1	7.94		−0.02		−0.13		3	B9 III	1770
	−57 04271	NGC 3532 - 385		11 04 16	−58 15.6	1	11.07		0.33		0.05		1		822
	−58 03107	NGC 3532 - 93		11 04 16	−58 25.5	2	9.34	.015	0.05	.015	0.00	.000	6	A1 V	822,1493
		NGC 3532 - 390		11 04 17	−58 18.3	1	11.36		0.45		−0.06		1		822
306012	−60 02525			11 04 17	−61 07.2	1	10.67		0.09		−0.31		2	B5 III	837
96548	−64 01629	LSS 2154	B	11 04 17	−65 14.3	1	10.82		0.25		0.07		1	WN	1732
96522	−57 04272	NGC 3532 - 453		11 04 18	−58 15.4	2	9.00	.045	−0.02	.020	−0.11	.025	7	A0 IV	822,1493
		NGC 3532 - 701		11 04 18	−58 24.	1	13.62		0.71		0.21		2		822
		NGC 3532 - 711		11 04 18	−58 24.	1	17.08		0.90				2		491
		NGC 3532 - 712		11 04 18	−58 24.	1	16.97		1.38				4		491
		NGC 3532 - 713		11 04 18	−58 24.	1	18.30		1.20				6		491
		NGC 3532 - 714		11 04 18	−58 24.	1	13.93		0.81		0.18		4		491
		NGC 3532 - 715		11 04 18	−58 24.	1	17.44		0.75				15		491
		NGC 3532 - 716		11 04 18	−58 24.	1	12.21		1.15		0.83		2		491
		NGC 3532 - 717		11 04 18	−58 24.	1	15.55		0.67		−0.12		3		491
		NGC 3532 - 718		11 04 18	−58 24.	1	16.92		0.75				3		491
		NGC 3532 - 719		11 04 18	−58 24.	1	16.92		0.99		−0.09		10		491
		NGC 3532 - 720		11 04 18	−58 24.	1	15.99		1.20				9		491
		NGC 3532 - 721		11 04 18	−58 24.	1	17.10		0.67		−0.19		3		491
		NGC 3532 - 722		11 04 18	−58 24.	1	13.63		0.69		0.27		4		491
		NGC 3532 - 723		11 04 18	−58 24.	1	17.15		0.88				3		491
		NGC 3532 - 724		11 04 18	−58 24.	1	16.46		0.98				4		491
		NGC 3532 - 725		11 04 18	−58 24.	1	17.04		0.66		0.06		5		491
	−58 03109	NGC 3532 - 104		11 04 18	−58 24.5	1	9.78		0.13		0.07		2	A0	822
96523	−58 03108			11 04 18	−59 22.3	3	9.29	.023	0.22	.036	−0.11	.045	7	B8 V	127,885,1087
96548	−64 01629	V385 Car	A	11 04 18	−65 14.4	2	7.70	.003	0.00	.000	−0.68	.014	5	WN	401,1732
96569	−68 01430			11 04 18	−68 30.7	3	7.30	.020	1.68	.009	2.04	.007	9	M0 III	1704,2012,3040
306007	−60 02526			11 04 19	−60 58.3	1	11.15		0.09		−0.05		2	B6	837
		Car sq 2 # 52		11 04 19	−61 11.2	1	13.67		0.82		−0.33		3		466
		G 254 - 15		11 04 20	+76 33.8	1	15.76		1.64				1		906
	−58 03111	NGC 3532 - 218		11 04 20	−58 26.0	1	10.58		0.19		0.10		1	A5 V	822
	−58 03110	NGC 3532 - 213		11 04 20	−58 29.6	3	8.99	.026	0.13	.008	0.06	.005	9	A2 V	822,1493,2026
306000	−60 02527			11 04 20	−60 51.4	1	10.78		0.24		0.20		2	B9	837
99685	−89 00034			11 04 20	−89 31.3	1	7.82		0.10		−0.07		6	A0 IV	826
96436	+2 02387	HR 4319	⋆ AB	11 04 21	+02 13.6	5	5.52	.008	0.97	.010	0.66	.003	25	G9 III	3,15,1080,1118,1417
96544	−58 03112	NGC 3532 - 221	⋆ A	11 04 21	−58 24.3	4	6.07	.114	1.20	.022			15	K2 II/III	15,1075,2006,2032
	−58 03114	NGC 3532 - 207		11 04 21	−58 32.1	4	8.79	.035	0.19	.013	−0.57	.021	11	B5	460,822,1493,2026
303758	−58 03115	LSS 2153		11 04 22	−59 05.3	1	10.38		0.47		−0.46		2	B0.5II	191
		Lod 306 # 24		11 04 22	−60 57.0	1	11.16		0.23		0.18		2	A0	837
	−57 04274	NGC 3532 - 226		11 04 23	−58 22.5	1	10.24		0.19		0.01		1	A3 V	822
96418	+26 02176			11 04 24	+25 48.5	1	6.87		0.48		0.01		3	F8 IV	833
	−57 04275	NGC 3532 - 231		11 04 24	−58 20.9	2	10.52	.015	0.33	.015	0.02		2	A9 V	822,2026
96546	−58 03116			11 04 24	−59 17.9	2	9.32	.000	0.03	.005	−0.02	.030	4	B9.5 V	885,1087
96568	−64 01630	HR 4326		11 04 24	−64 34.1	4	6.40	.011	0.11	.004	0.13	.005	21	A3 V	15,1075,1771,2038
		Cha T # 20	⋆ V	11 04 25	−76 02.0	1	13.59		1.47		1.00		1		437
	−58 03118	NGC 3532 - 203		11 04 27	−58 34.5	2	11.19	.044	0.24	.044	0.06		3		822,2026
306010	−60 02530			11 04 27	−61 01.9	1	10.62		0.09		−0.34		2	A3	837
96564	−58 03120	NGC 3532 - 215		11 04 28	−58 28.4	3	7.79	.011	−0.03	.008	−0.10	.000	68	B9.5III	822,1493,2026
303773	−58 03121			11 04 28	−59 15.8	1	10.52		0.06		−0.22		4		1087
96566	−61 02067	HR 4325		11 04 29	−62 09.2	5	4.61	.009	1.03	.016	0.82	.004	17	G8 III	15,1075,1754,2012,3077
96565	−58 03123	NGC 3532 - 202		11 04 30	−58 35.4	3	9.55	.017	0.01	.013	−0.01	.005	6	A0 IV	822,1493,2026
	−27 07867			11 04 31	−27 41.9	1	9.52		0.63		0.05		1	G5	78
		Car sq 2 # 53		11 04 31	−60 33.0	1	11.89		0.15		−0.62		3		466
		Car sq 2 # 54		11 04 32	−60 40.1	1	12.03		0.26		−0.59		3		466
96478	+11 02311	IDS11020N1127	A	11 04 35	+11 11.0	1	9.29		0.97		0.72		1	K0	3016
96478	+11 02311	IDS11020N1127	B	11 04 35	+11 11.0	1	9.52		0.94		0.61		1		3016
	−58 03125	NGC 3532 - 208		11 04 35	−58 31.9	1	10.16		0.18		0.11		1	A0	822
96584	−57 04279	NGC 3532 - 236		11 04 36	−58 18.8	3	8.23	.015	1.57	.008	1.85	.043	30	K5 III?	822,1671,2026
	−45 06692			11 04 37	−45 53.9	1	10.96		0.57		0.00		1		1696
	−58 03127	NGC 3532 - 212		11 04 38	−58 31.3	1	10.80	.068	0.35	.005	0.13		3		822,2026
	−57 04282	NGC 3532 - 382		11 04 39	−58 15.9	2	11.12	.024	0.36	.024	0.04		3	F0 IV	822,2026
	−57 04285	NGC 3532 - 222		11 04 40	−58 22.9	1	10.51		0.22		0.06		1		822
96457	+35 02211			11 04 41	+35 06.8	1	9.33		0.39		−0.02		2	F8	1601
96610	−58 03128	NGC 3532 - 216		11 04 41	−58 28.3	1	8.67		0.01		−0.06		3	A2	822
306027	−58 03129			11 04 41	−59 17.5	1	9.34		1.49		1.45		5	K5-7:	1087
96586	−59 03045			11 04 41	−59 26.7	2	8.98	.005	0.09	.010	0.01	.070	6	B9/A0 IV/V	127,1087
96587	−59 03046			11 04 41	−59 58.5	1	9.66		0.14		0.09		2	A1 Vn	432
96812	−67 01647			11 04 42	−67 39.2	1	7.60		0.08		0.05		2	A1 V	1771
		G 197 - 4		11 04 43	+60 14.8	2	13.79	.005	−0.02	.000	−0.79	.015	5		538,3028
		NGC 3532 - 371		11 04 43	−58 22.1	1	11.34		0.45		−0.03		1		822
96609	−58 03131	NGC 3532 - 363		11 04 43	−58 26.2	3	8.62	.025	0.01	.015	−0.04	.000	10	A0mA1III	822,1493,2026
	−58 03130	NGC 3532 - 336		11 04 43	−58 36.9	1	11.00		−0.01		−0.26		1		822
306035	−59 03047	LSS 2160		11 04 43	−59 34.5	1	8.72		0.11		−0.71		4	B1 V	1087

HD	DM	Other Id	N	Rem	α_{1950}	δ_{1950}	S	V	σ_V	B–V	σ_{B-V}	U–B	σ_{U-B}	n	Spectrum	References
96557	−31 08776	HR 4324			11 04 45	−32 18.9	3	6.58	.004	0.35	.005	0.00		8	F2 V	15,78,2012
96581	−46 06758				11 04 47	−46 53.1	1	9.68		0.17		0.18		1	A2 V	1770
	−58 03133	NGC 3532 - 349			11 04 47	−58 29.9	2	9.76	.009	0.07	.014	0.06	.005	5	A1 V	822,1493
96620	−58 03132	NGC 3532 - 345			11 04 47	−58 31.4	3	7.38	.017	0.01	.011	-0.01	.035	10	A0.5III	822,1493,2026
	−57 04288	NGC 3532 - 374		⋆ AB	11 04 48	−58 21.2	2	10.68	.063	0.22	.034	0.03		3		822,2026
	−57 04289	NGC 3532 - 367			11 04 48	−58 23.9	2	10.76	.015	0.25	.030	0.11		2		822,2026
96514	+13 02358				11 04 49	+13 17.1	1	7.70		1.02		0.88		2	K2 III	1648
96497	+22 02316				11 04 49	+22 19.4	1	8.20		0.63		0.15		2	G1 V	1648
	−58 03134	NGC 3532 - 342			11 04 49	−58 33.1	2	10.11	.014	0.12	.009	0.09		4		822,2026
96637	−57 04291	NGC 3532 - 370			11 04 51	−58 22.5	2	9.81	.042	0.18	.023	0.04		4	A2	822,2026
	−58 03136	NGC 3532 - 353			11 04 52	−58 29.5	1	10.04		0.14		0.08		1		822
		Cha T # 21			11 04 52	−77 05.7	2	11.24	.025	1.38	.035	0.64	.070	2		437,1633
	−58 03139	NGC 3532 - 362			11 04 53	−58 26.5	3	9.61	.034	0.15	.013	0.07	.050	11	A2 V	822,1493,2026
	−58 03137	NGC 3532 - 347			11 04 53	−58 30.8	2	10.72	.015	0.00	.035	-0.40		2	A0	822,2026
96622	−58 03140	LSS 2161			11 04 53	−59 23.8	3	8.90	.016	0.13	.023	-0.83	.065	10	O9.5 V	6,127,1087
	−30 08970				11 04 54	−30 32.7	1	10.56		1.18		1.17		1	K5	78
	−58 03138	NGC 3532 - 341			11 04 54	−58 33.6	3	9.78	.027	0.06	.014	0.04	.005	7	A1 V	822,1493,2026
305999	−60 02533	IDS11028S6033		AB	11 04 55	−60 49.1	1	9.44		0.89		0.21		2	F5	837
	−58 03142	NGC 3532 - 350			11 04 56	−58 30.0	1	10.06		0.20		0.09		1	A2	822
96616	−41 06343	HR 4327, V815 Cen		⋆ AB	11 04 58	−42 22.1	3	5.14	.004	0.03	.000	0.02		9	Ap SrCrEu	15,2012,3033
	−58 03143	NGC 3532 - 356			11 04 58	−58 29.1	1	10.35		0.39		0.00		1		822
306032	−59 03051				11 04 58	−59 25.0	1	10.57		0.03		-0.44		1	B2.5:IV:	1087
96638	−59 03052	LSS 2163			11 04 58	−59 31.3	3	8.56	.015	0.23	.044	-0.79	.058	13	O8 V	6,390,1087
		Lod 306 # 27			11 04 58	−60 44.9	1	11.70		0.03		-0.24		2	B8	837
96528	+24 02318	HR 4322			11 04 59	+23 35.7	4	6.49	.025	0.16	.005	0.12	.000	13	A5 m	70,1022,3058,8071
96574	+14 02345				11 05 00	+14 07.7	1	7.32		0.55		0.04		2	F9 V	1003
96651	−57 04295	NGC 3532 - 576			11 05 00	−58 07.2	2	8.98	.032	0.03	.018	0.00	.005	5	A0 V	822,1493
96653	−58 03144	NGC 3532 - 361			11 05 00	−58 26.8	2	8.38	.025	0.00	.005	-0.08	.015	7	A0 IIIn	822,1493
		Car sq 2 # 55			11 05 00	−60 45.6	1	12.67		0.21		-0.60		3		466
96391	+72 00515				11 05 01	+72 13.8	2	7.09	.005	0.35	.000	0.10	.005	4	F0	985,1733
96652	−57 04296	NGC 3532 - 380			11 05 01	−58 15.4	2	9.24	.018	0.05	.009	0.01	.009	5	A2	822,1493
96706	−70 01305	HR 4329		⋆ A	11 05 01	−70 36.4	4	5.57	.016	-0.06	.009	-0.63	.000	18	B3 IV/V	15,1075,2012,2038
	−57 04297	NGC 3532 - 373			11 05 02	−58 21.3	2	10.98	.000	0.30	.020	-0.69		2		822,2026
	−58 03146	NGC 3532 - 352			11 05 02	−58 29.4	1	10.37		0.19		0.09		1	A5 V	822
96667	−58 03145	NGC 3532 - 343			11 05 03	−58 32.7	2	9.59	.023	0.09	.009	0.01		5	A0	822,2026
96668	−58 03147	NGC 3532 - 337			11 05 03	−58 38.2	3	8.30	.018	-0.03	.014	-0.08	.025	8	A0 IIIs	822,1493,2026
	−58 03148	NGC 3532 - 339			11 05 05	−58 35.0	1	10.73		0.40		0.08		1		822
96632	−47 06521	IDS11029S4742		A	11 05 06	−47 58.5	1	8.31		0.22		0.15		1	A2mA5-A9	1770
96650		RW Cen			11 05 07	−54 51.0	1	8.78		2.87				2	N3	864
		NGC 3532 - 193			11 05 07	−58 32.2	1	8.34		-0.01		-0.08		3		822
96670	−59 03057	LSS 2167			11 05 07	−59 36.1	7	7.43	.032	0.14	.041	-0.78	.037	21	O7 V(f)n	6,127,164,191,540,976,1087
96685	−58 03150	NGC 3532 - 447		⋆ AB	11 05 08	−58 26.1	1	9.68		0.16		-0.22		1	B8	822
		NGC 3532 - 441			11 05 08	−58 35.7	2	10.94	.010	0.95	.015	0.54		3		822,2026
	−58 03151	NGC 3532 - 448			11 05 09	−58 24.6	2	9.92	.029	0.65	.010	0.22		3		822,2026
	−58 03152	NGC 3532 - 443			11 05 09	−58 34.4	1	10.90		0.28		0.06		1		822
		Car sq 2 # 56			11 05 11	−61 18.2	1	10.98		0.38		-0.71		3		466
96669	−59 03059	LSS 2168			11 05 12	−59 32.8	3	8.53	.027	0.20	.022	-0.68	.033	9	B1 II/III	540,976,1087
	−58 03154	NGC 3532 - 445			11 05 13	−58 32.0	2	10.85	.005	0.25	.010	0.03		3	A8 IV	822,2026
		Car sq 2 # 57			11 05 14	−61 12.8	1	12.89		0.49		-0.35		3		466
		Cha T # 22		⋆ V	11 05 15	−77 10.5	1	16.16		2.02		0.92		1		437,825
		NGC 3532 - 442			11 05 20	−58 35.1	1	11.87		0.12		-0.23		1		822
96684	−49 05846				11 05 21	−49 40.7	1	9.22		0.04		0.00		1	A0 V(n)	1770
		NGC 3532 - 446			11 05 21	−58 31.6	1	11.37		0.24		0.07		1		822
		NGC 3532 - 440			11 05 21	−58 36.2	2	10.53	.049	0.96	.015	0.71		3		822,2026
96714	−57 04306	NGC 3532 - 505			11 05 22	−58 12.2	1	9.47		0.00		-0.03		3	B9 V	1493
	−57 04310	NGC 3532 - 450			11 05 26	−58 22.2	2	10.81	.000	0.25	.029	0.06		3		822,2026
96715	−59 03064	LSS 2171			11 05 26	−59 41.6	4	8.25	.011	0.11	.013	-0.85	.020	12	O5	390,540,976,1087
96729	−58 03157	NGC 3532 - 449			11 05 27	−58 24.5	3	9.98	.043	0.00	.024	-0.35	.009	7	B9 IIIp	822,1493,2026
96716	−60 02539	IDS11034S6050		AB	11 05 27	−61 06.6	1	8.35		-0.01		-0.79		2	B1 II	401
		G 163 - 50			11 05 28	−04 52.9	5	13.06	.015	0.03	.009	-0.68	.008	93	DA	281,989,1620,1764,3060
96728	−55 04148				11 05 30	−56 23.4	1	8.22		-0.02		-0.33		2	B8 Vn	401
96730	−58 03160	NGC 3532 - 493			11 05 30	−58 31.1	2	9.96	.024	0.09	.005	0.04		3	A1 V	822,2026
306033	−59 03065				11 05 30	−59 30.4	1	9.36		1.16		1.17		4	K2 III:	1087
96495	+72 00516				11 05 31	+72 19.0	1	8.75		0.23		-0.50		2	F2	463
96700	−29 08875	HR 4328			11 05 31	−29 54.1	5	6.53	.011	0.60	.000	0.07	.013	14	G2 V	15,78,158,2012,3026
		Cha F # 25			11 05 33	−76 57.1	1	13.23		2.22				1	G8 III	1633
		G 163 - 51			11 05 34	−04 57.3	4	12.55	.021	1.51	.004	1.21	.025	42		281,1620,1764,3028
96755	−58 03161	NGC 3532 - 495			11 05 34	−58 29.1	1	8.40		0.02		-0.10		3	A0	822
96756	−58 03163				11 05 35	−59 24.0	1	7.49		1.36		1.50		7	K2:III:	1087
		Cha T # 23		⋆ V	11 05 35	−77 02.6	1	16.29		1.64		0.60		1		437
	−58 03164	NGC 3532 - 489			11 05 37	−58 36.2	1	11.33		0.10		-0.39		1		822
		NGC 3532 - 494			11 05 38	−58 29.4	1	11.01		0.31		0.06		1		822
	−58 03166	NGC 3532 - 488			11 05 38	−58 36.7	1	10.32		0.15		0.09		1		822
96772	−57 04315	NGC 3532 - 502			11 05 39	−58 12.9	1	9.61		0.07		0.05		3	A0 IV/V	1493
	−58 03167	NGC 3532 - 490			11 05 41	−58 35.5	1	10.64		0.21		0.05		1	A5 V	822
96774	−58 03168	NGC 3532 - 486			11 05 42	−58 41.0	1	9.93		0.08		0.10		3	A2 V	1493
	−27 07881				11 05 43	−27 59.8	5	9.32	.011	1.25	.015	1.18	.006	9	K7 V	78,158,912,1705,3072
96773	−58 03170	NGC 3532 - 492			11 05 43	−58 33.5	3	9.76	.019	0.08	.011	0.03	.000	7	A1 V	822,1493,2026
	−58 03171	NGC 3532 - 487			11 05 43	−58 37.6	2	11.00	.024	0.25	.049	0.06		3		822,2026
	+23 02309	G 58 - 44			11 05 44	+22 56.8	1	9.70		0.71		0.25		1	G8	333,1620

Table 1 589

HD	DM	Other Id	N Rem	α_{1950}	δ_{1950}	S	V	σ_V	B–V	σ_{B-V}	U–B	σ_{U-B}	n	Spectrum	References
		Cha T # 24	⋆ V	11 05 44	−76 16.1	2	14.98	.085	1.75	.035	0.76	.080	2		437,5005
96790	−57 04319	NGC 3532 - 499		11 05 45	−58 22.8	1	9.86		0.11		-0.72		1		822
306034	−59 03070			11 05 45	−59 30.4	1	10.30		0.40		0.18		2	Am	1087
96789	−57 04320	NGC 3532 - 670		11 05 48	−58 01.2	1	7.04		1.34		1.41		2	K2 II	1671
		LSS 2174		11 05 48	−63 34.8	1	11.62		0.44		-0.60		3	B3 Ib	1737
		Cha T # 25		11 05 48	−75 46.8	1	15.35		1.70		0.75		1	M3	437
96723	−29 08877	HR 4331		11 05 51	−29 42.1	4	6.48	.004	0.03	.000	0.05	.000	12	A1 V	15,78,1075,2012
96809	−58 03174	NGC 3532 - 491		11 05 51	−58 34.7	3	9.16	.030	0.07	.013	0.00	.012	5	A0 V	822,885,1493
		Cha F # 28		11 05 52	−77 09.9	1	15.14		2.50				2	K4 III	1633
96692	+16 02216	G 56 - 16		11 05 53	+16 02.6	2	9.76	.019	1.10	.019	1.01	.005	3	K0	333,1620,3077
96808	−57 04322	NGC 3532 - 501		11 05 55	−58 13.8	2	8.88	.019	0.06	.005	0.01	.023	4	A1 IV shell	822,1493
	−25 08473			11 05 57	−25 52.6	1	9.95		0.93		0.62		1	K3 V	3072
96826	−58 03175	NGC 3532 - 559		11 05 57	−58 43.9	2	9.61	.028	0.04	.019	0.03	.009	4	A0 V	885,1493
96656	+59 01351			11 05 58	+59 29.2	1	7.16		1.01				5	G5	1379
96810	−59 03075	LSS 2176		11 05 58	−59 57.5	2	8.68	.020	0.25	.000	-0.45	.035	5	B1 II	6,463
	−76 00652	HRC 245, DI Cha		11 05 58	−77 21.8	4	10.71	.027	1.18	.019	0.56	.054	4	G1Iab:pe	437,825,1588,1763
	−70 01310	LSS 2182		11 06 7	−70 38.3	1	11.07		0.12		-0.84		3		1737
96829	−60 02546	LSS 2177	⋆ AB	11 06 00	−60 17.2	7	7.31	.011	0.23	.012	-0.64	.026	12	B1/2 II	6,191,258,614,657,688,1075
		Car sq 2 # 58		11 06 00	−61 18.0	1	11.60		0.20		-0.70		3		466
		Cha T # 27	⋆ V	11 06 01	−76 35.9	1	14.80		1.26		-0.28		1	M1.5	437
96719	+17 02318			11 06 02	+17 28.5	1	6.81		1.02		0.79		2	G8 III	1648
	−44 07094			11 06 03	−44 33.5	1	9.47		0.38		-0.08		2		3077
96828	−59 03078	LSS 2179		11 06 04	−59 58.5	1	7.85		0.43		0.25		3	A0 II	463
		G 56 - 17		11 06 07	+08 17.9	1	16.08		0.31		-0.55		1		3060
96849	−58 03178	NGC 3532 - 564		11 06 07	−58 40.2	1	8.95		0.03		0.00		1	A0 IV	885
96738	+25 02344	HR 4332	⋆ AB	11 06 08	+24 55.8	2	5.68	.003	0.06		0.12		9	A3 IV	70,1022
		Car sq 2 # 59		11 06 14	−60 30.7	1	12.17		0.34		-0.42		3		466
96879	−58 03183	NGC 3532 - 566		11 06 16	−58 33.3	1	8.93		-0.07		-0.49		1	B9	822
303841	−58 03185			11 06 17	−59 01.4	1	10.50		0.17		0.14		2	B9	540
96880	−58 03184	LSS 2180		11 06 17	−59 08.5	4	7.57	.012	0.48	.012	-0.51	.030	9	B1 Iab/b	6,164,191,1737
306068	−59 03081			11 06 17	−60 14.6	1	10.80		0.11		-0.53		3	B5	2
96882	−60 02553			11 06 17	−61 00.8	1	9.05		-0.03		-0.86		2	B1 II/III	401
96864	−55 04156			11 06 18	−56 12.8	2	8.98	.006	0.10	.002	-0.65	.007	7	B3/5 III	540,976
96901	−63 01845	IDS11043S6404	AB	11 06 18	−64 19.8	2	8.86	.045	0.20	.000	-0.62	.005	2	B1 III/IV	55,8100
96819	−27 07886	HR 4334		11 06 19	−27 48.6	3	5.43	.004	0.07	.000	0.05	.005	11	A1 V	15,1075,2012
96781	+14 02347			11 06 20	+13 37.1	1	10.22		0.06		0.06		2	A0 V	272
96778	+30 02111			11 06 20	+30 18.7	1	7.08		0.87		0.48		2	G8 III	985
96898	−58 03186	NGC 3532 - 623		11 06 20	−58 37.1	1	7.93		0.40		0.22		4	F0 II	1493
		Cha T # 28		11 06 20	−77 23.4	1	15.34		1.71		0.46		1		437
	−43 06810			11 06 21	−43 58.9	2	9.81	.009	0.56	.005	-0.02		5	F8	2033,3077
96895	−57 04335	NGC 3532 - 633		11 06 21	−58 09.6	1	8.48		0.09		-0.82		1	B0 II/III	822
96883	−60 02554			11 06 22	−61 21.3	3	7.87	.028	0.39	.010	-0.18	.023	7	B8 Ib	191,540,976
96897	−58 03187	NGC 3532 - 629		11 06 24	−58 24.5	1	9.53		0.26		0.13		3	Ap	1493
96861	−48 06202			11 06 25	−49 23.4	1	9.06		0.12		0.10		1	A1 V	1770
96899	−60 02555			11 06 26	−60 42.6	3	9.16	.014	0.06	.008	-0.73	.019	7	B1 II/IIIn	2,540,976
		Steph 918		11 06 27	+03 16.5	1	11.91		1.41		1.21		1	K7	1746
96918	−58 03189	HR 4337, V382 Car		11 06 27	−58 42.2	9	3.90	.033	1.25	.015	0.98	.031	28	G2 Ia	15,549,1018,1034,1075*
		ApJS33,459 # 29		11 06 27	−65 31.0	1	15.12		1.41		-0.91		1		1753
96896	−57 04337	NGC 3532 - 631		11 06 28	−58 15.7	1	9.75		0.03		0.02		3	A0 V	1493
96919	−61 02075	HR 4338, V371 Car		11 06 29	−61 40.6	8	5.14	.018	0.23	.020	-0.45	.019	19	B9.5Ia	6,15,1018,1034,2001*
		CCS 1824		11 06 30	−81 31.3	1	9.01		3.42				2	N	864
96707	+68 00632	HR 4330, EP UMa		11 06 31	+67 28.9	1	6.06		0.22		0.13		3	F0p Sr	220
		Cha F # 29		11 06 31	−77 11.2	1	13.41		1.79		1.58		2	K4 III	1633
96917	−56 04238	LSS 2183		11 06 32	−56 44.7	7	7.08	.019	0.08	.004	-0.83	.024	18	O8	6,158,191,540,976*
96813	+37 02162	HR 4333, CO UMa		11 06 34	+36 34.9	2	5.79	.033	1.50	.011	1.53		7	M3.5 IIIab	70,3009
		Cha T # 31	⋆ V	11 06 38	−77 26.2	2	12.58	.065	1.24	.000	-0.14	.015	2	K2	437,825
	−28 08680			11 06 39	−29 53.8	1	10.42		0.38		-0.02		2	G5	1700
306050	−59 03082			11 06 39	−59 53.0	2	9.65	.005	0.10	.020	-0.57	.000	5	B5	2,463
96511	+82 00325			11 06 40	+82 00.3	1	7.20		0.66		0.26		2	G0	1502
96945	−59 03083	LSS 2184		11 06 40	−60 06.4	2	8.77	.036	0.22	.011	-0.51	.005	5	B1/2 III	540,976
97048	−76 00654	Cha T # 32	⋆ V	11 06 40	−77 23.0	6	8.45	.014	0.35	.010	0.25	.032	7	A0pe Shell	437,825,1588,1604*
96910	−47 06547			11 06 41	−48 07.9	1	8.12		0.04		0.00		2	Ap SiCr	1770
	−28 08681			11 06 42	−28 59.4	1	9.81		0.26		0.09		3	A5	1700
306060	−59 03084			11 06 42	−60 11.0	2	10.06	.020	0.13	.020	0.08	.010	6	A0	2,463
96944	−58 03198	NGC 3532 - 664		11 06 45	−58 33.6	1	9.12		-0.01		-0.05			A0 V	1493
96946	−60 02559	LSS 2185		11 06 45	−60 29.3	3	8.44	.017	0.21	.009	-0.71	.022	7	O7 f	2,540,976
	−41 06360			11 06 48	−41 35.0	1	10.12		0.91		0.58		1		78
96968	−56 04247	LSS 2186		11 06 49	−57 08.9	1	7.88		0.32		0.25		3	A8 II/III	8100
96969	−58 03199	NGC 3532 - 661		11 06 50	−58 41.3	1	9.06		0.02		0.00		1	A0 V	549
96970	−60 02561			11 06 50	−60 38.6	1	8.17		0.17		0.09		3	A1 V	2
		Cha T # 33		11 06 50	−77 17.5	1	12.83		1.47		1.12		5		1633
96834	+43 02083	HR 4336		11 06 51	+43 28.7	3	5.90	.005	1.56	.005	1.89	.005	7	M2 III	15,1003,3055
96833	+45 01897	HR 4335		11 06 52	+44 46.2	9	3.01	.008	1.14	.007	1.11	.008	35	K1 III	1,15,1077,1080,1355*
306112	−60 02562	LSS 2187		11 06 52	−60 48.1	1	9.70		0.14		-0.66		3	B3	2
		Cha T # 34		11 06 53	−77 28.3	1	16.45		1.95		1.58		1		437
96987	−59 03086			11 06 54	−59 56.2	1	9.53		0.17		0.13		2	A3/5 V	463
306053	−59 03087	LSS 2188		11 06 57	−60 06.0	2	10.24	.015	0.29	.015	-0.42	.000	5		2,463
		LSS 2189		11 06 57	−60 38.9	1	11.12		0.14		-0.68		3	B3 Ib	466
		Cha F # 30		11 06 57	−76 32.2	1	11.44		1.69				2	K3 III	1633
303831	−58 03202			11 07 00	−58 59.4	1	10.34		0.02		-0.10		1	A0	549

HD	DM	Other Id	N	Rem	α_{1950}	δ_{1950}	S	V	σ_V	B–V	σ_{B-V}	U–B	σ_{U-B}	n	Spectrum	References
97001	−60 02563				11 07 01	−60 33.3	1	9.24		0.18		0.10		3	A0 V	2
303810	−57 04346	NGC 3532 - 667			11 07 02	−58 16.0	1	9.02		1.49		1.61		3	K2	1493
97000	−58 03203	NGC 3532 - 665			11 07 02	−58 27.2	1	7.93		0.08		0.05		1	A2 IV	822
		AAS41,43 F1 # 1			11 07 02	−60 23.3	1	12.21		0.51		0.14		1	F4	820
96941	−25 08487	IDS11046S2527			11 07 03	−25 43.1	1	8.68		0.74		0.27		1	G6/8 V	78
97013	−59 03088				11 07 03	−60 03.0	1	8.88		0.23		−0.49		4	B2/3 II	463
		AAS41,43 F1 # 2			11 07 04	−60 25.0	1	12.28		0.37		−0.42		1		820
96981	−50 05743				11 07 05	−51 21.0	1	9.23		0.19		0.16		1	A2/3 IV	1770
		AAS41,43 F1 # 3			11 07 05	−60 22.6	1	11.25		0.55		0.04		1	F0	820
	−23 09765				11 07 06	−24 19.7	4	10.44	.005	1.54	.011	1.10		8	M2	158,912,1705,1746
		AAS41,43 F1 # 4			11 07 06	−60 24.1	1	11.42		0.34		−0.54		1	B3	820
96937	+3 02466	G 163 - 54			11 07 07	+02 43.6	2	7.68	.000	0.77	.005	0.40	.015	2	G5	1620,1658
306086	−59 03090	LSS 2190			11 07 08	−60 24.1	2	9.48	.005	0.27	.019	−0.55	.033	4	B5	2,820
		AAS41,43 F1 # 5			11 07 09	−60 18.0	1	11.27		0.39		0.31		2	A7	820
		AAS41,43 F1 # 6			11 07 09	−60 23.4	1	12.64		0.30		−0.22		1		820
		AAS41,43 F1 # 10			11 07 09	−60 26.7	1	12.86		0.83		0.29		1	G2	820
		AAS41,43 F1 # 8			11 07 10	−60 23.1	1	10.88		0.35		−0.52		1	B2	820
306085	−60 02566	LSS 2191			11 07 11	−60 26.4	1	9.75		0.34		−0.44		2	B2 Ve	820
303825	−58 03204				11 07 12	−58 48.0	1	10.08		1.01		0.80		1	K0	549
		AAS41,43 F1 # 9			11 07 12	−60 15.9	1	11.77		0.37		0.17		2	A0	820
		MCC 280			11 07 13	+21 53.8	1	10.99		1.27				1	K5	1017
96921	+33 02084				11 07 13	+32 48.1	1	7.16		1.07		1.04		2	K0	1625
		AAS41,43 F1 # 12			11 07 13	−60 25.2	1	12.81		0.44		−0.24		1		820
306111	−60 02567	LSS 2192			11 07 13	−60 49.8	3	10.15	.043	0.18	.010	−0.81	.048	6	B2 Ve	2,540,1737
97010	−52 04271				11 07 14	−52 57.9	1	8.95		0.11		0.08		1	A1/2 V	1770
303826	−58 03205				11 07 15	−58 50.2	2	10.43	.010	0.30	.000	0.22	.010	2	A3	287,549
	−41 06367				11 07 16	−42 11.4	2	9.73	.000	0.78	.000	0.34		8	G5	158,2033
		AAS41,43 F1 # 13			11 07 17	−60 25.4	1	11.79		0.31		−0.28		1	B6	820
		AAS41,43 F1 # 14			11 07 18	−60 21.7	1	12.61		0.38		−0.35		1	B5	820
		GD 128			11 07 19	+26 35.2	1	15.89		−0.07		−0.99		1		3060
96904	+56 01504	IDS11044N5615	A		11 07 19	+55 57.8	1	8.91		0.27		0.03		2	F0	1502
96953	+15 02297	IDS11047N1515			11 07 20	+14 58.9	1	8.54		0.86				3	K0	882
97026	−47 06560				11 07 20	−48 21.5	1	8.36		0.11		0.08		1	A0/1 V	1770
303824	−58 03209				11 07 20	−58 47.7	1	10.19		0.14		0.11		1	A2	549
		AAS41,43 F1 # 17			11 07 20	−60 32.6	1	12.20		0.45		0.12		1	A0	820
303830	−58 03210				11 07 21	−58 59.7	1	10.18		0.10		0.09		1	A0	549
		LP 214 - 47			11 07 22	+39 40.4	1	11.05		1.10		0.94		3		1723
97065	−58 03211				11 07 22	−58 54.9	1	9.50		−0.05		−0.32		1	B8/9 III	549
		AAS41,43 F1 # 18			11 07 22	−60 33.6	1	12.63		0.44		−0.24		1		820
97066	−59 03094				11 07 23	−59 59.0	1	9.34		0.08		0.08		3	A0 V	463
306071	−59 03093		A		11 07 23	−60 15.9	2	10.69	.005	0.21	.014	−0.48	.019	4	A0	287,820
306071	−59 03093		AB		11 07 23	−60 15.9	1	10.11		0.15		−0.39		2	A0	463
306071	−59 03093		B		11 07 23	−60 15.9	1	11.13		0.20		−0.34		3		820
		AAS41,43 F1 # 19			11 07 24	−60 26.9	1	12.07		0.33		−0.42		1	B5	820
		Car sq 2 # 61			11 07 25	−60 28.9	2	12.45	.019	0.76	.014	−0.08	.014	4		466,820
		AAS41,43 F1 # 20			11 07 25	−60 30.0	1	12.77		0.42		0.16		1		820
		AAS41,43 F1 # 22			11 07 26	−60 21.6	1	11.77		0.36		−0.18		1	B8	820
		AAS41,43 F1 # 23			11 07 26	−60 22.3	1	12.19		1.26		1.07		1		820
303827	−58 03214				11 07 27	−58 50.3	2	10.51	.025	0.19	.000	0.16	.025	2	A0	287,549
		AAS41,43 F1 # 24			11 07 27	−60 13.4	1	12.28		0.15		−0.44		1	B5	820
		G 56 - 19			11 07 28	+15 40.4	1	13.59		1.51		1.09		1		1620
	−9 03222				11 07 28	−10 00.9	1	10.54		1.40		1.27		1	M0 V	3072
		G 119 - 57			11 07 29	+29 13.4	2	13.22	.010	1.47	.010	1.10	.024	3		203,3078
96951	+36 02160				11 07 29	+35 36.4	1	7.86		0.04		0.06		2	A1 V	275
		G 10 - 3			11 07 29	−02 30.8	1	12.55		0.92		0.44		2		1658
97023	−31 08816	HR 4339			11 07 29	−32 05.7	2	5.80	.005	0.04	.000			7	A1 V	15,2027
306070	−59 03095				11 07 29	−60 15.3	4	10.29	.017	0.42	.016	−0.25	.016	9	B8	2,287,463,820
		WLS 1100 75 # 9			11 07 30	+74 34.0	1	13.28		0.66		0.20		2		1375
		AAS41,43 F1 # 26			11 07 30	−60 21.6	1	12.72		0.48		−0.12		1		820
306106	−60 02570				11 07 30	−60 46.5	1	10.88		0.07		−0.73		3	B5	466
303829	−58 03212				11 07 31	−58 53.3	1	9.84		1.11		0.99		1	K0	549
		AAS41,43 F1 # 27			11 07 31	−60 20.7	1	12.01		0.23		−0.16		1	B7	820
		AAS41,43 F1 # 28			11 07 31	−60 21.0	1	11.24		0.32		0.21		1	B9	820
97082	−58 03216	NGC 3532 - 480		⋆	11 07 32	−58 34.0	3	6.63	.094	0.75	.061	0.47	.025	3	F8	549,657,1484
		AAS41,43 F1 # 29			11 07 32	−60 25.5	1	12.52		0.46		0.13		1	A3	820
306116	−60 02571	LSS 2195			11 07 32	−61 00.6	2	9.71	.030	−0.03	.002	−0.83	.017	4	B0.5IV	540,976
		WLS 1112 50 # 10			11 07 34	+50 25.7	1	11.66		1.05		0.86		2		1375
306072	−59 03096				11 07 35	−60 17.7	3	9.86	.014	0.08	.015	−0.01	.027	7	B9	2,463,820
		AAS41,43 F1 # 31			11 07 35	−60 26.6	1	11.99		0.29		−0.45		1	B5	820
303822	−58 03221	NGC 3532 - 710			11 07 36	−58 42.5	1	10.55		0.20		0.14		1	A0	549
		AAS41,43 F1 # 32			11 07 36	−60 28.3	1	11.84		0.47		0.20		1	A7	820
303828	−58 03223				11 07 37	−58 50.8	2	10.81	.025	−0.02	.000	−0.27	.010	2	B9	287,549
96973	+48 01919				11 07 38	+48 04.1	1	7.27		0.47		0.02		2	F5	1601
		Car sq 2 # 63			11 07 38	−61 09.8	1	12.18		1.09		−0.12		3		466
		AAS41,43 F1 # 33			11 07 39	−60 28.0	1	12.03		0.57		0.04		1	F6	820
97005	+23 02313				11 07 41	+22 58.4	1	7.50		0.34		0.11		2	F0	1375
		AAS41,43 F1 # 34			11 07 41	−60 24.5	1	11.25		0.25		−0.50		1	B3	820
		AAS41,43 F1 # 35			11 07 41	−60 28.7	1	11.81		0.30		0.17		1	B9	820
306059	−59 03099				11 07 42	−60 11.9	2	9.42	.009	0.09	.019	−0.72	.066	4	B5	463,820
		AAS41,43 F1 # 39			11 07 42	−60 14.6	1	10.94		1.76		2.01		1		820

Table 1 591

HD	DM	Other Id	N Rem	α_{1950}	δ_{1950}	S	V	σ_V	B–V	σ_{B-V}	U–B	σ_{U-B}	n	Spectrum	References
		AAS41,43 F1 # 38		11 07 42	−60 16.7	1	12.18		0.13		-0.12		1		820
306073	−59 03098			11 07 42	−60 18.4	2	10.20	.010	0.15	.019	-0.47	.015	3	B9	463,820
		AAS41,43 F1 # 40		11 07 44	−60 17.4	1	11.26		0.84		0.41		4	F1	820
97136	−63 01852	LSS 2196		11 07 44	−63 31.4	2	9.27	.005	0.03	.011	-0.73	.008	5	B2 III	540,976
97436	−83 00396			11 07 44	−84 09.9	1	7.91		1.17		1.06		3	K0 III	1704
97073	−37 07067	IDS11054S3748	AB	11 07 45	−38 05.4	1	8.88		0.79		0.34		1	G8 V	78
		G 163 - 56		11 07 46	+04 35.7	1	14.10		0.73		0.04		2		1658
		AAS35,161 # 30		11 07 46	−77 12.9	1	13.17		1.53		1.19		1		5005
	−58 03224			11 07 47	−58 46.8	1	11.20		0.03		-0.50		1	B7	549
		AAS41,43 F1 # 41		11 07 47	−60 22.1	1	10.03		1.67		1.94		1	M2	820
		AAS41,43 F1 # 43		11 07 48	−60 32.5	1	12.32		0.43		-0.28		1	B6	820
		AAS41,43 F1 # 42		11 07 48	−60 37.4	1	11.56		0.29		-0.43		1	B2	820
97034	+31 02236			11 07 51	+30 37.5	1	8.65		0.10		0.10		2	A3 V	272
303823	−58 03225			11 07 51	−58 48.5	1	10.73		0.19		0.08		1	A3	549
306084	−60 02573			11 07 52	−60 27.9	1	11.13		0.11		0.08		1		820
306082	−60 02574			11 07 52	−60 30.3	2	10.05	.023	0.26	.023	-0.47	.047	4	B5	2,820
		Cha T # 40	⋆ V	11 07 52	−76 07.0	1	12.42		0.68		0.19		1	K6	437
97240	−77 00651	Cha F 31		11 07 52	−77 31.4	2	8.51	.000	0.45	.010	-0.02		6	F5 V	1633,2012
97117	−47 06572			11 07 53	−48 16.2	1	9.34		0.08		0.07		1	A0 V	1770
		AAS41,43 F1 # 46		11 07 53	−60 18.5	1	12.65		0.28		0.13		1	B9	820
		AAS41,43 F1 # 47		11 07 53	−60 37.1	1	11.57		0.31		-0.58		1	B0	820
97175	−69 01487	LSS 2204		11 07 53	−70 05.8	1	8.89		-0.07		-0.85		2	B0.5/1 III	401
97150	−58 03228			11 07 54	−58 35.1	1	9.53		-0.05		-0.27		1	B8/9 IV	549
97151	−59 03100	V353 Car		11 07 54	−59 49.4	3	7.71	.038	-0.09	.012	-0.81	.025	11	B2 Ve	2,164,463
		LP 552 - 10		11 07 55	+04 49.4	1	13.54		1.38		1.09		1		1696
		LB 1959		11 07 56	+60 13.1	1	12.39		-0.22		-0.93		1		832
		Lod 282 # 29		11 07 56	−58 43.4	1	11.48		0.67		0.12		1	F7	549
		AAS41,43 F1 # 48		11 07 56	−60 19.1	1	12.47		0.53		0.08		1	A8	820
		AAS41,43 F1 # 49		11 07 56	−60 22.6	1	12.08		1.61		1.65		1		820
306099	−60 02577	LSS 2197		11 07 56	−60 39.5	3	8.74	.017	0.17	.004	-0.73	.034	7	B3	2,466,820
		Tr 18 - 150		11 07 57	−60 23.9	1	13.64		0.55		0.24		3		1720
97152	−60 02578	V431 Car	⋆	11 07 57	−60 42.5	3	8.07	.039	-0.04	.090	-0.61	.067	7	WC	2,820,1096
		Lod 282 # 30		11 07 58	−58 43.5	1	11.49		0.57		0.01		1	F6	549
97166	−59 03102	NGC 3572 - 1		11 07 58	−59 58.6	4	7.89	.017	0.07	.011	-0.84	.057	13	O7.5III(f)	2,6,411,463
		Tr 18 - 140		11 07 58	−60 29.4	1	12.33		0.64		0.10		1		820
97132	−50 05759			11 07 59	−51 02.6	1	9.82		0.23		0.18		1	Ap SrEuCr	1770
	−28 08692			11 08 00	−29 08.5	1	9.86		1.07		0.92		1	K5 V	3072
97165	−58 03229	LSS 2200		11 08 00	−59 12.8	1	8.40		0.06		-0.01		2	B0 III	401
		AAS41,43 F1 # 53		11 08 00	−60 22.4	1	12.10		0.31		0.27		2	A0	820
306095	−60 02580			11 08 00	−60 33.3	2	10.37	.005	0.32	.019	-0.55	.009	4	B9	466,820
		NGC 3572 - 17		11 08 01	−59 59.5	1	11.50		0.19		-0.53		2		411
306083	−60 02579	Tr 18 - 139		11 08 02	−60 29.4	1	9.06		0.93		0.62		2	G5	820
		Tr 18 - 143		11 08 03	−60 29.0	1	11.49		0.44		0.19		1		820
		AAS41,43 F1 # 59		11 08 03	−60 30.8	1	12.50		0.59		0.14		1		820
		NGC 3572 - 16		11 08 04	−59 59.6	1	11.49		0.47		0.13		1		411
		AAS41,43 F1 # 58		11 08 04	−60 28.0	1	12.80		0.59		0.18		1		820
		NGC 3572 - 20		11 08 05	−59 58.6	1	12.16		0.16		-0.48		2		411
306096	−60 02582	LSS 2202		11 08 05	−60 35.5	2	9.28	.005	0.21	.028	-0.65	.019	4	B5	2,820
306122	−60 02583			11 08 05	−61 10.1	1	10.08		0.09		-0.68		3	B0 III	466
306054	−59 03103			11 08 06	−60 09.2	3	10.72	.005	0.16	.010	-0.57	.031	6	B9	2,463,820
306055				11 08 06	−60 10.8	1	11.33		0.34		0.10		1	A0	820
		NGC 3572 - 38		11 08 07	−59 57.2	1	11.65		0.23		-0.36		2		411
306081	−60 02585	Tr 18 - 138		11 08 07	−60 30.3	1	9.69		0.40		-0.62		1	B3	820
	−60 02584	Tr 18 - 137		11 08 07	−60 30.5	1	9.55		1.40		1.57		1		820
		NGC 3572 - 35		11 08 08	−59 56.3	1	12.02		0.23		-0.18		2		411
		AAS41,43 F1 # 64		11 08 08	−60 23.1	1	11.53		0.20		-0.39		1	B7	820
	−60 02588			11 08 10	−60 36.9	1	10.49		0.29		-0.50		1	B2	820
97188	−58 03233			11 08 11	−58 38.9	1	9.82		0.18		0.07		1	A2	549
306045	−59 03104			11 08 11	−59 50.4	2	9.61	.007	0.42	.030	0.00	.019	5	F2	463,540
	−60 02589			11 08 11	−60 34.3	1	9.81		1.49		1.34		1	G8	820
306098	−60 02587			11 08 11	−60 39.0	1	8.58		1.16		0.99		2	K0	820
		Feige 37		11 08 12	+07 36.	1	14.04		0.22		0.10		1		1298
		Lod 282 # 27		11 08 12	−58 45.1	1	11.26		0.10		-0.45		1	B6	549
		NGC 3572 - 45		11 08 12	−59 58.7	1	12.82		0.26		-0.38		2		411
		AAS41,43 F1 # 67		11 08 12	−60 27.9	1	13.25		0.23		0.14		1	A0	820
97100	+31 02239	IDS11056N3100	C	11 08 13	+30 45.8	1	9.08		0.89		0.49		2	G0	3024
		Tr 18 - 114		11 08 13	−60 25.7	1	12.87		0.21		0.02		1		820
306074	−59 03105			11 08 14	−60 15.0	1	10.84		0.22		-0.56		1	A2	820
	−75 00713	ApJ180,115 # 232		11 08 14	−76 12.7	1	10.53		0.72		0.50		2	F0 V	1633
97171	−51 05403			11 08 15	−52 17.0	1	8.80		0.00		-0.16		3	B9.5 IV	1770
		NGC 3572 - 68		11 08 15	−59 59.3	1	12.40		1.01		0.61		1		411
306056	−59 03106	LSS 2205		11 08 15	−60 09.4	3	10.05	.011	0.10	.012	-0.71	.039	7	B5	2,463,820
97033	+66 00704	G 236 - 60	⋆ A	11 08 16	+66 17.2	1	8.25		0.70		0.26		2	G5	1064
		NGC 3572 - 67		11 08 17	−59 59.0	1	10.35		0.17		-0.78		2		411
		NGC 3572 - 70		11 08 17	−59 59.6	1	11.90		0.17		-0.48		2		411
97206	−59 03109	NGC 3572 - 77		11 08 17	−60 00.3	3	9.63	.011	0.11	.008	-0.65	.021	8		2,411,463
		AAS41,43 F1 # 71		11 08 17	−60 21.1	1	12.69		0.21		0.18		2	A0	820
97101	+31 02240	G 119 - 59	⋆ B	11 08 18	+30 43.1	2	9.99	.022	1.50	.039	1.16	.035	6	M2 V	497,3072
		Lod 282 # 28		11 08 18	−58 45.0	1	11.35		0.06		-0.40		1	B7	549
97300	+68 00635	Cha T # 41		11 08 18	−76 20.5	3	9.02	.025	0.38	.017	0.16	.057	3	B9 V	437,1588,1604

HD	DM	Other Id	N Rem	α₁₉₅₀	δ₁₉₅₀	S	V	σ_V	B-V	σ_B-V	U-B	σ_U-B	n	Spectrum	References
97143	−5 03221			11 08 19	−05 43.8	1	9.69		0.43		-0.02		std	F2	1348
306048	−59 03110	NGC 3572 - 48		11 08 19	−59 57.4	1	9.58		0.13		-0.85		2		411
306080	−60 02590	Tr 18 - 136		11 08 19	−60 30.3	2	10.35	.005	0.27	.023	-0.60	.023	4		2,820
97101	+31 02240	G 119 - 60	★ A	11 08 20	+30 43.2	4	8.34	.029	1.35	.008	1.28	.010	14	K9 V	22,497,1197,3072
97159	−28 08695			11 08 20	−29 18.1	1	8.91		0.02		0.03		4	A0 V	1700
		NGC 3572 - 66		11 08 20	−59 58.7	1	11.31		0.18		-0.64		3		411
97207	−59 03111	NGC 3572 - 72	★ AB	11 08 21	−59 59.9	3	9.55	.016	0.12	.009	-0.63	.018	7		2,411,463
		LP 612 - 9		11 08 22	+01 26.2	1	11.58		0.62		0.01		2		1696
303920	−58 03236			11 08 22	−58 45.5	2	9.83	.000	0.05	.000	-0.48	.000	2	B5	287,549
		Tr 18 - 158		11 08 22	−60 31.6	1	13.44		0.48		0.25		1		1720
97185	−48 06240			11 08 23	−49 22.9	2	7.50	.005	-0.09	.005	-0.72	.000	5	B2/3 II/III	55,164
303918	−58 03237			11 08 23	−58 45.5	2	9.71	.000	0.07	.000	-0.47	.000	2	B5	287,549
		NGC 3572 - 63		11 08 23	−59 57.3	1	10.16		1.35		1.31		2		411
		NGC 3572 - 85		11 08 23	−60 01.2	1	12.37		0.20		-0.42		2		411
97223	−59 03113	Tr 18 - 112		11 08 23	−60 18.8	3	8.72	.010	0.04	.007	-0.44	.016	6	B3 V	2,463,820
		Tr 18 - 113		11 08 23	−60 21.5	1	12.04		0.28		-0.28		1		820
		Car sq 2 # 67		11 08 23	−60 55.0	1	12.55		0.94		-0.26		3		466
		Cha F # 33		11 08 23	−77 29.5	1	15.09		2.00				2	G8 III	1633
97184	−48 06241			11 08 24	−48 56.0	1	9.89		0.00		-0.05		1	A0 V	1770
97222	−59 03114	NGC 3572 - 124	★ AB	11 08 24	−59 52.3	7	8.79	.017	0.19	.007	-0.68	.036	17	B0.5IIn	2,6,191,411,463,540, 976
	−59 03112	NGC 3572 - 59		11 08 24	−59 56.8	1	10.35		0.23		-0.68		2		411
		NGC 3572 - 74		11 08 24	−59 59.6	1	11.89		0.18		-0.42		2		411
		AAS41,43 F1 # 76		11 08 24	−60 31.7	1	12.23		0.29		-0.43		1		820
303919	−58 03238			11 08 25	−58 45.1	2	10.24	.005	0.09	.005	-0.46	.010	2	B8	287,549
	−59 03115	NGC 3572 - 55		11 08 25	−59 55.8	1	10.69		0.28		-0.40		3		411
		AAS41,43 F1 # 77		11 08 25	−60 32.2	1	11.79		0.43		0.11		1	A9	820
		G 10 - 4		11 08 26	+06 41.8	3	11.41	.009	0.72	.000	-0.02	.012	5		1064,1620,3077
97050	+66 00705	IDS11052N6633	B	11 08 26	+66 18.1	1	8.18		1.37		1.61		2		1064
		Tr 18 - 132		11 08 26	−60 26.8	1	11.82		0.84		0.43		1		820
		Ho 10 - 12		11 08 27	−60 06.8	1	13.09		0.60		0.01		2		463
		AAS41,43 F1 # 78		11 08 27	−60 21.2	1	13.01		0.31		-0.20		1	B8	820
		Tr 18 - 92		11 08 28	−60 29.7	2	12.56	.020	0.61	.030	0.50	.035	4		411,820
		Cha T # 44	★ V	11 08 28	−76 18.6	1	13.29		1.45		0.44		1	K5:	437
		Tr 18 - 93		11 08 29	−60 27.7	2	12.45	.010	0.39	.005	-0.27	.010	3		411,820
97237	−60 02593	IDS11064S6023	AB	11 08 29	−60 38.9	1	8.43		1.21		1.06		2	G8 III	820
		NGC 3572 - 113		11 08 30	−59 57.0	1	11.52		0.99		0.66		1		411
		Ho 10 - 16		11 08 30	−60 05.8	1	13.33		0.45		0.16		2		463
		Ho 10 - 10		11 08 30	−60 07.6	1	12.75		0.29		-0.32		3		463
		NGC 3572 - 103		11 08 31	−59 59.6	1	12.98		0.32		-0.25		2		411
		Ho 10 - 13		11 08 31	−60 05.8	1	13.11		0.34		-0.21		3		463
		Ho 10 - 9		11 08 31	−60 06.2	2	12.70	.009	0.30	.005	-0.31	.000	4		411,463
		Ho 10 - 18		11 08 31	−60 06.7	1	13.49		0.33		-0.22		3		463
		Tr 18 - 91		11 08 31	−60 28.9	1	13.07		2.19				2		411
		NGC 3572 - 102		11 08 32	−59 59.7	1	12.49		1.29		1.20		1		411
		Ho 10 - 7		11 08 32	−60 06.9	2	12.52	.332	0.25	.005	-0.43	.074	5		411,463
		Ho 10 - 17		11 08 32	−60 07.5	1	13.40		0.49		0.21		2		463
		AAS41,43 F1 # 83		11 08 32	−60 10.2	1	11.96		0.62		0.12		1		820
		AAS41,43 F1 # 89		11 08 32	−60 37.0	1	11.90		0.26		0.15		1	B9	820
		Cha T # 45	★ V	11 08 32	−77 20.8	1	14.91		1.53		0.39		1	K7	437
		GD 129		11 08 33	+47 34.9	1	15.38		0.15		-0.63		1		3060
303921	−58 03240			11 08 33	−58 53.4	1	8.73		1.50		1.82		1	M0	549
		AAS41,43 F1 # 84		11 08 33	−60 09.3	1	12.67		0.17		-0.46		1		820
		AAS41,43 F1 # 85		11 08 33	−60 20.6	1	12.88		0.65		0.18		1	F6	820
		Tr 18 - 149		11 08 33	−60 23.5	1	13.58		1.27		0.65		2		1720
		Tr 18 - 95		11 08 33	−60 27.4	2	13.58	.034	0.28	.010	0.08	.019	3		411,820
	−60 02594	Car sq 2 # 68		11 08 33	−60 43.4	1	10.80		0.21		-0.53		3		466
	−59 03117	NGC 3572 - 95		11 08 34	−60 01.1	1	10.08		0.22		-0.68		2		411
97253	−59 03116	Ho 10 - 1		11 08 34	−60 06.8	4	7.11	.010	0.16	.020	-0.81	.027	14	O5.5III(f)	6,411,463,820
303917	−58 03241			11 08 35	−58 46.5	2	9.57	.025	0.62	.010	-0.09	.005	2	B5	287,549
		Ho 10 - 20		11 08 35	−60 05.6	1	13.79		0.66		0.26		3		463
		Ho 10 - 14		11 08 35	−60 07.1	2	13.36	.160	0.58	.030	0.19	.065	4		411,463
		Tr 18 - 111		11 08 35	−60 19.1	1	11.27		0.98		0.74		2		820
		Tr 18 - 65		11 08 35	−60 22.0	1	12.98		0.37		0.10		1		820
		Cha T # 46	★ V	11 08 35	−76 13.3	1	13.87		1.42		0.10		1	M0	437
		G 122 - 2	AB	11 08 36	+43 41.7	1	10.94		1.50		1.24		2	M4	7010
	−59 03118	Ho 10 - 6		11 08 36	−60 05.9	2	12.05	.069	0.65	.040	0.03	.030	5		411,463
		Ho 10 - 24		11 08 36	−60 06.	1	12.61		0.28		-0.22		2		411
		Tr 18 - 78		11 08 36	−60 20.2	1	12.25		0.36		0.01		2		820
		Tr 18 - 94		11 08 36	−60 27.5	2	12.55	.000	0.18	.000	-0.02	.010	3		411,820
		NGC 3572 - 122		11 08 37	−59 59.0	1	11.81		0.27		-0.46		2		411
		Ho 10 - 4		11 08 37	−60 06.8	2	11.32	.034	0.30	.024	-0.32	.010	6		411,463
		AAS41,43 F1 # 91		11 08 37	−60 19.6	1	13.18		0.48		0.01		1	B7	820
		Tr 18 - 63		11 08 37	−60 24.1	2	12.05	.025	0.25	.035	-0.04		2		820,2041
306077	−60 02595	Tr 18 - 133	★ V	11 08 37	−60 28.7	2	9.13	.059	0.93	.032	0.68	.036	5	G0	820,1720
		NGC 3572 - 96		11 08 38	−60 00.6	1	12.10		0.32		-0.43		2		411
		Tr 18 - 77		11 08 38	−60 20.3	1	12.53		0.48		-0.04		1		820
	−60 02596	Tr 18 - 97		11 08 38	−60 25.6	1	11.22		0.28		-0.27		1		820
		NGC 3572 - 114		11 08 39	−59 58.0	1	11.04		1.45		1.48		1		411
		Ho 10 - 22		11 08 39	−60 06.5	1	14.26		0.56		0.08		2		463
		Ho 10 - 5		11 08 39	−60 06.7	2	11.86	.020	0.29	.015	-0.32	.005	5		411,463

Table 1   593

HD	DM	Other Id	N Rem	α_{1950}	δ_{1950}	S	V	σ_V	B–V	σ_{B-V}	U–B	σ_{U-B}	n	Spectrum	References
	−59 03120	Ho 10 - 2		11 08 39	−60 07.0	3	10.25	.051	0.32	.013	-0.23	.021	10		411,463,820
		AAS41,43 F1 # 99		11 08 39	−60 10.1	1	11.93		0.13		0.07		2	B9	820
		WLS 1112 50 # 5		11 08 40	+52 11.8	1	13.44		0.85		0.45		2		1375
		Ho 10 - 21		11 08 40	−60 07.6	1	13.93		0.51		0.11		2		463
		Tr 18 - 62		11 08 40	−60 25.4	1	12.63		0.30		-0.25		1		820
		AAS41,43 F1 # 102		11 08 40	−60 34.0	1	12.06		0.47		0.05		2	B8	820
		AAS41,43 F1 # 103		11 08 40	−60 37.6	1	13.04		0.34		0.07		1		820
97272	−58 03242			11 08 41	−58 55.3	1	9.45		0.11		0.07		1	A1 IV/V	549
		Ho 10 - 15		11 08 41	−60 06.2	2	13.23	.023	0.53	.023	0.05	.033	4		411,463
		Tr 18 - 60		11 08 41	−60 24.6	1	12.61		0.71		0.19		1		820
306079		Tr 18 - 88		11 08 41	−60 30.4	2	11.13	.010	0.14	.010	0.08	.005	3	A0	411,820
		Tr 18 - 79		11 08 42	−60 19.7	1	12.96		0.36		-0.30		1		820
		Tr 18 - 66		11 08 42	−60 22.1	2	12.10	.023	0.22	.014	-0.19	.009	5	B8	820,1720
		Tr 18 - 89		11 08 42	−60 29.1	2	13.28	.034	0.27	.039	0.13	.024	3		411,820
97214	−10 03216			11 08 43	−10 41.0	3	9.24	.013	1.08	.016	0.93	.035	8	K5 V	803,1705,3078
		AAS41,43 F1 # 108		11 08 43	−60 17.0	1	12.94		0.66		0.02		1		820
97249	−40 06561			11 08 44	−41 19.0	1	8.49		0.52		0.01		1	F6 V	78
		AAS41,43 F1 # 111		11 08 44	−60 18.3	1	11.87		0.28		-0.26		1	B7	820
		Tr 18 - 59		11 08 44	−60 24.5	2	12.27	.026	0.17	.017	-0.19	.013	6		820,1720
97284	−60 02598	Tr 18 - 87		11 08 44	−60 28.2	5	8.87	.009	0.12	.013	-0.79	.012	60	B0 Iab/b	2,401,411,820,1720
97271	−57 04387	HR 4342		11 08 45	−58 11.0	3	6.88	.009	-0.09	.004	-0.45		9	B7 III	15,401,2012
		Ho 10 - 8		11 08 45	−60 06.1	2	12.23	.005	0.29	.020	-0.41	.010	5		411,463
		Ho 10 - 19		11 08 45	−60 06.5	1	13.56		0.70		0.38		2		463
	−59 03121			11 08 45	−60 09.1	1	11.89		0.26		-0.25		1	B8	820
		AAS41,43 F1 # 110		11 08 45	−60 18.2	1	12.55		1.46		1.20		1		820
97140	+59 01353			11 08 46	+59 10.3	1	7.34		0.61				5	G0	1379
		NGC 3572 - 91		11 08 46	−60 00.9	1	11.63		1.30		1.24		1		411
		Ho 10 - 23		11 08 47	−60 07.2	1	14.49		1.01		0.31		3		463
		Ho 10 - 11		11 08 47	−60 07.5	1	13.08		0.74		0.21		2		463
		Tr 18 - 67		11 08 47	−60 22.8	1	11.93		0.32		-0.01		1		820
		NGC 3572 - 116		11 08 48	−59 57.9	1	11.25		1.15		0.96		1		411
96571	+86 00161			11 08 49	+85 54.7	1	7.34		0.22		0.13		2	A2	985
		Tr 18 - 57		11 08 49	−60 23.9	2	12.37	.009	0.21	.013	-0.16	.022	6	B8	820,1720
		Tr 18 - 49		11 08 49	−60 26.1	1	13.40		0.90		-0.27		3		1720
		TV Leo		11 08 50	−05 37.2	1	11.44		0.12		0.16		1	A1	668
97283	−58 03245			11 08 50	−58 38.1	1	9.69		-0.04		-0.33		1	B8/9 IV	549
		Tr 18 - 68		11 08 50	−60 23.1	1	12.76		0.29		0.14		1		820
		Tr 18 - 129		11 08 50	−60 25.0	1	13.51		0.36		0.17		1		820
		Tr 18 - 73		11 08 51	−60 21.6	1	12.42		0.75		0.37		1		820
		Tr 18 - 56		11 08 51	−60 24.5	2	12.30	.014	1.25	.009	1.08	.009	4		820,1720
		Tr 18 - 48		11 08 51	−60 26.3	1	11.78		0.75		0.13		1		820
	−60 02602	Car sq 2 # 69		11 08 51	−60 53.7	1	10.30		0.93		-0.15		3		466
		G 163 - 59		11 08 52	−06 15.1	2	14.82	.023	1.64	.140	0.86		5		419,1663,3078
	−59 03125	NGC 3572 - 117		11 08 52	−59 58.1	1	11.28		0.28		0.20		1		411
		Tr 18 - 51		11 08 52	−60 25.6	1	13.33		0.43		0.25		3		1720
		Tr 18 - 47		11 08 52	−60 27.5	3	11.63	.021	0.24	.011	-0.34	.012	9	B7	411,820,1720
306057	−59 03126	Ho 10 - 3		11 08 53	−60 07.2	2	10.93	.023	0.16	.000	-0.06	.005	5	A0	463,820
97263	−46 06816			11 08 54	−47 19.6	1	9.29		0.00		-0.05		1	A0 V	1770
97281	−51 05415			11 08 54	−52 23.4	1	9.09		0.12		0.09		1	A2/3 V	1770
		Tr 18 - 69		11 08 54	−60 22.7	3	12.19	.016	0.22	.010	-0.04	.009	7	B8	820,1720,2041
97297	−59 03129			11 08 56	−60 02.2	1	8.95		0.03		-0.01		2	A0 IV	463
	−60 02603	Tr 18 - 134		11 08 56	−60 30.1	1	11.28		0.21		-0.13		1		820
306058	−59 03130			11 08 57	−60 14.3	3	9.94	.004	0.23	.020	-0.65	.022	6	B3	2,463,820
	−60 02604	Tr 18 - 50		11 08 57	−60 26.1	2	12.19	.122	0.22	.019	-0.02	.093	6	B8	820,1720
		AAS41,43 F1 # 129		11 08 57	−60 33.9	1	12.61		0.21		0.16		2	B9	820
97320	−64 01636			11 08 57	−65 09.1	5	8.16	.008	0.48	.013	-0.16	.010	14	F3 V	1097,1594,2033,3037,6006
97282	−52 04302			11 08 58	−52 39.5	1	10.01		0.05		0.04		1	A0 V	1770
		Tr 18 - 110		11 08 58	−60 14.2	1	11.50		0.66		0.14		1		820
		AAS41,43 F1 # 130		11 08 58	−60 16.6	1	12.50		0.34		-0.26		1	B6	820
		Tr 18 - 148		11 08 58	−60 28.2	1	13.54		0.29		0.09		4		1720
97230	+3 02470			11 08 59	+02 48.5	1	8.62		0.25		0.14		2	A3	1375
		AAS41,43 F1 # 131		11 08 59	−60 17.5	1	11.66		1.39		1.29		2		820
	−60 02607	Tr 18 - 53		11 08 59	−60 25.1	4	11.73	.019	0.21	.017	0.00	.015	8	B9	411,820,1720,2041
97319	−60 02606	LSS 2217		11 08 59	−60 50.8	6	8.53	.039	0.22	.018	-0.67	.033	15	O8	2,6,191,540,976,2021
97138	+69 00602	HR 4340		11 09 00	+68 32.6	1	6.43		0.13		0.11		2	A3 V	1733
		AAS41,43 F1 # 134		11 09 00	−60 18.1	1	12.76		0.31		-0.13		2	B7	820
306075	−59 03132	Tr 18 - 81		11 09 00	−60 19.1	2	10.70	.005	0.16	.042	-0.19	.005	4	A0	2,820
		Tr 18 - 71		11 09 00	−60 21.6	2	12.79	.000	0.29	.005	0.15	.005	3		820,1720
97233	−14 03277	MCC 602		11 09 01	−14 42.7	4	9.05	.014	1.21	.013	1.11	.021	7	K4 V	803,1705,1774,3072
	−59 03133	Tr 18 - 55		11 09 01	−60 23.7	3	11.85	.009	0.14	.007	-0.27	.004	7	B6	820,1720,2041
97227	+19 02426			11 09 02	+18 57.9	1	8.24		0.46		-0.05		2	F2	1648
		Tr 18 - 52		11 09 02	−60 25.4	3	13.22	.020	0.52	.012	0.10	.024	6		411,820,1720
306078	−60 02608	Tr 18 - 135		11 09 02	−60 31.0	1	10.43		0.19		-0.58		2	B9	820
	−60 02609			11 09 02	−60 35.4	1	11.62		0.48		0.24		1	F0	820
	−59 03136			11 09 03	−60 04.9	1	11.62		0.20		-0.46		1	B0	820
	−59 03135	Tr 18 - 70		11 09 03	−60 22.4	4	11.73	.010	0.20	.025	-0.05	.004	7	B8	411,820,1720,2041
		Tr 18 - 146		11 09 04	−60 18.9	1	11.71		0.23		-0.65		1		820
306173	−59 03137	Tr 18 - 145		11 09 04	−60 19.1	1	11.06		0.19		-0.39		1	B9	820
	−60 02611	Tr 18 - 54		11 09 04	−60 24.3	4	11.09	.012	0.23	.022	-0.26	.013	10		411,820,1720,2041
		vdB 47 # b		11 09 04	−61 05.3	1	12.99		0.47		-0.41		3		434

HD	DM	Other Id	N Rem	α_{1950}	δ_{1950}	S	V	σ_V	B–V	σ_{B-V}	U–B	σ_{U-B}	n	Spectrum	References
	−59 03139			11 09 05	−60 05.5	1	11.19		0.14		0.08		1	A0	820
306076	−59 03138	Tr 18 - 90		11 09 05	−60 23.2	4	9.32	.016	1.02	.003	0.68	.004	9	G2	411,820,1720,2041
		Tr 18 - 126		11 09 05	−60 25.8	1	13.84		0.36		0.22		1		1720
97244	+15 02301	HR 4341		11 09 06	+14 40.3	1	6.30		0.21		0.08		2	A5 V	252
		Tr 18 - 80		11 09 06	−60 20.0	1	13.01		0.32		0.09		1		820
		L 395 - 13		11 09 07	−40 48.3	1	13.24		1.48		1.20		3		3078
	−59 03141			11 09 07	−60 13.4	2	11.05	.005	0.17	.019	-0.58	.049	3	B2	463,820
		Tr 18 - 82		11 09 07	−60 19.0	1	11.59		0.58		0.17		1		820
97313	−49 05915			11 09 08	−49 53.8	1	7.97		0.21		0.13		1	A2/3 V	1770
		AAS41,43 F1 # 151		11 09 08	−60 36.4	1	12.35		0.39		0.19		1	A2	820
		Cha F # 34		11 09 08	−77 16.6	1	14.32		1.81				2	K3 III	1633
		Tr 18 - 86		11 09 10	−60 18.7	1	12.36		0.19		-0.02		1		820
		Tr 18 - 11		11 09 10	−60 23.1	2	13.10	.195	0.24	.073	-0.27	.005	3		820,1720
	−60 02612			11 09 10	−60 32.6	1	11.40		0.26		-0.42		2	B5	820
306145	−59 03142	LSS 2218	⋆ AB	11 09 11	−59 38.8	1	9.55		0.37				4	B2:Ven	2021
306097	−60 02613	LSS 2219		11 09 11	−60 38.8	6	8.91	.020	0.66	.013	-0.28	.042	17	O9 III	2,6,540,820,976,2021
97277	−22 03095	HR 4343		11 09 12	−22 33.1	7	4.47	.009	0.03	.004	0.05	.004	19	A1 V	15,369,1075,1705,2012*
	−59 03143	Tr 18 - 83		11 09 12	−60 18.0	1	11.80		0.28		-0.26		2		820
		AAS41,43 F1 # 124		11 09 12	−60 23.	1	13.11		0.32		-0.37		1		820
		AAS41,43 F1 # 136		11 09 12	−60 23.	1	12.42		0.25		-0.38		3	B7	820
		AAS41,43 F1 # 164		11 09 12	−60 23.	1	13.15		0.69		0.19		1		820
		AAS41,43 F1 # 175		11 09 12	−60 23.	1	13.62		0.42		0.31		1		820
		AAS41,43 F1 # 178		11 09 12	−60 23.	1	12.95		0.39		-0.12		1		820
	−60 02617			11 09 12	−60 34.6	1	11.82		0.28		-0.51		1	B5	820
97352	−60 02615			11 09 12	−60 52.6	2	9.57	.024	0.16	.005	-0.63	.002	6	B3 V	540,976
		Car sq 2 # 70		11 09 12	−61 14.1	1	14.75		0.80		0.03		3		466
		Tr 18 - 17		11 09 13	−60 20.5	2	12.77	.040	0.23	.030	0.06	.050	2		820,1720
		AAS41,43 F1 # 158		11 09 13	−60 30.2	1	12.61		0.61		-0.12		1		820
	−59 03145	Tr 18 - 108		11 09 14	−60 14.7	1	11.67		0.45		0.10		1		820
		Tr 18 - 85		11 09 14	−60 17.7	1	12.94		0.33		0.18		1		820
		Tr 18 - 16		11 09 14	−60 20.7	1	13.11		1.24		1.48		1		1720
	−59 03146	Tr 18 - 10		11 09 14	−60 22.7	3	11.50	.022	0.17	.019	-0.16	.012	9	B7	411,820,1720
		Tr 18 - 9		11 09 14	−60 23.5	2	12.66	.005	0.32	.005	0.16	.005	4		820,1720
		AAS41,43 F1 # 165		11 09 15	−60 16.8	1	12.78		0.38		0.08		1		820
		Tr 18 - 42		11 09 15	−60 25.9	2	12.83	.023	1.32	.000	1.21	.145	4		820,1720
		WLS 1100 5 # 8		11 09 16	+04 50.0	1	12.51		0.46		0.00		1		1375
		G 56 - 22		11 09 16	+16 06.3	2	13.61	.018	0.87	.009	0.40	.027	5		1620,1658
		Tr 18 - 121		11 09 16	−60 26.9	1	13.77		1.18		1.60		3		1720
97368	−58 03253	LSS 2220		11 09 17	−59 04.5	1	8.52		0.19		-0.66		2	B1 II	401
	−59 03148	Tr 18 - 84		11 09 17	−60 17.9	1	11.70		0.15		-0.23		3		820
97474	−75 00715	Cha F # 35		11 09 17	−75 50.3	1	8.69		0.65		0.23		2	G3 V	1633
	−59 03150	Tr 18 - 107		11 09 18	−60 16.4	1	11.08		0.31		-0.35		2		820
	−59 03149	Tr 18 - 8		11 09 18	−60 23.2	3	11.79	.015	0.18	.015	-0.18	.007	6	B8	411,820,1720
		Tr 18 - 43		11 09 18	−60 26.9	2	12.54	.027	0.20	.014	0.00	.032	5	B9	820,1720
		vdB 47 # a		11 09 18	−61 02.0	1	12.07		0.43		-0.55		3		434
303905	−58 03254			11 09 19	−58 38.6	1	9.87		0.99		0.73		1	K0	549
97367	−58 03255			11 09 19	−58 57.3	1	8.14		1.25		1.28		1	K2/3 III	549
306174	−59 03151	Tr 18 - 12		11 09 19	−60 21.9	4	9.86	.015	0.12	.013	-0.43	.003	10	B4	411,820,1720,2041
		Tr 18 - 7		11 09 19	−60 23.3	2	13.02	.033	0.30	.023	0.16	.033	4		820,1720
	−60 02619	Tr 18 - 1		11 09 19	−60 24.4	4	9.09	.011	0.11	.009	-0.42	.004	14	B4	411,820,1720,2041
		Tr 18 - 18		11 09 20	−60 20.3	1	12.58		0.34		0.07		1		820
		Tr 18 - 41		11 09 20	−60 25.1	3	11.85	.007	1.31	.012	1.25	.027	5		411,820,1720
		Cha T # 47		11 09 20	−77 01.5	1	15.54		1.26		-0.29		1	K7:	437
		Cha T # 48	⋆ V	11 09 21	−76 18.2	1	15.48		0.93		-1.02		1	M1:	437
	+25 02351			11 09 22	+24 59.0	1	10.89		0.47		-0.02		2		1375
	−59 03152	Ho 11 - 2		11 09 22	−60 06.6	1	11.38		0.13		-0.41		1		411
		AAS41,43 F1 # 179		11 09 22	−60 10.7	1	11.72		0.69		0.24		1	G5	820
		Tr 18 - 15		11 09 22	−60 21.6	3	12.49	.009	0.25	.010	0.03	.007	11	B9	411,820,1720
97382	−62 01912	LSS 2221		11 09 22	−62 56.0	1	9.70		0.25		-0.37		2	B5/7	540
	−28 08704			11 09 23	−29 10.2	1	10.45		1.10		0.95		1	K5 V	3072
	−59 03155			11 09 23	−60 08.5	2	10.33	.019	0.08	.009	-0.68	.033	4	B0	463,820
		AAS41,43 F1 # 180		11 09 23	−60 17.2	1	12.61		0.22		-0.14		1	B8	820
		Tr 18 - 6		11 09 23	−60 23.1	3	13.04	.021	0.26	.013	-0.33	.019	9		411,820,1720
		Tr 18 - 44		11 09 23	−60 27.4	1	13.26		0.39		0.26		1		820
97380	−59 03157			11 09 24	−59 54.8	1	9.22		0.05		-0.04		3	A0	463
97381	−59 03156	Ho 11 - 1		11 09 24	−60 06.3	4	8.32	.023	0.06	.007	-0.71	.028	7	B2	6,411,463,820
		Tr 18 - 14		11 09 24	−60 21.8	3	12.48	.018	0.28	.012	0.02	.025	7	B8	411,820,1720
306175	−59 03154	Tr 18 - 13	⋆ AB	11 09 24	−60 22.3	5	9.93	.013	0.33	.021	0.15	.022	12	F0	2,411,820,1720,2041
		Tr 18 - 3		11 09 24	−60 23.6	3	12.07	.019	0.22	.015	-0.42	.006	8		411,820,1720
97309	−19 03198			11 09 25	−19 52.6	1	8.01		0.22		0.14		2	A1mA3/5-A7	1375
306176	−59 03158	Tr 18 - 2		11 09 25	−60 23.9	3	10.13	.010	0.11	.004	0.06	.016	9	B8	411,820,1720
306182		Tr 18 - 45		11 09 25	−60 27.6	3	11.93	.022	0.53	.032	0.05	.022	7		2,820,1720
		AAS41,43 F1 # 189		11 09 25	−60 31.1	1	13.07		0.51		-0.13		1		820
		AAS41,43 F1 # 188		11 09 25	−60 37.5	1	12.40		0.54		-0.23		1	B6	820
		vdB 47 # c		11 09 25	−61 05.5	1	13.79		0.97		-0.24		3		434
		Tr 18 - 4		11 09 26	−60 24.1	2	12.45	.024	0.47	.015	0.15	.049	3		411,820
		Tr 18 - 39		11 09 26	−60 25.4	2	13.35	.044	0.39	.039	0.21	.039	3		820,1720
		Tr 18 - 46		11 09 26	−60 28.0	2	11.83	.019	1.33	.000	1.13	.010	3		820,1720
		Ho 11 - 3		11 09 27	−60 06.7	1	12.28		0.12		-0.46		1		411
	−59 03159			11 09 27	−60 13.0	2	10.44	.005	0.09	.049	-0.61	.054	3	B1	463,820

Table 1   595

HD	DM	Other Id	N	Rem	α_1950	δ_1950	S	V	σ_V	B–V	σ_B–V	U–B	σ_U–B	n	Spectrum	References
		Tr 18 - 106			11 09 27	−60 22.3	1	13.56		0.32		0.26		2		1720
		Tr 18 - 5			11 09 27	−60 24.2	2	12.41	.044	1.91	.054	1.70		3		411,820
		Tr 18 - 40			11 09 27	−60 25.9	2	12.07	.049	1.89	.000	1.90	.078	3		820,1720
		AAS41,43 F1 # 199			11 09 28	−60 14.5	1	12.27		0.21		-0.48		1	B4	820
	−59 03162				11 09 28	−60 14.5	1	12.04		0.27		0.21		1	B8	820
		Tr 18 - 105			11 09 28	−60 15.6	1	13.59		0.44		0.11		1		820
	−59 03161	Tr 18 - 19			11 09 28	−60 19.2	1	10.40		1.37		1.43		1		820
		Tr 18 - 25			11 09 28	−60 21.1	1	12.89		0.47		0.09		1		820
	−59 03163	Ho 11 - 4			11 09 29	−60 06.7	2	11.64	.010	0.69	.005	0.24	.035	2		411,820
		Tr 18 - 117			11 09 29	−60 23.5	1	13.56		0.27		0.33		1		820
		AAS41,43 F1 # 205			11 09 29	−60 33.8	1	12.16		0.29		-0.29		1	B7	820
97400	−59 03164	LSS 2222		⋆ A	11 09 31	−60 10.3	4	7.79	.014	0.09	.010	-0.58	.023	9		2,401,463,820
306189	−60 02622				11 09 31	−60 40.6	1	11.25		0.11		-0.09		1	B9	820
		LP 612 - 15			11 09 32	−02 16.9	1	11.45		0.96		0.73		1		1696
97396	−58 03261				11 09 32	−59 07.2	1	8.04		0.01		-0.04		2	A0 III/IV	401
97398	−59 03165				11 09 32	−60 02.3	2	6.69	.019	-0.05	.005	-0.30	.015	6	B8 V	401,463
		Ho 11 - 5			11 09 32	−60 06.8	2	11.61	.010	1.13	.005	0.89	.005	2		411,820
97399	−59 03166	LSS 2224		⋆ BC	11 09 32	−60 09.4	4	8.43	.016	0.04	.006	-0.76	.031	9		2,401,463,820
		Tr 18 - 104			11 09 32	−60 17.5	1	12.85		0.21		-0.32		1		820
		Tr 18 - 24			11 09 32	−60 20.1	1	13.38		0.34		0.21		1		820
97343	−25 08519				11 09 33	−25 51.8	4	7.05	.003	0.76	.004	0.37	.006	16	G8/K0 V	78,1075,1311,1775
		Tr 18 - 26			11 09 33	−60 20.9	1	12.66		0.68		-0.12		1		820
306181	−60 02623	Tr 18 - 37			11 09 33	−60 26.1	5	9.58	.016	0.12	.016	-0.71	.026	13	B0	2,411,820,1720,2041
306190	−60 02624				11 09 33	−60 45.9	1	9.68		0.16		-0.63		3	B5	2
		AAS41,43 F1 # 214			11 09 34	−60 14.7	1	13.69		0.48		0.20		1		820
97342	−20 03367				11 09 35	−21 03.9	1	8.29		0.63		0.16		1	G3/5 V	78
	−59 03167	Ho 11 - 6			11 09 35	−60 06.6	2	11.70	.050	0.25	.030	0.04	.050	2		411,820
		AAS41,43 F1 # 215			11 09 35	−60 16.3	1	13.24		0.31		-0.16		1		820
		AAS41,43 F1 # 218			11 09 35	−60 30.2	1	10.24		1.51		1.33		1	G5	820
306188	−60 02625				11 09 35	−60 40.2	1	11.01		0.22		0.00		1	A0	820
		Car sq 2 # 71			11 09 35	−60 48.0	1	11.81		0.37		-0.50		3		466
	−59 03168	Tr 18 - 103			11 09 36	−60 18.0	2	11.08	.055	0.23	.010	-0.44		2		820,2041
306172	−59 03168	Tr 18 - 96			11 09 36	−60 18.0	3	9.39	.011	1.24	.007	1.19	.008	9	K2	820,1720,2041
		Tr 18 - 28			11 09 36	−60 21.9	1	13.31		0.45		0.55		2		1720
		Tr 18 - 32			11 09 36	−60 23.7	4	11.29	.026	1.00	.014	0.72	.010	6		411,820,1720,2041
306177	−60 02627	Tr 18 - 34			11 09 36	−60 24.3	5	10.79	.017	0.11	.019	-0.40	.000	12	B6	2,411,820,1720,2041
97340	+0 02758				11 09 37	−00 08.3	1	8.14		0.23		0.10		2	A5	1776
		AAS41,43 F1 # 223			11 09 38	−60 32.1	1	12.14		0.44		-0.28		1		820
97472	−70 01336	HR 4349			11 09 38	−71 09.9	2	6.34	.005	1.37	.000	1.54	.000	7	K2/3 III	15,1075
		Tr 18 - 33			11 09 39	−60 24.1	2	13.46	.049	0.55	.005	0.05	.010	3		820,1720
		AAS41,43 F1 # 222			11 09 39	−60 29.0	1	13.25		0.53		0.30		1		820
		Cha F # 36			11 09 39	−77 15.2	1	13.76		2.22		2.40		2	K0 III	1633
97339	+13 02369				11 09 41	+13 29.6	1	8.23		1.30				2	K2	882
	+0 02759				11 09 41	−00 09.8	1	9.57		0.33		0.04		2	A5	1776
	−59 03169				11 09 41	−60 11.1	1	11.18		0.35		0.17		1	A1	820
97434	−60 02629	Tr 18 - 36		⋆ AB	11 09 41	−60 25.6	7	8.08	.009	0.16	.011	-0.79	.017	19	O9	2,6,411,463,820,1720,2041
		AAS41,43 F1 # 225			11 09 42	−60 08.0	1	12.31		0.27		-0.38		1	B6	820
		Tr 18 - 27			11 09 42	−60 21.5	2	12.74	.000	0.25	.023	0.10	.037	4		820,1720,2041
		Tr 18 - 152			11 09 42	−60 28.3	2	11.93	.014	0.58	.014	0.06	.019	4		820,1720
		Tr 18 - 29			11 09 43	−60 22.3	2	12.11	.032	0.29	.014	-0.45	.005	5		820,1720
		Leo II sq # 26			11 09 44	+22 14.8	1	13.25		0.61				1		988
		Leo II sq # 1			11 09 44	+22 23.3	1	11.87		0.54				4		988
		Steph 926			11 09 44	+30 31.2	1	11.27		1.12		1.02		1	K7	1746
		AAS41,43 F1 # 234			11 09 44	−60 15.7	1	12.82		1.38		1.10		1		820
		AAS41,43 F1 # 233			11 09 44	−60 33.6	1	12.67		0.53		0.10		1	B7	820
306153	−59 03171				11 09 45	−59 48.1	1	10.23		0.16		-0.47		3	B5	2
		Tr 18 - 102			11 09 45	−60 15.3	2	13.37	.080	0.43	.055	0.12	.195	2		820,820
		Tr 18 - 35			11 09 45	−60 24.7	1	13.12		0.65		0.45		2		820
		AAS41,43 F1 # 239			11 09 45	−60 31.9	1	12.50		0.60		-0.06		1		820
		Leo II sq # 3			11 09 46	+22 17.4	1	13.34		0.67				3		988
		Leo II sq # 2			11 09 46	+22 23.4	1	14.15		0.50				3		988
	−35 07049				11 09 46	−35 31.9	1	9.68		0.81		0.35		1	G8	78
	−59 03173				11 09 46	−60 10.9	1	11.71		0.21		-0.20		1	B8	820
	−59 03174	Tr 18 - 101			11 09 47	−60 15.2	1	12.24		0.33		0.19		1		820
		Tr 18 - 147			11 09 47	−60 20.6	2	13.55	.034	0.34	.019	0.23	.039	3		820,1720
		Car sq 2 # 72			11 09 47	−60 44.1	1	11.40		0.27		-0.59		3		466
	−60 02633				11 09 48	−60 32.5	1	10.42		1.18		1.10		1	G8	820
		WLS 1100 5 # 5			11 09 49	+06 24.3	1	11.12		0.52		-0.02		2		1375
97334	+36 02162	HR 4345		⋆ A	11 09 49	+36 05.3	5	6.40	.009	0.60	.010	0.12	.006	16	G0 V	70,101,196,245,1067
97333	+41 02168				11 09 49	+40 48.4	1	8.46		0.08		0.06		2	A2 V	272
		AAS41,43 F1 # 243			11 09 49	−60 23.1	1	13.44		0.50		0.26		1		820
97393	−31 08847	HR 4346			11 09 50	−32 09.7	3	6.35	.016	1.69	.046	1.90	.075	8	M1 III	1034,2032,3055
97431	−51 05435				11 09 50	−52 13.9	1	9.22		0.11		0.10		1	A0 V	1770
		AAS41,43 F1 # 245			11 09 50	−60 11.0	1	12.28		1.07		0.81		1	A7	820
		AAS41,43 F1 # 244			11 09 50	−60 21.6	1	13.61		0.47		0.06		1		820
97302	+55 01446	HR 4344			11 09 51	+55 10.0	3	6.65	.008	0.08	.000	0.08	.005	8	A4 V	1501,1502,1733
	−36 07031				11 09 51	−37 18.4	1	9.49		0.83		0.50		1	K0	78
97413	−45 06771				11 09 51	−45 59.7	1	6.28		0.16		0.04		2	A1 V	1770
306191	−60 02634				11 09 51	−60 50.0	1	10.65		0.26		-0.59		3	B8	466
		Leo II sq # 54			11 09 52	+22 15.2	1	15.37		0.97				1		988

HD	DM	Other Id	N	Rem	α_{1950}	δ_{1950}	S	V	σ_V	B–V	σ_{B-V}	U–B	σ_{U-B}	n	Spectrum	References
97450	−56 04292				11 09 52	−56 53.6	1	8.69		−0.06		−0.44		2	B5 V	401
		AAS41,43 F1 # 246			11 09 52	−60 13.2	1	12.77		0.15		0.03		1	B9	820
		AAS41,43 F1 # 249			11 09 54	−60 15.0	1	12.67		0.65		0.12		1	F4	820
		Tr 18 - 31			11 09 54	−60 24.7	1	12.80		0.25		0.17		1		820
		AAS41,43 F1 # 248			11 09 54	−60 29.2	1	12.67		0.22		0.00		1	B9	820
		Tr 18 - 100			11 09 55	−60 15.0	1	12.50		0.21		0.15		1	F4	820
	−60 02636				11 09 55	−60 28.7	1	10.97		1.33		1.24		2	K0	820
		Tr 18 - 156			11 09 56	−60 19.0	1	12.06		1.60		2.00		1		1720
97484	−60 02638	EM Car	4		11 09 56	−60 49.4	4	8.40	.018	0.31	.012	−0.65	.029	12	O8/9	6,540,976,2021
		Leo II sq # 13			11 09 57	+22 17.2	1	13.14		0.73				1		988
97471	−58 03268	LSS 2231	2		11 09 57	−58 31.9	2	9.30	.000	−0.01	.000	−0.87		6	B0 V	6,1021
		Tr 18 - 99			11 09 57	−60 17.1	1	12.41		0.29		−0.40		1		820
	+24 02328				11 09 58	+23 41.9	1	9.75		0.86		0.55		3		1723
97354	+47 01853				11 09 58	+46 56.8	1	9.14		0.52		−0.04		2	G0	1601
306180	−60 02640	Tr 18 - 151			11 09 58	−60 26.9	1	10.33		0.28		−0.55		4	B0	1720
		Tr 18 - 30			11 09 59	−60 23.3	2	12.19	.025	1.16	.005	1.12	.040	2		820,1720
		AAS41,43 F1 # 254			11 09 59	−60 38.1	1	12.28		0.55		−0.36		1		820
		Steph 928			11 10 00	+19 12.5	1	10.77		1.50		1.19		1	M1	1746
		Leo II sq # 14			11 10 00	+22 24.6	1	11.17		0.60				2		988
	−59 03179	Tr 18 - 98			11 10 00	−60 21.2	2	11.21	.023	0.18	.027	−0.59	.009	5	B0	820,1720
306180	−60 02640	EN Car			11 10 00	−60 27.1	2	10.30	.005	0.29	.014	−0.51	.033	4		2,820
		AAS41,43 F1 # 256			11 10 00	−60 32.2	1	12.32		1.21		1.26		1		820
97371	+36 02164	IDS11071N3621		B	11 10 01	+36 06.1	1	7.21		1.01		0.88		1	G8 III	245
97411	−17 03321	HR 4347		★AB	11 10 01	−18 13.6	4	6.12	.004	0.00	.017	−0.04		13	A0 V	15,176,252,2012
		Tr 18 - 153			11 10 01	−60 21.6	2	11.85	.026	0.16	.044	−0.54	.004	6	B2	820,1720
306206	−60 02641	Car sq 2 # 75			11 10 01	−61 00.3	2	10.95	.064	0.37	.005	−0.70	.015	5	B0 V	92,466
		Tr 18 - 157			11 10 02	−60 24.0	2	12.59	.045	0.37	.015	−0.06	.000	2	B8	820,1720
		AAS41,43 F1 # 259			11 10 02	−60 28.7	1	12.98		0.52		0.03		1		820
		AAS41,43 F1 # 260			11 10 02	−60 33.3	1	12.83		0.38		0.11		1	B8	820
		Car sq 2 # 74			11 10 02	−60 57.9	1	11.60		0.18		−0.63		3		466
97499	−60 02643	LSS 2233	2		11 10 02	−61 02.4	2	9.22	.005	0.06	.005	−0.78	.000	6	B1/2 IV/V	6,92
		AAS41,43 F1 # 261			11 10 03	−60 10.7	1	12.22		0.11		−0.52		1	B2	820
		Ho 12 - 5			11 10 03	−60 30.8	1	11.47		1.29		1.10		1		820
		AAS41,43 F1 # 263			11 10 03	−60 35.1	1	12.84		1.00		0.73		1		820
97485	−61 02103	IT Car			11 10 03	−61 29.0	1	7.93		0.93		0.66		1	F8 Iab/b	688
	+36 02165	G 119 - 64			11 10 04	+36 00.5	5	9.76	.004	0.43	.003	−0.20	.008	12	G0	979,1064,1620,1658,8112
		Leo II sq # 5			11 10 05	+22 14.2	1	12.38		0.65				2		988
		Leo II sq # 4			11 10 05	+22 17.1	1	15.25		0.95				3		988
		AAS41,43 F1 # 264			11 10 05	−60 10.9	1	12.78		0.20		−0.42		1	B2	820
		AAS41,43 F1 # 265			11 10 05	−60 18.0	1	13.21		0.34		0.21		1	A0	820
		Tr 18 - 155			11 10 05	−60 31.7	1	12.96		1.69		0.99		2		1720
		Cha T # 49		★V	11 10 05	−76 03.9	1	15.28		1.39		−0.19		1	M2	437
97428	−20 03374	HR 4348			11 10 06	−21 28.6	1	6.40		1.38				4	K3 III	2006
		AAS41,43 F1 # 269			11 10 06	−60 15.1	1	12.23		0.30		−0.25		1	B8	820
		AAS41,43 F1 # 268			11 10 06	−60 22.1	1	12.91		0.27		−0.30		1	B6	820
306183	−60 02647	Ho 12 - 1			11 10 06	−60 29.4	2	10.33	.009	0.19	.019	−0.71	.014	4		466,820
	−59 03182				11 10 07	−60 11.2	1	10.41		0.92		0.64		1	A7	820
	−60 02646	Ho 12 - 3			11 10 07	−60 30.1	1	10.80		0.32		−0.55		1		820
		LP 732 - 97			11 10 08	−15 12.8	1	9.77		0.61		0.05		2		1696
97498	−58 03279				11 10 08	−58 41.0	1	9.04		−0.05		−0.53		2	B3/5 III/V	401
	−59 03184				11 10 08	−60 10.2	1	9.85		1.86		1.97		1	K2	820
	−59 03183				11 10 08	−60 12.3	1	11.54		0.28		−0.56		1		820
	−59 03185				11 10 08	−60 14.0	1	11.36		0.38		−0.43		1	B0	820
		AAS41,43 F1 # 270			11 10 08	−60 19.3	1	12.09		0.57		0.14		1	F2	820
		Leo II sq # 15			11 10 09	+22 27.7	1	11.90		0.52				2		988
		Ho 12 - 4			11 10 09	−60 30.6	1	12.07		0.41		−0.20		1		820
		AAS41,43 F1 # 272			11 10 09	−60 30.9	1	13.14		0.34		0.12		1		820
	−59 03186				11 10 10	−60 17.3	1	11.58		1.06		0.88		1	F0	820
	−60 02648	Ho 12 - 2			11 10 10	−60 29.1	1	10.06		0.15		0.06		2		820
		AAS41,43 F1 # 278			11 10 11	−60 17.4	1	12.22		0.14		−0.13		1	B9	820
97370	+54 01428				11 10 12	+53 38.6	1	8.15		0.96		0.72		2	G5	1566
		Leo II sq # 11			11 10 13	+22 15.3	1	14.31		0.72				3		988
	−59 03187				11 10 13	−60 14.6	1	11.66		0.28		−0.45		1	B3	820
		AAS41,43 F1 # 280			11 10 13	−60 20.7	1	12.94		0.64		0.09		1		820
97522	−64 01641	LSS 2235	5		11 10 13	−64 56.8	5	7.74	.015	0.30	.008	−0.56	.030	11	B0.5II	6,191,540,976,8100
97535	−70 01338	IDS11084S7040	AB		11 10 13	−70 56.7	1	7.02		0.01		−0.36		2	B7 Vn	401
97496	−51 05447				11 10 14	−51 27.2	1	8.49		0.02		−0.01		1	A0 V	1770
306159	−59 03188				11 10 14	−60 06.3	1	8.40		2.07		2.03		1	M0	463
		AAS41,43 F1 # 282			11 10 14	−60 17.9	1	13.47		0.30		0.25		1		820
306178	−59 03189	Tr 18 - 154			11 10 15	−60 22.7	2	10.61	.015	0.21	.024	0.15	.010	3	B9	820,1720
306179	−60 02652				11 10 15	−60 27.5	2	10.39	.009	0.22	.009	−0.62	.028	4		2,820
97495	−48 06263	HR 4350			11 10 16	−48 49.8	2	5.35	.005	0.18	.000			7	A3 IV/V	15,2012
303922	−58 03282				11 10 16	−58 56.4	1	10.34		0.53				4	F7 V	2021
		AAS41,43 F1 # 285			11 10 16	−60 20.5	1	11.98		0.46		0.12		1	F2	820
		AAS41,43 F1 # 286			11 10 18	−60 22.7	1	11.76		0.51		0.10		1		820
	−60 02656	NGC 3590 - 69			11 10 18	−60 32.7	1	10.68		1.32		1.14		1		820
	−60 02655	NGC 3590 - 68			11 10 18	−60 33.2	1	10.82		0.41		0.08		1		820
		AAS41,43 F1 # 289			11 10 19	−60 25.3	1	13.33		0.37		−0.08		1		820
	−60 02657	LSS 2236			11 10 19	−60 36.7	1	10.19		0.29		−0.50		3		466
		AAS41,43 F1 # 292			11 10 20	−60 12.5	1	12.01		0.29		−0.43		2	B5	820

Table 1   597

HD	DM	Other Id	N	Rem	α_{1950}	δ_{1950}	S	V	σ_V	B–V	σ_{B-V}	U–B	σ_{U-B}	n	Spectrum	References
		AAS41,43 F1 # 290			11 10 20	−60 24.8	1	12.95		0.67		0.19		1		820
306187	−60 02658	AAS41,43 F1# 292			11 10 22	−60 31.4	2	10.19	.009	0.32	.019	−0.44	.028	4	B5	2,820
97521	−58 03286	St 13 - 19			11 10 23	−58 34.8	3	9.56	.034	0.03	.008	−0.78	.005	8	B2 V	6,411,2021
306184	−60 02658	NGC 3590 - 70		A	11 10 23	−60 31.3	1	10.26		0.17		−0.60		2	B1	820
306184	−60 02658	NGC 3590 - 70		B	11 10 23	−60 31.3	1	10.70		0.28		−0.54		3	B1	820
		AAS41,43 F1 # 295			11 10 25	−60 27.0	1	12.52		0.46		0.10		1		820
97440	+27 01992				11 10 26	+27 19.5	1	7.95		0.33		0.03		2	F0	1375
97533	−57 04420				11 10 26	−58 22.3	2	8.30	.027	0.03	.005	−0.84		5	B3/5 V	6,2021
	−71 01212	LSS 2241			11 10 26	−71 42.5	1	10.07		0.08		−0.78		2	B1 Ia	1737
		Leo II sq # 18			11 10 27	+22 36.8	1	12.88		0.98				2		988
97534	−59 03190	HR 4352, LSS 2237	⋆	A	11 10 27	−60 02.7	9	4.60	.007	0.54	.020	0.06	.054	31	A6 Iae	15,463,1018,1075,2001*
		AAS41,43 F1 # 296			11 10 27	−60 11.6	1	11.71		1.56		1.44		1		820
		AAS41,43 F1 # 297			11 10 27	−60 30.3	1	12.01		0.58		−0.17		1		820
97507	−33 07566				11 10 29	−34 08.0	1	8.59		0.62				4	G3 V	2012
		AAS41,43 F1 # 298			11 10 29	−60 22.2	1	13.22		0.34		0.26		1	A0	820
		Leo II sq # 6			11 10 30	+22 13.0	1	15.26		0.73				2		988
	−59 03192				11 10 31	−60 18.6	1	11.74		0.23		−0.35		1	B6	820
97420	+59 01355				11 10 34	+59 24.7	1	9.07		0.47				5	F6 V	1379
		LP 732 - 48			11 10 34	−12 31.6	1	12.51		0.41		−0.25		3		1696
97557	−59 03193	LSS 2238			11 10 34	−59 24.3	5	7.23	.007	0.00	.008	−0.69	.024	16	B2 III	6,158,191,540,976
		G 10 - 6			11 10 37	+00 30.7	2	10.28	.040	1.46	.000	1.24		2	M0	1620,1746
97488	+34 02206	IDS11079N3359	AB		11 10 37	+33 43.0	1	6.64		1.01		0.78		2	K0	1625
97503	+5 02463	G 10 - 7			11 10 39	+04 45.3	6	8.70	.010	1.18	.005	1.12	.014	28	K5 V	333,989,1003,1620,1705*
97438	+60 01314				11 10 39	+60 01.2	1	9.68		0.32				3	F0 III	1379
97581	−60 02665	NGC 3590 - 40			11 10 39	−60 27.9	5	8.84	.013	0.25	.013	−0.59	.020	13	B1 Iab	2,6,463,820,2021
97550	−49 05937	HR 4353			11 10 40	−49 27.8	3	6.10	.004	1.05	.005			11	G8 II/III	15,1075,2006
97551	−50 05793				11 10 40	−50 32.7	1	9.92		0.07		0.05		1	F7/8 V	1770
97455	+56 01508	IDS11078N5558	AB		11 10 41	+55 41.3	1	7.46		0.44		0.01		2	F2	1502
98042	−58 03389	St 13 - 13			11 10 41	−58 40.2	2	10.64	.025	0.26	.005	0.23	.045	2	A1 IV/V	287,411
	−60 02667				11 10 41	−60 26.5	1	11.15		0.23		−0.51		1	B3	820
	−60 02666	NGC 3590 - 31			11 10 41	−60 30.0	2	10.39	.000	0.27	.009	−0.54	.019	4		463,820
		NGC 3590 - 29			11 10 41	−60 31.3	1	13.53		0.55		0.83		2		463
		NGC 3590 - 64			11 10 41	−60 34.3	1	12.64		0.66		0.04		1		820
97583	−63 01860	HR 4355	⋆	A	11 10 41	−63 53.9	3	5.22	.015	−0.09	.016	−0.26	.005	7	B8 V	6,1771,2032
97502	+11 02333				11 10 42	+10 43.2	1	7.45		1.22		1.22		2	K2	1648
		NGC 3590 - 30			11 10 42	−60 30.6	1	13.59		0.45		0.07		2		463
		NGC 3590 - 65			11 10 42	−60 33.4	1	11.70		0.53		−0.30		3		820
97517	−17 03326				11 10 43	−17 53.4	1	8.11		0.61				4	G3 V	2033
		AAS41,43 F1 # 306			11 10 43	−60 17.5	1	12.66		0.38		0.13		1		820
		AAS41,43 F1 # 307			11 10 43	−60 27.4	1	12.80		0.32		−0.08		1		820
304026	−58 03391	St 13 - 9			11 10 44	−58 34.9	1	11.36		0.07		−0.20		1	B9	411
		St 13 - 17			11 10 44	−58 38.6	1	12.03		0.30		0.09		1		411
		NGC 3590 - 32			11 10 44	−60 30.2	1	12.79		0.35		−0.28		2		463
		NGC 3590 - 36			11 10 45	−60 29.1	1	12.38		0.34		−0.22		1		820
97528	−25 08531	TT Hya			11 10 46	−26 11.6	1	7.26		0.17		0.02		1	A1 III	1588
		NGC 3590 - 26			11 10 46	−60 31.4	1	13.39		0.60		0.40		3		463
97545	−32 07942				11 10 47	−33 00.2	1	9.12		0.70		0.24		1	G3/5 V	78
	−60 02669				11 10 47	−60 25.3	1	11.10		0.25		−0.48		3	B3	820
306165	−59 03197	LSS 2243			11 10 48	−60 09.7	2	9.82	.000	0.07	.005	−0.62	.000	5	B3	2,463
	−60 02670	NGC 3590 - 41			11 10 48	−60 28.5	2	11.43	.028	0.32	.042	0.20	.019	4		463,820
		NGC 3590 - 4			11 10 48	−60 30.7	2	12.79	.030	0.36	.000	−0.35	.035	4		411,463
		NGC 3590 - 24			11 10 48	−60 31.2	1	13.92		0.46		0.26		2		411
97617	−66 01550				11 10 48	−66 49.8	1	7.76		0.07		−0.34		2	B8/9 III	401
97580	−56 04307				11 10 49	−56 45.8	1	8.77		−0.01		−0.34		2	B8 V	401
303910	−58 03294	St 13 - 7			11 10 49	−58 37.7	2	10.83	.024	0.02	.000	−0.73	.019	3	B8	287,411
306158	−59 03199				11 10 49	−60 01.5	1	10.74		0.06		−0.17		3	B9	2
		NGC 3590 - 42			11 10 49	−60 28.6	2	11.46	.070	0.33	.033	0.14	.009	4		463,820
		NGC 3590 - 22			11 10 49	−60 31.2	2	12.24	.015	0.35	.005	−0.27	.020	4		411,463
306185	−60 02671	NGC 3590 - 23			11 10 50	−60 31.0	3	10.27	.015	0.29	.009	−0.50	.025	8		411,463,820
		NGC 3590 - 21			11 10 50	−60 31.2	2	12.55	.050	0.36	.000	−0.21	.025	4		411,463
		HBC 588	V		11 10 50	−76 20.8	2	11.55	.050	1.12	.005	0.75	.015	2	K0	1763,5005
		St 13 - 15			11 10 51	−58 36.8	1	12.20		0.36		0.06		1		411
	−58 03397	St 13 - 8			11 10 51	−58 37.7	2	11.17	.058	−0.03	.010	−0.68	.005	3		287,411
		NGC 3590 - 61			11 10 51	−60 33.3	1	12.63		0.66		0.03		1		820
		NGC 3590 - 7			11 10 52	−60 30.1	1	14.65		0.82		0.64		2		411
		NGC 3590 - 8			11 10 52	−60 30.3	2	14.16	.010	0.39	.020	0.10	.080	2		411,463
		NGC 3590 - 3			11 10 52	−60 31.0	3	11.94	.031	0.32	.015	−0.36	.007	7		411,463,820
		NGC 3590 - 20			11 10 52	−60 31.4	2	13.00	.060	0.42	.015	0.03	.020	4		411,463
304006	−58 03398	St 13 - 6			11 10 53	−58 35.2	1	10.62		−0.01		−0.78		2	B9	411
97597	−58 03299	St 13 - 1			11 10 53	−58 37.7	3	8.77	.007	−0.08	.015	−0.94	.021	12	B2 II/III	287,411,1732
	−60 02673	NGC 3590 - 1			11 10 53	−60 30.8	3	10.56	.022	0.30	.017	−0.51	.026	6		411,463,820
97501	+41 02170	HR 4351	⋆	AB	11 10 54	+41 21.6	1	6.33		1.15		1.12		1	K2 III	15
97576	−43 06872	HR 4354			11 10 54	−44 06.0	4	5.79	.009	1.66	.008	2.08		16	K5/M0 III	15,1075,2029,3005
97578	−47 06614				11 10 54	−47 56.9	1	10.17		1.46		1.11		1	R5	864
		St 13 - 11			11 10 54	−58 36.0	1	11.46		1.18		1.06		1		411
	−59 03203	LSS 2246			11 10 54	−60 17.3	1	11.18		0.23		−0.61		1	B2	820
		AAS41,43 F1 # 318			11 10 54	−60 22.1	1	12.15		0.19		−0.49		1	B4	820
	−60 02674	NGC 3590 - 19			11 10 54	−60 31.5	3	11.49	.086	0.30	.019	−0.41	.024	9		411,463,820
		HRC 247, CV Cha			11 10 54	−76 28.0	1	10.90		1.07		0.08		1	G8 V	1763
		Cha T # 52	V		11 10 54	−76 28.0	2	10.95	.025	1.04	.000	0.14	.195	2	K0	437,825

HD	DM	Other Id	N	Rem	α_{1950}	δ_{1950}	S	V	σ_V	B–V	σ_{B-V}	U–B	σ_{U-B}	n	Spectrum	References
		St 13 - 14			11 10 55	−58 36.5	1	12.12		0.05		-0.45		2		411
		St 13 - 18			11 10 55	−58 38.5	1	12.21		1.12		1.05		1		411
		NGC 3590 - 2			11 10 55	−60 30.6	2	13.28	.050	0.35	.010	-0.10	.085	4		411,463
306186	−60 02676	NGC 3590 - 18			11 10 55	−60 31.0	3	10.67	.057	0.29	.015	-0.43	.021	6	B8	411,463,820
		NGC 3590 - 80			11 10 55	−60 32.0	2	13.54	.015	0.37	.040	-0.14	.000	5		411,463
		Leo II sq # 17			11 10 56	+22 37.1	1	14.75		1.12				2		988
97616	−58 03301				11 10 56	−58 49.5	1	10.10		0.08		0.04		1	A2 V	1770
		NGC 3590 - 60			11 10 56	−60 33.6	1	12.81		0.46		-0.24		1		820
	−58 03302	St 13 - 12			11 10 57	−58 35.7	1	11.41		0.00		-0.59		1		411
		NGC 3590 - 46			11 10 57	−60 29.5	2	12.06	.014	0.25	.023	0.06	.009	4		463,820
		NGC 3590 - 10			11 10 57	−60 30.7	2	12.97	.075	0.37	.000	-0.09	.165	4		411,463
		Cha T # 53			11 10 57	−76 28.0	1	14.45		1.30		0.05		1		437
97513	+50 01803				11 10 58	+50 07.7	1	7.36		1.28		1.25		2	K0	1566
97629	−60 02677	LSS 2250			11 10 58	−60 50.4	2	9.53	.011	0.10	.024	-0.65	.032	5	B3/5 Ib/II	2,540
		Leo II sq # 16			11 10 59	+22 36.8	1	14.74		0.72				2		988
		NGC 3590 - 79			11 10 59	−60 31.1	2	12.36	.119	0.14	.129	-0.41	.119	5		411,463
97592	−49 05943	IDS11087S4935		A	11 11 00	−49 51.0	1	6.88		0.43		0.21		1	A1/2 V	1770
		AJ74,1125 T3# 4			11 11 00	−60 14.9	2	12.49	.030	1.15	.010	0.70	.090	2		2,820
		NGC 3590 - 11			11 11 00	−60 31.0	3	12.34	.082	0.37	.027	-0.28	.017	8		411,463,820
		AJ74,1125 T3# 3			11 11 01	−60 14.5	2	11.93	.010	0.33	.015	-0.38	.040	2	B3	2,820
306171	−59 03204	LSS 2251			11 11 01	−60 22.0	5	9.65	.013	0.15	.014	-0.69	.022	13	B0.5II-III	2,463,540,820,976
		AAS41,43 F1 # 324			11 11 01	−60 26.4	1	11.34		1.09		0.81		1		820
	−58 03304	St 13 - 16			11 11 02	−58 35.9	1	11.93		0.07		-0.32		1		411
303909	−58 03303	St 13 - 4			11 11 02	−58 36.7	2	9.84	.015	-0.01	.024	-0.71	.005	3	B5	287,411
		NGC 3590 - 82			11 11 02	−60 30.8	1	14.59		0.84		0.63		2		411
		NGC 3590 - 12			11 11 02	−60 31.0	1	14.43		0.61				1		463
		NGC 3590 - 5			11 11 03	−60 32.3	2	13.41	.020	0.34	.020	-0.13	.060	4		411,463
97561	+20 02572	IDS11084N2041		AB	11 11 04	+20 24.2	3	6.92	.005	0.75	.000	0.33	.000	7	G5 III	1003,1080,3077
		St 13 - 22			11 11 04	−58 37.1	1	13.62		0.30		0.13		2		411
		NGC 3590 - 16			11 11 04	−60 31.6	1	14.27		0.49		0.09		2		463
	−60 02680	NGC 3590 - 50			11 11 04	−60 32.3	3	10.98	.019	0.31	.017	-0.50	.027	7		411,463,820
		AAS41,43 F1 # 328			11 11 05	−60 17.2	1	11.71		1.14		0.84		1		820
		NGC 3590 - 13			11 11 05	−60 31.0	1	13.87		0.45		-0.01		3		463
		NGC 3590 - 83			11 11 05	−60 31.2	1	13.72		0.49		0.29		2		411
		LP 672 - 59			11 11 06	−04 59.2	1	13.51		0.73		0.01		2		1696
	−58 03308	St 13 - 10			11 11 07	−58 33.4	1	11.45		0.00		-0.69		1		411
		St 13 - 23			11 11 07	−58 36.0	1	13.44		0.52		0.07		2		411
303908	−58 03310	St 13 - 5			11 11 08	−58 35.4	1	10.68		0.04		-0.05		2	A3	411
		NGC 3590 - 81			11 11 08	−60 32.6	1	14.14		0.57		0.76		2		463
		AAS41,43 F1 # 330			11 11 09	−60 22.6	1	12.07		0.36		0.20		1	B8	820
		AAS41,43 F1 # 329			11 11 09	−60 23.9	1	11.85		0.73		0.17		1		820
		AAS41,43 F1 # 332			11 11 10	−60 26.4	1	11.82		1.08		0.81		1		820
		NGC 3590 - 49			11 11 10	−60 30.4	2	12.41	.049	0.50	.029	0.07	.005	3		463,820
97560	+40 02408				11 11 11	+40 15.1	1	7.92		0.66		0.13		2	G0	1601
306167	−59 03210				11 11 11	−60 16.0	1	10.83		0.20		-0.14		1	B9	820
97585	+0 02761	HR 4356			11 11 12	+00 12.2	2	5.41	.005	-0.03	.000	-0.05	.000	7	A0 V	15,1417
		AAS41,43 F1 # 334			11 11 14	−60 20.6	1	12.51		0.64		0.11		1		820
		St 13 - 21			11 11 15	−58 36.9	1	12.40		0.44		0.40		1		411
	−58 03312	St 13 - 20			11 11 15	−58 37.7	1	11.76		0.03		-0.55		2		411
	−60 02686				11 11 15	−60 27.7	1	11.74		0.30		0.14		1	A0	820
97674	−66 01552				11 11 16	−66 34.6	1	7.99		0.56		0.35		2	A1mA5-A8	1771
97672	−60 02688	LSS 2253			11 11 17	−60 47.3	1	9.61		0.22		-0.70		2	O9.5/B0 V	540
		Leo II sq # 7			11 11 18	+22 15.9	1	14.65		0.65				2		988
97654	−56 04317				11 11 18	−56 46.7	1	7.54		0.03		0.05		2	A0 V	401
304001	−58 03313	St 13 - 3			11 11 18	−58 39.3	1	10.06		0.01		-0.71		2	B8	411
	−60 02692				11 11 18	−60 28.3	1	11.66		0.60		0.07		1	F5	820
97669	−58 03314	St 13 - 2			11 11 20	−58 38.3	1	8.53		-0.02		-0.07		1	A0 IV/V	411
97670	−58 03315	HR 4361			11 11 20	−59 20.8	3	5.74	.009	-0.10	.005	-0.72		9	B2 III	6,15,2012
		NGC 3590 - 56			11 11 20	−60 33.6	1	11.79		0.73		0.19		1		820
97649	−45 06792				11 11 22	−46 12.1	1	9.85		0.12		0.16		1	A1 III/IV	1770
		AAS41,43 F1 # 338			11 11 22	−60 23.7	1	12.84		0.32		0.15		1	A0	820
306205	−60 02696				11 11 22	−60 59.6	2	9.98	.054	0.13	.015	-0.64		6	B1.5Ven	6,2021
306163	−59 03214				11 11 23	−60 09.5	1	11.23		0.40		0.07		3	A0	2
97651	−52 04350	HR 4360			11 11 24	−52 57.6	4	5.75	.007	1.31	.010	1.42		12	K2 III	15,1075,2006,3005
		NGC 3590 - 55			11 11 24	−60 33.4	1	11.21		0.43		0.20		1		820
97605	+8 02476	HR 4358		★ A	11 11 26	+08 20.1	5	5.79	.004	1.12	.004	1.13	.011	15	K3 III	13,15,1256,1355,3077
97605	+8 02476	IDS11088N0837		B	11 11 26	+08 20.1	1	11.83		0.74		0.46		4		13
306157	−59 03216	LSS 2254		AB	11 11 26	−60 01.2	4	10.31	.018	0.47	.017	-0.34	.044	12	B1 V	2,6,463,2021
		AAS41,43 F1 # 340			11 11 26	−60 24.4	1	12.42		0.34		-0.29		1	B6	820
306209	−60 02697	LSS 2255			11 11 26	−61 04.3	1	10.16		0.13		-0.73		2	B1 Ve	540
97603	+21 02298	HR 4357		★ A	11 11 27	+20 47.9	9	2.56	.007	0.13	.014	0.11	.009	33	A4 V	15,30,1007,1013,1119*
		MCC 605			11 11 28	+52 14.5	1	11.45		1.34				1	K4-5	1017
97707	−60 02698	LSS 2256			11 11 29	−60 28.1	9	8.10	.037	0.50	.020	-0.38	.033	22	B1/2 Ib	2,6,191,463,540,820*
97688	−51 05472	IDS11093S5115		A	11 11 32	−51 31.7	1	7.93		1.20		1.07		5	K0 II/III	1673
		Leo II sq # 8			11 11 33	+22 16.2	1	15.12		0.79				2		988
		WLS 1112 50 # 7			11 11 35	+47 48.3	1	11.19		1.04		0.75		2		1375
97687	−49 05956				11 11 35	−49 43.7	1	9.62		0.49		0.26		1	A0 III/IV	1770
97689	−52 04355				11 11 35	−52 35.0	2	6.82	.004	0.25	.001	0.12	.000	5	A0mA3/5-F0	355,1770
97633	+16 02234	HR 4359			11 11 37	+15 42.2	9	3.34	.023	-0.01	.014	0.04	.023	25	A2 V	15,1022,1119,1203*
97704	−50 05807				11 11 40	−51 18.5	1	9.96		0.24		0.11		1	A2/3 III/I	1770

Table 1 599

HD	DM	Other Id	N Rem	α₁₉₅₀	δ₁₉₅₀	S	V	σ_V	B–V	σ_B-V	U–B	σ_U-B	n	Spectrum	References
		Leo II sq # 10		11 11 43	+22 25.3	1	14.68		0.86				2		988
97643	+20 02573			11 11 46	+20 18.2	1	7.33		0.96		0.72		2	G5	1648
306162	−59 03224			11 11 46	−60 09.2	1	10.68		0.20		−0.26		3	B5	2
97746	−57 04454	GI Car		11 11 48	−57 38.3	2	8.21	.044	0.67	.039	0.40		3	F3/5 Iab/b	1484,8100
		Leo II sq # 23		11 11 49	+22 31.8	1	15.36		0.67				1		988
	−75 00719	VR h		11 11 51	−75 47.1	1	9.54		1.32		0.98		24	G9 III	1763
97658	+26 02184			11 11 53	+25 58.9	1	7.76		0.85		0.44		2	K1 V	3026
306166	−59 03196			11 11 56	−60 08.3	2	9.16	.005	1.52	.025	1.78	.075	2	K7	463,820
		NGC 3603 - 36		11 11 56	−61 01.1	2	14.75	.033	0.50	.042	0.42	.014	8		56,707
97584	+74 00456	G 254 - 18	⋆ AB	11 11 57	+73 44.8	2	7.68	.044	1.04	.005	0.92		5	K5	1197,3072
97718	−21 03268	IDS11095S2212	A	11 11 58	−22 28.4	1	10.06		0.68		0.14		1	G8 V	78
	−26 08434			11 11 58	−26 39.8	1	10.64		0.72		0.24		2		1696
97697	+17 02331			11 12 03	+16 36.6	1	7.98		1.58				2	K0	882
		I 119 354		11 12 04	+29 20.1	1	12.83		0.43		−0.29		3		1723
97714	+13 02373			11 12 05	+12 53.7	1	8.45		1.16		1.10		2	K5	3040
		GD 130		11 12 05	+29 20.0	1	12.80		0.45		−0.10		1		3060
	−23 09822			11 12 06	−23 38.8	1	10.95		0.32		0.10		2	A5	1776
97792	−55 04230			11 12 06	−55 46.5	1	8.04		−0.05		−0.63		2	B2/3 Vnne	401
	−17 03333			11 12 08	−17 38.3	1	10.07		1.71		1.72		2		1375
306247	−59 03236	FR Car		11 12 10	−59 46.8	2	9.35	.053	5.47	.493	0.63	.024	2	G5	689,1587
97805	−57 04465			11 12 12	−57 31.7	1	7.78		−0.03		−0.25		2	B9 IV	401
		NGC 3603 - 2		11 12 12	−61 00.5	2	12.90	.005	0.94	.034	0.42	.024	3		56,707
97851	−65 01646	LSS 2266		11 12 14	−65 37.1	2	8.51	.005	0.18	.004	−0.72	.027	5	B0 III	540,976
97849	−62 01937			11 12 18	−63 07.3	1	9.91		0.41		0.28		1	A7 IV/V	1771
306196	−60 02718	LSS 2265		11 12 20	−60 47.0	3	9.64	.019	0.33	.021	−0.39	.020	9	B0:Vn	2,6,2021
97782	−22 03102	IDS11099S2233	AB	11 12 21	−22 49.6	2	8.97	.024	1.13	.005	1.00	.066	3	K4 V	78,3072
97783	−22 03101			11 12 21	−23 22.5	3	9.07	.013	0.58	.008	0.00	.009	5	G1/2 V	78,803,2017
97848	−58 03351	LSS 2264		11 12 21	−58 45.1	2	8.68	.012	−0.01	.005	−0.90	.023	12	B0 III	6,540,976,1021
97843	−44 07168			11 12 27	−45 01.9	1	8.59		0.08		0.08		1	A1 V	1770
97823	−49 05974			11 12 28	−49 37.6	1	9.50		0.10		0.05		1	A1 V	1770
306168	−59 03241	GL Car		11 12 30	−60 23.2	3	9.60	.073	0.21	.018	−0.73	.009	8	B3:V	6,540,2021
	−39 06992			11 12 31	−39 49.5	2	10.28	.015	0.89	.005	0.61	.005	3		78,1696
97619	+79 00356			11 12 32	+78 34.9	1	6.91		1.30		1.46		2	K0	985
		G 56 - 26		11 12 33	+19 43.9	1	12.93		1.58		1.18		1		1620
97778	+23 02322	HR 4362		11 12 33	+23 22.1	5	4.63	.005	1.66	.005	1.84	.015	13	M3 IIb	15,369,1363,3016,8015
97866	−43 06899	HR 4364		11 12 33	−43 27.7	2	6.21	.005	1.61	.005	1.93		6	K4 III	2006,3005
306267	−59 03242	LSS 2270		11 12 34	−60 05.8	1	10.79		0.29		−0.59		3	B8	1737
		G 122 - 7		11 12 36	+43 33.3	1	13.55		0.80		0.36		3		7010
97840	−32 07964			11 12 38	−33 02.7	1	6.99		0.36				4	F3 V	2012
97763	+59 01360			11 12 40	+58 29.3	2	8.05	.000	1.22	.005	1.33	.019	9	K2	186,1775
97772	+60 01316	IDS11097N6019	AB	11 12 40	+60 03.1	1	8.63		0.43				2	F5	1379
	−59 03245			11 12 40	−60 09.7	1	9.12		0.06		−0.63		2		476
97913	−58 03366	LSS 2271		11 12 43	−58 54.1	5	8.80	.017	0.05	.012	−0.82	.026	14	O9/B0 V	6,191,540,976,1021
	+59 01361			11 12 47	+58 56.2	1	10.65		0.70		0.25		5		186
		NGC 3603 - 28		11 12 48	−60 59.3	1	13.46		0.52		0.23		2		707
97864	−18 03141			11 12 49	−19 21.9	1	6.88		0.08		0.08		7	A3 V	1628
97932	−60 02727			11 12 49	−61 10.7	1	9.43		−0.01		−0.65		2	B2 III	401
	−17 03336	IDS11104S1734	A	11 12 50	−17 51.7	4	10.02	.022	1.35	.021	1.19	.015	13	K7 V	694,803,1705,3073
	−17 03336	IDS11104S1734	C	11 12 50	−17 51.7	1	13.70		1.55		1.05		2		3073
		G 163 - 66		11 12 51	−05 54.5	1	12.65		0.87		0.22		2		1696
	−17 03337	IDS11104S1734	B	11 12 51	−17 51.6	4	9.96	.011	1.34	.012	1.21	.000	14	dM1	694,803,1705,3073
		NGC 3603 - 27		11 12 51	−60 59.2	1	15.08		1.02		−0.17		2		707
97911	−50 05828			11 12 52	−51 03.9	1	8.40		0.40		0.15		1	A2mA8-F3	1770
97632	+80 00350			11 12 53	+79 48.5	1	8.00		1.28		1.35		2	K0	1733
		LP 900 - 47		11 12 53	−28 20.4	1	12.35		0.87		0.50		2		1696
		NGC 3603 - 26		11 12 53	−60 59.1	2	14.36	.015	0.83	.035	0.12	.124	5		607,707
		NGC 3603 - 30		11 12 53	−60 59.3	1	12.58		0.17		0.08		2		707
97927	−57 04486			11 12 54	−57 59.2	1	7.71		−0.05		−0.25		2	B9/9.5IV	401
		NGC 3603 - 1		11 12 54	−60 59.	2	13.11	.023	1.14	.019	0.80	.070	8		56,707
		NGC 3603 - 6		11 12 54	−60 59.	1	14.16		1.35		0.20		1		56
		NGC 3603 - 11		11 12 54	−60 59.	1	13.84		0.68		0.22		1		707
		NGC 3603 - 15		11 12 54	−60 59.	1	12.32		0.20		−0.25		1		707
		NGC 3603 - 39		11 12 54	−60 59.	2	15.00	.035	0.66	.048			6		56,707
		NGC 3603 - 40		11 12 54	−60 59.	1	13.97		1.90		1.21		4		56
		NGC 3603 - 44		11 12 54	−60 59.	1	13.08		0.73		0.32		4		56
		NGC 3603 - 142		11 12 54	−60 59.	1	14.24		1.29		0.09		1		707
97878	−21 03273			11 12 55	−22 08.1	1	9.94		0.47		−0.08		2	G2/3wA/F0	1375
97895	−28 08735			11 12 55	−29 14.6	2	8.79	.008	−0.09	.004	−0.56	.000	9	B4 V	55,1775
97949	−60 02730			11 12 55	−60 24.7	1	9.48		0.27		0.15		16	A7 V	632
97859	+5 02468			11 12 56	+05 13.8	4	9.35	.009	−0.12	.004	−0.55	.015	30	A0	989,1026,1728,1729
		NGC 3603 - 54		11 12 56	−60 59.2	1	14.34		1.03		−0.30		1		607
97922	−51 05497			11 12 57	−52 15.1	1	10.03		−0.01		−0.08		1	A0 Vn	1770
		G 119 - 65		11 12 58	+32 18.5	2	13.10	.004	1.42	.000	1.13	.026	6		316,3062
		NGC 3603 - 25		11 12 58	−60 58.9	2	12.27	.005	1.37	.010	0.17	.075	9	B1.5Iab	607,707
97950	−60 02732	NGC 3603 - 70	⋆ AB	11 12 58	−60 59.	1	9.03		0.80		−0.19		1	WN5	1096
		NGC 3603 - 47		11 12 59	−60 59.7	1	12.67		1.04		−0.22		3	O4 V	607
97811	+59 01362			11 13 00	+59 23.6	4	7.94	.007	0.21	.008	0.15	.020	24	Am	186,1379,1775,8071
97966	−58 03372	LSS 2276		11 13 00	−59 08.6	4	8.85	.017	0.05	.016	−0.87	.020	11	B7 Iab/b	6,540,976,1021
97968	−59 03255			11 13 00	−59 46.1	1	9.36		0.04		−0.12		3	Ap Si	2
		NGC 3603 - 53		11 13 00	−60 59.4	2	14.03	.295	1.13	.045	−0.04	.025	2		607,707

HD	DM	Other Id	N Rem	α_{1950}	δ_{1950}	S	V	σ_V	B–V	σ_{B-V}	U–B	σ_{U-B}	n	Spectrum	References
		NGC 3603 - 18		11 13 00	−60 59.6	2	12.55	.061	1.16	.000	-0.04	.080	4	O6 If	607,707
		Steph 932		11 13 01	+55 36.3	1	11.22		1.41		1.06		1	M0	1746
		NGC 3603 - 19		11 13 02	−60 59.5	2	13.62	.064	0.98	.030	-0.19	.064	5		607,707
97969	−59 03259	LSS 2278		11 13 04	−59 53.5	3	7.70	.022	0.00	.017	-0.76	.045	9	B1/2II/IIIn	2,6,2021
		NGC 3603 - 24		11 13 04	−60 58.9	1	13.79		0.97		-0.39		1		607
		NGC 3603 - 23		11 13 06	−60 58.9	2	12.70	.055	1.11	.010	-0.14	.109	5	O9.5Iab	607,707
		NGC 3603 - 22		11 13 06	−60 59.0	2	13.18	.115	1.09	.030	-0.08	.105	6	O5 V(f)	607,707
		WD1113+413		11 13 08	+41 19.3	1	15.38		-0.17		-0.98		2	DA2	1727
233821	+51 01658			11 13 08	+50 54.8	1	9.21		0.40		-0.03		2	F8	1566
97959	−43 06907			11 13 08	−44 07.3	1	7.72		0.22		0.09		1	A1mA6-A9	1770
97918	−11 03063	SY Crt		11 13 09	−12 19.2	1	6.40		1.60		1.60		1	M3 III	3040
97957	−41 06438			11 13 10	−42 20.5	2	7.44	.013	0.51	.004	-0.04		6	F7 V	1311,2012
97943	−37 07126			11 13 11	−37 59.1	1	7.36		1.13		0.99		6	K0 III	1673
		NGC 3603 - 5		11 13 11	−60 59.2	2	12.74	.040	0.72	.040	-0.08	.015	2		56,707
97855	+53 01480	HR 4363	⋆ A	11 13 12	+53 02.7	1	6.50		0.43		-0.12		3	F6 V	3037
97855	+53 01480	IDS11103N5319	B	11 13 12	+53 02.7	1	8.03		0.60		-0.04		3		3032
97983	−53 04429			11 13 13	−53 45.5	1	9.40		0.10		0.12		1	A1/2 IV	1770
	+50 01805			11 13 14	+49 50.5	1	10.70		0.58		-0.01		2		1375
306199	−60 02735	NGC 3603 - 3		11 13 14	−60 58.7	2	11.06	.005	0.14	.020	-0.33	.010	10	A0	56,707
97907	+14 02367	HR 4365	⋆	11 13 15	+13 34.8	3	5.32	.005	1.19	.008	1.08	.032	5	K3 III	1080,4001,8087
		NGC 3603 - 14		11 13 17	−61 01.4	1	11.74		0.13		-0.29		1		707
97916	+2 02406			11 13 19	+02 21.6	4	9.21	.011	0.42	.007	-0.12	.009	8	F5 V	158,792,1003,3077
97941	−25 08553			11 13 19	−25 32.2	1	9.21		0.36		0.05		12	F0 V	588
97938	+13 02378			11 13 20	+12 53.2	2	6.82	.005	-0.02	.005	-0.08	.005	3	A0	272,8087
97937	+13 02379	HR 4366		11 13 21	+13 07.1	2	6.68	.015	0.28	.019	0.04	.015	3	F0 IV	254,8087
	−59 03264			11 13 22	−59 58.3	1	8.80		0.10		-0.82		2		476
97889	+60 01318			11 13 23	+60 12.9	2	6.92	.010	0.17	.005	0.10	.005	5	A3 V	985,1501
	−25 08554	LP 850 - 59		11 13 23	−25 25.6	1	10.60		1.24		1.05		1		3072
306200		NGC 3603 - 13		11 13 24	−61 02.0	2	10.35	.004	1.27	.009	1.31	.039	12	K2	56,707
98014	−57 04494	SY Car		11 13 26	−57 39.3	1	8.89		2.25		2.55		2	N3	864
97998	−38 07027			11 13 27	−39 03.2	2	7.36	.005	0.63	.000	0.06		2	G5 V	78,2039
98025	−56 04352			11 13 30	−57 05.0	1	6.43		0.10		0.14		2	A3 V	401
	−76 00656	VR i		11 13 31	−76 45.9	1	10.08		1.58		1.68		1	K3 III	1763
98022	−47 06664	IDS11113S4723	A	11 13 34	−47 38.8	1	6.54		-0.02		-0.08		2	A0 V	1770
	+49 02035			11 13 39	+48 53.4	1	9.58		0.20		0.10		2	A2	272
97991	−2 03312			11 13 39	−03 11.9	6	7.40	.006	-0.22	.005	-0.93	.012	70	B1 V	55,147,1509,1732,1775,6005
97935	+59 01363			11 13 40	+58 57.5	1	9.16		0.57		0.07		2	G5	186
97934	+59 01364			11 13 44	+59 13.1	1	8.24		1.01		0.84		std	K0 IV	186
98007	−2 03313			11 13 44	−03 29.3	4	8.94	.002	0.73	.006	0.32	.009	65	K0	147,1509,1775,6005
98050	−47 06670			11 13 45	−47 38.1	1	9.25		0.13		0.13		1	A2/3 V	1770
	+35 02227			11 13 47	+35 16.5	1	10.86		0.28		0.16		2		272
98051	−51 05513			11 13 47	−51 43.1	1	9.97		0.09		0.04		1	A0 IV/V	1770
306269	−59 03274	LSS 2280		11 13 48	−60 06.4	2	10.20	.014	0.05	.011	-0.72	.028	5	B0.5V	2,540
		Ton 1381		11 13 50	+32 11.	2	14.13	.005	0.04	.000	0.07	.005	3		286,313
		MCC 606		11 13 53	+26 44.0	1	10.52		1.28				1	K5	1017
97989	+50 01807	HR 4367		11 13 53	+49 45.0	1	5.88		1.08				3	gK0	70
	−69 01507			11 13 53	−69 42.5	1	10.72		0.68		0.21		1		697
310480	−69 01508	DI Car		11 13 55	−69 38.6	1	9.64		1.06		0.88		1	G5:	697
	−13 03333			11 13 57	−14 25.0	3	9.98	.026	1.41	.027	1.20	.000	6	K7 V	158,1017,3072
98072	−52 04399			11 13 58	−52 35.2	1	9.40		0.08		0.06		1	A0 V	1770
98031	+13 02380			11 13 59	+12 44.1	1	8.40		0.66				3	K0	882
98097	−62 01946			11 13 59	−63 08.9	2	9.84	.000	0.11	.007	-0.59	.015	8	B2/3 II	540,976
		L 192 - 72		11 14 03	−57 17.5	2	11.65	.010	1.48	.010	0.92		3		1705,3073
304027	−58 03399	LSS 2282		11 14 03	−59 06.9	1	10.04		0.20		-0.62		2		540
306249	−59 03283	LSS 2283		11 14 03	−59 44.9	1	10.10		0.02		-0.63		3	B8	2
306248	−59 03282			11 14 03	−59 45.6	1	10.63		0.14		-0.09		3	B9	2
		L 395 - 109		11 14 04	−43 44.3	1	14.15		1.61				1		3062
		G 45 - 45		11 14 06	+06 43.5	1	16.74		0.46		-0.38		1		3060
		AAS9,163 T5# 6		11 14 07	+58 38.6	1	10.84		1.21		1.27		7		186
98058	−2 03315	HR 4368	⋆ AB	11 14 07	−03 22.7	6	4.47	.005	0.21	.005	0.11	.023	20	A7 IVn	15,369,1075,1425,3023,8015
		Ton 1383		11 14 08	+20 36.	2	14.13	.025	-0.12	.010	-0.38	.005	8		286,313
98096	−45 06837	HR 4370	⋆ AB	11 14 08	−45 36.5	2	6.30	.005	0.41	.005			7	F3 V	15,2027
		Feige 38		11 14 12	+07 16.0	1	12.99		-0.22		-1.00		2	sdB	1298
		LP 318 - 355		11 14 17	+26 21.0	1	11.10		0.65		0.14		3		1723
	+23 02328			11 14 21	+22 45.5	1	10.24		0.23		0.07		2	A0	272
	+30 02129			11 14 22	+29 36.2	1	9.86		0.57		-0.09		3		1723
306258	−59 03290			11 14 22	−59 49.2	1	11.24		0.13		-0.24		3	B5	2
98088	−6 03344	HR 4369, SV Crt	⋆ AB	11 14 26	−06 51.7	3	6.13	.008	0.20	.003	0.14	.014	13	A2p	15,1202,1417
		L 395 - 108		11 14 26	−43 49.3	1	14.16		1.65				1		3062
98126	−37 07142			11 14 30	−38 07.6	1	6.88		1.10		1.05		7	K1/2 III	1628
306234	−58 03406	LSS 2285		11 14 32	−59 18.8	2	10.15	.005	0.28	.000	-0.41		6	B2 V	6,2021
98195	−69 01510			11 14 32	−69 38.0	1	7.79		0.10		0.02		2	A0 V	1771
98149	−52 04405			11 14 34	−53 13.8	1	9.55		0.03		0.03		1	A0 IV	1770
	+30 02130	G 147 - 36		11 14 35	+29 50.7	1	9.29		0.62		0.15		1	G0	1658
98117	+17 02337			11 14 38	+17 13.6	1	8.50		1.08				3	G5	882
98120	−14 03297			11 14 38	−14 48.9	1	8.82		0.06		-0.81		2	G5 IV/V	6
	−27 07978			11 14 40	−27 32.4	1	9.79		1.40		1.21		1	M0 V	3072
	−1 02505	G 10 - 10		11 14 42	−01 42.5	1	9.75	.036	1.17	.000	1.18		5	M0	158,1746
		IC 2714 - 183		11 14 42	−62 25.5	1	12.90		0.32				3		2042
98118	+2 02409	HR 4371		11 14 43	+02 17.1	8	5.17	.006	1.51	.006	1.84	.008	87	M0 III	3,15,369,1003,1417*

Table 1 601

HD	DM	Other Id	N Rem	α_{1950}	δ_{1950}	S	V	σ_V	B–V	σ_{B-V}	U–B	σ_{U-B}	n	Spectrum	References
		LP 906 - 51		11 14 44	−30 53.1	1	11.48		0.46		−0.18		2		1696
308351	−61 02168	IC 2714 - 180		11 14 44	−62 24.4	1	11.39		0.42				3		2042
98212	−66 01567			11 14 47	−66 39.0	1	9.03		0.16		−0.35		2	B5 III	401
98161	−37 07146	HR 4372		11 14 48	−37 44.5	3	6.25	.011	0.11	.010			11	A3 Vn	15,2018,2028
		LP 169 - 14		11 14 49	+46 39.9	1	11.12		0.68		0.14		2		1723
		IC 2714 - 179		11 14 49	−62 23.5	1	12.37		0.55				3		2042
98210	−60 02774	LSS 2291		11 14 52	−60 39.2	2	8.79	.000	0.05	.030	−0.72		7	B1/2 II	2,2021
		IC 2714 - 154		11 14 52	−62 26.7	1	13.60		0.12				3		2042
98294	−74 00789	Cha F # 41		11 14 54	−74 56.9	1	8.52		0.19		−0.11		2	B8 V	1633
98114	+59 01366			11 14 57	+59 17.8	1	9.73		0.38		−0.04		2	G0	186
98203	−50 05859			11 14 58	−51 08.0	1	8.71		0.03		0.05		1	A0 V	1770
	−62 01954	IC 2714 - 157		11 15 00	−62 25.5	1	10.53		0.35				3		2042
98154	+26 02189			11 15 01	+25 44.0	1	7.40		0.14		0.10		2	A3 V	105
		G 197 - 8		11 15·02	+57 17.7	1	12.11		0.63		0.07		1	sdF8	1658
98153	+28 01983			11 15 04	+27 38.1	1	6.94		0.20		0.10		2	A2	252
		G 163 - 70		11 15 05	+01 36.5	2	12.42	.002	0.52	.000	−0.15	.012	3	sdG0	1658,1696
98172	+12 02384			11 15 07	+12 40.7	1	7.60		1.25		1.46			K2	8087
98260	−60 02783	LSS 2292		11 15 09	−61 14.2	3	9.70	.009	0.26	.001	−0.63	.017	8	B1/2 Ib/II	6,540,2021
98152	+41 02176			11 15 10	+41 06.6	1	9.00		0.03		−0.02		1	A0 V	272
		GD 131		11 15 12	+03 21.2	1	15.39		0.10		−0.57		1		3060
98222	−37 07152			11 15 12	−38 06.5	1	8.18		0.67		0.21		1	G5 V	78
		Ton 1388		11 15 14	+21 46.	2	14.82	.005	0.11	.005	−0.97	.010	8		286,313
98220	−32 07996			11 15 14	−33 16.1	3	6.84	.005	0.50	.004	−0.01	.005	15	F7 V	1311,2012,3037
98221	−34 07345	HR 4373		11 15 14	−34 27.8	2	6.44	.005	0.41	.005			7	F3 V	15,2012
98278	−58 03424	IDS11130S5834	ABC	11 15 15	−58 49.9	1	6.70		−0.01		−0.28		2	B9.5/A0 III	401
98292	−67 01703	HR 4379	★ AB	11 15 16	−67 33.0	2	6.05	.005	1.76	.000	1.92	.005	7	M2 III	15,1075
		LP 214 - 66		11 15 17	+41 17.6	1	10.77		0.59		−0.02		3		1723
98233	−35 07111	HR 4376		11 15 19	−36 15.7	1	6.68		0.98				4	G8/K0 III	2006
98257	−52 04417			11 15 19	−53 12.8	1	9.40		0.16		0.15		1	A2 IV/V	1770
		IC 2714 - 106		11 15 19	−62 24.7	1	13.06		0.43				2		2042
98314	−62 01959	IDS11133S6256	AB	11 15 23	−63 12.2	2	7.46	.019	0.11	.010	−0.15	.015	3	B9 III	761,1771
	+58 01312			11 15 25	+58 24.2	1	9.92		0.40		−0.06		3		186
98310	−57 04536	LSS 2294		11 15 29	−57 43.1	2	9.49	.010	−0.03	.012	−0.80	.005	6	B4/5 V	540,976
98231	+32 02132	HR 4375	★ AB	11 15 29	+31 48.6	9	3.79	.018	0.59	.005	0.04	.009	35	G0 V	15,22,1013,1118,1193*
98284	−35 07113			11 15 34	−36 19.3	1	8.30		0.54				4	F6 V	2012
98329	−58 03431			11 15 35	−58 56.1	1	7.03		−0.08		−0.45		2	B9 II/III	401
306299	−60 02788	LSS 2295		11 15 35	−60 51.9	1	10.55		0.08		−0.52		2	B2 Ve	540
		Ton 1384		11 15 40	+20 08.	2	13.20	.010	−0.24	.005	−0.96	.005	8	sd	286,313
98214	+56 01511			11 15 40	+56 28.1	1	7.38		1.08		1.00		3	K0 III	37
306280	−59 03313	LSS 2297		11 15 42	−60 20.3	1	10.60		0.09		−0.62		2	B2 III	191
98247	+43 02102			11 15 43	+42 35.4	2	6.94	.005	1.01	.010	0.78	.005	5	K0 III	1501,1601
		G 10 - 11		11 15 44	−02 57.9	2	15.35	.050	0.07	.020	−0.75	.010	7		538,3060
98280	+12 02319	HR 4378		11 15 45	+12 15.5	3	6.66	.005	0.05	.014	0.05	.012	7	A2 V	1022,1733,3050
98262	+33 02098	HR 4377	★ AB	11 15 47	+33 22.0	9	3.48	.014	1.40	.008	1.55	.015	30	K3 III	1,15,1077,1080,1118*
98281	−4 03049	G 163 - 74		11 15 47	−04 47.5	6	7.29	.011	0.74	.007	0.27	.016	14	G8 V	22,1003,1311,1509,1705,3026
98325	−37 07159			11 15 47	−37 56.5	1	8.78		1.27		1.34		1	K2 III	78
98326	−49 06030			11 15 47	−50 22.0	1	9.46		−0.01		−0.02		1	B9.5 V	1770
	+51 01661			11 15 49	+51 02.2	1	10.77		0.59		0.01		3		1723
		G 236 - 63		11 15 50	+65 16.0	1	14.40		1.62				1		906
98363	−63 01876			11 15 50	−63 46.1	1	7.86		0.16		0.09		2	A2 V	1771
		LP 552 - 31		11 15 53	+04 18.2	1	12.83		1.33		1.30		1		1696
	+58 01313			11 15 54	+58 11.5	1	10.31		0.58		0.03		5	F8	186
		Cha T # 56	V	11 15 58	−76 48.2	2	13.63	.000	1.41	.000	0.44	.155	2	M0.5	437,825
98359	−48 06354			11 16 00	−48 30.8	1	7.31		0.06		0.05		3	B9.5 V	1770
98317	+36 02175			11 16 06	+35 45.8	1	6.68		1.13		1.03		2	K0	1601
98261	+62 01172			11 16 07	+62 27.4	1	6.94		1.11		1.02		3	K1 III	1501
98410	−62 01964	LSS 2299		11 16 09	−62 42.1	2	8.90	.025	0.20	.002	−0.53	.029	4	B2.5Ib/II	540,976
308354	−62 01965	IC 2714 - 30		11 16 11	−62 29.8	1	10.87		0.62				2	B9	2042
98383	−53 04460			11 16 14	−53 26.4	1	8.15		0.11		0.07		1	A1 V	1770
98345	+11 02343			11 16 15	+10 52.2	1	7.69		0.33		−0.02		3	F0	1648
98315	+59 01367			11 16 15	+59 09.6	1	7.96		1.36		1.48		4	K0	186
		Steph 933		11 16 16	+49 16.6	1	12.05		1.44		1.18		1	M1	1746
98392	−46 06946			11 16 16	−47 09.8	1	9.07		0.16		0.13		1	A1 V	1770
		G 147 - 43		11 16 18	+28 53.3	1	16.56		0.25		−0.58		1		3060
98366	+2 02411	HR 4381		11 16 21	+01 55.5	2	5.90	.005	1.04	.000	0.88	.005	7	gK0	15,1417
306294	−60 02800			11 16 21	−60 40.7	1	10.25		0.03		−0.65		2	B2 V	191
98353	+38 02225	HR 4380		11 16 25	+38 27.6	5	4.77	.017	0.11	.013	0.03	.005	13	A2 V	15,1022,1363,3023,8015
98408	−53 04461			11 16 25	−53 46.1	1	9.24		0.10		0.08		1	A1 V	1770
		LP 792 - 12		11 16 28	−16 59.4	1	10.92		0.99		0.84		1		1696
98436	−60 02801			11 16 29	−60 53.0	1	8.95		0.06		−0.56		2	B2/3 II/III	401
98388	+14 02374			11 16 30	+13 39.8	1	7.13		0.49		0.00		1	F8 V	8087
		L 37 - 4		11 16 30	−75 27.	2	12.20	.183	1.06	.023	0.74	.117	4		1696,3062
98418	−50 05887			11 16 34	−50 47.7	1	9.47		0.20		0.08		1	A1 IV	1770
98465	−62 01966	LSS 2302		11 16 34	−63 03.4	2	9.79	.001	0.31	.002	−0.67	.008	3	O/Be	540,1737
		LP 264 - 37		11 16 35	+36 46.3	1	11.78		1.32		1.29		4		1723
		GD 133		11 16 38	+02 37.0	1	14.57		0.19		−0.59		1		3060
98398	+6 02431	G 10 - 12		11 16 40	+05 57.2	2	9.29	.010	0.81	.005	0.33	.010	2	K0	333,1620,1658
98484	−60 02808			11 16 41	−61 22.3	1	8.32		0.22		0.17		2	A1/2 IV/V	1771
98496	−67 01712			11 16 44	−68 22.9	1	9.88		0.28		0.17		1	A7 IV/V	1771
	+14 02375			11 16 45	+14 02.9	1	9.51		0.96		0.72		1	K0	8087

HD	DM	Other Id	N Rem	α_{1950}	δ_{1950}	S	V	σ_V	B–V	σ_{B-V}	U–B	σ_{U-B}	n	Spectrum	References
		GD 134		11 16 46	+27 51.9	1	13.95		0.43		-0.26		1		3060
	+13 02387			11 16 47	+12 49.0	1	10.68		0.20		0.01		2	A2	272
98427	−0 02428	IDS11143S0106	A	11 16 50	−01 22.7	1	7.08		0.52		0.05		2	G0 IV-V	3026
98427	−0 02428	IDS11143S0106	B	11 16 50	−01 22.7	1	7.90		0.68		0.25		2		3032
98430	−13 03345	HR 4382		11 16 50	−14 30.5	8	3.56	.004	1.12	.007	0.98	.014	37	K0 III	3,15,1075,1311,1425*
98478	−47 06726			11 16 50	−47 54.5	1	8.18		0.67				4	F8/G0 IV	2012
98481	−57 04563			11 16 51	−57 57.1	2	9.17	.019	0.01	.005	-0.80		6	B2/3 III	6,2021
98477	−46 06957	IDS11146S4657	A	11 16 54	−47 13.2	1	8.58		0.08		0.01		3	B9.5 V	1770
98561	−74 00798	Cha F # 42		11 16 58	−75 08.5	1	8.33		0.45		0.30		2	A3.5IV	1633
98456	−27 08004			11 16 59	−28 12.1	1	7.32		0.92		0.61		6	G8 III	1657
		GD 135		11 17 02	−02 22.6	1	14.46		0.09		-0.66		1		3060
	+58 01314			11 17 03	+58 27.7	1	10.05		0.52		0.00		3		186
308395	−61 02198	LSS 2304		11 17 03	−61 47.1	2	10.37	.026	0.48	.002	-0.31	.037	4	B2 III	191,540
		AAS9,163 T5# 12		11 17 05	+58 10.6	1	10.68		0.48		-0.02		4		186
98617	−78 00638	HR 4385		11 17 06	−79 23.7	3	6.34	.004	0.26	.004	0.07	.010	15	A8 IIIm	15,1075,2038
308411	−62 01967			11 17 07	−62 58.2	1	10.29		0.33		-0.19		1	B8	761
98560	−63 01881	HR 4384		11 17 10	−64 18.6	4	5.98	.008	0.46	.011	-0.01	.009	17	F5 V	15,1075,2038,3062
98536	−60 02819			11 17 11	−60 39.9	1	9.32		0.03		-0.04		2	B9mA0/2-A5	127
		L 144 - 150		11 17 12	−64 20.	1	14.32		1.57				1		3062
		Mel 105 - 17		11 17 13	−63 08.2	1	14.53		0.54				4		56
		Mel 105 - 16		11 17 18	−63 08.6	1	11.54		0.36		-0.24		11		56
98627	−75 00722	Cha F # 43		11 17 18	−75 33.1	1	7.38		1.75		2.07		2	K5 III	1633
		LP 93 - 554		11 17 24	+60 04.5	1	10.63		0.74		0.19		2		1723
306352	−59 03349			11 17 25	−59 43.2	2	9.73		0.23		-0.63		2	A0	476
		Mel 105 - 10		11 17 25	−63 10.7	1	13.00		0.37		0.12		1		56
	+66 00717	SZ UMa	★	11 17 29	+66 07.0	5	9.31	.015	1.42	.009	1.08	.000	19	M1 V	1,694,1197,3078,8006
98556	−47 06740			11 17 30	−48 07.4	1	8.17		1.27		1.41		1	K3 III	1770
98555	−47 06741			11 17 31	−48 05.9	1	8.64		-0.03		-0.16		1	B9 IV	1770
98558	−52 04466	IDS11152S5256	A	11 17 32	−53 12.1	1	7.89		0.14		0.05		1	A1 V	1770
		Mel 105 - 18		11 17 32	−63 08.7	1	14.98		1.11				1		56
98488	+59 01368			11 17 34	+59 03.0	1	8.11		0.07		0.14		3	A2	186
		Lod 336 # 1		11 17 39	−58 35.0	1	11.27		1.82		1.49		1		287
98584	−57 04583			11 17 40	−57 52.8	1	7.59		-0.09		-0.49		2	B7 II	401
98598	−60 02827			11 17 40	−60 27.5	1	9.77		0.31		0.15		1	A7 IV/V	1771
		Lod 336 # 4		11 17 43	−58 36.1	1	12.64		0.44		0.39		1		287
		Mel 105 - 9		11 17 44	−63 13.2	1	12.09		1.32		1.21		2		56
		Mel 105 - 23		11 17 44	−63 13.9	1	14.08		0.49		0.33		3		56
98616	−63 01883			11 17 45	−63 28.5	1	8.73		0.28		0.09		1	A8/9 [IV]	1771
98526	+45 01912			11 17 47	+45 16.4	1	6.74		0.32		0.07		3	F0 III	1501
98596	−58 03475			11 17 47	−58 54.8	1	7.31		0.01		-0.18		2	B9 V	401
98672	−74 00801	HR 4387		11 17 48	−74 52.1	3	6.27	.004	-0.03	.000	-0.12	.000	11	B9.5/A0 V	15,1075,2038
98547	+18 02475			11 17 49	+17 35.1	1	7.15		0.12		0.08		2	A0	272
98614	−58 03477	LSS 2307		11 17 49	−59 17.8	3	8.46	.026	0.09	.007	-0.69	.015	7	B2 II	6,540,976
98499	+67 00692	HR 4383		11 17 52	+67 22.5	1	6.21		1.01				2	K0	71
98579	−27 08013	IDS11154S2747	A	11 17 52	−28 03.5	2	6.74	.039	1.12	.005	0.99	.068	3	K1 III	78,3062
98624	−60 02829	LSS 2308		11 17 53	−60 57.5	3	9.81	.082	0.26	.038	-0.72	.017	7	Be	6,540,2021
	+6 02436	G 45 - 48	★ A	11 17 55	+05 46.6	3	10.21	.027	0.75	.004	0.28	.008	4		333,1620,1658,1696
	+6 02436	G 45 - 49	★ B	11 17 55	+05 46.6	3	13.41	.008	1.42	.029	1.07	.024	3		333,1620,1658,1773
		LP 906 - 78		11 17 56	−28 03.9	1	14.29		1.39		1.10		2		3062
		Mel 105 - 24		11 17 57	−63 13.5	1	13.66		1.69				4		56
98562	+24 02344			11 17 58	+23 53.4	2	8.76	.005	0.64	.020	0.21		5	G2 V	20,3077
		Lod 336 # 5		11 17 58	−58 35.8	1	11.03		1.86		1.74		1		287
		Lod 336 # 3		11 17 58	−58 36.2	1	12.18		1.04		0.33		1		287
		LP 612 - 66		11 17 59	−01 25.5	1	10.51		0.76		0.34		2		1696
		Mel 105 - 25		11 18 00	−63 13.9	1	14.28		0.38		0.29		1		56
98639	−60 02832			11 18 01	−60 44.3	1	9.77		0.28		0.14		2	A2/3mA7-A7	127
98659	−66 01572	IDS11159S6630	AB	11 18 01	−66 46.6	1	7.50		0.11		-0.21		2	B8 II	401
	−34 07381			11 18 02	−34 33.8	2	10.39	.005	0.81	.005	0.45	.015	3	G0	78,1696
		Lod 336 # 2		11 18 04	−58 35.8	1	10.65		1.68		1.73		1		287
98657	−60 02833	LSS 2310		11 18 05	−61 07.3	1	9.62		0.51		-0.37		2	B1 Ib/II	540
		LOD 23 # 18		11 18 06	−60 46.1	1	12.00		0.34		0.06		2		127
		Ly 12		11 18 06	−62 28.	1	11.67		0.13		-0.51		3		279
		LOD 23 # 17		11 18 07	−60 48.1	1	12.21		0.22		0.04		2		127
98695	−71 01238	HR 4389		11 18 07	−71 43.2	2	6.40	.005	0.05	.000	-0.48	.000	7	B5 V	15,1075
98634	−40 06653			11 18 08	−40 41.0	1	9.41		0.06		0.03		1	A0 V	1770
98572	+52 01558			11 18 09	+52 02.2	1	7.32		0.39		-0.02		2	F0	985
98669	−60 02835			11 18 09	−60 49.4	1	7.70		0.23		0.14		2	F0/2 IV	127
		G 176 - 25		11 18 12	+37 55.2	2	11.34	.107	1.14	.004	1.08	.007	4	K7	1726,7010
98678		RS Cen		11 18 17	−61 36.0	1	8.12		1.71		1.48		1	M5 IIIe	975
98693	−59 03368	IDS11162S6002	A	11 18 21	−60 18.4	1	8.51		0.28		0.15		1	A2/3 V	1771
98649	−22 03121			11 18 24	−22 56.4	1	8.02		0.66		0.20		1	G3/5 V	78
98644	+13 02391			11 18 31	+12 49.5	2	8.59	.010	0.59	.010	0.07	.019	3	K0	1648,8087
	+28 01993			11 18 32	+27 45.9	1	9.27		1.18		1.16		2	K2	1569
98664	+6 02437	HR 4386		11 18 33	+06 18.2	7	4.04	.008	-0.06	.008	-0.10	.035	30	B9.5Vs	3,15,1363,1417,1509*
98618	+59 01369			11 18 36	+58 45.5	1	7.66		0.64		0.13		2	G5 V	186
98722	−57 04608			11 18 37	−58 12.7	1	8.02		0.39				4	F0/2 IV/V	2012
		AAS9,163 T5# 15		11 18 39	+58 58.0	1	10.29		0.55		0.00		3		186
98732	−58 03496			11 18 42	−58 25.8	1	7.02		1.00				4	K0 III	2033
	−62 01971	LSS 2312		11 18 42	−62 56.2	1	11.24		0.24		-0.57		2	B0	761
98718	−53 04498	HR 4390	★ AB	11 18 43	−54 13.0	7	3.89	.005	-0.16	.006	-0.60	.013	25	B5 Vn	15,26,1075,1637,1770*

Table 1 603

HD	DM	Other Id	N Rem	α_{1950}	δ_{1950}	S	V	σ_V	B–V	σ_{B-V}	U–B	σ_{U-B}	n	Spectrum	References
98733	−59 03376	LSS 2311		11 18 44	−59 53.2	6	7.95	.012	0.16	.013	−0.70	.018	12	B1/2 II	2,6,138,164,540,976
	−60 02841			11 18 45	−60 48.0	1	11.28		0.06		−0.35		2		127
	−60 02842			11 18 46	−60 46.7	2	10.85	.141	0.07	.039	−0.49		6		127,2021
98742	−60 02843			11 18 48	−60 43.8	1	10.00		0.31		0.14		2	A7/9 V	127
98662	+59 01370			11 18 50	+58 41.8	1	9.61		0.22		0.10		2	A5	186
		G 10 - 15		11 18 54	+08 26.6	2	13.13	.015	0.74	.005	0.11	.010	3		333,1620,1658
		Feige 40		11 18 54	+11 35.9	3	11.16	.044	−0.13	.014	−0.62	.015	4	B6	1026,1264,1298
98601	+77 00432			11 18 55	+77 20.7	1	8.61		0.68		0.21		2	G0	1375
		LP 792 - 18		11 18 56	−16 59.8	1	11.72		0.88		0.55		2		1696
		G 10 - 16		11 18 57	+05 34.0	1	12.26		1.17		1.13		1		1658
98712	−19 03242	SZ Crt	★ AB	11 18 57	−20 10.7	3	8.58	.121	1.23	.349	1.29	.135	6	K4/5 V	158,497,1705
		G 56 - 30		11 18 58	+18 28.5	2	10.96	.015	0.52	.005	−0.13	.005	3	G0	333,1620,1696
98673	+57 01316	HR 4388		11 18 58	+57 20.9	2	6.43	.005	0.16	.005	0.09	.005	4	A7 Vn	985,1733
304201	−58 03503			11 19 00	−59 01.6	1	9.55		0.04		−0.78		2	A2	476
308433	−61 02222			11 19 03	−61 46.0	1	10.46		0.61		−0.41		2	B0	1737
		Wolf 383		11 19 04	+13 47.9	1	12.41		0.59		−0.03		2		1696
		G 176 - 27		11 19 04	+50 54.1	1	11.29		0.83		0.44		2		1658
98767		CoD -55 03979		11 19 05	−55 29.3	1	10.59		1.81		2.09		2	N3	864
		G 10 - 17		11 19 06	+06 25.9	2	13.58	.025	1.62	.029	1.27		8		940,3078
98696	+46 01717	IDS11163N4553	AB	11 19 06	+45 36.4	1	8.11		0.25		0.05		2	A3 III	1601
	−55 04306			11 19 07	−55 51.0	1	9.96		0.63		−0.38		2	O9.5V	540
	−58 03508			11 19 09	−59 07.2	1	9.60		0.06		−0.75		3		476
98710	+35 02241			11 19 10	+35 08.1	1	8.73		0.26		0.10		2	Am	1601
98781	−49 06087			11 19 11	−50 14.6	1	9.66		0.11		0.07		1	A0 V	1770
98837	−73 00834			11 19 11	−74 11.4	1	8.54		0.40		0.36		2	A3 V	1776
98736	+19 02443	IDS11166N1844	AB	11 19 12	+18 27.9	1	7.94		0.89		0.64		3	G5	196
98763	−21 03292			11 19 17	−21 49.9	1	7.35		−0.06		−0.35		1	B8 V	55
98810	−62 01973			11 19 18	−63 02.1	1	9.77		0.20		−0.08		2	B8/9 III/IV	761
98818	−60 02850			11 19 21	−60 57.1	3	7.35	.007	1.12	.000	0.97		9	K0 III	258,1075,2012
		G 45 - 53		11 19 26	+12 19.8	1	12.48		1.36		1.17		1		333,1620
		GD 136		11 19 26	−04 55.2	1	13.71		0.45		−0.16		1		3060
98815	−48 06404			11 19 27	−48 40.2	1	8.22		−0.06		−0.36		3	B8/9 II/II	1770
98817	−60 02852			11 19 27	−60 42.8	1	8.30		2.14		1.97		2	M2 Iab/b	127
98804	−52 04495			11 19 28	−52 39.6	1	9.66		0.01		−0.08		2	B9.5 IV/V	1770
306403	−58 03516			11 19 28	−59 19.6	1	9.38		0.22		−0.55		3	B8	476
		A1 5		11 19 30	+14 05.9	1	16.37		0.45		−0.05		1		98
		WLS 1120-20 # 5		11 19 30	−17 34.2	1	12.90		0.80		0.13		2		1375
	−58 03518			11 19 31	−58 38.4	1	11.31		0.10		−0.58		2		476
306414	−59 03388			11 19 32	−59 33.8	2	9.99	.021	0.59	.007	−0.45	.002	3	B0	540,1737
98800	−24 09706	IDS11172S2414	AB	11 19 37	−24 30.2	1	9.08		1.26		1.07		1	K4 V	3072
98913	−74 00803	Cha F # 45		11 19 38	−75 10.2	1	9.54		0.51		−0.01		2	F6 V	1633
		WLS 1120-20 # 10		11 19 42	−19 15.4	1	13.30		0.73		0.09		2		1375
98846	−53 04507			11 19 42	−53 40.2	1	9.19		0.02		−0.02		1	A0 III/IV	1770
	−61 02238	LSS 2318		11 19 42	−61 33.4	1	10.94		0.72		−0.36		2	B0.5 Ia	1737
98876	−62 01977	IDS11176S6256	AB	11 19 43	−63 12.8	1	9.37		0.21		−0.28		1	B5 Vn	761
	+15 02325	G 56 - 34		11 19 44	+14 43.2	1	10.46		0.96		0.74		1	K2	333,1620
		A1 7		11 19 45	+14 16.8	1	15.00		0.51		0.04		1		98
306387	−60 02857	LSS 2319	★ AB	11 19 45	−60 45.3	2	9.58	.083	0.12	.049	−0.69		6	B1.5III	6,2021
	+31 02260			11 19 46	+31 10.8	1	10.27		1.01		0.83		1		280
98797	+30 02139			11 19 48	+29 32.4	1	8.79		0.94		0.72		1	G5	280
98824	+18 02481			11 19 52	+17 42.7	3	7.02	.014	1.07	.011	0.95	.004	5	K1 III	1003,1080,3035
98873	−52 04503			11 19 53	−53 04.9	1	9.37		0.05		−0.02		1	A0 IV	1770
	+24 02348			11 19 54	+23 54.5	1	9.71		0.31		0.10		2	A2	272
98772	+65 00828	HR 4391		11 19 54	+64 36.3	3	6.03	.008	0.08	.005	0.07	.008	7	A3 V	985,1501,1733
98867	−38 07092			11 19 56	−38 49.8	1	7.34		0.01		−0.05		3	B9.5 V	1770
		LP 850 - 23		11 19 57	−26 57.3	2	14.84	.138	0.99	.010	0.55	.010	7		1773,3060
98868	−42 06925			11 19 58	−43 03.1	2	9.63	.000	0.59	.009	0.04		5	G3 V	78,2033
	+31 02261			11 19 59	+30 44.0	1	10.58		0.49				3		280
		LP 850 - 24		11 20 00	−26 57.2	3	12.63	.117	0.56	.000	−0.17	.010	8		1696,1773,3060
98892	−43 07006	HR 4393		11 20 01	−44 22.3	2	6.12	.001	0.93	.005	0.68		6	G8 III	58,2007
	−60 02863			11 20 01	−60 48.1	1	11.62		0.21		−0.24		2		127
98911	−60 02865			11 20 03	−60 27.4	1	8.80		−0.01		−0.66		2	B2 IV	401
98841	+32 02141			11 20 04	+31 41.1	1	9.08		0.29		0.10		2	F0	1569
98866	−30 09150	IDS11176S3021	A	11 20 04	−30 37.7	2	8.15	.014	0.52	.005	0.03		7	F5 V	1279,2012
98866	−30 09150	IDS11176S3021	B	11 20 04	−30 37.7	1	11.52		1.24		1.26		3		1279
98839	+44 02083	HR 4392		11 20 05	+43 45.4	8	4.99	.013	1.00	.015	0.80	.010	33	G8 II	15,1080,1119,1363*
	+29 02151			11 20 06	+28 55.0	1	10.03		0.53				2		280
98927	−60 02867	LSS 2322		11 20 06	−60 47.9	6	9.16	.081	0.10	.026	−0.77	.024	13	B2/3 Ve	2,6,127,540,976,1021
98840	+34 02217			11 20 08	+33 30.3	1	8.47		0.21		0.03		2	A2 III-IV	272
		WLS 1120-20 # 7		11 20 08	−21 54.6	1	11.95		0.67		0.15		2		1375
	+27 02011			11 20 10	+26 57.1	1	9.88		1.49		1.97		1	K5	280
98851	+32 02142			11 20 11	+32 06.2	1	7.40		0.33		0.12		2	F2	1569
98922	−52 04507	IDS11179S5249	A	11 20 12	−53 05.7	1	6.78		0.04		−0.02		3	B9 Ve	1770
99015	−76 00662	HR 4397		11 20 15	−77 20.0	2	6.42	.005	0.20	.000	0.10	.010	7	A5 III/IV	15,1075
98882	+16 02249			11 20 17	+15 48.5	1	8.17		1.07				2	K0	882
98883	+15 02326			11 20 18	+14 35.9	2	8.30	.015	1.25	.000	1.31		4	K0	882,1648
	+27 02012			11 20 19	+26 53.6	1	10.45		1.35		1.54		1		280
98849	+59 01371			11 20 19	+58 51.8	1	9.00		1.05		0.85		4	G5	186
99014	−74 00804	Cha F # 46	★ AB	11 20 20	−74 50.9	1	9.72		0.66		0.10		2	G2 V	1633
98955	−58 03534	LSS 2324		11 20 21	−58 40.8	3	8.73	.023	−0.03	.005	−0.83	.004	6	B2 Ib/IIN	401,540,976

HD	DM	Other Id	N	Rem	α_{1950}	δ_{1950}	S	V	σ_V	B–V	σ_{B-V}	U–B	σ_{U-B}	n	Spectrum	References
308460	−62 01981				11 20 27	−62 49.4	1	11.23		0.28		0.15		2	A0	761
98986	−62 01982				11 20 27	−63 16.4	1	9.20		0.16		-0.26		1	Ap Si	761
		A1 11			11 20 28	+14 17.9	1	16.86		0.20		-0.28		1		98
		LP 374 - 39			11 20 29	+26 10.1	1	15.14		1.84				2		1663
237995	+59 01372				11 20 29	+58 52.2	1	9.78		0.45		-0.06		2	G0	186
	+32 02143				11 20 32	+32 12.9	1	9.98		0.97		0.72		1	K0	280
	+20 02594	G 56 - 36			11 20 40	+20 10.4	2	9.97	.034	0.48	.000	-0.14	.000	3	G0	333,1620,3077
98960	+0 02782	HR 4394			11 20 44	+00 24.4	2	6.04	.005	1.46	.000	1.78	.000	7	K3	15,1417
98993	−35 07163	HR 4396		★ A	11 20 47	−35 53.4	2	5.00	.005	1.46	.005	1.63		6	K4 III	2032,3005
99024	−59 03411	LSS 2327			11 20 48	−60 14.7	2	9.65	.005	0.02	.010	-0.77		6	B2 Ib/II	6,2021
		LOD 23 # 9			11 20 50	−60 48.2	1	12.17		0.25		0.08		2		127
99022	−56 04449	HR 4398			11 20 51	−56 30.3	2	5.78	.005	0.01	.000			7	A3/5p	15,2012
98991	−17 03367	HR 4395			11 20 53	−18 30.3	4	5.08	.004	0.43	.008	-0.05	.010	8	F3 IV	15,1008,2012,3026
98969	+39 02441				11 20 55	+38 58.1	1	8.87		0.37		-0.03		3	F8	1601
	+27 02014				11 20 56	+27 17.3	1	9.75		1.05				2	K2	280
98957	+33 02109				11 20 56	+32 43.0	1	8.96		0.33		0.00		2	F0	1569
306388	−60 02884	LSS 2329			11 20 57	−60 48.5	1	10.86		0.03		-0.61		2	B0	127
		GD 137			11 20 58	+24 09.8	1	16.25		0.28		-0.53		1		3060
		POSS 733 # 2			11 20 58	−11 24.7	1	15.33		1.41				2		1739
98967	+52 01561				11 21 00	+52 18.9	1	7.96		0.40		-0.02		2	F0	1375
98970	+29 02153				11 21 01	+28 59.6	2	8.33	.004	1.20	.008	1.14	.008	8	G5	280,1775
	+58 01316				11 21 03	+57 56.3	1	9.93		0.58		-0.01		2	G5	186
98989	+35 02246				11 21 05	+34 31.0	1	9.28		0.00		0.01		3	A0 V	272
308459	−62 01985				11 21 05	−62 48.8	1	11.08		0.18		-0.33		2	B5	761
	+30 02144				11 21 07	+29 44.9	1	10.84		0.51				2		280
		A1 14			11 21 10	+14 22.1	1	16.40		0.60		-0.30		1		98
99104	−64 01657	HR 4401		★ AB	11 21 11	−64 40.8	5	5.10	.005	-0.08	.006	-0.44	.014	19	B5 V	15,164,1075,1771,2038
		G 10 - 19			11 21 12	+08 50.1	2	11.18	.001	1.61	.073	1.18	.025	6	M0	3078,7009
	+25 02369				11 21 12	+25 13.6	1	9.02		1.29		1.55		1	K5	280
		AAS9,163 T5# 19			11 21 13	+58 42.1	1	10.79		1.11		1.08		7		186
		G 122 - 22			11 21 15	+45 49.1	1	11.93		1.17		0.93		3		7010
99066	−54 04508				11 21 15	−54 34.3	1	9.45		-0.06		-0.36		2	B8 III	401
99065	−52 04524				11 21 16	−52 54.7	1	8.98		0.01		-0.03		1	B9.5 V	1770
	+59 01373				11 21 17	+58 38.0	1	10.02		1.20		1.21		8		186
		POSS 733 # 1			11 21 17	−11 39.0	1	15.07		1.52				1		1739
99085	−58 03557				11 21 17	−58 48.3	1	8.10		-0.07		-0.43		2	B9 II	401
99161	−75 00724	Cha F # 48			11 21 17	−75 44.0	1	8.87		0.15		-0.05		2	B9.5V	1633
		LP 492 - 53			11 21 19	+09 54.3	1	12.66		0.44		-0.27		2		1696
99028	+11 02348	HR 4399		★ AB	11 21 19	+10 48.3	9	3.94	.009	0.41	.006	0.07	.004	31	F1 IV +G3 V	15,292,1007,1008,1013*
		RX Leo			11 21 20	+26 53.3	2	11.64	.083	0.25	.054	0.11	.016		F2	668,699
		AAS12,381 27 # A			11 21 21	+27 07.9	1	10.36		0.62		0.23		1		280
99077	−38 07117				11 21 21	−39 07.9	1	9.28		0.37		0.18		1	Fm δ Del	55
237997	+59 01374				11 21 26	+59 22.8	1	9.61		0.58		0.07		7	K0	186
99055	+2 02418	HR 4400			11 21 28	+01 41.0	2	5.38	.005	0.94	.000	0.66	.000	7	G8 III	15,1417
		BPM 20912			11 21 30	−50 42.	1	14.86		0.07		-0.74		4	DA	3065
99149	−65 01662				11 21 30	−65 33.6	1	8.15		0.13		0.08		2	A1 V	1771
		AAS12,381 34 # A			11 21 33	+33 56.9	1	10.28		0.96		0.85		1		280
99054	+13 02401				11 21 34	+12 44.2	2	8.24	.040	1.29	.000	1.33		4	K2	882,3077
	+24 02350				11 21 36	+23 52.4	2	10.10	.034	0.15	.015	0.11	.039	3	A2	272,280
99146	−58 03566	LSS 2332			11 21 37	−59 01.8	2	8.13	.044	0.23	.014	-0.71	.011	4	B3	540,976
		G 120 - 45			11 21 38	+21 38.1	3	14.25	.005	0.28	.025	-0.53	.010	8		203,1281,3078
308456	−62 01988				11 21 38	−62 38.5	1	11.04		0.22		0.07		2	A0	761
237998	+59 01375				11 21 39	+59 10.2	1	10.30		1.41		1.67		8	K5	186
99088	+14 02382				11 21 40	+14 10.6	1	8.34		0.60				3	K0	882
	+30 02148				11 21 41	+29 46.2	1	9.73		0.27		0.13		1	F0	280
		LP 672 - 76			11 21 41	−08 15.0	1	13.68		1.40		1.11		1		1696
99158	−60 02891	LSS 2333			11 21 41	−60 35.5	1	8.62		0.00		-0.64		2	B3 III	401
99159	−60 02892				11 21 42	−60 49.0	1	9.07		-0.04		-0.50		2	B3 V	127
99160	−60 02893	LSS 2334			11 21 44	−61 04.8	6	9.14	.022	0.21	.014	-0.72	.022	14	B7 Ib	2,6,540,976,1021,8100
		G 147 - 50			11 21 45	+35 45.4	1	13.19		0.63		-0.02		2		1658
		AAS9,163 T5# 24			11 21 45	+58 37.0	1	10.82		1.45		1.68		7		186
99175	−62 01990				11 21 45	−62 45.4	1	8.64		0.23		0.19		2	A0 V	761
	+33 02111				11 21 52	+33 20.2	1	10.02		1.01		0.88		1		280
	+34 02219				11 21 52	+33 42.6	1	10.13		0.92		0.56		1		280
99205	−69 01537	LSS 2336			11 21 52	−69 04.3	2	9.59	.000	0.01	.004	-0.72	.005	3	B1 III	540,976
99193	−62 01992	LSS 2335			11 21 53	−63 24.4	2	8.89	.012	0.21	.014	-0.55	.014	4	B2 III	540,976
		G 10 - 20			11 21 55	+09 40.9	1	12.87		1.01		0.75		1	K0	1696
99120	+22 02354				11 21 55	+22 03.2	1	9.53		0.09		0.05		1	A1 V	272
99119	+30 02149				11 21 57	+30 26.5	2	9.69	.004	0.31	.004	0.02	.000	10		280,1775
99171	−41 06529	HR 4403			11 21 58	−42 23.7	7	6.12	.009	-0.18	.005	-0.79	.009	22	B2 V	15,158,1075,1732,1770*
	+9 02487				11 22 01	+08 51.2	1	10.98		0.14		0.11		2	A2	1026
	+26 02207				11 22 02	+26 02.3	1	10.26		0.78		0.41		1		280
99167	−10 03260	HR 4402			11 22 05	−10 35.1	5	4.82	.011	1.55	.010	1.95	.050	21	K5 III	15,1024,2013,2029,3052
		A1 18			11 22 07	+14 20.7	1	16.21		0.51		-0.03		1		98
308458	−62 01994				11 22 08	−62 46.3	1	10.52		0.17		-0.67		2	B0.5V	540
99230	−66 01581				11 22 08	−66 43.6	1	8.44		0.23		0.12		2	A3 V	1771
		POSS 733 # 4			11 22 10	−12 07.7	1	17.10		1.56				2		1739
99264	−71 01248	HR 4406			11 22 11	−71 58.9	6	5.58	.006	0.05	.010	-0.59	.007	22	B2 IV- V	15,26,164,1075,2038,8023
99165	+30 02150				11 22 15	+30 23.7	1	9.25		1.03		0.96		1	G5	280
99164	+32 02146				11 22 15	+32 06.0	1	7.69		0.94		0.71		1	K0	280

Table 1 605

HD	DM	Other Id	N	Rem	α_{1950}	δ_{1950}	S	V	σ_V	B–V	σ_{B-V}	U–B	σ_{U-B}	n	Spectrum	References
99248	−62 01997				11 22 16	−63 04.0	1	8.91		0.13		−0.05		1	A0 IV	761
		MCC 39			11 22 17	+40 16.7	1	10.33		1.36		1.25		3	K4	3072
306480	−58 03578				11 22 17	−59 19.7	1	11.13		0.16				4	B5 III:	2021
	+27 02016				11 22 19	+27 27.6	1	10.24		0.94		0.69		1	K0	280
		AAS12,381 35 # A			11 22 19	+34 58.7	1	10.48		1.02		0.84		1		280
99196	+12 02335	HR 4404			11 22 23	+11 42.3	5	5.81	.018	1.38	.011	1.57	.013	12	K4 III	15,252,1003,1080,3035
99211	−16 03244	HR 4405		★ AB	11 22 23	−17 24.5	6	4.07	.012	0.21	.005	0.10	.005	21	A9 V	15,1075,1425,2012*
		Steph 941			11 22 24	+02 45.0	1	11.78		1.53		1.26		1	M0	1746
		V442 Cen			11 22 25	−35 37.3	1	11.88		−0.01		−0.65		1		1471
99263	−60 02906				11 22 27	−60 46.8	1	9.86		0.04		−0.43		2	B2/3	127
99279	−60 02911	IDS11204S6106		AB	11 22 29	−61 22.4	3	7.21	.014	1.26	.000	1.18		6	K5/M0 V	258,1075,1705
	−58 03581				11 22 30	−58 39.3	1	11.51		0.48		0.00		2		434
	+30 02152				11 22 31	+30 04.6	1	10.35		0.74		0.34		1	G0	280
99177	+50 01820				11 22 31	+49 39.0	1	8.21		0.52		0.10		2	F8	1733
	−42 06957	NGC 3680 - 79			11 22 31	−42 49.6	1	12.49		0.47		0.06		1		3017
	+24 02352				11 22 35	+24 11.0	1	10.37		0.96		0.71		1	G5	280
		G 163 - 80			11 22 35	−05 39.9	1	11.49		1.07		1.03		1		1658
		NGC 3680 - 69			11 22 35	−43 00.0	1	14.32		0.77		0.23		2		3017
99278	−60 02912				11 22 35	−60 51.0	2	9.87	.104	0.00	.020	−0.34	.000	5	B5/7 Ib/II	2,127
99194	+47 01871				11 22 36	+47 00.9	1	8.69		0.55		0.05		2	G0	1601
	−42 06958	NGC 3680 - 3			11 22 36	−42 58.9	1	11.60		0.55		0.02		3		3017
99207	+30 02153				11 22 37	+30 02.1	1	7.71		0.37		0.02		2	F0	1569
		G 147 - 54			11 22 37	+35 28.5	1	12.95		0.51		−0.19		2		1658
	−42 06959	NGC 3680 - 1			11 22 37	−42 53.9	1	11.98		0.47		0.00		2		3017
	+28 02002				11 22 40	+28 13.1	1	9.60		0.83		0.46		1	G5	280
		NGC 3680 - 70			11 22 40	−43 00.9	1	13.87		0.71		0.16		2		3017
		AAS12,381 35 # B			11 22 41	+34 48.3	1	10.19		1.16		1.21		1		280
	−19 03263				11 22 41	−20 28.3	1	10.39		0.75		0.30		2		1375
		NGC 3680 - 2			11 22 42	−42 55.4	1	14.75		0.69		0.20		2		3017
		NGC 3680 - 4			11 22 42	−42 59.3	1	12.95		0.50		0.04		3		3017
99316	−62 02003	LSS 2337			11 22 42	−63 09.4	6	7.40	.021	0.32	.008	−0.22	.030	14	B9.5Iab	164,191,540,761,976,8100
		NGC 3680 - 17			11 22 46	−42 53.0	1	12.05		1.01		0.81		2		3017
	+26 02209				11 22 47	+25 41.0	1	10.08		0.28		0.00		1	A5	280
99235	+35 02250				11 22 47	+35 26.5	1	8.77		0.43		0.05		2	G0	1569
		NGC 3680 - 72			11 22 48	−43 03.5	1	12.74		0.44		0.04		2		3017
99233	+43 02113				11 22 49	+42 54.4	1	7.87		0.54		−0.04		2	F8	1601
		NGC 3680 - 71			11 22 50	−43 02.8	1	12.58		0.79		0.42		2		3017
		NGC 3680 - 14			11 22 51	−42 57.0	1	12.40		0.48		0.01		3		3017
	−42 06962	NGC 3680 - 18			11 22 52	−42 52.5	1	12.10		0.78		0.33		3		3017
	−42 06963	NGC 3680 - 13			11 22 52	−42 57.9	1	10.90		1.15		1.06		2		3017
		NGC 3680 - 15			11 22 54	−42 55.3	1	12.54		0.51		0.05		2		3017
		NGC 3680 - 5			11 22 54	−42 59.9	1	12.86		0.49		0.00		3		3017
306449	−60 02918				11 22 54	−60 53.1	1	10.38		0.08		−0.32		2	B8	127
		POSS 733 # 5			11 22 55	−11 50.1	1	17.94		1.79				4		1739
	+29 02159				11 22 56	+29 12.9	1	9.31		1.18		1.29		1	K0	280
		NGC 3680 - 19			11 22 56	−42 52.6	1	12.49		0.51		0.06		3		3017
		NGC 3680 - 16			11 22 56	−42 54.0	1	13.35		0.55		−0.01		2		3017
		NGC 3680 - 6			11 22 56	−43 01.0	1	12.48		0.52		0.08		4		3017
	+27 02018				11 22 59	+26 50.2	1	10.79		0.95		0.80		2		1569
99267	+30 02154				11 22 59	+30 15.7	1	6.87		0.31		0.02		2	F0	1569
		KUV 11230+4240			11 22 59	+42 40.1	1	16.40		0.26		−0.64		1	DA	1708
99354	−60 02920	LSS 2338			11 22 59	−61 00.1	5	8.94	.006	0.10	.017	−0.73	.021	13	B1 IIIne	2,6,540,976,2021
99355	−60 02919	AZ Cen			11 22 59	−61 05.7	3	8.63	.097	0.69	.039	0.45		2	G0	688,1488,1490
99285	+17 02356	HR 4408		★ A	11 23 00	+16 43.9	3	5.61	.026	0.35	.009	−0.02	.015	7	F3 IV	70,254,3053
99232	+59 01377				11 23 00	+59 02.2	1	9.10		1.05		0.87		2	G5	186
		NGC 3680 - 12			11 23 00	−42 58.3	1	15.06		0.75		0.26		1		3017
	−42 06971	NGC 3680 - 20			11 23 02	−42 54.9	1	10.20		1.01		0.73		2		3017
	+58 01317				11 23 04	+58 27.3	1	10.33		0.51		−0.09		5		186
99322	−35 07189	HR 4409			11 23 04	−35 47.3	2	5.22	.005	1.00	.010	0.77		6	K0 III	2006,3069
	−42 06974	NGC 3680 - 11			11 23 05	−42 59.3	1	10.96		1.09		0.82		3		3017
99302	+27 02021				11 23 07	+27 01.3	2	7.30	.009	0.23	.000	0.09	.005	4	Am	280,8071
		POSS 733 # 6			11 23 07	−11 25.8	1	18.70		1.66				3		1739
99333	−37 07235	HR 4411		★ AB	11 23 07	−37 28.3	3	5.89	.008	1.55	.015	1.71	.005	8	M3 III	404,2007,3005
99349	−41 06540				11 23 07	−42 23.8	1	7.76		1.06		0.84		6	K0 III	1673
		NGC 3680 - 80			11 23 07	−42 54.8	1	14.69		0.73		0.26		3		3017
99266	+52 01563	IDS11204N5241		AB	11 23 08	+52 24.4	1	7.32		0.42		0.01		3	F0	1566
99283	+56 01518	HR 4407			11 23 08	+56 07.5	1	5.70		1.00		0.75		2	K0 III	1733
		NGC 3680 - 21			11 23 08	−42 55.3	1	14.57		0.65		0.16		3		3017
		NGC 3680 - 23			11 23 08	−42 58.7	1	14.06		0.64		0.02		1		3017
	+27 02019				11 23 09	+27 26.0	1	9.88		0.45				3	F2	280
		NGC 3680 - 22			11 23 09	−42 56.2	1	13.47		0.55		−0.02		3		3017
		NGC 3680 - 10			11 23 09	−43 00.3	1	12.91		0.44		0.03		2		3017
		NGC 3680 - 8			11 23 09	−43 01.5	1	12.80		1.32		1.30		2		3017
		NGC 3680 - 39			11 23 10	−42 54.3	1	12.41		0.52		0.07		2		3017
		NGC 3680 - 81			11 23 10	−42 58.7	1	13.84		0.48		0.01		1		3017
		NGC 3680 - 24			11 23 10	−42 58.9	1	13.92		0.56		−0.02		1		3017
		NGC 3680 - 9			11 23 10	−43 00.9	1	14.20		0.55		0.05		3		3017
		NGC 3680 - 73			11 23 10	−43 02.4	1	15.16		0.61		0.02		2		3017
		NGC 3680 - 40			11 23 11	−42 54.0	1	12.55		1.12		1.00		2		3017
		NGC 3680 - 7			11 23 11	−43 02.0	1	13.50		1.60				2		3017

HD	DM	Other Id	N Rem	α_{1950}	δ_{1950}	S	V	σ_V	B–V	σ_{B-V}	U–B	σ_{U-B}	n	Spectrum	References
		NGC 3680 - 38		11 23 12	−42 54.9	1	12.47		0.53		0.05		4		3017
		NGC 3680 - 33		11 23 12	−42 57.6	1	11.88		0.52		0.06		1		3017
		NGC 3680 - 25		11 23 12	−42 58.8	1	13.89		0.55		-0.06		1		3017
		L 144 - 68		11 23 12	−62 15.	1	13.74		1.27		1.17		2		3062
99391	−60 02923			11 23 13	−60 42.0	3	9.13	.103	-0.02	.013	-0.68	.012	6	B3 III/IV	127,540,976
99332	−26 08545			11 23 14	−27 16.4	1	7.93		0.32				4	F0 V	2012
		NGC 3680 - 36		11 23 14	−42 56.4	1	13.44		0.56		0.03		3		3017
		NGC 3680 - 35		11 23 14	−42 57.0	1	13.17		0.50		0.01		4		3017
	−42 06976	NGC 3680 - 34		11 23 14	−42 57.5	1	10.69		0.87		0.50		3		3017
	−42 06977	NGC 3680 - 26		11 23 14	−42 59.6	1	11.04		1.13		0.97		3		3017
306485	−59 03453	LSS 2340		11 23 15	−59 50.3	2	9.81	.030	0.12	.024	-0.65	.029	5	B3	2,540
99329	+4 02463	HR 4410		11 23 16	+04 08.1	2	6.36	.005	0.34	.005	0.03	.015	7	F1 IV	15,1417
		NGC 3680 - 37		11 23 17	−42 55.7	1	13.13		0.46		0.01		4		3017
		NGC 3680 - 30		11 23 17	−42 58.0	1	13.63		0.54		-0.02		2		3017
		Feige 41		11 23 18	+06 47.0	2	11.00		0.09	.000	-0.02	.000	2		1026,1264
		NGC 3680 - 31		11 23 18	−42 57.6	1	13.22		0.46		0.04		3		3017
	−42 06978	NGC 3680 - 27		11 23 18	−43 00.6	1	10.79		1.16		1.03		2		3017
	−42 06980	NGC 3680 - 29		11 23 19	−42 59.5	1	12.14		0.49		-0.03		2		3017
		NGC 3680 - 28		11 23 21	−43 00.0	1	15.24		0.88		0.53		1		3017
233832	+51 01664	G 122 - 23		11 23 22	+50 38.9	1	10.09		0.83		0.40		2	K0 V	1003
99363	−12 03423			11 23 23	−13 28.6	1	6.55		1.72		1.99		1	M1 III	3040
		NGC 3680 - 32		11 23 23	−42 57.7	1	12.95		0.49		0.02		3		3017
	−42 06981	NGC 3680 - 41		11 23 24	−42 53.4	1	11.01		1.15		0.97		2		3017
	+25 02374			11 23 25	+25 23.5	1	9.63		1.04		0.99		1	K5	280
99383	−38 07127			11 23 25	−38 35.9	8	9.07	.014	0.47	.008	-0.22	.011	23	F2 V	1088,1236,1311,1594*
99416	−59 03454			11 23 25	−59 54.4	4	8.88	.012	0.06	.017	-0.63	.023	9	B2 III	2,6,540,976
	−42 06983	NGC 3680 - 44		11 23 26	−42 55.8	1	10.06		1.23		1.16		3		3017
		NGC 3680 - 42		11 23 30	−42 53.5	1	13.04		0.47		0.04		1		3017
99415	−57 04688			11 23 30	−57 52.0	1	8.99		0.05		-0.55		2	B3 V	401
99359	+23 02349			11 23 31	+22 59.0	1	7.41		0.32		-0.01		2	F0	1569
	+26 02211			11 23 31	+25 56.3	1	10.17		1.10		1.11		1		280
	+59 01378			11 23 31	+58 34.6	1	9.89		0.64		0.20		3	G0	186
		NGC 3680 - 43		11 23 31	−42 53.5	1	12.50		0.45		-0.05		1		3017
		NGC 3680 - 46		11 23 31	−42 57.9	1	13.49		0.50		0.05		1		3017
		NGC 3680 - 45		11 23 32	−42 56.5	1	12.04		0.56		0.05		4		3017
99515	−77 00667	Cha F # 50		11 23 32	−77 25.5	1	9.37		0.39		0.14		2	A8 IV	1633
		NGC 3680 - 82		11 23 33	−42 59.0	1	13.50		0.50		0.03		5		3017
99453	−63 01893	HR 4413		11 23 33	−63 41.8	6	5.17	.008	0.50	.003	0.02	.002	14	F7 V	15,258,688,1488,2032,2033
	−14 03322			11 23 35	−15 26.5	3	10.44	.014	0.58	.005	-0.10	.010	6		516,1064,1696
99501	−75 00728	Cha F # 49		11 23 35	−75 37.0	1	9.36		0.46		-0.02		2	F2 V	1633
		G 56 - 39		11 23 37	+21 07.7	1	12.20		0.68		0.06		2	G3	333,1620
		NGC 3680 - 74		11 23 41	−43 04.6	1	13.81		0.01		0.15		1		3017
99469	−60 02931			11 23 44	−61 05.7	1	9.72		0.08		-0.44		3	B3/5 V	2
99373	+34 02222	HR 4412		11 23 45	+33 43.5	3	6.33	.014	0.43	.007	0.01	.009	7	F6 IV	70,1501,3053
		NGC 3680 - 58		11 23 45	−42 55.7	1	12.96		0.50		0.00		2		3017
	−42 06988	NGC 3680 - 48		11 23 45	−43 02.0	1	11.83		0.45		0.09		2		3017
		NGC 3680 - 75		11 23 45	−43 03.8	1	12.88		0.50		0.07		1		3017
99467	−58 03605			11 23 45	−59 04.7	1	7.51		-0.01		-0.28		2	B9 II/III	401
		NGC 3680 - 59		11 23 46	−42 57.1	1	12.56		1.19		1.13		2		3017
		NGC 3680 - 76		11 23 46	−43 05.5	1	12.43		1.24		1.27		1		3017
		NGC 3680 - 57		11 23 47	−42 53.7	1	12.56		0.44		0.08		2		3017
		NGC 3680 - 49		11 23 47	−43 00.9	1	13.08		0.86		0.35		2		3017
	+35 02252			11 23 48	+35 19.9	1	10.22		0.91		0.38		1		280
		NGC 3680 - 47		11 23 48	−43 01.7	1	13.75		0.86		0.39		1		3017
99419	+21 02318			11 23 50	+20 47.6	1	7.94		0.57		0.08		2	G5	1648
		NGC 3680 - 51		11 23 50	−42 57.9	1	14.70		0.68		0.08		3		3017
99404	+11 02353	G 56 - 40		11 23 51	+10 41.9	2	8.66	.000	0.68	.005	0.19	.025	2	G5	333,1620,1658
99487	−60 02934			11 23 52	−60 45.2	1	9.89		0.02		0.06		2	B9 V	127
306462	−60 02935			11 23 55	−61 18.2	1	11.43		0.14		-0.30		3	A0	2
		NGC 3680 - 60		11 23 56	−42 56.6	1	14.36		0.63		0.10		1		3017
		NGC 3680 - 52		11 23 56	−42 58.2	1	15.10		0.77		0.34		2		3017
99482	−51 05699			11 23 56	−51 56.4	1	9.69		0.00		-0.10		1	B9.5 V	1770
		NGC 3680 - 61		11 23 57	−42 57.0	1	14.46		0.79		0.31		2		3017
		NGC 3680 - 77		11 23 57	−43 04.0	1	12.72		0.49		0.03		1		3017
		W Crt		11 24 00	−17 39.0	2	10.77	.030	0.09	.025	0.09	.005	2	F4	597,700
		NGC 3680 - 50		11 24 01	−42 58.9	1	11.97		0.83		0.35		3		3017
	−42 06993	NGC 3680 - 56		11 24 03	−42 52.9	1	10.87		0.24		0.14		3		3017
99443	+26 02022			11 24 04	+26 40.2	1	8.24		0.97		0.74		1	K0	280
	−42 06994	NGC 3680 - 62		11 24 09	−43 06.0	1	9.75		0.50		0.01		1	G0	3017
99591	−75 00731			11 24 10	−75 27.1	2	9.36	.000	0.59	.000	0.08		7	F7 V	236,2012
		NGC 3680 - 53		11 24 12	−42 58.7	1	10.91		1.14		0.01		2		3017
99491	+3 02502	HR 4414	*A	11 24 13	+03 17.1	6	6.50	.008	0.79	.007	0.49	.014	23	K0 IV	13,22,1067,1080,1084,3077
99491	+3 02502	IDS11217N0333	AB	11 24 13	+03 17.1	3	6.15	.008	0.84	.020	0.62	.000	8	K0 IV	15,1417,1705
99547	−61 02315	LSS 2341		11 24 13	−61 44.6	1	10.50		0.11		-0.54		2		2
99492	+3 02503	G 10 - 23	*B	11 24 14	+03 16.7	6	7.58	.014	1.00	.012	0.93	.026	21	K2 V	13,22,1067,1084,1355,3077
99524	−42 06995	NGC 3680 - 54		11 24 15	−42 58.2	1	8.96		1.18		1.21		3	K2 III	3017
		NGC 3680 - 64		11 24 15	−43 04.6	1	12.41		0.90		0.57		1		3017
		NGC 3680 - 65		11 24 16	−43 01.8	1	13.09		0.59		0.00		2		3017
99508	−37 07243			11 24 17	−37 53.0	1	7.89		0.05		0.07		1	A1 IV/V	55
		NGC 3680 - 55		11 24 17	−42 53.8	1	12.02		0.87		0.50		2		3017

Table 1 607

HD	DM	Other Id	N Rem	α_{1950}	δ_{1950}	S	V	σ_V	B–V	σ_{B-V}	U–B	σ_{U-B}	n	Spectrum	References
		NGC 3680 - 78		11 24 17	−42 59.0	1	13.24		0.53		0.02		1		3017
		NGC 3680 - 63		11 24 17	−43 04.8	1	11.64		1.28		1.30		1		3017
99473	+29 02160			11 24 18	+28 41.0	1	7.64		0.30		0.02		2	A6 V	1569
99545	−58 03617			11 24 18	−59 08.	1	9.47		0.24		0.21		1	A7/8 V	1771
99546	−58 03620	LSS 2342		11 24 20	−59 09.7	4	8.29	.015	-0.04	.006	-0.94	.013	12	O7	6,540,976,2021
99556	−60 02941	HR 4415	★ A	11 24 20	−60 50.4	6	5.29	.041	-0.08	.008	-0.56	.014	17	B3 III/IV	2,6,26,127,2032,8023
	+29 02161			11 24 21	+29 23.8	1	10.38		1.04		0.94		1		280
99555	−58 03622			11 24 21	−59 22.6	1	8.85		-0.06		-0.60		2	B3 V	401
		NGC 3680 - 68		11 24 24	−42 54.8	1	13.30		0.58		0.08		1		3017
99520	−24 09749			11 24 25	−25 10.2	1	10.43		0.71		0.09		1	G6/8	78
		NGC 3680 - 66		11 24 25	−43 01.1	1	11.81		0.56		0.09		2		3017
99505	+22 02362			11 24 26	+22 07.8	2	7.61	.010	0.63	.000	0.10	.010	4	G5	1569,1648
		G 197 - 14		11 24 26	+59 50.0	2	11.40	.001	0.66	.000	-0.07	.006	7		1723,7010
99574	−52 04567	HR 4417	★ AB	11 24 27	−52 53.1	4	5.80	.004	0.51	.009	0.32		12	K0 III +	15,1075,1770,2009
		NGC 3680 - 67		11 24 28	−42 55.2	1	12.86		0.62		-0.05		1		3017
	+30 02157			11 24 30	+30 22.4	1	10.06		1.22				2		280
99518	+25 02376			11 24 31	+25 18.2	1	7.73		0.32		0.06		2	F0	1569
99504	+33 02116			11 24 32	+33 24.0	1	7.23		0.99		0.80		1	K0	280
	+24 02356			11 24 37	+24 24.9	1	9.80		1.10		0.94		1		280
99564	−11 03098	HR 4416	★ A	11 24 38	−12 04.9	3	5.94	.020	0.49	.007	0.01		8	F4 III-IV	15,254,2012
	+59 01379			11 24 40	+58 48.7	1	10.30		0.52		-0.05		3		186
99565	−14 03326	IDS11222S1506	AB	11 24 40	−15 22.2	2	7.64	.010	0.75	.000	0.26	.029	6	G8 V	214,3030
	−61 02325	LSS 2343		11 24 40	−61 44.5	1	11.14		0.74		-0.32		2	O7 III	1737
	+29 02163			11 24 44	+28 58.7	1	10.09		1.09				2		280
99619	−60 02948	V771 Cen		11 24 44	−61 05.6	1	6.81		1.98		2.11		1	M1/2 III	1490
99532	+57 01323			11 24 45	+56 51.0	1	8.10		0.46		-0.06		2	F8	3016
	−15 03267			11 24 46	−15 47.1	1	10.85		1.20		1.04		2	K7 V	3072
		G 120 - 51		11 24 47	+23 06.1	2	14.15	.005	1.22	.014	0.90	.005	5		419,3062
	+21 02321	G 120 - 50		11 24 48	+20 58.5	1	11.52		0.64		0.11		2	G0	1658
99548	+24 02357			11 24 48	+23 35.8	1	9.05		0.97		0.73		1	K0	280
		LP 672 - 38		11 24 48	−04 30.5	1	13.03		1.00		0.72		1		1696
308510	−62 02021			11 24 50	−62 39.9	1	10.16		0.30		-0.37		2	B5 II	540
	+25 02377			11 24 51	+24 45.8	1	9.84		0.93		0.84		1		280
	+28 02005			11 24 51	+28 21.7	1	10.17		0.50				3		280
238002	+58 01318			11 24 51	+58 01.0	1	9.06		1.06		0.94		3	K0	186
99644	−68 01506	IDS11227S6830	A	11 24 51	−68 46.3	1	8.10		0.19		0.02		2	A3 III/IV	1771
99531	+59 01380			11 24 52	+58 58.7	1	8.76		0.91		0.57		4	G5	186
99582	+24 02360	G 120 - 52	★ A	11 24 54	+24 19.1	1	9.27		0.77		0.53		1	K0	280
99582	+24 02360	IDS11223N2435	B	11 24 54	+24 19.1	1	9.74		1.14		1.20		1		280
	+30 02158			11 24 55	+29 43.9	1	9.73		1.08				2	K0	280
99594	+27 02024			11 25 00	+26 43.7	1	8.09		1.14		1.02		1	K2 III	280
99580	+35 02255			11 25 02	+34 31.9	1	8.68		1.09		0.84		1	K0	280
99593	+30 02159			11 25 03	+30 06.0	1	9.18		0.31		-0.01		2	F2 V	1569
	+4 02470			11 25 04	+04 15.3	2	10.66	.015	1.41	.010	1.25		2	K7	801,1746
	+13 02410	G 56 - 42		11 25 05	+12 55.2	1	10.60		1.13		1.10		1	K4.5	333,1620
	+29 02164			11 25 05	+28 58.4	1	10.30		0.55				2		280
99592	+45 01924	ST UMa		11 25 07	+45 27.6	2	6.19	.112	1.68	.005	1.70	.005	3	M4 III	1601,3001
99625	−25 08682			11 25 09	−25 35.1	1	6.76		1.08				1	K0 III	1705
	+23 02350			11 25 10	+23 18.2	1	10.43		0.38		-0.06		3		1569
99608	+36 02187			11 25 13	+35 38.8	1	8.53		1.18		1.27		1	K2	280
	+25 02380			11 25 15	+24 47.4	1	9.74		1.31		1.56		1		280
	+27 02026			11 25 16	+27 12.9	1	10.20		0.44		-0.04		1		280
	+34 02227			11 25 16	+33 41.4	1	9.61		1.03		0.89		1	K0	280
99682	−55 04384			11 25 17	−55 37.7	2	8.67	.005	0.50	.019	-0.06		4	F6 V	1594,6006
	+25 02381			11 25 19	+24 34.9	1	9.75		0.38		-0.02		2	F0	1569
99651	−0 02442	HR 4419	★ AB	11 25 20	−01 25.5	2	6.24	.005	1.04	.000	0.74	.005	7	K0	15,1256
99701	−52 04579			11 25 20	−53 00.2	1	9.89		0.10		0.09		1	A1 V	1770
99648	+3 02504	HR 4418	★ A	11 25 22	+03 07.9	10	4.95	.016	1.00	.004	0.79	.009	34	G8 II-III	3,15,1007,1013,1080*
	+27 02027			11 25 22	+27 12.9	1	9.98		0.75		0.46		1		280
99759	−75 00732	Cha F # 52		11 25 22	−75 39.6	1	8.48		0.27		0.06		2	B9.5V	1633
99620	+56 01521	IDS11226N5613	AB	11 25 24	+55 56.6	2	7.69	.005	0.20	.000	0.12	.005	5	Am	1502,8071
99647	+12 02338			11 25 26	+12 14.9	1	6.56		1.56		1.89		2	K2	1648
		LP 672 - 39		11 25 26	−08 53.7	3	12.38	.014	1.46	.040			10		158,1663,1705
99632	+48 01944			11 25 28	+47 43.9	2	8.10	.013	0.12	.009	0.08	.013	5	A3 V	1375,8071
	+37 02181	IDS11228N3651	A	11 25 30	+36 35.7	1	9.37		0.98		0.74		1	K0	280
99661	+33 02117			11 25 32	+33 13.5	1	9.02		1.03		0.95		1	K0	280
99786	−73 00855			11 25 32	−73 57.1	1	9.04		0.18		-0.03		2	B9 IV/V	1776
99734	−57 04716			11 25 33	−58 24.3	1	7.65		-0.12		-0.64		2	B3 V	401
99746	−52 04587			11 25 37	−52 50.6	1	8.79		-0.03		-0.28		1	B9 IV	1770
99757	−50 06032			11 25 41	−50 34.7	1	8.17		-0.11		-0.52		1	B7 II/III	1770
99758	−59 03488			11 25 41	−60 10.4	1	9.75		0.03		0.03		3	A0 IV	2
		Feige 42		11 25 42	+39 44.0	1	13.17		0.05		0.12		2		1298
306495	−59 03489			11 25 42	−59 52.5	1	10.06		0.08		-0.48		3	B5	2
99785	−65 01668			11 25 49	−65 54.8	1	7.70		0.11		-0.42		2	B5 V	401
304437	−57 04773			11 25 50	−57 28.8	1	10.99		0.22		0.12		1	A0	549
		G 10 - 25	A	11 25 53	+07 48.5	3	10.23	.010	1.19	.011	1.09	.016	5		333,1017,1620,3025
		G 10 - 25	B	11 25 53	+07 48.5	1	10.38		0.95		0.81		1		1620
306493	−59 03493	LSS 2348		11 25 59	−59 45.9	1	10.60		0.12		-0.53		3	B1 V	2
	+50 01822			11 26 00	+49 50.1	1	9.71		1.36		0.95		2	M8	1375
99751	+13 02415			11 26 01	+13 11.1	1	8.80		0.93				2	K0	882

HD	DM	Other Id	N Rem	α₁₉₅₀ δ₁₉₅₀	S	V	σ_V	B–V	σ_B–V	U–B	σ_U–B	n	Spectrum	References
306523	−60 02973	LSS 2349		11 26 07 −60 41.7	1	10.28		0.15		−0.60		3	B5	2
99825	−60 02975	IDS11240S6027	A	11 26 09 −60 43.5	1	9.43		0.31		0.18		1	A8/9 V	1771
99846	−66 01590			11 26 09 −67 19.3	1	9.65		0.41		0.20		1	A8 V	1771
99803	−41 06565	HR 4423	⋆ A	11 26 10 −42 23.9	2	5.16	.019	−0.03	.003	−0.12	.017	5	B9 V	1770,3017
99803	−41 06565	HR 4423	⋆ AB	11 26 10 −42 23.9	2	5.07	.005	−0.03	.000			7	B9 V	15,2012
99822	−55 04392			11 26 10 −55 36.7	1	9.13		0.47		−0.01		1	F5/6 V	6006
99824	−59 03496			11 26 10 −60 20.1	1	8.98		−0.03		−0.38		3	Ap Si	2
	−50 06040			11 26 11 −51 10.5	1	10.15		0.72				4		2013
99747	+62 01183	HR 4421		11 26 14 +62 03.0	2	5.86	.005	0.36	.009	−0.08	.009	4	F5 Va vw	254,3037
99823	−57 04730			11 26 15 −57 52.0	1	7.19		−0.03		−0.11		2	A0 V	1771
99857	−65 01669	LSS 2350		11 26 15 −66 12.8	5	7.47	.049	0.13	.026	−0.72	.051	9	B1 Ib	6,191,540,976,8100
99872	−71 01253	HR 4425	⋆ AB	11 26 15 −72 11.9	3	6.08	.004	0.14	.014	−0.43	.005	14	B4/5 V	15,1075,2038
	−50 06041			11 26 17 −50 31.7	1	10.15		0.72				4	F8	2033
99856	−64 01665	LSS 2351		11 26 17 −64 36.7	2	8.74	.016	0.22	.014	−0.71	.021	4	O8	540,976
99843	−52 04599			11 26 18 −53 07.2	1	9.64		0.08		0.10		1	A2 V	1770
99799	−21 03316			11 26 19 −22 07.2	1	7.66		1.39		1.58		2	K2/3 III	1375
99796	+26 02218			11 26 21 +25 39.3	1	8.74		0.29		0.08		2	F0	1569
		vdB 49 # b		11 26 21 −62 52.6	1	12.27		1.54		1.52		1		434
99787	+40 02433	HR 4422	⋆ AB	11 26 23 +39 36.7	3	5.29	.029	0.01	.012	0.01	.000	6	A2 V	1022,1363,3023
308570	−61 02348	LSS 2352		11 26 23 −62 06.4	3	10.00	.013	0.65	.007	−0.41	.008	4	O7 Ib(f)	2,540,1737
	+1 02570	G 10 - 26	⋆ AB	11 26 25 +01 22.2	2	9.82	.005	0.85	.005	0.58	.005	2	K3	333,1620,1658
99855	−60 02978			11 26 26 −60 48.1	1	9.38		0.13		−0.74		3	A1 IV/V	2
		AAS12,381 26 # A		11 26 29 +25 57.3	1	10.47		0.95		0.83		1		280
99869	−54 04559	IDS11242S5420	AB	11 26 29 −54 36.9	1	9.33		0.12		0.08		1	A1 V	1770
		GD 310		11 26 30 +38 25.4	2	11.89	.000	−0.13	.005	−1.00	.005	5		1298,3028
99898	−62 02033	LSS 2354	⋆ AB	11 26 37 −62 39.3	3	9.35	.013	0.34	.005	−0.62	.022	12	B2/5 III	92,191,390
99832	+31 02270	IDS11240N3059	AB	11 26 39 +30 42.1	1	7.20		0.47		−0.03		std	F5 IV-V	1292
99897	−61 02350	LSS 2355		11 26 39 −62 22.6	5	8.35	.020	0.16	.014	−0.79	.016	13	O6	2,6,540,976,2021
99831	+42 02209	IDS11239N4238		11 26 40 +42 22.1	1	8.91		0.30		0.10		2	Am	1240
		vdB 49 # a		11 26 40 −62 49.6	1	14.12		0.81		−0.09		3		434
306571	−59 03510			11 26 41 −59 41.0	1	11.30		0.18		−0.13		3		2
99894	−59 03513			11 26 45 −60 10.9	1	10.01		0.08		−0.10		3	A0	2
99861	+27 02030			11 26 46 +26 49.9	1	9.35		1.00		0.85		1	K0	280
99890	−55 04399	LSS 2356		11 26 46 −56 22.1	3	8.30	.015	−0.06	.001	−0.92	.017	6	B0.5V	6,540,976
99873	−0 02444			11 26 51 −00 34.4	1	7.03		1.38		1.63		6	K4 III	1657
308539	−63 01899			11 26 54 −63 45.3	1	10.11		0.17		−0.69		2	B0.5V	540
99859	+57 01324	HR 4424		11 26 56 +57 00.8	4	6.28	.008	0.15	.013	0.11	.005	12	A4 m	252,1379,3058,8071
	+28 02008			11 26 57 +28 21.9	1	8.80		1.56				2	M0	280
99944	−63 01900	IDS11247S6319	AB	11 26 57 −63 35.1	1	9.07		0.08		−0.75		2	B0/1 V	401
		LP 552 - 63		11 26 59 +06 55.7	1	12.96		1.43		1.13		2		1696
	−36 07209			11 27 00 −37 11.8	1	9.75		0.58		0.05		1	G0	78
99940	−59 03518			11 27 00 −59 58.4	1	9.99		0.28		0.17		3	A0	2
99953	−62 02039	LSS 2361		11 27 00 −63 16.7	6	6.45	.033	0.32	.007	−0.62	.015	13	B1/2 Iab/b	6,540,914,976,2012,8100
		KS 584		11 27 00 −63 17.	1	12.04		0.04		−0.36		2		540
99939	−57 04745	LSS 2360		11 27 02 −57 33.4	4	7.20	.023	0.05	.011	−0.79	.019	10	B1/2 Iab/b	6,540,976,2012
99935	−48 06546			11 27 03 −48 34.2	1	9.44		−0.01		−0.14		1	B9 III/IV	1770
		Wolf 399		11 27 04 +11 21.3	1	13.38		1.17		1.17		1		1696
	+33 02119	IDS11244N3335	AB	11 27 05 +33 18.8	1	9.64		0.74		0.37		1	G5	280
99902	+16 02266	HR 4426		11 27 06 +15 41.4	2	5.74	.000	1.34	.013	1.58		5	K2 III	70,1501
238006	+55 01467			11 27 07 +54 45.2	1	9.79		0.89				1	K0	1017
	+27 02032			11 27 08 +27 02.7	1	10.21		1.18		1.23		1		280
	+28 02009			11 27 08 +28 01.9	1	9.93		0.26		0.01		1		280
99922	−23 10009	HR 4428	⋆ AB	11 27 09 −24 11.3	2	5.75	.005	0.07	.000			7	A0 V	15,2027
99923	−27 08121			11 27 09 −27 45.3	1	6.67		1.35				1	K3 III	2008
	+25 02384			11 27 10 +25 04.0	1	9.99		0.27		0.07		2	A5	1569
	+30 02162	TU UMa		11 27 10 +30 20.6	2	9.32	.064	0.19	.036	0.11	.037	2	F2	668,699
99959	−52 04604			11 27 12 −53 21.0	1	8.01		0.05		0.07		1	A1 V	1770
99915	+25 02385			11 27 14 +25 13.4	1	7.90		0.48		0.01		2	F5	1569
99978	−55 04406			11 27 17 −55 37.2	2	8.65	.005	0.96	.014	0.44		4	G5/8 III	1594,6006
		AAS41,43 F2 # 1		11 27 17 −63 03.9	1	10.78		1.30		1.08		1	G6	820
308594	−62 02044			11 27 17 −63 07.8	1	11.06		0.42		0.29		1	A2	820
	+37 02183			11 27 18 +37 22.6	1	10.06		1.10		0.95		1		280
99992	−63 01902			11 27 20 −63 46.1	2	9.77	.009	0.13	.006	−0.68	.014	4	B0/5 IIIp	540,976
		G 10 - 28		11 27 22 +04 17.1	1	11.73		1.34		1.27		1		333,1620
99948	+17 02363			11 27 22 +17 15.7	1	8.80		0.00		−0.02		2	A2	1648
99947	+25 02386			11 27 24 +25 09.6	1	7.72		1.01		0.84		1	K0 III	280
	+32 02152			11 27 24 +32 15.2	1	10.80		0.13		0.13		2	A2	1026
99946	+30 02163	AW UMa	⋆ A	11 27 26 +30 14.6	3	6.94	.031	0.35	.014	0.00	.018	13	F0	280,1569,3016
99913	+55 01468	HR 4427		11 27 26 +54 38.3	2	6.51	.015	0.94	.005	0.68	.000	5	G9 III	1501,1733
		AAS41,43 F2 # 3		11 27 26 −63 04.3	1	11.53		0.12		−0.60			B2	820
99969	−16 03262			11 27 27 −17 29.6	1	7.77		1.61		1.84		2	M(3) III	1375
308593	−62 02046			11 27 27 −63 06.7	1	10.05		0.09		−0.69		1	B8	820
99946	+30 02164	IDS11248N3031	B	11 27 28 +30 15.5	5	9.47	.007	0.72	.022	0.26	.014	30		1292,1320,1388,1569,3032
100004	−50 06060			11 27 28 −51 23.3	3	7.39	.013	0.41	.000	−0.09	.002	9	F3 V	1311,2012,3037
99957	+26 02222			11 27 29 +25 34.8	1	7.71		1.37		1.56		1	K3 III	280
	+29 02170			11 27 29 +28 39.0	1	10.70		0.61		0.09		3		1569
	+23 02354			11 27 30 +23 16.2	1	10.58		0.76		0.36		1		280
	+14 02393			11 27 32 +14 23.4	1	9.03		0.48		0.07		2	F8	1648
100013	−53 04586			11 27 34 −54 21.4	1	9.12		0.08		0.10		1	A1/2 V	1770
99954	+48 01949			11 27 36 +47 31.2	1	7.46		0.96		0.73		2	K0 III-IV	172

Table 1   609

HD	DM	Other Id	N Rem	α₁₉₅₀	δ₁₉₅₀	S	V	σ_V	B–V	σ_B–V	U–B	σ_U–B	n	Spectrum	References
		AAS41,43 F2 # 5		11 27 36	−62 58.6	1	11.57		1.56		1.43		1	K0	820
	+36 02193	G 147 - 58		11 27 37	+36 06.1	2	9.89	.001	0.67	.000	0.11	.004	5	G2	1658,1723
	+37 02185			11 27 37	+36 55.5	1	10.12		1.09		1.02		1	K0	280
		AAS41,43 F2 # 6		11 27 39	−63 00.3	1	12.04		0.17		−0.29		1	B8	820
		AAS41,43 F2 # 8		11 27 40	−63 04.7	1	12.48		0.18		0.04		1	B9	820
308592	−62 02050			11 27 41	−63 04.7	1	10.62		0.14		−0.64		1	B5	820,8100
99967	+47 01880	HR 4430, EE UMa		11 27 42	+46 56.0	2	6.36	.010	1.26	.005	1.18	.000	4	K2 III	172,252
	+37 02186			11 27 43	+36 54.1	1	10.18		0.97		0.66		1	K0	280
100025	−59 03531	LSS 2362		11 27 44	−60 19.5	1	9.78		0.02		−0.64		3	B1/2 III	2
	+30 02165			11 27 45	+30 09.6	2	8.91		0.54				2	F8 V	280
		G 197 - 17		11 27 45	+62 09.4	1	10.67		0.71		0.27		1	G8	1658
99998	−2 03360	HR 4432		11 27 46	−02 43.6	4	4.77	.008	1.53	.007	1.80	.024	15	K4 III	15,1417,3016,6002
		AAS41,43 F2 # 9		11 27 46	−63 12.9	1	11.68		0.15		−0.50		1	B3	820
	+27 02033			11 27 47	+26 39.4	1	9.61		0.86		0.48		1	F8	280
99995	+44 02102			11 27 47	+43 51.2	1	6.67		0.91		0.63		3	K1 III	1501
100038	−53 04587	IDS11255S5407	A	11 27 47	−54 23.2	1	9.12		0.03		−0.04		1	B9 IV	1770
99984	+43 02122	HR 4431		11 27 49	+43 26.9	3	5.96	.019	0.49	.011	−0.03	.005	6	F7 III	70,254,3037
		AAS41,43 F2 # 10		11 27 50	−62 56.2	1	12.45		0.18		0.04		1	B9	820
308610	−62 02052	IDS11256S6300	AB	11 27 50	−63 17.0	1	10.35		0.32		−0.45		1	B5	820
100006	+19 02459	HR 4433		11 27 53	+18 41.1	2	5.53	.005	1.06	.011	0.82		4	K0 III	70,3051
	+20 02612			11 27 53	+20 06.2	1	9.78		0.97		0.67		1	G5	280
308609	−62 02055			11 27 55	−63 15.6	1	10.73		0.37		0.24		1	A0	820
	+48 01950			11 27 56	+48 10.9	1	9.54		0.28		0.02		2	A2	272
		Steph 950		11 27 57	+55 14.0	1	11.78		1.44		1.22		1	M0	1746
100065	−59 03537			11 27 58	−60 15.1	1	8.66		0.21		0.04		3	A5/7 III	2
		AAS41,43 F2 # 13		11 28 00	−62 52.4	1	12.21		0.37		0.31		1	A0	820
		AAS41,43 F2 # 14		11 28 00	−62 58.3	1	12.16		0.31		0.16		1	B8	820
	+28 02011			11 28 02	+28 01.9	1	10.19		1.04		0.88		1	G5	280
308591	−62 02058			11 28 02	−63 05.7	1	10.60		0.12		−0.66		1	B5	820
308588	−62 02059			11 28 04	−62 59.8	1	10.85		0.14		−0.47		3	A0	820
		AAS41,43 F2 # 17		11 28 04	−63 07.1	1	12.14		0.14		−0.42		2	B6	820
	+26 02223			11 28 05	+25 31.1	1	10.41		0.27		0.02		2		1569
		AAS41,43 F2 # 20		11 28 06	−63 08.5	1	12.12		0.15		−0.43		2		820
	+31 02272			11 28 07	+30 54.1	1	10.40		1.05		1.02		1	K2	280
		AAS41,43 F2 # 18		11 28 07	−62 55.0	1	12.11		0.45		0.25		1	A3	820
100098	−62 02062			11 28 07	−63 01.7	1	8.54		0.26		0.27		3	A1 III	820
100018	+42 02214	IDS11254N4150	AB	11 28 08	+41 33.8	2	6.96	.005	0.46	.010	−0.05	.015	3	F5	292,3030
100079	−59 03541			11 28 08	−59 44.7	1	8.75		0.31		0.17		1	A8/9 V	1771
100099	−63 01904	IC 2944 - 138		11 28 09	−63 32.5	5	8.07	.018	0.10	.014	−0.81	.014	11	O9 III	6,434,540,976,1312
100030	+48 01952	HR 4435		11 28 10	+48 12.4	3	6.42	.016	0.88	.001	0.53	.000	6	G9 IV	70,172,3016
	+29 02174			11 28 12	+29 03.4	1	10.28		1.13		1.09		1		280
304444	−57 04761			11 28 12	−58 13.1	1	9.42		1.49				1	K5	549
306592				11 28 12	−60 01.	1	11.59		0.25		0.17		3	B9	2
308595	−62 02064			11 28 13	−63 09.8	2	11.01	.005	0.06	.015	−0.62	.005	2	B9	287,820
		AAS41,43 F2 # 22		11 28 13	−63 14.1	1	12.04		0.18		−0.54		1	B3	820
		G 122 - 27		11 28 14	+44 01.8	1	11.13		1.21				2	K4	1726
100096	−61 02364			11 28 14	−61 53.6	1	8.48		0.20		0.09		2	A3mA5-A7	1771
304435	−57 04762			11 28 15	−58 00.6	1	11.15		0.19		0.09		1	A0	549
		AAS41,43 F2 # 23		11 28 15	−63 13.0	1	11.66		0.24		−0.54		1	B3	820
		AAS41,43 F2 # 24		11 28 16	−63 15.2	1	11.56		0.35		−0.41		1	B3	820
100056	+19 02462			11 28 17	+19 12.6	1	9.54		0.41		−0.04		2		1569
100041	+29 02176			11 28 17	+28 43.6	1	6.72		1.51		1.81		2	M3 III	1003
304391	−56 04554			11 28 18	−56 51.5	3	8.33	.000	1.06	.004	0.98	.014	9	K4V	158,657,2012
304456	−57 04763			11 28 18	−58 22.3	1	11.24		0.32		0.16		1	B9	549
306545	−60 02999			11 28 18	−61 10.7	1	10.54		0.10		−0.48		3	B5	2
308611				11 28 18	−63 20.9	1	11.35		0.31		−0.58		1	B8	820
	+35 02257			11 28 20	+34 48.7	1	10.09		0.96		0.75		1	K0	280
304443	−57 04764			11 28 20	−58 10.9	1	11.88		0.24		0.21		1		549
		Lod 372 # 19		11 28 20	−58 21.3	1	11.95		0.37		0.29		1	A0	549
		AAS41,43 F2 # 25		11 28 20	−63 09.3	1	11.57		1.25		1.15		1	G2	820
		vdB 50 # b		11 28 20	−63 32.1	1	12.52		0.73		0.20		3		434
304442	−57 04764			11 28 22	−58 10.9	2	11.05	.015	0.19	.005	−0.25	.010	2	B9	287,549
100201	−78 00657			11 28 22	−78 39.9	1	9.51		0.52		−0.01		19	F6 V	1763
99945	+81 00373	HR 4429		11 28 23	+81 24.2	2	6.13	.020	0.24	.005	0.12	.000	4	A2 m	985,1733
308590				11 28 23	−63 04.7	1	10.48		1.33		1.17		1		820
100126	−62 02065	LSS 2365		11 28 24	−62 30.6	1	7.93		0.15		−0.11		3	A0 Ib/II	8100
		Lod 372 # 17		11 28 25	−58 13.9	1	11.77		0.24		−0.20		1	B7	549
		Lod 372 # 16		11 28 25	−58 17.1	1	11.59		0.33		−0.17		1	B8	549
		AAS41,43 F2 # 29		11 28 25	−63 10.1	1	12.25		0.15		−0.01		1	B8	820
		AAS41,43 F2 # 28		11 28 26	−63 08.2	1	11.98		0.14		−0.57		1	B3	820
100055	+49 02062	HR 4436		11 28 27	+49 03.9	1	6.56		0.93		0.71		2	G9 III	172
	+63 00965			11 28 27	+63 25.0	2	9.95	.047	1.36	.022	1.28	.013	4	M0	3072,7008
100123	−59 03544			11 28 27	−59 45.1	2	8.52	.052	0.17	.019	0.07	.009	4	A1 V	2,1771
100029	+70 00665	HR 4434		11 28 28	+69 36.4	5	3.85	.016	1.61	.009	1.97	.005	19	M0 III	15,1025,1363,3016,8015
308597	−62 02066			11 28 28	−63 11.0	2	11.21	.010	0.13	.025	−0.47	.010	2	A0	287,820
100119	−41 06592			11 28 31	−41 39.0	2	7.21	.001	−0.06	.004	−0.21	.011	5	B8/9 IV	1770,3017
		Lod 372 # 15		11 28 31	−58 13.2	1	11.51		0.51		0.03		1	F6	549
		G 197 - 18		11 28 33	+59 28.2	1	13.06		1.00		0.76		3		7010
100135	−59 03545			11 28 33	−60 02.5	2	7.27	.005	−0.02	.030	−0.35	.010	5	B5/7 III	2,294
308596	−62 02068			11 28 33	−63 12.0	2	11.10	.025	0.11	.005	−0.58	.000	2	A0	287,820

HD	DM	Other Id	N Rem	α_{1950}	δ_{1950}	S	V	σ_V	B–V	σ_{B-V}	U–B	σ_{U-B}	n	Spectrum	References
100148	−56 04558	V419 Cen		11 28 34	−56 37.4	2	8.01	.020	0.69	.010	0.45		2	F7 II	657,1484
308598	−62 02070			11 28 34	−63 10.7	2	11.08	.010	0.07	.050	−0.62	.000	2	B9	287,820
		LP 732 - 35		11 28 35	−14 39.6	1	14.29		1.81				1		3078
		LTT 18170		11 28 37	+64 06.5	1	12.70		0.53		−0.09		4		1723
306619	−60 03006			11 28 37	−60 28.8	1	10.36		0.03		−0.53		5	B9	125
304445	−57 04768			11 28 38	−58 13.5	2	10.83	.026	0.19	.015	−0.15	.005	2	B9	287,549
		AAS41,43 F2 # 33		11 28 38	−62 51.6	1	11.04		1.58		1.76		2	G5	820
	+22 02369			11 28 39	+22 26.5	1	10.36		1.35		1.52		1		280
100162	−62 02071	IDS11264S6234	AB	11 28 39	−62 50.5	1	10.07		0.13		−0.12		2	B8/A0 V	820
		AAS41,43 F2 # 36		11 28 40	−63 15.7	1	11.95		0.26		0.07		1	B8	820
100163	−62 02072			11 28 41	−63 16.5	1	9.91		0.27		0.11		1	A1/2 IV/V	820
306593	−59 03549	LSS 2366		11 28 43	−59 58.5	2	9.94	.020	0.13	.050	−0.57	.015	5	B3	2,294
		AAS41,43 F2 # 37		11 28 43	−63 10.9	1	11.48		0.15		−0.46		1	B4	820
100128	+23 02358			11 28 44	+23 05.6	1	7.81		0.21		0.15		1	A3	280
	+84 00255			11 28 44	+84 16.8	1	9.45		1.25		0.98		2	K2	1733
308583	−62 02073			11 28 47	−62 48.2	1	10.93		0.14		−0.56		2	B8	820
		AAS41,43 F2 # 41		11 28 47	−63 10.5	1	12.74		0.22		−0.34		1	B6	820
		Lod 372 # 20		11 28 48	−58 13.1	1	12.14		0.19		−0.13		1	B9	549
304446	−57 04770			11 28 48	−58 14.0	2	10.95	.020	0.20	.010	−0.19	.005	2	B5	287,549
306582	−59 03551			11 28 48	−59 45.4	1	9.88		0.09		−0.08		3	B9	2
		AAS41,43 F2 # 39		11 28 48	−62 51.1	1	11.57		0.56		0.06		1	F6	820
		Lod 421 # 6		11 28 49	−63 05.3	2	11.88	.005	0.12	.030	−0.49	.015	2	B3	287,820
100199	−62 02075	LSS 2368		11 28 50	−62 40.2	7	8.19	.026	0.00	.007	−0.84	.017	18	B0/1 IIIne	2,191,401,540,976*
308589	−62 02074			11 28 50	−62 58.1	1	10.78		0.20		−0.14		1	A0	820
100113	+40 02436			11 28 51	+40 22.4	1	8.00		1.46		1.75		2	K2	1601
		AAS41,43 F2 # 43		11 28 51	−62 59.2	1	12.10		0.14		−0.41		1	B5	820
304447	−57 04771			11 28 52	−58 14.0	2	10.82	.015	0.19	.010	−0.22	.000	2	B8	287,549
		AAS41,43 F2 # 44		11 28 52	−63 09.5	1	12.11		0.60		0.13		1	F5	820
308545	−63 01908			11 28 52	−64 00.7	1	10.09		0.19		−0.61		2	B1 V	540
308581	−62 02076			11 28 54	−62 41.6	1	9.00		1.15		0.99		1	K0	401
		Lod 421 # 1		11 28 54	−63 00.5	1	12.29		0.41		−0.36		1		287
100150	+18 02505			11 28 55	+18 01.6	1	7.12		0.00		0.02		2	A1 III(V)	272
		Lod 372 # 14		11 28 55	−58 23.5	1	11.29		0.55		−0.02		1	F8	549
100200	−62 02077			11 28 55	−62 54.6	1	9.21		0.08		−0.28		2	B9 IV	820
	+38 02251			11 28 56	+38 04.0	1	9.72		1.43		1.56		1	K5	280
	+41 02201			11 28 56	+40 46.6	1	9.80		1.21		1.14		2	M0	3072
100213	−65 01675	TU Mus		11 28 56	−65 28.0	3	8.33	.044	0.05	.011	−0.86	.016	8	O8	401,540,976
		LP 613 - 2		11 28 57	+01 25.9	1	12.64		0.54		−0.18		2		1696
100198	−60 03011	HR 4438, V809 Cen		11 28 57	−61 00.1	7	6.37	.016	0.51	.008	−0.12	.019	18	A3 Ia	2,15,191,258,2013*
		AAS41,43 F2 # 46		11 28 59	−62 54.0	1	12.30		0.46		−0.48		3	B0	820
		AAS41,43 F2 # 47		11 28 59	−63 00.8	1	12.79		0.52		0.24		1		820
		Lod 421 # 5		11 28 59	−63 03.3	2	11.80	.010	0.43	.025	0.38	.130	2		287,820
		G 10 - 29		11 29 00	+02 30.6	1	12.49		1.44		1.12		1		3073
304441	−57 04772			11 29 00	−58 07.4	1	10.21		0.51		0.04		1	G0	549
100149	+31 02274			11 29 01	+31 14.9	1	8.03		0.89		0.59		1	G5 V	280
		Lod 421 # 2		11 29 03	−62 54.1	1	11.92		0.10		−0.34		1		287
		Lod 421 # 3		11 29 03	−62 54.1	2	12.03	.047	0.17	.056	−0.40	.070	4		287,820
	+30 02167			11 29 04	+29 29.0	1	9.90		1.02				2		280
304453	−57 04774			11 29 04	−58 19.1	1	9.35		0.84		1.10		1	K0	549
	+24 02364			11 29 07	+24 08.8	1	9.20		1.58		1.96		1	K5	280
		Lod 421 # 4		11 29 07	−63 01.3	1	11.83		0.15		−0.39		1		287
		Lod 421 # 7		11 29 07	−63 01.3	2	11.74	.095	0.19	.015	−0.39	.045	2		287,820
	+23 02359	G 120 - 57		11 29 09	+22 56.6	3	10.31	.031	1.47	.044	1.19	.036	6	M0	906,1569,7008
308599				11 29 09	−63 07.5	1	11.62		0.28		−0.43		1		820
100180	+15 02345	HR 4437	★ A	11 29 10	+14 38.6	2	6.27	.070	0.57	.001	0.07		4	G0 V	70,3024
100180	+15 02345	IDS11266N1455	B	11 29 10	+14 38.6	1	9.22		1.14		1.03		2	G5	3024
100231	−58 03688			11 29 10	−59 23.7	1	9.07		0.32		0.16		1	A7 V	1771
		AAS41,43 F2 # 52		11 29 11	−63 12.8	1	11.52		0.56		0.24		1	F0	820
304454	−57 04778			11 29 12	−58 20.6	1	10.01		0.18		−0.60		1	B3	549
		AAS41,43 F2 # 54		11 29 12	−63 13.4	1	11.80		1.24		0.97		1		820
		AAS41,43 F2 # 55		11 29 12	−63 21.7	1	12.31		0.24		−0.11		1	B8	820
		AAS41,43 F2 # 60		11 29 12	−63 21.7	1	12.26		0.41		0.22		1	A0	820
		G 254 - 24		11 29 13	+76 56.1	2	11.53	.000	0.66	.009	−0.05	.014	2		1064,1658
100242	−59 03556	LSS 2372		11 29 13	−60 22.4	6	8.37	.012	0.14	.026	−0.74	.023	21	B0/2 Ib/IIn	6,125,294,540,837,976
100243	−62 02080	LSS 2373		11 29 13	−62 57.3	4	9.94	.007	0.10	.009	−0.73	.020	8	B0/2	2,540,820,976
308613				11 29 13	−63 25.7	1	11.28		0.30		−0.41		1		820
100179	+25 02388			11 29 14	+24 35.3	2	7.13	.000	1.37	.010	1.60		4	K4 III	280,1569
		HA 193 # 2520		11 29 14	−60 23.9	1	13.24		0.43		0.12		8		125
	+31 02275			11 29 16	+31 19.1	1	11.00		0.29		0.03		2		1569
308600				11 29 17	−63 07.5	1	10.54		1.17		0.84		1		820
100308	−76 00673	Cha F # 53		11 29 17	−76 48.0	1	9.37		0.51		−0.04		2	F6 V	1633
100219	−19 03285	HR 4440		11 29 18	−20 30.0	2	6.23	.005	0.54	.000			7	F7 V	15,2012
		L 252 - 112		11 29 18	−53 46.	1	15.69		−0.12		−0.93		1	DAn	782
		BPM 21022		11 29 18	−56 15.	1	15.69		0.12		−0.93		1		3065
		AAS41,43 F2 # 57		11 29 18	−62 49.5	1	11.41		0.86		0.41		2	G8	820
		AAS41,43 F2 # 58		11 29 18	−62 57.7	1	12.88		0.32		−0.11		1	B8	820
	+23 02360			11 29 20	+23 05.1	2	9.96	.040	0.96	.010	0.79	.015	2	K2	280,1569
100204	+31 02276			11 29 21	+30 30.9	1	7.74		1.15				2	K1 IV	280
		LP 673 - 52		11 29 21	−08 42.6	1	12.58		0.81		0.37		3		1696
100250	−50 06090			11 29 21	−51 22.7	1	8.63		−0.01		−0.09		3	B9.5 IV/V	1770

Table 1

HD	DM	Other Id	N Rem	α₁₉₅₀	δ₁₉₅₀	S	V	σ_V	B–V	σ_B–V	U–B	σ_U–B	n	Spectrum	References
306608	−59 03559			11 29 22	−60 10.8	1	9.90		0.02		−0.35		64	B8	125
		L 396 - 7		11 29 23	−40 46.3	1	11.52		1.54				1		1705
100263	−62 02081			11 29 23	−63 08.9	2	9.67	.015	0.09	.015	−0.78	.005	3	B1/2 III/IV	401,820
100238	−5 03307			11 29 24	−06 11.6	1	6.76		1.05		0.96		3	K0	3077
	−60 03014			11 29 24	−60 27.8	1	10.47		0.12		0.08		2	A0	837
		L 144 - 39		11 29 24	−61 27.	1	13.45		1.37		0.98		1		3062
100264	−64 01668			11 29 24	−65 07.6	1	8.49		0.13				1	B9 V	178
		LP 396 - 7		11 29 25	−40 46.5	1	11.59		1.49		0.97		1		3073
306620	−60 03015			11 29 26	−60 26.8	2	9.91	.009	0.00	.005	−0.17	.023	8	A0	125,837
100261	−58 03692	HR 4441, o1 Cen	⋆ A	11 29 27	−59 10.0	7	5.13	.025	1.07	.027	0.77	.022	22	F7 Ia/ab	15,897,1018,1034,1417*
308606	−62 02082			11 29 27	−63 13.0	1	10.95		0.23		−0.52		1	B9	820
100279	−61 02379	IDS11272S6132	AB	11 29 28	−61 48.5	1	8.77		0.46		0.30		2	A2/3mA5-A7	1771
		AAS41,43 F2 # 62		11 29 28	−62 46.5	1	12.02		0.64		0.08		1	F6	820
		AAS41,43 F2 # 63		11 29 28	−63 03.8	1	13.25		0.24		−0.03		1	B4	820
100262	−58 03693	HR 4442, o2 Cen		11 29 29	−59 14.4	8	5.15	.021	0.48	.020	−0.01	.049	25	A2 Ia	15,540,976,1018,1034*
100276	−59 03562	LSS 2375		11 29 29	−60 19.8	6	7.22	.023	0.04	.006	−0.83	.023	15	B1 Ib	2,6,540,837,976,2012
		AAS41,43 F2 # 65		11 29 29	−63 05.3	1	13.32		0.66		0.15		1		820
306604	−59 03561			11 29 30	−60 14.9	1	9.58		0.00		−0.26		6	B9	125
100277	−60 03017			11 29 30	−60 25.0	3	7.89	.008	0.04	.026	−0.07	.021	13	B9.5V	125,294,837
		AAS41,43 F2 # 67		11 29 31	−63 02.3	1	13.66		0.24		−0.06		1		820
		AAS41,43 F2 # 66		11 29 31	−63 05.2	1	13.05		0.37		0.18		1	A0	820
100215	+39 02450			11 29 32	+39 12.1	2	7.99	.005	0.31	.000	0.03	.010	8	Am:	833,8071
100203	+61 01246	HR 4439	⋆ AB	11 29 32	+61 21.6	3	5.48	.000	0.50	.004	−0.01	.008	11	F4 V +G3 V	15,292,1008
		WLS 1120-20 # 8		11 29 32	−19 45.6	1	13.43		0.63		0.12		2		1375
306621	−60 03018			11 29 32	−60 27.2	1	10.47		0.25		0.04		2	A2	294
100235	+37 02192	IDS11269N3648	AB	11 29 33	+36 31.4	2	6.59	.010	0.96	.005	0.73		5	K0 III	280,1501
100278	−61 02380	LSS 2377		11 29 33	−61 29.8	3	7.69	.011	0.74	.009	0.49	.020	6	F5 Ib	657,2012,8100
		AAS41,43 F2 # 69		11 29 33	−62 49.9	1	11.90		0.14		−0.45		2	B4	820
308605				11 29 33	−63 11.8	1	11.46		0.24		0.12		1	A0	820
100295	−62 02084			11 29 34	−62 52.7	1	9.02		1.25		1.20		1	K0/1 III	820
		AAS41,43 F2 # 70		11 29 35	−63 26.8	1	11.07		1.74		2.01		1	M1	820
308601	−62 02085			11 29 36	−63 09.0	2	10.39	.006	0.11	.009	−0.76	.005	3	B0 V	540,820
		AAS41,43 F2 # 72		11 29 37	−62 57.4	1	13.16		0.59		0.40		2		820
		HA 193 # 2669		11 29 38	−60 22.6	1	12.25		0.62		−0.04		4		125
		AAS41,43 F2 # 74		11 29 38	−63 07.1	1	12.36		0.26		0.16		1		820
308608				11 29 38	−63 17.7	1	11.55		0.41		0.34		2	A2	820
100324	−67 01758	LSS 2380		11 29 38	−67 46.9	2	8.58	.014	0.16	.004	−0.70	.035	5	B2 IIIne	540,976
304440	−57 04783			11 29 39	−58 06.6	1	11.16		0.11		−0.48		1	B8	549
		AAS41,43 F2 # 77		11 29 40	−63 02.7	1	12.70		0.19		0.06		1	B8	820
		AAS41,43 F2 # 76		11 29 40	−63 03.0	1	13.05		0.57		0.21		1	F0	820
	+34 02228	IDS11270N3356	AB	11 29 41	+33 39.7	1	10.28		0.92		0.73		1	K0	280
306623	−60 03020			11 29 41	−60 30.8	2	10.27	.005	0.41	.005	−0.01	.042	8	F0	125,837
306634	−60 03021			11 29 41	−60 36.1	1	10.80		0.11		0.15		2	A2	837
		AAS41,43 F2 # 78		11 29 41	−63 09.3	1	13.10		0.26		−0.21		1		820
		AAS41,43 F2 # 79		11 29 42	−62 50.6	1	12.54		0.40		0.19		1	A0	820
		AAS41,43 F2 # 80		11 29 42	−63 08.9	1	13.15		0.12		−0.34		1		820
100359	−73 00864	IDS11277S7321	A	11 29 43	−73 37.6	2	6.88	.000	0.25	.015	−0.23		6	B7 IV	1776,2012
100287	−28 08928	HR 4444	⋆ AB	11 29 47	−28 59.2	2	4.99	.005	0.53	.000	0.02		4	F8 V	404,2008
		HA 193 # 2739		11 29 47	−60 21.8	1	12.40		0.47		0.16		10		125
100334	−57 04789			11 29 49	−58 13.7	1	8.90		0.48		0.00		1	F6 V	549
		AAS41,43 F2 # 81		11 29 49	−63 09.9	1	11.48		0.22		−0.61		1	B2	820
100323	−63 01911	LSS 2381		11 29 49	−63 38.0	3	8.58	.013	0.02	.007	−0.80	.016	6	B0.5/1 Ib	401,540,976
	+30 02168			11 29 50	+30 08.0	1	9.48		1.00				2	K0	280
100336	−65 11298	SY Mus		11 29 50	−65 08.5	1	10.64		1.21		−0.77		1		1753
306624	−60 03026			11 29 53	−60 31.9	1	10.66		0.30		0.18		2	A3	837
100369	−72 01139			11 29 53	−72 31.1	1	9.85		0.26				4	A0 V	2012
100266	+42 02217			11 29 54	+42 03.1	1	8.74		1.17		1.08		2	K0	1733
100307	−26 08620	HR 4445		11 29 54	−26 28.2	2	6.17	.010	1.66	.005	1.99		6	M2 III	2006,3055
		AAS41,43 F2 # 82		11 29 54	−63 03.2	1	13.37		0.64		0.01		1		820
100401	−75 00738			11 29 56	−76 01.6	1	8.36		1.15				4	G6 III/IV	2012
		A1 72		11 29 58	+17 18.5	1	15.55		0.54		−0.15		1		98
	−60 03027			11 29 58	−60 32.2	1	11.34		0.20		−0.46		2	B4	837
100335	−60 03028			11 29 58	−60 34.6	2	7.86	.015	−0.05	.010	−0.47	.015	5	B5 III	2,6
		AAS41,43 F2 # 83		11 29 58	−62 55.4	1	12.63		0.61		0.02		2		820
		AAS41,43 F2 # 84		11 29 58	−63 09.9	1	11.73		1.41		1.52		1		820
		A1 73		11 29 59	+17 35.9	1	15.90		0.58		−0.21		1		98
100355	−61 02391			11 29 59	−62 16.3	3	9.23	.009	0.00	.009	−0.67	.019	5	B2/3 III	2,540,976
		AAS41,43 F2 # 87		11 29 59	−63 05.4	1	12.47		0.67		0.06		1	F5	820
	+11 02369	G 57 - 7		11 30 00	+11 10.9	2	10.13	.000	0.59	.005	0.03	.000	3	G0	333,1620,1696
		AAS41,43 F2 # 85		11 30 00	−62 56.8	1	12.83		0.30		0.18		2	A0	820
100354	−60 03029			11 30 01	−60 53.0	1	9.78		0.01		−0.45		3	B3 II/III	2
		AAS41,43 F2 # 86		11 30 01	−62 47.8	1	11.91		0.20		0.06		2	B8	820
		AAS41,43 F2 # 88		11 30 01	−63 05.8	1	13.52		0.45		0.13		1		820
306622	−60 03030			11 30 02	−60 27.6	3	10.82	.010	0.03	.049	−0.14	.022	12	A0	125,294,837
		AAS41,43 F2 # 90		11 30 02	−63 01.3	1	12.11		1.14		0.80		1		820
		AAS41,43 F2 # 89		11 30 02	−63 02.8	1	12.66		0.43		0.23		1		820
		G 176 - 40		11 30 03	+44 16.3	1	15.22		1.56				4		1663,1705
	−60 03031			11 30 03	−60 25.9	1	11.67		0.29		−0.03		4		125
		AAS41,43 F2 # 91		11 30 03	−63 02.4	1	12.65		0.13		−0.30		1		820
100315	−6 03403			11 30 04	−06 38.2	2	9.62	.000	0.63	.000	0.17	.000	16	G5	830,1783

HD	DM	Other Id	N	Rem	α_{1950}	δ_{1950}	S	V	σ_V	B–V	σ_{B-V}	U–B	σ_{U-B}	n	Spectrum	References
		AAS41,43 F2 # 92			11 30 05	−62 55.3	1	13.14		0.62		0.03		2		820
		G 10 - 31			11 30 06	+05 29.7	1	13.77		1.49		1.22		1		333,1620
		G 10 - 32			11 30 06	+05 30.2	1	11.93		1.15		1.10		1		333,1620
		LP 850 - 46			11 30 06	−24 56.2	1	13.36		1.53						3062
100382	−66 01605	HR 4448			11 30 06	−66 41.1	4	5.89	.003	1.14	.005	1.12	.007	17	K1 III	15,1075,1642,2038
100301	+50 01832				11 30 07	+49 46.3	1	8.10		1.45		1.80		8	K3 V	7009
		AAS41,43 F2 # 93			11 30 07	−63 06.9	1	12.93		0.27		−0.36		1	B7	820
		AAS41,43 F2 # 94			11 30 07	−63 22.4	1	11.86		0.52		0.31		1	A7	820
	−60 03032	LSS 2383			11 30 09	−60 31.5	2	11.32	.005	0.29	.005	−0.73	.025	4	B2 IIIne2+	837,1737
100311	+39 02452				11 30 10	+39 08.4	1	7.89		0.13		0.11		3	A3 V	833
	−49 06265				11 30 12	−50 11.7	1	10.69		1.11		0.89		2		1730
100380	−59 03573				11 30 13	−59 26.5	1	6.77		0.16		0.16		2	A3 V	1771
		AAS41,43 F2 # 95			11 30 13	−63 00.8	1	13.39		0.43		0.06		1		820
		HA 193 # 2796			11 30 14	−60 22.7	1	11.54		0.12		−0.25		7		125
306615	−59 03574				11 30 14	−60 22.7	2	11.30	.015	0.26	.060	0.12	.020	4	A0	294,837
		AAS41,43 F2 # 96			11 30 14	−63 02.5	1	12.42		0.42		0.28		1	A0	820
100340	+6 02461				11 30 15	+05 33.2	5	10.12	.008	−0.24	.005	−0.96	.023	35	B9	989,1026,1728,1729,1732
100343	−7 03250	HR 4446			11 30 15	−07 33.1	4	5.94	.004	1.38	.000	1.64	.000	18	K4 III	15,1417,2013,2029
		AAS41,43 F2 # 97			11 30 15	−63 14.4	1	11.84		0.68		0.11		1	G0	820
100483	−78 00663				11 30 16	−79 16.6	1	8.79		1.44		1.48		18	K1/2 [III]	1763
100381	−60 03037				11 30 17	−60 28.3	4	8.75	.026	−0.02	.027	−0.63	.018	9	B2 III	2,6,294,837
306627	−60 03038				11 30 18	−60 36.6	2	10.07	.039	0.06	.019	−0.18	.019	3	A0	287,837
		AAS41,43 F2 # 98			11 30 18	−63 01.8	1	12.34		1.26		0.89		1		820
		AAS41,43 F2 # 99			11 30 18	−63 26.7	1	11.33		0.75		0.26		1	F2	820
100363	−11 03123	SU Crt			11 30 19	−11 45.4	7	8.64	.009	0.30	.006	0.03	.026	23	F2 V	158,1003,1311,1775,2033*
	+26 02230				11 30 20	+26 25.6	1	10.28		1.00		0.86		1	K5	280
	+35 02261				11 30 21	+34 34.0	1	9.88		0.98		0.76		1		280
		AAS41,43 F2 # 100			11 30 21	−63 04.0	1	11.71		0.63		0.10		1	F4	820
100378	−39 07168	HR 4447			11 30 22	−40 09.6	4	5.63	.009	1.58	.000	1.81		16	M1 III	15,1075,2027,3005
306614	−59 03577				11 30 22	−60 20.4	2	11.34	.025	0.16	.055	−0.22	.025	4	B8	294,837
		AAS41,43 F2 # 101			11 30 22	−62 45.6	1	12.04		1.27		1.18		1		820
100338	+34 02230				11 30 23	+34 19.7	1	7.20		1.32		1.54		1	K2	280
		AAS41,43 F2 # 102			11 30 23	−63 17.8	1	11.17		0.66		0.14		1	F7	820
100395	−35 07280	IDS11280S3539		AB	11 30 24	−35 56.0	1	6.68		0.60		0.15		1	G0 V	78
100398	−49 06268				11 30 24	−50 19.6	1	9.05		1.01		0.69		2	K0 III	1730
100373	+20 02616				11 30 25	+19 48.6	1	8.51		0.99		0.76		1	G5	280
100393	−30 09303	HR 4449			11 30 25	−30 48.7	6	5.11	.023	1.58	.007	1.95	.008	23	M2 IIIb	15,678,1075,2032,3005,8017
308680	−62 02092				11 30 25	−62 52.6	1	10.48		1.30		1.22		1	K2	820
308607	−62 02091				11 30 25	−63 18.7	1	10.54		0.26		−0.49		2	B9	820
100396	−43 07132	IDS11280S4309		A	11 30 27	−43 25.1	1	8.10		0.02		−0.07		1	A0 III	1770
100431	−65 01680	IDS11282S6519		AB	11 30 27	−65 35.2	1	8.18		0.14				1	B7/8 III/IV	178
	−60 03040				11 30 28	−60 26.1	2	11.32	.005	0.00	.000	−0.52	.015	5	B5	125,837
100427	−56 04583	IDS11282S5708		AB	11 30 29	−57 24.1	1	9.18		0.31		0.10		1	A[3]mA7-A9	1771
306613	−59 03580				11 30 31	−60 19.0	1	9.24		1.14		1.13		2	K0	837
100430	−61 02400				11 30 31	−61 32.5	1	8.22		0.19		0.18		1	A1 IV/V	1771
		AAS41,43 F2 # 106			11 30 31	−62 58.2	1	12.10		1.39		1.43		1		820
		AAS41,43 F2 # 105			11 30 31	−63 17.0	1	12.56		0.24		−0.40		1	B5	820
100360	+39 02453				11 30 32	+39 08.1	1	7.83		0.65		0.13		3	G2 V	1501
100407	−31 09083	HR 4450		★ A	11 30 32	−31 34.8	7	3.54	.007	0.95	.011	0.71	.009	25	G8 III	3,15,1075,2012,2032*
308604	−62 02093				11 30 32	−63 10.7	2	10.19	.028	0.09	.014	−0.71	.009	4	B0.5 V	434,820
	+33 02130				11 30 34	+33 28.1	1	10.23		0.92		0.68		1		280
		AAS41,43 F2 # 108			11 30 34	−63 04.7	1	12.56		1.27		1.16		1		820
		AAS41,43 F2 # 109			11 30 34	−63 12.0	1	11.99		0.39		0.26		1	A0	820
		AAS41,43 F2 # 111			11 30 34	−63 13.4	1	11.29		1.75		1.90		1		820
		AAS41,43 F2 # 112			11 30 34	−63 21.4	1	11.72		0.30		−0.54		1	B0	820
	+38 02256				11 30 35	+38 23.9	1	9.57		1.28		1.31		1	K5	280
308651	−61 02401				11 30 35	−62 11.6	1	10.08		0.00		−0.81		2	B0.5III	540
308602					11 30 35	−63 03.5	1	11.72		0.20		−0.62		1	B8	820
		AAS41,43 F2 # 110			11 30 36	−63 13.3	1	11.51		1.74		1.80		1		820
100444	−62 02094	LSS 2384			11 30 36	−63 22.2	5	8.43	.010	0.20	.010	−0.75	.018	9	O9 II	401,540,820,976,1737
100428	−59 03582				11 30 37	−60 00.3	1	9.15		0.53		0.05		2	F5/6 IV/V	294
		AAS41,43 F2 # 114			11 30 37	−62 59.7	1	12.43		0.20		−0.18		1	B8	820
	+35 02262				11 30 38	+35 27.1	1	10.25		0.22		0.09		1		280
	+51 01675				11 30 38	+51 01.1	1	9.77		1.03		0.84		4		1723
		HA 193 # 3137			11 30 39	−60 23.7	1	12.00		0.34		−0.02		3		125
308603					11 30 40	−63 11.6	1	10.09		1.52		1.47		1	K2	820
		HA 193 # 3143			11 30 41	−60 19.0	1	12.19		0.41		0.00		3		125
100442	−45 07079				11 30 42	−45 43.2	1	9.24		0.49		−0.02		2	F5 V	508
100454	−54 04607				11 30 42	−55 22.2	1	9.02		0.21		0.15		1	A7 V	1771
100405	+27 02035				11 30 43	+26 51.1	1	10.03		0.10		0.18		1	A2	280
100418	−15 03295	HR 4451			11 30 43	−16 00.2	2	6.05	.005	0.59	.014			5	F9 III	2006,6009
100414	+22 02374				11 30 44	+21 51.2	1	9.12		1.03		0.79		1	K0	280
		AAS41,43 F2 # 118			11 30 44	−63 03.9	1	13.50		0.37		0.28		1	A0	820
		AAS41,43 F2 # 117			11 30 45	−62 56.0	1	12.45		0.58		0.03		1		820
100464	−53 04614				11 30 46	−53 34.8	1	8.70		0.13		0.12		1	A2 V	1770
	−49 06273				11 30 47	−50 14.8	1	11.18		0.30		0.14		2		1730
		AJ74,1125 T5# 7			11 30 47	−62 42.7	1	13.64		0.18		−0.05		1		2
		AAS41,43 F2 # 119			11 30 50	−62 58.5	1	13.12		0.60		0.09		1		820
		AAS41,43 F2 # 120			11 30 50	−63 10.7	1	11.91		1.23		0.91		1		820
		LSS 2385			11 30 50	−64 25.5	1	11.33		0.72		−0.27		2	B3 Ib	1737

Table 1 613

HD	DM	Other Id	N Rem	α_{1950}	δ_{1950}	S	V	σ_V	B–V	σ_{B-V}	U–B	σ_{U-B}	n	Spectrum	References
100447	+20 02618			11 30 53	+19 57.4	1	7.01		1.28				2	K2	280
100434	+26 02231			11 30 53	+26 25.4	1	8.66		1.47		1.73		1	K0	280
		AAS41,43 F2 # 121		11 30 55	−62 52.2	1	11.40		0.97		0.66		1	F8	820
306628	−60 03048			11 30 57	−60 36.8	3	10.27	.051	0.06	.012	-0.02	.010	5	B9	287,294,837
		AAS41,43 F2 # 122		11 30 57	−63 17.2	1	11.80		0.55		0.12		1	F0	820
		Lod 402 # 34		11 30 58	−60 32.0	1	11.78		0.24		0.24		2	A2	837
		AAS41,43 F2 # 123		11 30 58	−63 09.3	1	10.98		1.29		1.19		1	K0	820
306705	−59 03588			11 30 59	−59 54.8	1	11.76		0.21		0.18		3	A0	2
306625	−60 03050			11 31 00	−60 29.1	2	10.21	.024	0.07	.015	0.06	.024	3	A0	287,837
100495	−62 02096	LSS 2386		11 31 00	−62 49.7	1	9.62		0.15		-0.70		2	B1 III	820
		AAS41,43 F2 # 125		11 31 01	−62 48.6	1	12.15		0.31		-0.13		2		820
		AAS41,43 F2 # 126		11 31 01	−63 04.8	1	13.02		0.31		0.37		1	A1	820
100508	−66 01609			11 31 01	−66 46.7	2	7.73	.009	0.83	.005	0.60		5	K0 IV/V	1642,2012
		AAS41,43 F2 # 127		11 31 03	−63 01.4	1	13.27		0.30		0.40		1	B9	820
		AAS41,43 F2 # 128		11 31 04	−63 06.8	1	11.09		0.89		0.52		1	F5	820
100516	−68 01524	LSS 2388		11 31 05	−68 33.8	1	8.49		0.19		-0.68		2	B2 II	401
	−59 03589			11 31 06	−60 18.5	1	11.04		0.51		0.01		2		837
306741	−59 03590			11 31 07	−60 16.4	1	11.01		0.28		0.18		2	B9	837
		Lod 402 # 31		11 31 07	−60 32.7	1	11.41		0.32		0.17		2	A3	837
		AAS41,43 F2 # 129		11 31 07	−63 21.3	1	12.50		0.26		-0.09		1	B8	820
		GD 138		11 31 08	+33 18.0	1	16.68		0.23		-0.64		1		3060
100507	−60 03053			11 31 08	−60 29.7	1	8.71		-0.04		-0.30		2	B8/9 II	837
306671	−58 03706	LSS 2387		11 31 09	−59 13.0	2	9.78	.010	0.15	.015	-0.65	.011	5	B0.5IV	540,976
		AAS41,43 F2 # 130		11 31 09	−62 48.9	1	11.79		0.67		0.16		2	F0	820
		AAS41,43 F2 # 131		11 31 09	−62 48.9	1	12.19		0.34		0.23		1	A0	820
		AAS41,43 F2 # 138		11 31 10	−62 50.0	1	12.21		0.13		-0.32		1	B7	820
100493	−39 07175	HR 4453	★ AB	11 31 11	−40 18.6	2	5.38	.005	0.12	.000			7	A2 IV/V	15,2012
100504	−45 07084	IDS11288S4546	A	11 31 11	−46 02.4	1	8.54		0.62				4	G1 V	2012
		AAS41,43 F2 # 133		11 31 11	−62 58.6	1	12.99		0.41		0.19		1		820
308702	−62 02097			11 31 11	−62 59.3	1	11.01		0.19		-0.50		1	A0	820
		AAS41,43 F2 # 134		11 31 12	−63 10.4	1	12.98		0.36		0.18		1	A2	820
	+2 02446			11 31 14	+02 03.1	1	10.14		1.39				1	R2	1238
308681				11 31 14	−62 56.6	1	11.72		0.14		-0.29		1	A0	820
308707	−62 02098			11 31 14	−63 18.7	1	9.97		1.19		1.06		1	K0	820
100546	−69 01557			11 31 14	−69 55.1	1	6.68		-0.01		-0.12		2	B9 Vne	1771
308618	−63 01918			11 31 15	−63 36.9	1	10.67		0.08		-0.62		1		401
304463	−58 03708			11 31 16	−58 39.4	1	10.48		0.27		-0.51		2	B1.5V	540
100530	−63 01919	IDS11290S6321	AB	11 31 16	−63 37.2	2	9.44	.113	0.06	.011	-0.68	.004	6	B1/2 III	540,976
100470	+37 02195	HR 4452		11 31 17	+37 05.6	4	6.39	.009	1.06	.005	0.92	.009	9	K0 III	15,1003,1080,3026
306740	−59 03597			11 31 17	−60 14.6	1	11.06		0.38		0.10		2	B9	837
		AJ74,1125 T5# 5		11 31 17	−62 47.0	1	12.40		0.20		0.23		1		2
100503	−30 09311			11 31 18	−30 48.7	1	8.74		1.69		1.84		1	G/Kp Ba	565
100528	−56 04596			11 31 18	−57 22.1	1	7.87		-0.02		-0.06		2	B9 V	1771
308703	−62 02099			11 31 18	−63 04.0	1	10.34		0.83		0.65		3	A3	820
100513	−45 07086	IDS11289S4559	AB	11 31 19	−46 15.4	1	8.15		0.20		0.10		2	A3 IV/V	508
		SS Leo		11 31 21	+00 14.5	2	10.42	.003	0.11	.011	0.10	.007	2	F1	668,699
100486	+14 02404			11 31 21	+13 49.7	1	7.89		1.14				2	K2	882
	+28 02014			11 31 21	+27 32.2	1	9.01		1.58		1.92		1	M0	280
306739	−59 03598			11 31 21	−60 14.1	1	11.36		0.24		0.24		2		837
306700	−59 03601			11 31 23	−59 46.9	1	10.26		0.11		0.01		2		837
306742	−59 03600			11 31 23	−60 20.0	2	10.50	.000	0.07	.005	0.01	.020	5	A0	2,837
308704	−62 02101			11 31 24	−63 08.9	1	11.19		0.10		-0.63		2	B8	820
		Lod 402 # 32		11 31 25	−60 31.7	1	11.49		0.21		0.16		2	A2	837
		AAS41,43 F2 # 140		11 31 25	−63 12.1	1	11.99		0.14		-0.49		1	B4	820
306645	−60 03057			11 31 26	−61 12.3	1	10.35		0.05		-0.40		3	B5	2
	+26 02232			11 31 27	+26 02.0	1	9.91		0.92		0.75		1	K7	280
308701				11 31 28	−62 59.5	1	11.71		0.23		0.09		1		820
		LP 793 - 5		11 31 29	−17 45.9	1	10.28		0.76		0.31		2		1696
100544	−60 03058			11 31 29	−60 36.6	2	8.47	.100	1.15	.025	0.91	.045	4	G3 II	294,837
		AAS41,43 F2 # 142		11 31 29	−63 02.5	1	13.32		0.33		0.36		1		820
		AAS12,381 35 # C		11 31 30	+35 19.0	1	10.66		1.10		1.08		1		280
		AAS41,43 F2 # 143		11 31 30	−63 01.5	1	13.08		0.37		0.32		1		820
308648	−61 02409			11 31 31	−61 48.8	1	11.35		0.15		-0.27		1		2
100537	−15 03297			11 31 32	−15 46.1	1	6.99		1.46		1.77		6	K4 III	1657
		AJ74,1125 T5# 6		11 31 32	−62 46.6	1	13.07		0.32		0.27		1		2
		AAS41,43 F2 # 144		11 31 32	−63 09.2	1	12.77		0.18		-0.07		1	B9	820
100555	−48 06618			11 31 33	−48 32.5	1	8.16		0.73				4	G6 V	2012
306654	−60 03060			11 31 33	−61 18.3	1	11.09		0.35		0.13		3	A2	2
	−74 00827			11 31 33	−75 15.5	2	9.82	.015	1.50	.015			7	K5	236,2012
308700				11 31 34	−63 00.6	1	11.67		0.16		-0.55		2		820
100518	+11 02372	HR 4454		11 31 35	+11 18.0	3	6.55	.004	0.18	.008	0.10	.006	6	A2 m	355,1022,3058
100517	+22 02375			11 31 35	+22 18.0	1	8.94		1.04		0.87		1	K2	280
		AAS41,43 F2 # 145		11 31 35	−63 03.1	1	13.12		0.15		-0.12		2		820
		AAS41,43 F2 # 147		11 31 36	−63 00.1	1	12.62		0.53		-0.26		2	B6	820
		AAS12,381 35 # D		11 31 37	+34 37.6	1	10.56		1.15		1.29		1		280
		LP 673 - 59		11 31 38	−08 13.1	1	12.21		0.97		0.60		2		1696
100592	−58 03713			11 31 40	−59 17.1	1	9.84		0.89		0.17		5	F6/8 V	897
		HA 55 # 198		11 31 41	+29 48.3	1	13.02		1.04		0.66		2		269
		AAS41,43 F2 # 148		11 31 41	−63 15.5	1	12.41		0.10		-0.37		1	B7	820
	−60 03061			11 31 42	−60 33.2	1	11.04		0.52		0.06		2	F8	837

HD	DM	Other Id	N Rem	α_{1950}	δ_{1950}	S	V	σ_V	B–V	σ_{B-V}	U–B	σ_{U-B}	n	Spectrum	References
308705				11 31 42	−63 10.3	1	10.83		0.72		0.16		1	F5	820
		LP 793 - 8		11 31 44	−17 03.1	1	11.78		0.58		−0.05		2		1696
100549	+23 02361			11 31 45	+23 09.8	1	9.29		−0.12		−0.44		1	B7 IV-III	280
		HA 55 # 114		11 31 46	+29 31.2	1	12.85		0.53		0.06		2		269
100563	+3 02521	HR 4455		11 31 48	+03 20.3	3	5.78	.017	0.45	.017	0.00	.009	9	F6 V	15,254,1417
		AAS41,43 F2 # 150		11 31 48	−63 21.6	1	12.29		1.19		0.94		1	F6	820
306746	−60 03063			11 31 50	−60 36.8	1	10.60		0.14		−0.27		2	B8	837
		NGC 3766 - 318		11 31 50	−61 25.2	1	9.63		−0.02		−0.57		1		51
		AAS41,43 F2 # 151		11 31 54	−62 56.3	1	11.85		1.10		0.70		1		820
		AAS41,43 F2 # 152		11 31 54	−63 01.3	1	12.59		0.35		0.35		2	B9	820
		AAS41,43 F2 # 153		11 31 54	−63 17.2	1	12.27		0.46		0.09		1	B9	820
100715	−79 00637			11 31 54	−79 58.5	1	8.04		0.24		0.11		2	A4 V	1730
		G 122 - 30		11 31 56	+40 26.2	1	11.24		0.78		0.21		4	K0	7010
100533	+67 00707			11 31 56	+66 34.0	1	9.01		0.59		0.16		2	G5	1733
		AAS41,43 F2 # 154		11 31 56	−63 19.3	1	12.06		0.43		−0.30		2	B6	820
100638	−64 01677			11 31 56	−65 08.0	2	7.16	.003	0.10	.003	−0.35	.001	27	B5/6 II/III	401,978
308679	−62 02108			11 31 57	−62 47.9	1	10.81		0.53		0.20		2	F2	820
308699	−62 02109			11 31 57	−63 01.8	1	11.25		0.08		−0.60		2	B8	820
		AAS41,43 F2 # 157		11 31 58	−63 05.6	1	11.52		1.26		1.08		2		820
100637	−56 04603			11 31 59	−57 21.0	1	7.50		1.27		1.29		6	K1 III	1673
		HA 55 # 116		11 32 01	+29 32.8	1	12.16		0.52		−0.01		2		269
100599	+24 02369			11 32 02	+23 37.6	1	8.91		1.11		1.15		1	K0	280
		AAS41,43 F2 # 158		11 32 02	−62 59.4	1	11.05		1.45		1.45		2	G8	820
	+34 02231			11 32 03	+34 06.2	1	9.42		1.22		1.42		1	K2	280
100623	−32 08179	HR 4458	★	11 32 03	−32 34.0	6	5.96	.007	0.81	.007	0.34	.015	23	K0 V	15,678,1013,2012,3078,8017
308706				11 32 03	−63 11.8	1	11.66		0.38		0.24		1		820
100625	−38 07209			11 32 04	−38 31.1	2	8.77	.005	1.11	.000	0.94		5	K0 III	78,2012
		AAS41,43 F2 # 159		11 32 04	−62 55.9	1	12.09		0.23		−0.03		1	B8	820
		AAS41,43 F2 # 161		11 32 05	−62 56.9	1	12.20		0.58		0.05		1		820
100600	+17 02374	HR 4456	★ AB	11 32 06	+17 04.4	21	5.94	.005	−0.15	.007	−0.64	.010	289	B4 V	15,30,130,154,369,667*
		G 122 - 31		11 32 06	+47 05.2	3	16.39	.006	0.02	.017	−0.93	.006	6	DA2	538,1727,3060
		HA 55 # 21		11 32 07	+29 12.1	1	11.11		0.72		0.46		2		269
		HA 55 # 168		11 32 07	+29 39.0	1	13.18		0.57		0.07		2		269
308697				11 32 07	−63 03.8	1	11.36		0.16		−0.14		3		820
	−23 10062			11 32 08	−23 35.6	1	11.17		1.53				1	M1	1746
308698	−62 02112			11 32 08	−63 01.9	1	10.67		0.24		0.16		2	A0	820
		AAS41,43 F2 # 165		11 32 09	−62 56.1	1	12.18		0.36		0.23		1	A2	820
		AAS41,43 F2 # 164		11 32 09	−63 01.1	1	11.25		1.72		2.00		3		820
100598	+32 02162			11 32 10	+32 26.8	1	8.51		0.32		0.01		2	A2	1569
		HA 55 # 22		11 32 12	+29 07.8	1	12.67		0.70		0.28		2		269
100651	−45 07105			11 32 13	−45 38.9	1	8.90		0.21		0.11		2	A5 V	508
306736	−59 03608			11 32 13	−60 22.6	1	10.37		0.07		−0.02		2	A0	837
100666	−63 01923			11 32 14	−63 40.7	1	8.77		0.31		0.16		1	A3mA7-A9	1771
308682	−62 02113			11 32 15	−62 56.7	1	10.74		0.56		0.07		2	B9	820
100663	−52 04682			11 32 16	−52 45.1	1	8.93		−0.01		−0.14		1	B9 V	1770
308696	−62 02114			11 32 16	−63 02.3	1	10.88		0.67		0.17		2	F5	820
		AAS41,43 F2 # 169		11 32 17	−62 57.1	1	12.75		1.20		0.45		2		820
		AAS41,43 F2 # 170		11 32 18	−62 57.2	1	12.47		0.61		0.02		1		820
		AAS41,43 F2 # 168		11 32 18	−63 11.7	1	11.96		0.58		0.06		1	F5	820
100615	+55 01473	HR 4457		11 32 20	+55 03.7	2	5.65	.007	1.03	.006	0.85		4	K0 III	71,1501
	−12 03458			11 32 20	−12 58.0	1	10.41		1.14		1.01		1	K7 V	3072
100674	−53 04636			11 32 20	−54 22.8	1	8.86		0.01		−0.03		1	A0 V	1770
		AAS41,43 F2 # 171		11 32 20	−62 57.5	1	11.91		1.25		0.79		1		820
100596	+67 00708			11 32 21	+67 27.7	1	8.98		0.23		0.06		10	A3	1228
100643	+31 02279			11 32 23	+30 47.1	2	7.40	.013	1.07	.002	1.03	.005	10	K0 IV	269,1569
100574	+71 00576	IDS11294N7121	AB	11 32 23	+71 04.8	1	8.85		0.57		0.14		3	G5	1723
100673	−53 04637	HR 4460		11 32 23	−53 59.3	6	4.62	.009	−0.08	.003	−0.20	.011	22	B9 Ve	15,1075,1637,1770,2012,3023
	+30 02175			11 32 26	+29 42.1	2	9.15	.061	0.61	.010	0.09		5	G0 V	269,280
100655	+21 02331	HR 4459		11 32 27	+20 43.1	2	6.46	.005	1.01	.000	0.85		6	G9 III	280,1501
100573	+75 00447			11 32 27	+74 31.1	1	9.26		1.05		0.80		2	K0	1375
100671	−45 07107			11 32 27	−46 10.6	1	7.70		1.56		1.90		1	K4 III	508
100712	−58 03727			11 32 28	−58 51.9	1	9.03		0.24		0.21		1	A8 V	1771
100689	−42 07086			11 32 30	−43 00.2	1	8.76		−0.03		−0.16		1	B9 III/IV	1770
		AAS41,43 F2 # 172		11 32 30	−63 01.2	1	12.48		0.16		−0.06		1	B8	820
100654	+36 02198	IDS11299N3558	AB	11 32 33	+35 40.9	1	8.63		0.37		0.14		3	F5	1569
100708	−48 06630	HR 4462		11 32 33	−48 51.7	4	5.50	.004	1.04	.000	0.92		16	K0 III	15,1075,2029,3005
		AAS41,43 F2 # 174		11 32 35	−62 57.0	1	12.84		0.28		0.07		1		820
		AAS41,43 F2 # 173		11 32 35	−63 08.5	1	12.40		0.28		0.08		1	B8	820
308694				11 32 36	−63 06.7	1	10.24		1.50		1.70		1	K5	820
308708				11 32 38	−63 14.1	1	11.64		0.33		0.18		1		820
	+23 02364			11 32 39	+23 06.0	1	9.84		1.60		1.95		1		280
308683	−62 02117			11 32 39	−62 51.4	1	10.28		1.20		1.08		1	K0	820
		HA 55 # 252		11 32 40	+29 56.6	1	11.44		0.60		0.08		1		269
100724	−52 04691			11 32 40	−52 57.9	1	6.89		1.29		1.35		1	K2 III	1770
308695	−62 02118			11 32 40	−62 59.8	1	10.04		1.78		1.91		1	K5	820
		HA 55 # 23		11 32 41	+29 05.2	1	11.93		0.50		−0.15		2		269
100725	−52 04692			11 32 41	−52 58.9	1	9.61		0.03		0.03		1	B9/A1 V	1770
100735	−55 04468			11 32 41	−55 50.2	1	6.97		0.92				4	G5 III	2012
100667	+54 01450			11 32 42	+53 51.0	1	8.04		0.66		0.21		2	G5	1566
100679	+35 02264			11 32 43	+34 47.2	2	9.01	.010	0.17	.000	0.14	.010	3	Am	272,280

Table 1   615

HD	DM	Other Id	N Rem	α₁₉₅₀	δ₁₉₅₀	S	V	σ_V	B–V	σ_B–V	U–B	σ_U–B	n	Spectrum	References
	+31 02280			11 32 44	+30 36.5	1	9.05		0.92		0.61		1	K0	280
		AAS41,43 F2 # 179		11 32 44	−63 00.3	1	10.91		1.69		1.96		1	K0	820
		HA 55 # 26		11 32 47	+29 09.7	1	11.29		0.45		-0.01		2		269
		AAS41,43 F2 # 180		11 32 47	−62 53.0	1	12.96		0.28		0.20		1		820
		AAS41,43 F2 # 181		11 32 47	−62 57.0	1	12.14		0.37		-0.40		1		820
100733	−46 07199	HR 4463, V763 Cen	⋆ AB	11 32 48	−47 05.7	3	5.72	.018	1.67	.010	1.93		13	M3 III	15,2012,3046
	+29 02183			11 32 49	+29 27.7	1	10.28		1.01		0.79		1	G0	280
		AAS41,43 F2 # 182		11 32 50	−62 58.1	1	11.78		0.72		0.27		1	G2	820
100698	+35 02265			11 32 53	+35 08.5	1	8.94		1.58		0.49		1	M3	280
308709				11 32 53	−63 12.6	1	11.62		0.28		0.17		1		820
100697	+54 01451			11 32 54	+54 10.0	1	8.79		0.09		0.09		2	A2	1733
100773	−60 03075			11 32 55	−60 37.1	3	6.59	.059	0.36	.024	0.02	.010	7	F0/2 V	294,2012,3077
		AAS41,43 F2 # 183		11 32 55	−63 07.7	1	12.20		0.17		-0.08		1	B8	820
100717	+11 02376			11 32 56	+11 27.9	1	6.60		1.05		0.91		2	K0	1648
		HA 55 # 171		11 32 56	+29 43.1	1	12.51		1.15		1.05		2		269
		HA 55 # 207		11 32 57	+29 45.4	1	11.98		0.50		-0.02		1		269
308687	−62 02120			11 32 57	−62 55.6	1	10.50		1.17		0.99		1	K2	820
		HA 55 # 30		11 32 58	+29 08.4	1	10.58		0.40		-0.09		3		269
	−31 09113			11 32 58	−32 15.1	4	9.82	.024	1.50	.017	1.17		9	M2 V	912,1705,2033,3073
		AAS41,43 F2 # 186		11 32 58	−62 56.7	1	12.39		0.69		0.06		1		820
		A1 91b		11 33 00	+16 15.1	1	15.92		0.45		-0.06		1		98
		HA 55 # 78		11 33 00	+29 18.3	1	10.94		0.61		0.20		1		269
		AAS41,43 F2 # 187		11 33 00	−63 00.2	1	11.40		0.91		0.45		1	F7	820
	+25 02392			11 33 01	+24 32.8	1	9.59		1.02		0.87		1	K2	280
		AAS41,43 F2 # 188		11 33 03	−62 57.8	2	12.33	.000	0.26	.009	0.19	.009	4	B9	2,820
		AAS41,43 F2 # 189		11 33 03	−63 02.2	1	11.89		0.44		0.27		1	B8	820
100726	+18 02510			11 33 05	+18 09.1	1	7.27		1.18		1.29		1	K0	280
		HA 55 # 31		11 33 05	+29 11.0	1	14.60		0.89		0.63		1		269
		SSII 17		11 33 06	+46 50.0	1	11.50		0.31		0.07		1	F0	1298
100791	−62 02121			11 33 06	−63 10.5	1	8.72		1.50		1.66		1	K2/3 III	820
100754	+26 02235			11 33 07	+26 16.8	1	8.91		1.36		1.69		1	K0	280
100740	+11 02377	HR 4464		11 33 08	+11 11.3	2	6.58		0.13	.000	0.08	.015	5	A4 Vn	985,1022
		HA 55 # 32		11 33 08	+29 05.3	1	14.16		0.97		0.53		1		269
	+36 02200			11 33 09	+36 12.1	1	8.75		1.33		1.51		1	K2	280
	+35 02266			11 33 10	+34 57.2	1	10.01		1.00		0.77		1		280
100764	−13 03407			11 33 11	−14 19.0	2	8.85	.075	1.10	.066	0.56		2	CH	1238,8005
	+25 02393			11 33 12	+24 53.3	1	9.32		1.00		0.76		2	K0	3072
100784	−45 07111			11 33 12	−45 54.4	1	8.72		0.38		-0.01		2	F2 V	508
100785	−46 07203	V785 Cen		11 33 12	−46 53.5	2	7.65	.077	1.22	.595	0.12		5	M3 II/III	508,2012
		AAS41,43 F2 # 191		11 33 12	−63 04.0	1	12.22		0.19		-0.25		1		820
		HA 55 # 35		11 33 13	+29 08.6	1	15.37		0.61		0.07		1		269
308693	−62 02123			11 33 13	−63 01.4	1	11.09		0.21		-0.61		1	B8	820
100807	−57 04839			11 33 14	−57 57.0	1	8.78		0.19		0.14		2	A5 V	1771
	−62 02124	IC 2944 - 1		11 33 15	−62 59.1	4	11.03	.013	0.09	.015	-0.76	.015	7	B2 III-IV	530,820,932,1737
	−60 03084	NGC 3766 - 111		11 33 16	−61 20.1	1	11.42		0.06		-0.48		2		56
		NGC 3766 - 103		11 33 16	−61 20.9	1	12.25		0.07		-0.32		2		56
	−62 02125	IC 2944 - 2		11 33 17	−62 59.2	2	11.35	.024	0.12	.005	-0.62	.010	3	B1.5Vn	530,820
	−36 07278			11 33 19	−37 11.5	1	9.48		1.31		1.42		3	K0	1700
308710				11 33 20	−63 13.2	1	11.56		0.12		0.03		1	A0	820
100826	−60 03090	IDS11310S6044	AB	11 33 22	−61 00.7	3	6.26	.024	0.14	.017	-0.25	.043	6	B9.5Iab/b	51,191,8100
	+17 02376			11 33 23	+17 14.7	1	9.55		1.10		0.97		2	K8	3072
		AAS41,43 F2 # 196		11 33 23	−63 03.9	1	11.62		0.75		0.26		1		820
	−60 03093	NGC 3766 - 114		11 33 24	−61 19.5	1	11.74		-0.01		-0.50		2		56
308692	−62 02130	IC 2944 - 3		11 33 24	−62 59.2	3	11.13	.029	0.11	.019	-0.54	.030	7	B3 V	2,530,1312
	+28 02020			11 33 25	+28 15.5	1	10.44		1.12		1.11		2		280
	−60 03094	NGC 3766 - 327		11 33 25	−61 18.1	5	7.17	.011	1.93	.024	2.18	.035	18	M1-Iab-Ib	51,56,91,1522,3034
308711	−62 02128			11 33 25	−63 11.4	1	9.15		1.42		1.52		1	K5	820
100775	+28 02021			11 33 26	+28 10.8	2	8.27	.015	0.53	.001	0.06	.008	14	F8 V	269,1569
		MN177,99 # 40		11 33 26	−46 29.5	1	11.70		1.43		1.22		1		508
		MN177,99 # 39		11 33 26	−46 34.5	1	12.64		0.72		0.29		2		508
		IC 2944 - 4		11 33 26	−62 57.9	1	11.98		0.38		-0.56		3		530
100840	−60 03095	NGC 3766 - 326		11 33 27	−61 17.0	4	8.18	.011	0.00	.027	-0.66	.044	14	B2 III	2,51,56,91
		IC 2944 - 5		11 33 27	−62 56.8	1	12.24		0.13		-0.52		3		530
100796	+31 02281			11 33 28	+30 59.8	2	8.42	.010	0.57	.005	0.08	.005	7	G0 V	1569,1625
100841	−62 02127	HR 4467	⋆ A	11 33 28	−62 44.6	6	3.13	.008	-0.04	.006	-0.15	.041	16	B9 III	6,15,1034,1075,2012,3023
306785	−60 03096	NGC 3766 - 317		11 33 29	−61 12.7	1	9.67		-0.04		-0.62		1	B2	51
100825	−46 07205	HR 4466		11 33 30	−47 21.9	4	5.24	.004	0.26	.004	0.10	.009	12	F0 V	15,1075,2012,3023
	−60 03097	NGC 3766 - 112		11 33 30	−61 19.5	1	10.36		0.59		0.09		8		56
308685	−62 02132			11 33 30	−62 52.9	1	10.72		0.69		-0.45		3	A2	2
		G 120 - 61		11 33 31	+29 06.6	3	13.55	.072	1.32	.017	1.09	.052	5		203,269,3062
	−24 09840			11 33 31	−24 35.1	1	11.92		0.49		-0.17		2		1696
	+36 02201			11 33 33	+36 24.0	1	9.28		1.10		0.99		1	K0	280
	−60 03098	NGC 3766 - 97		11 33 33	−61 21.5	1	9.62		-0.01				1	B2 V	91
	−46 07207			11 33 35	−46 42.4	1	11.68		0.51		0.11		2		508
	−60 03101	NGC 3766 - 137		11 33 35	−61 17.9	1	10.82		-0.01		-0.55		2		56
306794	−60 03102	NGC 3766 - 1	⋆ V	11 33 36	−61 19.6	2	8.60	.023	0.01	.005	-0.61	.005	4	B2 IVp	2,56
		MN177,99 # 38		11 33 37	−46 35.5	1	12.01		0.68		0.24		1		508
100901	−72 01143			11 33 37	−72 34.0	1	6.53		1.16				4	K0/1 III	2012
100809	+15 02352			11 33 39	+14 58.5	1	8.28		0.20		0.10		2	Am	272
	+22 02378			11 33 39	+22 10.6	1	10.48		0.60		0.16		1		280

HD	DM	Other Id	N Rem	α_{1950}	δ_{1950}	S	V	σ_V	B–V	σ_{B-V}	U–B	σ_{U-B}	n	Spectrum	References
100855	−52 04706			11 33 39	−53 01.3	1	9.28		0.36		0.18		1	A0/1 V	1770
100808	+28 02022	HR 4465	⋆AB	11 33 40	+28 03.5	4	5.81	.013	0.24	.007	0.07	.008	18	F0 V	70,269,1569,3024
100808	+28 02022	IDS11310N2820	C	11 33 40	+28 03.5	1	11.17		1.38		1.41		3		3024
		HA 55 # 37		11 33 40	+29 12.3	1	13.06		0.83		0.48		1		269
		HA 55 # 38		11 33 41	+29 05.8	1	15.59		0.84				1		269
	+47 01888			11 33 41	+46 55.4	1	11.08		0.08		0.24		2	Am:	1026
	−60 03108	NGC 3766 - 81		11 33 41	−61 20.7	1	10.00		-0.15				1	B2 Vnpe	91
		HA 55 # 39		11 33 42	+29 08.6	1	15.12		1.24				1		269
		NGC 3766 - 319		11 33 43	−61 30.5	1	10.08		0.33		0.20		1		51
308690	−62 02133	IC 2944 - 6		11 33 43	−62 57.1	2	10.27	.012	1.64	.016	1.75		4	K7	530,1312
308691	−62 02134	IC 2944 - 7		11 33 43	−62 59.2	1	10.22		1.27		1.22		3	K2 V	530
		KUV 352 - 9		11 33 45	+39 46.2	1	12.96		-0.13		-0.60		1		974
100865	−53 04644			11 33 45	−53 31.3	1	9.30		0.18		0.16		1	A2 V[m]	1770
306792	−60 03116	NGC 3766 - 7	⋆V	11 33 45	−61 18.2	1	10.11		0.01				1	B2 V	91
100856	−60 03112	NGC 3766 - 5		11 33 45	−61 18.8	2	8.06	.095	-0.10	.080	-0.85		2	B1.5III	2,91
	−46 07211			11 33 46	−46 41.8	1	10.10		1.22		1.23		2		508
306797	−61 02432	NGC 3766 - 240	⋆V	11 33 46	−61 25.5	2	9.60	.028	0.02	.023	-0.55	.037	4		2,56
	−60 03120	NGC 3766 - 8		11 33 48	−61 17.7	2	10.70	.015	-0.03	.000	-0.66		2	B4 V	56,91
		NGC 3766 - 322		11 33 48	−61 20.	1	12.54		0.07		-0.18		1		51
		NGC 3766 - 323		11 33 48	−61 20.	1	12.41		1.31		1.42		1		51
		NGC 3766 - 324		11 33 48	−61 20.	1	13.28		1.08		0.34		1		51
		NGC 3766 - 325		11 33 48	−61 20.	1	13.35		0.24		0.39		1		51
	−60 03122	NGC 3766 - 88	⋆V	11 33 48	−61 22.0	1	10.00		0.04				1	B3 npe	91
100851	−36 07288			11 33 49	−36 54.1	1	9.75		0.21		0.09		3	A2 V	1700
306820	−58 03739			11 33 49	−59 20.7	1	11.33		0.20		-0.25		2	B6 III	540
	−60 03125	NGC 3766 - 26		11 33 50	−61 19.0	2	9.16	.083	-0.02	.039	-0.66		3	B2 IVne	91,1737
	−60 03126	NGC 3766 - 63	⋆V	11 33 50	−61 21.0	1	9.26		0.01		-0.61		1	B1.5Vn	56
306798	−61 02434	NGC 3766 - 239		11 33 50	−61 25.1	2	9.44	.009	0.00	.019	-0.62	.000	4		2,56
100879	−62 02135	IC 2944 - 8	⋆V	11 33 50	−63 00.3	3	9.76	.015	0.01	.009	-0.78	.023	7	B0.5V	2,127,530
100844	+19 02477			11 33 51	+18 52.2	1	9.43		0.29		0.02		3	F0	1569
100843	+25 02394			11 33 51	+25 18.4	1	7.01		0.18		0.19		1	A7 V	280
	+26 02236			11 33 51	+26 07.1	1	9.60		1.13		1.14		1	K0	280
		HA 55 # 41		11 33 51	+29 11.3	1	15.58		0.54		0.01		1		269
		NGC 3766 - 9		11 33 51	−61 17.2	1	12.21		0.28		-0.17		1		56
	−60 03128	NGC 3766 - 27	⋆AB	11 33 52	−61 19.2	1	8.46		-0.04		-0.64		1	B2 IV-V	56
	−60 03133	NGC 3766 - 67	⋆V	11 33 54	−61 21.0	1	9.85		-0.12				1	B2 Vp	91
	+16 02272			11 33 55	+15 31.7	1	9.95		0.18		0.07		2	A2	272
		HA 55 # 42		11 33 55	+29 11.4	1	14.64		0.80		0.25		1		269
	−46 07214			11 33 55	−46 35.8	1	11.08		0.66		0.17		1		508
	+18 02512			11 33 56	+18 27.6	1	10.62		0.98		0.70		1		280
	−60 03134	NGC 3766 - 23		11 33 56	−61 18.4	1	10.42		0.04				1	B2 IV-V	91
	−60 03135	NGC 3766 - 30		11 33 56	−61 19.5	1	11.10		0.03		-0.41		1		56
100915	−60 03136	NGC 3766 - 316	⋆V	11 33 57	−61 11.4	2	8.57	.000	-0.04	.000	-0.68		2	B2 IVn	51,91
		NGC 3766 - 72		11 33 57	−61 23.2	1	13.02		0.19				1		56
	+36 02202			11 33 58	+35 31.2	1	9.42		1.06		1.02		1	K0	280
	−60 03137	NGC 3766 - 22		11 33 58	−61 19.1	1	9.99		-0.02				1	B2.5V	91
	+40 02442			11 33 59	+39 28.5	1	10.10		1.29		1.29		3	M0	1569
		KPS 352 - 203		11 33 59	+40 04.6	1	12.49		0.62		0.09		3		974
		G 56 - 48		11 34 01	+14 40.5	2	14.03	.010	0.80	.002	0.11	.015	4		333,1620,1658
306715	−59 03636			11 34 01	−59 59.2	1	10.62		0.04		-0.57		2	B2.5V	540
	−60 03147	NGC 3766 - 20		11 34 01	−61 18.6	1	9.43		0.05				1	B2 IV-Vn	91
	−60 03143	NGC 3766 - 52		11 34 01	−61 20.3	1	10.29		0.02				1	B1.5V	91
	−60 03145	NGC 3766 - 70		11 34 01	−61 22.4	1	9.07		0.01		-0.63		3	B2 IV-V	56
		LP 553 - 15		11 34 02	+05 16.2	1	12.17		0.90		0.55		4		1696
	+22 02379			11 34 02	+21 44.9	1	9.73		0.48		-0.08		1	F5	280
	−46 07216			11 34 02	−46 39.1	1	11.48		0.58		0.11		2		508
100929	−60 03140	HR 4472		11 34 02	−60 46.5	4	5.83	.016	-0.09	.007	-0.63	.001	12	B3 IV	6,15,26,2012
	−60 03149	NGC 3766 - 36	⋆V	11 34 02	−61 19.9	1	10.33		0.03				1	B4 Vne	91
	+23 02365			11 34 03	+22 39.9	1	9.31		1.01		0.80		1	K0	280
		LP 673 - 106		11 34 03	−08 13.2	1	12.37		0.49		-0.25		2		1696
		IC 2944 - 10		11 34 03	−62 54.9	1	11.04		1.28		0.89		3		530
308712		IC 2944 - 9		11 34 03	−63 09.7	2	11.69	.012	0.36	.024	0.28	.055	4	A5 V	530,1312
100869	+22 02381	IDS11315N2201	AB	11 34 05	+21 44.8	1	9.28		0.58		0.05		1	G5	280
	+34 02236			11 34 05	+33 43.1	1	8.59		1.46		1.80		1	K2	280
100893	−32 08199	HR 4469	⋆AB	11 34 06	−33 17.6	2	5.72	.010	1.02	.000	0.83		6	K0 III	404,2032
		MN177,99 # 44		11 34 06	−46 25.1	1	13.86		0.96		0.42		1		508
100930	−60 03152			11 34 06	−61 02.6	1	7.83		1.93				3	M1	1522
		NGC 3766 - 40		11 34 07	−61 19.3	1	12.72		0.16		-0.08		1		56
100943	−60 03155	NGC 3766 - 232		11 34 08	−61 23.3	3	7.15	.015	0.12	.009	-0.60	.010	6	B5 Ia	6,56,138
	+2 02451			11 34 09	+02 11.2	1	10.42		0.71				2		202
100889	−8 03202	HR 4468		11 34 09	−09 31.5	9	4.70	.024	-0.08	.006	-0.15	.026	31	B9.5Vn	3,15,1075,1079,1107*
100890	−11 03135			11 34 09	−12 05.5	1	9.20		0.42				4	F5	2033
100910	−33 07845			11 34 09	−34 22.3	2	7.57	.000	0.50	.005	-0.03		5	F6 V	78,2033
		NGC 3766 - 17		11 34 09	−61 18.6	1	12.86		0.14		-0.08		1		56
100912	−39 07207			11 34 10	−40 14.8	1	8.63		0.92		0.44		1	G8 III	78
100906	−18 03208			11 34 11	−18 41.6	2	9.67	.010	0.83	.010	0.51		3	G8(w) +F5	1594,6006
	−46 07218			11 34 11	−46 38.8	1	10.31		1.15		1.02		2		508
306791	−60 03157	NGC 3766 - 15	⋆V	11 34 11	−61 16.8	2	8.54	.020	0.06	.005	-0.55		2	B2 III	2,91
100926	−46 07219			11 34 12	−47 08.3	1	9.72		0.22		0.15		2	A3 III/IV	508
	−60 03158	NGC 3766 - 16		11 34 12	−61 18.2	1	10.01		0.00		-0.59		2		56

Table 1

HD	DM	Other Id	N	Rem	α₁₉₅₀	δ₁₉₅₀	S	V	σ_V	B–V	σ_B–V	U–B	σ_U–B	n	Spectrum	References
		LP 63 - 220	A		11 34 13	+67 21.5	1	10.93		0.63		0.15		3		1723
100911	−36 07291	HR 4470			11 34 13	−36 57.6	3	6.30	.004	0.06	.004	0.05		8	A1 V	15,78,2012
100942	−60 03159	IDS11319S6020	A		11 34 13	−60 37.0	1	7.77		−0.01		−0.25		2	B8/9 V	1771
306799	−60 03161	NGC 3766 - 48			11 34 14	−61 20.0	4	7.52	.022	1.84	.007	1.74	.151	14	M0 Ib	91,1522,1704,3034
		A1 100			11 34 17	+15 11.4	1	17.34		0.50		−0.70		1		98
100883	+50 01842				11 34 17	+49 52.5	1	9.34		0.20		0.04		2	F2	272
		LP 129 - 373			11 34 17	+52 38.0	1	10.45		1.15		1.16		3		1723
		MN177,99 # 41			11 34 17	−46 32.7	1	13.91		0.69		0.07		1		508
100940	−54 04652				11 34 17	−54 52.5	1	8.90		0.26		0.15		1	A9 V	1771
100922	−11 03136	IDS11318S1148	AB		11 34 18	−12 04.4	3	8.86	.007	0.80	.005	0.37		10	G5	176,214,2033
	−60 03164	NGC 3766 - 178			11 34 18	−61 17.5	1	11.38		0.03		−0.49		1		56
		IC 2944 - 126			11 34 18	−62 45.	1	14.99		0.73				1		1312
		IC 2944 - 127			11 34 18	−62 45.	1	14.86		0.81				1		1312
		IC 2944 - 128			11 34 18	−62 45.	1	14.63		0.91				1		1312
		IC 2944 - 129			11 34 18	−62 45.	1	14.10		0.87				1		1312
		IC 2944 - 130			11 34 18	−62 45.	1	14.56		0.36				1		1312
308688	−62 02139	IC 2944 - 12			11 34 18	−62 55.2	1	11.24		0.46		0.31		3	A0	530
308713		IC 2944 - 11			11 34 18	−63 07.8	2	11.41	.036	0.32	.040	0.17	.071	4	A3 V	530,1312
		MN177,99 # 42			11 34 20	−46 30.4	1	12.28		0.52		0.05		1		508
100969	−60 03168	NGC 3766 - 169			11 34 22	−61 15.0	1	9.17		−0.03		−0.67		3	B5	56
	+37 02199				11 34 23	+37 24.6	1	9.84		1.25		1.38		1	K2	280
100920	−0 02458	HR 4471			11 34 23	−00 32.9	10	4.30	.010	1.00	.008	0.75	.006	35	G9 III	3,15,1075,1075,1363*
	−79 00640				11 34 24	−80 03.5	1	8.99		1.43		1.44		2		1730
308689	−62 02140	IC 2944 - 13			11 34 26	−62 56.9	2	10.82	.000	0.12	.000	−0.68		4	B2 V	530,1312
		A1 102			11 34 28	+15 07.3	1	17.06		−0.46		−0.77		1		98
		GD 140			11 34 28	+30 04.6	1	12.50		−0.06		−0.98		2	DA	3060
	−48 06657				11 34 28	−48 43.4	2	10.90	.015	0.19	.000	0.09	.000	5	A7 V	1097,3077
100953	−32 08202	HR 4473			11 34 32	−32 42.6	3	6.29	.008	0.46	.007	0.04		8	F5 V	15,78,2012
100984	−50 06178				11 34 34	−50 38.8	1	9.16		−0.03		−0.19		1	B9 IV/V	1770
		NGC 3766 - 321			11 34 35	−61 29.3	1	11.44		0.04		−0.30		1		51
101008	−62 02142	IC 2944 - 14			11 34 35	−63 07.1	7	9.16	.011	−0.01	.016	−0.89	.025	18	B0 III	2,6,127,530,540,976,1312
100947	+28 02023				11 34 36	+28 03.1	3	7.66	.020	1.07	.007	0.97	.005	16	K1 III	20,269,1569
		NGC 3766 - 336			11 34 36	−60 54.3	1	13.98		1.15		0.43		3		56
	−76 00678				11 34 36	−77 13.6	1	10.46		0.34				4	A5	2012
100858	+78 00392				11 34 37	+77 52.4	1	6.49		1.61		1.94		4	K5	985
101007	−60 03178				11 34 37	−60 53.5	1	7.02		1.82		1.18		1	M3 Ib	51
100980	−38 07232				11 34 40	−39 11.5	2	8.72	.012	0.44	.002	−0.09		4	F3 V	1594,6006
	−46 07222				11 34 41	−46 50.1	1	10.50		1.36		1.36		2		508
101021	−60 03182	HR 4475			11 34 41	−61 04.0	6	5.15	.005	1.12	.010	1.07	.047	21	K0 III	15,51,1075,2013,2029,3005
	+48 01958	DF UMa	★ AB		11 34 43	+47 44.4	2	10.19	.124	1.17	.158	1.02		3	K4 V	1017,3072
	−45 07140				11 34 43	−46 15.6	1	10.90		0.26		0.14		1		508
	+19 02481				11 34 44	+18 36.8	1	10.24		0.93		0.79		1		280
		IC 2944 - 15			11 34 44	−63 02.4	1	11.29		1.15		0.76		3		530
	+36 02203				11 34 45	+36 19.4	1	10.60		1.09		1.05		1		280
101003	−46 07223				11 34 45	−47 06.9	1	8.85		0.43		−0.05		2	F5 V	508
101020	−54 04662				11 34 46	−55 19.1	1	8.96		0.17		0.15		1	A4 V	1771
		G 253 - 53			11 34 47	+82 05.2	1	11.77		1.50				1		906
		NGC 3766 - 320			11 34 47	−61 31.4	1	9.98		−0.03		−0.58		1		51
100973	+26 02237				11 34 48	+26 04.8	1	8.89		0.95		0.71		2	G5	280
101019	−53 04655				11 34 48	−53 42.3	1	8.94		0.18		0.12		1	A3 V	1771
100993	+26 02238				11 34 49	+25 41.8	1	8.17		0.54		0.02		2	F8 V	3026
	+34 02239				11 34 50	+34 10.6	1	9.02		0.96		0.65		1	K0	280
308816	−62 02143	IC 2944 - 16			11 34 51	−63 02.7	2	11.04	.018	0.09	.004	−0.67	.013	5	B1.5V	530,1312
310774	−65 01694				11 34 51	−65 31.5	1	10.18		0.07		−0.77		2	B8	540
		Feige 46			11 34 54	+14 27.	2	13.22	.033	−0.28	.019	−1.14	.014	4	sdO	1298,3028
		KPS 352 - 208			11 34 54	+39 42.3	1	11.60		1.00		0.77		3		974
		MN177,99 # 5			11 34 54	−46 26.1	1	14.25		1.35		1.56		1		508
308794	−62 02144	IC 2944 - 133			11 34 54	−62 41.8	1	9.16		0.01		−0.83		1	A2	1312
101030	−46 07225				11 34 55	−46 48.8	1	8.64		1.48		1.72		1	K4 III	508
100972	+45 01943				11 34 56	+44 59.6	1	6.85		0.04		0.07		2	B9	252
		IC 2944 - 17			11 34 56	−62 56.6	1	11.50		1.35		1.09		3		530
308802		IC 2944 - 18			11 34 57	−62 55.2	1	11.75		0.21		0.01		3		530
101015	+28 02024				11 34 58	+27 43.3	1	8.16		1.24		1.46		1	K0	280
101031	−46 07227				11 34 58	−46 52.8	1	9.27		0.91		0.55		2	G6/K0 (III)	508
		WLS 1100 75 # 5			11 34 59	+76 21.2	1	12.46		0.54		0.02		2		1375
		MN177,99 # 8			11 35 00	−46 29.5	1	12.49		0.78		0.39		1		508
101070	−62 02147	IC 2944 - 19			11 35 00	−62 52.2	3	8.94	.011	0.01	.005	−0.83	.035	8	B0.5IV	127,530,1312
308815	−62 02148	IC 2944 - 20			11 35 01	−62 59.7	3	9.65	.009	0.08	.016	−0.72	.026	9	B0.5V	127,530,1312
		Ton 1375			11 35 02	+21 46.	1	13.47		0.04		0.13		3		286
101051	−46 07230	IDS11326S4623	AB		11 35 03	−46 39.8	1	8.93		1.10		0.95		1	K0/1 III	508
309793	−62 02149	IC 2944 - 134			11 35 04	−62 45.6	1	10.98		0.08				1	B9	1312
		IC 2944 - 21			11 35 05	−62 54.0	1	11.21		1.63		2.09		3		530
101066	−46 07231				11 35 06	−47 01.7	1	7.81		0.31		0.03		2	F2 IV/V	508
100971	+68 00652	SU Dra			11 35 07	+67 36.5	1	9.27		0.10		0.15		1	A2	668
101083	−54 04665				11 35 07	−55 09.4	1	9.59		0.10		0.09		1	A1 Vn	1770
		IC 2944 - 22			11 35 07	−63 12.2	2	10.86	.004	1.46	.028	1.34	.130	4	K3 III	530,1312
101067	−47 06997	HR 4476			11 35 08	−47 28.2	3	5.44	.004	1.24	.005	1.30		10	K2 III	15,1075,3077
		NGC 3766 - 332			11 35 08	−61 05.4	1	9.89		0.02		−0.30		1		51
101085	−62 02150				11 35 08	−63 24.4	1	8.36		0.25		−0.21		2	B8 IV	127
	+2 02453				11 35 09	+02 27.1	1	10.43		1.47		1.73		1	M2	327

HD	DM	Other Id	N	Rem	α_{1950}	δ_{1950}	S	V	σ_V	B–V	σ_{B-V}	U–B	σ_{U-B}	n	Spectrum	References
		G 122 - 37			11 35 09	+42 06.7	1	14.47		1.48		1.08		1		3016
		MN177,99 # 1			11 35 09	−46 24.1	1	13.63		0.62		0.02		1		508
101084	−62 02151	IC 2944 - 23		★V	11 35 10	−63 04.1	4	9.23	.148	0.07	.009	−0.75	.029	9	B1 Ve	6,127,530,1312
101013	+51 01679	HR 4474			11 35 11	+50 53.7	2	6.15	.006	1.06	.010	0.76	.000	6	K0p	993,1080
101065	−46 07232	V816 Cen			11 35 11	−46 26.0	2	8.02	.010	0.77	.010	0.20		7	B5	508,2012
101105	−60 03195	IDS11328S6056		AB	11 35 11	−61 12.5	6	7.16	.017	0.00	.013	−0.63	.011	21	B2 III/IV	6,51,91,1075,1499,2012
101132	−75 00744	HR 4479			11 35 11	−75 37.2	4	5.64	.004	0.35	.008	−0.01	.009	16	A9 IV	15,278,1075,2012
101063	−28 08980				11 35 12	−28 34.4	3	9.46	.009	0.76	.005	0.10	.010	7	K5/M0 III	78,1594,3077
		MN177,99 # 2			11 35 12	−46 24.6	1	12.52		1.60		2.06		1		508
		IC 2944 - 24			11 35 13	−63 01.5	1	12.63		1.68		1.88		2		530
101058	+26 02239				11 35 15	+25 40.5	1	8.57		0.91		0.55		2	G5	280
		MN177,99 # 3			11 35 15	−46 24.7	1	14.03		1.03		0.83		1		508
		MN177,99 # 9			11 35 16	−46 19.5	1	13.04		0.75		0.26		2		508
		MN177,99 # 4			11 35 16	−46 25.5	1	12.78		0.62		0.10		1		508
	−46 07233				11 35 16	−46 32.5	1	10.81		1.01		0.75		1		508
101119	−60 03201				11 35 16	−60 42.4	1	7.34		0.01		−0.12		2	B9 III	401
		IC 2944 - 26			11 35 17	−63 02.3	1	12.76		0.42		0.07		2		530
		IC 2944 - 25			11 35 17	−63 13.5	2	12.29	.038	0.15	.014	−0.53	.014	2		530,1312
101059	+22 02384				11 35 18	+22 01.3	1	8.05		0.16		0.08		3	A3	1569
	−46 07235				11 35 18	−46 41.0	1	11.78		1.44		1.52		1		508
308800	−62 02152	IC 2944 - 27			11 35 19	−62 50.1	2	10.90	.018	0.12	.009	−0.21		5	B8 IV	530,1312
101040	+52 01576				11 35 20	+52 19.0	1	9.27		0.48		0.05		39	G5	588
		MN177,99 # 10			11 35 20	−46 18.8	1	12.53		1.50		1.49		1		508
	−47 07000				11 35 20	−48 20.5	2	10.84	.065	1.30	.005	1.29	.090	2	K4	78,3062
		IC 2944 - 28			11 35 20	−63 07.0	1	12.11		0.12		−0.49		2	O9.5Ve	530
101060	+16 02277				11 35 22	+16 03.2	1	8.83		−0.04		−0.14		2	A0 V	272
	−68 01538				11 35 22	−69 03.4	1	11.02		0.08		−0.86		2	O6	540
	+11 02382				11 35 23	+10 51.5	1	9.85		0.53		0.02		1	F8	289
	+20 02629				11 35 23	+19 56.0	1	9.62		0.95		0.64		1	G5	280
	−62 02153	IC 2944 - 29			11 35 23	−63 03.3	3	10.18	.018	0.04	.017	−0.81	.005	15	B1 V	91,530,1312
		Feige 47			11 35 24	+45 13.	1	16.50		−0.21		−0.69		1		1298
	−45 07146				11 35 24	−46 15.4	1	9.55		0.13		0.14		2		508
		MN177,99 # 26			11 35 24	−46 43.0	1	13.29		0.72		0.17		1		508
	+40 02445				11 35 25	+39 50.0	1	9.92		0.60				3	G0	280
101131	−62 02154	IC 2944 - 30			11 35 25	−63 02.7	5	7.14	.020	0.02	.013	−0.90	.015	18	O6.5	6,91,530,1312,2012
		IC 2944 - 31			11 35 26	−63 03.0	1	13.19		0.23		−0.01		2		530
101093	−0 02464	IDS11328S0103			11 35 27	−01 19.5	1	7.62		0.55		−0.01		2	F9 V	1003
101146	−61 02460				11 35 27	−62 11.5	1	8.07		0.01		−0.64		2	B1 IV	401
	+32 02170				11 35 28	+31 31.7	1	9.77		0.75		0.46		3	G5	1569
101091	+32 02169				11 35 28	+32 09.5	1	7.11		0.40		0.00		2	F2	1569
	+33 02134	IDS11328N3320	A		11 35 29	+33 02.9	1	10.84		0.52		−0.02		3	G0	1723
	+33 02134	IDS11328N3320	B		11 35 29	+33 02.9	1	12.13		0.72		0.39		3	G0	1723
		MN177,99 # 25			11 35 30	−46 41.2	1	12.30		0.53		0.18		1		508
	+38 02261				11 35 31	+38 12.0	1	10.72		1.20		1.06		1		280
101128	−46 07237				11 35 31	−46 33.4	1	9.28		0.06		0.07		3	A2 IV/V	508
101129	−51 05874				11 35 31	−52 04.5	1	9.97		0.14		0.09		1	A0 IV/V	1770
		IC 2944 - 32			11 35 31	−63 02.3	1	13.62		0.26		−0.07		2		530
		MN177,99 # 24			11 35 32	−46 40.6	1	13.42		1.23		0.91		1		508
308817	−62 02157	IC 2944 - 34			11 35 32	−63 04.2	3	10.67	.015	0.09	.019	−0.65	.016	13	B1.5V	91,530,1312
308818	−62 02156	IC 2944 - 33			11 35 32	−63 05.1	2	9.64	.014	0.04	.005	−0.83	.019	8	B0 V	530,1312
101162	−66 01629	HR 4485			11 35 32	−67 20.6	4	5.94	.007	1.02	.000	0.84	.005	17	K0 III	15,1075,1488,2038
		MN177,99 # 14			11 35 33	−46 26.6	1	13.40		0.87		0.39		1		508
308813	−62 02158	IC 2944 - 36			11 35 34	−63 02.3	2	9.28	.000	0.04	.000	−0.86	.000	9	O9.5V	530,1312
101112	+9 02523	HR 4478			11 35 35	+09 09.7	3	6.16	.004	1.08	.000	0.96	.004	9	K1 III	15,1417,3077
101110	+22 02385				11 35 35	+22 08.9	1	8.12		1.33		1.60		1	K2	280
		MN177,99 # 15			11 35 35	−46 29.1	1	14.01		0.46		0.01		1		508
		IC 2944 - 37			11 35 35	−63 03.6	1	13.37		0.22		−0.15		2		530
		IC 2944 - 35			11 35 35	−63 08.7	1	11.95		0.31		0.28		2	A2 V	530
		IC 2944 - 38			11 35 37	−63 05.2	2	12.68	.071	0.38	.042	0.29		2		530,1312
308819	−62 02160	IC 2944 - 39			11 35 37	−63 05.8	2	10.09	.005	0.15	.010	−0.23	.005	8	B9p(e)	530,1312
		MN177,99 # 13			11 35 38	−46 22.8	1	12.33		1.07		0.82		1		508
101108	+39 02458	IDS11330N3918		AB	11 35 40	+39 01.9	2	8.89	.039	0.17	.000	0.09	.039	3	A5p	272,280
308803	−62 02159	IC 2944 - 135			11 35 40	−62 54.4	1	11.83		0.15		−0.41		4		1312
101107	+44 02110	HR 4477			11 35 41	+43 54.2	2	5.60	.040	0.33	.012	−0.01		4	F2 III	70,254
		MN177,99 # 11			11 35 41	−46 19.3	1	12.43		0.97		0.73		1		508
		MN177,99 # 12			11 35 42	−46 20.1	1	12.63		1.10		0.93		1		508
		LP 613 - 92			11 35 43	−00 44.2	1	10.40		0.81		0.45		2		1696
308799		IC 2944 - 40			11 35 43	−62 48.7	1	10.50		1.09		0.92		3	K0	530
308814	−62 02162	IC 2944 - 41			11 35 43	−63 04.0	2	11.25	.005	0.16	.000	−0.39	.005	8	B5 Vn	530,1312
101190	−62 02163	IC 2944 - 43		★A	11 35 44	−62 55.2	8	7.31	.019	0.05	.014	−0.86	.024	21	O5	2,6,127,191,530,540*
		IC 2944 - 42			11 35 44	−62 56.6	1	11.18		2.23		2.88		3		530
		IC 2944 - 44			11 35 45	−62 55.5	1	12.45		0.17		−0.35		1		530
101174	−64 01682				11 35 45	−65 22.7	3	7.40	.007	0.10	.005	−0.45	.001	42	B4/5 IV	401,978,1642
101189	−61 02463	HR 4487			11 35 46	−61 33.0	5	5.15	.007	−0.02	.010	−0.17	.014	11	Ap HgMn	6,15,51,1771,2012
		KPS 352 - 214			11 35 48	+40 07.7	1	13.87		1.37		1.29		3		974
		IC 2944 - 45			11 35 48	−62 55.5	1	12.21		0.21		−0.10		2		530
101191	−62 02164	IC 2944 - 46			11 35 48	−63 06.8	6	8.49	.011	0.05	.020	−0.86	.015	16	O8 V	2,6,530,540,976,1312
		LOD 24 # 8			11 35 49	−61 47.5	1	12.11		0.10		−0.01		2		127
101154	−1 02546	HR 4484		★AB	11 35 51	−02 09.6	2	6.21	.005	1.12	.000	1.08	.005	7	G9 III	15,1417
308804	−62 02165	IC 2944 - 48			11 35 51	−62 54.0	2	11.28	.015	0.14	.010	−0.51	.010	5	O9.5Vn	530,1312

Table 1 619

HD	DM	Other Id	N	Rem	α_{1950}	δ_{1950}	S	V	σ_V	B–V	σ_{B-V}	U–B	σ_{U-B}	n	Spectrum	References
		IC 2944 - 49			11 35 51	−63 07.2	2	13.01	.079	0.14	.030	-0.37		6		530,1312
		IC 2944 - 47			11 35 51	−63 09.1	1	13.04		0.15		-0.29		2		530
		G 10 - 39			11 35 52	+03 30.7	1	11.55		1.37		1.11		1	K7	333,1620
101135	+20 02631				11 35 52	+20 22.7	1	8.69		0.21		0.01		2	A9 IV	1569
		IC 2944 - 50			11 35 52	−63 02.1	2	12.75	.687	1.24	.017	0.97		3		530,1312
	−62 02166	IC 2944 - 51			11 35 52	−63 06.8	3	10.82	.019	0.11	.020	-0.53	.025	8	B5 V	91,530,1312
101153	+8 02532	HR 4483, ω Vir			11 35 53	+08 24.7	5	5.36	.043	1.58	.015	1.57	.031	20	M4 III	3,15,1355,1417,3016
101133	+47 01894	HR 4480			11 35 53	+47 06.7	2	6.09	.015	0.37	.010	0.14		3	F1 III	39,70
		IC 2944 - 52			11 35 53	−63 04.5	2	13.17	.040	1.36	.094	1.40		6		530,1312
		IC 2944 - 53			11 35 53	−63 04.9	1	14.01		0.20		0.06		2		530
		IC 2944 - 54			11 35 53	−63 06.7	2	11.05	.019	0.11	.005	-0.58	.028	4	B5 V	530,1312
101151	+34 02242	HR 4482			11 35 54	+33 54.2	2	6.27	.010	1.32	.005	1.48		4	K2 III	252,280
		IC 2944 - 56			11 35 55	−62 55.1	1	13.04		0.24		-0.20		1		530
	−62 02167	IC 2944 - 57			11 35 55	−63 04.0	3	10.51	.129	0.16	.041	-0.51	.064	7	B2 V	91,530,1312
		IC 2944 - 55			11 35 55	−63 05.2	1	14.02		0.71		0.04		2		530
		IC 2944 - 58			11 35 55	−63 06.4	1	13.91		0.36		0.01		2		530
	+29 02191				11 35 56	+28 46.0	1	9.79		0.74		0.38		1	K0	280
306828	−59 03661				11 35 56	−59 28.4	1	9.90		0.03		-0.67		2	B2 IV	540
101205	−62 02168	IC 2944 - 59		⋆ AB	11 35 56	−63 05.7	8	6.46	.019	0.04	.018	-0.87	.015	23	O6	2,6,91,158,530,540*
		LP 733 - 14			11 35 57	−13 33.5	1	11.29		0.99		0.79		1		1696
		IC 2944 - 60			11 35 57	−63 06.1	1	13.32		0.22		-0.19		2		530
308820	−62 02170	IC 2944 - 61			11 35 57	−63 06.9	1	11.55		0.13		-0.41		4	B5 Vne	530
		ST Leo			11 35 58	+10 50.3	2	10.77	.006	0.07	.011	0.11	.020	2	F2	597,699
101223	−62 02171	IC 2944 - 62			11 35 58	−62 55.5	4	8.70	.011	0.16	.010	-0.75	.015	9	O8	2,530,540,1312
308828		IC 2944 - 63			11 35 58	−63 10.9	2	10.59	.020	0.75	.020	0.30	.015	6	G8 V	91,530
		MN177,99 # 18			11 35 59	−46 24.5	1	12.31		1.31		1.45		1		508
		IC 2944 - 64			11 35 59	−63 02.6	1	13.59		0.28		0.04		2		530
		IC 2944 - 65			11 35 59	−63 06.2	1	13.16		0.69		0.21		2		530
	+25 02399				11 36 00	+25 18.0	2	11.67	.004	-0.15	.011	-0.51	.007	6	A0p	275,1240
308812		IC 2944 - 66			11 36 00	−63 01.1	2	11.58	.004	0.26	.004	0.20	.034	3	A0 V	530,1312
101150	+65 00843	HR 4481		⋆ AB	11 36 01	+64 37.4	1	6.50		0.14		0.11		2	A5 IV	1733
308775	−61 02468				11 36 01	−62 24.4	2	10.56	.000	0.05	.010	-0.43	.030	5	B8	2,401
	+22 02386				11 36 02	+21 38.8	1	9.98		1.01		0.82		1	K0	280
101219	−53 04668				11 36 02	−53 58.7	1	9.45		0.34		0.19		1	A8 IV/V	1771
308781	−62 02169	IC 2944 - 136			11 36 02	−62 32.3	1	11.45		0.14		-0.43		3	B6 V	1312
		L 396 - 10			11 36 03	−41 05.6	1	13.78		1.72		1.20		2		3073
		IC 2944 - 67			11 36 03	−63 06.0	1	12.31		0.16		-0.32		2		530
	+55 01479				11 36 04	+55 23.0	1	9.96		0.82		0.53		4		1723
101163	+39 02459				11 36 05	+38 55.6	1	8.94		1.22		1.42		1	K2	280
		LOD 24 # 9			11 36 05	−61 47.6	1	11.57		0.18		0.08		2		127
		IC 2944 - 68			11 36 05	−63 02.9	1	13.52		0.58		0.32		2		530
		IC 2944 - 69			11 36 05	−63 07.1	1	13.94		0.70		0.20		2		530
		IC 2944 - 71			11 36 06	−63 05.1	2	12.40	.055	0.21	.064	-0.26	.000	3		530,1312
		IC 2944 - 70			11 36 06	−63 11.4	1	12.10		0.17		-0.44		2		530
101177	+45 01947	HR 4486		⋆ A	11 36 07	+45 23.1	2	6.46	.010	0.57	.007	0.03	.035	5	G0 V	22,3026
101177	+45 01947	IDS11335N4540	B		11 36 07	+45 23.1	2	8.37	.025	0.94	.020	0.67	.040	5	K2 V	22,3032
		IC 2944 - 72			11 36 07	−63 05.3	2	12.59	.004	1.27	.030	1.24	.098	3		530,1312
101179	+24 02374				11 36 08	+23 36.4	1	7.55		0.96				2	K0	280
101198	−12 03466	HR 4488		⋆ AB	11 36 08	−12 55.6	1	5.48		0.52				4	F7 V	2006
		IC 2944 - 73			11 36 08	−63 06.8	1	12.66		0.36		0.33		2		530
		IC 2944 - 74			11 36 08	−63 11.0	1	11.79		0.12		-0.52		2		530
101218	−50 06201				11 36 09	−50 59.3	1	9.85		0.11		0.12		1	A2 V	1770
		IC 2944 - 77			11 36 09	−63 05.4	2	12.90	.004	0.17	.009	-0.28	.013	3		530,1312
		IC 2944 - 75			11 36 09	−63 07.0	1	13.07		0.54		0.17		2		530
		IC 2944 - 76			11 36 09	−63 07.5	1	12.00		1.48		1.77		1		530
308806	−62 02173	IC 2944 - 78			11 36 10	−62 55.0	1	11.18		0.80		0.25		2	F5 V	530
101178	+39 02460	IDS11335N3944	A		11 36 11	+39 20.0	3	7.33	.022	1.61	.005	1.95	.030	8	M1 III	280,974,1501
	−46 07244				11 36 11	−46 43.4	1	11.07		0.38		0.12		1		508
		IC 2944 - 79			11 36 11	−63 12.0	1	11.87		0.26		-0.15		2		530
101215	−38 07247				11 36 12	−38 58.2	1	7.79		0.16		0.14		51	A1 IV/V	978
101235	−50 06358				11 36 13	−50 04.5	1	9.85		0.13		0.13		1	A1/2 V	1770
	+33 02136				11 36 14	+32 54.6	1	10.02		0.50		-0.06		1		280
		LOD 24 # 7			11 36 14	−61 51.1	1	12.65		0.02		-0.37		2		127
		IC 2944 - 81			11 36 14	−63 05.4	1	13.69		0.42		0.36		1		530
		IC 2944 - 80			11 36 14	−63 07.0	2	12.04	.009	0.13	.009	-0.38		3	B6 Vne	530,1312
	+34 02243				11 36 16	+33 51.1	1	9.39		0.96		0.70		1	K0	280
		IC 2944 - 82			11 36 16	−62 54.5	1	12.25		0.55		-0.38		2		530
101195	+17 02381				11 36 17	+17 22.7	1	8.53		0.99				2	K0	280
101194	+41 02218				11 36 17	+40 31.7	1	7.77		0.44		-0.08		3	F5	974
		IC 2944 - 84			11 36 17	−63 06.3	1	12.52		0.63		0.28		2		530
		IC 2944 - 83			11 36 17	−63 11.9	1	12.55		0.21		-0.22		2		530
	+34 02244				11 36 18	+34 04.0	1	10.13		0.28		0.01		1		280
101208	+24 02375				11 36 19	+24 05.0	1	9.15		0.30		-0.04		2	F5	1569
		Steph 961			11 36 19	+57 58.4	1	11.09		1.24		1.21		1	K7	1746
101206	+43 02135	G 122 - 38			11 36 21	+42 36.0	1	8.27		0.96		0.82		2	K5 V	1601
308811		IC 2944 - 85			11 36 21	−63 01.8	2	11.09	.015	0.38	.035	0.32	.015	6	A5 V	91,530
		IC 2944 - 86			11 36 22	−63 11.9	1	12.44		1.05		0.67		2		530
101266	−44 07473				11 36 24	−45 05.1	4	9.28	.005	0.66	.005	0.13		17	G3 V	116,1020,1075,2012
	+33 02137	IDS11338N3258	AB		11 36 25	+32 41.6	1	10.61		0.34		0.05		1		280
		IC 2944 - 87			11 36 25	−63 07.7	1	12.35		0.46		0.28		2		530

HD	DM	Other Id	N Rem	α_{1950}	δ_{1950}	S	V	σ_V	B–V	σ_{B-V}	U–B	σ_{U-B}	n	Spectrum	References
	+32 02171			11 36 26	+32 02.2	1	8.75		1.42		1.58		1	K5	280
101267	−52 04743			11 36 26	−53 08.3	1	9.36		0.03		0.04		1	A1 V	1770
101264	−38 07253			11 36 27	−39 03.0	1	9.34		0.40		0.06		49	F0 V	978
	−46 07248			11 36 27	−46 35.5	1	10.00		0.51		0.10		1		508
101242	+6 02478	G 57 - 11		11 36 28	+06 20.2	1	7.59		0.68		0.14		1	G5	333,1620
101227	+45 01949			11 36 28	+44 34.9	1	8.39		0.69		0.06		3	G0	3026
101265	−39 07229			11 36 28	−39 35.0	1	7.73		1.25		1.21		1	K1 III	1642
101259	−24 09867	HR 4489		11 36 29	−24 26.4	4	6.41	.006	0.82	.005	0.31		12	G6/8 V	15,78,2013,2028
308827	−62 02179	IC 2944 - 88		11 36 29	−63 10.1	3	10.77	.015	0.18	.015	-0.45	.005	8	B2.5V	530,996,1312
	+32 02172			11 36 31	+32 22.1	1	9.18		1.10		1.06		1	K0	280
	+20 02632			11 36 32	+19 45.0	1	9.54		1.13		1.17		1	K5	280
		LOD 24 # 11		11 36 32	−61 41.1	1	11.19		0.36		0.08		2		127
308829	−62 02184	IC 2944 - 89		11 36 33	−63 12.0	3	10.69	.052	0.21	.019	-0.21	.024	7	B6 Ve	530,996,1312
		IC 2944 - 90		11 36 35	−63 08.5	1	11.67		0.15		-0.42		2		530
	−26 08683			11 36 36	−27 24.9	1	10.03		1.22		1.15		2	K7 V	3072
308830	−62 02185	IC 2944 - 91		11 36 36	−63 13.5	3	10.12	.012	0.12	.009	-0.71	.038	8	B2 Vn	127,530,1312
101274	−37 07386	NGC 3783 sq1 1		11 36 37	−37 28.6	2	9.18	.005	0.06	.005	0.04	.010	7	A0	1687,1700
101298	−62 02186	IC 2944 - 92		11 36 38	−63 09.2	6	8.07	.010	0.07	.014	-0.83	.013	14	O6	2,530,540,976,996,1312
101317	−70 01404	IDS11344S7020	AB	11 36 38	−70 36.5	1	8.21		0.13		-0.09		2	B9 V	1771
		AA Leo		11 36 39	+10 36.2	2	11.67	.062	0.11	.024	0.13	.022	2		597,699
		LOD 24 # 13		11 36 40	−61 39.9	1	12.07		1.16		0.96		2		127
	−46 07252			11 36 41	−46 38.7	1	11.08		1.20		1.20		1		508
		IC 2944 - 93		11 36 41	−62 57.3	1	12.17		0.26		-0.24		2		530
101254	+52 01578	IDS11340N5244	A	11 36 42	+52 27.9	1	8.44		0.52		0.09		3	G0	1566
		NGC 3783 sq1 3		11 36 42	−37 33.5	1	11.03		0.59		0.11		3		1687
	+21 02340			11 36 43	+20 52.6	1	10.27		0.22		0.05		2	A2	272
		IC 2944 - 94		11 36 43	−63 04.6	1	12.06		0.49		0.42		2		530
101225	+73 00531			11 36 44	+72 49.6	1	9.22		0.08		0.05		2	A0	1375
101316	−60 03215			11 36 44	−61 07.1	1	8.92		0.17		0.23		1	A3 III/IV	51
		IC 2944 - 95		11 36 44	−62 56.3	1	12.08		0.43		0.38		2		530
308810	−62 02188	IC 2944 - 96	★ AB	11 36 45	−63 02.3	5	9.57	.026	0.08	.013	-0.74	.015	14	B1 V	91,191,401,530,1312
308826	−62 02189	IC 2944 - 97	★ V	11 36 45	−63 08.7	3	10.29	.071	0.12	.013	-0.67	.009	6	B1.5Vn	530,996,1312
101310	−51 05895			11 36 46	−51 37.2	1	8.45		0.64		0.36		1	G8/K1 III	1770
101312	−57 04894	IDS11344S5711	A	11 36 47	−57 27.7	1	7.28		-0.02		-0.08		2	A0 V	1771
101334	−65 01698	IDS11345S6527	A	11 36 47	−65 43.6	1	10.20		0.34		0.20		2	A0 V	401
101335	−68 01544	IDS11345S6822	AB	11 36 47	−68 38.4	1	8.93		0.42		0.16		1	A0 V	1771
101330	−61 02478			11 36 48	−62 08.7	1	7.31		-0.02		-0.38		2	B5/7 II	401
101333	−62 02190	IC 2944 - 98		11 36 48	−63 14.7	6	8.97	.016	0.09	.010	-0.75	.035	15	B0 III	127,191,530,540,976,1312
		NGC 3783 sq1 4		11 36 49	−37 29.1	1	11.99		0.51		-0.01		3		1687
		IC 2944 - 99		11 36 50	−63 06.5	1	12.63		0.35		0.29		2		530
101278	+17 02382			11 36 51	+16 48.9	1	8.38		1.14		1.13		1	K0	280
		LOD 24 # 15		11 36 51	−61 39.1	1	11.86		0.66		0.18		2		127
101289	+26 02241	IDS11343N2551	AB	11 36 52	+25 34.7	2	7.78	.045	0.58	.000	0.10		5	G0 V	20,3026
		KPS 352 - 217		11 36 52	+40 09.9	1	14.30		0.56		-0.03		3		974
306904	−60 03217			11 36 52	−61 08.1	1	8.58		1.31		1.52		1	K0	51
308822	−62 02193	IC 2944 - 100		11 36 53	−63 05.5	3	10.63	.009	0.40	.021	0.31	.019	7	A3 V	91,530,1312
		NGC 3783 sq1 5		11 36 54	−37 29.3	1	13.48		0.80		0.40		3		1687
101309	−38 07259	V829 Cen		11 36 54	−39 06.4	2	8.03	.071	0.95	.007	0.47	.002	2	G8 IV	1641,1642
		LOD 24 # 14		11 36 54	−61 42.7	1	11.39		0.13		0.04		2		127
101290	+22 02387	IDS11343N2152	A	11 36 55	+21 35.4	1	8.25		1.03		0.93		1	G5	280
		NGC 3783 sq1 2		11 36 55	−37 36.3	1	9.85		0.73		0.27		4		1687
101332	−62 02191	IC 2944 - 139		11 36 55	−62 39.4	7	7.63	.018	0.09	.008	-0.76	.021	16	B0 III/IV	2,6,138,191,540,976,8100
		IC 2944 - 101		11 36 55	−63 09.2	1	12.65		0.18		-0.42		2		530
		LOD 24 # 5		11 36 56	−61 53.9	1	11.61		0.56		0.13		2		127
308809	−62 02194	IC 2944 - 102		11 36 56	−63 00.3	2	9.98	.009	0.25	.000	-0.23	.004	5	B0.5V	530,1312
308823	−62 02195	IC 2944 - 104		11 36 56	−63 05.6	2	11.35	.004	0.22	.004	-0.47	.004	5	B6 V	530,1312
		IC 2944 - 103		11 36 56	−63 09.9	1	10.69		1.19		0.93		2	K1 III	530
		NGC 3783 sq1 7		11 36 57	−37 29.3	1	14.23		0.93		0.55		5		1687
308893	−61 02481			11 36 57	−62 12.8	1	9.49		1.75		1.90		6	K5	2
		LOD 24 # 18		11 36 58	−61 35.8	1	12.52		1.08		1.00		2		127
101301	+19 02482	IDS11344N1933	B	11 36 59	+19 16.6	1	8.50		0.35		0.08		3	F0 III-IV	3016
308860	−61 02482			11 36 59	−61 33.9	1	11.56		0.04		-0.27		2	A3	127
101304	+11 02384			11 37 00	+10 36.3	1	8.92		0.20		0.16		2	A6 m	1776
101327	−36 07316	IDS11345S3653	A	11 37 00	−37 09.2	1	7.42		0.46		-0.01		4	F5/6 V	1700
101349	−47 07021			11 37 00	−48 12.3	3	8.95	.004	0.87	.000	0.38		10	K1 (IV)	1075,2012,3008
101336	−61 02484			11 37 00	−61 39.1	1	9.90		0.09		-0.05		2	B9/A1 IV/V	127
101302	+19 02483	IDS11344N1933	A	11 37 01	+19 16.4	1	7.23		0.43		0.15		3	F0 IIIn	3016
101320	+27 02044	IDS11344N2731	AB	11 37 01	+27 14.0	1	8.19		0.67		0.31		4	A3 V+G0	1068
306932	−58 03774	LSS 2435		11 37 01	−59 05.4	1	9.64		0.10		-0.71		2		540
	−62 02198	IC 2944 - 105		11 37 01	−62 54.6	2	10.92	.015	0.70	.025	-0.37	.020	4	O9.5Ib	530,1737
		KUV 11370+4222		11 37 02	+42 22.0	1	16.56		0.22		-0.68		1	DA	1708
101303	+18 02515			11 37 03	+18 10.9	1	8.23		1.41				2	K5	280
308861	−61 02485			11 37 03	−61 37.4	1	10.42		0.42		0.19		2		127
308831	−62 02197	IC 2944 - 106		11 37 03	−63 10.4	1	11.04		0.21		-0.54		3	B1.5V	530
101300	+45 01951			11 37 04	+44 47.8	2	8.34	.035	0.94	.015	0.61	.015	4	G8 III	172,1733
		IC 2944 - 107		11 37 05	−63 10.0	1	13.81		0.23		-0.14		2		530
101348	−34 07607			11 37 08	−34 57.4	1	7.81		0.72		0.34		1	G2 V	78
101378	−61 02487			11 37 08	−62 21.8	1	9.42		0.02		-0.73		2	B3/5 III/V	401
		NGC 3783 sq1 6		11 37 09	−37 30.3	1	13.72		0.49		0.04		6		1687
101360	−50 06220	IDS11347S5108	AB	11 37 09	−51 24.5	1	8.20		0.73				4	G5 IV +F/G	2012

Table 1

HD	DM	Other Id	N Rem	α_{1950}	δ_{1950}	S	V	σ_V	B–V	σ_{B-V}	U–B	σ_{U-B}	n	Spectrum	References
	+3 02528			11 37 10	+03 18.6	1	10.72		1.55				1	K0	1017
101319	+45 01952	IDS11335N4540	AB	11 37 10	+45 26.0	1	7.51		0.40		0.01		2	F2	1601
		G 237 - 25		11 37 10	+59 58.4	1	11.59		1.36		1.34		2	M1	7010
		IC 2944 - 108		11 37 10	−63 04.4	1	13.42		0.35		0.31		2		530
101379	−64 01685	HR 4492, GT Mus	⋆ AB	11 37 10	−65 07.2	5	5.13	.043	0.80	.012	0.39	.022	10	G2 III+A0 V	15,1075,1641,1642,1771
	+32 02173			11 37 12	+32 09.	1	10.84		0.76		0.38		2	G0	1723
101359	−45 07176			11 37 12	−46 14.8	1	9.80		0.39		0.05		2	F0 V	508
101414		RR Mus		11 37 12	−72 16.9	1	8.50		2.87		3.51		2	Na	864
101390	−61 02488			11 37 13	−61 31.4	1	7.60		1.28		1.27		1	K0/1 III	51
	+41 02221			11 37 14	+40 41.2	1	9.78		0.70				2	G6 III	280
101373	−54 04692			11 37 14	−54 43.2	1	8.05		0.20		0.11		1	A5/7 IV	1771
308832	−62 02201	IC 2944 - 109		11 37 14	−63 12.3	1	11.37		0.17		−0.54		3	B1 V	530
308807	−62 02203	IC 2944 - 110		11 37 16	−62 58.2	1	11.52		0.25		−0.26		3	B1 Vn	530
		IC 2944 - 111		11 37 17	−63 10.9	1	11.59		0.17		−0.58		6	B0.5Vne	530
308837				11 37 17	−63 22.	1	11.32		0.20		−0.20		2		127
		IC 2944 - 112		11 37 18	−63 16.5	1	12.01		0.32		0.10		3		530
101369	−13 03420	HR 4490	⋆ AB	11 37 19	−14 11.5	2	6.20	.005	0.00	.000			7	A0 V	15,2029
101370	−15 03323	HR 4491		11 37 19	−16 20.6	2	6.18	.010	1.63	.005	1.75		8	M3 II/III	2006,3001
101388	−45 07178			11 37 19	−45 42.0	1	7.78		0.21		0.08		2	A7 IV/V	508
101413	−62 02205	IC 2944 - 113	⋆ B	11 37 21	−63 12.0	6	8.35	.007	0.07	.011	−0.80	.017	12	O9.5IV	2,401,530,540,976,1312
308834		IC 2944 - 114		11 37 21	−63 17.5	1	11.43		0.91		0.58		3	A0	530
101385	−19 03318	IDS11349S2011		11 37 22	−20 28.3	1	9.89		0.85		0.51		1	G8/K0 V	78
		IC 2944 - 115		11 37 22	−63 07.8	1	12.37		0.28		−0.20		2		530
		IC 2944 - 116		11 37 23	−63 07.0	1	12.88		0.31		−0.17		2		530
	−47 07032	V420 Cen		11 37 24	−47 41.1	1	9.63		0.60				1		688
		IC 2944 - 117		11 37 24	−63 12.4	1	11.96		0.19		−0.40		1		530
101424	−61 02491			11 37 25	−61 56.3	1	9.77		0.11		−0.29		1	B8 III	885
101436	−62 02206	IC 2944 - 118	⋆ A	11 37 25	−63 12.1	7	7.60	.023	0.06	.013	−0.83	.024	14	O7.5	2,191,401,530,540,976,1312
101366	+42 02233			11 37 26	+42 18.2	1	7.57		1.54		1.88		7	M2 III	172
101408	−45 07180			11 37 26	−45 31.5	3	7.25	.003	1.03	.005	0.76	.033	13	K0 III	1075,1499,3077
		IC 2944 - 119		11 37 26	−63 12.6	1	12.68		0.22		−0.36		1		530
101410	−49 06377	IDS11350S4956	A	11 37 28	−50 12.5	2	6.66	.003	0.00	.004	−0.01	.003	8	Ap EuCr	1628,1770
		G 236 - 80		11 37 30	+67 33.8	3	12.20	.024	1.45	.000	1.18	.014	8		940,1064,3078
101406	−37 07399	IDS11350S3733	B	11 37 30	−37 49.9	1	8.64		0.42		−0.03		1	G8 III	1700
	+30 02183			11 37 31	+29 46.2	1	10.68		1.05		1.06		1	K0	280
101437	−62 02208			11 37 32	−63 22.0	1	9.23		0.46		−0.01		2	A0 IV	127
	+19 02484			11 37 33	+18 42.8	1	9.93		1.53		1.80		1	K2	280
101397	+24 02378			11 37 33	+23 59.1	1	6.88		0.28		0.03		2	A3	1569
	+25 02401			11 37 33	+24 48.5	1	9.23		1.14		1.24		1	K2	280
	+36 02210			11 37 35	+36 27.3	1	10.01		1.20		1.22		1	K5	280
101396	+26 02243			11 37 36	+26 25.9	2	8.13	.025	0.93	.010	0.72		5	K1 V	20,3026
101394	+37 02205			11 37 36	+36 45.2	1	8.16		1.05		0.96		1	K0	280
		NGC 3783 sq1 11		11 37 36	−37 34.4	1	13.31		0.72		0.25		4		1687
		NGC 3783 sq1 10		11 37 36	−37 35.1	1	12.84		0.67		0.16		3		1687
308824		IC 2944 - 121		11 37 36	−63 07.6	1	11.24		0.16		−0.61		3	B5	530
308825	−62 02213	IC 2944 - 120		11 37 36	−63 10.2	1	10.64		0.16		−0.66		3	B1 IV	530
	+20 02636			11 37 38	+20 24.8	1	10.23		0.44				2	F5 IV	193
	+24 02379			11 37 39	+23 51.7	1	9.30		0.33		−0.02		2	A7	1569
	+20 02637			11 37 41	+19 48.6	1	9.53		1.07		0.96		1	K2	280
	+40 02449			11 37 42	+40 03.8	1	10.29		1.16		1.27		1	K1 III	280
		LP 907 - 412		11 37 43	−29 12.0	1	11.71		0.54		−0.17		2		1696
101431	−34 07610	HR 4494		11 37 43	−34 28.0	7	4.70	.009	−0.07	.006	−0.20	.023	30	B9 V	3,15,1075,1770,2012*
		AJ74,1125 T4# 8		11 37 43	−62 08.0	1	11.70		0.10		−0.46		2		
101391	+58 01331	HR 4493		11 37 44	+58 14.8	1	6.37		−0.12		−0.34		3	B9 p Hg(Mn)	220
101466	−60 03231			11 37 44	−60 54.5	1	7.35		0.05		−0.01		2	A0 IV	1771
101430	−27 08237			11 37 46	−28 13.1	1	7.07		1.67		2.02		5	K5/M0 III	1657
308833	−62 02216	IC 2944 - 122		11 37 46	−63 13.1	1	11.05		0.18		−0.62		3	B1 V	530
308874	−61 02496			11 37 47	−62 01.5	1	10.83		0.07		−0.21		3		2
		IC 2944 - 124		11 37 47	−63 05.3	1	11.64		0.19		−0.53		3		530
		IC 2944 - 123		11 37 47	−63 11.4	1	13.37		0.47		0.42		2		530
308896	−61 02497			11 37 49	−62 22.3	1	11.23		0.14		−0.23		3	A2	2
	+24 02380			11 37 50	+23 52.6	1	9.59		1.16		1.18		1	K2	280
101482	−61 02498			11 37 50	−61 56.5	1	9.69		0.08		−0.22		1	B8/9 III	885
101481	−60 03233			11 37 51	−61 08.0	1	9.13		0.32		0.08		1	F0 V	51
308841	−63 01939			11 37 51	−63 34.9	1	9.79		0.15		0.05		2	A0	127
		LP 793 - 20		11 37 52	−17 06.7	1	13.93		1.02		0.79		1		3062
101441	+12 02362	AI Leo		11 37 54	+11 28.4	1	8.72		1.71		1.78		1	M2	3001
101440	+17 02384			11 37 55	+17 04.2	1	9.52		0.30				3	F5	280
101479	−48 06703			11 37 55	−48 53.3	1	8.20		0.17		0.16		1	A1 V	1770
101459	+34 02245			11 37 59	+33 57.8	1	8.77		1.08		0.94		1	K5	280
101460	+13 02439			11 38 01	+13 05.0	1	6.65		0.99		0.73		1	F5	1700
308873	−61 02502			11 38 01	−61 57.2	1	9.97		0.04		−0.56		3	A0	2
	+23 02373			11 38 02	+22 38.9	1	9.47		0.23				4	A5	193
101469	+21 02342	IDS11354N2135	AB	11 38 03	+21 18.8	1	8.31		0.85		0.55		2	G5	1569
		LP 613 - 28		11 38 03	−01 44.4	1	13.25		0.81		0.35		2		1696
101511	−61 02504			11 38 04	−61 55.3	1	8.16		0.02		−0.24		2	B9 V	127
		G 10 - 42		11 38 05	+05 10.7	1	13.22		0.98		0.55		1		1696
	+32 02176			11 38 05	+31 36.9	1	9.26		0.27		0.11		1	G5	280
101493	−42 07144	IDS11356S4236	A	11 38 05	−42 52.7	1	8.97		0.46		−0.06		4	F3/5 V	1279
101493	−42 07144	IDS11356S4236	AB	11 38 05	−42 52.7	1	8.62		0.48				4	F3/5 V	2012

HD	DM	Other Id	N Rem	α₁₉₅₀	δ₁₉₅₀	S	V	σ_V	B–V	σ_B-V	U–B	σ_U-B	n	Spectrum	References
101493 −42 07144		IDS11356S4236	B	11 38 05	−42 52.7	1	9.98		0.53		−0.04		4		1279
	+38 02265			11 38 06	+37 56.3	1	10.32		1.47		1.76		1	K5	280
		AJ74,1125 T4# 10		11 38 06	−62 26.0	1	12.37		0.16		−0.42		3		2
	+30 02185			11 38 07	+30 19.9	1	9.43		1.58		1.63		1	M2	280
		LP 215 - 35		11 38 09	+39 37.1	1	12.92		0.86		0.48		2		1773
101485 +18 02517		G 121 - 5		11 38 11	+17 59.5	1	9.19		0.81				2	K0	280
101484 +22 02391		HR 4495		11 38 11	+21 37.8	2	5.27	.010	0.99	.005	0.79	.005	5	K1 III	1080,3016
		AJ74,1125 T4# 9		11 38 11	−62 11.3	1	12.05		0.51		0.14		3		2
		LP 169 - 76		11 38 12	+47 57.5	1	12.06		0.94		0.56		4		1723
		IC 2944 - 125		11 38 12	−63 14.2	1	11.42		1.28		1.33		3		530
		vdB 52 # a		11 38 12	−64 15.5	1	13.11		0.30		−0.36		3		434
	−79 00644			11 38 12	−80 14.6	1	9.62		1.15		0.75		2		1730
101487 +13 02440		AK Leo		11 38 13	+13 21.3	1	8.57		1.54		1.44		1	M2	3001
	+17 02386			11 38 14	+16 32.9	1	9.48		1.10				3	K0	280
	+40 02450			11 38 14	+40 20.8	1	10.10		1.05		0.89		1	G9 III	280
101543 −56 04683				11 38 14	−57 24.3	1	8.93		0.27		0.19		2	A5mA6-F0	1771
101545 −61 02508		IC 2944 - 132	★ AB	11 38 15	−62 17.5	7	6.37	.030	0.02	.009	−0.87	.021	13	B0 III	2,191,540,976,1312,1737,8100
101541 −53 04691		HR 4497		11 38 17	−53 41.5	3	5.98	.023	1.67	.005			11	M1 III	15,1075,2032
	+20 02638			11 38 19	+20 24.3	1	11.37		0.44				2	F5 IV-V	193
		AJ74,1125 T4# 11		11 38 20	−62 06.9	1	12.57		0.29		0.16		3		2
		CS 15625 # 3		11 38 21	+24 31.7	1	13.35		0.30		0.01		1		1744
101539 −42 07149				11 38 24	−42 27.5	1	8.55		0.01		−0.03		1	A0 V	1770
	+32 02177			11 38 25	+32 06.1	1	9.65		0.27		0.06		2	F0	1569
101501 +35 02270		HR 4496	★ A	11 38 25	+34 29.0	12	5.32	.015	0.72	.014	0.26	.018	56	G8 V	1,15,101,245,272,1080*
	+34 02247			11 38 27	+33 41.1	1	10.27		1.02		0.99		1		280
		CS 15625 # 22		11 38 28	+26 44.0	1	12.07		0.32		0.01		1		1744
101570 −61 02514		HR 4499		11 38 31	−61 48.8	5	4.93	.006	1.14	.019	0.80	.009	17	G3 Ib	15,127,1075,1488,2012
101582 −53 04693				11 38 34	−54 16.3	1	7.42		0.01		−0.01		1	A0 V	1770
101584 −54 04707				11 38 34	−55 17.8	1	7.02		0.39		0.46		2	F2pec shell	191
		KUV 352 - 6		11 38 36	+42 29.0	1	15.77		−0.17		−1.09		1		974
101581 −43 07228				11 38 37	−44 07.9	3	7.77	.010	1.06	.010	0.88	.025	8	K4/5 V	78,1075,3072
101558 +25 02403				11 38 38	+25 05.0	1	8.36		1.39		1.70		1	K0	280
		G 176 - 46		11 38 38	+39 35.9	2	12.61	.015	0.79	.005	0.28	.015	3	K2	1658,1773
	−79 00646			11 38 38	−79 58.5	1	9.92		1.23		0.91		2	K0	1730
101563 −28 09027		HR 4498		11 38 39	−28 55.3	3	6.43	.005	0.66	.007	0.17		8	G0 V	15,78,2012
	+22 02392			11 38 42	+22 13.6	1	9.95		0.99		0.80		1		280
		G 120 - 67		11 38 42	+25 22.4	2	13.32	.020	0.85	.010	0.40	.010	2		1658,3062
101599 −53 04696		IDS11363S5406	AB	11 38 43	−54 22.2	1	9.56		0.21		0.11		1	A4/5 V	1771
	+17 02388			11 38 45	+17 17.4	1	9.63		0.50		−0.01		2	F8	1569
		Ton 1421		11 38 45	+21 17.	1	14.96		0.28		0.11		1		286
		LP 433 - 38		11 38 46	+16 43.6	1	13.95		1.45		1.02		1		1696
101596 −46 07288				11 38 46	−46 37.0	1	9.12		0.31		0.17		2	F0 V	508
		LP 793 - 29		11 38 47	−20 57.9	1	11.42		0.60		−0.06		2		1696
308935 −62 02226		LSS 2450		11 38 49	−63 06.6	2	9.87	.025	0.13	.000	−0.69	.005	5		2,401
101638 −70 01410				11 38 50	−70 53.3	1	8.12		0.19		0.11		1	A4 V	1771
101615 −42 07155		HR 4502		11 38 51	−42 49.1	1	5.54		0.03		0.01		2	A0 V	1770
101612 −25 08823				11 38 52	−26 23.4	1	7.53		0.47		−0.06		1	F5/6 V	78
308885 −61 02518				11 38 52	−62 16.4	1	11.50		0.22		0.06		3	B9	2
308899 −61 02521		IDS11366S6203	A	11 38 56	−62 19.6	1	10.55		0.40		−0.57		1		1737
308899 −61 02521		IDS11366S6203	AB	11 38 56	−62 19.6	1	9.88		0.40		−0.56		1		1737
308899 −61 02521		IDS11366S6203	B	11 38 56	−62 19.6	1	10.86		0.43		−0.45		1		1737
	+41 02224			11 38 57	+40 54.6	1	11.69		−0.11		−0.17		1	Am:	280
101614 −40 06891				11 38 57	−40 44.3	1	6.86		0.58		0.02		1	G0 V	78
101606 +32 02179		HR 4501	★ AB	11 38 58	+32 01.4	3	5.75	.015	0.44	.012	−0.11	.015	7	F4 V	70,254,3037
		KUV 11390+4225		11 38 58	+42 25.3	1	16.20		−0.09		−1.11		1	DA	1708
101587 +20 02641				11 38 59	+19 48.5	1	8.77		1.16		1.23		1	K0	280
	+29 02196			11 39 01	+29 07.9	1	9.87		0.28				6	F0	193
101585 +45 01955				11 39 01	+44 28.4	1	7.93		1.58		1.91		2	M3 III	172
101604 +55 01481		HR 4500		11 39 02	+55 27.0	1	6.27		1.50				1	K5	71
101607 +21 02343				11 39 04	+20 41.0	1	9.03		0.26		0.12		2	F0	1569
101655 −61 02522				11 39 04	−62 01.9	1	9.84		0.41		−0.53		2	A2/3 IV	540
		G 122 - 40		11 39 06	+43 01.6	1	11.89		1.51		1.13		4		1663
101620 +42 02236				11 39 07	+41 30.9	1	6.93		0.43		0.00		3	F3 V	1501
	+17 02390			11 39 08	+16 57.1	1	9.16		1.17		1.24		1	K0	280
		G 197 - 27		11 39 08	+59 06.4	1	11.69		1.39		1.14		2	K5	7010
101666 −31 09181		HR 4503	★ A	11 39 14	−32 13.3	6	5.22	.022	1.47	.014	1.79	.013	21	K5 III	15,542,1075,2013,2029,3005
	+5 02529	G 10 - 43		11 39 15	+05 25.5	6	9.59	.008	1.25	.011	1.19	.011	29	K8	989,1017,1620,1705,1729,3077
101684 −63 01943		LSS 2455		11 39 16	−63 33.1	1	7.23		0.94		0.68		2	F5 Iab/b	127
101782 −82 00469		HR 4507	★ A	11 39 17	−82 49.3	3	6.31	.007	1.08	.000	0.89	.005	15	K0 III	15,1075,2038
	+16 02283			11 39 18	+16 04.7	2	9.04	.009	1.23	.014	1.41	.019	4	K2	272,280
101641 +35 02272				11 39 19	+34 35.1	1	8.06		1.15		1.10		1	K0	280
101657 +24 02383				11 39 21	+23 43.7	1	8.82		0.21		0.09		2	A3	1569
	+38 02268			11 39 22	+37 37.7	1	9.28		1.17				3	K1 III	280
101680 −39 07258				11 39 22	−40 17.7	1	9.21		0.87				4	K0 V(p)	2033
101727 −76 00686				11 39 23	−76 46.6	2	6.94	.004	0.40	.013	−0.05		13	F3 V	1499,2012
	+34 02249			11 39 25	+33 31.8	1	10.64		0.97		0.70		1		280
101713 −70 01413				11 39 26	−71 13.7	1	7.95		0.14		−0.10		2	B9 V	1771
	+22 02394			11 39 27	+21 56.9	1	9.68		1.08		0.96		1	K0	280
	+29 02197			11 39 27	+28 34.9	1	10.01		0.93				2		280
101679 −38 07276				11 39 27	−39 00.4	1	8.12		1.11		0.81		1	K0 III	1642

Table 1   623

HD	DM	Other Id	N Rem	α_{1950}	δ_{1950}	S	V	σ_V	B–V	σ_{B-V}	U–B	σ_{U-B}	n	Spectrum	References
101712	−62 02234	IC 2944 - 137	⋆ V	11 39 27	−63 08.2	3	7.93	.063	1.80	.015	0.77	.035	8	M3 I	138,1312,3034
		G 56 - 51		11 39 28	+15 03.0	1	12.58		1.48		1.24		1		333,1620
101700	−46 07304			11 39 28	−46 46.6	1	9.48		0.43		0.08		2	F3 IV/V	508
101676	+13 02443			11 39 29	+12 33.8	1	7.08		0.47		-0.04		2	F6 V	1648
101711	−59 03717	LSS 2456		11 39 29	−59 57.2	1	6.84		0.33		0.41		1	A8 III	1771
101688	+23 02375	HR 4505		11 39 30	+22 29.3	2	6.66	.005	0.35	.005	-0.05	.000	6	F2 IV-V	1501,1733
		G 120 - 68		11 39 32	+26 59.5	2	10.68	.000	1.52	.005	1.21	.019	3	M2	1744,3078
101695	−19 03326	HR 4506		11 39 32	−20 01.0	1	6.22		0.95				4	G8 IV	2006
	+16 02284			11 39 33	+16 23.4	1	9.96		1.08		1.16		1	K0	280
101724	−63 01944			11 39 33	−63 31.1	1	8.03		0.01		-0.41		2	B9 II/III	127
101675	+40 02451	IDS11369N4013	A	11 39 35	+39 56.3	1	8.78		0.71				2	G5 III	280
	+37 02208			11 39 40	+37 01.2	1	9.57		1.04				2	G7 III	280
		LP 375 - 23		11 39 42	+23 18.2	1	11.58		1.49		1.17		3		1723
101673	+67 00714	HR 4504		11 39 42	+67 01.3	2	5.29	.012	1.26	.013	1.24		4	K3 III	71,3016
101766	−74 00838			11 39 42	−74 50.4	1	8.15		1.53				4	K2 Ib/II	2012
	+30 02188			11 39 44	+29 57.1	1	9.49		0.96				2	K0	280
	+18 02519			11 39 49	+18 12.8	1	10.81		0.32		-0.02		2		1569
	+40 02453			11 39 49	+39 51.5	1	9.93		0.74				2	G6 III	280
	+41 02226			11 39 49	+41 19.7	1	9.41		1.33				2	K3 III	280
101763	−56 04706			11 39 51	−57 17.2	1	7.10		-0.02		-0.11		2	B9/9.5 V	1771
		Ross 116		11 39 55	+12 16.3	1	13.35		0.73		-0.01		2		1696
	+40 02454			11 39 58	+39 57.3	1	10.04		0.91		0.59		1	G7 III	280
	−61 02540	St 14 - 47		11 40 01	−62 17.9	2	9.56	.005	1.04	.019	0.69	.005	9	K0 III	2,1672
101805	−74 00839			11 40 01	−74 57.0	3	6.47	.005	0.53	.007	0.05	.005	10	F7 V	278,1075,2012
101794	−61 02541	St 14 - 13		11 40 02	−62 12.0	9	8.68	.021	0.05	.017	-0.74	.013	23	B1 Ibne	2,6,411,540,885,956*
101760	−35 07390			11 40 03	−35 43.0	1	8.14		0.94		0.65		1	G8 IV	78
101795	−62 02237			11 40 03	−62 51.9	1	7.33		0.06		-0.61		2	B2/3 III	401
101741	+41 02227			11 40 05	+41 10.7	1	8.43		1.03				2	K0 III	280
101753	+19 02492			11 40 08	+18 31.2	1	7.39		-0.15		-0.55		5	B6 V	1569
		CS 15625 # 4		11 40 11	+25 37.6	1	13.93		0.55		0.02		1		1744
	+35 02274			11 40 11	+34 37.8	1	8.88		1.39		1.66		1	K2	280
306962	−59 03727	LSS 2461		11 40 12	−59 42.0	5	9.70	.040	0.17	.017	-0.84	.019	10	Bpenn	6,540,976,1737,2021
		St 14 - 41		11 40 14	−62 13.6	1	13.73		0.20				2		1672
306922	−60 03272	LSS 2462		11 40 15	−61 12.4	3	10.28	.022	0.20	.008	-0.77	.003	7	Benn	6,540,1737,2021
		St 14 - 39		11 40 17	−62 10.7	1	14.00		1.02				2		1672
	−61 02546	St 14 - 46		11 40 18	−62 17.5	1	9.46		1.65		1.65		6		1672
		St 14 - 40		11 40 19	−62 11.0	1	13.25		0.29				2		1672
101827	−65 01702			11 40 19	−65 55.6	1	9.48		0.51		0.41		1	A7/9	1771
	+29 02198			11 40 20	+28 49.3	2	10.21	.024	0.09	.000	0.13	.014	2	A2	280,1026
		KK Cen		11 40 23	−58 42.8	1	10.84		1.04		0.63		1		689
101826	−60 03274			11 40 23	−61 13.7	1	8.73		-0.02		-0.62		2	B3 III	401
		St 14 - 42		11 40 23	−62 14.1	1	13.16		0.18		0.00		2		1672
		AAS12,381 40 # C		11 40 24	+40 15.9	1	10.88		0.86		0.61		1		280
		St 14 - 43		11 40 24	−62 15.8	1	13.59		0.49		0.18		3		1672
	+23 02376			11 40 26	+22 41.0	1	9.75		1.32		1.61		1		280
101837	−61 02551	St 14 - 12	⋆ V	11 40 26	−62 09.4	6	8.50	.022	-0.02	.009	-0.82	.011	19	B0.5IV	401,411,885,956,1672,2021
101838	−61 02550	St 14 - 14		11 40 26	−62 17.2	9	8.41	.011	0.02	.009	-0.82	.015	23	B1 II-III	2,6,411,540,885,956*
		St 14 - 44		11 40 27	−62 16.2	1	12.87		0.63				2		1672
101839	−63 01947			11 40 27	−63 48.4	2	8.05	.005	0.10	.005	-0.57	.000	6	B2 III	164,401
101807	+21 02346			11 40 28	+20 32.9	1	9.60		1.12		1.13		2		1569
		St 14 - 45		11 40 28	−62 16.5	1	14.15		1.16				2		1672
	−68 01554			11 40 28	−68 44.1	1	8.43		1.70				6		955
101823	−44 07525			11 40 30	−44 39.9	1	8.68		0.10		0.05		1	A0 V	1770
101834	−35 07393	IDS11381S3528	AB	11 40 34	−35 44.7	1	8.99		0.49		0.04		108	F5 V	264
		St 14 - 38		11 40 34	−62 10.8	1	14.19		0.27				2		1672
309018	−62 02241	LSS 2466		11 40 34	−62 31.7	2	10.27	.005	0.32	.024	-0.65	.073	6	O9 V	191,1747
101850	−59 03734			11 40 35	−60 22.5	3	9.23	.008	0.01	.016	-0.65	.008	6	B3 II/III	2,540,976
	+16 02285			11 40 37	+15 32.8	1	10.51		1.04		0.95		1	G5	280
		Steph 964		11 40 37	+47 23.3	1	11.70		1.16		1.11		1	K4	1746
		St 14 - 37		11 40 37	−62 10.9	1	12.73		0.81		0.42		3		1672
	+23 02377			11 40 38	+22 42.8	1	10.33		0.32		-0.03		2		1569
	+25 02405			11 40 39	+24 55.6	1	11.17		0.66		0.16		1	F0	1298
101817	+21 02347			11 40 40	+20 38.9	1	8.70		1.37		1.66		2	K0	1569
306989	−60 03278	V644 Cen		11 40 42	−60 27.4	3	9.63	.090	0.07	.014	-0.71	.025	5	B1 III	540,976,1588
	+12 02370			11 40 43	+12 21.6	1	9.70		0.39				7	A5	193
	−33 07915			11 40 43	−34 15.4	1	10.22		0.95		0.59		1	G5	6006
	−51 05974			11 40 44	−51 32.8	3	10.36	.015	1.53	.016	1.18	.009	6	K0	158,1705,3062
101841	+28 02033			11 40 46	+28 19.9	1	7.69		0.34				1	F3 II	6009
		St 14 - 36		11 40 46	−62 08.9	2	12.64	.017	0.23	.190	-0.16	.063	7		411,1672
101863	−35 07397			11 40 47	−35 27.6	1	8.88		1.05		0.86		8	G8 III	264
101917	−78 00677	HR 4509		11 40 47	−79 01.7	3	6.38	.004	0.89	.005	0.59	.000	14	K0 III/IV	15,1075,2038
		SSII 22		11 40 48	+44 05.0	1	12.48		0.44		-0.10		1	F2	1298
101903	−64 01690			11 40 49	−65 06.1	1	7.90		1.68				1	K3 II/III	1642
101856	+28 02034			11 40 51	+27 50.8	1	8.05		0.92		0.71		1	K0 III	280
101854	+38 02272			11 40 51	+37 31.1	1	8.38		1.32				2	K1 IV	280
101887	−54 04725			11 40 51	−55 18.4	1	8.78		0.23		0.15		1	A7 V	1771
101855	+32 02180			11 40 54	+32 02.4	1	6.77		1.09				2	K0	280
101883	−36 07371	HR 4508		11 40 57	−36 54.7	4	5.99	.014	1.45	.008	1.71		13	K3 III	15,1075,2032,3005
307007	−60 03283	LSS 2469		11 40 57	−60 54.7	2	10.10	.010	0.03	.008	-0.62	.032	5	B3 III	540,1747
101914	−65 01704			11 40 57	−65 58.8	1	8.22		0.24		-0.21		2	B9 II	401

HD	DM	Other Id	N	Rem	α₁₉₅₀	δ₁₉₅₀	S	V	σ_V	B–V	σ_B–V	U–B	σ_U–B	n	Spectrum	References
101853	+42 02241				11 40 59	+42 00.0	1	6.73		0.97		0.73		4	F8 IV	1501
		St 14 - 63			11 40 59	−62 11.0	1	13.39		1.36				1		1672
101872	+11 02393	IDS11384N1115	A		11 41 00	+10 58.3	1	8.58		0.33		0.03		2	A5	1776
	+16 02288				11 41 01	+16 00.5	1	10.38		1.04		0.97		1		280
		St 14 - 64			11 41 02	−62 11.6	1	14.50		1.35				1		1672
101898	−17 03453	IDS11385S1733	A		11 41 03	−17 49.2	1	8.59		0.42		0.02		4	F3 V	1700
		St 14 - 62			11 41 03	−62 10.9	1	13.93		0.76				1		1672
		G 57 - 15			11 41 04	+18 41.8	1	15.07		1.20		0.89		1		1620
101930	−57 04948				11 41 04	−57 44.0	1	8.21		0.91				4	K1 V	2012
		St 14 - 65			11 41 05	−62 11.3	1	14.69		0.34				1		1672
		St 14 - 34			11 41 05	−62 13.5	1	13.91		0.51		0.36		1		956
309002	−61 02558	St 14 - 9			11 41 05	−62 18.3	3	10.34	.006	0.09	.019	-0.34	.012	10	B6 V	287,411,1672
		St 14 - 35			11 41 06	−62 13.1	1	11.62		0.18		-0.37		4		956
	+42 02242				11 41 07	+41 44.9	1	9.29		0.99				2	K0 III	280
	+44 02120	IDS11385N4444	A		11 41 07	+44 27.6	1	11.47		0.48		-0.08		2		3032
	+44 02120	IDS11385N4444	B		11 41 07	+44 27.6	1	11.22		0.44		-0.07		2		3032
	+44 02120	IDS11385N4444	C		11 41 07	+44 27.6	1	11.10		1.00		0.66		2		3032
101871	+53 01506				11 41 07	+52 40.3	2	7.41	.000	0.15	.000	0.07	.010	6	A3 V	1501,1566
101947	−61 02559	St 14 - 4		★ V	11 41 07	−62 12.7	10	5.03	.020	0.79	.017	0.39	.044	34	G1 Ia +B?	15,411,956,1018,1034*
		St 14 - 10			11 41 07	−62 17.7	2	10.62	.009	1.31	.009	1.33	.009	9		411,1672
	+31 02287				11 41 08	+30 44.4	1	9.93		1.04		0.93		1	K0	280
101928	−53 04723				11 41 10	−53 33.3	1	9.09		0.23		0.09		1	A5 IV/V	1771
101906	+24 02386	IDS11386N2434	AB		11 41 11	+24 17.3	2	7.39	.024	0.87	.020	0.43	.033	12	G2 V	1499,3026
		MCC 40			11 41 11	+32 49.9	1	11.13		1.19		1.13		3	M0	1723
		AAS12,381 40 # B			11 41 11	+40 22.1	1	10.09		1.31		1.40		1		280
		St 14 - 33			11 41 11	−62 13.5	1	12.92		0.33		0.23		3		956
	+19 02495	IDS11386N1905	AB		11 41 12	+18 48.5	1	9.51		1.02		0.97		2		1569
101907	+17 02392				11 41 13	+16 36.0	2	8.03	.020	0.62	.015	0.11	.015	5	G0	1569,1648
101966	−67 01804				11 41 15	−68 12.1	1	7.01		0.35		0.19		1	A8/9 III/I	1771
101965	−62 02249				11 41 18	−62 57.4	1	8.62		-0.02		-0.48		2	B5 III	401
		St 14 - 58			11 41 19	−62 09.7	1	12.82		0.48		0.09		3		1672
309000	−61 02561	St 14 - 15			11 41 19	−62 11.3	3	10.75	.010	0.14	.006	0.05	.021	11		411,956,1672
		St 14 - 54			11 41 19	−62 12.1	1	14.55		0.39		0.04		2		1672
101964	−61 02560	St 14 - 3			11 41 19	−62 15.3	7	8.35	.013	-0.01	.011	-0.84	.019	17	B0 III	2,6,287,411,885,956,1672
		St 14 - 32			11 41 19	−62 16.8	2	12.62	.000	0.12	.010	-0.31	.020	7		956,1672
		St 14 - 55			11 41 20	−62 12.0	1	13.90		0.49				1		1672
		St 14 - 27			11 41 20	−62 13.2	2	12.73	.010	0.09	.000	-0.26	.010	7		956,1672
		St 14 - 30			11 41 20	−62 15.5	1	12.74		0.15		-0.12		3		956
		St 14 - 31			11 41 20	−62 15.7	1	13.98		0.26		0.11		4		956
		St 14 - 48			11 41 20	−62 17.1	1	14.12		0.56				2		1672
308976	−61 02564	LSS 2471, MP Cen			11 41 21	−61 27.9	1	10.18		0.20		-0.66		3	B8	1747
	+32 02182				11 41 22	+31 36.5	1	9.35		1.08		0.99		1	G5	280
101918	+40 02457				11 41 22	+40 21.6	1	8.75		1.55				2	K4 IV	280
101933	−5 03340	HR 4510			11 41 22	−06 23.9	2	6.06	.005	0.96	.000	0.72	.005	7	K0	15,1256
		St 14 - 59			11 41 22	−62 09.3	1	10.75		1.73		2.22		5		1672
		St 14 - 57			11 41 22	−62 10.1	1	11.70		0.29		-0.04		5		1672
		St 14 - 28			11 41 22	−62 14.3	1	13.02		0.35		-0.25		4		956
101828	+83 00336				11 41 23	+82 35.9	1	8.82		0.19		0.11		2	G5 II	8100
101993	−61 02565	St 14 - 2		★	11 41 23	−62 14.1	6	9.66	.021	0.01	.011	-0.74	.021	13	B2 IV-V	287,411,885,956,1672,1747
		St 14 - 29			11 41 23	−62 14.6	1	14.36		0.27		0.04		4		956
		St 14 - 56			11 41 24	−62 11.8	1	13.61		0.41				1		1672
		St 14 - 53			11 41 24	−62 12.7	1	14.73		0.51				1		1672
	+18 02522				11 41 26	+17 40.8	1	9.47		1.18		1.16		1		280
101994	−61 02567	St 14 - 1			11 41 26	−62 14.1	6	8.95	.022	-0.02	.013	-0.82	.046	13	B0 III-IV	191,287,411,885,956,1672
101959	−29 09302				11 41 27	−29 28.2	1	6.97		0.55		0.04		1	G0 V	78
309001	−61 02569	St 14 - 6		★	11 41 27	−62 16.8	5	10.16	.010	-0.01	.008	-0.74	.009	17	B2 IV	287,411,956,1672,1747
		St 14 - 60			11 41 28	−62 09.3	1	14.84		0.56				1		1672
309003	−61 02562	St 14 - 8			11 41 28	−62 21.1	4	10.41	.004	-0.01	.013	-0.74	.013	13	B2 IV	287,411,956,1672
101954	+19 02496				11 41 29	+19 27.1	1	9.23		1.26		1.50		1		280
		St 14 - 26			11 41 29	−62 13.2	2	12.90	.020	0.13	.010	-0.06	.055	7		956,1672
101995	−62 02250	HR 4513			11 41 29	−62 36.0	3	6.08	.013	0.06	.010	0.12	.005	7	A2 III/IV	401,1771,2032
		St 14 - 61			11 41 30	−62 09.5	1	15.81		0.34				1		1672
101953	+30 02189				11 41 31	+29 51.5	1	8.05		0.26		0.03		3	Am:	1569
101952	+60 01344				11 41 32	+59 54.2	1	10.02		0.05		0.09		1		1502
		St 14 - 52			11 41 33	−62 12.8	1	14.23		0.49				1		1672
		St 14 - 19			11 41 34	−62 14.3	2	12.39	.010	1.59	.000	1.26	.073	3		411,956
		St 14 - 50			11 41 34	−62 16.6	1	13.62		0.37				2		1672
		St 14 - 49			11 41 34	−62 17.0	1	12.69		0.56		-0.08		3		1672
101973	−32 08282				11 41 35	−33 06.5	1	10.18		0.83		0.44		1	K0 V	78
102006	−54 04734				11 41 35	−55 06.8	1	8.86		0.18		0.09		1	A3mA6-A9	1771
101968	+30 02191				11 41 36	+30 10.4	1	7.87		0.04		0.04		4	A2 V	1068
102009	−61 02571	St 14 - 5			11 41 36	−62 15.6	5	8.90	.015	-0.05	.010	-0.82	.017	14	B1 III-IV	287,411,885,956,1672
101980	+26 02250	HR 4512		★ A	11 41 37	+25 29.8	2	6.03	.010	1.53	.000	1.85		5	K5 III	70,3016
101980	+26 02250	IDS11390N2546	B		11 41 37	+25 29.8	1	10.82		0.63		0.19		3		3032
		LP 961 - 33			11 41 37	−34 24.8	1	12.33		0.66		0.01		2		1696
		St 14 - 18			11 41 38	−62 14.8	3	12.24	.023	0.49	.019	-0.32	.011	7		411,956,1672
		St 14 - 21			11 41 38	−62 15.1	2	13.47	.009	0.20	.018	0.00		5		956,1672
		St 14 - 25			11 41 39	−62 13.8	2	12.59	.019	0.71	.000	0.18	.039	3		956,1672
		St 14 - 24			11 41 39	−62 14.0	2	14.00	.000	0.26	.042	-0.04		4		956,1672
		G 122 - 43			11 41 40	+40 49.1	1	12.02		0.54		-0.20		2		1658

Table 1   625

HD	DM	Other Id	N Rem	α_{1950}	δ_{1950}	S	V	σ_V	B–V	σ_{B-V}	U–B	σ_{U-B}	n	Spectrum	References
		St 14 - 51		11 41 40	−62 15.2	1	14.98		0.90				2		1672
101977	+39 02465	IDS11390N3934	AB	11 41 41	+39 17.1	2	8.62	.015	0.35	.005	0.03	.005	5	A8 V	833,1569
		St 14 - 20		11 41 41	−62 15.5	3	12.49	.042	1.19	.012	1.04	.142	3		411,956,1672
		SSII 23		11 41 42	+40 50.	2	11.99	.040	0.50	.020	-0.22	.035	2	F2	1240,1298
102026	−53 04730			11 41 42	−53 28.2	1	8.52		0.08		0.03		1	A1 V	1771
		St 14 - 23		11 41 42	−62 14.3	1	13.05		0.19		0.03		1		956
101979	+38 02273			11 41 43	+37 48.4	1	9.20		1.04				2	K0 III	280
101978	+39 02466			11 41 43	+38 48.3	2	7.86	.024	1.48	.005	1.79		3	K2 IV	280,8041
		St 14 - 22		11 41 43	−62 14.9	1	13.10		0.27		-0.01		2		956
		LP 907 - 19		11 41 45	−30 39.5	1	12.56		0.48		-0.20		2		1696
		St 14 - 17		11 41 48	−62 15.3	3	12.21	.009	-0.01	.008	-0.55	.031	8	B6 V	411,956,1672
	+31 02289			11 41 50	+30 51.9	1	10.35		1.00		0.77		1		280
101998	+49 02079			11 41 50	+48 47.6	2	7.13	.010	1.45	.008	1.68	.065	10	K4 III	172,7008
		St 14 - 16		11 41 50	−62 14.4	2	12.66	.063	1.31	.054	1.17	.039	3		411,956
102000	+22 02396			11 41 52	+21 39.3	1	8.53		1.09		0.98		1	K0	280
102053	−61 02576	St 14 - 7		11 41 52	−62 17.1	5	8.89	.024	0.09	.010	-0.19	.015	13	B8 Vp	287,411,885,956,1672
	+25 02409			11 41 53	+24 39.9	4	9.78	.014	0.26	.006	0.08	.015	9	A2	193,272,280,1569
	+27 02049			11 41 53	+26 41.1	1	10.27		0.97		0.83		1		280
102014	+23 02380			11 41 59	+22 39.3	1	9.04		1.25		1.37		1	K2	280
	+26 02251	G 121 - 12		11 42 01	+25 48.9	3	10.37	.008	0.48	.014	-0.13	.007	4	F6	1064,1658,3077
	+29 02204			11 42 03	+28 42.4	1	9.81		1.19		1.12		1	K2	280
	+31 02290	G 121 - 13		11 42 04	+31 14.5	1	8.96		1.13		1.06		2	K8	3072
	+24 02388			11 42 05	+24 20.7	1	9.63		0.99		0.82		1	G5	280
102031	+21 02348			11 42 07	+21 12.8	1	9.39		1.04		0.93		2	K2	1569
	+35 02277			11 42 07	+34 31.2	1	9.22		0.89		0.59		3	G8 III	280
		G 121 - 14		11 42 08	+27 11.3	1	13.91		1.43		1.12		4		316
102056	+29 02206			11 42 08	+28 56.9	3	7.01	.013	0.00	.006	0.03	.006	16	Am:	1068,1499,8071
		GD 142		11 42 09	+36 58.5	1	15.07		0.06		-0.52		1		3060
102076	−48 06769			11 42 10	−48 52.0	1	7.12		0.99		0.75		25	K0 III	1621
102077	−48 06770	V838 Cen	⋆ AB	11 42 11	−49 08.4	1	8.98		0.90		0.50		25	K0/1 Vp	1621
310831	−66 01637	RT Mus		11 42 11	−67 01.6	1	8.70		0.73				1	F8	1488
102070	−17 03460	HR 4514		11 42 14	−18 04.4	9	4.72	.009	0.97	.006	0.74	.011	33	G8 IIIa	15,369,1007,1013,1075*
102101	−59 03765	LSS 2476		11 42 16	−60 07.4	1	7.62		0.08		-0.65		4	B3/5 Ib/II	158
	+35 02278			11 42 19	+35 17.2	1	10.10		0.94				3	G8 III	280
	+17 02393			11 42 20	+17 08.5	1	10.47		1.08		1.03		1	K0	280
		AAS9,163 T6# 1		11 42 21	+56 03.1	1	10.88		0.99		0.80		5		186
	−49 06451			11 42 21	−49 42.2	1	11.03		0.70		0.10		1		78
102118	−58 03821			11 42 21	−59 04.5	1	8.34		0.20		0.21		1	A3 V	1771
	+31 02291			11 42 23	+31 03.3	1	9.66		1.00				2	K0 III	280
	−40 06919			11 42 23	−40 25.8	1	10.36		1.24		1.11		2		1730
102113	−54 04743	IDS11400S5454	A	11 42 24	−55 11.0	1	7.61		0.11		0.03		1	A1 V	1771
102081	+20 02644			11 42 26	+19 46.7	1	8.49		0.28		0.01		2	A3	1569
308999	−61 02582			11 42 26	−62 12.2	2	9.66	.015	1.10	.015	0.96	.185	2	K0 III	137,287
	+23 02382			11 42 27	+22 47.3	1	9.56		0.96		0.68		1	G5	280
102093	+16 02289			11 42 29	+15 36.8	1	8.33		1.44		1.87		1	K2	280
102092	+17 02394			11 42 29	+17 26.3	1	8.50		1.08		1.09		1	K2	280
	+37 02211			11 42 29	+37 16.9	1	9.55		0.99				3	G8 III	280
102156	−73 00885			11 42 31	−73 39.9	1	7.52		1.76				4	K5 III	2012
102091	+22 02400			11 42 33	+21 49.4	1	9.54		1.24		1.27		1		280
	−40 06922			11 42 33	−40 28.3	1	10.32		0.65		0.17		2		1730
102125	−18 03229			11 42 34	−18 36.9	1	9.58		0.58		0.07		3	G1 V	1700
309004	−61 02585	St 14 - 11	⋆	11 42 35	−62 14.0	1	10.91	.018	0.00	.004	-0.68	.017	10	B2.5IV	287,411,956,1747
102155	−72 01157			11 42 35	−72 52.4	3	6.94	.010	1.38	.003	1.25	.003	17	K1 II	1499,1704,2012
102103	+15 02374			11 42 37	+14 32.5	1	6.49		1.15				2	K0	280
		G 57 - 17		11 42 38	+18 37.8	1	13.27		1.47		1.10		2		333,1620
	+29 02208			11 42 39	+28 52.4	1	10.02		1.12		1.04		1		280
309005	−61 02586			11 42 39	−62 13.4	1	10.88		0.09		-0.54		1	A0	287
	+41 02229			11 42 41	+41 04.9	1	9.38		0.95				2	G8 III	280
	+42 02245			11 42 41	+41 28.2	1	9.89		0.92				2	G8 III	280
102153	−62 02264	LSS 2478		11 42 41	−63 15.8	4	8.73	.004	-0.06	.011	-0.88	.007	9	B0 IV/V	401,540,976,1732
102124	+9 02545	HR 4515		11 42 42	+08 32.2	4	4.84	.008	0.18	.005	0.09	.010	10	A4 V	15,1256,1363,3023
		SSII 24		11 42 42	+40 26.	1	14.63		0.11		0.08		2		1298
102136	−40 06927			11 42 43	−40 28.1	1	9.36		0.78				4	G8/K0 V	2033
102150	−48 06777	HR 4516		11 42 44	−48 47.5	4	6.25	.007	1.17	.009			18	K1 III	15,1075,2013,2029
102122	+20 02465			11 42 46	+20 10.0	1	7.52		1.05				2	K0	280
102148	−34 07658			11 42 48	−34 31.9	1	8.90		1.13				4	K1/2 (III)	2012
102142	+28 02039			11 42 49	+27 29.8	2	7.24	.020	0.74	.015	0.34		5	G5 V	20,3026
102149	−35 07416			11 42 49	−36 07.5	1	7.86		1.13				4	K1 III/IV	2012
		G 122 - 44		11 42 51	+44 57.6	1	11.17		1.32		1.18		4	K4-5	7010
102191	−67 01810			11 42 51	−67 25.2	1	8.60		0.21		0.13		1	A2 V	1771
	+24 02389			11 42 55	+23 53.8	1	9.75		1.24		1.33		1		280
102158	+48 01964	G 122 - 45		11 42 55	+47 56.9	4	8.05	.018	0.60	.008	0.04	.019	8	G2 V	22,1003,1658,3026
	−61 02595	LSS 2480		11 42 55	−61 42.1	1	10.51		0.33		-0.60		3		1747
102161	+25 02411			11 42 56	+25 23.4	1	8.40		0.64		0.15		2	G0 V	3026
102160	+31 02293			11 42 56	+30 46.1	1	8.14		1.02				4	K0 III	280
	+36 02215			11 42 56	+35 33.4	1	10.28		1.20		1.28		1	K1 III	280
102159	+36 02216	TV UMa		11 42 58	+36 10.3	3	7.01	.085	1.54	.045	1.40	.044	5	M4 III	172,280,3001
102184	−36 07394			11 42 58	−37 20.7	1	7.88		0.43		-0.07			F5 V	78
		L 145 - 141		11 42 58	−64 33.5	4	11.50	.018	0.19	.009	-0.67	.022	17		1698,1765,2033,3073
	+40 02460			11 43 00	+39 30.2	1	10.54		0.94		0.69		1	K0 III	280

HD	DM	Other Id	N	Rem	α_{1950}	δ_{1950}	S	V	σ_V	B–V	σ_{B-V}	U–B	σ_{U-B}	n	Spectrum	References
		LP 553 - 70			11 43 01	+04 10.1	1	11.59		1.13		1.10		1		1696
102204	−57 04969				11 43 01	−57 51.0	1	9.29		0.22		0.16		1	A2mA7-A7	1771
102241	−74 00843	IDS11408S7419	AB		11 43 02	−74 35.8	2	8.39	.004	0.33	.006	0.18		6	A1 V	278,2012
102200	−45 07249				11 43 05	−45 47.0	2	8.76	.005	0.45	.010	-0.21		3	F2 V	1594,6006
		LSS 2481			11 43 05	−64 33.7	2	11.52	.010	0.18	.010	-0.69	.020	7	B3 Ib	1737,1747
	+72 00545				11 43 08	+72 22.3	1	9.02		1.17		1.12		2	K8	3072
310720	−63 01952	LSS 2482			11 43 10	−64 10.0	1	10.53		0.53		-0.38		4	B5	1747
		AAS9,163 T6# 2			11 43 11	+56 14.4	1	10.96		1.21		1.19		6		186
102220	−57 04970				11 43 11	−57 36.0	1	9.62		0.26		0.18		1	A3mA8-F0	1771
102217	−39 07291				11 43 12	−40 23.0	1	8.91		0.35		0.11		2	A9 V	1730
102249	−66 01640	HR 4520		⋆ A	11 43 14	−66 27.1	8	3.63	.015	0.16	.005	0.14	.016	27	A7 II/III	15,1034,1075,1409*
102232	−44 07564	HR 4519			11 43 15	−45 24.7	12	5.28	.004	-0.12	.007	-0.55	.007	67	B6 III	14,15,182,278,388,977*
102234	−48 06787				11 43 15	−48 56.9	1	9.35		0.37		0.13		1	A9/F0 V	1771
102212	+7 02479	HR 4517			11 43 17	+06 48.6	11	4.03	.011	1.51	.011	1.80	.025	41	M1 IIIab	3,15,369,1003,1119*
	+15 02375				11 43 17	+14 52.4	1	10.31		1.05		1.00		1		280
102208	+37 02212				11 43 17	+36 34.1	1	8.69		0.90				2	G8 III	280
	+25 02413				11 43 18	+25 05.9	1	9.80		1.55		1.83		1	M0	280
102248	−60 03314				11 43 19	−61 10.1	2	7.92	.000	0.01	.019	-0.43	.005	3	B6/8 II	2,401
	+27 02051				11 43 20	+26 41.3	1	9.92		0.93				2	G8 III	280
		G 148 - 6			11 43 20	+32 06.2	2	14.06	.030	1.61	.020	1.14		6	dM4	419,3028
309065	−60 03315				11 43 20	−61 11.2	2	8.83	.010	0.04	.030	-0.73	.010	2		2,401
		G 148 - 7			11 43 21	+32 06.2	2	13.66	.015	0.05	.000	-0.67	.010	8	DAs	419,3060
102226	+40 02461				11 43 21	+39 40.4	2	8.02	.015	1.23	.020	1.39		5	K2 III	280,1569
	+23 02383				11 43 23	+22 39.0	1	9.98		1.36		1.56		1		280
	+29 02210				11 43 23	+28 56.8	1	8.72		0.98				3	G9 III	280
		LP 319 - 21			11 43 23	+30 30.6	1	11.60		1.22		1.21		4		1723
	+33 02151				11 43 24	+33 12.6	2	9.83	.070	1.53	.000	1.17		2	M6 III	280,694
102247	−49 06464				11 43 24	−49 40.3	1	8.23		0.15		0.14		1	A1 IV/V	1771
102224	+48 01966	HR 4518			11 43 25	+48 03.4	6	3.71	.011	1.18	.008	1.15	.016	17	K2 III	15,172,1080,1363,4001,8015
		CS 15625 # 5			11 43 26	+24 28.3	1	13.89		0.84		0.11		1		1744
		AAS9,163 T6# 3			11 43 27	+55 52.3	1	12.20		0.82		0.59		7		186
	+27 02053				11 43 28	+27 07.3	1	9.49		1.05				2	K0 III	280
102293	−76 00688				11 43 31	−76 35.0	2	7.89	.006	0.08	.000	-0.01		6	B9.5IV	278,2012
102253	+7 02480				11 43 32	+07 27.1	1	6.87		1.64		1.95		2	M1	3012
102242	+40 02463				11 43 32	+39 40.8	1	8.64		0.81				2	G7 III	280
102243	+21 02353				11 43 34	+20 35.8	1	7.65		0.93		0.65		1	G6 III	280
		L 68 - 144			11 43 34	−73 55.4	1	14.52		1.43		1.19		2		3073
	+28 02040				11 43 36	+27 28.3	1	10.18		1.31		1.34		1	K2	280
102256	−37 07455				11 43 36	−37 52.1	2	8.51	.005	0.60	.005	0.01		2	G3 V	78,2039
102281	−61 02607				11 43 37	−61 42.2	1	9.42		0.33		0.17		1	A8 V	1771
	+29 02211				11 43 38	+29 17.7	1	10.09		1.37		1.58		2	K2	1569
	+40 02464				11 43 38	+39 36.5	1	10.08		0.66				2	G5 III	280
		AAS9,163 T6# 5			11 43 40	+55 47.2	1	12.45		0.87		0.65		7		186
309029	−62 02280	LSS 2484			11 43 40	−62 46.9	1	10.50		0.34		-0.43		4	B3	1747
238037	+56 01541				11 43 44	+56 22.9	1	10.10		0.51		0.02		3	F8	186
	+33 02152	IDS11411N3336	AB		11 43 45	+33 19.3	1	9.80		0.75				2	G6 III	280
102251	+56 01542				11 43 45	+55 37.2	1	8.25		1.11		0.95		3	K0 III	186
	+25 02416				11 43 48	+24 59.7	1	10.12		1.05		0.98		1		280
		AAS9,163 T6# 8			11 43 48	+55 54.4	1	12.05		0.57		-0.09		5		186
102272	+14 02434				11 43 49	+14 24.1	1	8.69		1.02		0.69		1	K2	280
		AAS9,163 T6# 6			11 43 49	+56 13.7	1	11.49		0.99		0.74		7		186
		LP 673 - 86			11 43 53	−09 18.0	1	13.17		1.20		1.13		1		1696
		SSII 26			11 43 54	+40 42.	2	11.87	.050	0.40	.060	-0.06	.000	2	F0	1240,1298
		vdB 53 # b			11 43 54	−65 17.	1	15.39		1.20		0.27		1		434
238038	+56 01543				11 43 55	+56 20.0	1	9.11		0.67		0.27		std	G0	186
102300	−24 09943				11 43 55	−24 41.9	1	7.68		0.57		0.02		1	F7 V	78
102322	−54 04764				11 43 55	−55 23.0	1	8.56		0.00		-0.11		1	B9 V	1770
		AAS9,163 T6# 10			11 43 56	+55 51.0	1	11.76		1.17		1.09		8		186
309008	−62 02283				11 43 59	−62 27.0	1	10.98		0.28		0.18		1	A2 V	137
	+51 01696	G 176 - 53			11 44 01	+51 10.0	2	9.92	.002	0.56	.009	-0.15		3	G0 VI	1064,1691
102320	−53 04756				11 44 01	−53 50.4	1	7.67		0.74		0.41		1	A3 V	1771
	+21 02354				11 44 04	+20 52.3	1	9.34		1.07		1.01		1	K2 II	280
102351	−62 02286				11 44 04	−62 52.0	1	9.68		0.08		-0.38		2	B8 II/III	401
102352	−63 01963	LSS 2485			11 44 04	−63 47.1	2	9.98	.049	0.05	.019	-0.78	.015	6	B2 Vne	401,1747
102350	−60 03325	HR 4522		⋆ A	11 44 05	−60 54.0	6	4.10	.007	0.90	.008	0.58	.007	24	G5 Ib/II	15,1075,1488,2012*
102333	−18 03235				11 44 06	−18 43.7	1	9.05		0.15		0.12		3	Ap EuCrSr	1700
		vdB 53 # a			11 44 06	−65 18.	1	13.07		0.45		-0.40		4		434
	+40 02465				11 44 08	+40 21.6	2	9.98	.005	0.75	.005	0.46		4	G7 III	280,1569
		LP 733 - 71			11 44 08	−13 43.5	4	11.69	.010	1.48	.012	1.05	.000	6	M3	912,1705,3073,3078
102365	−39 07301	HR 4523			11 44 08	−40 13.7	5	4.89	.011	0.67	.005	0.04	.000	19	G5 V	15,678,2012,3078,8017
309043	−62 02288	LSS 2486			11 44 08	−63 20.6	1	10.81		0.07		-0.59		4	B5	1747
102368	−61 02611	LSS 2487			11 44 10	−61 48.1	4	7.60	.024	0.02	.009	-0.86	.016	7	B0 III	2,401,540,976
	+29 02213				11 44 12	+29 06.9	1	9.31		1.21				3	G2-5 IIIp	280
102370	−64 01706				11 44 12	−64 29.3	2	6.89	.015	0.00	.090	-0.28	.000	4	B8 V	401,1771
	+37 02213				11 44 13	+36 31.9	1	9.75		1.05				2	K0 III	280
102343	+15 02378	IDS11417N1533	AB		11 44 16	+15 16.8	1	7.86		0.32		0.08		2	A5	1569
	+34 02258				11 44 16	+33 42.5	1	10.73		0.31		0.01		2		1569
102328	+56 01544	HR 4521			11 44 16	+55 54.4	2	5.29	.023	1.28	.005	1.49	.018	7	K3 III	186,1080
	−13 03442				11 44 18	−13 49.6	1	10.26		0.40		-0.20		2		1696
	+27 02054				11 44 19	+27 19.1	1	9.63		0.82				2	G8 III	280

Table 1 627

HD	DM	Other Id	N Rem	α_{1950}	δ_{1950}	S	V	σ_V	B–V	σ_{B-V}	U–B	σ_{U-B}	n	Spectrum	References
	+28 02041			11 44 19	+28 16.4	1	9.95		1.06		0.90		2		1569
	+31 02298			11 44 20	+30 53.3	1	9.06		1.06				2	K0 III	280
	+27 02055			11 44 21	+27 18.0	2	9.25	.050	1.58	.065	1.45		10	K3 V	280,7009
	+33 02153			11 44 21	+33 18.5	1	10.08		1.00				2	K0 III	280
102360	−18 03239			11 44 21	−18 38.1	1	9.94		1.25		1.29		1	K0 III/IV	1700
	+51 01697			11 44 23	+51 15.4	1	9.62		1.26		1.19		2	M0 V:p	3072
102400	−61 02616			11 44 23	−62 04.0	1	9.25		0.00		−0.65		2	B2 III	401
	+40 02466			11 44 24	+39 47.8	1	9.44		1.13				2	K0 III	280
	+17 02397			11 44 25	+16 29.3	1	9.05		0.64		0.14		2	G0	1569
	+32 02188			11 44 25	+32 06.7	3	10.74	.008	−0.07	.004	−0.27	.006	11		275,1026,1240
102399	−60 03330			11 44 26	−60 45.8	1	8.02		−0.04		−0.44		2	B7 V	401
	+27 02056			11 44 27	+27 12.7	1	9.79		1.02		0.80		2		1569
		G 197 - 30		11 44 27	+60 16.3	1	13.51		0.78		0.14		2		1658
102355	+62 01198			11 44 27	+61 40.8	2	6.61	.005	0.22	.000	0.10	.005	5	A7 V	985,1501
102415	−60 03333	LSS 2489		11 44 28	−61 11.0	3	9.15	.016	0.12	.015	−0.78	.016	5	B5	2,540,976
		AO 1128		11 44 29	+01 00.7	1	12.48		0.58		0.08		1		1748
	+19 02500			11 44 30	+19 20.4	1	10.20		0.98		0.80		1	G5	280
		Ton 592		11 44 30	+29 53.	1	14.03		0.14		0.11		3		1036
	+41 02231			11 44 32	+41 11.8	1	10.77		0.27		0.10		1	A5	280
102392	−11 03178	IDS11420S1116	AB	11 44 32	−11 32.7	3	9.02	.016	1.12	.028	1.09		6	K2	1017,2033,3072
309058	−63 01965	LSS 2490		11 44 34	−63 47.7	1	10.98		0.04		−0.59		4	A2	1747
310797	−65 01714		A	11 44 34	−65 35.3	1	10.61		0.85		0.44		4	K0	158
310797	−65 01714		B	11 44 34	−65 35.3	1	10.97		0.93				4		158
	+28 02042			11 44 35	+28 28.2	1	9.17		1.03				2	G8 III	280
		LB 2112		11 44 35	+61 32.3	2	13.38	.100	−0.20	.045	−0.85	.175	4		272,308
102389	+20 02648			11 44 36	+19 35.0	1	8.14		0.11		0.08		4	A2	1068
102326	+77 00440	IDS11417N7736	A	11 44 36	+77 19.1	1	8.75		0.73		0.45		2	G8 IV-V	1003
102397	−35 07438	HR 4524		11 44 36	−35 37.7	3	6.16	.004	0.95	.005			11	G8 III	15,1075,2032
	+37 02214			11 44 37	+37 06.0	1	10.44		0.72				2	G6 III	280
	+39 02468			11 44 38	+39 08.6	1	10.58		1.05		0.90		1	G8 III	280
		G 254 - 29		11 44 39	+78 57.8	2	10.79	.010	1.60	.005			2		979,3045
102405	+18 02529			11 44 41	+18 08.1	1	9.19		1.00		0.76		2	G5	1569
	+28 02043			11 44 41	+27 43.1	1	9.74		1.34		1.54		1	K0	280
	−61 02620	LSS 2491		11 44 41	−62 23.2	1	11.20		0.31		−0.66		4		1747
		CS 15625 # 6		11 44 44	+23 54.9	1	13.89		0.52		−0.13		1		1744
102404	+25 02418			11 44 44	+24 42.1	3	7.85	.009	1.25	.010	1.27	.021	7	K3 III	172,280,8041
102418	+28 02044			11 44 44	+27 40.5	1	8.48		0.42		0.02		2	F8	1569
307131	−58 03845			11 44 44	−58 55.9	1	9.48		1.06		0.81		1	K0 III	137
		AO 1129		11 44 45	+01 10.0	1	12.79		0.68		−0.07		1	G0 V	1748
102438	−29 09337	HR 4525		11 44 45	−30 00.3	4	6.48	.004	0.68	.000	0.22	.027	9	G5 V	15,1008,2012,3077
		G 122 - 46		11 44 46	+48 02.7	1	14.22		1.50		1.12		2		3016
	+22 02402			11 44 48	+21 50.4	1	10.45		0.95		0.76		1	K0	280
307160				11 44 49	−59 37.1	1	10.77		1.14		0.85		2	K0 III	137
102429	+14 02437			11 44 52	+13 39.8	1	9.05		0.30		0.03		2	F2	1569
102461	−57 04989	HR 4526		11 44 52	−57 25.1	6	5.40	.007	1.66	.007	1.98	.015	22	K5 III	15,1075,1311,2013*
102475	−61 02622	LSS 2493		11 44 52	−62 09.5	6	8.52	.013	−0.01	.019	−0.86	.014	16	B1 II	2,6,540,976,1021,8100
		LF14 # 5		11 44 52	−62 53.6	1	10.77		1.22		0.78		1	K0 III	137
	+27 02057			11 44 53	+26 41.5	1	9.53		0.91		0.57		2	G7 III	1569
102427	+40 02467			11 44 53	+40 12.3	2	8.66	.005	0.67	.029	0.32		3	G2 III	280,8038
102447	−18 03240			11 44 53	−18 39.3	1	9.10		1.46		1.74		2	K2/3 (III)	1700
102459	−52 04868			11 44 53	−53 22.5	1	8.59		0.20		0.13		1	A5 IV	1771
307077	−59 03801			11 44 53	−59 58.8	1	10.97		0.15		0.16		1	A0 V	137
307105	−60 03338			11 44 53	−60 55.1	1	9.13		1.10		0.91		1	K0 III	137
102476	−62 02294			11 44 55	−63 01.4	1	8.77		0.07		−0.29		2	B8/9 Ib	401
309037	−62 02295			11 44 56	−63 08.9	1	9.51		0.09		−0.79		2	B5	401
102506	−75 00754			11 44 57	−76 20.4	1	7.59		1.67				4	M6 III	2012
		LP 375 - 39		11 44 58	+24 17.6	1	9.71		0.61		0.14		2		1723
		LP 553 - 42		11 45 00	+07 02.2	1	12.20		0.49		−0.22		2		1696
	+23 02388			11 45 00	+23 15.8	1	9.93		1.00		0.87		1	K0	280
		Feige 49		11 45 00	−15 48.9	1	12.88		−0.30		−1.08		1	B1	1298
		LP 733 - 100		11 45 02	−11 29.0	1	12.31		0.67		0.18		2		1696
	+1 02611			11 45 06	+00 53.5	1	10.41		1.54		1.83		1	K2	1748
		G 10 - 50	⋆V	11 45 09	+01 05.9	6	11.12	.026	1.75	.015	1.36	.044	21		694,1017,1705,1764,3078,8039
		AO 1131		11 45 10	+01 03.2	1	11.41		0.64		0.20		1	G5 V	1748
		AO 1132		11 45 10	+01 05.4	1	11.20		1.70		1.52		1		1748
309036	−62 02300	LSS 2497		11 45 10	−63 05.9	3	10.64	.011	0.25	.014	−0.61	.025	8	B2 II	540,914,1747
102533	−65 01719	LSS 2498		11 45 13	−65 43.7	1	7.60		0.17		−0.12		3	B9 Ib	8100
		CCS 1917		11 45 15	−57 34.5	1	10.17		2.89		3.20		1		864
		G 236 - 82		11 45 17	+71 08.3	2	10.90	.075	0.63	.000	0.04	.007	3		1658,5010
		G 237 - 32		11 45 18	+71 38.1	1	11.38		1.37		1.34		5		7010
102493	+33 02156			11 45 19	+32 46.1	2	7.86	.019	1.25	.005	1.24		3	K2 III	280,8041
102510	+9 02549	HR 4528	⋆A	11 45 21	+08 31.4	4	5.31	.008	0.02	.009	0.04	.004	11	A0	15,1363,1417,3023
102494	+28 02046			11 45 21	+27 37.1	3	7.49	.012	0.87	.004	0.54	.014	6	G9 IV	172,280,8041
		AO 1134		11 45 22	+01 06.7	1	13.12		1.14		0.81		1	G0 I	1748
		G 57 - 19		11 45 22	+15 10.5	1	13.20		0.86		0.26		2		333,1620
	+56 01545			11 45 23	+56 27.3	1	10.00		1.01		0.87		7		186
102509	+21 02358	HR 4527, DQ Leo	⋆A	11 45 24	+20 29.8	6	4.53	.015	0.55	.007	0.29	.018	25	G5 III-IVe+	15,667,1008,1363,3026,8015
	+56 01546			11 45 24	+56 23.1	1	10.44		0.54		0.08		4		186
309083	−61 02632	LSS 2499		11 45 24	−61 42.3	1	10.02		0.02		−0.66		3	B8	1747
		LF14 # 8		11 45 24	−62 05.6	1	12.58		1.49		1.07		1		137

HD	DM	Other Id	N	Rem	α_{1950}	δ_{1950}	S	V	σ_V	B–V	σ_{B-V}	U–B	σ_{U-B}	n	Spectrum	References
		G 11 - 23			11 45 25	+08 04.5	1	15.93		0.15		-0.75		1		3060
102508	+41 02232				11 45 25	+40 57.2	1	8.99		0.93				2	G8 III	280
		AAS9,163 T6# 14			11 45 26	+56 20.3	1	11.20		0.57		0.06		5		186
307092	-60 03347	LSS 2500			11 45 26	-60 28.6	1	11.15		0.19		-0.68		4		1747
	+25 02419				11 45 28	+24 44.8	1	9.53		0.54		0.06		1		280
	-65 01721				11 45 28	-65 58.8	1	10.82		0.18		-0.27		2	B6 IV	540
		AAS9,163 T6# 15			11 45 30	+55 54.2	1	12.26		0.64		0.02		6		186
102552	-59 03809	SV Cen			11 45 30	-60 17.2	2	8.68	.013	0.05	.010	-0.71	.023	4	B1/3	540,976
309049	-63 01975				11 45 30	-63 42.1	1	10.01		0.72		0.64		2	A3	401
102540	-34 07692				11 45 33	-34 56.8	2	7.10	.005	0.76	.009	0.23		13	G5/6 V	1499,2012
102567	-61 02636	V801 Cen		⋆	11 45 34	-61 55.7	4	8.96	.060	0.20	.022	-0.85	.014	31	O/Bne	6,358,1737,1747
	+27 02060				11 45 36	+26 52.1	1	10.13		1.33		1.39		1		280
		AO 1135			11 45 37	+01 11.0	1	12.95		0.70		0.47		1	K2 V	1748
	+32 02189				11 45 38	+32 11.2	1	10.25		0.35		-0.02		1	F2	1298
102579	-65 01723	IDS11433S6533	AB		11 45 39	-65 50.3	1	8.33		0.93				4	K0 V	2033
309158	-60 03351				11 45 43	-61 07.8	1	10.84		0.47		0.05		1	F2 V	137
		AO 1136			11 45 44	+01 03.2	1	10.76		1.07		0.93		1	K0 III	1748
	+38 02278				11 45 44	+38 05.6	1	10.34		1.28		1.30		1	K1 III	280
102561	-49 06501				11 45 44	-49 42.9	1	7.89		0.22		0.15		2	A5 V	1771
102601	-76 00689				11 45 44	-76 37.2	1	8.05		0.88				4	G8 III/IV	2012
	+13 02456				11 45 46	+13 22.5	1	10.44		0.01		-0.01		2	A0	1026
	+43 02152				11 45 46	+42 41.6	1	9.18		0.86				2	G8 III	280
	+41 02233				11 45 49	+41 18.4	1	9.86		1.07		0.96		1	K1 III	280
102584	-66 01649	HR 4530, μ Mus			11 45 49	-66 32.2	4	4.74	.016	1.54	.008	1.90	.005	23	K4 III	1088,2006,2038,3055
102555	+29 02214				11 45 50	+28 41.7	2	7.23	.005	0.37	.005	0.02		5	F2	985,1733
	+20 02652				11 45 51	+19 39.0	1	9.40		1.18		1.13		1	K5	280
102574	-9 03366	HR 4529		⋆ A	11 45 51	-10 02.0	3	6.25	.010	0.58	.000	0.13	.005	14	F7 V	15,1417,2027
307118	-58 03858				11 45 51	-58 33.8	1	10.65		0.22		-0.04		1	B8 V	137
102582	-22 03220				11 45 55	-22 48.8	1	7.43		0.77		0.29		1	G6 IV	78
102569	+56 01547				11 45 56	+55 49.0	1	7.94		1.12		1.07		std	K1 III	186
		AO 1137			11 45 59	+01 02.3	1	11.43		1.12		1.22		1	K2 III	1748
102615	-72 01162				11 45 59	-73 12.1	1	8.57		0.40				4	F3 IV/V	2012
	-60 03361	LSS 2504			11 46 8	-60 56.8	1	10.97		0.31		-0.58		4		1747
		AO 1138			11 46 00	+01 02.6	1	12.07		0.73		0.21		1		1748
102589	+29 02216				11 46 01	+29 04.7	1	7.06		0.06		0.07		3	A2 V	1068
	+32 02190				11 46 01	+32 10.2	1	9.99		0.64				4	G5 III	280
102596	-46 07398				11 46 01	-47 16.6	1	8.99		1.41				4	K2 III	2012
102613	-55 04644				11 46 01	-55 41.3	1	8.09		0.21		0.13		1	A8 IV	1771
309035	-62 02312	LSS 2503			11 46 03	-63 06.9	3	10.33	.000	0.62	.005	-0.45	.038	8	O9 Ia	540,914,1747
102590	+15 02381	HR 4531		⋆ AB	11 46 04	+14 33.7	2	5.88	.020	0.29	.005	0.05		5	A8 V +G2 V	70,196
		AO 1139			11 46 05	+01 03.2	1	11.40		0.62		0.10		1	G1 V	1748
		CS 15625 # 8			11 46 05	+26 45.9	1	13.30		0.80		0.47		1		1744
		CS 15625 # 7			11 46 12	+26 12.0	1	14.58		0.02		0.18		1		1744
309165	-60 03363				11 46 12	-61 12.5	1	9.96		-0.01		-0.59		3	B3 V	434
102620	-26 08789	HR 4532, II Hya			11 46 13	-26 28.3	3	5.11	.014	1.59	.021	1.67	.020	14	M4 III	1024,2007,3052
		AAS12,381 15 # A			11 46 14	+14 40.4	1	9.95		1.06		0.97		1		280
307161	-59 03821				11 46 16	-59 33.5	1	10.96		0.48		0.11		2	F2 V	137
102617	+21 02361				11 46 17	+20 47.0	1	8.67		0.28		0.07		2	A7 IV	1569
309093	-61 02648				11 46 17	-62 02.9	1	11.15		0.17		-0.25		3	B8 V	434
	+34 02259	IDS11437N3415	AB		11 46 20	+33 58.8	1	9.70		0.30		0.01		2	Am:	1569
		G 254 - 31			11 46 20	+76 39.4	1	11.94		0.84		0.47		2		7010
		AAS12,381 42 # A			11 46 21	+41 48.6	1	10.46		1.54		1.91		1		280
309111	-62 02317	LSS 2505			11 46 21	-62 35.6	3	10.62	.069	0.11	.023	-0.73	.013	3		540,1737,1747
102626	+19 02504				11 46 23	+19 02.2	1	9.54		0.31		-0.01		2	F5	1569
		GD 143			11 46 24	+19 23.9	1	14.42		0.58		-0.16		1		3060
	-9 03371	X Crt			11 46 24	-10 10.1	2	10.38	.740	0.29	.060	0.11	.030	2	A2	668,700
	+15 02382				11 46 25	+14 47.1	1	8.44		0.57		0.05		1	G2 III	280
102627	+16 02296				11 46 27	+16 24.6	2	6.85	.000	1.07	.005	0.96		5	K2 III	280,1501
	+39 02470				11 46 28	+38 54.8	1	9.54		1.34		1.42		1	K3 IV	280
102616	+68 00662	IDS11437N6753	A		11 46 28	+67 36.4	1	7.43		0.72		0.38		2	F8	1726
102616	+68 00662	IDS11437N6753	B		11 46 28	+67 36.4	1	8.38		0.35		0.10		2		1726
102634	+0 02843	HR 4533			11 46 28	-00 02.4	2	6.14	.005	0.52	.000	0.07	.010	7	F7 V	15,1417
102657	-50 06375				11 46 30	-51 07.8	1	7.73		-0.03		-0.56		1	B3 II/III	55
102647	+15 02383	HR 4534, β Leo		⋆ A	11 46 31	+14 51.1	13	2.14	.012	0.09	.010	0.08	.021	52	A3 V	1,3,15,22,1006,1020*
102645	+37 02216				11 46 32	+37 23.7	1	8.25		1.13				2	K0 III	280
	-9 03372	RS Crt			11 46 33	-10 20.6	1	10.67		0.54		0.01		3	G0	588
102646	+28 02048				11 46 37	+28 24.0	3	7.37	.019	0.93	.007	0.71	.005	7	K0 III	20,172,8041
	+32 02192				11 46 37	+31 44.1	1	9.74		1.05				2	K0 III	280
102661	-6 03455				11 46 38	-07 05.0	1	6.93		1.54		1.84		4	K5	1657
102660	+17 02402	HR 4535			11 46 40	+16 31.3	5	6.04	.010	0.27	.008	0.14	.003	18	A3 m	70,270,1022,1199,3058
		CS 15625 # 9			11 46 40	+27 42.5	1	12.70		0.82		0.40		1		1744
	-60 03378	LSS 2506			11 46 40	-61 1.6	1	11.69		0.18		-0.46		4		1747
102677	-19 03352				11 46 41	-20 03.9	1	9.30		0.89		0.58		1	K2 V	3072
		SSII 29			11 46 42	+30 36.0	1	12.07		0.42		-0.10		1	F2	1298
102693	-56 04799	IDS11443S5708	A		11 46 44	-57 24.8	1	6.87		-0.03		-0.16		2	B9 V	1771
		LP 851 - 229			11 46 48	-26 15.0	2	12.12	.063	0.46	.010	-0.22	.010	2		1696,3060
	+17 02403				11 46 49	+17 09.5	1	10.26		0.30		0.00		2	A5	1569
102729	-68 01571				11 46 50	-68 25.2	1	8.50		0.25		0.17		1	A2 Vn	1771
		LP 553 - 46			11 46 51	+05 46.6	1	12.34		1.16		0.66		1		1696
102688	+19 02505				11 46 52	+18 30.8	1	8.44		1.55		1.86		1	M1	280

Table 1 629

HD	DM	Other Id	N Rem	α_{1950}	δ_{1950}	S	V	σ_V	B–V	σ_{B-V}	U–B	σ_{U-B}	n	Spectrum	References
102698	−15 03360			11 46 58	−16 19.6	1	10.03		1.11		1.00		1	K4 V	3072
102703	−45 07312	IDS11445S4531	AB	11 46 58	−45 47.4	5	7.18	.006	0.28	.006	0.12	.004	27	Fm δ Del	78,278,1075,1770,2011
102727	−54 04788	IDS11445S5457	A	11 46 58	−55 13.7	1	8.60		0.24		0.13		1	A3mA5-A7	1771
102686	+30 02194			11 46 59	+29 46.6	3	7.59	.020	0.93	.013	0.60	.000	6	G6 IIIp	172,280,8041
	+17 02404			11 47 00	+16 52.5	1	9.16		1.30		1.59		1	K2	280
		SSII 30		11 47 00	+32 55.9	1	12.86		0.25		0.07		1	F0:	1298
		LP 64 - 29		11 47 00	+67 36.0	1	11.05		0.97		0.85		4		1723
	+38 02282			11 47 03	+37 34.2	2	10.24	.000	0.29	.005	0.03	.024	3		280,1569
102715	+19 02506			11 47 04	+19 12.9	1	7.84		1.26		1.51		1	K0	280
102742	−63 01986			11 47 04	−64 20.7	1	9.79		0.13		-0.60		3	B3 Ve	540
102713	+35 02284	HR 4536		11 47 06	+35 12.6	2	5.71	.015	0.47	.006	0.03		5	F5 IV	70,3016
	+47 01906			11 47 06	+47 01.7	1	10.21		0.08		0.10		1	A0	1026
309104	−61 02661			11 47 08	−62 14.7	1	10.02		1.24		0.87		1	K0 III	137
309063	−63 01987	LSS 2508		11 47 11	−63 54.5	1	10.96		0.29		-0.70		4		1747
		SSII 32		11 47 12	+46 43.	1	12.25		0.35		0.08		1	F0	1298
	+21 02363			11 47 13	+20 58.6	1	10.11		0.88		0.64		1		280
		KUV 11472+3858		11 47 13	+38 57.6	1	17.46		-0.02		-0.82		1	DA	1708
		G 10 - 53		11 47 14	+05 15.0	1	12.48		1.02		0.60		2		333,1620
		G 10 - 54		11 47 14	+06 25.8	2	12.57	.015	0.61	.005	-0.12	.020	2		1620,1696
102776	−63 01988	HR 4537		11 47 14	−63 30.6	10	4.33	.013	-0.15	.006	-0.62	.008	52	B3 V	6,15,26,681,815,1075*
102770	−50 06388	IDS11448S5035	A	11 47 15	−50 51.4	1	8.81		0.20		0.11		1	A3 V	1771
102774	−61 02663			11 47 16	−61 8.2	1	9.98		0.15		0.04		1	B9/A0 IV	137
102769	−45 07316			11 47 18	−45 44.2	2	7.59	.009	1.30	.000			8	K3 III	1075,2033
102794	−53 04786			11 47 21	−54 13.9	1	8.19		0.25		0.14		1	A4 IV	1771
	+41 02238			11 47 23	+40 36.1	1	10.13		0.26		0.04		2		1569
309110	−62 02333	LSS 2509		11 47 28	−62 38.6	1	10.77		0.25		-0.58		4	B5	1747
	+21 02364			11 47 30	+21 22.7	1	9.71		1.29		1.13		1	K2	280
102781	+18 02532			11 47 31	+18 09.8	1	9.20		0.91		0.62		1	K0	280
	+56 01548			11 47 32	+55 32.8	1	10.44		0.58		0.08		3		186
102839	−69 01595	HR 4538		11 47 32	−69 56.9	5	4.96	.007	1.41	.009	1.23	.020	22	G6 Ib	15,1075,1075,2038,8100
102780	+21 02366			11 47 33	+20 33.9	1	8.14		1.59		1.97		1	K2	280
102779	+21 02365			11 47 33	+21 26.1	1	9.16		1.13		1.09		1	K2	280
307226	−60 03401			11 47 33	−60 56.7	1	10.94		0.13		-0.16		2	B9	807
102815	−51 06084			11 47 34	−51 25.3	1	7.46		0.02		0.05		1	A1 V	1771
102800	+5 02546	G 10 - 55		11 47 35	+05 11.3	1	9.18		0.78		0.32		1	K0	333,1620
102816	−52 04906			11 47 35	−52 48.3	1	8.61		0.18		0.12		1	A4 V	1771
		SSII 34		11 47 36	+46 50.	1	11.62		0.37		0.01		1	F0	1298
102807	+35 02285			11 47 37	+35 04.1	2	8.31	.029	1.13	.010	1.04		3	K1 III	280,8041
102814	−48 06872			11 47 37	−49 15.9	1	8.25		0.18		0.09		1	A3 V	1771
		vdB 56 # c		11 47 38	−64 35.6	1	14.03		0.70		0.42		1		434
		LF14 # 14		11 47 39	−59 30.0	1	11.65		1.33		1.00		1		137
102809	+13 02463			11 47 40	+13 17.5	1	8.47		0.23		0.10		3	A7 Vm	270
	+33 02159			11 47 40	+32 41.1	1	10.21		0.78		0.29		1	K0	280
307225	−60 03402			11 47 40	−60 54.8	1	11.65		0.19		0.06		2		807
		LP 319 - 34		11 47 41	+26 41.9	1	14.83		1.53		1.12		1		1773
307224	−60 03404			11 47 41	−60 54.2	1	10.31		0.25		0.23		2	A0	807
		Klemola 76		11 47 42	+04 01.	1	12.34		0.13		0.12		2		1026
	+40 02471			11 47 42	+40 05.8	1	10.18		1.04		0.88		1	K0 III	280
309119	−62 02336	LSS 2510		11 47 42	−62 45.8	1	10.69		0.19		-0.56		4	B9	1747
		vdB 56 # b		11 47 42	−64 35.7	1	12.16		0.42		-0.41		3	B1 V	434
		G 121 - 21		11 47 43	+25 35.0	1	15.55		1.41		0.73		1		3028
307228	−60 03406			11 47 44	−60 56.9	1	9.95		0.46		0.08		2	F8	807
	−62 02337	V350 Cen	*	11 47 44	−63 22.4	1	10.48		0.27		-0.63		4		1747
	−64 01716	LSS 2512		11 47 44	−64 33.7	1	10.79		0.26		-0.70		3		1747
		G 121 - 22		11 47 46	+25 35.3	1	15.55		0.23		-0.58		4		3060
		LP 265 - 581		11 47 47	+37 58.0	1	11.78		1.12		1.05		2		1773
102845	−15 03363	HR 4539		11 47 47	−15 35.2	1	6.13		0.95				4	G8 II/III	2006
102851	−50 06394			11 47 47	−51 11.6	1	8.77		1.07		0.90		2	K0/1 III	536
		NGC 3960 - 8		11 47 47	−55 24.0	1	13.86		0.56				2		905
		Feige 50		11 47 48	+52 50.0	1	11.96		0.19		0.13		1		3060
307178	−59 03843			11 47 48	−59 56.8	1	10.20		0.12		-0.03		1	A0 V	137
		LSS 2513		11 47 48	−64 35.7	1	10.47		0.43		-0.53		3	B3 Ib	1747
		LP 265 - 582		11 47 49	+37 58.1	1	11.30		1.19		1.17		2		1773
102843	−0 02498	G 10 - 56		11 47 49	−00 58.5	2	9.24	.015	0.78	.005	0.43	.025	2	G5	1620,1658
		NGC 3960 - 9		11 47 49	−55 24.2	1	15.37		0.58				2		905
307223	−60 03409			11 47 49	−60 49.5	2	10.00	.020	1.29	.005	1.30	.015	5	K2 III	137,807
		vdB 56 # a		11 47 49	−64 35.7	1	11.47		0.44		-0.50		4	B0.5V	434
		NGC 3960 - 6		11 47 50	−55 23.5	1	15.22		0.60				2		905
102879	−74 00854			11 47 50	−75 21.6	2	9.34	.015	1.18	.000	0.83		7	G8/K0 III	236,2012
		NGC 3960 - 5		11 47 51	−55 23.3	1	14.29		0.48				2		905
		NGC 3960 - 7		11 47 51	−55 23.4	1	14.52		0.61				2		905
102867	−62 02342			11 47 51	−62 29.1	1	9.82		0.12		0.00		1	B9 V	137
	−35 07477			11 47 52	−35 29.1	1	9.69		1.74		1.72		1		78
		NGC 3960 - 4		11 47 52	−55 23.2	1	14.99		0.60				2		905
		NGC 3960 - 10		11 47 52	−55 24.3	1	13.93		0.53				2		905
		NGC 3960 - 11		11 47 53	−55 24.4	1	13.45		0.57				2		905
		NGC 3960 - 15		11 47 53	−55 25.0	1	14.78		1.49				2		905
		NGC 3960 - 17		11 47 53	−55 25.7	1	14.26		0.61				2		905
	−60 03410			11 47 53	−60 53.3	1	11.00		0.27		0.38		2	A0	807
		NGC 3960 - 3		11 47 54	−55 23.2	1	12.54		0.60				2		905

HD	DM	Other Id	N	Rem	α_{1950}	δ_{1950}	S	V	σ_V	B–V	σ_{B-V}	U–B	σ_{U-B}	n	Spectrum	References
		NGC 3960 - 97			11 47 54	−55 23.5	1	16.36		0.77				2		905
		NGC 3960 - 12			11 47 54	−55 24.1	1	13.62		0.48				2		905
		NGC 3960 - 94			11 47 54	−55 24.5	1	16.00		0.74				2		905
		NGC 3960 - 16			11 47 54	−55 25.6	1	15.29		0.67				2		905
		NGC 3960 - 18			11 47 54	−55 25.7	1	12.65		1.62				2		905
102866	−60 03411				11 47 54	−60 45.2	1	9.43		1.01		0.71		2	K0	807
102865	−45 07326				11 47 55	−46 11.4	4	7.66	.004	0.39	.006	-0.02	.009	17	F2 V	78,977,1075,2011
		NGC 3960 - 13			11 47 55	−55 24.2	1	14.62		0.53				2		905
		NGC 3960 - 14			11 47 55	−55 24.5	1	14.99		0.81				2		905
		NGC 3960 - 22			11 47 55	−55 24.9	1	15.76		0.98				2		905
		NGC 3960 - 21			11 47 55	−55 25.3	1	14.93		0.53				2		905
		NGC 3960 - 20			11 47 55	−55 25.5	1	15.65		0.63				2		905
		NGC 3960 - 19			11 47 55	−55 26.0	1	15.34		0.69				2		905
		G 253 - 54			11 47 56	+85 22.3	1	15.73		1.61				1		906
		NGC 3960 - 30			11 47 56	−55 23.3	1	14.29		1.44				2		905
		NGC 3960 - 1			11 47 57	−55 22.5	1	14.31		0.48				2		905
		NGC 3960 - 2			11 47 57	−55 22.7	1	13.95		0.53				2		905
		NGC 3960 - 91			11 47 57	−55 23.2	1	13.32		1.17				2		905
		NGC 3960 - 23			11 47 57	−55 25.1	1	14.45		0.59				2		905
		Lod 481 # 28			11 47 57	−60 49.7	1	12.18		0.32		0.25		2	A0	807
		NGC 3960 - 92			11 47 58	−55 23.1	1	14.34		0.56				2		905
		NGC 3960 - 29			11 47 58	−55 23.5	1	15.50		0.72				2		905
		NGC 3960 - 28			11 47 58	−55 23.8	1	13.01		1.20				3		905
		NGC 3960 - 26			11 47 58	−55 24.4	1	14.93		0.86				2		905
102857	+44 02132				11 47 59	+43 56.2	1	7.92		0.78		0.40		2	G3 III	172
		NGC 3960 - 31			11 47 59	−55 22.9	1	15.14		0.57				2		905
		NGC 3960 - 32			11 47 59	−55 23.4	1	14.91		0.56				2		905
		NGC 3960 - 95			11 47 59	−55 24.8	1	15.99		0.73				2		905
		NGC 3960 - 24			11 47 59	−55 24.9	1	13.28		1.25				3		905
		NGC 3960 - 25			11 47 59	−55 24.9	1	13.94		0.63				2		905
		Lod 481 # 29			11 47 59	−60 45.6	1	12.29		0.16		-0.12		2	B8	807
		CS 15625 # 23			11 47 60	+28 01.1	1	13.13		0.48		-0.04		1		1744
		NGC 3960 - 34			11 48 00	−55 23.1	1	13.68		0.59				2		905
		NGC 3960 - 35			11 48 00	−55 23.7	1	15.21		0.71				2		905
		NGC 3960 - 27			11 48 00	−55 24.1	1	14.22		0.97				2		905
307222	−60 03414				11 48 00	−60 51.0	2	10.98	.015	0.23	.005	0.14	.015	3	A1 V	137,807
102878	−61 02677	HR 4541, LSS 2514			11 48 00	−62 22.3	7	5.69	.017	0.27	.009	-0.12	.057	18	A2 Iab	2,15,191,540,976,2012,8100
	+42 02250				11 48 01	+41 59.9	1	10.04		1.08		0.98		1	K1 III	280
		NGC 3960 - 39			11 48 03	−55 24.8	1	14.37		0.61				2		905
		NGC 3960 - 43			11 48 03	−55 25.8	1	13.57		1.33				3		905
		NGC 3960 - 100			11 48 04	−55 22.6	1	15.48		0.67				2		905
		NGC 3960 - 36			11 48 04	−55 23.4	1	13.74		0.48				2		905
		NGC 3960 - 37			11 48 04	−55 24.1	1	13.95		0.63				2		905
		NGC 3960 - 40			11 48 04	−55 25.0	1	13.89		0.59				2		905
102870	+2 02489	HR 4540		★ AB	11 48 05	+02 02.8	23	3.61	.009	0.55	.008	0.10	.011	276	F8 V	1,3,15,22,26,30,130*
102888	−26 08807	HR 4542			11 48 05	−27 00.0	1	6.48		0.98				4	G8 III	2006
		NGC 3960 - 42			11 48 05	−55 25.6	1	15.21		0.62				2		905
	+38 02284				11 48 06	+37 37.1	1	10.35		1.12		1.10		1	K1 III	280
		NGC 3960 - 99			11 48 06	−55 22.3	1	15.40		0.96				2		905
		NGC 3960 - 98			11 48 06	−55 22.8	1	15.86		0.87				2		905
		NGC 3960 - 45			11 48 06	−55 24.4	1	14.86		0.63				2		905
		NGC 3960 - 41			11 48 06	−55 25.3	1	13.15		1.30				3		905
	+41 02239				11 48 07	+41 04.1	1	9.37		1.13				2	K1 III	280
102885	−12 03505				11 48 07	−12 35.3	1	7.22		1.56		1.79		2	K4 III	3040
		NGC 3960 - 96			11 48 07	−55 23.4	1	15.97		0.82				2		905
		NGC 3960 - 50			11 48 07	−55 23.5	1	13.11		1.19				3		905
		NGC 3960 - 90			11 48 07	−55 23.7	1	14.15		0.53				2		905
		NGC 3960 - 44			11 48 07	−55 24.2	1	12.67		1.20				3		905
307229	−60 03417				11 48 07	−60 54.9	1	11.34		0.20		0.16		2		807
102907	−60 03416				11 48 07	−60 56.0	1	9.26		1.25		1.24		2	K0/1 III	807
		Steph 974		AB	11 48 08	+33 29.1	1	11.48		1.38		1.13		1	M1	1746
		NGC 3960 - 87			11 48 08	−55 24.4	1	13.13		0.56				2		905
		NGC 3960 - 72			11 48 08	−55 25.4	1	15.61		0.74				2		905
		NGC 3960 - 73			11 48 08	−55 25.8	1	15.67		0.66				2		905
		NGC 3960 - 88			11 48 09	−55 22.2	1	14.83		0.63				2		905
		NGC 3960 - 89			11 48 09	−55 22.2	1	13.45		0.49				2		905
		NGC 3960 - 46			11 48 09	−55 24.1	1	13.12		0.57				2		905
		NGC 3960 - 74			11 48 09	−55 26.0	1	13.56		0.51				2		905
309175	−60 03419				11 48 09	−61 23.1	1	9.70		0.28		0.19		1	A2 V	137
		NGC 3960 - 47			11 48 10	−55 24.1	1	13.65		1.17				3		905
102902	−32 08354				11 48 11	−32 51.8	1	7.34		0.74				4	G3 V	2033
		NGC 3960 - 52			11 48 11	−55 22.9	1	13.31		0.55				2		905
		NGC 3960 - 48			11 48 11	−55 23.8	1	15.12		0.61				2		905
		LF14 # 20			11 48 11	−59 45.7	1	13.08		0.16		0.15		1	A0 V	137
		SSII 35			11 48 12	+46 14.	1	12.70		0.35		0.01		1	F2	1298
		NGC 3960 - 51			11 48 12	−55 23.5	1	12.74		0.56				2		905
		NGC 3960 - 49			11 48 12	−55 23.9	1	13.41		0.79				2		905
		NGC 3960 - 61			11 48 12	−55 24.6	1	14.05		0.52				2		905
		NGC 3960 - 93			11 48 12	−55 25.9	1	16.12		0.67				2		905
	−60 03421				11 48 12	−60 57.3	1	11.21		0.56		0.15		2	F8	807

Table 1   631

HD	DM	Other Id	N Rem	α_{1950}	δ_{1950}	S	V	σ_V	B–V	σ_{B-V}	U–B	σ_{U-B}	n	Spectrum	References
102923	−61 02681			11 48 12	−62 16.4	1	8.51		0.00		-0.60		2	B3 II/III	401
102924	−62 02347			11 48 12	−62 55.7	1	9.93		0.03		-0.60		2	B3/5 Ib/II	401
		NGC 3960 - 81		11 48 13	−55 22.4	1	14.71		0.51				3		905
		NGC 3960 - 62		11 48 13	−55 24.8	1	13.79		0.65				2		905
102897	+19 02509			11 48 14	+18 54.2	1	9.93		0.28		0.03		2	F5	1569
		NGC 3960 - 82		11 48 14	−55 22.1	1	14.31		0.57				2		905
		NGC 3960 - 57		11 48 14	−55 23.4	1	14.31		0.52				2		905
		NGC 3960 - 58		11 48 14	−55 23.5	1	15.21		0.54				2		905
		NGC 3960 - 64		11 48 14	−55 26.0	1	14.09		0.57				2		905
		NGC 3960 - 53		11 48 15	−55 24.3	1	14.76		0.55				2		905
		NGC 3960 - 63		11 48 15	−55 26.2	1	13.85		0.52				3		905
		NGC 3960 - 54		11 48 16	−55 25.8	1	13.29		0.50				2		905
102955	−72 01165			11 48 16	−73 15.4	1	8.34		0.18				4	A0 IV	1594,2012
	−22 03230			11 48 18	−23 05.2	1	11.75		-0.25		-1.16		3		1732
307213	−60 03422			11 48 18	−60 42.1	1	10.06		1.09		0.92		2	K0	807
102940	−62 02350			11 48 18	−63 22.3	1	8.74		0.32		0.04		2	F0 V	401
	+17 02405			11 48 19	+16 39.7	1	9.42		1.01		0.78		2	K0	1569
		NGC 3960 - 55		11 48 19	−55 23.4	1	13.55		1.31				2		905
		NGC 3960 - 56		11 48 19	−55 23.5	1	14.92		0.58				2		905
		NGC 3960 - 101		11 48 19	−55 25.3	1	16.03		0.81				2		905
		NGC 3960 - 65		11 48 19	−55 26.2	1	14.07		0.41				2		905
		NGC 3960 - 80		11 48 20	−55 22.6	1	13.67		0.45				3		905
		NGC 3960 - 59		11 48 20	−55 23.2	1	14.52		0.51				2		905
102910	+13 02465	HR 4543	⋆ A	11 48 21	+12 33.4	2	6.38	.035	0.27	.006	0.08		4	A5 m	70,3058
102910	+13 02465	IDS11458N1250	B	11 48 21	+12 33.4	1	11.65		1.48		1.50		2		3024
	+23 02395			11 48 21	+22 30.5	1	11.15		0.16		0.13		2	A3	1026
	+30 02196			11 48 21	+30 06.2	1	10.21		0.33		0.03		2		1723
		NGC 3960 - 71		11 48 21	−55 24.9	1	13.97		1.84				2		905
		NGC 3960 - 66		11 48 21	−55 25.6	1	13.86		0.46				2		905
		LF14 # 22		11 48 21	−58 57.3	1	11.30		1.15		0.84		1	K1 III	137
		LF14 # 21		11 48 21	−62 18.0	1	12.96		0.00		-0.48		1		137
102909	+34 02262			11 48 22	+33 52.9	3	7.16	.016	1.03	.009	0.81	.010	6	G8 III	172,280,8041
		NGC 3960 - 79		11 48 22	−55 23.4	1	15.62		0.68				2		905
		NGC 3960 - 78		11 48 22	−55 23.6	1	15.92		0.67				2		905
		NGC 3960 - 60		11 48 23	−55 24.0	1	14.68		0.58				2		905
		NGC 3960 - 75		11 48 23	−55 25.9	1	15.63		0.64				2		905
	+37 02219	IDS11458N3727	AB	11 48 25	+37 10.1	1	8.87		0.99				2	K0 III	280
		NGC 3960 - 67		11 48 25	−55 25.8	1	13.98		0.54				2		905
		NGC 3960 - 69		11 48 25	−55 26.3	1	14.09		0.48				2		905
		G 122 - 49		11 48 27	+48 39.9	2	13.26	.005	1.84	.005	1.30		3	M7	316,3078
		NGC 3960 - 70		11 48 27	−55 24.8	1	13.42		0.45				2		905
		NGC 3960 - 68		11 48 27	−55 26.4	1	14.66		0.43				2		905
		NGC 3960 - 77		11 48 28	−55 24.3	1	14.57		0.48				2		905
		Lod 481 # 26		11 48 28	−60 56.8	1	11.75		0.32		-0.05		2	B8	807
102927	+19 02510	IDS11459N1924	A	11 48 29	+19 08.1	1	8.97		0.97		0.70		1	K0	280
102926	+23 02396			11 48 29	+23 01.3	1	8.92		1.02		0.88		1	K0	280
102928	−4 03152	HR 4544	⋆	11 48 29	−05 03.3	2	5.63	.005	1.06	.000	0.89	.005	7	K0 IV	15,1417
102969	−67 01836			11 48 30	−67 48.2	1	7.66		1.07		0.79		3	G8 III	1704
102941	+37 02220			11 48 31	+37 06.8	2	7.55	.020	1.10	.010	1.00		4	K1 III	172,280
		NGC 3960 - 76		11 48 32	−55 25.7	1	14.14		0.44				2		905
	+36 02219	G 148 - 13		11 48 33	+35 32.9	3	9.72	.057	1.51	.013	1.20	.012	8	M1 V	679,1569,7008
		NGC 3960 - 85		11 48 33	−55 24.8	1	13.88		0.45				2		905
102942	+34 02264	HR 4545	⋆ A	11 48 34	+33 39.2	2	6.25	.014	0.31	.005	0.14	.024	6	Am	1068,8071
		NGC 3960 - 84		11 48 34	−55 25.0	1	12.95		0.57				2		905
		NGC 3960 - 86		11 48 35	−55 25.0	1	15.09		1.12				2		905
	−70 01436			11 48 36	−70 31.4	1	9.34		0.91				2		1594
102964	−44 07614	HR 4546		11 48 38	−44 53.7	13	4.46	.007	1.30	.007	1.46	.007	52	K3 III	14,15,182,278,388*
307221	−60 03427	LSS 2516		11 48 38	−60 50.8	2	10.87	.015	0.19	.010	-0.51	.039	6		807,1747
	−60 03428			11 48 38	−60 57.9	1	11.24		0.11		-0.48		2	B5	807
	−60 03430	LSS 2517		11 48 41	−60 51.1	2	11.11	.019	0.20	.019	-0.49	.054	6	B4	807,1747
102981	−43 07333			11 48 42	−43 39.3	5	6.60	.007	-0.01	.004	-0.06	.006	29	A0 V	78,278,1770,2001,2011
	+33 02162			11 48 45	+33 28.7	1	9.58		0.97				2	G8 IV	280
102997	−61 02691	LSS 2518		11 48 45	−61 34.1	5	6.53	.016	0.30	.005	-0.54	.028	14	B5/7 Ia/ab	6,191,540,976,8100
		G 197 - 35		11 48 47	+54 28.5	1	16.72		0.20		-0.57		4		538
		LF14 # 23		11 48 47	−58 36.3	1	11.80		0.48		-0.09		1	F5 V	137
307219	−60 03433			11 48 48	−60 45.4	1	9.59		1.66		1.68		2	K7	807
102990	−11 03190	HR 4547	⋆ A	11 48 50	−11 54.6	2	6.34	.005	0.41	.005			7	F1 III-IV	15,2027
103007	−61 02696	LSS 2519		11 48 50	−61 43.0	2	9.28	.024	-0.01	.000	-0.81	.015	3	B2/3 III	401,1747
307266	−58 03880			11 48 52	−59 22.3	1	10.36		1.07		0.80		1	K0 III	137
102987	+43 02159			11 48 53	+43 12.5	1	9.03		1.06				4	K0 III	280
		KUV 11489+4052		11 48 55	+40 51.6	1	17.33		-0.08		-1.02		1	DB	1708
	−60 03434			11 48 56	−60 55.9	1	11.65		0.18		-0.14		2	B8	807
307230	−60 03435			11 48 59	−60 56.3	1	10.35		0.15		-0.44		2	A0	807
		LP 130 - 238		11 49 01	+55 46.2	1	11.00		0.57		-0.02		3		1723
	+35 02288			11 49 02	+35 28.1	1	9.64		0.78				2	G7 IV	280
		SSII 37		11 49 06	+31 33.9	1	12.88		0.33		-0.01		1	F0	1298
		KUV 11491+4104		11 49 06	+41 03.9	1	16.08		0.09		-0.67		1	DA	1708
103026	−30 09506	HR 4548		11 49 10	−30 33.2	4	5.84	.003	0.56	.005	0.04	.014	15	F8 V	15,1311,2012,3037
	−62 02358			11 49 10	−62 40.7	1	10.96		0.52		0.04		1	F3 V	137
	+15 02385			11 49 11	+15 06.9	1	9.78		1.08		1.00		1	K0	280

HD	DM	Other Id	N Rem	α_{1950}	δ_{1950}	S	V	σ_V	B–V	σ_{B-V}	U–B	σ_{U-B}	n	Spectrum	References
	−28 09134	LP 907 - 38		11 49 11	−28 35.8	2	10.45	.015	0.73	.005	0.25	.010	3		78,1696
		G 12 - 1		11 49 12	+14 34.6	1	12.15		1.22		1.22		1		333,1620
103010	+68 00664			11 49 13	+68 08.0	1	8.85		0.30		0.09		2	A2	1733
103043	−50 06415			11 49 14	−51 12.5	1	8.21		0.13		0.16		1	A3 IV/V	1771
	+27 02063			11 49 15	+26 38.6	2	11.10	.029	0.29	.000	0.09	.010	3	A2	272,1744
103036	−4 03155	TY Vir		11 49 17	−05 29.0	1	8.10		1.28		1.00		1	G3 Ibp	3069
		Lod 481 # 27		11 49 17	−60 51.2	1	11.93		0.23		-0.25		2	B8	807
		G 148 - 14		11 49 21	+33 23.9	1	10.97		1.06				1	K4:	1017
103066	−63 02003	IDS11469S6402	AB	11 49 21	−64 19.1	1	7.19		0.03		-0.46		2	B4 V	401
103046	+9 02552	IDS11468N0923	B	11 49 22	+09 06.1	1	7.93		0.49		0.02		3	K0 III-IV	3016
307298	−59 03869	LSS 2522		11 49 22	−60 00.8	2	10.35	.015	0.06	.006	-0.56	.014	5	B5	540,1747
		LF14 # 26		11 49 22	−61 01.7	1	11.52		0.55		0.11		2	F5 V	137
103047	+9 02553	IDS11468N0923	A	11 49 23	+09 06.5	1	7.40		0.97		0.70		3	K0	3016
	+35 02289			11 49 23	+34 30.7	1	10.07		1.06		0.92		1	K0 III	280
103079	−64 01724	HR 4549	★ AB	11 49 24	−64 55.7	7	4.89	.010	-0.12	.005	-0.56	.023	26	B4 V	6,15,26,1075,1637*
	+43 02160			11 49 25	+42 41.7	1	10.00		1.12				4	K0 III	280
		KUV 11495+3925		11 49 28	+39 25.1	1	15.47		-0.19		-1.16		1	sdB	1708
		KUV 352 - 1		11 49 28	+39 25.1	1	15.35		0.03		-0.46		1	sdB	974
307220	−60 03443			11 49 28	−60 48.1	1	11.17		0.24		0.16		2	A2	807
	−61 02713			11 49 30	−62 16.3	1	10.92		1.06		0.64		2	K0 III	137
307293	−59 03870	LSS 2523		11 49 31	−59 42.4	2	10.03	.043	0.05	.018	-0.64	.000	5	B3	540,1747
103078	−53 04799			11 49 32	−54 07.8	1	8.04		0.23		0.16		1	A4/5 V	1771
103077	−48 06899			11 49 33	−49 07.4	1	6.98		-0.10		-0.51		3	B5 V	1770
103072	+19 02511	G 57 - 27		11 49 34	+19 02.3	1	8.42		0.86		0.55		1	K2	333,1620
307231	−60 03444			11 49 34	−60 58.3	1	10.90		0.17		0.04		2	A0	807
103070	+34 02265			11 49 36	+34 07.6	1	8.61		1.08				2	K0 III	280
103069	+50 01871			11 49 38	+50 12.6	1	7.10		1.30		1.34		2	K1 III	172
		SW Crt		11 49 39	−24 12.5	1	16.41		0.40		0.22		1		1471
103100	−32 08370			11 49 39	−32 44.1	1	10.07		0.66		0.16		1	G3 V	78
103090	−53 04801			11 49 39	−53 29.6	1	9.52		0.05		0.05		1	B9.5 V	1770
103103	−58 03890			11 49 40	−58 46.4	1	8.10		0.24		0.17		2	A3 V	1771
	+31 02303			11 49 41	+31 01.1	1	9.59		0.85				2	G6 III	280
103101	−56 04836	HR 4551	★ A	11 49 41	−56 42.6	3	5.57	.008	0.07	.000	0.08		8	A2 V	15,1771,2012
103105	−60 03446			11 49 41	−60 46.5	1	10.00		0.14		-0.36		2	B5/7 III	807
	+14 02443			11 49 44	+13 45.9	1	8.77		1.46		1.66		1	K5	280
238052	+59 01406			11 49 44	+58 39.5	1	9.17		1.22		1.13		2	K5	1733
	+15 02386			11 49 45	+14 30.7	1	9.69		1.12		0.94		1	G0	280
	+25 02427			11 49 46	+24 45.6	1	10.28		1.08		1.01		1	K0 III	280
		CS 15625 # 25		11 49 47	+24 40.6	1	13.36		0.44		-0.09		1		1744
103112	+10 02347	IDS11472N1031		11 49 48	+10 13.5	1	7.54		1.05		0.89		2	K0	3008
	+13 02469			11 49 48	+13 09.6	1	10.16		1.18		1.22		1	K0	280
307232	−60 03450			11 49 49	−60 57.5	1	11.10		0.08		-0.44		2	A2	807
103137	−64 01725	UU Mus		11 49 50	−65 07.5	2	9.29	.109	0.93	.093	0.61	.011	2	F7/G0p	689,1587
	−60 03451			11 49 52	−61 05.1	1	11.26		0.17		-0.28		2	B6	807
309210	−61 02722	LSS 2524		11 49 53	−62 24.2	1	10.76		0.13		-0.50		4	B3	1747
		LF14 # 28		11 49 55	−62 07.3	1	12.07		0.97		0.60		1		137
103128	+23 02399			11 49 58	+22 59.4	1	9.20		0.93		0.59		1	K0	280
		CS 15625 # 11		11 49 59	+27 53.2	1	12.62		1.00		0.78		1		1744
		G 121 - 27		11 50 00	+27 47.8	3	12.29	.025	0.93	.010	0.48	.040	9		203,1663,3062
103146	−60 03454	VZ Cen		11 50 00	−61 14.8	1	8.32		0.02		-0.73		3	B2 III/IV	6
		CS 15625 # 24		11 50 01	+23 36.7	1	11.81		0.47		-0.06		1		1744
103140	+40 02472			11 50 03	+40 24.6	2	8.62	.000	0.78	.005	0.51		4	K0 III	280,1569
103170	−62 02373			11 50 04	−62 53.5	1	7.70		1.19				4	K1wG3 IV	2012
103167	−60 03457			11 50 05	−60 55.8	1	9.66		0.05		-0.58		2	B3/5 III	807
103095	+38 02285	HR 4550, CF UMa	★	11 50 06	+38 04.7	21	6.45	.008	0.75	.007	0.17	.011	233	G8 Vp	1,15,22,130,172,667*
103165	−54 04812	IDS11476S5423	A	11 50 06	−54 39.6	1	9.41		0.21		0.14		1	A4 V	1771
103169	−62 02374			11 50 06	−62 30.1	1	9.65		0.04		-0.58		1	B2/3 III	401
	+31 02306			11 50 08	+30 31.1	1	9.10		0.94				2	G6 III	280
103182	−61 02729			11 50 10	−61 57.4	1	7.22		-0.02		-0.51		2	B3 III	401
103152	+16 02307	IDS11476N1559	A	11 50 12	+15 42.9	1	6.84		0.25		0.12		3	A2	270
103151	+28 02052			11 50 12	+28 28.2	1	8.34		1.17				3	K2 III	280
		LP 673 - 114		11 50 12	−06 04.1	1	12.17		0.83		0.45		2		1696
103154	−6 03469	S Crt		11 50 12	−07 19.1	1	9.27		1.40		0.62		2	M3	1375
	−60 03459			11 50 12	−60 53.0	1	10.45		1.33		1.15		2	K6	807
		KS 643		11 50 12	−63 07.	1	11.42		1.16		0.73		2		540
	+22 02416			11 50 13	+21 36.0	1	10.25		1.02		0.90		1	G5	280
310918	−64 01727			11 50 15	−64 59.6	1	10.42		0.30		0.29		1	A2	1480
103191	−23 10243			11 50 16	−24 11.8	2	8.46	.036	0.71	.041	0.17		5	G5 IV	78,2033
	+15 02390			11 50 17	+14 39.4	1	10.26		1.00		0.83		1	G5	280
	+42 02252			11 50 17	+41 42.2	2	9.73	.000	0.74	.010	0.46		5	G7 III	280,1569
	+16 02308			11 50 18	+16 18.8	1	10.04		1.56		1.92		1		280
103171	+14 02444			11 50 19	+14 01.4	1	8.55		1.10		1.04		1	K2	280
	+16 02309			11 50 21	+16 01.7	1	9.16		1.24		1.46		1	K0	280
103188	+14 02445			11 50 22	+13 42.0	1	7.51		1.00		0.81		1	K0	280
		LF14 # 29		11 50 22	−61 12.5	1	12.99		1.06		0.80		1		137
103187	+47 01909			11 50 23	+46 50.2	2	8.15	.010	0.84	.005	0.51	.015	3	G6 III	172,8038
103192	−33 08018	HR 4552, β Hya	★ AB	11 50 23	−33 37.8	5	4.28	.004	-0.10	.004	-0.33	.005	21	Ap Si	15,1075,2012,3023,8015
103209	−50 06437			11 50 24	−50 48.7	1	8.97		0.31		0.17		1	A8/9 V	1771
	−49 06574			11 50 25	−50 24.8	1	10.34		1.08		0.90		1		444
	−60 03462			11 50 25	−60 55.1	1	10.74		1.09		0.88		1	K2 III	137

Table 1   633

HD	DM	Other Id	N Rem	α_{1950}	δ_{1950}	S	V	σ_V	B–V	σ_{B-V}	U–B	σ_{U-B}	n	Spectrum	References
	+38 02286			11 50 28	+37 51.1	1	9.69		1.15		1.06		2	K0 III	280
103208	−50 06438			11 50 28	−50 30.7	1	9.18		1.13		1.00		1	K1 III	444
103221	−49 06575			11 50 29	−49 43.6	1	8.45		0.18		0.11		2	A5 V	1771
		MCC 622		11 50 30	+19 13.0	1	11.70		1.51				1	M0	1017
	−49 06576			11 50 31	−50 24.0	1	9.81		1.19		1.30		1	K5	444
103224	−63 02021			11 50 31	−64 08.4	1	9.33		0.05		−0.50		2	B2/5	401
	+43 02163			11 50 36	+42 46.0	2	9.88	.000	0.21	.025	0.10	.010	2	A2	272,280
307260	−58 03899			11 50 36	−59 13.0	1	10.35		0.49		−0.06		1	F5 V	137
307300	−59 03894	LSS 2525		11 50 36	−60 06.6	2	10.27	.045	0.04	.010	−0.60	.020	5		540,1747
		AAS78,203 # 1		11 50 36	−64 08.	1	11.28		0.14		−0.57		3		1684
	+27 02065			11 50 38	+27 00.5	2	9.83	.000	1.52	.025	1.88		2	K4 III	280,694
103228	−14 03427	IDS11481S1507	AB	11 50 40	−15 23.7	1	8.15		0.30				4	F0 V	176
	−21 03407			11 50 40	−21 46.4	2	10.39	.005	1.12	.015	1.10	.045	2	K5	78,1696
	−50 06442	IDS11482S5018	AB	11 50 40	−50 34.9	1	9.49		1.57		1.80		1	K2	444
103243	−52 04976			11 50 40	−52 48.8	1	9.20		−0.01		−0.26		2	B9 IV	1770
103126	+87 00099	G 253 - 55		11 50 42	+86 30.3	1	8.28		0.72		0.25		1	G5	1658
		LP 673 - 42		11 50 43	−07 05.3	2	11.88	.014	1.55	.012			5		158,1705
	+21 02371	IDS11482N2053	AB	11 50 45	+20 35.9	1	10.08		0.52				4	F5	193
		LP 907 - 40		11 50 45	−31 07.1	1	13.61		1.70		1.11		1		3078
	+24 02404			11 50 47	+23 57.2	1	9.68		0.60		0.08		1		280,1569
		AA44,469 # 5		11 50 47	−50 33.1	1	12.29		0.66		0.10		1		444
	+37 02225			11 50 48	+37 00.6	1	8.75		1.13				2	K0 III	280
	+42 02253			11 50 51	+41 52.1	1	9.45		1.09				2	K2 III	280
	+43 02164			11 50 51	+42 52.1	1	9.73		0.87				3	G8 III	280
103270	−64 01729			11 50 52	−65 07.6	1	7.33		0.03		−0.52		2	B4 III/V	401
103271	−70 01439			11 50 52	−70 32.9	1	9.23		0.30		0.19		1	A3 V	1771
103266	−34 07760	HR 4553		11 50 55	−34 47.3	3	6.16	.005	0.08	.000	0.08		8	A2 V	15,78,2028
103267	−50 06446			11 50 56	−50 25.1	1	8.60		1.12		1.10		1	K1 III	444
	+34 02267			11 50 57	+33 59.2	1	10.16		1.25		1.37		1	K0 III	280
103306	−73 00903			11 50 57	−74 15.6	1	8.95		0.46				4	F3 IV/V	2012
	+30 02200			11 50 58	+29 36.0	1	9.23		1.15				2	K0 III	280
103262	+4 02537	G 12 - 2	★ A	11 50 59	+03 47.8	1	10.11		0.79		0.37		2	K0	1620
103262	+4 02537	IDS11486N0405	B	11 50 59	+03 47.8	1	11.09		0.97		0.80		2		1620
103283	−48 06922			11 50 59	−48 57.1	1	8.36		−0.08		−0.49		2	B9 II/III	1770
		LF14 # 32		11 50 59	−59 13.8	1	11.93		1.22		0.82		1		137
		Ly 94		11 51 00	−63 04.	1	11.60		0.50		−0.35		3		846
		HIC 57984		11 51 01	+49 52.7	1	11.24		1.32		1.21		5		1723
103246	+74 00476	IDS11483N7419	AB	11 51 02	+74 02.1	2	6.84	.029	0.57	.010	0.14	.010	3	F8	292,3030
	+0 02853			11 51 02	−00 02.3	1	10.09		0.49		−0.04		3	F8	281
	+42 02254			11 51 03	+42 12.6	1	10.58		1.53		1.65		1	K4 III	280
103281	−46 07474			11 51 04	−46 27.8	5	7.22	.004	1.03	.005	0.83	.005	13	K0 III	78,278,1075,2011,2024
309218	−62 02388	BB Cen		11 51 04	−62 34.5	1	9.85		0.88				1	F5	1488
	+37 02226			11 51 05	+37 04.9	1	10.11		1.10		1.03		1	K1 III	280
103295	−27 08365			11 51 05	−28 21.5	2	9.60	.000	0.80	.015	0.27		3	G5/6 III	1594,6006
103304	−61 02751			11 51 05	−62 12.7	2	10.83	.010	0.17	.005	−0.69	.000	6	B3 V	846,1737
103302	−48 06926			11 51 07	−49 02.5	1	8.30		0.06		0.02		1	Ap SrCrEu	1770
307261	−58 03904			11 51 08	−59 12.2	1	10.78		0.14		0.01		1	A1 V	137
		ApJS53,765 T3# 2		11 51 10	−64 57.9	1	10.94		0.78		0.30		1		1480
103289	+19 02514			11 51 11	+19 03.3	1	9.07		1.23		1.28		1	K0	280
103288	+34 02268			11 51 12	+33 53.6	1	7.03		0.31		0.14		1	F0	280
103287	+54 01475	HR 4554		11 51 13	+53 58.4	19	2.44	.005	0.00	.004	0.01	.009	172	A0 V	1,15,61,71,667,1006*
103309	+41 02244			11 51 15	+41 11.5	2	6.74	.010	1.06	.000	0.90		5	K1 III	172,280
		WLS 1200-5 # 6		11 51 15	−05 16.0	1	11.82		0.53		−0.03		2		1375
103338	−64 01731	LSS 2527		11 51 15	−65 00.7	3	7.50	.029	0.13	.004	−0.58	.009	6	B3 Ib	401,846,1684
	+32 02199			11 51 16	+31 30.2	1	9.02		1.15				2	K2 III	280
103313	+1 02624	HR 4555		11 51 17	+00 49.8	3	6.30	.008	0.20	.005	0.15	.007	9	F0 V	15,252,1256
103311	+14 02447			11 51 17	+14 18.5	1	7.74		0.26		0.05		2	F0	1569
	+32 02200			11 51 18	+31 57.2	1	10.00		1.03		0.86		1	G9 III	280
103312	+11 02409			11 51 19	+11 23.9	1	8.51		0.91		0.64		2	K0	1648
103348	−62 02399			11 51 20	−63 15.1	1	9.90		−0.02		−0.67		2	B2/3 III	401
	+33 02165			11 51 21	+33 09.2	1	10.14		0.95		0.64		1	G8 III	280
		AAS12,381 37 # A		11 51 21	+36 50.7	1	9.87		1.01				2		280
103324	+40 02474			11 51 22	+39 53.3	2	8.10	.015	0.99	.005	0.71		4	K0 IV	172,280
	+16 02313			11 51 24	+16 08.7	1	10.23		1.09		1.12		1		280
	+21 02373	CV Leo		11 51 24	+20 46.1	1	9.11		1.56		1.98		2	K7	280
		CS 15625 # 945		11 51 24	+25 31.9	1	12.82		0.56		−0.01		1		1744
	+30 02201			11 51 25	+29 50.4	1	10.89		1.23		1.20		2	K2 V	1723
103321	+72 00550	IDS11487N7229	AB	11 51 25	+72 12.1	2	7.60	.019	0.47	.005	0.02	.015	3	F5	292,3030
	+0 02854			11 51 26	−00 15.1	5	10.11	.003	0.57	.003	0.09	.007	45	G2 V	281,989,1728,1729,6004
		LF14 # 34		11 51 26	−59 59.8	1	12.54		1.28		1.22		1		137
309272	−61 02760	LSS 2528		11 51 26	−61 38.3	2	9.88	.000	0.10	.010	−0.67	.015	6	B5	846,1747
	+23 02402			11 51 27	+22 55.3	1	9.20		1.29		1.44		1	K5	280
103341	−0 02507			11 51 28	−00 45.7	3	8.27	.005	1.12	.005	0.95		13	K0 III	281,5006,6004
		HA 103 # 38		11 51 28	−00 55.8	2	10.70	.003	0.49	.026	−0.01		8	F8	281,6004
307350	−57 05086	LSS 2529		11 51 30	−58 14.6	2	9.62	.027	0.01	.007	−0.67	.003	6	B9	540,976
	+1 02625			11 51 31	+00 37.0	1	10.15		0.93				1	G5	1284
	+32 02201			11 51 31	+32 01.4	1	9.45		1.09				2	K0 III	280
		Ly 96		11 51 32	−61 12.6	1	11.51		0.15		−0.40		2		846
	−61 02763			11 51 32	−61 55.1	1	10.89		0.17		−0.32		4		1684
		G 12 - 4		11 51 33	+10 05.6	2	12.77	.035	1.71	.020			2		333,1620,1705

HD	DM	Other Id	N Rem	α_{1950}	δ_{1950}	S	V	σ_V	B–V	σ_{B-V}	U–B	σ_{U-B}	n	Spectrum	References
103354	+18 02539			11 51 33	+18 26.9	2	7.87	.005	0.47	.005	0.00	.010	6	F8	1569,1648
	+1 02626			11 51 35	+01 14.9	1	10.02		1.21		1.25		2		401
309298	−61 02764	LSS 2530		11 51 38	−62 11.1	2	11.06	.005	0.15	.010	-0.56	.035	5		846,1747
103391	−30 09530			11 51 39	−30 37.7	2	7.30	.010	1.63	.015	1.97		6	M1 III	2012,3040
		HA 103 # 42		11 51 41	−00 55.4	2	11.53	.010	0.75	.010	0.36		5	G5	281,6004
103400	−56 04861	HR 4556		11 51 41	−57 07.9	2	6.05	.005	0.05	.000			7	A0/1 III	15,2012
103376	+19 02515			11 51 42	+18 48.6	3	10.18	.017	-0.13	.019	-0.45	.000	10	A2	1026,1068,1298
		HA 103 # 44		11 51 42	−00 51.9	2	12.18	.002	0.73	.001	0.28		5	G8	281,6004
103396	−45 07373			11 51 42	−45 32.6	2	8.66	.000	0.36	.000	0.04		8	F2 V	1075,2011
	+0 02855			11 51 44	+00 12.5	2	10.42	.010	0.42	.015	-0.04		7	F5	281,6004
103401	−62 02403			11 51 44	−63 18.3	2	9.87	.000	0.02	.000	-0.44	.000	4	B5/7 V	401,401
		LF14 # 35		11 51 47	−59 42.9	1	11.65		1.75		2.10		1		137
		G 12 - 5		11 51 49	+09 05.0	1	12.72		0.86		0.43		1		333,1620
	+18 02540			11 51 49	+18 18.0	1	10.48		1.12		0.94		1		280
	+38 02289			11 51 49	+38 19.0	1	9.44		0.95				3	G9 III	280
	+19 02516			11 51 50	+19 11.8	1	9.49		1.36		1.66		1	K2	280
	+16 02316	IDS11493N1549	AB	11 51 51	+15 32.1	1	10.50		1.18		1.18		2	M0 V	1569
103406	+43 02166			11 51 51	+43 26.5	1	10.02		0.99				2	G9 III	280
309299	−61 02765	LSS 2531		11 51 51	−62 17.5	2	9.82	.015	0.19	.013	-0.69	.037	5	B1 II	540,1747
		LP 553 - 62		11 51 52	+03 14.6	1	11.68		0.40		-0.19		2		1696
103423	−45 07376			11 51 52	−46 12.8	4	7.97	.004	0.98	.005	0.67	.010	18	K0 III	768,977,1075,2011
		LP 793 - 107		11 51 53	−19 08.9	2	12.46	.063	0.49	.010	-0.17	.005	3		1696,3060
103407	+18 02541	IDS11494N1808	AB	11 51 55	+17 51.6	1	9.40		0.27		0.08		4		1068
	+33 02166			11 51 55	+32 44.2	1	10.18		1.18		1.13		1	K0 III	280
103437	−37 07536	HR 4557	⋆ AB	11 51 55	−37 28.3	3	6.45	.005	0.52	.004	-0.05		8	F7 V	15,258,2012
		LF14 # 36		11 51 55	−60 27.0	1	12.94		1.28		1.38		2		137
	−50 06462			11 51 56	−51 06.9	2	10.98	.010	0.92	.005	0.67	.019	3	G5	1696,3077
	−61 02767			11 51 56	−62 18.9	1	10.45		1.07		0.83		2	K0 III	137
		HA 103 # 376		11 51 58	−00 19.0	2	11.85	.001	0.94	.002	0.67		27	K2	281,6004
103432	+20 02658	G 121 - 29	⋆ A	11 51 59	+19 41.4	1	8.22		0.71		0.21		1	G0	3016
		G 12 - 6		11 52 00	+13 52.1	1	13.40		1.02		0.86		2		333,1620
103420	+14 02448			11 52 00	+14 25.8	1	7.80		0.44				1	F5	280
	+25 02431			11 52 01	+25 07.5	1	10.18		1.01		0.83		1		280
	+32 02203			11 52 01	+31 55.6	1	10.28		1.19		1.39		1	K1 III	280
103431	+20 02659	G 121 - 30	⋆ B	11 52 02	+19 42.4	1	8.43		0.76		0.27		1	G5	3016
	+37 02228			11 52 03	+37 24.8	1	9.99		1.47		0.95		1	M9	280
		IRC +40 229		11 52 03	+37 25.2	1	11.41		2.08				1		426
103457	−63 02031			11 52 03	−63 25.3	1	7.76		-0.01		-0.31		2	Ap Si	401
	+28 02059			11 52 04	+28 28.0	3	11.15	.093	0.23	.074	0.11	.033	5	A3	280,1026,1569
103453	−44 07648			11 52 04	−44 48.2	1	9.00		0.05		-0.03		1	A0 V	1770
		CS 15625 # 12		11 52 05	+24 23.2	1	12.98		0.59		0.03		1		1744
103430	+49 02092			11 52 05	+49 12.8	1	7.34		0.95		0.74		2	K0 IV wk1	172
103456	−53 04832	IDS11496S5407	AB	11 52 06	−54 23.8	1	9.03		0.30		0.13		1	A3/5	1771
	+0 02856			11 52 08	−00 09.2	2	10.75	.005	0.43	.015	-0.03		11		281,6004
	+26 02267			11 52 09	+26 04.0	1	10.11		0.96		0.68		1	G8 III	280
103447	+3 02569			11 52 10	+03 25.4	1	9.86		1.02		0.85		1	K2	3072
103462	−25 08930	HR 4558		11 52 10	−25 26.2	1	5.30		0.88				4	G8 III	2006
	+21 02374			11 52 12	+21 17.9	1	9.94		1.44		1.68		1	K7	280
	−13 03470			11 52 12	−13 44.9	1	10.68		1.12		1.08		1	K5 V	3072
309297	−61 02773	LSS 2532		11 52 12	−62 12.6	2	9.73	.000	0.19	.015	-0.60	.024	6		846,1747
		LSS 2533		11 52 14	−62 56.8	2	12.61	.035	1.64	.016	0.35	.019	10	B3 Ib	1684,1684
103465	−59 03922	LSS 2534		11 52 15	−60 24.4	1	8.39		0.47		0.36		5	A8 II/III	8100
103459	−0 02510			11 52 18	−01 10.1	2	7.60	.000	0.68	.005	0.28	.052	4	G5	803,3077
103481	−42 07295			11 52 18	−43 20.0	8	7.43	.005	1.28	.005	1.32	.005	44	K1/2 III	402,460,474,768,977*
		GD 145		11 52 19	+22 46.3	1	14.58		0.45		-0.22		1		3060
		AAS12,381 23 # A		11 52 20	+22 58.4	1	10.20		1.04		1.03		1		280
		GPEC 106		11 52 21	−26 00.5	1	11.08		0.79				2		1719
	+29 02228	G 121 - 31		11 52 22	+29 01.2	2	10.49	.051	1.37	.011	1.29	.039	6	M0 V	1569,7008
		AAS12,381 32 # A		11 52 23	+32 04.5	1	10.89		0.71		0.32		1		280
	+14 02450			11 52 24	+14 11.0	1	9.96		0.92		0.61		1		280
	+26 02268			11 52 25	+26 02.3	1	9.30		1.09		1.01		2	K0 III	280
103486	+0 02858			11 52 27	−00 16.7	4	8.35	.003	0.42	.003	0.09	.006	42	F2	281,989,1729,6004
	+0 02857			11 52 27	−00 28.3	2	10.60	.006	0.59	.014	0.09		8	G2 V	281,6004
103484	+9 02560	HR 4559		11 52 29	+08 43.3	2	5.57	.005	0.94	.000	0.67	.000	7	gK0	15,1417
103493	−55 04711	IDS11500S5532	AB	11 52 29	−55 48.9	1	6.69		0.64				4	G3 IV/V	2012
309266	−60 03496			11 52 29	−61 23.0	1	10.50		0.12		0.11		1	A2 V	137
		AAS12,381 37 # B		11 52 30	+36 48.6	1	11.00		0.17		0.18		1		280
103483	+47 01913	HR 4560, DM UMa	⋆ ABC	11 52 30	+46 45.3	1	6.54		0.11		0.08		5	A3 Vn	1084
103516	−62 02408	LSS 2535		11 52 30	−63 00.0	4	5.91	.014	0.20	.010	-0.06	.085	13	A2 Ib	191,401,2032,8100
		Lod 480 # 20		11 52 34	−58 06.9	1	11.92		0.20		0.11		1	A0	549
103498	+47 01914	HR 4561	⋆ D	11 52 36	+46 44.9	2	7.03	.000	0.01	.005	0.02	.005	7	A1 p CrEu	220,1084
	+33 02168			11 52 37	+32 40.6	1	10.68		1.02		0.93		1	K0 III	280
		HA 103 # 487		11 52 37	−00 45.6	2	11.85	.013	0.66	.001	0.19		27	G3	281,6004
103500	+37 02230	HR 4562		11 52 39	+37 02.1	3	6.48	.023	1.58	.010	1.78	.049	5	M3 III	172,280,3055
		S096 # 4		11 52 40	−64 57.3	1	10.97		1.60				2		1730
	+29 02229			11 52 41	+28 47.9	1	9.31		1.28		1.40		2	K2 III	1569
103522	−2 03438			11 52 42	−03 09.9	1	8.96		0.53		-0.03		2	F8	1375
	−74 00859			11 52 42	−74 44.0	1	10.83		0.39				4	A0	2012
103520	+39 02478			11 52 47	+39 02.1	3	7.02	.018	0.99	.007	0.74	.010	7	K0 III	172,280,833
103543	+26 02270			11 52 50	+25 48.0	4	6.97	.007	1.23	.008	1.24	.008	10	K1 III	105,172,280,8041

Table 1

635

HD	DM	Other Id	N	Rem	α_{1950}	δ_{1950}	S	V	σ_V	B–V	σ_{B-V}	U–B	σ_{U-B}	n	Spectrum	References
		Lod 480 # 23			11 52 50	−58 08.0	1	12.26		0.32		0.09		1	A0	549
		LF14 # 39			11 52 50	−59 42.6	1	12.51		0.10		−0.20		1	B9 V	137
103574	−63 02036				11 52 51	−63 25.5	1	7.98		-0.02		−0.58		2	B2 V	401
103544	+14 02451				11 52 53	+13 53.7	1	8.71		0.35		−0.04		2	F5	1569
103575	−71 01295	IDS11505S7116		A	11 52 53	−71 33.0	1	8.59		0.35		0.18		1	A1 V	1771
	+40 02477				11 52 55	+40 05.2	1	9.69		1.03				2	G8 III	280
103542	+35 02293				11 52 56	+34 48.1	2	8.49	.009	1.00	.005	0.76		4	G8 III	280,8041
	−21 03420				11 52 56	−22 06.3	3	10.17	.020	0.52	.004	-0.14	.000	5	F5	78,516,1064
		HA 103 # 603			11 52 58	−00 04.0	1	11.55		0.45		−0.04		21	F9	281
103576	+48 01985				11 53 02	+47 32.2	2	8.75	.030	0.29	.005	0.05	.005	7	A3	272,308
	−0 02511				11 53 03	−00 47.5	1	10.11		0.52		0.02		2	F9	281
	+22 02419				11 53 04	+22 26.3	1	9.38		1.52		1.62		1	K7	280
	+0 02859				11 53 05	+00 12.0	2	10.07	.008	0.51	.013	-0.01		7	F8	281,6004
		Lod 480 # 18			11 53 05	−58 09.7	1	11.80		0.21		0.13		1	A0	549
		LP 553 - 66			11 53 06	+05 32.4	1	11.91		1.36		1.26		1		1696
103578	+16 02319	HR 4564		⋆ A	11 53 06	+15 55.5	4	5.53	.009	0.11	.007	0.12	.005	15	A3 V	70,270,1022,3050
		LF14 # 40			11 53 06	−61 34.5	1	10.99		1.20		1.08		2	K2 III	137
103596	−27 08384	HR 4565			11 53 07	−28 11.9	4	5.93	.016	1.49	.010	1.88	.022	13	K4 III	15,1637,2029,3005
		BPM 36430			11 53 07	−48 24.0	1	12.85		-0.20		-1.01		2		3065
	+23 02403				11 53 08	+23 16.7	2	11.57	.030	0.28	.015	0.05	.005	3	B5	1026,1744
103577	+23 02404				11 53 09	+22 55.3	1	8.33		1.09		1.04		1	K0	280
	+35 02294				11 53 11	+34 51.0	1	10.20		0.72				2	G5	280
307321	−59 03937	LSS 2536			11 53 11	−60 20.8	2	9.93	.010	0.03	.010	-0.76	.019	6		846,1747
		G 12 - 9			11 53 12	+14 49.9	1	11.80		0.92		0.64		1		333,1620
		SSII 48			11 53 12	+31 42.	1	11.79		0.54		−0.17		1	F2	1298
	+36 02222				11 53 12	+35 43.0	1	9.40		1.41				2	K4 III	280
	+39 02479				11 53 12	+39 22.2	1	10.25		0.67				4		280
		LP 961 - 58			11 53 14	−37 59.9	1	12.00		1.57				1		3062
103619	−43 07370				11 53 14	−44 13.9	3	8.68	.002	0.34	.000	0.21	.012	16	A7/8 IV(m)	977,1075,2011
302005	−57 05106				11 53 15	−58 00.0	1	10.24		0.25		0.18		1	A0	549
302009	−57 05107				11 53 16	−58 00.2	1	11.04		0.25		0.22		1	A0	549
	+29 02231				11 53 18	+28 42.9	1	9.85		0.90		0.58		1	G7 III	280
		Lod 480 # 21			11 53 20	−58 02.9	1	12.08		0.31		0.23		1	A2	549
		AAS12,381 28 # A			11 53 22	+23 03.1	1	10.49		1.00		0.80		1		280
103605	+57 01343	HR 4566			11 53 22	+56 52.6	1	5.84		1.10				1	K1 III	71
103637	−38 07410	HR 4568			11 53 22	−39 24.6	3	6.12	.004	1.01	.004			11	K0 III	15,1075,2032
	+10 02357				11 53 23	+10 07.5	1	8.87		0.06		−0.53		3	A0	272
103639	−52 05024				11 53 23	−52 27.5	1	9.75		0.64		0.12		2	G8 (III) +G	1696
		Lod 480 # 16			11 53 23	−57 59.7	1	11.53		0.42		0.19		1	A5	549
		Lod 480 # 25			11 53 23	−57 59.7	1	12.43		0.58		0.10		1		549
		Lod 480 # 19			11 53 24	−58 06.7	1	11.81		0.68		0.08		1	F8	549
		LP 493 - 76			11 53 26	+12 42.9	1	11.70		0.78		0.41		2		1696
103614	+26 02271				11 53 26	+25 46.2	1	8.60		0.50		−0.03		2	F6 V	3026
	+32 02205				11 53 26	+31 28.8	1	8.95		1.15				2	K2 III	280
	+43 02167				11 53 26	+42 32.7	1	10.05		1.02				3	K0 III	280
		Lod 480 # 22			11 53 26	−58 06.6	1	12.15		0.40		0.22		1	A7	549
307378	−59 03939				11 53 26	−59 40.6	1	10.10		0.07		−0.01		2	A0 III	137
103628	+22 02421	IDS11509N2232		A	11 53 27	+22 15.6	1	8.06		0.51		0.07		2	F7 IV	3024
103628	+22 02421	IDS11509N2232		B	11 53 27	+22 15.6	1	9.64		0.58		0.08		2	F8	3024
309323	−62 02433				11 53 27	−62 36.3	1	10.09		0.04		−0.45		2	B8	401
103612	+41 02248				11 53 28	+40 55.6	2	6.91	.015	1.12	.005	1.02		4	K1 III	172,280
103632	−16 03358	HR 4567			11 53 28	−16 52.3	4	5.17	.003	-0.01	.027	0.01	.026	15	A0 V	3,15,2012,3023
		Lod 480 # 17			11 53 28	−58 06.8	1	11.76		1.48		1.25		1		549
103655	−63 02039				11 53 29	−64 21.3	1	7.82		1.18		0.98		2	G8 II/III	122
103673	−76 00697				11 53 30	−76 43.8	1	8.60		0.74				4	G5 V	2012
	+36 02224				11 53 31	+36 17.3	1	10.16		1.01		0.83		1	K0 III	280
103652	−59 03942	LSS 2537			11 53 31	−59 55.9	1	8.25		0.00		−0.83		2	B1 Iab	846
	+29 02232				11 53 32	+29 23.0	2	10.60	.004	0.21	.000	0.10	.030	7	A5	272,280
103646	+0 02860				11 53 32	−00 31.2	7	9.86	.003	0.37	.003	-0.06	.004	100	F5	281,989,1728,1729,1764*
	+15 02397				11 53 33	+14 30.8	1	10.09		1.60		1.83		1		280
103653	−62 02436				11 53 34	−62 39.4	1	10.11		0.07		−0.02		1	A0 V	137
309317	−62 02437				11 53 35	−62 30.8	1	9.84		-0.01		−0.68		2	B3	401
103643	+58 01345				11 53 36	+58 26.9	1	8.32		1.24		1.44		3	K3 II-III	37
103679	−49 06631				11 53 37	−49 35.0	1	8.25		-0.07		−0.34		4	B9 III(p)	1770
103661	+16 02320	WX Leo			11 53 38	+16 00.5	1	7.94		1.48		1.49		1	M2	280
103644	+37 02231				11 53 38	+36 31.6	3	7.91	.023	1.12	.014	1.03	.024	5	K1 III	172,280,8041
		Lod 480 # 14			11 53 38	−58 09.1	1	11.18		0.43		0.23		1	A1	549
		Lod 480 # 9			11 53 38	−58 09.1	1	11.19		0.37		0.26		1	A3	287
	−57 05111				11 53 38	−58 09.1	2	10.81	.030	0.37	.010	0.30	.020	2	A0	287,549
	+29 02233				11 53 39	+28 42.8	1	9.29		1.21		1.26		1	K0 III	280
	+33 02171				11 53 40	+33 11.0	3	10.59	.012	0.26	.009	-0.06	.006	6	A2	275,1026,1298
		LSS 2538			11 53 40	−62 56.6	1	12.51		1.85		0.37		5	B3 Ib	1684
103660	+29 02234				11 53 41	+29 08.2	3	7.30	.015	1.06	.005	0.95	.009	6	K1 III	172,280,8041
	−45 07400	IDS11512S4513		B	11 53 41	−45 30.5	1	9.88		0.58				4	F5	2033
		L 325 - 214			11 53 42	−48 24.	1	12.85		-0.20		-1.01		1	DA	782
		Lod 480 # 26			11 53 42	−58 09.7	1	12.76		0.48		0.16		1		549
103659	+36 02225	IDS11511N3600		AB	11 53 43	+35 43.6	1	6.64		0.38		0.07		2	F2	1569
103658	+43 02168				11 53 43	+43 18.4	1	10.30		0.74				2	F8	280
	−45 07401	IDS11512S4513		A	11 53 43	−45 29.7	1	9.70		0.55				4	F5	2033
309276	−61 02808				11 53 46	−61 43.9	1	10.45		0.23		0.10		2	A1 V	137

HD	DM	Other Id	N	Rem	α_{1950}	δ_{1950}	S	V	σ_V	B–V	σ_{B-V}	U–B	σ_{U-B}	n	Spectrum	References
103676	+27 02070				11 53 47	+26 57.4	1	6.78		0.35		0.04		2	F2	3077
103683	+42 02256				11 53 48	+42 17.5	2	7.03	.015	0.98	.010	0.77		5	G9 II-III	172,280
	−57 05113				11 53 48	−58 09.8	2	10.95	.020	0.34	.010	0.30	.020	2	A0	287,549
		CS 15625 # 17			11 53 50	+24 44.9	1	12.86		0.64		0.03		1		1744
	−57 05115				11 53 52	−58 02.7	1	11.41		0.23		-0.29		1	B6	549
103685	+14 02452				11 53 53	+14 27.9	1	6.72		1.46				2	K0	280
	−61 02807				11 53 53	−61 52.4	1	11.49		0.16		-0.32		2	B8 V	137
103704	−63 02044				11 53 53	−64 16.0	1	8.73		1.05		0.68		2	G5 III	122
	+25 02436				11 53 54	+25 16.0	1	9.88		0.90		0.56		1	G7 III	280
103681	+58 01346	Z UMa			11 53 54	+58 09.0	1	7.88		1.60		1.01		1	M5 III	3001
103684	+35 02295				11 53 55	+35 10.9	2	7.32	.015	1.12	.010	0.92		3	K1 III	280,8041
103691	+47 01915				11 53 55	+46 58.5	1	7.95		1.11		1.03		2	K0 IV	172
		HA 103 # 517			11 53 56	−00 07.6	2	11.10	.000	0.41	.005	-0.05		5	F5	281,5006
302006	−57 05116				11 53 56	−57 57.0	1	10.53		0.26		0.19		1	A2	549
309275	−61 02809				11 53 56	−61 40.0	1	9.53		1.15		0.91		2	G8 III	137
		HA 103 # 518			11 53 57	−00 15.9	1	11.22		0.64		0.04		4	G3	281
	+13 02473				11 53 59	+12 37.6	1	8.70		1.48		1.86		1	K5	280
103714	−63 02046				11 53 59	−63 30.8	1	8.24		0.07		-0.17		2	B9 II/III	401
103715	−70 01445	LSS 2539			11 53 59	−71 22.4	2	9.09	.029	0.18	.005	-0.77	.000	6	B0/3ne	400,846
		CS 15625 # 16			11 54 01	+25 10.3	1	14.44		0.01		0.08		1		1744
302007	−57 05117				11 54 01	−57 58.8	1	10.58		0.49		0.08		1	G5	549
103732	−57 05118				11 54 01	−58 09.2	1	9.75		0.20		0.18		1	A0 V	549
310911	−64 01741				11 54 01	−64 31.0	1	10.54		0.40		0.18		2	A0	122
103723	−20 03540				11 54 04	−21 08.4	2	10.06	.000	0.52	.005	-0.09	.010	2	F5/6 V	78,3077
302008	−57 05120				11 54 04	−58 00.5	1	10.40		0.55		0.05		1	K0	549
	+31 02313				11 54 05	+31 01.3	1	10.10		1.13		1.03		1	G9 III	280
103734	−62 02444	LSS 2540			11 54 05	−62 59.1	2	8.59	.005	0.50	.015	-0.13	.015	5	B9 Ibp Si	846,8100
	+28 02062				11 54 06	+28 20.7	1	9.92		1.18		1.28		1	K1 III	280
103707	+41 02250	IDS11515N4135	AB		11 54 06	+41 18.2	1	8.53		1.07				2	K0 III	280
103729	−44 07666				11 54 06	−45 16.1	9	8.53	.003	1.53	.003	1.85	.005	172	K3/4 (III)	863,977,1460,1628,1657,1673*
		CS 15625 # 938			11 54 07	+23 52.3	1	12.29		0.34		-0.08		1		1744
103720	−1 02594				11 54 08	−02 30.0	1	9.58		0.94		0.69		1	K3 V	3072
	+31 02314				11 54 09	+31 05.1	2	10.84	.037	0.21	.005	0.08	.023	4		1026,1068
103742	−31 09364	IDS11516S3143	B		11 54 10	−31 59.4	1	7.63		0.64		0.14		1	G3 V	78
		LF14 # 47			11 54 10	−61 10.7	1	12.87		0.27		-0.08		2		137
	+13 02474				11 54 11	+12 52.4	1	10.49		1.08		1.01		1		280
		G 57 - 29			11 54 11	+18 39.0	2	15.60	.044	0.26	.034	-0.59	.010	6		940,3060
103743	−31 09365	IDS11516S3143	A		11 54 11	−31 59.3	1	7.80		0.67		0.18		1	G3 V	78
103745	−45 07405				11 54 11	−45 25.5	1	9.26		0.34				4		2011
	−57 05121				11 54 11	−58 10.4	1	10.99		0.38		0.30		1	A6	549
103719	+33 02172				11 54 12	+32 28.9	3	7.98	.010	1.17	.007	1.21	.010	5	K2 III	172,280,8041
		HA 103 # 626			11 54 12	−00 06.6	2	11.84	.002	0.42	.003	-0.05	.003	34	G1	281,1764
103746	−46 07521	HR 4570			11 54 12	−46 47.7	12	6.26	.009	0.41	.006	0.04	.008	54	F5 V	14,15,78,182,278,388*
		Wray 912			11 54 12	−61 56.9	1	12.60		0.64		-0.12		4		1684
		LP 434 - 7			11 54 14	+18 17.2	1	12.49		0.78		0.37		2		1696
		Lod 480 # 24			11 54 14	−58 06.7	1	12.42		0.49		0.40		1		549
	+13 02475				11 54 15	+12 30.5	1	9.67		1.55		1.87		1		280
		LF14 # 48			11 54 15	−60 04.3	1	11.42		0.37		0.14		1	F0 V	137
	+0 02861				11 54 16	+00 20.4	1	11.03		0.97				1		1284
	+14 02453	G 12 - 11			11 54 17	+13 39.6	1	9.95		0.70		0.18		1	G5	333,1620
103736	+62 01204	HR 4569			11 54 17	+61 49.7	3	6.23	.019	0.96	.008	0.70	.000	7	G8 III	71,1501,1733
103738	+53 01517				11 54 18	+52 33.0	1	8.44		1.37		1.53		3	K0	1566
	+0 02862				11 54 18	−00 14.7	5	10.90	.005	1.09	.007	0.95	.007	36	K0 III	281,989,1728,1729,5006
307382	−59 03957				11 54 19	−59 53.0	1	9.74		1.11		0.90		1	K0 III	137
	+40 02480				11 54 20	+39 44.3	1	10.96		0.21		0.09		1		280
103777	−57 05122				11 54 21	−58 08.2	1	8.39		1.28		1.26		1	K1 III	549
309294	−61 02815				11 54 21	−62 06.4	2	10.52	.003	0.10	.006	-0.74	.015	3	B1.5II-Ib	540,1737
103764	−61 02814	Ru 97 - 34			11 54 21	−62 19.9	2	9.50	.020	-0.04	.020	-0.75	.050	4	B2 II/III	401,478
103762	−59 03959				11 54 23	−59 32.6	2	8.61	.016	-0.02	.011	-0.73	.006	4	B3 III	540,976
		CS 15625 # 13			11 54 24	+27 39.5	1	14.03		0.16		0.13		1		1744
103791	−57 05123				11 54 25	−58 02.4	1	8.86		0.05		-0.02		1	B9 V	549
		HA 103 # 528			11 54 26	−00 09.4	1	11.80		0.58		0.01		24	G8	281
103779	−62 02455	LSS 2543			11 54 26	−62 58.2	7	7.21	.015	0.00	.016	-0.85	.017	19	B0/1 II	6,158,191,540,976*
		CS 15625 # 15			11 54 27	+25 43.0	1	13.29		0.52		-0.06		1		1744
		HA 103 # 529			11 54 28	−00 10.4	1	12.25		0.99		0.81		26		281
307347	−60 03525	LSS 2544			11 54 28	−60 42.3	3	9.94	.005	0.13	.015	-0.52	.015	7		846,1684,1747
103769	+42 02257				11 54 30	+41 40.2	1	8.96		0.96				2	G8 III	280
309336	−62 02456				11 54 30	−62 39.0	2	10.69	.005	0.22	.015	-0.72	.020	6	B5	279,846
		HA 103 # 194			11 54 31	−00 45.4	1	12.30		0.70		0.23		2	G7	281
103789	−32 08413	HR 4571, LV Hya			11 54 31	−33 02.2	3	6.20	.004	-0.04	.006	-0.11		9	B9.5V	15,1770,2012
103770	+41 02252				11 54 32	+40 34.0	2	7.13	.015	1.04	.010	0.87		4	K1 III	172,280
103784	−8 03264				11 54 33	−09 08.5	1	6.17		-0.05		-0.10		1	A3	1770
103780	+34 02273				11 54 34	+33 37.9	1	8.80		1.15				2	K1 III	280
	−62 02458				11 54 34	−62 39.0	2	11.47	.015	0.29	.010	-0.37	.015	6		279,846
	−57 05127				11 54 36	−58 03.4	1	11.17		0.76		0.22		1	F8	549
103826	−57 05128				11 54 36	−58 10.6	2	8.48	.010	0.36	.005	0.27	.015	2	A2mA5-A7	549,1771
103807	−62 02459	LSS 2546			11 54 36	−63 23.9	1	7.59		0.65		0.34		2	F5 Ib	8100
103781	+27 02071				11 54 38	+27 26.3	1	8.83		1.02				2	G9 III	280
103799	+41 02253	HR 4572			11 54 40	+40 37.4	3	6.62	.012	0.47	.006	-0.03	.019	6	F6 V	272,1569,3037
103811	+44 02144				11 54 41	+43 38.6	1	9.41		1.07				2	K1 III	280

Table 1 637

HD	DM	Other Id	N	Rem	α_{1950}	δ_{1950}	S	V	σ_V	B–V	σ_{B-V}	U–B	σ_{U-B}	n	Spectrum	References
	−6 03481				11 54 42	−06 42.0	1	9.47		1.00		0.84		1	K8	3072
103822	−34 07803				11 54 42	−34 54.9	1	7.76		1.50		1.85		7	K5 III	1673
	−0 02513				11 54 43	−00 46.1	1	11.32		0.68		0.01		2	dG6	281
		Ru 97 - 18			11 54 43	−62 27.4	1	13.78		0.33		0.29		2		411
103800	+23 02407				11 54 44	+22 56.3	1	8.22		0.10		0.09		3	A2 V	1068
103836	−25 08950				11 54 44	−25 51.9	1	8.92		1.04		0.87		1	K3/4 V	78
		HA 103 # 646			11 54 46	−00 04.4	1	12.47		0.71		0.23		28	G5	281
	+39 02480				11 54 47	+39 04.5	1	9.44		0.97				2	G8 III	280
		Ru 97 - 37			11 54 48	−62 22.	1	12.00		1.10		0.74		2		478
		Ru 97 - 38			11 54 48	−62 22.	1	12.36		0.14		0.10		2		478
		Ru 97 - 39			11 54 48	−62 22.	1	12.82		0.70		0.12		2		478
		Ru 97 - 40			11 54 48	−62 22.	1	12.92		0.68		0.15		2		478
		Ru 97 - 41			11 54 48	−62 22.	1	14.35		0.46		0.26		1		478
		Ru 97 - 42			11 54 48	−62 22.	1	14.36		0.45		0.18		2		478
		Ru 97 - 43			11 54 48	−62 22.	1	14.56		0.84				1		478
		Ru 97 - 44			11 54 48	−62 22.	1	14.78		0.39				1		478
		Ru 97 - 45			11 54 48	−62 22.	1	14.94		0.45				1		478
		Ru 97 - 46			11 54 48	−62 22.	1	15.41		1.42				1		478
		Ru 97 - 47			11 54 48	−62 22.	1	16.11		0.73		0.14		1		478
		Ru 97 - 48			11 54 48	−62 22.	1	16.20		0.91		-0.17		1		478
		Ru 97 - 49			11 54 48	−62 22.	1	16.29		1.21				1		478
		Ru 97 - 51			11 54 48	−62 22.	1	16.97		0.74				1		478
		Ru 97 - 13			11 54 48	−62 25.0	2	13.61	.035	0.61	.010	0.08	.015	4		411,478
		Ru 97 - 17			11 54 48	−62 27.2	1	12.52		0.66		0.19		2		411
		HBC 591, T Cha		★	11 54 48	−79 05.1	3	10.48	.175	1.17	.062	0.70	.094	3	F5	776,825,1763
	−37 07566	LTT 4448			11 54 49	−37 54.4	1	10.41		1.65		1.92		1		78
103856	−48 06984	IDS11523S4848	AB		11 54 49	−49 04.9	1	6.67		0.36				4	F0 V	2012
		Ru 97 - 15			11 54 49	−62 26.0	2	13.18	.020	0.39	.005	0.20	.025	4		411,478
103833	+0 02864				11 54 50	+00 23.6	1	9.96		0.96				1	K5	1284
	+18 02544				11 54 50	+17 40.9	1	10.09		1.19		1.26		1		280
		HA 103 # 95			11 54 50	−00 48.9	1	10.72		1.18		1.18		2	gG8	281
103813	+27 02073				11 54 51	+27 02.6	2	7.47	.010	0.96	.015	0.65	.010	5	G7 IV-V	172,3026
		CCS 1937			11 54 51	−56 45.2	1	10.46		4.70				2		864
		Ru 97 - 10			11 54 51	−62 24.9	1	14.11		0.30		0.26		2		411
		Ru 97 - 16			11 54 51	−62 26.5	2	13.24	.015	0.40	.019	0.24	.054	3		411,478
	−0 02514				11 54 53	−00 40.2	2	11.18	.000	0.65	.011	0.10		5	G5	281,6004
		Ru 97 - 9			11 54 53	−62 24.7	1	13.30		0.41		0.18		2		411
103860	−63 02054				11 54 53	−64 19.4	1	9.14		0.22		0.14		2	A2 V	122
	+30 02205				11 54 54	+29 33.2	2	9.17	.010	1.38	.005	1.69		5	K3 III	196,280
103845	+48 01987				11 54 54	+48 04.0	2	8.30	.005	1.05	.010	0.93	.000	3	K0 II-III	172,8041
		Ru 97 - 36			11 54 54	−62 20.5	1	10.51		0.07		-0.34		2		478
		Ru 97 - 33			11 54 54	−62 26.9	1	13.79		0.39		0.34		2		411
103829	+54 01480				11 54 56	+53 50.0	1	9.24		0.67		0.25		3	F8	1566
		HA 103 # 97			11 54 56	−00 51.9	1	11.79		0.57		0.08		2	G2	281
		Ru 97 - 14			11 54 56	−62 25.4	2	12.41	.030	1.38	.000	1.08	.050	4		411,478
		Ru 97 - 19			11 54 57	−62 26.0	1	13.70		0.34		0.18		2		411
		Ru 97 - 11			11 54 59	−62 24.4	1	11.95		1.05		0.83		2		411
		Ru 97 - 12			11 55 00	−62 24.8	1	12.27		0.16		0.06		2		411
310910	−63 02057				11 55 00	−63 25.0	1	10.24		0.72		0.22		2	G5	122
103873	−64 01747				11 55 00	−64 28.0	1	9.70		0.27		0.15		2	A1 V	122
		G 12 - 12		★ V	11 55 01	+12 06.2	1	11.84		1.46		1.13		2		333,1620
		Ru 97 - 8			11 55 01	−62 23.3	2	12.29	.015	0.37	.030	0.26	.020	4		411,478
		Ru 97 - 24			11 55 02	−62 24.5	1	13.64		0.48		0.38		2		411
103872	−60 03532	LSS 2547			11 55 03	−61 16.3	1	8.89		0.10		-0.47		2	B5e	846
		Ru 97 - 25			11 55 03	−62 24.5	1	13.96		0.20		0.15		2		411
		Ru 97 - 26			11 55 03	−62 25.4	1	13.42		0.47		0.31		2		411
		Ru 97 - 20			11 55 03	−62 26.7	2	12.55	.015	0.99	.015	0.65	.088	3		411,478
		Ru 97 - 6			11 55 05	−62 22.3	1	13.73		0.31		0.23		2		411
		Ru 97 - 7			11 55 05	−62 22.9	1	13.02		0.48		0.37		2		411
		Ru 97 - 21			11 55 05	−62 27.2	1	12.36		1.76				1		411
		LF14 # 51			11 55 06	−59 26.2	1	12.69		1.52		1.36		1		137
		Ru 97 - 27			11 55 06	−62 25.3	1	13.29		0.43		0.27		2		411
309335	−62 02465				11 55 06	−62 41.5	1	10.79		0.52		0.06		1	F5V	137
	−58 03948				11 55 07	−58 49.6	1	10.81		0.98		0.77		1	K0 III	137
103884	−61 02829	HR 4573			11 55 08	−62 10.2	6	5.57	.007	-0.16	.007	-0.66	.010	18	B4 V	6,15,26,1732,2012,8023
		Ru 97 - 32			11 55 08	−62 22.4	1	12.47		0.26		0.24		2		411
103877	+18 02546				11 55 09	+17 44.8	2	6.78	.004	0.39	.000	0.18	.008	14	Am	1569,8071
103733	−60 03516				11 55 09	−60 38.9	1	10.07		0.15		0.09		1	A1 V	137
	−60 03535				11 55 09	−60 52.1	1	11.06		0.28		-0.54		2		846
		Ru 97 - 28			11 55 09	−62 24.9	2	13.18	.020	1.24	.005	0.81	.115	2		411,478
	+24 02410				11 55 10	+23 49.0	1	10.03		1.00		0.76		1		280
		Ru 97 - 22			11 55 10	−62 26.7	1	13.56		0.47		0.24		2		411
	+26 02277				11 55 11	+26 06.4	1	10.30		1.26		1.43		1	K1 III	280
		Ru 97 - 5			11 55 11	−62 21.1	2	13.22	.020	0.38	.010	0.26	.015	4		411,478
		Ru 97 - 29			11 55 11	−62 24.9	2	13.99	.010	0.49	.005	0.26	.015	3		411,478
		AAS12,381 42 # B	A		11 55 13	+41 54.5	1	11.61		0.04		0.18		1		280
		AAS12,381 42 # B	B		11 55 13	+41 54.5	1	11.91		0.59		-0.01		1		280
103899	−44 07677	IDS11527S4447	AB		11 55 13	−45 03.7	3	8.86	.007	0.29	.007	0.21	.004	16	A5mA7-A9	977,1075,2011
309293	−61 02830				11 55 14	−62 05.8	3	9.77	.019	0.36	.002	-0.59	.005	7	B0 IV	540,846,1684
103911	−46 07536				11 55 15	−46 30.5	3	8.83	.006	0.18	.005	0.17	.007	16	A5 V	977,1075,2011

HD	DM	Other Id	N Rem	α_{1950}	δ_{1950}	S	V	σ_V	B–V	σ_{B-V}	U–B	σ_{U-B}	n	Spectrum	References
		Ru 97 - 23		11 55 16	−62 26.5	1	13.55		0.22		0.12		2		411
103910	−43 07386	IDS11528S4310	AB	11 55 17	−43 26.5	4	7.98	.006	0.24	.004	0.15	.004	17	A2 III	977,1075,1770,2011
		Ru 97 - 4		11 55 17	−62 21.0	2	12.42	.035	1.25	.005	1.11	.055	4		411,478
103925	−74 00863			11 55 19	−75 05.9	2	8.81	.005	1.40	.005	1.31		7	K2/3 III	236,2012
	+25 02441	IDS11528N2531	AB	11 55 20	+25 14.7	1	9.60		0.91		0.61		2	K0	1569
	+41 02257			11 55 20	+41 06.8	1	10.04		0.36		−0.04		2		1569
		LP 734 - 54		11 55 20	−09 32.1	1	11.19		0.94		0.65		1		1696
		Ru 97 - 3		11 55 20	−62 20.4	2	12.11	.005	1.22	.015	0.96	.025	4		411,478
		Ru 97 - 1		11 55 21	−62 21.0	2	12.26	.022	0.23	.016	0.18	.013	47		411,478
		Ru 97 - 50		11 55 22	−62 22.0	1	16.91		0.58				1		478
		Ru 97 - 2		11 55 25	−62 21.4	2	13.02	.070	0.35	.020	0.22	.015	4		411,478
		Ru 97 - 30		11 55 25	−62 25.5	1	12.70		0.38		0.18		2		411
		LF14 # 53		11 55 25	−63 06.5	1	10.75		1.06		0.93		2	K0 III	137
309326	−62 02467			11 55 26	−62 32.7	1	10.56		0.11		−0.03		1	A0 V	137
103914	+14 02457			11 55 27	+14 18.9	2	8.43	.005	0.56	.005	0.05	.010	4	G5	1569,1648
103912	+49 02098	G 122 - 57		11 55 27	+48 29.0	2	8.38	.015	0.86	.000	0.43	.019	3	G7 IV wl1	172,1658
103932	−26 08883			11 55 27	−27 25.2	5	6.98	.012	1.13	.015	1.10	.030	19	K4 V	78,258,803,1075,3078
103924	−64 01751	LSS 2549		11 55 27	−65 02.6	1	7.48		0.52		0.31		2	F0 Ib/II	8100
		Ru 97 - 31		11 55 28	−62 25.0	1	12.94		0.31		0.10		2		411
	+13 02477			11 55 29	+13 10.2	1	10.32		0.97		0.75		1		280
103913	+25 02442			11 55 30	+25 25.0	2	8.28	.005	0.56	.005	0.05	.010	4	F8	105,1569
103939	−63 02062			11 55 31	−64 24.6	1	9.51		0.38		0.24		2	A2/3 IV	122
103928	+33 02174	HR 4574	★ AB	11 55 33	+32 33.2	1	6.42		0.29				2	A9 V	70
		LF14 # 55		11 55 34	−59 47.3	1	11.92		0.08		−0.22		1	B8 V	137
		Ru 98 - 16		11 55 34	−64 18.4	1	11.81		0.32		0.34		2		251
		Ru 97 - 35		11 55 38	−62 22.6	1	10.41		1.11		0.96		2		478
103949	−23 10299			11 55 39	−23 38.7	1	8.77		0.99		0.82		1	K3 V	3072
	+18 02547			11 55 41	+18 16.1	1	10.15		0.98		0.77		1		280
	+24 02414			11 55 42	+23 28.0	1	9.37		1.50		1.93		1	K7	280
	+30 02208			11 55 42	+30 03.2	1	10.16		0.57		0.06		3	F8	196
	−35 07576			11 55 42	−35 55.0	2	10.11	.002	0.52	.010	0.12		6	F5	1594,6006
103975	−47 07273			11 55 43	−47 41.9	3	6.76	.003	0.53	.004	−0.01	.013	10	F7 V	278,1075,2011
103961	−55 04751	HR 4576		11 55 43	−56 02.3	4	5.43	.008	−0.08	.000	−0.30		12	B8 III	15,1771,2013,2029
103955	+22 02425			11 55 44	+22 03.3	1	8.23		0.94		0.64		2	K0	1569
	+33 02175			11 55 45	+32 36.6	1	8.91		0.87				2	G5 III	280
	+35 02298			11 55 45	+35 27.1	1	9.56		0.45		−0.07		2	F5	1569
103953	+62 01206	HR 4575		11 55 45	+61 44.6	1	6.72		1.07		0.89		2	K0 III	1733
		CS 15625 # 20		11 55 46	+24 14.3	1	13.36		0.59		0.10		1		1744
103943	+42 02259			11 55 47	+42 06.6	1	9.77		0.64				2	G5	280
	−58 03965	LSS 2550		11 55 47	−59 09.3	3	10.38	.008	0.04	.015	−0.49	.011	9		846,1684,1747
103974	−40 07041	HR 4577	★ AB	11 55 48	−40 40.1	5	6.78	.011	0.96	.006	0.70		21	K1 III	15,404,1075,2013,2028
103979	−63 02067	Ru 98 - 15		11 55 48	−64 15.6	2	9.75	.020	0.19	.005	0.14	.030	4	A0	122,251
103962	−59 03971	LSS 2551		11 55 49	−59 28.6	2	7.35	.010	0.02	.000	−0.05	.045	4	B9 IIIp Si	1771,8100
311051	−63 02068	Ru 98 - 2		11 55 51	−64 20.2	1	9.86		0.25		0.22		2	A0	251
103966	+27 02076			11 55 53	+27 06.9	1	8.42		0.09		0.08		3	A3 V	1068
		CS 15625 # 21		11 55 54	+24 08.2	1	13.87		0.85		0.42		1		1744
103965	+32 02207			11 55 54	+32 12.0	3	7.89	.014	1.07	.005	0.94	.005	5	K1 III	172,280,8041
		LP 907 - 80		11 55 54	−27 23.3	1	12.98		0.82		0.31		2		1696
		Ru 98 - 3		11 55 54	−64 19.5	1	11.54		0.34		0.24		2		251
103994	−47 07275	IDS11534S4718	A	11 55 55	−47 34.4	1	8.60		0.03		−0.10		2	B9 V	1771
103994	−47 07275	IDS11534S4718	AB	11 55 55	−47 34.4	1	8.55		0.02		−0.10		1	B9 V	1770
103991	−22 03254			11 55 56	−23 21.6	2	9.50	.025	0.92	.007	0.66	.008	2	K2 V	78,3072
		CS 15625 # 29		11 55 57	+25 33.1	1	13.06		0.39		−0.06		1		1744
	−56 04918			11 55 57	−57 12.5	1	11.22		0.12		−0.28		4	B8	1684
103984	+48 01989			11 55 58	+48 02.6	2	6.94	.005	0.39	.005	−0.01	.000	5	F3 V	1501,1601
104004	−32 08429			11 55 58	−33 24.7	2	10.41	.010	0.56	.006	−0.06	.010	3	G3/5 V	78,1696
104006	−41 06879	IDS11535S4121	A	11 55 58	−41 38.4	4	8.92	.009	0.81	.004	0.34	.015	10	K1 V	1097,2012,2039,3077
309308	−61 02841			11 55 59	−62 23.0	1	11.50		0.12		−0.36		1	B8 V	137
		Ru 98 - 14		11 55 59	−64 12.5	1	11.72		0.13		0.01		2		251
104015	−70 01448			11 55 59	−70 27.7	2	6.98	.033	0.00	.009	−0.56		8	B2/3 IIIn	158,1075
		Feige 52		11 56 00	+12 31.	1	12.87		0.44		−0.07		1		3060
	+34 02276			11 56 01	+33 30.4	1	9.79		0.71				2	G5 III	280
103985	+40 02485			11 56 02	+40 27.5	1	8.21		1.37				2	K3 III	280
		G 197 - 38	AB	11 56 03	+59 50.0	1	11.03		1.34		1.15		3	M0	7010
311052	−63 02069	Ru 98 - 4		11 56 03	−64 19.2	2	9.93	.015	0.19	.005	0.16	.030	4	A2	122,251
104013	−63 02070	Ru 98 - 1		11 56 04	−64 21.1	1	9.55		0.10		−0.08		4	B9	251
104021	−52 05081			11 56 06	−52 29.4	1	7.49		0.22		0.08		2	A7 V	1771
104011	−59 03974			11 56 06	−60 22.6	2	9.47	.005	0.16	.014	−0.66	.009	4	B3/5e	846,1684
309428	−61 02843	LSS 2553		11 56 06	−62 10.7	3	10.76	.013	0.35	.005	−0.46	.011	12	B5	279,846,1747
104036	−77 00766			11 56 06	−77 32.8	1	6.73		0.22				4	A7 V	2012
104017	+38 02294			11 56 11	+38 09.3	2	7.84	.009	1.03	.009	1.02		4	K2 III	280,8041
		Ru 98 - 6		11 56 11	−64 18.2	1	10.88		0.44		0.15		2		251
	−63 02072	Ru 98 - 10		11 56 13	−64 16.0	1	10.08		0.31		0.20		2		251
	+31 02316			11 56 15	+31 24.7	1	9.00		1.26				1	K2 III	280
		Ru 98 - 11		11 56 15	−64 15.2	1	11.05		1.16		0.82		2		251
104033	−58 03978	LSS 2556		11 56 16	−59 22.2	1	8.49		0.02		−0.77		2	B1/2 III	846
104035	−63 02073	HR 4578, LSS 2555		11 56 16	−64 03.7	7	5.61	.017	0.17	.011	−0.07	.080	23	A1 Ib	15,122,191,1075,2032*
104043	−41 06881			11 56 18	−41 35.7	1	8.84		0.12		−0.09		1	A0 IV	1770
		LF14 # 57		11 56 18	−62 04.6	1	12.13		0.08		−0.45		2	B8 V	137
		He3 739		11 56 19	−62 17.	1	13.20		1.71		2.06		3		1684

Table 1   639

HD	DM	Other Id	N Rem	α_{1950}	δ_{1950}	S	V	σ_V	B–V	σ_{B-V}	U–B	σ_{U-B}	n	Spectrum	References
104047	−62 02476			11 56 19	−62 41.8	1	8.86		0.02		-0.72		2	B2 II	401
104049	−63 02077	Ru 98 - 5		11 56 20	−64 18.9	1	8.86		0.23		0.24		2	A2	251
	+16 02323			11 56 21	+15 32.7	1	8.81		1.47		1.84		1	K5	280
104039	−25 08963	HR 4579	⋆ AB	11 56 21	−25 37.8	4	6.42	.004	0.03	.002	0.03	.004	12	A1 IV/V	15,78,1637,2012
104059	−42 07341			11 56 21	−43 22.1	2	9.11	.007	0.35	.007	-0.09		5	F2 V	1594,6006
104048	−63 02078	Ru 98 - 13		11 56 23	−64 13.0	2	8.86	.005	0.22	.010	0.20	.030	4	A0	122,251
		Ru 98 - 7		11 56 23	−64 17.9	1	11.49		0.48		0.04		2		251
		LF14 # 58		11 56 25	−61 31.6	1	12.10		0.28		-0.12		1	B8 V	137
		Ru 98 - 9		11 56 25	−64 16.1	1	11.93		1.13		0.78		2		251
104062	−61 02852			11 56 26	−61 53.6	1	8.95		-0.03		-0.60		2	B3 IV/V	401
311048	−63 02079			11 56 27	−64 08.7	1	10.41		0.05		-0.22		2	A0	122
		LP 907 - 49		11 56 28	−28 01.8	1	11.88		1.23		1.17		1		1696
	+12 02405			11 56 29	+12 21.0	1	8.87		1.14		1.09		1	K0	280
104056	−3 03216	G 13 - 1		11 56 29	−04 29.8	3	9.00	.004	0.58	.000	-0.05	.005	5	G5	516,1620,1658
	−72 01184	LSS 2557		11 56 29	−73 09.1	3	10.69	.022	-0.08	.014	-0.93	.007	10	B0 III	391,400,846
104055	+1 02636	HR 4580	⋆ A	11 56 30	+00 48.5	4	6.17	.013	1.26	.007	1.40	.016	19	K2 IV	14,15,252,1417
104069	−47 07282			11 56 30	−48 03.8	1	8.60		0.17		0.14		2	A0 IV/V	1771
	+21 02377			11 56 31	+21 08.0	1	10.33		0.11		0.10		2		1026
104072	−73 00913			11 56 31	−73 43.8	1	7.71		1.40				4	K2 III	2012
104053	+24 02416			11 56 32	+24 11.1	1	9.04		1.21		1.18		1	K0	280
		Ru 98 - 8		11 56 32	−64 16.6	2	10.32	.020	1.73	.025	2.12	.205	4		122,251
	+37 02234			11 56 34	+37 11.3	1	10.38		1.10		0.98		1	K0 IV	280
311054	−63 02080	Ru 98 - 12		11 56 34	−64 14.1	1	10.48		0.30		0.20		2	A0	251
104078	−9 03408			11 56 36	−10 11.9	1	6.48		1.45		1.75		4	K2	1628
104067	−19 03382			11 56 36	−20 04.2	2	7.96	.029	0.98	.005	0.74	.044	3	K2 V	78,3072
104080	−45 07438			11 56 38	−45 33.2	4	6.35	.006	-0.07	.006	-0.23	.002	26	B8/9 V	278,1770,2001,2011
104081	−51 06236	HR 4582		11 56 38	−51 25.1	4	6.04	.004	1.28	.000			18	K1/2 III	15,1075,2013,2029
	−64 01758			11 56 39	−64 26.0	1	11.07		0.34		0.19		2		122
104076	+25 02445			11 56 40	+24 54.4	1	8.39		0.64		0.08		4	G0 V	3026
104084	−64 01759			11 56 40	−64 28.9	1	8.94		0.12		-0.06		2	B9 V	122
		MtW 56 # 21		11 56 41	+29 23.3	1	11.01		0.90		0.55		5		397
104075	+33 02176	HR 4581	⋆ AB	11 56 43	+33 26.7	3	5.96	.019	1.15	.000	1.11	.015	5	K1 III	70,172,8041
	+18 02550			11 56 44	+18 07.4	1	9.66		1.06		1.01		1	K0	280
	−58 03989			11 56 45	−58 39.9	1	10.53		0.12		0.09		1	A2 V	137
104088	+27 02080			11 56 46	+27 23.6	1	8.81		0.25		0.09		1	A5	280
	+13 02479			11 56 49	+13 10.4	1	9.36		1.04		0.91		1	K0	280
	+20 02662			11 56 50	+19 50.2	1	9.51		0.31		0.01		3	F0	1569
104125	−56 04926			11 56 50	−56 53.3	1	6.76		0.17		0.06		1	A2 V	1771
	+3 02582			11 56 51	+02 42.9	1	11.14		0.22		0.15		2		1026
104111	−62 02485	LSS 2558		11 56 53	−62 33.2	1	6.37		0.31		0.22		5	A8 II	8100
		AAS78,203 # 2		11 56 54	−59 03.	1	12.12		1.86		2.22		3		1684
	−59 03982	LSS 2559		11 56 54	−60 17.1	2	10.27	.005	0.13	.015	-0.63	.019	6		846,1747
	+12 02406			11 56 55	+12 07.8	1	9.10		0.95		0.68		1	G5	280
	+41 02262			11 56 55	+41 25.8	1	10.19		1.06		0.98		1	K0 III	280
104122	−48 07018			11 56 55	−48 51.7	1	7.39		-0.05		-0.31		4	B9 III	1770
	−75 00772			11 56 56	−75 46.3	1	10.90		0.34				4	A0	2012
		CS 15625 # 19		11 56 57	+25 21.7	1	14.09		0.54		-0.07		1		1744
		1156+295 # 13		11 56 58	+29 31.	1	15.36		0.66		0.15		4		1595
		1156+295 # 14		11 56 58	+29 31.	1	15.89		0.52		-0.03		4		1595
		1156+295 # 15		11 56 58	+29 31.	1	16.60		0.54		-0.07		4		1595
104113	+18 02552		A	11 56 59	+18 25.3	1	9.29		0.45		-0.02		2		1569
104113	+18 02552		B	11 56 59	+18 25.3	1	9.84		0.39		-0.06		2		1569
	+18 02551			11 57 00	+18 10.7	1	10.21		1.09		0.99		1		280
104141	−50 06554			11 57 00	−51 22.6	1	8.60		0.15		0.12		1	A3 V	1771
104138	−45 07444			11 57 01	−46 21.2	5	6.65	.005	0.56	.002	0.09	.009	20	F7 V	278,977,1075,2011,2024
		LP 64 - 405		11 57 03	+62 45.2	1	15.33		1.57				3		1663
104134	−29 09473			11 57 03	−30 20.3	1	10.07		0.52		0.04		2	F5 V	1730
104157	−54 04893	IDS11545S5435	A	11 57 03	−54 52.3	1	9.98		0.03		-0.06		1	B9/A2 V	1770
	+28 02066			11 57 04	+27 58.1	1	10.03		1.02		0.82		1	G9 III	280
104128	+40 02488			11 57 05	+40 16.7	1	8.40		1.14				1	K1 III	280
104174	−77 00772	HR 4583	⋆ AB	11 57 07	−77 56.6	4	4.90	.007	-0.06	.005	-0.15	.005	13	B9 Vn	15,1075,2016,2038
	+38 02295			11 57 08	+37 58.6	1	9.88		0.76				2	G6 III	280
309333	−62 02496	LSS 2560		11 57 08	−62 42.7	2	10.65	.005	0.64	.030	-0.37	.030	8		846,1747
104159	−58 03994			11 57 10	−58 54.8	1	10.10		0.06		-0.08		1	A0 V	137
		CS 15625 # 18		11 57 11	+25 22.8	1	13.76		0.50		-0.14		1		1744
	+34 02278			11 57 14	+34 14.3	1	9.11		1.10				2	G8 III	280
104148	+54 01482	IDS11547N5357	AB	11 57 15	+53 40.6	1	8.58		0.54		0.12		3	F8	1566
	+29 02243			11 57 16	+28 50.8	1	9.77		1.01		0.85		1	G9 III	280
233888	+52 01600	IDS11547N5209	A	11 57 17	+51 53.6	1	9.20		0.50		-0.06		2	G0	3032
233888	+52 01600	IDS11547N5209	B	11 57 17	+51 53.6	1	11.31		0.84		0.40		2		3032
		CS 15625 # 28		11 57 18	+27 35.1	1	13.59		0.47		-0.09		1		1744
	+31 02318			11 57 18	+30 49.5	1	10.24		1.06		0.88		1	G9 III	280
104181	+4 02556	HR 4585		11 57 23	+03 56.0	4	5.36	.009	0.00	.001	0.00	.000	10	A1 V	15,1256,1363,3023
104179	+34 02279	HR 4584		11 57 23	+34 18.8	3	6.50	.012	0.21	.011	0.15	.010	5	A9 III	39,70,254
104200	−55 04772	LSS 2561		11 57 23	−56 00.9	3	7.69	.019	-0.12	.009	-0.91	.014	7	B0.5/1 III	846,1684,1732
	+28 02068			11 57 24	+28 24.0	1	9.44		1.01		0.83		1	G8 III	280
104199	−54 04897			11 57 25	−55 18.4	1	8.78		0.00		-0.12		3	B9 III/IV	1770
104178	+44 02145			11 57 27	+43 37.3	1	8.52		1.13				2	K0 III	280
104214	−57 05169			11 57 27	−58 19.5	1	9.04		2.21		2.83		2	C3,2	864
	−17 03526			11 57 29	−18 08.1	2	10.75	.025	1.18	.015	1.16	.020	2	K7 V	1696,3072

HD	DM	Other Id	N	Rem	α_{1950}	δ_{1950}	S	V	σ_V	B–V	σ_{B-V}	U–B	σ_{U-B}	n	Spectrum	References
104207	+20 02664	GK Com			11 57 31	+19 41.9	2	6.99	.022	1.58	.002	1.57		25	M2	280,3042
		G 122 - 61			11 57 31	+43 52.4	1	15.71		0.30		-0.57		1		3060
104212	−46 07569				11 57 31	−46 30.3	5	8.37	.006	0.65	.004	0.18	.009	29	G3 V	460,977,1075,1770,2011
104208	+18 02557				11 57 32	+18 11.9	1	8.88		1.10		1.08		1		280
104211	−45 07451				11 57 32	−45 46.1	2	9.82	.004	0.48	.001	0.05		11	F3 V	863,2011
		MtW 56 # 24			11 57 34	+29 25.3	1	15.21		0.74		0.50		5		397
		MtW 56 # 23			11 57 34	+29 27.3	1	12.72		1.01		0.82		5		397
	−44 07694				11 57 34	−45 15.7	1	9.94		0.61		0.12		7	G5	863
104203	+41 02263				11 57 35	+40 34.8	1	9.14		0.81				3	G6 III	280
	−60 03566				11 57 35	−61 10.9	1	10.61		1.00		1.19		1	K1 III	137
104204	+37 02238	IDS11550N3717	A		11 57 36	+37 00.5	1	7.48		0.11		0.11		3	Am	1068
104226	−44 07695				11 57 36	−44 34.9	4	7.85	.004	0.54	.004	0.09	.011	20	F8 IV	977,1075,1770,2011
104222	−19 03385				11 57 39	−20 27.6	2	10.36	.010	0.56	.024	0.04		3	G2/3 V	1594,6006
104252	−54 04902				11 57 39	−54 40.7	1	8.50		0.56		0.09		2	F2 III	1770
104257	−65 01767	LSS 2562			11 57 39	−65 48.0	1	8.85		0.05		-0.73		2	B1/2 II/III	191
	−44 07697				11 57 41	−45 17.8	2	9.76	.002	0.50	.002	0.06		10	F2	863,2011
104243	+6 02536	G 12 - 13			11 57 42	+05 38.6	1	8.39		0.76		0.33		1	G5	333,1620
104204	+37 02239	IDS11550N3717	B		11 57 43	+37 00.7	1	8.29		1.27				3		280
104216	+81 00389	HR 4586			11 57 44	+81 07.9	3	6.19	.014	1.60	.005	1.90	.000	8	M2 III	15,1003,3035
	+14 02461				11 57 50	+14 06.5	1	11.03		0.26		-0.02		1		280
104241	+45 01983				11 57 50	+44 54.5	2	7.56	.009	0.07	.009	0.05	.005	4	A2 V	272,8071
104268	−29 09486				11 57 50	−30 19.4	1	7.97		0.98		0.74		2	K0 III	1730
307398	−60 03576				11 57 50	−60 35.3	1	10.91		0.28		0.18		2	A2 V	137
104276	−67 01882				11 57 50	−68 04.5	1	9.21		0.30		0.12		1	A7 III/IV	1771
104242	+43 02174				11 57 51	+42 52.0	1	10.25		0.58				2	K0	280
104262	+0 02875	IDS11553S0008	AB		11 57 51	−00 24.8	1	9.14		0.58				1	G5	1284
104274	−54 04904				11 57 51	−54 50.4	1	8.64		2.85		5.10		2	Nb	864
104259	+50 01881				11 57 53	+49 51.8	1	8.69		0.52		-0.02		1	G0	3016
	+52 01602				11 57 53	+52 14.0	1	10.07		1.21		1.28		2		1726
104283	−53 04891				11 57 53	−53 33.6	1	8.86		0.25		0.17		1	A3 V	1771
104285	−62 02512				11 57 53	−62 57.3	3	9.63	.004	0.01	.001	-0.72	.015	8	B1/2 V	401,540,976
		G 11 - 33			11 57 54	−03 48.	1	15.40		0.94		0.54		1		3060
104286	−69 01618				11 57 56	−69 52.5	1	9.12		0.29		0.19		1	A3 V	1771
		Wray 923			11 57 59	−62 45.2	1	14.38		0.79		-0.17		3		1684
		G 121 - 41			11 58 02	+28 44.0	1	12.46		1.36		1.05		2		3016
104290	+19 02521				11 58 04	+18 36.8	1	7.84		1.03		0.90		1	K0	280
		L 829 - 24			11 58 09	−13 32.5	1	12.74		1.55				1		1746
104307	−21 03443	HR 4588		⋆A	11 58 09	−21 33.5	1	6.28		1.22				4	K2 III	2006
104304	−9 03413	HR 4587			11 58 10	−10 09.7	10	5.56	.016	0.77	.011	0.43	.016	36	K0 IV	15,22,1008,1075,1197*
	−29 09489				11 58 10	−30 20.3	1	11.35		0.68		0.19		2		1730
	+32 02211				11 58 11	+32 10.2	1	9.98		0.98		0.73		1	G8 III	280
		DY Hya			11 58 13	−33 34.6	1	12.08		0.30		0.12		1		700
104332	−63 02096	LSS 2563			11 58 13	−63 34.2	1	8.47		-0.02		-0.81		3	B1 III	846
104319	+29 02245				11 58 14	+29 28.0	1	8.36		0.59		0.07		3	F8	196
104329	−48 07034				11 58 15	−49 17.4	1	7.86		1.19		1.07		5	K1/2 III	1673
104320	+13 02480				11 58 16	+12 47.6	1	9.52		0.97		0.76		1	K0	280
	+12 02412				11 58 17	+12 15.7	1	8.90		1.37		1.61		1	K5	280
104337	−18 03295	HR 4590			11 58 18	−19 22.8	1	5.26		-0.20				4	B2 IV	2007
104321	+7 02502	HR 4589			11 58 19	+06 53.6	7	4.66	.011	0.12	.008	0.11	.010	26	A5 V	15,369,1256,1363,3023*
104316	+71 00598				11 58 19	+70 31.0	1	6.78		0.08		0.08		3	A0	985
	+14 02463				11 58 22	+13 55.3	1	9.39		1.07		0.96		1	K0	280
104356	−0 02520	HR 4591			11 58 28	−01 29.3	3	6.32	.022	1.21	.005	1.06	.000	8	K0	15,1417,3077
104364	−68 01597				11 58 29	−68 54.5	1	9.13		0.12		-0.04		2	A0 III	1730
104350	+13 02481	AG Vir			11 58 30	+13 17.2	2	8.46	.060	0.25	.025	0.05	.005	2	A7 V	280,3002
104349	+28 02069				11 58 31	+28 14.5	2	8.76	.015	1.13	.005	1.27		3	K1 III	280,8041
		MtW 56 # 29			11 58 34	+29 23.3	1	15.66		0.65		-0.10		5		397
104366	+31 02321				11 58 34	+31 15.4	1	9.31		0.11		0.05		2	A3 V	1068
309504	−61 02882				11 58 34	−61 30.7	1	10.75		0.13		0.05		2	A0 V	137
		G 11 - 34			11 58 35	−01 27.3	3	10.95	.004	1.10	.011	0.99	.014	6		158,1658,1705
	−26 08918				11 58 35	−26 58.1	2	11.00	.028	0.48	.005	-0.05		4		1594,6006
	−61 02883	LSS 2564			11 58 35	−62 24.0	1	11.64		0.34		-0.50		5		846
	+35 02300				11 58 36	+35 13.4	1	9.46		1.18				3	K2 III	280
104388	−63 02098				11 58 38	−63 58.2	2	8.93	.005	0.04	.010	-0.24	.005	4	B8 V	122,401
		WLS 1200-5 # 10			11 58 39	−04 28.1	1	11.21		0.77		0.30		2		1375
104381	+13 02482				11 58 40	+12 39.4	2	6.92	.019	0.11	.005	0.08	.029	3	A1 V	272,280
	+29 02246				11 58 41	+29 02.4	1	10.07		1.04		0.89		1	G9 III	280
		LF14 # 64			11 58 42	−60 00.1	1	10.74		1.17		0.97		1	K2 III	137
104392	+25 02448				11 58 44	+24 29.8	1	8.17		1.33		1.48		1	K2 III	280
	+36 02229				11 58 44	+36 02.4	1	10.35		1.09		1.03		1	K0	280
104422	−63 02100				11 58 46	−63 48.2	1	9.04		-0.01		-0.63		2	B3 III	401
104418	−54 04917				11 58 49	−54 58.6	1	9.25		0.01		-0.18		2	B9.5 V	1770
104406	+26 02286				11 58 50	+25 59.7	2	8.97	.044	1.13	.010	1.04		3	K0 III	280,8041
104432	−61 02888				11 58 50	−62 19.0	1	8.47		0.04		-0.07		2	B9 V	401
	+44 02150				11 58 51	+43 47.5	1	10.78		0.89		0.55		1	G8 III	280
104391	+44 02151				11 58 51	+43 51.7	1	9.65		0.97				2	G9 III	280
	+17 02425				11 58 53	+17 12.4	1	9.44		1.08		0.96		1	K0	280
104420	−58 04020				11 58 53	−58 37.9	1	9.26		0.31		0.17		1	A8/9 IV/V	1771
		CS 15625 # 30			11 58 54	+25 14.8	1	11.68		1.33		1.23		1		1744
		LF14 # 65			11 58 54	−60 16.5	1	12.25		0.45		0.06		1	A2 V	137
104430	−56 04954	HR 4592			11 58 56	−57 13.5	4	6.15	.007	-0.01	.011	-0.03	.001	14	A0 V	15,1637,1771,2012

Table 1

641

HD	DM	Other Id	N	Rem	α_{1950}	δ_{1950}	S	V	σ_V	B–V	σ_{B-V}	U–B	σ_{U-B}	n	Spectrum	References
104425	+43 02177	IDS11570N4336		D	11 58 59	+43 19.2	1	8.83		0.48				5	F6 II	193
104445	−53 04897	IDS11565S5317		ABC	11 59 00	−53 33.4	1	8.45		0.27		0.22		1	A5 IV/V	1771
		LF14 # 66			11 59 02	−61 35.2	1	12.32		1.23		0.94		1		137
104465	−62 02526				11 59 03	−63 17.1	1	9.08		-0.05		-0.81		2	B2 II/III	401
104436	+65 00863				11 59 04	+65 13.1	2	7.26	.010	0.26	.005	0.04	.005	4	A3	1502,1733
104439	+11 02421				11 59 05	+11 20.9	1	8.81		1.26		1.37		1	K0	280
104438	+36 02230	HR 4593			11 59 06	+36 19.3	2	5.59	.005	1.01	.004	0.79		4	K0 III	70,172
		SSII 63			11 59 06	+42 17.	1	12.48		0.18		0.06		2	A0	1026
		L 829 - 10			11 59 06	−11 56.9	1	12.34		1.46				1		1746
104463	−51 06276				11 59 06	−51 31.7	1	8.43		0.18		0.16		1	A1mA4-A6	1771
104437	+40 02490				11 59 07	+39 52.0	1	8.60		0.67				2	G5 IV	280
309437	−61 02895	LSS 2565			11 59 07	−62 17.4	2	11.15	.000	0.04	.005	-0.60	.030	7		846,1747
		AAS12,381 22 # A			11 59 10	+21 51.3	1	10.40		0.86		0.64		1		280
104479	−68 01600				11 59 10	−68 55.2	1	6.96		1.28		1.23		2	K0 III	1730
104480	−69 01622				11 59 10	−70 07.7	1	8.34		0.19		0.10		1	A1 V	1771
104453	+14 02467				11 59 11	+13 41.3	1	8.95		0.96		0.70		1	G5	280
104452	+22 02430				11 59 11	+22 22.4	1	6.58		0.64				1	G0 II	6009
		LF14 # 67			11 59 11	−59 37.9	1	11.95		0.32		0.19		1	A2 V	137
104471	−33 08130	IDS11567S3406		AB	11 59 13	−34 22.3	3	6.90	.004	0.59	.000	0.08		9	G0 V	258,1075,2012
309460	−62 02528	LSS 2567			11 59 14	−62 46.7	2	10.28	.024	0.44	.034	-0.40	.024	6		846,1747
	+39 02485				11 59 15	+39 17.5	1	10.14		1.27		1.52		1	K2 III	280
104475	−53 04898				11 59 15	−53 41.3	1	9.22		0.31		0.21		1	A8 IV/V	1771
	+12 02414				11 59 16	+11 49.0	1	9.98		1.01		0.82		1		280
104481	−72 01194				11 59 17	−73 01.2	2	6.85	.004	0.06	.009	0.05		6	A1 V	278,2012
		CS 15625 # 31			11 59 18	+26 32.5	1	13.03		0.52		-0.19		1		1744
		KZ Cen			11 59 21	−46 00.0	1	11.75		0.25		0.15		1		1517
		LP 434 - 27			11 59 22	+19 38.1	1	12.32		1.07		1.00		1		1696
104484	+0 02880	IDS11569N0039		A	11 59 23	+00 22.8	1	7.79		0.30		-0.01		10	A2	14
104497	−19 03395				11 59 28	−20 31.8	1	7.79		0.16		0.15		5	A3 IV/V	1770
		LF14 # 68			11 59 28	−59 43.7	1	12.56		0.36		0.24		1	A2 V	137
104509	−51 06286				11 59 29	−51 32.5	1	9.05		0.21		0.09		3	A4 IV/V	1771
104495	+33 02185				11 59 30	+32 36.7	1	8.75		1.09				2	K0 IV	280
309467	−62 02531				11 59 30	−62 58.7	1	9.96		0.06		-0.50		2		1684
	−58 04026				11 59 32	−59 04.8	1	10.19		1.15		1.07		1	K2 III	137
		LF14 # 70			11 59 34	−60 17.4	1	11.50		1.41		1.08		1		137
104504	+32 02212				11 59 35	+31 31.7	1	8.74		0.52		0.06		4	F8	1371
104513	+43 02179	HR 4594, DP UMa		★ A	11 59 35	+43 19.4	3	5.24	.018	0.27	.011	0.08	.010	6	A7 m	1363,3023,8071
104523	−64 01772				11 59 35	−64 31.1	1	7.94		0.09		-0.19		2	B9 Ib/II	122
104522	−63 02105				11 59 36	−63 28.4	1	9.54		0.09		-0.02		2	B9 III/IV	401
		CS 15625 # 32			11 59 37	+27 47.0	1	12.29		0.47		-0.08		1		1744
	+40 02491				11 59 39	+39 49.6	2	9.53	.030	1.50	.015	1.41		2	M5 III	280,694
		LF14 # 71			11 59 39	−61 55.6	1	11.75		1.22		0.95		1		137
	−37 07613				11 59 40	−37 53.1	1	9.87		0.96		0.78		2		536
104532	−48 07070				11 59 40	−48 26.8	1	9.22		0.70				4	G3/5 V	2011
	−62 02535	LSS 2568			11 59 40	−62 55.7	2	11.02	.019	0.09	.024	-0.41	.015	6		846,1747
	−59 04018	LSS 2569			11 59 42	−60 03.3	2	10.59	.010	0.24	.010	-0.58	.025	5		846,1747
104555	−84 00371	HR 4595		★ A	11 59 45	−85 21.2	2	6.04	.005	1.29	.000	1.54	.000	7	K3 III	15,1075
104550	−54 04927				11 59 46	−55 17.1	1	9.47		0.36		0.08		1	A9 III/IV	1771
	−60 03619				11 59 46	−60 49.0	1	9.92		1.36		1.28		1	K2 III	137
104526	+43 02180				11 59 47	+43 25.1	1	8.38		1.15		1.01		1	K0 V	280
104552	−60 03621	LSS 2570			11 59 49	−61 10.4	3	9.47	.033	0.06	.024	-0.65	.008	7	B3/5e	846,1684,1747
	−59 04020				11 59 50	−60 21.5	1	10.00		0.60				2	F2 V	137
104553	−61 02906	LSS 2571			11 59 50	−62 08.1	3	7.33	.016	0.04	.008	-0.76	.017	6	B1 Ib	401,540,976
104565	−57 05199	LSS 2572			11 59 54	−57 57.9	5	9.25	.023	0.36	.013	-0.61	.030	14	O9/B0 Ia/b	191,540,976,1684,1737
104570	−70 01454	HR 4596			11 59 54	−71 12.6	4	6.41	.004	1.16	.009	1.14	.000	19	K1 III	15,1075,1075,2038
104568	−68 01603				11 59 55	−69 06.9	1	8.10		0.06		-0.16		2	B9 V	1075,1771
104540	+12 02415				11 59 56	+11 47.1	1	9.80		0.99		0.74		1	K0	280
104564	−55 04810				11 59 56	−56 16.9	1	7.38		0.04		0.07		2	A0 V	1771
104567	−62 02537	LSS 2573			11 59 56	−62 47.5	1	8.47		0.07		-0.77		2	B1 III	401
104556	+43 02182	IDS11570N4336		B	11 59 57	+43 22.2	4	6.64	.003	0.86	.005	0.48	.016	9	G8 V wk1	22,172,1003,1080
	+19 02524				12 00 00	+18 48.6	1	9.26		1.50		1.86		1	K5	280
		LF14 # 74			12 00 01	−61 40.0	1	12.79		0.17		-0.08		1		137
104571	−73 00922				12 00 01	−74 21.7	2	9.45	.005	0.23	.010	0.08		7	B8/9 V	236,2012
		G 11 - 35			12 00 02	+08 43.3	2	14.01	.009	1.55	.009	1.00	.000	5		316,3078
104585	−74 00867				12 00 02	−74 30.0	1	8.75		1.24				4	G8/K0 III	2012
	+14 02468				12 00 03	+13 37.8	1	10.39		0.92		0.70		1		280
		LF14 # 75			12 00 03	−58 50.5	1	11.60		0.22		0.10		2	A1 V	137
104600	−68 01604	HR 4597			12 00 03	−68 54.8	5	5.88	.005	-0.08	.000	-0.28	.005	20	B9 V	15,1075,1075,1771,2038
104574	+12 02416				12 00 05	+12 09.4	1	8.81		1.32		1.35		1	K0	280
	+36 02231				12 00 05	+36 25.3	1	10.32		0.79		0.49		1	G8 III	280
	+0 02884				12 00 06	−00 06.7	1	10.03		0.97				1	K0	1284
		LF14 # 76			12 00 06	−59 34.9	1	12.64		1.31		1.03		2		137
104573	+36 02232				12 00 07	+36 00.3	2	8.15	.006	0.23	.000	0.18	.006	9	A7 IV	1075,280
104572	+44 02152				12 00 07	+43 35.6	1	9.65		1.26				2	K2 III	280
309464	−62 02540				12 00 07	−62 53.0	1	9.72		0.50		0.17		2	F2 V	137
		G 122 - 67			12 00 08	+36 53.2	1	12.35		1.43				1		1705
104590	+25 02449				12 00 09	+24 43.6	3	7.75	.024	1.13	.007	1.13	.019	6	K2 III	172,280,8041
		LF14 # 78			12 00 09	−58 57.4	1	12.79		0.37		0.20		1	A2 V	137
	+16 02327				12 00 10	+16 11.7	1	10.23		1.01		0.78		1	K0	280
		NGC 4051 sq1 3			12 00 10	+44 47.8	1	14.16		0.54		-0.07		1		327

HD	DM	Other Id	N	Rem	α_{1950}	δ_{1950}	S	V	σ_V	B–V	σ_{B-V}	U–B	σ_{U-B}	n	Spectrum	References
104598	−45 07487	IDS11576S4551		AB	12 00 11	−46 07.6	2	9.74	.015	0.44	.008	0.14		11	F2 III/IV	863,2011
104589	+26 02288				12 00 13	+25 36.3	2	8.10	.019	1.09	.005	0.98		3	K1 III	280,8041
104586	+60 01370				12 00 14	+59 47.9	1	9.24		0.53		0.06		2	G0	1502
104614	−45 07488				12 00 16	−46 06.6	2	9.49	.004	1.04	.002	0.77		10	K0 III	863,2011
	+42 02266				12 00 17	+41 29.0	1	9.37		1.37				3	K3 III	280
104625	−6 03499	HR 4598			12 00 18	−07 24.3	2	6.21		1.49	.000	1.83	.000	7	K5	15,1256
104604	+15 02406				12 00 19	+15 21.5	1	9.99		1.02		0.84		1	K0	280
		G 122 - 68			12 00 19	+43 49.8	1	13.91		1.50		1.22		1		3062
104621	+27 02087	IDS11578N2733		A	12 00 22	+27 17.1	2	9.25	.025	0.79	.010	0.38	.020	2	G5 IV-V	280,8038
104631	−61 02914	LSS 2574			12 00 22	−61 53.8	7	6.76	.018	0.08	.010	−0.74	.020	19	O9.5 III/IV	6,158,191,540,976*
	+45 01986				12 00 24	+44 48.0	1	11.12		0.63		0.18		3	F8	327
104649	−61 02915			V	12 00 24	−62 23.6	2	7.91	.141	0.01	.005	−0.83	.073	6	B1 V	164,401
		LOD 25 # 21			12 00 25	−61 07.5	1	11.97		0.24		0.10		2		122
104635	+44 02153				12 00 26	+43 30.6	1	8.94		0.92				2	G8 III	280
104641	−21 03461				12 00 26	−21 52.4	1	7.85		1.62		2.02		4	K5 III	1657
104671	−62 02543	HR 4599		★ AB	12 00 28	−63 02.1	7	4.33	.009	0.27	.004	0.04	.013	24	A3mA8-A8	15,355,1075,1637,2012*
104666	−48 07081				12 00 29	−49 22.5	1	7.54		0.11		0.09		2	A1 V	1770
		G 122 - 69			12 00 31	+47 36.2	1	12.42		0.43		−0.28		2		1658
104663	−45 07491	IDS11580S4558		AB	12 00 32	−46 14.9	3	8.88	.008	0.57	.008	0.01	.006	16	F7 V	977,1075,2011
104664	−47 07335				12 00 32	−47 55.0	4	6.61	.002	0.24	.006	0.17	.000	33	A7 III	278,1075,1770,2011
		LF14 # 79			12 00 33	−62 15.6	1	12.12		0.80		0.34		1		137
		GD 146			12 00 34	+21 18.0	1	15.34		0.61		−0.09		1		3060
104685	−74 00870				12 00 35	−74 39.8	2	9.48	.010	0.51	.015	0.31		7	A3mA6-F0	236,2012
		G 121 - 46			12 00 37	+29 17.8	1	13.20		1.46		1.21		2		3032
309462	−62 02546				12 00 38	−62 46.5	1	10.14		0.08		−0.61		2	B5	401
	+29 02250	G 121 - 47		★ AB	12 00 39	+29 17.5	1	10.30		0.94		0.70		2	G8 V	3032
104683	−63 02115	LSS 2575			12 00 39	−64 04.4	5	7.94	.016	0.00	.011	−0.80	.022	10	B1 Ib/II	122,401,540,976,1684
311016	−68 01607				12 00 39	−68 49.4	1	9.77		0.78		0.36		2	G5	1730
104675	+37 02244				12 00 42	+37 00.1	1	8.94		1.31				2	K2 III	280
	+44 02154				12 00 43	+44 02.1	1	10.62		1.07		1.01		1	K1 III	280
	+30 02214				12 00 44	+29 49.5	1	10.14		0.80				3	G6 III	280
104688	+29 02251				12 00 45	+29 24.0	3	8.28	.011	1.48	.015	1.72	.008	9	K3 IV	280,397,8041
		WLS 1200-5 # 7			12 00 45	−06 45.7	1	13.84		0.67		0.30		2		1375
104695	−55 04820				12 00 45	−55 31.5	1	8.82		−0.03		−0.27		1	B9 III	1770
309457	−62 02548	LSS 2576			12 00 47	−62 36.3	4	9.94	.019	0.33	.015	−0.59	.025	11	B0	279,846,1684,1747
	−76 00708				12 00 47	−77 00.1	1	9.33		1.22				4	G5	2012
104705	−62 02549	LSS 2577			12 00 49	−62 25.0	6	7.79	.017	−0.01	.013	−0.88	.022	10	B0 III/IV	6,401,540,976,1684,1732
	+41 02266				12 00 50	+40 40.8	1	10.24		1.13		1.17		1	K1 III	280
		NGC 4051 sq1 1			12 00 51	+44 49.8	1	14.02		0.96		0.66		1		327
	+71 00600				12 00 54	+70 42.6	1	10.27		1.32				1	R2	1238
104720	−45 07495				12 00 55	−45 52.6	8	9.24	.006	1.51	.005	1.88	.011	65	K4/5 III	863,1075,1460,1628,1657*
104724	−71 01302				12 00 56	−71 54.9	2	8.01	.024	0.79	.010	0.33		6	G5 IV	2012,3008
104710	+30 02217				12 01 00	+29 57.5	2	7.39	.005	1.56	.015	1.76		4	M3 III	172,280
		WLS 1200-5 # 5			12 01 01	−03 00.8	1	12.70		0.47		−0.09		2		1375
104734	−56 04976				12 01 03	−57 18.8	1	7.95		0.21		0.09		1	A4 V	1771
	−62 02553				12 01 03	−62 52.7	1	9.85		0.04		−0.61		2		846
104747	−38 07479	IDS11585S3827		AB	12 01 04	−38 43.8	2	6.50	.005	0.50	.005			8	F7 V	1075,2012
104731	−41 06938	HR 4600			12 01 04	−42 09.2	14	5.15	.013	0.41	.008	−0.03	.008	82	F5 V	14,15,125,182,278,388*
		NGC 4051 sq1 4			12 01 05	+44 48.4	1	13.39		0.86		0.35		1		327
309514	−61 02923				12 01 06	−62 06.5	1	10.10		0.10		−0.12		3	A0	279
104752	−73 00924	HR 4601			12 01 08	−73 56.1	5	6.44	.006	1.22	.007	0.97	.006	21	G6 III	15,278,1075,2012,2038
309512	−61 02924				12 01 09	−62 00.2	1	10.57		0.21		0.01		5	A3	279
	+18 02563				12 01 10	+17 56.4	1	9.80		1.19		1.22		1	K0	1771
104762	−47 07341				12 01 11	−48 10.8	1	9.80		0.35		0.14		1	A2/3mA9-F0	1771
	+14 02470				12 01 14	+13 37.1	1	10.70		0.38		−0.05		2		1569
		LP 794 - 17			12 01 14	−16 15.2	1	10.76		1.29		1.17		1	K5	3062
		L 757 - 79			12 01 14	−16 32.2	1	10.77		1.26				1		1746
	−45 07500				12 01 14	−46 20.5	1	10.60		0.16				4		2011
	−60 03667				12 01 14	−61 05.1	1	9.65		0.64		0.22		2		122
	−44 07743				12 01 18	−45 13.0	3	10.52	.004	1.26	.002	1.31	.011	21		863,1460,2011
	−59 04034				12 01 18	−59 30.9	1	10.06		0.49		0.16		2	F5 V	137
		LOD 25 # 16			12 01 18	−61 02.2	1	10.93		1.13		0.77		2		122
	+42 02268				12 01 22	+42 18.8	1	10.53		0.86		0.55		1	G7 III	280
	−45 07501				12 01 22	−45 28.3	1	10.61		0.16		0.18		6		863
309515	−61 02926				12 01 23	−62 11.8	1	10.18		1.67				3	K5	279
104794	−51 06312				12 01 24	−52 22.6	1	9.59		0.30		0.15		1	A9 V	1771
	+39 02488				12 01 26	+38 29.4	1	9.66		1.27				2	K2 III	280
		LP 494 - 23			12 01 28	+11 25.5	1	13.94		1.20		1.11		1		1696
104785	+19 02526				12 01 28	+19 05.7	1	7.74		1.23		1.21		2	K2	1569
104784	+25 02454				12 01 29	+25 13.0	1	8.37		0.47		−0.01		3	F8 V	3026
104808	−47 07344				12 01 29	−47 49.7	1	9.11		0.28		0.14		1	A6 V	1771
104810	−63 02124				12 01 29	−64 15.6	2	7.36	.000	0.03	.000	−0.36	.010	4	Ap Si	122,401
		Ly 107			12 01 30	−58 49.	1	12.03		0.40		−0.45		2		846
	+16 02332				12 01 31	+16 08.6	1	10.50		1.08		0.96		1	K0	280
104806	−45 07506				12 01 31	−45 39.5	3	8.62	.000	1.43	.004	1.57	.002	27	K2 IV	863,977,2011
104800	+4 02568	G 11 - 36		★ B	12 01 32	+03 37.6	6	9.22	.035	0.59	.012	−0.05	.011	12	G0 V	22,158,202,516,1003*
104797	+28 02074				12 01 32	+28 20.3	1	8.59		0.36		−0.03		3	F5	1569
104783	+38 02297				12 01 32	+38 16.7	1	9.14		0.80				2	G5 III	280
104809	−61 02931				12 01 33	−62 01.4	1	9.75		0.05		−0.35		3	B7/9 II	279
104812	−68 01610				12 01 33	−69 14.5	1	6.80		1.25		0.31		1	A0 IV/V	1771

Table 1 643

HD	DM	Other Id	N Rem	α_{1950}	δ_{1950}	S	V	σ_V	B–V	σ_{B-V}	U–B	σ_{U-B}	n	Spectrum	References
	+37 02245			12 01 35	+37 23.5	1	9.59		1.14				2	K1 III	280
	−44 07746			12 01 35	−44 59.7	1	11.11		0.51				4		2011
	−74 00872			12 01 35	−75 18.3	2	10.99	.015	0.42	.015	0.24		7	A0	236,2012
104817	+2 02509			12 01 40	+01 44.4	3	7.68	.005	0.19	.009	0.09	.018	7	Am	355,1003,8071
104813	+45 01991			12 01 40	+44 43.7	1	8.51		1.05				2	K0 III	280
104822	−56 04984			12 01 40	−57 05.1	1	7.55		0.01		−0.02		2	A0/1 V[n]	1771
	+19 02527	R Com		12 01 42	+19 03.6	1	9.36		1.33		0.12		1	M8e	280
	+38 02298			12 01 42	+37 47.1	1	10.06		1.44		1.62		1	K2 IV	280
104839	−50 06630			12 01 42	−51 11.6	1	6.48		−0.05		−0.14		3	B9.5 V	1770
104827	+22 02437	HR 4602	★ AB	12 01 43	+21 44.3	2	5.88	.010	0.24	.013	0.11		5	F0 IV-V	70,8071
104843	−72 01197			12 01 43	−73 19.8	1	8.00		1.19				4	K1 II/III	2012
104841	−62 02561	HR 4603, θ2 Cru		12 01 44	−62 53.2	6	4.72	.012	-0.09	.008	−0.62	.010	21	B3 V	6,15,26,1075,2012,8023
104828	+10 02374			12 01 45	+09 28.3	1	9.92		1.09		0.93		1	K0	3072
	+27 02090			12 01 46	+27 17.4	1	10.24		1.15		1.17		1	K1 III	280
		POSS 377 # 31		12 01 49	+21 14.1	1	16.89		1.19				3		1739
104848	−60 03683	LSS 2578		12 01 49	−60 45.4	1	7.95		0.13		−0.60		2	B3 Ib/II	846
		LOD 25 # 14		12 01 49	−61 01.5	1	11.30		0.16		−0.35		2		122
	+34 02282			12 01 50	+33 53.4	1	10.06		1.10		1.12		1	K1 III	280
		POSS 377 # 15		12 01 54	+22 14.7	1	16.04		1.68				3		1739
104845	+37 02246			12 01 55	+36 38.0	1	8.73		1.09				3	K0 III	280
	−44 07749			12 01 56	−45 11.9	3	10.59	.007	0.59	.003	0.09	.007	20		863,1460,2011
	−60 03686			12 01 57	−61 15.6	1	9.00		1.37		1.25		1	K2 III	137
104869	−49 06768			12 01 59	−49 45.6	1	7.98		0.18		0.14		1	A4/5 IV	1771
	−59 04037			12 02 00	−59 37.0	1	10.00		0.54		0.08		1	F5 V	137
104876	−62 02564	LSS 2579		12 02 00	−62 37.5	4	8.72	.018	0.44	.010	−0.52	.020	9	B0/1 I/III	540,846,976,1684
104872	−54 04946			12 02 02	−55 09.1	1	7.72		−0.05		−0.25		3	B9 III	1770
104878	−67 01896	HR 4604		12 02 03	−68 03.0	7	5.35	.008	−0.01	.006	−0.18	.015	23	A0 IV	15,26,1075,1075,1771*
	+39 02489			12 02 05	+38 56.6	1	9.58		1.00				3	G9 III	280
		LB 2197		12 02 05	+60 48.4	2	13.61	.000	-0.34	.028	−1.24	.009	4		308,3060
		L 469 - 72		12 02 05	−37 58.8	1	11.78		1.52				1		3062
		POSS 377 # 32		12 02 07	+24 19.6	1	16.91		1.79				3		1739
	+31 02326			12 02 07	+31 24.7	1	9.94		1.07		1.02		1	K0 III	280
	+32 02216			12 02 07	+31 48.0	1	10.27		0.95		0.40		2		1569
104862	+36 02235			12 02 07	+35 50.7	3	7.40	.012	1.06	.008	0.91	.005	12	K1 III	172,280,1775
	+33 02188			12 02 08	+33 14.5	1	10.22		0.30		0.03		2		1569
104893	−28 09270			12 02 09	−28 54.4	2	9.22	.009	1.22	.019	0.94		4	F8/G2(e)	1594,6006
104904	+86 00176	HR 4606		12 02 10	+85 51.8	1	6.27		0.59				2	F6 V	71
104900	−58 04058			12 02 10	−58 58.5	2	6.26	.060	-0.07	.020	−0.16	.005	4	B9 Vn	122,1771
104902	−75 00777	HR 4605		12 02 10	−76 14.5	5	5.03	.007	1.49	.002	1.78	.005	21	K4 III	15,278,1075,2012,2038
104886	−4 03199	RX Vir	★ AB	12 02 11	−05 29.7	1	8.73		1.45		0.10		5	K0	897
	−58 04059			12 02 11	−59 18.2	2	10.04	.090	0.03	.025	−0.03	.005	4	B8 V	122,137
104901	−61 02933	BY Cru	★ A	12 02 11	−61 43.1	2	7.43	.010	0.19	.015	−0.20	.034	3	B8/9 Iab/b	279,8100
104901	−61 02934	IDS11596S6126	C	12 02 12	−61 42.7	2	10.19	.004	0.09	.017	−0.57	.009	12		279,846
	−26 08952	IK Hya		12 02 14	−27 23.6	1	9.96		0.32				1	A2:	1484
104920	−71 01306			12 02 15	−72 10.8	2	8.13	.003	0.18	.002	−0.09	.007	3	B9/A0 IV/V	953,1771
104917	−50 06635			12 02 16	−50 34.5	1	8.32		0.22		0.06		1	A9/F0 IV	1771
	−72 01201	BR Mus		12 02 16	−72 35.6	1	10.64		0.10		−0.53		1		953
104931	−51 06332	IDS11597S5151	AB	12 02 17	−52 07.8	1	9.92		0.32		0.08		1	A8/9 IV/V	1771
104907	+12 02420			12 02 18	+11 55.3	1	9.98		1.05		0.90		1	K0	280
		Sand 6		12 02 18	+23 26.	1	14.25		1.57				1	dM4	686
104906	+33 02189			12 02 18	+33 07.8	2	8.21	.015	1.00	.005	0.74		3	K0 III	280,8041
	+36 02236			12 02 18	+35 57.9	1	9.39		1.30				2	K2 III	280
		LOD 25 # 9		12 02 18	−60 59.5	1	11.34		0.54		0.05		2		122
309516	−61 02937			12 02 18	−62 09.0	1	10.28		0.12		−0.13		3	A0	279
	+35 02310	G 148 - 34		12 02 19	+34 54.5	1	10.66		1.02		0.93		1	K1 V	1658
		Cru sq 1 # 9		12 02 19	−62 19.5	1	12.63		0.28		0.18		2		279
104918	−61 02936			12 02 21	−62 19.2	1	10.04		0.06		−0.34		2	B5/7 V	279
104933	−60 03697	HR 4607		12 02 22	−60 41.4	5	5.96	.008	1.69	.016	2.04		20	M2 III	15,1075,1075,2006,3055
		G 13 - 6		12 02 23	+04 36.8	1	15.04		0.64		−0.12		1		202,3060
104936	−69 01626			12 02 23	−69 43.4	1	7.46		1.71		2.10		3	K5 III	1704
	+32 02217			12 02 24	+31 37.0	3	10.54	.035	0.28	.010	0.12	.016	7	Am	1026,1240,1298
104938	−74 00875			12 02 24	−75 24.5	2	7.72	.005	1.62	.010	1.76		7	K4 III	236,2012
		He3 748		12 02 28	−65 04.	1	13.17		1.21		0.02		5		1684
		E5 - o		12 02 29	−45 17.0	1	16.52		0.70				4		1460
		E5 - m		12 02 29	−45 17.5	1	15.81		0.61		−0.06		4		1460
		E5 - d		12 02 30	−45 14.	1	13.60		0.69				4		1075
		E5 - g		12 02 30	−45 22.	1	14.58		0.63				4		1075
		Cru sq 1 # 102		12 02 30	−62 08.	1	11.51		0.06		−0.48		2		846
	+42 02270			12 02 31	+41 55.3	1	10.06		1.20		1.29		1	K1 III	280
	−60 03702	LSS 2582		12 02 31	−60 35.4	2	10.28	.010	0.06	.015	−0.62	.015	5		846,1747
	+18 02565			12 02 32	+18 02.4	1	9.97		1.10		1.10		1	K0	280
		LOD 25 # 8		12 02 32	−61 00.5	1	11.73		0.17		−0.33		2		122
104957	+31 02327			12 02 33	+31 06.7	2	8.88	.005	0.26	.010	0.04	.005	8	Am	1068,1371
104969	−52 05197			12 02 33	−53 19.9	1	9.06		0.17		0.13		1	A2 V	1771
104971	−60 03704	IDS12000S6037	AB	12 02 33	−60 54.1	1	6.64		0.96		0.66		2	G8 III	122
	−58 04066			12 02 34	−59 09.9	1	10.29		1.45		1.35		2		122
104955	+48 01996			12 02 35	+47 40.8	1	8.72		0.84		0.45		1	G7 III	8034
104977	+14 02471			12 02 36	+13 45.1	1	7.62		1.47		1.75		1	K5	280
		E5 - k		12 02 36	−45 17.1	1	15.38		0.50		0.04		4		1460
		E5 - b		12 02 36	−45 19.	1	13.53		0.68				4		2012

HD	DM	Other Id	N Rem	α_{1950}	δ_{1950}	S	V	σ_V	B–V	σ_{B-V}	U–B	σ_{U-B}	n	Spectrum	References
		LOD 25+ # 14		12 02 36	−59 05.1	1	12.07		0.26		0.01		2		122
		MCC 632		12 02 38	+12 26.3	1	11.12		1.30				1	K5	1017
		LP 852 - 5		12 02 38	−25 17.0	1	12.94		0.26		-0.62		2		3060
104982	−28 09277			12 02 38	−28 26.0	2	7.78	.021	0.65	.000	0.14		6	G5 V	1311,2033
	−58 04067			12 02 38	−59 11.8	1	8.86		1.83		1.79		2	M0 III	122
104984	−60 03706	LSS 2583		12 02 38	−60 55.3	3	8.61	.010	0.08	.021	-0.57	.054	5	A0	122,846,1771
	+20 02679			12 02 39	+19 54.9	1	9.65		1.04		0.87		1	K2	280
104975	+39 02491	IDS12001N3925	AB	12 02 39	+39 08.1	1	9.10		0.30		0.07		1	A3	280
	−58 04069			12 02 39	−59 07.3	1	10.20		-0.01		-0.51		2	B4	122
	−58 04068			12 02 39	−59 10.6	1	11.50		0.26		0.14		2		122
	−60 03707			12 02 39	−61 01.2	1	10.09		0.12		-0.40		2	B3	122
104979	+9 02583	HR 4608		12 02 40	+09 00.6	8	4.12	.008	0.98	.009	0.63	.013	32	G8 III	3,15,667,1080,1256*
104976	+15 02408			12 02 40	+15 07.8	1	9.09		0.92		0.62		1	K0	1569
104988	−0 02532	G 11 - 37		12 02 40	−01 13.9	4	8.16	.017	0.76	.011	0.28	.030	12	G8 V	22,1003,2012,3077
309511	−61 02944			12 02 42	−61 59.0	1	9.60		1.18		1.01		3	K0	279
104992	−57 05234			12 02 43	−57 48.2	1	8.45		0.32		0.19		1	A4 V	1771
104994	−61 02945	LSS 2584		12 02 43	−61 46.4	2	10.88	.054	-0.05	.017	-0.84	.002	8	WR	279,1684
104985	+77 00461	HR 4609		12 02 44	+77 11.1	2	5.80	.000	1.01	.000	0.85	.000	4	G9 III	15,1003
		LOD 25+ # 13		12 02 44	−59 07.1	1	12.01		0.11				2		122
104993	−61 02946			12 02 45	−61 29.0	1	8.57		0.32		0.10		1	A4mA9-F3	1771
		E5 - h		12 02 47	−45 20.0	1	14.23		1.26		1.12		3		1460
		LOD 25+ # 17		12 02 47	−59 14.1	1	11.54		0.41		0.00		2		122
105016	−62 02573			12 02 47	−62 41.9	2	6.69	.010	0.29	.005	0.14	.010	3	A3mA7-F0	279,1771
		LP 434 - 42		12 02 48	+19 57.5	1	11.54		0.38		-0.14		2		1696
	+40 02496			12 02 48	+40 26.3	1	9.89		0.98		0.75		1	G9 III	280
		LF14 # 84		12 02 48	−59 41.4	1	12.53		0.32		0.21		1		137
		E5 - 59		12 02 49	−45 14.8	2	12.85	.019	0.07	.008	-0.02		10		1460,2011
	−60 03710	LSS 2585		12 02 49	−60 53.1	4	9.68	.011	0.46	.018	-0.56	.039	8	O9.5Ia	122,540,976,1684
		LOD 25 # 5		12 02 49	−61 01.2	1	11.89		0.38		0.04		2		122
104998	+32 02218			12 02 50	+31 29.3	3	8.28	.009	1.15	.020	1.18	.028	11	K0 III	37,1371,8100
105004	−25 09024	IDS12003S2602	AB	12 02 51	−26 18.8	6	10.26	.093	0.55	.024	-0.09	.004	13	F7/G0	176,1236,1696,2033*
	−58 04070			12 02 51	−59 05.9	1	8.95		1.57		1.71		2		122
	+16 02335			12 02 52	+15 37.8	1	10.67		0.60		0.14		2		1569
104999	+18 02566			12 02 52	+17 34.1	2	7.64	.000	0.41	.005	-0.03	.005	4	F2	1569,1648
		WLS 1200-5 # 9		12 02 53	−05 22.8	1	12.59		0.68		0.00		2		1375
		E5 - c		12 02 53	−45 15.8	2	13.40	.004	0.90	.008	0.43		8		1075,1460
		E5 - e		12 02 54	−45 19.	1	13.85		0.68				4		1075
		E5 - f		12 02 54	−45 19.	1	13.98		0.84				4		1075
105026	−63 02139	LSS 2586		12 02 55	−63 59.7	1	8.17		0.48		0.35		2	F0 Ib	122
		G 121 - 55		12 02 56	+22 20.2	1	13.20		0.73		0.10		1		1658
105020	+29 02252			12 02 57	+28 47.0	3	8.17	.014	1.33	.010	1.50	.003	23	K3 III	280,1613,8041
		LOD 25+ # 18		12 02 57	−59 16.9	1	11.68		1.86		2.07		2		122
	−60 03714			12 02 58	−60 34.5	1	10.33		0.14		-0.39		3		1684
		LF14 # 86		12 02 58	−61 45.0	1	10.93		1.23		0.96		2	K0 III	137
	−59 04047			12 02 59	−59 37.4	1	10.83		1.08		0.75		1	K0 III	137
105033	+43 02184			12 03 01	+43 26.6	1	8.65		1.16				3	K1 III	280
	−58 04074	LSS 2587		12 03 01	−58 25.0	1	11.51		0.34		-0.58		3		846
	+45 01994			12 03 03	+44 30.9	1	10.01		0.79				3	G0	280
105029	+69 00641	IDS12005N6921	E	12 03 03	+69 02.3	1	7.74		1.59				2	M1	70
105028	+69 00642	IDS12005N6921	A	12 03 03	+69 04.4	2	7.36	.010	1.06	.024	0.89		3	K0 III	70,3024
105028	+69 00642	IDS12005N6921	B	12 03 03	+69 04.4	1	10.98		0.63		0.10		3	G0	3024
	+16 02336			12 03 04	+15 44.2	1	9.97		0.72		0.21		2	G0	1569
		AAS12,381 34 # B		12 03 04	+33 49.3	1	10.48		1.07		1.03		1		280
		LP 674 - 65		12 03 05	−04 27.6	1	10.82		1.05		0.94		1		1696
		POSS 377 # 13		12 03 06	+25 03.4	1	15.97		1.45				3		1739
		L 254 - 2		12 03 06	−50 12.	1	12.25		0.79		0.27		2		1696
	+32 02219			12 03 07	+32 15.9	1	9.16		0.90				2	G7 IV	280
	−45 07522			12 03 07	−45 38.4	3	10.87	.007	0.99	.003	0.63	.014	22		863,1460,2011
	+14 02472			12 03 08	+14 17.7	1	10.80		0.53		-0.01		2	G5	1569
105054	−58 04075			12 03 08	−58 51.4	1	8.93		0.05		-0.05		2	A0 V	122
309522				12 03 08	−62 25.1	1	10.81		0.88		0.64		3	K5	279
105043	+63 00999	HR 4610	★ AB	12 03 09	+63 12.7	2	6.13	.000	1.17	.000	1.15	.000	4	K2 III	15,1003
	−61 02949			12 03 09	−61 50.9	1	10.36		1.22		1.05		3		279
105055	−62 02575	ZZ Cru		12 03 09	−63 13.5	1	9.61		0.04		-0.64		2	B3 V	401
	+37 02247			12 03 10	+36 32.9	1	10.16		1.04		0.97		1	K1 III	280
	−60 03718			12 03 11	−60 26.4	1	10.00		0.20		-0.16		2		122
		E5 - W		12 03 12	−45 29.	1	12.03		0.41				4		2011
	+23 02414	IDS12007N2343	AB	12 03 13	+23 27.0	1	10.69		0.75		0.32		1		280
		LF14 # 87		12 03 13	−62 26.3	1	11.49		0.78				1		137
105056	−68 01612	GS Mus		12 03 13	−69 17.7	4	7.42	.040	0.07	.009	-0.86	.021	11	O9/B0 Ia	164,191,846,1737
	+15 02409			12 03 14	+14 33.2	1	10.84		0.53		-0.02		2	F8	1569
105057	+54 01489			12 03 14	+53 45.7	2	8.20	.015	1.17	.020	1.18	.010	5	K2 III	37,1733
105061	−5 03419			12 03 15	−05 34.7	1	7.59		1.06		0.90		5	K0	897
	−59 04049			12 03 16	−60 06.2	1	10.83		0.19		0.14		1	A1 V	137
105065	−18 03319			12 03 17	−18 34.9	4	9.98	.018	1.34	.015	1.21	.011	14	K5 V	158,1705,1775,3072
105071	−64 01791	HR 4611		12 03 17	−65 16.1	7	6.31	.021	0.20	.012	-0.43	.030	20	B6 Iab/b	15,191,540,976,1075*
		E5 - a		12 03 18	−45 19.	1	13.04		0.61				4		2011
		E5 - Z		12 03 18	−45 30.	1	12.32		0.81				4		2011
		Cru sq 1 # 15		12 03 19	−62 06.6	1	12.44		0.56		0.18		2		279
	+27 02094			12 03 22	+26 56.8	1	9.78		1.09		1.01		2	G9 III	1569

Table 1

HD	DM	Other Id	N Rem	α_{1950}	δ_{1950}	S	V	σ_V	B–V	σ_{B-V}	U–B	σ_{U-B}	n	Spectrum	References
105078	−35 07694	HR 4612	★ A	12 03 22	−35 24.9	3	6.23	.005	−0.08	.001	−0.35		11	B7 V	15,1770,2012
105074	+26 02299			12 03 23	+25 56.7	1	9.65		0.94		0.69		1	G9 IV	8041
	+40 02497			12 03 23	+40 08.1	1	10.50		0.99		0.73		1	G9 III	280
		LOD 25+ # 20		12 03 23	−59 14.3	1	11.89		0.37		0.08		2		122
105075	+11 02425			12 03 24	+11 01.2	1	10.23		1.12		1.13		1	K2	280
		E5 - X		12 03 24	−45 28.	1	11.99		0.74				4		2011
		AAS12,381 36 # A		12 03 26	+36 10.5	1	10.56		0.89		0.66		1		280
105089	−2 03460	HR 4613		12 03 26	−02 51.2	2	6.36	.005	1.00	.000	0.80	.005	7	K0	15,1256
105087	+15 02410	G 12 - 16		12 03 27	+14 55.8	3	10.10	.014	0.83	.009	0.43	.015	5	K0	333,1569,1620,1658
	+69 00643	IDS12005N6921	D	12 03 27	+69 04.5	1	9.11		1.10				2	K0	70
105085	+24 02424			12 03 28	+23 29.0	1	7.49		0.33		0.02		1	F5	280
		LP 494 - 30		12 03 29	+12 35.2	1	13.25		0.53		−0.16		2		1696
105116	−45 07529			12 03 30	−45 36.7	4	8.53	.004	0.17	.004	0.10	.003	18	A2 V	977,1075,1770,2011
	−45 07528			12 03 31	−46 22.8	1	9.93		0.61				4		2011
	−60 03723	NGC 4103 - 1	★ V	12 03 31	−60 58.7	1	9.50		0.05		−0.63		2	B2 IVe	48
	+35 02311			12 03 32	+34 39.4	1	10.02		1.00				2	G8 III	280
	+41 02272			12 03 32	+41 12.9	1	10.51		0.70				3		280
105122	+69 00644	IDS12010N6915	ABC	12 03 32	+68 58.7	3	7.13	.023	0.44	.020	0.02	.020	5	F5	292,1381,3030
105102	+26 02300			12 03 33	+26 03.0	2	8.99	.015	1.24	.005	1.37	.015	2	K1 III	280,8041
105100	+32 02220			12 03 33	+32 25.5	2	8.95	.000	0.87	.005	0.49		5	G7 III	280,1569
	+48 01999			12 03 33	+47 32.9	1	9.75		0.71		0.27		1	G8 III	8034
		LF14 # 89		12 03 33	−60 07.6	1	12.45		0.16		0.20		1	A0 V	137
105121	−72 01204			12 03 33	−73 16.7	1	6.50		1.83				4	M2 III	2012
		NGP +48 051		12 03 34	+48 20.1	1	11.28		1.07		0.91		1	K0 III	8034
105101	+28 02077			12 03 35	+27 49.6	1	9.74		0.52		0.09		37	G5	1613
	+32 02221			12 03 35	+31 32.2	1	10.07		0.30		0.12		2	A5 IV	1569
		NGC 4103 - 15		12 03 35	−60 58.3	1	13.78		0.26		0.20		2		48
	+39 02493	IDS12010N3923	AB	12 03 36	+39 06.5	1	9.22		0.73				2	G6 IV	280
		NGC 4103 - 20		12 03 38	−60 56.3	1	13.88		1.35		1.20		2		48
		NGC 4103 - 21		12 03 39	−60 55.8	1	12.84		1.26		0.93		2		48
		NGC 4103 - 19		12 03 39	−60 57.1	1	14.24		1.21		0.80		2		48
		NGC 4103 - 16		12 03 39	−60 58.1	1	13.60		0.29		0.10		2		48
		Ross 636		12 03 40	+10 29.2	1	13.18		0.77		0.18		2		1696
	+28 02078	G 121 - 56		12 03 40	+27 46.6	3	10.13	.028	0.78	.018	0.00	.041	42	F6	308,1613,1658
	−60 03727	NGC 4103 - 2		12 03 40	−60 58.4	1	9.78		0.08		−0.52		3		48
	−44 07774			12 03 41	−45 16.1	2	10.82	.008	0.65	.000	0.19		11		863,2011
		LF14 # 90		12 03 41	−60 43.3	1	10.83		1.24		0.95		1	K0 III	137
		NGC 4103 - 17		12 03 41	−60 57.9	1	13.76		0.44		0.26		2		48
105139	−69 01632			12 03 41	−69 31.2	2	7.55	.005	0.01	.005	−0.65	.000	8	B2 II/III	164,400
	+16 02337			12 03 42	+15 40.2	1	10.08		1.26		1.36		1		280
	−58 04078			12 03 42	−58 43.9	1	9.93		0.47		0.00		1	F5 V	137
105138	−67 01903	HR 4614		12 03 43	−68 22.4	4	6.22	.004	1.24	.004	1.04	.000	19	G3 Ib	15,1075,1075,2038
105150	−58 04079			12 03 44	−58 52.0	2	7.66	.005	0.12	.010	0.12	.020	4	A0/1 V	122,1771
		NGC 4103 - 22		12 03 45	−60 55.4	1	13.75		1.31		0.90		2		48
		NGC 4103 - 18		12 03 45	−60 57.6	1	13.04		0.19		−0.12		2		48
	−60 03729	NGC 4103 - 3		12 03 45	−60 58.2	1	11.28		0.10		−0.40		2		48
	+68 00673			12 03 46	+67 56.6	1	9.17		1.15				1	K2	70
105140	+47 01926			12 03 47	+46 33.8	2	7.53	.010	1.51	.015	1.84	.015	4	K5 III	172,1733
105151	−65 01788	HR 4615	★ APB	12 03 47	−65 25.8	3	5.96	.057	0.61	.022	0.34	.044	11	G8/K0 III +	15,1075,1279
105154	−74 00877			12 03 47	−74 59.3	2	9.93	.005	0.53	.010	0.00		7	F6 V	236,2012
		NGC 4103 - 24		12 03 48	−60 55.5	1	12.52		0.14		−0.20		2		48
105152	−65 01788	IDS12012S6509	C	12 03 48	−65 25.8	2	8.01	.010	0.37	.005	0.04	.010	5	A3/5 V	1279,1771
105157	+13 02488			12 03 50	+12 47.6	1	8.61		0.93		0.55		1	G5	280
		NGC 4103 - 30		12 03 50	−60 56.9	1	13.34		0.25		0.00		2		48
		NGC 4103 - 106		12 03 50	−61 00.2	1	15.14		1.00				1		48
		G 11 - 39		12 03 51	+07 30.6	1	13.10		1.17		1.06		1		333,1620
		LF14 # 93		12 03 51	−59 15.9	1	12.27		0.28		0.02		1	A3 V	137
		NGC 4103 - 107		12 03 51	−61 00.3	1	14.85		0.62				1		48
	−59 04053			12 03 52	−60 14.7	1	11.48		0.12		−0.17		1	B8 V	137
	−60 03731	NGC 4103 - 4	★	12 03 55	−60 57.9	2	10.01	.015	0.08	.010	−0.58	.030	6		48,1747
105169	−64 01795			12 03 56	−65 12.8	1	8.59		0.30		0.19		1	A7 V	1771
	−44 07777			12 03 57	−45 10.4	4	9.99	.007	0.36	.003	0.17	.006	35	A2	863,1460,1704,2011
		NGC 4103 - 34		12 03 57	−60 59.3	1	12.10		0.14		−0.40		2		48
		POSS 377 # 28		12 03 58	+22 28.9	1	16.67		1.45				4		1739
	−60 03735	NGC 4103 - 330	★	12 03 58	−60 56.7	1	10.31		0.08		−0.61		4		1747
105163	+13 02490			12 03 59	+12 42.4	1	9.12		0.20		0.02		1	A8 IV	280
		Sand 12		12 04 00	+23 25.	1	13.97		1.62				1	dM3	686
		NGC 4103 - 42		12 04 01	−60 59.5	1	12.18		0.20		−0.25		2		48
		NGC 4103 - 38		12 04 02	−60 58.6	1	11.42		1.72		1.62		3		48
	+24 02425			12 04 03	+23 33.8	1	10.30		1.06		1.03		1		280
105178	−63 02143			12 04 03	−63 39.7	1	9.03		0.04		−0.26		2	B8 V	401
	−60 03738	NGC 4103 - 7	★	12 04 04	−60 57.0	2	10.25	.039	0.06	.010	−0.61	.005	6		48,1747
	−60 03739	NGC 4103 - 5		12 04 05	−60 58.6	1	9.92		0.14		−0.15		2		48
105183		Feige 56		12 04 06	+11 57.	4	11.07	.020	−0.12	.003	−0.57	.007	11	A0 V	272,1026,1298,1569
	−44 07778			12 04 06	−45 16.6	2	9.31	.009	1.67	.005	1.92		5		365,2011
	−60 03741	NGC 4103 - 333		12 04 06	−60 56.8	1	9.40		0.08		−0.62		3		1684
	−60 03742	NGC 4103 - 6		12 04 06	−60 57.7	1	9.82		0.10		−0.54		2		48
		NGC 4103 - 513		12 04 06	−60 58.	1	11.40		0.24				3		2042
		NGC 4103 - 514		12 04 06	−60 58.	1	12.90		0.13				3		2042
		NGC 4103 - 515		12 04 06	−60 58.	1	12.22		0.34				3		2042

HD	DM	Other Id	N	Rem	α_{1950}	δ_{1950}	S	V	σ_V	B–V	σ_{B-V}	U–B	σ_{U-B}	n	Spectrum	References
105181	+34 02285				12 04 07	+33 50.7	2	7.94	.019	1.52	.010	1.82		3	K4 III	280,8041
		POSS 377 # 30			12 04 09	+26 18.8	1	16.83		1.60				4		1739
105182	+30 02223				12 04 09	+29 46.1	1	8.27		1.34				2	K3 III	280
105193	−59 04054				12 04 09	−59 28.5	1	8.46		1.14		0.95		1	G8/K0 IV	137
105196	−72 01205				12 04 10	−72 47.6	1	8.32		0.13				4	B9 IV	2012
	−60 03745	NGC 4103 - 13		★	12 04 11	−61 00.0	3	10.46	.039	0.13	.017	−0.55	.045	10		48,1747,2042
105194	−60 03744	NGC 4103 - 8			12 04 12	−60 54.6	2	9.26	.020	0.12	.020	0.01		16	A0	48,2042
105200	+14 02473				12 04 14	+13 55.7	1	10.23		0.50		−0.08		2	F8	1569
	+23 02415				12 04 14	+23 23.3	1	10.35		1.06		0.92		1		280
105199	+31 02328				12 04 14	+30 32.4	1	9.83		0.18		0.09		2	A2 V	1068
105205	−11 03238				12 04 14	−11 57.8	1	6.77		0.32		0.06		5	F0	1628
	+24 02426				12 04 15	+24 25.1	1	10.08		0.90		0.51		1	G5	280
105210	−61 02958				12 04 15	−62 02.2	2	10.48	.033	0.17	.028	0.03	.009	4	B9/A1 V	137,279
105211	−63 02145	HR 4616		★ A	12 04 15	−64 20.1	4	4.14	.004	0.35	.007	0.01	.012	12	F0 IV	15,1075,2006,3026
105209	−58 04085				12 04 16	−59 18.8	1	8.68		0.21		0.13		1	A1 V	1771
	−60 03746	NGC 4103 - 12		★	12 04 16	−60 59.5	4	9.20	.028	0.18	.024	−0.73	.037	13	B2 IVe	48,1684,1737,1747
105216	+34 02286				12 04 20	+33 43.2	1	8.08		1.58				2	K5 III	280
	−60 03748	NGC 4103 - 11		★	12 04 21	−60 59.3	2	9.42	.005	0.11	.014	−0.60	.045	10		48,1747
105219	+6 02551	G 13 - 8		★ AB	12 04 22	+06 05.1	1	8.42		0.78				2	K0	202
		AAS12,381 39 # A			12 04 22	+38 37.8	1	11.63		0.08		0.08		1		280
105217	+21 02392	IDS12018N2103	AB		12 04 23	+20 46.3	2	8.90	.015	0.42	.005	0.15	.000	5	G5	1569,1648
	−60 03750				12 04 23	−60 53.8	1	9.97		0.23		−0.28		1	B6	549
309529	−62 02578				12 04 23	−62 39.2	2	9.73	.020	1.29	.020	1.18	.000	5	K2 III	137,279
105233	−58 04087				12 04 24	−59 25.0	1	8.55		0.23		0.04		2	F0 V	122
	−60 03749	NGC 4103 - 14		★	12 04 24	−61 00.9	3	10.02	.027	0.13	.022	−0.54	.045	8		48,1747,2042
105245	−61 02959				12 04 25	−62 14.2	1	9.73		0.03		−0.44		3	B8/9 II/III	279
	+28 02079				12 04 26	+28 04.5	1	10.05		0.97		0.66		1	G7 III	280
	−60 03754	NGC 4103 - 9			12 04 27	−60 56.9	2	10.94	.045	0.15	.000	−0.38		5		48,2042
	−61 02960				12 04 27	−61 58.7	1	9.27		1.37		1.22		3		279
105238	+12 02423				12 04 28	+11 48.4	1	10.21		0.36		−0.01		2	F5	1569
	+21 02393				12 04 28	+21 21.7	1	9.44		1.22		1.27		1	K2	280
	+27 02096				12 04 28	+27 17.7	1	10.37		0.73		0.44		1	G6 III	280
	−58 04088				12 04 28	−59 14.7	1	9.85		0.14		−0.33		2	B6	122
105246	−75 00780				12 04 29	−75 30.5	2	8.43	.005	1.75	.030	1.70		7	M0 III	236,2012
	+32 02223				12 04 32	+32 17.9	1	9.46		1.14				2	K1 III	280
	+22 02440				12 04 33	+21 35.2	1	10.56		1.03		0.97		1	K2	280
	−60 03756				12 04 33	−60 27.9	1	10.89		0.45		0.16		1	F2 V	137
	+25 02460				12 04 35	+24 53.1	1	10.41		1.11		1.09		1		280
105256	−60 03757	NGC 4103 - 10			12 04 35	−60 59.2	2	9.05	.015	0.12	.005	−0.52		6	B2 II/III	48,2042
	+23 02416				12 04 36	+22 31.9	1	9.76		0.96		0.78		1		280
105259	+48 02001				12 04 36	+47 34.6	1	8.29		1.36		1.56		1	K4 III	8034
105263	+10 02381				12 04 37	+09 56.4	1	7.38		0.32		0.10		2	A3	1733
105262	+13 02491				12 04 37	+13 15.9	2	7.08	.000	−0.01	.005	−0.08	.034	3	B9	280,3077
		POSS 377 # 23			12 04 37	+25 42.9	1	16.41		1.30				3		1739
105261	+15 02412				12 04 39	+15 22.7	1	9.63		1.37		1.40		1	K2	280
		G 237 - 44			12 04 39	+68 02.8	1	16.36		1.70				1		906
105274	−63 02147				12 04 40	−64 24.2	1	6.84		0.45		−0.05		2	F5 V	122
105260	+18 02568				12 04 41	+17 53.2	1	9.17		0.26				4	F0 III-IV	193
105266	−5 03424	RW Vir			12 04 41	−06 29.2	1	6.89		1.56		1.17		5	M5 III	897
		Cru sq 1 # 103			12 04 42	−62 18.	1	12.51		0.70		−0.29		4		846
	−4 03208	G 13 - 9			12 04 43	−05 26.9	5	10.02	.034	0.40	.008	−0.21	.012	13	A7	158,1064,1620,1696,3077
105284	−57 05260				12 04 45	−58 18.3	1	8.55		0.25		0.11		1	A2 V	1771
105281	−9 03434				12 04 48	−10 27.7	3	8.22	.009	0.38	.008	0.21	.012	11	Am	355,1003,1775
105283	−42 07433				12 04 48	−42 58.1	4	7.20	.003	0.41	.005	0.01	.012	18	F3 IV	768,977,1075,2011
105288	+43 02187	IDS12023N4339	A		12 04 49	+43 22.5	1	8.07		0.93				2	G9 III	280
105290	+20 02683				12 04 50	+19 37.9	1	8.19		1.14		1.09		1	K2	280
	−45 07547				12 04 50	−45 48.7	1	11.17		1.11				4		2011
105299	−62 02582				12 04 50	−63 17.3	1	10.14		0.00		−0.75		2	B2/3 V	401
	+16 02339				12 04 51	+16 24.9	1	9.93		0.96		0.71		2	K0	1569
	−44 07788				12 04 51	−44 36.6	1	9.32		1.68		2.07		6		863
105305	+12 02492				12 04 54	+12 56.8	1	9.66		0.47		−0.06		2	G0	1569
105302	+34 02288				12 04 54	+33 59.4	1	8.88		1.08				2	K0 III	280
105303	+23 02417				12 04 55	+23 19.2	1	7.61		0.28		0.09		1	F2	280
105304	+13 02493	G 12 - 17		★	12 04 56	+13 18.8	2	9.67	.024	0.96	.010	0.80	.005	3	K0	333,1569,1620
		Lod 565 # 13			12 04 56	−60 34.8	1	11.16		0.70		0.20		1	F8	549
		Lod 565 # 18			12 04 56	−60 34.8	1	12.10		0.60		0.25		1	F8	549
		POSS 377 # 24			12 04 57	+20 29.8	1	16.43		1.64				4		1739
105313	−44 07791				12 04 57	−45 18.2	13	7.31	.004	0.02	.003	−0.05	.008	225	B9 V	278,365,863,977,1075,1460*
105306	+12 02424				12 04 58	+12 13.1	1	9.02		1.01		0.84		1	K0	280
	−75 00781				12 04 59	−75 39.2	1	10.47		0.70				4	G0	2012
		Sand 14			12 05 00	+27 53.	1	15.09		1.51				1	dM1	686
105330	−30 09687				12 05 01	−31 07.8	2	6.73	.009	0.53	.004	0.00	.002	12	F7 V	1628,1770
		Cru sq 1 # 20			12 05 01	−62 11.6	1	13.23		1.36		1.16		4		279
	−45 07549				12 05 02	−45 41.2	1	9.86		1.44				4		1075
105322	+13 02494	IDS12025N1312	A		12 05 03	+12 55.6	1	8.80		0.36		0.04		2	F2	1569
105321	+14 02474				12 05 03	+13 47.7	1	7.65		1.01		0.85		1	K0	280
	+26 02304				12 05 03	+26 10.7	1	10.42		1.36				1	K2 III	694
105319	+42 02274				12 05 03	+42 20.9	1	7.63		1.58				2	K4 III	280
		G 13 - 10			12 05 03	−05 01.5	1	13.79		1.29		1.09		1		1658
	+49 02113				12 05 05	+48 30.1	1	10.42		0.93		0.60		1	G8 III	8034

Table 1   647

HD	DM	Other Id	N	Rem	α_{1950}	δ_{1950}	S	V	σ_V	B–V	σ_{B-V}	U–B	σ_{U-B}	n	Spectrum	References
		LSS 2598			12 05 05	−70 05.4	1	11.21		0.00		−0.79		3	B3 Ib	846
		GPEC 465			12 05 06	−29 47.6	1	11.01		0.89				1		1719
		Cru sq 1 # 21			12 05 06	−62 10.8	1	10.97		1.00		0.75		5		279
	+2 02516	G 13 - 11			12 05 07	+01 53.0	2	9.72	.010	1.05	.025	1.01	.040	2	K5	308,1620
	+15 02413				12 05 10	+15 24.6	1	10.59		0.90		0.67		1		1569
	+40 02499				12 05 10	+40 27.3	1	10.08		1.17		1.13		1	K1 III	280
105340	−74 00880	HR 4617			12 05 10	−75 05.3	5	5.17	.008	1.30	.002	1.37	.004	25	K2 II/III	15,278,1075,2012,2038
		AAS12,381 23 # B			12 05 11	+22 48.6	1	10.97		−0.16		−0.63		1		280
	+36 02238				12 05 11	+35 31.4	1	10.17		1.02		0.85		1	K0 III	280
105339	−60 03763				12 05 11	−60 27.6	1	8.89		0.11		−0.34		1	B5/7 III/V	549
		Cru sq 1 # 22			12 05 11	−62 07.5	1	14.73		0.79		0.06		6		279
		Sand 15			12 05 12	+25 23.	1	14.62		1.41				1	dM1	686
105341	+31 02329				12 05 12	+30 41.9	1	8.39		1.52				2	K3 III	280
	−60 03764				12 05 12	−60 33.4	2	10.02	.010	0.15	.000	−0.19	.015	2	B9	287,549
		LF14 # 98			12 05 12	−61 38.7	1	11.66		0.29		−0.12		1	B9 V	137
	+27 02099				12 05 14	+27 25.4	1	10.19		1.06		1.05		1	K0 III	280
105353	−58 04094				12 05 14	−59 00.9	1	8.02		−0.05		−0.39		2	B8 II	122
		NGP +48 057			12 05 15	+48 15.1	1	11.90		1.09		0.89		1	K0 III	8034
	+31 02330	IDS12027N3144	A		12 05 16	+31 27.2	1	10.09		0.76				3	K0	280
		Cru sq 1 # 23			12 05 16	−62 13.3	1	14.22		0.77		0.16		6		279
	−37 07677				12 05 17	−38 24.1	2	9.95	.007	0.86	.000	0.41		5		1594,6006
		Cru sq 1 # 24			12 05 18	−62 06.2	1	13.53		0.45		0.26		4		279
	+37 02252				12 05 19	+37 20.0	1	9.94		1.00				2	K0 III	280
105364	−45 07554				12 05 21	−45 38.8	2	9.37	.013	0.80	.001	0.41		11	G2 V	863,2011
	−34 07927				12 05 22	−35 14.6	2	10.34	.000	0.49	.000	−0.04		6	F8	1696,2033
	−60 03766				12 05 22	−60 38.1	2	10.04	.000	0.17	.000	0.04	.000	2	B9	287,549
105368	+43 02190				12 05 23	+43 02.7	1	9.82		1.25		1.34		1	K2 IV	280
105363	−44 07797				12 05 23	−45 06.5	1	9.06		1.05				4	G8/K0III/IV	2011
105365	−58 04095				12 05 23	−58 28.5	1	8.70		0.33		0.24		1	A8 III/IV	1771
		Cru sq 1 # 25			12 05 23	−62 13.3	1	12.24		1.48		1.46		7		279
		Cru sq 1 # 26			12 05 23	−62 19.0	1	11.89		0.21		−0.33		3		279
	+16 02340				12 05 24	+16 13.5	1	10.29		0.90		0.63		1		1569
105381	−44 07799			V	12 05 25	−44 32.5	3	8.07	.009	1.68	.008	1.99		13	M2 (III)	977,2001,2011
		Lod 565 # 17			12 05 26	−60 28.5	1	11.46		0.35		−0.06		1	B7	549
105372	+15 02414				12 05 27	+14 37.5	1	10.19		0.64		0.15		1	G5	1569
105369	+43 02191	IDS12029N4315	AB		12 05 27	+42 58.7	1	9.50		0.42				3	F8	280
	−62 02586	Cru sq 1 # 114			12 05 27	−62 37.1	1	10.75		0.11		−0.59		3		846
105371	+26 02307	Coma Ber 1			12 05 28	+25 39.4	1	8.50		0.47		−0.03		2	F6 V	105
105382	−49 06813	HR 4618, V863 Cen		⋆B	12 05 29	−50 23.0	9	4.46	.008	−0.16	.012	−0.68	.013	38	B6 IIIe	15,26,1075,1637,1770,2012*
105383	−50 06688	HR 4619		⋆C	12 05 29	−50 29.1	4	6.36	.006	−0.06	.011	−0.20		13	B9 V	15,1770,2012,2024
	+24 02428				12 05 30	+23 58.8	1	9.35		1.02		0.86		2	K0	1569
	+47 01929				12 05 30	+47 07.1	1	9.42		0.42		−0.05		2	F2	1601
	+16 02342				12 05 31	+15 43.2	1	10.60		0.90		0.70		1		1569
105400	−49 06815				12 05 31	−49 43.8	1	9.35		0.01		0.03		1	B9.5 V	1770
105401	−60 03771				12 05 32	−60 30.2	1	7.50		0.55		0.04		1	F8 V	549
105388	+31 02331				12 05 33	+31 19.7	1	7.45		−0.02		0.02		3	A0 V	1068
	+34 02289				12 05 34	+34 19.0	1	10.34		1.00		0.79		1	G9 III	280
105390	+0 02897				12 05 34	−00 07.2	1	9.26		0.45		−0.03		1	F5	289
105408	+13 02495				12 05 37	+12 53.2	1	10.45		0.40		−0.05		1	F5	1569
105421	+56 01568	IDS12031N5601	A		12 05 37	+55 44.6	2	7.79	.030	0.51	.020	−0.02	.005	5	F8	1502,3026
105415	−45 07558				12 05 37	−45 48.4	2	9.98	.007	0.17	.002	0.02		11	A0 V	863,2011
		Cru sq 1 # 27			12 05 37	−62 17.5	1	17.70		0.87		−0.02		4		279
105407	+14 02475				12 05 38	+13 35.8	1	9.55		0.95		0.75		1	K0	280
	+17 02434				12 05 38	+16 58.8	1	9.71		0.51		0.02		1	F8	1569
105409	−3 03239				12 05 39	−04 00.5	1	7.52		1.07		0.92		3	K0	1657
105416	−47 07396	HR 4620			12 05 39	−48 24.9	13	5.33	.005	−0.01	.006	−0.03	.008	81	B9 V	14,15,26,182,278,388*
	−61 02967				12 05 39	−62 09.2	1	10.34		1.10		0.86		4		279
	−61 02968				12 05 39	−62 23.9	2	10.07	.005	0.13	.000	−0.52	.009	13		279,846
105422	+56 01569	IDS12031N5601	B		12 05 40	+55 44.6	2	8.35	.000	0.60	.025	0.04	.000	5	F8	1502,3026
		Cru sq 1 # 30			12 05 40	−62 14.9	1	16.72		1.16				5		279
		Cru sq 1 # 31			12 05 40	−62 26.2	1	16.63		1.07		0.07		4		279
105418	−60 03774	LSS 2599			12 05 41	−61 21.4	1	9.00		0.59		0.54		3	F0 IV	1747
		Cru sq 1 # 32			12 05 41	−62 07.5	1	15.64		0.61		0.08		3		279
		LF14 # 99			12 05 42	−59 29.3	1	11.53		1.37		1.32		1	K3 III	137
105424	+31 02332				12 05 43	+30 33.2	2	7.59	.010	1.53	.005	1.83		3	K3 IV	280,8041
		Cru sq 1 # 33			12 05 43	−62 15.2	1	13.56		1.21		0.69		6		279
		G 13 - 12			12 05 45	+00 10.4	2	10.90	.005	0.95	.020	0.72	.005	2	K2	1620,1658
105435	−50 06697	HR 4621, δ Cen		⋆A	12 05 45	−50 26.6	11	2.58	.019	−0.13	.009	−0.89	.012	84	B2 IVne	15,26,198,681,815,1088*
		LF14 # 100			12 05 45	−60 53.5	1	11.33		0.45		0.22		1	F2 V	137
		Cru sq 1 # 34			12 05 46	−62 15.4	1	13.62		1.20		0.75		5		279
		Cru sq 1 # 35			12 05 46	−62 27.0	1	16.92		0.95				5		279
105437	−60 03777	HR 4622			12 05 47	−60 34.1	5	6.21	.008	1.73	.014	1.88		19	K3/4 II	15,549,1075,1075,2007
		Cru sq 1 # 36			12 05 47	−62 10.6	1	12.72		0.34		0.29		2		279
		Cru sq 1 # 37			12 05 47	−62 14.0	1	15.10		0.80		0.24		3		279
		LP 674 - 108			12 05 50	−03 46.4	1	12.37		0.43		−0.18		2		1696
105452	−24 10174	HR 4623			12 05 50	−24 27.0	11	4.02	.005	0.33	.011	−0.01	.014	52	F0 IV/V	3,15,418,1007,1008*
	−60 03778				12 05 50	−60 36.1	1	11.38		0.38		−0.26		1		549
	−58 04101				12 05 51	−58 46.0	1	11.24		0.42		0.04		2	F0 V	137
		Cru sq 1 # 39			12 05 51	−62 01.5	1	12.21		0.45		0.24		9		279
		G 11 - 40			12 05 52	−00 12.2	4	11.24	.016	1.42	.018	1.17		7	K4	158,1017,1705,3073

HD	DM	Other Id	N	Rem	α_{1950}	δ_{1950}	S	V	σ_V	B–V	σ_{B-V}	U–B	σ_{U-B}	n	Spectrum	References
105444	+15 02416				12 05 53	+15 24.1	1	9.66		1.02		0.83		1	K2	280
		Cru sq 1 # 41			12 05 53	−62 17.8	1	17.50		1.54				4		279
		Klemola 107			12 05 54	+12 37.	1	10.78		0.29		0.15		2		1026
105460	+13 02496				12 05 55	+13 23.6	1	9.72		0.92		0.63		1	K0	1569
	−45 07563				12 05 55	−45 36.9	1	9.88		1.44		1.52		6		863
	−60 03781				12 05 56	−60 33.9	2	10.24	.015	0.11	.000	-0.14	.000	2	B9	287,549
105457	−61 02972				12 05 56	−62 05.4	1	9.03		0.08		−0.27		2	Ap Si	279
	+35 02315				12 05 57	+34 57.9	1	9.54		0.70				2	G5 V	280
105464	−9 03439				12 05 57	−09 43.3	1	9.76		0.31		−0.47		3	F8	1732
	−62 02590				12 05 57	−62 25.5	1	9.78		1.12		0.93		15		279
	−58 04102				12 05 58	−59 24.8	1	10.30		0.19		0.07		1	A1 V	137
	−63 02154	LSS 2600			12 05 58	−63 30.3	2	10.36	.019	0.07	.005	−0.69	.024	6		846,1747
	+25 02465	Coma Ber 2			12 05 59	+25 04.7	1	9.10		1.08		1.06		1	K0 III	280
		Cru sq 1 # 44			12 05 59	−62 09.7	1	16.68		0.82		0.22		2		279
		LF14 # 103			12 06 00	−58 47.8	1	12.70		0.35		−0.02		1	F0 V	137
	+0 02900	UU Vir			12 06 01	−00 10.7	2	10.07	.174	0.14	.052	0.15	.014	2	F2	668,699
	−60 03783				12 06 01	−60 25.7	1	10.90		0.22		−0.14		1	B9	549
		Cru sq 1 # 45			12 06 02	−62 05.6	1	13.70		0.23		0.14		3		279
	−44 07807	E5 - V			12 06 03	−44 35.6	2	10.08	.002	0.38	.002	0.11		11	G0	863,2011
		Cru sq 1 # 46			12 06 03	−62 06.7	1	13.97		1.41		1.37		4		279
		Cru sq 1 # 47			12 06 03	−62 26.3	1	15.71		0.78		0.18		3		279
		Coma Ber 384			12 06 04	+26 46.4	1	15.07		1.38				2		913
		Coma Ber 385			12 06 04	+26 46.4	1	14.12		1.47				4		913
105475	+27 02100	Coma Ber 3			12 06 04	+26 46.4	4	6.99	.012	1.02	.010	0.84	.003	9	K0 III	172,1080,1355,3026
		Cru sq 1 # 48			12 06 04	−62 27.4	1	16.75		1.55				3		279
105486	+43 02192				12 06 05	+43 14.4	1	9.67		0.94				2	G9 III	280
	−61 02973				12 06 05	−62 19.3	1	10.14		1.28		1.40		2		279
105487	+13 02497				12 06 06	+13 04.1	1	9.44		0.78		0.32		1	K0	1569
	+41 02274				12 06 06	+40 55.5	1	9.45		0.87				2	G8 III	280
105474	+45 01997				12 06 06	+44 41.5	1	8.69		1.06				2	K0 III	280
105483	−67 01910				12 06 06	−68 21.6	1	7.94		0.25		0.14		1	A6 V	1771
105498	−44 07808				12 06 07	−44 38.9	9	8.06	.006	-0.03	.006	-0.24	.019	148	B8/9 V	116,278,460,1066,1075,1586*
		Lod 565 # 16			12 06 07	−60 38.9	1	11.38		1.75		2.02		1		549
	−60 03787				12 06 10	−60 34.9	2	10.21	.000	0.19	.005	0.12	.025	2	A0	287,549
		Lod 565 # 19			12 06 10	−60 37.9	1	12.12		0.63		−0.02		1	B8	549
	+17 02436				12 06 11	+17 10.8	1	9.94		1.30		1.52		1	K5	280
	−57 05272	LSS 2601			12 06 13	−58 23.5	4	9.98	.015	0.30	.011	−0.65	.029	7	O7	191,540,976,1684
	−60 03790				12 06 13	−60 34.9	2	10.09	.000	0.13	.000	0.10	.000	2	A0	287,549
	−60 03789				12 06 13	−60 44.5	1	11.35		0.15		−0.49		1	B3	549
		Cru sq 1 # 50			12 06 13	−62 07.6	1	13.02		0.51		0.02		4		279
105502	+15 02417				12 06 14	+15 18.9	1	9.83		0.86		0.46		1	K0	280
		Feige 57			12 06 18	+07 44.0	2	13.30		-0.03	.005	-0.06	.005	2		1264,3060
105516	+38 02304				12 06 18	+38 08.0	1	8.82		1.24				4	K1 III	280
105509	−43 07502	HR 4624, V788 Cen			12 06 18	−44 02.8	10	5.74	.007	0.24	.005	0.15	.004	52	A2mA5-F2	14,15,278,388,977,977*
105521	−40 07128	HR 4625, V817 Cen			12 06 19	−40 57.2	6	5.53	.050	-0.08	.011	−0.64	.018	25	B3 IVe	15,681,815,1637,1770,2012
	+17 02437				12 06 23	+16 59.3	1	10.49		1.09		0.97		1		1569
	+22 02442	G 59 - 1		★ A	12 06 23	+22 04.0	3	9.53	.021	0.65	.010	0.05	.004	5	G2 V	333,1003,1620,1773
105525	+49 02116				12 06 23	+49 27.8	1	7.45		1.03		0.88		2	K0 III	172
105523	−60 03792	LSS 2602			12 06 23	−61 12.3	2	9.40	.005	0.11	.005	−0.56	.020	4	B3 V	846,1747
		G 121 - 58			12 06 24	+22 04.1	1	14.02		1.38		0.83		2		1773
	+25 02466	Coma Ber 4			12 06 24	+24 58.8	1	9.36		1.08		1.06		1	K0 III	280
		Cru sq 1 # 51			12 06 24	−62 09.3	1	16.73		1.04		−0.21		2		279
105526	+13 02498				12 06 25	+12 57.9	1	8.92		0.33		−0.04		1	F0	1569
		Cru sq 1 # 52			12 06 27	−62 12.6	1	12.36		1.22		0.77		7		279
	+40 02503				12 06 29	+40 09.4	1	9.22		1.04				2	G9 III	280
105530	−50 06710				12 06 29	−50 52.1	1	8.61		0.18		0.06		1	A2/3 V	1771
105524	−75 00785				12 06 29	−76 21.5	1	9.10		0.72				4	F3 III/IV	2012
		LB 2229			12 06 32	+52 37.2	1	14.54		0.44		0.16		3		308
105533	−64 01802	IDS12039S6504		AB	12 06 32	−65 20.2	1	8.73		0.36		0.19		1	A9 V	1771
	−61 02978				12 06 33	−62 23.0	1	9.85		1.28		1.01		8		279
		Cru sq 1 # 54			12 06 33	−62 24.2	1	10.95		1.18		0.90		5		279
105547	+32 02227	IDS12040N3226		A	12 06 36	+32 12.0	1	9.15		0.34		−0.04		2	F2 V	1569
	+36 02241				12 06 36	+36 22.3	1	10.49		0.90		0.80		1	K0 III	280
105542	−61 02979				12 06 37	−61 45.2	1	9.44		0.08		−0.45		2	B7/8 III/IV	279
105544	−63 02158				12 06 37	−63 27.1	1	8.17		0.15		0.16		1	A3 V	1771
105540	−58 04111				12 06 38	−58 25.9	1	9.39		0.31		0.18		1	A3/4 IV/V	1771
	−61 02980				12 06 38	−61 34.3	1	10.05		0.23		0.01		19		279
		Cru sq 1 # 56			12 06 38	−62 24.5	1	16.88		1.21		0.64		4		279
		Cru sq 1 # 57			12 06 39	−62 22.4	1	16.05		0.23		0.16		4		279
105548	+17 02439				12 06 40	+17 27.9	2	7.23	.005	1.63	.010	2.03		4	M1 III	280,3040
		G 59 - 2			12 06 40	+19 45.0	1	11.04		1.29		1.24		1	K7	333,1620
105567	+15 02418				12 06 41	+15 06.5	1	7.81		0.49		0.03		2	F8	1569
	+27 02101				12 06 41	+27 01.3	1	10.50		1.13		1.20		1	K2 III	280
		LP 554 - 83			12 06 42	+05 45.1	1	10.71		0.98		0.84		1	K3	1696
105564	−63 02159	IDS12041S6319		AB	12 06 42	−63 35.7	1	9.02		0.27		0.19		1	A3/4 V	1771
		Cru sq 1 # 58			12 06 43	−62 22.9	1	14.58		0.68		0.01		2		279
	+22 02443				12 06 44	+22 27.2	1	10.28		0.94		0.78		1		280
	+36 02242				12 06 44	+35 59.4	1	9.99		-0.08		−0.26		1	B8	1026
		GPEC 515			12 06 44	−29 15.8	1	10.56		1.11				2		1719
	−61 02971				12 06 44	−61 56.1	1	9.60		0.62		0.17		3		279

Table 1   649

HD	DM	Other Id	N Rem	α_{1950}	δ_{1950}	S	V	σ_V	B–V	σ_{B-V}	U–B	σ_{U-B}	n	Spectrum	References
105562	−61 02981			12 06 44	−61 56.1	1	9.13		0.05		−0.28		3	B8/9 IV/V	279
105585	+43 02194			12 06 45	+42 45.6	3	8.50	.005	0.66	.005	0.21	.005	5	G5 IV	172,280,8038
105577	−42 07452			12 06 45	−42 33.6	1	7.93		0.53				4	F6 V	2012
	−61 02982	LSS 2603		12 06 45	−62 08.5	3	9.78	.012	0.08	.004	−0.71	.013	15		279,846,1747
105563	−63 02160	IDS12041S6316	AB	12 06 45	−63 32.5	1	7.13		1.17		−0.21		1	M1e	1684
105580	−59 04080			12 06 46	−59 29.5	3	7.14	.027	−0.06	.023	−0.52	.027	9	B3 IV	26,122,158
		Cru sq 1 # 61		12 06 46	−62 10.4	1	14.53		0.68		0.39		7		279
105566	+16 02343	IDS12042N1558	AB	12 06 47	+15 41.5	1	9.23		0.85		0.48		1	K0	280
105581	−60 03796			12 06 47	−60 40.5	1	8.66		0.08		−0.17		1	B9	549
		Cru sq 1 # 105		12 06 48	−63 51.	1	13.13		0.98		−0.09		4		846
105586	+30 02230			12 06 49	+29 44.3	3	8.02	.017	0.99	.005	0.76	.005	5	G9 III	172,280,8041
	+47 01931			12 06 49	+47 12.9	1	9.70		0.20		0.12		1	A2	272
	−60 03798			12 06 49	−61 08.0	1	10.41		0.19		0.13		2	A1 V	137
		POSS 377 # 17		12 06 52	+20 48.4	1	16.17		1.42				3		1739
105590	−11 03246	IDS12043S1118	AB	12 06 53	−11 34.6	3	6.56	.005	0.66	.000	0.17	.005	9	G0	158,258,1075
		G 11 - 41		12 06 56	+00 58.9	1	11.07		0.45		−0.22		4		158
105601	+39 02496			12 06 56	+38 54.7	3	7.38	.004	0.30	.010	0.10	.000	6	Am	1068,3002,8071
		Cru sq 1 # 62		12 06 56	−62 26.3	1	11.84		1.55		1.75		4		279
105606	−44 07824			12 06 57	−44 36.3	2	9.28	.009	0.34	.002	0.17		11	F0 IV/V	863,1075
	−61 02984			12 06 57	−61 57.0	1	10.01		1.38		1.64		4		279
	−61 02986			12 06 58	−61 54.9	1	10.38		0.09		−0.36		3		279
105615	−66 01707			12 06 58	−67 16.5	1	7.45		0.10		0.06		2	A0/1 V	1771
105603	+17 02441			12 06 59	+16 41.4	1	9.51		0.50		−0.10		1	F8	1569
	+17 02442			12 06 59	+16 47.7	1	11.08		0.66		0.29		1		1569
105617	+46 01773			12 06 59	+46 07.9	1	10.07		1.17		1.23		1	K0 III	8034
		G 197 - 45		12 06 59	+52 12.7	1	10.73		0.72		0.17		1	G5	1658
105610	−51 06417			12 06 59	−51 52.0	1	7.20		−0.07		−0.35		3	B8 II	1770
	+34 02290			12 07 00	+33 56.3	1	10.43		0.70		0.29		1		280
105613	−57 05277			12 07 01	−58 04.3	1	7.38		0.14		0.11		1	A3 V	1771
		LF14 # 105		12 07 01	−59 47.0	1	11.89		1.48		1.89		1		137
		G 11 - 42		12 07 03	+08 40.0	1	13.26		1.47		1.12		1		333,1620
105618	+12 02429			12 07 04	+11 29.4	1	8.57		0.71		0.26		1	G0	280
105625	−51 06418			12 07 04	−51 34.4	1	8.19		0.09		0.12		2	A1/2 V	1771
		AAS78,203 # 3		12 07 06	−61 05.	1	12.22		0.29		−0.13		2		1684
105627	−61 02987	LSS 2604		12 07 06	−62 18.2	6	8.14	.025	0.04	.018	−0.87	.022	25	O9 V	279,401,540,846,976,1684
105631	+41 02276	G 123 - 7		12 07 07	+40 31.9	3	7.46	.012	0.79	.004	0.43	.008	7	K0 V	172,1080,3026
105639	+2 02517	HR 4626	★ A	12 07 08	+02 10.7	3	5.94	.008	1.12	.000	1.14	.004	9	K3 III	15,1256,3077
105632	+33 02193			12 07 08	+33 21.9	2	7.57	.005	1.12	.005	1.21		5	K2 III	37,280
		Cru sq 1 # 66		12 07 08	−62 18.1	1	11.22		0.98				2		279
105635	+12 02430			12 07 10	+12 10.0	1	9.85		0.60		0.12		2	G5	1569
105633	+15 02419			12 07 10	+15 08.3	1	10.36		0.67		0.16		2	G5	1569
	−45 07577			12 07 11	−45 38.0	1	10.57		0.41		0.06		6		863
105650	−63 02162	LSS 2605		12 07 11	−63 38.2	4	8.16	.015	0.07	.008	−0.77	.014	8	B1 II	401,540,976,1684
105651	+45 01999			12 07 12	+45 08.9	1	10.15		0.73				2	G5	280
		NGP +46 054		12 07 12	+46 10.6	1	11.02		0.90		0.70		1	G9 II	8034
105634	+12 02431			12 07 13	+12 18.2	1	9.04		1.35		1.62		1	K2	280
	−45 07578			12 07 13	−45 42.4	1	10.55		0.42				4	F8	1075
	+11 02433			12 07 14	+10 37.3	1	10.31		1.01		0.79		1	G5	280
	−63 02164			12 07 14	−63 33.0	2	9.63	.010	0.04	.007	−0.72	.000	5	B1.5V	540,976
	+17 02443			12 07 18	+16 55.5	1	11.13		0.59		0.10		1		1569
		AAS12,381 22 # B		12 07 18	+21 43.1	1	10.77		0.41		0.06		1		280
	+29 02259			12 07 18	+29 12.4	1	9.47		1.08		0.97		1	K0 III	280
		AAS12,381 42 # C		12 07 18	+42 25.3	1	10.35		1.20		1.41		1		280
105658	−48 07194			12 07 18	−49 13.4	1	9.53		0.03		−0.02		1	B9.5/A0 IV	1770
		Cru sq 1 # 68		12 07 18	−62 17.1	1	10.47		1.87		2.05		3		279
	−62 02602			12 07 18	−62 33.9	1	9.94		1.04		0.68		2		279
105671	−45 07581	IDS12047S4539	A	12 07 20	−45 55.7	5	8.44	.008	1.14	.004	1.08	.007	24	K4 V	158,863,977,1705,2011
	−59 04083			12 07 20	−60 04.5	1	10.26		0.13		0.08		1	B9 V	137
105663	+40 02505			12 07 22	+39 42.7	1	8.17		1.56				3	K4 IV	280
105675	−63 02166	LSS 2606		12 07 22	−63 43.3	4	9.25	.013	0.21	.017	−0.73	.023	9	Be	401,540,976,1684
105678	+75 00469	HR 4627		12 07 23	+74 56.4	2	6.34	.004	0.50	.007	0.06		4	F6 IV	71,1733
		Sand 21		12 07 24	+21 45.	1	14.99		1.70				1	dM4	686
		NGC 4151 sq1 5		12 07 24	+39 45.1	1	11.22		1.02		0.83		3		327
	−59 04084			12 07 26	−60 09.3	1	9.56		1.13		0.91		1	K0 III	137
105680	+23 02423			12 07 27	+22 55.1	2	8.07	.005	0.30	.005	0.09	.010	7	Am	1068,8071
105686	−34 07956	HR 4628	★ AB	12 07 27	−34 25.6	3	6.16	.004	0.03	.000	−0.02		10	A0 V	15,404,2012
105688	−42 07457	IDS12049S4300	A	12 07 28	−43 16.8	1	8.77		0.53		0.10		4	F5 V	1075
105689	−46 07727	IDS12049S4705	AB	12 07 29	−47 22.1	1	9.57		0.32		0.12		2	A7 V	1771
105702	+6 02559	HR 4629		12 07 30	+06 05.1	4	5.72	.035	0.36	.014	0.17	.014	13	Am	3,15,355,1417
	−45 07583			12 07 30	−45 40.6	1	11.33		0.60				4		2011
		LF14 # 108		12 07 31	−59 42.4	1	12.56		0.28		0.11		2	A1 V	137
	+19 02536			12 07 32	+19 15.3	1	10.03		1.10		1.07		2	K5	280
	+22 02445			12 07 32	+21 34.8	1	9.64		1.60		1.92		1	M0	280
	−45 07584			12 07 32	−46 00.1	2	9.43	.008	1.38	.007	1.51		10		863,2011
	−62 02603			12 07 32	−62 31.9	1	10.15		0.20		0.07		3		279
	+23 02424			12 07 33	+22 29.5	1	10.31		0.86		0.53		1	G5	280
105695	+40 02506			12 07 33	+39 54.2	1	9.07		0.33		−0.07		3	F5	327
105707	−21 03487	HR 4630		12 07 33	−22 20.5	10	3.00	.011	1.33	.005	1.47	.004	173	K2 III	3,9,15,30,418,1020*
105698	+25 02468	Coma Ber 6		12 07 34	+24 32.3	1	8.79		0.50		−0.05		3	F9 V	3016
	−45 07585			12 07 34	−45 40.3	2	10.65	.009	0.45	.003	0.12		11		863,2011

HD	DM	Other Id	N Rem	α_{1950}	δ_{1950}	S	V	σ_V	B–V	σ_{B-V}	U–B	σ_{U-B}	n	Spectrum	References
	−61 02988			12 07 34	−61 54.9	1	10.50		0.14		0.05		1	B9 V	137
105699	+17 02444			12 07 35	+16 42.3	3	7.06	.005	1.33	.012	1.48		7	K2	280,882,1569
105714	−46 07731			12 07 35	−46 42.6	4	8.44	.004	0.27	.005	0.06	.007	29	A6 IV	460,977,1075,2011
		NGC 4147 - 3		12 07 36	+18 49.	1	17.06		0.12		0.14		1		82
		NGC 4147 - 4		12 07 36	+18 49.	1	17.48		0.04		-0.12		3		82
		NGC 4147 - 5		12 07 36	+18 49.	1	13.19		0.95		0.65		5		82
		NGC 4147 - 6		12 07 36	+18 49.	1	16.76		0.21				1		82
		NGC 4147 - 10		12 07 36	+18 49.	1	16.97		0.81		0.21		1		82
		NGC 4147 - 11		12 07 36	+18 49.	1	17.92		-0.22		-0.37		2		82
		NGC 4147 - 12		12 07 36	+18 49.	1	18.03		0.62		-0.39		1		82
		NGC 4147 - 16		12 07 36	+18 49.	1	17.16		0.12		-0.05		1		82
		NGC 4147 - 17		12 07 36	+18 49.	1	15.87		0.82		0.16		1		82
		NGC 4147 - 18		12 07 36	+18 49.	1	18.27		0.72		-0.51		1		82
		NGC 4147 - 104		12 07 36	+18 49.	1	14.44		0.63		0.02		1		82
		NGC 4147 - 105		12 07 36	+18 49.	1	17.34		0.73		-0.03		1		82
		NGC 4147 - 110		12 07 36	+18 49.	1	15.75		1.00				2		82
		NGC 4147 - 113		12 07 36	+18 49.	1	17.06		0.03		-0.01		1		82
		NGC 4147 - 114		12 07 36	+18 49.	1	15.22		0.93		0.48		1		82
		NGC 4147 - 115		12 07 36	+18 49.	1	17.28		-0.04		-0.13		1		82
		NGC 4147 - 203		12 07 36	+18 49.	1	16.97		0.69				1		82
		NGC 4147 - 204		12 07 36	+18 49.	1	16.96		0.47		-0.33		1		82
		NGC 4147 - 207		12 07 36	+18 49.	1	16.39		0.63		-0.07		1		82
		NGC 4147 - 209		12 07 36	+18 49.	1	16.55		0.80		0.34		1		82
		NGC 4147 - 313		12 07 36	+18 49.	1	15.03		1.09				1		82
		NGC 4147 - 314		12 07 36	+18 49.	1	17.54		-0.11				1		82
		NGC 4147 - 315		12 07 36	+18 49.	1	16.68		0.17		0.16		1		82
	+26 02311	Coma Ber 7		12 07 36	+26 10.4	1	9.25		1.53		1.88		1	K3 IV	280
	+5 02592	G 11 - 43		12 07 38	+04 35.5	3	9.20	.034	0.97	.018	0.79	.025	4	K5	202,308,1620
		G 148 - 36		12 07 39	+34 04.1	1	16.54		0.86		-0.07		1		782
105721	+43 02196			12 07 39	+43 11.2	1	10.44		0.79		0.46		3	G8 III	280
105724	+13 02499			12 07 40	+13 12.6	1	10.16		0.90		0.55		1	K0	280
105722	+15 02420			12 07 41	+14 48.3	1	8.93		0.44		0.01		2	F5	1569
	+19 02537	IDS12051N1931	AB	12 07 41	+19 14.8	1	9.84		0.47		-0.02		1	F8	280
		NGC 4151 sq1 4		12 07 41	+39 36.2	1	12.58		0.65		0.22		3		327
	−61 02989			12 07 41	−62 00.5	1	10.68		1.15		0.94		2	K0 III	137
105725	+12 02432			12 07 42	+12 09.0	1	10.08		0.60		0.06		2	G0	1569
105739	+49 02118			12 07 42	+48 48.6	1	8.35		0.91		0.59		2	G8 III	172
105740	+17 02445			12 07 44	+16 39.0	3	8.39	.021	0.97	.005	0.67		7	G5	280,882,1569
		V Com		12 07 44	+27 42.5	2	12.72	.237	0.04	.043	0.09	.043	2	A2	597,699
105736	−54 05020			12 07 44	−54 51.0	1	8.06		0.42				4	F3 V	2012
	−62 02604			12 07 44	−63 19.9	1	9.79		0.09		-0.51		2	B5	401
	+26 02312	Coma Ber 8		12 07 45	+25 30.4	1	9.56		1.41		1.48		1	K1 V	280
		NGC 4151 sq1 6		12 07 45	+39 53.0	1	10.90		1.20		1.16		2		327
	+20 02690	SY Com		12 07 47	+19 47.5	1	10.44		1.40		0.79		1	M4	3001
		Cru sq 1 # 106		12 07 48	−62 14.	1	13.78		0.85		-0.21		4		846
	+29 02261			12 07 49	+28 35.6	1	9.51		1.60		1.86		1	M1 III	280
105752	−61 02991			12 07 49	−61 56.2	1	9.16		0.06		-0.13		4	B8 V	279
	+34 02294			12 07 51	+33 43.7	1	9.79		1.11		1.07		1	K0 III	280
105753	−63 02172	LSS 2607		12 07 51	−63 49.3	4	9.16	.041	0.24	.018	-0.76	.023	11	B2/3e	401,540,976,1684
105756	+26 02313	Coma Ber 9		12 07 52	+25 36.6	1	9.45		0.56		0.06		3	G5	3016
		LP 554 - 79		12 07 53	+05 58.0	1	11.51		0.92		0.69		1		1696
105771	+29 02263			12 07 56	+29 20.8	2	7.50	.015	0.99	.005	0.74		5	G9 III	172,280
105776	−37 07714	HR 4631		12 07 58	−37 35.5	2	6.05	.005	0.21	.005			7	A5 V	15,2012
105777	−60 03805	IDS12053S6025	AB	12 07 58	−60 41.8	1	8.65		0.22		0.17		1	A1 V	549
105778	+17 02446	HR 4632		12 07 59	+17 05.2	4	6.39	.024	0.07	.017	0.10	.013	11	A4 V	252,280,1022,1068
	−62 02605			12 07 59	−62 39.6	2	11.05	.014	0.20	.000	0.07	.009	5	B9 V	125,137
		NGC 4151 sq1 2		12 08 00	+39 43.4	1	11.47		1.00		0.80		6		327
105791	+66 00748	G 237 - 48		12 08 00	+65 56.4	1	8.68		0.56		-0.02		2	F8 V	1003
		Cru sq 1 # 107		12 08 00	−62 06.	1	11.49		0.15		-0.45		2		846
		AAS61,331 # 7		12 08 01	+18 52.	1	11.33		-0.07		-0.23		3		1569
	+38 02307			12 08 03	+37 50.2	1	10.63		0.70		0.20		1		280
105784	−45 07592			12 08 03	−46 15.1	3	8.35	.004	1.03	.003	0.79	.001	18	G8 III	863,977,2011
105785	−55 04894			12 08 04	−56 09.8	1	6.96		0.10		0.12		1	A3 V	1771
105783	−41 07011			12 08 06	−41 45.2	1	9.14		1.15		0.92		2	K1/2 (III)p	536
		NGC 4151 sq1 3		12 08 07	+39 37.2	1	12.79		1.22		1.40		4		327
	−61 02996			12 08 07	−62 13.5	1	10.83		0.60		0.05		2	F8 V	137
	−61 02997			12 08 08	−62 03.1	1	10.21		0.60		0.19		3		279
	−62 02606	AJ77,216 T19# 109		12 08 08	−62 38.7	1	10.34		0.15		-0.64		2		846
105793	+13 02501			12 08 09	+12 33.3	1	10.07		1.06		0.89		1	K0	280
		NGP +33 054		12 08 11	+33 16.6	1	11.48		0.79		0.41		1	G7 III	8034
	+30 02232			12 08 12	+30 16.0	1	10.68		1.18		1.29		1	K1 III	280
	+33 02196			12 08 12	+33 21.4	1	10.49		0.94		0.61		1	G8 III	8034
		Cru sq 1 # 108		12 08 12	−62 08.	1	13.38		1.00		-0.17		3		846
		NGP +28 043		12 08 13	+28 11.2	1	10.52		1.10		1.05		1	K1 III	8041
	+40 02507			12 08 13	+39 36.1	1	9.83		0.59		-0.01		14	G0	327
105805	+28 02084	Coma Ber 10	⋆	12 08 14	+27 33.6	5	6.03	.011	0.11	.010	0.10	.010	19	A3 V	15,1022,1068,1127,8001
		NGP +28 044		12 08 14	+28 11.2	1	11.88		0.81		0.52		1	G8 III	8041
105806	+10 02386			12 08 16	+10 21.6	1	10.08		1.24		1.43		1	K2	280
105824	+40 02508	IDS12058N4027	AB	12 08 16	+40 10.2	1	6.85		0.28		0.05		1	A3	280
	−62 02609			12 08 16	−62 40.7	2	10.19	.005	0.32	.009	0.18	.005	4	A2 V	125,137

Table 1 651

HD	DM	Other Id	N Rem	α_{1950}	δ_{1950}	S	V	σ_V	B–V	σ_{B-V}	U–B	σ_{U-B}	n	Spectrum	References
	+37 02253			12 08 17	+37 26.0	2	9.89	.010	0.83	.030	0.43		5	G7 III	280,1569
105814	−27 08539			12 08 17	−27 47.3	1	7.32		1.09		0.96		5	K0 III	1657
	−44 07842			12 08 17	−45 17.0	2	10.21	.004	0.44	.007	0.17		11	F2	863,2011
105825	+33 02198			12 08 19	+32 41.3	1	9.22		0.27				2	A9 V	77
	+15 02421			12 08 20	+15 17.2	1	10.88		0.82		0.46		1		1569
105837	−45 07595			12 08 21	−46 02.8	3	7.52	.004	0.56	.000	-0.05	.009	16	G0/1 V	977,1075,2011
		G 11 - 44		12 08 22	+00 41.0	4	11.11	.024	0.43	.008	-0.22	.015	6		1064,1620,1696,3077
105843	+33 02199			12 08 23	+33 17.2	2	8.70	.034	1.05	.015	0.91		3	K0 III	77,8034
		MCC 636		12 08 26	+41 20.1	1	10.53		1.34				1	M0	1017
105861	+44 02166	G 123 - 9		12 08 26	+44 16.9	1	10.50		0.62		-0.02		2	G5	1658
105852	−44 07845	HR 4636	⋆ AB	12 08 26	−45 08.7	14	6.61	.007	1.07	.007	0.79	.012	155	K0 III	14,15,116,125,278,388*
105845	+15 02422			12 08 27	+14 42.1	2	8.49	.020	1.47	.025			5	K2	280,882
105841	−60 03812	HR 4634		12 08 27	−60 59.9	5	6.07		0.40	.004			19	F2 III/IV	15,1075,1488,2012,2018
105857	−55 04903			12 08 28	−56 07.3	1	7.34		0.12		0.08		8	A2 V	1770
		G 12 - 20		12 08 29	+12 25.3	2	12.10	.025	0.81	.000	0.38	.005	2		333,1620,1658
	+28 02085			12 08 29	+28 11.8	1	10.44		1.02		0.91		1	G9 III	280
105862	+36 02243			12 08 29	+35 39.5	1	9.17		0.57				2	F9 V	77
105850	−22 03305	HR 4635		12 08 29	−23 19.4	3	5.46	.006	0.05	.008	0.06		11	A1 V	15,1637,2012
105865	+19 02539			12 08 30	+18 54.2	1	8.03		1.14				2	K2	280
		LP 320 - 157		12 08 30	+29 35.	1	14.96		1.59				1	dM4	686
	+33 02200			12 08 30	+33 14.2	1	10.34		0.47				2		77
		Cru sq 1 # 109		12 08 30	−62 13.	1	11.55		1.23		0.00		3		846
105864	+25 02470	Coma Ber 11		12 08 31	+24 35.8	1	9.15		0.66		0.17		4	G5	3016
	−61 03000			12 08 32	−62 11.5	1	10.55		0.14		-0.34		3		279
	−61 03001			12 08 33	−61 54.6	2	9.88	.000	0.18	.000	-0.42	.000	5		279,846
105880	+43 02197			12 08 34	+42 37.4	1	9.97		1.35		1.57		1	K3 III	280
105866	+13 02502			12 08 35	+12 45.0	1	8.71		0.51		0.05		2	F5	1569
105873	−45 07598			12 08 35	−46 06.9	2	9.47	.006	0.26	.008	0.14		10	A5 IV	863,2011
		CCS 1968, Z Cru		12 08 35	−64 09.5	1	8.97		4.33				2	Nb	864
	+18 02575			12 08 37	+17 35.2	1	10.50		1.07		0.90		2		1569
105874	−51 06448	IDS12060S5140	A	12 08 37	−51 56.3	1	7.77		0.23		0.09		1	A6 V	1771
105881	+39 02500			12 08 38	+39 24.5	1	8.04		0.40		-0.06		3	F3 V	833
105877	−68 01624			12 08 38	−68 50.3	1	9.72		0.36		0.19		1	A8/9 V	1771
105878	−71 01318	LSS 2608		12 08 39	−71 59.6	1	9.89		0.06		-0.74		2	B1 V	846
	−61 03002			12 08 41	−62 22.5	1	10.18		1.58				3		279
105882	+14 02477			12 08 42	+13 49.9	1	9.32		0.38		-0.03		2	F2	1569
	+29 02264			12 08 42	+29 23.2	2	9.59	.025	0.55	.010	-0.04		5	G0	77,3016
105892	−63 02178	LSS 2609		12 08 42	−63 41.4	3	8.21	.025	0.12	.019	-0.70	.021	6	B1/2 Ib/III	401,540,976
105898	+25 02471	Coma Ber 13		12 08 43	+25 01.9	1	7.54		0.86		0.33		4	G2 V	3026
105897	+57 01357			12 08 43	+56 56.4	1	8.09		1.11		1.14		3	K1 III	37
		WLS 1200-5 # 8		12 08 45	−05 27.9	1	12.13		0.29		0.11		2		1375
105901	−5 03444			12 08 45	−05 38.8	1	8.20		0.62		0.13		2	G0	1064
105905	−44 07848			12 08 45	−44 35.0	2	9.32	.005	0.89	.005	0.58		10	K2 V	863,2011
105902	−7 03360	IDS12062S0720	AB	12 08 46	−07 36.7	2	8.75	.000	1.08	.005	0.68	.005	2	K5	554,565
105943	+82 00356	HR 4639	⋆ A	12 08 47	+81 59.3	1	6.00		1.62				2	gK5	71
105910	+23 02429			12 08 49	+22 52.3	1	7.50		0.26		0.18		1	A9 III	280
	−62 02611			12 08 49	−62 34.0	1	10.52		0.14		-0.27		5		125
		NGP +26 056		12 08 50	+26 08.3	1	11.47		0.72		0.53		1	G7 III	8041
105926	+32 02228			12 08 51	+32 15.6	1	8.23		0.46				2	F5	77
105919	−43 07524			12 08 51	−44 00.3	6	6.58	.003	0.46	.003	-0.03	.005	32	F5 V	278,977,977,1075,1770,2011
105924	+44 02168			12 08 53	+43 58.9	1	9.50		1.04				2	K0 III	280
	+16 02345			12 08 55	+16 18.0	1	10.41		0.45		0.02		2	F8	1569
105920	−50 06752	HR 4637		12 08 55	−51 04.8	4	6.22	.005	0.82	.004	0.49		16	G6 III (+G)	15,1075,2027,3005
	−61 03003			12 08 55	−62 11.8	2	10.53	.000	0.16	.000	-0.44	.000	5		279,846
		CCS 1971		12 08 55	−63 29.0	1	10.41		2.81				3		864
	−62 02613			12 08 56	−62 31.5	1	11.19		0.18		-0.42		5		125
105944	+45 02002			12 08 58	+44 31.7	1	10.50		0.65		-0.04		1	G7 III	280,8034
		AAS12,381 28 # B		12 08 59	+28 13.8	1	10.50		1.11		1.05		1		280
105963	+54 01499	IDS12065N5359	A	12 08 59	+53 42.1	1	8.03		0.88		0.56		1	K0	1080
		LP 734 - 39		12 08 59	−15 19.2	2	12.38	.005	0.72	.020	0.03	.015	5		1696,3060
105946	+35 02321			12 09 00	+34 32.3	1	7.30		0.35		-0.05		2	F4IV	1733
		Cru sq 1 # 110		12 09 00	−62 32.	1	11.19		0.18		-0.44		2		846
		L 103 - 27		12 09 00	−69 19.	1	10.69		1.14		1.03		1		1696
105945	+36 02246	IDS12065N3639	A	12 09 02	+36 22.0	2	8.84	.009	0.27	.005	0.02	.000	4	F0 IV	1068,1569
105937	−51 06455	HR 4638		12 09 02	−52 05.4	8	3.95	.006	-0.16	.006	-0.61	.014	37	B3 V	15,26,1075,1637,1770,2012*
105947	+15 02423			12 09 03	+14 54.3	1	9.75		0.77		0.38		1	G5	1569
	+25 02472	Coma Ber 14		12 09 03	+25 03.9	1	9.45		1.06		1.06		1	K0 III	280
105964	+26 02315	Coma Ber 15		12 09 03	+26 00.9	1	8.85		0.59		-0.02		3	G0 V	3026
	+36 02247	IDS12065N3639	B	12 09 03	+36 22.5	1	9.72		0.68		0.28		2		1569
105956	−46 07751			12 09 03	−46 57.7	2	7.81	.008	1.13		0.92		4	K1/2 III	861,2012
		POSS 377 # 40		12 09 04	+25 47.1	1	17.36		1.56				4		1739
	+22 02446			12 09 06	+21 35.5	1	10.76		0.26		0.04		1		280
105967	+4 02583			12 09 07	+04 20.0	1	6.93		0.15		0.14		3	A0	355
105958	−62 02616			12 09 07	−63 05.7	2	9.16	.015	0.20	.005	-0.33	.010	5	B5 II/III	125,401
105973	−50 06758			12 09 09	−50 40.7	1	7.68		-0.03		-0.16		3	B9 III/IV	1770
105965	+15 02424			12 09 10	+15 09.7	1	9.85		1.31		1.41		2	K2	1569
		AAS12,381 41 # A		12 09 11	+40 32.1	1	11.04		0.06		0.07		1		280
	+22 02447			12 09 13	+21 50.4	1	10.05		1.07		1.04		1	K2	280
105983	+14 02478			12 09 14	+13 51.0	1	9.99		0.64		0.06		2	G5	1569
105982	+24 02436			12 09 14	+23 36.0	1	6.71		1.04				2	K2	280

HD	DM	Other Id	N	Rem	α_{1950}	δ_{1950}	S	V	σ_V	B–V	σ_{B-V}	U–B	σ_{U-B}	n	Spectrum	References
105993	−44 07855				12 09 14	−44 50.2	4	8.55	.004	0.15	.007	0.15	.007	18	A1 IV/V	768,977,1075,2011
		G 121 - 63			12 09 15	+25 52.4	1	13.54		0.84		0.47		3		1658
106002	+57 01359	HR 4641			12 09 16	+57 20.0	1	6.31		1.53		1.87		2	gK5	1733
	+41 02277				12 09 17	+41 06.2	1	9.87		0.96				2	G9 III	280
		LF14 # 114			12 09 18	−60 21.6	1	11.59		0.62		0.11		1	F6 V	137
		Ly 109			12 09 18	−62 50.	2	11.67	.000	0.18	.000	−0.50	.005	8		125,846
105981	+26 02316	Coma Ber 16		⋆	12 09 19	+26 08.9	2	5.66	.000	1.41	.015	1.60		5	K2 V	70,172
	+21 02396	IDS12068N2057	A		12 09 20	+20 40.5	1	10.23		0.37				4	F3 V	193
105999	−62 02619				12 09 20	−63 06.1	1	7.43		0.36		0.07		2	Ap SrCrEu	1771
106000	−63 02185	LSS 2611			12 09 20	−64 13.9	2	6.54	.000	0.27	.005	0.13		6	A2/3 II	1075,8100
106003	+39 02503	IDS12068N3919	A		12 09 21	+39 02.8	2	8.52	.000	0.64	.005	0.20		5	G2 V	172,280
106001	−74 00885				12 09 22	−74 39.7	2	8.46	.005	1.26	.000	1.15		5	K1 III	657,2012
	−62 02620				12 09 23	−62 25.1	1	10.78		0.23		0.14		4		125
106025	+13 02503	IDS12069N1333	AB		12 09 24	+13 16.7	1	8.54		0.43		−0.03		2	F8	1569
106004	+14 02479				12 09 24	+14 17.7	2	9.52	.005	1.00	.000	0.79		4	K0	882,1569
106015	−58 04153				12 09 24	−58 44.3	1	8.91		0.25		0.15		1	A3 V	1771
106016	−61 03006				12 09 25	−61 50.8	1	8.25		−0.01		−0.28		2	B8 II/III	279
	−62 02621	AJ77,216 T19# 108			12 09 25	−62 39.2	1	10.98		0.30		−0.40		2		846
106026	+12 02433				12 09 26	+11 34.2	1	8.32		1.04		0.97		1	K0	280
106024	+14 02480				12 09 26	+14 20.9	2	9.49	.015	1.22	.015	1.32		4	K2	882,1569
		Cru sq 1 # 77			12 09 28	−61 49.1	1	12.21		0.82		0.39		2		279
106038	+14 02481	G 12 - 21			12 09 29	+13 32.7	7	10.17	.015	0.46	.014	−0.18	.006	10	F6 V-VI	333,979,1003,1569*
106022	+29 02265	HR 4642		⋆ A	12 09 29	+28 48.9	2	6.52	.030	0.36	.003	0.01		4	F5 V	77,3016
106053	+78 00411				12 09 29	+77 43.1	2	6.73	.015	0.04	.020	0.06		10	A0	252,1258
106036	−62 02622				12 09 31	−63 10.6	2	6.99	.020	0.09	.000	0.07	.000	5	A2 V	125,1771
106030	−1 02633				12 09 32	−01 33.7	1	9.89		0.72				1	G0	1284
106048	−53 04986				12 09 33	−53 57.1	1	8.64		0.31		0.02		1	A9 V	1771
238087	+59 01428				12 09 36	+59 12.3	1	10.04		1.28		1.23		1	K5	3003
	−60 03825				12 09 36	−60 28.5	1	10.77		0.13		0.10		1	A0 V	137
106057	+21 02398	HR 4643			12 09 37	+20 49.2	1	5.57		0.95				2	K0 II-III	70
106046	−45 07609				12 09 37	−46 08.2	3	8.43	.001	0.33	.001	0.14	.004	16	A2mA5-F0	977,1075,2011
	−58 04158				12 09 39	−58 32.1	1	9.99		1.21		1.13		1	K1 III	137
	+37 02254				12 09 42	+37 15.8	1	9.22		0.89				2	G8 III	77
		Cru sq 1 # 78			12 09 42	−62 33.0	1	12.72		0.51		0.22		3		279
106068	−62 02624	HR 4644, LSS 2612			12 09 42	−62 40.4	8	5.92	.010	0.30	.007	−0.28	.151	29	B8 Ia/ab	15,401,1020,1075,1684*
	+26 02317	Coma Ber 17			12 09 43	+26 18.3	1	9.95		0.50		0.00		2	dF8	3016
	+33 02202				12 09 44	+33 22.5	1	11.01		0.92		0.58		1	G8 III	8034
		WLS 1224 50 # 6			12 09 44	+50 21.7	1	11.69		0.73		0.26		2		1375
106207	+87 00104				12 09 44	+87 12.6	1	8.03		1.14		1.07		2	K5	1733
		WLS 1220 30 # 6			12 09 46	+30 23.5	1	11.01		0.63		0.03		2		1375
106086	−59 04097				12 09 46	−59 47.4	1	6.84		0.23		0.15		2	A8 IV	1771
	−62 02625				12 09 46	−62 31.5	1	9.49		1.51		1.67		5		279
106074	+13 02504				12 09 47	+13 07.4	1	10.16		0.47		0.04		1	F5	1569
106073	+15 02425				12 09 48	+15 06.8	2	9.52	.020	0.91	.005	0.61		4	K0	882,1569
	−5 03449				12 09 48	−05 41.4	1	10.60		0.60		0.10		2		1064
		Cru sq 1 # 111			12 09 48	−61 57.	1	11.22		0.15		−0.44		2		846
		Cru sq 1 # 112			12 09 48	−63 50.	1	13.71		1.20		−0.01		4		846
	+38 02310				12 09 49	+37 46.4	1	9.71		0.90				4	G5	280
		NGP +26 059			12 09 50	+26 08.3	1	11.26		0.99		0.70		1	G8 III	8041
238090	+55 01519	G 197 - 49		⋆ AB	12 09 50	+54 45.7	1	9.86		1.42		1.16		2	K5	3072
106102	+56 01575				12 09 50	+55 59.1	1	7.21		1.20		1.32		3	K2 III-IV	37
106083	−43 07533				12 09 50	−43 43.3	4	7.92	.000	1.16	.004	1.01	.003	27	K0 III	460,977,1075,2011
106090	+14 02482				12 09 51	+13 52.2	1	10.65		0.46		−0.06		1	F5	1569
106089	+15 02426				12 09 51	+15 25.3	1	10.44		0.49		0.04		2	G0	1569
	+41 02278				12 09 51	+40 41.9	2	10.12	.015	0.25	.000	0.11	.010	3	A2	272,280
106092	−5 03450	G 13 - 18			12 09 51	−06 04.1	4	9.98	.016	0.99	.015	0.79	.009	14	K0	158,1620,1746,1775
	+26 02318	Coma Ber 18			12 09 53	+25 51.1	1	9.31		1.09		1.03		1	K0 III	280
106103	+28 02087	Coma Ber 19		⋆ V	12 09 53	+27 39.5	5	8.10	.021	0.41	.017	−0.04	.014	25	F5 V	1127,1375,1613,3016,8001
106112	+78 00412	HR 4646			12 09 53	+77 53.6	2	5.13	.009	0.32	.008	0.10		4	A5 m	1363,3058
		NGP +26 063			12 09 55	+25 33.6	1	10.87		1.09		0.86		1	G8 III	8041
	−56 05091				12 09 56	−56 36.7	1	10.47		0.18		0.12		2	A0 III	403
106116	−2 03481	G 11 - 45		⋆ AB	12 09 57	−02 48.7	9	7.43	.012	0.71	.008	0.27	.019	22	G4 V	22,1003,1375,1509*
106101	−52 05332				12 09 57	−52 29.9	1	7.38		0.12		0.07		1	A1/2 V	1770
	+26 02319				12 09 59	+25 33.6	1	10.03		1.11		1.01		1	K0 III	8041
	+28 02088				12 09 59	+27 30.0	2	10.20	.005	1.04	.005	0.97	.030	2	K0 III	280,8034
		AAS78,203 # 4			12 10 00	−64 33.	1	13.09		1.74		1.71		4		1684
	+37 02255				12 10 01	+36 32.0	1	9.21		1.49				2	K3 III	77
106107	−44 07859				12 10 02	−45 15.1	3	8.89	.005	0.92	.005	0.53	.003	18	G6 IV	863,977,2011
	−62 02626	LSS 2613			12 10 02	−63 19.3	4	9.61	.043	0.21	.009	−0.49	.022	12	B3 V:ne	540,976,1684,1747
106114	+36 02248	IDS12075N3619	AB		12 10 03	+36 02.4	1	8.55		0.41				2	F5 V	77
106110	−59 04102				12 10 03	−59 45.7	1	7.62		0.07		0.05		2	B9.5/A0 IV	1771
106111	−69 01646	HR 4645, S Mus			12 10 04	−69 52.4	2	5.93	.015	0.78	.005	0.44		2	F6 Ib	657,1484
		LF14 # 117			12 10 05	−58 59.3	1	12.80		0.28		−0.21		1		137
		Steph 997			12 10 06	+13 19.9	1	11.80		1.36		1.26		1	K7	1746
	+30 02238				12 10 09	+30 16.5	1	9.40		0.57				2	G0	77
		POSS 377 # 25			12 10 12	+25 14.2	1	16.58		1.48				4		1739
	+40 02511				12 10 13	+39 40.9	1	9.38		1.41				2	K4 III	280
106132	−58 04164				12 10 13	−59 22.6	1	7.13		1.38		1.48		1	K0/1 III	137
106136	+11 02436				12 10 14	+11 23.4	1	8.38		1.61		1.78		1	M1	280
106150	+52 01617				12 10 16	+52 13.1	1	7.25		0.91		0.57		2	K0	1375

Table 1 653

HD	DM	Other Id	N Rem	α_{1950}	δ_{1950}	S	V	σ_V	B–V	σ_{B-V}	U–B	σ_{U-B}	n	Spectrum	References
	+18 02578			12 10 17	+17 38.0	1	9.78		0.36		-0.05		2	F2	1569
106155	+13 02505			12 10 18	+13 11.1	1	9.69		0.53		0.01		1	F8	1569
		Sand 27		12 10 18	+27 12.	1	14.98		1.35				1	dM3	686
		MU Cen		12 10 18	-44 11.	1	13.19		0.17		-0.64		1		1471
		Cru sq 1 # 113		12 10 18	-62 08.	1	11.29		0.17		-0.46		2		846
106153	+15 02427			12 10 19	+14 47.6	2	9.41	.005	0.93	.005	0.66		3	K0	280,882
106145	-58 04165			12 10 19	-59 23.2	1	8.56		1.21		1.17		2	K0 III	137
	+28 02089	Coma Ber 20		12 10 20	+27 50.3	1	9.17		1.24		1.43		1	K1 III	280
106152	+32 02229			12 10 20	+32 04.5	2	7.77	.000	0.54	.005	0.06		4	F5	77,1375
106146	-61 03022	LSS 2614		12 10 22	-62 02.1	1	7.80		0.29		-0.12		3	B9 Ib	125
106154	+13 02506			12 10 23	+13 25.0	1	10.26		1.10		1.02		1	K0	280
106156	+10 02391	G 12 - 22		12 10 24	+10 19.2	4	7.92	.008	0.79	.002	0.44	.011	7	G8 V	333,861,1003,1620,3026
	+22 02450			12 10 24	+21 29.5	1	8.45		1.00		0.83		1	K0	280
		SSII 96		12 10 24	+49 21.	1	11.56		0.47		-0.22		2	F2	1240
	+24 02438			12 10 25	+24 26.1	1	10.02		0.28		0.15		1		280
106167	-55 04921	IDS12078S5527	A	12 10 25	-55 43.6	1	8.59		0.22		0.08		1	A3 V	1771
		POSS 377 # 21		12 10 26	+21 20.1	1	16.35		1.74				3		1739
106171	+41 02280			12 10 26	+40 31.6	1	8.70		1.27				2	K2 III	280
	-62 02628			12 10 27	-62 29.3	1	9.91		0.80		0.60		4		125
		WLS 1200 15 # 6		12 10 28	+14 43.7	1	12.82		0.61		0.18		2		1375
	+15 02428			12 10 30	+15 28.0	1	11.01		0.50		-0.02		1		1569
106184	+29 02267	Coma Ber 22		12 10 31	+28 54.8	1	7.68		1.59				2	K4 III	280
		Wray 948		12 10 31	-60 36.3	1	12.72		0.83		-0.28		3		1684
		G 12 - 23		12 10 33	+08 59.7	2	13.38	.015	0.64	.007	0.01	.005	3		1620,1658
106187	+12 02434			12 10 33	+12 13.0	1	9.66		0.75		0.35		1	K0	280
106186	+13 02507			12 10 33	+13 20.4	2	9.27	.030	1.03	.020	0.87		4	K0	882,1569
	+23 02432			12 10 34	+22 54.9	1	9.80		0.86		0.46		1	G5	280
		POSS 377 # 34		12 10 34	+23 15.6	1	17.06		1.76				4		1739
106202	-55 04926			12 10 35	-55 47.2	1	8.24		0.31		0.18		1	A8 V	1771
106198	-33 08252	HR 4647		12 10 37	-33 50.8	1	6.50		1.62				4	M4 III	2006
106210	+11 02439	G 12 - 24		12 10 40	+11 06.5	5	7.56	.010	0.67	.007	0.16	.013	28	G3 V	22,333,861,1003,1620,3026
106209	+12 02435			12 10 41	+11 47.7	1	7.52		1.56		1.95		1	K5	280
106211	-6 03528			12 10 42	-07 26.3	1	9.68		0.67		0.18		2	G5	1375
106224	+22 02451			12 10 45	+22 18.3	1	7.52		0.11		0.10		2	A3 V	1068
106223	+31 02340			12 10 45	+30 33.7	4	7.44	.009	0.29	.008	-0.08	.000	16	Fp	77,275,1068,1240
106236	+46 01775			12 10 45	+46 15.8	1	10.02		0.24				5	A9 V	193
		He3 763		12 10 45	-60 45.	1	11.85		0.31		-0.46		3		1684
	-61 03026	LSS 2615		12 10 45	-62 02.3	2	10.42	.005	0.21	.019	-0.53	.019	6		846,1747
	+21 02401			12 10 46	+21 10.2	1	10.49		1.11		1.04		1	K0	280
106233	-44 07871			12 10 46	-45 20.9	3	8.90	.008	0.67	.005	0.12	.001	18	G3/5 V	863,977,2011
		AJ77,733 T19# 11		12 10 46	-62 33.1	1	12.17		0.18		0.05		4		125
		LF14 # 120		12 10 46	-62 36.8	1	11.47		1.14		0.73		1	G8 III	137
106238	+30 02241			12 10 48	+29 32.1	2	9.43	.019	1.10	.005	1.03		3	K1 III	77,8034
	-60 03836	LSS 2616		12 10 48	-61 19.7	2	9.73	.010	0.26	.010	-0.44	.019	6	B8	846,1747
	-63 02194	AAS78,203 # 5		12 10 48	-64 02.5	1	10.41		0.66		0.17		6		1684
106231	-38 07581	HR 4648		12 10 49	-38 39.1	3	5.75	.004	-0.14	.007	-0.60		10	B4 III	15,1770,2012
	+35 02325			12 10 51	+35 12.2	1	9.91		1.60				1	M1 III	694
	+20 02695			12 10 52	+19 37.9	1	9.96		1.04		0.98		1	F8	280
106251	+11 02440	HR 4650		12 10 53	+10 32.4	5	5.85	.014	0.26	.010	0.10	.012	13	A2 m	70,355,1022,3058,8071
		WD1210+533		12 10 55	+53 20.6	1	14.12		-0.34		-1.23		3	DAO1	1727
106252	+10 02392			12 10 56	+10 19.4	1	7.36		0.64				2	G0	280
106250	+17 02451			12 10 57	+17 27.6	1	7.58		1.28		1.37		1	K2	280
	+24 02439	Coma Ber 23		12 10 57	+23 32.5	2	9.41	.015	0.59	.015	0.01	.015	2	dG0	333,1620,1658
		G 11 - 46		12 10 59	+07 49.0	2	11.57	.010	0.86	.015	0.50		3	K3	202,333,1620
	+27 02103	Coma Ber 24		12 10 59	+27 19.3	1	9.94		0.60		0.13		3	dG0	3016
		NGP +27 065		12 10 59	+27 24.3	2	10.31	.005	0.93	.005	0.69	.015	2	G9 III	280,8034
		Sand 28		12 11 00	+31 30.	1	13.32		1.51				1	dM3	686
		G 123 - 14		12 11 00	+47 27.0	1	12.91		0.63		-0.08		2		1658
		L 326 - 11		12 11 00	-45 23.	1	12.52		1.32		1.22		1		3073
106260	-60 03840			12 11 00	-60 55.0	1	8.77		0.23		0.22		1	A3 V	1771
106257	-33 08257	HR 4651	⋆ AB	12 11 01	-33 30.9	1	6.33		0.08				4	A0 V	2032
106279	+31 02341			12 11 03	+30 50.4	2	8.39	.024	0.92	.005	0.67		3	G8 III	77,8034
106278	+31 02342			12 11 04	+31 14.5	1	8.81		1.00				2	K0 III	77
	+27 02104	Coma Ber 25		12 11 05	+27 24.6	2	9.47	.029	1.58	.019	1.95		3	M2 III	280,694
		G 198 -B6	A	12 11 05	+39 17.6	1	14.89		1.45		1.25		1		3062
		G 198 -B6	B	12 11 05	+39 17.6	1	16.48		-0.07		-0.82		1		3062
106261	-62 02633	LSS 2617		12 11 05	-62 39.2	4	7.94	.014	0.13	.009	-0.65	.018	12	B2 III	164,540,976,1684
106248	-77 00804	HR 4649		12 11 05	-78 17.7	3	6.34	.004	1.23	.005	1.41	.000	12	K2/3 III	15,1075,2038
106247	-75 00792			12 11 07	-75 44.6	2	8.68	.000	0.56	.003	0.26		6	F0 IV	278,2012
	-62 02634			12 11 08	-62 31.1	1	10.33		0.16		0.06		4		125
		G 11 - 47		12 11 09	+00 55.6	2	12.92	.014	0.76	.007	0.12	.002	4		1620,1658
106282	+10 02393			12 11 12	+10 01.2	1	9.22		1.21		1.17		1	K0	280
	+22 02453	IDS12087N2151	AB	12 11 12	+21 34.4	1	9.00		0.71		0.33		1	G5	280
106292	+37 02257			12 11 14	+37 00.9	1	8.93		0.48				2	F7 V	77
		AJ77,733 T19# 12		12 11 15	-62 30.3	1	12.87		0.52		0.35		6		125
106294	+16 02348			12 11 16	+15 44.5	2	9.27	.000	0.68	.005	0.21		4	K0	882,1569
106304	-40 07168			12 11 17	-40 35.6	3	9.07	.013	0.02	.007	-0.02	.014	3	B9 V	55,1311,3077
106308	-57 05338			12 11 18	-57 40.0	2	7.91	.010	0.12	.005	0.05	.000	3	A0 V	401,1771
	+16 02349			12 11 19	+15 47.2	1	10.85		0.56		0.10		2		1569
106306	-52 05364			12 11 19	-52 51.9	1	8.37		0.18		0.05		1	A2 IV/V	1771

HD	DM	Other Id	N Rem	α_{1950}	δ_{1950}	S	V	σ_V	B–V	σ_{B-V}	U–B	σ_{U-B}	n	Spectrum	References
106330 +29 02268	Coma Ber 26			12 11 21	+28 34.4	1	8.53		1.50				2	K4 IV	280
106309 −58 04174	LSS 2618			12 11 22	−59 07.1	3	7.82	.027	−0.03	.011	−0.83	.007	7	B2 III/Vne	164,401,1684
106307 −54 05065				12 11 23	−55 12.3	1	7.20		1.38		1.55		1	K3 III	897
		Feige 58, IK Vir		12 11 24	+02 17.	2	11.80		0.18	.000	0.07	.000	2		1026,1264
		NGP +27 072		12 11 24	+26 46.1	1	10.40		0.62		0.13		1	G5	8038
		EG 86, HZ 21		12 11 25	+33 13.1	1	14.22		−0.36		−1.25		1	DA:	3028
106321 −45 07630	HR 4652		★ AB	12 11 25	−45 26.8	8	5.30	.003	1.43	.005	1.58	.007	47	K3 III	14,15,278,388,977,977*
		LF14 # 121		12 11 25	−59 40.8	1	11.83		0.48		0.15		1	F2 V	137
106329 +40 02512				12 11 26	+40 15.9	1	8.46		0.97				2	G9 III	280
106323 −50 06802				12 11 27	−51 02.1	1	8.49		0.02		−0.05		3	B9.5 V	1770
	−59 04119	LSS 2620		12 11 27	−60 19.4	2	10.26	.053	0.26	.005	−0.57	.043	7	O9.5Ve	94,846
106325 −61 03036	LSS 2619			12 11 27	−62 00.8	4	8.57	.027	0.23	.008	−0.62	.019	13	B1/2 Ib/II	125,191,540,976
		G 121 - 68		12 11 29	+29 51.5	1	14.96		1.64				1		686
106337 −35 07804	IDS12089S3600		A	12 11 30	−36 16.7	2	7.86	.006	−0.10	.002	−0.50	.005	2	B7 (III)	55,1770
	−60 03847			12 11 30	−60 41.9	1	9.86		1.34		1.32		2	K2 III	137
		LF14 # 122		12 11 30	−60 59.1	1	12.92		1.06		0.55		1		137
		AJ77,216 T19# 111		12 11 30	−62 46.	1	12.58		0.99		−0.10		4		846
106350 +5 02600	G 13 - 21			12 11 31	+05 16.5	2	8.68	.039	0.80	.010	0.41		3	K0	202,1620
106349 +12 02436	IDS12090N1242		A	12 11 31	+12 25.5	1	7.93		1.16		1.21		1	K0	280
		LF14 # 124		12 11 33	−59 28.8	1	12.87		0.71		0.19		1		137
	−62 02637			12 11 34	−62 30.5	1	10.21		0.14		−0.30		4		125
106360 −53 05009				12 11 35	−53 38.5	1	7.78		0.08		0.07		1	A2 V	1770
106343 −63 02203	HR 4653			12 11 35	−64 07.8	11	6.22	.016	0.11	.014	−0.77	.037	41	B1.5Ia	15,540,976,1018,1075*
106344 −65 01819				12 11 35	−66 16.3	2	7.13	.000	−0.01	.005	−0.47		8	B4 V	164,2012
106365 +33 02205	IDS12091N3320		A	12 11 36	+33 03.7	3	6.85	.019	1.14	.010	1.06	.005	10	K2 III	1003,1084,3035
106365 +33 02206	IDS12091N3320		B	12 11 36	+33 03.7	3	8.77	.005	0.59	.016	0.07	.013	7	F8	1003,1084,3024
106352 −0 02550				12 11 36	−01 21.8	1	9.77		0.71				1	G5	1284
106366 +31 02343				12 11 37	+31 05.9	1	8.25		0.64				2	G0	77
106361 −59 04125				12 11 39	−59 45.9	1	8.62		0.26		0.27		1	A3mA5-A7	1771
	+34 02298			12 11 41	+33 46.1	1	9.47		1.08				2	K0 III	77
106381 +66 00751				12 11 41	+66 23.2	1	6.76		1.31		1.52		2	K0	985
106384 −4 03235	FG Vir		★ A	12 11 42	−05 26.4	1	6.57		0.26		0.05		4	A9 III	3024
106384 −4 03235	IDS12091S0510		B	12 11 42	−05 26.4	1	13.51		0.38		−0.17		3		3024
106384 −4 03235	IDS12091S0510		C	12 11 42	−05 26.4	1	11.85		1.33		1.35		3		3024
	+32 02230			12 11 43	+31 30.3	1	11.42		0.82		0.53		1	G8 III	8034
106383 +34 02299				12 11 43	+33 43.7	2	8.04	.005	0.86	.005	0.47		3	G8 IV	77,8034
	+36 02249		AB	12 11 43	+35 55.0	1	9.68		0.80				2	G5	77
	+41 02281			12 11 43	+41 06.7	1	10.18		0.64				2	G6 IV	280
	−59 04126			12 11 43	−59 39.2	1	10.45		1.17		1.08		1	K1 III	137
		LF14 # 126		12 11 44	−61 15.3	1	11.96		0.31		−0.16		2	B8 V	137
106362 −65 01820	LSS 2621			12 11 44	−66 14.8	4	7.45	.037	0.00	.011	−0.84	.019	10	B1 Ib	164,191,401,1684
		G 13 - 22		12 11 45	+00 54.2	2	13.43	.014	1.59	.005	1.00		5		419,906
106398 +27 02105	Coma Ber 27			12 11 46	+26 47.0	2	7.41	.015	0.92	.005	0.62		5	G8 III	172,280
	+54 01502	G 197 - 51		12 11 46	+53 53.1	1	9.63		0.56		−0.01		1	sdF8	1658
	−60 03849	LSS 2622		12 11 46	−60 49.1	1	11.22		0.21		−0.47		3		846
106400 +12 02437	AH Vir		★ AB	12 11 48	+12 05.9	2	9.12	.115	0.76	.009	0.34	.022	11	K2 V	1569,3077
106396 +45 02008				12 11 48	+45 01.4	1	9.51		1.12				3	K0	280
	+28 02091	Coma Ber 28		12 11 49	+28 05.8	1	10.10		0.47		−0.01		3	F8 V	3016
	+42 02282			12 11 49	+42 13.3	1	10.07		0.82				2		280
106397 +44 02171				12 11 50	+43 40.1	1	9.58		1.19				2	K2 III	280
106391 −61 03039	LSS 2623			12 11 50	−61 37.4	4	8.62	.021	0.77	.012	−0.15	.035	11	B2 Ia/ab	191,540,976,1684
106399 +13 02509				12 11 52	+12 47.4	1	10.74		0.86		0.45		2	G5	1569
	+32 02231			12 11 52	+31 29.9	2	10.83	.010	0.98	.020	0.81	.005	2	K0 III	280,8034
106420 +47 01940				12 11 52	+47 19.6	1	8.21		−0.14		−0.45		2	B8 V	1240
	−59 04127	LSS 2624		12 11 52	−60 08.0	1	10.73		0.39		−0.57		5	B1 V	94
	−28 09374			12 11 53	−29 18.3	2	10.25	.005	0.82	.014	0.33		5		1594,6006
106414 −50 06814	TV Cen			12 11 53	−51 15.2	1	7.93		2.82		3.30		1	Nb	864
		G 59 - 7		12 11 54	+24 52.4	1	12.18		1.48		1.12		1		333,1620
106411 −43 07555				12 11 55	−43 49.5	3	9.27	.014	0.57	.011	−0.06		8	F7 V	1594,2033,6006
	+27 02106			12 11 56	+26 30.6	2	10.26	.015	1.07	.010	1.02	.000	2	K0 III	280,8034
	−60 03850			12 11 56	−61 20.0	1	11.05		0.55		0.10		1	F5 V	137
106393 −66 01723				12 11 56	−67 22.6	1	8.80		−0.03		−0.81		2	B2 V	401
106428 +45 02009				12 11 57	+45 19.3	1	10.32		0.67				3	G5	280
106422 +11 02441				12 11 59	+11 22.5	1	10.31		1.09		0.97		1	K0	280
		Sand 33		12 12 00	+24 24.	1	13.62		1.52				1	dM4	686
	−59 04128	LSS 2625		12 12 00	−60 02.7	2	11.00	.000	0.40	.012	−0.59	.001	7	B1 Ve	94,1684
106425 −53 05013				12 12 03	−53 33.8	1	8.99		0.28		0.13		1	A1mA7-A9	1771
	−60 03853			12 12 05	−61 15.0	1	11.20		0.29		−0.31		3		1684
106449 +40 02513				12 12 06	+39 37.2	2	6.97	.020	1.59	.010	1.96		4	K7 III	172,280
106450 +15 02430				12 12 07	+14 55.9	1	10.53		0.57		0.06		2	G5	1569
	−59 04130	LSS 2626		12 12 07	−60 02.5	5	10.42	.017	0.37	.014	−0.66	.011	13	O6 V	94,540,976,1684,1737
106453 −24 10236	IDS12096S2413		AB	12 12 08	−24 29.8	1	7.46		0.72				4	K0/1 V +(G)	2033
		LF14 # 129		12 12 09	−61 45.0	1	11.53		0.26		0.14		1	A1 V	137
106426 −62 02642				12 12 09	−62 29.3	1	9.30		0.45		0.24		1	A9 V	1771
	−62 02643			12 12 10	−62 48.3	1	10.87		0.83		0.46		1	G8 III	137
	+28 02092			12 12 11	+28 10.1	1	10.33		0.77		0.38		1	G6 III	8034
106462 +46 01777				12 12 11	+45 54.4	1	9.99		0.34		0.16		1	F2	280
	+23 02435			12 12 13	+23 22.5	1	10.55		1.05		1.00		1		280
106456 −43 07560				12 12 14	−43 50.0	7	7.55	.006	1.11	.006	0.97	.004	36	K0 II/III	402,863,977,1075,2001*
	+30 02244			12 12 15	+30 18.5	1	10.42		1.15		1.20		1	K2 III	280

Table 1 655

HD	DM	Other Id	N Rem	α_{1950}	δ_{1950}	S	V	σ_V	B–V	σ_{B-V}	U–B	σ_{U-B}	n	Spectrum	References
106478 +54 01504	HR 4654			12 12 15	+53 42.8	1	6.16		1.04				2	gK0	70
106460 −63 02207				12 12 15	−64 10.9	1	8.28		0.03		-0.64		2	B2 III	401
106480 +15 02431				12 12 20	+15 23.8	1	10.49		0.65		0.23		2	K0	1569
106479 +29 02269	Coma Ber 30			12 12 20	+28 54.5	2	8.67	.028	1.04	.009	0.87		4	G9 III	280,8034
+43 02200				12 12 20	+43 12.2	1	10.48		1.12		1.12		1	K1 III	280
106474 −54 05075	V369 Cen			12 12 20	−54 32.5	2	8.18	.106	1.68	.023	1.46	.009	22	M5 II	897,3042
+20 02698				12 12 22	+19 31.1	1	9.09		1.43		1.61		1	K5	280
106489 −40 07182				12 12 22	−40 51.6	1	7.48		0.65				4	G5 V	2012
	LF14 # 130			12 12 23	−61 36.7	1	11.55		1.30		1.29		1		137
	LP 376 - 36			12 12 24	+23 41.	1	13.51		1.41				1	dM3	686
106485 −20 03606	HR 4655			12 12 24	−20 34.0	3	5.82	.004	1.05	.005			11	K0 III	15,2013,2029
	POSS 377 # 27			12 12 28	+22 10.2	1	16.64		1.49				4		1739
106509 +34 02301				12 12 28	+33 32.0	1	8.15		0.47				2	F5	77
106495 +15 02432				12 12 29	+14 32.6	2	9.42	.019	1.04	.029	0.72		3	K0	280,882
106490 −58 04189	HR 4656, δ Cru			12 12 29	−58 28.2	5	2.79	.009	-0.24	.004	-0.91	.014	17	B2 IV	26,1034,1088,2035,8023
106510 +30 02245				12 12 30	+30 25.1	1	8.28		0.53				2	F8 V	77
106511 +13 02510				12 12 33	+13 01.9	1	10.72		0.91		0.54		2	G5	1569
106515 −6 03532	IDS12100S0642	AB		12 12 33	−06 58.7	3	7.35	.004	0.82	.003	0.45	.004	14	G5	258,1499,2012
	HA 80 # 154			12 12 35	+14 48.7	1	11.90		0.62		0.14		1		269
106525 +33 02207				12 12 36	+32 29.7	1	9.54		0.56				2		77
106516 −9 03468	HR 4657	★ A		12 12 36	−10 01.2	10	6.11	.010	0.47	.013	-0.13	.020	25	F5 V	15,22,254,742,1003*
106541 +33 02209				12 12 39	+33 07.1	1	8.19		0.51				2	G0	77
	HA 80 # 155			12 12 40	+14 49.7	1	10.84		0.55		-0.01		1		269
106521 −60 03861				12 12 40	−60 45.1	1	8.97		0.47		0.37		1	A9 IV	1771
106526 +13 02511				12 12 41	+13 25.0	1	10.70		0.63		0.06		1	G0	1569
106542 +17 02454				12 12 41	+17 11.1	2	6.82	.007	1.19	.013	1.17	.002	11	K2	1375,1499
+49 02126	G 198 - 37			12 12 41	+49 00.0	1	10.50		1.46		1.34		3	M2 V	7008
+23 02436				12 12 42	+22 31.4	1	9.34		0.32		0.14		1	F0	280
+38 02315				12 12 42	+38 10.5	1	10.00		0.36				2	F5	77
106522 −60 03862				12 12 42	−61 19.9	1	9.12		0.33		0.27		1	A7 IV/V	1771
	AAS78,203 # 6			12 12 42	−61 40.	1	13.89		1.14		0.05		3		1684
	POSS 377 # 33			12 12 43	+26 14.0	1	17.02		1.59				5		1739
106543 +15 02433				12 12 44	+14 50.6	1	10.90		0.55		-0.02		2	G5	1569
−54 05079	IDS12101S5426	AB		12 12 44	−54 42.4	1	9.91		0.79		0.36		2		1696
106556 +47 01943				12 12 46	+47 23.6	1	7.32		1.01		0.75		2	G5 II	172
106574 +71 00610	HR 4659			12 12 46	+70 28.7	1	5.71		1.19				2	gK2	71
	G 59 - 8			12 12 47	+19 41.4	2	13.15	.005	0.71	.005	0.08	.000	3		333,1620,1658
+29 02270				12 12 47	+29 22.8	2	9.70	.020	1.04	.005	0.89	.015	2	G9 III	280,8034
106576 +56 01579				12 12 47	+55 47.3	1	8.70		0.91		0.62		2	G0	1502
−59 04136				12 12 47	−59 37.1	1	10.59		0.13		0.07		1	A1 V	137
106558 +32 02210				12 12 50	+32 55.7	1	9.00		0.43				2	F5 IV	77
106557 +39 02505				12 12 50	+39 00.8	1	8.14		0.05		0.07		2	A2 V	1068
106549 −30 09789	IDS12102S3032	AB		12 12 50	−30 48.9	2	9.79	.044	1.45	.041	1.20		6	K5 V	173,3072
	HA 80 # 119			12 12 51	+14 40.9	1	11.21		1.05		0.83		2		269
106577 +38 02316				12 12 53	+37 44.5	1			0.91				2	G9 V	77
	POSS 377 # 39			12 12 54	+22 56.0	1	17.26		1.46				3		1739
	LP 494 - 66			12 12 55	+10 07.2	1	12.16		1.31		1.32		1		1696
	HA 80 # 193			12 12 55	+14 59.4	1	13.05		0.94		0.74		1		269
+36 02250				12 12 55	+35 50.7	2	9.19	.005	0.97	.025	0.78		5	K0 III	77,1569
106572 −41 07056	HR 4658			12 12 55	−41 38.0	6	6.25	.008	1.00	.005	0.72	.045	24	K0 III	15,1075,1311,2013*
+24 02442				12 12 56	+23 56.0	1	10.42		1.05		0.87		1		280
106592 +40 02514	G 198 - 38	★ AB		12 12 57	+40 24.3	1	8.82		0.91				3	G9 IV	280
	IBVS2612 # 3			12 12 57	+52 32.1	1	12.27		0.55		0.03		1		1548
	NGC 4216 sq1 8			12 12 58	+13 20.5	1	13.96		0.69				2		492
106605 +53 01535				12 12 58	+52 59.7	1	7.54		1.06		0.87		1	K0	1548
106591 +57 01363	HR 4660	★ A		12 12 58	+57 18.6	24	3.31	.007	0.08	.005	0.07	.004	315	A3 V	1,15,61,71,667,985*
106593 +39 02506				12 12 59	+38 56.3	3	7.69	.005	0.29	.005	0.12	.007	5	F0 V	105,280,1569
	NGC 4216 sq1 6			12 13 00	+13 29.4	1	12.55		0.67				2		492
106589 −47 07490				12 13 00	−48 08.4	1	8.94	.004	0.67	.004	0.14	.001	18	G5 V	863,977,2011
	Ly 114			12 13 00	−60 30.	1	11.48		0.48		-0.51		5		94
	HA 80 # 81			12 13 02	+14 25.9	1	11.49		0.62		0.07		1		269
106594 +12 02438				12 13 04	+11 34.2	1	10.01		1.14		1.16		1	K0	280
	NGC 4216 sq1 7			12 13 04	+13 26.5	1	15.14		0.67				2		492
106590 −60 03863	LSS 2627			12 13 04	−60 51.1	4	7.93	.025	0.23	.009	-0.62	.026	8	B1 III	191,540,976,1684
	NGP +27 075			12 13 05	+26 33.9	1	11.96		1.03		0.95		1	K1 III	8034
	NGP +27 079			12 13 07	+26 34.3	1	11.21		0.75		0.19		1	G5	8038
106602 −57 05365	IDS12105S5751	AB		12 13 07	−58 08.3	2	9.75	.020	0.07	.030	-0.23	.025	4	B9 V	401,3039
106606 +13 02512				12 13 08	+12 44.8	1	9.99		0.93		0.58		2	K0	1375
106607 +11 02442				12 13 09	+11 08.7	1	8.40		1.12		1.13		1	K0	280
+28 02094	Coma Ber 32			12 13 09	+28 10.5	2	9.81	.000	1.03	.000	0.85	.020	2	G9 III	280,8034
−60 03864	LSS 2628			12 13 09	−61 03.2	5	9.43	.013	0.64	.009	-0.39	.038	14	O9.5Ia	191,540,976,1684,1737
106619 +33 02211				12 13 10	+33 04.2	1	9.32		0.58				2		77
	G 12 - 26			12 13 11	+14 44.5	1	14.38		1.55				1		906
	AAS12,381 39 # B			12 13 11	+38 42.6	1	10.46		1.11		1.05		1		280
106618 +44 02175				12 13 11	+44 12.3	1	9.56		1.48				2	M2 III	280
	IBVS2612 # 4			12 13 11	+52 55.8	2	12.55	.005	1.53	.015	1.00		3	dMe:	1548,1746
106612 −22 03322	HR 4661	★ AB		12 13 11	−23 04.5	2	6.53	.005	0.46	.010			8	F5/6 V	176,2007
	IBVS2612 # 7			12 13 12	+52 37.4	1	13.24		0.75		0.15		1		1548
	NGC 4216 sq1 1			12 13 13	+13 14.7	1	12.39		0.62				1		492
+27 02108	Coma Ber 33			12 13 14	+26 34.4	2	9.48	.020	1.01	.015	0.85	.000	2	K0 III	280,8034

HD	DM	Other Id	N	Rem	α_{1950}	δ_{1950}	S	V	σ_V	B–V	σ_{B-V}	U–B	σ_{U-B}	n	Spectrum	References
106625	−16 03424	HR 4662			12 13 14	−17 15.9	17	2.58	.009	−0.11	.005	−0.32	.034	232	B8 IIIp	3,9,15,26,198,1006*
106621	+9 02611	IDS12107N0935		AB	12 13 16	+09 18.8	1	8.85		0.28		0.03		2	A2	1569
		EG 87, EG UMa		V	12 13 16	+52 47.5	2	13.29	.068	0.53	.005	−0.46	.029	3	DA:	1548,3028
		POSS 377 # 6			12 13 18	+24 28.3	1	14.54		1.45				4		1739
	+29 02271	Coma Ber 34			12 13 18	+29 04.6	3	9.31	.014	1.50	.005	1.78	.016	16	M3 III	280,1613,8034
106640	+36 02254				12 13 18	+35 31.5	1	9.16		0.53				2	G0 V	77
106639	+43 02202				12 13 19	+43 00.6	1	10.09		1.01				3	G8 III	280
106616	−64 01835	LSS 2629			12 13 19	−64 54.9	4	7.97	.024	0.01	.017	−0.77	.015	9	B1 Ib/II	401,540,976,1684
106631	−42 07516	IDS12107S4300		A	12 13 20	−43 17.0	1	9.26		0.36		0.25		2	A1mA5-F0	1776
106677	+73 00549	HR 4665, DK Dra			12 13 21	+72 49.8	1	6.29		1.14				2	K0 III	71
106633	−54 05083	IDS12107S5432		A	12 13 21	−54 48.9	1	8.35		0.00		−0.13		3	B9 V	1770
	−60 03867	LSS 2630			12 13 23	−60 34.7	2	10.89	.029	0.30	.029	−0.43	.005	6		846,1684
	−61 03055				12 13 23	−62 21.1	1	9.97		0.55		0.13		2	F5 V	137
		POSS 377 # 5			12 13 24	+23 56.2	1	14.32		0.61				3		1739
		AAS78,203 # 7			12 13 24	−59 24.	1	14.84		1.51		1.28		4		1684
	+15 02435				12 13 25	+15 19.2	1	9.57		0.66		0.19		2	K0	1569
106655	−46 07804				12 13 25	−47 07.2	1	9.66		0.03		0.01		1	A0 V	1770
106635	−63 02217				12 13 25	−64 10.6	1	8.35		−0.02		−0.58		2	B3 V	401
106637	−64 01837	LSS 2631			12 13 25	−65 23.4	2	8.22	.005	−0.03	.005	−0.81	.000	4	B2/3 III	401,846
	+6 02573	G 12 - 27	★	AB	12 13 26	+05 51.1	3	9.45	.017	1.22	.005	0.96		8	K8	176,202,3072
		IBVS2612 # 8			12 13 27	+52 46.7	1	13.83		0.71		0.11		1		1548
106661	+15 02436	HR 4663			12 13 28	+15 10.6	6	5.10	.016	0.06	.007	0.08	.014	31	A3 V	269,1022,1068,1363*
	−60 03868	LSS 2632			12 13 28	−60 47.1	1	10.86		0.29		−0.45		2		846
		Sand 37			12 13 30	+28 22.	1	15.32		1.69				1	dM3	686
	+45 02011				12 13 30	+45 24.4	1	10.72		1.02		0.75		1	G8 III	8034
	−59 04147	LSS 2633	★	AB	12 13 30	−60 15.2	5	9.41	.029	0.42	.013	−0.54	.021	19	B0 II	94,540,976,1684,8100
106678	+28 02095	Coma Ber 35			12 13 31	+28 19.6	1	8.37		0.41		0.03		3	F6 V	3016
		NGC 4216 sq1 4			12 13 32	+13 35.2	1	11.76		0.66				1		492
106680	−2 03487				12 13 33	−02 44.2	1	8.81		0.62		0.19		2	G5	1064
106670	−49 06954				12 13 33	−49 37.8	2	8.86	.000	0.83	.015	0.30		3	G3/5 Vw	1594,6006
		NGC 4216 sq1 2			12 13 34	+13 14.8	1	11.72		0.94				1		492
106659	−68 01637				12 13 34	−68 53.9	1	9.82		0.22		0.17		2	A2/3 V	1771
106674	−62 02658				12 13 35	−62 42.9	2	9.21	.004	0.11	.004	0.00	.004	17	A0 V	125,137
106691	+26 02321	Coma Ber 36			12 13 37	+26 02.3	4	8.14	.022	0.40	.003	−0.04	.005	7	F3 V	1127,1148,8001,8023
106690	+41 02284	HR 4666	★	A	12 13 37	+40 56.3	3	5.69	.005	1.59	.005	1.95	.010	5	M1 III	172,542,3001
106690	+41 02284	HR 4666	★	AB	12 13 37	+40 56.3	2	5.62	.019	1.49	.005	1.45		3	M1 III+F7 V	70,542
106690	+41 02284	IDS12111N4113		B	12 13 37	+40 56.3	1	8.77		0.57		0.09		1	F7 V	542
106676	−71 01323	HR 4664			12 13 37	−72 20.2	4	6.21	.007	−0.01	.003	−0.06	.000	22	A0 V	15,1075,1075,2038
106672	−60 03870				12 13 38	−60 45.7	1	9.77		0.21		0.22		1	A3/5 V	1771
		LF14 # 134			12 13 38	−62 24.3	1	12.17		0.31		0.24		1	A2 V	137
		BH Cru			12 13 39	−56 00.3	1	7.41		2.71		2.91		1	SC	864
106712	+42 02287	IDS12112N4227		AB	12 13 41	+42 10.6	1	8.71		1.50				3	M0 III	280
		G 13 - 25			12 13 41	−04 53.5	1	11.57		1.13		1.14		1		1696
	+18 02582				12 13 42	+17 36.6	1	10.41		0.79		0.33		2		1569
		IBVS2612 # 2			12 13 42	+52 49.4	1	11.62		0.66		0.20		1		1548
		AAS78,203 # 8			12 13 42	−64 07.	1	13.41		0.64		−0.45		3		1684
	+27 02109	Coma Ber 38			12 13 44	+27 15.3	1	10.24		1.13		1.24		1	K1 III	280
106713	+29 02272	Coma Ber 37			12 13 44	+29 06.9	3	9.40	.021	0.56	.005	0.05	.004	24	G5	77,1613,3016
	+27 02110				12 13 47	+27 13.5	2	10.28		1.01		0.84		2	G8 III	280,8041
		Steph 1001			12 13 47	+27 21.2	2	12.19	.009	1.46	.019	1.25	.009	4	M0	685,1746
106714	+24 02443	Coma Ber 39	★		12 13 49	+24 13.4	3	4.97	.030	0.96	.008	0.69	.005	6	K0 III	1080,1363,3016
106708	−62 02659	LSS 2634			12 13 49	−63 15.2	5	8.10	.028	0.09	.014	−0.78	.016	12	B0/1 III	125,401,540,976,1684
		HA 80 # 203			12 13 50	+14 54.1	1	11.00		1.16		1.26		1		269
		LF14 # 135			12 13 53	−59 10.9	1	11.36		0.51		0.04		1	F5 V	137
106734	+12 02439				12 13 54	+12 15.9	1	10.34		0.44		−0.01		2	F5	1569
106730	−63 02219	LSS 2636			12 13 54	−64 14.3	6	8.50	.064	0.10	.040	−0.92	.029	16	B0/1e	401,540,820,976,1684,1737
		LSS 2635			12 13 54	−64 49.3	1	11.27		0.02		−0.61		3	B3 Ib	846
106731	−64 01839				12 13 54	−65 13.7	1	8.53		0.03		−0.16		2	B9 V	1771
		NGC 4216 sq1 10			12 13 55	+13 28.0	1	15.15		0.44				3		492
	+18 02583				12 13 55	+17 36.9	2	9.92	.029	0.68	.000	0.27	.015	3		280,1569
106746	+16 02351				12 13 56	+16 18.4	1	9.83		0.39		0.01		2	F5	1569
		CS 16026 # 14			12 13 56	+28 07.3	1	14.73		0.23		0.07		1		1744
106742	−62 02661				12 13 57	−62 55.2	2	6.97	.014	0.60	.014			8	G0 V	1075,2033
106747	+13 02514				12 13 58	+12 47.3	2	9.71	.030	1.08	.005	1.01		4	K2	280,882
		NGC 4216 sq1 9			12 13 59	+13 27.3	1	13.89		0.60				3		492
106760	+33 02213	HR 4668			12 14 00	+33 20.4	6	5.00	.009	1.14	.004	1.07	.009	12	K0.5 IIIb	77,1080,1355,1363*
106748	+12 02440				12 14 01	+11 50.3	1	10.31		1.06		0.90		2	G5	1569
106761	+14 02483				12 14 01	+14 26.2	1	9.91		1.17		1.07		2	K2	280
	−59 04152	LSS 2637			12 14 01	−60 07.6	4	10.72	.009	0.29	.011	−0.69	.025	13	O9.5II	94,191,540,1684
	+25 02478	Coma Ber 40			12 14 02	+24 53.2	5	10.65	.022	−0.03	.006	−0.02	.010	20	A0 V	275,1026,1068,1240*
	+44 02176				12 14 02	+43 57.0	1	9.23		0.97				2	G9 III	280
	−64 01840				12 14 02	−64 59.9	1	10.15		0.05		−0.42		1	B5	820
	+30 02249				12 14 03	+29 59.0	1	10.76		0.69		0.26		1	G5 III	8034
	+28 02096	Coma Ber 41			12 14 04	+27 55.9	1	9.75		1.01		0.75		1	G8 III	8034
		AAS41,43 F3 # 2			12 14 04	−64 29.3	1	11.46		0.22		0.05		1	B9	820
106762	+10 02398				12 14 05	+09 54.2	1	10.39		0.66		0.22		1	G5	280
		HA 80 # 128			12 14 05	+14 40.3	1	14.23		0.73		0.30		1		269
		Sand 38			12 14 06	+22 25.	1	13.48		1.48				1	dM1	686
106758	−61 03062				12 14 07	−61 51.1	1	10.10		0.13		0.05		1	A0 V	137
106774	+14 02484				12 14 08	+13 34.5	3	8.84	.017	1.26	.016	1.32	.000	6	K0	280,492,882

Table 1 657

HD	DM	Other Id	N Rem	α_{1950}	δ_{1950}	S	V	σ_V	B–V	σ_{B-V}	U–B	σ_{U-B}	n	Spectrum	References
106773 +15 02438				12 14 10	+15 04.0	1	10.50		1.00		0.83		1	G5	280
		POSS 377 # 41		12 14 10	+24 47.5	1	17.42		0.97				2		1739
		POSS 377 # 42		12 14 10	+24 51.9	1	18.46		0.90				3		1739
106775 −1 02639				12 14 10	−02 27.7	2	8.04	.008	1.00	.003	0.81		5	K0	1284,1657
		POSS 377 # 10		12 14 11	+24 47.3	1	15.42		1.04				2		1739
106759 −73 00968				12 14 11	−73 51.9	3	7.83	.001	0.37	.013	0.27	.010	7	A2 V	278,657,2012
106784 +40 02516		IDS12117N4009	AB	12 14 12	+39 52.2	1	7.22		0.17		0.08		3	A3 IV	1068
106771 −51 06534				12 14 12	−51 48.0	1	9.83		0.30		0.11		1	A8/9 V	1771
		LP 852 - 59		12 14 13	−22 12.4	1	11.07		1.36		1.22		1		1773
	−60 03874	LSS 2638		12 14 13	−61 16.1	3	10.44	.023	0.66	.015	−0.33	.019	8		279,846,1684
	+38 02320			12 14 14	+37 54.3	1	9.38		1.23				2	K2 III	280
106783 +45 02014				12 14 14	+44 40.8	1	9.18		1.50		1.76		1	K4 V	280
106802 +45 02015				12 14 14	+44 51.9	1	9.65		0.99				2	G8 III	280
	+30 02250			12 14 15	+29 51.9	2	10.08	.005	0.35	.015	−0.03		3	F2	77,1744
	+48 02011			12 14 15	+47 53.1	1	10.36		1.07		0.91		2	K0 III	1375
	−63 02227			12 14 15	−64 15.5	1	10.13		1.27		1.13		2	K0	820
106786 +12 02441				12 14 16	+11 46.6	1	10.43		0.59		0.04		2	G0	1569
106785 +13 02515				12 14 16	+12 49.0	2	9.01	.005	1.13	.015	1.11		3	K0	280,882
	+30 02251	IDS12117N3037	AB	12 14 16	+30 20.8	1	9.48		0.88				2	G8 IV	77
106782 −72 01233				12 14 18	−73 15.4	2	8.64	.009	0.10	.011	−0.41		6	B5 V	278,2012
		POSS 377 # 8		12 14 19	+24 51.6	1	14.60		0.70				2		1739
	+25 02479			12 14 19	+25 07.4	2	10.51	.030	0.80	.005	0.52	.050	2	G8 III	280,8041
		CS 16026 # 11		12 14 19	+29 12.7	1	13.49		0.11		0.14		1		1744
106793 −55 04964				12 14 19	−56 07.8	3	9.68	.018	0.02	.017	−0.35	.033	6	B8/9II/IIIe	401,1684,3039
		G 13 - 26	AB	12 14 20	+03 14.6	2	13.27	.014	1.49	.181			7		202,538
		LP 734 - 101		12 14 20	−11 32.4	1	10.26		0.75		0.36		2		1696
	+36 02256			12 14 21	+36 12.9	1	9.34		0.61				2	G0	77
		CS 16026 # 7		12 14 22	+30 37.9	1	13.89		0.69		0.16		1		1744
106781 −64 01843				12 14 22	−64 58.4	1	9.31		−0.02		−0.70		1	B3/7	820
		POSS 377 # 43		12 14 23	+24 54.4	1	19.58		0.26				3		1739
		AAS41,43 F3 # 9		12 14 23	−64 31.6	1	11.75		0.54		0.06		1	F5	820
106796 −64 01845				12 14 23	−65 01.8	1	9.21		0.17		0.11		1	A1/2 Vn	820
106797 −64 01844		HR 4669		12 14 23	−65 24.9	4	6.06	.010	0.03	.007	−0.06	.012	12	A0 V	15,1075,1771,2038
106804 +15 02439				12 14 24	+15 05.9	2	9.54	.000	1.37	.029	1.45		3	K2	280,882
		Feige 59		12 14 24	+15 50.9	1	11.55		−0.06		−0.10		1	B9	1298
	+16 02354			12 14 24	+16 06.0	2	10.16	.020	1.04	.005	0.88	.015	6	K0	1569,1569
		SSII 116		12 14 24	+39 34.	1	13.17		0.51		−0.09		1	A0	1298
	+45 02016			12 14 24	+45 22.0	2	9.06	.019	1.50	.005	1.82		3	K4 III	280,8034
		Ly 118		12 14 24	−64 37.	2	11.84	.024	0.06	.005	−0.53	.005	3	B3	820,846
106814 +28 02097		Coma Ber 43		12 14 25	+28 01.1	1	8.96		1.61				2	M2 III	280
	+38 02321			12 14 25	+37 45.7	1	9.86		0.69				2	G0	77
	−64 01846			12 14 25	−64 28.6	1	10.19		0.06		−0.43		2	B5	820
		CS 16026 # 6		12 14 26	+30 59.4	1	14.35		0.36		−0.21		1		1744
		IBVS2612 # 5		12 14 26	+52 40.9	1	12.80		0.58		−0.04		1		1548
	−59 04156			12 14 27	−60 23.8	1	10.74		0.28		−0.39		4		1684
		Coma Ber 201		12 14 28	+26 46.2	1	11.16		0.47		−0.02		2		1148
106819 −15 03442		HR 4670		12 14 28	−16 25.0	3	6.03	.080	0.10	.000	0.12		9	A2 V	15,252,2029
	−60 03877			12 14 28	−60 48.0	1	10.24		1.09		0.79		2	K0 III	137
		AAS41,43 F3 # 10		12 14 28	−64 39.6	1	12.05		0.10		−0.39		1	B4	820
106815 +12 02442				12 14 30	+12 03.4	1	9.32		1.40		1.60		1	K0	280
		Sand 39		12 14 30	+31 26.	1	14.15		1.62				1	dM4	686
106810 −59 04158				12 14 30	−59 55.0	1	9.35		0.44		0.26		1	A8/9 V	1771
		POSS 377 # 3		12 14 31	+24 45.5	1	14.20		0.61				4		1739
106833 −3 03258				12 14 32	−03 58.1	1	10.08		0.88		0.59		4	K2	158
		AAS41,43 F3 # 11		12 14 32	−64 12.1	1	10.84		1.14		0.83		1	K0	820
		IBVS2612 # 6		12 14 33	+52 56.2	1	13.00		0.59		0.03		1		1548
	−64 01849			12 14 33	−64 45.0	1	10.89		0.07		−0.33		1	B6	820
		HA 80 # 129		12 14 35	+14 37.0	1	12.16		0.73		0.26		2		269
		CS 16026 # 5		12 14 35	+31 02.4	1	14.16		0.71		0.22		1		1744
106851 +49 02128				12 14 35	+48 42.6	1	8.04		1.22		1.18		2	K2 III	172
		Coma Ber 277		12 14 36	+26 32.4	1	12.65		0.98		0.88		3		1148
		LP 852 - 60		12 14 36	−21 39.2	1	13.05		0.70		0.17		1		1696
		Steph 1003		12 14 38	+11 40.3	1	12.06		1.51		1.14		1	M1	1746
		HA 80 # 167		12 14 38	+14 44.9	1	13.73		0.69		0.03		1		269
106842 +25 02481		Coma Ber 44		12 14 39	+24 54.2	2	8.35	.010	1.05	.000	0.89		3	G8 III	280,8034
	+27 02112	Coma Ber 45		12 14 40	+26 29.2	2	9.86	.005	0.64	.010	0.20	.005	3		280,1148
		CS 16026 # 3		12 14 40	+31 45.9	1	13.95		0.65		0.05		1		1744
106837 −54 05095		IDS12120S5420	A	12 14 40	−54 36.9	1	8.31		0.22		0.16		1	A3 V	1771
		Coma Ber 294		12 14 41	+26 05.8	1	14.27		0.71		0.26		1		1148
	−7 03382			12 14 41	−07 32.0	1	10.50		1.12		1.10		1	K5 V	3072
106840 −62 02667				12 14 41	−62 57.0	1	8.97		0.70				4	G5/6 IV/V	2033
		LP 376 - 40		12 14 42	+25 37.	1	14.40		1.57				1		686
106839 −59 04159				12 14 43	−59 47.8	1	9.85		0.13		−0.34		1	B8 II	137
		Coma Ber 278		12 14 44	+26 22.7	1	12.32		0.60		−0.01		1		1148
107192 +88 00071		HR 4686		12 14 45	+87 58.6	1	6.28		0.35		−0.01		5	F2 V	15
106853 +9 02612				12 14 46	+09 25.2	1	9.72		1.46		1.81		1	K2	280
		POSS 377 # 4		12 14 46	+25 17.3	1	14.23		0.77				4		1739
		ApJS33,459 # 232		12 14 46	−63 11.7	1	11.93		0.50		−0.46		4		1684
		HA 80 # 130		12 14 47	+14 34.0	1	14.58		0.90		0.65		1		269
106846 −49 06970				12 14 47	−50 10.7	1	10.02		0.27		0.18		1	A7/8 V	1771

HD	DM	Other Id	N	Rem	α_{1950}	δ_{1950}	S	V	σ_V	B–V	σ_{B-V}	U–B	σ_{U-B}	n	Spectrum	References
		HA 80 # 170			12 14 48	+14 46.9	1	13.67		0.63		0.09		2		269
106858	+23 02441				12 14 48	+22 29.2	1	8.13		0.13		0.09		3	A2	1068
		Coma Ber 328			12 14 48	+25 36.9	1	12.32		0.51		-0.01		1		1148
		Coma Ber 308			12 14 48	+25 59.4	1	11.26		1.17		0.98		2	K0 III	1148
106857	+29 02274	Coma Ber 46			12 14 48	+29 00.2	2	8.99	.018	0.42	.005	-0.05		9	F5 V	77,3026
106855	-20 03615	IDS12122S2030		AB	12 14 49	-20 46.7	1	9.39		0.94		0.43		1	K1 V	3072
		HA 80 # 172			12 14 51	+14 49.8	1	14.78		0.92		0.24		1		269
		POSS 377 # 38			12 14 51	+24 47.6	1	17.23		0.52				2		1739
106849	-67 01931	HR 4671, ϵ Mus			12 14 51	-67 41.0	4	4.04	.036	1.58	.000	1.64	.053	17	M4 III	678,1088,3053,8017
		CS 16026 # 2			12 14 52	+32 06.8	1	13.69		0.67		0.09		1		1744
	+39 02507				12 14 52	+39 09.6	1	9.54		1.25		1.43		1	K2	280
	+43 02205				12 14 52	+43 04.4	1	9.74		1.00				2	G9 III	280
	+33 02214				12 14 53	+32 43.3	1	10.70		0.93		0.59		1	G9 III	8034
311161					12 14 53	-68 29.8	1	9.78		1.50				1	A0	906
106877	+7 02529	IDS12124N0709		AB	12 14 54	+06 52.8	1	7.91		0.39				4	F2	176
106869	-48 07307				12 14 54	-48 38.8	2	6.82	.005	0.58	.009	0.08		8	F8 V	158,1075
		AAS41,43 F3 # 14			12 14 54	-64 37.1	1	11.22		0.97		0.55		1	G0	820
106886	+39 02508				12 14 55	+38 55.8	1	8.53		1.03				2	K0 III	280
		AAS41,43 F3 # 13			12 14 55	-64 12.9	1	10.78		1.09		0.79		2	G8	820
		AAS41,43 F3 # 23			12 14 55	-64 12.9	1	11.97		0.33		0.23		2	A2	820
		HA 80 # 132			12 14 56	+14 37.7	1	14.82		0.47		-0.18		1		269
106871	-57 05375	AB Cru			12 14 56	-57 53.2	3	8.54	.182	0.14	.010	-0.80	.015	9	B1/2 III/V	540,976,1684
		HA 80 # 133			12 14 57	+14 37.9	1	12.23		0.62		0.02		2		269
		AAS41,43 F3 # 15			12 14 57	-64 34.4	1	12.26		0.28		0.21		1	A2	820
106885	+44 02177				12 14 58	+44 17.8	1	8.12		1.32				2	K3 III	280
		LP 434 - 79			12 14 59	+15 52.2	1	12.59		0.87		0.52		2		1696
106887	+29 02275	Coma Ber 47		★ AB	12 15 00	+29 12.9	5	5.71	.012	0.16	.008	0.12	.008	13	A4 m	77,1022,1068,3052,8071
106874	-64 01850				12 15 00	-64 39.0	1	9.19		1.26		1.19		1	G8/K1 III	820
	+36 02257				12 15 02	+35 48.6	1	9.09		1.32				2	K0 III	77
106884	+54 01510	HR 4672			12 15 02	+53 28.2	2	5.80	.015	1.32	.020	1.52		4	K5 III	70,3001
		AAS41,43 F3 # 17			12 15 02	-64 35.9	1	11.66		0.38		0.23		1	A0	820
		G 148 -B4	A		12 15 03	+32 21.9	1	15.12		1.58				1	DAn	3016
		G 148 -B4	B		12 15 03	+32 21.9	2	16.98	.024	0.36	.015	-0.48	.015	3		3028,3060
106888	+15 02441				12 15 04	+14 43.3	2	8.17	.010	0.54	.002	0.04	.002	27	F8	269,1569
		CS 16026 # 12			12 15 04	+28 42.6	1	13.95		0.59		0.06		1		1744
106912	+12 02443				12 15 05	+11 29.6	1	9.11		1.09		1.01		1	K0	280
		Coma Ber 279			12 15 05	+26 25.2	1	12.43		0.48		-0.05		5		1148
	+31 02347				12 15 06	+31 24.6	1	10.78		0.93		0.71		1	K0 III	8034
		AAS78,203 # 9			12 15 06	-56 21.	1	9.46		1.51		1.35		4		1684
		POSS 377 # 37			12 15 07	+24 43.0	1	17.20		1.18				2		1739
		AAS12,381 34 # C			12 15 07	+34 04.1	1	10.51		1.13		1.20		1		280
	+45 02017				12 15 07	+44 59.9	1	9.99		0.61				2	G6 III	280
		LSS 2640			12 15 07	-67 43.5	1	11.61		0.22		-0.52		3	B3 Ib	846
	+45 02018	G 123 - 18			12 15 08	+45 26.8	2	9.64	.005	0.68	.005	0.23		4	G5	280,1658
106903	-49 06972	SW Cen			12 15 08	-49 27.4	1	10.18		0.15		-0.02		2	B9 V	1770
	+31 02348				12 15 09	+30 39.4	1	10.66		1.12		1.13		1	K1 III	280
106902	-44 07914				12 15 09	-45 06.9	11	7.95	.006	-0.09	.005	-0.56	.005	63	B3/5 II/III	402,474,768,863,977,1075*
		Coma Ber 327			12 15 10	+25 45.8	1	14.40		0.62		0.29		1		1148
106922	-35 07842	HR 4675		★ AB	12 15 10	-35 49.0	2	6.14	.005	0.01	.000			7	A0 V	15,2012
106926	+15 02442	HR 4676			12 15 12	+15 25.4	1	6.34		1.37				2	K4 III	280
	-61 03072	LSS 2641			12 15 12	-61 48.0	2	10.60	.005	0.45	.005	-0.32	.005	5		125,846
		Coma Ber 329			12 15 13	+25 40.9	1	12.80		1.10		0.63		1		1148
		CS 16026 # 4			12 15 13	+31 34.6	1	13.69		0.66		0.08		1		1744
106906	-55 04978				12 15 13	-55 41.9	1	7.78		0.46				4	F5 V	2012
	+40 02519				12 15 15	+39 51.2	1	9.90		0.72				2	G5	280
		POSS 377 # 2			12 15 16	+24 48.5	1	13.83		0.37				2		1739
		Coma Ber 295			12 15 16	+26 07.5	1	14.31		0.95		0.46		1		1148
		LF14 # 139			12 15 16	-61 18.9	1	12.79		0.18		-0.27		1	B8 V	137
106929	+10 02402				12 15 17	+10 23.9	3	9.81	.024	-0.12	.012	-0.51	.018	5	B7 IV	272,280,308
		CS 16026 # 9			12 15 17	+30 31.8	1	14.16		0.61		0.01		1		1744
	+45 02019				12 15 17	+45 18.6	2	10.07	.010	1.23	.005	1.20		3	K0 III	280,8034
	-61 03073				12 15 17	-62 01.8	1	10.14		1.04		0.75		1	K0 III	137
106909	-63 02230				12 15 17	-63 54.5	1	8.30		0.36		0.05		2	F0 V	122
106949	+15 02443				12 15 19	+15 17.8	2	8.36	.011	0.53	.002	0.03	.003	22	F8	269,1569
106955	-23 10471	IDS12127S2327		ABC	12 15 19	-23 44.1	1	6.95		0.10		-0.06		3	B9 V	1770
106946	+26 02323	Coma Ber 49			12 15 20	+25 50.9	5	7.87	.026	0.36	.008	0.02	.043	9	F3 V	105,1127,1148,8001,8023
106940	-49 06975				12 15 20	-49 52.8	1	8.78		0.25		0.11		1	A4 V	1771
	+25 02483	Coma Ber 350			12 15 21	+25 23.3	3	9.93	.012	1.49	.015	1.81	.047	5	K5 III	280,694,1148
		ON 325 # 1			12 15 21	+30 23.	1	13.70		0.69		0.26		2		326
		ON 325 # 2			12 15 21	+30 23.	1	14.95		0.56		-0.07		2		326
		ON 325 # 3			12 15 21	+30 23.	1	15.17		0.49		-0.12		2		326
		ON 325 # 4			12 15 21	+30 23.	1	15.89		0.82		0.54		2		326
	+36 02258				12 15 21	+35 36.4	1	9.74		0.61				2	G5	77
	-63 02232				12 15 21	-64 13.1	1	9.57		1.07		0.79		2	G3	820
106911	-78 00741	HR 4674			12 15 22	-79 02.1	6	4.24	.010	-0.13	.005	-0.52	.011	31	B4 V	15,681,815,1034,1075,2038
106950	+12 02444	IDS12128N1221		AB	12 15 23	+12 03.3	1	10.26		0.37		-0.01		2	F0	1569
	+32 02233				12 15 23	+32 16.2	2	9.58	.044	0.43	.019	-0.01		3	F5	77,1744
	+21 02405				12 15 24	+21 11.8	1	10.00		1.06		0.90		1	G5	280
		Coma Ber 296			12 15 24	+26 07.5	1	14.43		0.78		0.63		2		1148
106943	-60 03890				12 15 24	-61 11.5	2	7.51	.005	-0.02	.000	-0.42		6	B7 IV	401,1075

Table 1 659

HD	DM	Other Id	N Rem	α_{1950}	δ_{1950}	S	V	σ_V	B–V	σ_{B-V}	U–B	σ_{U-B}	n	Spectrum	References
106964	+12 02445			12 15 25	+12 04.9	2	9.01	.049	1.33	.005	1.54	.010	3	K0	280,1569
106963	+13 02517			12 15 25	+12 48.1	1	10.21		0.57		0.07		1	F8	1569
107113	+87 00107	HR 4683		12 15 25	+86 42.8	1	6.33		0.43		-0.08		5	F4 V	15
		SW Dra		12 15 26	+69 47.2	1	9.94		0.16		0.15		1	F4	668
		AAS41,43 F3 # 19		12 15 26	-64 24.8	1	12.36		0.24		0.02		2	A3	820
		CS 16026 # 13		12 15 27	+28 08.0	1	12.53		1.42		1.21		1		1744
106971	+33 02217			12 15 28	+33 00.7	1	9.14		0.53				2		77
	+23 02443			12 15 29	+22 33.6	1	10.10		0.94		0.60		2		1569
	+29 02276			12 15 29	+28 55.5	1	10.02		0.37				2	F4	77
106961	-61 03075	IDS12128S6137	AB	12 15 29	-61 53.9	2	8.94	.005	0.16	.005	0.05	.005	5	A0 V	125,137
106970	-51 06557	IDS12129S5145	A	12 15 31	-52 01.7	1	6.84		-0.06		-0.31		3	B8 III	1770
		AAS41,43 F3 # 20		12 15 31	-64 26.8	1	11.44		0.34		0.18		3	A0	820
	+22 02458			12 15 32	+22 27.1	1	10.33		0.98		0.72		1	K0	280
		Coma Ber 271		12 15 33	+26 48.6	1	11.80		0.55		-0.02		3		1148
	+33 02218			12 15 33	+32 46.0	1	10.39		0.95		0.62		1	G8 III	8034
		LF14 # 142	V	12 15 33	-63 00.1	3	9.52	.032	1.51	.042	1.06	.050	3	K4 V	137,689,1587
		POSS 377 # 1		12 15 34	+24 44.5	1	12.02		0.51				2		1739
	+29 02277			12 15 34	+29 18.3	2	10.30	.020	1.10	.005	1.01	.015	2	K0 III	280,8041
		CS 16026 # 8		12 15 35	+30 34.1	1	14.74		0.10		0.16		1		1744
		AAS41,43 F3 # 21		12 15 35	-64 24.3	1	11.48		0.13		-0.46		2	B2	820
	-64 01853			12 15 35	-65 03.5	1	10.86		0.08		-0.34		3		1684
		HA 80 # 136		12 15 36	+14 39.5	1	12.22		1.34		1.54		2		269
106976	-3 03263	HR 4678	★ A	12 15 36	-03 40.3	1	6.56		0.33		0.01		2	F0	13
106976	-3 03263	HR 4678	★ AB	12 15 36	-03 40.3	3	5.98	.004	0.37	.012	0.00	.013	9	F2 V	13,15,1417
		AAS41,43 F3 # 22		12 15 36	-64 41.9	1	11.70		0.23		0.12		1	A0	820
106987	+13 02518			12 15 37	+13 06.8	1	10.35		0.48		-0.04		1	G0	1569
	+41 02285			12 15 37	+41 23.9	1	10.58		1.04		0.83		1	G9 III	280
		HA 80 # 137		12 15 39	+14 36.2	1	12.67		0.59		0.02		1		269
		HA 80 # 175		12 15 39	+14 52.0	1	15.09		1.05		0.91		1		269
	-62 02675			12 15 39	-62 43.7	1	9.00		0.93		0.71		1		1696
		He3 775	★	12 15 39	-62 44.	1	9.31		0.09		-0.52		2		1684
		AAS41,43 F3 # 24		12 15 39	-64 26.1	1	12.05		0.42		0.21		2	B7	820
106988	+10 02404			12 15 40	+10 08.8	2	8.88	.058	0.16	.026	0.14		3	A5 IVm	193,280
106999	+28 02100	Coma Ber 52		12 15 40	+27 34.7	5	7.47	.013	0.16	.019	0.09	.018	8	A5 V	1068,3016,8001,8023,8071
107028	+69 00657	IDS12133N6921	A	12 15 40	+69 04.5	1	7.66		0.84		0.45		3	G5	3016
106979	-50 06874			12 15 40	-50 58.0	1	7.56		1.49		1.84		6	K4 III	1673
	-63 02234			12 15 40	-63 46.0	1	11.18		0.16		-0.86		2		122
	-63 02233			12 15 40	-63 47.2	1	9.11		0.71		0.31		2		122
		AAS41,43 F3 # 25		12 15 41	-64 18.0	1	11.97		0.21		0.06		2	B8	820
	+16 02356	HY Com		12 15 42	+16 25.8	1	10.26		0.21		0.09		1		280
	+35 02330			12 15 42	+34 51.4	1	10.30		0.38				2	F6	77
		AAS41,43 F3 # 26		12 15 42	-64 36.5	1	11.79		0.50		0.33		1	A3	820
106983	-63 02235	HR 4679	★ A	12 15 43	-63 43.5	7	4.04	.013	-0.17	.009	-0.68	.020	21	B2/3 V	15,26,122,1034,1075*
	+28 02101			12 15 44	+27 47.8	1	10.71		0.88		0.53		1	G7 III	8034
		CS 16026 # 76		12 15 44	+28 48.2	1	14.38		0.45		-0.18		1		1744
106984	-64 01854			12 15 44	-64 26.2	1	8.13		1.65		1.86		2	K3/5	820
		HA 80 # 19		12 15 46	+14 05.6	1	10.69		1.16		1.19		2		269
		POSS 377 # 7		12 15 47	+24 29.1	1	14.57		2.03				2		1739
		Ton 76		12 15 47	+30 09.7	1	14.21		0.51		-0.23		1		1744
106997	-60 03894			12 15 47	-60 38.2	1	8.43		1.02		0.70		1	G8/K0 III	137
	+19 02549			12 15 48	+19 17.4	1	10.14		0.99		0.67		1	K0	280
	+25 02484			12 15 48	+24 44.6	1	10.73		0.72		0.34		1		280
		L 104 - 107		12 15 48	-69 12.	1	12.20		0.84		0.37		2		1696
		Coma Ber 371		12 15 49	+25 07.4	1	12.07		0.59		0.09		1		1148
	+31 02349			12 15 49	+31 11.7	1	10.70		1.03		0.81		1	K0 III	8034
107029	+42 02291			12 15 49	+42 18.2	1	8.72		1.08				3	K0 III	280
		LP 734 - 102		12 15 49	-12 35.4	1	12.92		0.39		-0.14		2		1696
107016	+13 02519			12 15 50	+12 44.8	1	10.24		0.92		0.52		1	G5	1569
	-64 01855			12 15 50	-64 30.8	1	10.12		0.13		-0.30		2	B3	820
		AAS41,43 F3 # 29		12 15 50	-64 38.2	1	10.47		1.45		1.56		1	G8	820
		CS 16026 # 16		12 15 51	+28 17.9	1	14.21		0.59		-0.08		1		1744
	+19 02550			12 15 52	+19 25.5	2	8.76	.000	0.98	.005	0.72	.010	7	K0	1569,1648
		Coma Ber 326		12 15 52	+25 49.0	1	14.61		0.94		0.54		1		1148
		Coma Ber 297		12 15 52	+26 09.9	1	12.58		1.61		1.46		1		1148
		Coma Ber 276		12 15 52	+26 30.8	1	12.38		1.04		0.91		1		1148
	+45 02021			12 15 54	+44 37.0	1	9.61		1.17				2	K1 III	280
		AAS41,43 F3 # 30		12 15 54	-64 36.2	1	12.22		0.42		0.29		2	A4	820
107031	+17 02460			12 15 57	+16 28.9	2	9.80	.015	1.12	.010	1.19		4	K2	882,1569
107022	-43 07592			12 15 57	-43 52.0	4	8.41	.004	0.69	.002	0.19	.002	32	G5 V	863,977,1075,2011
		AAS41,43 F3 # 31		12 15 58	-64 31.3	1	12.34		0.32		0.18		1	B7	820
107053	+33 02219			12 15 59	+33 01.6	1	6.72		0.22				2	A5 V	77
		LF14 # 144		12 15 59	-58 53.9	1	11.94		1.29		1.18		2		137
		Coma Ber 348		12 16 00	+25 33.8	1	13.71		0.95		0.77		1		1148
107054	+31 02350	HR 4680		12 16 01	+30 31.7	3	6.24	.009	0.30	.004	0.02	.005	4	A9.5III	77,1068,3037
	-62 02679			12 16 01	-62 31.3	1	11.10		0.45		0.06		1	F5 V	137
		POSS 377 # 16		12 16 02	+21 23.3	1	16.14		1.23				3		1739
107047	-52 05457			12 16 02	-52 29.1	1	9.42		0.03		-0.05		1	B9.5 V	1770
		AAS41,43 F3 # 32		12 16 03	-64 29.5	1	11.66		0.24		-0.13		1	B7	820
107067	+23 02447	Coma Ber 53		12 16 05	+23 23.9	3	8.74	.008	0.52	.006	0.02	.000	4	F7 V	1127,8001,8023
	+33 02220			12 16 05	+33 07.7	1	10.12		0.37				2	F3	77

HD	DM	Other Id	N Rem	α_{1950}	δ_{1950}	S	V	σ_V	B–V	σ_{B-V}	U–B	σ_{U-B}	n	Spectrum	References
107070	+0 02920	HR 4681		12 16 06	−00 30.6	2	5.89	.005	0.17	.000	0.12	.010	7	A5 Vn	15,1417
		Coma Ber 309		12 16 07	+25 49.8	1	13.53		0.50		0.03		1		1148
107085	+41 02287			12 16 07	+41 11.6	2	8.38	.025	0.91	.020	0.65		4	G9 III-IV	172,280
107050	−63 02239			12 16 07	−64 24.5	1	9.26		0.34		0.16		2	A3 III/IV	820
		Coma Ber 307		12 16 10	+25 58.6	1	13.53		1.10		1.02		1		1148
107086	+26 02324	Coma Ber 55		12 16 10	+26 28.0	2	7.60	.020	0.44	.005	0.01	.005	5	F8 V	1148,3016
		LF14 # 146		12 16 10	−61 03.1	1	11.33		0.19		0.11		1	A0 V	137
	+21 02408			12 16 13	+20 49.3	1	8.88		1.18		1.31		1	K2	280
	+22 02460	TW Com		12 16 13	+22 22.0	1	9.50		1.44		1.43		1	K5	280
		LF14 # 147		12 16 13	−58 41.8	1	10.72		1.33		1.45		1	K2 III	137
		AAS41,43 F3 # 34		12 16 13	−64 12.3	1	11.55		0.27		0.18		1	B9	820
		Coma Ber 346		12 16 14	+25 31.4	1	10.89		0.71		0.20		2		1148
		NGP +28 068		12 16 14	+28 19.7	1	11.22		0.92		0.65		1	K0 III	8041
	+37 02264			12 16 14	+37 15.1	1	10.34		0.40				2	F8	77
107087	+19 02552			12 16 15	+19 26.8	1	8.03		0.56		0.10		2	G0 V	1569
		LP 734 - 90		12 16 16	−11 18.4	1	12.51		0.44		-0.18		3		1696
		LF14 # 148		12 16 16	−58 57.5	1	11.52		1.24		0.72		1		137
107088	+15 02444			12 16 17	+15 05.7	1	9.97		0.81		0.37		1	K0	280
		Coma Ber 275		12 16 17	+26 35.7	1	10.82		0.95		0.74		3	K1 V	1148
		AAS41,43 F3 # 35		12 16 17	−64 35.8	1	10.19		1.22		1.08		3	G5	820
	+25 02485	Coma Ber 56		12 16 18	+24 34.1	1	9.48		0.88		0.48		3	G7 V	3016
		Coma Ber 349		12 16 18	+25 26.6	1	14.74		0.64		0.00		1		1148
		Coma Ber 347		12 16 18	+25 33.0	1	14.19		0.82		0.26		1		1148
107079	−54 05113	HR 4682	⋆ A	12 16 19	−54 51.9	4	5.00	.005	1.60	.008	1.94	.005	19	M1 III	15,1075,2012,3055
		Wray 959		12 16 20	−61 34.6	1	12.59		0.50		-0.26		7		1684
107100	+13 02520			12 16 21	+13 16.6	1	10.77		0.91		0.61		1	G5	1569
		Coma Ber 325		12 16 21	+25 43.4	1	12.89		0.59		0.10		1		1148
	−58 04209	LSS 2642		12 16 21	−59 10.8	1	11.15		0.29		-0.44		2		846
	+16 02358			12 16 22	+16 09.8	1	9.97		1.20		1.28		1		280
		Coma Ber 293		12 16 24	+26 20.4	1	11.70		0.94		0.55		2	G8 III	1148
	+31 02351			12 16 24	+30 44.8	1	9.59		0.48				2	F8	77
		Coma Ber 323		12 16 25	+25 47.4	1	13.51		0.59		0.07		1		1148
	+28 02102			12 16 25	+28 07.8	1	10.73		0.62		0.16		1	G5	8038
		AAS41,43 F3 # 37		12 16 25	−64 53.9	1	12.16		0.36		0.19		1	A2	820
	+18 02587			12 16 26	+17 36.4	1	8.80		0.47		0.02		2	F8	1569
		Coma Ber 1023		12 16 26	+26 09.9	1	11.51		1.03				3		913
107130	+43 02207			12 16 26	+43 21.3	1	9.13		0.97				2	G8 III	280
107114	+24 02445	Coma Ber 57		12 16 27	+23 50.8	1	8.36		0.96		0.75		1	K0	280
107115	+14 02486			12 16 28	+13 38.7	2	9.34	.005	0.98	.019	0.73		3	K0	280,882
		Coma Ber 324		12 16 28	+25 42.6	1	11.29		0.71		0.29		2		1148
		Coma Ber 322		12 16 28	+25 47.4	1	12.25		1.03		0.82		1		1148
		LP 734 - 96		12 16 28	−15 09.2	1	11.56		1.02		0.91		1		1696
		AAS41,43 F3 # 36		12 16 28	−64 28.4	1	12.24		1.33		1.20		1	F5	820
107097	−60 03898			12 16 29	−60 51.5	1	7.76		0.04				4	B9 IV	1075
107116	+13 02521			12 16 30	+13 11.9	2	10.38	.050	0.22	.005	0.09	.025	4	A3	272,280
		Feige 60		12 16 30	+15 24.0	1	12.89		0.07		0.13		1		3060
107132	+25 02486	Coma Ber 58		12 16 30	+25 07.4	4	8.83	.014	0.50	.004	0.01	.012	6	F7 V	1127,1148,8001,8023
		Coma Ber 1027		12 16 30	+25 47.3	1	12.32		1.03				1		913
		Ton 607		12 16 30	+26 03.	1	16.49		0.10		-0.72		2		1036
		SSII 122		12 16 30	+47 54.	2	12.42	.000	0.41	.000	-0.23	.000	2	A7	1026,1240
		G 12 - 30	⋆ V	12 16 31	+11 24.0	2	13.80	.014	1.85	.023	1.03	.026	4		419,3078
107131	+26 02326	Coma Ber 60	⋆ V	12 16 31	+26 17.1	6	6.48	.008	0.18	.005	0.09	.006	18	A4 IV-V	15,252,1127,8001,8015,8052
		Coma Ber 280		12 16 31	+26 27.6	1	11.49		0.90		0.57		2	G8 III	1148
	+28 02103	Coma Ber 59	⋆	12 16 31	+28 19.6	2	8.98	.035	0.75	.005	0.28		4	G5 IV	280,3016
	−63 02240			12 16 31	−63 42.0	1	11.14		0.20		-0.18		2		122
107111	−64 01858			12 16 31	−64 50.6	1	9.86		0.19		0.12		2	A1/2 IV/V	820
		G 12 - 31		12 16 32	+06 11.3	1	15.94		1.55		1.25		2		3032
		G 12 - 32		12 16 32	+06 11.6	1	14.97		1.45		1.14		3		3032
		POSS 377 # 22		12 16 32	+23 06.4	1	16.39		1.41				3		1739
		AAS41,43 F3 # 40		12 16 32	−64 40.2	1	12.84		0.28		0.17		2	A0	820
		AAS41,43 F3 # 39		12 16 33	−64 10.6	1	10.88		0.44		0.16		1	B9	820
		AAS41,43 F3 # 41		12 16 33	−64 29.1	1	12.25		0.58		0.04		1	F4	820
107134	+11 02448			12 16 34	+10 52.2	1	8.72		1.04		0.90		1	K0	280
	−63 02241			12 16 34	−63 43.7	1	10.82		0.12		-0.48		2		122
107112	−65 01838			12 16 34	−65 32.9	1	9.21		0.38		0.17		1	A8 V	1771
107146	+17 02462			12 16 35	+16 49.7	3	7.03	.009	0.61	.005	0.08	.003	12	G2 V	1355,1499,3026
		Coma Ber 330		12 16 35	+25 36.2	1	14.09		0.90		0.49		1		1148
	+33 02221			12 16 35	+33 01.9	1	9.63		0.40				2	F3	77
		AAS41,43 F3 # 43		12 16 35	−64 11.9	1	12.42		0.22		0.12		1	B9	820
		AAS41,43 F3 # 44		12 16 35	−64 16.9	1	10.94		0.93		0.53		1	F9	820
	−64 01859			12 16 35	−64 37.3	1	9.90		0.97		0.61		2	G5	820
		G 148 - 43		12 16 36	+32 07.4	1	12.31		1.50		1.18		1		1744
107193	+75 00470	HR 4687		12 16 36	+75 26.3	1	5.38		-0.02		-0.05		2	A1 V	3023
107137	−10 03442			12 16 36	−11 14.7	1	8.81		0.51		-0.01		1	K5	3026
107158	+41 02288			12 16 37	+40 35.4	2	7.18	.010	0.99	.010	0.75		4	K0 III	172,280
107143	−47 07530			12 16 37	−47 33.1	1	7.86		0.11		0.07		1	A1 V	1770
107126	−58 04212	IDS12140S5844	A	12 16 38	−59 00.8	1	8.61		0.12		0.03		1	A1 V	1771
		Coma Ber 351		12 16 39	+25 24.2	1	12.33		1.00		0.57		1		1148
		Coma Ber 298		12 16 39	+26 10.7	1	13.00		0.62		0.12		3		1148
		AAS41,43 F3 # 42		12 16 39	−64 27.8	1	11.82		1.69		1.90		1		820

Table 1 661

HD	DM	Other Id	N	Rem	α_1950	δ_1950	S	V	σ_V	B–V	σ_B–V	U–B	σ_U–B	n	Spectrum	References
	+38 02323				12 16 40	+37 40.0	1	10.56		1.06		0.93		1	K0 III	280
	−60 03901				12 16 40	−60 33.0	1	11.17		0.20		0.13		1	A0 V	137
		AAS41,43 F3 # 46			12 16 40	−64 40.8	1	10.84		1.69		1.72		1	K5	820
		Coma Ber 264			12 16 41	+27 07.3	1	12.41		0.63		0.11		1		1148
	+28 02104	Coma Ber 202			12 16 41	+27 44.8	1	10.81		0.53		0.05		2	G0	1375
	+30 02254				12 16 41	+29 28.7	1	10.10		0.45				2	F6	77
		AAS41,43 F3 # 47			12 16 41	−64 36.6	1	12.84		0.20		0.06		1	A2	820
		LOD 26- # 7			12 16 43	−63 44.7	1	11.09		0.53		0.18		2		122
		AAS41,43 F3 # 48			12 16 43	−64 24.5	1	12.47		0.45		0.29		1	A4	820
	−64 01860				12 16 43	−64 40.0	1	10.57		0.23		0.19		2	A0	820
		Coma Ber 1040			12 16 44	+28 12.5	1	12.58		1.12				1		913
	−64 01861				12 16 44	−64 34.8	1	10.09		0.06		−0.59		2	B3	820
107159	+25 02487	Coma Ber 61			12 16 45	+25 20.3	4	7.79	.022	0.30	.005	0.08	.020	6	A5 V	1148,3016,8001,8023
		Coma Ber 203			12 16 45	+25 42.8	3	10.80	.017	0.87	.016	0.56	.006	5	K0	1127,8001,8023
		G 13 - 29			12 16 46	+02 43.4	2	11.08	.019	0.91	.005	0.62		3	K3	202,1620
107161	−8 03323	IDS12142S0821		AB	12 16 46	−08 38.2	1	6.84		1.08				1	K0 III	2008
107153	−54 05115				12 16 47	−55 02.5	1	9.66		0.32		0.12		1	A3mA8-A8	1771
107170	+15 02445				12 16 48	+14 49.2	3	6.54	.009	1.12	.006	0.99	.004	27	K0	269,1375,1569
107168	+23 02448	Coma Ber 62		★	12 16 48	+23 18.7	6	6.26	.007	0.17	.004	0.15	.008	24	A5 m	15,1127,3016,8001*
		Coma Ber 263			12 16 48	+27 07.2	1	13.47		1.47		1.28		2		1148
	+31 02352				12 16 48	+31 01.2	1	10.11		0.67				3	G5	280
107152	−52 05467				12 16 49	−52 44.9	1	8.58		−0.05		−0.30		2	B8 V	1770
		LF14 # 150			12 16 49	−61 33.1	1	12.58		1.21		0.89		1		137
		Coma Ber 306			12 16 50	+25 59.4	1	10.34		0.41		−0.03		3		1148
	+28 02105	Coma Ber 63			12 16 50	+28 12.2	1	10.14		0.58		0.14		2	dG1	3016
	+36 02259				12 16 50	+35 47.3	2	9.72	.015	1.15	.044	1.04		3	K1 III	77,280
107154	−64 01862				12 16 50	−65 02.7	1	9.75		0.09		−0.36		1	B8/9 Ib	820
107175	−1 02648				12 16 51	−01 57.2	1	9.34		0.70				1	G5	1284
107145	−76 00726				12 16 51	−76 31.4	4	6.83	.005	0.44	.007	−0.01	.004	14	F5 IV/V	278,1075,1075,2012
	−64 01863				12 16 52	−64 26.6	1	9.57		1.53		1.81		1	K3	820
		NGP +29 080			12 16 54	+28 39.4	1	10.83		0.63		0.20		1	G5	8038
107196	+14 02487				12 16 56	+14 12.4	1	10.16		0.55		0.02		2	G5	1569
		Coma Ber 269			12 16 56	+26 53.5	1	11.68		0.52		0.05		2		1148
		Coma Ber 270			12 16 56	+26 53.5	1	12.82		0.60		−0.05		2		1148
107211	+40 02524				12 16 56	+39 54.1	2	8.38	.015	0.66	.005	0.26		4	G2 IV	172,280
107214	+25 02488	Coma Ber 65			12 16 57	+24 33.7	3	9.03	.006	0.58	.006	0.04	.004	4	F5 V	1127,8001,8023
	+29 02279	G 121 - 77			12 16 57	+28 39.5	3	10.62	.005	1.45	.015	1.25	.008	4	M2 V	679,1744,3016
		AAS41,43 F3 # 53			12 16 57	−64 32.3	1	12.36		0.40		0.23		1	B9	820
		AAS41,43 F3 # 54			12 16 58	−64 29.6	1	12.42		0.29		0.10		1	A0	820
		Coma Ber 262			12 16 59	+27 09.7	1	12.66		0.62		0.20		1		1148
107213	+28 02106	Coma Ber 66		★	12 16 59	+28 26.2	2	6.38	.033	0.52	.015	0.08		8	F8 V	70,3016
	−36 07775				12 16 59	−36 25.9	1	10.18		1.15		1.08		1	K5 V	3072
		AAS78,203 # 10			12 17 00	−61 47.	1	13.42		1.11		−0.08		5		1684
107212	+30 02255				12 17 01	+30 26.7	2	8.78	.024	0.88	.005	0.61		3	G8 III	77,8034
		WLS 1220 30 # 10			12 17 01	+30 39.4	1	12.31		0.82		0.41		2		1375
	+45 02022				12 17 01	+45 12.8	1	10.74		0.85		0.44		1	G6 V	280
		LF14 # 151			12 17 01	−59 25.4	1	10.60		1.49		1.00		2	K3 III	137
107199	−18 03367	R Crv			12 17 02	−18 58.7	2	10.98	.335	1.53	.110	0.16	.330	2	M6 e	635,817
		AAS41,43 F3 # 55			12 17 02	−64 33.8	1	12.28		0.52		0.26		1		820
		AAS12,381 9 # A			12 17 03	+08 57.6	1	10.98		0.28		0.10		1		280
107224	+41 02289				12 17 03	+41 13.0	1	7.86		0.49				2	F5	280
107215	+11 02449				12 17 04	+11 16.1	1	10.26		1.03		0.87		1	K0	280
	+45 02023				12 17 04	+45 09.5	1	9.46		0.52		0.04		1	G0	280
		G 12 - 33			12 17 05	+06 56.2	1	12.36		1.23		1.18		1	K5	333,1620
107216	+9 02615				12 17 06	+09 05.5	1	9.89		1.07		0.91		1	K0	280
		Sand 50			12 17 06	+21 47.	1	14.46		1.46				1		686
		NGP +26 079			12 17 06	+25 29.5	1	11.29		0.68		0.30		1	G5	8038
107190	−63 02243				12 17 06	−64 07.7	1	8.65		1.77		2.13		2	K5/M0 III	820
		Coma Ber 204			12 17 07	+25 34.6	1	11.35		0.44		−0.06		4		1148
		Coma Ber 321			12 17 07	+25 37.0	1	13.37		0.89		0.50		1		1148
		Coma Ber 274			12 17 07	+26 35.7	1	12.46		0.89		0.49		3	G8 IV	1148
		AAS41,43 F3 # 58			12 17 07	−64 46.6	1	12.09		0.29		0.19		1	B9	820
		AAS41,43 F3 # 59			12 17 07	−64 48.2	1	11.65		0.26		−0.08		1	B9	820
		AAS41,43 F3 # 60			12 17 08	−64 24.5	1	12.39		0.14		−0.16		1	B5	820
	−64 01864				12 17 09	−64 32.0	1	10.71		0.22		0.14		2	A0	820
107209	−62 02688	LSS 2643			12 17 10	−62 34.6	5	6.80	.005	0.25	.007	−0.29	.009	15	B9/A0 Iab/b	285,657,1020,1075,2012
		Coma Ber 352			12 17 11	+25 23.4	1	12.19		0.61		0.03		2		1148
107222	−57 05400				12 17 11	−57 54.8	1	9.43		0.31		0.09		1	A7 IV/V	1771
		LF14 # 152			12 17 11	−62 00.2	1	11.81		0.71		0.35		1		137
		LSS 2644			12 17 11	−63 28.5	1	11.29		0.61		−0.15		4	B3 Ib	846
107223	−64 01865				12 17 13	−64 44.2	1	8.18		0.09		0.04		2	B9.5V	820
		Coma Ber 353			12 17 14	+25 24.2	1	12.51		0.78		0.33		1		1148
		AAS41,43 F3 # 63			12 17 14	−64 30.6	1	11.96		0.15		0.00		3	B9	820
		Coma Ber 320			12 17 15	+25 42.6	1	13.17		0.91		0.73		1		1148
		AAS41,43 F3 # 64			12 17 16	−64 47.4	1	12.22		0.26		−0.10		1	B8	820
	+17 02463				12 17 17	+17 05.3	1	7.89		0.68		0.32		2	K0	1569
	+40 02525				12 17 17	+39 31.7	1	10.04		1.10		1.08		1	K1 III	280
		AAS41,43 F3 # 57			12 17 17	−64 53.3	1	12.26		0.17		−0.06		1	B8	820
		Coma Ber 370			12 17 18	+25 11.4	1	14.14		0.87		0.53		1		1148
	+26 02327	Coma Ber 205			12 17 18	+25 30.6	3	10.02	.029	1.14	.007	1.11	.020	8	K1 III	280,1148,8041

HD	DM	Other Id	N Rem	α_{1950}	δ_{1950}	S	V	σ_V	B–V	σ_{B-V}	U–B	σ_{U-B}	n	Spectrum	References
		Coma Ber 261		12 17 18	+27 06.4	1	12.25		0.54		0.11		1		1148
		G 197 - 54		12 17 18	+59 07.4	1	11.85		1.27				1	K5-M0	1017
		AAS41,43 F3 # 65		12 17 18	−64 15.1	1	11.85		0.43		0.13		1	A8	820
107256	+12 02448			12 17 19	+11 52.6	1	8.30		1.49		1.43		1	M2	280
107254	+15 02446			12 17 19	+14 39.6	2	8.95	.013	0.46	.003	0.04	.003	22	F5	269,1569
107276	+29 02280	Coma Ber 68		12 17 20	+28 44.5	6	6.68	.018	0.17	.012	0.09	.004	13	Am	77,1068,1127,8001*
		CCS 1986		12 17 20	−58 39.4	1	11.06		3.12				1	N3	864
		AAS41,43 F3 # 66		12 17 20	−64 38.3	1	12.04		0.17		0.01		1	B8	820
107255	+14 02488			12 17 21	+13 55.3	1	10.40		0.71		0.29		1	K0	1569
107274	+49 02130	HR 4690		12 17 21	+49 15.7	3	5.28	.015	1.63	.031	1.97	.005	6	M0 III	70,172,3001
107259	+0 02926	HR 4689	⋆	12 17 21	−00 23.3	9	3.89	.010	0.02	.005	0.07	.023	40	A2 IV	3,15,1075,1075,1363*
107257	+11 02450			12 17 22	+10 45.7	1	8.39		0.57		0.06		2	G0	1569
107253	+15 02447			12 17 22	+15 15.9	1	9.96		0.62		0.21		2	G5	1569
		Steph 1007		12 17 22	+53 03.4	1	11.18		1.44		1.19		1	K7-M0	1746
107260	−14 03493			12 17 22	−14 47.1	1	7.94		1.07		0.92		3	K0 III	1657
107264	−41 07116			12 17 24	−41 40.1	1	8.65		1.48		1.53		1	Kp Ba	565
107250	−56 05169			12 17 24	−56 26.4	1	7.87		0.01		−0.18		2	B9 III	401
		POSS 377 # 12		12 17 25	+24 21.1	1	15.91		1.48				3		1739
		Coma Ber 372		12 17 25	+25 02.7	1	14.16		1.11		0.89		2		1148
		He3 776		12 17 25	−62 32.	1	12.86		0.86		−0.37		4		1684
		LSS 2645		12 17 25	−66 45.9	1	11.72		0.14		−0.73		5	B3 Ib	846,1737
107269	−63 02244			12 17 26	−64 03.9	1	8.72		1.68		1.80		2	M3 III	820
		AAS41,43 F3 # 68		12 17 26	−64 31.9	1	11.00		1.53		1.78		2	G8	820
107278	+10 02409			12 17 27	+09 53.4	1	10.00		1.53		1.95		1	K5	280
		CS 16026 # 15		12 17 27	+27 54.1	1	13.38		0.63		0.05		1		1744
		Coma Ber 369		12 17 28	+25 10.6	1	14.38		0.67		−0.03		1		1148
107287	+31 02353			12 17 28	+30 45.8	2	8.01	.005	1.20	.019	1.38		3	K1 III	77,8034
		LF14 # 153		12 17 28	−58 54.0	1	12.66		0.38		0.00		1		137
		AAS41,43 F3 # 69		12 17 28	−64 51.2	1	11.16		0.41		0.12		1	A8	820
107252	−69 01654	IDS12147S6913	A	12 17 28	−69 29.9	1	8.98		0.27		0.29		1	A3 V	1771
107277	+12 02449			12 17 29	+11 54.2	1	10.23		0.77		0.43		2	G5	1569
	+31 02354			12 17 29	+30 32.0	1	10.40		1.04		0.89		1	G8 III	8034
107286	+44 02180			12 17 30	+43 53.2	2	7.68	.020	0.84	.000	0.48		4	G8 III	172,280
107268	−63 02245	IDS12148S6333	AB	12 17 30	−63 49.6	1	8.63		1.66		1.64		2	K2/3 III	122
	+18 02590			12 17 31	+17 33.8	1	8.49		1.57		1.88		1	M0	280
107270	−63 02246			12 17 31	−64 22.2	1	7.12		1.10		0.77		1	G5p Ba	820
107288	+14 02489	IDS12150N1425	AB	12 17 33	+14 08.0	1	6.94		1.08				3	K0	280
		Coma Ber 319		12 17 33	+25 46.6	1	14.72		0.68		0.21		1		1148
		Coma Ber 222		12 17 33	+26 11.5	1	11.26		0.98		0.80		2	G8 III	1148
		Coma Ber 292		12 17 33	+26 20.3	1	13.34		0.88		0.45		1		1148
107303	+16 02360			12 17 34	+15 59.6	1	8.05		0.49		−0.01		3	F8	1569
	+28 02107	Coma Ber 69		12 17 35	+28 11.1	1	10.30		0.68		0.03		2	dG5	3016
107295	−21 03511	HR 4691	⋆ AB	12 17 35	−21 53.9	2	5.96	.005	0.82	.000			7	K0 III +F/G	15,2027
		AAS41,43 F3 # 71		12 17 35	−64 10.1	1	12.40		0.24		0.12		1	B9	820
107305	+10 02410			12 17 36	+10 10.2	1	7.57		1.35		1.63		1	K0	280
		Coma Ber 281		12 17 36	+26 28.4	1	10.38		0.62		0.03		3		1148
107304	+15 02448			12 17 37	+15 19.6	1	10.22		0.67		0.23		2	G5	1569
	+16 02361			12 17 38	+16 11.6	1	10.04		0.41		0.01		2		1569
		CS 16026 # 18		12 17 39	+29 01.1	1	13.87		0.45		−0.21		1		1744
107298	−50 06906			12 17 39	−50 56.4	1	7.70		0.16		0.08		1	A6 V	1771
	−59 04180			12 17 40	−59 37.0	1	10.38		0.32		0.22		3	A2 V	137
		AAS41,43 F3 # 72		12 17 40	−64 41.8	1	11.83		0.12		−0.40		1	B6	820
		ApJS33,459 # 235		12 17 41	−63 04.9	1	12.01		0.18		−0.46		3		1684
		AAS41,43 F3 # 73		12 17 41	−64 43.4	1	12.60		0.10		−0.12		1	B9	820
107324	+42 02292			12 17 42	+41 41.1	1	8.93		0.27		0.05		3	F0 V	1068
107301	−65 01842	HR 4692		12 17 42	−65 33.9	4	6.20	.005	−0.04	.009	−0.15	.008	13	B9 V	15,1075,1771,2038
		AAS41,43 F3 # 74		12 17 43	−64 44.1	1	11.67		0.19		0.03		1	B9	820
107341	+38 02324	IDS12152N3827	A	12 17 44	+38 10.8	1	6.60		1.00				1	K1 III	3024
107341	+38 02324	IDS12152N3827	AB	12 17 44	+38 10.8	2	6.71	.025	1.00	.000	0.85		4	K1 III	77,172
107341	+38 02324	IDS12152N3827	B	12 17 44	+38 10.8	1	10.28		0.57		0.08		4	K0	3032
107300	−64 01867			12 17 44	−65 02.5	1	9.98		0.07		−0.59		1	B3/7	820
		G 12 - 34		12 17 45	+16 48.3	1	13.69		1.38		1.11		1		1620
		Lod 615 # 2		12 17 45	−64 29.8	2	11.80	.015	0.25	.005	0.15	.015	3	B9	287,820
107314	−51 06593			12 17 46	−51 41.8	1	8.64		0.20		0.17		1	A4 V	1771
		Coma Ber 305		12 17 47	+25 59.4	1	12.00		0.58		0.01		1		1148
107326	+26 02329	Coma Ber 70	⋆	12 17 47	+26 16.7	6	6.14	.010	0.30	.003	0.07	.010	21	F0 IV	15,374,3013,8001,8015,8023
		AAS12,381 35 # E		12 17 47	+35 10.5	1	10.37		0.63				3		280
	−64 01868			12 17 47	−65 22.2	2	10.41	.019	0.03	.000	−0.65	.022	4	B2.5IV	540,976
	+34 02305			12 17 48	+33 44.2	1	10.15		0.41				2	F5	77
107328	+4 02604	HR 4695	⋆ A	12 17 49	+03 35.5	11	4.97	.012	1.17	.010	1.15	.005	55	K1 III	3,15,1003,1080,1363*
107325	+27 02114	Coma Ber 71	⋆	12 17 49	+26 53.9	4	5.54	.000	1.09	.013	1.06	.020	7	K1 III	172,252,8034,8053
		Wray 961		12 17 49	−60 28.2	1	12.50		0.75		−0.31		5		1684
		AAS41,43 F3 # 78		12 17 49	−64 25.6	1	12.42		0.20		−0.07		1	B8	820
		AAS41,43 F3 # 77		12 17 49	−64 55.6	1	11.02		1.03		0.70		1	G5	820
	+1 02684	IDS12153N0108	AB	12 17 50	+00 51.6	2	10.09	.084	1.52	.018	1.20		5	M0 V	1705,7008
		Coma Ber 373		12 17 50	+25 04.3	1	11.83		0.56		0.09		2		1148
	+34 02306			12 17 50	+34 00.6	1	10.06		0.45				2	F8	77
		LSS 2648		12 17 51	−64 31.0	3	11.72	.022	0.10	.018	−0.53	.012	5	B3 Ib	287,820,846
		AAS41,43 F3 # 80		12 17 52	−64 25.0	1	10.61		1.85		2.16		1	K2	820
	−64 01869			12 17 52	−65 20.3	2	10.25	.010	0.00	.000	−0.70	.009	4	B2 III	540,976

Table 1 663

HD	DM	Other Id	N Rem	α_{1950}	δ_{1950}	S	V	σ_V	B–V	σ_{B-V}	U–B	σ_{U-B}	n	Spectrum	References
		Coma Ber 368		12 17 53	+25 11.4	1	11.84		0.55		0.00		6		1148
		Coma Ber 223		12 17 53	+26 19.4	1	10.87		0.67		0.21		1	G5 V	1748
107346	+10 02412			12 17 54	+10 14.9	1	10.34		0.83		0.10		1	K0	280
	+15 02449			12 17 54	+15 20.7	1	10.65		0.53		0.01		2		1569
		Coma Ber 331		12 17 54	+25 38.6	1	13.40		0.55		0.03		1		1148
	+36 02261			12 17 54	+35 58.9	1	10.04		1.21		1.28		1	K2 III	280
		AAS41,43 F3 # 81		12 17 54	−64 51.5	1	12.16		0.33		0.16		1	B8	820
		LSS 2647		12 17 54	−70 35.2	1	10.96		0.09		-0.67		1	B3 Ib	540
107316	−72 01238			12 17 54	−73 07.4	1	8.99		0.24				4	B5 V	2012
107347	+9 02617			12 17 55	+09 27.2	1	9.41		1.28		1.37		1	K2	280
		LF14 # 155		12 17 56	−61 52.5	1	13.12		0.37		0.26		1		137
		AAS41,43 F3 # 82		12 17 56	−64 29.6	2	11.64	.019	0.24	.015	0.06	.000	3	B9	287,820
107343	+13 02522			12 17 57	+12 34.5	1	10.54		1.26		1.41		1	K5	1569
	−63 02247	IDS12152S6407	AB	12 17 57	−64 24.0	1	10.03		0.76		0.47		1	A3	820
		AAS41,43 F3 # 83		12 17 57	−64 31.0	2	11.68	.025	0.22	.000	0.05	.000	2	B8	287,820
107348	−21 03514	HR 4696	⋆ AB	12 17 58	−21 56.3	5	5.21	.022	-0.10	.004	-0.37	.026	13	B8 V	15,1079,1637,2029,3023
		LP 555 - 77		12 17 59	+07 02.7	1	12.11		0.85		0.50		2		1696
107344	+12 02450			12 17 59	+11 39.2	1	9.40		1.49		1.84		1	K5	280
		AAS41,43 F3 # 84		12 17 59	−64 37.5	1	11.22		0.14		-0.29		1	B8	820
		AAS41,43 F3 # 87		12 17 59	−64 49.8	1	11.86		0.52		0.22		1		820
107364	+12 02451			12 18 00	+12 23.6	1	10.64		0.48		-0.05		1	G0	1569
		AAS41,43 F3 # 86		12 18 00	−64 24.2	1	12.19		0.35		0.12		1	A0	820
107338	−64 01871			12 18 00	−64 52.3	1	9.26		0.21		0.15		1	A2 V	820
	−58 04223			12 18 01	−58 47.5	1	11.16		0.32		0.20		1	A3 V	137
107363	+14 02490			12 18 02	+14 01.4	2	9.88	.010	1.00	.005	0.59		4	K0	280,882
107381	+35 02331			12 18 02	+34 32.4	2	9.15		1.01	.045	0.85		4	K0 III	77,1569
107357	−64 01872			12 18 03	−64 41.0	1	7.99		0.93		0.60		1	G5 III	820
107397	+62 01224	RY UMa		12 18 04	+61 35.2	1	6.97		1.80		2.12		1	M3:III	3001
107380	+44 02182			12 18 05	+43 41.2	1	8.45		1.25				2	K0 IV	280
	+39 02513			12 18 06	+38 35.9	1	10.69		0.93		0.70		1	K0 III	280
		AAS41,43 F3 # 90		12 18 06	−64 49.6	1	11.60		0.78		0.30		1	G0	820
107382	+20 02708			12 18 07	+20 20.6	1	8.35		1.02		0.89		1	K0	280
107356	−58 04226			12 18 07	−58 49.4	1	9.07		0.10		0.15		2	A0 V[n]	1771
		Coma Ber 310		12 18 08	+25 53.8	1	11.70		0.85		0.42		3	G7 III	1148
		BSD 5 # 560		12 18 08	+75 23.1	1	11.92		0.28		0.10		1	A5	1058
		AAS41,43 F3 # 91		12 18 08	−64 46.0	1	11.56		0.38		0.29		1	A5	820
107384	+15 02451			12 18 09	+15 25.3	2	9.04	.009	1.13	.014	1.05		4	K0	280,882
		Coma Ber 1101		12 18 09	+28 02.7	1	11.22		0.92				2		913
		AAS41,43 F3 # 92		12 18 09	−64 40.7	1	10.77		1.08		0.78		1	G5	820
107383	+18 02592	HR 4697	⋆ AB	12 18 11	+18 04.1	7	4.74	.015	1.01	.005	0.79	.012	30	G8 III	15,667,1080,1363,3016*
107398	+27 02115	Coma Ber 73	⋆ A	12 18 11	+27 20.0	1	7.08		0.35		0.01		1	F3 V	280
107398	+27 02115	Coma Ber 72	⋆ B	12 18 11	+27 20.0	1	7.18		0.38		-0.01		1	F3 V	280
		AAS41,43 F3 # 94		12 18 11	−64 05.9	1	11.66		0.38		0.23		1	A4	820
		AAS41,43 F3 # 93		12 18 11	−64 29.5	1	12.43		0.44		0.15		1	A8	820
107359	−72 01239			12 18 11	−72 32.3	1	9.24		0.49		0.00		5	F6 V	518
	+45 02027			12 18 12	+44 55.3	1	10.08		1.18		1.17		1	K0 III	280
107387	−19 03465			12 18 12	−19 33.6	1	10.38		0.67		0.12		1	G5	3072
	−59 04184	LSS 2649		12 18 12	−60 12.7	3	10.15	.039	0.21	.008	-0.55	.032	6		540,976,1684
		AAS41,43 F3 # 95		12 18 12	−64 40.6	1	11.77		0.21		-0.08		1	B9	820
	+35 02332	IDS12157N3533	AB	12 18 13	+35 16.3	2	9.33		1.03	.055	0.80		4	K0 III	77,1569
		LSS 2650		12 18 13	−60 48.1	1	11.39		0.40		-0.50		3	B3 Ib	846
		Coma Ber 345		12 18 15	+25 30.6	1	11.72		0.55		0.03		4		1148
107399	+26 02330	Coma Ber 76	⋆ A	12 18 15	+26 02.6	6	9.07	.009	0.54	.010	0.02	.006	32	G0 V	1127,1148,1613,1748*
107399	+26 02330	Coma Ber 76	⋆ B	12 18 15	+26 02.6	1	13.13		1.05		0.90		3		3016
107392	−46 07866			12 18 15	−47 10.6	6	7.11	.008	0.25	.003	0.13	.003	29	A7 III	278,977,1075,1770,2001,2011
107401	+14 02491			12 18 16	+13 35.3	2	9.86	.015	1.11	.020	1.12		2	K2	280,882
	+25 02492	Coma Ber 77		12 18 16	+24 29.6	1	9.01		1.02		0.90		1	gK1	280
		Coma Ber 1106		12 18 16	+26 33.2	1	13.28		0.85				1		913
		AAS41,43 F3 # 96		12 18 16	−64 41.9	1	12.27		0.33		0.28		2	A0	820
		AAS41,43 F3 # 97		12 18 16	−64 44.1	1	12.06		0.56		0.12		1	F5	820
107415	+16 02362			12 18 17	+15 49.1	2	6.46	.005	1.02	.005	0.79		4	K0	252,280
	−59 04186			12 18 17	−60 23.6	1	10.92		0.56		0.06		2	F5 V	137
		AAS41,43 F3 # 100		12 18 17	−64 19.7	1	12.19		0.46		0.11		1		820
107416	+12 02452			12 18 18	+11 57.1	1	10.37		0.54		0.03		2	G0	1569
		Coma Ber 367		12 18 18	+25 15.4	1	13.71		0.58		0.00		1		1148
	−63 02249			12 18 18	−63 43.8	1	10.52		0.24		0.08		2		122
	−63 02248			12 18 18	−64 18.4	1	9.97		0.51		0.05		1	F5	820
		AAS41,43 F3 # 102		12 18 18	−64 35.9	1	11.90		0.80		0.31		1	G2	820
		AAS41,43 F3 # 98		12 18 18	−64 53.3	1	11.86		0.21		0.12		1	A0	820
	+24 02448	Coma Ber 78		12 18 19	+24 22.8	1	9.25		1.18		1.33		1	gK2	280
		Coma Ber 354		12 18 19	+25 23.4	1	12.56		0.59		0.06		2		1148
107395	−61 03095			12 18 19	−61 46.0	1	9.00		0.01		-0.48		2	B5 III	401
		AAS41,43 F3 # 101		12 18 19	−64 19.7	1	11.98		0.25		-0.05		1	B9	820
		BSD 5 # 561		12 18 20	+75 13.6	1	12.16		0.48		-0.04		1	F7	1058
107394	−58 04230			12 18 20	−58 33.3	1	9.21		1.01		0.61		1	G5 III	137
		AAS41,43 F3 # 103		12 18 20	−64 51.8	1	11.18		1.16		0.99		1		820
107418	−12 03614	HR 4699	⋆ A	12 18 21	−13 17.3	3	5.15	.012	1.05	.007	0.92	.010	14	K0 III	3,2035,3016
		AAS41,43 F3 # 104		12 18 21	−64 51.2	1	12.25		0.32		0.26		2	A0	820
		Coma Ber 332		12 18 22	+25 37.8	1	13.44		1.26		1.02		2		1148
107427	+26 02331	Coma Ber 79		12 18 22	+26 12.2	7	9.09	.011	0.12	.006	0.12	.015	13	A2 V	1068,1148,1240,1748,3016*

HD	DM	Other Id	N Rem	α_{1950}	δ_{1950}	S	V	σ_V	B–V	σ_{B-V}	U–B	σ_{U-B}	n	Spectrum	References
		AAS41,43 F3 # 105		12 18 22	−64 26.1	1	12.45		0.42		0.31		1	A5	820
	+17 02464			12 18 25	+17 16.0	1	8.99		0.95		0.85		1	K2	280
		AAS41,43 F3 # 106		12 18 25	−64 40.7	1	13.16		0.24		-0.01		1		820
		AAS41,43 F3 # 107		12 18 25	−64 40.7	1	12.22		0.20		-0.21		1		820
107465	+58 01371	HR 4701		12 18 26	+58 08.5	2	5.53	.013	1.44	.012	1.71		3	K5 III	71,3001
107422	−41 07125	IDS12158S4200	AB	12 18 28	−42 17.1	6	6.82	.008	0.04	.007	0.04	.007	43	A0 V	278,1075,1637,1770,2001,2011
		Coma Ber 366		12 18 29	+25 16.2	1	14.28		0.45		-0.29		1		1148
107424	−56 05184			12 18 29	−56 29.8	1	7.76		-0.01		-0.35		2	B8/9 III	401
		Coma Ber 333		12 18 30	+25 35.3	1	12.24		0.64		0.13		3		1148
		AAS41,43 F3 # 108		12 18 31	−64 53.6	1	12.58		0.19		0.02		1	B8	820
	+16 02363			12 18 32	+16 04.5	1	10.29		0.60		0.11		2		1569
	+34 02307			12 18 32	+34 16.7	1	9.89		0.57				2	G0	77
107469	+25 02493	Coma Ber 81	⋆	12 18 33	+25 18.4	5	7.31	.009	0.88	.005	0.59	.010	11	G9 III	172,1080,1148,3026,8034
107448	+10 02414			12 18 34	+09 38.9	1	9.26		1.33		1.43		1	K2	280
107468	+26 02332	Coma Ber 80		12 18 34	+25 59.9	7	8.06	.006	1.06	.009	0.93	.011	27	K0 III	1148,1613,1748,3040,8001*
107452	−10 03452	IDS12160S1055	A	12 18 34	−11 11.9	1	8.25		0.29		0.10		2	F0p	1003
107485	+38 02326			12 18 35	+38 18.1	2	7.48	.020	0.98	.025	0.72		5	G9 III	77,172
107467	+46 01778			12 18 35	+45 47.5	2	7.39	.010	0.97	.005	0.66		5	G8 III	172,280
	+15 02452			12 18 36	+15 18.6	1	10.13		0.47		-0.02		2		1569
		Coma Ber 355		12 18 36	+25 23.4	1	13.12		0.62		0.18		1		1148
107447	−61 03100	T Cru		12 18 37	−62 00.2	2	6.32	.010	0.77	.005	0.52		2	G2 Ib	657,688
	+14 02493			12 18 38	+14 07.3	1	10.36		1.31		1.47		1		280
107470	+12 02454			12 18 39	+12 24.7	1	10.31		0.33		0.16		1	A5	280
	−48 07360			12 18 39	−49 02.5	1	10.46		0.40		-0.55		2	B0 I	403
107442	−53 05086			12 18 39	−54 13.5	1	8.44		0.28		0.11		1	A8 V	1771
107446	−59 04188	HR 4700		12 18 39	−60 07.5	4	3.58	.005	1.42	.003	1.63	.000	19	K3/4 III	15,1075,2012,3077
		AAS41,43 F3 # 109		12 18 39	−64 16.2	1	11.55		0.54		0.02		2	F5	820
		Coma Ber 365		12 18 40	+25 13.0	1	14.25		0.73		0.09		1		1148
		Coma Ber 364		12 18 40	+25 17.8	1	14.69		0.71		0.35		1		1148
107487	+32 02234			12 18 40	+31 34.8	1	8.68		0.55				2	F8 V	77
	+44 02183			12 18 40	+43 58.8	1	9.81		1.03				2	K0 III	280
		Coma Ber 283		12 18 41	+26 22.7	1	12.26		0.65		0.16		3		1148
		Ton 730 # 5		12 18 41	+28 28.5	1	14.85		1.15		0.84		1		157
107486	+35 02333			12 18 41	+34 57.9	2	7.18		1.08	.005	1.06		4	K1 III	77,1569
	+40 02526			12 18 41	+39 47.3	1	9.51		1.01				2	K0 III	280
		AAS41,43 F3 # 110		12 18 41	−64 52.0	1	12.29		0.29		0.22		2	B9	820
107484	+42 02295			12 18 42	+41 45.6	3	7.72	.011	1.16	.005	1.17	.015	7	K0 III	37,172,280
107461	−56 05189			12 18 42	−57 16.6	1	8.86		-0.01		-0.44		2	B6 III	401
		Ton 730 # 2		12 18 43	+28 29.6	1	14.23		0.77		0.34		1		157
		UV Vir		12 18 44	+00 38.7	2	11.35	.005	0.09	.014	0.16	.018	2	A4	597,699
		Coma Ber 282		12 18 45	+26 25.9	2	11.47	.010	1.06	.040	0.84	.005	4	K3 V	1148,1748
		Coma Ber 260		12 18 45	+27 05.5	1	12.01		0.64		0.04		1		1148
		LOD 26- # 11		12 18 45	−63 47.2	1	10.61		1.14		0.91		2		122
		AAS41,43 F3 # 111		12 18 46	−64 30.6	1	12.71		0.48		0.22		1	A1	820
107496	+16 02364			12 18 47	+16 06.3	1	9.39		0.57		0.16		2	G5	1569
		Coma Ber 224		12 18 47	+25 22.6	1	11.51		0.46		-0.01		3		1148
107495	+32 02235			12 18 47	+32 22.4	1	9.12		0.96				2	K0 III	77
	+39 02515			12 18 47	+38 52.6	1	10.33		1.13		1.19		1	K2 III	280
107497	+15 02453			12 18 48	+15 02.2	1	9.81		0.45		0.01		2	G0	1569
		AAS41,43 F3 # 112		12 18 48	−64 38.8	1	12.10		0.70		0.20		1		820
	+44 02184			12 18 49	+44 20.1	1	10.19		1.32		1.48		1	K1 IV	280
		AAS41,43 F3 # 113		12 18 49	−64 38.4	1	11.53		0.09		-0.05		1	B9	820
		Coma Ber 225		12 18 50	+25 17.0	1	11.67		0.48		-0.06		3		1148
		Coma Ber 226		12 18 51	+25 23.3	1	11.50		0.57		0.04		2		1148
107512	+32 02236			12 18 51	+32 27.2	2	9.14	.000	0.53	.005	0.10		4	G0	77,1375
	+26 02333	Coma Ber 228		12 18 52	+25 56.9	3	10.31	.011	0.52	.008	-0.02	.003	24	F8 V	1148,1613,1748
		Coma Ber 207		12 18 52	+26 10.6	2	10.95	.005	0.53	.005	-0.02	.010	8	G0	1148,1748
	+26 02334	Coma Ber 227		12 18 52	+26 12.2	2	10.24	.010	0.57	.010	0.04	.010	5	F7 V	1148,1748
107514	+13 02523			12 18 53	+12 57.4	2	9.63	.015	0.74	.010	0.23		3	G5	882,1569
107494	−48 07365			12 18 53	−49 20.9	1	9.38		0.25		0.14		1	A7 V	1771
		Wray 963		12 18 53	−61 27.5	1	12.34		0.25		-0.28		4		1684
107515	+12 02455			12 18 54	+12 06.4	1	10.61		0.65		0.15		2	G0	1569
		LOD 26- # 10		12 18 54	−63 45.7	1	11.06		0.65		0.10		2		122
		AAS41,43 F3 # 114		12 18 54	−64 31.8	1	12.61		0.35		0.19		1		820
		Steph 1009		12 18 55	+34 19.4	1	10.95		1.26		1.16		6	K4	1746
	−64 01873			12 18 55	−64 29.5	1	9.94		0.10		-0.23		1	B8	820
		AAS41,43 F3 # 115		12 18 55	−64 34.5	1	12.47		0.19		0.12		1	B9	820
107513	+25 02495	Coma Ber 82		12 18 56	+25 16.5	3	7.43	.009	0.28	.002	0.02	.003	5	Am	1127,8001,8023
		AAS41,43 F3 # 117		12 18 56	−64 27.5	1	12.19		0.20		-0.35		1		820
		Coma Ber 380		12 18 57	+24 56.3	1	15.20		0.46		-0.15		2		1148
		Coma Ber 356		12 18 58	+25 25.7	1	12.77		0.37		-0.01		1		1148
	−64 01874	LSS 2651		12 18 58	−65 05.0	3	10.16	.019	0.00	.008	-0.80	.017	3	B0.5V	540,820,976
		ON 231 # 1		12 19 1	+28 30.	1	12.08		0.59		0.08		2		326
		ON 231 # 2		12 19 1	+28 30.	1	13.08		0.60		0.02		1		326
		ON 231 # 3		12 19 1	+28 30.	1	13.51		0.96		0.84		2		326
		ON 231 # 4		12 19 1	+28 30.	1	14.82		1.55		1.26		2		326
		ON 231 # 5		12 19 1	+28 30.	1	15.60		0.85		0.78		1		326
		ON 231 # 6		12 19 1	+28 30.	1	16.63		0.35		0.09		1		326
		ON 231 # 7		12 19 1	+28 30.	1	17.05		0.64				1		326
107531	+25 02496	Coma Ber 83		12 19 01	+24 47.4	2	8.94	.028	0.47	.005	-0.02	.009	4	F8 V	1148,3016

Table 1 665

HD	DM	Other Id	N Rem	α_{1950}	δ_{1950}	S	V	σ_V	B–V	σ_{B-V}	U–B	σ_{U-B}	n	Spectrum	References
		BSD 5 # 566		12 19 01	+75 11.5	1	12.54		0.52		0.00		1	G0	1058
		AAS41,43 F3 # 119		12 19 01	−64 34.5	1	12.79		0.17		0.12		1	A0	820
107532	+13 02524			12 19 02	+12 45.0	1	9.86		0.59		0.08		1	G5	1569
107549	+34 02308			12 19 02	+33 44.8	1	9.28		0.59				2	G0	77
107509	−60 03925			12 19 02	−61 15.6	1	7.91		0.57				4	F8 V	1075
		AAS41,43 F3 # 120		12 19 02	−64 27.5	1	12.47		0.53		0.07		1	F0	820
		AO 1148		12 19 03	+26 23.7	1	11.41		0.62		0.14		1	F7 V	1748
		Ton 730 # 1		12 19 03	+28 29.6	1	13.55		0.80		0.46		1		157
		AJ77,733 T20# 4		12 19 05	−64 47.6	2	11.76	.037	0.10	.033	-0.14	.014	4	B8	125,820
		Coma Ber 254		12 19 06	+27 11.9	1	12.58		0.53		-0.05		1		1148
		Ton 730 # 3		12 19 06	+28 37.1	1	14.09		1.21		1.34		1		157
		SSII 131		12 19 06	+48 05.	1	11.96		0.02		0.01		1	A2	3060
		LP 675 - 3		12 19 06	−05 10.2	1	12.84		0.85		0.46		2		1696
		AAS78,203 # 11		12 19 06	−61 39.	1	13.11		0.81		0.00		4		1684
	−61 03106			12 19 06	−62 23.2	1	10.05		1.10		0.89		1	K0 III	137
		AAS41,43 F3 # 122		12 19 06	−64 50.2	1	10.50		0.64		0.07		1	G0	820
		AJ77,733 T20# 5		12 19 06	−65 03.4	1	13.12		1.64		2.01		3		125
107526	−53 05092			12 19 07	−54 07.4	1	9.09		0.27		0.18		2	A5/7 V[m]	1771
		AJ77,733 T20# 3		12 19 07	−64 51.1	2	11.64	.014	0.80	.000	0.40	.005	4		125,820
107582	+62 01227	G 197 - 56		12 19 08	+62 01.7	2	8.25	.033	0.62	.005	0.02	.005	4	G2 V	1003,3026
		AAS41,43 F3 # 124		12 19 08	−64 17.1	1	10.11		1.67		1.88		2	K3	820
	+17 02466			12 19 09	+17 05.8	1	10.90		0.60		0.05		2		1375
107527	−64 01876			12 19 09	−64 41.5	2	9.41	.009	-0.01	.009	-0.60	.005	5	B3/5 II	125,820
		Coma Ber 291		12 19 10	+26 18.6	1	12.08		0.59		0.01		2		1148
	+31 02356			12 19 10	+30 40.9	1	10.72		0.65		0.22		1	G6 III	8034
107541	−34 08098			12 19 10	−34 30.2	1	9.39		1.06		0.55		1	Kp Ba	565
	+20 02712			12 19 11	+20 06.5	1	10.28		1.27		1.34		1	K5	280
		NGP +28 059		12 19 11	+27 34.7	1	11.27		0.85		0.46		1	G7 III	8041
		AAS41,43 F3 # 126		12 19 11	−64 45.0	1	11.52		0.26		0.22		2	A0	820
		AAS78,203 # 12		12 19 12	−61 31.	1	13.00		0.31		-0.38		5		1684
107568	+27 02116	Coma Ber 84		12 19 13	+26 53.2	3	8.58	.045	0.95	.008	0.71	.008	6	G9 III	280,1148,8034
		BSD 5 # 567		12 19 13	+74 55.2	1	12.78		0.68		0.19		2	G1	1058
107546	−57 05420			12 19 14	−58 08.3	1	8.47		0.07		-0.21		2	B9 V	401
		AAS41,43 F3 # 127		12 19 14	−64 30.1	1	12.63		0.23		-0.33		1		820
107543	−55 05019	HR 4702		12 19 15	−56 05.8	2	5.91	.010	1.56	.005	1.64		6	K4 III	2035,3005
		CS 16026 # 25		12 19 16	+28 56.9	1	14.27		0.66		0.32		1		1744
107565	−64 01877	LSS 2652		12 19 16	−64 25.0	4	8.69	.006	0.15	.009	-0.69	.018	10	B1 II/III	540,820,976,1684
		CS 16026 # 22		12 19 17	+31 56.6	1	14.72		0.56		0.10		1		1744
	+46 01779			12 19 17	+45 44.5	1	9.33		1.12				2	K0 III	280
107583	+27 02117	Coma Ber 85		12 19 18	+26 49.6	4	9.31	.015	0.58	.005	0.06	.004	6	G0 V	1127,1148,8001,8023
		AAS41,43 F3 # 129		12 19 18	−64 10.5	1	11.25		0.68		0.19		1	F8	820
		Coma Ber 357		12 19 19	+25 20.1	1	13.02		0.99		0.26		1		1148
107566	−66 01747	HR 4703	★ A	12 19 19	−67 14.7	4	5.14	.004	0.19	.005	0.15	.013	12	A5 V	15,1075,2038,3023
	+32 02237			12 19 20	+32 08.1	1	9.87		0.48				2	F8	77
		AAS41,43 F3 # 130		12 19 20	−64 22.5	1	10.23		0.96		0.64		1	F6	820
		CS 16026 # 27		12 19 21	+28 02.2	1	13.66		0.77		0.37		1		1744
107597	+41 02290			12 19 21	+40 59.2	2	7.98	.025	0.98	.005	0.72		5	G9 III	172,280
		AAS41,43 F3 # 131		12 19 21	−64 25.0	1	11.74		0.42		0.20		1	A2	820
		AAS41,43 F3 # 132		12 19 21	−64 40.4	1	12.13		0.39		0.24		1	A5	820
		Coma Ber 374		12 19 22	+25 04.2	1	14.87		0.35		-0.22		1		1148
	+28 02108			12 19 22	+28 25.6	2	9.90	.005	0.88	.005	0.60	.010	2	G8 III	280,8041
107581	−64 01878			12 19 22	−65 02.2	1	9.43		0.03		-0.77		1	B3 III	820
107547	−72 01240	IDS12165S7257	AB	12 19 22	−73 13.6	3	6.76	.007	0.17	.010	0.14		10	A4/5 V	278,1075,2012
		Coma Ber 358		12 19 23	+25 24.9	1	12.33		0.51		-0.06		1		1148
107579	−50 06926			12 19 23	−51 06.5	1	8.60		0.22		0.21		1	A4 IV	1771
107567	−67 01939	HR 4704		12 19 23	−68 01.8	4	5.73	.004	1.04	.000	0.82	.005	19	K0 III	15,1075,1075,2038
107598	+15 02454			12 19 24	+15 07.4	1	9.53		0.46		-0.05		2	G0	1569
	+26 02335	Coma Ber 208		12 19 24	+25 53.7	2	10.67	.005	0.35	.005	-0.07	.000	3	F2 V	1148,1748
	+29 02282	IDS12169N2906	A	12 19 24	+28 50.8	1	9.77		0.54				2		77
	−61 03108			12 19 24	−62 07.2	1	10.01		0.17		0.12		1	A2 V	137
		Coma Ber 259		12 19 25	+27 03.8	1	14.50		0.28		-0.10		2		1148
107596	+42 02296	G 123 - 26		12 19 25	+42 25.1	3	9.44	.036	1.36	.018	1.23	.015	4	M0 V	679,1017,3072
	−63 02250			12 19 25	−64 09.0	1	10.78		0.16		-0.16		1	B7	820
107611	+28 02109	Coma Ber 86		12 19 26	+27 35.2	4	8.53	.020	0.46	.004	-0.02	.000	5	F7 V	1127,3026,8001,8023
		AAS41,43 F3 # 134		12 19 26	−64 28.9	1	12.13		0.22		-0.26		1	B4	820
	+26 02336	Coma Ber 87		12 19 27	+25 57.8	4	9.59	.011	0.43	.011	0.03	.005	27	dF6	1148,1613,1748,3016
		CS 16026 # 23		12 19 27	+31 23.3	1	12.80		0.17		0.03		1		1744
	−62 02707	LSS 2653		12 19 27	−63 04.2	2	10.06	.004	0.08	.006	-0.68	.019	7	B2 III	540,976
238106	+56 01589			12 19 28	+55 41.4	1	9.68		0.40		0.02		2	G0	1502
107610	+47 01955			12 19 29	+47 27.6	1	6.35		1.11		1.08		2	K2 III	172
107612	+17 02469			12 19 30	+17 01.3	3	6.64	.040	0.04	.018	0.04	.004	12	A2p	1063,1068,3033
		Coma Ber 344		12 19 30	+25 33.7	1	14.97		0.81		0.15		1		1148
107613	+12 02456			12 19 31	+12 08.0	1	10.17		1.16		1.03		2	K0	1569
		BSD 5 # 572		12 19 31	+75 15.4	1	12.46		0.52		-0.01		1	G0	1058
107634	+33 02225			12 19 32	+33 22.1	1	7.48		1.04				2	K0 III	77
107592	−63 02251			12 19 32	−64 24.3	1	7.54		1.29		1.27		1	K1 III	820
	−64 01880			12 19 32	−64 31.6	1	10.69		0.32		-0.05		3	B3	820
	+25 02497	Coma Ber 88		12 19 33	+25 22.5	3	10.32	.016	0.90	.005	0.53	.013	5	G6 III	280,1148,8034
		CS 16026 # 26		12 19 33	+28 20.6	1	13.99		0.04		-0.04		1		1744
107633	+35 02334			12 19 33	+34 32.0	1	8.80		0.63				2	G6 V	77

HD	DM	Other Id	N	Rem	α_{1950}	δ_{1950}	S	V	σ_V	B–V	σ_{B-V}	U–B	σ_{U-B}	n	Spectrum	References
	+44 02188				12 19 33	+43 40.7	1	10.58		1.00		0.83		1	G8 III	280
107593	−64 01879	LSS 2654			12 19 33	−65 19.8	3	7.75	.008	0.04	.005	−0.79	.016	7	B1 Ib	401,540,976
	−59 04192				12 19 34	−60 09.3	2	11.14	.018	0.44	.022	0.16	.008	4	B9 IV	540,976
107632	+43 02212	G 123 - 27			12 19 35	+42 29.9	1	9.65		0.65		0.00		2	G5	1569
107620	−37 07842				12 19 36	−38 10.4	1	7.54		1.00		0.73		4	G8 III	1673
		AAS41,43 F3 # 138			12 19 36	−64 29.7	1	12.58		0.33		−0.11		3	B8	820
		AAS78,203 # 13			12 19 36	−67 22.	1	10.41		1.84		1.91		6		1684
107637	+12 02457				12 19 37	+11 51.1	1	9.97		1.04		1.02		2	K0	1569
107636	+13 02525				12 19 37	+12 42.5	1	10.75		0.67		0.23		1	G5	1569
		BSD 5 # 571			12 19 37	+75 11.4	1	11.50		0.65		0.21		1	G2	1058
		vdB 57 # a			12 19 37	−63 00.9	1	13.08		0.80		0.15		3		434
107654	+33 02226				12 19 38	+32 38.5	1	9.34		0.59				2	G0	77
		BSD 5 # 274			12 19 38	+74 06.1	1	12.44		0.73		0.08		2	G2	1058
	+32 02238				12 19 39	+32 22.2	1	9.97		0.82				2	G8 III	77
107626	−44 07958				12 19 39	−44 53.1	1	8.83		0.14		0.09		1	A0 V	1770
	−61 03111	LSS 2655			12 19 39	−62 17.7	2	10.83	.045	0.14	.024	−0.59	.018	5		846,1684
	+18 02597				12 19 40	+17 37.3	1	9.44		1.25		1.36		1	K2	280
107655	+25 02498	Coma Ber 89		★	12 19 40	+25 03.1	6	6.19	.015	−0.01	.009	−0.03	.024	24	A0 V	15,70,1022,1068,8001,8015
107656	+10 02416				12 19 42	+10 07.4	1	8.47		1.34		1.45		1	K0	280
		AAS41,43 F3 # 139			12 19 43	−64 24.1	1	10.86		0.21		−0.17		2	B4	820
	+19 02556				12 19 44	+18 52.6	1	9.75		0.93		0.67		1	K0	280
		Coma Ber 363			12 19 44	+25 12.1	1	14.73		0.87		0.37		1		1148
	−63 02252				12 19 45	−64 18.8	1	10.70		0.15		−0.48		1	B5	820
		AAS41,43 F3 # 140			12 19 45	−64 33.1	1	11.99		0.28		0.17		2	B9	820
	+42 02297				12 19 46	+42 03.8	1	10.66		1.02		0.90		1	K0 III	280
	−63 02253				12 19 46	−64 17.7	1	10.46		0.13		−0.20		1	B8	820
		Klemola 123			12 19 48	+37 28.	1	12.27		−0.04		0.01		2		1026
		MCC 651			12 19 48	+48 43.0	1	10.39		1.03				1	K8	1017
107672	+10 02417				12 19 49	+10 13.2	1	8.61		1.24		1.32		1	K0	280
		Coma Ber 375			12 19 50	+25 02.6	1	11.59		0.44		−0.05		5		1148
107650	−52 05532	IDS12172S5233	AB		12 19 50	−52 49.7	1	9.83		0.20		0.15		1	A6 V	1771
		Coma Ber 359			12 19 51	+25 24.9	1	10.87		0.59		0.13		2		1148
		Coma Ber 343			12 19 51	+25 26.5	2	11.30	.026	1.39	.021	1.15	.004	3		333,1148
107652	−61 03112				12 19 51	−61 56.3	1	8.32		0.03		−0.31		2	B9 V	401
		vdB 57 # c			12 19 51	−63 01.2	1	11.37		0.34		0.27		4	A1 V	434
	−62 02711	LSS 2656			12 19 51	−63 08.8	3	10.29	.018	0.11	.010	−0.72	.011	5	B0.5IV	540,846,976
		POSS 377 # 20			12 19 52	+25 44.4	1	16.35		1.30				3		1739
		CS 16026 # 24			12 19 52	+31 04.2	1	14.87		0.45		−0.07		1		1744
	−59 04193	LSS 2657			12 19 52	−59 59.2	3	9.42	.018	0.31	.004	−0.59	.036	7	B0.5II	191,540,976
	+33 02227	G 148 - 51			12 19 53	+32 53.6	1	9.59		0.94				2	K0 V	77
107665	−50 06934				12 19 53	−50 31.5	1	8.21		0.27		0.16		1	A6 IV/V	1771
107685	+23 02453	Coma Ber 90			12 19 54	+22 44.5	3	8.55	.009	0.47	.004	−0.04	.002	4	F6 V	1127,8001,8023
		Coma Ber 362			12 19 54	+25 14.5	1	14.60		0.58		0.12		1		1148
		AAS12,381 41 # B			12 19 54	+41 06.1	1	11.67		−0.18		−0.62		1		280
		AAS78,203 # 14			12 19 54	−58 36.	1	12.16		0.26		−0.39		3		1684
		Coma Ber 334			12 19 55	+25 36.0	1	11.61		0.88		0.54		2	G8 III	1148
	+41 02291				12 19 55	+41 03.8	1	9.32		1.04				2	K0 III	280
		BSD 5 # 278			12 19 55	+74 27.8	1	11.50		0.62		−0.10		1	G0 p	1058
107663	−48 07380				12 19 55	−49 02.9	1	7.94		0.14		0.13		1	A2 IV/V	1770
107667	−64 01883	LSS 2658			12 19 55	−65 09.6	2	8.32	.016	0.01	.006	−0.76	.021	4	B1 III/V	540,976
		vdB 57 # d			12 19 56	−63 01.7	1	10.50		0.97		0.61		3		434
	−64 01884	LSS 2659			12 19 56	−64 54.0	1	10.28		0.02		−0.66		1	B0	820
	+39 02516				12 19 57	+38 59.8	1	10.02		0.86				2	G8 III	280
107705	+6 02599	HR 4708		★ A	12 19 59	+05 35.0	2	6.42		0.57	.015	0.06	.005	5	F8 V	13,3077
107705	+6 02599	HR 4708		★ AB	12 19 59	+05 35.0	3	6.40	.004	0.60	.000	0.08	.004	10	F8 V	15,1417,8015
107705	+6 02599	IDS12174N0552	B		12 19 59	+05 35.0	2	9.39	.040	1.10	.025	1.00	.045	5		13,3077
107700	+26 02337	Coma Ber 91		★ A	12 20 00	+26 07.4	11	4.80	.018	0.49	.005	0.26	.009	52	A4 V + F6	15,542,667,1008,1084*
107700	+26 02337	Coma Ber 91		★ B	12 20 00	+26 07.4	1	12.52		0.89		0.56		3		3024
		HA 5 # 399			12 20 00	+75 06.4	1	12.67		0.56		0.09		1		1058
		BSD 5 # 577			12 20 00	+75 24.2	1	11.04		0.56		−0.03		1	G0	1058
	+17 02470				12 20 01	+16 38.2	1	9.53		0.80		0.44		1	K0	1569
		Coma Ber 273			12 20 01	+26 38.6	1	12.67		0.57		0.02		1		1148
		Coma Ber 229			12 20 01	+26 54.8	1	12.10		0.64		0.05		1		1148
107701	+26 02338	Coma Ber 92		★ C	12 20 02	+26 07.7	6	8.55	.015	0.52	.011	0.01	.007	11	F8 V	542,1084,1127,1748,8001,8023
	+18 02600				12 20 03	+18 04.0	1	9.72		1.51		1.90		1	K5	280
		POSS 377 # 29			12 20 03	+26 20.4	1	16.68		0.89				4		1739
		CS 16026 # 21			12 20 03	+32 16.1	1	13.92		0.74		0.35		1		1744
107725	+27 02118	Coma Ber 93			12 20 04	+26 53.9	3	7.86	.011	1.36	.022	1.61	.004	6	K3 III	280,1148,8034
107760	+74 00493	AS Dra		★	12 20 04	+73 31.4	2	8.07	.047	0.74	.005	0.23	.024	4	G7 V	1003,3016
107684	−52 05539				12 20 04	−53 20.1	1	8.71		0.26		0.11		1	A5 IV	1771
	−64 01885				12 20 04	−64 54.0	2	10.88	.024	0.06	.015	−0.60	.005	3	B1	820,846
	+33 02228				12 20 05	+32 34.6	1	9.64		1.33				2	K3 III	77
		AO 1152			12 20 06	+25 46.5	1	11.10		0.58		0.03		1	F8 V	1748
107696	−56 05202	HR 4706			12 20 06	−57 23.9	4	5.38	.008	−0.10	.003	−0.42	.006	11	B7 Vn	15,26,1771,2012
107697	−58 04245				12 20 06	−59 20.7	1	8.86		0.63		0.60		1	A8 IV	1771
	−62 02716	LSS 2660			12 20 06	−62 24.8	1	10.31		0.31		−0.60		3		846
		Coma Ber 382			12 20 07	+24 46.7	1	15.12		0.60		−0.10		2		1148
107726	+16 02365				12 20 08	+15 56.9	2	9.77	.045	0.64	.015	0.09		4	K0	882,1569
		NGP +25 058			12 20 08	+25 20.8	1	11.30		1.01		0.76		1	G8 III	8034
		AAS41,43 F3 # 145			12 20 08	−64 32.5	1	11.71		0.34		0.20		1	B5	820

Table 1 667

HD	DM	Other Id	N	Rem	α_{1950}	δ_{1950}	S	V	σ_V	B–V	σ_{B-V}	U–B	σ_{U-B}	n	Spectrum	References
107742	+27 02119	Coma Ber 94			12 20 09	+27 24.6	3	8.47	.045	1.40	.032	1.71	.004	4	K2 III	280,1148,8041
		Coma Ber 381			12 20 10	+24 43.5	1	12.62		0.55		0.03		1		1148
		Coma Ber 311			12 20 10	+25 56.8	2	10.47	.005	0.71	.000	0.27	.005	6	G8 V	1148,1748
107743	+20 02713				12 20 11	+20 25.6	1	7.89		1.25		1.29		1	K2	280
	+26 02339	Coma Ber 95			12 20 11	+26 05.0	2	10.05	.024	0.45	.005	-0.02	.010	3	F5 I-II	1748,3016
107744	+9 02620				12 20 12	+08 32.9	1	9.55		1.02		0.84		2	K0	1569
107741	+32 02239				12 20 12	+31 31.6	2	8.25	.024	1.10	.010	1.04		3	K0 III	77,8034
		LP 908 - 50			12 20 12	−28 22.7	1	14.30		1.36		1.02		2		3062
107720	−64 01886				12 20 13	−64 27.2	2	7.12	.001	1.12	.002	1.05	.006	6	K1 III	820,1704
		AAS41,43 F3 # 148			12 20 15	−64 12.8	1	11.11		0.27		-0.08		2	B6	820
	+31 02357				12 20 16	+31 04.7	1	8.92		1.35				2	K4 III	77
	−64 01887				12 20 17	−64 47.2	2	10.49	.066	0.48	.215	-0.03	.155	4	B7	125,820
		G 59 - 15		A	12 20 18	+17 41.6	1	15.76		1.48				1		906
		G 59 - 15		B	12 20 18	+17 41.6	1	15.66		1.62				1		906
		Coma Ber 285			12 20 18	+26 23.3	1	12.64		0.53		0.05		1		1148
		Coma Ber 284			12 20 18	+26 28.1	1	12.50		0.66		-0.02		2		1148
		LP 320 - 474			12 20 18	+27 28.	1	13.84		1.55				1	dM2	686
107752	+12 02458				12 20 20	+11 52.9	1	10.05		0.75		0.18		2	G5	1569
	+25 02500	Coma Ber 210			12 20 20	+25 20.8	3	10.56	.009	1.07	.018	1.04	.011	5	K0 III	280,1148,8034
		LF14 # 161			12 20 21	−60 58.5	1	10.63		1.24		1.05		2	K2 III	137
		AAS41,43 F3 # 149			12 20 21	−64 30.5	1	11.66		0.28		0.12		1	A1	820
107699		RS Mus			12 20 21	−75 13.6	1	8.81		4.04				2	Nb	864
107763	+40 02529				12 20 22	+39 33.2	1	8.16		1.13				2	K2 III	280
107762	+44 02190				12 20 22	+44 05.1	2	7.90	.015	0.94	.005	0.74		4	G9 III	172,280
		AAS41,43 F3 # 150			12 20 22	−64 32.2	1	12.00		0.42		0.24		1	A0	820
		Coma Ber 342			12 20 24	+25 34.4	1	13.71		0.82		0.59		1		1148
		HA 5 # 198			12 20 25	+74 35.7	1	13.26		0.54		0.02		1		1058
		Ross 693			12 20 25	−17 23.2	1	11.69		0.81		0.36		2		1696
		Coma Ber 253			12 20 26	+27 19.8	1	11.47		0.54		0.05		2		1148
		He3 781			12 20 26	−62 04.	1	13.65		1.14		0.03		3		1684
107751	−65 01852				12 20 26	−65 49.0	2	9.10	.000	0.39	.005	0.12	.008	4	A9 V	1688,1771
107764	+17 02472				12 20 27	+16 31.2	3	8.28	.012	1.05	.010	0.88		5	K0	280,882,1569
		AAS41,43 F3 # 151			12 20 27	−64 46.3	1	12.40		0.16		0.07		1	B9	820
	+29 02283				12 20 28	+28 36.7	1	10.80		0.73		0.49		1		8038
		Coma Ber 1245			12 20 29	+25 15.6	1	11.32		1.05				3		913
107775	+12 02459				12 20 30	+11 40.9	1	9.48		1.23		1.38		1	K0	280
	+34 02309				12 20 30	+33 34.9	1	10.07		0.95				2	G8 III	280
		L 326 - 61			12 20 30	−46 21.	1	13.63		1.60		1.04		1		3073
107773	−66 01752	HR 4710			12 20 31	−67 21.5	6	6.35	.006	0.89	.000	0.52	.026	27	G8/K0 IV	15,1075,1311,2012*
		CS 16026 # 28			12 20 32	+27 43.9	1	13.68		0.02		0.07		1		1744
		GD 477			12 20 32	+76 23.1	1	14.53		0.58		-0.17		3		308
107774	+13 02526				12 20 33	+12 57.7	3	9.24	.015	1.10	.019	0.95	.019	4	K0	280,882,3040
107759	−63 02255				12 20 33	−63 48.7	1	8.74		1.72		1.84		2	K2/3	122
	−61 03121	LSS 2661			12 20 34	−61 27.0	3	10.09	.009	0.15	.007	-0.73	.027	9	O9.5III	540,846,976
	+19 02559				12 20 35	+19 25.3	1	9.32		1.11		1.10		1	K0	280
		Coma Ber 335			12 20 35	+25 40.8	1	12.18		0.59		-0.06		1		1148
	+31 02358				12 20 35	+30 53.2	1	9.29		0.53				2	F8	77
		POSS 377 # 26			12 20 37	+21 17.1	1	16.60		1.56				4		1739
		AAS41,43 F3 # 152			12 20 37	−64 12.3	1	10.88		0.60		0.06		2	F5	820
107793	+26 02340	Coma Ber 97			12 20 38	+26 07.7	4	9.10	.012	0.56	.015	0.01	.008	5	F8 V	1127,1748,8001,8023
	+33 02229				12 20 38	+33 12.8	1	9.90		0.49				2	F8	77
	−63 02256				12 20 38	−64 14.4	1	10.88		0.27		0.00		2	B8	820
		LSS 2662, BI Cru			12 20 40	−62 21.7	3	10.68	.112	1.30	.039	0.21	.071	6	B3 Ib	846,1684,1737,1753
		AAS41,43 F3 # 154			12 20 40	−64 17.9	1	10.44		1.78		2.06		1	K5	820
		Coma Ber 1257			12 20 41	+25 50.7	1	12.66		1.32				2		913
	+13 02527				12 20 42	+13 22.5	1	10.70		0.57		0.10		1	G5	1569
		NGC 4340 sq1 7			12 20 42	+17 02.8	1	13.59		0.77				1		492
		Ton 610			12 20 42	+23 27.	1	15.64		-0.07		-0.94		3	DA	1036
		Coma Ber 312			12 20 42	+25 50.3	1	12.71		1.21		1.32		1		1148
		HA 5 # 199			12 20 42	+74 39.2	1	11.42		0.48		-0.04		1		1058
		AAS78,203 # 15			12 20 42	−57 28.	1	14.29		1.52		0.89		5		1684
	−63 02257				12 20 42	−64 24.0	1	10.69		0.18		0.02		1	B8	820
107772	−65 01854				12 20 42	−66 24.1	1	9.78		0.22		0.01		2	A0 Vn	1688
	+33 02230				12 20 43	+33 13.3	1	9.54		0.38				2	F4	77
107788	−52 05552				12 20 43	−53 22.2	1	8.71		-0.10		-0.85		4	B2 III/IV	400
		NGC 4349 - 304			12 20 43	−61 32.4	1	11.34		0.19				1		2041
107790	−64 01889	IDS12180S6419		AB	12 20 43	−64 36.0	1	9.06		0.00		-0.56		1	B2 IV/V	820
107811	+14 02495			A	12 20 44	+13 54.0	1	8.55		0.33		0.06		1	F5	1569
107814	−11 03291	TT Crv			12 20 44	−11 32.1	1	6.50		1.64		1.90		8	M1	3040
		AAS41,43 F3 # 158			12 20 44	−64 18.6	1	10.80		1.44		1.28		1	G8	820
	−64 01890				12 20 44	−64 48.7	1	10.38		0.03		-0.69		1	B2	820
107810	+14 02496				12 20 45	+14 14.0	1	10.27		0.85		0.54		2	K0	1569
107815	−24 10314	HR 4711			12 20 45	−24 33.8	4	5.67	.009	1.16	.005			18	K1 III	15,1075,2013,2028
		Coma Ber 304			12 20 46	+26 02.3	1	12.40		0.49		-0.03		1		1148
107823	+43 02215				12 20 46	+43 10.6	1	8.80		0.43		-0.07		3	F5	1569
		AAS41,43 F3 # 159			12 20 46	−64 23.7	1	11.23		1.30		1.14		1	G8	820
107791	−65 01855				12 20 47	−65 35.0	1	10.02		0.15		-0.10		2	B8/9 IV/V	1688
		NGC 4340 sq1 9			12 20 48	+16 56.6	1	14.32		0.99				2		492
		NGC 4372 sq2 7			12 20 48	−72 26.0	1	10.63		0.37		0.18		1		518
107825	+20 02715				12 20 49	+19 56.3	1	8.26		1.20		1.18		1	K2	280

HD	DM	Other Id	N Rem	α_{1950}	δ_{1950}	S	V	σ_V	B–V	σ_{B-V}	U–B	σ_{U-B}	n	Spectrum	References
		Coma Ber 1266		12 20 50	+26 50.0	1	12.77		0.83				1		913
107813	−6 03559			12 20 50	−06 46.5	3	9.39	.012	0.36	.004	−0.10	.007	7	F2 V	158,1003,3037
		Coma Ber 251		12 20 52	+27 34.3	1	11.75		0.66		0.10		2		1148
107805	−60 03938	R Cru		12 20 52	−61 21.1	2	6.45	.065	0.64	.040	0.45		2	F7 Ib/II	657,688
	−64 01891			12 20 52	−64 37.8	1	10.82		0.03		−0.66		1	B2	820
	+43 02217			12 20 53	+43 21.7	1	10.34		0.96		0.87		1	K0 III	280
	−63 02258	LSS 2663	⋆ AB	12 20 54	−64 02.5	1	10.75		0.09		−0.62		2		846
		AAS41,43 F3 # 162		12 20 54	−64 34.8	1	11.63		0.34		0.20		1	A0	820
		AAS41,43 F3 # 161		12 20 54	−64 38.5	1	12.01		0.23		0.16		1	A2	820
	+44 02191			12 20 55	+43 50.5	1	10.78		0.93		0.66		1	G8 III	280
107821	−63 02259			12 20 56	−63 35.6	2	7.40	.029	0.17	.015	0.08	.000	3	A3 V	122,1771
		AAS41,43 F3 # 163		12 20 56	−64 26.2	1	11.61		0.17		−0.41		1	B5	820
	+21 02415	G 59 - 17		12 20 57	+20 34.1	1	10.02		1.07		0.95		1	K7	3072
		NGP +26 094		12 20 57	+26 19.4	1	11.99		0.95		0.75		1	K0 III	8034
107832	−34 08117	HR 4712		12 20 57	−35 08.1	3	5.31	.004	−0.08	.000	−0.23		10	B9 III	15,1770,2029
	+23 02456			12 20 58	+23 07.5	1	10.42		1.07		0.90		1		280
		Coma Ber 1278		12 20 58	+26 10.3	1	12.13		1.20				2		913
		NGP +26 098		12 20 58	+26 19.4	1	10.45		0.80		0.38		1	G5	8038
107833	−38 07700	HR 4713		12 20 58	−39 01.5	2	6.39	.005	0.30	.005			7	F2 V	15,2012
		AAS41,43 F3 # 164		12 20 58	−64 15.6	1	10.47		1.81		2.10		1	K8	820
		NGC 4340 sq1 4		12 20 59	+17 06.0	1	13.40		0.79				2		492
107854	+25 02501	Coma Ber 98		12 20 59	+24 52.2	6	7.40	.010	1.06	.012	0.94	.011	10	K0 III	172,280,1148,8001*
107853	+27 02120	Coma Ber 99		12 20 59	+26 50.6	5	9.09	.019	0.48	.003	0.03	.013	22	F7 III	830,1148,1783,8001,8023
		Coma Ber 252		12 20 59	+27 23.0	1	12.44		0.62		0.04		1		1148
		AAS41,43 F3 # 165		12 20 59	−64 19.9	1	10.22		1.44		1.42		1	K0	820
	+17 02473	G 59 - 18		12 21 00	+17 10.9	3	10.19	.021	0.72	.005	0.10	.005	6	G5	516,1620,1696
		NGC 4349 - 5		12 21 00	−61 35.6	1	11.51		1.34		0.95		3		1695
		NGC 4349 - 53		12 21 00	−61 38.0	1	11.33		1.12				1		2041
		LOD 26- # 14		12 21 00	−63 44.4	1	9.80		1.89		2.07		2		122
		WLS 1200 15 # 7		12 21 01	+13 10.9	1	12.18		1.14		1.03		2		1375
107855	+14 02497	IDS12185N1429	AB	12 21 01	+14 12.0	1	8.98		0.64		0.12		2	G5	1569
		NGC 4349 - 9		12 21 01	−61 34.3	1	11.59		1.33		1.04		2		1695
		NGC 4340 sq1 5		12 21 02	+17 07.6	1	13.38		0.73				2		492
107838	−63 02261	IDS12183S6326	A	12 21 02	−63 42.5	2	8.13	.005	0.27	.019	0.09	.000	3	A4mA7-A7	122,1771
		LP 555 - 11		12 21 03	+07 34.4	1	11.81		0.62		−0.04		2		1696
	+20 02717			12 21 03	+19 39.5	1	9.58		1.12		1.02		1		280
	−57 05439	NGC 4337 - 3		12 21 04	−57 48.6	1	10.85		0.66		0.21		4		251
		AO 1156		12 21 05	+26 21.4	1	10.33		1.21		1.28		1	K2 III	1748
		NGC 4340 sq1 6		12 21 06	+17 08.4	1	15.29		0.70				2		492
		Coma Ber 272		12 21 06	+26 43.3	1	12.86		0.48		−0.01		1		1148
		WLS 1224 50 # 10		12 21 06	+49 54.9	1	13.86		0.82		0.44		2		1375
		HA 5 # 130		12 21 06	+74 29.0	1	11.56		1.04		0.87		1		1058
107860	−38 07701	HR 4714		12 21 06	−38 38.0	3	5.79	.005	−0.08	.002	−0.21		10	B9 V	15,1770,2029
	−63 02262			12 21 06	−64 22.9	1	10.86		0.13		−0.54		1	B3	820
		Coma Ber 376		12 21 08	+25 03.2	1	14.58		0.37		−0.17		1		1148
	−64 01893			12 21 08	−64 37.0	1	10.90		0.18		−0.17		1	B6	820
		NGC 4349 - 25		12 21 09	−61 31.9	1	12.18		0.23				1		2041
107865	+12 02462			12 21 10	+12 11.9	1	8.85		1.03		0.83		1	K0	280
		Coma Ber 255		12 21 10	+27 12.4	1	12.89		1.43		1.22		2		1148
107869	−29 09709	IDS12186S2947	A	12 21 10	−30 03.5	2	6.39	.030	1.61	.015	1.92	.025	5	K5 III	803,3012
		NGC 4337 - 14		12 21 10	−57 50.1	1	13.69		1.19		0.83		2		251
	+27 02121	Coma Ber 102		12 21 11	+26 52.7	3	9.34	.015	0.61	.010	0.08	.000	5	G0 V	1127,8001,8023
107877	+27 02122	Coma Ber 101		12 21 11	+27 15.4	3	8.37	.023	0.45	.007	−0.03	.003	4	F6 V	1127,8001,8023
		HA 5 # 2		12 21 12	+74 03.7	1	12.98		0.62		0.11		2		1058
	+16 02368			12 21 13	+15 31.2	1	10.92		0.85		0.54		1		1569
		AAS41,43 F3 # 169		12 21 13	−64 40.2	1	12.06		0.34		0.06		1	B8	820
		NGC 4337 - 13		12 21 15	−57 51.3	1	11.86		1.57		1.81		2		251
		He3 784, Ly 124		12 21 16	−64 34.	3	11.11	.067	0.20	.037	−0.71	.032	6	O7	820,846,1684
	−57 05441	NGC 4337 - 2		12 21 17	−57 48.4	1	10.93		0.75		0.38		3		251
		NGC 4372 sq3 5		12 21 17	−72 23.2	1	13.34		0.89		0.31		4		518
		HA 5 # 3		12 21 18	+74 10.7	1	11.82		0.65		0.14		1		1058
		Ly 125		12 21 18	−62 44.	1	11.76		0.28		−0.53		3		846
		Coma Ber 377		12 21 19	+25 03.2	1	13.32		0.33		−0.05		1		1148
		NGC 4372 sq2 10		12 21 19	−72 20.6	1	12.01		0.87		0.36		8		518
		NGC 4340 sq1 1		12 21 20	+16 52.6	1	13.51		0.93				2		492
107904	+43 02218	HR 4715, AI CVn		12 21 20	+42 49.2	3	6.07	.019	0.34	.011	0.18	.000	7	F3 IV	39,70,3013
107888	+13 02529			12 21 21	+12 51.6	2	10.40	.010	1.42	.016			3	K5	1625,1705
		AAS61,331 # 17		12 21 21	+37 59.	1	11.09		0.31		−0.02		3		1569
		NGC 4337 - 4		12 21 21	−57 47.2	1	11.53		0.70		0.21		2		251
		NGC 4349 - 113		12 21 21	−61 30.9	1	12.61		0.14				1		2041
107887	+14 02499			12 21 22	+14 24.9	1	10.10		0.67		0.21		1		1569
		CS 16026 # 29		12 21 22	+31 43.9	1	14.15		0.57		−0.02		1		1744
		NGC 4337 - 12		12 21 22	−57 51.3	1	13.02		1.17		0.94		2		251
		NGC 4337 - 11		12 21 22	−57 52.0	1	12.16		0.18		0.13		4		251
107907	+16 02369			12 21 23	+15 53.7	2	8.82	.010	0.99	.010	0.74		2	K0	280,882
107908	+14 02500			12 21 24	+14 20.1	1	9.48		0.63		0.12		1	G5	1569
107906	+17 02475			12 21 25	+16 41.8	3	8.94	.005	0.81	.019			3	K2	280,882
107905	+41 02292			12 21 26	+40 59.8	1	8.05		1.65				3	M1 III	280
107885	−57 05442			12 21 26	−57 38.4	2	7.75	.028	0.55	.009	−0.02		4	F8 Vw	1594,6006
		NGC 4349 - 308		12 21 26	−61 29.8	1	11.89		0.29				1		2041

Table 1   669

HD	DM	Other Id	N Rem	α_{1950}	δ_{1950}	S	V	σ_V	B–V	σ_{B-V}	U–B	σ_{U-B}	n	Spectrum	References
107863	−76 00729			12 21 26	−76 56.3	1	6.95		1.54				4	K3 III	2012
107739	−85 00343	HR 4709		12 21 26	−85 52.4	3	6.32	.004	1.07	.010	0.90	.005	14	K0 III	15,1075,2038
		NGC 4337 - 5		12 21 27	−57 46.4	1	12.52		0.23		0.16		2		251
	+17 02476			12 21 28	+16 46.9	1	10.69		0.93		0.88		1		280
	+18 02604			12 21 28	+18 02.1	1	9.63		0.94		0.77		1	G5	280
	+45 02030			12 21 29	+45 18.7	1	9.63		0.81				3	G7 III	280
		NGC 4337 - 15		12 21 29	−57 50.7	1	13.26		0.36		0.43		2		251
	−59 04201	LSS 2664		12 21 29	−60 02.5	1	9.73		0.18		−0.49		2		846
		Coma Ber 1328		12 21 30	+27 09.6	1	14.09		1.05				1		913
		BSD 5 # 293		12 21 30	+73 59.2	1	12.19		0.66		0.09		2	G1	1058
	−57 05444	NGC 4337 - 1		12 21 31	−57 49.0	1	10.53		0.24		0.20		5		251
	−61 03135	NGC 4349 - 203		12 21 31	−61 39.3	1	11.53		1.29		0.92		2		1695
107936	+24 02452	Coma Ber 103		12 21 33	+24 15.6	1	8.82		1.08		0.97		1	K0	280
107935	+26 02343	Coma Ber 104		12 21 33	+26 07.7	5	6.71	.012	0.23	.007	0.05	.012	7	A7 V	280,1127,1148,8001,8023
108149	+86 00180			12 21 33	+85 35.0	1	8.90		0.12		0.09		6	A0	1219
		PHI2/2 10		12 21 33	−32 05.8	1	11.62		1.13				1		1719
		NGC 4337 - 7		12 21 33	−57 52.0	1	11.91		0.47		0.08		2		251
107901	−66 01755			12 21 33	−66 47.0	1	8.72		0.38		0.04		2	F0 V	1688
		NGC 4337 - 8		12 21 34	−57 46.0	1	12.98		0.52		0.09		2		251
	−72 01243			12 21 34	−72 38.0	1	10.30		0.42		0.11		7		518
		HA 5 # 403		12 21 35	+75 07.2	1	13.31		0.46		−0.05		1		1058
		NGC 4372 sq1 8		12 21 35	−72 22.5	2	14.31	.039	0.64	.025	0.27	.010	8		372,518
	+19 02562			12 21 36	+19 09.9	1	10.49		1.13		1.26		1		280
	+26 02342	Coma Ber 213		12 21 36	+26 24.4	4	10.49	.013	0.77	.008	0.38	.008	6	K0 V	1127,1748,8001,8023
107950	+52 01626	HR 4716		12 21 36	+51 50.3	7	4.79	.012	0.88	.010	0.61	.011	26	G7 III	15,667,1080,1118,4001*
		PHI2/2 8		12 21 37	−31 32.9	1	13.43		1.08				2		1719
107937	+6 02606	FK Vir		12 21 38	+06 14.9	1	7.65		1.62		1.70		11	M2	3042
	−59 04204	LSS 2665		12 21 38	−60 12.9	1	11.18		0.38		−0.55		4		846
		NGC 4372 sq1 10		12 21 38	−72 25.2	1	14.62		1.10		0.39		3		372
		NGC 4340 sq1 2		12 21 39	+16 55.4	1	13.40		0.52				2		492
	+37 02271			12 21 40	+36 36.4	1	9.78		0.42				2	F6	77
107934	+40 02533			12 21 40	+40 08.4	1	8.99		1.20				2	K1 III	280
107931	−46 07903			12 21 40	−47 05.7	1	6.61		−0.05		−0.22		3	B9 V	1770
		NGC 4337 - 16		12 21 40	−57 49.5	1	13.60		0.70		0.24		2		251
		Coma Ber 1344		12 21 41	+27 09.9	1	13.89		1.48				3		913
		NGC 4337 - 10		12 21 41	−57 51.5	1	13.42		0.51		0.15		2		251
		NGC 4372 sq1 5		12 21 41	−72 23.1	1	12.92		1.49		1.14		3		372
		NGC 4349 - 212		12 21 42	−61 37.	1	14.14		0.50		0.14		3		1146
		NGC 4349 - 214		12 21 42	−61 37.	1	14.26		0.44		−0.35		3		1146
		NGC 4349 - 216		12 21 42	−61 37.	1	14.70		0.33		−0.13		1		1146
		NGC 4349 - 217		12 21 42	−61 37.	1	13.47		0.26		−0.21		3		1146
		NGC 4349 - 222		12 21 42	−61 37.	1	14.03		0.55		−0.01		2		1146
		NGC 4349 - 228		12 21 42	−61 37.	1	13.79		0.63		0.29		2		1146
		NGC 4349 - 229		12 21 42	−61 37.	1	14.09		0.84				1		1146
		NGC 4349 - 230		12 21 42	−61 37.	1	13.97		0.50		0.31		1		1146
		NGC 4349 - 236		12 21 42	−61 37.	1	13.73		0.58		0.10		1		1146
		NGC 4349 - 238		12 21 42	−61 37.	1	11.68		0.34		0.10		3		1146
		NGC 4349 - 239		12 21 42	−61 37.	1	12.12		0.37		0.09		2		1146
		POSS 377 # 19		12 21 43	+22 37.9	1	16.34		1.71				3		1739
	+29 02284	Coma Ber 105		12 21 43	+28 41.0	1	10.08		0.50		0.01		2	dG0	3016
107952	+10 02418			12 21 44	+09 47.1	1	8.96		1.16		1.16		1	K0	280
		NGC 4372 sq1 9		12 21 45	−72 22.3	1	14.51		1.21		0.47		3		372
		NGC 4372 sq1 7		12 21 45	−72 25.5	1	14.05		1.26		0.65		3		372
	+25 02502	Coma Ber 214		12 21 47	+24 36.2	3	10.01	.009	0.79	.004	0.33	.001	5	G9 V	1127,8001,8023
	+36 02264			12 21 47	+36 18.9	1	9.49		0.89				2	G8 III	280
	+42 02300			12 21 47	+41 57.5	1	9.84		1.04		1.03		3	K1 III	280
		NGC 4337 - 6	A	12 21 47	−57 51.2	1	12.23		0.41		0.23		1		251
		NGC 4337 - 6	AB	12 21 47	−57 51.2	1	12.08		0.55		0.23		1		251
	+20 02718			12 21 48	+19 35.0	1	10.04		1.16		1.04		1	K0	280
107966	+26 02344	Coma Ber 107	★ V	12 21 48	+26 22.5	8	5.18	.014	0.08	.004	0.10	.005	23	A3 IV	15,1022,1068,1127*
107944	−61 03138	NGC 4349 - 188		12 21 48	−61 38.1	4	8.34	.008	0.74	.005	0.51	.005	10	F2 III	285,657,2012,2041
107945	−64 01895			12 21 48	−64 39.8	1	8.05		1.28		1.28		1	K1/2 III	820
		NGC 4349 - 127		12 21 49	−61 32.6	1	10.83		1.56		1.48		3		1695
	−61 03139	NGC 4349 - 167		12 21 49	−61 36.3	1	10.06		0.25				1	B7 III	2041
		NGC 4372 sq1 3		12 21 49	−72 23.3	2	12.69	.029	1.37	.010	0.98	.038	10		372,518
107967	+17 02477			12 21 51	+16 37.2	2	8.97	.010	1.00	.049			3	K2	280,882
		Coma Ber 378		12 21 51	+25 09.5	1	11.21		0.51		0.06		3		1148
		Coma Ber 299		12 21 52	+26 11.0	1	12.38		1.14		1.21		2		1148
107957	−48 07401	S Cen		12 21 52	−49 09.8	2	7.53	.046	1.78	.046	2.28	.124	9	C4,5	864,3038
		NGC 4337 - 9		12 21 52	−57 50.4	1	13.66		0.48		0.17		2		251
		NGC 4349 - 185		12 21 52	−61 38.3	1	11.97		1.34		0.95		2		1695
		Coma Ber 1358		12 21 53	+26 11.6	1	12.39		1.18				1		913
		HA 5 # 133		12 21 53	+74 29.5	1	11.75		0.70		0.20		1		1058
107986	+15 02459			12 21 55	+14 46.7	1	9.65		0.61		0.08		1	G5	1569
	+28 02111	Coma Ber 108		12 21 55	+27 45.0	1	9.99		0.60		0.10		2	dG5	3016
	−61 03143	NGC 4349 - 131		12 21 55	−61 32.8	1	11.07		0.17				1		2041
		NGC 4349 - 168		12 21 55	−61 35.4	1	11.48		1.30		0.92		3		1695
		Coma Ber 360		12 21 56	+25 24.6	1	12.49		0.45		−0.14		2		1148
108007	+26 02345	Coma Ber 109	★ AB	12 21 56	+25 51.6	6	6.41	.018	0.27	.006	0.07	.005	16	F0 V	15,938,1127,1748,8001,8015
	+38 02328			12 21 56	+38 10.5	1	10.33		0.67				2	G6 III	280

HD	DM	Other Id	N	Rem	α_{1950}	δ_{1950}	S	V	σ_V	B–V	σ_{B-V}	U–B	σ_{U-B}	n	Spectrum	References
107959	−51 06659				12 21 56	−51 55.2	1	9.16		0.30		0.14		1	A7 V	1771
107988	+9 02622				12 21 57	+08 37.4	1	10.46		0.67		0.20		1	G5	280
108005	+43 02220	IDS12195N4339	AB		12 21 57	+43 21.9	1	7.92		0.33		0.17		std	F0	1370
107979	−60 03955		AB		12 21 57	−60 37.6	1	7.11		1.25				4	K2 III +A/F	1075
107947	−71 01326				12 21 57	−72 19.6	2	6.60	.009	0.00	.000	−0.04	.005	13	A0 V	518,1771
		NGC 4349 - 312			12 21 58	−61 40.8	1	10.73		1.08				1		2041
107981	−60 03956	NGC 4349 - 211			12 21 59	−61 24.1	2	9.19	.012	0.03	.002	−0.39		6	A0	1146,2041
		PHI2/2 7			12 21 60	−31 37.6	1	13.37		1.02				2		1719
108009	+13 02530				12 22 00	+13 09.1	2	10.26	.049	1.02	.029	0.75		3	K0	492,1569
		NGC 4374 sq1 8			12 22 00	+13 11.3	1	13.44		0.59				1		492
108021	+32 02241				12 22 01	+31 33.4	2	8.81	.005	0.74	.005	0.36		5	G6 IV	77,196
		LSS 2667			12 22 01	−60 33.2	1	12.30		0.33		−0.19		2	B3 Ib	846
108006	+38 02329	IDS12196N3817			12 22 02	+37 59.0	2	9.47	.005	0.47	.005	−0.01		4	G5	77,1569
107982	−65 01859				12 22 03	−65 34.5	1	9.15		0.12		−0.39		1	B6 II	1688
		NGC 4372 sq1 2			12 22 03	−72 22.5	1	12.16		2.03		2.62		3		372
		NGC 4372 sq1 11			12 22 03	−72 26.8	1	14.69		1.06		0.58		3		372
	+38 02330	IDS12196N3817	AB		12 22 04	+37 59.7	1	10.90		0.50				5	F9 V	193
108022	+19 02563				12 22 05	+18 56.7	1	8.20		1.52		2.02		1	K2	280
107998	−40 07281	HR 4718	⋆ AB		12 22 05	−41 06.4	6	6.24	.003	1.17	.014	1.10	.028	22	K2/3 III	15,404,1075,2013,2030,3005
107976	−57 05451	NGC 4337 - 17	⋆ A		12 22 05	−57 50.5	1	7.72		0.98		0.65		3	F8 IV/V	251
107976	−57 05451	NGC 4337 - 18	⋆ B		12 22 05	−57 50.5	1	8.00		0.36		0.18		2	G0	251
108056	+60 01398				12 22 07	+59 47.7	1	9.00		0.44		−0.04		2	F5	1502
		Coma Ber 265			12 22 08	+26 56.0	1	12.56		0.65		0.19		1		1148
	−60 03960	LSS 2669			12 22 08	−61 06.4	1	11.45		0.26		−0.20		3		846
		NGC 4349 - 314			12 22 08	−61 38.1	1	12.41		0.32				1		2041
		NGC 4349 - 313			12 22 08	−61 39.4	1	12.86		0.42				1		2041
108002	−64 01898	LSS 2668			12 22 08	−64 56.1	7	6.93	.017	0.11	.010	−0.73	.030	14	B2 Ia/ab	138,158,191,401,540*
107983	−70 01481				12 22 08	−70 27.1	1	8.10		0.16		−0.03		2	B9/9.5 V	1771
108001	−64 01899				12 22 09	−64 40.9	1	9.66		0.09		−0.07		2	B9/A0 V	1688
108020	+42 02302				12 22 10	+41 36.6	2	7.21	.005	1.32	.005	1.38		4	K2 III	172,280
		HA 5 # 408			12 22 10	+75 08.0	1	12.59		0.49		−0.09		1		1058
		PHI2/2 4			12 22 10	−32 09.8	1	13.37		1.14				2		1719
		NGC 4349 - 174			12 22 10	−61 35.5	1	11.02		1.48		1.34		2		1695
		NGC 4349 - 175			12 22 10	−61 35.5	1	10.97		1.44				1		2041
		Coma Ber 383			12 22 12	+24 54.4	1	14.55		−0.12				2		1148
		SSII 143			12 22 12	+39 25.	2	11.89	.004	−0.04	.005	−0.16	.014	6	B8	275,1026
	−64 01900				12 22 12	−64 48.5	1	9.94		0.13		−0.19		2	B8	1688
	−64 01901				12 22 12	−64 55.2	1	9.88		0.09		−0.66		1	B5	1688
		Ross 695			12 22 13	−17 56.0	3	11.32	.016	1.60	.012	1.11	.029	15	M4 V:	158,1705,3078
		Coma Ber 341			12 22 14	+25 31.7	1	12.59		0.58		0.00		1		1148
108041	+11 02455				12 22 15	+10 59.7	1	9.21		0.76		0.44		1	G5	280
	−66 01758				12 22 16	−66 47.6	1	9.82		0.17		−0.07		2	B8	1688
		NGP +33 096			12 22 17	+33 01.5	1	10.78		1.08		1.01		1	K1 III	8034
		Coma Ber 336			12 22 18	+25 39.7	1	12.65		0.40		−0.17		1		1148
108035	−49 07074	IDS12196S4938	A		12 22 18	−49 54.9	1	8.21		0.32		0.04		1	A8/9 V	1771
		NGC 4349 - 213			12 22 18	−61 26.1	1	12.49		1.45		1.11		4		1146
108038	−63 02267				12 22 18	−64 07.3	1	9.39		0.23		0.23		2	B9.5III/IV	1688
		Coma Ber 318			12 22 19	+25 44.5	1	13.81		0.31		0.03		2		1148
		NGC 4349 - 215			12 22 19	−61 25.4	1	13.30		0.57		−0.07		3		1146
		NGC 4372 sq1 1			12 22 19	−72 19.5	3	11.27	.019	1.34	.010	1.09	.000	21		372,518,518
		NGC 4374 sq1 7			12 22 20	+13 15.0	1	13.02		0.87				1		492
108078	+31 02362				12 22 20	+31 18.6	3	7.56	.012	1.13	.011	1.19	.000	7	K2 III	37,172,280
108076	+39 02519	G 123 - 29			12 22 20	+38 35.7	3	8.02	.015	0.56	.004	−0.05	.004	6	G0 V	22,1003,1658
		POSS 377 # 11			12 22 21	+25 14.0	1	15.60		1.07				3		1739
108077	+32 02243				12 22 22	+32 06.0	2	8.03		0.96	.000	0.67		4	G7 III	77,172
108079	+17 02479				12 22 23	+16 28.3	2	8.09	.010	1.13	.000	1.17		3	K2	280,882
		NGC 4374 sq1 1			12 22 24	+13 06.1	1	11.91		0.49				3		492
		NGC 4349 - 219			12 22 24	−61 24.5	1	13.30		0.63		0.10		3		1146
		NGC 4349 - 220			12 22 26	−61 24.8	1	13.36		0.37		0.09		4		1146
108053	−64 01904				12 22 26	−64 50.2	1	9.31		0.14		−0.16		2	B8/9 III	1688
		NGC 4374 sq1 2			12 22 27	+13 04.2	1	13.67		0.49				3		492
108090	+34 02311				12 22 27	+33 48.4	1	9.33		0.44				2	F5 V	77
	+46 01784				12 22 27	+45 51.7	1	9.73		1.53				2	M5 III	280
		NGC 4372 sq1 4			12 22 27	−72 20.5	1	12.88		1.53		1.53		3		372
		NGC 4374 sq1 3			12 22 28	+13 03.2	1	14.26		0.59				1		492
		HA 5 # 72			12 22 29	+74 16.3	1	12.95		0.62		0.11		1		1058
108063	−41 07163	HR 4721			12 22 29	−42 14.2	3	6.10	.005	0.65	.004			14	G5 III +F/G	15,1075,2027
		NGC 4349 - 311			12 22 29	−61 42.2	1	12.07		0.17				1		2041
108054	−65 01862	HR 4720			12 22 29	−65 29.5	3	6.29	.008	0.97	.010	0.75	.005	11	G8/K0 IV	15,1075,2038
	+26 02346	Coma Ber 110			12 22 30	+25 56.4	3	10.24	.023	0.28	.003	0.06	.006	22		1148,1569,1613
108100	+43 02221				12 22 30	+43 07.9	2	7.15	.014	0.37	.004	0.04	.010	11	F2	245,1370
		NGC 4349 - 218			12 22 30	−61 24.0	1	12.32		0.20		−0.15		4		1146
108039	−75 00812				12 22 30	−75 45.9	1	8.40		1.02				4	K2 IV	2012
		Coma Ber 361			12 22 31	+25 11.8	1	11.85		0.40		−0.03		2		1148
108102	+26 02347	Coma Ber 111	⋆ C		12 22 32	+25 50.3	5	8.09	.036	0.51	.003	0.01	.011	25	F8 V	1127,1148,1613,8001,8023
108092	+12 02463				12 22 33	+12 12.9	1	9.85		0.98		0.76		1	K0	280
		Coma Ber 286			12 22 35	+26 26.9	1	12.34		0.48		0.02		1		1148
	+49 02137				12 22 35	+49 25.1	2	10.65	.010	−0.14	.005	−0.60	.000	5	B7 V	1026,1240
108101	+35 02337				12 22 36	+35 02.5	3	9.11	.006	0.23	.013	0.11	.010	4	Am	77,275,1068
	+33 02234				12 22 37	+33 01.6	2	10.47	.005	1.00	.010	0.81	.005	2	K0 III	280,8034

Table 1   671

HD	DM	Other Id	N Rem	α_{1950}	δ_{1950}	S	V	σ_V	B−V	σ_{B-V}	U−B	σ_{U-B}	n	Spectrum	References
108107 −10 03467		HR 4722		12 22 37	−11 20.0	2	5.94	.005	0.03	.000			7	A1 V	15,2028
108104 +15 02460				12 22 38	+14 41.0	1	9.98		0.58				2	K0	882
108122 +38 02331				12 22 38	+37 29.9	1	8.25		1.13		1.14		1	K1 III	280
108073 −68 01650		IDS12198S6855	A	12 22 38	−69 12.0	1	7.51		−0.02		−0.42		2	B5/7 V	401
		NGC 4372 sq1 6		12 22 39	−72 18.8	1	13.14		1.55		1.16		3		372
		NGC 4374 sq1 4		12 22 40	+13 02.0	1	14.26		0.70				1		492
108135 +57 01373		HR 4726		12 22 40	+57 03.3	2	5.85	.035	1.62	.002	1.83		3	M3 IIIb	71,3055
		PHI2/2 121		12 22 40	−31 20.0	1	12.51		0.87				1		1719
108110 −27 08670		HR 4723		12 22 41	−27 28.3	1	6.09		1.27				4	K3 III	2032
		PHI2/2 2		12 22 41	−31 28.7	1	12.13		1.23				1		1719
		NGC 4349 - 221		12 22 41	−61 25.4	1	11.26		1.28		1.30		4		1146
108088 −65 01863				12 22 41	−65 32.9	1	10.05		0.19		0.05		1	B9/A0 IV/V	1688
108134 +61 01294				12 22 43	+60 58.5	1	7.36		0.57		0.00		4	G0p	3026
108114 −34 08146		HR 4724		12 22 43	−34 54.6	2	5.71	.008	−0.07	.008	−0.24		8	B9 IV/V	1770,2032
		NGC 4349 - 225		12 22 43	−61 29.0	1	13.10		0.23		0.17		2		1146
		NGC 4374 sq1 5		12 22 44	+13 02.7	1	11.64		0.54				2		492
108123 +24 02455		Coma Ber 113	⋆	12 22 44	+24 12.2	6	6.04	.009	1.09	.009	1.00	.007	26	K1 III	15,70,1007,1013,3016,8015
+30 02267				12 22 44	+30 13.7	2	9.63	.009	0.47	.019	0.07		4	F8	77,1375
		NGC 4349 - 309		12 22 45	−61 37.9	1	11.23		0.19				1		2041
108105 +1 02694		SS Vir		12 22 46	+01 04.5	3	6.82	.494	4.19	.106	3.24	.436	4	C6,3e	635,817,864
		NGC 4349 - 223		12 22 46	−61 26.8	1	12.15		0.19		0.00		4		1146
		Coma Ber 317		12 22 47	+25 45.2	1	12.82		0.61		0.11		1		1148
+37 02273				12 22 47	+36 35.0	1	9.53		0.52				2	F8	77
108150 +64 00896		HR 4727	⋆ A	12 22 47	+64 04.8	1	6.32		0.91				2	G8 III	71
		PHI2/1 69		12 22 48	−31 23.7	1	13.21		1.02				2		1719
108153 +32 02244		G 148 - 55		12 22 50	+32 08.7	2	8.74	.039	0.89	.024	0.67		3	K0 V	77,3072
		AAS12,381 44 # A		12 22 51	+43 44.9	1	10.49		0.93		0.66		1		280
108119 −61 03158				12 22 51	−62 01.6	1	9.06		0.57		0.14		2	F7/8 V	122
108154 +24 02457		Coma Ber 114		12 22 52	+23 30.4	3	8.58	.014	0.45	.001	−0.05	.003	4	F7 V	1127,8001,8023
		Coma Ber 303		12 22 52	+26 01.2	1	12.58		0.54		−0.02		1		1148
		NGP +27 098		12 22 52	+27 17.8	1	11.31		0.70		0.15		1	G5	8038
		G 148 - 56		12 22 52	+31 36.2	2	13.62	.020	0.93	.005	0.58	.010	2		1658,1744
+36 02268				12 22 52	+36 15.4	3	10.42	.025	−0.22	.031	−0.79	.000	21	B3	77,1026,1068
		HA 5 # 205		12 22 52	+74 33.8	1	12.74		0.55		0.06		1		1058
108120 −62 02732				12 22 52	−63 20.9	1	9.22		0.26		0.21		2	A0/1 V	1688
108132 −61 03160		NGC 4349 - 224		12 22 53	−61 28.2	2	9.05	.015	0.05	.011	−0.30		6	B9	1146,2041
		NGC 4349 - 226		12 22 53	−61 29.3	1	12.77		0.35		0.06		3		1146
+29 02287				12 22 54	+28 55.0	1	10.13		0.97		0.70		1	G9 III	8041
108152 +40 02536				12 22 54	+39 30.5	1	9.57		0.76				2	G8 IV	280
		Coma Ber 256		12 22 55	+27 09.6	1	11.26		0.57		0.00		3		1148
108144 −40 07288				12 22 55	−40 34.4	2	9.72	.005	0.60	.005	−0.01		6	G2/3 V	1696,2033
108118 −56 05225				12 22 55	−57 03.4	1	8.18		0.13		0.04		2	B9.5V	401
108155 +16 02370				12 22 56	+16 26.9	2	8.21	.005	0.88	.010	0.62		3	K0	280,882
+22 02470				12 22 56	+21 32.6	2	9.35	.009	0.30	.009	0.08		5	F1 III	193,280
		NGP +27 099		12 22 56	+27 22.1	1	10.84		0.69		0.37		1	G8	8038
		NGP +27 100		12 22 56	+27 23.3	2	9.68	.035	1.08	.010	0.98	.015	2	K0 III	8034,8041
		Malm +29 157		12 22 56	+28 58.4	1	12.95		1.41		1.23		3	dM2	685
+31 02363				12 22 56	+30 52.0	1	9.86		0.57				2	F8	77
−61 03161				12 22 56	−62 07.9	1	11.66		0.34		0.29		2	A0	122
		Coma Ber 1454		12 22 58	+27 30.7	1	13.43		1.06				1		913
−61 03162				12 22 58	−62 08.3	1	11.44		0.06		0.04		2	A0	122
		NGC 4372 sq2 9		12 22 58	−72 15.7	1	12.10		1.51				1		518
108176 +10 02421				12 22 59	+10 01.7	1	8.12		1.53		1.86		1	K5	280
108174 +39 02520				12 22 59	+38 37.5	4	7.66	.008	1.12	.011	1.10	.011	9	K2 III	37,105,172,280
238116 +56 01592				12 22 59	+55 47.2	1	8.44		1.24		1.15		2	K0	1502
108147 −63 02270				12 22 59	−63 44.7	1	6.98		0.53				4	F8/G0 V	1075
108177 +2 02538		G 13 - 35		12 23 01	+01 34.0	8	9.66	.009	0.43	.008	−0.23	.012	14	F5 VI	202,792,979,1003,1620*
108175 +38 02333				12 23 01	+37 40.6	1	8.78		0.57				2	G5	77
108187 +19 02565				12 23 02	+19 08.6	1	8.77		1.18		1.21		2	K2	1569
		Coma Ber 258		12 23 02	+27 06.3	1	12.81		0.57		−0.01		2		1148
+27 02127		Coma Ber 115		12 23 02	+27 24.2	1	9.67		1.09		0.99		6	K0 III	1148
108186 +49 02138				12 23 02	+48 38.9	1	7.78		1.03		0.76		2	G9 III-IV	172
+26 02348				12 23 03	+26 23.3	2	10.30	.015	1.00	.005	0.86	.010	2	K0 III	280,8034
		NGC 4349 - 227		12 23 03	−61 29.7	1	12.41		0.36		0.11		3		1146
+22 02471				12 23 04	+22 06.7	2	10.07	.049	1.13	.000	1.10	.019	3	K7	1569,3072
		NGC 4349 - 310		12 23 05	−61 40.8	1	11.87		0.36				1		2041
		NGC 4374 sq1 6		12 23 07	+13 06.6	1	13.46		0.77				1		492
+32 02245				12 23 07	+31 33.2	1	10.12		0.63				2		77
108170 −61 03166		LSS 2674		12 23 07	−61 58.3	6	8.70	.017	0.38	.018	−0.56	.037	13	O9.5/B0	122,191,540,976,1684,1737
108172 −65 01866				12 23 07	−65 26.5	1	9.27		0.05		−0.45		2	B3/5 V	1688
		Coma Ber 1467		12 23 08	+26 03.4	1	12.77		0.97				1		913
		PHI2/1 66		12 23 08	−31 52.6	1	13.32		0.83				3		1719
108201 +24 02458		Coma Ber 116		12 23 09	+23 37.7	1	7.62		1.19		1.21		1	K0	280
−61 03167				12 23 09	−62 03.3	1	9.65		1.07		0.90		2		122
108133 −72 01244				12 23 09	−72 40.1	1	9.50		1.13		0.82		6	G5/8 III/IV	518
		POSS 377 # 35		12 23 10	+24 09.5	1	17.08		2.19				4		1739
		Coma Ber 302		12 23 10	+25 59.5	1	12.63		0.49		0.01		1		1148
−61 03169				12 23 10	−62 13.6	1	11.63		0.29		−0.11		2		122
+45 02032				12 23 11	+45 20.5	1	9.95		0.68				2	G5 III	280
−60 03977		LSS 2675		12 23 11	−61 11.4	2	10.26	.026	0.05	.002	−0.56	.009	4	B5 II	540,976

HD	DM	Other Id	N	Rem	α_{1950}	δ_{1950}	S	V	σ_V	B–V	σ_{B-V}	U–B	σ_{U-B}	n	Spectrum	References
108148	−72 01245				12 23 11	−72 35.1	1	9.89		0.58		0.05		7	G2 V	518
	−61 03171				12 23 16	−62 11.2	1	10.10		1.16		0.92		2		122
		NGC 4372 sq3 6			12 23 16	−72 29.1	1	13.56		1.27		0.77		2		518
		LP 852 - 47			12 23 17	−24 16.6	1	12.82		0.99		0.49		4		3078
108202	−0 02573				12 23 18	−00 50.0	1	10.29		1.18		1.17		1	K7 V	3072
		Coma Ber 287			12 23 19	+26 27.5	1	12.92		0.65		0.18		3		1148
		G 13 - 36			12 23 19	−05 54.9	1	11.41		1.02		0.92		1		1658
108196	−48 07413	IDS12206S4902		A	12 23 19	−49 18.7	1	6.91		0.10		0.08		1	A0 V	1770
		Coma Ber 268			12 23 20	+26 50.1	1	11.89		0.59		0.03		3		1148
108211	+14 02501				12 23 21	+14 02.8	1	10.01		1.11		0.87		1	K0	280
		WLS 1200 15 # 9			12 23 22	+14 28.4	1	14.03		0.70		0.14		2		1375
108227	+26 02349	Coma Ber 117			12 23 22	+26 14.1	2	9.32	.015	0.46	.000	−0.04	.005	6	G5	1148,3016
108226	+27 02129	Coma Ber 118			12 23 22	+27 03.2	5	8.35	.007	0.46	.009	−0.04	.004	28	F6 V	1127,1148,1613,8001,8023
		LOD 26 # 20			12 23 22	−62 14.5	1	12.45		0.47		0.27		2		122
108225	+39 02521	HR 4728			12 23 23	+39 17.7	2	5.01	.010	0.96	.007	0.73		4	G8 III-IV	1080,1363
	−48 07414				12 23 23	−48 34.4	4	10.73	.012	1.13	.011	0.96	.005	16		1279,2013,2033,3073
		V373 Cen			12 23 24	−45 33.0	1	13.91		0.08		−0.80		1		1471
	+8 02599				12 23 26	+08 21.2	2	10.36	.000	1.26	.021	1.26		2	M0	280,1705
	+45 02034				12 23 27	+44 54.2	1	10.94		0.96		0.66		1	G8 III	280
		HA 5 # 12			12 23 28	+74 07.4	1	11.95		0.61		0.15		1		1058
	−61 03172				12 23 28	−61 35.9	1	10.72		0.22		0.01		1	B9	768
108238	+26 02350	Coma Ber 119			12 23 30	+25 57.5	3	8.24	.008	1.10	.004	1.08	.016	6	K0 III	280,1148,8034
	+39 02522				12 23 31	+38 53.7	1	10.20		0.74		0.38		1	G7 III	280
	−59 04222				12 23 31	−59 49.2	1	11.49		0.30		0.13		1	A0	549
		Lod 623 # 30			12 23 31	−59 49.4	1	12.10		0.49		0.33		1	F8	549
108239	+16 02371				12 23 32	+16 08.4	1	6.73		1.11				3	K0	280
		NGC 4349 - 231			12 23 32	−61 32.6	1	13.29		0.26		0.30		2		1146
	+30 02268				12 23 33	+30 24.1	1	9.75		0.44				2	F4	77
108252	+43 02224				12 23 33	+43 23.7	1	8.65		1.16				2	K0 III	280
108230	−31 09692				12 23 33	−32 02.7	2	9.35	.004	−0.15	.004	−0.83	.008	8	B2 III/IV	55,1775
108222	−64 01914				12 23 33	−64 44.7	1	9.73		0.09		−0.33		1	B8/9 II/III	1688
	−59 04223				12 23 34	−59 49.5	1	10.95		0.24		0.09		1	A0	549
	−61 03174				12 23 34	−61 27.2	1	11.30		0.54		0.16		1	F6	768
	−61 03173				12 23 34	−62 09.3	1	11.00		0.25		0.21		2	A0	122
		LP 555 - 78			12 23 35	+05 04.3	1	10.94		0.67		0.12		2		1696
	+27 02130	Coma Ber 120			12 23 35	+27 01.3	3	9.76	.009	0.81	.008	0.31	.013	21	dG7	1148,1613,3016
	+30 02269	Coma Ber 121			12 23 35	+29 30.1	1	9.66		1.01				2	K0 III	77
		CS 16026 # 32			12 23 35	+31 16.8	1	13.59		0.76		0.39		1		1744
		CS 16026 # 30			12 23 36	+31 38.5	1	14.39		0.50		−0.04		1		1744
	+16 02372				12 23 37	+15 36.9	1	10.19		1.04		1.07		1	K2	280
		LOD 26 # 10			12 23 40	−61 48.8	1	11.76		0.16		0.02		2		122
		Coma Ber 1504			12 23 41	+26 38.9	1	12.77		0.92				1		913
	+33 02235				12 23 41	+33 15.3	1	9.92		0.56				2	F8	77
		NGC 4406 sq1 4			12 23 42	+13 17.6	1	13.79		0.66				1		492
	+39 02524				12 23 42	+39 15.1	1	9.02		1.17				4	K1 III	280
	+37 02275				12 23 43	+36 38.1	1	10.24		1.11		0.99		1	K0 III	280
108271	+53 01547				12 23 43	+52 57.2	1	8.30		0.46		−0.03		2	F8	1733
		LOD 26 # 21			12 23 43	−62 13.3	1	12.52		1.29		1.29		2		122
108250	−62 02742	HR 4729		⋆ C	12 23 43	−62 50.7	5	4.85	.009	−0.14	.012	−0.59	.007	20	B4 IV	15,1075,1637,2006,2012
108237	−57 05474				12 23 44	−57 49.9	1	8.39		0.02		−0.40		2	B8 II/III	401
		NGC 4349 - 232			12 23 44	−61 33.4	1	12.48		0.26		−0.09		3		1146
		LOD 26 # 22			12 23 44	−62 12.8	1	12.74		0.80		0.12		2		122
		G 59 - 20			12 23 46	+20 07.5	1	12.29		1.22		1.13		1		1620
		L 255 - 27			12 23 48	−53 16.	2	12.28	.094	0.58	.012	−0.14	.015	5		1696,3065
	−61 03178				12 23 48	−61 32.7	2	10.48	.024	0.07	.019	−0.42	.041	4	B6	768,1146
108248	−62 02745	HR 4730		⋆ AB	12 23 48	−62 49.3	3	0.76	.012	−0.25	.008	−0.96		6	B0.5IV	721,1034,2006
		BPM 7543			12 23 48	−65 56.	1	13.97		0.40		−0.54		3	DA	3065
108257	−50 06975	HR 4732		⋆ A	12 23 49	−51 10.4	6	4.81	.005	−0.14	.005	−0.64	.013	18	B3 V(n)	15,26,1770,2012,8015,8023
	+28 02114	Coma Ber 123			12 23 50	+27 39.9	1	10.10		0.46		−0.01		2	dG0	3016
		HA 5 # 411			12 23 50	+75 06.6	1	12.98		0.51		0.05		1		1058
		NGC 4406 sq1 1			12 23 51	+13 10.5	1	12.84		0.53				3		492
	−61 03179				12 23 51	−61 56.7	1	9.79		0.08		−0.22		2	B6	122
108284	+17 02483				12 23 53	+16 37.9	1	9.22		1.11		1.09		1	K2	280
	−61 03180				12 23 53	−61 28.4	1	11.26		0.40		0.30		1	A0	768
	+23 02459				12 23 54	+22 56.7	1	9.75		1.30		1.51		1	K5	280
		Coma Ber 1516			12 23 54	+25 32.3	1	11.96		0.93				1		913
108283	+28 02115	Coma Ber 125		⋆	12 23 54	+27 32.7	8	4.95	.017	0.27	.010	0.18	.001	17	A9 IVnp	15,379,853,1127,1363*
108285	+13 02534				12 23 55	+12 54.3	1	10.96		0.19		0.09		2	A2	1026
		Coma Ber 290			12 23 55	+26 24.2	1	12.50		0.62		0.04		1		1148
108299	+37 02276				12 23 56	+37 23.4	1	8.20		1.40				3	K2 IV	280
	+45 02035				12 23 56	+44 59.3	1	9.94		0.84				2	G8 III	280
108298	+40 02538				12 23 57	+40 10.0	1	8.52		1.20				3	K1 IV	280
108301	+15 02463				12 23 58	+14 50.9	1	8.04		1.04		0.86		1	K0	280
	−61 03181				12 23 59	−61 37.6	2	10.59	.002	0.15	.002	−0.08	.088	4	B9	768,1146
	+23 02460				12 24 00	+23 23.0	1	9.91		0.99		0.78		1	K0	280
	−75 00813				12 24 00	−75 28.7	2	9.50	.025	1.15	.005	0.88		7	K0	236,2012
108357	+75 00471				12 24 01	+75 18.1	1	10.49		0.71		0.34		1	G5	1058
	+15 02464				12 24 04	+15 23.7	1	10.22		1.00		0.86		1		280
		POSS 377 # 18			12 24 06	+25 12.8	1	16.28		0.85				3		1739
		Coma Ber 1525			12 24 06	+27 03.0	1	13.91		0.98				1		913

Table 1

HD	DM	Other Id	N Rem	α_{1950}	δ_{1950}	S	V	σ_V	B–V	σ_{B-V}	U–B	σ_{U-B}	n	Spectrum	References
108295	−60 03989	Cr 258 - 2		12 24 06	−60 28.6	1	9.41		−0.02		−0.63		4	B2 Vn	251
	+21 02418			12 24 07	+21 13.6	3	10.26	.029	0.07	.012	0.10	.019	4	A0	272,280,1569
	+21 02419			12 24 07	+21 17.2	2	10.88	.005	0.31	.005	−0.05	.019	3		1569,1569
		Lod 623 # 16		12 24 07	−59 48.5	1	11.14		1.18		0.99		1	K3	549
108294	−43 07674			12 24 08	−44 12.6	2	8.28	.001	−0.05	.001	−0.27	.003	6	B8/9 V	1732,1770
108346	+55 01533			12 24 09	+55 26.2	3	7.05	.014	0.10	.000	0.10	.005	8	Am	1501,1502,8071
108309	−48 07426	HR 4734		12 24 09	−48 38.1	5	6.26	.018	0.68	.003	0.22	.018	18	G3/5 V	15,1279,1311,2012,3077
	−59 04226			12 24 11	−59 39.4	1	11.89		0.21		−0.28		1	B8	549
		Lod 623 # 29		12 24 11	−59 48.4	1	11.94		0.24		−0.15		1	B8	549
108296	−65 01871			12 24 11	−65 32.7	1	9.72		0.24		0.12		2	B9.5V	1688
		Coma Ber 215		12 24 12	+24 53.2	1	10.61		1.02		0.81		2	G9 III	1148
	+47 01961		V	12 24 13	+46 44.6	1	11.50		0.36		−0.23		2		1569
108323	−32 08713	HR 4735		12 24 13	−32 33.2	2	5.56	.006	0.01	.003	0.03		7	B9 V	1770,2006
	−61 03184			12 24 13	−61 48.2	1	10.59		0.22		−0.31		2	B3	122
108399	+72 00565	HR 4740		12 24 14	+72 12.4	2	6.27	.035	1.06	.005	0.85	.005	5	K0	1733,3016
		HA 5 # 213		12 24 14	+74 36.3	1	13.31		0.49		0.06		1		1058
	−59 04228			12 24 15	−59 38.7	1	11.39		0.33		0.23		1	A0	549
	−60 03993	Cr 258 - 3		12 24 15	−60 28.7	2	9.77	.041	1.31	.024	1.36	.041	4	K2 III	251,1685
108347	+26 02351	Coma Ber 126		12 24 16	+26 25.6	1	9.22		0.91		0.65		2	G8 III	1569
	+34 02314			12 24 16	+33 52.7	1	10.54		1.04		0.88		1	K0 III	280
		Cr 258 - 26		12 24 16	−60 26.2	1	11.37		1.08		0.80		3		248
		Cr 258 - 18		12 24 16	−60 30.4	1	12.58		0.93		0.59		2		251
		Cr 258 - 36		12 24 16	−60 31.5	1	14.84		0.77		0.26		3		248
108310	−51 06692			12 24 17	−52 06.6	1	9.85		0.30		0.13		1	A9 V	1771
		Cr 258 - 33		12 24 17	−60 32.5	1	13.75		0.59		0.16		3		248
		NGC 4372 sq3 4		12 24 17	−72 30.0	1	12.93		1.00		0.54		5		518
108358	+27 02133	Coma Ber 127	⋆ A	12 24 18	+27 06.2	1	8.80		0.54		0.04		2	G5	3016
		Cr 258 - 24		12 24 18	−60 29.4	2	13.05	.094	0.28	.005	0.21	.005	5		248,251
		Cr 258 - 17		12 24 18	−60 30.5	1	12.71		0.23		0.21		2		251
	−61 03185			12 24 18	−61 39.7	1	11.47		0.15		−0.02		1	B9	768
		Cr 258 - 16		12 24 19	−60 30.6	1	12.47		0.18		0.25		2		251
		Cr 258 - 32		12 24 19	−60 31.8	1	13.75		0.55		0.15		3		248
		Cr 258 - 34		12 24 20	−60 26.2	1	14.02		0.61		0.15		3		248
	−60 03994			12 24 20	−61 24.6	1	10.53		0.15		−0.07		1	B9	768
	−61 03186			12 24 20	−61 24.9	1	10.52		0.16		−0.07		1	B1 V	768
108348	+14 02502			12 24 21	+14 19.3	1	8.88		0.99		0.91		1	K0	280
		NGP +25 067		12 24 21	+25 09.4	1	10.66		0.84		0.42		1	G5 III	8041
		Coma Ber 1537		12 24 21	+26 32.6	1	11.94		1.15				3		913
	+16 02375			12 24 22	+15 45.3	1	9.85		1.14		1.20		1	K2	280
		Cr 258 - 30		12 24 22	−60 28.0	1	13.68		0.37		0.12		3		248
108329	−65 01873			12 24 22	−65 44.6	1	8.53		0.24		−0.30		2	B5 V	1688
	−61 03188			12 24 23	−61 47.8	1	10.27		0.38		0.16		2		122
		NGC 4372 sq3 1		12 24 23	−72 18.7	1	12.30		1.76		2.01		3		518
108380	+49 02139			12 24 24	+48 56.3	2	7.92	.025	1.08	.005	0.89	.025	4	K1 III	172,1733
108420	+74 00496			12 24 24	+74 00.4	1	9.82		0.53		0.00		1	G0	1058
		Cr 258 - 23		12 24 24	−60 29.5	1	12.98		0.31		0.25		2		251
	+9 02625			12 24 25	+08 50.4	1	11.16		0.32		0.01		2	F0	1569
		Cr 258 - 25		12 24 25	−60 30.0	1	13.58		0.48		0.15		2		251
		Cr 258 - 15		12 24 25	−60 31.6	1	12.48		0.18		0.14		2		251
		LOD 26- # 20		12 24 25	−64 05.6	1	13.70		0.76		0.24		2		122
	+2 02541	G 13 - 38		12 24 26	+01 50.8	1	10.48		0.70		0.10		2	F8	333,1620
108330	−65 01874			12 24 26	−65 58.6	1	9.80		0.27		0.08		2	B9 V	1688
		G 148 - 59		12 24 27	+27 17.8	1	14.88		1.73				1	dM4	686
108381	+29 02288	Coma Ber 129	⋆	12 24 27	+28 32.8	9	4.35	.012	1.13	.004	1.15	.002	25	K1 III	15,37,172,379,667*
		Cr 258 - 27		12 24 27	−60 27.2	1	12.13		0.19		0.13		3		248
		Cr 258 - 28		12 24 27	−60 32.6	1	13.02		1.36		1.09		3		248
		Coma Ber 313		12 24 28	+25 52.8	1	11.94		0.46		0.00		2		1148
		Cr 258 - 7		12 24 28	−60 31.3	1	11.62		0.13		0.01		2		251
108382	+27 02134	Coma Ber 130	⋆	12 24 29	+27 06.1	10	5.00	.014	0.08	.008	0.13	.004	25	A3 IV	15,379,853,1022,1068*
108353	−60 03998	Cr 258 - 14	⋆ AB	12 24 29	−60 30.2	3	8.41	.030	0.05	.012	−0.27	.005	10	B9 III	251,401,1075
	−60 03999	Cr 258 - 8		12 24 29	−60 31.2	1	11.76		0.28		0.15		2		251
	−60 03997	Cr 258 - 1		12 24 29	−60 33.6	3	9.04	.030	1.49	.033	1.57	.052	7	K3 II	248,251,1685
		LOD 26 # 7		12 24 29	−61 54.5	1	12.18		0.35		0.00		1		122
		Cr 258 - 11		12 24 30	−60 29.1	1	12.21		0.18		0.15		2		251
	−60 04000	Cr 258 - 12		12 24 30	−60 29.7	1	11.09		0.06		−0.17		4	B8 III	251
		Cr 258 - 13		12 24 30	−60 30.0	1	10.93		0.07		−0.28		2	B7 V	251
108383	+12 02467			12 24 31	+12 05.4	1	9.93		1.16		1.17		1	K2	280
108354	−61 03191			12 24 31	−61 56.1	1	9.56		0.04		−0.30		2	B8 II	122
	−61 03190	LSS 2678		12 24 31	−62 15.5	1	11.00		0.35		−0.53		3		1684
108374	−55 05063	IDS12218S5511	AB	12 24 32	−55 27.5	3	7.14	.009	0.00	.001	−0.31	.002	9	B8/9 III	243,401,1770
	−60 04002	Cr 258 - 9		12 24 33	−60 30.8	1	12.08		0.20		0.10		2		251
		CS 16026 # 33		12 24 34	+29 38.3	1	14.48		0.55		0.01		1		1744
108342	−71 01327			12 24 34	−72 10.7	1	8.90		0.54		0.05		9	F8 V	518
		POSS 377 # 14		12 24 35	+20 15.1	1	16.01		1.58				3		1739
108370	−49 07103			12 24 35	−50 04.5	1	8.93		0.21		0.13		1	A3 V	1771
	−60 04004	Cr 258 - 4		12 24 35	−60 28.3	1	11.43		0.12		−0.01		2		251
		Coma Ber 132		12 24 36	+27 07.4	3	9.87	.020	0.68	.005	0.18	.004	4	G5 V	1127,8001,8023
108376	−57 05487			12 24 36	−57 44.2	1	8.87		0.10		−0.46		2	B7 Ib/II	401
		Cr 258 - 22		12 24 36	−60 30.6	1	13.22		0.40		0.35		2		251
		Cr 258 - 10		12 24 36	−60 31.1	1	11.51		1.23		1.06		2		251

HD	DM	Other Id	N Rem	α_{1950}	δ_{1950}	S	V	σ_V	B–V	σ_{B-V}	U–B	σ_{U-B}	n	Spectrum	References
108355	−63 02283	HR 4736		12 24 36	−63 30.7	5	6.01	.015	0.08	.013	−0.29	.005	18	B8 IV	15,1075,1688,1771,2012
108343	−74 00917			12 24 36	−75 11.8	2	8.03	.003	0.24	.008	0.08		6	A8 IV/V	278,2012
	+21 02421			12 24 37	+21 14.7	1	9.78		0.99		0.84		2	K2	1569
	−60 04005	Cr 258 - 5		12 24 38	−60 29.2	1	10.48		0.09		−0.15		2	B8 V	251
		Cr 258 - 6		12 24 39	−60 29.9	1	12.52		0.38		0.33		2		251
108373	−53 05146			12 24 40	−54 11.6	1	9.04		0.29		0.15		2	A7 IV/V	1771
108436	+70 00698			12 24 41	+70 00.5	1	8.46		0.64		0.09		2	G0	1733
	−63 02285			12 24 42	−64 02.7	1	11.21		0.26		−0.04		2		122
		Coma Ber 217		12 24 43	+26 43.2	1	11.08		0.44		−0.02		2	F5	1148
		Coma Ber 267		12 24 43	+26 51.3	1	12.90		0.52		−0.09		1		1148
108396	−58 04289	HR 4739, BL Cru		12 24 43	−58 42.9	3	5.48	.065	1.54	.011	1.51	.047	8	M4/5 III	803,2007,3055
108421	+27 02135	Coma Ber 133	⋆AB	12 24 44	+27 18.3	3	8.26	.029	1.02	.010	0.91	.020	4	K2 V	1569,8001,8023
108408	+37 02278			12 24 44	+36 39.1	3	7.66	.012	0.16	.010	0.15	.024	7	Am	77,1068,8071
108401	−32 08720			12 24 44	−32 30.0	1	8.27		1.29		1.31		6	K1 III	1673
	+25 02506	Coma Ber 134		12 24 45	+25 09.8	3	10.09	.008	0.79	.015	0.39	.010	6	G6 III	1148,3016,8041
	+15 02465			12 24 46	+14 44.0	1	10.26		1.44		1.63		1		280
		NGP +25 068		12 24 46	+25 16.4	1	10.56		0.98		0.62		1	G9 III	8041
108406	−55 05067			12 24 46	−55 32.8	2	8.78	.002	−0.03	.005	−0.40	.004	3	B8 V	401,1770
	−60 04007	Cr 258 - 19		12 24 46	−60 27.9	2	11.24	.010	0.19	.020	−0.25	.005	5		248,251
	−61 03194			12 24 46	−61 27.3	1	11.31		0.14		−0.04		1	B9	768
	+20 02721			12 24 47	+20 21.7	1	9.55		1.07		0.93		2		1569
		Lod 623 # 33		12 24 47	−59 52.7	1	12.34		0.30		0.22		1	A0	549
108398	−63 02288			12 24 47	−63 56.8	1	9.22		0.12		−0.03		2	B9/A0III/IV	1688
		NGC 4431 sq1 8		12 24 48	+12 39.0	2	12.72	.061	0.82	.084	0.30		4		492,1531
108405	−47 07625			12 24 48	−47 51.1	4	9.28	.008	0.58	.007	−0.06	.000	10	G3 V	1594,1696,2033,6006
	−59 04231			12 24 49	−59 53.0	1	10.80		0.59		0.08		1	F8	549
108397	−62 02760			12 24 49	−63 17.6	1	9.80		0.23		0.18		2	A0/3 V	1688
	+22 02474			12 24 50	+22 16.9	1	10.35		0.67		0.28		1		280
	+45 02037			12 24 50	+44 31.8	1	10.20		0.62				2		280
		Cr 258 - 35		12 24 50	−60 30.1	1	14.64		0.68		0.30		3		248
	+27 02136	Coma Ber 135		12 24 51	+27 26.8	3	10.11	.016	0.92	.008	0.60	.009	3	G8 III	280,8001,8023
108416	−55 05069			12 24 51	−56 02.0	1	9.36		0.40		0.05		1	A8/F0 V	1771
108418	−55 05069			12 24 51	−56 02.0	1	9.36		0.27		−0.27		2	A8/F0 V	1688
		Cr 258 - 20		12 24 52	−60 28.4	1	12.34		0.51		0.12		2		251
		NGC 4431 sq1 1		12 24 54	+12 28.0	2	10.32	.025	0.68	.015	0.23	.005	13		492,1531
		NGC 4431 sq1 11		12 24 54	+12 29.8	2	14.44	.032	0.69	.036	0.19		5		492,1531
		Coma Ber 1571		12 24 55	+26 30.3	1	13.29		1.53				2		913
108414	−49 07109			12 24 55	−50 20.2	1	9.98		0.27		0.13		1	A8 V	1771
		Cr 258 - 29		12 24 55	−60 30.6	1	13.03		0.31		0.29		3		248
	−60 04010			12 24 55	−61 16.4	1	11.18		0.31		−0.34		1	B6	768
108440	+8 02604			12 24 56	+08 17.4	1	8.06		1.12		1.13		1	K2	280
		Coma Ber 379		12 24 56	+25 03.4	1	11.65		1.26		1.24		2	K2 V	1148
		CS 16026 # 35		12 24 56	+28 01.5	1	13.78		0.36		−0.11		1		1744
	+44 02193			12 24 56	+43 46.1	1	10.49		1.00		0.55		2	G0	1569
108449	+25 02507	Coma Ber 136		12 24 57	+24 56.0	1	8.24		0.30		0.05		2	F8 V	3016
	+38 02337			12 24 57	+37 31.4	1	10.65		0.40				2		280
108417	−63 02292			12 24 57	−63 36.5	1	8.88		0.25		0.16		2	A1 V	1688
		AAS12,381 11 # A		12 24 59	+11 12.2	1	10.43		0.89		0.57		1		280
108453	+11 02465			12 24 59	+11 18.2	1	9.23		1.12		1.00		1	K2	280
		Cr 258 - 31		12 24 59	−60 30.1	1	13.69		2.27		1.79		3		248
		NGC 4431 sq1 12		12 25 00	+12 40.2	2	14.60	.009	0.60	.023	−0.01		4		492,1531
108452	+12 02470			12 25 01	+11 33.4	2	10.77	.066	0.14	.005	0.08	.005	4	Am	272,280
		NGC 4431 sq1 7		12 25 01	+12 40.9	2	11.91	.009	0.50	.007	−0.02	.007	6		492,1531
	−60 04011			12 25 01	−61 16.8	1	10.39		1.03		0.73		1	K0	768
108468	+18 02611			12 25 02	+18 06.7	1	7.44		0.96				3	G5 III	280
		PHI2/2 97		12 25 02	−30 48.4	1	12.95		0.93				3		1719
108434	−61 03198	LSS 2680		12 25 02	−61 45.2	3	8.93	.007	0.17	.025	−0.49	.012	7	B2/3 II/III	122,540,976
	−63 02293			12 25 02	−64 02.5	1	10.47		0.19		−0.05		2		122
		LP 435 - 136		12 25 03	+15 28.0	1	11.73		0.57		−0.03		2		1696
		Coma Ber 288		12 25 03	+26 23.1	1	11.92		0.57		−0.04		1		1148
108466	+28 02116	Coma Ber 137		12 25 03	+28 23.2	3	7.04	.021	1.21	.013	1.24	.005	7	K1 III	172,280,8034
	−61 03199			12 25 03	−61 46.0	1	9.78		0.14		−0.44		2		122
	−60 04012	Cr 258 - 21		12 25 04	−60 28.4	2	9.89	.006	0.02	.015	−0.33	.005	5	B9 Ib	248,251
108467	+25 02508	Coma Ber 138		12 25 05	+24 30.2	1	8.79		1.06				2	G8 III	280
	−61 03200			12 25 05	−61 28.4	1	11.23		0.31		0.23		1	A3	768
108465	+42 02307			12 25 06	+41 37.9	1	6.54		0.52		0.29		2	A3	1569
	+46 01787			12 25 06	+45 53.1	1	10.28		1.10		0.96		1	K0 III	280
	−14 03519			12 25 06	−15 29.2	1	10.67		1.13		1.12		1	K5 V	3072
108469	+13 02536			12 25 07	+12 29.8	2	9.19	.025	0.63	.010	0.11	.008	15	G5	492,1531
		NGC 4431 sq1 4		12 25 07	+12 36.1	2	12.67	.018	0.87	.007	0.59	.030	4		492,1531
		HA 5 # 285		12 25 07	+74 45.8	1	12.85		0.53		−0.06		2		1058
108486	+26 02352	Coma Ber 139		12 25 08	+26 11.3	5	6.69	.024	0.17	.009	0.09	.007	7	A5 m	346,1127,1148,8001,8023
		PHI2/1 52		12 25 08	−31 58.6	1	13.06		0.87				2		1719
108473	+2 02542	3C273 # 10		12 25 09	+02 06.8	1	9.23		0.98		0.69		3	G5	1687
108471	+9 02628	HR 4741		12 25 09	+08 53.2	3	6.37	.013	0.94	.010	0.68	.000	9	G8 III	15,1417,3016
108470	+10 02426			12 25 09	+09 37.5	1	9.85		1.36		1.40		1	K2	280
		NGC 4431 sq1 3		12 25 09	+12 31.0	2	13.36	.016	0.54	.006	0.01	.010	3		492,1531
	−59 04233			12 25 10	−60 00.0	1	11.37		0.32		0.26		1	A3	549
		Lod 624 # 45		12 25 11	−61 29.6	1	12.48		0.62		0.33		1	A8	768
108502	+56 01598	HR 4745		12 25 13	+55 59.4	2	5.70	.000	1.55	.007	1.92		5	M2 IIIb	70,1501

Table 1   675

HD	DM	Other Id	N Rem	α_{1950}	δ_{1950}	S	V	σ_V	B–V	σ_{B-V}	U–B	σ_{U-B}	n	Spectrum	References
108477	−15 03471	HR 4742		12 25 13	−16 21.3	2	6.31	.029	0.86	.019	0.50		6	G5 II	252,2007
		Coma Ber 316		12 25 15	+25 48.6	1	12.61		1.50		1.33		1		1148
	+27 02137	Coma Ber 218		12 25 15	+26 43.1	1	11.30		0.46		-0.05		2	dG5	1148
	−59 04234	NGC 4439 - 15		12 25 15	−59 50.0	2	11.28	.015	0.46	.034	0.21	.005	3	A8	251,549
108344	−83 00461			12 25 16	−83 31.5	1	6.66		0.04		-0.29		6	B8 V	1628
108519	+28 02118	Coma Ber 140		12 25 17	+27 42.0	4	7.71	.015	0.27	.005	0.07	.016	6	A9 V	1375,3016,8001,8023
108506	−3 03298	HR 4746, FT Vir		12 25 17	−04 20.3	3	6.21	.004	0.44	.005	0.09	.020	10	F4 III-II	15,1256,3026
		G 255 - 11		12 25 18	+75 40.7	1	10.56		1.20				1		906
108485	−59 04236	IDS12226S5929	B	12 25 18	−59 45.2	1	9.94		1.16		1.02		1		549
	+29 02290	Coma Ber 141		12 25 19	+28 28.3	1	9.64		0.68		0.16		3	dK0	3016
	−60 04018			12 25 19	−61 18.3	1	10.13		0.18		0.20		1	B9	768
	−61 03201			12 25 19	−61 54.0	1	11.21		0.18		0.02		2	B9	122
108447	−69 01657	IDS12224S6959	AB	12 25 19	−70 16.2	1	7.64		1.44				4	K3 III	1075
		3C 273 # 11		12 25 20	+02 08.8	1	10.45		1.48		1.77		4		1687
108483	−49 07115	HR 4743		12 25 20	−49 57.2	7	3.91	.006	-0.20	.008	-0.78	.016	29	B2 V	15,26,1034,1075,1770*
108485	−59 04237	LSS 2681	★ A	12 25 20	−59 45.7	2	8.03	.019	0.38	.005	0.28	.042	4	A2 II	549,8100
		NGC 4439 - 22		12 25 20	−59 46.8	1	13.70		0.48		0.29		3		248
		G 12 - 37		12 25 21	+05 29.1	1	13.76		1.39				2		202
108505	+9 02629			12 25 21	+09 10.5	1	8.90		0.90		0.63		2	G5	1569
	+48 02022			12 25 21	+47 46.2	1	10.19		1.08		0.97		2	K0 III	1375
108503	+33 02238			12 25 22	+32 45.1	1	9.11		0.74				2	G7 IV	77
		Coma Ber 219		12 25 23	+25 22.3	2	11.19	.010	0.82	.000	0.39	.000	2	G9 V	8001,8023
	+47 01963			12 25 24	+47 05.5	1	9.29		0.68		0.19		2	G0	1601
		NGC 4439 - 16		12 25 24	−59 49.1	2	12.75	.109	0.49	.015	0.28	.055	5		248,251
	−60 04021			12 25 24	−61 23.5	1	10.98		0.40		0.31		1	A3	768
	−61 03203			12 25 24	−61 54.4	1	10.61		0.24		-0.39		2		122
	+19 02570	TV Com		12 25 27	+19 04.7	1	9.05		1.85		2.04		1	M2	280
		CS 16026 # 39		12 25 27	+31 48.8	1	14.79		0.44		-0.20		1		1744
108500	−61 03204	IDS12227S6112	ABC	12 25 28	−61 29.2	1	6.82		0.67		0.18		1	G3 V	768
		G 60 - 6		12 25 29	+07 43.9	2	11.07	.005	0.92	.020	0.69	.005	2	K2	1620,1658
108523	−14 03521			12 25 29	−15 22.4	1	8.30		0.72				4	G6 V	2033
	−61 03205			12 25 29	−61 25.0	1	11.21		0.10		-0.04		1	B8	768
	−61 03206			12 25 29	−61 31.9	1	9.49		0.01		-0.48		1	B7	768
108501	−63 02297	HR 4744		12 25 29	−64 03.9	9	6.04	.014	0.02	.010	-0.04	.010	35	A0 V	15,122,1075,1075,1688,1771*
108534	+36 02269			12 25 32	+35 38.8	2	9.36		1.08	.080	0.88		4	K0 III	77,1569
		PHI2/2 111		12 25 32	−30 34.0	1	12.31		0.86				3		1719
		PHI2/1 49		12 25 33	−31 26.8	1	12.59		1.07				1		1719
		NGC 4439 - 11		12 25 33	−59 50.8	1	12.93		0.25		-0.10		2		251
108547	+19 02571			12 25 34	+18 36.4	1	8.58		1.26		1.33		1	K2	280
	+23 02463	Coma Ber 142		12 25 34	+23 25.1	1	9.96		0.65		0.13		1	G5	280
108545	+30 02273		AB	12 25 34	+30 05.6	2	9.12	.005	0.90	.015	0.58		4	G8 III	77,1569
		NGC 4439 - 18		12 25 34	−59 48.7	1	13.64		1.45		1.46		2		251
	−59 04238	NGC 4439 - 10		12 25 34	−59 50.9	2	11.14	.015	1.13	.010	0.87	.029	3		251,549
108546	+21 02422			12 25 35	+20 37.8	1	7.80		1.28		1.38		1	K0	280
		POSS 377 # 36		12 25 35	+23 39.2	1	17.13		1.62				4		1739
		Coma Ber 257		12 25 35	+27 03.1	1	12.25		0.55		-0.01		1		1148
	−59 04239	NGC 4439 - 1		12 25 35	−59 47.9	3	10.74	.017	0.15	.027	-0.40	.004	6	B2 III	251,287,549
		NGC 4439 - 17		12 25 35	−59 49.3	1	12.95		1.40		1.47		2		251
		HA 5 # 493		12 25 36	+75 13.5	1	11.91		0.68		0.25		1		1058
	+19 02572			12 25 37	+19 18.6	1	10.67		1.02		1.10		1		280
		G 255 - 12		12 25 37	+75 40.2	2	15.73	.041	1.72	.023			5		906,1663
108530	−61 03209	HR 4747		12 25 37	−61 31.1	7	6.22	.012	1.25	.005	1.30	.012	23	K2 III	15,285,768,1075,1637*
108531				12 25 37	−61 47.9	2	8.22	.001	-0.01	.042	-0.25	.003	4	B8 V	122,1688
	−60 04027			12 25 38	−61 20.9	1	11.04		0.16		0.08		1	A0	768
108548	+16 02377			12 25 39	+15 53.7	1	8.60		0.42		-0.05		2	F8	1569
108574	+45 02038	IDS12232N4521	A	12 25 39	+45 04.2	1	7.50		0.58		0.08		2	F5	3026
	−59 04240	NGC 4439 - 12		12 25 39	−59 49.7	3	11.54	.106	0.22	.012	-0.33	.009	7	B5 V:	248,251,549
		NGC 4439 - 9		12 25 39	−59 51.0	1	12.79		0.21		-0.07		2		251
108559	+16 02378			12 25 40	+15 31.7	1	8.80		1.12		1.13		1	K0	280
	+42 02308			12 25 40	+41 43.3	1	9.91		0.69				3	G5	280
108575	+45 02038	IDS12232N4521	B	12 25 40	+45 04.1	1	8.12		0.68		0.20		2		3026
108528	−59 04241			12 25 40	−60 06.7	1	7.74		0.09		0.11		1	A1 V	1771
		L 68 - 28	★ A	12 25 40	−71 12.2	2	13.70	.033	1.59	.002	1.11		4		1705,3078
		Coma Ber 266		12 25 41	+26 47.8	1	12.42		0.44		-0.10		1		1148
	−59 04242	NGC 4439 - 8		12 25 41	−59 50.8	2	11.64	.005	0.19	.024	-0.35	.010	3	B7 V:	251,549
		HA 5 # 417		12 25 42	+75 05.4	1			0.80		0.49		1		1058
108541	−38 07753	HR 4748		12 25 42	−38 45.9	2	5.45	.006	-0.08	.003	-0.21		8	B8 V	1770,2006
		NGC 4439 - 19		12 25 42	−59 49.1	1	14.78		1.72		1.02		3		251
	−59 04243	NGC 4439 - 13		12 25 42	−59 49.8	2	12.16	.018	0.23	.018	-0.21	.005	5		251,549
		NGC 4439 - 20		12 25 42	−59 51.5	1	15.57		0.62		0.40		3		251
	−60 04029			12 25 42	−61 19.9	1	11.25		0.32		0.28		1	A3	768
108560	+12 02474			12 25 43	+12 06.2	1	7.84		1.36		1.66		1	K0	280
	+37 02279	IDS12232N3714	AB	12 25 43	+36 57.5	1	9.42		0.46				2	F5	77
	−59 04244	Ho 14 - 7		12 25 43	−59 34.2	1	11.34		0.31		0.17		1		549
		L 68 - 27	★ B	12 25 43	−71 12.3	1	13.72		1.59		1.11		3		3078
108576	+25 02510	Coma Ber 143		12 25 44	+25 18.6	1	8.89		0.56		0.00		2	G5	3016
	+41 02296			12 25 44	+41 02.5	1	10.56		0.84		0.58		1	G8 IV	280
108544	−60 04031			12 25 44	−60 37.1	1	8.94		1.75		1.97		4	K0 III	322
108577	+13 02539			12 25 45	+12 37.3	3	9.60	.016	0.69	.013	0.13	.012	12	G5	492,1531,1569
		NGC 4431 sq1 6		12 25 46	+12 39.2	2	12.53	.019	0.44	.050	-0.19	.000	5		492,1531

HD	DM	Other Id	N Rem	α_{1950}	δ_{1950}	S	V	σ_V	B–V	σ_{B-V}	U–B	σ_{U-B}	n	Spectrum	References
108564	−16 03469			12 25 46	−16 38.1	4	9.46	.015	0.98	.020	0.67	.021	9	K2 V	158,1064,1696,3072
	−59 04245	NGC 4439 - 2	★ AB	12 25 46	−59 48.4	4	10.25	.077	0.17	.010	-0.34	.013	12	B2 n	248,251,287,549
		NGC 4439 - 14		12 25 47	−59 49.6	1	12.76		0.50		0.18		2		251
		Lod 624 # 38		12 25 47	−61 41.3	1	11.66		0.34		0.22		1	A3	768
	+38 02339			12 25 48	+38 17.5	1	10.56		0.87		0.55		1	G8 III	280
108554	−55 05082			12 25 48	−56 10.8	1	7.92		0.15		0.09		2	A0/1 V	1771
		Ho 14 - 10		12 25 48	−59 32.	1	13.32		0.52		0.23		2		251
		Ho 14 - 11		12 25 48	−59 32.	1	14.13		0.46		0.31		2		251
		Lod 623 # 32		12 25 48	−59 49.9	1	12.31		0.25		-0.14		1	B8	549
		NGC 4439 - 7		12 25 48	−59 50.5	2	12.73	.058	0.27	.058	-0.04	.000	3	B9	251,549
	−61 03211			12 25 48	−61 46.4	1	11.37		0.26		-0.48		1	B3	768
		Coma Ber 337		12 25 49	+25 34.9	1	15.58		1.36		0.53		2		1148
		HA 5 # 149		12 25 49	+74 27.2	1	12.02		0.31		0.11		1		1058
108570	−55 05084	HR 4749	★ A	12 25 49	−56 07.7	6	6.15	.007	0.92	.000	0.64	.009	19	K0/1 III	15,243,1075,1311,2007,3077
	−59 04246	NGC 4439 - 3		12 25 49	−59 49.0	5	10.43	.077	0.15	.020	-0.33	.023	10	B3 IV	248,251,287,401,549
		NGC 4439 - 4		12 25 49	−59 49.4	1	12.32		1.13		0.86		2		251
		NGC 4439 - 5		12 25 49	−59 49.9	1	12.32		0.22		-0.14		2		251
	−59 04247	NGC 4439 - 6		12 25 49	−59 50.2	2	11.03	.009	0.76	.014	0.18	.019	4		251,549
	−61 03212	LSS 2684		12 25 49	−61 34.6	4	9.44	.017	0.21	.014	-0.55	.018	11	B2.5II	540,768,976,1684
		3C 273 # 2		12 25 50	+02 16.8	3	12.24	.014	0.44	.004	-0.04	.014	6		327,1463,1687
		Coma Ber 314		12 25 51	+25 52.4	1	12.84		0.62		0.02		1		1148
	−59 04248	Ho 14 - 1		12 25 51	−59 31.4	1	10.49		0.36		0.22		1		549
	+45 02039			12 25 52	+45 24.3	1	9.30		0.73				3	G5 III-IV	280
108593	+12 02475			12 25 53	+12 19.0	1	9.76		0.82		0.57		1	K0	280
	+42 02309			12 25 55	+41 55.9	2	10.78	.016	0.02	.004	0.12	.016	4	A0	275,1026
108581	−17 03632			12 25 55	−18 01.1	3	9.23	.024	1.23	.017	1.15		7	K4 V	1705,2033,3072
	+30 02276			12 25 56	+30 10.5	1	10.67		0.87		0.52		1	G8 III	8034
	+37 02281			12 25 56	+37 27.8	1	10.22		0.88				2	G9 IV	280
	+74 00497			12 25 56	+74 14.8	1	10.70		0.68		0.18		1		1058
		Lod 623 # 36		12 25 56	−59 31.8	1	12.75		0.33		0.20		1		549
	+32 02247			12 25 57	+32 12.5	1	9.94		1.06		0.92		1	K0 III	280
108612	+35 02342			12 25 57	+34 54.1	2	8.46		1.13	.050	1.03		4	K1 III	77,1569
		NGC 4439 - 24		12 25 57	−59 50.1	1	14.53		0.55		0.30		3		248
		Lod 623 # 34		12 25 58	−59 30.8	1	12.63		0.38		0.27		1	A3	549
		NGC 4439 - 21		12 25 58	−59 49.8	1	13.47		1.67		1.40		3		248
		LSS 2686		12 25 58	−61 33.1	1	11.37		0.28		-0.47		4	B3 Ib	1684
	+30 02277	IDS12235N3026	AB	12 25 59	+30 09.3	2	10.06	.019	0.72	.014	0.31		4	G7 III	280,8034
	+38 02340			12 25 59	+37 38.7	1	10.27		1.17		1.24		1	K1 III	280
		Coma Ber 338		12 26 00	+25 36.4	1	12.55		0.44		-0.20		1		1148
	+42 02310			12 26 00	+42 21.0	1	10.37		1.55		1.75		1	M2 III	280
		NGC 4439 - 23		12 26 00	−59 50.3	1	14.48		0.84		0.61		3		248
108591	−60 04035			12 26 00	−60 38.5	1	9.34		1.09		0.68		1	G1/3 III/IV	401
	−60 04034			12 26 00	−61 11.6	1	10.26		0.19		0.19		1	Ap	768
	+32 02248			12 26 01	+31 57.0	1	9.94		0.78				2	G9 III	280
		BONN82 T13# 13		12 26 01	−60 37.4	1	13.01		0.74		0.30		3	G7	322
		Lod 624 # 39		12 26 01	−61 41.4	1	11.67		0.16		0.11		1	A0	768
108614	+12 02477			12 26 02	+12 23.6	1	7.95		1.02		0.87		1	K0	280
108613	+14 02505			12 26 02	+14 13.6	1	9.70		1.53		1.85		1	K5	280
	−61 03216			12 26 02	−61 30.8	1	11.91		0.20		-0.41		1	B5	768
108599	−15 03475			12 26 03	−15 43.8	1	7.67		1.51		1.87		1	K5 III	1746
		Lod 624 # 44		12 26 03	−61 30.9	1	12.11		0.31		0.07		1	B8	768
108558	−74 00919			12 26 03	−75 20.6	1	8.01		1.52				4	K3/4 III	2012
		3C 273 # 9		12 26 04	+02 13.9	1	13.84		1.41		1.13		5		1687
108629	+30 02279			12 26 04	+29 52.0	1	8.58		1.66				2	M0 III	77
	+17 02488			12 26 05	+17 02.7	1	8.97		1.53		1.67		1	M0	280
		3C 273 # 1		12 26 06	+02 34.7	2	11.86	.013	0.90	.144	0.42	.166	6		327,1595
108610	−61 03218	IDS12233S6119	A	12 26 06	−61 35.7	4	6.93	.006	-0.05	.007	-0.55	.008	9	B3 IV/V	158,401,768,1688
		NGC 4472 sq1 13		12 26 07	+08 23.6	1	11.53		0.91				1		492
108604	−51 06712			12 26 07	−52 02.0	1	9.50		0.32		0.07		1	A9 V	1771
		Lod 623 # 27		12 26 07	−59 36.9	1	11.65		0.57		0.03		1	F9	549
108642	+27 02138	Coma Ber 144	★	12 26 08	+26 30.2	5	6.53	.009	0.18	.007	0.11	.009	14	A7 m	15,1127,1148,8001,8015
		LOD 26+ # 16		12 26 08	−59 12.6	1	11.03		0.44		0.33		2		122
		LOD 26+ # 17		12 26 08	−59 14.4	1	11.70		0.49				2		122
		BONN82 T13# 11		12 26 08	−60 36.9	1	12.30		1.43		1.32		3	M0	322
108608	−60 04037			12 26 08	−60 39.2	2	9.40	.009	-0.05	.012	-0.47	.015	3	B9 III/IV	401,1688
108609	−60 04038			12 26 09	−60 49.7	1	9.81		0.13		-0.53		1	B3 III/V	1688
	−61 03219			12 26 10	−61 34.1	1	11.27		0.22		0.23		1	A0	768
		NGC 4458 sq1 9		12 26 11	+13 29.8	1	13.78		0.61		0.14		1		492
		NGC 4458 sq1 8		12 26 11	+13 30.2	1	14.78		0.59		0.09		1		492
		Coma Ber 339		12 26 11	+25 35.6	1	14.77		1.34		0.84		1		1148
108603	−51 06713	IDS12235S5138	A	12 26 11	−51 54.1	1	8.07		0.25		0.12		1	A4/5 V	1771
		3C 273 # 3		12 26 12	+02 18.8	3	11.87	.020	1.00	.004	0.56	.015	9		327,1463,1687
		NGC 4458 sq1 2		12 26 13	+13 22.6	2	13.39	.010	1.51	.022	1.27	.107	3		492,1531
108627	−60 04042			12 26 13	−60 59.4	1	9.55		0.14		0.09		1	A0 V	1688
		HZ 26		12 26 14	+28 37.2	1	13.90		-0.08		-0.29		1	B5	1744
	+33 02240			12 26 14	+33 12.0	1	9.87		0.39				2	F4	77
	−6 03580	G 13 - 42		12 26 14	−07 14.6	2	10.86	.010	1.12	.015	1.09	.005	2	K5	1620,1658
108651	+26 02353	Coma Ber 145	★ BC	12 26 15	+26 10.6	6	6.64	.014	0.22	.006	0.08	.006	30	Am	15,346,1084,1127,8001,8015
	+13 02540			12 26 16	+13 21.9	2	9.41	.021	0.50	.005	0.05	.000	14	F8	492,1531
108625	−49 07123			12 26 16	−50 22.3	1	7.80		0.16		0.14		2	A4 V	1771

Table 1   677

HD	DM	Other Id	N Rem	α_{1950}	δ_{1950}	S	V	σ_V	B–V	σ_{B-V}	U–B	σ_{U-B}	n	Spectrum	References
	−59 04250			12 26 16	−59 46.7	1	10.33		1.13		0.85		1	K2	549
		Ross 948		12 26 17	−10 23.2	1	10.96		1.51				1	M2	1746
	−60 04044			12 26 17	−60 40.2	1	9.34		1.06		0.71		7	gG8	322
		3C 273 # 4		12 26 18	+02 23.1	4	12.65	.023	0.53	.023	-0.02	.015	12		327,1463,1595,1687
	+22 02476			12 26 18	+22 16.8	1	9.13		0.97		0.75		1	G5	280
		HA 5 # 153		12 26 18	+74 30.8	1	12.36		0.38		0.03		1		1058
		HA 5 # 220		12 26 18	+74 34.9	1	12.32		0.61		0.03		1		1058
		BONN82 T13# 23		12 26 18	−60 35.0	1	14.64		1.55				3		322
		Coma Ber 289		12 26 19	+26 27.5	1	13.83		0.32		-0.13		2		1148
		Coma Ber 315		12 26 20	+25 51.5	1	12.96		0.52		0.01		1		1148
	+45 02040			12 26 20	+44 29.3	1	9.69		1.18				2	G9 III	280
108639	−60 04047	LSS 2687		12 26 21	−60 31.7	7	7.81	.008	0.08	.011	-0.82	.015	23	B1 III	322,401,540,976,1075*
	+10 02429			12 26 22	+10 15.9	1	10.67		1.11		1.17		1	K0	900
		BONN82 T13# 10		12 26 22	−60 37.6	1	12.21		0.27		0.24		3	A5	322
		G 198 - 54		12 26 23	+49 52.1	1	15.37		1.53				1		906
		HA 5 # 354		12 26 23	+74 59.2	1	11.56		0.80		0.50		1		1058
		HA 5 # 154		12 26 24	+74 31.6	1	12.04		0.66		0.15		1		1058
108662	+26 02354	Coma Ber 146	⋆ A	12 26 25	+26 11.4	9	5.29	.014	-0.06	.009	-0.12	.009	61	A0p	15,1022,1068,1084*
108661	+38 02341			12 26 25	+38 20.6	1	9.46		0.43				2	F6 V	77
		LP 735 - 13		12 26 25	−15 07.3	1	13.69		1.56		1.20		1		1696
		3C 273 # 6		12 26 26	+02 13.4	2	12.92	.013	1.10	.004	0.76	.009	6		327,1687
	+33 02242			12 26 26	+33 12.0	2	8.94	.015	0.99	.005	0.76		3	G9 III	77,8034
	+33 02241			12 26 26	+33 27.0	2	9.56	.005	1.23	.015	1.33		3	K2 III	77,8034
		G 12 - 38		12 26 27	+08 42.3	1	12.06		1.60				1	M2:	1705
108675	+29 02292	Coma Ber 147		12 26 27	+29 09.8	2	8.59	.065	0.55	.050	0.03		4	F6 IV-V	77,3026
	−60 04048			12 26 27	−61 12.9	1	10.74		0.09		-0.19		1	B8	768
		NGC 4472 sq1 11		12 26 28	+08 15.2	1	11.84		0.56				4		492
	+27 02139	Coma Ber 220		12 26 28	+26 49.1	1	10.78		0.89		0.54		3	G9 IV	1148
		WLS 1224 50 # 5		12 26 28	+52 20.9	1	13.86		0.94		0.67		2		1375
		3C 273 # 80		12 26 29	+02 18.8	1	14.92		1.11		0.95		4		1687
		3C 273 # 7		12 26 30	+02 19.9	4	13.53	.014	0.61	.019	0.09	.006	41		327,1463,1595,1687
		NGC 4472 sq1 12		12 26 30	+08 18.4	1	12.02		1.04				2		492
108674	+38 02342			12 26 30	+37 52.3	1	9.10		0.38				2	F4 V	77
108658	−56 05255			12 26 31	−56 44.3	1	8.38		0.05		-0.37		2	B7 III	401
		NGC 4458 sq1 3		12 26 32	+13 21.1	2	12.98	.000	0.61	.009	0.10	.005	4		492,1531
108676	+14 02507			12 26 34	+14 13.6	1	8.72		1.63		2.05		1	K5	280
		Coma Ber 1655		12 26 34	+28 11.5	1	13.18		1.48				2		913
		3C 273 # 5		12 26 35	+02 16.9	4	12.70	.017	0.67	.016	0.09	.015	14		327,1463,1595,1687
		NGC 4458 sq1 5		12 26 35	+13 19.2	2	11.58	.002	0.90	.025	0.53	.008	8		492,1531
		NGC 4458 sq1 6		12 26 35	+13 31.6	2	10.88	.010	0.63	.010	0.07	.027	10		492,1531
		SSII 156		12 26 36	+32 34.	1	11.62		0.05		0.09		1	A2	275
	+37 02282			12 26 36	+36 41.4	1	9.90		0.64				2		77
108680	−1 02674	FZ Vir		12 26 36	−02 09.2	1	6.90		1.63		1.86		10	M1	3040
		3C 273 # 8		12 26 37	+02 11.3	2	14.62	.416	1.16	.082	1.06	.101	16		327,1687
		NGC 4458 sq1 4		12 26 37	+13 20.4	2	13.87	.029	0.63	.040	0.22	.049	3		492,1531
		PHI2/2 100		12 26 37	−30 34.8	1	12.59		0.96				3		1719
108693	+32 02250	IDS12242N3157	AB	12 26 38	+31 40.0	3	7.92	.045	0.60	.008	0.00	.005	8	G0	77,938,3016
		Lod 624 # 40		12 26 38	−61 37.2	1	11.76		0.38		-0.23		1	B6	768
108659	−66 01783	LSS 2689		12 26 38	−66 35.1	2	7.30	.015	0.32	.005	-0.34	.017	4	B5 Iab/Ib	191,1684
		RR CVn		12 26 39	+34 55.3	2	11.73	.436	0.13	.039	0.15	.051	2	A9.7	668,699
108682	−30 09942			12 26 39	−30 33.6	4	9.06	.008	0.79	.003	0.32	.021	19	K0 V	158,1775,2033,3077
	+29 02293	Coma Ber 148		12 26 40	+28 34.2	1	8.91		1.49				2	K4 IV	280
		Coma Ber 1658		12 26 41	+26 38.4	1	13.33		1.08				2		913
		HA 5 # 422		12 26 41	+75 10.6	1	13.29		0.60		0.07		2		1058
108683	−37 07905			12 26 41	−37 59.2	1	9.38		3.09				3	C4-5,4-5	864
108686	−53 05169			12 26 41	−53 41.4	1	8.29		0.96		0.47		1	G8/K1 III	1770
108687	−53 05169			12 26 41	−53 41.4	1	8.29		0.95		0.47		1	B/A V	1770
	−60 04053			12 26 42	−61 16.1	1	10.80		0.52		0.24		1	A9	768
		Vir sq 1 # 15		12 26 43	+10 36.4	1	13.23		0.70		0.22		1		900
		Coma Ber 340		12 26 43	+25 33.8	1	11.97		0.43		-0.05		1		1148
		HA 5 # 24		12 26 43	+74 09.1	1	13.30		0.42		-0.02		1		1058
108691	−59 04255			12 26 43	−59 57.3	1	9.54		0.09		-0.33		1	B8 II	549
		He3 794		12 26 43	−64 33.	1	11.97		0.46		-0.07		6		1684
		Coma Ber 1661		12 26 44	+27 02.9	1	13.72		0.61				1		913
		G 148 - 63		12 26 44	+27 39.2	1	13.86		1.54				1	dM4	686
		CS 16026 # 37		12 26 44	+29 24.3	1	15.07		0.36		-0.19		1		1744
108711	+43 02227			12 26 44	+43 17.5	1	8.45		1.17				2	K1 III	280
108690	−58 04323			12 26 44	−59 12.6	1	8.75		1.88		2.09		2	M0/1	122
	−59 04256	IDS12240S5932	AB	12 26 44	−59 48.4	1	10.56		0.31		-0.13		1	B8	549
	+41 02297			12 26 45	+41 05.6	1	10.86		0.20		0.11		1		280
	−60 04054			12 26 45	−61 21.1	1	10.71		0.15		-0.10		1	B8	768
		LP 615 - 196		12 26 46	−01 57.1	1	11.46		0.68		0.06		2		1696
108671	−65 01882			12 26 46	−66 19.3	1	9.26		0.42		0.01		2	F0/2 V	1688
108713	+35 02343			12 26 47	+34 58.8	1	7.56		0.56				2	F9 V	77
108712	+36 02270			12 26 47	+35 48.5	1	8.55		0.43				2	G0	77
		Coma Ber 1667		12 26 48	+27 21.6	1	12.68		0.89				1		913
	+2 02547			12 26 49	+02 15.2	1	10.25		0.31		-0.02		1	F5	272
108714	+18 02614			12 26 49	+17 35.9	2	7.71	.007	0.11	.008	0.01	.008	17	A3 IV	861,1068
		L 194 - 11		12 26 49	−55 42.8	1	13.27		1.62		1.24		1		3078
	−64 01939	NGC 4463 - 2		12 26 49	−64 31.5	1	11.68		0.32		-0.15		2		251

HD	DM	Other Id	N	Rem	α_{1950}	δ_{1950}	S	V	σ_V	B–V	σ_{B-V}	U–B	σ_{U-B}	n	Spectrum	References
		NGC 4472 sq1 2			12 26 52	+08 14.8	1	14.02		0.63				2		492
		LOD 26+ # 14			12 26 52	−59 10.7	1	12.85		0.66				2		122
		NGC 4458 sq1 12			12 26 53	+13 24.1	1	14.03		0.88		0.61		2		1531
	−64 01940	NGC 4463 - 3			12 26 53	−64 31.7	1	10.46		0.25		−0.24		2	B5 III:	251
	+20 02727				12 26 54	+19 35.5	1	9.89		0.29				4	F0 III	193
		NGC 4472 sq1 1			12 26 55	+08 16.0	1	12.88		1.05				2		492
	−58 04325				12 26 55	−59 11.6	1	10.70		0.06		−0.29		2		122
	−64 01941	NGC 4463 - 4			12 26 55	−64 31.1	1	11.26		0.25		−0.22		2		251
	+11 02467				12 26 56	+11 24.9	1	9.25		1.38		1.66		1	K5	280
		Coma Ber 301			12 26 56	+25 56.8	1	11.03		0.51		0.02		3		1148
	+33 02243				12 26 56	+32 38.9	1	9.87		0.54				2		77
		G 199 - 17			12 26 56	+53 49.2	2	14.19	.005	1.54	.005			4		419,3078
		PHI2/1 43			12 26 56	−32 16.6	1	12.61		0.98				3		1719
108722	+24 02464	Coma Ber 149		★	12 26 57	+24 23.1	4	5.48	.010	0.43	.010	0.09	.004	14	F5 V	15,254,1008,3026
		LP 853 - 74			12 26 57	−21 51.1	1	10.86		0.88		0.60		2		1696
		Vir sq 1 # 11			12 26 58	+10 21.8	1	12.45		0.99		0.93		2		900
	+31 02370				12 26 58	+30 29.4	1	9.76		1.52		1.90		1	K5 III	280
		HA 5 # 423			12 26 58	+75 10.4	2	10.98	.038	0.39	.013	0.00	.000	6		271,1058
	−60 04060				12 26 58	−61 18.9	1	10.65		0.12		−0.26		1	B7	768
		Klemola 135			12 27 00	+08 38.	2	11.46	.030	0.05	.010	0.08	.025	3		280,1026
		Vir sq 1 # 24			12 27 01	+10 30.1	1	14.60		0.86		0.54		1		900
	+30 02280				12 27 01	+29 46.1	2	9.41	.000	0.56	.017	0.11		23	G5	77,1613
108736	+30 02281	IDS12245N3004		AB	12 27 01	+29 47.3	2	9.39	.009	0.54	.001	0.07		28	G5	77,1613
	−59 04257				12 27 01	−59 58.4	1	10.44		1.14		0.86		1	K2	549
	−59 04258				12 27 03	−59 54.3	1	10.48		0.33		0.23		1	A0	549
	+19 02574				12 27 04	+19 06.3	1	10.86		0.24		0.11		2	A5	1569
		CS 16026 # 38			12 27 04	+29 50.3	1	12.95		0.36		−0.05		1		1744
108753	+31 02371				12 27 04	+30 54.5	1	9.42		0.51				2	G5	77
		HA 5 # 499			12 27 04	+75 21.0	1	12.94		0.65		0.19		1		1058
	−59 04259				12 27 04	−59 58.7	1	11.25		0.62		0.15		1	G8	549
108719	−64 01942	NGC 4463 - 5			12 27 04	−64 30.8	1	8.41		0.21		−0.63		4	B1 III	251
	−64 01943	NGC 4463 - 1			12 27 04	−64 32.1	2	8.28	.008	0.79	.022	0.75	.016	5	F5 Iab	251,1688
	+30 02282				12 27 06	+30 21.2	1	10.22		0.93				3	G8 III	280
		NGC 4463 - 6			12 27 06	−64 31.	1	12.24		0.33		−0.19		2		251
		G 60 - 8			12 27 07	+00 06.9	1	11.40		1.02		0.92		1	K3	1620
	+14 02508				12 27 07	+14 24.6	1	10.44		1.00		0.88		1	G5	280
		NGC 4463 - 7			12 27 07	−64 30.6	1	11.17		0.26		−0.21		2		251
		NGC 4463 - 8			12 27 07	−64 30.6	1	11.91		0.31		−0.10		2		251
	−64 01944	NGC 4463 - 9			12 27 07	−64 30.6	1	12.04		0.46		0.43		2		251
		NGC 4472 sq1 3			12 27 08	+08 11.2	1	14.57		0.82				2		492
108732	−55 05097	HR 4754			12 27 08	−56 14.9	5	5.79	.011	1.57	.009	1.79	.039	14	M1 III	15,1034,1075,2007,3005
		NGC 4472 sq1 4			12 27 09	+08 10.3	1	14.18		0.95				2		492
108752	+34 02317				12 27 09	+33 49.5	1	9.55		0.52				2	K0	77
108754	−2 03528	G 13 - 43			12 27 10	−03 03.0	5	9.03	.012	0.70	.007	0.14	.020	9	G7 V	22,742,1003,1620,3077
108730	−48 07460				12 27 10	−48 35.6	1	9.84		0.04		0.02		1	B9.5 V	1770
	+25 02511	Coma Ber 150		★ V	12 27 11	+24 47.8	3	9.80	.016	0.78	.006	0.31	.002	4	G9 V	1127,8001,8023
	+25 02512				12 27 11	+25 13.0	1	10.02		0.43		−0.11		2	F8	3016
		Coma Ber 1684			12 27 11	+27 56.6	1	14.60		0.81				1		913
	+40 02540				12 27 12	+39 52.7	1	9.42		0.70				2	G0	280
108733	−59 04261				12 27 12	−59 33.9	1	9.29		1.50		1.42		1	K0	549
	−61 03228				12 27 12	−61 37.6	1	10.98		0.33		0.22		1	A5	768
		Vir sq 1 # 14			12 27 13	+10 27.0	1	13.07		0.92		0.52		1		900
108765	+21 02424	HR 4756			12 27 13	+21 10.4	4	5.68	.014	0.07	.007	0.09	.007	17	A3 V	70,1022,1068,3052
	+42 02312				12 27 13	+42 16.1	1	9.98		1.58		1.85		1	M1 III	280
	−64 01945	NGC 4463 - 13			12 27 13	−64 34.2	1	9.90		0.37		−0.12		2	B6 III	251
	+30 02283				12 27 14	+30 17.9	1	9.47		0.52				2	G0	77
108750	−60 04066				12 27 14	−61 10.8	2	8.38	.003	0.31	.008	0.08	.001	4	A8 V	1688,1771
108775	+15 02469				12 27 15	+14 55.6	1	7.26		1.25				2	K0	280
	−60 04067				12 27 15	−61 22.7	1	11.24		0.41		0.26		1	A5	768
		Coma Ber 1686			12 27 16	+25 49.2	1	11.23		1.04				2		913
108767	−15 03482	HR 4757		★ A	12 27 16	−16 14.2	12	2.95	.010	−0.05	.005	−0.08	.017	33	B9.5V	3,15,1020,1034,1075*
108767	−15 03481	IDS12247S1558		B	12 27 16	−16 14.2	2	8.40	.081	0.87	.062	0.64	.081	7	K0 V	1752,3024
108759	−41 07219	HR 4755			12 27 17	−41 27.6	1	6.02		1.52				4	M2 II/III	2006
		LOD 26+ # 12			12 27 17	−59 10.9	1	11.56		0.52				2		122
108735	−72 01256				12 27 17	−72 42.5	1	7.07		0.28				4	A6 IV	2012
	+13 02543				12 27 20	+13 05.0	2	10.61	.039	0.29	.000	0.01	.010	3	A2	272,280
		Coma Ber 1689			12 27 20	+27 24.4	1	14.02		0.53				1		913
	−33 08445				12 27 20	−34 18.0	1	9.63		1.35		1.56		2	K5	1730
	−60 04068	LSS 2694			12 27 20	−61 1.9	1	11.84		0.37		−0.53		3	B1 V	1737
	+28 02122				12 27 21	+27 51.5	1	9.66		1.43		1.84		1	K4 III	280
		Vir sq 1 # 12			12 27 22	+10 38.9	1	12.55		0.58		−0.01		2		900
		Vir sq 1 # 22			12 27 23	+10 40.0	1	14.30		0.70		0.10		2		900
	−64 01946	NGC 4463 - 11			12 27 23	−64 31.0	1	10.08		0.29		−0.20		4	B6 IV	251
		Vir sq 1 # 19			12 27 24	+10 22.2	1	14.00		0.72		0.40		2		900
108769	−33 08447				12 27 24	−34 13.4	2	9.04	.007	−0.15	.001	−0.77	.008	3	B2 III	55,1730
108771	−52 05638				12 27 24	−53 19.1	1	8.19		0.00		−0.02		2	A0 IV	1771
		NGC 4463 - 10			12 27 24	−64 31.	1	13.08		0.61		0.56		2		251
	+12 02480				12 27 25	+12 06.2	1	9.58		1.37		1.57		1	M0	280
		NGC 4472 sq1 8			12 27 26	+08 20.5	1	13.40		0.86				1		492
108807	+25 02513	Coma Ber 152			12 27 26	+24 36.9	3	7.86	.011	0.44	.008	−0.04	.005	4	F7 V	3016,8001,8023

Table 1 679

HD	DM	Other Id	N Rem	α_{1950}	δ_{1950}	S	V	σ_V	B–V	σ_{B-V}	U–B	σ_{U-B}	n	Spectrum	References
	−61 03230	LSS 2695		12 27 26	−62 12.3	3	10.47	.005	0.73	.017	−0.37	.031	7	O7 III	191,1684,1737
108789	−52 05639			12 27 27	−53 19.8	1	8.63		0.06		0.04		2	A1/2 V	1771
108805	+26 02356	Coma Ber 153		12 27 28	+26 23.8	2	8.04	.010	0.97	.005	0.74		3	G8 III	280,8041
		Coma Ber 1695		12 27 29	+27 35.4	1	13.31		1.11				1		913
		HA 5 # 224		12 27 29	+74 37.5	1	13.29		0.57		−0.02		1		1058
108806	+25 02514	Coma Ber 154		12 27 30	+24 49.5	1	8.55		1.31				2	K1 III	280
108799	−12 03647	HR 4758	⋆ AB	12 27 30	−13 07.0	4	6.38	.012	0.58	.012	0.05	.005	11	G0 V	292,1414,2006,3016
	−59 04264			12 27 31	−59 35.5	1	11.50		0.28		0.21		1	A2	549
108773	−60 04070	LSS 2696		12 27 31	−60 42.8	3	6.67	.005	0.45	.004	0.39	.019	8	F0 Ib	322,329,1075
	−64 01947	NGC 4463 - 12		12 27 31	−64 31.6	1	11.77		0.28		−0.09				251
		NGC 4472 sq1 5		12 27 32	+08 12.1	1	12.22		1.02				3		492
108751	−71 01331	IDS12246S7128	A	12 27 32	−71 44.8	1	8.03		0.35		0.10		1	A8/9 III/I	1771
		Malm +27 233		12 27 34	+26 56.9	1	11.78		1.37		1.25		3	dM1	685
	−61 03233			12 27 35	−61 44.3	1	9.93		0.64		0.38		1	G0	768
	+10 02430			12 27 36	+10 22.5	1	10.41		0.61		0.05		2	G5	900
108815	+18 02617			12 27 37	+18 10.3	1	7.45		1.58		1.61		1	M2	280
	−61 03235			12 27 37	−61 41.0	1	11.84		0.12		−0.27		1	B7	768
	+20 02728			12 27 38	+19 37.1	1	9.08		1.14		1.13		1	K5	280
108844	+59 01444	HR 4760		12 27 38	+58 40.8	2	5.34	.015	0.19	.002	0.13	.002	5	A5 δ Del	3058,8071
108821	−22 03383	HR 4759		12 27 40	−23 25.2	3	5.64	.005	1.66	.011	2.03	.020	13	M0 III	1024,2007,3005
	+42 02314			12 27 41	+41 37.4	1	9.97		0.80				2	G5	280
108845	+52 01631	HR 4761	⋆ A	12 27 41	+51 48.7	3	6.22	.015	0.51	.011	0.03	.030	6	F8 V	70,254,272
108791	−72 01258			12 27 41	−72 48.8	1	6.93		1.15				4	G8/K0 II	2012
		NGC 4472 sq1 7		12 27 42	+08 19.3	1	12.52		0.62				3		492
	+29 02294	Coma Ber 155		12 27 42	+28 41.0	2	9.55	.073	0.87	.000	0.45	.034	3	G5 III-IV	3016,8038
108804	−61 03237			12 27 42	−61 32.5	2	8.40	.006	0.42	.018	0.14	.011	3	F2 V	768,1688
	−61 03236			12 27 42	−61 36.0	1	10.91		0.26		0.09		1	B9	768
108833	+32 02250	T CVn		12 27 44	+31 46.8	1	9.90		1.47		0.73		1	M3	3001
108861	+59 01446	HR 4762		12 27 44	+59 02.6	1	6.08		0.98				1	G8 III-IV	71
		Coma Ber 221		12 27 45	+25 18.3	3	10.37	.009	0.82	.009	0.42	.001	5	G7 V	1127,8001,8023
		NGP +28 092		12 27 45	+27 50.1	1	10.82		0.74		0.32		1	G5	8038
108834	+28 02123	Coma Ber 156		12 27 45	+27 50.1	1	8.62		1.17		1.04		2	K0 IV	1569
108871	+63 01017			12 27 45	+63 07.4	1	9.28		0.52				2		1733
108812	−59 04266			12 27 45	−60 20.5	1	9.76		0.40		0.17		1	A8/9 IV/V	1771
108846	+40 02542			12 27 46	+39 31.5	2	7.68	.010	0.45	.005	−0.03	.015	5	F5 V	1501,1569
		NGC 4472 sq1 6		12 27 47	+08 15.3	1	11.77		1.31				2		492
	−61 03239			12 27 47	−61 30.8	1	10.39		0.36		0.02		1	B8	768
108849	+5 02634	BK Vir		12 27 48	+04 41.6	1	7.46		1.54		0.78		1	M7	3001
		Vir sq 1 # 17		12 27 48	+10 31.9	1	13.77		0.73		0.26		2		900
108813	−60 04073			12 27 48	−61 07.3	1	8.53		0.50		−0.02		2	F5 V	1688
108847	+31 02373			12 27 49	+31 16.2	3	9.12	.022	1.02	.010	0.87		5	G9 V	77,1017,3072
108863	+22 02478			12 27 50	+22 13.5	1	7.71		0.99				3	K0	280
		Coma Ber 1710		12 27 50	+24 14.5	1	12.44		0.76				1		913
	+38 02343			12 27 50	+37 47.3	1	9.57		0.51				2	F8	77
108792	−74 00922			12 27 50	−75 07.5	3	7.51	.004	0.13	.005	−0.06	.032	10	B9 V	278,1075,2012
	+10 02431			12 27 51	+10 19.7	1	10.88		0.57		0.07		2		900
		Vir sq 1 # 21		12 27 51	+10 21.8	1	14.19		0.60				1		900
		Vir sq 1 # 9		12 27 51	+10 38.3	1	11.96		0.75		0.36		1		900
108862	+23 02465			12 27 51	+22 42.2	1	8.63		1.31		1.48		1	K2	280
	−25 09286	SV Hya		12 27 53	−25 46.3	2	9.84	.065	0.13	.005	0.08	.005	2	A5	668,700
		LOD 26+ # 11		12 27 53	−59 04.7	1	11.55		0.20		0.20		2		122
		Vir sq 1 # 25		12 27 54	+10 22.3	1	14.65		1.15				1		900
		Vir sq 1 # 26		12 27 54	+10 37.9	1	14.82		0.72		0.21		1		900
108840	−43 07710			12 27 54	−44 05.4	1	8.29		0.00		−0.07		3	B9 V	1770
	−61 03240			12 27 54	−61 43.1	1	10.58		0.29		0.10		1	B8	768
108872	+41 02298			12 27 55	+41 06.4	1	8.60		1.07				2	K1 III	280
	+42 02315			12 27 55	+41 32.7	1	9.66		1.07				2	K0 III	280
108907	+70 00700	HR 4765, CQ Dra		12 27 56	+69 28.7	2	4.96	.004	1.62	.001	1.81		6	M3 IIIa	1363,3053
		CS 16026 # 40		12 27 57	+31 11.4	1	13.37		0.40		−0.19		1		1744
	+36 02274			12 27 57	+36 02.2	1	10.08		0.54				2	G0	77
	+37 02284			12 27 57	+37 20.5	2	9.76	.005	0.21	.005	0.08		4	Am	77,1240
		HA 5 # 85		12 27 59	+74 20.8	1	12.00		0.58		0.14		1		1058
	+6 02622			12 28 00	+06 20.0	1	10.35		0.85		0.50		2	K0	1696
108873	+34 02318			12 28 00	+34 22.5	1	9.57		0.39				2	G5	77
		He3 795		12 28 00	−61 01.	1	12.28		0.64		−0.47		3		1684
		AAS12,381 9 # C		12 28 01	+09 01.3	1	11.25		0.25		0.13		1		280
108875	+10 02432	IDS12255N1016	AB	12 28 01	+09 59.6	1	7.46		0.47		0.00		2	F2	1569
		Steph 1020	A	12 28 02	−10 25.0	1	11.66		1.29		1.08		2	K6	1746
	−61 03241			12 28 02	−61 41.4	1	10.02		1.09		0.88		1	K0	768
		Vir sq 1 # 20		12 28 03	+10 27.4	1	14.17		0.86		0.43		3		900
		Coma Ber 1721		12 28 03	+26 43.4	1	14.07		0.36				1		913
		LP 735 - 53		12 28 04	−12 52.6	1	12.93		1.37		1.31		1		1696
108891	+35 02346			12 28 06	+34 47.3	2	8.84	.010	0.70	.019	0.22		6	G5	77,1569
	+10 02433			12 28 08	+09 35.5	1	10.05		1.11		0.90		1	K0	280
108909	+27 02140	Coma Ber 159		12 28 08	+27 19.8	1	9.18		0.60		−0.05		2	G0	3016
108908	+34 02319			12 28 09	+34 01.5	1	8.50		0.33				2	F2 V	77
108885	−33 08456			12 28 10	−34 15.5	1	8.98		0.62		0.10		2	G2 V	1730
	+22 02479			12 28 11	+21 30.3	1	9.74		1.52		1.92		1	K2	280
		Coma Ber 1732		12 28 11	+27 42.4	1	13.17		0.82				1		913
108869	−60 04083			12 28 12	−61 02.9	1	9.37		0.27		0.18		2	A1/2 V	1688

HD	DM	Other Id	N	Rem	α_{1950}	δ_{1950}	S	V	σ_V	B–V	σ_{B-V}	U–B	σ_{U-B}	n	Spectrum	References
108870	−62 02798				12 28 12	−62 28.6	1	9.59		0.55		0.07		2	F3/5	1688
108888	−53 05183				12 28 13	−54 01.7	1	9.43		0.12		0.10		1	B9.5 V	1770
		LOD 26+ # 10			12 28 15	−59 01.1	1	11.82		0.50		0.10		2		122
	−64 01950				12 28 15	−65 22.9	1	9.65		0.63		0.19		3	F8	1688
		HA 5 # 294			12 28 16	+74 49.1	1	12.66		0.62		0.05		2		1058
108910	−3 03309				12 28 16	−03 47.1	1	6.88		1.43		1.64		2	K3	1003
108915	+13 02545				12 28 17	+12 45.8	1	8.65		0.91		0.59		2	K0	1375
108889	−58 04335	IDS12255S5859	AB		12 28 17	−59 15.9	1	8.86		0.14		0.04		2	B9 V	122
	−65 01887	IDS12254S6559	AB		12 28 17	−66 15.5	1	10.40		0.24		0.18		2	A0	1688
	+9 02635				12 28 18	+08 43.3	1	9.98		1.21		1.21		1	K2	280
	+45 02041				12 28 19	+45 00.3	1	10.16		1.08		1.03		1	K0 III	280
	+20 02729				12 28 20	+20 10.1	1	11.14		0.24		0.08		2	A0	272
108903	−56 05272	HR 4763		★ A	12 28 23	−56 50.0	5	1.62	.011	1.60	.008	1.76	.012	13	M4 III	15,507,1034,1075,2012
108904	−61 03247				12 28 23	−61 37.9	1	7.99		0.50		0.02		2	F6 V	1688
		LOD 26+ # 9			12 28 25	−58 58.0	1	12.07		0.57				2		122
		Coma Ber 1743			12 28 26	+26 35.4	1	13.97		0.88				1		913
108905	−61 03249				12 28 26	−62 23.8	1	9.64		0.51		0.03		3	F5 IV/V	1688
	−58 04337				12 28 27	−59 06.6	1	10.80		0.57		0.18		2		122
108943	+43 02233				12 28 28	+43 26.1	1	9.16		0.45		−0.05		2	F8	1569
	+44 02196				12 28 28	+44 27.9	1	9.04		1.32				2	K2 III	280
108954	+53 01554	HR 4767			12 28 28	+53 21.0	3	6.21	.018	0.55	.013	0.02	.035	6	F8 V	70,254,3026
		Vir sq 1 # 10			12 28 30	+10 17.0	1	12.03		0.47		−0.08		1		900
		Vir sq 1 # 6			12 28 30	+10 30.7	1	10.90		0.80		0.42		1		900
108925	−56 05274	HR 4764		★ B	12 28 30	−56 48.3	3	6.42	.011	0.15	.016	0.18		8	A3 V	15,1771,2012
108906	−65 01889				12 28 30	−66 06.7	1	9.78		0.17		0.05		2	A0 V	1688
108945	+25 02517	Coma Ber 160		★ V	12 28 31	+24 50.6	6	5.45	.007	0.05	.005	0.10	.003	26	A2p	15,1127,1263,3033*
		CS 16026 # 43			12 28 32	+30 41.2	1	14.19		0.48		0.04		1		1744
108944	+32 02252				12 28 32	+31 42.0	1	7.34		0.50				2	F9 V	77
		Vir sq 1 # 7			12 28 34	+10 26.1	1	11.72		0.95		0.67		2		900
108976	+28 02125	Coma Ber 162			12 28 34	+28 00.4	4	8.59	.022	0.48	.002	−0.05	.004	5	F7 V	1127,3026,8001,8023
		HA 5 # 87			12 28 34	+74 20.8	1	11.67		0.46		−0.01		2		1058
		HA 5 # 230			12 28 34	+74 37.6	1	12.69		0.60		0.04		1		1058
	+16 02384				12 28 35	+16 21.1	1	9.94		1.12		1.12		1	K2	280
108957	+17 02489				12 28 35	+16 53.4	1	7.45		1.13		1.08		2	K2	1375
108955	+38 02344				12 28 35	+38 26.4	2	8.52	.029	0.91	.005	0.62		3	K0 III	280,8041
		LOD 26+ # 4			12 28 35	−59 05.1	1	12.24		0.34		0.28		2		122
	+34 02320				12 28 36	+34 09.9	1	9.55		1.09	.020	1.02		4	K0 III	77,1569
	−58 04339				12 28 36	−59 06.6	1	11.60		0.16		−0.07		2		122
		Vir sq 1 # 13			12 28 37	+10 35.0	1	12.99		0.92		0.68		1		900
		Coma Ber 1753			12 28 37	+27 17.6	1	13.98		0.74				1		913
108973	+40 02543				12 28 38	+39 51.5	2	6.80	.005	1.04	.000	0.92		5	K1 III	172,280
108935	−42 07705				12 28 38	−43 06.4	2	9.78	.029	1.14	.010	1.01		6	K3 V	2033,3072
		CS 16026 # 44			12 28 39	+27 45.6	1	14.15		0.55		−0.07		1		1744
108975	+36 02276				12 28 39	+36 05.6	2	8.87		1.18	.030	1.21		4	K1 III	77,1569
	+39 02529				12 28 41	+38 53.1	1	9.88		0.71				2	G7 III	280
		Vir sq 1 # 8			12 28 42	+10 34.2	1	11.87		0.55		0.04		1		900
		LB 11236			12 28 42	+31 08.0	1	14.05		0.11		0.13		1		1744
	+32 02253				12 28 42	+31 35.1	1	10.12		0.48				2	F8	77
108939	−60 04090				12 28 42	−60 36.5	2	8.07	.008	0.01	.003	−0.27	.008	4	B8 III	401,1688
	+11 02469				12 28 43	+10 36.2	1	9.70		0.63		0.10		1	G5	900
	+26 02341	Coma Ber 211			12 28 43	+26 18.5	3	10.31	.027	1.19	.012	1.24	.025	4	K1 III	280,1148,8034
	−24 10403				12 28 43	−25 20.2	1	10.70		0.46		−0.03		1		6006
	+9 02636	G 12 - 39			12 28 46	+09 05.6	4	9.73	.049	1.45	.017	1.22	.005	6	M0:	1003,1017,1705,3072
108953	−58 04341				12 28 47	−58 56.0	1	9.19		0.80		0.55		2	G8/K0 IV/V	122
108985	+8 02609	HR 4770			12 28 49	+07 52.8	2	6.04	.005	1.52	.000	1.88	.000	7	K5	15,1417
		Vir sq 1 # 16			12 28 50	+10 25.4	1	13.29		0.67		0.16		1		900
		Vir sq 1 # 23			12 28 50	+10 25.4	1	14.53		0.85		0.44		2		900
		CS 16026 # 42			12 28 50	+30 50.4	1	13.81		0.59		0.04		1		1744
	−24 10404				12 28 51	−25 21.1	1	10.65		0.96		0.75		1		6006
108967	−48 07477				12 28 51	−48 42.6	1	8.05		0.23		0.15		1	A2/3 IV	1771
108968	−58 04344	HR 4768, BG Cru			12 28 52	−59 08.9	4	5.47	.063	0.63	.035	0.41		5	F7 Ib/II	122,138,1488,2008
		Coma Ber 1759			12 28 53	+26 43.1	1	13.00		0.66				1		913
	−64 01952				12 28 53	−64 55.0	2	11.02	.005	0.93	.003	0.22	.002	5		1684,1737
	+18 02621				12 28 54	+17 58.1	1	9.14		1.03		0.89		2	K0	1569
	−60 04095				12 28 54	−60 45.8	1	9.63		0.09		−0.62		1	B8	1688
108969	−60 04094				12 28 55	−61 04.5	1	8.92		0.03		−0.40		2	B6/7 IV	1688
		Coma Ber 1761			12 28 56	+25 03.6	1	12.54		0.59				1		913
	−42 07709				12 28 56	−43 23.9	1	10.78		1.00		0.81		1	K5 V	3072
	+27 02142	Coma Ber 163			12 28 57	+26 41.5	1	9.08		1.33		1.51		2	K2 IV	1569
109011	+55 01536				12 28 57	+55 23.7	4	8.12	.026	0.93	.013	0.64	.007	9	K2 V	1007,1013,7009,8023
		Coma Ber 1768			12 28 58	+24 16.2	1	13.39		0.50				1		913
		Coma Ber 1766			12 28 58	+24 25.9	1	13.01		0.80				1		913
		Coma Ber 1769			12 28 59	+26 35.2	1	14.41		0.52				1		913
108983	−61 03255				12 29 00	−61 57.4	1	9.86		0.25		0.21		2	A1 IV/V	1688
		MN177,99 # 7			12 29 01	+31 06.8	1	12.47		0.90		0.66		1		508
109005	−10 03487	IDS12264S1031	A		12 29 01	−10 47.8	1	7.97		0.22		0.06		3	A7	3016
109005	−10 03486	IDS12264S1031	B		12 29 01	−10 48.1	1	8.40		0.32		0.04		3		3016
108997	−54 05231				12 29 01	−55 04.9	1	8.48		1.08		0.75		4	G8 III	243
108998	−55 05115	IDS12263S5534	A		12 29 02	−55 50.6	1	8.89		0.10		0.00		3	B9.5 V	1770
	−60 04101				12 29 02	−61 08.5	1	10.19		0.08		−0.36		3		1684

Table 1

HD	DM	Other Id	N Rem	α_{1950}	δ_{1950}	S	V	σ_V	B–V	σ_{B-V}	U–B	σ_{U-B}	n	Spectrum	References
109029	+33 02244			12 29 03	+33 17.5	1	7.45		0.38				2	F4 V	77
	+25 02518	Coma Ber 164	⋆ A	12 29 04	+24 53.8	1	9.49		0.45		-0.01		2	F6 V	3016
	+25 02518	Coma Ber 164	⋆ B	12 29 04	+24 53.8	1	10.56		0.61		0.11		2		3016
109030	+26 02359	Coma Ber 166		12 29 04	+26 08.8	3	7.89	.027	0.04	.017	0.06	.011	6	A0p	1068,1240,3016
109012	+27 02143	Coma Ber 165		12 29 04	+27 20.5	3	7.51	.013	1.19	.014	1.24	.000	7	K1 III	172,280,8041
109014	−4 03296	HR 4772		12 29 04	−04 46.6	2	6.18	.005	1.04	.000	0.87	.005	7	K0	15,1256
108999	−61 03256			12 29 04	−61 33.0	1	9.27		0.01		-0.20		2	B9 IV/V	1688
		AAS12,381 44 # B		12 29 05	+44 11.7	1	9.94		1.22		1.42		1		280
		LOD 26+ # 6		12 29 05	−59 02.1	1	11.48		0.16		0.10		2		122
109000	−62 02805	HR 4771		12 29 05	−63 13.8	5	5.95	.011	0.27	.009	0.14	.003	17	A8 III	15,1075,1688,1771,2012
109033	+11 02470			12 29 06	+10 32.7	1	8.54		0.71		0.29		2	G5	900
	+25 02519			12 29 06	+24 30.3	1	11.31		0.07		0.09		2		1026
		Coma Ber 1780		12 29 06	+25 20.1	1	12.93		0.80				1		913
108995	−48 07478			12 29 06	−49 00.6	1	9.10		0.26		0.18		1	A1/2mA5-A9	1771
		Coma Ber 1782		12 29 07	+27 54.6	1	12.78		0.90				1		913
	+32 02254	IDS12267N3225	AB	12 29 07	+32 17.9	2	9.67	.009	0.44	.014	-0.02		4		77,1375
109052	+43 02235			12 29 07	+42 48.4	1	8.65		0.99				2	K0 III	280
108970	−72 01261	HR 4769		12 29 07	−72 43.5	5	5.88	.005	1.10	.008	1.05	.005	21	K1 III	15,278,1075,2012,2038
109032	+12 02484			12 29 08	+12 24.2	1	8.09		0.32		0.01		2	F0	1569
		Coma Ber 1784		12 29 08	+24 53.9	1	12.13		0.89				1		913
	+29 02296	Coma Ber 167		12 29 08	+28 43.0	1	9.39		0.56		0.08		2	dF8	3016
		HA 5 # 296		12 29 09	+74 51.8	1	13.21		0.68		0.17		2		1058
109017	−31 09743	IDS12265S3122	AB	12 29 10	−31 38.5	2	8.24	.015	0.88	.005	0.46		6	G8 IV	2012,3008
109055	+22 02483			12 29 11	+22 24.0	2	8.85	.000	-0.02	.000	0.05	.000	17	A0 V	1068,1613
109053	+34 02322			12 29 11	+34 11.6	2	9.24		0.91	.005	0.56		4	G8 III	77,1569
		CS 16026 # 47		12 29 13	+28 28.8	1	14.12		0.66		0.14		1		1744
109042	−49 07170			12 29 13	−49 37.7	1	8.97		0.02		-0.26		1	B8 IV	1770
109068	+46 01791			12 29 14	+45 30.1	1	7.66		0.20		0.08		1	A7 V	1240
	+16 02385			12 29 15	+16 10.0	1	9.64		1.27		1.36		1	K2	280
		Coma Ber 1794		12 29 15	+26 41.1	1	13.16		0.62				1		913
		HA 5 # 232		12 29 15	+74 38.3	2	10.87	.019	0.56	.007	0.12	.030	5	F8	271,1058
109024	−55 05117			12 29 15	−56 00.7	2	7.43	.009	0.05	.004	-0.28	.005	5	B8 IV	401,1770
	+23 02468			12 29 16	+22 48.1	1	10.45		1.07		1.05		1		280
109070	+23 02467			12 29 16	+23 14.7	1	8.03		1.06		1.00		1	K0	280
	+44 02198			12 29 16	+43 29.6	1	10.23		0.61				2	G0	280
109044	−52 05663			12 29 16	−53 02.4	1	7.78		0.03		0.04		2	A0 IV	1771
		AAS12,381 13 # A		12 29 17	+12 38.4	1	11.01		0.25		0.05		1		280
	+34 02323			12 29 17	+33 45.8	2	9.10	.041	1.54	.012	2.00		10	K4 V	77,7009
	+38 02346			12 29 18	+38 02.4	1	9.92		0.87		0.57		1	G8 III	280
		Coma Ber 1802		12 29 19	+26 10.1	1	13.17		0.64				1		913
109116	+75 00472			12 29 19	+74 33.3	2	9.33	.017	0.53	.009	0.08	.013	6	G5	271,1058
		PHI2/2 71		12 29 19	−30 31.8	1	12.56		1.03				1		1719
	+28 02126	Coma Ber 169		12 29 20	+28 15.3	1	9.76		0.74		0.19		2	G5	3016
	+11 02471			12 29 22	+10 57.1	1	10.13		0.88		0.63		1	K0	280
109069	+30 02287			12 29 22	+29 35.4	3	7.54	.006	0.33	.012	0.00	.011	17	F0 V	77,1613,3016
109045	−55 05120			12 29 22	−55 51.2	1	6.55		1.68		0.95		4	K5 III	243
		Wray 985		12 29 22	−61 43.9	1	11.61		0.28		-0.05		4		1684
		Coma Ber 1804		12 29 24	+26 41.0	1	13.90		0.78				1		913
109047	−61 03261			12 29 24	−62 13.4	1	9.47		0.26		0.21		2	A1 IV	1688
109074	−31 09746	HR 4774		12 29 25	−32 15.5	2	6.45	.005	0.20	.000			7	A3 V	15,2012
109065	−61 03263			12 29 27	−61 27.2	2	8.19	.003	0.14	.006	0.11	.001	3	A1 Vn	1688,1771
109026	−71 01336	HR 4773		12 29 27	−71 51.4	6	3.86	.008	-0.15	.005	-0.61	.017	17	B5 V	15,26,1034,1075,1637,8023
109085	−15 03489	HR 4775		12 29 29	−15 55.2	8	4.30	.013	0.38	.007	0.02	.013	42	F2 V	3,15,1008,1075,1425*
		PHI2/2 49		12 29 29	−30 00.0	1	12.03		1.03				3		1719
	+8 02612			12 29 33	+08 01.7	1	11.01		0.26		0.03		2		1569
		HA 5 # 93		12 29 33	+74 21.4	1	12.01		0.60		0.10		1		1058
		Coma Ber 1815		12 29 34	+25 23.2	1	12.30		0.84				1		913
	+44 02199			12 29 34	+43 45.5	1	9.70		1.48				2	M6 III	280
	−61 03264	LSS 2702		12 29 34	−61 26.0	1	11.23		0.58		-0.47		3	O7 III	1737
	−62 02808			12 29 34	−62 34.1	1	10.17		1.02		-0.19		4		1684
		WLS 1200 15 # 8		12 29 36	+14 59.4	2	11.19	.045	0.32	.010	-0.01	.015	3		280,1375
109080	−61 03265			12 29 36	−61 41.8	1	9.72		0.24		0.28		2	A0 IV	1688
	+11 02472			12 29 38	+10 44.8	1	9.96		1.30		1.47		1	K2	280
	+30 02288			12 29 39	+30 12.0	1	10.03		1.28		1.43		2	K3 III	1375
109118	+21 02428			12 29 40	+21 11.7	1	8.58		1.27		1.23		2	K2	1569
109091	−59 04296	IDS12269S5921	AB	12 29 41	−59 37.2	1	7.75		0.07		-0.32		2	B8 V	401
		L 327 - 29		12 29 42	−46 07.	1	11.60		0.85		0.48		2		1696
	+12 02485			12 29 43	+12 03.8	1	10.80		0.70		0.28		1		280
109067	−74 00927			12 29 43	−75 09.5	2	7.78	.002	0.41	.012	0.00		6	F3 IV/V	278,2012
		HA 5 # 298		12 29 44	+74 49.2	1	12.96		0.66		0.24		2		1058
	−39 07674			12 29 45	−39 49.3	3	10.92	.019	0.50	.007	-0.02	.000	7	F6	1696,2033,6006
109111	−49 07174			12 29 45	−49 40.0	1	8.68		0.28		0.07		1	A9 III/IV	1771
		G 12 - 40		12 29 46	+12 27.4	2	10.93	.005	1.17	.075	1.03	.060	2	K5	1620,1658
109117	+26 02360	Coma Ber 171		12 29 46	+26 14.8	1	9.13		0.62		0.14		2	G5	3016
		WLS 1224 50 # 9		12 29 47	+49 57.2	1	10.71		0.58		0.02		1		1375
		G 12 - 41		12 29 48	+12 27.0	2	12.49	.010	1.40	.015	1.26	.015	2		1620,1658
		Coma Ber 1830		12 29 48	+23 58.8	1	12.68		0.44				1		913
	+34 02325			12 29 49	+34 03.7	1	10.02		0.88				2	G8 III	280
	+46 01792			12 29 50	+45 46.4	1	8.68		1.52				2	M1 III	280
109102	−62 02810			12 29 50	−62 26.9	1	10.04		0.26		0.16		2	B9/A0 V	1688

HD	DM	Other Id	N Rem	α_{1950}	δ_{1950}	S	V	σ_V	B–V	σ_{B-V}	U–B	σ_{U-B}	n	Spectrum	References
109127	+27 02145	Coma Ber 172		12 29 51	+27 03.4	1	8.96		0.34		0.03		2	F3 V	3016
	+13 02547			12 29 52	+12 28.6	1	10.43		0.98		0.58		1		280
		CS 16026 # 51		12 29 52	+28 50.5	1	14.04		0.60		-0.08		1		1744
109156	+44 02202			12 29 52	+43 28.6	2	8.14	.015	1.08	.005	0.96		4	K0 III wk1	172,280
109103	-63 02336			12 29 53	-63 57.1	1	10.06		0.21		-0.53		1	B3/5 III	1688
		CS 16026 # 55		12 29 54	+30 06.8	1	14.48		0.59		-0.01		1		1744
109114	-60 04116			12 29 54	-60 35.8	1	9.86		0.12		0.09		2	A0 V	1688
		Steph 1021		12 29 55	+20 40.0	1	12.90		1.56		1.12		2	M4	1746
		PHI2/2 46		12 29 55	-31 15.9	1	12.72		1.48				3		1719
		CS 16026 # 58		12 29 56	+32 09.9	1	13.46		0.66		0.12		1		1744
109132	-20 03667			12 29 56	-20 56.1	1	6.40		1.11		1.05		3	K0 III	1628
	+21 02429	IDS12275N2139	AB	12 29 58	+21 22.4	1	9.50		0.47		0.00		1		1569
109157	+28 02128	Coma Ber 173		12 29 59	+28 21.7	1	9.11		0.83		0.47		2	G7 IV	3016
109124	-61 03269			12 29 59	-61 36.0	1	8.99		0.18		0.05		2	B8/9 III	1688
		GD 317, EG 90		12 30 01	+41 45.9	2	15.74	.015	-0.05	.015	-0.83	.010	2	DA:	1727,3028
109141	-13 03552	HR 4776		12 30 01	-13 35.0	2	5.75	.010	0.38	.010	-0.04		6	F3 IV/V	254,2035
	+33 02246			12 30 02	+32 40.2	1	10.16		1.01		0.81		1	G9 III	280
	+31 02380			12 30 03	+30 47.5	1	10.63		1.10		0.95		1	K0 III	280
	+33 02247			12 30 03	+33 00.2	1	10.52		0.99		0.78		1	G8 III	280
		AAS12,381 36 # B		12 30 03	+36 02.7	1	10.21		1.15		1.09		1		280
109213	+75 00473	IDS12280N7522	A	12 30 03	+75 05.2	1	8.56		1.00		0.97		1	G9 III	8100
109213	+75 00473	IDS12280N7522	AB	12 30 03	+75 05.2	7	7.50	.100	1.10	.035	1.05	.095	7	G9 III	271,8100
	+36 02278			12 30 04	+35 36.4	1	9.71		0.63				2	G0	77
	+37 02289			12 30 04	+37 19.3	1	9.04		0.48				2	F7	77
		Coma Ber 1848		12 30 05	+27 45.6	1	12.36		0.96				1		913
	+16 02388			12 30 06	+15 30.1	1	10.40		0.90		0.66		1		280
		Coma Ber 1849		12 30 07	+27 48.6	1	13.02		0.76				1		913
		CS 16026 # 49		12 30 07	+28 32.5	1	12.30		1.27		1.10		1		1744
109137	-61 03272			12 30 07	-61 51.0	1	9.83		0.17		0.12		3	B9.5V	1688
	+24 02468	Coma Ber 174		12 30 08	+23 32.2	1	9.37		0.75		0.41		1	dK0	280
109179	+36 02279			12 30 08	+36 06.4	1	8.82		0.53				2	G5	77
109246	+75 00474			12 30 08	+74 45.9	2	8.75	.010	0.65	.005	0.19	.005	32	G0	271,1058
	-53 05198		V	12 30 08	-54 22.4	1	10.16		1.77		1.90		4		243
109202	+57 01378			12 30 11	+57 15.2	1	8.11		0.60		0.17		2	F8/9V	1733
		AAS12,381 8 # A		12 30 12	+08 12.0	1	10.54		0.22		0.13		1		280
		HA 5 # 506		12 30 12	+75 18.2	1	13.12		0.72		0.35		1		1058
		Coma Ber 1855		12 30 13	+23 55.9	1	12.18		1.04				1		913
		G 59 - 22		12 30 13	+26 53.7	2	13.25	.020	0.64	.002	-0.03	.007	4		1620,1658
109180	+9 02637			12 30 14	+09 22.2	1	9.25		0.59		0.15		3	K0	8088
109185	+24 02470	Coma Ber 176		12 30 15	+23 41.7	1	7.45		0.29		0.08		1	F0	280
		S Com		12 30 17	+27 18.2	2	10.91	.023	0.06	.017	0.10	.003	2	F4	668,699
		CS 16026 # 48		12 30 17	+28 27.6	1	13.33		0.90		0.54		1		1744
	+35 02349			12 30 18	+34 51.6	2	9.60		1.34	.045	1.60		4	K2 III	77,1569
		Coma Ber 1861		12 30 19	+24 02.4	1	12.56		0.65				1		913
		Coma Ber 1860		12 30 19	+24 54.1	1	13.02		1.08				1		913
	+27 02146	Coma Ber 177		12 30 20	+27 25.4	1	10.15		0.29		0.01		1		280
109164	-60 04128	LSS 2705	★ AB	12 30 20	-60 40.3	2	7.96	.119	0.04	.010	-0.79		6	B2 II	401,1075
109138	-74 00932			12 30 22	-75 06.6	1	9.47		0.76				4	K0 V	2012
109203	+28 02129	Coma Ber 178		12 30 23	+27 49.7	2	8.98	.024	0.80	.000	0.41		3	G7 III	280,8041
		CS 16026 # 57		12 30 24	+31 18.6	1	14.40		0.43		-0.14		1		1744
		SSII 167		12 30 24	+37 32.	1	13.96		0.20		0.05		1	A5	1298
	-60 04130	LSS 2706		12 30 24	-60 40.1	1	9.96		0.03		-0.76		1		401
109152	-68 01682			12 30 24	-68 33.6	1	9.33		0.38		0.11		1	A8/F0 V	1771
		Coma Ber 1867		12 30 25	+26 19.8	1	14.46		0.05				1		913
		G 12 - 42		12 30 26	+12 28.3	1	13.28		1.57		1.25		1		1620
		Coma Ber 1870		12 30 26	+26 20.1	1	13.76		0.53				1		913
109195	-51 06792			12 30 26	-51 48.4	1	6.55		-0.02		-0.09		3	B9.5 V	1770
109216	+22 02485			12 30 27	+21 58.5	1	9.01		0.31		0.10		1	F2 V	280
		Coma Ber 1872		12 30 27	+27 49.1	1	13.38		0.63				1		913
	+30 02290			12 30 27	+29 50.4	2	11.49	.025	0.23	.025	0.05		5	A2	77,272
		HA 5 # 32		12 30 27	+74 04.2	1	12.33		0.62		0.13		1		1058
		Coma Ber 1874		12 30 29	+26 26.5	1	13.05		0.86				1		913
109217	+11 02473	HR 4777		12 30 31	+10 34.3	2	6.28	.020	0.95	.005	0.71		4	G8 III	280,3077
109192	-43 07738			12 30 31	-44 09.7	1	10.23		0.71		-0.28		1	G8/K0 III	1737
109214	+26 02361	Coma Ber 180		12 30 32	+26 14.5	2	8.78	.05	1.12	.005	1.11		4	K1 III	280,1569
		Coma Ber 1877		12 30 32	+27 59.3	1	11.25		1.00				3		913
109198	-57 05547			12 30 32	-57 26.4	1	7.75		-0.01		-0.45		2	B8 III	401
109197	-54 05241			12 30 33	-54 42.3	2	7.61	.000	0.07	.010	0.07	.010	6	A2 V	243,1771
109182	-62 02818			12 30 34	-62 45.6	1	10.74		0.09		-0.45		2	B5 V	1688
109193	-45 07836			12 30 35	-45 35.8	1	8.38		0.00		-0.12		3	B9/A0 V	1770
109236	+8 02614	CI Vir		12 30 36	+07 31.6	1	9.21		1.49		1.22		1	M3	3001
		CS 16026 # 54		12 30 36	+29 54.0	1	14.34		0.54		-0.09		1		1744
		HA 5 # 507		12 30 36	+75 21.2	1	13.50		0.45		-0.03		1		1058
		Coma Ber 1888		12 30 39	+25 43.0	1	11.73		0.94				1		913
109210	-53 05202			12 30 39	-54 14.3	1	9.90		0.02		-0.12		4	B9/A0 Vn	243
109199	-65 01905			12 30 39	-65 50.4	1	7.69		0.09		-0.53		2	B3/4 IV	1688
109200	-68 01684			12 30 39	-68 28.5	4	7.13	.018	0.84	.013	0.46	.019	11	K1 V	258,2012,2033,3077
	+14 02514			12 30 42	+14 16.5	1	9.73		1.17		1.18		1	K2	280
109315	+75 00475			12 30 44	+74 44.0	2	10.12	.003	0.66	.010	0.25	.000	7	K0	271,1058
109238	-18 03416	HR 4778		12 30 45	-19 31.0	3	6.25	.004	0.30	.007			11	F0 IV/V	15,2013,2029

Table 1   683

HD	DM	Other Id	N Rem	α_{1950}	δ_{1950}	S	V	σ_V	B–V	σ_{B-V}	U–B	σ_{U-B}	n	Spectrum	References
		Coma Ber 1894		12 30 46	+25 12.1	1	12.78		1.06				1		913
109268	+46 01793			12 30 46	+45 36.8	1	9.06		0.00		0.00		1	A2 V	1240
	+44 02203			12 30 48	+44 00.5	1	10.15		0.77				2		280
	+17 02496			12 30 49	+16 48.7	1	9.85		1.60		1.67		1		280
	+34 02329			12 30 49	+33 46.6	2	9.46		1.44	.030	1.77		4	K4 III	77,1569
	+34 02330			12 30 49	+34 17.7	1	9.87		0.61				2		77
		G 12 - 43	⋆ AB	12 30 50	+09 17.6	6	12.46	.021	1.84	.013	1.21	.066	22	M7	1620,1705,1764,8006*
109242	−53 05205			12 30 51	−53 34.4	1	9.02		0.33		0.18		1	A5 V	1771
109270	+8 02616			12 30 52	+08 13.3	1	7.58		1.10		1.03		4	K5	280
		PHI2/2 40		12 30 53	−30 41.4	1	12.75		1.00				3		1719
		SSII 169		12 30 54	+34 31.0	1	12.52		0.32		−0.01		1	F2	1298
109252	−54 05246			12 30 54	−54 38.8	1	9.37		0.94		0.68		4	K2 V	243
		Coma Ber 1901		12 30 55	+24 08.3	1	13.08		0.63				1		913
	+10 02438	IDS12284N1044	A	12 30 56	+10 28.2	2	10.19	.085	0.81	.015	0.54	.020	2	G5	280,280
	+31 02381			12 30 56	+31 22.4	1	9.17		1.18				2	K1 III	77
109280	+34 02331	IDS12285N3443	AB	12 30 57	+34 26.3	1	8.41		0.71				2	G2 V	77
	−33 08486	GPEC 1172		12 30 57	−34 16.8	1	11.58		0.72				2		1719
109243	−62 02821			12 30 57	−62 51.7	1	10.16		0.17		−0.11		2	B8/9 V	1688
		Coma Ber 1903		12 30 58	+24 40.8	1	12.93		1.01				1		913
109272	−12 03659	HR 4779		12 30 58	−12 33.3	3	5.58	.006	0.86	.001	0.47	.002	24	G8 III/IV	3,418,2007
	+28 02132			12 30 59	+27 40.1	1	10.11		1.13		1.12		1	K1 III	280
109285	+8 02617			12 31 00	+07 57.3	1	6.91		1.04				2	K0	280
	+39 02533			12 31 00	+39 24.2	1	9.12		1.12				2	K1 III	280
	+23 02473			12 31 01	+22 45.7	1	10.19		0.91		0.58		1		280
		Coma Ber 1908		12 31 01	+26 26.6	1	11.12		1.05				3		913
		Coma Ber 1909		12 31 01	+27 44.2	1	13.67		0.49				1		913
109281	+30 02291			12 31 01	+30 19.0	1	8.67		1.03				2	K0 III	77
	+36 02280			12 31 01	+35 31.7	2	9.44		1.07	.085	0.75		4	G8 III	77,1569
	+41 02303	G 123 - 38		12 31 01	+40 40.8	1	10.11		0.99		0.83		2	K0 V	1569
109304	+41 02302			12 31 01	+41 00.2	1	8.50		0.50				2	G0	280
	+43 02238			12 31 01	+42 34.7	1	10.64		0.84		0.53		1	G7 III	280
	+19 02581			12 31 02	+19 08.2	1	10.31		1.22		1.11		1		280
109282	+25 02522	Coma Ber 182		12 31 03	+24 43.5	2	7.34	.063	1.64	.010	1.87	.015	3	M1	280,1569
		G 59 - 23		12 31 04	+24 06.6	1	10.71		1.19		1.15		1	K4	1620
109305	+38 02347			12 31 04	+38 20.7	3	6.52	.012	1.04	.008	0.89	.019	5	K0 III-IV	172,280,8041
109307	+25 02523	Coma Ber 183	⋆	12 31 05	+24 33.5	6	6.29	.009	0.11	.005	0.09	.005	17	A3 IV-Vs	15,1022,1068,1127*
109266	−61 03283			12 31 05	−61 29.8	2	8.97	.007	0.00	.002	−0.40	.006	4	B8 V	401,1688
109253	−63 02344			12 31 05	−64 06.1	1	9.60		0.17		−0.31		1	B8 II/III	1688
	+40 02548			12 31 10	+39 34.0	1	8.79		1.04				2	G9 III	280
109276	−55 05133			12 31 10	−56 17.7	1	7.56		0.11		0.10		1	A2 V	1771
109317	+34 02332	HR 4783		12 31 11	+33 31.4	4	5.41	.010	1.01	.011	0.83	.005	17	K0 III	77,1080,3016,8034
109309	−8 03372	HR 4781		12 31 12	−09 10.6	5	5.48	.007	−0.03	.010	−0.05	.034	22	A0 V	3,15,1417,2012,3023
	+25 02524			12 31 14	+25 27.1	1	9.81		1.09		1.04		1	K0 III	280
		CS 16026 # 52		12 31 14	+29 47.3	1	13.04		1.02		0.90		1		1744
	+37 02292			12 31 14	+36 30.9	2	10.28	.025	0.96	.000	0.67	.005	2	G8 III	280,1748
	+41 02304			12 31 14	+41 14.1	1	9.09		1.19				3	K2 III	280
	+10 02439			12 31 15	+09 53.8	1	10.08		0.97		0.78		1		280
109312	−49 07195	HR 4782	⋆ AB	12 31 15	−49 38.0	3	6.38	.012	0.45	.015	0.02		9	F3 III/IV	15,2012,3053
		G 148 - 68		12 31 16	+34 55.4	1	13.68		1.24		1.07		1		3062
	−63 02345			12 31 16	−64 14.8	1	11.02		0.22		−0.55		9		1684
	+10 02440			12 31 18	+09 48.5	1	9.72		1.52		1.83		1		280
109313	−53 05209			12 31 19	−53 29.9	1	7.97		0.02		−0.07		3	B9.5 V	1770
109345	+34 02333	HR 4784		12 31 20	+33 39.6	3	6.24	.006	1.05	.000	0.90	.008	14	K0 III	77,1355,8034
109358	+42 02321	HR 4785	⋆	12 31 22	+41 37.7	13	4.26	.007	0.59	.005	0.05	.011	249	G0 V	1,15,22,130,667,1077*
109387	+70 00703	HR 4787, κ Dra		12 31 22	+70 03.8	6	3.85	.028	−0.13	.011	−0.57	.019	18	B6 IIIpe	15,154,1203,1212,1363,8015
109346	+27 02147	Coma Ber 185		12 31 23	+27 08.2	1	9.68		0.38		−0.03		12	F3 V	1613
109324	−49 07197			12 31 23	−49 47.1	1	7.30		0.14		0.11		1	A3 V	1771
		CS 16026 # 50		12 31 25	+28 33.0	1	13.62		0.59		0.00		1		1744
109333	−13 03557			12 31 25	−14 21.8	2	9.09	.045	1.11	.000	1.01		7	K3/4 V	2033,3072
109314	−60 04147	LSS 2708		12 31 25	−60 54.8	2	9.59	.049	0.12	.013	−0.74	.003	2	B5	1577,1688
		Coma Ber 1938		12 31 26	+27 24.6	1	11.72		1.00				2		913
		Coma Ber 1939		12 31 27	+26 38.0	1	13.65		0.45				1		913
109334	−22 03399			12 31 27	−23 16.2	1	7.44		1.13		0.90		4	K0 III	1657
		NGC 4526 sq1 8		12 31 29	+08 04.5	1	14.46		0.61		0.03		1		1531
		GPEC 1206		12 31 29	−34 37.2	1	11.99		0.91				2		1719
		NGC 4526 sq1 9		12 31 30	+07 57.1	2	12.58	.040	1.08	.045	0.97		2		492,1531
109359	+27 02149	Coma Ber 186	⋆ AB	12 31 30	+27 06.8	3	9.08	.022	0.56	.008	0.03	.030	16	G0 V	853,1613,3016
		NGC 4526 sq1 2		12 31 31	+07 49.5	1	13.37		0.60		0.00		1		1531
		NGC 4526 sq1 7		12 31 31	+08 04.2	2	11.98	.005	0.68	.170	0.02		2		492,1531
	+14 02515			12 31 31	+14 09.1	1	10.15		1.40		1.66		1	K0	280
		CS 16026 # 46		12 31 31	+28 13.5	1	14.03		0.46		−0.16		1		1744
		HA 5 # 98		12 31 31	+74 18.1	1	11.31		0.35		0.02		1		1058
	−60 04151			12 31 31	−60 53.5	1	10.43		1.14		0.99		1		1577
		Coma Ber 1947		12 31 32	+26 38.9	1	13.26		0.74				1		913
		ApJS57,743 T4# 3		12 31 32	−60 53.5	1	12.85		0.45		0.11		1		1577
		Coma Ber 1948		12 31 33	+24 11.7	1	12.18		0.99				2		913
		VX Cru		12 31 33	−60 57.5	1	11.42		1.39		1.02		1		689
		Coma Ber 1956		12 31 39	+26 07.1	1	12.78		0.96				1		913
109390	+32 02256			12 31 39	+32 18.2	1	8.30		0.52				2	F8 V	77
109371	−50 07088			12 31 39	−51 22.2	1	9.67		0.22		0.15		1	A6 V	1771

HD	DM	Other Id	N	Rem	α₁₉₅₀	δ₁₉₅₀	S	V	σ_V	B–V	σ_B–V	U–B	σ_U–B	n	Spectrum	References
		ApJS57,743 T4# 5			12 31 39	−60 56.9	1	12.59		0.15		0.03		1		1577
		NGC 4526 sq1 1			12 31 40	+07 48.4	1	13.25		0.76		0.46		1		1531
109388	+43 02239				12 31 40	+42 39.9	1	9.00		0.47		-0.01		2	F5	1569
109400	+47 01969				12 31 40	+47 01.4	2	7.31	.010	1.06	.005	0.86		4	G8 III	172,280
		ApJS57,743 T4# 6			12 31 40	−60 56.8	1	12.43		1.86				1		1577
		Coma Ber 1958			12 31 43	+24 26.3	1	13.67		1.13				1		913
		CS 16026 # 53			12 31 44	+29 45.4	1	14.80		0.64		0.06		1		1744
109389	+33 02248				12 31 44	+32 28.3	1	8.75		1.03				2	G9 III	77
		ApJS57,743 T4# 7			12 31 44	−60 57.9	1	11.69		0.32		0.24		1		1577
109379	−22 03401	HR 4786			12 31 45	−23 07.2	8	2.65	.006	0.90	.004	0.59	.009	159	G5 II	3,9,15,1020,1034,1075*
	+31 02384				12 31 46	+30 55.9	4	11.03	.023	0.18	.011	0.09	.008	7	A3 V	77,1026,1240,1744
	+10 02441				12 31 47	+09 56.9	1	9.40		1.46		1.81		1	M0	280
		HA 5 # 238			12 31 48	+74 40.4	2	12.06	.007	0.39	.003	-0.07	.003	7	F7	271,1058
		AAS12,381 45 # A			12 31 49	+44 52.5	1	10.12		1.51		1.87		1		280
		Coma Ber 1965			12 31 51	+23 54.7	1	13.73		0.62				1		913
		NGC 4526 sq1 6			12 31 52	+07 59.6	2	11.28	.006	0.67	.000	0.23	.007	6		492,1531
109415	+27 02150	Coma Ber 187			12 31 52	+26 33.9	1	8.23		0.40		-0.08		2	F3 V	3016
		G 59 - 24			12 31 55	+15 33.4	4	12.01	.009	0.42	.010	-0.25	.007	8		1064,1620,1658,3077
		G 13 - 46			12 31 55	−01 57.3	1	12.95		1.34		1.10		1		1658
		NGC 4526 sq1 5			12 31 56	+07 58.1	2	13.34	.006	0.61	.035	0.28	.025	4		492,1531
109397	−57 05563				12 31 56	−57 25.9	1	8.76		0.27		0.25		1	A3 V	1771
109372	−67 02005	BO Mus			12 31 57	−67 28.9	1	6.00		1.71		1.47		1	M6 II/III	3076
109417	+8 02619				12 31 58	+08 00.6	3	6.74	.017	1.09	.007	1.01	.009	12	K2	280,492,1531
	+30 02295				12 31 58	+30 03.2	1	10.10		1.25		1.30		1	K2 III	280
	+33 02249				12 31 58	+32 29.0	1	10.33		1.14		1.08		1	K0 III	280
109409	−43 07755	HR 4788		★ AB	12 31 59	−44 23.7	4	5.77	.007	0.69	.010	0.28		18	G1 V	15,1075,1311,2027
	+35 02353				12 32 00	+35 20.4	1	10.07		0.90				2	G7 III	280
	+44 02207				12 32 00	+43 39.7	1	10.65		0.21		0.09		2	A7 V	1240
		Wray 988, RT Cru			12 32 00	−64 17.4	1	12.80		1.84		1.53		3		1684
109438	+21 02431				12 32 02	+21 10.5	1	8.45		1.16		1.12		2	K0	1569
		LP 615 - 197			12 32 02	−01 17.3	1	13.63		0.57		-0.21		2		1696
		CS 16026 # 63			12 32 03	+30 52.9	1	12.81		1.17		1.00		1		1744
		PHI2/2 33			12 32 03	−30 44.5	1	12.60		1.02				2		1719
	+17 02498				12 32 04	+16 42.1	1	9.69		1.12		1.03		2	K0	1569
109422	−31 09777				12 32 04	−31 35.9	1	8.91		0.72				4	G5 V	2033
		LP 555 - 89			12 32 05	+04 56.8	1	13.05		1.43		1.27		1		1696
		G 123 - 39			12 32 05	+39 30.0	1	11.04		0.92		0.72		1	K1	1658
109439	+12 02488				12 32 06	+11 34.6	1	8.15		0.90		0.58		1	G5	280
		Malm +25 283			12 32 06	+25 03.2	1	12.19		1.31		1.23		3		685
		Malm +25 282			12 32 06	+25 03.3	1	12.21		1.33		1.23		3	dM1	685
		CS 16026 # 62			12 32 06	+31 01.1	1	13.53		0.60		0.04		1		1744
	+42 02322				12 32 06	+41 56.8	1	9.76		1.20		1.36		1	K2 III	280
	+20 02737				12 32 08	+19 42.2	1	9.35		1.00		0.84		1	K2	280
		NGC 4526 sq1 4			12 32 09	+07 55.7	2	13.35	.033	0.61	.015	0.10	.018	3		492,1531
109461	+41 02305				12 32 10	+40 48.9	1	8.81		0.89				2	G8 III	280
109443	−22 03402				12 32 10	−23 12.0	1	9.25		0.36		-0.09		1	F0 V	3016
109399	−72 01275	LSS 2710			12 32 12	−72 26.5	6	7.62	.018	0.00	.010	-0.84	.036	12	B0.5III	191,391,400,540,976,1034
		NGC 4526 sq1 3			12 32 13	+07 52.9	2	12.60	.023	0.69	.002	0.25	.014	4		492,1531
		NGC 4526 sq1 12			12 32 13	+07 55.3	2	14.63	.019	0.75	.000	0.27	.019	3		492,1531
		HA 5 # 239			12 32 13	+74 28.8	1	11.76		1.51		1.81		6	K2	271
109451	+16 02390				12 32 14	+16 05.1	1	8.75		0.84		0.66		1	K0	280
		Coma Ber 1987			12 32 15	+26 05.9	1	13.25		1.12				1		913
109463	+25 02526	Coma Ber 189			12 32 16	+24 30.0	1	7.70		1.39				2	K5 III	280
		ON 254 # 7			12 32 16	+29 41.6	1	14.19		0.93		0.77		1		930
		Coma Ber 1991			12 32 17	+26 50.5	1	12.41		1.04				3		913
		CS 16026 # 64			12 32 17	+30 40.3	1	13.80		0.54		0.01		1		1744
109462	+37 02293	Up 1 - 5			12 32 17	+36 35.6	4	9.13	.039	0.43	.041	-0.08	.010	6	G0	77,948,1141,1748
		Coma Ber 1993			12 32 18	+24 26.2	1	12.01		1.13				2		913
		HA 5 # 308			12 32 18	+74 44.2	1	12.04		0.66		0.13		1		1058
		BPM 21487			12 32 18	−51 27.	1	14.78		0.60		-0.20		2		3065
		Coma Ber 1994			12 32 19	+25 10.4	1	12.88		0.48				2		913
	+27 02151	Coma Ber 190			12 32 19	+27 06.3	2	10.12	.005	0.46	.013	0.02	.031	24	dF5	1613,3016
		LP 555 - 40			12 32 20	+05 20.4	1	16.62		1.66				4		1663
109486	+11 02475				12 32 20	+11 06.7	1	7.88		0.28		0.08		1	F0	280
109482	+29 02301				12 32 20	+29 21.8	4	8.06	.021	0.90	.016	0.65	.019	10	G8 III	44,77,172,8100
	+42 02323				12 32 20	+42 02.1	2	9.80	.019	0.28	.005	-0.03	.010	3	F0 V	1026,1240
		PHI2/2 31			12 32 21	−31 09.3	1	12.91		1.14				2		1719
109485	+23 02475	HR 4789		★ AB	12 32 22	+22 54.3	7	4.81	.009	-0.01	.009	-0.01	.011	35	A0 IV	15,1022,1068,1363*
109484	+25 02528	Coma Ber 191			12 32 24	+24 47.4	1	8.42		0.97		0.74		1	K0	280
		EG 91, AM CVn		★	12 32 29	+37 54.6	2	14.18	.000	-0.22	.010	-1.02	.010	2	DBp	348,3028
	+15 02479				12 32 26	+14 46.2	1	10.30		1.17		1.16		1	G5	280
		Coma Ber 2000			12 32 26	+24 33.4	1	14.07		0.58				1		913
109483	+28 02134	Coma Ber 192			12 32 26	+27 43.9	1	9.96		0.57		0.02		2	G0	3016
	+37 02295				12 32 27	+36 51.7	3	9.41	.016	1.11	.020	1.10	.000	4	K1 III	77,280,1748
109498	+23 02476	G 59 - 25			12 32 28	+23 25.6	1	8.75		0.59		0.08		1	G5	333,1620
	+9 02644				12 32 29	+09 04.6	1	10.28		1.07		0.94		1	K0	280
	+37 02296	Up 1 - 6			12 32 29	+36 40.6	3	9.87	.009	0.49	.030	0.01	.000	4	F7	77,948,1141
		Compar HZ29			12 32 29	+37 52.5	2	12.40	.010	0.65	.007	0.03	.007	8		348,446
		G 12 - 44			12 32 30	+10 06.5	2	11.41	.005	1.45	.005	1.12		2	M1	333,1620,1705
		CS 16026 # 65			12 32 30	+30 32.1	1	13.42		0.66		0.11		1		1744

Table 1 685

HD	DM	Other Id	N Rem	α_{1950}	δ_{1950}	S	V	σ_V	B–V	σ_{B-V}	U–B	σ_{U-B}	n	Spectrum	References
		HA 5 # 172		12 32 30	+74 29.2	1	13.12		0.79		0.40		1		1058
		ON 254 # 2		12 32 32	+29 34.7	1	14.84		1.28		1.16		1		930
		ON 254 # 4		12 32 33	+29 32.9	1	13.35		0.78		0.54		1		930
		ON 254 # 6		12 32 33	+29 34.	1	15.73		0.80		0.32		1		930
109497	+30 02296			12 32 33	+30 28.1	1	8.63		0.49				2	F6 IV	77
109509	+37 02297	Up 1 - 2		12 32 33	+36 36.1	4	8.13	.017	0.49	.019	0.02	.017	6	F8	77,948,1141,1748
		GD 148		12 32 33	+47 54.2	1	14.52		0.06		-0.68		2	DA	3060
109448	-67 02009			12 32 33	-68 04.5	1	7.52		0.03		0.03		2	A0 V	1771
109510	+19 02584	HR 4791	⋆ B	12 32 36	+18 39.1	2	6.58	.015	0.26	.015	0.11	.005	6	A7 m	3024,8071
		NGP +26 119		12 32 36	+25 33.2	1	11.91		0.91				1	K2 IV	44
		ON 254 # 1		12 32 36	+29 36.4	1	12.09		0.61		0.25		1		930
109511	+19 02584	HR 4792	⋆ A	12 32 37	+18 39.1	3	5.03	.024	1.15	.018	1.11	.000	6	K2 III	1080,1363,3024
	+26 02364			12 32 37	+25 34.5	3	10.16	.013	0.98	.008	0.84	.010	4	K0 III	44,280,8041
109519	+22 02490	HR 4793		12 32 38	+22 09.4	2	5.87	.015	1.24	.019	1.33		7	K1 III	70,1501
		ON 254 # 3		12 32 38	+29 32.9	1	12.73		0.85		0.84		1		930
109551	+70 00705	HR 4795		12 32 38	+70 17.8	2	4.90	.022	1.32	.003	1.15		5	K2 III	1363,3016
109475	-61 03296			12 32 38	-61 27.9	2	6.61	.003	0.47	.005	0.04		7	F2/3 III	1688,2029
109520	+12 02489			12 32 39	+11 49.4	1	7.50		1.04		0.93		1	K0	280
109491	-57 05574			12 32 39	-57 40.3	1	8.81		-0.03		-0.42		2	B8 III/IV	401
109492	-61 03298	HR 4790		12 32 39	-61 33.9	6	6.22	.009	0.73	.004	0.27	.019	23	G3 III	15,329,1034,1075,1075,2029
	-63 02356	LSS 2711		12 32 39	-63 56.7	2	10.41	.005	0.19	.014	-0.55	.048	5	B2 Ve	191,1684
		HA 5 # 514		12 32 40	+75 14.0	1	13.37		0.52		0.02		1		1058
		HA 5 # 513		12 32 40	+75 14.7	1	11.76		0.55		-0.01		1		1058
109477	-62 02834			12 32 40	-62 50.0	1	9.42		0.20		0.12		2	B9 V	1688
	+31 02385	IDS12303N3134	A	12 32 41	+31 20.4	1	9.79		0.70				2	G5 IV	77
		SSII 176		12 32 42	+40 25.	1	13.07		-0.07		-0.27		1	A0	1298
109493	-62 02835			12 32 42	-62 35.8	1	9.85		0.10		-0.29		1	B9 III	1688
109476	-61 03297			12 32 43	-62 04.9	1	9.78		0.10		0.02		2	A0 V	1688
109478	-64 01966			12 32 44	-65 00.1	1	8.96		0.38		0.04		2	F0 IV	1688
109530	+37 02298	Up 1 - 1		12 32 45	+36 42.0	4	7.28	.023	0.46	.025	0.05	.008	6	F5	77,948,1141,1748
		Coma Ber 2011		12 32 46	+23 50.8	1	13.53		0.89				1		913
109504	-60 04168			12 32 47	-61 13.6	1	9.27		0.06		-0.27		2	B8/9 V	1688
		ON 254 # 5		12 32 48	+29 33.3	1	14.51		1.10		1.47		1		930
	+26 02366			12 32 49	+26 18.5	1	11.20		0.91				2	G9 III	44
		AAS12,381 10 # A		12 32 50	+09 43.6	1	11.04		-0.02		0.00		1		280
	+36 02284	Up 1 - 7		12 32 50	+36 24.1	3	9.38	.043	0.54	.017	0.08	.044	5	F8	77,948,1748
	+48 02038			12 32 50	+47 31.1	2	10.97	.030	0.17	.005	0.03	.005	5	A3 V	1026,1240
109543	+33 02251			12 32 51	+33 19.4	2	8.80	.005	0.68	.015	0.20		3	G2 IV	77,8038
	+45 02048			12 32 52	+44 42.6	1	9.34		1.28				2	K2 III	280
	+36 02285			12 32 53	+35 32.7	2	9.51	.018	0.51	.000	0.04		11	F8	77,1569
109542	+37 02299	Up 1 - 3		12 32 53	+36 36.8	4	8.23	.021	0.41	.028	-0.02	.032	6	F2	77,948,1141,1748
		Coma Ber 2018		12 32 54	+24 38.6	1	12.68		0.83				1		913
		CS 16026 # 61		12 32 54	+31 34.6	1	14.25		0.15		0.07		1		1744
		HA 5 # 174		12 32 54	+74 27.2	2	12.43	.010	0.63	.010	0.10	.003	7		271,1058
109524	-34 08280	IDS12302S3420	A	12 32 54	-34 36.3	1	7.91		1.03		0.83		3	K4 V	3072
109524	-34 08280	IDS12302S3420	B	12 32 54	-34 36.3	1	11.91		1.60		1.28		2		3072
109505	-60 04170	LSS 2712		12 32 54	-61 17.9	4	8.00	.014	0.23	.007	-0.67	.018	8	B2 II	401,540,976,1688
		AO 705		12 32 56	+36 29.5	1	11.68		1.03		0.77		1	K1 V	1748
		NGP +29 120		12 32 57	+29 03.8	1	11.91		0.67		0.35		1	G8 III	8034
109517	-61 03303	IDS12301S6117	AB	12 32 58	-61 33.9	2	8.72	.002	0.12	.002	0.06	.001	4	B9 V	401,1688
109552	+29 02303			12 32 59	+29 06.8	1	8.10		0.59				2	F8 IV	77
	+10 02444			12 33 00	+09 56.0	1	9.70		1.14		1.19		1	K0	280
		HA 5 # 38		12 33 00	+74 04.1	1	12.69		0.61		0.04		1		1058
	-59 04330	LSS 2713		12 33 00	-60 14.9	3	10.27	.017	0.44	.021	-0.59	.040	8	O9.5Iab	191,401,1684
	+45 02050			12 33 01	+45 06.0	1	9.43		1.07				2	K0 III	280
109528	-61 03305			12 33 01	-62 13.0	1	10.32		0.21		0.00		2	A0 III/IV	1688
109553	+16 02392			12 33 02	+15 54.7	1	8.09		1.37		1.60		1	K5	280
		Coma Ber 2021		12 33 02	+24 26.0	1	14.25		0.48				1		913
	+36 02286			12 33 03	+36 01.2	1	9.54		0.53		0.02		13	F8	1569
		HA 5 # 515		12 33 04	+75 16.3	1	12.51		0.39		-0.03		1		1058
109536	-40 07376	HR 4794		12 33 04	-40 44.8	6	5.12	.011	0.22	.017	0.03	.030	16	A5 IV/V(+F)	15,2013,2016,2029*
	+44 02208			12 33 06	+43 59.0	1	10.49		0.96		0.83		1	G9 III	280
	+12 02490			12 33 07	+11 36.1	3	9.77	.015	1.54	.026	1.84	.069	8	K5	280,492,1531
	+16 02393			12 33 07	+15 33.6	1	10.58		1.13		1.22		1		280
	+18 02631			12 33 12	+17 30.0	1	10.55		0.64		0.08		2		1569
109581	+34 02336			12 33 12	+34 20.1	1	8.46		1.60				2	K3 IV	77
		NGP +26 121		12 33 14	+26 11.9	2	11.93	.029	1.00	.019			3	K0 III	44,492
		PHI2/2 23		12 33 14	-30 51.5	1	12.75		1.19				2		1719
109550	-62 02838			12 33 14	-62 34.1	1	8.80		0.10		-0.29		1	B9 II	1688
		NGC 4565 sq1 2		12 33 15	+26 15.4	1	11.04		0.51				3		492
	-45 07872			12 33 15	-45 38.7	4	11.09	.022	1.49	.015	1.13	.014	8	M1	158,912,1705,3073
		NGC 4565 sq1 3		12 33 16	+26 18.5	1	12.85		1.23				1		492
	+47 01973			12 33 17	+46 29.7	1	9.94		1.32		1.42		1	K2	280
	+30 02298			12 33 19	+30 28.1	2	9.56	.005	0.96	.015	0.76		3	G9 III	77,8034
109573	-39 07717	HR 4796	⋆ AB	12 33 19	-39 35.6	2	5.79	.005	0.01	.000			7	A0 V	15,2027
109575	-49 07226			12 33 20	-50 22.7	1	8.45		0.16		0.18		1	A3 (V)	1771
109585	-19 03521	HR 4797, TU Crv		12 33 21	-20 15.1	3	6.20	.005	0.33	.005	0.07		9	F0 V	15,401,2029
	+21 02433			12 33 22	+20 57.3	1	11.23		0.22		0.11		2		1569
109615	+40 02551			12 33 22	+39 57.5	2	7.30	.005	-0.05	.005	-0.06	.015	4	A0 V	1068,1240
		HA 5 # 244		12 33 22	+74 41.7	1	11.36		0.51		0.05		1		1058

HD	DM	Other Id	N Rem	α_{1950}	δ_{1950}	S	V	σ_V	B−V	σ_{B-V}	U−B	σ_{U-B}	n	Spectrum	References
109590	−41 07285			12 33 23	−42 02.1	1	7.97		0.44				4	F5 V	2033
	+12 02491			12 33 24	+11 45.3	2	10.09	.009	0.52	.002	0.03	.020	10		492,1531
		NGC 4564 sq1 5		12 33 24	+11 46.5	2	12.34	.064	0.52	.023	-0.07	.000	5		492,1531
	+38 02349			12 33 24	+37 48.0	1	9.04		0.48				2		77
	+14 02519			12 33 26	+13 46.0	1	10.13		0.97		0.59		1	K0	280
	+23 02477			12 33 26	+22 32.3	1	9.69		0.88		0.69		1	K2	280
		Coma Ber 2030		12 33 26	+24 30.3	1	13.55		0.82				1		913
109593	−51 06827			12 33 26	−52 08.5	1	6.56		1.63		2.01		5	K5/M0 III	1628
109628	+12 02492	G 60 - 20	⋆ AB	12 33 27	+11 40.9	3	8.43	.022	0.60	.001	0.05	.007	12	G2 V	492,1003,1531
	+25 02530			12 33 27	+25 03.1	1	10.53		1.03				2	K0 III	44
109616	+29 02305			12 33 27	+28 58.2	3	9.46	.027	0.93	.007	0.69		5	G8 III	44,77,8034
		CS 16026 # 60		12 33 27	+31 58.7	1	14.26		0.60		-0.16		1		1744
		SSII 177		12 33 27	+42 39.2	3	12.02	.026	-0.24	.013	-0.95	.022	6	sdB	1026,1240,1298
	+38 02350			12 33 28	+38 03.7	1	9.57		0.40				2	F8	77
109563	−66 01834			12 33 29	−66 53.7	1	7.26		-0.03		-0.11		2	B9 IV	1688
		NGC 4564 sq1 3		12 33 30	+11 43.4	2	12.31	.041	0.70	.015	0.20	.011	4		492,1531
109626	+30 02299			12 33 30	+30 12.1	2	8.91	.005	1.16	.019	1.15		3	K1 III	77,8034
109625	+32 02258			12 33 30	+32 00.5	1	7.97		0.55				std	G0	77
		SV CVn		12 33 30	+37 29.2	1	12.20		0.16		0.10		1	A0	597
109655	+46 01797			12 33 30	+46 03.3	2	7.29	.020	1.60	.010	1.93		4	K5 III	172,280
109647	+52 01638			12 33 30	+51 29.8	1	8.49		0.95		0.70		2	K0	3016
		PHI2/2 19		12 33 30	−30 46.8	1	12.11		1.00				2		1719
	+26 02368			12 33 32	+26 01.8	1	10.07		1.05				2	G8 III	44
		NGC 4569 sq1 4		12 33 33	+13 27.4	1	10.95		0.49		-0.02		3		492
109627	+26 02369	Coma Ber 194		12 33 33	+25 42.0	5	7.72	.018	1.05	.012	0.95	.013	9	K0 III	44,105,172,280,8034
109649	+32 02259			12 33 34	+32 16.8	1	7.37		1.35				std	K3 III	77
109648	+37 02300	Up 1 - 4		12 33 34	+36 32.0	3	8.77	.013	0.50	.024	-0.01	.019	5	F8	77,948,1748
109595	−61 03307			12 33 36	−61 51.7	1	10.12		0.23		0.22		2	A1 IV/V	1688
		Coma Ber 2036		12 33 37	+24 27.0	1	14.35		0.65				1		913
		CS 16026 # 66		12 33 38	+27 56.8	1	14.12		0.37		-0.10		1		1744
109650	+30 02300			12 33 38	+29 43.7	1	8.80		1.26				2	K0 III	77
	+45 02051			12 33 42	+44 57.4	1	10.44		0.97		0.79		1	G9 III	280
	+35 02355			12 33 43	+35 03.6	2	9.50		1.02	.015	0.89		4	K0 III	77,1569
109638	−42 07765			12 33 43	−43 12.8	1	7.62		1.43		1.67		5	K3 III	1673
		NGC 4565 sq1 4		12 33 44	+26 12.3	1	13.61		0.75				4		492
109614	−66 01837	IDS12308S6640	AB	12 33 44	−66 56.3	1	8.57		0.00		-0.34		2	B8 IV/V	1688
	−59 04337			12 33 45	−59 52.7	1	11.40		0.21		-0.32		4		1684
		L 38 - 15		12 33 45	−76 40.6	4	10.99	.035	1.30	.027	1.06		8		158,912,1705,3073
		NGC 4564 sq1 6		12 33 46	+11 44.8	2	12.23	.011	0.99	.000	0.76	.030	5		492,1531
	+13 02555			12 33 46	+13 18.0	1	10.28		0.42		0.02		3	F5	492
109681	+41 02308			12 33 48	+40 51.7	2	7.61	.000	1.09	.020	1.06		4	K2 III	172,280
	+44 02210			12 33 48	+44 10.6	1	10.44		1.11		1.11		1		280
109680	+49 02151			12 33 48	+49 27.5	1	8.78		0.29		0.05		2	F0 V	1240
		BSD 5 # 354		12 33 48	+74 02.3	1	12.35		0.64		0.16		1	G0	1058
	−65 01917			12 33 48	−65 46.9	1	9.59		0.14		-0.67		2	B5	1688
	+46 01798			12 33 49	+45 57.5	1	9.88		0.71				2	G5	280
	+10 02449			12 33 50	+09 40.9	1	10.43		0.97		0.86		2		1569
		NGP +26 124		12 33 50	+25 30.8	1	11.92		0.72				1	G6	44
109683	+34 02337			12 33 50	+34 11.9	1	8.83		0.47				2	G0	77
109702	+57 01382	IDS12316N5723	AB	12 33 52	+57 06.1	1	7.23		1.18		1.27		3	K2 III	37
		CS 16026 # 67		12 33 53	+27 33.1	1	13.39		-0.01		0.04		1		1744
		HA 5 # 107		12 33 53	+74 21.0	2	11.98	.011	0.50	.011	0.05	.004	5		271,1058
109642	−60 04183			12 33 53	−60 27.0	1	9.46		0.23		0.17		2	A2 IV/V	1771
109644	−62 02843			12 33 53	−62 37.7	1	8.69		0.17		0.13		2	A0 IV	1688
		Coma Ber 2042		12 33 55	+24 02.2	1	12.40		1.21				3		913
109623	−66 01841			12 33 55	−66 50.5	1	9.70		0.03		-0.14		2	B9/A0 V	1688
		Coma Ber 2044		12 33 57	+25 03.6	1	13.13		0.75				1		913
		NGC 4565 sq1 5		12 33 57	+26 16.9	1	13.47		0.84				1		492
109691	+32 02260			12 33 57	+32 28.3	3	8.87	.006	-0.04	.000	-0.07	.009	14	A0 V	77,1068,1240
		Coma Ber 2045		12 33 58	+24 02.1	1	14.30		0.54				1		913
	+46 01799			12 33 58	+46 25.1	1	9.78		1.01				2	G9 III	280
		HA 5 # 313		12 33 58	+74 49.1	1	13.04		0.78		0.44		1		1058
109664	−54 05270			12 33 58	−54 24.9	1	9.96		0.33		0.13		1	A9 V	1771
		HA 5 # 178		12 33 59	+74 25.1	2	10.65	.022	0.39	.007	0.00	.004	5		271,1058
109675	−49 07233			12 34 00	−50 03.6	1	6.40		0.03		-0.04		1	A0 V	1770
	+32 02261			12 34 04	+31 51.9	1	9.75		0.56				2	F8	77
		NGC 4564 sq1 7		12 34 05	+11 39.9	1	12.19		0.84		0.59		2		492
		NGC 4569 sq1 5		12 34 05	+13 32.5	1	12.08		0.47				2		492
109693	−4 03312			12 34 05	−05 08.8	1	9.75		0.45		-0.01		3	F8	1731
109667	−58 04402			12 34 05	−59 13.9	1	9.28		0.01		-0.39		2	B7 IV/V	401
		CS 16026 # 59		12 34 06	+32 20.0	1	14.48		0.22		-0.02		1		1744
	+33 02254			12 34 06	+32 37.9	1	10.36		1.12		1.07		1	G8 III	8034
	+45 02052			12 34 06	+45 10.5	2	8.61	.020	0.93	.010	0.65		4	G9 IV	172,280
		NGC 4564 sq1 8		12 34 07	+11 53.0	2	10.74	.049	0.61	.015	0.00	.054	3		492,1531
109729	+60 01406	HR 4800, T UMa		12 34 07	+59 45.7	1	8.11		1.40		0.77		1	M4	3001
	+20 02740			12 34 08	+19 54.1	1	10.15		1.03		0.90		1	K2	280
	+12 02493			12 34 09	+12 12.8	1	10.13		1.09		1.00		1	K0	280
		NGP +33 123		12 34 09	+32 52.8	1	10.92		0.95		0.71		1	K0 III	8034
109697	−32 08814			12 34 09	−33 07.8	1	9.97		-0.17		-0.89		4	B2 III	1732
		Coma Ber 2047		12 34 10	+24 58.9	1	13.69		0.74				1		913

Table 1   687

HD	DM	Other Id	N Rem	α_{1950}	δ_{1950}	S	V	σ_V	B–V	σ_{B-V}	U–B	σ_{U-B}	n	Spectrum	References
	+45 02053			12 34 10	+45 22.7	1	10.40		1.04		0.95		1	K0 III	280
109678	−61 03312			12 34 10	−62 23.2	1	9.48		0.24		0.07		2	B8 IV/V	1688
109668	−68 01702	HR 4798, α Mus	★ A	12 34 11	−68 51.6	4	2.69	.009	-0.21	.006	-0.84	.018	13	B2 IV	26,1034,1088,8023
	+28 02137	G 59 - 27		12 34 12	+27 45.1	4	10.90	.020	0.40	.008	-0.23	.013	7	sdF2	516,1298,1620,1744
	+16 02394			12 34 13	+15 32.3	1	9.27		1.11		1.12		1	K0	280
109704	−5 03535	HR 4799		12 34 13	−05 33.4	1	5.86		0.07		0.08		8	A3 V	1088
		NGC 4565 sq1 9		12 34 14	+26 08.9	1	14.73		0.57				1		492
		NGC 4569 sq1 6		12 34 15	+13 32.8	1	11.67		0.27				1		492
		NGC 4569 sq1 9		12 34 18	+13 21.7	1	14.54		0.78				1		492
	+33 02255			12 34 18	+33 11.1	2	9.55	.034	1.00	.019	0.86		3	G9 III	77,8034
		L 19 - 83		12 34 18	−85 20.	1	12.77		0.52		-0.09		2		1696
	+16 02395			12 34 19	+15 29.7	1	10.76		0.27		0.10		1		280
109730	+30 02302			12 34 20	+29 58.0	1	9.31		0.52				2	G0	77
109740	+41 02309			12 34 20	+41 21.5	1	8.79		1.16				2	K1 III	280
		Coma Ber 2052		12 34 22	+23 58.4	1	13.95		0.59				1		913
	−20 03682			12 34 23	−21 04.3	1	10.78		0.48		-0.06		2		1696
		Coma Ber 2053		12 34 24	+24 17.9	1	11.98		1.13				2		913
		Ton 624		12 34 24	+28 56.	1	15.64		0.17		0.13		1		1036
		AAS78,203 # 16		12 34 24	−63 05.	1	11.70		2.05		2.51		1		1684
	+27 02152			12 34 25	+27 20.4	1	10.12		1.79				2	M7	44
	+40 02554			12 34 27	+40 02.2	1	9.69		1.15				3	K1 III	280
109724	−56 05346			12 34 27	−56 30.8	1	8.67		-0.02		-0.50		2	B5 V	401
109742	+17 02504	HR 4801		12 34 28	+17 21.9	3	5.69	.008	1.43	.014	1.73	.015	7	K5 III	70,1501,3001
		NGC 4569 sq1 10		12 34 29	+13 22.0	1	14.72		0.57				1		492
	+15 02482			12 34 30	+15 16.0	1	10.42		1.03		1.00		1		280
	+33 02256			12 34 30	+32 52.5	1	10.15		1.51		0.81		1	M6	280
	+15 02483			12 34 31	+14 31.6	2	8.93	.020	0.52	.005	0.05	.010	5	G0	1569,1648
		NGC 4569 sq1 8		12 34 35	+13 28.1	1	12.93		0.48				2		492
109764	+9 02648			12 34 36	+09 04.3	1	6.59		0.25		0.21		1	A2	280
		CS 16026 # 75		12 34 36	+32 11.0	1	14.08		0.51		-0.12		1		1744
109762	+33 02257			12 34 36	+33 18.1	2	8.59	.005	0.26	.005	0.11		6	Am	77,1068
109763	+12 02494			12 34 37	+12 27.4	1	9.11		0.90		0.50		2	K0	1569
		NGC 4569 sq1 7		12 34 38	+13 31.6	1	13.09		0.95				1		492
		NGC 4565 sq1 7		12 34 38	+26 11.6	1	13.93		0.62				1		492
		CS 16026 # 69		12 34 39	+29 02.5	1	13.73		0.60		0.08		1		1744
109738	−67 02025			12 34 39	−67 35.4	1	8.26		0.19		0.05		1	A4/5 [IV]w	1771
		NGC 4565 sq1 8		12 34 40	+26 11.2	1	13.92		0.62				1		492
109781	+47 01976			12 34 40	+46 57.4	2	8.25	.020	0.83	.010	0.36		4	G8 III	172,280
109771	+12 02495			12 34 41	+12 05.6	1	7.98		1.00		0.84		1	K0	280
		Coma Ber 2059		12 34 43	+24 39.7	1	14.07		0.37				1		913
		NGC 4565 sq1 6		12 34 43	+26 16.7	1	12.78		0.66				1		492
109752	−57 05600	IDS12319S5732	AB	12 34 43	−57 48.8	1	7.98		0.01		-0.35		2	B8 III	401
	+31 02387			12 34 47	+30 41.8	1	9.97		0.62				2	G0	77
109793	+13 02561			12 34 51	+13 15.5	1	8.22		0.89		0.59		1	G5	280
		CS 16026 # 74		12 34 52	+32 01.8	1	14.45		0.56		-0.11		1		1744
		AAS12,381 29 # A		12 34 53	+28 58.7	1	10.30		0.94		0.63		1		280
	+25 02534			12 34 55	+25 20.5	7	10.52	.011	-0.27	.014	-1.10	.018	29	sdO	272,830,1026,1240,1298,1783*
		HA 5 # 316		12 34 57	+74 43.5	1	12.23		0.75		0.27		1		1058
109787	−47 07745	HR 4802		12 34 57	−48 16.0	5	3.86	.003	0.05	.005	0.03	.003	21	A2 V	15,1075,2012,3023,8015
	+40 02555			12 34 58	+40 22.0	1	9.16		1.13				3	K0 III	280
		HA 5 # 43		12 34 58	+74 09.9	1	12.00		0.64		0.19		1		1058
		GPEC 1434		12 35 00	−34 46.5	1	11.46		1.04				2		1719
109761	−68 01706			12 35 00	−69 19.7	1	7.39		0.93				4	G6 III	1075
109777	−57 05602			12 35 01	−57 35.4	1	7.56		0.03		-0.22		2	B8 V	401
109804	+29 02306			12 35 03	+28 30.0	3	9.85	.024	1.00	.005	0.80	.010	4	G9 III	44,280,8034
	+43 02241			12 35 03	+43 19.7	1	10.13		1.00		0.86		1	K0 III	280
109799	−26 09233	HR 4803	★ AB	12 35 03	−26 51.8	7	5.44	.017	0.33	.016	0.03	.006	19	F0 V	15,1007,1008,1013*
		NGP +28 100		12 35 05	+27 35.0	1	11.80		0.70				2	G8 III	44
	+8 02624			12 35 06	+07 42.1	1	9.44		0.99		0.72		1	G5	280
	+10 02452			12 35 06	+10 03.7	1	10.64		0.23		0.16		1		280
109815	+11 02478			12 35 06	+10 51.5	1	7.58		1.34		1.61		1	K2	280
109814	+13 02562			12 35 07	+12 56.7	1	8.99		1.35		1.44		1	K0	280
		CS 16026 # 72		12 35 10	+30 49.6	1	13.54		0.62		0.08		1		1744
	+38 02351			12 35 10	+38 05.3	1	8.82		1.16				2	K1 III	280
109838	+46 01802			12 35 10	+45 31.7	1	8.04		0.31		0.00		2	F2 V	1240
		G 164 - 5		12 35 11	+38 12.2	1	12.10		0.58		-0.09		2		1658
		LSS 2715		12 35 12	−62 28.6	2	10.91	.000	0.80	.011	-0.41	.028	7	B3 Ib	1684,1737
	+16 02396			12 35 13	+15 50.6	2	10.18	.034	0.23	.019	0.12	.024	3		280,1569
109839	+39 02540			12 35 14	+38 57.6	1	7.35		0.48		-0.02		2	F5 V	105
	+46 01803			12 35 14	+45 38.0	1	10.77		0.94		0.69		2	G5	280
	−51 06859		V	12 35 16	−51 43.6	5	10.66	.012	1.53	.039	1.15	.010	8	M3	912,1696,1705,2033,3078
		Coma Ber 2070		12 35 18	+24 54.2	1	13.41		0.54				1		913
109808	−55 05161	IDS12325S5523	AB	12 35 18	−55 39.4	2	6.94	.008	0.16	.006	0.05	.002	3	A1/2 V	401,1770
109801	−65 01926	IDS12324S6517	AB	12 35 19	−65 33.6	4	8.45	.016	0.38	.013	0.03	.014	7	A9 V	1594,1688,1771,6006
		KUV 12353+2925		12 35 20	+29 24.9	1	17.29		-0.13		-0.84		1	DA	1708
109845	+33 02258			12 35 20	+32 48.4	2	8.74	.005	0.33	.000	0.02		4	F3 V	77,1240
	+10 02453			12 35 21	+10 16.6	1	9.73		0.86		0.51		1	K0	280
	+19 02591			12 35 21	+19 10.3	1	9.42		1.48		1.85		1	K5	280
109810	−61 03317			12 35 22	−61 31.9	1	9.86		0.10		-0.23		1	B8/9 II/III	1688
	+41 02310			12 35 23	+41 27.3	1	10.11		1.05		0.94		1	G9 III	280

HD	DM	Other Id	N	Rem	α₁₉₅₀	δ₁₉₅₀	S	V	σ_V	B–V	σ_B-V	U–B	σ_U-B	n	Spectrum	References
109827	−49 07250				12 35 23	−49 49.1	1	9.21		−0.05		−0.30		1	B9 IV	1770
		G 164 - 6			12 35 24	+32 12.5	1	15.76		1.28		1.10		1		3062
109811	−64 01973				12 35 24	−65 04.3	1	9.66		0.43		0.00		2	F0 V	1688
	+12 02497	G 60 - 21			12 35 25	+12 11.5	1	11.09		1.07		0.92		1		1620
	+26 02373				12 35 25	+26 04.9	2	10.28	.013	0.25	.004	0.09	.000	11	F0 IV	1026,1068
	+37 02303				12 35 27	+37 18.7	1	9.29		1.00				2	K0 III	77
		GD 266			12 35 27	+37 47.1	1	15.62		0.60		−0.11		1		3060
	+12 02498				12 35 28	+11 43.1	1	10.24		0.80		0.40		1	G5	280
		CS 16032 # 7			12 35 31	+03 01.8	1	14.67		0.32		0.09		1		1744
109860	+4 02631	HR 4805			12 35 31	+03 33.4	3	6.33	.005	0.02	.010	0.03	.024	9	A1 V	15,252,1417
109842	−46 08050				12 35 31	−46 50.4	2	8.07	.000	0.51	.000			8	F5 V	1075,2012
109851	−44 08126				12 35 33	−45 11.7	1	10.32		0.39		0.14		2	A9 V	1730
	+36 02290				12 35 34	+35 53.7	1	10.36		0.67				2	G6 III	280
	+45 02055				12 35 34	+45 01.1	1	11.62		1.31		1.24		2	K7	1746
		Steph 1027			12 35 37	+45 00.8	1	11.48		1.28		1.26		2	K5	1746
	+22 02494	G 59 - 29			12 35 38	+22 22.9	1	10.97		0.99		0.86		1	K5	1620
109813	−75 00827				12 35 38	−75 47.5	1	6.82		1.68		1.99		5	K4 III	1628
	−66 01856				12 35 39	−67 17.4	1	11.28		0.08		0.00		3		1684
		LP 377 - 29			12 35 40	+23 58.0	1	11.35		1.49				1		1017
		BSD 5 # 361			12 35 40	+74 03.4	1	13.19		0.42		−0.03		1	F5	1058
		He3 801			12 35 40	−62 15.	1	11.41		0.69		−0.35		4		1684
		AAS78,203 # 17			12 35 42	−62 05.	1	11.64		0.48		−0.28		3		1684
109832	−68 01709				12 35 42	−68 29.3	1	8.11		0.35		−0.01		1	A9 V	1771
	−60 04204	LSS 2717			12 35 43	−60 58.7	2	10.35	.030	0.55	.005	−0.26	.022	5	B2 Ve	191,1684
110010	+80 00389				12 35 44	+79 29.4	2	6.98	.015	0.61	.015	0.17	.050	4	G0	1733,3026
109875	−10 03512	IDS12331S1058	A		12 35 44	−11 14.5	1	7.64		1.48		1.78		3	K0	1657
	−15 03508				12 35 44	−15 42.3	1	10.37		1.53		1.87		2	M0	1746
		PHI4/2 35			12 35 45	−29 43.7	1	13.48		0.87				3		1719
	+21 02437				12 35 46	+20 44.4	1	9.26		1.09		0.91		2	K5	1569
	+39 02541				12 35 46	+38 57.2	1	10.11		0.30		0.07		2	F0 V	1240
109880	−47 07755				12 35 46	−47 40.1	1	7.76		0.01		−0.04		3	B9 (IV)	1770
	+27 02154	Coma Ber 199			12 35 47	+26 37.9	4	10.15	.016	0.88	.002	0.51	.032	10	G7 III	44,280,1613,8034
109895	+32 02262				12 35 47	+32 12.2	1	9.72		0.72				2	K0	77
		NGP +28 102			12 35 48	+28 15.9	1	11.67		1.04				2	K0 III	44
109896	+2 02560	HR 4807, FW Vir			12 35 49	+02 07.8	4	5.67	.015	1.59	.004	1.83	.005	73	M3+IIIb	15,1256,3055,8015
		PHI4/2 34			12 35 49	−29 36.6	1	13.56		0.93				2		1719
		CS 16032 # 4			12 35 50	+03 01.4	1	12.84		1.22		1.19		1		1744
	−66 01860	LSS 2718			12 35 50	−66 35.7	2	10.31	.020	0.10	.001	−0.70	.009	4	B1 III	540,976
109867	−66 01861	HR 4806, LSS 2720	⋆ A		12 35 53	−66 55.1	12	6.25	.017	0.05	.008	−0.84	.014	45	B1 Ia	15,116,540,976,1020,1075*
	+27 02155	Coma Ber 200			12 35 55	+26 36.4	1	9.94		0.66		0.20		6	dK0	1613
109928	+32 02263				12 35 55	+32 21.7	2	9.53	.015	0.51	.015	−0.03		4	G5	77,1569
109890	−52 05754				12 35 56	−53 20.5	1	8.96		0.14		0.10		1	A2 V	1771
	−64 01975				12 35 56	−64 51.7	1	10.52		0.63		−0.05		2	B8	1688
109914	+7 02561	HR 4808, R Vir			12 35 58	+07 15.8	8	7.06	.268	1.46	.058	1.18	.072	18	M4	70,635,814,817,3051*
	+36 02292				12 35 59	+36 27.1	2	11.04	.005	0.25	.015	−0.01		4		77,1569
109857	−74 00955	HR 4804	⋆ AB		12 36 00	−75 05.7	6	6.49	.012	0.09	.005	−0.24	.006	29	B8 Vn	15,278,1075,1637,2012,2038
		NGC 4590 sq1 4			12 36 01	−26 24.2	1	11.99		0.99				2		438
109953	+44 02212				12 36 02	+44 22.7	1	8.85		0.98				2	G9 III	280
		L 187 - 386			12 36 02	−49 32.5	2	13.78	.000	0.18	.010	−0.68	.024	8	DA6	1698,1765
		L 327 - 186			12 36 02	−49 32.5	1	13.96		0.18		−0.70		2	DA	3065
109941	+29 02308				12 36 03	+28 50.9	3	8.70	.011	1.09	.010	1.06		5	K0 III	44,280,8034
109942	+14 02523				12 36 04	+14 04.8	1	7.18		1.17				2	K0	280
	+32 02264				12 36 04	+31 45.9	2	9.93	.005	0.63	.010	0.11		4		77,1569
		G 59 - 30			12 36 05	+20 56.2	1	12.47		1.18		1.06		1	K7	1620
	+21 02438				12 36 06	+20 59.7	1	9.13		1.01		0.79		2	K2	1569
109891	−61 03319	IDS12332S6159	AB		12 36 06	−62 15.3	2	8.62	.010	0.24	.007	0.13	.033	4	A0 V	1688,1771
		G 164 - 8			12 36 07	+35 29.8	1	14.85		1.31		1.26		1		3062
		HA 5 # 385			12 36 07	+74 55.3	1	12.24		0.63		0.09		1		1058
		NGC 4590 sq2 7			12 36 07	−26 32.4	1	14.14		1.30		1.24		3		548
109930	−4 03319	G 13 - 50	⋆ A		12 36 08	−05 03.1	1	8.55		0.90		0.64		5	K0	1731
109930	−4 03319	G 13 - 50	⋆ A		12 36 08	−05 03.1	2	9.03	.020	0.85	.005	0.55	.005	2	K0	1620,1658
109930	−4 03319	G 13 - 50	⋆ B		12 36 08	−05 03.1	2	9.66	.010	0.98	.025	0.79	.000	2		1620,1658
109931	−17 03668	HR 4809			12 36 08	−17 58.6	3	6.00	.012	0.30	.004	0.10		9	F0 V	15,2029,3037
		CS 16032 # 5			12 36 09	+03 01.3	1	14.39		0.55		0.09		1		1744
109944	−3 03329	IDS12335S0350	A		12 36 09	−04 05.9	1	6.67		1.60		1.97		2	K5	3040
	+30 02309				12 36 10	+29 35.9	1	10.35		1.05		0.98		1	K1 III	280
		L 471 - 42			12 36 10	−38 04.9	1	12.74		1.69		1.25		1		3078
109955	+9 02649				12 36 11	+09 02.6	1	9.51		0.65		0.23		1	K0	280
	+25 02537				12 36 11	+25 28.2	1	10.92		1.29				2	K2 III	44
	+36 02293				12 36 11	+35 35.5	1	10.16		0.67				2		280
109979	+46 01805				12 36 11	+45 29.5	1	7.23		0.38		−0.07		2	F2	1569
109892	−68 01713				12 36 11	−69 16.1	2	8.44	.020	0.03	.005	−0.08	.010	4	B9.5V	401,1771
109954	+24 02480				12 36 12	+23 54.6	1	8.42		0.86		0.57		1	K0	280
	+29 02309				12 36 12	+29 12.8	1	10.29		0.81				2	G7 III	44
		NGC 4590 sq2 6			12 36 13	−26 28.9	1	13.83		0.68		0.08		3		548
109885	−70 01502				12 36 13	−71 20.9	3	9.03	.020	0.13	.016	−0.66	.018	9	B2 III	400,540,976
	+75 00476				12 36 14	+74 45.6	1	10.05		0.46		0.00		2	F5	1058
		NGP +28 103			12 36 15	+27 56.6	1	11.88		0.97				2	K2 III	44
		CS 16026 # 70			12 36 15	+29 49.0	1	14.31		0.57		0.02		1		1744
		G 13 - 51			12 36 15	−04 02.8	1	13.52		1.60		1.28		5		203

Table 1   689

HD	DM	Other Id	N Rem	α_{1950}	δ_{1950}	S	V	σ_V	B–V	σ_{B-V}	U–B	σ_{U-B}	n	Spectrum	References
		NGC 4590 sq2 9		12 36 15	−26 29.1	1	14.66		0.88		0.21		3		548
		NGC 4590 sq1 3		12 36 16	−26 33.3	2	11.71	.005	0.90	.025	0.53		7		438,548
110042	+72 00575			12 36 17	+72 24.0	1	8.34		-0.02		-0.06		2	A0	1733
	+39 02542			12 36 18	+39 13.8	1	10.20		0.99		0.83		1	G8 III	280
		NGC 4590 sq1 10		12 36 18	−26 24.4	1	12.48		0.61				3		438
		CS 16032 # 1		12 36 19	+03 03.4	1	13.46		0.48		-0.13		1		1744
		NGP +25 086		12 36 19	+24 42.7	1	11.75		1.69				2	M2 III	44
109981	+35 02359			12 36 19	+34 51.1	4	7.98	.018	1.14	.030	1.14	.007	8	K2 III	37,77,172,8041
233963	+51 01772			12 36 19	+51 08.9	1	9.32		0.39		-0.04		2	F2	1566
	+12 02499			12 36 20	+11 34.5	1	10.53		0.93		0.50		1	G5	280
109983	+14 02525			12 36 20	+14 04.7	1	8.86		1.60		1.41		1	M2	280
	+46 01806			12 36 20	+46 24.6	1	10.43		0.95		0.73		1	K0 III	280
		HA 5 # 116		12 36 20	+74 22.5	1	11.92		0.85		0.51		2		1058
	−59 04362	LSS 2723		12 36 20	−60 10.9	3	10.01	.009	0.14	.018	-0.78	.005	6	B0.5 Vne1	976,1684,1737
		BSD 5 # 366		12 36 21	+73 58.6	1	13.10		0.44		-0.04		1	F8	1058
	+17 02509			12 36 22	+16 29.9	1	10.15		1.32		1.40		1	K2	280
109980	+41 02312	HR 4811		12 36 22	+41 09.0	1	6.37		0.18		0.07		2	A7 Vn	1240
	+20 02747			12 36 23	+20 24.2	1	10.38		0.32		0.01		2		1569
109995	+40 02558			12 36 23	+39 35.1	4	7.61	.009	0.05	.010	0.13	.018	13	A0p	275,1068,1240,3025
109960	−29 09845	HR 4810		12 36 23	−30 08.9	3	5.88	.007	1.21	.005			11	K2/3 III	15,2013,2029
109962	−44 08136			12 36 23	−45 17.2	1	9.60		0.38		0.00		2	F2 V	1730
		G 60 - 24		12 36 24	+11 58.4	2	11.52	.005	1.48	.011	1.19		2		1705,3078
110012	+47 01977			12 36 24	+47 02.9	2	9.86	.019	0.05	.000	0.07	.010	3	A2 V	1026,1240
109949	−53 05249			12 36 24	−54 16.5	1	8.84		0.33		0.20		1	A9 IV	1771
109937	−62 02862			12 36 24	−63 05.2	1	9.42		0.40		-0.39		2	B2/3 III	1688
		HA 5 # 320		12 36 25	+74 51.9	1	13.03		0.70		0.22		1		1058
		NGC 4590 sq1 18		12 36 25	−26 34.8	1	13.69		1.03				2		438
		GD 149		12 36 28	+31 46.1	1	14.52		0.64		-0.05		1		3060
		NGP +26 125		12 36 29	+25 58.0	3	11.11	.025	1.05	.005	1.02	.000	3	K0 III	44,8034,8041
109997	+14 02528			12 36 30	+14 03.7	1	9.20		0.25		0.10		1	F8	280
	+14 02527			12 36 30	+14 08.5	1	9.47		1.11		1.04		1	K0	280
		NGP +28 104		12 36 30	+28 09.2	1	12.07		0.73				2	G8 III	44
109964	−51 06877			12 36 30	−52 17.8	1	9.15		1.61				2	K4/5 (III)	44
109974	−51 06878			12 36 31	−51 30.9	1	9.02		0.04		-0.06		1	B9 IV/V	1770
		CS 16032 # 6		12 36 32	+03 05.1	1	13.42		0.34		-0.09		1		1744
	+18 02637			12 36 32	+17 42.7	1	9.46		0.29		0.03		2	F0	1569
		CS 16026 # 73		12 36 32	+30 59.2	1	13.38		0.33		-0.08		1		1744
		NGC 4590 sq3 5		12 36 32	−26 28.8	1	14.62		0.83		0.18		6		1638
		NGC 4590 sq1 6		12 36 32	−26 39.7	1	12.36		0.80				4		438
		CS 16032 # 2		12 36 33	+03 03.0	1	13.03		0.82		0.52		1		1744
109996	+23 02479	HR 4812	⋆ A	12 36 33	+22 56.1	2	6.39	.010	1.10	.000	1.06		5	K1 III	280,1501
109986	−26 09254			12 36 34	−26 39.4	1	9.27		0.93				5	G8 IV	438
		NGP +26 126		12 36 36	+26 12.0	1	10.91		0.96				1	K0 III	44
		NGC 4590 sq3 7		12 36 36	−26 29.7	1	14.71		0.63		0.01		6		1638
		NGC 4590 sq3 2		12 36 37	−26 28.8	1	13.36		1.13		0.66		5		1638
		NGC 4590 sq3 3		12 36 37	−26 31.3	1	13.71		0.99		0.45		5		1638
110024	+21 02439	HR 4815		12 36 38	+21 20.2	1	5.46		0.96				2	G9 III	70
		HA 5 # 188		12 36 38	+74 25.2	1	12.34		0.56		0.08		2		1058
		NGC 4590 sq3 4		12 36 38	−26 29.2	1	14.14		0.91		0.30		2		1638
110026	+15 02491	IDS12341N1453	AB	12 36 39	+14 36.1	2	8.01	.025	0.37	.000	0.14	.015	5	Am	1569,8071
110013	−4 03321			12 36 39	−05 16.3	1	9.05		0.85		0.54		5	K0	1731
		NGC 4590 sq3 6		12 36 39	−26 30.2	1	14.68		0.83		0.24		2		1638
	+16 02399			12 36 40	+15 42.6	1	9.80		1.46		1.76		1	K2	280
110014	−7 03452	HR 4813	⋆ A	12 36 40	−07 43.2	7	4.65	.011	1.24	.010	1.38	.016	35	K2 III	3,15,1075,1088,2012*
		NGC 4590 sq1 17		12 36 40	−26 33.2	1	13.57		0.98				2		438
	−64 01976			12 36 40	−64 52.1	1	10.88		0.66		0.38		2	A0	1688
109952	−77 00859			12 36 40	−77 34.4	2	9.05	.040	1.19	.005	1.05	.005	5	K5 V	803,3073
		NGC 4590 sq3 1		12 36 42	−26 30.4	1	12.74		1.34		1.14		2		1638
		PHI4/2 20		12 36 42	−30 23.4	1	13.63		0.95				1		1719
109977	−61 03326			12 36 43	−61 59.5	1	10.28		0.23		-0.32		2	B7/9	1688
109978	−61 03327	LSS 2726		12 36 43	−62 09.5	5	8.81	.022	0.41	.009	-0.59	.033	10	O9 III	191,540,976,1684,1737
	+28 02139			12 36 44	+27 41.9	2	11.08	.009	0.74	.004	0.40		3	G6 III	44,1019
		NGP +31 108		12 36 44	+30 51.3	1	10.99		0.93		0.68		1	K0 III	8034
110065	+41 02313			12 36 44	+41 15.5	1	8.25		0.91				2	K0 III	280
		PHI4/2 15		12 36 45	−29 38.7	1	13.20		1.03				2		1719
		CCS 2021		12 36 45	−45 06.5	1	10.22		3.39				2	R	864
110043	+31 02390			12 36 46	+30 51.5	2	8.66	.015	1.28	.029	1.25		3	K0 III	77,8034
110044	+30 02310	G 164 - 9		12 36 47	+29 30.8	1	9.01		0.87				2	G8 IV	77
		PHI4/3 58		12 36 48	−29 20.8	1	13.42		1.15				2		1719
	+25 02541			12 36 49	+25 26.2	3	10.48	.008	1.13	.012	1.14	.020	4	K0 III	44,280,8041
		NGP +26 127		12 36 49	+26 14.8	1	12.30		0.85				1	G8	44
		NGC 4590 sq1 12		12 36 49	−26 22.2	1	12.90		1.12				4		438
		NGC 4590 sq2 8		12 36 49	−26 38.9	1	14.49		0.89		0.31		3		548
		CS 16032 # 8		12 36 51	+02 05.3	1	13.70		0.07		0.16		1		1744
110066	+36 02295	HR 4816, AX CVn		12 36 51	+36 13.6	7	6.44	.032	0.06	.014	0.02	.011	21	A0 p SrCrEu	77,275,1022,1068,1202*
		NGC 4594 sq1 1		12 36 51	−11 23.7	1	10.49		0.87				4		492
110006	−57 05617			12 36 51	−57 37.8	1	9.74		0.64				4	G5/6 V	2034
109993	−66 01866			12 36 52	−66 45.9	1	8.16		0.12		-0.15		2	B8 V	1688
		CS 16032 # 3		12 36 54	+03 03.5	1	14.22		0.78		0.38		1		1744
110085	+35 02362			12 36 54	+34 29.2	3	9.13	.010	0.20	.005	0.07	.009	7	A7 V	77,1068,1240

HD	DM	Other Id	N Rem	α_{1950}	δ_{1950}	S	V	σ_V	B–V	σ_{B-V}	U–B	σ_{U-B}	n	Spectrum	References
	+17 02514			12 36 55	+17 04.8	1	9.46		1.02		0.91		1	K0	280
	+25 02542			12 36 55	+24 35.7	1	9.80		1.17		1.29		1		280
		NGC 4590 sq1 8		12 36 56	−26 37.0	2	12.45	.005	0.52	.015	-0.06		7		438,548
110020	−65 01941	HR 4814, FH Mus		12 36 57	−66 14.2	5	6.26	.006	-0.05	.002	-0.20	.006	18	B8 V	15,1075,1688,1771,2038
	−9 03530			12 36 58	−09 53.9	1	10.30		0.91		0.57		2		401
110035	−36 07990	IDS12343S3644	AB	12 36 58	−37 00.5	3	9.05	.010	1.06	.020	0.91		12	K3 V	173,2012,3077
	+48 02045			12 36 59	+48 03.5	1	11.10		0.26		0.08		2	F2 V	1240
110058	−48 07576			12 36 59	−48 55.4	1	7.99		0.15		0.04		2	A0 V	1771
	+11 02479	NGC 4596 sq1 1		12 37 01	+10 34.0	2	10.18	.055	1.11	.015	0.97		2	K0	280,492
		NGC 4590 sq1 7		12 37 01	−26 30.9	1	12.38		2.08				2		438
	+13 02566			12 37 02	+12 30.8	1	10.08		1.43		1.86		1		280
		NGC 4594 sq1 2		12 37 03	−11 22.4	1	12.87		0.75		0.46		3		492
110021	−67 02034			12 37 03	−67 32.7	1	9.11		0.43		0.25		1	A8/F0 V	1771
		NGC 4590 sq1 14		12 37 04	−26 21.1	1	13.23		0.47				2		438
	+31 02392			12 37 05	+31 16.6	2	9.37	.005	0.96	.019	0.65		3	G8 III	77,8034
		PHI4/3 53		12 37 05	−28 36.5	1	13.47		1.05				1		1719
		HA 5 # 324		12 37 06	+74 53.5	1	12.98		0.50		-0.02		1		1058
110039	−60 04218			12 37 06	−61 21.2	1	10.17		0.25		0.12		2	B9	1688
110087	−26 09259			12 37 08	−26 38.9	2	8.88	.005	1.18	.005	1.19		8	K2 III	438,548
	+37 02305			12 37 09	+36 34.8	1	9.45		0.50				2	F8	77
	+31 02393			12 37 10	+31 05.3	1	10.10		0.39				2	F5	77
		NGC 4594 sq1 8		12 37 10	−11 14.8	1	12.82		0.75				2		492
110086	−10 03524			12 37 10	−11 22.5	1	10.12		0.67		0.13		6	G0	492
110073	−39 07748	HR 4817		12 37 10	−39 42.7	7	4.64	.008	-0.09	.007	-0.40	.023	30	B8p	15,1075,1202,1770,2012*
		NGC 4590 sq1 16		12 37 12	−26 35.3	2	13.42	.015	1.10	.000	0.65		5		438,548
110041	−69 01686			12 37 12	−69 27.3	1	8.83		0.08		-0.28		2	B8 V	401
110062	−62 02865			12 37 13	−63 10.7	1	8.91		0.29		0.18		3	B9 V	1688
		NGC 4590 sq1 13		12 37 14	−26 36.6	2	12.95	.020	0.45	.010	-0.20		7		438,548
	+17 02517			12 37 15	+16 58.8	1	10.21		1.21		1.33		1		280
		LP 377 - 36		12 37 16	+25 47.4	1	14.82		1.64				1		686
	+28 02140			12 37 18	+27 51.7	3	10.31	.012	1.48	.025	1.79		4	K5 III	44,694,1019
		NGP +29 129		12 37 19	+28 35.5	1	11.13		0.85		0.51		1	G8 III	8034
110134	+35 02363			12 37 19	+35 21.7	1	8.19		0.42				2	F5 V	77
	+32 02267			12 37 20	+32 15.2	1	9.88		1.29		1.47		1	K2 III	280
		NGC 4590 sq1 11		12 37 20	−26 23.9	1	12.67		0.91				2		438
		NGP +26 128		12 37 22	+26 03.7	2	10.94	.035	0.93	.005	0.57		2	G8 III	44,8034
	+16 02400			12 37 23	+16 17.4	2	9.49	.010	0.99	.010	0.96	.005	3		280,1569
	+45 02057			12 37 23	+45 09.4	1	8.91		1.18				2	K1 III	280
		CCS 2023		12 37 23	−57 06.4	1	10.43		4.51				3		864
110079	−65 01946			12 37 23	−65 35.3	1	9.15		0.15		-0.06		2	B9 V	1688
110135	+27 02157			12 37 25	+26 46.6	2	9.36	.000	0.97	.010	0.69		3	G8 III	44,8034
		HA 5 # 241		12 37 25	+74 28.9	1	11.86		0.92		0.71		4	K0	271
		NGC 4596 sq1 3		12 37 26	+10 26.0	1	12.97		0.56				2		492
	+20 02752			12 37 26	+19 51.6	1	9.12		1.22		1.35		1	K2	280
		NGC 4590 sq1 5		12 37 26	−26 23.4	1	12.07		1.02				2		438
110080	−69 01689			12 37 27	−70 16.3	2	7.39	.005	0.26	.020	0.20		5	A5 V	1075,1771
		NGC 4596 sq1 6		12 37 28	+10 18.5	1	10.89		0.79				3		492
	+13 02567	G 60 - 26		12 37 28	+12 55.3	3	9.83	.023	0.65	.005	0.00	.017	6	G0	516,1620,1696
	+31 02394			12 37 28	+31 01.3	1	10.29		1.00		0.70		1	G9 III	8034
		NGC 4596 sq1 2		12 37 29	+10 31.1	1	12.72		0.63				6		492
	−66 01872			12 37 30	−67 13.0	2	10.42	.022	0.06	.001	-0.67	.005	4	B2.5II-III	540,976
110182	+52 01644			12 37 31	+52 00.0	1	7.63		0.05		0.07		2	A0	1375
110166	+37 02306			12 37 32	+37 16.4	5	8.77	.015	-0.14	.015	-0.38	.010	12	B8 V	77,275,1068,1240,1298
		HA 5 # 54		12 37 32	+74 07.7	1	12.85		0.62		0.10		1		1058
		NGC 4594 sq1 9		12 37 32	−11 11.5	1	10.54		0.70				1		492
		NGC 4596 sq1 4		12 37 33	+10 25.3	1	12.05		0.60				6		492
110102	−62 02869			12 37 33	−63 00.6	1	10.20		0.55		0.33		2	B9	1688
110103	−63 02387			12 37 33	−64 21.7	1	10.24		0.51		0.33		2	A2/3 V	1688
110081	−71 01361			12 37 33	−71 57.1	1	9.54		0.42		0.16		1	A7 V	1771
		NGC 4596 sq1 5		12 37 34	+10 24.4	1	10.85		0.42				6		492
		U Com		12 37 36	+27 46.3	1	11.49		0.15		0.12		1	A9	668
110193	+43 02249			12 37 36	+43 02.9	1	8.79		1.14				2	K2 III	280
	+49 02153			12 37 36	+49 06.2	1	9.64		0.29		0.12		2	Am:	1240
	+21 02442	G 59 - 32		12 37 37	+21 05.3	2	9.00	.025	0.79	.025	0.49	.010	2	K0	333,1620,1658
	+25 02544			12 37 37	+25 15.8	3	9.55	.004	1.04	.021	0.92		5	K0 III	44,280,8041
110117	−61 03335			12 37 37	−61 43.2	1	9.43		0.29		-0.07		2	B9 IV	1688
110183	+28 02141			12 37 39	+28 11.1	1	10.03		0.52		0.02		3	F9 V	1019
110148	−52 05782			12 37 39	−53 17.4	1	9.82		0.27		0.16		1	A8 V	1771
	−61 03336			12 37 39	−61 27.3	1	10.44		0.51		-0.52		4		1684
	−10 03527			12 37 40	−11 26.7	1	9.69		1.08				5	K2	492
110184	+9 02653			12 37 42	+08 48.1	2	8.31	.035	1.17	.005	0.76	.015	2	G5	280,1569
		Klemola 152		12 37 42	+27 31.	1	11.80		0.15		0.08		2		1026
	+23 02482			12 37 43	+22 34.7	1	10.83		0.66		0.24		1		280
110194	+34 02341			12 37 43	+34 26.4	2	7.08		1.40	.000	1.66		4	K4 III	77,1569
		HA 5 # 56		12 37 43	+74 09.1	1	10.85		0.49		-0.01		2		1058
		NGC 4594 sq1 4		12 37 44	−11 21.9	1	13.19		0.40				3		492
	+14 02530			12 37 45	+13 42.3	1	10.13		0.49				4	F8 IV	193
110195	+29 02313			12 37 46	+28 29.8	2	10.06	.043	0.74	.000	0.34		3	G3 V	44,1019
		SSII 189		12 37 48	+29 39.9	1	12.89	.000	0.44	.000	-0.04		1	F0:	1298
	+42 02329			12 37 48	+41 54.9	1	10.10		1.13		1.09		1	K1 III	280

Table 1   691

HD	DM	Other Id	N Rem	α_{1950}	δ_{1950}	S	V	σ_V	B–V	σ_{B-V}	U–B	σ_{U-B}	n	Spectrum	References
110151	−60 04225			12 37 48	−60 38.3	1	8.72		0.14		0.02		2	B9.5IV	1688
	+26 02378			12 37 49	+25 55.3	3	9.98	.020	1.24	.004	1.35	.045	4	K1 III	44,280,8034
	+17 02519			12 37 50	+16 53.6	1	10.72		0.51		−0.02		2		1569
	+75 00477	IDS12360N7514	A	12 37 50	+74 57.2	1	10.45		0.68		0.25		1		1058
	+75 00477	IDS12360N7514	B	12 37 50	+74 57.2	1	10.88		0.72		0.33		1		1058
	+28 02142			12 37 51	+28 02.9	1	10.80		0.59		0.12		1	G0 V	1019
	+19 02592			12 37 52	+18 36.8	1	10.16		0.28		0.01		2	A5	1569
	+17 02520			12 37 54	+17 20.7	1	9.60		1.17		1.27		1	K2	280
		HA 5 # 257		12 37 54	+74 36.4	1	12.87		0.55		0.06		1		1058
	+23 02483			12 37 55	+23 17.7	1	9.79		0.45		−0.04		46	F5	1613
		NGC 4596 sq1 8		12 37 57	+10 19.6	1	13.04		0.67				2		492
110163	−61 03340			12 37 57	−61 42.9	1	9.05		0.25		0.02		2	B9 III	1688
		GD 318		12 37 59	+48 56.5	1	12.53		0.41		−0.13		3		308
		HA 5 # 258		12 37 59	+74 42.2	1	10.91		0.94		0.69		1		1058
		HA 5 # 389		12 37 59	+74 56.0	1	11.67		0.65		0.10		1		1058
		Ton 632		12 38 00	+30 19.	1	15.61		0.43		−0.16		1		1036
		Ton 102		12 38 00	+51 36.	1	13.54		−0.25		−0.88		2		272
110221	+8 02626			12 38 03	+07 58.4	1	8.00		0.89		0.60		1	G5	280
110313	+69 00671	G 237 - 70		12 38 03	+69 04.6	2	7.87	.005	0.61	.000	0.08	.028	4	F8	1658,3026
110312	+75 00479	IDS12362N7458	A	12 38 04	+74 41.7	1	8.13		1.50		1.76		1	K2	1058
		NGC 4594 sq1 7		12 38 04	−11 25.8	1	12.52		0.67				2		492
	−46 10865	LSS 3670		12 38 05	−46 50.7	1	10.03		0.82		0.00		1		954
110250	+11 02483			12 38 08	+11 12.3	1	9.07		1.05		0.97		1	K2	280
110248	+31 02396			12 38 08	+30 39.1	5	7.65	.011	0.31	.011	0.17	.009	24	Am	77,275,1068,1240,8071
		L 399 - 68		12 38 08	−43 18.4	2	12.24	.000	1.74	.011	1.39		4		1705,3073
		NGC 4596 sq1 9		12 38 09	+10 28.9	1	13.17		0.78				1		492
	+8 02627			12 38 10	+08 23.9	1	10.77		0.98		0.66		1		280
		NGP +28 107		12 38 10	+27 35.0	1	12.09		1.02				2	K0	44
		NGC 4594 sq1 10		12 38 10	−11 22.1	1	13.78		1.01				2		492
110225	−27 08810	IDS12355S2710	A	12 38 11	−27 26.3	1	7.57		−0.04		−0.23		3	B9 IV/V	1770
		WLS 1224 50 # 8		12 38 12	+49 38.2	1	13.56		0.78		0.31		2		1375
110249	+13 02569			12 38 13	+12 59.4	1	7.90		1.53		1.93		1	K2	280
	+23 02484			12 38 13	+23 14.7	2	9.68	.007	1.08	.001	0.98	.006	16	K5	280,1613
110277	+52 01646			12 38 13	+51 45.4	1	8.85		0.46		0.08		1	F8	272
		BSD 5 # 376		12 38 13	+74 03.4	1	12.85		0.65		0.15		1	G0	1058
	+46 01811			12 38 16	+45 29.7	1	9.57		1.17				2	K0 III	280
110207	−67 02040			12 38 17	−68 18.0	1	8.52		0.15		0.12		1	A1 V	1771
	+32 02268			12 38 18	+32 07.7	1	10.12		0.52				2	F8	77
110191	−71 01365			12 38 19	−71 49.1	1	9.31		0.50		0.29		1	A7mA9-F2	1771
	+32 02269			12 38 21	+32 23.7	1	9.84		0.49				2	F8	77
110253	−43 07813			12 38 25	−43 49.6	1	6.72		1.27				4	K2 III	2012
110231	−60 04232			12 38 25	−60 40.5	1	9.85		0.10		−0.02		4	B9	1688
110232	−60 04231			12 38 25	−60 48.0	1	10.18		0.10		−0.29		1	B8/9 II/III	1688
110314	+41 02318			12 38 26	+40 47.8	3	8.29	.022	0.65	.005	0.23	.029	5	G2 V	172,280,8038
110533	+84 00286			12 38 26	+83 55.1	1	7.35		0.59		0.10		2	G0	3026
110296	+34 02342			12 38 27	+34 22.7	2	7.80	.018	1.55	.001	1.90		5	K4 V	77,7008
		SW CVn		12 38 27	+37 21.7	2	12.16	.132	0.06	.027	0.10	.068	2	A0	597,699
110297	+27 02158	IDS12360N2741	AB	12 38 28	+27 24.8	1	7.82		0.44		0.04		3	F5 V	1019
110281	+0 02969			12 38 30	−00 20.8	5	9.36	.009	1.59	.005	1.66	.012	137	K2 V	281,989,1729,1764,6004
		NGC 4609 - 39		12 38 30	−62 44.9	1	13.18		0.33		0.22		1		139
		AAS78,203 # 18		12 38 30	−65 35.	1	13.38		1.85		1.26		3		1684
110270	−48 07596			12 38 31	−48 32.6	1	7.92		−0.05		−0.23		4	B8/9 III	1770
	+18 02642			12 38 32	+18 23.4	1	9.41		1.14		0.99		1	G5	280
110258	−59 04388	AG Cru		12 38 33	−59 31.2	3	7.82	.058	0.59	.027	0.40	.005	4	F8 Ib/II	122,1484,1490
110245	−66 01881			12 38 33	−66 41.7	1	8.37		0.73		0.44		2	F8/G0 III	1688
110315	+16 02404	G 59 - 33		12 38 35	+15 39.4	3	7.93	.025	1.11	.016	1.04	.014	6	K2	333,1620,1758,3072
		CS 16032 # 12		12 38 36	+03 02.2	1	14.30		0.50		−0.08		1		1744
		Ton 633		12 38 36	+28 31.	1	15.78		0.04		0.05		1		1036
		Ton 634		12 38 36	+28 35.	1	15.38		0.33		−0.24		1		1036
	+26 02379			12 38 37	+25 58.8	1	10.24		0.76				2	G7 V	44
	+29 02315			12 38 38	+28 38.8	3	10.81	.020	1.07	.009	0.90	.005	4	K0 III	44,1019,8034
	+0 02970			12 38 38	−00 27.5	1	10.94		0.83		0.22		2	K2 V	281
110287	−45 07944	HR 4818		12 38 38	−45 52.3	3	5.84	.009	1.51	.009	1.72		9	K4 III	15,2012,3005
110318	−12 03676	HR 4822	★ AB	12 38 40	−12 44.4	4	5.25	.057	0.43	.009	0.09	.010	4	F3 IV	15,1008,2008,3053
110326	+31 02397			12 38 41	+30 42.7	5	6.95	.018	0.27	.012	0.05	.010	21	Am	77,275,1068,1240,8071
110289	−52 05798			12 38 42	−52 34.5	1	8.35		2.85				4	K0 III	3039
110290	−53 05275			12 38 42	−54 14.6	3	8.93	.004	0.92	.032	0.54	.023	40	G8 III	1589,1699,1730
		CS 16032 # 13		12 38 43	+03 03.4	1	14.52		0.68		0.25		1		1744
	+24 02484			12 38 44	+23 45.3	1	9.92		1.06		1.11		1		280
	+38 02355			12 38 44	+37 29.2	1	10.46		1.59				2	M4 III	694
110304	−48 07597	HR 4819	★ AB	12 38 45	−48 41.1	7	2.16	.006	−0.01	.006	−0.01	.005	17	A0 IV	15,507,1034,1075,2012*
		S172 # 3		12 38 45	−54 23.8	1	10.79		1.11		0.87		2		1730
110322	−42 07831			12 38 46	−43 22.3	1	9.39		0.10		0.08		1	A0 V	1770
		NGC 4609 - 36		12 38 46	−62 44.2	1	12.44		0.32		0.06		1		139
		NGC 4609 - 45		12 38 48	−62 46.8	1	13.41		1.21		1.33		1		139
		CS 16032 # 9		12 38 49	+02 05.8	1	14.61		0.47		−0.10		1		1744
		CS 16032 # 18		12 38 49	+03 00.8	1	14.11		0.45		−0.16		1		1744
		CS 16032 # 11		12 38 50	+02 02.8	1	13.83		0.03		0.09		1		1744
	+37 02309			12 38 50	+36 44.4	1	9.98		1.09				3	K0 III	280
	+0 02971			12 38 50	−00 27.8	1	10.45		0.89		0.50		4	G8 V	281

HD	DM	Other Id	N	Rem	α_{1950}	δ_{1950}	S	V	σ_V	B–V	σ_{B-V}	U–B	σ_{U-B}	n	Spectrum	References
		S172 # 4			12 38 52	−54 19.7	1	11.44		1.06				2		1730
		NGC 4609 - 49			12 38 52	−62 45.3	1	14.62		0.96		0.45		1		139
		NGC 4609 - 44			12 38 52	−62 45.6	1	13.52		2.55		1.80		2		139
110294	−67 02044				12 38 52	−67 31.1	1	9.08		0.21		0.18		1	A4 IV/V	1771
		NGC 4609 - 18			12 38 53	−62 41.2	1	12.22		0.34		0.28		2		139
		NGC 4609 - 35			12 38 53	−62 44.2	1	13.49		0.39		0.44		1		139
		SSII 191			12 38 54	+29 27.	1	13.88		0.06		0.13		3	A0	275
110310	−64 01983				12 38 54	−64 27.3	1	7.37		0.44		0.10		2	F0/2 IV	1688
		PHI4/3 23			12 38 55	−28 37.3	1	11.76		0.98				1		1719
		NGC 4609 - 40			12 38 55	−62 45.1	1	13.47		0.42		0.34		3		139
110324	−55 05187				12 38 56	−55 38.2	1	8.97		0.09		0.01		3	B9/A0 IV/V	1770
	+37 02310				12 38 58	+37 17.5	1	9.91		0.45				2	F5	77
		HA 5 # 329			12 38 58	+74 46.1	1	12.05		0.50		0.04		1		1058
		NGC 4609 - 28			12 38 58	−62 43.2	1	11.98		0.43		0.32		1		139
	+35 02365				12 38 59	+35 26.9	1	10.01		0.95				2	G9 III	280
110333	−52 05802	IDS12362S5220		A	12 38 59	−52 36.0	1	8.23		0.29		0.19		1	A6 IV/V	1771
	+8 02628				12 39 00	+07 57.5	1	10.26		1.01		0.87		1		280
		NGP +26 131			12 39 00	+26 27.7	1	10.94		1.31				1	K2 III	44
110363	+27 02159				12 39 00	+26 59.3	4	9.75	.011	1.14	.006	1.10	.010	5	K0 III	44,280,1019,8034
		Ton 635			12 39 00	+29 10.	1	12.85		-0.18		-1.06		1		1036
110311	−68 01731	HR 4820, R Mus			12 39 00	−69 08.0	1	5.89		0.75				1	F7 Ib	1484
		G 59 - 35			12 39 01	+27 20.9	1	11.95		1.28		1.12		1		1620
110375	+39 02544				12 39 01	+38 39.8	1	8.33		0.38		-0.05		2	F2 IV	105
		CS 16032 # 10			12 39 02	+02 01.4	1	14.48		0.48		-0.07		1		1744
		HA 104 # 423			12 39 02	−00 14.8	1	15.60		0.63		0.05		1		1764
110377	+11 02484	HR 4824, GG Vir	★	A	12 39 03	+10 42.0	2	6.24	.068	0.19	.006	0.10		3	A7 Vn	70,3058
110392	+41 02320				12 39 03	+40 51.2	2	7.63	.010	0.92	.000	0.67		4	K0 III	172,280
110335	−59 04393	HR 4823, CH Cru			12 39 03	−59 24.7	6	4.93	.016	-0.04	.010	-0.41	.021	15	B6 IVe	26,122,1637,1771,2006,8023
		NGC 4609 - 16			12 39 03	−62 40.1	1	12.42		0.60		0.14		4		139
		NGC 4609 - 19			12 39 03	−62 42.2	1	13.00		0.35		0.16		2		139
	+42 02330				12 39 04	+41 45.2	1	10.58		0.32		0.16		1		280
		PHI4/3 20			12 39 04	−28 04.9	1	13.34		0.86				2		1719
110378	+8 02630				12 39 05	+08 03.5	1	9.35		0.98		0.79		1	K2	280
		HA 104 # 425			12 39 05	−00 15.2	1	12.31		0.68		0.25		13		281
		CS 16032 # 14			12 39 06	+03 05.3	1	14.44		0.65		0.04		1		1744
		HA 104 # 428			12 39 07	−00 10.0	1	12.63		0.99		0.75		16		1764
		HA 104 # 427			12 39 07	−00 16.0	1	12.45		0.55		0.05		7	F8	281
110379	−0 02601	HR 4825	★	AB	12 39 07	−01 10.5	17	2.74	.010	0.36	.006	-0.04	.012	104	F0 V	15,22,1008,1013,1020*
	−62 02880	NGC 4609 - 13			12 39 07	−62 40.4	2	11.14	.029	0.27	.000	-0.05	.000	7		139,807
		NGC 4609 - 38			12 39 07	−62 46.9	1	13.40		0.52		0.25		2		139
110409	+50 01941				12 39 09	+49 33.9	1	6.91		1.30		1.34		2	K2 III	172
110356	−46 08104				12 39 09	−47 19.9	1	7.83		1.05		0.90		5	K0 III	1673
		LP 217 - 54		B	12 39 10	+41 21.7	1	11.04		0.76		0.30		1		1773
		104 L1			12 39 10	−00 06.3	1	14.61		0.63		0.06		1		1764
		NGC 4609 - 41			12 39 10	−62 45.3	1	13.29		0.43		0.25		1		139
	+19 02595				12 39 11	+18 36.4	1	9.42		1.39		1.67		1	K5	280
	+27 02160				12 39 11	+27 12.6	2	10.72	.021	1.04	.005	0.85		3	K0 III	44,1019
	−62 02883	NGC 4609 - 8			12 39 11	−62 42.9	2	11.47	.010	0.19	.000	-0.22	.005	7		139,807
110385	−18 03442	HR 4827			12 39 12	−19 29.1	4	6.03	.006	0.40	.003	0.07	.005	11	F2 V	15,401,2012,3077
		NGC 4609 - 29			12 39 12	−62 41.0	1	13.05		0.38		0.20		2		139
		NGC 4609 - 17			12 39 12	−62 42.1	1	13.37		0.30		0.08		2		139
110410	+31 02398				12 39 13	+30 38.8	1	9.32		0.38				2	F5 V	77
	+38 02357				12 39 13	+38 05.9	1	9.78		1.00		0.90		1	K0 III	280
		NGC 4609 - 47			12 39 13	−62 44.4	1	13.67		0.42		0.36		1		139
		NGC 4609 - 48			12 39 14	−62 44.4	1	13.21		0.37		0.24		1		139
	−62 02885	NGC 4609 - 7			12 39 15	−62 41.8	2	10.55	.010	0.13	.000	-0.44	.000	6		139,807
		HA 104 # 430			12 39 16	−00 09.4	1	13.86		0.65		0.13		8		1764
		HA 104 # 134			12 39 16	−00 39.4	1	10.92		0.99		0.85		2	K2	281
	−62 02886	NGC 4609 - 9			12 39 16	−62 45.4	2	11.40	.025	0.24	.020	-0.18	.020	5		139,807
110412	+10 02459	IDS12368N1026		A	12 39 17	+10 09.6	1	6.99		0.32		0.10		2	F0	1569
		LP 217 - 54		A	12 39 17	+41 19.5	1	15.47		1.46		0.99		1		1773
		NGC 4609 - 5			12 39 17	−62 41.4	1	12.09		0.21		-0.04		4		139
	−62 02884	NGC 4609 - 6			12 39 17	−62 41.7	2	10.42	.000	0.16	.015	-0.43	.010	6		139,807
		NGC 4609 - 26			12 39 17	−62 45.7	1	13.08		0.68		0.34		1		139
	+8 02631				12 39 18	+07 40.3	1	8.88		1.20		1.27		1	F8	280
	+30 02313				12 39 18	+29 54.9	1	10.26		0.43				2	F6	77
110360	−59 04396	LSS 2732			12 39 18	−60 22.7	5	9.32	.013	0.17	.010	-0.77	.040	15	O7/8	191,401,540,976,1684
		NGC 4609 - 12			12 39 18	−62 48.1	1	11.33		1.46		1.61		3		139
		NGC 4609 - 25			12 39 18	−62 51.2	2	11.73	.052	1.18	.005	0.00	.014	4		139,496
	+47 01982				12 39 19	+46 41.4	1	9.30		1.11				2	K1 III	280
	−62 02887	NGC 4609 - 1			12 39 19	−62 44.6	2	9.59	.014	0.70	.000	0.26	.005	8		139,807
		NGC 4609 - 14			12 39 20	−62 42.3	1	11.82		0.64		0.28		3		139
110411	+11 02485	HR 4828, ρ Vir			12 39 21	+10 30.7	6	4.88	.007	0.08	.012	0.04	.018	24	A0 V	15,1022,1363,3023*
	+18 02647				12 39 22	+17 49.0	5	11.86	.007	-0.33	.009	-1.21	.014	19	sdO	1026,1026,1036,1264*
		HA 5 # 263			12 39 22	+74 41.2	1	11.87		1.06		0.89		1		1058
		HA 5 # 334			12 39 22	+74 45.2	1	12.02		0.59		0.04		1		1058
		HA 104 # 540			12 39 22	−00 06.4	1	10.82		0.87		0.52		2	dK0	281
110418	−6 03626				12 39 22	−07 13.5	1	6.78		1.58		1.88		2	K5	3040
		NGC 4609 - 46			12 39 22	−62 44.8	1	13.02		0.33		0.26		1		139
	+27 02162				12 39 23	+26 32.1	2	10.18	.009	1.02	.004	0.87		3	K0 III	44,1019

Table 1   693

HD	DM	Other Id	N Rem	α_{1950}	δ_{1950}	S	V	σ_V	B–V	σ_{B-V}	U–B	σ_{U-B}	n	Spectrum	References
	+27 02161			12 39 23	+27 06.2	1	10.54		0.58		-0.08		1	F8	1019
	+30 02314	IDS12369N3016	AB	12 39 23	+29 59.6	1	10.19		0.38				2	F5	77
110462	+63 01026	HR 4833		12 39 23	+62 59.2	2	6.05	.025	0.03	.015	0.08	.005	4	A2 III	985,1733
		HA 5 # 197		12 39 23	+74 32.0	1	11.07		0.48		0.01		1		1058
	−62 02888	NGC 4609 - 2		12 39 23	−62 43.2	3	9.72	.071	0.18	.005	-0.34	.007	8		139,807,1688
		Steph 1031		12 39 24	+43 59.3	1	11.11		1.32		1.27		1	K7	1746
		NGC 4609 - 51		12 39 24	−62 42.	1	14.11		1.98		2.59		1		139
		NGC 4609 - 52		12 39 24	−62 42.	1	13.10		0.60		0.38		1		139
		NGC 4609 - 37		12 39 24	−62 42.6	1	13.44		0.83		0.38		1		139
		NGC 4609 - 27		12 39 24	−62 47.0	1	13.69		0.41				1		139
110423	+7 02568	HR 4829	⋆ AB	12 39 25	+07 04.8	5	5.58	.005	0.00	.006	0.01	.015	20	A2 V	3,15,814,1417,3023
	−62 02891	NGC 4609 - 20		12 39 26	−62 37.9	2	10.88	.131	0.51	.141	0.32	.019	3		139,807
	−62 02890	NGC 4609 - 10		12 39 26	−62 44.4	2	10.43	.009	0.20	.009	-0.33	.009	8		139,807
	−62 02889	LOD 670 # 8		12 39 26	−62 44.9	1	10.28		0.22		-0.14		2	B8	807
	−62 02889	NGC 4609 - 21		12 39 26	−62 44.9	2	10.91	.005	0.23	.005	-0.24	.020	5		139,807
110463	+56 01618			12 39 27	+55 59.9	3	8.27	.007	0.95	.004	0.76	.005	5	K3 V	1007,1013,8023
110373	−62 02892	NGC 4609 - 4	⋆ A	12 39 27	−62 42.0	3	9.57	.026	0.15	.008	-0.40	.009	8	B8	139,807,1688
	−62 02893	NGC 4609 - 3		12 39 27	−62 43.3	2	10.68	.019	0.20	.015	-0.21	.010	6		139,807
		NGC 4609 - 24		12 39 27	−62 44.8	1	12.17		0.39		0.30		2		139
		NGC 4609 - 11		12 39 27	−62 45.1	1	11.28		0.22		-0.19		3		139
110438	+35 02366			12 39 28	+35 19.4	1	9.17		0.59				2	G5	77
		HA 5 # 264		12 39 28	+74 42.7	1	12.39		0.63		0.18		1		1058
		NGC 4609 - 50	⋆ B	12 39 28	−62 42.1	1	12.88		0.24		0.08		1		139
		CS 16032 # 79		12 39 29	+02 01.7	1	14.76		0.58		-0.04		1		1744
110439	+9 02661			12 39 29	+09 16.6	1	8.80		1.08		1.02		1	K0	280
		HA 104 # 325		12 39 29	−00 25.2	1	15.58		0.69		0.05		1		1764
		PHI4/3 15		12 39 31	−28 13.5	1	13.53		0.94				1		1719
		NGC 4609 - 15		12 39 31	−62 41.5	1	12.12		0.30		0.12		2		139
	+14 02533			12 39 32	+14 26.0	1	8.76		1.59		1.77		1	M0	280
	+18 02648			12 39 32	+17 37.9	1	9.91		1.05		1.00		1	K0	280
110390	−60 04244			12 39 32	−60 44.7	1	7.19		0.17		0.11		2	A1 V	1688
		PHI4/3 13		12 39 33	−28 14.8	1	13.00		0.89				2		1719
		NGC 4609 - 42		12 39 33	−62 47.3	1	13.64		0.39		0.21		1		139
		NGC 4609 - 23		12 39 33	−62 48.6	1	12.13		0.24		-0.02		1		139
		NGC 4609 - 22		12 39 35	−62 42.1	1	12.88		0.32		0.25		3		139
311829	−60 04242			12 39 36	−61 19.5	1	10.70		0.59		-0.29		3	B0	1684
110464	+33 02261			12 39 38	+33 08.2	1	9.07		0.51				2	G0	77
		HA 104 # 330		12 39 38	−00 24.2	1	15.30		0.59		-0.03		6		1764
110466	+11 02486			12 39 39	+11 09.4	3	9.50	.023	0.18	.024	0.11	.015	5	A2 V	193,272,280
	+36 02299			12 39 39	+35 54.6	1	9.80		1.23				2	K0 III	77
		HA 104 # 439		12 39 39	−00 13.8	1	11.33		0.79		0.45		5	dG8	281
		HA 104 # 440		12 39 40	−00 08.4	1	15.11		0.44		-0.23		5		1764
	+36 02300			12 39 41	+36 04.8	1	9.62		0.59				2	F8	77
		NGC 4636 sq1 6		12 39 42	+02 56.2	1	13.43		0.67		0.22		2		492
		CS 16032 # 15		12 39 42	+03 01.1	1	13.18		0.00		0.05		1		1744
		SSII 193		12 39 42	+31 12.	2	13.20	.004	0.02	.004	0.03	.008	4	A0	275,1298
		HA 104 # 237		12 39 43	+01 07.7	1	15.40		1.09		0.92		1		1764
110465	+27 02163	IDS12372N2654	AB	12 39 43	+26 38.0	4	9.28	.009	1.07	.027	1.01	.003	7	K2 V +K3 V	44,1019,1381,3072
		NGP +30 109		12 39 43	+30 00.9	1	11.30		0.90		0.63		1	G8 III	8034
311846	−61 03349			12 39 43	−61 44.6	1	10.26		1.13		0.12		3	B0	1684
110469	+0 02972			12 39 44	−00 02.1	2	8.00	.004	0.51	.009	0.01		15	F5	281,6004
		PASP86,394 # 207		12 39 44	−62 50.5	1	12.57		1.30		0.24		3		496
		NGC 4636 sq1 5		12 39 45	+02 55.3	1	12.77		0.59		-0.03		2		492
		CS 16032 # 17		12 39 46	+03 01.2	1	11.76		0.12		0.15		1		1744
	+30 02315			12 39 46	+30 01.0	2	9.94	.005	1.08	.005	1.09		3	K0 III	280,8034
		HA 104 # 443		12 39 46	−00 08.9	1	15.37		1.33		1.28		2		1764
		104 L2		12 39 46	−00 18.8	1	16.05		0.65		-0.17		2		1764
	+29 02316			12 39 47	+28 52.7	3	10.03	.020	0.97	.009	0.74	.004	4	G9 III	44,1019,8034
		HA 104 # 444		12 39 47	−00 16.1	1	13.27		0.70		0.34		2		281,1764
		HA 104 # 334		12 39 47	−00 24.0	1	13.48		0.52		-0.07		16		1764
		Feige 68		12 39 48	+16 41.0	1			0.05		0.10		1		1264
		SSII 194		12 39 48	+32 13.	1			0.14		0.15		1	A2	275
		HA 104 # 335		12 39 48	−00 16.7	2	11.69	.013	0.62	.001	0.15	.002	25	G3	281,1764
110445	−53 05285			12 39 48	−54 15.9	2	7.09	.001	0.05	.001	0.02	.000	16	A0 V	1699,1770
110434	−65 01966			12 39 48	−66 11.0	1	7.58		-0.02		-0.51		2	B8/9 III	1688
		HA 104 # 239		12 39 49	+01 03.0	1	13.94		1.36		1.29		3		1764
	+26 02380			12 39 49	+26 13.7	2	10.01	.000	1.41	.008	1.65		3	K2 V	44,1019
		NGP +34 104		12 39 49	+33 57.9	1	11.40		0.91		0.50		1	G6 III	8034
110458	−48 07608	HR 4831		12 39 49	−48 32.3	5	4.66	.004	1.10	.008	1.01	.005	22	K0 III	15,1075,2012,3077,8015
		HA 104 # 336		12 39 51	−00 23.5	1	14.40		0.83		0.50		6		1764
110501	+34 02344			12 39 53	+33 57.8	3	6.62	.000	1.07	.017	1.04	.010	5	K2 III-IV	77,985,8034
110432	−62 02898	HR 4830, BZ Cru		12 39 53	−62 47.1	10	5.31	.025	0.24	.018	-0.83	.014	41	B2pe	164,681,807,815,1034*
110500	+46 01814			12 39 54	+46 09.1	2	7.03	.009	0.24	.000	0.10	.004	5	Am	1240,8071
	+46 01815			12 39 54	+46 18.6	1	9.66		0.94				2	G8 III	280
		HA 104 # 337		12 39 54	−00 17.9	4	11.21	.006	0.77	.005	0.34	.009	53	dG8	1728,281,989,1729
110433	−62 02899	NGC 4609 - 31		12 39 55	−62 54.7	3	9.04	.012	0.25	.008	-0.32	.004	5	B2 IV/V	139,807,1688
	+28 02143			12 39 56	+28 13.2	1	9.99		0.78		0.33		1	G4 V	1019
		NGC 4609 - 33		12 39 56	−62 43.6	1	12.58		0.31		0.12		2		139
	+10 02460			12 39 57	+10 08.6	1	10.24		1.05		0.94		1		280
		HA 104 # 338		12 39 57	−00 22.1	1	16.06		0.59		-0.08		5		1764

HD	DM	Other Id	N Rem	α_{1950}	δ_{1950}	S	V	σ_V	B–V	σ_{B-V}	U–B	σ_{U-B}	n	Spectrum	References
	−27 08817			12 39 58	−27 52.6	2	10.45	.010	0.44	.005	−0.01		3		1594,6006
110461	−55 05194	HR 4832		12 39 58	−55 40.4	3	6.06	.017	−0.03	.006	−0.19	.007	9	B9 V	401,1770,2032
110449	−60 04249			12 39 58	−60 42.1	1	9.36		0.13		−0.62		1	B2 III	1688
		NGC 4609 - 30		12 39 58	−62 42.1	1	12.09		0.24		0.05		1		139
		NGC 4636 sq1 4		12 40 00	+02 50.0	1	13.71		0.79		0.23		2		492
		HA 104 # 339		12 40 00	−00 25.2	1	15.46		0.83		0.71		1		1764
		HA 104 # 244		12 40 01	+01 02.2	1	16.01		0.59		−0.15		2		1764
110487	−4 03335			12 40 01	−05 11.6	1	7.74		0.45		−0.01		3	F2	1731
		NGC 4609 - 32		12 40 01	−62 50.4	2	11.28	.033	1.34	.009	0.17	.005	4		139,496
		CS 16032 # 80		12 40 02	+03 03.1	1	13.13		0.82		0.49		1		1744
110483	−47 07796			12 40 02	−47 39.2	1	8.48		1.14		0.88		1	G8 Ib/II	565
		NGC 4609 - 34		12 40 02	−62 43.5	1	12.54		0.32		−0.04		1		139
	+28 02144			12 40 04	+27 42.5	2	10.89	.004	1.01	.002	0.69		3	G6 III	44,1019
	+36 02301			12 40 04	+36 13.4	1	9.57		0.69				2	G0	77
	+29 02317			12 40 05	+29 23.0	1	10.47		0.61		0.10		1	G2 V	1019
		SSII 195		12 40 06	+30 05.	1	13.28		0.32		−0.10		3	A0	275
	+39 02548			12 40 06	+39 26.6	1	10.16		0.95		0.62		1	G7 III	280
		NGC 4636 sq1 7		12 40 07	+02 59.8	1	13.80		1.00		0.76		1		492
		CS 16032 # 16		12 40 09	+03 04.4	1	14.81		0.45		−0.05		1		1744
110523	+29 02318			12 40 09	+29 26.2	2	9.33	.009	0.59	.004	0.16		3	G1 V	77,1019
110522	+34 02345			12 40 09	+33 51.9	1	9.44		0.59				2	G5	77
	+43 02251			12 40 09	+42 47.3	1	9.92		0.32		0.00		2	A5	1569
	+73 00566	G 237 - 72		12 40 09	+73 14.3	3	9.27	.023	0.62	.012	−0.03	.029	6	G0	308,1658,3026
110477	−60 04251			12 40 09	−60 52.5	3	7.76	.005	0.44	.004	0.00		10	F3 IV/V	1075,1688,2033
	−62 02901			12 40 10	−62 55.7	1	11.35		0.48		0.38		2	A2	807
110478	−62 02902	NGC 4609 - 43		12 40 11	−62 38.3	3	9.39	.027	1.43	.007	1.45	.015	7	K0 III	139,807,1685
110524	+29 02319			12 40 12	+28 38.1	1	7.51		0.38		−0.02		4	F4 V	1019
	+39 02549			12 40 12	+39 22.0	1	10.42		1.02		0.83		1	G8 III	280
		HA 104 # 453		12 40 12	−00 09.1	1	12.24		0.77		0.27		16	dG8	281
		LSS 2736		12 40 13	−61 40.9	2	10.55	.020	1.78	.007	0.58	.024	6	B3 Ib	1684,1737
	+0 02973			12 40 14	+00 04.3	2	10.01	.004	0.42	.011	−0.02		15	F5	281,6004
311885	−62 02903	Ho 15 - 6		12 40 14	−62 52.4	2	10.58	.034	0.28	.005	−0.23	.005	3		139,807
		G 13 - 53		12 40 15	+03 01.2	3	11.31	.010	1.11	.010	1.06	.012	5		492,1620,1658
110535	+34 02347			12 40 15	+33 42.8	1	9.02		1.05				2	G9 III	77
	−59 04401			12 40 15	−59 40.3	1	11.22		0.24		−0.34		2		1684
		PASP86,394 # 204		12 40 16	−62 50.9	1	13.21		1.11		0.09		5		496
		NGP +25 089		12 40 16	+25 15.0	1	12.46		0.75				2	G8	44
		NGP +28 110		12 40 16	+27 54.2	1	12.27		0.82				2	G7	44
		NGP +28 109		12 40 16	+27 59.8	1	11.19		0.77				2	G8 III	44
		Ho 15 - 17		12 40 17	−62 50.8	1	11.79		1.58		1.78		1		139
		SSII 196		12 40 18	+30 25.9	1	12.64		0.35		0.03		1	F2	1298
		HA 104 # 455		12 40 18	−00 07.9	1	15.10		0.58		−0.02		7		1764
110506	−55 05197	HR 4834		12 40 18	−55 54.1	4	5.99	.012	−0.08	.004	−0.30	.005	12	B7/8 V	15,401,1770,2012
		HA 104 # 454		12 40 19	−00 09.4	1	11.94		0.75		0.25		19	dG8	281
110498	−60 04253	LSS 2737		12 40 19	−61 22.5	4	9.67	.007	0.50	.009	−0.50	.013	11	B0.5III	540,1684,1688,2021
		NGC 4636 sq1 3		12 40 20	+02 48.8	1	12.66		0.51		0.04		2		492
		NGC 4636 sq1 2		12 40 20	+02 54.6	1	12.76		0.63		0.07		3		492
		HA 104 # 457		12 40 20	−00 12.4	1	16.05		0.75		0.52		4		1764
		MtW 104 # 30		12 40 20	−00 15.6	1	12.34		0.64		0.19		3		397
		HA 104 # 456		12 40 20	−00 15.7	1	12.36		0.62		0.14		2		1764
		Lod 670 # 27		12 40 20	−62 43.8	1	12.02		0.28		−0.05		2	B8	807
	+45 02058			12 40 21	+44 53.6	1	9.56		0.86				2	G8 III	280
		MtW 104 # 31		12 40 21	−00 12.4	1	16.07		0.85		0.50		3		397
		MtW 104 # 32		12 40 22	−00 13.0	1	14.81		0.83		0.54		3		397
	+21 02448			12 40 23	+20 37.7	1	9.07		1.38		1.52		1	K5	280
110536	−1 02716	IDS12378S0143	A	12 40 23	−01 58.7	1	9.09		0.94		0.63		2	K2 V	3072
110536	−1 02716	IDS12378S0143	B	12 40 23	−01 58.7	1	13.92		1.49		0.84		1		3072
		Ho 15 - 43		12 40 23	−62 49.4	1	14.34		0.84		0.06		2		393
110480	−71 01370			12 40 23	−72 10.4	1	8.77		0.16		0.15		1	A2 V	1771
110517	−53 05291			12 40 24	−54 08.2	1	8.58		1.15		0.99		12	G8/K0 III	1589
110537	−3 03341	G 13 - 54		12 40 25	−03 46.4	1	7.84		0.66		0.19		1	G0	1620
		LSS 2738		12 40 25	−61 16.6	1	11.81		0.60		−0.22		3	B3 Ib	1684
110509	−65 01973			12 40 25	−66 20.9	1	9.40		0.07		−0.05		2	B9.5V	1688
		G 164 - 13		12 40 26	+30 29.2	1	15.36		1.64				1		686
	−53 05293	UW Cen		12 40 26	−54 15.3	2	10.18	.071	0.51	.159	0.07	.045	2		1589,1699
		Ho 15 - 42		12 40 26	−62 50.4	1	14.07		0.94		0.61		2		393
	+27 02164			12 40 27	+26 50.8	3	10.61	.019	1.01	.007	0.75	.021	4	G8 III	44,1019,8041
110571	+26 02382	IDS12380N2627	AB	12 40 28	+26 11.0	3	8.77	.019	1.01	.004	0.74	.000	3	G9 IV	44,1019,8034
	+37 02312			12 40 28	+37 06.4	1	9.82		0.71				2	G7 III	77
		LSS 2739		12 40 28	−61 45.7	1	12.31		1.01		0.01		3	B3 Ib	1737
		HA 104 # 460		12 40 29	−00 11.9	1	12.89		1.29		1.24		16		1764
		MtW 104 # 40		12 40 29	−00 11.9	1	12.91		1.28		1.31		3		397
		NGC 4636 sq1 8		12 40 30	+02 55.9	1	14.62		0.99				1		492
		CS 16032 # 21		12 40 32	+03 02.5	1	13.04		−0.05		−0.11		1		1744
110572	+0 02975			12 40 32	−00 15.9	1	9.69		0.47		0.00		3	G0	397
		G 59 - 36		12 40 33	+21 33.9	1	10.83		0.92		0.69		1	K0	333,1620
		PHI4/3 4		12 40 34	−28 40.0	1	11.63		0.93				2		1719
		Ho 15 - 30		12 40 34	−62 49.0	1	14.84		0.92		−0.05		2		393
110551	−53 05294			12 40 35	−54 22.4	1	9.33		−0.04		−0.36		2	F0 V	1730
110532	−58 04453	HR 4835	★ AB	12 40 35	−58 37.7	5	6.39	.022	1.09	.016	1.00		21	K0/1 III	15,404,1075,1075,2006

Table 1 695

HD	DM	Other Id	N Rem	α_{1950}	δ_{1950}	S	V	σ_V	B–V	σ_{B-V}	U–B	σ_{U-B}	n	Spectrum	References
		LSS 2740		12 40 35	−61 23.5	1	11.09		0.66		-0.19		3	B3 Ib	1684
		Ho 15 - 2		12 40 35	−62 49.5	3	12.20	.023	0.84	.013	-0.14	.033	9		139,393,496
		Ho 15 - 1	⋆ V	12 40 35	−62 49.8	4	12.15	.149	0.94	.068	-0.23	.039	11		139,393,496,496
		Ho 15 - 29		12 40 35	−62 50.0	1	14.64		0.94		0.11		2		393
	+0 02976	IDS12381N0033	A	12 40 36	+00 16.3	3	9.71	.006	0.48	.002	-0.03	.003	65	F8	281,989,1729
		SSII 197		12 40 36	+32 19.	1	13.05		-0.03		-0.09		4	A0	275
		Ly 153		12 40 36	−62 50.	1	11.91		0.89		-0.16		2		496
110583	+13 02572			12 40 37	+13 01.0	1	9.40		1.07		0.95		1	K2	280
110582	+25 02548			12 40 37	+24 34.4	1	8.00		1.01		0.86		1	K0	280
		LSS 2741		12 40 37	−61 14.6	1	11.43		0.56		-0.23		2	B3 Ib	1684
	+47 01984			12 40 38	+47 25.7	1	11.15		0.73		0.29		2	G8 V	1375
		Ho 15 - 27		12 40 38	−62 49.6	1	14.46		1.60		0.40		2		393
		Ho 15 - 28		12 40 38	−62 49.7	1	14.35		1.29		0.64		2		393
		Ho 15 - 25	AB	12 40 38	−62 49.7	1	12.21		0.92		-0.20		2		393
	+33 02263			12 40 39	+33 27.6	1	10.02		0.47				2	F6	77
		Ho 15 - 26	AB	12 40 39	−62 49.6	1	12.82		0.86		-0.10		2		393
		Ho 15 - 31		12 40 40	−62 49.4	1	15.68		0.86		0.02		2		393
		Ho 15 - 34		12 40 40	−62 49.8	1	14.81		0.98		0.39		2		393
		Ho 15 - 16	V	12 40 40	−62 51.2	2	13.94	.024	0.63	.024	-0.08	.394	3		139,393
	+24 02487			12 40 41	+23 32.7	1	9.43		1.27		1.47		1	K2	280
	+38 02360			12 40 41	+38 22.2	2	9.71	.004	0.35	.013	-0.02	.004	5	F0p	1240,1569
		HA 104 # 350		12 40 41	−00 16.9	1	12.63		0.67		0.17		7		1764
		PHI4/3 1		12 40 41	−28 10.9	1	12.74		1.11				3		1719
		Ho 15 - 32		12 40 41	−62 49.6	1	15.11		0.96		-0.09		2		393
	+18 02651			12 40 42	+17 29.4	1	10.31		1.14		1.12		1		280
110575	−39 07785	HR 4836		12 40 42	−39 54.2	4	6.44	.011	0.24	.022	0.05	.016	14	A7 III	15,1637,2012,8071
		Ho 15 - 33		12 40 42	−62 49.7	1	14.72		0.90		0.30		2		393
		NGC 4647 sq1 3		12 40 43	+11 49.3	1	13.68		0.83				2		492
110612	+10 02461			12 40 44	+10 22.5	2	7.42	.023	1.58	.014	1.80	.007	5	M1	280,3025
		Ho 15 - 41		12 40 44	−62 49.4	1	14.68		0.93		-0.18		2		393
	+17 02527			12 40 45	+17 23.9	2	10.09	.024	0.68	.000	0.25	.005	3	G0	280,1569
	+38 02361			12 40 45	+38 20.1	3	9.81	.046	0.28	.036	0.08	.010	7	Am	77,1240,1569
	+17 02528			12 40 46	+17 22.3	2	9.59	.024	0.94	.010	0.73	.019	3	G0	280,1569
110593	−40 07456			12 40 46	−41 02.9	2	7.13	.010	1.60	.010	1.83		6	M2 III	2012,3012
		Ho 15 - 39		12 40 46	−62 49.9	1	14.70		0.85		-0.02		2		393
		Ho 15 - 35	AB	12 40 46	−62 50.4	1	14.94		1.38		1.25		2		393
		Ho 15 - 13		12 40 46	−62 53.3	1	12.93		1.43		1.08		1		139
		Ho 15 - 37		12 40 47	−62 50.7	1	14.73		0.70		0.64		2		393
		NGC 4647 sq1 4		12 40 48	+11 43.3	1	14.23		0.56				2		492
		HA 104 # 470		12 40 48	−00 13.5	1	14.31		0.73		0.10		1		1764
110591	−25 09394			12 40 48	−26 12.6	1	9.33		1.05		0.67		1	G8 II	565
		Ho 15 - 40		12 40 48	−62 49.5	1	14.77		0.87		0.05		2		393
		AAS78,203 # 19		12 40 48	−66 12.	1	12.54		0.69		-0.16		5		1684
311881	−62 02905			12 40 49	−62 43.5	1	11.23		0.23		-0.16		2	A0	807
110628	+26 02383			12 40 51	+26 24.0	2	6.69	.004	0.35	.004	0.11	.015	3	F2 III	1019,1569
110627	+27 02165			12 40 51	+26 41.6	1	9.68		0.62		0.16		1	G0	1019
		BSD 5 # 391		12 40 51	+74 34.2	1	12.78		0.91		0.55		1	G1	1058
		Ho 15 - 38		12 40 51	−62 50.6	1	14.77		1.89		0.90		2		393
110605	−30 10073			12 40 52	−31 06.7	1	9.12		0.73				4	G8 V	2012
		Ho 15 - 11		12 40 52	−62 45.9	1	11.91		0.68		0.26		1		139
		NGC 4647 sq1 14		12 40 53	+11 51.3	1	14.62		0.57				2		492
		AAS12,381 13 # B		12 40 53	+12 44.7	1	10.46		0.92		0.72		1		280
	+18 02652			12 40 53	+18 06.1	1	9.43		1.01		0.80		2	K0	1569
110678	+61 01312	HR 4840	⋆	12 40 53	+61 25.7	2	6.39	.009	1.27	.002	1.39		8	K2 III	71,1501
311884		Ho 15 - 3	⋆ V	12 40 53	−62 48.8	5	10.80	.045	0.82	.023	-0.18	.031	12	WN6+O5	139,393,496,807,1684
110642	+31 02401			12 40 55	+31 04.5	1	9.67		0.58				2	G5	77
		G 255 - 17		12 40 55	+78 09.6	1	15.84		1.86				1		906
110569	−68 01740			12 40 55	−69 09.7	1	9.32		0.25		0.12		1	A6 V	1771
	+36 02303			12 40 58	+36 26.8	2	9.58	.010	1.02	.005	0.70		6	G7 III	280,1569
		AAS12,381 41 # C		12 40 58	+41 09.8	1	10.67		1.09		1.07		1		280
110621	−43 07834			12 40 58	−44 24.1	4	9.91	.012	0.46	.005	-0.16	.000	13	Fpw	1236,1594,2033,3077
		Ho 15 - 4		12 40 59	−62 48.5	3	12.28	.037	1.02	.014	-0.14	.036	8		139,393,496
	+40 02566			12 41 01	+39 54.8	1	9.11		0.41		0.00		2	F5	1733
		Ho 15 - 15		12 41 01	−62 49.9	1	13.28		0.68		0.28		1		139
	+12 02507			12 41 02	+12 27.9	1	10.04		1.13		1.13		2	K2	1569
		GPEC 1834		12 41 02	−32 47.3	1	11.78		0.85				2		1719
110619	−37 08082			12 41 02	−37 25.9	1	7.52		0.67				4	G5 V	2012
311880	−62 02906			12 41 02	−62 41.6	1	11.05		0.28		0.13		2	B9	807
		BSD 5 # 392		12 41 03	+74 19.5	1	12.01		0.66		0.18		1	G1	1058
	−7 03470			12 41 03	−08 26.1	1	11.10		0.86		0.55		2		1696
110646	−0 02603	HR 4837		12 41 04	−01 18.1	7	5.93	.007	0.86	.007	0.45	.015	18	G8 IIIp	15,1007,1013,1417*
		Ho 15 - 5		12 41 04	−62 47.1	1	12.04		0.59		0.31		2		139
110679	+26 02384			12 41 08	+26 03.3	3	9.19	.013	0.86	.007	0.45		6	G5 III	44,280,8034
		G 59 - 37		12 41 09	+25 22.7	1	12.96		1.63				1	dM4	686
		LSS 2747		12 41 09	−61 22.6	1	11.08		0.83		-0.10		1	B3 Ib	1684
110610	−63 02406			12 41 09	−63 56.3	2	8.58	.000	0.45	.010	0.18	.006	3	A9 IV/V	1688,1771
110680	+26 02385			12 41 10	+25 57.2	3	7.63	.013	0.30	.005	0.04	.004	7	A9 III	833,1569,1569
		NGC 4647 sq1 5		12 41 11	+11 41.7	1	13.26		0.59				2		492
110687	+42 02334			12 41 11	+41 32.1	1	7.90		1.63				3	M3 III	280
110652	−31 09860			12 41 11	−31 46.5	1	9.51		0.80		0.37		1	K0 V	3072

HD	DM	Other Id	N Rem	α_{1950}	δ_{1950}	S	V	σ_V	B–V	σ_{B-V}	U–B	σ_{U-B}	n	Spectrum	References
311879	−62 02907			12 41 11	−62 41.8	1	10.33		0.20		−0.24		2	B8	807
		Ho 15 - 10		12 41 11	−62 46.6	1	11.15		1.35		1.14		1		139
		Ho 15 - 12		12 41 11	−62 53.6	1	12.80		0.78		0.60		1		139
110635	−53 05303			12 41 12	−53 31.8	1	9.28		0.34		0.13		1	A9 V	1771
	+13 02573			12 41 13	+13 25.0	1	9.52		1.31		1.48		1	K2	280
110688	+36 02304			12 41 13	+35 35.0	1	9.21		0.51				2	F6 V	77
		HA 104 # 364		12 41 13	−00 18.1	1	15.80		0.60		−0.13		1		1764
	+44 02220			12 41 15	+44 07.3	1	9.82		1.09		1.09		1	K0 III	280
		PHI4/1 178		12 41 15	−29 38.1	1	12.72		0.95				3		1719
		NGC 4647 sq1 6		12 41 16	+11 43.6	1	12.58		0.66				2		492
110653	−35 08155	HR 4838		12 41 16	−36 04.5	3	6.38	.006	−0.06	.000	−0.23		10	B8/9 V	15,1770,2012
110625	−61 03352			12 41 16	−62 12.9	2	9.79	.017	0.35	.005	−0.11	.024	4	B8/9 III	432,1688
		SSII 199, AQ CVn		12 41 18	+32 38.	1	13.14		0.09		0.13		11	A0	275
	+45 02059			12 41 19	+45 00.4	1	10.05		0.71				2	G6 III	280
		NGC 4647 sq1 12		12 41 20	+11 54.2	1	13.86		0.56				2		492
		NGP +28 111		12 41 20	+28 09.7	1	11.63		0.98				2	K0 III	44
		HA 104 # 366		12 41 20	−00 18.5	1	12.91		0.87		0.42		1		1764
110666	−27 08832	HR 4839		12 41 20	−28 03.0	2	5.48	.000	1.34	.005	1.50		6	K3 III	2006,3005
		NGC 4647 sq1 7		12 41 21	+11 42.5	1	13.72		0.59				2		492
110703	+15 02500	IDS12388N1540	AB	12 41 21	+15 23.1	1	8.78		0.25		0.10		1	F2	280
110659	−55 05201			12 41 21	−55 57.5	2	8.27	.020	0.00	.000	−0.39	.004	3	B7 III	401,1770
110639	−60 04259	LSS 2748		12 41 21	−61 07.3	2	8.43	.030	0.69	.003	−0.29	.032	5	B1 Ib-II	191,1684
311883	−62 02908	Ho 15 - 7		12 41 21	−62 50.0	2	10.77	.010	0.66	.015	0.15	.000	4		139,807
		HA 104 # 479		12 41 22	−00 16.4	1	16.09		1.27		0.67		1		1764
		NGC 4647 sq1 11		12 41 23	+11 53.9	1	12.18		0.60				2		492
		PHI4/1 176		12 41 23	−30 21.9	1	13.52		0.87				1		1719
110640	−60 04260			12 41 23	−61 15.1	2	9.01	.009	0.26	.012	0.20	.004	3	A1 V	1688,1771
		NGC 4647 sq1 13		12 41 24	+11 56.7	1	14.82		0.58				1		492
		SSII 200		12 41 24	+32 15.9	1	12.51		0.40		−0.08		1		1298
		HA 104 # 367		12 41 25	−00 17.2	1	15.84		0.64		−0.13		1		1764
	+0 02978			12 41 25	−00 21.6	1	10.88		0.96		0.61		2	G8 III	281
110660	−63 02407			12 41 26	−63 46.9	1	9.93		0.53		−0.32		2	B3/5	1688
		ApJS33,459 # 238		12 41 27	−61 18.6	1	12.29		0.72		−0.26		3		1684
		PHI4/1 170		12 41 28	−29 45.4	1	12.36		1.06				3		1719
110744	+33 02264			12 41 29	+33 11.7	1	9.34		1.02				2	G8 III	77
		Ho 15 - 9		12 41 29	−62 47.4	1	12.12		0.76		0.22		1		139
311886	−62 02909			12 41 29	−62 56.7	1	10.96		0.41		−0.30		2	B8	807
110722	+11 02487			12 41 30	+11 11.7	2	8.64	.000	0.32	.004	0.00	.004	12	F2	1569,1648
		NGP +26 135		12 41 30	+26 21.5	1	12.06		0.90				1	K0 III	44
110676	−58 04466			12 41 30	−58 46.0	1	8.82		0.17		0.17		1	A2 V	1771
110721	+21 02451	IDS12390N2143	AB	12 41 31	+21 26.8	1	7.98		0.25		0.20		1	F0	280
110743	+35 02369	G 164 - 15		12 41 31	+34 37.1	1	8.79		0.83				2	G8 III	77
		NGC 4647 sq1 10		12 41 32	+11 52.0	1	14.62		0.46				2		492
110745	+28 02145			12 41 32	+27 36.0	1	8.68		0.55		0.06		3	G0 V	1019
		NGP +28 112		12 41 34	+27 57.0	1	11.43		0.88				2	G8 III	44
		NGP +26 136		12 41 35	+26 20.0	1	12.26		0.79				1	G8	44
311882	−62 02910	Ho 15 - 8		12 41 35	−62 47.7	2	11.19	.024	0.33	.010	0.05	.024	3		139,807
	−60 04263			12 41 38	−60 52.9	2	11.11	.000	0.56	.018	−0.31	.002	5		1684,1737
		LP 736 - 2		12 41 39	−15 29.5	1	12.35		0.40		−0.20		2		1696
		PHI4/1 167		12 41 39	−29 38.9	1	13.30		1.09				2		1719
		Ho 15 - 14		12 41 39	−62 46.6	1	12.93		0.74		0.33		1		139
	+30 02318			12 41 42	+30 25.5	2	9.79	.010	1.00	.000	0.85		3	G9 III	280,8034
110698	−56 05410	IDS12389S5644	AB	12 41 42	−57 00.8	2	6.65	.010	0.08	.005	−0.01	.005	4	A0 IV/V	401,1771
110699	−58 04468			12 41 42	−58 53.1	1	10.17		0.23		−0.40		3	B9/A0 Vn	1684
		HA 104 # 483		12 41 43	−00 11.2	1	12.08		0.67		0.18		14	dG8	281
110700	−59 04410			12 41 43	−60 03.1	1	9.50		0.39		0.32		1	A8/F0 V	1771
110787	+36 02305	IDS12393N3619	AB	12 41 44	+36 02.5	3	7.10	.005	0.30	.007	0.10	.015	7	Am	77,1240,8071
		LP 676 - 6		12 41 44	−07 12.7	1	11.34		1.11		0.93		2		1696
	+9 02665			12 41 45	+08 29.4	1	9.57		1.27		1.52		1	K2	280
	+45 02060			12 41 45	+44 58.9	1	9.84		1.08				2	K0 III	280
	−7 03477	HW Vir		12 41 45	−08 24.0	1	10.83		−0.23		−0.96		24	sdB	1732
		NGC 4647 sq1 9		12 41 46	+11 47.2	1	14.01		0.68				2		492
	+26 02386			12 41 46	+26 12.8	3	10.71	.024	1.04	.005	0.78	.008	4	G8 III	44,1019,8034
110813	+61 01313	S UMa		12 41 46	+61 22.0	3	8.23	.251	2.13	.012	1.78	.144	3	S1.5 e	817,8005,8022
		HA 104 # 484		12 41 46	−00 14.6	1	14.41		1.02		0.73		1		1764
	−69 01698			12 41 46	−70 15.6	1	9.75		0.48		−0.05		1		742
		NGP +26 137		12 41 48	+26 20.5	1	12.52		0.87				1	G8	44
110715	−64 01990			12 41 48	−64 41.4	2	8.65	.000	0.39	.011	0.19	.026	5	B9 V	122,1688
		HA 104 # 485		12 41 50	−00 13.9	1	15.02		0.84		0.49		1		1764
110733	−53 05313			12 41 50	−53 48.7	1	7.40		1.41		1.64		5	K2/3 III	1673
		G 60 - 29		12 41 52	+06 46.9	1	13.06		1.42		1.07		1		1620
110801	+37 02316			12 41 52	+36 45.4	2	8.68	.010	1.16	.015	1.23		4	K2 III	37,77
		LOD 27 # 18		12 41 52	−61 54.3	1	11.06		2.54				2		122
	+8 02637			12 41 53	+08 16.2	1	8.89		1.44		1.60		2	K2	280,1569
110736	−62 02914	CE Cru		12 41 53	−62 41.8	2	9.30	.017	0.22	.003	−0.31	.012	4	B6 V	807,1688
110789	+22 02506			12 41 54	+22 16.2	2	7.61	.005	0.46	.005	−0.06	.005	5	F6 V	1569,1648
110788	+28 02146			12 41 54	+28 15.3	4	8.05	.005	0.97	.009	0.70	.005	8	G9 IV	44,172,280,1019
110717	−68 01744			12 41 54	−68 45.4	2	7.35	.015	0.11	.020	0.12		5	A3 V	1075,1771
		PASP86,394 # 208		12 41 55	−62 48.7	1	11.93		1.38		0.27		3		496
	+14 02534			12 41 56	+13 45.1	1	9.82		1.29		1.47		1	K2	280

Table 1 697

HD	DM	Other Id	N Rem	α_{1950}	δ_{1950}	S	V	σ_V	B–V	σ_{B-V}	U–B	σ_{U-B}	n	Spectrum	References
	+25 02555			12 41 56	+25 01.9	3	9.16	.019	0.92	.008	0.50		5	G7 III	44,280,8034
110734	−61 03353			12 41 56	−61 28.9	1	10.26		0.21		-0.06		2	B8/9 III	1688
	−64 01991			12 41 56	−64 48.6	2	11.20	.010	0.74	.004	-0.33	.007	4		1684,1737
110716	−68 01745	HR 4841		12 41 56	−68 33.4	4	6.15	.007	0.69	.007	0.38	.005	19	F5 Ib	15,1075,1075,2038
110737	−64 01992			12 41 57	−65 02.7	4	8.48	.005	0.38	.025	0.15	.020	4	B9.5V	122,1688
	+39 02554			12 41 58	+38 41.5	1	9.52		0.71				2	G6 III	280
	−65 01986			12 41 58	−66 21.3	1	10.57		0.06		-0.56		2		1684
	+28 02147	IDS12395N2745	AB	12 41 59	+27 29.0	1	10.22		0.59		0.12		1	G0 V	1019
110833	+52 01650	G 199 - 36		12 41 59	+52 05.1	4	7.04	.014	0.94	.006	0.75	.018	17	K2 V	1067,1080,1197,3016
110735	−61 03355			12 41 59	−61 56.4	1	8.85		0.90		0.62		2	F6/7 V	122
		NGP +25 091		12 42 00	+25 14.5	1	11.65		1.00				2	K0 III	44
		HA 104 # 490		12 42 00	−00 09.5	2	12.58	.004	0.53	.007	0.06	.006	26	G3	281,1764
	+13 02574			12 42 02	+13 08.4	1	9.60		1.08		1.01		1	K0	280
110815	+14 02536			12 42 03	+13 37.3	1	8.84		1.06		0.97		1	K0	280
	+15 02502			12 42 03	+14 31.9	1	9.72		1.02		0.81		1	G5	280
110814	+35 02370	IDS12397N3457	A	12 42 03	+34 41.4	1	9.49		1.04				2	G8 III	280
	+37 02317	IDS12397N3650	AB	12 42 05	+36 33.0	1	10.13		0.76				3	G7 III	280
110803	+0 02979			12 42 05	−00 06.0	2	9.93	.001	0.48	.002	0.01		21		281,6004
110784	−52 05849			12 42 05	−52 42.0	1	8.98		0.21		0.16		2	A3/4 V	1771
	+40 02568			12 42 06	+39 52.5	1	10.44		0.67				2		280
110834	+44 02221	HR 4843		12 42 06	+44 22.6	2	6.35	.014	0.44	.009	0.13		8	F6 IV	70,1501
		CS 16032 # 20		12 42 08	+03 03.9	1	13.31		0.61		0.05		1		1744
110773	−64 01993			12 42 08	−64 38.8	2	9.64	.042	0.34	.005	0.01	.009	4	B9 III	122,1688
110772	−59 04418			12 42 11	−59 48.2	1	7.23		1.33		1.34		1	K0 III	1490
	+39 02555			12 42 12	+38 52.2	1	10.40		1.13		1.07		1	K0 III	280
110835	+43 02253			12 42 12	+43 24.0	2	6.96	.005	1.32	.005	1.52		4	K3 III	172,280
110785	−58 04474			12 42 15	−59 21.7	1	9.98		0.06		-0.76		4	B3/5 II	1684
110786	−61 03356	LSS 2752	★ AB	12 42 16	−61 56.7	1	7.68		1.25		0.49		2	A3/4 Ib	122
		CS 16032 # 19		12 42 17	+03 04.3	1	12.72		0.17		0.09		1		1744
110844	+29 02320			12 42 17	+29 07.9	4	8.31	.030	1.13	.001	1.07	.000	6	K0 III	44,77,1019,8034
		SSII 202		12 42 18	+32 46.	1	12.76		0.17		0.12		6	A0	275
		SSII 203		12 42 18	+35 16.	2	12.30	.034	0.50	.000	-0.15	.015	3	F2	1240,1298
	+37 02318	TX CVn		12 42 18	+37 02.2	2	9.62	.050	0.55	.005	-0.26	.015	2		280,1753
110798	−55 05208			12 42 18	−56 03.3	2	8.70	.006	0.05	.003	-0.13	.011	3	B8 V	401,1770
	+17 02529			12 42 19	+16 35.7	1	10.40		1.39		1.53		1	K5	280
110800	−64 01994			12 42 19	−65 10.9	2	9.03	.007	0.35	.006	0.22	.014	4	B9.5V	122,1688
		CS 16032 # 81		12 42 20	+03 04.3	1	14.01		0.52		-0.05		1		1744
110809	−55 05209			12 42 20	−55 27.1	1	9.86		0.40		0.21		1	A8/9 V	1771
110855	+14 02537	G 61 - 11		12 42 23	+13 45.8	2	9.06	.015	0.66	.000	0.20	.010	2	G0	333,1620,1658
110854	+36 02306			12 42 24	+35 55.0	4	8.29	.006	-0.04	.010	-0.09	.004	16	A0 V	77,270,1068,1240
	+13 02575			12 42 25	+12 50.9	1	10.48		0.87		0.55		1		280
	+35 02371			12 42 27	+34 29.5	1	9.81		0.63				2	G0	77
	+37 02319			12 42 27	+36 31.7	1	9.95		0.68				2	G6 III	280
110811	−61 03358			12 42 28	−61 30.4	1	10.00		0.42		0.32		2	A1 IV	1688
110871	+33 02266			12 42 29	+33 21.4	1	7.78		0.40				2	F6 V	77
		LP 853 - 71		12 42 30	−23 09.2	1	12.30		0.89		0.56		3		1696
	−64 01996			12 42 30	−65 09.4	2	10.06	.004	0.35	.014	0.30	.028	4	A0	122,1688
311815	−60 04271			12 42 31	−60 58.9	1	10.77		0.47				4	B5 V	2021
110821	−60 04272			12 42 34	−60 43.3	2	9.66	.001	0.14	.002	-0.27	.002	3	B8 V	761,1688
		Wray 1008		12 42 34	−61 21.0	1	12.72		1.41		0.47		3		1684
110872	+24 02489			12 42 36	+23 52.0	1	7.75		1.63		1.99		1	M1	280
110883	+28 02148	IDS12402N2756	A	12 42 36	+27 40.0	3	7.45	.014	1.08	.013	0.95	.011	4	K2 III	44,1019,8034
110838	−47 07830			12 42 36	−47 53.8	2	6.64	.010	1.15	.010			8	K0/1 III	2012,2033
110882	+28 02149			12 42 38	+28 01.2	1	8.91		0.60		0.07		1	G1 V	1019
110897	+40 02570	HR 4845		12 42 38	+39 33.0	8	5.96	.011	0.55	.005	-0.04	.014	25	F7 V	15,254,667,1003,1008*
	+48 02052			12 42 39	+48 18.7	1	10.92		0.26		0.04		2	A7 V	1240
		NGP +26 139		12 42 40	+25 27.3	1	12.13		1.08				1	K0	44
110829	−60 04273	HR 4842	★ A	12 42 41	−60 42.4	2	4.70	.001	1.05	.002	0.99	.017	8	K0 III	1279,1637
110829	−60 04273	HR 4842	★ AB	12 42 41	−60 42.4	4	4.68	.015	1.05	.006	0.93	.012	12	K0 III	15,1034,1075,2012
110829	−60 04273	IDS12398S6026	B	12 42 41	−60 42.4	1	10.24		0.10		-0.34				1279
110830	−61 03359			12 42 41	−61 55.1	2	9.71	.018	0.67	.030	0.22	.000	4	F2/3 IV/V	122,1688
		G 60 - 30		12 42 43	+06 30.7	1	11.04		1.13		1.09		1	K7	1620
110884	+28 02150			12 42 43	+27 36.4	1	9.14		0.59		0.10		1	G1 V	1019
	+30 02320			12 42 43	+29 37.5	1	10.77		0.87		0.49		2	G6 III	1019
		HA 104 # 598		12 42 43	−00 00.2	4	11.48	.004	1.11	.010	1.05	.013	116	K2	281,989,1729,1764
	−64 01997			12 42 43	−65 20.9	3	9.73	.004	0.13	.012	-0.60	.019	9	B2 III	122,540,976
	+31 02402	IDS12403N3139	AB	12 42 44	+31 22.1	1	9.91		0.70				2	G5	77
110831	−65 01993			12 42 44	−66 05.3	1	9.51		0.10		0.06		2	A0 V	1688
	−57 05684			12 42 46	−57 25.0	1	11.60		0.20		-0.69		3		1684
		NGP +25 092		12 42 47	+25 14.5	2	11.62	.019	0.96	.005	0.67		3	K0 III	44,8034
110914	+46 01817	HR 4846, Y CVn		12 42 47	+45 42.8	6	5.04	.117	2.56	.067	6.29	.436	9	C7 I	15,172,3001,8015,8022,8027
110898	+5 02673	G 60 - 31		12 42 50	+04 38.2	2	9.50	.019	0.61	.010	0.02		3	G5	202,333,1620
	+13 02577			12 42 50	+13 11.3	1	10.16		1.18		1.09		1		280
	+44 02224			12 42 50	+43 28.6	1	9.73		0.66				2	G2 V	280
110862	−57 05686			12 42 52	−57 56.2	1	9.31		0.33		0.23		2	A3/5 V	1771
		LOD 27 # 14		12 42 52	−61 56.4	1	13.49		1.19				2		122
311774	−59 04422			12 42 53	−60 17.2	2	10.95	.020	0.28	.020	-0.35	.080	4	B8	761,837
110930	+29 02321			12 42 54	+28 57.3	3	9.80	.011	0.89	.009	0.52	.006	6	G7 III	44,1019,8034
		SSII 206		12 42 54	+41 02.	1	11.14		0.14		0.11		2	A2	1240
110902	−19 03560			12 42 55	−19 52.3	1	7.57		0.02		-0.01		2	A0 V	401

HD	DM	Other Id	N	Rem	α_{1950}	δ_{1950}	S	V	σ_V	B–V	σ_{B-V}	U–B	σ_{U-B}	n	Spectrum	References
110865	−62 02917				12 42 55	−62 25.2	1	10.06		0.54		0.47		2	A1/2 IV/V	1688
110933	+14 02538				12 42 56	+13 41.2	1	8.14		1.02		0.81		1	G5	280
110932	+15 02503	IDS12404N1455		B	12 42 56	+14 38.2	2	7.72	.019	0.43	.014	-0.10	.033	2	G5	280,1068
110932	+15 02504	IDS12404N1455		A	12 42 56	+14 38.8	2	7.22	.005	-0.11	.000	-0.37	.005	3	A0	280,1068
	+26 02388				12 42 56	+26 03.0	1	10.92		0.65		0.21		1	G2 V	1019
110891	−47 07833				12 42 56	−48 02.3	1	8.64		0.32		0.07		1	A8 IV	1771
		NGP +25 093			12 42 57	+25 13.1	1	12.30		0.88				2	K1 III	44
110878	−57 05690	LSS 2757			12 42 58	−58 13.5	1	7.91		0.51		0.44		3	A8/9 III	8100
110863	−59 04423	LSS 2756		★ AB	12 42 58	−60 16.7	4	9.06	.059	0.30	.013	-0.52	.011	8	B2 II/III	540,837,976,1684
	+12 02509				12 42 59	+12 10.6	1	9.15		1.27		1.49		1	K2	280
110950	+30 02321				12 42 59	+30 03.0	1	8.21		0.56				2	G2 V	77
	+16 02413				12 43 00	+15 29.4	1	9.96		0.88		0.56		1		280
		NGP +28 115			12 43 00	+28 15.0	1	11.58		0.81				2	G8 III	44
110864	−60 04274				12 43 00	−60 29.9	1	9.57		0.13		-0.43		1	B3 II/III	761
110934	+11 02489				12 43 01	+10 51.0	1	9.34		0.84		0.59		1	K0	280
	+47 01990				12 43 01	+46 45.2	1	9.14		1.24				2	K2 III	280
		LOD 27 # 15			12 43 01	−61 53.5	1	12.00		0.79		0.10		2		122
		LOD 27 # 11			12 43 01	−62 00.1	1	12.80		0.84		0.10		2		122
	+29 02322				12 43 03	+28 52.8	3	10.27	.016	0.91	.006	0.55	.005	4	G8 III	44,1019,8034
110949	+30 02322				12 43 04	+30 28.2	1	8.47		0.45				2	F8 V	77
110951	+8 02639	HR 4847, FM Vir			12 43 05	+07 56.8	6	5.22	.012	0.33	.009	0.15	.011	17	A8 m	15,355,1363,1417,3016,8071
110963	+29 02323				12 43 06	+29 05.5	2	9.43	.015	0.35	.027	-0.05		5	F3 V	77,1019
	+25 02556				12 43 07	+25 11.7	2	9.57	.015	1.02	.005	0.81		3	G9 III	44,8034
		Lod 682 # 19			12 43 08	−60 13.8	1	11.35		0.54		0.44		2	A3	837
311854	−61 03362	LOD 27- # 13			12 43 09	−61 56.2	1	11.04		0.37		0.25		2	A2 V	122
	+27 02166				12 43 10	+26 31.5	1	11.16		0.52		-0.03		1		1019
110964	+27 02167				12 43 10	+27 24.1	2	9.09	.017	1.61	.013	1.68		3	M4 III	694,1019
111112	+81 00402	HR 4852			12 43 10	+80 53.7	2	6.37	.030	0.14	.015	0.14	.005	4	A5 m	985,1733
110879	−67 02064	HR 4844		★ AB	12 43 11	−67 50.1	8	3.04	.006	-0.19	.007	-0.75	.013	22	B2 V +B3 V	15,26,507,1034,1075*
		G 61 - 12			12 43 12	+14 10.6	1	13.03		0.98		0.83		1		1658
110910	−55 05214				12 43 12	−56 12.4	1	9.36		0.29		0.14		1	A7 V	1771
	+32 02273				12 43 13	+32 27.1	1	9.94		1.24		1.34		1	K2 III	280
110986	+40 02571				12 43 15	+40 04.1	1	8.68		0.94				2	G8 III	280
	+15 02505				12 43 16	+15 20.8	1	10.49		0.94		0.69		1	K0	280
	+22 02509				12 43 19	+22 18.8	1	9.91		0.42				5	F2 V	193
110988	+34 02350				12 43 19	+33 48.9	3	7.51	.029	0.98	.016	0.76	.010	5	G9 III-IV	77,172,8034
		Lod 682 # 23			12 43 20	−60 20.9	1	11.49		0.37		0.29		2	A0	837
110912	−61 03363				12 43 20	−62 01.6	1	9.76		0.23		-0.11		2	B8 II/III	122
	+10 02466	IDS12409N1042		A	12 43 23	+10 25.2	1	10.01		1.05		0.96		1	K0	280
110925	−60 04277	IDS12405S6017		AB	12 43 23	−60 33.9	2	8.27	.019	0.57	.008	0.23	.020	4	F2 IV/V	837,1688
110995	+14 02539				12 43 25	+13 51.4	1	8.32		0.99		0.78		1	K0	280
	+15 02507				12 43 27	+14 41.8	1	10.07		1.29		1.40		1		280
110946	−64 02001				12 43 27	−64 38.7	2	9.14	.025	0.26	.007	-0.58	.056	3	B1 III	122,1688
	+34 02352				12 43 28	+33 52.9	1	10.21		0.45				2		77
		Lod 682 # 28			12 43 28	−60 21.8	1	12.20		0.33		-0.11		2	B9	837
	+30 02323				12 43 29	+29 36.3	1	11.33		0.50		-0.03		1		1019
	+11 02490				12 43 30	+10 50.3	1	9.21		1.29		1.46		1	K2	280
110956	−55 05215	HR 4848		★ A	12 43 30	−56 12.9	8	4.64	.010	-0.16	.004	-0.64	.013	25	B3 V	15,26,1075,1637,1770,2012*
110956	−55 05216	IDS12406S5556		B	12 43 30	−56 12.9	1	8.79		0.22		0.05		5		321
311792	−60 04280				12 43 30	−60 33.8	1	11.23		0.22		-0.49		2	B8	837
	+19 02606				12 43 32	+19 18.9	1	9.99		1.03		0.83		1	K0	280
	−59 04432				12 43 33	−60 19.0	1	10.97		0.73		0.20		2		837
	+18 02659	IDS12411N1748		AB	12 43 34	+17 31.7	1	9.80		0.53		0.10		3	F8	1569
110959	−59 04431				12 43 34	−60 15.9	2	9.26	.010	0.09	.010	-0.36	.029	3	B7 Ib	761,837
311791	−60 04281				12 43 34	−60 33.3	1	11.40		0.51		0.22		2	A2	837
		LOD 27 # 6			12 43 34	−62 04.5	1	12.47		0.72		0.16		2		122
111004	+30 02324				12 43 35	+29 52.4	2	9.18	.196	0.38	.022	0.03		3	F6 V	77,1019
		MN120,163 # 15			12 43 35	−62 39.0	1	12.49		1.20		1.00		2		375
		Lod 682 # 21			12 43 37	−60 19.2	1	11.44		0.80		0.39		2	G0	837
110975	−62 02919				12 43 37	−62 52.3	1	9.81		0.27		-0.29		1	B7 III	1688
110972	−58 04494	LSS 2760			12 43 39	−59 04.7	1	9.70		0.48		-0.03		74	B3	1764
110974	−59 04434				12 43 39	−59 55.9	1	9.54		0.25		0.16		2	A1 Vn	122
110960	−61 03365				12 43 39	−62 04.2	1	10.19		0.53		0.47		2	A1/2 V	1688
111013	+29 02325				12 43 40	+29 15.7	3	9.48	.017	1.01	.009	0.84	.009	4	G8 III	44,1019,8034
311924	−60 04435				12 43 42	−60 19.8	1	11.20		0.47		0.26		2	B9	837
	+29 02324				12 43 43	+28 57.3	1	10.55		0.41		-0.04		1	F8	1019
311795	−60 04282				12 43 43	−60 51.9	2	10.27	.011	0.40	.001	-0.43	.015	3	B3	540,976
311926	−59 04438				12 43 45	−60 23.6	1	11.16		0.24		0.03		2	A0	837
110984	−60 04285	LSS 2762			12 43 47	−60 54.8	4	8.94	.014	0.46	.011	-0.47	.022	8	B1 II/III	540,761,976,1684
		SSII 208			12 43 48	+27 27.	1	13.79		0.18		0.15		4	A0	275
		Ton 642			12 43 48	+27 34.	1	14.12		-0.23		-0.89		2		1036
		LOD 27 # 8			12 43 48	−62 03.6	1	12.61		0.83		0.21		2		122
111028	+10 02468	HR 4849		★ A	12 43 50	+09 49.1	5	5.66	.007	0.99	.007	0.78	.017	12	K1 III-IV	15,22,1003,1080,3077
	+26 02390				12 43 50	+26 08.8	2	10.51	.004	1.08	.004	0.98		3	K0 III	44,1019
	+22 02510				12 43 53	+21 28.4	1	9.53		1.15		1.12		1	K2	280
		LOD 27 # 5			12 43 54	−62 10.3	1	12.40		0.42		-0.03		2		122
111031	−11 03361				12 43 55	−11 32.4	2	6.87	.010	0.69	.005	0.31		9	G5	1075,1311
	+35 02372				12 43 58	+35 07.5	1	9.58		0.97				2	G9 III	280
111003	−64 02003	LSS 2763			12 43 58	−64 46.1	3	9.86	.022	0.35	.009	-0.46	.008	5	B3/5 II/III	540,976,1688
111057	+17 02532				12 43 59	+16 47.7	1	8.46		0.23		0.10		2	A2 m	1068

Table 1 699

HD	DM	Other Id	N Rem	α_{1950}	δ_{1950}	S	V	σ_V	B–V	σ_{B-V}	U–B	σ_{U-B}	n	Spectrum	References
	+30 02325			12 43 59	+29 58.2	1	10.66		0.79		0.50		1	G9 III	1019
111044	−0 02608			12 43 59	−00 32.9	2	8.30	.006	1.01	.001	0.78	.022	7	K0	1371,1657
311927	−59 04439			12 44 00	−60 22.3	1	11.18		0.26		0.14		2	A0	837
111056	+21 02455			12 44 02	+21 19.5	1	8.58		0.21		0.09		2	A7 IV-III	1068
111065	+27 02168			12 44 02	+26 37.1	1	9.64		0.53		0.01		1	G0	1019
111069	+12 02511			12 44 03	+11 39.1	1	8.63		0.63		0.17		1	G5	280
	+24 02492			12 44 03	+23 59.0	1	9.54		1.03		1.04		1	K0	280
	+33 02267			12 44 03	+33 20.2	2	9.20	.000	1.13	.005	1.21		3	K1 III	280,8034
111032	−32 08912	HR 4850	⋆ A	12 44 04	−33 02.5	4	5.87	.023	1.33	.010	1.60		12	K3 III	15,1075,2032,3005
111011	−59 04440			12 44 04	−60 08.0	1	8.05		1.15		0.84		2	G8 III/IV	122
111066	+24 02493	G 61 - 14	⋆ A	12 44 05	+24 25.3	1	6.86		0.53		0.05		1	F8 V	333,1620
111068	+13 02580			12 44 06	+13 14.2	1	8.64		1.49		1.56		1	K5	280
		NGP +26 141		12 44 06	+26 18.6	1	11.76		1.07				1	K0	44
	−64 02006			12 44 06	−65 22.6	2	10.13	.000	0.09	.015	-0.63	.034	3	B8	122,1688
311963	−60 04287			12 44 07	−60 34.8	1	10.46		0.35		0.27		2	A2	837
111024	−62 02921			12 44 07	−62 48.8	4	9.01	.010	0.22	.013	-0.20	.025	12	B5 III	125,375,1271,1688
		LSS 2764		12 44 08	−61 39.4	1	12.08		0.83		-0.03		2	B3 Ib	1737
111067	+17 02533	HR 4851		12 44 09	+16 51.0	2	5.16	.029	1.36	.005	1.53	.005	5	K3 III	1080,3016
111025	−64 02007			12 44 09	−65 12.8	1	9.58		0.16		-0.05		2	B8 V	1688
	+31 02404			12 44 10	+30 58.5	1	10.25		1.15		1.15		1	K1 III	280
311923	−59 04443			12 44 10	−60 18.0	1	11.21		0.22		-0.06		2	A0	837
		AAS12,381 22 # C		12 44 11	+22 12.1	1	10.95		0.19		0.19		1		280
311888				12 44 12	−62 55.	2	11.80	.034	0.52	.059	0.33		8		375,1271
		G 123 - 58		12 44 13	+40 45.2	1	13.96	.020	0.80	.005	0.20	.015	2		1658,3062
311928	−59 04444			12 44 13	−60 23.9	1	10.80		0.32		0.26		2	A2	837
111037	−60 04288			12 44 14	−60 57.0	1	8.77		0.38		0.01		2	F0 V	1688
111113	+37 02324			12 44 16	+37 07.3	1	9.17		0.46				2	F2 V	77
	−64 02009			12 44 17	−65 21.8	1	9.61		0.58		0.19		2	G0	122
		SSII 209		12 44 18	+27 44.	1	12.89		0.09		0.15		2	A2	275
	+34 02354			12 44 18	+34 05.2	1	10.00		0.62				2		77
		PHI4/2 121		12 44 18	−29 12.3	1	12.18		1.27				3		1719
	+17 02534			12 44 19	+17 26.1	1	10.50		1.23		1.34		1		1569
	+26 02391			12 44 19	+26 08.5	1	10.77		0.59		0.09		1		1019
		NGP +25 095		12 44 20	+25 20.1	1	11.52		0.99				2	G9 IV	44
		Lod 682 # 29		12 44 20	−60 24.3	1	12.33		0.42		-0.20		2	B6	837
	+33 02268			12 44 21	+32 53.6	1	9.64		1.06				2	K0 III	280
		NGP +26 142		12 44 24	+25 53.4	1	11.95		0.71				1	G6	44
	+20 02759			12 44 25	+19 52.9	1	9.99		0.57		-0.03		1	G5	280
111131	+27 02169			12 44 25	+27 14.6	1	9.44		0.41		-0.01		3	F5 V	1019
		Lod 682 # 22		12 44 25	−60 08.5	1	11.45		0.33		0.24		2	A0	837
	+20 02760			12 44 26	+19 41.4	1	9.78		1.12		1.10		1	K5	280
		Lod 682 # 27		12 44 26	−60 24.2	1	12.02		0.66		0.37		2	A2	837
	+23 02491	G 59 - 38		12 44 27	+22 43.8	2	9.52	.025	0.60	.000	0.07	.010	2	G0	1620,1658
111153	+43 02258			12 44 27	+43 25.5	1	7.95		0.44		0.04		2	F8	1569
111054	−61 03367			12 44 27	−62 16.3	2	9.70	.025	0.38	.013	0.02	.007	4	B7 III/V	122,1688
	+26 02392			12 44 28	+25 34.4	1	10.49		0.83				2	G8 V	44
		CCS 2031		12 44 28	−59 25.1	1	8.86		5.80				1	N	864
		Wolf 436		12 44 29	+01 54.6	1	12.35		0.50		-0.22		2		1696
111133	+6 02660	HR 4854, EP Vir		12 44 30	+06 13.5	9	6.33	.012	-0.05	.005	-0.06	.009	69	A0p SrEuCr	15,147,1063,1079,1202*
		NGP +26 143		12 44 30	+25 42.7	1	12.09		0.77				1	G8 III	44
		AAS78,203 # 20		12 44 30	−64 19.	1	10.98		0.38		-0.35		2		1684
111132	+9 02669			12 44 31	+09 20.2	1	6.86		1.34				2	K0	280
		AAS12,381 45 # B		12 44 32	+45 26.1	1	10.48		1.09		1.03		1		280
328811	−46 08179			12 44 32	−46 43.6	1	10.89		0.28		-0.47		3		1081
111077	−57 05703	LSS 2767		12 44 32	−57 41.9	3	10.36	.015	0.26	.008	-0.64	.013	5	Be	540,976,1684
	+10 02471			12 44 33	+09 49.0	1	9.15		1.11		1.08		1	K0	280
111155	+14 02544			12 44 34	+13 53.4	1	7.91		0.40		-0.02		2	F5	1569
111087	−53 05329			12 44 34	−54 11.2	1	9.33		0.10		0.00		2	B9.5 V	1770
111086	−52 05889			12 44 35	−52 48.0	1	8.73		0.15		0.10		2	A2 V(n)	1771
111091	−62 02926			12 44 35	−62 41.3	4	10.42	.027	0.29	.007	0.22	.031	11	A1/3 V	125,375,1271,1688
		PHI4/1 99		12 44 36	−30 25.2	1	13.22		1.22				3		1719
111078	−63 02413			12 44 36	−64 23.7	4	9.69	.021	0.34	.012	0.26	.044	4	A0 V	122,1688
		AAS61,331 # 11		12 44 37	+22 59.	1	11.18		0.28		0.06		2		1569
		AAS12,381 46 # A		12 44 38	+45 40.8	1	10.39		1.09		0.99		1		280
311966	−60 04290			12 44 38	−60 46.8	1	10.93		0.32		-0.56		2	B8	761
111105	−59 04448			12 44 39	−59 48.5	2	7.25	.005	0.18	.019	0.15	.029	3	A2	122,1771
111106	−59 04449	IDS12417S5937	AB	12 44 39	−59 53.5	1	9.17		0.14		-0.12		2	B9	122
		NGP +25 096		12 44 40	+24 57.7	1	11.11		1.02				2	K0 III	44
	+27 02170			12 44 40	+27 18.7	1	10.47		0.66		0.17		1	K2	1019
	+30 02326			12 44 40	+29 43.6	2	10.19	.009	0.96	.008	0.77	.025	2	G8 III	1019,8034
		Lod 682 # 24		12 44 40	−60 16.1	1	11.51		0.37		0.20		2	A2	837
		Lod 682 # 30		12 44 40	−60 23.2	1	12.49		0.73		0.68		2		837
111163	+16 02420			12 44 41	+15 51.9	1	6.64		0.89				2	K0	280
311922				12 44 41	−60 16.1	1	12.80		0.41		0.27		1		761
		NGP +28 116		12 44 42	+27 57.6	1	11.10		0.95				2	G8 III	44
111180	+33 02269			12 44 42	+32 50.5	1	7.73		1.49		1.88		1	K3 V	280
		GPEC 2045		12 44 42	−28 06.6	1	10.79		0.96				2		1719
111079	−70 01522			12 44 42	−71 19.4	1	8.44		0.07		-0.42		4	B6 V	400
111165	+7 02575			12 44 43	+07 16.8	4	8.46	.005	1.17	.002	1.22	.010	45	K0	147,1509,1531,6005
111164	+12 02512	HR 4855	⋆ A	12 44 43	+12 13.9	2	6.09	.024	0.12	.002	0.13		3	A3 V	70,3050

HD	DM	Other Id	N Rem	α_{1950}	δ_{1950}	S	V	σ_V	B–V	σ_{B-V}	U–B	σ_{U-B}	n	Spectrum	References
		G 59 - 39		12 44 44	+14 58.4	1	13.48		1.47		1.10		2		3028
	+40 02572			12 44 44	+39 58.0	1	9.46		0.51		0.02		2	F8	1569
111102	−57 05707			12 44 44	−58 01.5	1	6.95		0.24				4	F0 III	1075
311929	−60 04291			12 44 44	−60 25.5	2	10.44	.010	0.15	.010	-0.24	.005	3	B9	761,837
111107	−64 02012			12 44 44	−64 28.7	2	10.07	.042	0.17	.010	-0.53	.016	3	B2/3	122,1688
		G 61 - 17		12 44 45	+14 58.8	2	15.86	.000	0.21	.010	-0.54	.010	6	DAs	1663,3060
111092	−64 02010			12 44 45	−65 21.7	2	9.68	.006	0.12	.002	-0.18	.009	4	B9/8 IV	122,1688
	+9 02671			12 44 47	+09 17.2	1	9.21		1.35		1.58		1	K5	280
111123	−59 04451	HR 4853, β Cru	⋆ A	12 44 47	−59 24.9	7	1.25	.016	-0.24	.006	-1.00	.016	19	B0.5III	15,26,507,1034,1075*
111125	−62 02927			12 44 47	−62 49.4	4	9.84	.013	0.14	.017	-0.41	.019	9	A3 V	125,375,1271,1688
	+34 02356			12 44 48	+33 40.8	1	9.97		0.91				3	G9 III	280
		Lod 682 # 25		12 44 50	−60 17.3	1	11.58		0.19		-0.47		2	B5	837
		Lod 682 # 31		12 44 50	−60 17.3	1	12.50		0.36		0.05		2	A0	837
		LP 736 - 33		12 44 51	−12 32.3	1	14.81		0.14		-0.57		3		3060
311962	−60 04293			12 44 51	−60 38.5	1	11.20		0.21		0.08		2	A0	761
111124	−62 02928	LSS 2768	A	12 44 51	−62 43.4	8	9.36	.034	0.69	.020	-0.40	.019	22	B1/3 I	125,375,540,976,1271,1684*
111124	−62 02928		B	12 44 51	−62 43.4	1	10.55		1.05		0.01		3		375
	+34 02357	IDS12424N3429	AB	12 44 52	+34 13.0	1	10.05		0.49				2	F8	77
		AAS12,381 15 # C		12 44 54	+14 50.7	1	11.11		0.18		0.12		1		280
		A2 23b		12 44 54	+28 55.4	1	15.65		0.56		-0.10		1		98
	+32 02274	G 164 - 20		12 44 54	+31 29.4	1	9.84		1.29		1.24		1	K4 V	3072
311969				12 44 54	−60 50.3	1	10.99		0.29		0.13		2		761
111150	−62 02929			12 44 55	−63 10.6	1	10.17		0.50		0.37		2	A3/5 V	1688
111151	−64 02013			12 44 57	−64 34.9	1	10.07		0.31		0.05		2	B8 IV	1688
	+43 02259			12 44 58	+42 40.7	1	10.10		1.08		0.96		1	K0 III	280
111199	−5 03569	HR 4856	⋆ A	12 44 58	−06 01.7	4	6.27	.017	0.54	.008	0.06	.012	16	F7 V	15,254,1417,3053
311904	−59 04453			12 45 00	−60 02.7	1	11.22		0.47		0.22		2		122
311921	−59 04455			12 45 00	−60 18.1	2	11.30	.015	0.24	.024	-0.06	.000	3	A2	761,837
		SOS 9081		12 45 00	−60 37.	1	11.63		0.21		-0.42		2	B5	761
		MN120,163 # 14		12 45 00	−62 40.7	1	12.64		0.55		0.23		5		375
	−63 02414			12 45 00	−64 07.0	2	10.53	.003	0.31	.001	-0.50	.001	3	B1.5III	540,976
	−59 04456			12 45 01	−59 46.6	1	10.30		0.21		-0.40		2	B2	122
	+33 02270			12 45 02	+33 02.6	2	10.24	.000	1.19	.020	1.34	.030	2	K2 III	280,8034
	+16 02422			12 45 04	+15 52.3	1	9.90		0.92		0.64		1	G5	280
111224	+13 02582			12 45 05	+12 49.5	1	9.41		0.90		0.59		1	K0	280
	+30 02328			12 45 05	+30 00.5	1	10.80		0.39		-0.02		1	F4	1019
		Lod 682 # 26		12 45 05	−60 15.3	1	11.87		0.36		0.27		2	A4	837
111205	−45 08024			12 45 08	−46 13.0	1	9.15		0.06		0.05		1	A0 V	1770
	−59 04458	IDS12422S5929	AB	12 45 08	−59 45.7	1	10.66		0.24		-0.02		2		122
111194	−59 04459			12 45 08	−60 04.4	1	8.88		0.14		-0.64		2	K2/3 II	191
111175	−62 02931			12 45 08	−63 23.6	1	10.64		0.46		0.32		2	A0/1 V	1688
111233	+28 02151			12 45 10	+27 40.9	1	9.18		0.55		0.17		1	G0 V	1019
111174	−62 02932			12 45 10	−62 45.9	5	8.11	.019	0.39	.007	0.31	.040	18	A3 V	125,375,1271,1688,1771
	+33 02271			12 45 11	+33 07.0	2	9.02	.005	1.35	.019	1.68		4	K2 III	280,8034
111270	+63 01034	HR 4859		12 45 11	+63 03.2	1	5.89		0.20		0.10		2	A9 V	1733
111161	−66 01944	IDS12422S6635	A	12 45 11	−66 51.5	1	7.60		0.19		0.07		1	A3 III/IV	1771
		G 14 - 2		12 45 12	−08 19.8	1	13.24		0.73		-0.09		2		3060
		SOS 9118		12 45 12	−60 46.	1	11.59		0.32		-0.42		2	B5	761
111176	−63 02415			12 45 12	−64 22.8	1	9.71		0.15		-0.21		2	Ap Si	122
111193	−59 04460	LSS 2769		12 45 13	−59 56.2	6	7.95	.018	0.22	.008	-0.68	.026	11	A2/3 IV	122,401,540,976,1684,1771
111226	−24 10540	HR 4857		12 45 14	−24 34.8	4	6.43	.008	-0.06	.006	-0.41	.035	11	B8 V	15,1079,1770,2012
		NGP +28 118		12 45 16	+27 58.0	1	11.20		1.10				2	K1 III	44
	+33 02272			12 45 16	+33 16.5	1	10.14		1.09		0.91		1	G9 III	8034
111209	−59 04463			12 45 16	−60 05.9	1	9.62		0.27		0.20		2	A2 V	122
		MN120,163 # 12		12 45 16	−62 59.1	1	12.41		1.60		1.63		4		375
311931	−59 04461			12 45 17	−60 23.1	2	10.63	.019	0.12	.005	-0.06	.005	3	A0	761,837
	+31 02407			12 45 18	+30 29.6	1	10.33		1.14		1.06		1	K0 III	280
		L 105 - 2		12 45 18	−65 06.	1	13.10		0.69		-0.04		4		3062
111239	+4 02653	HR 4858		12 45 19	+03 50.7	4	6.40	.005	1.60	.004	1.82	.004	17	M3 IIIb	15,1088,1417,3055
111253	+27 02171			12 45 19	+27 06.1	1	10.11		0.39		-0.04		1	F5 V	1019
111256	+14 02545			12 45 21	+13 41.8	1	8.05		0.99		0.78		1	K0	280
		PHI4/2 105		12 45 21	−28 32.6	1	13.61		0.86				1		1719
311906	−59 04464			12 45 21	−59 59.6	1	9.61		1.28		1.12		2	K0	122
311932	−60 04296			12 45 21	−60 31.9	1	10.25		0.56		0.23		2	A3	837
111240	−0 02610			12 45 22	−00 44.3	3	9.21	.005	0.28	.006	0.04	.009	24	A5	830,1371,1783
111255	+22 02512			12 45 23	+21 54.6	1	8.02		1.38		1.53		1	K0	280
111271	+30 02329			12 45 23	+29 48.3	2	7.26	.010	0.34	.016	0.06		6	F0	77,1019
	+10 02473			12 45 24	+10 23.9	1	8.85		1.56		1.90		1	K5	280
311919	−59 04465			12 45 24	−60 18.1	1	9.96		0.60		0.07		2	G0	837
		SOS 9119		12 45 24	−60 43.	1	11.68		0.30		-0.28		2	B6	761
111254	+23 02494			12 45 25	+22 59.5	1	8.15		1.32		1.47		1	K2	280
	+31 02408			12 45 27	+31 05.7	1	9.83		0.55				2	F8	77
		MN120,163 # 13		12 45 27	−62 48.7	1	12.53		0.36		0.23		2		375
		G 60 - 32		12 45 28	+10 01.7	2	11.40	.005	1.56	.006	1.16		4		1705,3078
		LSS 2771		12 45 28	−60 58.6	1	10.93		0.82		-0.18		3	B3 Ib	1737
111235	−44 08235			12 45 29	−44 26.7	1	6.88		-0.02		-0.11		3	B9 V	1770
		NGP +28 119		12 45 30	+27 59.0	1	11.37		0.89				2	G9 V	44
111272	+19 02608			12 45 31	+19 06.6	1	6.91		0.94				2	K0	280
111306	+50 01948			12 45 31	+50 25.8	1	6.81		0.39		-0.03		3	F0	985
111285	+24 02495			12 45 32	+24 22.1	1	7.18		0.96				2	G8 III	20

Table 1 701

HD	DM	Other Id	Rem	α_{1950}	δ_{1950}	S	V	σ_V	B–V	σ_{B-V}	U–B	σ_{U-B}	n	Spectrum	References
111335	+67 00764	HR 4863		12 45 32	+67 03.8	1	5.43		1.56				2	gK5	71
	+33 02273			12 45 33	+33 18.4	1	10.90		0.98		0.66		1	G8 III	8034
111275	+0 02983			12 45 33	−00 05.2	1	7.97		0.55		0.13		4	F8	1371
		MN120,163 # 18		12 45 33	−62 37.3	1	13.49		0.93		0.64		1		375
111248	−49 07389			12 45 35	−50 11.5	1	10.03		0.29		0.13		1	A9 V	1771
111284	+27 02172			12 45 36	+27 22.8	3	8.14	.010	1.09	.010	1.05	.013	4	K1 III	44,1019,8034
111276	−5 03571			12 45 36	−05 41.0	1	9.16		0.95		0.58		4	G5	492
111262	−31 09900			12 45 36	−32 03.1	1	7.41		1.43		1.71		7	K4 III	1673
	+35 02373	IDS12432N3453	AB	12 45 38	+34 36.5	1	9.58		0.99				2	G9 III	280
		MN120,163 # 16		12 45 43	−63 01.5	1	12.96		0.89		0.36		1		375
		Wray 1015		12 45 43	−63 44.7	1	13.18		0.92		−0.07		5		1684
111237	−64 02018			12 45 43	−64 29.7	1	10.22		0.28		0.04		2	B8 V	1688
111232	−67 02079			12 45 43	−68 09.3	1	7.59		0.70				4	G8 V	2012
111308	+14 02546	HR 4861		12 45 44	+13 49.6	3	6.56	.005	0.01	.011	0.04	.026	7	A1 V	1022,1068,3050
		NGC 4697 sq1 2		12 45 44	−05 38.2	1	14.02		0.71				2		492
	−59 04468	LSS 2772		12 45 45	−59 38.2	2	10.29	.040	0.21	.008	−0.66		6	B2	540,2021
111295	−26 09340	HR 4860		12 45 46	−27 19.4	4	5.66	.003	0.95	.006			18	G8 III	15,1075,2018,2027
111269	−53 05339			12 45 46	−53 24.5	1	9.38		0.06		0.02		1	B9.5 IV	1770
	+21 02457			12 45 48	+21 09.1	1	10.16		0.32		0.11		2	A3m	1026
		NGC 4697 sq1 3		12 45 48	−05 37.3	1	11.69		1.33				2		492
111307	+20 02761			12 45 49	+19 35.7	1	7.60		1.61		1.75		1	M1	280
311934	−60 04298			12 45 49	−60 27.4	2	10.72	.010	0.18	.005	−0.44	.000	4	B9	761,837
	+30 02330			12 45 50	+29 37.7	1	10.50		0.45		0.01		1	F6	1019
111318	+31 02409			12 45 50	+30 40.1	1	7.66		1.12				2	K0 III	77
	+25 02564			12 45 51	+25 24.5	1	9.60		0.97				2	K2 III	44
	−63 02417			12 45 51	−64 10.4	2	11.01	.005	0.24	.006	−0.19	.000	4		122,1684
	+37 02325			12 45 52	+36 36.1	1	9.27		0.92				2	G8 III	77
111312	−14 03581			12 45 54	−15 26.9	1	7.94		0.96		0.55		2	K2 V	3072
111319	+21 02458			12 45 55	+20 54.6	1	8.16		1.54		1.82		1	K2	280
111347	+31 02410			12 45 57	+31 02.4	1	8.03		0.44				2	F7 V	77
	+16 02424			12 45 58	+16 14.2	1	10.09		1.20		1.38		1	K0	280
111336	+18 02663			12 45 59	+17 51.2	1	8.32		1.02		0.79		1	K0	280
111348	+28 02152			12 45 59	+28 03.4	1	8.73		0.52		−0.02		1	F8 V	1019
111289	−56 05448			12 45 59	−56 34.9	2	8.62	.000	0.45	.000	−0.07		4	F3/5 V	1594,6006
111283	−64 02020	IDS12430S6503	AB	12 45 59	−65 19.2	1	7.35		−0.01		−0.43		2	B6 III/V	122
111346	+42 02340			12 46 00	+42 16.0	1	8.64		1.15				2	G9 III	280
		NGC 4697 sq1 4		12 46 01	−05 37.2	1	10.34		0.70		−0.11		4		492
	−45 08032			12 46 01	−45 53.2	2	9.71	.019	0.47	.014	−0.03		4		1594,6006
110945	−58 04490	X Cru		12 46 03	−19 14.5	1	8.15		0.87				1	G0/2 Ib	1484
		NGC 4697 sq1 5		12 46 04	−05 36.6	1	13.99		0.60				2		492
	+41 02331			12 46 05	+40 42.0	1	10.89		0.87		0.74		1	G8 III	280
	+43 02260			12 46 05	+43 24.4	1	9.22		0.29		0.02		2	F3 V	1240
111420	+71 00630			12 46 05	+71 13.2	2	7.31	.064	1.26	.002	1.50	.000	5	K3 II-III	3025,8100
		MN120,163 # 17		12 46 05	−62 47.8	1	13.12		0.55		0.39		3		375
	+25 02567			12 46 06	+25 00.4	1	10.37		0.90		0.60		2		1569
	+25 02566	IDS12436N2519	AB	12 46 06	+25 02.2	1	9.93		0.46		−0.01		2	F6	1569
111366	+37 02326			12 46 06	+36 36.9	1	8.30		1.16				2	K0 III	77
	+17 02541			12 46 07	+17 14.1	2	9.85	.015	1.63	.010	1.99	.005	2		280,1569
111367	+27 02173	G 149 - 10		12 46 08	+26 52.2	2	8.79	.014	0.59	.005	0.09	.000	2	G1 V	1019,1658
111349	+17 02542			12 46 09	+17 12.6	1	8.64		1.14		1.15		1	K2	280
		NGC 4697 sq1 6		12 46 09	−05 34.6	1	13.70		0.51				1		492
111303	−60 04301			12 46 09	−60 47.7	1	9.04		0.53		0.30		2	F0 IV	1688
		G 14 - 3		12 46 10	−01 15.9	1	14.22		1.05		0.76		1		1658
		NGC 4697 sq1 11		12 46 10	−05 30.0	1	11.67		0.63				1		492
		GD 480		12 46 11	+61 25.9	1	13.27		0.40		−0.19		4		308
111350	−12 03699			12 46 12	−13 07.0	1	8.74		1.20		1.23		3	K2 III	1700
111369	+9 02676			12 46 13	+09 27.8	1	8.83		1.16		1.23		1	K0	280
		NGC 4697 sq1 7		12 46 13	−05 34.0	1	13.86		0.64				1		492
	+12 02516			12 46 15	+12 26.9	2	10.15	.037	−0.08	.005	−0.17	.005	4	A0	272,280
111332	−59 04474			12 46 15	−60 23.4	1	8.14		1.44		0.23		3	K2/K3 III	1684
111290	−71 01389	LSS 2773		12 46 15	−71 27.5	3	7.76	.012	0.00	.008	−0.78	.021	9	B1 Ib/II	400,540,976
		GPEC 2156		12 46 16	−34 02.1	1	11.18		0.95				2		1719
111382	+14 02547			12 46 18	+13 35.8	1	7.91		1.30		1.36		2	K2	1569
	+33 02274			12 46 18	+32 51.7	1	10.19		1.04		0.94		1	K0 III	280
312050	−62 02939			12 46 18	−62 53.4	3	11.02	.014	0.27	.011	0.09	.055	9	A0	125,375,1271
		A2 44		12 46 20	+29 10.5	1	15.83		0.51		−0.14		1		98
		NGC 4697 sq1 8		12 46 20	−05 36.0	1	12.30		0.50				1		492
111398	+12 02518	IDS12438N1238	A	12 46 21	+12 22.2	1	7.07		0.66		0.19		3	G5 V	280
111395	+25 02568	HR 4864		12 46 21	+25 06.8	5	6.30	.038	0.70	.004	0.25	.008	11	G7 V	15,1008,1019,1119,3077
111342	−52 05919			12 46 21	−53 14.3	1	9.23		−0.04		−0.17		1	B9 V	1770
	+8 02642			12 46 22	+07 49.3	1	9.67		0.90		0.75		2	K2	1569
111397	+14 02549	HR 4865		12 46 24	+14 23.7	9	5.70	.014	0.02	.009	0.09	.032	36	A1 V	3,15,70,1007,1013*
	+31 02411			12 46 24	+31 21.2	1	9.94		1.01				2	G9 III	280
111421	+49 02163	HR 4866		12 46 24	+48 44.3	2	6.27	.000	0.18	.009	0.17	.005	4	A6 m	252,1240
111357	−50 05920			12 46 24	−51 09.3	1	8.90		0.24		0.20		1	A4/5 IV	1771
		PHI4/1 65		12 46 25	−30 58.1	1	13.27		0.77				1		1719
311958	−60 04303			12 46 25	−60 39.5	1	10.74		0.10		−0.18		2	A0	761
111343	−63 02418			12 46 25	−63 55.8	1	9.22		0.86		0.69		2	A2 III	1688
	+29 02326			12 46 27	+29 25.4	1	10.58		0.44		−0.06		1	F8	1019
111359	−54 05340			12 46 28	−54 27.0	1	9.26		0.30		0.13		1	A1mA6-A7	1771

Catalogue of mean UBV data

HD	DM	Other Id	N	Rem	α_{1950}	δ_{1950}	S	V	σ_V	B–V	σ_{B-V}	U–B	σ_{U-B}	n	Spectrum	References
111423	+42 02341				12 46 29	+41 50.5	1	8.74		0.51		0.04		2	F8 V	172
111456	+61 01320	HR 4867			12 46 29	+60 35.5	5	5.85	.000	0.46	.005	-0.04	.004	11	F5 V	15,254,1007,1013,8015
111363	-60 04306				12 46 29	-60 25.6	2	9.09	.000	0.09	.000	-0.49	.025	5	B3 III	401,761
111315	-71 01391	HR 4862			12 46 29	-71 42.8	4	5.56	.015	1.17	.003	0.95	.002	13	G8 Ib/II	15,565,1075,1637
111425	+35 02375				12 46 30	+35 26.8	1	9.30		0.58				2	G5	77
111424	+37 02328				12 46 30	+37 06.7	1	9.24		0.39				2	F3	77
	+39 02564	G 123 - 62			12 46 30	+38 31.2	1	10.43		1.00		1.04		1	K0 III	280
111422	+44 02225				12 46 30	+44 25.6	1	9.80		0.17		0.10		2	A5 V	1240
111365	-61 03374				12 46 30	-62 15.6	2	10.26	.025	0.25	.012	-0.35	.018	3	B2/5	122,1688
	+11 02499				12 46 31	+10 31.4	1	10.52		1.10		1.13		1	K0	280
	+20 02762				12 46 31	+20 27.6	1	9.85		1.22		1.36		1	K2	280
		NGC 4697 sq1 10			12 46 32	-05 32.4	1	11.61		0.89				1		492
		LOD 27+ # 15			12 46 32	-60 12.3	1	10.64		1.08		0.68		2		122
		CCS 2035, RX Cru			12 46 32	-61 29.8	1	9.89		5.51				3	N?	864
111364	-60 04307				12 46 33	-60 58.2	1	10.34		0.29		0.25		2	A1/2 V	1688
111426	+35 02376				12 46 34	+34 51.8	2	9.34	.015	0.51	.030	0.03		4	K0	77,1569
	+36 02308				12 46 34	+36 12.7	2	10.97	.000	0.37	.005	-0.06		4	F2	77,1569
		G 123 - 63			12 46 34	+40 59.8	1	13.61		1.52				1		3062
111444	+42 02342				12 46 35	+42 21.0	1	8.68		1.00				2	K0 III	280
	+23 02498				12 46 36	+23 16.9	1	10.19		0.37				2	F2 V	193
	+9 02678				12 46 37	+09 14.5	1	9.58		1.45		1.78		1		280
111457	+34 02358				12 46 38	+34 03.2	1	8.20		1.40				2	K2 IV	77
		PHI4/1 63			12 46 38	-30 41.8	1	13.39		0.90				2		1719
111377	-60 04309	LSS 2775			12 46 39	-60 41.6	3	9.53	.013	0.13	.012	-0.67	.011	11	B2/5 II/III	540,761,976
		A2 49			12 46 40	+29 10.3	1	16.32		0.76		-0.27		1		98
	+47 01994				12 46 40	+46 44.5	1	9.66		0.64				2	G0 IV	280
111468	+55 01555				12 46 40	+54 31.2	1	8.52		0.21		0.09		2	A3/5V/IV	1733
111427	-5 03577				12 46 40	-05 33.0	1	9.40		0.67		0.09		4	G5	492
		SOS 9152			12 46 42	-60 27.	1	11.89		0.20		-0.35		2	B2	761
		PHI4/1 62			12 46 43	-30 57.6	1	13.08		0.99				3		1719
111417	-45 08040				12 46 44	-45 33.2	2	8.30	.014	1.40	.000			8	K3 III	1075,2033
	+27 02174				12 46 45	+27 09.4	3	9.70	.013	1.07	.006	0.84	.011	4	K0 III	44,1019,8034
		LP 853 - 77			12 46 45	-26 03.7	1	12.25		0.47		-0.20		2		1696
311908	-59 04479				12 46 45	-60 06.5	1	10.93		0.31		0.22		2	A2	122
	+29 02327				12 46 47	+28 44.2	1	10.80		0.46		-0.06		1	F5	1019
	+43 02261	IDS12444N4246	A		12 46 47	+42 29.8	2	8.50	.000	0.99	.000	0.74		4	G9 III	280,1569
111513	+62 01257	G 199 - 41			12 46 47	+61 39.0	1	7.35		0.62		0.20		3	G1 V	22
111408	-59 04480				12 46 47	-60 14.5	1	9.17		0.13		-0.30		2	B7 IV	122
		Ton 651			12 46 48	+29 07.	1	16.00		0.40		0.06		1		98
111393	-65 02027				12 46 49	-66 15.3	1	9.09		0.12		-0.27		2	B6 IV	1688
	+43 02262	IDS12444N4246	B		12 46 50	+42 30.3	1	8.83		0.37				2	F0	280
111469	+28 02153	HR 4869	A	★	12 46 51	+27 49.5	5	5.78	.014	0.04	.012	0.06	.006	27	A2 V	275,1019,1022,1068,3050
111469	+28 02153	IDS12444N2806	B		12 46 51	+27 49.5	1	13.45		0.71		0.21		4		3024
	+45 02064				12 46 51	+45 11.4	1	10.45		1.06		0.96		1	K0 III	280
311957	-60 04311				12 46 51	-60 40.6	1	11.13		0.13		-0.12		2	B9	761
	+17 02543				12 46 52	+16 31.5	1	10.50		1.11		1.14		3	K0	1569
111483	+36 02309				12 46 52	+35 35.6	2	8.06	.010	1.24	.005	1.32		4	K3 III	77,172
111409	-63 02420				12 46 52	-64 20.5	3	7.62	.004	0.19	.005	0.07	.018	7	B9.5IV	122,1688,1771
	+11 02500				12 46 53	+11 15.7	1	10.02		1.15		1.26		1	K2	280
	-63 02421				12 46 55	-63 34.1	1	10.40		0.46		0.13		2	A0	1688
	+39 02566				12 46 56	+39 19.0	1	10.31		0.56		0.04		2	G0	1569
		AAS12,381 42 # D			12 46 56	+42 00.4	1	11.16		0.16		0.17		1		280
111436	-50 07300				12 46 57	-51 07.2	1	7.78		0.19		0.09		2	A3/5 V	1771
	+31 02414				12 47 02	+30 42.5	1	10.73		0.90		0.67		1	G8 III	280
111498	+10 02474				12 47 03	+09 38.7	1	7.65		0.90		0.60		1	K0	280
		G 237 - 78	V	★	12 47 03	+66 23.0	1	10.92		1.64				1	M3	694
	-63 02422				12 47 04	-64 16.9	1	9.81		0.51		0.05		2		122
	+34 02359				12 47 05	+33 36.5	1	10.26		0.56				2		77
111514	+27 02175				12 47 06	+26 29.7	2	8.77	.005	1.13	.005	1.12		4	K2 III	44,1019
		SOS 9151			12 47 06	-60 27.	1	11.89		0.22		-0.24		2	B6	761
	-62 02940				12 47 09	-62 28.8	2	10.55	.015	1.06	.012	-0.05	.007	4		1684,1737
	+18 02666				12 47 10	+17 38.0	1	9.30		0.93		0.69		2	K0	1569
	+34 02360				12 47 10	+33 47.4	1	9.97		0.96				2	K0 III	280
111499	-14 03587	SV Crv			12 47 10	-14 48.4	1	6.90		1.67		1.65		1	M3/4 III	3046
111515	+2 02585	G 14 - 5	A	★	12 47 12	+01 28.1	5	8.15	.009	0.69	.006	0.13	.020	9	G8 V	22,333,1003,1620,3077,6006
111525	+42 02343				12 47 12	+42 19.5	2	8.59	.005	0.19	.005	0.10	.005	4	A7 V	1068,1240
111446	-61 03376				12 47 13	-62 22.2	3	6.63	.005	1.41	.026	1.49	.052	12	K3 III	122,375,1271
111463	-59 04483	HR 4868, LSS 2779			12 47 14	-60 07.7	5	6.73	.021	0.34	.009	0.25	.032	16	A3 II	122,1034,1505,2035,8100
		GD 481			12 47 16	+73 41.6	1	14.53		0.25		0.01		3		308
111465	-63 02424				12 47 17	-63 54.3	2	10.20	.006	0.50	.006	0.32	.033	3	A6/8 V	1688,1771
111526	+19 02610				12 47 18	+18 36.9	1	8.76		0.90		0.66		1	K0	280
111540	+29 02328				12 47 18	+29 25.8	2	9.52	.009	0.61	.007	0.20		3	G1 V	77,1019
	-58 04539				12 47 19	-58 44.1	1	11.21		0.07				4		2021
111539	+39 02568				12 47 20	+38 38.4	3	8.07	.032	1.07	.008	0.94	.015	5	K0 III	172,280,8041
311953					12 47 20	-60 43.2	1	11.78		0.21		0.07		3		761
	+35 02380				12 47 21	+34 30.9	1	9.25		1.09				2	K0 III	280
111573	+49 02164				12 47 21	+48 30.5	1	9.16		0.32		0.00		2	F0 V	1240
111572	+49 02165				12 47 21	+49 00.7	2	6.50	.035	1.13	.005	1.12	.020	4	K1 III	172,252
111542	+22 02517				12 47 22	+21 54.8	2	8.47	.009	0.25	.005	0.12	.014	4	A7 IV	1068,1569
111541	+27 02176				12 47 22	+26 42.2	4	6.88	.014	1.06	.009	0.93	.006	8	K0 III	44,172,1019,8041

Table 1   703

HD	DM	Other Id	N Rem	α_{1950}	δ_{1950}	S	V	σ_V	B–V	σ_{B-V}	U–B	σ_{U-B}	n	Spectrum	References
311999	−60 04312			12 47 23	−61 18.8	2	10.77	.034	0.67	.014	−0.33	.004	6	O9.5V	92,1684
	+14 02551			12 47 24	+14 02.5	1	9.98		1.10		1.10		1	K0	280
		NGP +25 098		12 47 24	+25 18.9	1	11.35		1.12				2	K2 III	44
111478	−57 05738			12 47 24	−57 46.9	1	7.78		0.46				4	F2 II	1075
111503	−45 08049			12 47 25	−45 54.1	2	9.53	.009	0.50	.000	0.04		4	F5 V	1594,6006
		Za CVn		12 47 26	+44 02.7	2	11.49	.082	0.17	.041	0.06	.045	2	A5	597,699
		G 60 - 38		12 47 28	+04 14.2	1	16.71		−0.02		−1.13		1		3060
	+30 02331			12 47 28	+29 42.8	1	11.20		0.80		0.41		1	G8 III	1019
111505	−59 04485 LSS 2781			12 47 29	−60 23.5	3	9.11	.018	0.11	.024	−0.69	.015	6	B2/3 III	401,761,1505
111562	+9 02681			12 47 30	+08 50.5	1	9.44		0.34		0.00		2	F8	1569
	−47 11039 LSS 3742			12 47 30	−48 11.3	1	10.62		0.51		−0.35		1		954
111519	−47 07893 HR 4871			12 47 30	−48 11.3	3	6.24	.010	0.05	.007	0.04		8	A0 V	15,1770,2012
312000				12 47 30	−61 20.3	1	10.86		0.69		−0.17		2		1684
		AAS78,203 # 21		12 47 30	−64 33.	1	12.59		0.22		−0.24		4		1684
111480	−67 02092			12 47 30	−68 06.0	1	8.32		0.18		0.12		1	A3 V	1771
		NGP +25 099		12 47 31	+25 07.7	1	12.13		0.97				2	G8	44
111574	+40 02578			12 47 31	+40 07.1	1	8.68		1.01				2	K0 III	280
		LP 909 - 66		12 47 33	−27 57.9	1	11.21		0.96		0.76		4		1696
		LSS 2782		12 47 33	−61 51.3	1	10.53		0.99		−0.18		3	B3 Ib	1737
	+8 02647			12 47 35	+08 22.5	1	10.28		0.94		0.73		1		280
	−47 11040 LSS 3743			12 47 35	−47 43.8	1	10.98		0.42		−0.38		1		954
311951				12 47 35	−60 36.9	1	12.50		0.35		0.30		1		761
		SSII 216		12 47 36	+32 31.0	1	11.92		0.35		0.09		1	F2	1298
	−16 03543			12 47 36	−17 07.0	2	10.67	.023	1.26	.009	1.18		5	K5	158,1746
311950				12 47 36	−60 36.9	1	12.10		0.19		−0.36		1		761
		G 198 - 80		12 47 37	+47 23.2	1	15.55		1.53		1.29		1		1773
		PHI4/2 73		12 47 37	−28 29.5	1	12.61		0.81				3		1719
111535	−46 08203			12 47 37	−46 57.0	2	7.96	.009	0.48	.000			8	F3/5 V	1075,2033
	+30 02332			12 47 38	+29 52.9	1	10.38		1.07		0.92		1	K0 III	1019
	+17 02544			12 47 40	+17 26.6	1	10.30		0.24		0.14		1		280
111564	−29 09950			12 47 40	−30 18.3	1	7.61		0.61				4	G0 V	2012
111536	−51 07041			12 47 41	−51 57.7	1	7.67		0.18		0.11		1	A1/2 IV/V	1771
111538	−57 05740			12 47 43	−57 53.7	1	7.06		1.53				4	K5 III	1075
	+40 02579			12 47 45	+40 09.9	1	9.49		0.99				2	G8 III	280
	+47 01996 G 123 - 64			12 47 46	+47 23.7	1	10.12		0.53		−0.08		1	G0	1658
	+36 02310			12 47 47	+35 29.3	1	10.20		0.55				2	F6	77
		AAS12,381 36 # C		12 47 47	+35 53.9	1	10.54		0.92		0.70		2		280
111604	+38 02373 HR 4875			12 47 48	+37 47.3	4	5.88	.039	0.15	.011	0.08	.009	10	A3 V	77,1022,1068,1240
111591	+23 02502 HR 4873			12 47 49	+23 08.2	2	6.44	.015	1.00	.005	0.79	.010	5	K0 III	280,3016
		AAS12,381 25 # A		12 47 49	+24 59.7	1	10.91		0.22		0.17		1		280
	+37 02331			12 47 50	+37 08.5	1	9.39		0.54				2	F8	77
	+32 02279			12 47 51	+32 27.3	1	10.64		0.99		0.88		1	G9 III	280
111605	+26 02393			12 47 52	+26 25.9	1	9.61		0.66		0.28		2	G5 V	1019
		GD 319		12 47 52	+55 22.3	1	12.31		0.03		−0.93		2		308
111557	−62 02942			12 47 52	−62 43.3	3	9.64	.048	0.53	.011	0.12	.027	9	F3/5 IV/V	375,1271,1688
	+44 02227			12 47 53	+44 27.9	1	10.34		1.22		1.33		1	K1 III	280
	−50 07311			12 47 53	−51 14.3	1	10.50		0.69		0.16		2	G0	1696
311916	−59 04487			12 47 53	−60 12.5	1	11.17		0.17		−0.55		2	B8	761
		SSII 217		12 47 54	+29 48.	1	13.62		0.08		0.07		5	A0	275
	+34 02362			12 47 55	+33 58.8	1	10.15		0.50				2	F6	77
111628	+36 02311			12 47 55	+35 59.9	1	8.52		1.45				2	K4 III	77
	+39 02570 IDS12456N3919		AB	12 47 58	+39 04.3	1	9.58		0.96				2	G8 III	280
111597	−33 08653 HR 4874		⋆A	12 47 58	−33 43.6	4	4.90	.006	−0.04	.003	−0.12	.003	13	B9 V	15,26,1770,2029
111578	−59 04488 LSS 2785			12 47 58	−60 10.5	1	9.15		0.16		−0.57		2	B2/3 II	761
111558	−68 01777 LSS 2784			12 48 00	−69 22.4	4	7.25	.022	0.11	.030	−0.45	.021	10	B7 Ib	158,191,540,976
		He3 828		12 48 02	−57 34.5	1	14.29		0.90		−0.67		1		1753
		G 60 - 40		12 48 03	+09 17.6	1	14.79		1.64				1		906
	+71 00632 G 237 - 79			12 48 04	+71 28.6	1	10.08		0.97		0.90		3		1625
		PHI4/2 68		12 48 04	−28 10.6	1	12.46		1.05				3		1719
	+27 02177			12 48 05	+26 29.4	1	9.83		0.61		0.12		2	G0	1019
111588	−52 05947 HR 4872			12 48 05	−52 30.9	2	5.72	.005	0.13	.000			7	A5 V	15,2012
111571	−68 01778			12 48 05	−68 28.9	1	9.81		0.33		0.13		1	A8 V	1771
	+8 02649			12 48 07	+07 47.6	1	10.16		0.95		0.54		1	K0	280
	+27 02178			12 48 08	+27 25.4	2	10.41	.005	1.00	.005	0.83		2	K1 III	44,8041
111580	−64 02038 IDS12451S6422		AB	12 48 08	−64 38.9	1	9.03		0.51		0.03		2	F5 V	1688
311981	−60 04323			12 48 09	−60 58.0	1	9.10		0.66		−0.32		2	A0	540
	+17 02546			12 48 10	+17 03.1	1	10.20		0.89		0.56		1		280
111659	+33 02276			12 48 10	+32 51.0	1	10.01		0.55				2		77
111631	+0 02989 G 14 - 6			12 48 10	−00 29.4	14	8.49	.019	1.41	.012	1.26	.009	51	M0.5V	1,116,694,1006,1075,1088*
		PHI4/2 33		12 48 11	−30 55.3	1	12.49		1.00				1		1719
111599	−54 05346			12 48 11	−55 08.2	1	8.13		0.15		0.07		3	B9/A0 V	1770
111589	−61 03379			12 48 11	−62 07.3	1	9.93		0.27		0.05		2	B9/A0 V	1688
	+23 02504			12 48 13	+23 22.1	1	9.68		0.97		0.73		1	K5	280
	−12 03708			12 48 13	−13 10.2	1	11.13		0.58		−0.10		2		1696
	+21 02462 IDS12458N2105		AB	12 48 14	+20 48.4	2	9.20	.000	1.02	.000	0.86		5	K5	1381,3030
	+41 02332			12 48 14	+40 33.6	1	10.64		1.12		1.03		1	K0 III	280
111610	−48 07720			12 48 14	−48 31.3	1	9.48		0.12		−0.04		1	B9.5 V	1770
311913	−59 04493			12 48 15	−60 05.7	1	10.14		0.15		0.07		1	A0	761
	+36 02312			12 48 16	+36 12.0	1	10.47		0.86		0.41		1	G6 III	280
	+23 02505			12 48 17	+22 29.0	1	9.54		0.99		0.70		1	G5	280

HD	DM	Other Id	N	Rem	α_{1950}	δ_{1950}	S	V	σ_V	B–V	σ_{B-V}	U–B	σ_{U-B}	n	Spectrum	References
111660	+8 02650				12 48 18	+08 22.3	1	8.29		0.95		0.76		1	G5	280
111622	−49 07426				12 48 18	−49 49.0	1	8.73		0.03		−0.02		3	B9.5 V	1770
111612	−58 04550				12 48 18	−58 48.0	2	8.11	.000	−0.04	.010	−0.56	.007	8	B3 IV/V	401,1732
111689	+46 01824	IDS12460N4637	A		12 48 19	+46 20.3	1	8.15		0.95				2	G8 III	280
111613	−59 04494	NGC 4755 - 20		⋆	12 48 19	−60 03.5	10	5.74	.019	0.38	.022	−0.09	.026	36	A1 Ia	15,26,1018,1034,1075*
	+27 02179				12 48 20	+27 05.9	1	10.74		0.64		0.09		1	G2 III	1019
111690	+33 02277				12 48 21	+33 20.3	1	9.64		0.73				2	G6 IV	280
	−63 02431				12 48 21	−63 48.3	1	10.99		0.81		−0.12		2		1684
		ApJS33,459 # 38			12 48 22	−64 43.7	1	15.00		1.34		−0.36		1		1753
111639	−52 05957				12 48 24	−52 27.7	1	9.56		0.35		0.16		1	A7 IV	1771
111651	−28 09735				12 48 25	−28 54.3	1	8.01		0.24				1	A3/5 III	1473
	−28 09736				12 48 28	−28 49.8	1	11.23		0.48				1		1473
111937	+83 00365				12 48 30	+82 42.1	1	9.12		0.32		0.03		2	A3	1733
		SOS 9255			12 48 30	−60 10.	1	11.37		0.25		−0.33		1	B8	761
		BPM 7855			12 48 30	−61 02.	1	15.86		−0.14		−1.07		1		3065
	+30 02333				12 48 33	+29 30.7	1	11.33		0.60		0.05		1		1019
	+42 02345				12 48 33	+41 31.9	1	9.58		1.16				2	K2 III	280
311941	−60 04325				12 48 33	−60 25.5	1	10.72		0.20		0.10		2	A0	761
111717	+32 02281				12 48 35	+31 44.9	1	7.88		0.37				2	F2 V	77
	+41 02334				12 48 36	+40 31.4	1	10.52		1.20		1.28		2	G9 III	280
111656	−59 04498				12 48 38	−60 13.5	2	9.94	.005	0.15	.005	−0.53	.010	5	B3 III	761,1505
		G 164 - 25			12 48 39	+29 34.1	1	14.95		1.64				1		686
111732	+33 02278				12 48 39	+33 01.1	1	8.75		1.35				2	K1 III	77
112014	+84 00289	HR 4892		⋆ B	12 48 39	+83 41.4	2	5.84	.010	−0.03	.035	−0.07	.045	4	A0 V	1733,3016
111654	−57 05749				12 48 39	−58 12.8	2	7.37	.007	1.66	.006	1.97		5	K3 III	1075,1490
	+27 02180				12 48 40	+26 41.0	3	9.90	.012	0.84	.005	0.43	.007	4	G5 III	44,1019,8041
111862	+17 02549	IDS12472N1737	C		12 48 41	+17 18.7	1	8.76		1.43		1.78		1		280
	+38 02374				12 48 42	+37 39.5	2	9.34	.014	0.29	.005	0.03		4	F0 V	77,1240
	+34 02363				12 48 43	+33 38.3	1	10.01		0.48				2	F5	77
	−63 02432				12 48 45	−63 44.2	1	11.19		0.78		−0.18		2		1684
	+30 02334				12 48 46	+29 34.4	1	10.48		1.30		1.24		1	K0 IV	1019
112028	+84 00290	HR 4893		⋆ A	12 48 46	+83 41.1	2	5.31	.030	0.00	.035	−0.06		4	A1 IIIShell	1733,3023
111720	−9 03569	HR 4877		⋆ A	12 48 47	−10 04.0	1	6.50		1.02		0.82		1	G8 III	13
111720	−9 03569	HR 4877		⋆ AB	12 48 47	−10 04.0	2	6.40	.005	1.02	.000	0.80	.005	7	G8 III	15,1417
111720	−9 03568	IDS12462S0948	B		12 48 47	−10 04.0	2	9.78		0.89		0.45		2	G0	13
111687	−60 04329				12 48 48	−60 48.4	4	9.19	.008	0.06	.009	−0.30	.013	11	B8 III	401,761,1505,1688
111733	+9 02683				12 48 49	+08 28.9	1	7.63		1.09		1.02		1	G5	280
111721	−12 03709				12 48 49	−13 12.9	6	7.97	.004	0.80	.009	0.16	.025	21	G6 V	742,861,1594,1700,2033,3077
		NGP +25 100			12 48 50	+25 16.1	1	12.46		0.64				2	G5	44
111743	+28 02154				12 48 50	+27 36.8	3	9.71	.010	1.03	.016	0.84		5	G9 III	44,694,1019
		Hya sq 1 # 9			12 48 50	−28 56.1	1	12.65		0.71				1		1473
	+14 02554				12 48 52	+13 52.8	1	9.85		0.97		0.88		1	K0	280
111742	+29 02329				12 48 52	+28 37.3	3	9.22	.020	0.92	.009	0.68	.017	4	G8 III	44,1019,8034
111674	−63 02433				12 48 52	−64 24.2	2	8.80	.021	0.26	.000	0.09	.001	4	A5/7 V	1688,1771
111744	+13 02596				12 48 53	+12 38.7	2	8.88	.029	0.01	.005	−0.02	.024	3	A0 V	272,280
	+32 02283				12 48 53	+32 12.0	4	10.23	.029	0.12	.019	0.07	.010	10	A2 V	77,275,1026,1240
	+41 02335				12 48 53	+41 02.8	1	9.54		0.91				2	G7 III	280
111699	−59 04502				12 48 53	−60 08.1	1	9.84		0.13		−0.55		3	B2/3 III	1505
		NGP +25 101			12 48 54	+25 20.5	1	11.36		1.08				2	K0 III	44
		Steph 1036			12 48 54	+42 11.4	1	11.90		1.33		1.28		1	K4	1746
111763	+29 02330	G 164 - 27			12 48 55	+29 07.9	2	9.11	.009	0.84	.001	0.53		3	G8 V	44,1019
	+26 02396				12 48 57	+25 50.6	1	10.46		0.89				2	G6 IV	44
111688	−62 02944				12 48 57	−62 52.2	1	9.90		0.61		−0.04		1	B8/9 II/III	1688
		AAS12,381 8 # B			12 48 58	+08 17.1	1	10.27		0.98		0.70		1		280
111709	−51 07065				12 48 58	−51 51.4	1	9.26		0.34		0.18		1	A(3)mA5-A7	1771
		LP 377 - 78			12 49 00	+22 24.	1	12.99		1.60				1	dM4	686
		LP 377 - 79			12 49 00	+22 25.	1	14.33		1.72				1	dM3	686
	+43 02264				12 49 00	+42 37.5	1	10.54		0.89		0.49		1	G5 III	280
		GD 482			12 49 01	+83 13.9	1	13.34		0.65		0.10		3		308
		PHI4/1 21			12 49 01	−30 29.7	1	12.80		0.83				1		1719
111782	+28 02155	IDS12466N2747	AB		12 49 02	+27 30.3	1	9.30		0.41		−0.01		4	F5 V	1019
111714	−59 04504				12 49 02	−60 16.7	2	9.31	.019	0.16	.010	−0.59	.019	6	B2/3 II/III	761,1505
312110	−59 04503	IDS12461S6004	AB		12 49 02	−60 20.7	1	10.42		0.16		0.01		3	A0	761
111765	+3 02703	HR 4878			12 49 04	+03 19.7	2	6.01	.005	1.29	.000	1.44	.000	7	gK4	15,1256
111711	−54 05352				12 49 04	−55 05.4	1	9.80		0.39		0.24		1	A8/9 V	1771
	+23 02507				12 49 05	+22 31.7	1	10.02		1.53		1.90		3	gM0	685
111796	+43 02265				12 49 05	+42 44.7	1	8.32		1.13				3	K0 III	280
		PHI4/1 19			12 49 05	−30 43.1	1	12.47		0.96				2		1719
	−28 09744				12 49 06	−29 00.8	1	10.45		0.04				1		1473
111797	+29 02331				12 49 08	+29 05.2	3	9.83	.000	0.37	.003	−0.04		3	G0	77,1019
111748	−30 10169				12 49 08	−30 55.8	1	6.96		−0.06		−0.19		4	B8 V(p Si)	1770
	+17 02550				12 49 11	+17 01.9	1	8.48		1.02		0.85		2	K0	1569
	+42 02347				12 49 11	+42 01.0	1	10.73		0.52		0.15		2		1569
111774	−39 07879	HR 4879			12 49 11	−39 24.5	4	5.97	.013	−0.10	.002	−0.40		15	B7/8 V	15,1770,2013,2028
111813	+26 02397				12 49 12	+25 46.9	2	9.12	.000	0.87	.010			4	K1 V	44,280
		AAS78,203 # 22			12 49 12	−62 52.	1	11.45		0.75		−0.08		2		1684
		POSS 737 # 7			12 49 13	−11 49.8	1	18.54		1.86				3		1739
	−28 09745				12 49 14	−28 47.9	1	9.65		1.06				1	K0	1473
111741	−61 03381				12 49 15	−62 24.4	1	10.59		0.32		0.05		2	A0	1688
111812	+28 02156	HR 4883			12 49 16	+27 48.7	11	4.94	.007	0.67	.005	0.20	.004	54	G0 IIIp	15,1007,1008,1019*

Table 1   705

HD DM	Other Id	N Rem	α_{1950}	δ_{1950}	S	V	σ_V	B–V	σ_{B-V}	U–B	σ_{U-B}	n	Spectrum	References
111775 −47 07917	HR 4880		12 49 16	−47 49.4	3	6.32	.006	0.03	.003	0.00		8	A0 V	15,1770,2012
111814 +20 02768			12 49 17	+19 30.0	1	8.11		0.71		0.33		2	G5	1569
111786 −26 09369	HR 4881		12 49 17	−26 28.0	3	6.14	.007	0.24	.004			11	A0 III	15,2013,2028
+39 02571			12 49 18	+38 38.5	1	10.28		0.65				2	G0	280
	Lod 694 # 30		12 49 18	−60 31.0	1	12.25		0.18		−0.22		2	B8	837
+23 02508			12 49 19	+22 49.0	1	10.50		1.17		1.00		3	dM0	685
	Lod 694 # 28		12 49 19	−60 30.9	1	12.03		0.44		0.32		2	A5	837
111777 −55 05262			12 49 20	−56 18.0	4	8.48	.021	0.61	.011	−0.02	.011	8	G0/3	657,1696,2012,6006
111758 −56 05468			12 49 20	−56 26.1	1	9.28		0.08		−0.60		2	B8	401
111832 +8 02652			12 49 24	+08 10.6	1	7.78		1.01		0.81		1	G5	280
	NGP +26 157		12 49 25	+25 32.4	1	10.70		0.97				1	K0 III	44
+30 02335			12 49 25	+30 12.7	1	11.24		0.60		0.06				1019
+32 02286			12 49 25	+32 06.0	1	10.97		1.10		1.05		1	K0 III	8034
−28 09748			12 49 25	−28 46.5	1	11.01		0.47						1473
111845 +19 02613	IDS12470N1943	A	12 49 26	+19 26.4	2	7.27	.059	0.37	.045	0.04	.045	5	A2	3024,8071
111845 +19 02613	IDS12470N1943	AB	12 49 26	+19 26.6	1	6.85		0.36		0.07		2	F2	1068
111844 +19 02613	IDS12470N1943	B	12 49 26	+19 26.6	1	7.87		0.56		0.06		3	F2	3024
111842 +26 02399	IDS12470N2613	A	12 49 27	+25 56.8	4	7.55	.008	1.46	.016	1.76	.005	9	K4 III	44,172,280,833
+27 02181			12 49 27	+27 03.1	2	11.27	.000	1.03	.006	0.81		4	K0 V	44,1019
+42 02348			12 49 27	+41 57.8	1	10.18		1.16		1.03		1	K0 III	280
312113 −60 04332			12 49 29	−60 27.6	1	11.21		0.16		−0.26		2		837
+28 02157			12 49 30	+28 23.7	1	10.00		0.58		0.10		1		1019
111482 −84 00407	HR 4870	★ AB	12 49 30	−84 51.1	3	5.44	.012	1.00	.012	0.80	.005	12	K0 III	15,1075,2038
+29 02332			12 49 31	+29 08.3	2	11.15	.004	0.98	.009	0.77		3	K0 III	44,1019
+31 02415			12 49 31	+30 38.5	1	9.87		0.55				2	G5	77
+37 02332			12 49 31	+36 43.2	1	9.71		1.15				2	K1 III	77
111790 −53 05359	HR 4882	★ AB	12 49 31	−53 33.5	1	6.24		1.13				4	G8 Ib/II	2035
111808 −48 07737			12 49 32	−49 19.6	3	7.86	.004	−0.05	.007	−0.34	.000	13	B8 IV	1732,1770,2014
111779 −62 02947			12 49 32	−62 46.9	1	9.71		0.46		−0.10		2	B5/7 V	1688
+43 02266			12 49 34	+43 09.8	1	9.25		1.16				2	K1 III	280
	LP 909 - 45		12 49 34	−32 10.7	1	11.84		1.44		1.25		2		1696
	Lod 694 # 29		12 49 34	−60 35.1	1	12.20		0.23		−0.09		2	B9	837
+13 02598			12 49 35	+13 13.9	1	10.46		1.15		1.27		1		280
	PHI4/1 7		12 49 35	−29 45.6	1	11.91		1.36				2		1719
+26 02401			12 49 36	+25 43.7	2	9.91	.010	1.18	.010	1.12		3	K2 V	44,280
312116 −60 04333			12 49 36	−60 32.5	1	10.50		0.59		0.05		2	G0	837
111860 +32 02287			12 49 37	+32 02.8	2	8.92	.044	1.15	.054	1.10		3	K0 III	77,8034
111859 +41 02337			12 49 37	+40 30.8	1	8.10		0.36		0.01		2	F3 V	1240
311979 −60 04334			12 49 37	−60 58.6	2	10.44	.011	0.31	.009	−0.47	.029	3	B3	540,976
111822 −51 07074			12 49 38	−52 23.7	2	7.86	.000	−0.05	.010	−0.93		7	B0.5/1 Ib	400,2012
111823 −52 05976			12 49 38	−52 46.5	1	8.50		0.14		0.13		1	A3 V	1771
111861 +28 02158			12 49 39	+28 24.1	4	8.80	.018	1.45	.020	1.76	.011	6	K5 III	44,694,1019,8034
312096 −59 04515			12 49 39	−60 04.0	1	10.89		0.27		−0.60		2	A0	1684
	EX Hya		12 49 42	−28 58.6	2	13.41	.180	−0.05	.025	−1.03		2	M5/M6:	1471,1473
111821 −50 07336			12 49 42	−51 11.8	1	9.37		0.30		0.11		1	A9 V	1771
111824 −58 04569			12 49 42	−58 58.5	1	9.28		0.21		0.22		1	A3 IV/V	1771
111862 +17 02551	HR 4884	★ A	12 49 43	+17 20.7	2	6.34	.020	1.60	.010	1.99		4	M0 III	280,3001
	G 59 - 41		12 49 43	+23 00.4	1	11.47		1.43		1.26		1	M25	1620
+30 02336			12 49 43	+29 55.3	2	11.22	.033	0.99	.010	0.85	.013	2	K0 III	1019,8034
111827 −64 02050			12 49 45	−64 32.7	1	8.34		0.17		−0.26		2	B5 III	1688
111825 −60 04335			12 49 46	−60 32.3	2	9.37	.039	0.05	.019	−0.67	.015	6	B1/3 Vnn	837,1505
	Lod 694 # 21		12 49 46	−60 42.7	1	11.41		0.22		0.19		2	A0	837
	GD 150		12 49 47	+16 01.0	1	14.63		−0.27		−0.96		1		3060
+17 02552			12 49 47	+16 48.2	1	8.67		0.96		0.73		1	G5	280
+31 02416			12 49 47	+31 20.2	1	10.92		1.16		1.10		1	K0 III	8034
	NGP +26 160		12 49 48	+25 53.4	1	11.46		0.71				1	G6 III	44
	G 59 - 42		12 49 49	+23 04.0	1	11.20		1.26		1.23		1	K7	1620
	Hya sq 1 # 1		12 49 49	−28 57.7	1	13.48		0.64				1		1473
	LSS 2795		12 49 49	−61 14.7	1	11.47		0.86		−0.01		2	B3 Ib	1684
111826 −62 02948			12 49 51	−63 03.8	1	9.40		0.35		0.25		1	A0 V	1688
−28 09749			12 49 52	−28 56.8	1	11.35		0.87				1		1473
312115 −60 04336			12 49 52	−60 27.3	1	11.41		0.31		0.30		2	A2	837
111891 +27 02182			12 49 53	+26 45.7	1	9.35		0.48		0.00		3	F6 V	1019
+27 02183			12 49 53	+27 16.0	3	10.54	.012	1.04	.013	0.84		5	G4 III	44,694,1019
111892 +17 02553	IDS12472N1737	B	12 49 54	+17 22.8	1	6.97		0.57		0.11		2	F8	3026
111838 −59 04516			12 49 54	−60 17.5	2	10.29	.014	0.09	.000	−0.03	.000	4	B8	761,1505
	GD 151		12 49 55	+18 13.1	1	15.48		−0.08		−0.85		1		3060
	Lod 694 # 20		12 49 55	−60 40.4	1	11.41		0.19		−0.01		2	B9	837
	NGP +31 138		12 49 56	+31 23.5	1	10.95		0.78		0.35		1	G7 IV	8034
	WLS 1300-5 # 6		12 49 57	−05 09.6	1	13.28		0.55		−0.12		2		1375
	Hya sq 1 # 5		12 49 57	−28 58.7	1	12.70		1.18				1		1473
111908 +8 02654	IDS12474N0745	A	12 49 58	+07 28.7	2	9.39	.040	1.38	.005	1.63	.020	2	R0	280,8005
111908 +8 02654	IDS12474N0745	B	12 49 58	+07 28.7	1	9.47		1.06		0.68		1	G0	280
111908 +8 02654	IDS12474N0745	C	12 49 58	+07 28.7	1	9.99		1.01				1		1238
111893 +16 02430	HR 4886		12 49 58	+16 23.6	4	6.30	.035	0.17	.013	0.11	.014	10	A7 V	70,1022,1068,3016
111868 −28 09754			12 49 58	−28 44.7	1	9.62		0.47				1	F2 V	1473
111919 +29 02333			12 49 59	+29 06.7	3	8.82	.015	1.04	.009	0.83	.005	4	G9 III	44,1019,8034
	Hya sq 1 # 3		12 49 59	−28 55.4	1	12.17		0.60				1		1473
312075 −59 04513			12 49 61	−60 02.0	1	10.42		0.47		−0.26		3	B0	1684,1684
+12 02522			12 50 00	+11 43.1	1	9.76		1.04		0.92		1	K2	280

HD	DM	Other Id	N Rem	α_{1950}	δ_{1950}	S	V	σ_V	B–V	σ_{B-V}	U–B	σ_{U-B}	n	Spectrum	References
	+30 02337			12 50 00	+29 52.8	1	10.51		0.76		0.37		1	G3 V	1019
	−51 07080			12 50 00	−51 37.2	1	10.87		0.50		0.00		1	F6 III	55
111938	+47 01997			12 50 01	+47 02.9	1	8.49		0.62		0.12		2	G0	1569
	+32 02288	IDS12477N3228	AB	12 50 02	+32 11.7	1	10.09		0.51				2		77
	−63 02441			12 50 02	−64 08.2	1	9.41		0.62		0.18		1	F8	1688
	+38 02376			12 50 05	+37 30.7	1	10.06		0.43				2	F6	77
312107	−60 04337			12 50 05	−60 26.6	1	10.48		0.14		−0.17		2	B5	837
		Hya sq 1 # 4		12 50 06	−28 54.4	1	12.89		1.29				1		1473
111874	−64 02053			12 50 06	−64 31.3	1	9.12		0.17		−0.01		1	B8 III/IV	1688
111939	+34 02365			12 50 07	+34 06.4	1	8.91		1.16				2	K1 III	77
		Hya sq 1 # 15		12 50 07	−28 46.6	1	11.74		0.53				1		1473
		Hya sq 1 # 6		12 50 08	−28 59.4	1	12.02		0.94				1		1473
111872	−56 05476			12 50 08	−56 31.9	1	9.00		0.19		0.14		5	B9.5 V	1770
	+9 02686			12 50 09	+08 34.5	1	9.90		0.94		0.72		1	G5	280
	+27 02184			12 50 09	+26 50.4	1	10.74		0.48		−0.03		2	G0 III	1019
	+31 02417			12 50 09	+31 23.3	2	9.31	.010	1.39	.034	1.52		3	K2 III	77,8034
312106	−60 04338			12 50 09	−60 29.2	1	10.72		1.12		0.86		2	K0	837
		Lod 694 # 25		12 50 09	−60 32.2	1	11.62		0.70		0.33		2	K0	837
111873	−60 04339			12 50 09	−60 36.6	2	8.76	.015	0.57	.015	0.06	.019	6	G0 V	837,1505
111926	+11 02505			12 50 10	+11 28.8	1	9.42		0.58		0.10		3	G5	196
		AS Vir		12 50 10	−09 58.7	2	11.51	.109	0.18	.026	0.12	.031	2	A6	668,699
111884	−54 05360	HR 4885		12 50 10	−54 40.9	7	5.92	.007	1.31	.006	1.40	.016	26	K2 III	15,657,1075,2013,2020*
312079	−59 04523	NGC 4755 - 452		12 50 10	−60 09.7	1	10.04		0.18		−0.56		3	B5	1684
	+36 02313			12 50 11	+36 17.3	1	10.26		0.37				2	F5	77
	+30 02338			12 50 12	+30 05.9	2	10.03	.030	0.39	.011	0.00		3	F3	77,1019
111886	−60 04341	IDS12473S6012	AB	12 50 13	−60 27.9	5	9.22	.011	0.06	.016	−0.63	.011	15	B3 III/V	540,761,837,976,1505
111958	+44 02231			12 50 14	+44 02.5	1	8.82		1.37				2	K3 III	280
111915	−48 07753	HR 4888		12 50 16	−48 40.3	4	4.33	.006	1.37	.005	1.58	.000	20	K3/4 III	15,1075,2012,8015
312117	−60 04342			12 50 16	−60 33.5	1	10.45		0.11		−0.58		2	B8	837
	+27 02185			12 50 17	+26 32.4	1	10.34		0.30		0.02		2	F2	1019
	+20 02770			12 50 18	+20 14.1	1	10.21		1.14		1.03		1	K0	280
111900	−52 05989			12 50 18	−52 57.2	1	7.09		0.05		0.05		1	A1 V	1770
		Lod 694 # 15		12 50 18	−60 28.8	1	10.87		1.53		1.66		2	K5	837
		NGC 4755 - 13		12 50 19	−60 02.6	3	12.52	.102	0.32	.019	−0.01	.015	8	A0 V	458,1505,2002
	−28 09761			12 50 20	−28 48.5	1	11.29		0.48				1		1473
		POSS 737 # 4		12 50 21	−12 22.9	1	17.39		1.63						1739
	−59 04528	NGC 4755 - 7	⋆ V	12 50 21	−60 07.0	4	9.79	.030	0.13	.017	−0.68	.013	13	B0.5V	458,1505,1737,2002
	+27 02186			12 50 22	+26 41.2	1	10.51		0.47		−0.08		2		1019
111904	−59 04529	NGC 4755 - 1	⋆	12 50 22	−60 03.4	7	5.77	.013	0.33	.017	−0.37	.026	28	B9 Ia	15,458,1018,1034,1505*
111876	−70 01533			12 50 22	−71 00.1	1	7.12		0.05		−0.04		2	B9.5/A0 V	1771
		NGC 4755 - 12		12 50 23	−60 02.3	3	12.00	.035	0.27	.018	−0.38	.010	10	B8 V	458,1505,2002
		NGC 4755 - 424		12 50 23	−60 06.9	1	13.20		0.31		−0.04		3		1505
	−59 04530	NGC 4755 - 11		12 50 23	−60 07.5	3	11.41	.009	0.24	.032	−0.32	.000	10	B8 V	458,1505,2002
		POSS 737 # 5		12 50 24	−12 28.6	1	17.61		1.17				3		1739
		NGC 4755 - 140		12 50 24	−60 02.8	1	13.42		1.75		1.71		2		1505
	−59 04531	NGC 4755 - 117		12 50 24	−60 05.2	2	10.84	.015	0.20	.020	−0.58	.020	6		458,1505
	−59 04536	NGC 4755 - 138		12 50 25	−60 02.9	1	11.25		0.20		−0.53		3		1505
	−59 04533	NGC 4755 - 410		12 50 25	−60 06.1	2	10.72	.005	0.21	.025	−0.38	.005	6	B3 V	458,1505
		NGP +28 123		12 50 26	+27 51.3	1	11.85		0.78				2	G8 IV	44
111996	+34 02366			12 50 26	+34 00.5	2	9.45	.024	1.10	.015	0.89		3	K1 V	77,3072
	−59 04537	NGC 4755 - 10		12 50 26	−60 04.5	3	10.98	.067	0.18	.018	−0.53	.019	12	B1 V	458,1505,2002
	−59 04532	NGC 4755 - 113		12 50 26	−60 05.7	2	10.17	.000	0.16	.024	−0.59	.010	6	B1 V	458,1505
		NGC 4755 - 413		12 50 26	−60 06.5	1	11.66		0.16		−0.43		3		458
	−59 04535	NGC 4755 - 414		12 50 26	−60 06.7	2	10.25	.015	0.15	.015	−0.59	.015	5	B1 V	458,1505
111830	−77 00880			12 50 26	−77 45.9	1	7.78		1.27		1.21		3	K0 III	1704
111916	−59 04538	NGC 4755 - 0		12 50 27	−59 55.7	3	9.31	.013	0.25	.015	−0.63	.027	12	B1 II	540,976,1505
		NGC 4755 - 139		12 50 27	−60 02.7	1	11.68		0.25		−0.42		4		1505
	+43 02269			12 50 29	+43 02.8	1	9.62		1.28		1.40		1	K2 III	280
		NGC 4755 - 146		12 50 29	−60 02.5	1	14.08		0.53		0.32		3		1505
		NGC 4755 - 423		12 50 29	−60 06.9	1	13.76		0.42		0.22		3		1505
		NGP +25 102		12 50 30	+24 42.0	1	11.05		1.08				2	K0 III	44
111945	−12 03712			12 50 30	−13 02.2	1	9.17		0.90		0.52		3	G8 III/IV	1700
		NGC 4755 - 145		12 50 30	−60 02.8	1	13.80		0.40		0.13		3		1505
	−59 04539	NGC 4755 - 158		12 50 30	−60 05.0	1	10.98		0.20		−0.37		3		458
		Lod 694 # 26		12 50 30	−60 37.7	1	11.64		0.18		−0.21		2	B8	837
		Lod 694 # 31		12 50 30	−60 37.7	1	12.44		0.07		−0.36		2	B7	837
312119	−60 04343			12 50 30	−60 43.8	1	10.75		0.12		−0.55		2	B5	837
		LSS 2804		12 50 30	−61 35.3	2	10.97	.074	1.22	.071	−0.21	.046	5	B3 Ib	1684,1737
	−28 09765			12 50 31	−28 52.8	1	10.70		0.57				1		1473
		Lod 694 # 23		12 50 31	−60 29.3	1	11.50		0.30		0.19		2	A2	837
		NGC 4755 - 408		12 50 32	−60 06.4	1	11.63		0.19		−0.39		2		458
		Lod 694 # 18		12 50 32	−60 36.4	1	11.18		1.16		0.88		2	K1	837
	−59 04540	NGC 4755 - 8	⋆ AB	12 50 33	−60 08.3	3	9.78	.161	0.20	.018	−0.63	.040	8		458,1505,2002
	+17 02554			12 50 34	+17 06.3	1	10.21		1.11		1.06		2		1569
	−59 04541	NGC 4755 - 115	⋆ AB	12 50 34	−60 04.7	2	10.32	.080	0.16	.020	−0.60	.015	6	B1 V	458,1505
111980	−17 03723	IDS12479S1758	AB	12 50 35	−18 14.4	6	8.36	.013	0.54	.006	−0.11	.006	21	F7 V	22,1003,1594,1775,2012,3077
	−59 04542	NGC 4755 - 418	⋆ V	12 50 35	−60 07.5	2	9.74	.090	0.13	.015	−0.62	.025	6	B2 V	458,1505
		Lod 694 # 17		12 50 35	−60 33.4	1	10.96		1.11		0.76		2	K1	837
112001	+27 02187			12 50 36	+27 04.0	2	7.69	.004	0.62	.005	0.13	.022	3	G0 IV	1019,3016
112000	+28 02159			12 50 36	+28 06.2	1	9.75		0.49		0.02		1	G0	1019

Table 1 707

HD	DM	Other Id	N	Rem	α_{1950}	δ_{1950}	S	V	σ_V	B–V	σ_{B-V}	U–B	σ_{U-B}	n	Spectrum	References
		SSII 226			12 50 36	+28 28.	1	12.68		0.11		0.10		3	A2	275
		NGC 4755 - 2619			12 50 36	−60 04.	1	14.53		0.92		0.31		3		1505
		NGC 4755 - 2623			12 50 36	−60 04.	1	14.30		1.92		1.96		4		1505
		NGC 4755 - 2629			12 50 36	−60 04.	1	15.82		0.92		0.15		3		1505
		NGC 4755 - 2631			12 50 36	−60 04.	1	15.38		0.86		0.32		5		1505
		NGC 4755 - 4814			12 50 36	−60 04.	1	14.15		0.42		0.23		3		1505
		NGC 4755 - 4824			12 50 36	−60 04.	1	14.98		0.52		0.23		3		1505
		NGC 4755 - 537			12 50 36	−60 04.	1	14.86		0.44		0.29		3		1505
		NGC 4755 - 116			12 50 36	−60 04.3	1	12.52		0.22		−0.39		3		1505
	+11 02508				12 50 37	+11 29.4	1	9.59		1.43		1.75		1	M0	280
111998	−2 03593	HR 4891			12 50 37	−03 16.9	4	6.13	.020	0.50	.009	0.02	.013	11	F5 V	15,254,1256,3077
		NGC 4755 - 406			12 50 37	−60 06.6	1	11.56		0.15		−0.47		2		458
	+40 02586				12 50 38	+40 21.7	1	10.46		0.96		0.71		1	G8 III	280
		NGC 4755 - 137			12 50 38	−60 03.2	1	12.65		0.31		−0.33		3		1505
111934	−59 04543	NGC 4755 - 106		⋆V	12 50 38	−60 05.2	4	6.91	.021	0.22	.014	−0.67	.004	17	B1.5Ib	158,458,1505,8100
		NGC 4755 - 107			12 50 38	−60 05.5	2	11.31	.135	0.25	.000	−0.31	.010	6		458,1505
	−59 04544	NGC 4755 - 405			12 50 38	−60 06.4	2	10.21	.010	0.10	.010	−0.65	.010	6	B2 V	458,1505
	+14 02557				12 50 39	+14 16.3	1	10.30		1.25		1.26		1	K2	280
	−28 09767				12 50 39	−29 02.7	1	10.96		0.48				1		1473
		NGC 4755 - 105		⋆V	12 50 39	−60 04.9	2	8.66	.038	0.15	.005	−0.71	.005	10	B1 III	458,1505
		NGC 4755 - 403			12 50 39	−60 06.3	1	11.30		0.13		−0.47		2		458
	−59 04546	NGC 4755 - 417			12 50 39	−60 07.4	3	9.81	.101	0.21	.032	−0.74	.026	7	B2 IVne	91,458,1505
112002	+13 02599				12 50 40	+12 44.2	1	7.88		0.14		0.12		3	A2	270
111968	−39 07893	HR 4889			12 50 40	−39 54.4	5	4.27	.005	0.21	.005	0.12	.007	21	A4 IV	15,1075,2012,3023,8015
		NGC 4755 - 411			12 50 40	−60 07.2	1	11.38		0.17		−0.44		2		458
	+10 02480				12 50 41	+09 44.2	1	9.74		1.31		1.46		1	K2	280
	−59 04547	NGC 4755 - 4			12 50 41	−60 04.7	6	7.50	.098	2.19	.027	2.45	.068	21	M2-Iab	91,458,1505,1522,2002,3034
		NGC 4755 - 104			12 50 41	−60 05.0	1	11.03		1.57		1.39		4		1505
312137	−60 04344				12 50 41	−60 56.5	2	9.92	.016	0.43	.019	−0.47	.012	3	B3	540,976
	+9 02688				12 50 42	+09 02.1	1	8.98		1.30		1.42		1	K2	280
		NGC 4755 - 103			12 50 42	−60 05.2	1	13.25		0.31		−0.05		3		1505
		NGC 4755 - 402			12 50 42	−60 06.6	1	11.48		0.14		−0.52		2		458
		NGC 4755 - 427			12 50 42	−60 08.3	1	11.80		0.16		−0.39		2		458
111935	−63 02446				12 50 42	−63 28.8	1	10.35		0.50		0.35		1	A2/3 V	1688
112029	+46 01828				12 50 43	+46 06.8	1	9.25		1.04				2	G9 III	280
		NGC 4755 - 214			12 50 43	−60 04.5	1	11.17		0.20		−0.44		3		458
		NGC 4755 - 315			12 50 43	−60 07.6	1	11.92		0.24		−0.47		2		458
	−59 04548	NGC 4755 - 316			12 50 43	−60 07.8	1	11.44		0.17		−0.44		2		458
		POSS 737 # 6			12 50 44	−12 41.6	1	17.83		1.51				2		1739
		NGC 4755 - 213			12 50 44	−60 04.7	1	11.65		0.18		−0.38		3		458
	−59 04549	NGC 4755 - 301			12 50 44	−60 06.2	2	9.73	.129	0.11	.010	−0.66	.000	5	B1 V	458,1505
	+21 02464				12 50 45	+20 59.9	1	10.14		1.12		1.08		1	K5	280
111971	−56 05482	IDS12478S5707		AB	12 50 45	−57 23.2	2	8.02	.024	0.52	.015	−0.10		3	F7 V	1594,6006
		NGC 4755 - 207			12 50 45	−60 04.8	2	12.10	.035	0.24	.015	−0.33	.015	5		458,1505
312104	−60 04345				12 50 45	−60 26.7	1	10.60		0.10		−0.58		2	B5	837
112030	+33 02280				12 50 46	+32 33.4	2	8.69	.005	0.89	.010	0.60		3	G7 III	77,8034
		NGC 4755 - 224			12 50 46	−60 03.5	2	10.31	.005	0.18	.020	−0.64	.020	6	B0 V	458,1505
	−59 04553	NGC 4755 - 18			12 50 46	−60 06.0	2	9.60	.024	0.19	.029	−0.59		7	B2 IV	91,1505
	−59 04552	NGC 4755 - 5			12 50 46	−60 07.9	4	8.35	.018	0.13	.006	−0.70	.014	11	B1 III	91,458,1505,2002
	−59 04550	NGC 4755 - 9			12 50 47	−60 02.3	3	10.01	.009	0.22	.026	−0.58	.015	10	B1 Vn	458,1505,2002
	−59 04551	NGC 4755 - 223		⋆V	12 50 47	−60 03.6	2	7.96	.022	0.20	.000	−0.67	.003	45	B0 V	458,1505
		NGC 4755 - 19		A	12 50 47	−60 06.0	2	9.66	.078	0.17	.034	−0.62	.010	6	B1.5Vpne	91,1505
		NGC 4755 - 19		AB	12 50 47	−60 06.0	1	8.86		0.16		−0.58		3		458
		POSS 437 # 5			12 50 48	+15 39.1	1	17.67		1.03				2		1739
	−59 04556	NGC 4755 - 203			12 50 48	−60 05.6	2	10.28	.015	0.14	.015	−0.62	.020	7	B1 V	458,1505
111952	−60 04346	LSS 2812			12 50 48	−60 28.0	4	9.46	.020	0.05	.010	−0.71	.016	9	B2/5	401,761,837,1505
112031	+29 02334				12 50 49	+29 15.7	2	8.01	.010	0.33	.004	0.04		5	F0 V	77,1019
		NGC 4755 - 216			12 50 49	−60 04.4	1	12.48		0.25		−0.26		3		1505
		NGC 4755 - 222			12 50 49	−60 04.4	1	13.50		0.31		0.05		3		1505
		NGC 4755 - 215			12 50 49	−60 04.7	2	11.68	.040	0.17	.010	−0.39	.015	6		458,1505
111973	−59 04555	NGC 4755 - 2		⋆	12 50 49	−60 06.3	6	5.96	.024	0.23	.016	−0.58	.015	21	B3 Ia	458,1018,1034,1505*
	−59 04557	NGC 4755 - 305		⋆AB	12 50 49	−60 06.8	2	8.52	.060	0.11	.005	−0.71	.005	6	B0.5IVn	458,1505
112033	+22 02519	HR 4894		⋆AB	12 50 50	+21 31.0	6	4.92	.020	0.90	.007	0.65	.011	13	K0 III+F3 V	938,1080,1084,1363*
112033	+22 02519	HR 4894		⋆C	12 50 50	+21 31.0	2	9.76	.015	0.73	.010	0.30	.029	5	G0	1084,3024
	+30 02339				12 50 50	+29 47.9	1	11.01		0.69		0.23		1	G0 V	1019
	+44 02232				12 50 50	+43 32.1	1	10.15		0.91		0.76		1	G9 III	280
	−59 04559	NGC 4755 - 306			12 50 51	−60 07.0	3	10.04	.077	0.24	.023	−0.66	.009	8	D2 IVne	91,458,1505
		NGC 4755 - 225			12 50 52	−60 03.2	1	13.74		0.35		0.06		4		1505
		NGC 4755 - 202			12 50 52	−60 05.7	2	10.05	.010	0.16	.025	−0.62	.005	6	B1 V	458,1505
	−59 04558	NGC 4755 - 201			12 50 52	−60 06.3	3	9.41	.014	0.14	.015	−0.65	.012	11	B0.5V	458,1505,1737
	−59 04561	NGC 4755 - 313			12 50 52	−60 07.6	1	11.64		0.12		−0.51		2	B2 IVne	458
312120	−60 04347				12 50 52	−60 41.4	1	10.64		0.09		−0.58		2	B5	837
	−22 03467	LW Hya			12 50 53	−22 36.1	1	9.63		0.90		0.35		3	G8 III-IV	6008
	−59 04562	NGC 4755 - 210			12 50 53	−60 05.3	1	10.12		0.15		−0.56		3	B2 Vn	458
	−59 04560	NGC 4755 - 307			12 50 53	−60 05.3	2	9.61	.040	0.16	.025	−0.62	.005	6	B1.5V	458,1505
		NGC 4755 - 312			12 50 53	−60 07.7	1	12.08		0.23		−0.37		2		458
		Lod 694 # 24			12 50 54	−60 30.3	1	11.60		0.28		0.24		2	A2	837
	−47 11104	LSS 3784			12 50 56	−47 47.2	1	10.70		0.74		−0.05		1		954
		NGC 4755 - 208			12 50 56	−60 05.4	1	12.10		0.20		−0.32		2		458
	−59 04563	NGC 4755 - 332			12 50 56	−60 10.3	1	11.02		0.13		−0.47		3		458

HD	DM	Other Id	N Rem	α_{1950}	δ_{1950}	S	V	σ_V	B–V	σ_{B-V}	U–B	σ_{U-B}	n	Spectrum	References
		NGC 4755 - 231		12 50 57	−60 02.5	1	13.58		0.40		0.09		4		1505
	−59 04564	NGC 4755 - 6	⋆V	12 50 57	−60 08.7	4	9.07	.026	0.08	.015	-0.70	.000	13	B2 III	91,458,1505,2002
		NGC 4755 - 217		12 50 58	−60 05.1	1	12.04		0.22		-0.33		3		458
	−59 04565	NGC 4755 - 311		12 50 58	−60 07.1	2	10.78	.000	0.18	.025	-0.44	.010	8	B2 V	458,1505
111991	−61 03385			12 50 59	−61 30.0	1	9.14		0.47		0.42		2	A2 IV	1688
	+13 02601			12 51 00	+13 02.6	1	10.26		0.93		0.77		1		280
111990	−59 04566	NGC 4755 - 3	⋆AB	12 51 00	−60 03.8	5	6.79	.015	0.24	.006	-0.58	.008	30	B2 Ib	458,1505,2001,2012,8100
112035	−7 03503			12 51 01	−07 34.1	1	8.10		0.25		0.18		2	A0	1375
312122	−60 04348			12 51 01	−60 33.1	1	10.67		0.16		0.04		2		837
312123	−60 04349			12 51 01	−60 33.4	1	10.89		0.38		0.22		2		837
112082	+47 01998			12 51 02	+46 55.7	2	7.41	.100	1.58	.040	1.66		4	M3 III	172,280
112039	−22 03470			12 51 02	−22 50.3	1	8.42		0.49				4	F5/6 V	2012
112048	−3 03373	HR 4896		12 51 03	−03 57.1	2	6.39	.055	1.10	.005	1.01	.015	10	K0	252,1088
112009	−51 07101	IDS12482S5144	A	12 51 03	−52 00.3	1	8.38		0.30		0.07		1	A8/F0 V	1771
112060	+20 02772			12 51 04	+19 45.3	1	6.44		0.78				3	G5 IV	280
112070	+34 02367			12 51 04	+34 10.6	1	8.95		0.98				2	G9 III	77
		NGC 4755 - 229		12 51 04	−60 02.6	1	13.56		1.47		1.33		4		1505
111992	−62 02954	IDS12480S6237	AB	12 51 04	−62 53.7	1	8.56		0.53		0.45		2	A1 V	1688
	+9 02689			12 51 05	+08 35.6	1	10.56		0.96		0.74		1		280
312039	−62 02955			12 51 05	−62 40.4	1	10.57		0.35		-0.22		1	A0	1688
111955	−71 01400			12 51 05	−71 29.8	1	9.64		0.47		0.12		1	A9/F0 IV/V	1771
	+43 02271			12 51 07	+43 25.4	1	10.30		0.34		-0.01		2	F2 V	1240
	−59 04569	NGC 4755 - 343		12 51 07	−60 06.9	1	11.37		0.16		-0.39		3		458
		NGC 4755 - 230		12 51 09	−60 02.7	1	13.87		0.57		0.15		3		1505
		NGC 4755 - 218		12 51 09	−60 05.9	1	12.30		0.28		-0.37		3		1505
	+30 02340			12 51 10	+30 18.8	1	10.68		0.75		0.32		1	G6 III	1019
		NGC 4755 - 14		12 51 11	−60 04.3	3	12.89	.067	0.74	.026	0.11		7		458,1505,2002
		AAS12,381 8 # C		12 51 12	+07 47.0	1	10.93		0.51		-0.01		1		280
	+28 02160			12 51 12	+27 54.2	1	11.31		0.55		-0.01		1	G2 III	1019
		NGP +28 124		12 51 12	+28 24.0	1	11.92		0.74				2	G6 IV	44
	+30 02341			12 51 12	+30 23.1	1	9.86		0.85		0.48		1	G6 V	1019
		NGC 4755 - 17		12 51 12	−60 03.4	2	13.32	.042	0.58	.009	0.42		4		1505,2002
112083	+38 02377	IDS12488N3831	AB	12 51 13	+38 14.6	2	9.17	.080	0.54	.045	0.05		4	G0	77,1569
		POSS 737 # 2		12 51 13	−12 37.1	1	14.68		1.24				1		1739
		NGC 4755 - 15		12 51 14	−60 04.2	3	13.27	.030	0.49	.005	0.25	.005	7		458,1505,2002
312084	−59 04571	NGC 4755 - 228		12 51 14	−60 05.5	2	11.03	.020	0.19	.010	-0.50	.005	6	B8	458,1505
		NGC 4755 - 227		12 51 15	−60 06.1	1	13.70		0.37		0.24		3		1505
112013	−61 03386			12 51 15	−62 22.2	1	10.33		0.36		0.08		1	B9	1688
	+41 02339			12 51 16	+41 09.7	1	8.73		1.08				2	K0 III	280
	+31 02419			12 51 17	+31 29.0	3	11.65	.026	0.11	.008	0.09	.011	7	A2 V	275,1026,1240
		NGC 4755 - 16		12 51 17	−60 03.6	3	13.43	.021	0.30	.033	0.02	.005	6		458,1505,2002
112027	−60 04350	LSS 2819		12 51 17	−60 48.4	5	9.22	.013	0.59	.014	-0.27	.022	17	B1 Ib	540,837,976,1505,1684
	+21 02465			12 51 18	+20 55.3	1	9.52		0.32		0.01		2	F5	1569
	+27 02188			12 51 18	+27 16.0	1	9.92		0.55		-0.04		1	G0 V	1019
112026	−60 04351			12 51 18	−60 37.4	6	8.68	.018	0.02	.007	-0.81	.015	13	B0/1 IV	401,540,837,976,1505,1688
112097	+13 02602	HR 4900		12 51 19	+12 41.4	5	6.24	.010	0.27	.007	0.04	.015	14	A7 III	15,355,1007,1013,8015
	+32 02289			12 51 19	+31 29.5	2	11.17	.015	0.47	.005	-0.06	.010	3	F5 V:p	1240,1298
112084	+19 02614			12 51 20	+19 19.7	1	7.06		1.14				2	K0	280
		AAS12,381 8 # D		12 51 21	+07 45.4	1	10.74		0.97		0.79		1		280
	+12 02523			12 51 21	+11 48.6	1	9.97		0.88		0.47		1		280
112114	+36 02314			12 51 22	+36 00.5	2	8.13	.020	1.00	.000	0.73		4	K0 IV wk1	172,280
112074	−17 03726	IDS12487S1730	A	12 51 22	−17 46.0	1	6.80		0.28				4	A3/5 +F	2006
112045	−61 03388	IDS12483S6116	AB	12 51 22	−61 31.9	1	9.59		0.46		0.26		1	A1 IV/V	1688
	+16 02434			12 51 23	+16 06.4	1	9.57		1.08		1.03		1	K0	280
		SSII 227		12 51 24	+29 52.	1	14.03		0.08		0.11		3	A0	275
		SSII 228		12 51 24	+32 32.	1	13.30		0.20		0.12		3	A0	275
112044	−57 05776	HR 4895, S Cru		12 51 24	−58 09.6	3	6.24	.028	0.57	.034	0.41	.031	4	F7 Ib/II	657,1484,1490
112126	+33 02282			12 51 25	+32 46.7	2	8.73	.005	1.38	.030	1.31		4	G9 IV	77,1569
		LSS 2820		12 51 25	−61 09.0	1	12.20		1.08		0.11		2	B3 Ib	1684
112115	+25 02571			12 51 29	+24 54.4	2	8.11	.015	1.03	.005			4	K1 III	44,280
112127	+27 02189			12 51 30	+27 03.1	4	6.90	.013	1.26	.002	1.43	.005	8	K2 III	37,44,1019,8034
112160	+60 01426			12 51 30	+59 47.8	1	7.84		0.30		0.05		2	F0	1502
	+25 02572			12 51 32	+25 02.3	1	10.07		0.30		0.02		2	F5	1569
	+28 02161			12 51 32	+28 25.0	1	10.68		0.51		0.00		1	G0	1019
112057	−61 03389			12 51 33	−61 33.4	1	10.17		0.26		-0.02		1	A0/1 V	1688
		NGP +27 165		12 51 34	+27 07.3	1	10.68		0.72		0.33		1	G5 III	8034
		NGP +26 163		12 51 35	+25 53.4	1	11.62		0.79				1	G5	44
112100	−24 10592	IDS12489S2459	AB	12 51 35	−25 15.8	1	9.90		0.92		0.60		2	K1 V	3072
112128	+16 02435			12 51 36	+15 36.1	1	8.76		1.15		1.20		1	K0	280
		SSII 229		12 51 36	+29 10.	1	12.18		-0.02		0.03		3	A0	275
112139	+13 02603			12 51 37	+12 36.2	1	8.71		1.03		0.88		1	K0	280
		Lod 694 # 27		12 51 37	−60 37.0	1	11.80		0.11		-0.31		2	B8	837
112092	−56 05487	HR 4898	⋆A	12 51 38	−56 54.4	7	4.03	.014	-0.18	.009	-0.76	.017	20	B2 IV-V	15,26,1034,1075,2012*
	+10 02483			12 51 40	+09 38.7	1	10.01		1.17		1.24		1		280
	+18 02673			12 51 40	+17 39.1	1	8.78		0.91		0.58		2	K0	1569
	+28 02162			12 51 40	+27 35.0	2	11.06	.004	1.12	.009	0.94		3	K0 IV	44,1019
112091	−56 05487	HR 4899, μ2 Cru	⋆B	12 51 40	−56 53.8	10	5.09	.059	-0.09	.020	-0.57	.036	41	B5 Vne	15,26,681,815,1034*
112078	−58 04584	HR 4897, λ Cru		12 51 40	−58 52.5	9	4.63	.007	-0.16	.005	-0.62	.010	45	B4 Vne	15,26,681,815,1034*
112152	+28 02163			12 51 41	+27 56.8	3	9.08	.011	0.20	.006	0.10	.014	8	A3 V	275,1019,1068
112131	−10 03570	HR 4901		12 51 42	−11 22.7	2	5.99	.005	0.08	.000			7	A2 V	15,2028

Table 1 709

HD	DM	Other Id	N	Rem	α_{1950}	δ_{1950}	S	V	σ_V	B–V	σ_{B-V}	U–B	σ_{U-B}	n	Spectrum	References
112140	+12 02525				12 51 43	+11 35.0	1	9.24		0.82		0.39		3	G5	196
		POSS 437 # 1			12 51 43	+15 49.3	1	14.69		1.40				2		1739
312037	−62 02959				12 51 43	−62 31.8	1	10.63		0.67		−0.23		2	B5	1684
	+31 02420				12 51 44	+30 57.8	1	10.12		0.45				2	F6	77
	+28 02164			A	12 51 45	+28 09.7	2	10.43	.000	0.48	.002	0.03	.002	3	G1 V	1003,1019
	+28 02164			B	12 51 45	+28 09.7	3	10.56	.009	0.74	.000	0.39	.024	4		44,1019,8041
	+36 02316				12 51 45	+36 12.1	1	10.08		1.19		1.39		1	K0 III	280
112151	+37 02334				12 51 45	+37 15.5	1	9.18		0.37				2	G5	77
112142	−8 03449	HR 4902, ψ Vir			12 51 45	−09 16.1	7	4.79	.012	1.59	.012	1.56	.019	39	M3 III	3,15,418,1256,3055*
	+30 02342	IDS12494N3003		A	12 51 47	+29 46.0	1	10.70		0.66		0.18		1	G2 V	1019
112171	+34 02369	HR 4904			12 51 50	+33 48.3	5	6.26	.010	0.20	.024	0.09	.010	12	A7 IV	77,252,1022,1068,1240
112185	+56 01627	HR 4905, ϵ UMa		⋆	12 51 50	+56 13.9	9	1.76	.012	−0.02	.007	0.02	.012	31	A0 p Cr	15,667,1007,1013,1263*
112144	−19 03597				12 51 51	−19 49.6	1	6.73		1.61		1.99		5	M0 III	1628
		LP 854 - 2			12 51 51	−21 53.8	1	12.72		1.40		1.23		2		1696
	+29 02335				12 51 52	+29 28.6	2	11.02	.000	0.78	.004	0.41		2	G6 V	44,1019
		GD 152			12 51 52	+39 00.6	1	14.24		0.47		−0.21		1		3060
	+27 02190				12 51 53	+27 21.6	2	10.44	.004	0.95	.003	0.78		3	K2 III	44,1019
112172	+29 02336				12 51 53	+28 43.2	3	9.14	.071	1.57	.029	1.60		6	M3 III	44,694,1019
		HA 32 # 267			12 51 53	+44 39.9	1	14.13		0.58		−0.01		2		269
		LSS 2823			12 51 54	−61 57.0	1	11.73		1.26		0.27		3	B3 Ib	1737
	+29 02337				12 51 55	+29 02.5	3	10.40	.009	0.96	.001	0.66	.019	3	G8 III	44,1019,8041
	+9 02690				12 51 58	+08 46.5	1	9.24		1.12		1.04		2	K0	1569
	+42 02350				12 51 59	+42 00.5	1	10.14		1.17		1.27		1	K2 III	280
		POSS 737 # 3			12 51 59	−12 03.3	1	17.33		1.77				1		1739
112109	−62 02960				12 51 59	−63 22.2	1	8.06		0.34		0.02		1	F0 V	1688
112123	−61 03391				12 52 00	−62 17.3	3	7.74	.002	0.07	.019	0.08	.008	5	A0/1 V	401,1688,1771
112186	+12 02527				12 52 03	+12 10.7	1	8.78		1.16		1.28		1	K0	280
	+43 02274	IDS12497N4321		A	12 52 03	+43 04.0	1	9.50		1.03				2	G9 III	280
		Malm +25 461			12 52 04	+25 16.4	1	11.08		1.16		1.15		3	dM0	685
112147	−58 04589				12 52 09	−58 44.9	2	9.13	.035	0.26	.022	−0.82	.004	5	Be	1684,1737
112197	+22 02521				12 52 10	+21 51.3	1	9.01		0.11		0.10		1	A2 V	1068
	+36 02317				12 52 11	+36 10.0	1	10.06		0.75				2	K2 III	280
112220	+47 02000				12 52 11	+47 02.9	2	7.47	.025	1.04	.005	0.92		4	K0 III	172,280
112164	−43 07953	HR 4903			12 52 11	−43 52.7	3	5.88	.007	0.64	.005	0.23		11	G1 V	15,158,2012
	−45 10884	LSS 3727			12 52 11	−46 20.1	1	10.98		0.50		−0.20		2		954
		LSS 2826			12 52 13	−61 18.1	2	10.70	.010	1.10	.007	0.04	.007	5	B3 Ib	1684,1737
		LSS 2824			12 52 13	−66 06.7	1	11.82		0.30		−0.61		4	B3 Ib	1737
	+8 02656				12 52 14	+07 31.8	1	9.53		1.07		1.00		1	K0	280
	+37 02335				12 52 15	+36 55.9	1	9.78		0.60				2	G0	77
	+47 02001				12 52 15	+46 39.1	1	9.98		0.04		0.04		2	A2 V	1240
	+13 02605				12 52 16	+13 09.0	1	9.96		0.94		0.78		1		280
	+29 02338				12 52 16	+29 14.3	2	10.20	.000	0.74	.004	0.33		2	G7 V	44,1019
112166	−52 06021				12 52 17	−53 08.8	1	9.79		0.09		0.04		1	B9.5/A0 V	1770
112192	−41 07470				12 52 19	−42 01.2	2	6.82	.000	−0.13	.000	−0.66	.005	5	B2/3 Vn	55,164
112150	−68 01795				12 52 20	−68 43.6	1	7.53		0.23		0.19		1	A3/5 IV/V	1771
	+18 02675				12 52 21	+17 56.5	1	9.34		1.01		0.88		1	K0	280
	+39 02579				12 52 21	+38 56.8	1	10.10		1.32		1.48		1	K2 III	280
	+52 01661				12 52 21	+52 02.6	1	11.26		−0.08		−0.14		2		1026
112168	−59 04579				12 52 21	−60 01.1	3	9.17	.020	0.11	.016	−0.20	.004	8	B9 V	540,976,1505
112221	+12 02528				12 52 23	+11 45.1	1	9.10		0.45				3	G5	196
	−5 03596	G 14 - 12			12 52 23	−06 03.7	5	10.43	.020	1.18	.019	1.09	.009	8	K5	158,1017,1620,1658,1705
	+31 02421				12 52 24	+31 16.2	1	10.13		0.53				2	F8	77
112169	−62 02961				12 52 24	−62 51.7	1	9.62		0.47		−0.03		1	F2 V	1688
	+28 02165				12 52 25	+28 11.0	2	9.78	.014	1.36	.002	1.53		2	K1 IV	44,1019
	+10 02485				12 52 27	+10 13.3	1	9.60		1.16		1.08		1	K2	280
	+14 02561				12 52 27	+14 04.7	1	9.38		1.26		1.36		1	K5	280
	+31 02422				12 52 28	+31 18.3	1	10.09		0.65				2	G0	77
	+31 02423				12 52 30	+30 45.3	2	10.38	.009	1.17	.020	0.98		2	K0 III	694,1019
	+38 02380				12 52 30	+38 03.0	1	9.40		1.10		1.02		4	G9 III	280
	+40 02590				12 52 30	+39 59.6	1	8.83		1.16				2	K1 III	280
312143	−61 03392				12 52 30	−61 34.8	1	10.19		0.78		−0.32		2	B5	1684
112233	+28 02166				12 52 31	+28 15.8	1	9.60		0.28		0.10		1	F0 V	1019
112213	−42 07975	HR 4906			12 52 31	−42 38.7	4	5.48	.010	1.68	.008	2.05		13	M0 III	15,1075,2032,3005
112181	−59 04580	LSS 2829			12 52 31	−60 22.8	5	8.81	.016	0.09	.013	−0.71	.019	15	B2/3 II/III	401,540,976,1505,2021
	+32 02291				12 52 32	+31 39.1	1	10.42		0.53				2		77
		HA 32 # 269			12 52 32	+44 40.3	1	13.96		0.78		0.48		2		269
112170	−63 02450				12 52 32	−64 05.8	1	9.98		0.32		0.04		1	B9 V	1688
		G 60 - 46			12 52 33	+08 04.3	2	11.66	.019	0.66	.005	−0.01	.015	3		1620,1696
		AT Vir			12 52 35	−05 11.0	3	10.71	.040	0.11	.013	0.11	.004	3	A2	668,699,700
	−20 03740				12 52 35	−21 05.5	1	10.61		0.68		0.16		1		6006
		POSS 437 # 3			12 52 37	+15 52.5	1	15.66		1.52				1		1739
	+17 02559				12 52 37	+17 21.3	1	10.62		0.22		0.07		2		1569
112234	+20 02773				12 52 37	+19 53.2	1	6.68		1.15				3	K0	280
	+29 02339				12 52 37	+29 00.7	2	9.62	.009	0.64	.003	0.19		3	G3 V	77,1019
		NGP +28 128			12 52 38	+28 19.4	1	12.04		0.81				2	G9 III	44
112248	+29 02340				12 52 38	+28 30.3	1	8.41		0.41		0.04		4	F5 V	1019
112202	−58 04593				12 52 38	−58 45.0	2	10.05	.019	0.13	.003	−0.52	.008	3	B3/5 V	540,976
	+45 02069				12 52 39	+45 22.8	1	9.40		1.26				2	K0 III	280
112195	−62 02962				12 52 39	−63 20.9	1	9.09		0.20		0.03		2	B9 V	1688
		HA 32 # 273			12 52 40	+44 42.9	1	13.70		1.28		0.87		1		269

HD	DM	Other Id	N	Rem	α_{1950}	δ_{1950}	S	V	σ_V	B–V	σ_{B-V}	U–B	σ_{U-B}	n	Spectrum	References
112264	+47 02003	HR 4909, TU CVn			12 52 40	+47 28.1	3	5.84	.027	1.58	.026	1.54	.029	5	M5 III	70,172,3001
		POSS 737 # 1			12 52 41	−12 13.9	1	14.58		1.14				2		1739
	+29 02341				12 52 42	+29 16.2	1	10.78		0.67		0.17		1	G8 V	1019
	−20 03741				12 52 42	−21 05.5	2	10.65	.015	0.50	.005	0.12		3		1594,6006
112257	+28 02167				12 52 43	+28 02.3	1	7.84		0.66		0.18		3	G2 V	3026
112238	−20 03742				12 52 44	−21 28.7	1	8.64		0.64				4	G3/5 V	2012
	+8 02658	G 60 - 47			12 52 45	+08 07.7	2	9.93	.000	0.80	.000	0.40	.010	2	K0 V	1620,1658
	+40 02591				12 52 45	+39 54.8	1	10.00		1.08		0.95		1	K0 III	280
	+43 02278	IDS12506N4309		B	12 52 46	+42 52.8	1	10.82		0.54		0.07		2	F8	1569
	+10 02486				12 52 47	+10 26.7	1	9.88		1.01		0.81		1		280
	+18 02677				12 52 48	+17 57.8	1	9.53		1.11		1.15		1	K2	280
		POSS 437 # 7			12 52 50	+15 29.6	1	18.20		1.29				4		1739
112275	+33 02284				12 52 50	+33 08.2	2	8.27	.010	1.08	.015	1.10		3	K0 III	77,8034
		HA 32 # 209			12 52 50	+44 25.5	1	15.51		0.79		0.32		1		269
112241	−29 10014				12 52 50	−29 47.9	1	6.90		0.07		0.09		5	A2/3 V	1628
		NGP +28 129			12 52 52	+28 10.5	1	11.63		0.85				2	G6 III	44
112297	+45 02070				12 52 52	+44 33.8	1	8.97		0.47		−0.12		1	F8 V	269
	+46 01830				12 52 52	+45 51.7	1	10.81		0.96				2	G8 V	1569
	+20 02775				12 52 53	+20 01.8	1	8.60		1.57		1.89		1	M0	280
	+27 02191				12 52 53	+26 38.3	2	10.27	.004	0.75	.003	0.39		3	G6 V	44,1019
		EG 93, HZ 34			12 52 53	+37 48.8	1	15.66		−0.28		−1.26		2	DA	3028
	+29 02342				12 52 54	+29 22.3	1	11.09		0.60		0.12		1	G5 V	1019
112225	−60 04357	IDS12499S6022		AB	12 52 56	−60 38.7	2	8.01	.003	0.30	.007	0.11	.007	3	A8 V	1688,1771
112279	+11 02513				12 52 58	+10 59.2	1	9.03		1.14		1.19		1	K0	280
		NGP +28 130			12 52 58	+27 39.9	1	10.74		1.02				2	K0 III	44
		NGP +28 131			12 52 58	+28 05.0	1	11.95		0.91				2	G8	44
112277	+13 02607				12 52 59	+12 58.4	1	7.98		0.96		0.71		1	K0	280
	−7 03509				12 52 59	−08 27.1	2	9.66	.000	0.72	.000	0.20	.000	4	G5	516,1064
112244	−56 05498	HR 4908		⋆ A	12 52 59	−56 33.9	5	5.37	.027	0.02	.011	−0.86	.041	11	O9 Ia/ab	191,1034,1684,2007,8100
112278	+12 02529	IDS12505N1202		A	12 53 00	+11 46.0	2	6.91	.010	1.54	.005	1.65		6	M2	280,3040
112298	+32 02293				12 53 00	+32 20.4	1	8.20		1.18				3	K0 III	280
112243	−52 06033	IDS12501S5234		A	12 53 00	−52 50.1	1	8.26		0.27		0.08		1	A8 III/IV	1771
112299	+26 02404				12 53 02	+26 00.6	1	8.42		0.52		0.06		1	F8 V	3026
	+36 02319				12 53 02	+35 45.4	1	10.31		1.14		0.99		1	G8 III	280
112281	+0 03002				12 53 05	+00 19.6	1	6.78		0.98		0.70		2	K0	985
112300	+4 02669	HR 4910		⋆ A	12 53 05	+03 40.1	11	3.38	.007	1.57	.014	1.80	.017	49	M3 III	3,15,369,418,1075*
		NGP +26 169			12 53 06	+26 07.6	1	11.62		0.73				1	G8 III	44
112313	+26 02405	IN Com			12 53 08	+26 09.7	1	8.81		0.81		0.24		1	G5	280
	+27 02192			A	12 53 08	+27 20.1	1	11.00		0.99		0.63		1	K0 IV	1019
112263	−47 07967				12 53 09	−47 45.8	1	8.07		0.32		0.08		3	A8/9 III	1771
112219	−71 01404	HR 4907			12 53 09	−71 54.9	3	5.93	.004	1.12	.005	0.90	.005	11	G8 III	15,1075,2038
	+22 02523				12 53 11	+22 27.1	1	9.63		0.64		0.19		1	G2	280
		G 60 - 48			12 53 12	+12 49.9	3	11.32	.020	0.49	.004	−0.22	.013	4		333,1620,1696,3077
	+27 02194				12 53 15	+27 28.2	1	10.89		0.50		−0.05		1	F8	1019
112304	−14 03605	HR 4911			12 53 15	−15 03.4	1	6.17		0.00				4	A0 V	2007
	+27 02193				12 53 16	+27 12.6	1	10.51		0.70		0.25		1	G3 V	1019
		AAS35,161 # 47			12 53 17	−76 30.9	1	14.32		0.56		−0.32		1		5005
	+12 02530				12 53 18	+12 21.4	1	10.16		1.02		0.78		1		280
	+32 02295				12 53 18	+31 57.5	1	9.44		1.15				3	K1 III	280
		HA 32 # 160			12 53 18	+44 15.5	1	14.36		0.81		0.29		1		269
112339	+36 02320				12 53 19	+36 13.2	1	9.45		0.43				2	G5	77
112353	+32 02296				12 53 21	+32 16.3	1	6.84		1.09				2	K1 III	280
		HA 32 # 377			12 53 23	+44 56.8	1	10.61		0.62		0.07		1		269
		HA 32 # 279			12 53 24	+44 42.3	1	15.23		0.70		−0.03		1		269
		BPM 21641			12 53 24	−56 09.	1	15.01		0.68		−0.17		2	DF	3065
112254	−66 02001				12 53 26	−67 21.0	1	9.24		0.25		0.14		1	A4/6 IV/V	1771
112394	+57 01400				12 53 27	+57 22.6	4	9.20	.084	0.71	.006	0.22	.019	9	G5 IV-V	1007,1013,3016,8023
312126	−60 04362				12 53 27	−60 58.9	1	10.05		0.10		−0.75		2	A2	540
112272	−63 02454	LSS 2834			12 53 27	−64 05.4	6	7.36	.036	0.81	.018	−0.27	.030	12	B1 Ia/ab	138,191,540,976,1684,1737
112253	−66 02002				12 53 27	−67 16.2	1	9.32		0.26		0.17		1	A5	1771
112340	+17 02562				12 53 28	+16 29.9	1	9.35		0.09		0.06		3	A2	1068
	+45 02072				12 53 29	+44 57.9	1	10.64		0.61				2		280
112429	+66 00778	HR 4916			12 53 29	+65 42.6	4	5.21	.025	0.28	.004	0.02	.009	11	F0 V	15,254,1008,3023
		HA 32 # 99			12 53 30	+44 07.9	1	11.53		0.64		0.15		1		269
112295	−60 04363				12 53 30	−61 03.9	1	9.48		0.63		0.21		1	B8/9 II/III	1688
		HA 32 # 281			12 53 31	+44 37.0	1	13.67		0.56		−0.16		1		269
	+29 02343				12 53 33	+29 22.9	2	10.77	.000	0.75	.003	0.36		2	G6 V	44,1019
112319		V Cru			12 53 34	−57 37.8	1	9.81		4.09		5.20		3	Ne	864
	+14 02562				12 53 35	+14 20.1	1	9.86		1.06		0.82		2	K0	1569
		HA 32 # 283			12 53 36	+44 36.8	1	13.22		0.54		−0.05		1		269
	+24 02508				12 53 37	+23 50.7	2	9.23	.019	1.48	.000	1.83	.009	4	gM1	280,685
		NGP +28 132			12 53 38	+28 23.8	1	11.32		1.16				2	K2 III	44
	+26 02407				12 53 39	+26 17.5	1	9.67		0.34				4	F1 IV-V	193
	+27 02195				12 53 39	+26 44.4	2	10.01	.009	1.14	.006	1.17		3	K1 III	44,1019
112346	−45 08114				12 53 39	−45 56.3	1	8.35		0.06		0.05		1	A0 V	1770
		SSII 232			12 53 40	+28 23.5	5	12.79	.027	0.11	.022	−0.74	.018	7	sdB	98,275,974,1240,1298,3027
112412	+39 02580	HR 4914		⋆ B	12 53 40	+38 35.1	7	5.61	.013	0.34	.009	0.03	.013	13	F0 V	1,15,1028,1240,6001*
112413	+39 02580	HR 4915, α2 CVn		⋆ A	12 53 41	+38 35.3	12	2.89	.012	−0.11	.007	−0.32	.006	26	A0p SiEuHg	1,15,1022,1028,1068*
	+30 02343				12 53 43	+30 03.8	1	11.57		0.53		−0.05		1		1019
		HA 32 # 215			12 53 43	+44 28.7	1	15.49		0.70		0.26		1		269

Table 1   711

HD	DM	Other Id	N Rem	α_{1950}	δ_{1950}	S	V	σ_V	B–V	σ_{B-V}	U–B	σ_{U-B}	n	Spectrum	References
	+43 02281	IDS12514N4252	A	12 53 44	+42 36.2	1	9.66		1.05		0.93		1	K0 III	280
	+45 02074			12 53 45	+44 31.7	1	9.32		1.15				2	K1 III	280
112350	−55 05288			12 53 45	−55 30.5	1	8.14		0.06		−0.23		3	B9 II/III	1770
		POSS 437 # 6		12 53 46	+16 07.6	1	17.91		1.51				3		1739
	+28 02168			12 53 46	+28 23.0	1	11.32		1.17		1.24		1	G8 IV	1019
	+36 02321			12 53 46	+35 37.2	1	9.28		1.58				3	M2 III	280
		SSII 233		12 53 48	+32 44.	1	14.54		0.01		−0.07		2	A0	275
112374	−25 09508	HR 4912, LN Hya		12 53 48	−26 11.4	1	6.62		0.68				4	F3 Ia	2006
	+30 02344			12 53 49	+29 36.1	4	10.46	.006	1.38	.009	1.62	.016	18	K3 III	694,830,1019,1783
	−24 10619			12 53 49	−24 39.3	1	10.15		1.24		1.15		2	K4	3072
112361	−47 07972	IDS12510S4709	AB	12 53 49	−47 24.9	1	6.75		0.45				5	F5 IV/V	1414
112414	+30 02345			12 53 50	+29 53.9	1	9.36		0.23		0.07		1	A7 V	1019
		HA 32 # 105		12 53 50	+44 10.3	1	13.38		0.62		0.11		1		269
	+38 02382			12 53 51	+38 03.4	1	10.17		1.07		0.92		1	K0 III	280
112431	+40 02596			12 53 51	+40 01.4	2	8.92	.015	0.21	.010	0.10	.000	4	Am	1068,1240
		HA 32 # 107		12 53 53	+44 10.8	1	13.66		0.73		0.34		1		269
112445	+33 02286			12 53 55	+32 48.1	2	8.44	.005	1.05	.010	0.92		3	G9 III	280,8034
		G 61 - 21		12 53 56	+15 58.8	2	13.82	.017	1.29	.009	1.07	.024	6		203,3078
112364	−59 04600			12 53 59	−59 28.3	7	7.38	.025	0.21	.007	−0.69	.022	17	B1 III	164,191,401,540,976*
		SSII 234		12 54 00	+27 45.	1	13.45		0.07		0.20		2	A0	275
112463	+43 02282	IDS12517N4333	AB	12 54 01	+43 16.3	1	9.12		0.76				2	G5 V	280
112381	−53 05397	V823 Cen		12 54 02	−54 19.0	2	6.50	.003	−0.08	.010	−0.15	.006	4	Ap SiCr	401,1770
112446	+10 02487			12 54 04	+09 35.2	1	7.58		1.06		0.98		1	K0	280
112447	+9 02694			12 54 05	+09 25.6	1	8.47		1.25		1.45		1	K5	280
112366	−62 02965	LSS 2837		12 54 05	−63 11.6	2	7.54	.070	0.76	.019	−0.02	.025	4	B6 Iab/b	191,1684
112486	+54 01556	HR 4917	★ AB	12 54 06	+54 22.2	2	5.84	.020	0.20	.010	0.09		5	A5 m	70,8071
112390	−56 05506			12 54 06	−56 52.4	1	9.77		0.05		−0.59		2	B3 III	401
	+29 02344			12 54 09	+29 25.2	1	9.86		1.02		0.84		1	G9 III	1019
112407	−48 07807			12 54 10	−48 41.5	1	8.70		0.31		0.07		1	A7/9 (V)	1771
		POSS 437 # 2		12 54 11	+15 34.2	1	15.66		1.61				2		1739
112406	−48 07808			12 54 11	−48 38.8	1	8.78		0.38		0.02		2		1771
112409	−50 07394	HR 4913		12 54 11	−50 55.7	3	5.17	.005	−0.07	.006	−0.32		10	B8/9 V	1770,2014,2032
	+8 02659			12 54 12	+07 49.2	1	9.71		1.02		0.88		1	K0	280
	+27 02196			12 54 12	+26 38.7	1	10.91		1.18				2	K0 III	44
112369	−68 01802			12 54 12	−69 04.5	1	8.23		0.29		0.17		1	A7 V	1771
112383	−67 02129			12 54 13	−67 41.4	2	6.78	.005	0.03	.030	0.09		5	A2 IV/V	1075,1771
	+37 02337			12 54 14	+37 07.6	1	10.10		1.06		1.03		2	K2 V	1569
112476	+11 02515			12 54 15	+10 46.8	2	8.33	.000	0.17	.005	0.09	.010	3	A1 V	272,280
	+32 02300			12 54 15	+32 24.1	1	10.08		1.04		0.93		1	K0 III	280
		LSS 2838		12 54 17	−61 07.3	1	11.56		1.00		0.12		2	B3 Ib	1684
		SSII 235		12 54 18	+47 35.	1	11.45		0.15		0.05		2	A5	1240
		L 328 - 123		12 54 18	−49 41.	1	14.25		1.08		0.62		3		3062
112425	−49 07506			12 54 18	−49 48.2	1	8.81		0.18		0.12		1	A2/3 V	1771
		LSS 2839		12 54 18	−61 07.4	1	11.42		1.00		−0.12		2	B3 Ib	1684
112487	+32 02301			12 54 19	+32 13.2	4	9.66	.024	0.12	.011	0.11	.013	21	A3 V	275,1026,1068,1240
112501	+44 02234	IDS12520N4406	AB	12 54 20	+43 49.3	3	6.97	.004	0.11	.004	0.14	.007	18	Am	269,1240,1569
112437	−46 08281			12 54 20	−46 56.2	2	8.18	.005	1.24	.000			8	K2 III	1075,2033
		HA 32 # 109		12 54 21	+44 12.7	1	11.88		0.50		−0.02		1		269
	−64 02087	NGC 4815 - 7		12 54 26	−64 42.5	1	10.29		1.16		0.78		2		251
	+28 02169			12 54 27	+28 04.4	1	9.71		0.59		0.05		1	G1 V	1019
112515	+46 01832	IDS12522N4609	A	12 54 28	+45 53.0	2	8.51	.009	0.34	.005	0.11		4	F2 Vp:	1240,3024
112515	+46 01832	IDS12522N4609	B	12 54 28	+45 53.0	2	9.53	.019	0.45	.005	0.00		4	F2	1240,3024
112559	+66 00780	RY Dra		12 54 28	+66 15.9	3	6.37	.060	3.24	.075	8.34	.112	3	C7 I	8005,8022,8027
		HA 32 # 174		12 54 29	+44 21.3	1	12.41		0.72		0.15		1		269
112502	+30 02346			12 54 30	+30 19.6	1	9.86		0.66		0.22		1	G3 V	1019
	−57 05809			12 54 30	−58 18.0	3	10.63	.066	0.13	.018	−0.54	.048	5	B4 II-Ib	540,976,1684
		AAS78,203 # 23		12 54 30	−62 44.	1	11.09		0.72		−0.31		2		1684
	+38 02385			12 54 33	+37 37.4	1	10.49		0.62				2	G8 IV	280
	+25 02577			12 54 34	+24 40.0	1	9.23		1.18		1.33		2	K5	280
		HA 32 # 176		12 54 34	+44 16.7	1	10.20		1.06		1.01		1		269
		HA 32 # 338		12 54 34	+44 50.6	1	10.94		0.49		−0.04		1		269
112469	−56 05510			12 54 34	−57 10.7	1	8.33		0.27		0.22		1	A3/4 IV	1771
		GD 153		12 54 35	+22 18.2	1	13.42		−0.25		−1.18		1	DAwk	3060
		V485 Cen		12 54 39	−32 57.0	1	17.55		0.42		−1.20		1		1471
112459	−62 02969			12 54 40	−63 23.2	1	8.70		0.57		0.27		1	A3mA8-F0	1771
		NGC 4815 - 9		12 54 41	−64 42.6	1	13.03		1.00		0.24		1		251
	+9 02695			12 54 42	+09 22.0	1	9.99		1.13		1.09		1		280
		HA 32 # 292		12 54 42	+44 36.1	1	12.78		0.53		−0.03		1		269
112471	−60 04369	LSS 2840		12 54 42	−60 37.0	1	8.82		0.20		−0.66		2	B1 II/III	401
	−64 02088	NGC 4815 - 2		12 54 42	−64 40.8	1	10.12		1.27		1.09		3		251
112503	+9 02696	IDS12522N0850	AB	12 54 43	+08 33.8	1	6.82		0.46		0.08		2	F5	985
		HA 32 # 177		12 54 43	+44 17.2	1	11.40		0.56		0.01		1		269
112484	−57 05810	LSS 2841		12 54 43	−58 07.1	4	9.05	.016	0.04	.004	−0.74	.011	10	B2 II/III	401,540,976,2021
	+44 02235			12 54 44	+44 17.1	1	10.24		1.05		1.03		1	K0 III	280
112481	−49 07513	V856 Cen		12 54 44	−49 30.6	1	8.36		0.04		−0.74		1	B2 II/III	55
112485	−60 04370			12 54 49	−60 32.7	2	9.48	.011	0.10	.016	−0.72	.028	3	B1/3 III	401,1688
		NGC 4815 - 3		12 54 49	−64 41.3	1	12.38		0.53		0.07		2		251
312174	−62 02971			12 54 50	−63 02.6	2	9.82	.010	0.73	.016	−0.29	.019	3	B2	540,976
		NGC 4815 - 8		12 54 50	−64 42.4	1	12.95		2.02		1.50		1		251
112570	+46 01833	HR 4919		12 54 51	+46 26.9	4	6.12	.008	1.01	.004	0.80	.005	17	K0 III-IV	70,172,269,1569

HD	DM	Other Id	N	Rem	α_{1950}	δ_{1950}	S	V	σ_V	B–V	σ_{B-V}	U–B	σ_{U-B}	n	Spectrum	References
112491	−53 05402				12 54 51	−53 50.5	2	9.62	.010	0.04	.000	−0.65	.005	5	B2/3 II/III	400,401
		NGC 4815 - 4		AB	12 54 51	−64 41.1	1	13.16		0.70		0.70		2		251
112519	−21 03635	HR 4918			12 54 53	−22 29.0	3	6.31	.004	1.07	.000			11	K0 III	15,1075,2035
	−64 02090	NGC 4815 - 1			12 54 54	−64 41.1	1	9.63		0.87		0.67		3		251
		NGC 4815 - 5			12 54 54	−64 41.8	1	13.09		0.42		0.03		2		251
312242	−60 04371				12 54 56	−60 43.8	2	9.29	.001	0.22	.003	−0.63	.013	3	B8	401,1688
	+8 02660				12 54 58	+08 07.5	1	9.81		1.20		1.24		1		280
112497	−60 04372	LSS 2843			12 54 58	−60 47.1	2	8.43	.033	0.40	.001	−0.46	.031	5	B1/2 II/III	540,976
112541	+14 02565				12 54 59	+14 19.7	1	9.50		1.37		1.54		1	K2	280
112571	+31 02425	IDS12526N3055		AB	12 55 00	+30 38.4	1	8.65		0.38		−0.02		2	F5 V	1019
		NGC 4815 - 6			12 55 00	−64 42.	1	13.40		1.01		0.66		2		251
		AAS12,381 12 # A			12 55 01	+12 03.3	1	10.45		1.01		0.81		1		280
		G 61 - 23			12 55 01	+18 56.7	2	9.83	.008	0.70	.000	0.17	.015	4	G8	1620,1658
112572	+25 02578	IDS12526N2530		AB	12 55 02	+25 13.3	2	8.53	.035	0.79	.010	0.41	.025	4	G5	1569,1569
112510	−54 05385				12 55 02	−54 30.5	2	9.33	.005	0.01	.005	−0.47	.005	6	B8 Ib/II	400,401
112573	+19 02618	G 61 - 24		⋆A	12 55 05	+18 57.7	2	8.99	.020	0.67	.005	0.14	.025	2	K0	1620,1658
112575	−13 03627				12 55 07	−14 11.6	3	9.11	.007	1.13	.009	0.98		7	K4 V	1705,2033,3072
	+13 02609				12 55 09	+12 37.6	1	9.71		1.01		0.79		1	K2	280
112549	−36 08224				12 55 10	−36 31.7	1	8.57		−0.05		−0.21		2	B9 IV	1770
	+29 02346				12 55 12	+29 14.9	1	10.25		0.75		0.29		1	G4 V	1019
112610	+44 02238				12 55 12	+44 05.5	2	8.11	.002	0.43	.002	−0.03	.011	15	F5	269,1569
112532	−53 05405	IDS12523S5339		A	12 55 13	−53 54.9	1	7.37		0.07		0.04		2	A0 V	1770
112600	+11 02518				12 55 14	+10 30.2	1	9.44		1.09		1.04		1	K0	280
	+29 02345				12 55 14	+28 42.0	1	9.86		0.56		0.11		1	G0 V	1019
		G 123 - 74			12 55 14	+41 13.1	1	12.79		1.48		1.17		1		3062
112532	−53 05405	IDS12523S5339		AB	12 55 15	−53 54.9	1	7.17		0.15		0.04		2	A0 V	401
	+33 02287	IDS12529N3321		AB	12 55 15	+33 04.7	1	9.97		0.64				3	G5 IV	280
	−63 02461				12 55 15	−63 56.4	1	10.61		0.68		0.22		2		1730
112535	−58 04616				12 55 16	−58 34.4	1	9.09		0.35		0.26		1	A7 IV/V	1771
312190	−59 04614	AAS78,203 # 24			12 55 16	−59 48.6	1	11.03		0.20		−0.53		2	B8	1684
112536	−60 04375				12 55 16	−60 44.8	1	10.02		0.23		−0.05		1	B7/8 V	1688
	+29 02347				12 55 17	+29 05.8	2	9.68	.005	0.83	.005	0.40		2	G8 V	44,1019
	+36 02322	BF CVn		⋆A	12 55 18	+35 29.8	1	10.60		1.42		1.13		1	M0 V	3016
	+36 02322	G 164 - 31		⋆B	12 55 18	+35 29.8	2	13.16	.000	1.63	.014	1.14		4		538,3016
112611	+26 02409				12 55 19	+26 12.4	1	8.91		1.10		1.05		1	K2	280
112565	−46 08291				12 55 19	−46 47.7	1	6.86		0.04		0.03		1	A1 III/IV	1770
112612	+10 02490				12 55 21	+09 36.5	1	7.16		0.24		0.17		1	A3	280
	+27 02197				12 55 21	+27 18.4	1	10.78		0.46		0.01		1	F5	1019
112527	−63 02462				12 55 21	−64 08.2	1	9.26		0.16		−0.23		1	B9 IV	1688
312157	−61 03399				12 55 22	−62 12.5	1	10.49		0.71		−0.21		2	B3	1684
	−58 04617				12 55 23	−58 58.8	1	9.64		0.33		−0.41		2	B8	1684
	+34 02371				12 55 26	+33 55.2	1	9.49		0.91				3	G7 III	280
		HA 32 # 230			12 55 26	+44 25.0	1	11.90		0.51		−0.13		1		269
	+31 02427				12 55 29	+30 41.4	1	11.03		0.78		0.44		1	K0 III	1019
112641	+37 02341				12 55 29	+37 00.0	1	8.94		0.95				2	G9 III	280
112537	−64 02093				12 55 29	−64 57.4	1	9.75		0.28		0.25		1	A1 V	1688
112625	+8 02661				12 55 30	+07 50.6	1	9.40		0.81		0.45		2	G5	1569
	+23 02514				12 55 31	+23 06.8	1	9.76		0.91		0.60		1	G5	280
112652	+33 02288				12 55 31	+32 37.5	2	9.67	.019	0.67	.014	0.30		4	G5 V	280,8034
112663	+53 01580				12 55 31	+52 39.0	1	9.55		0.48		−0.02		2	K0	1566
112583	−57 05823				12 55 31	−57 59.1	1	7.82		0.24		0.13		1	A7 V	1771
	+22 02528				12 55 32	+22 10.4	1	10.38		0.56		0.05		1		280
	+16 02441				12 55 33	+16 09.8	1	10.78		0.95		0.87		1	K0	280
		Wray 1038			12 55 36	−63 59.3	1	12.68		0.55		−0.24		5		1684
	+18 02680				12 55 37	+17 42.7	1	9.75		1.08		0.90		2		1569
112604	−45 08134				12 55 37	−45 41.8	1	8.86		0.01		−0.11		1	A0 IV	1770
112642	+18 02681				12 55 38	+18 02.3	1	8.36		1.63		1.94		1	K5	280
112595	−53 05408				12 55 39	−54 01.4	2	8.74	.005	−0.03	.001	−0.36	.005	3	B8 II/III	401,1770
	+37 02342				12 55 42	+36 48.6	1	10.23		1.18		1.21		1	K0 III	280
		vdB 58 # a			12 55 42	−66 01.9	1	12.36		0.40		−0.19		3	B4 V	434
	+30 02347				12 55 43	+29 52.9	1	10.75		0.58		0.11		1	G1 V	1019
112568	−69 01739				12 55 46	−69 53.6	1	9.01		0.34		0.28		1	A3mA5-A7	1771
	+25 02580	IDS12533N2504		AB	12 55 47	+24 47.6	1	10.36		0.70		0.34		2	G5	280
	−62 02975	IDS12527S6258		AB	12 55 47	−63 14.3	1	10.43		0.66		−0.37		2		1684
		AAS12,381 28 # C			12 55 50	+27 55.6	1	11.28		0.22		0.11		1		280
112607	−62 02977				12 55 54	−63 22.3	2	8.06	.023	0.18	.009	−0.23	.017	4	B7/8 III	401,1688
	+37 02343				12 55 57	+37 13.1	1	10.41		1.15		1.17		1	K0 III	280
		Sand 178			12 56 00	+26 46.	1	14.08		1.51				1		686
112679	−6 03705				12 56 00	−06 40.7	1	7.76		1.09		1.02		3	K0	1657
		NGC 4833 - 1			12 56 00	−70 36.	1	12.52		1.56				2		107
		NGC 4833 - 3			12 56 00	−70 36.	1	12.75		1.50				2		107
		NGC 4833 - 4			12 56 00	−70 36.	1	13.73		1.07				2		107
		NGC 4833 - 5			12 56 00	−70 36.	1	15.59		1.17				1		107
		NGC 4833 - 6			12 56 00	−70 36.	1	13.83		1.25				1		107
		NGC 4833 - 8			12 56 00	−70 36.	1	14.42		1.03				1		107
		NGC 4833 - 9			12 56 00	−70 36.	1	13.08		1.21				1		107
		NGC 4833 - 14			12 56 00	−70 36.	1	14.15		1.12				1		107
		NGC 4833 - 15			12 56 00	−70 36.	1	13.33		1.30				1		107
		NGC 4833 - 16			12 56 00	−70 36.	1	13.56		1.28				1		107
		NGC 4833 - 22			12 56 00	−70 36.	1	13.67		0.70				1		107

Table 1   713

HD	DM	Other Id	N	Rem	α_{1950}	δ_{1950}	S	V	σ_V	B–V	σ_{B-V}	U–B	σ_{U-B}	n	Spectrum	References
		NGC 4833 - 31			12 56 00	−70 36.	1	13.85		1.28				1		107
		NGC 4833 - 33			12 56 00	−70 36.	1	13.94		1.24				1		107
		NGC 4833 - 35			12 56 00	−70 36.	1	12.52		1.65				2		107
		NGC 4833 - 49			12 56 00	−70 36.	1	13.80		1.03				1		107
		NGC 4833 - 68			12 56 00	−70 36.	1	15.18		0.59				1		107
		NGC 4833 - 78			12 56 00	−70 36.	1	15.35		0.67				1		107
		NGC 4833 - 79			12 56 00	−70 36.	1	12.64		1.60				2		107
		NGC 4833 - 81			12 56 00	−70 36.	1	15.57		0.32				1		107
		NGC 4833 - 82			12 56 00	−70 36.	1	12.40		1.73				2		107
		NGC 4833 - 84			12 56 00	−70 36.	1	14.86		0.78				1		107
		NGC 4833 - 85			12 56 00	−70 36.	1	15.73		0.41				1		107
		NGC 4833 - 86			12 56 00	−70 36.	1	13.77		0.66				1		107
		NGC 4833 - 87			12 56 00	−70 36.	1	14.66		1.07				1		107
		NGC 4833 - 90			12 56 00	−70 36.	1	13.61		1.21				1		107
		NGC 4833 - 92			12 56 00	−70 36.	1	13.80		1.15				1		107
		NGC 4833 - 93			12 56 00	−70 36.	1	15.79		0.34				1		107
		NGC 4833 - 94			12 56 00	−70 36.	1	13.16		1.43				1		107
		NGC 4833 - 97			12 56 00	−70 36.	1	13.95		1.02				1		107
		NGC 4833 - 106			12 56 00	−70 36.	1	13.32		1.29				1		107
		NGC 4833 - 114			12 56 00	−70 36.	1	13.99		0.98				1		107
		NGC 4833 - 115			12 56 00	−70 36.	1	13.69		1.23				1		107
		NGC 4833 - 117			12 56 00	−70 36.	1	13.56		1.32				1		107
		NGC 4833 - 126			12 56 00	−70 36.	1	14.12		1.15				1		107
		NGC 4833 - 139			12 56 00	−70 36.	1	13.99		1.34				2		107
		NGC 4833 - 140			12 56 00	−70 36.	1	12.74		1.48				2		107
		NGC 4833 - 159			12 56 00	−70 36.	1	13.91		1.31				2		107
		NGC 4833 - 161			12 56 00	−70 36.	1	13.77		1.25				2		107
		NGC 4833 - 163			12 56 00	−70 36.	1	14.23		1.25				2		107
		NGC 4833 - 164			12 56 00	−70 36.	1	13.25		1.32				2		107
		NGC 4833 - 165			12 56 00	−70 36.	1	13.80		1.16				2		107
		NGC 4833 - 168			12 56 00	−70 36.	1	13.00		0.68				2		107
		NGC 4833 - 171			12 56 00	−70 36.	1	14.21		1.41				1		107
		NGC 4833 - 177			12 56 00	−70 36.	1	13.70		0.97				2		107
		NGC 4833 - 178			12 56 00	−70 36.	1	15.09		1.16				1		107
		NGC 4833 - 179			12 56 00	−70 36.	1	13.42		1.10				2		107
		NGC 4833 - 261			12 56 00	−70 36.	1	14.09		0.97				1		107
	+45 02076				12 56 01	+44 49.9	1	10.46		1.64				2	M3 III	694
	+45 02077				12 56 01	+45 12.8	1	10.24		0.56		−0.01		2		1569
	+13 02610				12 56 02	+12 39.9	1	9.60		1.08		0.99		1	K0	280
	+30 02348				12 56 02	+29 43.6	1	10.33		0.64		0.18		1	G3 V	1019
112706	+26 02410	IDS12536N2627		A	12 56 03	+26 11.5	1	8.47		0.98		0.80		1	K0	280
	+12 02536				12 56 04	+11 54.2	1	10.14		1.22		1.23		1	K0	280
	+31 02429				12 56 05	+30 55.8	1	9.98		0.75		0.35		1	G6 V	1019
112696	−2 03605				12 56 05	−02 38.0	1	6.64		1.45		1.69		2	K2	1375
	+30 02349				12 56 06	+30 17.4	1	9.53		1.16		1.14		1	K0 V	1019
		KUV 12562+2839			12 56 07	+28 39.0	1	16.76		0.38		−0.25		1	DA?	1708
112637	−62 02979				12 56 09	−63 02.1	1	9.61		0.35		−0.27		1	B2 III	1688
112734	+29 02348				12 56 10	+28 35.3	2	6.97	.000	0.23	.001	0.05	.003	2	A5	1019,1690
	+27 02199				12 56 11	+27 27.9	1	10.21		0.42		−0.01		1	F8	1019
	+39 02585				12 56 11	+38 40.8	1	10.14		1.09		0.91		1	G8 III	280
		T Com			12 56 13	+23 24.5	1	13.65		1.94		0.36		1	gM8	685
	+44 02240				12 56 13	+43 59.5	1	9.80		1.07		1.01		1	K0 III	280
112735	+22 02531				12 56 14	+22 19.0	2	7.23	.010	0.56	.010	0.10		5	G0	280,3026
112672	−55 05311				12 56 14	−55 38.1	1	9.56		0.07		−0.05		1	B9.5 V	1770
112622	−70 01543				12 56 14	−70 33.9	1	8.80		0.31		0.15		1	A5 IV/V	1771
		AAS12,381 8 # E			12 56 15	+08 05.4	1	10.38		1.20		1.08		1		280
	+8 02664				12 56 16	+07 44.8	1	10.26		1.10		1.07		1	K0	280
112753	+28 02170	IDS12539N2801		AB	12 56 16	+27 44.7	2	7.98	.004	0.64	.006	0.16	.012	3	G0 V	1019,3026
112661	−61 03401	LSS 2849		★AB	12 56 16	−62 01.1	3	9.19	.014	0.65	.011	−0.27	.024	5	B0/1 III/IV	540,976,1688
112737	+9 02700	CN Vir			12 56 17	+08 28.8	1	8.24		1.63		1.66		1	M1	3001
112685	−45 08144				12 56 17	−45 41.5	1	7.86		0.32				4	F0 V	2012
112754	+27 02201	IDS12539N2740		AB	12 56 18	+27 23.3	1	8.64		0.86		0.49		1	G4 II	1019
112689	−54 05395				12 56 18	−54 26.5	1	9.07		0.04		−0.30		2	B7 III/IV	401
112662	−63 02468				12 56 19	−63 54.8	1	9.73		0.49		0.25		1	A0	1688
	+30 02350				12 56 20	+30 20.5	1	11.19		0.64		0.09		1	G1 V	1019
112736	+17 02568				12 56 21	+16 28.8	1	9.19		0.92		0.56		1	G5	280
	+44 02241				12 56 22	+44 04.5	1	10.13		0.63				2	G5 V	280
	+45 02078				12 56 22	+45 03.4	1	10.31		0.83				2	K1 III	280
	+34 02374	G 164 - 34			12 56 23	+33 30.4	1	10.24		0.64		0.10		1		1658
112690	−59 04629	LSS 2851			12 56 24	−59 35.7	2	9.00	.006	0.20	.003	−0.66	.027	4	B1/3 III	540,976
	+23 02517				12 56 25	+23 12.8	1	10.47		1.01		0.79		1		280
		LP 616 - 93			12 56 26	+00 27.6	1	10.92		0.67		0.04		2		1696
	+10 02493				12 56 26	+09 29.9	1	10.69		0.88		0.51		1		280
112768	+20 02781				12 56 26	+19 49.5	1	8.30		1.17		1.23		1	K2	280
112769	+18 02682	HR 4920			12 56 27	+17 40.7	6	4.77	.023	1.57	.007	1.94	.010	25	M0 III	3,15,1363,3001,6001,8015
112758	−9 03595	IDS12539S0918		AB	12 56 28	−09 34.0	4	7.55	.013	0.79	.012	0.35	.025	13	K0 V	22,2033,3008,8025
112799	+45 02079				12 56 30	+45 05.5	1	9.68		1.02				2	K0 III	280
		Wray 1039			12 56 31	−59 19.8	1	11.26		0.26		−0.52		1		1684
112703	−63 02472				12 56 31	−64 06.1	2	7.64	.003	0.49	.008	−0.02	.006	3	F5 V	1688,1730
	+21 02473				12 56 32	+21 08.8	1	9.70		1.39		1.68		1	K5	280

HD	DM	Other Id	N Rem	α_{1950}	δ_{1950}	S	V	σ_V	B–V	σ_{B-V}	U–B	σ_{U-B}	n	Spectrum	References
	+29 02349			12 56 32	+28 56.2	1	10.80		0.53		0.06		1		1019
112716	−55 05314			12 56 32	−55 55.2	1	8.51		0.37		0.19		1	A2mA4-A8	1771
		NGC 4833 sq1 20		12 56 35	−70 36.0	2	12.52	.015	1.56	.015	1.15	.050	7		290,1638
112800	+35 02387			12 56 37	+34 48.9	1	8.08		1.61		1.75		2	M1	1569
112814	+40 02598			12 56 37	+40 06.8	3	6.86	.018	0.95	.008	0.62	.049	5	G9 II-III	172,280,8041
112802	+8 02665	IDS12541N0826	A	12 56 38	+08 10.2	1	7.87		1.08		0.99		1	K0	280
		LSS 2854		12 56 38	−61 21.9	2	10.52	.090	1.19	.014	0.09	.011	9	B3 Ib	1684,1737
		G 164 - 36		12 56 42	+32 15.2	1	14.05		1.63				1	dM4	686
	+42 02360			12 56 43	+42 26.4	1	9.19		1.33		1.61		1	K3 III	280
	−63 02473	LSS 2857		12 56 47	−63 45.3	2	10.50	.029	0.57	.011	-0.37	.028	3	B0.5V	540,976
112859	+47 02007			12 56 48	+47 25.3	1	7.98		0.93		0.54		2	G8 III-IVp	172
112860	+47 02006			12 56 49	+46 39.5	1	8.34		1.09				2	K1 III	280
112764	−55 05316	IDS12539S5522	A	12 56 49	−55 38.5	1	8.23		0.05		-0.33		2	B6/8 IV/V	1770
112764	−55 05316	IDS12539S5522	AB	12 56 50	−55 38.5	1	7.80		0.04		-0.26		2	B6/8 IV/V	401
112781	−55 05317	IDS12539S5522	A	12 56 51	−55 38.7	1	8.99		0.08		-0.04		1	B9	1770
		NGC 4833 sq2 1		12 56 51	−70 35.8	1	12.40		1.68		1.77		5		1638
112828	+11 02520			12 56 52	+10 56.1	1	8.06		1.15		1.23		1	K0	280
	+40 02599			12 56 52	+40 19.5	1	10.35		1.12		1.07		1	K1 III	280
		NGC 4833 sq1 1		12 56 54	−70 34.7	2	12.40	.025	1.64	.005	1.49	.025	10		290,1638
	−5 03606			12 56 55	−06 07.4	1	10.03		0.60		0.06		2		1375
112780	−53 05416			12 56 56	−54 06.6	1	7.97		1.01				4	G8 III/IV	2033
	+11 02522			12 56 58	+10 36.1	1	9.62		1.32		1.14		1	K0	280
	+40 02600			12 56 58	+39 34.8	1	10.70		0.98		0.70		1	G6 III	280
	+43 02285	IDS12547N4317	A	12 56 58	+43 01.1	1	9.32		1.75		2.03		1	K3 V	280
112751	−63 02474	LSS 2858		12 56 58	−64 01.7	3	9.27	.006	0.40	.003	-0.47	.020	11	B2 Ib/II	540,976,1730
112845	+15 02527			12 56 59	+15 00.2	1	9.04		1.58		1.77		1	K2	280
	+20 02783			12 56 59	+20 01.2	1	9.73		1.10		1.09		1		280
	+27 02202			12 56 59	+27 25.1	1	10.62		0.61		0.12		1		1019
	+35 02388			12 57 01	+35 23.8	1	10.30		0.90		0.64		2	K2 IV	1569
112872	+31 02431			12 57 02	+30 42.3	1	9.71		0.80		0.46		1	G6 III:	1019
112869	+38 02389	TT CVn		12 57 02	+38 05.2	3	9.21	.035	1.84	.020	1.82		5	R6p	280,1238,8005
112956	+69 00681	G 237 - 84		12 57 02	+69 03.1	2	8.07	.000	0.67	.005	0.13	.009	4	G0	308,1658
112784	−59 04634	LSS 2861		12 57 02	−60 19.4	3	8.26	.010	0.06	.009	-0.83	.020	6	O9.5III	540,976,1684
	+6 02685			12 57 03	+06 17.6	1	9.78		1.00		0.72		3	K0	8088
		AAS12,381 8 # F		12 57 03	+08 05.4	1	10.49		1.34		1.57		1		280
112873	+30 02351	IDS12547N3036	AB	12 57 04	+30 19.8	1	9.23		0.83		0.44		1	G7 V	1019
112785	−61 03406	LSS 2860		12 57 04	−62 08.9	3	9.69	.009	0.50	.011	-0.45	.022	5	B8	540,976,1688
112846	−3 03384	HR 4921	★ A	12 57 05	−03 32.6	4	5.80	.015	0.18	.000	0.07	.008	10	A3 V	12,13,15,1417
112846	−3 03384	IDS12545S0316	B	12 57 05	−03 32.6	1	10.15		0.55		0.29		4		12,13
312228	−60 04379			12 57 05	−60 31.6	1	11.15		0.32		-0.57		2	B5	1684
		WEIS 60454		12 57 06	+29 15.7	1	16.04		1.41				1		1692
		WEIS 61090		12 57 07	+28 50.7	1	16.62		0.55				1		1692
112887	+28 02171			12 57 08	+28 20.1	2	7.19	.000	0.42	.001	-0.02	.001	14	F4 V	1019,1690
112886	+29 02350			12 57 09	+28 30.5	2	8.16	.000	0.44	.002	-0.04	.001	8	F7 V	1019,1690
112848	−19 03612			12 57 09	−20 01.3	1	7.23		0.50				2	F7 IV/V	961
		G 60 - 53		12 57 11	+01 48.7	1	11.42		1.37		1.32		1	K5	1696
112786	−63 02475			12 57 11	−64 20.9	1	9.53		0.19		-0.14		1	B9 V	1688
	+40 02601			12 57 12	+40 08.3	1	9.08		1.01		0.83		1	G9 III	280
		WEIS 60626		12 57 13	+29 06.5	1	12.99		0.76				1		1692
112788	−68 01814			12 57 13	−68 37.9	1	9.62		0.29		0.17		1	A6 V	1771
	−69 01743	LSS 2859		12 57 13	−69 56.4	3	9.42	.027	0.02	.001	-0.79	.022	11	B2 Vn	400,540,976
		WEIS 60551		12 57 14	+29 12.2	1	12.09		0.55				1		1692
112888	+10 02496			12 57 15	+09 47.7	1	6.71		1.05				2	K0	280
	+15 02529			12 57 15	+14 58.3	1	9.56		1.51		1.82		1		1569
112874	+17 02570			12 57 15	+16 32.4	2	10.73	.024	0.04	.000	0.03	.000	3	A3 IV-V	1026,1068
112914	+42 02361	G 123 - 79		12 57 15	+42 15.2	1	8.64		0.94		0.72		2	G9 V	3072
312258	−60 04380			12 57 16	−61 23.4	2	10.33	.005	1.26	.017	0.19	.033	5		1684,1737
		GD 267		12 57 18	+04 47.7	1	15.05		-0.09		-0.88		1	DA	3060
113049	+76 00473	HR 4927		12 57 19	+75 44.5	2	6.00	.039	0.99	.015	0.75		13	K0 III	252,1258
112812	−61 03408			12 57 19	−62 01.0	1	10.06		0.46		0.37		1	A1/2 V	1688
	+13 02612			12 57 20	+13 12.8	1	9.98		1.58		1.87		1	M0	280
112940	+47 02008			12 57 21	+46 43.5	1	9.04		0.31		0.09		2	Am	1240
112825	−59 04639			12 57 21	−59 24.9	1	9.71		0.32		-0.58		2	B1/2 IV/V	1684
		WEIS 62312		12 57 22	+28 25.9	1	15.54		0.61				1		1692
		WEIS 60523		12 57 22	+29 10.4	1	14.60		0.74				1		1692
112789	−71 01415			12 57 22	−72 09.6	1	8.10		0.14		0.08		2	A0 V	1771
112790	−71 01414			12 57 22	−72 21.3	1	9.17		0.37		0.20		1	A7/8[m]	1771
		G 61 - 26		12 57 23	+20 34.5	1	14.60		1.54		1.21		3		316
		G 149 - 28		12 57 23	+27 50.2	1	15.40		0.28		-0.68		1		3060
	+43 02287			12 57 23	+42 52.8	1	9.90		1.05		0.94		1	K0 III	280
	+23 02519			12 57 25	+23 02.4	1	8.80		1.40		1.60		1	K5	280
		WEIS 61005		12 57 25	+28 52.0	1	12.27		1.01				1		1692
	−19 03613			12 57 27	−19 44.8	1	9.37		0.94				2	G5	961
112900	+15 02530			12 57 28	+14 33.3	1	9.61		1.47		1.79		1	K2	280
		WEIS 63891		12 57 28	+27 53.4	1	12.25		0.58				3		1692
	+37 02346			12 57 28	+37 15.2	1	10.19		0.21		0.10		1		280
112842	−59 04640			12 57 28	−60 06.4	7	7.05	.012	0.22	.006	-0.45	.026	24	B4 II	116,540,976,1020,1075*
		LSS 2866		12 57 28	−61 24.4	1	11.58		1.11		0.09		2	B3 Ib	1737
		WEIS 62047		12 57 30	+28 40.1	1	13.17		0.62				2		1692
	+29 02351			12 57 30	+28 54.1	3	9.35	.000	0.83	.000	0.44	.002	37	G6 III	588,1019,1692

Table 1   715

HD	DM	Other Id	N Rem	α_{1950}	δ_{1950}	S	V	σ_V	B–V	σ_{B-V}	U–B	σ_{U-B}	n	Spectrum	References
238179	+55 01569			12 57 30	+54 56.8	1	9.75		0.71		0.31		3	G8 V	3016
	+23 02520			12 57 32	+23 26.0	1	10.04		1.11		1.04		1	K0	280
		WEIS 63350		12 57 34	+28 24.8	1	14.09		0.83				1		1692
	+47 02010			12 57 34	+46 54.8	1	10.05		0.63				2	G0	280
112853	−60 04382			12 57 34	−60 35.4	1	9.57		0.17		0.05		1	A0 III	1688
	+25 02582			12 57 36	+25 22.6	1	10.23		1.45		1.77		2	K5	280
		G 60 - 54		12 57 37	+03 45.3	4	15.88	.046	0.64	.000	-0.09	.000	4		1,1281,1705,3028
		WEIS 63432		12 57 39	+28 12.5	1	13.30		0.91				1		1692
		WEIS 62013		12 57 40	+28 40.1	1	13.22		1.48				2		1692
		WEIS 62012		12 57 40	+28 40.8	1	12.37		0.89				2		1692
		MN201,849 # 42		12 57 41	−20 05.5	1	11.71		0.99				2		961
112941	+6 02688			12 57 42	+05 56.6	1	9.08		0.48		-0.05		3	G0	8088
	+28 02172			12 57 43	+27 29.7	1	10.33		0.98		0.72		1		1019
		WEIS 63484		12 57 43	+28 05.6	1	12.44		0.89				2		1692
	−58 04640			12 57 43	−59 19.6	1	10.33		0.27		-0.56		4		1684
		WEIS 61039		12 57 44	+28 55.9	1	13.81		0.66				2		1692
		MN201,849 # 106		12 57 44	−20 07.5	1	13.18		1.00				2		961
		WEIS 63433		12 57 45	+28 11.9	1	12.09		0.45				2		1692
112943	−1 02754	G 14 - 19		12 57 45	−02 26.1	7	9.77	.010	1.19	.010	1.16	.011	15	K7 V	202,1017,1197,1620,1658*
		WEIS 63425		12 57 46	+28 14.8	1	13.78		1.18				3		1692
112973	+35 02391			12 57 46	+35 11.3	1	7.64		0.90		0.62		2	G5	1569
112843	−71 01416	LSS 2865		12 57 46	−72 21.6	3	9.55	.013	0.09	.004	-0.69	.018	13	B1/2 II/III	400,540,976
112962	+8 02666			12 57 47	+08 28.2	1	9.12		1.23		1.51		1	K5	280
	−19 03614			12 57 47	−20 06.6	1	10.03		0.42				2	F8	961
112935	−32 09083	HR 4922		12 57 48	−33 14.1	2	6.01	.005	0.39	.005			7	F3 V	15,2012
	+10 02498			12 57 49	+10 10.4	1	10.46		0.28		0.06		1		280
	+21 02476			12 57 49	+20 59.6	1	9.90		1.00		0.90		1	K0	280
112988	+35 02392			12 57 49	+35 16.1	1	7.78		0.91		0.61		2	G0	1569
		WEIS 63340		12 57 51	+28 22.3	1	15.46		1.11				1		1692
		WEIS 61034		12 57 51	+28 54.7	1	10.74		0.51				1		1692
	+29 02352		AB	12 57 51	+28 55.8	1	10.75		0.49		-0.06		1		1019
	+31 02433			12 57 51	+30 30.6	1	11.89		0.86		0.36		1	G8 IV	1019
		MN201,849 # 156		12 57 51	−20 10.4	1	14.10		0.51				2		961
112961	+14 02567			12 57 52	+14 06.9	2	7.92	.005	0.85	.005	0.46	.005	3	G5	1569,1648
		WEIS 61064		12 57 52	+28 57.4	1	11.04		0.56				1		1692
112989	+31 02434	HR 4924	★ AB	12 57 53	+31 03.3	6	4.89	.014	1.16	.014	1.05	.005	11	K1 IIIp	1019,1080,1363,1690*
		WEIS 63925		12 57 54	+27 59.1	1	15.03		1.01				1		1692
		BPM 7961		12 57 54	−72 18.	1	15.18		0.01		-0.73		1		3065
		WEIS 61032		12 57 55	+28 54.4	1	10.71		0.48				1		1692
	+27 02203			12 57 56	+27 02.1	1	10.70		0.57		0.11		1	G1 V	1019
		WEIS 63254		12 57 56	+28 21.7	1	13.85		0.81				1		1692
		MN201,849 # 127		12 57 56	−20 10.3	1	13.67		0.85				2		961
112898	−66 02030			12 57 57	−67 20.2	1	9.39		0.33		0.24		1	A3mA4-A9	1771
112937	−45 08162			12 57 58	−46 08.3	1	9.34		0.02		-0.02		1	B9.5 V	1770
		WEIS 62019		12 58 00	+28 39.8	1	11.90		0.68				1		1692
		WEIS 61094		12 58 00	+28 49.6	1	11.66		1.09				1		1692
113020	+43 02290			12 58 00	+42 44.0	1	9.39		0.97				2	G8 III	280
	−61 03413	LSS 2868		12 58 00	−62 19.9	1	11.49		0.57		-0.27		2		779
112992	−2 03609	HR 4925	★ AB	12 58 01	−03 06.0	2	5.98	.005	1.12	.000	1.10	.005	7	K2 III	15,1417
113002	+20 02784			12 58 02	+19 54.1	2	8.75	.010	0.75	.005	0.21	.010	5	K0	1569,1648
		WEIS 62345		12 58 02	+28 28.0	1	14.35		0.87				1		1692
	+29 02353			12 58 02	+28 42.7	2	10.36	.000	0.53	.001	0.03		2	F8	1019 ,1692
113021	+32 02311			12 58 02	+32 02.9	2	6.67	.015	0.92	.005	0.58		5	G5	196,280
113092	+67 00773	HR 4928		12 58 02	+66 52.0	3	5.36	.040	1.30	.010	1.29	.000	6	K2 III	15,1003,3016
112911	−62 02981			12 58 03	−62 45.2	1	7.45		1.77		1.91		3	K2 III	1704
		G 60 - 55	★ V	12 58 04	+05 57.1	3	13.40	.024	1.75	.007	1.10	.066	7		419,1705,8088
	+23 02521			12 58 04	+22 53.5	1	9.78		0.97		0.64		1		280
		WEIS 63251		12 58 04	+28 23.1	1	15.32		0.54				1		1692
	−19 03615			12 58 04	−20 05.5	1	9.73		0.49				2	F5	961
113001	+36 02328	IDS12557N3618	AB	12 58 05	+36 01.5	2	9.57	.019	0.08	.014	-0.87	.005	6	sdO+F	963,1280
113036	+42 02362			12 58 05	+42 12.0	1	8.73		0.30		0.02		2	F0 V	1569
		MN201,849 # 243		12 58 05	−20 10.9	1	14.72		0.96				2		961
		MN201,849 # 49		12 58 06	−20 07.0	1	12.23		0.65				2		961
		Wray 1046		12 58 06	−59 23.7	1	12.01		0.30		-0.23		2		1684
		BC Cha	★	12 58 06	−77 35.2	1	14.49		1.23		-0.19		1		5005
113064	+55 01571			12 58 07	+54 30.4	1	8.56		1.18		1.25		3	K2 III	37
		MN201,849 # 285		12 58 07	−20 10.4	1	15.28		0.67				2		961
	−19 03616			12 58 07	−20 11.3	1	10.31		1.53				2		961
		MN201,849 # 80		12 58 07	−20 13.2	1	12.94		0.65				2		961
	+38 02391			12 58 08	+37 59.2	1	9.29		1.54		1.88		1	K3 III	280
	+8 02668			12 58 10	+07 55.6	1	10.28		1.04		0.92		1	K0	280
		WEIS 61074		12 58 10	+28 59.6	1	12.54		0.81				2		1692
		MN201,849 # 132		12 58 11	−20 05.4	1	13.90		0.87				2		961
113022	+19 02622	HR 4926	★ A	12 58 12	+18 38.5	4	6.20	.018	0.43	.015	0.00	.007	10	F6 Vs	70,254,1084,3077
113022	+19 02622	IDS12557N1855	B	12 58 12	+18 38.5	2	9.48	.015	0.53	.005	0.04	.000	5	G5	1084,1569
		AAS78,203 # 25		12 58 12	−66 37.	1	11.36		2.16		2.65		5		1684
113023	+13 02616			12 58 13	+13 04.8	1	10.01		0.94		0.75		1	G5	280
112953	−60 04387	LSS 2869		12 58 14	−60 47.2	3	8.88	.014	0.75	.012	-0.23	.014	6	B0/1 Ib/II	540,976,1684
		MN195,678 # 11		12 58 14	−61 21.3	1	12.58		1.04		0.73		5		1474
	+40 02604			12 58 16	+39 41.8	1	9.18		1.22		1.39		1	K2 III	280

HD	DM	Other Id	N	Rem	α_{1950}	δ_{1950}	S	V	σ_V	B–V	σ_{B-V}	U–B	σ_{U-B}	n	Spectrum	References
112954	−62 02982				12 58 16	−62 39.4	1	8.37		0.49		0.26		2	B9 IV	1688
		WEIS 63308			12 58 17	+28 07.1	1	11.89		0.68				4		1692
	+19 02623				12 58 18	+19 23.8	1	9.61		1.24		1.27		1	K5	280
	+29 02354	WEIS 61955			12 58 18	+28 46.8	2	9.84	.004	0.59	.002	0.13		5	G0 V	1019,1692
	+13 02618	DT Vir			12 58 20	+12 38.7	2	9.72	.005	1.46	.019	1.10		3	M2	1705,3072
112938	−67 02157	IDS12551S6719	AB		12 58 20	−67 35.3	1	7.93		0.14		0.10		1	A2 V	1771
		MN201,849 # 28			12 58 21	−20 28.8	1	11.83		0.56				2		961
	+23 02522				12 58 22	+22 59.1	1	10.43		0.86		0.71		1		280
114282	−88 00076				12 58 22	+87 55.1	2	7.47	.005	1.43	.007	1.54	.039	12	K2	1332,1733
	−59 04650				12 58 22	−59 39.2	1	11.00		0.23		−0.41		1	B5	768
		WEIS 63301			12 58 24	+28 04.5	1	11.65		1.14				1		1692
		MN201,849 # 223			12 58 24	−20 12.9	1	14.86		0.62				2		961
113009	−44 08388				12 58 24	−44 29.0	3	7.66	.004	−0.08	.001	−0.47	.003	6	B5 III	55,1770,2014
		AAS78,203 # 26			12 58 24	−62 28.	1	11.39		0.89		0.14		3		1684
		WEIS 61100			12 58 26	+28 54.3	1	15.27		1.32				1		1692
112998	−57 05852	IDS12555S5736	A		12 58 27	−57 52.1	1	8.95		0.22		0.22		1	A5/7	1771
113076	+31 02436				12 58 28	+30 47.0	2	9.04	.000	0.67	.000	0.17	.000	2	G4 V	1019,1690
	+48 02073				12 58 28	+48 28.5	1	10.68		0.34		−0.04		2	F0 V	1240
113075	+39 02589				12 58 29	+39 11.7	1	8.69		1.17		1.13		1	K0 V	280
		WEIS 61101			12 58 30	+28 55.0	1	14.47		0.54				1		1692
		MN201,849 # 18			12 58 30	−19 48.5	1	11.28		0.77				2		961
	+24 02524				12 58 31	+24 23.6	1	10.32		1.08		1.01		1		280
		BSD 57 # 10			12 58 31	+26 47.2	1	10.68		0.78		0.33		2	G2	97
112999	−59 04651				12 58 31	−60 24.1	2	7.39	.010	0.06	.003	−0.42	.029	4	B6 IIIn	540,976
113093	+32 02312				12 58 32	+31 53.2	1	9.54		0.48		−0.06		3	G5	196
		LB 2520			12 58 32	+59 20.4	1	15.31		0.13		−0.63		4	DA	308
113052	−19 03617				12 58 33	−20 30.5	1	6.92		1.31				2	K2 III	961
113139	+57 01408	HR 4931	★	AB	12 58 35	+56 38.1	13	4.93	.004	0.37	.005	0.01	.004	267	F1 V +G6 V	1,15,130,292,1006*
113079	+5 02702				12 58 36	+04 37.7	1	7.24		0.25		0.09		2	A5	401
	+37 02350				12 58 36	+37 14.7	2	9.99	.015	0.11	.019	0.09	.029	3	A2	272,280
113094	+25 02583				12 58 38	+24 35.1	1	7.81		1.13				4	K1 III	20
113012	−59 04654	LSS 2872			12 58 38	−59 48.4	7	8.14	.017	0.11	.006	−0.78	.015	17	B0 III	401,540,768,976,1684*
113140	+46 01836				12 58 39	+46 13.1	2	8.14	.003	1.14	.001	1.13	.001	10	K1 III	269,1569
113044	−44 08391	IDS12558S4439	A		12 58 39	−44 55.2	1	9.18		0.14		0.05		1	A0 V	1770
	−63 02484	LSS 2870			12 58 39	−63 29.0	2	10.49	.000	0.31	.014	−0.59	.014	4	B0.5V	540,976
113077	+16 02445				12 58 41	+16 04.5	1	9.20		0.90		0.52		1	G5	280
113096	+17 02572				12 58 41	+16 55.8	1	9.72		0.88		0.61		1	G5	280
113095	+17 02573	HR 4929			12 58 41	+17 23.5	1	5.96		0.96				2	K0 III	70
113014	−61 03419	LSS 2873	★	AB	12 58 42	−61 54.7	2	9.05	.006	0.32	.002	−0.54	.003	3	B2 III	779,1688
	+8 02669				12 58 44	+08 14.6	1	10.56		0.19		0.17		1	A2	280
	+24 02525				12 58 44	+23 38.3	1	10.45		1.03		0.74		1	K0	280
113101	−7 03525	G 14 - 22			12 58 45	−08 10.2	5	8.97	.014	0.73	.008	0.26	.014	11	G8 V	1003,1620,1658,2033,3077
113083	−26 09470				12 58 45	−27 06.2	5	8.04	.008	0.54	.007	−0.11	.005	13	G1 V	1594,1696,2012,2017,3077
113016	−63 02485	RZ Cen			12 58 46	−64 21.5	4	9.13	.091	0.19	.007	−0.62	.023	12	B1/2 III/V	401,540,976,1688
	+28 02173				12 58 47	+27 44.0	1	10.63		0.58		0.07		1	G0 V	1019
		WEIS 61147			12 58 47	+29 05.1	1	12.88		0.79				1		1692
		LSS 2875			12 58 47	−61 15.0	1	11.24		1.52		0.46		2	B3 Ib	1737
113034	−61 03421	LSS 2874	★	AB	12 58 47	−61 33.8	5	9.23	.025	1.04	.021	0.02	.026	10	B0/1 III	191,540,976,1684,1688
		WEIS 61202			12 58 48	+28 51.3	1	10.82		0.77				1		1692
		WEIS 60362			12 58 48	+29 10.5	1	14.06		0.70				1		1692
112985	−70 01548	HR 4923			12 58 48	−71 16.8	4	3.61	.007	1.18	.000	1.25	.010	14	K2 III	15,1075,2038,3005
		WEIS 32977			12 58 49	+30 34.5	1	11.49		0.58				1		1692
	+9 02707				12 58 51	+08 44.1	1	10.31		1.00		0.63		1		280
113116	−8 03470	G 14 - 23			12 58 53	−09 11.1	2	9.64	.010	0.55	.010	−0.05	.005	2	G0	1620,1658
		G 123 - 82			12 58 54	+41 13.6	1	13.43		1.45				1		3062
113124	+13 02619				12 58 55	+13 26.3	1	8.09		0.11		0.11		2	A1 V	1068
	+26 02417				12 58 55	+25 36.8	1	10.09		1.22		1.33		1	K2	280
		WEIS 61130			12 58 55	+29 00.4	1	13.15		0.65				1		1692
		MN201,849 # 192			12 58 55	−20 08.0	1	14.54		0.72				2		961
	−59 04664				12 58 55	−59 39.4	1	11.61		0.30		0.24		1	A0	768
	−59 04663				12 58 56	−59 42.4	1	11.04		0.27		−0.22		1	B8	768
	+19 02625				12 58 56	+18 49.3	1	9.64		1.20		1.19		2		1569
		WEIS 63175			12 58 57	+28 06.9	1	14.33		1.54				1		1692
		WEIS 63200			12 58 58	+28 21.8	1	13.87		0.82				1		1692
113090	−54 05409				12 58 58	−55 18.2	1	8.71		0.26		0.23		1	A5 V	1771
312207	−59 04665				12 58 59	−59 55.4	1	11.15		0.19		−0.34		1	B9	768
312253	−60 04390				12 58 59	−60 44.0	2	9.91	.001	0.42	.007	−0.43	.013	3	B3	540,976
113168	+38 02394				12 59 00	+38 19.0	1	7.87		−0.03		−0.04		2	A1 IV	1068
		MN201,849 # 169			12 59 00	−20 05.8	1	14.17		0.76				2		961
	+30 02353				12 59 02	+29 53.7	1	10.04		1.01		0.75		1	G5 III:	1019
113059	−66 02038				12 59 02	−66 26.3	1	9.31		0.38		0.26		1	A7/F0 III	1771
	+8 02671				12 59 04	+08 13.2	1	9.77		1.12		1.08		1	K0	280
113170	+22 02537				12 59 08	+21 32.3	1	7.07		1.46				2	K5	280
	+29 02355	WEIS 61196			12 59 08	+28 53.8	2	10.02	.024	0.85	.003	0.36		2	G7 III:	1019,1692
113109	−56 05541				12 59 09	−56 58.1	3	9.19	.013	0.03	.008	−0.69	.011	7	B2/3 IV	401,540,976
		WEIS 64075			12 59 10	+27 53.8	1	12.07		1.02				1		1692
	−19 03618				12 59 10	−20 25.3	1	10.01		1.66				2		961
		WEIS 64025			12 59 11	+27 52.4	1	12.10		1.00				2		1692
	+30 02354				12 59 11	+30 20.3	1	11.05		0.49		−0.04		1	K0	1019
	+31 02437				12 59 12	+31 24.1	1	10.41		1.30		1.54		1	K3 V	280

Table 1   717

HD	DM	Other Id	N	Rem	α_{1950}	δ_{1950}	S	V	σ_V	B–V	σ_{B-V}	U–B	σ_{U-B}	n	Spectrum	References
113158	−18 03528	UY Vir			12 59 13	−19 30.3	1	7.98		0.36		0.12		1	A9 IV	1768
	−59 04668				12 59 13	−59 42.8	1	11.49		0.64		0.12		1	F0	768
		LSS 2878			12 59 13	−62 25.8	1	11.31		1.00		0.19		2	B3 Ib	779
113117	−47 08040				12 59 14	−48 18.8	1	8.71		0.33		0.08		1	A9/F0 IV	1771
		WEIS 61924			12 59 15	+28 38.1	1	12.44		0.61				1		1692
		GPEC 2643			12 59 15	−32 53.0	1	10.40		1.34				1		1719
113134	−51 07200	IDS12564S5115	AB		12 59 19	−51 31.2	1	9.12		0.00		−0.49		4	B5 IV/V	400
	+31 02438				12 59 20	+30 57.7	1	10.19		0.98		0.80		1	K2 V	1019
	+31 02439				12 59 20	+31 25.8	1	9.83		1.08				2	K2 V	280
	+13 02621				12 59 21	+12 57.6	1	10.11		1.22		1.33		1	G0	280
		WEIS 63150			12 59 21	+28 14.1	1	11.89		1.07				2		1692
	−60 04392	LSS 2880			12 59 21	−61 7.2	1	11.43		1.38		1.15		2		1737
	+37 02354				12 59 22	+36 51.8	1	10.61		1.03		0.77		1		280
		MN201,849 # 59			12 59 23	−19 43.9	1	12.55		0.74				2		961
113174	−19 03619				12 59 23	−20 22.9	1	9.21		1.08				2	K0/1 III	961
	−19 03620				12 59 24	−19 50.0	1	9.98		0.54				2	F5	961
	+47 02011				12 59 25	+46 32.3	1	10.38		0.27		0.02		1		280
		AAS61,331 # 8			12 59 27	+20 45.	1	11.07		0.31		0.14		2		1569
	+28 02174				12 59 27	+27 36.4	1	10.70		0.59		0.10		1	G1 V	1019
		G 14 - 24			12 59 27	−01 49.0	4	12.81	.016	0.74	.015	0.02	.016	9		419,1658,1705,3028
	+30 02355				12 59 29	+29 51.1	4	10.65	.022	−0.07	.007	−0.10	.024	18	A0	1019,1026,1068,1298
		WEIS 34589			12 59 30	+29 47.8	1	11.62		0.53				1		1692
		WEIS 63119			12 59 32	+28 12.1	1	11.63		0.61				2		1692
		WEIS 63057			12 59 32	+28 22.9	1	13.76		0.67				1		1692
		WEIS 60267			12 59 32	+29 17.9	1	11.50		0.70				1		1692
	+29 02356				12 59 32	+29 27.6	1	11.35		0.84		0.36		1		1019
		MN201,849 # 118			12 59 32	−19 50.6	1	13.45		0.95				2		961
		GPEC 2650			12 59 32	−34 05.4	1	11.02		0.87				6		1719
113151	−59 04674				12 59 32	−59 39.9	1	9.46		0.23		0.11		1	A0 IV	768
		MN201,849 # 87			12 59 33	−19 42.3	1	13.15		0.55				2		961
113213	+19 02628				12 59 34	+18 53.7	1	8.69		1.17		1.13		1	K2	280
113241	+44 02246				12 59 35	+43 59.7	1	9.10		1.17				2	K1 III	280
		He3 851			12 59 35	−62 02.	1	12.53		0.52		−0.25		5		1684
	+30 02356				12 59 36	+30 19.6	1	11.23		0.94		0.66		1	K3 III	1019
	−59 04676				12 59 36	−59 56.0	1	10.50		0.45		0.30		1	A5	768
		WEIS 62447			12 59 37	+28 27.6	1	13.42		0.64				1		1692
		MN201,849 # 23			12 59 37	−19 51.8	1	11.55		0.63				2		961
		WEIS 63081			12 59 38	+28 17.7	1	12.09		0.74				2		1692
		WEIS 63059			12 59 38	+28 21.8	1	12.41		0.88				1		1692
113152	−60 04395				12 59 38	−61 03.8	1	7.84		0.41		−0.03		1	F2 V	1688
	−63 02495	LSS 2883			12 59 38	−63 34.0	4	10.04	.017	0.73	.012	−0.47	.017	7		540,976,1684,1737
	+27 02205				12 59 39	+26 59.4	1	10.36		0.50		0.07		1	F6	1019
113194	−26 09476				12 59 39	−26 30.9	1	8.81		0.94		0.65		1	K4 V	3072
113120	−70 01553	HR 4930, LSS 2879			12 59 39	−71 12.4	9	6.00	.029	0.03	.017	−0.87	.019	47	B1.5 IIIne	15,540,681,815,976,1075*
113203	−27 08984				12 59 40	−28 01.1	1	6.71		1.52		1.82		9	K3 III	1628
113226	+11 02529	HR 4932		⋆ A	12 59 41	+11 13.6	9	2.83	.019	0.93	.015	0.73	.015	30	G8 III	3,15,1080,1118,1194*
113242	+29 02357	WEIS 60243			12 59 41	+29 15.6	4	8.93	.019	0.57	.005	0.10	.015	7	F8 V	1019,1690,1692,3026
	+31 02440				12 59 42	+31 26.1	1	10.06		1.06		0.97		1	G8 III	280
	−59 04679				12 59 42	−59 38.0	3	10.29	.061	0.31	.010	−0.33	.003	5	B8 Iab	540,768,976
		WEIS 61225			12 59 43	+28 58.8	1	14.21		0.58				1		1692
	−19 03621				12 59 43	−19 46.4	1	10.71		1.21				2		961
113163	−60 04396	LSS 2885			12 59 43	−60 28.5	4	7.78	.011	0.28	.023	−0.59	.017	7	B1 III	540,976,1034,1684
	+28 02175				12 59 44	+28 04.8	2	11.34	.005	0.57	.000	0.08		2		1019,1692
	−59 04680				12 59 44	−59 40.7	1	11.18		0.32		0.01		1	B9	768
113225	+13 02622				12 59 46	+13 04.7	1	9.95		0.97		0.66		2	G5	1569
113243	+17 02577				12 59 46	+17 03.8	1	9.32		1.00		0.76		1	K0	280
113153	−68 01824				12 59 47	−68 31.6	2	8.56	.019	0.54	.000	0.10		4	F7 V	1594,6006
	+12 02539				12 59 48	+11 29.8	1	9.34		0.98		0.71		1	K0	280
		Lod 733 # 30			12 59 48	−59 40.8	1	12.43		1.70		1.69		1		768
		WEIS 33436			12 59 49	+30 17.8	1	11.48		0.60				1		1692
	+42 02366				12 59 49	+41 43.9	1	10.68		0.73		0.45		1		280
	−19 03622				12 59 49	−19 51.9	1	10.91		0.72				2		961
	−59 04681				12 59 49	−59 40.8	1	11.37		0.26		−0.48		1	B4	768
113164	−63 02497				12 59 49	−63 48.1	1	10.48		0.37		0.24		1	A1/2 V	1688
	+30 02357				12 59 50	+30 17.9	1	10.89		0.51		0.04		1	G2	1019
113337	+64 00927	HR 4934		⋆ A	12 59 50	+63 52.7	2	6.07	.039	0.42	.012	−0.01	.010	3	F5 V	254,3016
113191	−64 02124				12 59 51	−64 31.0	2	8.91	.001	0.45	.008	0.23	.012	2	A9 IV	1688,1771
		WEIS 63056			12 59 53	+28 23.9	1	11.92		0.71				1		1692
113166	−70 01555				12 59 54	−70 44.2	1	9.57		0.45		0.25		1	A9 V[p]	1771
		MN201,849 # 20			12 59 55	−19 47.7	1	11.24		0.65				2		961
	−59 04682				12 59 55	−59 42.4	1	10.30		0.26		0.10		1	A0	768
113269	+21 02480				12 59 56	+20 50.9	1	8.68		1.02		0.78		2	K0	1569
113199	−59 04683				12 59 56	−59 43.8	1	8.81		0.21		0.17		1	A2 V	768
113284	+31 02442				12 59 58	+30 37.5	2	7.94	.000	0.35	.001	0.03	.000	6	F1 IV	1019,1690
	−59 04684	LSS 2887			12 59 58	−59 54.6	2	10.42	.014	0.17	.009	−0.55		5	B3	768,2021
		WEIS 62467			12 59 59	+28 35.1	1	13.89		0.65				1		1692
113198	−59 04685				12 59 59	−59 41.4	1	9.46		0.24		0.12		1	A1 IV/V	768
113271	+16 02449				13 00 00	+16 08.9	1	9.31		1.02		0.84		1	K0	280
113299	+43 02293	IDS12577N4251	A		13 00 00	+42 35.2	1	9.53		0.25		0.11		4	A6 m	270
113211	−58 04659				13 00 00	−59 20.4	1	8.53		0.30		0.06		1	A9 V	1771

HD	DM	Other Id	N	Rem	α_{1950}	δ_{1950}	S	V	σ_V	B–V	σ_{B-V}	U–B	σ_{U-B}	n	Spectrum	References
113270	+16 02450				13 00 01	+16 18.6	1	9.03		0.91		0.72		1	K0	280
		WEIS 60225			13 00 02	+29 16.1	1	12.52		0.59				2		1692
113302	+38 02396				13 00 02	+37 36.9	1	8.18		0.50		0.05		1	F8 V	619
113301	+39 02592				13 00 02	+39 05.5	2	8.97	.005	0.43	.000	0.03	.020	5	F6 V	1501,1569
		WEIS 62483			13 00 03	+28 30.4	1	15.80		0.61				1		1692
		WEIS 60232			13 00 03	+29 12.7	1	11.31		0.80				1		1692
	+30 02358				13 00 03	+29 59.0	1	11.15		0.69		0.23		1	G5	1019
113300	+39 02593				13 00 03	+39 09.0	1	9.25		0.47		-0.07		2	G5	1569
	-63 02501	LSS 2888			13 00 03	-63 26.8	2	10.10	.003	0.54	.005	-0.43	.012	3	B0.5II-III	540,976
113048	-80 00684				13 00 03	-80 52.9	1	7.20		1.03				4	K0 III	2012
	+29 02358	IDS12577N2930		AB	13 00 04	+29 14.1	1	11.21		0.49		0.00		1	F5	1019
		WEIS 60228			13 00 04	+29 14.3	1	10.20		0.48				1		1692
		MN201,849 # 39			13 00 04	-20 10.2	1	11.64		1.08				2		961
	-59 04687				13 00 04	-59 40.2	1	11.64		0.34		0.27		1	A2	768
113285	+5 02708	RT Vir			13 00 06	+05 27.3	1	7.41		1.66		1.00		1	M1	3076
	+17 02579				13 00 07	+17 15.1	1	9.87		1.00		0.81		1		280
		WEIS 35605			13 00 07	+29 24.8	1	12.63		1.04				1		1692
		WEIS 60191			13 00 08	+29 23.4	1	13.93		0.53				1		1692
113200	-63 02503				13 00 08	-63 51.3	1	9.71		0.40		0.27		1	A2/3 V	1688
	+19 02630				13 00 09	+18 30.5	1	10.24		0.98		0.80		1		280
113320	+23 02528	IDS12577N2330		A	13 00 09	+23 12.9	1	8.15		0.97		0.81		1	K0	280
		WEIS 61440			13 00 09	+28 56.8	1	14.95		0.34				1		1692
		WEIS 34280			13 00 09	+30 01.4	1	12.23		0.97				1		1692
	+46 01838				13 00 09	+46 02.7	1	9.21		1.40		1.65		1	K5	280
113236	-48 07882				13 00 09	-49 16.4	3	8.12	.007	0.00	.002	-0.42		9	B7 III/IV	1770,2014,2014
		AJ82,345 # 192			13 00 09	-62 58.1	1	13.46		0.83		0.04		2		779
	+14 02569				13 00 10	+13 44.3	1	10.34		0.91		0.67		1	K0	280
		WEIS 61305			13 00 10	+29 02.2	1	13.89		0.63				1		1692
	-59 04689				13 00 10	-59 36.2	1	10.78		0.52		0.34		1	A5	768
113237	-55 05348	IDS12572S5534		AB	13 00 11	-55 50.5	2	8.03	.007	-0.02	.004	-0.33	.008	4	B8 V	401,1770
113319	+33 02296				13 00 12	+32 42.1	1	7.55		0.65		0.13		3	G8 IV	196
		MN201,849 # 54			13 00 12	-20 08.5	1	12.34		0.65				2		961
		FQ Com			13 00 14	+31 34.4	1	12.77		1.89		1.63		1	gM7	685
	+14 02571				13 00 15	+14 04.3	1	10.37		0.72		0.33		1	G5	280
113339	+31 02443				13 00 15	+30 33.0	2	9.22	.000	0.59	.000	0.08	.001	2	G0 V	1019,1690
113338	+31 02444				13 00 16	+31 20.9	1	9.54		0.72				2	G2 V	1692
		LB 27			13 00 17	+27 56.8	1	14.18		-0.15		-1.00		1	sdO	974
113321	+17 02580				13 00 18	+16 44.1	1	9.37		1.04		0.90		1	K0	280
		HZ 37			13 00 18	+38 43.	1	12.63		-0.09		-0.21		2		1026
		G 14 - 25			13 00 18	-02 24.4	2	11.51	.044	1.31	.005	1.16		3	K5	202,1620,3077
113309	-19 03624				13 00 18	-20 13.3	1	8.65		0.52				2	F6 V	961
113291	-26 09486				13 00 18	-27 14.1	1	8.59		1.26		1.13		1	Kp Ba	565
	-59 04690				13 00 18	-59 41.2	1	11.66		0.35		-0.04		1	Ap	768
113238	-61 03432				13 00 18	-62 01.6	1	10.23		0.27		0.14		1	B8/9 V	1688
		WEIS 33098			13 00 20	+30 38.7	1	11.73		1.13				1		1692
	-19 03625	IDS12577S1939		AB	13 00 20	-19 54.7	1	10.31		0.53				2	F8	961
		AJ82,345 # 195			13 00 21	-61 57.9	1	12.56		0.86		-0.20		2		779
113239	-63 02505				13 00 22	-63 41.9	1	10.23		0.22		-0.02		1	B9/A0 V	1688
	-59 04691				13 00 23	-59 31.3	1	10.53		0.26		0.13		1	A1	768
		A2 368			13 00 24	+30 39.5	1	16.47		0.42		-0.21		1		98
234004	+53 01589				13 00 24	+52 30.9	1	9.70		0.33		0.00		2	F2	1566
	+27 02207				13 00 26	+27 06.1	1	11.17		0.52		0.02		1	F8	1019
		WEIS 62495			13 00 27	+28 27.0	1	12.59		0.96				1		1692
	+27 02208				13 00 29	+26 39.5	1	10.00		0.99		0.76		1	G8 III	280
		WEIS 62479			13 00 29	+28 33.4	1	14.14		1.19				1		1692
113379	+46 01839				13 00 29	+45 39.2	1	7.63		1.64		1.95		1	M1	280
113261	-59 04693				13 00 29	-60 06.7	1	8.89		0.22				4	B8 Ib	2021
113349	+12 02541				13 00 30	+12 08.9	1	9.20		1.40		1.61		1	K2	280
		WEIS 61833			13 00 31	+28 39.3	1	11.70		0.44				3		1692
		WEIS 64308			13 00 32	+28 02.2	1	10.97		0.72				2		1692
		G 123 - 84			13 00 32	+41 47.3	1	12.95		1.50				1		3062
113365	+23 02530	IDS12582N2311		A	13 00 33	+22 54.4	1	6.97		0.08		0.09		2	A0	1068
	+40 02611				13 00 33	+39 43.8	1	10.52		0.33		0.01		2		1569
		WLS 1300-5 # 7			13 00 34	-07 11.0	1	11.66		1.08		0.84		2		1375
113352	-5 03618	G 14 - 26			13 00 36	-05 51.3	2	9.73	.010	0.59	.020	0.05	.010	2	G5	1620,1658
113380	+23 02531				13 00 37	+23 28.2	1	7.46		1.12				2	K0	280
	+25 02585				13 00 37	+25 28.1	1	10.33		0.91		0.62		1	K2	280
113436	+60 01439	HR 4936			13 00 37	+59 59.1	2	6.53	.005	0.05	.000	0.06	.000	7	A3 Vn	1501,8071
		WEIS 64298			13 00 38	+27 54.5	1	12.02		1.03				1		1692
113314	-48 07887	HR 4933			13 00 39	-49 15.5	4	4.84	.007	0.02	.005	0.02	.006	11	A0 V	15,26,1770,2029
113281	-63 02506				13 00 39	-63 25.2	1	9.23		0.28		0.11		2	B9 IV	1688
		WEIS 62941			13 00 40	+28 19.6	1	13.98		0.79				1		1692
		G 149 - 31			13 00 41	+24 14.2	1	15.56		1.61				1		906
	-59 04696				13 00 41	-59 32.8	1	11.61		0.37		0.27		1	A0	768
	-59 04698				13 00 41	-59 50.2	1	11.02		0.28		0.18		1	A1	768
113381	+14 02574				13 00 42	+14 22.9	1	10.48		0.94		0.88		1	K0	280
113423	+51 01802				13 00 42	+51 18.4	1	8.29		1.00		0.74		2	K0	1733
	+18 02631				13 00 45	+18 40.4	1	10.11		0.93		0.69		2	K0	1569
113406	+24 02531				13 00 45	+24 05.7	1	6.98		1.59				2	M1 III	280
		G 62 - 9			13 00 46	+04 23.4	1	11.31		0.78		0.36		1		1658

Table 1 719

HD	DM	Other Id	N	Rem	α_{1950}	δ_{1950}	S	V	σ_V	B–V	σ_{B-V}	U–B	σ_{U-B}	n	Spectrum	References
	−23 10879				13 00 46	−24 09.1	2	9.97	.028	0.49	.009	0.01		4	F0	1594,6006
113383	−5 03619	IDS12582S0553	AB		13 00 47	−06 10.1	1	9.02		1.31		1.17		1	F0	768
		Ton 141			13 00 48	+28 58.	1	15.11		0.07		0.08		2		1036
	+40 02612				13 00 48	+40 10.6	1	10.66		1.08		1.13		1	K1 III	280
	+16 02452				13 00 49	+15 49.1	1	10.64		0.84		0.37		1		280
	+29 02359	IDS12585N2909	AB		13 00 51	+28 52.5	1	10.32		0.63		0.09		1	G2 V	1019
		WEIS 62990			13 00 52	+28 10.7	1	14.57		0.68				2		1692
		WEIS 61403			13 00 54	+29 09.9	1	13.68		0.84				1		1692
113446	+43 02294				13 00 56	+42 33.8	1	9.07		1.12		0.94		1	K0 III	280
113407	+15 02535				13 00 57	+15 03.4	1	9.10		1.38		1.58		1	K3 III	280
		POSS 219 # 3			13 00 58	+39 48.2	1	15.86		1.42				2		1739
113348	−60 04406				13 00 59	−61 11.6	2	9.19	.003	0.49	.015	0.25	.013	2	A9 V	1688,1771
		WEIS 62575			13 01 00	+28 29.6	1	13.98		0.66				1		1692
113347	−59 04700	LSS 2894			13 01 00	−59 55.3	1	8.37		0.53		0.41		3	A8 III/IV	8100
	+40 02613				13 01 01	+39 54.3	1	10.30		1.42		1.24		1	K3 III	280
	−63 02511	LSS 2893			13 01 02	−63 41.4	1	10.38		0.60		−0.35		2	B0.5II-III	540
		WEIS 62651			13 01 04	+28 31.9	1	16.14		0.75				1		1692
113415	−19 03629	HR 4935	★ AB		13 01 05	−20 18.9	5	5.58	.004	0.56	.005	0.06	.005	15	F7 V	15,196,258,1075,2035
		L 195 - 10			13 01 06	−55 21.	1	10.67		1.00		0.89		1		1696
113374	−59 04702				13 01 08	−59 38.7	1	10.06		0.28		−0.17		1	B8/9 Ib/II	768
	−60 04408				13 01 08	−61 02.9	2	10.42	.035	1.19	.008	0.18	.016	5		1684,1737
	+21 02483				13 01 09	+20 48.6	1	9.25		1.02		0.77		1	K0	280
		WEIS 62981			13 01 09	+28 11.5	1	13.18		0.80				1		1692
		G 149 - 33			13 01 10	+23 45.1	1	15.61		1.79				1		906
		WEIS 62550			13 01 11	+28 36.7	1	13.52		1.30				1		1692
	+28 02176	WEIS 64343			13 01 13	+27 51.6	2	9.78	.004	0.98	.002	0.75		4	G8 III:	1019,1692
113428	−23 10882				13 01 14	−24 15.0	1	7.50		1.36		1.37		7	K0 III	1657
	−60 04412	LSS 2896			13 01 14	−60 52.0	2	10.26	.025	0.74	.002	−0.27	.038	4	B0 Ib	191,1684
	+28 02177				13 01 15	+28 23.2	1	10.71		0.96		0.70		1	G5 V	1019
113468	+29 02360				13 01 15	+28 43.1	3	9.14	.021	0.58	.002	0.17	.002	8	F7 V	1019,1690,1692
	−59 04703				13 01 15	−59 49.2	1	10.48		0.28		0.17		1	A1	768
113390	−59 04707				13 01 16	−59 41.0	1	9.61		0.30		−0.20		1	B9	768
		WEIS 61772			13 01 17	+28 41.4	1	14.06		0.84				1		1692
113399	−55 05359				13 01 17	−55 54.5	1	8.90		0.10		−0.21		3	B7 II	1770
113376	−65 02141				13 01 17	−65 39.2	2	8.48	.000	0.73	.002	0.26	.007	35	G3 V	1589,1699
		WEIS 62620			13 01 18	+28 23.6	1	10.69		0.96				1		1692
	−63 02512	LSS 2897			13 01 18	−63 39.8	2	9.91	.009	0.49	.009	−0.45	.031	4	B2 III	540,976
	+46 01841				13 01 19	+45 44.0	1	10.27		0.77				2	K0 III	280
113459	−2 03622	HR 4937	★ AB		13 01 20	−03 23.7	2	6.58	.005	0.30	.005	0.09	.015	7	F0	15,1417
	+29 02361				13 01 21	+28 54.2	1	10.80		0.87		0.41		1	G5 V:	1019
		AJ82,345 # 201			13 01 21	−62 29.8	1	13.53		1.06		0.21		2		779
		Ton 143			13 01 22	+27 02.6	1	15.64		-0.41		-1.24		1	sdO	974
		WEIS 61473			13 01 23	+28 53.1	1	10.70		0.85				1		1692
	+7 02608	CO Vir			13 01 24	+07 20.1	1	9.54		1.24		1.30		1	M8	3001
	−59 04710				13 01 25	−59 37.1	1	10.85		0.35		0.21		1	A0	768
312287	−62 02992				13 01 25	−62 49.9	1	10.65		0.92		−0.20		2	B0	779
	−63 02513	LSS 2898			13 01 25	−63 41.5	3	9.83	.045	0.62	.049	−0.36	.060	6	B0 III	191,540,1684
		BF Cha	★		13 01 25	−77 22.9	1	12.37		1.48		1.09		1		5005
	+30 02359				13 01 26	+30 24.7	1	12.00		0.89		0.53		1	K0	1019
113493	+31 02445				13 01 26	+31 16.2	3	7.33	.015	1.05	.005	0.93	.001	4	K0 III	1019,1690,1692
113401	−61 03437				13 01 26	−62 07.5	1	10.39		0.54		0.08		1	F5	1688
113421	−59 04711	LSS 2902			13 01 27	−59 33.4	4	9.34	.006	0.20	.009	−0.66	.012	13	B0 Vn	540,768,976,2021
	+85 00215				13 01 28	+84 46.0	1	10.74		0.14		0.12		7	B8	1219
		WEIS 60122			13 01 29	+29 12.6	1	11.64		0.58				1		1692
113494	+31 02446				13 01 29	+31 01.5	2	8.51	.000	0.40	.000	0.02	.002	6	F2 IV	1019,1690
312256	−61 03438				13 01 29	−61 24.2	1	9.76		0.82		−0.22		2		1684
		AJ82,345 # 203			13 01 29	−62 07.0	1	12.36		0.65		−0.13		2		779
113496	+12 02545				13 01 30	+11 29.9	1	7.39		1.51		1.48		1	M2	280
		WEIS 64436			13 01 30	+27 50.1	1	14.29		0.56				1		1692
		WEIS 62610			13 01 30	+28 28.7	1	11.17		1.03				1		1692
113422	−61 03439	LSS 2901			13 01 30	−61 26.5	6	8.23	.011	0.82	.020	−0.21	.036	11	B1/2 Ia/ab	138,191,540,914,976,1684
113492	+33 02298				13 01 31	+33 11.3	1	9.23		0.92		0.68		1	G5	280
		WEIS 62925			13 01 32	+28 11.9	1	14.44		0.84				1		1692
113495	+16 02454				13 01 34	+16 23.9	1	9.28		1.06		1.08		1	K0	280
	+31 02447				13 01 34	+30 55.8	1	10.32		0.49		-0.02		1	F6	1019
	+38 02397				13 01 34	+37 40.2	1	10.29		0.64		0.08		1	G5	619
	+43 02295				13 01 35	+43 11.3	1	10.60		0.52		-0.01		2		1569
	+12 02546				13 01 36	+11 34.5	1	9.34		1.15		1.03		1	K0	280
	+23 02532				13 01 36	+22 33.6	1	8.94		1.15		1.12		1	K5	280
		WEIS 62638			13 01 36	+28 26.4	1	14.57		0.60				1		1692
113451	−47 08064				13 01 36	−47 51.9	2	7.62	.004	-0.03	.002	-0.32		7	B9 III	1770,2014
113515	+31 02448				13 01 37	+30 44.1	2	9.55	.005	1.00	.000	0.76		2	G8 III	1019,1692
113545	+43 02296				13 01 37	+43 16.5	1	6.81		1.63				2	M1	280
		WEIS 60144			13 01 39	+29 24.7	1	13.60		1.40				1		1692
113432	−62 02993	LSS 2906			13 01 39	−63 13.4	4	8.84	.020	0.78	.015	−0.20	.021	8	B1 Iab/b	191,540,976,1684
		WEIS 64390			13 01 46	+27 59.7	1	10.35		0.45				1		1692
		POSS 219 # 1			13 01 46	+39 50.1	1	14.76		1.56				1		1739
113561	+30 02360				13 01 49	+30 29.3	2	9.38	.000	0.71	.005	0.36		2	G5 III	1019,1692
	−12 03747	IDS12592S1243	AB		13 01 49	−12 59.7	1	9.66		0.71				6	G0	176
113456	−59 04714				13 01 49	−59 49.7	1	9.31		0.22		0.07		1	B9 III/IV	768

HD	DM	Other Id	N Rem	α_{1950}	δ_{1950}	S	V	σ_V	B–V	σ_{B-V}	U–B	σ_{U-B}	n	Spectrum	References
		POSS 219 # 7		13 01 50	+40 01.0	1	17.96		1.39				3		1739
	+40 02616			13 01 50	+40 07.5	1	9.39		0.96		0.68		1	K0	280
113457	−63 02515			13 01 51	−64 10.4	3	6.65	.004	0.00	.004	-0.04	.008	5	A0 V	401,1688,1771
		WEIS 64441		13 01 52	+27 49.9	1	15.17		0.54				1		1692
113578	+44 02250	G 123 - 85		13 01 53	+44 23.3	1	8.95		1.05				2	K3 V	280
		He3 856		13 01 53	−61 15.	1	13.21		1.00		0.16		4		1684
		WEIS 62719		13 01 54	+28 21.5	1	14.81		0.75				1		1692
113562	+28 02179			13 01 54	+28 22.5	3	9.34	.005	0.43	.006	0.03	.000	10	F0 V	1019,1690,1692
113504	−40 07660			13 01 56	−40 24.8	1	7.92		0.44		0.02		5	F5/6 V	1731
	+27 02209			13 01 57	+26 34.5	1	10.31		0.96		0.64		1	G8 III	280
		WEIS 61759		13 01 59	+28 37.2	1	13.75		0.31				1		1692
		WEIS 61792		13 01 59	+28 46.5	1	11.99		0.70				2		1692
113523	−40 07662	HR 4938, V789 Cen		13 01 59	−40 55.7	3	6.27	.013	1.67	.010			11	M3/4 III	15,1075,2006
	+39 02597			13 02 02	+38 42.1	1	9.18		1.05		0.92		1	G8 III	280
		AJ82,345 # 212		13 02 02	−62 12.5	1	13.04		0.97		0.12		1		779
		G 60 - 60		13 02 03	+10 44.1	2	13.67	.000	0.62	.002	-0.04	.000	4		333,1620,1658
	+20 02790			13 02 04	+19 54.5	1	10.28		0.99		0.69		1		280
		WEIS 61577		13 02 06	+28 52.0	1	12.04		0.63				1		1692
		WEIS 61529		13 02 06	+29 10.2	1	14.04		0.74				1		1692
		WEIS 25550		13 02 06	+29 32.	1	16.32		1.49				1		784
		WEIS 60037		13 02 07	+29 16.8	1	15.65		0.94				1		1692
		WLS 1300-5 # 5		13 02 07	−03 09.1	1	12.27		0.64		0.01		2		1375
		WEIS 61750		13 02 08	+28 40.3	1	12.24		0.51				1		1692
113538	−51 07244			13 02 08	−52 09.2	4	9.05	.019	1.37	.023	1.27	.012	11	K5/M0 V	158,1705,2012,3078
113511	−63 02519	LSS 2912		13 02 09	−63 47.0	4	9.05	.006	0.46	.013	-0.52	.026	10	B0 IV/V	540,976,1684,1688
	+20 02791			13 02 10	+19 36.5	1	9.88		0.63		0.15		2	G5	1569
		WEIS 64426		13 02 10	+27 52.8	1	10.89		0.95				1		1692
	+27 02210			13 02 11	+26 32.1	1	9.31		1.01		0.80		1	G8 III	280
	+30 02361	WEIS 24026		13 02 11	+30 22.5	2	10.57	.009	0.56	.007	0.03		2	G0 V	1019,1692
	+57 01416			13 02 12	+56 32.0	1	10.52		1.09		0.89		2		1375
113540	−53 05455			13 02 12	−54 00.7	1	9.52		0.48		0.05		1	A9/F2 V	1771
	+31 02449			13 02 13	+30 34.1	1	10.84		0.41		0.01		1	F3	1019
113609	+10 02509			13 02 14	+09 43.1	1	8.16		1.05		0.87		1	K0	280
113537	−46 08383			13 02 14	−46 50.9	2	6.43	.000	0.42	.000			8	F3 III	1075,2012
		Wray 1060		13 02 14	−62 56.8	1	12.37		0.94		0.13		4		1684
		KUV 13023+3145		13 02 17	+31 45.1	1	16.99		-0.22		-1.15		1	DB	1708
113608	+15 02539			13 02 19	+14 59.7	1	9.82		1.34		1.54		1	K0	280
	+35 02402			13 02 20	+35 17.7	1	10.04		1.09		1.02		2	K0 III	1569
113649	+40 02618			13 02 21	+39 35.5	1	7.72		0.96		0.71		1	G8 V	280
	−60 04430			13 02 23	−61 24.0	1	9.99		1.08		-0.11		2		1684
		WEIS 24092		13 02 24	+30 10.	1	16.78		1.46				1		784
		WEIS 23055		13 02 24	+30 26.	1	16.01		1.52				1		784
	+10 02510			13 02 25	+10 11.8	1	10.52		1.01		0.94		1		280
		EG 97, HZ 39		13 02 25	+28 23.5	2	15.47	.034	-0.30	.019	-1.16	.000	3	DAs	974,3028
	+21 02484			13 02 26	+20 51.7	1	9.46		0.95		0.72		2	K5	1569
113623	+16 02458			13 02 27	+16 20.8	1	9.26		1.28		1.51		1	K2	280
	+19 02634			13 02 27	+19 12.4	1	9.58		0.95		0.83		2	K0	1569
		LB 2539		13 02 27	+59 42.9	1	14.52		-0.13		-1.08		4	DBp	308
113612	−27 09018			13 02 27	−28 15.0	1	7.69		0.86				4	G8 (V) (+F)	2012
	+46 01843			13 02 28	+46 03.2	1	9.93		0.78				2	K0	280
113640	+11 02532			13 02 29	+10 36.9	1	9.35		0.24		0.02		1	F0	280
		WEIS 25462		13 02 30	+29 09.	1	16.90		1.46				1		784
		WEIS 61605		13 02 30	+29 09.	1	13.52		1.42				1		784
113573	−56 05566			13 02 31	−56 35.0	1	8.77		0.08		-0.57		2	B3/5 V	401
113541	−62 02997			13 02 31	−63 06.2	1	9.77		0.49		0.28		1	B8/9 III	1688
		G 61 - 28		13 02 32	+16 39.9	1	12.34		1.00		0.82		1	K3	1620
113602	−51 07248	HR 4939		13 02 34	−51 50.9	4	6.42	.006	1.69	.005	1.99		13	M1 III	15,1075,2032,3005
		G 149 - 36		13 02 35	+24 32.6	1	15.67		1.68				1		906
	−64 02149	Y Mus		13 02 35	−65 14.7	2	10.32	.043	0.95	.014	0.41	.049	2	Fp	1589,1699
	+30 02362			13 02 38	+30 25.9	1	11.16		0.56		0.05		1	G2	1019
		WEIS 64401		13 02 39	+27 58.4	1	13.97		0.63				1		1692
113672	+38 02398			13 02 39	+37 32.4	1	9.21		0.73		0.34		1	G8 V	619
		WEIS 60040		13 02 40	+29 19.8	1	13.57		1.08				1		1692
		WEIS 64501		13 02 41	+27 51.5	1	12.67		0.71				1		1692
		MCC 684	AB	13 02 41	+56 10.2	1	10.41		1.49				2	M0	1619
		RY Com		13 02 42	+23 32.7	2	11.92	.236	0.09	.032	0.04	.009	2	A5	597,699
		WEIS 25472		13 02 42	+29 41.	1	14.22		1.48				1		784
113673	+32 02318			13 02 43	+32 17.6	1	8.67		1.52		1.90		1	M0 III	280
		G 60 - 62		13 02 44	+03 14.0	2	10.96	.020	0.98	.000	0.82	.015	2		1620,1658
	+24 02536			13 02 44	+24 28.3	1	9.79		1.46		1.74		1	K5	280
113576	−69 01757			13 02 44	−69 26.7	1	9.24		0.32		0.17		1	A6 IV/V	1771
		CS 16027 # 16		13 02 46	+30 26.1	1	14.08		0.54		-0.07		1		1744
		WEIS 60031		13 02 47	+29 12.1	1	11.11		0.42				1		1692
	+30 02363			13 02 47	+29 30.3	1	11.25		0.55		-0.05		1		1019
		WEIS 25391		13 02 47	+29 31.8	1	11.29		0.56				1		1692
		WEIS 22864		13 02 48	+30 46.	1	13.06		1.54				1		784
	−60 04432			13 02 48	−61 05.7	2	9.89	.005	1.21	.023	0.04	.000	4		1684,1737
		WEIS 62776		13 02 49	+28 20.4	1	11.36		0.67				2		1692
113587	−64 02151	IDS12596S6505	AB	13 02 49	−65 20.9	1	8.90		0.33		0.19		1	A8 V	1771
	+29 02362	WEIS 61735		13 02 50	+28 39.7	2	9.89	.000	0.66	.002	0.24		2	G3 V	1019,1692

Table 1   721

HD	DM	Other Id	N Rem	α_{1950}	δ_{1950}	S	V	σ_V	B–V	σ_{B-V}	U–B	σ_{U-B}	n	Spectrum	References
113605 −61 03445		LSS 2916		13 02 50	−61 58.2	4	9.72	.022	0.32	.010	-0.45	.016	13	B2/3ne	540,779,976,1684
+11 02533				13 02 51	+11 16.5	1	10.15		1.14		1.17		1	K2	280
		WLS 1300-5 # 9		13 02 51	−04 45.9	1	13.51		1.06		1.17		2		1375
113606 −63 02527		LSS 2914		13 02 51	−63 58.5	4	8.62	.008	0.45	.017	-0.55	.025	7	O9.5III	540,976,1684,1737
		LP 854 - 23		13 02 52	−27 08.8	1	12.96		1.45		1.06		2		1696
113618 −59 04726				13 02 52	−59 27.4	1	9.50		0.34		0.16		2	A0 V	122
		AAS12,381 25 # B		13 02 53	+25 10.8	1	10.20		0.95		0.65		1		280
113589 −66 02060				13 02 53	−66 46.7	1	9.49		0.42		0.16		1	A1mA7-F0	1771
113730 +37 02360				13 02 54	+37 20.3	1	9.15		0.20		0.13		2	Am	1068
		WEIS 23108		13 02 55	+30 26.7	1	10.68		1.40				1		1692
+38 02400				13 02 55	+38 16.5	1	9.82		0.80		0.45		1	G8 III	280
		WEIS 64568		13 02 56	+27 48.8	1	14.63		0.90				1		1692
+30 02364				13 02 56	+30 26.2	1	10.67		1.38		1.58		1		1019
+25 02591				13 02 57	+25 14.9	1	9.45		1.09		1.10		1	K0	280
113716 +11 02535				13 02 58	+10 31.5	1	8.72		1.00		0.79		1	G5	280
−60 04434		LSS 2917		13 02 58	−60 56.3	3	9.59	.008	0.91	.009	-0.18	.015	5	O9.5IV	540,976,1684
		Feige 72		13 03 00	+06 15.	1	11.62		0.17		0.13		2		3060
113713 +15 02541		IDS13005N1516	A	13 03 00	+14 59.6	1	7.94		0.45		-0.06		2	F5 V	1648
		WEIS 61723		13 03 00	+28 37.8	1	13.47		0.67				1		1692
		POSS 219 # 5		13 03 00	+39 41.8	1	16.32		1.65				3		1739
		WEIS 52202		13 03 03	+28 05.8	1	12.52		0.83				1		1692
+28 02180				13 03 03	+28 09.8	1	10.78		0.45		0.09		1	F3	1019
113731 +14 02578				13 03 05	+13 29.6	1	7.46		0.97		0.71		1	K0	280
		WEIS 23131		13 03 05	+30 27.1	1	10.73		0.65				1		1692
		CS 16027 # 9		13 03 05	+31 13.4	1	14.11		0.65		0.15		1		1744
+22 02542				13 03 06	+21 53.0	1	10.34		1.16		1.12		1		280
113679 −37 08363				13 03 06	−38 14.8	3	9.70	.014	0.60	.005	0.00		7	G2/3 V	1696,2033,3025
−61 03446		LSS 2918		13 03 06	−61 37.5	2	10.11	.008	0.53	.007	-0.39	.018	3	B1 III	540,976
113682 −40 07678				13 03 07	−40 46.4	1	10.07		1.00		-0.17		3	K1 III	1737
		PASP75,256 # 44		13 03 10	+37 45.7	1	13.78		0.56		0.10		1		619
		PASP75,256 # 38		13 03 10	+37 50.1	1	11.44		0.66		0.18		1		619
113770 +42 02370		IDS13009N4213	AB	13 03 10	+41 56.9	1	8.67		0.48		-0.06		3	F5	1569
113658 −60 04436				13 03 11	−60 52.5	1	9.81		0.15		-0.14		1	B9 III	1688
+26 02426				13 03 12	+26 11.8	1	10.23		0.96		0.79		1	K0	280
113635 −67 02178				13 03 12	−67 50.2	1	8.23		0.16		0.15		1	A3 V	1771
		BK Cha	⋆	13 03 12	−77 14.5	1	14.63		1.48		0.34		1		5005
		PASP75,256 # 43		13 03 13	+37 47.8	1	13.24		0.79		0.28		1		619
113733 −13 03651				13 03 13	−13 50.6	1	7.20		1.50		1.81		9	K4 III	1628
		G 61 - 29	⋆ V	13 03 16	+18 17.0	1	15.69		-0.10		-0.97		1		3060
113771 +27 02212				13 03 17	+26 51.2	1	7.50		0.93		0.64		1	K0 III	280
113889 +73 00583		HR 4950	⋆ AB	13 03 17	+73 17.5	1	6.41		0.20		0.08		2	F0 V	1733
		WEIS 50527		13 03 18	+29 03.9	1	12.06		0.79				1		1692
		WEIS 25352		13 03 18	+29 28.	1	17.42		1.43				1		784
		WEIS 23930		13 03 18	+30 10.	1	16.99		1.54				1		784
		POSS 219 # 4		13 03 18	+39 37.7	1	15.87		0.71				4		1739
113659 −64 02160				13 03 19	−64 48.8	1	8.03		-0.01		-0.92		4	O8/9 III	1684
−60 04438		LSS 2920		13 03 21	−60 44.7	3	10.07	.021	0.38	.015	-0.48	.013	6		540,976,1684
113703 −47 08088		HR 4940	⋆ A	13 03 22	−48 11.7	8	4.70	.006	-0.14	.001	-0.58	.012	32	B4 V	15,26,1075,1770,2012,2014*
113703 −47 08088		IDS13005S4756	B	13 03 22	−48 11.7	1	10.80		0.82				6		321
+28 02181				13 03 23	+27 41.5	1	10.77		0.55		0.00		1	F8	1019
		PASP75,256 # 40		13 03 23	+37 42.5	1	12.31		0.43		-0.07		1		619
+39 02598				13 03 23	+39 11.8	1	10.48		1.54		2.01		1	K3 III	280
		POSS 219 # 8		13 03 23	+39 40.9	1	18.25		0.98				4		1739
		CS 16027 # 12		13 03 24	+31 04.4	1	14.52		0.47		-0.17		1		1744
113797 +36 02337		HR 4943		13 03 24	+36 04.0	6	5.22	.037	-0.08	.008	-0.21	.004	12	B9 V	1022,1068,1079,1363*
113688 −58 04683				13 03 24	−59 19.7	1	9.24		0.16		-0.70		2	B2/3 Ib/II	122
113723 −51 07262				13 03 25	−52 17.2	1	8.90		0.26		0.23		1	A4mA5-F0	1771
113689 −63 02535		IDS13002S6402	AB	13 03 25	−64 17.6	2	9.09	.077	0.42	.008	0.13	.006	2	A9 V	1688,1771
		CS 16027 # 6		13 03 26	+31 53.5	1	13.72		0.63		0.12		1		1744
−66 02071				13 03 26	−67 20.1	1	11.21		0.21		-0.42		4		1684
		CS 16027 # 10		13 03 27	+31 12.7	1	14.11		0.76		0.28		1		1744
		PASP75,256 # 39		13 03 28	+37 42.3	1	11.49		0.71		0.26		1		619
−64 02161				13 03 28	−64 47.9	1	8.06		0.01		-0.88		2	O9 IV	540
		CS 16027 # 13		13 03 29	+31 02.1	1	14.69		0.51		-0.03		1		1744
113811 +40 02621				13 03 29	+39 52.3	1	7.31		1.51				2	K5 III	280
113827 +50 01979				13 03 29	+49 44.2	1	9.30		1.17		1.04		1	K0	3072
		AJ82,345 # 221		13 03 29	−62 35.8	1	12.73		1.53		0.44		2		779
				13 03 29	−64 41.9	1	9.15		0.07		-0.12		1	B8/9 V	1688
113690 −64 02162		WEIS 25329		13 03 30	+29 30.	1	16.96		1.53				1		784
		WEIS 23888		13 03 30	+30 04.	1	17.46		1.48				1		784
		WEIS 23962		13 03 30	+30 17.	1	15.84		1.52				1		784
113784 +16 02460				13 03 31	+16 07.0	1	9.56		1.02		0.84		1	G5	280
+25 02594				13 03 32	+24 38.5	1	10.00		0.48		-0.14		1		3025
+29 02363				13 03 33	+29 12.3	2	9.91	.009	0.69	.005	0.21	.009	3	G4 V	97,1019
113708 −64 02163		LSS 2921		13 03 34	−64 56.8	3	8.16	.008	0.00	.056	-0.84	.032	8	B1/2 IIIn	122,1688,2012
		LP 322 - 809		13 03 36	+26 38.	1	14.77		1.58				1		686
113751 −53 05464				13 03 36	−54 01.5	1	7.94		0.27		0.09		1	A5/7 V	1771
		WEIS 23961		13 03 37	+30 17.4	1	11.03		0.61				1		1692
113847 +46 01847		HR 4945	⋆ AB	13 03 37	+45 32.1	1	5.63		1.13				2	K1 III	70
113829 +39 02599				13 03 38	+38 40.2	1	8.43		1.06		0.90		1	K0 III	280

HD	DM	Other Id	N Rem	α_{1950}	δ_{1950}	S	V	σ_V	B–V	σ_{B-V}	U–B	σ_{U-B}	n	Spectrum	References
	+29 02364			13 03 39	+28 32.0	2	9.51	.005	0.63	.002	0.13	.004	3	G4 V	1019,1744
		CS 16027 # 15		13 03 39	+30 31.6	1	14.06		0.53		-0.05		1		1744
		PASP75,256 # 41		13 03 39	+37 53.3	1	12.25		0.93		0.64		1		619
113862	+54 01570			13 03 40	+53 42.6	1	8.93		0.41		-0.04		2	F5	1566
	+25 02595			13 03 41	+25 27.0	1	9.80		0.52		0.03		2	G0	1569
113767	-49 07643			13 03 41	-49 25.1	2	8.48	.023	0.82	.014	0.38		5	G8 V	2012,3008
	-61 03452	LSS 2923		13 03 41	-61 31.0	4	9.27	.016	0.77	.012	-0.27	.025	7	O9.5II	191,540,976,1684
113726	-61 03451			13 03 41	-61 50.6	2	8.72	.023	0.47	.019	0.01	.022	3	F5 V	122,1688
		Sand 212		13 03 42	+29 35.	1	15.92		1.57				1		784
		WEIS 23878		13 03 42	+30 04.	1	17.21		1.42				1		784
		WEIS 23965		13 03 42	+30 20.	1	16.53		1.41				1		784
		PASP75,256 # 42		13 03 42	+37 57.0	1	13.31		0.57		0.00		1		619
113740	-58 04688			13 03 42	-59 10.6	2	7.92	.000	0.28	.005	0.05	.005	3	A7/8w	122,1771
113801	-19 03634			13 03 44	-19 47.5	2	8.50	.033	1.18	.019	1.05		2	K0 III	1238,8005
113778	-40 07682	HR 4941		13 03 45	-41 19.3	1	5.59		1.05				4	K0 II/III	2006
113742	-61 03454			13 03 45	-61 40.6	1	9.17		0.36		-0.50		1	B1/2 III	1688
		CS 16027 # 17		13 03 46	+30 26.7	1	14.51		0.71		0.14		1		1744
		CS 16027 # 2		13 03 46	+32 53.9	1	14.19		0.64		0.15		1		1744
113864	+42 02372			13 03 46	+41 48.6	1	8.71		0.22		0.07		2	A8 V	1068
113865	+29 02365	HR 4948	★ ABC	13 03 47	+29 17.8	6	6.54	.012	0.07	.018	0.08	.024	10	A3 IV	97,1019,1022,1068*
113753	-59 04735			13 03 47	-60 00.7	1	8.77		0.22				4	B5 III	2014
		BPM 88744		13 03 48	+28 51.	1	14.29		1.60				1		784
		WEIS 25301		13 03 48	+29 43.	1	17.04		1.50				1		784
	-62 03006	LSS 2926		13 03 48	-62 33.1	1	10.27		0.63		-0.38		2		779
	+21 02486	IDS13014N2116	AB	13 03 49	+20 59.8	1	9.43		1.29		1.22		1	K4	3072
		CS 15622 # 36		13 03 50	+28 20.4	1	11.97		0.96		0.80		1		1744
113755	-62 03010	IDS13008S6304	AB	13 03 50	-63 20.1	1	9.80		0.54		0.35		1	A3/8 III/V	1688
		CS 16027 # 14		13 03 52	+30 55.6	1	14.65		0.67		0.14		1		1744
		CS 16027 # 1		13 03 52	+32 58.4	1	12.81		0.34		-0.01		1		1744
		WEIS 25327		13 03 54	+29 34.	1	17.46		1.70				1		784
		WEIS 21909		13 03 54	+30 59.	1	13.83		1.44				1		784
		CS 16027 # 5		13 03 54	+31 56.6	1	14.91		0.46		-0.14		1		1744
		PASP75,256 # 45		13 03 54	+37 51.8	1	14.02		0.85		0.57		1		619
113754	-62 03007	LSS 2927		13 03 54	-62 43.4	4	9.48	.019	0.62	.014	-0.46	.015	7	O6/7	540,779,976,1684
113848	+21 02487	HR 4946	★ AB	13 03 55	+21 25.3	2	6.01	.030	0.38	.010	-0.02		6	F4 V	254,1118
	+30 02365			13 03 56	+29 41.4	3	11.45	.015	0.47	.009	-0.05	.003	5	G5	97,1019,1692
113892	+41 02355	IDS13016N4128	A	13 03 56	+41 11.4	1	7.25		1.65				2	M1 III	280
113891	+47 02017			13 03 56	+47 12.9	1	9.80		0.28		0.03		2	G5	1569
	-31 10079			13 03 56	-31 59.2	1	10.99		0.75		0.23		2		1696
113781	-56 05583			13 03 56	-56 44.0	2	8.47	.007	-0.05	.007	-0.75	.004	8	B2 III	401,1732
113867	+23 02537			13 03 57	+22 32.8	1	6.82		0.29				9	F0	1118
113866	+23 02538	HR 4949, FS Com		13 03 57	+22 53.0	5	5.60	.009	1.58	.008	1.63	.004	17	M5 III	15,3042,8011,8015,8022
		CS 16027 # 21		13 03 58	+29 01.5	1	13.94		0.66		0.14		1		1744
113791	-49 07644	HR 4942	★ A	13 03 59	-49 38.3	7	4.27	.011	-0.19	.004	-0.78	.019	28	B3 V	15,26,1075,1770,2012*
113791	-49 07644	IDS13011S4922	B	13 03 59	-49 38.3	1	9.46		0.51		0.02		5		321
		WEIS 25299		13 04 00	+29 44.	1	17.27		1.72				1		784
	-61 03456			13 04 00	-61 47.0	1	11.08		0.40		-0.29		2	B3	122
	+29 02366			13 04 01	+28 33.0	2	11.76	.010	0.28	.003	0.04	.005	3		1019,1744
	-59 04737			13 04 01	-59 34.4	1	10.18		0.66		0.32		2		122
	+38 02402			13 04 04	+37 31.5	1	9.64		0.46		-0.02		1	F8	619
		CS 16027 # 88		13 04 05	+28 35.3	1	12.68		0.72		0.09		1		1744
	+39 02601			13 04 05	+38 48.6	1	10.06		1.54		1.84		1	K5	280
113693	-76 00747			13 04 05	-77 02.5	1	8.88		1.03		0.85		1	K3 V	3072
		BM Cha	★	13 04 05	-77 39.1	1	14.89		1.40		0.35		1		5005
		WEIS 23830		13 04 06	+30 18.	1	14.84		1.42				1		784
113893	+11 02538	G 61 - 30		13 04 07	+11 18.8	1	8.52		0.60		0.09		1	G0	333,1620
113880	+15 02544			13 04 07	+15 17.7	1	10.11		1.02		0.92		1	K0	1569
	+28 02182			13 04 07	+28 20.5	2	9.36	.000	0.42	.002	-0.02	.001	4	F0 IV	1019,1690
		CS 16027 # 11		13 04 07	+31 02.7	1	14.70		0.39		-0.20		1		1744
	+47 02018			13 04 07	+46 47.3	1	9.30		0.97		0.84		1	K0	280
113852	-35 08441	HR 4947		13 04 07	-35 35.6	3	5.64	.007	0.03	.000			11	A0 V	15,2013,2029
113795	-63 02540			13 04 07	-63 29.8	1	9.76		0.31		0.14		1	A1 V	1688
		AJ82,345 # 225		13 04 08	-61 52.4	1	13.53		0.79		-0.03		1		779
		CS 16027 # 4		13 04 09	+32 19.4	1	14.55		0.51		-0.07		1		1744
		CS 16027 # 7		13 04 10	+31 49.7	1	14.05		0.68		0.20		1		1744
		GD 485		13 04 10	+70 51.4	1	13.47		0.35		-0.15		3		308
	+33 02301			13 04 11	+33 18.1	1	9.91		1.47		1.82		1		280
	+29 02367			13 04 12	+29 07.5	1	10.84		0.61		0.11		3	G0 V	1019
113921	+34 02389			13 04 15	+34 17.2	1	8.15		1.14		1.20		1	K0 III	280
		AJ82,345 # 227		13 04 16	-62 49.2	1	12.37		1.31		0.36		2		779
		HA 57 # 115		13 04 17	+29 31.4	2	11.28	.005	0.51	.015	-0.04		2		97,1692
113841	-55 05381			13 04 17	-55 32.3	1	9.67		0.35		0.17		2	A9 V	1771
113823	-59 04740	HR 4944	★ AB	13 04 18	-59 35.6	5	5.99	.008	0.48	.007	0.11	.010	15	B9 IV+F8/G2	15,122,401,1637,2012
	-59 04741			13 04 18	-59 45.1	1	9.85		1.26		1.02		2		122
	-59 04742			13 04 19	-59 47.8	1	10.96		0.26		-0.12		2		122
		HA 57 # 169		13 04 22	+29 40.0	1	12.81		0.64		0.14		2		97
		WEIS 23870		13 04 23	+30 02.6	1	11.51		0.97				1		1692
	+30 02366			13 04 23	+30 02.6	1	11.50		0.94		0.81		1	K0	1019
	-61 03462	LSS 2931		13 04 23	-62 06.3	5	9.18	.015	0.55	.013	-0.40	.026	13	B0.5II	540,779,976,1684,1688
113807	-65 02156			13 04 23	-65 36.7	1	7.58		0.10		0.11		1	A1 V	1771

Table 1　　　　　　　　　　　　　　　　　　　　　　　　　　　　　　　　　　　　　723

HD	DM	Other Id	N	Rem	α₁₉₅₀	δ₁₉₅₀	S	V	σ_V	B–V	σ_B–V	U–B	σ_U–B	n	Spectrum	References
	+34 02390				13 04 24	+34 01.5	1	10.37		0.97		0.84		1	G5 IV	280
113994	+62 01275	HR 4953			13 04 25	+62 18.6	2	6.13	.005	0.98	.010	0.65		7	G7 III	252,1379
113958	+28 02184	IDS13020N2844	A		13 04 26	+28 28.5	2	8.99	.000	0.53	.001	0.10	.000	8	F7 V	1019,1690
	+29 02368	IDS13021N2908	A		13 04 26	+28 52.5	3	10.82	.035	0.56	.017	0.03	.010	4		1019,1019,1692
		HA 57 # 171			13 04 26	+29 39.6	1	14.18		0.60		0.08		1		1744
		CS 16027 # 20			13 04 28	+29 19.8	1	14.95		0.44		0.04		1		1744
113873	−52 06187				13 04 28	−52 29.0	1	7.68		0.19		0.19		1	A6 V	1771
	−61 03463				13 04 28	−62 08.8	1	10.30		0.35		−0.13		2	B6	122
	+9 02718				13 04 30	+08 34.3	1	9.71		1.23		1.44		1	K5	280
		WEIS 23825			13 04 30	+30 18.	1	16.33		1.51				1		784
		WEIS 52002			13 04 31	+28 14.2	1	12.76		0.85				1		1692
113888	−52 06189				13 04 31	−53 24.0	2	8.63	.001	0.01	.001	−0.21	.011	5	B9 III	401,1770
113859	−59 04748				13 04 32	−59 32.8	1	9.70		0.22		0.00		2	A0 IV/V	122
113860	−63 02543				13 04 32	−63 44.7	1	9.07		0.47		0.16		1	F0 V	1688
	−63 02544	LSS 2932			13 04 34	−63 56.1	3	9.98	.019	0.47	.016	−0.44	.040	5	B0.5Iab	191,540,976
		LP 378 - 541			13 04 36	+21 05.	2	12.60	.032	1.64	.044	1.19		6		686,7009
		G 149 - 40			13 04 36	+28 07.2	1	13.84		1.65				1		784
		WEIS 25225			13 04 36	+29 34.	1	16.80		1.44				1		784
	+31 02450				13 04 37	+31 22.4	1	9.68		1.60		1.88		1	M3 III	1019
113902	−52 06194	HR 4951			13 04 39	−53 11.6	3	5.70	.004	−0.07	.001	−0.27		10	B8 V	15,1770,2012
		Steph 1044			13 04 42	−22 12.1	1	11.13		1.56		1.90		2	M0	1746
		WEIS 52724			13 04 43	+27 49.5	1	12.82		1.05				1		1692
	+12 02555				13 04 44	+11 32.7	1	8.89		1.45		1.79		1	K5	280
		Comp to F74			13 04 44	+19 52.6	1	11.98		0.91		0.47		3		308
		POSS 219 # 6			13 04 44	+39 32.1	1	17.88		0.64				4		1739
113984	+1 02789	IDS13022N0107	AB		13 04 46	+00 51.2	2	7.06	.009	0.48	.007	−0.05	.002	4	F5 V	1003,3016
		WEIS 51212			13 04 46	+28 43.8	1	11.20		1.33				2		1692
113914	−49 07653				13 04 46	−49 27.5	1	7.81		0.08		0.14		1	A1/2 IV/V	1771
113996	+28 02185	HR 4954			13 04 47	+27 53.6	8	4.80	.020	1.48	.015	1.85	.015	38	K5 III	15,1019,1080,1363*
113995	+29 02369				13 04 47	+28 58.8	1	8.40		1.37		1.54		3	K4 III	1019
		CS 16027 # 18			13 04 47	+30 02.5	1	15.00		0.38		−0.17		1		1744
113931	−46 08415				13 04 47	−47 01.4	1	9.43		0.04		0.03		1	A0 V	1770
	+0 03022				13 04 48	+00 22.2	1	11.40		0.01		0.01		2	A0	1026
		WEIS 24391			13 04 48	+29 52.	1	17.36		1.48				1		784
	+28 02186				13 04 49	+27 35.0	1	10.56		0.66		0.23		2	G4 V	1019
		He3 863			13 04 49	−47 44.4	1	12.55		1.57		0.28		1		1753
113997	+17 02589				13 04 51	+16 29.4	1	9.25		1.09		1.00		1	K0	280
	+31 02451	WEIS 21939			13 04 52	+31 03.4	2	10.13	.009	0.67	.003	0.18		2	G2 IV:	1019,1692
		CS 16027 # 8			13 04 52	+31 46.9	1	14.52		0.62		0.19		1		1744
113904	−64 02183	HR 4952, θ Mus	⋆ AB		13 04 52	−65 02.4	6	5.50	.017	−0.02	.009	−0.86	.025	19	WC +O9.5 Ia	15,404,1034,1075,2038,8100
	+29 02370	IDS13025N2917	A		13 04 54	+29 01.2	2	10.68		0.77		0.34		2	G3 V	1019
		CS 16027 # 3			13 04 54	+32 33.4	1	13.79		0.45		−0.20		1		1744
113949	−51 07282				13 04 54	−52 15.4	1	8.78		0.02		−0.07		1	B9 V	1770
		GD 268			13 04 56	+15 00.5	1	17.06		−0.05		−0.99		1		3060
		G 14 - 30			13 04 56	−04 09.4	1	11.20		1.24		1.22		1	K5	1658
		G 164 - 44			13 04 57	+41 33.8	1	13.92		1.54				1		3062
113935	−59 04756				13 04 57	−59 24.2	1	9.49		0.20		−0.26		2	B8 Ib/II	122
		LSS 2936			13 04 57	−63 15.6	1	11.85		0.73		−0.22		4	B3 Ib	1684
113918	−64 02186				13 04 58	−65 16.4	2	8.67	.001	0.36	.002	0.11	.001	20	A8 V	1699,1771
	−65 02163				13 04 58	−65 44.7	1	9.84		0.17		−0.76		4		1684
113953	−55 05389				13 04 59	−56 07.1	2	7.30	.005	−0.03	.000	−0.34	.004	5	B8 IV	401,1770
113954	−56 05593	IDS13020S5609	AB		13 04 59	−56 25.0	2	7.36	.112	0.11	.005	0.02	.005	3	A1 V	401,1771
		Comp to F74			13 05 00	+19 52.9	1	12.75		0.92		0.59		2		272
		Feige 74			13 05 00	+19 52.9	1	12.04		−0.08		−0.34		3		272
113974	−39 08062				13 05 00	−39 46.6	1	8.17		1.33		1.48		1	K3 III	1731
114037	+27 02217	IDS13028N2701	A		13 05 01	+26 47.4	2	7.60	.005	1.09	.000	0.99	.024	3	K1 III	280,3077
114016	+13 02634	IDS13026N1250	A		13 05 05	+12 32.4	1	9.14		0.90		0.55		1	G5	280
	+31 02452	WEIS 21967			13 05 06	+30 55.6	2	10.40	.009	0.57	.003	0.09		2	F8	1019,1692
		HA 57 # 175			13 05 07	+29 36.6	1	13.03		0.95		0.71		3		97
114058	+47 02020				13 05 08	+46 43.0	2	9.65	.014	0.23	.005	0.14		5	A7 Vm	193,280
		LOD 28 # 11			13 05 08	−62 07.6	1	12.04		0.37		−0.03		2		122
113955	−63 02556	IDS13019S6333	AB		13 05 09	−63 49.1	1	10.08		0.16		−0.04		1	B9 IV	1688
		AAS12,381 10 # B			13 05 10	+10 20.3	1	10.64		0.92		0.70		2		280
	+12 02557				13 05 11	+12 21.0	1	9.47		1.06		0.88		1	K0	280
	−60 04460				13 05 11	−60 58.0	1	10.50		0.49		−0.51		6		1684
		POSS 219 # 2			13 05 12	+39 43.2	1	15.85		1.41				4		1739
	−61 03469	LSS 2937			13 05 12	−61 54.7	3	9.85	.013	0.31	.018	−0.58	.041	5	B5	122,779,1688
114060	+24 02539	G 149 - 41	⋆ A		13 05 15	+24 16.5	2	8.16	.013	0.70	.004	0.18	.013	6	G5 V	1003,3032
	+35 02406				13 05 15	+34 40.1	1	9.34		1.18		1.10		2	K3 III	3072
114060	+24 02540	G 149 - 42	⋆ B		13 05 16	+24 17.1	2	8.54	.004	0.78	.004	0.31	.009	6		1003,3032
114059	+30 02367	IDS13029N2959	AB		13 05 16	+29 42.9	3	9.21	.010	0.91	.005	0.60	.005	8	G8 V	97,1019,1690
114038	−9 03628	HR 4955			13 05 16	−10 28.4	7	5.19	.017	1.14	.004	1.12	.000	20	K2 III	15,1007,1013,2006*
113979	−62 03020				13 05 17	−63 14.1	1	10.08		0.59		0.37		1	A2	1688
113968	−60 04461				13 05 18	−60 58.6	1	8.80		0.02		−0.50		1	B2/3 IV	1688
	+38 02405				13 05 19	+38 09.8	1	10.09		0.76		0.32		1	G8 III	619
	+47 02021				13 05 19	+46 55.1	1	9.82		1.15		1.12		1	K2	280
114071	+30 02368				13 05 20	+29 44.3	3	9.62	.000	0.48	.005	−0.04	.030	7	F7 V	97,1019,1690
	−60 04463	IDS13022S6040	AB		13 05 20	−60 56.1	1	8.82		0.62		−0.39		4		1684
		AJ82,345 # 238			13 05 21	−62 41.8	1	13.79		1.03		0.12		2		779
	+39 02605				13 05 23	+39 24.7	1	10.02		1.19		1.17		1	K1 III	280

HD	DM	Other Id	N Rem	α_{1950}	δ_{1950}	S	V	σ_V	B–V	σ_{B-V}	U–B	σ_{U-B}	n	Spectrum	References
114023	−49 07659			13 05 23	−50 11.6	1	8.65		0.25		0.19		1	A5 V	1771
113991	−60 04465			13 05 23	−60 40.0	1	9.39		0.25		−0.54		1	B3 III	1688
		WEIS 29436		13 05 24	+29 53.	1	17.47		1.51				1		784
	+37 02365			13 05 26	+37 20.9	1	9.66		1.14		1.16		1	K2 III	280
114072	+20 02796			13 05 28	+20 12.0	2	7.88	.005	0.41	.005	0.02	.000	4	F4 V	1569,1648
	+24 02541			13 05 28	+23 53.8	1	10.43		1.50		0.94		1		280
	+37 02366			13 05 29	+36 31.5	1	10.38		1.06		0.91		1	K0 V	280
114012	−62 03022			13 05 29	−62 56.7	1	9.10		0.50		0.36		1	A0 V	1688
114092	+28 02187	HR 4956	⋆	13 05 30	+27 49.4	5	6.19	.007	1.36	.006	1.63	.013	24	K4 III	252,1019,1690,1692,3077
		HA 57 # 122		13 05 30	+29 32.1	1	14.93		0.80		0.30		2		97
114084	+33 02304			13 05 31	+33 14.4	2	9.90	.004	0.15	.003	0.14		6	A2 III	193,280
	+70 00727			13 05 31	+70 22.4	1	9.17		0.48		0.02		2	F2	1733
113992	−64 02196			13 05 31	−64 31.1	1	9.32		0.03		−0.12		1	B9 III	1688
		HA 57 # 25		13 05 33	+29 11.0	1	10.93		0.53		0.06		3		97
	+39 02606			13 05 33	+39 24.1	1	9.10		0.90		0.56		1	G8 V	280
		AAS12,381 40 # C		13 05 34	+39 52.5	1	10.52		1.00		1.00		1		280
114024	−59 04766	LSS 2941		13 05 34	−59 24.0	5	9.78	.021	0.26	.014	−0.67	.023	13	O9.5	122,403,540,1684,2021
114011	−60 04466	LSS 2940		13 05 34	−60 55.2	4	9.25	.026	0.76	.005	−0.16	.040	13	B3/5 Ib/II	191,540,976,8100
		vdB 59 # a		13 05 35	−62 00.3	2	12.55	.020	0.82	.000	−0.05		5		122,434
		WEIS 23712		13 05 36	+30 19.	1	16.86		1.51				1		784
	+32 02320			13 05 36	+32 00.9	1	10.46		1.00		0.75		1	G8 III	280
114124	+43 02300			13 05 36	+42 38.4	1	7.98		1.25		1.42		1	K0 III	280
		LOD 28 # 10		13 05 36	−62 03.9	1	12.00		0.59				2		122
114108	+37 02367			13 05 37	+37 14.5	2	8.69	.005	1.09	.005	0.97	.010	3	K1 III	619,1569
114042	−53 05486			13 05 37	−54 20.8	1	7.83		0.22		0.22		1	A2 IV	1771
114026	−59 04769	LSS 2942		13 05 37	−60 04.3	3	8.26	.016	0.08	.006	−0.77	.021	6	B1/2 IV/V	401,540,976
114093	+25 02599			13 05 38	+25 05.8	1	6.83		0.91				3	G8 III	280
		G 14 - 31		13 05 39	+01 36.1	1	13.00		0.97		0.73		1		1658
		POSS 323 # 3		13 05 39	+29 38.3	1	13.09		0.60				1		1739
		POSS 323 # 5		13 05 39	+29 41.5	1	14.72		0.57				2		1739
114051	−39 08066			13 05 39	−39 38.9	1	8.60		0.50		0.04		3	F5 V	1731
	+29 02371			13 05 40	+28 45.4	1	11.88		0.52		0.01		1		1019
		CS 16027 # 27		13 05 40	+29 06.5	1	14.61		0.34		−0.12		1		1744
		HA 57 # 124		13 05 40	+29 32.2	1	14.36		0.65		0.09		2		97
		HA 57 # 182		13 05 40	+29 38.2	1	13.08		0.62		0.01		2		97
113283	−86 00283			13 05 40	−87 17.6	1	7.12		0.71				4	G3/5 V	2012
		POSS 323 # 2		13 05 41	+29 35.6	1	12.90		0.53				2		1739
		HA 57 # 183		13 05 41	+29 35.7	1	12.93		0.51		−0.06		1		97
114034	−62 03023			13 05 41	−63 04.2	1	10.20		0.21		−0.17		1	B9 II/III	1688
114094	+4 02696	G 60 - 66		13 05 42	+04 02.6	6	9.67	.011	0.69	.012	0.20	.013	29		333,830,1003,1620,1658*
	+28 02188			13 05 44	+28 29.1	1	10.46		0.44		−0.02		2	F6 IV	1019
		MtW 57 # 14		13 05 44	+29 43.8	1	16.61		0.66		0.11		2		97
114066	−39 08067			13 05 44	−39 53.9	1	8.28		1.04		0.88		2	G8/K0 III	1731
	+29 02372			13 05 45	+29 03.3	1	9.95		1.15		1.16		2	K0 III	1019
	−63 02565	LSS 2943		13 05 45	−64 16.2	2	10.16	.013	0.11	.008	−0.69	.013	3	B2 II	540,976
		HA 57 # 184		13 05 46	+29 38.6	2	15.75	.015	0.63	.005	0.05		4		97,1739
		HA 57 # 226		13 05 46	+29 52.9	1	12.30		0.97		0.75		2		97
114076	−40 07705			13 05 46	−41 22.6	1	9.39		0.82				4	K0 V	2033
		POSS 323 # 12		13 05 47	+29 35.9	1	16.94		0.55				3		1739
		AAS35,161 # 62	V	13 05 47	−77 41.4	1	15.29		1.47		−0.39		1		5005
	+27 02220			13 05 48	+27 26.4	1	10.28		1.27		1.50		1		280
	+37 02368			13 05 48	+37 18.0	1	10.44		0.46		−0.03		1		619
		L 545 - 61		13 05 48	−33 50.	1	13.53		0.58		−0.08		3		3060
	−61 03472	LSS 2944		13 05 48	−61 50.9	2	9.89	.015	0.33	.010	−0.54	.060	4	B1	122,779
	+20 02798			13 05 49	+20 23.1	2	10.31	.015	0.30	.011	−0.02		6	F0 V	193,1569
		HA 57 # 125		13 05 49	+29 26.5	1	12.47		0.57		0.07		2		97
		POSS 323 # 14		13 05 49	+29 41.2	1	18.96		0.73				4		1739
114095	−6 03742	G 14 - 32		13 05 50	−07 02.6	3	8.35	.004	0.96	.020	0.56	.000	8	G5	516,1620,2033
		LOD 28 # 5		13 05 51	−61 56.0	1	12.79		0.52				2		122
		Z Com		13 05 52	+18 49.0	2	13.31	.168	0.12	.030	0.14		2		597,699
		AJ82,345 # 242		13 05 52	−62 11.5	1	11.96		1.40		0.25		2		779
		WEIS 25046		13 05 54	+29 32.	1	16.76		1.56				1		784
	+32 02321			13 05 54	+32 18.8	2	10.05	.015	0.26	.000	0.05	.024	3	F0	280,1569
	+10 02513			13 05 55	+09 53.8	1	8.75		1.51		1.80		1	K5	280
		G 149 - 43		13 05 55	+14 27.7	1	13.50		0.92		0.56		1		1658
		HA 57 # 186		13 05 55	+29 43.7	2	15.03	.005	0.71	.024	0.19		6		97,1739
		AJ82,345 # 243		13 05 55	−62 11.4	1	11.58		1.66		0.49		2		779
114044	−64 02199			13 05 55	−64 38.8	1	9.63		0.04		−0.44		1	B5 V	1688
114146	+39 02607	IDS13036N3917	AB	13 05 56	+39 00.4	1	8.39		0.71				2	G5 p	280
114113	−8 03491	HR 4957		13 05 56	−08 43.0	2	5.54	.005	1.18	.000	1.18	.000	7	K3 III	15,1417
		HA 57 # 229		13 05 57	+29 48.2	1	12.95		0.71		0.35		2		97
114131	+16 02466	IDS13035N1602	AB	13 05 58	+15 45.5	2	7.83	.023	0.94	.005	0.74		4	K1 IV	280,882
	+10 02514			13 05 59	+09 42.1	1	9.94		1.24		1.33		1	K5	280
	+30 02369	POSS 323 # 1		13 05 59	+29 45.1	3	11.64	.005	0.84	.013	0.46	.006	8	G0	97,1019,1739
		CS 16027 # 31		13 05 60	+30 45.0	1	14.65		0.65		0.07		1		1744
		WEIS 25041		13 06 00	+29 36.	1	17.50		1.63				1		784
		WEIS 25016		13 06 00	+29 43.	1	17.50		1.47				1		784
		HA 57 # 231		13 06 00	+29 52.7	1	12.38		0.46		−0.02		2		97
114159	+43 02301			13 06 00	+43 26.8	1	7.64		1.58		1.87		1	M1	280
114171	+54 01573			13 06 00	+53 49.6	1	8.26		0.34		−0.05		3	F5	1566

Table 1   725

HD	DM	Other Id	N	Rem	α_{1950}	δ_{1950}	S	V	σ_V	B–V	σ_{B-V}	U–B	σ_{U-B}	n	Spectrum	References
		LP 322 - 706			13 06 01	+30 36.	1	13.88		1.51				1		784
		CS 16027 # 33			13 06 01	+32 41.8	1	14.07		0.32		-0.04		1		1744
		AJ82,345 # 244			13 06 01	−61 53.0	1	13.37		1.02		-0.06		1		779
		LSS 2945			13 06 02	−62 40.4	1	11.72		0.52		-0.41		2	B3 Ib	779
	+24 02544				13 06 03	+24 05.7	1	10.46		0.65		0.12		1		280
		WEIS 50807			13 06 03	+29 01.8	1	11.27		0.45				1		1692
	+40 02626				13 06 03	+39 59.9	1	9.71		0.98		0.73		1	G8 V	280
	−58 04709	AAS78,203 # 28			13 06 03	−58 34.5	1	10.56		0.18		-0.44		2		1684
		HA 57 # 187			13 06 04	+29 36.6	1	14.93		0.98		0.70		2		97
		POSS 323 # 8			13 06 04	+29 36.6	1	14.99		0.91				2		1739
		HA 57 # 232			13 06 04	+29 44.2	2	14.93	.005	0.47	.010	-0.22		3		97,1739
		HA 57 # 129			13 06 05	+29 26.5	1	12.43		0.58		-0.01		2		97
		HA 57 # 188			13 06 05	+29 34.6	1	14.06		0.59		0.01		4		97
		CS 16027 # 32			13 06 05	+30 47.3	1	14.24		0.66		0.18		1		1744
		WEIS 23737			13 06 06	+30 15.	1	15.55		1.48				1		784
114102	−48 07952				13 06 09	−49 13.7	1	7.84		0.07		0.10		1	A1 V	1771
		LOD 28 # 8			13 06 09	−61 59.8	1	12.51		0.40				2		122
		CS 16027 # 34			13 06 10	+32 59.6	1	14.03		0.12		0.16		1		1744
		G 62 - 15			13 06 11	+08 20.4	1	15.09		1.68				1		906
		HA 57 # 233			13 06 11	+29 48.6	1	13.49		0.77		0.37		2		97
	+30 02370			A	13 06 12	+29 36.2	2	12.28	.012	1.25	.019	1.16	.013	4		97,1019
	+30 02370			B	13 06 12	+29 36.2	1	14.00		0.64		0.16		1		1019
		G 164 - 46			13 06 13	+40 43.2	1	14.00		0.80		0.22		2		3060
		HA 57 # 191			13 06 14	+29 35.5	1	13.94		0.68		0.14		2		97
		G 14 - 33			13 06 14	−03 42.4	2	11.17	.010	0.66	.005	0.01	.039	3		1620,1658
114172	+30 02371				13 06 15	+29 39.0	4	8.56	.008	0.61	.007	0.00	.027	32	G0 V	97,1019,1690,3026
		HA 57 # 193			13 06 15	+29 42.1	2	13.99	.000	0.83	.000	0.44		5		97,1739
	+31 02453	WEIS 22366			13 06 15	+30 44.6	2	10.53	.000	0.61	.002	0.16		2	G3 V	1019 ,1692
		WEIS 21066			13 06 15	+31 22.0	1	11.52		0.54				1		1692
		HA 57 # 130			13 06 16	+29 30.8	1	14.13		0.67		0.14		3		97
		MtW 57 # 60			13 06 16	+29 33.9	1	16.95		0.63		-0.10		1		97
	−61 03474				13 06 16	−61 57.0	1	10.47		0.99		0.38		2		122
		HA 57 # 131			13 06 17	+29 32.5	1	15.91		0.77		0.21		2		97
114190	+23 02541				13 06 18	+23 14.3	1	8.85		1.00		0.79		1	K0	280
		G 14 - 34			13 06 18	−01 15.2	1	12.75		1.51				2		202
114174	+6 02697	G 60 - 67			13 06 19	+05 29.0	4	6.80	.004	0.66	.007	0.19	.033	8	G5 IV	22,333,1003,1620,3026
	+28 02189				13 06 20	+27 52.4	1	8.87		0.99		0.80		2	K0 III-IV	1019
114136	−39 08076				13 06 20	−39 36.4	1	8.06		0.48				4	F5 V	2012
	−64 02205				13 06 20	−64 47.6	1	10.82		0.12		-0.25		1	B8	549
		WEIS 22087			13 06 21	+30 57.2	1	10.19		1.06				2		1692
		CS 16027 # 35			13 06 21	+33 08.1	1	11.69		0.01		0.04		1		1744
114149	−22 03515	HR 4958			13 06 21	−22 51.1	3	4.95	.014	1.05	.000	0.94	.010	14	K0 III	244,2035,3016
114218	+37 02369				13 06 22	+37 29.4	1	8.68		1.07		1.00		2	K1 III	1569
114122	−62 03028	LSS 2946			13 06 22	−62 36.3	8	8.57	.006	0.58	.018	-0.45	.018	20	B0 III	92,403,540,779,976,1684*
	+31 02455				13 06 23	+30 57.1	2	10.17	.000	1.05	.006	0.85	.131	2	K2 III:	280,1019
		WEIS 24966			13 06 24	+29 38.	1	15.85		1.51				1		784
		HA 57 # 194			13 06 24	+29 42.2	2	14.98	.005	0.43	.029	-0.22		3		97,1739
		WEIS 24550			13 06 24	+29 49.	1	16.77		1.40				1		784
		POSS 323 # 13			13 06 25	+29 39.9	1	17.47		0.92				2		1739
	+14 02580				13 06 26	+14 28.2	1	11.34		0.29		0.05		1		280
	−64 02207				13 06 26	−64 56.2	1	10.97		0.20		0.12		1	A0	549
		L 472 - 66			13 06 27	−39 52.3	1	12.86		1.60		1.10		1		3078
		HA 57 # 134			13 06 28	+29 30.7	1	13.19		0.78		0.31		3		97
		MtW 57 # 74			13 06 28	+29 35.5	2	16.42	.015	0.72	.019	0.20		3		97,1739
		MtW 57 # 78			13 06 29	+29 38.2	1	15.86		1.54		1.17		1		97
		LP 854 - 65			13 06 30	−24 55.7	1	11.26		0.86		0.51		3		1696
	−64 02209				13 06 30	−64 47.0	1	9.86		1.15		0.87		1	K2	549
	−64 02208				13 06 30	−65 05.2	1	10.98		0.56		0.23		1	F2	549
	+35 02410				13 06 31	+35 22.8	1	10.08		1.08		1.03		2	K0	1569
	−64 02210	LOD 757 # 13			13 06 32	−65 06.7	1	10.28		1.25		1.01		1	K3	549
114220	+13 02637				13 06 35	+12 55.3	1	10.16		0.89		0.55		1	K0	280
		HA 57 # 135			13 06 35	+29 32.5	2	15.21	.000	0.35	.005	-0.19	.015	3		97,1744
		AAS12,381 34 # D			13 06 35	+34 13.4	1	11.18		0.26		0.08		1		280
		KUV 13066+2800			13 06 36	+28 00.3	1	16.74		0.59		-0.17		1		1708
		WEIS 24911			13 06 36	+29 27.	1	16.63		1.62				1		784
	+14 02581				13 06 37	+14 22.5	1	10.20		0.93		0.63		1	G5	280
114203	−8 03495	HR 4959			13 06 37	−09 16.3	2	6.31	.005	1.02	.000	0.80	.005	7	K0	15,1417
114141	−62 03029				13 06 38	−62 43.1	1	9.64		0.49		0.43		1	A5/7 V	1688
114241	+19 02642				13 06 39	+18 53.5	1	6.91		1.49				2	K2	280
		LP 268 - 35			13 06 39	+34 09.2	1	11.54		1.22		1.23		1		1773
114168	−56 05609				13 06 41	−56 30.1	2	8.59	.002	0.00	.004	-0.18	.003	5	B9 III/IV	401,1770
114256	+10 02516	HR 4960			13 06 42	+10 17.3	1	5.78		1.00				2	K0 III	70
	+17 02592				13 06 43	+17 17.6	1	9.48		1.03		0.89		1	K0	280
	−60 04480	LSS 2947			13 06 43	−60 51.9	2	9.56	.015	0.25	.019	-0.58		6	B1 V	403,2021
		CS 16027 # 28			13 06 44	+29 15.8	1	12.81		0.64		0.13		1		1744
114254	+31 02456				13 06 44	+30 30.6	2	8.26	.000	0.42	.001	-0.03	.001	6	F3 V	1019,1690
114283	+47 02023				13 06 44	+47 07.2	1	8.71		0.50		0.07		1	F8	1569
		Steph 1049			13 06 45	+49 20.8	1	11.04		1.54		1.19		2	M0	1746
		WEIS 50994			13 06 46	+28 56.5	1	11.33		0.53				2		1692
		L 545 - 73			13 06 46	−34 35.0	1	12.42		1.51				1		3062

HD	DM	Other Id	N	Rem	α_{1950}	δ_{1950}	S	V	σ_V	B–V	σ_{B-V}	U–B	σ_{U-B}	n	Spectrum	References
114157	−63 02577				13 06 46	−64 17.7	2	9.20	.005	0.26	.007	0.25	.008	2	A7/8 V	1688,1771
114284	+29 02373	IDS13044N2913		A	13 06 48	+28 56.7	1	9.08		0.71		0.33		1	G0 V	1019
		CS 16027 # 23			13 06 49	+28 16.6	1	13.83		0.77		0.41		1		1744
	+40 02627				13 06 53	+40 15.9	1	10.45		1.14		1.13		1	K0 IV	280
	−8 03497				13 06 53	−09 19.9	1	10.30		1.12		1.14		2	K5 V	3072
114169	−64 02212				13 06 54	−64 59.9	5	9.61	.011	-0.03	.002	-0.79	.019	11	B1/2 III	122,540,549,976,1688
	+28 02190				13 06 55	+28 05.8	1	10.83		0.63		0.13		2	G5 III	1019
		CS 16027 # 24			13 06 55	+28 38.0	1	14.54		0.55		0.05		1		1744
		CS 16027 # 25			13 06 57	+28 41.5	1	14.71		0.50		-0.18		1		1744
	+39 02609				13 06 57	+38 36.2	1	9.58		1.18		1.29		2	K2 III	280
		Wray 1069			13 06 57	−63 13.3	1	12.38		0.93		-0.24		4		1684
	−64 02215				13 06 58	−64 52.6	1	11.12		0.14		-0.14		1	B9	549
114260	−21 03660				13 07 00	−21 55.3	3	7.36	.000	0.73	.000	0.25	.010	7	G6 V	258,2012,3026
		L 147 - 173			13 07 00	−64 05.	1	14.80		0.84		0.42		3		3062
	+25 02602				13 07 01	+24 35.4	2	10.13	.029	0.04	.015	0.12	.058	6	A0	1026,1068
		WEIS 21415			13 07 01	+31 10.3	1	11.38		0.65				1		1692
		WEIS 21159			13 07 01	+31 21.8	1	10.40		1.46				1		1692
114187	−64 02216				13 07 01	−64 32.0	1	10.06		0.08		-0.41		1	B6/8 III	1688
114300	+18 02696				13 07 03	+17 45.0	2	8.26	.000	1.49	.007	1.84	.000	3	K2	280,3040
114199	−63 02579	LSS 2948			13 07 05	−64 11.4	3	9.47	.008	0.30	.011	-0.65	.012	5	B1 Ia	540,976,1684
	−64 02218				13 07 05	−65 09.2	1	10.84		0.20		0.09		1	B9	549
	+25 02603				13 07 06	+25 03.0	1	9.71		1.08		1.04		1	K0	280
	+28 02191	WEIS 52483			13 07 07	+27 58.8	3	10.42	.008	0.73	.006	0.34	.005	4	K0 IV	280,1019,1692
		CS 16027 # 26			13 07 07	+28 48.4	1	14.21		0.64		-0.07		1		1744
114311	+31 02458				13 07 07	+30 42.1	2	9.04	.000	0.50	.001	0.02	.002	6	F6 V	1019,1690
114287	−9 03636	HR 4961			13 07 08	−10 03.8	2	5.93	.005	1.49	.000	1.69	.000	7	K5 III	15,1417
		CS 16027 # 30			13 07 09	+30 07.4	1	15.00		0.57		-0.01		1		1744
114214	−62 03033				13 07 09	−62 25.5	1	9.31		0.71		0.42		1	F0 V	1688
114286	−6 03750				13 07 10	−07 23.3	2	7.34	.004	1.09	.016	0.91	.017	5	K0	1375,1657
114213	−60 04485	LSS 2950			13 07 12	−61 12.4	5	8.97	.018	0.93	.014	-0.01	.037	17	B1 Ib	403,540,976,1684,2021
		G 164 - 47			13 07 13	+29 15.2	2	14.18	.025	1.72	.010			2	dM4	686,784
114325	+20 02802				13 07 16	+20 24.0	1	8.32		0.25		0.05		2	A3	1569
	+24 02548				13 07 18	+23 46.4	1	10.62		0.80		0.59		1		280
	+8 02679				13 07 19	+08 27.0	1	10.59		1.01		0.87		1	K0	280
114252	−61 03483				13 07 19	−61 34.0	1	8.63		0.37		0.04		1	F0 V	1688
114326	+17 02595	HR 4962			13 07 20	+17 06.9	2	5.91	.020	1.45	.005	1.65		4	K5 III	252,280
114357	+38 02407	HR 4964			13 07 20	+37 41.3	1	6.02		1.15				2	K3 III	70
114327	+17 02596				13 07 21	+16 44.6	1	8.26		1.07				3	K0	882
114330	−4 03430	HR 4963		⋆ A	13 07 21	−05 16.4	9	4.38	.007	-0.01	.005	0.00	.013	30	A1 V	15,53,542,1075,1107*
114330	−4 03430	IDS13048S0500		B	13 07 21	−05 16.4	1	8.00		0.20		-0.09		2		542
114200	−70 01567	LSS 2949			13 07 23	−70 32.6	4	8.47	.021	0.09	.007	-0.90	.019	9	B0/2 Ve	400,540,976,1737
		Sand 226			13 07 24	+29 20.	1	15.98		1.48				2		694
114376	+39 02611	HR 4967		⋆ BC	13 07 24	+38 48.0	5	6.28	.035	-0.13	.007	-0.48	.005	11	B7 III	70,1022,1068,1079,1690
114343	+16 02468				13 07 25	+16 07.9	1	8.98		0.46		0.01		2	F8	1569
		GPEC 2904			13 07 26	−33 26.6	1	10.91		1.11				2		1719
		WEIS 21166			13 07 29	+31 23.1	1	10.96		0.72				1		1692
		AAS12,381 34 # E			13 07 29	+34 22.0	1	10.32		1.40		0.92		1		280
114277	−62 03035				13 07 29	−62 45.6	1	9.81		0.32		0.08		1	B8/9 V	1688
114263	−63 02585				13 07 29	−64 15.6	2	8.39	.005	0.01	.005	-0.24	.002	3	B8 Vn	401,1688
		WEIS 22274			13 07 30	+30 52.1	1	10.46		1.04				1		1692
114377	+32 02324	IDS13052N3154		AB	13 07 30	+31 38.0	1	9.38		0.71				2	K5/M0	280
114446	+57 01419				13 07 30	+57 05.9	1	7.01		0.60		0.08		2	F8	3016
		AAS78,203 # 29			13 07 30	−60 14.	1	10.05		0.22		-0.38		6		1684
238198	+59 01489				13 07 31	+58 49.9	1	9.78		0.88		0.53		2	K0	3032
	+31 02459				13 07 32	+30 52.3	1	10.43		1.04		0.86		1	K0 V	1019
114380	+16 02469				13 07 33	+16 24.8	1	7.86		0.41		0.02		2	F2	1569
114378	+18 02697	HR 4968		⋆ AB	13 07 33	+17 47.6	7	4.32	.012	0.45	.005	-0.06	.007	17	F5 V	15,292,1008,1197,1363*
114278	−62 03037				13 07 33	−63 18.8	1	10.26		0.30		0.22		1	A0/1 V	1688
114401	+29 02374				13 07 34	+29 04.2	2	8.62	.007	1.21	.007	1.26	.012	8	K1 II	1019,8100
114381	+13 02638				13 07 35	+12 35.2	3	9.53	.032	0.02	.002	0.02		7	A0 V	193,272,280
	−64 02222				13 07 35	−65 00.7	2	10.70	.000	0.08	.000	-0.25	.015	3	B8	122,549
114279	−64 02223				13 07 36	−64 28.2	1	10.30		0.10		-0.07		1	A0	1688
		G 61 - 35			13 07 39	+22 46.5	1	12.45		1.10		0.89		1		3078
		GD 154			13 07 39	+35 25.7	1	15.33		0.18		-0.59		1		3060
	−63 02589				13 07 39	−64 16.3	1	10.05		0.20		-0.57		1	B8	1688
114306	−59 04795				13 07 42	−59 27.6	2	8.63	.015	0.22	.005	-0.25		6	B8 II	122,2014
	−64 02224				13 07 43	−64 57.1	1	10.50		0.07		-0.30		1	B7	549
114402	+16 02470				13 07 44	+15 36.3	1	9.04		0.61		0.08		2	G5	1569
	+25 02604			V	13 07 44	+24 52.0	2	9.40	.209	1.44	.024	0.88	.141	3	M5	280,1569
114428	+37 02371				13 07 44	+37 25.1	1	8.63		0.50		0.07		1	G5	619
114427	+39 02613				13 07 44	+38 59.4	2	7.18	.000	0.17	.005	0.17	.014	4	A7 IV	280,833
		LP 557 - 10			13 07 45	+03 54.1	1	11.66		0.47		-0.16		2		1696
	+15 02548				13 07 45	+15 25.2	1	10.54		0.56		0.07		1	F8	1569
	+27 02222				13 07 45	+26 55.2	1	10.02		1.20		1.09		1		280
		B2 1308+326 # 6			13 07 45	+32 35.3	2	14.32	.009	0.37	.014	-0.04	.063	5		1540,1595
	+39 02612				13 07 45	+39 20.4	1	8.91		1.54		1.89		2	M0	280
	+10 02518				13 07 46	+09 48.2	1	9.37		1.01		0.90		2	K5	1569
	+31 02460				13 07 46	+30 59.0	2	9.95	.009	0.69	.003	0.19		2	G4 V	1019,1692
114447	+39 02614	HR 4971		⋆ A	13 07 46	+38 45.8	1	5.91		0.29				2	A9 III-IV	70
		B2 1308+326 # 5			13 07 47	+32 36.2	1	16.74		0.57		0.02		1		1540

Table 1   727

HD	DM	Other Id	N Rem	α₁₉₅₀	δ₁₉₅₀	S	V	σ_V	B–V	σ_B-V	U–B	σ_U-B	n	Spectrum	References
114319	−61 03487			13 07 48	−61 49.3	1	8.23		0.41		0.01		1	F0/2 IV/V	1688
		Wray 1072		13 07 49	−62 41.4	1	13.06		0.77		-0.18		4		1684
		B2 1308+326 # 4		13 07 50	+32 36.1	1	13.85		0.67		0.40				1540
114318	−60 04493			13 07 50	−61 06.8	1	10.11		0.15		-0.12		1	B9/A0 III	1688
114320	−63 02595			13 07 51	−63 33.0	1	10.11		0.17		-0.02		1	B9.5V	1688
114321	−63 02592			13 07 51	−64 04.1	1	8.89		0.17		0.14		1	A0 V	1688
		CS 16027 # 40		13 07 53	+32 13.3	1	14.97		0.91		0.60		1		1744
114317	−60 04494	IDS13048S6026	AB	13 07 53	−60 41.6	1	9.80		0.09		-0.25		1	B9 III	1688
		CS 16027 # 37		13 07 54	+33 08.9	1	13.71		0.74		0.37		1		1744
114504	+63 01056	HR 4974	⋆ A	13 07 54	+62 29.7	2	6.53	.010	-0.02	.005	-0.02	.005	5	A1 V	985,1733
114341	−59 04801	LSS 2957		13 07 54	−59 51.4	2	9.06	.013	0.12	.007	-0.72	.003	7	B0 IIInn	540,976
114463	+26 02431			13 07 56	+25 49.3	1	8.80		0.73		0.30		1	K0	280
114340	−59 04804	LSS 2958		13 07 56	−59 28.8	9	8.03	.023	0.54	.017	-0.49	.025	20	B1 Ia	122,138,403,540,914,976*
110994	−89 00037	BQ Oct		13 07 57	−89 31.3	1	6.80		1.70				4	M3 III	2012
114448	+30 02372			13 07 58	+29 42.6	3	8.45	.009	1.05	.009	0.90	.004	14	K0 III	97,1019,1690
		WEIS 23508		13 07 58	+30 27.4	1	10.59		0.52				1		1692
114365	−51 07329	HR 4965, V824 Cen		13 07 59	−52 18.1	3	6.06	.001	-0.10	.004	-0.38	.012	7	Ap Si	401,1770,2035
114405	−34 08700			13 08 00	−34 43.5	1	9.34		1.08		0.81		3	K0/1 III	3072
	+30 02373			13 08 01	+30 25.9	1	10.61		0.51		-0.05		1	F8	1019
	+27 02223			13 08 02	+26 33.0	1	10.45		0.89		0.60		1	G0	280
114391	−53 05500			13 08 05	−53 54.0	1	8.63		0.01		-0.35		2	B7 V	401
	−64 02228			13 08 06	−64 49.0	1	10.66		0.30		0.23		1	A3	549
	+24 02549			13 08 07	+23 45.3	1	9.47		0.98		0.75		1	G5	280
	+38 02408			13 08 07	+38 17.7	1	10.24		1.44		1.74		1	K5	280
114449	−4 03432	IDS13055S0424	AB	13 08 07	−04 40.6	1	7.61		0.40				4	F2	1351
		CS 16027 # 42		13 08 08	+31 14.1	1	13.28		1.27		1.25		1		1744
114392	−54 05469	IDS13051S5450	AB	13 08 08	−55 06.2	1	9.58		0.40		0.11		1	A8/F0 V	1771
	+31 02461	WEIS 22210		13 08 11	+31 04.0	2	10.53	.000	0.26	.030	0.13		3		1019,1692
114480	+13 02640			13 08 13	+13 16.7	1	10.10		0.32		-0.08		2	F5	1569
114394	−58 04723	LSS 2959		13 08 13	−59 18.7	4	8.21	.016	0.13	.033	-0.73	.024	8	B1 III	122,401,540,976
		B2 1308+326 # 3		13 08 14	+32 33.8	2	13.64	.090	0.70	.009	0.31	.005	5		1540,1595
	+45 02090			13 08 15	+45 10.3	1	9.39		1.24		1.36		1	K2 III	280
114423	−46 08455	IDS13054S4645	A	13 08 16	−47 01.4	1	8.21		0.10		-0.02		1	A0 V	1770
114439	−48 07988			13 08 16	−48 44.7	1	8.61		0.30		0.11		1	A6 IV/V	1771
114369	−63 02603			13 08 16	−64 21.3	1	10.24		0.16		-0.23		1	A0	1688
	−64 02229			13 08 16	−65 03.8	1	10.90		0.40		0.27		1	A7	549
		B2 1308+326 # 7		13 08 17	+32 33.5	1	16.44		0.62		-0.14				1540
114435	−41 07648	HR 4970		13 08 17	−41 58.0	2	5.78	.005	0.52	.000			7	F5 IV	15,2012
114493	+14 02584	IDS13058N1350	AB	13 08 18	+13 34.4	2	7.11	.005	1.12	.015	0.98		4	K0	280,3040
114519	+36 02344	RS CVn		13 08 18	+36 12.0	2	8.11	.055	0.58	.000	0.07	.015	5	K2 III	196,1569
114370	−68 01843			13 08 20	−68 39.1	1	7.99		0.38		0.29		1	A6 V	1771
114495	−1 02781			13 08 23	−02 27.8	1	8.67		0.48		-0.05		2	G5	1375
114371	−69 01772	HR 4966	⋆ A	13 08 23	−69 40.6	3	5.91	.004	0.41	.005	-0.04	.010	11	F3 IV/V	15,1075,2038
		LP 617 - 8		13 08 24	+00 27.8	3	12.74	.028	0.47	.032	-0.19	.059	4		1064,1696,3060
114507	+13 02641			13 08 25	+13 05.9	2	9.15	.005	1.07	.005	1.00		3	K0	280,882
	+29 02375			13 08 25	+28 48.6	1	9.73		0.46		0.09		1	F9 V	1019
114424	−62 03042			13 08 25	−62 35.6	1	9.58		0.31		-0.19		4	B8 IIne	1684
114520	+22 02550	IDS13060N2146	AB	13 08 26	+21 30.0	1	6.86		0.42		0.03		1	F2 II	8100
114537	+35 02414			13 08 26	+34 34.0	1	6.92		0.45		0.07		3	F5 V	1501
	+38 02409			13 08 26	+38 03.0	1	10.53		1.25		1.34		1		280
	+9 02724			13 08 27	+09 03.8	1	8.99		1.04		0.88		1	K0	280
	+29 02376			13 08 27	+28 56.0	1	10.00		0.44		0.02		1	F8	1019
114441	−54 05472	LSS 2962		13 08 27	−55 05.5	2	8.03	.005	0.13	.005	-0.76	.000	6	B2 Vnne	400,401
		CS 16027 # 38		13 08 28	+32 44.2	1	11.74		-0.01		0.06		1		1744
	−64 02232			13 08 28	−64 52.3	1	10.20		1.20		1.13		1	K3	549
	+15 02550			13 08 29	+14 50.7	1	10.60		0.68		0.26		1		280
	+29 02377			13 08 29	+29 15.2	2	11.09	.004	0.69	.001	0.28	.005	4	G0	97,1019
114425	−63 02604			13 08 29	−64 18.1	1	9.56		0.26		0.21		1	A5 V	1688
114556	+29 02378			13 08 31	+28 37.3	2	8.81	.000	0.27	.000	0.19	.001	2	A9 III	1019,1690
114474	−42 08175	HR 4973		13 08 31	−43 06.2	2	5.26	.015	1.05	.005	0.92		6	K1/2 III	2006,3077
		B2 1308+326 # 1		13 08 33	+32 34.6	2	13.04	.014	0.49	.018	-0.04	.009	5		1540,1595
114460	−61 03496			13 08 35	−62 01.9	1	9.58		0.15		-0.20		2	B8	1688
		G 61 - 36		13 08 36	+12 36.0	1	14.22		0.71		0.09				1658
		B2 1308+326 # 2		13 08 37	+32 37.1	2	12.72	.014	0.65	.005	0.26	.005	5		1540,1595
	+30 02374			13 08 38	+30 25.6	1	11.51		1.18		1.13		1	K8	1019
114538	+14 02585			13 08 39	+14 10.6	2	7.94	.002	0.39	.000	0.04	.000	17	F2 IV:	269,1569
114443	−64 02234			13 08 39	−65 01.9	3	9.71	.004	1.08	.000	0.82	.000	4	K0	122,287,549
114461	−62 03046	HR 4972, LSS 2964		13 08 40	−63 02.2	4	6.33	.013	0.42	.015	0.30	.031	13	A8 II/III	15,1688,2012,8100
	−64 02235			13 08 40	−65 08.9	3	9.52	.004	1.12	.000	0.81	.008	4	K2	122,287,549
		WEIS 13144		13 08 42	+30 26.4	1	11.52		1.18				1		1692
114703	+68 00714	G 238 - 26		13 08 42	+67 45.6	2	8.88	.000	0.92	.000	0.70	.018	5	K2 V	22,1003
		WEIS 15847		13 08 43	+29 23.2	1	11.47		0.76				1		1692
	+37 02374			13 08 43	+36 51.6	1	9.80		1.17		1.15		2		1569
		G 14 - 38		13 08 44	−04 33.7	1	10.87		0.74		0.28		1		1658
		GD 155		13 08 45	+38 11.0	1	15.54		0.56		-0.09		1		3060
114489	−60 04508	LSS 2966		13 08 45	−60 51.1	2	6.76	.015	0.46	.000	0.34		9	A9/F0 II	2012,8100
114478	−62 03048	LSS 2965		13 08 45	−62 33.0	4	8.67	.016	0.49	.011	-0.50	.018	14	B1 Ib/II	92,403,1684,8100
		GD 486		13 08 46	+73 00.8	1	14.87		0.53		-0.15		3		308
	−64 02238			13 08 46	−65 08.1	3	9.53	.000	1.12	.008	0.83	.004	4	K2	122,287,549
		KUV 13088+3139		13 08 47	+31 38.8	1	16.79		0.16		-0.60		1	DA	1708

HD	DM	Other Id	N Rem	α_{1950}	δ_{1950}	S	V	σ_V	B–V	σ_{B-V}	U–B	σ_{U-B}	n	Spectrum	References
		CS 22877 # 13		13 08 47	−09 07.8	1	14.64		0.51		−0.23		1		1736
		Ton 146		13 08 48	+28 53.	1	15.31		0.18		0.14		1		1036
	+42 02379			13 08 48	+41 41.6	1	9.88		1.04		0.87		1	G8 III	280
		WLS 1300-5 # 8		13 08 49	−05 00.8	1	11.14		0.49		0.00		2		1375
114479	−67 02208			13 08 50	−67 28.3	1	9.82		0.26		0.18		1	A7 V	1771
114604	+31 02462	IDS13065N3122	AB	13 08 51	+31 05.5	1	9.36		0.71		0.34		1	G5	1019
		CS 16027 # 36		13 08 52	+33 14.5	1	15.27		0.30		−0.14		1		1744
114512	−52 06276			13 08 52	−52 32.8	1	7.14		1.71		1.82		7	M2 III	1673
114499	−60 04509			13 08 52	−61 19.5	1	9.67		0.19		−0.13		1	A7 IV/V	1688
114606	+10 02519	G 63 - 5	★ A	13 08 53	+09 53.3	9	8.74	.011	0.62	.005	0.05	.009	18	G1 V	22,333,908,1003,1013,1620*
114592	+15 02551			13 08 53	+14 35.7	1	9.17		0.66		0.19		2	G0	1569
	+30 02375			13 08 53	+30 04.5	3	10.25	.017	0.43	.008	−0.03	.006	4	G0	97,1019,1692
		G 63 - 6		13 08 54	+09 51.9	1	12.31		1.31		1.12		2		1773
114606	+10 02519	G 63 - 6	★ B	13 08 54	+09 51.9	2	12.32	.005	1.28	.005	1.14	.065	2		1658,3032
114622	+34 02400			13 08 54	+33 43.9	1	7.94		0.30		−0.01		2	F0	1569
		CS 16027 # 39		13 08 56	+32 16.9	1	13.78		0.66		0.10		1		1744
114576	−25 09653	HR 4978	★ AB	13 08 56	−26 17.2	3	6.49	.011	0.19	.005			11	A5 V	15,176,2027
		AAS12,381 18 # A		13 08 57	+17 39.9	1	10.36		1.16		1.16		1		280
114490	−63 02608			13 08 57	−63 41.0	1	9.93		0.09		−0.06		1	B9 IV	1688
	+9 02725			13 08 58	+08 58.1	1	9.90		0.96		0.64		1		280
114513	−56 05628			13 08 58	−56 41.5	1	8.18		0.02		−0.15		2	B9 IV	401
	+20 02805			13 09 01	+20 06.0	1	9.38		1.53		1.78		1	M0	280
114623	+16 02474			13 09 03	+15 59.8	1	9.88		0.52		0.01		2	G0	1569
	+28 02192			13 09 04	+27 50.9	1	9.67		0.84		0.49		2	G8 III	1019
	−63 02609	AAS78,203 # 30		13 09 04	−64 22.0	1	10.89		0.53		−0.45		2		1684
		Lod 757 # 29		13 09 04	−64 55.2	1	13.02		0.30		0.04		1	A8	549
114636	+27 02224			13 09 05	+26 39.1	1	8.38		1.02		0.81		1	K1 III	280
114530	−60 04512	LSS 2968		13 09 05	−60 48.3	2	9.07	.003	0.31	.002	−0.59	.017	9	O9.5III	403,540,976
		Lod 757 # 15		13 09 05	−64 57.6	1	10.51		1.77		2.02		1		549
114637	+22 02552	IDS13067N2227	A	13 09 07	+22 11.1	1	6.82		0.88				2	G5 III	280
114635	+29 02379			13 09 07	+29 08.0	3	8.85	.000	0.50	.002	0.02	.025	8	F6 V	97,1019,1690
		Lod 757 # 26		13 09 07	−64 52.7	1	11.75		0.92		0.61		1	K2	549
114517	−64 02239			13 09 07	−65 01.7	3	9.29	.000	1.40	.000	1.38	.000	4	K0/1 III	122,287,549
	+25 02607			13 09 08	+25 00.5	1	9.61		0.89		0.68		1	K0	280
114658	+33 02307			13 09 08	+32 33.6	1	8.94		1.10		1.17		1	K0	280
114516	−63 02612	LSS 2967		13 09 08	−63 32.8	1	8.68		0.19		−0.69		2	B0 Ve	401
	−64 02241			13 09 08	−64 52.8	1	11.50		0.19		0.09		1	B9	549
114639	+16 02475			13 09 09	+16 00.3	1	9.64		1.05		0.93		1	F8	280
114638	+20 02806			13 09 09	+20 04.9	1	8.34		0.94		0.62		2	G5	1569
		CS 16027 # 41		13 09 09	+32 03.0	1	14.70		0.08		0.14		1		1744
114529	−59 04815	HR 4975, V831 Cen	★ ABC	13 09 09	−59 39.3	7	4.57	.024	−0.08	.006	−0.38	.007	19	B7 V	15,122,1075,1771,2012*
		Lod 757 # 27		13 09 09	−64 56.4	1	11.78		0.59		0.05		1	F5	549
114597	−23 10950			13 09 11	−24 18.2	1	7.97		1.34				4	K1/2 III	2012
		G 164 - 50		13 09 12	+29 51.3	1	14.42		1.55				1		784
		WEIS 13910		13 09 13	+30 11.5	1	11.43		0.69				1		1692
114674	+41 02364			13 09 13	+41 03.5	1	7.16		0.92				2	K0	280
114444	−74 01029	LSS 2963		13 09 13	−75 02.9	2	10.31	.005	−0.02	.010	−0.80	.010	3	B1 II/III	55,400
	+36 02345			13 09 14	+36 12.2	1	9.96		1.06		0.88		1	G8 III	280
114613	−37 08437	HR 4979		13 09 15	−37 32.3	5	4.85	.003	0.70	.003	0.31	.005	21	G3 V	15,1075,2012,3077,8015
	−64 02245			13 09 16	−64 53.4	1	9.92		1.29		1.27		1	K3	549
114659	+13 02643			13 09 20	+12 59.8	2	8.32	.010	0.97	.005			3	K0	280,882
		Lod 757 # 28		13 09 20	−64 55.1	1	11.96		0.11		−0.34		1	B7	549
	+19 02646	G 61 - 38	★ A	13 09 21	+18 38.8	3	9.89	.005	0.64	.000	0.04	.009	3	G0	927,1620,1658,3024
	+19 02646	G 61 - 38	★ B	13 09 21	+18 38.8	1	10.96		1.06		0.87		1		3024
114587	−57 05944			13 09 22	−58 23.1	1	9.15		0.36		0.24		1	A7 V	1771
	+47 02028			13 09 23	+46 56.0	1	10.15		1.04		0.93		1		280
	+10 02520			13 09 24	+09 46.7	1	9.26		1.47		1.80		1	K5	280
114676	+15 02552			13 09 24	+14 57.7	1	9.70		0.98		0.70		1	G5	280
114642	−15 03613	HR 4981	★ A	13 09 24	−15 55.8	7	5.04	.011	0.46	.008	0.02	.011	26	F6 V	3,15,244,254,1008*
114553	−64 02248			13 09 24	−64 57.1	2	9.48	.000	1.03	.000	0.90	.000	2	K2	287,549
114589	−61 03501			13 09 25	−61 45.8	1	9.12		0.54		0.06		1	F7 V	1688
		WEIS 14901		13 09 26	+29 47.0	1	10.77		0.58				1		1692
114569	−64 02249	IDS13062S6446	AB	13 09 26	−65 01.6	2	8.09	.015	−0.01	.010	−0.42	.019	3	B8 III	122,549
114688	+15 02553			13 09 28	+14 56.3	1	10.10		0.36		−0.06		2	F5	1569
114662	−20 03787			13 09 28	−20 37.2	2	7.61	.015	1.18	.002	1.30	.280	6	K0 III	3040,8100
	−61 03502			13 09 28	−62 09.4	1	10.31		0.75		−0.30		3		1684
	+45 02092			13 09 29	+45 16.4	1	10.57		1.10		1.00		1	G8 IV	280
114570	−65 02201	HR 4977		13 09 29	−65 57.7	3	5.89	.008	0.05	.016	0.04	.005	8	A0 Vn	15,1075,1771
	+30 02376			13 09 30	+29 46.1	1	10.74		0.56		0.00		1	G0 V	1019
		CCS 2065		13 09 30	−58 18.4	1	11.07		2.82				2		864
114603	−60 04517			13 09 30	−61 12.0	1	9.00		0.15		−0.08		1	B9.5V	1688
	−60 04518	LSS 2972		13 09 31	−60 45.2	2	10.41	.005	0.38	.002	−0.55	.052	5	B1 ne	403,1684
114710	+28 02193	HR 4983	★ A	13 09 32	+28 07.9	19	4.26	.014	0.57	.006	0.07	.011	292	G0 V	1,15,22,30,101,130*
	+12 02563			13 09 33	+12 16.3	1	10.02		1.24		1.37		1	K0	280
		CS 16027 # 43		13 09 34	+29 05.0	1	14.95		0.39		−0.25		1		1744
		LP 677 - 57		13 09 36	−08 12.2	1	13.26		0.88		0.53		1		1696
	−60 04519			13 09 36	−60 40.0	2	9.98	.007	0.22	.005	−0.45	.012	3	B2.5V	540,976
114590	−64 02256	IDS13063S6506	AB	13 09 37	−65 21.9	1	8.96		0.39		0.21		1	A3mA8-A8	1771
114620	−60 04520			13 09 39	−61 16.8	2	7.27	.026	0.35	.015	0.13	.016	2	A9/F0III/IV	1688,1771
		WEIS 13211		13 09 41	+30 19.0	1	10.61		0.91				1		1692

Table 1 729

HD	DM	Other Id	N Rem	α_{1950}	δ_{1950}	S	V	σ_V	B–V	σ_{B-V}	U–B	σ_{U-B}	n	Spectrum	References
	+34 02402			13 09 41	+34 02.3	1	9.74		1.01		0.82		1		280
		AAS78,203 # 31		13 09 42	−61 53.	1	11.94		0.41		0.25		4		1684
	+28 02194			13 09 43	+28 27.8	1	10.63		0.58		0.10		2	G0 V	1019
114724	+25 02610	HR 4984		13 09 44	+24 31.4	1	6.33		0.98				3	K1 III	280
	+25 02611			13 09 44	+24 52.5	1	9.36		1.24		1.17		1	K2	280
	+30 02377			13 09 44	+30 18.5	1	10.59		0.88		0.61		1	G6 V	1019
	−9 03642			13 09 45	−09 50.8	1	12.80		0.96		0.53		2	K7 V	3072
114692	−34 08720	IDS13070S3412	A	13 09 45	−34 28.8	1	7.76		0.53				4	F5/6 V	2012
114649	−53 05512			13 09 45	−53 39.9	2	8.53	.035	0.51	.005	0.04		5	F5 V	1594,6006
114725	+16 02476	IDS13073N1640	AB	13 09 46	+16 23.7	1	7.46		0.40		-0.04		2	F5 V	1569
	+43 02308			13 09 46	+43 18.5	1	10.14		0.95		0.74		1	G8 V	280
114630	−59 04827	HR 4980	⋆ A	13 09 47	−59 33.0	4	6.18	.032	0.59	.011	0.10	.005	10	G0 V	15,122,2012,3077
	+18 02699			13 09 48	+18 15.4	1	9.65		1.01		0.95		1	K0	280
		G 61 - 40		13 09 49	+18 35.3	1	15.58		1.64				1	K5	3024
		CS 22877 # 14		13 09 49	−09 04.0	1	13.57		0.09		0.16		1		1736
	+36 02348			13 09 50	+35 39.1	1	10.30		1.03		0.85		1		280
	−60 04525			13 09 51	−61 00.5	1	9.78		0.17		-0.09		1	A0	1688
114761	+30 02378			13 09 53	+30 14.5	1	8.45		1.52		1.88		1	K3 V	1019
114778	+35 02420			13 09 53	+35 04.7	2	8.42	.000	0.46	.000	-0.04	.000	11	F5	1229,1769
114653	−62 03064			13 09 53	−63 20.6	1	8.61		0.16		-0.08		1	B8 III	1688
	+11 02544			13 09 55	+10 40.1	1	10.34		0.96		0.67		1		280
114762	+18 02700	G 63 - 9		13 09 55	+17 46.9	6	7.31	.013	0.53	.007	-0.07	.010	11	F9 V	22,333,792,1003,1620*
		LP 617 - 88		13 09 56	+01 05.2	1	11.92		0.63		0.03		2		1696
	+23 02547			13 09 56	+23 06.7	1	10.26		1.26		1.32		1		280
		G 149 - 55		13 09 57	+27 08.1	1	14.51		1.43				1		686
114707	−42 08196	HR 4982		13 09 58	−42 26.1	3	6.21	.007	1.06	.000			11	K0 III	15,2013,2029
		G 256 - 7		13 09 59	+85 18.6	1	15.97		0.80		0.13		1	DG?	782
114729	−31 10156			13 09 59	−31 36.2	4	6.68	.006	0.62	.006	0.05		14	G0 V	1075,1311,2012,2020
114671	−63 02628			13 09 59	−63 44.0	1	10.24		0.09		-0.47		1	B4 III	1688
		WEIS 11310		13 10 00	+31 06.5	1	10.91		0.49				1		1692
	−64 02264			13 10 00	−64 36.0	1	10.64		0.05		-0.38		2	B5 IV	540
114654	−64 02262			13 10 00	−64 52.1	3	9.88	.004	0.14	.008	0.11	.020	4	A1 V	122,549,1688
	−60 04528	LSS 2976		13 10 01	−60 49.5	5	8.75	.012	0.22	.014	-0.65	.022	10	B1 III	540,976,1684,1688,2021
114670	−62 03066			13 10 02	−63 02.1	2	9.21	.003	0.45	.009	0.23	.015	2	A9 V	1688,1771
114655	−67 02213			13 10 02	−68 14.0	1	7.93		0.16		0.23		1	A3 V	1771
	+43 02310			13 10 03	+42 40.6	1	10.26		0.99		0.95		1	K1 III	280
	+12 02564			13 10 04	+11 34.7	1	9.74		1.11		0.97		1		280
114780	+12 02565	HR 4986		13 10 04	+11 49.3	1	5.77		1.51				2	M0 III	70
	+74 00526	G 255 - 28		13 10 04	+74 07.2	2	9.34	.004	0.58	.000	0.02	.000	5	G2	196,1064
114533	−77 00890	HR 4976		13 10 04	−78 11.0	2	5.84	.005	1.07	.000	0.73	.005	7	G2 Ib	15,1075
		CS 16027 # 49		13 10 05	+30 37.2	1	13.40		0.41		0.03		1		1744
		CS 16027 # 52		13 10 06	+31 28.6	1	14.97		0.24		-0.10		1		1744
114685	−62 03067			13 10 06	−63 07.2	1	9.25		0.23		0.24		1	A0 V	1688
114672	−64 02267			13 10 06	−65 00.9	3	9.48	.000	1.14	.005	0.98	.000	4	K0/2 III	122,287,549
	+28 02195		AB	13 10 07	+27 57.7	1	10.60		0.49		-0.01		2	F9 IV	1019
	−64 02268			13 10 07	−64 52.1	2	10.64	.010	0.39	.034	0.21	.015	3	A8	122,549
114793	+19 02648	HR 4987		13 10 09	+19 01.0	2	6.52	.005	0.86	.015	0.50		3	G8 III	280,1569
	−28 09945			13 10 11	−29 23.1	2	9.88	.023	0.87	.018	0.52		5		1594,6006
		CS 22877 # 8		13 10 12	−10 16.6	1	14.42		0.05		0.12		1		1736
114698	−63 02630			13 10 14	−63 50.3	1	10.42		0.10		0.04		1	B9/A0 V	1688
114656	−70 01573			13 10 15	−71 15.6	1	9.48		0.48		0.27		1	A0mA7-F0	1771
		Sand 240		13 10 18	+27 49.	1	13.43		1.53				1	dM4	686
		CS 16027 # 89		13 10 18	+30 01.2	1	14.59		0.62		0.06		1		1744
		WEIS 12451		13 10 18	+30 36.	1	13.58		1.42				1		784
	+74 00528			13 10 18	+74 23.0	1	10.48		0.46		-0.05		1	G0	1064
	+30 02379	WEIS 13025		13 10 19	+30 26.9	2	10.99	.014	0.57	.003	0.06		2	G0	1019,1692
114718	−62 03073			13 10 21	−62 50.3	1	8.44		0.17		-0.06		1	B9 III/IV	1688
114813	+9 02727			13 10 23	+08 39.5	1	8.48		0.81		0.36		1	G5	280
	+21 02499			13 10 23	+20 54.5	1	9.76		0.98		0.76		1	K2	280
114733	−57 05952	LSS 2980	⋆ AB	13 10 23	−58 05.8	2	9.53	.018	0.27	.008	-0.56	.015	4	B2/3 III	540,976
	−59 04839			13 10 23	−59 47.3	1	10.74		0.20		0.18		2	A2	837
114735	−61 03511			13 10 23	−61 26.7	1	9.41		0.21		-0.02		1	B9 III	1688
114720	−63 02632	SS Cen		13 10 23	−63 53.1	1	9.59		0.09		-0.18		1	B8 V	1688
	+26 02434			13 10 24	+25 51.3	1	10.04		0.98		0.77		1	K2	280
114839	+27 02227			13 10 24	+27 08.8	1	8.47		0.29		0.07		1	Am	280
114719	−63 02633			13 10 24	−63 25.1	1	9.79		0.28		0.16		1	A2/3 III	1688
	+24 02553			13 10 25	+24 11.8	1	10.02		0.97		0.70		1	K0	280
		CS 16027 # 46		13 10 25	+28 57.5	1	14.92		0.44		-0.28		1		1744
114772	−50 07589	HR 4985	⋆ AB	13 10 25	−50 26.1	2	5.90	.006	-0.02	.004	-0.08		7	B9.5V	1770,2035
114773	−51 07359			13 10 25	−52 08.6	1	8.35		0.26		0.19		1	A6 IV	1771
114819	+20 02809			13 10 27	+20 11.0	1	7.37		1.25		1.26		1	K0	280
	+28 02196			13 10 27	+27 37.1	1	9.40		1.02		0.90		2	K2	280
	+29 02380			13 10 27	+29 12.5	3	10.26	.032	0.18	.008	0.11	.009	7	A2	1019,1026,1068
	+12 02566			13 10 28	+12 03.5	1	9.61		1.08		0.94		1	K0	280
	−40 07755			13 10 28	−41 20.4	1	10.96		1.16		1.08		2	G5	1696
	−61 03512	LSS 2978		13 10 28	−62 20.0	2	10.02	.009	0.34	.011	-0.51	.011	3	B1 III	540,976
114737	−62 03079	LSS 2981		13 10 30	−63 19.3	4	8.00	.018	0.17	.010	-0.76	.019	7	O9 III	401,540,976,1688
		CS 16027 # 90		13 10 31	+29 51.9	1	14.49		0.79		0.47		1		1744
	−60 04535	LSS 2982		13 10 31	−60 26.5	2	10.34	.010	0.22	.008	-0.54	.000	3	B1.5V	540,976
114738	−63 02636			13 10 31	−64 02.9	2	7.83	.012	0.14	.020	0.15	.003	2	A1 V	1688,1771

HD	DM	Other Id	N	Rem	α_{1950}	δ_{1950}	S	V	σ_V	B–V	σ_{B-V}	U–B	σ_{U-B}	n	Spectrum	References
		CS 16027 # 48			13 10 32	+30 01.1	1	14.18		0.63		0.09		1		1744
		WEIS 13284			13 10 32	+30 16.8	2	11.23	.005	0.31	.000	0.09		2		280,1692
114877	+43 02311				13 10 35	+42 53.5	1	6.78		0.93				2	G5	280
		CS 16027 # 54			13 10 36	+31 57.0	1	15.77		-0.09		-0.05		1		1744
114879	+35 02423				13 10 36	+34 32.8	1	8.95		0.17		0.06		2	A2	1068
114739	-64 02275				13 10 36	-64 46.1	1	8.48	.011	0.41	.005	0.19	.009	3	F0 IV	122,1688
114757	-62 03083				13 10 37	-63 19.3	2	9.27	.014	0.16	.020	-0.32	.001	2	B6/8 Ve	401,1688
		CS 16027 # 56			13 10 38	+32 53.9	1	13.94		0.46		-0.19		1		1744
114913	+59 01493	IDS13086N5859		A	13 10 38	+58 42.8	1	8.66		0.50		0.01		2	F7 V	1502
		WLS 1320-20 # 7			13 10 39	-22 00.8	1	11.07		0.52		0.05		2		1375
114865	+16 02477				13 10 40	+15 34.9	2	7.64	.005	1.12	.000			4	K0	280,882
	+11 02545				13 10 41	+10 53.5	1	10.37		1.10		1.01		1	K0	280
		G 61 - 41			13 10 41	+20 27.2	1	13.02		1.58				1		686
		FW Com			13 10 41	+31 17.4	1	15.20		0.23		0.08		1		1744
	+45 02094				13 10 41	+44 40.0	1	9.30		1.13		1.17		1	K1 III	280
114864	+18 02702	IDS13083N1833		A	13 10 42	+18 17.4	1	9.25		1.50		1.62		1	K0	280
		AAS12,381 21 # A			13 10 42	+21 00.4	1	10.69		0.58		-0.01		1		280
	+37 02380				13 10 42	+37 26.2	1	9.78		0.27		0.07		2	F0	1569
	-60 04539				13 10 43	-60 54.3	2	10.54	.065	0.74	.014	-0.34	.010	4		1684,1737
		AAS12,381 28 # D			13 10 44	+27 31.3	1	10.06		1.04		0.85		2		280
		WEIS 14114			13 10 44	+29 57.0	1	11.05		0.89				1		1692
114775	-64 02281	IDS13075S6421		AB	13 10 44	-64 36.8	1	9.00		0.08		-0.16		1	B9 IV	1688
		LP 910 - 33			13 10 45	-32 11.2	1	14.00		1.46		1.03		1		3062
114889	+19 02649	HR 4992			13 10 46	+18 59.5	2	6.11	.010	1.20	.006			3	G8 III	70,280
	+29 02381	IDS13084N2910		AB	13 10 46	+28 54.0	1	10.58		0.53		0.03		1	G0	1019
		WEIS 11277			13 10 46	+31 10.7	1	11.21		1.01				1		1692
114866	-2 03651				13 10 46	-03 13.8	1	7.28		1.48		1.82		5	K2	897
114846	-18 03562	HR 4990		★ AB	13 10 46	-18 33.7	2	6.26	.010	0.09	.005	0.00		8	A0 V	244,2007
	-62 03086				13 10 46	-62 37.0	1	10.71		0.36		-0.61		3		1684
	+38 02412	S CVn			13 10 48	+37 38.5	1	9.75		1.32		1.51		1	K3 III	280
114792	-61 03516	LSS 2985		★ AB	13 10 48	-62 23.4	3	6.82	.014	0.90	.008	0.60	.044	10	F5/6 Ib	1688,2012,8100
		CS 16027 # 50			13 10 49	+31 00.9	1	14.20		0.03		0.11		1		1744
114883	+14 02586				13 10 50	+13 39.4	1	7.41		1.22				1	K0	280
		CS 16027 # 47			13 10 50	+29 11.1	1	13.97		0.35		-0.07		1		1744
		GPEC 3038			13 10 51	-31 39.5	1	11.31		0.91				2		1719
		HZ 42			13 10 52	+31 37.9	1	14.61		-0.12		-0.56		1	B5	1744
	+9 02730				13 10 53	+09 03.7	1	8.65		1.57		1.90		1	M0	280
		Sand 243			13 10 54	+26 05.	1	14.05		1.50				1	dM3	686
		CS 22877 # 12			13 10 56	-09 19.4	1	13.53		0.14		0.18		1		1736
114800	-62 03090	LSS 2986			13 10 57	-63 06.5	4	7.96	.045	0.11	.012	-0.84	.021	10	B2 III/Vne	401,540,976,1684
		WEIS 12460			13 10 58	+30 39.6	1	11.89		0.85				2		1692
	+37 02381				13 10 59	+37 14.2	1	9.96		0.97		0.67		2	K0 III	1569
114833	-51 07363				13 10 59	-51 43.2	1	9.29		0.03		-0.03		1	B9.5 III	1770
		WEIS 12965			13 11 00	+30 29.0	1	11.78		0.94				1		1692
		LP 910 - 73			13 11 02	-33 35.3	1	11.29		0.97		0.78		1		1696
114914	+29 02382				13 11 03	+28 35.4	2	9.56	.000	0.39	.001	0.02	.002	2	F3 V	1019,1690
	+30 02380			A	13 11 03	+30 29.1	1	11.79		0.93		0.72		1		1019
	+30 02380			B	13 11 03	+30 29.1	1	12.14		0.61		0.11		1		1019
114957	+47 02031				13 11 03	+46 45.5	1	8.41		0.53		0.03		2	G5	1733
		CS 16027 # 55			13 11 04	+32 46.7	1	13.85		0.81		0.43		1		1744
114941	+41 02367				13 11 04	+40 49.5	1	8.63		1.54		1.57		1	K5	280
114835	-58 04738	HR 4988			13 11 04	-58 25.1	1	5.89		1.08				4	K0/1 III	2006
114817	-61 03519				13 11 04	-61 52.5	2	10.03	.000	0.12	.005	-0.25	.009	4	B9 V	125,1688
114873	-42 08213	HR 4991			13 11 05	-42 52.5	8	6.16	.017	1.37	.010	1.63	.010	26	K4 III	15,1075,2012,2013*
114855	-54 05489	LSS 2988			13 11 05	-54 25.7	1	8.39		0.84		0.60		3	F5 Ia/ab	8100
	+12 02567				13 11 06	+11 31.0	1	9.87		0.28		0.09		2		1569
114929	+32 02330				13 11 06	+32 29.3	1	9.28		0.95		0.63		1	G5	280
		WLS 1320-20 # 6			13 11 07	-19 30.2	1	11.94		0.64		0.12		2		1375
	-62 03092				13 11 07	-62 52.8	1	10.60		0.42		-0.52		3		1684
114818	-63 02640				13 11 07	-64 04.1	1	10.02		0.36		-0.34		1	B3/5 III	1688
114930	+32 02331				13 11 08	+31 56.2	1	9.03		0.29		0.03		2	F1 III-IVm	1569
		CS 22877 # 15			13 11 08	-08 59.9	1	13.23		0.40		-0.19		1		1736
	-44 08508				13 11 08	-44 43.4	1	10.79		0.02		-0.40		7		1770
114837	-58 04740	HR 4989		★ AB	13 11 08	-58 50.2	4	4.91	.007	0.48	.000	-0.06		10	F5 V	15,1488,2012,3037
114931	+16 02480				13 11 09	+15 45.9	1	10.35		0.53		0.01		1	K0	1569
	+29 02383				13 11 11	+28 55.3	2	9.60	.009	1.27	.020	1.41		2	K2 III	694,1019
	+36 02349				13 11 11	+35 32.2	1	9.48		0.89		0.65		1	K0 III	280
	+30 02381	WEIS 12971			13 11 12	+30 28.5	2	11.28	.005	0.57	.004	0.04		2	F8	1019,1692
	+26 02438				13 11 16	+26 07.0	1	10.01		0.95		0.73		1	G5	280
114959	+21 02502				13 11 17	+20 40.9	1	8.30		1.06		0.90		1	K0	280
114975	+37 02383				13 11 18	+37 09.1	1	6.48		1.60				2	M1	280
115061	+68 00717	IDS13101N6749		B	13 11 18	+67 34.4	1	6.98		1.14		1.08		2	K0	3016
114875	-56 05656				13 11 18	-57 17.1	1	9.54		0.33		0.26		1	A2mA6-A8	1771
114958	+30 02382				13 11 19	+30 02.7	1	8.59		0.74		0.31		1	G6 V	1019
114976	+30 02383	IDS13090N3021		AB	13 11 20	+30 05.0	1	7.28		1.10		1.04		1	G9 III	1019
		G 149 - 58			13 11 21	+22 34.3	1	13.80		1.51				1	dM3	686
	+38 02413				13 11 21	+38 28.7	1	8.91		1.16		1.26		1	K0 III	280
		WEIS 13781			13 11 22	+30 04.0	1	10.71		0.51				3		1692
115003	+48 02098				13 11 22	+48 03.8	1	9.32		0.20				6	A8 V	193
	+24 02554				13 11 23	+24 00.6	1	10.08		1.05		0.93		1		280

Table 1 731

HD	DM	Other Id	N	Rem	α_{1950}	δ_{1950}	S	V	σ_V	B–V	σ_{B-V}	U–B	σ_{U-B}	n	Spectrum	References
114858	−64 02286				13 11 23	−64 27.9	1	9.23		0.07		−0.20		1	B8 III	1688
	+31 02463				13 11 24	+30 38.6	2	9.94	.009	0.59	.001	0.06		2	G0 V	1019,1692
114988	+33 02311				13 11 24	+32 47.7	2	6.68	.010	0.78	.005	0.47		4	G2 II	280,1569
114945	−16 03620				13 11 24	−17 09.7	1	9.58		0.59		0.07		12	G3 V	1603
114921	−40 07764				13 11 24	−40 56.8	1	7.60		1.03				4	K0 III	2012
114960	+2 02646				13 11 25	+01 43.3	4	6.57	.017	1.41	.007	1.72	.014	15	K5 III	1003,1509,1531,3077
114987	+43 02312				13 11 25	+43 16.6	2	9.73	.018	0.17	.018	0.15		5	A5	193,280
115019	+57 01424	IDS13095N5714		B	13 11 26	+56 56.7	2	8.03	.065	1.18	.015	1.10	.005	6	K2 II	1625,8100
115004	+40 02633	HR 4997			13 11 27	+40 25.0	3	4.93	.012	1.05	.012	0.92	.005	7	K0 III	1080,1363,3016
	+9 02731				13 11 28	+08 48.2	1	9.27		1.29		1.42		1	K2	280
114989	+26 02439				13 11 29	+25 57.8	1	7.32		0.42		0.00		3	F7 V	833
	−60 04545				13 11 29	−60 50.5	1	9.63		0.16		−0.39		1	B9	1688
114886	−62 03096	LSS 2989		⋆ AB	13 11 29	−63 19.0	5	6.87	.019	0.12	.010	−0.80	.015	13	B1/2 Ib/II	401,540,976,1684,2012
	+28 02199				13 11 30	+28 05.7	1	9.92		0.50		−0.04		1		1019
114961	−2 03653	SW Vir			13 11 30	−02 32.5	1	7.02		1.55		0.72		5	M7 III:	897
114946	−19 03651	HR 4995			13 11 30	−19 40.1	2	5.32	.015	0.88	.010	0.44		6	G6 V	2007,3008
		WEIS 14189			13 11 32	+29 58.2	1	11.28		1.00				1		1692
114991	+10 02523				13 11 33	+10 02.3	1	8.60		0.83		0.45		2	K0	1569
115043	+57 01425	IDS13095N5714		A	13 11 34	+56 58.4	8	6.83	.006	0.60	.002	0.09	.008	191	G1 V	130,1007,1008,1013*
114990	+16 02481				13 11 36	+16 12.7	1	9.05		0.77		0.44		2	G5	1569
114909	−61 03524	IDS13084S6130		AB	13 11 37	−61 45.9	2	9.53	.021	0.16	.003	−0.07	.019	4	B8/9 V	125,1688
		CS 16027 # 91			13 11 40	+29 53.0	1	14.92		0.58		0.02		1		1744
		CS 16027 # 92			13 11 41	+29 55.2	1	14.13		0.79		0.41		1		1744
	+14 02587				13 11 43	+14 16.2	1	10.40		1.36		1.71		1	K5	280
		CS 16027 # 93			13 11 44	+29 56.1	1	14.42		0.72		0.16		1		1744
114993	−23 10974	IDS13090S2345		AB	13 11 44	−24 01.2	1	6.74		0.37		−0.04		1	F0 V	1776
	+24 02556				13 11 45	+23 34.8	1	9.49		1.04		1.03		1	K0	280
		CS 16027 # 45			13 11 45	+28 48.6	1	14.30		0.28		−0.04		1		1744
115136	+68 00720	IDS13101N6749		A	13 11 46	+67 33.1	1	6.54		1.13		1.06		2	K2 III	3016
114971	−48 08050	HR 4996			13 11 46	−48 41.5	4	5.87	.011	1.06	.000	0.93		12	K1 III	15,2013,2028,3009
114981	−37 08461				13 11 50	−38 23.3	1	7.11		−0.06		−0.64		1	B3/5 (V)ne	55
114911	−67 02224	HR 4993, η Mus		⋆ A	13 11 50	−67 37.8	6	4.79	.004	−0.09	.005	−0.35	.008	20	B7 V	15,1075,1771,2012,2038,3023
115045	+31 02464				13 11 51	+31 19.9	2	9.56	.000	0.06	.000	0.07	.000	4	A3 V	1068,1690
114928	−63 02651				13 11 52	−64 00.1	1	10.10		0.04		−0.18		1	B9 V	1688
114938	−61 03529				13 11 53	−61 52.9	1	10.11		0.12		−0.33		1	B5/7 V	1688
		WEIS 13796			13 11 54	+30 09.0	1	11.09		0.69				1		1692
		G 14 - 39			13 11 54	−03 50.1	4	12.86	.033	0.92	.019	0.38	.047	10		202,419,1658,3060
114912	−69 01784	HR 4994			13 11 56	−69 24.9	2	6.36	.005	1.21	.000	1.33	.000	7	K2/3 III	15,1075
		HA 81 # 45			13 11 57	+13 59.5	1	12.60		0.51		−0.06		1		269
115337	+81 00416	HR 5009		⋆ AB	13 11 57	+80 44.1	1	6.25		0.94				2	K0 Ib	71
115023	−16 03624	IDS13093S1713		AB	13 11 57	−17 29.3	1	9.53		0.54		0.04		2	F3/5 V	1375
	+10 02527				13 11 59	+09 36.9	1	10.14		0.92		0.60		1		280
	−60 04551	LSS 2991			13 11 59	−60 42.1	3	9.81	.015	0.53	.009	−0.35	.024	11	B1 IIIne	540,976,1688
		He3 881			13 11 59	−61 58.	1	12.62		0.71		−0.17		4		1684
	+44 02256				13 12 00	+44 10.7	1	10.17		1.10		0.99		1	K0 III	280
114952	−62 03103				13 12 00	−62 41.2	1	9.96		0.39		0.30		1	A3	1688
115046	+12 02572	HR 4998		⋆ AB	13 12 02	+11 35.8	2	5.67	.000	1.50	.000	1.86	.000	4	M0 III	15,1003
		CS 22877 # 11			13 12 03	−09 20.0	1	13.86		0.60		−0.04		1		1736
115227	+73 00587	HR 5003			13 12 06	+73 03.8	2	6.59	.005	0.08	.000	0.07	.005	5	A2 V	985,1733
115062	−9 03646				13 12 08	−10 06.3	1	6.94		1.58		1.88		2	K5	3071
115074	+16 02483				13 12 09	+15 33.0	1	8.68		0.44		−0.06		2	F5	1569
	+47 02035				13 12 09	+46 44.0	1	10.28		1.35		1.61		1		280
	+39 02621				13 12 10	+39 04.0	1	10.70		1.02		0.82		1	G8 III	280
	+28 02200				13 12 11	+28 15.4	1	10.83		0.55		0.09		1	G0	1019
	+14 02588				13 12 14	+14 09.3	1	10.51		1.04		0.86		1	K0	280
	+25 02614				13 12 15	+24 41.8	1	10.34		0.43		−0.08		2		1569
115103	+30 02384				13 12 15	+29 40.0	3	8.61	.060	0.49	.010	0.05	.005	3	F6 V	1019,1690,3026
326343	−41 07688				13 12 15	−42 11.9	3	10.61	.021	0.41	.026	−0.40	.019	7		1406,1583,1707
115000	−61 03532				13 12 15	−62 07.3	1	9.44		0.21		0.06		1	B9 III	761
		CS 22877 # 2			13 12 17	−11 52.7	1	14.37		−0.04		−0.12		1		1736
115031	−54 05505				13 12 18	−55 04.5	1	8.34		0.64				4	G2/3 V	2033
115032	−54 05503	IDS13093S5451		AB	13 12 18	−55 07.1	1	9.40		0.02		−0.31		1		1771
115080	−10 03635	IDS13097S1050		AP	13 12 19	−11 06.1	2	7.04	.005	0.68	.010	0.12		6	G0	2033,3077
	+33 02314				13 12 20	+33 11.3	1	9.79		0.53		0.01		2	G0 III	1569
		CS 16027 # 63			13 12 21	+28 39.0	1	15.00		0.37		−0.20		1		1744
115050	−35 08547	HR 4999			13 12 21	−36 06.4	4	6.18	.011	0.97	.006			18	K1 III	15,1075,2013,2028
		CS 16027 # 59			13 12 22	+31 47.0	1	13.93		0.09		0.03		1		1744
		G 62 - 22			13 12 25	+05 41.9	1	11.15		1.12		1.05		1		1620
115067	−36 08422				13 12 25	−37 09.7	3	8.06	.008	−0.10	.009	−0.48	.003	12	B8 II/III	55,1770,1775
	+28 02202				13 12 27	+28 26.6	1	10.50		0.46		−0.08		1		1019
		WEIS 14246			13 12 27	+29 58.7	1	11.75		1.04				1		1692
	−60 04557				13 12 29	−60 53.9	2	10.98	.000	0.71	.001	−0.38	.021	5		1684,1737
		G 199 - 64			13 12 30	+59 20.1	1	11.22		1.39				2	K6	1746
		WEIS 14250			13 12 31	+30 00.9	1	11.01		0.69				2		1692
	+40 02634				13 12 32	+40 09.3	1	9.03		1.18		1.29		1	K1 III	280
	−61 03538	AAS78,203 # 32			13 12 32	−61 33.9	1	10.13		0.85		0.42		2		1684
115042	−61 03539	LSS 2995			13 12 32	−61 36.0	6	9.07	.012	0.78	.018	−0.25	.032	18	B0	125,403,540,976,1684,1737
	+30 02385				13 12 33	+29 48.4	1	11.70		1.08		0.91		1	K0 IV	1019
		HA 81 # 8			13 12 34	+13 51.9	1	11.17		0.53		−0.06		1		269
	−61 03540	LSS 2996			13 12 34	−61 43.5	1	10.18		0.26		−0.53		1	B1.5IV	540

HD	DM	Other Id	N Rem	α_{1950}	δ_{1950}	S	V	σ_V	B–V	σ_{B-V}	U–B	σ_{U-B}	n	Spectrum	References
115034	−63 02662			13 12 34	−63 37.2	2	8.78	.018	0.08	.003	-0.76	.014	4	B1/2 V	540,976
	+25 02615			13 12 35	+25 28.7	1	10.26		0.99		0.81		1	K0	280
		CS 16027 # 94		13 12 35	+31 38.1	1	14.28		0.43		-0.17		1		1744
	−61 03541			13 12 36	−61 40.6	1	10.42		1.20		1.00		3		125
115165	+23 02551			13 12 37	+23 18.9	1	8.83		1.20		1.24		1	K2	280
		Ross 466		13 12 37	−10 45.1	1	11.75		0.81		0.34		2		1696
	+8 02688			13 12 39	+08 27.8	1	8.77		1.20		1.26		1	K2	280
	+19 02653			13 12 39	+19 15.4	1	9.68		0.96		0.74		1		280
115122	−16 03627	UW Vir		13 12 40	−17 12.4	1	9.05		0.30		0.16		3	A8 IV/V	3077
	−36 08429			13 12 42	−36 43.6	1	10.40		0.58		-0.08		3		3009
	−60 04558	LSS 2997		13 12 42	−60 33.7	5	9.48	.015	0.54	.021	-0.48	.034	12	O6 f	403,540,976,1684,1737
	−61 03542			13 12 43	−62 02.3	1	11.19		0.21		-0.16		3		125
115166	+19 02654			13 12 44	+19 11.0	1	6.57		1.47				2	K0	280
	+22 02562			13 12 44	+22 03.6	1	10.51		1.17		1.13		1		280
238208	+57 01428			13 12 45	+57 16.9	2	9.72	.028	0.82	.008	0.40		2	K2 V	1007,8023
115095	−49 07760			13 12 45	−50 13.6	2	8.51	.010	-0.04	.010	-0.38		5	B7 IV	1770,2014
	+34 02406			13 12 46	+33 52.9	1	9.58		1.06		1.00		1	K0	280
	+35 02427			13 12 47	+34 51.8	1	9.22		1.32		1.58		2	K2	1569
		WEIS 44162		13 12 48	+27 46.	1	13.04		1.46				1		784
115182	+30 02386			13 12 49	+30 01.3	2	9.35	.005	1.13	.011	1.04		2	G9 III	694,1019
115197	+36 02352			13 12 50	+35 42.6	1	6.85		0.15		0.11		3	A2	1068
		CS 22877 # 5		13 12 50	−11 09.7	1	14.81		0.06		0.12		1		1736
115071	−61 03544	LSS 2998		13 12 50	−62 19.2	4	7.96	.016	0.24	.014	-0.71	.023	10	O9/B0 V	94,401,540,976,1684
115183	+16 02486			13 12 51	+16 27.5	1	7.49		1.49		1.86		1	K2	280
	+41 02372			13 12 51	+40 39.7	1	9.88		1.01		0.88		1	K0	280
		CS 16027 # 60		13 12 55	+29 27.5	1	14.60		0.66		0.12		1		1744
	+43 02314	TV CVn		13 12 57	+42 31.8	1	9.79		1.48		1.04		1	M6	280
115268	+60 01456			13 12 57	+59 33.6	2	8.21	.020	0.24	.010			9	A3	1118,1379
	−62 03111			13 13 00	−62 27.0	2	10.19	.005	0.18	.003	-0.63	.009	2	B1 V	540,976
115114	−61 03546	LSS 3000		13 13 03	−61 29.9	7	9.65	.154	0.30	.036	-0.65	.053	19	Be	92,94,403,540,976*
	+29 02384			13 13 04	+28 45.3	1	10.63		0.99		0.83		1	K0 III	1019
115143	−50 07622			13 13 04	−50 37.9	1	8.49		-0.04		-0.26		1	B8 V	1770
115231	+9 02736	G 62 - 23		13 13 08	+09 16.9	1	8.45		0.66		0.17		1	G5	333,1620
115157	−46 08519			13 13 08	−46 25.6	2	8.45	.001	-0.07	.000	-0.36		5	B8 IV/V	1770,2014
		LSS 3002		13 13 10	−61 46.3	1	11.93		0.32		-0.39		4	B3 Ib	1684
	+13 02647			13 13 11	+13 02.4	1	9.00		0.38		0.00		1	F2	1569
	−61 03549	LSS 3003		13 13 11	−62 17.7	2	10.11	.045	1.07	.007	-0.12	.014	4		94,1684
	−36 08436			13 13 12	−36 44.4	1	11.14		0.76		-0.36		1		1753
115269	+53 01608			13 13 13	+52 32.6	1	9.05		0.60		0.03		2	G0	1566
	+12 02575			13 13 14	+12 24.5	1	10.71		0.48		-0.06		1		1569
	+12 02576	G 61 - 45	★ AB	13 13 15	+12 09.5	3	9.31	.012	0.86	.012	0.60	.017	3	K0	333,1569,1620,1658
115256	+29 02385			13 13 15	+29 00.3	1	7.29		1.31		1.53		1	K3 III	1690
115174	−50 07626			13 13 15	−50 41.5	1	8.59		0.29		0.06		1	A9 III/IV	1771
115202	−19 03653	HR 5001		13 13 16	−19 40.7	6	5.22	.011	1.02	.007	0.88	.015	19	K1 III	15,244,803,1075,2007,4001
115271	+41 02374	HR 5004		13 13 17	+41 07.1	2	5.79	.015	0.19	.001	0.12		6	A7 V	70,253
	+71 00646			13 13 17	+70 34.5	2	10.13	.000	0.75	.000	0.17	.000	4	G0	516,1064
115244	+15 02557			13 13 18	+15 13.8	1	8.60		0.38		-0.05		2	F2	1569
		WEIS 13424		13 13 18	+30 17.9	1	11.59		0.57				1		1692
		CS 16027 # 58		13 13 18	+31 55.0	1	15.07		0.44		-0.21		1		1744
115148	−60 04563			13 13 18	−61 17.0	1	9.12		0.08		-0.41		2	B5 III	401
	+41 02375			13 13 22	+40 51.8	2	10.79	.054	0.23	.005	0.03	.000	3	A0	272,280
	+44 02259			13 13 22	+44 10.5	1	10.22		0.82		0.54		1	K0 III	280
	+12 02577			13 13 23	+11 42.9	1	10.19		0.93		0.68		1		280
234024	+54 01583			13 13 23	+53 48.0	1	9.23		0.31		0.02		2	A5	1566
	+30 02387			13 13 26	+29 36.7	3	9.58	.005	0.47	.006	0.13	.002	4	F3 V	1019,1690,1692
115206	−49 07772			13 13 26	−49 55.6	1	9.92		0.29		0.16		1	A8/F0	1771
115149	−64 02316	HR 5000		13 13 26	−64 52.4	4	6.06	.008	0.43	.007	-0.02	.019	16	F5 V	15,122,1075,2038
		G 149 - 65		13 13 29	+25 48.9	1	13.67		1.39				1		686
115289	+9 02737			13 13 30	+09 28.4	1	8.28		0.94		0.63		1	G5	280
		Feige 75		13 13 30	+13 14.	1	14.51		-0.21		-0.93		1		3060
		NGC 5053 - 2		13 13 30	+17 57.	1	16.08		0.69		0.24		1		86
		NGC 5053 - 14		13 13 30	+17 57.	1	15.19		0.80		0.26		1		589
		NGC 5053 - 56		13 13 30	+17 57.	1	16.43		0.73		0.08		1		86
		NGC 5053 - 63		13 13 30	+17 57.	1	16.54		0.20		0.09		1		86
		NGC 5053 - 68		13 13 30	+17 57.	1	15.10		0.76		0.28		1		86
		NGC 5053 - 85		13 13 30	+17 57.	1	16.00		0.77		0.40		2		86
		NGC 5053 - 86		13 13 30	+17 57.	1	16.33		0.79		0.23		1		86
		NGC 5053 - 88		13 13 30	+17 57.	1	15.26		0.86		0.43		1		86
		NGC 5053 - 90		13 13 30	+17 57.	1	15.96		0.76		0.35		1		86
		NGC 5053 - 94		13 13 30	+17 57.	1	16.43		0.69		0.23		1		86
		NGC 5053 - 97		13 13 30	+17 57.	1	15.13		0.88		0.50		1		86
		NGC 5053 - 1002		13 13 30	+17 57.	2	12.89	.025	0.71	.010	0.28	.005	5		86,589
		NGC 5053 - 1003		13 13 30	+17 57.	2	14.00	.004	1.14	.008	0.79	.027	11		86,589
		NGC 5053 - 1004		13 13 30	+17 57.	1	14.44		1.02		0.68		1		589
		NGC 5053 - 1005		13 13 30	+17 57.	1	14.53		0.87		0.46		1		589
		NGC 5053 - 1006		13 13 30	+17 57.	2	14.61	.010	0.88	.015	0.42	.015	2		86,589
		NGC 5053 - 1007		13 13 30	+17 57.	2	14.53	.073	1.04	.015	0.70	.131	3		86,589
		NGC 5053 - 1008		13 13 30	+17 57.	1	14.64		0.70		0.20		1		589
		NGC 5053 - 1009		13 13 30	+17 57.	2	14.80	.005	0.95	.010	0.59	.030	2		86,589
		NGC 5053 - 1010		13 13 30	+17 57.	1	14.82		1.02		0.41		1		589

Table 1   733

HD	DM	Other Id	N	Rem	α_{1950}	δ_{1950}	S	V	σ_V	B−V	σ_{B-V}	U−B	σ_{U-B}	n	Spectrum	References
		NGC 5053 - 1011			13 13 30	+17 57.	2	14.93	.017	0.66	.038	0.04	.008	14		589,589
		NGC 5053 - 1012			13 13 30	+17 57.	2	14.89	.050	1.00	.015	0.62		4		86,589
		NGC 5053 - 1013			13 13 30	+17 57.	1	14.97		1.00		0.36		3		589
		NGC 5053 - 1015			13 13 30	+17 57.	1	15.54		0.91		0.44		1		589
		NGC 5053 - 1016			13 13 30	+17 57.	1	15.61		0.86		0.28		2		589
		NGC 5053 - 1017			13 13 30	+17 57.	1	15.62		0.57		0.12		1		589
		NGC 5053 - 1018			13 13 30	+17 57.	1	16.05		0.77		0.30		2		589
		NGC 5053 - 1019			13 13 30	+17 57.	1	16.21		0.77		0.21		1		589
		NGC 5053 - 1020			13 13 30	+17 57.	1	16.36		0.81		0.34		1		589
		NGC 5053 - 1021			13 13 30	+17 57.	1	16.40		0.72		0.23		1		589
		NGC 5053 - 1022			13 13 30	+17 57.	1	16.50		0.54		-0.03		1		589
		NGC 5053 - 1023			13 13 30	+17 57.	1	16.52		0.50				1		589
		NGC 5053 - 1024			13 13 30	+17 57.	1	16.56		0.74		0.12		2		589
		NGC 5053 - 1025			13 13 30	+17 57.	1	16.57		0.25				1		589
		NGC 5053 - 1026			13 13 30	+17 57.	1	16.61		0.49				1		589
		NGC 5053 - 1027			13 13 30	+17 57.	1	16.64		0.24				1		589
		NGC 5053 - 1028			13 13 30	+17 57.	1	16.76		0.66		0.03		1		589
		NGC 5053 - 1029			13 13 30	+17 57.	1	16.81		0.12		0.06		1		589
		NGC 5053 - 1030			13 13 30	+17 57.	1	16.92		0.03		-0.02		1		589
		NGC 5053 - 1031			13 13 30	+17 57.	1	16.92		-0.01		0.06		1		589
		NGC 5053 - 1032			13 13 30	+17 57.	1	17.01		-0.08				1		589
		NGC 5053 - 1033			13 13 30	+17 57.	1	17.01		0.03		0.03		1		589
		NGC 5053 - 1034			13 13 30	+17 57.	1	17.04		0.59		0.06		1		589
		NGC 5053 - 1035			13 13 30	+17 57.	1	17.11		0.66				2		589
		NGC 5053 - 1036			13 13 30	+17 57.	1	17.14		-0.01		0.11		1		589
		NGC 5053 - 1050			13 13 30	+17 57.	1	15.83		0.65		0.22		1		86
		NGC 5053 - 1051			13 13 30	+17 57.	1	16.62		0.59		0.09		2		86
		NGC 5053 - 9508			13 13 30	+17 57.	1	16.60		0.24		0.05		1		86
		WEIS 41679			13 13 30	+29 05.	1	13.32		1.66				1		784
		HA 81 # 106			13 13 32	+14 09.8	1	13.15		0.57		0.00		1		269
115301	+22 02564				13 13 32	+21 38.8	1	7.43		0.02		-0.01		3	A3 V	1068
	−2 03660	G 14 - 41			13 13 32	−03 05.7	2	10.19	.015	0.71	.010	0.23	.020	2		1620,1658
		G 61 - 46			13 13 33	+14 31.8	1	13.02		1.39		1.09		1		1620
		WEIS 15570			13 13 34	+29 26.2	1	12.14		1.22				2		1692
115236	−44 08539	UY Cen			13 13 37	−44 26.4	2	6.70	.253	2.73	.100	4.06		6	SC	864,3039
		LSS 3006			13 13 38	−63 40.1	1	11.00		0.88		-0.05		2	B3 Ib	1737
		HA 81 # 157			13 13 40	+14 15.7	1	12.37		0.56		0.00		1		269
		WEIS 12828			13 13 41	+30 31.6	1	10.96		0.52				2		1692
115223	−59 04878	LSS 3008			13 13 41	−60 11.6	1	8.66		0.59		0.30		2	A0 Ib/II	403
		WEIS 13685			13 13 42	+30 06.3	1	11.09		0.57				1		1692
		HA 81 # 210			13 13 43	+14 32.5	1	14.44		0.70		0.36		1		269
	−62 03121				13 13 43	−63 05.7	1	10.69		0.52		-0.49		4		1684
		CS 16027 # 64			13 13 44	+28 21.4	1	14.23		0.48		0.01		1		1744
	+26 02443				13 13 47	+25 58.7	1	9.56		1.04		0.82		1	K0	280
	+40 02640				13 13 47	+39 31.5	1	9.91		0.99		0.77		1	G8 IV	280
		HA 81 # 211			13 13 48	+14 26.0	1	13.74		0.87		0.69		1		269
115319	+19 02655	HR 5007			13 13 48	+19 18.9	2	6.45	.005	0.97	.010	0.75		4	G8 III	280,3016
115321	+15 02558				13 13 49	+15 17.3	2	8.29	.015	0.37	.020	0.04	.045	5	F0	1569,1648
		WEIS 13695			13 13 49	+30 11.1	1	12.14		0.89				1		1692
115211	−66 02142	HR 5002			13 13 49	−66 31.2	4	4.86	.009	1.49	.009	1.58	.005	13	K2 Ib/II	15,1075,2038,3005
		G 164 - 55			13 13 50	+32 02.8	1	12.64		0.94		0.63		1	K3	1658
	+41 02376				13 13 50	+40 32.3	1	9.79		1.10		1.08		1	G8 V	280
	+41 02377				13 13 50	+41 20.5	1	10.44		1.16		1.10		1	K2 V	280
115338	+31 02466				13 13 51	+31 09.5	1	8.69		1.19				2	G5	280
115308	−0 02674	HR 5005, DK Vir			13 13 51	−01 07.6	2	6.68	.009	0.31	.009	0.06	.014	9	F1 IV	39,1088
115349	+36 02354	G 164 - 56			13 13 52	+36 09.1	1	8.28		0.64		0.12		1	G2 V	1658
115322	+7 02627	FH Vir			13 13 53	+06 46.1	1	7.15		1.60		1.55		11	M6 III	3071
		CS 16027 # 57			13 13 53	+32 12.7	1	14.60		0.01		0.05		1		1744
		AAS12,381 17 # A			13 13 55	+17 17.1	1	10.35		1.29		1.46		1		280
		HA 81 # 213			13 13 56	+14 28.2	1	15.14		0.50		0.17		1		269
	−18 03574				13 13 56	−18 51.9	1	10.54		-0.13				1		6009
115339	+28 02205	G 149 - 68			13 13 57	+28 00.2	2	8.24	.055	0.72	.015	0.27		7	G8 V	20,3026
		HZ 43			13 13 57	+29 21.5	1	12.67		-0.13		-1.21		1	DA	1744
	+45 02099				13 13 58	+45 30.1	1	10.13		0.94		0.69		1	K0 III	280
	+41 02378				13 13 59	+40 33.8	1	10.04		1.09		1.17		1	K1 III	280
	−61 03558	V593 Cen			13 13 59	−62 21.7	1	10.24		0.30		-0.52		2		94
		EG 98, HZ 43			13 14 00	+29 22.	2	12.73	.078	-0.11	.009	-1.16	.013	6	DAw	940,3028
	−61 03559	St 16- 117			13 14 01	−62 11.6	1	10.58		0.26		-0.51		2		1576
115352	+18 02708				13 14 03	+17 33.8	1	7.70		0.69		0.31		2	G0	1648
	−61 03560	St 16- 116			13 14 04	−62 10.6	1	10.32		0.43		0.34		1		1576
115402	+45 02100				13 14 05	+44 59.2	1	9.22		1.07		0.94		2	K1 III	280
115311	−33 08934	IDS13113S3404		AB	13 14 05	−34 20.1	2	8.05	.010	0.73	.005	0.25		9	G5 V	214,1414
115280	−54 05516	IDS13110S5417		A	13 14 05	−54 33.3	1	9.25		0.36		0.13		1	A7/8 V	1771
115310	−30 10457	HR 5006			13 14 06	−31 14.5	1	5.10		0.96				4	K0 III	2006
115365	+20 02814	HR 5010		⋆AB	13 14 07	+20 02.9	2	6.44	.005	0.26	.010	0.09	.005	5	F0 V	252,3016
115281	−59 04881	IDS13110S5915		AB	13 14 08	−59 30.7	1	8.88		0.03		-0.38		2	B9 II	401
115381	+19 02657				13 14 11	+19 10.0	1	6.73		0.97				1	K0	280
		LP 378 - 897			13 14 12	+23 31.	1	13.52		1.55				1		686
115282	−60 04569	IDS13110S6033		AB	13 14 12	−60 48.4	2	8.97	.010	0.16	.000	-0.59		6	B2 III	401,2014
		G 149 - 70			13 14 13	+28 08.2	2	13.27	.027	1.66	.027	1.32		5		203,686

HD	DM	Other Id	N	Rem	α_{1950}	δ_{1950}	S	V	σ_V	B–V	σ_{B-V}	U–B	σ_{U-B}	n	Spectrum	References
		CS 16027 # 62			13 14 14	+28 47.2	1	13.85		0.76		0.27		1		1744
	+44 02262				13 14 15	+43 31.7	1	9.44		0.98		0.84		1	K1 III	280
		LP 557 - 41			13 14 16	+08 26.4	1	11.13		0.87		0.55		2		1696
115383	+10 02531	HR 5011		★ A	13 14 18	+09 41.1	11	5.21	.011	0.58	.007	0.10	.008	26	F8 V	15,101,245,1008,1119*
		LP 378 - 932			13 14 18	+21 49.	1	13.98		1.26				1		686
	+44 02263				13 14 18	+43 51.9	1	10.03		1.07		1.00		1	K0 III	280
115382	+13 02648				13 14 19	+12 40.9	1	8.39		0.63		0.12		1	G5	1569
115403	+20 02815				13 14 19	+20 04.7	2	7.60	.005	0.26	.000	0.02	.019	6	F0 IV	1068,3032
115331	−43 08165	HR 5008			13 14 19	−43 43.0	5	5.84	.008	0.19	.013	0.08	.056	14	A2mA4-A9/F0	15,355,1770,2012,8071
115368	−12 03785	IDS13117S1237	A		13 14 20	−12 53.7	1	7.77		1.06		0.88		3	K0 III	1657
115404	+17 02611	G 63 - 18		★ AB	13 14 22	+17 17.0	3	6.54	.024	0.92	.021	0.62	.020	6	K2 V	22,1080,1620
	+18 02709	NGC 5053 - 1			13 14 24	+17 55.9	2	9.69	.014	1.16	.004	1.06		14	K2	86,589
		RT Com			13 14 25	+17 23.	2	13.56	.165	0.15	.035	0.07	.060	2		597,668
115444	+37 02388				13 14 25	+36 38.7	2	8.97	.009	0.78	.000	0.17	.005	8	K0	1569,1569
115390	−17 03808				13 14 26	−17 38.4	1	9.42		0.47		-0.03		11	F5 V	1603
115286	−67 02237	IDS13110S6758	A		13 14 26	−68 13.9	1	6.68		0.09		-0.16		2	B9 IV	1771
		CS 22877 # 28			13 14 27	−08 42.3	1	15.16		0.28		0.13		1		1736
115460	+44 02264				13 14 28	+44 06.2	1	8.46		1.08		1.12		1	K1 III	280
115425	+14 02590				13 14 30	+14 26.1	2	8.95	.010	1.28	.000	1.36	.090	2		269,280
115424	+15 02560				13 14 30	+14 44.5	1	9.11		1.08		0.98		1	G5	280
115316	−61 03566	St 16- 113		★	13 14 30	−62 11.4	6	9.33	.013	0.21	.020	-0.69	.023	20	B3/6 III/IV	540,761,976,1576*
		HA 81 # 166			13 14 31	+14 18.8	1	13.10		1.07		1.18		1		269
		CS 22877 # 31			13 14 31	−09 24.6	1	14.14		0.12		0.19		1		1736
	−60 04571				13 14 31	−60 28.7	1	10.14		0.55		-0.38		3		1684
		HA 81 # 167			13 14 32	+14 19.9	1	12.89		0.70		0.22		1		269
		AAS12,381 38 # A			13 14 33	+37 55.4	1	11.14		0.25		0.14		1		280
		HA 81 # 168			13 14 35	+14 14.8	1	12.29		0.88		0.50		1		269
115462	+29 02386				13 14 35	+29 18.5	1	8.45		1.31				1	K3 III	280
		Cr 268 - 1			13 14 35	−66 48.5	1	11.57		1.00		0.62		6		412
		Cr 268 - 3			13 14 35	−66 48.9	1	12.65		0.26		0.06		2		412
		Cr 268 - 2			13 14 36	−66 48.8	1	11.89		0.15		-0.39		2		412
115464	+20 02816				13 14 38	+19 33.7	1	8.53		1.03		0.86		1	K0	280
115463	+21 02509				13 14 38	+21 07.8	1	8.19		1.42		1.67		1	K0	280
115360	−53 05541				13 14 38	−54 19.8	1	9.37		0.35		0.15		1	A9 V	1771
		HA 81 # 170			13 14 41	+14 19.2	1	12.26		0.82		0.51		1		269
		CS 16027 # 66			13 14 41	+30 06.5	1	14.51		0.59		-0.05		1		1744
115396	−42 08271				13 14 41	−43 02.1	1	7.46		0.11		0.09		1	A1/2 V	1770
115345	−61 03568	St 16- 112			13 14 42	−62 11.5	3	9.06	.015	0.23	.026	-0.37	.010	7	B5/7 III	92,761,1576
	+10 02532				13 14 43	+10 06.6	1	9.68		1.18		1.22		1	K0	280
		Cr 268 - 10			13 14 45	−66 48.2	1	13.94		0.44		0.28		2		412
		G 164 - 61			13 14 46	+40 04.9	2	13.49	.010	0.60	.007	-0.10	.022	3		1658,3062
	+45 02101				13 14 46	+45 16.5	1	9.86		1.51		1.65		1	K2	280
		CS 22877 # 32			13 14 46	−09 42.2	1	14.45		0.23		0.05		1		1736
		LSS 3013		★ V	13 14 46	−62 10.3	1	11.86		0.86		-0.06		4	B3 Ib	1684
115478	+14 02591	HR 5013			13 14 47	+13 56.3	4	5.33	.002	1.30	.008	1.50	.010	20	K3 III	269,1080,1569,3016
	+23 02555				13 14 47	+22 50.6	1	10.13		1.20		1.27		1		280
115415	−46 08538				13 14 47	−47 19.6	2	6.65	.011	0.00	.004	-0.06		8	B9 V	1770,2014
	+3 02751	IDS13123N0332	A		13 14 48	+03 16.5	1	9.38		1.20		1.07		2		3032
	+3 02751	IDS13123N0332	B		13 14 48	+03 16.5	1	9.56		1.00		0.71		2		3032
		Feige 76			13 14 48	+13 03.	1	15.18		-0.12		-0.65		2		1298
115363	−63 02684	LSS 3014			13 14 49	−63 25.4	6	7.75	.021	0.62	.015	-0.44	.037	11	B1 Ia	138,403,540,976,1684,1737
115612	+69 00694	HR 5018			13 14 50	+68 40.3	2	6.20		-0.06	.000	-0.15	.009	3	B9.5V	985,1079
		Cr 268 - 12			13 14 50	−66 48.8	1	12.85		0.30		0.13		2		412
	+13 02649				13 14 52	+12 44.3	1	10.30		0.54		-0.11		2	F0	1569
	+33 02319				13 14 53	+32 35.4	1	9.50		1.27		1.33		1	K3 III	280
		vdB 60 # c			13 14 53	−62 25.1	1	12.02		0.49		-0.32		3		434
		Cr 268 - 9			13 14 54	−66 47.8	1	13.08		0.35		0.13		2		412
		Cr 268 - 4			13 14 54	−66 49.0	1	12.20		0.36		0.17		2		412
115488	+0 03040	HR 5014		★ AB	13 14 56	−00 24.8	2	6.36	.005	0.27	.005	0.03	.010	7	F0 V	15,1256
		Cr 268 - 13			13 14 57	−66 48.8	1	13.39		0.60		0.15		2		412
		HA 81 # 259			13 14 58	+14 41.4	1	11.87		0.99		0.67		1		269
		Cr 268 - 5			13 14 58	−66 49.4	1	12.68		0.28		0.06		2		412
		Cr 268 - 6			13 14 59	−66 49.9	1	11.75		1.20		0.99		2		412
		HA 81 # 125			13 15 00	+14 13.9	1	12.04		0.60		0.00		1		269
115519	+33 02320				13 15 00	+32 45.1	1	7.66		0.53		0.00		2	F8	1569
		AAS78,203 # 33			13 15 00	−62 12.	1	10.10		0.26		-0.58		2		1684
	+27 02230				13 15 01	+27 25.6	1	9.72		0.94		0.65		1	K0	280
115520	+31 02468				13 15 01	+30 52.6	1	8.42		0.19		-0.02		2	F0	1569
		CS 22877 # 36			13 15 01	−11 02.8	1	14.32		0.05		0.15		1		1736
115400	−62 03137	LSS 3016			13 15 02	−63 11.2	3	6.82	.019	0.75	.037	0.51	.019	7	F2 Ib	403,2012,8100
		Cr 268 - 7			13 15 02	−66 50.5	1	12.70		0.48		0.01		2		412
115521	+6 02722	HR 5015			13 15 05	+05 44.0	6	4.81	.017	1.65	.013	1.93	.015	27	M2 IIIa	3,15,1075,1363,2029,3055
115538	+18 02710				13 15 06	+17 33.3	1	7.51		1.00		0.74		2	G5	1569
		AAS78,203 # 34			13 15 06	−62 34.	1	11.10		1.46		0.46		4		1684
115556	+24 02559				13 15 07	+23 54.5	2	9.55	.010	0.16	.005	0.09	.055	3	A5 III	280,1068
115419	−61 03573				13 15 07	−61 57.2	1	9.30		0.16		0.02		1	B9/9.5III	761
115537	+19 02658				13 15 09	+19 28.7	1	7.94		1.44		1.75		1	K2	280
115346	−74 01044				13 15 09	−74 34.5	1	8.38		0.86				4	G8/K0 V	2033
	+13 02650				13 15 10	+13 04.5	1	10.05		0.52		-0.05		2	F8	1569
115539	+14 02593				13 15 10	+14 01.6	6	7.22	.008	0.95	.007	0.63	.004	27	G8 III-IV	269,882,1003,1080*

Table 1 735

HD	DM	Other Id	N Rem	α_{1950}	δ_{1950}	S	V	σ_V	B–V	σ_{B-V}	U–B	σ_{U-B}	n	Spectrum	References
115572	+37 02391	IDS13129N3721	AB	13 15 10	+37 04.7	1	8.58		1.16		1.21		2	K0 III	1569
115470	−43 08175			13 15 10	−43 47.5	1	6.80		−0.02		−0.09		3	B9.5 V	1770
115436	−56 05698			13 15 10	−57 02.2	1	7.08		0.13		0.06		2	B9 III	1771
115558	+19 02659			13 15 11	+19 17.1	1	8.40		0.92		0.58		1	G5	280
		St 16 - 64		13 15 11	−62 14.5	1	12.68		0.37		−0.19		4		1576
115626	+59 01499			13 15 12	+59 00.8	2	9.00	.015	0.38	.010	−0.05		9	F5	1118,1371
		St 16- 124		13 15 12	−62 13.7	1	13.19		0.96		0.48		2		1576
		St 16 - 97		13 15 13	−62 13.4	1	13.55		0.36		0.15		5		1576
	+36 02357			13 15 14	+35 58.8	1	9.64		0.29		0.07		2	A2	272
	−62 03141	vdB 60 # d		13 15 14	−62 24.0	2	10.57	.019	0.45	.000	−0.46	.032	6		434,1684
	+14 02594			13 15 16	+14 07.0	1	10.25		0.49		0.02		2		1569
		G 61 - 49		13 15 18	+17 48.1	1	11.43		0.58		−0.09		2		1696
		CS 16027 # 67		13 15 18	+31 57.4	1	14.12		0.57		0.00		1		1744
115604	+41 02380	HR 5017, AO CVn		13 15 18	+40 50.1	9	4.73	.010	0.30	.007	0.21	.010	32	F3 III	15,39,253,254,1008*
115473	−57 05981			13 15 18	−57 52.9	1	9.08		0.27		−0.26		2	WC	1684
		St 16- 100		13 15 19	−62 12.7	1	13.84		1.41		1.24		2		1576
115495	−49 07797			13 15 20	−49 39.0	1	9.11		−0.07		−0.49		1	B9 II/III	1770
115455	−61 03575	St 16- 111		13 15 20	−62 13.7	6	7.97	.015	0.19	.015	−0.78	.016	29	O7.5III((f))	92,94,540,976,1576,1684
		St 16- 121		13 15 21	−62 16.1	1	12.32		1.78				3		1576
	−62 03144			13 15 22	−62 33.7	1	11.55		0.23		0.17		3		144
	−61 03576	St 16- 107		13 15 23	−62 16.0	4	9.50	.018	0.23	.025	−0.68	.015	18	B0.5V	92,94,761,1576
		CS 22877 # 38		13 15 24	−11 43.0	1	13.55		0.02		0.10		1		1736
	−61 03577	St 16- 114		13 15 24	−62 04.9	1	9.45		1.71		1.95		1		1576
		Cr 268 - 8		13 15 24	−66 56.	1	13.28		0.33		0.20		2		412
		Cr 268 - 11		13 15 24	−66 56.	1	13.52		0.35		0.12		2		412
		Cr 268 - 14		13 15 24	−66 56.	1	14.05		0.32		0.13		2		412
		Cr 268 - 15		13 15 24	−66 56.	1	14.14		1.02		0.54		1		412
		Cr 268 - 16		13 15 24	−66 56.	1	13.32		0.87		0.30		2		412
		Cr 268 - 17		13 15 24	−66 56.	1	13.78		0.35		0.19		2		412
		Cr 268 - 18		13 15 24	−66 56.	1	14.31		0.35		0.16		2		412
		Cr 268 - 19		13 15 24	−66 56.	1	14.49		0.88		0.47		2		412
		Cr 268 - 20		13 15 24	−66 56.	1	14.41		0.53		0.29		2		412
		Cr 268 - 21		13 15 24	−66 56.	1	13.63		0.32		0.19		2		412
		Cr 268 - 22		13 15 24	−66 56.	1	13.16		0.32		0.26		2		412
		ST Com		13 15 25	+21 02.7	1	10.91		0.18		0.20		1	F5	597
	−61 03579	St 16 - 29		13 15 25	−62 15.0	3	10.46	.012	0.23	.017	−0.67	.019	9	B2 V	92,94,1576
		St 16- 126		13 15 26	−62 05.6	1	13.40		1.57				1		1576
		Cen sq 1 # 109		13 15 26	−62 32.7	1	11.80		0.42		0.38		3		144
		RV Com		13 15 27	+18 57.	1	13.92		0.20		0.09		1		597
115507	−42 08282			13 15 28	−43 20.7	1	8.60		−0.05		−0.18		3	B9.5 V	1770
115527	−45 08372			13 15 28	−45 30.1	1	6.86		0.00		−0.06		2	A0 V	1770
		St 16- 119		13 15 28	−62 07.9	1	12.20		1.08		1.08		1		1576
	−61 03581	St 16 - 27		13 15 29	−62 14.3	3	10.11	.012	0.25	.015	−0.64	.022	12	B2 V	94,144,1576
	−61 03580	St 16 - 26		13 15 29	−62 15.0	2	11.60	.008	0.25	.004	−0.48	.016	9		94,1576
		St 16 - 31		13 15 29	−62 15.6	1	14.37		0.45		0.32		2		1576
	+13 02652			13 15 30	+12 30.5	1	8.81		1.11		1.03		2	K0	280
		Sand 262		13 15 30	+21 13.	1	13.67		0.98				1		686
		St 16- 122		13 15 30	−62 06.2	1	12.54		1.35		1.04		1		1576
		Cen sq 1 # 113		13 15 30	−62 31.0	1	13.00		0.35		−0.15		3		144
	−59 04894			13 15 32	−59 32.8	2	10.27	.018	0.04	.001	−0.71	.009	3	B1 V	540,976
		St 16 - 9		13 15 32	−62 13.1	1	13.49		0.97		0.50		1		1576
		St 16 - 10		13 15 32	−62 13.3	1	13.50		1.04		0.57		3		1576
		St 16 - 11		13 15 32	−62 13.6	1	10.18		0.24		−0.61		4		92
115677	+58 01437			13 15 33	+58 05.1	2	7.80	.010	1.49	.010			8	K0	1118,1379
	−61 03583	St 16- 118		13 15 33	−62 15.2	2	11.53	.052	0.32	.008	−0.30	.000	9		94,1576
115606	+13 02653			13 15 34	+13 15.8	4	8.55	.009	0.31	.006	0.11	.006	24	A2	193,269,270,1569
		CS 22877 # 27		13 15 34	−08 35.3	1	15.02		0.06				1		1736
115529	−50 07660	HR 5016		13 15 34	−51 01.4	3	6.18	.005	0.01	.007	−0.05		8	A0 V	15,1770,2012
		St 16- 125		13 15 34	−62 15.1	1	13.35		0.50		−0.18		2		1576
115613	+28 02207			13 15 35	+27 43.0	2	8.57	.025	0.57	.010	0.05		6	F8 V	20,3026
		St 16 - 12		13 15 35	−62 13.6	1	13.38		0.45		0.28		5		1576
		St 16 - 33		13 15 35	−62 15.9	1	13.84		1.00		0.88		1		1576
		HA 81 # 126		13 15 36	+14 04.9	1	13.90		0.76		0.29		1		269
		L 196 - 30		13 15 36	−55 52.	1	12.94		0.96		0.72		2		3062
		St 16 - 25		13 15 36	−62 14.9	2	12.49	.009	0.31	.014	−0.18	.014	4		94,1576
	+28 02208	IDS13133N2836	A	13 15 37	+28 19.7	1	9.36		0.96		0.70		2	K1 III	280
		St 16 - 34		13 15 37	−62 15.8	1	13.51		0.38		0.04		4		1576
115439	−71 01458	HR 5012		13 15 37	−71 46.3	3	6.03	.003	1.35	.000	1.46	.000	16	K3 III	15,1075,2038
		HA 81 # 78		13 15 38	+14 01.0	1	14.79		0.57		0.11		1		269
115577	−27 09145			13 15 38	−28 04.1	2	6.80	.005	0.98	.000	0.65		5	K0 III	2012,3077
115512	−55 05483			13 15 38	−56 19.6	1	10.01		0.14		0.10		1	A9 V	1771
		HA 81 # 82		13 15 42	+14 01.9	1	13.28		0.62		0.04		1		269
	+28 02209			13 15 42	+28 24.5	1	9.04		0.48		−0.06		1	G0	280
115514	−61 03585	V378 Cen		13 15 42	−62 07.2	2	8.32	.050	0.92	.010	0.62		2	F5 Iab/b	688,8100
		St 16- 127		13 15 42	−62 13.8	1	13.76		0.36		0.17		4		1576
115564	−49 07803			13 15 43	−49 36.9	2	7.35	.001	−0.02	.003	−0.22		7	B8 III	1770,2014
115533	−61 03586	LSS 3026	★ AB	13 15 46	−61 37.9	1	10.08		0.08		−0.76		2	B2/5 Ib/II	401
		St 16 - 23		13 15 46	−62 15.2	1	12.80		0.37		0.02		5		1576
115720	+59 01500	IDS13138N5917	AB	13 15 47	+59 00.8	1	8.04		1.05				5	G9 IV	1118
115617	−17 03813	HR 5019	★ A	13 15 47	−18 02.0	23	4.74	.008	0.71	.007	0.26	.011	254	G5 V	1,3,15,22,30,116,125*

HD	DM	Other Id	N Rem	α_{1950}	δ_{1950}	S	V	σ_V	B–V	σ_{B-V}	U–B	σ_{U-B}	n	Spectrum	References
		LP 378 - 974		13 15 48	+21 55.	1	13.88		1.55				1		686
	−61 03587	St 16- 109		13 15 48	−62 18.4	3	10.73	.005	0.29	.010	-0.54	.020	9	O9.5V	92,94,761
		vdB 60 # b		13 15 48	−62 19.2	1	12.70		0.35		-0.28		3		434
	+40 02641			13 15 49	+39 34.5	1	9.66		1.08		1.01		1	K0 III	280
		CS 22877 # 30		13 15 49	−08 51.4	1	15.26		0.10				1		1736
		St 16- 120		13 15 50	−62 06.4	1	12.32		0.49		0.37		5		1576
		St 16- 123		13 15 50	−62 06.6	1	13.18		0.66		0.18		5		1576
		St 16 - 21		13 15 50	−62 15.3	1	13.74		0.58		0.40		3		1576
		HA 81 # 84		13 15 52	+13 59.7	1	14.28		0.53		-0.06		1		269
		HA 81 # 85		13 15 52	+14 02.0	1	13.39		0.88		0.44		1		269
		St 16- 115		13 15 52	−62 03.8	1	10.13		1.23		0.84		1	F5 Ib	1576
	+42 02386			13 15 53	+42 16.7	1	9.25		1.13		1.09		1	K0	280
		G 238 - 30		13 15 54	+64 31.0	1	12.91		0.47		-0.28		3		1658
		HA 81 # 86		13 15 55	+13 57.3	1	15.67		0.51		-0.06		1		269
		LTT 13872		13 15 55	+64 31.1	2	13.29	.010	0.60	.010	-0.12	.020	3		1064,3025
		CS 22877 # 20		13 15 56	−07 33.5	1	13.86		0.13		0.17		1		1736
	+25 02618			13 15 57	+25 08.1	1	10.36		1.08		1.03		1	G5	280
	+12 02580			13 15 58	+12 11.8	1	10.00		0.96		0.74		2		1569
115706	+51 01829			13 15 58	+50 42.6	1	7.95		0.43		-0.06		2	F2	1566
	+24 02560			13 16 00	+23 36.5	1	9.22		1.51		1.75		1	K7	280
		Ru 107 - 8		13 16 00	−64 41.1	1	13.64		0.48		0.31		2		412
115598	−55 05486			13 16 05	−56 01.0	1	8.62		-0.04		-0.44		2	B7 III	401
		Ru 107 - 7		13 16 05	−64 40.9	1	13.92		0.31		0.02		2		412
115735	+50 01994	HR 5023, BK CVn		13 16 07	+49 56.7	2	5.14	.010	-0.03	.037	-0.20		3	A0 V	1363,3023
115569	−64 02343			13 16 07	−64 56.7	1	9.65		0.34		0.16		1	A9 IV/V	1771
	+11 02553			13 16 08	+10 59.4	1	10.54		1.11		1.08		1		280
115678	+13 02655			13 16 08	+13 10.5	1	8.33		1.56				1	M1	280
115722	+36 02360			13 16 08	+36 13.0	1	9.54		0.31		0.08		1	Am	280
115723	+34 02410	HR 5022		13 16 09	+34 21.6	1	5.82		1.35				2	K4.5III	70
		Ru 107 - 6		13 16 09	−64 41.3	1	12.67		0.54		0.27		2		412
	+11 02554			13 16 10	+11 04.6	1	9.93		0.99		0.70		1	K0	280
	+45 02102			13 16 12	+44 50.9	1	10.35		0.96		0.69		1	G8 IV	280
115659	−22 03554	HR 5020	★ A	13 16 12	−22 54.5	12	2.99	.010	0.92	.003	0.65	.010	57	G8 III	3,15,244,1007,1013*
115708	+27 02234	HH Com		13 16 14	+26 37.7	2	7.82	.005	0.25	.001	0.05	.007	10	A2p	1068,1202
		CS 16027 # 65		13 16 14	+29 10.1	1	14.04		0.50		0.02		1		1744
115644	−35 08592			13 16 15	−35 50.8	1	10.24		0.47		0.01		2	F3/F5 V	1696
115622	−58 04801	IDS13131S5814	AB	13 16 16	−58 30.0	1	9.44		0.50		0.19		1	A9 V	1771
115583	−66 02164			13 16 16	−67 06.1	1	7.27		0.00		-0.19		2	B9 V	1771
	+12 02581			13 16 17	+11 54.8	2	9.20	.029	0.17	.010	0.11	.010	3	A2	272,280
		Ru 107 - 5		13 16 17	−64 41.0	1	12.53		0.37		-0.02		2		412
115585	−70 01589			13 16 17	−70 35.5	1	7.42		0.75				4	G5/6 IV/V	2012
	+12 02582			13 16 18	+12 05.7	1	7.98		1.55		1.93		1	K5	280
115751	+43 02318			13 16 18	+43 17.9	1	8.26		0.94		0.78		1	K0 III	280
	−61 03598	LSS 3031		13 16 18	−62 23.0	3	10.30	.012	0.83	.010	-0.15	.012	8	B5 Iap	92,94,1684
		Ru 107 - 4		13 16 18	−64 40.8	1	13.13		0.45		0.06		2		412
115709	+4 02721	HR 5021		13 16 19	+03 57.0	2	6.61	.005	0.06	.000	0.02	.005	7	A1 IV	15,1256
		Ru 107 - 1		13 16 20	−64 41.1	2	11.19	.010	1.44	.004	0.87	.009	8		412,1685
		Ru 107 - 2		13 16 21	−64 41.4	1	11.92		0.64		0.25		2		412
		Ru 107 - 17		13 16 22	−64 41.7	1	13.95		1.04		0.77		2		412
	+26 02447			13 16 23	+26 03.3	1	10.13		1.51		1.44		1	M5	280
	+27 02235			13 16 24	+26 34.8	1	8.97		1.09		1.04		1	K2	280
	+28 02210			13 16 24	+28 24.8	1	9.58		1.05		0.91		1	K0	280
		Feige 77		13 16 24	+29 02.	1	15.11		0.56		-0.15		1		3060
		G 14 - 45		13 16 24	−02 48.5	6	10.85	.014	1.01	.009	0.69	.023	16	K4:	516,1064,1620,1705,3003,3073
	+28 02211	IDS13140N2839	AB	13 16 25	+28 23.3	1	9.65		0.92		0.77		1	K2	280
		Ru 107 - 3		13 16 25	−64 41.1	1	12.85		0.38		-0.11		2		412
		Ru 107 - 16		13 16 25	−64 41.4	1	13.67		0.88		0.30		2		412
115752	+32 02338			13 16 26	+32 26.0	1	8.16		0.27		0.06		5	F0	1569
	+47 02046			13 16 26	+46 33.8	1	9.55		1.01		0.91		1	G5	280
	−61 03599			13 16 26	−62 03.8	1	10.48		0.48		0.30		3		144
115625	−62 03159			13 16 27	−62 46.4	1	9.06		0.14		0.12		1	B9 IV	761
		Ru 107 - 15		13 16 27	−64 40.0	1	14.37		0.43		0.20		2		412
115737	+13 02656			13 16 28	+12 38.2	2	8.29	.010	0.98	.000	0.75		3	K2	280,882
		Ru 107 - 14		13 16 28	−64 40.2	1	12.98		0.36		0.15		2		412
115762	+25 02619	G 149 - 77	★ A	13 16 29	+24 52.4	1	8.52		0.71		0.28		2	G2 V	3026
		Ru 107 - 10		13 16 29	−64 41.1	1	14.50		0.36		0.30		2		412
		Ru 107 - 9		13 16 30	−64 41.2	1	13.29		0.31		-0.13		2		412
	−22 09861			13 16 31	−22 45.2	1	10.26		0.83		0.39		2		1696
115736	+15 02561			13 16 32	+14 44.1	3	8.98	.029	1.28	.014	1.28	.015	4	K0	280,882,3040
		Ru 107 - 11		13 16 32	−64 40.6	1	12.53		0.42		-0.04		2		412
115781	+34 02411	BL CVn		13 16 33	+33 42.1	1	8.13		1.14		0.94		1	K1 V	280
		Ru 107 - 12		13 16 33	−64 40.4	1	13.49		0.83		0.72		2		412
		Cen sq 1 # 112		13 16 34	−62 33.3	1	12.09		0.25		0.02		3		144
		Ru 107 - 13		13 16 34	−64 40.2	1	13.86		1.13		0.25		2		412
115652	−61 03600			13 16 35	−61 42.4	1	7.97		0.04		-0.32		3	B8 II/III	401
	−62 03166			13 16 39	−62 30.6	1	10.77		0.16		-0.09		3		144
		Cen sq 1 # 108		13 16 43	−62 17.7	1	11.56		0.69		0.22		3		144
115764	+10 02535			13 16 44	+09 42.8	1	8.42		1.29		1.37		1	K2	280
116459	+85 00222			13 16 45	+85 00.9	2	7.27	.010	0.51	.010	0.03	.005	4	G0	1733,3026
115810	+35 02435	HR 5025	★ D	13 16 46	+35 23.4	1	6.02		0.24				2	F0 IV	70

Table 1 737

HD	DM	Other Id	N	Rem	α_{1950}	δ_{1950}	S	V	σ_V	B–V	σ_{B-V}	U–B	σ_{U-B}	n	Spectrum	References
115715	−52 06393				13 16 47	−52 34.7	1	9.06		0.15		0.12		1	B9.5/A0 V	1770
	+27 02236				13 16 52	+27 26.6	1	10.34		0.98		0.76		1		280
		Cen sq 1 # 111			13 16 52	−62 02.6	1	12.95		0.69		0.20		3		144
		HA 81 # 93			13 16 55	+13 55.3	1	11.72		0.66		0.14		1		269
	+32 02340				13 16 55	+31 43.7	1	10.58		1.17		1.22		1		280
	+42 02388				13 16 56	+42 03.9	1	9.70		0.31		0.00		2	F2	1569
115770	−29 10263				13 16 56	−30 04.2	1	8.53		-0.11		-0.58		6	B5 III	1732
115704	−61 03608	LSS 3033		⋆ AB	13 16 57	−61 44.6	5	8.07	.021	0.46	.021	-0.54	.033	13	B0.5Iab	401,403,540,976,1684
	+25 02620				13 16 59	+25 23.5	1	9.30		0.92		0.56		1	K0	280
	+40 02645				13 17 00	+39 58.6	1	9.30		1.10		1.10		1	K1 III	280
115831	+9 02742				13 17 02	+08 54.4	1	8.21		1.50		1.79		1	K2	280
		GD 156			13 17 03	+20 33.9	1	16.29		0.48		-0.17		1		3060
	+27 02237				13 17 03	+27 10.0	1	9.34		1.10		0.99		1	K0	280
		G 177 - 31			13 17 04	+45 20.8	1	14.13		0.03		-0.56		1	DAs	3060
115671	−69 01808				13 17 04	−69 25.1	1	7.47		0.25		0.29		1	A2 V	1771
		CS 16027 # 74			13 17 06	+28 19.1	1	15.21		0.41		-0.21		1		1744
115883	+45 02104				13 17 06	+45 21.8	1	8.73		0.46		0.04		2	F5	1569
115773	−40 07824				13 17 06	−40 55.7	1	6.74		0.52				4	F6 V	1075
	−62 03173				13 17 06	−62 29.0	1	10.79		0.30		-0.54		2	B2 V	434
	+32 02341				13 17 07	+32 11.6	1	10.56		1.10		1.11		1	K3 III	280
115744	−55 05498	IDS13140S5514	A		13 17 07	−55 30.2	1	8.64		0.26		0.17		1	A3 V	1771
115772	−39 08187				13 17 08	−39 40.6	2	9.66	.019	0.84	.000	0.38		3	G6wF7/G0	1594,6006
115760	−52 06400				13 17 08	−53 12.7	1	8.57		-0.01		-0.18		1	B9/9.5 V	1770
115830	+10 02536	IDS13146N1013	AB		13 17 09	+09 57.4	1	7.86		0.34		0.05		2	F0	1569
326338	−41 07748				13 17 10	−41 28.2	1	9.96		0.47		0.33		2	F0 III	1707
	+35 02436	G 164 - 64		⋆ A	13 17 11	+35 24.2	1	9.52		1.47		1.18		3	M1	3072
	+35 02436	G 164 - 64		⋆ AB	13 17 11	+35 24.2	1	9.49		1.49				9	M1	1197
	+35 02436	G 164 - 65		⋆ B	13 17 11	+35 24.2	2	12.09	.000	1.59	.000	1.52		4		3072,8105
115856	+22 02570				13 17 12	+22 29.0	1	9.71		1.08		1.10		1	K2	280
		Ross 470			13 17 16	−09 34.7	1	11.88		1.28		1.25		1		1696
115898	+46 01862	V CVn			13 17 17	+45 47.4	1	7.54		1.51		1.16		2	M1	1569
	−61 03612				13 17 17	−62 05.6	1	10.93		0.77		-0.33		2		1684
115746	−62 03174				13 17 17	−63 09.0	1	9.41		0.31		-0.65		4	B2/4 III/V	1684
	+23 02561				13 17 18	+23 19.5	1	9.55		0.70		0.22		1	K0	280
115673	−73 01134	T Mus			13 17 19	−74 10.8	1	7.84		2.84		6.90		3	Np	864
		G 164 - 66			13 17 22	+33 36.6	1	10.66		1.50		1.22		1	M2	679
115778	−59 04912	HR 5024		⋆ A	13 17 23	−59 30.6	1	6.18		0.43				4	F3/5 II	2032
		Feige 78			13 17 24	+29 40.	2	13.77	.019	0.06	.015	0.02	.000	3		1744,3060
	+7 02634	G 62 - 30			13 17 25	+07 07.2	2	9.78	.000	0.64	.000	0.02	.000	3	G0	927,1620
115928	+36 02362				13 17 26	+36 20.7	1	7.18		0.97				2	G5	280
115885	+9 02743				13 17 27	+08 45.0	1	7.08		1.13		1.12		1	K0	280
	+46 01863				13 17 27	+45 45.5	1	10.35		0.80		0.24		2	G0	1569
115820	−48 08118				13 17 27	−48 57.7	1	7.98		0.22		0.06		1	A7/8 V	1771
115900	+17 02616				13 17 30	+16 55.4	1	9.54		1.10		0.96		2		1569
		CS 16027 # 72			13 17 30	+28 33.9	1	14.29		0.57		-0.05		1		1744
115927	+37 02396				13 17 30	+36 38.1	1	7.62		1.04				2	G5	280
115929	+28 02213				13 17 32	+28 22.0	2	8.24	.020	0.45	.010	0.00		5	F6 V	20,3026
	+39 02630				13 17 35	+39 27.8	1	9.88		1.41		1.40		1	K2 III	280
115823	−52 06405	HR 5026			13 17 35	−52 29.1	6	5.47	.017	-0.13	.004	-0.53	.002	17	B5 III/IV	15,26,1770,2007,8015,8023
115805	−59 04914	LSS 3036			13 17 35	−59 43.7	3	9.74	.023	0.31	.026	-0.70	.012	9	B1 Vnne	540,1684,2021
115792	−62 03178				13 17 35	−63 05.7	1	10.24		0.34		0.23		1	A8/F0 [V]	1771
115901	+10 02537				13 17 36	+10 15.7	1	8.79		1.05		0.90		1	K0	280
		Feige 80			13 17 36	+12 19.9	2	11.37		-0.10	.000	-1.02	.019	3	sdO+A?	1026,1264
115953	+48 02108	IDS13154N4818	AB		13 17 36	+48 02.4	3	8.45	.018	1.48	.002	1.19	.033	5	K0	938,1758,3072
		G 14 - 46			13 17 38	−01 24.8	1	13.49		1.56				2		202
		CS 16027 # 68			13 17 39	+33 14.2	1	14.94		0.48		-0.17		1		1744
		AV Vir			13 17 40	+09 27.7	1	11.42		0.23		0.12		1	A4	597
		V 840 Cen			13 17 40	−55 34.9	1	14.14		1.30		0.30		1		1753
115954	+39 02631				13 17 41	+38 37.8	1	8.34		0.64		0.25		6	G5	1569
115842	−55 05504	HR 5027, LSS 3038			13 17 41	−55 32.3	10	6.02	.011	0.30	.009	-0.71	.013	48	B0.5Ia/ab	15,285,540,681,815,976*
115794	−64 02356				13 17 41	−65 17.2	1	9.55		0.40		0.27		1	A9 V	1771
	+12 02586				13 17 42	+11 30.5	1	10.19		1.11		1.11		1		280
		LP 437 - 59			13 17 42	+15 08.3	1	12.39		0.72		0.18		2		1696
115942	+16 02493	IDS13153N1605	A		13 17 42	+15 49.6	1	7.25		1.08		1.03		1	G5	280
115903	−10 03655				13 17 42	−11 02.5	1	6.73		1.00		0.79		3	K0	1311
		CS 16027 # 73			13 17 45	+28 18.8	1	14.19		0.48		-0.18		1		1744
115968	+38 02431	G 164 - 67			13 17 47	+38 25.3	2	8.00	.005	0.71	.010	0.31	.005	3	G8 III	1569,1658
115892	−36 08497	HR 5028			13 17 47	−36 26.9	6	2.75	.014	0.04	.009	0.03	.020	17	A2 V	15,1020,1075,2012*
		LP 617 - 36			13 17 48	−02 45.8	1	11.04		0.58		-0.10		1		1064
		G 14 - 47			13 17 50	−01 23.6	1	11.61		1.51				2		202
	+42 02390				13 17 52	+41 38.5	1	10.49		0.85		0.52		1	G8 III	280
	+16 02494				13 17 56	+15 42.9	1	9.31		1.55		1.82		1	M0	280
		WLS 1320-20 # 10			13 17 56	−19 42.2	1	12.21		1.20		1.20		2		1375
115981	+18 02717				13 18 00	+17 50.5	2	7.30	.005	0.39	.000	-0.06	.010	4	F2 V	1569,1648
115912	−46 08580	HR 5029			13 18 00	−46 37.1	3	5.76	.004	1.11	.010			11	K1 III	15,1075,2007
115846	−66 02171				13 18 00	−67 16.5	2	7.05	.010	-0.04	.009	-0.54	.012	7	B3 IV	26,158
		CS 16027 # 69			13 18 01	+31 13.7	1	14.83		0.29		-0.01		1		1744
		Cen sq 1 # 106			13 18 01	−62 06.8	1	11.39		1.56		1.79		3		144
		Cen sq 1 # 105			13 18 01	−62 27.7	1	11.28		1.09		0.76		3		144
		Ross 471			13 18 02	−12 10.4	1	12.92		0.89		0.31		2		1696

HD	DM	Other Id	N Rem	α_{1950}	δ_{1950}	S	V	σ_V	B–V	σ_{B-V}	U–B	σ_{U-B}	n	Spectrum	References
	+13 02657			13 18 03	+12 30.6	1	9.48		0.50		-0.05		1	F8	280
	+27 02238			13 18 04	+26 54.6	1	9.05		1.44		1.78		1	K7	280
116010	+40 02647	HR 5032		13 18 05	+40 24.7	1	5.60		1.21				3	K1 III	70
	+18 02718			13 18 06	+17 34.2	1	9.65		1.16		1.21		1		280
	+22 02572			13 18 08	+21 52.9	1	10.07		0.99		0.86		1	K0	280
115915	-51 07453			13 18 08	-51 24.6	3	7.55	.009	-0.05	.007	-0.31	.012	9	B8/9 IV	401,1770,2014
115862	-62 03186			13 18 08	-62 57.7	1	10.73		0.09		-0.71		2	F5 V	434
115995	+3 02758	HR 5031	⋆ AB	13 18 09	+03 12.2	2	6.25	.005	0.10	.000	0.09	.005	7	A3 V	15,1417
115994	+9 02745			13 18 09	+09 11.1	1	8.97		1.00		0.78		1	K5	280
	+12 02589			13 18 09	+12 27.7	1	10.20		1.02		0.74		1		280
	+13 02658			13 18 10	+12 48.0	2	9.29	.000	1.41	.005	1.66	.010	2	K5	272,280
116044	+45 02105			13 18 12	+45 22.3	1	8.66		0.94		0.60		1	G8 IV	280
116012	+4 02729	G 62 - 33		13 18 13	+04 23.5	4	8.58	.009	0.94	.015	0.75	.011	8	K2 V	22,333,1003,1620,3008
116091	+60 01458			13 18 13	+60 07.4	1	8.01		0.63				5	F8	1118
	-7 03595			13 18 14	-07 38.1	1	10.75		1.00		0.78		1		1696
115879	-65 02279			13 18 14	-65 46.3	1	8.35		0.10		0.11		1	A1 V	1771
	+10 02539			13 18 15	+10 18.0	1	9.81		0.46				4	F5 V	193
		G 63 - 21		13 18 15	+11 08.4	1	11.62		1.49				1	M1	1746
116029	+25 02623			13 18 16	+24 54.7	1	7.86		1.00		0.90		1	K1 III	280
116056	+43 02321			13 18 16	+43 22.4	1	8.13		0.81		0.51		2	K2 III	1569
	-61 03623			13 18 16	-62 23.6	1	10.44		0.20		-0.06		3		144
		G 63 - 22		13 18 17	+11 09.3	1	15.95		1.56				1		906
		CS 22877 # 26		13 18 18	-08 23.1	1	13.52		0.22		0.13		1		1736
	-60 04609			13 18 18	-60 32.3	1	11.07		0.41		-0.44		4		1684
		Cen sq 1 # 110		13 18 18	-62 14.8	1	12.15		1.86		2.12		3		144
	+20 02826			13 18 20	+20 06.0	1	9.78		1.14		1.12		1		280
115985	-21 03700			13 18 20	-22 07.3	1	7.27		0.96		0.51		2	G6/8 III	1375
	+10 02540	G 63 - 23		13 18 21	+10 11.8	2	9.89	.010	1.14	.000	1.06	.039	3	K8	1696,3072
	+12 02590			13 18 22	+11 53.4	1	10.02		0.26		0.12		1	A5	280
	+36 02363			13 18 23	+35 58.0	1	9.54		1.55		1.87		1	K2	280
	+14 02600			13 18 25	+13 38.1	1	9.11		1.44		1.72		1	K5	280
	+10 02541			13 18 28	+09 34.2	1	9.17		1.04		0.87		1	K0	280
	-44 08587			13 18 28	-45 10.9	1	11.49		0.54		-0.17		2	G5	1696
115987	-39 08202			13 18 30	-39 43.7	2	7.48	.003	-0.01	.001	-0.06	.002	4	B9.5V	55,1770
115988	-42 08333			13 18 31	-43 22.3	2	6.69	.003	-0.02	.004	-0.11		7	B9 V	1770,2014
	+35 02439	G 164 - 68		13 18 34	+34 33.3	1	10.62		1.51		1.16		10	M0	7009
		G 199 - 66		13 18 34	+56 50.5	1	12.68		0.50		-0.21		2		1658
	+29 02394			13 18 36	+28 54.2	1	9.96		1.36		1.55		1		280
116188	+65 00927			13 18 37	+65 22.0	1	8.67		0.17		0.08		2	A2	1733
115951	-63 02716			13 18 37	-64 22.4	1	8.80		0.18		-0.13		1	A3/5 V	1771
116127	+44 02265			13 18 40	+44 15.1	2	6.52	.005	0.41	.000	-0.03	.000	7	F3 V	985,1501
	+42 02392			13 18 41	+42 00.8	1	9.07		1.09		1.11		1	K0 III	280
115990	-54 05559	IDS13156S5454	A	13 18 41	-55 09.3	1	7.76		0.00		-0.13		3	B9 IV/V	1770
116002	-48 08130			13 18 43	-49 13.3	1	9.63		0.29		0.18		1	A2mA7-A8	1771
	+16 02496			13 18 44	+15 33.2	1	10.10		1.10		1.03		1		280
	+34 02414			13 18 44	+33 36.6	1	10.13		1.11		1.07		1	K0 III	280
	+24 02570			13 18 45	+23 45.5	1	8.88		0.97		0.69		1	K0	280
116170	+54 01588			13 18 47	+54 07.3	1	9.07		0.95		0.56		2	K0	1733
	+14 02601			13 18 48	+14 06.6	1	10.16		0.79		0.36		1	K3	280
116061	-18 03587	HR 5033		13 18 48	-19 13.6	4	6.20	.008	0.10	.009	0.08		15	A2 V	15,244,252,2013,2029
		Wray 1093		13 18 50	-63 19.5	1	12.80		0.80		-0.37		4		1684
115977	-63 02720	UX Cen		13 18 50	-63 57.5	1	7.62		2.50		4.68		2	Nb	864
116048	-34 08828			13 18 51	-34 38.1	1	8.18		0.96				4	G8 III	2012
116110	+14 02602			13 18 54	+14 24.7	1	7.28		1.49				1	K2	280
116038	-50 07708			13 18 54	-51 01.2	1	7.09		0.21		0.07		2	A3/4 III/I	1771
116003	-59 04926			13 18 54	-59 55.4	6	6.92	.024	0.01	.011	-0.83	.011	14	B1 II	158,401,403,540,976*
	+38 02436			13 18 55	+38 04.5	1	9.10		0.54		0.02		2	M2	1569
116156	+38 02435			13 18 55	+38 07.1	1	6.92		0.53		0.07		2	F8 V	1569
116064	-38 08457			13 18 55	-39 03.1	6	8.80	.015	0.47	.010	-0.21	.010	17	F/Gw(A)	1594,2012,2013,2017*
	+46 01864			13 18 56	+45 46.3	1	9.97		1.04		0.77		1	K0 IV	280
		CS 22877 # 45		13 18 57	-11 13.6	1	13.31		0.28		0.15		1		1736
	+42 02394			13 18 58	+41 55.1	1	8.88		1.13		1.14		1	K0 III	280
116053	-49 07856			13 18 58	-49 39.1	2	7.76	.005	-0.05	.010	-0.46		7	B3 III	2014,3036
		Ross 472		13 19 00	-11 37.6	1	12.18		0.75		0.25		2		1696
116158	+34 02415			13 19 02	+33 33.3	1	8.94		1.15		1.20		1	K0	280
	+44 02267	AV CVn		13 19 03	+44 13.0	2	9.80	.042	1.89	.009	2.15	.033	4	S2.9:	280,1569
	+12 02591			13 19 04	+12 05.8	1	9.42		0.96		0.68		1	K0	280
		G 177 - 34		13 19 06	+46 39.0	1	14.55		0.00		-0.66		1		3060
	+28 02218			13 19 08	+28 22.8	1	9.58		1.14		1.06		1	K5	280
115967	-71 01467	HR 5030	⋆ AB	13 19 08	-71 53.1	5	6.05	.016	0.08	.008	-0.34	.005	23	B5 V	15,1075,1637,2012,2038
116160	+2 02664	HR 5037		13 19 09	+02 21.0	3	5.69	.006	0.06	.008	0.03	.006	8	A2 V	15,695,1256
		AAS12,381 14 # A		13 19 09	+13 47.0	1	10.61		0.75		0.23		1		280
116159	+18 02721			13 19 09	+18 29.2	1	8.72		1.10		1.04		1	K0	280
		CS 22877 # 34		13 19 09	-10 06.6	1	12.62		0.07		0.16		1		1736
116084	-51 07465	HR 5036		13 19 13	-51 55.3	5	5.84	.015	0.11	.004	-0.72	.013	14	B2 Ib	400,401,1770,2007,8100
116173	+16 02498			13 19 14	+16 20.6	1	7.10		1.36		1.60		1	K0	280
		G 14 - 48		13 19 16	-02 21.9	1	12.11		1.40				2		202
116204	+39 02635	BM CVn		13 19 17	+39 08.5	3	7.19	.027	1.16	.005	1.02	.020	6	K2	105,280,1569
	+44 02268			13 19 21	+44 24.8	1	9.26		0.92		0.58		1	G8 IV	280
116272	+61 01353	G 199 - 67		13 19 21	+61 20.7	1	8.46		0.76		0.34		2	G5	1733

Table 1

739

HD	DM	Other Id	N Rem	α_{1950}	δ_{1950}	S	V	σ_V	B–V	σ_{B-V}	U–B	σ_{U-B}	n	Spectrum	References
116072 −60 04627	HR 5034, V790 Cen	★ C	13 19 21	−60 42.6	4	6.22	.056	0.01	.012	−0.59	.010	14	B2.5 Vn	13,26,1088,1771
+9 02750			13 19 23	+08 53.2	1	10.26		1.00		0.89		1	K2	280
116087 −60 04627	HR 5035	★ AB	13 19 23	−60 43.6	5	4.53	.024	−0.14	.007	−0.61	.020	13	B3 V	13,15,26,1075,8023
	Cen sq 1 # 308		13 19 25	−61 45.9	1	12.17		0.69		0.21		3		144
+37 02400			13 19 26	+36 44.5	1	9.45		1.18		1.24		2	K0	1569
116231 +43 02323			13 19 26	+43 14.4	1	8.61		1.56		1.94		1	K5	280
116175 −11 03498			13 19 29	−12 19.1	1	6.88		1.61		2.06		7	K5 III	1024
116134 −49 07862			13 19 30	−50 08.1	1	9.60		0.26		0.14		1	A7/9 V	1771
116117 −53 05572			13 19 30	−54 14.7	1	7.94		0.69		−0.10		3	B8/9 II/III	8100
	He3 894		13 19 32	−64 17.	1	10.87		0.98		0.34		4		1684
116232 +26 02450			13 19 34	+26 15.6	1	7.61		0.97		0.71		1	G8 III	280
+28 02220	IDS13175N2810	A	13 19 34	+27 53.4	1	10.50		0.60		0.11		3		1569
116246 +40 02649			13 19 35	+39 38.5	2	8.76	.025	−0.06	.020	−0.11	.015	4	A1 V	280,1068
116233 +25 02625			13 19 37	+25 08.6	1	7.07		0.22		0.14		1	A9 III	280
116235 +5 02737	HR 5040		13 19 38	+05 25.0	3	5.86	.004	0.11	.010	0.07	.032	10	A2 m	15,1417,8071
116119 −61 03638	LSS 3050		13 19 39	−61 45.1	3	7.89	.017	0.71	.008	−0.07	.015	8	B9.5Ia	403,1684,2012
+21 02517			13 19 42	+20 33.2	1	9.89		1.07		0.96		1	K0	280
	CS 16027 # 76		13 19 42	+29 41.8	1	14.11		0.70		0.14		1		1744
−59 04936			13 19 43	−59 29.6	1	9.96		1.02		0.77		2	K2	837
+41 02386			13 19 44	+40 35.9	1	9.44		1.09		1.12		1	K1 III	280
	Cen sq 1 # 306		13 19 44	−61 44.5	1	11.58		1.85		2.22		3		144
−61 03639	LSS 3052		13 19 44	−62 10.8	4	10.13	.027	0.84	.011	−0.24	.024	9	B1 Ia	403,549,1684,1737
116273 +36 02366	IDS13175N3609	A	13 19 46	+35 53.5	1	8.25		0.44		−0.02		2	F8 V	1569
	CS 22877 # 46		13 19 46	−11 01.8	1	14.05		0.07		0.17		1		1736
	BC Vir		13 19 50	+06 08.7	1	12.02		0.23		0.08		1		699
+19 02663			13 19 51	+19 07.6	1	9.68		1.09		0.97		2	K0	1569
+27 02243	IDS13175N2741	AB	13 19 51	+27 25.5	1	9.16		1.35		1.62		1	K7	280
−61 03642			13 19 51	−62 18.5	1	11.34		0.49		0.20		1	A5	549
	CS 16027 # 77		13 19 52	+31 37.2	1	15.22		0.04		0.10		1		1744
116303 +44 02269	HR 5045		13 19 53	+44 09.9	3	6.35	.016	0.25	.009	0.08	.009	7	A7 m	280,1569,8071
116197 −47 08260	HR 5038	★ AB	13 19 53	−47 40.9	2	6.17	.009	0.18	.000	0.17		5	A4 V	1771,2007
	Lod 821 # 20		13 19 54	−59 25.1	1	11.53		0.18		0.18		2	A0	837
116287 +31 02477			13 19 55	+30 33.2	1	6.77		1.34				2	K2 III	280
116168 −60 04630			13 19 55	−60 53.7	1	9.21		0.12		−0.58		2	B3 II	401
+27 02245	IDS13176N2703	AB	13 19 56	+26 46.8	1	8.77		0.77		0.36		2	G0	1569
116289 +9 02752			13 19 59	+09 08.7	1	9.05		0.17		0.18		1	A3	280
−62 03217	LSS 3055		13 20 3	−62 52.9	1	11.69		0.37		−0.63		2	B0 Vn	1737
	G 164 - 69		13 20 00	+29 09.1	1	12.65		0.64		−0.03		2		1658
116288 +21 02519			13 20 01	+20 57.5	1	7.28		1.37		1.61		1	K0	280
116199 −58 04835			13 20 01	−58 43.6	1	9.40		0.31		0.19		1	A8 IV/V	1771
116186 −59 04938			13 20 02	−59 58.0	1	8.09		1.11		0.87		4	G5 II	1673
−61 03647			13 20 02	−62 04.6	1	11.48		0.34		−0.32		1	B3	549
−61 03646			13 20 02	−62 17.2	1	11.23		0.29		−0.44		1	B5	549
116226 −47 08261	HR 5039		13 20 03	−48 18.1	4	6.37	.006	−0.08	.007	−0.45	.012	11	B5 V	26,1770,2007,3036
+18 02724			13 20 04	+17 57.2	1	10.28		1.03		0.96		2		280
−59 04941			13 20 05	−59 25.6	1	10.47		0.98		0.72		2	K2	837
−61 03648			13 20 05	−62 20.8	1	10.32		0.66		0.17		1	F8	549
+24 02574			13 20 06	+23 43.4	1	10.44		1.22		1.45		1		280
116275 −12 03802	IDS13175S1240	A	13 20 07	−12 55.5	1	7.77		0.15		0.14		2	A2 V	3032
116275 −12 03801	IDS13175S1240	B	13 20 07	−12 55.5	1	10.82		0.50		−0.04		2		3032
+25 02627			13 20 10	+24 42.2	1	9.12		1.15		1.08		1	K5	280
+30 02391			13 20 11	+30 11.6	1	9.76		1.34		1.58		1	K2	280
	Ton 704		13 20 12	+32 25.6	1	14.67		0.51		−0.08		1		1744
+29 02401			13 20 13	+28 46.9	1	9.55		0.96		0.60		1	K0	280
+31 02478	IDS13178N3103	A	13 20 13	+30 46.9	1	11.05		0.22		0.00		2		1569
	CS 22877 # 49		13 20 14	−11 43.0	1	13.70		0.00		0.05		1		1736
	Lod 807 # 37		13 20 14	−62 11.7	1	12.28		0.55		0.15		1	F0	549
	Lod 807 # 42		13 20 16	−62 16.6	1	12.97		0.41		0.26		1		549
+28 02221			13 20 18	+28 09.8	1	9.39		0.93		0.60		1	K0	280
116329 +26 02453			13 20 19	+26 06.0	1	9.10		0.44		0.20		3	F7 V	3026
116228 −59 04946			13 20 19	−59 29.0	1	8.73		1.71		1.90		2	K2 III	837
116229 −59 04945			13 20 20	−60 06.9	1	8.29		0.24		0.10		1	A5 V	1771
116292 −16 03650	HR 5044		13 20 20	−17 28.4	3	5.36	.030	0.99	.005	0.74	.015	12	K0 III	244,2006,3016
116278 −32 09322	HR 5043		13 20 20	−32 55.7	2	6.19	.024	1.64	.005	1.96		6	M2 III	2007,3055
116269 −49 07875			13 20 22	−49 47.1	1	10.20		0.02		−0.21		2	B9 III	3036
	Lod 807 # 43		13 20 23	−62 14.7	1	13.00		0.38		0.23		1		549
	Lod 821 # 13		13 20 24	−59 24.8	1	10.75		1.10		1.03		2	K4	837
	Lod 821 # 23		13 20 24	−59 28.3	1	11.84		0.21		0.08		2	A0	837
−59 04947			13 20 26	−59 31.0	1	10.61		0.17		−0.15		2	B9	837
+28 02222			13 20 27	+28 01.3	1	10.16		1.34		1.48		1		280
+29 02402			13 20 27	+29 07.4	1	9.82		1.24		1.34		2	K0	1569
116344 +18 02725			13 20 28	+18 02.2	1	8.68		0.23		0.07		2	A7 III	1569
+19 02664			13 20 28	+19 13.2	1	10.78		0.32		0.04		3		1569
116345 +17 02620			13 20 29	+17 19.6	1	9.22		0.92		0.72		1		280
	G 238 - 32		13 20 29	+68 56.1	1	12.66		0.60		−0.09		2	F9	1658
	Lod 807 # 28		13 20 29	−62 11.9	1	11.35		1.70		1.75		1	K9	549
+21 02521			13 20 30	+20 41.5	1	9.92		1.53		1.86		1	K5	280
	CS 16027 # 75		13 20 31	+28 59.0	1	14.62		0.59		0.06		1		1744
116363 +17 02621			13 20 32	+16 31.9	1	8.84		0.18		0.11		3	A2	1068
116364 +16 02501			13 20 33	+15 30.7	3	7.59	.005	1.26	.016	1.31	.029	5	K0	280,882,3040

HD	DM	Other Id	N Rem	α_{1950}	δ_{1950}	S	V	σ_V	B–V	σ_{B-V}	U–B	σ_{U-B}	n	Spectrum	References
		AAS61,331 # 3		13 20 35	+13 33.	1	10.54		0.29		0.08		2		1569
116405	+45 02106			13 20 35	+44 58.5	3	8.35	.010	-0.07	.007	-0.17	.053	5	A0 V	272,280,1569
		G 149 - 81		13 20 36	+24 44.1	1	12.93		1.62		1.12		1		3078
	-61 03652			13 20 36	-61 49.4	1	11.26		0.53		0.31		1	F0	820
116379	+19 02665			13 20 37	+18 32.3	1	7.92		0.16		0.09		3	A5 Vm	1068
	-59 04950	IDS13174S5914	AB	13 20 37	-59 29.2	1	10.36		0.07		-0.45		2	B5	837
116243	-63 02732	HR 5041		13 20 38	-64 16.5	4	4.52	.004	0.84	.006	0.48	.014	13	G6 II	15,122,1075,2006
116394	+29 02403			13 20 39	+28 47.0	1	8.88		1.40		1.69		1	K5 V	280
		AAS41,43 F4 # 3		13 20 39	-62 16.1	1	12.00		0.48		0.31		1	A3	820
		G 255 - 32		13 20 40	+74 28.2	2	11.64	.009	0.48	.002	-0.22	.009	3		1064,1658
		He3 876		13 20 40	-61 51.	1	11.95		0.61		0.14		4		1684
		Lod 807 # 40		13 20 40	-62 10.2	1	12.48		0.31		0.19		1	A0	549
	+32 02344	IDS13184N3213	A	13 20 41	+31 56.5	1	9.76		0.82		0.50		1	G8 V	1773
116406	+39 02639			13 20 41	+38 37.3	1	9.16		1.02		0.84		1	K0 III	280
	-61 03653			13 20 41	-61 57.8	2	10.63	.000	0.86	.005	0.30	.025	2	F8	549,820
116282	-59 04951	LSS 3059		13 20 42	-59 32.9	4	9.66	.013	0.39	.008	-0.60	.028	13	B0 IV	540,837,976,1684
116407	+29 02404			13 20 43	+29 20.1	1	8.89		1.32		1.56		1	K5 III	280
116365	-4 03469	HR 5047		13 20 43	-04 39.8	2	5.88	.005	1.43	.000	1.66	.000	7	K3 III	15,1256
		AAS61,331 # 4		13 20 44	+13 59.	1	10.87		0.52		0.08		1		1569
		Lod 807 # 31		13 20 44	-62 05.0	1	11.59		1.52		1.42		1		549
		Ton 706		13 20 45	+32 08.4	1	15.33		-0.11		-0.25		1		1744
		LP 323 - 56		13 20 46	+31 59.0	1	14.35		1.54		1.20		1		1773
		Lod 821 # 19		13 20 48	-59 33.0	1	11.42		0.16		-0.06		2	B9	837
		Lod 807 # 39		13 20 48	-62 07.1	1	12.46		0.57		0.24		1		549
	-63 02736			13 20 48	-63 25.4	2	10.46	.024	0.74	.008	-0.39	.020	6		1684,1737
	-59 04953			13 20 49	-59 25.8	1	10.99		0.32		0.17		2	A2	837
116300	-59 04954			13 20 50	-59 37.1	1	10.60		0.09		-0.42		2	A0	837
116338	-49 07884	HR 5046		13 20 51	-49 33.7	3	6.49	.013	0.96	.000			11	G8 III	15,1075,2032
		AM Vir		13 20 52	-16 24.7	2	11.14	.070	0.27	.015	0.11	.005	2	A6	597,700
		Lod 807 # 32		13 20 54	-62 19.1	2	11.80	.040	0.19	.025	0.00	.000	2	B9	549,820
		CS 16027 # 80		13 20 55	+33 01.1	1	14.77		0.61		0.05		1		1744
116326	-52 06454			13 20 55	-53 19.8	1	8.88		0.03		-0.03		1	B9.5 V	1770
	-59 04955			13 20 56	-59 30.0	1	9.85		0.49		0.02		2	F8	837
	-61 03656			13 20 56	-62 11.4	3	11.26	.021	0.15	.017	-0.28	.021	3	B8	287,549,820
	+44 02270			13 20 57	+44 17.7	1	9.34		1.18		1.10		1	G8 V	280
116313	-61 03658			13 20 57	-61 59.9	2	9.44	.005	1.45	.010	1.44	.030	2	K2	549,820
		Lod 807 # 35		13 20 57	-62 17.2	2	11.91	.020	0.40	.020	0.27	.010	2	A0	549,820
	-61 03657			13 20 57	-62 18.4	2	11.88	.010	0.31	.015	0.26	.020	2	A0	549,820
	+39 02640			13 20 58	+38 57.2	1	9.21		1.00		0.76		1	G8 III	280
		He3 878		13 20 58	-61 47.	1	11.80		0.41		-0.39		2		1684
	+17 02624			13 21 00	+17 11.4	1	9.76		1.16		1.28		1		280
116441	+18 02726			13 21 00	+17 33.4	1	7.53		1.16		1.23		1	K0	280
116476	+40 02653			13 21 00	+39 53.9	1	8.38		1.18		1.14		1	K0 III	280
	-59 04956			13 21 00	-59 33.2	1	10.92		0.07		-0.46		2	B6	837
	-59 04957			13 21 01	-59 28.7	1	10.61		1.22		1.08		2	K5	837
116356	-53 05586			13 21 02	-53 31.9	1	7.41		0.07		0.03		1	A1 V[n]	1771
116314	-61 03659			13 21 03	-62 19.3	2	9.89	.005	0.04	.015	-0.47	.000	2	B5 V	549,820
	-61 03660			13 21 04	-62 13.9	3	10.70	.017	0.24	.005	0.03	.005	3	A0	287,549,820
	-61 03661			13 21 05	-62 16.4	2	11.50	.020	0.19	.005	-0.26	.020	2	B8	549,820
116442	+3 02765	IDS13186N0314	A	13 21 06	+02 58.9	1	7.06		0.78		0.38		5	G5	1084
		Ross 985		13 21 07	-12 46.8	2	11.84	.045	1.41	.005			2	K7	1705,1746
116443	+3 02766	IDS13186N0314	B	13 21 08	+02 59.0	1	7.36		0.83		0.51		5	G5	1084
116244	-74 01057	HR 5042		13 21 08	-74 37.5	2	5.04	.005	1.10	.005	1.02	.005	7	K0 III	15,1075
116444	+0 03050	IDS13186S0012	AB	13 21 09	-00 27.5	1	8.82		0.62				2	G5	1414
116428	-17 03841			13 21 09	-17 59.6	1	8.86		0.43		-0.01		2	F2 V	1375
116358	-59 04959			13 21 09	-59 38.0	1	10.26		0.26		0.26		2	A7 III	837
		CS 22877 # 21		13 21 10	-07 45.6	1	15.32		0.09				1		1736
		Lod 807 # 33		13 21 10	-62 16.0	1	11.76		1.16		0.79		1		549
		AAS41,43 F4 # 12		13 21 11	-62 14.4	1	11.87		0.46		0.16		1	A2	820
		Lod 807 # 38		13 21 13	-62 13.6	2	12.38	.015	0.20	.020	-0.19	.005	2	B7	549,820
	+13 02661			13 21 14	+13 12.5	1	9.26		0.48		0.00		1	F5	1569
116495	+29 02405	G 164 - 71	★ AB	13 21 14	+29 29.7	2	8.90	.001	1.36	.023	1.20	.005	6	K5	938,3072
	+41 02390			13 21 14	+41 11.8	1	10.42		0.91		0.65		1	G8 III	280
	-61 03663			13 21 14	-61 43.4	1	10.97		0.67		0.08		3		144
	+14 02607			13 21 15	+14 18.5	1	9.94		0.98		0.66		1	K0	280
116445	-13 03696			13 21 15	-14 16.4	1	10.35		-0.01		-0.72		4	K1 IV	400
116388	-53 05591			13 21 15	-53 44.3	2	7.58	.002	0.07	.003	0.05	.003	36	A1 V	1699,1771
116359	-60 04645			13 21 16	-61 03.6	1	9.81		0.15		-0.61		2	B3 II/III	401
		Lod 821 # 24		13 21 17	-59 33.8	1	11.95		0.23		0.18		2	A2	837
116479	+17 02625			13 21 18	+17 10.9	1	8.76		1.40		1.67		1	K0	280
		HZ 44		13 21 18	+36 23.	1	11.71		-0.27		-1.16		3	sdO	1026
	-59 04960			13 21 18	-59 26.2	1	11.28		0.52		0.03		2	F8	837
		AAS41,43 F4 # 15		13 21 19	-62 02.6	1	11.66		0.85		0.24		1	B6	820
	-61 03664			13 21 19	-62 11.6	3	11.16	.013	0.32	.009	0.22	.011	3	A0	287,549,820
		Lod 807 # 36		13 21 20	-62 18.8	2	12.07	.050	0.27	.050	0.16	.000	2	A0	549,820
116499	+18 02728			13 21 21	+18 28.6	1	8.54		1.56		1.83		1	K2	280
		Lod 821 # 22		13 21 21	-59 34.6	1	11.57		0.50		0.19		2	F6	837
116373	-59 04961			13 21 21	-60 00.2	1	9.26		0.35		0.15		1	A8 V	1771
	-61 03667			13 21 21	-62 01.6	2	9.73	.024	0.67	.015	0.12	.034	3	F2	549,820
	-61 03666			13 21 21	-62 02.9	2	9.40	.015	1.25	.005	1.11	.015	2		549,820

Table 1 741

HD	DM	Other Id	N	Rem	α_{1950}	δ_{1950}	S	V	σ_V	B–V	σ_{B-V}	U–B	σ_{U-B}	n	Spectrum	References
	+46 01866				13 21 22	+45 55.4	1	8.98		1.05		0.94		1	K0	280
		Lod 821 # 25			13 21 22	−59 28.7	1	12.04		0.15		−0.31		2	B8	837
		Lod 807 # 41			13 21 22	−62 04.8	1	12.65		0.57		0.18		1		549
	+47 02056				13 21 23	+47 28.2	1	10.34		0.14		0.06		2	A2	272
	−61 03668				13 21 24	−62 19.4	2	10.22	.005	0.67	.000	0.17	.015	2	F5	549,820
		G 63 - 24			13 21 25	+11 29.5	2	11.23	.005	1.41	.010	1.18		2	M0	1620,1746
116516	+19 02668				13 21 25	+18 38.2	1	8.58		0.97		0.66		2	G5	1569
116515	+26 02455				13 21 25	+25 48.5	2	7.38	.015	1.04	.015	0.84	.019	3	K0	105,280
		G 14 - 49			13 21 25	−02 28.7	1	12.79		0.53		−0.17		2		1696
116400	−54 05577				13 21 25	−54 57.4	1	9.15		0.24		0.14		1	A3/5 V	1771
		AAS41,43 F4 # 20			13 21 25	−61 56.9	1	11.69		0.34		0.23		1	A0	820
238224	+58 01441				13 21 26	+58 10.0	3	9.75	.015	1.26	.017	1.15	.026	4	M0 V	1007,3072,8023
116374	−61 03669				13 21 26	−62 05.7	2	8.99	.010	0.22	.000	0.13	.005	2	B9 IV	549,820
	−58 04849				13 21 29	−59 18.4	1	11.56		0.25		0.18		2	A1	837
		AAS41,43 F4 # 22			13 21 29	−62 01.0	1	12.09		0.38		0.29		2	B9	820
	−59 04963				13 21 30	−59 38.6	1	11.05		0.30		0.21		2	A5	837
		AAS78,203 # 35			13 21 30	−63 42.	1	10.81		0.15		−0.50		4		1684
		AAS78,203 # 36			13 21 30	−63 43.	1	11.84		0.59		0.11		4		1684
	−65 02315				13 21 30	−65 38.2	1	10.51		0.29		−0.31		4		1684
	+13 02662				13 21 31	+12 32.7	1	10.66		0.78		0.38		1	K0	1569
		LP 677 - 74			13 21 31	−08 30.4	1	11.09		0.77		0.32		2		1696
116448	−37 08596				13 21 31	−37 46.4	2	7.14	.001	1.47	.002	1.81		14	K3 III	1673,2012
	−61 03670				13 21 31	−62 13.1	3	10.96	.021	0.21	.017	0.10	.013	3	A0	287,549,820
116500	−3 03460				13 21 32	−03 56.6	2	9.59	.006	0.04	.000	0.01	.009	29	A0	1255,1308
116434	−50 07747	IDS13185S5059	A		13 21 32	−51 14.6	1	7.02		0.08		0.06		1	A2 V	1771
		Lod 821 # 26			13 21 32	−59 25.1	1	12.13		0.53		0.34		2	A8	837
		Cen sq 1 # 317			13 21 32	−61 44.7	1	14.43		0.85		0.18		3		144
	−61 03671				13 21 32	−62 05.7	3	11.08	.005	0.21	.005	−0.21	.021	3	B8	287,549,820
116452	−49 07898				13 21 33	−49 40.9	1	9.60		−0.03		−0.31		2	B7 III/IV	3036
		Cen sq 1 # 309			13 21 34	−61 43.7	2	12.28	.000	0.50	.005	0.24	.015	5		144,820
		Cen sq 1 # 313			13 21 34	−61 43.7	1	13.06		0.63		0.18		3		144
	−61 03672				13 21 34	−62 19.0	2	10.23	.000	1.21	.000	1.02	.015	2	G8	549,820
116403	−60 04649				13 21 35	−60 46.1	1	9.00		0.02		−0.39		2	B5 II/III	401
	−61 03673				13 21 36	−62 19.5	2	9.80	.015	1.12	.005	0.89	.010	2	G8	549,820
		AAS78,203 # 37			13 21 36	−64 24.	1	12.42		1.42		0.52		4		1684
	−59 04966				13 21 37	−59 26.1	1	10.89		0.16		0.15		2	A2	837
	−61 03674				13 21 37	−62 06.5	3	11.09	.012	0.37	.009	0.24	.040	3	A2	287,549,820
116581	+37 02404	HR 5052			13 21 38	+37 17.7	3	6.12	.041	1.65	.015	1.92	.059	9	M3 III	70,1501,3001
116455	−50 07749				13 21 38	−50 37.3	1	10.31		0.00		−0.70		2	B2 II	3036
		AAS41,43 F4 # 29			13 21 38	−61 45.5	1	11.99		0.31		−0.14		2	B8	820
	−61 03675				13 21 38	−62 18.7	3	10.46	.012	0.17	.005	−0.32	.017	3	B6	287,549,820
	−58 04852				13 21 40	−59 14.3	1	10.63		0.86		0.39		2	K3	837
	−61 03677				13 21 40	−61 43.6	2	10.91	.005	0.19	.010	0.08	.005	6	B8	144,820
		AAS41,43 F4 # 35			13 21 41	−61 43.6	1	11.73		0.38		−0.05		2	B8	820
	+14 02608				13 21 43	+14 29.4	1	10.04		0.58		0.08		2	G0	1569
	+45 02107				13 21 43	+45 22.0	1	10.70		1.00		0.84		1	K1 III	280
116438	−60 04651	LSS 3063			13 21 43	−60 42.7	4	8.06	.021	0.19	.008	−0.65	.016	9	B1 III	401,540,976,1684
		NGC 5139 - 1853			13 21 44	−47 22.8	1	15.51		0.01		−0.53		2		249
116487	−46 08622				13 21 45	−47 20.3	2	9.09	.005	1.15	.025	1.13		8	K2 III	517,2001
116420	−63 02741	LSS 3061			13 21 45	−64 22.8	1	9.09		0.65		0.49		1	F0 II	8100
116421	−65 02317				13 21 45	−65 38.9	1	9.29		0.07		−0.28		4	B7 Vn	1684
116608	+44 02271				13 21 46	+43 51.6	3	9.44	.012	0.15	.011	0.05	.005	7	A1 V	193,272,280
		AAS41,43 F4 # 32			13 21 46	−61 48.1	1	12.47		0.46		0.22		2	A8	820
		G 63 - 25			13 21 47	+15 45.9	1	12.26		1.37		1.28		1		1620
	+14 02609				13 21 48	+14 14.9	1	10.21		0.97		0.72		1		280
	+42 02397				13 21 48	+42 07.9	1	10.30		1.04		0.87		1	K0 III	280
		NGC 5128 sq1 1			13 21 48	−43 09.3	1	12.12		0.62		0.08		2		99
		Cen sq 1 # 310			13 21 48	−61 43.0	1	12.63		0.81		0.27		3		144
		LSS 3064			13 21 48	−63 16.7	1	11.67		0.78		−0.28		2	B3 Ib	1737
116422	−65 02318				13 21 49	−66 11.6	1	9.48		0.20		−0.32		4	B5ep Shell	1684
	+41 02391				13 21 50	+40 57.4	1	10.01		1.52		1.91		1	K2	280
	+26 02457				13 21 51	+25 48.4	1	10.12		0.97		0.77		1	K2	280
116457	−63 02743	HR 5048		★ A	13 21 52	−64 13.5	6	5.30	.007	0.40	.009	−0.01	.018	21	F3 IV/V	15,122,1075,2012,2038,3026
116617	+35 02445				13 21 55	+34 56.7	1	8.10		1.03		0.81		2	K0 III	1569
116656	+55 01598	HR 5054		★ AB	13 21 55	+55 11.2	6	2.05	.009	0.03	.028	0.03	.006	20	A2 V	15,1007,1013,1363*
		WLS 1320-20 # 9			13 21 56	−20 36.9	1	13.30		0.72		0.30		2		1375
	−59 04969				13 21 56	−60 11.6	1	11.05		0.38		−0.24		2		1684
	+16 02503				13 21 57	+15 56.7	1	10.26		1.00		0.73		1		280
	+40 02656				13 21 57	+40 28.6	1	8.86		1.10		1.12		1	K0 III	280
116568	−4 03472	HR 5050			13 21 57	−04 54.2	3	5.74	.046	0.41	.012	−0.06	.009	14	F3 V	254,285,1088
116526	−48 08169				13 21 57	−49 02.3	1	10.80		0.07		−0.18		2	B9 Ib/II	3036
116456	−61 03678				13 21 57	−62 17.3	2	10.09	.005	0.14	.010	−0.05	.005	2	B8 IV/V	549,820
	−61 03679				13 21 58	−62 22.2	1	10.94		0.31		0.29		2	A3	820
116491	−58 04858	LSS 3068			13 21 59	−58 39.4	2	8.80	.007	0.10	.013	−0.62	.014	4	B3 III	540,976
	+9 02754				13 22 00	+09 09.1	1	10.59		0.98		0.78		1	K0	280
		NGC 5139 - 15			13 22 00	−47 06.2	2	10.56	.010	1.63	.010	1.98		4		319,1643
		NGC 5139 - 5567			13 22 01	−47 06.9	2	15.51	.005	0.79	.000	0.15		4		319,1643
		AAS41,43 F4 # 36			13 22 01	−61 57.5	1	12.57		0.64		0.18		1	F0	820
116594	+13 02663	HR 5053			13 22 02	+12 41.5	2	6.44	.001	1.06	.001	0.91	.004	18	K0 III	269,1569
	+13 02664				13 22 03	+12 37.2	1	9.98		0.96		0.65		1		280

HD	DM	Other Id	N	Rem	α_{1950}	δ_{1950}	S	V	σ_V	B–V	σ_{B-V}	U–B	σ_{U-B}	n	Spectrum	References
116553	−43 08256				13 22 04	−43 39.7	2	8.65	.024	0.28	.019	0.05	.010	7	A9 V	99,1492
	−61 03681				13 22 04	−61 45.7	2	10.79	.010	0.55	.000	0.28	.010	5	F0	144,820
116634	+32 02346				13 22 05	+32 20.7	1	7.74		0.06		0.10		3	A2	1068
	+43 02327			V	13 22 05	+42 44.0	1	10.15		0.59		0.06		2		1569
	−62 03242				13 22 05	−62 50.5	1	11.21		0.56		−0.45		2		1684
		LSS 3067			13 22 05	−63 07.4	1	11.24		0.54		−0.31		3	B3 Ib	1737
		G 63 - 26			13 22 06	+20 43.1	3	12.19	.010	0.45	.010	−0.17	.015	4		1064,1620,3029
116635	+31 02483				13 22 07	+30 48.9	1	9.70		0.17		0.10		2	F2	1569
116636	+23 02568				13 22 08	+23 26.9	1	8.69		1.33		1.63		1	K2	280
116538	−51 07500				13 22 08	−51 34.9	2	7.91	.005	−0.06	.009	−0.88	.014	4	B2 Ib/II	55,400
		NGC 5139 - 3641			13 22 09	−47 13.8	2	14.28	.005	0.87	.005			3		931,1643
		NGC 5139 - 3642			13 22 09	−47 14.2	2	13.86	.010	0.87	.021	0.36		5		441,1643
116508	−61 03682	IDS13189S6111		A	13 22 09	−61 26.1	1	8.46		0.32		−0.01			A9 V	1771
116458	−69 01838	HR 5049			13 22 09	−70 22.0	4	5.66	.004	−0.03	.000	−0.17	.016	15	Ap HgMn	15,1075,1771,2038
		G 63 - 27			13 22 10	+15 07.0	2	11.10	.015	1.12	.005	1.05	.005	2	K5	1620,1658
116557	−49 07906				13 22 10	−49 34.2	1	10.38		−0.06		−0.34		2	Ap Si	3036
116619	+15 02568				13 22 11	+14 43.8	1	8.90		0.75		0.38		1	K0	280
		CS 16027 # 81			13 22 11	+32 15.4	1	14.37		0.49		0.02				1744
		AU Vir			13 22 11	−06 42.7	2	11.39	.005	0.17	.010	0.11	.010	2	A0	668,700
		NGC 5139 - 4412			13 22 11	−47 10.8	1	15.38		0.75				1		1643
116620	+12 02593				13 22 12	+11 39.4	1	8.12		0.96		0.77		1	K0	280
		NGC 5139 - 3383			13 22 12	−47 15.4	2	14.91	.015	1.01	.005			2		931,1643
	−61 03683				13 22 12	−62 08.4	3	11.12	.025	0.37	.009	0.25	.017	3	A0	287,549,820
		NGC 5139 - 4415			13 22 14	−47 11.0	2	14.02	.005	1.14	.010	1.08		2		319,1643
116558	−49 07908				13 22 14	−49 49.6	1	9.80		−0.03		−0.21		2	B8 V	3036
		Ton 708			13 22 15	+28 33.6	1	15.24		−0.32		−1.14		1		1744
		NGC 5139 - 5015			13 22 15	−47 09.2	3	14.43	.010	0.90	.005	0.45		5		319,931,1643
	+40 02657				13 22 16	+39 51.2	1	10.22		0.35		0.02		2		1569
	+25 02631				13 22 17	+24 59.2	1	10.39		1.02		0.83		1		280
		NGC 5139 - 5760			13 22 17	−47 06.2	2	14.99	.023	0.81	.009	0.18		4		319,1643
		NGC 5139 - 3386			13 22 17	−47 15.7	2	15.10	.015	0.82	.015			2		931,1643
116560	−51 07504				13 22 17	−51 37.8	2	7.16	.005	0.04	.015	0.04	.020	4	A0 V	401,1771
116562	−53 05600	IDS13192S5343		AB	13 22 17	−53 58.7	1	9.42		0.27		0.11		35	A2 IV	1699
		Ru 166 - 3			13 22 17	−63 11.5	1	12.70		0.49		0.44		2		251
		NGC 5139 - 483			13 22 18	−47 17.8	2	12.81	.010	1.07	.005	0.59		6		319,1643
		NGC 5139 - 399			13 22 19	−47 21.0	4	12.69	.026	0.43	.008	0.00	.000	6		319,517,931,1643
		NGC 5139 - 3889			13 22 21	−47 13.6	1	14.90		0.89				2		1643
		NGC 5139 - 5952			13 22 22	−47 05.4	3	14.79	.009	0.86	.017	0.35	.054	8		319,441,1643
		AAS41,43 F4 # 39			13 22 22	−62 20.8	1	10.95		1.36		1.38		1	K0	820
		NGC 5139 - 3388			13 22 23	−47 15.3	2	14.84	.010	0.90	.020			2		931,1643
116586	−46 08628	NGC 5139 - 3			13 22 23	−47 22.4	6	8.57	.012	1.29	.013	1.43	.005	53	K0	319,517,931,1643,2001,3000
	+11 02561				13 22 24	+11 00.0	3	10.12	.005	1.10	.014	1.03	.010	17	M1 V	280,830,1783
	+34 02419				13 22 24	+34 25.8	1	9.06		1.52		1.91		1	M0	280
		NGC 5139 - 1595			13 22 24	−47 23.7	1	13.49		0.98				1		1643
		NGC 5139 - 4418			13 22 25	−47 11.6	2	14.53	.010	0.97	.025			2		931,1643
116587	−51 07508				13 22 25	−52 06.1	1	7.68		0.22		0.09		1	A5 IV/V	1771
		NGC 5139 - 2586			13 22 26	−47 18.0	2	14.92	.040	0.73	.010			2		931,1643
		Ru 166 - 2			13 22 26	−63 11.6	1	12.79		0.61		0.13		2		251
	+20 02833				13 22 27	+20 18.3	1	10.20		0.98		0.83		1		280
		NGC 5139 - 5325			13 22 27	−47 08.0	1	15.52		0.00				1		1643
		NGC 5139 - 1597			13 22 27	−47 24.1	1	15.44		0.92				1		1643
		NGC 5139 - 1598			13 22 27	−47 24.4	1	15.25		1.06				1		1643
		DY Cen			13 22 27	−53 59.2	2	12.44	.082	0.38	.032	−0.61	.028	2		1589,1699
116694	+36 02370				13 22 28	+35 31.6	1	9.70		0.38		−0.01		2	G5	1569
		NGC 5139 - 5761			13 22 28	−47 05.6	1	15.45		0.06				3		1643
		NGC 5139 - 2588			13 22 28	−47 18.5	3	13.50	.012	1.13	.012			6		546,931,1643
		NGC 5139 - 5326			13 22 29	−47 08.4	1	15.08		0.08				1		1643
		AAS41,43 F4 # 40			13 22 29	−62 08.5	1	13.19		0.48		0.36		1		820
		NGC 5139 - 2231			13 22 30	−47 20.2	1	15.51		0.04				1		1643
	−61 03684				13 22 30	−61 48.5	3	10.20	.006	1.10	.002	0.80	.011	6	K2	144,820,1730
		AAS41,43 F4 # 42			13 22 30	−61 54.9	1	12.03		0.46		0.21		1	A7	820
		LP 911 - 81			13 22 31	−28 25.5	1	11.55		1.20		1.19		1		1696
		Ru 166 - 1			13 22 31	−63 11.4	1	11.29		1.10		0.93		2		251
	+38 02441				13 22 32	+37 36.5	1	10.36		1.00		0.78		1	G8 IV	280
		NGC 5139 - 2589			13 22 32	−47 18.7	1	15.13		0.07				1		1643
116658	−10 03672	HR 5056, α Vir		★A	13 22 33	−10 54.1	18	0.98	.010	−0.23	.008	−0.94	.008	125	B1 III-IV +	1,15,26,53,198,667*
		AAS41,43 F4 # 43			13 22 33	−62 19.1	1	11.14		1.11		1.06		1	G8	820
		NGC 5139 - 2812			13 22 35	−47 17.0	1	14.69		0.20				1		1643
116615	−46 08635				13 22 35	−47 22.3	3	10.07	.018	1.19	.015	1.10	.005	8	G8/K1 (III)	517,2001,3021
		NGC 5128 sq1 3			13 22 37	−43 10.0	1	13.71		1.07		0.75		1		99
116616	−50 07758				13 22 37	−51 17.1	1	9.07		0.25		0.14		2	A5 V	1771
	−61 03685				13 22 37	−62 01.1	1	10.78		0.71		0.16		1	F5	820
		Cen sq 1 # 314			13 22 38	−61 45.6	1	13.41		0.74		0.17		3		144
		AAS41,43 F4 # 46			13 22 38	−62 12.3	1	12.18		0.55		0.28		1	A8	820
116625	−46 08636				13 22 39	−47 18.6	3	10.04	.012	1.14	.022	0.84	.014	12	G5/8	517,2001,3021
		AAS41,43 F4 # 45			13 22 39	−61 59.6	1	12.52		0.32		−0.11		1		820
	+42 02398				13 22 40	+41 46.1	1	10.44		1.02		0.85		1	G8 III	280
		AAS41,43 F4 # 47			13 22 40	−62 12.5	1	12.20		0.18		−0.02		1	B9	820
		NGC 5139 - 6120			13 22 41	−47 03.8	1	14.85		0.64				3		1643
		NGC 5139 - 2035			13 22 42	−47 20.9	1	15.28		0.02				1		1643

Table 1   743

HD	DM	Other Id	N	Rem	α₁₉₅₀	δ₁₉₅₀	S	V	σ_V	B–V	σ_B-V	U–B	σ_U-B	n	Spectrum	References
	+22 02577				13 22 43	+21 33.0	1	10.42		1.14		1.14		1		280
		CS 16027 # 84			13 22 43	+30 01.9	1	14.12		-0.04		0.11		1		1744
116706	+24 02578	HR 5057			13 22 44	+24 06.9	3	5.77	.028	0.06	.010	0.08	.009	8	A3 IV	70,1022,1068
		AAS41,43 F4 # 48			13 22 44	-61 41.8	1	12.48		0.52		0.30		1	A2	820
		AAS41,43 F4 # 49			13 22 44	-62 07.3	1	11.36		0.14		-0.12		1		820
		NGC 5139 - 1609			13 22 45	-47 24.6	1	14.33		0.98				1		1643
116649	-47 08295	NGC 5139 - 1			13 22 45	-47 33.6	1	7.58		0.10				2	A0	1643
116589	-61 03686				13 22 45	-62 08.5	2	9.40	.005	0.09	.010	-0.28	.005	2	B7/9 II/III	549,820
		NGC 5139 - 5964			13 22 46	-47 05.1	1	13.98		0.91				3		1643
	-60 04663				13 22 46	-61 08.5	1	10.61		0.11		-0.29		2	B6	122
		NGC 5128 sq1 4			13 22 47	+43 27.4	1	10.33		0.30		0.22		2		99
	-61 03687				13 22 47	-62 16.5	2	11.21	.000	0.11	.015	-0.31	.005	2	B6	549,820
		CS 16027 # 83			13 22 48	+30 22.1	1	14.48		0.65		0.00		1		1744
		NGC 5128 sq2 11			13 22 49	-42 31.6	1	12.10		1.08		0.86		2		5013
116663	-46 08638	NGC 5139 - 4			13 22 49	-47 15.9	5	8.78	.033	0.00	.015	-0.12	.014	19	B9 V	319,517,1770,2001,3036
		AAS41,43 F4 # 52			13 22 49	-62 01.5	1	12.96		0.48		0.46		1	A8	820
		G 62 - 40			13 22 50	-01 06.2	2	13.57	.028	0.73	.025	0.09		4		202,1658
		LOD 29 # 16			13 22 50	-61 10.8	1	11.76		0.68		0.19		2		122
		AAS41,43 F4 # 54			13 22 50	-62 07.4	1	11.46		1.26		1.01		1		820
116607	-61 03688				13 22 50	-62 18.0	2	8.98	.010	0.12	.005	0.02	.000	2	B9 III	549,820
	+13 02665				13 22 51	+13 08.5	1	10.33		0.69		0.21		1	G5	1569
		AAS41,43 F4 # 56			13 22 51	-61 51.1	1	12.55		0.51		0.16		1		820
	-61 03689				13 22 51	-62 05.3	2	10.41	.010	0.13	.005	-0.47	.005	2	B4	549,820
116651	-50 07764				13 22 52	-50 55.0	1	7.87		0.22		0.08		1	A5 III/IV	1771
	-61 03690				13 22 52	-61 48.5	1	11.64		0.33		0.23		2	B9	820
	+32 02347				13 22 53	+32 14.7	1	10.37		1.12		1.11		1	K2 III	280
116725	+11 02563				13 22 54	+11 01.0	1	8.15		1.05		0.85		1	K0	280
	+39 02643				13 22 54	+39 11.0	1	10.12		1.02		0.82		2	K3 III	3072
		NGC 5139 - 347			13 22 55	-47 24.3	1	12.54		0.74				2		1643
	+60 01460				13 22 56	+59 56.9	1	9.29		1.56		1.86		4	M1	1371
		LOD 29 # 18			13 22 56	-61 12.0	1	12.83		2.08				2		122
	-61 03691				13 22 57	-61 47.3	2	11.64	.005	0.36	.010	0.28	.025	5	A0	144,820
	+34 02422				13 22 58	+34 10.4	1	9.51		1.19		1.22		1	K0	280
116783	+41 02392				13 22 58	+40 42.7	1	8.45		1.53		2.02		1	K3 III	280
116664	-50 07765				13 22 58	-51 13.2	1	7.12		0.18		0.09		1	A5 V	1771
		G 164 - 73			13 22 59	+32 32.0	2	12.15	.000	1.32	.000	1.14		2	K7	1658,1759
		GD 157			13 22 59	+37 01.0	1	13.24		0.44		-0.21		1		3060
116665	-56 05767				13 22 60	-56 46.8	1	9.48		0.41		0.26		1	A8 V	1771
		NGC 5139 - 53		V	13 23 00	-47 21.	3	11.57	.032	1.69	.014	1.62	.136	7		319,517,931
116753	+26 02460				13 23 02	+25 48.5	2	8.79	.005	0.25	.000	0.09	.009	4	F2	280,833
		PG1323-086			13 23 02	-08 33.7	1	13.48		-0.14		-0.68		8	sdB-O	1764
		L 257 - 47			13 23 02	-51 25.7	1	14.60		0.00		-0.79		2	DA	3073
		LOD 29 # 14			13 23 02	-61 09.4	1	12.65		0.38		0.28		2		122
		AAS41,43 F4 # 59			13 23 02	-62 00.0	1	11.70		0.33		-0.14		1	B7	820
	-42 08403	NGC 5128 sq1 5			13 23 03	-42 55.7	1	10.54		1.40		1.65		2		99
	-27 09225				13 23 04	-28 06.8	2	11.03	.018	1.54	.000	1.16		5	M2.5	158,1705
116687	-42 08404				13 23 04	-42 25.3	2	9.70	.013	0.26	.004	0.11	.000	11	A9 V	1492,5013
		AAS41,43 F4 # 60			13 23 04	-62 12.7	1	11.92		0.69		0.18		1	F4	820
116677	-57 06048				13 23 05	-57 49.0	1	8.42		0.42		0.15		2	A9 III/IV	1771
	-60 04667				13 23 05	-61 13.9	1	11.24		0.27		-0.12		2		122
		AAS41,43 F4 # 62			13 23 05	-61 57.6	1	12.02		0.24		0.03		1	A0	820
		AAS41,43 F4 # 63			13 23 05	-62 01.9	1	12.68		0.56		0.51		1	A0	820
		LP 617 - 58			13 23 06	+02 39.0	1	11.24		0.80		0.39		2		1696
116699	-43 08266				13 23 06	-43 37.0	3	8.58	.016	0.03	.020	0.00	.008	7	A0 V	99,1492,1770
		NGC 5139 - 162			13 23 06	-47 17.	1	12.14		1.70		1.89		10		319
		NGC 5139 - 491			13 23 06	-47 23.8	1	12.84		0.93				2		1643
		AAS41,43 F4 # 61			13 23 06	-62 14.6	1	13.03		0.11		-0.27		1		820
		AAS41,43 F4 # 64			13 23 07	-62 20.8	1	11.04		1.19		0.98		1	G8	820
		NGC 5128 sq2 2			13 23 09	-42 34.7	2	11.52	.025	0.58	.005	0.02	.020	5		1492,5013
		LOD 29 # 13			13 23 09	-61 14.7	1	12.62		0.56				2		122
		AAS41,43 F4 # 66			13 23 09	-61 39.2	1	12.65		0.41		0.31		1	A1	820
	-61 03692	LSS 3072			13 23 09	-61 47.6	6	9.69	.012	1.32	.010	0.27	.021	15	B2.5 Ia	144,403,820,1684,1730,1737
		AAS41,43 F4 # 65			13 23 09	-62 20.6	1	11.88		0.67		0.35		1	G0	820
		OO Cen			13 23 09	-62 54.0	1	11.62		1.60		1.33		1		689
	+20 02834				13 23 10	+20 09.6	1	10.50		1.02		0.77		1	K0	280
		PH1323-086A			13 23 12	-08 34.8	1	13.59		0.39		-0.02		12		1764
		AAS41,43 F4 # 68			13 23 12	-62 09.2	1	12.69		0.44		0.08		1	F2	820
116784	+16 02508				13 23 13	+15 49.4	1	6.87		1.20		1.35		1	K0	280
116842	+55 01603	HR 5062			13 23 13	+55 14.9	18	4.01	.005	0.17	.004	0.08	.004	213	A5 V	1,15,401,985,1006,1007*
		PG1323-086C			13 23 13	-08 33.0	1	14.00		0.71		0.25		13		1764
		PG1323-086B			13 23 13	-08 36.3	1	13.41		0.76		0.27		13		1764
116742	-26 09707				13 23 13	-27 12.6	1	8.46		0.33		0.14		1	B9/A0mA8	1770
116713	-39 08246	HR 5058			13 23 13	-39 29.7	4	5.11	.007	1.18	.013	0.95	.035	15	K0.5 III	565,993,1311,3005
		NGC 5128 sq2 10			13 23 13	-42 29.2	1	16.58		0.69		0.17		1		1492
		NGC 5128 sq2 5			13 23 13	-42 35.1	2	13.06	.015	1.37	.020	1.47	.000	5		1492,5013
		NGC 5128 sq1 7			13 23 13	-43 02.0	1	13.34		1.51		1.53		1		99
		NGC 5128 sq2 3			13 23 15	-42 32.3	2	11.73	.025	0.67	.005	0.18	.015	11		1492,5013
		NGC 5128 sq2 8			13 23 15	-42 35.3	1	14.25		0.64		-0.10		3		1492
116717	-48 08188				13 23 15	-48 31.6	1	6.71		0.03		0.01		2	A0 V	1771
116690	-58 04867				13 23 15	-59 03.5	1	9.78		0.25		0.13		2	A5 IV/V	122

744

Catalogue of mean UBV data

HD	DM	Other Id	N Rem	α_{1950}	δ_{1950}	S	V	σ_V	B–V	σ_{B-V}	U–B	σ_{U-B}	n	Spectrum	References
		AAS41,43 F4 # 69		13 23 15	−62 06.3	1	13.03		0.50		0.41		1		820
	−61 03693			13 23 16	−62 03.4	1	11.23		0.27		−0.26		1	B7	820
116926	+68 00724	G 238 - 36		13 23 17	+68 26.0	1	9.58		0.70		0.13		2	G5	3026
116579	−74 01059	HR 5051		13 23 17	−74 25.9	3	6.62	.004	−0.05	.005	−0.23	.005	12	B9 V	15,1075,2038
116689	−58 04869	NGC 5138 - 319		13 23 18	−58 38.5	3	9.72	.027	0.04	.025	0.04	.028	7	B9 V	122,185,847
		AAS41,43 F4 # 71		13 23 18	−62 09.8	1	11.11		1.28		1.14		2	G8	820
116678	−63 02750			13 23 18	−63 37.2	1	9.90		0.43		0.15		1	A7/F0	1771
		NGC 5128 sq2 12		13 23 19	−42 28.4	1	13.00		1.25		1.22		4		5013
	−61 03694			13 23 19	−61 44.9	1	10.94		0.80		0.19		2	G0	820
	+24 02579			13 23 20	+23 57.3	1	10.15		1.12		1.17		1		280
	−42 08410	NGC 5128 sq1 8		13 23 20	−43 22.6	1	10.98		1.24		1.44		2		99
		AAS41,43 F4 # 72		13 23 20	−62 09.9	1	12.41		0.98		0.66		1		820
		NGC 5128 sq2 9		13 23 21	−42 31.2	1	15.22		0.76		0.21		4		1492
		NGC 5138 - 125		13 23 21	−58 45.4	1	13.24		0.17		−0.08		3		847
		NGC 5138 - 121		13 23 21	−58 47.0	1	13.37		0.54		0.05		2		847
		AAS41,43 F4 # 74		13 23 21	−61 58.0	1	12.54		0.64		0.12		1		820
		NGC 5128 sq2 6		13 23 22	−42 32.3	2	13.54	.015	1.36	.000	1.48	.035	7		1492,5013
	−58 04870			13 23 22	−58 51.7	1	10.04		0.36		0.20		2		122
		NGC 5128 sq2 7		13 23 23	−42 34.6	1	13.75		1.02		0.69		4		1492
		NGC 5128 sq1 9		13 23 23	−43 27.7	1	14.87		0.89		0.47		1		99
		NGC 5128 sq2 4		13 23 24	−42 33.7	1	12.63		0.56		0.07		2		1492
	−61 03695			13 23 24	−62 07.1	1	10.09		0.67		0.17		1	F5	820
	−61 03696			13 23 25	−61 50.5	1	11.06		0.62		0.12		1	F5	820
		NGC 5139 - 24		13 23 26	−47 00.9	2	10.81	.009	0.36	.005	0.33		4		319,1643
		NGC 5138 - 124		13 23 26	−58 46.5	1	13.26		0.45		0.11		2		847
		AAS41,43 F4 # 77		13 23 26	−62 05.4	1	12.38		0.38		0.20		1	A0	820
116802	−2 03683	W Vir		13 23 27	−03 07.1	1	10.02		0.88				1	F2 Ib	688
	+14 02612			13 23 28	+13 56.1	1	9.49		1.52		1.51		1	M2	280
		PG1323-086D		13 23 28	−08 35.0	1	12.08		0.59		0.01		6		1764
		NGC 5128 sq1 10		13 23 28	−43 29.3	1	11.85		1.27		1.28		2		99
116721	−58 04871	NGC 5138 - 223		13 23 28	−58 40.4	3	8.69	.024	1.66	.020	1.81	.047	6	M1 III	122,185,847
		AAS41,43 F4 # 79		13 23 28	−61 52.2	1	12.27		0.37		0.22		1	B9	820
		AAS41,43 F4 # 78		13 23 28	−62 06.0	1	11.99		0.22		0.09		1	B9	820
		NGC 5138 - 115		13 23 29	−58 47.9	1	13.95		0.63		0.15		2		847
	−42 08413	NGC 5128 sq1 11		13 23 31	−43 00.2	1	10.49		0.61		−0.07		2		99
	−62 03260			13 23 31	−62 24.4	1	10.96		0.34		0.21		1	A2	820
116769	−22 03589	IDS13208S2243	AB	13 23 32	−22 58.9	1	8.64		0.34		0.17		4	A6(m)A6-A9	164
	+21 02525			13 23 33	+21 09.5	1	9.42		1.09		0.95		2	K2	1569
		NGC 5128 sq1 12		13 23 33	−43 24.8	1	11.77		0.42		0.23		2		99
		NGC 5138 - 47		13 23 33	−58 45.9	2	13.40	.000	0.50	.005	0.09	.005	5		185,847
	+22 02578			13 23 35	+22 20.9	1	10.82		−0.03		−0.01		1		280
	−61 03698			13 23 35	−61 49.3	1	11.43		0.41		−0.25		1	B2	820
	−61 03697			13 23 35	−62 07.2	1	10.92		0.26		−0.02		1	B7	820
		AAS41,43 F4 # 82		13 23 35	−62 11.0	1	12.42		0.46		0.35		1	A8	820
		AAS41,43 F4 # 81		13 23 35	−62 25.1	1	12.23		0.29		0.23		1	A2	820
		NGC 5138 - 134		13 23 36	−58 41.3	1	13.31		0.25		0.12		3		847
		NGC 5138 - 51		13 23 36	−58 44.2	1	13.69		0.29		0.17		2		847
		AAS41,43 F4 # 86		13 23 36	−61 42.6	1	12.37		0.25		−0.35		1		820
		NGC 5138 - 135		13 23 37	−58 41.0	1	12.62		1.69		1.77		2		847
		AAS41,43 F4 # 85		13 23 37	−61 49.1	1	12.40		0.62		0.38		1	A5	820
116878	+45 02108	IDS13215N4501	AB	13 23 38	+44 45.2	2	8.28	.000	0.72	.005	0.24		4	G8 V +G8 V	1381,3030
116831	−0 02686	HR 5059		13 23 38	−00 56.0	5	5.96	.021	0.19	.012	0.13	.009	14	A7 III	15,53,252,285,1256
		NGC 5138 - 53		13 23 38	−58 43.9	1	13.96		0.67		0.11		3		847
		AAS41,43 F4 # 87		13 23 38	−62 08.7	1	12.72		0.46		0.29		1		820
	+24 02581			13 23 40	+24 08.9	1	10.32		1.21		1.26		1	K2	280
116775	−48 08195			13 23 40	−49 21.2	1	8.26		0.13		0.14		1	A2 V	1771
		AAS41,43 F4 # 88		13 23 40	−62 05.8	1	12.19		0.43		0.36		2	A0	820
	+14 02613			13 23 41	+13 44.9	1	10.38		1.42		1.64		1	M0 III	280
	+43 02330			13 23 41	+43 16.3	1	9.89		1.12		1.01		1	K1 V	280
		AAS41,43 F4 # 91		13 23 41	−61 38.0	1	11.38		0.60		0.15		2	F0	820
	−61 03699			13 23 41	−62 07.4	1	10.80		0.56		0.06		1	F5	820
		AAS41,43 F4 # 92		13 23 43	−61 56.2	1	10.94		2.70		2.20		8	M5	820
	−61 03700			13 23 43	−62 08.1	1	11.44		0.19		−0.44		1	B5	820
		AAS41,43 F4 # 90		13 23 43	−62 18.9	1	11.70		0.30		0.23		2	A1	820
	−58 04873	NGC 5138 - 54		13 23 44	−58 45.6	2	10.62	.010	0.38	.010	0.14	.005	5	F0 III	185,847
		AAS41,43 F4 # 93		13 23 44	−62 02.5	1	11.92		0.36		0.27		1	B8	820
		AAS41,43 F4 # 98		13 23 45	−61 37.7	1	12.24		0.35		0.23		1	A0	820
116734	−62 03265			13 23 45	−62 24.0	1	8.39		0.48		0.03		1	F5 IV/V	820
116866	+21 02526			13 23 46	+21 04.3	2	8.62	.015	0.50	.005	0.00	.005	4	F8	1569,1648
116789	−46 08649			13 23 46	−47 04.1	1	8.43		0.09		0.06		1	A0 V	1770
		NGC 5138 - 38		13 23 46	−58 49.1	1	13.32		0.41		0.26		3		847
		NGC 5138 - 37		13 23 46	−58 49.2	1	13.28		0.31		0.20		3		847
		AAS41,43 F4 # 96		13 23 46	−62 13.2	1	12.84		0.45		0.37		1		820
116765	−58 04874			13 23 47	−59 04.7	1	9.71		0.08		−0.17		2	B9 III/IV	122
		AAS41,43 F4 # 97		13 23 47	−62 21.2	1	12.27		0.34		0.16		1		820
		NGC 5139 - 11		13 23 48	−47 13.	1	10.04		1.11		0.87		1		319
		NGC 5139 - 12		13 23 48	−47 13.	1	10.10		1.16		1.18		1		319
		NGC 5139 - 14		13 23 48	−47 13.	1	10.36		1.32				1		931
		NGC 5139 - 21		13 23 48	−47 13.	1	10.48		0.22				2		931
		NGC 5139 - 36		13 23 48	−47 13.	3	11.17	.024	1.59	.013	1.85	.124	8		319,517,4002

Table 1   745

HD	DM	Other Id	N	Rem	α_{1950}	δ_{1950}	S	V	σ_V	B–V	σ_{B-V}	U–B	σ_{U-B}	n	Spectrum	References
		NGC 5139 - 40			13 23 48	−47 13.	3	11.36	.009	1.49	.009	1.32	.038	7		319,1468,3036
		NGC 5139 - 43			13 23 48	−47 13.	3	11.61	.022	1.71	.015	1.82	.014	7		319,1468,3036
		NGC 5139 - 46			13 23 48	−47 13.	1	11.54		1.58		1.55		2		319
		NGC 5139 - 48			13 23 48	−47 13.	1	11.51		1.56		1.44		2		319
		NGC 5139 - 49			13 23 48	−47 13.	1	11.58		1.59		1.41		2		319
		NGC 5139 - 55			13 23 48	−47 13.	2	11.51	.045	1.67	.010	1.25	.040	5		319,3036
		NGC 5139 - 56			13 23 48	−47 13.	2	11.68	.020	1.64	.015	1.72	.010	5		319,3036
		NGC 5139 - 58			13 23 48	−47 13.	2	11.66	.029	1.43	.017	1.21	.034	6		319,3036
		NGC 5139 - 61		V	13 23 48	−47 13.	3	11.58	.059	1.63	.013	1.67	.005	8		319,931,3036
		NGC 5139 - 62			13 23 48	−47 13.	1	11.50		1.65		1.71		2		319
		NGC 5139 - 65			13 23 48	−47 13.	1	11.61		1.50		1.37		3		3036
		NGC 5139 - 66			13 23 48	−47 13.	1	11.53		1.62		1.58		2		3036
		NGC 5139 - 67			13 23 48	−47 13.	1	11.64		1.50		1.39		3		3036
		NGC 5139 - 70			13 23 48	−47 13.	1	11.61		1.80		1.68		7		319
		NGC 5139 - 73			13 23 48	−47 13.	1	11.70		1.64		1.62		2		3036
		NGC 5139 - 74			13 23 48	−47 13.	2	11.78	.040	1.38	.020	1.16	.020	4		319,3036
		NGC 5139 - 83			13 23 48	−47 13.	2	11.65	.005	1.04	.029	0.84		3		517,931
		NGC 5139 - 84			13 23 48	−47 13.	1	11.87		1.66		1.79		2		319
		NGC 5139 - 85			13 23 48	−47 13.	1	11.84		1.61		1.62		2		319
		NGC 5139 - 90			13 23 48	−47 13.	1	11.66		1.64		1.66		9		319
		NGC 5139 - 91			13 23 48	−47 13.	2	11.83	.024	1.39	.005	1.22	.033	7		319,3036
		NGC 5139 - 94			13 23 48	−47 13.	1	11.80		1.39		1.18		3		3036
		NGC 5139 - 95			13 23 48	−47 13.	1	11.77		1.62		1.61		3		3036
		NGC 5139 - 96			13 23 48	−47 13.	2	11.77	.019	1.48	.002	1.39	.019	7		319,3036
		NGC 5139 - 102			13 23 48	−47 13.	1	11.68		1.46		1.33		5		319
		NGC 5139 - 109			13 23 48	−47 13.	1	11.78		1.40		1.14		2		3036
		NGC 5139 - 113			13 23 48	−47 13.	1	11.76		0.39		0.07		6		319
		NGC 5139 - 118			13 23 48	−47 13.	1	11.90		0.65		0.14		3		517
		NGC 5139 - 124			13 23 48	−47 13.	1	11.83		1.38		1.25		5		319
		NGC 5139 - 128			13 23 48	−47 13.	1	11.83		1.05				4		546
		NGC 5139 - 132			13 23 48	−47 13.	2	11.74	.000	1.71	.000	1.92	.000	3		687,1468
		NGC 5139 - 134			13 23 48	−47 13.	1	11.80		1.52		1.17		1		1468
		NGC 5139 - 139			13 23 48	−47 13.	1	11.94		1.51		1.48		2		1468
		NGC 5139 - 143			13 23 48	−47 13.	3	11.95	.008	0.45	.004	0.08	.005	10		319,517,931
		NGC 5139 - 146			13 23 48	−47 13.	1	12.04		1.38				1		931
		NGC 5139 - 150			13 23 48	−47 13.	1	12.00		1.70		1.90		10		319
		NGC 5139 - 155			13 23 48	−47 13.	1	12.00		1.42		1.25		5		319
		NGC 5139 - 156			13 23 48	−47 13.	2	11.94	.005	1.55	.000	1.64	.136	4		319,517
		NGC 5139 - 159			13 23 48	−47 13.	2	12.03	.024	1.33	.015	1.06	.024	9		319,3036
		NGC 5139 - 161			13 23 48	−47 13.	2	11.97	.005	1.39	.005	1.24		5		319,931
		NGC 5139 - 171			13 23 48	−47 13.	1	12.07		1.48		1.46		1		319
		NGC 5139 - 173			13 23 48	−47 13.	1	11.99		0.60				2		931
		NGC 5139 - 179			13 23 48	−47 13.	2	12.03	.000	1.72	.000	2.05	.000	8		687,1468
		NGC 5139 - 180			13 23 48	−47 13.	1	12.07		1.24				1		931
		NGC 5139 - 183			13 23 48	−47 13.	1	12.07		1.33				1		931
		NGC 5139 - 209			13 23 48	−47 13.	2	12.12	.025	1.33	.005	0.80		2		517,931
		NGC 5139 - 212			13 23 48	−47 13.	1	12.30		1.23				1		931
		NGC 5139 - 213			13 23 48	−47 13.	1	12.22		1.14		0.72		5		319
		NGC 5139 - 219			13 23 48	−47 13.	1	12.20		1.58		1.66		5		319
		NGC 5139 - 234			13 23 48	−47 13.	3	12.26	.008	1.17	.009	0.81	.005	11		319,546,931
		NGC 5139 - 246			13 23 48	−47 13.	1	12.44		1.21		0.90		2		3036
		NGC 5139 - 253			13 23 48	−47 13.	2	12.34	.008	1.37	.004	1.28	.004	9		319,1468
		NGC 5139 - 270			13 23 48	−47 13.	1	12.42		1.50		1.46		2		319
		NGC 5139 - 276			13 23 48	−47 13.	1	12.47		1.33		1.20		2		1468
		NGC 5139 - 279			13 23 48	−47 13.	1	12.32		1.38		1.08		4		319
		NGC 5139 - 287			13 23 48	−47 13.	2	12.41	.000	1.42	.000	1.30	.019	3		319,1468
		NGC 5139 - 289			13 23 48	−47 13.	1	12.44		1.26		0.88		4		319
		NGC 5139 - 297			13 23 48	−47 13.	1	12.44		1.31		1.02		6		319
		NGC 5139 - 305			13 23 48	−47 13.	2	12.57	.010	1.30	.019	0.93	.010	3		1468,3036
		NGC 5139 - 312			13 23 48	−47 13.	2	12.44	.000	1.30	.004	0.96		6		319,931
		NGC 5139 - 313			13 23 48	−47 13.	1	12.55		1.24		0.97		2		3036
		NGC 5139 - 320			13 23 48	−47 13.	2	12.79	.049	1.69	.054	1.58	.010	3		319,1468
		NGC 5139 - 325			13 23 48	−47 13.	1	12.49		1.17				4		546
		NGC 5139 - 331			13 23 48	−47 13.	1	12.54		1.65		1.94		3		319
		NGC 5139 - 357			13 23 48	−47 13.	1	12.29		1.09		1.18		1		1468
		NGC 5139 - 359			13 23 48	−47 13.	1	12.61		1.27				2		931
		NGC 5139 - 361			13 23 48	−47 13.	1	12.61		1.21				4		546
		NGC 5139 - 364			13 23 48	−47 13.	1	12.50		1.11		0.63		3		319
		NGC 5139 - 367			13 23 48	−47 13.	1	12.56		1.50		1.65		1		1468
		NGC 5139 - 370			13 23 48	−47 13.	1	12.82		1.02		0.48		5		319
		NGC 5139 - 371			13 23 48	−47 13.	1	12.68		1.60		1.76		2		319
		NGC 5139 - 378			13 23 48	−47 13.	1	12.66		1.28				1		931
		NGC 5139 - 379			13 23 48	−47 13.	1	12.62		1.19		0.70		3		319
		NGC 5139 - 380			13 23 48	−47 13.	2	12.66	.029	1.24	.019	0.88		3		931,3036
		NGC 5139 - 384			13 23 48	−47 13.	1	12.65		1.62		1.73		3		319
		NGC 5139 - 394			13 23 48	−47 13.	1	12.66		1.35		1.19		4		319
		NGC 5139 - 398			13 23 48	−47 13.	1	12.80		1.18		0.78		2		3036
		NGC 5139 - 402			13 23 48	−47 13.	2	12.70	.030	1.08	.017	0.73	.089	7		319,517
		NGC 5139 - 407			13 23 48	−47 13.	1	12.73		1.22				1		931
		NGC 5139 - 415			13 23 48	−47 13.	1	12.66		1.10		0.60		3		319

HD	DM	Other Id	N Rem	α_{1950}	δ_{1950}	S	V	σ_V	B–V	σ_{B-V}	U–B	σ_{U-B}	n	Spectrum	References
		NGC 5139 - 423		13 23 48	−47 13.	2	12.67	.023	1.14	.019	0.73		4		319,931
		NGC 5139 - 425		13 23 48	−47 13.	2	12.78	.000	1.74	.000	1.97	.000	3		687,1468
		NGC 5139 - 430		13 23 48	−47 13.	1	12.77		1.13				3		546
		NGC 5139 - 434		13 23 48	−47 13.	1	12.76		1.16		0.84		15		319
		NGC 5139 - 442		13 23 48	−47 13.	1	12.83		1.18		0.79		2		4002
		NGC 5139 - 447		13 23 48	−47 13.	2	12.89	.000	1.74	.000	1.94	.000	3		687,1468
		NGC 5139 - 450		13 23 48	−47 13.	1	12.94		1.10		0.65		4		3036
		NGC 5139 - 462		13 23 48	−47 13.	1	12.76		1.01		0.54		4		319
		NGC 5139 - 463		13 23 48	−47 13.	1	12.85		1.45				1		1468
		NGC 5139 - 464		13 23 48	−47 13.	1	12.85		1.18		0.82		3		319
		NGC 5139 - 470		13 23 48	−47 13.	1	12.83		1.02				4		546
		NGC 5139 - 472		13 23 48	−47 13.	1	12.92		1.36		1.17		1		1468
		NGC 5139 - 473		13 23 48	−47 13.	2	12.84	.029	1.11	.010	0.62		6		546,3036
		NGC 5139 - 474		13 23 48	−47 13.	1	12.81		1.06		0.58		3		319
		NGC 5139 - 476		13 23 48	−47 13.	1	12.79		1.07		0.64		4		319
		NGC 5139 - 481		13 23 48	−47 13.	1	12.88		1.11		0.66		5		319
		NGC 5139 - 482		13 23 48	−47 13.	1	12.93		1.31		1.40		6		319
		NGC 5139 - 487		13 23 48	−47 13.	1	12.89		1.22		1.25		1		1468
		NGC 5139 - 497		13 23 48	−47 13.	2	13.00	.000	1.21	.005	0.98	.164	5		1468,3036
		NGC 5139 - 509		13 23 48	−47 13.	1	12.94		1.00		0.56		2		319
		NGC 5139 - 537		13 23 48	−47 13.	1	12.90		1.24		0.85		4		319
		NGC 5139 - 541		13 23 48	−47 13.	1	12.91		1.16				4		546
		NGC 5139 - 542		13 23 48	−47 13.	1	12.92		0.04		-0.09		2		319
		NGC 5139 - 543		13 23 48	−47 13.	1	12.98		1.05		0.67		2		319
		NGC 5139 - 547		13 23 48	−47 13.	1	12.98		0.65				2		931
		NGC 5139 - 557		13 23 48	−47 13.	1	12.68		1.43		0.93		1		517
		NGC 5139 - 558		13 23 48	−47 13.	1	13.06		1.07		0.75		2		3036
		NGC 5139 - 566		13 23 48	−47 13.	1	13.00		1.10		0.72		2		3036
		NGC 5139 - 569		13 23 48	−47 13.	1	13.04		1.07		0.56		2		3036
		NGC 5139 - 571		13 23 48	−47 13.	1	13.00		1.08				2		931
		NGC 5139 - 575		13 23 48	−47 13.	1	13.04		0.98		0.51		4		319
		NGC 5139 - 591		13 23 48	−47 13.	3	13.06	.021	1.07	.014	0.55		10		319,546,931
		NGC 5139 - 595		13 23 48	−47 13.	1	13.05		1.20		0.78		5		319
		NGC 5139 - 600		13 23 48	−47 13.	1	13.04		1.26				4		546
		NGC 5139 - 601		13 23 48	−47 13.	1	12.98		1.04		0.51		3		319
		NGC 5139 - 605		13 23 48	−47 13.	1	13.01		1.16		0.60		3		319
		NGC 5139 - 970		13 23 48	−47 13.	1	12.16		1.50		1.10		2		424
		NGC 5139 - 1031		13 23 48	−47 13.	1	14.82		0.95				3		441
		NGC 5139 - 1032		13 23 48	−47 13.	1	14.93		0.91				3		441
		NGC 5139 - 1048		13 23 48	−47 13.	1	15.75		-0.05		-0.19		2		249
		NGC 5139 - 1118		13 23 48	−47 13.	1	14.67		0.65		0.10		2		249
		NGC 5139 - 1137		13 23 48	−47 13.	1	14.92		0.46		0.08		2		249
		NGC 5139 - 1152		13 23 48	−47 13.	1	14.17		1.08		0.79		3		441
		NGC 5139 - 1155		13 23 48	−47 13.	1	15.12		1.13		0.91		2		441
		NGC 5139 - 1156		13 23 48	−47 13.	1	14.40		0.91		0.28		2		441
		NGC 5139 - 1158		13 23 48	−47 13.	1	14.84		0.20		0.22		2		249
		NGC 5139 - 1250		13 23 48	−47 13.	1	14.53		0.89		0.31		3		441
		NGC 5139 - 1256		13 23 48	−47 13.	1	14.77		0.22		0.21		2		249
		NGC 5139 - 1263		13 23 48	−47 13.	1	14.79		1.07		0.72		3		441
		NGC 5139 - 1271		13 23 48	−47 13.	1	15.61		0.08		-0.18		2		249
		NGC 5139 - 1280		13 23 48	−47 13.	1	15.34		0.10		0.03		2		249
		NGC 5139 - 1362		13 23 48	−47 13.	1	15.38		0.11		-0.02		2		249
		NGC 5139 - 1365		13 23 48	−47 13.	1	15.44		0.08		-0.09		2		249
		NGC 5139 - 1385		13 23 48	−47 13.	1	14.89		0.15		0.27		2		249
		NGC 5139 - 1400		13 23 48	−47 13.	1	14.08		1.13		1.09		3		441
		NGC 5139 - 1403		13 23 48	−47 13.	1	15.42		0.09		-0.53		2		249
		NGC 5139 - 1404		13 23 48	−47 13.	1	14.58		0.72		0.15		3		441
		NGC 5139 - 1432		13 23 48	−47 13.	1	14.39		0.91		0.26		3		441
		NGC 5139 - 1455		13 23 48	−47 13.	2	13.97	.010	0.99	.013	0.49		6		441,546
		NGC 5139 - 1458		13 23 48	−47 13.	1	13.92		1.06				4		546
		NGC 5139 - 1493		13 23 48	−47 13.	1	13.37		1.06				4		546
		NGC 5139 - 1590		13 23 48	−47 13.	1	14.76		1.26		1.29		4		441
		NGC 5139 - 1615		13 23 48	−47 13.	1	13.88		0.97				1		546
		NGC 5139 - 1651		13 23 48	−47 13.	1	14.05		1.01				4		546
		NGC 5139 - 1692		13 23 48	−47 13.	1	13.64		0.88				4		546
		NGC 5139 - 1860		13 23 48	−47 13.	1	13.60		0.82		0.20		2		517
		NGC 5139 - 1865		13 23 48	−47 13.	1			1.04		0.66		1		517
		NGC 5139 - 1871		13 23 48	−47 13.	1	13.83		0.95		0.22		1		4002
		NGC 5139 - 1890		13 23 48	−47 13.	2	13.50	.009	0.94	.027			5		546,931
		NGC 5139 - 1898		13 23 48	−47 13.	1	13.12		1.08				4		546
		NGC 5139 - 1978		13 23 48	−47 13.	1	13.44		1.02				4		546
		NGC 5139 - 2024		13 23 48	−47 13.	2	14.79	.024	0.18	.029	0.25	.029	3		249,319
		NGC 5139 - 2042		13 23 48	−47 13.	1	13.10		1.04				1		931
		NGC 5139 - 2050		13 23 48	−47 13.	1	14.17		0.95				4		546
		NGC 5139 - 2219		13 23 48	−47 13.	1	14.67		0.93		0.33		3		441
		NGC 5139 - 2225		13 23 48	−47 13.	1	14.33		0.89		0.27		2		517
		NGC 5139 - 2229		13 23 48	−47 13.	1	13.04		0.86		0.31		2		517
		NGC 5139 - 2367		13 23 48	−47 13.	1	13.53		1.07				4		546
		NGC 5139 - 2379		13 23 48	−47 13.	2	14.82	.044	0.21	.024	0.23	.015	3		249,319
		NGC 5139 - 2801		13 23 48	−47 13.	1	14.53		0.96		0.46		4		441

Table 1   747

HD	DM	Other Id	N	Rem	α_{1950}	δ_{1950}	S	V	σ_V	B–V	σ_{B-V}	U–B	σ_{U-B}	n	Spectrum	References
		NGC 5139 - 3091			13 23 48	−47 13.	1	13.81		0.97				4		546
		NGC 5139 - 3598			13 23 48	−47 13.	1	14.36		0.87		0.31		2		319
		NGC 5139 - 3631			13 23 48	−47 13.	1	14.87		0.84		0.21		4		441
		NGC 5139 - 3639			13 23 48	−47 13.	1	14.45		0.91		0.28		4		441
		NGC 5139 - 3847			13 23 48	−47 13.	1	13.58		0.67				1		931
		NGC 5139 - 3881			13 23 48	−47 13.	1	13.09		1.22				2		931
		NGC 5139 - 3898			13 23 48	−47 13.	1	13.95		0.96				4		546
		NGC 5139 - 4135			13 23 48	−47 13.	1	14.05		0.60		0.00		2		319
		NGC 5139 - 4146			13 23 48	−47 13.	2	14.20	.010	0.97	.005	0.37		2		319,931
		NGC 5139 - 4153			13 23 48	−47 13.	1	14.63		0.22		0.24		2		319
		NGC 5139 - 4369			13 23 48	−47 13.	2	13.09	.009	1.19	.023			5		546,931
		NGC 5139 - 4373			13 23 48	−47 13.	1	13.96		0.76		0.25		2		319
		NGC 5139 - 4402			13 23 48	−47 13.	1	14.79		0.20		0.15		1		319
		NGC 5139 - 4403			13 23 48	−47 13.	2	14.67	.000	0.95	.011	0.41	.027	5		319,441
		NGC 5139 - 4404			13 23 48	−47 13.	1	14.92		0.09		0.19		2		319
		NGC 5139 - 4405			13 23 48	−47 13.	2	14.45	.005	0.95	.014	0.44	.020	5		319,441
		NGC 5139 - 4407			13 23 48	−47 13.	1	15.41		0.93		0.61		1		319
		NGC 5139 - 4408			13 23 48	−47 13.	1	13.21		0.58				2		931
		NGC 5139 - 4409			13 23 48	−47 13.	1	15.70		0.84		0.23		3		319
		NGC 5139 - 4419			13 23 48	−47 13.	2	14.68	.045	0.20	.020	0.24		2		319,931
		NGC 5139 - 4465			13 23 48	−47 13.	1	14.69		0.61				1		931
		NGC 5139 - 4654			13 23 48	−47 13.	1	14.89		0.78		0.28		1		319
		NGC 5139 - 4655			13 23 48	−47 13.	1	16.14		0.62		0.01		2		319
		NGC 5139 - 4658			13 23 48	−47 13.	1	16.38		0.75				1		319
		NGC 5139 - 4659			13 23 48	−47 13.	1	16.27		0.95				1		319
		NGC 5139 - 4660			13 23 48	−47 13.	1	15.85		0.84		0.05		1		319
		NGC 5139 - 4661			13 23 48	−47 13.	1	16.31		0.77		0.38		1		319
		NGC 5139 - 4663			13 23 48	−47 13.	2	14.88	.049	0.90	.034	0.24		3		319,931
		NGC 5139 - 4665			13 23 48	−47 13.	1	14.63		0.60		0.06		1		319
		NGC 5139 - 4686			13 23 48	−47 13.	1	13.32		1.05				4		546
		NGC 5139 - 5000			13 23 48	−47 13.	1	15.08		0.13		0.18		2		249
		NGC 5139 - 5002			13 23 48	−47 13.	1	14.56		0.79		0.40		2		319
		NGC 5139 - 5003			13 23 48	−47 13.	1	14.29		0.63		0.07		2		319
		NGC 5139 - 5005			13 23 48	−47 13.	1	14.91		0.17		0.23		2		319
		NGC 5139 - 5006			13 23 48	−47 13.	3	14.37	.000	0.98	.015	0.58	.019	5		319,441,931
		NGC 5139 - 5007			13 23 48	−47 13.	1	15.76		0.88		0.15		2		319
		NGC 5139 - 5013			13 23 48	−47 13.	1	15.46		0.87				2		319
		NGC 5139 - 5021			13 23 48	−47 13.	2	14.42	.020	0.25	.020	0.17		2		319,931
		NGC 5139 - 5022			13 23 48	−47 13.	1	14.41		1.01		0.53		2		319
		NGC 5139 - 5044			13 23 48	−47 13.	1	13.73		1.03				4		546
		NGC 5139 - 5268			13 23 48	−47 13.	1	13.49		1.04				4		546
		NGC 5139 - 5312			13 23 48	−47 13.	1	16.12		0.83				1		319
		NGC 5139 - 5315			13 23 48	−47 13.	1	16.42		0.73				3		319
		NGC 5139 - 5318			13 23 48	−47 13.	1	14.22		1.00		0.62		4		441
		NGC 5139 - 5319			13 23 48	−47 13.	1	16.01		0.78		0.02		2		319
		NGC 5139 - 5321			13 23 48	−47 13.	1	15.76		0.76		0.11		1		319
		NGC 5139 - 5324			13 23 48	−47 13.	2	14.83	.040	1.08	.020	0.55		2		319,931
		NGC 5139 - 5585			13 23 48	−47 13.	1	13.20		1.12				4		546
		NGC 5139 - 5752			13 23 48	−47 13.	2	14.38	.018	0.97	.001	0.50	.036	5		319,441
		NGC 5139 - 5759			13 23 48	−47 13.	2	13.08	.015	0.94	.005	0.47		3		319,931
		NGC 5139 - 5891			13 23 48	−47 13.	1	14.38		1.46				1		931
		NGC 5139 - 5938			13 23 48	−47 13.	1	14.75		0.18		0.16		2		249
		NGC 5139 - 5941			13 23 48	−47 13.	1	13.50		0.88		0.32		1		319
		NGC 5139 - 6107			13 23 48	−47 13.	1	14.50		0.90		0.33		3		441
		NGC 5139 - 6110			13 23 48	−47 13.	1	14.92		0.93		0.57		4		441
		NGC 5139 - 6113			13 23 48	−47 13.	1	13.65		1.00		0.50		1		319
		NGC 5139 - 6119			13 23 48	−47 13.	1	13.45		0.93		0.54		1		319
		NGC 5139 - 6258			13 23 48	−47 13.	1	14.13		0.91		0.35		1		319
		NGC 5139 - 6259			13 23 48	−47 13.	1	13.46		1.11		0.70		1		1468
		NGC 5139 - 6304			13 23 48	−47 13.	1	14.87		0.73		0.14		5		3036
		NGC 5139 - 6305			13 23 48	−47 13.	1	14.55		0.94		0.32		5		3036
		NGC 5139 - 6327			13 23 48	−47 13.	1	13.15		1.03				5		546
		NGC 5139 - 6355			13 23 48	−47 13.	1	13.31		1.11				3		546
		NGC 5139 - 6400			13 23 48	−47 13.	1	14.24		0.86		0.27		1		319
		NGC 5139 - 6442			13 23 48	−47 13.	1	13.15		0.88				1		931
		NGC 5139 - 6460			13 23 48	−47 13.	1	13.30		1.07				4		546
		NGC 5139 - 6465			13 23 48	−47 13.	1	13.19		0.95				4		546
		NGC 5139 - 6477			13 23 48	−47 13.	1	13.78		0.90		0.14		4		3036
		NGC 5139 - 6497			13 23 48	−47 13.	1	13.50		1.03		0.38		3		3036
		NGC 5139 - 6523			13 23 48	−47 13.	1	13.94		1.04				5		546
		NGC 5139 - 6595			13 23 48	−47 13.	1	14.03		1.07		0.63		3		441
		NGC 5139 - 6642			13 23 48	−47 13.	1	14.94		0.86		0.24		4		441
		NGC 5139 - 6747			13 23 48	−47 13.	1	13.89		0.73		0.17		3		441
		NGC 5139 - 6840			13 23 48	−47 13.	1	14.52		0.32		0.22		2		249
		NGC 5139 - 6903			13 23 48	−47 13.	1	15.44		0.04		-0.04		2		249
		NGC 5139 - 6912			13 23 48	−47 13.	1	15.22		0.07		0.07		2		249
		NGC 5139 - 6920			13 23 48	−47 13.	1	14.73		0.18		0.22		2		249
		NGC 5139 - 6938			13 23 48	−47 13.	1	14.98		0.08		0.08		2		249
		NGC 5139 - 6941			13 23 48	−47 13.	1	15.25		0.04		0.03		2		249
		NGC 5139 - 6956			13 23 48	−47 13.	1	14.81		0.16		0.16		2		249

HD	DM	Other Id	N	Rem	α_{1950}	δ_{1950}	S	V	σ_V	B–V	σ_{B-V}	U–B	σ_{U-B}	n	Spectrum	References
		NGC 5139 - 6980			13 23 48	−47 13.	1	14.71		0.18		0.26		2		249
		NGC 5139 - 6986			13 23 48	−47 13.	1	15.28		0.10		0.01		2		249
		NGC 5139 - 9001			13 23 48	−47 13.	1	10.49		0.59		0.32		1		1622
		NGC 5139 - 9006		V	13 23 48	−47 13.	1	12.13		1.80		1.52		1		1468
		NGC 5139 - 9017		V	13 23 48	−47 13.	1	12.51		1.75		1.89		1		1468
		NGC 5139 - 9102			13 23 48	−47 13.	1	14.30		0.92				4		546
		NGC 5139 - 9198			13 23 48	−47 13.	1	13.71		1.10				3		546
		NGC 5138 - 35			13 23 48	−58 49.0	2	11.47	.015	0.08	.015	-0.40	.015	4	B9 V	185,847
		AAS41,43 F4 # 99			13 23 48	−62 10.3	1	12.50		0.37		0.22		1		820
		NGC 5138 - 42			13 23 49	−58 47.6	1	12.83		0.27		0.17		4		847
	−61 03701				13 23 49	−61 43.2	1	11.41		0.38		0.21		1	B8	820
116956	+57 01438	G 223 - 4			13 23 50	+57 13.8	1	7.28		0.81		0.45		2	G9 IV-V	1733
		AAS41,43 F4 # 100			13 23 50	−62 01.1	1	11.35		0.85		0.27		1	G0	820
		AAS41,43 F4 # 101			13 23 50	−62 05.8	1	12.37		0.75		0.62		1		820
	+25 02634				13 23 51	+25 03.8	1	9.54		0.96		0.86		1	K2	280
		AAS41,43 F4 # 103			13 23 51	−61 52.3	1	12.64		0.60		0.23		1		820
116748	−64 02408				13 23 51	−64 49.3	1	8.82		0.32		0.21		1	A3mA5-A9	1771
	+35 02450	IDS13216N3540	AB		13 23 52	+35 24.1	2	9.65		0.68		0.23		2	F8	1569
		NGC 5138 - 63		V	13 23 52	−58 42.9	1	10.30		1.50		1.74		5	K3	847
		NGC 5138 - 6			13 23 52	−58 46.7	1	11.76		0.11		-0.35		3	B9	847
		AAS41,43 F4 # 104			13 23 52	−62 15.6	1	11.35		0.19		-0.09		2	B8	820
		AAS41,43 F4 # 105			13 23 52	−62 22.7	1	12.57		0.31		0.22		1		820
		CS 16027 # 82			13 23 53	+31 48.9	1	14.50		0.55		-0.06		1		1744
		NGC 5138 - 61			13 23 53	−58 43.6	2	13.34	.009	0.23	.005	0.11	.000	9		185,847
		NGC 5138 - 13			13 23 53	−58 44.4	2	13.76	.000	0.35	.005	0.07	.005	7		185,847
		NGC 5138 - 7			13 23 53	−58 45.8	1	13.66		0.45		0.21		3		847
		AAS41,43 F4 # 107			13 23 53	−61 50.2	1	12.72		0.68		0.15		1		820
		AAS41,43 F4 # 108			13 23 54	−61 54.5	1	11.00		1.76		1.56		1	M3	820
		AAS41,43 F4 # 106			13 23 54	−62 09.0	1	12.78		0.51		0.23		1		820
116749	−66 02224				13 23 54	−66 29.2	1	7.80		0.14		0.08		2	A0/1 V	1771
		AAS41,43 F4 # 109			13 23 55	−61 45.5	1	12.33		0.28		0.08		1	B9	820
116858	−23 11071	IDS13211S2346	A		13 23 56	−24 02.0	2	8.79	.009	0.93	.005	0.59		5	K3 V	2033,3072
		NGC 5138 - 12			13 23 56	−58 44.6	2	13.69	.014	0.40	.010	0.18	.005	7		185,847
		NGC 5138 - 8	AB		13 23 56	−58 45.6	2	12.27	.005	0.20	.005	-0.06	.010	7	B8	185,847
116779	−58 04878				13 23 56	−59 12.8	1	8.73		1.04		0.78		2	K0 III	122
	−63 02756	LSS 3076			13 23 56	−63 30.1	1	10.92		0.27		-0.62		2		496
		NGC 5138 - 9			13 23 58	−58 45.2	2	12.01	.005	0.66	.010	0.31	.010	5		185,847
		NGC 5138 - 3			13 23 58	−58 47.7	2	12.85	.010	0.18	.000	-0.07	.010	7		185,847
	+22 02579				13 23 59	+22 00.9	1	9.84		0.95		0.64		1	K2	280
		NGC 5138 - 11			13 23 59	−58 44.6	2	11.63	.005	0.09	.010	-0.40	.010	6	B7	185,847
116845	−21 03712				13 24 00	−22 02.3	2	8.27	.005	0.89	.005	0.55	.010	12	G3 III	164,1775
		NGC 5138 - 1			13 24 00	−58 46.4	1	13.96		0.69		0.20		3		847
		NGC 5138 - 2			13 24 00	−58 46.6	1	13.75		0.32		0.16		3		847
		CS 22877 # 52			13 24 01	−12 17.3	1	14.29		0.04		0.10		1		1736
116835	−40 07894	HR 5060			13 24 01	−41 14.3	1	5.69		1.47				4	K3 III	2006
		NGC 5138 - 144			13 24 01	−58 41.6	1	13.58		1.19		0.62		1		847
116869	−3 03468				13 24 02	−04 11.2	1	9.47		1.04		0.70		1	K1 III-Ba2	565
		NGC 5138 - 143			13 24 03	−58 40.6	1	13.47		0.78		0.23		2		847
		NGC 5138 - 27			13 24 03	−58 47.4	1	13.79		0.57		0.15		3		847
		AAS41,43 F4 # 110			13 24 03	−61 51.6	1	11.92		0.88		0.49		1		820
	−61 03702				13 24 03	−61 56.5	1	10.46		0.33		0.25		1	A2	820
116927	+24 02583				13 24 04	+23 49.0	1	8.38		1.03		1.00		1	K0	280
116911	+28 02226				13 24 04	+28 27.5	1	9.68		0.33		0.00		2	G5	1569
116870	−11 03516	HR 5064			13 24 04	−12 26.9	4	5.24	.030	1.53	.038	1.74	.035	10	K5 III	15,252,1003,3035
116836	−48 08202	HR 5061	★ AB		13 24 04	−48 53.1	1	6.32		0.01		-0.07		1	A0 III/IV	1771
		NGC 5138 - 99			13 24 04	−58 49.7	1	11.86		0.12		-0.32		3	B8	847
116795	−61 03704	V380 Cen	★ AB		13 24 04	−61 36.9	1	9.09		0.09		-0.48		2	B3/5 V	820
116781	−62 03270	LSS 3078	★ AB		13 24 04	−62 23.4	3	7.60	.050	0.13	.014	-0.92	.009	15	O9/B1 Ie	540,820,976
		NGC 5138 - 23			13 24 05	−58 46.7	1	12.85		0.26		0.16		3		847
		AAS41,43 F4 # 115			13 24 05	−61 48.2	1	11.74		0.29		-0.34		1	B5	820
	−61 03703				13 24 05	−62 05.4	1	11.00		0.32		0.26		2	A1	820
	−58 04880	NGC 5138 - 22			13 24 06	−58 46.5	2	10.70	.000	1.47	.010	1.46	.214	6	K0	185,847
		NGC 5138 - 26			13 24 06	−58 47.1	1	13.95		0.29		0.19		3		847
116957	+46 01868	HR 5067			13 24 08	+46 17.3	1	5.88		0.97				2	gK0	70
		NGC 5138 - 69			13 24 08	−58 43.2	1	12.73		0.19		-0.10		3		847
		AAS41,43 F4 # 116			13 24 08	−62 08.5	1	12.31		0.33		0.16		1	B9	820
		NGC 5139 - 6509			13 24 09	−47 01.6	1	14.72		0.79				1		1643
		NGC 5138 - 95			13 24 09	−58 48.5	1	12.80		0.20		0.08		3	A0	847
116859	−34 08881				13 24 10	−35 21.6	1	8.48		0.00		-0.13		3	B9.5 III/I	1770
		NGC 5138 - 18	AB		13 24 10	−58 45.7	1	13.90		0.60		0.16		3		847
		NGC 5138 - 94			13 24 10	−58 49.1	1	12.38		0.16		-0.18		4	B9	847
		AAS41,43 F4 # 118			13 24 10	−61 58.8	1	11.65		0.22		-0.31		1	B6	820
		AAS41,43 F4 # 117			13 24 10	−62 05.5	1	12.79		0.32		0.22		1	A5	820
116796	−62 03271	LSS 3079			13 24 10	−63 16.8	6	8.46	.020	0.07	.012	-0.87	.029	16	O9 II	125,403,496,540,976,1684
		NGC 5138 - 15			13 24 11	−58 45.0	2	11.82	.000	0.17	.015	-0.17	.015	4	B8	185,847
		NGC 5138 - 17			13 24 11	−58 45.6	1	13.29		0.23		0.08		3		847
	−61 03705				13 24 11	−62 02.5	1	11.38		0.15		-0.45		1		820
		BPM 22038			13 24 12	−59 57.	1	14.01		0.60		-0.08		1		3065
		AAS41,43 F4 # 120			13 24 12	−61 59.1	1	13.09		0.27		0.02		1	A0	820
		AAS41,43 F4 # 122			13 24 12	−62 02.2	1	11.39		1.46		1.51		1		820

Table 1   749

HD	DM	Other Id	N Rem	α_{1950}	δ_{1950}	S	V	σ_V	B–V	σ_{B-V}	U–B	σ_{U-B}	n	Spectrum	References
		AAS41,43 F4 # 123		13 24 13	−62 05.0	1	13.29		0.29		0.19		1		820
		AAS41,43 F4 # 119		13 24 13	−62 22.3	1	11.89		0.15		−0.07		1	B8	820
		NGC 5138 - 19		13 24 14	−58 45.7	1	12.37		0.17		−0.10		3	B9	847
		AAS41,43 F4 # 124		13 24 14	−62 03.7	1	12.48		0.30		−0.02		1		820
238233	+59 01511			13 24 15	+59 09.5	1	9.19		1.03				3	K0	1379
116827	−61 03706			13 24 16	−61 51.6	1	9.51		0.11		−0.57		1	B2/5 Vn	820
	−61 03707			13 24 16	−62 02.6	1	11.08		0.15		−0.47				820
117043	+64 00949	HR 5070		13 24 17	+63 31.0	2	6.50	.000	0.74	.000	0.32	.029	6	G6 V	1733,3026
		NGC 5139 - 6540		13 24 17	−47 02.4	1	14.98		0.60				1		1643
		NGC 5139 - 296		13 24 17	−47 04.1	2	12.43	.010	1.19	.015			3		931,1643
		NGC 5138 - 78	AB	13 24 17	−58 44.9	1	12.24		0.15		−0.04		3	B9	847
		AAS41,43 F4 # 129		13 24 17	−62 02.4	1	13.29		0.14		0.09		1		820
		AAS41,43 F4 # 128		13 24 17	−62 05.8	1	12.76		0.70		0.02		1		820
	−61 03708			13 24 17	−62 20.5	1	11.33		0.22		0.14		1	B9	820
	+43 02334	IDS13221N4305	B	13 24 18	+42 48.5	1	11.09		0.10		0.09		2	A0	1026
		SX UMa		13 24 18	+56 31.0	1	10.58		0.12		0.14		1	A8.5	668
		NGC 5139 - 6208		13 24 18	−47 03.6	1	14.66		0.14				2		1643
		NGC 5138 - 79		13 24 18	−58 45.5	1	11.38		1.63		1.80		6		847
		AAS41,43 F4 # 131		13 24 18	−62 04.7	1	11.60		0.26		0.16		1	B9	820
	+23 02569			13 24 19	+23 05.1	1	9.62		0.63		0.11		1	G5	280
116920	−23 11076	IDS13211S2346		13 24 19	−24 01.8	3	8.72	.009	0.91	.009	0.63		9	K3 V	2012,2013,3072
		NGC 5139 - 6342		13 24 19	−47 03.2	1	13.29		0.98				1		1643
		NGC 5139 - 6210		13 24 19	−47 03.7	1	14.95		0.09				2		1643
116862	−48 08206	HR 5063		13 24 19	−49 07.3	6	6.27	.009	−0.13	.006	−0.67	.012	24	B3 III/IV	15,1075,1770,2014,2029,3036
116848	−53 05615			13 24 19	−53 53.7	1	8.38		0.19		0.09		1	A3 IV	1771
116826	−61 03709			13 24 19	−61 46.7	1	10.14		0.20		−0.06		1	A0	820
		AAS41,43 F4 # 133		13 24 19	−61 58.0	1	11.78		0.81		0.27		1	G5	820
		AAS41,43 F4 # 130		13 24 19	−62 12.5	1	12.67		0.42		0.23		1	A7	820
116873	−39 08260	HR 5065	⋆ A	13 24 20	−39 54.2	1	6.40		0.98				4	G8/K0 III	2009
		AAS41,43 F4 # 134		13 24 20	−62 00.1	1	11.85		1.44		1.60		1		820
116940	+18 02736			13 24 21	+17 54.5	1	9.10		1.17		1.29		1	K0	280
		AAS61,331 # 14		13 24 21	+28 41.	1	10.78		0.31		−0.05		2		1569
		G 223 - 5		13 24 21	+60 03.1	1	12.26		1.47		1.18		1		1658
		NGC 5138 - 74		13 24 21	−58 43.1	1	13.24		0.21		−0.05		3		847
		NGC 5139 - 6343		13 24 22	−47 03.3	1	14.71		0.17				1		1643
		NGC 5139 - 6062		13 24 22	−47 04.9	2	13.77	.015	0.93	.015			3		931,1643
		NGC 5138 - 88		13 24 23	−58 47.8	1	12.01		0.12		−0.31		3	A0	847
116898	−21 03714	IDS13217S2151	AB	13 24 24	−22 06.3	1	8.09		0.51		0.02		4	F6 V	244
		AAS41,43 F4 # 135		13 24 24	−62 03.2	1	11.54		0.49		0.22		1	F2	820
		NGC 5139 - 6219		13 24 25	−47 03.6	1	14.77		1.18				1		1643
		NGC 5139 - 6217		13 24 25	−47 03.8	1	14.70		0.33				1		1643
		AAS41,43 F4 # 136		13 24 25	−62 05.9	1	12.54		0.50		−0.36		1		820
116959	+15 02578	IDS13220N1454	AB	13 24 26	+14 38.0	1	8.25		0.63		0.15		1	G5	1569
		NGC 5139 - 2342		13 24 26	−47 20.1	1	13.82		0.97				1		1643
	−58 04883	NGC 5138 - 151		13 24 26	−58 42.3	2	10.46	.010	1.06	.010	0.69	.029	6	G8	185,847
		NGC 5138 - 87		13 24 26	−58 47.7	1	13.16		0.19		0.04		3		847
		NGC 5139 - 6075		13 24 27	−47 05.2	2	14.06	.010	0.79	.000			3		931,1643
		NGC 5138 - 83		13 24 27	−58 46.2	1	13.99		0.47		0.21		3		847
	−61 03711			13 24 27	−61 46.8	1	10.87		0.47		0.26		1	A2	820
116841	−61 03710			13 24 27	−62 01.8	1	8.51		1.64		1.96		1	K3 III	820
		NGC 5139 - 1980		13 24 28	−47 22.3	1	14.80		1.08				1		1643
		NGC 5139 - 269		13 24 28	−47 22.7	3	12.36	.012	1.22	.011	0.86		6		319,931,1643
		AAS41,43 F4 # 138		13 24 28	−61 58.1	1	10.86		1.50		1.46		1	G8	820
		AAS41,43 F4 # 141		13 24 29	−61 58.4	1	12.14		1.20		0.97		1		820
		AAS41,43 F4 # 139		13 24 29	−62 14.7	1	12.28		0.64		0.18		1		820
116960	+12 02594			13 24 30	+12 10.1	2	8.00	.024	−0.02	.000	−0.19	.034	3	A0 V	272,280
	+21 02530			13 24 30	+21 03.7	1	9.62		1.06		0.98		1	K5	280
	−60 04683			13 24 30	−61 08.8	1	10.68		0.16		−0.06		2	B8	122
		AAS41,43 F4 # 142		13 24 30	−62 00.7	1	13.34		0.47		0.12		1		820
	+19 02674			13 24 31	+18 36.3	1	9.82		0.93		0.68		1		280
		NGC 5138 - 168		13 24 32	−58 46.4	1	11.86		0.09		−0.35		3	B7	847
		AAS41,43 F4 # 143		13 24 32	−62 03.0	1	12.09		0.15		−0.28		1	B8	820
		AAS41,43 F4 # 144		13 24 32	−62 03.3	1	12.26		0.25		−0.04		1	B9	820
	+24 02584			13 24 33	+23 44.9	1	8.98		1.39		1.65		1	K7	280
		NGC 5139 - 2174		13 24 33	−47 21.7	1	14.79		0.82				1		1643
		NGC 5139 - 13		13 24 33	−47 27.3	1	10.18		1.23				5		1643
	−60 04685			13 24 33	−61 09.9	1	11.04		0.38		0.24		2		122
		AAS41,43 F4 # 145		13 24 33	−62 04.3	1	12.97		0.21		−0.08		1		820
		AAS41,43 F4 # 146		13 24 34	−61 56.6	1	12.80		0.62		0.30		1		820
		NGC 5139 - 2176		13 24 35	−47 20.9	1	14.70		0.18				1		1643
		AAS41,43 F4 # 147		13 24 35	−62 01.5	1	13.12		0.26		−0.07		1		820
		Ton 711		13 24 36	+23 18.	1	14.99		0.02		−0.31		2		1036
	−61 03712	LSS 3085		13 24 36	−62 03.6	1	11.32		0.16		−0.44		1	B5	820
116929	−30 10579			13 24 37	−30 28.8	1	9.61		0.88		0.50		2	G8 IV	1730
		NGC 5139 - 2770		13 24 37	−47 18.6	2	14.27	.055	0.88	.005			2		931,1643
	−61 03714			13 24 37	−61 38.0	1	10.57		0.17		−0.46		2	B5	820
		NGC 5139 - 2771		13 24 38	−47 18.5	1	14.77		0.91				1		1643
	−61 03713			13 24 38	−62 04.1	1	11.46		0.17		−0.35		1	B6	820
116849	−65 02335			13 24 38	−66 01.2	1	9.23		0.17		−0.81		4	B1 Vpe	1684
		NGC 5128 sq1 13		13 24 40	−43 15.4	1	13.06		0.59		−0.04		1		99

HD	DM	Other Id	N	Rem	α_{1950}	δ_{1950}	S	V	σ_V	B–V	σ_{B-V}	U–B	σ_{U-B}	n	Spectrum	References
	−58 04886				13 24 40	−58 38.7	1	10.56		0.07		−0.08		2		122
	−61 03715				13 24 40	−61 53.3	1	11.10		0.18		−0.55		1	B3	820
	−46 08659				13 24 41	−47 03.3	1	10.66		1.33		1.46		3		517
		NGC 5139 - 2181			13 24 41	−47 21.3	1	14.64		1.12				1		1643
117008	+22 02580				13 24 42	+21 49.8	1	9.23		0.20		0.16		1	A3	280
	−61 03716				13 24 42	−62 06.5	1	11.08		0.36		−0.24		1	B5	820
		AAS41,43 F4 # 152			13 24 42	−62 10.3	1	11.97		0.64		0.16		1		820
		AAS41,43 F4 # 153			13 24 42	−62 12.5	1	12.02		0.34		0.24		2	A0	820
		NGC 5139 - 199			13 24 43	−47 24.0	1	12.20		1.38				2		1643
116875	−60 04687	Ho 16 - 80			13 24 43	−60 48.2	2	7.31	.014	0.04	.000	−0.32	.014	9	B8 V	401,1740
	−39 08265				13 24 45	−40 07.2	1	10.84		0.33		0.08		2		1730
		NGC 5139 - 2360			13 24 45	−47 20.3	1	14.57		0.90				1		1643
	−61 03718				13 24 45	−61 52.9	1	10.06		0.21		−0.33		1	B6	820
116976	−15 03668	HR 5068			13 24 47	−15 42.9	9	4.75	.009	1.09	.005	1.06	.007	35	K1 III	3,15,30,1075,1088*
		NGC 5139 - 418			13 24 47	−47 18.6	2	12.72	.023	1.11	.009	0.66		4		319,1643
		NGC 5139 - 2187			13 24 47	−47 20.8	1	15.35		0.07				1		1643
		NGC 5139 - 2186			13 24 47	−47 21.1	1	15.21		0.86				1		1643
	+57 01439				13 24 48	+56 34.4	1	10.83		0.21				1		1213
	−61 03719				13 24 48	−61 30.4	1	10.66		0.77		0.48		2		122
		AAS41,43 F4 # 156			13 24 48	−61 57.3	1	13.01		0.24		−0.12		1		820
		NGC 5139 - 2781			13 24 49	−47 17.9	2	13.66	.225	0.89	.045			2		931,1643
		NGC 5139 - 2189			13 24 49	−47 21.6	1	14.74		0.97				1		1643
116887	−60 04689				13 24 49	−61 05.5	1	8.53		1.12		1.03		2	K0 III	122
		AAS41,43 F4 # 160			13 24 49	−61 51.0	1	10.89		1.16		0.78		1	F8	820
		AAS41,43 F4 # 157			13 24 49	−61 58.6	1	10.63		1.56		0.80		3	G5	820
		Ho 16 - 58			13 24 50	−60 54.1	1	12.44		0.40		0.20		2		1740
		AAS41,43 F4 # 158			13 24 50	−62 12.1	1	11.39		1.01		0.63		2	G3	820
116865	−67 02286	IDS13213S6721		A	13 24 50	−67 36.6	1	7.06		0.41		0.25		1	A5 III/IV	1771
117028	+29 02408				13 24 51	+29 08.0	1	8.24		0.96		0.69		1	G8 III	280
117187	+73 00592	HR 5073		★ A	13 24 51	+72 39.0	2	5.79	.012	1.62	.009	1.90		3	M1 IIIab	71,3001
		LOD 29 # 9			13 24 51	−61 27.2	1	12.38		0.24		0.08		2		122
		AAS41,43 F4 # 159			13 24 51	−62 14.4	1	12.30		0.23		−0.17		1	B8	820
	−46 08661	IDS13219S4655		AB	13 24 52	−47 10.8	1	10.39		1.30		1.47		5		517
		AAS41,43 F4 # 164			13 24 52	−61 41.2	1	11.18		0.82		0.27		2	G2	820
		AAS41,43 F4 # 161			13 24 52	−61 51.7	1	11.94		0.42		0.06		2	B9	820
	+14 02615				13 24 53	+14 12.3	1	10.59		0.48		−0.02		2	F9 III	1569
		Ho 16 - 62			13 24 53	−60 59.0	1	11.60		0.41		0.25		2		1740
116888	−61 03721				13 24 53	−62 03.5	1	10.12		0.24		0.18		2	A0/1 IV	820
		AAS41,43 F4 # 163			13 24 53	−62 09.2	1	12.29		0.60		0.04		1		820
	−63 02766				13 24 53	−63 25.3	1	9.82		1.13		0.92		3		125
		G 63 - 28			13 24 55	+12 18.2	1	11.42		1.33		1.22		1	K7	1620
117030	+21 02532				13 24 55	+20 31.9	1	8.17		1.32		1.54		1	K0	280
		Ho 16 - 55			13 24 55	−60 54.8	1	12.57		0.51		0.27		2		1740
116987	−30 10583				13 24 57	−30 31.6	1	8.32		0.46		0.00		3	F3 V	1730
116969	−39 08267				13 24 57	−40 08.1	1	8.92		0.61		0.17		2	G0 V	1730
		NGC 5139 - 446			13 24 57	−47 21.1	3	12.77	.014	1.11	.012	0.69		10		319,546,1643
		Ho 16 - 63			13 24 57	−61 01.7	1	12.39		0.47		−0.09		2		1740
	+9 02757	G 63 - 29			13 24 58	+09 29.3	1	10.76		1.00		0.85		1	K5	1696
	+11 02569				13 24 58	+11 10.7	1	10.00		0.93		0.72		1		280
	+14 02616				13 24 58	+13 49.5	1	9.85		1.02		0.83		2	K0 III	1569
		LOD 29 # 8			13 24 58	−61 27.2	1	11.38		0.23		−0.24		2		122
	−61 03723				13 24 58	−61 27.3	1	10.03		0.18		−0.01		2		122
116947	−53 05622				13 24 59	−53 54.3	1	8.39		0.08		−0.03		3	B9 V	1770
		AAS41,43 F4 # 165			13 24 59	−62 12.5	1	12.82		0.46		0.19		1		820
		AAS41,43 F4 # 166			13 25 00	−61 53.1	1	12.91		0.15		−0.13		1	B9	820
116908	−62 03275	V402 Cen			13 25 00	−63 19.5	1	9.20		0.10		−0.03		3	A0 IV	125
	+38 02448				13 25 01	+38 28.1	1	9.58		1.02		0.78		2	G8 IV	1569
117077	+39 02646				13 25 01	+38 47.6	1	8.72		1.16		1.10		1	K0 III	280
	−30 10585				13 25 01	−30 25.5	1	11.11		1.06		0.88		2		1730
116979	−46 08663				13 25 01	−47 07.8	1	9.41		0.68		0.33		5	K0 III +A5	517
		NGC 5139 - 2194			13 25 01	−47 20.8	1	13.89		0.97				1		1643
117044	+14 02617				13 25 03	+14 10.4	1	8.10		0.34		0.09		1	F0 II	280
116980	−47 08344	NGC 5139 - 5			13 25 04	−47 26.8	2	8.70	.000	0.56	.005			18	G0	931,1643
	−61 03725				13 25 04	−61 55.1	1	10.26		0.61		0.34		1	F0	820
		GD 324			13 25 05	+57 09.8	1	13.07		0.47		−0.15		2		308
116948	−58 04889				13 25 06	−59 16.4	1	7.07		0.96		0.66		2	K0 III	122
		AAS41,43 F4 # 169			13 25 06	−61 47.8	1	11.80		0.35		−0.14		1	B8	820
117062	+25 02637				13 25 07	+24 34.2	1	8.92		0.33		−0.05		7	F2 V	3026
116972	−53 05624				13 25 07	−54 15.0	1	8.91		0.07		0.08		1	B9.5 V	1770
	−61 03726	LSS 3087			13 25 07	−62 08.3	2	10.30	.000	0.46	.000	−0.50	.035	4	B0.5Ia-Iab	403,820
116890	−68 01929	HR 5066, EZ Mus			13 25 07	−69 22.1	4	6.20	.012	0.01	.009	−0.41	.012	13	Ap Si	15,1075,1771,2038
		Ho 16 - 54			13 25 08	−60 56.2	1	11.99		1.09		0.93		2		1740
		AAS41,43 F4 # 170			13 25 08	−62 07.5	1	11.56		1.13		0.58		1		820
117064	+16 02510				13 25 09	+16 15.3	2	9.10	.000	0.95	.015	0.77		3	K0	280,882
116991	−43 08297	IDS13222S4315		AB	13 25 09	−43 30.6	2	7.21	.005	1.08	.000	0.87	.010	15	K1 III	99,1492
		AAS41,43 F4 # 171			13 25 09	−62 06.9	1	12.70		0.62		0.08		1		820
		CS 16027 # 86			13 25 10	+29 07.1	1	13.89		0.58		−0.03		1		1744
	+39 02647				13 25 10	+39 24.5	1	8.94		1.55		1.70		1	M0	280
		AAS41,43 F4 # 173			13 25 10	−61 54.8	1	12.81		0.47		0.29		1		820
	−61 03727				13 25 10	−62 11.1	1	9.81		0.87		0.39		2	G5	820

Table 1   751

HD	DM	Other Id	N Rem	α_{1950}	δ_{1950}	S	V	σ_V	B–V	σ_{B-V}	U–B	σ_{U-B}	n	Spectrum	References
	+25 02638			13 25 11	+25 16.1	1	9.21		1.02		0.84		1	K2	280
	+47 02060			13 25 11	+47 04.7	1	9.15		0.64		0.11		2	F8	1601
	−27 09236			13 25 11	−28 01.7	1	10.19		1.21		1.14		4		158
	−61 03728			13 25 11	−61 55.8	1	10.36		1.41		1.41		1	K1	820
117099	+36 02375	G 164 - 74		13 25 12	+35 59.4	1	8.79		0.99		0.88		2	K5	1569
	−22 03595			13 25 12	−23 15.4	1	10.39		0.77		0.30		4	K0	164
		Ho 16 - 38		13 25 12	−60 58.6	1	13.34		0.42		0.00		1		1740
		AAS41,43 F4 # 174		13 25 12	−62 01.2	1	12.52		0.55		0.05		1		820
		Ho 16 - 52		13 25 13	−60 52.5	1	12.26		0.57		0.22		2		1740
		LP 557 - 79		13 25 14	+02 57.6	1	12.30		0.50		-0.12		2		1696
	+11 02571			13 25 14	+11 13.0	1	9.48		1.60		1.94		1	M2	280
	+27 02254			13 25 14	+27 21.8	1	11.03		0.17		0.10		1		280
117014	−32 09376			13 25 14	−33 01.0	1	8.38		-0.05		-0.19		1	B8 V	1770
		NGC 5139 - 2003		13 25 14	−47 21.9	1	15.21		0.79				1		1643
		NGC 5139 - 272		13 25 14	−47 22.1	2	12.38	.005	1.35	.014	1.19		7		319,1643
		AAS41,43 F4 # 176		13 25 14	−61 55.5	1	12.59		0.50		0.24		1	A8	820
		13H25M15S		13 25 15	+17 57.2	1	11.43		0.81		0.44		2		1696
	−62 03280	LSS 3088		13 25 16	−63 19.6	1	10.84		0.31		-0.37		3		125
		NGC 5139 - 2204		13 25 17	−47 21.4	1	15.00		0.72				2		1643
		NGC 5139 - 2005		13 25 17	−47 22.5	2	15.04	.044	0.10	.005	0.14		3		249,1643
116994	−50 07794			13 25 17	−51 02.0	1	8.79		0.31		0.09		1	A9 V	1771
116951	−61 03729			13 25 17	−62 04.2	1	9.62		0.26		0.23		2	A3 IV/V	820
		AAS41,43 F4 # 177		13 25 17	−62 11.1	1	11.67		0.56		0.07		1	F5	820
117122	+45 02110			13 25 18	+44 57.9	1	8.43		0.66		0.17		1	G5	280
	−60 04691			13 25 18	−61 16.3	1	10.76		0.16		-0.15		2		122
	−61 03730			13 25 18	−61 54.4	1	10.62		0.67		0.18		1	F8	820
	+41 02398			13 25 19	+40 54.6	1	9.91		1.39		1.58		1	K0	280
		AAS41,43 F4 # 178		13 25 19	−62 15.7	1	12.08		0.67		0.28		1	B1	820
		Ho 16 - 41		13 25 20	−60 56.9	1	11.10		1.56		1.78		3		1740
116973	−63 02773			13 25 20	−63 41.9	1	9.10		1.21		1.20		4	G8/K0 III	125
		NGC 5128 sq1 15		13 25 21	−43 37.5	2	15.21	.052	0.89	.047	0.46	.005	4		99,1492
117101	+25 02639			13 25 22	+24 39.8	1	8.64		0.31		0.05		1	F5	280
		Ho 16 - 51		13 25 22	−60 51.0	1	12.83		0.60		0.44		3		1740
		AAS41,43 F4 # 181		13 25 22	−62 07.9	1	12.31		0.48		0.30		2	B8	820
		NGC 5128 sq1 16		13 25 23	−42 53.9	1	12.48		0.28		0.19		2		99
		Ho 16 - 70		13 25 23	−60 47.1	1	11.72		0.37		0.28		3		1740
117200	+65 00935	HR 5074	⋆ A	13 25 24	+64 59.6	1	6.66		0.38		0.00		4	F0	1084
		Ho 16 - 42		13 25 24	−60 56.1	1	11.80		0.25		-0.41		3		1740
		AAS41,43 F4 # 182		13 25 24	−61 46.2	1	11.67		0.64		-0.30		1	B3	820
	+33 02340			13 25 25	+33 15.4	1	9.13		1.03		0.88		1	G8 V	280
116909	−60 04692	Ho 16 - 71		13 25 26	−60 45.9	1	9.49		0.08		-0.18		2	B8/B8 III	1740
		Ho 16 - 39		13 25 26	−60 58.1	1	12.06		0.21		-0.36		3		1740
	+6 02746			13 25 27	+06 28.1	1	11.05		0.11		0.13		2		1026
117123	+34 02424			13 25 27	+33 53.1	1	9.18		1.10		1.12		1	K3 V	280
		NGC 5128 sq1 17		13 25 27	−43 33.4	2	15.13	.083	0.82	.000	0.46	.010	3		99,1492
	+18 02738			13 25 28	+18 12.3	1	10.32		0.93		0.66		1		280
		Ho 16 - 40		13 25 29	−60 57.5	1	11.53		0.20		0.20		3		1740
117201	+65 00936	HR 5075	⋆ B	13 25 30	+64 58.7	1	7.04		0.40		-0.02		4	F0	1084
117083	−21 03718			13 25 30	−22 08.5	1	7.43		1.24		1.32		4	K1 III	244
		Ho 16 - 33		13 25 30	−60 58.7	1	11.32		0.66		0.11		3		1740
		Ho 16 - 73		13 25 30	−61 03.1	1	12.21		0.35		0.17		3		1740
		Ho 16 - 74		13 25 30	−61 03.6	1	12.66		0.31		0.27		4		1740
117137	+35 02453			13 25 31	+35 24.7	1	8.72		0.99		0.77		2	G5	1569
117018	−52 06511			13 25 31	−52 56.9	1	7.97		0.20		0.18		2	A3 IV	1771
		NGC 5139 - 25		13 25 32	−47 26.6	2	10.79	.005	0.51	.000			6		931,1643
	+14 02620			13 25 34	+13 47.9	1	9.59		1.11		1.06		1	K2 II	280
	+14 02619			13 25 34	+14 13.1	1	10.37		0.96		0.61		1	G8 II	280
117655	+84 00311			13 25 35	+83 34.0	1	7.25		0.70		0.28		3	G5	3026
		NGC 5128 sq1 18		13 25 35	−43 42.5	2	11.31	.015	0.29	.015	0.08	.005	6		99,1492
		AAS41,43 F4 # 185		13 25 35	−61 45.5	1	12.34		0.29		0.05		1	A0	820
117000	−61 03732	LSS 3091		13 25 35	−61 50.6	6	6.64	.017	1.06	.017	0.54	.029	6	F0/2 Ia	403,820,8100
	−61 03733			13 25 35	−61 51.2	1	11.50		0.29		0.16		2		820
117055	−47 08353			13 25 36	−48 09.0	2	9.38	.002	-0.08	.002	-0.47	.009	3	B9 III	1770,3036
	−58 04893			13 25 37	−58 35.0	1	9.85		0.22		0.15		2	A1	122
		Ho 16 - 45		13 25 37	−60 53.8	1	13.23		0.62		0.06		3		1740
116925	−74 01063			13 25 37	−74 38.8	1	8.48		1.14		0.97		2	K0 III	1730
		AAS41,43 F4 # 186		13 25 38	−61 56.0	1	11.25		1.14		0.30		1	B3	820
		NGC 5128 sq1 19		13 25 39	−43 32.3	2	10.90	.029	1.65	.019	1.81	.029	3		99,1492
117114	+10 02548	IDS13232N0959	A	13 25 40	+09 43.2	1	7.61		1.17		1.24		2	K0	280
		Ho 16 - 64		13 25 40	−60 55.9	1	12.23		0.62		0.00		2		1740
		AAS41,43 F4 # 187		13 25 40	−62 06.5	1	12.36		0.92		0.61		1	G8	820
117138	+19 02677			13 25 41	+18 32.9	1	8.53		0.30		0.02		2	F2	1569
117024	−63 02778			13 25 41	−63 36.9	7	7.10	.015	0.05	.013	-0.71	.018	27	B2 Ib	125,403,496,540,976*
117157	+33 02341	IDS13234N3306	AB	13 25 42	+32 50.4	1	8.68		0.26		0.09		4	F0	1569
		Ho 16 - 50		13 25 42	−60 52.1	1	11.56		0.54		0.21		4		1740
117025	−64 02418	HR 5069		13 25 42	−64 25.0	6	6.10	.011	0.12	.007	0.04	.007	19	Ap SrEuCr	15,122,355,1075,1771,2038
		NGC 5128 sq1 20		13 25 43	−43 25.7	2	13.43	.005	0.62	.019	0.04	.049	3		99,1492
117086	−43 08308			13 25 43	−43 42.5	2	9.60	.009	0.45	.014	-0.04	.037	8	F2 V	99,1492
	−61 03734			13 25 43	−62 17.2	1	11.13		0.43		0.15		1	A5	820
117126	−0 02691	G 14 - 54		13 25 44	−00 34.6	3	7.43	.008	0.65	.008	0.17	.019	5	G5	803,1620,1658

HD	DM	Other Id	N Rem	α_{1950}	δ_{1950}	S	V	σ_V	B–V	σ_{B-V}	U–B	σ_{U-B}	n	Spectrum	References
116852	−78 00813			13 25 44	−78 35.8	1	8.47		−0.09		−0.09		1	O9 III	400
117105	−26 09740			13 25 45	−27 08.2	3	7.20	.000	0.58	.000	0.02	.016	7	G1 V	861,1311,2012
		Ho 16 - 32		13 25 45	−60 57.9	1	13.17		0.34		0.00		2		1740
	+41 02399			13 25 46	+41 22.4	1	8.96		0.96		0.70		1	K0 III	280
		G 14 - 55	AB	13 25 47	−02 05.8	2	11.24	.000	1.54	.015			6	M2	158,202
	−60 04695	Ho 16 - 19		13 25 47	−60 55.8	2	11.35	.005	0.24	.009	−0.37	.000	9	B1 IV	251,1740
	−60 04694	Ho 16 - 31		13 25 47	−60 57.6	2	11.41	.005	0.25	.010	−0.38	.010	4	B2 V	251,1740
		Ho 16 - 29		13 25 47	−61 00.4	1	13.20		0.36		0.01		4		1740
		NGC 5128 sq1 22		13 25 49	−43 34.1	2	11.93	.010	0.39	.015	0.08	.010	3		99,1492
117057	−58 04894			13 25 49	−58 41.0	1	8.17		−0.04		−0.43		2	Ap Si	122
		Ho 16 - 48		13 25 49	−60 52.1	1	12.78		0.31		−0.02		3		1740
		Ho 16 - 46		13 25 49	−60 53.8	1	13.62		0.31		0.03		2		1740
		Ho 16 - 3		13 25 49	−60 56.7	1	13.11		0.84		0.48		2		1740
	−60 04697	Ho 16 - 20		13 25 49	−60 56.7	2	10.49	.000	0.22	.005	−0.46	.005	12	B2 V	251,1740
	−60 04698	Ho 16 - 18		13 25 50	−60 55.5	2	11.05	.005	0.20	.000	0.14	.000	6		251,1740
		Ho 16 - 28		13 25 50	−61 00.9	1	13.25		0.45		0.39		3		1740
		AAS41,43 F4 # 189		13 25 50	−62 05.4	1	12.30		0.46		0.32		1		820
117141	−8 03562			13 25 51	−09 29.1	1	8.20		1.17		1.09		6	K0	1657
117058	−60 04700	LSS 3093		13 25 51	−60 27.0	1	9.16		0.33		0.37		3	A3 II/III	8100
		Ho 16 - 49		13 25 51	−60 50.2	1	10.29		0.09		−0.11		3		1740
		Ho 16 - 47		13 25 51	−60 53.7	1	12.29		1.39		1.21		2		1740
		NGC 5128 sq1 23		13 25 52	−43 19.2	1	13.90		0.78		0.38		2		99
		Ho 16 - 61		13 25 52	−60 55.7	1	13.03		0.48		0.20		4		1740
	−60 04701	Ho 16 - 21		13 25 52	−60 56.5	2	10.29	.005	0.22	.010	−0.46	.000	6	B2 III	251,1740
		Ho 16 - 30		13 25 52	−60 57.5	2	12.71	.110	1.19	.019	0.77	.005	7		251,1740
		AJ77,733 T22# 10		13 25 52	−63 20.7	1	11.77		0.18		−0.17		3		125
		AAS41,43 F4 # 190		13 25 53	−62 17.8	1	11.81		0.58		0.15		1		820
117059	−61 03735			13 25 56	−62 06.4	1	9.23		1.23		1.14		2	K0/2	820
		AAS41,43 F4 # 192		13 25 57	−62 16.7	1	10.73		1.32		1.16		1	K0	820
	−62 03290	LSS 3094		13 25 57	−63 15.0	3	10.65	.011	0.93	.011	−0.16	.013	10	B2 Ia	125,496,1737
	+16 02512			13 25 58	+16 01.7	1	9.13		1.16		1.24		1	K0	280
117188	+18 02739			13 25 58	+18 06.3	1	8.96		1.19		1.28		1	K2	280
	−30 10597			13 25 58	−30 43.5	1	10.47		0.82		0.29		2		1730
117119	−43 08311			13 25 58	−43 28.7	2	8.07	.005	1.33	.000	1.57	.014	9	K3 III	99,1492
		Ho 16 - 16		13 25 58	−60 54.9	2	12.91	.025	0.32	.005	−0.09	.025	5		251,1740
		Steph 1071		13 25 59	+12 37.3	1	12.54		1.23		1.07		1	K7	1746
117176	+14 02621	HR 5072	★ A	13 25 59	+14 02.7	18	4.97	.009	0.71	.007	0.26	.009	300	G2.5 Va	1,3,15,22,30,116,130*
117242	+53 01622	HR 5076		13 25 59	+53 00.3	2	6.35	.005	0.23	.002	0.12		4	A7 III	70,254
		Ho 16 - 1		13 25 59	−60 56.5	1	10.91		0.76		−0.31		3		1737
	−60 04703	Ho 16 - 22		13 25 59	−60 56.5	2	11.05	.161	0.68	.049	−0.28	.019	6	B2e	251,1740
		Ho 16 - 5		13 25 60	−60 51.4	1	13.06		0.61		0.08		2		1740
117203	+31 02493			13 26 00	+31 24.5	1	6.92		1.38				2	K2	280
	−60 04705	Ho 16 - 86		13 26 00	−60 44.6	1	11.32		0.36		0.22		2		1740
		Ho 16 - 17		13 26 00	−60 55.8	2	13.06	.005	1.56	.015	1.72	.125	4		251,1740
		Ho 16 - 23		13 26 00	−60 57.4	2	12.91	.023	0.33	.000	0.27	.009	8		251,1740
		Ho 16 - 26		13 26 00	−60 58.0	2	12.89	.009	0.29	.018	−0.14	.005	9		251,1740
	+20 02836			13 26 01	+20 05.2	1	10.34		0.96		0.81		1		280
		G 165 - 6		13 26 01	+29 12.8	1	12.94		0.92		0.68		2		1658
		Ho 16 - 6		13 26 01	−60 52.3	1	12.78		0.83		0.38		3		1740
	−61 03736			13 26 01	−61 46.7	3	9.81	.074	0.25	.016	−0.63	.020	11	B2 IVe	540,820,976
117189	+16 02514			13 26 02	+16 22.6	2	8.23	.010	1.11	.005	1.12		3	K0	280,882
		Cr 271 - 9		13 26 03	−63 55.6	1	12.10		0.36		0.26		2		251
		AAS12,381 21 # B		13 26 04	+21 11.2	1	10.47		1.03		0.88		1		280
		Ho 16 - 15		13 26 04	−60 55.0	2	13.10	.010	0.61	.015	0.08	.090	4		251,1740
		GPEC 3672		13 26 05	−31 05.2	1	11.21		0.87				2		1719
		AAS41,43 F4 # 195		13 26 05	−61 42.8	1	11.48		0.84		0.42		1		820
		AAS41,43 F4 # 194		13 26 05	−62 08.9	1	11.33		1.50		1.46		1		820
117074	−63 02779			13 26 05	−64 12.1	1	8.64		0.02		−0.50		2	B9 V	122
117204	+28 02228			13 26 07	+27 36.4	1	8.67		1.04		0.90		1	K2	280
	−22 03598			13 26 07	−23 14.7	1	10.49		0.57		0.01		4	G0	164
	−61 03737			13 26 07	−62 00.6	1	10.54		0.53		0.10		1	F2	820
		LP 497 - 95		13 26 09	+14 27.1	1	11.65		1.02		0.88		1		1696
117281	+51 01846	HR 5079		13 26 09	+50 50.7	1	6.98		0.26		0.08		2	F1 IV	164
		Ho 16 - 7		13 26 09	−60 53.2	1	13.60		0.38		0.16		2		1740
		Ho 16 - 79		13 26 11	−60 51.1	1	13.17		0.39		−0.18		2		1740
	−60 04706	Ho 16 - 60		13 26 11	−60 56.5	2	11.64	.005	0.23	.015	−0.44	.000	6		251,1740
		Ho 16 - 24		13 26 12	−60 57.5	1	12.21		1.83		2.05		1		1740
117176	+14 02622	IDS13236N1419	B	13 26 13	+13 59.6	1	8.52		1.00		0.76		1		280
		AAS41,43 F4 # 197		13 26 13	−62 14.1	1	12.19		0.25		−0.16		1	B8	820
		Cr 271 - 10		13 26 13	−63 54.9	1	11.72		0.73		0.23		2		251
117261	+41 02400	HR 5077		13 26 14	+40 59.3	2	6.49	.015	0.93	.005	0.61		5	G8 III	280,1501
	−60 04707			13 26 14	−61 20.1	1	10.51		0.15		−0.34		2	B5	122
		Ho 16 - 10		13 26 15	−60 56.0	1	12.46		0.23		−0.29		4		1740
		Ho 16 - 4		13 26 15	−60 56.0	1	10.01		0.25		−0.52		2		540
	−60 04708	Ho 16 - 14		13 26 15	−60 56.1	3	10.09	.049	0.22	.024	−0.52	.013	15	B3 V	251,976,1740
		AAS41,43 F4 # 198		13 26 16	−62 01.5	1	10.74		1.42		1.55		1	K2	820
	+24 02587			13 26 17	+23 32.8	1	8.85		1.57		1.91		1	M0	280
117145	−44 08669			13 26 17	−45 16.7	2	6.61	.009	−0.04	.003	−0.12		7	B9 IV/V	1770,2014
		AAS41,43 F4 # 200		13 26 18	−61 45.6	1	11.58		0.34		−0.21		2	B8	820
		S040 # 4		13 26 18	−74 37.3	1	10.55		1.06		0.71		2		1730

Table 1 753

HD	DM	Other Id	N Rem	α_{1950}	δ_{1950}	S	V	σ_V	B–V	σ_{B-V}	U–B	σ_{U-B}	n	Spectrum	References
117262	+34 02426			13 26 19	+33 55.5	1	8.14		1.09		1.01		1	G8 IV	280
117133	−53 05635			13 26 19	−53 47.4	1	8.81		0.00		-0.18		1	B9 IV/V	1770
117110	−60 04710	Ho 16 - 9		13 26 20	−60 52.2	1	9.14		0.53		0.00		3	G0 V	1740
117150	−50 07812	HR 5071		13 26 21	−50 54.4	3	5.05	.004	0.06	.010			11	A1 V	15,2012,2014
117149	−48 08237			13 26 22	−48 37.2	1	10.10		0.27		0.23		1	A9 V	1771
		Ho 16 - 67		13 26 22	−60 48.6	1	12.88		0.41		0.32		3		1740
		AAS41,43 F4 # 199		13 26 22	−62 20.2	1	11.34		0.62		0.12		1	F4	820
	+28 02229	IDS13241N2828	A	13 26 24	+28 10.2	1	10.24		0.67		0.12		2		3060
	+28 02229	IDS13241N2828	B	13 26 24	+28 10.2	1	16.41		0.45		-0.20		1		3060
		Cr 271 - 5		13 26 24	−63 57.5	1	12.10		0.20		-0.16		2		251
		CCS 2093		13 26 25	−54 56.9	1	9.35		4.49				2		864
	−60 04711	Ho 16 - 85		13 26 25	−60 45.5	1	10.72		0.27		0.27		3		1740
117151	−55 05583			13 26 26	−56 14.1	1	8.88		0.01		-0.39		2	B6 III	401
	−63 02780			13 26 26	−63 25.1	1	11.14		0.56		0.12		2		125
117111	−64 02428	LSS 3097		13 26 26	−65 14.6	3	7.67	.023	0.08	.014	-0.76	.005	7	B2 Vne	540,976,1684
117282	+33 02342			13 26 27	+33 28.1	1	8.45		1.42		1.79		2	K0	1569
	+43 02336			13 26 27	+43 01.1	1	9.42		1.40		1.63		1	K2 III	280
		Ho 16 - 13		13 26 28	−60 56.5	1	12.17		1.91		2.12		2		1740
117263	+29 02411			13 26 30	+28 32.7	1	8.81		1.02		0.85		1	K0	280
117566	+79 00422	HR 5091		13 26 30	+78 54.1	2	5.76	.010	0.77	.001	0.35		5	G2.5 IIIb	71,3016
117205	−22 03600			13 26 31	−22 54.7	2	9.13	.005	1.05	.005	0.75	.005	10	K0 III	164,1775
	+12 02596			13 26 32	+11 58.6	1	8.44		1.16		1.16		1	K2	280
		AJ77,733 T22# 12		13 26 32	−63 18.1	1	12.33		0.25		0.07		6		125
		Ho 16 - 83		13 26 33	−61 01.2	1	11.68		1.96		2.20		4		1740
		Cr 271 - 4		13 26 33	−63 57.7	1	11.79		0.17		-0.24		2		251
117193	−47 08365			13 26 34	−47 37.0	1	6.93		0.09		0.02		1	A1 V	1771
		AAS41,43 F4 # 201		13 26 34	−62 07.2	1	10.99		1.16		0.87		1	F5	820
117135	−63 02781	Cr 271 - 1		13 26 34	−63 59.6	1	9.50		0.56		0.16		4	F3 (IV)	251
117265	+16 02516			13 26 35	+16 10.1	2	8.84	.005	1.02	.015	0.97		3	K0	280,882
	−63 02782	Cr 271 - 6		13 26 35	−63 56.0	1	11.27		0.24		0.22		4		251
117211	−44 08674			13 26 36	−45 13.4	1	8.62		-0.04		-0.10		1	B9 V	1770
117170	−53 05639			13 26 36	−53 44.7	3	7.64	.008	-0.01	.009	-0.71	.005	10	B3 II/III	164,400,401
117134	−61 03741			13 26 36	−62 09.6	1	9.51		0.17		-0.74		2	B0/1 IIIn	820
117283	+19 02678			13 26 37	+19 08.1	1	8.68		0.22		0.08		2	F0	1569
117376	+60 01461	HR 5085	⋆ AB	13 26 37	+60 12.2	1	5.40		-0.01		-0.03		2	A1 Vn	3023
	+12 02597	G 63 - 30	⋆ AB	13 26 38	+11 43.8	1	8.46		0.82		0.52		1	K0	280
	+14 02624			13 26 38	+14 10.6	1	10.38		1.03		0.75		1	G8 III	280
		AJ77,733 T22# 11		13 26 39	−63 27.2	1	12.09		0.11		-0.25		4		125
		Cr 271 - 8		13 26 39	−63 55.0	1	12.36		0.22		-0.22		2		251
117301	+32 02356			13 26 40	+32 16.7	1	8.49		0.26		0.07		1	F0 V	280
	−55 05588	LSS 3099		13 26 40	−55 51.3	3	10.68	.023	0.29	.015	-0.65	.005	5		540,976,1737
		AAS41,43 F4 # 203		13 26 40	−62 07.2	1	11.91		0.32		0.00		1	B9	820
117267	−0 02694	HR 5078		13 26 41	−01 06.3	6	6.42	.014	1.12	.009	1.11	.014	17	K0 III	15,252,1256,1509,1531,3077
		Ho 16 - 66		13 26 41	−60 50.5	1	11.58		0.34		0.36		2		1740
117361	+51 01847	HR 5083		13 26 42	+50 58.7	2	6.43	.005	0.37	.000	0.03		4	F5 V Sr	70,254
117220	−36 08607			13 26 43	−37 18.1	2	9.01	.014	0.84	.000	0.34		4	G5w	1594,6006
		Ho 16 - 76		13 26 43	−60 55.4	1	12.81		0.57		0.17		3		1740
		Ho 16 - 82		13 26 43	−61 01.1	1	12.06		0.26		-0.35		4		1740
	−61 03742	LSS 3098		13 26 43	−62 02.1	1	9.58		1.25		0.86		3	A8	820
117304	+11 02575	HR 5081		13 26 44	+11 04.6	1	5.65		1.05				2	K0 III	70
		Cr 271 - 7		13 26 44	−63 56.2	1	12.45		0.18		-0.18		2		251
238241	+59 01514			13 26 46	+59 18.8	1	9.96		0.49				3	G0	1379
117303	+15 02581			13 26 47	+15 04.4	1	8.16		1.18		1.29		1	K2	280
117246	−17 03862			13 26 47	−18 28.2	2	6.90	.005	1.40	.016	1.56	.005	4	K2/3 III	1003,3040
117347	+37 02411			13 26 50	+36 34.2	1	9.10		1.03		0.83		2	K5	1569
		Ho 16 - 78		13 26 50	−60 56.6	1	13.77		0.45		0.07		1		1740
117185	−61 03745			13 26 50	−61 54.8	1	9.83		0.12		-0.29		1	B8 III	820
	−63 02784	Cr 271 - 2		13 26 51	−63 59.8	1	11.35		0.07		-0.02		2		251
		G 63 - 32		13 26 52	+11 42.7	2	12.12	.015	1.46	.002	0.97		2	M6r	1705,3078
		Ho 16 - 65		13 26 52	−60 51.6	1	11.09		0.54		0.20		3		1740
		Cr 271 - 3		13 26 52	−63 59.3	1	11.80		0.49		0.05		2		251
117378	+42 02403			13 26 53	+42 29.8	1	7.64		0.56				2	G0	280
		CCS 2094, YY Mus		13 26 53	−68 36.6	1	11.26		3.14				2		864
		Ton 714		13 26 54	+23 05.	1	14.90		0.09		0.05		2		1036
		Ho 16 - 75		13 26 54	−60 55.2	1	12.90		0.44		0.36		2		1740
	+29 02413			13 26 55	+29 02.9	2	10.98	.015	0.12	.005	0.13	.034	3	A0	272,280
117286	−21 03723			13 26 57	−22 06.6	1	8.67		0.62		0.05		2	G2 V	1375
		LP 855 - 76		13 26 57	−26 24.8	1	12.55		1.25		1.17		1		1696
117195	−61 03746			13 26 57	−62 05.3	1	9.50		0.62		0.12		2	G8/K0	820
117287	−22 03601	HR 5080, R Hya	⋆ A	13 26 58	−23 01.4	8	5.57	.199	1.58	.053	0.56	.075	30	M7 IIIe	15,635,814,3051,8005*
		Ho 16 - 77		13 26 59	−60 57.5	1	13.09		0.51		0.31		2		1740
		AAS41,43 F4 # 207		13 26 59	−62 03.7	1	11.55		0.35		0.09		1	B8	820
	+38 02450			13 27 00	+38 12.8	1	9.12		1.35		1.37		1	K1 III	280
117196	−61 03747			13 27 01	−62 09.3	1	9.80		0.32		0.28		2	A2 III/IV	820
	+26 02465			13 27 03	+25 37.3	1	10.00		1.02		0.88		1	K0	280
		Ho 16 - 84		13 27 03	−61 03.3	1	11.54		0.23		-0.42		4		1740
	−60 04718	LSS 3100		13 27 03	−61 05.9	2	10.10	.013	0.28	.007	-0.46	.001	9	B3 II	540,976
	+25 02640			13 27 05	+24 47.3	1	10.60		1.16		1.13		2	K0	280
	−60 04719	Ho 16 - 72		13 27 05	−60 59.4	1	11.02		0.23		-0.48		2		1740
		G 63 - 33		13 27 06	+17 20.3	1	12.55		1.38		1.10		1		1696

HD	DM	Other Id	N	Rem	α1950	δ1950	S	V	σV	B-V	σB-V	U-B	σU-B	n	Spectrum	References
	+31 02496				13 27 06	+31 00.6	1	9.98		1.03		0.86		2	G8 III	1569
	-60 04720	Ho 16 - 81			13 27 07	-61 00.8	1	10.64		0.46		0.23		3		1740
	-61 03749				13 27 07	-62 05.2	1	11.02		0.24		0.14		1	B9	820
117253	-58 04909				13 27 10	-58 24.4	1	6.77		1.01		0.85		2	K0 III	122
117389	+33 02343				13 27 11	+33 25.5	1	8.69		1.45		1.77		2	M0 III	1569
		AAS12,381 43 # A			13 27 11	+43 06.9	1	10.50		1.13		1.08		1		280
117239	-60 04721	Ho 16 - 68			13 27 11	-60 55.0	1	10.27		0.19		0.18		2	B9 V	1740
117154	-74 01065				13 27 12	-74 31.9	1	8.68		1.53		1.76		2	K3 III	1730
	+16 02518				13 27 13	+15 37.3	1	9.74		0.88		0.46		1	G5	280
	+10 02551				13 27 16	+10 18.6	1	9.59		1.13		1.10		2	K0	280
117256	-60 04724	Ho 16 - 69			13 27 17	-60 57.6	1	10.16		0.11		-0.17		2	B9 V	1740
		AAS41,43 F4 # 210			13 27 18	-62 00.6	1	11.72		0.31		0.18		1	A0	820
		AAS12,381 41 # D			13 27 21	+41 20.9	1	9.32		1.62		1.57		1		280
		He3 904			13 27 23	-57 42.8	1	14.16		1.64		-0.82		1		1753
117325	-48 08255				13 27 25	-49 00.4	1	9.29		0.25		0.15		1	A9 V	1771
117405	+6 02750	HR 5087			13 27 27	+06 16.2	2	6.50	.005	0.96	.000	0.60	.005	7	K0	15,1256
	+11 02576	G 63 - 34			13 27 27	+10 39.0	8	9.05	.007	1.50	.014	1.23	.017	22	M1 V	1,1017,1197,1389,1620,1705*
	+28 02231				13 27 27	+27 56.0	1	8.76		1.49		1.33		1	M5	280
117295	-57 06090				13 27 27	-58 22.5	1	9.82		0.37		0.10		1	A8/9 V	1771
	+29 02416				13 27 29	+28 54.2	1	9.60		0.92		0.65		1	G8 III	280
		G 14 - 57			13 27 29	-08 26.7	2	14.23	.070	1.67	.009	1.23		4		419,1705
117404	+7 02655	HR 5086			13 27 30	+07 26.2	2	6.16	.005	1.47	.000	1.80	.000	7	K5	15,1256
	+38 02451				13 27 30	+37 51.6	1	9.26		1.08		1.04		1	K0 III	280
		NGC 5168 - 51			13 27 34	-60 41.1	1	12.48		0.33		0.22		2		251
117434	+19 02680				13 27 36	+19 19.1	1	7.23		1.02		0.89		2	K0	1569
		BPM 22046			13 27 36	-55 15.	1	13.90		0.86		0.48		1		3065
	+41 02402				13 27 38	+40 49.3	1	9.69		1.28		1.52		1	K2 III	280
117343	-55 05598				13 27 38	-55 45.8	1	9.48		0.35		0.12		1	A8 V	1771
117435	+16 02519				13 27 39	+15 34.3	1	8.70		0.28		0.12		1	A3	280
117380	-22 03602				13 27 39	-22 59.5	1	9.84		0.50		0.04		4	F5 V	164
		NGC 5168 - 158			13 27 39	-60 40.6	1	13.29		1.33		1.11		2		251
117476	+35 02456				13 27 40	+34 46.9	1	7.72		0.23		0.08		2	A3	1569
	-7 03632			A	13 27 40	-08 18.6	6	12.32	.016	0.08	.011	-0.61	.010	17	DA	203,1036,1281,1698,1705,3078
	-7 03632			B	13 27 40	-08 18.6	1	14.34		1.63		1.48		3		3078
		NGC 5168 - 55			13 27 40	-60 42.2	1	12.12		0.24		-0.16		2		251
117392	-22 03603				13 27 41	-23 14.2	1	9.49		1.00		0.75		4	K0 III	164
	+20 02839				13 27 43	+19 34.3	1	9.95		1.19		1.27		1	K0	280
	+30 02405				13 27 44	+30 00.7	1	9.98		1.17		1.26		1	K3 III	280
	+31 02497				13 27 45	+30 37.2	1	9.64		0.93		0.66		2	K0 IV	1569
117408	-22 03604	SS Hya		★AB	13 27 45	-23 23.5	1	7.88		0.00		-0.06		4	A0 V	244
		NGC 5168 - 144			13 27 45	-60 39.4	1	12.66		0.29		0.12		2		251
117298	-68 01938	IDS13242S6836		AB	13 27 46	-68 52.0	1	8.68		0.28		0.14		2	A0 V	1771
	+16 02521				13 27 48	+15 53.0	1	10.52		1.22		1.24		1		280
117464	+17 02637				13 27 48	+16 38.7	1	9.17		1.04		1.03		1	K0	280
		NGC 5168 - 61			13 27 48	-60 42.6	1	13.59		0.38		0.26		2		251
117436	-5 03706	HR 5088		★A	13 27 49	-06 12.8	2	6.10	.020	0.32	.010	0.05	.010	10	F2 V	13,1088
117436	-5 03706	IDS13252S0557		B	13 27 49	-06 12.8	1	11.05		0.94		0.47		2		13
117326	-63 02789				13 27 50	-64 12.8	3	9.53	.012	0.13	.004	-0.62	.012	10	B2 II	122,540,976
117356	-60 04734	IDS13246S6013		A	13 27 51	-60 28.2	1	7.85		0.17		0.09		1	A3 V	1771
	+11 02578				13 27 53	+11 11.3	1	9.77		1.29		1.40		1	K2	280
	+47 02064				13 27 53	+46 45.5	1	10.92		0.02		0.05		3	A2	272
	-59 05011				13 27 54	-60 13.0	2	10.67	.009	0.13	.003	-0.55	.014	3	B2.5V	540,976
		NGC 5168 - 159			13 27 54	-60 41.	1	13.37		1.43		1.32		2		251
	-60 04735	NGC 5168 - 156		★AB	13 27 55	-60 40.5	1	10.36		0.13		-0.40		5		251
		NGC 5168 - 157		★B	13 27 55	-60 40.6	2	10.54	.055	1.23	.022	0.95	.009	5		251,1685
117357	-61 03760	LSS 3103			13 27 55	-61 28.5	3	9.01	.050	0.24	.035	-0.71	.022	9	O9.5/B0 V	401,540,976
	+17 02638				13 27 56	+16 59.1	1	9.78		1.34		1.34		1		280
117345	-65 02356				13 27 56	-65 59.5	1	9.40		0.46		0.19		2		1771
117155	-78 00816				13 27 56	-78 49.1	1	7.69		1.32		1.30		3	K2 II	1704
117478	+8 02721	IDS13255N0800		AB	13 27 59	+07 44.6	1	8.42		0.33				4	A5	176
		NGC 5168 - 11			13 27 59	-60 39.3	1	12.59		0.22		-0.39		2		251
		NGC 5168 - 130			13 27 59	-60 40.0	1	12.11		0.26		0.09		2		251
		NGC 5168 - 26			13 27 59	-60 41.2	1	12.09		0.27		0.05		2		251
		LSS 3104			13 28 00	-63 10.1	1	11.36		0.71		-0.22		3	B3 Ib	1737
		NGC 5168 - 20			13 28 01	-60 40.7	1	13.89		0.68		0.60		2		251
		NGC 5168 - 27			13 28 04	-60 41.3	1	13.16		0.32		0.18		2		251
		LP 911 - 24			13 28 05	-32 22.1	1	11.48		1.34		1.14		1		3062
117911	+82 00395				13 28 07	+81 31.4	1	8.72		0.26		0.06		2	A5	1502
		G 149 - 94			13 28 08	+19 25.7	2	14.75	.019	1.80	.010	1.86		7		316,3078
117440	-38 08592	HR 5089		★AB	13 28 08	-39 09.0	5	3.89	.005	1.19	.014	1.04	.012	16	G8 II/III	15,1075,2012,4001,8015
	+15 02584				13 28 09	+15 25.3	1	9.71		0.97		0.69		1		280
		BV Cen			13 28 10	-54 43.1	1	13.05		0.77		-0.22		9	G5:V	860
117399	-60 04739	V659 Cen			13 28 13	-61 19.5	2	6.49	.005	0.70	.000	0.38	.005	3	F6/7 Ib	122,657
		AAS61,331 # 13			13 28 18	+27 56.	1	11.32		0.24		0.05		2		1569
	-64 02439				13 28 18	-64 36.8	1	10.33		0.06		-0.57		2	B3	807
117316	-73 01158				13 28 19	-74 21.2	1	8.17		0.45		0.10		2	F2 IV	1730
	+40 02660				13 28 21	+39 40.2	1	9.17		1.41		1.73		1	K5	280
		WLS 1320-20 # 11			13 28 21	-17 44.9	1	11.39		0.52		0.18		2		1375
		AAS12,381 44 # C			13 28 23	+43 46.2	1	10.24		1.04		0.91		1		280
117555	+24 02592	FK Com		★C	13 28 25	+24 29.4	4	8.15	.030	0.87	.008	0.41	.036	5	G5 II	280,1512,1569,8090

Table 1   755

HD	DM	Other Id	N Rem	α_{1950}	δ_{1950}	S	V	σ_V	B–V	σ_{B-V}	U–B	σ_{U-B}	n	Spectrum	References
		Tr 21 - 22		13 28 25	−62 32.8	1	12.37		0.14		0.05		1		1672
117415	−64 02440			13 28 26	−64 35.3	2	9.61	.020	0.08	.000	0.03	.015	4	A0 V	122,807
117588	+44 02274			13 28 28	+44 02.9	1	8.50		0.40		0.01		2	F8	1569
	+10 02554			13 28 29	+10 29.0	1	9.40		1.62		1.87		2	M0	280
	+13 02672			13 28 29	+12 32.9	1	10.65		0.74		0.30		2	G0	1696
	+29 02418			13 28 29	+28 51.3	1	10.92		0.63		0.29		2		1569
117567	+24 02593	IDS13261N2445	A	13 28 30	+24 29.7	4	7.62	.006	0.43	.004	-0.03	.010	6	F2	1512,1569,1625,8090
	+38 02454			13 28 30	+38 02.0	1	8.82		1.56		1.79		1	M2	280
	−3 03488			13 28 30	−03 51.0	1	10.41		0.92		0.69		1	K2 V	3072
117484	−46 08696			13 28 30	−46 28.7	1	7.54		0.01		-0.04		3	B9.5 V	1770
117470	−51 07566	V701 Cen	⋆ A	13 28 31	−51 30.8	1	8.93		0.02		-0.24		1	B9 V(n)	1770
		Ru 108 - 11		13 28 33	−58 13.8	1	11.75		0.62		-0.01		2		251
	−57 06103	Ru 108 - 3		13 28 34	−58 10.3	1	11.20		0.16		0.00		2		251
	−62 03325	Tr 21 - 10		13 28 35	−62 33.3	2	11.08	.020	0.07	.005	-0.37	.020	5		251,1672
		Tr 21 - 19		13 28 38	−62 30.3	1	12.88		0.62		0.13		2		1672
117471	−57 06105	Ru 108 - 1	⋆ AB	13 28 39	−58 10.9	1	8.47		0.41		0.09		2	F5	251
		Tr 21 - 14		13 28 39	−62 33.3	2	11.98	.005	0.13	.010	-0.04	.030	5		251,1672
117589	+27 02262			13 28 40	+26 38.8	1	7.39		0.29		0.08		1	A5	280
		Tr 21 - 9		13 28 40	−62 33.0	2	11.46	.025	1.16	.000	0.97	.015	5		251,1672
117460	−62 03326	LSS 3106	⋆ A	13 28 40	−62 47.1	1	7.43		0.07		-0.62		2	B0/1 III	976
117460	−62 03326	LSS 3106	⋆ AB	13 28 40	−62 47.1	1	7.13		0.06		-0.66		2	B0/1 III	540
		G 165 - 7		13 28 41	+30 45.2	1	16.04		0.72		0.78		5		1663
		Ru 108 - 10		13 28 41	−58 13.4	1	11.70		0.48		0.05		2		251
	−62 03327	Tr 21 - 8		13 28 41	−62 31.2	2	11.77	.010	0.14	.000	-0.05	.030	4		251,1672
	−64 02443			13 28 41	−64 37.0	2	10.34	.010	0.01	.015	-0.64	.015	4	B2	122,807
	+18 02742			13 28 42	+18 23.1	1	10.09		1.20		1.30		1		280
	−17 03874			13 28 42	−17 48.6	1	9.95		1.07		0.90		1	K5 V	3072
		L 196 - 36		13 28 42	−55 58.	2	12.19	.095	1.37	.015	1.10	.020	4		1696,3062
117488	−57 06106	Ru 108 - 2		13 28 42	−58 11.5	1	11.04		0.13		0.03		2		251
		Tr 21 - 11		13 28 42	−62 31.8	2	12.41	.025	0.24	.005	0.11	.035	5		251,1672
		Lod 848 # 31		13 28 42	−64 27.3	1	11.64		0.54		0.10		2		807
117510	−52 06544			13 28 43	−52 45.9	1	9.11		-0.04		-0.32		1	B8 V	1770
	−63 02800			13 28 43	−64 14.4	1	10.19		0.43		0.00		2	F3	807
	−64 02445			13 28 44	−64 25.9	1	10.03		0.62		0.14		2	F9	807
	+21 02543			13 28 45	+21 11.7	1	10.02		1.06		0.92		1	K2	280
		Tr 21 - 20		13 28 45	−62 30.1	1	13.02		0.28		0.19		2		1672
117558	−27 09278	HR 5090		13 28 46	−27 51.3	2	6.46	.005	0.10	.000			7	A1 V	15,2012
117610	+22 02589			13 28 47	+21 35.4	1	8.84		1.09		1.05		1	K0	280
117492	−62 03329	Tr 21 - 1		13 28 47	−62 31.4	2	8.92	.020	0.02	.000	-0.56	.020	8	B5 III	251,1672
		Tr 21 - 16		13 28 47	−62 33.1	2	12.92	.000	0.40	.000	0.18	.025	5		251,1672
	−64 02446			13 28 47	−64 40.1	2	10.84	.015	0.32	.005	0.23	.045	4	A5	122,807
	−62 03330	Tr 21 - 3		13 28 48	−62 32.3	2	10.46	.010	0.04	.000	-0.46	.019	6	B2 V	251,1672
117474	−64 02447		A	13 28 48	−64 28.3	2	8.89	.010	1.13	.000	0.92	.010	4	G8 III	122,807
117474	−64 02447		B	13 28 48	−64 28.3	1	9.86		0.70		0.17		2		122
	+37 02416	IDS13265N3657	AB	13 28 49	+36 41.9	1	9.70		1.03		0.92		2		1569
117490	−60 04744	LSS 3110		13 28 49	−60 33.5	4	8.90	.022	0.03	.006	-0.84	.026	8	O9/B0 Vn	401,403,540,976
117475	−64 02448			13 28 49	−64 42.0	3	9.13	.021	0.01	.004	-0.23	.010	8	B9 V	122,807,2014
	−57 06110	Ru 108 - 4		13 28 50	−58 11.5	1	11.08		0.18		0.20		2		251
		Tr 21 - 21		13 28 50	−62 30.9	1	14.07		0.62		0.27		1		1672
		Tr 21 - 13		13 28 50	−62 32.5	2	13.24	.050	0.37	.030	0.14	.065	4		251,1672
117473	−63 02803			13 28 50	−64 17.7	2	9.42	.005	0.34	.010	0.16	.005	3	A8/9 V	807,1771
	−63 02802			13 28 50	−64 18.6	1	10.56		0.28		-0.61		2	B1	807
117360	−76 00767	HR 5082, S Cha	⋆ A	13 28 50	−77 18.6	3	6.50	.016	0.47	.025	-0.08	.015	9	F5 V	13,2033,3062
117360	−76 00769	IDS13246S7703	B	13 28 50	−77 18.6	2	9.32	.025	0.92	.015	0.68	.015	5		13,3062
117360	−76 00767	IDS13246S7703	C	13 28 50	−77 18.6	1	11.34		1.27		1.26		4		3062
	−62 03332	Tr 21 - 4		13 28 51	−62 31.8	2	11.19	.024	0.09	.005	-0.31	.029	6	B8 V	251,1672
117624	+18 02743			13 28 52	+17 50.0	1	8.39		0.28		0.14		2	A8 Vm	1068
	+40 02661			13 28 52	+40 10.9	2	10.89	.000	0.16	.005	-0.29	.015	3	A5	280,1569
		G 14 - 59		13 28 53	−07 44.5	1	12.41		1.43		1.16		1		1696
117513	−62 03335	Tr 21 - 2		13 28 53	−62 32.3	2	8.95	.015	0.02	.005	-0.52	.020	7	B2 III	251,1672
117656	+33 02344			13 28 54	+32 46.3	1	8.35		1.02		0.82		2	K1 III	1569
	−58 04927			13 28 54	−58 43.8	1	10.66		0.12		0.07		1	A0	890
		G 149 - 96		13 28 55	+22 03.5	1	13.66		0.80		0.15		2		1658
	+23 02576			13 28 56	+23 22.0	1	10.04		0.66		0.15		1		280
	+29 02421	IDS13266N2928	A	13 28 56	+29 12.7	1	9.17		1.24		1.45		1	K3 III	280
		CS 22889 # 10		13 28 56	−08 39.0	1	14.47		-0.03						1736
		Tr 21 - 12		13 28 56	−62 31.8	2	12.62	.040	0.30	.015	0.11	.050	5		251,1672
		Tr 21 - 15		13 28 56	−62 33.2	2	13.40	.015	0.38	.015	0.22	.055	4		251,1672
117527	−57 06112	Ru 108 - 5		13 28 57	−58 12.8	1	9.33		0.10		-0.06		2	B9	251
	−57 06113	Ru 108 - 6		13 28 58	−58 14.2	1	10.40		0.07		-0.10		2		251
	+30 02408			13 29 00	+29 49.9	1	10.11		1.18		1.34		1	K2 IV	280
	−45 08522			13 29 00	−46 04.3	1	10.49		0.64		0.13		4	M0	158
		Tr 21 - 23		13 29 00	−62 29.8	1	13.09		0.20		0.18		2		1672
		Lod 848 # 29		13 29 00	−64 33.1	1	11.24		0.16		0.02		2	B9	807
		Tr 21 - 5		13 29 01	−62 32.0	2	11.85	.039	0.19	.015	0.08	.024	6		251,1672
		AAS12,381 12 # B		13 29 02	+11 47.5	1	10.49		1.06		0.94		1		280
117673	+37 02417			13 29 03	+36 44.3	2	7.56	.029	1.55	.010	1.75		3	M1	280,1569
		Tr 21 - 18		13 29 03	−62 32.7	1	13.39		0.61		0.15		2		1672
	+32 02358			13 29 04	+31 52.6	1	10.21		0.78		0.48		1	G8 III	280
117710	+42 02405	HR 5096		13 29 06	+42 21.8	2	6.07	.010	1.07	.018	1.08		4	K2 III	70,1569

HD	DM	Other Id	N	Rem	α_{1950}	δ_{1950}	S	V	σ_V	B–V	σ_{B-V}	U–B	σ_{U-B}	n	Spectrum	References
	+45 02114				13 29 06	+44 54.7	1	8.88		1.00		0.79		1	G8 III	280
	−64 02451				13 29 07	−64 48.4	1	11.01		0.27		0.20		2	A0	807
117635	−1 02832	G 62 - 44			13 29 08	−02 03.8	8	7.34	.014	0.78	.011	0.31	.012	20	G9 V	22,158,202,1003,1509*
	−57 06116	Ru 108 - 7			13 29 09	−58 12.9	1	10.05		0.04		-0.42		2		251
		Ru 108 - 9			13 29 09	−58 13.2	1	11.78		0.57		0.26		2		251
		Lod 915 # 15			13 29 09	−59 08.9	1	11.42		1.14		0.82		1	K3	890
		Tr 21 - 6			13 29 09	−62 31.8	2	12.10	.010	0.23	.010	0.12	.030	5		251,1672
	+11 02579				13 29 11	+10 36.5	1	10.42		1.19		1.23		1		280
117597	−37 08695	HR 5092	4		13 29 11	−38 08.5		6.16	.008	1.03	.008			18	K1 III	15,1075,2013,2029
117598	−42 08503				13 29 11	−42 26.8	1	8.33		-0.03		-0.06		3	B9 IV	1770
		Tr 21 - 17			13 29 11	−62 32.1	2	12.91	.010	0.53	.000	-0.11	.010	4		251,1672
		Lod 915 # 14			13 29 12	−59 08.9	1	11.36		0.40		0.25		1	A5	890
		Lod 915 # 35			13 29 12	−59 08.9	1	12.39		0.36		0.17		1	A6	890
		Lod 915 # 26			13 29 12	−59 11.7	1	11.93		0.17		0.14		1	A3	890
	+11 02580				13 29 15	+11 27.0	1	9.90		1.02		0.82		1	K0	280
		Tr 21 - 7			13 29 15	−62 31.1	2	11.69	.015	0.14	.015	-0.16	.025	7		251,1672
		Lod 848 # 34			13 29 15	−64 46.1	1	12.73		0.47		-0.35		2	WR	807
	+24 02595				13 29 16	+23 37.9	1	9.93		1.05		0.96		2	K0	280
117562	−57 06117	Ru 108 - 8			13 29 16	−58 12.3	1	9.87		0.10		-0.07		2	B9	251
	−64 02453				13 29 16	−64 34.6	1	11.31		0.14		-0.03		2	B9	807
117696	+19 02682				13 29 17	+19 13.4	1	7.87		1.12		1.14		1	K0	280
		Tr 21 - 24			13 29 17	−62 30.3	1	13.33		0.43		0.19		1		1672
		WLS 1320-20 # 8			13 29 18	−20 00.9	1	13.29		0.55		-0.10		2		1375
117616	−43 08356				13 29 19	−43 47.3	1	8.64		0.11		0.10		1	A0 V	1770
117661	−17 03877	HR 5094, HX Vir	⋆	AB	13 29 21	−18 28.3	3	6.00	.008	0.19	.008	0.14	.000	11	A7 IV/V	244,2007,8071
	−64 02454				13 29 21	−64 36.4	2	10.95	.010	0.13	.010	0.06	.020	4	A1	122,807
117728	+37 02418				13 29 22	+37 22.2	1	7.68		1.07		0.89		2	K0	1569
117675	−5 03714	HR 5095	8		13 29 22	−05 59.9	8	4.69	.007	1.61	.017	1.94	.016	37	M2 III	15,369,1075,1088,1509*
		Lod 915 # 21			13 29 22	−59 11.1	1	11.66		0.20		0.19		1	A3	890
		Feige 81			13 29 24	+15 55.9	1	13.48		-0.22		-1.02		2	sdBp	1298
117563	−63 02808				13 29 24	−64 13.9	1	8.98		0.95		0.62		2	G8 III/IV	807
117711	+17 02641				13 29 25	+17 23.3	2	8.66	.010	0.36	.000	0.02	.002	5	F5	1733,8088
117729	+30 02411				13 29 26	+30 03.3	1	9.47		0.32		0.03		2	K0	1569
	−63 02809				13 29 26	−64 23.0	1	10.52		0.12		-0.02		2	A0	807
		G 165 - 8			13 29 28	+29 32.0	1	11.95		1.57				1	M4e	1746
117697	+9 02773				13 29 29	+09 14.0	1	8.03		0.56		0.08		3	F8	3026
	+20 02843				13 29 29	+19 58.3	1	10.16		1.11		1.09		1	G5	280
117730	+27 02264				13 29 29	+26 51.5	1	7.68		1.55		1.88		1	K5	280
		Lod 915 # 24			13 29 31	−59 00.3	1	11.73		0.49		0.05		1	F7	890
	+21 02544				13 29 33	+21 13.4	1	10.14		1.02		0.83		2	K5	1569
117619	−53 05658				13 29 33	−53 34.4	1	8.49		1.91		2.22		2	C4,5	864
	−64 02456				13 29 33	−64 31.4	1	10.75		0.20		0.08		2	A0	807
117606	−58 04929				13 29 35	−58 59.0	1	9.11		0.18		0.14		1	A1 V	890
	+32 02359				13 29 37	+32 02.2	2	10.47		0.00	.009	0.01	.014	5	A1 IV	1026,1264
		WLS 1340 35 # 6			13 29 37	+35 11.8	1	13.27		0.77		0.29		2		1375
117745	+15 02588				13 29 38	+15 25.3	1	9.10		0.94		0.61		1	K0	280
	+31 02500	IDS13274N3140		AB	13 29 38	+31 25.0	1	10.41		1.43				3	M0 V	1625
	+38 02456				13 29 38	+38 13.6	1	11.46		0.18		0.10		1		280
	−64 02457				13 29 38	−64 35.4	2	9.75	.040	0.58	.005	0.11	.010	4	F8	122,807
		Lod 915 # 19			13 29 39	−59 05.5	1	11.52		1.00		0.72		1	K0	890
	+42 02407				13 29 41	+41 48.7	1	9.89		1.14		1.10		1	K0	280
	−64 02458				13 29 41	−64 38.0	2	9.99	.020	0.51	.000	0.11	.005	4	F8	122,807
117746	+10 02556				13 29 42	+10 24.3	1	8.10		1.53		1.85		1	K2	280
117845	+59 01517				13 29 43	+59 12.6	1	8.08		0.56				5	G2 V	1379
117777	+29 02423	VZ CVn			13 29 44	+28 50.5	1	9.25		0.37		-0.01		6	F2 V	588
		Lod 915 # 7			13 29 44	−59 04.6	1	10.57		1.60		1.84		1		890
117718	−28 10127	HR 5098	3		13 29 46	−29 18.5	3	6.45	.005	0.43	.009	0.00		9	F5 IV	15,2027,3053
	+37 02420	IDS13276N3718		A	13 29 48	+37 02.7	1	10.91		0.92		0.54		2	G5	1375
	+44 02277				13 29 48	+43 44.0	1	9.37		0.29		0.03		2	A5	280
117716	−28 10128	HR 5097			13 29 48	−28 26.2	1	5.69		0.04				4	A1 Vn	2018
	−64 02460				13 29 49	−64 49.1	1	10.96		0.18		0.07		2	A0	807
	+23 02578				13 29 50	+23 26.1	1	9.24		0.82		0.60		1	K0	280
117815	+47 02066				13 29 50	+47 29.5	1	7.08		0.21		0.06		3	A3 V	1501
	−64 02461				13 29 50	−64 35.7	2	9.45	.025	1.08	.010	0.78	.005	4	K3	122,807
	+39 02654				13 29 54	+39 09.2	1	10.79		0.25		0.06		1		280
	−58 04930				13 29 57	−59 04.1	1	10.61		0.47		0.14		1	F3	890
		Cen sq 1 # 207			13 29 57	−62 08.4	1	11.33		0.67		0.13		3		144
	−64 02463				13 29 58	−64 30.5	1	10.38		1.13		0.95		2	K3	807
117630	−64 02464				13 29 59	−64 36.3	2	9.85	.350	0.19	.010	0.11	.005	4	B8/9 IV/V	122,807
		Lod 915 # 25			13 30 00	−59 03.3	1	11.77		0.62		0.19		1	G0	890
	−58 04931				13 30 00	−59 11.3	1	10.36		0.49		0.11		1	F2	890
	+15 02590	IDS13276N1459		A	13 30 01	+14 43.0	1	9.51		1.13		1.08		2	K0	280
117650	−63 02814				13 30 01	−64 18.9	2	10.08	.000	0.10	.000	0.13	.005	3	A0	287,807
117816	+21 02545				13 30 02	+21 08.1	1	8.41		1.56		1.92		1	K5	280
117831	+31 02503				13 30 02	+30 43.8	1	8.77		1.13		1.06		2	K0 III	1569
117668	−60 04756				13 30 02	−60 02.5	1	9.17		0.13		-0.61		2	B1/2 Vn	401
117738	−47 08401				13 30 04	−47 30.9	1	7.77		0.18		0.14		1	A3 III/IV	1771
	+36 02381				13 30 05	+36 31.4	1	9.80		0.35		-0.07		2	F2	1569
117651	−64 02465	HR 5093	4		13 30 05	−65 22.6	4	6.36	.003	-0.01	.005	-0.08	.008	17	A0 V	15,1075,1771,2038
117830	+35 02460				13 30 07	+35 08.7	1	8.33		0.82		0.39		2	G0	1569

Table 1 757

HD	DM	Other Id	N	Rem	α_{1950}	δ_{1950}	S	V	σ_V	B–V	σ_{B-V}	U–B	σ_{U-B}	n	Spectrum	References
		G 62 - 45			13 30 08	+08 08.1	1	10.93		0.99		0.87		1	K7	1658
117846	+37 02421	IDS13279N3720	A		13 30 08	+37 04.5	1	7.28		1.00		0.80		1	G8 III	3032
117846	+37 02421	IDS13279N3720	AB		13 30 08	+37 04.5	1	6.81		0.75				2	G8 III	280
117846	+37 02421	IDS13279N3720	B		13 30 08	+37 04.5	1	8.17		0.42		-0.01		1		3032
117800	-8 03580				13 30 09	-08 55.0	1	10.03		0.03		-0.13		1	G5	890
117687	-60 04759	LSS 3115			13 30 09	-61 11.3	2	9.32	.018	0.12	.002	-0.71	.011	7	B2 Ib/II	540,976
117670	-64 02466				13 30 09	-64 30.4	1	7.64		0.48		0.04		2	F5 V	807
		Lod 848 # 32			13 30 10	-64 31.3	1	11.67		0.13		-0.16		2	B8	807
		G 150 - 9			13 30 11	+24 27.6	1	12.92		1.24		0.98		3		316
117789	-14 03739	HR 5099		⋆ A	13 30 11	-15 06.4	3	5.54	.013	1.23	.007	1.19	.024	10	K1.5 IIIb	244,2007,3005
		Lod 915 # 28			13 30 11	-59 05.8	1	12.02		0.11		-0.12		1	B9	890
		Feige 83			13 30 12	+53 38.	1	12.72		0.03		0.03		1		3060
	+26 02466				13 30 13	+26 06.5	1	9.66		1.12		1.00		1	K0	280
	+31 02505				13 30 13	+31 03.6	1	10.18		0.16		0.08		2		1569
117901	+60 01470				13 30 13	+60 06.1	1	8.95		0.45		-0.03		4	F8	1371
		Lod 915 # 10			13 30 14	-59 04.9	1	10.80		1.45		1.25		1	K5	890
117704	-61 03793	LSS 3116			13 30 14	-62 03.7	3	8.90	.014	0.21	.012	-0.59	.024	8	B1/2 III	92,540,976
		Lod 915 # 31			13 30 15	-59 04.5	1	12.22		0.15		0.13		1	A0	890
117858	+36 02382	G 165 - 11			13 30 16	+36 17.7	2	7.98	.014	0.56	.009	0.00	.005	4	G0	1569,1658
		G 62 - 46			13 30 17	+01 32.7	1	17.11		0.38		-0.47		1		3060
	+23 02581				13 30 17	+22 44.5	1	9.67		1.03		0.82		2		1569
		G 63 - 36		⋆ AB	13 30 18	+17 04.2	2	11.40	.004	1.54	.006	1.22	.023	4	M4 Ve	8088,3003
117875	+40 02663				13 30 19	+39 32.9	2	8.38	.015	0.31	.000	0.09	.010	5	F1 V	1501,1569
117770	-48 08290				13 30 19	-48 41.1	1	9.08		0.27		0.15		1	A1mA5-A9	1771
116877	-86 00300				13 30 19	-86 28.2	1	7.66		1.08		0.87		5	G8/K0 III	1704
117818	-9 03711	HR 5100			13 30 20	-09 54.5	2	5.20	.005	0.96	.000	0.61	.005	7	K0 III	15,1256
	+23 02582	IDS13280N2313	A		13 30 22	+22 56.9	1	10.16		1.09		0.98		1		280
117707	-64 02468	LSS 3118			13 30 23	-64 55.3	4	9.41	.013	0.55	.020	-0.41	.015	11	B1 Ia	138,403,540,976
	+43 02339				13 30 24	+42 59.2	1	8.15		1.60		1.95		2	M0	280
	+41 02408				13 30 25	+41 07.7	1	10.46		0.99		0.93		1	G8 V	280
117876	+25 02643	HR 5102		⋆ AB	13 30 26	+24 36.3	6	6.10	.009	0.96	.008	0.70	.005	10	G8 III	15,1003,1080,1512*
	+31 02506				13 30 29	+30 34.7	1	9.22		1.30		1.56		2	K3 III	1569
117893	+31 02507				13 30 30	+31 00.3	1	7.50		1.06		0.94		2	K0	1569
		L 106 - 26			13 30 30	-66 32.	1	11.30		1.05		0.71		2		3062
		L 106 - 29			13 30 30	-66 32.	1	11.48		0.98		0.73		1		1696
		WD1330+473			13 30 31	+47 19.1	1	15.29		-0.11		-0.92		2	DA3	1727
		VW Cen			13 30 31	-63 47.8	1	10.29		1.50		1.43		1		1587
		LP 558 - 2			13 30 32	+08 12.8	1	12.68		0.42		-0.23		2		1696
		LP 678 - 89			13 30 32	-04 43.4	1	10.19		0.85		0.51		2		1696
	+19 02683				13 30 33	+19 29.5	1	10.53		1.20		1.34		2		1569
		Cen sq 1 # 209			13 30 35	-62 40.6	1	11.40		0.10		-0.24		3		144
117902	+35 02462	IDS13283N3525	AB		13 30 36	+35 09.8	1	6.80		0.22		0.09		6	F0 III	3030
	+40 02664				13 30 36	+40 18.3	1	9.94		0.93		0.60		1	K0 III	280
	+27 02265				13 30 37	+26 37.6	1	9.20		1.15		1.24		1	K2	280
117848	-27 09298				13 30 38	-27 25.3	1	9.48		0.64				4	G0 V	2033
		Cen sq 1 # 210			13 30 38	-62 08.7	1	12.17		1.52		1.67		3		144
		Lod 915 # 18			13 30 41	-59 00.1	1	11.50		0.55		0.16		1	F8	890
		Cen sq 1 # 213			13 30 41	-62 23.6	1	13.02		0.27		-0.19		3		144
		Cen sq 1 # 206			13 30 42	-62 35.8	1	11.22		0.51		0.05		3		144
	-58 04938				13 30 43	-59 06.2	1	10.47		0.50		0.10		1	F5	890
		Lod 915 # 27			13 30 43	-59 06.9	1	11.98		0.34		0.27		1	A0	890
		Lod 915 # 34			13 30 43	-59 06.9	1	12.31		0.29		-0.34		1	B5	890
117787	-60 04765				13 30 43	-60 55.5	1	9.04		0.40		0.07		1	A9 V	1771
117935	+44 02278				13 30 45	+44 23.2	1	8.54		0.32		0.09		2	F2	1569
		GD 269			13 30 46	+03 36.4	1	15.86		0.01		-0.79		1		3060
		Lod 915 # 32			13 30 46	-58 56.8	1	12.28		0.41		0.20		1	A6	890
117912	+26 02468				13 30 47	+26 08.2	1	8.24		1.07		0.96		1	G5	280
	+16 02525				13 30 48	+16 22.6	1	10.44		0.11		0.13		1		280
117880	-17 03883				13 30 48	-18 15.4	6	9.06	.008	0.07	.011	0.10	.028	31	B9 IV/V	308,830,1003,1775,1783,3077
	-61 03802	Cen sq 1 # 205			13 30 48	-62 22.4	1	10.75		0.18		-0.20		3		144
117797	-61 03803	LSS 3119			13 30 49	-62 09.6	2	9.19	.005	0.49	.011	-0.55	.010	5	O8f	92,540
117867	-32 09445				13 30 50	-33 15.4	1	9.34		0.38		-0.02		27	A9 V	978
117850	-37 08720	IDS13280S3745	A		13 30 50	-38 00.1	1	8.13		0.07		0.13		1	A3/7	1771
	-58 04940				13 30 50	-58 59.0	1	11.11		0.54		0.14		1	G0	890
	-58 04941				13 30 53	-59 00.5	1	11.09		0.12		0.13		1	A0	890
	+44 02279				13 30 57	+43 31.6	1	9.40		1.08		0.91		2		3072
		Lod 915 # 22			13 30 58	-59 04.3	1	11.66		0.28		0.15		1	A4	890
		Lod 848 # 23			13 30 58	-64 42.5	1	10.65		1.35		1.50		2		807
		Lod 915 # 17			13 31 00	-59 02.1	1	11.47		0.25		0.13		1	A0	890
		Lod 848 # 33			13 31 00	-64 41.9	1	11.90		0.26		0.20		2	A0	807
	+17 02642				13 31 01	+16 57.6	1	9.87		1.12		0.98		2	K2	1569
	+26 02469	G 150 - 11			13 31 02	+26 21.1	1	9.89		0.82		0.46		1	K2	1658
117881	-44 08727				13 31 02	-44 57.4	1	9.20		0.10		0.05		1	A0 V(n)	1770
		Lod 915 # 29			13 31 02	-58 54.6	1	12.14		0.37		0.18		1	A4	890
117936	+9 02776	G 62 - 49			13 31 04	+08 50.5	2	8.01	.010	1.02	.013	0.94	.028	5	K0	265,7009
		Cen sq 1 # 212			13 31 04	-62 16.3	1	12.75		0.64		0.16		3		144
		Lod 915 # 23			13 31 05	-58 56.5	1	11.69		0.56		0.13		1	G0	890
	-64 02471				13 31 05	-64 39.4	1	9.47		1.70		1.91		2	K8	807
117981	+34 02430				13 31 06	+33 54.8	1	6.66		1.18				2	K2 III	280
		Cen sq 1 # 211			13 31 06	-62 19.5	1	12.56		0.33		0.28		3		144

HD	DM	Other Id	N	Rem	α_{1950}	δ_{1950}	S	V	σ_V	B–V	σ_{B-V}	U–B	σ_{U-B}	n	Spectrum	References
	−63 02830				13 31 06	−64 20.5	1	10.43		0.24		−0.22		2	B7	807
		Cen sq 1 # 217			13 31 07	−62 17.3	1	13.45		1.66		1.49		3		144
117828	−63 02831				13 31 07	−64 17.3	2	9.63	.000	0.11	.000	0.13	.015	3	A0 V	287,807
118019	+48 02127				13 31 08	+48 00.9	1	7.44		1.08		1.01		2	K0	1733
	−55 05624				13 31 08	−55 23.6	1	10.77		0.18		−0.72		2		540
117854	−58 04943				13 31 09	−59 15.2	3	7.53	.012	0.64	.000	0.06	.025	7	G0 V	657,1311,2012
	+44 02280				13 31 10	+44 01.0	1	10.53		1.04		0.83		1	G8 III	280
117827	−63 02832				13 31 10	−64 16.3	2	10.00	.000	0.10	.000	0.07	.000	3	B9/A0 IV/V	287,807
	−58 04944				13 31 11	−58 50.3	1	10.36		0.99		0.74		1	K0	890
	+28 02237				13 31 12	+27 36.5	1	8.74		1.46		1.81		1	K5	280
117997	+45 02117				13 31 12	+44 32.9	1	8.16		1.64		1.96		1	K3 III	280
	+37 02423				13 31 15	+37 25.5	1	10.04		0.80		0.44		2	K0	1569
117983	+19 02686				13 31 16	+18 48.5	1	9.69		0.28		0.07		2	F5	1569
117980	+34 02431				13 31 16	+34 19.2	1	9.04		1.29		1.52		1	K4 III	280
	−58 04946				13 31 16	−58 59.1	1	10.35		0.98		0.52		1	K2	890
		Lod 915 # 16			13 31 16	−59 04.6	1	11.46		0.40		0.32		1	A5	890
117856	−62 03374	LSS 3120			13 31 17	−63 04.8	4	7.38	.033	0.21	.006	−0.69	.022	9	B0/1 III	403,540,976,8100
	+42 02411				13 31 18	+41 34.2	1	10.36		1.12		1.18		1	K2 III	280
	+33 02348				13 31 19	+32 39.1	1	9.13		1.57		1.86		2	K5	1569
	+54 01607	RV UMa			13 31 21	+54 14.7	2	10.24	.131	0.10	.044	0.13	.020	2	F0	668,699
117916	−36 08672				13 31 21	−37 03.1	1	10.13		0.21		0.14		1	A8 V	1771
		Lod 915 # 20			13 31 21	−58 57.1	1	11.60		0.61		0.11		1	G0	890
	+22 02595				13 31 22	+22 13.7	1	9.60		1.12		0.94		1	K5	280
118001	+17 02645				13 31 24	+16 31.8	1	8.07		1.11		1.12		1	K0	280
117927	−33 09161				13 31 24	−34 07.9	1	7.55		1.07		0.94		3	K0 III	1731
117919	−47 08417	HR 5103		★ AB	13 31 26	−48 01.0	3	6.32	.006	−0.05	.001	−0.37		10	B8 III	15,1770,2029
118005	+4 02763				13 31 27	+04 10.6	1	8.89		0.14		0.13		1	A3 III	695
118002	+16 02527				13 31 28	+15 47.5	1	9.26		1.03		0.95		1	K2	280
	+47 02068	IDS13294N4719		AB	13 31 28	+47 03.2	1	8.82		1.02		0.82		2	G5	1601
	+22 02596				13 31 30	+21 33.6	1	10.03		1.14		1.04		1		280
118021	+13 02680				13 31 32	+12 31.1	1	9.18		0.26		0.12		2	A8 Vm	1375
	+22 02597				13 31 32	+21 39.5	1	9.47		1.17		1.23		1	K5	280
117970	−24 10977	RW Hya			13 31 32	−25 07.5	2	8.85	.058	1.50	.010	0.33		3	M III pe	867,1753
118051	+33 02349				13 31 33	+32 35.7	1	7.80		0.69		0.24		2	G5 III	1375
118022	+4 02764	HR 5105, CW Vir			13 31 36	+03 54.9	15	4.93	.011	0.03	.007	0.01	.020	68	A1 p SrCrEu	3,15,667,695,1007*
		WLS 1340 15 # 7			13 31 36	+15 12.6	1	11.39		0.47		−0.04		2		1375
117939	−38 08635				13 31 36	−38 38.8	4	7.29	.007	0.67	.011	0.12	.018	8	G3 V	742,1311,2012,6006
118052	+32 02361				13 31 37	+31 48.4	1	8.91		1.32		1.51		1	K5 III	280
	−49 08031				13 31 37	−49 37.3	1	9.10		0.27		−0.62		3		976
	+35 02464				13 31 39	+34 51.2	1	9.98		0.99		0.82		2	K0 V	1569
117987	−26 09804				13 31 40	−27 14.9	2	9.27	.018	0.99	.007	0.74		5	K2 V	2033,3072
		Lod 915 # 33			13 31 40	−58 48.0	1	12.29		0.35		0.24		1	A2	890
	+19 02687				13 31 41	+18 35.9	1	9.94		0.99		0.71		2		1569
		CS 22889 # 6			13 31 42	−09 40.1	1	14.65		0.39		−0.19		1		1736
118036	+0 03075	IDS13292N0012		AB	13 31 43	−00 03.5	2	7.38	.024	0.91	.005	0.67		11	K4 III	176,938
	+45 02118				13 31 46	+45 09.8	1	9.70		1.00		0.77		1	G8 III	280
	−58 04951				13 31 46	−58 56.5	1	11.04		0.15		0.16		1	A2	890
118066	+19 02688				13 31 47	+18 40.3	1	8.40		0.26		0.07		3	F0	1569
118096	+33 02350				13 31 48	+33 28.8	2	9.30	.058	1.10	.019	0.90		3	K5 IV	906,3072
		Lod 915 # 30			13 31 48	−58 52.2	1	12.17		0.38		0.24		1	A2	890
	+5 02767	G 62 - 50			13 31 49	+04 55.5	1	9.97		1.38		1.21		1	M0.5V:	3072
	+44 02284	IDS13297N4348		AB	13 31 50	+43 32.7	1	10.11		0.89		0.60		1	G8 III	280
	+38 02457				13 31 52	+37 53.7	1	10.05		1.10		1.00		2	K8	3072
	+43 02340				13 31 52	+43 24.1	1	9.25		1.07		0.99		1	G8 III	280
	+75 00510	G 255 - 34			13 31 52	+75 15.7	1	10.23		1.30		1.23		3	K8	196
	+41 02409				13 31 53	+40 29.7	1	10.48		1.10		1.03		1	K2	280
118010	−32 09459	HR 5104			13 31 53	−33 03.3	1	6.44		1.26				4	K2 III	2035
118068	+14 02636				13 31 54	+14 22.0	1	7.20		1.11		1.02		1	K0	280
117947	−59 05043				13 31 54	−59 54.1	1	8.21		0.35		0.05		1	A8 IV	1771
118097	+33 02351				13 31 55	+32 43.8	1	9.87		0.22		0.08		2		272
		Lod 915 # 36			13 31 56	−58 49.8	1	12.55		0.24		0.11		1	A2	890
118055	−15 03695				13 31 59	−16 04.0	2	8.87	.010	1.29	.000	1.07		3	K0wF8	1594,6006
		He3 916			13 31 59	−64 30.4	1	13.26		1.60		1.32		1		1753
118143	+42 02413				13 32 00	+41 31.5	1	9.09		1.52		1.86		1	K0	280
118054	−12 03843	HR 5106		★ AB	13 32 01	−12 57.5	3	5.91	.004	0.02	.004	−0.09		11	A0 V	1414,2006,3030
	−56 05819				13 32 01	−57 22.7	2	10.97	.010	0.38	.002	−0.21	.008	3	B8 II-Ib	540,976
117959	−61 03821				13 32 03	−61 41.8	1	8.06		0.22		0.19		1	A1mA5-A6	1771
		LSS 3122			13 32 03	−61 55.9	1	11.93		0.82		−0.20		2	B3 Ib	1737
	+22 02599				13 32 04	+21 55.4	1	10.15		1.20		1.33		1		280
		G 150 - 15			13 32 06	+27 32.2	1	14.42		1.39				1		1759
118014	−51 07601				13 32 06	−52 23.1	1	8.57		0.03		−0.08		1	B9.5 V	1770
118100	−7 03646	EQ Vir			13 32 07	−08 05.1	3	9.33	.033	1.19	.014	1.04	.000	10	K7 V	158,1705,3072
	+25 02648				13 32 08	+25 10.4	1	9.82		1.18		1.16		1	K2	280
	+25 02647	IDS13297N2532		A	13 32 08	+25 18.1	1	10.55		0.99		0.78		1		280
118156	+39 02658	HR 5108		★ A	13 32 09	+39 02.7	2	6.38	.005	0.21	.017	0.11		4	F0 IV	70,105
118098	+0 03076	HR 5107			13 32 09	−00 20.5	9	3.37	.008	0.11	.005	0.12	.028	43	A3 V	3,15,1020,1075,1075*
	+36 02384				13 32 12	+36 14.6	1	8.79		0.58		0.14		2	G5 III	1569
118214	+56 01667	HR 5109			13 32 12	+55 36.3	3	5.58	.019	−0.04	.004	−0.07	.013	7	A0 V	70,752,3023
	−62 03383	Cen sq 1 # 203			13 32 15	−62 30.9	1	10.24		1.13		0.88		3		144
	+13 02682				13 32 16	+13 01.5	1	9.62		0.67		0.14		1	G0	280

Table 1   759

HD	DM	Other Id	N	Rem	α_{1950}	δ_{1950}	S	V	σ_V	B–V	σ_{B-V}	U–B	σ_{U-B}	n	Spectrum	References
	+23 02586				13 32 16	+23 17.3	1	9.92		1.25		1.42		1	K2	280
118016	−61 03825	LSS 3125			13 32 17	−61 36.7	2	7.94	.035	0.11	.008	-0.67	.013	4	B0/1 V	540,976
118158	+21 02547				13 32 18	+20 54.1	1	8.30		0.33		0.06		3	F0	1569
118102	−24 10984				13 32 19	−25 02.9	1	9.00		0.54				1	F7 V	867
117949	−68 01952				13 32 19	−69 18.5	1	9.03		0.21		-0.04		2	A0 V	1771
		LP 558 - 8			13 32 25	+05 07.2	1	11.56		0.56		-0.08		2		1696
118232	+49 02227	HR 5112			13 32 25	+49 16.3	5	4.69	.021	0.13	.014	0.12	.015	14	A5 V	15,1363,3023,6001,8015
		Cen sq 1 # 218			13 32 25	−62 33.3	1	13.68		0.77		0.20		3		144
117374	−85 00384	HR 5084			13 32 25	−85 31.9	2	5.57	.005	0.18	.000	0.15	.010	7	A2mA5-A8	15,1075
	+37 02425				13 32 26	+37 26.3	1	9.56		1.13		1.01		1	K0	1569
118129	−6 03843				13 32 27	−06 42.7	3	8.18	.006	1.07	.007	0.95	.001	33	K2	147,1509,6005
118105	−34 08989				13 32 28	−34 23.1	1	8.96		0.55		-0.01		1	F7 V	1731
118204	+28 02241				13 32 29	+27 45.9	1	8.53		0.82		0.40		1	G5	280
	+11 02582				13 32 30	+11 03.0	1	10.95		1.26		1.22		2	K5	1569
		Cen sq 1 # 216			13 32 30	−62 33.2	1	13.30		0.44		0.27		3		144
118017	−64 02481				13 32 31	−64 39.2	1	7.76		0.96		0.67		2	G8 III	122
	−62 03388				13 32 32	−62 29.5	1	9.68		1.58		1.74		3		144
118217	+32 02363				13 32 33	+31 32.3	1	9.45		0.94		0.72		1		280
118090	−47 08437				13 32 33	−48 15.7	1	7.73		0.18		0.10		1	A5 V	1771
118180	+13 02683				13 32 34	+12 46.2	1	6.56		1.02		0.94		1	K0	280
	+33 02355	IDS13303N3345		A	13 32 34	+33 28.9	1	9.73		1.01		0.85		2	G8 III	1569
118216	+37 02426	HR 5110, BH CVn			13 32 34	+37 26.3	8	4.97	.009	0.40	.008	0.06	.005	41	F2 IV	15,374,667,1008,1363*
118131	−33 09171				13 32 37	−33 45.7	1	7.75		1.16		1.14		27	K0 III	978
118047	−60 04783				13 32 37	−61 08.2	1	8.11		0.35		0.22		1	A6 V	1771
118328	+68 00730	IDS13312N6817		AB	13 32 39	+68 01.4	1	8.56		0.61		0.07		2	G1 V	1003
118145	−21 03736			V	13 32 40	−22 21.8	1	7.30		1.66		1.90		4	M1 III	244
118243	+43 02341				13 32 41	+42 56.9	1	8.97		0.70		0.34		2	G5	1569
118205	+12 02605				13 32 42	+12 13.3	1	8.47		1.36		1.46		1	K5	280
118107	−49 08048				13 32 42	−50 03.4	1	10.15		0.34		0.18		1	A9 IV/V	1771
		G 62 - 51			13 32 43	−01 22.8	1	13.13		1.47		1.24		1		1658
118234	+21 02548				13 32 45	+21 02.3	1	7.57		1.08		0.93		1	K0	280
	−48 08321				13 32 45	−49 09.3	1	10.82		0.69		0.15		2	G0	1696
		Cen sq 1 # 208			13 32 45	−62 06.2	1	11.37		1.25		1.05		3		144
	+42 02416				13 32 46	+42 10.5	1	9.85		1.17		1.15		1	K2 III	280
		Feige 84			13 32 48	+13 43.0	2	11.85	.009	-0.17	.005	-0.72	.023	4	sdB3	1026,1298
	+25 02649				13 32 48	+25 03.7	1	8.83		0.70		0.25		5	G8 V	3026
118187	−21 03738				13 32 48	−21 46.1	1	7.10		0.48		0.01		4	F7/8 V	244
118162	−31 10483				13 32 48	−31 45.7	1	9.45		0.17		0.12		1	A2 III/IV	1770
118244	+23 02587				13 32 50	+22 45.2	1	6.91		0.43		-0.10		3	F5 V	3016
118109	−57 06151				13 32 50	−57 27.7	1	9.57		0.38		0.31		1	A8 V	1771
	+0 03077				13 32 51	−00 08.2	5	10.28	.018	1.44	.016	1.21	.017	10	K7 V	158,281,1764,3072,5006
	+18 02755	IDS13305N1820		AB	13 32 54	+18 04.8	1	10.20		0.81		0.42		1	K0	280
118265	+21 02549				13 32 54	+20 51.7	1	9.06		1.23		1.27		1	K5	280
		Ton 165			13 32 54	+28 09.	2	15.47	.047	-0.34	.061	-1.04	.000	2		98,1036
	+39 02659				13 32 55	+38 33.7	1	9.66		1.28		1.43		1	K2	280
118219	−4 03515	HR 5111			13 32 55	−05 08.5	2	5.72	.005	0.95	.000	0.66	.005	7	G6 III	15,1417
118094	−62 03394				13 32 55	−62 53.4	1	7.95		0.00				4	B8 Vn	2014
	+25 02650				13 32 57	+24 50.6	1	10.11		0.95		0.67		1		280
	+29 02432				13 32 57	+29 23.4	1	9.05		0.90		0.55		1	G8 V	280
	+36 02385				13 32 57	+36 20.9	1	9.73		0.44		-0.01		2	F5	1569
		Steph 1077			13 32 58	+30 26.4	1	11.72		1.44		1.20		1	M0	1746
118134	−52 06597				13 32 59	−52 27.5	1	8.31		0.20		0.08		1	A6 V	1771
118264	+23 02588				13 33 02	+23 03.6	1	7.97		1.02		0.80		1	G5	280
118136	−56 05828				13 33 02	−57 07.2	1	8.74		0.30		0.16		1	A5 V	1771
		KN Cen			13 33 02	−64 18.2	3	9.41	.236	1.47	.140	1.04	.151	3		657,689,1587
118165	−47 08445	IDS13300S4711		A	13 33 03	−47 26.5	1	8.57		0.23		0.16		1	A7 V	1771
118266	+10 02565	HR 5114		★ A	13 33 04	+10 27.6	3	6.49	.014	1.04	.015	0.88	.016	7	K1 III+F6 V	280,1084,3077
118266	+10 02566	IDS13306N1043		B	13 33 04	+10 27.6	2	8.95	.005	0.48	.000	-0.01	.010	6	F6 V	1084,3077
118288	+37 02428				13 33 04	+36 38.4	1	7.91		1.11		1.14		2	K0	1569
118123	−62 03397				13 33 05	−62 31.3	1	9.89		0.12		-0.16		3	B9 II/III	144
118295	+44 02285	HR 5116			13 33 06	+44 27.1	2	6.84	.000	0.20	.000	0.17	.063	3	F0 III	280,1569
118246	−5 03730	GP Vir			13 33 07	−05 54.1	8	8.07	.012	-0.14	.007	-0.62	.015	92	B8	147,728,989,1509,1728,1729*
118193	−48 08328				13 33 07	−48 32.8	1	8.08		0.10		0.12		1	A1 V	1771
	+38 02460				13 33 08	+37 39.7	1	8.85		1.39		1.61		1	K5	280
	+35 02465				13 33 09	+35 29.0	1	10.89		0.23		0.08		2	A2	272
	+44 02286				13 33 09	+44 22.8	1	10.92		1.07		0.81		2		1569
	−64 02483				13 33 09	−64 43.7	1	11.01		0.16		-0.35		2		122
		HA 105 # 307			13 33 10	−00 27.3	1	12.05		0.69		0.22		1	G0	281
118296	+35 02466				13 33 13	+35 14.0	1	7.92		1.00		0.76		2	K0 III	1569
118239	−33 09179				13 33 14	−34 07.1	1	8.95		0.47		0.02		5	F3 V	1731
		G 177 - 43			13 33 15	+51 43.7	1	12.04		0.98		0.83		1	K3	1658
118238	−32 09477	V764 Cen			13 33 17	−33 13.4	3	8.94	.092	1.24	.007	1.10	.018	27	K2 IIIp	1641,2012,3049
		AAS12,381 44 # D			13 33 18	+43 57.6	1	10.27		1.32		1.51		1		280
118280	−0 02708				13 33 19	−00 42.6	5	8.80	.004	1.36	.004	1.61	.010	58	dK4	281,989,1729,5006,6004
118311	+38 02461				13 33 21	+37 57.5	1	8.18		1.07		1.00		1	K2	280
118555	+76 00492				13 33 21	+76 03.1	1	7.85		0.06		0.08		2	A0	1733
118289	+9 02785	FP Vir			13 33 22	+08 32.9	1	7.10		1.62		1.57		15	M2	3042
118196	−57 06156	LSS 3128			13 33 22	−58 14.6	1	7.69		0.72		0.44		3	F5 II	8100
		HA 105 # 513			13 33 24	−00 06.1	1	12.14		0.67		0.21		3	G0	281
118290	+0 03079				13 33 26	−00 19.4	5	8.31	.007	1.52	.004	1.90	.016	52	M0 III	281,989,1657,1729,6004

HD	DM	Other Id	N Rem	α_{1950}	δ_{1950}	S	V	σ_V	B−V	σ_{B-V}	U−B	σ_{U-B}	n	Spectrum	References
		Pis 18 - 6		13 33 28	−61 50.1	1	12.17		1.35		1.07		2		251
	+1 02831	G 62 - 52		13 33 29	+01 27.7	3	10.88	.019	0.67	.004	0.01	.008	4	F8	333,927,1620,1696
		Ton 720		13 33 30	+24 41.	1	14.57		0.02		0.01		4		1036
118198	−63 02856	LSS 3127		13 33 32	−63 23.5	2	8.47	.012	0.18	.012	-0.74	.018	4	O9/9.5III	540,976
		HA 105 # 18		13 33 34	−00 56.2	1	11.66		0.58		0.17		1	G2	281
118256	−52 06607			13 33 34	−52 56.2	1	8.05		-0.06		-0.50		2	B5 IV	401
	−62 03403	Cen sq 1 # 204		13 33 34	−62 36.5	1	10.72		0.11		-0.08		3		144
	−45 08570			13 33 38	−45 27.4	1	10.55		0.18		0.05		2		1770
		Pis 18 - 5		13 33 38	−61 48.8	1	13.36		0.56		0.14		2		251
	+30 02422			13 33 40	+29 52.6	1	10.03		0.82		0.57		1	K0 IV	280
	+37 02429			13 33 40	+37 18.4	1	10.09		1.05		0.93		2		1569
118330	−0 02710			13 33 41	−00 40.5	5	7.06	.008	0.53	.003	-0.01	.004	42	F8 V	281,989,1729,3037,6004
		Pis 18 - 3		13 33 41	−61 49.5	1	10.95		1.35		1.27		2		251
118258	−55 05647	IDS13305S5539	AB	13 33 42	−55 54.1	1	7.97		0.85				4	G6 V	2012
		Pis 18 - 4		13 33 42	−61 48.6	1	13.48		0.25		0.17		2		251
118361	+24 02604			13 33 44	+24 11.3	1	8.06		0.92		0.66		1	G5	280
		HA 105 # 216		13 33 44	−00 38.9	1	11.52		0.96		0.68		5		281
	−61 03840	Pis 18 - 1		13 33 44	−61 48.2	2	10.87	.025	0.14	.013	-0.46	.009	3		251,540
	+30 02423			13 33 45	+29 37.9	1	10.31		1.18		1.14		1	K0 III	280
	+45 02119			13 33 45	+44 36.9	1	9.74		0.94		0.70		1	G8 III	280
		Pis 18 - 2		13 33 45	−61 48.8	1	11.66		0.57		0.20		2		251
	+21 02553			13 33 46	+20 53.7	1	9.81		1.11		1.08		1	K2	280
	+32 02364			13 33 46	+31 35.1	1	10.30		1.06		0.97		1	G8 III	280
		WD1333+524		13 33 48	+52 28.2	1	16.33		-0.02		-0.74		2	DA3	1727
	+0 03080			13 33 48	−00 18.7	1	10.62		0.95		0.62		2	G8 IV	281
118261	−61 03841	HR 5113	⋆ AB	13 33 48	−61 26.1	2	5.62	.005	0.50	.000			7	F7 V	15,2012
	+36 02386			13 33 54	+36 26.6	1	9.42		0.41		0.00		2	F2	1569
		CS 22889 # 13		13 33 54	−07 60.0	1	13.71		0.16		0.16		1		1736
118242	−64 02487	IDS13304S6425	AB	13 33 54	−64 40.9	1	7.55		-0.04		-0.30		2	Ap Si	122
118389	+20 02848	IDS13316N2031	A	13 33 56	+20 14.7	1	8.73		1.48		1.79		1	K5	280
		NGC 5236 sq1 6		13 33 57	−29 48.7	1	14.29		0.44		-0.09		3		5013
	+36 02387			13 33 58	+36 07.0	1	8.81		1.52		1.92		1	M0	280
	−0 02711			13 33 58	−00 43.2	1	11.20		0.43		-0.03		1	F8	281
118319	−33 09189	HR 5117		13 33 58	−34 12.8	2	6.48	.015	1.03	.000	0.85		8	K0 III	244,2035
		GD 325		13 33 59	+48 44.0	2	14.01	.010	0.02	.010	-0.93	.010	6	DB + dM	308,940
	+14 02638			13 34 00	+14 03.3	1	9.59		0.96		0.67		2	K0	1569
	+22 02606			13 34 00	+21 47.7	1	9.45		1.01		0.82		1	K5	280
118686	+77 00516	HR 5131		13 34 02	+76 48.1	1	6.46		1.59		1.94		2	K5 III:	1733
118349	−25 09900	HR 5120	⋆ A	13 34 02	−26 14.4	2	5.78	.000	0.22	.000	0.20	.000	4	A7 V	1279,3030
118349	−25 09900	HR 5120	⋆ AB	13 34 02	−26 14.4	2	5.38	.005	0.23	.000	0.17		6	A7 V +A	404,2007
118349	−25 09900	IDS13313S2559	B	13 34 02	−26 14.4	2	6.72	.000	0.24	.000	0.09	.000	4		1279,3030
	+75 00511	G 255 - 38	⋆ B	13 34 03	+74 45.3	1	9.78		1.14		1.11		2	K5	1773,3003
	−0 02712			13 34 06	−00 41.5	1	11.03		0.55		0.06		1	F9	281
118335	−36 08713			13 34 06	−36 50.7	1	7.64		0.03		0.00		3	A0 V	1770
118338	−43 08418	HR 5118		13 34 06	−43 53.3	3	5.98	.008	0.94	.005			14	G8/K0 III	15,1075,2029
	+18 02757			13 34 07	+18 24.3	2	9.80	.005	0.75	.005	0.16	.005	4	F8	1569,1569
118350	−31 10503			13 34 07	−32 01.7	1	9.87		0.27		0.11		1	A2mA9-A9	1770
		G 62 - 53		13 34 10	+03 56.7	5	14.68	.022	0.95	.005	0.37	.004	7	DK	1,203,1281,1705,3078
		HA 105 # 228		13 34 10	−00 37.2	1	11.20		1.12		1.03		1	G3	281
118390	−0 02713			13 34 10	−01 00.4	4	8.35	.003	1.04	.004	0.87	.004	40	G9 III	281,989,1729,6004
118337	−42 08580			13 34 10	−42 31.6	1	7.41		1.57		1.97		8	K5 III	1673
		HA 105 # 30		13 34 13	−00 58.3	1	11.26		0.57		0.08		1	G2	281
		G 165 - 18		13 34 18	+36 39.1	1	16.31		0.40		-0.51		1		3060
		LDS 4551		13 34 18	−16 04.	1	13.85		1.54		1.20		4		3028
		LDS 4552		13 34 18	−16 04.	1	15.55		-0.06		-0.83		2	DA	3028
118356	−46 08753			13 34 19	−47 17.0	1	8.38		-0.08				4	B7 II/III	2014
	+13 02687			13 34 20	+12 32.1	3	9.93	.030	0.23	.004	0.08	.015	7	A7 V	193,272,280
118379	−40 08022			13 34 20	−40 38.6	1	7.57		0.17		0.10		1	A3 IV/V	1771
118322	−55 05650	RV Cen		13 34 21	−56 13.3	1	7.37		2.87		3.72		2	Nb	864
118354	−45 08578	HR 5121		13 34 22	−46 10.4	2	5.90	.006	-0.11	.002	-0.41		7	B8 V	1770,2014
118458	+26 02473			13 34 23	+26 08.0	1	8.18		0.26		0.09		1	A9 IV	280
		G 63 - 40		13 34 25	+19 20.8	2	11.94	.005	0.85	.000	0.29		3		1696,1759
		HA 105 # 535		13 34 25	−00 11.8	2	11.45	.003	0.39	.003	-0.23	.005	23		989,1729
118412	−22 03630	TV Hya		13 34 25	−23 21.7	1	7.13		0.27		0.04		4	A9 V	244
	−61 03847	LSS 3133		13 34 26	−61 28.7	2	9.73	.001	0.30	.002	-0.43	.013	3	B3 II-III	540,976
	+8 02735	G 62 - 54		13 34 29	+08 01.6	2	10.01	.010	1.08	.002	0.92		2	M0	1620,1705
	+36 02391			13 34 30	+35 55.1	1	10.71		0.43		-0.09		1	F5	280
118368	−50 07917			13 34 31	−50 26.6	1	10.23		0.37		0.12		1	A8/9 V	1771
	+36 02392			13 34 33	+35 55.1	1	10.93		0.57		0.00		1	K5	280
118369	−55 05653	IDS13314S5529	A	13 34 33	−55 43.8	1	8.64		0.02		-0.20		2	B9 V	1770
	−61 03848			13 34 34	−61 45.1	2	10.44	.003	0.09	.014	-0.57	.009	2	B2.5IV	540,976
		NGC 5236 sq1 7		13 34 36	−29 42.2	1	14.59		0.99		0.66		2		5013
	+15 02601			13 34 37	+14 46.8	1	10.58		0.97		0.90		1	K0	280
118508	+25 02652	HR 5123		13 34 38	+24 52.1	3	5.72	.013	1.57	.009	1.91	.048	9	M2 III	70,1501,3001
118536	+50 02014	HR 5126		13 34 38	+49 44.5	1	6.46		1.20		1.21		2	K1 III	1733
118418	−35 08861			13 34 38	−36 19.1	1	7.44		0.11		0.10		1	A2 IV	1770
		L 106 - 73		13 34 38	−67 49.4	1	15.57		0.30		-0.59		4		3073
	+20 02849			13 34 39	+20 06.1	1	9.62		0.93		0.66		1	G5	280
	+35 02469			13 34 41	+35 18.4	1	8.90		0.51		-0.01		3	F8	1569
118370	−57 06166			13 34 41	−57 57.2	1	8.13		0.08				4	B9.5V	2033

Table 1 761

HD	DM	Other Id	N Rem	α_{1950}	δ_{1950}	S	V	σ_V	B–V	σ_{B-V}	U–B	σ_{U-B}	n	Spectrum	References
	+17 02650			13 34 42	+16 39.0	1	9.73		1.24		1.33		1	K2	280
	+17 02651			13 34 42	+17 27.9	1	9.66		0.79		0.35		2	G0	1375
		HA 105 # 131		13 34 42	−00 54.8	1	12.30		0.64		0.09		1	G0	281
118434	−34 09016			13 34 42	−34 47.7	1	7.25		0.09		0.08		4	A2/3 V	243
118383	−55 05654			13 34 42	−56 22.6	2	8.30	.014	0.31	.009	0.26	.009	5	A3 IV	243,1771
		HA 105 # 437		13 34 43	−00 22.7	2	12.54	.007	0.24	.004	0.07	.006	9		281,1764
		HA 105 # 436		13 34 43	−00 24.2	1	11.58		0.61		0.00		17	F8	281
	+52 01720			13 34 45	+52 18.1	1	9.82		1.58		2.00		6		1404
118384	−57 06169	HR 5122	⋆ AB	13 34 50	−58 09.6	4	6.42	.009	1.12	.005	0.93		14	K1 III	15,404,1075,2007
		NGC 5236 sq1 3		13 34 51	−29 46.1	1	12.04		1.29		1.48		3		5013
	−30 10744			13 34 52	−30 53.0	1	9.89		1.24		1.36		4	F8	119
118285	−75 00882	HR 5115		13 34 52	−75 25.8	3	6.34	.004	0.00	.005	-0.26	.000	12	B8 IV	15,1075,2038
118468	−41 07995			13 34 53	−42 15.6	1	7.82		0.10		0.08		1	A1 V	1770
	+26 02474			13 34 55	+26 23.7	1	9.33		1.11		1.11		1	K0	280
118576	+30 02428	IDS13326N3036	A	13 34 55	+30 20.3	3	9.30	.031	0.63	.006	0.12	.017	9	G8 V	196,1064,7009
118576	+30 02428	IDS13326N3036	B	13 34 55	+30 20.3	3	10.51	.009	0.84	.019	0.56	.011	8	G0	196,1064,7009
118465	−34 09020	IDS13321S3433	AB	13 34 55	−34 48.6	1	7.19		0.69		0.20		4	G3 V	244
118423	−55 05658	IDS13317S5543	AB	13 34 55	−55 58.1	2	8.04	.001	0.01	.001	-0.07	.011	7	B9 III	243,1770
118326	−69 01899	IDS13312S6941	AB	13 34 56	−69 55.9	1	9.23		0.54		0.21		1	A8/F0 [IV]	1771
118344	−69 01898	HR 5119		13 34 56	−70 11.4	3	6.09	.004	1.41	.005	1.73	.000	14	K3 III	15,1075,2038
118526	+0 03082			13 34 57	+00 02.0	4	8.77	.007	0.35	.006	0.03	.005	39	A5	281,989,1729,6004
		NGC 5236 sq1 5		13 34 59	−29 44.1	1	12.98		0.68		0.10		3		5013
		NGC 5236 sq1 2		13 34 59	−29 45.1	1	12.04		0.87		0.59		3		5013
118385	−63 02869	LSS 3134		13 34 59	−63 48.1	1	7.40		0.14		0.00		2	B9.5II	403
118558	+13 02692			13 35 01	+13 12.1	1	9.39		0.33		-0.06		2	F2	1569
		HA 105 # 135		13 35 01	−00 48.9	1	11.28		0.65		0.05		1	G2	281
	+25 02655			13 35 02	+25 19.8	1	9.07		1.14		1.21		1	K2	280
	+41 02413			13 35 02	+41 15.3	1	9.81		1.59		1.88		1	M2	280
		LSS 3135		13 35 02	−63 31.8	1	10.93		0.84		-0.28		3	B3 Ib	1737
	+25 02656			13 35 03	+24 59.0	1	9.68		0.88		0.49		1	G5	280
118450	−49 08074			13 35 03	−50 05.7	3	6.64	.004	-0.08	.006	-0.45	.003	9	B5 II	401,1770,2014
146171	+0 03083			13 35 05	+00 05.3	1	10.54		0.63		0.10		1	G5 V	281
	−45 08585			13 35 05	−46 17.8	1	10.41		1.12		0.86		6		863
118487	−47 08466			13 35 06	−47 42.6	1	8.24		0.24		0.20		1	A5/7 V	1771
118436	−57 06175			13 35 06	−58 03.4	1	7.98		0.14		0.14		2	A0 IV/V	1771
		HA 105 # 447		13 35 08	−00 21.6	1	13.13		0.56		0.06		1	G3	281
		LSS 3136		13 35 10	−63 23.5	1	11.17		1.00		-0.04		1	B3 Ib	1737
118427	−66 02274			13 35 12	−66 24.4	1	9.91		0.33		0.18		1	A8/9 V	1771
	+36 02393	G 165 - 19		13 35 13	+35 58.4	1	9.07		1.40		1.16		3	M9	3072
118579	+0 03084			13 35 13	−00 22.3	7	9.18	.004	0.25	.004	0.04	.011	56	A2	281,989,1728,1729,1775*
		NGC 5236 sq1 1		13 35 13	−29 35.2	1	11.64		0.87		0.50		3		5013
118623	+37 02433	HR 5127	⋆ AB	13 35 14	+36 32.9	7	4.83	.016	0.23	.011	0.11	.018	19	A6 III+F0 V	15,292,938,1363,3030*
118517	−36 08726	IDS13324S3658	A	13 35 14	−37 13.8	1	8.66		0.17		0.11		2	A0 IV(m)	1770
118495	−49 08077			13 35 17	−49 30.6	2	9.02	.006	-0.02	.006	-0.31	.003	3	B8 V	1730,1770
118496	−49 08076			13 35 18	−49 48.1	1	7.91		1.34		1.49		5	K2 III	897
118489	−56 05853			13 35 18	−56 28.6	1	8.26		0.56		0.22		4	A0 V G/KIII	243
118497	−50 07928			13 35 19	−51 17.9	1	8.11		-0.02				4	B2 III	2014
	+44 02288			13 35 20	+43 49.4	1	9.32		1.54		1.80		1	M0	280
	−59 05072			13 35 22	−60 03.1	1	11.15		0.14		-0.32		2	B6 V	540
118643	+34 02435			13 35 23	+33 59.5	2	7.22	.018	1.36	.005	1.55	.014	7	K3 II	1080,8100
	+23 02590			13 35 24	+23 18.3	1	10.04		0.93		0.65		2		1569
	+26 02476			13 35 25	+25 44.2	1	9.71		1.46		1.74		1		280
118624	+14 02643			13 35 26	+13 42.1	1	8.34		1.41		0.93		1	M2	280
	+36 02394			13 35 26	+36 24.1	1	9.46		1.00		0.84		1	K0	280
118500	−57 06179			13 35 27	−57 54.1	1	9.08		0.32		0.06		1	A8/F0 [IV]	1771
		HA 105 # 563		13 35 28	−00 14.0	1	11.22		0.53		-0.05		2	F8	281
	−0 02715	IDS13329S0020	A	13 35 28	−00 34.9	1	10.32		1.16		1.22		1	K2	281
	+26 02477			13 35 30	+25 40.9	1	9.50		1.24		1.46		1	K2	280
118669	+42 02424			13 35 30	+42 27.3	1	7.59		1.52		1.69		1	M1	280
118520	−56 05856	HR 5124	⋆ AB	13 35 32	−57 22.1	1	6.01		1.14				4	G5 Ib	2006
118386	−71 01495			13 35 32	−72 15.0	1	7.97		0.31		0.25		1	A3mA5-A8	1771
118658	+27 02273			13 35 33	+27 03.6	1	8.58		0.95		0.72		1	K0 III	280
	+52 01722			13 35 33	+52 11.8	1	10.10		0.42		-0.06		6		1404
	+19 02691			13 35 34	+19 19.0	1	9.63		1.53		1.92		1	K5	280
		CS 22889 # 23		13 35 35	−11 54.2	1	14.55		0.02		0.00		1		1736
118546	−52 06633			13 35 35	−52 53.8	1	9.45		0.06		-0.02		1	B9.5/A0 V	1770
118659	+19 02692	G 63 - 44		13 35 36	+19 24.4	3	8.84	.008	0.68	.004	0.07	.016	5	G5	516,1620,1658
	+12 02610			13 35 37	+11 32.7	1	9.71		0.98		0.68		1	K0	280
118475	−67 02325			13 35 37	−67 24.9	3	6.96	.004	0.62	.004	0.18	.005	8	G2/3 IV/V	657,1311,2012
	+0 03085			13 35 38	−00 20.4	3	10.26	.004	0.41	.001	-0.03		16	F2	281,5006,6004
118670	+23 02591			13 35 39	+22 47.0	1	7.02		0.94				3	G5	280
118660	+15 02602	HR 5129		13 35 41	+14 33.4	1	6.54		0.24		0.09		2	A9 Vs	254
	−0 02716			13 35 41	−01 02.0	2	9.96	.004	0.86	.007	0.53		13	dG9	281,6004
118567	−53 05704			13 35 41	−53 39.4	1	9.42		0.12		0.15		1	A3/5 V	1771
118548	−54 05680			13 35 41	−54 46.2	1	7.37		0.19		0.16		1	A2mA5-A7	1771
118741	+51 01856	HR 5133	⋆ AB	13 35 43	+50 58.1	3	6.49	.056	1.56	.004	1.70	.043	12	M2 II-III +	1733,3001,8100
	+27 02274			13 35 45	+27 05.6	1	10.09		1.09		1.05		2	K5	1569
		NGC 5236 sq1 4		13 35 45	−29 31.5	1	12.40		0.98		0.70		3		5013
118742	+39 02662	G 165 - 21	⋆ C	13 35 47	+39 25.8	1	9.15		0.92		0.62		4		3024
118591	−49 08082			13 35 48	−49 32.2	1	9.52		1.22		1.27		2	K0 III	1730

HD	DM	Other Id	N Rem	α_{1950}	δ_{1950}	S	V	σ_V	B−V	σ_{B-V}	U−B	σ_{U-B}	n	Spectrum	References
	+48 02138	G 177 - 46	⋆ AB	13 35 49	+48 23.5	1	9.76		1.38		1.18		2	M0	3072
		HA 105 # 256		13 35 49	−00 38.7	1	11.82		0.61		0.18		19	F8	281
118603	−51 07640			13 35 49	−51 57.2	1	8.09		0.27		0.15		1	A8 III/IV	1771
118701	+34 02436			13 35 50	+34 00.6	1	8.35		1.52		1.92		1	K5	280
	+41 02415			13 35 50	+41 28.2	1	9.37		0.94		0.67		1	K0 III	280
118671	−0 02717			13 35 50	−00 44.8	2	9.13	.007	0.49	.002	0.02		11	G0	281,6004
118742	+39 02663	G 165 - 22	⋆ AB	13 35 52	+39 26.0	2	7.76	.009	0.70	.017	0.22	.009	6	G5 V	1003,3024
118506	−67 02327			13 35 52	−68 21.4	1	9.70		0.44		0.16		1	A8/9 V	1771
118936	+76 00493	G 255 - 40		13 35 53	+76 09.9	1	8.17		0.67		0.19		2	G5	3026
118646	−28 10181	HR 5128		13 35 53	−29 18.4	4	5.83	.005	0.40	.004	0.00		16	F3 V	15,1075,2027,3053
118571	−60 04836			13 35 53	−60 43.8	3	8.79	.023	-0.02	.010	-0.84	.013	7	B1 III	401,540,976
	+34 02437			13 35 56	+34 23.8	1	11.03		0.11		0.13		1		280
	−0 02718			13 35 57	−00 34.1	1	10.75		0.67		0.21		1	G0	281
118551	−63 02883			13 35 58	−63 43.8	1	9.37		0.28		0.14		1	A7 IV/V	1771
118904	+71 00659	HR 5139		13 35 59	+71 29.8	1	5.51		1.20				2	gK2	71
118648	−32 09509			13 35 59	−32 51.5	2	6.98	.002	-0.05	.007	-0.12	.001	7	B9.5V	244,1770
118570	−58 05004			13 35 59	−59 11.9	1	10.14		0.41		0.06		1	A9/F2	1771
	−63 02886	LSS 3140		13 36 6	−63 24.5	1	10.32		0.89		-0.19		3	O9.5 Ib	1737
	+19 02696			13 36 01	+18 35.4	1	10.05		1.02		0.84		1		280
118743	+28 02245	BZ Boo		13 36 02	+27 32.5	1	8.18		0.30		0.09		2	F0 III	1569
	−61 03871	LSS 3139		13 36 02	−61 55.4	3	9.61	.020	1.19	.026	0.21	.035	5	B5 Ia	540,976,1737
118704	−7 03663			13 36 03	−08 10.0	2	8.49	.002	0.54	.000			36	F8	130,1351
	+43 02342			13 36 04	+42 36.0	1	10.24		1.14		1.20		1		280
	+45 02122			13 36 04	+44 38.6	1	10.56		0.93		0.64		1	K0	280
118809	+51 01858			13 36 04	+51 12.8	1	7.84		0.98		0.74		2	G5	1566
118705	−7 03664			13 36 04	−08 13.4	2	9.17	.001	0.46	.000			36	G5	130,1351
	−63 02884			13 36 05	−64 02.0	1	9.73		0.96		0.76		1		287
		LP 558 - 18		13 36 07	+06 33.1	1	13.56		1.12		0.99		1		1696
	+10 02572			13 36 07	+10 22.5	1	9.81		0.34		0.14		3	A2	272
	+30 02431			13 36 07	+29 37.2	7	10.06	.017	-0.14	.011	-0.64	.012	37	B6	280,830,979,1026,1068*
		HA 105 # 155		13 36 07	−00 51.9	1	12.29		0.63		0.12		1	G0	281
118572	−63 02885			13 36 07	−64 12.0	1	10.34		0.52		-0.01		1	F7/8 V	287
	+24 02609			13 36 08	+23 49.0	1	10.31		1.14		1.21		1		280
	−0 02719			13 36 08	−00 59.0	2	9.97	.001	0.66	.009	0.15		15	gG2	281,6004
118522	−70 01653	HR 5125		13 36 08	−70 32.1	3	6.58	.004	1.29	.005	1.09	.000	11	K0 III	15,1075,2038
	−0 02720			13 36 09	−00 44.1	1	11.58		0.70		0.24		1	F9	281
118638	−55 05666			13 36 10	−56 02.3	1	8.80		1.51		1.76		4	K3 III	243
	−63 02887	LSS 3141		13 36 11	−64 04.7	1	10.34		0.57		-0.04		1	F3 II	287
		AAS12,381 42 # E		13 36 13	+42 08.6	1	10.32		1.07		0.99		1		280
	−49 08088			13 36 13	−49 35.0	1	10.76		1.29		1.21		2		1730
		LSS 3142		13 36 13	−60 04.9	1	11.56		0.81		-0.15		2	B3 Ib	1737
118709	−24 11011			13 36 15	−24 36.5	1	7.51		1.08		0.93		5	K0 III	1657
		Lod 894 # 4		13 36 16	−64 04.7	1	10.81		1.10		0.85		1		287
		HA 105 # 359		13 36 17	−00 31.1	1	11.73		0.55		-0.02		20	F8	281
		G 150 - 26		13 36 19	+23 19.1	1	15.52		1.56				1		906
	−63 02889			13 36 20	−64 01.0	1	10.59		0.14		-0.05		1		287
		HA 105 # 788		13 36 22	+00 14.4	1	11.28		0.46		0.00		1	F2	281
	−63 02891			13 36 24	−64 07.4	1	10.21		0.46		-0.46		1		287
	+24 02610			13 36 27	+23 47.3	2	10.01	.015	0.29	.009	0.02		6	A7	193,1569
		HA 105 # 792		13 36 28	+00 05.6	1	10.73		1.34		1.55		1	dK4	281
118791	+11 02586			13 36 28	+11 30.3	1	8.53		1.17		1.23		1	K5	280
		HA 105 # 577		13 36 30	−00 14.5	1	11.49		0.72		0.26		7	G0	281
118728	−35 08884			13 36 31	−35 59.2	1	7.48		0.05		0.05		1	A1 V	1770
118823	+24 02611			13 36 32	+24 30.2	1	8.26		1.08		1.07		1	K2 III	280
118849	+39 02665	IDS13343N3853		13 36 32	+38 38.0	1	8.71		0.90		0.64		1	K0	280
		HA 105 # 165		13 36 32	−00 51.8	1	11.63		0.46		-0.02		7	F8	281
		HA 105 # 693		13 36 33	−00 04.6	1	11.88		0.61		0.10		2	G0	281
118697	−53 05712			13 36 34	−53 53.7	2	7.34	.010	-0.01	.006	-0.09	.023	5	B9.5V	401,1770
118937	+67 00792			13 36 35	+66 51.5	1	7.87		0.08		0.08		3	A2	1733
118824	+24 02612			13 36 36	+24 23.1	2	7.79	.005	0.37	.010	-0.04	.010	4	F0	1569,1625
118839	+19 02697	HR 5137		13 36 38	+18 31.1	1	6.48		1.20		1.27		1	K3 III	280
118797	−12 03861			13 36 38	−13 03.7	1	10.15		0.96		0.75		1	K1/2 V	3072
	+34 02438			13 36 39	+34 10.0	1	9.60		1.25		1.49		1	K2 III	280
		WLS 1340 35 # 10		13 36 39	+34 52.8	1	12.23		0.51		-0.13		2		1375
118666	−63 02896	HR 5130		13 36 39	−64 19.4	3	5.74	.024	0.39	.004	0.01	.010	15	F2 III/IV	15,1075,2038
	+22 02613			13 36 40	+21 53.5	1	9.84		0.98		0.74		1	K0	280
118716	−52 06655	HR 5132, ϵ Cen	⋆ A	13 36 42	−53 12.8	9	2.30	.012	-0.23	.010	-0.92	.010	28	B1 III	15,26,507,1020,1034*
	+19 02698			13 36 44	+18 58.3	1	9.92		1.03		0.92		1	K0	280
118781	−39 08390	HR 5135, V765 Cen		13 36 44	−39 29.7	1	6.27		1.65				4	M4 III	2007
118840	+11 02588			13 36 45	+10 45.8	1	6.64		1.44		1.79		2	M3 III	1648
		HA 105 # 697		13 36 45	−00 03.7	1	11.84		0.91		0.73		4	G0	281
118871	+44 02290			13 36 46	+43 58.3	1	9.08		0.80		0.44		2	K0	280
	+36 02397			13 36 49	+36 13.1	1	10.47		1.10		1.08		2	K8	1569
118798	−31 10533			13 36 50	−32 14.2	1	9.78		0.29		0.15		1	A0mA8-F0	1770
118799	−39 08392	HR 5136		13 36 52	−39 47.9	1	5.60		1.30				4	K2/3 III	2007
	+0 03087			13 36 53	−00 07.9	1	10.58		0.54		-0.08		4	G0	281
		CS 22889 # 25		13 36 53	−12 11.8	1	14.95		0.19				1		1736
	+30 02432			13 36 54	+30 17.1	1	9.50		1.22		1.38		1	K5 III	280
118767	−49 08095	HR 5134, V744 Cen		13 36 54	−49 41.8	2	5.88	.397	1.58	.013	1.34	.092	6	M5 III	897,3002
118873	+27 02275			13 36 55	+26 48.8	1	9.46		0.11		0.09		3	A3 IV-V	1068

Table 1

763

HD	DM	Other Id	N Rem	α_{1950}	δ_{1950}	S	V	σ_V	B–V	σ_{B-V}	U–B	σ_{U-B}	n	Spectrum	References
118887	+35 02471			13 36 55	+34 31.0	1	8.48		1.31		1.61		2	K3 IV	1569
	+12 02612			13 37 00	+11 46.5	1	10.02		1.08		0.92		1	K0	280
118874	+18 02762			13 37 01	+17 50.9	1	8.86		1.03		0.96		1	K0	280
118850	−0 02721			13 37 01	−00 58.3	4	9.42	.004	0.99	.013	0.76	.002	52	G8 III	281,989,1729,6004
118769	−56 05865	XX Cen		13 37 01	−57 21.6	3	7.46	.148	0.80	.046	0.60	.030	3	F7/8 II	657,689,1484
118684	−69 01904			13 37 01	−69 44.3	1	8.80		0.33		0.28		1	A3 V	1771
118768	−55 05676			13 37 02	−56 16.4	1	9.67		0.28		0.18		4	A2 III	243
118553	−76 00774			13 37 03	−77 15.7	1	8.53		0.03		-0.47		6	B3 III	1732
238263	+56 01671			13 37 04	+55 54.8	1	9.22		0.89		0.58		2	K2	1502
118914	+32 02370			13 37 05	+32 01.6	1	8.88		0.66		0.23		1	G0	280
118770	−57 06200			13 37 05	−57 40.3	1	7.63		0.48		0.44		1	A8 III	1771
118905	+27 02276			13 37 06	+26 56.3	1	7.16		1.03				1	K1 III	280
		HA 105 # 174		13 37 06	−00 53.1	1	12.04		0.94		0.60		1	G3	281
118889	+11 02589	HR 5138	★ AB	13 37 07	+11 00.0	3	5.58	.011	0.33	.000	0.05	.002	5	F0 IV+F2 IV	292,1381,3030
118699	−69 01905			13 37 10	−69 54.6	1	9.73		0.35		0.20		1	A7/8 V	1771
118816	−53 05717			13 37 15	−53 36.4	1	7.82		-0.01		-0.33		2	Ap Si	401
	+16 02535			13 37 17	+15 51.8	1	10.17		1.31		1.49		1	K2	280
	+46 01889	G 177 - 50		13 37 19	+46 26.7	1	10.23		1.45		1.08		4	M2	7008
		AJ74,1000 M3 # 2		13 37 22	+28 14.8	1	11.53		1.04				3		209
		HA 105 # 595		13 37 24	−00 13.6	1	11.62		0.69		0.19		12	G0	281
	+40 02674			13 37 26	+40 19.9	1	9.82		1.04		0.86		1	K0	280
		G 64 - 12		13 37 29	+00 12.9	6	11.45	.024	0.38	.006	-0.23	.000	26	sdF0	281,516,1620,1658*
		HA 105 # 599		13 37 29	−00 12.3	1	11.70		0.62		0.08		13	F8	281
		HA 105 # 815		13 37 30	+00 12.9	1	11.45		0.38		-0.24		14	A5p	1764
	+13 02698	G 63 - 46		13 37 32	+12 50.8	3	9.38	.017	0.58	.000	-0.06	.020	5	F9 V	1003,1620,1658
118926	−3 03508	G 62 - 60		13 37 33	−03 56.4	4	9.62	.021	1.39	.023	1.19	.019	9	M0 V	1620,1705,2034,3072
	+32 02373			13 37 34	+32 23.1	1	10.83		0.29		0.05		1		280
119024	+53 01640	HR 5142		13 37 34	+53 10.4	2	5.43	.030	0.10	.006	0.04		4	A3 Vn	70,3023
		CS 22889 # 39		13 37 34	−07 49.4	1	14.59		-0.01		-0.15		1		1736
	+29 02444			13 37 35	+28 36.1	2	9.67	.014	1.16	.000	1.23		4	K0	209,280
118805	−64 02511			13 37 35	−65 01.1	1	9.24		0.33		0.21		1	A7 V	1771
		CS 22889 # 37		13 37 36	−08 58.9	1	13.78		0.07		0.16		1		1736
118878	−43 08458			13 37 36	−44 04.6	1	6.57		0.05		-0.05		2	A0 V	1770
		G 238 - 44		13 37 37	+70 32.4	2	12.79	.000	-0.09	.000	-0.84	.000	5		1281,3028
118894	−33 09233			13 37 38	−34 06.3	1	7.36		0.78		0.39		4	G2 IV/V	244
118896	−42 08641			13 37 38	−42 51.8	1	9.68		0.26		0.09		1	A9 V	1771
	+21 02559			13 37 39	+21 13.5	1	9.80		1.12		1.09		3	K5	1569
		RU UMi		13 37 39	+70 03.4	1	10.00		0.29		0.01		1	A2	588
		LP 498 - 31		13 37 40	+13 25.4	1	12.22		1.43		1.19		1		1696
118971	+26 02481			13 37 40	+26 10.8	1	7.83		0.92		0.56		1	G8 III	280
		HA 105 # 600		13 37 41	−00 14.2	1	11.84		0.56		0.08		13	F8 p	281
	+29 02445			13 37 42	+28 52.5	1	10.71		1.01				3	G0	209
	+30 02434			13 37 42	+30 17.5	1	9.52		1.34		1.60		1	K2	280
		BPM 8436		13 37 42	−67 55.	1	14.34		1.12		0.68		1		3065
		GD 326		13 37 44	+61 03.9	1	13.83		0.50		-0.22		2		308
	+40 02675			13 37 45	+40 21.1	1	10.22		1.10		1.04		1	K0	280
	+19 02700			13 37 47	+18 42.3	1	9.94		0.36		-0.01		2		1569
119227	+75 00512	V UMi		13 37 47	+74 33.8	1	7.74		1.53		1.44		1	M1	3001
		HA 105 # 496		13 37 47	−00 20.3	1	11.33		0.31		0.06		2	A1	281
118870	−58 05038			13 37 49	−58 58.5	1	7.69		0.09		-0.03		2	B9.5 V	1771
	+43 02344			13 37 51	+43 29.5	1	10.94		0.04		0.14		1		280
	+23 02597			13 37 52	+22 43.8	1	9.71		0.93		0.62		1	K0	280
118981	+2 02705	G 62 - 61		13 37 54	+02 24.4	2	8.21	.000	0.57	.000	0.03	.020	2	G0	333,1620,1658
	+40 02676			13 37 54	+40 27.3	1	11.04		0.13		0.10		1		280
		ApJ144,259 # 36		13 37 54	−19 38.	1	11.51		-0.33		-1.23		4		1360
		LP 855 - 50		13 37 55	−24 21.6	2	13.20	.065	0.76	.015	-0.03	.005	4		1696,3060
	+23 02598			13 37 56	+22 46.7	1	9.32		1.27		1.37		1	K5	280
	+35 02472			13 37 57	+35 18.4	1	9.07		1.56		1.90		2	M0 V	1569
119025	+29 02446			13 37 58	+29 16.8	1	8.61		1.00		0.84		1	K0 IV	280
118931	−44 08809			13 37 58	−44 60.0	1	9.24		0.32		0.13		1	A1mA5/8-F0	1770
119007	+22 02616			13 37 59	+22 29.1	1	7.68		1.45		1.79		1	K2	280
119035	+31 02526	HR 5143		13 37 59	+31 15.8	2	6.22	.010	0.96	.002	0.68		5	G8 III	70,1501
	−33 09240			13 38 01	−33 26.8	1	10.70		0.68		0.15		4		244
	+37 02438			13 38 04	+37 15.4	1	9.11		1.09		1.00		2	K2 III	1569
		G 177 - 51		13 38 06	+44 01.4	3	12.76	.011	1.64	.022			6		538,1705,1759
		Feige 87		13 38 06	+61 07.0	1	11.63		-0.07		-0.90		2		3060
118910	−56 05874			13 38 06	−57 11.7	1	8.95		0.49		-0.06		60	F5 V	1593
	+25 02659			13 38 07	+25 06.0	1	10.46		1.13		1.06		1		280
	−61 03907			13 38 08	−61 56.2	1	9.56		1.25		0.22		2	B5 Ia	403
118972	−33 09242			13 38 10	−34 12.6	2	6.94	.027	0.86	.000	0.47	.050	5	K1 V	244,3072
119036	+19 02701			13 38 11	+18 50.7	1	8.79		0.96		0.69		2	K0	1569
118986	−29 10516			13 38 12	−29 28.7	1	7.86		0.98		0.70		4	G8 III	1657
118961	−44 08812			13 38 13	−44 23.8	1	9.37		0.48		0.03		1	F5 V(W)	1770
		G 177 - 52		13 38 15	+47 27.7	1	15.30		1.73				1		1663
119055	+20 02858	HR 5144	★ AB	13 38 17	+20 12.5	4	5.72	.026	0.01	.011	0.01	.009	12	A1 V	70,542,1022,1068
118987	−33 09245			13 38 17	−33 24.2	2	9.17	.005	1.42	.005	1.62	.020	10	K3 III	244,1775
119082	+20 02859			13 38 18	+20 15.9	1	7.39		0.24		0.07		3		1068
118964	−51 07668			13 38 19	−52 21.3	1	8.44		1.18				4	K1 III	2012
119081	+28 02248	HR 5145	★ A	13 38 21	+28 19.1	3	6.23	.008	1.29	.008	1.50		7	K3 III	70,209,3016
119112	+40 02677			13 38 21	+40 14.6	1	8.53		1.04		0.92		1	K0	280

HD	DM	Other Id	N	Rem	α_{1950}	δ_{1950}	S	V	σ_V	B–V	σ_{B-V}	U–B	σ_{U-B}	n	Spectrum	References
118965	−52 06679				13 38 22	−53 21.5	1	8.69		0.06		−0.15		1	B8/9 III	1770
119083	+18 02766				13 38 23	+18 03.2	1	8.17		1.08		0.97		2	K0	1569
119124	+51 01859	HR 5148		⋆ A	13 38 24	+50 46.3	2	6.33	.010	0.53	.010	−0.02	.010	5	F7 V	1733,3026
119124	+51 01859	IDS13364N5101		B	13 38 24	+50 46.3	2	10.50	.008	1.38	.007	1.22	.125	4		3016,7009
		Steph 1083			13 38 25	+50 46.1	1	10.46		1.36		1.11		1	K7	1746
	+32 02374				13 38 27	+31 55.5	1	10.00		0.56				2		77
	+40 02678				13 38 30	+39 58.9	1	9.63		1.14		1.16		1	K0	280
119213	+57 01456	HR 5153, CQ UMa			13 38 32	+57 27.6	3	6.27	.014	0.11	.011	0.02	.000	11	A4 p SrCrEu	252,1379,3016
118991	−53 05725	HR 5141		⋆ A	13 38 32	−54 18.4	1	5.26		−0.10		−0.28		3	B9 Vn	1770
118991	−53 05725	HR 5141		⋆ AB	13 38 32	−54 18.4	1	4.99	.011	−0.06	.004	−0.23	.011	11	B9 Vn+A0 V	401,404,1770,2006
	+40 02679				13 38 34	+39 47.4	1	9.03		1.31		1.28		1	K2	280
119125	+32 02375				13 38 35	+32 01.1	1	9.37		0.92		0.66		1	K0	280
118913	−68 01981				13 38 39	−68 59.7	1	7.71		0.09		0.01		2	Ap EuCrSr	1771
119126	+23 02600	HR 5149			13 38 40	+22 44.9	2	5.64	.015	1.01	.001	0.82		5	G9 III	70,1501
		AJ74,1000 M3 # 4			13 38 40	+28 33.3	1	11.91		0.74				3		209
118978	−58 05059	HR 5140			13 38 41	−58 32.1	5	5.39	.019	−0.03	.008	−0.25	.005	15	B9 III	15,26,1075,1771,2009
		Klemola 179			13 38 42	+15 49.	1	11.65		−0.02		0.10		2		1026
	+32 02376				13 38 42	+32 04.6	1	9.75		1.57		1.84		1	M0	280
	+24 02619				13 38 44	+23 33.7	1	10.51		1.04		0.86		1	K0	280
119127	+16 02539				13 38 45	+16 10.9	1	8.33		0.93		0.66		1	K0	280
119171	+34 02439				13 38 45	+33 39.5	1	8.72		0.92		0.68		1	K0 III	280
119086	−22 03645	HR 5146		⋆ AB	13 38 45	−23 11.9	4	6.59	.010	0.06	.012	0.06	.005	13	A1 V	13,15,244,2030
119086	−22 03645	IDS13360S2257		C	13 38 45	−23 11.9	1	10.34		1.37		0.99		3		13
119042	−43 08467				13 38 49	−43 25.0	1	9.69		0.39		0.01		1	A9/F2	1771
119031	−48 08413				13 38 49	−48 52.1	1	8.33		0.13		0.15		1	A3 V	1771
	+29 02447				13 38 50	+28 57.8	1	10.23		0.76				3	K0	209
119228	+55 01625	HR 5154			13 38 51	+54 56.0	6	4.66	.018	1.63	.009	1.95	.023	21	M2 IIIab	15,1119,1363,3016*
119090	−32 09549	HR 5147, T Cen			13 38 53	−33 20.7	1	6.75		1.58		1.62		5	M3e	897
119091	−33 09252				13 38 53	−33 30.8	2	8.92	.010	0.98	.010	0.65	.035	9	G8 III	244,897
119067	−42 08653				13 38 53	−42 55.4	1	8.57		0.26		0.05		1	A9 V	1771
119069	−45 08635				13 38 53	−45 36.0	3	8.43	.004	−0.20	.003	−0.97	.009	10	B3 II	26,55,1732
		LP 323 − 239			13 38 54	+30 17.1	1	15.83		1.75				2		1663
	−19 03718				13 38 54	−19 43.3	1	11.10		1.09		0.85		9		1732
	+18 02767				13 38 58	+18 13.7	1	9.99		0.99		0.77		2		1569
119020	−57 06211				13 38 58	−58 22.1	1	7.93		0.24		0.11		1	A6 V	1771
119149	−7 03674	HR 5150			13 38 59	−08 27.1	1	5.00	.004	1.62	.006	1.93	.017	22	M2 III	3,1088,3055
119070	−47 08526				13 38 59	−48 12.7	3	9.13	.013	0.78	.005	0.31		11	G8 V	1075,2033,3008
119103	−42 08656				13 39 01	−42 53.7	3	7.14	.011	−0.06	.004	−0.17	.005	8	B9 IV	55,1770,2014
	+37 02439				13 39 04	+37 05.9	1	10.00		1.15		1.03		2	K2 III	1569
119046	−56 05884	IDS13358S5649		A	13 39 04	−57 04.3	1	8.06		0.18		0.17		2	A1/2 V	1771
		G 63 − 47			13 39 07	+15 04.6	1	12.09		1.50		1.23		1		1620
119121	−51 07677				13 39 14	−51 24.1	3	8.55	.012	0.05	.005	−0.11	.006	9	B9 V	401,1770,2014
	−61 03926	LSS 3148			13 39 14	−61 29.2	2	6.84	.029	0.26	.005	−0.33		6	B8 I	403,2012
	+19 02704				13 39 15	+19 30.4	1	9.05		1.45		1.81		1	K5	280
	+39 02670				13 39 15	+39 07.5	1	9.95		1.45		1.79		1	K5	280
	+30 02436				13 39 16	+29 54.7	1	9.88		1.03		0.71		1	K0	280
	+20 02863				13 39 17	+19 58.2	1	10.56		0.31		0.00		2		1569
		AJ74,1000 M3 # 5			13 39 17	+28 27.3	1	11.49		0.65				3		209
119185	−12 03870				13 39 17	−12 46.6	1	8.91		1.00		0.64		1	G8 III(pBa)	565
119049	−62 03478	LSS 3147			13 39 17	−63 16.0	2	9.69	.000	0.18	.001	−0.71	.010	4	B1 II/III	540,976
119076	−61 03927	IDS13359S6123		AB	13 39 18	−61 38.4	1	6.87		1.16		0.88		27	K0 III +A	978
119217	+0 03090	G 62 − 62			13 39 22	+00 07.7	3	9.75	.018	1.29	.011	1.32	.030	5	M0 V	265,1705,3072
	−19 03719				13 39 24	−20 13.9	1	10.62		0.80		0.36		2		1696
	−59 05128				13 39 27	−60 15.6	1	10.77		0.43		0.14		1	F0	549
	+15 02608				13 39 28	+15 23.7	1	10.06		1.08		0.93		1		280
	+29 02448				13 39 28	+28 42.4	1	9.84		1.13		1.08		1	K2	280
	+29 02449				13 39 28	+29 00.9	1	11.00		0.80				3	K0	209
119191	−33 09259	IDS13366S3328		AB	13 39 28	−33 43.7	1	6.66		0.42		−0.02		4	F0/3 +(F/G)	244
	+13 02701				13 39 30	+12 32.6	1	9.72		1.36		1.48		1	G0	280
		GD 158			13 39 32	+26 13.3	1	12.82		0.16		0.01		1		3060
		G 238 − 46			13 39 33	+68 22.8	1	11.94		1.16		1.15		1	K3	1658
		WLS 1340 15 # 9			13 39 34	+15 03.5	1	13.76		0.86		0.37		2		1375
		LSS 3150			13 39 38	−56 31.0	1	10.27		1.01		0.65		2	B3 Ib	1737
119159	−56 05891	HR 5151, LSS 3151			13 39 39	−56 31.0	1	6.00		−0.08				4	B0.5 III	2007
119144	−59 05132				13 39 40	−60 17.7	1	10.09		0.09		−0.18		1	B8/9 V	549
		AJ74,1000 M3 # 6			13 39 42	+28 19.5	1	11.85		0.79				3		209
		AJ74,1000 M3 # 16			13 39 43	+28 55.0	1	11.93		0.60				3		209
119221	−42 08664				13 39 43	−42 56.0	1	7.28		0.18		0.08		3	A4 V	1771
119287	+23 02601				13 39 44	+22 52.3	1	8.23		1.17		1.33		1	K0	280
119288	+9 02798	HR 5156		⋆	13 39 45	+08 38.5	6	6.16	.014	0.41	.009	−0.06	.010	19	F3 Vp	15,254,333,1417,1620,3037
	+2 02711				13 39 46	+01 45.4	6	10.37	.005	−0.16	.003	−0.70	.012	77	B5	989,1026,1728,1729,1732,1764
119193	−50 07983	HR 5152			13 39 46	−50 32.3	1	6.41		1.67				4	M0 III	2007
		DM 0202711E			13 39 48	+01 45.4	1	10.57		0.52		0.03		4		1732
119162	−60 04901				13 39 48	−60 41.0	2	10.15	.025	0.25	.003	−0.49	.001	3	Be	540,976
119164	−61 03933				13 39 49	−61 27.4	5	7.20	.006	1.29	.003	1.13	.005	57	G8 II/III	657,657,978,1621,2012
119222	−43 08480				13 39 49	−43 56.3	2	7.01	.007	−0.11	.007	−0.45		7	B6 III	1770,2014
		WLS 1340 35 # 7			13 39 50	+32 41.6	1	11.76		0.45		−0.03		2		1375
	+39 02671				13 39 50	+39 11.7	1	10.16		0.78		0.45		1	K0	280
	−33 09264				13 39 50	−33 48.0	1	9.19		1.59		1.85		4	K5	244
119333	+29 02450				13 39 52	+28 37.8	1	10.62		0.60				3		209

Table 1   765

HD	DM	Other Id	N	Rem	α_{1950}	δ_{1950}	S	V	σ_V	B–V	σ_{B-V}	U–B	σ_{U-B}	n	Spectrum	References
119291	−0 02725	G 62 - 64			13 39 52	−01 26.0	1	9.26		1.18		1.20		1	K7 V	3072
		WLS 1340 15 # 6			13 39 53	+17 25.7	1	13.89		0.88		0.62		2		1375
		WLS 1340 35 # 5			13 39 53	+37 28.6	1	10.69		1.28		1.42		2		1375
		NGC 5272 - 297			13 39 54	+28 38.	1	12.89		1.42		1.42		1		8014
		NGC 5272 - 1127			13 39 54	+28 38.	1	16.14		-0.02		0.01		2		645
		NGC 5272 - 1130			13 39 54	+28 38.	1	15.82		0.01		0.01		2		645
		NGC 5272 - 1140			13 39 54	+28 38.	1	15.68		0.39		0.00		1		645
		NGC 5272 - 1142			13 39 54	+28 38.	1	15.58		0.44		0.02		1		645
		NGC 5272 - 1147			13 39 54	+28 38.	1	16.37		-0.10		-0.22		1		645
		NGC 5272 - 1151			13 39 54	+28 38.	1	15.82		0.07		0.13		1		645
		NGC 5272 - 1156			13 39 54	+28 38.	1	16.28		-0.07		-0.16		1		645
		NGC 5272 - 1158			13 39 54	+28 38.	1	17.10		-0.12		-0.55		1		645
		NGC 5272 - 1213			13 39 54	+28 38.	1	15.76		0.09		0.16		1		645
		NGC 5272 - 1215			13 39 54	+28 38.	1	15.61		0.45		-0.03		1		645
		NGC 5272 - 1233			13 39 54	+28 38.	1	15.57		0.14		0.08		1		645
		NGC 5272 - 1245			13 39 54	+28 38.	1	15.64		0.12		0.10		1		645
		NGC 5272 - 1257			13 39 54	+28 38.	1	14.94		-0.30		-1.11		2		645
		NGC 5272 - 1306			13 39 54	+28 38.	1	15.71		0.08		0.09		1		645
		NGC 5272 - 1309			13 39 54	+28 38.	1	15.61		0.48		-0.01		5		645
		NGC 5272 - 1312			13 39 54	+28 38.	1	15.71		0.08		0.10		1		645
		NGC 5272 - 1314			13 39 54	+28 38.	1	15.76		0.09		0.11		1		645
		NGC 5272 - 1319			13 39 54	+28 38.	1	15.46		0.42		0.00		1		645
		NGC 5272 - 1322			13 39 54	+28 38.	1	15.62		0.12		0.05		1		645
		NGC 5272 - 1328			13 39 54	+28 38.	1	12.81		1.37		1.26		1		8014
		NGC 5272 - 1336			13 39 54	+28 38.	1	17.47		0.19		0.01		1		645
		NGC 5272 - 1351			13 39 54	+28 38.	1	15.46		0.50		-0.01		1		645
		NGC 5272 - 1357			13 39 54	+28 38.	1	16.95		-0.21		-0.47		1		645
		NGC 5272 - 1418			13 39 54	+28 38.	1	15.73		0.12		0.05		1		645
		NGC 5272 - 1429			13 39 54	+28 38.	1	15.68		0.42		0.03		1		645
		NGC 5272 - 1506			13 39 54	+28 38.	1	17.41		0.21		0.00		1		645
		NGC 5272 - 1604			13 39 54	+28 38.	1	17.00		0.71		0.07		1		645
		NGC 5272 - 1626			13 39 54	+28 38.	1	15.76		0.01		0.05		1		645
		NGC 5272 - 1648			13 39 54	+28 38.	1	15.90		0.02		0.06		1		645
		NGC 5272 - 2174			13 39 54	+28 38.	1	18.45		0.20		-0.10		1		645
		NGC 5272 - 2251			13 39 54	+28 38.	1	18.46		0.33		-0.14		1		645
		NGC 5272 - 2309			13 39 54	+28 38.	1	18.46		0.31		-0.10		1		645
		NGC 5272 - 5182			13 39 54	+28 38.	1	15.70		0.08		0.11		25		645
		NGC 5272 - 5193			13 39 54	+28 38.	1	14.80		0.89		0.36		6		645
		NGC 5272 - 5206			13 39 54	+28 38.	1	9.85		1.15		1.06		47		645
		NGC 5272 - 5216			13 39 54	+28 38.	1	14.09		1.01		0.62		2		645
		NGC 5272 - 5235			13 39 54	+28 38.	1	15.76		0.08		0.13		1		645
		NGC 5272 - 5297			13 39 54	+28 38.	1	12.89		1.42		1.40		1		645
		NGC 5272 - 6402			13 39 54	+28 38.	1	12.66		0.64		0.09		2		645
	+22 02619				13 39 55	+21 37.8	1	10.58		0.95		0.58		1		280
		G 165 - 24			13 39 55	+37 05.2	1	12.01		0.77		0.26		2	K0	1658
119334	+24 02622				13 39 56	+24 27.6	1	8.92		0.97		0.77		1	G5	280
119476	+65 00953	HR 5162			13 39 56	+65 04.5	2	5.85	.000	0.08	.005	0.08	.005	4	A2 V	985,1733
119250	−40 08096	HR 5155			13 39 56	−41 08.9	1	5.98		1.02				4	K0 III	2006
		G 63 - 48			13 39 57	+19 02.3	1	12.36		1.48		1.22		1		1620
	+36 02401				13 39 57	+35 32.4	1	9.87		1.11		1.11		1	K2 III	280
		LP 678 - 81			13 39 57	−07 11.5	1	12.35		0.78		0.27		2		1696
119267	−32 09566	IDS13371S3210	A		13 39 58	−32 25.1	1	8.07		1.12		0.88		4	G8/K0 III	244
		Lod 1010 # 50			13 39 58	−60 20.1	1	11.89		0.15		0.07		1	B9	549
119237	−52 06695				13 40 02	−53 15.9	1	9.35		0.02		-0.25		1	B9 III	1770
	+19 02706				13 40 03	+19 18.1	1	10.12		1.15		1.20		1		280
		Lod 1010 # 32			13 40 03	−60 00.8	1	11.33		1.05		0.53		1	K3	549
	−59 05137				13 40 03	−60 16.6	1	11.13		0.27		0.19		1	A0	549
	+38 02471				13 40 04	+38 04.6	1	8.86		1.21		1.25		1	K2 V	280
119410	+39 02672				13 40 05	+39 12.0	2	8.34	.019	0.90	.010	0.57	.000	3	G9 III	105,280
119335	+11 02593				13 40 06	+11 27.8	1	8.29		1.42		1.71		1	K5	280
	+28 02250				13 40 06	+28 07.6	1	9.37		0.94				3	G5	209
119211	−58 05080				13 40 06	−58 52.3	1	10.09		0.27		0.08		3	A8/F0 [V]	1771
	−59 05139				13 40 06	−59 58.7	1	11.13		0.34		0.26		1	A3	549
	−59 05138				13 40 06	−60 02.7	1	9.64		1.77		1.98		1	M4	549
119350	+15 02609				13 40 07	+15 23.9	1	7.25		1.30				1	K2	280
	−59 05141				13 40 07	−60 11.4	1	9.87		1.62		1.70		1	K3	549
119268	−44 08832				13 40 08	−44 23.3	1	8.61		0.07		0.06		2	A0 V	1770
119109	−73 01184				13 40 11	−73 23.1	2	7.46	.010	0.00	.005	-0.51	.000	9	B3 V	164,400
119392	+24 02624				13 40 13	+23 34.3	1	7.33		1.55				2	M1	280
	+43 02347				13 40 13	+43 30.3	1	10.08		1.07		1.05		1	K2	280
		LP 855 - 54			13 40 13	−22 50.7	1	11.16		0.53		-0.13		2		1696
		Lod 1010 # 57			13 40 14	−60 01.6	1	12.58		0.45		-0.04		1		549
119243	−59 05142				13 40 14	−60 11.0	1	8.67		1.13		0.80		1	G8 III	549
	+40 02683				13 40 15	+39 41.2	1	9.88		0.59		0.10		2		1569
119278	−45 08650				13 40 15	−45 47.7	1	8.29		0.03		-0.04		3	B9 V	1770
	+28 02252				13 40 16	+27 31.4	1	9.70		1.01		0.88		1	K2	280
		CS 22889 # 40			13 40 16	−07 34.0	1	15.26		0.09				1		1736
119308	−34 09094				13 40 18	−35 02.3	1	7.85		0.04		-0.01		1	Ap SrCrEu	1770
119256	−56 05898	IDS13370S5705	AB		13 40 18	−57 20.3	1	7.31		1.13		0.95		2	K1 II	536
		Lod 1010 # 54			13 40 19	−60 09.3	1	12.26		0.47		0.26		1		549

HD	DM	Other Id	N Rem	α_{1950}	δ_{1950}	S	V	σ_V	B–V	σ_{B-V}	U–B	σ_{U-B}	n	Spectrum	References
119258	−59 05144			13 40 20	−60 08.3	2	10.49	.000	0.18	.000	0.05	.000	2	A0/1 V	287,549
119445	+42 02431	HR 5160		13 40 21	+41 55.5	1	6.30		0.86				2	G6 III	70
	−59 05145			13 40 22	−59 54.7	1	10.90		0.67		0.16		1	F8	549
		CS 22889 # 35		13 40 26	−09 20.0	1	14.95		-0.03				1		1736
	+44 02296			13 40 27	+44 09.7	1	10.41		0.99		0.84		1		280
		G 165 - 25		13 40 29	+33 32.9	1	11.97		1.64				1	M7	1746
119338	−37 08847			13 40 29	−38 00.3	2	8.94	.005	-0.09	.020	-0.53	.011	4	B5 V	26,55
	+16 02541			13 40 30	+15 44.9	1	9.76		1.08		1.20		1	K2	280
119446	+34 02444			13 40 30	+34 05.8	1	7.86		1.21		1.18		1	K0	280
119458	+35 02474	HR 5161		13 40 30	+35 14.4	2	5.99	.010	0.86	.003	0.53		5	G5 III	70,1501
119425	+4 02775	HR 5159	★ AB	13 40 33	+03 47.4	6	5.35	.011	1.12	.008	1.04	.007	15	K1 III	15,542,1003,1080,1417,3077
		Lod 1010 # 35		13 40 33	−60 24.1	1	11.43	-	0.65		0.12		1	F6	549
	+35 02475			13 40 34	+35 21.6	1	9.11		1.30		1.50		2	K3 III	1569
119283	−58 05092	IDS13373S5844	A	13 40 35	−58 51.1	1	6.52		0.01		-0.22		2	B8 V	1771
119285	−60 04913	V851 Cen		13 40 35	−61 06.9	2	7.82	.005	1.07	.011	0.82	.006	26	K1 Vp	1621,1641
	+40 02685			13 40 36	+39 49.8	2	10.00	.005	0.34	.005	0.16	.000	3		280,1569
		EG 103		13 40 36	+57 15.	1	16.73		0.55		-0.27		1	DA	3028
119360	−37 08850	IDS13377S3709	AB	13 40 36	−37 25.2	1	9.65		0.11		0.02		1	A3/5 V + A	1771
119329	−50 07991			13 40 36	−51 22.4	1	8.25		0.74				4	G5 IV	2033
119359	−33 09275			13 40 38	−33 59.2	1	9.29		0.51		-0.02		4	F5 V	244
119361	−41 08089	HR 5157	★ AB	13 40 40	−41 49.0	4	5.97	.011	-0.08	.005	-0.33	.016	18	B8 III	1637,1770,2006,2014
119477	+28 02253			13 40 41	+27 42.3	1	8.78		1.22		1.26		1	K0	280
		WLS 1340 15 # 5		13 40 43	+12 51.6	1	13.89		0.75		0.22		2		1375
		LP 438 - 44		13 40 43	+14 25.1	1	11.01		0.59		0.07		2		1696
		G 150 - 31		13 40 43	+28 32.2	1	11.96		1.20		1.18		1		1658
119496	+30 02439			13 40 46	+29 55.8	1	9.13		0.06		0.04		3	A2 V	1068
	−59 05147			13 40 46	−60 03.8	1	11.51		0.27		0.19		1	A1	549
	+20 02864			13 40 51	+20 12.9	1	9.51		1.25		1.19		1	K0	280
119515	+32 02378	IDS13387N3231	A	13 40 51	+32 16.3	1	9.30		1.20		1.24		1	G8 III	280
	+41 02418			13 40 51	+40 52.4	1	9.30		0.91		0.55		2	K0	1569
	+42 02433			13 40 52	+41 35.7	1	10.33		1.06		0.93		1		280
119402	−41 08092			13 40 52	−42 19.9	1	9.12		0.27		0.17		1	A8 IV	1771
		AAS12,381 19 # A		13 40 53	+18 48.3	1	10.69		0.58		0.05		1		280
		CCS 2117		13 40 53	−70 21.	1	9.88		4.52				2	N	864
		Lod 1010 # 53		13 40 54	−60 13.7	1	12.09		0.51		0.07		1	F3	549
119461	−3 03522	IDS13383S0346	A	13 40 55	−04 01.4	1	7.10		1.29		1.31		3	K2 III	3024
119461	−3 03522	IDS13383S0346	B	13 40 55	−04 01.4	1	10.12		0.53		0.04		4		3024
119461	−3 03522	IDS13383S0346	C	13 40 55	−04 01.4	1	14.16		0.75		0.17		1		3024
	+23 02604			13 40 57	+22 40.3	1	11.04		0.29		0.01		1	A5	280
	−59 05148			13 40 57	−60 10.4	2	10.42	.000	0.09	.000	-0.14	.000	2	B9	287,549
	−59 05149			13 40 58	−60 08.6	2	11.12	.000	0.20	.000	-0.20	.000	2	B9	287,549
119534	+32 02379			13 40 59	+32 15.6	1	9.24		1.13		1.11		2	K2 V	280
119583	+50 02021	IDS13390N5032	A	13 40 59	+50 16.7	1	7.73		0.92		0.68		2	K0	1566
	−59 05151			13 41 02	−59 58.4	1	11.44		0.29		0.09		1	B9	549
	−59 05150			13 41 02	−60 11.7	1	11.30		0.39		0.20		1	F5	549
119503	−8 03624	IDS13386S0859	A	13 41 04	−09 13.4	1	9.61		0.98		0.80		2	K2 V	3072
119503	−8 03624	IDS13386S0859	B	13 41 04	−09 13.4	1	11.98		1.40		1.32		2	K5	3072
		Lod 1010 # 41		13 41 06	−59 57.4	1	11.57		1.24		0.91		1		549
	+30 02441			13 41 07	+29 45.4	1	9.48		1.17		1.28		1	K0	280
119430	−42 08685			13 41 07	−43 11.4	1	7.09		-0.02		-0.02		2	A0 V	1770
119419	−50 07998	HR 5158, V827 Cen		13 41 07	−50 45.7	3	6.46	.004	-0.14	.012	-0.29		8	Ap SiCr	15,1771,2027
119386	−59 05154			13 41 07	−59 59.7	2	10.35	.000	0.28	.000	0.22	.000	2	A2	287,549
119387	−59 05156			13 41 08	−60 06.1	2	10.47	.000	0.10	.000	-0.11	.000	2	B8/9 IV	287,549
	+12 02615			13 41 09	+12 03.9	1	9.87		1.14		1.22		1	K0	280
119550	+15 02614	G 63 - 51		13 41 10	+14 37.0	1	6.94		0.63		0.16		1	G2 V	333,1620
		AJ74,1000 M3 # 15		13 41 10	+29 05.0	1	11.40		0.56				3		209
	+14 02654			13 41 12	+14 14.5	1	9.70		1.05		0.73		1		280
	+14 02653			13 41 12	+14 25.5	1	8.95		1.42		1.67		1	K2	280
		AJ74,1000 M3 # 8		13 41 13	+28 13.9	1	10.73		0.86				3		209
	+39 02675			13 41 13	+39 30.1	1	9.28		1.10		0.98		2	K8	3072
	+14 02655			13 41 14	+13 48.3	1	9.30		1.08		0.89		1	K0	280
	+36 02404			13 41 15	+36 05.8	1	9.97		1.05		0.95		1	K1 III	280
	−59 05158			13 41 15	−60 01.7	1	11.44		0.51		0.08		1	F5	549
		Lod 1010 # 58		13 41 16	−59 57.3	1	12.68		0.35		0.22		1	A7	549
119537	−4 03540	HR 5163		13 41 18	−05 14.9	2	6.50	.005	0.05	.000	0.01	.000	7	A1 V	15,1417
	−59 05159			13 41 20	−60 19.9	1	10.70		0.38		0.31		1	A5	549
	−60 04924			13 41 20	−60 26.4	1	10.50		0.13		-0.06		1	B8	549
	+43 02348			13 41 21	+42 44.5	1	10.10		1.04		0.92		1	K0	280
119617	+36 02405			13 41 22	+35 35.8	1	8.57		1.10		1.05		1	K2 III	280
		Lod 1010 # 55		13 41 22	−60 05.5	1	12.38		0.27		0.16		1		549
		Lod 1010 # 52		13 41 22	−60 06.3	1	12.08		0.17		0.06		1		549
	+6 02784	IDS13389N0559	AB	13 41 23	+05 43.2	1	10.85		0.20		0.04		1	A4	1026
119584	+23 02606	HR 5164		13 41 23	+22 57.1	2	6.13	.005	1.43	.016	1.70		4	K4 III	70,3016
119585	+18 02773			13 41 24	+17 39.5	2	7.88	.005	1.02	.010	0.82	.005	4	K0	1569,1648
	+21 02566			13 41 24	+20 59.0	2	9.70	.019	0.19	.006	0.19	.000	7	A2	193,272
		Ton 730		13 41 24	+25 55.	1	15.91		0.57		-0.84		3		1036
119616	+36 02406			13 41 26	+35 58.0	1	7.96		0.98		0.82		1	G5	280
	+14 02656			13 41 27	+13 50.3	1	9.48		0.99		0.73		1	G5	280
	+40 02686			13 41 28	+40 02.1	1	10.42		1.15		1.20		1		280
	+33 02366			13 41 34	+33 15.9	1	8.89		1.68		2.01		2	M0	1569

Table 1   767

HD	DM	Other Id	N Rem	α_{1950}	δ_{1950}	S	V	σ_V	B–V	σ_{B-V}	U–B	σ_{U-B}	n	Spectrum	References
119586	−7 03685			13 41 35	−08 14.2	1	9.17		0.87		0.51		1	K2 V	3072
119618	+22 02624			13 41 37	+22 04.1	1	8.27		1.65		1.83		1	K5	280
		Lod 1010 # 59		13 41 38	−60 00.1	1	12.75		0.31		0.21		1	A7	549
		G 62 - 66		13 41 40	+05 07.8	1	11.45		1.31		1.31		2	K4-5	265
	+29 02453			13 41 40	+28 54.5	1	11.58		0.96				3		209
		Lod 1010 # 39		13 41 40	−60 02.6	1	11.51		1.10		1.41		1		549
119525	−48 08447			13 41 41	−48 29.2	1	10.24		0.29		0.10		1	A9/F0 V	1771
		Lod 1010 # 56		13 41 41	−60 04.3	1	12.58		0.37		0.21		1	A7	549
	+15 02615			13 41 42	+15 13.1	1	9.94		0.45		-0.05		1	F5 III	1722
		L 258 - 146		13 41 42	−53 52.	2	12.49	.025	1.55	.031	1.38		2		1705,3062
	+25 02672			13 41 43	+25 02.1	1	10.28		0.31				4	F2 IV	193
119488	−59 05160			13 41 45	−60 20.6	2	9.98	.000	0.21	.000	0.12	.000	2	B9/A2	287,549
119487	−59 05161			13 41 46	−59 58.6	2	10.12	.000	0.14	.000	0.09	.000	2	B8/9 V	287,549
119605	−15 03731	HR 5165		13 41 48	−15 55.7	8	5.59	.013	0.81	.005	0.41	.010	27	G0 Ib-IIa	15,1007,1013,2006*
119608	−17 03918			13 41 48	−17 41.2	3	7.54	.025	-0.07	.015	-0.82	.015	10	B1 Ib	399,400,8100
	−61 03968			13 41 48	−61 44.5	1	11.48		0.16		-0.16		1		287
		CS 22889 # 44		13 41 49	−08 18.4	1	15.27		0.18				1		1736
119490	−61 03970	IDS13384S6125	AB	13 41 49	−61 39.6	1	8.26		0.04				4	B6/7 III	2014
119665	+26 02488			13 41 50	+25 32.1	1	8.80		0.52		-0.02		3	F6 V	3026
		Lod 1010 # 49		13 41 50	−59 54.3	1	11.77		0.40		0.18		1	F0	549
119514	−59 05162			13 41 51	−59 58.0	2	10.54	.000	0.25	.000	0.18	.000	2	A1/2 V	287,549
119686	+28 02254			13 41 52	+27 50.2	1	7.12		0.26		0.09		1	A9 IV	280
119704	+41 02420			13 41 52	+41 30.0	1	8.35		1.52		1.91		1	K2	280
	−61 03974			13 41 52	−61 46.5	1	11.26		0.23		-0.10		1		287
	−61 03973	LSS 3153		13 41 52	−62 10.5	1	10.70		0.70		-0.38		3	O7 III	1737
		AAS61,331 # 9		13 41 54	+20 44.	1	10.69		0.28		0.14		2		1569
119573	−44 08854			13 41 54	−44 58.3	1	8.17		0.18		0.06		1	A5/7 IV/V	1771
119666	+22 02625			13 41 55	+22 07.2	1	8.93		1.01		0.84		1	K2	280
120103	+80 00421			13 41 55	+80 27.3	1	7.07		1.56		1.94		2	K5	1502
		Lod 1010 # 31		13 41 55	−59 56.9	1	11.30		1.16		0.77		1	K2	549
	+13 02703			13 41 56	+13 28.0	1	9.98		1.09		0.95		1		280
		G 177 - 57		13 41 56	+54 05.9	1	9.91		1.28		1.25		2		7010
119765	+52 01733	HR 5169		13 41 58	+52 18.9	2	6.02	.000	0.01	.004	-0.01		5	A1 V	70,3016
119623	−24 11057	HR 5166		13 41 58	−25 15.0	1	6.21		1.38				4	K3 III	2007
		AAS12,381 43 # B		13 42 01	+42 59.6	1	10.19		1.02		0.88		1		280
		G 62 - 67		13 42 01	−02 53.5	1	13.17		0.55		-0.21		1		1620
	−59 05164			13 42 01	−60 22.5	1	11.24		0.58		-0.02		1	B6	549
		Lod 1010 # 37		13 42 02	−59 53.7	1	11.44		0.70		0.10		1	K0	549
		AAS61,331 # 2		13 42 03	+11 55.	1	10.35		0.36		-0.04		2		1569
119638	−13 03761			13 42 03	−13 58.3	2	6.91	.001	0.54	.001			32	G2 V	130,1351
		Lod 1010 # 45		13 42 03	−60 21.5	1	11.67		0.54		-0.15		1	B5	549
	+37 02447			13 42 04	+37 07.9	1	10.83		0.23		-0.01		1		1569
		AAS12,381 13 # C		13 42 07	+12 49.5	1	10.26		0.10		0.16		1		280
		AAS61,331 # 15		13 42 07	+30 42.	1	10.25		0.20		0.03		2		1569
119558	−59 05165			13 42 07	−59 55.7	2	10.86	.000	0.29	.000	0.15	.000	2	A0	287,549
	+37 02448			13 42 08	+36 34.3	1	9.17		1.06		0.92		2	K0	1569
		Lod 1010 # 34		13 42 08	−59 57.0	1	11.37		1.19		1.05		1	M0	549
		Lod 1010 # 51		13 42 09	−60 13.4	1	11.90		0.31		0.23		1	B9	549
	+15 02616			13 42 10	+15 19.0	1	11.15		0.92		0.45		1	G5 III	1722
	−59 05166			13 42 10	−60 10.5	1	11.34		0.19		0.06		1	B8	549
	−61 03981			13 42 11	−61 47.4	1	11.50		0.12		0.01		1		287
	−58 05131			13 42 12	−58 24.0	1	9.59		1.30		1.04		3		803
119731	+24 02629	IDS13399N2410	AB	13 42 13	+23 55.1	2	8.96	.005	0.32	.005	0.07	.005	3	A5	280,1569
119748	+29 02454			13 42 13	+29 13.6	2	8.32	.023	1.03	.005	0.86		4	K1 III	209,280
		AJ74,1000 M3 # 9		13 42 14	+28 15.1	1	11.25		0.64				3		209
	+31 02533			13 42 14	+30 54.0	1	10.18		0.69		0.31		2	G2 IV	1569
		Lod 1010 # 22		13 42 16	−59 47.4	1	10.93		1.59		1.80		1	K4	549
119578	−59 05167			13 42 16	−59 55.5	2	10.53	.000	0.18	.000	0.19	.000	2	B9 IV/V	287,549
	+13 02705			13 42 18	+12 36.4	1	10.26		1.11		1.01		1	K0	280
119547	−63 02942	LSS 3154		13 42 18	−63 42.5	2	8.28	.027	-0.03	.006	-0.89	.012	4	B1 III/Vn	540,976
119643	−41 08119			13 42 19	−42 09.4	1	9.33		0.25		0.06		1	A7/8 III/I	1771
119644	−44 08861			13 42 19	−45 05.9	2	8.11	.009	-0.09	.003	-0.53	.005	4	B3/5 III	26,55
	−60 04938			13 42 19	−60 23.6	1	11.62		0.03		0.02		1	B8	549
	+37 02449			13 42 20	+36 50.7	1	9.59		1.10		1.01		2	G5	1569
		Lod 1010 # 44		13 42 20	−60 19.0	1	11.65		0.45		0.30		1	A8	549
	+15 02617			13 42 21	+15 11.1	1	9.57		0.51		0.04		1	F8 III	1722
119822	+54 01620			13 42 21	+53 54.3	1	8.99		0.44		0.06		2	F8	1566
119629	−48 08458			13 42 21	−48 32.5	2	6.76	.000	0.54	.000			8	F7 V	1075,2033
119632	−48 08459	IDS13393S4840	A	13 42 23	−48 54.7	1	8.97		0.02		-0.20		2	B9 IV/V	1770
	+15 02618			13 42 24	+14 41.4	1	9.56		1.10		1.04		1		280
120084	+78 00466	HR 5184		13 42 25	+78 18.9	1	5.91		1.01				2	K0	71
		Lod 1010 # 24		13 42 25	−60 04.4	1	10.98		1.16		0.98		1	K3	549
	−50 08008	NGC 5286 sq2 4		13 42 26	−51 06.0	1	11.03		0.43				3		485
	+34 02447			13 42 27	+33 34.7	1	9.27		1.26		1.46		1	K5 III	280
		Steph 1091		13 42 29	+02 20.5	1	10.90		1.39		1.17		2	K5	1746
119768	+21 02570			13 42 29	+20 43.4	1	8.18		1.53		1.81		1	K5	280
	−52 06731			13 42 29	−52 44.8	1	9.73		0.86				3	G3	955
	−3 03527			13 42 30	−04 22.1	3	10.53	.028	1.36	.003	1.24	.005	6	M2	158,679,1705
	+26 02490			13 42 31	+25 47.1	1	10.01		1.03		0.98		1		280
	−62 03544			13 42 31	−63 06.0	1	10.65		0.49		-0.46		2		94

HD	DM	Other Id	N	Rem	α_{1950}	δ_{1950}	S	V	σ_V	B–V	σ_{B-V}	U–B	σ_{U-B}	n	Spectrum	References
		AJ74,1000 M3 # 10			13 42 33	+28 16.2	1	11.47		1.00				3		209
119634	−59 05168				13 42 33	−59 53.8	1	10.98		0.22		0.10		1	B9/A0 V	549
	+27 02279				13 42 34	+27 26.8	1	10.02		1.18		1.16		1		280
119658	−48 08461				13 42 34	−48 50.0	1	9.19		0.44		−0.02		2	F5 V	1770
		Lod 1010 # 47			13 42 34	−59 52.5	1	11.70		0.64		0.23		1		549
119750	−2 03727	IDS13400S0249	A		13 42 35	−03 03.9	1	8.25		0.28		0.04		2	A5	3016
119750	−2 03727	IDS13400S0249	B		13 42 35	−03 03.9	1	11.00		0.75		0.28		2		3032
	−59 05169				13 42 35	−59 53.7	1	11.27		0.23		0.16		1	B8	549
	−59 05170				13 42 37	−59 46.6	1	8.96		1.40				1	K3	549
		Lod 1010 # 46			13 42 38	−60 03.3	1	11.67		0.74		0.24		1		549
		Lod 1010 # 60			13 42 38	−60 03.3	1	12.98		0.48		0.18		1		549
	+18 02776	IDS13402N1821	A		13 42 39	+18 03.7	4	9.79	.045	1.42	.012	1.06	.115	9	M1	1003,1197,3078,8105
	−59 05171	IDS13393S5933	AB		13 42 39	−59 48.1	1	10.30		0.23		0.14		1	A2	549
		MCC 52			13 42 42	+02 32.9	1	10.84		1.40				1	M0	1017
		WLS 1340 35 # 9			13 42 42	+35 04.5	1	14.40		0.57		0.03		2		1375
		Lod 1010 # 48			13 42 42	−60 00.4	1	11.72		0.47		0.20		1	A3	549
		NGC 5286 sq1 5			13 42 45	−50 59.5	1	13.62		0.61		0.02		3		371
119659	−59 05173				13 42 46	−59 53.7	1	9.31		0.47		0.23		1	A8 III/IV	1771
		Lod 1010 # 43			13 42 46	−59 57.0	1	11.64		0.43		0.20		1		549
	−62 03555	NGC 5281 - 17			13 42 47	−62 40.1	1	10.98		0.16		−0.32		2		251
	−62 03554	NGC 5281 - 18			13 42 47	−62 41.2	1	11.10		0.08		−0.35		2		251
		G 150 - 34			13 42 48	+25 16.2	1	13.64		1.51		1.12		4		203
119646	−61 03987	LSS 3156			13 42 48	−62 12.1	5	6.60	.020	0.12	.007	−0.69	.022	14	B1 Ib/II	92,403,540,976,8100
	+16 02548				13 42 49	+16 20.4	1	9.23		1.45		1.72		1	K5	280
	+28 02256				13 42 49	+28 18.7	2	9.88	.019	1.04	.014	0.83		4	K0	209,280
	+28 02257				13 42 49	+28 26.3	2	9.43	.019	1.30	.005	1.44		4	K2	209,280
		RZ CVn			13 42 49	+32 54.2	1	10.88		0.12		0.21		1	A0	597
119752	−25 09972	HR 5167			13 42 49	−25 51.9	3	5.80	.004	0.02	.000			11	A0 V	15,2018,2028
		AO 970			13 42 50	+14 55.3	1	11.71		0.89		0.68		1		1722
119756	−32 09603	HR 5168			13 42 50	−32 47.5	6	4.23	.005	0.38	.003	0.00	.007	25	F3 V	15,244,1075,2012,3026,8015
	−62 03557	NGC 5281 - 10			13 42 52	−62 39.4	1	10.58		0.09		−0.33		2	B8 V	251
		AO 971			13 42 53	+14 51.6	1	11.72		1.06		0.67		1		1722
119786	−15 03735	HR 5170	★ A		13 42 53	−15 31.0	1	6.19		0.05				4	A0 V	2006
		NGC 5281 - 9			13 42 53	−62 39.6	1	10.89		0.13		−0.29		2	B8 V	251
	+37 02450				13 42 54	+36 53.3	1	9.64		1.06		0.97		2	K1 III	1569
	−50 08013	NGC 5286 sq1 1			13 42 55	−51 04.3	2	11.07	.020	0.09	.000	0.04		6	A2	371,485
		NGC 5286 sq1 4			13 42 55	−51 05.7	2	13.47	.005	0.80	.005	0.29		4		371,485
119726	−51 07724				13 42 55	−52 15.8	1	7.59		0.22		0.12		1	A5/6 V	1771
	−62 03558	NGC 5281 - 8			13 42 55	−62 39.6	1	10.04		0.08		−0.39		2	B7 V	251
	−62 03559	NGC 5281 - 11			13 42 55	−62 40.9	2	8.39	.005	1.67	.003	1.84	.020	4	K2 II/III	251,1685
		NGC 5286 sq1 6			13 42 56	−51 10.6	1	14.59		1.55		1.21		2		371
		NGC 5281 - 15			13 42 56	−62 38.9	1	12.73		0.32		0.28		2		251
	−62 03560	NGC 5281 - 7			13 42 56	−62 40.0	1	11.06		0.51		0.03		2		251
119805	−11 03587				13 42 57	−12 08.1	1	7.99		1.18		1.18		3	K2	1657
	+27 02283				13 42 59	+27 22.0	1	9.77		1.60		1.88		1		280
119727	−54 05725				13 42 59	−54 26.0	1	6.43		0.10		0.05		2	A1 V	1771
	−62 03561	NGC 5281 - 6			13 42 59	−62 40.6	1	10.43		0.07		−0.38		2	B4 V	251
		NGC 5281 - 5			13 43 00	−62 40.9	1	11.44		0.14		−0.25		2		251
	+22 02626				13 43 01	+22 02.2	1	9.02		0.95		0.50		1	K0	280
		NGC 5281 - 16			13 43 01	−62 38.8	1	13.29		0.27		0.23		2		251
119682	−62 03565	NGC 5281 - 3			13 43 01	−62 40.4	3	8.00	.027	0.18	.036	−0.88	.032	7	O9e	251,403,1737
		NGC 5281 - 4			13 43 01	−62 41.5	1	11.29		0.07		−0.40		2	B8 V	251
119757	−42 08713				13 43 02	−43 00.1	1	9.53		0.18		0.12		1	A1/2 IV	1770
	−62 03567	NGC 5281 - 14			13 43 03	−62 38.4	1	10.21		0.05		−0.46		2		251
	−62 03568	NGC 5281 - 2			13 43 03	−62 40.1	1	8.80		0.08		−0.33		3	B9 III	251
119875	+32 02385				13 43 04	+31 44.5	1	8.75		1.14		1.17		1	K2 III	280
	+12 02623				13 43 06	+11 40.9	1	9.46		0.25		0.08		3	A0	272
119699	−62 03570	NGC 5281 - 1			13 43 06	−62 39.6	1	6.61		0.18		0.20		7	A2 Ib	251
	−62 03569	NGC 5281 - 12			13 43 06	−62 40.7	1	11.28		0.12		−0.33		2		251
	+23 02609				13 43 07	+22 58.8	1	10.32		1.15		1.21		1		280
119683	−66 02314				13 43 07	−66 30.3	1	9.44		0.33		0.22		1	A7 IV	1771
	+33 02370				13 43 08	+33 06.2	1	10.05		1.01		0.88		2	G8 III	1569
		NGC 5286 sq1 2			13 43 08	−51 11.5	1	12.52		0.71		0.15		3		371
		OP 471 # 3			13 43 09	+45 16.5	1	15.45		0.64		−0.01		1		930
		NGC 5286 sq1 3			13 43 09	−51 11.8	1	13.34		0.70		0.25		2		371
	+18 02777				13 43 11	+17 59.5	1	9.37		1.43		1.75		1	K5	280
		GD 327			13 43 11	+45 18.1	1	14.03		0.57		−0.11		1		308
	−33 09314				13 43 11	−33 41.5	3	9.97	.013	0.79	.005	0.20	.005	7	Gp	158,1594,6006
119780	−45 08687				13 43 11	−46 06.9	1	7.56		0.09		0.09		1	A1 V	1771
119728	−59 05175				13 43 11	−60 13.0	1	9.79		0.24		0.17		1	A0 V	549
		NGC 5281 - 13			13 43 11	−62 39.4	1	11.31		1.14		0.79		2		251
119865	+12 02624				13 43 12	+12 30.7	1	8.60		1.50		1.86		1	K5	280
119850	+15 02620				13 43 12	+15 09.7	6	8.46	.014	1.45	.016	1.10	.020	21	M3 V	22,1003,1197,1722,3078,8006
119876	+16 02551				13 43 14	+16 12.5	1	8.19		0.91		0.62		1	K0	280
	+17 02663				13 43 15	+16 40.1	1	9.52		1.61		1.91		1	M0	280
		OP 471 # 2			13 43 15	+45 17.4	1	14.28		0.66		0.23		1		930
119853	−11 03591	HR 5173	★ AB		13 43 16	−12 10.6	1	5.51		0.90				4	G8 III	2035
119869	−7 03694				13 43 17	−07 32.6	1	8.90		0.46				8	F5	1351
	−62 03574	LSS 3159			13 43 17	−62 29.0	2	10.26	.020	0.53	.015	−0.44	.025	5	B1 Ib	403,1737
119914	+35 02480				13 43 21	+34 53.9	1	7.48		0.96		0.69		2	G5	1569

Table 1   769

HD	DM	Other Id	N Rem	α_{1950}	δ_{1950}	S	V	σ_V	B–V	σ_{B-V}	U–B	σ_{U-B}	n	Spectrum	References
119762	−59 05176			13 43 21	−60 18.5	1	9.06		1.09		0.76		1	K0/1 II/III	549
119992	+56 01683	HR 5177		13 43 22	+56 08.1	2	6.50	.020	0.47	.010	-0.08	.005	4	F7 V	254,3026
119915	+31 02536			13 43 26	+30 51.5	1	8.58		1.38		1.55		2	K3 V	1569
		CS 22889 # 52		13 43 29	−10 11.8	1	13.73		-0.27		-0.95		1		1736
119834	−50 08017	HR 5172	⋆ A	13 43 29	−51 11.0	4	4.64	.004	0.96	.008	0.73	.005	17	G8/K0 III	15,485,1075,2012
		AO 973		13 43 30	+14 58.6	1	11.96		0.66		0.18		1		1722
		Ton 736		13 43 30	+29 20.	1	14.61		0.17		0.21		1		313
119848	−49 08188			13 43 32	−49 48.5	1	9.05		0.13		0.12		1	A1 V	1770
119944	+27 02285			13 43 34	+27 28.6	1	8.00		1.24		1.13		1	K2 III	280
119700	−70 01670			13 43 34	−70 58.9	1	6.85		0.00		-0.19		2	B9 V	1771
		AO 974		13 43 35	+14 53.4	1	11.23		0.65		0.17		1		1722
		NGC 5286 sq2 1		13 43 35	−51 11.7	1	10.76		0.24				3		485
119931	+5 02794	HT Vir	⋆ AB	13 43 36	+05 21.9	1	7.16		0.55				4	G0	176
119815	−60 04961			13 43 39	−60 26.9	2	8.84	.011	0.04	.003	-0.62	.003	7	B5 V	540,976
119796	−61 04003	HR 5171, V766 Cen	⋆ AB	13 43 40	−62 20.4	5	6.60	.072	2.12	.098	1.16		14	G8 Ia/0	15,164,403,2006,8036
120565	+83 00397	HR 5203		13 43 41	+83 00.2	1	5.98		1.01				2	G9 III	71
	+15 02622			13 43 42	+14 33.8	1	9.43		0.95		0.58		1	K0	280
	+15 02623			13 43 42	+15 18.5	1	10.70		0.71		0.25		1		1722
119932	+0 03098	G 64 - 23		13 43 47	−00 12.5	1	9.36		0.93		0.75		1	K0	1620
	+15 02624			13 43 48	+15 10.	1	9.91		0.65		0.12		1	G3 III	1722
120004	+38 02477			13 43 50	+37 46.5	1	7.59		0.55		0.11		2	F8	1569
		OP 471 # 1		13 43 50	+45 15.2	1	12.54		0.84		0.65		1		930
119857	−52 06740			13 43 50	−53 02.0	1	7.23		1.17		1.18		4	K1 IV	1673
120022	+41 02423	IDS13418N4132	AB	13 43 55	+41 17.3	1	7.56		0.98		0.74		1	K0	1569
	−50 08025	NGC 5286 sq2 2		13 43 55	−51 18.7	1	10.98		0.30				3		485
	+27 02287			13 43 57	+26 49.6	1	10.15		1.19		1.39		1	K0	280
		OP 471 # 4		13 43 58	+45 17.5	1	13.96		0.65		0.24		1		930
120005	+31 02540			13 43 59	+31 08.9	1	6.53		0.47		0.00		3	F8 V	3016
119884	−51 07732			13 43 59	−52 01.2	3	6.94	.011	0.04	.021	-0.04	.020	7	A0 V	401,1771,3039
119898	−51 07733			13 44 00	−52 14.2	1	8.58		0.23		0.14		1	A5 V	1771
120006	+31 02539			13 44 01	+30 42.6	1	8.33		1.42		1.72		2	K0	1569
		Steph 1094		13 44 01	−03 52.7	1	11.65		1.34		1.16		1	K7	1746
119921	−35 08995	HR 5174	⋆ A	13 44 01	−36 00.1	4	5.14	.009	-0.02	.002	-0.05		13	A0 V	15,1770,2013,2029
119896	−47 08606			13 44 01	−47 46.7	3	8.21	.017	0.37	.010	0.00	.005	6	A9/F0 V	1594,1771,6006
	+14 02660			13 44 02	+14 28.6	1	9.53		1.43		1.76		1	M0	280
119980	+0 03099			13 44 04	−00 19.1	1	10.90		0.76		0.30		1	K3 V	3072
	−52 06741	AM Cen		13 44 04	−53 06.5	2	8.41	.295	2.93	.025	3.71		2	SC	864,8051
	+40 02689			13 44 06	+40 20.3	1	9.73		1.12		1.19		1	K5	280
120047	+41 02424	HR 5179		13 44 06	+41 20.3	1	5.87		0.21		0.06		2	A5 V	252
	+77 00521			13 44 06	+77 28.9	2	9.45	.000	0.66	.000	0.08	.000	4	G5	516,1064
119862	−62 03593			13 44 06	−62 54.8	1	9.95		0.29		0.18		1	A5/8	1771
		AO 977		13 44 09	+15 25.9	1	9.89		0.48		-0.06		1		1722
120048	+39 02678	HR 5180		13 44 09	+38 45.2	1	5.94		0.94				2	G9 III	70
	+35 02485			13 44 16	+35 28.9	1	9.42		1.23		1.43		2	K2 III	1569
	+36 02408			13 44 16	+35 48.6	1	9.14		0.97		0.75		1	K0	280
120049	+28 02259			13 44 17	+28 11.5	1	8.34		0.26		0.09		2	Am	1068
		NGC 5286 sq2 5		13 44 17	−51 06.4	1	12.18		0.51				2		485
119938	−49 08194	HR 5175		13 44 18	−50 00.0	3	5.91	.004	0.29	.008			11	A3mA5-F0	15,1075,2009
119954	−44 08888			13 44 19	−44 30.4	1	9.45		0.29		0.11		1	A2mA7-A7	1770
	+17 02669			13 44 21	+17 07.6	1	10.01		1.04		0.96		1		280
		NGC 5286 sq2 6		13 44 21	−51 04.5	1	12.41		0.56				2		485
120064	+26 02494	HR 5182		13 44 24	+25 57.1	3	5.98	.014	0.51	.007	0.11	.010	8	G5 III+A7V:	70,833,1501
		LP 558 - 40		13 44 27	+05 58.6	1	15.17		1.59				1		1759
120066	+7 02690	HR 5183	⋆ A	13 44 28	+06 36.1	5	6.33	.012	0.62	.011	0.15	.013	19	G0 V	15,22,1417,1620,3077
120066	+7 02690	IDS13420N0651	B	13 44 28	+06 36.1	1	10.03		1.20		1.16		3		3032
120025	−18 03681			13 44 29	−19 00.4	1	6.74		0.32		0.17		3	A2/3mA7-F3	355
119971	−49 08198	HR 5176		13 44 30	−50 04.3	4	5.46	.010	1.35	.010	1.35		13	K2 III	15,1075,2006,3077
119970	−49 08199			13 44 31	−49 52.5	1	7.28		0.10		0.13		1	A1/2 V	1771
	−33 09330			13 44 34	−34 14.7	1	10.47		1.01		0.85		1	K0	861
120033	−8 03639	HR 5178	⋆ AB	13 44 35	−09 27.6	3	6.04	.005	1.42	.000	1.66	.000	8	K5 III	15,1417,3002
119911	−62 03600	LSS 3161		13 44 35	−63 12.0	2	8.19	.032	0.06	.003	-0.78	.011	4	B1 Ib/II	540,976
119985	−45 08701			13 44 36	−45 48.9	2	8.55	.014	0.63	.005			8	G0 V	1075,2033
120015	−40 08161			13 44 38	−41 01.6	1	9.22		0.16		0.12		1	A2/3 V	1770
	+20 02870			13 44 39	+19 42.3	2	10.19	.015	0.08	.000	0.11	.005	3	A2	272,280
119974	−55 05733			13 44 40	−55 42.7	1	7.20		0.01		-0.31		3	B5/7 III	1770
120052	−17 03932	HR 5181		13 44 42	−17 36.6	6	5.44	.015	1.62	.009	1.96	.026	20	M2 III	15,31,1003,1024,2007,3055
120198	+55 01634	HR 5187, CR UMa	⋆ A	13 44 44	+54 40.9	4	5.68	.021	-0.06	.012	-0.11	.026	12	B9 p EuCr	70,752,1063,3033
120086	−1 02858			13 44 44	−02 11.7	5	7.87	.008	-0.19	.004	-0.80	.004	16	B2 V	55,1509,1531,1732,2033
120000	−48 08487	IDS13416S4840	A	13 44 44	−48 54.9	1	8.09		0.05		-0.01		2	A0 V	1771
120036	−31 10649	IDS13419S3156	A	13 44 45	−32 11.1	1	8.94		1.32		1.23		2	K5/M1 V	3072
120036	−31 10649	IDS13419S3156	B	13 44 45	−32 11.1	1	9.12		1.36		1.24		2	K5	3072
120017	−46 08860			13 44 46	−47 09.3	1	8.75		0.11		0.08		1	A0/1 V	1770
	−5 03763			13 44 48	−05 53.0	1	10.28		1.02		0.86		2	K5 V	1017,3072
120164	+39 02680	HR 5186	⋆ A	13 44 50	+38 47.5	2	5.48	.015	1.06	.024	0.84		5	K0 III	70,3024
120164	+39 02679	IDS13427N3903	B	13 44 50	+38 47.5	1	8.98		0.58		0.01		2	F8	3024
	+41 02426			13 44 51	+41 12.0	1	8.69		1.31		1.48		2	K5	280
120053	−29 10599			13 44 51	−30 04.5	1	10.36		0.48		-0.01		3	F5 V	1700
120136	+18 02782	HR 5185	⋆ AB	13 44 53	+17 42.3	11	4.50	.007	0.48	.006	0.05	.009	184	F7 V + F8 V	1,3,15,130,254,1077*
120119	+11 02601			13 44 54	+11 11.2	1	8.59		1.48		1.81		2	K5	1648
	+32 02387			13 44 57	+31 50.8	1	10.11		1.31		1.51		1		280

HD	DM	Other Id	N Rem	α_{1950}	δ_{1950}	S	V	σ_V	B–V	σ_{B-V}	U–B	σ_{U-B}	n	Spectrum	References
	+36 02410			13 44 57	+35 38.8	1	9.99		1.39		1.76		1	M0	280
		G 63 - 54		13 44 59	+10 36.6	3	15.09	.010	0.35	.047	-0.48	.013	6	DAwk	316,1620,3060
	+7 02692	G 65 - 2		13 45 01	+06 34.0	2	10.04	.000	1.21	.009	1.18	.000	4	M0 V	333,1620,3072
		CS 22889 # 53		13 45 02	-11 59.4	1	14.22		0.34				1		1736
		G 64 - 25		13 45 08	+03 45.3	1	13.97		0.93		0.50		1		333,1620
	+17 02670			13 45 08	+17 10.2	1	9.73		1.32		1.40		1	K2	280
	+43 02353			13 45 10	+42 49.1	1	9.55		1.09		1.01		1	K0	280
	+25 02679			13 45 11	+24 48.4	1	9.95		1.08		1.01		1	K2	280
120095	-41 08160			13 45 11	-41 56.2	1	7.97		0.03		-0.04		3	B9/A0 V	1770
120077	-46 08865			13 45 11	-47 14.0	1	8.47		0.16		0.12		1	A0 V	1770
120094	-40 08168			13 45 12	-40 45.0	1	9.38		0.12		0.09		1	A1 V	1770
		G 65 - 4		13 45 15	+06 54.1	1	14.48		1.54		1.29		1		333,1620
120042	-58 05185			13 45 15	-59 14.8	1	7.32		0.09		0.11		1	A1/2 V	1771
	+60 01492			13 45 16	+60 25.0	1	9.00		1.36		1.52		1	K2	1746
120184	+20 02872			13 45 18	+19 33.4	1	8.71		1.08		0.94		1	G5	280
	+22 02632			13 45 19	+21 41.2	1	11.58		1.51				1	M0	1017
120170	-8 03641			13 45 20	-08 32.4	1	9.04		0.93		0.67		2	K0	536
120245	+38 02479			13 45 21	+38 08.5	1	6.95		1.13				2	K0 III	280
120265	+43 02354			13 45 22	+43 02.6	1	8.09		0.31				2	A5	193
120186	-7 03704			13 45 24	-07 46.5	2	7.71	.001	0.55	.000			32	G0	130,1351
	+34 02452			13 45 31	+34 18.5	1	10.06		1.13		1.11		1	K0 III	280
120232	+19 02716			13 45 34	+19 11.6	1	7.69		1.66		1.94		1	K5	280
		AAS12,381 38 # B		13 45 34	+37 43.2	1	10.46		1.02		0.88		1		280
120315	+50 02027	HR 5191		13 45 34	+49 33.7	11	1.86	.009	-0.19	.007	-0.67	.017	41	B3 V	1,15,154,667,1068*
	+39 02683			13 45 37	+39 08.5	1	9.09		1.08		0.98		1	K0	280
120020	-70 01677			13 45 37	-70 23.0	1	9.17		0.40		0.24		1	A6/7 V	1771
120247	+14 02664			13 45 38	+13 35.1	1	7.55		1.10		1.02		1	K0	280
		G 65 - 5		13 45 39	+06 56.9	1	14.80		1.47		1.13		1		333,1620
	+13 02710			13 45 39	+12 46.0	1	10.48		1.05		0.88		1		280
	+14 02663			13 45 39	+13 53.2	1	8.81		1.08		0.95		1	K0	280
120278	+32 02389			13 45 39	+32 22.4	1	9.03		1.21		1.26		1	K2 III	280
		G 65 - 6		13 45 41	+05 46.1	1	11.22		1.23		1.18		1	K5	333,1620
	+31 02544			13 45 42	+30 56.3	1	9.66		0.69		0.03		3	G0	1569
120175	-43 08554	XY Cen		13 45 42	-44 15.8	1	7.41		0.82		0.55		1	M5/6 III	1593
	+36 02411			13 45 45	+36 15.4	1	10.31		0.96		0.66		1	G6 V	280
120132	-60 04998			13 45 47	-60 30.2	1	7.54		0.12		0.05		2	B9 IV/V	1771
		LP 380 - 5		13 45 48	+23 49.6	2	15.64	.005	1.10	.005	0.47	.000	14		470,940
	+21 02574			13 45 50	+21 30.5	1	10.64		1.16		1.13		2		1569
120176	-45 08718			13 45 50	-46 18.5	1	10.34		0.19		0.12		1	A(3) (V)	1770
	+40 02690			13 45 52	+39 41.0	1	10.04		1.46		1.74		1		280
	+40 02691			13 45 54	+39 52.9	1	9.86		1.05		0.99		1	K2	280
		LP 380 - 6		13 45 58	+23 51.6	2	15.31	.019	1.96	.005			12		470,940
	+24 02640			13 45 59	+24 19.9	1	9.98		1.00		0.79		2	K0	1569
120348	+42 02440			13 45 59	+42 17.8	3	6.58	.007	1.09	.013	1.01	.014	6	K1 III	280,1003,1569
		IC 4329a sq1 10		13 46 00	-30 07.1	1	12.63		1.00		0.67		4		1687
120237	-35 09019	HR 5189	★A	13 46 02	-35 27.2	1	6.58		0.55		0.05		3	G0 V	1279
120237	-35 09019	HR 5189	★AB	13 46 02	-35 27.2	5	6.53	.007	0.56	.004	0.04	.000	19	G0 V	15,244,1075,2012,3077
120237	-35 09019	IDS13432S3512	B	13 46 02	-35 27.2	1	10.13		1.23		1.24		3		1279
120224	-43 08558			13 46 02	-43 40.9	1	9.88		0.11		0.11		2	A0/1 V	1770
	-71 01520			13 46 02	-71 44.6	1	9.89		0.80		0.38		2		1696
		Wolf 507		13 46 04	+17 57.3	1	12.80		0.72		0.07		2		1696
120334	+24 02641			13 46 04	+23 41.9	1	7.97		1.16		1.20		1	K0	280
120223	-43 08559			13 46 04	-43 29.1	1	8.97		0.97				4	G8 IV	2012
		G 63 - 55		13 46 05	+13 14.1	1	11.28		0.94		0.68		2		333,1620
120349	+32 02391			13 46 05	+31 39.0	1	7.54		0.28		0.08		1	A9 III	280
		IC 4329a sq1 1		13 46 07	-30 07.2	1	10.34		0.73		0.26		3		1687
	-60 05005			13 46 07	-61 12.4	1	11.00		0.46		-0.54		1		1737
		SS CVn		13 46 08	+40 09.0	1	11.52		0.11		0.16		1	A0	597
	+39 02685			13 46 09	+39 26.3	1	9.94		0.96		0.81		1	K0	280
120299	-8 03642			13 46 09	-09 18.7	1	9.35		0.34		0.01		2	A3	1776
120270	-30 10912			13 46 09	-30 27.8	1	7.66		1.30		1.46		4	K2/3 III	1700
120253	-40 08181			13 46 09	-40 37.9	2	8.78	.005	0.90	.012	0.38		6	G8 (III/IV)	2012,3008
	-59 05209	LSS 3166		13 46 13	-60 21.5	3	9.82	.001	0.78	.018	-0.32	.014	4	O9.5Ia	540,976,1737
		G 63 - 56		13 46 15	+15 45.4	1	13.47		0.99		0.51		1		1620
		G 65 - 8		13 46 16	+09 57.9	2	13.99	.015	0.95	.010	0.72	.005	3		333,1620
120350	+13 02711			13 46 16	+13 25.9	2	8.61	.049	0.26	.004	0.10		3	A9 IV	193,280
120364	+30 02449			13 46 16	+30 06.6	1	7.70		1.01		0.88		1	K0	280
	+24 02644			13 46 20	+24 18.3	1	9.32		1.37		1.65		1	K5	280
	+25 02682			13 46 20	+25 27.6	1	10.68		0.72		0.24		1		280
120256	-47 08637			13 46 21	-47 36.1	1	9.24		0.22		0.12		1	A0 V(m)	1770
120273	-43 08566	IDS13433S4343	A	13 46 22	-43 57.8	1	8.03		-0.01		-0.01		3	B9/A0 IV	1770
120420	+31 02547	HR 5195		13 46 23	+31 26.3	1	5.62		1.01				2	K0 III	70
	-64 02576			13 46 23	-64 32.6	1	11.50		0.33		0.28		2	A0	807
120435	+40 02693			13 46 25	+40 29.7	1	7.76		1.53		1.87		1	M1	280
120211	-61 04046			13 46 27	-62 01.6	2	7.92	.015	0.11	.006	-0.66	.006	5	B2 III	540,976
		G 64 - 27		13 46 28	+03 02.6	2	11.18	.005	1.57	.020	1.21		2	M2	333,1620,1746
120421	+28 02262			13 46 28	+28 07.8	1	7.15		1.07				3	K1 III	20
120322	-33 09355			13 46 28	-33 34.5	1	7.69		0.12		0.08		1	A2 V	1770
	+27 02295			13 46 29	+26 52.5	1	10.28		0.96		0.77		1		280
120307	-41 08171	HR 5190, ν Cen		13 46 30	-41 26.4	9	3.41	.009	-0.23	.008	-0.88	.015	45	B2 IV	15,26,681,815,1034*

Table 1

HD	DM	Other Id	N	Rem	α_{1950}	δ_{1950}	S	V	σ_V	B–V	σ_{B-V}	U–B	σ_{U-B}	n	Spectrum	References
	+7 02699				13 46 31	+07 03.3	1	10.08		0.63		0.00		2	G5	1696
120323	−33 09358	HR 5192, V806 Cen			13 46 32	−34 12.1	10	4.19	.016	1.50	.011	1.44	.014	37	M5 III	15,678,2032,3053,4001*
120197	−66 02338				13 46 34	−66 34.5	1	8.57		0.39		−0.03		1	F2 IV	6006
		CS 22889 # 60			13 46 35	−10 07.5	1	13.07		−0.01		−0.05		1		1736
120340	−38 08828				13 46 35	−38 24.6	1	8.93		0.12		0.09		1	A1 V	1770
		RX CVn			13 46 36	+41 38.0	1	12.19		0.17		0.10		1		597
120324	−41 08172	HR 5193, μ Cen		★ A	13 46 36	−42 13.5	8	3.35	.049	−0.17	.020	−0.89	.039	40	B2 IV/Vne	26,681,815,1212,1637*
120324	−41 08172	HR 5193, μ Cen		★ AB	13 46 36	−42 13.5	1	2.94		−0.16				3	B2 IV/Vn	15
120324	−41 08172	IDS13436S4159	B		13 46 36	−42 13.5	1	12.96		0.56		0.07		1		321
120463	+43 02357				13 46 37	+42 35.2	2	7.11	.005	1.10	.005	1.04		5	K1 III	280,1501
	+43 02358				13 46 38	+42 56.5	1	9.91		1.38		1.68		1	K2	280
		G 63 - 57			13 46 39	+14 34.9	1	16.68		0.20		−0.56		1		3060
		IC 4329a sq1 8			13 46 39	−30 08.7	1	11.71		1.27		1.24		3		1687
	+22 02635				13 46 41	+21 48.8	1	9.86		0.34		0.02		2	F0	1569
	−64 02578				13 46 41	−64 31.0	1	11.50		0.39		0.22		2	A4	807
120408	−1 02860	CE Vir			13 46 42	−01 40.9	2	9.60	.045	1.18	.205	1.44		2	K2	793,3001
	−48 08516				13 46 43	−49 21.1	1	11.42		0.65		−0.17		3		1649
120327	−50 08057	IDS13436S5032	A		13 46 43	−50 46.5	1	8.47		0.11		0.08		1	A1 V	1770
	+13 02712				13 46 47	+13 08.3	1	9.24		0.89		0.61		1	K0	280
	+14 02665				13 46 47	+13 54.4	1	10.26		1.05		0.88		1		280
120476	+27 02296	G 150 - 41		★ AB	13 46 47	+27 13.7	2	7.06	.010	1.12	.001	1.01	.019	5	K4 V	938,3072
120528	+53 01658				13 46 47	+53 30.6	4	8.57	.014	0.66	.007	0.22	.009	7	G5 V	1007,1013,3016,8023
		IC 4329a sq1 2			13 46 47	−30 03.2	1	11.10		1.58		1.83		5		1687
		AAS12,381 18 # B			13 46 49	+17 56.6	1	9.97		1.15		1.19		1		280
120499	+40 02694	HR 5199, R CVn			13 46 49	+39 47.4	1	8.55		1.22		0.41		1	M4	3001
		IC 4329a sq1 5			13 46 53	−30 04.9	1	14.07		0.57		−0.04		3		1687
120439	−7 03708				13 46 54	−08 00.3	1	9.38		0.37		−0.01		2	A3	1776
120083	−77 00914				13 46 54	−77 58.2	1	9.60		1.01		0.93		1	K3 V	3072
	+14 02666				13 46 56	+14 02.9	1	9.87		1.04		0.93		1	K0	280
		IC 4329a sq1 3			13 46 56	−30 01.7	1	13.00		0.65		0.14		3		1687
		IC 4329a sq1 11			13 46 57	−30 04.1	1	14.84		0.77		0.30		6		1687
120329	−56 05956				13 46 57	−57 00.6	2	8.34	.000	0.74	.000	0.37		5	G3/5 V	657,2012
		IC 4329a sq1 6			13 46 59	−30 04.2	1	14.57		0.63		−0.01		5		1687
	+26 02498				13 47 00	+26 07.8	1	9.79		0.86		0.39		1	G5	280
		IC 4329a sq1 4			13 47 00	−30 01.6	1	13.44		0.61		0.11		5		1687
	+18 02787				13 47 02	+18 07.6	1	8.45		1.26		1.33		1	K2	280
	+20 02875				13 47 02	+20 11.2	1	9.91		1.12		1.10		1	K2	280
120477	+16 02564	HR 5200			13 47 04	+16 02.7	7	4.05	.017	1.52	.012	1.89	.017	26	K5 III	3,15,31,1080,1363*
		IC 4329a sq1 7			13 47 04	−30 03.5	1	14.62		0.81		0.51		4		1687
	+25 02683				13 47 05	+24 32.2	1	10.02		1.05		0.88		1		280
120467	−21 03781				13 47 05	−21 51.4	4	8.16	.025	1.26	.017	1.24	.069	10	K4/5 V	22,2017,2033,3078
120529	+35 02488				13 47 07	+34 36.8	1	9.27		0.39		0.01		2	F8	1569
		IC 4329a sq1 9			13 47 08	−30 05.2	1	13.91		0.57		−0.02		3		1687
120452	−17 03937	HR 5196			13 47 09	−17 53.2	5	4.97	.010	1.06	.005	0.91	.017	14	K0 III	196,219,244,2007,3077
120361	−58 05207				13 47 11	−59 13.6	1	9.87		0.38		0.20		2	A7/F0	1771
	−64 02580				13 47 11	−64 41.6	1	11.15		0.05		−0.21		2	B7	807
	+27 02298				13 47 12	+26 41.0	1	9.59		1.13		1.14		1	K2	280
120413	−47 08649				13 47 12	−48 01.6	1	8.08		0.95		0.61		4	G6 III	1673
	+20 02877				13 47 14	+20 30.9	1	9.98		1.00		0.83		2	K0	1569
120500	+9 02814				13 47 15	+08 39.4	1	6.59		0.13				4	A0	2006
120455	−28 10277	HR 5197			13 47 16	−28 50.0	2	6.18		−0.02	.004	−0.13		5	A0 V	1079,2035
	−64 02581				13 47 16	−64 29.4	1	11.30		0.28		0.24		2	A2	807
120581	+50 02029				13 47 17	+49 39.0	1	8.21		0.88		0.47		2	G5	1566
	−64 02582				13 47 18	−64 37.5	1	11.04		0.55		0.12		2	F5	807
120566	+38 02482				13 47 19	+37 32.1	1	9.03		0.56		0.00		2	G2 V	1569
120457	−39 08501	HR 5198			13 47 20	−39 39.2	1	6.44		0.99				4	K1 III	2006
120539	+21 02578	HR 5201			13 47 21	+21 30.7	3	4.91	.011	1.44	.013	1.65	.000	6	K4 III	1080,1363,3053
120430	−51 07765				13 47 21	−52 02.7	1	8.64		0.80		0.41		2	K1 + A0 V	1770
120400	−56 05960	V381 Cen			13 47 22	−57 20.0	2	7.55	.275	0.77	.145	0.42		2	F8 Ib/II	657,1484
	−64 02583				13 47 22	−64 39.6	1	10.56		0.30		0.26		2	A5	807
120541	+13 02715				13 47 26	+13 26.4	1	6.66		0.19		0.14		1	A2	280
120567	+28 02265				13 47 27	+27 49.0	1	8.72		1.00		0.81		1	G5	280
120540	+20 02878				13 47 28	+19 35.7	1	7.64		1.28		1.52		1	K0	280
		G 150 - 43			13 47 28	+28 10.3	1	16.32		0.33		−0.59		1		3060
120486	−38 08840				13 47 29	−38 51.0	1	7.78		0.88		0.04		3	A0 V	1770
	+16 02565	G 63 - 59			13 47 33	+15 36.9	2	10.16	.005	1.02	.000	0.95	.005	2	K8	333,1620,3072
120600	+37 02457	HR 5204			13 47 34	+36 52.8	2	6.42	.005	0.24	.000	0.07	.000	4	A7 IV-V	1569,1733
120487	−40 08195	IDS13446S4031	A		13 47 35	−40 45.6	1	9.01		0.22		0.08		1	A8 V	1771
	+19 02718				13 47 37	+19 04.1	1	9.18		0.99		0.73		2	K0	1569
120569	+14 02669				13 47 39	+13 44.1	1	8.57		1.05		0.88		1	K2	280
	+20 02879				13 47 40	+19 41.1	1	9.44		1.42		2.01		1		280
		G 63 - 60			13 47 41	+12 49.8	1	14.52		1.55		1.28		1		333,1620
	−64 02587				13 47 42	−64 34.8	1	11.20		0.59		0.09		2	F7	807
120617	+35 02490				13 47 44	+35 01.7	1	8.51		1.46		1.84		2	K5	1569
120650	+33 02378				13 47 45	+32 46.2	1	7.46		1.37		1.58		2	K3 III	1375
120682	+55 01636				13 47 45	+54 36.0	1	8.65		1.27		1.41		3	K2 III	37
120459	−56 05963				13 47 45	−57 13.6	2	7.99	.009	0.21	.005	0.22		5	A2 IV/V	1771,2012
120533	−29 10620				13 47 47	−30 00.9	1	9.31		0.37		−0.05		3	F6 V	1700
120618	+30 02451				13 47 48	+29 31.6	1	8.96		0.98		0.77		1		280
120544	−19 03754	HR 5202			13 47 50	−19 39.0	1	6.53		0.51				4	F6 IV/V	2007

HD	DM	Other Id	N	Rem	α_{1950}	δ_{1950}	S	V	σ_V	B–V	σ_{B-V}	U–B	σ_{U-B}	n	Spectrum	References
		Lod 995 # 20			13 47 50	−64 46.2	1	12.30		0.19		−0.16		2	B8	807
	+31 02553				13 47 52	+31 28.8	1	9.68		0.99		0.87		2	K2 V	1569
120404	−68 02014	HR 5194			13 47 52	−69 09.3	3	5.74	.003	1.72	.005	2.01	.005	18	K5/M0 III	15,1075,2038
120602	+6 02800	HR 5205			13 47 54	+05 44.7	2	6.00	.005	0.90	.000	0.59	.000	7	K0	15,1417
	−64 02589				13 47 54	−64 38.3	1	10.33		0.24		0.18		2	A1	807
120377	−70 01684				13 47 54	−70 30.5	1	9.12		0.12		−0.07		3	Ap Si	400
120571	−19 03755				13 47 57	−20 19.4	1	9.29		1.15		0.90		1	K0 II	565
120636	+21 02579	IDS13456N2145		B	13 47 59	+21 30.2	1	7.32		0.88				2	G0	280
		AAS61,331 # 12			13 48 00	+25 09.	1	11.44		−0.09		−0.27		2		1569
120651	+21 02580	IDS13456N2145		A	13 48 02	+21 31.4	1	6.84		0.93				3	G0	280
120702	+43 02359				13 48 03	+42 48.3	1	7.01		0.37		0.05		3	F2 IV	1501
	+23 02619				13 48 04	+23 02.7	1	9.93		1.06		0.94		1		280
120494	−61 04074	NGC 5316 - 135			13 48 04	−61 52.9	1	8.43		1.75		1.70		2	M4 II	311,1695
120664	+27 02299				13 48 05	+26 32.2	1	8.85		0.84		0.41		2	G5: V:	1733
120620	−3 03537				13 48 05	−04 01.3	1	9.62		1.10		0.77		1	K2	565
	+27 02300				13 48 07	+27 19.4	1	10.28		1.00		0.80		1		280
120787	+62 01318	HR 5213			13 48 08	+61 44.3	2	5.96	.000	0.97	.016	0.64		4	G3 V	71,3016
120521	−57 06339	LSS 3168			13 48 10	−58 17.5	4	8.55	.030	0.23	.013	−0.69	.027	7	O8	138,403,540,976
120771	+55 01637				13 48 11	+55 07.0	2	7.59	.005	1.51	.024	1.56	.063	6	M1	1733,3016
	−64 02590				13 48 12	−64 36.7	1	10.87		0.27		0.15		2	A0	807
	+24 02648				13 48 13	+23 53.4	1	10.23		0.79		0.27		2		280
120495	−64 02591				13 48 13	−64 42.8	1	9.08		0.17		0.19		2	A2 IV	807
		Lod 995 # 17			13 48 14	−64 44.2	1	11.76		0.65		0.11		2	G0	807
121623	+85 00234				13 48 17	+84 45.7	1	7.74		1.22		1.26		2	K0	1733
120607	−29 10630				13 48 17	−30 19.1	1	8.19		1.63		1.68		4	M3 III	1700
120535	−56 05969				13 48 18	−56 30.9	2	10.60	.002	0.13	.006	−0.63	.010	4	B2 III	540,976
	+17 02672				13 48 19	+17 15.9	1	9.54		0.48		−0.02		2	F8	1375
120496	−67 02407				13 48 20	−67 31.3	1	8.50		0.27		0.19		1	A3mA4-A6	1771
120559	−56 05970				13 48 21	−57 11.0	4	7.98	.012	0.66	.006	0.01	.024	8	G5wF8 V	657,742,2012,3077
120591	−44 08934				13 48 23	−44 41.3	1	9.43		0.09		0.08		1	A1 V	1770
120592	−47 08662	IDS13453S4748		A	13 48 25	−48 02.7	2	7.37	.005	0.80	.000	0.33	.005	7	G6 (III)	158,1279
120611	−44 08935				13 48 27	−44 49.7	1	9.54		0.26		0.04		1	A9 V	1771
120593	−47 08664	IDS13453S4748		B	13 48 27	−48 03.1	2	7.47	.005	0.49	.005	−0.06	.009	7	F5 V	164,1279
	+23 02621				13 48 30	+23 23.6	1	9.74		1.08		0.95		1	K0	280
120562	−59 05239				13 48 30	−59 26.1	1	9.24		0.49		0.22		1	F2 IV	549
		Lod 1095 # 17			13 48 31	−59 29.0	1	11.84		0.64		0.07		1		549
		LP 856 - 53			13 48 33	−27 19.0	1	15.00		0.10		0.10		1		3028
	+19 02721				13 48 34	+18 42.1	1	9.67		1.11		1.02		1		280
		CS 22889 # 58			13 48 34	−10 48.7	1	13.55		−0.11		−0.41		1		1736
120690	−23 11329	HR 5209			13 48 35	−24 08.4	5	6.43	.010	0.70	.008	0.27	.008	16	G5 V	15,1008,1075,2033,3026
		SZ CVn			13 48 36	+37 51.9	1	12.96		0.31		0.07		1		699
120595	−56 05973	IDS13453S5658		AB	13 48 37	−57 12.6	1	8.57		−0.01				4	B3 V	2012
120753	+21 02582				13 48 38	+21 30.4	1	8.54		1.04		0.88		2	G5	1569
120752	+24 02649				13 48 38	+23 53.4	1	9.21		1.12		1.10		2		280
120788	+36 02414	IDS13465N3624		AB	13 48 40	+36 09.6	1	8.32		1.02		0.87		1	K0	280
120672	−35 09054	HR 5208			13 48 41	−36 11.1	3	6.35	.011	0.48	.004	0.01		8	F6 V	15,941,2029
120640	−46 08909	HR 5206			13 48 41	−46 39.1	3	5.77	.012	−0.16	.006	−0.75	.016	12	B2 Vp	26,1637,2032
120751	+24 02650	IDS13464N2416		A	13 48 43	+24 00.6	1	9.79		1.02		0.94		2		1569
	+33 02380				13 48 45	+33 30.7	1	9.47		1.15		1.16		2	K2 III	1569
120874	+59 01533	HR 5216			13 48 45	+58 47.2	2	6.46	.001	0.08	.003	0.07	.003	3	A3 V	752,1733
120817	+42 02444				13 48 46	+42 24.1	1	7.67		0.16		0.12		1	A2	280
120691	−30 10960				13 48 47	−31 04.3	2	7.15	.004	0.52	.000	0.02		6	F7/8 V	1311,2033
120578	−62 03688				13 48 48	−62 31.6	1	7.99		0.02		−0.38		3	B5 II	1732
120641	−52 06787	IDS13456S5219		A	13 48 49	−52 33.8	3	7.52	.028	0.28	.009	0.06	.014	6	B8 V	13,401,1770
		Lod 995 # 18			13 48 50	−64 31.5	1	11.87		0.29		0.24		2	A4	807
120642	−52 06787	HR 5207		★A	13 48 51	−52 33.9	2	5.26	.006	−0.09	.004	−0.33	.007	4	B8 V	13,1770
120642	−52 06787	IDS13456S5219		★AB	13 48 51	−52 33.9	1	5.14		−0.06				4	B8 Vn	2006
	+16 02570				13 48 52	+16 03.0	1	9.49		1.31		1.49		1	K5	280
120818	+35 02492	HR 5214			13 48 52	+35 01.2	6	6.65	.006	0.12	.005	0.07	.000	19	A5 IV	15,1007,1013,1022*
		Lod 995 # 19			13 48 52	−64 46.0	1	12.06		0.26		0.12		2	A0	807
120802	+27 02301				13 48 54	+27 22.5	1	8.18		1.16				4	K1 III	20
120830	+36 02415				13 48 55	+36 30.4	2	9.32	.017	0.22	.003	0.10		6	A7 Vm	193,280
120730	−28 10288				13 48 56	−28 40.6	1	10.00		0.81		0.47		2	G8 V	1696
120709	−32 09676	HR 5210		★A	13 48 56	−32 44.8	1	4.57		−0.16		−0.71		2	B4 III	1770
120709	−32 09676	HR 5210		★AB	13 48 56	−32 44.8	5	4.31	.005	−0.13	.000	−0.60	.005	16	B4 III	15,1020,1075,2012,8015
		Lod 1095 # 30			13 48 56	−59 32.8	1	12.61		0.19		−0.13		1	B8	549
120803	+25 02689				13 48 57	+24 56.6	2	7.59	.015	1.17	.005	1.06	.027	3	K1 III	280,3077
120819	+35 02493	HR 5215			13 48 57	+34 54.7	2	5.88	.005	1.61	.010	1.96		7	M2 III	1118,3001
120710	−32 09676	HR 5211		★B	13 48 57	−32 44.9	2	6.04	.023	−0.02	.017	−0.17	.027	4	B9 V	321,1770
		Lod 995 # 15			13 48 57	−64 31.5	1	11.56		0.41		0.31		2	A4	807
120631	−61 04088	NGC 5316 - 7			13 48 58	−61 34.7	1	8.52		0.43		−0.06		1	F5 V	311
120773	+7 02707				13 48 59	+06 43.1	1	9.77		0.43		−0.02		1	F2	289
120734	−36 08903	V757 Cen			13 49 00	−36 22.6	1	8.40		0.65		0.04		1	F7/G0	941
120564	−69 01945				13 49 00	−69 48.0	1	7.16		0.10		0.12		1	A1 V[n]	1771
		MCC 701			13 49 04	+26 34.4	1	11.00		1.37		1.32		1	K6	679
120711	−46 08917				13 49 05	−47 02.4	1	8.83		0.35		0.17		1	A1mA7-F0	1770
		LP 798 - 42			13 49 06	−20 16.6	1	13.01		0.70		0.16		4		3062
120213	−82 00585	HR 5188			13 49 06	−82 25.2	3	5.94	.004	1.46	.005	1.58	.010	9	K2/3 IIIp	15,536,1075
	+14 02672				13 49 08	+13 39.5	1	10.28		1.10		1.03		2	K0	1569
	+19 02723				13 49 08	+19 03.8	1	9.08		1.22		1.21		1	K0	280

Table 1   773

HD	DM	Other Id	N Rem	α_{1950} δ_{1950}	S	V	σ_V	B–V	σ_{B-V}	U–B	σ_{U-B}	n	Spectrum	References
120759	−31 10706	HR 5212	⋆ AB	13 49 08 −31 22.3	3	6.11	.007	0.48	.000			11	F7 V	15,2020,2029
		Lod 1095 # 28		13 49 09 −59 25.3	1	12.45		0.21		-0.08		1	B6	549
	+7 02708	BB Vir		13 49 10 +06 40.1	2	10.70	.005	0.09	.005	0.12	.025	2	A2	597,668
	−59 05248			13 49 10 −59 29.7	1	11.39		0.42		0.22		1	A0	549
	−64 02602			13 49 11 −64 47.1	1	11.73		0.22		0.17		2	A2	807
120847	+23 02623			13 49 12 +23 00.7	1	8.68		1.19		1.22		1	K0	280
	−58 05237			13 49 13 −59 20.1	1	11.45		0.38		0.17		1	A6	549
120647	−64 02603			13 49 13 −64 34.5	1	9.07		1.00		0.58		2	G1/3 III	807
120893	+38 02485			13 49 18 +37 57.9	1	8.37		0.54		0.11		2	F6 V	1569
	−59 05250			13 49 18 −59 42.6	1	11.23		0.45		-0.03		1	B8	549
120715	−56 05980			13 49 19 −57 06.8	1	9.86		0.39		0.23		1	A9 V	1771
	−59 05251			13 49 19 −59 39.3	1	9.82		1.09		0.73		1	K3	549
	+14 02674			13 49 22 +14 18.2	1	9.70		1.17		1.24		2	K0	1569
120777	−33 09383			13 49 22 −33 36.9	1	9.99		0.27		0.12		1	A2mF0-F0	1770
120678	−62 03703	LSS 3170		13 49 23 −62 28.5	4	7.80	.062	0.13	.032	-0.91	.027	13	O/Be	403,540,976,1732,1737
120807	−30 10966	IDS13466S3017	A	13 49 25 −30 32.3	1	7.76		0.21		0.09		5	A3 IV	1700
	+20 02887			13 49 26 +20 26.1	1	9.95		1.41		1.74		1		280
120763	−45 08770	IDS13464S4522	AB	13 49 27 −45 36.5	1	9.38		0.32		0.05		1	A9 V	1771
120780	−50 08092	IDS13463S5026	AB	13 49 27 −50 40.5	3	7.37	.018	0.90	.005	0.54	.014	9	K3 V	258,2012,3008
	+32 02397			13 49 28 +32 23.7	1	10.37		1.18		1.18		1	K2 III	280
		LP 498 - 73		13 49 30 +09 19.8	1	12.05		0.62		0.01		2		1696
120895	+25 02691			13 49 30 +24 56.1	1	8.08		1.37		1.67		1	K3 III	280
	+36 02416			13 49 30 +36 20.6	1	9.01		1.46		1.77		1	K5	280
		Lod 1095 # 33		13 49 30 −59 36.5	1	13.16		0.32		0.04		1	B8	549
120809	−36 08911			13 49 32 −36 35.4	1	7.92		1.52		1.85		1	M0 III	941
120739	−61 05056			13 49 32 −61 06.5	4	7.85	.013	0.33	.014	-0.32	.016	10	B3 III	403,540,976,2014
	−61 04096			13 49 32 −61 44.5	2	10.49	.014	0.14	.003	-0.61	.011	3	B2.5II-III	540,976
120916	+33 02381			13 49 33 +33 32.0	1	9.14		1.09		1.07		2	K2 III	1569
120680	−65 02503			13 49 33 −66 16.0	3	7.09	.012	0.10	.004	-0.56	.012	8	B2 II	540,976,2012
120933	+35 02496	HR 5219, AW CVn		13 49 35 +34 41.5	5	4.75	.016	1.65	.012	1.93	.025	26	K5 III	15,1363,3055,6001,8015
		LSE # 44		13 49 36 −47 55.	1	12.45		-0.24		-1.18		1	sdO	1650
	−59 05255			13 49 38 −59 33.3	2	10.56	.005	0.16	.010	0.04	.015	2	B9	287,549
	−64 02606			13 49 38 −64 33.6	1	11.49		0.21		-0.11		2	B8	807
120950	+40 02701			13 49 39 +39 55.0	1	7.40		1.56				2	M2	280
		CS 22889 # 57		13 49 42 −10 53.1	1	14.15		0.29		0.04		1		1736
		WLS 1340 35 # 8		13 49 43 +34 57.5	1	11.85		0.80		0.34		2		1375
	+36 02418			13 49 43 +35 57.7	1	8.93		1.08		1.04		1	G8 III	280
		Lod 1095 # 32		13 49 43 −59 37.7	1	12.78		0.04		-0.74		1	O9	549
120768	−59 05258			13 49 44 −59 46.2	1	9.93		0.68		0.19		1		549
120767	−59 05257			13 49 44 −59 46.3	1	9.68		0.51		0.17		1		549
	+27 02303			13 49 45 +26 52.6	1	10.83		0.75		0.32		1		679
121146	+69 00724	HR 5227	⋆ A	13 49 45 +68 33.8	2	6.35	.059	1.10	.083			3	K2 IV	71,3009
	−61 04098	NGC 5316 - 8		13 49 45 −61 31.5	1	10.75		0.13		-0.03		1		311
	+42 02446			13 49 46 +42 17.8	1	10.15		1.38		1.60		1		280
		NGC 5316 - 137		13 49 46 −61 39.5	1	11.83		0.17		0.00		1		311
	+18 02791			13 49 48 +17 36.7	1	9.26		1.40		1.74		1	K2	280
		Lod 1095 # 18		13 49 48 −59 26.0	1	11.86		0.25		-0.03		1	B6	549
	−59 05260			13 49 50 −59 36.2	1	11.48		0.32		0.22		1	B9	549
120934	+12 02635	HR 5220		13 49 51 +12 24.7	4	6.07	.036	0.04	.001	0.01	.013	8	A1 V	70,1022,1221,3050
		Lod 1095 # 24		13 49 51 −59 28.0	1	12.12		1.18		0.83		1	K2	549
120966	+26 02501			13 49 52 +26 22.6	1	9.07		0.26		0.01		4	F1 V	1068
	−63 03028	LSS 3171		13 49 52 −64 13.5	1	10.44		0.64		-0.36		3	B1 Ib	1737
	+16 02572			13 49 53 +15 50.9	1	9.78		1.01		0.74		1	K0	280
	+30 02457			13 49 53 +30 25.3	1	9.84		1.00		0.80		1	K0	280
120797	−56 05989			13 49 54 −56 27.1	2	9.07	.033	0.52	.019	0.00		4	F7 V	1594,6006
	−59 05262			13 49 54 −59 35.2	2	10.93	.000	0.17	.010	0.08	.010	2	B9	287,549
	−61 04100	NGC 5316 - 136		13 49 54 −61 39.9	2	10.64	.015	0.11	.035	-0.17	.015	2		311,460
120783	−61 04102	NGC 5316 - 1		13 49 55 −61 23.0	3	7.89	.025	1.50	.022	1.60	.026	11	K4 III	311,1695,1704
	+29 02462			13 49 57 +29 21.3	1	9.89		1.00		0.80		1	K0	280
	+40 02702			13 49 57 +39 50.0	1	10.16		1.00		0.77		1	K0	280
121130	+65 00963	HR 5226, CU Dra	⋆ A	13 49 58 +64 58.2	4	4.66	.019	1.57	.010	1.86	.025	15	M3.5III	15,1363,3016,8015
		Lod 1095 # 12		13 49 58 −59 31.0	1	11.29		1.42		1.37		1	K5	549
		NGC 5316 - 6		13 49 58 −61 31.7	1	12.54		0.40		0.12		1		311
		LSE # 153		13 50 00 −46 30.	1	11.35		-0.26		-1.24		1		1650
		Lod 1095 # 19		13 50 00 −59 40.4	1	11.88		0.46		0.34		1	A0	549
		NGC 5316 - 33		13 50 00 −61 36.0	1	12.43		0.28		0.12		2		460
	−61 04103	NGC 5316 - 35		13 50 00 −61 37.1	5	9.43	.027	1.49	.027	1.41	.025	11		311,460,1653,1695,4002
		NGC 5316 - 77		13 50 00 −61 38.0	1	12.77		0.57		0.37		1		311
234078	+50 02030	G 177 - 62		13 50 01 +50 11.9	1	8.90		1.33		1.20		2	K5	3072
120785	−64 02609			13 50 01 −64 57.9	1	9.99		0.47		0.20		1	F0 V	549
	+22 02642			13 50 02 +22 25.6	1	10.17		1.04		0.87		1	K0	280
	−59 05263			13 50 02 −59 34.1	2	10.65	.005	0.16	.010	0.11	.000	2	B9	287,549
120798	−61 04105	NGC 5316 - 2		13 50 02 −61 24.4	1	8.95		-0.01		-0.12		1	B9 IV/V	311
	+34 02462			13 50 03 +33 58.3	1	9.61		1.23		1.35		1	K2 III	280
		Lod 1095 # 25		13 50 03 −59 22.9	1	12.21		0.98		0.48		1		549
	−61 04106	NGC 5316 - 31		13 50 03 −61 35.4	4	9.38	.027	1.43	.020	1.21	.025	10		311,460,1695,4002
		NGC 5316 - 38		13 50 04 −61 36.9	3	12.63	.047	0.25	.004	0.08	.029	6		311,460,1653
120905	−31 10722			13 50 05 −31 33.2	1	8.94		0.26		0.06		1	A2 II/III	1770
		Lod 1095 # 20		13 50 05 −59 26.8	1	11.96		0.21		-0.16		1	B7	549
		Lod 1095 # 29		13 50 05 −59 37.3	1	12.59		0.23		0.14		1		549

HD	DM	Other Id	N	Rem	α₁₉₅₀	δ₁₉₅₀	S	V	σ_V	B–V	σ_B–V	U–B	σ_U–B	n	Spectrum	References
	−61 04108	NGC 5316 - 75			13 50 07	−61 37.2	2	11.22	.005	0.21	.000	-0.08	.000	5		460,1653
120997	+17 02676				13 50 08	+16 58.5	1	6.65		1.52		1.86		1	K5	280
121457	+79 00431				13 50 09	+79 14.6	1	6.59		1.15		1.07		3	G5	985
		Lod 1095 # 26			13 50 09	−59 31.1	1	12.39		1.26		1.06		1		549
		NGC 5316 - 160			13 50 09	−61 40.3	1	13.97		0.66				1		311
120799	−62 03725				13 50 10	−63 15.4	2	9.11	.011	0.12	.002	-0.56	.001	5	B2 III	540,976
		G 150 - 46			13 50 11	+14 40.1	1	11.63		1.50				1	K7	1746
120814	−64 02611				13 50 11	−64 36.3	1	10.83		0.30		0.20		2	A7/F0	807
		Lod 995 # 11			13 50 11	−64 36.4	1	11.30		0.30		0.20		2	A2	807
	+25 02693				13 50 12	+25 16.4	1	9.82		0.35				7	F1 IV	193
	+36 02419				13 50 12	+35 52.6	1	9.73		0.90		0.52		2	K0	1569
120859	−57 06368				13 50 12	−58 11.6	1	8.09		0.10		-0.05		2	B9 V	1771
		NGC 5316 - 113			13 50 12	−61 39.1	1	11.67		0.44		0.31		3		1653
	−62 03729				13 50 12	−62 35.8	2	11.03	.004	0.11	.009	-0.41	.011	3	B5 III	540,976
	−61 04110	NGC 5316 - 112			13 50 13	−61 38.4	1	11.15		0.17		-0.24		3		1653
		NGC 5316 - 159			13 50 13	−61 40.6	1	13.51		0.29		0.22		1		311
121049	+35 02497				13 50 15	+35 20.4	1	8.69		1.21		1.27		2		1569
		Lod 1095 # 9			13 50 15	−59 28.4	1	11.01		1.17		0.84		1	K3	549
		NGC 5316 - 83			13 50 15	−61 38.3	1	13.08		0.52		0.24		3		1653
120941	−31 10726				13 50 16	−31 28.5	1	8.15		-0.07		-0.23		3	B9 V	1770
121048	+36 02420				13 50 17	+35 55.5	2	8.76	.044	0.25	.001	0.12		3	A8 Vm	193,280
330928	−48 08563				13 50 17	−48 53.3	1	10.82		0.59		0.13		2		1081
120886	−51 07813				13 50 17	−52 15.3	1	8.26		0.10		0.08		1	A0 V	1770
	−59 05266				13 50 17	−59 41.5	1	10.93		0.29		0.17		1	B8	549
	+23 02624				13 50 18	+23 29.1	1	9.68		1.11		1.00		1	K0	280
		Lod 1095 # 31			13 50 18	−59 31.2	1	12.74		0.74		0.21		1	F8	549
		Lod 1095 # 21			13 50 18	−59 40.2	1	12.01		0.24		0.15		1	A0	549
		NGC 5316 - 15			13 50 18	−61 33.5	1	13.00		0.44		0.14		1		1653
120955	−31 10729	HR 5221		⋆ A	13 50 19	−31 40.9	2	4.74	.005	-0.11	.039	-0.55	.021	3	B4 IV	542,1770
120955	−31 10729	IDS13475S3126		⋆ AB	13 50 19	−31 40.9	6	4.73	.015	-0.14	.005	-0.56	.006	21	B4 IV +Am	15,26,1075,2012,8015,8052
120955	−31 10727	IDS13475S3126	B		13 50 19	−31 40.9	3	8.42	.048	0.30	.031	0.03	.043	7	Am	321,542,1770
		Lod 1095 # 34			13 50 19	−59 38.5	1	13.36		0.44		0.33		1	A0	549
120841	−61 04113	NGC 5316 - 118			13 50 19	−61 39.9	2	9.85	.035	0.16	.040	-0.05	.025	2	B8 III/IV	311,460
120861	−60 05062	IDS13469S6032	AB		13 50 20	−60 46.8	1	8.30		0.42				4	F2 V	2012
		Lod 1095 # 23			13 50 21	−59 34.1	1	12.06		0.56		0.00		1	F8	549
	−61 04114	NGC 5316 - 156			13 50 21	−61 40.8	1	11.53		0.28		0.02		3		1653
	+21 02584	IDS13480N2145	AB		13 50 22	+21 30.6	1	11.08		0.94		0.69		1	K0	280
120907	−51 07814				13 50 22	−52 12.0	1	8.82		0.15		0.16		1	A1 V	1770
	−61 04117	NGC 5316 - 72			13 50 22	−61 36.9	4	9.85	.008	1.58	.023	1.65	.016	10		311,460,1653,1695,4002
		Lod 1002 # 12			13 50 22	−64 59.7	1	11.52		1.18		0.84		1		549
	−61 04118	NGC 5316 - 42			13 50 23	−61 35.5	2	11.71	.009	0.20	.033	-0.09	.009	4		311,1653
		NGC 5316 - 9			13 50 24	−61 33.4	1	12.10		0.28		0.08		3		1653
		NGC 5316 - 5			13 50 25	−61 32.2	3	12.98	.016	0.36	.017	0.22	.030	5		311,460,1653
120908	−52 06805	HR 5217			13 50 27	−53 07.6	6	5.88	.015	0.01	.002	-0.42	.004	19	B4 III	15,26,1770,2006,8015,8023
	−61 04122	NGC 5316 - 71			13 50 27	−61 37.3	1	10.79		0.21		-0.24		3		1653
		NGC 5316 - 122			13 50 27	−61 39.3	1	13.34		0.41		0.31		3		1653
		Lod 1002 # 16			13 50 27	−65 04.6	1	11.99		1.37		1.37		1		549
		LP 498 - 78			13 50 29	+14 21.4	1	12.56		0.65		0.01		2		1696
120958	−38 08883	V774 Cen			13 50 29	−38 48.7	2	7.60	.000	-0.10	.005	-0.78	.005	5	B3 Vnne	55,400
120959	−38 08882	IDS13475S3852	A		13 50 29	−39 07.1	1	8.75		0.12		0.13		1	A2/3 IV	1770
	−64 02613				13 50 29	−65 09.2	2	11.47	.000	0.15	.000	-0.01	.000	2	B9	287,549
121063	+24 02656				13 50 30	+23 34.8	1	8.01		1.52		1.88		2	K2	280
120960	−41 08231	IDS13475S4108	A		13 50 30	−41 23.3	1	7.86		0.30		0.04		1	A8/9 IV	1771
	−61 04125	NGC 5316 - 45			13 50 30	−61 36.0	2	9.56	.011	1.60	.017	1.65	.041	8		311,460,1695
	−61 04124	NGC 5316 - 124			13 50 30	−61 39.4	1	10.87		0.26		-0.05		3		1653
		Lod 1095 # 22			13 50 32	−59 35.5	1	12.02		0.67		0.11		1	G0	549
	−61 04126	NGC 5316 - 4			13 50 33	−61 33.7	3	11.54	.013	0.29	.025	-0.01	.011	6		311,460,1653
		GD 159			13 50 34	+18 33.1	1	12.81		0.20		-0.02		1		3060
120974	−37 08978				13 50 34	−38 01.2	1	7.60		0.14		0.09		1	A4 III/IV	1771
		NGC 5316 - 110			13 50 35	−61 38.1	1	13.70		0.49		0.21		2		1653
	+18 02794				13 50 36	+17 45.8	1	10.60		0.30		0.00		2	F2	1569
		LP 678 - 43			13 50 36	−09 00.5	1	14.70		0.33		-0.63		3		3060
	−64 02615				13 50 36	−65 03.7	2	11.47	.000	0.16	.000	-0.02	.000	2	B9	287,549
120987	−35 09090	HR 5222		⋆ AB	13 50 37	−35 25.1	2	5.53	.000	0.44	.000	-0.03		8	F3 V	244,2007
		NGC 5316 - 67			13 50 38	−61 36.5	1	13.84		0.56		0.22		3		1653
		NGC 5316 - 86			13 50 38	−61 36.9	1	12.65		0.16		0.08		1		311
		NGC 5316 - 154			13 50 38	−61 40.9	2	13.59	.090	0.62	.010	0.12	.065	2		311,460
	−59 05272				13 50 39	−59 39.3	1	11.64		0.19		0.06		1	B8	549
		NGC 5316 - 108			13 50 40	−61 38.4	1	12.32		0.74		0.25		1		1653
		Lod 1002 # 13			13 50 40	−65 10.8	1	11.59		0.69		0.16		1	K0	549
	+22 02643				13 50 41	+22 22.5	1	9.25		1.11		1.06		1	K0	280
	+23 02626				13 50 41	+23 10.7	1	10.30		0.84		0.62		1		280
121147	+41 02433				13 50 41	+40 32.8	2	9.12	.027	0.00	.009	0.05		5	A2 III	193,280
120863	−67 02423				13 50 41	−67 59.5	1	7.89		0.24		0.07		1	A7 V	1771
		NGC 5316 - 63			13 50 42	−61 35.9	2	13.23	.005	0.57	.015	0.11	.005	6		311,460,1653
		Lod 1002 # 18			13 50 42	−65 12.1	1	12.10		0.28		-0.25		1	B3	549
		NGC 5316 - 87			13 50 43	−61 37.8	2	13.75	.054	0.46	.058	0.38	.015	1		311,460
		Lod 1095 # 27			13 50 44	−59 26.7	1	12.41		0.24		-0.02		1	B8	549
		FY Hya			13 50 46	−29 19.7	1	11.90		0.15		0.06		1	F2	597
	−61 04129	NGC 5316 - 103			13 50 46	−61 37.5	1	10.41		0.29		0.08		1		1653

Table 1

HD	DM	Other Id	N	Rem	α_{1950}	δ_{1950}	S	V	σ_V	B−V	σ_{B-V}	U−B	σ_{U-B}	n	Spectrum	References
121131	+28 02272	G 150 - 47			13 50 47	+28 03.5	2	8.36	.020	0.82	.010	0.38		8	K1 V	20,3026
121148	+35 02498				13 50 47	+35 07.5	1	8.99		0.50		0.03		2	G0 III	1569
121003	−34 09222				13 50 47	−34 25.1	1	7.81		0.06		0.09		1	A0/1 V	1770
121149	+28 02273				13 50 48	+27 53.6	1	8.54		0.59		0.11		2	G0 V	3026
	−56 05997	LSS 3172			13 50 49	−56 50.4	2	10.51	.001	0.27	.011	-0.64	.019	3	B0.5II-Ib	540,976
121109	+13 02720				13 50 50	+12 59.3	1	7.19		0.28		0.09		2	A3	1375
121107	+18 02795	HR 5225			13 50 50	+18 10.7	2	5.70	.005	0.85	.005	0.50		4	G5 III	70,1569
	+20 02890			V	13 50 50	+20 24.8	1	10.40		0.38		-0.02		2		1569
120991	−46 08931	HR 5223, V767 Cen		⋆ A	13 50 50	−46 52.9	6	6.07	.046	-0.07	.017	-0.92	.023	39	B2 IIne	13,681,815,1637,1770,2006
120991	−46 08931	IDS13477S4638		B	13 50 50	−46 52.9	1	11.02		0.07		-0.48		4		13
		G 65 - 11			13 50 51	+06 14.0	1	13.01		1.45		1.12		1		333,1620
	−61 04130	NGC 5316 - 95			13 50 51	−61 36.8	1	10.87		0.22		-0.13		3		1653
		Lod 1002 # 23			13 50 51	−65 14.1	1	13.03		0.46		0.24		1		549
	+23 02628				13 50 54	+23 00.7	1	9.72		1.46		1.58		1	K5	280
121164	+29 02464	HR 5229			13 50 54	+28 53.6	2	5.89	.015	0.20	.010	0.14		4	A7 V	252,280
121004	−45 08786				13 50 54	−46 17.6	5	9.04	.006	0.60	.007	-0.05	.013	9	G2 V	742,1696,2012,2017,3077
	+37 02459				13 50 55	+37 10.6	1	10.00		1.02		0.80		2	K0 III	1569
	−48 08574				13 50 55	−48 46.9	1	11.22		0.60		-0.30		2		1649
	+22 02644				13 50 56	+21 59.8	1	10.78		0.41		-0.08		2	F5	1569
121005	−47 08699				13 50 56	−48 11.5	1	9.92		0.09		0.03		1	A0 V	1770
		Lod 1095 # 10			13 50 56	−59 29.1	1	11.14		1.25		0.97		1	K3	549
		CoD -77 00622		⋆ V	13 50 56	−77 33.4	1	8.76		1.43		0.86		1	M3	817
121056	−34 09223	HR 5224			13 50 57	−35 04.1	4	6.18	.003	1.03	.008	0.84	.019	14	K0 III	15,244,2027,3008
	+25 02696				13 50 58	+25 11.6	1	9.18		1.11		1.09		1	K0	280
	+26 02503				13 50 58	+25 55.1	1	10.19		0.73		0.28		1	G5	280
	−61 04133	NGC 5316 - 101			13 50 58	−61 37.1	3	10.02	.010	0.18	.028	-0.12	.015	6		311,460,1653
		Lod 1002 # 20			13 50 58	−65 14.1	1	12.62		0.59		0.18		1		549
120913	−67 02426	HR 5218			13 50 59	−67 24.4	3	5.71	.004	1.49	.000	1.67	.000	15	K2 III	15,1075,2038
121007	−50 08104				13 51 00	−51 09.3	2	8.20	.001	0.07	.002	0.03		6	B9.5V	1770,2014
121183	+27 02306				13 51 01	+27 19.9	1	8.41		0.89				4	K0 IV	20
121197	+41 02434				13 51 01	+40 35.0	1	6.53		1.55				2	K5	280
121111	−5 03775				13 51 01	−05 56.5	1	7.69		0.55				12	G0	1351
	+13 02721	G 65 - 12			13 51 02	+13 12.1	3	9.80	.000	1.39	.031	1.27	.010	4	M0	333,1620,1705,3072
121184	+24 02658				13 51 03	+24 24.3	2	7.94	.000	1.47	.000	1.75	.000	4	K3 III	280,280
	+37 02460				13 51 03	+37 10.3	1	9.93		0.98		0.77		2	K0 III	1569
	−64 02621				13 51 04	−65 06.5	2	11.41	.000	0.17	.000	0.03	.000	2	B9	287,549
	+24 02659				13 51 07	+24 23.0	1	9.12		1.28		1.39		2	K2	280
		Lod 1002 # 19			13 51 07	−65 08.1	1	12.50		0.63		0.15		1		549
121211	+41 02435				13 51 08	+40 57.7	1	8.36		1.13		1.14		1	K2	280
		Lod 1002 # 17			13 51 08	−65 15.9	1	12.03		0.42		0.10		1		549
	−61 04149	NGC 5316 - 3			13 51 09	−61 33.0	3	9.93	.006	1.73	.008	1.99	.096	6		311,460,1695,4002
121081	−27 09474				13 51 10	−27 53.6	1	8.44		0.00		-0.16		2	B9.5(p Si)	1770
121057	−48 08578				13 51 10	−48 26.8	1	7.18		0.19		0.14		2	A4 V	1771
121212	+34 02467				13 51 11	+34 01.9	1	6.90		1.48				2	K5 III	280
121058	−49 08284				13 51 12	−50 03.3	1	9.52		0.16		0.14		1	A1 V	1770
121213	+28 02274				13 51 13	+27 44.4	1	8.15		1.24		1.29		1	K5	280
120978	−61 04140				13 51 14	−62 07.3	1	9.83		0.38		0.13		1	A8/F0 [V]	1771
121247	+42 02449	IDS13492N4241		AB	13 51 16	+42 25.8	1	6.88		0.05		0.12		1	A0	280
	−44 08967				13 51 16	−44 29.3	1	9.66		0.90		0.60		2		3008
121198	+13 02723				13 51 18	+12 55.6	1	8.55		0.48		0.00		2	F8	1648
	+40 02703				13 51 18	+40 05.6	1	8.65		1.01		0.86		1	K0	280
121040	−55 05779				13 51 18	−55 29.6	1	7.27		0.28		0.03		1	A9 V	1771
121023	−60 05076				13 51 20	−60 39.2	1	9.50		0.45		0.21		1	A8 IV/V	1771
		Lod 1002 # 11			13 51 20	−65 12.7	1	11.49		0.62		0.10		1	F8	549
	+27 02308				13 51 21	+26 52.1	1	10.14		0.29		0.14		1		280
		AAS61,331 # 5			13 51 24	+16 37.	1	11.47		0.17		0.12		2		1569
	−64 02626				13 51 25	−65 07.7	1	9.78		0.58		0.11		1	F8	549
121156	−27 09478	HR 5228			13 51 26	−28 19.4	1	6.04		1.13				4	K2 III	2035
	+36 02423				13 51 27	+35 37.9	1	9.40		1.03		0.89		1	K0	280
121297	+53 01667	EH UMa			13 51 27	+52 34.1	2	6.78	.005	1.55	.019	1.53	.010	9	M2	1501,1566
	−64 02627				13 51 29	−65 07.3	1	11.38		0.89		0.43		1	G0	549
	+28 02275				13 51 30	+28 12.7	1	10.53		1.15		1.17		1		280
	+29 02468				13 51 30	+29 16.8	1	10.71		0.40		-0.03		2		1569
121094	−51 07828				13 51 30	−51 22.6	1	8.05		0.22		0.20		1	A3 V	1771
	+18 02797	XZ Boo			13 51 32	+17 31.6	1	9.53		1.46		1.29		2	M7	1569
121117	−44 08969				13 51 32	−45 08.9	1	8.60		0.18		0.16		1	A3 V	1771
	−64 02628				13 51 32	−65 07.7	1	11.34		0.42		0.10		1	F0	549
	+29 02469				13 51 33	+29 01.1	1	10.19		1.11		1.10		1		280
		G 238 - 55			13 51 33	+67 04.1	1	15.69		1.90				1		906
121141	−47 08714				13 51 33	−47 53.4	3	7.17	.009	0.36	.012	0.01		10	F3 V	1075,2033,3077
	+41 02436				13 51 34	+40 40.0	1	9.16		0.98		0.79		1	K0	280
121248	+14 02678				13 51 36	+14 04.8	1	7.76		0.38		0.16		1	A5	280
121096	−54 05788				13 51 36	−55 13.7	1	8.55		0.32		0.10		1	A2mA9-F0	1771
		AAS61,331 # 18			13 51 41	+41 20.	1	10.86		0.36		0.01		2		1569
121249	+10 02605	G 65 - 13			13 51 42	+10 29.8	2	9.02	.005	0.73	.005	0.28	.019	3	K0	333,1620,1658
		G 65 - 14			13 51 43	+07 01.4	1	14.73		1.38		1.03		1		333,1620
	+35 02499				13 51 43	+34 47.7	1	9.76		1.18		1.23		2	K2	1569
121073	−61 04156	IDS13482S6144		A	13 51 43	−61 59.1	1	8.71		0.34		0.14		1	A8 V	1771
	+32 02404	IDS13496N3238		AB	13 51 45	+32 24.4	1	10.14		1.04		1.06		1		280
	+32 02405				13 51 47	+32 18.4	1	10.24		0.85		0.41		1		280

HD	DM	Other Id	N	Rem	α_{1950}	δ_{1950}	S	V	σ_V	B–V	σ_{B-V}	U–B	σ_{U-B}	n	Spectrum	References
121319	+29 02471				13 51 53	+28 34.4	1	7.73		1.01				3	K0 III	20
	+39 02695				13 51 54	+38 59.6	1	9.14		1.53		2.02		1	M0	280
		AAS12,381 24 # A			13 51 55	+24 31.4	1	10.94		0.18		0.19		1		280
		Lod 1002 # 15			13 51 56	−65 08.7	1	11.89		1.35		1.16		1	M0	549
	+34 02468				13 51 58	+33 57.1	1	10.64		0.93		0.72		1	K0 III	280
		LP 798 - 56			13 51 58	−20 19.4	1	11.12		1.31		1.27		1		1696
121190	−51 07832	HR 5230			13 51 58	−51 54.9	3	5.68	.020	−0.07	.004	−0.31	.005	8	B9 V	26,1770,2014
		Lod 1002 # 5			13 51 59	−65 09.1	1	11.25		0.57		0.05		1	F8	549
	−64 02632				13 51 59	−65 09.1	1	10.58		1.85		2.08		1	M4	549
	+24 02660				13 52 00	+23 45.1	1	9.78		0.97		0.76		1	K0	280
	+40 02704				13 52 00	+39 52.7	1	9.84		1.02		0.78		1	G5	280
	+34 02469				13 52 01	+34 27.0	1	10.45		0.25		0.06		1		280
121409	+54 01630	HR 5238			13 52 01	+53 58.4	3	5.70	.008	−0.04	.008	−0.07	.001	7	A0 V	70,752,3016
	−64 02633				13 52 01	−65 04.1	1	10.30		1.08		0.86		1	K3	549
	+25 02697				13 52 02	+25 27.8	1	10.54		0.49		0.00		2	F8	1569
		G 65 - 15			13 52 03	+09 57.2	1	14.86		1.54		1.16		1		333,1620
121367	+38 02489				13 52 04	+38 29.5	1	9.15		0.25		0.07		1	F5	280
121368	+35 02501				13 52 05	+34 51.1	1	7.66		1.02		0.85		2	K0	1569
121226	−46 08944				13 52 05	−46 50.8	1	7.44		0.08		0.08		1	A1 V	1770
121344	+25 02698				13 52 07	+25 27.1	1	9.17		0.55		0.00		2	F8	1569
121299	−0 02758	HR 5232			13 52 08	−01 15.5	4	5.15	.009	1.09	.018	1.04	.035	11	K2 III	15,1256,3005,4001
121261	−35 09113				13 52 08	−35 53.6	2	9.26	.035	1.24	.015	1.05		4	G2/3 II	1594,6006
121369	+30 02461	IDS13499N3024	AB		13 52 10	+30 09.6	1	7.47		0.29		0.08		1	F2	280
121225	−46 08945	IDS13491S4611	A		13 52 10	−46 25.4	1	9.97		0.04		−0.14		1	B9.5 V	1770
121271	−28 10318	IDS13494S2836	AB		13 52 12	−28 50.7	1	9.57		1.45		1.22		1	K5/M0 V	3072
121239	−49 08300				13 52 13	−49 26.1	1	9.23		0.11		0.07		1	A1 V	1770
		Lod 1002 # 14			13 52 14	−65 09.3	1	11.68		0.54		0.10		1	F5	549
	−44 08976				13 52 15	−44 52.0	1	9.66		0.90				4	K2 V	2033
	+35 02502				13 52 16	+35 01.1	1	9.64		1.47		1.85		1	K5 III	1569
121370	+19 02725	HR 5235	⋆ A		13 52 18	+18 38.9	25	2.68	.009	0.58	.008	0.20	.007	457	G0 IV	1,3,15,61,71,130,667⋆
121208	−57 06394				13 52 18	−58 12.2	1	9.37		0.32		0.21		1	A9 V	1771
121371	+16 02577				13 52 19	+15 31.4	1	7.93		0.97		0.68		1	K0	280
121410	+30 02462				13 52 21	+30 27.0	1	8.57		0.35		−0.04		2	F0	1371
121325	−7 03728	HR 5233	⋆ AB		13 52 21	−07 48.8	3	6.19	.011	0.53	.005	−0.01		20	F8 V + G0	15,1256,1351
121241	−52 06840				13 52 21	−52 26.5	1	9.96		0.23		0.18		1	A2 IV	1770
121263	−46 08949	HR 5231			13 52 25	−47 02.6	6	2.54	.012	−0.23	.011	−0.90	.006	16	B2.5IV	15,507,1075,1770,2012,8015
121228	−58 05282	LSS 3174			13 52 25	−59 07.6	3	7.81	.015	0.18	.009	−0.65	.022	7	B1 Ib	403,540,976
121264	−46 08950				13 52 26	−47 11.5	1	9.20		0.36		0.04		1	A8/F2	1771
121266	−48 08591				13 52 26	−49 07.1	1	8.01		0.16		0.18		1	A1 IV	1770
121209	−63 03053				13 52 27	−63 38.8	1	7.24		0.10		−0.04		2	B9.5 III/I	1771
		Wolf 518			13 52 31	+17 19.1	1	12.86		1.26		1.23		1		1696
121288	−42 08856				13 52 32	−42 22.8	1	7.70		1.22				4	K1 III	2012
121290	−42 08854				13 52 32	−42 52.9	1	9.54		0.27		0.15		1	A1/2mA9-A9	1770
	+21 02591				13 52 33	+20 57.8	1	9.61		1.25		1.47		1	K5	280
121292	−47 08730				13 52 33	−47 37.3	1	9.03		−0.03		−0.31		3	B8 V	26
		LP 380 - 23			13 52 34	+24 03.3	1	13.25		0.72		0.17		1		1773
		LP 380 - 24			13 52 36	+23 59.0	1	15.66		1.68		1.43		1		1773
121291	−44 08979				13 52 36	−44 24.3	1	7.86		0.70		0.29		2	A0 Vn + K2	1770
	+24 02661				13 52 43	+23 36.6	1	9.96		1.30		1.42		1	K2	280
	+37 02469				13 52 49	+37 29.7	1	10.60		0.90		0.48		2	K0	1375
121358	−42 08860				13 52 49	−42 57.2	1	8.13		0.12		0.12		1	A2 IV/V	1770
121397	−30 11015	HR 5237			13 52 52	−31 02.4	1	6.51		0.90				4	G8 III	2007
331016	−48 10996				13 52 53	−49 05.2	1	9.63		1.18		0.44		1		954
121476	+19 02728				13 52 54	+18 56.1	1	7.90		1.49		1.90		1	M0 III	280
121230	−68 02035				13 52 54	−68 57.8	1	8.80		0.21		0.16		1	A3 V	1771
121399	−38 08917				13 52 58	−39 09.9	1	7.19		0.38		0.21		3	A0 IV/V	1770
121492	+20 02897				13 52 59	+19 37.9	1	7.88		1.42		1.63		1	K0	280
121360	−51 07844				13 52 59	−51 33.4	1	9.07		0.16		0.10		1	A0 V	1770
	+18 02801				13 53 01	+17 55.1	1	10.30		0.92		0.60		1		280
121400	−41 08288				13 53 02	−42 09.6	1	8.11		0.20		0.16		1	A4 IV	1771
121336	−53 05805	HR 5234	⋆ AB		13 53 02	−53 53.3	5	6.14	.016	0.07	.010	0.03	.011	12	A1 V	15,401,404,1770,2029
121447	−17 03961				13 53 03	−18 00.3	3	7.83	.010	1.74	.013	1.96	.013	16	Kp Ba	565,993,2033
121481	−8 03667				13 53 09	−09 18.9	1	6.73		1.55		1.92		7	K5	1628
	+18 02802				13 53 13	+17 32.4	1	8.89		1.49		1.84		1	K5	280
121416	−45 08815	HR 5239			13 53 13	−46 20.8	5	5.82	.007	1.14	.000	1.11		20	K1 III	15,1075,2013,2029,3009
	+30 02463				13 53 14	+30 20.2	1	10.31		1.12		1.09		1		280
121496	−9 03804				13 53 14	−09 30.7	1	6.85		0.47				7	F6 IV-V	1351
121384	−54 05806	HR 5236	⋆ A		13 53 14	−54 27.4	3	6.00	.009	0.78	.000	0.23		11	G6 IV/V	2006,2018,3077
121384	−54 05806	IDS13499S5412	B		13 53 14	−54 27.4	1	12.00		1.42		1.23		2		3008
		POSS 192 # 4			13 53 16	+48 19.2	1	16.80		1.05				2		1739
121533	+13 02728	G 65 - 16			13 53 22	+12 41.3	1	8.56		0.62		0.06		2	G5	333,1620
	−48 08610				13 53 22	−48 53.2	1	11.41		0.60		−0.26		1		1649
	−48 10982	LSS 3677			13 53 22	−48 53.2	1	11.19		0.50		−0.34		2		954
121364	−60 05103				13 53 22	−61 21.7	1	10.08		0.37		0.19		1	A6/9	1771
121560	+14 02680	HR 5243			13 53 25	+14 18.0	2	6.17	.010	0.50	.001	−0.09		4	F6 V	70,3026
		LP 799 - 96			13 53 26	−15 59.0	1	13.98		0.70		0.20		3		3062
121559	+23 02631				13 53 28	+22 55.5	1	7.57		1.49		1.87		1	K2	280
121604	+25 02702				13 53 38	+24 46.5	1	8.78		0.96		0.86		1	G5	280
	+34 02471				13 53 38	+34 17.1	1	10.03		1.10		1.05		1	K0 V	280
	+46 01913				13 53 39	+46 00.4	1	9.93		0.09		0.08		8	Am	1202

Table 1   777

HD	DM	Other Id	N	Rem	α_{1950}	δ_{1950}	S	V	σ_V	B–V	σ_{B-V}	U–B	σ_{U-B}	n	Spectrum	References
121483	−45 08822				13 53 41	−46 08.5	3	6.95	.001	-0.13	.004	-0.68		13	B2/3 IV	1732,2012,2014
	+41 02441				13 53 44	+40 44.0	1	8.96		1.21		1.25		1	K2	280
121550	−15 03774				13 53 45	−15 57.1	1	9.34		0.81		0.35		4	G8/K0 V	3062
121647	+40 02706				13 53 46	+39 42.5	1	7.36		1.22		1.28		1	K2 III	280
121626	+29 02473				13 53 49	+28 55.0	2	7.00	.005	0.04	.005	0.05	.039	3	A0	272,280
		WD1353+409			13 53 49	+40 55.2	1	15.55		-0.14		-0.85		2	DA3	1727
121648	+26 02508	ZZ Boo			13 53 52	+26 09.8	2	6.79	.005	0.36	.005	-0.01	.000	5	F2 V	588,1569
121800	+66 00824				13 53 54	+66 21.7	1	9.11		-0.17		-0.75		3	G7	963
121607	+1 02865	HR 5244			13 53 55	+01 17.7	2	5.90	.005	0.21	.005	0.13	.010	7	A8 V	15,1417
121682	+32 02411	HR 5245			13 53 57	+32 16.6	2	6.34	.025	0.36	.012	0.04		4	F4 IV-V	70,254
121608	−9 03807				13 53 58	−09 47.4	2	7.68	.003	0.54	.000			34	F8	130,1351
	+25 02703				13 53 59	+24 52.0	1	9.82		1.01		0.87		1	K2	280
		G 255 - 43			13 54 00	+79 05.7	1	10.58		1.45				2	M2 V	1625
		L 259 - 146			13 54 00	−54 39.	1	14.07		1.31		1.00		1		3062
	−61 04200				13 54 00	−61 36.8	1	10.60		0.65		0.15		1	F7	549
121474	−63 03070	HR 5241			13 54 00	−63 26.6	3	4.70	.005	1.11	.004	1.04	.005	14	K0 III	15,1075,2012
121579	−27 09503	IDS13512S2710		A	13 54 02	−27 25.0	2	7.88	.030	0.50	.004	0.05		5	F3 V	1279,2012
121579	−27 09503	IDS13512S2710		B	13 54 02	−27 25.0	1	11.92		1.02		1.12		1		1279
121518	−57 06410	V412 Cen		⋆ A	13 54 03	−57 28.0	1	7.03		1.74		1.69		5	M3 Iab/b	897
121387	−71 01537				13 54 04	−72 04.8	2	8.62	.023	0.43	.005	-0.11		4	F5wA9 V	1594,6006
	+26 02510				13 54 05	+26 27.9	1	9.98		1.19		1.36		1		280
121696	+32 02412				13 54 07	+32 07.5	1	7.84		0.22		0.11		1	F0	280
	+35 02505	IDS13519N3507		AB	13 54 07	+34 52.6	1	9.65		0.29		0.07		2	F0	1569
121629	−6 03907				13 54 07	−07 03.8	1	9.48		0.97		0.75		1	K2 V	3072
121553	−49 08323				13 54 08	−50 00.9	1	9.28		0.03		-0.09		1	B9.5 V	1770
	−61 04203				13 54 10	−61 35.3	1	11.45		0.43		0.13		1	B8	549
121611	−22 03685				13 54 11	−22 51.6	1	7.59		-0.01		-0.07		2	B9.5 V	1770
121596	−34 09257				13 54 12	−34 39.6	1	7.41		1.47		1.82		6	K3 III	1673
121683	+16 02583				13 54 13	+16 08.1	1	6.83		0.94		0.60		1	K0	280
121710	+28 02278	HR 5247			13 54 17	+27 44.2	5	5.00	.007	1.43	.009	1.72	.006	18	K3 III	269,1080,1363,1569,3053
121632	−26 10004	IDS13515S2702		AB	13 54 17	−27 16.8	1	8.24		0.84				5	G8/K0 V	173
		Lod 1101 # 19			13 54 17	−61 37.2	1	11.97		0.57		0.09		1	F6	549
121711	+21 02592				13 54 20	+20 53.9	1	8.81		1.11		1.07		1	G5	280
121615	−36 08965				13 54 20	−36 49.3	1	9.04		0.14		0.09		1	A1/2 V	1770
121570	−52 06865				13 54 20	−52 30.6	1	10.34		0.29		0.23		1	A2 IV/V	1770
121920	+70 00762				13 54 22	+70 10.7	1	7.27		1.16		1.06		2	K0	1733
121581	−50 08146				13 54 22	−50 55.7	1	9.66		0.26		0.19		1	A0 V	1770
	+38 02493				13 54 23	+37 43.0	1	10.32		1.01		0.83		1	K0 III	280
121725	+29 02475				13 54 25	+28 34.9	1	8.76		1.01		0.82		1	K0	280
121726	+24 02666				13 54 26	+23 47.7	1	9.22		1.13		1.08		1		280
	+30 02464	IDS13522N3040		AB	13 54 27	+30 25.1	1	9.25		0.61		0.08		2	G0	1371
121556	−59 05310				13 54 28	−59 38.5	2	8.83	.032	-0.03	.000	-0.69	.009	4	B5 Vn	540,976
	+16 02584				13 54 30	+16 16.0	1	10.56		0.95		0.64		1		280
	+34 02472				13 54 30	+34 10.8	1	8.88		1.56		1.83		1	M0	280
121780	+49 02253				13 54 30	+49 15.4	1	7.32		0.99		0.76		3	K0	1601
121617	−46 08973				13 54 32	−46 45.9	1	7.30		0.08		0.04		1	A1 V	1770
121598	−51 07861				13 54 32	−51 37.9	1	10.39		0.18		0.15		1	A1 V	1770
	−62 03784				13 54 32	−62 56.8	1	11.32		0.29		-0.38		2	B4 II-III	540
121544	−65 02551	IDS13509S6509		AB	13 54 35	−65 23.8	1	7.93		0.18		0.15		1	A2/3 V	1771
121781	+36 02427				13 54 36	+35 42.0	2	8.95	.015	0.04	.015	0.02	.000	3	A1 V	272,280
121521	−68 02047				13 54 36	−68 59.2	1	9.82		0.38		0.10		1	A9 IV/V	1771
		G 150 - 50			13 54 37	+25 43.4	1	12.47		0.70		0.00		2		1658
	+35 02508	IDS13525N3541		A	13 54 38	+35 25.7	1	9.07		1.19		1.37		3	K3 III	1569
121620	−53 05812				13 54 39	−53 27.7	1	7.08		0.96				4	G6/8 III	2033
		Lod 1101 # 1			13 54 39	−61 28.9	1	10.58		1.69		1.29		1	K3	287
	+17 02683				13 54 40	+17 18.4	1	9.79		1.04		0.89		1		280
		GD 160			13 54 40	+26 26.0	1	15.20		0.63		-0.13		1		3060
121699	−22 03687	HR 5246			13 54 40	−22 46.8	1	6.14		1.43				4	K2/3 III	2007
	−61 04207				13 54 40	−61 32.7	2	10.45	.010	1.11	.005	0.82	.050	2	K5	287,549
	+25 02705				13 54 42	+25 20.7	1	10.30		1.21		1.37		1		280
		NGC 5367 - 5			13 54 42	−39 43.	2	12.81	.023	0.90	.027	0.57	.032	5		556,734
		NGC 5367 - 6			13 54 42	−39 43.	2	13.69	.272	0.47	.019	0.14	.000	4		556,734
		NGC 5367 - 7			13 54 42	−39 43.	1	13.31		0.64		0.18		1		556
		NGC 5367 - 8			13 54 42	−39 43.	1	12.91		0.68		0.48		1		556
		NGC 5367 - 10			13 54 43	−39 52.2	1	12.11		0.53		0.07		1		556
	−48 11005	LSS 3687			13 54 43	−48 50.3	1	11.08		0.44		-0.26		2		954
121657	−48 08632	IDS13516S4836		A	13 54 43	−48 50.3	1	8.78		0.08		0.03		1	A0 V	1770
121754	+11 02613				13 54 44	+10 33.0	1	8.91		0.37		-0.05		3	F0	1648
	−39 08581	NGC 5367 - 1		⋆ A	13 54 44	−39 43.7	1	10.26		0.20		-0.15		1		556
	−39 08581	NGC 5367 - 1		⋆ AB	13 54 44	−39 43.7	2	9.75	.030	0.32	.004	-0.19	.008	7		556,734
	−39 08581	NGC 5367 - 1		⋆ B	13 54 44	−39 43.7	1	10.70		0.51		-0.19		1		556
121764	+21 02593				13 54 45	+21 11.9	2	6.70	.010	1.56	.010	1.94		6	K4 III	280,1501
121639	−52 06876				13 54 45	−52 25.1	2	8.32	.009	-0.03	.004	-0.30	.004	4	B9 III	401,1770
	−57 06417				13 54 45	−58 11.3	1	11.70		0.17		0.06		1	A0	549
	−61 04208				13 54 45	−61 33.4	2	10.23	.005	1.07	.000	0.84	.050	2	K3	287,549
		Lod 1101 # 24			13 54 46	−61 24.5	1	12.37		0.36		0.04		1	B8	549
121557	−65 02553	HR 5242		⋆ AB	13 54 46	−65 33.4	3	6.19	.004	1.04	.005	0.82	.005	14	K0 III	15,1075,2038
	−57 06418				13 54 48	−58 07.6	1	11.32		0.14		0.10		1	A0	549
	+7 02721	G 65 - 17			13 54 49	+07 13.6	1	10.01		0.88		0.63		1	K2	333,1620
	+25 02707				13 54 49	+24 55.4	1	9.99		1.21		1.40		1	K7	280

HD	DM	Other Id	N Rem	α_{1950}	δ_{1950}	S	V	σ_V	B–V	σ_{B-V}	U–B	σ_{U-B}	n	Spectrum	References
	+39 02700			13 54 50	+38 42.9	1	10.93		0.60		0.11		2	G8 III	1569
	+15 02649			13 54 52	+15 27.2	1	9.09		0.28		0.05		1	F2	280
121825	+44 02312			13 54 53	+44 31.5	1	7.67		0.58		0.07		2	G0	1601
	−61 04210			13 54 53	−61 31.1	1	11.10		0.75		0.19		1	F6	549
	+34 02473	G 165 -B5	B	13 54 54	+34 03.4	1	16.16		0.08		-0.70		4	DA	3060
	−39 08583	NGC 5367 - 2		13 54 54	−39 49.7	2	10.10	.013	0.12	.004	-0.26	.038	7		556,734
	+34 02473	G 165 -B5	A	13 54 55	+34 02.9	1	9.09		0.54		-0.04		4	F8	3029
121658	−55 05802			13 54 55	−56 06.6	1	7.53		2.82		5.50		2	Na	864
		Lod 1171 # 8		13 54 56	−58 13.5	1	11.02		1.32		1.35		1	K3	549
		G 150 - 53		13 55 00	+24 07.4	1	15.77		1.65				1		906
121733	−37 09044			13 55 01	−37 55.5	1	7.63		0.23		0.09		2	A9 V	1771
121692	−49 08335			13 55 01	−50 17.5	1	10.20		0.19		0.15		1	A1 V	1770
	+24 02669			13 55 02	+23 52.3	1	9.85		1.21		1.27		1	K0	280
121953	+66 00825			13 55 02	+65 35.9	1	7.59		0.66		0.14		5	F5	3026
121979	+67 00812			13 55 02	+67 11.3	1	8.47		0.77		0.32		2	K0	3026
121730	−30 11049			13 55 02	−30 43.3	1	7.58		1.07				4	K0 III	2012
	−61 04213			13 55 02	−61 39.1	1	10.75		1.22		1.04		1	K4	549
121827	+27 02312			13 55 03	+26 39.9	1	8.48		0.98		0.79		1	K0	280
	−39 08586	NGC 5367 - 4		13 55 03	−39 43.2	2	9.14	.009	1.01	.009	0.81	.027	5	K0	556,734
		NGC 5367 - 9		13 55 03	−39 52.5	1	11.96		0.50		0.08		1		556
	+29 02477			13 55 04	+29 02.0	1	9.98		1.08		1.16		1	K0	280
		NGC 5367 - 3		13 55 04	−39 44.5	2	12.44	.009	0.56	.023	0.04	.037	4		556,734
121703	−50 08152			13 55 04	−50 54.9	1	10.11		0.37		0.28		1	A2/3 IV	1770
		Lod 1171 # 24		13 55 05	−58 12.6	1	12.04		0.49		0.14		1	A7	549
	−61 04215			13 55 06	−61 29.5	1	10.96		0.49		0.13		1	F4	549
121574	−68 02049			13 55 08	−69 20.0	1	8.99		1.08				4	K0 III	2012
		Lod 1101 # 15		13 55 09	−61 39.1	1	11.53		0.52		0.02		1	F2	549
121844	+25 02708			13 55 10	+25 14.5	1	7.89		1.11		1.06		1	K1 III	280
	+35 02511			13 55 11	+34 57.7	1	9.86		0.28		0.10		2	A2	272
	−57 06421			13 55 11	−58 11.1	1	10.78		1.16		0.81		1	K3	549
121785	−19 03782			13 55 12	−20 24.1	1	8.76		0.30		0.10		3	A9/F0 IV/V	803
121743	−41 08329	HR 5248		13 55 13	−41 51.4	7	3.82	.007	-0.22	.006	-0.83	.010	20	B2 V	15,26,1020,1075,1770*
	−61 04219			13 55 16	−61 28.2	2	10.12	.000	1.22	.000	1.01	.000	2	K5	287,549
121735	−48 08640	IDS13521S4823	ABV	13 55 17	−48 37.9	1	9.89		0.21		0.02		1	A0 V	1770
121746	−47 08772			13 55 18	−48 13.3	2	7.14	.014	0.48	.000			8	F5 IV/V	1075,2033
		Lod 1101 # 23		13 55 18	−61 19.8	1	12.26		0.20		0.09		1	B9	549
121439	−77 00922	HR 5240		13 55 18	−78 20.9	3	6.10	.019	0.03	.003	-0.16	.005	11	B9 III	15,1075,1637
121705	−57 06422			13 55 19	−57 39.2	2	9.41	.000	0.09	.000	0.03	.000	2	A0 V	287,885
		Lod 1101 # 18		13 55 19	−61 21.9	1	11.78		0.38		-0.32		1	B6	549
		G 65 - 18		13 55 20	+08 03.9	1	15.75		1.48		1.10		1		333,1620
	+23 02633			13 55 20	+22 41.7	1	10.28		0.87		0.61		1		280
	+30 02469	ST CVn		13 55 20	+30 06.0	2	11.14	.083	0.24	.019	0.10	.054	3	A1	597,1569
	+33 02389			13 55 20	+33 03.1	1	9.13		1.17		1.21		2	K2 IV	1569
		Lod 1101 # 5		13 55 20	−61 27.8	1	10.58		1.69		1.29		1	M0	549
121747	−49 08342			13 55 22	−50 04.9	1	10.27		0.18		0.13		1	A1 V	1770
121880	+16 02588			13 55 28	+16 26.7	2	7.60	.000	0.05	.000	0.08	.029	3	A0 V	272,280
	+31 02564			13 55 28	+31 00.8	1	10.00		1.03		0.82		2	G8 III	1569
121719	−58 05306			13 55 29	−59 08.8	1	10.61		0.21		0.08		1	B8/9 III/IV	287
122020	+63 01105			13 55 34	+63 02.1	1	7.32		0.51		0.09		2	F5	1502
		Lod 1171 # 18		13 55 34	−58 12.1	1	11.69		0.58		0.04		1	F8	549
121790	−44 09010	HR 5249		13 55 35	−44 33.6	7	3.86	.007	-0.21	.006	-0.80	.012	29	B2 IV/V	15,26,1075,1637,1770*
	−57 06423			13 55 35	−58 18.6	1	11.66		0.23		-0.04		1	B9	549
121789	−42 08907			13 55 36	−42 45.4	1	9.41		0.28		0.12		1	A9 V	1771
	+75 00523	G 255 - 45		13 55 39	+74 57.5	2	9.73	.009	0.61	.014	0.03	.009	4	G5	308,1658
	−49 08351			13 55 39	−49 33.6	1	10.22		0.88		0.64		3	G5	803
	−57 06425			13 55 41	−58 08.8	2	11.27	.005	0.25	.005	0.18	.050	2	A0	287,549
121907	+17 02687			13 55 42	+16 38.8	2	7.41	.034	0.18	.005	0.12	.005	3	A0	272,280
121847	−24 11202	HR 5250		13 55 42	−24 43.7	4	5.20	.019	-0.09	.009	-0.38	.020	8	B8 Vp Shell	1079,1770,2007,3023
121849	−33 09467			13 55 43	−33 45.2	3	8.17	.000	0.69	.007	0.17	.005	16	G5 V	158,1775,2012
121934	+33 02390			13 55 44	+32 50.6	1	7.20		0.94		0.73		1	K0	280
121865	−11 03642			13 55 46	−11 48.9	1	7.05		0.98				2	G5	1351
121995	+52 01757	G 223 - 39	⋆ AB	13 55 47	+52 14.2	2	8.27	.010	0.66	.005	0.17		4	G5	1381,3030
		Lod 1101 # 12		13 55 47	−61 35.8	1	11.22		1.23		1.16		1	K5	549
121708	−67 02449			13 55 47	−67 36.3	1	8.25		1.71		2.06		8	K3 III	1704
121908	+2 02752	IDS13533N0243	AB	13 55 48	+02 27.8	2	8.57	.005	0.52	.002	0.00		10	G0	176,3026
		G 65 - 19		13 55 48	+12 48.9	1	12.25		1.66		1.29		1		333,1620
	+34 02475			13 55 48	+34 05.1	1	10.36		0.95		0.81		1	K1 III	280
		Lod 1171 # 23		13 55 48	−58 08.0	1	11.99		0.46		0.17		1	A8	549
	+28 02279			13 55 49	+28 29.2	1	10.03		1.10		1.18		1	K2	280
121909	−0 02769	BH Vir		13 55 50	−01 25.1	1	10.20		0.54		0.03		1	F8 V	588
		G 65 - 20		13 55 51	+01 07.5	1	14.28		1.47		1.40		1		333,1620
		LP 499 - 9		13 55 51	+09 33.6	1	13.42		0.51		-0.20		2		1696
		Lod 1171 # 21		13 55 51	−58 07.1	1	11.78		0.66		0.14		1	G0	549
		Lod 1101 # 11		13 55 51	−61 23.9	1	11.12		1.30		1.20		1	K5	549
		Lod 1171 # 14		13 55 52	−58 16.0	1	11.43		1.29		1.09		1	K4	549
	−57 06428			13 55 53	−58 07.4	1	9.97		1.14		0.91		1	K3	549
	−58 05308			13 55 53	−58 22.7	1	11.57		0.24		-0.10		1	B7	549
	−66 02395	LSS 3178		13 55 53	−67 15.2	2	10.45	.005	0.10	.010	-0.78	.045	5	B0 III	403,1737
	+30 02470			13 55 55	+29 58.3	1	10.47		0.48		-0.01		12	F5	272
121821	−50 08159			13 55 56	−50 42.8	1	8.98		0.06		0.02		1	B9 V	1770

Table 1   779

HD	DM	Other Id	N Rem	α₁₉₅₀	δ₁₉₅₀	S	V	σ_V	B–V	σ_B–V	U–B	σ_U–B	n	Spectrum	References
	−57 06429			13 55 56	−58 11.0	2	11.16	.005	0.09	.000	−0.16	.000	2	B9	287,549
		Lod 1171 # 34		13 55 56	−58 11.3	1	13.04		0.60		0.02		1		549
		Lod 1171 # 16		13 55 57	−58 09.2	1	11.63		0.53		0.01		1	F6	549
		Lod 1171 # 25		13 55 57	−58 14.7	1	12.14		0.29		0.18		1	A1	549
121956	+19 02736			13 55 58	+19 11.2	1	8.43		1.28		1.35				280
121794	−59 05324			13 55 58	−60 19.4	1	7.90		0.27		0.23		1	A1 V	1771
122064	+62 01325	HR 5256	⋆	13 55 59	+61 44.0	1	6.52		1.01		0.98		2	K3 V	1733
121807	−57 06431			13 55 59	−57 35.7	2	9.38	.000	0.05	.000	−0.15	.000	2	B8/9 III	287,885
122343	+79 00435			13 56 00	+78 56.3	1	8.21		0.29		0.12		2	A5	1733
		Lod 1101 # 16		13 56 00	−61 27.0	1	11.59		1.07		0.68		1		549
		Lod 1101 # 22		13 56 00	−61 33.3	1	12.08		0.32		−0.41		1	B5	549
	−61 04233			13 56 00	−61 36.4	1	10.57		1.12		0.88		1	K2	549
	+39 02704			13 56 01	+38 49.8	1	10.26		1.38		1.67		1		280
		Lod 1101 # 20		13 56 01	−61 26.2	1	12.05		0.32		0.24		1		549
121852	−44 09013			13 56 03	−45 13.5	2	7.38	.010	0.50	.015	0.01		6	F8 V	1075,3037
121853	−49 08356	HR 5251		13 56 04	−50 07.6	1	5.91		0.96				4	G8/K0 III	2007
		Lod 1171 # 22		13 56 07	−57 57.8	1	11.91		0.20		0.10		1	A0	549
121822	−57 06432			13 56 07	−58 09.3	1	9.71		0.17		0.05		1	B9.5V	549
121808	−60 05131			13 56 07	−61 08.3	1	8.09		0.25		0.15		1	A3 IV	1771
	−61 04237			13 56 08	−61 33.1	1	11.68		0.32		0.13		1	B8	549
	+20 02905			13 56 09	+19 50.3	1	9.63		0.98		0.67		2	G5	1569
		Lod 1171 # 26		13 56 09	−58 20.8	1	12.15		0.22		0.13		1	A0	549
	−58 05313			13 56 09	−59 06.6	1	10.63		0.26		−0.19		1		287
		G 177 - 66		13 56 10	+51 01.9	1	13.15		0.69		0.04		1		1658
121854	−51 07880	IDS13529S5159	AB	13 56 10	−52 13.9	1	10.01		0.40		0.24		1	A2mA7-A8	1770
121836	−53 05820			13 56 10	−53 56.2	1	9.56		0.17		0.15		1	A1 V	1770
	−58 05314			13 56 10	−58 24.9	1	10.66		0.57		0.00		1	G0	549
		Lod 1171 # 31		13 56 12	−57 59.5	1	12.63		0.53		0.04		1		549
121796	−64 02663			13 56 13	−65 18.0	1	7.58		0.16		0.17		1	A1 IV	1771
	+33 02392			13 56 14	+33 05.8	1	10.54		1.08		0.97		1		280
121980	+15 02651	HR 5254		13 56 15	+14 53.6	2	6.01	.010	1.42	.019	1.71	.005	3	K5 III	252,3001
121968	−2 03766			13 56 16	−02 40.3	4	10.26	.007	−0.18	.004	−0.91	.014	55	B5	989,1026,1729,1764
		AAS12,381 38 # C		13 56 17	+37 55.1	1	10.77		1.00		0.93		1		280
122037	+41 02445			13 56 17	+41 05.3	1	8.47		1.08		1.04		1	K0	280
		Lod 1171 # 11		13 56 17	−58 17.5	1	11.24		1.12		0.87		1	K3	549
	−61 04239			13 56 17	−61 36.0	1	11.48		0.32		−0.16		1	B6	549
		Ly 1 - 18		13 56 17	−61 54.7	1	14.36		0.61		0.29		2		1672
		LSS 3181		13 56 17	−62 20.2	1	11.05		0.92		−0.25		3	B3 Ib	1737
121996	+22 02650	HR 5255		13 56 18	+21 56.4	3	5.67	.154	−0.03	.016	0.02	.000	7	A0 Vs	70,1022,3023
122008	+23 02636			13 56 18	+23 07.0	2	9.60	.009	0.20	.000	0.09		6	A5 Vm	193,280
		L 197 - 13		13 56 18	−57 47.	1	13.29		1.22		0.98		1		3062
		UY Boo		13 56 20	+13 11.7	2	10.44	.193	0.16	.058	0.09	.021	2	A2	597,699
	+40 02710			13 56 20	+40 07.2	1	8.66		1.54		1.83		1	K3 V	280
		Lod 1171 # 33		13 56 20	−57 58.4	1	12.74		0.64		0.12		1	G0	549
		Lod 1171 # 32		13 56 20	−58 06.4	1	12.67		0.37		0.14		1	A5	549
		Ly 1 - 17		13 56 20	−61 54.9	1	14.80		0.80		0.39		2		1672
		Ly 1 - 16		13 56 20	−61 55.4	1	14.37		0.45		0.24		2		1672
		LP 856 - 26		13 56 21	−23 18.9	1	15.01		0.27		−0.61		4		3062
	−57 06434	IDS13529S5751	A	13 56 21	−58 06.0	2	11.09	.015	0.25	.020	0.26	.110	2	A0	287,549
121981	−6 03911			13 56 22	−06 40.9	1	6.96		1.00				2	G5	1351
121899	−48 08653			13 56 23	−48 46.5	1	8.92		−0.02		−0.33		1	B8 V	1770
121875	−57 06435			13 56 23	−57 35.3	2	9.29	.000	0.19	.000	0.21	.000	2	A2 IV	287,885
		Ly 1 - 19		13 56 23	−61 54.5	1	12.50		0.33		−0.11		3		1672
		Ly 1 - 15		13 56 23	−61 55.3	1	13.92		1.23		0.90		2		1672
		Ly 1 - 20		13 56 24	−61 54.7	1	11.74		0.45		0.07		3		1672
	+20 02906			13 56 25	+19 40.1	1	9.83		1.29		1.31		1		280
	+28 02281			13 56 25	+28 29.1	1	10.72		0.50		0.02		2		1569
		Ly 1 - 21		13 56 25	−61 54.6	1	13.39		0.70		0.13		1		1672
		Ly 1 - 14		13 56 26	−61 55.5	1	10.52		1.71		1.78		2		1672
		Ly 1 - 11		13 56 27	−61 56.3	1	14.01		0.68		0.09		1		1672
121877	−58 05317			13 56 28	−59 07.7	1	10.61		0.17		−0.05		1	B9/A0 IV/V	287
		Ly 1 - 23		13 56 28	−61 54.4	1	14.39		0.59		0.36		1		1672
		Ly 1 - 13		13 56 28	−61 55.4	1	14.15		0.91		0.44		1		1672
		Ly 1 - 12		13 56 28	−61 56.0	1	14.06		0.42		0.16		2		1672
	+39 02706			13 56 29	+38 55.5	1	9.56		1.12		1.08		1	K0	280
		Ly 1 - 24		13 56 29	−61 54.3	1	14.40		0.53		0.38		1		1672
		Ly 1 - 22		13 56 29	−61 54.8	1	12.97		0.34		0.01		3		1672
		Lod 1171 # 29		13 56 31	−58 10.6	1	12.48		0.66		0.14		1	F9	549
	+19 02739			13 56 33	+19 05.0	1	9.38		1.05		0.89		1	K0	280
	+25 02711			13 56 33	+25 08.8	1	10.36		1.08		0.77		1		280
	+38 02497			13 56 33	+38 17.2	1	9.77		1.20		1.33		1	K4 III	280
		Lod 1171 # 30		13 56 33	−58 11.6	1	12.54		0.50		0.10		1	F0	549
		Ly 1 - 9		13 56 33	−61 55.1	1	14.11		0.37		0.18		2		1672
		Ly 1 - 10		13 56 33	−61 56.1	1	13.46		0.47		0.09		2		1672
		Ly 1 - 7		13 56 35	−61 54.4	1	13.17		0.36		−0.07		2		1672
		Ly 1 - 8		13 56 35	−61 55.3	1	12.50		0.28		0.17		3		1672
122052	+25 02712			13 56 36	+24 56.1	3	7.27	.011	1.21	.005	0.94	.008	20	G0 III	280,1569,1569
		LP 739 - 5		13 56 37	−13 27.9	2	12.74	.020	0.66	.020	−0.14	.015	4		1696,3062
		Ly 1 - 6		13 56 38	−61 54.1	1	13.03		0.31		0.22		3		1672
		Ly 1 - 5		13 56 39	−61 54.1	1	11.08		0.16		0.08		3		1672

HD	DM	Other Id	N Rem	α_{1950}	δ_{1950}	S	V	σ_V	B–V	σ_{B-V}	U–B	σ_{U-B}	n	Spectrum	References
		Ly 1 - 4		13 56 40	−61 54.4	1	14.85		1.27		0.49		2		1672
		Ly 1 - 2		13 56 40	−61 54.9	1	12.46		2.37		2.32		3		1672
	−61 04249			13 56 41	−61 34.0	1	11.11		0.37		0.26		1	A3	549
		Ly 1 - 3		13 56 41	−61 54.7	1	16.22		0.87		0.17		1		1672
		Lod 1171 # 20		13 56 44	−58 11.3	1	11.76		0.31		-0.05		1	B8	549
121901	−60 05135	HR 5252		13 56 44	−61 14.3	3	6.49	.009	0.33	.009			11	F0/2 III/IV	15,2020,2027
	+29 02480			13 56 45	+29 09.1	1	10.04		0.88		0.61		1	K0	280
	+27 02315			13 56 46	+27 15.1	1	9.27		0.99		0.81		1	K0	280
		Ly 1 - 1		13 56 46	−61 55.2	1	12.41		0.50		0.27		3		1672
122080	+26 02511	IDS13545N2618	AB	13 56 47	+26 03.5	1	7.11		0.18		0.12		1	A5	280
122132	+47 02108			13 56 47	+46 50.3	1	7.01		1.65		1.99		2	M2 III	1601
122149	+54 01636			13 56 47	+53 49.3	1	7.94		0.59		0.17		2	G2 IV	1566
121983	−32 09762			13 56 48	−33 17.8	3	8.11	.005	-0.10	.009	-0.75	.008	12	B3 II	26,55,1775
		LP 619 - 73		13 56 49	−02 34.2	1	11.55		0.70		0.08		2		1696
		Lod 1171 # 28		13 56 49	−58 13.7	1	12.40		0.55		0.36		1	A0	549
122081	+23 02638			13 56 50	+22 42.9	1	8.35		1.05		0.77		1	G5	280
	+23 02639			13 56 50	+22 58.4	1	9.85		1.54		1.90		1	M0	280
	+18 02806			13 56 51	+18 05.4	1	9.71		1.03		0.89		1	K0	280
	−57 06438			13 56 54	−57 55.9	1	10.92		0.34		-0.06		1	B8	549
	+34 02476	G 165 - 39		13 56 57	+34 06.7	5	10.05	.010	0.40	.014	-0.21	.013	10	sdF5	792,1003,1064,1658,3025
122120	+23 02640			13 57 00	+23 06.7	1	9.00		1.16		1.10		2		3072
		Lod 1101 # 21		13 57 01	−61 29.8	1	12.08		0.55		-0.01		1		549
	+37 02478			13 57 02	+37 15.4	1	9.45		0.98		0.72		2	K0	1569
121965	−57 06439			13 57 02	−58 19.3	1	10.31		0.20		-0.24		1	B8/9 IV/V	549
121891	−67 02459			13 57 02	−68 08.7	1	9.67		0.34		0.19		1	A7 IV/V	1771
		G 150 - 54		13 57 05	+25 29.0	1	10.73		1.26				1	K6	1746
121976	−56 06072			13 57 05	−57 17.0	1	10.33		0.13		-0.26		1	A3/5 V	885
121932	−65 02573	HR 5253	⋆ A	13 57 05	−66 01.6	3	5.96	.003	0.35	.004	0.05	.015	17	F2 III	15,1075,2038
	+16 02597			13 57 06	+16 05.4	1	10.82		0.44		-0.07		2		1569
122133	+25 02713	IDS13548N2447	AB	13 57 07	+24 32.6	1	8.15		0.24		0.21		1	F0	280
	−58 05322			13 57 07	−58 22.5	1	10.31		0.48		0.06		1	F2	549
		CCS 2138		13 57 07	−61 22.0	1	11.60		3.02				1	CS	864
		Lod 1171 # 27		13 57 08	−58 13.7	1	12.16		0.35		0.16		1	A2	549
	+38 02498			13 57 09	+38 07.1	1	9.93		1.16		1.17		1	K2	280
	+35 02513			13 57 10	+34 37.8	1	9.88		1.03		0.87		2	K0	1569
122015	−45 08869			13 57 10	−45 53.1	1	6.60		0.19		0.09		1	A7 V	1771
122066	−24 11215	HR 5257		13 57 12	−24 46.0	2	5.78	.010	0.48	.002	0.05		6	F6 V	2007,3053
	−64 02670	LSS 3183		13 57 12	−64 47.3	1	11.37		0.38		-0.64		3	O9.5 III	1737
122106	−2 03768	HR 5258		13 57 14	−03 18.4	1	6.33	.086	0.48	.014	0.05	.009	9	F8 V	15,254,1417
		G 64 - 34	⋆ V	13 57 16	−05 08.4	2	12.01	.030	1.02	.015	0.72	.035	2		1620,1658
122017	−53 05828			13 57 17	−53 57.9	1	8.39		0.18		0.16		1	A2 V	1770
122002	−56 06076			13 57 18	−57 13.4	1	10.14		0.21		-0.32		1	B9 II/III	885
		RU CVn		13 57 20	+31 53.5	1	11.36		0.10		0.11		1	A2	668
122167	+23 02642			13 57 21	+23 10.4	1	8.63		0.57		0.08		2	G0	1569
	+37 02480	RW CVn		13 57 25	+37 26.4	1	9.46		1.51		1.03		3	gM7	1569
	+25 02715			13 57 26	+25 28.8	1	9.74		1.02		0.91		1	K0	280
122109	−35 09181			13 57 31	−35 32.6	1	8.02		0.03		0.05		3	A0 V	1770
	−26 10045			13 57 32	−26 53.6	1	10.72		0.71		0.12		2		1696
	−58 05324			13 57 33	−59 14.4	2	10.41	.029	0.28	.003	-0.40	.001	4	B2.5IV	540,976
	+25 02716			13 57 34	+25 04.8	1	10.02		1.19		1.34		1	K7	280
234096	+51 01885			13 57 34	+50 45.5	1	9.29		0.39		0.03		2	F5	1566
122201	+24 02675			13 57 36	+23 32.5	1	9.21		0.96		0.62		1	G5	280
121992	−66 02405			13 57 36	−66 34.4	1	8.25		0.21		0.19		1	A1/2 V	1771
122253	+42 02455			13 57 38	+42 17.5	1	7.80		0.90		0.61		3	K0	1733
122093	−48 08672			13 57 39	−49 17.1	1	8.28		0.26		0.06		2	A9 V	1771
	+19 02740			13 57 40	+18 43.9	1	9.95		1.47		1.84		1		280
122076	−52 06912			13 57 41	−53 11.8	1	9.54		0.15		0.06		1	A0 IV	1770
	+42 02456			13 57 43	+41 35.5	1	9.41		1.02		0.81		1	G5	280
	+18 02808			13 57 44	+18 31.2	1	9.73		1.35		1.47		1	K2	280
122137	−37 09079			13 57 45	−37 52.9	1	8.08		0.26		0.16		1	A1/2 + (F)	1770
122128	−41 08367			13 57 45	−42 06.4	1	9.33		0.30		0.14		1	A9 V	1771
		LSS 3184		13 57 48	−65 55.6	1	12.60		0.03		-0.84		2	B3 Ib	1737
122096	−56 06082			13 57 53	−56 59.2	1	6.93		1.76		1.92		5	K4 II/III	897
122174	−37 09080			13 57 55	−37 46.3	2	8.55	.021	-0.05	.007	-0.18	.006	3	B9 Vn	1770,6006
	+34 02477			13 58 00	+34 00.2	1	9.34		1.42		1.67		1	K3 III	280
	+18 02809			13 58 01	+17 52.4	1	9.25		1.08		0.99		1	K0	280
122158	−43 08723			13 58 02	−43 57.3	1	6.70		0.14		0.14		1	A4/5 V	1771
122285	+35 02516			13 58 03	+34 36.9	1	8.36		1.11		0.95		3	K0 III	1569
122196	−37 09083			13 58 04	−37 48.5	7	8.73	.012	0.46	.008	-0.19	.018	20	F0 (V)w	1088,1236,1594,1696*
122159	−47 08803	IDS13549S4806	AB	13 58 04	−48 20.7	3	8.58	.010	-0.02	.004	-0.28	.003	8	B8 IV/V	26,1770,2014
122242	−7 03751			13 58 13	−07 49.8	1	8.04		0.91		0.52		4	K0	1657
122316	+38 02501			13 58 15	+38 06.8	1	8.80		1.49		0.77		1	M3	280
		G 150 - 56		13 58 16	+21 47.6	1	15.85		1.68				1		906
122208	−38 08981			13 58 16	−39 00.7	1	7.95		0.13		0.11		1	Ap SrCrEu	1770
122326	+39 02708			13 58 17	+39 16.3	2	6.63	.000	1.30	.005	1.43	.014	4	K3 III	280,833
122210	−39 08628	HR 5259	⋆ A	13 58 17	−39 58.8	1	6.13		1.25				4	K1 III	2032
122345	+41 02449			13 58 19	+41 03.1	1	8.72		1.31		1.54		2	K2	280
122301	+16 02598			13 58 20	+15 37.1	1	8.07		1.25		1.32		1	K0	280
122101	−66 02407	IDS13545S6632	ABC	13 58 21	−66 46.8	1	9.45		0.43		0.27		1	A5/7 IV/V	1771
122142	−60 05151			13 58 22	−60 33.1	1	7.97		0.11				4	B4 V	2014

Table 1   781

HD	DM	Other Id	N Rem	α_{1950}	δ_{1950}	S	V	σ_V	B–V	σ_{B-V}	U–B	σ_{U-B}	n	Spectrum	References
	+18 02811	G 135 - 6		13 58 23	+18 20.2	2	10.31	.044	1.21	.024	1.22	.019	3	K6	1569,1696
	+34 02479			13 58 25	+33 35.6	1	9.88		1.01		0.87		1		280
	+21 02597			13 58 27	+21 27.7	1	10.28		1.09		1.02		1		280
122384	+41 02450			13 58 31	+40 47.6	4	8.49	.006	0.46	.003	-0.01	.006	39	G0	1502,1569,1569,1648
122303	−1 02892	G 64 - 35		13 58 31	−02 25.4	3	9.71	.012	1.48	.011	1.13	.009	6	K5	158,1705,3073
	+20 02910			13 58 32	+20 23.6	1	9.57		1.04		0.84		2	K0	1569
122223	−44 09040	HR 5260		13 58 36	−45 21.7	7	4.34	.011	0.60	.007	0.26	.007	35	F7 II/III	15,1075,1637,2012*
	+36 02433			13 58 37	+35 38.0	1	9.80		1.54		1.99		1	K5 III	280
122364	+22 02651			13 58 43	+22 13.1	1	7.11		0.91				2	K0 IV	280
122259	−38 08988			13 58 44	−39 11.8	1	7.53		0.13		0.11		1	A3 IV/V	1771
122232	−55 05836			13 58 47	−55 54.3	2	9.22	.025	0.45	.005	-0.07		4	F3/5 V	1594,6006
	+18 02812			13 58 48	+17 37.8	1	9.86		1.31		1.50		1		280
	+36 02434			13 58 48	+35 38.3	1	9.88		1.16		1.14		1		280
122404	+31 02574			13 58 49	+30 55.7	1	8.37		1.36		1.57		2	K0	1371
122260	−44 09043			13 58 49	−44 47.7	1	9.65		0.29		0.14		2	A9 V	1771
122385	+22 02652			13 58 50	+21 38.4	1	9.15		0.93		0.61		1		280
122422	+33 02395			13 58 51	+32 43.9	1	7.74		0.22		0.09		1	A8 IV	280
122365	+9 02835	HR 5262		13 58 52	+09 08.2	2	5.98	.005	0.09	.000	0.09	.005	7	A2 V	15,1417
122274	−42 08960	IDS13558S4256	AB	13 58 53	−43 10.4	1	9.77		0.12		0.08		1	A1 V	1770
122275	−46 09027	IDS13557S4652	AB	13 58 53	−47 06.9	1	9.30		0.27		0.13		1	A7/8 V	1771
122386	+14 02686			13 58 54	+13 58.4	1	7.14		0.88		0.51		2	G8 III-IV	1648
122405	+28 02287	HR 5263		13 58 54	+27 37.6	1	6.23		0.17				2	A7 III	70
122292	−38 08990			13 58 54	−38 43.8	1	8.97		0.18		0.09		1	A4/5 IV/V	1771
	+41 02451			13 59 02	+40 37.0	2	9.81	.005	1.23	.005	1.34	.061	4	K5	280,1569
122180	−68 02066			13 59 02	−68 58.6	1	8.13		-0.08		-0.59		2	B2/3 III	400
	+40 02716			13 59 03	+40 19.3	1	10.14		1.04		0.81		1	K0	280
122442	+29 02483			13 59 04	+28 39.0	3	8.17	.010	0.18	.007	0.17	.019	17	A2	269,1569,1775
122408	+2 02761	HR 5264	⋆ A	13 59 06	+01 47.1	16	4.25	.008	0.10	.004	0.12	.020	93	A3 V	1,3,15,272,1007,1013*
122408	+2 02761	IDS13566N0202	B	13 59 06	+01 47.1	1	9.48		0.45				2		3024
		L 547 - 141		13 59 06	−34 49.	1	12.69		0.87		0.23		3		3062
122456	+32 02419			13 59 09	+31 48.3	1	6.88		1.42				2	K3 III	280
	+23 02643			13 59 10	+23 26.8	1	9.43		1.31		1.50		2	K5	280
	+32 02420			13 59 10	+32 31.9	1	10.32		1.12		1.09		1	G8 III	280
	+19 02745			13 59 11	+19 04.4	1	9.44		1.42		1.72		1	K5	280
122457	+17 02691			13 59 14	+16 59.9	1	7.39		0.50		0.04		2	F8	1569
122443	+18 02813			13 59 14	+17 54.6	2	6.36	.005	1.22	.005	0.86		4	G5	280,1569
	+19 02746			13 59 14	+18 43.8	1	10.20		0.96		0.76		1	K0	280
		G 65 - 22		13 59 16	+09 10.3	5	11.57	.009	0.74	.010	0.09	.014	10		158,333,538,1064,1620*
122503	+33 02396			13 59 21	+32 49.4	2	8.52	.009	1.26	.012	1.22	.021	8	K2	280,1775
122322	−51 07914			13 59 21	−51 50.0	1	10.41		0.31		0.22		1	A2 IV	1770
122546	+50 02041			13 59 23	+50 22.8	2	8.46	.015	1.19	.000	1.32	.010	5	K2 III	37,1733
	−66 02415			13 59 23	−66 23.7	1	9.25		0.88		0.45		3		3056
	+39 02712			13 59 24	+39 16.3	1	10.61		1.10		1.20		1	K3 III	280
122394	−35 09210			13 59 24	−35 56.3	2	7.05	.002	0.04	.001	0.05	.000	5	A0 V	401,1770
122324	−55 05841	LSS 3188		13 59 25	−55 45.8	5	9.07	.019	0.37	.018	-0.58	.010	14	B0.5Ia	26,138,403,540,976
122297	−59 05356			13 59 25	−59 56.0	1	8.01		0.30		0.28		1	A2 IV	1771
122517	+36 02435			13 59 26	+36 27.9	1	7.53		0.84				2	K0	280
	+33 02398			13 59 29	+33 04.2	1	10.19		0.98		0.75		1	K0 III	280
	+34 02480			13 59 29	+34 28.8	1	9.29		1.24		1.44		1	K3 III	280
122336	−51 07915			13 59 29	−51 39.6	1	10.06		0.27		0.18		1	A2 V	1770
	+30 02475			13 59 30	+30 20.1	1	9.11		1.01		0.88		1	K0	280
		HA 58 # 137		13 59 31	+29 13.6	1	12.24		0.59		0.09		2		269
122430	−26 10060	HR 5265		13 59 32	−27 11.4	2	5.47	.005	1.33	.005	1.44		6	K3 III	2007,3005
122375	−49 08402			13 59 33	−49 53.9	1	9.09		0.18		0.14		1	A1 V	1770
		Lod 1202 # 13		13 59 33	−58 27.5	1	10.90		1.40		1.28		1	K5	549
		MCC 149	AB	13 59 34	+15 44.1	1	10.61		1.44				2	M0	1625
		HA 58 # 230		13 59 34	+29 33.0	1	11.47		0.63		0.17		1		269
		HA 58 # 231		13 59 35	+29 26.9	1	12.17		0.64		0.24		1		269
122313	−61 04286	LSS 3187		13 59 35	−62 01.5	2	9.47	.009	0.29	.006	-0.65	.020	8	O5/7	540,976
		MCC 149	A	13 59 36	+15 45.5	1	10.58		1.46				1	M2	1705
	−59 05362	LSS 3189		13 59 36	−60 09.6	1	9.52		0.81		-0.30		2		1737
	+28 02288			13 59 37	+27 57.9	1	10.17		1.02		0.78		1	K2	280
122518	+22 02654			13 59 38	+21 47.8	1	7.94		0.59				1	G0	280
122338	−58 05335			13 59 38	−58 42.5	1	8.80		0.57		0.39		1	A9 V	549
		Lod 1202 # 18		13 59 39	−58 40.4	1	11.58		0.48		0.35		1	A0	549
122547	+33 02399			13 59 44	+33 04.0	4	9.48	.028	1.24	.035	1.02	.029	7	R2	280,1238,1569,8005
122574	+41 02454			13 59 44	+40 39.6	2	6.80	.010	1.17	.000	1.24		6	K2	280,1569
		AAS12,381 22 # D		13 59 45	+22 29.0	1	11.27		0.20		0.02		1		280
	+33 02400			13 59 45	+33 04.7	2	8.89	.005	0.95	.015	0.83	.005	3	K0	280,1569
	+39 02713			13 59 46	+39 08.6	1	10.38		1.23		1.42		1	K3 III	280
122314	−66 02419		V	13 59 47	−66 29.8	1	7.87		0.35		0.19		1	A5 IV/Vs	1771
122447	−39 08652			13 59 51	−40 20.0	1	8.81		0.37		0.06		1	A2mA8-A9	1770
	+16 02602			13 59 52	+16 30.7	1	10.34		1.10		1.08		1	K0	280
		G 64 - 37		13 59 53	−05 24.4	3	11.12	.016	0.37	.009	-0.22	.023	5		1620,1696,6006
	+32 02422			13 59 54	+32 07.6	1	9.10		1.14		1.21		1	K0	280
122449	−46 09040			13 59 55	−46 34.4	4	8.13	.008	-0.07	.007	-0.43	.001	10	B8 Ib/II	26,55,1770,2014
122548	+11 02620			13 59 56	+10 34.0	1	7.09		1.07		1.00		2	K0	1648
122562	+21 02598			14 00 02	+21 07.4	2	7.66	.024	0.96	.010	0.89		3	G5	280,1569
122464	−48 08712			14 00 02	−48 33.1	1	9.98		0.19		0.06		1	A1 V	1770
122438	−55 05846	HR 5266		14 00 03	−55 58.4	3	5.92	.008	1.21	.010			11	K2 III	15,1075,2032

HD	DM	Other Id	N Rem	α_1950	δ_1950	S	V	σ_V	B–V	σ_B-V	U–B	σ_U-B	n	Spectrum	References
	+38 02506			14 00 04	+37 51.3	1	9.33		0.90		0.73		2	K2 III	280
122563	+10 02617	HR 5270		14 00 05	+09 55.6	7	6.20	.008	0.91	.009	0.36	.020	15	F8 IV	15,1003,1064,1509*
122651	+44 02319			14 00 06	+43 48.0	1	7.68		0.90		0.59		2	G5	1601
	+38 02507			14 00 07	+37 44.8	1	9.91		1.12		1.08		2	K2 III	280
122510	−31 10859	HR 5268	⋆ AB	14 00 07	−31 26.7	2	6.17	.005	0.48	.000			7	F6 V	15,2012
122511	−39 08653			14 00 09	−39 27.7	1	8.00		-0.04		-0.13		2	B9 IV	1770
122637	+36 02436			14 00 11	+36 21.2	1	7.23		0.46		-0.02		2	F5	1569
		BPM 22431		14 00 12	−58 23.	1	14.55		0.95		0.27		3		3065
	−70 01704			14 00 12	−70 33.3	1	9.52		0.07		-0.38		4		400
122675	+46 01922	HR 5271		14 00 13	+45 59.7	2	6.26	.030	1.34	.025	1.45		6	K4 III	252,1118
	−61 04292	LSS 3190		14 00 13	−62 04.3	2	8.76	.027	0.19	.002	-0.64	.012	4	B1.5IIp	540,976
	+20 02913			14 00 14	+20 02.1	2	9.84	.020	0.29	.005	0.06	.020	5	A3	1569,1776
122479	−51 07921			14 00 14	−51 49.2	2	7.37	.010	-0.07	.003	-0.64		7	B2 III	26,2014
	+27 02321			14 00 17	+27 03.3	1	9.86		1.34		1.64		1	M0	280
	+40 02721			14 00 17	+40 25.2	2	9.82	.015	0.25	.005	0.11	.010	3	A2	272,280
122512	−40 08366			14 00 17	−40 26.0	1	9.81		0.98		0.68		2	A2mA7-F0	1770
	−58 05339			14 00 17	−58 31.7	1	10.40		0.81		0.53		1	F0	549
122451	−59 05365	HR 5267, β Cen	⋆ AB	14 00 17	−60 08.0	2	0.61	.000	-0.25	.011	-0.98		5	B1 III	1034,2006
		Lod 1202 # 17		14 00 18	−58 31.9	1	11.56		1.48		1.55		1		549
122740	+57 01478	IDS13586N5742	AB	14 00 19	+57 27.9	1	7.76		0.47		0.00		2	F5	3030
122740	+57 01478	IDS13586N5742	C	14 00 19	+57 27.9	1	10.42		1.56		1.79		1		3030
122577	−16 03785			14 00 20	−17 07.6	1	6.35		1.38		1.55		9	K2/3 III	1628
	+41 02455			14 00 21	+40 43.9	1	10.58		1.11		1.06		1	K0	280
122496	−51 07922	IDS13571S5111	A	14 00 21	−51 25.2	1	9.31		0.11		-0.01		1	A0 V	1770
122450	−58 05340	LSS 3191		14 00 21	−59 13.4	3	9.26	.027	0.44	.010	-0.50	.018	6	B1/5N	403,540,976
	+35 02517			14 00 23	+35 11.7	1	9.18		1.00		0.76		2	K0 III	1569
122250	−76 00799	HR 5261, θ Aps		14 00 23	−76 33.4	1	5.48		1.48		1.07		5	M7 III	897
122579	−16 03786	IDS13577S1707	A	14 00 24	−17 22.0	1	8.72		0.68				4	G3 V	2033
122532	−40 08373	HR 5269, V828 Cen		14 00 24	−41 11.0	4	6.10	.003	-0.12	.007	-0.38	.009	13	Ap Si	15,1770,2027,8071
	−58 05341			14 00 24	−58 22.0	1	10.97		0.59		0.24		1	F7	549
122653	+18 02815			14 00 28	+18 13.7	1	8.04		1.50		1.88		1	K0	280
		G 165 - 42		14 00 28	+33 14.5	1	15.06		1.23		0.95		1		3062
122578	−16 03787	IDS13577S1707	B	14 00 28	−17 19.9	1	9.08		0.68				4	G3/5 V	2033
		L 197 - 165		14 00 30	−59 09.	1	13.96		1.59				1		3062
122535	−47 08837			14 00 31	−47 58.3	1	9.86		0.11		0.08		1	A1 V	1770
122676	+15 02658			14 00 32	+15 12.9	1	7.14		0.72		0.29		2	G5	1648
	+47 02112	G 200 - 16	⋆ AB	14 00 32	+46 34.9	4	9.16	.007	1.49	.018	1.10	.078	13	M1	497,906,1197,3072
122693	+25 02722	IDS13583N2503	AB	14 00 34	+24 48.2	1	8.06		0.57		0.07		2	F8 V	3031
	−58 05342			14 00 34	−58 29.5	1	9.69		1.47		1.28		1	K3	549
122716	+31 02578			14 00 38	+30 34.0	1	9.13		0.79		0.32		2	G8 V	1371
	+36 02437			14 00 38	+36 31.4	1	9.44		1.37		1.68		1	K5	280
122569	−39 08661			14 00 38	−39 48.6	1	9.02		0.05		-0.01		1	Ap CrEuSr	1770
		AAS61,331 # 6		14 00 39	+17 22.	1	10.29		0.33		0.09		2		1569
122557	−46 09053			14 00 39	−47 09.1	1	8.52		0.18		0.16		1	A5 V	1771
122525	−53 05856			14 00 39	−53 28.8	1	8.75		0.24		0.21		1	Ap SrCrEu	1770
	−58 05343			14 00 40	−58 28.0	2	10.49	.010	0.41	.005	0.20	.035	2	A0	287,549
	−33 09515			14 00 41	−34 10.8	1	10.08		1.08		0.95		1		1730
		GD 161		14 00 43	+21 00.6	1	14.96		0.46		-0.24		1		3060
		AAS12,381 29 # B		14 00 43	+28 56.0	1	10.72		1.70		1.32		1		280
122909	+69 00733	HR 5282		14 00 44	+68 55.1	1	6.34		1.40				2	K5	71
	+17 02696			14 00 46	+16 45.6	1	10.28		1.16		1.39		1		280
122589	−42 08994			14 00 47	−42 45.3	1	9.14		0.05		0.05		2	A0 V	1770
		G 135 - 12		14 00 48	+14 56.5	1	12.40		1.50		1.16		1		1696
122730	+21 02600			14 00 49	+20 32.5	1	8.84		1.02		0.80		1	G5	280
	−33 09517			14 00 50	−34 20.4	1	11.32		0.95				2		1730
		Lod 1202 # 26		14 00 50	−58 27.7	1	12.14		0.52		0.34		1	A5	549
	−58 05345			14 00 50	−58 38.1	1	10.94		0.25		0.07		1	B9	549
	+40 02722			14 00 51	+39 39.2	1	9.47		0.95		0.70		1	K0	280
122540	−57 06479			14 00 53	−58 19.4	1	10.26		0.27		0.05		1	A0	549
	−58 05346			14 00 55	−58 27.3	2	10.47	.000	0.46	.000	0.30	.035	2	A0	287,549
	+16 02518			14 00 56	+26 16.8	1	9.43		1.19		1.30		1	K7	280
122767	+25 02723			14 00 58	+24 50.2	1	7.96		1.33		1.43		1	K3 III	280
122784	+33 02402			14 01 00	+33 03.2	1	7.22		1.49		1.88		1	K5 III	280
	+20 02914			14 01 01	+20 07.1	1	9.14		1.61		1.89		1	K5	280
	+37 02489			14 01 01	+37 07.2	1	10.05		0.98		0.77		2	K0	1569
122768	+23 02644			14 01 02	+22 44.1	1	6.85		0.31		0.14		2	F0	280
	−33 09522			14 01 02	−34 10.7	1	10.52		1.09				2		1730
	+24 02680	IDS13587N2438	AB	14 01 03	+24 23.9	1	9.94		1.08		0.93		1	K0	280
		Lod 1202 # 23		14 01 03	−58 33.9	1	11.81		0.34		0.13		1	B9	549
122617	−50 08231			14 01 04	−50 54.1	1	9.67		0.29		0.21		1	A7 V	1771
122742	+11 02625	HR 5273		14 01 05	+11 01.8	5	6.29	.028	0.73	.010	0.30	.009	15	G8 V	101,245,1067,1080,3077
122703	−21 03824	HR 5272		14 01 05	−22 10.9	1	6.30		0.45				4	F3 IV	2006
122664	−36 09061			14 01 05	−37 02.6	1	8.34		0.22		0.09		1	A7 III/IV	1771
		Lod 1202 # 24		14 01 06	−58 35.1	1	11.96		0.69		0.40		1		549
122560	−61 04312			14 01 06	−62 04.0	1	9.95		0.39		0.17		1	A8/9	1771
122744	+8 02810	HR 5274		14 01 08	+07 47.2	2	6.25	.005	0.94	.000	0.70	.000	7	G9 III	15,1417
	+21 02601			14 01 08	+21 15.2	1	10.14		1.36		1.82		1	K2	280
122866	+51 01889	HR 5280		14 01 08	+51 12.7	2	6.16	.005	0.01	.000	0.00		8	A2 V	1118,1501
		G 65 - 23		14 01 10	+13 33.6	1	12.12		1.31		1.24		1		1620
	+29 02484			14 01 11	+28 52.1	1	10.48		0.90		0.52		1	G0	280

Table 1　　　　　　　　　　　　　　　　　　　　　　　　　　　　783

HD	DM	Other Id	N	Rem	α_{1950}	δ_{1950}	S	V	σ_V	B−V	σ_{B-V}	U−B	σ_{U-B}	n	Spectrum	References
122595	−58 05347	V413 Cen			14 01 11	−58 39.2	1	9.78		0.15		-0.17		1	B8 IV/V	549
122796	+28 02291				14 01 12	+27 45.0	2	7.28	.145	1.06	.000	0.91	.020	2	K1 III	280,3040
122618	−53 05863				14 01 12	−53 47.6	1	9.36		0.21		0.17		1	A2 V	1770
122769	+9 02842	IDS13588N0858	AB		14 01 15	+08 43.6	3	7.58	.005	0.46	.015	0.00	.005	6	F5	292,1381,3030
122685	−41 08429				14 01 16	−41 31.5	1	7.85		0.15		0.12		1	A5 V	1771
	−58 05348				14 01 17	−58 28.5	2	10.88	.010	0.32	.005	0.19	.010	2	A0	287,549
122722	−33 09526				14 01 19	−34 16.9	1	9.15		0.56		0.08		2	F7 V	1730
122647	−52 06969				14 01 19	−52 55.1	1	10.09		0.23		0.17		1	A1 V	1770
122687	−47 08847				14 01 21	−48 18.0	1	9.18		1.04		0.68		1	Kp Ba	565
	+33 02404				14 01 22	+33 18.9	1	10.37		0.30		0.02		2		1569
	−58 05350				14 01 22	−58 34.7	1	10.01		1.75		2.10		1	M0	549
	+40 02724				14 01 24	+39 54.8	1	9.43		1.09		0.97		1	K0	280
122797	+5 02836	HR 5275			14 01 25	+05 08.4	2	6.23	.005	0.40	.005	-0.03	.010	7	F4 V	15,1417
122705	−49 08424				14 01 27	−49 49.9	1	7.66		0.12		0.07		1	A2 V	1770
		HA 58 # 195			14 01 29	+29 19.7	1	11.04		0.51		0.01		2		269
	+20 02915				14 01 30	+19 55.5	1	9.84		1.01		0.88		2		1569
		HA 58 # 242			14 01 30	+29 29.0	1	12.89		1.36		1.25		1		269
122723	−44 09068				14 01 30	−45 02.9	1	8.34		0.23		0.09		1	A5 V	1771
		L 197 - 49			14 01 30	−55 23.	1	11.06		0.93		0.56		2		1696
		HA 58 # 196			14 01 32	+29 19.5	1	12.18		1.28		1.22		1		269
		G 65 - 24			14 01 34	+07 50.1	1	12.35		1.17		1.05		1		1620
		HA 58 # 243			14 01 34	+29 26.3	1	14.60		0.55		-0.12		1		269
122724	−49 08427				14 01 36	−49 57.9	1	8.03		0.20		0.08		1	A1 III/IV	1770
		HA 58 # 197			14 01 37	+29 16.8	1	13.84		0.61		0.17		1		269
		Lod 1202 # 20			14 01 37	−58 34.7	1	11.66		0.99		0.36		1	G9	549
122815	−4 03614	HR 5276			14 01 38	−05 08.5	1	6.38		1.32		1.57		8	K0	1088
123154	+73 00615				14 01 40	+73 10.6	1	8.62		0.92		0.64		3	K0	3025
	+23 02645	G 135 - 16			14 01 41	+22 47.2	1	10.16		0.67		0.10		1	G2	1658
122736	−48 08738				14 01 41	−49 01.8	1	9.79		0.19		0.11		1	A1 IV/V	1770
122757	−40 08392	IDS13587S4059	A		14 01 42	−41 13.3	1	8.67		0.11		0.10		1	A2/3 V	1770
122669	−61 04317	LSS 3193			14 01 43	−62 16.1	4	8.97	.014	0.36	.016	-0.72	.015	11	O/Be	92,540,976,1737
122837	−14 03863	HR 5277			14 01 44	−14 43.9	2	6.32	.029	1.08	.000	0.95		6	K1 III +G	252,2007
	+35 02520				14 01 45	+34 53.8	1	10.50		0.35		0.22		1		280
122691	−61 04318	LSS 3194			14 01 46	−62 20.9	4	9.20	.007	0.35	.008	-0.75	.022	12	O/Be	92,540,976,1737
	−59 05381				14 01 47	−59 24.8	1	11.02		0.41		0.25		1		287
		Lod 1202 # 11			14 01 48	−58 24.4	1	10.86		1.62		1.41		1	K4	549
		CCS 2142, U Cir			14 01 48	−66 46.6	1	9.58		2.78		3.66		2	N	864
	+21 02602	IDS13595N2114	A		14 01 50	+20 59.9	2	10.19	.010	1.09	.019	1.01	.015	3	K5	1569,1746
	+21 02602	IDS13595N2114	BC		14 01 50	+20 59.9	1	11.66		1.48		1.16		2	K4	1746
		Lod 1202 # 16			14 01 51	−58 38.2	1	11.15		1.29		1.02		1	K0	549
	+29 02485				14 01 54	+29 08.1	1	10.30		0.99		0.74		1	K0	280
122780	−53 05866				14 01 54	−53 59.8	1	9.43		0.08		-0.04		1	A0 III/IV	1770
122739	−58 05352				14 01 56	−58 24.0	1	9.41		1.60		1.80		1	K2/3	549
122711	−62 03895				14 01 56	−62 55.0	1	8.70		0.09				4	B3 V	2014
122807	−38 09035				14 01 57	−39 07.6	1	9.35		0.27		0.11		1	A9 V	1771
		Lod 1202 # 25			14 01 57	−58 24.8	1	11.97		0.60		0.25		1	G8	549
122779	−52 06981				14 02 03	−53 00.3	1	9.54		0.18		0.16		1	A1/2 V	1770
122910	+2 02768	HR 5283			14 02 05	+02 32.2	2	6.27	.005	1.02	.000	0.88	.005	7	K0	15,1417
	−17 03996				14 02 05	−18 07.4	1	9.91		1.19		1.14		5		627
	−59 05384				14 02 07	−59 34.6	1	10.87		0.33		0.26		1		287
122945	+22 02659				14 02 10	+21 37.6	2	8.53	.010	0.17	.005	0.10	.005	3	A0 V	272,280
122781	−58 05354				14 02 10	−58 21.8	1	9.84		1.43		1.41		1	K5	549
		Lod 1202 # 21			14 02 10	−58 41.0	1	11.67		0.45		0.39		1	A2	549
122843	−42 09011				14 02 11	−42 37.3	1	9.41		0.16		0.19		1	A3/4 V	1771
122762	−59 05385				14 02 12	−60 13.3	1	10.09		0.32		0.30		2	A5/8	1771
122992	+29 02486				14 02 13	+29 22.9	3	7.91	.010	1.55	.017	1.81	.040	10	M1	269,280,397
	+32 02425				14 02 14	+32 00.4	1	9.90		1.00		0.80		2	G5	1569
		HA 58 # 157			14 02 16	+29 12.1	1	13.53		0.56		0.12		2		269
		HA 58 # 200			14 02 17	+29 20.0	1	14.70		0.96		0.31		1		269
		HA 58 # 248			14 02 18	+29 29.9	1	14.14		0.85		0.35		1		269
122857	−42 09015				14 02 20	−43 08.8	1	9.02		0.28		0.13		1	A7/8 III/I	1771
	+24 02683				14 02 23	+24 19.4	1	9.99		1.00		0.95		1		280
		Lod 1202 # 22			14 02 23	−58 32.1	1	11.68		0.28		0.28		1	B9	549
		HA 58 # 202			14 02 24	+29 17.7	1	14.83		0.62		0.07		1		269
123048	+35 02521	IDS14002N3454	AB		14 02 24	+34 39.7	2	8.05	.010	0.18	.000	0.12	.029	3	A1 V	272,280
		Lod 1202 # 19			14 02 24	−58 40.9	1	11.61		0.70		0.12		1	F9	549
122844	−54 05887	HR 5278			14 02 25	−54 25.8	2	6.16	.005	0.24	.005			7	A5 III/IV	15,2029
		HA 58 # 249			14 02 26	+29 28.0	1	15.56		0.60		0.19		1		269
122875	−47 08859	NGC 5460 - 9			14 02 26	−48 13.3	1	8.54		0.91		0.37		3	G6 IV	111
123033	+26 02521	IDS14002N2618	A		14 02 29	+26 03.5	1	6.95		0.44		-0.06		2	F6 V	3026
	+38 02511				14 02 29	+37 35.1	1	10.51		0.25		0.08		2		280
122958	−15 03805	HR 5284			14 02 30	−16 05.8	1	6.56		0.10				4	A1/2 V	2009
122956	−14 03867				14 02 31	−14 37.1	1	7.22		1.01				2	G6 IV/VwF6	1594
122917	−35 09249	IDS13596S3604	A		14 02 31	−36 18.4	1	8.20		0.33		0.08		1	A9 V	1771
		MtW 58 # 81			14 02 32	+29 17.7	1	14.76		0.69		0.07		5		397
	−59 05390				14 02 32	−59 29.0	1	11.16		0.23		0.26		1		287
	+24 02684				14 02 33	+24 31.3	1	9.71		1.11		0.99		1	K0	280
		MtW 58 # 86			14 02 35	+29 24.2	1	11.87		0.78		0.27		5		397
	+29 02488				14 02 39	+29 16.7	2	9.37	.008	0.98	.017	0.70	.000	6	K0	280,397
		HA 58 # 207			14 02 39	+29 17.6	1	16.00		0.77		0.30		1		269

HD	DM	Other Id	N Rem	α_{1950}	δ_{1950}	S	V	σ_V	B–V	σ_{B-V}	U–B	σ_{U-B}	n	Spectrum	References
		HA 58 # 208		14 02 40	+29 23.0	1	11.43		0.63		0.13		1		269
123133	+53 01685			14 02 40	+52 39.3	1	7.82		1.25		1.31		2	K0	1566
		AAS12,381 22 # E		14 02 41	+21 44.8	1	10.41		0.99		0.80		1		280
		MtW 58 # 93		14 02 41	+29 22.9	1	13.79		0.82		0.46		5		397
		MtW 58 # 94		14 02 42	+29 17.5	1	16.11		0.56		0.25		4		397
	+18 02819			14 02 43	+17 36.4	1	10.52		1.15		1.25		1		280
	+21 02603			14 02 43	+21 26.9	1	9.08		1.15		1.14		1	K2	280
122920	−45 08923			14 02 44	−46 15.4	1	9.87		0.35		0.11		1	A9 V	1771
	−47 08861	NGC 5460 - 38		14 02 44	−48 08.0	1	9.77		1.28		1.24		3		111
	−47 08862	NGC 5460 - 65		14 02 45	−47 55.5	1	11.42		0.15		0.13		2		111
		G 255 - 47		14 02 47	+73 29.5	1	13.51		0.65		-0.07		1		1658
	−47 08863	NGC 5460 - 66		14 02 47	−47 55.0	1	11.60		0.19		0.16		2		111
	+36 02440			14 02 49	+36 03.2	1	8.98		0.94		0.68		1	G8 V	280
	+20 02917			14 02 51	+19 58.0	1	11.00		0.24		0.11		3		1569
123000	−33 09539			14 02 51	−33 57.5	1	9.22		0.17		0.10		1	A2/3 V	1770
122925	−53 05871			14 02 51	−53 58.1	1	8.11		-0.03		-0.40		3	B7 III	26
	+30 02478			14 02 52	+29 38.0	1	10.41		1.16		1.20		2		280
122879	−59 05395	HR 5281, LSS 3197		14 02 52	−59 28.6	10	6.41	.014	0.12	.013	-0.79	.019	33	B0 Ia	15,244,403,540,914*
	−47 08865	NGC 5460 - 58		14 02 53	−48 18.2	1	10.86		0.63		0.17		4		111
123037	−17 04002	IDS14001S1736	A	14 02 54	−17 50.0	1	8.15		0.58		0.02		5	F7/G0 +A2/5	627
123003	−37 09146			14 02 56	−37 29.4	1	7.72		-0.01		-0.05		3	A0 IV/V	1770
122980	−40 08405	HR 5285, khi Cen		14 02 59	−40 56.5	7	4.35	.007	-0.20	.006	-0.77	.011	26	B2 V	15,26,1020,1075,1770*
	−47 08868	NGC 5460 - 36		14 02 59	−48 08.0	1	10.82		0.11		0.04		3	A0	111
122982	−44 09078			14 03 00	−44 27.0	1	7.24		0.01		-0.01		3	A0 V	1770
	−47 08869	NGC 5460 - 67		14 03 00	−48 01.5	1	11.24		0.16		0.13		3		111
122983	−47 08870	NGC 5460 - 35		14 03 01	−48 08.9	1	9.94		0.04		-0.13		3	B9 IV/V	111
123299	+65 00978	HR 5291		14 03 02	+64 36.9	15	3.66	.007	-0.05	.005	-0.09	.006	94	A0 III	1,15,667,985,1077,1203*
	+33 02410			14 03 05	+32 38.0	1	9.04		1.15		1.19		1	K2 III	280
123004	−42 09027	HR 5286		14 03 05	−42 51.2	1	6.20		0.98				4	G8 III	2032
	−48 08756	NGC 5460 - 57		14 03 05	−48 22.1	1	10.72		0.29		0.24		4	A2	111
123021	−40 08407			14 03 06	−40 56.2	2	8.34	.000	0.25	.014	0.06	.005	4	A7 III/IV	321,1771
	+26 02523			14 03 08	+26 23.2	1	9.35		1.03		0.83		1	K0	280
122905	−61 04332			14 03 08	−62 08.2	1	7.25		1.80		2.13		9	M1 III	1704
	−47 08871	NGC 5460 - 37		14 03 09	−48 09.3	1	10.66		0.19		0.12		3	A0	111
	−47 08872	NGC 5460 - 103		14 03 10	−48 03.8	1	11.38		0.55		0.10		2		111
		NGC 5466 - 1411		14 03 12	+28 46.	1	16.59		0.11				1		1634
		NGC 5466 - 2506		14 03 12	+28 46.	1	17.21		0.57				2		1634
		NGC 5466 - 2508		14 03 12	+28 46.	1	15.86		0.62				3		1634
		NGC 5466 - 3410		14 03 12	+28 46.	1	17.31		0.66				1		1634
		NGC 5466 - 3501		14 03 12	+28 46.	1	16.40		0.34				1		1634
		NGC 5466 - 4409		14 03 12	+28 46.	1	17.44		0.61				1		1634
	−47 08873	NGC 5460 - 96		14 03 14	−47 52.4	1	10.72		0.28		0.23		3	A0	111
		NGC 5466 sq1 27		14 03 15	+28 45.	1	13.71		1.26				2		1634
		NGC 5466 sq1 7		14 03 15	+28 45.	1	13.69		1.36				1		1634
122987	−55 05876			14 03 16	−55 56.9	1	10.20		0.30		0.27		2	A2/3 IV	401
		HA 58 # 401		14 03 18	+29 56.7	1	11.44		0.55		0.06		1		269
	−64 02731	LSS 3198		14 03 18	−65 15.4	1	10.62		0.30		-0.70		3	O5 IIIn	1737
		NGC 5460 - 68		14 03 19	−48 02.2	1	12.49		0.36		0.17		3		111
	−47 08877	NGC 5460 - 98		14 03 20	−47 48.3	1	10.25		0.52		0.08		3		111
123190	+23 02647			14 03 22	+22 58.2	1	7.67		1.16				1	K0	280
	−47 08879	NGC 5460 - 39		14 03 25	−47 59.8	1	10.67		0.15		-0.01		3	A2	111
	−47 08880	NGC 5460 - 40		14 03 25	−48 03.7	1	11.49		0.16		0.13		3		111
	−47 08881	NGC 5460 - 64		14 03 26	−47 55.6	1	10.84		0.08		-0.07		3	A2	111
		HA 58 # 67		14 03 28	+28 54.2	1	13.06		0.90		0.83		1		269
123079	−45 08931			14 03 30	−46 21.0	1	8.70		0.10		0.10		1	A0 V	1770
	+18 02821			14 03 31	+18 24.0	1	10.34		1.03		0.94		2		1569
123123	−26 10095	HR 5287		14 03 31	−26 26.5	6	3.27	.014	1.12	.009	1.04	.005	18	K2 III	15,1075,2012,4001*
	+27 02329			14 03 32	+27 07.6	1	9.79		1.52		1.97		1	M0	280
		G 64 - 42		14 03 35	+02 25.2	1	14.90		1.46				1		1620
	−61 04337			14 03 35	−61 26.6	1	10.96		0.22		-0.47		2		432
		Feige 89		14 03 36	+07 02.	1	14.36		-0.22		-0.91		1		3060
123338	+57 01487			14 03 37	+56 45.5	2	7.11	.005	1.04	.005	0.90	.010	5	K0 III-IV	1501,1502
123096	−47 08883	NGC 5460 - 8		14 03 38	−48 02.5	1	9.36		0.04		-0.21		4	B9 IV/V	111
123081	−50 08259			14 03 38	−51 03.5	1	9.64		0.20		0.17		1	A1 V	1770
123177	−8 03696			14 03 39	−08 39.2	2	6.53	.000	0.02	.002	0.03		9	A0	1509,2006
123097	−47 08886	NGC 5460 - 7		14 03 41	−48 03.6	1	9.29		0.03		-0.21		4	B9 IV/V	111
123233	+10 02630			14 03 42	+10 08.0	1	8.64		0.08		0.04		2	A0 V	272
		HA 58 # 171		14 03 42	+29 13.0	1	12.17		0.51		-0.07		1		269
123180	−15 03808			14 03 43	−15 58.6	1	10.10		1.04		0.85		1	K3 V	3072
123139	−35 09260	HR 5288	⋆ A	14 03 44	−36 07.5	7	2.05	.011	1.01	.013	0.88	.024	16	K0 IIIb	15,1020,1034,1075*
		NGC 5460 - 102		14 03 44	−48 02.2	1	10.93		0.22		0.18		4		111
123008	−63 03134	LSS 3199		14 03 44	−64 13.9	5	8.83	.021	0.38	.016	-0.61	.034	8	B0 Ia/b	138,403,540,976,1737
	+31 02581			14 03 46	+31 02.0	1	10.56		0.31		-0.04		3		1569
123301	+35 02523			14 03 46	+34 37.3	1	8.38		1.08		1.02		1	K1 III	280
		G 64 - 43		14 03 46	−01 05.3	1	15.80		-0.05		-0.99		1	DC	3060
123112	−46 09087			14 03 46	−47 21.1	1	7.01		0.04		0.00		1	A(p SiCr)	1770
	−61 04341			14 03 46	−61 28.8	1	10.41		0.23		-0.36		2		432
	+3 02855	G 64 - 44	⋆ AB	14 03 47	+03 28.4	1	10.28		1.19		1.24		1	M0	333,1620
		HA 58 # 258		14 03 47	+29 27.7	1	13.17		0.65		0.09		1		269
123179	−14 03872		V	14 03 47	−15 16.2	1	9.97		0.24		0.15		2	A2 III	1768

Table 1   785

HD	DM	Other Id	N	Rem	α_{1950}	δ_{1950}	S	V	σ_V	B–V	σ_{B-V}	U–B	σ_{U-B}	n	Spectrum	References
		GD 271			14 03 48	+12 21.6	1	14.36		0.67		-0.03		1		3060
	−47 08888	NGC 5460 - 33			14 03 48	−48 16.3	1	11.18		0.59		0.20		3		111
122862	−74 01142	HR 5279			14 03 48	−74 36.9	6	6.02	.004	0.58	.006	0.05	.005	26	G2/3 IV	15,1075,2012,2020*
	−59 05406				14 03 50	−59 32.8	1	10.62		0.23		0.06		1	B9	549
123009	−65 02610				14 03 50	−65 54.8	1	8.03		0.38		0.19		1	A5mA8-F2	1771
123056	−59 05404	LSS 3200			14 03 51	−60 14.0	2	8.14	.023	0.14	.013	-0.75	.018	4	B2/4 III	540,976
	−32 09847				14 03 53	−33 17.5	1	10.79		0.44		0.02		2		1696
		NGC 5460 - 32			14 03 53	−48 07.5	1	12.04		0.28		0.19		3		111
123282	+22 02661				14 03 54	+22 24.3	1	7.41		1.05				1	G5	280
	−47 08889	NGC 5460 - 49			14 03 58	−47 55.8	1	10.91		0.07		0.04		3	A0	111
		NGC 5460 - 47			14 04 02	−47 59.0	1	11.95		0.33		0.20		3		111
		Lod 1225 # 21			14 04 02	−59 32.7	1	11.71		0.33		0.21		1		549
	−47 08891	NGC 5460 - 73			14 04 03	−48 05.1	1	9.67		1.12		0.95		5	K5	111
	+22 02664				14 04 04	+21 49.7	1	9.84		1.19		1.08		1		280
123255	−8 03697	HR 5290			14 04 04	−09 04.5	7	5.46	.011	0.35	.005	0.07	.022	53	F2 IV	15,130,254,285,688*
123265	−4 03616	G 64 - 46			14 04 06	−05 16.7	2	8.37	.019	0.84	.019	0.56	.005	4	K0	803,1620
123130	−52 07001				14 04 06	−53 17.1	1	8.74		0.01		-0.31		3	B8 V	26
		NGC 5460 - 104			14 04 08	−48 00.3	1	13.07		0.45		0.29		2		111
	−47 08892	NGC 5460 - 76			14 04 08	−48 03.7	1	11.56		0.26		0.19		4		111
123222	−34 09376				14 04 09	−35 15.2	1	7.19		0.96				4	G8/K0 III	2012
123339	+28 02292				14 04 10	+27 57.8	2	8.10	.001	0.37	.004	-0.06	.005	12	F2	269,1569
	+29 02490				14 04 10	+29 07.4	1	10.00		1.26		1.35		1	K2	280
123323	+18 02824				14 04 11	+18 09.0	1	7.70		1.06		0.94		1	K0	280
123182	−47 08893	NGC 5460 - 74			14 04 11	−48 04.8	1	9.94		0.07		-0.03		4	B9 (V)	111
123183	−47 08895	NGC 5460 - 75			14 04 12	−48 04.5	1	9.84		0.09		-0.04		1	A0 V	1770
123183	−47 08895	NGC 5460 - 86			14 04 12	−48 04.5	1	10.11		0.06		-0.09		4	A0	111
123184	−47 08894	NGC 5460 - 54		⋆ AB	14 04 12	−48 11.2	2	9.52	.051	0.05	.002	-0.04	.022	6	A0	111,1770
	−47 08896	NGC 5460 - 11			14 04 12	−48 14.8	1	9.36		1.20		1.18		4	K2	111
123351	+31 02582			V	14 04 13	+31 05.2	1	7.58		1.03		0.81		8	K0	269
123148	−56 06156				14 04 13	−56 39.2	1	7.94		0.02		-0.35		2	A2 V	401
	−47 08897	NGC 5460 - 34			14 04 16	−48 16.0	1	9.69		1.04		0.88		4	K0	111
123201	−47 08899	NGC 5460 - 1		⋆ AB	14 04 18	−48 04.8	2	8.70	.090	0.03	.014	-0.20	.087	5	A0	111,1770
	−61 04350				14 04 18	−61 27.1	1	9.48		1.59		1.80		1		1577
		ApJS57,743 T3# 12			14 04 18	−61 28.3	1	13.08		0.65		0.14		1		1577
		LundsII141 # 24			14 04 18	−61 37.	1	11.10		0.60		-0.24		2		432
		NGC 5460 - 48			14 04 19	−48 00.6	1	12.31		0.33		0.21		2		111
123202	−47 08900	NGC 5460 - 12			14 04 19	−48 13.7	3	8.95	.030	0.02	.009	-0.17	.032	6	B9 IV/V	111,400,1770
123061	−67 02498				14 04 20	−68 06.3	1	9.15		0.35		0.17		1	A8 V	1771
	+38 02514	W CVn			14 04 21	+38 04.0	3	10.28	.298	0.24	.094	0.13	.047	3	F2 III	280,668,699
123399	+39 02720	IDS14022N3854	A		14 04 21	+38 39.3	1	7.97		0.89		0.58		2	K0	985
123223	−43 08803				14 04 21	−43 42.3	1	9.11		0.32		0.08		1	A9/F0 V	1771
		NGC 5460 - 77			14 04 22	−48 05.1	1	10.21		0.11		-0.01		4		111
		NGC 5460 - 53			14 04 22	−48 08.1	1	11.47		1.33		1.35		3		111
	−47 08903	NGC 5460 - 31		⋆ AB	14 04 22	−48 08.8	1	10.55		0.19		0.14		4		111
123225	−47 08901	NGC 5460 - 5			14 04 23	−48 05.6	2	8.88	.023	0.00	.005	-0.33	.001	5	B9	111,1770
	+27 02332				14 04 24	+27 20.4	1	9.51		1.07		1.03		1		280
123224	−47 08904	NGC 5460 - 2			14 04 25	−47 56.9	3	7.99	.016	0.02	.003	-0.22	.002	10	B8 II	111,1770,2014
123226	−47 08902	NGC 5460 - 13			14 04 25	−48 14.0	1	9.09		0.03		-0.24		4	B8 IV/V	111
	−63 03138	LSS 3201			14 04 25	−63 58.0	2	10.29	.005	0.38	.009	-0.65	.000	4	O7 IIIn	390,1737
	−47 08905	NGC 5460 - 56			14 04 26	−48 10.1	1	10.60		0.16		0.14		5	A0	111
	+29 02491				14 04 27	+28 44.4	1	10.71		0.97		0.67		2		280
		NGC 5460 - 55			14 04 27	−48 10.2	1	10.94		0.13		0.09		5		111
123247	−48 08775	NGC 5460 - 3			14 04 28	−48 28.0	3	6.45	.034	-0.01	.008	-0.03	.010	9	B9.5V	111,1770,1771
123168	−59 05410				14 04 30	−59 37.3	1	8.81		0.12		-0.18		1	B8 IV	549
		ApJS57,743 T3# 7			14 04 31	−61 30.3	1	13.28		0.44		0.32		1		1577
123151	−62 03941	HR 5289			14 04 31	−62 58.3	3	6.40	.004	1.01	.005			11	G8/K0 III	15,1075,2006
123408	+35 02525	IDS14024N3515	A		14 04 32	+35 01.0	1	7.00		1.00				2	K0	280
123710	+75 00526				14 04 32	+74 48.6	1	8.21		0.59		-0.07		4	G5	3026
	−59 05411	LSS 3202			14 04 32	−59 32.4	1	10.83		0.74		-0.27		1	B3	549
		ApJS57,743 T3# 11			14 04 33	−61 25.9	1	11.88		1.16		0.82		1		1577
123204	−58 05372				14 04 34	−58 24.8	1	8.76		0.18				4	B7 III	2014
		GD 490			14 04 35	+64 04.5	1	15.07		0.42		-0.17		3		308
		BPM 38119			14 04 36	−49 51.	1	14.14		0.73		0.10		1		3065
		ApJS57,743 T3# 3			14 04 36	−61 28.0	1	12.64		1.33		1.04		1		1577
	−47 08906	NGC 5460 - 61			14 04 37	−47 53.3	1	10.70		0.01		-0.08		2	A0	111
123269	−47 08907	NGC 5460 - 14			14 04 39	−48 17.1	1	9.54		0.01		-0.19		4	B8 V	111
		ApJS57,743 T3# 5			14 04 39	−61 29.8	1	13.66		0.61		0.32		1		1577
		QY Cen			14 04 39	−61 29.8	2	11.87	.630	2.12	.290	1.37		2		689,1577
123270	−48 08776				14 04 40	−48 43.9	1	9.76		0.26		0.19		1	A2/5 (V)	1770
	+23 02649				14 04 41	+23 26.8	1	10.26		1.04		0.84		1	K0	280
123409	+29 02493				14 04 41	+28 40.5	2	6.90	.003	1.00	.006	0.78	.005	12	K0	269,1569
		ApJS57,743 T3# 4			14 04 42	−61 27.2	1	13.49		0.63		0.19		1		1577
	+20 02921				14 04 43	+20 25.8	1	9.64		1.45		1.83		1	K5	280
	−61 04357				14 04 44	−61 27.9	1	11.79		0.25		-0.40		1		1577
	−61 04358				14 04 44	−61 28.2	1	10.56		0.11		-0.22		1		1577
		OQ 208 # 1			14 04 46	+28 41.	1	10.74		0.96		0.68		2		326
		OQ 208 # 2			14 04 46	+28 41.	1	12.11		0.98		0.71		1		326
		OQ 208 # 3			14 04 46	+28 41.	1	12.97		0.90		0.62		2		326
		OQ 208 # 4			14 04 46	+28 41.	1	14.32		1.24				1		326
		OQ 208 # 5			14 04 46	+28 41.	1	15.00		0.69		0.13		1		326

HD	DM	Other Id	N Rem	α_{1950}	δ_{1950}	S	V	σ_V	B–V	σ_{B-V}	U–B	σ_{U-B}	n	Spectrum	References
		OQ 208 # 6		14 04 46	+28 41.	1	15.30		0.70				1		326
		OQ 208 # 7		14 04 46	+28 41.	1	15.62		0.62		-0.04		1		326
		OQ 208 # 8		14 04 46	+28 41.	1	16.04		0.67		-0.08		1		326
		OQ 208 # 9		14 04 46	+28 41.	1	17.02		0.87		0.04		1		326
123291	-43 08808			14 04 47	-43 45.4	2	8.24	.000	0.01	.003	-0.08		5	B9.5/A0 IV	1770,2014
	-48 11137	LSS 3731		14 04 47	-48 33.8	1	10.58		0.72		-0.15		2		954
		G 165 - 47		14 04 49	+38 51.7	4	14.55	.027	1.72	.025	2.05		8		316,1705,1759,3078
123344	-34 09382			14 04 49	-34 42.7	1	7.35		0.02		0.00		1	A0 V	1770
123229	-61 04359	LSS 3203		14 04 50	-62 05.3	1	9.05		1.12		0.70		3	F0/2 Ib/II	8100
	-47 08908	NGC 5460 - 60		14 04 51	-47 51.5	1	10.68		0.06		-0.07		2	A0	111
		Lod 1225 # 27		14 04 52	-59 29.8	1	12.41		0.39		0.26		1		549
		Lod 1225 # 5		14 04 53	-59 35.8	1	10.49		1.64		1.97		1		549
123439	+13 02750			14 04 59	+12 49.9	1	9.16		0.60		0.17		2	F8	1648
	-59 05417			14 05 01	-59 24.3	1	11.09		0.42		0.28		1	A5	549
	-59 05418			14 05 02	-59 29.8	1	11.31		0.19		0.16		1	A0	549
123471	+23 02652			14 05 03	+22 42.0	2	9.28	.063	0.27	.122	0.01	.104	5	B9	280,1569
	-61 04363			14 05 03	-61 26.6	1	9.88		1.07		0.76		1		1577
123802	+75 00527			14 05 04	+74 57.3	1	7.48		1.16		1.32		2	K0	3016
123333	-51 07976			14 05 04	-52 03.1	1	9.50		0.95				4	K2 V	2012
		Lod 1225 # 18		14 05 04	-59 34.0	1	11.50		0.26		0.15		1		549
		ApJS57,743 T3# 14		14 05 04	-61 27.0	1	11.20		1.26		1.19		1		1577
	-47 08912	NGC 5460 - 62	⋆ AB	14 05 05	-47 54.5	1	10.08		0.60		0.20		2	F8	111
	-47 08910	NGC 5460 - 43		14 05 06	-48 17.6	1	10.50		0.08		-0.02		3	B9	111
		NGC 5460 - 42		14 05 07	-48 06.7	1	12.16		0.28		0.17		3		111
		Lod 1225 # 22		14 05 07	-59 38.0	1	11.71		0.38		0.20		1		549
	+24 02688			14 05 08	+23 46.2	1	9.70		0.95		0.77		1		280
123360	-47 08914	NGC 5460 - 4		14 05 08	-48 01.4	1	8.77		0.46		0.01		5	F3 V	111
	-47 08913	NGC 5460 - 41		14 05 08	-48 07.5	1	11.65		0.52		0.08		2		111
123189	-69 02007			14 05 08	-69 24.6	2	9.81	.001	0.54	.019	-0.02	.000	3	F3 V	540,976
	-48 08787	NGC 5460 - 45		14 05 09	-48 21.9	1	10.89		0.10		0.04		3	A0	111
		Lod 1225 # 25		14 05 09	-59 17.4	1	12.23		0.32		-0.17		1		549
		G 64 - 47		14 05 11	+04 10.9	2	14.57	.010	1.60	.015			2		1620,1759
123296	-59 05420			14 05 11	-59 28.8	1	8.63		0.10		-0.17		1	B8/9 V	549
	+30 02484			14 05 15	+29 39.4	1	9.85		1.28		1.48		2	K7	280
		Lod 1225 # 19		14 05 15	-59 18.0	1	11.50		0.65		0.17		1		549
	-60 05207			14 05 15	-60 32.5	1	11.17		0.26		-0.39		2	B3 V	540
123573	+47 02118			14 05 16	+47 04.4	1	8.43		0.45		0.00		2	F8	1601
	+60 01512	IDS14037N6015	B	14 05 16	+59 59.8	1	9.26		0.44		-0.05		2	F5	1502
123389	-41 08492			14 05 16	-41 48.6	1	8.44		0.36		0.21		1	A2/3 V	1770
		Lod 1225 # 24		14 05 18	-59 23.6	1	11.92		0.43		0.08		1		549
	+18 02827			14 05 20	+17 58.3	1	9.94		1.33		1.43		1		280
	-58 05381			14 05 20	-59 18.4	1	11.48		0.46		0.25		1	A8	549
		G 165 - 48		14 05 21	+39 37.8	1	15.30		1.29				1		1759
	-47 08916	NGC 5460 - 44	⋆ AB	14 05 21	-48 19.8	1	10.31		0.35		0.13		3	A2	111
	-58 05382			14 05 21	-59 17.4	1	10.90		0.24		0.12		1	A0	549
123453	-12 03966	IDS14027S1227	AB	14 05 22	-12 41.4	1	7.63		0.58				15	G8 (V) +A/F	1351
	-59 05421			14 05 22	-59 36.6	1	11.23		0.17		-0.31		1	B6	549
123533	+31 02584			14 05 23	+31 12.9	1	8.43		1.09				2	K0 III	280
		Feige 90		14 05 24	+05 22.	1	14.04		-0.12		-0.48		1		3060
	+24 02689			14 05 24	+23 58.5	1	9.34		1.55		1.99		1	K5	280
123335	-58 05383	HR 5292		14 05 24	-59 02.4	6	6.34	.011	0.05	.006	-0.50	.008	20	B3 III/V	26,244,540,976,1637,2032
123428	-32 09871			14 05 25	-32 49.0	1	7.26		1.02				4	K0 III	2012
		Lod 1225 # 15		14 05 25	-59 25.1	1	11.41		0.21		0.00		1		549
	+29 02496			14 05 26	+29 27.6	1	10.08		1.12		1.14		2	K2	280
	+42 02466	G 165 - 49		14 05 26	+41 49.1	2	9.92	.000	0.68	.000	0.18	.020	2	G8	1658,3060
	-59 05424	Lod 1225 # 12		14 05 26	-59 29.0	2	11.27	.000	0.28	.000	0.19	.000	2	A0	287,549
	-59 05425	Lod 1225 # 20		14 05 28	-59 28.4	2	11.71	.040	0.28	.020	0.17	.020	2	A0	287,549
	+34 02491			14 05 29	+33 38.4	1	9.54		0.97		0.72		1	G5	280
123519	+17 02704			14 05 30	+17 06.2	1	7.78		1.49				2	K2	882
		NGC 5460 - 78		14 05 30	-48 02.2	1	12.33		0.35		0.19		2		111
	-59 05426			14 05 30	-59 36.1	1	11.43		0.21		0.16		1	A0	549
	-47 08918	NGC 5460 - 79		14 05 31	-48 01.7	1	11.96		0.56		0.14		2		111
		Lod 1225 # 23		14 05 31	-59 36.0	1	11.73		0.26		0.20		1		549
		Lod 1225 # 28		14 05 32	-59 30.2	2	12.46	.000	0.35	.000	0.23	.000	2		287,549
	+19 02757			14 05 33	+19 05.4	1	10.22		0.96		0.59		1		280
		AAS12,381 28 # E		14 05 35	+28 20.2	1	10.09		1.10		1.04		1		280
	-47 08920	NGC 5460 - 51		14 05 37	-48 16.5	1	11.56		0.23		0.17		3		111
123376	-59 05427	IDS14020S5915	AB	14 05 37	-59 29.4	1	9.21		1.92		1.88		1	K5	549
123431	-41 08497			14 05 39	-42 21.7	2	8.72	.003	0.00	.002	-0.03		6	A0 V	1770,2014
		NGC 5460 - 80		14 05 40	-48 00.4	1	12.88		0.68		0.13		2		111
		Lod 1225 # 29		14 05 41	-59 30.3	2	12.81	.000	0.50	.000	0.25	.000	2		287,549
123432	-47 08923	NGC 5460 - 6		14 05 42	-48 08.8	1	7.56		1.12		1.06		4	K2 III	111
123445	-42 09065	HR 5294	⋆ A	14 05 45	-43 14.0	3	6.19	.013	-0.06	.003	-0.27		9	B9 V	1770,2014,2032
123485	-44 09112	IDS14027S4506	A	14 05 52	-45 20.3	1	8.80		0.11		0.12		1	A1 V	1770
123611	+27 02336			14 05 54	+27 00.8	1	8.36		1.14		1.14		1	K0	280
123657	+44 02325	HR 5299, BY Boo		14 05 56	+44 05.5	6	5.26	.028	1.58	.012	1.66	.013	19	M4 III	15,1118,3001,8003*
	-47 08925	NGC 5460 - 59		14 05 57	-48 01.0	1	10.66		0.38		0.14		2	A5	111
123612	+25 02733			14 05 58	+24 33.1	1	6.61		1.39				1	K5 III	280
	+39 02724			14 05 58	+39 01.3	1	9.06		1.28		1.40		1	K2 II	280
123628	+23 02653			14 05 59	+22 48.0	1	8.38		0.99		0.73		1	G5	280

Table 1 787

HD	DM	Other Id	N	Rem	α_{1950}	δ_{1950}	S	V	σ_V	B–V	σ_{B-V}	U–B	σ_{U-B}	n	Spectrum	References
		Steph 1124			14 05 59	−16 17.8	1	10.77		1.59		1.94		2	M0	1746
123614	+17 02706				14 06 00	+17 20.7	1	8.27		0.97				1	K0	882
123462	−51 07991				14 06 00	−51 52.8	1	10.40		0.17		0.14		1	A2 V	1770
	−59 05430				14 06 00	−59 22.1	1	11.13		0.31		-0.03		1	B8	549
		LP 619 - 35			14 06 01	+00 10.4	1	12.08		1.05		0.87		1		1696
	+20 02930				14 06 02	+19 33.7	1	9.24		1.18		1.28		1	K0	280
	+19 02758				14 06 07	+19 05.7	1	10.20		1.01		0.82		1	K0	280
123545	−29 10849	IDS14032S3006	AB		14 06 08	−30 20.1	1	9.56		0.31		0.16		1	A1/2mA7-A8	1770
		Lod 1225 # 14			14 06 09	−59 27.7	1	11.38		1.14		0.80		1		549
123514	−46 09112				14 06 11	−47 06.7	1	9.12		0.20		0.17		1	A4 V	1771
		CS 15623 # 1			14 06 12	+23 35.0	1	13.75		0.14		0.19		1		1744
123598	−18 03757	FR Vir			14 06 14	−19 00.5	2	7.10	.000	1.59	.007	1.78	.019	9	M2 III	803,3071
		Lod 1225 # 26			14 06 14	−59 29.2	1	12.24		0.35		-0.20		1		549
123465	−59 05432				14 06 14	−59 46.4	1	9.31		0.05		-0.51		4	B3 V	244
	−47 08930	NGC 5460 - 108			14 06 17	−47 59.0	1	10.57		1.06		0.90		2		111
123515	−50 08294	HR 5296, V869 Cen	⋆	A	14 06 17	−51 16.1	5	5.97	.016	-0.05	.004	-0.31		16	B8 V	15,1075,1770,2006,2014
		G 64 - 48			14 06 19	+04 47.0	3	11.07	.004	1.16	.008	1.00	.054	4	K5	265,1620,1746
123377	−69 02012	HR 5293	⋆	ABC	14 06 19	−70 04.2	3	6.04	.007	1.73	.010	1.83	.005	13	K4 II	15,1075,2038
123630	−9 03865	HR 5298			14 06 20	−10 05.9	3	6.46	.001	1.00	.005	0.72	.004	13	G8 III	285,688,1088
123528	−50 08295	IDS14031S5023	AB		14 06 20	−50 36.9	1	9.17		0.11		0.08		1	A1 V	1770
		Steph 1125			14 06 22	+24 05.1	1	12.33		1.46		1.21		1	M1	1746
123549	−46 09114	IDS14032S4627	AB		14 06 22	−46 41.2	1	9.15		0.13		0.09		1	A0 V	1770
		G 135 - 20			14 06 23	+17 46.2	1	11.62		0.73		0.21		2	G8	1696
		BPM 8752			14 06 24	−66 27.	1	14.37		0.34		0.09		1		3065
123782	+50 02047	HR 5300, CF Boo	⋆	A	14 06 25	+49 41.6	2	5.26	.005	1.64	.010	1.92		13	M1.5III	1118,3055
123583	−37 09183				14 06 25	−38 11.8	1	7.86		0.12		0.08		1	A3 IV	1771
123585	−43 08827				14 06 27	−44 07.8	2	9.28	.015	0.51	.005	-0.04	.010	2	F7 Vwp Sr	565,3077
123505	−60 05215				14 06 27	−61 16.6	4	9.68	.011	0.78	.007	0.26	.016	12	G8 V	158,742,2012,3077
	+22 02668				14 06 29	+21 39.8	1	10.39		1.00		0.73		1	K0	280
124063	+75 00529	HR 5305			14 06 32	+74 49.8	2	6.45	.005	0.14	.000	0.12	.000	4	A7 V	985,1733
		GD 162			14 06 33	+13 56.6	1	15.40		0.46		-0.25		1		3060
	−30 11195				14 06 34	−30 41.5	1	11.81		1.50				1	M0	1705
123569	−52 07028	HR 5297	⋆	A	14 06 34	−53 12.1	6	4.74	.007	0.94	.004	0.71	.008	28	G8 III	15,1075,1311,1637*
123757	+34 02495				14 06 36	+34 02.2	1	8.48		0.98		0.90		1	K2 III	280
123467	−66 02463	IDS14028S6608	AB		14 06 37	−66 22.0	1	8.42		0.25		0.15		1	A1 V	1771
123821	+52 01776				14 06 38	+51 49.7	2	8.68	.030	1.06	.020	0.89	.045	7	G8 IIIp	37,8100
	+31 02586				14 06 40	+30 50.6	1	9.48		0.58		0.07		7	K0	1569
123783	+32 02435				14 06 43	+32 15.3	1	8.14		1.33		1.54		1	K3 III	280
123724	+18 02831				14 06 44	+17 35.5	1	8.50		1.07				2	K0	882
123624	−47 08933	NGC 5460 - 10			14 06 45	−48 10.6	1	8.66		0.31		0.06		3	F2 V	111
123622	−42 09081				14 06 46	−43 01.0	1	9.68		0.11		0.09		1	A1 V	1770
		G 15 - 8			14 06 47	−04 43.8	1	11.87		1.46		1.30		1		1620
123635	−43 08831	IDS14037S4348	AB		14 06 47	−44 02.7	2	7.72	.010	-0.02	.004	-0.36		5	B9 II	1770,2014
		AAS61,331 # 16			14 06 48	+31 06.	1	11.33		-0.01		-0.06		3		1569
123570	−59 05435				14 06 48	−59 38.9	1	9.99		0.17		0.05		4	B9/A0 V	244
		G 178 - 5			14 06 50	+44 28.7	1	11.93		1.14		1.07		1	K5	3060
123492	−69 02014	HR 5295			14 06 53	−69 29.0	4	6.06	.008	0.18	.004	0.13	.010	19	A6 IV	15,1075,1771,2038
		Feige 91			14 06 54	+59 55.0	1	13.38		-0.28		-1.07		1	DAs	3060
123651	−45 08971				14 06 55	−46 01.9	2	8.17	.000	0.57	.000			8	G0/1 V	1075,2012
		LP 499 - 41			14 06 58	+12 33.4	1	13.08		0.96		0.79		2		1696
123760	+10 02637				14 07 00	+10 28.9	1	8.00		0.65		0.15		2	G5 V	3026
123664	−45 08972				14 07 00	−45 40.7	1	7.64		0.10		0.14		1	A1 V	1770
123701	−29 10862				14 07 01	−30 12.4	1	8.50		1.04		0.66		1	Kp Ba	565
123898	+51 01892				14 07 02	+51 01.7	1	8.72		0.55		0.02		2	G0	1566
123590	−61 04382	LSS 3204			14 07 02	−62 14.6	1	7.62		0.14		-0.83		2	O7/8	432
123822	+26 02529				14 07 03	+25 40.7	1	8.52		0.98		0.71		1	G8 III	280
123899	+50 02048				14 07 03	+49 36.6	1	8.92		0.34				2	F8	1118
	−35 09296				14 07 04	−35 24.3	1	9.69		0.51		-0.01		2		401
123652	−50 08308				14 07 06	−50 30.3	1	10.19		0.20		0.15		1	A2/3 V	1770
123682	−44 09127	IDS14040S4430	A		14 07 09	−44 45.0	5	8.29	.011	0.69	.009	0.14	.005	15	G3/5 V	1075,1279,2012,2013,3008
123844	+21 02618				14 07 12	+20 42.5	2	7.46	.005	0.44	.005	-0.01	.015	5	F2	1569,1648
123682	−44 09130	IDS14040S4430	BC		14 07 12	−44 43.9	2	9.62	.010	0.91	.002	0.64	.005	4	A0/1 V	1696,3008
123977	+60 01516	HR 5302			14 07 13	+59 34.4	2	6.46	.000	1.03	.005	0.82	.010	5	K0 III	1003,1080
123859	+27 02340				14 07 15	+27 07.8	1	8.20		1.30		1.41		1	K0	280
123846	+15 02670				14 07 16	+14 39.1	2	8.37	.005	1.28	.000	1.34		3	K0	882,1648
	+35 02528				14 07 18	+35 21.5	2	10.59	.000	1.08	.013	0.90	.004	7	K0	280,1775
123746	−30 11209				14 07 18	−30 11.2	2	6.76	.004	0.11	.014	0.08		2	A2 IV/V	925,1770
123878	+23 02657				14 07 22	+23 23.4	1	8.42		1.15		1.20		1	K0	280
		BPM 38165			14 07 24	−47 31.	1	14.31		-0.07		-0.89		1	DA	3065
123747	−38 09130	IDS14044S3855	AB		14 07 25	−39 09.3	1	8.70		0.06		0.03		2	A0 V + (A)	1770
123879	+19 02933				14 07 26	+19 58.5	1	7.75		1.29		1.45		1	K0	280
	+21 02619				14 07 26	+21 32.4	1	10.35		0.29		-0.02		2		1569
123877	+26 02530				14 07 26	+26 04.9	1	7.93		1.49		1.81		1	K5 III	280
123706	−53 05895				14 07 27	−53 48.8	1	10.13		0.13		0.10		1	A1 V	1770
123829	−12 03980				14 07 28	−13 06.6	1	9.08		0.60				31	G1/2 V	6011
123731	−46 09127				14 07 28	−46 58.3	1	10.14		0.32		0.21		1	A9 V	1771
		AAS12,381 32 # B			14 07 29	+31 38.2	1	10.45		1.05		1.08		1		280
		AAS12,381 30 # B			14 07 30	+30 01.9	1	11.53		0.33		0.15		1		280
123929	+31 02588				14 07 33	+31 29.4	2	7.27	.005	0.85	.000	0.47		5	K0	280,1569
	+33 02417				14 07 36	+32 38.2	1	10.56		1.08		1.11		1	K2 III	280

HD	DM	Other Id	N Rem	α_{1950}	δ_{1950}	S	V	σ_V	B–V	σ_{B-V}	U–B	σ_{U-B}	n	Spectrum	References
123944	+38 02518			14 07 38	+37 33.9	1	7.58		0.35		-0.04		3	F0	1569
123773	-42 09090			14 07 39	-43 05.0	1	9.79		0.38		0.14		1	A2mF0-F3/5	1770
	+24 02694			14 07 41	+24 05.3	2	8.92	.010	1.02	.000	0.80	.024	3	K0	280,1569
123721	-59 05443			14 07 41	-59 32.2	1	8.76		0.04				4	B9 II	2014
	-12 03984			14 07 44	-12 55.5	1	10.67		0.73				29	G5	6011
	-13 03834			14 07 46	-13 41.6	4	10.68	.009	0.60	.007	-0.11	.015	8	F7	158,1064,3077,6006
123930	+18 02834			14 07 47	+17 57.1	1	9.04		0.49				2	K0	882
123962	+35 02529			14 07 48	+35 19.0	2	8.94	.004	0.27	.004	0.07	.004	8	A2	280,1775
123884	-17 04022			14 07 48	-17 45.3	6	9.32	.011	0.05	.012	-0.26	.020	30	B8/9 Ia/abp	55,830,1311,1775,1783,8100
123797	-48 08820	NGC 5460 - 109		14 07 48	-48 32.8	2	6.79	.219	0.59	.419	0.03		6	G5 IV	111,2012
123798	-48 08819			14 07 49	-49 02.2	1	8.55		0.30		0.02		2	A8/9 IV	1771
		PG1407-013		14 07 51	-01 16.1	1	13.75		-0.26		-1.12		2	sdB	1764
123945	+20 02934			14 07 52	+19 35.5	2	8.75	.005	0.58	.000	-0.02	.010	5	G0	1569,1648
	+29 02502			14 07 53	+28 37.7	1	10.24		0.43		-0.03		2		1569
		L 332 - 60		14 07 54	-46 26.	1	11.54		1.58		1.77		1		1696
124018	+41 02471			14 07 57	+41 00.8	1	7.15		1.21		1.31		2	K2	1601
	-59 05445			14 08 01	-59 24.5	2	9.40	.000	1.85	.000	1.66	.000	2		287,549
123978	+15 02673			14 08 04	+14 39.0	1	8.48		1.11				2	G5	882
	-59 05447			14 08 05	-59 23.9	1	9.21		1.92		1.88		1		287
123934	-15 03817	HR 5301, ET Vir		14 08 06	-16 04.0	4	4.92	.030	1.68	.010	2.05	.100	24	M1 III	1024,2006,3055,6002
	+20 02936			14 08 07	+20 00.4	1	9.92		1.06		0.90		1		280
123999	+25 02737	HR 5304		14 08 07	+25 19.7	7	4.83	.004	0.53	.007	0.07	.004	32	F9 IV	15,667,1008,1363,3053*
	+31 02590			14 08 07	+31 11.3	1	9.29		1.08				2	G5 III	280
	+32 02437			14 08 07	+31 45.9	1	9.34		1.26		1.42		1	G5	280
124019	+28 02294			14 08 12	+27 52.1	1	8.56		0.65		0.12		2	G2 V	3026
	+32 02438			14 08 12	+32 26.1	1	9.32		0.98		0.78		1	G8 IV	280
124033	+36 02446			14 08 12	+35 41.0	1	8.85		0.46		-0.01		2	G5	1569
124101	+52 01777			14 08 13	+52 29.3	1	7.88		0.96		0.67		2	K0	1566
123949	-18 03763			14 08 13	-18 54.5	1	8.73		1.37		1.30		1	K1p Ba	565
		GD 163		14 08 16	+32 22.7	2	14.00	.042	0.00	.019	-0.78	.023	4	DA	419,3060
124034	+15 02674			14 08 20	+15 31.6	1	6.97		1.45				2	K2	882
123923	-44 09145	IDS14052S4459	AB	14 08 23	-45 13.3	1	9.74		0.46		0.25		1	A(3)mF0-F3	1770
123840	-58 05403			14 08 23	-59 03.9	1	9.16		0.53		0.04		4	F5 V	244
123951	-29 10879			14 08 25	-29 33.0	1	7.58		1.20		1.26		5	K0/1 III	1657
		G 65 - 29		14 08 26	+07 07.1	1	12.01		0.92		0.64		1	K2	1620
	+37 02502			14 08 27	+36 44.6	1	9.74		0.42		-0.02		5	F2	1569
123984	-12 03993	AL Vir		14 08 27	-13 04.5	2	9.12	.015	0.39	.015	0.32		2	F3/5 III	688,6011
	-61 04409			14 08 27	-61 42.9	1	11.21		0.17		-0.46		2		432
124102	+38 02521			14 08 28	+37 43.7	1	8.51		0.64		0.21		5	G3 III	1569
124066	+15 02675	IDS14062N1541	A	14 08 32	+15 27.1	1	8.28		0.69		0.26		1	G0	3029
124066	+15 02675	IDS14062N1541	B	14 08 32	+15 27.1	1	16.83		0.75		0.35		2		3029
		He2 104, V852 Cen		14 08 34	-51 12.3	1	13.69		0.90		-0.27		1		1753
	-33 09619			14 08 36	-33 27.1	1	9.96		0.45				1	G0	925
		AAS12,381 33 # A		14 08 37	+32 37.9	1	11.55		0.07		0.13		1		280
		G 124 - 4		14 08 38	+01 46.4	2	13.45	.010	0.57	.010	-0.15	.012	3		1658,1696
		WLS 1424 55 # 6		14 08 40	+55 12.2	1	12.05		0.58		0.03		2		1375
		LP 799 - 29		14 08 41	-16 56.2	1	12.21		0.74		0.14		2		1696
		G 65 - 30		14 08 44	+07 04.0	1	14.05		1.54		0.96		1		1620
124214	+57 01492			14 08 45	+57 00.7	1	8.95		0.33		0.09		2	F5	1375
		L 476 - 18		14 08 48	-36 28.	2	13.56	.010	0.63	.005	-0.07	.010	3		1696,3062
	-32 09927			14 08 49	-32 49.2	2	10.44	.000	0.34	.001	0.12	.001	8	A0	1698,1765
124170	+42 02468			14 08 51	+42 24.1	1	8.64		0.16		0.12		3	A2	1307
	+29 02504			14 08 52	+29 08.8	1	10.20		1.31		1.45		1		280
124115	+2 02783	HR 5307		14 08 59	+01 35.8	4	6.44	.012	0.48	.005	0.03	.010	10	F6 V	15,254,1250,1256
	+37 02503	IDS14069N3656	AB	14 09 01	+36 41.9	1	10.12		0.55		0.04		2	G0 V	1569
124547	+78 00478	HR 5321		14 09 01	+77 46.9	2	4.87	.031	1.37	.003	1.39		5	K3 III	1363,3053
124012	-47 08963			14 09 02	-48 09.4	1	8.20		-0.03		-0.22		1	B8 V	1770
124186	+32 02443	HR 5310		14 09 04	+32 31.8	2	6.11	.010	1.26	.000	1.43	.015	3	K4 III	252,280
	+81 00465	G 255 - 51		14 09 05	+80 51.3	1	10.34		1.47		1.14		4	M2	7008
124106	-11 03684			14 09 05	-12 22.5	2	7.93	.010	0.87	.005	0.58		6	K1 V	2034,3072
	+30 02489			14 09 08	+29 49.8	1	10.11		1.06		0.98		2	K0	280
124045	-47 08965			14 09 11	-47 54.6	1	8.81		0.21		0.15		1	A1mA4-A7	1770
	+30 02490	G 165 - 51		14 09 14	+30 19.3	1	10.32		1.17		1.17		2	K6:	1569
123995	-58 05411			14 09 14	-59 09.3	1	8.09		0.52		0.26		4	A9 III/IV	244
		LP 799 - 72		14 09 17	-18 34.5	1	12.07		0.56		-0.15		2		1696
	+26 02536			14 09 18	+26 25.5	1	10.52		1.16		1.21		2	M0	1569
	+33 02422			14 09 20	+33 07.4	1	10.00		1.19		1.14		1	G8 III	280
124046	-49 08547			14 09 20	-50 13.0	1	10.13		0.28		0.18		1	A2/3 III	1770
	+10 02643	G 65 - 31		14 09 21	+09 41.9	1	10.58		1.08		1.08		1	K5	333,1620
		AAS61,331 # 10		14 09 23	+21 16.	1	10.82		0.33		0.13		2		1569
124094	-42 09117	IDS14063S4253	A	14 09 25	-43 07.5	1	8.19		0.19		0.18		1	A3 V	1771
	+20 02940			14 09 26	+20 32.8	1	9.43		1.35		1.56		1	K5	280
124243	+36 02450			14 09 26	+35 38.6	1	9.00		0.60		0.01		2	G5 III	1569
		G 65 - 32		14 09 28	+00 59.3	1	13.94		1.02		0.63		1		1620
	+33 02423			14 09 28	+32 46.6	1	10.63		0.95		0.61		1	G8 III	280
123781	-75 00939			14 09 28	-76 13.4	1	7.84		0.04		-0.09		5	B9 IV	897
		G 65 - 33		14 09 33	+07 20.3	1	12.94		1.04		0.74		1		1620
124162	-23 11551	HR 5309		14 09 34	-24 07.8	1	6.34		1.35				4	K2 III	2035
124082	-54 05925			14 09 34	-55 05.9	1	9.29		0.20		-0.26		2	B7/8 III	401
		CS 22883 # 9		14 09 38	+11 04.7	1	15.32		0.06				1		1736

Table 1   789

HD	DM	Other Id	N Rem	α_{1950}	δ_{1950}	S	V	σ_V	B–V	σ_{B-V}	U–B	σ_{U-B}	n	Spectrum	References
		AAS12,381 33 # B		14 09 39	+33 05.4	1	10.95		0.26		0.04		1		280
	+34 02502			14 09 39	+33 53.1	1	9.99		0.96		0.82		1	K0	280
		G 64 - 52		14 09 39	−00 21.2	2	12.98	.020	1.63	.005	1.19	.010	5		203,3073
123396	−82 00593			14 09 40	−83 18.9	1	8.97		1.19		0.78		1	G8 II/IIIp	565
124144	−39 08769			14 09 41	−39 33.4	1	8.62		0.11		0.10		1	A1 V	1770
124224	+3 02867	HR 5313, CU Vir	★ A	14 09 44	+02 38.6	10	5.01	.011	−0.12	.006	−0.40	.015	93	B9p Si	15,112,667,1063,1075*
124145	−42 09126			14 09 47	−42 30.6	1	9.27		0.17		0.14		1	A3 V	1771
		BPM 38217		14 09 48	−48 25.	1	12.79		0.74		0.18		2		3065
124176	−40 08492			14 09 50	−40 36.0	1	6.75		−0.01		−0.07		1	B9 V	1770
		G 65 - 34		14 09 51	+11 51.5	1	11.43		1.29		1.28		1	M2	333,1620
124270	+21 02625			14 09 51	+20 52.6	1	7.60		0.24		0.08		2	A5	1569
124320	+39 02731			14 09 53	+39 03.9	2	8.85	.005	0.13	.010	0.11	.005	5	A2 V	272,833
		EG 105		14 09 54	+15 47.0	1	18.12		0.96		0.30		2	DA:	3028
124206	−26 10158	HR 5312		14 09 54	−27 01.6	2	5.10	.025	1.15	.010	1.13		7	K2 III	2007,3016
124147	−53 05912	HR 5308		14 09 54	−53 25.9	1	5.56		1.44				4	K5 III	2009
		Steph 1131		14 09 58	+18 34.4	1	11.98		1.33		1.16		1	K7-M0	1746
		LP 559 - 87		14 09 59	+05 31.7	1	13.37		0.71		0.13		2		1696
		Feige 92		14 10 00	+50 21.0	2	11.53	.085	−0.13	.010	−0.61	.010	4	B9	1026,1298
		Lod 1289 # 37		14 10 02	−57 35.8	1	12.60		0.53		0.33		2	A0	807
	+38 02525			14 10 03	+37 44.7	1	9.42		0.97		0.70		1	K0	280
		LP 739 - 33		14 10 03	−14 30.7	1	13.65		0.73		−0.07		2		3062
		G 65 - 35		14 10 05	+10 22.9	1	14.36		1.54		1.34		1		1620
	+27 02345			14 10 07	+27 32.9	1	10.44		1.09		0.98		1		280
124152	−58 05420			14 10 07	−59 07.8	1	8.24		0.21		0.09		4	A0 IV/V	244
124292	−2 03804	G 64 - 54		14 10 10	−03 04.9	1	7.03		0.74		0.24		1	G0	1620
124179	−51 08038			14 10 10	−51 26.5	1	10.01		0.24		0.23		1	A1 V	1770
124150	−57 06541			14 10 10	−57 36.3	1	10.31		0.18		−0.26		2	B8/9 IV/V	807
	+28 02300			14 10 12	+28 17.5	1	9.73		1.18		1.31		1	K5	280
124346	+29 02505	IDS14080N2911	AB	14 10 13	+28 57.1	1	7.80		0.38		0.09		2	F2	3032
124346	+29 02505	IDS14080N2911	C	14 10 13	+28 57.1	1	9.78		0.50		0.01		2		3032
124294	−9 03878	HR 5315		14 10 13	−10 02.5	7	4.18	.013	1.34	.007	1.46	.006	36	K3 III	3,15,1075,1088,2012*
124195	−54 05933	HR 5311, V716 Cen		14 10 15	−54 23.5	5	6.12	.043	0.07	.011	−0.39	.003	14	B5 Ve	15,401,1075,1770,2006
		Lod 1289 # 34		14 10 15	−57 40.7	1	12.27		0.47		0.40		2	A1	807
124399	+44 02336	IDS14083N4439	AB	14 10 16	+44 25.4	1	7.76		0.34		0.02		2	F2	3016
124228	−39 08775			14 10 18	−39 44.7	1	7.87		0.16		0.14		1	A2/3 III	1770
124347	+23 02662			14 10 19	+22 36.6	1	7.97		0.43		0.04		2	F0	1375
124263	−32 09948	IDS14074S3302	A	14 10 19	−33 16.5	1	8.52		0.03		−0.04		1	A0 V	1770
124281	−26 10163	HR 5314		14 10 21	−26 22.7	1	6.24		1.08				4	K0 III	2007
	−62 04007			14 10 21	−62 47.7	2	9.95	.018	0.15	.006	−0.68	.012	4	B1.5V	540,976
		He2 106		14 10 23	−63 11.8	1	12.83		0.97		−0.23		1		1753
	+24 02700			14 10 24	+24 02.9	2	8.88	.010	0.96	.000	0.83	.050	4	K0	1569,1569
	+25 02740			14 10 25	+24 55.0	1	9.94		1.16		1.21		1	K5	280
		Ross 845, GQ Vir		14 10 26	−11 47.1	2	13.87	.024	1.52	.005	0.83		3	M5	1705,3073
124283	−32 09951			14 10 26	−33 17.1	1	7.82		0.30				1	F0 V	925
124254	−45 09019			14 10 27	−45 45.1	1	7.43		0.23		0.10		2	A4 V	1771
		G 64 - 55		14 10 28	−05 14.7	2	11.74	.000	1.23	.015	1.15	.045	2		1620,1658
124255	−46 09164			14 10 29	−47 16.7	1	9.13		0.11		0.09		2	A0/1 V	1770
		Lod 1289 # 22		14 10 30	−57 50.1	1	11.23		0.59		0.25		2	F3	807
	+38 02526			14 10 36	+37 36.7	1	10.66		1.11		1.14		1	K0 III	280
	−57 06544			14 10 37	−57 37.3	2	11.29	.000	0.37	.000	0.12	.000	3	B9	287,807
	−58 05425			14 10 37	−58 24.9	1	10.45		0.66		0.09		2	B6 III	540
	−57 06545			14 10 39	−57 40.2	2	11.09	.000	0.35	.000	0.10	.000	3	B9	287,807
124182	−65 02652	IDS14068S6541	AB	14 10 39	−65 55.3	2	6.94	.010	−0.02	.011	−0.57	.008	7	B3 II/III	26,158
		WLS 1420 0 # 6		14 10 40	+00 23.1	1	13.32		0.43		−0.05		2		1375
124358	−11 03691			14 10 40	−11 55.4	2	9.48	.035	0.92	.015	0.45		5	G5	742,1594
		Lod 1289 # 13		14 10 41	−57 46.6	1	10.73		1.35		1.32		2	K5	807
124268	−53 05915			14 10 44	−53 41.9	1	7.29		2.92		6.20		2	Nb	864
	−57 06547			14 10 44	−57 38.4	2	11.20	.000	0.33	.000	0.22	.063	3	B9	287,807
		Lod 1289 # 16		14 10 44	−57 40.5	2	10.95	.010	1.40	.010	1.36	.010	3	M0	287,807
124440	+35 02531			14 10 45	+35 08.3	1	8.77		1.20		1.39		1	K2 III	280
124197	−65 02655	IDS14069S6514	AB	14 10 45	−65 28.1	4	6.71	.017	−0.03	.011	−0.51	.011	12	B3 V	26,158,540,976
124373	−11 03692			14 10 48	−12 00.5	1	10.21		0.45		0.01		1	F8	742
124459	+36 02451			14 10 50	+35 50.1	1	8.04		0.48		0.02		1	F8	272
		Lod 1289 # 36		14 10 50	−57 35.4	1	12.45		0.57		0.43		2	A1	807
124374	−12 04001			14 10 51	−12 42.5	1	9.23		0.55		0.07		15	F7/8 V	1308
	+24 02704			14 10 53	+23 34.2	1	9.44		1.05		0.80		1	K0	280
	+37 02505			14 10 53	+36 47.5	2	9.73	.000	0.29	.010	0.07	.010	7	A2	272,1569
	−6 03950	G 124 - 12		14 10 53	−06 43.6	1	10.14		1.44		1.22		4	M0	158
124334	−43 08895			14 10 58	−43 47.5	1	10.05		0.12		0.11		1	A2 III/IV	1770
	−57 06549			14 10 59	−57 58.1	1	9.81		1.20		1.09		2	K5	807
	+24 02705			14 11 00	+23 57.9	1	9.13		1.14		1.18		1	K2	280
124442	+13 02762	IDS14086N1303	AB	14 11 01	+12 48.4	2	8.26	.043	0.55	.014	0.07		7	G0	173,3016
	+32 02446			14 11 01	+32 21.6	1	9.88		1.11		1.04		1	K0 III	280
	+28 02302			14 11 02	+27 54.5	1	8.93		1.02				1	G5	280
124401	−11 03693			14 11 02	−11 36.3	2	6.98	.002	1.02	.001			48	G5	130,1351
		Lod 1289 # 17		14 11 04	−57 52.0	1	10.96		1.39		1.37		2	M6	807
		Lod 1289 # 33		14 11 04	−57 52.0	1	12.23		0.54		0.13		2	F0	807
	+21 02627			14 11 05	+21 26.6	1	9.54		1.56		1.90		1	K7	280
		Ton 755		14 11 06	+27 12.	1	14.39		0.03		0.12		3		313
124425	−0 02796	HR 5317		14 11 06	−00 36.6	4	5.91	.018	0.47	.004	0.02	.005	11	F7 V	15,1003,1417,3053

HD	DM	Other Id	N	Rem	α_{1950}	δ_{1950}	S	V	σ_V	B–V	σ_{B-V}	U–B	σ_{U-B}	n	Spectrum	References
		WLS 1420-20 # 6			14 11 06	−19 39.7	1	11.11		0.63		0.18		2		1375
124568	+51 01899				14 11 07	+50 56.8	1	7.82		0.12		0.11		2	A0	1566
124730	+70 00778	HR 5334			14 11 08	+69 40.0	1	5.24		1.58		1.84		5	M2 IIIab	3001
		CS 22883 # 11			14 11 09	+12 07.5	1	13.07		0.09		0.19		1		1736
		Lod 1282 # 5			14 11 09	−59 00.1	1	10.89		1.42		1.38		1	K5	549
	+31 02594				14 11 10	+31 13.8	1	9.73		1.20		1.04		2	K0	1569
	+26 02540				14 11 13	+26 05.9	1	10.42		1.06		0.95		1		280
		Lod 1282 # 21			14 11 17	−58 59.8	1	12.32		0.43		0.29		1	A5	549
	+22 02674				14 11 18	+22 27.0	1	9.79		0.26		0.04		1		280
124604	+56 01717				14 11 18	+55 47.4	1	9.69		0.31		0.03		2	F8	1502
	−57 06552				14 11 18	−58 20.4	1	10.57		0.22		0.16		2	A5	807
124314	−61 04431	LSS 3212		⋆ AB	14 11 20	−61 28.4	1	6.64		0.21				4	O7	2012
124752	+68 00771	IDS14103N6803		AB	14 11 21	+67 49.2	3	8.54	.005	0.81	.007	0.52	.005	6	K0 V	1007,1013,8023
124299	−63 03183				14 11 21	−63 40.0	1	8.22		1.22		1.27		5	K1 III	897
124548	+25 02744				14 11 23	+25 00.7	1	8.03		0.05		0.06		2	A2 V	1375
	+35 02534				14 11 24	+35 15.1	1	10.65		0.40		−0.19		3		1569
	+37 02506				14 11 26	+37 06.4	1	10.68		0.53		−0.02		2		1569
124462	−17 04036				14 11 26	−17 48.0	1	9.10		1.15		1.00		2	K0/1 III	1375
	−57 06556				14 11 26	−57 58.5	1	10.38		0.30		0.23		2	A5	807
	+26 02542				14 11 27	+26 19.9	1	9.75		1.10		1.07		1	K7	280
124367	−56 06206	HR 5316, V795 Cen		⋆ A	14 11 27	−56 51.2	9	5.03	.026	−0.06	.015	−0.63	.010	48	B5 Vne	15,26,540,681,815,976*
		Lod 1282 # 33			14 11 27	−58 46.5	1	13.11		0.40		0.25		1		549
	+26 02541				14 11 28	+25 42.3	1	10.28		1.06		0.95		1		280
	+35 02535				14 11 28	+35 04.2	1	9.80		1.04		0.82		1	K0	280
		Lod 1282 # 29			14 11 29	−58 46.0	1	12.97		0.70		0.07		1		549
		G 135 - 29			14 11 31	+15 44.3	1	16.20		0.07		−0.65		4		3060
		LP 739 - 88			14 11 32	−13 19.3	1	14.90		0.49		−0.06		2		3062
		Lod 1282 # 30			14 11 32	−58 45.6	1	13.04		0.57		0.38		1		549
124099	−77 00940	HR 5306			14 11 34	−77 25.9	3	6.48	.012	1.40	.018	1.38	.036	11	K2 IIp	15,1075,1637
124586	+31 02595				14 11 35	+31 25.7	1	7.52		−0.08		−0.24		2	A0	272
	−57 06557				14 11 35	−58 00.2	1	10.40		0.11		−0.01		2	B9	807
	+31 02596	IDS14094N3128		AB	14 11 37	+31 13.5	1	10.12		1.01		0.78		2	M0	1569
124498	−14 03902	IDS14089S1453		AB	14 11 37	−15 07.1	1	10.46		1.28		0.96		2	K7 V	3072
124498	−14 03902	IDS14089S1453		C	14 11 37	−15 07.1	1	14.02		1.60		0.99		3		3072
124433	−41 08589	HR 5318			14 11 37	−41 36.3	3	5.60	.004	0.93	.000	0.58		8	G8 III	15,2027,3005
124466	−29 10924				14 11 38	−30 17.9	1	9.86		0.34				1	F0 V	925
124467	−32 09960				14 11 39	−33 19.6	1	9.35		0.33		0.11		1	A2mA8-F2	1770
	−45 09031				14 11 39	−46 01.5	1	10.52		1.22		1.10		1		55
124674	+52 01782	HR 5328		⋆ A	14 11 40	+52 01.3	3	6.70	.014	0.39	.005	−0.04	.000	6	F1 V	1,1028,3024
124570	+13 02764	HR 5323			14 11 41	+13 11.6	4	5.55	.021	0.54	.016	0.09	.013	13	F7 IV	3,70,254,3026
124508	+29 02508	IDS14095N2934		AB	14 11 41	+29 20.3	1	6.80		0.34		0.15		1	A2	280
124675	+52 01782	HR 5329, κ2 Boo		⋆ A	14 11 42	+52 01.4	3	4.52	.020	0.20	.000	0.14	.000	8	A8 IV	1013,1028,3024
124675	+52 01782	HR 5329, kappa2		⋆ AB	14 11 42	+52 01.4	5	4.39	.018	0.20	.005	0.10	.005	9	A8 IV +F1 V	15,1007,1118,1363,8015
		AAS12,381 32 # C			14 11 43	+32 30.9	1	10.73		0.91		0.57		1		280
124469	−37 09260				14 11 43	−38 08.9	1	8.69		0.31		0.01		1	A9 V	1771
		Lod 1282 # 23			14 11 44	−58 54.5	1	12.38		0.90		0.50		1	K2	549
	+37 02507				14 11 45	+36 49.5	1	9.22		0.47		−0.01		2	F8	1569
124553	−5 03837	HR 5322			14 11 45	−05 43.0	4	6.36	.006	0.59	.005	0.15	.012	12	F9 V	15,1256,1311,3026
124642	+30 02494	G 165 - 56			14 11 46	+30 26.9	1	8.06		1.06		0.99		2	K5	1569
	+35 02536				14 11 46	+34 45.3	1	9.42		1.42		1.72		1	K2 III	280
		Lod 1282 # 9			14 11 46	−58 54.6	1	11.29		0.93		0.47		1	K1	549
	+23 02666				14 11 47	+22 40.0	1	10.36		1.24		1.39		1		280
124448	−45 09033	V821 Cen			14 11 47	−46 03.4	6	9.99	.009	−0.09	.009	−0.80	.006	29	(B)p	55,158,400,843,1112,2012
	−58 05432				14 11 47	−58 56.5	1	10.82		0.19		−0.21		1	B8	549
123998	−80 00706	HR 5303			14 11 49	−80 46.5	4	4.92	.014	0.25	.009	0.10	.010	14	A2mA7-F2	15,1075,2016,2038
	−34 09470				14 11 50	−34 30.8	1	9.37		0.54				1	F8	925
		Lod 1282 # 34			14 11 50	−58 53.2	1	13.12		0.48		0.32		1		549
		CS 22883 # 4			14 11 52	+09 45.8	1	14.22		0.43		−0.20		1		1736
	−61 04441	LSS 3213			14 11 52	−61 25.0	1	10.72		0.27		−0.50		2	B1.5V	540
	+28 02304				14 11 53	+27 49.3	1	8.65		1.23		1.34		1	K0	280
	+28 02305				14 11 54	+28 10.0	1	9.29		1.11		1.09		1	K0	280
	+33 02429				14 11 54	+33 09.7	1	9.22		0.98		0.80		1	K0	280
		Lod 1282 # 22			14 11 54	−58 53.7	1	12.38		0.72		0.23		1	F8	549
		Lod 1282 # 15			14 11 56	−58 49.8	1	11.74		0.99		0.44		1	K0	549
124694	+46 01944				14 11 57	+46 33.5	1	7.19		0.52		0.00		3	G0	196
124538	−31 11016				14 11 58	−32 04.3	1	8.27		1.68				1	M1 III	925
124454	−52 07087	HR 5319			14 11 58	−53 16.6	1	6.39		1.57				4	K3 III	2006
		CS 22883 # 5			14 11 60	+09 48.4	1	14.75		−0.04		−0.07		1		1736
	−58 05434				14 12 02	−58 44.0	1	11.42		0.37		−0.16		1	B8	549
124696	+36 02453				14 12 04	+35 50.2	2	6.91	.010	1.60	.000	2.01		7	K7 III	280,1501
124575	−20 03989				14 12 04	−20 49.9	1	7.47		1.47		1.72		3	K3/4 III	8100
		Lod 1282 # 16			14 12 05	−59 08.2	1	11.77		0.84		0.33		1	K0	549
124628	+2 02794				14 12 06	+02 01.8	1	8.69		0.78		0.30		2	G5	1375
		G 64 - 60			14 12 06	−03 03.3	2	14.21	.028	1.55	.014	1.45	.173	4		203,3073
	+35 02537				14 12 07	+35 27.1	1	10.24		0.99		0.79		1	K0	280
	−58 05435				14 12 07	−58 45.0	1	10.66		0.16		−0.29		1	B8	549
124524	−46 09185				14 12 08	−47 00.3	1	8.99		0.08		0.02		1	A0 V	1770
124485	−50 08387				14 12 09	−50 38.0	1	9.68		0.28		0.17		1	A1 III/IV	1770
	−57 06560				14 12 11	−57 36.0	1	10.49		0.79		0.25		2	F8	807
	+33 02431				14 12 12	+33 16.4	1	8.83		0.92		0.62		1	G8 III	280

Table 1 791

HD	DM	Other Id	N	Rem	α_{1950}	δ_{1950}	S	V	σ_V	B–V	σ_{B-V}	U–B	σ_{U-B}	n	Spectrum	References
124540 −39 08797					14 12 12	−40 03.9	1	9.02		0.14		0.16		1	A2/3 V	1770
		Lod 1282 # 17			14 12 15	−58 53.9	1	12.02		1.30		1.03		1	K4	549
−58 05436					14 12 15	−58 53.9	2	11.18	.015	0.52	.010	0.26	.005	2	B8	287,549
124578 −34 09476	IDS14093S3440		A		14 12 16	−34 53.8	1	9.50		0.16		0.14		1	A1/2 IV	1770
−58 05437					14 12 16	−59 01.3	1	10.42		1.11		0.82		1	K3	549
		G 200 - 24			14 12 17	+46 13.2	1	13.18		1.49				1		1759
124677 +16 02627	G 135 - 31		⋆ A		14 12 19	+16 14.7	1	8.58		0.82		0.51		2	G5	1648
124525 −50 08390					14 12 19	−50 38.9	1	10.07		0.86		0.54		4	G8 III/IV	158
−57 06562	Lod 1289 # 15				14 12 19	−58 00.3	1	10.87		0.48		0.26		2	A8	807
		Lod 1282 # 28			14 12 19	−58 57.9	1	12.90		0.62		0.42		1		549
124609 −21 03869					14 12 20	−22 19.9	1	8.25		0.07		0.04		2	A0 V(n)	1375
124470 −57 06561					14 12 20	−58 09.8	1	9.75		0.60		0.13		2	A7/F0 IV	807
−58 05438					14 12 20	−58 55.9	2	11.36	.065	0.23	.050	-0.04	.085	2	B8	287,549
124681 +4 02841	HR 5331, FS Vir				14 12 22	+03 34.1	2	6.45	.005	1.60	.000	1.80	.061	16	M4 IIIab	1088,3046
124714 +22 02677					14 12 22	+21 38.8	1	8.29		1.48		1.80		2	K2	280
124713 +22 02678	HR 5333				14 12 22	+22 06.4	2	6.39		0.18	.015	0.09	.005	5	A7 V	280,1022
124755 +42 02472	HR 5335				14 12 23	+41 45.2	1	6.24		1.04				2	K2	70
124679 +10 02654	HR 5330		⋆ AB		14 12 24	+10 20.1	3	5.29	.013	1.00	.005	0.77	.000	7	K1 III	70,1080,3077
−1 02926					14 12 24	−02 04.2	1	9.50		1.01		0.70		2	K0	1375
−57 06563	Lod 1289 # 18				14 12 24	−57 58.5	1	11.00		0.63		0.05		2	F9	807
124486 −57 06564					14 12 25	−58 09.1	1	9.06		1.11		0.85		2	G6/8 III	807
		Lod 1282 # 19			14 12 26	−59 06.3	1	12.20		0.44		0.21		1	A0	549
+25 02746					14 12 27	+25 01.0	2	9.66	.005	0.31	.010	0.04	.005	4	F8	1569,1569
		Lod 1282 # 27			14 12 27	−58 49.9	1	12.84		0.46		0.25		1		549
124580 −44 09181	HR 5325, GQ Vir				14 12 28	−44 46.0	3	6.30	.004	0.60	.005			11	G1 V	15,2018,2029
124796 +39 02739					14 12 30	+39 30.5	1	7.40		1.08		0.98		4	K2 III	1501
124615 −36 09218					14 12 30	−37 18.6	1	9.31		0.20		0.11		1	A2/3 IV(m)	1770
−58 05439					14 12 30	−58 54.2	2	11.36	.020	0.25	.025	-0.07	.080	2	B9	287,549
+23 02668					14 12 31	+23 02.2	1	9.36		1.16		1.24		1	K5	280
+30 02497					14 12 31	+30 00.1	1	9.83		1.12		1.05		2	K0	1569
		Lod 1289 # 32			14 12 31	−57 59.1	1	12.11		0.50		0.46		2	A2	807
		Lod 1282 # 31			14 12 31	−58 52.9	1	13.06		0.37		0.24		1	A2	549
		G 65 - 36			14 12 32	+10 38.9	1	12.37		1.52		1.21		1	M2	1620
		Lod 1289 # 27			14 12 33	−57 53.5	1	11.59		0.40		0.29		2	A0	807
		Lod 1282 # 24			14 12 33	−58 59.2	1	12.63		0.53		0.33		1	F0	549
		G 65 - 37			14 12 35	+04 20.3	1	13.95		1.16				1		1620
		Lod 1282 # 32			14 12 35	−58 56.2	1	13.06		0.37		0.26		1	A2	549
124597 −46 09194					14 12 37	−46 48.0	1	9.77		0.04		-0.23		1		1770
124597 −46 09194					14 12 37	−46 48.0	1	9.94		0.00		-0.28		1		1770
124581 −51 08076					14 12 37	−51 45.9	1	9.94		0.26		0.18		1	A0 V	1770
		Lod 1282 # 20			14 12 37	−58 47.0	1	12.30		0.44		0.11		1	A8	549
+36 02456					14 12 38	+36 29.5	1	10.03		1.06		0.98		1	G8 III	280
124683 −17 04046	HR 5332				14 12 38	−17 58.1	4	5.52	.081	-0.02	.017	-0.01	.040	14	A0 V	3,252,1079,2009
		Lod 1282 # 1			14 12 38	−59 08.7	1	10.25		2.09		2.42		1	M5	549
+25 02747					14 12 39	+25 14.6	1	9.79		1.10		0.97		1	K0	280
		Lod 1289 # 19			14 12 40	−57 54.6	1	11.00		1.42		1.22		2	K2	807
		Lod 1282 # 26			14 12 40	−58 50.7	1	12.69		0.76		0.12		1		549
124471 −66 02490	HR 5320		⋆ A		14 12 40	−66 21.4	5	5.75	.006	-0.07	.009	-0.85	.029	25	B1.5 III	15,1075,1732,2038,8100
124715 −2 03811					14 12 41	−03 13.0	3	9.24	.004	0.57	.008	0.03	.007	7	G0	516,742,1064
124317 −75 00949					14 12 41	−75 51.4	1	7.35		1.71		1.85		9	M3 III	1704
		Lod 1289 # 29			14 12 42	−57 58.0	1	11.86		0.28		-0.12		2	B8	807
+26 02544					14 12 43	+25 42.4	1	9.67		1.22		1.32		1	K2	280
124797 +24 02707					14 12 44	+23 55.1	1	6.75		0.33		0.12		2	F0	1569
		G 178 - 9			14 12 45	+42 03.5	2	13.76	.015	0.85	.012	0.23	.129	3		1658,3060
−1 02928					14 12 45	−02 15.8	1	10.47		0.37		0.08		2		1375
124490 −66 02491					14 12 46	−66 56.2	1	9.16		0.34		0.24		1	A7 V	1771
124757 +3 02874	IDS14103N0336		AB		14 12 48	+03 21.7	4	7.06	.011	0.55	.011	0.01	.011	11	F8	292,938,1381,3030
124773 +18 02851					14 12 49	+17 45.0	1	9.00		-0.03		-0.09		3	A0 V	272
−31 11026					14 12 50	−31 55.0	1	10.74		0.84		0.44		2		1696
−57 06566					14 12 51	−58 02.4	1	11.50		0.21		0.12		2	A1	807
		Lod 1282 # 14			14 12 52	−58 48.2	1	11.43		1.07		0.62		1	K2	549
		Lod 1282 # 18			14 12 52	−59 00.4	1	12.10		0.60		0.23		1	F5	549
−57 06567					14 12 55	−57 32.1	1	10.40		0.70		0.16		2	F8	807
−57 06568					14 12 57	−57 31.5	1	10.64		0.62		0.07		2	F8	807
124601 −59 05476	HR 5326, R Cen		⋆ A		14 12 57	−59 40.9	1	7.55		2.10				1	M5e	8051
124635 −52 07103					14 12 58	−53 11.5	1	10.19		0.25		0.19		1	A2/4 V	1770
		Lod 1282 # 25			14 12 59	−58 58.1	1	12.64		0.33		0.18		1	A0	549
		G 65 - 38			14 13 00	+00 50.5	1	13.06		1.40		1.33		1		1620
124620 −56 06215	IDS14095S5650		AB		14 13 01	−57 03.9	1	7.17		0.20				4	B9 IV	1075
124621 −57 06569					14 13 01	−57 39.3	1	10.13		0.38		0.30		2	A2 IV	807
		G 65 - 39			14 13 04	+04 53.9	3	14.32	.018	1.73	.005	1.26	.100	7		419,1620,3078
		Lod 1282 # 6			14 13 05	−59 06.8	1	10.96		0.81		0.28		1	G0	549
124759 −15 03837					14 13 06	−15 51.0	1	7.98		1.23		1.18		5	K0 III	1657
		CS 22883 # 7			14 13 08	+10 35.6	1	12.38		0.07		0.20		1		1736
+24 02708					14 13 08	+24 18.0	1	9.92		0.82		0.44		1	G5	280
+31 02597					14 13 11	+31 31.5	1	9.76		0.79		0.45		2	G5 III	1569
124883 +28 02309					14 13 12	+27 57.9	1	7.29		0.25		0.16		1	A3 mF2	280
124721 −44 09188					14 13 14	−44 57.5	1	9.85		1.04		0.84		2	K1/2 (III)	536
−57 06571					14 13 16	−57 47.9	1	10.77		0.20		-0.38		2	B6	807
+62 01340	G 223 - 47				14 13 18	+61 36.0	1	10.00		0.96		0.74		3	K0	308

HD	DM	Other Id	N	Rem	α_{1950}	δ_{1950}	S	V	σ_V	B−V	σ_{B-V}	U−B	σ_{U-B}	n	Spectrum	References
124740	−40 08541				14 13 18	−41 00.1	1	7.86		−0.02		−0.18		2	A(p SiCr)	1770
124780	−32 09982	HR 5337			14 13 20	−33 00.6	1	6.56		0.30				4	F0 V	2007
		CS 22883 # 15			14 13 22	+11 26.1	1	16.07		−0.36		−1.13		1		1736
124884	+13 02770				14 13 22	+12 58.9	1	8.81		1.66				2	K0	882
		G 178 - 10			14 13 22	+45 15.0	1	11.82		1.46				1		1759
124851	−6 03957	IDS14107S0636	AB		14 13 22	−06 49.8	2	8.96	.000	0.83	.018	0.55		7	K0	214,1414
124864	+10 02657	IDS14109N1046	AB		14 13 23	+10 32.1	1	9.22		0.87		0.60		4	K0	3030
124864	+10 02657	IDS14109N1046	C		14 13 23	+10 32.1	1	9.75		0.84		0.46		1	K0	3032
124897	+19 02777	HR 5340			14 13 23	+19 26.5	14	−0.05	.013	1.23	.008	1.27	.011	89	K1 IIIb	1,15,22,198,667,1034*
124850	−5 03843	HR 5338			14 13 23	−05 45.8	11	4.08	.010	0.51	.007	0.03	.018	41	F7 V	3,15,254,1008,1020*
124766	−42 09181	IDS14103S4249	A		14 13 23	−43 02.6	1	8.93		0.09		0.08		1	A1 V	1770
124689	−57 06572	RR Cen			14 13 25	−57 37.4	1	7.29		0.36		0.05		2	A9/F0 V	807
		G 166 - 14			14 13 26	+23 10.6	1	16.41		−0.10		−0.95		1		3060
	+27 02353		A		14 13 32	+27 30.9	1	10.78		0.73		0.35		1		280
	+27 02353		B		14 13 32	+27 30.9	1	11.13		0.29		0.06		1		280
125019	+53 01699	HR 5345			14 13 32	+52 46.1	2	6.58	.006	0.10	.002	0.08	.009	6	A4 V	374,1375
	−58 05443				14 13 32	−58 51.5	1	11.27		0.27		0.03		1	B9	549
124914	+13 02771				14 13 34	+12 41.2	1	8.16		0.94				1	G5	882
		Lod 1289 # 31			14 13 35	−57 38.7	1	12.09		0.28		−0.08		2	B9	807
	+23 02670				14 13 36	+23 12.7	1	10.17		0.96		0.60		1		280
		Lod 1289 # 28			14 13 37	−57 37.6	1	11.85		0.66		0.37		2	A0	807
		GD 164			14 13 38	+28 21.6	1	15.71		0.62		−0.10		1		3060
		Lod 1289 # 35			14 13 39	−57 36.0	1	12.39		0.26		0.03		2	A0	807
	−58 05444				14 13 42	−59 04.7	1	11.35		0.38		0.17		1	B9	549
124953	+19 02779	HR 5343, CN Boo			14 13 43	+19 08.6	5	5.98	.012	0.26	.011	0.05	.000	12	A8 III	15,252,1007,1013,8015
124915	−5 03845	HR 5341			14 13 44	−06 23.4	1	6.43		0.28		0.06		8	A9 III	1088
124786	−51 08093				14 13 44	−51 31.1	1	9.71		0.24		0.20		1	A2 IV	1770
	−57 06575				14 13 44	−57 52.2	1	11.32		0.20		0.13		2	A0	807
	−21 03873				14 13 46	−21 31.9	1	10.67		1.37		0.07		1		1753
		Lod 1289 # 30			14 13 46	−57 34.9	1	12.04		0.28		0.07		2	A0	807
	−57 06576				14 13 47	−57 52.6	1	11.35		0.30		0.12		2	A0	807
124997	+35 02539				14 13 50	+34 40.1	1	8.16		1.68		1.90		1	K0	280
124931	−2 03812	HR 5342			14 13 54	−02 57.9	1	6.14		−0.01		−0.01		8	A1 V	1088
	+36 02461				14 13 57	+36 10.8	1	10.11		0.93		0.51		2	G8 III	1569
124812	−55 05964	LSS 3216			14 13 57	−55 32.0	2	6.63	.010	0.23	.010	−0.01		7	A1 II	1075,8100
	+33 02433	G 165 - 59		⋆ A	14 14 04	+32 49.0	1	9.57		1.10		1.08		1	K5	280
	+33 02434	IDS14119N3303		B	14 14 04	+32 50.3	1	9.99		0.38		−0.03		3		1569
	+35 02540				14 14 05	+34 45.9	1	10.04		1.12		1.19		1	K2	280
124919	−31 11050				14 14 06	−31 29.2	1	9.90		0.48				1	F5 V	925
125076	+46 01948				14 14 07	+45 47.8	1	6.53		1.03		0.79		3	K0	196
125140	+57 01496	IDS14125N5708	AB		14 14 08	+56 54.2	1	9.21		0.80				3	G5	3024
125140	+57 01496	IDS14125N5708	C		14 14 08	+56 54.2	1	12.08		0.63				3		3024
	+26 02546				14 14 10	+25 53.1	1	8.94		1.10		1.06		1	K0	280
		LP 97 - 674	AB		14 14 12	+59 41.1	1	11.68		1.52				1		1746
125040	+20 02954	HR 5346		⋆ AB	14 14 13	+20 21.2	1	6.25		0.49				2	F8 V	70
		XX Vir			14 14 13	−06 03.0	1	11.59		0.12		0.10		1	F6	597
	−60 05274				14 14 14	−61 15.0	1	11.40		0.42		−0.43		1		287
		CS 22883 # 20			14 14 16	+09 23.4	1	14.58		0.40				1		1736
124990	−17 04053				14 14 18	−18 21.3	1	6.22		0.98				4	K0 III	2009
125111	+40 02760	HR 5347			14 14 22	+39 58.5	3	6.39	.009	0.36	.006	−0.06	.028	6	F3 V	70,254,3037
	+26 02547				14 14 23	+26 10.7	1	9.05		1.03		0.87		1	G5	280
125193	+57 01498	IDS14125N5708			14 14 23	+56 55.3	1	6.68		0.58		0.15		2	F8	1502
125113	+32 02451				14 14 24	+32 26.4	1	6.73		0.93				2	G8 III	280
125161	+52 01784	HR 5350, ι Boo		⋆ A	14 14 24	+51 35.8	5	4.75	.006	0.21	.010	0.06	.000	20	A9 V	15,1363,3024,6001,8015
124940	−41 08626				14 14 25	−41 56.5	1	9.67		0.34		0.03		1	A9 V	1771
125161	+52 01785	IDS14126N5150	B		14 14 26	+51 36.4	2	8.24	.038	0.83	.024	0.41	.052	4	A2	1003,3016
		CS 22883 # 13			14 14 27	+12 27.2	1	14.44		0.17		0.18		1		1736
125162	+46 01949	HR 5351			14 14 29	+46 19.0	4	4.18	.004	0.08	.000	0.06	.015	12	A0 Shell	15,1118,3023,8015
	−83 00553				14 14 29	−83 46.6	1	9.18		0.66		0.20		2	G5	1730
	+29 02516				14 14 31	+29 10.3	1	9.57		1.02		0.72		2	K0	1569
125122	+32 02452				14 14 32	+32 23.8	1	8.20		1.14		1.17		1	K3 IV	280
124961	−40 08559				14 14 33	−40 37.8	1	7.88		0.00		−0.13		2	B9 V	1770
124654	−74 01161				14 14 33	−74 57.0	1	7.70		0.07		−0.04		4	B9/9.5V	158
125006	−30 11301				14 14 34	−31 06.6	1	9.70		0.27		0.19		1	A1mA3-F0	1770
		Lod 1256 # 1			14 14 36	−61 11.1	1	11.67		0.41		−0.32		1		287
	+35 02541				14 14 37	+34 42.5	1	10.66		1.07		1.00		1		280
		TV Boo			14 14 37	+42 35.5	1	10.70		0.10		0.13		1	B9	668
125405	+74 00573				14 14 37	+73 33.9	1	8.80		0.51		0.10		2	F8	1733
124941	−50 08421				14 14 38	−51 17.2	1	9.54		0.17		0.12		1	A0 V	1770
		G 65 - 40			14 14 39	+10 49.6	2	11.53	.010	1.51	.010	1.27		2	M2	333,1620,1746
124879	−62 04049	IDS14109S6233	AB		14 14 40	−62 46.4	2	8.39	.010	0.48	.000	−0.04		3	F6 V	742,1594
125007	−42 09195				14 14 45	−42 21.6	2	7.04	.007	−0.04	.003	−0.18		5	B9 V	1770,2014
124909	−60 05276				14 14 46	−60 40.1	1	9.08		0.28		−0.63		4	O9/B0 V	92
	−60 05277				14 14 50	−61 11.3	1	11.32		0.34		−0.46		1		287
	+30 02501				14 14 51	+30 10.8	1	10.71		0.05		0.08		1		280
125260	+55 01669				14 14 51	+55 31.8	1	6.90		1.13		1.16		3	K1 III	37
124979	−50 08427	LSS 3219			14 14 51	−51 16.4	2	8.52	.014	0.10	.014	−0.82	.046	9	O8	400,403
	−59 05488				14 14 52	−59 40.3	1	10.25		0.23		0.01		4		244
125031	−36 09253				14 14 53	−37 06.5	1	9.51		0.29		0.10		1	A9 IV	1771
		G 165 - 61			14 14 54	+31 56.6	1	13.15		2.00				1		1759

Table 1   793

HD	DM	Other Id	N Rem	α_{1950}	δ_{1950}	S	V	σ_V	B–V	σ_{B-V}	U–B	σ_{U-B}	n	Spectrum	References
	−59 05491			14 14 54	−59 24.5	1	9.66		0.14		-0.24		4	A0	244
		KS 885		14 14 54	−63 02.	1	11.49		0.36		-0.47		3		540
	+27 02357			14 14 57	+27 30.9	1	9.29		0.54		-0.04		2	G0	1375
124995	−53 05938			14 14 58	−54 05.5	1	8.39		0.09		-0.34		1	B8 III	1770
	+35 02542			14 15 00	+35 01.7	1	10.50		0.98		0.82		1		280
125180	+15 02690	HR 5352		14 15 05	+15 29.6	3	5.81	.030	1.68	.048	2.00	.069	7	M3 IIIa	70,97,3055
125011	−53 05940			14 15 05	−53 53.9	1	7.57		1.11		0.85		5	K0 III	1673
		L 260 - 53		14 15 06	−52 10.1	1	13.27		1.47		1.10		1		3078
125034	−50 08431			14 15 07	−50 40.1	1	9.58		0.29		0.17		1	A(7)	1770
	−83 00555			14 15 08	−83 58.3	1	11.31		0.38		0.21		2		1730
125230	+34 02513			14 15 10	+34 07.5	1	8.78		0.56		0.05		2	G0	1569
125194	+12 02677			14 15 11	+11 33.9	1	7.60		1.31		1.54		2	K0	1648
		MCC 714	AB	14 15 12	+32 58.6	2	11.43	.044	1.17	.005	1.17	.024	3	K4	497,1746
125070	−46 09229			14 15 14	−47 07.2	1	9.24		0.04		0.03		1	B9 IV/V	1770
	−60 05283	LSS 3220		14 15 14	−60 48.0	1	9.87		0.39		-0.37		2		432
124834	−73 01267			14 15 14	−73 44.3	1	6.57		0.03		-0.50		11	B3 III/IV	1628
		Feige 93, EG 107		14 15 15	+13 15.7	3	15.31	.019	-0.22	.012	-1.10	.011	4	DAwk	1264,1298,1727
		GD 493		14 15 16	+70 27.5	1	14.62		0.51		-0.03		3		308
125015	−58 05464	IDS14117S5813	AB	14 15 16	−58 27.4	1	7.27		0.01		-0.48		4	B7 Vn	244
		G 64 - 62		14 15 18	+00 23.3	1	11.73		1.37		1.22		1	M1	1620
125038	−58 05465			14 15 18	−58 44.5	1	7.76		0.36		0.07		4	F0 IV	244
125213	+12 02678			14 15 19	+12 24.7	1	9.04		1.64				2		882
	+34 02514			14 15 19	+33 59.7	1	9.77		0.44		-0.05		2	F5	1569
	+32 02453			14 15 21	+31 48.9	1	9.22		1.50		1.76		2	K4 III	280
125184	−6 03964	HR 5353	⋆ A	14 15 21	−07 18.5	4	6.47	.008	0.72	.004	0.34	.005	17	F9 V	15,1088,2027,3077
		G 124 - 20		14 15 22	−06 26.2	1	16.20		0.07		-0.86		1		3060
125247	+14 02722			14 15 26	+14 08.1	1	8.13		1.55				2	K2	882
		G 64 - 63		14 15 26	−00 17.6	1	12.76		1.59		1.17		1		1620
125150	−32 10005	HR 5348		14 15 26	−32 59.4	3	6.53	.007	0.28	.007			11	A9 IV/V	15,2013,2029
	+46 01951			14 15 27	+45 40.1	2	10.09	.085	1.45	.003	1.22		8	M0	1619,7008
		NGC 5548 sq1 4		14 15 28	+25 30.0	1	11.88		0.92		0.60		3		327
125072	−58 05467			14 15 31	−59 08.3	6	6.65	.011	1.03	.009	0.94	.012	22	K3 V	244,285,1075,2012*
	+15 02691			14 15 33	+15 07.5	1	9.50		0.43		0.02		3	F2	97
125349	+51 01908	HR 5360		14 15 34	+51 32.3	1	6.20		0.04				2	A2 IV	70
		CS 22883 # 14		14 15 35	+12 16.4	1	14.77		0.04		0.15		1		1736
125171	−35 09404			14 15 35	−35 40.0	1	8.99		0.29		0.06		1	A9 V	1771
	+28 02312			14 15 36	+27 48.1	1	9.19		1.44		1.75		1	M0	280
125016	−66 02505			14 15 36	−66 34.4	1	8.09		0.24		0.25		1	A1 IV/V	1771
		NGC 5548 sq1 2		14 15 38	+25 18.1	1	15.38		0.69		0.17		1		327
		CCS 2153		14 15 38	−56 45.8	1	10.50		3.22				2		864
	+23 02673			14 15 40	+23 21.1	1	10.35		1.09		0.96		1		280
125154	−41 08649			14 15 40	−42 00.6	1	9.19		0.06		0.04		2	A0 V	1770
125104	−55 05976			14 15 40	−55 39.1	2	7.31	.015	0.11	.005	-0.56		6	B4 IV/V	401,1075
	+24 02714			14 15 43	+23 50.3	1	9.39		1.47		1.86		1	M0	280
		NGC 5548 sq1 1		14 15 43	+25 19.4	1	13.80		0.65		0.07		1		327
	+29 02521			14 15 46	+28 55.3	1	9.69		1.22		1.26		1	K5	280
	+13 02777	G 65 - 43		14 15 47	+12 58.5	1	10.57		0.88		0.50		1	K3	333,1620
125200	−36 09259			14 15 48	−37 06.0	1	8.03		0.26		0.07		1	A8 IV	1771
		NGC 5548 sq1 3		14 15 50	+25 29.0	1	13.05		0.76		0.33		3		327
125248	−18 03789	HR 5355, CS Vir		14 15 52	−18 29.1	3	5.89	.005	0.00	.013	-0.08	.049	12	Ap Si(Cr)	1063,1202,2035
	+25 02756			14 15 53	+25 27.6	1	10.50		0.52		0.25		2		1569
125320	+27 02360			14 15 53	+27 01.6	1	8.12		0.77				4	G5 IV	20
125351	+36 02468	HR 5361		14 15 53	+35 44.4	7	4.82	.016	1.06	.003	0.91	.013	23	K1 III	15,667,1080,1363,3016*
		HA 82 # 128		14 15 55	+14 53.1	1	11.60		0.58		-0.01		2		97
125117	−59 05501			14 15 55	−59 31.4	1	7.95		0.09		-0.18		4	B8/9 III/IV	244
	+29 02523	IDS14137N2851	AB	14 15 56	+28 36.7	1	10.08		0.72		0.51		1	K0	280
125406	+48 02188	HR 5363		14 15 57	+48 13.9	2	6.33	.005	0.45	.009	0.05		5	F5 V	70,1501
	+49 02287	IDS14141N4913	AB	14 15 57	+48 59.0	1	9.12		0.32		0.04		2	A2	272
125237	−31 11077			14 15 57	−31 28.5	1	8.71		1.26				1	K0 III	925
		CS 22883 # 18		14 16 01	+10 06.3	1	14.64		0.03		0.15		1		1736
		HA 82 # 187		14 16 02	+15 02.9	1	12.26		1.08		0.85		2		97
		HA 82 # 186		14 16 02	+15 03.7	1	15.33		1.00		0.72		1		97
		HA 82 # 189		14 16 04	+15 02.4	1	14.81		0.77		0.39		1		97
125334	+13 02779			14 16 06	+12 40.7	1	9.55		0.99				2		882
125390	+39 02744			14 16 06	+39 11.8	1	8.23		0.97		0.73		4	G7 III	1501
		HA 82 # 236		14 16 07	+15 10.4	1	14.45		0.70		0.30		2		97
	+27 02361			14 16 07	+27 22.7	1	9.89		1.52		1.30		1	M7	280
125335	+11 02662			14 16 09	+10 44.4	2	7.22	.010	0.32	.015	0.11	.005	6	Am	355,8071
	+39 02745			14 16 10	+39 19.1	1	10.47		0.25		0.06		2	A2	272
125276	−25 10271	HR 5356	⋆ AB	14 16 10	−25 35.4	4	5.86	.008	0.50	.004	-0.12	.005	13	F7(w)F3 V	15,285,2012,3037
125158	−60 05294	HR 5349		14 16 10	−61 02.5	6	5.22	.009	0.28	.009	0.15	.005	19	A5/7mA7-F2	15,244,355,1075,2006,2016
125238	−45 09084	HR 5354		14 16 11	−45 49.7	5	3.54	.009	-0.18	.005	-0.72	.009	23	B2.5IV	15,26,1075,2012,8015
125353	+13 02780			14 16 12	+12 40.0	1	9.30		1.16				2	G5	882
124771	−79 00755	HR 5336		14 16 13	−79 52.8	5	5.06	.009	-0.11	.007	-0.57	.018	18	B3 V	15,26,1075,1637,2038
	+30 02504			14 16 14	+29 40.3	1	10.43		0.98		0.81		2		1569
	−60 05295	LSS 3224		14 16 14	−60 35.0	1	11.07		0.35		-0.44		2	B1.5V	540
125301	−31 11081	IDS14133S3123	A	14 16 15	−31 36.4	1	9.00		0.37		0.14		1	A1mA9-F2/3	1770
		LP 439 - 450		14 16 16	+19 52.0	1	11.93		0.53		-0.13		2		1696
125253	−45 09086			14 16 17	−45 52.2	1	7.10		0.07		0.07		1	A1 V	1770
125354	−5 03853	G 124 - 22		14 16 20	−06 21.9	3	9.11	.025	1.30	.006	1.18	.017	7	K7 V	158,1705,3072

HD	DM	Other Id	N Rem	α_{1950}	δ_{1950}	S	V	σ_V	B–V	σ_{B-V}	U–B	σ_{U-B}	n	Spectrum	References
	−62 04068	LSS 3223		14 16 20	−62 35.3	2	10.70	.015	0.63	.031	−0.41	.002	5	B0.5 Ia	540,1737
	+23 02677			14 16 21	+23 17.9	1	9.28		1.01		0.77		2	K0	1375
	+25 02757			14 16 21	+25 32.1	1	9.38		1.08		0.97		1	K0	280
		G 200 - 29		14 16 21	+49 53.4	1	14.44		1.58				1		1759
125391	+18 02861			14 16 22	+17 35.0	1	7.94		0.96					K0	882
125283	−36 09268	HR 5357		14 16 22	−36 46.4	4	5.93	.010	0.07	.008	0.04	.005	11	A2 Vn	15,401,1771,2012
125337	−12 04018	HR 5359		14 16 24	−13 08.5	10	4.52	.006	0.13	.005	0.12	.016	79	A1 V	15,130,355,1075,1088*
	+15 02692			14 16 26	+15 26.6	1	10.18		1.08		0.96		2		97
		HA 82 # 284		14 16 27	+15 17.9	1	14.47		0.67		0.24		1		97
	+15 02693			14 16 27	+15 27.7	1	11.19		0.55		−0.03		2		97
125206	−60 05298	LSS 3226		14 16 27	−60 51.1	1	7.92		0.26		−0.71		4	O9.5V	92
125409	+20 02957			14 16 29	+20 05.1	1	8.14		1.17		1.17		2	K0	1648
125159	−65 02681			14 16 31	−65 53.5	2	10.01	.017	0.19	.003	−0.50	.005	7	B8	540,976
		LSS 3225		14 16 34	−63 39.6	1	11.43		0.32		−0.56		4	B3 Ib	92
		G 124 - 23		14 16 35	−07 03.9	4	13.41	.025	1.63	.013	1.09	.005	8		203,1705,1774,3078
125342	−34 09535			14 16 37	−34 51.5	1	9.17		0.29		0.10		1	A6 III/IV	1771
	+33 02441			14 16 41	+32 55.6	1	9.85		1.49		1.82		1	K2	280
125241	−60 05300	LSS 3227		14 16 41	−60 39.6	7	8.28	.014	0.49	.013	−0.55	.021	21	O9/9.5	125,138,219,403,432*
	+15 02694			14 16 43	+14 42.8	1	9.93		0.58		0.02		2	G0	97
	+32 02456			14 16 43	+32 20.3	1	9.99		1.18		1.30		1	K2 III	280
125287	−55 05983			14 16 45	−55 33.7	1	7.10		0.96				4	G6 III	1075
	+30 02505			14 16 46	+30 04.5	1	9.22		1.55		1.96		1	M0	280
125557	+52 01788			14 16 46	+52 15.8	1	6.92		0.14		0.10		2	A2	1566
125244	−63 03218			14 16 48	−64 13.9	1	9.51		0.88		0.63		4	K1/2 V	158
124639	−82 00601	HR 5327		14 16 48	−82 37.2	3	6.42	.009	0.03	.010	−0.33	.004	11	B8 Ve	15,1075,1637
125288	−55 05984	HR 5358		14 16 49	−56 09.4	7	4.34	.013	0.12	.007	−0.44	.007	30	B5 Ib/II	15,26,1075,1637,2012*
		HA 82 # 238		14 16 50	+15 08.7	1	12.57		0.85		0.42		2		97
	+30 02506			14 16 50	+30 25.4	1	9.92		0.62		0.14		1	G0	280
125451	+13 02782	HR 5365	★ A	14 16 51	+13 14.0	7	5.41	.012	0.38	.011	−0.03	.014	28	F5 IV	3,15,1007,1008,1013*
		HA 82 # 359		14 16 52	+15 27.3	1	13.64		0.68		0.15		1		97
125538	+39 02749	HR 5369		14 16 52	+38 59.8	1	6.87		1.06		0.89		2	K0 III	105
125304	−57 06602	LSS 3228		14 16 52	−57 45.3	1	8.42		0.80		0.48		3	F5 Ib	8100
		HA 82 # 239		14 16 53	+15 15.7	1	12.74		0.62		0.02		3		97
		MtW 82 # 57		14 16 53	+15 16.3	1	15.54		1.21		0.85		2		97
125316	−52 07150			14 16 53	−53 20.2	1	10.15		0.37		0.25		1	A1 V	1770
125547	+42 02481			14 16 57	+42 14.1	2	7.10	.000	1.04	.010	0.83	.005	7	K1 III	1501,1601
		HA 82 # 287		14 16 58	+15 23.7	1	12.68		0.45		−0.02		2		97
		G 165 - 62		14 16 58	+38 52.3	1	11.51		1.39				2		1759
125454	−1 02938	HR 5366		14 16 58	−02 02.1	3	5.17	.052	1.02	.012	0.84	.017	14	G9 III	37,1088,3016
		AJ77,733 T23# 5		14 16 58	−60 43.8	1	13.22		−0.03		−0.68		1		125
125455	−4 03665	G 124 - 24	★ AB	14 17 00	−04 55.2	4	7.56	.018	0.85	.009	0.53	.004	14	K1 V	22,1197,2034,3073
		HA 82 # 288		14 17 01	+15 16.7	1	15.54		1.06		0.81		1		97
125383	−42 09235	HR 5362	★ AB	14 17 01	−42 49.8	2	5.55	.000	0.91	.010	0.60		6	G8 III	404,2007
125504	+15 02695			14 17 03	+15 09.9	2	8.30	.023	1.50	.015	1.83		11	K0	97,882
		HA 82 # 289		14 17 03	+15 17.0	1	13.90		0.71		0.22		3		97
		HA 82 # 243		14 17 05	+15 13.0	1	11.64		0.62		0.08		2		97
125366	−52 07152			14 17 07	−52 48.7	1	9.00		0.12		0.01		1	B9 V	1770
125489	+1 02913	HR 5368		14 17 08	+00 36.8	2	6.18	.005	0.20	.000	0.08	.010	7	A7 V	15,1256
		WLS 1420 0 # 10		14 17 08	−00 08.7	1	12.60		0.63		0.16		2		1375
125418	−35 09423			14 17 08	−36 01.6	1	8.73		0.10		0.08		1	A2 V	1770
	+30 02507			14 17 09	+29 53.1	1	9.55		1.13		1.03		1	K0	280
		MtW 82 # 82		14 17 10	+15 05.1	1	15.24		0.91		0.37		1		97
	+27 02364			14 17 12	+27 31.2	1	10.36		1.08		1.05		2	K4	1569
		HA 82 # 198		14 17 13	+15 03.6	1	13.85		1.06		0.92		2		97
		HA 82 # 244		14 17 14	+15 10.1	1	14.13		0.66		0.21		2		97
125632	+55 01678	HR 5372		14 17 16	+55 05.6	1	6.48		0.18		0.06		2	A5 Vn	1733
125560	+16 02637	HR 5370		14 17 23	+16 32.1	6	4.84	.016	1.23	.004	1.39	.014	24	K3 III	3,15,1080,1363,3051,8015
125332	−63 03224	V418 Cen		14 17 26	−64 00.8	1	7.10		1.74		1.92		5	K5 III	897
	+26 02552			14 17 28	+25 52.5	1	9.98		1.02		0.93		1	K2	280
125522	−10 03876			14 17 28	−10 49.9	1	10.10		0.98		0.85		1	K3 V	3072
		G 124 - 25		14 17 29	−09 23.0	2	12.91	.045	1.59	.009	1.14		2		1705,3078
125473	−37 09336	HR 5367	★ A	14 17 30	−37 39.4	5	4.04	.005	−0.03	.001	−0.11	.007	22	A0 IV	15,1075,1770,2012,8015
125442	−44 09236	HR 5364		14 17 30	−44 57.4	3	4.77	.012	0.31	.009	0.05		9	F0 IV	15,401,2012
	−60 05308			14 17 30	−60 35.9	1	10.63		0.13		−0.18		std		125
		HA 82 # 245		14 17 31	+15 09.5	1	14.77		0.55		−0.07		1		97
125443	−48 08962			14 17 34	−48 35.9	1	9.86		0.17		0.12		1	A1 V	1770
		HA 82 # 365		14 17 36	+15 33.6	1	11.42		0.98		0.67		2		97
	+31 02604			14 17 36	+31 12.8	1	10.14		1.60		1.80		1	K2	280
125444	−48 08963			14 17 37	−49 08.3	1	7.53		0.18		0.08		1	A6 V	1771
		HA 82 # 246		14 17 38	+15 14.3	1	15.36		0.73		0.49		1		97
		HA 82 # 292		14 17 38	+15 24.5	1	12.05		1.10		1.11		2		97
125509	−34 09551			14 17 38	−34 26.9	2	7.70	.000	−0.03	.001	−0.09	.007	4	A0 V	401,1770
125508	−33 09736			14 17 40	−33 27.4	2	9.12	.023	0.19	.016	0.10		2	A2 V	925,1770
	+15 02696			14 17 41	+15 33.1	1	9.94		1.15		1.00		4	G5	97
125446	−53 05952			14 17 43	−53 47.7	1	9.45		1.08		0.71		4	G5 III	119
125642	+39 02750	HR 5373		14 17 45	+39 01.4	6	6.34	.012	0.05	.004	0.05	.004	16	A2 V	15,1007,1013,1022*
		G 135 - 40		14 17 46	+22 55.8	1	12.47		0.76		0.12		1		1658
	+26 02553			14 17 46	+26 13.2	1	8.88		1.24		1.28		1	K2	280
126629	+84 00322			14 17 46	+84 10.5	1	8.38		1.55		1.89		2		1733
		1418+54 # 8		14 17 48	+54 40.6	1	11.35		0.69		0.31		5		1595

Table 1   795

HD	DM	Other Id	N	Rem	α_{1950}	δ_{1950}	S	V	σ_V	B–V	σ_{B-V}	U–B	σ_{U-B}	n	Spectrum	References
		HA 82 # 370			14 17 51	+15 35.8	1	13.34		0.90		0.61		3		97
		LP 620 - 67			14 17 52	+02 11.0	1	12.91		1.34		1.14		1		1696
		Ton 194			14 17 52	+25 43.0	1	13.70		-0.19		-0.65		2	sdB	1036
125608	+9 02878	IDS14154N0903	A		14 17 53	+08 48.7	1	7.29		0.98		0.42		1	G5	1545
		HA 82 # 372			14 17 53	+15 35.3	1	11.77		0.64		0.09		2		97
		AJ77,733 T23# 4			14 17 56	-60 39.0	1	11.97		0.29		-0.12		1		125
125658	+31 02605	HR 5374			14 17 57	+30 39.5	5	6.45	.007	0.15	.004	0.11	.004	17	A5 III	985,1022,1199,3016,8071
		G 65 - 44			14 17 58	+06 18.4	1	11.13		1.12		1.01		1	K5	333,1620
		WLS 1500 75 # 6			14 17 58	+75 12.7	1	11.12		0.94		0.54		2		1375
125498	-51 08132				14 17 58	-52 05.6	2	8.56	.004	0.11	.002	-0.03		5	B9.5V	1770,2014
		Steph 1145			14 18 02	+39 16.7	1	12.31		1.51		1.00		2	M0:e:	1746
	+29 02526				14 18 03	+28 43.5	1	9.47		0.97		0.59		2	K0	1569
		1418+54 # 7			14 18 03	+54 32.5	1	14.94		0.53		-0.02		1		1540
125554	-36 09293				14 18 03	-37 17.3	1	9.64		0.13		0.10		1	A2/3 IV	1770
125541	-41 08691				14 18 03	-41 28.7	1	8.87		0.29		0.03		1	A9 V	1771
		HA 82 # 249			14 18 05	+15 12.1	1	14.19		0.84		0.38		1		97
	+32 02458				14 18 05	+31 56.4	1	9.28		1.06		1.12		2	K3 III	280
125465	-60 05315	V339 Cen			14 18 05	-61 19.2	1	8.58		1.11		0.76		1	F7 II	1587
	+27 02366				14 18 06	+27 13.0	1	9.55		1.03		0.84		1	G5	280
		L 107 - 107			14 18 06	-68 49.	1	12.64		0.77		0.06		2		3062
125659	+14 02725				14 18 08	+14 08.1	1	8.60		1.32				1	K0	882
		LP 500 - 6			14 18 10	+13 26.9	1	11.99		0.47		-0.16		2		1696
		HA 82 # 252			14 18 10	+15 11.5	1	14.33		0.73		0.19		1		97
	+54 01662				14 18 13	+54 33.2	1	10.55		0.40		-0.02		1		1540
125582	-42 09253				14 18 14	-43 04.3	1	9.15		0.34		0.11		1	A3/5mA8-	1770
	+31 02608				14 18 15	+31 22.7	1	9.53		1.14		1.21		1	K0	280
		G 124 - 26			14 18 16	-08 51.5	1	15.48		0.31		-0.59		1		3060
125565	-40 08607				14 18 16	-41 06.5	1	9.25		0.03		-0.02		1	B9.5 V	1770
125569	-46 09263				14 18 16	-46 26.1	1	8.71		0.25		0.09		1	A5 V	1771
	+38 02548	G 165 - 63			14 18 17	+38 10.7	1	10.37		0.82		0.51		1	K2	1658
125709	+27 02367	IDS14160N2701	AB		14 18 18	+26 48.1	1	7.58		0.16		0.07		2	A2	1569
125595	-39 08857				14 18 18	-40 09.8	3	9.01	.018	1.10	.003	1.05		9	K3/4 V	158,1705,2012,3072
	+15 02697				14 18 19	+15 07.1	1	8.96		1.64		1.97		3	M0	97
125616	-33 09745				14 18 19	-34 01.5	1	8.73		1.41				1	K2/3 III	925
		CS 22883 # 28			14 18 20	+09 28.9	1	13.55		0.03		0.14		1		1736
		1418+54 # 5			14 18 22	+54 36.8	1	15.89		0.80		1.14		1		1540
		AAS12,381 28 # F			14 18 23	+27 45.7	1	10.65		1.05		0.89		1		280
		HA 82 # 100			14 18 24	+14 44.8	1	10.96		0.66		0.19		2		97
		Ton 765			14 18 24	+28 17.	1	14.96		0.08		0.05		1		313
	+31 02610				14 18 25	+30 55.3	1	9.85		1.10		1.08		1	K0 III	280
125583	-45 09113				14 18 25	-45 38.6	1	9.17		0.21		0.13		1	A5 V	1771
125570	-48 08973				14 18 25	-48 50.0	1	9.04		0.23		0.22		1	A4 V	1771
125545	-57 06616	LSS 3229			14 18 27	-58 03.8	5	7.40	.020	0.15	.004	-0.70	.027	14	B1 Iab/b	158,403,540,976,8100
125796	+49 02294	IDS14166N4858	AB		14 18 28	+48 44.1	4	7.42	.018	0.55	.006	0.06	.006	10	F8	292,938,1381,3030
		G 239 - 12			14 18 28	+73 28.1	1	11.61		0.41		-0.22		2		1658
125728	+26 02554				14 18 30	+26 18.2	2	6.79	.019	0.92	.019	0.51		4	G8 II	20,8100
125573	-52 07174				14 18 30	-52 46.2	1	9.63		0.16		-0.04		1	B9 V	1770
125586	-50 08480				14 18 31	-50 42.7	1	9.65		0.08		-0.03		1	B9.5 V	1770
		HA 82 # 302			14 18 33	+15 24.1	1	12.71		0.46		-0.04		2		97
	+28 02314				14 18 35	+27 50.6	1	10.12		0.99		0.78		1	G5	280
		1418+54 # 3			14 18 38	+54 37.9	2	14.33	.026	0.64	.013	0.17	.022	6		1540,1595
238341	+55 01679				14 18 38	+54 44.1	1	9.27		0.55		0.05		1	G5	1540
124882	-83 00557	HR 5339			14 18 39	-83 26.5	3	4.31	.000	1.31	.000	1.45	.000	11	K2 III	15,1075,2038
		G 124 - 27			14 18 40	-00 53.2	1	13.12		1.67				1		1759
		1418+54 # 4			14 18 42	+54 35.1	2	13.70	.026	0.56	.004	-0.07	.022	6		1540,1595
125679	-19 03859				14 18 42	-19 27.1	1	8.84		0.20		-0.09		1	A1/2 V	1776
		1418+54 # 2			14 18 44	+54 42.9	1	12.81		0.88		0.56		1		1540
125647	-38 09309				14 18 44	-38 27.1	1	7.96		0.10		0.06		1	A1/2 IV	1770
125602	-53 05956				14 18 44	-53 31.9	1	9.67		0.42		0.27		1	A2 IV	1770
125798	+37 02519				14 18 45	+36 37.2	1	7.36		0.18		0.10		2	A0	272
	+9 02879	G 65 - 45			14 18 47	+09 12.1	1	10.59		0.97		0.87		1	K0	333,1620
125649	-44 09259				14 18 52	-44 25.9	1	7.74		1.22				4	K1 III	2012
		CS 22883 # 33			14 18 57	+11 59.0	1	15.36		-0.08				1		1736
	+31 02611				14 18 58	+30 58.1	1	10.80		0.87		0.51		1		280
	+15 02698				14 18 59	+15 07.3	1	9.42		0.46		-0.05		std	F8	97
125698	-35 09449				14 18 59	-35 43.5	1	9.79		0.20		0.17		1	A2 III	1770
		Steph 1146			14 19 00	+14 14.7	1	11.83		1.23		1.15		1	K7	1746
		GD 334			14 19 01	+57 40.6	1	14.18		0.50		-0.21		4		308
	+77 00539				14 19 01	+77 22.2	1	10.08		1.39		1.05		2	M8	1375
125651	-49 08685	T Lup			14 19 01	-49 37.3	1	8.77		3.19				2	Nb	864
125628	-57 06619	HR 5371		★ A	14 19 01	-58 13.9	1	4.92		0.86		0.50		3	G8 III	1279
125628	-57 06619	HR 5371		★ AB	14 19 01	-58 13.9	2	4.78	.010	0.79	.005	0.39		8	G8 III+F5 V	244,2009
125628	-57 06619	IDS14154S5800	B		14 19 01	-58 13.9	1	7.00		0.41		-0.03		3	F5 V	1279
125785	+12 02684				14 19 02	+11 52.3	1	7.93		0.48		0.02		2	F8	1648
125858	+39 02754				14 19 02	+39 12.9	2	7.93	.000	0.41	.005	-0.02	.010	6	F5 V	833,1601
125918	+57 01504				14 19 02	+57 08.8	1	8.85		1.01		0.90		4	G9 II	8100
125699	-36 09303				14 19 04	-37 15.2	1	9.72		0.29		0.04		1	A9 V	1771
125816	+13 02784				14 19 08	+12 59.6	2	8.66	.029	0.72	.005	0.34		3	G5	882,1648
		CS 22883 # 25			14 19 10	+08 07.0	1	15.10		-0.16		-1.01		1		1736
	+1 02916				14 19 12	+01 00.7	2	9.61	.005	1.34	.014	1.18		5	K0	1594,6006

HD	DM	Other Id	N Rem	α_{1950}	δ_{1950}	S	V	σ_V	B–V	σ_{B-V}	U–B	σ_{U-B}	n	Spectrum	References
125669	−54 05975			14 19 14	−55 12.2	1	7.20		0.11				4	B8/9 III	1075
125759	−31 11119			14 19 16	−31 26.7	1	9.28		0.98				1	K0 III	925
125745	−34 09570	HR 5376		14 19 19	−34 33.6	2	5.56	.002	-0.09	.004	-0.41		5	B8 V	1770,2006
125721	−47 09082	HR 5375, LSS 3231	⋆ AP	14 19 22	−48 05.6	6	6.09	.010	-0.14	.005	-0.91	.010	18	B2 II/III	15,26,404,1770,2013,2029
125737	−46 09278			14 19 23	−46 45.8	1	9.74		0.17		0.10		1	A1 V	1770
125788	−19 03863			14 19 25	−20 15.9	1	8.97		0.46		0.00		2	F3 V	1375
125630	−66 02519	BS Cir		14 19 29	−66 25.0	1	6.84		0.10		-0.06		1	Ap Si[Cr]	1771
		G 178 - 19		14 19 31	+41 40.1	1	15.47		1.47				2		1663
		CS 22883 # 30		14 19 35	+11 13.6	1	14.51		0.40		-0.19		1		1736
125777	−45 09127			14 19 43	−46 03.1	1	8.52		0.20		0.11		1	A3 V	1771
125764	−51 08149			14 19 45	−51 32.8	2	9.64	.025	0.49	.002	-0.04		5	F5 V	1594,6006
	+20 02972	G 135 - 42		14 19 48	+20 29.2	1	10.45		0.62		0.08		1		1658
	+30 02512	G 166 - 22		14 19 48	+29 51.7	3	8.58	.021	1.26	.023	1.22	.020	6	K5 V	22,1733,3072
125789	−46 09282			14 19 49	−46 26.3	2	8.37	.009	-0.01	.007	-0.12		5	B9/A0 V	1770,2014
		G 178 - 20		14 19 51	+37 53.6	1	12.29		1.50				1		906
125790	−50 08497			14 19 51	−50 43.2	1	10.45		0.22		0.15		1	A1/2 V	1770
125883	−6 03983			14 19 52	−07 22.9	1	9.35		0.47		0.10		2	F8	1064
125823	−38 09329	HR 5378, V761 Cen		14 19 57	−39 17.1	7	4.38	.044	-0.19	.005	-0.74	.016	38	B2 V	15,26,747,1075,1770*
125825	−40 08626			14 19 57	−40 44.7	2	8.26	.004	0.00	.002	-0.19		5	B8/9 III	1770,2014
		WLS 1420 0 # 5		14 19 58	+02 35.9	1	12.18		1.34		1.31		2		1375
		CS 22883 # 27		14 19 58	+09 00.4	1	15.09		0.15				1		1736
	+15 02700			14 19 59	+14 57.7	1	10.36		0.96		0.77		2		97
125810	−50 08501	HR 5377	⋆ AB	14 19 59	−50 32.7	1	6.02		1.36				4	K2 III	2032
126009	+30 02513	CI Boo		14 20 02	+29 35.8	1	6.60		1.60		1.46		1	M4 III	3040
125924	−7 03835			14 20 04	−08 01.3	3	9.67	.014	-0.20	.003	-0.85	.013	11	B2 IV	55,272,1732
	−60 05331	Ly 2 - 70		14 20 05	−61 10.6	1	10.29		0.14		-0.13		2	B6	837
		Ly 2 - 73		14 20 06	−61 12.1	1	13.43		0.56		0.21		2		1667
125886	−30 11378			14 20 07	−30 49.8	1	9.89		0.56				1	G2 V	925
	+15 02701			14 20 08	+14 41.1	1	10.07		0.28		0.16		2	F2	97
126029	+36 02478			14 20 08	+35 52.5	1	8.15		0.58		0.06		2	G0	1601
125866	−41 08720			14 20 08	−42 01.6	2	7.09	.001	-0.06	.001	-0.37		5	B9 IV	1770,2014
125867	−45 09131			14 20 08	−45 41.3	1	9.05		0.32		0.14		1	A2mA7-A8	1770
125829	−49 08699			14 20 08	−49 37.5	2	8.15	.018	0.06	.006	-0.36		5	B7 III	1770,2014
		WLS 1420 25 # 10		14 20 09	+24 25.6	1	12.32		0.37		-0.07		2		1375
	−60 05332			14 20 09	−60 44.6	1	11.17		0.18		-0.01		2	B9	837
		Ly 2 - 37		14 20 09	−61 09.8	1	12.10		0.54		0.10		2		1667
	−61 04512			14 20 10	−61 21.4	1	10.85		0.74		0.38		2	K0	837
125849	−46 09289	RS Lup		14 20 12	−47 17.7	1	9.65		2.65		5.00		2	Nb	864
125932	−27 09803	HR 5381		14 20 13	−27 31.5	7	4.78	.016	1.31	.006	1.53	.012	34	K3 III	3,15,418,1075,2012*
125792	−58 05515	LSS 3232		14 20 13	−59 11.4	2	9.08	.029	0.54	.001	-0.43	.027	4	B1 Iab/b	540,976
125779	−60 05333	Ly 2 - 30	⋆ AB	14 20 15	−61 04.6	1	9.17		0.16		-0.23		2	B9	837
		Ly 2 - 36		14 20 15	−61 09.5	2	11.71	.025	0.24	.005	0.15	.005	4	B9	837,1667
		Lod 1296 # 16		14 20 16	−61 15.2	1	10.65		0.98		0.71		2	K4	837
126030	+26 02559	UV Boo		14 20 17	+25 46.6	2	8.15	.014	0.40	.000	-0.05	.014	4	F5 V	105,1375
126138	+54 01668			14 20 18	+53 44.9	2	7.58	.010	-0.13	.010	-0.48	.000	5	B9	963,1566
		Ly 2 - 28		14 20 18	−61 04.4	1	14.79		4.24		1.70		2		1667
		Ly 2- 105		14 20 18	−61 10.	2	15.78	.000	1.06	.000	1.02	.000	4		1667,1667
		Ly 2- 106		14 20 18	−61 10.	1	14.73		1.48		0.31		2		1667
		Ly 2- 107		14 20 18	−61 10.	1	11.51		1.20		1.07		2		1667
		Ly 2- 108		14 20 18	−61 10.	1	14.31		0.17		-0.93		2		1667
		Ly 2 - 10		14 20 19	−61 08.0	1	12.35		1.65		1.94		2		1667
	−60 05336			14 20 20	−60 54.9	1	10.55		0.64		0.20		2	G0	837
	−60 05337			14 20 22	−60 57.2	1	11.13		0.44		0.09		2	B9	837
	−60 05335	Ly 2 - 27		14 20 22	−61 05.0	4	9.65	.016	0.17	.013	-0.21	.012	8	B5	837,1101,1667,1667
		Steph 1151		14 20 23	+49 03.7	1	12.15		1.32		1.09		1	M0	1746
125869	−52 07195	HR 5380		14 20 23	−52 57.0	3	5.99	.007	1.10	.000			11	K1 III	15,2013,2029
		Ly 2 - 13		14 20 24	−61 10.7	1	12.45		0.30		0.30		2		1667
		G 65 - 46		14 20 25	+11 40.1	1	14.46		1.57		1.61		1		333,1620
126031	+15 02702			14 20 26	+15 10.0	1	7.54		0.34		0.14		2	F1 Vm	97
	−60 05338	Ly 2 - 6		14 20 26	−61 06.6	2	10.43	.045	0.24	.075	0.01	.090	4	B6	837,1667
		Ly 2 - 8		14 20 26	−61 07.0	1	12.37		0.29		0.23		2		1667
126051	+18 02870			14 20 28	+17 40.8	1	8.28		1.04				1	K0	882
	−60 05339	Ly 2 - 7		14 20 29	−61 06.8	2	9.66	.100	0.35	.035	-0.49	.040	4	B2	837,1667
125968	−27 09806			14 20 30	−27 35.5	2	7.76	.005	0.66	.000			8	G3/5 V	1075,2033
		Lod 1296 # 28		14 20 30	−61 03.1	1	11.72		0.34		0.21		2	A0	837
		Lod 1296 # 29		14 20 30	−61 03.1	1	12.01		0.24		0.14		2	A0	837
		Ly 2 - 14		14 20 31	−61 10.9	1	12.82		0.63		0.38		2		1667
		Steph 1150		14 20 35	+37 50.6	1	12.20		0.96		0.82		1	K7	1746
	−7 03837			14 20 35	−07 53.2	1	9.92		1.13		1.10		2	K5 V	3072
125937	−45 09134			14 20 35	−45 26.2	1	8.11		0.22		0.09		1	A8/9 (V)	1771
	−60 05340	Ly 2 - 4		14 20 35	−61 05.6	1	11.25		0.19		-0.10		2	B8	1667
	+46 01959			14 20 36	+45 53.2	1	9.48		0.25		0.09		2	A0	272
		Ly 2 - 2		14 20 36	−61 05.8	1	11.45		1.41		0.81		2		837
	−71 01584			14 20 38	−71 22.0	1	10.54		0.46		0.26		3	B5	400
	+39 02758			14 20 39	+38 42.5	1	8.87		0.27		0.09		2	F0 IV	105
125852	−60 05341	Ly 2 - 1		14 20 39	−61 07.4	3	9.21	.094	0.13	.022	-0.15	.018	7	B9	837,1101,1667
		Ly 2 - 15		14 20 40	−61 10.5	1	12.34		0.46		0.12		2		1667
		LP 800 - 51		14 20 41	−17 04.5	1	11.90		0.74		0.16		2		1696
126053	+1 02920	HR 5384	⋆	14 20 42	+01 28.5	10	6.27	.011	0.64	.007	0.08	.013	37	G1 V	15,22,333,908,1008*
126100	+24 02728			14 20 43	+24 20.0	1	8.53		1.02		0.85		3	K0	1625

Table 1 797

HD	DM	Other Id	N Rem	α_{1950}	δ_{1950}	S	V	σ_V	B–V	σ_{B-V}	U–B	σ_{U-B}	n	Spectrum	References
126035	−11 03729	HR 5383		14 20 44	−11 29.2	1	6.21		0.99				4	G7 III	2006
125996	−33 09774			14 20 47	−33 28.9	1	9.76		0.73				1	G3 V	925
125952	−51 08163			14 20 47	−52 11.8	1	8.71		0.17		0.09		1	A0 V	1770
	+27 02372			14 20 48	+27 25.9	1	10.32		0.15		0.10		2	A3	1375
125880	−60 05342	Ly 2 - 16		14 20 48	−61 09.8	3	7.68	.013	0.24	.004	0.16	.047	7	A0	837,1101,1667
125833	−64 02832	IDS14169S6502	AB	14 20 48	−65 15.5	1	8.13		1.12				4	K0 III	2012
126141	+25 02770	HR 5387		14 20 52	+25 33.8	4	6.23	.017	0.37	.011	-0.03	.005	8	F5 V	70,105,254,3016
125901	−60 05344			14 20 54	−61 20.4	1	9.09		0.04		-0.12		2	B9.5V	837
126128	+9 02882	HR 5385	★ BC	14 20 55	+08 40.3	1	6.86		0.43		-0.01		4	F0 V +F2 V	3030
126129	+9 02882	HR 5386	★ A	14 20 55	+08 40.4	1	5.14		0.01		-0.01		2	A0 m	3030
126129	+9 02882	IDS14185N0854	★ ABC	14 20 55	+08 40.4	3	4.86	.010	0.05	.007	-0.03	.000	9	A0m+F0V+F2V	15,1363,1417
125881	−63 03246			14 20 56	−63 28.7	1	7.25		0.60				4	G2 V	2012
125835	−67 02574	HR 5379, LSS 3233		14 20 56	−67 58.2	4	5.58	.010	0.48	.015	-0.01	.007	19	A1 Iab/b	15,1075,2038,8100
126001	−40 08646			14 20 59	−41 00.7	1	7.81		0.25		0.20		1	A5 IV	1771
		Feige 94		14 21 00	+21 03.	1	13.87		-0.05		-0.06		1		3060
		Ly 2 - 42		14 21 00	−61 10.8	1	12.94		0.52		0.01		2		1667
	−60 05346	Ly 2 - 55		14 21 01	−61 03.8	2	10.26	.005	0.15	.005	-0.33	.090	4	B5	837,1667
		Ly 2 - 41		14 21 01	−61 11.6	1	14.48		2.79		1.92		2		1667
	−60 05345	Ly 2 - 79		14 21 01	−61 14.2	1	11.02		0.19		-0.02		2	B5	837
126056	−27 09811			14 21 04	−27 31.2	1	8.17		-0.03		-0.16		1	B9 V	1770
	−61 04522			14 21 04	−61 22.4	1	9.71		1.54		1.81		2		837
	−60 05347	Ly 2 - 52		14 21 05	−61 05.5	2	11.01	.015	0.19	.015	-0.30	.065	4	B3	837,1667
		Ly 2 - 23		14 21 05	−61 05.9	1	12.24		0.30		0.12		2		1667
		Ly 2 - 22		14 21 06	−61 06.3	1	12.15		0.62		-0.04		2		1667
		G 65 - 48		14 21 12	+08 55.9	1	14.62		1.50				1		333,1620
		G 135 - 47		14 21 12	+21 14.7	2	13.75	.004	1.53	.004	1.08		6		316,1759
126289	+54 01671	S Boo		14 21 12	+54 02.2	1	10.56		1.52		0.50		1	M4	635
		CS 22883 # 26		14 21 13	+08 12.2	1	13.65		0.03		0.12		1		1736
126131	−15 03862			14 21 13	−15 52.5	1	6.72		0.09		0.08		11	A1 V	1628
126091	−33 09784			14 21 15	−33 42.1	1	9.22		0.21		0.13		1	A1(m)	1770
125975	−61 04524			14 21 15	−61 21.5	1	8.42		1.64		1.92		2	K5/M0 III	837
	−60 05348	Ly 2 - 44		14 21 17	−61 09.5	2	11.03	.000	0.23	.015	0.16	.025	4	A2	837,1667
		Ly 2 - 43		14 21 18	−61 10.3	1	14.44		2.42		-0.49		2		1667
125855	−70 01750			14 21 19	−70 42.0	1	8.90		0.38		0.17		1	A7 V	1771
126062	−46 09302			14 21 21	−46 57.1	1	7.45		0.07		0.04		1	A1 V	1770
	−56 06260	LSS 3236		14 21 21	−56 55.0	2	9.92	.000	1.25	.025	0.38	.025	4	B8 Ia	403,1737
126108	−32 10084			14 21 24	−32 45.3	1	8.74		0.05		0.02		1	A0 V	1770
125987	−61 04525	IDS14177S6108	AB	14 21 24	−61 22.8	1	9.86		0.05		-0.03		2	A0 V	837
		Lod 1378 # 20		14 21 25	−57 54.5	1	11.60		0.33		0.23		2	A0	807
		WLS 1420-20 # 9		14 21 27	−19 59.6	1	11.27		0.45		0.07		2		1375
126265	+39 02760			14 21 29	+39 33.4	1	7.24		0.56		0.10		2	F8	1601
126004	−60 05353	Ly 2 - 90	★ AB	14 21 29	−61 05.7	1	9.41		0.21		-0.36		2	B5	837
		Ton 197		14 21 30	+31 50.	2	15.32	.020	-0.12	.005	-1.01	.015	3	DAwk	1036,3028
126201	+6 02874			14 21 31	+06 02.9	1	7.20		1.01		0.70		2	K0	3008
126200	+8 02857	HR 5388		14 21 33	+08 28.2	3	5.92	.053	0.05	.013	0.08	.017	8	A3 V	15,1417,1545
126110	−44 09299			14 21 34	−44 58.7	2	7.90	.000	0.02	.002	-0.08		7	B9 V	1770,2014
126135	−40 08656			14 21 37	−40 31.7	2	6.97	.001	-0.06	.001	-0.29		7	B8 V	1770,2014
	+25 02773			14 21 38	+24 46.2	1	9.67		0.27		0.14		2	A2	272
125990	−65 02718	HR 5382		14 21 38	−65 56.8	3	6.35	.008	0.14	.005	0.05	.005	12	A3 V	15,1075,2038
126174	−29 11045			14 21 40	−29 23.6	1	8.30		0.08		0.05		1	A0 V	1770
	−60 05354			14 21 41	−61 15.0	1	11.38		0.19		0.02		2	B9	837
126248	+6 02875	HR 5392		14 21 42	+06 02.8	3	5.11	.025	0.12	.004	0.09	.000	12	A5 V	1088,1363,3016
	−61 04531			14 21 42	−61 21.6	1	10.57		0.48		0.15		2	F6	837
		LP 560 - 15		14 21 44	+08 16.9	1	11.58		1.35		1.29		1	K4	1696
		WLS 1420-20 # 7		14 21 46	−22 26.7	1	12.80		0.53		0.17		2		1375
	−61 04532			14 21 46	−61 26.1	1	11.36		0.19		0.10		2	B9	837
126270	+16 02642	IDS14194N1644		14 21 49	+16 30.0	1	6.90		0.65		0.39		2	Am	272
126307	+28 02318			14 21 49	+27 38.3	1	6.38		1.61				4	K4 III	20
126271	+8 02858	HR 5394		14 21 51	+08 18.7	4	6.19	.005	1.20	.008	1.27	.015	14	K4 III	15,1417,1509,3077
		LundsII139 # 106		14 21 51	−60 00.8	1	11.19		0.62		0.15		3		1101
		CS 22883 # 37		14 21 54	+11 43.0	1	14.73		0.55		0.04		1		1736
126218	−24 11469	HR 5390		14 21 57	−24 34.8	1	5.32		0.96				4	K0 III	2006
126194	−36 09338			14 21 58	−37 13.2	2	6.70	.002	0.11	.003	0.01	.005	4	A1 V	401,1770
126251	−11 03736	HR 5393	★ AB	14 21 59	−11 26.6	4	6.49	.003	0.42	.004	0.02		51	F4 III	15,130,1256,1351
126273	−1 02951			14 22 01	−02 07.0	2	7.18	.000	1.63	.005	1.88	.075	4	M1	1375,3040
126178	−46 09309			14 22 04	−46 47.1	1	8.90		0.25		0.17		1	A5 V	1771
126116	−59 05555			14 22 05	−60 16.1	1	8.47		0.29		0.02		3	A8 IV	1101
126221	−38 09353			14 22 09	−39 08.7	1	8.54		0.20		0.12		1	A5 IV	1771
	−60 05357			14 22 11	−60 43.0	1	9.66		1.14		1.00		3		1101
126117	−60 05356			14 22 11	−60 47.1	1	8.47		1.26		1.23		3	K2 III	1101
		CS 22883 # 42		14 22 12	+09 30.8	1	14.31		0.15				1		1736
		GD 165, CX Boo		14 22 12	+09 30.9	1	14.32		0.14		-0.59		1	DA	3060
126470	+56 01731			14 22 13	+55 48.3	1	8.52		0.26		0.07		2	F2	1502
126198	−52 07213			14 22 17	−53 13.7	1	8.02		0.04		-0.44		2	B8/9 II	1770
		Lod 1378 # 14		14 22 18	−57 55.0	1	11.19		0.70		0.37		2		807
		Lod 1378 # 17		14 22 18	−57 55.0	1	11.35		0.36		0.35		2	A0	807
126238	−43 09039	IDS14191S4311	AB	14 22 19	−43 25.0	3	7.68	.017	0.81	.004	0.19	.025	4	G3/8wF0/2	742,1594,6006
		G 65 - 49		14 22 27	+09 06.7	1	12.26		1.63		1.23		1		333,1620
		G 65 - 50		14 22 28	+04 38.2	1	15.53		1.41				1		333,1620
		CS 22883 # 40		14 22 29	+10 14.2	1	15.33		0.35		-0.10		1		1736

HD	DM	Other Id	N	Rem	α_{1950}	δ_{1950}	S	V	σ_V	B–V	σ_{B-V}	U–B	σ_{U-B}	n	Spectrum	References
126313	−32 10097			A	14 22 32	−32 44.8	2	7.58	.000	1.14	.000	0.89	.015	2	Kp Ba	565,871
126313	−32 10097			B	14 22 32	−32 44.8	1	10.29		0.29		0.12		1		871
	−57 06636				14 22 36	−57 53.2	1	10.78		0.26		0.18		2	A0	807
126298	−44 09315				14 22 38	−44 42.6	1	8.71		0.03		−0.11		3	B9.5 V	1770
		Lundsll139 # 110			14 22 39	−59 42.8	1	10.85		1.54		1.80		3		1101
126366	−19 03879	IDS14199S1931	BC		14 22 40	−19 44.4	2	6.99	.010	0.14	.015	0.08	.005	5	A2 V	13,1084
126297	−44 09318				14 22 40	−44 26.6	1	9.46		0.32		0.19		1	Ap CrEuSr	1770
126367	−19 03880	HR 5397		⋆ A	14 22 42	−19 44.7	2	6.63	.005	0.14	.005	0.10	.015	5	A1/2 V	13,1084,2009
	−59 05562	Lundsll139 # 108			14 22 46	−59 44.1	1	11.09		0.35		0.01		3		1101
	−59 05?	Lundsll139 # 105			14 22 46	−59 57.7	1	11.26		0.29		0.10		3		1101
126400	−26 10280	HR 5399			14 22 54	−26 37.6	1	6.48		0.94				4	K0 III	2006
		Lundsll139 # 104			14 22 54	−60 03.6	1	11.89		2.08				3		1101
		WLS 1424 55 # 5			14 22 55	+57 31.0	1	13.92		0.66		0.25		2		1375
126341	−44 09322	HR 5395, τ Lup		⋆ A	14 22 55	−44 59.8	8	4.55	.005	−0.16	.005	−0.79	.021	24	B2 IV	15,26,1075,1770,2001,2011*
126354	−44 09323	HR 5396		⋆ AB	14 22 57	−45 09.3	10	4.35	.007	0.42	.006	0.20	.018	55	F4 IV + A7:	15,278,1020,1075,1637,1770*
	−7 03842				14 22 58	−08 11.6	1	11.26		0.52		−0.15		2	G2	1696
	−57 06638				14 23 03	−57 48.1	1	11.14		0.42		0.31		2	A0	807
	−57 06637				14 23 03	−57 52.5	1	11.05		0.28		0.25		2	A0	807
126386	−41 08757	HR 5398			14 23 04	−42 05.6	5	6.32	.015	1.18	.026	1.30		20	K2 III	15,2012,2013,2029,3077
125873	−79 00761		AB		14 23 04	−79 31.4	1	8.95		0.57				4	F7/8 V	2012
		GD 166			14 23 05	+21 54.7	1	15.25		0.41		−0.23		1		3060
	−57 06639				14 23 06	−58 00.1	2	10.83	.002	0.23	.004	−0.18	.006	5	B9	807,1586
126241	−65 02732	HR 5391			14 23 06	−65 35.8	3	5.85	.004	1.50	.000	1.76	.000	13	K3 III	15,1075,2038
	−59 05564	NGC 5606 - 164			14 23 07	−59 27.1	1	10.48		0.12		−0.33		5		1725
	−59 05565	NGC 5606 - 165			14 23 08	−59 26.6	1	10.92		0.12		0.08		3		1725
	−64 02848				14 23 08	−64 38.4	1	10.40		0.35		−0.32		2	B4 II-III	540
126892	+77 00541				14 23 10	+76 53.4	1	8.66		0.93		0.60		2	G5	1502
126285	−60 05366				14 23 10	−60 31.7	1	7.83		1.10		0.92		3	G8/K0 III	1101
126512	+21 02649	G 166 - 25			14 23 11	+20 49.4	2	7.29	.015	0.57	.025	−0.01	.030	5	F9 V	22,3077
126320	−59 05568	AS Cir			14 23 21	−60 03.9	1	8.08		1.70		1.84		3	M2 III	1101
126515	+1 02927	FF Vir			14 23 23	+01 13.1	1	7.11		0.01		−0.02		3	A2p	1202
	−57 06642	LSS 3237			14 23 23	−57 56.5	1	10.60		0.17		−0.50		2	B3	807
	−57 06641				14 23 23	−58 00.2	1	10.86		0.25		−0.15		2	B9	807
	+24 02733	G 166 - 27		⋆ A	14 23 25	+23 51.1	2	9.72	.014	1.42	.019	1.26		4	M3	694,3078
	+24 02733	G 166 - 27		⋆ AB	14 23 25	+23 51.1	1	9.14		1.47				1	M3	3025
126458	−31 11182				14 23 25	−32 00.3	1	9.54		0.01		−0.08		1	A0 V	1770
126597	+39 02764	HR 5402			14 23 27	+38 37.1	2	6.28	.005	1.23	.009	1.26		4	K3 III	70,105
		G 124 - 38			14 23 27	−03 51.9	1	12.97		0.65		−0.07		4		1696
126459	−32 10114				14 23 27	−33 01.5	1	7.16		0.66		0.41		1	A(5) V	1770
		NGC 5606 - 134			14 23 27	−59 30.0	1	12.12		0.35		−0.27		2		1725
	+24 02733	G 166 - 28		⋆ B	14 23 28	+23 51.3	2	9.98	.010	1.44	.012	1.27		3		694,3078
126357	−59 05569	NGC 5606 - 107		⋆	14 23 28	−59 30.7	2	9.06	.005	0.22	.000	−0.65	.027	5	B1 III	1725,8100
126660	+52 01804	HR 5404		⋆ A	14 23 30	+52 04.9	9	4.05	.011	0.50	.005	0.01	.009	68	F7 V	1,15,254,1077,1118*
	−59 05570	NGC 5606 - 133			14 23 30	−59 29.7	1	11.32		0.25		0.21		2		1725
	+24 02734	IDS14211N2407	C		14 23 32	+23 50.8	1	9.50		1.44		1.18		2	K7	3016
126358	−60 05370				14 23 34	−60 45.0	1	8.83		1.24		1.13		1	F3 IV/V	6006
126583	+14 02733	G 65 - 51			14 23 35	+13 39.6	2	8.06	.005	0.74	.009	0.33		4	G5	333,882,1620
	−57 06643				14 23 35	−57 48.3	1	11.22		0.60		0.33		2		807
	−59 05571	NGC 5606 - 139			14 23 35	−59 32.4	1	10.58		0.09		−0.30		2		1725
126582	+14 02734				14 23 37	+14 18.5	1	8.61		1.21				1	K0	882
	−59 05572				14 23 37	−60 05.9	1	10.16		1.12		0.98		2		1101
		Lod 1378 # 24			14 23 38	−57 46.3	1	11.74		0.74		0.52		2		807
		NGC 5606 - 149			14 23 38	−59 31.7	1	13.35		0.68		0.09		2		1725
		CS 22883 # 39			14 23 40	+10 52.2	1	13.75		0.02		0.09		1		1736
126598	+26 02569				14 23 41	+26 29.4	1	7.40		1.31				3	K4 III	20
		LP 800 - 11			14 23 41	−20 57.2	1	11.26		1.42		1.60		1		1696,3062
		Lod 1378 # 22			14 23 41	−57 48.1	1	11.69		0.57		0.21		2		807
		Lod 1378 # 23			14 23 41	−57 48.1	1	11.72		0.67		0.26		2	G0	807
		NGC 5606 - 148			14 23 41	−59 31.9	1	12.50		1.13		0.69		3		1725
		NGC 5606 - 132			14 23 42	−59 27.8	1	13.02		0.35		0.02		3		1725
		NGC 5606 - 147			14 23 42	−59 31.4	1	12.51		0.49		0.14		5		1725
		NGC 5606 - 167			14 23 42	−59 33.8	1	12.73		0.51		0.16		2		1725
126475	−39 08918	HR 5400			14 23 43	−39 39.0	4	6.34	.005	−0.09	.009	−0.32		13	B9 V	15,1770,2013,2029
126795	+65 00995				14 23 44	+65 23.0	1	9.28		0.59		0.00		2	K0	3026
	−57 06644				14 23 44	−57 59.6	2	10.63	.007	0.27	.010	−0.07	.002	5	B9	807,1586
126500	−26 10290				14 23 45	−26 38.0	2	8.16	.003	1.46	.001	1.64	.000	14	K2 III	1657,1732
126599	+15 02711				14 23 46	+14 33.9	1	9.02		0.13		0.07		2	A1 V	272
126476	−39 08919				14 23 47	−40 12.5	1	8.03		0.18		0.12		1	A4 V	1771
	−59 05574	NGC 5606 - 128			14 23 49	−59 27.3	2	10.56	.010	0.42	.005	0.20	.005	10		1725,1725
		NGC 5606 - 168			14 23 49	−59 32.9	1	12.45		0.36		0.23		3		1725
	−57 06645				14 23 50	−57 50.7	1	10.10		1.07		0.74		2	K0	807
		Lundsll139 # 101			14 23 50	−60 06.4	1	10.44		1.35		1.50		3		1101
	−57 06646				14 23 52	−57 51.2	1	11.39		0.40		0.34		2	A1	807
	−59 05574	NGC 5606 - 20			14 23 55	−59 26.1	1	11.34		1.32		1.27		2		1725
		NGC 5606 - 131			14 23 55	−59 27.8	1	13.07		0.45		0.17		2		1725
126504	−45 09188	HR 5401		⋆ A	14 23 58	−45 54.5	10	5.82	.005	0.31	.004	0.14	.004	73	A1mA5/7-F2	15,278,355,977,977,1075*
		NGC 5606 - 156			14 23 60	−59 24.9	1	12.68		0.52		−0.07		1		1725
		NGC 5606 - 109			14 24 00	−59 23.3	1	13.34		0.37		−0.09		1		1725
		NGC 5606 - 111			14 24 00	−59 23.8	1	13.22		0.37		−0.08		1		1725
		NGC 5606 - 130			14 24 00	−59 28.0	1	12.50		0.51		0.14		2		1725

Table 1 799

HD	DM	Other Id	N Rem	α_{1950}	δ_{1950}	S	V	σ_V	B–V	σ_{B-V}	U–B	σ_{U-B}	n	Spectrum	References
		NGC 5606 - 14	⋆E	14 24 01	−59 24.7	2	11.73	.015	0.31	.005	-0.33	.010	4		251,1725
	−59 05577	NGC 5606 - 2	⋆B	14 24 03	−59 24.6	1	9.71		0.23		-0.66		2	B1 IV	251
		NGC 5606 - 15	⋆D	14 24 04	−59 24.7	1	10.58		0.20		-0.53		1	B1.5IV	251
		NGC 5606 - 10		14 24 04	−59 25.1	1	14.32		0.59		0.34		2		251
		NGC 5606 - 12	⋆C	14 24 04	−59 25.1	3	12.16	.010	0.46	.000	0.28	.007	8		251,1101,1725
	−57 06647			14 24 05	−57 55.1	1	10.65		0.61		0.12		2	F6	807
		NGC 5606 - 160		14 24 05	−59 26.2	1	13.16		0.35		-0.10		1		1725
126449	−59 05578	NGC 5606 - 1	⋆A	14 24 06	−59 24.5	2	8.78	.010	0.25	.005	-0.68	.000	4	B0 IV	251,1725
		NGC 5606 - 13	⋆B	14 24 06	−59 25.5	3	12.83	.005	0.34	.011	-0.26	.010	6		251,1101,1725
		NGC 5606 - 11		14 24 07	−59 25.1	3	12.91	.016	0.32	.012	-0.08	.014	5		251,1101,1725
126661	+19 02810	HR 5405		14 24 08	+19 27.0	4	5.41	.029	0.23	.009	0.21	.034	15	F0 m	3,70,3058,8071
126523	−46 09328			14 24 08	−47 17.7	1	9.16		0.00		-0.19		1	B9 IV	1770
		NGC 5606 - 5	⋆A	14 24 08	−59 24.2	3	10.96	.027	0.23	.013	-0.52	.017	9		251,1101,1725
		LP 680 - 15		14 24 10	−04 36.1	1	11.35		0.74		0.15		2		1696
126525	−51 08206			14 24 10	−51 42.6	2	7.82	.000	0.70	.000	0.21		7	G5 V	2012,3026
		NGC 5606 - 6		14 24 10	−59 24.4	3	11.92	.012	0.35	.004	-0.20	.023	7		251,1101,1725
126587	−21 03903			14 24 11	−22 01.2	3	9.13	.005	0.82	.012	0.16	.000	7	GwA/F	742,1594,6006
		NGC 5606 - 9		14 24 12	−59 23.9	3	12.78	.008	0.37	.023	-0.10	.019	9		251,1101,1725
		NGC 5606 - 7		14 24 13	−59 24.5	3	12.48	.016	0.28	.011	-0.32	.012	8		251,1101,1725
		NGC 5606 - 8		14 24 13	−59 24.5	3	13.10	.044	0.28	.025	-0.10	.013	6		251,1101,1725
		NGC 5606 - 151		14 24 13	−59 31.2	1	12.14		1.16		0.90		3		1725
126546	−43 09060			14 24 14	−44 19.2	1	9.23		0.11		0.08		3	A0 V	1770
126548	−46 09329			14 24 14	−46 52.5	2	7.66	.005	-0.03	.005	-0.25		6	B9 V	1770,2014
	−59 05579	NGC 5606 - 4		14 24 14	−59 24.1	3	9.91	.053	0.25	.007	-0.64	.035	6	B0 IV	251,1101,1725
		NGC 5606 - 150		14 24 15	−59 30.4	1	12.59		0.44		0.37		3		1725
	−59 05580	NGC 5606 - 3		14 24 16	−59 24.7	3	9.86	.008	0.27	.007	-0.63	.023	10	B0 IV	251,1101,1725
		NGC 5606 - 155		14 24 16	−59 28.4	1	12.19		1.60		0.00		2		1725
126209	−76 00826	HR 5389		14 24 16	−76 30.3	2	6.06	.005	1.18	.000	1.11	.000	7	K0/1 III	15,1075
126477	−58 05542	NGC 5606 - 166		14 24 17	−59 14.9	1	9.88		0.03		-0.32		5	B8 III/V	1725
126561	−45 09190			14 24 18	−45 59.3	1	7.23		0.00		-0.04		2	A0 V	1770
126676	+9 02890			14 24 20	+08 36.4	1	7.56		0.00		0.02		1	A0	1545
		WLS 1424 55 # 7		14 24 20	+52 54.9	1	11.88		0.55		-0.02		2		1375
	+29 02534			14 24 21	+29 20.6	1	10.06		0.20		0.10		2	A0	272
		NGC 5606 - 154		14 24 21	−59 29.4	1	13.11		0.44		0.38		2		1725
126695	+17 02737	IDS14220N1652	AB	14 24 22	+16 38.3	1	8.23		0.46		-0.04		2	F8	3030
		Ton 772		14 24 24	+28 26.	1	15.18		0.37		-0.04		2		313
		LP 620 - 79		14 24 24	−02 54.8	1	12.86		0.68		0.02		2		1696
126575	−45 09191			14 24 24	−45 58.7	1	7.69		-0.03		-0.23		1	B9 V	1770
		NGC 5606 - 153		14 24 24	−59 30.5	1	11.18		1.57		1.32		3		1725
	−59 05586	NGC 5606 - 21		14 24 25	−59 22.6	1	11.02		0.35		-0.50		3		1725
		NGC 5606 - 136		14 24 27	−59 23.2	1	12.31		0.34		-0.31		4		1725
126605	−40 08691			14 24 28	−41 07.8	1	9.35		0.12		0.09		1	A1 V	1770
		NGC 5606 - 141		14 24 30	−59 20.0	1	12.87		0.62		0.21		2		1725
		NGC 5606 - 137		14 24 30	−59 23.1	1	12.63		0.72		0.23		3		1725
		Lod 1378 # 21		14 24 31	−58 03.3	1	11.65		0.33		-0.14		2	B7	807
		NGC 5606 - 135		14 24 32	−59 22.0	1	12.34		0.30		0.20		2		1725
126528	−59 05587			14 24 32	−59 54.3	1	8.87		0.62		0.23		3	G3 IV	1101
126606	−43 09065			14 24 33	−44 06.3	1	7.00		0.10		0.09		2	A1 V	1770
		Lod 1378 # 5		14 24 36	−57 47.2	1	10.61		1.24		1.00		2	K3	807
		Lod 1378 # 19		14 24 37	−58 03.1	1	11.52		0.55		0.13		2	F0	807
126681	−17 04092			14 24 38	−18 11.0	4	9.30	.010	0.60	.004	-0.10	.010	15	G3 V	1003,1775,2033,3077
	−59 05589	NGC 5606 - 23		14 24 41	−59 22.6	1	10.91		0.31		-0.44		6		1725
126778	+29 02535			14 24 42	+28 48.9	2	8.16	.015	0.91	.005	0.66	.005	3	K0 III	1003,3077
		Ton 199		14 24 44	+33 11.1	1	13.93		-0.15		-1.04		2	sdB	1036
		NGC 5606 - 143		14 24 44	−59 26.8	1	13.15		0.33		0.29		2		1725
		Lod 1378 # 25		14 24 46	−58 02.5	1	11.83		0.36		0.29		2	A3	807
126722	−5 03880	HR 5406		14 24 47	−05 53.7	2	6.18	.020	0.09	.002	0.09	.007	10	A2 IV	1088,3016
	−59 05590			14 24 47	−60 02.4	1	10.72		0.42		0.16		3		1101
126549	−61 04556			14 24 47	−61 59.1	1	8.59		-0.05				4	B5/6 V	2014
	+30 02523			14 24 48	+30 23.3	1	9.56		0.26		0.10		2	A2	272
126764	+13 02798			14 24 49	+12 36.4	1	8.41		0.84				3	G5	882
		NGC 5606 - 142		14 24 49	−59 23.3	1	12.33		0.44		0.18		5		1725
126877	+53 01711			14 24 51	+52 51.0	1	8.12		1.42		1.55		3	K0	1566
126669	−36 09372			14 24 52	−37 03.1	1	10.17		0.32		0.12		1	A9 V	1771
126592	−57 06652			14 24 54	−57 55.1	1	8.89		1.28		0.89		2	K2/3 III	807
		NGC 5617 - 4		14 24 55	−60 26.8	1	10.86		1.12		0.78		2		1685
		NGC 5606 - 127		14 24 59	−59 29.2	1	11.20		0.11		-0.05		2		1725
126766	−12 04055			14 25 02	−13 08.1	2	6.65	.000	0.42	.000			36	F5 V	130,1351
		NGC 5606 - 140		14 25 02	−59 21.1	1	11.70		0.33		-0.44		2		1725
126610	−58 05549	HR 5403		14 25 03	−58 58.5	3	6.45	.008	0.14	.000			11	A0 II/III	15,1075,2009
		ST Vir		14 25 05	−00 40.5	1	10.84		0.07		0.08		1	A7	668
	+23 02698			14 25 06	+22 56.4	1	8.97		0.99		0.75		2	K5	1375
126812	+15 02714			14 25 07	+14 58.7	1	7.27		1.29		1.38		2	K0	1648
126593	−59 05593	LSS 3250	⋆AB	14 25 07	−60 19.0	2	8.58	.012	0.48	.011	-0.42	.034	4	B1/2 Ib	540,976
	−57 06654			14 25 08	−57 54.3	1	11.30		0.20		0.01		2	B9	807
126710	−45 09199			14 25 11	−45 38.9	1	8.56		0.29		0.07		1	A8/9 IV	1771
	−64 02866			14 25 12	−64 58.4	1	11.15		0.18		-0.25		2	B7 II-III	540
126769	−28 10712	HR 5407	⋆ABC	14 25 14	−29 16.1	7	4.96	.008	-0.07	.006	-0.40	.011	25	B8 V	15,1075,1079,1088,1770*
	−69 02055	LSS 3247		14 25 15	−69 54.0	1	10.08		0.09		-0.79		3	B2 III	400
		NGC 5617 - 41		14 25 17	−60 31.6	1	14.74		0.92		0.35		2		746

HD	DM	Other Id	Rem	α₁₉₅₀	δ₁₉₅₀	S	V	σ_V	B-V	σ_B-V	U-B	σ_U-B	n	Spectrum	References
		Ton 202		14 25 18	+26 46.	3	15.75	.140	0.15	.106	-0.76	.016	4	DC	98,1036,3028
126656	-57 06655			14 25 20	-57 59.2	1	9.62		0.18		-0.02		2	B9 III	807
126772	-33 09825			14 25 21	-33 38.2	1	8.73		1.29				1	K2 III	925
126640	-60 05381	NGC 5617 - 55		14 25 22	-60 25.4	4	8.90	.046	1.83	.010	1.84	.031	7	M2 II?	412,746,1685,4002
		G 124 - 43		14 25 23	-00 09.2	1	13.96		1.65		1.32		4		316
	-57 06656			14 25 23	-57 53.1	1	10.96		0.25		0.24		2	A1	807
		G 124 - 44		14 25 24	-00 09.1	1	14.04		1.68		1.24		3		316
		EG 110, MY Aps		14 25 24	-81 07.	1	13.00		0.00				1	DA	3028
		G 124 - 45		14 25 27	-00 55.2	1	11.19		0.72		0.22		2		1658
		LP 620 - 75		14 25 27	-00 55.2	1	13.75		0.25		-0.53		1		3065
126829	-15 03877			14 25 27	-15 53.2	1	9.85		1.12		1.01		1	K4 V	3072
126692	-55 06023			14 25 27	-55 54.8	1	6.90		0.52				4	A3 II	1075
126943	+41 02504	HR 5411		14 25 29	+41 14.9	2	6.65	.000	0.35	.005	-0.05	.018	4	F2 IV	254,3016
		SW Boo		14 25 31	+36 16.0	1	11.76		0.08		0.12		1		668
	+24 02735	G 166 - 29		14 25 32	+24 03.7	2	10.84	.035	1.27	.002	1.08		4	M0	1017,7008
126759	-47 09162			14 25 34	-47 46.1	8	6.39	.003	-0.11	.005	-0.35	.007	44	Ap Si	278,977,977,1075,1770,2011*
		NGC 5617 - 82		14 25 35	-60 31.3	1	14.55		0.52		0.39		2		746
126868	-1 02957	HR 5409	*AB	14 25 37	-02 00.3	7	4.81	.011	0.70	.015	0.20	.004	26	G2 III	15,1075,1088,1118*
126914	+17 02739			14 25 38	+17 08.5	1	8.56		0.91				2	K0	882
	-74 01182			14 25 38	-74 32.1	1	10.24		-0.01		-0.58		4	B5 V	400
127029	+53 01714			14 25 39	+53 32.0	1	7.99		0.57		0.03		2	G0	1566
		LP 440 - 26		14 25 40	+16 10.7	1	11.07		0.89		0.68		2		1696
		AF Vir		14 25 41	+06 46.0	1	10.94		0.14		0.06		1	A2	597
126760	-48 09083			14 25 41	-48 28.4	1	8.74		0.28		0.05		1	A9 V	1771
	-60 05401	NGC 5617 - 347		14 25 41	-60 26.4	1	10.13		1.33		1.10		2	K5	1685
		NGC 5617 - 106		14 25 41	-60 32.8	1	13.84		1.47		1.13		2		746
		NGC 5617 - 116		14 25 44	-60 28.9	3	10.68	.009	1.75	.013	1.72	.048	12		746,1685,4002
		NGC 5617 - 114		14 25 44	-60 31.3	1	12.48		0.44		-0.09		2		746
		G 65 - 52		14 25 48	+04 36.6	2	12.17	.013	0.60	.002	-0.10	.002	4		333,1620,1658
126970	+29 02538			14 25 48	+29 29.2	1	7.58		0.88		0.53		2	G5 IV	3016
126803	-46 09347			14 25 48	-46 31.2	4	8.94	.011	0.68	.008	0.09	.009	20	G5 V	863,977,2011,3073
		NGC 5617 - 135		14 25 48	-60 30.5	1	11.87		0.39		0.01		2		412
	-60 05384	NGC 5617 - 137		14 25 48	-60 34.1	2	9.77	.000	0.21	.005	-0.33	.010	5	B3	403,746
		NGC 5617 - 138		14 25 48	-60 34.8	1	13.36		0.42		0.15		3		746
126693	-63 03268			14 25 48	-63 56.3	2	10.24	.022	0.21	.010	-0.45	.010	7	B3/5 Vne	540,976
		G 65 - 53		14 25 49	+05 32.1	1	13.12		1.65		1.29		1		333,1620
126788	-52 07254			14 25 52	-52 40.3	1	8.56		0.10		-0.03		2	B9 V	1770
		NGC 5617 - 158		14 25 52	-60 29.7	1	11.75		0.44		0.07		2		412
		NGC 5617 - 155		14 25 52	-60 29.9	1	11.70		0.32		0.23		2		412
		G 65 - 54		14 25 53	+05 32.4	1	12.53		1.57		1.24		1		333,1620
126945	+16 02652			14 25 53	+16 20.7	1	7.44		0.66				2	G5	882
126990	+36 02493			14 25 53	+35 48.1	1	7.45		0.95		0.74		2	K0	1601
		NGC 5617 - 164		14 25 54	-60 30.3	1	11.87		0.43		0.08		2		412
		NGC 5617 - 163		14 25 54	-60 30.5	1	11.77		0.40		0.01		2	B5	412
		NGC 5617 - 168		14 25 55	-60 28.9	1	13.34		1.52		1.73		1		412
		NGC 5617 - 170		14 25 55	-60 30.9	1	12.54		0.47		0.06		2		412
	-61 04565	LSS 3251		14 25 57	-61 45.8	1	10.62		0.19		-0.45		2	B3 IV	540
	-60 05386	NGC 5617 - 185		14 25 58	-60 30.6	2	10.24	.020	0.35	.000	-0.54	.020	4		412,1737
	-60 05385	NGC 5617 - 182		14 25 58	-60 31.9	2	11.36	.005	0.42	.015	-0.52	.020	5		412,746
		G 200 - 40		14 25 59	+54 01.0	1	11.97		1.38				5		538
		G 200 - 39		14 25 59	+54 01.6	1	15.04		-0.07		-0.98		3		538
		NGC 5617 - 187		14 25 59	-60 29.4	1	12.08		0.41		0.00		2		412
		NGC 5617 - 191		14 25 59	-60 30.5	1	12.82		0.40		-0.04		2		412
		NGC 5617 - 188		14 25 59	-60 31.6	1	12.15		0.41		0.08		2		412
		NGC 5617 - 199		14 26 00	-60 28.1	1	15.11		0.56		0.45		2		746
		NGC 5617 - 198		14 26 00	-60 29.0	1	12.57		0.40		0.10		2		412
	-60 05387	NGC 5617 - 195		14 26 00	-60 29.5	1	10.90		0.35		-0.27		2		412
	-60 05388	NGC 5617 - 202		14 26 00	-60 31.2	2	10.31	.000	0.42	.009	-0.36	.004	21		412,746
126991	+25 02782	G 166 - 31		14 26 01	+24 44.3	1	7.90		0.80		0.24		3	G2 V	3026
		NGC 5617 - 205		14 26 01	-60 29.6	1	12.74		0.43		0.00		2		412
		NGC 5617 - 207		14 26 02	-60 29.8	1	13.03		0.47		0.14		2		412
		Steph 1161	AB	14 26 03	+63 13.8	1	11.13		1.45		1.20		2	M2	1746
126927	-6 04009	HR 5410		14 26 03	-06 40.6	2	5.41	.000	1.49	.000	1.75	.000	10	K5 III	1088,3005
	-45 09206			14 26 03	-46 14.1	2	10.35	.000	1.44	.004	1.26		5	K5	158,1705
		NGC 5617 - 215		14 26 03	-60 32.6	2	11.69	.014	1.96	.014	2.30	.070	4		412,746
		NGC 5617 - 219		14 26 04	-60 29.5	1	13.37		0.42		0.10		2		412
	-60 05390	NGC 5617 - 216		14 26 04	-60 29.6	1	10.95		0.35		-0.33		2		412
		NGC 5617 - 218		14 26 04	-60 30.0	1	13.68		0.47		0.14		2		412
		NGC 5617 - 227		14 26 05	-60 27.9	3	10.42	.001	1.68	.030	1.57	.029	4		412,1685,4002
		NGC 5617 - 226		14 26 05	-60 30.3	2	12.05	.000	0.42	.010	-0.07	.015	5		412,746
		NGC 5617 - 225		14 26 05	-60 31.3	1	13.24		0.45		0.12		2		412
	-44 09372			14 26 06	-45 04.3	1	10.19		0.44		0.20		2	F2	1730
		NGC 5617 - 240		14 26 07	-60 30.2	1	13.50		0.63		0.17		2		412
		NGC 5617 - 244		14 26 08	-60 29.7	1	13.39		0.45		0.16		2		412
		NGC 5617 - 255		14 26 10	-60 31.0	1	13.24		0.70		0.51		2		412
	-60 05397	NGC 5617 - 291		14 26 11	-60 30.1	1	11.59		0.27		-0.21		4	B6	412
	-60 05394	NGC 5617 - 258		14 26 11	-60 30.2	1	10.54		0.77		-0.26		2	B5	412
		Feige 95		14 26 12	+21 19.3	2	13.22		-0.24	.005	-1.07	.005	2	sdB2	1264,3060
127065	+36 02495	HR 5416		14 26 12	+36 25.2	2	6.21	.010	1.18	.014	1.19		4	K0 III	70,1733
126920	-21 03915			14 26 14	-22 24.2	1	9.11		0.56				4	G1 V	2033

Table 1   801

HD	DM	Other Id	N	Rem	α_{1950}	δ_{1950}	S	V	σ_V	B–V	σ_{B-V}	U–B	σ_{U-B}	n	Spectrum	References
126807	−59 05606				14 26 16	−59 33.2	1	9.02		-0.02				4	B8 III	2014
126808	−60 05395	NGC 5617 - 283			14 26 18	−60 37.5	3	8.83	.037	1.25	.008	1.14	.024	6	K1 III	742,746,1685
		CS 22883 # 51			14 26 19	+10 39.0	1	15.03		0.18				1		1736
127043	+28 02331	HR 5414	⋆ B		14 26 19	+28 30.7	1	7.62		0.00		-0.01		4	A1 V	1084
		LTT 5721	⋆ V		14 26 19	−62 28.1	5	11.11	.022	1.88	.018	1.48	.055	31	M5.5 eV	621,1494,1705,3045,8042
127067	+28 02332	HR 5415	⋆ A		14 26 21	+28 30.8	1	7.12		-0.02		-0.04		4	A1 V	1084
		NGC 5617 - 290			14 26 23	−60 31.2	1	13.12		0.64		0.14		3		746
126859	−55 06025	V853 Cen			14 26 25	−55 54.5	1	6.96		0.23				4	A6 V	1075
		NGC 5617 - 304			14 26 26	−60 31.5	1	14.95		0.93		0.21		3		746
		G 178 - 27			14 26 27	+38 12.8	1	11.22		0.43		-0.22		1	sdF5	3060
126922	−44 09375				14 26 29	−44 56.8	1	7.93		0.27		0.06		1	A9 V	1771
126842	−60 05399	NGC 5617 - 319			14 26 29	−60 35.7	4	9.02	.012	0.43	.008	-0.06	.008	7	F3 V	742,746,1594,6006
127093	+26 02575				14 26 32	+26 04.6	1	6.69		1.59				4	M III	20
		LP 174 - 340			14 26 36	+46 07.9	1	16.99		2.22				1		1663
127069	+11 02684				14 26 37	+11 15.9	1	7.58		1.25		1.35		2	K0	1648
		LP 500 - 93			14 26 39	+13 53.8	1	12.17		0.66		-0.02		2		1696
126936	−44 09377				14 26 39	−44 52.9	2	9.56	.008	0.45	.001	0.14	.002	3	Ap CrEuSr	1730,1770
		NGC 5617 - 344			14 26 39	−60 30.2	1	14.06		1.46		1.17		2		746
		CS 22883 # 50			14 26 41	+10 15.5	1	14.30		0.24		0.08		1		1736
		NGC 5617 - 348			14 26 43	−60 28.2	1	13.72		1.49		1.14		2		746
	+45 02184	G 178 - 28			14 26 44	+44 37.1	1	9.74		0.87		0.54		1	K2	3060
126964	−38 09413				14 26 47	−39 02.5	1	9.61		0.13		0.11		1	A1 V	1770
127013	−29 11100				14 26 48	−29 42.5	1	9.60		1.03		0.67		5	G8 III/IV	897
		NGC 5617 - 362			14 26 49	−60 34.9	1	12.68		1.32		1.01		3		746
126905	−60 05406				14 26 50	−60 38.4	1	9.26		1.12		0.87		1	G5/8 III	742
127243	+50 02084	HR 5420	⋆		14 26 53	+50 04.1	2	5.59	.000	0.85	.001	0.44		6	G3 IV	70,3016
126981	−44 09383	HR 5412	⋆ AB		14 26 54	−45 05.9	8	5.50	.007	-0.08	.003	-0.26	.004	29	B8 Vn	15,26,278,1075,1770,2011*
126997	−43 09098				14 26 57	−43 38.5	1	6.91		0.13		0.02		3	A0/1 V	1770
126124	−83 00558				14 26 60	−83 55.7	1	7.15		1.57		1.79		2	K3 III	1730
126983	−48 09098	HR 5413	⋆ A		14 27 00	−49 17.8	3	5.37	.008	0.05	.003	0.03		9	A0 V	15,1770,2012
126999	−46 09361				14 27 03	−46 43.0	1	10.13		1.20				4	K5/M0 V	2011
126862	−67 02595	HR 5408	⋆ A		14 27 04	−67 29.7	3	5.84	.008	1.01	.005	0.79	.005	15	K1 III	15,1075,2038
127263	+47 02150				14 27 06	+47 22.4	2	8.14	.020	0.24	.005	0.09	.010	4	Am	272,1733
		BPM 22719			14 27 06	−51 49.	1	14.56		0.76		0.13		1		3065
127017	−48 09102				14 27 07	−49 09.0	1	8.59		0.04		-0.12		2	B9 V	1770
126986	−51 08240				14 27 07	−52 02.0	1	9.67		0.29		-0.23		3	B9/A0 V(nn)	1586
126940	−59 05612	LSS 3257			14 27 09	−60 05.1	1	9.33		0.44		-0.45		1	B2 Ib/II	8100
	+16 02658	G 135 - 56			14 27 13	+15 43.7	1	10.68		1.47		1.17		3	M3	3078
		WLS 1420 0 # 7			14 27 16	−01 55.8	1	12.23		0.68		0.17		2		1375
127167	+1 02941	HR 5418			14 27 17	+01 03.0	2	5.93	.005	0.16	.000	0.10	.005	7	A5 IV	15,1417
		CS 22874 # 26			14 27 17	−23 45.1	1	11.54		0.23		0.15		1		1736
		CS 22874 # 27			14 27 19	−23 44.9	1	15.29		0.25		0.19		1		1736
127168	−3 03634	IDS14248S0348	A		14 27 24	−04 01.5	1	7.06		0.40		-0.01		1	F3 V	3077
127168	−3 03634	IDS14248S0348	B		14 27 24	−04 01.5	1	11.16		1.12		1.00		4		3024
		CS 22874 # 12			14 27 28	−25 16.0	1	15.13		0.09		0.16		1		1736
		L 260 - 93			14 27 30	−53 52.5	2	11.63	.018	1.51	.001			5		158,1705
127227	+16 02659				14 27 32	+16 25.9	2	7.49	.020	1.53	.040	1.84		5	K5 III	882,3077
		CS 22874 # 25			14 27 33	−23 46.9	1	13.33		0.12		0.18		1		1736
127084	−47 09184				14 27 33	−48 03.8	1	9.05		0.19		0.15		1	A1 V	1770
		G 178 - 30			14 27 35	+39 45.6	2	13.52	.010	0.87	.010	0.30	.025	2		1658,3060
127700	+76 00527	HR 5430	⋆ A		14 27 36	+75 55.1	4	4.27	.041	1.43	.007	1.68	.020	11	K4 III	15,1363,3016,8015
127334	+42 02508	HR 5423			14 27 39	+42 01.2	2	6.35	.005	0.70	.005	0.26	.055	5	G5 V	1733,3016
127174	−28 10740				14 27 40	−29 05.3	1	7.66		1.44		1.71		5	K3 III	897
127304	+32 02482	HR 5422	⋆ A		14 27 41	+32 00.8	4	6.06	.004	-0.03	.005	-0.09	.008	9	A0 Vs	70,1022,1079,3032
127304	+32 02482	IDS14256N3214	B		14 27 41	+32 00.8	2	11.32	.020	0.86	.027	0.52	.002	5		1752,3032
127110	−48 09109				14 27 42	−48 36.9	1	8.66		0.01		-0.22		1	B9 V	1770
127335	+39 02773	V Boo			14 27 44	+39 05.0	1	7.89		1.56		1.13		1	M4	817
	−65 02755	LSS 3259			14 27 46	−66 6.8	1	10.82		0.13		-0.79		2	O9.5 V	1737
127152	−40 08729	HR 5417			14 27 48	−40 37.4	1	6.39		1.44				4	K3 III	2006
		G 166 - 33			14 27 49	+29 47.6	1	14.66		1.65				1		1759
127208	−21 03917	HL Lib			14 27 50	−22 14.4	2	6.97	.024	0.19	.004	-0.33	.014	13	B8 Ve	1628,1770
127123	−49 08810				14 27 50	−50 14.7	1	9.72		0.22		0.19		1	A2 IV/V	1770
127141	−47 09188				14 27 57	−48 04.0	1	8.61		0.10		0.06		1	A0 V	1770
		Feige 96			14 28 00	+21 30.0	1	13.00		0.08		0.18		1		1298
127154	−46 09380				14 28 01	−46 51.1	2	7.29	.007	-0.03	.002	-0.12		6	Ap (Si)	1770,2014
127233	−29 11116	Y Cen			14 28 02	−29 52.6	1	8.58		1.44		0.72		5	M7/8 III	897
127193	−38 09430	HR 5419			14 28 04	−38 38.9	1	5.97		1.06				4	K1 III	2006
127215	−40 08735				14 28 05	−40 26.1	1	7.49		0.14		0.04		1	A1 V	1770
127112	−58 05576				14 28 05	−58 40.1	1	9.15		0.10		-0.35		3	B7 III	1586
	−16 03882	UW Lib			14 28 08	−16 35.3	2	9.20	.110	1.06	.420	1.30		2	K0	793,3076
127254	−34 09694				14 28 11	−34 27.6	1	7.76		0.07		0.05		3	A0 V	1770
127339	−7 03856	IDS14257S0811	A		14 28 12	−08 25.3	6	9.41	.015	1.39	.018	1.24	.021	21	K7 V	1003,1197,1705,1775*
127216	−45 09225				14 28 13	−45 29.6	2	9.03	.002	1.04	.000	0.75		10	G8 III	863,2011
127386	+25 02786				14 28 14	+25 18.7	1	8.61		0.47		-0.04		2	G5	3016
127337	+5 02886	HR 5424	⋆ A		14 28 15	+04 59.6	1	6.01		1.42		1.66		8	gK4	1088
127255	−39 08975				14 28 18	−40 08.9	1	9.15		0.05		0.00		3	A(0) V	1770
		LTT 5734			14 28 20	−12 04.2	2	11.94	.005	1.60	.018			5		158,1705
		LP 560 - 32			14 28 21	+05 20.8	1	11.89		0.64		0.01		2		1696
127352	−5 03896	IDS14258S0522	AB		14 28 24	−05 34.9	2	7.69	.009	0.78	.024	0.38		8	G5	214,1414
		V Cen sq # 6			14 28 24	−56 38.6	1	12.77		1.84		1.33		2		973

HD	DM	Other Id	N Rem	α1950	δ1950	S	V	σV	B−V	σB−V	U−B	σU−B	n	Spectrum	References
		V Cen sq # 9		14 28 28	−56 38.1	1	13.90		0.48		0.26		2		973
127294	−42 09398			14 28 30	−42 52.0	4	7.85	.003	0.50	.002	0.02	.009	17	F7/8 V	977,1075,1770,2011
		V Cen sq # 3		14 28 31	−56 37.8	1	12.03		0.61		0.29		2		973
127356	−14 03970	IDS14258S1511	AB	14 28 34	−15 24.8	2	7.98	.009	0.71	.004	0.17		6	G5 V	1003,2033
127340	−26 10340			14 28 36	−26 22.3	2	8.88	.005	0.77	.005	0.27		6	G6/8 V	1696,2012
		CS 22871 # 15		14 28 37	−19 28.0	1	14.74		0.23		0.17		1		1736
		V Cen sq # 1		14 28 38	−56 37.0	1	10.62		0.26		0.06		2		973
127277	−49 08822			14 28 39	−50 12.3	1	9.57		0.18		0.10		1	A1 V	1770
		V Cen sq # 2		14 28 40	−56 41.9	1	11.03		1.50		1.60		2		973
127506	+36 02500	G 178 - 31		14 28 42	+35 40.3	3	8.70	.000	1.06	.031	0.93	.047	5	K3 V	196,1746,3072
		V Cen sq # 10		14 28 45	−56 38.7	1	13.99		0.79		0.23		2		973
127326	−45 09236	IDS14256S4539	AB	14 28 48	−45 52.7	1	8.54		0.30		0.04		1	A8/9 V	1771
		CS 22871 # 11		14 28 53	−19 37.4	1	15.06		0.11		0.17		1		1736
127377	−30 11494			14 28 53	−30 41.8	1	10.03		0.17				1	A2 IV	925
127392	−30 11497			14 28 55	−30 58.8	2	9.68	.020	0.68	.000	0.04	.030	2	Gp Ba	565,3077
		G 65 - 55		14 28 56	+12 30.3	1	11.53		1.36		1.21		1	K7	333,1620
		G 124 - 49		14 28 57	−00 29.0	1	11.19		0.95		0.76		2		3072
127297	−56 06296	HR 5421, V Cen		14 28 57	−56 40.0	4	6.67	.275	0.81	.155	0.48	.006	7	F5 Ib/II	657,1484,1490,2006
		V Cen sq # 8		14 28 57	−56 41.7	1	12.79		0.47		0.32		2		973
127296	−54 06044			14 28 58	−55 14.6	2	7.38	.000	0.29	.005	0.10		5	A8 IV	657,1075
		V Cen sq # 4		14 28 59	−56 42.6	1	12.19		1.07		0.94		2		973
		CS 22874 # 20		14 29 00	−24 18.6	1	15.52		0.21		0.22		1		1736
127346	−51 08270			14 29 03	−51 45.0	1	7.27		0.03		−0.10		2	B9 III	1770
		CS 22874 # 13		14 29 05	−25 05.1	1	15.57		0.16		0.23		1		1736
		CS 22874 # 29		14 29 08	−23 20.4	1	15.09		0.18		0.23		1		1736
		V Cen sq # 7		14 29 08	−56 39.5	1	12.78		0.68		0.23		2		973
		POSS 741 # 3		14 29 09	−12 25.7	1	16.00		1.59				2		1739
127419	−27 09889			14 29 11	−27 29.2	2	7.47	.000	0.29	.000			8	A5/7 IV	1075,2012
		LP 913 - 32		14 29 11	−31 38.8	1	11.13		0.75		0.20		2		1696
		V Cen sq # 5		14 29 12	−56 38.7	1	12.46		0.80		0.28		2		973
		LP 440 - 36		14 29 13	+16 49.0	1	11.47		0.92		0.73		1		1696
		WLS 1424 55 # 9		14 29 14	+55 21.7	1	13.70		0.55		0.13		2		1375
		CS 22871 # 8		14 29 14	−20 21.1	1	14.79		0.05		0.08		1		1736
127381	−49 08831	HR 5425		14 29 14	−50 14.2	4	4.41	.009	−0.19	.005	−0.85	.014	16	B2 III/IV	15,1075,1770,2012
		POSS 741 # 2		14 29 15	−12 40.3	1	14.78		1.55				1		1739
		Ton 774		14 29 18	+28 31.	1	14.42		0.67		−0.39		1		313
		WLS 1420 0 # 8		14 29 19	−00 01.8	1	13.58		0.58		0.10		2		1375
		G 224 - 9		14 29 27	+63 52.7	1	14.71		1.33				1		1759
		WLS 1420-20 # 8		14 29 27	−20 09.4	1	12.29		1.58		1.31		2		1375
127441	−44 09420			14 29 27	−44 21.0	1	8.73		0.27		0.23		1	A5/6 IV	1771
127493	−22 03804			14 29 31	−22 26.2	3	10.03	.012	−0.25	.009	−1.18	.017	11	sdO8p	400,1732,1775
		CS 22874 # 14		14 29 31	−24 59.8	1	14.95		−0.34		−1.12		1		1736
127442	−44 09421			14 29 32	−44 53.2	1	8.64		1.75		1.95		5	M2/3 III	1673
		Lod 1339 # 4		14 29 32	−61 48.8	1	11.94		0.27		−0.15		1	F5	287
127760	+59 01589			14 29 34	+58 50.4	1	8.29		1.19		1.32		3	K2 III	37
127464	−41 08861			14 29 34	−41 42.5	1	9.60		0.31		0.12		1	A9 V	1771
127445	−46 09400			14 29 34	−46 58.9	2	8.40	.006	0.00	.001	−0.16		6	B9 V	1770,2014
127821	+63 01136	HR 5436		14 29 35	+63 24.4	2	6.11	.000	0.40	.002	−0.06	.002	3	F5 IV	254,3016
		POSS 741 # 4		14 29 35	−12 28.6	1	16.38		1.69				2		1739
127522	−17 04110			14 29 35	−17 39.7	1	8.53		0.21		0.17		2	A2 IV	1375
		NGC 5662 - 457		14 29 37	−56 20.2	1	12.39		0.74		0.21		2		1715
	−56 06305	NGC 5662 - 508		14 29 37	−56 36.3	1	11.12		0.29		0.09		2		1715
127317	−67 02603			14 29 37	−67 23.9	1	6.93		−0.05				4	B2/3 Vn	2012
	−61 04597			14 29 38	−61 46.4	1	11.10		0.50		0.27		1		287
	−61 04596			14 29 38	−61 48.8	1	11.24		0.30		0.13		1		287
	+73 00633			14 29 39	+72 57.2	1	10.08		1.48		1.84		2	K5	1375
127617	+19 02817			14 29 40	+18 59.2	1	8.78		−0.11		−0.55		4	A3	8079
127665	+31 02628	HR 5429	★ A	14 29 40	+30 35.4	10	3.58	.009	1.30	.007	1.44	.006	49	K3 III	1,15,667,1007,1077*
		CS 22871 # 17		14 29 40	−19 12.6	1	14.18		0.04		0.01		1		1736
	−61 04598			14 29 40	−61 46.1	1	11.41		0.24		0.20		1		287
		NGC 5662 - 509		14 29 41	−56 35.4	1	12.52		0.41		0.35		4		1715
		NGC 5662 - 455		14 29 42	−56 22.0	1	10.78		1.29		1.14		2		1715
		CS 22874 # 4		14 29 44	−27 10.0	1	14.31		0.21		0.22		1		1736
127427	−55 06046	NGC 5662 - 510		14 29 44	−55 59.8	1	7.76		1.48		1.30		2	G8 II/III	1715
	−27 09894			14 29 45	−28 02.3	1	10.67		1.24		1.26		1	K7	1696
	−56 06306	NGC 5662 - 448		14 29 45	−56 27.3	1	11.30		0.27		0.06		2		1715
	−61 04600			14 29 45	−61 46.4	1	10.86		0.17		−0.40		1		287
127667	+19 02818			14 29 46	+19 03.3	1	7.84		0.50		−0.08		4	F8	8079
127498	−41 08864			14 29 46	−42 11.1	1	7.78		0.24		0.06		1	A7/8 IV	1771
		NGC 5662 - 458		14 29 46	−56 19.6	1	13.06		0.54		0.14		2		1715
	+11 02687	G 135 - 59		14 29 47	+11 34.2	1	9.69		1.21		1.17		3	M0 V:p	3072
		NGC 5662 - 456		14 29 47	−56 20.8	1	11.44		1.89		1.97		2		1715
		CS 22874 # 23		14 29 49	−23 55.2	1	15.13		0.29		0.12		1		1736
		CS 22874 # 6		14 29 50	−26 32.4	1	13.49		0.46		0.08		1		1736
		POSS 741 # 6		14 29 51	−12 52.2	1	17.77		1.81				3		1739
		GD 336		14 29 54	+37 19.7	2	15.28	.033	−0.20	.023	−1.15	.009	4	DA	308,1727
127332	−69 02066			14 29 54	−69 37.9	1	10.13		0.46		0.20		1	A8/9 V	1771
		NGC 5662 - 506		14 29 55	−56 39.6	1	12.30		0.45		0.45		2		1715
		CS 22874 # 10		14 29 56	−26 00.1	1	14.19		0.37		−0.41		1		1736
127449	−58 05594			14 29 56	−58 36.1	1	7.72		−0.01		−0.60		3	B2/3 Vn	1586

Table 1   803

HD	DM	Other Id	N Rem	α_{1950}	δ_{1950}	S	V	σ_V	B–V	σ_{B-V}	U–B	σ_{U-B}	n	Spectrum	References
		G 65 - 56		14 30 01	+05 59.6	2	10.65	.009	1.21	.014	1.17	.042	4	K4	333,1620,3072
127486	−54 06053	HR 5426		14 30 01	−54 46.7	4	5.86	.005	0.48	.000	0.04		16	F5 IV	15,1075,2012,3053
127501	−52 07301	HR 5427	⋆ A	14 30 02	−52 27.6	3	5.86	.005	1.09	.005			14	K0 III	15,1075,2030
127530	−48 09140			14 30 03	−48 53.3	1	8.27		0.00		−0.25		2	B8/9 II/II	1770
127487	−56 06308	NGC 5662 - 450		14 30 03	−56 24.1	1	10.55		0.20		−0.02		2	A2 V	1715
127762	+38 02565	HR 5435, γ Boo	⋆ A	14 30 04	+38 31.6	11	3.04	.013	0.19	.006	0.12	.012	33	A7 III	1,15,39,254,667,1077*
		NGC 5662 - 454		14 30 04	−56 20.0	1	13.05		0.76		0.19		2		1715
	−56 06309	NGC 5662 - 507		14 30 04	−56 38.3	1	10.97		0.69		0.31		2		1715
		NGC 5662 - 453		14 30 05	−56 20.4	1	13.14		0.60		0.12		2		1715
		NGC 5662 - 451		14 30 05	−56 22.5	1	12.27		0.40		0.31		2		1715
		NGC 5662 - 505		14 30 06	−56 41.2	1	10.19		1.44		1.44		2		1715
127726	+27 02388	HR 5433	⋆ AB	14 30 07	+26 53.8	3	5.95	.065	0.21	.004	0.07	.005	12	A7 IV-V	1022,1381,3030
		CS 22874 # 15		14 30 07	−24 55.8	1	13.87		0.03		−0.08		1		1736
		NGC 5662 - 449		14 30 07	−56 24.8	1	12.61		0.47		0.26		2		1715
127605	−31 11280			14 30 09	−31 26.6	1	9.19		−0.02		−0.11		1	B9.5 V	1770
		GD 495		14 30 10	+63 59.8	1	14.46		0.10		0.12		3		308
127624	−30 11519	HR 5428	⋆ AB	14 30 11	−30 29.7	2	6.07	.010	1.04	.005	0.84		7	K0 III	404,2007
	−56 06311	NGC 5662 - 441		14 30 11	−56 37.9	1	11.34		0.24		−0.01		2		1715
		G 135 - 61		14 30 12	+13 58.8	1	14.08		1.41		1.20		3		316
		CS 22874 # 32		14 30 12	−23 16.9	1	14.23		0.14		0.23		1		1736
		NGC 5662 - 459		14 30 13	−56 17.6	1	11.94		0.50		0.29		2		1715
		NGC 5662 - 452		14 30 13	−56 21.5	1	12.99		0.40		0.19		2		1715
	−72 01542	IDS14254S7305	AB	14 30 13	−73 18.8	1	10.17		−0.02		−0.91		3	B1 III	400
127531	−51 08278			14 30 14	−52 09.4	1	8.88		0.13		0.11		1	A0 V	1770
127788	+33 02471			14 30 15	+33 24.8	1	8.69		0.04		0.03		2	A3 V(III)	272
127739	+22 02715	HR 5434		14 30 16	+22 28.8	3	5.92	.021	0.35	.019	0.04	.010	6	F2 IV	70,254,3016
127515	−56 06312	NGC 5662 - 442		14 30 16	−56 31.6	2	9.88	.011	0.21	.009	−0.20	.006	6	B7/8 Vne	1586,1715
		CS 22874 # 19		14 30 19	−24 35.2	1	14.95		0.12		0.19		1		1736
127534	−55 06050	NGC 5662 - 511		14 30 20	−55 59.6	1	10.81		0.27		0.19		2	A0 IV	1715
	+32 02487			14 30 21	+31 56.7	1	10.63		0.50				1	G5	1213
127929	+60 01547	HR 5437		14 30 21	+60 26.7	2	6.27	.005	0.24	.000	0.13	.013	4	F0 III	1733,3016
		CS 22874 # 22		14 30 21	−24 12.6	1	15.10		0.13		0.18		1		1736
		CS 22874 # 30		14 30 23	−23 21.7	1	15.34		0.09		0.14		1		1736
127568	−51 08282			14 30 23	−52 08.4	1	9.55		0.21		0.15		1	A0 V	1770
		NGC 5662 - 512		14 30 24	−55 57.5	1	10.88		0.25		0.11		2		1715
		NGC 5662 - 460		14 30 24	−56 10.8	1	12.20		0.38		0.25		2		1715
	+2 02835	IDS14279N0244	AB	14 30 25	+02 30.7	1	9.83		0.61		0.08		2	G0	1375
		LP 440 - 71		14 30 25	+16 14.3	1	13.28		0.57		−0.21		2		1696
		CS 22874 # 21		14 30 25	−24 19.6	1	13.89		0.09		0.15		1		1736
		NGC 5662 - 447		14 30 25	−56 28.9	1	12.20		0.90		0.30		2		1715
		CS 22874 # 11		14 30 26	−25 27.3	1	15.64		0.08		0.17		1		1736
		NGC 5662 - 461		14 30 28	−56 05.3	1	12.09		0.33		−0.04		2		1715
		NGC 5662 - 343		14 30 29	−56 26.1	1	12.93		0.42		0.09		2		1715
		NGC 5662 - 1		14 30 30	−56 28.8	1	11.74		1.77		1.70		2		1715
127690	−21 03924			14 30 31	−22 06.4	1	10.37		0.60		0.15		2	F7 V	1375
127535	−59 05631	V841 Cen		14 30 31	−60 11.3	1	8.71		1.07		0.79		30	K1 IV/Ve	1621
127489	−63 03303	LSS 3263		14 30 31	−63 59.0	2	9.03	.086	0.18	.018	−0.67	.023	7	B9.5V	540,976
		NGC 5662 - 346		14 30 32	−56 19.4	1	12.52		0.30		−0.06		3		1715
		CS 22874 # 5		14 30 34	−26 46.0	1	14.73		0.08		0.18		1		1736
		NGC 5662 - 357		14 30 34	−56 11.8	1	12.41		0.75		0.25		2		1715
		NGC 5662 - 355		14 30 36	−56 15.2	1	11.78		0.30		0.22		2		1715
		NGC 5662 - 354		14 30 36	−56 16.0	1	12.42		0.73		0.25		2		1715
127369	−73 01302			14 30 36	−73 28.5	1	6.93		1.61		1.89		9	K4 III	1628
		NGC 5662 - 344		14 30 37	−56 24.9	1	13.02		0.57		0.29		2		1715
		WD1430+427		14 30 38	+42 43.4	1	14.23		−0.14		−0.66		2	DA2	1727
	−56 06313	NGC 5662 - 445		14 30 40	−56 31.6	1	11.37		0.26		0.15		2		1715
		NGC 5662 - 443		14 30 40	−56 33.1	1	11.60		0.29		0.15		2		1715
		NGC 5662 - 440		14 30 41	−56 35.8	1	13.05		0.49		0.27		2		1715
127764	−6 04025			14 30 42	−06 43.0	1	7.77		1.53		1.85		3	K0	1657
		NGC 5662 - 504		14 30 42	−56 42.3	1	11.22		1.19		0.84		2		1715
		NGC 5662 - 436		14 30 45	−56 41.3	1	12.28		0.49		0.27		4		1715
	+27 02391			14 30 46	+27 16.1	1	9.22		0.57		0.07		2	G5	1375
127679	−48 09149			14 30 47	−48 50.2	1	9.84		0.17		0.11		1	A2 V	1770
		NGC 5662 - 444		14 30 47	−56 32.3	1	12.02		0.57		0.31		2		1715
		NGC 5662 - 352		14 30 49	−56 18.5	1	12.62		0.80		0.40		2		1715
127694	−44 09436			14 30 51	−45 18.6	2	9.39	.005	0.50	.005	−0.05		8	F5 V	116,2011
	−56 06315	NGC 5662 - 5		14 30 51	−56 29.7	1	10.99		0.27		0.02		2		1715
127717	−41 08887			14 30 52	−41 56.5	1	9.04		0.29		0.08		1	A9 V	1771
		LP 680 - 30		14 30 54	−04 07.4	1	11.37		1.27		1.17		1	K4	1696
	−9 03964			14 30 54	−09 42.4	1	10.54		1.27		1.20		4	K4	158
		NGC 5662 - 348		14 30 55	−56 20.8	1	11.93		0.52		0.32		2		1715
		CS 22874 # 17		14 30 56	−24 37.0	1	15.12		0.10		0.12		1		1736
		NGC 5662 - 349		14 30 56	−56 20.4	1	12.54		0.66		0.41		2		1715
		NGC 5662 - 9		14 30 56	−56 23.8	1	12.05		0.35		0.27		2		1715
		NGC 5662 - 503		14 30 56	−56 43.1	1	12.80		1.09		0.53		1		1715
127610	−59 05633	LSS 3264		14 30 56	−59 33.9	1	7.84		0.64		0.41		3	F0 II	8100
128000	+56 01746	HR 5442		14 30 57	+55 37.1	1	5.76		1.50				1	gK5	71
127716	−41 08890	HR 5431		14 30 57	−41 52.8	2	6.60	.001	0.06	.001	0.14		6	A2 IV	1770,2032
		NGC 5662 - 464		14 30 57	−55 59.8	1	13.19		0.54		0.25		2		1715
		NGC 5662 - 347		14 30 57	−56 21.4	1	11.61		0.72		0.33		2		1715

HD	DM	Other Id	N Rem	α_{1950}	δ_{1950}	S	V	σ_V	B−V	σ_{B-V}	U−B	σ_{U-B}	n	Spectrum	References
127655	−56 06317			14 30 59	−56 53.5	2	8.10	.000	0.23	.005			8	A0 III	1075,2014
		NGC 5662 - 358		14 31 00	−56 10.9	1	12.21		0.26		0.02		3		1715
		CS 22874 # 16		14 31 02	−24 43.9	1	14.68		0.12		0.06		1		1736
	−56 06319	NGC 5662 - 439		14 31 03	−56 36.4	1	10.64		0.22		0.09		2		1715
127684	−56 06318	NGC 5662 - 437		14 31 03	−56 38.6	1	10.26		0.20		-0.17		2	B9	1715
127537	−68 02133			14 31 03	−68 27.9	2	8.55	.019	0.43	.000	-0.02		4	F3 IV/V	742,1594
		NGC 5662 - 463		14 31 04	−56 00.6	1	11.95		0.49		0.30		2		1715
128642	+81 00482			14 31 05	+81 01.9	1	6.91		0.76		0.34		2	G5	1502
		NGC 5662 - 359		14 31 06	−56 09.6	1	12.03		0.33		0.25		2		1715
		NGC 5662 - 350		14 31 06	−56 20.7	1	13.43		0.43		0.33		3		1715
		CS 22871 # 3		14 31 07	−20 59.8	1	14.50		0.11		0.17		1		1736
127871	+10 02703	G 66 - 5	⋆ A	14 31 08	+09 33.6	5	8.82	.018	0.92	.005	0.68	.024	10	K2 V	22,265,1003,1658,3008
		NGC 5662 - 438		14 31 08	−56 38.4	1	11.14		1.30		1.12		2		1715
	−59 05634	LSS 3265		14 31 08	−60 12.5	3	9.50	.010	0.75	.005	-0.27	.018	9	O9 I	403,540,976
		NGC 5662 - 19		14 31 09	−56 32.0	1	12.33		0.39		0.33		2		1715
	−55 06057	NGC 5662 - 353		14 31 11	−56 17.6	1	11.37		0.35		-0.06		2		1715
128002	+46 01969			14 31 12	+46 05.9	1	9.36		0.29		0.02		2	A5 V	272
		NGC 5662 - 22		14 31 12	−56 27.9	1	11.63		0.29		0.15		4		1715
	−56 06320	NGC 5662 - 21		14 31 12	−56 29.6	1	11.19		0.24		0.08		4		1715
127871	+10 02703	G 66 - 6	⋆ B	14 31 13	+09 33.6	1	14.60		1.85		1.65		1		3008
		NGC 5662 - 25		14 31 13	−56 27.4	1	13.32		0.91		0.44		2		1715
		NGC 5662 - 24		14 31 13	−56 32.5	1	13.11		0.65		0.26		1		1715
127685	−60 05433			14 31 13	−60 57.9	1	9.33		0.19		0.12		3	A1 V	976
		CS 22874 # 3		14 31 15	−27 20.0	1	13.97		0.11		0.19		1		1736
		NGC 5662 - 29		14 31 15	−56 20.1	1	12.64		0.44		0.32		3		1715
	−56 06321	NGC 5662 - 27		14 31 15	−56 28.5	1	10.18		0.22		-0.13		2		1715
127750	−45 09269			14 31 16	−46 05.1	1	7.61		0.04		0.01		2	A0 V	1770
128127	+62 01352	G 223 - 69		14 31 17	+61 39.4	2	9.31	.009	0.74	.009	0.34	.019	4	K0	308,1658
127722	−55 06058	NGC 5662 - 356		14 31 17	−56 12.8	1	10.29		0.10		-0.24		2	B8 V	1715
127986	+37 02545	HR 5441, CP Boo		14 31 18	+37 10.8	2	6.40	.005	0.51	.005	0.07	.005	5	F8 IV	1501,1733
127721	−55 06060	NGC 5662 - 462		14 31 20	−56 02.7	1	10.01		0.13		-0.21		2	B8 V	1715
		NGC 5662 - 360		14 31 20	−56 09.1	1	12.23		0.38		0.26		2		1715
128091	+59 01591			14 31 21	+58 38.6	1	8.12		1.29		1.54		3	K3 III	37
		CS 22871 # 4		14 31 24	−20 36.6	1	14.15		0.31		0.18		1		1736
	+32 02489	RS Boo		14 31 25	+31 58.4	1	9.73		0.10		0.11		1	A5	668
127733	−56 06322	NGC 5662 - 40		14 31 25	−56 35.1	3	8.61	.023	1.70	.012	1.86	.044	7	K2 III	251,746,1715
		CS 22871 # 13		14 31 26	−19 24.3	1	14.89		0.16				1		1736
		NGC 5662 - 46		14 31 26	−56 20.4	1	13.27		0.89		0.31		2		1715
		CS 22874 # 9		14 31 29	−26 04.5	1	13.58		0.34		0.05		1		1736
	−55 06061	NGC 5662 - 47		14 31 29	−56 18.5	2	10.74	.018	0.17	.004	0.00	.022	3	B9 V	251,1715
128039	+44 02368			14 31 30	+44 33.0	1	9.26		0.18		0.10		2	A0 V	272
		CS 22871 # 22		14 31 30	−17 36.0	1	14.23		0.20		0.20		1		1736
127799	−48 09159			14 31 30	−48 37.1	1	9.54		0.14		0.19		1	A2 IV/V	1770
		NGC 5662 - 435		14 31 31	−56 43.7	1	12.10		0.51		0.33		2		1715
	−59 05640			14 31 31	−59 41.3	2	10.76	.022	0.35	.007	-0.41	.023	6	B7 Ia	540,1586
127724	−59 05642	HR 5432		14 31 32	−59 47.8	3	6.39	.004	1.25	.005			11	K2 III	15,1075,2035
128041	+36 02505	IDS14295N3601	AB	14 31 33	+35 48.2	1	7.63		0.75		0.35		2	G5	1601
		CS 22874 # 2		14 31 33	−27 37.6	1	13.50		-0.08		-0.50		1		1736
127753	−56 06325	NGC 5662 - 49		14 31 33	−56 20.7	5	7.06	.018	1.86	.013	2.04	.054	11	K3 II	251,746,1075,1490,1715
127754	−56 06324	NGC 5662 - 48		14 31 33	−56 30.2	3	9.54	.049	0.23	.039	-0.18	.013	6	B8 V	251,746,1715
	−11 03759			14 31 35	−12 18.5	3	11.34	.011	1.63	.009	1.13	.045	6	M4	158,1705,8006
		NGC 5662 - 54		14 31 37	−56 23.5	1	12.39		1.53		1.30		3		746
		NGC 5662 - 55		14 31 38	−56 30.6	1	12.11		0.49		0.26		2		1715
127987	+11 02691			14 31 40	+10 44.5	1	8.02		0.16		0.07		2	A2	1648
		NGC 5662 - 58		14 31 40	−56 29.7	1	11.91		0.34		0.25		2		1715
127782	−56 06326	NGC 5662 - 61		14 31 40	−56 33.2	4	9.41	.014	0.18	.011	-0.23	.015	13	B8 V	251,746,1586,1715
		CS 22871 # 5		14 31 41	−20 37.8	1	15.30		0.14		0.20		1		1736
	−55 06062	NGC 5662 - 362		14 31 41	−56 10.7	1	11.66		0.23		0.13		2		1715
		CS 22871 # 19		14 31 42	−19 08.3	1	13.25		-0.28		-1.10		1		1736
		NGC 5662 - 70		14 31 43	−56 24.8	2	11.44	.033	0.37	.013	0.31	.019	3		251,1715
127847	−44 09450			14 31 44	−45 07.9	1	8.61		0.16		0.19		1	A3 V	1771
		NGC 5662 - 361		14 31 45	−56 11.2	1	12.71		0.62		0.25		2		1715
127756	−60 05437	LSS 3266		14 31 45	−60 47.4	5	7.58	.017	0.12	.014	-0.76	.026	12	B1/2 Vne	164,403,540,976,1586
	−66 02568			14 31 47	−66 37.6	1	10.94		0.16		-0.55		2	B2 V	540
		NGC 5662 - 363		14 31 48	−56 08.8	1	11.46		1.47		1.31		2		1715
		NGC 5662 - 76		14 31 48	−56 20.1	1	15.27		0.93		0.35		3		746
234152	+51 01929			14 31 49	+50 58.3	1	9.45		0.50		0.01		2	F8	1566
127817	−56 06328	NGC 5662 - 81		14 31 50	−56 24.2	3	9.16	.021	0.20	.014	-0.26	.006	6	B7 IV	251,746,1715
		G 66 - 7		14 31 51	+12 48.0	1	10.79		1.10		1.04		1	K5	333,1620
128165	+53 01719	G 200 - 49		14 31 51	+53 07.4	1	7.23		0.99		0.84		3	K3 V	1080
		NGC 5662 - 86		14 31 51	−56 14.3	2	11.58	.017	0.22	.010	0.11	.023	3		251,1715
	−55 06063	NGC 5662 - 78		14 31 51	−56 14.3	2	11.13	.043	1.86	.018	2.11	.108	5		746,1715
	−56 06330	NGC 5662 - 85		14 31 51	−56 23.6	3	10.61	.035	0.57	.014	-0.11	.004	9	A2 IIIp	251,746,1715
127818	−56 06329	NGC 5662 - 82		14 31 51	−56 33.3	2	9.52	.022	0.57	.005	0.17	.002	3	F6 IV/V	251,1715
127879	−43 09167			14 31 52	−43 20.1	1	7.83		0.24		0.07		1	A9 V	1771
127864	−45 09276			14 31 53	−46 14.6	9	6.89	.004	-0.02	.004	-0.13	.008	71	B9 V	116,278,863,977,977,1075*
		NGC 5662 - 365		14 31 53	−56 06.9	1	11.78		0.30		0.17		2		1715
		NGC 5662 - 91		14 31 54	−56 21.4	1	13.28		0.66		0.30		2		1715
		NGC 5662 - 92		14 31 54	−56 23.4	2	13.37	.017	0.75	.025	0.26	.021	4		746,1715
127835	−55 06064	NGC 5662 - 97		14 31 55	−56 15.3	2	9.39	.014	0.11	.009	-0.28	.000	4	B7 V	251,1715

Table 1   805

HD	DM	Other Id	N Rem	α_{1950}	δ_{1950}	S	V	σ_V	B–V	σ_{B-V}	U–B	σ_{U-B}	n	Spectrum	References
	−56 06332	NGC 5662 - 94		14 31 55	−56 24.4	3	11.56	.013	0.31	.022	0.22	.014	6		251,746,1715
		LP 680 - 33		14 31 56	−05 20.5	1	13.32		1.44		1.02		1		1696
	−56 06334	NGC 5662 - 104	⋆ AB	14 31 56	−56 20.8	3	9.89	.024	0.25	.020	-0.06	.008	6	B8 V	251,746,1715
	−56 06333	NGC 5662 - 99		14 31 56	−56 23.9	3	9.86	.011	0.20	.013	-0.24	.004	18	B7 IV	251,746,1715
	−60 05441			14 31 56	−61 18.1	1	9.85		0.11		-0.43		2	B8	1586
		NGC 5662 - 465		14 31 57	−56 02.0	1	12.09		0.31		0.03		2		1715
127836	−56 06335	NGC 5662 - 107		14 31 57	−56 30.0	3	9.80	.019	0.26	.015	-0.12	.009	6	B8 V	251,746,1715
		NGC 5662 - 102		14 31 57	−56 31.7	1	11.36		1.39		1.23		2		1715
		NGC 5662 - 433		14 31 57	−56 41.3	1	13.62		0.69		0.16		2		1715
	−1 02966			14 32 02	−01 45.2	1	10.02		1.01		0.87		1	K5 V	3072
127964	−19 03903	HR 5438		14 32 02	−20 13.3	1	6.48		0.14				4	A3 V	2007
		NGC 5662 - 117		14 32 03	−56 18.5	1	13.02		0.89		0.30		2		1715
		NGC 5662 - 116		14 32 03	−56 24.9	2	11.71	.027	0.29	.019	0.25	.032	3		251,1715
128093	+33 02474	HR 5445	⋆ A	14 32 04	+32 45.2	4	6.33	.016	0.40	.004	-0.02	.011	8	F4 V	70,254,272,289,3037
		NGC 5662 - 364		14 32 04	−56 09.7	1	11.43		0.72		0.16		2		1715
		NGC 5662 - 118		14 32 04	−56 22.8	3	10.66	.019	1.30	.000	1.09	.014	13		251,746,1715
		NGC 5662 - 122		14 32 04	−56 26.9	1	14.98		0.99		0.42		3		746
		CS 22874 # 36		14 32 05	−23 60.0	1	13.38		0.10		0.16		1		1736
		NGC 5662 - 121		14 32 05	−56 31.9	1	11.86		0.42		0.00		2		1715
	−56 06336	NGC 5662 - 120		14 32 05	−56 32.2	1	10.99		0.25		-0.09		2		1715
128078	+25 02796	IDS14298N2450	A	14 32 06	+24 36.6	1	8.40		0.04		0.00		2	A1 V	272
		NGC 5662 - 466		14 32 06	−56 00.0	1	11.84		0.29		0.16		2		1715
	−56 06337	NGC 5662 - 126		14 32 06	−56 23.6	2	10.68	.024	0.23	.002	-0.01	.004	3	A1 V	251,1715
127838	−60 05443			14 32 07	−60 35.4	4	9.12	.054	0.35	.009	-0.50	.021	8	B1 Ib/II	403,540,976,8100
		SV Boo		14 32 08	+39 19.7	2	12.41	.374	0.16	.011	0.18	.017	2		668,699
		NGC 5662 - 131		14 32 08	−56 26.8	1	13.37		1.47		1.06		3		746
		NGC 5662 - 434		14 32 08	−56 38.0	1	12.32		0.38		-0.12		2		1715
		NGC 5662 - 467		14 32 10	−55 57.3	1	10.98		1.36		1.09		2		1715
		NGC 5662 - 135		14 32 10	−56 11.3	1	12.07		0.44		-0.05		2		1715
		NGC 5662 - 133		14 32 10	−56 22.7	2	12.48	.023	1.19	.005	0.85	.027	6		746,1715
	−56 06339	NGC 5662 - 432		14 32 10	−56 41.3	1	10.89		0.21		0.02		2		1715
		NGC 5662 - 369		14 32 11	−56 04.9	1	12.61		1.05		0.52		2		1715
		NGC 5662 - 137		14 32 11	−56 21.2	1	13.55		0.66		0.26		2		1715
127866	−56 06340	NGC 5662 - 136		14 32 11	−56 24.0	3	8.29	.022	0.16	.012	-0.24	.004	7	B7 III	251,746,1715
		NGC 5662 - 141		14 32 12	−56 18.2	2	11.78	.061	0.55	.103	0.38	.013	5		251,1715
		NGC 5662 - 142		14 32 12	−56 21.3	1	12.48		0.49		0.35		4		1715
	−56 06341	NGC 5662 - 146		14 32 13	−56 20.3	3	10.83	.024	0.31	.015	0.13	.020	8	A1 V	251,1586,1715
		CS 22874 # 7		14 32 14	−26 39.0	1	15.21		0.14		0.21		1		1736
		NGC 5662 - 366		14 32 14	−56 07.2	1	10.69		1.17		0.86		2		1715
128184	+47 02158			14 32 16	+47 00.3	1	6.51		0.05		0.09		2	A0	3016
		NGC 5662 - 156		14 32 16	−56 24.1	1	12.74		1.04		0.56		1		1715
	−56 06342	NGC 5662 - 155		14 32 16	−56 30.8	2	10.19	.030	1.35	.012	1.21	.003	5		746,1715
127968	−35 09635			14 32 17	−35 55.2	1	9.58		0.09		0.10		1	A2 IV/V	1770
		NGC 5662 - 367		14 32 18	−56 08.4	1	12.40		0.59		0.31		2		1715
127972	−41 08917	HR 5440, η Cen		14 32 19	−41 56.4	10	2.36	.035	-0.22	.017	-0.84	.019	50	B1.5Vne	15,26,681,507,815*
127954	−43 09174			14 32 20	−44 13.9	1	9.46		0.14		0.11		1	A1 V	1770
		NGC 5662 - 368		14 32 20	−56 06.8	1	12.98		0.63		0.17		2		1715
		G 124 - 52		14 32 21	−02 11.6	1	14.41		1.65				3		538
127971	−40 08794	HR 5439	⋆ AB	14 32 21	−41 17.9	3	5.88	.005	-0.08	.003	-0.41		9	B7 V	1770,2007,2014
		NGC 5662 - 169		14 32 21	−56 13.8	1	11.52		1.75		1.82		2		1715
	−56 06344	NGC 5662 - 172		14 32 22	−56 24.9	2	11.36	.027	0.40	.004	0.25	.038	3		251,1715
		NGC 5662 - 431		14 32 22	−56 40.5	1	11.91		0.31		0.19		2		1715
	−55 06066	NGC 5662 - 177		14 32 23	−56 19.6	2	11.15	.022	0.24	.017	0.10	.031	3	B9 V	251,1715
		NGC 5662 - 181		14 32 23	−56 23.2	1	14.28		0.93		0.45		3		746
		NGC 5662 - 179		14 32 24	−56 19.3	1	12.90		0.48		0.24		2		1715
127924	−55 06068	NGC 5662 - 187		14 32 25	−56 14.5	2	9.22	.019	0.11	.000	-0.31	.004	4	B8 IV	251,1715
127900	−55 06067	NGC 5662 - 184		14 32 25	−56 18.0	3	8.82	.021	0.15	.017	-0.28	.012	6	B8 IV	251,746,1715
		NGC 5662 - 186		14 32 25	−56 20.6	2	11.59	.028	0.37	.002	0.30	.032	4		251,1715
		NGC 5662 - 188		14 32 25	−56 22.7	1	14.33		1.12		0.73		3		746
127992	−41 08919			14 32 26	−41 58.1	1	9.12		0.20		0.14		1	A5 V	1771
		NGC 5662 - 190		14 32 26	−56 19.3	1	13.17		0.67		0.70		3		1715
127975	−46 09428	IDS14292S4615	A	14 32 27	−46 28.1	1	8.33		0.17		0.17		1	A2/3 V	1771
127975	−46 09428	IDS14292S4615	AB	14 32 27	−46 28.1	1	8.12		0.17		0.15		1	A2/3 V	1770
		POSS 741 # 7		14 32 28	−12 29.7	1	18.04		1.50				3		1739
		CS 22874 # 54		14 32 28	−26 23.9	1	14.98		0.36		0.05		1		1736
		NGC 5662 - 198		14 32 28	−56 24.5	1	14.71		0.77		0.32		3		746
	+17 02757			14 32 29	+16 38.9	1	9.35		0.47		0.03		2	F5	1648
		NGC 5662 - 430		14 32 29	−56 38.0	1	12.57		1.53		1.51		2		1715
128167	+30 02536	HR 5447	⋆ A	14 32 30	+29 57.7	11	4.46	.004	0.36	.005	-0.08	.002	219	F3 V	15,130,374,1008,1013*
128197	+38 02570			14 32 30	+38 13.8	1	8.10		0.20		0.13		2	Am	272
		NGC 5662 - 202		14 32 30	−56 24.8	2	12.49	.037	0.65	.027	0.27	.001	5		746,1715
		NGC 5662 - 199		14 32 31	−56 15.2	1	13.02		0.56		0.29		2		1715
		NGC 5662 - 429		14 32 31	−56 38.9	1	13.18		0.57		0.31		2		1715
127850	−63 03315			14 32 32	−63 55.7	1	8.05		0.46				4	F5 V	2012
128385	+66 00855			14 32 32	+65 36.8	1	6.61		0.49		-0.01		2	F5	1502
127976	−46 09431			14 32 32	−46 35.9	5	6.78	.008	0.96	.006	0.67	.008	19	G8 III	977,1075,2001,2011,2024
128185	+29 02555			14 32 33	+28 36.8	1	7.89		0.56		0.06		2	F8 V	3026
		NGC 5662 - 206		14 32 33	−56 19.0	1	15.12		0.97		0.31		3		746
		NGC 5662 - 205		14 32 34	−56 24.4	2	12.24	.009	0.41	.001	0.31	.003	6		746,1715
		NGC 5662 - 210		14 32 35	−56 24.4	1	13.64		0.71		0.19		3		1715

HD	DM	Other Id	N	Rem	α₁₉₅₀	δ₁₉₅₀	S	V	σ_V	B–V	σ_B–V	U–B	σ_U–B	n	Spectrum	References
		Ton 207			14 32 36	+25 04.	1	14.72		-0.01		0.11		2		1036
		G 166 - 36			14 32 36	+30 18.9	1	14.78		1.47				1		1759
128198	+37 02551	HR 5448			14 32 36	+36 50.7	2	6.02	.010	1.38	.001	1.59		6	K4 III	70,1501
		POSS 741 # 1			14 32 36	-12 43.0	1	13.86		1.27				1		1739
		CS 22874 # 37			14 32 36	-24 10.9	1	13.42		0.29		0.09		1		1736
		NGC 5662 - 323			14 32 36	-56 32.0	1	13.52		0.83		0.24		3		1715
		NGC 5662 - 324			14 32 36	-56 32.0	1	12.76		0.81		0.50		3		1715
		G 166 - 37			14 32 37	+25 23.3	1	12.66		0.69		0.03		2		1658
		NGC 5662 - 214			14 32 37	-56 27.5	1	11.44		0.38		0.12		3		1715
127926	-60 05449	LSS 3267			14 32 37	-60 50.2	2	9.17	.017	0.37	.014	-0.47	.024	5	B1/2 II	540,976
		G 66 - 8			14 32 38	+04 34.6	2	13.42	.005	0.95	.010	0.65	.019	3		1620,1658
128052	-38 09493				14 32 38	-39 15.7	1	9.63		0.18		0.10		1	A2 III/IV	1770
		CS 22871 # 49			14 32 39	-22 14.8	1	15.36		0.10		0.13		1		1736
128033	-42 09465	Z Lup			14 32 39	-43 09.0	1	8.34		3.13				2	C6,3	864
	-55 06071	NGC 5662 - 222			14 32 39	-56 19.8	2	10.59	.024	0.31	.002	-0.16	.005	3	B7 V	251,1715
		NGC 5662 - 425			14 32 40	-56 44.6	1	12.56		0.59		0.26		2		1715
		NGC 5662 - 322			14 32 41	-56 30.6	1	13.23		1.41		0.98		2		1715
		CS 22871 # 9			14 32 42	-20 14.8	1	11.92		0.01		-0.17		1		1736
	-55 06072	NGC 5662 - 228			14 32 42	-56 17.9	2	11.32	.023	0.57	.003	0.21	.024	3		251,1715
		CS 22874 # 53			14 32 43	-26 18.2	1	13.58		0.02		-0.09		1		1736
		NGC 5662 - 470			14 32 43	-55 58.5	1	12.67		0.93		0.60		2		1715
		NGC 5662 - 428			14 32 43	-56 39.2	1	12.41		1.41		1.07		2		1715
		NGC 5662 - 468			14 32 44	-56 00.3	1	13.02		0.44		0.14		2		1715
		NGC 5662 - 232			14 32 44	-56 14.5	1	12.37		0.38		0.28		2		1715
		NGC 5662 - 231			14 32 44	-56 20.6	1	12.74		0.73		0.18		2		1715
		NGC 5662 - 427			14 32 44	-56 40.5	1	13.09		0.53		0.13		3		1715
128332	+57 01519	HR 5451			14 32 45	+57 17.2	2	6.47	.010	0.50	.002	-0.03		3	F6 V	71,3026
		NGC 5662 - 370			14 32 45	-56 05.3	1	11.76		0.26		0.15		2		1715
		NGC 5662 - 371			14 32 47	-56 04.1	1	12.66		0.59		0.28		2		1715
	-56 06346	NGC 5662 - 239			14 32 47	-56 21.9	3	11.08	.017	0.30	.011	0.12	.016	6	B9 V	251,746,1715
		NGC 5662 - 469			14 32 48	-56 00.1	1	12.96		0.93		0.29		1		1715
		NGC 5662 - 247			14 32 48	-56 14.8	1	12.16		0.64		0.43		2		1715
		NGC 5662 - 319			14 32 48	-56 29.4	1	13.92		0.70		0.20		2		1715
		G 66 - 9			14 32 49	+12 26.6	2	12.00	.024	0.50	.015	-0.27	.010	3		333,1620,1696
		NGC 5662 - 320			14 32 50	-56 28.8	1	13.66		0.75		0.27		2		1715
		CS 22871 # 6			14 32 51	-20 34.5	1	13.26		0.36		0.17		1		1736
		NGC 5662 - 201			14 32 51	-56 16.0	1	11.86		0.49		0.31		2		1715
		NGC 5662 - 318			14 32 51	-56 29.5	1	12.32		0.79		0.33		2		1715
		G 135 - 67			14 32 52	+17 07.2	2	15.03	.019	1.65	.005	1.01	.009	4		316,3078
	-55 06073	NGC 5662 - 249			14 32 52	-56 19.5	2	11.04	.022	0.24	.001	0.06	.016	3	A0 V	251,1715
		G 66 - 10			14 32 53	+10 12.7	1	13.38		1.42				1		1759
		NGC 5662 - 372			14 32 53	-56 07.4	1	12.34		1.36		0.98		2		1715
		NGC 5662 - 426			14 32 53	-56 40.3	1	13.67		0.83		0.32		2		1715
		Ton 208			14 32 54	+29 01.	1	15.98		-0.09		-1.17		1		1036
	+34 02541	G 166 - 38			14 32 55	+33 57.7	2	9.57	.019	1.29	.030	1.17		5	K8	1017,3072
128333	+50 02095	HR 5452, CH Boo			14 32 55	+49 35.1	2	5.74	.000	1.56	.000	1.88		4	M1 IIIab	70,3055
		CS 22874 # 45			14 32 55	-24 41.1	1	15.50		-0.19		-0.90		1		1736
		NGC 5662 - 373			14 32 56	-56 09.8	1	12.34		1.25		0.88		2		1715
128220	+19 02824				14 32 57	+19 26.0	4	8.51	.017	0.23	.013	-0.79	.025	7	sdO+G	963,1026,1418,8088
		NGC 5662 - 471			14 32 57	-55 56.3	1	12.07		0.29		-0.03		2		1715
		GD 167			14 32 58	+14 41.7	1	14.19		0.53		-0.03		1		3060
	-64 02936				14 32 58	-64 36.4	1	11.27		0.70		0.15		4	B8 II-Ib	540
		Ton 209			14 33 00	+24 00.0	1	12.54		-0.20		-1.01		2		1036
		NGC 5662 - 261			14 33 01	-56 14.3	1	12.71		0.60		0.35		4		1715
128068	-45 09293	HR 5444		★ A	14 33 02	-46 01.7	10	5.54	.005	1.49	.010	1.72	.008	52	K3 III	15,116,278,863,977*
128386	+56 01749				14 33 03	+55 46.2	1	8.11		0.99		0.70		2	G9 III	1502
		NGC 5662 - 316			14 33 03	-56 31.2	1	12.39		0.68		0.18		2		1715
128120	-34 09774				14 33 04	-34 41.3	1	7.58		1.67		1.96		7	M1/2 III	1673
		NGC 5662 - 317			14 33 04	-56 32.4	1	13.19		0.40		0.22		2		1715
128188	-10 03931				14 33 05	-11 11.0	2	10.00	.029	0.99	.019	0.50		3	K0	1594,6006
		CS 22874 # 35			14 33 07	-23 32.9	1	14.33		0.20		0.20		1		1736
		NGC 5662 - 264			14 33 07	-56 24.4	1	12.65		0.45		0.34		2		1715
	-56 06347	NGC 5662 - 423			14 33 07	-56 39.3	1	9.95		0.56		0.00		2		1715
128254	+19 02827	IDS14308N1942		AB	14 33 08	+19 28.4	1	8.43		1.06		0.89		5	K0	8088
	-64 02939	LSS 3268			14 33 08	-64 35.0	3	10.90	.014	0.58	.012	-0.36	.005	9		540,1586,1737
128132	-35 09645				14 33 10	-35 41.9	1	8.93		0.03		0.02		3	A0 V	1770
		NGC 5662 - 269			14 33 10	-56 23.4	1	11.30		0.25		0.09		2		1715
		CS 22874 # 57			14 33 12	-27 00.1	1	15.03		-0.19		-0.94		1		1736
	-56 06349	NGC 5662 - 314			14 33 12	-56 30.3	1	11.10		0.46		-0.08		2		1715
		NGC 5662 - 424			14 33 12	-56 41.3	1	12.34		0.44		-0.09		2		1715
128233	+0 03206	IDS14307N0041		AB	14 33 14	+00 27.6	1	8.61		0.58		0.11		3	F8	1371
		NGC 5662 - 321			14 33 14	-56 34.6	1	13.18		0.74		0.23		2		1715
128152	-39 09047	HR 5446			14 33 16	-39 22.8	1	6.13		1.05				4	K1 III	2007
		NGC 5662 - 273			14 33 16	-56 22.5	1	11.75		1.01		0.58		2		1715
		NGC 5662 - 374			14 33 17	-56 07.1	1	11.60		1.57		1.48		2		1715
	-55 06075	NGC 5662 - 275			14 33 17	-56 13.6	1	10.69		0.43		0.36		3		1715
	-56 06351	NGC 5662 - 300			14 33 17	-56 25.9	1	10.57		0.68		0.23		2		1715
		NGC 5662 - 315			14 33 18	-56 31.1	1	12.32		1.54		1.36		2		1715
		NGC 5662 - 313			14 33 20	-56 30.4	1	13.20		0.39		0.24		2		1715
128205	-29 11183				14 33 21	-29 58.8	1	9.47		0.16		0.13		1	A1 V	1770

Table 1

HD	DM	Other Id	N	Rem	α_{1950}	δ_{1950}	S	V	σ_V	B–V	σ_{B-V}	U–B	σ_{U-B}	n	Spectrum	References
		NGC 5662 - 473			14 33 21	−56 01.1	1	12.50		0.79		0.31		2		1715
		NGC 5662 - 422			14 33 21	−56 38.7	1	12.19		1.69		1.53		2		1715
128204	−24 11586				14 33 22	−24 54.3	2	10.00	.014	0.47	.014	0.01		4	F3 V	1594,6006
		CS 22874 # 55			14 33 22	−26 34.4	1	15.25		0.34		0.00		1		1736
		NGC 5662 - 281			14 33 22	−56 18.0	1	12.69		0.93		0.43		2		1715
127943	−69 02071				14 33 24	−69 47.9	1	10.01		0.48		0.17		1	A9/F0 [V]	1771
		NGC 5662 - 284			14 33 25	−56 19.8	1	12.38		0.75		0.22		2		1715
128272	+0 03207	IDS14307N0041			14 33 26	+00 26.2	2	7.71	.015	0.98	.001	0.76	.006	8	G8 III	861,1371
		NGC 5662 - 285			14 33 26	−56 20.1	1	11.19		0.38		0.24		2		1715
128368	+35 02581				14 33 27	+34 54.2	1	8.67		1.24		1.28		2	K0	1601
		POSS 741 # 5			14 33 27	−12 38.8	1	16.75		1.66				2		1739
		G 239 - 21			14 33 29	+67 21.4	1	14.06		1.53				1		1759
		NGC 5662 - 472			14 33 29	−55 57.5	1	12.59		0.44		0.33		2		1715
		NGC 5662 - 474			14 33 29	−56 01.9	1	12.58		0.70		0.33		2		1715
		NGC 5662 - 312			14 33 29	−56 30.4	1	13.19		0.45		0.20		2		1715
		NGC 5662 - 421			14 33 29	−56 38.1	1	13.26		1.05		0.57		1		1715
128089	−57 06712	LSS 3269			14 33 29	−58 13.1	2	8.62	.048	0.40	.001	−0.51	.033	4	B0.5/1 III	540,976
		NGC 5662 - 290			14 33 30	−56 18.7	1	12.44		0.53		0.36		2		1715
		NGC 5662 - 289			14 33 30	−56 19.8	1	13.38		0.79		0.15		2		1715
		NGC 5662 - 419			14 33 31	−56 42.1	1	11.75		1.00		0.52		2		1715
		NGC 5662 - 310			14 33 32	−56 24.3	1	12.34		0.70		0.32		2		1715
128020	−67 02616	HR 5443			14 33 32	−67 42.7	6	6.03	.006	0.50	.000	−0.03	.008	28	F7 V	15,1075,2012,2020*
127962	−69 02072				14 33 32	−70 12.3	1	9.90		0.35		0.21		1	A7/8 V	1771
128311	+10 02710	G 66 - 11			14 33 33	+09 58.0	1	7.49		0.98		0.78		1	K0	333,1620
128206	−34 09778				14 33 34	−35 09.5	1	9.88		0.34		0.13		1	A9 V	1771
		NGC 5662 - 418			14 33 34	−56 42.4	1	13.43		0.61		0.42		2		1715
128207	−39 09050	HR 5449			14 33 35	−39 59.7	2	5.74	.005	−0.12	.004	−0.44		6	B8 V	1770,2006
128238	−28 10823				14 33 36	−28 38.8	1	9.39		0.25		0.19		1	A1/2(m)	1770
	+60 01552				14 33 38	+59 54.5	1	9.87		0.53		0.02		3	F8	1502
128239	−30 11560				14 33 39	−30 49.1	1	8.59		0.02		0.00		2	B9.5/A0 II	1770
		NGC 5662 - 420			14 33 39	−56 40.5	1	12.16		0.95		0.51		2		1715
	+24 02743				14 33 40	+24 26.2	1	8.95		0.19		0.10		2	A0	272
128224	−36 09491	IDS14306S3614		A	14 33 41	−36 26.9	1	7.68		−0.08		−0.46		2	B8(p Si)	1770
		NGC 5662 - 309			14 33 42	−56 26.3	1	13.36		0.86		0.47		2		1715
		NGC 5662 - 414			14 33 43	−56 37.0	1	13.83		0.57		0.38		2		1715
		NGC 5662 - 415			14 33 43	−56 37.1	1	13.06		1.39		0.86		2		1715
		NGC 5662 - 475			14 33 44	−56 04.3	1	9.97		1.94		2.12		3		1715
	−55 06078	NGC 5662 - 377			14 33 44	−56 12.8	1	10.98		0.22		0.05		3		1715
128210	−45 09297				14 33 45	−45 39.6	1	9.16		0.26		0.14		1	A2mA3-A7	1770
		NGC 5662 - 378			14 33 45	−56 13.6	1	11.66		1.51		1.38		3		1715
		NGC 5662 - 376			14 33 46	−56 12.0	1	13.37		0.58		0.40		2		1715
		CS 22874 # 46			14 33 47	−25 08.2	1	14.63		0.06		0.08		1		1736
		Feige 97			14 33 48	+30 20.	2	12.38	.073	0.03	.024	0.15	.010	3		1026,1298
128263	−34 09781				14 33 48	−34 34.6	1	8.03		0.08		0.11		1	A1 V	1770
		NGC 5662 - 311			14 33 48	−56 27.9	1	11.85		1.37		1.25		2		1715
		LP 800 - 71			14 33 49	−18 06.6	1	10.69		0.60		−0.02		2		1696
		NGC 5662 - 412			14 33 49	−56 35.9	1	12.87		1.44		1.27		2		1715
128211	−46 09449				14 33 50	−46 47.9	4	8.78	.005	0.12	.006	0.10	.009	18	A1 V	977,1075,1770,2011
128402	+23 02710	HR 5454			14 33 51	+23 28.0	2	6.37	.005	1.05	.005	0.90	.000	5	K0	985,1733
128279	−28 10826				14 33 51	−28 53.5	3	8.01	.024	0.64	.003	−0.01	.029	6	Gw	462,742,1594
		NGC 5662 - 413			14 33 52	−56 36.7	1	12.79		0.60		0.24		3		1715
		NGC 5662 - 417			14 33 52	−56 41.3	1	12.78		0.53		0.37		2		1715
128137	−59 05660	IDS14301S5955		AB	14 33 52	−60 08.4	2	9.28	.278	0.30	.292	0.04		6	B8 IV/V	1696,2014
		NGC 5662 - 375			14 33 54	−56 09.2	1	13.67		0.56		0.38		2		1715
		NGC 5662 - 305			14 33 54	−56 17.1	1	13.55		0.88		0.17		1		1715
	−56 06357	NGC 5662 - 411			14 33 54	−56 35.7	1	11.35		0.29		0.09		2		1715
	+20 02992	IDS14316N1951		AB	14 33 56	+19 37.4	1	9.80		0.65		0.16		2	G0	8088
128316	−22 03817				14 33 56	−22 56.9	1	10.63		0.36		0.25		2	K0 IV	837
		NGC 5662 - 416			14 33 56	−56 39.2	1	12.57		1.37		1.15		2		1715
		NGC 5662 - 308			14 33 59	−56 25.6	1	12.32		0.58		0.28		2		1715
		NGC 5662 - 410			14 34 00	−56 35.8	1	13.71		0.58		0.30		2		1715
128266	−45 09302	HR 5450		* AB	14 34 03	−45 55.0	6	5.40	.013	1.00	.018	0.74	.050	17	G8 III+A1 V	15,278,1075,1279,2001,2011
		NGC 5662 - 306			14 34 03	−56 17.6	1	12.25		0.71		0.29		2		1715
128284	−36 09499	IDS14310S3706		AB	14 34 04	−37 18.9	1	8.05		0.04		−0.06		1	A1 V + A	1771
128267	−45 09303	IDS14308S4542		C	14 34 04	−45 54.7	2	8.49	.017	0.01	.005	−0.09	.005	4		1279,1770
		NGC 5662 - 304			14 34 04	−56 20.2	1	12.10		0.55		0.07		2		1715
128214	−56 06358				14 34 05	−57 14.5	1	8.56		0.72		0.36		4	G5 IV	1101
		CS 22874 # 56			14 34 08	−26 43.6	1	14.30		0.07		0.09		1		1736
		CS 22871 # 39			14 34 09	−20 27.1	1	15.49		0.11		0.19		1		1736
128357	−27 09940				14 34 10	−28 01.5	1	9.39		0.38		0.17		1	A0mA9-F2	1770
128356	−25 10441				14 34 11	−25 35.0	1	10.31		0.69		0.21		1	K3 V	3072
		NGC 5662 - 513			14 34 12	−55 56.2	1	12.21		0.61		0.13		2		1715
		NGC 5662 - 476			14 34 13	−56 03.3	1	12.22		0.33		0.12		2		1715
234162	+51 01936				14 34 14	+51 06.5	1	9.09		1.01		0.75		2	K2	1566
128319	−36 09502				14 34 14	−36 29.9	1	8.80		0.14		0.15		1	A2 V	1770
		NGC 5662 - 380			14 34 14	−56 11.4	1	12.67		1.41		1.22		2		1715
		Steph 1170			14 34 15	+14 34.8	1	11.82		1.47		1.22		1	K7-M0	1746
		NGC 5662 - 409			14 34 16	−56 37.8	1	13.76		0.73		0.19		2		1715
		CS 22871 # 33			14 34 17	−18 59.9	1	13.70		0.14		0.19		1		1736
	−55 06984	NGC 5662 - 382			14 34 17	−56 15.1	1	10.89		0.48		0.34		2		1715

HD	DM	Other Id	N Rem	α_{1950}	δ_{1950}	S	V	σ_V	B–V	σ_{B-V}	U–B	σ_{U-B}	n	Spectrum	References
	−56 06359			14 34 17	−57 13.4	1	9.94		1.39		1.41		4		1101
128428	−3 03648	G 124 - 54		14 34 18	−04 03.8	3	7.78	.013	0.76	.013	0.37		8	G0	2017,2034,3077
		CS 22874 # 33		14 34 18	−22 43.3	1	13.95		0.35		0.03		1		1736
		CS 22874 # 47		14 34 18	−25 14.3	1	13.63		0.10		0.19		1		1736
128321	−43 09204			14 34 18	−43 27.3	1	9.90		0.14		0.12		1	A1 V	1770
128429	−11 03770	HR 5455		14 34 20	−12 05.6	9	6.20	.019	0.46	.007	-0.03	.012	37	F5 V	15,22,253,254,742*
		CS 22874 # 44		14 34 20	−24 43.3	1	14.71		0.13		0.21		1		1736
128360	−36 09503			14 34 20	−36 30.3	1	7.98		0.15		0.08		1	A1 V	1770
	−55 06085	NGC 5662 - 477		14 34 21	−56 02.0	1	10.98		0.21		0.11		2		1715
		NGC 5662 - 379		14 34 21	−56 10.4	1	13.72		0.50		0.40		4		1715
		NGC 5662 - 408		14 34 21	−56 36.8	1	12.16		0.74		0.29		2		1715
		NGC 5662 - 502		14 34 21	−56 42.5	1	11.53		0.45		0.00		2		1715
128227	−59 05665			14 34 21	−59 55.6	2	8.32	.006	1.06	.003	0.81	.001	31	K0 III	1621,1642
		NGC 5662 - 303		14 34 22	−56 20.9	1	11.92		0.43		-0.08		3		1715
		NGC 5662 - 307		14 34 22	−56 25.8	1	13.08		0.67		0.34		2		1715
		GD 168		14 34 24	+15 54.6	1	16.58		0.00		-0.74		1		3060
		CS 22874 # 52		14 34 24	−26 14.8	1	15.13		0.13		0.24		1		1736
		NGC 5662 - 302		14 34 24	−56 19.3	1	13.41		0.56		0.17		3		1715
		NGC 5662 - 479		14 34 29	−56 06.3	1	12.20		0.39		-0.14		2		1715
		NGC 5662 - 407		14 34 29	−56 39.1	1	11.36		0.48		0.18		2		1715
128391	−33 09939			14 34 30	−34 03.4	1	9.28		0.25		0.14		1	A1/2 II/II	1770
		KS 907		14 34 30	−68 48.	1	10.98		1.19		0.80		3		540
128345	−48 09198	HR 5453		14 34 31	−49 12.5	5	4.04	.006	-0.15	.001	-0.56	.008	15	B5 V	15,1075,1770,2012,8015
		NGC 5662 - 381		14 34 31	−56 14.0	1	12.00		0.76		0.24		2		1715
		NGC 5662 - 478		14 34 32	−56 02.6	1	12.44		0.37		0.22		2		1715
		NGC 5662 - 402		14 34 32	−56 26.2	1	13.02		0.74		0.32		2		1715
		CS 22871 # 23		14 34 33	−17 29.4	1	13.08		-0.18		-0.87		1		1736
	−56 06361	NGC 5662 - 403		14 34 33	−56 26.4	1	10.93		0.59		0.39		3		1715
128392	−35 09667			14 34 34	−35 30.9	1	9.87		0.41		0.12		1	A9 V	1771
		NGC 5662 - 501		14 34 34	−56 41.8	1	12.52		0.51		0.11		2		1715
128344	−47 09279			14 34 35	−47 48.4	4	6.65	.010	-0.02	.002	-0.26	.002	32	B8 V	278,1075,1770,2011
	−55 06087	NGC 5662 - 301		14 34 35	−56 18.6	1	11.65		0.28		0.18		3		1715
		G 201 - 5		14 34 36	+55 46.2	2	11.51	.015	0.41	.000	-0.24	.010	5	sdF6	308,1658
		CS 22874 # 43		14 34 36	−24 48.6	1	14.28		0.11		0.07		1		1736
		NGC 5662 - 401		14 34 38	−56 12.9	1	11.70		0.34		0.25		2		1715
		NGC 5662 - 405		14 34 41	−56 28.3	1	12.48		0.38		0.33		2		1715
	−56 06362	NGC 5662 - 406		14 34 43	−56 36.5	1	10.89		0.48		0.34		2		1715
		NGC 5662 - 514		14 34 44	−56 01.1	1	11.61		0.40		0.21		2		1715
		NGC 5662 - 481		14 34 46	−56 10.6	1	12.01		1.53		1.17		2		1715
		NGC 5662 - 483		14 34 46	−56 12.5	1	11.39		1.30		1.09		2		1715
128393	−46 09457			14 34 47	−46 39.8	1	9.61		0.41		0.06		1	A9 V	1771
	+61 01443			14 34 48	+61 26.8	1	9.20		0.98		0.84		2	K0	1733
		NGC 5662 - 482		14 34 49	−56 11.3	1	13.52		0.70		0.14		1		1715
		CS 22874 # 51		14 34 50	−26 22.4	1	14.46		0.12		0.19		1		1736
		LP 500 - 69		14 34 51	+09 39.3	1	15.68		1.81				1		1759
	−60 05471	LSS 3272		14 34 52	−60 51.1	1	10.19		0.37		-0.33		2		1586
		CS 22874 # 58		14 34 54	−27 02.4	1	13.65		0.06		-0.01		1		1736
128218	−69 02074			14 34 55	−70 04.5	1	8.52		0.32		0.31		1	A3 V	1771
128413	−45 09312			14 34 56	−45 39.3	13	6.82	.031	1.17	.005	1.21	.003	196	K2 III	116,278,460,474,768*
129245	+80 00448	HR 5479		14 34 57	+79 52.6	3	6.26	.000	1.30	.000	1.46	.000	5	K3 III	15,1003,3002
128580	+11 02700	G 66 - 15	⋆ A	14 34 58	+11 14.6	2	9.58	.025	0.66	.010	0.13	.000	2		333,1620
		G 66 - 16		14 34 58	+11 15.1	3	10.75	.012	0.83	.008	0.54	.009	3		1620,1658,1773
		Steph 1172		14 34 58	+55 56.4	1	11.51		1.40		1.24		1	K7	1746
128549	+1 02957			14 34 59	+00 39.6	1	9.04		0.90		0.46		2	G5	1371
128609	+27 02400	R Boo		14 34 59	+26 57.2	2	8.18	.635	1.41	.025	0.71	.120	2	M4	635,3001
128512	−17 04128			14 35 00	−17 35.8	1	9.83		1.14		1.03		26	K0 III/IV	588
	−55 06089	NGC 5662 - 480		14 35 02	−56 08.1	1	10.91		0.33		0.27		1		1715
128381	−56 06363			14 35 04	−57 11.4	1	8.42		1.87		2.05		4	M2 III	1101
128661	+36 02509	BW Boo		14 35 06	+36 08.8	1	7.14		0.12		0.04		4	F0 V	1501
		NGC 5662 - 404		14 35 07	−56 27.6	1	11.10		0.65		0.14		2		1715
		CS 22874 # 42		14 35 08	−24 45.8	1	13.91		0.43		-0.16		1		1736
128718	+48 02222	IDS14334N4839	AB	14 35 10	+48 26.2	2	7.74	.015	0.46	.011	-0.07		4	F2	938,1381
		CS 22874 # 59		14 35 12	−27 05.7	1	15.71		0.08		-0.02		1		1736
128488	−38 09529	HR 5456		14 35 12	−38 34.7	1	6.02		1.43				4	K3 III	2007
128293	−67 02622	LSS 3271	⋆ AB	14 35 13	−67 59.3	2	6.73	.030	-0.03	.022	-0.88		7	B2 Vne	26,2012
		CS 22871 # 32		14 35 18	−18 58.4	1	14.17		0.06		0.02		1		1736
		He3 1016		14 35 18	−61 11.	1	12.11		0.47		-0.32		3		1586
128532	−34 09807	IDS14323S3450	A	14 35 21	−35 03.6	1	6.80		0.16		0.13		3	A3 V	1771
		CS 22871 # 45		14 35 22	−21 32.9	1	14.46		0.08		0.04		1		1736
128517	−42 09509			14 35 22	−43 09.1	1	7.71		0.12		0.11		1	A2 V	1770
		G 223 - 74		14 35 25	+58 34.2	1	11.78		1.52				1		1759
		LP 680 - 44		14 35 25	−08 53.1	1	11.48		0.74		0.20		2		1696
128596	−12 04104			14 35 25	−12 41.6	1	7.48		0.65				19	G2 V	1351
128476	−55 06092			14 35 27	−56 16.9	3	7.44	.011	1.05	.000	0.74	.007	10	G8 III	657,1075,4001
128597	−12 04105			14 35 28	−12 51.2	1	8.78		0.72				4	G6 V	2012
		G 66 - 17		14 35 31	+05 30.0	1	14.26		1.53		1.06		1		3016
		CS 22874 # 34		14 35 35	−23 01.8	1	14.49		0.25		0.10		1		1736
		CS 22874 # 50		14 35 36	−26 08.9	1	14.73		0.16		0.25		1		1736
128568	−37 09580			14 35 37	−37 36.5	1	8.76		0.21		0.14		1	A1/2mA7-A8	1770
		G 66 - 18		14 35 39	−00 37.8	1	13.10		0.86		0.48		2		1620

Table 1

809

HD	DM	Other Id	N	Rem	α_{1950}	δ_{1950}	S	V	σ_V	B–V	σ_{B-V}	U–B	σ_{U-B}	n	Spectrum	References
128456	−63 03349				14 35 40	−63 19.8	1	7.63		1.38		1.49		6	K2/3 III	1704
		Steph 1171		A	14 35 42	+18 01.4	1	12.41		1.33		1.22		1	K7-M0	1746
		Steph 1171		B	14 35 42	+18 01.4	1	14.16		1.53				2	K7-M0	1746
		CS 22874 # 39			14 35 43	−24 14.6	1	14.89		0.15		0.21		1		1736
		NGC 5694 sq1 8			14 35 44	−26 15.6	1	11.58		0.60				2		438
		CS 22874 # 41			14 35 46	−24 42.2	1	14.80		0.12		0.17		1		1736
		LP 680 - 45			14 35 49	−07 25.6	1	12.02		0.70		0.08		2		1696
128615	−38 09539				14 35 50	−38 36.0	1	9.11		0.06		0.00		2	A0 V	1770
128582	−46 09469	HR 5457			14 35 53	−46 22.0	10	6.06	.004	0.51	.004	0.03	.014	51	F6 V	14,15,116,278,977,1075*
128750	+18 02906	HR 5462			14 35 54	+18 30.9	2	5.90	.005	1.10	.000	0.99		7	gK2	1080,1118
		G 124 - 57			14 35 55	−03 02.1	1	13.84		1.49				1		1759
		NGC 5694 sq1 13			14 35 57	−26 14.6	1	12.25		1.12				2		438
		WLS 1424 55 # 8			14 35 59	+57 27.9	1	12.02		0.74		0.41		2		1375
		CS 22871 # 31			14 35 60	−18 41.1	1	14.50		0.15		0.22		1		1736
	−26 10405				14 36 03	−26 30.9	1	10.23		0.40				3	F0	438
128617	−48 09218	HR 5458			14 36 03	−48 50.3	3	6.39	.009	0.44	.002	0.01		9	F5 IV/V	15,2012,3077
128585	−50 08713				14 36 03	−50 56.6	1	9.27		0.06		-0.49		2	B3 IV	400
128799	+20 03000				14 36 09	+20 04.3	1	8.08		0.48		-0.04		2	G0	1733
		CS 22871 # 28			14 36 10	−18 13.6	1	13.44		0.35		0.14		1		1736
128649	−42 09520				14 36 10	−42 32.9	1	9.55		0.05		-0.08		2	Ap CrSi	1770
128620	−60 05483	HR 5459		⋆ A	14 36 11	−60 37.8	4	0.00	.005	0.65	.027	0.23		8	G2 V	507,621,1705,3078
128620	−60 05483	IDS14328S6025		⋆ AB	14 36 11	−60 37.8	1	-0.29		0.72				1	G2 V +K1 V	507
128621	−60 05483	HR 5460		⋆ B	14 36 11	−60 37.8	3	1.35	.010	0.87	.026	0.63		7	K1 V	621,507,3078
		Feige 98			14 36 12	+27 42.9	1	11.76		-0.02		-0.05		1		1298
128522	−64 02961	LSS 3273			14 36 14	−64 37.1	3	8.95	.020	0.25	.025	-0.61	.027	6	B1 Ia/ab	403,540,976
128752	−9 03975				14 36 19	−10 20.4	1	6.73		1.00				11	G5	1351
128588	−59 05679				14 36 19	−59 43.0	2	9.07	.005	0.18	.003	-0.38	.004	10	B3 Vne	567,1586
128902	+44 02376	HR 5464			14 36 20	+43 51.4	4	5.71	.011	1.48	.000	1.66	.035	8	K2 III	15,252,1003,1080
	−58 05654				14 36 20	−58 34.5	1	10.45		0.26		-0.39		2		1586
		L 198 - 35			14 36 25	−56 41.2	1	11.33		1.42		1.16		3		3060
	+0 03214				14 36 26	−00 07.0	2	9.66	.001	0.58	.006	0.03		10	G2 V	281,6004
		GD 498			14 36 27	+68 16.8	1	13.79		0.34		-0.14		2		308
		CS 22874 # 78			14 36 27	−23 07.7	1	14.97		0.08		0.05		1		1736
128636	−52 07393	IDS14330S5234		AB	14 36 27	−52 47.2	2	8.30	.010	0.48	.000	-0.08		3	G0/2 +F	742,1594
		CS 22871 # 40			14 36 30	−20 37.6	1	12.72		0.26		0.08		1		1736
128736	−33 09968				14 36 30	−34 02.7	1	10.39		0.33		0.20		1	A1mA8-F3	1770
		CS 22871 # 34			14 36 32	−19 03.3	1	15.47		0.24		0.19		1		1736
		NGC 5694 sq1 16			14 36 34	−26 22.4	1	12.96		1.06				1		438
		CS 22874 # 79			14 36 35	−22 57.1	1	14.76		0.16		0.18		1		1736
		NGC 5694 sq1 7			14 36 37	−26 20.6	1	10.50		1.46				3		438
		CS 22871 # 78			14 36 39	−17 44.8	1	14.75		0.22		0.24		1		1736
		CS 22874 # 69			14 36 39	−25 12.5	1	14.88		0.14		0.26		1		1736
128998	+54 01693	HR 5467			14 36 40	+54 14.3	2	5.84	.010	-0.01	.005	0.00		5	A1 V	985,1733
	−25 10470				14 36 41	−26 18.9	1	10.31		1.48				3		438
		NGC 5694 sq1 12			14 36 41	−26 26.8	1	12.08		0.53				2		438
128757	−38 09551	IDS14336S3807		AB	14 36 42	−38 20.2	1	8.61		0.08		0.02		1	A1 IV/V	1770
	−62 04233	LSS 3274			14 36 43	−62 19.6	1	10.76		0.38		-0.37		2	B3 II	34,540
128726	−45 09332				14 36 44	−45 31.7	12	8.06	.004	0.02	.004	-0.07	.010	182	A0 V	116,125,278,460,768,863*
	−26 10413				14 36 45	−26 27.1	1	10.04		0.51				5		438
128787	−26 10415	IDS14340S2616		A	14 36 46	−26 30.5	1	6.99		0.45				3	F5 V	438
128806	−28 10858				14 36 47	−28 24.2	1	8.96		0.38		0.17		1	A1mA7-F2	1770
128674	−56 06368	IDS14331S5635		A	14 36 47	−56 48.6	4	7.38	.019	0.67	.008	0.15		13	G5/6 V	1020,1075,2012,3026
128674	−56 06368	IDS14331S5635		B	14 36 47	−56 48.6	1	11.30		1.42		1.17		4		3008
		NGC 5694 sq1 11			14 36 50	−26 20.5	1	12.06		1.21				3		438
	−26 10417				14 36 52	−26 29.0	2	9.79	.041	1.08	.002	0.87		5	K5 V	438,3072
	+22 02725				14 36 54	+21 41.0	1	10.12		1.56		1.94		2	K5	1746
128788	−40 08857				14 36 54	−40 41.1	1	8.28		0.21		0.11		1	A5 V	1771
	−45 09334				14 36 54	−45 24.9	2	9.94	.004	0.51	.000	-0.05		10	G5	863,2011
		BPM 9040			14 36 54	−62 45.	1	14.76		0.56		-0.19		2		3065
128866	+0 03216				14 36 55	−00 02.0	5	9.09	.005	0.71	.007	0.30	.006	39	G5 IV	281,989,1728,1729,6004
128713	−55 06107	HR 5461			14 36 55	−56 13.6	4	6.29	.013	1.17	.005			18	K0/1 II	15,1075,1075,2035
		NGC 5694 sq1 17			14 36 56	−26 25.6	1	14.76		0.85				1		438
128760	−45 09336				14 36 56	−45 19.9	5	8.12	.005	0.57	.004	0.08	.008	22	F8/G0 V	861,977,1075,1770,2011
		G 135 - 79			14 36 58	+20 22.1	1	12.51		0.96		0.72		1		1658
129002	+45 02204	HR 5468			14 36 59	+44 37.2	2	5.35	.040	0.00	.007	-0.04		4	A1 V	70,3023
	−62 04234	LSS 3275			14 36 59	−62 43.1	4	10.47	.048	0.32	.027	-0.46	.027	7	B2	34,540,976,1586
	−0 02851				14 37 02	−00 58.9	1	10.19		0.76		0.30		2		1696
128775	−45 09337	IDS14338S4522		A	14 37 02	−45 34.8	4	6.62	.014	-0.12	.008	-0.37	.008	16	Ap Si	278,1075,1770,2011
	+0 03217				14 37 03	+00 05.5	1	11.04		0.57		0.02		3	G2 V	281
	+0 03218				14 37 05	+00 25.6	1	10.51		0.90		0.67		2	K0 III	281
128967	+27 02404	IDS14349N2714		AB	14 37 07	+27 01.3	1	8.53		0.39		0.10		2	F2	1733
128819	−40 08860	IDS14340S4025		AB	14 37 07	−40 37.6	2	6.66	.004	-0.07	.000	-0.27		6	B8/9 V	1770,2014
129004	+33 02482	RV Boo			14 37 09	+32 45.3	1	8.06		1.47		0.68		1	M2	3001
		CS 22871 # 42			14 37 10	−21 07.3	1	12.94		0.16		0.22		1		1736
		NGC 5694 sq1 14			14 37 10	−26 19.8	1	12.34		1.30				2		438
	+20 03004				14 37 13	+19 38.6	3	10.12	.050	-0.13	.008	-0.52	.007	6	B6 IV	1026,1264,1298
	−62 04237				14 37 18	−62 31.4	1	10.81		0.56		0.15		1		287
128855	−40 08864				14 37 20	−40 23.6	1	7.34		0.05		0.09		1	A1/2 IV	1770
	−62 04239				14 37 23	−62 37.4	1	10.71		0.60		-0.02		1		287
		LundsII141 # 134			14 37 24	−62 00.	1	11.77		0.42		-0.45		2		34

HD	DM	Other Id	N Rem	α₁₉₅₀	δ₁₉₅₀	S	V	σ_V	B–V	σ_B V	U–B	σ_U–B	n	Spectrum	References
	−74 01221			14 37 24	−74 39.0	1	10.17		0.04		−0.31		3		400
128856	42 09541			14 37 25	−43 09.4	1	8.57		0.03		−0.01		2	A0 V	1770
	+0 03220			14 37 26	+00 23.3	1	10.13		0.90		0.58		2	G8 III	281
	+19 02836			14 37 26	+19 20.5	2	11.53	.035	0.19	.030	0.12	.030	2		1026,1298
		CS 22871 # 69		14 37 28	−18 34.1	1	15.86		0.02		−0.30		1		1736
128931	−28 10870			14 37 28	−29 05.2	2	7.81	.000	0.46	.000			8	F6 V	1075,2033
		LP 740 - 41		14 37 31	−11 58.2	1	12.29		0.63		−0.07		2		1696
		NGC 5694 sq1 10		14 37 31	−26 17.6	1	11.62		0.73				4		438
128908	−33 09984			14 37 32	−33 26.3	1	9.73		0.33		0.15		1	A9 V	1771
		HA 106 # 1024		14 37 33	+00 14.6	2	11.59	.014	0.33	.008	0.08	.016	29	A6	281,1764
	−62 04243			14 37 33	−62 47.7	1	11.15		0.44		0.08		2	B7	34
		CS 22874 # 70		14 37 34	−24 39.4	1	15.15		0.22		0.22		1		1736
		CS 22874 # 67		14 37 38	−25 44.1	1	15.60		0.07		0.03		1		1736
128947	−26 10425			14 37 38	−26 25.2	1	9.92		0.48				4	F3 V	438
		CS 22874 # 68		14 37 44	−25 25.4	1	14.86		0.35		0.06		1		1736
128840	−57 06739			14 37 45	−57 28.6	1	9.10		0.07		−0.30		2	Ap Si	567
129061	+14 02767			14 37 48	+14 04.3	1	9.19		0.88		0.49		1	G5	1748
129207	+54 01695			14 37 48	+53 39.3	1	7.12		0.97		0.74		2	G5	1566
128986	−13 03944			14 37 48	−13 50.0	2	6.98	.001	1.63	.001			31	K4/5 III	130,1351
128959	−26 10427			14 37 48	−26 31.0	3	9.22	.023	0.60	.005	0.04	.000	7	F7 V	742,1594,6006
		HA 106 # 863		14 37 51	−00 00.4	2	10.60	.003	0.54	.001	0.03		15	F7	281,6004
		CS 22874 # 61		14 37 55	−26 17.1	1	14.66		0.06		0.04		1		1736
		CS 22874 # 60		14 37 55	−26 42.0	1	14.08		0.43		0.14		1		1736
129333	+64 01017			14 37 56	+64 30.4	1	7.54		0.61		0.03		3	F8	3016
128859	−57 06742			14 37 56	−57 55.2	1	10.15		0.26		−0.27		1	B7/8	567
128974	−35 09702	HR 5466		14 37 57	−35 55.3	4	5.66	.005	−0.09	.009	−0.32		13	Ap Si	15,1770,2013,2029
128948	−45 09344			14 37 58	−45 29.5	3	10.00	.007	0.13	.003	0.11	.005	12	A0 IV/V	863,1770,2011
		Ton 780		14 37 59	+23 20.4	1	14.91		−0.24		−0.99		4	sdB	1036
129170	+38 02578			14 38 00	+38 19.5	1	7.59		1.45		1.72		2	K0	1601
		CS 22874 # 75		14 38 00	−23 35.5	1	14.91		0.12		0.17		1		1736
		HA 106 # 869		14 38 02	+00 02.5	1	12.06		0.82		0.51		9	F8	281
		HA 106 # 1042		14 38 04	+00 07.4	1	11.61		0.84		0.41		7	G2	281
	−62 04245			14 38 04	−62 40.3	1	10.69		0.64		0.07		1		287
		AO 1019		14 38 05	+13 40.3	1	11.61		0.66		0.15		1	F7 V	1748
129132	+22 02731	HR 5472	⋆	14 38 06	+22 11.3	2	6.13	.025	0.40	.002	0.03	.007	4	G0 V	1733,3016
129267	+56 01754			14 38 06	+55 55.0	1	8.78		0.93		0.57		2	K0 III-IV	1502
128960	−46 09493		AB	14 38 07	−47 13.9	2	8.69	.046	0.06	.035	−0.40		5	B8 III/IV	1770,2014
	+0 03221			14 38 12	−00 19.7	2	10.57	.004	0.51	.009	0.03		13		281,6004
128917	−58 05672	HR 5465		14 38 12	−58 24.1	3	6.22	.005	0.45	.005			14	F5 V	15,1075,2012
		G 178 - 41		14 38 15	+45 30.7	2	12.66	.000	0.49	.005	−0.22	.005	2		1658,3060
		HA 106 # 545		14 38 16	−00 17.1	1	14.05		0.64		0.20		6		281
128918	−58 05673			14 38 17	−58 50.8	1	9.84		0.24		−0.13		2	B8 IV/V	567
129153	+14 02769	HR 5473		14 38 19	+13 44.9	2	5.91	.006	0.22	.007	0.05		4	F0 V	70,3016
		HA 106 # 700		14 38 19	−00 10.8	1	9.78		1.36		1.58		44		1764
		CS 22874 # 72		14 38 19	−24 01.7	1	14.31		0.21		0.12		1		1736
129018	−47 09337			14 38 19	−47 44.4	1	8.55		0.30		0.11		1	A7 III/IV	1771
		HA 106 # 702		14 38 21	−00 04.4	2	10.60	.000	0.48	.007	−0.02		13	F7	281,6004
		HA 106 # 548		14 38 21	−00 16.4	1	13.48		1.05		0.87		18		281
	−62 04248			14 38 21	−62 42.3	1	10.83		0.54		0.10		1		287
129174	+17 02768	HR 5475	⋆ A	14 38 22	+16 37.9	1	5.00		−0.11		−0.41		3	B9 p MnHg	3023
129174	+17 02768	IDS14360N1651	⋆ AB	14 38 22	+16 37.9	4	4.53	.013	−0.03	.010	−0.32	.008	10	B9 p + A6 V	15,1022,1363,8015
	+0 03222			14 38 22	−00 04.4	4	9.79	.004	1.36	.007	1.56	.017	42	K0 III	281,989,1729,6004
		HA 106 # 550		14 38 22	−00 15.1	1	13.21		0.99		0.70		28		281
128951	−57 06746			14 38 22	−57 44.4	1	10.19		0.33		−0.20		5	B8	567
	+15 02743			14 38 23	+14 51.4	1	12.00		1.02		0.80		1	G0	1748
129175	+17 02768	HR 5476	⋆ B	14 38 23	+16 37.9	1	5.88		0.25		0.10		2	A6 V	3016
		HA 106 # 553		14 38 24	−00 14.9	1	13.34		0.68		0.21		26		281
129109	−8 03813			14 38 24	−09 17.4	1	9.45		0.95		0.59		1	K3 V	3072
128898	−64 02977	HR 5463, α Cir	⋆ A	14 38 26	−64 45.5	1	3.18		0.26				1	Ap SrEuCr	1279
128898	−64 02977	HR 5463, α Cir	⋆ AB	14 38 26	−64 45.5	8	3.18	.010	0.24	.005	0.13	.049	23	Ap SrEuCr	15,1020,1075,2001*
128898	−64 02977	IDS14344S6432	B	14 38 26	−64 45.5	2	8.46	.009	1.23	.057	1.42		4		1279,3003
		CS 22871 # 70		14 38 28	−18 23.9	1	14.76		0.45		−0.15		1		1736
		CS 22871 # 77		14 38 30	−17 50.6	1	15.39		0.08		0.12		1		1736
		UKST 867 # 10		14 38 31	+00 09.1	1	14.40		0.75				2		1584
129055	−43 09248			14 38 31	−43 43.1	2	8.80	.024	0.46	.005	−0.09		3	F3 V	742,1594
	+18 02910			14 38 32	+18 08.0	2	11.52		−0.05	.000	−0.20	.014	5	A0	1026,1264
		CS 22871 # 75		14 38 34	−18 06.9	1	13.28		0.17		0.22		1		1736
129157	−13 03947			14 38 35	−13 42.9	1	8.41		0.87				8	K0 III/IV	1351
	−62 04250			14 38 35	−63 07.1	1	10.70		0.56		0.11		1	F8	549
		AO 1013		14 38 36	+13 58.8	1	10.96		1.01		0.76		1	G8 V	1748
129056	−46 09501	HR 5469, α Lup	⋆ A	14 38 36	−47 10.5	10	2.30	.015	−0.21	.009	−0.88	.005	24	B1.5Vn	15,26,278,507,1020*
		UKST 867 # 21		14 38 37	−00 01.3	1	16.67		0.70				1		1584
129068	−41 09049			14 38 37	−42 11.4	1	9.45		0.15		0.13		1	A2 V	1770
		UKST 867 # 11		14 38 38	+00 24.0	1	14.41		0.57				3		1584
129269	+34 02551			14 38 39	+34 29.6	1	8.83		0.30		0.04		3	A3	1625
		POSS 176 # 4		14 38 41	+47 33.5	1	16.57		1.54				1		1739
		CS 22874 # 77		14 38 42	−23 26.6	1	14.57		0.19		0.25		1		1736
		POSS 176 # 1		14 38 43	+47 59.8	1	13.74		1.19				1		1739
		UKST 867 # 16		14 38 44	+00 30.8	1	15.45		0.73				2		1584
129070	−45 09355			14 38 44	−46 05.1	5	7.35	.010	−0.01	.008	−0.16	.022	22	F5 V	116,977,1075,2001,2011

Table 1   811

HD	DM	Other Id	N Rem	α_{1950}	δ_{1950}	S	V	σ_V	B–V	σ_{B-V}	U–B	σ_{U-B}	n	Spectrum	References
129247	+14 02770	HR 5478	★ AB	14 38 45	+13 56.5	9	3.78	.010	0.04	.016	0.06	.021	36	A2 III	15,938,1007,1013,1022*
		CS 22871 # 68		14 38 45	−19 44.9	1	15.75		0.45		0.18		1		1736
	−62 04252			14 38 45	−63 15.4	1	11.30		0.53		0.05		1	F6	549
	−62 04253			14 38 46	−63 15.2	1	10.73		0.61		0.12		1	F8	549
129116	−37 09618	HR 5471		14 38 51	−37 34.8	7	3.99	.005	−0.17	.004	−0.69	.013	28	B2.5V	15,26,1020,1075,1637*
		CS 22874 # 76		14 38 52	−23 34.1	1	15.02		0.05		−0.06		1		1736
129161	−30 11624	HR 5474	★ AB	14 38 52	−30 43.2	2	6.37	.007	−0.09	.003	−0.38		5	Ap Si	1770,2007
129230	+0 03223			14 38 53	+00 19.0	4	8.12	.005	1.03	.008	0.84	.007	35	G8 III	281,989,1729,6004
		GD 169		14 38 53	+15 57.9	1	14.16		0.51		−0.19		1		3060
		CS 22871 # 60		14 38 53	−20 37.7	1	14.17		0.40		0.12		1		1736
		Ton 781		14 38 54	+25 59.	1	14.83		0.03		0.03		2		1036
		Lod 1375 # 19		14 38 55	−63 15.4	1	11.14		0.79		0.24		1	G2	549
		CS 22874 # 73		14 38 56	−23 47.5	1	15.71		0.35		0.17		1		1736
	−57 06753			14 38 56	−58 09.1	1	10.50		0.45		0.02		3		1586
129178	−28 10881			14 38 57	−29 08.7	3	8.13	.009	0.48	.007	−0.06		10	F5/6 V	1075,2033,3037
		Lod 1409 # 124		14 38 57	−61 29.7	1	12.69		2.01		1.65		2		1667
		G 201 - 9		14 38 58	+56 04.1	1	11.95		1.44				1	M3	1746
129162	−32 10301	IDS14360S3223	AB	14 38 58	−32 36.1	1	8.22		−0.03		−0.15		2	B9/A0 V	1770
		Lod 1375 # 27		14 39 00	−63 02.4	1	12.06		0.34		0.10		1	B9	549
		HA 106 # 573		14 39 02	−00 22.1	1	11.02		1.02		0.80		7	G3	281
129196	−28 10884			14 39 02	−28 36.3	1	8.10		0.22		0.17		1	A1mA1-F0	1770
129163	−39 09130			14 39 02	−39 38.2	1	9.71		0.37		0.04		1	A9 V	1771
		G 66 - 20		14 39 05	+06 40.8	1	12.05		1.50		1.19		1	M0	3016
	+0 03224			14 39 05	−00 13.3	6	9.34	.005	1.31	.005	1.47	.015	50	K0 III	281,397,989,1584,1729,6004
129355	+32 02504	RW Boo		14 39 06	+31 47.1	1	7.75		1.34		0.89		1	M2	3001
		CS 22871 # 56		14 39 06	−21 02.4	1	14.24		0.17		0.07		1		1736
		CS 22874 # 80		14 39 06	−22 39.0	1	14.96		0.13		0.23		1		1736
129040	−61 04673			14 39 06	−61 29.7	1	10.23		0.29		0.05		1	B8/9 IV/V	549
129290	+14 02771	G 135 - 80	★ A	14 39 07	+13 48.9	3	8.39	.010	0.60	.008	0.06	.005	5	G2 V	1003,1748,3016
129290	+14 02771	G 135 - 81	★ B	14 39 07	+13 48.9	1	13.48		1.45		1.11		2		3016
		Lod 1409 # 125		14 39 08	−61 38.3	1	11.31		0.46		0.39		2		1667
		CS 22871 # 53		14 39 09	−21 48.7	1	13.70		0.08		0.00		1		1736
129090	−57 06756			14 39 09	−57 52.4	1	7.35		0.12		−0.28		3	B7 II	567
		UKST 867 # 25		14 39 10	+00 25.5	1	17.09		0.31				1		1584
		MtW 106 # 63		14 39 10	−00 15.7	1	15.42		0.62		0.17		3		397
	−0 02856			14 39 10	−00 30.1	1	10.44		0.72		0.39		1	G2	281
129312	+8 02903	HR 5480		14 39 11	+08 22.5	6	4.86	.007	1.00	.004	0.76	.004	39	G8 III	3,1080,1088,1363,1509,4001
		AO 1016		14 39 11	+14 13.7	1	12.12		0.49		−0.02		1	B7 V	1748
		HA 106 # 902		14 39 12	+00 06.1	1	11.77		0.99		0.80		9	G7	281
		UKST 867 # 12		14 39 12	−00 03.0	1	14.96		1.22				1		1584
129216	−32 10306			14 39 12	−32 33.3	1	6.64		0.92				4	G8 IV	1075
		Ton 211		14 39 12	−32 38.	1	15.41		−0.02		0.06		1		1036
129357	+29 02568			14 39 13	+29 16.5	1	7.81		0.63		0.12		2	G2 V	3031
		HA 106 # 728		14 39 13	−00 08.5	1	12.27		0.55		0.00		12	F8	281
	−28 10889			14 39 14	−29 17.7	1	12.46		0.56		−0.12		2		1696
		CS 22871 # 62		14 39 16	−20 26.0	1	14.94		0.14				1		1736
129218	−34 09850			14 39 16	−34 31.8	1	9.11		0.27		0.14		1	A9 V	1771
129220	−38 09581			14 39 16	−39 16.1	1	9.09		0.10		0.13		1	A2 IV/V	1770
129426	+46 01981			14 39 17	+45 37.8	1	7.72		0.94		0.54		2	G5	1601
129201	−41 09059			14 39 17	−42 14.4	1	9.42		0.05		0.08		1	A1 IV	1770
		CS 22871 # 80		14 39 18	−17 19.1	1	15.31		0.18		0.23		1		1736
129336	+12 02729	HR 5481		14 39 19	+11 52.5	4	5.55	.008	0.94	.000	0.67	.007	8	G8 III	15,1003,1080,3077
129271	−13 03957			14 39 19	−13 38.5	1	8.05		0.82				11	G6/8 V	1351
		Lod 1409 # 123		14 39 19	−61 30.5	1	12.30		0.68		0.19		2		1667
		Lod 1375 # 23		14 39 19	−63 05.1	1	11.36		0.67		0.14		1	F8	549
129123	−55 06131			14 39 20	−56 10.3	1	8.84		0.14		−0.36		4	B3 III/V	567
		MtW 106 # 82		14 39 22	−00 18.6	1	14.59		0.62		0.10		2		397
	+14 02772			14 39 23	+13 47.3	1	9.72		0.56		0.00		1	G0	1748
129500	+52 01819			14 39 24	+52 27.0	1	7.17		1.15		1.15		2	K0	1566
129313	−0 02858			14 39 25	−00 26.3	1	9.27		0.41		0.08		1	F8	281
129124	−57 06759			14 39 26	−57 40.5	1	9.11		0.05		−0.26		3	B8 V	567
		Lod 1375 # 18		14 39 27	−63 14.5	1	11.00		1.33		1.02		1	K3	549
	−59 05699			14 39 28	−59 38.3	2	10.35	.005	0.29	.005	−0.13	.009	4		567,1586
		Lod 1409 # 31		14 39 29	−61 38.4	2	11.62	.024	0.46	.000	−0.09	.010	3	B7 p	549,1667
129093	−62 04255			14 39 29	−63 08.7	1	9.31		0.23		−0.37		1	B5 V	549
		MtW 106 # 92		14 39 30	−00 16.2	1	13.12		0.66		0.14		3		397
129092	−62 04257	IDS14356S6232	AB	14 39 30	−62 45.3	1	6.41		−0.08				4	B3 IV	2012
		MtW 106 # 94		14 39 31	−00 16.7	1	15.49		0.85		0.44		3		397
129221	−45 09366			14 39 31	−45 23.1	3	8.12	.004	0.49	.001	0.24	.023	16	F2 III	977,1075,2011
129412	+25 02816			14 39 32	+24 45.1	2	7.66	.005	0.50	.005	0.01	.005	5	F7 V	1733,3026
	−62 04256			14 39 32	−63 15.3	1	10.92		0.25		−0.32		1	B6	549
		CS 22871 # 64		14 39 33	−20 18.3	1	14.65		0.25		0.20		1		1736
		Lod 1409 # 122		14 39 33	−61 28.2	1	11.82		0.71		0.19		2		1667
		Lod 1375 # 32		14 39 33	−63 10.5	1	12.74		0.70		0.30		1	F0	549
		UKST 867 # 8		14 39 34	+00 09.7	1	13.45		0.50				3		1584
		vdB 62 # a		14 39 34	−58 31.6	1	14.12		0.75		−0.03		3		434
129580	+58 01523	G 223 - 79	★ A	14 39 35	+58 10.4	1	7.43		0.93		0.63		3	G8 IV	3024
129580	+58 01523	IDS14382N5823	B	14 39 35	+58 10.4	1	8.34		0.72		0.25		4		3024
		LP 440 - 58		14 39 36	+17 11.3	1	16.67		1.54				1		1663
		HA 106 # 588		14 39 36	−00 13.3	2	11.78	.004	1.09	.011	0.94		27	G5	281,1584

HD	DM	Other Id	N Rem	α_{1950}	δ_{1950}	S	V	σ_V	B−V	σ_{B-V}	U−B	σ_{U-B}	n	Spectrum	References
		vdB 62 # c		14 39 36	−58 27.4	1	12.40		0.67		−0.27		3		434
		Lod 1409 # 30		14 39 36	−61 39.6	2	11.57	.019	0.45	.015	0.08	.034	3	B7 p	549,1667
129430	+21 02674	HR 5483		14 39 37	+21 20.2	1	6.40		0.93		0.58		3	G8 III	1733
		G 136 - 4		14 39 38	+10 35.3	1	15.13		1.07		0.85		1		1658
		Lod 1375 # 31		14 39 40	−63 10.4	1	12.40		0.46		0.30		1	A1	549
		Wolf 538		14 39 43	+03 55.7	1	14.01		1.01		0.76		1		1696
129562	+53 01728	IDS14381N5304	AB	14 39 43	+52 51.6	1	8.50		0.49		0.00		2	F8	1375
		vdB 62 # b		14 39 43	−58 30.7	1	11.63		0.58		−0.24		3		434
	−62 04259			14 39 43	−63 10.5	2	10.15	.025	0.50	.045	0.09	.035	2	F5	287,549
129167	−61 04676			14 39 46	−61 39.9	1	9.69		0.42		0.31		1	A8 IV/V	549
		CS 22871 # 81		14 39 49	−17 12.8	1	13.85		0.60		0.17		1		1736
129281	−47 09355			14 39 54	−47 21.5	2	6.91	.009	−0.05	.000	−0.31		6	B9 II/III	1770,2014
129379	−18 03882			14 39 56	−19 06.0	1	6.90		1.20		1.16		10	K0 III	1628
128780	−78 00884			14 39 56	−79 11.6	1	9.56		0.65		0.16		1	F8/G2	1658
		UKST 867 # 19		14 39 58	+00 06.4	1	16.32		0.76				2		1584
	−59 05700	LSS 3277		14 39 59	−60 00.0	2	10.73	.030	0.39	.004	−0.47	.012	5	B3 Iab	540,1586
	−0 02861			14 40 00	−00 26.2	1	10.28		0.68		0.22		1	dG2	281
		BPM 22844		14 40 00	−52 08.	1	13.95		0.59		−0.06		1		3065
		UKST 867 # 22		14 40 01	−00 09.7	1	16.77		0.64				2		1584
		Lod 1375 # 26		14 40 01	−63 10.0	1	11.93		0.55		0.36		1	B9	549
		SZ Boo		14 40 02	+28 25.0	1	11.92		0.11		0.10		1	A3	668
		CS 22871 # 52		14 40 02	−22 11.8	1	13.96		0.05		0.05		1		1736
		POSS 176 # 5		14 40 03	+48 29.1	1	17.15		1.55				2		1739
129415	−16 03922			14 40 03	−17 05.6	1	10.30		0.26		0.18		1	A2 III	1770
129243	−58 05686			14 40 05	−58 34.9	1	8.55		0.10		−0.50		3	B3 IV/V	567
		CS 22874 # 71		14 40 06	−24 16.4	1	14.28		0.74		0.21		1		1736
		UKST 867 # 7		14 40 08	+00 32.7	1	13.45		0.64				1		1584
		POSS 176 # 2		14 40 08	+47 58.6	1	14.54		0.84				1		1739
129398	−28 10895			14 40 08	−28 46.7	1	9.21		0.16		0.15		1	A1 V	1770
	+20 03009	G 135 - 82		14 40 09	+19 43.1	1	10.08		1.34		1.28		2	M0	3072
	−0 02862			14 40 10	−00 27.2	1	10.15		0.71		0.15		1	dG3	281
		CS 22871 # 63		14 40 11	−20 30.2	1	13.74		0.22		0.24		1		1736
		CS 22871 # 65		14 40 12	−20 14.0	1	14.79		0.40		0.02		1		1736
		CS 22874 # 62		14 40 13	−26 06.8	1	14.08		0.30		0.14		1		1736
129418	−29 11261			14 40 13	−29 35.0	1	8.70		0.23		0.20		1	A2/3mA3-A9	1770
129350	−45 09374			14 40 13	−46 12.9	2	8.36	.001	1.19	.004	1.16		12	K1 III	977,2011
129225	−62 04263	LSS 3278		14 40 13	−63 03.4	4	9.42	.018	0.32	.005	−0.60	.018	9	B1/2 II	34,540,549,976
		UKST 867 # 14		14 40 14	+00 37.1	1	15.39		0.82				1		1584
129601	+40 02808			14 40 15	+39 37.2	1	8.19		1.01		0.83		2	K0	1601
129285	−55 06136			14 40 15	−56 08.7	1	9.79		0.06		−0.27		3	B9 V	567
	+20 03010	G 135 - 83	★ AB	14 40 16	+19 41.7	1	9.12		1.29		1.27		2	K4:	3072
129417	−24 11636			14 40 16	−24 53.9	1	8.16		0.46				4	F3 V	2012
		CS 22874 # 66		14 40 16	−25 34.4	1	15.41		0.17		0.20		1		1736
129517	+7 02830	IDS14379N0701	AB	14 40 19	+06 48.1	1	7.21		0.63				6	F8	173
		CS 22871 # 57		14 40 20	−20 58.8	1	13.61		0.13		0.18		1		1736
129433	−24 11637	HR 5484		14 40 20	−24 47.1	3	5.68	.004	0.00	.003	−0.05	.021	7	B9.5V	1079,1770,2007
	−62 04264			14 40 21	−63 12.0	1	10.88		0.35		0.17		1	A9	549
	−61 04677			14 40 22	−61 31.3	3	9.69	.029	1.10	.005	0.90	.063	4	K0	287,549,1667
	−62 04265			14 40 22	−62 54.2	1	10.99		0.26		−0.11		1	B8	549
130235	+80 00451			14 40 23	+80 00.1	1	7.28		1.62		1.95		2	K0	1733
		LP 914 - 60		14 40 24	−28 01.8	1	11.43		0.81		0.44		2		1696
		Lod 1375 # 8		14 40 24	−62 51.4	1	10.63		1.23		0.80		1	K2	549
129502	−5 03936	HR 5487		14 40 25	−05 26.5	9	3.88	.010	0.38	.005	−0.03	.011	33	F2 III	15,253,1008,1075,1088*
		Lod 1409 # 32		14 40 25	−61 37.4	2	11.76	.136	0.51	.029	0.12	.049	3	F6	549,1667
		Lod 1409 # 129		14 40 27	−61 40.5	1	10.77		1.89		2.59		2		1667
		UKST 867 # 13		14 40 30	+00 10.3	1	15.02		0.85				1		1584
		LundsII141 # 145		14 40 30	−61 47.	1	11.72		0.45		−0.44		2		34
		GD 170		14 40 33	+09 58.4	1	13.95		0.44		−0.13		1		3060
		Lod 1409 # 115		14 40 33	−61 26.5	1	12.75		0.49		0.39		2		1667
129329	−58 05692			14 40 34	−58 50.0	1	9.74		0.23		−0.53		6	B3 V	567
129456	−34 09868	HR 5485		14 40 35	−34 57.6	5	4.04	.005	1.35	.000	1.53	.000	16	K3 III	15,1075,2012,3051,8015
		Lod 1409 # 120		14 40 36	−61 32.0	1	12.92		2.42		3.15		2		1667
129288	−62 04269			14 40 37	−63 17.3	2	10.00	.000	0.46	.000	0.11	.000	2	A9/F0 V	287,549
	−61 04680			14 40 38	−61 39.4	2	10.87	.063	0.55	.005	0.05	.019	3	F8	549,1667
		CS 22874 # 64		14 40 40	−25 41.1	1	15.25		0.44		−0.03		1		1736
		UKST 867 # 18		14 40 42	+00 35.9	1	16.26		0.58				1		1584
		UKST 867 # 17		14 40 43	−00 14.2	1	15.90		1.00				2		1584
		Lod 1409 # 28		14 40 44	−61 30.3	2	11.52	.000	0.74	.010	0.24	.029	3	G0	549,1667
		POSS 176 # 6		14 40 46	+48 25.7	1	17.81		1.60				2		1739
		UKST 867 # 20		14 40 47	+00 00.9	1	16.62		0.84				1		1584
	−61 04682			14 40 47	−61 30.0	3	10.25	.025	1.14	.005	0.98	.039	4	B9	287,549,1667
		Lod 1375 # 20		14 40 47	−63 06.2	1	11.23		1.75				1	M0	549
129798	+61 01451	HR 5492, DL Dra	★ AB	14 40 48	+61 28.5	5	6.25	.006	0.41	.008	0.00	.009	23	F2 V	15,1007,1013,1334,8015
129387	−56 06405			14 40 48	−57 11.2	1	9.78		0.07		−0.31		4	B8 III/IV	567
	−61 04683			14 40 48	−61 31.2	3	10.09	.034	1.36	.005	1.17	.083	4	K3	287,549,1667
	+0 03225			14 40 50	+00 18.5	3	10.81	.003	0.45	.006	−0.04		11	F5	281,1584,6004
129474	−46 09535			14 40 50	−47 02.6	9	8.83	.003	1.62	.005	1.96	.009	121	K4 (III)	863,977,1075,1460,1628,1657*
		Lod 1375 # 30		14 40 50	−63 09.9	1	12.38		0.44		0.25		1	A0	549
129779	+55 01704			14 40 51	+55 00.9	1	7.62		1.61		1.57		2	M2	1375
	+6 02932	G 66 - 22	A	14 40 53	+06 02.4	4	10.46	.011	0.71	.005	0.10	.003	9	G6 V	516,1003,1620,3077

Table 1

HD	DM	Other Id	N Rem	α_{1950}	δ_{1950}	S	V	σ_V	B–V	σ_{B-V}	U–B	σ_{U-B}	n	Spectrum	References
	+6 02932		B	14 40 53	+06 02.4	1	13.36		1.57				1		3032
		POSS 176 # 7		14 40 53	+47 45.0	1	17.99		1.80				2		1739
129369	−62 04271			14 40 53	−63 10.2	2	10.15	.025	0.47	.010	0.10	.020	2	F0/2 IV	287,549
		Lod 1409 # 113		14 40 54	−61 27.7	1	12.41		0.53		0.08		2		1667
		Lod 1409 # 114		14 40 54	−61 27.7	1	10.84		1.82		2.38		2		1667
	−61 04686			14 40 54	−61 32.5	3	9.85	.034	1.15	.005	0.81	.068	4	K3	287,549,1667
		LP 500 - 87		14 40 55	+11 33.7	1	14.41		1.64		1.30		1		1773
	+0 03226			14 40 55	−00 09.4	2	9.66	.002	0.57	.010	0.04		11	G0 V	281,6004
		CS 22874 # 89		14 40 56	−24 31.6	1	14.61		0.14		0.21		1		1736
	+0 03227			14 40 57	+00 10.0	2	9.91		0.62	.005	0.07		3	G5 V	281,1584
		HA 106 # 459		14 40 57	−00 27.8	3	11.59	.008	0.33	.004	0.08	.004	27		1728,989,1729
		CS 22874 # 88		14 40 57	−24 11.1	1	14.00		0.04		-0.15		1		1736
129570	−26 10468			14 40 58	−26 28.0	1	9.92		0.34		0.11		1	A(2)(m)	1770
	−61 04687			14 40 58	−61 37.0	2	10.77	.010	0.42	.000	-0.03	.019	3	B7	549,1667
		Lod 1409 # 112		14 40 59	−61 27.7	1	12.66		0.71		0.21		2		1667
129569	−23 11878			14 41 00	−23 52.0	1	8.81		0.02		-0.08		1	B9 IV/V	1770
		Lod 1375 # 28		14 41 00	−62 52.6	1	12.09		0.68		0.31		1	A8	549
	−59 05702	LSS 3280		14 41 01	−59 33.5	2	10.73	.010	0.68	.001	-0.38	.007	5	O6	540,1737
		CS 22871 # 85		14 41 03	−18 30.6	1	14.36		0.13		0.22		1		1736
		Lod 1375 # 14		14 41 03	−62 51.9	1	10.89		1.22		0.93		1	K3	549
129421	−58 05696			14 41 04	−58 41.7	1	9.39		0.00		-0.36		2	Ap Si	567
129405	−60 05502			14 41 05	−61 18.6	1	7.01		2.01		2.15		1	K4 III	549
129546	−33 10022			14 41 06	−33 56.6	2	10.12	.032	0.43	.010	-0.04		4	F2 V	1594,6006
		L 261 - 41		14 41 06	−52 13.	1	12.43		1.19		1.13		1		1696
	−52 07463			14 41 06	−52 19.9	1	10.92		0.41		-0.13		2		540
		POSS 176 # 3		14 41 08	+47 32.5	1	14.58		1.29				1		1739
		Lod 1375 # 29		14 41 08	−63 17.5	1	12.33		0.42		0.23		1	A0	549
	+27 02411	G 166 - 42		14 41 09	+26 57.7	1	9.67		0.73		0.26		3	G8	196
		Lod 1409 # 21		14 41 10	−61 28.6	3	10.79	.025	1.62	.005	1.30	.073	4		287,549,1667
	−61 04689			14 41 10	−61 29.8	2	11.27	.005	0.46	.015	0.12	.015	3	A9	549,1667
129462	−57 06772	HR 5486		14 41 11	−58 16.0	4	6.11	.005	1.01	.009			18	K0 III	15,1075,1075,2006
		Lod 1375 # 15		14 41 12	−63 21.2	1	10.91		1.63		1.66		1	K9	549
129666	+0 03228			14 41 13	+00 15.6	1	9.10		0.62		0.10		3	G5 V	281
129712	+27 02413	HR 5490, W Boo		14 41 13	+26 44.4	6	4.81	.010	1.66	.010	1.93	.022	17	M3-III	15,1118,3055,8015*
	−61 04690			14 41 14	−61 30.2	3	10.49	.273	1.18	.351	0.85	.414	4	K1	287,549,1667
		LP 740 - 48		14 41 15	−11 39.7	2	11.50	.030	0.60	.023	-0.19	.007	4		1696,3062
		BPM 22855		14 41 18	−59 17.	1	12.87		0.53		-0.18		1		3065
129422	−62 04275	HR 5482	★ A	14 41 18	−62 39.8	5	5.36	.009	0.29	.007	0.07	.005	13	A9 III/IV	15,1637,1754,2012,2016
		CS 22874 # 86		14 41 20	−23 43.8	1	14.85		0.35		0.19		1		1736
		CS 22874 # 106		14 41 20	−26 27.6	1	13.13		0.25		0.20		1		1736
		UKST 867 # 15		14 41 22	+00 20.4	1	15.44		1.46				1		1584
	−44 09571	E6 - 58		14 41 22	−45 14.4	2	10.41	.010	0.27	.005	0.17		8	A2	116,2011
		Lod 1409 # 108		14 41 22	−61 31.6	1	12.52		0.54		0.38		2		1667
		Feige 102		14 41 24	+51 56.	1	16.13		-0.18		-0.78		1		3060
129464	−61 04691			14 41 25	−61 39.5	1	9.44		0.51		-0.04		1	F8/G0 V	549
130043	+69 00765			14 41 26	+69 19.6	1	9.20		0.56		0.00		3	K0	3026
129667	−15 03941			14 41 27	−15 30.7	1	9.40		0.40		0.17		1	A5/7mA6-F3	1770
		CS 22874 # 81		14 41 27	−22 37.9	1	14.38		0.36		0.00		1		1736
		CS 22874 # 135		14 41 28	−26 04.6	1	12.98		0.48		0.24		1		1736
129511	−56 06414			14 41 28	−56 54.4	1	8.53		-0.01		-0.39		3	B8 III	567
129465	−62 04276			14 41 29	−63 13.6	1	9.45		1.15		0.96		1	G8/K0 III	549
		CS 22874 # 90		14 41 31	−24 31.9	1	15.52		0.08		0.11		1		1736
129078	−78 00893	HR 5470		14 41 33	−78 50.1	3	3.83	.017	1.43	.004	1.68	.004	8	K3 III	15,1034,1075
129557	−55 06150	HR 5488, BU Cir	★ A	14 41 34	−55 23.5	7	6.10	.010	-0.07	.008	-0.80	.004	93	B2 III	15,540,567,976,1075*
		CS 22874 # 85		14 41 35	−23 38.4	1	15.29		0.10		0.06		1		1736
	−44 09574			14 41 36	−45 16.1	1	9.92		1.17				4		2011
		CS 22871 # 88		14 41 37	−19 07.8	1	14.44		0.09		0.14		1		1736
		Lod 1409 # 106		14 41 39	−61 31.6	1	13.00		0.60		0.47		2		1667
129727	+0 03229			14 41 40	−00 24.5	6	9.48	.006	0.38	.001	-0.04	.003	50	F0	281,989,1728,1729,1775,6004
129578	−55 06152	IDS14380S5511	B	14 41 41	−55 23.8	1	7.51		1.39		1.48		6	K2 III	1673
		Lod 1409 # 19		14 41 41	−61 44.7	1	10.72		1.20		0.90		1	K2	549
		BPM 9085		14 41 42	−66 10.	1	13.58		1.38				1		3065
129623	−44 09576			14 41 43	−44 27.3	6	7.52	.004	0.94	.010	0.60	.008	31	G6 III	116,460,977,1075,1770,2011
	−61 04695			14 41 43	−61 29.8	3	9.97	.022	1.32	.004	1.28	.080	4	K3	287,549,1667
129715	−21 03954	IDS14390S2149		14 41 44	−22 02.3	1	9.32		1.17				1	K2/3 V	1746
		UKST 867 # 24		14 41 45	+00 12.2	1	16.90		0.64				1		1584
		Lod 1409 # 101		14 41 46	−61 29.1	1	11.86		4.31		3.56		2		1667
		CS 22874 # 83		14 41 47	−23 19.2	1	15.17		0.15		0.19		1		1736
129846	+41 02523	HR 5493		14 41 48	+40 40.2	2	5.73	.000	1.40	.002	1.62		6	K4 III	70,1501
129642	−49 09033			14 41 48	−49 42.0	1	8.39		0.95				4	K3 V	2012
		Lod 1409 # 23		14 41 48	−61 21.7	1	10.92		0.68		0.17		1	F8	549
		CS 22874 # 84		14 41 51	−23 27.9	1	12.26		0.41		0.14		1		1736
	−62 04278			14 41 51	−62 55.3	1	10.39		0.32		-0.13		1	B8	549
129685	−34 09888	HR 5489		14 41 55	−34 58.9	6	4.91	.007	0.01	.004	-0.02	.008	17	A0 V	15,1075,1770,2012,3023,8015
	−62 04279			14 41 55	−62 56.2	1	10.76		0.40		-0.08		1	B8	549
		UKST 867 # 9		14 41 57	+00 08.4	1	13.47		0.67				1		1584
	+22 02742			14 41 57	+22 23.5	1	9.92		1.25				1	M0	1017
129660	−44 09580			14 41 57	−45 12.9	8	9.38	.002	0.27	.004	0.16	.009	90	A7 V	1075,1460,1628,1657,1673*
		Lod 1375 # 21		14 41 58	−63 06.8	1	11.28		0.63		0.04		1	F8	549
		Lod 1409 # 38		14 42 00	−61 16.7	1	12.63		0.74		0.05		1		549

HD	DM	Other Id	N Rem	α_{1950}	δ_{1950}	S	V	σ_V	B–V	σ_{B-V}	U–B	σ_{U-B}	n	Spectrum	References
	−60 05505			14 42 01	−61 17.4	1	11.03		0.31		0.22		1	A2	549
129686	−42 09625			14 42 03	−42 31.8	1	9.31		0.17		0.18		2	A2 III/IV	1770
		UKST 867 # 5		14 42 05	+00 35.0	1	11.21		1.66				1		1584
		CS 22871 # 83		14 42 05	−17 43.4	1	15.15		0.14		0.21		1		1736
		Lod 1375 # 24		14 42 05	−63 08.0	1	11.41		1.37		1.36		1		549
	−60 05506			14 42 06	−61 17.4	2	10.13	.000	0.96	.000	0.36	.000	2	K0	287,549
		Lod 1409 # 103		14 42 06	−61 29.6	1	12.12		0.46		0.22		2		1667
		Lod 1409 # 105		14 42 06	−61 33.1	1	12.42		0.54		0.41		2		1667
	+0 03231			14 42 08	+00 22.7	1	10.37		0.98				1	K0	1584
129688	−45 09393			14 42 09	−45 27.3	8	9.22	.003	0.04	.003	0.02	.005	129	A0/1 V	1075,1460,1628,1657,1673*
		Lod 1409 # 104		14 42 09	−61 29.9	1	12.89		0.82		0.41		2		1667
		UKST 867 # 23		14 42 10	+00 28.7	1	16.89		0.54				1		1584
	−45 09394			14 42 11	−45 37.2	1	10.40		0.40				4		2011
		Lod 1409 # 7		14 42 11	−61 45.0	1	9.66		1.88		2.08		1	M5	549
		CS 22871 # 94		14 42 12	−21 13.6	1	12.65		0.33		0.19		1		1736
		CS 22874 # 105		14 42 13	−26 19.4	1	12.62		0.04		−0.04		1		1736
	−44 09582		AB	14 42 13	−44 55.9	1	11.53		0.53				4		2011
		Lod 1409 # 29		14 42 13	−61 44.4	1	11.52		1.96		2.14		1	M5	549
		Lod 1375 # 25		14 42 13	−63 03.1	1	11.79		0.53		0.18		1	A8	549
		Boo sq 1 # 1		14 42 14	+22 11.2	1	13.18		0.95		0.75		4		625
129732	−35 09765	IDS14392S3543	A	14 42 14	−35 56.2	1	7.36		0.10		0.07		1	A1 IV/V	1770
		Boo sq 1 # 7		14 42 16	+22 12.7	1	14.42		0.64		0.16		4		625
		Lod 1409 # 34		14 42 18	−61 11.3	1	11.89		0.51		0.10		1	F2	549
		Lod 1409 # 33		14 42 18	−61 22.7	1	11.61		0.74		0.10		1	F8	549
		Lod 1409 # 36		14 42 19	−61 24.5	1	12.16		0.64		0.37		1	A5	549
		Lod 1409 # 37		14 42 20	−61 21.9	1	12.27		0.51		0.35		1	A5	549
129735	−46 09556			14 42 26	−46 35.6	5	8.66	.006	0.45	.007	0.02	.010	21	F3/5 V	977,1075,2001,2011,2013
	−61 04699			14 42 27	−61 20.3	1	10.82		0.43		0.33		1	A2	549
129645	−62 04284			14 42 27	−63 09.2	1	10.68		0.22		0.18		1	A3/5 II/IV	549
		CS 22871 # 87		14 42 28	−18 46.0	1	14.25		0.07		0.15		1		1736
129747	−45 09398			14 42 28	−45 40.2	5	8.47	.009	0.69	.014	0.16	.020	22	G5 V	116,768,977,1075,2011
	−61 04698			14 42 28	−61 38.7	3	11.11	.013	0.45	.016	−0.27	.005	5	B6	34,549,1667
		Boo sq 1 # 8		14 42 29	+22 12.8	1	15.18		0.62		0.14		4		625
		CS 22874 # 101		14 42 31	−25 11.4	1	14.79		0.29		0.18		1		1736
		CS 22871 # 89		14 42 34	−19 11.8	1	13.70		0.03		−0.07		1		1736
	−44 09587			14 42 34	−45 02.9	2	10.71	.004	0.34	.002	0.06		14		1460,2011
		Boo sq 1 # 6		14 42 36	+22 00.5	1	14.99		0.69		0.27		4		625
129831	−26 10493	IDS14397S2631	A	14 42 36	−26 43.9	1	8.07		0.29		0.09		1	A2/3 V	1770
129902	−0 02867	HR 5496		14 42 37	−01 12.4	2	6.05	.015	1.61	.000	1.93	.000	10	M1 III	1088,3035
	−60 05510	LSS 3281		14 42 37	−60 34.7	2	10.52		0.60		−0.36		2	B0.5V	540
129672	−61 04700			14 42 37	−61 37.6	1	9.74		0.15		−0.04		1	B8/9 V	549
130044	+45 02214			14 42 38	+45 23.8	1	6.74		0.28		0.01		4	F0 IV	1501
129791	−44 09590	IDS14394S4427	A	14 42 40	−44 39.4	4	6.91	.004	0.05	.007	−0.01	.005	19	A0 V	977,1075,1771,2011
129792	−46 09559			14 42 43	−46 34.5	1	9.57		0.44				4	F2 V	2011
		Lod 1409 # 35		14 42 44	−61 24.7	1	12.01		0.61		0.14		1	F2	549
		Boo sq 1 # 4		14 42 47	+22 00.2	1	13.62		0.89		0.41		4		625
129989	+27 02417	HR 5506	★ AB	14 42 48	+27 17.0	6	2.38	.012	0.97	.004	0.72	.024	16	K0 II-III	15,1080,1363,3016*
129708	−60 05511	BP Cir		14 42 48	−61 15.1	1	7.61		0.75		0.34		1	F2/3 II	549
129972	+17 02780	HR 5502		14 42 54	+17 10.5	5	4.64	.108	0.97	.007	0.74	.010	18	K0 III	15,1080,1118,3016,8015
129956	+1 02972	HR 5501		14 42 57	+00 55.6	7	5.68	.005	−0.02	.005	−0.07	.018	78	B9.5V	3,147,252,1079,1088*
		E6 - r		14 42 57	−45 12.5	2	14.17	.004	0.98	.003	0.55		6		1075,1460
129793	−53 06094	NGC 5749 - 112		14 42 58	−54 07.8	1	8.50		0.11		0.11		2	A0 IV	1761
		Boo sq 1 # 2		14 42 59	+21 59.5	1	12.94		0.74		0.15		4		625
129773	−56 06423			14 42 59	−57 01.5	1	7.95		0.04		−0.29		3	B7/8 III	567
		E6 - p		14 43 00	−45 11.	1	13.61		0.70				4		1075
		E6 - q		14 43 00	−45 15.	1	13.60		1.10				4		1075
		E6 - h		14 43 00	−45 20.	1	11.72		1.16				4		2011
		Boo sq 1 # 3		14 43 01	+22 07.7	1	13.36		0.60		0.00		4		625
130004	+14 02779	G 136 - 12		14 43 02	+14 03.5	1	7.87		0.88		0.68		2	K0	1648
130025	+19 02854	HR 5507		14 43 02	+19 05.7	1	6.13		0.83				2	K0	70
129975	+0 03234			14 43 04	−00 09.3	5	8.37	.003	1.51	.012	1.86	.023	83	K3 III	147,989,1509,1729,6005
		E6 - g		14 43 05	−45 22.7	2	12.18	.001	0.55	.004	0.09		7		1460,2011
		CS 22871 # 93		14 43 06	−20 24.4	1	13.42		0.43		0.19		1		1736
129926	−24 11661	HR 5497	★ AB	14 43 06	−25 13.9	4	4.93	.008	0.35	.006	0.07	.004	11	F0 III +dF9	158,219,404,2007
		E6 - s		14 43 06	−45 12.	1	14.59		0.77				4		1075
		E6 - u		14 43 06	−45 13.	1	15.11		0.85				4		1075
		E6 - l		14 43 07	−45 16.2	2	12.34	.007	1.32	.004	1.34		7		1075,1460
129857	−45 09404			14 43 07	−45 33.5	8	9.58	.007	1.10	.006	0.94	.008	68	K0 III	116,1460,1628,1657,1673*
130084	+33 02489	HR 5510		14 43 08	+33 00.0	2	6.26	.020	1.58	.001	1.90		4	M1 IIIb	70,3001
129858	−46 09562	HR 5494		14 43 08	−47 13.9	8	5.73	.009	0.07	.009	0.06	.009	53	A1 V	14,15,116,278,1075,1770*
	−44 09592			14 43 09	−45 08.9	1	10.67		1.49						1075
		TW Boo		14 43 10	+41 14.3	2	10.91	.283	0.20	.082	0.12	.004	2	F5	668,699
129978	−14 04023	HR 5503	★ AB	14 43 12	−15 15.0	1	6.33		1.19				4	K2 III	2007
		E6 - v		14 43 12	−45 13.	1	15.67		1.23				4		1075
		E6 - o		14 43 12	−45 16.	1	13.73		0.44				4		1075
129958	−18 03891			14 43 14	−18 46.2	1	8.15		0.25		0.09		1	A3 III/IV	1770
129944	−22 03844	HR 5499		14 43 14	−22 56.6	1	5.81		0.98				4	K0 III	2035
129740	−65 02865			14 43 15	−65 58.2	1	7.35		−0.07				4	B3 V	2012
		CS 22874 # 82		14 43 16	−22 48.9	1	14.49		0.14		0.23		1		1736
	−45 09408			14 43 16	−45 22.4	2	10.70	.002	1.08	.002	0.81		6		861,1075

Table 1   815

HD	DM	Other Id	N Rem	α_{1950}	δ_{1950}	S	V	σ_V	B–V	σ_{B-V}	U–B	σ_{U-B}	n	Spectrum	References
		E6 - n		14 43 17	−45 11.7	2	12.82	.005	1.06	.005	0.71		6		1075,1460
		CS 22871 # 92		14 43 18	−20 17.6	1	14.35		0.13		0.16		1		1736
		CS 22874 # 100		14 43 18	−25 19.0	1	15.40		0.15		0.23		1		1736
		E6 - m		14 43 18	−45 14.	1	12.31		1.33				4		1075
129794	−61 04704			14 43 18	−61 24.1	2	9.65	.000	1.30	.000	1.17	.000	2	K1/2 III	287,549
129929	−36 09605			14 43 19	−37 00.8	1	8.09		−0.18		−0.87		1	B2	55
129980	−20 04087	HR 5504	★ AB	14 43 20	−20 57.9	3	6.42	.012	0.61	.014	0.12		9	G2 V	1414,2007,3077
	−44 09593			14 43 20	−44 54.8	2	9.95	.003	0.45	.000	0.03		10	F2	863,2011
		LP 621 - 75		14 43 22	−00 18.9	1	13.13		0.53		−0.18		4		1696
	−44 09594			14 43 22	−44 59.5	2	10.17	.020	1.24	.001	1.17		10		863,2011
		LP 501 - 12		14 43 24	+14 29.1	1	12.17		0.73		0.14		2		1696
		E6 - t		14 43 24	−45 12.	1	14.68		0.75				4		1075
		LundsII141 # 155		14 43 24	−62 54.	1	11.34		0.59		−0.08		2		34
		L 477 - 3		14 43 26	−35 08.8	2	11.88	.020	0.87	.010	0.29	.015	4		1696,3073
	−44 09595			14 43 26	−45 10.9	2	10.15	.005	0.81	.002	0.37		10	K0	863,2011
	−3 03668			14 43 27	−04 17.1	1	10.77		0.97		0.43		1		565
129896	−53 06101	NGC 5749 - 111		14 43 29	−54 07.4	1	9.57		1.09		0.96		2	K1 II/III	1761
129824	−61 04705			14 43 29	−61 35.1	1	9.21		1.05		0.74		1	G6/8 III	549
	+17 02782			14 43 30	+16 54.4	1	9.90		0.67		0.23		3		291
		LundsII141 # 158		14 43 30	−61 53.	1	11.53		0.36		−0.27		2		34
129893	−51 08457	HR 5495	★ AB	14 43 31	−52 10.4	1	5.21		0.98				4	G8 III	2006
		LP 681 - 90		14 43 32	−06 00.7	1	13.28		0.84		0.53		2		1696
129981	−31 11449	V553 Cen		14 43 32	−31 57.7	1	8.24		0.62				1	G3 Ib/II	688
		NGC 5749 - 110		14 43 32	−54 10.0	1	10.02		2.02		1.61		1		1761
130108	+17 02783			14 43 34	+17 00.5	1	7.35		1.20		1.22		11	K2	291
130188	+42 02531	IDS14417N4248	AB	14 43 36	+42 35.5	2	7.31	.018	0.46	.009	0.01	.002	5	F5	292,3030
130188	+42 02531	IDS14417N4248	C	14 43 36	+42 35.5	1	12.46		1.34		1.18		3		3024
		CS 22871 # 90		14 43 37	−19 21.9	1	15.09		0.26		0.09		1		1736
	−44 09598			14 43 37	−45 01.4	3	10.14	.011	0.04	.009	−0.53	.004	8	B6 Ve	55,1586,2011
129964	−44 09599			14 43 40	−44 32.4	3	9.07	.005	1.20	.004	1.18	.007	14	K1/2 III	116,863,2011
129932	−51 08461	HR 5498	★ AB	14 43 42	−51 59.8	2	6.07	.004	0.09	.001	0.15		5	A1 III/IV	1770,2007
130109	+2 02862	HR 5511		14 43 43	+02 06.1	20	3.74	.012	−0.01	.007	−0.02	.015	198	A0 V	1,3,15,30,125,130,667*
	+17 02784			14 43 43	+16 59.2	1	9.86		0.51		0.04		4	F8	291
130144	+15 02758	HR 5512		14 43 44	+15 20.5	2	5.75	.093	1.53	.033	1.26		6	M5 IIIab	70,3025
130010	−30 11697			14 43 44	−31 15.2	1	9.74		0.11		0.12		1	A1 IV/V	1770
		NGC 5749 - 109		14 43 47	−54 09.4	1	12.28		0.23		−0.27		1		1761
130069				14 43 48	−19 13.	1	11.22		0.17		0.15		1	A0/1 IV	1770
		E6 - i		14 43 48	−45 16.	1	11.99		1.17				4		1075
130145	+10 02739	IDS14414N1005	AB	14 43 49	+09 51.6	5	7.27	.014	0.62	.009	0.10	.011	10	G1 V +G4 V	292,861,938,1003,3030
		NGC 5749 - 108		14 43 49	−54 11.6	1	13.10		0.50		0.00		1		1761
129881	−61 04707			14 43 49	−61 33.6	1	10.36		0.29		0.18		1	A6/8 III	549
129842	−65 02871			14 43 49	−65 31.9	1	8.44		0.03		−0.34		5	B8/9 IV/V	1586
	−42 09658			14 43 51	−42 41.2	1	10.26		0.05		−0.62		1	B2 III-II	55
130215	+28 02365			14 43 52	+27 43.3	1	7.98		0.83		0.52		2	K2 V	3026
		E6 - k		14 43 54	−45 18.	1	11.74		1.41				4		1075
129897	−61 04708			14 43 54	−61 22.8	1	8.36		1.32		1.22		1	K0 II/III	549
130095	−26 10505			14 43 56	−27 02.3	5	8.12	.008	0.08	.005	0.10	.022	11	B9 V	55,164,1311,1770,3077
130055	−37 09686	HR 5508		14 43 56	−38 04.8	1	5.94		1.33				4	K3 III	2006
		NGC 5749 - 107		14 43 57	−54 12.0	1	11.92		0.25		0.00		2		1761
		NGC 5749 - 106		14 43 57	−54 15.0	1	12.77		0.44		0.25		1		1761
		NGC 5749 - 105		14 43 59	−54 15.7	1	12.27		0.73		0.25		2		1761
		NGC 5749 - 104		14 43 59	−54 17.0	1	11.87		0.35		0.14		1		1761
	+17 02785	G 136 - 13		14 44 03	+16 43.1	2	9.26	.015	1.27	.010	1.18	.015	3	K5 V	1003,3072
		CS 22874 # 103		14 44 03	−26 01.9	1	15.36		0.12		0.21		1		1736
130014	−46 09574			14 44 03	−47 17.5	1	9.97		0.03		−0.18		2	B8 V	400
130035	−43 09322			14 44 04	−44 14.7	5	9.37	.006	0.37	.001	0.08	.006	24	F0 V	460,863,1075,1770,2011
129811	−70 01807			14 44 04	−70 23.4	1	7.37		1.34		1.31		7	K1 III	1704
	−44 09602			14 44 05	−45 07.6	1	11.74		0.53				4		2011
		E6 - e		14 44 06	−45 11.	1	11.95		0.58				4		2011
		TY Aps		14 44 06	−71 07.2	1	11.24		0.32		0.24		1	F2	700
		NGC 5749 - 94	AB	14 44 07	−54 22.4	1	11.81		1.83		2.18		1		1761
		NGC 5749 - 103		14 44 08	−54 18.8	1	13.45		0.60		0.09		1		1761
		NGC 5749 - 93		14 44 09	−54 22.7	1	11.97		1.27		0.95		2		1761
		NGC 5749 - 69		14 44 11	−54 09.9	1	11.90		0.38		0.28		2		1761
		G 200 - 58		14 44 12	+46 46.2	1	14.77		1.85				1		1759
		NGC 5749 - 62		14 44 14	−54 06.1	1	12.14		0.46		0.08		2		1761
130156	−15 03949			14 44 15	−16 15.8	1	9.31		0.44		0.20		1	A3mA7-F3	1770
		NGC 5749 - 60		14 44 15	−54 03.8	1	11.78		0.93		0.53		1		1761
130073	−43 09326	HR 5509		14 44 16	−43 20.9	7	6.30	.008	1.08	.006	0.89	.013	28	G8 III	14,15,278,977,1075*
		E6 - f		14 44 16	−45 12.	1	11.57		1.10				4		1075
		NGC 5749 - 68		14 44 17	−54 12.4	1	12.84		0.57		0.08		3		1761
		NGC 5749 - 61		14 44 19	−54 05.2	1	11.39		0.44		0.13		2		1761
		G 239 - 26		14 44 20	+71 39.9	1	12.25		0.48		−0.24		2		1658
130037	−53 06105	NGC 5749 - 70		14 44 20	−54 14.1	1	10.17		0.17		−0.48		4	B4 V	1761
		NGC 5749 - 99		14 44 21	−54 19.0	1	13.52		0.72		0.15		1		1761
		NGC 5749 - 102		14 44 22	−54 16.9	1	10.70		1.17		0.97		3		1761
		LTT 5870		14 44 23	−12 31.7	2	12.09	.005	1.54	.007			3	K7	940,1705
130157	−20 04093	HR 5513		14 44 23	−21 07.0	2	6.05	.010	1.64	.005	1.98		6	K5 III	2007,3005
		NGC 5749 - 67		14 44 23	−54 11.9	1	12.04		0.65		0.18		3		1761
		CS 22871 # 97		14 44 25	−21 25.6	1	12.27		0.24		0.16		1		1736

HD	DM	Other Id	N	Rem	α_{1950}	δ_{1950}	S	V	σ_V	B–V	σ_{B-V}	U–B	σ_{U-B}	n	Spectrum	References
	−44 09603				14 44 26	−45 03.9	4	10.53	.004	0.26	.004	0.14	.005	24		861,863,1460,2011
		CS 22874 # 110			14 44 27	−27 23.6	1	15.79		0.21		0.18		1		1736
		NGC 5749 - 98			14 44 27	−54 19.8	1	13.45		0.83		0.22		2		1761
130158	−25 10534	HR 5514			14 44 28	−25 24.9	2	5.61	.000	-0.05	.006	-0.18		3	B9 IV/V	1770,2008
130133	−40 08977				14 44 28	−41 03.0	1	8.44		0.14		0.12		1	A2/3 V	1771
		NGC 5749 - 91		V	14 44 28	−54 23.8	1	12.57		1.46		1.68		2		1761
	−44 09604				14 44 29	−45 07.4	1	10.97		0.49				4		2011
129954	−66 02645	HR 5500		⋆ A	14 44 29	−66 23.1	5	5.90	.013	-0.10	.011	-0.70	.012	23	B2.5 Ve	15,26,1075,1637,2038
		G 178 - 49			14 44 31	+39 11.4	1	13.60		0.98		0.50		1		3060
		NGC 5749 - 101			14 44 31	−54 16.8	1	13.60		0.65		0.04		2		1761
		NGC 5749 - 95			14 44 31	−54 20.7	1	13.73		0.49		0.09		2		1761
130039	−59 05717				14 44 33	−59 47.0	1	10.35		0.25		-0.16		3	B5 III	567
		CS 22874 # 93			14 44 34	−24 23.2	1	11.76		0.27		0.18		1		1736
		NGC 5749 - 100			14 44 35	−54 16.3	1	13.42		0.48		0.32		3		1761
		CS 22874 # 99			14 44 36	−24 58.8	1	14.86		0.47		-0.16		1		1736
130161	−34 09923				14 44 36	−34 31.5	1	7.85		0.34		0.06		1	A9 V	1771
130163	−39 09222				14 44 36	−39 43.0	1	6.93		0.02		0.01		1	A0 V	1770
130119	−45 09419				14 44 36	−45 56.9	6	8.09	.003	-0.03	.003	-0.25	.006	52	B8 III	474,977,1075,1586,1770,2011
		CS 22874 # 87			14 44 39	−23 51.0	1	15.40		0.21		0.22		1		1736
		CS 22874 # 107			14 44 40	−26 34.8	1	13.35		0.27		0.26		1		1736
		G 136 - 14			14 44 41	+17 17.7	1	11.84		1.41				1		1759
		NGC 5749 - 97			14 44 41	−54 19.2	1	13.51		0.58		0.05		3		1761
		NGC 5749 - 92			14 44 41	−54 21.8	1	12.96		0.38		0.20		3		1761
130149	−46 09586				14 44 42	−47 03.2	1	8.73		0.04		-0.12		1	B9 V	1770
		NGC 5749 - 96			14 44 42	−54 19.3	1	12.66		0.68		0.13		2		1761
130020	−61 04713				14 44 42	−61 54.2	1	9.01		0.18				4	B5/7 III	2014
		NGC 5749 - 64			14 44 43	−54 08.9	1	13.44		0.61		0.34		2		1761
	−54 06165	NGC 5749 - 90			14 44 43	−54 29.8	1	10.94		0.62		0.07		1		1761
130307	+3 02938	G 66 - 24			14 44 45	+02 54.8	1	7.78		0.90		0.57		1	G8 V	333,1620
		NGC 5749 - 63			14 44 45	−54 08.1	1	12.84		0.41		0.26		2		1761
		NGC 5749 - 87			14 44 45	−54 25.1	1	13.07		1.12		0.93		2		1761
		NGC 5749 - 88			14 44 45	−54 25.4	1	12.74		0.50		0.20		2		1761
		NGC 5749 - 85			14 44 46	−54 20.9	1	13.91		0.64		0.05		2		1761
		NGC 5749 - 89			14 44 48	−54 27.5	1	13.18		1.14		1.04		1		1761
130259	−25 10537	HR 5516			14 44 49	−25 52.7	4	5.25	.016	0.94	.012	0.65	.022	15	G8/K0 III	3,418,2007,3016
		CS 22874 # 132			14 44 52	−23 14.2	1	13.82		0.28		0.14		1		1736
		NGC 5749 - 73			14 44 52	−54 15.6	1	13.92		0.89		0.46		1		1761
		NGC 5749 - 79			14 44 53	−54 17.7	1	12.56		0.29		0.13		3		1761
		NGC 5749 - 80			14 44 53	−54 18.7	1	12.96		0.38		0.21		3		1761
		NGC 5749 - 71		AB	14 44 54	−54 15.3	1	13.02		0.70		0.32		2		1761
		NGC 5749 - 82			14 44 54	−54 19.1	1	11.84		0.50		0.16		3		1761
		NGC 5749 - 86			14 44 55	−54 24.0	1	12.52		0.55		0.01		1		1761
130152	−53 06108	NGC 5749 - 72			14 44 56	−54 15.9	2	9.66	.018	0.18	.016	-0.33	.006	11	B6 Ib/II	1101,1761
		NGC 5749 - 74		V	14 44 57	−54 16.3	1	13.02		0.42		0.25		3		1761
	−60 05521				14 44 58	−60 32.1	1	10.77		0.43		-0.15		2		1586
		NGC 5749 - 57			14 44 60	−54 02.5	1	10.92		1.33		1.20		1		1761
		NGC 5749 - 66			14 44 60	−54 11.2	1	11.16		1.01		0.65		4		1761
		He3 1103			14 45 00	−44 09.8	1	13.20		0.80		-0.17		1		1753
		NGC 5749 - 81			14 45 00	−54 19.8	2	11.76	.022	0.49	.003	0.03	.010	5		1101,1761
		CS 22874 # 95			14 45 01	−24 45.2	1	13.05		0.31		0.14		1		1736
130201	−45 09421				14 45 01	−45 27.7	2	10.10	.007	0.07	.002	0.11		10	A0 IV/V	863,2011
130274	−26 10519	HR 5517			14 45 02	−26 26.3	3	5.77	.002	-0.01	.008	-0.07	.015	7	B9.5V	1079,1770,2007
	−53 06110	NGC 5749 - 59		AB	14 45 02	−54 05.7	1	10.70		0.31		0.17		2		1761
130242	−36 09641				14 45 03	−36 42.5	1	9.35		0.16		0.11		1	A2 IV	1770
		NGC 5749 - 76			14 45 03	−54 18.0	1	14.08		0.64		0.20		1		1761
		NGC 5749 - 78			14 45 03	−54 18.4	1	13.95		0.67		0.25		2		1761
		NGC 5749 - 65			14 45 04	−54 11.4	1	12.43		1.38		1.16		2		1761
130385	+18 02933				14 45 05	+17 57.9	1	8.41		0.54		0.05		2	G5	1648
		CS 22874 # 96			14 45 06	−24 52.4	1	14.05		0.01		-0.20		1		1736
	−57 06791	LSS 3283			14 45 06	−57 21.8	2	10.33	.015	0.95	.046	-0.16	.019	4	B1 Iab	540,1737
		CS 22874 # 131			14 45 07	−23 44.1	1	15.44		0.12		0.11		1		1736
130079	−64 03016				14 45 07	−65 02.8	1	10.49		0.32		0.16		2	B9 V	434
130021	−68 02185				14 45 08	−68 43.7	2	6.50	.002	-0.10	.002	-0.64		9	B2 V	1732,2012
	−53 06112	NGC 5749 - 77			14 45 09	−54 19.1	2	10.85	.015	0.24	.009	-0.27	.015	6		1101,1761
		NGC 5749 - 83			14 45 09	−54 21.5	1	13.08		0.71		0.26		1		1761
		NGC 5749 - 49			14 45 10	−54 13.6	1	12.18		1.45		1.30		2		1761
		NGC 5749 - 35			14 45 10	−54 16.0	1	14.18		0.67		0.08		1		1761
130325	−12 04134	HR 5518			14 45 11	−12 37.8	2	6.34	.010	1.10	.002	0.95		6	K0 III	253,2007
130278	−38 09655				14 45 11	−39 11.8	1	9.52		0.30		0.16		1	A2mA8-F2	1771
		NGC 5749 - 33			14 45 11	−54 16.6	1	13.04		0.42		0.20		2		1761
		NGC 5749 - 75			14 45 11	−54 18.4	1	12.39		0.35		0.07		2		1761
		CS 22871 # 96			14 45 12	−21 20.4	1	13.71		0.11		0.16		1		1736
130277	−38 09657				14 45 14	−38 49.2	1	8.17		0.20		0.14		1	A4/5 IV	1771
		NGC 5749 - 84			14 45 15	−54 23.4	1	13.51		0.62		0.08		2		1761
		NGC 5749 - 56			14 45 16	−54 07.1	1	11.66		0.28		0.13		2		1761
		NGC 5749 - 36			14 45 17	−54 15.1	1	12.76		0.43		0.25		3		1761
		NGC 5749 - 37			14 45 17	−54 15.1	1	13.64		0.64		0.10		3		1761
		NGC 5749 - 34			14 45 18	−54 16.2	1	13.03		0.36		0.24		4		1761
		NGC 5749 - 32			14 45 19	−54 17.2	1	13.37		0.63		0.02		2		1761
		NGC 5749 - 31			14 45 19	−54 18.4	1	11.64		2.04		2.36		2		1761

Table 1   817

HD	DM	Other Id	N Rem	α_{1950}	δ_{1950}	S	V	σ_V	B–V	σ_{B-V}	U–B	σ_{U-B}	n	Spectrum	References
	−45 09428			14 45 20	−45 57.3	1	9.60		1.04				4		2011
		NGC 5749 - 30		14 45 20	−54 18.6	1	12.30		1.35		1.09		1		1761
		G 124 - 72		14 45 21	−02 57.4	2	13.28	.000	1.63	.000	1.20		4		203,1759
		NGC 5749 - 38		14 45 23	−54 15.5	2	11.58	.015	0.26	.005	-0.07	.014	7		1101,1761
		NGC 5749 - 12		14 45 23	−54 26.1	1	13.33		0.45		0.34		2		1761
		NGC 5749 - 51		14 45 24	−54 11.9	1	13.69		0.78		0.26		3		1761
	−53 06113	NGC 5749 - 17		14 45 24	−54 15.5	2	12.13	.022	0.50	.010	0.27	.036	5		1101,1761
130185	−59 05721			14 45 24	−59 41.8	1	9.29		0.32		-0.09		4	B7 II	567
	−53 06114	NGC 5749 - 29		14 45 25	−54 18.5	2	10.26	.032	1.26	.000	1.31	.012	6		1101,1761
		NGC 5749 - 23		14 45 25	−54 19.6	2	10.84	.020	0.27	.009	-0.20	.012	6		1101,1761
		NGC 5749 - 16		14 45 25	−54 21.6	1	13.71		0.62		0.10		3		1761
		CS 22874 # 97		14 45 26	−25 00.5	1	15.01		0.46		-0.18		1		1736
		NGC 5749 - 50		14 45 26	−54 12.1	1	12.57		0.51		0.03		2		1761
130244	−53 06115	NGC 5749 - 39		14 45 26	−54 15.3	2	10.38	.022	0.12	.013	0.03	.045	7	A0 V	1101,1761
130205	−58 05719	IDS14416S5859	AB	14 45 26	−59 12.0	1	6.62		1.54				4	K3/4 III	1075
		NGC 5749 - 18		14 45 27	−54 20.6	1	13.91		0.73		0.14		2		1761
130227	−56 06441	HR 5515		14 45 27	−56 27.5	7	6.22	.007	1.12	.006	1.06	.000	28	K1 III	15,1075,2013,2018*
		NGC 5749 - 47		14 45 28	−54 13.6	1	13.71		0.68		0.13		2		1761
		G 66 - 26		14 45 29	+04 10.0	1	10.96		0.99		0.86		1		1620
	−45 09429			14 45 29	−45 24.5	1	9.60		1.05		0.81		5		863
		NGC 5749 - 27		14 45 29	−54 17.7	1	12.53		0.33		0.19		3		1761
		NGC 5749 - 26		14 45 29	−54 18.0	2	10.87	.025	0.23	.012	-0.26	.004	5		1101,1761
	−53 06117	NGC 5749 - 25		14 45 29	−54 18.1	2	11.13	.016	0.31	.006	-0.12	.013	6		1101,1761
		NGC 5749 - 22		14 45 29	−54 19.7	2	11.29	.022	0.25	.026	-0.27	.003	6		1101,1761
		NGC 5749 - 14		14 45 29	−54 23.4	1	13.85		0.50		0.33		2		1761
130328	−36 09645	HR 5519, V768 Cen		14 45 31	−36 25.6	1	6.10		1.52		1.28		10	M5 III	3042
		NGC 5749 - 53		14 45 31	−54 10.6	1	13.63		0.75		0.34		2		1761
		NGC 5749 - 48		14 45 31	−54 13.2	1	14.34		0.71		0.09		2		1761
		NGC 5749 - 54		14 45 32	−54 06.6	1	12.77		0.41		0.21		1		1761
		NGC 5749 - 24		14 45 32	−54 19.3	1	13.31		0.60		0.21		3		1761
		NGC 5749 - 55		14 45 33	−54 05.6	1	12.44		0.66		0.05		2		1761
	−53 06119	NGC 5749 - 15		14 45 33	−54 15.4	2	12.13	.015	0.33	.019	-0.02	.016	6		1101,1761
130834	+72 00652			14 45 34	+72 10.5	1	7.50		0.15		0.16		2	A5	1733
		NGC 5749 - 40		14 45 34	−54 15.4	2	11.49	.016	0.26	.002	-0.05	.025	7		1101,1761
		NGC 5749 - 28		14 45 34	−54 18.3	2	11.94	.004	1.44	.010	1.33	.051	4		1101,1761
		NGC 5749 - 20		14 45 35	−54 20.4	1	14.02		0.69		0.08		2		1761
		NGC 5749 - 19		14 45 35	−54 21.2	1	13.28		0.62		0.04		3		1761
		NGC 5749 - 11		14 45 36	−54 29.4	1	13.20		0.60		0.42		2		1761
		NGC 5749 - 41		14 45 37	−54 16.3	1	13.28		0.46		0.29		1		1761
130500	+26 02603			14 45 38	+25 40.8	1	8.30		0.98		0.72		4	G8 II-III	8100
		NGC 5749 - 45		14 45 38	−54 14.7	1	13.21		0.64		0.35		1		1761
		NGC 5749 - 46		14 45 39	−54 13.1	1	11.87		1.29		0.99		3		1761
		CS 22874 # 130		14 45 40	−23 52.8	1	13.77		0.22		0.22		1		1736
		NGC 5749 - 42		14 45 40	−54 16.3	1	13.59		0.68		0.09		2		1761
130264	−56 06445			14 45 40	−57 13.8	1	7.76		0.61				4	F5 V	2012
130247	−59 05724			14 45 41	−59 30.2	1	9.28		0.37		-0.16		1	B7 II	567
		NGC 5749 - 13		14 45 42	−54 24.6	1	12.34		0.38		0.10		3		1761
		NGC 5749 - 21		14 45 43	−54 20.5	1	14.16		0.80		0.20		2		1761
130372	−32 10387			14 45 44	−32 34.9	1	9.82		0.39		0.01		1	A9 V	1771
130265	−58 05724			14 45 44	−58 55.4	2	8.54	.015	0.65	.000	0.15		7	G3 V	803,1075
130684	+60 01567			14 45 45	+59 55.1	1	7.35		0.25		0.05		2	A2	1502
		CS 22871 # 115		14 45 47	−17 24.9	1	14.11		0.10		0.18		1		1736
130555	+36 02531			14 45 48	+35 51.7	1	7.57		0.41		-0.04		2	F5	1601
130411	−16 03940			14 45 48	−16 48.9	1	10.03		0.40		0.07		1	A9/F0 V	1770
		NGC 5749 - 52		14 45 48	−54 10.2	1	13.65		0.66		0.17		1		1761
		NGC 5749 - 10		14 45 48	−54 29.3	1	13.16		0.71		0.26		2		1761
		CS 22874 # 117		14 45 50	−25 26.8	1	14.81		0.26		0.28		1		1736
		NGC 5749 - 44		14 45 50	−54 14.3	1	13.13		0.57		0.23		1		1761
		NGC 5749 - 43		14 45 51	−54 14.1	1	13.94		0.54		0.34		1		1761
130298	−55 06191	LSS 3284		14 45 53	−56 13.2	5	9.25	.017	0.43	.014	-0.56	.022	12	O5/6	403,540,567,976,1586
		CS 22871 # 109		14 45 54	−18 35.5	1	14.46		0.08		0.08		1		1736
130365	−45 09435			14 45 54	−45 40.0	3	9.08	.003	0.20	.001	0.12	.002	10	A1 V	1075,1770,2011
130388	−35 09820			14 45 56	−35 20.5	1	7.62		0.20		0.14		1	A2 III	1770
130364	−45 09436			14 45 56	−45 35.5	1	9.29		0.51				4	F7/8 IV	2011
		NGC 5749 - 6		14 45 59	−54 15.2	1	13.64		0.51		0.35		2		1761
130430	−19 03963			14 45 60	−19 49.9	1	10.44		0.07		0.00		1	B9.5 V	1770
130348	−53 06122	NGC 5749 - 7		14 46 02	−54 18.3	2	9.69	.025	1.03	.002	0.77	.003	8	G8 III/IV	1101,1761
	+39 02801			14 46 03	+38 40.5	1	9.70		1.31		1.14		2	M2	3072
		NGC 5749 - 9		14 46 04	−54 28.3	1	12.02		0.39		0.18		2		1761
130463	−12 04137			14 46 05	−12 54.7	1	7.64		1.06		0.91		5	K0 III	1657
		NGC 5749 - 8		14 46 06	−54 24.0	1	12.56		0.47		0.28		2		1761
		NGC 5749 - 5	V	14 46 09	−54 16.7	1	13.39		0.55		0.29		2		1761
130603	+24 02779	HR 5524	★ AB	14 46 10	+24 34.4	2	6.16	.015	0.49	.010	0.08		4	F2 V	70,3016
130414	−39 09247			14 46 10	−39 51.6	1	9.80		0.39		0.01		1	A9 V	1771
130505	−14 04034			14 46 14	−14 30.2	1	10.65		0.24		0.18		1	A1/2 IV	1770
130377	−52 07544			14 46 14	−52 32.2	1	7.16		0.09		-0.02		2	B9 IV	1770
130350	−58 05725	LSS 3285		14 46 15	−59 07.5	3	9.02	.021	0.56	.010	-0.34	.032	13	B2 Ib/II	540,567,976
		G 239 - 29		14 46 16	+68 26.3	1	13.46		1.63							1759
	+42 02535	G 178 - 50		14 46 18	+41 43.8	2	10.56	.005	0.75	.005	0.26	.015	2	G8	1658,3060
		CS 22874 # 116		14 46 18	−25 47.1	1	15.73		0.11		0.18		1		1736

HD	DM	Other Id	N Rem	α_{1950}	δ_{1950}	S	V	σ_V	B–V	σ_{B-V}	U–B	σ_{U-B}	n	Spectrum	References
130416	−44 09627			14 46 18	−44 19.5	2	9.47	.000	0.35	.005			5	F3 IV/V	2001,2011
130417	−46 09605			14 46 18	−47 18.1	1	8.71		0.03		0.01		2	B9.5 V	1770
		CS 22871 # 111		14 46 19	−18 27.8	1	13.20		0.34		0.15		1		1736
130415	−43 09360			14 46 19	−43 30.9	1	8.65		0.31		0.08		1	A8/9 IV	1771
		NGC 5749 - 4		14 46 19	−54 12.2	1	13.13		0.55		0.41		2		1761
130557	−0 02886	HR 5522		14 46 20	−00 38.4	1	6.13		−0.04		−0.10		8	B9 V(SiCr)	1088
		NGC 5749 - 3		14 46 23	−54 13.2	1	13.01		0.51		0.12		2		1761
130604	+6 02946	IDS14439N0622	AB	14 46 24	+06 09.8	2	6.85	.010	0.45	.010	−0.01		6	F6 V	1381,3030
130435	−44 09628			14 46 24	−44 39.8	1	10.02		0.43				4	F0 IV/V	2011
		Wolf 546		14 46 25	+00 51.8	1	12.40		0.82		0.38		2		1696
130529	−23 11916	HR 5521	⋆A	14 46 25	−24 02.7	1	5.68		1.30				4	K3 III +G	2007
130380	−58 05726	LSS 3286		14 46 28	−58 46.2	1	6.85		1.19		0.77		3	F8 Ib/II	8100
		GD 171		14 46 29	+18 47.0	1	12.57		0.41		−0.23		3		308
		Steph 1185		14 46 29	−05 43.9	1	11.53		1.34		1.22		1	K7	1746
		CS 22874 # 127		14 46 30	−23 50.3	1	13.64		0.10		0.16		1		1736
130559	−13 03986	HR 5523	⋆AB	14 46 34	−13 56.5	2	5.32		0.07	.000	−0.05		7	Ap SrEuCr	1063,2007
		CS 22874 # 129		14 46 34	−23 44.0	1	14.86		0.13		0.13		1		1736
155717	+0 03246			14 46 35	+00 28.3	1	9.92		0.94		0.75		1	K3 V	3072
130563	−19 03966			14 46 37	−19 41.8	1	7.20		0.23		0.18		11	A5/7 IV	1770
		NGC 5749 - 1		14 46 38	−54 24.0	1	11.90		0.46		0.18		1		1761
		CS 22874 # 126		14 46 39	−23 51.2	1	14.32		0.23		0.22		1		1736
		NGC 5749 - 2		14 46 39	−54 17.6	1	12.35		0.75		0.17		1		1761
130540	−32 10407			14 46 46	−32 46.8	1	8.64		0.03		0.01		2	A0 V	1770
130454	−51 08500			14 46 46	−52 10.7	1	7.77		0.03		−0.24		2	B8 II	1770
	−25 10553	IDS14439S2540	A	14 46 47	−25 52.7	2	11.70	.020	1.49	.006	1.05		2		1705,3078
	−25 10553	IDS14439S2540	B	14 46 47	−25 52.7	2	12.09	.020	1.50	.022	1.05		2		1705,3078
130669	+10 02747	G 136 - 18	⋆AB	14 46 49	+10 25.4	3	8.44	.007	0.87	.007	0.60	.028	8	K2 V +K2 V	1003,1381,3030
130532	−40 09007			14 46 49	−40 26.0	1	9.35		0.10		0.08		1	A0 V	1770
	+26 02606	G 166 - 45		14 46 50	+25 54.9	6	9.73	.008	0.43	.005	−0.23	.013	11	sdF9	308,979,1003,1658*
		CS 22871 # 110		14 46 50	−18 29.0	1	15.47		0.08		0.09		1		1736
		G 223 - 82		14 46 51	+63 08.5	1	11.26		0.85		0.51		1		1658
131020	+68 00801			14 46 53	+68 10.1	1	9.59		0.65		0.10		3	K0	3026
	−61 04732	LSS 3287		14 46 53	−61 20.5	1	9.84		0.70		−0.32		2	B0 Ia	403
129752	−81 00679			14 46 53	−82 01.7	1	9.09		0.48		0.04		3	F5 V	156
130543	−41 09194	IDS14437S4126	A	14 46 54	−41 38.4	1	7.82		0.24		0.14		1	A9 V	1771
		CS 22871 # 106		14 46 57	−20 15.3	1	14.99		0.13		0.16		1		1736
130623	−19 03968			14 46 58	−19 24.7	1	10.29		0.50		0.11		1	A7/F0	1770
130437	−59 05734	LSS 3288		14 46 58	−60 04.7	4	9.91	.034	0.71	.021	−0.39	.031	8	Be	403,567,1586,1737
		CS 22874 # 113		14 46 60	−26 17.1	1	14.89		0.13		0.17		1		1736
	+59 01609	G 223 - 83		14 47 00	+59 07.0	2	9.76	.014	0.62	.000	0.07	.005	4	G3	308,1658
130705	+10 02748			14 47 01	+10 15.1	1	6.64		1.26		1.46		3	K4 II-III	37
130766	+25 02839			14 47 02	+25 21.5	1	6.71		1.34		1.53		4	K3 II	8100
	−51 08507	Ho 18 - 3		14 47 05	−52 05.5	1	10.03		1.20		1.05		7		128
130476	−59 05737			14 47 05	−59 27.8	1	9.59		0.41		0.02		6	B9 Ib/II	567
		He3 1032		14 47 06	−59 39.	1	12.31		1.52		0.46		4		1586
		Ho 18 - 4		14 47 07	−52 04.0	1	12.22		0.37		−0.30		4		128
		Ho 18 - 5		14 47 08	−52 03.2	1	12.83		0.62		0.04		3		128
130817	+38 02593	HR 5529		14 47 09	+38 01.0	3	6.17	.009	0.36	.010	−0.09	.000	6	F3 V	70,254,3037
130534	−51 08508	Ho 18 - 1		14 47 10	−52 03.5	2	8.80	.044	0.37	.024	−0.17	.016	17	B3	128,1586
		Ho 18 - 14		14 47 10	−52 04.4	1	13.45		0.39		0.07		2		128
		Ho 18 - 2		14 47 11	−52 04.1	1	11.26		0.35		−0.13		4		128
		Ho 18 - 12		14 47 12	−52 03.	1	13.74		1.35		0.48		3		128
		Ho 18 - 15		14 47 12	−52 03.	1	13.68		0.88		0.05		2		128
		Ho 18 - 16		14 47 12	−52 03.	1	13.62		1.52		0.92		2		128
		CS 22874 # 134		14 47 15	−22 48.9	1	15.72		0.10		0.04		1		1736
		Ho 18 - 6		14 47 15	−52 03.6	1	11.68		0.41		−0.01		2		128
		Ho 18 - 7		14 47 16	−52 03.5	1	12.38		0.49		0.01		3		128
		Ho 18 - 13		14 47 18	−52 02.3	1	13.16		1.48		1.04		2		128
		G 124 - 76		14 47 19	−04 03.1	1	14.17		1.31		1.07		3		316
130694	−27 10073	HR 5526		14 47 21	−27 45.2	6	4.41	.003	1.39	.006	1.48	.007	16	K3 III	15,1075,2012,3077*
		Ho 18 - 8		14 47 21	−52 03.9	1	12.49		0.44		−0.05		3		128
130788	+11 02730			14 47 23	+10 42.0	1	7.09		0.10		0.05		3	A2	1648
130550	−55 06200			14 47 23	−56 14.4	1	9.48		0.54		0.25		2	A5/7 V	403
130656	−34 09964			14 47 25	−34 53.2	1	8.07		0.15		0.10		1	A2 IV	1770
	−51 08512	Ho 18 - 9		14 47 26	−52 03.9	2	11.20	.040	0.33	.018	−0.19	.031	5		128,540
		Ho 18 - 10		14 47 27	−52 03.3	1	12.24		0.39		−0.02		4		128
130818	+23 02744	RY Boo		14 47 28	+23 14.3	1	7.14		0.42		−0.07		3	F6 IV	196
		CS 22874 # 115		14 47 29	−25 52.2	1	15.12		0.22		0.22		1		1736
130945	+46 01993	HR 5533		14 47 32	+46 19.4	2	5.75	.018	0.48	.007	0.00	.001	55	F7 IV	254,274
		CS 22871 # 99		14 47 33	−21 30.8	1	14.99		0.06		0.11		1		1736
		G 66 - 30		14 47 35	+01 02.9	4	11.05	.015	0.40	.003	−0.18	.008	10	sdF5	516,1620,1696,3077
		CS 22871 # 107		14 47 35	−20 08.8	1	14.02		0.46		−0.14		1		1736
		L 478 - 87		14 47 36	−37 20.	1	13.24		0.93		0.37		2		3062
130551	−60 05530			14 47 37	−60 43.6	5	7.18	.015	0.45	.010	−0.10	.004	10	F5 V	742,1075,1594,3037,6006
	+45 02224		AB	14 47 39	+45 28.9	1	10.70		1.11		0.80		2		497
	−51 08516	Ho 18 - 11		14 47 39	−52 03.8	1	11.34		0.36		−0.18		3		128
130697	−42 09722			14 47 43	−42 37.0	7	6.83	.006	0.13	.002	0.14	.005	57	A2 V	278,977,977,1075,1770*
		CS 22874 # 124		14 47 44	−24 10.5	1	14.48		0.15		−0.91		1		1736
		CS 22874 # 114		14 47 46	−25 49.3	1	14.80		0.05		0.01		1		1736
130987	+47 02177	IDS14460N4700	A	14 47 48	+46 47.0	1	7.55		0.50		−0.01		2	F8	1601

Table 1 819

HD	DM	Other Id	N Rem	α_{1950}	δ_{1950}	S	V	σ_V	B–V	σ_{B-V}	U–B	σ_{U-B}	n	Spectrum	References
130917	+29 02581	HR 5532	⋆A	14 47 49	+28 49.3	2	5.80		0.05	.004	0.08		6	A4 V	70,1022
130612	−59 05739			14 47 51	−60 17.6	1	8.94		0.49		-0.12		4	B3 II/III	567
130731	−35 09847			14 47 54	−36 10.8	1	9.22		0.32		0.13		1	A9 IV/V	1771
130871	+7 02850	G 66 - 31		14 47 55	+07 01.3	4	9.08	.007	0.96	.005	0.72	.009	8	K2 V	265,1003,1705,3008
131040	+51 01957	HR 5537	⋆A	14 47 55	+51 34.8	2	6.49	.015	0.39	.010	-0.01	.010	3	F4 V	1733,3016
131040	+51 01957	IDS14463N5147	B	14 47 55	+51 34.8	1	9.86		0.98		0.62		3		3016
130819	−15 03965	HR 5530	⋆B	14 47 55	−15 47.4	15	5.15	.010	0.40	.007	-0.03	.014	70	F4 IV	3,15,1006,1008,1034*
130895	+17 02790			14 47 56	+17 00.7	1	9.20		0.37		-0.02		3	F8	1648
130779	−36 09679	IDS14448S3641	AB	14 47 56	−36 53.4	1	9.06		0.81				5	G8 V	173
130758	−40 09026			14 47 57	−40 36.0	1	9.11		0.27		0.13		1	A7 V	1771
131041	+49 02326	HR 5538	⋆AB	14 47 59	+48 55.5	1	5.69		0.47				1	F6 V + F5 V	71
130948	+24 02786	HR 5534		14 48 02	+24 07.0	2	5.86	.005	0.56	.000	0.01		4	G2 V	70,3077
		CS 22871 # 104		14 48 02	−20 36.2	1	14.91		0.42		-0.09		1		1736
		CS 22871 # 103		14 48 02	−20 40.9	1	15.28		0.37		0.10		1		1736
130760	−45 09466			14 48 05	−46 16.7	1	9.11		0.24		0.09		4	A1 V	244
130841	−15 03966	HR 5531	⋆A	14 48 06	−15 50.1	16	2.75	.007	0.15	.007	0.10	.018	79	A3 IV	3,15,30,1006,1020,1034*
130761	−46 09630			14 48 09	−46 52.7	5	7.99	.008	0.44	.009	-0.03	.014	21	F5 V	116,977,1075,2001,2011
130458	−72 01604	HR 5520	⋆AB	14 48 11	−72 59.2	3	5.59	.004	0.82	.000	0.42	.000	14	G5 III	15,1075,2038
		GD 172		14 48 16	+18 56.6	1	15.34		0.60		-0.16		1		3060
		CS 22874 # 133		14 48 17	−23 08.2	1	15.09		0.11		0.15		1		1736
130900	−14 04045			14 48 18	−14 49.3	1	7.19		0.58				28	K1 III +A/F	130
		G 136 - 21		14 48 19	+19 38.4	1	14.54		1.56				1		906
130807	−43 09391	HR 5528	⋆AB	14 48 22	−43 22.2	11	4.32	.006	-0.15	.006	-0.62	.009	52	B5 IV	15,26,125,278,1020,1075*
130792	−44 09663			14 48 22	−45 15.2	1	9.73		0.33				4	F0 IV/V	2011
		POSS 327 # 3		14 48 23	+29 26.8	1	17.01		1.41				2		1739
		G 66 - 32		14 48 24	+07 46.2	2	15.46	.005	0.03	.009	-0.67	.000	5		1,316
		LP 801 - 63		14 48 24	−17 47.7	1	10.83		0.52		-0.09		2		1696
		CS 22874 # 125		14 48 24	−24 04.9	1	14.99		0.13		0.15		1		1736
	−58 05734			14 48 24	−59 12.3	1	10.64		0.60		-0.06		5		567
130952	−1 02991	HR 5535		14 48 25	−02 05.5	5	4.94	.013	0.99	.003	0.72	.015	20	G8 III-IV	15,1003,1256,3016,4001
130970	+0 03253	HR 5536		14 48 26	−00 03.1	2	6.16	.010	1.40	.005	1.68	.005	10	K3 III	252,1088
130902	−27 10082			14 48 28	−27 25.9	1	8.57		0.16		0.09		1	A0 V	1770
130921	−16 03949	IDS14457S1634	A	14 48 30	−16 47.2	1	9.33		0.38		0.13		1	A8/F0 IV	1770
130701	−63 03436	HR 5527, AX Cir		14 48 30	−63 36.3	3	5.81	.114	0.71	.037	0.27		6	G3 II + B8:	1484,1754,2006
131111	+37 02580	HR 5541		14 48 31	+37 28.6	4	5.48	.005	1.03	.007	0.84	.004	9	K0 III-IV	15,1003,1080,3035
130957	−17 04192			14 48 34	−18 00.2	1	9.28		0.36		0.04		1	A9 V	1770
131023	+10 02752	IDS14462N1008	AB	14 48 37	+09 55.7	1	7.40		0.76		0.34		2	K0	3016
130956	−17 04193	IDS14458S1728	A	14 48 37	−17 40.3	1	8.74		0.33		0.10		1	A2/3 + (G)	1770
		CS 22871 # 113		14 48 37	−19 19.5	1	14.85		0.15		0.23		1		1736
131194	+53 01741			14 48 39	+52 37.1	1	7.15		0.52		0.05		2	F5	1566
		L 406 - 53		14 48 41	−40 58.5	1	13.36		1.39				1		3062
130861	−42 09734			14 48 41	−43 02.7	1	9.06		0.33		0.10		1	A8 V	1771
		PSD 195 # 107		14 48 42	−58 26.	1	13.99		0.69		0.09		2	F4	1101
130862	−44 09665			14 48 43	−45 15.0	4	9.41	.008	0.22	.006	0.16	.007	18	A3/5 III	116,863,1075,2011
130795	−56 06466			14 48 43	−57 15.6	1	8.86		0.04		-0.45		3	B4 V	567
130903	−40 09037			14 48 45	−40 36.0	1	7.93		-0.08				4	B(2)p	2014
130989	−17 04196			14 48 47	−17 35.0	1	6.53		0.46		-0.04		4	F6 V	3077
130811	−56 06467			14 48 47	−56 23.1	2	10.21	.002	0.31	.002	-0.09	.230	4	B2 II/III	540,567
130992	−23 11940			14 48 50	−24 05.6	3	7.83	.013	1.01	.010	0.87	.005	11	K3 V	1311,2012,3078
130904	−44 09667			14 48 57	−44 43.3	8	7.20	.008	0.54	.003	0.05	.008	47	F6 V	278,863,977,977,1075,1770*
130881	−51 08533	IDS14455S5115	A	14 48 59	−51 27.1	1	9.35		0.07		-0.06		1	B9.5 V	1770
	−60 05540			14 49 00	−60 44.4	1	15.07		1.19		1.08		2		1101
130930	−43 09401			14 49 01	−43 36.7	2	8.70	.002	0.93	.003	0.73		12	K2 V	977,2011
130830	−59 05742			14 49 01	−59 33.9	1	12.06		0.55		0.05		2	A9 IV	1101
130931	−46 09638			14 49 02	−46 25.6	4	9.00	.004	0.54	.003	0.06	.006	22	F(3) (V)	863,977,1075,2011
130932	−46 09639			14 49 02	−46 33.3	3	6.86	.007	1.66	.013	1.95		13	M1/2 III	977,2001,2011
131156	+19 02870	HR 5544, ξ Boo	⋆A	14 49 05	+19 18.5	3	4.56	.051	0.76	.016	0.24	.005	61	G8 V	150,1080,1351
131156	+19 02870	HR 5544, ξ Boo	⋆AB	14 49 05	+19 18.5	11	4.56	.025	0.77	.011	0.28	.018	44	G8 V +K5Ve	15,150,938,1007,1013*
131156	+19 02870	IDS14468N1931	B	14 49 05	+19 18.5	2	6.95		1.10		1.15			K5V e	150,3030
131133	+8 02927	G 66 - 33		14 49 07	+07 56.9	3	9.51	.009	0.64	.009	0.40	.040	3	F8	333,1620,1658,1696
131047	−19 03974			14 49 09	−20 13.6	1	10.43		0.41		0.05		1	A9 V	1770
	−31 11530			14 49 12	−31 31.1	1	12.09		0.73		0.05		2		1696
		Feige 103		14 49 18	+11 51.9	1	12.38		-0.16		-0.55		1		1298
130998	−46 09642			14 49 21	−46 48.6	5	8.18	.009	0.05	.006	0.06	.008	20	A0 Vn	768,977,1075,1770,2011
130997	−45 09484			14 49 23	−45 23.2	2	9.14	.006	1.09	.001	0.91		10	K0/1 III/IV	863,2011
131077	−38 09726			14 49 33	−38 44.4	1	8.51		0.15		0.17		1	A7 V	1771
131117	−30 11780	HR 5542		14 49 34	−30 22.3	4	6.29	.017	0.60	.008	0.10		13	G1 V	15,1075,2007,3077
131096	−31 11536			14 49 34	−31 53.0	1	9.46		0.45		0.04		1	A9 V	1771
		G 66 - 34		14 49 38	+05 36.1	1	12.13		1.42		1.10		1		1620
130912	−63 03442			14 49 40	−63 34.2	1	8.91		0.02		-0.25		4	A3/5 V	1586
131301	+39 02806			14 49 41	+39 18.5	2	7.87	.010	0.44	.005	-0.05	.010	5	F5 V	833,1601
131120	−37 09760	HR 5543		14 49 42	−37 35.9	4	5.02	.003	-0.17	.003	-0.73	.008	12	B7 II/III	15,26,1770,2029
131162	−23 11946			14 49 44	−23 54.8	1	7.57		-0.01		-0.20		2	B9 IV	1770
131078	−46 09649	IDS14464S4613	AB	14 49 44	−46 25.9	3	8.15	.001	0.70	.001	0.23	.008	16	G5 V	977,1075,2011
131079	−46 09648			14 49 44	−46 39.3	1	9.39		1.70		1.91		4	M1/2 (III)	244
131140	−33 10135			14 49 47	−33 25.4	1	8.05		0.38		0.15		1	A9 V	1771
131141	−34 09996			14 49 47	−35 10.6	1	9.48		0.33		0.20		1	A(p SrEu)	1770
		Feige 104		14 49 48	+55 49.	1	13.85		-0.03		-0.08		1		3060
131099	−43 09416			14 49 48	−43 31.5	4	8.14	.003	0.24	.007	0.15	.008	27	A5 V	977,1075,1770,2011
		L 199 - 37		14 49 48	−56 13.	1	13.85		1.42				1		3062

HD	DM	Other Id	N	Rem	α_{1950}	δ_{1950}	S	V	σ_V	B–V	σ_{B-V}	U–B	σ_{U-B}	n	Spectrum	References
		WD1449+168			14 49 51	+16 50.2	1	15.44		−0.09		−0.89		1	DA3	1727
131197	−25 10584				14 49 51	−25 50.3	1	9.72		0.90				4	K2 III/IV	2012
131198	−26 10568				14 49 54	−27 02.7	1	9.13		0.19		0.14		1	A2 IV	1770
131015	−57 06819				14 49 55	−57 43.8	1	10.12		0.29		−0.21		1	B5/7 III	567
130940	−65 02914	IDS14458S6600		AB	14 49 56	−66 12.9	1	6.98		0.58				4	G0 V	2012
131250	+0 03259	G 66 - 35			14 49 57	+00 19.1	1	8.68		0.62		0.08		2	G5 V	333,1620
131165	−37 09766				14 49 57	−38 03.1	1	6.58		0.01		0.00		1	A0 V	1770
		G 66 - 36			14 49 58	+00 22.5	1	12.58		1.58		1.12		1		1620
131168	−45 09492	IDS14467S4527		A	14 50 02	−45 39.1	2	7.05	.014	−0.09	.024	−0.73	.005	4	B2/3 Vnne	401,1279
131168	−45 09492	IDS14467S4527		AB	14 50 02	−45 39.1	3	6.96	.036	−0.06	.024	−0.75		10	B2/3 Vnne	1586,2011,2014
131168	−45 09492	IDS14467S4527		B	14 50 02	−45 39.1	1	9.62		0.98		0.72		2	B8	1279
131168	−45 09492	IDS14467S4527		C	14 50 02	−45 39.1	1	13.40		0.56				2		1279
129289	−85 00415				14 50 04	−86 16.4	1	7.71		1.11		0.99		6	K0 III	1704
		G 66 - 37			14 50 05	+12 36.0	1	11.61		1.52		1.21		1		1620
131169		S Lup			14 50 05	−46 24.7	1	9.03		1.94		1.30		1	S e	817
		G 66 - 38			14 50 07	+07 10.0	2	13.00	.015	0.54	.008	−0.17	.008	4		1620,1658
131167	−43 09424				14 50 07	−43 42.3	1	9.31		0.33		0.08		1	A8/9 V	1771
131223	−27 10096				14 50 08	−27 37.3	1	9.19		0.09		0.04		1	A0 V	1770
131213	−32 10441				14 50 08	−33 05.5	1	9.53		0.34		0.10		1	A8 V	1771
131445	+54 01710				14 50 09	+53 57.2	1	8.48		1.24		1.42		3	K3 III	37
131507	+59 01615	HR 5552			14 50 10	+59 29.8	3	5.45	.010	1.37	.010	1.60	.000	7	K4 III	15,1003,1080
131334	+19 02874	IDS14479N1909		AB	14 50 12	+18 56.4	2	8.28	.005	0.61	.005	0.03	.009	4	G0 V	1003,3016
130942	−69 02146				14 50 16	−69 39.4	1	6.38		−0.06		−0.48		5	B5 V	1732
	−60 05552	LSS 3292			14 50 17	−60 56.0	2	10.58	.057	0.90	.013	−0.23	.021	3		1586,1737
		POSS 327 # 2			14 50 18	+29 59.8	1	15.71		1.38				1		1739
	−44 09682				14 50 18	−45 00.4	1	10.05		0.47				4	F5	2011
		PSD 195 # 156			14 50 18	−59 59.	2	11.86	.015	0.23	.010	−0.03	.029	3		1101,1101
131124	−58 05742				14 50 19	−58 42.4	1	8.93		0.14		−0.07		1	B9 III/IV	567
		POSS 327 # 1			14 50 24	+29 52.0	1	14.65		1.55				1		1739
131058	−65 02918	HR 5539			14 50 27	−65 47.3	5	6.08	.005	−0.06	.005	−0.60	.010	25	B3 Vn	15,26,1075,1637,2038
131226	−44 09685				14 50 32	−45 06.5	3	9.09	.000	0.28	.007	0.16		9	A7 III/IV	1075,2001,2011
131243	−45 09496				14 50 35	−46 03.3	5	8.55	.074	0.01	.021	−0.06	.035	23	B9.5V	244,977,1075,1770,2011
	−29 11374				14 50 36	−29 36.9	1	11.07		1.06				2		1730
131451	+32 02523				14 50 38	+32 07.7	1	6.87		1.06		0.89		2	K0	1625
131205	−55 06235	IDS14470S5507		AB	14 50 39	−55 19.8	1	9.45		0.52		0.13		1	F5/6 V	96
131295	−33 10146				14 50 43	−33 29.3	1	7.87		0.19		0.16		1	A2 IV	1770
131258	−43 09432				14 50 44	−43 23.7	3	8.22	.003	1.06	.007	0.79	.009	24	G8 II	863,977,2011
131873	+74 00595	HR 5563		⋆ AB	14 50 50	+74 21.6	5	2.07	.008	1.47	.006	1.77	.007	19	K4 III	15,1194,1363,3016,8015
		LOD 34+ # 5			14 50 50	−55 16.1	1	12.32		0.56		0.03		1		96
		LP 621 - 41			14 50 52	−00 29.1	1	12.58		0.51		−0.19		3		1696
		LOD 34+ # 10			14 50 56	−55 26.1	1	13.24		0.76		0.03		1		96
		LOD 34+ # 9			14 50 57	−55 23.7	1	12.34		0.63		0.10		1		96
	−54 06208				14 50 58	−55 16.6	1	11.32		0.19		0.09		1		96
		LOD 34+ # 2			14 50 59	−55 14.4	1	12.59		0.40		0.23		1		96
131369	−26 10585				14 51 01	−26 31.7	1	8.40		0.26		0.11		1	A2 III	1770
		LOD 34+ # 3			14 51 01	−55 15.2	1	12.49		0.43		0.02		1		96
131473	+16 02705	HR 5550		⋆ AB	14 51 03	+15 54.5	3	6.40	.011	0.56	.008	0.12	.010	7	F4 IV +G1IV	938,1733,3016
131509	+29 02592				14 51 03	+28 42.6	1	7.96		0.86		0.52		4	K0 V	3026
		POSS 327 # 4			14 51 03	+29 57.3	1	17.13		0.99				3		1739
		LOD 34+ # 15			14 51 06	−55 31.7	1	12.95		0.70		0.36		1		96
131511	+19 02881	HR 5553			14 51 07	+19 21.2	4	6.01	.008	0.83	.004	0.51	.013	13	K2 V	22,1080,1197,3077
131299	−50 08899				14 51 07	−51 12.2	2	8.83	.023	0.61	.005	0.11		5	G3 IVw	742,1594
	−55 06237				14 51 08	−55 33.2	1	10.37		0.65		0.18		1		96
		AJ85,265 # 1			14 51 09	−60 16.6	1	13.14		1.39		0.28		2		779
	−55 06239				14 51 10	−55 24.6	1	9.89		1.60		1.71		1		96
131476	+6 02957				14 51 11	+06 26.7	1	6.63		1.17				1	K2 III	2008
		LP 858 - 46			14 51 12	−24 07.9	1	11.30		0.73		0.20		2		1696
		G 136 - 27			14 51 13	+11 46.8	2	15.31	.023	1.87	.075			4		419,1759
		LOD 34+ # 16			14 51 13	−55 30.9	1	13.26		0.93				1		96
		GD 173			14 51 17	+00 37.7	1	15.29		−0.14		−0.96		1	DA	3060
131300	−55 06240				14 51 18	−55 22.5	1	9.33		0.95				1	K5/M0	96
131399	−33 10153	IDS14483S3344		A	14 51 21	−33 56.4	1	7.65		0.27		0.18		1	A1 V	1771
131399	−33 10153	IDS14483S3344		AB	14 51 21	−33 56.4	1	7.05		0.12		0.06		1	A1 V	1770
		LOD 34+ # 13			14 51 23	−55 34.2	1	12.38		0.64		0.16		1		96
131552	+25 02845				14 51 26	+25 01.9	1	9.00		0.48		−0.01		3	F8	1733
131430	−24 11735	HR 5548			14 51 26	−24 26.3	2	5.28	.020	1.33	.010	1.54		7	K3 III	2007,3016
		S448 # 4			14 51 26	−29 43.3	1	11.12		1.19				2		1730
131325	−55 06242				14 51 28	−55 31.7	1	9.15		0.07		−0.31		1	B5 V	96
131582	+23 02751	G 166 - 52		⋆ A	14 51 31	+23 32.9	2	8.67	.018	0.93	.022	0.72	.013	5	K3 V	22,1003
131374	−48 09449				14 51 31	−48 29.6	1	7.91		−0.04		−0.38		2	B9 II	1770
		Wolf 558			14 51 33	+02 07.8	1	12.83		0.62		−0.12		2		1696
131432	−32 10457	HR 5549		⋆ AB	14 51 34	−33 05.9	1	5.82		1.42				4	K2 III	2007
		LOD 34+ # 12			14 51 34	−55 34.8	1	12.01		0.41		0.18		1		96
131484	−18 03927				14 51 35	−19 16.9	1	10.16		0.38		0.13		1	A3/7 II	1770
131485	−27 10114				14 51 37	−28 04.3	1	8.89		0.28		0.14		1	A1 III/IV	1770
131459	−31 11563				14 51 39	−31 31.7	1	9.30		0.27		0.17		1	A2 IV/V	1770
		G 166 - 53			14 51 40	+23 45.5	1	11.69		1.59				1	M5	1759
131530	−11 03827	HR 5554			14 51 40	−11 41.7	2	5.80	.005	0.98	.010	0.72		6	G7 III	2007,3077
131342	−59 05753	HR 5546			14 51 42	−59 54.6	5	5.20	.005	1.16	.004	1.15		20	K2 III	15,1075,2013,2028,3005
		LP 858 - 47			14 51 46	−26 19.2	1	11.92		1.03		0.87		2		1696

Table 1 821

HD	DM	Other Id	N Rem	α_{1950}	δ_{1950}	S	V	σ_V	B–V	σ_{B-V}	U–B	σ_{U-B}	n	Spectrum	References
131461	−35 09890	IDS14487S3601		14 51 47	−36 13.7	1	7.27		0.05		0.01		1	A0/1 V	1770
131461	−35 09890	IDS14487S3601	B	14 51 47	−36 13.7	1	7.24		0.08		0.03		2	A0/1 V	1770
	−29 11389			14 51 48	−29 29.4	1	10.11		0.66		0.20		2	G0	1730
131355	−58 05756			14 51 48	−58 38.4	1	9.72		0.15		-0.13		4	B9 IV	567
131712	+43 02443			14 51 54	+43 11.1	1	8.67		1.00		0.80		2	K0	1733
131488	−40 09088			14 51 54	−40 55.0	1	8.02		0.11		0.09		1	A1 V	1771
131502	−35 09892			14 51 55	−35 57.8	1	8.98		0.27		0.17		1	A9 V	1771
131464	−46 09672			14 51 57	−46 25.7	4	7.31	.011	0.97	.006	0.64	.002	20	G6 III	977,1075,2011,4001
		AJ85,265 # 2		14 51 59	−59 20.6	1	12.98		1.23		0.09		2		779
	+26 02621	G 166 - 54		14 52 00	+25 46.3	1	11.00		0.40		-0.21		2	F0	3016
		AJ85,265 # 3		14 52 03	−59 14.4	1	12.06		1.09		0.54		2		779
131503	−43 09452			14 52 07	−44 07.9	4	7.98	.004	0.24	.006	0.16	.007	17	A4 V	977,1075,1771,2011
131109	−76 00924	HR 5540, R Aps		14 52 07	−76 27.7	2	5.35	.005	1.44	.005	1.70		10	K4 III	2038,3005
	+16 02708	CE Boo	⋆	14 52 08	+16 18.3	1	10.20		1.48		1.15		2	M3	3016
131466	−54 06219	IDS14486S5459	AB	14 52 14	−55 11.4	1	9.59		0.56		0.15		1	F5 V	96
131518	−43 09455			14 52 16	−44 11.4	1	9.13		0.33		0.12		1	A8/9 V	1771
131588	−29 11397			14 52 17	−29 19.7	1	8.41		0.63		0.06		2	G1 V	1730
131424	−60 05565			14 52 19	−61 03.9	1	7.71		0.41				4	F3 V	2012
131746	+35 02624			14 52 20	+35 26.9	1	8.43		1.11		0.96		2	K0	1601
131545	−45 09522			14 52 24	−45 46.3	1	8.11		0.25		0.10		1	A8/9 IV	1771
		G 178 - 55		14 52 27	+35 45.6	1	12.36		1.51				1		1759
131653	−8 03858	G 151 - 10		14 52 27	−08 53.4	4	9.52	.004	0.72	.000	0.16	.000	7	G5	516,1064,1620,3077
		G 66 - 41		14 52 29	+10 09.0	1	11.31		1.40		1.11		1	M1	333,1620
131544	−44 09709			14 52 29	−44 29.7	2	7.94	.002	1.55	.007	1.90		12	K3 III	977,2011
131635	−16 03964			14 52 30	−17 14.4	1	9.92		0.21		0.17		1	A2 III/IV	1770
131623	−25 10610			14 52 31	−25 24.8	1	8.31		1.67		2.06		5	K5 III	1657
131670	−6 04102			14 52 34	−07 05.5	1	8.00		1.20		1.07		1	G7 III-Ba1	565
131624	−30 11815			14 52 37	−30 58.8	1	9.36		0.19		0.19		1	A2 IV	1770
131491	−61 04768	IDS14486S6157	AB	14 52 37	−62 09.8	1	6.37		-0.04		-0.48		4	B4 III	158
	−61 04769	LSS 3294		14 52 38	−61 38.5	1	10.01		0.58		-0.32		2	B1 V	403
131625	−33 10169	HR 5558	⋆ A	14 52 40	−33 39.2	2	5.32	.001	0.04	.000	0.00		6	A0 V	1770,2007
131492	−62 04337	HR 5551, θ Cir	⋆ AB	14 52 41	−62 34.8	6	5.24	.146	-0.01	.036	-0.73	.079	36	B4 Vnpe	26,681,815,1586,1637,2035
		He3 1039		14 52 42	−59 14.	1	11.50		0.97		0.27		3		1586
131562	−52 07634	HR 5556		14 52 43	−52 36.5	1	5.38		0.13				4	A1 IV/V	2006
130650	−82 00629	HR 5525		14 52 49	−83 01.7	2	5.64	.005	0.95	.000	0.75	.005	7	G8/K0 III	15,1075
131749	+0 03273	IDS14503N0034	AB	14 52 50	+00 21.9	1	8.88		0.44		-0.10		4	F8	3016
		LP 741 - 20		14 52 52	−15 20.4	2	14.66	.010	1.60	.015			3		940,3078
131609	−44 09716			14 52 52	−44 54.9	1	9.34		0.54				4	F6 V	2011
131698	−23 11971			14 52 54	−24 04.7	1	8.17		0.23		-0.25		1	A0 II/III	1770
131627	−43 09464			14 52 56	−43 53.4	3	7.89	.005	0.16	.000	0.12	.001	16	A3 IV	977,1075,2011
131719	−26 10605			14 52 58	−26 55.5	1	9.00		1.00		0.79		1	K3 V	3072
131638	−44 09722			14 53 02	−44 42.0	1	8.34		0.02		-0.11		2	B9 V	1770
131826	+13 02866			14 53 05	+12 38.0	1	6.78		0.05		0.02		2	A2	1648
131637	−44 09723			14 53 05	−44 30.2	4	6.76	.002	0.05	.007	0.07	.005	17	A1 V	977,1075,1771,2011
		Feige 105		14 53 07	+12 57.0	1	12.99		0.02		0.05		1		1298
131657	−47 09543	HR 5559	⋆ AB	14 53 07	−47 40.7	7	5.63	.003	-0.04	.004	-0.26	.002	51	B9 V	15,278,1075,1637,1770*
131789	−13 04015			14 53 08	−13 41.9	1	7.59		0.32				26	F2 IV	130
131678	−46 09685			14 53 11	−46 45.0	5	7.09	.011	0.58	.005	0.33	.010	20	K0III +A1 V	977,1075,1771,2001,2011
		G 178 - 56		14 53 12	+43 13.7	2	11.91	.030	0.70	.005	0.16	.025	2		1658,3060
131612	−57 06846			14 53 15	−57 59.4	1	8.44		0.02		-0.48		3	B5 III	567
131750	−30 11826			14 53 19	−30 40.5	1	8.57		0.22		0.07		1	Ap SrCrEu	1770
		G 66 - 42		14 53 21	+07 30.1	1	11.32		1.43		1.31		1	M2	333,1620
		G 166 - 55		14 53 24	+33 30.5	2	13.02	.005	0.62	.005	-0.09	.015	2		1658,3060
131752	−38 09785	HR 5561		14 53 24	−39 12.9	2	6.36	.005	0.06	.005	-0.02		5	A0/1 V	1771,2007
131643	−55 06256			14 53 24	−55 42.7	1	7.44		0.39		0.00		1	F2/3 V	96
131751	−34 10047	IDS14504S3414	AB	14 53 26	−34 25.8	1	7.35		0.53				3	F8 V	1414
		LP 621 - 48		14 53 28	−03 07.2	2	13.11	.070	0.57	.025	-0.15	.005	4		1696,3060
131774	−32 10480	HR 5562		14 53 28	−32 26.1	1	6.06		1.41				4	K3 III	2007
131705	−50 08939	HR 5560		14 53 30	−51 14.8	2	6.62	.019	1.69	.005	2.04		6	M2 III	2007,3055
133002	+83 00431	HR 5596		14 53 34	+82 43.1	2	5.65	.004	0.68	.001	0.17		7	F9 V	71,3026
135294	+87 00143			14 53 35	+87 25.3	1	6.91		1.46		1.67		3	K0	1502
131777	−41 09307			14 53 40	−42 15.6	2	8.14	.005	0.09	.005	0.07		5	A1 V	1771,2014
131684	−58 05767			14 53 44	−59 00.2	1	9.43		0.13		-0.09		1	B9 IV/V	567
132142	+54 01716	G 200 - 62		14 53 46	+53 52.5	6	7.77	.021	0.79	.007	0.33	.010	9	K1 V	22,792,908,1003,1658*
132028	+38 02603			14 53 47	+38 10.5	1	8.74		0.29		0.10		2	A0	1601
131835	−35 09917			14 53 47	−35 29.6	1	7.89		0.16		0.09		1	A2 IV	1770
		G 66 - 43		14 53 51	+09 04.9	1	9.89		0.92		0.67		2		333,1620
131951	+15 02796	HR 5567		14 53 51	+14 38.8	2	5.77		-0.06	.000	-0.08	.000	5	A0 V	1022,1221
132027	+40 02829	IDS14519N4003	AB	14 53 51	+39 51.0	1	8.03		0.40		-0.07		2	F0	3030
		MCC 727		14 53 54	+17 56.9	1	10.97		1.44				8	K4	1619
132029	+32 02531	HR 5569	⋆ AB	14 53 54	+32 30.1	2	6.12		0.10	.011	0.07		6	A2 V	70,1022
	−49 09199			14 53 54	−49 53.0	1	11.76		0.49		-0.22		2		1649
132388	+72 00661			14 53 57	+71 39.0	1	9.05		0.96		0.73		3	K0	196
131954	+3 02957	IDS14515N0319	AB	14 54 00	+03 07.2	1	7.55		1.09		0.99		3	K0	3016
131885	−25 10619			14 54 00	−26 05.0	1	6.91		0.02		-0.07		1	A0 V	1770
		L 262 - 45		14 54 00	−52 00.	1	13.80		0.92		0.31		2		3062
		AJ85,265 # 4		14 54 00	−59 18.1	1	12.35		0.93		-0.05		2		779
131918	−10 03989	HR 5564		14 54 03	−11 12.5	2	5.46	.000	1.49	.000	1.70		6	K4 III	2006,3005
131867	−41 09317			14 54 10	−42 04.4	1	8.72		0.28		0.12		1	A7 IV	1771
132031	+17 02803			14 54 11	+17 18.3	1	7.77		0.55		0.04		3	G0	1648

HD	DM	Other Id	N	Rem	α_{1950}	δ_{1950}	S	V	σ_V	B–V	σ_{B-V}	U–B	σ_{U-B}	n	Spectrum	References
		LP 441 - 33			14 54 11	+18 06.6	1	15.74		1.98				1		1759
131887	−36 09771				14 54 11	−36 18.6	1	8.65		0.37		0.20		1	A2mA8-F2	1771
131802	−57 06850				14 54 12	−57 33.9	1	8.99		0.11		-0.24		6	B8 III	567
131919	−28 11055	HR 5565			14 54 14	−28 57.4	3	6.29	.006	-0.02	.007	-0.18	.022	7	B8/9 V	1079,1770,2007
131425	−76 00931	HR 5547			14 54 15	−76 57.7	3	5.92	.008	1.05	.000	0.82	.005	11	G8 II	15,1075,2038
131901	−32 10490				14 54 16	−32 38.0	1	7.20		0.09		0.09		1	A1 V	1770
132032	+13 02867				14 54 21	+13 21.0	1	8.11		0.60		0.14		2	G5	1648
132051	+9 02966	G 66 - 44			14 54 23	+08 47.8	1	8.65		0.83		0.45		1	K0	333,1620
		vdB 64 # b			14 54 24	−59 33.9	1	11.25		0.41		0.17		4	B9 V	434
	+21 02704				14 54 25	+21 16.3	2	9.41	.017	1.03	.009	0.80		3	K7	1017,3072
	−49 09208				14 54 26	−50 10.0	1	11.88		0.51		-0.09		1		1649
131889	−46 09696				14 54 28	−46 52.8	1	9.25		0.27		0.14		1	A7/8 V	1771
131551	−74 01281	HR 5555		⋆ A	14 54 29	−74 50.0	3	6.19	.007	-0.04	.000	-0.19	.005	13	B9 V	15,1075,2038
131979	−27 10139				14 54 30	−27 51.4	1	9.08		0.38		0.20		1	A0mA7-A9	1770
131976	−20 04123	IDS14516S2058	B		14 54 31	−21 11.3	3	8.01	.035	1.52	.025	1.21	.021	14	K5V	22,3078,8006
131903	−41 09322				14 54 31	−42 17.8	1	7.41		0.33		0.05		1	A9/F0 V	1771
131977	−20 04125	HR 5568		⋆ A	14 54 32	−21 11.5	7	5.76	.035	1.11	.008	1.06	.005	28	K4 V	15,22,164,1013,3077*
131977	−20 04125	IDS14516S2058		⋆ AB	14 54 32	−21 11.5	3	5.64	.011	1.12	.015	1.08		10	K4 V	1013,1075,2007
131941	−33 10191				14 54 33	−34 13.6	1	8.64		0.39		0.16		1	A3mA7-F2	1771
132052	−3 03696	HR 5570			14 54 34	−04 08.6	10	4.49	.011	0.32	.006	0.04	.019	56	F0 V	3,15,253,254,1008*
	−49 09212				14 54 34	−49 55.2	1	12.08		0.75		0.09		2		1649
131856	−56 06514				14 54 34	−56 22.3	1	10.07		0.29		-0.21		3	B7/8 II	567
131992	−24 11772				14 54 36	−25 14.5	1	6.94		0.21		0.12		1	A2/3 V	1770
		vdB 64 # a			14 54 36	−59 27.9	1	10.63		0.32		-0.22		3		434
131980	−32 10492				14 54 37	−32 37.8	1	7.67		0.25		0.20		1	A4 V	1771
131923	−48 09494	HR 5566			14 54 42	−48 39.5	8	6.35	.009	0.71	.008	0.25	.019	30	G3/5 V	15,116,158,1020,1075*
132254	+50 02126	HR 5581			14 54 43	+49 49.9	2	5.64	.010	0.50	.005	0.00	.015	5	F5 V	254,3016
132054	−17 04219				14 54 46	−17 43.8	1	9.67		0.31		0.21		1	A3/5(m)	1770
132145	+22 02764	HR 5574			14 54 48	+21 45.4	3	6.49	.004	0.00	.012	0.01	.004	7	A1 V	70,1022,3050
		L 478 - 144			14 54 48	−38 24.	1	14.87		0.84		0.45		1		3062
		AJ85,265 # 5			14 54 50	−59 53.3	1	12.63		0.99		-0.07		2		779
	−63 03464	LSS 3296			14 54 50	−63 23.5	1	11.30		0.56		-0.22		3		1586
132146	+16 02715	HR 5575			14 54 52	+16 35.3	1	5.71		0.94				1	K0	71
		LP 136 - 39			14 54 58	+14 11.0	1	14.73		1.68				3		1663
132132	+0 03277	HR 5573		⋆ A	14 54 59	+00 02.0	1	5.52		1.13		1.11		8	K1 III	1088
132016	−44 09742	IDS14517S4443	A		14 55 01	−44 54.8	1	8.35		0.00		-0.13		2	B9.5 V	1770
132112	−11 03841	FY Lib			14 55 03	−12 14.2	1	9.67		1.60		1.30		1	M5 III	3040
		AJ85,265 # 6			14 55 06	−59 43.3	1	11.22		1.23		0.15		2		779
132041	−35 09930				14 55 07	−35 51.9	2	7.80	.001	-0.07	.004	-0.38	.001	2	B8 II/III	55,1770
132018	−46 09704	IDS14517S4702	B		14 55 08	−47 14.6	2	8.44	.019	0.05	.005	-0.39	.005	3		401,1771
132256	+25 02853				14 55 12	+25 31.4	1	7.32		0.66		0.22		2	G2 IV	105
132094	−36 09786				14 55 15	−37 09.7	1	7.26		-0.03		-0.12		1	B9 V	1770
132058	−42 09853	HR 5571			14 55 15	−42 56.0	9	2.68	.013	-0.23	.005	-0.87	.020	23	B2 IV	15,26,278,507,1020,1075*
132406	+53 01752	G 201 - 25			14 55 24	+53 35.2	1	8.45		0.65		0.20		2	G0	1566
132096	−39 09402	HR 5572			14 55 24	−39 42.4	2	6.16	.005	1.22	.010	1.24		6	K1 III	2007,3005
132296	+19 02891				14 55 28	+19 11.2	1	6.61		0.93		0.63		2	K0	1648
	+18 02959	U Boo			14 55 29	+17 43.1	1	10.94		1.61		1.19		1		635
132304	+25 02855				14 55 29	+24 52.4	1	6.94		1.39		1.51		2	K3 III	3040
		G 166 - 57			14 55 29	+31 36.4	1	11.08		1.32		1.02		2	K5	3078
132230	−10 03994	HR 5578			14 55 31	−10 57.3	1	6.60		0.01				4	A1 V	2035
		AJ85,265 # 7			14 55 31	−59 26.1	1	12.17		0.84		-0.16		2		779
	−16 03974				14 55 34	−17 20.5	1	10.23		0.75		0.23		3	K1	1696
132119	−45 09559				14 55 35	−45 50.7	2	8.48	.000	0.21	.003	0.18	.004	7	A5 V	1673,1771
		He3 1043			14 55 36	−59 18.	1	11.48		0.58		-0.18		3		1586
132219	−27 10148	HR 5577		⋆ AB	14 55 41	−27 27.5	2	5.65	.000	0.26	.015	0.14		7	A4 V + A6 V	2007,8071
		AJ85,265 # 8			14 55 41	−60 46.1	1	11.43		0.90		-0.20		3		779
132101	−51 08631				14 55 46	−51 22.6	1	6.78		-0.04		-0.47		2	B5 V	1770
		AJ85,265 # 9			14 55 49	−59 38.1	1	10.47		1.35		0.14		2		779
131246	−82 00636	HR 5545			14 55 51	−82 50.5	3	5.64	.004	1.29	.005	1.15	.000	12	G8 Ib	15,1075,2038
	−58 05773				14 55 52	−58 44.6	1	10.12		0.66		-0.08		1		567
132466	+51 01970				14 55 53	+51 03.2	1	8.53		0.82		0.41		2	G0	1566
132343	+14 02812				14 55 54	+14 14.2	1	6.77		1.32		1.52		2	K3 III	1648
132200	−41 09342	HR 5576		⋆ AB	14 55 54	−41 54.3	9	3.13	.007	-0.21	.007	-0.78	.011	23	B2 IV	15,26,278,507,1075,1770*
132203	−44 09758				14 55 58	−45 14.6	3	8.82	.007	1.56	.010	1.76	.015	17	K2/3 (III)	863,977,1075,1770
		G 166 - 58			14 55 59	+29 49.8	2	15.60	.009	0.28	.019	-0.53	.042	4	DA,F	316,3060
132202	−42 09867	IDS14528S4304	AB		14 56 02	−43 16.4	3	7.04	.007	0.36	.003	0.04	.009	16	F2 III	977,1075,2011
132238	−37 09836	HR 5579			14 56 04	−37 40.9	2	6.45	.004	-0.08	.000	-0.34		5	B8 V	1770,2007
131891	−72 01644				14 56 04	−72 31.3	1	8.07		-0.01		-0.17		3	B9p Shell	1586
132136	−56 06523	LSS 3297			14 56 05	−57 11.5	1	8.25		0.99		0.61		3	F0 II/III	8100
132181	−52 07681	V Lup			14 56 08	−53 12.5	1	9.16		2.15		2.80		2	C5,5	864
132137	−57 06866				14 56 09	−58 07.1	1	8.69		0.16		-0.27		3	B8 II	567
132242	−42 09871	HR 5580			14 56 10	−42 57.7	6	6.10	.005	0.60	.003	0.29	.006	27	F7 II	15,278,977,1770,2011,2038
	−57 06867	LSS 3298			14 56 10	−57 52.4	2	10.25	.014	0.84	.027	-0.25	.012	4	O9.5IV	540,1737
132345	−10 03999	HR 5582		⋆ A	14 56 11	−10 56.7	3	5.84	.021	1.26	.000	1.37	.064	9	K3 III	13,2035,3024
132345	−10 03999	IDS14535S1044	B		14 56 11	−10 56.7	2	9.91	.195	0.88	.035	0.62	.085	8		13,3024
	−43 09510				14 56 15	−43 53.3	4	10.16	.019	1.32	.009	1.17	.041	13	K7 V	863,1075,1705,3072
132375	−4 03783	HR 5583		⋆ AB	14 56 16	−04 47.3	5	6.09	.008	0.50	.005	0.04	.007	19	F8 V	15,253,1075,1417,3053
	−0 02911	EH Lib			14 56 22	−00 45.0	1	9.46		0.19		0.13		1	A5	668
132545	+41 02543	IDS14546N4130	AB		14 56 26	+41 17.5	1	8.49		0.50		-0.01		2	F5	1601
132347	−30 11870	IDS14535S3019	AB		14 56 27	−30 30.7	1	7.01		0.25		0.15		1	A3mA5-F0	1771

Table 1   823

HD	DM	Other Id	N	Rem	α1950	δ1950	S	V	σV	B–V	σB–V	U–B	σU–B	n	Spectrum	References
132301	−43 09514				14 56 27	−43 36.7	4	6.58	.009	0.47	.005	−0.03	.006	18	F7 IV/V	977,1075,2011,3077
132285	−45 09566				14 56 27	−45 57.7	1	7.14		0.19		0.14		1	A3 V	1771
132447	+10 02770				14 56 29	+10 33.2	1	8.89		0.51		0.06		2	F8	1733
132004	−70 01891				14 56 29	−71 17.6	1	6.68		−0.02		−0.50		9	B5 V	1628
		G 201 - 27			14 56 30	+56 51.7	1	15.52		1.95				5		538
132302	−45 09567				14 56 32	−46 02.9	3	7.16	.009	−0.01	.007	−0.09	.015	7	A0 V	401,1771,2014
132380	−33 10215				14 56 46	−34 11.7	1	9.86		0.32		0.11		1	A9 V	1771
132127	−67 02719	IDS14523S6735		AB	14 56 46	−67 47.1	1	6.98		−0.06				4	B4 V	1075
132813	+66 00878	HR 5589, RR UMi			14 56 47	+66 07.9	7	4.59	.016	1.59	.017	1.59	.017	15	M5 III	15,1363,3052,8015*
132411	−19 03997				14 56 49	−19 26.3	1	10.20		1.08		0.93		1	K3 V	3072
132395	−33 10217				14 56 52	−33 33.5	1	9.14		0.27		0.21		1	A0 V	1770
		vdB 65 # a			14 56 52	−63 04.9	1	13.86		1.84		1.26		5		434
132525	+5 02954	HR 5584			14 56 53	+04 46.0	2	5.91	.010	1.60	.005	1.90	.010	10	M2 III	1088,3005
132453	−20 04134				14 56 54	−20 58.8	1	8.84		0.27		0.17		1	A7 IV	1770
132414	−32 10523				14 56 57	−32 38.1	1	7.68		0.61		0.34		1	A1 V + (G)	1770
132475	−21 04009				14 56 59	−21 48.4	7	8.55	.021	0.56	.012	−0.12	.025	15	F5/6 V	22,742,1064,1594,1658*
132432	−34 10103				14 57 07	−34 54.4	1	9.44		0.26		0.19		1	A7 V	1771
		Steph 1197			14 57 08	+49 10.0	1	11.90		1.32		1.25		1	K7	1746
133086	+75 00547	IDS14575N7518		A	14 57 08	+75 06.0	1	6.84		1.00		0.81		2	K0	1375
132492	−25 10644				14 57 11	−25 29.5	1	10.22		0.34		0.16		1	A2	1770
132320	−59 05775				14 57 11	−59 41.8	1	9.40		0.31		−0.53		3	B1 III	567
132599	+20 03051				14 57 12	+20 34.2	1	8.56		1.25		1.38		2	K0	1733
132128	−71 01707				14 57 13	−71 53.0	1	8.15		0.78		0.44		32	A2 III/IV	1699
	−60 05591	LundsII139 # 314			14 57 29	−60 50.4	1	10.38		1.08		0.69		2		1101
132478	−45 09583				14 57 31	−45 50.3	1	8.49		−0.05		−0.40		2	B8/9 II	401
132569	−15 04007				14 57 32	−16 20.5	1	9.84		0.89		0.56		1	K1 V	3072
132571	−18 03945				14 57 33	−18 25.8	1	7.81		1.02		0.70		8	G8 IV	1657
		BPM 39096			14 57 36	−48 39.	1	12.70		0.47		−0.15		1		3065
	+47 02190				14 57 37	+47 00.9	1	9.11		0.60		0.13		2	F8	1601
132772	+39 02820	HR 5588			14 57 42	+39 27.8	2	5.66	.015	0.31	.002	0.01		4	F2 III	70,254
		AJ85,265 # 10			14 57 43	−56 35.9	1	11.63		0.58		0.09		2		779
132574	−36 09827				14 57 45	−36 28.3	1	9.04		0.30		0.15		1	A9 V	1771
132529	−41 09376				14 57 47	−41 38.6	1	9.27		0.24		0.15		1	A8/9 V	1771
	−11 03853				14 57 48	−12 15.1	2	10.19	.058	0.53	.015	−0.07		3		1594,6006
		AJ85,265 # 11			14 57 50	−59 29.6	1	12.30		1.75		0.50		1		779
	+11 02751	G 66 - 47			14 57 51	+11 13.4	1	10.00		0.90		0.65		1	K2	333,1620
132418		CoD -60 05413			14 57 51	−60 48.4	1	10.00		1.81		1.47		3	M2/3	1101
132481	−55 06296				14 57 54	−56 03.8	4	6.87	.016	−0.07	.012	−0.78	.012	11	B1/2 Vn	540,567,976,2012
	−29 11465				14 57 57	−29 51.0	1	11.21		0.56		−0.14		2		1696
132683	−10 04011	G 151 - 18			14 58 01	−10 55.9	3	9.48	.028	1.39	.025	1.27	.010	7	K7 V	158,1705,3072
132604	−37 09863	HR 5585			14 58 02	−37 51.6	1	5.89		1.24				4	K2/3 III	2006
	−60 05592	LundsII139 # 313			14 58 02	−61 17.1	2	10.20	.005	0.26	.005	−0.06	.029	3		1101,1101
132910	+54 01724	IDS14566N5416		A	14 58 04	+54 04.1	1	6.84		0.32		0.03		3	F0	1084
132909	+54 01725	IDS14566N5416		B	14 58 05	+54 03.5	1	7.64		0.30		0.02		3	F0	1084
		LundsII139 # 312			14 58 06	−61 18.0	1	9.90	.005	0.60	.000	0.10	.054	2		1101,1101
132756	+9 02983	G 66 - 49			14 58 09	+08 48.1	1	7.29		0.69		0.19		1	G0	333,1620
132519	−57 06881				14 58 09	−58 11.4	1	9.95		0.24		−0.15		1	B8/9 II	567
		WLS 1500 75 # 5			14 58 13	+77 32.2	1	11.02		0.45		−0.04		2		1375
132742	−7 03938	HR 5586, δ Lib		★AB	14 58 18	−08 19.3	3	4.92	.010	0.00	.000	−0.10	.005	14	B9.5V	15,1088,8015
132742	−7 03938	IDS14556S0807		C	14 58 18	−08 19.3	1	12.80		1.12		0.96		3		627
132742	−7 03938	IDS14556S0807		D	14 58 18	−08 19.3	1	12.13		0.70		0.18		3		627
	−60 05597	LSS 3301			14 58 18	−60 55.8	1	10.45		0.63		−0.15		2	B5 II-Ib	540
		G 66 - 51			14 58 19	+02 19.6	2	10.62	.005	0.71	.000	0.10	.009	4	K0	333,1620,1696
132686	−32 10542				14 58 22	−32 30.1	2	9.28	.054	0.41	.000	0.03		3	F0wA	742,1594
		IC 4499 sq1 2			14 58 22	−82 02.1	1	12.09		0.63		0.18		2		156
	+68 00813	G 239 - 36			14 58 25	+67 56.8	1	9.77		0.58		0.04		2	G5	1064
132776	−2 03927				14 58 26	−02 40.4	1	8.84		1.15		1.13		2	G5	536
		IC 4499 sq1 3			14 58 26	−82 04.4	1	11.35		1.72		1.34		2		156
132538	−61 04797				14 58 28	−61 25.8	1	12.62		0.64		0.10		1	A7 III	1101
132797	−3 03707	IDS14560S0342		AB	14 58 33	−03 54.0	1	9.05		1.09		0.86		4	K2	158
132688	−40 09201				14 58 33	−40 51.9	1	9.49		1.38		1.29		1	G5/8	3008
132879	+22 02772	HR 5592			14 58 38	+22 14.6	1	6.26		1.19		1.20		2	K0	1733
132743	−36 09839	IDS14555S3632		A	14 58 39	−36 43.9	1	7.96		0.21		0.12		1	A6 V	1771
	−53 06199	NGC 5822 - 450			14 58 40	−53 58.9	1	11.40		0.36		0.25		2		807
		NGC 5822 - 451			14 58 40	−53 59.0	1	11.45		0.39		0.23		2		807
		Steph 1198			14 58 44	+13 53.4	1	11.03		1.39		1.22		1	K7-M0	1746
132833	−2 03928	HR 5590			14 58 44	−02 33.5	2	5.49	.020	1.67	.010	1.98	.025	10	MO III	1088,3005
132799	−24 11805				14 58 45	−24 52.3	1	8.80		0.30		0.20		1	A5 IV/V	1770
132761	−31 11646				14 58 46	−31 20.8	1	7.72		0.25		0.12		1	A2/3 IV	1770
	−53 06200	NGC 5822 - 447			14 58 48	−54 01.1	1	10.96		0.48		0.20		2		807
		PSD 195 # 418			14 58 48	−60 46.	2	10.51	.010	0.15	.010	0.07	.015	3	F2	1101,1101
		G 15 - 2			14 58 51	+05 45.0	1	12.15		1.45		0.89		1		333,1620
		G 167 - 11			14 58 51	+22 00.0	1	14.64		0.93		0.66		2		1658
132763	−33 10244	HR 5587			14 58 52	−34 09.7	3	6.21	.009	0.24	.004	0.14	.014	8	A8 IV	1771,2007,8071
		NGC 5822 - 449			14 58 52	−54 03.0	1	11.36		0.38		0.24		2		807
	−53 06201	NGC 5822 - 448			14 58 52	−54 03.0	1	11.03		0.43		0.26		2		807
133029	+47 02192	HR 5597, BX Boo		★A	14 58 56	+47 28.5	4	6.36	.006	−0.14	.005	−0.28	.007	122	B9 p SiSrCr	274,1063,1202,3024
133029	+47 02192	IDS14572N4740		B	14 58 56	+47 28.5	1	9.56		0.46		−0.04		4	F6 V	3024
	−53 06203	NGC 5822 - 444			14 58 56	−54 06.3	1	10.73		0.37		0.27		2		807
		NGC 5822 - 452			14 58 58	−53 56.8	1	11.52		0.57		0.09		2		807

HD	DM	Other Id	N	Rem	α_{1950}	δ_{1950}	S	V	σ_V	B–V	σ_{B-V}	U–B	σ_{U-B}	n	Spectrum	References
		IC 4499 sq1 4			14 58 58	−82 02.4	1	12.36		0.66		0.27		2		156
132883	−2 03930	IDS14564S0246		A	14 58 59	−02 58.0	2	6.68	.004	1.21	.013	1.21	.004	6	K1:IV-V:	1028,3024
132883	−2 03930	IDS14564S0246		B	14 58 59	−02 58.0	2	10.13	.026	0.80	.013	0.37	.004	6		1028,3024
		G 224 - 39			14 59 02	+60 33.4	1	13.63		1.54				1		1759
		AJ85,265 # 12			14 59 03	−58 11.0	1	12.83		0.77		−0.22		1		779
133030	+40 02838				14 59 07	+40 16.9	1	7.09		1.02		0.77		2	K0 III	1601
132851	−27 10183	HR 5591			14 59 07	−27 51.8	3	5.84	.007	0.16	.005	0.16		10	A4 IV	15,2027,8071
		NGC 5822 - 373			14 59 07	−54 06.9	1	12.03		0.39		0.26		2		807
132713	−57 06887				14 59 08	−57 41.3	1	8.11		0.39		0.14		1	Fm δ Del	96
132950	+16 02722				14 59 09	+16 04.1	2	9.14	.014	1.04	.005	0.91	.014	4	K2	679,3072
	+45 02247	G 179 - 6			14 59 09	+45 37.1	2	9.14	.017	1.44	.017	1.18		3	M0	1017,3072
132933	+0 03297	HR 5594		⋆ AB	14 59 15	+00 03.4	2	5.70	.000	1.51	.010	1.55	.005	10	M0.5 IIb	1088,3005
	−53 06207	NGC 5822 - 443			14 59 15	−54 01.1	1	9.72		1.22		1.07		2		807
		vdB 65 # b			14 59 15	−63 11.1	1	12.75		1.13		0.16		3		434
132785	−48 09559				14 59 16	−48 20.5	1	9.31		0.39				4	F0/2 V	2012
	−53 06208	NGC 5822 - 371			14 59 17	−54 06.8	1	11.21		0.33		0.23		2		807
234200	+52 01842				14 59 19	+52 35.9	1	9.28		0.35		0.08		2	F0	1566
		NGC 5822 - 364			14 59 19	−54 08.0	1	11.56		0.78		0.24		2		807
		NGC 5822 - 348			14 59 21	−54 16.5	1	10.91		1.02		0.72		2		1685
	−11 03861				14 59 24	−11 58.5	1	10.08		0.79		0.40		2		1696
132953	−7 03944				14 59 29	−07 22.8	1	6.40		0.18		0.10		8	A3	1210
	−57 06893	LSS 3302			14 59 29	−57 54.3	1	9.95		0.47		−0.38		4		567
	−50 09020				14 59 31	−50 48.6	1	11.94		0.57		0.00		2		1649
	−53 06209	NGC 5822 - 380			14 59 31	−54 03.6	1	11.23		0.27		0.21		2		807
	−53 06211	NGC 5822 - 384			14 59 32	−54 02.3	1	10.73		0.39		0.25		2		807
	−53 06210	NGC 5822 - 374			14 59 32	−54 05.6	1	11.04		0.37		0.24		2		807
		AJ85,265 # 13			14 59 32	−59 50.1	1	12.14		1.41		0.27		2		779
132655	−67 02733				14 59 33	−67 27.6	1	10.11		0.02		−0.23		2	B8 II/III	400
	−53 06212	NGC 5822 - 375			14 59 36	−54 05.0	2	9.71	.019	1.22	.004	1.10	.000	4		807,1685
		NGC 5822 - 453			14 59 37	−53 54.6	1	11.91		0.71		0.17		2		807
	−53 06213	NGC 5822 - 446			14 59 37	−53 59.6	1	10.85		0.42		0.30		2		807
		NGC 5822 - 215			14 59 38	−54 11.4	1	10.32		0.56		0.17		2		807
	−53 06214	NGC 5822 - 216			14 59 38	−54 11.4	2	10.67	.024	0.37	.017	0.22	.004	3		807,1780
	−32 10555				14 59 39	−32 58.9	1	9.89		1.21				2	K2	438
	−45 09610				14 59 43	−46 05.8	3	9.87	.006	1.19	.011	1.08		11	K5 V	863,1705,2012
		NGC 5822 - 224			14 59 43	−54 14.8	2	10.83	.020	1.04	.007	0.71	.012	4		1685,1780
		NGC 5822 - 226			14 59 44	−54 16.1	1	12.59		0.42		0.15		2		1780
132825	−56 06554				14 59 44	−57 10.4	1	9.59		1.16		0.81		1	G8/K0 III	96
	−53 06215	NGC 5822 - 219			14 59 45	−54 12.0	1	11.21		0.35		0.25		1		1780
		NGC 5822 - 454			14 59 47	−53 54.9	1	12.00		0.40		0.24		2		807
	−53 06217	NGC 5822 - 385			14 59 48	−54 02.8	1	11.17		0.35		0.25		2		807
		PSD 195 # 442			14 59 48	−59 05.	2	12.15	.015	1.16	.029	0.83		3	G0	1101,1101
	−59 05788	LSS 3303			14 59 48	−60 10.2	1	10.30		0.59		−0.24		2	B3 II-Ib	540
	−53 06218	NGC 5822 - 393			14 59 49	−54 00.1	3	10.23	.022	0.95	.007	0.51	.005	8		807,1597,1780
133008	−7 03946				14 59 50	−07 38.7	2	6.59	.025	0.18	.010	0.10	.015	4	A2	627,1210
132977	−28 11110				14 59 50	−28 25.2	1	9.23		0.14		0.09		1	A0 V	1770
		NGC 5822 - 456			14 59 51	−53 51.4	1	12.37		0.60		0.13		2		807
	−53 06219	NGC 5822 - 204			14 59 51	−54 05.2	2	11.30	.007	0.43	.009	0.21	.019	3		807,1780
132920	−42 09940				14 59 54	−42 24.7	3	10.00	.002	1.17	.009	1.11	.014	13	K3 V	158,863,1075
		NGC 5822 - 205			14 59 54	−54 05.4	2	11.92	.003	0.37	.015	0.21	.019	3		807,1780
131596	−84 00490	HR 5557			14 59 54	−84 35.7	3	5.88	.022	−0.04	.015	−0.12	.005	11	B9.5V	15,1075,2038
133124	+25 02861	HR 5600			14 59 55	+25 12.3	6	4.82	.009	1.50	.003	1.83	.005	18	K4 III	15,1080,1118,3016*
132955	−32 10560	HR 5595		⋆ A	14 59 55	−32 26.9	3	5.44	.003	−0.13	.003	−0.63	.004	8	B3 V	26,1770,2006
		NGC 5822 - 455			14 59 55	−53 54.2	1	12.22		0.44		0.18		2		807
		NGC 5822 - 110			14 59 55	−54 12.2	1	12.80		0.44		0.11		1		1780
		NGC 5822 - 118			14 59 56	−54 16.0	1	13.63		0.62		0.06		1		1780
133013	−20 04141				14 59 57	−21 13.8	1	9.69		0.34		0.15		1	A9 V	1770
	−56 06558				14 59 57	−57 15.7	1	11.01		0.37		0.20		1		96
	−53 06220	NGC 5822 - 190			14 59 59	−54 06.2	2	10.65	.003	0.43	.002	0.25	.023	3		807,1780
		NGC 5822 - 114			14 59 59	−54 14.4	1	13.41		1.31		0.90		1		1780
		NGC 5822 - 115			14 59 59	−54 14.8	1	12.06		0.37		0.19		1		1780
		NGC 5822 - 119			14 59 59	−54 16.0	1	13.11		0.53		0.08		2		1780
	−53 06221	NGC 5822 - 120			14 59 59	−54 16.5	2	10.65	.020	0.37	.005	0.24	.001	4		807,1780
		NGC 5824 sq1 6			15 00 00	−32 52.4	1	11.89		0.65				2		438
	−52 10251				15 00 00	−52 22.7	1	10.79		0.12		−0.57		2		954
	−54 06294	NGC 5822 - 328			15 00 00	−54 23.3	1	11.05		0.41				1		1780
	−53 06222	NGC 5822 - 2			15 00 01	−54 10.9	5	9.50	.029	1.05	.003	0.72	.009	11		95,807,1597,1780,4002
132842	−60 05607				15 00 01	−60 37.5	2	8.46	.009	0.09	.013	−0.18		6	B8/9 II	1101,2014
		NGC 5822 - 229			15 00 03	−54 18.0	1	11.72		0.42		0.17		2		1780
133208	+40 02840	HR 5602			15 00 04	+40 35.2	5	3.51	.026	0.96	.013	0.73	.018	17	G8 IIIa	15,1080,1363,8015,8100
132960	−40 09218				15 00 05	−41 04.6	3	7.39	.004	−0.15	.005	−0.95	.004	7	B1 IbN	164,400,401
133209	+39 02826				15 00 06	+39 07.4	2	8.16	.005	0.53	.005	0.07	.005	5	F7 V	833,1601
		NGC 5822 - 20			15 00 06	−54 12.5	1	13.67		0.85				1		95
		G 136 - 50			15 00 07	+10 25.1	1	11.40		1.13		1.10		2	K4	1696
133112	−2 03933	HR 5599			15 00 09	−02 50.1	2	6.59	.005	0.20	.000	0.16	.005	11	A5 m	1088,8071
133073	−9 04058				15 00 09	−10 11.8	1	7.62		1.11		0.99		5	K0	1657
132996	−36 09862				15 00 09	−36 43.4	1	7.77		0.61				4	G3 V	2033
		NGC 5822 - 102			15 00 10	−54 08.5	4	10.84	.016	1.02	.012	0.70	.011	12		807,1597,1780,4002
	−53 06224	NGC 5822 - 201			15 00 11	−54 02.7	4	10.26	.015	1.06	.011	0.65	.014	10		807,1597,1780,4002
		NGC 5822 - 21			15 00 11	−54 12.2	1	13.77		0.68				1		95

Table 1   825

HD	DM	Other Id	N Rem	α_{1950}	δ_{1950}	S	V	σ_V	B–V	σ_{B-V}	U–B	σ_{U-B}	n	Spectrum	References
		NGC 5822 - 3144		15 00 11	−54 13.4	1	14.66		1.68		1.64		1		1780
	−57 06896			15 00 11	−57 21.0	1	10.69		0.26		0.16		1		96
	−53 06225	NGC 5822 - 101		15 00 13	−54 08.1	2	11.06	.006	0.31	.000	0.24	.018	5		807,1780
		NGC 5822 - 121		15 00 13	−54 16.3	1	12.92		0.47		0.09		1		1780
133161	+16 02725			15 00 14	+16 14.9	1	7.02		0.59		0.17		2	G2 V	985
		NGC 5822 - 184		15 00 14	−54 04.8	1	12.70		0.80		0.29		2		1780
		NGC 5822 - 96		15 00 14	−54 06.7	1	14.09		0.60		0.06		2		1780
132922	−54 06297	NGC 5822 - 323		15 00 14	−54 22.0	2	10.09	.000	0.08	.000	−0.09		3	B9 IV/V	807,1780
	−53 06226	NGC 5822 - 200		15 00 15	−54 02.3	1	10.63		0.34		0.38		2		807
133388	+60 01582	HR 5608		15 00 17	+60 24.0	2	5.93	.005	0.10	.010	0.10	.000	4	A4 V	985,1733
		NGC 5822 - 43		15 00 17	−54 13.6	1	12.02		0.35		0.19		2		1780
		AJ85,265 # 14		15 00 17	−56 50.6	1	12.22		0.93		−0.16		2		779
		NGC 5824 sq1 8		15 00 18	−32 50.5	1	12.19		1.33				2		438
133163	+14 02823			15 00 19	+13 39.7	1	8.45		0.92				1	K2 III	882
		NGC 5822 - 183		15 00 19	−54 04.9	1	12.95		1.06		0.93		2		1780
		NGC 5824 sq1 10		15 00 20	−32 51.7	1	12.80		1.12				2		438
		NGC 5822 - 23		15 00 20	−54 11.0	1	13.54		0.92				1		95
		NGC 5822 - 22		15 00 20	−54 13.1	2	12.63	.019	0.46	.018	0.13	.008	3		95,1780
		NGC 5822 - 235		15 00 20	−54 20.5	1	12.81		0.47		0.13		2		1780
		NGC 5822 - 198		15 00 21	−54 02.0	1	11.68		0.29		0.21		2		1780
		NGC 5822 - 94		15 00 21	−54 06.3	1	11.41		0.32		0.24		1		1780
		NGC 5823 - 326		15 00 21	−55 26.5	1	11.33		1.64		1.54		1		4002
133165	+2 02905	HR 5601		15 00 22	+02 17.2	6	4.40	.005	1.04	.009	0.88	.007	25	M0.5 IIIb	15,1075,1080,1088*
133621	+72 00664	G 239 - 38		15 00 22	+71 57.7	3	6.66	.014	0.62	.020	0.04	.011	8	G0	196,308,3026
		NGC 5824 sq1 17		15 00 22	−32 52.1	1	14.40		0.66				2		438
132944	−53 06229	NGC 5822 - 1		15 00 22	−54 08.7	5	9.09	.018	1.28	.006	1.17	.009	12	G8 III	95,807,1597,1780,4002
133130	−12 04192			15 00 23	−12 39.7	1	7.69		0.04		0.01		2	B9.5 IV/V	1770
		NGC 5822 - 404		15 00 23	−53 58.3	1	12.01		0.40		0.20		1		1780
		NGC 5822 - 182		15 00 23	−54 05.4	1	12.14		0.40		0.23		2		1780
		NGC 5822 - 24		15 00 23	−54 10.5	1	13.70		1.02				1		95
	−58 05796			15 00 23	−58 35.4	2	11.86	.029	1.40	.049	1.07		3		1101,1101
		NGC 5822 - 405		15 00 24	−53 58.1	2	11.31	.001	0.44	.015	0.21	.063	5		807,1780
		G 136 - 51		15 00 26	+18 22.2	1	11.27		1.01		0.73		1	K5	1696
133095	−28 11118			15 00 26	−28 51.1	1	8.74		0.24		0.21		1	A1mA5-A8	1770
133077	−35 09996			15 00 26	−35 44.8	1	6.64		0.23		0.12		1	A4 IV	1771
		NGC 5822 - 44		15 00 28	−54 15.3	1	12.38		0.43		0.18		1		1780
		NGC 5822 - 181		15 00 29	−54 04.1	1	12.85		0.56		0.08		2		1780
		NGC 5822 - 45		15 00 29	−54 15.0	1	13.47		0.62		0.40		1		1780
		NGC 5822 - 42		15 00 30	−54 09.4	1	12.56		0.42		0.16		2		1780
		NGC 5823 - 22		15 00 31	−55 35.5	2	13.76	.034	0.77	.029	0.36	.029	3		95,534
	−60 05613	LSS 3304		15 00 32	−60 39.1	1	9.98		0.42		−0.57		2	B0 II	540
	−53 06230	NGC 5822 - 4		15 00 33	−54 13.3	3	9.94	.008	0.73	.005	0.44	.013	6		95,807,1780
		NGC 5823 - 24		15 00 33	−55 33.0	2	14.17	.010	1.59	.030	1.75	.745	2		95,534
		NGC 5823 - 21		15 00 33	−55 34.4	2	13.01	.019	0.57	.014	0.01	.005	4		95,534
		NGC 5824 sq1 12		15 00 34	−32 52.7	1	12.98		0.56				3		438
	−53 06231	NGC 5822 - 3		15 00 34	−54 14.1	5	10.31	.023	1.07	.012	0.76	.008	12		95,807,1597,1780,4002
132965	−55 06325			15 00 34	−56 13.9	1	8.94		0.02		−0.26		1	B7 III	567
		G 66 - 55		15 00 35	+08 53.1	1	11.39		0.80		0.31		1		333,1620
		NGC 5824 sq1 14		15 00 35	−32 52.0	1	13.98		1.14				4		438
		NGC 5822 - 3111		15 00 36	−54 14.0	1	14.85		0.83		0.43		3		1780
132905	−63 03493	HR 5593		15 00 36	−63 50.2	2	5.17	.005	0.94	.009	0.63		5	G8 III	1754,2009
		WLS 1500 75 # 7		15 00 37	+73 12.3	1	10.96		1.06		0.86		2		1375
	−53 06233	NGC 5822 - 177		15 00 37	−54 03.9	2	11.08	.006	0.21	.002	0.13	.004	5		807,1780
		NGC 5822 - 34		15 00 37	−54 10.5	1	14.15		0.82		0.21		3		1780
	−53 06234	NGC 5822 - 5		15 00 37	−54 12.0	3	10.27	.010	0.40	.012	0.25	.061	5		95,807,1780
	−53 06232	NGC 5822 - 51		15 00 37	−54 16.8	3	10.47	.017	1.02	.012	0.73	.039	5		807,1685,1780
		NGC 5822 - 232		15 00 37	−54 19.1	1	11.43		0.42		0.22		2		1780
		NGC 5822 - 233		15 00 37	−54 19.8	1	13.83		1.21		1.22		1		1780
		NGC 5822 - 223		15 00 37	−54 19.8	2	11.02	.002	0.33	.007	0.25	.003	5		807,1780
132984	−56 06567			15 00 37	−56 43.3	5	7.40	.019	−0.03	.010	−0.77	.003	16	B2 III/IV	158,540,567,976,1732
		NGC 5822 - 49		15 00 38	−54 15.9	1	12.57		0.50		0.11		3		1780
132946	−59 05795			15 00 38	−60 09.6	1	9.55		0.24		−0.22		4	B6/7 Ib/II	567
		NGC 5822 - 408		15 00 39	−53 59.0	1	12.32		0.48		0.24		2		1780
	−53 06235	NGC 5822 - 50		15 00 39	−54 15.9	2	10.83	.060	0.37	.010	0.19	.017	3		807,1780
		NGC 5823 - 25		15 00 39	−55 32.5	2	13.73	.028	0.56	.019	0.33	.047	4		95,534
133057	−43 09585			15 00 40	−43 32.7	1	8.51		0.23		0.06		1	A8 V	1771
		NGC 5822 - 6		15 00 41	−54 11.4	1	10.78		1.02		0.70		2		807
	−53 06236	NGC 5822 - 28		15 00 41	−54 12.4	2	11.24	.011	0.34	.012	0.23	.001	5		807,1780
		NGC 5822 - 242		15 00 41	−54 21.2	1	13.24		1.22		0.85		2		1780
	−55 06327	NGC 5823 - 20		15 00 41	−55 34.6	2	11.08	.019	0.66	.005	0.12	.005	4		95,534
132985	−57 06900			15 00 41	−57 30.2	1	7.40		0.08		−0.30		1	B8 II	96
		PSD 195 # 475		15 00 42	−61 37.	2	10.22	.019	1.07	.024	0.73	.107	3		1101,1101
		NGC 5822 - 208		15 00 43	−54 01.4	1	13.15		0.58		0.05		2		1780
		NGC 5822 - 125		15 00 43	−54 18.2	1	11.92		0.36		0.21		2		1780
		NGC 5822 - 407		15 00 44	−53 59.6	1	12.20		2.15		2.77		2		1780
	−53 06237	NGC 5822 - 174		15 00 44	−54 04.8	1	11.40		0.32		0.25		2		807
		NGC 5822 - 243		15 00 44	−54 21.1	1	12.99		0.73		0.39		2		1780
		NGC 5822 - 25		15 00 45	−54 11.7	1	12.91		0.59		0.09		2		1780
		NGC 5822 - 54		15 00 45	−54 14.0	1	14.56		0.71		0.16		1		1780
		LOD 34 # 20		15 00 45	−57 21.8	1	12.51		0.75				1		96

HD	DM	Other Id	N	Rem	α_{1950}	δ_{1950}	S	V	σ_V	B–V	σ_{B-V}	U–B	σ_{U-B}	n	Spectrum	References
		NGC 5822 - 86			15 00 46	−54 07.1	1	12.89		1.73		1.89		2		1780
		NGC 5822 - 237			15 00 46	−54 18.9	1	13.06		0.49		0.10		2		1780
		Steph 1204			15 00 47	+49 28.8	1	11.90		1.41		1.21		1	K7	1746
		NGC 5822 - 26			15 00 47	−54 11.9	1	12.90		0.61		0.08		2		1780
	−53 06238	NGC 5822 - 27			15 00 47	−54 13.0	2	11.31	.028	0.32	.005	0.22	.003	4		807,1780
	−54 06302	NGC 5822 - 238			15 00 47	−54 20.0	2	10.66	.013	0.45	.002	0.27	.003	4		807,1780
	−55 06328	NGC 5823 - 19			15 00 47	−55 34.1	2	11.47	.024	0.33	.005	0.16	.034	3		95,534
		NGC 5824 sq1 11			15 00 48	−32 47.9	1	12.95		0.59				3		438
		NGC 5822 - 173			15 00 48	−54 05.3	1	13.74		0.60		0.05		1		1780
	−53 06241	NGC 5822 - 83			15 00 48	−54 06.0	1	10.91	.010	0.35	.006	0.25	.004	5		807,1780
133000	−58 05798				15 00 48	−58 23.2	1	9.86		0.57		0.09		3	B8 II	567
		NGC 5823 - 26			15 00 49	−55 32.3	2	12.03	.010	1.69	.000	1.83	.185	3		95,534
		LOD 34 # 19			15 00 49	−57 10.4	1	11.98		0.42				1		96
		NGC 5822 - 410			15 00 50	−53 59.0	2	11.29	.017	0.33	.007	0.23	.012	3		807,1780
		NGC 5822 - 239			15 00 50	−54 19.8	1	11.09		0.44		0.24		2		1780
	−54 06304	NGC 5822 - 240			15 00 50	−54 20.1	3	9.49	.018	1.34	.010	1.33	.020	6		807,1685,1780
	−54 06303	NGC 5822 - 316			15 00 50	−54 24.1	1	10.52		1.03		0.76		2		1685
	−55 06329				15 00 51	−55 24.4	1	11.00		0.52		0.10		3		1586
	−53 06242	NGC 5822 - 57			15 00 53	−54 15.1	1	10.96		0.40		0.23		3		1780
	−55 06330	NGC 5823 - 16			15 00 53	−55 37.0	3	11.07	.021	0.44	.005	0.22	.018	6		95,534,1730
		NGC 5824 sq1 16			15 00 54	−32 48.7	1	14.18		0.77				2		438
133289	+15 02811				15 00 55	+15 27.8	1	9.62		0.77				2	G5	882
		NGC 5822 - 82			15 00 55	−54 05.9	1	14.42		0.76		0.15		3		1780
		NGC 5822 - 77			15 00 55	−54 07.4	1	12.88		0.47		0.11		2		1780
		NGC 5822 - 7			15 00 55	−54 11.3	2	12.20	.025	0.40	.015	0.16	.016	3		95,1780
		NGC 5822 - 130			15 00 55	−54 17.8	1	11.09		0.64		0.20		1		1780
		NGC 5823 - 17			15 00 55	−55 36.0	1	13.52		1.22		0.78		1		95
		G 15 - 4			15 00 56	+03 58.3	2	12.10	.020	1.47	.007	1.13	.012	4		333,1620,3078
		NGC 5822 - 241			15 00 56	−54 19.4	1	12.40		0.40		0.23		2		1780
	−54 06305	NGC 5822 - 312			15 00 56	−54 21.9	2	9.91	.014	0.85	.020	0.51		3		807,1780
		NGC 5823 - 285			15 00 56	−55 18.3	1	10.82		1.23		1.16		1		4002
133215	−15 04017				15 00 57	−16 09.1	1	9.67		0.04		-0.11		1	B9 III/IV	1770
		NGC 5824 sq1 5			15 00 57	−32 48.5	1	11.68		0.61				4		438
		NGC 5822 - 59			15 00 57	−54 14.6	1	12.72		0.46		0.11		2		1780
133330	+28 02391				15 00 58	+28 27.7	1	7.03		0.12		0.13		2	A2 V	252
		NGC 5822 - 416			15 00 58	−53 58.1	1	12.73		0.72		0.18		1		1780
133045	−57 06901				15 00 58	−57 27.3	1	10.42		0.34		0.13		1	A3 III/IV	96
		NGC 5822 - 81			15 00 59	−54 06.0	1	13.27		0.60				2		1780
		NGC 5822 - 1136			15 00 59	−54 08.5	1	15.27		0.93		0.48		1		1780
	−53 06243	NGC 5822 - 8			15 00 59	−54 09.3	5	10.39	.017	1.05	.007	0.78	.012	11		95,807,1597,1780,4002
		NGC 5823 - 18			15 00 59	−55 34.7	2	11.56	.023	1.63	.014	1.68	.023	4		95,534
		LOD 34 # 18			15 00 59	−57 12.3	1	11.55		1.02				1		96
		NGC 5822 - 172			15 01 01	−54 04.3	1	11.30		0.62		0.05		3		1780
133025	−60 05620	IDS14571S6021		AB	15 01 01	−60 33.4	1	8.65		0.02				4	B8/9 IV	2014
	−53 06244	NGC 5822 - 80			15 01 02	−54 06.6	4	10.36	.024	1.00	.013	0.62	.035	11		807,1597,1780,4002
133065	−57 06902				15 01 04	−57 28.9	1	9.44		0.79		0.35		1	G5 IV/V	96
		NGC 5822 - 412			15 01 05	−53 59.9	1	12.89		0.60		0.42		1		1780
133392	+35 02642	HR 5609			15 01 06	+35 24.0	1	5.51		1.02				2	K0	70
133216	−24 11834	HR 5603, σ Lib			15 01 08	−25 05.2	13	3.30	.032	1.68	.026	1.95	.016	41	M3 IIIa	3,15,418,1024,1034*
		NGC 5822 - 418			15 01 08	−53 58.1	1	11.77		0.34		0.15		1		1780
		NGC 5822 - 73			15 01 08	−54 11.3	1	14.19		0.71		0.14		1		1780
	−53 06247	NGC 5822 - 156			15 01 09	−54 07.6	2	11.03	.006	0.43	.002	0.15	.010	5		807,1780
		NGC 5822 - 62			15 01 09	−54 14.5	1	12.68		0.60		0.11		2		1780
	−54 06309	NGC 5822 - 246			15 01 09	−54 18.8	2	10.20	.000	0.38	.009	0.24	.013	3		807,1780
		NGC 5822 - 61			15 01 10	−54 15.0	1	12.99		0.47		0.10		2		1780
		NGC 5823 - 15			15 01 10	−55 36.0	2	14.08	.063	0.57	.039	0.38	.078	3		95,534
		NGC 5822 - 131			15 01 11	−54 16.8	1	12.75		0.43		0.12		2		1780
		NGC 5822 - 2264			15 01 11	−54 20.1	1	13.38		0.47		0.34		2		1780
	−53 06249	NGC 5822 - 158			15 01 12	−54 07.5	1	10.96		0.34		0.27		2		1780
	−53 06248	NGC 5822 - 9			15 01 12	−54 08.6	3	10.79	.012	0.33	.008	0.22	.088	5		95,807,1780
133119	−52 07785				15 01 13	−53 09.9	2	9.25	.042	0.67	.005	0.21		4	G0 IV/V	742,1594
		NGC 5822 - 18			15 01 13	−54 12.7	2	12.10	.002	0.40	.011	0.16	.020	4		95,1780
		NGC 5822 - 155			15 01 14	−54 07.8	1	11.94		0.39		0.19		1		1780
		G 179 - 9			15 01 15	+35 24.5	2	14.16	.025	1.40	.020			5		1663,1759
		NGC 5822 - 419			15 01 16	−53 58.2	1	12.49		0.44		0.17		1		1780
		NGC 5823 - 94			15 01 16	−55 26.9	1	13.30		0.53				2		905
		NGC 5823 - 14			15 01 16	−55 35.8	2	14.09	.068	0.65	.034	0.10	.024	3		95,534
133217	−35 10007				15 01 17	−36 05.4	1	7.89		0.24		0.13		1	A5 V	1771
		NGC 5822 - 17			15 01 17	−54 11.7	2	12.36	.006	0.48	.025	0.09	.005	4		95,1780
		G 66 - 58			15 01 18	+09 52.5	1	12.41		1.30		1.30		1		333,1620
		NGC 5823 - 89			15 01 18	−55 28.3	1	13.60		0.55				2		905
	−55 06338	NGC 5823 - 27			15 01 18	−55 31.0	5	9.71	.012	1.91	.014	2.23	.029	10		95,534,905,1730,4002
132907	−70 01920				15 01 18	−70 56.3	2	7.68	.005	-0.04	.005	-0.42	.000	7	B6/7 III	164,400
133218	−36 09869				15 01 19	−36 39.1	2	10.34	.015	0.39	.005	0.06		6	F2 V	1696,2033
	−53 06251	NGC 5822 - 10			15 01 19	−54 06.2	5	9.63	.019	1.17	.007	1.07	.006	11		95,807,1597,1780,4002
		NGC 5822 - 16			15 01 19	−54 10.9	2	12.11	.015	0.40	.023	0.21	.006	4		95,1780
		NGC 5822 - 133			15 01 19	−54 15.4	1	14.15		0.63		0.06		3		1780
		NGC 5823 - 13			15 01 19	−55 36.2	2	12.12	.042	0.51	.009	0.12	.037	4		95,534
133484	+45 02251	HR 5612			15 01 21	+44 50.3	2	6.64	.025	0.45	.002	0.01	.005	4	F5 IV-V	254,3053
		NGC 5823 - 93			15 01 21	−55 27.4	1	13.84		0.75				2		905

Table 1   827

HD	DM	Other Id	N Rem	α_{1950}	δ_{1950}	S	V	σ_V	B–V	σ_{B-V}	U–B	σ_{U-B}	n	Spectrum	References
		LOD 34 # 17		15 01 21	−57 20.7	1	11.72		0.45		−0.07		1		96
133294	−14 04102			15 01 22	−15 16.0	1	8.23		−0.05		−0.48		2	B8 II	1770
133271	−21 04019	IDS14585S2142	AB	15 01 22	−21 53.9	1	9.77		0.43		0.10		1	A3/5 II	1770
		NGC 5824 sq1 4		15 01 22	−33 00.0	1	11.36		0.67				3		438
		NGC 5822 - 137		15 01 22	−54 14.4	1	11.24		0.41		0.23		2		1780
		NGC 5822 - 19		15 01 23	−54 13.1	1	14.31		0.80		0.23		1		1780
		NGC 5823 - 98		15 01 23	−55 25.8	1	11.41		0.53				2		905
		G 66 - 59		15 01 24	+10 56.2	3	13.19	.010	0.66	.014	−0.09	.009	6		316,333,1620,1658
		NGC 5822 - 11		15 01 24	−54 07.5	3	11.02	.021	1.01	.008	0.66	.017	8		95,1597,1780
		NGC 5822 - 69		15 01 24	−54 11.5	1	13.71		1.61				2		1780
	−56 06570			15 01 24	−56 32.3	1	10.50		0.36		−0.22		1		1586
133332	−2 03939			15 01 25	−02 39.6	2	10.34	.165	1.49	.099	1.28		2	R5	1238,8005
	−53 06253	NGC 5822 - 423		15 01 25	−54 02.8	1	10.72		0.40		0.21		1		1780
		NGC 5823 - 103		15 01 25	−55 24.9	1	13.50		0.60				2		905
		NGC 5823 - 10		15 01 25	−55 34.2	3	13.51	.005	0.71	.010	0.20	.000	5		95,534,905
		NGC 5822 - 168		15 01 26	−54 04.3	1	12.12		0.44		0.13		2		1780
		NGC 5822 - 15		15 01 26	−54 10.1	2	12.22	.006	0.43	.008	0.13	.031	4		95,1780
		NGC 5823 - 111		15 01 26	−55 23.4	1	13.60		0.64				2		905
		NGC 5823 - 97		15 01 26	−55 26.2	2	11.90	.880	1.41	.465	0.20		4		905,4002
		NGC 5823 - 82		15 01 26	−55 26.5	1	12.74		0.53				2		905
		NGC 5823 - 83		15 01 26	−55 27.1	1	12.04		1.98				3		905
		NGC 5823 - 9		15 01 26	−55 33.9	2	13.06	.005	0.51	.010	0.41		3		95,534
129723	−87 00235	HR 5491, BP Oct		15 01 26	−87 56.9	4	6.48	.008	0.30	.005	0.09	.015	18	A2mA7-F2	15,1075,2012,2038
		NGC 5822 - 422		15 01 27	−54 02.2	1	13.04		0.87		0.48		1		1780
		NGC 5822 - 424		15 01 27	−54 03.1	1	12.02		0.38		0.16		1		1780
133220	−40 09243	HR 5604, GM Lup		15 01 28	−40 40.0	3	6.39	.011	1.46	.000			11	M6 III	15,2013,2029
		NGC 5822 - 165		15 01 28	−54 06.2	1	11.41		0.33		−0.22		3		1780
		NGC 5822 - 135		15 01 28	−54 06.2	2	10.22	.085	0.08	.004	0.05	.259	5		807,1780
	−55 06341	NGC 5823 - 126		15 01 28	−55 19.4	1	10.51		1.07		0.76		1		4002
133198	−43 09598			15 01 29	−43 36.6	1	8.93		0.33		0.18		1	A8 V	1771
		NGC 5823 - 118		15 01 29	−55 22.3	1	13.56		0.59				2		905
		NGC 5823 - 1016		15 01 29	−55 24.3	1	14.44		1.70				2		905
		NGC 5823 - 73		15 01 29	−55 25.7	1	12.60		0.54				2		905
		NGC 5822 - 139		15 01 30	−54 14.9	1	13.37		0.90		0.36		3		1780
		NGC 5823 - 74		15 01 30	−55 25.6	1	13.94		0.70				2		905
		NGC 5823 - 29		15 01 30	−55 29.3	1	12.95		0.53				2		905
		NGC 5823 - 8		15 01 30	−55 32.2	2	13.45	.024	0.72	.005	0.25		3		95,905
		NGC 5824 sq1 3		15 01 31	−32 58.5	1	11.23		1.70				2		438
		NGC 5822 - 425		15 01 31	−54 04.2	1	12.32		0.44		0.15		1		1780
		NGC 5823 - 1014		15 01 31	−55 25.1	1	15.55		0.98				2		905
	−49 09301			15 01 33	−50 16.1	1	9.74		0.74		−0.28		2	B0 Ia	403
		NGC 5822 - 14		15 01 33	−54 09.9	1	12.70		1.77				1		95
		NGC 5823 - 120		15 01 33	−55 22.2	1	13.62		0.51				2		905
		NGC 5823 - 1002		15 01 33	−55 26.1	1	15.11		0.65				2		905
		NGC 5822 - 144		15 01 34	−54 13.4	1	11.67		0.34		0.24		3		1780
		NGC 5822 - 143		15 01 34	−54 14.0	1	12.22		1.85		2.24		2		1780
		NGC 5823 - 72		15 01 34	−55 25.1	1	12.40		0.89				2		905
		NGC 5823 - 30		15 01 34	−55 28.6	1	13.93		0.60				2		905
		NGC 5823 - 117		15 01 35	−55 22.5	1	13.72		0.57				2		905
		NGC 5823 - 7		15 01 35	−55 33.0	3	13.78	.018	1.44	.010	1.03	.029	5		95,534,905
133485	+35 02644	HR 5613		15 01 36	+34 45.6	1	6.59		1.02				2	G8 III-IV	70
		NGC 5823 - 122		15 01 36	−55 21.6	1	13.87		0.57				2		905
		NGC 5823 - 1001		15 01 36	−55 27.1	1	15.02		0.74				2		905
133408	+6 02983	HR 5610	★ A	15 01 37	+05 41.2	1	7.13		0.29		0.03		2	F0 IV	3016
133408	+6 02983	HR 5610	★ AB	15 01 37	+05 41.2	1	6.49		0.31		−0.01		8	F0 IV +F1 V	1088
133408	+6 02983	IDS14592N0553	B	15 01 37	+05 41.2	1	7.36		0.31		0.00		2	F1 V	3072
	+0 03300			15 01 38	−00 04.6	1	9.70		0.91		0.60		1	K3 V	3072
		NGC 5822 - 13		15 01 38	−54 08.6	1	13.34		0.66		0.10		3		1780
		NGC 5823 - 121		15 01 38	−55 22.1	1	13.76		0.60				2		905
133432	+11 02762			15 01 39	+10 55.8	1	7.88		1.49		1.90		2	K5	1648
133241	−45 09629			15 01 40	−45 23.6	1	8.75		0.19		0.19		1	A4 V	1771
		NGC 5822 - 310		15 01 41	−54 22.9	1	12.09		0.45		0.19		2		1780
133181	−54 06312	NGC 5822 - 311		15 01 41	−54 23.3	1	9.89		0.10				1	A0 III/IV	1780
		G 179 - 10		15 01 42	+35 28.5	1	14.43		1.44				3		1663
133242	−46 09773	HR 5605	★ AB	15 01 42	−46 51.4	7	3.89	.006	−0.14	.004	−0.59	.015	35	B5 V	15,278,1020,1075,1770*
		NGC 5823 - 1003		15 01 43	−55 25.6	1	14.38		0.60				2		905
134879	+84 00339			15 01 43	+84 13.5	1	8.22		1.29		1.41		4	K0	1502
		NGC 5822 - 147		15 01 44	−54 12.2	1	12.06		0.39		0.19		2		1780
		NGC 5823 - 37		15 01 44	−55 26.6	1	11.58		0.48				2		905
		NGC 5823 - 33		15 01 44	−55 28.1	1	12.47		0.52				2		905
		NGC 5823 - 4		15 01 44	−55 30.1	3	13.02	.009	0.52	.025	0.15	.033	6		95,534,905
		NGC 5824 sq1 7		15 01 45	−32 45.9	1	12.10		0.58				3		438
		NGC 5823 - 32		15 01 45	−55 27.8	1	13.78		0.54				2		905
	−53 06256	NGC 5822 - 148		15 01 46	−54 11.5	3	10.09	.017	1.63	.005	1.93	.038	9		1597,1780,4002
		NGC 5823 - 62		15 01 46	−55 23.2	1	12.52		0.49				2		905
		NGC 5823 - 31		15 01 46	−55 27.4	1	13.66		0.74				2		905
		NGC 5823 - 34		15 01 46	−55 28.7	2	12.56	.005	0.46	.015	0.02		6		534,905
		NGC 5823 - 6		15 01 46	−55 33.1	3	12.96	.014	0.46	.008	0.10	.040	3		95,534,905
133319	−37 09920			15 01 47	−37 29.7	1	8.08		0.20		0.08		1	A5 IV	1771
		NGC 5823 - 1009		15 01 48	−55 23.6	1	14.46		1.03				2		905

HD	DM	Other Id	N	Rem	α_{1950}	δ_{1950}	S	V	σ_V	B–V	σ_{B-V}	U–B	σ_{U-B}	n	Spectrum	References
		NGC 5823 - 1008			15 01 48	−55 24.7	1	14.33		0.55				2		905
		NGC 5823 - 1004			15 01 48	−55 26.5	1	14.83		0.93				2		905
		NGC 5823 - 35			15 01 49	−55 28.5	1	12.76		0.52				2		905
133461	+15 02814				15 01 50	+14 57.2	1	9.02		1.16				1	K2 III	882
	−55 06347	NGC 5823 - 3			15 01 50	−55 29.6	3	10.50	.007	0.14	.010	0.02	.009	6		95,534,905
		NGC 5823 - 5			15 01 50	−55 33.4	3	13.25	.019	1.59	.018	1.51	.044	5		95,534,905
	−53 06258	NGC 5822 - 261			15 01 51	−54 13.2	2	10.81	.002	0.34	.005	0.23	.005	2		768,1780
		NGC 5822 - 2233			15 01 51	−54 14.6	1	10.96		1.04		0.72		2		1780
		NGC 5822 - 438			15 01 51	−54 14.7	2	10.96	.020	1.04	.002	0.74	.007	8		807,1597
		NGC 5823 - 1006			15 01 51	−55 25.3	1	15.72		1.07				2		905
133564	+45 02253	G 179 - 11			15 01 52	+45 10.8	1	8.86		0.63		0.00		3	G5	308
		NGC 5823 - 1005			15 01 52	−55 26.1	1	14.35		1.46				2		905
		NGC 5823 - 45			15 01 52	−55 26.2	1	12.72		0.47				2		905
		NGC 5822 - 260			15 01 53	−54 13.6	1	12.34		0.43		0.18		1		1780
		NGC 5823 - 64			15 01 53	−55 23.6	1	13.06		0.44				2		905
		NGC 5823 - 1007			15 01 53	−55 25.4	1	16.31		0.95				2		905
		NGC 5823 - 42			15 01 55	−55 25.8	1	12.07		0.42				2		905
		NGC 5823 - 1011			15 01 56	−55 22.7	1	16.22		1.10				2		905
133339	−36 09878				15 01 58	−37 11.6	1	8.66		0.16		0.15		1	A3 IV	1771
	−53 06260	NGC 5822 - 265			15 01 58	−54 11.4	2	11.24	.091	0.40	.050	0.18	.049	2		768,1780
	−55 06348	NGC 5823 - 2			15 01 58	−55 29.0	3	10.68	.018	0.44	.009	0.07	.028	6		95,534,905
133666	+56 01780				15 01 59	+56 13.8	1	6.86		1.13		1.11		2	K2	1502
		NGC 5822 - 275			15 01 59	−54 14.3	4	10.93	.026	1.23	.003	1.05	.022	10		807,1597,1780,4002
		NGC 5823 - 58			15 01 59	−55 23.0	1	12.70		0.47				2		905
		NGC 5823 - 57			15 01 59	−55 23.3	1	11.68		0.50				2		905
		NGC 5823 - 56			15 01 59	−55 23.5	1	13.98		0.47				2		905
		NGC 5823 - 43			15 01 59	−55 25.2	1	12.70		0.42				2		905
		NGC 5823 - 44			15 01 59	−55 26.2	1	13.81		0.53				2		905
		NGC 5823 - 36			15 01 59	−55 26.5	1	13.52		0.49				2		905
		NGC 5822 - 266			15 02 00	−54 11.4	1	11.61		0.44		0.24		1		768,1780
	−53 06261	NGC 5822 - 264			15 02 00	−54 11.9	2	11.09	.010	0.49	.024	0.23	.028	2		768,1780
		NGC 5823 - 11			15 02 00	−55 14.	1	14.10		0.58		0.43		1		95
		NGC 5823 - 23			15 02 00	−55 14.	1	13.93		1.97				1		95
		NGC 5822 - 299			15 02 01	−54 19.7	2	11.40	.002	0.51	.000	0.07	.008	2		768,1780
	−53 06262	NGC 5822 - 445			15 02 02	−54 06.0	2	10.76	.088	0.36	.015	0.30	.034	3	A6	768,807
		NGC 5822 - 285			15 02 02	−54 15.5	1	12.86		0.88		0.23		1		1780
	−55 06349	NGC 5823 - 1			15 02 02	−55 28.4	3	10.84	.010	0.44	.018	0.08	.023	6		95,534,905
		NGC 5822 - 296			15 02 03	−54 18.4	1	13.01		1.60		1.30		1		1780
133340	−40 09257	HR 5607		⋆ A	15 02 04	−40 52.4	1	5.15		1.01				4	G8 III	2007
133412	−18 03965				15 02 05	−18 23.7	1	9.59		1.20		1.17		2	K4 V	3072
		NGC 5823 - 59			15 02 05	−55 23.3	1	13.84		0.49				2		905
133341	−44 09842				15 02 06	−45 11.3	1	8.78		0.27		0.17		1	A7 V	1771
		NGC 5822 - 430			15 02 06	−54 06.8	2	12.04	.027	0.41	.010	0.19	.027	2		768,1780
	−53 06264	NGC 5822 - 276			15 02 06	−54 14.0	4	10.55	.015	0.73	.010	0.44	.025	6		768,807,1780,4002
		NGC 5822 - 292			15 02 07	−54 16.7	1	12.19		0.41		0.17		1		1780
		NGC 5822 - 290			15 02 07	−54 17.4	1	12.35		0.51		0.20		1		1780
133640	+48 02259	HR 5618, I Boo		⋆ AB	15 02 08	+47 50.9	4	4.77	.027	0.65	.000	0.10	.006	8	F9/G1 Vn	15,938,1013,3026
	−53 06255	NGC 5822 - 151			15 02 08	−54 13.8	3	10.83	.023	1.06	.008	0.73	.006	10		1597,1780,4002
133279	−53 06266	NGC 5822 - 277			15 02 08	−54 13.9	3	10.31	.006	0.21	.023	0.14	.034	4	A2 IV/V	768,807,1780
	−53 06265	NGC 5822 - 284			15 02 08	−54 15.0	3	10.95	.012	0.41	.010	0.25	.005	4		768,807,1780
		NGC 5822 - 291			15 02 08	−54 17.2	1	12.69		0.45		0.16		1		1780
133049	−71 01729	HR 5598			15 02 08	−71 42.7	5	6.52	.006	1.59	.000	1.90	.016	39	K4 III	15,1075,1589,2012,2038
		LP 859 - 5			15 02 09	−23 09.2	1	13.79		1.03		0.75		2		3062
		NGC 5823 - 217			15 02 09	−55 23.2	1	13.69		0.54				2		905
	−8 03901				15 02 10	−08 37.3	2	9.48	.005	0.92	.005	0.37		5	G0	742,1594
	−53 06269	NGC 5822 - 279			15 02 13	−54 11.4	3	11.01	.040	0.17	.021	0.13	.010	4		768,807,1780
		LOD 34 # 16			15 02 14	−57 33.4	1	11.13		1.26				1		96
	−63 03505	LSS 3307			15 02 16	−63 30.6	2	9.54	.030	1.16	.025	0.05	.005	5	B0 Ia+	403,1737
133582	+27 02447	HR 5616			15 02 18	+27 08.5	4	4.54	.016	1.24	.012	1.33	.009	11	K2 III	15,1080,1118,8015
133342	−50 09071				15 02 19	−50 21.2	1	10.51		0.34		-0.36		2	K2 III	1737
		NGC 5822 - 282			15 02 19	−54 12.0	1	11.57		1.87		2.09		1		1780
		NGC 5822 - 281			15 02 20	−54 11.2	3	11.42	.026	0.37	.011	0.28	.023	4		768,807,1780
	−53 06271	NGC 5822 - 280			15 02 20	−54 11.3	3	11.20	.037	0.49	.017	0.26	.015	4		768,807,1780
133466	−22 03897				15 02 22	−22 49.2	1	7.56		0.06		0.00		1	A0 V	1770
	+4 02959	CG Vir			15 02 22	+04 35.9	1	10.56		0.47		-0.02		6	F8	588
	+10 02786	G 66 - 60			15 02 25	+10 25.9	1	10.78		0.82		0.58		1	K2	333,1620
	+6 02986	G 15 - 5			15 02 26	+05 50.3	2	9.82	.015	1.31	.013	1.14		2	K5	333,1620,1705
		NGC 5823 - 274			15 02 26	−55 16.0	1	11.51		0.39		0.14		3		534
	−57 06909				15 02 26	−57 32.6	1	11.03		0.40		0.24		1		96
		Lod 2115 # 17			15 02 28	−54 10.2	1	12.00		0.40		0.26		1	A5	768
		LOD 34 # 15			15 02 29	−57 30.5	1	11.63		0.49				1		96
		G 15 - 6			15 02 30	+04 16.8	1	9.87		0.70		0.13		1	G4	1620
		NGC 5823 - 273			15 02 30	−55 16.0	1	13.85		0.64		0.00		2		534
	−53 06276				15 02 32	−54 12.8	1	10.87		0.63		0.15		1	F0	768
133365	−54 06320	NGC 5822 - 437			15 02 33	−54 18.8	1	9.52		0.61		0.14		1	G0	768
133399	−48 09606				15 02 34	−48 41.4	2	6.49	.005	-0.14	.000	-0.67	.005	6	B2/3 V	158,400
133641	+26 02647				15 02 35	+25 39.3	1	8.18		0.48		-0.05		2	F8	105
133366	−54 06321	NGC 5822 - 440			15 02 36	−54 24.2	1	10.20		0.12		0.04		1	B9 V	768
133421	−48 09607				15 02 42	−49 04.6	1	8.28		0.94				4	K0 IV/V	2012
133624	+14 02830	G 136 - 62			15 02 44	+14 17.1	1	8.95		0.76				2	G5	882

Table 1   829

HD	DM	Other Id	N	Rem	α_{1950}	δ_{1950}	S	V	σ_V	B–V	σ_{B-V}	U–B	σ_{U-B}	n	Spectrum	References
133725	+39 02832				15 02 48	+38 47.9	1	7.50		0.48		-0.05		2	F8 V	105
133529	-25 10710	HR 5614			15 02 51	-25 35.8	5	6.66	.007	-0.01	.001	-0.37	.015	14	B7 V(n)	15,1079,1770,2013,2029
	+23 02769				15 02 56	+23 10.5	1	10.40		-0.02		-0.66		3		963
133569	-19 04019				15 02 57	-20 13.3	1	8.66		0.03		0.02		1	A0 V	1770
133423	-57 06911				15 03 01	-57 23.5	1	9.72		0.34		0.23		1	A2 III	96
133531	-36 09896				15 03 03	-37 05.8	1	8.29		0.24		0.06		1	A5 III/IV	1771
	-54 06322				15 03 03	-54 18.3	1	10.79		0.34		-0.14		1	B7	768
133550	-35 10035	HR 5615			15 03 05	-36 04.3	1	6.27		1.65				4	K5 III	2006
133909	+60 01586				15 03 06	+59 43.7	1	7.43		0.17		0.09		2	A2	1502
		LP 859 - 7			15 03 07	-22 04.7	1	11.28		0.53		-0.08		2		1696
133440	-56 06579				15 03 07	-57 13.0	1	8.64		0.09		-0.11		1	B9 III	96
	-58 05808	IDS14593S5816		AB	15 03 08	-58 27.6	1	10.70		0.67		0.15		2	F8	807
133574	-34 10172				15 03 09	-35 12.8	1	8.70		0.34		0.05		1	A9/F0 V	1771
		GD 175			15 03 10	-07 03.1	2	15.90	.009	0.39	.009	-0.43	.009	4	DAwk	940,3060
133994	+66 00887	HR 5629			15 03 11	+66 06.8	1	6.23		0.05		0.07		2	A2 Vs	1733
		G 224 - 44			15 03 13	+60 34.6	2	10.99	.010	1.49	.005			3	M0	1746,1759
133454	-58 05809				15 03 17	-58 46.6	1	10.35		0.28		0.24		2	A0 III/IV	807
133518	-51 08745				15 03 21	-51 50.2	3	6.36	.028	-0.09	.015	-0.72	.002	8	B2 IVp He	540,976,2012
	-58 05811				15 03 23	-58 37.2	1	10.48		0.63		0.15		2	F6	807
133507	-56 06582				15 03 25	-57 14.8	1	7.20		1.31		1.22		1	K2 III	96
133650	-23 12060	IDS15006S2335		A	15 03 27	-23 46.7	1	9.03		0.25		0.20		1	A2 III	1770
133508	-57 06913				15 03 29	-57 19.9	1	9.83		0.20		0.02		1	A1 V	96
	-57 06914				15 03 29	-57 22.1	1	10.02		0.84		0.48		1		96
	-59 05816				15 03 30	-59 56.6	1	10.28		0.95		0.41		4		567
133495	-60 05634				15 03 30	-60 35.1	1	8.52		0.02		-0.45		2	B3 Vn	1586
133672	-25 10718				15 03 31	-25 49.6	1	9.45		0.39		0.22		1	A8/9 V	1770
133652	-30 11960	HR 5619			15 03 31	-30 43.5	2	5.96	.000	-0.08	.001	-0.44		5	Ap Si	1770,2007
133670	-21 04030	HR 5620			15 03 34	-21 50.3	1	6.17		1.05				4	K0 III	2007
		Steph 1211			15 03 35	+48 17.2	1	11.59		1.26		1.18		1	K7	1746
		Lod 2045 # 16			15 03 36	-58 32.8	1	11.82		1.23		0.13		2	K0	807
133456	-64 03095	HR 5611			15 03 38	-65 05.0	4	6.16	.004	1.47	.000	1.70	.000	23	K3 III	15,1075,1075,2038
		Ross 1046			15 03 40	+00 01.8	1	12.10		0.70		0.12		2		1696
133673	-31 11717	IDS15007S3204		AB	15 03 43	-32 15.5	1	8.14		1.65		1.79		8	K4 III	1673
		LSS 3308		AB	15 03 43	-49 45.3	1	11.56		1.34		0.61		3	B3 Ib	1737
		LSS 3308		A	15 03 43	-49 45.4	1	12.33		1.77		1.36		1	B3 Ib	1737
		LSS 3308		B	15 03 43	-49 45.4	1	12.73		0.83		0.34		1	B3 Ib	1737
		Steph 1210			15 03 45	+15 24.1	1	11.60		1.52		1.19		1	M0	1746
133962	+48 02262	HR 5627		★ AB	15 03 46	+48 20.6	2	5.57	.008	0.00	.005	-0.06	.012	69	A1 V	274,3016
133612	-47 09703				15 03 47	-47 47.7	2	8.90	.010	0.93	.005	0.54		6	G8 IV	2012,3008
		Lod 2045 # 14			15 03 47	-58 31.7	1	11.11		0.99		0.58		2	K0	807
133774	-15 04026	HR 5622			15 03 50	-16 03.9	2	5.17	.005	1.59	.005	1.92		7	K5 III	2035,3016
133594	-53 06288				15 03 54	-54 08.4	1	8.88		0.16				1	B5 III/V	96
133716	-37 09948				15 03 55	-37 45.2	2	7.18	.005	0.07	.015	0.06	.005	3	A1 V	401,1771
133631	-48 09630	HR 5617			15 03 56	-48 53.8	3	5.75	.012	0.92	.000			11	G8 III	15,1075,2035
133385	-71 01738				15 03 56	-72 01.0	1	6.73		-0.05				4	B2 III/IV	2012
	-58 05816				15 03 57	-58 22.4	1	10.59		0.56		0.24		2	F6	807
	+29 02621				15 03 58	+29 11.5	1	9.39		0.89		0.58		3	K0	1733
	+65 01033	G 224 - 46			15 03 58	+65 00.3	2	9.47	.010	0.82	.005	0.44	.025	6	G5	196,308
		BPM 39210			15 04 00	-47 00.	1	13.22		0.62		0.00		1		3065
133656	-47 09706	LSS 3309			15 04 00	-48 06.4	1	7.51		0.37		0.21		2	A1/2 Ib/II	403
133800	-15 04028	IDS15012S1606		A	15 04 01	-16 17.5	1	6.39		0.16		0.11		3	A1 V	1311
	-58 05817				15 04 01	-58 21.5	1	9.53		1.23		0.98		2	A8	807
133750	-32 10600				15 04 08	-32 43.1	1	7.18		0.07		0.04		1	A0 V	1770
133802	-26 10714				15 04 09	-27 17.1	1	7.87		0.35		0.21		1	A9 V	1770
	-58 05818				15 04 09	-58 33.5	1	10.56		0.41		0.29		2	A3	807
		LP 741 - 65			15 04 10	-10 57.7	1	12.23		1.18		1.09		1		1696
		Ton 787			15 04 12	+29 28.	1	16.38		-0.20		-0.69		1		313
133733	-42 10029	IDS15009S4243		A	15 04 12	-42 54.9	1	8.50		0.24		0.13		1	A3 V	1771
133803	-29 11538				15 04 14	-29 18.7	1	8.15		0.35		0.03		1	A9 V	1771
133681	-53 06292				15 04 14	-54 02.5	1	9.02		0.39				1	F0 III/IV	96
		LOD 35+ # 5			15 04 18	-54 08.2	1	12.86		0.56				1		96
133618	-59 05821				15 04 18	-59 46.6	1	9.18		0.12				4	B5/7 V	2033
133633	-59 05823				15 04 19	-59 45.6	1	8.79		0.65				4	G5/8	2033
133444	-71 01743	S Aps			15 04 20	-71 52.0	2	9.71	.152	1.25	.016	0.62	.041	2	C	1589,1699
133697	-53 06293				15 04 22	-54 07.6	1	9.97		0.08				1	B8/9 IV/V	96
133660	-58 05820				15 04 23	-58 30.5	2	8.87	.005	0.08	.000	-0.13	.010	6	B8/9 III	567,807
133723	-53 06294				15 04 25	-54 01.4	1	8.17		0.19				1	A3 IV/V	96
133735	-49 09345				15 04 26	-49 37.2	1	12.32		0.59		0.02		3	G6/8 IV	1649
234206	+54 01730				15 04 28	+53 53.4	1	9.19		0.37		-0.03		2	F2	1566
133682	-58 05821				15 04 30	-58 30.4	2	8.83	.005	0.05	.005	-0.39	.010	6	B6 V	567,807
		G 66 - 63			15 04 33	+12 01.2	1	10.38		0.90		0.61		1	K0	333,1620
133928	-1 03021	IDS15020S0154		AB	15 04 33	-02 05.4	1	7.66		0.57				4	F8	173
133967	+14 02837				15 04 35	+14 17.5	1	8.79		0.95				1	K0	882
133822	-45 09664	HS Lup			15 04 35	-45 23.2	2	7.73	.000	0.73	.000			8	G6 V	1075,2012
134044	+37 02608	HR 5630			15 04 37	+36 38.8	1	6.35		0.52				1	F8 V	71,254
133985	+13 02894				15 04 40	+13 32.2	1	8.34		0.93				1	G5	882
133896	-25 10728				15 04 40	-25 50.3	1	9.80		0.39		0.10		1	A9 V	1770
133820	-43 09641				15 04 41	-43 31.2	2	7.03	.005	0.04	.000	0.01	.020	4	A0 V	401,1771
		OR 109 # 6			15 04 42	+10 32.7	1	14.27		0.64		0.16		1		930
133785	-53 06297				15 04 45	-54 09.1	1	9.49		0.38				1	F0 V	96

HD	DM	Other Id	N	Rem	α_{1950}	δ_{1950}	S	V	σ_V	B–V	σ_{B-V}	U–B	σ_{U-B}	n	Spectrum	References
		OR 109 # 7			15 04 46	+10 35.0	1	15.12		0.54		0.14		1		930
	+72 00668				15 04 46	+71 57.5	1	9.27		1.03		0.86		3	G5	196
133995	+10 02789				15 04 48	+10 29.3	1	8.09		0.40		0.30		1	F0	930
	+30 02611				15 04 48	+30 12.1	2	9.12	.005	1.24	.000	1.12	.005	4	G8 III	1003,3077
		LOD 35+ # 7			15 04 48	−54 09.3	1	12.73		0.77				1		96
	−58 05823				15 04 50	−58 21.4	1	10.52		1.15		0.93		2	K0	807
134190	+55 01730	HR 5635			15 04 51	+54 44.9	3	5.25	.008	0.96	.006	0.63	.005	6	G8 III	1080,1363,3016
	−53 06298				15 04 52	−54 07.5	1	10.66		0.52				1		96
	−58 05824				15 04 53	−58 23.2	1	10.86		0.26		0.18		2	A5	807
		OR 109 # 2			15 04 56	+10 30.	1	14.55		0.59		0.19		1		930
		OR 109 # 3			15 04 56	+10 30.	1	15.80		0.68		0.30		1		930
		OR 109 # 4			15 04 56	+10 30.	1	15.98		0.57		0.16		1		930
		OR 109 # 5			15 04 56	+10 30.	1	14.97		0.72		0.28		1		930
134319	+64 01046				15 04 57	+64 14.2	1	8.42		0.68		0.11		3	G5	3026
133880	−40 09305	HR 5624, HR Lup			15 04 57	−40 23.6	5	5.79	.010	-0.16	.016	-0.41		20	Ap Siλ4200	15,401,1075,2013,2029
133738	−61 04838	LSS 3311			15 04 59	−61 41.8	4	6.89	.037	0.00	.012	-0.80	.016	11	B1/2III/Vne	540,976,1586,2012
133638	−68 02307	T TrA		★ AB	15 04 59	−68 31.8	1	6.84		-0.02				4	B9 IV	1075
		BPM 9323			15 05 00	−77 00.	1	15.17		0.04		-0.91		1		3065
133810	−58 05826				15 05 01	−58 25.1	1	9.75		0.57		0.13		2	F7/G0	807
133683	−66 02725	HR 5621, LSS 3310			15 05 01	−66 53.6	6	5.76	.011	0.68	.007	0.41	.040	23	F5 Ib	15,403,1075,1075,2038,8100
134063	+23 02775				15 05 02	+22 45.4	1	7.82		0.92		0.54		2	G5	1003
134064	+19 02924	HR 5633		★ AB	15 05 03	+18 38.0	2	6.02		0.06	.005	0.06		7	A1 V +A1 V	1022,1381
		LOD 35+ # 10			15 05 04	−54 03.7	1	11.81		0.60				1		96
		LOD 35+ # 9			15 05 04	−54 05.5	1	13.19		0.60				1		96
134083	+25 02873	HR 5634		★ A	15 05 06	+25 03.8	10	4.93	.007	0.43	.006	-0.03	.009	70	F5 V	15,667,1007,1008,1013*
134224	+52 01850				15 05 06	+52 28.6	1	8.63		0.46		0.03		2	G5	1566
134066	+9 03000	IDS15027N0937		AB	15 05 08	+09 25.0	1	6.72		0.68		0.19		2	G5	3026
		Lod 2045 # 15			15 05 10	−58 36.9	1	11.52		0.70		0.16		2	F8	807
134047	+6 03001	HR 5631			15 05 11	+05 41.4	4	6.16	.006	0.94	.003	0.70	.010	13	K0 III	15,1256,1509,8100
133792	−63 03518	HR 5623			15 05 14	−63 27.1	1	6.28		0.06				4	A0p SrCrEu	2006
133900	−50 09112				15 05 15	−50 29.9	1	8.62		0.02		-0.04		2	B9.5 V	1770
	+25 02874	G 167 - 19			15 05 16	+25 07.2	3	10.00	.038	1.39	.022	1.12	.105	10	K7 V	1003,1197,3078
134085	+12 02791				15 05 17	+12 25.0	1	9.33		0.88				4	G5	882
133936	−40 09309				15 05 18	−41 07.6	1	9.36		0.30		0.09		1	A7/8 V	1771
133901	−50 09113	LSS 3312			15 05 18	−50 59.0	1	9.27		0.32		-0.02		2	B8/9 Iab	403
133835	−58 05829				15 05 18	−58 37.5	1	9.99		0.48		0.05		2	F0 V	807
133937	−42 10050	HR 5625			15 05 20	−42 46.8	7	5.84	.011	-0.12	.011	-0.47	.011	17	B7 V	15,26,400,1770,2029*
134113	+9 03001	G 66 - 65			15 05 22	+09 04.3	4	8.26	.016	0.57	.007	-0.04	.023	10	G0	22,333,1003,1620,3077
	−53 06305	LOD 35+ # 11			15 05 22	−54 09.6	1	10.95		0.76				1		96
	−58 05831				15 05 22	−58 21.8	1	11.04		0.20		0.15		2	A2	807
	+7 02898				15 05 24	+07 29.6	1	10.07		0.45		0.02		3	F8	1732
133851	−59 05829				15 05 25	−59 42.8	1	8.97		0.22		-0.35		5	B3 III	567
133971	−39 09578				15 05 26	−40 05.7	1	8.25		0.32		0.02		1	A9 V	1771
		LOD 35+ # 13			15 05 26	−54 05.0	1	12.15		0.42				1		96
133955	−44 09889	HR 5626		★ AB	15 05 28	−45 05.3	9	4.05	.008	-0.18	.004	-0.68	.010	28	B3 V +B3 V	15,26,1020,1075,1586,1770*
134088	−7 03963	G 151 - 28			15 05 33	−07 43.0	2	8.00	.000	0.58	.004	-0.05	.000	7	G0 V:	1003,1775
	−53 06307				15 05 34	−54 01.5	1	10.54		0.48				1		96
133823	−65 02993				15 05 34	−65 19.0	2	9.63	.015	0.04	.001	-0.66	.011	8	B3 II	540,976
134152	+14 02839				15 05 35	+14 34.0	1	9.77		0.67				2	G0 V	882
134151	+15 02819				15 05 35	+15 08.9	1	9.33		1.26				4	K0	882
133904	−56 06596				15 05 35	−56 52.6	1	8.00		1.17		1.06		6	K0 III	1673
	+32 02547	G 167 - 20			15 05 39	+32 36.7	4	11.13	.005	0.68	.007	0.13	.014	7	G6 V	22,1003,1064,3025
134417	+62 01385				15 05 42	+62 00.8	2	8.80	.010	0.33	.003	0.13	.026	2	F0	762,969
134585	+72 00670				15 05 42	+72 04.5	1	7.45		1.16		1.13		3	K2	3026
		LP 502 - 3			15 05 44	+14 34.4	1	13.56		0.80		0.39		2		1696
134169	+4 02969				15 05 48	+04 07.3	2	7.69	.010	0.53	.015	-0.08	.005	2	G1 V wlm	792,6006
134073	−36 09937				15 05 59	−36 56.0	1	9.11		0.42		0.03		1	A9 V	1771
134140	−26 10730				15 06 00	−26 18.4	2	6.93	.024	1.72	.005	1.89		6	M1 III	2012,3040
134115	−30 11988				15 06 03	−30 29.4	1	8.58		0.31		0.28		1	A1 IV	1770
134282	+27 02457				15 06 05	+26 54.1	1	7.87		1.00		0.82		4	G8 II	8100
134141	−27 10236				15 06 05	−28 10.3	2	7.64	.023	0.31	.008	0.08		5	A9 V	1770,2012
134320	+26 02656	HR 5638			15 06 14	+26 29.5	3	5.67	.005	1.24	.000	1.25	.010	7	K2 III	1118,1501,8031
134352	+34 02604				15 06 17	+33 53.8	1	6.69		0.98		0.77		2	K0	1625
	−3 03732	G 15 - 7			15 06 19	−03 35.4	1	10.49		0.72		0.13		3		1620
134119	−44 09901				15 06 20	−44 22.3	2	8.63	.000	0.01	.010	-0.38		6	B6 V(n)	401,2014
134058	−52 07898				15 06 20	−52 40.8	1	8.00		0.59		-0.05		1	A0 III/IV	3077
134019	−57 06922				15 06 20	−57 43.3	1	9.38		0.04		-0.34		3	B7 III	567
		Ross 1035			15 06 21	+12 14.4	1	12.25		0.55		-0.17		2		1696
134305	+13 02899				15 06 22	+12 40.7	1	7.25		0.20		0.13		2	A7p	1648
		G 136 - 68			15 06 24	+17 34.8	1	15.01		1.59		1.57		4		419
134095	−50 09126				15 06 24	−50 38.9	1	8.42		-0.01		-0.29		1	B8 V	1770
	−55 06394				15 06 24	−56 05.0	1	10.19		0.25		-0.20		3		1586
134335	+25 02876	HR 5640		★ A	15 06 25	+25 17.9	1	5.81		1.24				2	K1 III	1118
		L 108 - 85			15 06 30	−67 48.	2	11.22	.058	0.53	.000	-0.14	.015	3		1696,3060
134323	+13 02901	HR 5639			15 06 31	+13 25.5	2	6.12	.020	0.96	.005	0.65	.030	4	K0	252,3016
		UMi sq # 2			15 06 33	+67 18.2	1	13.54		0.66				1		1618
134336	+9 03003	G 136 - 41			15 06 36	+09 23.1	1	10.58		0.56		-0.04		3	F8	1696
134216	−33 10350	IDS15035S3326		A	15 06 37	−33 37.5	1	8.50		0.39		0.10		1	A7 V	1771
	−36 09947				15 06 41	−37 07.4	1	11.08		0.61		0.18		2	G0	1696
	−56 06605	LSS 3313			15 06 43	−56 44.3	1	10.13		1.20		0.02		3		1737

Table 1 831

HD	DM	Other Id	N Rem	α_{1950}	δ_{1950}	S	V	σ_V	B–V	σ_{B-V}	U–B	σ_{U-B}	n	Spectrum	References
134060	−60 05656	HR 5632		15 06 43	−61 14.0	5	6.30	.007	0.62	.008	0.15	.004	20	G2 V	15,285,688,1637,2012
134493	+50 02146	HR 5648		15 06 44	+50 14.7	1	6.39		1.03				1	K0 III	71
134218	−40 09336			15 06 44	−40 59.4	2	7.49	.006	−0.05	.008	−0.22	.012	3	B9 V	401,1770
134646	+63 01167	IDS15059N6330	A	15 06 53	+63 18.5	1	6.89		0.39		0.03		4	F4 III	1502
	+24 02824	G 167 - 22		15 06 54	+24 12.2	1	9.30		1.07		0.94		2	K8 V:	3077
134255	−38 10020	HR 5636		15 06 54	−38 36.2	1	5.97		0.87				1	G6 III	2008
134405	+15 02823			15 06 56	+14 59.8	1	9.74		1.00				2	G5	882
		LP 682 - 7		15 06 56	−08 30.2	1	11.46		0.70		0.12		2		1696
134607	+58 01552			15 06 59	+57 50.9	1	7.72		1.52		1.84		4	K5	1733
	−57 06926	LSS 3314		15 07 9	−57 31.4	1	10.31		1.23		−0.06		2		1737
		CS 22890 # 9		15 07 00	+02 00.4	1	12.66		0.13		0.20		1		1736
134288	−36 09955			15 07 00	−36 22.1	1	8.65		0.41				4	F0 V	2012
134290	−37 09990			15 07 03	−37 17.1	1	8.36		0.12		0.08		1	A1/2 IV/V	1771
		G 15 - 9		15 07 06	+03 21.1	1	11.45		1.51		1.17		1	M2	333,1620
134586	+53 01766			15 07 06	+53 27.1	1	7.46		0.29		0.05		2	F0	1566
134038	−67 02780			15 07 06	−67 31.3	1	7.45		0.04				4	B9 IV	1075
		G 15 - 10		15 07 08	−04 33.5	3	12.02	.023	0.65	.005	−0.11	.024	6		308,1620,1696
	−59 05842			15 07 08	−59 25.6	1	9.82		1.33		1.50		2		1101
133980	−71 01757			15 07 11	−71 54.6	1	10.59		0.55		0.04		10	G5	1589
		UMi sq # 10		15 07 16	+67 39.8	1	14.52		1.61				1		1618
	−42 10084			15 07 19	−42 46.8	1	10.63		1.43		1.19		4	M0	158
134373	−25 10758	HR 5641		15 07 21	−26 08.6	1	5.76		1.05				4	K0 III	2007
134331	−43 09677	IDS15040S4321	A	15 07 21	−43 32.4	4	7.01	.003	0.62	.003	0.13		15	G2 V	1075,1279,2012,2013
133981	−72 01714	HR 5628		15 07 22	−72 34.9	2	6.00	.005	0.00	.000	−0.24	.000	7	B8/9 III	15,1075
134374	−27 10248	IDS15044S2805	AB	15 07 23	−28 15.6	1	9.53		0.70		0.21		4	G(5) V +F/G	158
134330	−43 09678	IDS15040S4321	B	15 07 23	−43 31.6	2	7.60	.005	0.72	.000	0.30		7	G6/8 V	1279,2012
	−56 06610	LSS 3316		15 07 23	−57 09.9	1	10.44		0.88		−0.17		2	B0 II-III	540
134439	−15 04042	IDS15047S1553	A	15 07 28	−16 08.5	12	9.07	.025	0.77	.012	0.17	.018	27	K0/1 V	22,158,742,908,1003,1097*
134440	−15 04041	IDS15047S1553	D	15 07 28	−16 13.5	11	9.44	.015	0.85	.009	0.34	.016	26	G8/K1 (V)	158,742,908,1097,1658,1705*
134258	−56 06611	IDS15037S5705	AB	15 07 29	−57 16.8	1	9.06		0.09		−0.27		1	B8 IV	567
	−60 05663	LSS 3315		15 07 29	−61 00.6	1	10.76		0.41		−0.49		2	B0.5V	540
134270	−54 06367	HR 5637	⋆ AB	15 07 31	−55 09.4	2	5.51	.033	1.14	.007	0.76		54	G2 Ib/II	1567,2006
134259	−57 06930			15 07 33	−57 43.9	1	8.09		0.02		−0.42		5	B7 Ib/II	567
134011	−72 01717			15 07 34	−72 41.4	1	9.23		0.20		0.18		32	F8/G0 IV	1699
134392	−37 09999			15 07 37	−37 56.4	1	9.29		0.38		0.11		1	A9/F0 IV/V	1771
		GD 176		15 07 46	−10 34.1	1	15.42		0.23		−0.50		1		3060
134569	+16 02745			15 07 48	+15 47.5	2	9.02	.005	1.24	.010	1.34		5	K2	882,3077
	−60 05667			15 07 53	−60 24.9	1	11.16		0.56		−0.38		2	B1 II-III	540
134411	−39 09619			15 07 54	−39 40.5	1	9.56		−0.18		−0.82		1	B2 III/IV	55
134825	+62 01388			15 07 55	+61 55.9	1	8.62		1.34		1.36		1	K5	969
134679	+39 02838	IDS15061N3921	A	15 08 00	+39 09.9	1	8.36		0.54		−0.04		3	F8 V	833
134679	+39 02838	IDS15061N3921	AB	15 08 00	+39 09.9	1	8.21		0.58		0.02		3	F8	833
134499	−21 04047			15 08 01	−21 51.5	1	9.80		0.37		0.13		1	A9 IV	1770
134627	+12 02796			15 08 08	+11 51.7	1	6.88		1.65		1.96		3	M1	1648
134443	−44 09921	HR 5643	⋆ C	15 08 08	−45 05.4	2	7.38	.005	1.08	.005	0.95	.005	6	K0 III	12,164
		CS 22890 # 3		15 08 09	−00 43.5	1	15.39		0.10				1		1736
		UMi sq # 11		15 08 12	+67 39.4	1	14.17		0.94				1		1618
134444	−44 09922	HR 5642	⋆ AB	15 08 12	−45 05.4	2	6.44	.015	1.04	.000	0.84	.010	6	K1 III	12,164
134694	+25 02879			15 08 16	+25 38.0	1	8.48		0.49		0.02		3	F8	833
134518	−35 10101	HP Lup	⋆ A	15 08 17	−36 03.0	1	9.26		0.33		0.18		1	A8 V	1771
134345	−62 04411			15 08 20	−62 50.8	1	6.84		1.41				4	K3 III	2012
134809	+52 01858			15 08 21	+52 36.0	1	8.34		1.23		1.23		3	K0	1566
		UMi sq # 47		15 08 21	+67 37.4	1	14.38		0.44				1		1618
134481	−48 09704	IDS15050S4822	⋆ A	15 08 27	−48 33.0	7	3.86	.013	−0.05	.006	−0.11	.015	18	B9.5Vne	12,15,1075,1279,1637*
134481	−48 09704	IDS15050S4822	⋆ AB	15 08 27	−48 33.0	3	3.69	.015	−0.03	.004	−0.12		14	B9.5Vne+A5V	1075,2012,8015
134630	−12 04214			15 08 28	−12 52.0	1	7.34		1.23		1.21		5	K0 III	1657
134482	−48 09705	HR 5647	⋆ B	15 08 28	−48 33.3	5	5.69	.011	0.14	.006	0.10	.011	15	A5 V	12,15,1075,1279,1637
		UMi sq # 42		15 08 33	+67 09.6	1	12.57		0.78				1		1618
134483	−51 08827	IDS15051S5143	B	15 08 33	−51 55.1	3	6.69	.013	0.50	.011	0.00		8	F6 V	1279,2033,3024
134540	−40 09357			15 08 34	−40 48.4	1	9.33		0.30		0.10		1	A0mA8-A8	1771
134772	+33 02550			15 08 37	+33 16.0	1	6.92		1.11		0.97		3	K1 III	1501
		Steph 1215		15 08 38	+33 52.7	1	11.09		1.43		1.22		2	M0	1746
134631	−20 04163			15 08 38	−20 30.0	1	10.13		0.59		0.04		4	F6 V	1731
134484	−53 06337			15 08 39	−53 28.2	1	8.29		0.01		−0.19		3	B9 III	1586
134632	−20 04164			15 08 40	−21 07.3	1	9.02		0.55		0.06		4	F5 V	1731
134557	−44 09928	IDS15053S4438	AB	15 08 41	−44 49.5	1	7.89		0.04		−0.11		2	A0 IV	401
134505	−51 08830	HR 5649	⋆ A	15 08 41	−51 54.6	5	3.41	.008	0.92	.004	0.66	.000	19	G8 III	15,1075,1279,2012,3024
134716	+12 02799			15 08 42	+12 38.9	1	9.57		1.00				3	G5	882
134633	−23 12121			15 08 42	−23 26.6	1	10.35		0.33		0.15		1	A2 II/III	1770
134591	−34 10235			15 08 43	−34 34.5	2	8.36	.004	0.06	.000	−0.36	.004	8	B5 III	55,1775
134522	−48 09707			15 08 44	−49 14.0	1	9.51		0.06		−0.45		1	B5 III	400
134401	−65 03001	LSS 3317		15 08 48	−65 46.9	3	8.99	.017	0.06	.016	−0.63	.019	9	B2 Vne	540,976,1586
134613	−32 10658	IDS15058S3227	B	15 08 50	−32 38.7	1	9.03		0.29		0.25		2	A2/3 (V)	1770
134614	−32 10658	IDS15058S3227	A	15 08 51	−32 38.5	1	8.71		0.27		0.24		1	A0/1 (V)	1770
134508	−56 06627			15 08 52	−56 44.7	1	9.41		0.21		−0.32		3	B7 II	567
134752	+13 02906			15 08 53	+13 13.2	1	8.74		1.17				1	K2	882
		UMi sq # 15		15 08 53	+67 40.0	1	14.88		0.86				1		1618
134468	−61 04856	HR 5645		15 08 56	−61 33.4	1	6.32		1.90				4	K4 Ib	2035
	−56 06629	LSS 3318		15 08 58	−57 1.6	1	10.13		1.73		0.45		2		1737
		UMi sq # 54		15 08 59	+67 21.4	1	13.26		0.78				1		1618

HD	DM	Other Id	N	Rem	α_{1950}	δ_{1950}	S	V	σ_V	B–V	σ_{B-V}	U–B	σ_{U-B}	n	Spectrum	References
134597	−47 09779	HR 5650			15 09 02	−48 01.8	3	6.32	.004	1.11	.005			11	K2 III	15,1075,2007
		GD 177			15 09 03	+07 25.8	1	12.56		0.48		−0.13		3	sdG	308,3060
134526	−57 06943				15 09 05	−58 05.7	1	9.60		0.02		−0.21		1	B8 IV	567
134793	+9 03006	LV Ser			15 09 08	+08 42.3	1	7.56		0.14		0.06		6	A3p	1202
134942	+47 02207	IDS15071N4714	A		15 09 09	+47 02.2	1	9.85		0.51		0.03		2	K0	1601
134942	+47 02207	IDS15071N4714	B		15 09 09	+47 02.2	1	9.84		0.71		0.26		2		1601
134719	−20 04167				15 09 09	−20 56.4	1	9.83		0.62		0.13		4	F6/7 V	1731
		G 151 - 34			15 09 11	−10 02.9	1	14.26		1.66		1.02		2		3078
134685	−34 10242				15 09 11	−34 46.8	1	7.67		0.13		0.09		1	A0 V	1771
134639	−46 09872				15 09 11	−46 34.6	1	9.33		0.16		0.08		1	B9.5 V	1770
134920	+39 02841				15 09 17	+39 08.3	1	8.83		0.59		0.06		2	F8 V	105
134759	−19 04047	HR 5652		⋆ AP	15 09 22	−19 36.2	8	4.53	.005	−0.09	.007	−0.35	.021	28	A0p Si	15,1063,1075,1088*
134759	−19 04047	IDS15066S1925		⋆ B	15 09 22	−19 36.2	1	9.82		0.74		0.31		2	G5 IV	3024
		GD 178			15 09 25	+32 15.5	2	14.12	.005	0.07	.014	−0.63	.014	5	DA	419,3060
134760	−22 03911				15 09 26	−23 09.4	1	9.70		0.35		0.11		1	A3 II	1770
134687	−44 09932	HR 5651			15 09 27	−44 18.8	7	4.82	.006	−0.18	.012	−0.68	.015	25	B3 IV	15,26,1075,1770,2012*
134671	−50 09177				15 09 28	−50 32.9	1	9.77		0.12		−0.23		3	B7 II	1586
134453	−69 02267	HR 5644, X TrA			15 09 29	−69 53.6	2	5.38	.135	3.43	.078	6.69		6	C5,5	864,897
		CS 22890 # 10			15 09 31	+02 20.8	1	13.73		−0.04		−0.02		1		1736
134796	−15 04048				15 09 32	−16 13.4	1	7.39		−0.01		−0.13		4	B9 V	1770
134870	+12 02802				15 09 34	+12 25.9	1	8.47		1.51				1	K2	882
		G 179 - 18			15 09 34	+45 00.8	1	14.95		0.97		0.51		1		1658
		UMi sq # 19			15 09 34	+67 35.0	1	13.12		0.77				1		1618
		G 136 - 76			15 09 39	+18 08.8	2	13.46	.018	1.52	.009	1.08		5		203,1759
		CCS 2221		V	15 09 39	−60 08.9	1	9.82		4.85				3		864
135075	+51 01987				15 09 42	+50 54.5	1	9.53		0.42		−0.03		2	G5	1566
134833	−20 04170				15 09 43	−20 45.5	1	9.25		0.86		0.57		4	K0/1 III/I	1731
134672	−55 06428				15 09 43	−55 32.7	1	7.50		0.80		0.41		2	G2 IV	1567
		LSS 3319			15 09 45	−59 00.5	1	13.66		1.93		0.66		2	B3 Ib	779
134922	+14 02845				15 09 46	+14 33.8	1	9.21		1.53		1.32		2	M2	271
		UMi sq # 20			15 09 47	+67 35.3	1	13.06		0.55				1		1618
134943	+19 02935	HR 5654, FL Ser		⋆ AB	15 09 48	+19 09.8	2	5.85	.048	1.54	.010	1.45		3	M4 IIIab	71,3001
134963	+22 02801				15 09 51	+22 30.1	1	6.62		1.62		1.88		2	M2	3040
134799	−36 09984	IDS15067S3652	A		15 09 52	−37 03.4	1	8.04		0.27		0.07		1	A3/4 V	1771
134799	−36 09984	IDS15067S3652	AB		15 09 52	−37 03.4	1	7.32		0.23		0.11		3	A3/4 V +F/G	404
	−63 03534				15 09 54	−64 11.4	2	10.35	.033	0.32	.024	−0.38	.006	5	B2.5V	540,976
134657	−60 05680	IDS15059S6058	AB		15 09 55	−61 09.4	2	6.35	.005	−0.04	.005			8	B5 III	2012,2014
		UMi sq # 40			15 09 58	+67 14.0	1	14.16		0.77				1		1618
		GD 179			15 09 58	−09 35.1	1	16.16		0.23		−0.54		1		3060
134837	−35 10119	HR 5653			15 09 58	−35 54.3	3	6.09	.004	−0.08	.004	−0.36		8	B8 V	15,1771,2028
135078	+37 02616				15 10 02	+36 56.5	1	7.21		0.79		0.30		2	G5	252
	+62 01390				15 10 02	+62 02.4	1	10.50		0.50		−0.02		1	G0	969
		G 15 - 13			15 10 05	+06 13.4	3	12.31	.013	0.75	.006	0.08	.012	10		333,1620,1663,3073
134815	−43 09711				15 10 05	−44 00.6	1	7.96		0.13		0.12		2	A0 IV	1771
	−59 05859				15 10 05	−59 35.6	1	10.81		0.39		−0.17		1		115
	−54 06394				15 10 07	−54 20.5	1	10.01		0.37		−0.27		5		567
135384	+68 00823	HR 5672			15 10 11	+67 58.1	1	6.18		0.15		0.12		2	A8 Vn	1733
		KS 936			15 10 12	−60 52.	1	11.49		0.51		−0.47		2		540
134985	−0 02941	G 15 - 14			15 10 14	−00 58.5	3	9.29	.013	0.78	.010	0.37	.009	6	K0	158,1620,1705
		LP 859 - 19			15 10 15	−21 47.4	1	11.27		0.90		0.51		1		1696
134783	−53 06354	LSS 3321			15 10 17	−53 56.8	1	9.46		0.22		−0.59		2	B3 II/III	540
135025	+7 02914				15 10 22	+07 35.2	1	8.74		0.47		0.02		2	F5	1375
134946	−23 12133	HR 5655			15 10 22	−23 49.3	4	6.47	.006	−0.04	.007	−0.41	.010	10	B8 III	15,1079,1770,2029
134874	−41 09635				15 10 22	−41 16.4	1	7.67		−0.05		−0.17		2	Ap Si	401
134606	−70 01985				15 10 23	−70 19.9	1	6.86		0.74				5	G5 IV	897
	−59 05863				15 10 24	−59 26.4	2	10.82	.005	0.85	.015	−0.23	.005	3		779,1737
	−59 05864	LSS 3320			15 10 24	−59 27.6	3	10.74	.049	0.86	.019	−0.26	.009	6	O7	103,540,779
134967	−19 04055	HR 5656			15 10 28	−19 27.6	3	6.07	.004	0.12	.005			11	A3 Vn	15,2018,2028
135101	+19 02939	HR 5659		⋆ A	15 10 29	+19 28.1	5	6.69	.015	0.68	.000	0.26	.015	12	G5 V	15,22,308,1028,3077
135101	+19 02939	G 136 - 78		⋆ B	15 10 29	+19 28.5	4	7.53	.013	0.73	.005	0.34	.011	10	G7 V	22,308,1028,3077
135162	+39 02846				15 10 30	+39 10.4	2	7.73	.010	1.00	.010	0.76	.010	7	K0 III	833,1601
134733	−62 04423				15 10 30	−62 34.8	2	8.62	.003	0.34	.015	−0.44	.019	5	B3 III	540,976
134709	−65 03007				15 10 30	−65 58.3	1	7.86		0.16				4	A1 V	1075
134986	−24 11927				15 10 33	−24 41.1	1	9.19		0.19		0.18		1	A2/3 V	1770
134987	−24 11928	HR 5657			15 10 33	−25 07.3	3	6.44	.004	0.70	.000			11	G5 V	15,1075,2006
		G 201 - 39			15 10 34	+56 36.2	1	16.24		0.28		−0.53		1		3062
		G 201 - 40			15 10 36	+56 36.3	1	15.74		1.56		1.56		1		3062
135103	+15 02831				15 10 37	+14 59.4	2	9.07	.030	1.04	.050	0.74		4	K0	882,1648
134950	−35 10127				15 10 38	−35 38.9	1	8.33		0.27		0.21		1	A6 V	1771
134930	−43 09716	IDS15073S4325	A		15 10 38	−43 36.4	1	7.36		0.21		0.10		1	A4/5 IV/V	1771
135145	+28 02412	IDS15086N2818	A		15 10 40	+28 06.9	1	8.37		0.59		0.07		4	G0 V	3026
135145	+28 02411	IDS15086N2818	B		15 10 40	+28 06.9	1	9.35		0.70		0.22		1		3024
	+3 02996				15 10 41	+02 50.1	1	9.73		0.49		0.00		2	F8	1375
134929	−40 09410	IDS15074S4105	B		15 10 42	−41 16.1	2	8.69	.017	0.83	.013	0.46	.026	6	G8/K0 III	1279,3077
134928	−40 09409	IDS15074S4105	A		15 10 42	−41 16.1	2	8.48	.031	0.79	.013	0.37	.022	6	G8 V	1279,3077
134928	−40 09409	IDS15074S4105	AB		15 10 42	−41 16.4	1	7.81		0.81				4	G8 V + G0	2012
134889	−50 09194				15 10 42	−51 07.1	2	9.86	.039	0.19	.007	−0.38	.020	5	B5 Ib/II	540,1586
134864	−55 06439				15 10 44	−55 52.8	1	8.08		0.08		−0.34		3	B7 II	567
	+83 00442				15 10 47	+82 56.7	1	9.99		1.15				1	R0	1238
134844	−57 06957				15 10 47	−58 09.6	3	9.15	.012	0.24	.009	−0.57	.015	9	B1 IV/V	115,540,976

Table 1　　　　　　　　　　　　　　　　　　　　　　　　　　　　　　　　　833

HD	DM	Other Id	N	Rem	α_{1950}	δ_{1950}	S	V	σ_V	B–V	σ_{B-V}	U–B	σ_{U-B}	n	Spectrum	References
		LundsII141 # 217			15 10 48	−58 59.	1	11.51		0.61		0.05		1		115
		CS 22890 # 11			15 10 49	+02 26.2	1	14.61		0.38		-0.17		1		1736
		WLS 1520-15 # 6			15 10 49	−14 36.0	1	11.92		1.23		1.29		2		1375
135421	+62 01393	IDS15099N6214	A		15 10 51	+62 02.5	1	7.88		0.53		-0.03		1	F7 V	762
135421	+62 01393	IDS15099N6214	B		15 10 51	+62 02.5	1	8.59		0.63		0.06		1		762
135030	−28 11228				15 10 52	−28 50.6	1	9.90		0.47		0.30		1	A8 III	1770
135050	−25 10800				15 10 53	−25 18.1	1	9.19		0.26		0.15		1	A6 IV	1770
		vdB 66 # a			15 10 54	−62 34.9	1	11.64		0.62		-0.11		4	B3 V	434
135051	−25 10801	HR 5658	★ AB		15 10 55	−26 00.5	1	5.84		1.14				4	G8/K0 II	2007
134990	−38 10086	IDS15078S3807	A		15 10 56	−38 18.3	1	7.06		0.15		0.08		1	A1/2 V	1771
135066	−23 12144				15 10 58	−24 11.6	1	7.76		0.93		0.64		3	K0 III	1657
135067	−25 10802				15 10 58	−25 51.9	1	8.23		0.30		0.25		1	A5 III	1770
134877		LSS 3322			15 10 59	−59 39.3	1	11.34		0.85		-0.06		2	Of	779
135052	−27 10274				15 11 00	−27 21.2	1	9.40		0.31		0.24		1	A5 IV	1770
	−8 03922				15 11 04	−08 25.9	1	9.56		0.60		0.03		3	G0	196
134735	−69 02281				15 11 05	−70 13.7	1	6.71		0.46		0.05		11	F5 IV	1628
		UMi sq # 36			15 11 06	+67 22.1	1	12.81		0.60				1		1618
		Cir sq 1 # 4			15 11 06	−59 39.	1	11.34		0.87		-0.08		1		115
		UMi sq # 32			15 11 07	+67 27.6	1	12.35		0.72				1		1618
135201	+15 02834				15 11 08	+14 58.5	1	10.34		0.64		0.18		2	F8	271
135202	+14 02847				15 11 10	+14 36.5	1	10.40		0.61		0.08		2	G0	271
135034	−42 10157	IDS15078S4301	A		15 11 10	−43 11.9	1	7.30		0.15		0.16		1	A1 V + B/A	1771
		Pis 20 - 14			15 11 10	−58 52.6	1	12.75		0.90		0.06		1		103
		UMi sq # 35			15 11 11	+67 22.9	1	14.56		1.25				1		1618
135221	+16 02752				15 11 12	+16 06.5	2	8.76	.035	1.06	.025	0.87		4	K0	882,1648
	−59 05870	LSS 3324			15 11 12	−59 17.6	4	10.13	.038	0.82	.006	-0.23	.012	7	B0 II-III	103,115,540,779
		HA 83 # 287			15 11 14	+14 36.7	1	11.92		0.46		-0.04		2	G0	271
135151	−8 03923				15 11 15	−09 18.4	1	8.02		0.51		-0.04		3	G0	196
134957	−56 06647				15 11 18	−56 49.8	1	9.73		0.19		-0.17		1	B8 III/IV	567
135263	+23 02789	HR 5665			15 11 19	+23 10.1	4	6.32	.008	0.06	.006	0.06	.005	10	A2 V	985,1022,1501,3050
		Pis 20 - 13			15 11 19	−58 53.7	5	11.29	.008	0.89	.020	-0.19	.032	11		103,251,779,1672,1737
135204	−0 02944	IDS15088S0058	AB		15 11 20	−01 09.6	7	6.60	.012	0.77	.010	0.36	.013	29	K0 V	22,1003,1013,1197*
135323	+39 02847				15 11 21	+38 51.1	1	8.08		1.07		0.88		2	K1 III	105
134958	−57 06960	LSS 3326			15 11 22	−57 59.3	6	8.08	.041	0.34	.018	-0.58	.023	14	B2 IIne	92,115,540,567,976,1586
135247	+12 02809				15 11 23	+12 13.6	1	8.53		1.56		1.79		4	K5	1648
		Pis 20 - 1			15 11 23	−58 52.3	3	12.57	.255	1.00	.029	-0.14	.045	8	B3 Iab	103,251,1672
	−58 05864	LSS 3325			15 11 23	−59 02.8	1	10.72		0.77		-0.28		2	B0.5Ia	779
134999	−55 06445				15 11 24	−55 37.1	1	9.43		0.05		-0.28		3	B9 IV	567
		Pis 20 - 12			15 11 24	−58 52.7	1	14.19		0.98		-0.29		2		103
		Pis 20 - 10			15 11 24	−58 53.4	3	12.92	.041	0.85	.027	-0.29	.052	6		103,251,1672
		Cir sq 1 # 7			15 11 24	−59 00.	1	11.03		0.52		-0.30		1		115
	−3 03746	G 15 - 15			15 11 25	−03 36.9	6	9.84	.013	1.12	.008	0.98	.021	27	K4 V	158,830,1003,1620,1783,3077
		Pis 20 - 2			15 11 25	−58 52.7	3	12.49	.015	0.99	.021	0.08	.014	8	B1 II/Ib	103,251,1672
135265	+14 02849				15 11 26	+14 17.3	1	10.00		0.72		0.30		3	G0	271
135264	+14 02850	IDS15091N1449	AB		15 11 26	+14 14.0	1	8.66		0.54		-0.01		3	G0 V	271
135205	−1 03036	Y Ser			15 11 26	−01 42.0	1	8.43		1.80		1.83		1	M5 IIIe	3076
		Pis 20 - 11	★ B		15 11 26	−58 53.2	4	10.74	.040	0.95	.017	-0.23	.052	10	WN6	92,103,251,1672
		Pis 20 - 9	★ D		15 11 26	−58 53.4	3	11.38	.032	0.88	.009	-0.20	.077	6		103,251,1672
		Pis 20 - 3			15 11 27	−58 52.9	3	11.95	.009	0.94	.026	-0.18	.048	7	B0 Ib	103,251,1672
134959	−58 05866	Pis 20 - 8	★ A		15 11 27	−58 53.1	11	8.12	.037	1.01	.033	-0.08	.025	24	B0	92,103,251,403,540,567*
		Pis 20 - 7	★ C		15 11 28	−58 53.2	3	10.48	.032	0.89	.089	-0.24	.032	6	O7	103,251,1672
135694	+72 00674	G 257 - 9			15 11 29	+72 01.6	1	9.09		0.65		0.11		3	K0	3026
135088	−41 09658				15 11 30	−42 08.1	1	7.49		0.01		-0.02		2	B9.5 IV	1770
		Pis 20 - 26			15 11 30	−58 53.	1	13.40		0.45		0.02		2		251
		Pis 20 - 5			15 11 30	−58 53.0	1	11.85		0.88		-0.20		1		103
		Pis 20 - 6			15 11 30	−58 53.0	1	11.66		0.90		-0.16		1		103
	−39 09668				15 11 31	−39 30.3	1	11.43		0.33		0.10		2		1730
135153	−31 11813	HR 5660			15 11 33	−31 20.0	5	4.90	.005	0.38	.005	0.28	.013	15	F3 III	15,1075,2012,3026,8015
		WLS 1520-15 # 11			15 11 34	−17 08.3	1	12.90		0.57		-0.05		2		1375
		Pis 20 - 4			15 11 34	−58 52.8	3	12.91	.019	0.68	.012	0.10	.072	6		103,251,1672
		AP Ser			15 11 35	+10 10.3	1	10.86		0.19		0.13		1	A9.5	668
135365	+35 02663	IDS15096N3515	AB		15 11 36	+35 04.1	1	8.64		0.43		0.01		2	F5	1601
		L 263 - 272			15 11 36	−53 47.	1	14.04		0.77		0.00		1		3065
		Pis 20 - 25			15 11 36	−58 51.4	4	11.91	.052	0.71	.028	-0.11	.022	8		103,251,779,1672
		HA 83 # 291			15 11 37	+14 34.4	1	12.83		0.76		0.35		3		271
135038	−54 06410				15 11 37	−54 23.1	2	8.39	.025	0.01	.002	-0.39	.001	5	B8 II	540,976
135208	−17 04283	IDS15088S1804	A		15 11 38	−18 14.6	2	6.74	.000	0.44	.002	-0.02		6	F5 V	2033,3037
135208	−17 04284	IDS15088S1804	B		15 11 38	−18 14.6	3	8.58	.011	0.51	.008	-0.08		10		2013,2033,3037
135248	−8 03927				15 11 40	−08 39.5	1	9.33		1.11		0.99		3	K2	196
	−58 05869	LSS 3330			15 11 40	−59 12.9	4	10.25	.036	0.80	.023	-0.19	.023	7	B0.5III	103,115,540,779
135402	+38 02629	HR 5673			15 11 41	+38 27.1	1	6.20		1.21				1	gK2	71
135230	−17 04285	HR 5662			15 11 44	−17 35.0	4	6.18	.005	-0.02	.026	-0.26	.028	8	B9 III	252,1079,1770,2007
		LP 562 - 17			15 11 46	+04 09.4	1	13.72		0.53		-0.20		2		1696
135297	+0 03322	FI Ser			15 11 48	+00 33.3	2	8.00	.022	0.00	.006	-0.05	.012	6	A0p	861,1202
135171	−38 10105				15 11 48	−39 14.9	1	8.84		1.17		1.09		2	K2 IV	1730
135232	−24 11946				15 11 51	−25 01.2	1	10.05		0.38		0.13		1	A3/5(m)	1770
135113	−52 08049				15 11 54	−52 20.4	2	8.98	.006	0.15	.001	-0.44	.004	5	B4 II	540,976
		CS 22890 # 27			15 11 55	−00 18.6	1	13.25		0.96		0.72		1		1736
		CS 22890 # 26			15 11 56	−00 17.5	1	13.97		0.06		0.09		1		1736
134583	−78 00972	IDS15055S7806	A		15 11 57	−78 17.3	1	7.01		0.33		0.08		10	F0 V	1628

HD	DM	Other Id	N Rem	α_{1950}	δ_{1950}	S	V	σ_V	B–V	σ_{B-V}	U–B	σ_{U-B}	n	Spectrum	References
135190	−39 09675			15 11 58	−39 32.2	1	8.70		0.18		0.19		2	A3 III	1730
135174	−43 09739	IDS15086S4347	A	15 11 58	−43 57.9	2	6.70	.034	−0.06	.018	−0.30	.002	4	B9 IV	401,1770
135093	−56 06653			15 11 59	−56 27.1	1	10.25		0.37		−0.07		4	B7/A0	567
135438	+32 02561	HR 5674	⋆ A	15 12 03	+31 58.4	1	5.99		1.52				1	K5	71
		LP 802 - 51		15 12 03	−18 26.4	2	11.43	.079	0.56	.007	−0.22	.010	5		1696,3060
		HA 83 # 368		15 12 07	+14 45.7	1	11.40		1.05		0.90		2		271
		HA 83 # 300		15 12 08	+14 38.7	1	11.62		0.48		0.10		2		271
	+14 02851			15 12 11	+14 32.7	1	10.60		0.66		0.12		2		271
135367	−4 03840	HR 5669		15 12 12	−05 19.1	1	6.27		1.47		1.80		8	K2	1088
135517	+43 02475			15 12 13	+43 14.0	2	6.65	.010	0.19	.005	0.11	.015	4	A5	252,1733
135302	−27 10289			15 12 14	−27 40.3	1	7.50		0.19		0.16		1	A1 V	1770
		Ton 788		15 12 18	+24 20.9	1	13.21		−0.18		−1.00		3	sdB	1036
		WLS 1520 5 # 6		15 12 19	+05 13.8	1	13.23		0.48		−0.06		2		1375
135137	−57 06973			15 12 20	−57 52.4	1	9.18		0.11		−0.13		1	B9 III	115
135530	+42 02577	HR 5677		15 12 22	+42 21.4	2	6.14	.010	1.63	.021	1.93		4	M2 IIIa	71,3055
135502	+29 02640	HR 5676		15 12 24	+29 20.9	3	5.26	.000	0.05	.018	0.07	.010	8	A2 V	1022,1118,3050
135235	−47 09824	HR 5663	⋆ AB	15 12 24	−47 53.4	2	5.95	.010	0.21	.000	0.09		6	A2 IIIs	404,2007
	+33 02560	G 179 - 22		15 12 25	+33 11.9	1	10.71		0.73		0.08		3	G3	308
	+66 00897			15 12 26	+65 57.9	1	8.95		1.14		1.15		2	K2	1733
		BPM 39332		15 12 30	−49 31.	1	14.27		0.55		−0.14		1		3065
234219	+54 01736	G 201 - 44		15 12 32	+54 03.2	1	10.52		0.45		−0.17		3	sdF8	308
135307	−43 09744			15 12 33	−43 00.1	1	7.98		0.14		0.12		1	A2 III/IV	1771
135160	−60 05698	HR 5661	⋆ AB	15 12 33	−60 43.2	6	5.73	.008	−0.08	.010	−0.89	.020	25	B0.5 Ve	15,26,1075,1586,1637,2030
	−39 09686			15 12 34	−39 28.7	1	10.87		0.26		0.14		2		1730
135482	+5 02985	HR 5675		15 12 42	+05 07.4	4	5.32	.005	1.09	.011	0.91	.005	19	K0 III	1080,1088,3016,4001
135346	−41 09682	HR 5667		15 12 46	−41 18.4	2	5.17	.010	0.57	.010	0.08		6	G5 Ia + B	401,2007
135139	−65 03013			15 12 47	−65 54.5	1	6.83		0.01				4	B5/7 V	1075
		Pis 21 - 2		15 12 48	−59 28.	1	12.00		1.50		0.42		2		524
		Pis 21 - 3		15 12 48	−59 28.	1	13.25		1.52		0.40		2		524
		Pis 21 - 4		15 12 48	−59 28.	1	13.92		1.32		0.26		2		524
		Pis 21 - 6		15 12 48	−59 28.	1	14.06		1.09		0.48		2		524
		Pis 21 - 7		15 12 48	−59 28.	1	13.68		1.45		0.34		2		524
		Pis 21 - 9		15 12 48	−59 28.0	1	12.12		0.50		0.00		1		524
135348	−43 09749	HR 5668		15 12 49	−43 18.0	2	6.04	.000	−0.13	.005	−0.64		6	B3 IV	401,2007
		Pis 21 - 1		15 12 49	−59 28.7	1	9.51		2.07		1.82		3		524
		Pis 21 - 5		15 12 49	−59 29.0	1	13.28		1.31		0.37		1		524
135240	−60 05701	HR 5664, LSS 3331	⋆ A	15 12 53	−60 46.4	5	5.07	.009	−0.07	.009	−0.93	.017	19	O8.5 V	15,1075,1732,2012,2035
		Pis 21 - 11		15 12 54	−59 28.9	1	11.37		1.21		0.14		2		524
135241	−60 05703	IDS15089S6045	A	15 12 55	−60 56.2	1	8.05		−0.04		−0.71		5	B3 V	1732
135486	−19 04061			15 12 57	−19 21.6	1	8.96		0.00		−0.12		1	B9.5 III	1770
135289	−55 06464			15 12 57	−55 43.8	1	10.57		0.17		−0.22		1	B8/9 V	567
135485	−14 04160			15 12 58	−14 30.5	3	8.17	.016	−0.08	.005	−0.54	.016	6	B3 V	55,399,8100
		Pis 21 - 8		15 12 59	−59 29.0	1	12.56		0.86		0.30		1		524
		Pis 21 - 10		15 12 59	−59 29.9	1	11.23		0.45		−0.25		1		524
135446	−30 12072			15 13 00	−30 32.7	1	8.17		0.66		0.21		1	G1/2 V	1770
135276	−59 05879			15 13 01	−59 48.4	1	10.49		0.59		0.07		1	K5	1732
135631	+38 02631	IDS15112N3840	AB	15 13 03	+38 29.1	1	7.15		0.35		0.06		2	F0	3016
135449	−32 10698			15 13 05	−32 42.5	3	9.39	.025	0.44	.000	−0.02	.010	5	F0	742,1594,6006
135445	−30 12075	IDS15101S3012	A	15 13 06	−30 23.6	1	9.44		0.42		0.25		1	A0 III	1770
135415	−43 09755			15 13 06	−43 46.0	1	7.90		−0.10		−0.44		2	Ap Si	401
135451	−35 10164			15 13 08	−36 03.3	1	9.75		0.35		0.21		1	A8/9 V	1771
135414	−43 09756			15 13 08	−43 40.3	1	8.37		0.34		0.25		1	A2/3mA7-F3	1771
135633	+23 02791	IDS15110N2255	A	15 13 15	+22 44.0	1	8.48		0.57		0.09		2	G0 V	3016
135633	+23 02791	IDS15110N2255	B	15 13 15	+22 44.0	1	12.05		1.25		1.07		2		3032
135316	−57 06978			15 13 15	−58 08.7	1	9.87		0.36		0.18		4	A2 III	1101
135559	+0 03327	HR 5679		15 13 16	+00 33.4	2	5.62	.005	0.18	.000	0.08	.010	7	A4 V	15,1256
	−50 09232			15 13 16	−50 55.3	1	11.85		0.49		−0.13		2		1649
135454	−41 09693			15 13 18	−42 11.2	2	6.76	.000	−0.03	.000	−0.11	.013	3	B9.5V	401,1770
135354	−56 06667			15 13 18	−56 50.7	2	9.20	.010	0.01	.001	−0.30	.014	6	B8 Vn	567,1586
		G 136 - 85		15 13 22	+19 23.3	2	13.58	.005	1.30	.025	1.10		7		316,1759
135578	−4 03847	IDS15108S0432	AB	15 13 24	−04 42.8	1	7.20		0.51		−0.04		3	F8	3030
135291	−63 03544	HR 5666		15 13 24	−63 25.6	3	4.85	.004	1.25	.000	1.32	.000	11	K2 III	15,1075,2012
		LP 562 - 21		15 13 26	+05 49.0	1	13.32		0.60		−0.15		2		1696
135355	−57 06980			15 13 26	−58 11.1	1	7.70		1.64		1.92		4	K2/5	1101
135722	+33 02561	HR 5681	⋆ A	15 13 29	+33 30.2	10	3.48	.010	0.95	.005	0.67	.011	51	G8 III	1,15,1028,1077,1080*
135722	+33 02562	IDS15115N3341	B	15 13 29	+33 30.0	4	7.84	.009	0.59	.000	0.01	.004	10	G0 Vvar	1,1028,1084,3024
135534	−21 04065	HR 5678		15 13 29	−22 12.9	2	5.51	.005	1.36	.015	1.48		6	K2 III	2007,3005
135508	−33 10418	IDS15104S3331	AB	15 13 29	−33 42.3	1	9.34		0.57		0.44		1	A2 IV	1770
135378	−57 06982			15 13 29	−58 15.9	1	9.63		1.06		0.76		4	K3 III	1101
		CS 22890 # 24		15 13 32	+00 07.1	1	13.41		0.57		−0.06		1		1736
135679	+26 02670			15 13 32	+25 49.6	2	6.98	.000	−0.08	.005	−0.29	.010	5	A0	105,3016
135379	−58 05875	HR 5670		15 13 35	−58 37.0	5	4.06	.011	0.09	.008	0.08	.005	18	A3 V	15,1075,2012,2024,3023
		CS 22890 # 15		15 13 38	+02 20.3	1	12.46		0.14		0.16		1		1736
	−52 08088			15 13 44	−52 46.7	1	10.97		0.17		−0.20		3		1586
	−61 04882			15 13 44	−61 18.6	2	10.55	.006	0.32	.003	−0.23	.008	5	B6 II-III	540,976
135697	+14 02853			15 13 46	+13 39.2	1	7.13		0.89				1	G5	882
135637	−12 04225			15 13 46	−13 02.0	1	8.01		0.28		0.09		2	A7 III/IV	1375
135698	+11 02786			15 13 47	+11 00.6	1	8.73		0.39		0.10		4	F5	1648
135601	−24 11965			15 13 47	−24 51.1	1	9.98		0.34		0.10		1	A(2) (II)	1770
135337	−65 03020			15 13 47	−66 16.7	1	8.82		0.00				4	B7 IV	1075

Table 1

HD	DM	Other Id	N	Rem	α_{1950}	δ_{1950}	S	V	σ_V	B–V	σ_{B-V}	U–B	σ_{U-B}	n	Spectrum	References
		CS 22890 # 19			15 13 50	+01 26.1	1	13.42		0.15		0.18		1		1736
135617	−23 12173				15 13 51	−23 49.6	1	7.30		0.21		0.21		1	A3 V	1770
135602	−26 10788				15 13 53	−26 58.0	1	9.36		0.48		0.14		1	A9 V	1770
135462	−57 06984				15 13 53	−57 40.4	1	10.36		0.16		-0.11		3	B9 II/III	567
		NGC 5897 sq1 2			15 13 57	−20 45.8	2	12.42	.030	0.79	.075	0.35	.050	7		514,514
		Steph 1227			15 14 00	+07 13.5	1	10.48		1.17		1.14		1	K7	1746
		G 224 - 57			15 14 00	+64 44.5	1	14.04		1.86				1		1759
135619	−34 10292	IDS15109S3412	A		15 14 02	−34 23.6	1	7.99		0.17		0.10		1	A2 IV	1771
136064	+67 00876	HR 5691			15 14 03	+67 32.2	5	5.14	.016	0.53	.006	0.07	.021	10	F9 IV	15,1003,1008,1363,3053
		Steph 1228			15 14 03	+74 19.3	1	11.48		1.03		0.93		1	K5	1746
		NGC 5897 sq1 8			15 14 03	−20 47.7	1	14.45		0.73		0.28		1		514
135583	−42 10218				15 14 06	−43 12.0	1	8.48		0.21		0.06		1	A9 IV/V	1771
135477	−59 05887				15 14 06	−59 54.7	2	8.21	.009	-0.04	.010	-0.69	.019	6	B2 V	115,1732
135792	+17 02843	IDS15118N1710	AB		15 14 07	+16 58.8	1	7.65		0.59		0.04		2	G0	1648
		BF Ser			15 14 10	+16 37.7	1	11.05		0.06		-0.02		1	A9.5	597
135774	+10 02818				15 14 11	+09 53.8	1	6.71		0.34		0.16		3	A2	8071
		LSS 3332			15 14 11	−62 38.3	1	11.32		0.94		-0.17		3	B3 Ib	1737
135725	−7 03992	G 152 - 3	⋆AB		15 14 12	−08 06.0	1	7.86		0.74		0.31		4	G5	158
135382	−68 02383	HR 5671			15 14 13	−68 29.8	5	2.88	.008	0.01	.005	-0.01	.010	16	A0 IV	15,1034,1075,2038,3023
	−59 05891				15 14 15	−59 48.2	1	10.75		0.13		0.02		13		1732
		NGC 5897 sq1 3			15 14 16	−20 42.2	1	13.13		1.15		0.96		1		514
135944	+51 01990	IDS15128N5118	A		15 14 19	+51 07.3	1	6.48		0.85		0.46		2	G5	1566
135742	−8 03935	HR 5685			15 14 19	−09 12.0	21	2.61	.007	-0.11	.003	-0.36	.011	159	B8 V	1,3,15,26,198,1006*
		NGC 5897 sq1 1			15 14 20	−20 44.7	1	11.34		0.86		0.63		1		514
135624	−45 09805				15 14 23	−45 31.9	1	7.46		1.66		2.01		8	M1 III	1673
135540	−58 05880				15 14 24	−58 30.7	1	9.33		0.14		-0.03		3	B9.5V	567
135891	+37 02625				15 14 27	+37 15.1	1	7.12		0.58		0.05		2	F8	1601
136174	+69 00789	HR 5693			15 14 28	+69 07.7	1	6.52		0.02		0.02		2	A1 Vn	1733
135667	−39 09720				15 14 28	−39 37.8	2	7.33	.005	1.25	.001	1.33		6	K1 III	861,2012
135926	+39 02858				15 14 34	+38 58.4	1	8.36		1.03		0.77		2	K0 III	105
135728	−30 12099				15 14 34	−31 16.6	1	8.62		0.39		0.25		1	Ap SrEuCr	1770
135840	+14 02856				15 14 35	+14 24.4	1	8.38		1.02				2	G5	882
		G 151 - 41			15 14 35	−06 08.9	2	13.89	.074	1.53	.078			6		1663,1759
135794	−8 03937				15 14 35	−08 26.6	1	9.33		0.39		0.03		5	F0	1306
		NGC 5897 - 9			15 14 36	−20 50.	1	13.48		1.43		1.08		2		3036
		NGC 5897 - 160			15 14 36	−20 50.	1	13.63		1.31		0.94		2		3036
		NGC 5897 - 209			15 14 36	−20 50.	1	13.77		1.30		0.84		2		3036
		NGC 5897 - 255			15 14 36	−20 50.	1	13.42		1.63		1.54		3		3036
		NGC 5897 - 263			15 14 36	−20 50.	1	13.31		1.54		1.25		1		3036
		NGC 5897 - 302			15 14 36	−20 50.	1	14.13		1.21		0.87		2		3036
		NGC 5897 - 316			15 14 36	−20 50.	1	13.94		1.27		0.88		2		3036
		NGC 5897 - 366			15 14 36	−20 50.	1	13.98		1.25		0.90		2		3036
		NGC 5897 sq1 6			15 14 37	−20 43.3	1	14.10		0.77		0.43		1		514
	−59 05894				15 14 38	−59 56.3	1	10.88		0.39		-0.16		2		115
	−59 05895	LSS 3334			15 14 40	−59 46.0	5	10.21	.033	0.87	.019	-0.21	.024	15	B0.5Ia	103,403,1101,1732,1737
		GD 440			15 14 43	+03 21.4	1	14.00		-0.01		-0.95		3		1764
135591	−60 05720	HR 5680, LSS 3336	⋆AB		15 14 46	−60 18.9	4	5.45	.023	-0.09	.009	-0.93	.032	14	O8 IIIp	138,1732,2006,8100
135758	−29 11630	HR 5686			15 14 47	−29 58.0	4	4.34	.005	1.10	.000	1.07	.005	12	K1 II/III	15,1075,2012,8015
	+17 02844				15 14 48	+17 15.0	1	8.50		1.60				1	M0	882
	−60 05718	LSS 3335			15 14 48	−61 14.4	2	10.27	.009	0.30	.002	-0.55	.009	4	B1.5II-III	540,976
		CS 22890 # 17			15 14 52	+02 15.4	1	13.61		0.10		0.19		1		1736
135730	−40 09481	HR 5682			15 14 52	−40 52.7	5	6.28	.011	0.18	.016	0.14	.000	15	A2mA2-A7	15,355,1771,2012,8071
		L 200 - 41	A		15 14 54	−56 17.	1	11.60		0.81		0.29		1		3062
		L 200 - 42	B		15 14 54	−56 17.	1	13.85		1.19		1.02		1		3062
		LP 562 - 25			15 14 55	+06 54.8	1	11.94		1.04		0.89		1		1696
135812	−24 11978				15 14 56	−24 48.0	1	8.50		0.31		0.22		1	A5/7 IV	1770
135965	+29 02648	IDS15129N2913	A		15 15 00	+29 01.8	1	8.75		0.60		0.03		1	K0	3016
135767	−44 10006				15 15 02	−44 59.5	1	9.10		0.01		-0.05		1	B9.5/A0 V	1770
135689	−53 06408				15 15 02	−54 10.9	1	7.00		1.75				4	M1/2 III	2012
		CS 22890 # 64			15 15 03	+02 56.5	1	14.70		0.39		-0.24		1		1736
136324	+72 00677				15 15 03	+71 56.0	1	9.12		1.37		1.52		7	F8	1655
		NGC 5897 sq1 4			15 15 03	−20 50.6	1	13.65		0.98		0.65		1		514
135734	−47 09860	HR 5683	⋆AB		15 15 03	−47 41.6	7	4.26	.015	-0.09	.016	-0.37	.012	18	B8 Ve	15,404,1075,1637,1770*
		ID5006/98			15 15 06	−02 24.4	1	13.37		0.49		0.14		1		1748
		UU Boo			15 15 10	+35 17.2	2	11.74	.235	0.09	.069	-0.03	.047	2		597,699
135714	−56 06685				15 15 11	−56 23.7	1	10.12		0.03		-0.22		1	Ap Si	567
135570	−66 02752				15 15 12	−67 06.6	1	8.50		0.05		-0.38		3	B6/7 V	1586
135592	−66 02753	R TrA			15 15 16	−66 18.9	4	6.38	.061	0.59	.020	0.37	.006	4	F7 Ib/II	657,688,1484,1490
135814	−40 09490				15 15 20	−40 39.2	1	8.73		0.29		0.07		1	A8/9 V	1771
135692	−59 05903				15 15 20	−60 11.6	1	7.89		0.00		-0.25		1	B8 V	567
	+60 01598				15 15 22	+59 57.2	1	9.55		0.44				5	F5 II	1379
135978	+12 02818				15 15 26	+11 59.4	2	8.88	.005	1.14	.000	1.02		6	K2	882,3077
135992	+13 02919				15 15 27	+13 21.3	1	8.63		1.45				2		882
	+30 02637				15 15 27	+29 47.1	1	9.73		1.08				1	R0	1238
136046	+29 02649				15 15 29	+29 09.0	1	9.20		0.62		0.00		1	G5	3032
135859	−38 10180				15 15 32	−38 41.9	1	9.32		0.34		0.22		1	A9 IV/V	1771
135753	−59 05905				15 15 33	−59 19.8	1	9.76		0.02		-0.58		1	B3 III	115
136136	+44 02444	IDS15139N4410	A		15 15 37	+43 58.5	1	9.37		1.19		1.12		1	G5	3016
136136	+44 02444	IDS15139N4410	B		15 15 37	+43 58.5	1	9.58		1.01		0.73		1		3016
135896	−30 12117	HR 5688	⋆A		15 15 37	−31 01.7	1	6.18		1.23				4	G6/8 III	2007

HD	DM	Other Id	N Rem	α_{1950}	δ_{1950}	S	V	σ_V	B–V	σ_{B-V}	U–B	σ_{U-B}	n	Spectrum	References
		CS 22890 # 31		15 15 39	−01 08.0	1	13.30		0.23		0.22		1		1736
135876	−40 09496	HR 5687, GG Lup		15 15 39	−40 36.4	4	5.59	.008	−0.10	.005	−0.47	.003	10	B7 V	15,401,1770,2012
135951	−24 11984			15 15 40	−24 49.1	1	7.69		0.19		0.12		1	A1/2 V	1770
		TV Lib		15 15 41	−08 16.8	2	11.23	.005	0.08	.010	0.07	.010	2	F0	597,700
135897	−39 09739			15 15 45	−39 18.4	1	9.50		0.30		0.18		1	A9 V	1771
135878	−41 09748			15 15 47	−42 15.3	1	7.86		0.12		0.10		1	A2 IV/V	1771
135802	−58 05886			15 15 47	−59 06.4	1	10.06		0.21		−0.27		3	B8 Ib/II	567
	−18 04031			15 15 48	−18 26.6	1	10.34		1.22		1.03		4	K5p	158
135786	−60 05736			15 15 51	−60 22.8	1	7.98		−0.07		−0.62		1	B2/3 IV	567
136028	+0 03337	HR 5690		15 15 52	−00 16.8	2	5.88	.000	1.51	.000	1.81	.000	10	K5 III	1088,3005
135803	−59 05907			15 15 53	−59 41.2	1	9.57		0.02		−0.43		3	B7 IV	567
	−56 06692			15 15 54	−57 00.1	1	10.97		0.29		−0.13		3		567
	+60 01600			15 15 56	+60 00.2	1	9.66		1.43				3	K5	1379
135971	−31 11861			15 15 58	−31 41.9	1	9.55		0.21		0.13		1	A2/3 V	1770
		NGC 5904 - 1061		15 16 00	+02 17.	1	13.40		1.13		0.60		2		3074
		NGC 5904 - 1068		15 16 00	+02 17.	1	12.48		1.45		1.41		4		3074
		NGC 5904 - 3003		15 16 00	+02 17.	1	12.42		1.48		1.48		4		3074
		NGC 5904 - 3078		15 16 00	+02 17.	1	12.56		1.31		1.18		1		3074
		NGC 5904 - 4019		15 16 00	+02 17.	1	12.60		1.34		1.30		2		3074
		NGC 5904 - 4047		15 16 00	+02 17.	1	12.43		1.42		1.34		4		3074
		NGC 5904 - 4059		15 16 00	+02 17.	1	12.70		1.34		1.23		2		3074
	−61 04896	LSS 3339		15 16 00	−62 04.2	5	9.64	.040	0.51	.006	−0.39	.018	14	B2 Ve	403,540,976,1586,1737
		CS 22890 # 58		15 16 04	+02 25.8	1	15.12		0.34				1		1736
135737	−67 02836	HR 5684	⋆ AB	15 16 05	−67 18.0	6	6.27	.010	−0.09	.006	−0.61	.015	27	B2 V	15,26,1075,1075,1637,2038
135936	−45 09830			15 16 07	−45 17.1	1	9.18		0.20		0.16		1	A5 V	1771
135885	−54 06445			15 16 09	−54 54.8	3	9.68	.005	0.07	.004	−0.58	.008	10	B4/5	540,567,976
136138	+21 02755	HR 5692		15 16 10	+20 45.3	1	5.70		0.97				1	G8 IIIa	71
135984	−36 10060	IDS15130S3620	AB	15 16 10	−36 31.1	1	7.88		0.33		0.06		1	A9 V	1771
135997	−31 11865	IDS15131S3140	A	15 16 11	−31 50.9	1	8.90		0.29		0.21		1	A3 IV	1770
	+25 02894	IDS15140N2522	A	15 16 12	+25 10.0	1	9.99		1.21				1	K8	1017
136176	+27 02477	IDS15140N2712	AB	15 16 12	+27 01.2	5	6.60	.019	0.56	.009	0.00	.012	11	G0 V +G0 V	292,938,1118,1381,3030
136032	−23 12202	IDS15133S2354	A	15 16 12	−24 05.0	1	7.16		0.40		0.13		1	K1 III + A	1771
135901	−54 06447	SY Nor		15 16 12	−55 00.9	1	9.76		1.53		0.79		1	K0/2 III	657
136013	−33 10441			15 16 14	−33 51.1	1	7.76		0.08		0.06		1	A0 V	1771
135739	−69 02315			15 16 16	−69 48.5	1	7.77		1.41				5	K3 III	897
136160	+10 02823	IDS15139N1048	A	15 16 18	+10 36.5	2	7.11	.020	0.50	.005	−0.01	.005	5	F8 V	1028,3016
136160	+10 02823	IDS15139N1048	B	15 16 18	+10 36.5	2	8.04	.019	0.62	.005	0.11	.009	4	G5	1028,3032
	−48 09835	LSS 3341		15 16 18	−49 2.7	1	10.45		0.39		−0.57		3		1737
		LP 802 - 56		15 16 19	−20 49.9	1	11.28		0.79		0.17		2		3062
136014	−36 10062	HR 5689		15 16 20	−36 54.8	2	6.19	.005	0.96	.000			7	G8 III/IV	15,2027
		CS 22890 # 42		15 16 21	+00 56.7	1	12.74		0.19		0.17		1		1736
		G 167 - 32		15 16 26	+31 48.3	1	14.57		1.48				3		316,1759
135916	−56 06697			15 16 28	−56 35.3	1	9.80		−0.04		−0.44		3	Ap Siλ4200	567
	−60 05744			15 16 28	−60 56.1	2	10.45	.019	0.14	.004	−0.41	.011	9	B4 III	540,976
136290	+46 02052			15 16 33	+45 48.1	1	6.65		1.62		2.00		5	K5 III	1501
135917	−59 05917			15 16 33	−59 21.8	4	7.32	.009	−0.09	.006	−0.88	.007	17	B1 III	540,567,976,1732
		CS 22890 # 56		15 16 34	+02 21.7	1	15.20		0.08				1		1736
136231	+26 02676			15 16 34	+25 57.2	2	8.56	.040	0.57	.012	0.00		5	G0 V	20,3026
136083	−25 10874	IDS15136S2547	AB	15 16 34	−25 58.2	1	9.18		0.21		0.12		1	A0 V	1770
135918	−59 05918			15 16 37	−60 07.1	1	9.52		0.03		−0.38		1	B8/9 V	567
136342	+52 01865			15 16 38	+51 47.9	1	8.20		1.00		0.75		2	K0	1566
136161	−1 03046			15 16 39	−01 59.1	2	8.86	.012	0.33	.004	0.24	.025	7	A3 V	1003,1775
		LP 859 - 35		15 16 39	−24 19.9	2	11.67	.010	0.61	.000	−0.09	.010	4		1696,3060
136140	−8 03947	FZ Lib		15 16 40	−08 57.9	1	6.90		1.55		1.44		17	M2	3040
	+15 02842			15 16 41	+15 20.9	1	8.86		1.14		1.06		2	K2	1648
		CS 22890 # 32		15 16 41	−00 18.9	1	14.09		0.36		0.11		1		1736
		Steph 1229		15 16 44	+29 25.9	1	11.65		1.47		1.19		1	M0	1746
136202	+2 02944	HR 5694, MQ Ser	⋆ A	15 16 45	+01 57.2	13	5.06	.007	0.54	.009	0.06	.011	55	F8 III-IV	1,15,22,116,1028,1034*
136202	+2 02944	IDS15142N0209	B	15 16 45	+01 57.2	2	10.11	.000	1.34	.000			4	K4	1,1028
	+86 00228		A	15 16 46	+85 56.5	1	10.97		1.12		1.06		1	K8	679
	+86 00228		B	15 16 46	+85 56.5	1	11.07		1.18		1.21		1		679
		CS 22890 # 55		15 16 47	+02 18.0	1	15.24		0.05				1		1736
	−7 04003			15 16 50	−07 32.4	6	10.57	.009	1.60	.010	1.21	.013	24	M5	158,694,1006,1705,3078,8006
136274	+26 02677	G 167 - 33	⋆	15 16 52	+25 52.4	4	7.96	.018	0.74	.005	0.30	.035	10	G8 V	20,22,1003,3077
136003	−55 06509	LSS 3343		15 16 53	−55 57.1	7	6.79	.020	0.20	.010	−0.63	.023	17	B1 Iab/b	96,158,403,540,567*
136004	−55 06508			15 16 54	−56 13.5	1	9.26		0.47		0.05		1	F3 V	96
	−30 12135			15 16 54	−30 39.1	1	11.59		0.53		−0.09		1		1696
		CS 22890 # 54		15 16 56	+02 17.1	1	15.84		−0.02				1		1736
136142	−31 11877			15 16 58	−31 31.7	1	9.23		0.33		0.27		1	A7 V	1770
136163	−29 11639			15 17 01	−29 44.9	1	9.52		0.44		0.32		1	A0 V	1770
136037	−55 06510			15 17 03	−56 06.8	1	8.55		1.26		1.30		1	K1 III	96
136217	−14 04175			15 17 04	−15 10.3	1	8.71		1.11		0.81		2	G8/K0 III	1375
136164	−34 10322			15 17 04	−34 44.7	1	7.79		0.17		0.09		1	A2 V	1771
	−56 06700			15 17 06	−57 11.0	2	10.79	.009	0.79	.001	0.13	.007	8	B5 III	540,567
136726	+72 00678	HR 5714		15 17 07	+72 00.3	2	5.01	.005	1.37	.002	1.60		4	K4 III	1363,3016
136144	−38 10203			15 17 07	−38 27.2	1	8.72		0.13		0.13		1	A3 V	1771
		LP 859 - 37		15 17 13	−24 06.4	1	12.72		0.52		−0.08		2		1696
		CS 22890 # 63		15 17 14	+02 46.9	1	13.84		0.40		−0.18		1		1736
	+29 02654	G 167 - 35		15 17 16	+29 23.0	1	10.25		1.26				1	K7	1746
136234	−24 12001	IDS15143S2437	A	15 17 16	−24 47.8	1	9.19		0.31		0.24		1	A5/6 IV	1770

Table 1 837

HD	DM	Other Id	N Rem	α_{1950}	δ_{1950}	S	V	σ_V	B–V	σ_{B-V}	U–B	σ_{U-B}	n	Spectrum	References
		WLS 1520 5 # 10		15 17 17	+04 35.0	1	12.88		0.81		0.45		2		1375
136328	+16 02765			15 17 17	+16 08.2	1	8.43		0.86				3	K2	882
		LOD 35 # 12		15 17 17	−56 08.4	1	12.87		0.80		0.27		1		96
136418	+42 02587			15 17 18	+41 55.0	1	7.88		0.93		0.67		2	G5	1601
136257	−8 03949			15 17 18	−08 28.7	2	7.54	.013	0.54	.004	0.03	.013	6	F9 V	158,1003
136168	−44 10038			15 17 21	−44 37.3	1	9.45		0.08		−0.14		1	B9 IV/V	1770
		CS 22890 # 38		15 17 22	+00 27.1	1	15.26		0.21		0.19		1		1736
		KS 956		15 17 24	−59 19.	1	11.61		0.91		−0.05		2		540
		WLS 1500 75 # 9		15 17 28	+74 58.7	1	11.82		0.71		0.29		2		1375
		LOD 35 # 14		15 17 28	−56 12.7	1	13.42		0.58				1		96
136403	+33 02574	HR 5702		15 17 29	+32 41.7	3	6.32		0.23	.010	0.09	.005	11	A2 m	1022,1199,8071
136074	−59 05926			15 17 29	−59 52.4	1	9.37		0.01		−0.64		1	B2/3 IV	115
136246	−27 10330			15 17 30	−28 06.4	1	7.17		0.09		0.06		14	A1 V	1770
136207	−42 10288			15 17 32	−42 54.2	1	9.04		0.28		0.23		1	A8/9 V	1771
136377	+16 02767			15 17 33	+16 38.6	1	8.33		1.28				3	K5	882
		LP 742 - 69		15 17 33	−15 09.0	1	12.20		0.79		0.29		2		1696
136129	−55 06516			15 17 36	−56 01.4	1	9.17		0.43		0.07		1	F3 V	96
136188	−48 09860			15 17 38	−48 37.2	1	10.47		0.02		−0.05		2	B9 V	1097
		LOD 35 # 16		15 17 38	−56 05.8	1	12.09		0.52		0.12		1		96
136617	+60 01603		V	15 17 40	+59 42.1	1	8.20		1.60				6	K5 V	1379
136247	−38 10215			15 17 42	−39 09.7	1	10.00		0.28		0.12		1	A9 V	1771
		LOD 35 # 17		15 17 43	−56 07.4	1	10.78		2.10		1.61		1		96
		LOD 35 # 11		15 17 43	−56 12.5	1	12.49		0.73		0.33		1		96
		LP 859 - 39		15 17 45	−23 01.8	1	12.01		0.55		−0.10		2		1696
136404	+14 02845			15 17 46	+14 43.7	1	7.53		1.65		1.95		2	M1	3040
136919	+74 00609			15 17 48	+74 13.6	1	6.67		1.02		0.87		2	K0	985
		CS 22890 # 43		15 17 50	+01 27.1	1	14.24		0.04		0.12		1		1736
136378	+0 03346	G 15 - 17	⋆ A	15 17 52	+00 25.7	1	9.29		0.80		0.35		1	G5	333,1620
136562	+50 02165			15 17 52	+50 23.7	1	7.62		0.10		0.09		2	A0	1566
		G 15 - 18		15 17 54	+00 22.5	1	12.15		1.44				1		333,1620
136061	−66 02764			15 17 54	−66 25.2	3	7.92	.015	0.69	.005	0.18	.000	9	G2 V	285,657,1075
	−55 06520			15 17 56	−56 05.5	1	10.87		0.22		0.21		1		96
	−45 09856			15 17 58	−45 29.7	2	10.70	.006	0.93	.005	0.62		11		143,1770
136224	−55 06521			15 18 01	−55 30.2	1	9.27		−0.05		−0.52		1	B6 V	567
		CS 22890 # 39		15 18 02	+00 27.5	1	15.05		0.22				1		1736
136366	−17 04312	HR 5701		15 18 03	−17 58.7	3	6.16	.005	1.02	.007	0.73		8	G8 II/III	15,2028,3077
		LOD 35 # 10		15 18 03	−56 10.1	1	11.84		0.78		0.41		1		96
136093	−65 03038			15 18 03	−66 02.0	4	7.05	.005	0.12	.008	0.09	.000	10	A3 V	285,657,1075,1490
		CS 22890 # 66		15 18 04	+03 19.4	1	14.79		0.28				1		1736
136512	+30 02647	HR 5709	⋆ A	15 18 04	+29 47.8	6	5.51	.016	1.01	.010	0.77	.022	14	K0 III	15,252,1003,1080,1118,4001
136298	−40 09538	HR 5695, δ Lup	⋆	15 18 05	−40 28.1	5	3.22	.005	−0.23	.004	−0.88	.016	17	B1.5IV	15,26,1075,2012,8015
		LOD 35 # 8		15 18 08	−56 04.7	1	12.05		0.73		0.31		1		96
136526	+31 02722	IDS15161N3104	A	15 18 10	+30 52.5	1	9.85		0.48		−0.03		2	G5	3016
136526	+31 02722	IDS15161N3104	B	15 18 10	+30 52.5	1	10.10		0.50		−0.03		2		3016
136300	−44 10048			15 18 10	−44 45.5	2	6.69	.005	0.00	.025	−0.04	.020	4	A0 V	401,1771
136528	+28 02420			15 18 11	+27 55.8	1	8.45		1.08		0.87		3	G5	186
136442	−1 03047	HR 5706	⋆	15 18 12	−02 13.9	2	6.34	.000	1.06	.002	1.07	.000	12	K0 V	1088,3016
136407	−15 04083	HR 5703	⋆ A	15 18 13	−15 22.2	5	6.18	.086	0.39	.005	0.03		5	F2 V	254,2007
		LOD 35 # 9		15 18 13	−56 06.6	1	12.57		1.52				1		96
		CS 22890 # 36		15 18 15	−00 05.5	1	14.62		0.21		0.14		1		1736
		G 151 - 49		15 18 15	−12 59.9	1	11.37		1.24				1		1759
136282	−48 09871			15 18 15	−48 36.0	1	9.22		−0.02		−0.33		2	B8 II	401
136250	−54 06469			15 18 15	−55 03.6	1	10.27		0.13		−0.35		5	B7 Vn	567
136346	−34 10328			15 18 16	−35 00.4	1	9.29		0.79		0.32		4	F3 V	119
136347	−37 10171	HR 5697	⋆ A	15 18 16	−38 02.4	1	6.60		−0.10		−0.32		1	Ap Si	1771
136347	−37 10171	HR 5697	⋆ AB	15 18 16	−38 02.4	2	6.48	.000	−0.06	.000	−0.29		7	A0p Si	404,2007
136334	−40 09539	HR 5696		15 18 18	−40 34.2	2	6.20	.000	0.08	.010	0.10		6	A1 V	1771,2032
136239	−58 05897	LSS 3345		15 18 21	−58 58.1	6	7.80	.018	0.93	.020	−0.15	.022	12	B2 Iae	115,138,403,540,567*
136267	−55 06522			15 18 24	−56 05.9	1	9.65		1.01		0.78		1	K1/2 IV/V	96
		CS 22890 # 41		15 18 25	+00 59.0	1	12.61		0.35		0.16		1		1736
136352	−47 09919	HR 5699		15 18 25	−48 08.1	6	5.65	.016	0.64	.008	0.07	.017	20	G3/5 V	15,1075,1637,2024*
	+28 02421			15 18 26	+27 49.7	1	10.70		1.23		1.16		5		186
136514	+1 03067	HR 5710	⋆ AB	15 18 29	+00 53.8	4	5.35	.007	1.19	.000	1.21	.008	11	K3 III	15,542,1417,3077
136479	−5 04057	HR 5707	⋆ A	15 18 29	−05 38.7	2	5.54	.009	1.05	.006	0.98	.015	13	K1 III	1088,1509
		Steph 1234		15 18 30	+39 17.7	1	10.79		1.56		1.90		3	M2	1746
136480	−6 04181			15 18 30	−06 26.0	1	7.35		1.16		1.11		26	K2	978
		GD 180		15 18 31	−06 54.2	1	14.57		0.58		−0.14		1		3060
136458	−19 04084	S Lib		15 18 31	−20 12.5	1	9.10		1.86		1.49		1	M1/3	3076
136445	−27 10333			15 18 34	−27 44.7	1	8.17		0.52		0.00		4	F5 V	55,2012
136316	−52 08213			15 18 36	−53 03.5	1	7.61		1.21		0.80		4	G0/2 IV/V	119
		LundsII141 # 231		15 18 36	−59 16.	1	10.78		0.81		−0.20		1		115
136729	+52 01869	HR 5715		15 18 37	+52 08.3	1	5.65		0.12		0.11		2	A4 V	3052
136422	−35 10236	HR 5705	⋆ A	15 18 38	−36 04.9	5	3.56	.010	1.53	.005	1.87	.008	16	K5 III	15,1075,2012,3077,8015
	−59 05934			15 18 38	−59 35.6	1	10.46		0.29		−0.27		1		567
136423	−37 10176			15 18 39	−37 22.2	1	8.10		0.24		0.08		1	A7 IV	1771
136351	−47 09922	HR 5698		15 18 39	−47 44.8	3	5.00	.009	0.50	.000	0.04		9	F8 V	15,2012,3053
	−45 09862			15 18 41	−46 13.9	1	10.64		0.52		0.07		2		1770
	+28 02422			15 18 44	+27 50.3	1	10.53		0.69		0.10		7		186
136303	−58 05901			15 18 45	−58 57.9	1	8.09		−0.02		−0.39		1	B7 III	567
136654	+31 02724			15 18 48	+31 39.4	2	6.89	.004	0.53	.030	0.01		26	F8 V	3016,8097

HD	DM	Other Id	N Rem	α_{1950}	δ_{1950}	S	V	σ_V	B–V	σ_{B-V}	U–B	σ_{U-B}	n	Spectrum	References
		LundsII141 # 233		15 18 48	−59 24.	1	10.51		0.42		−0.37		1		115
		WLS 1520-15 # 5		15 18 49	−12 39.0	1	11.60		0.68		0.19		2		1375
136393	−50 09318			15 18 52	−50 22.5	1	11.46		0.78		−0.20		2	A0 IV	1649
136356	−55 06524			15 18 52	−56 09.1	1	10.10		0.10		0.00		1	B9.5V	96
136643	+25 02902	HR 5711		15 18 57	+25 08.2	1	6.39		1.23				1	K0	71
136751	+44 02453	HR 5716		15 18 58	+44 36.9	2	6.18	.025	0.37	.013	0.03	.008	4	F3/4 IVs	254,3053
136482	−37 10181			15 18 58	−37 27.4	1	6.66		−0.07		−0.23		1	B8/9 V	1770
		GD 181		15 18 59	+15 41.7	1	13.40		0.62		0.00		2		3060
		G 224 - 59		15 18 59	+63 40.4	1	16.63		−0.04		−0.94		3		538
136655	+26 02681	MCC 739	A	15 19 01	+25 44.9	1	8.92		0.95		0.75		1	K0	679
136655	+26 02681	MCC 739	B	15 19 01	+25 44.9	2	10.85	.042	1.31	.009	1.27		2	K7	679,1017
	−43 09842			15 19 05	−43 41.0	1	11.19		0.94		0.68		2	G5	1696
136359	−60 05760	HR 5700		15 19 06	−60 28.7	5	5.65	.009	0.48	.003	0.00		17	F5 V	15,2013,2020,2028,3053
	−51 09025			15 19 07	−51 57.4	1	11.75		0.40		−0.35		3		1649
136466	−47 09926	IDS15157S4733	AB	15 19 08	−47 44.4	2	7.68	.005	0.69	.005	0.10		8	G6 V	285,2012
	+59 01641			15 19 09	+59 12.3	1	9.36		0.57				5		1379
136486	−43 09845			15 19 09	−44 06.9	2	6.74	.010	0.06	.020	0.06	.015	4	A1/2 V	401,1771
		G 151 - 52		15 19 10	−13 53.2	1	13.54		0.90		0.38		3		3060
	−50 09325			15 19 10	−51 07.8	1	10.31		0.79		0.44		2	F8	1696
		LP 915 - 27		15 19 11	−27 39.5	1	13.28		1.70				1		3062
	−51 09026			15 19 14	−51 28.7	1	12.03		0.55		−0.32		2		1649
136504	−44 10066	HR 5708	★ AB	15 19 17	−44 30.7	5	3.37	.016	−0.18	.006	−0.74	.020	17	B2 IV/V	15,26,1075,2012,8015
136398	−59 05941			15 19 17	−60 03.8	1	8.38		0.00		−0.42		1	B5 III	115
136549	−33 10481	IDS15162S3348	A	15 19 20	−33 58.8	1	7.66		0.04		0.01		1	A0 V	1771
136753	+31 02725	S CrB		15 19 22	+31 32.8	2	8.11	.140	1.30	.050	0.18	.180	2	M4	3001,8027
136415	−58 05908	HR 5704	★ AB	15 19 23	−59 08.6	3	4.50	.004	0.20	.004	−0.36	.010	11	B5 IV + F8	15,1075,2012
136711	+18 03008			15 19 27	+18 37.1	2	7.73	.009	1.21	.009	1.25	.005	4	K3 II-III	3077,8100
	−4 03873	G 15 - 19		15 19 27	−04 35.9	4	9.47	.024	1.32	.015	1.23	.010	8	K7 V	158,1620,1705,3072
136732	+11 02794			15 19 33	+10 40.9	1	8.30		0.21		0.10		2	A3	1648
		CS 22890 # 35		15 19 33	−00 11.0	1	15.12		0.05				1		1736
		Steph 1235		15 19 37	−18 35.3	1	11.64		1.75		2.33		3	M2	1746
136471	−58 05910	LSS 3347		15 19 37	−58 57.3	5	8.98	.015	0.94	.017	0.03	.040	11	B5 Ia	115,403,540,976,8100
		WLS 1520 5 # 7		15 19 38	+02 50.5	1	13.73		0.66		0.10		2		1375
	+27 02481			15 19 43	+27 34.6	1	9.25		1.63		1.77		7	K7	186
		AAS9,163 T7# 4		15 19 43	+27 57.3	1	11.29		0.69		0.16		7		186
136556	−49 09578	LSS 3350		15 19 45	−49 56.3	2	9.07	.063	0.29	.015	−0.64	.013	6	B2/3 (V)ne	540,1586
	+28 02423			15 19 48	+28 13.2	1	9.49		0.97		0.72		4	K0	186
136849	+33 02581	HR 5718		15 19 48	+33 06.7	4	5.37	.001	−0.07	.006	−0.20	.014	8	B9 Vn	1022,1079,1629,3023
	−52 08240			15 19 50	−52 57.8	1	9.51		0.15		−0.08		1		96
136557	−52 08243			15 19 52	−52 58.9	1	8.96		−0.01		−0.35		1	B4 V	96
136713	−10 04088			15 19 54	−10 28.8	1	7.99		0.95		0.79		1	K2 V	3072
136647	−37 10189			15 19 54	−38 12.5	1	9.60		0.30		0.13		1	A3mA8-A8	1771
		AAS9,163 T7# 7		15 19 57	+28 06.0	1	9.86		1.53		1.78		9		186
136703	−26 10842			15 19 57	−26 30.7	1	6.52		1.12				4	K0 III	2006
136664	−36 10103	HR 5712		15 19 57	−36 40.8	1	4.55		−0.16		−0.65		2	B4 V	1770
136664	−36 10103	HR 5712		15 19 57	−36 40.8	6	4.53	.004	−0.15	.004	−0.63	.019	19	B4 V	15,26,1075,1586,2012,8015
136537	−56 06729			15 19 57	−57 09.2	1	6.76		1.22		0.95		9	G2 II	1628
	−59 05952			15 19 57	−59 59.6	1	11.01		0.22		−0.37		1		115
		LP 442 - 73		15 20 00	+19 35.6	1	10.50		0.49		−0.06		1		1696
		LOD 36+ # 16		15 20 00	−52 32.9	1	11.59		0.57		0.10		1		96
136489	−62 04469	IDS15158S6240	AB	15 20 00	−62 50.8	1	9.02		1.05		0.78		1	G6 III	1642
		BPM 9483		15 20 00	−68 47.	1	14.68		0.57		−0.32			SD	3065
136831	+13 02928	HR 5717		15 20 01	+12 44.7	3	6.28	.005	0.00	.013	−0.02	.014	9	A0 V	252,1022,3016
136538	−59 05953			15 20 03	−59 17.0	1	9.49		0.10		−0.36		1	B8 II	567
		L 551 - 74		15 20 06	−34 01.	1	15.55		0.14		−0.76		5		3060
136834	+1 03071	G 15 - 20		15 20 11	+01 36.1	4	8.29	.026	0.99	.005	0.86	.015	8	K3 V	22,333,1003,1620,3077
		AAS9,163 T7# 9		15 20 12	+27 53.2	1	11.33		1.10		0.86		5		186
	+28 02424			15 20 14	+28 13.5	1	10.08		0.58		0.01		4		186
136801	−14 04188			15 20 14	−14 57.4	1	6.47		1.62				4	K4 III	2007
136901	+26 02685	UN CrB	★ A	15 20 16	+25 48.1	2	7.23	.012	1.24	.005	1.09		8	K1 III	20,833
136735	−31 11915			15 20 19	−31 37.7	1	8.82		0.40		0.25		1	A9 V	1770
		CS 22890 # 76		15 20 22	+00 28.6	1	15.68		0.02				1		1736
	+27 02482			15 20 22	+27 37.7	1	10.04		0.86		0.38		4	G0	186
	−52 08259			15 20 22	−52 35.2	1	10.44		0.32		0.25		1		96
136592	−59 05954	IDS15164S5910	AB	15 20 22	−59 20.9	4	10.33	.020	0.27	.006	−0.35	.014	7	B3/5	115,125,540,567
	−63 03580	LSS 3349		15 20 23	−63 27.4	3	9.72	.015	0.19	.004	−0.71	.011	5	B0.5IVn	540,976,1737
		L 263 - 144		15 20 24	−52 27.	1	14.27		1.54				1		3062
		LP 742 - 71		15 20 25	−12 28.9	1	12.17		1.11		0.94		1		1696
136924	+16 02773	G 136 - 99		15 20 26	+16 26.3	1	8.24		0.59		0.03		2	G5	308
136954	+32 02575			15 20 26	+32 10.9	1	8.16		1.35				2	K0	8097
136779	−29 11665			15 20 27	−29 48.8	1	9.67		0.54		0.31		1	A7/8 V	1770
		AAS9,163 T7# 11		15 20 29	+27 56.2	1	10.56		0.87		0.58		7		186
		LOD 36+ # 12		15 20 29	−52 35.2	1	13.21		0.59				1		96
136923	+19 02961	G 136 - 101		15 20 32	+19 05.7	1	7.20		0.76		0.34		3	K0	308
		WLS 1520-15 # 9		15 20 32	−15 02.3	1	12.80		0.71		0.19		2		1375
		LOD 36+ # 18		15 20 33	−52 27.5	1	12.60		0.63				1		96
	−58 05914	LSS 3351		15 20 34	−59 13.8	3	10.83	.019	0.47	.000	−0.37	.013	9	B2.5II	115,125,540
		WLS 1520 5 # 5		15 20 37	+07 17.5	1	12.71		0.48		0.02		2		1375
136707	−52 08265			15 20 39	−52 32.3	1	7.78		1.58		0.97		1	K3 III	96
	−60 05770			15 20 39	−61 12.4	1	8.97		1.80		1.95		2		540

Table 1

HD	DM	Other Id	N Rem	α_{1950}	δ_{1950}	S	V	σ_V	B–V	σ_{B-V}	U–B	σ_{U-B}	n	Spectrum	References
137003	+28 02425			15 20 40	+28 14.1	2	7.37	.014	0.98	.009	0.70		13	G8 III	20,186
136135	−78 00994			15 20 41	−79 07.9	1	7.64		1.07		1.07		6	K1 III	1704
136866	−16 04070			15 20 42	−16 23.2	1	7.63		1.39		1.56		3	K2/3 III	8100
136838	−27 10349			15 20 42	−27 46.5	1	8.94		0.34		0.30		1	A1 V	1770
		LOD 36+ # 17		15 20 42	−52 27.6	1	12.42		0.39		0.25		1		96
136669	−56 06739			15 20 43	−57 11.6	2	10.00	.003	0.35	.011	-0.22	.007	5	B5/6	540,567
137071	+40 02877	HR 5726		15 20 46	+39 45.5	1	5.50		1.60				1	K5	71
137422	+72 00679	HR 5735, γ UMi		15 20 47	+72 00.7	5	3.05	.005	0.06	.016	0.09	.019	12	A3 II-III	15,1363,3023,8015,8100
	−52 08270			15 20 47	−52 32.3	1	11.26		0.60		0.22		1		96
137050	+32 02577	UU CrB		15 20 51	+31 43.9	1	8.63		0.55				14	F8	8097
	+28 02426			15 20 53	+27 42.4	1	10.41		1.08		0.75		7	K0	186
		WLS 1520-15 # 7		15 20 55	−17 05.8	1	11.59		0.77		0.27		2		1375
136739	−52 08272	GH Lup		15 20 57	−52 40.6	3	7.78	.089	1.23	.039	0.98	.038	4	G2 Iab	96,403,1587
		CS 22890 # 71		15 20 58	+03 01.1	1	11.74		0.46		0.68		1		1736
136852	−31 11920			15 20 58	−31 50.2	1	9.68		0.22		0.16		1	A0 IV	1770
		LOD 36+ # 9		15 20 58	−52 39.6	1	11.77		0.78		0.50		1		96
137004	+14 02866			15 21 00	+14 13.3	1	8.36		1.33				3	K2	882
136636	−65 03051			15 21 02	−65 51.6	1	8.64		1.15		0.85		1	G6/8 II	565
136807	−48 09912	IDS15175S4814	AB	15 21 04	−48 24.3	2	7.52	.014	0.07	.009	0.02	.008	3	B9.5/A0 V	401,1770
	+65 01050			15 21 06	+65 22.5	1	10.14		0.55		-0.04		2	G0	1064
	+28 02427			15 21 07	+27 44.9	1	10.03		0.46		-0.07		4	F5	186
136956	−11 03940	HR 5720		15 21 07	−12 11.5	1	5.72		1.04				4	G8 III	2007
136857	−43 09874			15 21 07	−43 45.0	1	8.00		0.10		-0.01		2	A0 V	1771
136788	−52 08278	IDS15175S5222	AB	15 21 07	−52 33.3	1	9.40		0.47		0.18		1	F0/2 V	96
137107	+30 02653	HR 5727	★ AB	15 21 08	+30 28.0	10	4.98	.018	0.57	.012	0.04	.006	32	G0 V	15,292,938,1007,1013*
137107	+30 02653	IDS15191N3039	C	15 21 08	+30 28.0	1	13.35		0.58		-0.04		2		3030
136957	−13 04151			15 21 08	−13 36.0	1	8.55		0.04		-0.10		1	B8/9 V	1770
		WLS 1536 45 # 6		15 21 09	+45 26.0	1	11.35		0.69		0.21		2		1375
137006	−0 02961	HR 5721		15 21 09	−00 50.7	8	6.13	.018	0.26	.011	0.06	.011	29	F0 V	15,254,1007,1013,1088*
		AJ77,733 T24# 6		15 21 10	−59 18.1	1	11.75		0.30		0.22		5		125
		AAS9,163 T7# 14		15 21 12	+28 11.4	1	11.56		1.08		0.85		6		186
		CS 22890 # 77		15 21 12	−00 33.8	1	10.99		1.15		0.75		1		1736
	−52 08283			15 21 13	−52 37.3	1	11.77		0.24		0.23		1		96
		LOD 36+ # 6		15 21 15	−52 44.7	1	13.05		0.58				1		96
137147	+32 02578		V	15 21 20	+31 59.3	2	8.04	.033	0.24	.195			8	F0	1247,8097
		AAS9,163 T7# 16		15 21 21	+27 41.3	1	11.49		1.23		1.20		5		186
136933	−39 09827	HR 5719	★ AB	15 21 28	−39 32.0	2	5.37	.005	-0.11	.005	-0.39		5	Ap Siλ4200	1771,2007
137052	−9 04138	HR 5723		15 21 29	−10 08.6	6	4.93	.005	0.45	.004	0.03	.022	22	F5 IV	3,15,418,1075,2012,3026
	−52 08289			15 21 29	−52 45.7	1	10.63		0.34		0.21		1		96
		AJ77,733 T24# 10		15 21 29	−59 24.3	1	13.34		1.34		0.99		2		125
136672	−67 02864	HR 5713	★ A	15 21 30	−68 08.0	4	5.88	.004	1.01	.008	0.77	.005	20	G8/K0 III	15,1075,1075,2038
136935	−43 09884			15 21 31	−43 58.7	1	8.07		0.07		-0.28		2	B6 II	401
136874	−50 09371			15 21 31	−50 57.8	1	8.28		1.11		0.92		2	K0 III	1730
137257	+47 02219			15 21 32	+47 05.0	1	8.43		0.52		0.04		2	G0	1601
		CS 22890 # 68		15 21 34	+03 20.2	1	14.10		0.08		0.23		1		1736
		GD 182		15 21 34	−12 41.4	3	13.56	.012	0.62	.020	-0.07	.025	8		1696,3060,7009
136959	−31 11931	IDS15185S3121	AB	15 21 34	−31 32.2	1	9.52		0.20		0.08		1	A0 V	1770
136934	−40 09591			15 21 34	−40 25.2	1	9.17		0.30		0.22		1	A9 V	1771
136810	−59 05974			15 21 34	−60 04.4	1	9.65		0.08		-0.16		1	B8 IV	115
		G 136 - 103		15 21 35	+17 39.3	3	13.67	.012	1.79	.021	1.20	.032	11		203,1705,3078
137034	−23 12264			15 21 36	−24 11.6	1	8.61		0.40		0.17		2	A3mA7-A9	1771
137126	+13 02935			15 21 37	+13 32.5	1	8.37		0.96				6	G5	882
	+31 02730			15 21 37	+31 01.1	1	10.45		1.44				1	K7	1017
	−5 04065			15 21 38	−05 26.1	1	10.18		0.90		0.69		1	K3 V	3072
	+28 02428			15 21 39	+27 44.6	1	10.00		1.58		1.67		8		186
		LOD 36+ # 4		15 21 39	−52 49.2	1	13.73		0.65				1		96
	−52 08293			15 21 40	−52 50.1	1	10.71		0.49		0.31		1		96
137389	+62 01410	HR 5731		15 21 41	+62 13.5	4	5.97	.011	-0.03	.016	-0.07	.019	10	A0p Si	252,1079,1379,3016
		CS 22890 # 69		15 21 42	+03 15.7	1	12.71		0.22		0.07		1		1736
136846	−59 05978			15 21 42	−59 21.8	1	10.04		0.09		-0.01		4	A0 II/III	125
137181	+25 02908			15 21 43	+25 27.7	1	8.25		0.42		0.05		2	F5	105
		LP 859 - 44		15 21 44	−24 17.0	1	13.51		1.24				1		3062
		AJ77,733 T24# 8		15 21 47	−59 20.8	1	12.04		0.68		0.10		2		125
137443	+63 01192	HR 5737		15 21 48	+63 31.2	1	5.79		1.27				1	K2	71
	−52 08297			15 21 48	−52 49.7	1	11.81		0.27		0.22		1		96
137182	+11 02800			15 21 51	+10 43.9	1	7.36		0.18		0.11		3	A0	1648
137015	−37 10225	HR 5722	★ AB	15 21 52	−37 59.6	2	7.03	.005	0.10	.018	0.05		5	A1/2 V	1771,2032
136899	−56 06746			15 21 52	−56 50.3	1	7.48		0.07		-0.41		3	B8 III	567
	+28 02429			15 21 53	+28 05.3	1	9.48		1.01		0.69		4	K0	186
		BPM 77842		15 21 54	−13 11.	1	10.34		1.22		1.21		1	K7 V	3072
	−50 09378			15 21 54	−51 06.3	1	10.24		1.83				3		1730
	−46 10046			15 21 55	−46 27.4	1	10.71		0.75		0.27		2	G0	1696
	−6 04196			15 21 57	−06 52.8	1	9.88		1.07		0.88		2	K3 V	3072
136543	−75 01158	U Aps		15 22 01	−75 45.3	1	8.07		3.81				2	Nb	864
136947	−52 08303			15 22 02	−52 53.0	1	6.71		0.29		0.16		1	F0 IV/V	96
	+24 02856			15 22 05	+24 11.9	1	8.95		1.05		0.80		2	K2	1625
137058	−38 10289	HR 5724	★ A	15 22 05	−38 33.5	4	4.60	.000	0.00	.000	-0.05	.019	12	A0 V	15,1075,2012,3023
137077	−32 10782			15 22 06	−32 50.4	1	9.35		0.55		0.35		1	A8 V	1770
		L 335 - 31		15 22 06	−46 12.	1	13.68		0.68		-0.10		4		3065
		L 335 - 67		15 22 06	−46 59.	1	12.67		0.57		-0.15		3		1696

HD	DM	Other Id	N Rem	α_{1950}	δ_{1950}	S	V	σ_V	B–V	σ_{B-V}	U–B	σ_{U-B}	n	Spectrum	References
136968	−53 06483			15 22 11	−53 35.7	1	8.44		−0.03		−0.63		3	B5 Vne	1586
137061	−45 09906			15 22 12	−45 31.3	1	9.37		0.07		0.01		1	B9.5 V	1770
	+37 02635	YZ Boo		15 22 13	+37 02.0	1	10.36		0.18		0.08		1	F0	668
		AJ77,733 T24# 7		15 22 14	−59 36.4	1	11.81		0.43		0.32		2		125
137119	−35 10279			15 22 18	−36 01.4	1	7.59		0.11		0.07		1	A2 V	1771
136972	−57 07044			15 22 20	−57 23.4	1	9.32		0.42		−0.08		1	B5/6 II	567
		WLS 1520 5 # 9		15 22 21	+04 37.9	1	12.84		0.69		0.15		2		1375
137390	+45 02284	HR 5732		15 22 24	+45 26.8	1	6.01		1.20		1.23		2	K2 III	252
137444	+54 01744			15 22 24	+53 51.7	1	9.02		0.42		0.02		2	F8	1566
137241	−5 04069			15 22 25	−06 04.2	1	7.36		1.14		1.08		25	K0	978
137063	−50 09390			15 22 25	−51 10.4	1	9.50		1.03		0.75		3	G8/K0 III	1730
		Steph 1240	B	15 22 29	+56 19.6	1	12.03		1.40		1.19		2	M0	1746
137082	−50 09395			15 22 29	−50 37.4	1	8.51		0.52				4	F7 V	2012
137187	−27 10361			15 22 31	−27 45.2	1	9.83		0.40		0.30		1	A3/5 III/I	1770
137229	−23 12276			15 22 33	−23 30.7	1	8.56		0.29		0.14		1	A7 IV	1771
137391	+37 02636	HR 5733	⋆ A	15 22 36	+37 33.1	7	4.31	.011	0.31	.008	0.07	.013	22	F0 V	1,15,1028,1084,1118*
137392	+37 02637	HR 5734	⋆ BC	15 22 37	+37 31.3	5	6.52	.022	0.60	.005	0.12	.009	13	G1 V	15,938,1028,1084,3030
		Steph 1240	A	15 22 37	+56 20.0	1	11.77		1.39		1.03		2	M0	1746
		AJ77,733 T24# 9		15 22 38	−59 28.4	1	12.66		0.50		0.35		2		125
136999	−62 04481			15 22 41	−62 35.5	1	7.74		0.06		−0.08		30	B9.5V	978
137044	−59 05986			15 22 42	−59 41.3	2	9.39	.008	0.01	.020	−0.51	.016	10	B5/7 III	115,125
		G 152 - 21		15 22 47	−15 02.1	1	14.31		1.05		0.73		1		3062
		LP 915 - 32		15 22 47	−31 17.9	1	10.44		0.66		0.09		2		1696
137424	+30 02656			15 22 55	+30 14.0	1	9.27		0.60		0.15		2	K0	1733
137243	−33 10515	IDS15198S3323	A	15 22 55	−33 33.6	1	9.70		0.46		0.29		1	A0 IV	1770
136977	−65 03059			15 22 56	−65 56.9	1	6.91		−0.01				4	B8 II/III	1075
137370	+15 02855			15 22 58	+14 45.7	1	8.44		1.16				4	K0	882
137260	−32 10792			15 22 58	−32 52.4	1	9.85		0.35		0.25		1	A0 V	1770
137084	−60 05786			15 23 00	−60 16.2	1	9.25		0.12		−0.34		1	B5 II	567
137303	−26 10870			15 23 02	−26 31.9	3	8.83	.023	1.05	.008	0.95	.025	13	K3/4 V	158,1775,3072
137245	−37 10240			15 23 03	−37 32.9	1	8.92		0.14		0.12		1	A2 V	1771
137304	−28 11339			15 23 06	−28 31.2	1	9.94		0.50		0.33		1	A9 V	1770
137066	−64 03178	HR 5725		15 23 10	−64 21.4	4	5.70	.005	1.65	.008	1.98	.000	23	K5/M0 III	15,1075,1075,2038
		G 15 - 22		15 23 14	−03 56.9	1	14.85		1.05		0.92		1		3062
137264	−40 09622			15 23 16	−40 42.6	1	9.15		−0.02		−0.17		1	B9 V	1770
		BPM 39513		15 23 18	−45 38.	1	13.75		0.70		0.06		3		3065
		G 151 - 59		15 23 19	−01 53.8	1	10.41		0.82		0.51		1		1658
	−24 12066			15 23 19	−24 48.8	1	8.77		0.45		0.20		19	A2	1699
		Steph 1241		15 23 25	+39 45.9	1	10.54		1.22		1.16		1	K7	1746
137471	+15 02858	HR 5739		15 23 28	+15 36.2	4	5.17	.028	1.65	.008	1.96	.012	19	M1 III	3,15,1003,3001
	+37 02639			15 23 29	+37 23.2	1	9.02		1.38				1	M2	1213
137309	−44 10123			15 23 30	−44 25.9	1	8.59		0.00		−0.12		1	B9 IV(p Si)	1770
137164	−62 04482	LS TrA		15 23 30	−62 50.8	2	7.37	.023	1.05	.005	0.82	.017	2	K1/2 IVp	1641,1642
137629	+47 02224			15 23 31	+47 14.2	1	6.79		0.60		0.07		2	F8 IV	1375
137323	−43 09908			15 23 33	−43 56.0	1	7.76		0.38		0.18		1	A3mF2-F2	1771
137826	+67 00887	G 225 - 10		15 23 34	+66 43.8	1	8.72		0.69		0.15		2	G6 V	1003
137350	−35 10286			15 23 35	−35 36.4	1	7.60		0.10		0.06		1	A2 V(n)	1771
137510	+19 02966	HR 5740		15 23 38	+19 39.3	2	6.28	.004	0.61	.011	0.20		3	G0 V	71,3016
137717	+56 01804			15 23 45	+55 47.1	1	7.76		1.14		1.14		2	K1 III	37
137759	+59 01654	HR 5744	⋆ A	15 23 49	+59 08.4	13	3.29	.021	1.17	.007	1.22	.006	58	K2 III	1,15,667,1007,1077*
137531	+13 02943			15 23 51	+13 06.9	1	8.25		0.89		0.58		2	G5	1648
		BPM 77860		15 23 54	−01 05.	1	10.74		0.90		0.65		2	K5 V	3072
		CS 22890 # 82		15 23 58	+00 22.2	1	14.13		0.28		0.18		1		1736
137630	+32 02581			15 23 59	+32 38.9	1	7.02		1.20		1.28		3	K2 III	1501
	−59 06006			15 23 59	−59 31.5	3	10.15	.028	0.22	.012	−0.39	.025	9	B3 IV	115,125,540
137569	+15 02862			15 24 01	+14 52.1	2	7.90	.045	0.10	.154	−0.37		7	B5 III:	2033,3016
137718	+51 02003			15 24 02	+50 46.1	1	7.66		1.60		1.98		2	K5	1566
137570	+10 02853			15 24 06	+10 12.6	2	6.92	.030	1.62	.020	1.89	.005	4	M1	1733,3040
137432	−36 10161	HR 5736	⋆ A	15 24 06	−36 35.6	6	5.45	.005	−0.15	.003	−0.61	.004	17	B5 V	15,26,1770,2007,8015,8023
137327	−57 07057			15 24 07	−57 58.9	2	8.91	.005	0.18	.010	−0.29	.020	2	B8 IV	115,567
137719	+44 02464			15 24 16	+44 28.6	1	7.17		1.40		1.64		2	K5 III	1601
		CS 22890 # 81		15 24 17	−00 32.6	1	13.27		0.05		−0.03		1		1736
	−23 12296			15 24 18	−23 25.5	1	10.83		0.49		−0.19		2		1696
		L 72 - 91		15 24 18	−74 55.	1	15.93		−0.06		−0.93		1		3065
137704	+34 02645	HR 5741		15 24 20	+34 30.5	4	5.44	.011	1.41	.007	1.65	.005	9	K4 III	15,1003,1080,3035
137688	+28 02432			15 24 24	+28 18.1	1	7.45		1.35				3	K3 III	20
		NGC 5927 sq1 11		15 24 26	−50 37.6	1	11.42		1.49		1.43		3		89
		NGC 5927 sq3 2		15 24 26	−50 37.6	1	11.37		1.49		1.42		5		780
137610	−5 04079			15 24 29	−05 38.6	1	7.81		1.66		2.02		3	K5	1657
137514	−29 11699			15 24 30	−29 21.3	1	9.48		0.25		0.00		1	Ap Si	1770
	−57 07059	LSS 3355		15 24 31	−57 56.8	2	9.66	.035	1.50	.071	0.61	.035	4	B9 Ib	540,1737
		WLS 1536 45 # 11		15 24 39	+42 47.7	1	11.12		0.51		−0.03		2		1375
137380	−60 05810			15 24 40	−60 32.1	1	9.44		0.04		−0.13		3	B9 II/III	1586
137561	−32 10811			15 24 46	−32 21.6	1	9.63		0.51		0.09		1	F2 V	1770
137405	−60 05814	LSS 3356		15 24 47	−60 37.7	3	9.35	.013	0.12	.008	−0.66	.014	10	B0.5/1 Iab	540,976,8100
137465	−51 09132	HR 5738	⋆ AB	15 24 48	−51 25.5	2	6.08	.005	1.07	.015	0.76		7	G2 II	404,2006
137438	−56 06765			15 24 49	−56 52.7	1	9.34		0.07		−0.24		3	B6/7 V	567
137613	−24 12084	HM Lib		15 24 50	−24 59.8	8	7.48	.022	1.19	.013	0.84	.019	13	C2,2Hd	158,842,864,1238,1589,1699*
		G 152 - 25		15 24 51	−02 25.3	1	14.85		0.96		0.36		1		3062
137518	−44 10140	HV Lup, LSS 3359		15 24 51	−44 57.7	5	7.82	.053	0.11	.014	−0.68	.027	16	B1 (I/IIIn)	400,403,540,976,1586

Table 1

841

HD	DM	Other Id	N	Rem	α₁₉₅₀	δ₁₉₅₀	S	V	σ_V	B–V	σ_B-V	U–B	σ_U-B	n	Spectrum	References
137723	+10 02854	IDS15225N1003	AB		15 24 53	+09 52.5	1	7.99		0.55		0.06		2	G0	3030
137595	−33 10526				15 24 54	−33 22.3	2	7.49	.003	0.02	.007	−0.74	.000	2	B2 II/III	55,1770
137457	−58 05951				15 24 54	−59 01.1	1	9.48		0.13		−0.25		3	B8 II/III	567
137482	−51 09133				15 24 55	−51 51.8	1	10.38		0.93		−0.13		2	K3/4 III	1737
		LP 502 - 51			15 24 58	+12 17.9	1	12.21		1.10		0.94		1		1696
137594	−32 10814				15 24 58	−32 18.6	1	10.66		0.11		−0.02		1	B9/A0 IV	1770
	−58 05953	LSS 3357			15 24 58	−58 54.9	1	10.71		0.61		−0.36				115
137439	−59 06020				15 24 59	−59 56.2	3	9.50	.009	0.19	.004	−0.46	.017	8	B2 II	115,540,976
137597	−36 10169	GO Lup	⋆ AB		15 25 00	−37 11.1	2	7.05	.020	1.64	.015	1.73		7	M4 III	2012,3042
		G 151 - 61			15 25 03	−08 51.1	2	15.41	.043	1.83	.005			7		940,3032
137792	+18 03024	IDS15228N1759	AB		15 25 04	+17 48.5	1	8.08		0.49		0.02		2	F8	3016
137502	−53 06519				15 25 06	−53 16.4	3	8.79	.022	0.48	.005	−0.06	.000	5	F8 V	742,1594,6006
	−58 05955				15 25 06	−58 57.0	1	10.47		0.29		−0.45		1		115
137522	−55 06572				15 25 09	−55 27.3	2	7.80	.015	0.31	.010	−0.02		7	B8/9 II	567,2014
		CS 22890 # 80			15 25 10	−00 56.3	1	14.83		−0.07				1		1736
	+3 03032				15 25 11	+02 46.2	3	10.17	.012	1.37	.012	1.23		5	M0	1017,1705,7008
137384	−67 02885				15 25 12	−67 18.8	1	7.35		−0.09				4	B2/3 IV	1075
		LundsII141 # 249			15 25 16	−57 54.2	1	10.10		0.78		−0.12		1		115
		CS 22890 # 86			15 25 18	+00 58.4	1	14.11		0.31		−0.02		1		1736
137693	−26 10890				15 25 19	−26 56.7	1	10.17		0.25		0.20		1	A2/3 III	1770
137620	−44 10143				15 25 19	−44 43.2	2	7.75	.001	−0.04	.004	−0.31	.009	4	B8 III	401,1770
		NGC 5927 sq1 4			15 25 24	−50 31.2	1	11.29		0.40		0.22		3		89
		NGC 5927 sq3 1			15 25 24	−50 31.2	1	11.23		0.34		0.22		5		780
137725	−20 04239				15 25 25	−21 02.2	1	7.17		1.05		0.72		4	K0 III/IV	1657
137744	−16 04089	HR 5743			15 25 26	−16 32.6	2	5.65	.010	1.54	.010	1.84		7	K4 III	2006,3016
		NGC 5927 sq3 5			15 25 26	−50 32.2	1	13.03		0.56		0.15		5		780
137763	−8 03981	G 151 - 62	⋆ A		15 25 27	−09 10.2	3	6.91	.026	0.81	.015	0.47	.050	14	K2 V	1067,2033,3072
138230	+73 00678				15 25 28	+73 01.3	1	8.60		0.91		0.64		2	K0	1733
		NGC 5927 sq3 3			15 25 28	−50 33.0	1	12.50		0.29		0.26		4		780
137853	+25 02916	HR 5745			15 25 30	+25 16.5	1	6.02		1.62				1	K5	71
137778	−8 03983	G 151 - 63	⋆ B		15 25 30	−09 10.8	3	7.59	.023	0.91	.011	0.66	.009	12	K2 V	1509,2033,3072
		PG1525-071			15 25 31	−07 06.1	1	15.05		−0.20		−1.15		1	sdO	1764
137707	−30 12256				15 25 31	−30 55.5	1	9.54		0.32		0.24		1	A5 IV	1770
137543	−59 06033				15 25 31	−59 32.4	2	9.97	.001	0.15	.001	−0.36	.004	6	B9	540,976
137505	−62 04517				15 25 31	−63 05.7	1	8.71		1.47		1.66		1	K2/3 III	1642
		IRC +20 281			15 25 32	+19 44.1	1	13.90		1.71				2		8019
		PG1525-071D			15 25 32	−07 06.2	1	16.30		0.56		0.31		1		1764
		PG1525-071A			15 25 33	−07 05.5	1	13.51		0.76		0.26		1		1764
		PG1525-071B			15 25 34	−07 05.8	1	16.40		0.73		0.14		1		1764
		PG1525-071C			15 25 36	−07 04.0	1	13.53		1.11		1.13		1		1764
		NGC 5927 sq1 1			15 25 38	−50 37.5	1	11.31		0.41		0.10		3		89
137728	−31 11981	IDS15226S3108	A		15 25 40	−31 18.2	1	7.14		0.28		0.25		1	A3 V	1770
137844	+3 03034	IDS15232N0312	AB		15 25 42	+03 01.7	2	8.12	.020	0.30	.020	0.04		5	A5	173,3016
137854	+12 02838				15 25 42	+12 02.0	1	6.92		0.38		0.07		2	F0	1648
		BSD 15 # 421			15 25 42	+59 34.	1	11.96		0.07		0.09		5	B9	272
137676	−49 09653				15 25 44	−49 46.8	3	7.67	.011	0.77	.010	0.26	.023	9	G8 V	285,2012,3026
		NGC 5927 sq1 2			15 25 44	−50 34.6	1	11.16		1.42		1.64		3		89
137603	−58 05962	LSS 3362			15 25 45	−58 24.5	1	9.71		1.24		0.03		1	WR	115
137909	+29 02670	HR 5747, β CrB	⋆ AB		15 25 46	+29 16.6	8	3.68	.013	0.28	.010	0.11	.008	23	F0p	15,39,1008,1202,1263*
137809	−8 03985				15 25 46	−08 59.3	1	8.33		1.46		1.71		1	K2	1746
		CS 22890 # 94			15 25 49	+02 25.1	1	15.40		−0.23		−0.91		1		1736
137678	−50 09468				15 25 51	−50 33.8	2	7.73	.030	0.30	.010	0.24	.030	7	A6 V	89,676
137679	−50 09467	IDS15222S5102	AB		15 25 52	−51 13.1	1	7.66		1.14		1.18		3	K2/3 III	676
		MCC 744			15 25 53	+25 57.8	1	11.07		1.61				1	M0	1017
137697	−51 09150				15 25 54	−51 16.3	1	8.29		−0.02		−0.33		4	B7 III	676
137709	−46 10100	HR 5742			15 25 55	−46 33.7	4	5.23	.007	1.75	.009			12	K4 III	15,138,1075,2006
138004	+43 02500	G 179 - 32	⋆ B		15 25 56	+43 03.3	1	9.86		1.23				1		1746
138004	+43 02500	G 179 - 32	⋆ A		15 25 56	+43 03.5	1	7.47		0.67		0.14		2	G5	1601
137640	−57 07068				15 25 56	−57 18.1	1	9.69		0.19		0.04		4	B9 V	567
		G 137 - 8			15 25 58	+16 53.6	3	13.69	.008	1.36	.017	1.14	.063	6		316,1759,1774
	−58 05968	LSS 3363			15 25 58	−58 48.0	2	10.13	.025	0.90	.013	−0.11	.034	2	B2 II-Ib	115,540
137387	−72 01802	HR 5730, κ1 Aps	⋆ A		15 26 01	−73 13.1	6	5.48	.032	−0.13	.006	−0.79	.025	34	B3 Ib/IIIne	15,26,681,815,1075,1637
		CS Ser			15 26 05	+03 15.7	1	12.13		0.18		0.09		1	F1	699
137898	+2 02965	HR 5746			15 26 07	+02 00.9	3	5.15	.008	0.23	.004	0.08	.004	11	A8 IV	1088,1363,3023
137911	+7 02968				15 26 10	+07 05.6	1	8.75		0.34		0.10		2	A5	1375
	−61 04983				15 26 10	−61 49.3	1	10.53		1.02		0.95		1		1696
137712	−58 05976				15 26 24	−58 55.8	1	9.51		0.13		0.08		1	B8 III/IV	567
	−56 06776				15 26 27	−56 33.7	1	11.17		0.59		0.04		6		567
137733	−59 06046				15 26 33	−59 29.0	1	9.17		0.25		−0.34		1	B5 II	115
		LP 562 - 45			15 26 36	+03 43.8	1	13.90		1.45		1.22		1		1696
137877	−37 10284				15 26 37	−37 38.8	1	9.29		0.34		0.15		1	A9 V	1771
138100	+39 02875				15 26 40	+38 53.7	1	6.68		0.36		0.00		2	F3 V	105
137949	−16 04093	GZ Lib			15 26 45	−17 16.2	1	6.68		0.38		0.14		4	Ap SrEuCr	1202
	−58 05983				15 26 46	−59 15.6	2	10.04	.003	0.24	.007	−0.46	.026	3	B3 III	115,540
138245	+62 01414	HR 5754			15 26 47	+62 26.9	2	6.51	.010	0.14	.000	0.07	.030	4	A5 IV	1733,3016
137951	−23 12326				15 26 49	−23 55.2	1	8.13		0.42		0.20		1	A2 III/IV	1770
137901	−38 10389				15 26 51	−38 24.9	1	7.63		0.21		0.14		3	A2 IV	1771
138265	+61 01509	HR 5755			15 26 52	+60 50.5	3	5.89	.020	1.44	.005	1.65	.019	9	K5 III	252,1379,1733
138085	+16 02789				15 26 54	+16 34.0	1	6.38		1.00		0.64		2	G8 III-IV	3016
		GD 345			15 26 55	+57 21.5	1	14.11		0.39		−0.16		3		308

HD	DM	Other Id	N	Rem	α_{1950}	δ_{1950}	S	V	σ_V	B–V	σ_{B-V}	U–B	σ_{U-B}	n	Spectrum	References
137978	−23 12329				15 26 55	−23 52.4	1	8.16		0.41		0.18		2	F0 IV	1771
137919	−41 09965	IDS15237S4134		A	15 27 00	−41 44.9	1	6.37		-0.03		-0.18		1	B9.5 V	1770
137919	−41 09965	IDS15237S4134		AB	15 27 00	−41 44.9	1	6.23		0.01		-0.16		1	B9.5 V	1770
	−59 06058				15 27 03	−59 39.2	1	9.78		0.14		-0.36		3	B5 V	540
		G 15 - 23			15 27 04	+06 19.4	2	10.96	.000	0.70	.005	0.04	.000	4		516,1620
137838	−57 07072				15 27 05	−57 18.9	1	8.27		0.16		-0.27		2	B7 II	567
		GD 502			15 27 06	+67 34.4	1	13.67		0.56		-0.12		3		308
138213	+47 02227	HR 5752			15 27 08	+47 22.4	1	6.15		0.10		0.12		3	AmA3-F0 V?	8071
138008	−24 12118				15 27 09	−24 21.5	1	9.39		0.21		0.22		1	A3 III	1770
137891	−51 09171				15 27 11	−51 59.7	2	9.80	.005	0.85	.010	-0.25	.020	4	B9 V	403,1737
	+65 01055				15 27 13	+65 34.5	2	10.27	.099	1.52	.080	1.50		2	R5	1238,8005
137935	−45 09964				15 27 17	−45 32.7	1	8.41		0.01		-0.16		1	B8/9 V	1770
138026	−28 11377				15 27 18	−28 27.0	1	9.98		0.59		0.31		1	A9/F0 V	1770
		Lod 2158 # 11			15 27 19	−57 56.9	2	11.29	.000	0.73	.000	0.00	.000	3	B6	768,837
	−47 10044				15 27 21	−47 49.0	1	10.98		0.70		0.15		2	G5	1696
137990	−37 10298				15 27 22	−37 31.1	1	9.57		0.51		0.07		1	A7/8 V	1771
		WD1527+090			15 27 25	+09 05.9	1	14.29		-0.10		-0.90		1	DA1	1727
137628	−73 01590				15 27 29	−74 10.6	2	8.97	.029	1.02	.005	0.83		6	K3VCNIII/IV	2012,3008
		LP 915 - 40			15 27 30	−29 29.0	1	12.83		0.52		-0.10		3		3062
137937	−52 08437				15 27 31	−52 43.4	1	8.67		-0.02		-0.40		3	B6 III	567
137963	−51 09180				15 27 34	−51 51.1	2	9.85	.010	0.91	.000	-0.16	.000	3	B9 IV	403,1737
	−65 03082				15 27 37	−65 20.8	2	10.75	.010	-0.04	.019	-0.58	.004	3	B4 IV	540,976
138338	+55 01756	HR 5759			15 27 39	+55 21.9	2	6.44	.010	0.08	.005	0.08	.010	6	A3 m	1501,8071
137925	−57 07075				15 27 39	−58 05.9	2	10.44	.010	0.09	.000	0.03	.000	3	Ap Si	768,837
138367	+57 01590	IDS15266N5747		A	15 27 42	+57 36.8	1	6.87		0.49		0.01		2	F6 IV-V	1003
138105	−20 04246	HR 5749			15 27 43	−20 33.5	3	6.21	.013	0.18	.006	0.09	.013	9	A3 V	1770,2007,8071
		G 239 - 56			15 27 47	+70 58.8	1	12.44		0.98		0.75		1		1658
138302	+48 02300	IDS15262N4803		A	15 27 48	+47 53.0	1	6.87		0.04		0.04		3	A1 V	1501
138124	−21 04129				15 27 49	−22 01.2	1	8.72		0.46		0.19		2	A1mA9-F3	1771
	−57 07078				15 27 49	−58 09.2	3	10.73	.004	0.34	.008	0.00	.024	4	B8	115,768,837
138232	+26 02694	IDS15257N2551		AB	15 27 50	+25 40.7	1	7.63		1.43		1.61		2	K4 III	1625
		CCS 2254, RU TrA			15 27 50	−62 23.0	1	9.02		4.72				2	Np	864
138137	−16 04099	HR 5750			15 27 51	−16 26.4	1	5.82		1.06				4	K0 III	2007
		G 179 - 33		AB	15 27 58	+43 03.4	2	14.19	.024	1.74	.014			7		940,1759
		CS 22884 # 12			15 27 58	−09 45.5	1	14.74		0.37				1		1736
		WLS 1520-15 # 8			15 28 02	−15 04.3	1	12.84		1.03		0.74		2		1375
138138	−33 10564	IDS15250S3329		ABC	15 28 08	−33 39.0	1	6.86		0.17		0.07		1	A2/3 V	1770
138188	−18 04083				15 28 11	−18 40.0	1	7.66		0.28		0.13		2	A2 III/IV	1771
		PG1528+062B			15 28 12	+06 11.4	1	11.99		0.59		0.01		1		1764
		G 15 - 24			15 28 17	+08 34.1	5	11.44	.009	0.57	.000	-0.11	.008	8		516,1064,1620,1696,3077
	−60 05868	LSS 3364			15 28 18	−60 22.0	2	9.83	.016	0.24	.005	-0.58	.027	7	B1 II-III	540,976
		PG1528+062A			15 28 21	+06 11.6	1	15.55		0.83		0.36		1		1764
138341	+31 02742	HR 5760			15 28 21	+31 27.4	3	6.46	.008	0.19	.005	0.13	.008	8	A4 IV	1022,1629,8071
		PG1528+062			15 28 22	+06 11.1	1	14.77		-0.25		-1.09		1	sdOB	1764
		G 15 - 25			15 28 24	+07 28.5	1	11.66		1.24		1.11		1		333,1620
		PG1528+062C			15 28 28	+06 10.3	1	13.48		0.64		0.07		1		1764
138524	+62 01416	HR 5768			15 28 28	+62 16.2	2	6.31	.020	1.38	.015	1.61	.005	5	K4 III	1501,1733
138279	+2 02972	VY Ser			15 28 30	+01 51.2	2	9.75	.002	0.25	.005	0.09	.005	2	F5	668,699
138290	+9 03055	HR 5758			15 28 30	+08 44.9	1	6.56		0.37		-0.06		8	F4 V w	1088
138421	+46 02074				15 28 31	+46 33.4	1	7.59		1.30		1.34		2	K2	1733
138322	+17 02873				15 28 32	+17 36.6	1	8.99		0.18		0.13		2		1648
138383	+37 02651	HR 5761			15 28 34	+36 58.4	1	6.52		1.13		1.18		2	K0	1733
	−53 06547				15 28 35	−54 07.2	1	10.52		0.28		-0.28		2	B4 IV	540
		Lod 2158 # 17			15 28 36	−57 54.4	2	11.69	.000	0.70	.000	0.29	.000	3		768,837
		Lod 2158 # 19			15 28 36	−57 54.4	2	11.80	.019	0.60	.005	0.15	.005	3	F3	768,837
		Lod 2159 # 23			15 28 40	−57 56.9	1	11.90		0.49		0.15		2	F3	837
138221	−32 10868	HR 5753			15 28 42	−32 42.7	2	6.47	.014	0.09	.003	-0.34		6	B7 V	1770,2006
		ST Boo			15 28 44	+35 57.2	2	10.70	.212	0.19	.071	0.09	.015	2	F2	668,699
138323	+3 03045				15 28 46	+02 53.8	1	8.87		0.69		0.25		2	G5	1375
		Lod 2159 # 10			15 28 46	−57 57.4	1	11.18		1.68		1.68		2	K7	837
138191	−44 10190				15 28 47	−44 53.9	2	6.82	.001	-0.02	.004	-0.17	.011	3	B9 IV/V	401,1770
138269	−23 12356				15 28 48	−24 10.5	1	9.81		0.25		0.23		1	A1 III/IV	1770
138204	−38 10425	HR 5751			15 28 48	−38 27.2	2	6.24	.005	0.21	.005			7	A3 III	15,2027
138422	+34 02655				15 28 49	+34 38.1	1	6.83		0.01		-0.01		3	A0 V	1501
138282	−16 04104				15 28 49	−16 51.2	1	8.21		1.69		1.96		2	M1/2 III	1375
		Lod 2159 # 21			15 28 49	−57 55.9	1	11.78		1.43		0.76		2	K9	837
138268	−19 04128	HR 5756		⋆ A	15 28 50	−19 59.7	2	6.26	.002	0.21	.006	0.14	.007	3	A5 V	542,1770
138268	−19 04128	IDS15260S1949		⋆ AB	15 28 50	−19 59.7	4	6.21	.010	0.22	.004	0.12		17	A5 V +G0 V	15,404,2028,2033
138268	−19 04128	IDS15260S1949		B	15 28 50	−19 59.7	1	8.94		0.55		0.07		3	G0 V	542
138112	−59 06095				15 28 50	−59 38.3	3	9.53	.010	0.16	.007	-0.51	.018	7	B5	115,540,976
138151	−54 06561				15 28 51	−54 45.5	1	9.32		0.06		-0.22		3	Ap Si	567
	−75 01197				15 28 53	−75 30.0	1	9.51		-0.07		-0.80		3	B1 Vn	400
		Lod 2158 # 14			15 28 57	−58 03.6	2	11.58	.005	0.23	.000	0.13	.000	3	A0	768,837
		Lod 2158 # 16			15 28 57	−58 03.6	2	11.68	.000	0.47	.005	0.22	.005	3	A9	768,837
	−40 09712				15 28 58	−41 05.6	3	9.30	.004	1.52	.005	1.12	.014	6	M0	158,1705,3078
138145	−58 06019				15 28 58	−58 29.5	1	10.24		0.17		-0.21		1	B8/9 II/III	567
	−57 07080				15 28 59	−58 07.4	2	11.56	.000	0.28	.000	0.09	.005	3	B9	768,837
138356	+2 02973				15 29 01	+02 02.1	1	8.81		0.41		0.03		1	G0	289
138131	−60 05874				15 29 02	−60 44.4	1	9.15		0.09		-0.28		3	B6 Vne	1586
	+73 00679				15 29 03	+72 43.6	1	9.08		1.20		1.19		2	K0	1375

Table 1 843

HD	DM	Other Id	N	Rem	α₁₉₅₀	δ₁₉₅₀	S	V	σ_V	B–V	σ_B-V	U–B	σ_U-B	n	Spectrum	References
137179	−82 00654				15 29 03	−83 04.0	1	8.75		−0.08		−0.69		1	B2/3 II/III	55
		CS 22884 # 15			15 29 04	−09 18.4	1	13.23		0.13		0.23		1		1736
138295	−27 10419				15 29 05	−27 59.8	1	7.93		0.25		0.23		1	A0 V	1770
138167	−57 07082				15 29 06	−57 37.2	1	8.77		0.14		−0.34		1	Ap Si	115
138168	−57 07081	IDS15252S5751	A		15 29 06	−58 01.5	2	8.78	.000	0.44	.005	0.03	.005	3	F3 IV	768,837
138271	−37 10316				15 29 07	−37 43.0	1	9.14		0.40		0.17		1	A8/9 V	1771
138481	+41 02609	HR 5763			15 29 08	+41 00.2	7	5.03	.009	1.59	.006	1.89	.025	22	K5 III	15,1007,1013,1080*
138554	+54 01750				15 29 09	+54 08.4	1	8.82		0.51		−0.01		2	F8	1566
	−12 04271				15 29 10	−12 37.7	1	9.93		1.21		1.13		2	G5	1375
138181	−57 07083	IDS15252S5751	B		15 29 10	−58 01.8	2	8.71	.000	0.44	.005	0.02	.010	3	F3/5 V	768,837
138297	−32 10873				15 29 12	−33 06.6	1	8.62		0.30		0.24		1	A3 III/IV	1770
		Lod 2159 # 26			15 29 12	−57 55.5	1	12.31		0.47		0.38		2	A0	837
		Lod 2158 # 18			15 29 13	−58 04.8	2	11.73	.005	0.48	.010	0.36	.000	3	A0	768,837
138312	−29 11754				15 29 14	−30 01.5	1	9.52		0.13		0.06		1	A0 V	1770
138372	−7 04044				15 29 15	−08 16.0	1	9.41		1.08				5	G5	8088
138343	−21 04135				15 29 15	−21 47.9	1	7.23		0.04		−0.16		2	B9 V	1770
		G 152 - 35			15 29 17	−11 12.6	1	12.12		0.56		−0.22		2		3062
138344	−23 12359	GG Lib			15 29 18	−23 42.7	2	6.85	.016	1.59	.012	1.48		8	M4 III	219,6002
138345	−23 12360	IDS15264S2352	AB		15 29 19	−24 02.2	1	9.73		0.39		0.14		1	A1 III	1770
138467	+23 02822				15 29 21	+22 41.9	1	9.03		0.53		0.02		2	G5	1375
	−58 06026	LSS 3367			15 29 22	−58 28.8	1	8.71		2.05		1.44		1	B1 Ia	1737
138193	−57 07084				15 29 23	−58 12.6	4	9.70	.006	0.24	.007	−0.36	.027	6	A0 IV	115,540,768,837
138389	−8 03998				15 29 25	−08 51.6	1	8.97		1.16				3	K0	8088
138525	+37 02653	HR 5769			15 29 28	+36 47.2	2	6.39	.004	0.50	.000	0.05		3	F6 III	71,1733
		WLS 1520 5 # 8			15 29 29	+04 29.1	1	12.72		0.98		0.56		2		1375
		Ly 4 - 2			15 29 30	−55 03.	1	13.77		1.54				1		412
		Ly 4 - 3			15 29 30	−55 03.	1	14.57		1.51				1		412
		Ly 4 - 4			15 29 30	−55 03.	1	14.49		1.00		0.03		1		412
		Ly 4 - 5			15 29 30	−55 03.	1	12.81		2.30		2.47		1		412
		Ly 4 - 6			15 29 30	−55 03.	1	13.83		1.78				1		412
		Lod 2158 # 21			15 29 31	−57 52.6	2	12.00	.005	0.44	.000	0.30	.005	3	A5	768,837
138225	−57 07085				15 29 32	−58 01.5	2	10.51	.005	0.12	.000	0.00	.005	3	B9 V	768,837
138346	−31 12033				15 29 35	−31 44.6	1	9.09		0.12		0.09		1	A0 V	1770
		Ly 4 - 1			15 29 35	−55 04.4	1	12.48		0.59		0.17		1		412
	−57 07086				15 29 36	−57 53.2	2	10.63	.005	0.50	.000	0.21	.000	3	A8	768,837
	+64 01073				15 29 37	+64 11.0	1	9.44		0.32		0.00		10	F0	1603
		GD 183			15 29 40	+11 08.0	1	15.40		0.56		−0.18		1		3060
		Lod 2158 # 20			15 29 40	−57 48.7	2	11.91	.010	0.40	.000	0.28	.000	3	A5	768,837
	−8 03999	UZ Lib			15 29 41	−08 22.0	1	9.16		1.04				1	K0 IIIe	8088
138426	−18 04088				15 29 41	−19 14.1	1	8.53		0.14		0.12		1	Ap SrCr(Eu)	1770
138413	−19 04135	HR 5762			15 29 44	−19 30.1	4	5.50	.005	0.16	.004	0.13	.022	11	A2 IV	1771,2007,3058,8071
138415	−21 04137				15 29 46	−21 37.6	1	9.97		0.40		0.22		1	B9/A0 II	1770
		GD 184			15 29 50	+14 06.5	1	15.74		0.52		−0.15		1		3060
138274	−58 06034				15 29 50	−58 21.2	1	8.80		0.14		0.09		1	A0 III/IV	768
138527	+16 02797	HR 5770			15 29 51	+16 13.5	3	6.23		−0.04	.005	−0.20	.020	8	B9 V	1022,1079,3016
138416	−33 10583				15 29 55	−33 58.0	1	9.66		0.38		0.30		1	A1 V	1770
		Lod 2159 # 27			15 29 57	−57 46.4	1	12.37		0.39		0.18		2	A3	837
	−58 06039	LSS 3368			15 29 57	−59 04.4	3	10.56	.074	0.38	.026	−0.56	.025	8		115,540,1586
138629	+41 02611	HR 5774		★ AB	15 29 59	+41 04.1	4	5.00	.022	0.09	.017	0.13	.025	11	A5 V	15,1363,3023,8015
138170	−68 02493				15 29 59	−68 43.3	1	7.44		1.68		1.85		7	M2 III	1704
		CS 22884 # 5			15 30 00	−11 40.9	1	13.75		0.67		0.11		1		1736
138485	−16 04110	HR 5764			15 30 05	−16 41.1	8	5.50	.006	−0.14	.003	−0.76	.005	29	B2 Vn	15,26,1732,1770,2007,6001*
138316	−57 07089				15 30 11	−58 13.9	1	10.63		0.36		0.25		1	A8/F0	768
	−57 07090				15 30 11	−57 52.6	1	11.20		0.54		0.13		2	F5	837
		Lod 2159 # 25			15 30 11	−57 52.7	1	12.13		0.54		0.09		2	F5	837
138488	−24 12155	HR 5765		★ A	15 30 12	−24 19.3	2	7.00	.029	0.24	.019	0.06	.024	3	A3/5 V	1771,3030
138488	−24 12155	HR 5765		★ ABC	15 30 12	−24 19.3	1	6.27		0.23				4	A3 V +A9/F2	3030
138488	−24 12155	IDS15272S2409		★ BC	15 30 12	−24 19.3	1	7.12		0.30		0.00		2	A9/F2	2007
138460	−29 11765				15 30 12	−29 33.2	1	9.97		0.18		0.12		1	B9/A0 V	1770
138852	+64 01074	HR 5785			15 30 13	+64 22.6	1	5.80		0.96				1	K0 III-IV	71
138430	−43 10002				15 30 15	−43 55.7	1	8.89		0.05		−0.04		1	B9.5 V	1770
138428	−39 09965				15 30 16	−39 50.4	1	9.19		0.27		0.11		1	A5 IV/V	1771
138503	−24 12156				15 30 20	−24 51.5	1	9.10		−0.03		−0.73		1	B2/3(IV)(n)	55
138491	−33 10590				15 30 22	−34 11.5	1	8.00		0.38		0.31		2	A5 V	1770
138562	−0 02982	HR 5772			15 30 23	−01 01.1	1	5.50		1.09		1.02		8	K0 III	1088
138363	−56 06795	IDS15265S5644	AB		15 30 27	−56 54.3	1	7.12		1.12		0.97		30	K1 II	1621
138641	+20 03118	IDS15283N2005	AB		15 30 34	+19 54.7	1	8.06		0.49		0.03		5	F5	3030
		LP 623 - 3			15 30 37	−00 02.0	1	14.07		1.03		0.54		1		1696
		CS 22884 # 13			15 30 37	−09 34.9	1	14.33		0.33		0.13		1		1736
	−58 06053	LSS 3371			15 30 40	−58 32.7	1	9.80		1.62		0.03		4	VV Cephi	1737
	−54 06576				15 30 41	−54 30.8	1	9.72		0.55		0.05		2	B9 II-Ib	540
	−57 07091				15 30 41	−57 54.9	2	10.84	.005	0.83	.000	0.47	.005	3	K0	768,837
		PG1530+057A			15 30 42	+05 43.8	1	13.71		0.83		0.41		1		1764
		PG1530+057			15 30 42	−05 42.5	1	14.21		0.15		−0.79		1	sdB	1764
138505	−39 09970	HR 5767			15 30 43	−39 53.9	2	5.82	.000	1.69	.009	1.93		5	M2 III	2035,3055
	−57 07092				15 30 45	−57 54.3	2	11.69	.015	0.22	.005	0.21	.010	3	A1	768,837
138685	+16 02799				15 30 46	+16 10.8	2	6.52	.024	1.26	.005	1.40	.029	7	K2	897,1648
138574	−20 04260				15 30 48	−20 22.8	1	8.61		0.29		0.18		1	A3 IV/V	1770
138549	−30 12336				15 30 48	−30 51.1	2	7.96	.009	0.71	.005	0.28		8	G8 V	164,1075
		PG1530+057B			15 30 50	+05 43.8	1	12.84		0.75		0.33		1		1764

HD	DM	Other Id	N Rem	α_{1950}	δ_{1950}	S	V	σ_V	B–V	σ_{B-V}	U–B	σ_{U-B}	n	Spectrum	References
		Lod 2158 # 12		15 30 50	−58 08.0	2	11.33	.000	0.41	.000	−0.07	.000	3	B7	768,837
		WLS 1540 25 # 6		15 30 55	+24 59.5	1	13.21		0.56		0.03		2		1375
138749	+31 02750	HR 5778	⋆AB	15 30 55	+31 31.6	6	4.13	.012	−0.12	.010	−0.55	.010	32	B6 Vnne	15,154,1118,1212,1629,8015
138600	−24 12161			15 30 56	−24 56.6	1	7.62		0.58		0.11		1	G0/1 V	55
138615	−18 04092			15 30 57	−19 11.6	1	9.90		0.26		0.15		1	A1 III/IV	1770
		AR Ser		15 30 58	+02 56.3	1	11.69		0.22		0.09		1	F3	668
138575	−32 10898			15 31 01	−33 00.2	1	6.98		0.02		−0.01		2	A0 V	1770
138564	−38 10464	HR 5773		15 31 03	−39 10.9	3	6.36	.001	−0.03	.003	−0.11	.014	7	B9 V	401,1770,2009
138648	−16 04112			15 31 07	−16 49.9	2	8.14	.000	0.84	.002	0.46		6	K0 V	2033,3008
	−33 10593			15 31 07	−33 46.4	1	9.56		0.50				3		1594
138565	−43 10012	IDS15278S4307	AB	15 31 09	−43 17.6	2	8.46	.003	0.02	.007	−0.13	.007	3	B9 V	401,1770
138477	−56 06798			15 31 09	−56 49.0	2	8.51	.008	0.27	.001	−0.19	.006	7	B7 III	567,1586
138651	−20 04262			15 31 14	−21 08.6	1	8.15		0.21		0.15		2	A0/1 V	1771
234242	+52 01879			15 31 16	+52 29.0	1	9.70		0.42		0.03		2	F5	1566
138566	−45 10019	IDS15279S4604	A	15 31 20	−46 14.1	1	9.34		0.05		−0.10		1	B9.5 IV	1770
138652	−27 10440			15 31 23	−27 57.2	1	10.32		0.20		0.16		1	B9 V	1770
138672	−22 03977	EI Lib	⋆A	15 31 25	−22 50.0	1	9.88		0.64		0.20		1	A3/7	1770
138672	−22 03977	IDS15285S2240	B	15 31 25	−22 50.0	1	10.19		0.44		0.18		1	F5	1770
138584	−46 10175			15 31 25	−47 04.4	1	8.69		−0.03		−0.43		1	B7 II/III	1770
138716	−9 04171	HR 5777		15 31 26	−09 53.7	8	4.62	.011	1.01	.005	0.85	.012	34	K1 IV	15,30,1003,1075,1088*
137333	−84 00510	HR 5729		15 31 26	−84 18.2	2	5.56	.005	0.11	.000	0.07	.005	7	A2 V	15,1075
138583	−46 10176			15 31 27	−46 38.0	1	8.68		−0.03		−0.43		2	B6 III	401
	−56 06804			15 31 29	−56 24.2	1	10.78		0.38		−0.07		3		1586
138519	−56 06802			15 31 29	−57 15.1	2	7.92	.008	0.07	.000	−0.39	.004	33	Ap Si	567,978
		GD 185		15 31 30	−02 17.2	1	14.03		0.00		−0.81		1	DA	3060
		GD 186		15 31 33	+18 29.0	1	16.20		0.16		−0.58		1		3060
138536	−58 06074			15 31 33	−58 42.0	1	7.99		0.03		−0.32		1	B8 III	567
138721	−21 04145			15 31 34	−21 45.1	1	8.55		0.28		0.19		1	A1/2 V	1770
138688	−27 10443	HR 5775		15 31 35	−27 52.8	2	5.12	.025	1.30	.005	1.22		8	K4 III	2007,3016
138803	+17 02880	HR 5783		15 31 36	+17 18.3	2	6.50	.005	0.31	.000	0.04	.000	5	F3 III	1648,1733
	−56 06807	He3 1081		15 31 39	−56 28.0	1	10.46		1.11		0.81		3		1586
		CS 22884 # 20		15 31 41	−08 23.7	1	14.95		0.46		−0.14		1		1736
138763	−5 04100	HR 5779		15 31 42	−05 31.7	2	6.50	.005	0.58	.000	0.03	.005	7	F7 V	15,1256
		CS 22884 # 2		15 31 42	−12 02.8	1	12.80		0.30		0.29		1		1736
		LP 916 - 51		15 31 43	−28 11.9	1	12.22		0.89		0.52		2		1696
138764	−8 04010	HR 5780		15 31 44	−09 01.0	7	5.17	.009	−0.09	.002	−0.45	.013	75	B6 IV	15,26,154,1256,3047*
138498	−65 03100	HR 5766	⋆AB	15 31 46	−65 26.8	4	6.50	.007	0.35	.007	0.02	.015	24	F3 IV/V	15,1075,1075,2038
138690	−40 09760	HR 5776	⋆AB	15 31 48	−41 00.0	5	2.78	.013	−0.21	.009	−0.82	.010	17	B2 IV	15,26,1075,2012,8015
	−51 09261			15 31 49	−51 33.7	1	11.31		1.22		0.18		2		1649
	−58 06077			15 31 49	−58 42.1	1	10.86		0.41		−0.23		2	B3 IV	540
138753	−26 10944			15 31 52	−26 59.2	1	8.51		0.19		0.17		1	A0 V	1770
138403	−71 01889			15 31 54	−71 45.0	1	10.48		−0.23		−1.05		3		1586
	+27 02509			15 31 58	+26 40.5	1	8.15		1.62		2.00		2	M0	1375
	−62 04634			15 31 58	−62 34.7	2	10.43	.004	0.10	.013	−0.46	.006	3	B4 IV	540,976
		LP 623 - 6		15 32 00	+02 26.9	1	15.23		1.61				1		1759
139777	+80 00480	HR 5829	⋆A	15 32 00	+80 37.0	3	6.57	.015	0.67	.005	0.13	.000	6	G8 IV-V	15,1028,3026
		MCC 746		15 32 03	+38 04.9	1	11.37		1.48				2	K5	1619
		CS 22884 # 21		15 32 04	−08 06.7	1	14.39		0.14		0.18		1		1736
138539	−66 02798			15 32 05	−66 22.9	1	8.91		−0.04		−0.40		2	B7 IV/V	1586
138538	−65 03102	HR 5771	⋆A	15 32 07	−66 09.1	3	4.09	.007	1.16	.003	1.16	.000	16	K1/2 III	15,1075,2038
138790	−24 12179			15 32 08	−24 56.4	1	8.04		0.44		−0.02		1	F0/2 V	55
		G 15 - 26		15 32 10	+02 22.8	3	13.43	.030	1.48	.028	1.24		5		333,538,1620,1759
		G 137 - 22		15 32 12	+14 26.3	1	13.83		1.56		1.06		4		316
139813	+80 00481	IDS15350N8047	B	15 32 13	+80 37.0	2	7.30	.005	0.80	.009	0.38	.005	4	G5	1028,3030
138678	−55 06631			15 32 14	−55 41.6	1	8.84		0.08		−0.26		3	B7/8 IV	567
138766	−36 10286			15 32 15	−36 59.7	1	9.51		0.42		0.28		1	A8 V	1771
138813	−25 10982			15 32 16	−25 34.1	1	7.32		0.06		0.00		2	A0 V	1771
138821	−20 04266			15 32 17	−20 50.5	1	8.77		0.29		0.17		1	A1(M)A5-A8	1770
138791	−28 11429			15 32 17	−28 53.0	1	7.68		0.32		0.11		1	A2mA7-F0	1770
		CG Lib		15 32 18	−24 10.2	1	11.18		0.36		0.20		1	F1	700
		G 15 - 27		15 32 19	+01 58.7	1	13.51		1.48		1.28		1		333,1620
138743	−49 09787	R Nor	⋆AB	15 32 21	−49 20.5	1	7.69		1.77		1.33		1	M3e	817
138918	+11 02821	HR 5789, δ Ser	⋆AB	15 32 25	+10 42.3	3	3.80	.000	0.26	.007	0.12	.005	8	F0 IV	15,1363,8015
138769	−44 10239	HR 5781	⋆AB	15 32 26	−44 47.6	7	4.54	.004	−0.18	.007	−0.69	.017	27	B3 IVp	15,26,1020,1075,1770*
	−58 06087	LSS 3372		15 32 28	−58 20.7	2	10.44	.002	0.68	.004	−0.37	.016	4	O6	540,1737
		CS 22884 # 6		15 32 29	−11 30.9	1	14.77		0.03		−0.43		1		1736
		GD 346		15 32 30	+57 41.0	1	13.89		0.53		−0.05		3		308
		BPM 23550		15 32 30	−59 35.	1	14.17		0.77		0.13		1		3065
138679	−60 05926			15 32 31	−60 23.1	2	8.89	.039	−0.01	.002	−0.68	.000	4	B2 III/IV	540,976
138936	+2 02977	HR 5791		15 32 33	+01 50.1	1	6.55		0.28		0.02		8	F0 III	1088
139006	+27 02512	HR 5793, α Crb		15 32 34	+26 52.9	9	2.23	.009	−0.02	.005	−0.03	.007	47	A0 V	1,15,667,1007,1013*
138593	−68 02520			15 32 35	−68 20.0	1	8.39		0.49				4	F5 V	2012
138289	−77 01134	HR 5757		15 32 35	−77 45.1	5	6.18	.007	1.21	.012	1.36	.009	21	K2 III	15,1075,1637,2012,8100
138871	−24 12183			15 32 38	−24 59.0	1	8.67		0.50		0.28		2	A9 V	1771
139007	+25 02932	IDS15305N2520	A	15 32 40	+25 10.0	1	8.26		0.52		0.04		3	F8 V	3031
138872	−25 10985			15 32 40	−26 08.1	1	8.81		0.13		0.05		1	B9 III	1770
138729	−58 06089	LSS 3373		15 32 41	−58 24.7	3	8.96	.009	0.44	.010	−0.45	.024	9	B1/2 Ib	540,567,976
138905	−14 04237	HR 5787	⋆AP	15 32 43	−14 37.5	5	3.90	.004	1.01	.008	0.74	.007	20	K0 III	15,1075,1088,2012,8015
		G 137 - 24		15 32 45	+12 57.8	1	15.07		0.24		−0.63		1		3060
138816	−43 10036	HR 5784		15 32 46	−44 13.9	4	5.43	.013	1.50	.005	1.82		13	K4/5 III	15,1075,2006,3005

Table 1 845

HD	DM	Other Id	N	Rem	α₁₉₅₀	δ₁₉₅₀	S	V	σ_V	B−V	σ_B−V	U−B	σ_U−B	n	Spectrum	References
		LP 623 - 62			15 32 50	+01 44.0	1	13.59		1.28		1.20		1		1696
139951	+81 00517				15 32 50	+80 56.3	1	6.94		0.96		0.67		2	K0	1502
139669	+77 00592	HR 5826			15 32 51	+77 31.0	1	4.96		1.58		1.89		2	K5 III	3053
		CS 22884 # 33			15 32 54	−10 05.2	1	14.52		0.50		−0.13		1		1736
138860	−43 10040				15 33 01	−43 50.7	2	7.28	.007	−0.05	.002	−0.37	.003	4	B8 II/III	401,1770
138923	−32 10930	HR 5790		⋆ AB	15 33 03	−32 55.6	2	6.25	.003	−0.09	.002	−0.29		4	B8 V	1770,2008
138922	−27 10456				15 33 04	−27 46.6	1	10.06		0.34		0.27		1	A3/5 IV	1770
	−62 04664				15 33 06	−62 18.7	1	11.17		0.48		−0.36		2	B2	34
		G 137 - 26			15 33 07	+17 52.8	2	12.41	.010	1.59	.010	1.20	.020	4		316,3078
		G 137 - 25			15 33 07	+17 53.1	2	14.96	.029	1.79	.010	1.79		7		316,3078
138846	−49 09800			AB	15 33 09	−50 00.1	1	8.38		0.67				4	F8/G0 V	2012
	−53 06587				15 33 09	−54 14.3	1	9.87		0.53		0.08		1		96
		LP 683 - 12			15 33 14	−04 48.8	1	13.62		1.02		0.76		1		1696
139074	+18 03044	HR 5795			15 33 17	+17 49.3	1	6.12		0.94				1	K0	71
		LOD 36 # 6			15 33 18	−54 12.7	1	12.50		0.39		0.09		1		96
		L 480 - 69			15 33 21	−37 43.4	2	12.79	.058	1.60	.010			3		1705,3062
138940	−40 09787				15 33 23	−40 53.7	1	7.62		−0.03		−0.13		2	B9 V	1770
139153	+39 02889	HR 5800			15 33 25	+39 10.5	2	5.12	.005	1.63	.010	2.03	.015	4	M2 IIIab	105,3001
138862	−54 06599	IDS15296S5410	A		15 33 25	−54 19.9	1	8.71		0.73		0.45		1	K0 V	96
		G 167 - 50			15 33 27	+28 01.1	1	13.55		0.70		−0.02		1		1658
138878	−54 06601				15 33 29	−54 18.3	2	7.96	.000	0.02	.014	−0.24	.014	5	B8 II/III	96,567
139087	+11 02826	HR 5796			15 33 30	+11 25.8	1	6.07		1.09				1	G5	71
138942	−44 10253				15 33 34	−44 23.5	1	9.19		0.07		0.00		1	B9 V	1770
139010	−32 10938				15 33 35	−32 28.3	1	8.02		0.10		0.10		2	A0/1 V	1770
138895	−53 06592	IDS15298S5402	AB		15 33 37	−54 12.0	1	7.79		0.91		0.53		1	G3/5 III	96
	−53 06594				15 33 38	−54 08.6	1	10.35		0.96		0.72		1		96
		LOD 36 # 8			15 33 39	−54 09.8	1	12.91		0.83		0.34		1		96
139020	−28 11439	IDS15306S2831	A		15 33 40	−28 41.1	1	8.93		0.42		0.29		1	A4 III	1770
		Steph 1257			15 33 41	+15 00.1	1	12.09		1.06		0.89		1	K6	1746
	+35 02705				15 33 42	+35 14.9	1	8.97		1.07		0.97		2	K0	1733
138997	−41 10104				15 33 44	−41 40.2	1	8.76		0.24		0.06		1	A6 V	1771
139307	+50 02195				15 33 45	+49 51.8	1	7.33		1.56		1.90		2	K5	1566
139012	−41 10105	IDS15304S4108	A		15 33 46	−41 17.8	1	9.08		0.35		−0.03		1	A(8)	1771
139320	+53 01793				15 33 47	+53 35.9	1	8.66		0.82		0.38		2	G5	1566
	+16 02806				15 33 49	+16 00.3	1	8.65		1.34		1.85		5	K2	897
		CS 22884 # 28			15 33 50	−09 03.6	1	14.12		0.11		0.02		1		1736
139035	−32 10942				15 33 50	−32 39.1	1	8.19		0.05		−0.05		3	B8/9 III	1770
139022	−40 09794				15 33 50	−40 51.2	1	7.13		1.43		1.59		10	K2/3 III	1673
		LOD 36 # 10			15 33 51	−54 08.2	1	10.99		1.01		0.80		1		96
139357	+54 01756	HR 5811			15 33 57	+54 05.2	1	5.97		1.18				1	gK4	71
139284	+38 02678	HR 5808			15 33 58	+38 32.3	1	6.28		1.30		1.42		2	K2	1733
	−51 09307				15 33 58	−51 52.5	1	12.06		0.41		−0.31		2		1649
		LOD 36 # 11			15 33 58	−54 08.9	1	11.63		1.35		1.31		1		96
138910	−59 06182				15 33 58	−59 42.8	1	8.06		0.08		−0.52		4	B3 III	567
139063	−27 10464	HR 5794		⋆ AB	15 33 59	−27 58.3	6	3.59	.016	1.38	.006	1.58	.003	27	K3 III	3,15,418,1075,2012,8015
139137	−0 02988	HR 5799			15 34 00	−00 23.8	4	6.51	.008	0.72	.002	0.44	.026	19	G8 III +A	97,281,1088,6004
		WLS 1536 45 # 9			15 34 03	+45 54.6	1	12.47		0.54		0.00		2		1375
139195	+10 02884	HR 5802			15 34 05	+10 10.6	5	5.26	.010	0.94	.006	0.67	.006	20	K0 III:	993,1080,1509,3016,4001
139094	−26 10958				15 34 06	−26 19.7	4	7.39	.009	0.08	.003	−0.28	.011	11	B8 IV/V	26,861,1054,1770
	−53 06600				15 34 07	−54 06.8	1	9.90		1.29		1.31		1		96
139216	+15 02890	tau4 Se			15 34 09	+15 15.9	3	6.52	.085	1.46	.069	1.03	.050	12	M4 III	897,1501,3040
139095	−31 12108				15 34 09	−31 53.6	1	7.91		0.32		0.06		1	A9/F0 V	1770
139323	+40 02903	G 179 - 37		⋆ C	15 34 10	+39 59.7	1	7.56		0.97		0.78		3	K3 V	3030
139225	+16 02807	HR 5804			15 34 11	+16 17.0	1	5.95		0.28		0.02		2	F3 V	254
		G 152 - 40			15 34 12	−13 57.2	1	12.70		1.57				1		1705
		L 480 - 93			15 34 12	−38 20.4	1	14.00		0.87		0.37		2		3062
139000	−54 06607				15 34 13	−54 21.5	1	6.99		0.14		0.11		1	A1/2 V	96
139197	+0 03379				15 34 14	−00 05.3	6	9.04	.004	0.40	.003	0.15	.010	57	A5	97,281,989,1729,5006,6004
139586	+68 00842				15 34 15	+67 58.4	1	7.00		0.79		0.43		3	G5	3026
139341	+40 02905	G 179 - 38		⋆ AB	15 34 16	+39 58.0	4	6.80	.019	0.91	.010	0.70	.006	15	K2 V	938,1080,1381,3030
139002	−56 06827				15 34 16	−57 06.7	1	8.27		1.74		2.01		3	M1 III	978
139324	+36 02623				15 34 18	+35 52.2	1	7.48		0.62		0.16		2	G5	1601
		CS 22884 # 23			15 34 18	−08 04.3	1	14.18		0.18		0.22		1		1736
		LOD 36 # 14			15 34 18	−54 07.5	1	12.72		0.44				1		96
139182	−17 04380				15 34 21	−17 52.3	1	9.60		−0.02		−0.34		1	B8/9 II	1770
		LOD 36 # 15			15 34 21	−54 08.1	1	11.17		1.05		0.87		1		96
139268	+15 02891				15 34 23	+15 05.4	2	7.09	.005	0.05	.005	0.03	.010	5	A0	1733,3016
	−54 06609				15 34 24	−54 18.9	1	9.77		1.09		0.86		1		96
		Steph 1259			15 34 25	+40 59.8	1	11.34		1.39		1.22		1	M0	1746
139549	+64 01078				15 34 25	+64 15.6	1	9.11		0.40		−0.06		18	F8	1603
		LP 623 - 63			15 34 25	−03 13.8	2	13.23	.005	1.12	.015	1.05	.030	2		1696,3062
		POSS 564 # 1			15 34 26	+05 44.0	1	14.00		1.13				1		1739
238480	+56 01814				15 34 27	+55 41.4	1	9.05		0.44		0.03		2	F8	1502
139160	−25 11000	HR 5801			15 34 28	−26 07.0	5	6.19	.013	−0.01	.009	−0.43	.007	15	B9 IV	26,1054,1079,1770,2006
		LOD 36 # 16			15 34 29	−54 11.4	1	11.80		0.54		0.08		1		96
	−74 01448				15 34 29	−74 55.0	1	7.70		0.96		0.69		2		278
	+0 03380				15 34 31	+00 10.6	1	10.62		0.61		0.11		2	G0	97
	−50 09646				15 34 31	−51 02.5	1	11.12		1.16		0.66		1		96
141155	+85 00263				15 34 33	+84 59.7	1	6.85		1.01		0.78		3	K0	985
		AAS35,161 # 71		V	15 34 33	−34 21.3	1	13.83		1.24		−0.11		1		5005

HD	DM	Other Id	N Rem	α_{1950}	δ_{1950}	S	V	σ_V	B–V	σ_{B-V}	U–B	σ_{U-B}	n	Spectrum	References
139202	−21 04152			15 34 35	−21 57.1	1	6.97		0.29		0.24		1	A1mA3-F0	1770
		LP 916 - 12		15 34 38	−30 46.4	1	12.56		0.57		-0.14		2		1696
139493	+55 01766	HR 5818		15 34 39	+54 47.7	1	5.79		0.05		0.04		2	A2 V	1733
139478	+52 01886	HR 5817		15 34 40	+52 14.0	1	6.76		0.31		0.02		2	F4 III	254
139040	−56 06833			15 34 40	−56 29.7	1	10.50		0.32		-0.08		4	B7/9 III	567
139127	−42 10601	HR 5797	⋆ A	15 34 41	−42 24.3	6	4.33	.009	1.41	.013	1.72	.010	21	K4.5 III	15,1075,1279,1311*
139127	−42 10601	IDS15313S4214	B	15 34 41	−42 24.3	1	11.16		1.10		1.10		2		1279
	+45 02308			15 34 42	+45 00.3	1	10.51		1.06		0.87		2		1375
		LOD 36 # 17		15 34 43	−54 10.2	1	12.58		0.45		0.31		1		96
139287	+0 03381			15 34 44	+00 00.0	5	8.43	.006	0.62	.008	0.08	.008	37	G3 V	97,281,989,1729,6004
		LOD 37+ # 11		15 34 44	−51 04.2	1	11.50		0.45		0.02		1		96
	+74 00622			15 34 45	+74 30.5	1	9.68		0.62		0.15		2		1375
138800	−73 01625	HR 5782	⋆ A	15 34 46	−73 17.1	4	5.65	.007	-0.04	.002	-0.37	.008	16	B7 III/IV	15,1075,1637,2038
		GD 347		15 34 47	+50 23.7	1	15.71		0.25		-0.44		2	DAwk	308
		LOD 36 # 18		15 34 48	−54 10.1	1	11.98		0.67		0.18		1		96
238482	+56 01815	IDS15336N5551	AB	15 34 50	+55 41.1	1	8.66		0.45		-0.03		2	F5	1502
139250	−18 04112			15 34 50	−19 04.7	1	8.95		0.15		0.15		1	B9 V	1770
139389	+30 02682	HR 5813		15 34 51	+30 09.3	2	6.47	.005	0.40	.000	-0.07	.005	5	F5 V	1501,1733
139254	−22 03989	HR 5806		15 34 51	−22 58.6	3	5.81	.026	1.07	.014	0.92	.026	10	K0 III	1637,2007,3008
		HA 107 # 969		15 34 52	+00 25.5	1	12.60		0.59		0.04		1	F8	281
		HA 107 # 970		15 34 52	+00 28.3	2	10.94	.011	1.60	.004	1.75	.015	48	K3	281,1764
139308	−0 02990			15 34 54	−00 43.3	5	7.77	.008	1.28	.004	1.31	.005	37	K0 III	281,989,1657,1729,6004
		CS 22884 # 36		15 34 54	−11 20.8	1	13.40		0.20		0.22		1		1736
139162	−42 10605			15 34 55	−42 50.4	1	9.90		0.29		0.10		1	A8/9 V	1771
	−50 09658	LSS 3376		15 34 56	−50 58.9	1	11.30		0.50		-0.42		1		540
		LOD 36 # 19		15 34 56	−54 08.4	1	11.27		0.26		0.18		1		96
139084	−57 07121	V343 Nor		15 34 59	−57 32.6	2	8.12	.039	0.81	.001	0.35	.013	31	K0 V	1621,1641
		LP 803 - 53		15 35 00	−18 27.4	1	13.34		0.54		-0.23		3		3062
		KS 987		15 35 00	−66 22.	1	10.26		1.17		0.93		3		540
	−57 07122			15 35 03	−58 08.5	1	9.98		1.18		0.23		2	B1.5V	540
139255	−29 11831	IDS15320S2939	AB	15 35 04	−29 48.0	1	9.69		0.26		0.18		2	A2 V	1770
139129	−51 09324	HR 5798	⋆ A	15 35 06	−52 12.6	3	5.43	.004	0.00	.000	0.00		8	A0 V	15,2012,3023
138965	−69 02422	HR 5792		15 35 08	−70 03.9	5	6.43	.005	0.08	.004	0.06	.007	17	A1 V	15,285,688,1075,2038
139290	−27 10478	HR 5809		15 35 13	−28 02.6	1	6.32		1.17				4	K1 III	2007
139206	−45 10078			15 35 13	−45 32.8	2	6.97	.005	-0.08	.002	-0.42	.002	3	B5 V	401,1770
139233	−38 10532	HR 5805		15 35 15	−38 59.8	3	6.58	.005	-0.07	.002	-0.20	.001	7	B9 V	401,1770,2009
		HA 107 # 568		15 35 20	−00 07.5	1	13.05		1.15		0.86		7		1764
		L 201 - 12		15 35 21	−54 58.	1	12.82		1.37		1.21		1		3073
139329	−20 04285	HR 5810		15 35 22	−20 51.2	1	5.84		1.08				4	K0 III	2007
139347	−24 12207			15 35 23	−24 29.5	1	7.93		0.22		0.16		2	A0 V	1771
139208	−50 09665			15 35 23	−51 04.8	1	8.61		1.55		1.80		1	K2 III	96
		LP 563 - 16		15 35 24	+04 44.3	1	13.14		1.28		1.17		1		1696
139271	−38 10536	HR 5807	⋆ AB	15 35 24	−38 57.9	2	6.04	.000	0.21	.000	0.17		6	A3/5mA3-F2	1771,2006
	−48 10153	LSS 3378		15 35 24	−48 26.2	3	11.48	.019	0.45	.012	-0.31	.004	12	He Star	223,1514,1737
139406	+0 03382			15 35 27	+00 29.0	2	9.55	.004	0.48	.005	0.00		14	F5	281,6004
		HA 107 # 435		15 35 29	−00 15.4	2	10.92	.010	1.60	.008	1.74	.071	27		989,1729
		HA 107 # 991		15 35 33	+00 26.1	1	12.10		1.08		0.79		10	G8	281
139457	+10 02886			15 35 35	+10 24.5	3	7.10	.005	0.51	.007	-0.03	.012	8	F8 V	1003,1733,3037
139365	−29 11837	HR 5812		15 35 35	−29 36.9	7	3.66	.011	-0.18	.005	-0.68	.016	29	B2.5V	15,26,1054,1075,1088*
139309	−40 09811			15 35 35	−40 56.1	1	8.90		0.49		0.29		1	A1mA7-F2	1771
		CS 22884 # 25		15 35 37	−08 01.0	1	15.82		0.07				1		1736
		HA 107 # 442		15 35 38	−00 16.0	2	11.14	.000	1.30	.000	1.38		2	gG8	281,5006
139381	−21 04157			15 35 39	−21 26.2	1	8.27		0.39		0.06		2	F0 V	1771
		LOD 37+ # 8		15 35 42	−51 02.1	1	11.59		0.44		0.13		1		96
138867	−75 01222	HR 5786		15 35 42	−75 55.2	4	5.94	.005	-0.04	.001	-0.15	.005	16	B9 V	15,1075,1637,2038
	+0 03383			15 35 43	+00 25.2	2	10.43	.006	0.63	.002	0.16		10	G3 IV	281,6004
		LOD 37+ # 4		15 35 43	−50 57.2	1	12.55		1.74		1.01		1		96
139210	−58 06132			15 35 45	−58 58.7	1	9.09		0.08		-0.03		1	B8 Ib	567
	+0 03384			15 35 46	+00 17.8	1	10.25		1.07		0.81		2	G9 III	281
139275	−50 09676			15 35 46	−50 54.9	1	8.39		0.71		0.21		1	B9/A0 IV/V	96
	−52 08681	IDS15321S5216	AB	15 35 47	−52 25.4	1	10.10		0.23		-0.20		2	B7	837
		LOD 37+ # 5		15 35 48	−50 57.5	1	13.21		0.57		0.05		1		96
		LOD 37+ # 7		15 35 49	−50 59.7	1	12.29		1.46		1.73		1		96
		LOD 37+ # 3		15 35 50	−50 55.8	1	11.83		0.24		0.04		1		96
139211	−59 06206	HR 5803		15 35 50	−59 44.6	4	5.95	.004	0.49	.009	-0.01		16	F6 V	15,1075,2012,3037
		HA 107 # 446		15 35 51	−00 20.3	1	12.76		0.71		0.22		2		97
139350	−41 10145			15 35 51	−41 31.8	1	9.72		0.36		0.13		1	A9 V	1771
	−52 08685			15 35 51	−52 22.6	1	10.99		0.46		0.24		2	F3	837
		LOD 37+ # 6		15 35 52	−50 59.5	1	13.16		0.72		0.18		1		96
139237	−57 07128	LSS 3380		15 35 52	−58 05.4	2	9.89	.004	0.48	.008	-0.45	.011	4	B0/1	540,567
		LP 916 - 15		15 35 57	−28 25.7	2	11.83	.058	0.49	.015	-0.21	.010	3		1696,3060
139461	−8 04032	HR 5816	⋆ A	15 35 58	−08 37.7	1	6.48		0.52		-0.02		2	F6 V	3032
139461	−8 04032	IDS15333S0828	⋆ AB	15 35 58	−08 37.7	1	5.78		0.50		-0.03		8	F6 V +	1088
		HA 107 # 1006		15 36 00	+00 24.1	2	11.71	.003	0.77	.002	0.28	.007	54	G3	281,1764
	+0 03385			15 36 00	−00 09.0	2	10.59	.024	1.22	.024	1.21	.024	3	K0 III	97,281
		HA 107 # 450		15 36 00	−00 17.4	2	11.71	.005	0.76	.008	0.29	.004	22		989,1729
139641	+40 02907	HR 5823		15 36 02	+40 30.9	3	5.26	.015	0.89	.005	0.49	.029	7	G8 III-IV	15,1003,3016
139513	−0 02991			15 36 02	−00 26.2	4	9.44	.007	1.30	.005	1.30	.009	30	K0 III	97,281,989,1729
139446	−18 04118	HR 5814		15 36 02	−19 08.3	4	5.37	.006	0.88	.009	0.52	.009	17	G8 III/IV	3,15,1003,2007
		HA 107 # 720		15 36 03	+00 12.1	1	13.12		0.60		0.09		1		1764

Table 1   847

HD	DM	Other Id	N Rem	α₁₉₅₀	δ₁₉₅₀	S	V	σ_V	B–V	σ_B-V	U–B	σ_U-B	n	Spectrum	References
		HA 107 # 452		15 36 03	−00 13.6	2	11.01	.024	1.09	.015	0.88	.029	3	G5	97,281
		CS 22884 # 32		15 36 03	−09 44.5	1	14.18		0.29		0.22		1		1736
139689	+47 02252			15 36 04	+47 28.5	1	9.06		0.48		-0.04		2	G5	1375
		HA 107 # 453		15 36 05	−00 23.0	1	13.79		0.78		0.38		2		97
139515	−5 04128			15 36 06	−06 09.3	1	8.37		0.41		0.20		4	A5	1776
139486	−19 04169			15 36 07	−19 34.1	3	7.64	.008	0.03	.009	-0.08	.006	9	B9 V	26,1054,1770
139608	+24 02901	LY Ser		15 36 08	+24 41.1	1	6.80		1.50		0.94		1	M III	3040
		HA 107 # 456		15 36 09	−00 29.5	2	12.92	.000	0.92	.004	0.61	.040	12		97,1764
	−52 08695			15 36 09	−52 19.3	1	11.04		0.26		0.03		2	B9	837
139538	+0 03386			15 36 11	+00 06.2	1	9.89		0.61		0.10		2	G5	97
		HA 107 # 351		15 36 11	−00 22.4	3	12.34	.003	0.56	.003	0.01	.009	31	G1	97,281,1764
	−50 09681			15 36 11	−51 07.4	1	11.29		0.27		0.07		1	A0	96
		HA 107 # 457		15 36 13	−00 29.1	1	14.90	.010	0.82	.029	0.31	.035	2		97,1764
139354	−51 09346			15 36 13	−51 47.9	1	9.52		0.44				4	F3 IV/V	2012
	+6 03077			15 36 14	+06 32.7	1	10.50		0.55		-0.10		4		1696
139621	+23 02838	IDS15341N2300	AB	15 36 14	+22 50.2	1	7.44		1.11		1.01		2	K0	1625
	−52 08696			15 36 14	−52 51.8	1	11.04		0.10		-0.14		2	B8	837
139778	+54 01758	HR 5828		15 36 15	+54 40.3	1	5.87		1.08				1	K0	71
		HA 107 # 592		15 36 16	−00 07.5	2	11.84	.008	1.33	.016	1.40	.025	8		97,1764
		AAS35,161 # 65	V	15 36 16	−34 36.6	1	12.59		1.43		1.04		1		5005
139314	−57 07133	LSS 3382		15 36 16	−57 54.1	3	10.08	.082	0.39	.014	-0.59	.034	6	O/Be	403,540,1586
		HA 107 # 458		15 36 17	−00 14.7	1	11.68		1.21		1.19		1		1764
		HA 107 # 459		15 36 18	−00 12.8	1	12.28		0.90		0.43		1		1764
139497	−20 04291			15 36 19	−21 03.9	2	9.64	.005	0.55	.010	0.00		3	F5 V	1594,6006
	−50 09685			15 36 19	−51 04.1	1	11.52		0.52		0.20		1		96
	−52 08699			15 36 20	−52 23.7	1	11.86		0.33		0.14		2	B9	837
		HA 107 # 862		15 36 22	+00 19.5	2	11.21	.015	0.55	.000	0.08	.015	4	G0	97,281
139609	+12 02870			15 36 22	+12 24.9	1	7.44		-0.09		-0.34		3	B9	3016
139431	−42 10636		V	15 36 22	−42 36.3	4	7.44	.083	0.02	.021	-0.74	.017	17	B2 Vne	164,400,401,1586
139368	−53 06623			15 36 22	−53 27.3	1	9.88		0.07		-0.31		3	B9 Ib/II	567
		HA 107 # 215		15 36 24	−00 33.4	1	16.05		0.12		-0.08		2		1764
		HA 107 # 213		15 36 24	−00 34.5	1	14.26		0.80		0.26		1		1764
		HA 107 # 212		15 36 24	−00 35.8	1	13.38		0.68		0.14		1		1764
139518	−22 03996	HR 5819		15 36 25	−22 59.3	2	6.34	.003	0.03	.008	0.00		6	B9.5V	1770,2007
	−52 08700			15 36 25	−52 24.0	1	11.20		0.56		0.41		2	A9	837
		HA 107 # 1014		15 36 26	+00 25.3	1	12.11		0.55		0.00		6	F7	281
139610	+11 02834			15 36 26	+10 46.7	1	7.08		0.98		0.68		2	G5	1648
	−51 09352			15 36 26	−51 49.6	1	9.85		0.64				4		2012
139432	−43 10091			15 36 27	−43 27.4	4	7.59	.004	0.06	.000	-0.38	.003	13	B7 II/III	158,400,401,1770
139590	+0 03387			15 36 28	−00 08.9	6	7.50	.009	0.55	.005	0.06	.009	50	G0 V	97,281,989,1729,3026,6004
139400	−50 09690			15 36 28	−51 10.7	1	9.34		0.47		0.18		1	F0 III	96
	−52 08702	IDS15328S5212	AB	15 36 28	−52 21.7	1	10.59		1.02		0.73		2	K4	837
		HA 107 # 596		15 36 29	−00 04.1	2	11.39	.000	0.56	.005	0.05	.015	4	F8	97,281
139172	−69 02425			15 36 30	−70 02.2	1	10.53		0.05		-0.06		2	B8/9 IV/V	1730
		HA 107 # 1020		15 36 31	+00 24.8	2	11.84	.000	0.48	.000	0.00		11	F3	281,5006
		HA 107 # 357		15 36 32	−00 29.5	1	14.42		0.68		0.03		1		1764
		Lod 2313 # 25		15 36 32	−52 22.2	1	12.36		0.47		0.11		2		837
139519	−27 10492	IDS15335S2719	A	15 36 34	−27 28.9	1	7.65		0.36		0.24		1	A2/3 IV/V	1771
	−52 08703			15 36 34	−52 20.4	1	10.50		0.28		-0.09		2	B8	837
	−52 08705			15 36 34	−52 30.1	1	11.28		0.12		0.11		2	A0	837
		HA 107 # 599		15 36 35	+00 04.7	2	14.68	.007	0.70	.006	0.25	.003	9		97,1764
139521	−33 10631	HR 5820		15 36 35	−34 15.0	4	4.66	.005	0.99	.010	0.73	.004	14	G8/K0 III	15,1075,2012,8015
	−52 08704			15 36 35	−52 31.7	1	12.05		0.25		0.23		2		837
		HA 107 # 359		15 36 36	−00 25.8	1	12.80		0.58		-0.12		3		1764
	−50 09693			15 36 36	−51 06.8	1	10.47		1.39		1.23		1		96
		Lod 2313 # 24		15 36 37	−52 24.2	1	12.29		0.60		0.06		2	G2	837
		HA 107 # 600		15 36 38	−00 06.1	1	14.88		0.50		0.05		4		1764
		WLS 1540 25 # 10		15 36 39	+24 51.6	1	12.36		0.94		0.65		2		1375
	−9 04191	IDS15339S0957	A	15 36 39	−10 06.8	1	10.18		1.08		0.88		2	K5 V	3072
139798	+47 02253	HR 5830		15 36 40	+46 57.7	6	5.76	.018	0.35	.011	-0.02	.005	19	F2 V	15,254,1007,1013,3016,8015
		HA 107 # 464		15 36 40	−00 23.0	1	12.32		0.58		0.02		2		97
139747	+39 02895			15 36 41	+39 22.7	1	8.77		0.91		0.58		3	K0 III	833
139780	+44 02493			15 36 41	+43 46.0	1	6.82		0.16		0.13		5	A3 V	1501
		HA 107 # 601		15 36 41	−00 03.7	1	14.65		1.41		1.26		13		1764
139465	−44 10297			15 36 41	−44 51.5	1	7.40		1.26				4	K3 III	2012
139724	+32 02607			15 36 42	+31 42.8	1	7.20		0.23		0.09		1	A3	1629
		HA 107 # 871		15 36 43	+00 18.6	1	12.48		0.65		0.12		10	G5	281
		HA 107 # 602		15 36 45	−00 05.7	3	12.12	.003	0.98	.009	0.60	.013	32	G7	97,281,1764
	−52 08708			15 36 45	−52 34.0	1	10.68		0.32		0.18		2	A4	837
		HA 107 # 465		15 36 46	−00 15.7	2	11.61	.016	0.55	.000	0.07	.012	19	G1	97,281
	−52 08710			15 36 47	−52 18.3	2	10.36	.020	0.32	.034	-0.28	.015	5	B6	837,1350
	−52 08709			15 36 47	−52 31.5	1	11.67		0.47		0.42		2	A2	837
		HA 107 # 604		15 36 48	+00 09.1	1	14.47		1.10		1.08		1		97
		CS 22884 # 65		15 36 48	−07 41.8	1	13.83		0.48		-0.14		1		1736
	−52 08711			15 36 49	−52 27.0	1	10.63		1.31		1.17		2	K6	837
139559	−30 12423	IDS15338S3104	A	15 36 52	−31 13.4	1	8.33		0.40		0.07		1	A7/9	1770
139761	+35 02711	HR 5827		15 36 53	+34 50.2	1	6.11		1.02				1	K0	71
	−45 10108	LSS 3383		15 36 53	−45 37.1	3	10.72	.011	-0.01	.004	-0.80	.021	7	B0.5V	540,976,1021
		HA 107 # 365		15 36 54	−00 19.8	1	11.60		1.10		0.83		21	G7	281
139749	+26 02712	IDS15348N2604	AB	15 36 55	+25 54.5	1	8.16		0.55		-0.04		2	G0 V	3026

HD	DM	Other Id	N	Rem	α_{1950}	δ_{1950}	S	V	σ_V	B–V	σ_{B-V}	U–B	σ_{U-B}	n	Spectrum	References
		HA 107 # 467			15 36 55	−00 26.4	1	13.62		0.82		0.18		1		97
139627	−19 04176				15 36 55	−19 33.2	1	8.38		0.19		0.14		2	A2 V	1771
139545	−39 10074				15 36 55	−40 02.0	1	9.15		0.56		0.24		1	A9 V	1771
	+0 03388				15 36 57	−00 07.3	2	10.49	.020	1.12	.015	0.93	.010	4	G8 III	97,281
139525	−43 10101				15 36 58	−44 07.6	1	6.98		-0.07		-0.37		2	Ap Si	401
139452	−54 06629				15 36 58	−54 55.8	1	8.19		0.00		−0.39		3	B7 III/IV	567
		HA 107 # 610			15 36 59	+00 04.3	1	13.31		1.07		0.87		2		97
139468	−51 09361				15 37 00	−52 04.2	1	8.83		0.06		−0.32		3	B7 III	1350
	−51 09360				15 37 00	−52 14.0	1	11.26		0.15		0.05		2	B9	837
		Lod 2313 # 17			15 37 00	−52 17.9	1	11.37		0.77		0.32		2		837
		HA 107 # 612			15 37 01	+00 05.4	1	13.95		0.63		0.02		1		97,1764
		Lod 2313 # 22			15 37 01	−52 27.8	1	12.13		0.62		0.53		2	A5	837
139612	−28 11478				15 37 02	−29 03.8	1	9.20		0.14		0.08		1	A0 V	1770
139505	−52 08715				15 37 02	−52 18.8	1	8.37		1.17		1.04		2	K0 II	837
		HA 107 # 611			15 37 03	−00 02.9	1	14.33		0.89		0.46		2		1764
	−50 09709				15 37 03	−50 57.7	1	11.12		0.29		−0.07		1	A0	96
		HA 107 # 469			15 37 04	−00 16.9	2	12.18	.013	0.58	.004	0.06	.004	13	G0	97,281
139906	+50 02206	HR 5835			15 37 06	+50 35.1	1	5.84		0.83				1	G8 III	71
	−52 08722				15 37 06	−52 20.5	1	10.71		0.07		0.00		2	B9	837
	−52 08718				15 37 06	−52 28.1	1	11.88		0.31		0.23		2	A4	837
		POSS 564 # 2			15 37 08	+05 14.1	1	14.97		1.54				2		1739
		GD 348			15 37 08	+65 11.3	1	14.64		0.18		−0.49		2		308
		HA 107 # 614			15 37 09	−00 03.5	1	13.93		0.62		0.03		2		1764
139613	−30 12431	HR 5822			15 37 09	−31 03.1	2	6.33	.010	1.41	.015	1.65		6	K3 III	2007,3005
139579	−42 10647				15 37 09	−42 51.1	2	8.36	.007	-0.04	.002	−0.28	.001	8	B8 V	401,1732
	−50 09710				15 37 10	−50 57.5	1	11.59		0.59		0.23		1		96
	−57 07143				15 37 14	−58 11.1	1	11.24		0.52		−0.18		2		1586
139648	−29 11857				15 37 16	−29 48.4	1	8.64		0.22		0.13		1	A3 V	1770
		HA 107 # 619			15 37 17	−00 08.4	1	12.12		0.62		0.06		19	F7	281
		HA 107 # 473			15 37 19	−00 10.2	3	10.20	.012	1.09	.007	0.90	.010	14	G6	97,281,6004
139663	−23 12458	HR 5824	7		15 37 19	−23 39.4	7	4.97	.011	1.32	.006	1.51	.004	35	K3 III	15,1637,2007,3016*
	−52 08724				15 37 19	−52 20.1	1	10.75		0.28		-0.02		2	B8	837
139649	−33 10640				15 37 21	−34 05.0	1	9.44		0.38		0.22		1	A0 V	1770
139614	−42 10650	IDS15340S4210	A		15 37 22	−42 20.2	1	8.27		0.23		0.08		1	A7 V	1771
139599	−47 10210	HR 5821	⋆ A		15 37 25	−47 34.5	2	6.23	.000	1.63	.000	1.98		6	K5/M0 III	2035,3005
139530	−55 06663				15 37 26	−55 41.9	1	10.32		0.26		0.18		1	A1 III/IV	96
139892	+37 02665	HR 5834	⋆ AB		15 37 30	+36 47.8	4	4.67	.008	-0.12	.003	−0.47	.003	11	B7 V	938,1022,1079,1363
140227	+69 00806	HR 5844			15 37 30	+69 26.7	2	5.61	.011	1.36	.008	1.58		4	K5 IIIb	71,3001
		LOD 37+ # 20			15 37 30	−50 54.4	1	12.36		0.59		−0.01		1		96
139937	+45 02317				15 37 33	+45 16.7	1	8.12		0.33		0.07		2	F0	1601
		HA 107 # 626			15 37 33	−00 07.8	1	13.47		1.00		0.73		12		1764
139676	−35 10436				15 37 34	−35 15.9	1	7.57		0.28		0.08		1	A9 V	1770
		HA 107 # 627			15 37 35	−00 07.7	1	13.35		0.78		0.23		12		1764
	−52 08731				15 37 37	−52 16.5	1	10.84		0.36		−0.01		2	B7	837
139938	+43 02519				15 37 41	+42 47.2	1	8.84		1.18		1.18		2	K0	1375
	−55 06667				15 37 41	−55 51.8	1	10.62		0.19		0.05		1		96
		HA 107 # 484			15 37 43	−00 11.6	3	11.32	.004	1.24	.003	1.28	.010	25	G8	281,1764,5006
139664	−44 10310	HR 5825	7		15 37 45	−44 29.8	7	4.63	.010	0.40	.005	-0.03	.014	25	F3/5 V	15,1075,1586,2012*
139634	−50 09717				15 37 45	−51 01.6	1	8.91		0.07		−0.16		1	A0 V	96
139862	+12 02875	HR 5831	⋆ A		15 37 48	+12 12.8	3	6.19	.042	0.97	.015	0.72	.009	10	G8 II	71,542,8100
139862	+12 02875	IDS15354N1223	B		15 37 48	+12 12.8	1	10.33		0.49		-0.01		3	F5 V	542
139600	−55 06668				15 37 48	−55 56.4	1	8.83		0.15		-0.01		1	B9 IV	96
139764	−20 04298				15 37 49	−21 01.7	1	9.81		0.41		0.21		1	A9 V	1770
139840	−0 02997				15 37 51	−00 45.1	4	8.07	.002	1.12	.007	0.95	.003	28	G8 III	281,989,1729,6004
238488	+56 01821				15 37 58	+55 53.6	1	9.62		0.44		−0.02		2	F8	1502
139766	−32 10988				15 37 59	−32 29.5	1	8.67		0.33		0.21		1	A7/8 V	1770
139787	−27 10498				15 38 00	−27 54.3	1	9.77		0.20		0.16		1	A4 III	1770
139819	−16 04138				15 38 02	−17 08.3	1	9.60		0.34		0.18		1	A9 V	1770
140117	+58 01583	HR 5841			15 38 03	+58 05.1	1	6.45		1.09		1.04		2	K1 III	37
		L 624 - 50			15 38 05	−28 32.7	2	11.80	.006	0.49	.001	−0.22	.009	9		1698,1765
	−52 08746				15 38 05	−52 20.6	1	10.77		0.46		0.16		2	F0	837
		CS 22884 # 60			15 38 07	−08 42.6	1	14.26		0.37		−0.30		1		1736
		HA 107 # 636			15 38 08	−00 05.3	1	14.87		0.75		0.12		1		1764
139820	−19 04183				15 38 08	−20 15.4	1	10.18		0.28		0.18		1	A3 II/III	1770
		HA 107 # 639			15 38 12	−00 07.6	1	14.20		0.64		−0.03		6		1764
		POSS 564 # 3			15 38 14	+05 01.3	1	15.69		1.63				2		1739
		Ton 245	⋆		15 38 17	+26 58.2	2	13.84	.020	-0.19	.020	−0.93	.010	14	DA1	1036,1727
		HA 107 # 640			15 38 17	−00 07.2	1	15.05		0.75		0.09		3		1764
		G 179 - 43	A		15 38 18	+43 39.4	1	12.20		1.70				1	M2	419
		G 179 - 43	AB		15 38 18	+43 39.4	1	11.91		1.57				1	M2	1746
		HA 107 # 641			15 38 20	−00 06.4	2	11.08	.000	0.49	.005	0.06	.010	3	F9	97,281
139808	−36 10338				15 38 21	−36 58.2	1	10.57		0.30		0.24		1	A0 III/IV	1770
		LSS 3386			15 38 22	−54 01.8	2	10.93	.033	1.61	.000	0.43	.019	4	B3 Ib	962,1737
139508	−69 02426				15 38 25	−70 06.4	1	7.67		0.01		0.00		3	A0 V	1730
		LP 683 - 23			15 38 28	−05 30.5	1	12.20		1.04		0.85		2		1696
139909	−13 04226				15 38 28	−13 48.6	1	6.88		0.06		−0.03		2	B9.5 V	1770
139717	−54 06636	U Nor			15 38 28	−55 09.1	2	9.09	.065	1.55	.006	1.13	.024	2	F8 Ib/II	689,1587
139867	−28 11499				15 38 29	−29 08.4	2	7.29	.006	1.74	.021	2.03		9	M2/3 III	1657,2012
140048	+29 02697				15 38 33	+29 37.6	1	10.23		0.45		−0.06		31	G0	588
140027	+16 02816	HR 5840	5		15 38 41	+16 11.1	5	6.01	.005	0.91	.015	0.61	.005	15	G8 III	15,1007,1013,3016,8015

Table 1 849

HD	DM	Other Id	N	Rem	α_{1950}	δ_{1950}	S	V	σ_V	B–V	σ_{B-V}	U–B	σ_{U-B}	n	Spectrum	References
139790	−49 09901	LSS 3387			15 38 41	−49 50.3	3	8.48	.025	0.13	.012	-0.65	.014	7	B2 (III)n	540,976,1586
		CS 22884 # 47			15 38 43	−11 18.5	1	13.37		0.19		0.19		1		1736
139975	−18 04132				15 38 53	−18 35.9	1	9.62		0.34		0.22		1	A3 II/III	1770
		CS 22884 # 49			15 38 56	−10 42.2	1	14.25		0.02		-0.28		1		1736
139900	−40 09847				15 38 56	−40 39.7	1	9.04		0.77				4	G6 V	2012
139977	−23 12472				15 38 59	−24 08.4	1	7.42		0.33		0.10		2	F0/2 V	1771
139871	−49 09909	HR 5832			15 38 59	−49 19.8	1	6.04		1.31				4	K2 III	2009
		G 137 - 35			15 39 00	+13 59.4	1	14.98		1.56		1.09		5		203
139997	−19 04188	HR 5838	⋆	A	15 39 04	−19 31.1	6	4.72	.012	1.57	.009	1.94	.018	29	K5 III	15,1024,1075,1088*
139831	−55 06675				15 39 06	−55 41.0	1	9.87		0.14		-0.25		1	B8 IV	96
139976	−22 04009				15 39 07	−22 52.6	1	8.38		0.26		0.22		1	A1 IV	1770
140228	+41 02622				15 39 09	+41 06.3	1	8.56		0.69		0.23		2	G0	1733
		CS 22884 # 62			15 39 10	−07 59.3	1	15.02		0.15				1		1736
		LSS 3388			15 39 10	−59 30.7	1	12.05		0.47		-0.48		3	B3 Ib	1737
		G 167 - 53			15 39 11	+25 22.2	1	14.13		0.97		0.66		1		1658
140229	+38 02688				15 39 11	+38 26.4	1	8.22		0.42		-0.02		2	F8	1601
139943	−37 10439				15 39 11	−37 25.7	1	9.73		0.23		0.18		1	A1 V	1770
139720	−66 02815				15 39 11	−66 21.7	1	8.18		0.52				4	F6 IV/V	2012
140007	−21 04172				15 39 15	−21 44.4	3	8.16	.010	1.25	.002	1.11	.000	4	K0 III	742,979,8112
139658	−69 02429				15 39 16	−70 08.2	1	8.17		0.66		0.22		3	G3 V	1730
		S068 # 4			15 39 18	−70 03.3	1	11.06		0.26		0.11		2		1730
140159	+20 03138	HR 5842	⋆	AB	15 39 19	+19 49.8	8	4.52	.013	0.05	.009	0.04	.010	32	A1 V +A1 V	1,15,1022,1203,1363*
		Steph 1264			15 39 21	+18 37.7	1	12.31		1.48		1.11		1	M0	1746
		LP 803 - 60			15 39 21	−19 17.9	3	11.86	.012	1.61	.005	1.05		7		158,1705,3073
139980	−37 10441	HR 5837			15 39 23	−37 16.0	1	5.24		0.98				4	G8/K0 III	2035
139961	−44 10333				15 39 25	−44 47.1	4	8.85	.011	0.09	.005	0.16	.004	11	A0 V	158,1075,2017,3077
140160	+13 02982	HR 5843, khi Ser			15 39 26	+13 00.4	7	5.32	.018	0.04	.007	0.05	.004	20	A0 p Sr	15,112,1007,1013,1022*
		GD 188			15 39 27	+25 33.3	1	15.50		0.14		-0.79		1		3060
140122	+0 03389	IDS15369N0047		AB	15 39 28	+00 37.0	2	7.24	.050	0.20	.022	0.16	.005	5	A0	3030,8071
		WLS 1540 25 # 5			15 39 28	+27 18.7	1	13.37		0.84		0.45		2		1375
140008	−34 10494	HR 5839			15 39 29	−34 33.1	6	4.74	.010	-0.14	.006	-0.55	.018	19	B5 V	15,26,1075,1770,2012,8015
140037	−31 12192				15 39 31	−32 01.6	1	7.52		-0.07		-0.51		1	B5 III	1770
140038	−32 11013				15 39 33	−33 02.2	1	10.53		0.30		0.22		1	A3/5 III	1770
	−59 06252				15 39 34	−59 30.1	2	10.28	.001	0.52	.002	0.23	.001	3	B9 IV	540,976
139794	−64 03251				15 39 35	−65 13.9	1	8.31		-0.07		-0.47		6	B5 IV/V	1732
140054	−31 12194				15 39 36	−31 35.8	1	10.04		0.16		0.11		1	A2 III/IV	1770
		GD 189			15 39 37	−03 32.1	1	15.20		0.25		-0.55		1	DA	3060
140055	−32 11016				15 39 37	−32 41.5	1	9.78		0.26		0.17		1	A7 IV	1770
139912	−55 06676				15 39 37	−56 00.2	1	7.35		1.14		0.82		1	G6 III	96
140232	+18 03059	HR 5845			15 39 40	+18 37.3	2	5.81	.000	0.20	.000	0.11	.005	6	A2 m	3058,8071
	−59 06254				15 39 40	−59 42.6	2	10.54	.001	0.27	.005	-0.29	.004	3	B4 IV	540,976
139913	−57 07163				15 39 41	−57 17.6	1	7.23		0.46		0.39		27	A4 II/III	978
		WLS 1540 25 # 7			15 39 42	+22 54.6	1	12.45		0.69		0.13		2		1375
		LSE # 125			15 39 42	−39 10.	1	12.38		-0.14		-1.13		1		1650
		POSS 444 # 3			15 39 43	+17 38.4	1	14.96		1.58				3		1739
		GD 349			15 39 43	+59 48.5	1	15.38		0.38		-0.08		2		308
140056	−36 10355				15 39 44	−36 52.7	1	10.03		0.39		0.30		1	A8/9 V	1770
140075	−34 10497	IDS15366S3423		A	15 39 45	−34 32.2	1	8.74		0.51		0.42		1	A5/7 III	1770
		Lod 2326 # 1			15 39 45	−52 17.6	1	13.15		0.51		0.44		1	A1 V	889
139915	−59 06257	HR 5836			15 39 45	−60 07.8	4	6.49	.010	1.04	.005			18	G0 Ib	15,1075,2013,2030
	−52 08775				15 39 47	−52 18.8	1	12.13		0.32		0.08		1	B9 V	889
141218	+81 00523				15 39 48	+80 46.5	1	6.90		0.97		0.78		2	G5	985
		Lod 2326 # 3			15 39 49	−52 18.8	1	12.52		0.42		0.16		1	B9 V	889
	−52 08779				15 39 50	−52 18.0	1	10.82		1.39		1.46		1	K2 III	889
	−52 08778				15 39 50	−52 19.2	1	11.88		0.71		0.31		1	F2 III	889
	−58 06214				15 39 51	−58 47.5	1	10.68		0.26		-0.23		2	B5 V	540
140164	−15 04165	IDS15372S1541		AB	15 39 57	−15 51.3	1	7.07		0.51		0.03		3	F7 V	938
140105	−30 12478				15 39 59	−30 50.0	1	9.77		0.45		0.15		1	A2 IV/V	1770
140191	−18 04137				15 40 02	−18 50.0	1	10.40		0.32		0.19		1	A8 V	1770
140168	−24 12242				15 40 03	−24 59.5	1	8.96		0.33		0.30		1	A3 V	1770
140127	−34 10499				15 40 05	−35 05.1	1	8.51		0.46		-0.01		2	F5 V	1730
140106	−37 10453				15 40 05	−37 52.4	1	10.19		0.08		0.01		1	B9 V	1770
	−54 06650				15 40 06	−54 28.2	1	10.25		0.56		-0.12		2	B3 IV	540
		He3 1088			15 40 06	−56 56.	1	12.36		1.03		0.04		3		1586
140611	+66 00916				15 40 09	+66 00.1	1	9.45		0.65		0.13		3		3026
	−52 08782				15 40 10	−52 23.7	1	11.30		0.48		0.27		1	A7	820
	−58 06225				15 40 10	−58 21.4	2	10.34	.065	0.31	.015	-0.43	.019	7		540,567
	−52 08783				15 40 11	−52 28.8	1	11.23		0.63		0.08		1	F8	820
140041	−54 06651	T Nor			15 40 12	−54 49.7	1	10.59		1.78		0.56		1	M4/6e	817
140385	+30 02695				15 40 14	+29 47.0	1	8.54		0.70		0.10		2	G2 V	3026
140172	−33 10664				15 40 15	−33 14.9	1	7.45		1.15		0.88		1	K0 III	1770
140487	+49 02417				15 40 18	+49 36.2	1	8.61		0.61		0.10		2	G5	1733
140212	−24 12245				15 40 18	−24 29.8	1	9.42		0.27		0.25		1	A2 III/IV	1770
140043	−56 06872				15 40 18	−56 57.3	1	8.20		0.15		-0.18		1	B8/9 IV	567
140194	−30 12484				15 40 20	−30 22.6	1	7.43		0.45		0.15		1	A2 V	1770
140079	−51 09421				15 40 20	−51 44.8	1	7.23		-0.02		-0.38		3	B8 III	1350
140216	−31 12207				15 40 21	−31 51.0	1	9.57		0.31		0.16		1	A9 V	1770
140195	−33 10666				15 40 21	−33 19.7	1	9.21		0.20		0.12		2	B9.5 V	1770
	−51 09423				15 40 21	−52 07.3	1	10.69		0.15		0.14		1	A2	820
140042	−55 06679				15 40 21	−56 00.5	1	7.06		0.01		-0.44		1	B8/9 II/III	96

HD	DM	Other Id	N Rem	α_{1950}	δ_{1950}	S	V	σ_V	B–V	σ_{B-V}	U–B	σ_{U-B}	n	Spectrum	References
140283	−10 04149			15 40 23	−10 46.3	16	7.21	.013	0.49	.010	-0.20	.015	51	F3 VI	22,158,742,861,908,1003*
	−52 08789			15 40 23	−52 30.7	1	12.04		0.42		0.13		1	F0	820
140238	−27 10521			15 40 25	−27 35.5	1	9.15		0.32		0.06		2	A9 V	1771
		LP 803 - 50		15 40 26	−20 05.0	1	13.11		1.65		1.15		2	M1	3078
	−51 09427			15 40 26	−51 19.4	1	11.51		0.82		-0.21		2	A0	1649
	−52 08790			15 40 26	−52 26.1	1	12.00		0.50		0.02		1	F0	820
	−52 08792			15 40 29	−52 28.6	1	11.82		0.44		0.13		1	A0	820
140217	−34 10504			15 40 30	−34 51.5	1	9.61		0.71		0.61		1	A1/2 IV	1770
140197	−39 10116	IDS15372S3909	A	15 40 30	−39 18.2	1	7.59		0.14		0.10		3	A2 V	1771
140130	−49 09937	IDS15369S4954	A	15 40 31	−50 03.5	1	6.82		0.36		0.06		1	F2 IV/V	96
140109	−51 09429			15 40 31	−51 33.4	1	8.15		-0.05		-0.46		3	B7/8 III	1350
		AJ85,265 # 123		15 40 31	−53 45.3	1	12.87		1.38		0.20		1		962
140198	−41 10236			15 40 33	−41 51.2	1	8.19		0.07		0.04		2	A0 V	1771
140301	−14 04266	HR 5847		15 40 36	−14 53.1	2	6.30	.014	1.13	.018	0.90		5	K0 III	2007,3077
140436	+26 02722	HR 5849, γ Crb	⋆ AB	15 40 38	+26 27.2	7	3.83	.010	0.00	.010	-0.04	.007	32	B9 IV +A3 V	15,245,1022,1203,1363*
140259	−28 11519			15 40 38	−28 26.1	1	8.37		0.32		0.01		2	A9 V	1771
	−49 09940			15 40 38	−50 05.8	1	11.44		0.58		0.04		1		96
	−4 03956			15 40 39	−05 05.6	1	9.98		1.11		0.96		1	K5 V	3072
140766	+66 00917			15 40 47	+66 32.7	1	8.75		1.50		1.84		2	K2	1733
	−51 09434			15 40 47	−51 54.3	1	12.26		0.34		0.12		4		820
	−51 09435			15 40 48	−51 55.3	1	10.54		0.32		0.22		2	A2	820
140175	−52 08796	IDS15370S5246	AB	15 40 48	−52 55.8	1	8.14		0.01		-0.29		3	B8 III/IV	1350
140096	−59 06274			15 40 48	−59 35.6	1	8.77		0.09		-0.27		1	B8 V	567
140438	+14 02922	IDS15385N1359	⋆ AB	15 40 50	+13 49.5	2	6.49	.010	0.87	.007	0.51	.005	4	G5 III	1733,3016
	−51 09436			15 40 52	−51 53.0	1	10.91		0.28		0.17		2	A0	820
140097	−60 06053			15 40 52	−61 08.4	1	7.08		0.40				4	F3/5 IV/V	1075
	−52 08801			15 40 53	−52 18.6	1	12.12		0.36		0.22		1	B9	820
	−52 08799			15 40 53	−52 24.1	1	10.86		0.48		0.12		1	F3	820
	+26 02723	IDS15388N2636	AB	15 40 54	+26 26.4	2	10.72	.019	1.29	.009	1.30		7	K7	1746,7008
140372	−17 04408			15 40 56	−17 47.2	1	9.93		0.32		0.15		1	A3/5 III	1770
140222	−51 09438			15 40 57	−52 12.3	1	9.34		-0.02		-0.31		1	B9 III	820
140285	−41 10245	HR 5846	⋆ AB	15 41 00	−41 39.6	4	5.93	.005	0.00	.007	-0.20	.008	12	A0 V +B	15,1075,1771,2012
140221	−51 09439			15 41 00	−52 03.5	1	9.44		0.03		-0.06		1	B9 V	820
	−57 07185			15 41 01	−57 34.7	1	10.83		0.50		-0.15		3		1586
140150	−57 07184			15 41 01	−57 44.1	1	9.29		0.17		-0.29		3	B8 Ib/II	567
		Steph 1268		15 41 03	+50 36.5	1	12.49		1.43		1.17		1	K7-M0	1746
140330	−34 10510			15 41 04	−35 07.3	1	8.61		1.88				2	M0 III	1730
140349	−28 11523	IDS15380S2831	A	15 41 05	−28 40.5	1	9.42		0.48		0.23		1	A9/F0	1771
140349	−28 11523	IDS15380S2831	AB	15 41 05	−28 40.5	1	9.44		0.44		0.22		1	A9/F0 + G	1770
		POSS 444 # 1		15 41 06	+17 33.9	1	14.05		1.42				1		1739
		LOD 38+ # 4		15 41 06	−50 03.9	1	12.71		0.64		0.04		1		96
		AAS41,43 F5 # 14		15 41 06	−52 21.7	1	13.13		0.67		0.25		1		820
140223	−53 06674			15 41 06	−54 13.2	1	8.79		0.20		-0.45		4	B2/3	567
140387	−16 04151			15 41 07	−16 42.7	1	7.26		1.45		1.68		4	K3 III	1657
140201	−55 06683			15 41 08	−55 37.1	1	9.48		0.29		-0.16		1	A2 IV	96
140287	−44 10356			15 41 10	−44 25.3	1	8.81		0.01		-0.22		2	B9 IV	1770
		CpD -51 08499		15 41 10	−52 01.7	1	11.46		0.34		0.17		1	A0	820
	−52 08807			15 41 10	−52 21.1	1	11.29		1.04		0.71		3	G1	820
140401	−18 04140			15 41 11	−18 47.1	1	9.56		0.67		0.18		1	A9 V	1770
		AAS41,43 F5 # 17		15 41 11	−52 19.6	1	11.66		1.46		1.42		3		820
140202	−55 06684			15 41 11	−55 55.4	1	10.27		0.10		-0.27		1	B8/9 II	96
140331	−38 10623			15 41 14	−38 14.8	1	10.37		0.13		0.08		1	A0 V	1770
	−52 08809			15 41 14	−52 34.4	1	11.13		0.46		0.17		1	F0	820
140417	−15 04171	HR 5848		15 41 15	−15 30.9	3	5.41	.005	0.23	.005	0.07		10	F0 IV	15,2027,8071
	−52 08812			15 41 15	−52 18.7	1	11.95		0.40		0.18		1	A7	820
	−52 08814			15 41 17	−52 17.3	1	10.80		1.27		1.62		1	G8	820
		ApJ139,442 Anon	A	15 41 18	+26 25.	1	10.70		1.28		1.32		1		1064
		ApJ139,442 Anon	B	15 41 18	+26 25.	1	11.68		0.67		0.16		1		1064
		G 239 - 58		15 41 19	+68 34.3	2	14.37	.009	1.53	.039			6		940,1759
140419	−23 12487			15 41 19	−24 14.3	1	7.62		0.26		0.14		2	A2mA2-F0	1771
		CpD -51 08501		15 41 19	−52 09.0	1	11.86		0.86		0.40		1		820
	−51 09441			15 41 20	−51 50.6	1	9.95		0.28		0.18		1	A1	820
	−51 09442			15 41 20	−52 06.8	1	10.42		0.32		0.27		1	A2	820
140440	−15 04173			15 41 21	−15 44.4	1	8.73		0.15		0.14		1	A1 V	1770
	−51 09445			15 41 21	−51 59.4	1	10.79		0.41		0.33		1	B8	820
140289	−52 08816			15 41 22	−52 16.3	1	8.51		1.21		1.11		1	K1 III	820
140288	−51 09446			15 41 23	−52 02.9	1	9.42		0.17		-0.37		1	B3/5 III	820
	−55 06686	LOD 36+ # 8		15 41 23	−55 33.9	1	10.19		0.52		0.15		1		96
	−57 07190	LSS 3390		15 41 24	−57 15.6	1	9.92		0.48		-0.44		1	B0 Ia	540,1737
140047	−69 02441			15 41 24	−69 42.4	1	8.79		0.44		0.03		1	F2 V	742
		CpD -51 08508		15 41 26	−51 50.3	1	11.11		0.31		0.19		1	A4	820
140114	−67 02947			15 41 26	−67 17.3	1	9.55		-0.03		-0.37		1	B8/9 III	400,1586
	−49 09956			15 41 27	−50 05.1	1	10.39		0.44		0.14		1	F2	96
	−51 09447			15 41 28	−52 08.3	1	9.67		1.41		1.47		1	K0	820
140728	+52 01898	HR 5857, BP Boo		15 41 29	+52 31.1	1	5.51		-0.07		-0.10		2	B9 p SiCr	3033
	−52 08819			15 41 29	−52 14.5	1	11.33		0.12		-0.22		2	B8	820
140538	+2 02989	HR 5853	⋆ ABC	15 41 31	+02 40.4	4	5.87	.011	0.69	.011	0.23	.010	14	G5 V	1080,1088,1118,3077
	−51 09448			15 41 31	−52 01.5	1	10.99		0.14		0.12		1	A0	820
		AAS41,43 F5 # 32		15 41 31	−52 06.9	1	12.77		0.70		0.20		1		820
		AAS41,43 F5 # 30		15 41 31	−52 14.2	1	13.22		0.66		-0.03		1		820

Table 1   851

HD	DM	Other Id	N Rem	α_{1950}	δ_{1950}	S	V	σ_V	B–V	σ_{B-V}	U–B	σ_{U-B}	n	Spectrum	References
140536	+8 03073			15 41 32	+07 45.7	1	8.33		0.07		0.13		3	A0	1776
		AAS41,43 F5 # 34		15 41 33	−52 09.2	1	12.74		0.56		0.01		1		820
		AAS41,43 F5 # 35		15 41 33	−52 14.1	1	12.64		0.54		0.16		1		820
	−52 08822			15 41 33	−52 18.1	1	11.25		0.21		0.14		1	A0	820
140557	+13 02993			15 41 34	+12 58.5	1	8.15		1.48		1.83		2	K5	1648
		AAS41,43 F5 # 36		15 41 34	−52 12.1	1	12.76		0.73		0.21		1	F2	820
		AAS41,43 F5 # 38		15 41 36	−52 00.8	1	12.71		0.60		0.38		1		820
	−52 08823			15 41 37	−52 37.9	1	11.81		0.26		0.22		1	A0	820
		AAS41,43 F5 # 39		15 41 38	−52 09.2	1	12.74		0.79		0.53		1	A8	820
140290	−55 06687			15 41 39	−55 51.0	1	8.99		0.33		0.16		1	A7 V	96
140205	−64 03257			15 41 39	−64 44.2	1	9.15		−0.01		−0.25		2	B9 II/III	400
140493	−15 04174			15 41 40	−16 14.7	1	9.59		0.35		0.13		1	A9 IV/V	1770
140442	−30 12498			15 41 40	−30 31.8	1	7.45		0.09		0.08		1	A1 V	1770
		AAS41,43 F5 # 42		15 41 40	−52 15.4	1	12.83		0.71		0.08		1	F8	820
		CpD -51 08515		15 41 41	−51 50.2	1	10.83		1.40		1.08		1		820
		AAS41,43 F5 # 40		15 41 41	−52 18.7	1	12.76		1.14		1.06		1		820
140354	−52 08827			15 41 41	−52 24.6	1	8.31		1.80		2.04		1	K4 III	820
		AAS41,43 F5 # 44		15 41 42	−52 01.6	1	12.17		1.32		0.75		1		820
		AAS41,43 F5 # 43		15 41 42	−52 15.1	1	12.14		0.45		0.38		1	B7	820
140494	−23 12491			15 41 43	−24 03.2	1	9.09		0.47		0.30		1	A8 V	1770
329937	−48 10268			15 41 43	−48 52.7	1	10.88		0.27		−0.23		2	A5	1586
	−51 09452			15 41 44	−51 57.0	1	10.87		0.49		0.05		1	F5	820
140311	−55 06689	IDS15378S5540	AB	15 41 44	−55 49.2	1	8.14		0.10		−0.16		1	B8 V	96
	+13 02995			15 41 45	+13 08.1	1	11.90		0.62		−0.07		2		1696
		AAS41,43 F5 # 47		15 41 45	−52 11.8	1	13.49		0.70		0.46		1		820
140463	−33 10681	IDS15386S3353	AB	15 41 46	−34 02.5	1	9.88		0.78		0.27		20	G6/8 IV/V	1763
		AAS41,43 F5 # 48		15 41 47	−52 16.2	1	12.70		0.53		0.08		1	F5	820
140573	+6 03088	HR 5854	⋆ A	15 41 48	+06 34.9	28	2.64	.011	1.17	.005	1.25	.010	169	K2 IIIb	1,3,15,26,30,198,667*
		POSS 444 # 2		15 41 48	+17 23.4	1	15.27		0.86				1		1739
	−52 08833			15 41 48	−52 28.3	1	10.99		1.14		0.86		1	G3	820
		AAS41,43 F5 # 50		15 41 49	−52 25.1	1	12.51		0.42		0.28		1	B8	820
		AJ85,265 # 126		15 41 49	−54 33.5	1	13.09		1.28		0.24		1		962
		NGC 5986 sq2 8		15 41 50	−37 42.2	1	13.12		0.91		0.38		2		1538
		AAS41,43 F5 # 51		15 41 51	−52 18.5	1	11.88		0.91		0.24		1	G8	820
140475	−34 10514	IDS15387S3447	A	15 41 53	−34 56.7	1	7.71		0.09		0.05		1	A2 V	1770
		AAS41,43 F5 # 53		15 41 54	−52 04.4	1	12.42		0.48		0.42		1	A0	820
140770	+47 02264			15 41 56	+47 03.6	1	8.38		0.05		0.06		2	A0	1601
		AAS41,43 F5 # 54		15 41 56	−52 12.9	1	12.87		0.43		0.16		1	F0	820
		AAS41,43 F5 # 55		15 41 56	−52 20.0	1	12.62		0.58		0.13		1	F0	820
	−52 08839			15 41 57	−52 16.1	1	11.33		0.39		0.14		1	A3	820
	−52 08835			15 41 57	−52 25.1	1	11.49		0.70		0.09		1	G2	820
		NGC 5986 sq2 6		15 41 58	−37 23.7	1	12.61		0.79		0.17		3		1538
	−51 09454			15 41 58	−51 54.3	1	10.49		1.24		1.17		1	K0	820
	−52 08836			15 41 58	−52 36.8	1	10.07		1.88		2.08		1	M0	820
140336	−58 06277	LSS 3392		15 41 58	−58 47.8	3	9.73	.010	0.26	.005	−0.68	.022	11	O/Be	540,976,1586
140497	−33 10684			15 41 59	−33 59.9	2	9.31	.054	0.97	.010	0.68	.027	5	A9/F0 V	119,1770
	−52 08840			15 41 59	−52 26.7	1	11.78		0.69		0.16		1	F2	820
		AAS41,43 F5 # 61		15 42 00	−52 02.6	1	11.63		1.15		0.69		1	F0	820
		AAS41,43 F5 # 60		15 42 00	−52 20.4	1	12.41		0.38		0.38		1	A1	820
140716	+32 02621	HR 5855		15 42 01	+32 40.4	1	5.56		1.06				1	K0	71
140543	−21 04180			15 42 01	−21 39.5	3	8.89	.011	−0.01	.010	−0.89	.002	7	B1 Iab/b(n)	26,1054,1770
	−33 10685	HRC 248, HT Lup		15 42 01	−34 08.1	1	10.30		1.26		0.93		1	K2 V	1763
	−33 10685	HT Lup		15 42 01	−34 08.1	1	10.31	.019	1.26	.009	0.90	.052	3	Ge	776,825,5005
		NGC 5986 sq2 5		15 42 02	−37 26.4	1	11.78		0.48		0.26		4		1538
		NGC 5986 sq2 7		15 42 02	−37 31.0	1	12.87		1.54		1.60		3		1538
		AAS41,43 F5 # 64		15 42 02	−52 04.4	1	13.01		0.49		0.22		1		820
		AAS41,43 F5 # 63		15 42 02	−52 04.8	1	12.42		0.39		0.02		1		820
	−52 08841			15 42 02	−52 26.2	1	11.12		0.43		0.10		1	A5	820
140665	+15 02906	IDS15397N1537	A	15 42 03	+15 27.4	1	8.20		0.65		0.14		4	G0	3016
140665	+15 02906	IDS15397N1537	B	15 42 03	+15 27.4	1	10.51		1.13		1.04		3		3032
		AAS41,43 F5 # 65		15 42 03	−52 05.2	1	13.05		0.82		0.66		1		820
		GD 190		15 42 04	+18 16.1	1	14.72		−0.10		−1.00		1	DB	3060
		NGC 5986 sq1 8		15 42 04	−37 40.9	1	11.76		1.52				2		485
		AAS41,43 F5 # 67		15 42 04	−52 04.0	1	13.20		1.10		0.52		1		820
	−51 09456			15 42 05	−52 03.9	1	11.31		0.65		0.16		2	F5	820
	−37 10474			15 42 06	−37 43.6	2	10.43	.010	1.81	.000	2.10	.035	6		156,1538
	−52 08844			15 42 06	−52 27.8	1	10.56		0.11		0.05		1	B9	820
	−52 08845			15 42 07	−52 20.1	1	11.40		0.22		0.14		1	B9	820
	−51 09458			15 42 08	−52 05.8	1	11.09		1.33		1.40		1	G8	820
		AAS41,43 F5 # 70		15 42 08	−52 10.9	1	12.48		0.29		0.21		1	B9	820
140517	−37 10479	IDS15389S3719	AB	15 42 09	−37 28.5	2	8.59	.000	0.58	.000	0.27	.005	7	F2 III/IV	156,1538
		AAS41,43 F5 # 73		15 42 10	−52 12.2	1	12.62		1.67		1.80		1		820
	−52 08846			15 42 10	−52 37.7	1	11.15		0.53		0.23		1	F3	820
140277	−64 03262			15 42 10	−64 56.8	1	9.87		−0.05		−0.43		2	B5/7 V	400
140700	+16 02822			15 42 12	+16 40.0	1	7.43		1.51		1.86		4	K5 II-III	8100
140447	−52 08847			15 42 12	−52 26.6	1	8.17		0.09		0.04		1	A0 IV	820
		AAS41,43 F5 # 75		15 42 13	−51 52.9	1	12.20		0.49		0.02		1		820
	−51 09459			15 42 13	−52 05.5	1	10.54		0.23		0.19		2	A5	820
		AAS41,43 F5 # 78		15 42 15	−52 12.8	1	12.87		0.50		0.45		1	A1	820
	−44 10370			15 42 17	−44 20.2	2	10.86	.004	0.04	.005	−0.72		6		540,1021

HD	DM	Other Id	N	Rem	α_{1950}	δ_{1950}	S	V	σ_V	B–V	σ_{B-V}	U–B	σ_{U-B}	n	Spectrum	References
	−58 06283				15 42 17	−58 50.5	1	10.71		0.20		−0.31		2	B5 IV	540
		G 257 - 20			15 42 18	+76 09.4	4	12.22	.009	1.66	.016	1.20	.032	7		538,801,906,1705
	−52 08851				15 42 18	−52 39.7	1	11.81		0.64		0.18		1	F8	820
140520	−43 10195				15 42 19	−43 38.0	1	8.48		0.02		−0.43		2	B8 II	1770
	−51 09462	IN Nor			15 42 19	−51 44.6	1	10.02		0.41		−0.21		3		1350
		AAS41,43 F5 # 77			15 42 20	−51 59.9	1	12.89		0.43		0.21		1		820
	−33 10690				15 42 22	−34 02.7	1	9.48		1.59		1.80		21	K2	1763
	−44 10371				15 42 22	−44 22.0	1	10.82		0.56		0.00		2		540
		AAS41,43 F5 # 80			15 42 22	−51 51.7	1	12.26		0.47		0.22		1		820
140449	−56 06885				15 42 25	−57 07.6	1	10.25		0.32		−0.15		3	B8/A0	1586
140729	+17 02906	HR 5858			15 42 26	+17 25.2	1			0.00		−0.03		4	A0 V	1022
140600	−28 11549	IDS15394S2811		A	15 42 27	−28 20.6	1	10.65		0.22		0.18		1	A1 V	1770
140600	−28 11549	IDS15394S2811		B	15 42 27	−28 20.6	1	10.69		0.27		0.20		1	A1	1770
	−51 09465				15 42 27	−52 10.5	1	11.24		0.17		−0.10		1	B8	820
140869	+45 02325				15 42 28	+45 28.3	1	8.30		0.82		0.48		2	G5	1601
		LOD 38+ # 9			15 42 29	−50 03.1	1	12.58		0.37		0.09		1		96
140652	−18 04152				15 42 30	−18 57.1	1	7.56		0.23		0.14		2	A5 IV	1771
		AAS41,43 F5 # 82			15 42 30	−52 21.8	1	12.79		0.42		0.20		1	A2	820
		He3 1092			15 42 30	−66 20.	1	13.52		0.86		−0.65		1		1753
140730	+17 02907				15 42 31	+16 53.6	1	6.99		1.20		1.29		2	K2	1648
140601	−31 12241				15 42 31	−31 48.2	1	9.13		0.05		0.07		1	A0 V	1770
		AAS41,43 F5 # 83			15 42 31	−52 09.0	1	11.66		1.38		1.19		1	G5	820
		WLS 1540 25 # 9			15 42 33	+24 58.7	1	10.94		0.89		0.45		2		1375
		AAS41,43 F5 # 85			15 42 36	−52 05.3	1	11.81		1.60		1.66		1		820
		CS 22884 # 90			15 42 37	−11 45.5	1	13.54		0.20		0.19		1		1736
	−52 08856				15 42 37	−52 17.2	1	11.85		0.58		0.35		1	B9	820
	−52 08857				15 42 38	−52 16.6	1	10.29		0.84		0.44		1	K0	820
		AAS41,43 F5 # 88			15 42 38	−52 19.6	1	12.76		0.58		0.00		1	F0	820
		AAS41,43 F5 # 86			15 42 38	−52 37.2	1	11.36		1.25		1.10		1	G5	820
140408	−62 04857				15 42 38	−62 38.6	1	7.10		0.49				4	F6/7 V	1075
	−37 10486				15 42 39	−37 47.2	1	11.36		0.67		0.26		3		156
		NGC 5986 sq2 4			15 42 40	−37 48.2	1	11.35		0.70		0.14		2		1538
	−51 09466				15 42 40	−51 51.0	1	11.20		1.30		1.19		3		820
		AAS41,43 F5 # 89			15 42 40	−52 00.6	1	13.23		0.49		0.20		1		820
140481	−57 07203				15 42 40	−57 48.4	1	9.97		0.17		−0.18		3	B7/9 III/IV	567
		G 179 - 50			15 42 41	+42 32.6	2	12.63	.009	1.24	.005	1.23		4		419,906
		G 179 - 49			15 42 41	+42 32.8	2	12.94	.014	1.30	.000	1.23		4		419,906
		G 152 - 46			15 42 42	−13 39.8	2	12.32	.027	1.11	.023	1.06		5		158,3062
		AAS41,43 F5 # 91			15 42 43	−52 12.1	1	12.75		0.46		0.16		1	A8	820
		AAS41,43 F5 # 92			15 42 43	−52 18.9	1	12.29		0.39		0.10		1	A7	820
140812	+21 02813				15 42 44	+21 35.9	1	7.49		0.44		−0.05		4	F5 V	1501
140635	−29 11933				15 42 45	−29 42.4	1	9.23		0.29		0.08		2	A9 V	1771
		LOD 38+ # 10			15 42 46	−50 04.6	1	12.88		0.72		0.18		1		96
		CpD -51 08536			15 42 46	−52 08.1	1	11.17		0.78		0.27		1	G0	820
		AAS41,43 F5 # 96			15 42 47	−51 57.7	1	12.53		0.60		−0.04		1		820
		AAS41,43 F5 # 94			15 42 48	−52 11.9	1	12.86		0.61		0.42		1	A5	820
		AAS41,43 F5 # 98			15 42 50	−52 07.0	1	12.29		1.45		1.70		1		820
		CpD -51 08537			15 42 50	−52 11.4	1	12.05		0.56		0.06		1	F5	820
	−52 08859				15 42 50	−52 29.1	1	11.78		0.32		0.28		1	B9	820
		AAS41,43 F5 # 99			15 42 51	−52 06.7	1	12.88		0.44		0.07		1	B9	820
	−52 08860				15 42 53	−52 18.5	1	11.94		0.30		0.12		1	B9	820
	−51 09471				15 42 54	−52 12.4	1	10.68		0.51		0.05		1	F5	820
	−52 08861				15 42 54	−52 18.6	1	12.12		0.33		0.13		1	B9	820
140775	+5 03072	HR 5859			15 42 55	+05 36.2	7	5.57	.007	0.03	.004	0.06	.012	85	A1 V	15,1007,1013,1088*
		POSS 444 # 4			15 42 57	+17 30.4	1	17.35		2.23				1		1739
		AAS41,43 F5 # 103			15 42 57	−52 01.9	1	12.81		0.65		0.52		1		820
		AAS41,43 F5 # 104			15 42 58	−52 01.4	1	12.78		0.59		0.18		1		820
140582	−52 08866	IDS15392S5205	AB		15 42 58	−52 14.2	1	9.84		0.13		0.06		2	B9 V	820
		AAS41,43 F5 # 109			15 42 59	−51 57.2	1	12.74		0.42		−0.02		1		820
		AAS41,43 F5 # 106			15 42 59	−52 14.4	1	11.51		0.36		0.06		2		820
	−52 08865				15 42 59	−52 21.0	1	10.88		0.48		0.04		1	F3	820
		AAS41,43 F5 # 108			15 43 00	−52 18.5	1	12.83		0.72		0.49		3		820
140532	−57 07208				15 43 00	−57 57.6	1	10.31		0.09		−0.36		2	B6/8 Vp Si	540
140605	−51 09474				15 43 02	−51 59.5	2	7.06	.004	−0.03	.004	−0.47		24	B5/7 IV	820,2012
		AAS41,43 F5 # 110			15 43 02	−52 19.5	1	12.52		0.31		0.22		1	A3	820
141039	+53 01806				15 43 03	+53 08.3	1	7.40		0.62		0.09		2	G5	3026
		AAS41,43 F5 # 111			15 43 03	−52 10.5	1	12.33		0.63		0.65		1	A0	820
140913	+28 02469				15 43 04	+28 37.5	1	8.08		0.54		0.09		2	G0 V	3026
		G 16 - 8			15 43 04	−02 06.3	2	11.62	.020	0.71	.015	0.06	.000	5		1620,3062
		AAS41,43 F5 # 114			15 43 05	−52 07.0	1	12.61		1.58		2.20		1		820
		AAS41,43 F5 # 115			15 43 05	−52 10.5	1	11.54		1.34		1.34		1	A0	820
		AAS41,43 F5 # 113			15 43 06	−52 24.7	1	12.01		0.55		0.01		1	F4	820
		vdB 67 # a			15 43 06	−56 53.	1	11.89		0.70		0.33		3		434
140815	+1 03125	HR 5861			15 43 07	+01 02.8	3	6.31	.012	1.19	.005	1.17	.038	9	K0 III	15,252,1256
		CpD -51 08544			15 43 08	−52 08.6	1	12.22		0.44		0.04		1		820
140606	−51 09475				15 43 08	−52 08.6	1	8.19		1.54		1.75		5	K2	820
		AAS41,43 F5 # 116			15 43 08	−52 20.0	1	12.51		0.74		0.32		1	G2	820
		CpD -51 08546			15 43 09	−52 11.1	1	11.26		0.41		0.19		1		820
140620	−51 09476				15 43 09	−52 11.1	1	8.86		−0.02		−0.46		4	B9 II/III	820
		AAS41,43 F5 # 121			15 43 09	−52 16.4	1	12.43		0.62		0.05		1		820

Table 1

HD	DM	Other Id	N	Rem	α_{1950}	δ_{1950}	S	V	σ_V	B–V	σ_{B-V}	U–B	σ_{U-B}	n	Spectrum	References
140722	−27 10550	HR 5856		⋆ AB	15 43 10	−27 54.4	5	6.50	.012	0.33	.007	0.04		19	F0 V	15,1637,2013,2020,2028
		NGC 5986 sq1 11			15 43 10	−37 41.0	1	12.13		0.71				2		485
		CpD -51 08547			15 43 10	−51 54.0	1	11.81		0.45		0.19		1		820
		AAS41,43 F5 # 122			15 43 10	−52 28.9	1	11.89		0.70		0.29		1	F5	820
	−52 08868				15 43 11	−52 33.3	1	11.73		0.41		0.12		1	A2	820
		LP 443 - 58			15 43 12	+15 19.0	1	11.71		1.14		1.14		1		1696
329973	−49 09986				15 43 12	−49 59.7	1	11.17		0.19		-0.07		1	A0	96
140621	−51 09477				15 43 12	−52 09.9	1	8.37		0.38		-0.02		4	F6 V	820
		AAS41,43 F5 # 126			15 43 12	−52 17.1	1	12.56		0.42		0.24		1	A2	820
327413	−45 10194	LSS 3396			15 43 13	−45 54.3	1	10.74		0.19		-0.73		2	O9.5IV	540
		CpD -51 08549			15 43 14	−51 57.8	1	11.08		1.21		1.07		1	F2	820
140755	−20 04316				15 43 15	−20 21.1	1	10.06		0.42		0.22		1	A2/3 Ib/II	1770
		AAS41,43 F5 # 130			15 43 15	−52 03.8	1	12.93		0.53		-0.06		2		820
140622	−51 09478	IDS15395S5203		AB	15 43 15	−52 12.2	1	9.11		0.06		-0.06		4	A0 V	820
		AAS41,43 F5 # 128			15 43 15	−52 23.0	1	11.18		1.25		1.13		1	K0	820
	−51 09480				15 43 17	−51 54.9	1	10.83		0.59		0.12		1	F4	820
140690	−42 10755	IDS15399S4255		AB	15 43 18	−43 04.8	2	8.08	.005	0.66	.000			8	G5 V	1075,2033
140484	−65 03139	IDS15388S6508		⋆ AB	15 43 18	−65 17.3	3	5.53	.004	0.22	.005	0.09	.010	16	A5 III/IV	15,1075,2038
		AAS41,43 F5 # 131			15 43 19	−52 26.3	1	11.17		1.22		0.97		1	G8	820
	−58 06297				15 43 19	−58 35.7	1	10.70		0.17		-0.41		2	B4 IV	540
140693	−46 10356				15 43 20	−46 36.2	1	9.03		-0.01		-0.13		1	B9 V	1770
		LOD 38+ # 13			15 43 20	−50 02.6	1	11.69		1.11		0.92		1		96
		AAS41,43 F5 # 134			15 43 21	−52 15.1	1	12.75		0.58		0.27		1		820
		AAS41,43 F5 # 132			15 43 21	−52 23.6	1	11.59		1.03		0.67		1	G8	820
	−52 08872				15 43 22	−52 16.1	1	10.40		1.09		0.82		1	G8	820
140850	−0 03005			V	15 43 23	−01 17.4	5	8.82	.005	1.67	.007	2.04	.031	74	K2 III	147,989,1509,1729,6005
140734	−30 12525				15 43 23	−30 30.2	1	9.61		0.21		0.19		1	A3/5 IV	1770
	+5 03080	G 16 - 9			15 43 24	+05 11.7	3	9.14	.022	0.84	.009	0.39	.004	7	K2	333,1658,1696,3032
		NGC 5986 sq1 7			15 43 27	−37 43.5	1	11.31		1.35				3		485
	−58 06304				15 43 27	−58 14.4	1	11.29		0.32		-0.32		2	B3 V	540
140678	−52 08874				15 43 28	−52 27.7	1	9.62		0.42		0.08		1	F0 V	820
		AAS41,43 F5 # 137			15 43 29	−52 04.4	1	12.64		0.50		0.14		1	F0	820
		AAS41,43 F5 # 138			15 43 29	−52 10.5	1	12.62		0.64		-0.06		1	F2	820
140873	−1 03092	HR 5863			15 43 30	−01 38.9	7	5.39	.006	-0.03	.004	-0.41	.008	51	B8 III	15,147,1079,1256,1509*
		HRC 249, GW Lup			15 43 30	−34 21.	1	13.54		1.01		-0.45		1	M0	776
140783	−31 12256				15 43 31	−31 45.5	1	9.39		0.25		0.15		1	A7 III	1770
	−37 10496				15 43 31	−37 36.8	2	10.42	.000	1.24	.010	1.02	.035	5		156,1538
		CpD -51 08555			15 43 31	−52 01.5	1	11.61		0.47		0.15		1	F0	820
140784	−34 10524	HR 5860		⋆ AB	15 43 32	−34 31.7	4	5.60	.005	-0.12	.004	-0.42	.011	13	B7 Vn	15,1075,1770,2012
		NGC 5986 sq1 10			15 43 32	−37 41.7	1	11.95		1.37				2		485
		AAS41,43 F5 # 139			15 43 32	−52 11.2	1	13.89		0.23		-0.26		1		820
		AJ85,265 # 128			15 43 32	−53 51.4	1	11.58		1.39		0.24		1		962
140757	−36 10396				15 43 36	−36 46.9	1	7.96		0.14		0.08		1	A2 IV/V	1770
	−52 08878				15 43 36	−52 27.6	1	10.54		0.33		0.17		1	A5	820
		AAS41,43 F5 # 142			15 43 38	−52 06.9	1	12.57		0.49		0.12		1	F0	820
141173	+58 01591				15 43 39	+58 35.3	1	7.38		0.36		-0.02		2	F0	1502
140758	−40 09921				15 43 40	−41 03.6	1	7.82		0.20		0.20		1	A2 V	1771
		AAS41,43 F5 # 144			15 43 40	−52 12.4	1	12.49		0.53		0.06		1		820
		AAS41,43 F5 # 143			15 43 40	−52 21.8	1	12.50		0.80		0.37		1	G0	820
		AAS41,43 F5 # 145			15 43 41	−52 11.2	1	12.12		0.57		0.17		1	F0	820
		CpD -51 08561			15 43 44	−52 06.9	1	12.14		0.41		0.21		1	B9	820
140680	−55 06703				15 43 44	−55 55.2	1	9.74		0.09		-0.21		5	Ap Si	567
	−51 09486				15 43 45	−51 58.9	1	10.67		2.08		2.00		2		820
	−51 09485				15 43 45	−52 10.6	1	10.51		0.54		0.06		3	F0	820
		AAS41,43 F5 # 147			15 43 45	−52 14.5	1	12.48		1.07		0.77		1		820
		GD 350			15 43 46	+60 26.2	1	12.74		0.47		-0.12		2		308
140681	−57 07217				15 43 46	−57 39.2	2	8.95	.029	0.65	.000	0.17		3	F6 IV/V	742,1594
		AAS41,43 F5 # 149			15 43 47	−52 20.9	1	12.24		0.49		0.21		1	B8	820
		LOD 38+ # 16			15 43 48	−50 03.3	1	13.28		0.71		0.37		1		96
140855	−24 12286				15 43 49	−25 05.3	1	8.90		0.38		0.22		1	A5 IV	1770
140817	−35 10503	IDS15407S3512		AB	15 43 51	−35 21.3	1	6.83		0.03		-0.10		2	A0 V	1770
		GD 351			15 43 52	+62 52.7	1	14.82		-0.22		-1.23		2		308
140840	−35 10504	IDS15407S3512		C	15 43 52	−35 21.8	1	7.35		0.01		-0.04		2	B9/A0 V	1770
	−51 09488				15 43 52	−51 58.6	1	10.78		0.13		-0.26		1	B6	820
141003	+15 02911	HR 5867		⋆ A	15 43 53	+15 34.6	16	3.67	.010	0.06	.008	0.09	.019	74	A3 V	1,3,15,30,542,1006*
141003	+15 02911	IDS15416N1544		B	15 43 53	+15 34.6	3	9.96	.005	1.00	.005	0.79	.015	8	K3 V	1,542,1028,3024
140818	−37 10498				15 43 53	−37 36.4	2	10.68	.040	0.32	.024	0.21	.024	5	A0 V	156,1770
		AAS41,43 F5 # 154			15 43 53	−52 01.8	1	12.48		0.37		0.22		2		820
140741	−51 09491				15 43 54	−52 02.4	1	9.94		-0.02		-0.31		1	B9 II/III	820
	−52 08884				15 43 55	−52 18.7	1	12.40		0.37		0.21		1	B9	820
	−56 06894				15 43 55	−57 07.9	1	10.50		0.40		-0.08		5		567
	−51 09492				15 43 56	−51 53.8	1	11.15		0.26		0.20		3	A0	820
	−52 08883				15 43 56	−52 36.1	1	11.77		0.44		0.20		1	B9	820
	−52 08885				15 43 58	−52 17.5	1	9.66		1.40		1.50		1	K0	820
		Ton 803			15 44 00	+25 18.	1	14.11		-0.25		-1.21		3		1036
141004	+7 03023	HR 5868			15 44 01	+07 30.5	24	4.43	.008	0.60	.007	0.10	.009	109	G0 V	1,3,15,101,116,125*
		LOD 38+ # 17			15 44 01	−50 03.1	1	13.28		0.67		-0.09		1		96
140936	−16 04158				15 44 03	−16 26.1	1	10.07		0.29		0.22		1	A3/5 II	1770
140937	−18 04163	IDS15412S1849		AB	15 44 04	−18 59.5	1	8.71		0.34		0.12		1	A5 IV	1771
140875	−30 12535				15 44 04	−30 38.6	1	9.43		0.21		0.16		1	A2 IV	1770

HD	DM	Other Id	N	Rem	α_{1950}	δ_{1950}	S	V	σ_V	B–V	σ_{B-V}	U–B	σ_{U-B}	n	Spectrum	References
140861	−39 10157	HR 5862			15 44 05	−40 02.4	1	6.42		0.88				4	G8III/IV(p)	2032
140986	−5 04161	HR 5866			15 44 06	−05 57.9	1	6.23		1.17		1.09		8	K0	1088
141335	+63 01222	G 224 - 81			15 44 08	+63 00.8	2	8.94	.014	0.58	.005	−0.04	.014	4	G5	308,1658
141025	+1 03129	IDS15417N0103	A		15 44 11	+00 54.7	1	9.78		0.53		−0.02		2	G0	3028
141025	+1 03129	IDS15417N0103	B		15 44 11	+00 54.7	1	15.27		−0.32		−1.19		2	DAwk	3028
	−51 09494				15 44 11	−52 04.9	1	10.74		0.30		0.21		1	A0	820
141142	+39 02911				15 44 13	+39 15.9	2	8.76	.005	0.38	.000	−0.04	.000	6	F0 V	1501,1601
	−4 03974				15 44 13	−05 06.0	1	10.23		0.95		0.64		1	K2 V	3072
140955	−21 04187				15 44 14	−21 22.7	1	9.78		0.35		0.30		1	A2/3 (IV)	1770
140901	−37 10500	HR 5864	⋆ A		15 44 14	−37 45.6	5	6.01	.012	0.72	.004	0.30	.005	14	G6 IV	15,156,2012,2024,3077
140901	−37 10500	IDS15410S3736	B		15 44 14	−37 45.6	2	12.82	.024	0.28	.024	−0.45	.068	3	DAs	832,3060
	−51 09495				15 44 16	−51 55.1	1	10.00		0.22		0.22		3	A0	820
329905	−48 10320	LSS 3399			15 44 18	−48 28.5	4	10.43	.032	0.64	.010	−0.36	.045	9	O9 Ia	403,540,1021,1737
	−52 08894				15 44 20	−52 28.2	1	11.06		0.24		0.17		1	A2	820
		G 179 - 54			15 44 21	+39 23.4	2	13.51	.005	0.62	.005	−0.12	.020	2		1658,3060
140968	−25 11095				15 44 22	−25 42.8	1	8.20		0.18		0.10		2	A2 V	1771
140921	−36 10409				15 44 22	−36 57.0	1	10.00		0.52		0.33		1	A2/5(m)	1770
	−52 08897				15 44 23	−52 33.1	1	10.66		1.25		0.96		1	G8	820
141006	−17 04421				15 44 28	−17 56.2	1	8.26		0.23		0.16		2	A3 IV	1771
329976	−49 10004				15 44 28	−50 02.9	1	10.60		0.36		0.01		1	A2	96
	−51 09498				15 44 28	−52 10.8	1	11.15		0.49		0.31		1	A3	820
141007	−19 04215				15 44 29	−19 47.1	1	10.58		0.30		0.18		1	A0 III	1770
		G 137 - 49			15 44 32	+09 33.1	1	12.67		0.92		0.66		1		1696
141186	+36 02645	IDS15427N3645	A		15 44 32	+36 36.0	1	7.57		0.48		0.05		2	F5	1601
	−6 04279				15 44 33	−06 21.8	1	10.48		0.85		0.49		1	K3 V	3072
140992	−27 10560				15 44 33	−28 07.2	1	10.40		0.31		0.14		1	A7/8 II	1770
140958	−37 10502				15 44 33	−38 06.3	1	8.06		0.21		0.08		1	A6 V	1770
140882	−49 10007				15 44 33	−49 35.4	1	7.45		1.58				4	K3/4 III	2012
		LOD 38+ # 20			15 44 34	−49 59.7	1	13.22		0.42		0.21		1		96
		CS 22884 # 112			15 44 35	−08 39.7	1	14.49		0.39				1		1736
	−52 08902				15 44 36	−52 31.5	1	10.46		1.31		1.41		1	K0	820
		AAS35,161 # 72	V		15 44 37	−35 19.4	1	14.79		1.12		−0.53		1		5005
		CS 22884 # 97			15 44 38	−10 57.8	1	14.86		0.60		−0.07		1		1736
140843	−56 06903				15 44 39	−56 57.0	1	8.52		0.19		−0.28		1	B5 III/IV	567
141072	−2 04044				15 44 40	−02 27.8	1	8.23		1.40		1.55		1	K3 III	1746
	−52 08905				15 44 43	−52 25.3	2	10.51	.009	0.48	.004	−0.06	.017	6	B7	567,820
140826	−59 06332				15 44 45	−60 12.9	1	6.64		0.42				4	F5 V	1075
	−53 06701	LSS 3401			15 44 49	−54 05.4	3	10.40	.037	0.70	.011	−0.32	.009	7	B0 V	540,962,1350
	+15 02915	G 137 - 52			15 44 51	+15 20.2	1	9.61		0.99		0.72		1	K0	1696
		LP 916 - 28			15 44 51	−27 43.9	2	12.17	.030	0.83	.018	0.12	.013	4		1696,3062
		AAS35,161 # 74	V		15 44 52	−35 06.7	1	14.18		1.55		0.22		1		5005
140976	−43 10241				15 44 52	−43 18.3	1	8.35		0.02		−0.11		2	B9/9.5 V	1770
	−51 09506				15 44 53	−52 10.8	1	10.61		0.25		0.16		1	B9	820
141063	−25 11098				15 44 55	−25 50.0	1	7.00		0.17		0.13		2	A1 IV	1771
140926	−54 06687	LSS 3402			15 44 56	−54 14.6	5	7.90	.015	−0.02	.006	−0.75	.010	9	B2/3 Vnne	540,567,962,976,1586
141187	+14 02939	HR 5870			15 44 58	+14 16.1	2	5.70	.005	0.09	.005	0.07	.015	5	A3 V	1022,3050
141372	+52 01903				15 45 03	+52 18.1	1	8.02		1.07		0.97		2	K0	1566
141091	−24 12296				15 45 03	−25 03.7	1	7.16		0.14		0.02		2	A0 V	1771
140979	−52 08912	HR 5865	⋆ AB		15 45 04	−52 17.1	4	6.05	.010	1.38	.005	1.45		12	K2/3 III	15,820,2013,2030
141051	−39 10168				15 45 05	−39 21.8	1	10.62		0.39		0.27		1	A1/2 (IV)	1770
	−51 09508				15 45 06	−52 06.2	1	9.96		0.73		0.29		1	G8	820
		CS 22884 # 111			15 45 07	−08 36.6	1	15.16		0.22				1		1736
	−51 09511				15 45 08	−52 07.0	1	11.79		0.50		0.10		1	F3	820
	−51 09509				15 45 08	−52 12.8	1	11.33		0.60		0.14		1	B8	820
140945	−57 07242				15 45 10	−57 31.2	1	9.76		0.53		0.05		1	F2 IV	1577
141132	−26 11043				15 45 17	−26 21.2	1	8.84		0.32		0.28		1	A2 III/IV	1770
	−52 08917				15 45 17	−52 54.3	1	10.36		0.45		−0.15		5		567
141107	−28 11585				15 45 18	−28 38.2	2	7.69	.010	0.42	.010	−0.09		3	F2 V	742,1594
		ApJS57,743 T5# 3			15 45 18	−57 32.5	1	12.76		0.61		0.14		1		1577
141079	−36 10421				15 45 19	−37 10.2	1	9.37		0.39		0.19		1	A8 III	1770
141399	+47 02267				15 45 20	+47 08.4	1	7.21		0.77		0.42		2	K0	1375
		AA Nor			15 45 20	−57 31.0	1	13.24		1.03		0.93		1		689
141108	−32 11130				15 45 22	−32 48.7	1	9.82		0.05		−0.10		1	B9.5 V	1770
140946	−59 06345				15 45 22	−59 59.9	2	9.27	.003	0.15	.007	−0.62	.007	7	B2/3 Vnne	540,976,1586
141472	+55 01777	HR 5878			15 45 23	+55 37.6	1	5.92		1.39				1	gK3	71
		ApJS57,743 T5# 2			15 45 24	−57 31.9	1	11.70		1.29		1.21		1		1577
		ApJS57,743 T5# 6			15 45 25	−57 30.1	1	12.98		0.41		0.15		1		1577
141164	−23 12525				15 45 28	−23 40.9	1	6.72		0.10		0.06		1	A2 IV/V	1771
141112	−36 10424				15 45 28	−36 56.3	1	10.89		0.29		0.23		1	A2 III/IV	1770
141054	−51 09521				15 45 28	−52 03.8	1	7.35		1.70		1.85		8	M2/3 III	1673
		ApJS57,743 T5# 7			15 45 28	−57 29.8	1	13.40		0.40		0.27		1		1577
		G 224 - 83			15 45 29	+62 35.9	1	12.76		0.60		−0.17		2		1658
141017	−56 06912				15 45 34	−56 44.0	1	9.49		0.16		0.06		1	B9.5III	567
		ApJS57,743 T5# 4			15 45 35	−57 32.9	1	12.29		0.34		0.27		1		1577
	−53 06706	LSS 3404			15 45 36	−54 08.4	2	10.85	.055	0.66	.020	−0.25	.015	4		962,1350
		ApJS57,743 T5# 10			15 45 36	−57 30.2	1	13.41		0.42		0.26		1		1577
141352	+28 02475				15 45 38	+28 37.3	1	7.48		0.44		0.00		4	F6 V	8089
141135	−39 10174				15 45 38	−39 14.6	1	10.10		0.61		0.35		1	A5/7 (IV)	1770
141135	−39 10174				15 45 38	−39 14.6	1	10.78		0.71		0.26		1	A5/7 (IV)	1770
	−52 08925				15 45 38	−53 08.6	1	9.15		1.32		1.14		1		96

Table 1 855

HD	DM	Other Id	N	Rem	α_{1950}	δ_{1950}	S	V	σ_V	B-V	σ_{B-V}	U-B	σ_{U-B}	n	Spectrum	References
141067	-52 08927				15 45 39	-52 44.3	1	8.00		0.12		0.08		1	A0 V	96
	-1 03099				15 45 40	-01 37.1	1	10.33		1.65		2.03		2	M1	1746
141180	-27 10575				15 45 40	-27 25.8	2	8.27	.006	-0.03	.012	-0.25	.007	3	B9 III	1054,1770
141083	-52 08928				15 45 45	-52 39.6	1	8.23		-0.02		-0.33		1	B6 III/IV	96
141190	-29 11976				15 45 46	-29 19.9	1	7.98		0.26		0.07		1	A7 IV	1770
	-52 08930				15 45 47	-52 43.4	1	10.46		0.41		0.24		1		96
142105	+78 00527	HR 5903			15 45 48	+77 57.0	4	4.31	.010	0.04	.005	0.06	.016	16	A3 Vn	15,1363,3023,8015
141249	-17 04428				15 45 48	-18 15.4	1	10.19		0.38		0.19		1	Ap SrEu(Cr)	1770
141166	-41 10325				15 45 51	-41 38.4	2	7.40	.010	-0.01	.003	-0.31	.005	3	B8 III	401,1770
	-53 06708				15 45 51	-53 16.4	1	10.14		0.54		0.13		1		96
141353	+14 02940	HR 5874			15 45 53	+13 56.6	2	6.00	.000	1.26	.011	1.49		3	gK2	71,3077
	-57 07255				15 45 53	-57 30.8	1	7.86		2.82		0.02		1		1577
141653	+63 01225	HR 5886			15 45 54	+62 45.2	2	5.19	.000	0.05	.008	0.05		4	A2 IV	1363,3050
	-53 06709				15 45 54	-53 21.8	1	11.36		0.28				1		96
		LOD 37 # 7			15 45 54	-53 22.3	1	12.02		1.83				1		96
		LOD 37 # 4			15 45 55	-53 21.4	1	11.65		2.57				1		96
	-35 10525	GQ Lup			15 45 56	-35 30.0	3	11.72	.158	0.96	.050	0.03	.086	3	K7 V	776,825,5005
	-53 06710				15 45 56	-53 22.9	1	10.86		0.26		0.00		1		96
	-35 10525	HRC 250, GQ Lup			15 45 58	-35 29.9	1	11.50		1.18		0.21		1	K7 V	1763
	-61 05295	LSS 3403			15 45 58	-62 12.2	1	11.28		0.15		-0.75		2		1737
141211	-35 10526				15 45 60	-35 52.6	1	9.34		0.32		0.17		1	A0 V	1770
141192	-38 10677				15 46 00	-38 54.8	1	10.17		0.44		0.32		1	A0 IV	1770
141456	+32 02631	HR 5877			15 46 03	+31 53.3	1	6.39		1.51		1.91		2	K5	1733
141253	-29 11984				15 46 07	-30 03.4	1	8.21		0.21		0.07		1	A5 V	1770
141212	-37 10520				15 46 08	-37 41.9	1	9.31		0.16		0.07		1	A0 V	1770
	-52 08940				15 46 10	-52 42.7	1	8.99		1.15		0.86		1	G5	96
	-52 08939				15 46 10	-52 57.2	1	10.85		0.62		0.08		1		96
		LOD 37 # 13			15 46 10	-52 57.7	1	12.17		1.40				1		96
		AJ85,265 # 131			15 46 10	-53 41.9	1	11.69		1.15		0.06		1		962
142123	+77 00609				15 46 11	+77 18.2	1	7.98		0.99		0.69		2	G5	1375
	-52 08942				15 46 15	-52 46.9	1	10.69		0.56		0.08		1		96
141324	-15 04194	SS Lib			15 46 16	-15 23.0	1	10.44		0.37		0.20		1	A8/9 II	1770
		LOD 37 # 10			15 46 17	-52 58.8	1	12.65		0.57				1		96
		AJ77,733 T25# 8			15 46 17	-56 05.4	1	12.49		0.52		0.06		5		125
		LP 503 - 53			15 46 19	+13 06.2	1	12.94		1.09		0.98		1		1696
141378	-3 03829	HR 5875			15 46 19	-03 40.0	4	5.54	.019	0.12	.008	0.10	.013	12	A5 IV	15,252,1417,8071
		LOD 37 # 11			15 46 19	-52 59.0	1	12.27		0.60				1		96
141168	-52 08944	HR 5869		*A	15 46 19	-53 03.5	7	5.77	.011	-0.08	.006	-0.23	.008	24	B9 V(n)	15,26,96,567,1075*
141194	-48 10349	HR 5871			15 46 20	-48 45.6	3	5.83	.004	0.07	.000	0.05	.005	11	A1 V	15,1075,2012
141675	+55 01779	HR 5887		*AB	15 46 25	+55 31.8	2	5.86	.000	0.25	.000	0.16	.000	6	A3 m	3058,8071
141152	-57 07264				15 46 25	-57 30.4	1	8.72		0.08		-0.27		1	B7 III	567
	-52 08950				15 46 26	-52 40.0	1	8.86		1.82		1.94		1		96
141199	-52 08951				15 46 27	-52 39.1	1	9.86		0.09		-0.28		1	A0	96
	-54 06706	LSS 3405			15 46 28	-54 17.6	3	10.72	.041	0.66	.013	-0.28	.010	6	B0.5IV	540,962,1350
141477	+18 03074	HR 5879			15 46 29	+18 17.7	7	4.10	.008	1.61	.012	1.98	.023	28	M0.5IIIab	3,15,369,814,1363*
141458	+13 03012				15 46 30	+12 52.6	1	6.81		0.03		-0.01		2	A0	401
141527	+28 02477	HR 5880, R CrB			15 46 31	+28 18.5	5	5.90	.290	0.60	.032	0.17	.096	8	G0 Iep	817,1481,1549,1565,8089
	-59 06361				15 46 31	-59 30.5	1	10.68		0.59		0.28		2	B9 III	540
141380	-16 04162				15 46 32	-17 14.9	1	10.94		0.29		0.17		1	A7 II	1770
141337	-29 11989				15 46 33	-29 16.4	1	9.09		0.31		0.20		1	A8/9 III	1770
141327	-32 11150				15 46 33	-32 39.4	1	7.48		0.01		-0.10		2	B9 V	1770
141729	+60 01635				15 46 34	+59 43.3	2	6.53	.005	1.39	.005	1.61	.005	5	K3 III	1501,1502
141381	-19 04229				15 46 34	-19 28.0	1	10.72		0.30		0.23		1	A0/1 III	1770
141423	-2 04051	IDS15440S0256	A		15 46 35	-03 04.3	1	8.23		0.45		0.21		2	A0	1776
141403	-20 04331				15 46 40	-20 18.8	1	9.06		0.17		0.12		1	A0 V	1770
141529	+23 02862				15 46 44	+23 23.6	1	7.83		0.49		-0.01		3	F8	1625
141404	-20 04332				15 46 45	-20 37.6	3	7.71	.005	0.13	.004	0.05	.014	8	A0 III/IV	26,1054,1770
141405	-25 11116				15 46 45	-25 35.1	1	10.57		0.33		0.17		1	B8/A1 V	1770
141296	-45 10251	HR 5872		*ABC	15 46 46	-45 15.0	4	6.10	.007	0.30	.010			15	A9 III	15,2013,2020,2028
	-57 07268	LSS 3407			15 46 47	-57 33.0	1	10.68		0.27		-0.46		2	B2 V	540
141384	-33 10743				15 46 51	-33 58.5	1	9.60		0.22		0.17		1	A0 V	1770
141280	-52 08965				15 46 54	-52 51.1	1	9.88		0.18		0.09		1	A3 III	96
	-45 10255				15 46 56	-46 06.8	1	10.98		0.57		0.23		3	B9 III	540
		G 16 - 10			15 46 58	+07 53.4	1	14.87		1.45		1.26		4		316
	-55 06727				15 46 58	-56 10.9	1	10.27		1.16		0.86		4		125
141677	+42 02641				15 46 59	+42 28.9	1	8.81		0.85		0.45		2	G5	1375
141441	-24 12314				15 46 59	-24 60.0	1	9.25		0.47		0.27		1	A2/3 V	1770
141171	-63 03697				15 46 59	-64 01.9	1	7.74		0.45				4	F5 V	1075
141513	-2 04052	HR 5881			15 47 00	-03 16.7	7	3.54	.007	-0.04	.005	-0.09	.037	30	A0 V	3,15,1075,1088,1203*
141442	-25 11122				15 47 01	-25 15.5	1	8.74		0.19		0.14		1	A0 V	1770
141341	-44 10434				15 47 02	-44 33.0	1	8.84		0.03		-0.25		2	B8 V	1770
		MCC 740			15 47 03	+43 04.2	1	11.17		1.42				1	K5	1017
141406	-36 10447				15 47 03	-36 50.0	1	9.94		0.16		0.05		1	B9 V	1770
		LP 443 - 65			15 47 05	+14 47.5	1	12.31		0.43		-0.18		2		1696
141467	-24 12315				15 47 06	-24 58.0	1	9.23		0.47		0.29		1	A3 III	1770
141444	-28 11605				15 47 06	-28 33.2	1	8.96		0.10		0.09		1	B9.5 V	1770
141445	-29 11999				15 47 09	-29 16.3	1	9.69		0.41		0.25		1	A0mF0-F0	1770
141318	-54 06711	HR 5873		*A	15 47 13	-54 54.3	3	5.77	.013	0.01	.016	-0.72	.031	13	B2 III	26,567,8100
141318	-54 06711	IDS15433S5445		*AB	15 47 13	-54 54.3	2	5.72	.005	0.06	.000			7	B2 III	15,2012
		CS 22884 # 108			15 47 14	-09 05.1	1	14.24		0.50		-0.13		1		1736

HD	DM	Other Id	N Rem	α_{1950}	δ_{1950}	S	V	σ_V	B–V	σ_{B-V}	U–B	σ_{U-B}	n	Spectrum	References
141479	−26 11056			15 47 14	−27 12.6	1	10.20		0.26		0.19		1	A7/8 IV	1770
141480	−29 12003			15 47 18	−29 15.7	1	9.02		0.20		0.22		1	A3 III/IV	1770
141446	−40 09978			15 47 18	−40 22.8	1	10.11		0.48		0.23		1	A2	1770
141569	−3 03833	IDS15447S0337	ABC	15 47 20	−03 46.2	1	7.18		0.08		0.02		4	B9	1776
		GD 191		15 47 21	−12 41.2	1	13.11		0.49		-0.05		1		3060
	−50 09903			15 47 21	−50 32.1	2	10.71	.023	1.19	.003	1.13		5		158,1705
141516	−21 04199			15 47 22	−22 06.9	1	10.19		0.30		0.24		1	A0 IV	1770
		AJ77,733 T25# 7		15 47 22	−56 07.0	1	12.40		0.57		0.14		3		125
	+46 02115	G 179 - 56		15 47 24	+45 56.6	2	9.20	.009	0.65	.005	0.12	.014	4	G5	308,1658
	−55 06728			15 47 25	−56 04.3	1	11.49		0.47		0.10		4		125
	−55 06729			15 47 25	−56 06.5	1	10.34		0.52		0.11		4		125
141690	+25 02973	IDS15453N2546	A	15 47 26	+25 36.7	2	8.68	.015	0.63	.015	0.17	.025	6	G0 IV	833,3026
141690	+25 02973	IDS15453N2546	B	15 47 26	+25 36.7	1	10.66		0.81		0.46		3		3026
141482	−39 10194			15 47 26	−39 42.3	1	8.83		0.58		0.43		1	A3 V	1770
141714	+26 02737	HR 5889, δ Crb	★	15 47 30	+26 13.2	5	4.62	.011	0.80	.008	0.36	.020	13	G3.5 III-IV	15,1363,3026,4001,8015
141575	−20 04335			15 47 35	−20 41.8	1	8.19		0.27		0.20		1	A2 V	1770
141553	−23 12544			15 47 35	−23 59.8	1	10.55		0.30		0.21		1	A1/2 III/I	1770
		G 16 - 12		15 47 36	−03 30.7	1	13.60		1.38		1.04		1		1696
141595				15 47 36	−17 34.	1	10.63		0.42		0.17		1	A8/F2(m)	1770
141498	−38 10694			15 47 36	−39 11.0	1	9.91		0.52		0.36		1	A5 II/III	1770
141614	−13 04262			15 47 40	−14 15.1	1	10.44		0.37		0.27		1	A1/2 II	1770
141519	−37 10537			15 47 40	−37 42.6	1	9.40		0.15		0.09		1	A1 IV	1770
141769	+27 02545			15 47 43	+27 27.4	1	9.06		0.42		0.03		2	F8	1375
		Cn1-1		15 47 43	−48 36.	1	10.93		0.91		0.48		1		1753
141826	+40 02929	V CrB		15 47 44	+39 43.4	3	9.27	.033	4.41	.000	7.44	.607	3	N2	8005,8022,8027
141988	+62 01431			15 47 44	+62 29.7	1	8.54		0.33		0.09		3	F1 IV-V	8071
141680	+2 03007	HR 5888		15 47 46	+02 20.9	8	5.22	.009	1.02	.004	0.81	.007	49	G8 III	15,1007,1013,1080*
		G 179 - 57		15 47 46	+34 57.5	1	13.26		1.62				1		1759
141556	−33 10754	HR 5883		15 47 46	−33 28.6	7	3.95	.008	-0.05	.006	-0.13	.009	27	B9.5III-IV	15,26,1075,1770,2012*
141657	−2 04055	IDS15452S0245	A	15 47 47	−02 54.3	1	8.51		0.69		0.22		1	F2	1776
141484	−47 10376			15 47 47	−48 06.7	1	8.89		0.18				4	A1/2 IV	2012
141576	−29 12011			15 47 48	−29 21.2	1	9.09		0.13		0.09		1	A0 V	1770
	−55 06732			15 47 48	−55 55.8	1	10.82		0.68		0.17		7		125
143802	+85 00269			15 47 49	+85 26.5	1	6.89		0.20		0.12		2	A5	1733
141637	−25 11131	HR 5885		15 47 58	−25 36.0	10	4.64	.019	-0.05	.010	-0.73	.008	30	B1.5Vn	15,26,1020,1054,1075*
141578	−36 10465			15 47 58	−36 59.9	1	9.79		0.39		0.25		1	A9 V	1770
141544	−46 10430	HR 5882		15 47 58	−46 54.6	4	6.00	.005	1.15	.010	1.12		12	K2 III	15,1075,2006,3077
141559	−41 10369			15 48 02	−42 09.0	1	8.27		0.07		-0.12		3	B9 V	1770
141486	−55 06735			15 48 05	−56 12.1	1	9.19		0.10		0.00		9	B9 IV	125
141641	−31 12337			15 48 06	−31 21.4	1	9.03		-0.07		-0.54		1	Ap Si	1770
141989	+56 01833			15 48 07	+56 23.3	1	7.29		1.28		1.44		3	K2 III	1501
141639	−29 12016			15 48 08	−30 09.7	1	10.48		0.23		0.18		1	A5/7 II/II	1770
	−55 06737			15 48 08	−55 51.2	1	11.37		0.48		0.27		6		125
141695	−19 04239			15 48 10	−19 51.5	1	9.43		0.32		0.17		1	A6 II/III	1770
	−48 10380	LSS 3412		15 48 11	−48 58.4	2	10.30	.001	1.11	.039	0.02	.013	4	B2.5 Ia	540,1737
141696	−20 04338			15 48 12	−20 49.7	1	9.86		0.36		0.21		1	A3/5 III	1770
		G 202 - 16		15 48 13	+51 11.8	1	12.05		1.53				1	M3:	1746
		L 338 - 92		15 48 18	−47 55.	1	14.10		0.80		0.25		2		3062
141795	+4 03069			15 48 19	+04 37.6	7	3.71	.009	0.15	.005	0.13	.031	36	A2 m	3,15,355,1088,1363*
		Nor sq 3 # 314		15 48 20	−54 32.2	1	12.47		1.68		0.48		2		336
		LSS 3410		15 48 21	−54 53.4	1	11.73		0.97		0.12		1	B3 Ib	962
141413	−64 03286	HR 5876		15 48 21	−65 00.2	3	6.53	.004	0.18	.005	0.11	.010	17	A5 IV	15,1075,2038
	−61 05327	LSS 3408		15 48 22	−61 49.4	2	10.19	.009	0.11	.003	-0.51	.008	7	B3 IV	540,976
141850	+15 02918	HR 5894, R Ser		15 48 23	+15 17.0	1	7.10		1.39		0.39		1	M4	3001
142124	+63 01228			15 48 24	+63 17.6	1	8.20		0.35		0.03		3	F0	1733
		AJ77,733 T26# 10		15 48 24	−54 30.0	2	11.81	.005	0.91	.000	0.05	.009	8		125,336
	−54 06722			15 48 25	−55 12.3	1	10.51		0.45		-0.08		2	B6 IV	540
141666	−39 10197			15 48 26	−39 25.3	1	9.16		0.33		-0.16		1	B6 II/III	1770
143173	+83 00453			15 48 28	+83 06.1	1	7.47		0.20		0.14		3	A2	985
		VY Lib		15 48 29	−15 36.7	1	11.11		0.30		0.22		1	F2:	700
		He3 1108		15 48 30	−54 00.	1	10.67		0.11		-0.25		3		1586
141772	−18 04180			15 48 32	−18 56.7	1	10.29		0.41		0.21		1	A8/9 V	1770
		AAS35,161 # 77		15 48 32	−35 47.8	1	12.46		1.30		0.73		1		5005
141739	−31 12343			15 48 33	−31 48.9	1	9.72		0.24		0.16		1	A5 III/IV	1770
	+8 03095	G 16 - 13		15 48 34	+08 34.5	3	10.00	.016	0.59	.000	-0.04	.008	6	G2	516,1620,1696
		AJ85,265 # 134		15 48 34	−53 59.8	1	11.02		1.68		0.51		1		962
141774	−20 04343			15 48 35	−20 26.3	3	7.70	.004	0.09	.003	-0.10	.013	9	B9 V	26,1054,1770
141522	−59 06394	LSS 3409		15 48 35	−59 57.3	3	9.78	.007	0.23	.004	-0.65	.029	8	B0 Ia/ab	403,540,976
141059	−77 01170			15 48 36	−77 35.1	1	7.35		0.99		0.75		5	K0 III	1704
141773	−19 04241			15 48 37	−19 31.5	1	10.44		0.32		0.19		1	A3/5 II	1770
	−54 06725			15 48 38	−54 17.0	1	11.07		0.50		-0.08				125
141851	−2 04058	HR 5895		15 48 39	−02 56.4	2	5.10	.005	0.12	.000	0.06	.005	7	A3 Vn	15,1417
		LSS 3413		15 48 42	−56 08.1	1	11.90		0.76		-0.31		2	B3 Ib	336
141740	−36 10472			15 48 44	−36 33.2	1	9.10		0.33		0.24		1	A5 IV/V	1770
141702	−42 10847			15 48 44	−42 26.5	6	8.15	.017	0.53	.008	0.00	.009	14	F6/7 V	742,1594,2012,2013*
141829	−14 04290			15 48 47	−14 37.4	1	9.30		0.29		0.17		1	A2 III/IV	1770
	−54 06727			15 48 47	−54 24.1	1	10.45		0.69		0.26		5		125
141853	−13 04269	HR 5896		15 48 51	−13 59.0	1	6.19		1.00				4	G8 III	2035
141760	−39 10201			15 48 52	−39 31.8	1	8.68		0.39		0.25		1	A5 IV/V	1770
	−52 09018			15 48 52	−53 03.0	2	10.30	.011	0.46	.008	-0.32	.008	4	B2 V	540,567

Table 1

857

HD	DM	Other Id	N Rem	α_{1950}	δ_{1950}	S	V	σ_V	B–V	σ_{B-V}	U–B	σ_{U-B}	n	Spectrum	References
		AJ77,733 T26# 12		15 48 52	−54 32.2	1	12.30		0.75		−0.05		4		125
141972	+22 02889			15 48 54	+22 28.1	1	7.87		0.44		0.03		2	F0	1375
		Wolf 594		15 48 58	+08 12.3	1	12.60		1.42		1.23		1		1696
141779	−39 10202			15 48 58	−39 22.0	1	8.11		0.20		0.07		1	A5/6 IV/V	1770
141761	−40 10016	IDS15456S4018	AB	15 48 58	−40 27.2	1	9.40		0.37		0.29		1	A2 IV	1770
141585	−62 04990	HR 5884		15 49 00	−62 27.5	5	6.20	.008	1.47	.008	1.70		47	K3 III	15,978,1075,1075,2032
141992	+21 02829	HR 5899		15 49 04	+21 07.6	5	4.77	.016	1.53	.005	1.89	.018	12	K5 III	15,1080,1363,3016,8015
141814	−32 11195			15 49 04	−32 55.8	1	10.84		0.22		0.18		1	A1 III/IV	1770
	−31 12353			15 49 06	−31 54.0	2	10.12	.019	0.86	.014	0.37		4		1594,6006
141832	−29 12030	HR 5893	⋆ A	15 49 07	−29 44.2	4	6.39	.006	0.98	.000	0.85		13	K0 III	15,2013,2030,3005
141724	−50 09939	HR 5890		15 49 09	−50 28.0	1	6.60		1.53				4	K3 III	2007
	−54 06728			15 49 11	−54 31.9	1	11.51		0.58		0.11		4		125
141882	−21 04207			15 49 13	−21 32.0	1	8.28		0.30		0.20		1	A3 IV/V	1770
141836	−36 10478	IDS15460S3625	A	15 49 13	−36 34.2	1	9.45		0.25		0.20		1	A2/3 III	1770
		Nor sq 3 # 319		15 49 13	−54 37.2	1	13.38		1.70		0.71		2		336
141919	−9 04249			15 49 14	−09 19.3	1	8.96		0.89		0.47		2	K2 V	3072
141833	−34 10568			15 49 14	−35 08.4	1	10.65		0.46		0.24		1	A2	1770
141834	−35 10550			15 49 15	−35 17.0	1	10.32		0.29		0.22		1	B9.5 IV	1770
	+45 02342			15 49 17	+45 41.4	1	10.49		1.30		1.34		2	K0	1375
		WLS 1540 25 # 8		15 49 19	+24 55.0	1	13.71		0.81		0.34		2		1375
141837	−40 10023			15 49 19	−40 50.2	1	10.37		0.34		0.23		1	A8/F0 V	1770
		G 168 - 14		15 49 21	+29 40.3	1	13.01		1.58				1	M7	1746
142091	+36 02652	HR 5901	⋆ A	15 49 21	+35 48.7	5	4.81	.009	1.00	.005	0.87	.005	15	K0 III-IV	15,1118,3016,4001,8015
142053	+25 02981			15 49 22	+25 27.2	2	7.47	.010	1.09	.000	0.95	.050	5	K1 II-III	105,8100
		G 152 - 58		15 49 24	−14 21.6	2	11.79	.015	0.56	.019	−0.13	.015	3		1696,3062
141900	−22 04043			15 49 27	−22 52.5	1	10.01		0.29		0.28		1	A1 IV	1770
141689	−61 05351			15 49 27	−61 31.0	1	9.90		0.18		−0.27		4	B2/3ne	1586
	−54 06731			15 49 30	−54 23.9	1	11.42		0.45		0.24		4		125
		G 16 - 14		15 49 31	−03 46.4	1	12.71		1.34		1.17		1		1620
	−10 04182	G 152 - 60		15 49 32	−10 39.9	1	10.48		0.64		0.21		2		1696
141886	−39 10205			15 49 32	−39 23.3	1	10.45		0.25		0.10		1	B9 V	1770
		AJ77,733 T26# 13		15 49 32	−54 34.4	1	12.62		0.60		0.15		5		125
141782	−53 06732			15 49 35	−54 03.2	1	8.73		0.00		−0.43		1	B1/2 Vn	567
	−54 06733			15 49 35	−54 30.3	1	9.25		0.10		−0.20		15	B9	125
		GD 192		15 49 36	−11 54.9	1	14.71		0.61		−0.10		1		3060
	−50 09946			15 49 38	−50 30.1	1	11.28		0.50		−0.04		2	B5 V	540
141939	−26 11080			15 49 40	−26 56.5	1	8.24		0.22		0.15		1	A3 III/IV	1770
141905	−39 10207			15 49 41	−39 44.7	1	8.31		0.24		0.19		1	A2 V	1770
142093	+15 02923			15 49 42	+15 23.2	1	7.35		0.61		0.02		5	G2 V	814
142039	+4 03073			15 49 43	+04 08.3	1	9.06		0.58		0.07		1	F5	1776
	−54 06735			15 49 43	−54 46.7	1	10.84		0.69		0.13		4		125
141904	−37 10566			15 49 44	−38 03.2	1	11.04		0.43		0.35		1	A8/F0 V	1770
		AJ77,733 T26# 20		15 49 44	−54 27.5	1	14.36		0.82		0.49		5		125
141783	−57 07296			15 49 44	−57 46.1	1	9.09		0.45		−0.16		1	B6 Ib	567
	+11 02874	G 137 - 59		15 49 46	+11 01.6	2	9.37	.015	1.41	.015	1.27		3	M2	1705,3072
141974	−23 12560			15 49 47	−23 18.8	1	10.27		0.39		0.23		1	A2 III/IV	1770
		GD 193		15 49 48	+20 06.8	1	16.19		0.07		−0.83		1		3060
141960	−29 12044			15 49 50	−29 41.6	1	8.53		0.28		0.09		2	A9 V	1770
141923	−40 10031			15 49 51	−40 16.8	1	10.60		0.39		0.29		1	A7/9 V	1770
141961	−34 10575			15 49 53	−35 04.4	1	10.68		0.32		0.31		1	A3 II	1770
141976	−29 12045			15 49 55	−30 06.2	1	9.44		0.30		0.11		1	A(7) (II)	1770
		Nor sq 3 # 309		15 49 56	−54 26.3	1	11.99		1.98		0.79		3		336
		BK Nor		15 49 56	−59 05.2	1	9.77		2.74				1	CS	864
		AJ77,733 T26# 9		15 49 57	−54 37.3	1	11.72		0.61		0.12		5		125
	−54 06745			15 49 59	−54 30.4	1	11.60		0.87		0.47		5		125
	+33 02642			15 50 02	+33 05.9	2	10.84	.007	−0.17	.011	−0.85	.004	14	B2	1026,1036
142016	−30 12626			15 50 05	−30 37.8	1	7.26		0.20		0.08		2	A4 IV/V	1771
	−61 05361			15 50 07	−61 36.3	2	10.41	.007	0.12	.007	−0.51	.003	8	B3 IV	540,976
141998	−34 10580			15 50 08	−35 02.0	1	9.32		0.45		0.24		1	A9 V	1770
141979	−38 10718			15 50 10	−38 26.1	1	10.42		0.58		0.21		1	A8/F0	1770
		GD 194		15 50 12	+18 19.1	1	14.83		0.11		−0.84		1	DAs	3060
		AJ77,733 T26# 11		15 50 12	−54 29.8	1	11.98		0.91		0.49		5		125
	−35 10567	R Lup		15 50 14	−36 09.0	1	10.16		1.92		1.18		1	M5	817
		L 153 - 157		15 50 18	−69 17.	1	13.50		0.86		0.15		3		3060
		LP 683 - 47		15 50 19	−05 25.9	1	11.97		1.47		1.24		1		1696
142041	−31 12369			15 50 19	−31 29.1	1	9.19		0.26		0.08		2	A8/9 V	1771
		LP 860 - 42		15 50 20	−25 16.8	1	13.00		0.62		−0.06		3		1696
142097	−21 04214			15 50 25	−21 49.4	1	8.44		0.38		0.23		2	A5 V	1771
	−55 06764	LSS 3416		15 50 25	−55 37.0	1	11.56		0.66		−0.31		2		1737
142096	−19 04249	HR 5902		15 50 26	−20 01.1	7	5.03	.006	−0.02	.005	−0.58	.007	20	B3 V	15,26,1054,1770,2035*
141926	−54 06759	LSS 3417	⋆ AB	15 50 26	−55 10.9	2	8.60	.014	0.55	.005	−0.43	.032	4	B2nne	540,976
142130	−13 04276			15 50 29	−13 26.7	1	9.50		0.34		0.16		1	A7/8 III	1770
		LP 503 - 58		15 50 30	+12 36.6	1	12.94		0.99		0.88		1		1696
141767	−68 02585	HR 5891		15 50 31	−68 27.4	3	5.08	.004	1.12	.005	0.94	.005	15	G5 IIa	15,1075,2038
142146	−13 04277			15 50 32	−13 23.1	1	9.69		0.26		0.26		1	A5 IV	1770
142114	−24 12352	HR 5904	⋆ AB	15 50 36	−25 10.8	8	4.59	.007	−0.07	.008	−0.66	.014	29	B2.5Vn	15,26,1020,1054,1075*
		AJ77,733 T26# 17		15 50 37	−54 28.5	1	13.47		0.68		0.39		6		125
141913	−60 06191	HR 5898	⋆ AB	15 50 38	−60 35.8	3	6.13	.017	0.10	.000			11	B9 II	15,1075,2032
142244	+17 02926	HR 5909		15 50 40	+17 33.1	2	6.29	.005	1.24	.000	1.21	.005	5	K0	1648,1733
		CW Ser		15 50 41	+06 14.2	1	11.59		0.23		0.18		1		597

HD	DM	Other Id	N Rem	α_{1950}	δ_{1950}	S	V	σ_V	B–V	σ_{B-V}	U–B	σ_{U-B}	n	Spectrum	References
142098	−34 10584			15 50 43	−35 07.0	1	9.73		0.47		0.38		1	A3/5 II	1770
141891	−63 03723	HR 5897	⋆ A	15 50 43	−63 16.7	7	2.85	.006	0.30	.005	0.04	.012	29	F0 III/IV	15,1020,1075,1075*
141891	−63 03723	IDS15463S6307	B	15 50 43	−63 16.7	1	13.22		0.83		0.15		3		1518
142197	−2 04064			15 50 46	−02 52.9	1	7.50		0.44		−0.10		1	F5	1776
142099	−38 10727			15 50 49	−38 56.9	1	10.29		0.46		0.25		1	A9 V	1770
	−54 06774	IDS15470S5416	⋆ AB	15 50 49	−54 25.1	1	9.04		1.20		0.74		1		689
142147	−27 10629			15 50 50	−28 02.6	1	10.12		−0.03		−0.36		1	B9 II	1770
142267	+13 03024	HR 5911	⋆ AB	15 50 52	+13 21.1	6	6.10	.009	0.59	.013	0.00	.024	11	G0 IV	15,22,792,908,1003,3026
142268	+12 02909			15 50 53	+12 30.0	1	7.36		1.03		0.86		2	K2	1648
142165	−24 12354	HR 5906		15 50 54	−24 23.1	7	5.38	.012	−0.01	.009	−0.41	.009	20	B6 IVn	15,26,1054,1770,2006*
327503	−44 10492			15 50 56	−44 52.4	1	11.71		1.30		1.19		1	K0	1696
142373	+42 02648	HR 5914		15 50 57	+42 35.4	10	4.62	.015	0.56	.012	0.01	.015	48	F9 V	1,15,22,254,879,1077*
142184	−23 12569	HR 5907		15 50 57	−23 49.8	8	5.41	.013	−0.03	.009	−0.61	.007	27	B2.5 Vne	15,26,1054,1637,1770,2007*
142198	−16 04174	HR 5908		15 50 58	−16 35.0	7	4.14	.016	1.02	.005	0.82	.007	28	K0 III	15,1003,1075,1088*
	−50 09971			15 51 00	−50 47.0	1	10.48		0.48		−0.06		2	B9	1586
142186	−26 11091			15 51 02	−27 06.8	1	10.07		0.23		0.18		1	A1 III	1770
142029	−54 06778			15 51 02	−54 27.9	1	10.00		0.13		−0.15		5	B8/9 IV/V	125
142150	−37 10580			15 51 06	−37 30.6	1	10.58		0.51		0.27		1	A9/F0 V	1770
142531	+56 01838	HR 5922		15 51 07	+55 58.4	1	5.81		0.96				1	K0	71
		Wolf 600		15 51 08	+03 56.5	1	11.43		0.49		−0.01		4		1696
		AN Ser		15 51 11	+13 07.0	2	10.45	.043	0.19	.021	0.23	.009	2	F0	668,699
	−53 06768	LSS 3420		15 51 11	−53 51.3	3	9.18	.021	0.60	.025	−0.36	.022	7	B1 Ib	403,962,1350
	+72 00699	RS UMi		15 51 13	+72 21.6	1	10.07		0.74		0.23		3	G0	588
142081	−52 09067			15 51 13	−52 59.5	2	8.58	.015	0.47	.000	−0.04		3	F6 V	742,1594
142188	−36 10503			15 51 15	−37 01.1	1	10.71		0.39		0.25		1	A8/9 (IV)	1770
142357	+16 02840	HR 5913	⋆ AB	15 51 17	+16 13.3	1	6.22		0.41		0.03		2	F5 II-II	1733
141969	−65 03171	LSS 3418		15 51 19	−66 00.4	1	11.93		0.33		0.11		3	Be	1586
142049	−59 06428	HR 5900	⋆ AB	15 51 20	−60 01.8	5	5.76	.005	0.35	.006	0.11		19	G5II/III+A3	15,404,1075,2006,2012
142232	−27 10640			15 51 22	−27 24.9	1	8.81		0.45		0.24		1	A3/5mF0-F2	1770
142233	−31 12384			15 51 26	−31 58.2	1	10.37		0.38		0.23		1	A3/5 IV	1770
142201	−39 10225			15 51 26	−40 00.1	1	10.51		0.20		0.10		1	B9 III	1770
142250	−26 11096	HR 5910		15 51 27	−27 11.5	4	6.14	.012	−0.06	.004	−0.45	.008	14	B6 Vp	26,1054,1770,2006
142200	−38 10733			15 51 28	−38 41.3	1	9.81		0.21		−0.35		1	B5 II/III	1770
142202	−40 10051			15 51 30	−40 50.0	1	9.94		0.28		0.25		1	A1 V	1770
		He3 1112		15 51 30	−55 16.	1	12.06		1.16		0.27		3		1586
142532	+46 02124			15 51 34	+46 11.2	2	7.26	.010	0.96	.005	0.68	.005	5	G8 III	1501,1601
142288	−25 11183			15 51 38	−25 51.6	1	9.28		1.24		1.23		1	K5 V	3072
142301	−24 12365	HR 5912, V927 Sco	⋆	15 51 39	−25 05.8	5	5.87	.008	−0.06	.004	−0.60	.008	16	B8 IIIp	26,1054,1079,1770,2007
		LP 803 - 58		15 51 40	−16 24.0	1	11.50		1.05		0.93		1		1696
142152	−54 06788	LSS 3421		15 51 40	−54 37.8	5	9.63	.021	0.38	.013	−0.55	.027	13	B0 III	403,540,962,976,1350
		NGC 6005 - 5		15 51 40	−57 17.4	1	11.77		1.83		1.85		1		412
	−53 06782			15 51 42	−53 49.1	1	10.63		0.09		−0.23		2	B7 V	540
142315	−22 04046			15 51 44	−22 37.2	3	6.86	.005	0.04	.003	−0.21	.014	10	B9 V	26,1054,1770
		NGC 6005 - 3		15 51 45	−57 18.2	1	12.78		1.13		0.92		1		412
327623	−44 10499			15 51 46	−45 08.7	1	10.19		0.38		−0.32		2	B7 Iab	540
	−54 06791	LSS 3423		15 51 46	−54 29.8	3	10.40	.012	0.56	.004	−0.47	.009	8	O9.5IV	92,336,540
		NGC 6005 - 6		15 51 46	−57 16.1	1	13.43		0.61		0.41		1		412
142276	−36 10510			15 51 47	−36 37.9	1	9.70		0.47		0.28		1	A9 V	1770
		NGC 6005 - 1		15 51 47	−57 17.6	1	12.37		0.90		0.58		1		412
		NGC 6005 - 2		15 51 47	−57 17.6	1	13.28		1.05		0.83		1		412
		NGC 6005 - 4		15 51 47	−57 17.9	1	13.98		0.96		0.62		1		412
	−55 06802	LSS 3422		15 51 48	−56 09.3	1	10.88		0.52		−0.28		2	B2 IV	540
		KS 1027		15 51 48	−66 01.	1	11.08		0.61		0.08		2		540
		Nor sq 3 # 117		15 51 49	−54 30.3	1	12.90		0.66		−0.16		1		336
142737	+62 01435			15 51 50	+62 27.9	1	9.53		1.24		1.34		2		1375
142254	−42 10901			15 51 50	−42 27.5	2	6.68	.000	0.40	.000	−0.06		12	F3 V	1088,2012
		Nor sq 3 # 119		15 51 51	−54 35.4	1	12.97		1.31		0.23		2		336
		G 16 - 15	A	15 51 52	−04 08.1	3	11.49	.005	0.94	.017	0.72	.000	7		1064,1620,1658
		G 16 - 15	B	15 51 52	−04 08.1	1	14.24		1.34		1.01		1		1064
142139	−60 06208	HR 5905		15 51 52	−60 20.2	3	5.76	.013	0.07	.008	0.05		10	A3 V	15,26,2012
142291	−39 10230			15 51 53	−40 07.0	1	9.59		0.20		−0.18		1	Ap Si	1770
142256	−44 10501			15 51 53	−44 22.8	1	6.97		0.00		−0.28		3	B8 V	1770
		AJ77,733 T27# 7		15 51 53	−54 06.6	1	12.03		0.32		0.16		5		125
142277	−42 10903			15 51 54	−42 22.5	1	7.60		1.08				4	K0 II +G	2012
	+14 02955			15 51 57	+14 25.2	1	9.70		1.07		0.88		2	K5	3072
142317	−33 10804			15 51 57	−34 05.6	1	9.06		0.43		0.32		1	A0 V	1770
142302	−38 10741			15 52 00	−39 07.5	1	9.65		0.43		0.29		1	A5 III	1770
	−45 10331			15 52 00	−45 23.6	2	11.08	.001	0.88	.008	0.43		16		143,863
142318	−35 10585			15 52 02	−35 28.4	1	9.99		0.45		0.34		1	A2/3 III	1770
		AJ77,733 T27# 5		15 52 03	−54 09.6	1	11.51		1.30		0.96		6		125
142378	−18 04195	HR 5915	⋆ AB	15 52 07	−19 14.2	7	5.94	.012	−0.01	.007	−0.53	.011	20	B5 V	15,26,1054,1770,2007*
142422	−12 04363			15 52 08	−13 16.0	1	10.15		0.31		0.19		1	A5 II	1770
142381	−21 04224			15 52 11	−21 34.2	1	10.41		0.38		0.35		1	A5 II/III	1770
142237	−54 06802			15 52 11	−54 48.4	6	8.80	.030	0.11	.011	−0.51	.020	19	B2 Vne	403,540,567,976,1350,1586
142404	−23 12576			15 52 12	−24 11.6	1	9.14		0.19		0.14		1	A1 V	1770
142304	−44 10503			15 52 13	−44 23.2	1	6.83		0.02		−0.44		1	B2/5 V	1770
		Nor sq 3 # 38		15 52 14	−54 04.7	1	12.18		1.03		−0.02		2		336
142500	+9 03116	HR 5919, FP Ser		15 52 15	+08 43.6	2	6.28	.005	0.18	.000	0.08	.010	7	A7 Vn	15,1256
	−54 06807	LSS 3424		15 52 16	−54 33.6	3	9.95	.031	0.42	.008	−0.50	.019	7	B0.5V	336,540,1350
142403	−21 04225			15 52 18	−22 07.8	1	9.95		0.17		0.08		1	B9.5 III/I	1770

Table 1   859

HD	DM	Other Id	N Rem	α_{1950}	δ_{1950}	S	V	σ_V	B–V	σ_{B-V}	U–B	σ_{U-B}	n	Spectrum	References
142424 −22 04052				15 52 19	−23 13.3	1	8.41		0.37		0.23		1	A4 IV/V	1770
142574 +20 03166	HR 5924			15 52 22	+20 27.4	6	5.44	.005	1.59	.005	1.94	.007	12	M0 III	15,1003,1080,3009*
142407 −30 12663	HR 5916			15 52 22	−30 56.3	1	6.21		1.37				4	K4 III	2009
142261 −56 07043				15 52 22	−57 02.2	1	9.86		0.16		−0.28		2	B5/7 III	540
	G 202 - 19			15 52 23	+47 38.2	1	13.41		0.68		0.06		1		1658
142362 −37 10588				15 52 26	−38 03.0	1	10.86		0.44		0.36		1	A7 V	1770
142363 −40 10065				15 52 26	−40 30.1	1	9.59		0.21		0.13		1	A8 V	1770
−54 06810				15 52 26	−54 29.5	1	9.93		0.45				4	B0	1021
−55 06813				15 52 26	−56 06.5	2	10.44	.011	0.55	.004	−0.22	.006	7	B2.5IV	540,567
142553 +11 02883				15 52 27	+11 39.6	1	7.69		0.22		0.07		2	A2	1648
142445 −25 11190	HR 5917			15 52 28	−26 07.2	2	5.63	.010	0.14	.005	0.18		6	A3 V	401,2035
+18 03087				15 52 30	+18 10.6	1	9.55		0.40		0.05		3	F2	1648
	Nor sq 3 # 10			15 52 31	−54 01.7	1	11.32		0.51		−0.20		2		336
	AAS35,161 # 81			15 52 32	−37 52.8	1	15.61		1.19		0.74		1		5005
141864 −76 01123				15 52 33	−76 28.4	2	9.00	.000	0.67	.000	0.12		3	F6/7 V	742,1594
142575 +5 03113				15 52 35	+05 12.9	4	8.61	.005	0.37	.005	−0.07	.004	6	F0 V	979,1003,3077,8112
142679 +39 02930				15 52 35	+38 52.6	1	8.92		0.89		0.57		2	G5	1601
142429 −35 10590				15 52 36	−35 31.9	1	7.76		1.08		0.75		10	G8 III	1673
142502 −14 04309				15 52 37	−14 53.7	1	9.52		0.43		0.29		1	Ap SrEuCr	1770
142427 −34 10610				15 52 37	−34 38.6	1	10.17		0.40		0.31		1	A1/2	1770
142576 +5 03114				15 52 40	+05 10.9	1	9.06		0.50		−0.02		4	F2	158
142456 −26 11106	IDS15496S2627		AB	15 52 40	−26 36.1	1	7.32		0.49				2	F7 V	173
142430 −36 10518				15 52 40	−36 17.1	1	9.84		0.39		0.25		1	A7 IV	1770
142431 −40 10070				15 52 42	−40 33.5	1	7.07		0.13		0.09		1	A2 V	1770
	WLS 1612 60 # 6			15 52 44	+60 20.6	1	13.88		0.57		0.04		2		1375
142554 −4 04000				15 52 44	−04 19.9	1	9.79		0.19		0.11		2	A0	1776
142448 −39 10237	HR 5918		★ A	15 52 45	−39 43.1	2	6.03	.000	0.15	.003	−0.07		6	B9 V	1770,2006
−53 06816				15 52 47	−54 13.3	1	10.30		0.69		0.05		6		125
−53 06818	LSS 3426			15 52 48	−53 55.8	2	10.51	.020	1.22	.025	0.11	.010	5	B1 Ia	336,1737
−54 06818	LSS 3425			15 52 49	−54 54.0	1	10.42		0.71		−0.35		2		181
142432 −42 10916				15 52 50	−42 20.1	1	9.03		0.30		−0.08		3	B9 III	1770
	HBC 605			15 52 51	−37 47.4	2	11.86	.005	1.34	.000	1.09	.040	2	M0	1763,5005
142459 −35 10592				15 52 52	−35 38.9	1	10.26		0.24		−0.10		1	Ap Si	1770
142460 −38 10753				15 52 56	−39 07.0	1	10.07		0.49		0.40		1	A5/7 II	1770
142461 −40 10073				15 52 56	−40 23.2	1	8.90		1.21		1.10		2	K0 III	1763
−53 06825	LSS 3428			15 52 57	−53 56.5	1	10.70		0.59		−0.44		2		336
142780 +43 02542	HR 5932			15 52 58	+43 17.0	7	5.37	.016	1.64	.012	1.96	.011	35	M3 III	15,667,1118,3055,8011*
142364 −55 06823				15 52 58	−55 51.5	1	8.65		0.11		−0.16		2	B9 III	181
142484 −36 10524				15 53 01	−36 26.2	1	9.07		0.21		0.11		2	A0 V	1770
142742 +34 02709	IDS15511N3439		A	15 53 02	+34 30.5	1	7.16		0.16		0.15		3	A2	3024
142742 +34 02709	IDS15511N3439		B	15 53 02	+34 30.5	1	10.43		0.79		0.32		4		3024
142541 −26 11113				15 53 02	−27 02.4	1	9.36		0.27		0.19		2	A1/2 V	1771
142524 −32 11257				15 53 02	−32 56.5	1	9.61		0.15		0.14		1	A0/1 V	1770
142540 −25 11192	IDS15500S2539		A	15 53 03	−25 47.4	1	7.62		0.22		0.14		1	A2/3 III	1770
	AJ85,265 # 143			15 53 03	−54 08.2	1	11.70		1.32		0.19		1		962
142349 −57 07363				15 53 04	−57 26.1	1	8.80		0.08		−0.51		1	B5 IV	567
142542 −31 12407	HR 5923			15 53 05	−31 38.5	1	6.29		0.43				4	F6 V	2009
142577 −21 04227				15 53 08	−21 22.8	1	8.55		0.34		0.28		1	A1 V	1770
	LP 274 - 24			15 53 10	+33 59.6	1	15.95		1.76				2		1663
142578 −23 12585				15 53 10	−23 19.9	1	8.63		0.23		0.19		1	A2 IV/V	1770
142579 −23 12584				15 53 10	−23 44.9	1	8.27		0.38		0.06		1	F0 V	1771
142580 −25 11193	IDS15502S2531		A	15 53 11	−25 39.6	1	8.56		0.18		0.17		1	A2 IV	1770
142413 −53 06836				15 53 14	−54 12.6	2	9.23	.003	0.38	.010	0.25	.014	18	A5 III	125,219
	AT Ser			15 53 16	+08 03.3	2	11.05	.048	0.19	.015	0.09	.015	2	F2	668,699
142600 −25 11197				15 53 17	−25 17.8	1	9.31		0.43		0.05		1	F2 V	1771
142559 −36 10528				15 53 24	−36 24.6	1	10.37		0.23		0.14		1	A0 IV/V	1770
142560 −37 10602	HRC 251, RU Lup			15 53 24	−37 40.7	4	11.03	.225	0.56	.133	−0.78	.130	4	G5 V:e	776,1588,1763,5005
142763 +19 03036	HR 5931			15 53 25	+18 45.9	3	6.27		−0.10	.009	−0.42	.009	9	B8 III	1022,1079,1501
142640 −13 04290	HR 5927			15 53 26	−14 15.2	2	6.31	.000	0.49	.007	0.03		6	F6 V	2007,3053
142451 −54 06837				15 53 26	−54 23.0	1	9.23		0.02		−0.54		5	B3 V	567
142529 −47 10456	HR 5921			15 53 27	−48 01.0	3	6.30	.009	0.38	.004	−0.06		15	F2 V	15,1088,2012
142617 −30 12674				15 53 28	−30 29.0	1	9.24		0.23		0.25		1	A5 IV	1770
142415 −59 06464				15 53 28	−60 03.3	3	7.32	.008	0.62	.004	0.12	.005	9	G1 V	285,657,1075
142602 −34 10622	IDS15503S3408		A	15 53 29	−34 17.2	1	8.94		0.42		0.36		1	A2 V	1770
142468 −53 06845	LSS 3429			15 53 31	−54 11.3	10	7.90	.018	0.59	.011	−0.42	.020	30	B1 Ia/Iab	125,138,181,219,336*
142452 −56 07084				15 53 31	−56 53.8	2	9.78	.017	0.08	.008	−0.38	.009	9	B9 II/III	540,976
142663 −13 04291				15 53 32	−13 32.8	1	10.79		0.30		0.28		1	A5/7 II	1770
−53 06850	LSS 3430			15 53 36	−53 49.0	2	10.95	.049	0.50	.010	−0.39	.024	5		336,1350
142630 −33 10826	HR 5926		★ B	15 53 41	−33 49.2	1	5.60		0.01		0.11		1	B9 V	1770
142629 −33 10826	HR 5925		★ A	15 53 41	−33 49.3	1	5.11		0.12		0.07		1	A3 V	1770
142629 −33 10826	HR 5925		★ AB	15 53 41	−33 49.3	4	4.58	.004	0.10	.005	0.07	.005	12	A3 V	15,1075,2012,2016
142666 −21 04228				15 53 43	−21 53.0	1	8.89		0.56		0.33		2	A8 V	1771
142703 −14 04314	HR 5930			15 53 44	−14 41.1	3	6.12	.005	0.24	.004	−0.06		8	A2 Ib/II	15,1771,2027
	G 137 - 68			15 53 45	+17 59.8	1	14.47		1.63				1		906
142687 −23 12591				15 53 46	−23 57.1	1	10.59		0.21		0.18		1	B9/A0 III	1770
	Nor sq 3 # 90			15 53 46	−54 23.8	1	12.71		0.67		−0.13		2		336
142725 −16 04180				15 53 48	−16 40.8	1	9.14		0.34		0.28		1	A3 III	1770
142669 −28 11714	HR 5928		★ A	15 53 48	−29 04.2	10	3.88	.008	−0.20	.007	−0.81	.013	44	B2 IV/V	3,15,26,1020,1034,1054*
142926 +42 02652	HR 5938			15 53 49	+42 42.6	3	5.77	.019	−0.12	.006	−0.39	.025	6	B9 pe	879,1079,1118
142705 −22 04057				15 53 49	−23 02.4	1	7.77		0.17		0.12		1	A0 V	1771

HD	DM	Other Id	N Rem	α_{1950}	δ_{1950}	S	V	σ_V	B–V	σ_{B-V}	U–B	σ_{U-B}	n	Spectrum	References
142643	−35 10607			15 53 49	−35 47.0	1	6.91		0.11		0.12		1	A3 V	1770
142670	−30 12679			15 53 50	−31 07.7	1	10.11		0.25		0.22		1	A3 (III)	1770
	−55 06837	LSS 3431		15 53 50	−55 36.1	1	10.67		0.36		−0.52		2		181
142548	−55 06838			15 53 52	−55 55.7	2	9.20	.020	0.11	.005	−0.29	.010	5	B8 Ib/II	181,567
142688	−25 11202			15 53 53	−26 01.8	1	10.31		0.32		0.17		1	A9 V	1770
		Wolf 610		15 53 54	+08 05.2	1	11.64		0.51		−0.15		2		1696
		vdB 68 # a		15 53 54	−53 50.	1	11.00		0.49		−0.40		4	B0.5V	434
142846	+15 02933			15 53 56	+14 53.7	1	8.28		0.35		0.05		2	F2 IV	1776
		G 16 - 16		15 53 56	−04 40.6	1	12.00		1.20		1.09		1		1620
142726	−24 12397			15 53 56	−24 58.9	1	9.63		0.41		0.16		1	A8 III/IV	1770
142646	−40 10087			15 53 56	−40 47.0	1	8.65		1.58		1.61		11	K2 III	1763
	−54 06846	LSS 3432		15 53 56	−54 24.9	5	10.35	.019	0.44	.017	−0.52	.015	13	B0 III	181,336,540,1021,1350
142645	−39 10246			15 53 57	−40 06.6	1	9.45		0.32		0.21		1	A0 V	1770
142908	+38 02712	HR 5936	★ A	15 53 58	+38 05.4	4	5.44	.008	0.33	.016	0.02	.008	8	F0 IV	15,254,1008,3053
		AAS35,161 # 126		15 53 58	−42 31.3	1	14.10		1.34		−0.38		1		5005
142565	−54 06849	LSS 3433		15 53 58	−54 14.9	7	8.99	.019	0.47	.020	−0.49	.024	16	B1 II	138,181,336,403,540*
		AJ85,265 # 146		15 53 58	−54 39.4	1	11.69		1.46		0.37		1		962
		AJ77,733 T27# 6		15 54 00	−54 15.4	1	11.98		0.57		−0.10		4		125
142691	−35 10611	HR 5929		15 54 06	−36 02.5	2	5.80	.005	1.02	.016	0.50		6	K0/1 III	1770,2035
142584	−56 07105			15 54 06	−56 15.0	1	7.25		1.24		1.03		3	G6 Ib/II	8100
142860	+16 02849	HR 5933	★ A	15 54 08	+15 49.4	15	3.85	.016	0.48	.006	−0.02	.009	72	F6 V	1,3,15,22,30,667,1006*
	+44 02522	G 180 - 12		15 54 08	+44 22.5	1	10.04		0.65		0.11		2	G5	308
143031	+54 01780			15 54 08	+54 10.6	1	8.21		0.97		0.65		2	K0	1566
142800	−12 04373			15 54 08	−12 36.3	1	10.11		0.43		0.18		1	A9 V	1770
142751	−30 12683			15 54 10	−30 38.5	1	8.14		1.33		1.22		1	K0/1 II	565
142784	−24 12403	IDS15512S2458	A	15 54 14	−25 06.6	1	8.48		0.34		0.15		2	A5 IV	1771
142804	−15 04221	IDS15514S1545	A	15 54 15	−15 53.4	1	6.59		1.80		2.11		21	M1 III	3040
142804	−15 04221	IDS15514S1545	B	15 54 15	−15 53.4	1	13.73		0.70		0.14		2		3024
142805	−21 04233			15 54 16	−21 20.5	3	7.14	.000	0.17	.008	0.06	.015	10	A0 IV	26,1054,1771
142709	−42 10934			15 54 16	−42 28.7	3	8.05	.019	1.13	.012	1.00		10	K4 V	1075,2033,3077
	−53 06867	LSS 3434		15 54 17	−53 42.8	2	11.05	.000	0.78	.012	−0.19	.013	5		181,1586
142929	+25 02994			15 54 18	+25 19.2	1	8.41		0.51		0.03		3	F8 V	3026
142769	−30 12686			15 54 18	−31 00.7	1	9.66		0.27		0.19		1	A5 IV/V	1770
		Nor sq 3 # 96		15 54 18	−54 23.3	1	12.28		0.57		−0.31		2		336
142729	−39 10251			15 54 19	−39 31.4	1	9.73		0.52		0.38		1	A0	1770
142978	+39 02933			15 54 20	+39 34.4	1	7.85		0.40		−0.04		2	F2	1601
142634	−53 06868	LSS 3435		15 54 21	−54 11.9	7	8.72	.027	0.67	.012	−0.38	.018	23	B0.5Ia/Iab	125,181,219,336,403*
142514	−64 03320	HR 5920	★ AB	15 54 22	−64 53.7	4	5.74	.004	−0.07	.005	−0.40	.005	22	B7 III	15,1075,1075,2038
143078	+56 01844			15 54 23	+55 57.7	1	8.28		1.42		1.71		1	K5	1746
	−54 06858	LSS 3436		15 54 23	−54 48.0	1	10.71		0.60		−0.28		1		181
	−53 06872	LSS 3437		15 54 25	−53 46.2	2	11.89	.010	0.56	.010	−0.36	.020	5		181,336
142863	−2 04077			15 54 26	−02 50.7	1	8.97		0.40		0.29		2	A0	1776
142806	−26 11122	IDS15514S2653	AB	15 54 26	−27 00.8	1	10.22		0.35		0.17		1	A9 V	1770
142754	−40 10094	IDS15511S4042	AB	15 54 26	−40 51.2	4	8.60	.007	0.17	.011	−0.66	.010	12	B2 III	400,540,567,976
142786	−34 10630			15 54 32	−35 02.1	1	9.19		0.56		0.19		1	A9 V	1770
	−54 06867			15 54 40	−55 03.0	1	10.87		0.46		−0.21		5		567
142787	−38 10772			15 54 41	−38 28.1	1	9.48		0.44		0.32		1	A3 III	1770
		AAS35,161 # 128		15 54 42	−41 43.2	1	14.63		1.49		0.69		1		5005
		Nor sq 3 # 99		15 54 42	−54 20.5	1	12.45		0.59		0.04		1		336
142930	+3 03104	IDS15522N0342	A	15 54 44	+03 32.9	1	7.29		0.21		0.16		3	A0	1776
		G 16 - 17		15 54 44	+05 15.5	2	15.35	.040	1.59	.005	1.22		5		538,3078
142883	−20 04364	HR 5934		15 54 45	−20 50.4	5	5.85	.007	0.02	.005	−0.49	.009	14	B3 V	26,55,1054,1770,2007
		G 16 - 18		15 54 46	+05 16.6	2	13.39	.035	1.47	.000	1.14		5		538,1774,3078
	+21 02852	G 137 - 69		15 54 48	+20 44.5	2	9.98	.020	0.82	.005	0.51	.040	5	K0	308,1658
142884	−23 12597	V928 Sco		15 54 50	−23 23.0	3	6.78	.007	0.02	.004	−0.49	.013	6	B8/9 III	26,1054,1770
142851	−31 12426			15 54 50	−31 35.1	1	7.02		−0.04		−0.10		1	A0 V	1770
143187	+59 01691	HR 5949		15 54 51	+59 03.3	1			−0.02		−0.19		1	A0 V	1079
		VY Aps		15 54 53	−74 20.1	1	9.18		5.34				1	SC	864
142914	−16 04184			15 54 54	−16 50.4	1	10.22		0.49		0.29		2	A5	1770
142821	−41 10468			15 54 55	−41 31.0	1	7.90		0.18		−0.10		2	B9 V	1770
142969	+14 02969	HR 5940		15 54 56	+14 33.4	3	5.54	.012	1.14	.004	1.11	.074	5	K1 IV	1003,1080,4001
143062	+35 02749	IDS15531N3547	AB	15 54 57	+35 38.8	1	8.94		0.68		0.26		2	K0	1601
	−54 06873			15 54 57	−54 45.1	1	9.69		0.57		−0.05		1		78
142835	−39 10258			15 54 58	−39 54.1	1	10.56		0.39		0.26		1	A3/5 V	1770
142902	−22 04058			15 55 02	−22 24.2	1	9.03		0.42		0.20		2	A9 V	1770
		Nor sq 3 # 45		15 55 02	−54 03.1	1	13.27		0.79		0.19		1		336
142887	−33 10835	IDS15519S3347	A	15 55 03	−33 55.7	1	9.10		0.36		0.09		1	A0 III/IV	1770
142774	−53 06888			15 55 04	−53 55.3	2	9.17	.000	0.12	.010	−0.44	.015	5	B3 III/IV	181,567
		G 152 - 67		15 55 08	−02 22.7	2	12.31	.010	0.66	.005	−0.05	.029	3		1696,3062
142775	−54 06878	LSS 3440		15 55 08	−54 16.2	6	9.12	.013	0.71	.008	−0.32	.019	18	B0/1 III	181,336,403,540,567,976
142699	−62 05073			15 55 11	−62 40.4	1	7.02		1.57				4	K3 III	1075
142889	−37 10620	HR 5935		15 55 13	−37 21.6	1	6.31		1.01				4	K0 III	2007
142758	−58 06543	LSS 3439		15 55 16	−58 35.1	5	7.05	.018	0.22	.023	−0.66	.027	11	B1 Ia	403,540,976,1737,2012
142915	−34 10635			15 55 17	−34 40.2	1	10.65		0.18		0.07		1	A0 III/IV	1770
142872	−45 10362			15 55 18	−45 29.8	1	10.05		0.47		−0.05		3	B8 Ib/II	567
	−16 04187	G 153 - 15		15 55 19	−16 27.5	1	10.92		0.83		0.16		2	sdK1:	3062
142904	−38 10777	IDS15520S3853	A	15 55 19	−39 01.9	1	11.26		0.56		0.35		2	A2/3 V	1770
142903	−38 10777	IDS15520S3853	B	15 55 19	−39 01.9	1	10.92		0.60		0.45		1	A2	1770
143090	+30 02726			15 55 21	+29 54.4	1	10.09		0.93		0.62		7	K0	272
		G 152 -B4	A	15 55 22	−08 59.5	1	14.35		1.56				3		3016

Table 1 861

HD	DM	Other Id	N	Rem	α_{1950}	δ_{1950}	S	V	σ_V	B–V	σ_{B-V}	U–B	σ_{U-B}	n	Spectrum	References
		G 152 -B4	B		15 55 22	−08 59.5	1	14.80		0.09		-0.63		3	DA	3060
142983	−13 04302	HR 5941, FX Lib			15 55 23	−14 08.2	8	4.88	.031	-0.09	.010	-0.22	.045	45	B5 IIIpe	15,26,379,681,815,1770*
		Nor sq 2 # 14			15 55 24	−53 40.	1	11.87		0.88		-0.06		2		181
		Nor sq 3 # 48			15 55 24	−54 10.9	1	13.31		0.88		0.08		2		336
142984	−15 04226				15 55 25	−15 20.1	1	8.19		0.35		0.29		3	A7 IV/V	1770
142944	−29 12138				15 55 25	−29 20.5	1	10.16		0.17		0.13		1	A0 V	1770
142759	−62 05077				15 55 26	−62 33.7	1	8.90		0.43		-0.04		1	F3 V	1487
142985	−16 04188				15 55 27	−16 47.8	1	10.28		0.50		0.21		1	A3/5 (III)	1770
		Nor sq 3 # 106			15 55 27	−54 26.4	1	12.25		0.73		0.24		1		336
142986	−18 04209				15 55 30	−18 20.0	1	10.01		0.48		0.34		2	A1 III	1770
143107	+27 02558	HR 5947		⋆ AB	15 55 31	+27 01.3	25	4.14	.014	1.23	.006	1.28	.007	191	K2 IIIab	1,15,245,667,985,1006*
143016	−17 04461				15 55 33	−17 36.4	1	8.50		0.64				4	G3 V	2012
142988	−20 04368				15 55 34	−20 45.0	1	7.28		0.25		0.18		1	A3 III	1770
		Nor sq 3 # 124			15 55 34	−54 33.8	1	12.58		1.21		0.32		2		336
142990	−24 12427	HR 5942, V913 Sco			15 55 35	−24 41.3	5	5.42	.009	-0.09	.005	-0.66	.005	15	B5 IV	26,1054,1079,1770,2007
143007	−23 12605				15 55 38	−23 45.8	1	8.96		0.64				4	G3 V	2012
142591	−71 01939				15 55 38	−71 46.8	1	8.47		0.77				4	F8 II	2012
	−51 09671				15 55 41	−51 40.8	1	10.67		0.37		0.23		1	A7	96
		Nor sq 2 # 15			15 55 42	−53 49.	1	11.57		1.06		-0.09		2		181
143209	+40 02948	HR 5950			15 55 44	+39 50.2	1	6.31		1.08				1	K0	71
143433	+64 01102	G 225 - 35			15 55 44	+63 57.4	1	9.54		0.89		0.58		3	G5	196
	−50 10041	LSS 3441			15 55 44	−50 50.0	3	10.47	.035	0.46	.011	-0.42	.036	8		540,1021,1586
		Nor sq 3 # 108			15 55 46	−54 21.2	1	12.44		0.85		0.08		2		336
		G 16 - 20			15 55 47	+02 11.8	2	10.80	.000	0.61	.014	-0.08	.014	4		333,1620,1696
	−51 09672				15 55 47	−51 38.9	1	9.92		0.65		0.20		1		96
142918	−51 09674				15 55 48	−51 45.4	1	9.33		1.16		1.10		1	K(0) (III)	96
143018	−25 11228	HR 5944		⋆ A	15 55 49	−25 58.3	7	2.89	.008	-0.19	.005	-0.91	.010	79	B1 V +B2 V	3,15,26,1054,1075*
	−51 09675				15 55 50	−51 41.6	1	9.24		1.34		1.47		1		96
142994	−38 10783				15 55 51	−38 36.4	1	7.18		0.30		0.07		1	A3/5 II	1770
		AAS35,161 # 129			15 55 53	−41 48.5	1	13.01		1.28		0.24		1		5005
	−50 10045	LSS 3442			15 55 54	−50 17.5	2	10.24	.008	0.55	.004	-0.15		6	B3 III	540,1021
142946	−49 10219				15 55 58	−49 28.1	1	9.67		0.29		-0.39		3	B3 III	567
142947	−51 09678				15 55 58	−51 34.7	1	9.12		0.04		-0.41		1	B4 III	96
142919	−53 06911	HR 5937			15 56 01	−53 52.8	2	6.11	.005	0.00	.000	-0.48		8	B5 III	567,2006
143162	+13 03037				15 56 04	+12 58.6	1	8.10		1.00		0.81		2	K0	1648
143067	−20 04370				15 56 04	−20 52.5	1	8.43		0.34		0.28		1	A4 III	1770
143008	−36 10559				15 56 04	−37 03.4	1	10.08		0.40		0.31		1	A1/2 III/I	1770
		HRC 252, RY Lup			15 56 04	−40 13.6	2	10.52	.037	1.04	.032	0.55	.002	2	G0	776,1588
		HRC 252, RY Lup			15 56 05	−40 13.6	1	11.50		1.23		0.91		1	G0 V:	1763
143009	−41 10478	HR 5943			15 56 05	−41 36.2	3	4.98	.004	0.99	.010			11	K0 II/III	15,1075,2006
143022	−40 10112				15 56 07	−40 43.4	1	8.19		0.16		-0.02		3	B9.5 V	1770
		LOD 38 # 5			15 56 08	−51 40.2	1	11.54		0.58		0.00		1		96
143641	+72 00703				15 56 09	+72 32.1	1	7.23		1.03		0.86		2	K0	1733
143069	−26 11137				15 56 09	−26 26.5	1	8.11		0.29		0.12		2	A3 V	1771
143051	−32 11309				15 56 09	−32 52.2	1	6.97		0.02		-0.15		2	B9 IV/V	1770
142948	−53 06916				15 56 10	−53 42.6	3	8.02	.029	0.97	.010	0.55	.005	6	G5 III/IV	742,1594,6006
	−54 06898	LSS 3443			15 56 10	−54 13.4	3	10.95	.017	1.10	.000	0.03	.017	6		181,336,1737
143093	−19 04274	IDS15533S1939	B		15 56 12	−19 47.6	1	8.52		0.42		0.09		2	A8/9	1771
143094	−19 04275	IDS15533S1939	A		15 56 13	−19 47.8	1	8.11		0.29		0.14		2	A5/7 V	1771
		G 180 - 17			15 56 18	+39 11.1	1	14.69		1.59				1		1759
142996	−51 09684				15 56 21	−52 12.2	1	9.26		0.10		-0.35		1	B8 III	567
143112	−26 11140				15 56 25	−26 52.5	1	7.01		1.11				4	G8 III	2012
		G 180 - 18			15 56 27	+35 32.7	1	12.69		1.59				1	M6	1746
143113	−28 11755				15 56 28	−28 50.8	1	9.09		0.22		0.21		2	A3/4 V	1771
	−51 09690				15 56 28	−51 47.3	1	11.87		0.34		0.31		1		96
143133	−24 12439				15 56 29	−24 58.5	1	9.14		0.05		0.01		1	B9 V	1770
143272	+26 02762				15 56 30	+26 41.1	1	8.18		1.04		0.80		1	K0 II-III	8100
	−51 09691				15 56 31	−51 43.8	1	10.77		0.33		0.22		1		96
143291	+28 02503	G 168 - 20			15 56 32	+27 52.7	3	8.02	.009	0.76	.007	0.30	.005	7	K0 V	22,1003,3026
143115	−29 12159				15 56 32	−29 56.5	1	7.19		0.26		0.09		2	A2mA7-F0	1771
	−51 09692				15 56 32	−51 39.8	1	9.70		1.68		1.89		1		96
		LOD 38 # 14			15 56 34	−51 43.5	1	12.70		0.73				1		96
		LOD 38 # 9			15 56 34	−51 48.4	1	12.23		0.31				1		96
	−53 06923	LSS 3444			15 56 34	−53 46.2	3	9.84	.014	0.95	.004	-0.07	.028	4	O7 III	181,540,1737
143163	−17 04465				15 56 35	−18 06.8	1	10.00		0.56		0.18		1	A9 V	1770
143084	−40 10113	HR 5945			15 56 35	−40 30.7	1	6.49		1.52				4	K1 II/III	2007
143466	+55 01793	HR 5960, CL Dra			15 56 36	+54 53.4	6	4.96	.007	0.27	.005	0.06	.013	28	F0 IV	15,253,1363,3023,6001,8015
143313	+25 03003	MS Ser			15 56 38	+25 42.8	2	8.28	.064	1.00	.006	0.73	.014	3	K2 V	3026,8088
143099	−38 10794	IDS15535S3807			15 56 39	−38 16.1	1	9.27		0.54		0.01		4	F5 V	321
		LOD 38 # 10			15 56 39	−51 48.8	1	11.41		0.72		0.28		1		96
143036	−51 09695				15 56 39	−51 49.4	1	8.81		0.44		0.13		1	F2 IV/V	96
143148	−31 12442				15 56 40	−31 41.6	1	7.41		0.27		0.06		2	A7 III/IV	1771
142941	−63 03765	HR 5939, S TrA			15 56 40	−63 38.2	2	6.03	.019	0.56	.007	0.38	.017	2	F8 II	657,1490
		G 152 - 69			15 56 42	−05 50.7	1	14.18		1.59		1.00		6		316
143116	−35 10644				15 56 42	−36 09.6	1	9.77		0.93		0.66		2	K3 V	3072
		LOD 38 # 17			15 56 42	−51 40.8	1	11.49		0.63		0.23		1		96
		Steph 1294			15 56 43	+49 35.7	1	11.84		1.41		1.08		1	K7	1746
143149	−33 10848				15 56 43	−33 14.6	1	6.83		0.03		0.03		2	A0 V	1771
	−51 09697				15 56 44	−51 35.6	1	10.37		0.21		0.16		1	A0	96
143118	−38 10797	HR 5948		⋆ AB	15 56 48	−38 15.3	5	3.41	.008	-0.23	.004	-0.85	.018	18	B2.5IV	15,26,1075,2012,8015

HD	DM	Other Id	N Rem	α₁₉₅₀	δ₁₉₅₀	S	V	σ_V	B–V	σ_B-V	U–B	σ_U-B	n	Spectrum	References
143118	−38 10797	IDS15535S3807	B	15 56 48	−38 15.3	1	7.87		0.16		0.05		4		321
		LOD 38 # 15		15 56 48	−51 43.7	1	12.90		0.48				1		96
143392	+39 02936			15 56 49	+39 22.9	1	9.17		0.36		0.03		3	F2 V	1501
	+25 03005			15 56 51	+24 45.8	1	8.79		0.98		0.73		3	K0	1733
143120	−45 10373			15 56 52	−45 18.6	4	7.51	.015	0.74	.005	0.38		16	G5 IV	158,1075,2012,2013
143193	−24 12443			15 56 53	−24 24.7	1	10.08		0.38		0.17		2	A9 V	1770
		G 225 - 36		15 56 54	+59 24.7	1	10.55		1.39				2	K5	1619
		AJ84,127 # 2227		15 56 54	−08 10.	1	11.88		0.65		0.13		1		801
143226	−14 04324			15 56 55	−14 28.5	1	10.68		0.31		0.21		1	A5/7 II	1770
143192	−22 04065			15 56 55	−22 25.6	1	9.56		0.40		0.28		1	A8/9 V	1770
143294	+4 03092			15 56 59	+03 54.3	1	9.41		0.35		0.18		2	A2	1776
143195	−29 12164			15 57 00	−29 50.8	1	9.08		0.11		0.11		1	A0 V	1770
143181	−38 10800			15 57 01	−38 51.2	1	7.28		−0.01		−0.06		2	B9 V	1770
143295	−4 04017			15 57 04	−04 56.1	1	9.08		0.94		0.69		2	K2 V	3072
143393	+29 02748			15 57 05	+29 34.4	1	7.10		1.13		1.07		2	K2 III	1003
		AAS35,161 # 130	V	15 57 06	−41 35.2	1	14.52		1.22		−0.15		1		5005
143138	−47 10489			15 57 06	−47 45.0	1	8.70		1.39				4	G5 III	2012
		S100 # 3		15 57 06	−63 42.5	1	11.32		1.37				2		1730
143435	+37 02695	HR 5957		15 57 07	+36 47.1	1	5.62		1.49				1	gK5	71
	−51 09700			15 57 08	−51 38.5	1	10.72		0.71		0.22		1		96
	−7 04156			15 57 09	−08 06.5	4	10.49	.006	1.52	.011	1.22	.035	10	M1	158,801,1705,7008
	−53 06932	LSS 3445		15 57 10	−54 09.7	2	10.30	.010	0.86	.010	−0.22	.005	4		181,475
	−63 03768			15 57 10	−63 42.2	1	10.75		0.17		0.14		2		1730
143665	+65 01092	G 225 - 37		15 57 11	+65 32.5	1	9.04		0.86		0.55		3	G5	196
143122	−52 09162			15 57 12	−52 36.6	1	9.63		0.18		−0.31		3	B7 Ib/II	567
143101	−54 06922	HR 5946		15 57 12	−54 26.2	2	6.12	.005	0.26	.005			7	A5 V	15,2027
143039	−62 05104			15 57 16	−62 53.6	1	9.62		0.24				4	A5 IV/V	1075
143453	+30 02733			15 57 18	+30 31.2	1	8.78		0.36		0.12		3	F0	1371
143275	−22 04068	HR 5953	★	15 57 22	−22 28.9	10	2.32	.009	−0.12	.006	−0.91	.007	219	B0.3 IV	3,15,26,1034,1054*
143247	−31 12458			15 57 22	−31 57.6	1	8.63		0.20		0.16		1	A3 V	1770
143455	+25 03009			15 57 23	+25 34.5	2	7.79	.000	1.48	.007	1.73	.022	10	K2	833,8088
143454	+26 02765	HR 5958, T CrB	★ AB	15 57 25	+26 03.7	3	9.90	.068	1.40	.024	0.59	.059	23	gM3 + sdBe	866,1753,3001
143300	−15 04229			15 57 26	−16 01.2	1	9.72		0.33		0.27		1	A1 IV	1770
143232	−38 10803			15 57 26	−38 56.8	1	6.67		0.24		0.14		2	A5mA5-F2	1770
	−59 06536	NGC 6025 - 56		15 57 26	−59 15.5	1	11.04		0.24		0.23		1		162
143058	−63 03772			15 57 26	−63 38.3	2	8.69	.000	0.52	.003	0.00	.005	3	F7 V	1490,1730
143333	−16 04196	HR 5954	★ A	15 57 31	−16 23.3	6	5.47	.014	0.52	.007	0.04	.022	14	F7 V	15,22,254,1003,1075,3077
143248	−40 10120	HR 5952	★ AB	15 57 31	−40 17.7	4	6.21	.010	0.01	.004	0.00	.016	12	A0 V	15,1075,1770,2012
143356	−12 04391			15 57 33	−12 58.5	1	8.44		0.33		0.11		1	A9 V	1770
		G 137 - 74		15 57 34	+15 08.7	2	14.01	.018	1.57	.018	1.13		5		316,1759
		AA50,429# 8		15 57 34	−54 00.1	1	12.67		0.75		−0.11		2		475
		AA50,429# 9		15 57 36	−54 00.7	1	11.09		1.98		2.16		1		475
143234	−45 10378			15 57 37	−45 13.3	1	8.69		0.26				4	B9.5V	2012
143123	−60 06294	NGC 6025 - 43		15 57 37	−60 22.2	1	9.34		0.02		−0.36		2	B5 II/III	162
	−60 06296	NGC 6025 - 66		15 57 38	−60 26.0	1	10.87		0.14		0.02		1		162
143584	+50 02239	HR 5964		15 57 39	+50 01.4	2	6.07	.035	0.29	.009	0.03	.000	5	F0 IV	253,254
		AA50,429# 3		15 57 39	−53 59.	1	13.17		0.66		−0.11		3		475
		AA50,429# 4		15 57 39	−53 59.	1	12.79		0.70		−0.18		3		475
		AA50,429# 10		15 57 39	−53 59.	1	13.47		0.61		−0.07		2		475
		AA50,429# 11		15 57 39	−53 59.	1	13.75		0.76		0.02		2		475
		AA50,429# 12		15 57 39	−53 59.	1	13.89		0.70		−0.01		2		475
		AA50,429# 13		15 57 39	−53 59.	1	14.62		0.73		0.40		2		475
		AA50,429# 7	A	15 57 39	−53 59.	1	11.47		0.87		−0.10		4		475
		AA50,429# 7	B	15 57 39	−53 59.	1	11.00		1.10		−0.29		4		475
143183	−53 06947	AA50,429# 1	★ A	15 57 39	−53 59.7	1	7.75		2.71		2.71		6	K2	475
143183	−53 06947	AA50,429# 1	★ AB	15 57 39	−53 59.7	3	8.16	.160	2.16	.113	0.82	.119	29	K2	475,540,8051
	−1 03125			15 57 40	−01 39.1	1	10.33		1.25		1.24		1	K7 V	3072
143373	−13 04311			15 57 40	−13 18.9	1	9.48		0.25		0.25		1	A2 IV	1770
		AA50,429# 6		15 57 40	−54 00.2	1	12.86		0.78		−0.11		3		475
		AA50,429# 5		15 57 41	−54 00.4	1	12.54		0.68		−0.15		3		475
	−53 06948			15 57 45	−53 28.9	1	10.27		0.57		−0.23		2		181
	−59 06540	NGC 6025 - 57		15 57 46	−60 06.8	1	12.07		0.29		0.27		1		162
		NGC 6025 - 55		15 57 46	−60 10.6	1	12.59		0.55		0.17		1		162
143184	−55 06932			15 57 47	−55 52.2	1	8.82		0.11		−0.38		2	B4 III	567
	−59 06541	NGC 6025 - 54		15 57 48	−60 10.3	1	11.11		0.16		0.11		1		162
	−53 06950	AA50,429# 2		15 57 50	−54 00.3	4	9.76	.008	0.52	.007	−0.51	.017	11	B0 Ia	181,403,475,540
	−60 06297	NGC 6025 - 64		15 57 53	−60 22.2	1	11.71		0.19		0.20		2		162
143424	−15 04230			15 57 54	−15 16.4	1	9.22		0.90		0.61		2	K1 V	3072
143376	−24 12455			15 57 54	−24 39.2	1	8.52		0.31		0.20		1	A5 IV/V	1770
	−60 06298	NGC 6025 - 65		15 57 55	−60 24.4	1	10.86		0.15		0.07		1		162
		G 225 - 40		15 57 56	+58 28.1	1	15.22		1.54				1		1759
		NGC 6025 - 78		15 57 56	−60 18.4	1	12.56		0.36		0.23		1		162
	−60 06300	NGC 6025 - 42		15 57 57	−60 20.2	1	10.88		0.56		0.29		1		162
143377	−26 11153			15 57 58	−27 04.3	1	9.06		0.35		0.15		1	A5 III	1770
143378	−27 10707			15 57 58	−27 50.3	1	9.15		0.17		0.17		2	A2 V	1771
	−53 06955			15 57 58	−53 45.5	1	9.75		0.52				4		1021
	−53 06954	LSS 3448		15 57 58	−53 54.6	2	11.20	.010	0.62	.020	−0.33	.010	4		181,475
143218	−54 06938			15 57 58	−54 42.5	4	9.60	.012	0.38	.015	−0.35	.013	13	B2 II/III	181,540,567,976
143104	−66 02875	IDS15532S6616	AB	15 57 58	−66 25.1	1	9.32		−0.10		−0.72		6	B2 IIIp	400
		NGC 6025 - 53		15 58 04	−60 14.2	1	12.34		0.52		0.16		1		162

Table 1 863

HD	DM	Other Id	N Rem	α₁₉₅₀	δ₁₉₅₀	S	V	σ_V	B–V	σ_B–V	U–B	σ_U–B	n	Spectrum	References
		MtW 60 # 63		15 58 05	+29 53.5	1	12.91		0.75		0.36		6		397
143459	−7 04162	HR 5959		15 58 05	−08 16.3	2	5.54	.005	0.05	.000	−0.05	.000	7	A0 Vs	15,1256
143359	−36 10589			15 58 05	−36 50.1	1	10.02		0.49		0.21		1	A5/7 II/II	1770
143403	−28 11789			15 58 07	−28 51.7	1	9.90		0.24		0.16		2	A4 V	1771
		MtW 60 # 68		15 58 08	+29 52.3	1	15.63		0.79		0.25		7		397
	−60 06303	NGC 6025 - 63		15 58 08	−60 19.1	1	11.27		0.14		0.10		2		162
143156	−63 03776			15 58 09	−63 45.1	1	8.12		−0.09		−0.53		2	B3/5 III	400
143404	−31 12470	HR 5956		15 58 10	−31 45.0	1	6.33		1.45				4	K4 III	2007
143586	+30 02735			15 58 11	+29 46.4	2	8.58	.006	1.06	.004	0.93	.004	10	K0	397,1371
		MtW 60 # 77		15 58 11	+29 51.7	1	14.12		0.91		0.54		7		397
143483	−11 04044			15 58 12	−12 11.8	1	8.69		0.26		0.18		2	A0 III/IV	1771
	−53 06964			15 58 12	−53 44.8	1	11.71		0.41		−0.47		2		181
143427	−27 10710			15 58 13	−27 55.9	1	10.57		0.22		0.16		1	A1/2 IV	1770
	−54 06948	LSS 3449		15 58 13	−54 41.4	3	11.32	.014	0.63	.004	−0.44	.006	7		181,1586,1737
143220	−60 06305	NGC 6025 - 37		15 58 13	−60 14.7	1	10.15		0.26		0.28		3	A2 III	162
143304	−49 10258			15 58 14	−49 49.4	1	8.37		0.17		−0.29		3	B5 III	567
		MtW 60 # 83		15 58 16	+29 52.2	1	16.90		0.71		0.41		4		397
143405	−38 10816			15 58 16	−38 18.3	1	9.22		0.35		0.23		1	A7 V	1770
143321	−50 10074	IDS15546S5050	AB	15 58 19	−50 58.9	1	6.37		0.02		−0.41		3	B5 V	567
	−60 06308	NGC 6025 - 41		15 58 19	−60 21.3	1	11.29		0.12		−0.34		1		162
143361	−44 10569			15 58 21	−44 17.6	1	9.20		0.77				4	G6 V	2012
142022	−83 00593	IDS15470S8357	A	15 58 21	−84 05.8	2	7.67	.010	0.79	.010	0.43		8	G8/K0 V	285,2012
143553	+4 03096	HR 5963		15 58 22	+04 34.0	1	5.82		1.00		0.79		8	gK0	1088
		G 202 - 25		15 58 22	+45 52.8	1	11.04		0.87		0.57		2		1658
	−60 06310	NGC 6025 - 62		15 58 22	−60 18.3	1	12.65		0.36		0.24		1		162
	−60 06311	NGC 6025 - 38		15 58 22	−60 27.5	1	10.68		0.12		−0.01		2		162
143221	−62 05120			15 58 24	−62 59.5	2	7.26	.014	1.64	.014			8	K5 III	1075,2012
143472	−24 12458			15 58 25	−25 03.5	1	7.78		0.35		0.24		1	A2mA7-F2	1771
	−59 06553	NGC 6025 - 28		15 58 26	−60 12.3	1	11.19		0.34		0.28		2		162
143238	−62 05122	HR 5951		15 58 28	−62 24.2	3	6.23	.012	−0.04	.000			11	B9.5V	15,1075,2006
	−52 09179			15 58 29	−53 01.7	1	10.03		0.25		−0.18		2	B8	1586
143588	+12 02933			15 58 31	+12 37.0	1	8.10		0.39		0.11		2	F2	1648
	−60 06313	NGC 6025 - 29		15 58 31	−60 15.5	1	10.69		0.09		−0.08		4		162
142878	−76 01130			15 58 33	−77 06.0	1	7.53		1.35		1.50		5	K3 III	1704
144061	+71 00762	G 240 - 2	⋆ AB	15 58 35	+71 01.9	2	7.26	.015	0.65	.005	0.06	.015	6	G5	308,3026
143818	+57 01628			15 58 36	+57 33.4	1	9.63		0.49		−0.04		2	F8	1375
143487	−30 12753			15 58 36	−30 46.6	1	9.44		0.65		0.22		1	Apec	1770
		NGC 6025 - 52		15 58 36	−60 14.7	1	12.69		0.42		0.19		2		162
	−60 06316	NGC 6025 - 5		15 58 37	−60 20.3	2	10.21	.025	0.09	.010	−0.15	.013	6		162,1730
143287	−59 06555	NGC 6025 - 11		15 58 38	−60 11.8	1	8.34		0.04		−0.29		4	B8 Ib/II	162
143288	−60 06317	NGC 6025 - 6		15 58 38	−60 18.3	1	8.96		0.04		−0.26		5	B8	162
143473	−37 10654			15 58 40	−37 23.7	1	7.43		0.09		−0.29		40	Ap Si	1770
143463	−42 10980			15 58 43	−42 33.5	2	6.91	.000	0.48	.005			8	F6 V	1075,2012
143687	+31 02805			15 58 44	+31 42.5	2	6.61	.015	1.09	.000	0.96	.005	6	K1 III	1501,1625
		NGC 6025 - 69		15 58 44	−60 15.7	1	13.68		0.68		0.13		1		162
		NGC 6025 - 76		15 58 44	−60 19.2	1	13.04		0.41		0.32		1		162
	−57 07490			15 58 45	−57 54.3	1	11.62		0.22		−0.14		2	B7 IV	540
		NGC 6025 - 77		15 58 45	−60 19.8	1	12.96		0.62		0.28		1		162
	−59 06556	NGC 6025 - 58		15 58 46	−60 04.9	1	11.40		0.19		0.18		1		162
143309	−59 06557	NGC 6025 - 12	⋆ V	15 58 46	−60 11.9	1	9.30		0.02		−0.34		4	B8 Ib/II	162
143488	−36 10595			15 58 47	−36 36.3	1	7.00		0.00		−0.01		1	A0 V	1770
	−60 06319	NGC 6025 - 13		15 58 47	−60 17.0	1	9.62		0.05		−0.24		4		162
		NGC 6025 - 36		15 58 47	−60 17.4	1	11.02		1.81		2.11		2		162
	−60 06321	NGC 6025 - 30		15 58 49	−60 23.0	1	10.88		0.16		0.08		2		162
	−60 06322	NGC 6025 - 14		15 58 50	−60 16.0	1	9.85		0.04		−0.33		5		162
143464	−44 10577			15 58 51	−44 32.5	1	10.02		0.62				4	G5 (V)	2012
		NGC 6025 - 70		15 58 51	−60 11.4	1	13.63		0.61		0.06		1		162
143557	−22 04069			15 58 52	−22 40.1	1	9.34		0.37		0.27		1	A7 IV/V	1770
143504	−38 10823			15 58 52	−38 47.2	1	10.36		0.21		−0.09		1	A(p Si)	1770
	−60 06323	NGC 6025 - 32		15 58 52	−60 12.9	1	11.87		0.24		0.22		2		162
		NGC 6025 - 67		15 58 52	−60 15.9	1	12.71		0.42		0.15		2		162
		NGC 6025 - 75		15 58 52	−60 17.8	1	12.86		0.44		0.19		1		162
	−59 06558	NGC 6025 - 27		15 58 53	−60 12.3	1	11.04		0.18		0.14		3		162
		NGC 6025 - 31		15 58 53	−60 13.8	1	12.48		0.48		0.14		2		162
143289	−63 03778	IDS15544S6305	AB	15 58 53	−63 13.7	4	7.84	.006	0.53	.001	0.01	.004	10	F6 V	285,657,1075,1487
		G 16 - 25		15 58 54	+05 32.2	3	13.34	.006	0.59	.011	−0.14	.013	7		1064,1620,1658
143535	−29 12193	IDS15558S2926	AB	15 58 54	−29 33.6	1	9.30		0.44		0.12		1	A9 V + (F)	1771
143408	−53 06978			15 58 54	−53 17.4	1	8.81		0.09		−0.37		5	B5 II/III	567
143688	+24 02957			15 58 55	+24 35.5	1	8.82		0.47		0.01		1	F6 V	3026
143705	+29 02752			15 58 55	+29 05.2	3	7.96	.008	0.60	.008	0.08	.004	6	G0 V	1371,1733,3026
	−60 06325	NGC 6025 - 8		15 58 57	−60 14.5	1	8.86		0.04		−0.30		7		162
143567	−21 04255			15 58 58	−21 50.5	3	7.19	.004	0.09	.007	−0.11	.004	12	B9 V	26,1054,1770
	−53 06980	LSS 3451		15 58 58	−54 01.5	2	10.81	.000	0.87	.005	−0.19	.005	4		181,475
143340	−60 06326	NGC 6025 - 7		15 58 58	−60 16.1	1	8.05		0.03		−0.36		7	B9 V	162
143666	+18 03101	HR 5966		15 58 59	+17 57.3	3	5.12	.000	0.99	.014	0.74	.000	7	G8 III	15,1003,1080
	−60 06327	NGC 6025 - 15		15 59 01	−60 18.7	1	9.63		0.40		0.26		3		162
		AR Her		15 59 02	+47 03.7	2	10.65	.059	0.08	.003	0.10	.023	2	A9.5	668,699
	−60 06328	NGC 6025 - 19		15 59 03	−60 26.1	1	12.09		0.34		0.33		2		162
143538	−34 10692			15 59 04	−35 07.9	1	8.62		0.37		0.00		1	F0 V	1770
		NGC 6025 - 50		15 59 04	−60 14.6	1	13.11		0.57		0.07		1		162

HD	DM	Other Id	N Rem	α_{1950}	δ_{1950}	S	V	σ_V	B–V	σ_{B-V}	U–B	σ_{U-B}	n	Spectrum	References
	−60 06329	NGC 6025 - 35		15 59 04	−60 21.9	1	10.97		0.15		0.09		2		162
	−60 06330	NGC 6025 - 49		15 59 05	−60 15.0	1	11.20		0.21		0.19		1		162
143540	−37 10658	IDS15558S3746	AB	15 59 06	−37 54.6	1	8.63		0.19		0.13		1	A0 V	1770
	−60 06332	NGC 6025 - 4		15 59 06	−60 22.0	1	9.73		0.08		−0.22		3		162
	−59 06562	NGC 6025 - 24		15 59 07	−60 11.3	1	10.06		0.06		−0.16		5		162
143388	−60 06334	NGC 6025 - 22		15 59 07	−60 14.6	1	9.15		0.03		−0.36		7	B8	162
143761	+33 02663	HR 5968	⋆ A	15 59 08	+33 27.2	7	5.42	.018	0.60	.010	0.09	.008	18	G2 V	15,22,254,1003,1008*
143806	+39 02942	IDS15574N3927	A	15 59 08	+39 19.0	2	6.78	.005	0.12	.000	0.07	.000	6	A0	833,1601
	+23 02881	G 168 - 22		15 59 09	+23 13.0	1	10.66		0.71		0.22		2		1658
143914	+54 01788	IDS15580N5448	AB	15 59 09	+54 39.7	1	7.83		0.20		0.14		2	A0	1566
143569	−33 10881			15 59 09	−33 31.0	1	8.35		0.17		0.13		1	A0 V	1770
	−53 06984			15 59 09	−53 39.9	1	11.32		0.73		−0.19		2		181
	−60 06333			15 59 09	−60 41.3	1	11.73		0.70		0.22		2		540
		NGC 6025 - 74		15 59 10	−60 18.2	1	13.54		0.63		0.12		1		162
	−60 06335	NGC 6025 - 10		15 59 10	−60 24.6	1	10.80		0.19		0.15		2		162
	−60 06336	NGC 6025 - 23		15 59 11	−60 13.1	1	10.59		0.16		0.04		5		162
		NGC 6025 - 45		15 59 11	−60 23.1	1	13.47		1.82		2.06		1		162
		L 265 - 6		15 59 12	−49 59.	1	11.69		0.96		0.74		1		1696
143873	+51 02050			15 59 13	+50 57.5	1	8.49		0.41		0.01		2	F5	1566
143560	−37 10659			15 59 13	−37 40.5	1	9.56		0.26		0.07		1	A0 III/IV	1770
143413	−60 06338	NGC 6025 - 3		15 59 13	−60 20.9	3	8.41	.014	0.05	.008	−0.30	.017	12	B7 IV	162,1730,1732
		NGC 6025 - 71		15 59 14	−60 15.5	1	12.97		0.43		0.15		1		162
	−60 06339	NGC 6025 - 16		15 59 14	−60 18.1	1	9.88		0.05		−0.18		4		162
143600	−22 04071			15 59 16	−22 32.9	3	7.33	.005	0.10	.001	−0.07	.003	13	B9.5V	26,1054,1770
		NGC 6025 - 51		15 59 16	−60 12.7	1	13.02		1.43		1.49		3		162
143412	−59 06563	NGC 6025 - 25		15 59 17	−60 11.9	1	9.70		0.06		−0.20		5	B8	162
	−60 06341	NGC 6025 - 21	⋆ AB	15 59 17	−60 16.1	1	9.98		0.62		0.10		3		162
		NGC 6025 - 44		15 59 17	−60 22.3	1	13.35		0.51		0.09		1		162
	−60 06340	NGC 6025 - 20		15 59 17	−60 23.5	2	11.23	.025	0.24	.005	0.13	.020	5		162,1730
143510	−50 10084			15 59 19	−51 11.4	1	9.66		0.46		−0.12		3	B7 II	567
	−60 06342	NGC 6025 - 39		15 59 19	−60 27.5	1	11.24		0.40		0.22		1		162
	−60 06343	NGC 6025 - 17		15 59 20	−60 18.3	1	11.17		0.09		0.05		1		162
143343	−65 03208			15 59 21	−65 31.9	1	10.41		1.09		0.94		3	B8 Vnp	1586
		NGC 6025 - 68		15 59 22	−60 09.9	1	12.08		1.41		1.39		1		162
143578	−34 10700			15 59 23	−35 06.9	2	8.75	.002	0.10	.003	−0.19	.004	4	B8 IV/V	1586,1770
	−60 06344	NGC 6025 - 9		15 59 23	−60 22.8	1	11.83		0.22		0.19		2		162
143414	−62 05141	LT TrA		15 59 23	−62 33.3	1	10.10		−0.07		−0.71		5	WN	400
143447	−59 06568	NGC 6025 - 59		15 59 24	−60 03.6	1	10.69		0.09		−0.03		1	B9	162
	−60 06345	NGC 6025 - 46		15 59 25	−60 19.4	1	11.09		0.18		0.12		1		162
143807	+30 02738	HR 5971		15 59 26	+29 59.4	9	4.98	.009	−0.06	.009	−0.19	.012	45	A0 p Hg	15,1022,1119,1202*
		G 180 - 22		15 59 27	+37 22.8	1	13.00		0.89		0.37		1	K7	1658
143572	−43 10485			15 59 27	−43 18.2	1	7.35		0.11		−0.10		3	B9.5 V	1770
143474	−57 07500	HR 5961	⋆ AB	15 59 27	−57 38.2	6	4.63	.009	0.24	.012	0.08	.010	19	A5 IVs	15,1075,1279,2012*
143474	−57 07500	IDS15554S5730	C	15 59 27	−57 38.2	1	8.02		0.73		0.27		1		1279
143592	−34 10703			15 59 28	−35 04.1	1	8.63		0.06		−0.20		1	Ap Si	1770
143448	−60 06348	NGC 6025 - 1		15 59 28	−60 21.6	6	7.27	.033	−0.04	.013	−0.78	.017	27	B1.5V	162,540,976,1586,1732,1737
143449	−60 06349	NGC 6025 - 2		15 59 28	−60 22.5	2	8.11	.001	0.03	.005	−0.37	.008	15		162,1732
143635	−24 12473	IDS15565S2418	AB	15 59 29	−24 26.5	1	8.74		0.20		0.13		1	A1 IV	1770
143545	−48 10511			15 59 30	−48 35.1	4	9.20	.029	0.46	.008	−0.40	.020	14	B1/2ne	540,567,976,1586
	−60 06350	NGC 6025 - 33		15 59 30	−60 14.6	1	11.26		0.16		0.10		1		162
143546	−48 10512	HR 5962		15 59 32	−49 05.5	4	4.65	.003	0.92	.003	0.63	.005	20	G8 III	15,1075,2012,8015
		G 180 - 23		15 59 33	+36 57.2	2	14.36	.000	0.17	.000	−0.56	.000	2		1,3028
143619	−28 11817	HR 5965	⋆ A	15 59 33	−28 59.9	1	6.03		1.31				4	K3 III	2007
	−53 06992	LSS 3453		15 59 35	−53 32.7	1	10.27		0.56		−0.21		2	B2.5V	540
	−60 06352	NGC 6025 - 47		15 59 35	−60 19.8	1	11.01		0.31		0.22		1		162
		HRC 253, EX Lup		15 59 42	−40 10.2	2	13.18	.045	1.20	.000	−0.01	.110	2		776,825
	−60 06355	NGC 6025 - 18		15 59 43	−60 23.2	1	12.26		0.52		0.29		1		162
143714	−17 04472			15 59 45	−18 02.0	1	8.94		0.43		0.26		1	A2 III	1770
143692	−22 04072			15 59 46	−23 02.6	1	8.01		0.21		0.17		1	A2 IV/V	1770
	−54 06977			15 59 46	−54 21.6	1	10.80		0.74		0.15		2	B6 II-III	540
143652	−33 10890			15 59 48	−33 55.8	1	10.27		0.31		0.26		1	A0/1 IV	1770
143715	−24 12481	IDS15569S2444	A	15 59 51	−24 52.6	1	7.04		0.18		−0.01		1	A0 V	1771
143715	−24 12481	IDS15569S2444	AB	15 59 51	−24 52.6	1	7.07		0.15		−0.01		1	A0 V	1770
143511	−59 06576	NGC 6025 - 26		15 59 51	−60 04.7	1	8.31		0.21		0.18		4	A0 IV/V	162
	−60 06361	NGC 6025 - 40		15 59 52	−60 23.9	1	11.15		0.20		0.06		1		162
	+62 01446	G 225 - 43		15 59 53	+61 48.0	2	9.99	.015	1.29	.015	1.23		3	M0	906,3072
	−56 07221	LSS 3454		15 59 53	−56 25.0	3	10.12	.012	0.44	.010	−0.51	.018	6		181,540,1586
	−60 06362	NGC 6025 - 34		15 59 53	−60 15.4	1	10.51		0.10		−0.07		3		162
143918	+35 02755			15 59 55	+35 00.0	1	8.83		0.28		0.04		3	A3	1601
		NGC 6025 - 72		15 59 56	−60 14.6	1	13.19		0.66		0.19		1		162
143696	−33 10891			15 59 58	−33 58.3	1	10.10		0.36		0.24		2	A0 IV	1770
143675	−34 10708			15 59 58	−35 09.0	1	8.06		0.20		0.09		1	A5 IV/V	1770
143694	−29 12209			15 59 59	−29 31.7	1	10.09		0.34		0.18		1	A1 III	1770
143781	−8 04136			16 00 00	−08 21.4	1	7.70		1.56		1.90		4	K5	1657
143746	−22 04075			16 00 00	−23 01.6	1	10.29		0.26		0.17		2	A1/2 IV	1770
	−59 06578	NGC 6025 - 60		16 00 00	−60 06.5	1	10.48		0.10		−0.01		1		162
	−59 06579	NGC 6025 - 73		16 00 00	−60 12.3	1	10.49		1.14		0.96		1		162
	−60 06366	NGC 6025 - 48		16 00 03	−60 15.5	1	10.87		1.10		1.03		3		162
143699	−38 10832	HR 5967		16 00 04	−38 27.9	7	4.89	.005	−0.14	.002	−0.58	.016	28	B6 IV	15,26,1020,1075,1770*
143728	−27 10730			16 00 06	−28 00.6	1	9.51		0.03		−0.18		4	B9.5 IV	1770

Table 1 865

HD	DM	Other Id	N	Rem	α_{1950}	δ_{1950}	S	V	σ_V	B–V	σ_{B-V}	U–B	σ_{U-B}	n	Spectrum	References
143747	−26 11170				16 00 07	−26 29.5	1	8.45		0.12		0.08		2	A0 V	1770
143894	+23 02886	HR 5972			16 00 08	+22 56.5	6	4.83	.004	0.07	.006	0.08	.022	33	A3 V	15,1022,1363,3023*
143785	−19 04295				16 00 12	−19 42.2	1	7.32		0.27		0.26		1	A2/3 III	1770
143786	−24 12488				16 00 13	−24 31.6	1	8.52		0.33		0.17		2	A1mA6-A8	1770
143766	−26 11171				16 00 13	−26 34.6	1	6.96		0.52		-0.02		1	F7 V	1770
	−60 06368	NGC 6025 - 61			16 00 13	−60 16.8	1	11.40		0.16		0.15		1		162
143748	−31 12500				16 00 14	−31 52.3	1	9.59		0.36		0.25		1	B8 V	1770
143605	−56 07228				16 00 15	−56 18.2	3	9.07	.011	0.14	.004	-0.48	.011	8	B3 V	181,540,976
144015	+40 02962				16 00 17	+40 09.6	1	6.86		1.29		1.40		3	K1 III	1501
143731	−37 10669				16 00 18	−37 27.4	1	10.32		0.51		0.21		4	A9 V	1770
330345	−49 10290	LSS 3456			16 00 18	−49 44.1	2	10.27	.003	0.63	.016	-0.34		6	B0.5V	540,1021
143787	−25 11295	HR 5969			16 00 19	−25 43.6	2	4.98	.024	1.23	.005	1.23		6	K3 III	2007,3052
143733	−38 10837				16 00 19	−38 35.4	1	10.74		0.23		-0.12		1	B8 IV/V	1770
143346	−72 01902	HR 5955			16 00 19	−72 16.0	3	5.69	.004	1.17	.000	1.26	.000	14	K1 III	15,1075,2038
	−6 04346	G 153 - 21			16 00 22	−06 19.3	1	10.19		0.71		0.12		2		1658
143788	−28 11827				16 00 22	−28 21.5	1	9.43		0.16		0.16		3	A2 III	1770
		Ly 6 - 21			16 00 22	−51 44.7	1	10.72		0.12		-0.06		2		477
143790	−31 12505	HR 5970			16 00 24	−31 51.8	1	6.01		0.47				4	F4 IV	2007
143700	−47 10516				16 00 25	−47 20.3	4	9.31	.027	0.40	.016	-0.42	.024	12	B2/3(III)ne	540,567,976,1586
	−55 07003	LSS 3455			16 00 27	−55 15.7	3	10.48	.010	0.75	.009	-0.29	.009	7	O9.5V	181,540,1737
144283	+65 01095				16 00 28	+65 05.3	1	7.07		0.09		0.10		2	A2	985
143767	−34 10715				16 00 30	−35 05.6	1	9.58		0.32		0.19		1	A1 IV	1770
143822	−28 11831				16 00 35	−28 15.7	1	9.42		0.14		0.10		4	A0 V	1770
		Ly 6 - 25			16 00 39	−51 48.6	1	12.93		0.98		0.20		2		477
		G 137 - 78			16 00 42	+20 44.4	2	12.55	.005	1.63	.000	1.15	.059	5		203,3078
143792	−41 10507				16 00 42	−41 34.4	1	8.22		0.40		0.25		3	B9 V	1770
143738	−51 09746	Ly 6 - 2			16 00 43	−51 48.3	5	9.20	.008	0.11	.011	-0.55	.012	13	B5	412,477,540,567,976
	−54 06990	LSS 3457			16 00 43	−55 06.9	1	11.33		0.59		-0.40		1		181
143549	−68 02610				16 00 43	−68 18.0	1	7.69		-0.11		-0.61		1	B3 III	400
143812	−37 10676				16 00 47	−38 06.2	1	8.80		0.16		0.10		1	A0 V	1770
330344	−49 10293				16 00 48	−49 53.1	3	9.86	.011	0.45	.004	-0.31	.009	10	B2 IV	540,567,976
144204	+53 01834	HR 5981			16 00 49	+53 03.2	1	5.93		1.48				1	K2	71
143899	−19 04296				16 00 49	−19 26.3	1	8.30		1.08		0.65		1	G6/8 II	565
143751	−51 09748				16 00 49	−51 30.4	1	9.66		0.45		-0.14		3	B3 II/III	567
143824	−40 10163				16 00 51	−40 42.5	1	7.57		0.09		0.06		1	A1 V	1770
143846	−36 10618				16 00 53	−37 10.2	1	7.86		0.60				4	G2 V	2033
143900	−24 12499	HR 5973			16 00 54	−24 35.4	1	6.21		1.38				4	K2/3 III	2007
	−54 06991	LSS 3458			16 00 55	−54 58.7	1	11.54		0.46		-0.35		1		181
		Ly 6 - 3			16 00 56	−51 46.2	2	11.40	.000	0.52	.025	-0.14	.020	4		412,477
144284	+58 01608	HR 5986			16 00 57	+58 41.9	17	4.00	.007	0.52	.003	0.10	.004	242	F8 IV-V	1,15,61,71,985,1077*
		Ly 6 - 28			16 00 57	−51 47.8	1	13.56		0.94		0.52		2		477
		Ly 6 - 32			16 00 57	−51 49.8	1	14.76		1.28				2		477
144302	+60 01645				16 00 58	+59 46.1	1	7.84		0.63		0.21		2	G5	3016
		Ly 6 - 35			16 00 58	−51 48.5	1	16.44		1.43				1		477
		Ly 6 - 12			16 00 59	−51 50.0	2	13.30	.030	1.13	.005	0.51	.070	4		412,477
		Ly 6 - 27			16 01 00	−51 47.	1	13.39		2.24				2		477
		Ly 6 - 29			16 01 00	−51 47.	1	13.86		0.86				2		477
		Ly 6 - 30			16 01 00	−51 47.	1	13.90		2.58				3		477
		Ly 6 - 31			16 01 00	−51 47.	1	14.45		1.35				2		477
		Ly 6 - 38			16 01 01	−51 48.7	1	17.36		1.30				1		477
		Ly 6 - 24			16 01 02	−51 45.7	1	12.66		0.85		0.20		2		477
		Ly 6 - 4			16 01 02	−51 46.6	2	11.64	.015	0.61	.010	0.15	.035	4		412,477
143752	−54 06993				16 01 02	−55 00.2	1	8.37		0.19		-0.34		5	B3 III	567
		Ly 6 - 1			16 01 03	−51 49.1	3	10.59	.029	0.21	.011	0.07	.010	44		412,435,477
		Ly 6 - 11			16 01 03	−51 50.1	3	13.45	.011	1.14	.041	0.42	.086	5		412,435,477
143815	−48 10528				16 01 04	−48 41.2	1	9.20		0.38		0.01		3	B9 III	567
143956	−19 04298				16 01 05	−19 37.8	1	7.80		0.14		-0.02		1	B9 V	1770
		Ly 6 - 19			16 01 05	−51 49.4	2	14.74	.015	1.16	.005	0.24	.049	3		435,477
		Ly 6 - 20			16 01 05	−51 49.5	2	15.18	.024	1.22	.019	0.43		3		435,477
143495	−72 01903				16 01 05	−72 34.8	2	9.47	.005	0.07	.000	-0.37	.005	3	B9/A0 IIpSi	55,400
		G 16 - 28			16 01 06	+02 45.4	1	12.09		0.72		0.07		2		333,1620
143902	−32 11386	HR 5974			16 01 06	−33 04.6	3	6.09	.007	0.35	.004			11	F3 III	15,2013,2029
		Ly 6 - 37			16 01 06	−51 50.1	1	16.48		1.23				1		477
143937	−21 04264	IDS15582S2139		AB	16 01 07	−21 47.4	1	8.64		0.91		0.49		4	K0 V	158
		Ly 6 - 36			16 01 07	−51 48.1	1	16.45		1.04				1		477
		Ly 6 - 75		⋆ V	16 01 07	−51 49.1	2	11.42	.180	1.94	.130	1.57	.205	2		477,689
		Ly 6 - 7			16 01 07	−51 49.4	3	12.54	.018	2.26	.037	1.59	.102	10		412,435,477
		Ly 6 - 17			16 01 08	−51 47.6	2	12.80	.025	0.83	.000	0.31	.040	5		412,477
		Ly 6 - 8		AB	16 01 08	−51 49.6	3	12.45	.015	1.17	.019	0.50	.014	6		412,435,477
		Ly 6 - 9			16 01 08	−51 49.8	3	13.48	.025	1.13	.042	0.43	.033	7		412,435,477
		Ly 6 - 10			16 01 08	−51 49.9	3	13.16	.038	1.08	.007	0.06	.014	8		412,435,477
		Ly 6 - 18			16 01 09	−51 47.9	2	14.43	.070	0.86	.055	0.21	.050	4		412,477
		Ly 6 - 34			16 01 09	−51 48.8	1	16.03		1.27		0.59		1		477
	−54 06995	LSS 3459			16 01 09	−54 46.9	1	10.82		0.65		-0.36		2		181
143682	−62 05172				16 01 09	−62 34.4	1	8.42		1.28				4	K1 III	1075
		Ly 6 - 16			16 01 10	−51 47.4	2	12.42	.020	0.94	.015	0.29	.005	5		412,477
		Ly 6 - 14			16 01 11	−51 49.4	3	14.06	.037	1.17	.023	0.37	.034	6		412,435,477
		Ly 6 - 13			16 01 11	−51 49.8	2	14.85	.085	1.03	.055	0.76	.265	4		412,477
		Ly 6 - 33			16 01 12	−51 49.1	1	14.87		1.24				2		477
143864	−46 10541				16 01 13	−46 16.8	1	9.67		0.24		-0.23		5	B8	567

HD	DM	Other Id	N	Rem	α_{1950}	δ_{1950}	S	V	σ_V	B–V	σ_{B-V}	U–B	σ_{U-B}	n	Spectrum	References
		Ly 6 - 15			16 01 13	−51 49.0	2	14.35	.085	0.80	.020	0.23	.050	4		412,477
144206	+46 02142	HR 5982			16 01 14	+46 10.5	6	4.74	.012	-0.10	.012	-0.32	.004	18	B9 III	15,1079,1363,3033*
143926	−33 10903				16 01 14	−33 41.7	1	8.51		0.59		0.38		1	A7mF0-F0	1770
143796	−55 07018				16 01 14	−56 12.2	3	7.48	.213	1.91	.017	1.93	.016	13	M2 Iab/b	1673,2012,3040
143974	−23 12669				16 01 15	−23 26.1	1	10.43		0.56		0.10		1	F3/5 V	1770
143959	−27 10742	IDS15582S2720	A		16 01 15	−27 28.4	1	9.78		0.23		0.19		1	A2 IV	1770
		Ly 6 - 6			16 01 16	−51 46.5	2	11.53	.010	1.67	.010	1.88	.015	5		412,477
	−53 07041				16 01 16	−53 12.7	1	10.38		0.36		-0.35		2	B2.5V	540
144046	+5 03131	HR 5976		★ A	16 01 17	+05 07.4	2	6.08	.005	0.96	.000	0.72	.027	11	G9 III	1088,3077
143975	−24 12506				16 01 18	−25 08.3	1	8.24		0.25		0.18		2	A5 V	1770
143928	−37 10680	HR 5975		★ A	16 01 18	−37 43.5	2	5.90	.000	0.40	.000			8	F3 V	2006,2018
		AV Ser			16 01 19	+00 44.0	2	11.04	.203	0.29	.056	0.21	.004	2	F2	668,699
143960	−32 11390				16 01 19	−32 16.2	1	10.01		0.39		0.17		2	A8/9 V	1770
143817	−55 07025				16 01 19	−55 40.5	1	7.51		0.84		0.53		3	F7 II	8100
143994	−23 12670				16 01 20	−23 36.7	1	10.71		1.16		0.92		1	F2 V	1770
	+60 01647	IDS16005N6021	AB		16 01 22	+60 12.5	1	10.49		0.68		0.18		2	G5	1375
143939	−39 10298	IDS15580S3910	A		16 01 22	−39 17.9	1	6.96		-0.08		-0.19		2	Ap SiCr	1770
	+67 00922	AG Dra			16 01 23	+66 56.4	5	9.57	.160	1.11	.217	-0.57	.323	6	K1 IIpe	867,1591,1753,3025,8094
143927	−37 10681				16 01 23	−37 40.8	1	7.08		0.00		-0.35		3	B8/9 V	1770
143938	−36 10626				16 01 25	−36 47.5	1	10.14		0.27		-0.11		1	B8 II/III	1770
	−51 09751	Ly 6 - 5			16 01 25	−51 46.2	2	10.33	.005	0.24	.005	0.16	.005	4		412,477
143830	−54 07001				16 01 26	−54 55.3	1	10.29		0.25		-0.18		5	B8	567
143756	−62 05177				16 01 28	−62 17.5	1	9.25		-0.02		-0.51		3	B5 III	400
144208	+37 02708	HR 5983			16 01 29	+36 46.1	1	5.83		0.56				2	F7 III+A2 V	71
144033	−13 04330				16 01 29	−13 38.4	1	10.20		0.50		0.31		1	A2/3mF2-F3	1770
143961	−37 10682				16 01 32	−37 46.4	1	9.21		0.34		0.23		1	A1 V	1770
		Ly 6 - 26			16 01 32	−51 48.3	1	13.16		0.68		0.20		2		477
		Ly 6 - 22			16 01 32	−51 50.3	1	11.66		0.59		0.18		2		477
143867	−52 09224				16 01 32	−52 49.2	1	10.75		0.31		-0.16		5	B2/3 (III)	567
	+42 02667	G 180 - 24			16 01 33	+42 23.2	4	9.85	.009	0.47	.021	-0.20	.009	10	sdF8	308,792,1064,3025
		Ly 6 - 23			16 01 33	−51 49.2	1	11.99		0.48		0.28		2		477
144149	+18 03108				16 01 35	+17 56.3	1	6.75		1.21		1.27		3	K0	1648
		G 153 - 23			16 01 35	−03 40.7	1	14.84		1.02		0.44		2		3062
144070	−10 04237	HR 5978		★ AB	16 01 37	−11 14.2	7	4.16	.013	0.45	.007	0.01	.009	26	F6 IV-V+G8V	15,938,1075,1088,2012*
144070	−10 04237	IDS15589S1106	C		16 01 37	−11 14.2	1	7.30		0.75				2		3030
143977	−37 10684				16 01 37	−37 43.4	1	9.99		0.39		0.28		1	A0 V	1770
		LP 648 - 17			16 01 40	−06 07.9	1	15.37		1.69				1		1759
144071	−11 04056				16 01 40	−11 43.3	1	7.51		1.82		2.16		2	K5 III	262
143512	−74 01516				16 01 40	−74 42.3	1	9.29		0.62		0.04		2	G0 V	1117
144087	−11 04057	IDS15589S1110	A		16 01 41	−11 18.8	2	7.44	.013	0.75	.004	0.31	.011	6	G8 V	1028,3072
144087	−11 04057	IDS15589S1110	AB		16 01 41	−11 18.8	1	6.95		0.77		0.38		4	G8 V	158
144088	−11 04058	IDS15589S1110	B		16 01 41	−11 18.8	2	8.02	.004	0.85	.006	0.53	.011	6	K2 V	1028,3072
144090	−12 04407				16 01 47	−13 14.5	1	9.02		0.40		0.23		1	A1/2 III	1770
		LP 444 - 18			16 01 50	+18 50.2	1	12.47		0.98		0.69		1		1696
	−50 10122				16 01 53	−50 55.6	2	10.26	.016	0.56	.002	-0.24	.020	7	B2.5IV	540,567
144050	−28 11851	IDS15589S2814	A		16 01 57	−28 22.2	1	9.76		0.32		0.23		3	A1 III	1770
144035	−35 10710				16 01 58	−36 07.9	1	8.84		0.28		0.20		1	A0 III/IV	1770
143832	−62 05179				16 01 58	−63 12.4	1	7.50		0.09				4	A1/2 V	1075
144287	+25 03020				16 01 59	+25 22.9	4	7.10	.021	0.77	.000	0.34	.021	8	G8 V	22,1003,1080,3026
144113	−20 04392				16 01 59	−20 45.8	1	8.11		0.31		0.20		2	A3 IV/V	1770
144405	+53 01835				16 02 00	+53 21.2	1	7.90		1.22		1.21		2	K5	1566
144074	−31 12533				16 02 03	−31 19.2	1	8.45		0.22		0.16		1	A5 V	1770
	+45 02368	G 202 - 28			16 02 08	+45 25.6	1	10.32		1.25				1	M0	1017
144098	−32 11399	IDS15590S3237	AB		16 02 08	−32 45.5	1	9.20		0.30		0.20		1	A1 V	1770
143888	−61 05539				16 02 09	−61 58.1	1	9.40		0.01		-0.41		3	B7 III	400
144135	−23 12682				16 02 10	−23 54.3	1	10.09		1.13		0.93		2	K1 II	1770
144051	−37 10688				16 02 10	−37 40.3	1	10.75		0.64		0.22		1	A9/F0 V	1770
144542	+59 01697	HR 5995			16 02 14	+59 32.9	2	6.19	.001	1.58	.007	1.91		5	M1 III	71,3035
144115	−31 12535				16 02 14	−31 19.2	1	7.08		0.25		0.15		1	A2mA2-F2	1770
143983	−55 07051				16 02 14	−55 38.5	1	7.91		1.57		1.88		3	K4 III	8100
144174	−16 04213				16 02 15	−16 29.8	1	10.58		0.14		0.02		1	B9 V	1770
144116	−33 10915				16 02 15	−33 52.1	1	8.00		0.24		0.19		1	A2 IV/V	1770
144117	−34 10746				16 02 18	−34 24.4	1	9.86		0.20		0.10		1	A0 V	1770
144214	−11 04060				16 02 19	−11 30.2	1	9.09		0.50		0.70		2	F5	262
144175	−23 12687				16 02 19	−23 32.0	1	7.67		0.08		-0.01		3	B9 V	1770
	−51 10368	LSS 3684			16 02 19	−51 30.2	1	10.51		0.25		-0.49		1		954
144359	+34 02731				16 02 20	+34 18.9	1	6.73		0.06		0.02		2	A0	252
144137	−31 12537	IDS15592S3158	AB		16 02 20	−32 06.2	1	8.93		0.52				4	F5 V	2012
144360	+32 02671				16 02 23	+32 38.0	1	7.92		1.27		1.32		1	K2	1746
144118	−39 10302	IDS15590S3934	A		16 02 23	−39 42.5	1	7.82		0.24		0.09		1	A5 V	1770
	−71 01956				16 02 26	−71 13.6	1	10.44		0.41		-0.20		2		1696
	−52 09235	LSS 3461			16 02 27	−52 40.7	1	10.90		0.51		-0.42		2		181
144179	−32 11405	IDS15593S3235	ABC		16 02 29	−32 43.5	2	7.84	.000	0.82	.000	0.37		8	K0 V	285,2012
		L 153 - 57			16 02 30	−61 22.	1	14.09		1.40		1.12		1		3073
144190	−23 12690				16 02 31	−23 57.3	1	10.19		0.16		0.13		2	A0 V	1770
144177	−28 11861				16 02 31	−28 26.6	1	8.89		0.53		-0.02		1	F5 V	1770
144218	−19 04308	HR 5985		★ C	16 02 32	−19 40.0	6	4.91	.013	-0.02	.007	-0.68	.014	16	B2 V	1,15,1034,1586,1770,6002
144217	−19 04307	HR 5984		★ AB	16 02 32	−19 40.2	11	2.56	.049	-0.07	.008	-0.86	.015	47	B0.5V	1,15,1034,1054,1075,1088*
144191	−25 11323				16 02 32	−25 59.0	1	9.96		0.29		0.23		2	A1/2mA5-A7	1770
144221	−22 04079				16 02 33	−23 10.8	1	11.25		0.75		0.30		4	A9 V	1770

Table 1 867

HD	DM	Other Id	N	Rem	α_{1950}	δ_{1950}	S	V	σ_V	B–V	σ_{B-V}	U–B	σ_{U-B}	n	Spectrum	References
144178	−29 12259				16 02 33	−29 17.6	1	10.59		0.22		0.22		1	A1 III	1770
143967	−62 05184				16 02 35	−62 37.7	1	7.69		−0.04				4	B7 III	1075
144223	−25 11327				16 02 38	−26 05.9	1	10.45		0.30		0.16		1	A8 V	1770
144255	−22 04080				16 02 43	−22 50.1	1	10.77		0.46		0.07		1	F5	1770
144253	−20 04399				16 02 44	−20 18.6	5	7.40	.025	1.05	.019	0.93	.028	14	K3/4 V	158,196,1067,1705,3008
144254	−21 04269				16 02 46	−21 42.2	1	7.79		0.18		0.12		4	A0 V	1770
144194	−35 10716				16 02 46	−35 54.3	1	9.23		0.35		0.24		1	A7 IV	1770
144195	−36 10635				16 02 47	−36 23.3	1	9.97		0.49		0.29		1	A9 V	1770
	−53 07073	LSS 3462			16 02 49	−54 05.5	2	10.52	.000	0.99	.000	−0.11	.000	4		181,1737
144273	−19 04309				16 02 50	−19 32.8	1	7.53		0.12		−0.04		2	B9 V	1770
143999	−62 05187	U TrA			16 02 51	−62 46.6	1	7.64		0.49				1	F8 Ib/II	1484
144224	−35 10720				16 02 52	−35 38.1	1	9.75		0.44		0.19		1	A9 V	1770
144274	−22 04081				16 02 53	−23 14.5	1	9.66		0.58		0.16		1	F3 V	1770
144257	−30 12822				16 02 55	−31 03.9	1	11.61		0.63		0.14		2	A1 III/IV	1770
144289	−20 04402				16 02 56	−20 20.8	1	9.31		0.27		0.21		2	A1 IV/V	1770
144197	−44 10625	HR 5980			16 02 57	−45 02.4	6	4.72	.008	0.23	.005	0.15	.009	24	A1mA7-F(3)	15,355,1075,2012,3023,8015
		L 201 - 29			16 03 00	−55 56.	1	14.28		1.49		1.13		1		3062
144291	−25 11335				16 03 02	−25 33.4	1	9.87		0.48		0.10		1	F2 V	1770
144362	−5 04234	HR 5989		★ APB	16 03 04	−06 09.4	3	6.35	.066	0.43	.022	−0.03	.029	10	F2 IV	254,1088,3032
144362	−5 04234	IDS16004S0601		C	16 03 04	−06 09.4	1	11.31		1.09		0.85		1		3032
144362	−5 04234	IDS16004S0601		D	16 03 04	−06 09.4	1	11.87		1.32		1.12		1		3032
	+68 00861	G 240 - 5			16 03 05	+68 39.6	1	9.74		0.73		0.20		3	G5	3026
		G 180 - 27			16 03 06	+39 17.6	2	14.22	.033	1.72	.000			4		538,906
144334	−23 12700	HR 5988, V929 Sco			16 03 07	−23 28.3	5	5.91	.004	−0.08	.006	−0.57	.004	17	B8 V	26,1054,1079,1770,2035
144144	−53 07081				16 03 07	−53 41.6	1	8.82		0.22		0.04		5	B9 III	567
	−11 04062				16 03 10	−11 28.7	1	10.91		1.52		1.48		2	K3 III	262
	−53 07082	LSS 3463			16 03 10	−54 02.9	1	9.80		0.83		−0.17		2	B0.5II-III	540
144277	−38 10874				16 03 11	−39 07.5	1	7.87		0.10		0.06		1	A1 V	1770
144198	−48 10548				16 03 11	−48 38.9	1	8.68		0.35		0.01		3	B8/9 III	567
	−52 09243	LSS 3464			16 03 11	−52 55.6	2	10.30	.063	1.53	.010	0.55	.036	3		1586,1737
144426	+8 03134	HR 5992			16 03 12	+08 13.8	3	6.29	.004	0.08	.005	0.10	.004	10	A3 m	15,1417,8071
144579	+39 02947	IDS16013N3926	A		16 03 13	+39 17.4	7	6.66	.013	0.73	.004	0.21	.008	27	G8 V	22,908,1013,1080,1197*
	−11 04063				16 03 15	−11 39.1	1	10.29		1.33		1.18		2	G8 III	262
144294	−36 10642	HR 5987			16 03 18	−36 40.1	7	4.22	.010	−0.18	.011	−0.69	.018	21	B2.5Vn	15,26,1075,1770,2012*
144390	−5 04235	HR 5990		★ AB	16 03 20	−06 00.3	1	6.40		1.03		0.80		8	K0	1088
144393	−11 04064				16 03 20	−12 10.2	1	8.51		0.59		0.05		2	F7/8 V	262
144296	−40 10192	IDS16000S4021	A		16 03 20	−40 29.7	1	8.45		0.25		0.23		1	A3 V	1770
144366	−19 04311	IDS16005S1931	A		16 03 21	−19 38.9	1	9.27		0.28		0.18		1	A1 IV/V	1770
144293	−36 10643				16 03 21	−36 20.4	1	10.49		0.45		0.31		1	A3 IV	1770
144335	−30 12833				16 03 22	−31 01.5	1	10.43		0.25		0.17		1	A1 IV	1770
		G 153 - 24			16 03 23	−10 25.9	1	14.44		0.92		0.22		2		3062
	−53 07090	LSS 3466			16 03 23	−54 03.7	3	9.34	.016	0.80	.012	−0.27	.027	6	B0.5Iab	181,403,540
144183	−55 07079	HR 5979			16 03 23	−56 03.5	2	6.16	.000	0.58	.005	0.34		8	F2 II	2035,8100
144411	−10 04243				16 03 27	−10 51.7	1	9.69		0.98		0.81		2	K2 V	3072
144516	+10 02955				16 03 28	+10 04.2	1	6.96		0.04		−0.04		2	A2	1733
144315	−40 10193				16 03 28	−40 34.0	1	10.07		0.15		−0.31		3	B2/3 III	567
144412	−18 04235				16 03 31	−18 31.3	1	8.58		0.30		0.22		1	A2 V	1770
144394	−20 04404				16 03 32	−21 06.3	1	9.95		0.30		0.22		2	A1/2 V	1770
144515	+11 02910	NQ Ser		★	16 03 33	+10 49.2	2	8.27	.005	0.79	.005	0.27	.016	4	G8 V	1003,3026
144368	−33 10924				16 03 33	−33 48.5	1	11.04		0.32		0.17		1		1770
144379	−26 11209				16 03 34	−26 59.1	1	10.43		0.26		0.17		2	A1 IV	1770
144369	−33 10924				16 03 34	−33 48.7	1	10.74		0.33		0.21		1	A1/2 V	1770
		NGC 6031 - 7			16 03 34	−53 53.0	1	13.04		0.39		0.03		2		412
144396	−26 11210				16 03 36	−26 17.0	1	9.27		0.41		0.23		1	A1mA7-F0	1770
	−53 07141	NGC 6031 - 12			16 03 36	−53 53.7	1	10.86		0.29		−0.15		4		412
144449	−11 04067				16 03 37	−12 00.3	1	8.80		0.62		0.16		2	F8	262
		LP 861 - 23			16 03 37	−23 20.3	1	13.26		0.76		0.13		3		1696
		NGC 6031 - 5			16 03 37	−53 52.6	1	14.31		0.91		0.32		2		412
		NGC 6031 - 13			16 03 38	−53 53.4	1	12.71		0.38		0.13		2		412
144413	−19 04313				16 03 39	−20 05.8	1	10.14		0.30		0.27		2	Ap SrEu(Cr)	1770
144395	−21 04272				16 03 39	−22 01.2	1	9.90		0.48		0.12		1	F0 V	1770
		AAS35,161 # 88		V	16 03 39	−38 54.3	1	12.93		0.70		−0.47		1		5005
144243	−54 07045				16 03 39	−54 31.2	1	8.16		0.48		−0.04		1	F7 V	96
144431	−22 04084				16 03 40	−22 45.5	1	9.32		0.50		−0.01		1	F3 V	1770
		NGC 6031 - 4			16 03 40	−53 52.7	1	12.90		0.37		0.10		2		412
		NGC 6031 - 1			16 03 40	−53 53.1	1	13.87		0.59		0.49		2		412
	−56 07301				16 03 40	−56 20.6	1	11.26		0.30		−0.23		2	B5 IV	540
		NGC 6031 - 31			16 03 43	−53 52.5	1	12.21		0.52		0.19		2		412
		NGC 6031 - 29	A		16 03 43	−53 52.7	1	13.25		0.37		0.20		2		412
		NGC 6031 - 29	B		16 03 43	−53 52.7	1	13.56		0.52		0.43		2		412
		NGC 6031 - 30			16 03 44	−53 52.6	1	12.40		0.36		−0.02		2		412
		NGC 6031 - 24			16 03 44	−53 53.0	1	13.15		0.59		0.36		2		412
144351	−42 11037				16 03 45	−43 09.3	1	7.70		0.21		0.20		1	A1 V	1770
		NGC 6031 - 22			16 03 45	−53 53.4	1	12.30		0.41		0.14		2		412
144397	−33 10925				16 03 47	−34 09.1	1	9.60		0.44		0.33		2	A3 III	1770
144350	−41 10526				16 03 47	−41 56.1	1	7.18		1.60		1.69		9	K2/3 III	1673
		NGC 6031 - 28			16 03 47	−53 52.6	1	13.12		0.50		0.38		2		412
		NGC 6031 - 26			16 03 47	−53 52.9	1	13.88		0.80		0.67		2		412
144494	−12 04415				16 03 48	−12 21.4	1	8.94		1.10		0.82		2	G8/K0 III	262
		NGC 6031 - 38			16 03 48	−53 52.5	1	12.55		0.43		0.10		2		412

HD	DM	Other Id	N Rem	α₁₉₅₀	δ₁₉₅₀	S	V	σ_V	B–V	σ_B–V	U–B	σ_U–B	n	Spectrum	References
		AAS35,161 # 90		16 03 49	−39 03.0	1	14.84		1.66		0.62		1		5005
	−53 07149	NGC 6031 - 74	⋆ V	16 03 49	−53 53.1	1	10.96		0.34		-0.15		3		412
144493	−11 04068			16 03 50	−11 46.3	1	7.56		0.30		0.08		2	F0	262
144262	−56 07305			16 03 51	−56 59.4	1	9.70		0.23		-0.26		2	B6 Ib/II	540
144009	−70 02163			16 03 51	−70 55.9	2	7.23	.005	0.73	.005	0.31		8	G8 V	158,2012
144452	−21 04274			16 03 52	−21 47.2	1	9.32		0.68		0.16		1	F2 V	1770
144470	−20 04405	HR 5993		16 03 53	−20 32.1	8	3.95	.011	-0.04	.008	-0.82	.005	33	B1 V	15,26,401,1054,1075,1770*
144414	−33 10926			16 03 53	−34 01.7	1	9.64		0.43		0.28		1	A1mA5/7-F0	1770
144451	−21 04275	IDS16010S2136	A	16 03 54	−21 43.8	1	7.82		0.34		0.05		1	F0 V	1770
144432	−27 10778	IDS16008S2727	AB	16 03 54	−27 35.2	1	8.20		0.36		0.13		1	A9/F0 V	1770
144472	−23 12709			16 03 55	−24 01.1	1	8.74		0.25		0.18		2	A2/3 V	1770
145742	+81 00541			16 03 58	+80 45.8	1	7.57		1.06		0.95		3	K0	196
144433	−32 11425			16 03 58	−33 00.8	1	7.61		0.24		0.08		1	A6 IV(m)	1770
144723	+36 02689			16 03 59	+36 39.7	1	7.38		0.87		0.47		2	G5	1601
144471	−23 12711			16 03 59	−23 45.3	1	9.89		0.45		0.12		1	F2 V	1770
144415	−36 10648	HR 5991	⋆ A	16 03 59	−36 37.3	3	5.72	.007	0.30	.004			11	F1 IV	15,2018,2028
144497	−22 04086			16 04 01	−22 50.5	1	9.23		0.95		0.64		1	K2 V	3072
144523	−15 04246			16 04 02	−16 00.7	1	8.06		0.34		0.17		1	A6/7 IV	1770
	−53 07124	LSS 3468		16 04 02	−53 22.5	1	10.52		0.58		-0.33		2		181
144320	−54 07060			16 04 02	−54 59.8	5	9.19	.024	0.24	.010	-0.55	.013	16	B1 Vne	181,540,567,976,1586
144455	−33 10931	IDS16009S3326	A	16 04 04	−33 34.5	1	7.95		0.15		0.04		1	A0 IV/V	1770
144231	−62 05193	LL TrA		16 04 04	−62 50.9	1	6.88		-0.08				4	Ap Si	1075
	−11 04069			16 04 06	−12 11.0	1	11.51		0.68		0.24		2	G5	262
144473	−31 12570	IDS16010S3119	AB	16 04 06	−31 27.2	1	9.48		0.35		0.15		1	A9 V	1770
144416	−42 11042			16 04 08	−42 20.5	1	9.12		0.24		0.10		3	B9.5 V	1770
144499	−26 11215			16 04 09	−26 58.0	1	8.53		0.43		0.24		1	A9 V	1770
	−52 09255	LSS 3469		16 04 09	−52 19.0	1	11.00		0.68		-0.25		2		181
		AW Ser		16 04 12	+15 29.8	1	12.39		0.18		0.12		1	F6	699
144569	−16 04219			16 04 13	−16 48.6	1	7.90		0.16		0.08		2	B9.5 V	1770
144525	−25 11355			16 04 13	−25 36.6	1	9.42		0.22		0.16		2	A2 III	1770
144584	−11 04071			16 04 16	−12 09.9	1	9.86		0.42		0.30		2	A7 V	262
144585	−13 04342	HR 5996		16 04 16	−13 56.3	6	6.32	.017	0.66	.003	0.19	.031	16	G5 V	15,253,254,2020,2028,3077
		G 16 - 29		16 04 18	+08 31.3	3	11.57	.020	1.44	.020	1.70		4		265,1705,1759
144527	−28 11883			16 04 18	−28 44.0	1	9.06		0.39		0.10		1	F0 V	1770
		L 265 - 128		16 04 18	−52 49.	1	14.24		1.50				1		3062
		PSD 204 # 356		16 04 18	−75 22.	1	10.87		0.36				2	F0	1117
	−11 04072			16 04 19	−11 58.7	1	10.16		1.40		1.26		2	K2 III	262
144476	−37 10706			16 04 21	−37 17.3	1	8.82		0.43		0.25		1	A7 V	1770
144586	−17 04491			16 04 22	−17 48.1	1	7.79		0.12		0.00		2	B9 V	1770
144501	−35 10736			16 04 23	−35 47.2	1	10.13		0.38		0.25		1	A0 V	1770
144439	−47 10556	LSS 3471		16 04 24	−47 53.6	3	9.33	.012	0.85	.007	-0.09	.023	8	B2 Iab	540,567,976
144608	−20 04408	HR 5997		16 04 28	−20 44.1	7	4.32	.011	0.84	.007	0.52	.015	23	G3 II-III	3,15,418,1075,1586*
144502	−37 10708	IDS16012S3728	AB	16 04 29	−37 36.6	2	9.41	.005	0.49	.004	0.12	.008	8	B9 II	567,1770
144587	−23 12717			16 04 30	−23 49.1	1	8.35		0.39		0.14		2	A9 V	1770
144478	−43 10539			16 04 33	−43 41.8	1	8.49		0.12		0.00		3	B9 V	1770
144638	−12 04422			16 04 35	−12 43.9	1	9.86		0.42		0.15		1	A9 V	1770
144530	−40 10213	IDS16012S4022	A	16 04 37	−40 30.0	1	8.85		0.29		0.23		1	A3 V	1770
144660	−7 04195			16 04 39	−08 02.2	1	9.25		0.92		0.59		1	K4 V	3072
144612	−30 12852			16 04 40	−30 28.6	1	8.39		0.04		-0.18		1	Ap Si	1770
144801	+22 02924			16 04 41	+21 49.9	1	9.03		0.33		0.12		2	A5	1776
145368	+73 00707			16 04 41	+73 16.9	1	6.85		0.47		0.01		2	F5	3016
144872	+39 02950	G 180 - 31		16 04 42	+38 46.4	3	8.61	.010	0.95	.005	0.70	.016	13	K3 V	22,1003,1775
144613	−30 12853			16 04 42	−30 34.8	1	9.12		0.33		0.17		1	A8/9 V	1770
	−54 07082			16 04 42	−55 10.6	1	9.79		0.85		-0.21		2		181
	+35 02774	G 180 - 33		16 04 45	+34 47.1	1	10.41		1.20		1.27		3	M0	7008
144591	−35 10742			16 04 47	−36 05.9	1	6.76		-0.07		-0.22		3	B9 V	1770
144708	−12 04425	HR 6002	⋆ AB	16 04 50	−12 36.8	3	5.74	.008	0.02	.018	-0.11	.008	8	B9 V	1770,1771,2007
		AAS35,161 # 96	V	16 04 51	−39 00.6	1	14.06		1.48		0.91		1		5005
144661	−24 12552	HR 5998		16 04 52	−24 19.8	5	6.33	.011	-0.06	.005	-0.52	.007	20	B8 IV/V	26,1054,1079,1770,2035
	−49 10336			16 04 53	−49 31.9	1	11.38		0.69		0.05		3		1586
	+24 02972			16 04 54	+24 33.5	1	8.92		1.30		1.39		22	K2	1222
144741	−5 04242			16 04 56	−05 27.9	1	9.86		0.58		0.14		2	G0	1064
144662	−26 11224			16 04 58	−26 44.1	1	9.80		0.39		0.18		1	A5/7 III	1770
144479	−54 07093			16 04 58	−54 24.0	4	7.94	.025	0.05	.009	-0.41	.009	6	B6 II	96,540,567,976
	−60 06419			16 04 58	−60 37.6	1	9.69		1.03		0.68		2		540
145309	+70 00863	IDS16052N7032	A	16 04 59	+70 23.7	1	6.86		0.12		0.08		2	A0	985
144726	−11 04074			16 04 59	−11 47.3	1	8.56		0.53		0.11		2	F5	262
144727	−14 04354			16 05 00	−14 20.2	1	9.71		0.33		0.19		1	A3 IV	1770
		AAS35,161 # 97		16 05 00	−38 56.4	1	15.75		1.51		0.35		1		5005
144710	−15 04251			16 05 01	−15 58.7	1	9.42		0.51		0.15		2	F0 V	1771
		AAS35,161 # 98	V	16 05 01	−38 56.8	1	12.88		1.40		0.60		1		5005
145048	+52 01941			16 05 02	+52 27.1	1	9.03		0.23		0.10		2	A0	1566
144663	−27 10796			16 05 03	−27 35.9	2	7.18	.007	1.59	.006	1.92		8	K4/5 III	1657,2012
144665	−33 10943			16 05 03	−33 54.1	1	8.77		0.09		-0.16		3	B9 V	1770
	+0 03459			16 05 04	+00 09.9	1	9.77		1.00		0.85		2	K3 V	3072
144690	−25 11369	HR 6001		16 05 04	−26 11.7	4	5.37	.013	1.64	.022	1.99	.053	13	M2 III	1024,2007,2017,3055
144505	−53 07156			16 05 04	−54 00.2	1	9.38		-0.03		-0.56		4	B3 III	567
145046	+53 01842			16 05 05	+53 39.0	1	8.35		0.58		0.06		2	G5	1566
	+67 00925	G 225 - 47		16 05 05	+66 55.3	2	9.87	.009	0.56	.005	-0.04		5	G2	867,3026
330379	−48 10565			16 05 05	−48 21.2	1	10.65		0.40		-0.36		5	B3	567

Table 1   869

HD	DM	Other Id	N Rem	α₁₉₅₀	δ₁₉₅₀	S	V	σ_V	B–V	σ_B–V	U–B	σ_U–B	n	Spectrum	References
144839	+13 03069			16 05 06	+13 28.1	1	7.34		0.32		0.01		4	F3 III	3016
144713	−24 12556	IDS16021S2439	AB	16 05 06	−24 48.2	1	9.17		0.63		0.11		1	F3/5 V	1770
144555	−51 09806	LSS 3472		16 05 07	−51 49.8	4	9.40	.020	0.38	.002	−0.47	.016	9	B2 (Ib)ne	540,567,976,1586
145622	+77 00616	HR 6034		16 05 08	+76 55.7	1	5.73		0.05		0.08		2	A3 V	1733
		G 168 - 32		16 05 09	+26 58.5	2	13.33	.000	1.62	.023	1.25		4		419,1759
	+47 02298			16 05 10	+47 37.9	1	9.90		1.17		1.08		2	K4	3072
144729	−23 12723			16 05 11	−23 43.1	1	9.18		0.46		0.09		1	F0 V	1770
144480	−57 07613	HR 5994		16 05 11	−57 48.1	1	5.57		−0.04				4	B9.5V	2032
144889	+22 02926	HR 6005		16 05 12	+21 57.3	2	6.14	.000	1.37	.000	1.51	.000	4	K4 III	15,1003
144730	−24 12557			16 05 12	−24 28.5	1	10.50		0.72		0.22		1	F0 V	1770
144667	−38 10894	HR 6000	⋆ A	16 05 13	−38 57.6	4	6.65	.014	−0.07	.004	−0.43	.004	7	A0/3 III	12,164,1588,1770
144668	−38 10893	HR 5999, V856 Sco	⋆ CD	16 05 13	−38 58.4	3	6.89	.085	0.32	.023	0.23	.030	3	A7/F0	12,1588,1770
144874	+10 02958	HR 6004		16 05 14	+10 01.5	2	5.64	.008	0.20	.001	0.16	.039	10	A7 V	3,3016
144556	−54 07099			16 05 14	−55 02.9	2	7.27	.005	1.79	.009	2.10		5	K5 III	657,2012
144691	−34 10789			16 05 16	−35 01.9	1	8.34		0.28		0.05		1	A9 V	1770
144784	−12 04427			16 05 17	−12 52.2	1	8.42		0.46		−0.01		2	F3 V	262
144748	−24 12560			16 05 20	−24 59.7	1	8.65		0.16		0.17		2	Ap EuCrSr	1770
144921	+25 03031	SX Her		16 05 21	+25 02.5	2	8.26	.110	1.55	.010	1.38	.055	2	K2	793,3001
		AAS35,161 # 108		16 05 21	−38 58.4	1	13.14		1.48		1.28		1		5005
144575	−55 07124			16 05 22	−55 49.5	2	8.05	.030	0.08	.004	−0.49	.005	4	B3 IV	540,976
144805	−16 04222			16 05 25	−17 00.2	1	8.61		0.81		0.45		1	K0 + A0 V	1770
144558	−58 06653			16 05 25	−58 29.3	1	10.42		0.07		−0.41		2	B5/7 Ib	540
	−55 07127			16 05 26	−55 35.3	1	10.08		0.54		0.26		2	F0 V	540
	−11 04076			16 05 27	−12 07.6	1	10.89		0.50		0.30		2		262
144904	+16 02885			16 05 28	+15 50.6	1	7.82		1.23				2	K2	882
145082	+47 02300			16 05 30	+47 38.2	1	6.61		−0.01		−0.05		2	A0	252
		AAS35,161 # 110	V	16 05 30	−38 55.4	1	14.89		1.01		−0.67		1		5005
144481	−62 05202			16 05 30	−62 50.1	1	6.48		0.23				4	A3mA7-F2	1075
		GD 195		16 05 31	+17 45.3	1	16.68		0.17		−0.49		1		3060
		G 153 - 27		16 05 31	−10 17.6	2	14.73	.036	1.72	.061	0.99		5		940,3078
144751	−35 10752			16 05 33	−35 24.9	1	9.82		0.22		0.16		1	A0 III/IV	1770
144768	−28 11900			16 05 34	−28 43.4	1	9.45		0.37		0.15		1	Fm δ Del	1770
144647	−49 10345	LSS 3475		16 05 34	−49 28.3	3	10.01	.035	0.63	.049	−0.38	.006	8	O7	33,540,1021
144769	−28 11901			16 05 35	−29 01.8	1	9.49		0.33		0.24		2	A4/5 V	1771
145069	+39 02954			16 05 37	+39 17.2	2	8.42	.010	0.54	.005	0.03	.025	5	G0 III	833,1601
144840	−12 04429	G 153 - 28	⋆ A	16 05 37	−13 00.0	1	8.73		0.97		0.78		4	K3 V	3072
144840	−12 04429	G 153 - 28	⋆ B	16 05 37	−13 00.0	1	13.24		1.59		1.23		3		3072
144822	−20 04414			16 05 39	−20 37.6	1	8.40		0.61		0.19		1	F3 V	1770
	−11 04077			16 05 40	−12 14.5	1	10.45		0.45		0.35		2	A5	262
144628	−56 07345			16 05 41	−56 19.1	4	7.11	.010	0.86	.007	0.47	.010	11	K1/2 V	285,657,2012,3008
145120	+46 02148			16 05 42	+46 21.3	1	9.72		0.20		0.16		1	G5	1770
144823	−25 11379			16 05 42	−25 14.7	1	8.77		0.54		0.10		1	F3 V	1770
144937	+10 02959			16 05 43	+10 12.8	1	6.74		0.16		0.14		3	A3	3016
144892	−9 04305	IDS16030S0950	AB	16 05 43	−09 58.1	2	6.69	.009	0.50	.002	−0.02		8	F6 V +F8 V	173,3016
144841	−22 04098			16 05 43	−22 25.2	1	10.60		0.38		0.20		1	F2	1770
144788	−34 10795			16 05 43	−34 38.8	1	10.26		0.30		0.23		1	A1/2 IV	1770
		AJ74,882 T1# 6		16 05 43	−51 08.1	1	12.87		0.99		−0.05		2		33
144844	−23 12731	HR 6003		16 05 44	−23 33.2	5	5.88	.005	0.01	.004	−0.32	.003	17	B9 V	26,1054,1079,1770,2007
	−57 07626	LSS 3474		16 05 45	−57 33.5	1	10.87		0.09		−0.65		2	B2.5II	540
		AAS35,161 # 134	V	16 05 46	−41 32.5	1	13.88		0.61		−0.52		1		5005
144695	−49 10348	LSS 3476		16 05 47	−49 48.9	2	9.73	.015	0.58	.005	−0.43	.019	5	9.5 (Ib/II)	540,976
		LP 564 - 77		16 05 48	+04 01.5	1	12.77		0.61		−0.13		2		1696
145001	+17 02964	HR 6008	⋆ A	16 05 49	+17 10.7	2	5.00	.000	0.94	.004	0.62	.009	5	G8 III	3,1080
145000	+17 02965	HR 6009	⋆ B	16 05 49	+17 11.2	1	6.25		1.14		1.11		2	K1 III	1080
144235	−74 01520			16 05 49	−74 55.9	1	8.23		0.02		−0.16		3	A1 V	1117
	−11 04078			16 05 50	−11 56.0	1	10.81		1.34		1.22		2	K0 III	262
144907	−12 04431	IDS16031S1228	AB	16 05 52	−12 36.1	1	9.73		0.63		0.10		2	F6 V	262
144847	−31 12601			16 05 54	−31 57.9	1	9.28		0.41		0.08		2	A9 V	1771
		L 481 - 66		16 05 54	−37 57.	1	13.71		1.57				1		3062
144908	−12 04432			16 05 55	−12 50.9	1	10.36		0.48		0.25		1	A8 II/III	1770
144845	−27 10808			16 05 56	−28 03.2	1	9.28		0.43		0.25		1	F0 V	1770
144596	−62 05207			16 05 58	−62 47.5	1	7.44		−0.03				4	B9/A0 III	1075
		AJ74,882 T2# 1		16 06 00	−48 57.4	1	10.45		0.61		−0.30		1		33
	−50 10177	LSS 3477		16 06 00	−50 56.8	2	10.45	.009	0.80	.018	−0.06		9		567,1021
144880	−31 12603			16 06 01	−31 57.9	3	7.45	.005	0.53	.004	−0.04	.005	9	F7 V	1311,2012,3037
145002	+8 03141	HR 6010, FS Ser		16 06 03	+08 40.0	4	5.72	.008	1.60	.011	1.71	.023	17	M3.5 IIIa	3,1088,3016,4001
144774	−50 10178			16 06 03	−50 19.7	1	9.31		0.17		−0.34		3	B8 Ib	567
144925	−18 04240			16 06 08	−18 51.8	1	7.79		0.21		0.07		2	A0 V	1770
330386	−48 10572	LSS 3480	⋆ AB	16 06 08	−48 56.8	2	10.86	.574	0.63	.003	−0.25	.026	5	B5	540,1101
144939	−15 04255			16 06 09	−16 03.2	1	9.83		0.43		0.29		1	A4 V	1770
144881	−34 10802			16 06 09	−34 50.2	1	8.27		0.31		0.10		1	A1mA6-F2	1770
144300	−74 01522			16 06 09	−75 09.8	1	7.89		0.25		0.02		3	A7 V	1117
	−53 07190	LSS 3478		16 06 10	−53 00.7	1	10.93		0.26		−0.52		2		181
145454	+68 00864	HR 6025		16 06 11	+67 56.5	1	5.44		−0.02		−0.08		2	A0 Vn	3023
145050	+9 03153	FQ Ser		16 06 12	+08 44.7	1	6.45		1.54		1.48		12	M2	3042
144882	−36 10668			16 06 12	−37 02.8	1	8.68		0.49		0.41		1	A5 III/IV	1770
		AJ74,882 T2# 2		16 06 12	−48 57.8	1	11.67		0.63		−0.28		1		33
	−49 10355			16 06 12	−49 36.2	1	10.42		0.74		0.42		2		33
144851	−44 10674			16 06 16	−44 40.6	1	8.73		0.08		−0.35		4	B8 V	567
144980	−11 04079			16 06 17	−12 12.6	1	9.55		0.25		0.18		2	A0 V	262

HD	DM	Other Id	N	Rem	α₁₉₅₀	δ₁₉₅₀	S	V	σ_V	B–V	σ_B–V	U–B	σ_U–B	n	Spectrum	References
144926	−30 12880	IDS16032S3047	A		16 06 17	−30 55.2	1	7.14		0.33		0.08		1	A7 IV	1771
144940	−21 04284				16 06 18	−21 48.6	1	9.34		0.55		0.21		1	F2 IV	1770
144927	−32 11456	HR 6006		⋆AB	16 06 20	−32 31.1	2	6.18	.005	0.79	.000	0.45		8	K0/2 III	404,2007
144813	−48 10576	V Nor			16 06 20	−49 06.4	1	8.50		2.11				1	M6 (III)	8051
		Ly 7 - 7			16 06 20	−55 09.5	1	13.62		1.90				3		1101
144941	−26 11229				16 06 21	−27 08.6	2	10.13	.015	0.04	.008	-0.71	.002	4	B8	1054,1770
	−37 10718				16 06 21	−37 56.6	2	10.89	.000	0.90	.010	0.66	.019	6		158,1696,3060
		AAS35,161 # 116		V	16 06 21	−39 11.8	1	13.74		1.44		1.07		1		5005
		G 16 - 31			16 06 24	+01 59.3	1	10.16		0.85		0.47		1		333,1620
144984	−22 04102				16 06 24	−23 02.8	1	8.45		0.23		0.18		2	A1 V	1770
		AJ74,882 T1# 2			16 06 24	−49 33.0	1	10.83		0.60		-0.20		2		33
144981	−19 04322				16 06 26	−19 19.5	1	8.07		0.20		0.10		2	A0 V	1770
144897	−40 10236				16 06 26	−41 01.6	2	8.62	.005	0.22	.002	-0.05	.005	2	Ap EuCr	567,1770
144982	−20 04416				16 06 27	−20 48.1	1	10.36		0.62		0.15		1	F3/5 V	1770
144942	−27 10813				16 06 28	−27 57.6	1	9.93		0.38		0.14		2	A(3) II	1770
145085	+3 03132	HR 6011			16 06 29	+03 35.1	1	5.90		1.47		1.81		8	gK5	1088
145122	+17 02967	HR 6013			16 06 31	+17 20.2	1			0.00		-0.07		3	A0 Vnn	1022
144815	−53 07205				16 06 31	−53 18.4	2	9.15	.010	-0.03	.005	-0.57	.019	3	B8 Ib	181,567
144964	−33 10959				16 06 33	−33 13.6	1	9.42		0.27		0.20		1	A3 V	1770
	−48 10579	LSS 3483			16 06 33	−48 49.8	1	11.68		0.69		-0.24		3		1101
144816	−55 07159	Ly 7 - 1		⋆AB	16 06 33	−55 13.0	3	7.60	.020	0.08	.005	-0.50	.027	9	B3	540,976,1101
145056	−11 04081				16 06 36	−12 14.7	2	7.58	.026	0.07	.025	-0.05	.042	3	B9 V	262,1770
145057	−12 04437				16 06 36	−12 55.3	2	7.12	.010	0.22	.004	0.14	.003	3	A5 IV	262,1770
144986	−28 11917	IDS16035S2842	AB		16 06 36	−28 50.1	1	10.06		0.55		0.19		1	F3 V	1770
145059	−16 04230				16 06 38	−17 03.4	1	7.88		0.64		0.18		3	G2/3 V	803
144899	−47 10573				16 06 39	−47 47.0	1	8.97		0.66				4	G2 V	2012
	−55 07161	Ly 7 - 2			16 06 39	−55 25.0	1	10.29		1.03		0.64		4	G5 III	1101
144987	−33 10961	HR 6007			16 06 40	−33 24.9	2	5.51	.004	-0.09	.006	-0.29		7	B8 V	1770,2007
145030	−20 04418				16 06 41	−20 42.3	1	10.39		0.58		0.16		1	F5	1770
144900	−48 10581	LSS 3484		⋆AB	16 06 41	−48 50.2	4	9.64	.029	0.74	.009	-0.28	.024	10	O7/8	33,540,1021,1101
		EG 116			16 06 42	+42 13.	1	13.85		0.06		-0.54		2	DA	3028
145031	−26 11236				16 06 42	−26 18.5	1	8.18		0.16		0.11		2	A0 V	1770
145010	−31 12614	IDS16035S3120	A		16 06 42	−31 28.4	1	9.13		0.48		0.07		1	A8/F0 V	1770
145010	−31 12614	IDS16035S3120	B		16 06 42	−31 28.4	1	10.03		0.56		0.06		1		1770
144855	−53 07214				16 06 42	−53 44.4	1	7.89		1.84		2.02		3	M0/2	8100
145148	+6 03169	HR 6014		⋆	16 06 43	+06 31.2	8	5.95	.014	0.99	.014	0.85	.026	17	K0 IV	15,22,333,1003,1067*
144857	−53 07213				16 06 44	−54 11.6	2	8.86	.010	0.09	.010	-0.22	.025	2	B9 Ib/II	96,567
		Ly 7 - 6			16 06 45	−55 10.2	1	12.33		0.83		0.30		3		1101
144858	−55 07162				16 06 46	−55 56.5	4	7.14	.021	0.01	.008	-0.45	.005	9	B4/5 III	158,540,567,976
144989	−36 10674				16 06 47	−36 30.8	1	9.39		0.40		-0.07		3	B8 II	1770
144965	−39 10338	IDS16034S3952	AB		16 06 47	−39 59.9	5	7.06	.006	0.15	.007	-0.50	.011	15	B3 Vne	158,400,567,1586,1770
144946	−42 11073				16 06 47	−42 41.1	1	8.74		0.17		-0.10		3	B9 V	1770
144918	−48 10583	LSS 3485		⋆AB	16 06 47	−48 55.0	5	9.93	.031	0.78	.021	-0.30	.020	13	O5/7	33,432,540,1021,1101
145035	−32 11466				16 06 52	−32 36.6	1	9.93		0.51		0.15		4	F2 V	119
	−58 06665				16 06 54	−58 37.2	1	10.67		0.65		-0.11		2	B8 Ia	540
		AAS35,161 # 122			16 06 55	−39 00.3	1	14.37		1.47		1.08		1		5005
	−0 03060				16 06 57	−00 02.6	1	10.65		1.08		0.85		1	K5 V	3072
		LSS 3486			16 06 57	−48 57.8	2	11.98	.079	0.87	.000	-0.09	.010	3	B3 Ib	33,1101
145101	−22 04104				16 06 59	−22 16.7	1	7.70		0.29		0.20		1	A3 IV/V	1770
	−54 07145	Ly 7 - 4			16 07 00	−55 09.7	1	11.00		0.27		0.03		4		1101
		Ly 7 - 39			16 07 00	−55 10.	1	15.92		0.83		0.30		1		1101
		Ly 7 - 40			16 07 00	−55 10.	1	13.57		0.61		0.30		1		1101
		Ly 7 - 41			16 07 00	−55 10.	1	16.35		1.10				1		1101
		Ly 7 - 42			16 07 00	−55 10.	1	14.38		1.36		0.30		2		1101
		Ly 7 - 43			16 07 00	−55 10.	1	15.20		1.59		0.40		3		1101
		Ly 7 - 44			16 07 00	−55 10.	1	13.91		1.42		1.10		1		1101
		Ly 7 - 45			16 07 00	−55 10.	1	16.46		1.33				1		1101
		Ly 7 - 46			16 07 00	−55 10.	1	16.32		1.46				1		1101
		Ly 7 - 47			16 07 00	−55 10.	1	15.87		0.85		0.30		1		1101
		LP 744 - 50			16 07 01	−13 53.7	1	10.63		0.77		0.31		2		1696
145100	−17 04502	HR 6012			16 07 02	−18 12.6	3	6.46	.007	0.44	.000			11	F3 V	15,2013,2030
144966	−46 10590	LSS 3490			16 07 02	−46 51.0	2	10.02	.005	0.59	.008	-0.39		6	O9.5V	540,1021
145152	−11 04082				16 07 05	−12 11.3	1	9.56		1.23		0.97		2	K1 III	262
145153	−12 04441				16 07 06	−12 45.3	1	7.47		1.18		1.04		2	K0 III	262
144970	−48 10585	LSS 3489			16 07 06	−48 53.5	6	9.88	.041	0.76	.029	-0.31	.032	16	B0	33,540,1021,1101,1586,1737
145328	+36 02699	HR 6018		⋆AB	16 07 08	+36 37.0	7	4.76	.010	1.01	.006	0.87	.008	39	K0 III-IV	15,1003,1080,1363*
144969	−48 10587	LSS 3491			16 07 08	−48 39.9	7	8.29	.018	0.94	.013	-0.13	.025	16	B1 Ia	33,403,540,976,1101*
144932	−55 07171	Ly 7 - 3			16 07 08	−55 31.0	1	8.90		0.18		-0.24		5	A0	1101
145127	−24 12577				16 07 09	−24 27.1	1	6.65		0.02		-0.02		2	A0 V	1770
145389	+45 02376	HR 6023			16 07 12	+45 03.9	7	4.22	.025	-0.07	.005	-0.25	.009	46	B9 p Mn	15,879,1079,1363,3033*
145155	−22 04105				16 07 12	−22 19.8	1	9.74		0.61		0.06		2	F3 V	1770
145129	−25 11398				16 07 12	−25 17.6	1	9.50		0.27		0.13		2	A2 IV	1770
145102	−26 11240	V952 Sco			16 07 12	−26 46.7	2	6.59	.010	0.07	.003	-0.18	.005	8	Ap Si	26,1054
145086	−34 10815				16 07 12	−34 15.3	1	10.26		0.45		0.23		1	A2 II/III	1770
145206	−3 03884	HR 6016			16 07 13	−03 20.2	3	5.39	.007	1.45	.002	1.45	.004	66	K4 III	15,1417,3047
145074	−35 10777				16 07 14	−35 52.1	1	10.36		0.52		0.32		1	Ap (EuCrSr)	1770
144991	−50 10193				16 07 14	−51 10.0	1	9.74		0.27		-0.25		4	Ap Si	567
		HA 204 # 2533			16 07 14	−75 23.8	1	12.01		0.94		0.62		11		1499
	−54 07152	LSS 3488			16 07 15	−54 17.9	1	10.49		0.26		-0.66		2		181
	−54 07151	LSS 3487			16 07 15	−54 39.3	2	10.30	.004	0.57	.009	-0.44	.011	4	B0 II-III	181,540

Table 1   871

HD	DM	Other Id	N	Rem	α_{1950}	δ_{1950}	S	V	σ_V	B–V	σ_{B-V}	U–B	σ_{U-B}	n	Spectrum	References
	−48 10588	LSS 3493			16 07 18	−48 46.5	2	11.31	.050	0.61	.000	-0.32	.040	4		33,1101
144973	−54 07155				16 07 18	−54 21.4	1	9.59		0.19		-0.55		2	B2 II/III	181
	−55 07178	Ly 7 - 5			16 07 19	−55 12.3	1	9.84		1.18		0.98		3	G7 III	1101
144951	−55 07175				16 07 19	−56 07.4	3	8.12	.024	0.01	.004	-0.56	.012	5	B3 V	540,567,976
		HA 204 # 2545			16 07 21	−75 26.0	1	12.06		0.45		-0.01		11		1499
145188	−21 04287				16 07 22	−22 01.6	1	7.05		0.16		-0.05		1	A0 V	1770
145274	+11 02926				16 07 23	+10 59.7	1	8.51		0.24		0.08		2	A0	1648
		LP 917 - 16			16 07 24	−28 54.0	1	12.90		0.68		-0.16		3		3062
144972	−54 07159	QZ Nor			16 07 24	−54 13.5	2	8.78	.052	0.87	.033	0.51	.028	4	F6 I	96,8100
	−50 10196	LSS 3494			16 07 25	−50 29.2	1	10.60		0.89		-0.15		3		33
145189	−22 04106				16 07 26	−22 58.6	1	7.78		0.20		0.12		1	A3 III/IV	1770
145038	−48 10590				16 07 26	−48 50.1	1	10.46		2.11		1.80		1	K1 III	1101
145373	+30 02759				16 07 29	+29 58.8	1	8.36		1.47		1.67		1	K5	1746
330370	−47 10578				16 07 29	−48 02.1	1	10.51		0.65		-0.05		3	B8	1586
		AAS35,161 # 123		V	16 07 30	−38 45.5	1	15.08		1.15		-0.89		1		5005
144887	−64 03406				16 07 30	−64 30.0	1	9.08		-0.08		-0.46		3	B7 II/III	400
		L 42 - 20			16 07 30	−76 38.	1	13.92		0.89		0.30		3		3062
		G 138 - 3			16 07 32	+10 16.5	2	12.14	.000	0.91	.000	0.69	.000	3		1696,3060
		LP 444 - 28			16 07 32	+17 04.9	1	13.41		0.70		0.13		2		1696
145134	−40 10248	IDS16041S4010		AB	16 07 32	−40 18.4	1	9.31		0.24		-0.04		6	B8 V	567
		G 240 - 8			16 07 34	+64 31.4	1	12.40		1.41		1.30		1	M1	1658
	−59 06643				16 07 35	−59 52.7	1	11.52		0.18		-0.38		1		1586
145435	+41 02673				16 07 36	+41 13.5	1	6.72		0.51		0.02		2	G0 V	1601
		Ly 7 - 8			16 07 36	−55 15.3	1	12.19		0.54		0.00		3		1101
	−56 07381				16 07 36	−56 15.0	1	9.86		1.20		1.04		3		1586
145209	−24 12585				16 07 38	−24 50.7	1	8.48		0.42		0.10		1	F0 V	1770
145210	−25 11407				16 07 39	−25 44.5	1	9.69		0.20		0.16		1	A1 IV	1770
145105	−46 10592				16 07 41	−47 01.7	1	9.55		0.32		-0.10		5	B8 IV	567
145236	−21 04290				16 07 42	−21 31.1	1	10.07		0.46		0.16		2	A7/9 II	1770
145041	−54 07168				16 07 42	−54 15.1	1	8.80		1.29		1.23		1	K0/2 III	96
145404	+26 02791				16 07 43	+26 07.6	1	8.51		0.57		0.05		1	G0 V	3026
145249	−23 12748				16 07 45	−23 18.2	1	10.11		0.55		0.12		1	F0/2 V	1770
145531	+51 02061				16 07 46	+50 51.8	1	8.17		0.40		0.05		2	F2	1566
145158	−45 10500				16 07 46	−45 12.2	1	6.64		0.48				4	F6 V	2012
	−19 04326				16 07 47	−19 15.8	1	9.42		1.18		0.98		35	K0	1763
		LP 861 - 31			16 07 49	−25 05.6	1	15.35		0.16		-0.65		1		3060
145190	−35 10784				16 07 50	−35 46.3	1	9.62		0.49		0.34		1	A3mA6-F2	1770
145107	−51 09847				16 07 51	−51 29.4	3	9.71	.014	0.33	.003	-0.25	.017	8	B5 II/IIIe	540,567,976
		G 137 - 86			16 07 52	+09 16.1	1	13.06		0.48		-0.20		3		3060
145263	−25 11411				16 07 53	−25 23.6	1	8.98		0.41		0.05		1	F0 V	1770
145191	−40 10251	HR 6015			16 07 53	−40 59.3	2	5.85	.005	0.28	.005			7	F2 III (+A)	15,2012
145250	−29 12343	HR 6017			16 07 55	−29 17.2	2	5.11	.015	1.13	.010	1.02		8	K0 III	2007,3016
145293	−13 04357				16 07 56	−13 22.6	2	10.02	.028	0.66	.014	0.22		4	F7/8wA8	1594,6006
	−52 09317				16 07 57	−52 47.0	3	10.48	.018	1.17	.015	1.13	.013	8		158,1696,3062
145457	+27 02595				16 08 00	+26 52.3	1	6.60		1.06		0.83		2	K0 III	3016
		vdB 69 # a			16 08 00	−51 08.	1	11.27		2.56		2.23		6	B3 V	434
145458	+25 03039				16 08 01	+25 37.0	2	7.40	.010	0.94	.000	0.62	.025	5	G8 II-III	105,8100
145251	−32 11476				16 08 01	−32 30.7	1	10.21		0.27		0.17		1	A0 IV	1770
145109	−53 07265				16 08 01	−54 11.6	1	8.71		0.26		0.15		1	A2	96
	−54 07178				16 08 01	−54 43.4	1	10.53		0.16		-0.69		2		181
145110	−54 07185				16 08 02	−54 13.5	1	6.53		0.02		-0.12		1	B9 IV	96
	−55 07198	Ly 7 - 9			16 08 02	−55 12.6	1	12.53		0.36		0.30		3		1101
145674	+58 01622	HR 6036		★ AB	16 08 03	+58 04.0	1	6.33		0.04		-0.02		2	A1 V	1733
145278	−31 12634				16 08 05	−31 39.4	1	8.25		0.18		0.17		1	A1 V	1770
		Nor sq 2 # 45			16 08 06	−52 20.	1	12.08		0.90		0.02		2		181
145297	−25 11415				16 08 07	−25 17.2	1	9.54		0.45		0.16		2	A9 V	1770
	−55 07201	Ly 7 - 10			16 08 07	−55 13.1	1	11.04		0.34		0.26		3	F0 V	1101
145459	+25 03039	RU Her			16 08 09	+25 12.0	2	12.23	.000	1.58	.000			3	M4	8019,8027
	−55 07202	LSS 3495			16 08 10	−55 44.9	2	10.62	.003	0.50	.008	0.06	.006	5	B7 Ib	181,1586
		He3 1154			16 08 12	−51 21.	1	12.05		0.53		-0.35		2		1586
		HBC 630			16 08 15	−18 57.1	1	11.98		1.36		0.90		1	K2 IV	1763
145320	−30 12914				16 08 15	−30 36.5	1	10.71		0.61		0.06		1	A1 III/IV	1770
145217	−49 10383	LSS 3499			16 08 15	−50 10.6	5	9.95	.010	0.83	.027	-0.25	.022	17	B2	33,540,567,1021,1649
145333	−22 04110				16 08 16	−23 04.5	1	9.02		0.44		0.14		1	F0 V	1770
145280	−35 10793				16 08 16	−35 34.4	1	10.06		0.23		0.13		1	A2 III/IV	1770
		G 138 - 5			16 08 18	+16 39.3	3	10.71	.022	0.91	.015	0.70	.014	6	K2	316,1696,3060
145332	−21 04292				16 08 18	−21 18.7	1	10.36		0.62		0.16		1	F5/6	1770
145175	−53 07288	NGC 6067 - 229			16 08 18	−54 00.6	1	8.66		1.28		1.56		1	K3 III	1407
145694	+56 01867	HR 6038			16 08 20	+55 57.5	2	6.42	.020	1.07	.010	0.94	.005	5	K0 III	1501,1733
	−49 10387	LSS 3500			16 08 21	−49 14.5	1	10.72		0.93		-0.15		3		1649
145335	−31 12637				16 08 23	−31 39.5	1	9.42		0.37		0.26		1	A8 V	1770
	−53 07284	NGC 6067 - 227			16 08 23	−54 06.8	1	11.86		0.29				1		1407
	−53 07285	NGC 6067 - 228			16 08 23	−54 11.0	1	11.86		0.41				1		1407
145351	−26 11246				16 08 25	−26 36.5	1	9.14		0.34		0.19		1	A7 V	1770
	−53 07292	NGC 6067 - 230			16 08 28	−54 10.6	1	11.66		0.54				1		1407
	−53 07293	NGC 6067 - 231			16 08 29	−53 59.1	1	11.10		0.35		0.02		1		1407
145353	−26 11247				16 08 30	−27 01.3	3	6.95	.022	0.14	.013	-0.08	.013	7	B9 V	26,1054,1771
		AAS35,161 # 124			16 08 32	−38 54.6	1	13.33		1.48		1.27		1		5005
	−53 07294	NGC 6067 - 232			16 08 32	−53 57.0	1	11.33		1.28				1		1407
	−53 07296	NGC 6067 - 233			16 08 34	−54 02.3	1	11.58		0.29				1		1407

HD	DM	Other Id	N Rem	α_{1950}	δ_{1950}	S	V	σ_V	B–V	σ_{B-V}	U–B	σ_{U-B}	n	Spectrum	References
		G 137 - 87		16 08 35	+13 24.2	4	12.87	.022	0.50	.021	-0.21	.023	9		1658,1696,3060,7009
145515	+15 02963			16 08 35	+15 18.2	1	8.43		1.03				2	K0	882
	-53 07298	NGC 6067 - 234		16 08 35	-54 10.8	1	11.30		0.18				1		1407
	-53 07299	NGC 6067 - 235		16 08 37	-53 57.0	1	10.95		0.27				1		1407
	-53 07300	NGC 6067 - 236		16 08 37	-54 02.1	1	11.23		1.23				1		1407
	-53 07302	NGC 6067 - 237		16 08 38	-54 07.4	1	12.10		0.27				1		1407
145427	-16 04238			16 08 40	-17 09.5	1	9.98		0.43		0.20		1	A8/9 V	1770
		NGC 6067 - 215		16 08 40	-54 00.4	1	15.27		0.97		0.66		1		1407
		HRC 254, V866 Sco		16 08 42	-18 31.	1	11.30		1.06		-0.45		1	K5 V	1763
		NGC 6067 - 216		16 08 42	-54 00.3	1	14.55		0.80		0.26		1		1407
	-53 07307	NGC 6067 - 239		16 08 43	-54 08.0	1	11.33		0.38		0.04		1		1407
	-53 07308	NGC 6067 - 240		16 08 43	-54 10.1	1	9.99		1.70		1.72		1		1407
	-53 07310	NGC 6067 - 241		16 08 44	-54 08.5	1	11.66		0.17				1		1407
145408	-27 10836	IDS16057S2718	AB	16 08 46	-27 25.0	1	7.87		0.36		0.05		1	A5/7 (III)	1770
145675	+44 02549	G 180 - 35		16 08 47	+43 57.0	3	6.65	.009	0.88	.005	0.67	.011	13	K0 V	1067,1080,3016
145357	-42 11100	IDS16054S4207	A	16 08 48	-42 14.8	1	8.15		0.31		0.07		1	A5 V	1770
	-53 07316	NGC 6067 - 243		16 08 48	-54 10.0	1	12.22		0.23				1		1407
	-53 07318	NGC 6067 - 245		16 08 49	-54 01.4	1	12.45		0.32		0.04		1		1407
	-53 07319	NGC 6067 - 246		16 08 49	-54 05.0	1	11.68		0.30				1		1407
145727	+49 02469			16 08 50	+48 44.4	1	7.42		0.95		0.62		3	G7 III	1501
	-52 09334			16 08 50	-52 39.5	1	10.35		0.43		-0.15		2		1586
		NGC 6067 - 3		16 08 50	-54 00.5	1	13.36		1.45				1		1407
	-53 07324	NGC 6067 - 247	AB	16 08 52	-54 03.3	2	10.09	.005	1.48	.042	1.28		2	K2 II	1407,4002
	-53 07325	NGC 6067 - 248		16 08 52	-54 04.7	1	11.34		0.19				1		1407
	-49 10393	LSS 3503		16 08 53	-50 00.2	1	11.30		0.75		-0.19		3		1649
145468	-22 04113			16 08 54	-22 25.0	1	8.21		0.32		0.19		2	A3 V	1770
	-53 07326	NGC 6067 - 249		16 08 54	-54 05.9	1	11.66		0.26				1		1407
	-53 07327	NGC 6067 - 250		16 08 54	-54 08.7	1	12.25		0.26				1		1407
	-53 07334	NGC 6067 - 254		16 08 56	-54 03.2	1	10.70		0.26		-0.18		1	B8 III	1407
	-53 07335	NGC 6067 - 255		16 08 56	-54 04.1	1	11.77		0.24				1		1407
145302	-52 09337			16 08 57	-53 05.1	1	9.49		1.05		0.82		1	G8 (III)	8100
327922	-44 10698	LSS 3507		16 08 58	-44 40.6	2	10.10	.003	1.00	.031	-0.07	.003	4	B1.5 Iab	540,1737
		NGC 6067 - 17		16 08 58	-54 03.0	1	12.24		1.46				1		1407
	-53 07338	NGC 6067 - 256		16 08 58	-54 08.1	1	11.89		0.22				1		1407
145467	-20 04429			16 08 59	-20 58.6	1	8.98		0.50		0.16		1	F0 V	1770
	-53 07341	NGC 6067 - 258		16 08 59	-54 08.4	1	12.18		0.24				1		1407
	-53 07343	NGC 6067 - 260		16 09 00	-54 02.0	1	10.42		0.24				1		1407
	-53 07344	NGC 6067 - 261		16 09 01	-54 05.6	2	8.78	.005	1.77	.004	1.93	.021	2	K2 Ib	1407,4002
	-53 07346	NGC 6067 - 263		16 09 01	-54 06.9	1	11.64		0.29				1		1407
	-53 07347	NGC 6067 - 264		16 09 02	-54 09.8	1	10.34		0.13		-0.40		1	B5 V	1407
145142	-66 02911			16 09 02	-66 56.9	1	7.84		1.08		0.95		5	K0 III	1704
145304	-53 07350			16 09 03	-53 56.3	5	8.82	.011	0.13	.009	-0.56	.012	11	B1/2 III/IV	96,181,540,567,976
		NGC 6067 - 49		16 09 03	-54 06.3	1	12.92		1.35				1		1407
145501	-19 04332	HR 6026	★ CD	16 09 04	-19 19.3	4	6.27	.021	0.13	.008	-0.36	.013	14	B8/9 V	1054,1084,1088,1770
	-53 07353	NGC 6067 - 267		16 09 04	-54 04.0	1	9.03		0.18		-0.51		1	B2 III	1407
	-53 07352	NGC 6067 - 266		16 09 04	-54 07.9	1	12.05		0.22				1		1407
145886	+60 01658	IDS16086N6001	A	16 09 05	+60 00.0	1	6.77		0.05		0.07		3	A2 V	1501
		HBC 633		16 09 05	-18 59.2	1	11.66		1.17		0.71		1	K1 IV	1763
145502	-19 04333	HR 6027	★ AB	16 09 05	-19 19.9	10	4.00	.011	0.04	.010	-0.64	.009	43	B2 IV	15,15,26,1054,1075,1084*
145443	-32 11492			16 09 05	-32 15.6	1	10.16		0.21		0.18		1	A0 V	1770
	-53 07358	NGC 6067 - 269		16 09 06	-54 07.9	1	12.36		0.25				1		1407
145589	+10 02971	HR 6032	★ AB	16 09 06	+09 50.4	2	6.52	.005	0.25	.005	0.09	.010	7	F0 IV	15,1256
		G 138 - 8		16 09 06	+13 30.3	2	15.09	.005	0.23	.000	-0.64	.005	4		419,3060
	-53 07361	NGC 6067 - 272		16 09 06	-54 03.1	1	11.09		0.21				1		1407
	-53 07360	NGC 6067 - 271		16 09 06	-54 05.6	1	10.64		0.22		-0.24		1	B5 III	1407
	-53 07359	NGC 6067 - 270		16 09 06	-54 05.9	1	11.64		0.26				1	B8 V	1407
145324	-54 07226	LSS 3504		16 09 06	-54 13.9	1	7.32		0.37				1	A5 Ib/II	138
145504	-22 04114			16 09 07	-23 07.0	1	9.31		0.52		0.11		1	F0 V	1770
	-53 07363	NGC 6067 - 274		16 09 07	-54 04.3	1	10.90		0.27				1		1407
	-53 07362	NGC 6067 - 273		16 09 07	-54 06.3	1	11.18		0.23				1		1407
		NGC 6067 - 4		16 09 08	-54 00.8	1	14.57		0.75				1		1407
	-53 07364	NGC 6067 - 275		16 09 08	-54 04.7	2	9.16	.019	1.83	.019	2.03		2	K4 Ib	1407,4002
145505	-23 12769	IDS16062S2340	A	16 09 09	-23 47.6	1	8.46		0.26		0.17		2	A2 IV	1770
	-53 07366	NGC 6067 - 276	★ AB	16 09 09	-54 06.8	1	9.69		1.87				1	K3 II+K3Ib	1407
145503	-21 04297			16 09 10	-21 15.4	1	9.78		0.45		0.22		3	A3/7 (II)	1770
145483	-28 11962	HR 6029	★ A	16 09 10	-28 17.3	1	5.88		-0.03		-0.17		1	B9 V	1770
145483	-28 11962	HR 6029	★ AB	16 09 10	-28 17.3	5	5.68	.010	0.01	.009	-0.20	.009	14	B9 V +F2 V	15,1054,1079,1770,2030
	-51 09872	LSS 3506		16 09 10	-51 34.8	1	10.24		0.59		0.34		3	A2 II	1649
	-53 07370	NGC 6067 - 277	AB	16 09 10	-54 06.4	1	11.00		0.23				1		1407
145469	-28 11961			16 09 11	-28 52.4	1	8.59		0.42		0.05		1	F2 V	1770
145444	-37 10729	IDS16059S3751	AB	16 09 11	-37 58.7	1	9.88		1.21		1.13		2	K3/4 V	3072
	-53 07371	NGC 6067 - 278	AB	16 09 11	-54 04.9	1	12.26		0.37				1		1407
	-53 07372	NGC 6067 - 279		16 09 11	-54 07.3	1	10.92		0.18		-0.26		1		1407
145519	-18 04243			16 09 12	-18 56.0	4	7.98	.006	0.25	.004	0.00	.007	27	B9 V	26,1054,1763,1770
	-53 07376	NGC 6067 - 281		16 09 12	-54 04.7	1	12.02		0.29				1		1407
145647	+17 02982	HR 6035		16 09 13	+16 47.6	3	6.07	.026	0.03	.004	0.03	.020	6	A0 V	252,1022,3050
145482	-27 10841	HR 6028		16 09 13	-27 47.9	9	4.58	.007	-0.16	.006	-0.75	.012	34	B2 V	15,26,1020,1054,1075,1586*
	-50 10226	LSS 3508		16 09 13	-50 55.5	1	9.26		0.83		0.43		3	A3 II	1649
	-53 07381	NGC 6067 - 283		16 09 14	-54 03.5	1	11.62		0.26				1		1407
145646	+17 02983			16 09 15	+17 06.3	2	7.71	.005	1.31	.010	1.42		3	K5	882,3040

Table 1   873

HD	DM	Other Id	N Rem	α_{1950}	δ_{1950}	S	V	σ_V	B–V	σ_{B-V}	U–B	σ_{U-B}	n	Spectrum	References
145506	−24 12602			16 09 15	−24 58.1	1	10.07		0.33		0.24		2	A2/3 III/I	1770
	−53 07382	NGC 6067 - 284		16 09 15	−54 03.4	1	11.52		0.22		−0.19		1		1407
		NGC 6067 - 19		16 09 15	−54 03.6	1	15.10		0.84				1		1407
	−53 07384	NGC 6067 - 286		16 09 15	−54 09.8	1	10.15		0.18		−0.10		1	B8 IV	1407
145570	−9 04324	HR 6031		16 09 16	−09 56.2	9	4.93	.005	0.09	.008	0.10	.008	35	A3 IV	15,355,1075,1088,2012*
145991	+66 00937	G 225 - 49	⋆ A	16 09 17	+65 57.7	2	9.25	.009	0.73	.009	0.32	.009	5		308,1658
		NGC 6067 - 5		16 09 18	−54 05.	1	14.67		0.42				1		1407
		NGC 6067 - 214		16 09 18	−54 05.	1	13.98		0.30		0.28		1		1407
	−53 07389	NGC 6067 - 289		16 09 18	−54 05.5	1	11.98		0.24		−0.23		1	B8 III	1407
	−53 07388	NGC 6067 - 288		16 09 18	−54 09.6	1	10.55		0.25		−0.14		1	B8 V	1407
145394	−47 10593			16 09 19	−47 56.5	1	10.53		0.36		−0.13		5	B8 Ib/II	567
	−53 07398	NGC 6067 - 295		16 09 20	−54 07.6	1	11.18		0.21		−0.19		1	B8 III	1407
		vdB 70 # a		16 09 21	−50 15.7	1	12.82		0.95		−0.09		3		434
	−53 07400	NGC 6067 - 297	⋆ A	16 09 21	−54 06.5	2	8.26	.009	1.09	.006	0.80	.016	2	G0 Ib	1407,4002
	−53 07400	NGC 6067 - 298	⋆ AB	16 09 21	−54 06.5	1	9.01		0.48		0.42		1	F0 Ib	1407
	−53 07407	NGC 6067 - 301		16 09 21	−54 10.2	1	11.10		0.30				1		1407
	−53 07399	NGC 6067 - 296		16 09 21	−54 11.6	1	12.39		0.18				1		1407
145384	−53 07413	HR 6022		16 09 22	−53 32.6	2	5.90	.054	1.95	.005	2.21		6	M0Ib/II+F/G	2007,3005
	−53 07402	NGC 6067 - 299		16 09 22	−54 02.5	1	10.23		0.23		−0.14		1	B8 III	1407
		NGC 6067 - 217		16 09 23	−54 09.3	1	15.26		0.89		0.40		1		1407
145361	−55 07229	HR 6019		16 09 23	−55 24.8	2	5.80	.005	0.35	.005			7	F2 IV/V	15,2027
	−49 10400			16 09 24	−49 47.0	1	10.85		0.61		0.13		1		96
145607	−8 04180	HR 6033		16 09 25	−08 25.2	1	5.42		0.12		0.12		8	A4 V	1088
	−53 07416	NGC 6067 - 303		16 09 25	−54 02.3	2	10.00	.024	1.17	.441	1.40		2	K2 II-Ib	1407,4002
	−53 07419	NGC 6067 - 306		16 09 26	−54 06.2	2	10.05	.024	1.55	.032	1.53	.016	2	K3 II	1407,4002
145554	−19 04334			16 09 27	−19 27.1	3	7.65	.010	0.15	.007	−0.09	.009	13	B9 V	26,1054,1771
145412	−49 10402			16 09 28	−49 45.4	1	6.85		0.17		0.09		1	A0 V	96
	−51 09885	LSS 3509		16 09 29	−51 16.0	3	11.62	.013	0.54	.011	−0.46	.010	8		1586,1649,1737
		NGC 6067 - 218		16 09 29	−54 09.8	1	14.33		0.40		0.27		1		1407
	−53 07427	NGC 6067 - 311		16 09 29	−54 10.2	1	10.88		0.42				1		1407
145713	+23 02909	HR 6039, LQ Her		16 09 30	+23 37.4	1	5.58		1.57		1.62		1	M4.5 IIIa	3001
145397	−54 07245	HR 6024	⋆ A	16 09 31	−54 30.2	2	4.95	.005	1.03	.009	0.81		5	G8 III	96,2007
145595	−16 04244			16 09 33	−17 13.5	1	9.86		0.44		0.26		1	A5 II/III	1770
145535	−31 12661			16 09 33	−31 27.3	1	9.95		0.45		0.25		1	A8 V	1770
	−49 10405			16 09 34	−49 48.6	1	11.74		0.32		0.19		1		96
	−53 07434	NGC 6067 - 314		16 09 34	−54 04.8	1	11.83		0.26		−0.12		1	B8 V	1407
145629	−10 04267			16 09 35	−11 04.4	1	8.51		0.42		0.12		2	A5	1771
145556	−27 10846			16 09 38	−28 11.7	2	8.90	.003	0.07	.001	−0.36	.030	3	B4 II/III	1054,1770
145537	−34 10850			16 09 39	−34 31.4	2	10.41	.003	0.09	.005	−0.78	.009	3	B2 Ib/II	55,540
145572	−26 11259			16 09 40	−26 45.2	1	9.74		0.31		0.16		1	A0 III/IV	1770
145415	−53 07442	NGC 6067 - 316	⋆ AB	16 09 40	−54 00.0	1	8.86		0.93		0.10		1	B3	1407
	−53 07443	NGC 6067 - 317		16 09 42	−54 11.0	1	12.26		0.28				1		1407
	−53 07446	NGC 6067 - 318		16 09 43	−54 11.8	1	12.26		0.24				1		1407
145654	−10 04268			16 09 44	−10 21.3	1	8.53		1.70		2.02		1	K5	1746
145573	−27 10851			16 09 44	−28 00.3	1	7.98		0.44		−0.01		1	F3 V	1770
		G 16 - 34		16 09 45	+05 37.5	1	11.88		0.94		0.70		2		333,1620
145802	+33 02696	HR 6043	⋆ AB	16 09 45	+33 28.2	1	6.29		1.21				2	K2 III	71
145730	+12 02970	IDS16074N1210	A	16 09 46	+12 02.3	1	8.50		0.31		0.09		2	A3	3016
145730	+12 02970	IDS16074N1210	B	16 09 46	+12 02.3	1	9.67		0.50		0.00		2		3016
		HBC 634		16 09 46	−18 51.8	1	10.86		1.17		0.81		1	K0 IV	1763
145417	−57 07690			16 09 47	−57 25.5	7	7.53	.012	0.82	.012	0.29	.017	18	G8/K0 Vw	285,742,1075,2017*
	−53 07453	NGC 6067 - 322		16 09 48	−54 01.6	1	12.33		0.24		−0.14		1		1407
	−53 07454	NGC 6067 - 323		16 09 48	−54 09.3	1	10.24		1.54				1		1407
		G 138 - 12		16 09 49	+11 31.5	2	13.39	.023	1.19	.014	0.97	.009	4		419,3060
145631	−19 04337			16 09 49	−19 22.5	4	7.59	.008	0.15	.012	−0.05	.017	13	B9 V	26,1054,1770,1776
	−49 10409			16 09 49	−49 40.4	1	11.46		0.58		0.18		1		96
145772	+14 03012	IDS16076N1448	A	16 09 50	+14 40.6	1	8.42		1.45		1.69		2	K2	3077
		WLS 1612 60 # 7		16 09 50	+57 29.0	1	12.95		0.51		−0.13		2		1375
145612	−21 04299			16 09 50	−21 42.9	1	10.17		0.61		0.13		1	F0 V	1770
	−53 07458	NGC 6067 - 325		16 09 50	−54 07.3	1	10.36		0.32		−0.04		1	B8 III	1407
145488	−50 10239			16 09 52	−50 28.4	1	10.29		0.17		−0.29		5	8 II(p Si)	567
146044	+60 01659			16 09 55	+59 48.3	1	8.13		0.68		0.23		2	G0	1502
145522	−49 10410			16 09 56	−49 46.1	1	8.62		0.24		−0.22		1	B7 III	96
145492	−51 09893	LSS 3511		16 09 56	−52 03.8	5	9.15	.015	0.51	.019	−0.38	.014	12	B2 Ib	181,403,540,567,976
		NGC 6067 - 220		16 09 56	−54 05.9	1	14.80		0.75		0.25		1		1407
	−35 10818			16 09 57	−35 52.6	1	10.60		1.15		1.13		2	K5 V	3072
	−50 10241	LSS 3512		16 09 57	−50 58.7	2	10.81	.092	0.57	.005	−0.34		7		33,1021
		NGC 6067 - 219		16 09 57	−54 06.3	1	13.12		0.27		0.01		1		1407
145849	+36 02706	HR 6046	⋆ AB	16 09 58	+36 33.2	1	5.63		1.34				2	K3 III	71
145596	−33 11001			16 09 58	−34 10.8	1	7.42		1.82		2.21		6	M2 III	1673
145682	−14 04368			16 09 59	−15 12.4	1	9.71		0.37		0.18		1	A7/8 V	1770
	−54 07260	LSS 3510		16 10 00	−54 47.3	1	10.54		0.17		−0.69		2	B1 II-Ib	540
		Nor sq 2 # 50		16 10 00	−55 01.	1	12.68		0.39		−0.37		2		181
145657	−29 12373			16 10 01	−29 31.2	1	8.92		0.16		0.01		2	B9.5 V	1770
145521	−48 10614			16 10 01	−48 23.8	2	9.99	.007	0.40	.006	−0.30	.014	7	B2 III	540,567
	−49 10412			16 10 04	−49 45.1	1	11.68		0.58		0.31		1		96
330462	−49 10413			16 10 05	−49 42.1	1	10.93		0.40		0.31		1	A0	96
145633	−34 10860	He3 1158		16 10 07	−34 28.0	1	11.12		1.12		0.79		1	K1 III:	1586
145931	+42 02683	HR 6050	⋆ A	16 10 08	+42 30.1	2	5.87	.007	1.47	.014	1.74		4	K4 II+F7 V	71,542
145931	+42 02683	IDS16085N4238	B	16 10 08	+42 30.1	1	10.73		0.58		0.01		2	F8 IV	542

HD	DM	Other Id	N	Rem	α₁₉₅₀	δ₁₉₅₀	S	V	σ_V	B–V	σ_B–V	U–B	σ_U–B	n	Spectrum	References
145632	−33 11004				16 10 09	−33 43.2	1	10.40		0.09		−0.38		2	B8 Ib/II	1770
	−54 07264	LSS 3513			16 10 09	−54 58.0	2	10.12	.008	0.38	.001	−0.51	.009	4	B0.5V	181,540,1737
145774	−1 03144				16 10 10	−01 35.7	2	7.48	.004	−0.12	.001	−0.81	.006	10	B8	1732,3016
		Nor sq 2 # 52			16 10 12	−52 03.	1	11.51		1.33		0.22		2		181
145718	−22 04119	V718 Sco			16 10 13	−22 21.5	1	8.90		0.49		0.34		1	A8 III/IV	1770
145614	−41 10572				16 10 13	−41 13.3	1	9.58		0.43		−0.10		4	B9 III	567
145542	−51 09899				16 10 13	−51 46.8	1	7.77		1.56		1.79		2	K3 III	33
145523	−53 07484				16 10 15	−54 02.7	1	7.82		0.58		0.07		1	G2 V	96
145660	−33 11006				16 10 16	−33 52.6	1	9.90		0.28		0.22		2	A2/3 IV	1770
	−51 10475	LSS 3734			16 10 16	−51 33.1	1	12.05		0.39		−0.30		2		954
145788	−3 03891	HR 6041			16 10 19	−04 05.6	1	6.24		0.13		0.04		8	A1 V	1088
		Pis 22 - 2			16 10 19	−51 44.2	1	12.61		2.63		1.17		1		33
		Pis 22 - 5			16 10 19	−51 44.6	1	13.35		2.12		0.86		1		33
		Pis 22 - 6			16 10 19	−51 44.9	1	12.95		2.19		0.86		1		33
		Pis 22 - 7			16 10 19	−51 45.1	1	14.01		2.37		0.73		1		33
		Pis 22 - 1			16 10 19	−51 45.5	1	11.71		1.99		1.90		2		33
		Pis 22 - 3			16 10 20	−51 44.3	1	13.15		2.27		0.92		1		33
		Pis 22 - 4			16 10 20	−51 44.7	1	13.19		2.30		1.12		1		33
		Pis 22 - 10			16 10 20	−51 46.3	1	11.31		1.30		1.14		1		33
145957	+39 02961				16 10 21	+39 11.0	1	6.40		1.26		1.29		3	K2 III	833
145579	−51 09904				16 10 21	−51 20.0	1	9.74		0.24		−0.27		3	B7/8 Ib/II	567
		Pis 22 - 8			16 10 22	−51 45.2	1	13.54		0.72		0.13		2		33
145719	−27 10856	IDS16073S2709	A		16 10 23	−27 16.6	1	8.41		0.30		0.09			A6 IV	1770
145598	−50 10252				16 10 23	−50 39.2	1	8.66		0.66				4	G5 V (+F/G)	2033
145578	−49 10419				16 10 24	−49 38.4	1	10.16		0.33		0.18		1	B8/A0 V	96
	−50 10251	LSS 3514			16 10 24	−51 07.3	2	10.53	.049	0.36	.007	−0.36	.027	4	B2.5 IV	33,540
		G 16 - 35			16 10 25	+06 51.7	2	13.12	.015	0.68	.015	0.01	.025	5		333,1620,1658,3060
145851	+12 02971				16 10 26	+12 42.5	1	8.21		1.11				1	K2	882
145700	−35 10825				16 10 28	−35 40.3	1	9.10		0.24		0.03		2	B9 IV/V	1770
		EE Her			16 10 29	+18 08.8	1	13.55		0.49		0.11		1	a2	699
145749	−20 04438				16 10 29	−21 03.8	1	9.98		0.57		0.11		1	F3 V	1770
	−49 10423				16 10 29	−49 49.8	1	10.11		1.05		0.65		1		96
		LOD 39 # 8			16 10 36	−49 40.5	1	11.62		1.09		0.54		1		96
145891	+13 03089				16 10 38	+12 55.6	1	7.02		0.23		0.10		3	A3	3016
145637	−50 10259				16 10 38	−51 03.5	1	9.54		0.20		−0.36		1	B3 II/III	567
145600	−53 07506				16 10 40	−53 26.0	1	8.96		0.18		−0.35		1	B8 Ib/II	567
145976	+27 02603	HR 6052	★ AB		16 10 41	+26 47.9	1	6.52		0.38		−0.05		1	F3 V	254
		LSS 3515			16 10 42	−52 04.7	2	11.39	.000	1.17	.000	0.13	.000	6	B3 Ib	181,1737
145780	−31 12681				16 10 43	−31 54.5	1	9.95		0.28		0.16		1	A2/3 III/I	1770
145809	−21 04305				16 10 44	−21 16.4	1	6.69		0.61		0.08		3	G3 V	3026
145792	−24 12623	HR 6042	★ AB		16 10 45	−24 17.7	5	6.40	.009	0.03	.008	−0.46	.008	16	B6 IV	26,1054,1079,1770,2007
		GD 196			16 10 46	+16 39.6	2	15.73	.073	0.06	.015	−0.63	.000	3		1663,3060
	−55 07263				16 10 46	−55 57.8	1	10.37		0.16		−0.16		3		1586
145892	+5 03165	HR 6047			16 10 47	+05 08.8	1	5.47		1.47		1.77		8	K0	1088
145810	−23 12786				16 10 47	−24 07.7	1	8.35		0.33		0.18		2	A1/2 IV	1770
	−51 09912				16 10 47	−51 36.6	1	11.31		0.53		−0.12		2		33
145835	−12 04453				16 10 49	−12 32.1	1	9.79		0.72		0.21		2	G8/K0 +(F)	1375
145934	+13 03090				16 10 50	+13 21.9	1	8.56		0.98				1	K0	882
145854	−12 04454				16 10 50	−12 54.5	1	8.07		0.25		0.16		1	A0 III	1770
	−55 07266				16 10 50	−55 18.0	1	10.25		0.53		0.22		2		540
145684	−49 10429				16 10 51	−49 40.7	1	10.18		0.30		0.16		1	A7/F0 (II)	96
145664	−52 09393				16 10 51	−52 12.7	4	8.29	.013	0.32	.017	−0.45	.018	7	B2 II/III	115,181,540,976
145793	−27 10858	IDS16078S2800	A		16 10 52	−28 07.6	1	7.94		0.24		0.13		1	A2 V + (A)	1770
145544	−63 03854	HR 6030	★ A		16 10 52	−63 33.6	4	3.84	.005	1.10	.005	0.86	.005	21	G2 Ib-IIa	15,1075,1075,2012
	−50 10263	LSS 3518			16 10 53	−50 40.8	2	10.25	.078	0.37	.005	−0.40	.001	5	B5	33,1586
	−49 10430				16 10 54	−49 55.7	1	11.13		0.72		0.15		1		96
330398	−46 10611				16 10 55	−47 10.2	1	10.86		0.72		0.00		4	F2	567
		G 202 - 32			16 10 57	+45 31.0	1	12.94		1.41				1		1759
145958	+13 03091	IDS16086N1348	AB		16 10 58	+13 39.6	3	6.68	.008	0.76	.005	0.34	.000	12	G8 V	1080,1197,3026
146010	+21 02886				16 10 58	+21 41.5	2	6.70	.005	0.15	.000	0.09	.015	4	A2	985,1776
145877	−15 04271				16 10 58	−16 11.0	1	9.84		0.43		0.19		1	A2mA8-F0	1770
145856	−22 04120				16 10 58	−22 18.8	1	9.29		0.32		0.22		2	A2 III	1770
145894	−7 04233				16 11 03	−07 59.5	1	6.84		1.11		0.89		2	K0	1375
145857	−23 12791				16 11 03	−24 02.3	1	9.14		0.37		0.21		2	A0mA7-A8	1770
146100	+39 02963	G 180 - 39	★ AB		16 11 04	+39 28.9	1	8.56		0.71		0.18		6	G0	3016
145897	−11 04096	HR 6048			16 11 05	−11 42.7	2	5.22	.000	1.40	.025	1.54		9	K3 III	2007,3005
	−52 09398				16 11 05	−53 01.5	1	11.13		0.42		−0.34				115
145879	−21 04306				16 11 06	−21 57.7	1	9.90		0.28		0.15		2	A2 IV	1770
330465	−49 10433				16 11 06	−49 56.4	1	11.63		0.56		0.16		1	A5	96
	−50 10270	LSS 3519			16 11 06	−51 06.0	1	11.12		0.52		−0.15		2		33
145837	−27 10860				16 11 08	−27 40.4	1	10.04		0.43		0.29		1	A3 III	1770
	−49 10434				16 11 08	−50 07.7	1	11.20		0.51		0.17		2		33
	−52 09404				16 11 09	−52 14.6	2	11.90	.005	0.52	.000	−0.19	.030	4		1696,3065
145838	−32 11525	HR 6044			16 11 11	−32 53.1	1	5.92		1.02				4	K0 III	2009
145826	−39 10364				16 11 15	−39 27.6	1	8.18		0.53		0.44		2	A3 V	1770
	−49 10438	LSS 3522			16 11 15	−50 04.9	1	10.73		0.57		−0.32		2		33
		GD 197			16 11 16	−04 40.7	1	15.81		0.70		−0.02		1		3060
		LP 917 - 1			16 11 17	−28 22.6	1	12.90		1.50				2		3062
145939	−13 04383	LS IV -13 001			16 11 18	−13 50.8	1	10.14		0.00		−0.62		1	B2/3 V	1770
	−51 09925	LSS 3521			16 11 18	−52 04.1	2	10.34	.049	0.93	.005	−0.13	.005	3		115,181

Table 1   875

HD	DM	Other Id	N Rem	α_{1950}	δ_{1950}	S	V	σ_V	B–V	σ_{B-V}	U–B	σ_{U-B}	n	Spectrum	References
145917	−18 04247			16 11 19	−18 42.6	1	8.68		0.68		0.10		21	G3 V	1763
145840	−38 10924	IDS16080S3853	A	16 11 20	−39 00.3	1	7.70		0.32		0.04		1	A9 V	1770
145942	−21 04308			16 11 22	−22 00.6	1	7.73		0.27		0.18		2	A2 IV	1770
	−55 07285			16 11 24	−55 35.9	1	10.61		0.44		-0.13		3		1586
145814	−49 10441			16 11 27	−49 45.0	1	8.27		1.58		1.62		1	K2/3 III	96
		LOD 39 # 14		16 11 28	−49 57.0	1	12.13		0.53		0.13		1		96
	−50 10277			16 11 31	−50 48.1	1	10.46		1.05		0.54		1		33
145964	−20 04444	HR 6051		16 11 32	−20 58.9	2	6.41	.006	-0.01	.005	-0.16		6	B9 V	1770,2007
		LOD 39 # 15		16 11 33	−49 57.3	1	12.13		0.67		0.01		1		96
145880	−39 10367			16 11 34	−39 30.1	1	7.16		0.01		-0.02		4	B9.5 V	1770
145794	−52 09416	V349 Nor		16 11 34	−52 47.8	3	8.74	.004	0.22	.007	-0.62	.016	6	B2 II/III	115,540,976
	−54 07309	Nor sq 2 # 58		16 11 34	−54 51.6	1	10.45		0.21		-0.52		2		181
146081	+16 02908			16 11 36	+16 33.6	1	7.77		1.67		1.78		3	M1	1648
145966	−23 12794			16 11 36	−24 01.	1	10.32		0.47		0.21		2	A3/5 Ib/II	1770
145842	−47 10611	HR 6045		16 11 37	−47 14.8	2	5.13	.005	-0.12	.005	-0.40		8	B8 V	567,2007
	−50 10282			16 11 38	−51 01.8	1	10.85		0.48		-0.18		2	B9	33
145944	−31 12695			16 11 40	−32 09.9	1	9.88		0.37		0.09		1	A8 V	1770
		LOD 39 # 16		16 11 40	−49 57.6	1	12.26		0.61		0.17		1		96
		LOD 39 # 16s		16 11 40	−49 57.6	1	12.41		0.92				1		96
145828	−50 10284			16 11 40	−50 51.5	2	9.71	.049	0.35	.024	-0.62	.024	3	B0.5 III	33,115
145999	−21 04312			16 11 41	−21 45.8	1	10.28		0.40		0.29		2	A7 III	1770
145782	−57 07716	HR 6040		16 11 41	−57 47.2	2	5.62	.005	0.14	.000			7	A3 III	15,2030
146051	−3 03903	HR 6056	⋆ A	16 11 43	−03 34.0	6	2.73	.014	1.58	.010	1.95	.007	23	M1 III	15,1075,1088,3016*
146084	+6 03184	HR 6057		16 11 46	+06 01.7	1	6.30		1.15		1.17		8	K2 III	1088
145997	−18 04249	HR 6053	⋆	16 11 46	−18 24.5	1	6.32		1.09				4	K1 III	2007
	−52 09419			16 11 47	−52 55.2	1	11.33		0.29		-0.21		1		115
145815	−53 07544			16 11 47	−53 55.3	1	7.97		0.46		0.02		1	F5 V	96
146000	−22 04125			16 11 48	−23 12.0	1	10.59		0.21		0.19		1	A0 IV	1770
145846	−52 09422	LSS 3524		16 11 49	−52 14.8	5	8.79	.017	0.38	.014	-0.57	.025	9	B1/2ne	115,181,540,976,1586
146001	−25 11453	HR 6054		16 11 51	−25 21.1	7	6.06	.011	0.04	.008	-0.37	.009	21	B8 V	26,540,976,1054,1079*
146029	−22 04127			16 11 55	−22 15.3	3	7.38	.005	0.08	.007	-0.07	.010	11	B9 V	26,1054,1770
145921	−42 11132	HR 6049		16 11 55	−42 46.5	1	6.14		1.11				4	K2 III	2006
	−56 07487			16 11 56	−56 47.2	1	11.59		1.10		0.61		3		1586
		WLS 1620-10 # 6		16 11 57	−09 30.4	1	11.95		0.97		0.79		2		1375
145907	−49 10451			16 11 57	−49 57.6	1	9.83		0.42		0.18		1	F0 IV/V	96
		LP 624 - 55		16 12 00	+02 22.8	1	14.80		1.61				5		940
		Ton 256		16 12 00	+26 13.	1	15.41		0.65		-0.78		1		1036
145865	−52 09426			16 12 00	−52 16.6	1	10.54		0.24		-0.28		1	A2 II/III	115
146002	−33 11032			16 12 01	−34 01.9	1	10.13		0.26		0.21		1	A1/2 IV	1770
145766	−63 03863			16 12 02	−64 02.4	1	8.06		0.95		0.64		1	G8 III	1487
146116	−0 03082			16 12 04	−00 16.4	1	7.68		1.07		0.90		2	G8 III	536
146053	−19 04344			16 12 05	−19 55.3	1	10.07		0.54		0.24		1	A8/F3(m)	1770
145983	−40 10296			16 12 05	−40 16.4	1	9.48		0.24		-0.18		6	B7 Ib/II	567
		LundsII141 # 410		16 12 05	−52 11.7	2	10.98	.060	1.01	.010	-0.19	.010	2		115,181
145689	−67 03054	HR 6037		16 12 05	−67 49.0	3	5.87	.087	0.14	.005	0.08	.005	13	A3 V	15,1075,2038
146055	−24 12637			16 12 06	−24 27.8	1	8.40		0.33		0.11		2	B9 V	1770
		Nor sq 1 # 11		16 12 06	−51 30.	1	12.21		0.53		-0.25		1		115
	−51 09935	Nor sq 1 # 9		16 12 06	−52 11.4	2	10.59	.025	0.37	.015	-0.19	.015	2		115,181
146385	+56 01875			16 12 07	+55 51.5	1	9.90		-0.13		-0.41		2	B9	1502
146088	−18 04251			16 12 09	−18 17.0	1	9.02		0.37		0.21		1	A3 III	1770
146069	−23 12803			16 12 09	−23 38.1	1	9.54		0.51		0.09		1	F2 V	1770
		G 168 - 36		16 12 10	+20 49.0	2	11.41	.012	1.33	.020	1.24	.034	3		1658,5010
146087	−14 04379			16 12 10	−14 30.8	2	8.14	.043	0.32	.010	0.03	.002	4	A8/9 IV/V	1770,3063
145925	−51 09938			16 12 12	−51 41.1	1	9.51		0.32		-0.17		2	B7 III	567
146603	+67 00930	HR 6069		16 12 13	+67 16.2	2	6.20	.005	0.99	.002	0.72		4	G8 III	71,3016
146926	+76 00594	HR 6079		16 12 13	+76 00.3	2	5.48		-0.12	.016	-0.38	.040	4	B8 V	985,1079
146118	−11 04101			16 12 13	−11 35.4	1	8.73		0.35		0.07		2	A3	1771
146089	−21 04315			16 12 14	−22 00.0	1	10.20		0.63		0.09		1	F5 V	1770
146030	−34 10891			16 12 15	−35 08.4	1	9.53		0.43		0.31		1	A3 V	1770
	−52 09436			16 12 18	−52 11.7	1	10.35		0.23		-0.39		1		115
146119	−12 04460			16 12 19	−13 06.0	1	8.65		0.40		0.21		1	A2 IV	1770
146360	+47 02318			16 12 25	+47 00.1	1	8.91		0.36		0.10		2	F5	1601
	−51 09944			16 12 25	−51 36.0	1	11.37		0.32		-0.28		1		115
145947	−53 07579			16 12 26	−53 29.4	1	10.23		0.15		-0.55		1		115
146121	−21 04316	IDS16095S2131	AB	16 12 27	−21 39.7	1	10.14		0.61		0.09		1	F5/6 V	1770
	−52 09440			16 12 27	−52 39.0	1	11.05		0.25		-0.35		1		115
146173	+6 04391			16 12 29	−07 14.2	1	8.09		0.69		0.07		1	F0	1776
		LP 444 - 35		16 12 30	+19 13.	1	12.90		1.54				3		419
		AJ74,882 T1# 11		16 12 30	−50 18.0	1	11.90		0.46		-0.09		2		33
146120	−21 04317			16 12 31	−21 27.8	1	8.96		0.42		0.24		2	A2mA8-F2	1770
146140	−19 04346			16 12 33	−20 06.9	1	9.94		0.54		0.29		1	F0 V	1770
145619	−72 01920			16 12 34	−72 55.2	1	6.56		0.02		0.00		11	B9 V	1628
	−50 10301	LSS 3526		16 12 36	−50 18.9	1	10.77		0.56		-0.25		2	B9	33
	−57 07744	NGC 6087 - 110		16 12 37	−57 48.3	1	11.07		0.26				2		1385
		GD 198		16 12 38	−11 11.1	1	15.53		-0.11		-0.94		1	DB	3060
	−52 09444			16 12 38	−52 57.6	1	11.07		0.36		-0.12		1		115
325496	−42 11143			16 12 39	−43 02.7	1	10.32		0.38				4	A0	1021
327996	−45 10543			16 12 39	−45 32.2	1	10.72		0.36		0.09		4	B8	567
	−51 09166	LSS 3525		16 12 39	−52 09.1	2	11.61	.060	0.81	.010	-0.25	.045	2		115,181
146279	+11 02947			16 12 40	+11 36.9	1	7.54		0.32		0.02		3	F0	1648

HD	DM	Other Id	N Rem	α₁₉₅₀	δ₁₉₅₀	S	V	σ_V	B–V	σ_B-V	U–B	σ_U-B	n	Spectrum	References
	−53 07591			16 12 42	−53 50.4	1	13.29		0.58		0.34		1		885
		NGC 6087 - 113		16 12 42	−57 57.9	1	11.90		0.50				2		1385
146175	−22 04129			16 12 43	−22 41.0	1	9.28		0.21		0.17		1	A1/A2 V	1770
		NGC 6087 - 111		16 12 43	−57 52.0	1	12.47		0.35				2		1385
	−57 07747	NGC 6087 - 109		16 12 45	−57 46.9	1	11.31		0.14				2		1385
	−9 04337			16 12 46	−09 45.4	1	9.66		1.70		1.67		1		565
146214	−12 04463			16 12 46	−12 33.4	2	7.50	.028	0.15	.001	0.14	.024	3	A1 V	1770,3063
	−53 07592			16 12 46	−53 40.3	1	10.36		0.08		−0.44		1		115
		NGC 6087 - 112		16 12 46	−57 52.8	1	11.80		1.21				2		1385
146361	+34 02750	HR 6063, TZ CrB	⋆ AB	16 12 48	+33 59.0	3	5.27	.128	0.58	.029	0.04	.021	5	G0 V	254,292,938
146362	+34 02750	HR 6064	⋆ C	16 12 48	+33 59.1	1	12.31		1.40		1.28		3	G0	7009
	−51 09949	LSS 3527		16 12 48	−51 54.2	5	9.87	.018	0.86	.007	−0.22	.022	9	B0.5 Ia	115,181,540,1021,1737
146003	−53 07594	HR 6055		16 12 48	−53 41.3	4	5.46	.019	1.72	.007	1.96	.027	8	M2 III	96,885,2007,3055
145366	−78 01092	HR 6020, δ1 Aps	⋆ A	16 12 48	−78 34.4	4	4.73	.033	1.68	.011	1.67	.007	16	M5 IIIb	15,1075,1279,3052
	−51 09950	LSS 3528		16 12 49	−52 01.9	3	9.80	.021	0.83	.011	−0.21	.017	7	B1 Iab	115,540,1021
	−52 09447			16 12 49	−53 07.5	1	10.67		0.32		−0.07		1		115
146195	−21 04319			16 12 50	−21 49.3	1	10.41		0.44		0.25		2	A3/5 II	1770
		G 153 - 38		16 12 51	−07 21.3	1	10.36		0.75		0.29		2		7010
146058	−50 10309	LSS 3529		16 12 51	−50 29.0	3	9.57	.012	0.46	.008	−0.47	.007	7	B0/2 I	33,540,976
146018	−53 07599	IDS16090S5329	AB	16 12 52	−53 36.3	2	8.52	.010	0.15	.010	−0.04	.020	2	A0 V	96,885
	−51 09951	Nor sq 1 # 23		16 12 53	−51 42.0	1	10.82		0.46		−0.11		1	A0	115
146233	−7 04242	HR 6060	⋆ A	16 12 54	−08 14.3	11	5.50	.015	0.65	.009	0.17	.017	27	G2 Va	15,22,101,245,254*
146216	−19 04348			16 12 55	−19 57.1	1	8.08		0.31		0.08		1	A8 III	1770
145388	−78 01093	HR 6021	⋆ B	16 12 55	−78 32.7	4	5.27	.012	1.41	.012	1.61	.011	12	K3 III	15,1075,1279,3046
	−52 09455			16 12 58	−52 27.8	2	10.76	.010	0.29	.000	−0.43	.005	2		115,181
146197	−29 12406			16 12 59	−29 41.3	1	9.90		0.32		0.11		1	B9 IV/V	1770
	−2 04146			16 13 01	−02 16.1	1	10.49		0.86				1	K5 V	3072
	−51 09955	LSS 3530		16 13 01	−51 50.6	4	9.81	.022	0.91	.019	−0.16	.023	6	B1 Iab	115,181,432,540
		NGC 6087 - 108		16 13 01	−57 50.4	1	12.29		0.55				2		1385
146254	−14 04383	HR 6061		16 13 02	−14 43.5	4	6.09	.008	0.09	.013	0.02	.021	9	A0 III	252,1770,2007,3063
146059	−53 07607	IDS16092S5327	AB	16 13 04	−53 34.3	3	6.39	.008	0.79	.008	0.32	.005	6	G6 IV +G2 V	96,885,2012
330554	−49 10469			16 13 05	−49 38.8	1	10.05		0.70		0.30		1	G0	96
		Nor sq 2 # 67		16 13 06	−51 50.	1	11.74		0.47		−0.19		1		181
		Nor sq 1 # 27		16 13 06	−52 31.	1	11.71		0.55		−0.11		1		115
	−53 07610			16 13 07	−53 17.2	1	11.41		0.23		−0.27		1		115
		NGC 6087 - 115		16 13 07	−57 59.4	1	12.76		0.34				2		1385
146124	−49 10471			16 13 08	−49 44.0	1	7.69		0.77		0.50		1	G6 V	96
		G 138 - 16		16 13 10	+07 29.1	2	15.31	.012	0.83	.002	0.23	.023	4		1658,3060
	−51 09958	LSS 3531		16 13 11	−51 41.2	2	10.82	.078	0.97	.015	−0.02	.015	3		33,115
		Nor sq 1 # 30		16 13 12	−51 55.	2	11.88	.005	0.64	.060	−0.34	.065	2		115,181
	−57 07759	NGC 6087 - 107		16 13 12	−57 52.1	1	11.88		0.43				2		1385
	−51 09959			16 13 15	−51 35.4	1	12.01		0.50		−0.34		1		115
		NGC 6087 - 116		16 13 15	−58 01.3	1	12.80		0.53				2		1385
146388	+19 03075	HR 6065		16 13 16	+18 56.0	2	5.70	.010	1.13	.005	1.07	.050	5	K3 III	252,3077
146236	−27 10877			16 13 16	−28 02.4	1	9.14		0.47		0.25		2	A1mA9-F2	1770
146143	−49 10474	HR 6058		16 13 16	−49 56.7	8	4.98	.019	0.80	.010	0.50	.014	24	F7 Ib	15,96,1018,1034,1075*
146470	+32 02697			16 13 19	+32 16.3	1	8.43		1.35		1.51		2	K4 III	1003
146282	−22 04131			16 13 19	−23 00.3	2	9.46	.005	0.37	.005	0.16		5	A5	370,1770
146283	−22 04132			16 13 19	−23 02.3	1	8.42		1.27				3	G8/K1 (III)	370
		NGC 6093 sq1 7		16 13 20	−22 59.3	1	10.55		1.75				3		370
146266	−24 12644			16 13 21	−24 56.3	1	8.41		0.30		0.19		2	A1 V	1770
146077	−57 07762	NGC 6087 - 19		16 13 21	−57 58.0	1	9.90		0.04				1	B8 Ib/II	1385
	−53 07618			16 13 22	−53 19.6	1	10.66		0.18		−0.34		1		115
	−51 09960			16 13 23	−52 00.3	1	10.75		0.38		−0.10		1		115
146285	−24 12645			16 13 24	−24 51.9	3	7.93	.007	0.23	.003	−0.10	.007	10	B8 V	26,1054,1770
		NGC 6093 sq1 10		16 13 25	−22 46.9	1	11.51		1.14				2		370
146284	−23 12816	IDS16104S2402	AB	16 13 26	−24 09.5	3	6.71	.004	0.15	.006	−0.16	.009	9	B9 III/IV	26,1054,1770
146145	−52 09469	HR 6059	⋆ A	16 13 28	−52 57.8	5	6.31	.008	0.28	.012	0.11		19	A8 IV	15,285,2013,2020,2028
		Mark 876 # 2		16 13 30	+65 48.0	1	14.09		0.55		0.05		1		829
		Nor sq 1 # 34		16 13 30	−51 20.	1	11.68		0.52		−0.27		1		115
146413	+7 03125	G 17 - 8	⋆ AB	16 13 31	+07 29.2	3	8.74	.001	1.11	.024	1.02	.017	7	K3 V +K5 V	265,938,3030
		NGC 6093 sq1 12		16 13 31	−22 56.3	1	12.29		1.19				2		370
146144	−52 09471			16 13 31	−52 30.9	1	10.37		0.28		0.29		1	A5/7 V	115
	−51 09964	LSS 3533		16 13 32	−51 38.8	5	9.87	.032	0.93	.014	−0.12	.033	10	B0.5 Ia	33,115,540,1021,1737
	−53 07629			16 13 32	−53 29.2	1	11.92		0.31		0.38		3		1586
146286	−32 11559			16 13 33	−32 37.1	1	9.79		0.29		0.24		1	A0 V	1770
		CS 22872 # 5		16 13 34	−04 55.5	1	13.33		0.47		0.14		1		1736
		CS 22872 # 16		16 13 35	−02 56.2	1	15.37		0.24		0.27		1		1736
		NGC 6087 - 157		16 13 38	−57 45.6	1	12.78		0.49		0.18		4		1614
		NGC 6087 - 103		16 13 39	−57 39.4	1	12.33		0.32				2		1385
		Mark 876 # 4		16 13 40	+65 49.0	1	14.89		0.42		−0.09		1		829
	−52 09475	LSS 3534		16 13 40	−52 18.9	1	11.80		0.46		−0.35		1		115
	−57 07767	NGC 6087 - 104		16 13 40	−57 40.7	1	11.65		0.70				2		1385
		CS 22872 # 24		16 13 41	−02 19.6	1	13.40		0.14		0.05		1		1736
146330	−24 12646	IDS16107S2459	A	16 13 42	−25 06.4	1	8.86		0.40		0.25		3	A5 IV	1770
	−57 07768	NGC 6087 - 106		16 13 43	−57 51.0	2	11.87	.047	0.19	.028	0.07		2		69,1385
146148	−57 07769	NGC 6087 - 12		16 13 44	−57 49.0	3	7.48	.012	1.18	.004	0.92	.028	8	G4 III	69,460,1614
146537	+27 02613	HR 6068		16 13 45	+27 32.8	1	6.14		1.31				2	K2	71
146332	−29 12411	IDS16106S2930	AB	16 13 45	−29 37.3	7	7.62	.021	0.19	.011	−0.34	.013	21	B3 III	26,173,400,1054,1770*
147142	+75 00586	HR 6082		16 13 46	+75 20.1	2	6.34	.000	1.29	.005	1.38	.015	4	K2 IV	1502,1733

Table 1   877

HD	DM	Other Id	N Rem	α_{1950}	δ_{1950}	S	V	σ_V	B–V	σ_{B-V}	U–B	σ_{U-B}	n	Spectrum	References
146331	−25 11464			16 13 47	−25 44.4	1	8.36		0.34		0.08		2	B9 V	1770
146671	+48 02380			16 13 49	+48 27.1	1	8.99		1.33		1.50		3	K5	196
	−57 07774	NGC 6087 - 61		16 13 49	−57 33.3	1	11.56		0.14				1		1385
146161	−57 07771	NGC 6087 - 18		16 13 49	−57 58.9	1	9.98		0.10				1	B9 V	1385
	−57 07776	NGC 6087 - 60		16 13 50	−57 34.3	1	11.79		0.68				1		1385
146160	−57 07773	NGC 6087 - 16		16 13 50	−57 51.9	2	8.70	.013	1.73	.014	2.10		5	M3 III	69,1385
	−57 07772	NGC 6087 - 117		16 13 51	−58 04.9	1	11.79		0.37				2		1385
147620	+81 00543			16 13 52	+81 16.2	1	7.90		1.10		0.97			G5	1733
146333	−31 12732			16 13 52	−31 44.3	1	10.29		0.34		0.22		1	A8/F0 (III	1770
146187	−57 07775	NGC 6087 - 17		16 13 52	−57 55.8	2	8.78	.001	0.09	.030	-0.28		2	B8 V	60,1385
		WLS 1612 60 # 9		16 13 53	+59 55.9	1	11.26		0.53		-0.07		2		1375
146367	−22 04133			16 13 54	−22 59.2	2	8.68	.009	0.45	.003	0.18		5	A9 V	370,1770
146224	−53 07650			16 13 54	−53 25.9	4	7.51	.021	-0.03	.012	-0.53	.014	6	B3 III	96,115,540,976
		CS 22872 # 19		16 13 55	−02 46.7	1	13.63		0.40		0.12		1		1736
146314	−36 10720			16 13 55	−36 26.5	1	10.41		0.51		0.05		1	B8/9 (Ib)	1770
	−52 09482			16 13 55	−52 40.6	1	10.59		0.18		-0.22		1		115
		NGC 6087 - 102		16 13 56	−57 40.2	1	12.65		0.14				2		1385
	−57 07778	NGC 6087 - 118		16 13 56	−58 05.3	1	11.35		0.16				2		1385
	−51 09969	LSS 3536		16 13 57	−51 32.8	2	10.95	.068	0.49	.019	-0.20	.029	3	B2	33,115
146204	−57 07779	NGC 6087 - 15		16 13 57	−57 43.7	2	10.17	.014	0.05	.006	-0.38		4	B8 V	69,1385
146242	−52 09483			16 13 58	−52 52.5	2	8.82	.005	0.15	.005	-0.56	.005	2	B3/5 III	115,181
	−55 07345			16 13 58	−55 26.8	1	10.65		0.34		0.19		3		1586
		NGC 6093 sq1 14		16 13 59	−22 43.5	1	12.81		0.69				2		370
		Mark 876 # 3		16 14 01	+65 50.2	1	14.46		0.59		0.06		1		829
	−57 07781	NGC 6087 - 59		16 14 01	−57 36.1	1	12.11		0.13				1		1385
		NGC 6087 - 2407		16 14 01	−57 48.8	1	12.28		1.40				2		69
146639	+36 02714			16 14 02	+36 40.6	1	6.99		1.34		1.53		2	K5	1733
146416	−20 04454	HR 6066		16 14 02	−21 10.9	5	6.61	.009	0.01	.008	-0.17	.005	15	B9 V	26,1054,1079,1770,2007
	−57 07780	NGC 6087 - 45		16 14 02	−57 50.1	1	11.06		0.18				2		1385
		Mark 876 # 1		16 14 03	+65 51.7	1	11.67		1.53		1.73		1		829
146397	−27 10884			16 14 04	−27 23.3	1	10.00		0.52		0.42		1	A1 IV	1770
	−52 09485			16 14 04	−52 13.6	1	10.73		0.14		-0.24		1		115
	−57 07783	NGC 6087 - 105		16 14 04	−57 45.9	1	12.34		0.18				1		1385
		G 180 - 45		16 14 05	+35 56.5	1	9.62		1.23		1.36		2	K5	7010
	−57 07784	NGC 6087 - 119		16 14 06	−58 06.9	1	11.38		0.60				2		1385
	−57 07786	NGC 6087 - 44		16 14 07	−57 48.4	2	10.30	.016	1.40	.006	1.61		5		69,1385
146490	−5 04266			16 14 08	−05 22.4	1	7.44		0.54		0.09		1	A2	1776
	−57 07788	NGC 6087 - 101		16 14 08	−57 41.4	1	11.26		0.06				1		1385
	−57 07787	NGC 6087 - 123		16 14 08	−57 58.1	1	12.24		0.41				1		1385
		NGC 6087 - 122		16 14 08	−58 00.8	1	12.74		0.61				1		1385
146604	+23 02916			16 14 09	+23 14.7	1	6.56		1.01		0.74		2	K0	985
146813	+56 01879			16 14 09	+55 55.4	3	9.08	.011	-0.20	.016	-0.84	.017	12	B8	272,308,1775
		NGC 6087 - 48		16 14 09	−57 49.0	2	12.75	.033	0.24	.002	0.14		4		69,1385
	−57 07789	NGC 6087 - 46		16 14 09	−57 50.5	1	11.44		0.23				1		1385
146868	+61 01574	G 225 - 51	⋆A	16 14 10	+60 47.2	1	7.69		0.66		0.08		3	G5	3016
146335	−42 11164			16 14 10	−42 53.5	1	8.20		0.19		-0.25		6	B6 III	567
		NGC 6087 - 120		16 14 10	−58 03.9	1	11.20		1.43				2		1385
145869	−73 01713			16 14 10	−74 11.1	1	7.27		1.02		0.79		5	K0 III	1704
		CS 22872 # 15		16 14 12	−03 25.0	1	15.25		0.34		0.29		1		1736
	−13 04391			16 14 12	−13 14.9	1	12.64		1.35		1.19		1	K5 V	3072
		Nor sq 1 # 43		16 14 12	−51 31.	1	14.44		0.49		-0.24				115
	−52 09487	LSS 3538		16 14 13	−53 02.7	3	10.54	.016	0.23	.008	-0.59	.006	5		115,181,1586
		NGC 6087 - 121		16 14 13	−58 02.7	1	12.32		0.75				1		1385
	−57 07791	NGC 6087 - 14		16 14 14	−57 44.2	2	9.69	.020	0.09	.008	-0.26		4	B8 V	69,1385
		G 168 - 40		16 14 15	+31 36.4	1	14.41		0.88		0.42		2		1658
146437	−23 12827			16 14 15	−24 06.0	1	8.94		0.32		0.24		2	A1 III	1770
	−57 07792	NGC 6087 - 43		16 14 15	−57 46.4	1	12.13		0.28				2		1385
146247	−58 06733			16 14 15	−58 16.2	1	7.22		1.23		1.41		1	K2 III	669
146457	−22 04136			16 14 16	−22 48.0	2	8.45	.011	0.32	.004	0.17		5	A5 III/IV	370,1770
146261	−57 07793	NGC 6087 - 13		16 14 16	−57 42.2	3	9.29	.008	0.04	.017	-0.40	.000	5	B8 II	60,69,1385
	−57 07852	NGC 6087 - 34		16 14 16	−57 48.9	2	11.09	.010	0.28	.014	0.17		2		1385,1614
146514	−3 03910	HR 6067		16 14 17	−03 49.9	2	6.21	.057	0.32	.014	0.04	.066	9	A9 Vn	254,1088
		CS 22872 # 18		16 14 18	−02 51.5	1	13.94		0.27		0.27		1		1736
		NGC 6093 sq1 11		16 14 18	−22 58.1	1	12.17		0.83				2		370
		Nor sq 1 # 45		16 14 18	−52 25.	1	11.64		0.32		-0.29		1		115
146564	+1 03191	IDS16118N0126	AB	16 14 20	+01 20.6	1	9.41		0.61				5	G0	173
	−51 09977	LSS 3539		16 14 20	−51 49.1	4	10.71	.024	0.78	.029	-0.30	.008	5		115,181,432,1737
		NGC 6087 - 2301		16 14 20	−57 49.2	1	12.65		0.29		0.26		2		1614
	−57 07796	NGC 6087 - 41		16 14 21	−57 45.4	2	11.37	.025	0.12	.009	-0.07		2		69,1385
		NGC 6087 - 2302		16 14 21	−57 49.4	1	12.56		0.29		0.24		2		1614
		HA 84 # 332		16 14 22	+14 47.5	1	11.03		1.05		0.84		8		1499
		NGC 6093 sq1 13		16 14 22	−22 59.6	1	12.64		0.82				2		370
		L 410 - 21		16 14 24	−40 51.	1	11.31		0.93		0.70		1		1696
		NGC 6087 - 2303		16 14 24	−57 49.6	1	12.80		1.32		1.36		2		1614
146439	−33 11058			16 14 25	−33 39.2	1	9.54		0.30		0.27		1	A1 IV/V	1770
146305	−53 07668			16 14 25	−53 20.8	1	10.29		0.06		-0.20		1	B8/9 II	115
		CS 22872 # 25		16 14 26	−02 00.0	1	14.80		0.22		0.28		1		1736
		NGC 6093 sq1 16		16 14 26	−22 57.0	1	13.51		1.49				2		370
146337	−48 10684			16 14 26	−49 05.5	1	8.10		0.00		-0.42		1	B5/7 II	567
	−52 09498			16 14 26	−52 23.6	1	11.56		0.36		-0.28		1		115

HD	DM	Other Id	N	Rem	α_{1950}	δ_{1950}	S	V	σ_V	B–V	σ_{B-V}	U–B	σ_{U-B}	n	Spectrum	References
	−52 09497				16 14 26	−52 31.9	1	10.47		0.26		−0.17		1		115
	−57 07798	NGC 6087 - 42			16 14 26	−57 45.9	1	11.85		0.26				1		1385
		NGC 6087 - 125			16 14 26	−58 05.8	1	12.32		0.23				1		1385
146494	−22 04137				16 14 27	−23 13.1	1	9.05		0.40		0.16		1	A7 III	1770
	−51 09979				16 14 28	−51 54.8	1	10.72		0.33		−0.16		2	B9	432
	−57 07802	NGC 6087 - 62			16 14 28	−57 39.2	1	12.09		0.25				1		1385
		NGC 6087 - 3302			16 14 28	−57 43.8	1	11.41		0.18		0.06		3		69
	−57 07801	NGC 6087 - 40			16 14 28	−57 43.8	1	11.39		0.27				1		1385
146294	−57 07803	NGC 6087 - 11			16 14 28	−57 49.0	3	9.40	.010	0.03	.005	−0.36	.013	7	B8 III/IV	69,1385,1614
	−57 07799	NGC 6087 - 51			16 14 28	−57 58.7	1	11.27		0.25				1		1385
	−57 07800	NGC 6087 - 124			16 14 28	−58 02.0	1	11.67		0.43				1		1385
146493	−22 04138				16 14 29	−23 05.1	1	10.34		0.64		0.22		1	F3/5	1770
146271	−57 07804	NGC 6087 - 1			16 14 29	−57 34.5	2	8.31	.002	0.02	.005	−0.43		4	B8 Ib	69,1385
	−57 07805	NGC 6087 - 57			16 14 31	−57 31.9	1	11.60		0.14				1		1385
		NGC 6087 - 69			16 14 33	−57 47.6	3	12.17	.036	0.28	.010	0.21	.010	7		69,1385,1614
		NGC 6087 - 70			16 14 33	−57 48.3	3	11.95	.038	0.77	.402	1.82	.104	6		69,1385,1614
		HA 84 # 334			16 14 34	+14 49.1	1	12.00		0.97		0.63		8		1499
146870	+53 01856				16 14 34	+53 21.6	1	6.95		0.38		0.01		2	F0	1566
	−51 09984	LSS 3540			16 14 34	−51 19.1	3	10.32	.012	0.38	.018	−0.51	.005	7		33,115,1021
	−57 07808	NGC 6087 - 63			16 14 35	−57 40.8	1	12.08		0.27				1		1385
146544	−15 04291				16 14 36	−15 19.5	1	9.55		0.49		0.16		1	A8/9 V	1770
146343	−53 07679				16 14 37	−53 30.6	1	8.93		−0.01		−0.38		1	B9 II	115
146517	−23 12832				16 14 38	−23 44.3	1	10.26		0.46		0.39		2	A1 III/IV	1770
	−57 07811	NGC 6087 - 129		⋆AB	16 14 38	−57 49.5	3	9.73	.020	0.18	.005	0.12	.026	8	A0 V	69,1385,1614
146306	−57 07809	NGC 6087 - 20			16 14 38	−58 00.8	3	8.46	.011	0.04	.020	−0.38	.009	5	B7 III	60,69,1385
146324	−57 07816	NGC 6087 - 10			16 14 39	−57 48.6	3	7.94	.006	0.07	.008	−0.36	.015	8	B5 V	69,1385,1614
	−57 07812	NGC 6087 - 71			16 14 39	−57 53.0	1	12.38		0.31				1		1385
	−22 04140	R Sco			16 14 40	−22 49.4	1	10.36		1.45		0.94		1	M3	817
	−57 07817	NGC 6087 - 25			16 14 40	−57 47.2	4	10.01	.074	0.10	.023	−0.26	.015	11		60,69,1385,1614
		G 153 - 40			16 14 41	−12 50.3	1	15.00		0.10		−0.75		5	DA	3060
146441	−44 10756				16 14 41	−45 01.1	1	9.91		0.37		−0.23		2	B5 II	567
	−57 07818	NGC 6087 - 128			16 14 41	−57 49.4	3	8.72	.006	0.03	.008	−0.42	.007	15	B5 V	69,1385,1614
	−22 04141	S Sco			16 14 42	−22 46.3	1	10.46		1.49		1.26		1	M3	817
	−57 07819	NGC 6087 - 39			16 14 42	−57 44.7	2	10.79	.065	0.85	.014	0.46		3		1385,1614
146323	−57 07821	NGC 6087 - 155		⋆V	16 14 42	−57 46.7	2	6.16	.015	0.79	.006	0.53	.011	2	F8 Ib	669,688
		NGC 6087 - 130			16 14 42	−57 47.	3	9.17	.026	0.04	.005	−0.38	.004	7		69,1385,1614
		NGC 6087 - 131			16 14 42	−57 47.	1	11.21		0.12				2		1385
145782	−57 07716	NGC 6087 - 132		⋆	16 14 42	−57 47.	1	5.61		0.12				2	A3 III	1385
		NGC 6087 - 133			16 14 42	−57 47.	1	12.79		0.85				2		1385
		NGC 6087 - 134			16 14 42	−57 47.	1	12.56		0.70				2		1385
		NGC 6087 - 135			16 14 42	−57 47.	1	10.72		1.67				2		1385
		NGC 6087 - 136			16 14 42	−57 47.	1	12.15		0.99				2		1385
		NGC 6087 - 137			16 14 42	−57 47.	1	13.65		1.32				2		1385
		NGC 6087 - 138			16 14 42	−57 47.	1	13.86		0.57				2		1385
		NGC 6087 - 139			16 14 42	−57 47.	1	12.47		0.35				2		1385
		NGC 6087 - 140			16 14 42	−57 47.	1	12.45		0.32				2		1385
		NGC 6087 - 141			16 14 42	−57 47.	1	14.29		1.42				2		1385
		NGC 6087 - 142			16 14 42	−57 47.	1	12.71		1.23				2		1385
		NGC 6087 - 143			16 14 42	−57 47.	1	14.20		1.07				2		1385
		NGC 6087 - 144			16 14 42	−57 47.	1	13.86		0.94				2		1385
		NGC 6087 - 145			16 14 42	−57 47.	1	13.70		1.40				2		1385
		NGC 6087 - 146			16 14 42	−57 47.	1	11.48		1.44				2		1385
		NGC 6087 - 147			16 14 42	−57 47.	1	14.47		0.86				2		1385
		NGC 6087 - 148			16 14 42	−57 47.	1	11.83		0.64				2		1385
		NGC 6087 - 149			16 14 42	−57 47.	1	12.12		0.26				2		1385
		NGC 6087 - 150			16 14 42	−57 47.	1	12.17		0.21				2		1385
		NGC 6087 - 151			16 14 42	−57 47.	1	12.94		0.70				2		1385
		NGC 6087 - 152			16 14 42	−57 47.	1	12.39		0.75				2		1385
		NGC 6087 - 153			16 14 42	−57 47.	1	13.02		0.44				2		1385
		NGC 6087 - 154			16 14 42	−57 47.	1	10.60		0.38				2		1385
		NGC 6087 - 2206			16 14 42	−57 47.	1	12.84		0.34		0.06		2		69
		NGC 6087 - 73			16 14 42	−57 55.3	1	11.40		0.39				1		1385
	−51 09985				16 14 43	−52 09.1	1	10.42		0.35		−0.19		1		115
146698	+19 03077				16 14 44	+18 58.1	1	7.42		1.54		1.84		1	K2	1648
146738	+29 02803	HR 6074		⋆A	16 14 44	+29 16.4	5	5.78	.019	0.06	.013	0.10	.008	14	A3 V	15,1007,1013,1022,8015
146518	−30 13013				16 14 45	−30 21.5	1	10.31		0.18		0.07		1	B9 III/IV	1770
	−50 10354				16 14 47	−51 08.6	1	10.87		0.50		−0.36		1		115
	−57 07823	NGC 6087 - 65			16 14 47	−57 48.3	3	12.08	.013	0.22	.005	0.16	.036	6		69,1385,1614
	−57 07822	NGC 6087 - 64			16 14 47	−57 49.2	3	11.46	.016	0.15	.005	0.03	.026	7		69,1385,1614
146442	−45 10556				16 14 48	−45 43.1	1	9.04		0.30		−0.51		5	B2 III	567
	−57 07827	NGC 6087 - 99			16 14 48	−57 37.1	1	11.69		0.24				2		1385
	−57 07826	NGC 6087 - 127			16 14 48	−57 44.2	3	10.19	.012	0.09	.002	−0.31	.013	8	B8 V	69,1385,1614
146499	−39 10388				16 14 49	−39 37.6	2	9.71	.015	0.52	.005	−0.01		6	G0 V	1696,2012
		NGC 6087 - 4201			16 14 49	−57 44.3	1	11.50		0.34		0.22		2		1614
	−57 07824	NGC 6087 - 74			16 14 49	−57 55.8	1	12.19		0.28				1		1385
146590	−22 04144				16 14 50	−22 46.6	2	9.90	.020	0.48	.017	0.22		4	A9 V	370,1770
328007					16 14 50	−46 08.1	1	11.19		0.36		−0.40		5	A5	567
	−57 07828	NGC 6087 - 50			16 14 51	−57 59.6	2	10.62	.024	0.06	.004	−0.22		5		69,1385
146423	−50 10357				16 14 52	−51 05.0	1	8.67		0.52				4	F3 IV/V	2012
	−57 07831	NGC 6087 - 38			16 14 52	−57 42.2	2	11.44	.012	0.21	.011	0.12		4		69,1385

Table 1   879

HD	DM	Other Id	N	Rem	α_{1950}	δ_{1950}	S	V	σ_V	B–V	σ_{B-V}	U–B	σ_{U-B}	n	Spectrum	References
146569	−31 12749				16 14 54	−31 31.8	1	8.24		0.11		0.06		1	A0 V	1770
	−52 09514	LSS 3543			16 14 54	−52 11.4	2	10.80	.045	0.77	.020	-0.29	.010	2		115,181
	−57 07832	NGC 6087 - 75			16 14 54	−57 54.5	1	11.25		0.13				1		1385
146444	−49 10513				16 14 55	−49 17.6	5	7.67	.057	0.04	.020	-0.68	.013	18	B2 Vne	164,540,567,976,1586
	−51 09993	LSS 3544			16 14 55	−51 20.5	2	10.29	.024	1.32	.015	0.39	.049	3		33,115
	−57 07835	NGC 6087 - 37			16 14 55	−57 43.1	1	11.02		0.13		-0.08		3		69
	−57 07833	NGC 6087 - 49			16 14 55	−57 57.5	1	10.38		1.08				2		1385
	−49 10514	LSS 3545			16 14 58	−49 46.1	1	10.80		0.75		-0.52		2	B0:V:r	1649
	−57 07836	NGC 6087 - 76			16 14 58	−57 54.5	1	11.26		0.26				1		1385
146568	−31 12750				16 14 59	−31 27.4	1	9.18		0.23		0.14		1	A2/3 IV	1770
	−57 07838	NGC 6087 - 97			16 14 59	−57 31.2	1	11.86		0.22				2		1385
	−57 07837	NGC 6087 - 66			16 14 59	−57 49.5	3	12.09	.015	0.25	.006	0.20	.018	6		69,1385,1614
		NGC 6087 - 67			16 14 59	−57 50.3	1	12.57		0.29				1		1385
		NGC 6087 - 1303			16 14 59	−57 50.3	1	12.56		0.37		0.17		1		69
		CS 22872 # 3			16 14 60	−05 21.8	1	15.40		0.26		0.32		1		1736
		AJ74,882 T1# 13			16 15 00	−50 54.2	1	11.18		0.17		-0.10		3		33
	−50 10364				16 15 01	−50 53.8	1	10.33		0.50		-0.43		2	A0	33
		CS 22872 # 9			16 15 03	−04 40.8	1	13.75		0.30		0.29		1		1736
	−57 07841	NGC 6087 - 55			16 15 03	−57 34.5	1	11.23		0.01				1		1385
146547	−38 10942				16 15 04	−38 40.1	1	9.83		0.42		0.09		1	B9 III	1770
	−57 07839	NGC 6087 - 126			16 15 04	−57 59.1	1	11.57		0.28				1		1385
		CS 22872 # 14			16 15 05	−03 22.9	1	14.30		0.68		0.19		1		1736
		G 153 - 41			16 15 05	−15 28.5	3	13.42	.013	-0.22	.006	-1.11	.033	53		281,1764,3062
146607	−30 13018				16 15 06	−30 51.	1	10.31		0.11		-0.03		2	A(p Si)	1770
146594	−33 11068	IDS16119S3324		AB	16 15 06	−33 31.0	1	9.49		0.48		0.15		1	A9 V	1770
	−57 07845	NGC 6087 - 68			16 15 06	−57 50.0	2	12.08	.048	0.25	.010	0.25		3		1385,1614
		NGC 6087 - 98			16 15 07	−57 36.1	1	12.48		0.16				2		1385
146404	−57 07842	NGC 6087 - 21			16 15 07	−57 58.2	3	7.65	.023	1.78	.010	2.17	.005	9	M0 III	69,460,1385
	−52 09521				16 15 08	−52 45.1	1	10.58		0.29		-0.17		1		115
		CS 22872 # 10			16 15 09	−04 33.4	1	14.62		0.52		0.15		1		1736
147231	+71 00775	G 240 - 12			16 15 10	+71 03.4	1	7.90		0.63		0.22		2	G5	3026
146647	−22 04145				16 15 10	−22 35.0	1	9.79		0.60		0.30		1	F3 IV/V	1770
146606	−27 10893				16 15 10	−27 55.2	1	7.06		0.00		-0.07		4	A0 V	1771
		CS 22872 # 17			16 15 11	−02 56.3	1	14.78		0.16		0.09		1		1736
	−57 07850	NGC 6087 - 36			16 15 11	−57 45.1	3	10.38	.006	0.07	.001	-0.27	.022	7	B8 V	69,1385,1614
		WLS 1612 60 # 5			16 15 12	+62 29.2	1	11.60		0.81		0.34		2		1375
146624	−28 12037	HR 6070			16 15 12	−28 29.5	7	4.78	.009	0.02	.007	0.00	.017	17	A0 V	15,1075,1586,1770,2012*
	−52 09525	LSS 3546			16 15 12	−52 18.5	2	10.70	.055	0.31	.000	-0.45	.025	2		115,181
	−57 07851	NGC 6087 - 35			16 15 12	−57 47.2	3	10.09	.014	0.39	.001	0.13	.040	7	A2 V:	69,1385,1614
		NGC 6087 - 77			16 15 12	−57 55.1	1	12.33		0.31				1		1385
	−57 07853	NGC 6087 - 54			16 15 13	−57 35.0	1	11.52		0.43				1		1385
		LP 624 - 70			16 15 15	+01 44.4	1	15.11		1.74				2		1759
146463	−54 07452				16 15 15	−54 50.4	4	8.08	.022	-0.04	.017	-0.65	.012	17	B3 Vnne	540,976,1586,1732
146427	−56 07552				16 15 15	−57 06.1	2	9.21	.010	0.10	.009	-0.63	.005	6	B1/2 II/III	540,976
328004	−45 10561				16 15 16	−45 54.2	1	10.95		0.32		-0.18		3	B8	1586
		MN137,275 # 8			16 15 16	−53 25.0	1	12.22		1.20		1.13		5		656
	−57 07856	NGC 6087 - 53			16 15 16	−57 39.1	1	11.62		0.35				1		1385
146448	−57 07854	NGC 6087 - 8			16 15 16	−57 45.9	6	8.99	.032	0.05	.015	-0.38	.016	12	B5 II/III	60,69,540,976,1385,1614
		NGC 6087 - 4301	A		16 15 16	−57 45.9	1	10.60		0.40		0.23		2		1614
		NGC 6087 - 4301	B		16 15 16	−57 45.9	1	11.32		0.15		0.04		2		1614
146610	−38 10944				16 15 17	−38 31.9	1	9.06		0.55		0.05		1	F7 V	1770
	−50 10372	LSS 3548			16 15 17	−50 46.6	2	11.04	.019	0.46	.034	-0.29	.005	6	B0.5IV	33,92
	−53 07709	MN137,275 # 13			16 15 17	−53 30.0	1	10.50		1.10		0.87		2		656
146625	−33 11073				16 15 18	−33 60.0	1	10.11		0.30		-0.13		1	B8/9 III	1770
146501	−49 10525				16 15 19	−49 17.1	1	7.27		0.06		-0.34		4	B5/7 (Vn)	164
		MN137,275 # 7			16 15 19	−53 25.3	1	11.52		1.40		1.60		2		656
146678	−22 04147				16 15 20	−23 09.7	1	9.69		0.67		0.25		1	F5/6 V	1770
146349	−64 03442	IDS16107S6424		AB	16 15 20	−64 31.7	1	6.70		0.00				4	B9/9.5V	1075
147321	+73 00713	HR 6088			16 15 21	+73 31.0	2	5.99	.010	0.08	.000	0.15		14	A3 V	252,1258
		LundsII141 # 438			16 15 22	−51 46.7	1	11.03		0.47		-0.13		1		115
146649	−33 11075				16 15 24	−33 60.0	1	9.93		0.24		-0.18		1	B8/9 III/I	1770
		NGC 6087 - 78			16 15 24	−57 59.3	1	11.96		0.77				1		1385
146664	−30 13023				16 15 27	−31 00.3	1	9.70		0.15		0.12		1	A0 IV	1770
		NGC 6087 - 92			16 15 27	−57 48.3	2	12.47	.034	0.34	.000	0.24		3		1385,1614
146484	−57 07858	NGC 6087 - 9			16 15 27	−57 48.9	4	9.43	.012	0.06	.013	-0.27	.013	5	B9 V	60,69,1385,1614
		MN137,275 # 10			16 15 28	−53 23.9	1	12.24		1.47		1.65		2		656
		MN137,275 # 9			16 15 28	−53 24.3	1	12.48		0.42		0.29		3		656
146706	−22 04148				16 15 29	−23 09.2	2	7.55	.004	0.13	.016	-0.09	.004	4	B9 V	40,1770
146483	−57 07859	NGC 6087 - 7			16 15 29	−57 47.1	4	8.28	.001	0.06	.013	-0.23	.038	7	B9 Ib	60,69,1385,1614
		CS 22872 # 23			16 15 30	−02 20.5	1	14.93		0.16				1		1736
		NGC 6087 - 1504			16 15 31	−57 48.3	1	13.10		0.44		0.24		2		1614
		NGC 6087 - 1503			16 15 33	−57 48.5	1	12.43		0.79		0.43		1		1614
146381	−64 03446				16 15 33	−64 22.6	1	8.35		-0.06		-0.49		2	B6 III	1586
146815	+6 03198				16 15 34	+06 12.0	1	7.27		1.03		0.88		4	G7 II	8100
	−51 10000	LSS 3550			16 15 34	−51 52.1	1	10.42		0.58		0.00		1	B7 II	115
	−53 07725	MN137,275 # 11			16 15 35	−53 29.9	1	10.35		1.31		1.27		2		656
		NGC 6087 - 52			16 15 37	−57 48.1	1	10.99		1.68				2		1385
	−57 07862	NGC 6087 - 33			16 15 37	−57 49.0	1	11.98		0.21				1		1385
	−51 10005	LSS 3551			16 15 39	−51 43.8	2	11.10	.024	0.77	.015	-0.24	.044	3	B1 III	115,1649
146791	−4 04086	HR 6075		⋆A	16 15 40	−04 34.3	9	3.23	.006	0.97	.014	0.75	.009	30	G9.5 IIIb	15,1020,1075,1088*

HD	DM	Other Id	N	Rem	α_{1950}	δ_{1950}	S	V	σ_V	B–V	σ_{B-V}	U–B	σ_{U-B}	n	Spectrum	References
	−57 07864	NGC 6087 - 32			16 15 41	−57 48.7	1	12.10		0.24				1		1385
		CS 22872 # 28			16 15 42	−01 40.7	1	13.67		0.13		0.09		1		1736
		CS 22872 # 20			16 15 42	−03 01.4	1	13.24		0.17		0.16		1		1736
146743	−21 04330				16 15 42	−21 28.3	1	9.01		0.56		0.08		1	F3 V	1770
	−51 10008	LSS 3552			16 15 42	−51 20.8	3	11.05	.051	0.36	.011	-0.46	.015	4	B1 III	33,115,1649
146577	−52 09555				16 15 43	−52 57.8	1	9.25		0.06		-0.31		1	B7/8 III	115
		MN137,275 # 5			16 15 43	−53 24.3	1	11.94		0.39		0.10		2		656
146915	+23 02918				16 15 44	+23 43.7	1	7.20		1.46		1.75		2	K2	1625
	−57 07865	NGC 6087 - 6			16 15 44	−57 56.0	1	10.37		0.02				1	A0 V	1385
		CS 22872 # 13			16 15 45	−03 45.2	1	13.64		0.32		0.24		1		1736
146721	−30 13028				16 15 45	−30 22.6	1	8.39		0.12		0.10		1	A0 V	1770
146628	−49 10531	LSS 3553			16 15 45	−49 38.4	2	9.96	.020	0.65	.010	-0.37		6	B1/2 Ia	33,1021
	−57 07869	NGC 6087 - 93			16 15 45	−57 38.3	1	11.69		0.53				2		1385
146531	−57 07866	NGC 6087 - 22			16 15 45	−58 02.8	2	9.78	.041	0.04	.002	-0.46		4	B5 III	1385,1586
136509	−88 00133				16 15 46	−89 09.0	1	9.29		1.41		1.54		3	K3/4	826
146597	−53 07737				16 15 47	−53 23.4	1	8.39		0.28		0.25		3	A4 IV	656
		HA 132 # 801			16 15 48	−15 01.6	1	10.84		1.81		2.10		10		1499
146667	−42 11188	HR 6071		★ AB	16 15 48	−42 33.2	2	5.44	.005	0.10	.000			7	A3 V	15,2012
		AJ74,882 T1# 14			16 15 48	−51 20.4	1	11.70		0.52		0.02		2		33
146946	+32 02702	G 180 - 47			16 15 49	+31 55.3	2	6.86	.000	0.57	.005	0.04	.013	4	G0 V	1733,3016
146681	−40 10309				16 15 49	−40 42.0	1	8.94		0.44		0.15		1	A7 II/III	1770
325484	−41 10607				16 15 49	−42 10.8	1	11.26		0.40		0.14		4	B8	567
		MN137,275 # 81			16 15 49	−53 24.0	1	12.12		0.16		-0.23		2		656
146596	−52 09561				16 15 50	−52 39.1	2	7.98	.001	0.08	.002	-0.42	.022	3	B5 IV/V	115,1586
	−53 07739				16 15 50	−53 25.7	1	11.16		0.13		-0.35		9		656
146555	−57 07872	NGC 6087 - 5			16 15 50	−57 47.3	3	10.29	.010	0.00	.012	-0.36	.010	3	Ap	60,69,1385
146296	−70 02196				16 15 50	−71 09.6	1	9.77		0.60		-0.06		2	G3 V	3077
146795	−16 04265				16 15 51	−16 19.4	1	9.00		0.39		0.15		1	A5/6 III	1770
	−57 07873	NGC 6087 - 91			16 15 51	−57 44.8	1	11.62		0.61				2		1385
		CS 22872 # 22			16 15 53	−02 18.9	1	15.87		0.21		0.18		1		1736
146614	−53 07744				16 15 54	−53 28.0	1	9.46		0.06		-0.22		2	B8/9 Ib	656
		HA 132 # 813			16 15 55	−15 02.6	1	11.49		0.82		0.34		10		1499
146761	−30 13030				16 15 55	−30 18.2	1	8.45		0.20		0.17		1	A3 V	1770
		MN137,275 # 14			16 15 55	−53 28.6	1	12.28		1.71		1.05		2		656
146707	−40 10312				16 15 56	−40 49.4	1	8.88		0.46		0.14		2	B9 IV	1770
		MN137,275 # 6			16 15 56	−53 25.1	1	11.62		0.52		0.01		1		656
		MN137,275 # 15			16 15 57	−53 29.1	1	11.79		1.34		1.06		4		656
		NGC 6087 - 96			16 15 57	−57 36.6	1	12.38		0.23				2		1385
		LP 624 - 71			16 15 58	+00 07.4	1	12.95		0.66		-0.07		2		1696
	+55 01823	CR Dra		★	16 15 58	+55 23.7	1	9.96		1.46		1.08		2	M0	3016
		MN137,275 # 16			16 15 58	−53 32.1	1	11.62		1.22		1.10		2		656
		HRC 255	V		16 15 59	−23 09.3	1	12.89		1.02		0.52		1		776
	−53 07751	MN137,275 # 21			16 15 59	−53 26.4	1	11.05		0.10		-0.42		2		656
		MN137,275 # 22			16 15 59	−53 26.6	1	11.67		1.26		1.13		2		656
		NGC 6087 - 89			16 15 59	−57 43.0	1	12.64		0.42				2		1385
		NGC 6087 - 31			16 16 00	−57 48.7	1	12.39		0.61				1		1385
146599	−57 07876	NGC 6087 - 23			16 16 00	−58 03.8	1	9.82		0.00				1	B8 III	1385
147094	+52 01956				16 16 01	+52 29.7	1	9.06		0.64		0.17		2	K0	1566
146775	−27 10902				16 16 02	−28 10.2	4	7.68	.017	0.62	.004	0.10	.000	18	G5 V	158,1075,1775,2012
	−50 10387	LSS 3555			16 16 02	−50 19.2	2	10.16	.010	0.72	.010	-0.12		6		33,1021
		NGC 6087 - 79			16 16 02	−58 00.5	1	12.42		0.32				1		1385
146878	−10 04292				16 16 04	−10 39.9	1	9.59		0.44		0.26		2	F0	1375
		MN137,275 # 19			16 16 04	−53 29.9	1	11.85		1.40		1.44		3		656
		MN137,275 # 20			16 16 04	−53 30.4	1	12.18		0.22		-0.02		3		656
		AJ74,20 # 37			16 16 05	−15 41.0	1	11.16		0.73		0.16		3		3063
146686	−49 10536	HR 6072		★ A	16 16 05	−50 02.1	7	4.02	.011	1.08	.002	1.16	.020	23	K0 III	15,1034,1075,1649*
	−52 09572				16 16 06	−52 17.9	1	9.89		0.22		-0.14		1		115
	−52 09571				16 16 06	−52 27.4	1	10.11		1.12		1.03		1		3072
	−57 07878	NGC 6087 - 27			16 16 07	−57 44.3	1	11.29		0.47				1		1385
		NGC 6087 - 81			16 16 07	−58 00.2	1	12.75		0.82				2		1385
	−57 07879	NGC 6087 - 80			16 16 08	−58 00.7	1	11.88		0.39				1		1385
	−53 07758	MN137,275 # 12			16 16 10	−53 28.3	1	10.75		0.18		0.09		3		656
146850	−14 04398	HR 6078			16 16 11	−14 45.1	2	5.97	.024	1.52	.000	1.95		6	K3 III	252,2007
146834	−19 04357	HR 6076			16 16 12	−20 05.9	5	6.29	.022	1.08	.023	0.78	.029	20	K0 III	15,206,1007,1013,2007,8015
		L 202 - 179			16 16 12	−59 09.	1	15.08		0.10		-0.84		1		3065
		MN137,275 # 17			16 16 13	−53 31.8	1	11.82		1.28		1.27		3		656
146412	−69 02521				16 16 14	−69 36.2	1	8.54		0.42				4	F3 IV/V	2012
145344	−83 00602				16 16 15	−83 21.5	1	8.42		0.48				1	F5 V	1486
		CS 22872 # 27			16 16 17	−01 56.0	1	13.82		0.74		0.06		1		1736
146879	−16 04267				16 16 17	−16 54.3	2	9.48	.027	0.35	.000	0.25	.005	2	A2 III	1770,3063
	−57 07885	NGC 6087 - 29			16 16 18	−57 50.3	1	11.51		0.16				1		1385
		NGC 6087 - 85			16 16 18	−57 53.4	1	11.31		1.61				2		1385
147025	+26 02817				16 16 19	+26 01.0	1	6.68		0.91		0.56		2	G8 III	985
		MN137,275 # 18			16 16 19	−53 32.2	1	11.88		0.44		0.10		2		656
	−57 07887	NGC 6087 - 82			16 16 19	−57 57.2	1	11.57		0.18				2		1385
		CS 22872 # 29			16 16 21	−01 13.1	1	12.88		0.25		0.13		1		1736
		CS 22872 # 6			16 16 21	−04 58.5	1	10.89		0.26		0.12		1		1736
146835	−30 13040	IDS16132S3040	B		16 16 21	−30 46.9	2	7.10	.010	0.56	.015	0.01	.020	5	G0/3	13,1279
		CS 22872 # 12			16 16 22	−04 01.6	1	14.92		0.32		0.28		1		1736
	−57 07890	NGC 6087 - 28			16 16 22	−57 49.2	1	11.31		0.08				1		1385

Table 1 881

HD	DM	Other Id	N Rem	α_{1950}	δ_{1950}	S	V	σ_V	B–V	σ_{B-V}	U–B	σ_{U-B}	n	Spectrum	References
146836	−30 13041	HR 6077	⋆ A	16 16 23	−30 47.2	3	5.50	.024	0.47	.015	-0.01	.024	6	F5 IV	13,1279,3026
146836	−30 13041	IDS16132S3040	AB	16 16 23	−30 47.2	1	5.30		0.48				4	F5 IV	2006
146931	−6 04399			16 16 24	−07 11.9	1	7.81		0.27		0.22		3	A0	1776
147232	+60 01665	HR 6086, AT Dra		16 16 25	+59 52.5	3	5.44	.047	1.60	.022	1.71	.004	9	M4 IIIa	71,1501,3001
146690	−54 07493	HR 6073		16 16 25	−55 01.2	1	5.77		0.97				4	K0 III	2035
	−55 07422			16 16 25	−55 15.7	1	11.18		0.12		-0.07		2	B9 V	540
	−57 07893	NGC 6087 - 90		16 16 25	−57 38.2	1	11.51		0.53				2		1385
		NGC 6087 - 86		16 16 27	−57 53.1	1	12.57		0.81				2		1385
	−57 07892	NGC 6087 - 83		16 16 27	−57 56.8	1	11.92		0.27				2		1385
147006	+16 02924			16 16 29	+16 44.1	2	8.16	.019	1.12	.019	0.99		3	K0	882,1648
146748	−50 10398			16 16 29	−50 21.8	1	10.07		0.03		-0.15		3	B8/9 II/III	567
	−50 10397			16 16 29	−50 52.0	1	10.67		0.41		-0.15		2	B5	92
146934	−13 04400			16 16 30	−13 44.	1	9.56		0.41		0.19		1	A3 II	1770
146936	−15 04298			16 16 30	−16 12.	1	10.29		0.35		0.30		1	A2 IV	1770
146897	−21 04334			16 16 32	−21 17.0	1	9.10		0.49		0.03		1	F2/3 V	1770
146898	−24 12666			16 16 34	−24 43.1	1	9.57		0.91		0.35		2	F5 V	1770
146899	−26 11287			16 16 34	−26 44.9	2	10.23	.015	0.63	.001	0.38	.019	5	A5 II	40,1770
146854	−33 11097			16 16 34	−34 01.8	1	9.08		0.31		0.22		1	A2 IV	1770
146935	−15 04300			16 16 35	−15 25.6	3	8.51	.015	0.25	.002	0.22	.012	5	A1 V	515,1770,3063
147379	+67 00935	G 225 - 57	⋆ A	16 16 37	+67 21.5	7	8.61	.034	1.41	.010	1.24	.022	24	M0 V	1,22,1077,1118,1197*
		CS 22872 # 21		16 16 37	−02 50.4	1	15.02		0.19		0.27		1		1736
		NGC 6087 - 88		16 16 37	−57 42.2	1	11.52		0.34				2		1385
147379	+67 00935	G 225 - 58	⋆ B	16 16 38	+67 22.6	4	10.69	.017	1.50	.008	1.18	.004	8		1,694,3072,8006
146880	−34 10940			16 16 39	−34 23.9	1	9.16		0.30		0.20		1	A0 V	1770
146800	−47 10664			16 16 39	−48 05.7	2	8.92	.010	0.96	.005	0.61		6	K2 V	2012,3077
146952	−18 04260			16 16 40	−18 42.5	1	7.10		0.27		0.18		1	A1 IV	1770
146950	−15 04301			16 16 41	−15 25.6	3	9.88	.014	0.35	.009	0.28	.016	4	A2 IV	515,1770,3063
	−50 10405	LSS 3558		16 16 43	−51 06.4	1	11.22		0.38		-0.15		2	B3 III	1649
146938	−29 12450			16 16 45	−29 14.3	1	10.64		0.25		-0.03		1	B9 III/IV	1770
145621	−82 00682			16 16 45	−82 26.6	2	6.78	.018	1.10	.027	0.95	.007	14	K0 III	1628,1704
	−37 10765			16 16 46	−37 25.4	1	10.60		1.57				1	M4	1705
	−37 10765			16 16 46	−37 25.4	1	14.15		1.79				1		1705
	−37 10765		AB	16 16 47	−37 25.4	1	10.56		1.58		1.20		4		158
	−57 07903	NGC 6087 - 4		16 16 47	−57 52.0	1	10.47		0.04				1		1385
146905	−39 10410			16 16 48	−39 22.8	1	7.06		0.08		-0.19		2	B9 V	1770
146903	−36 10738			16 16 50	−36 26.8	1	9.61		0.43		0.07		1	B9 II/III	1770
146840	−49 10549	LSS 3559		16 16 50	−49 23.3	3	9.87	.012	0.34	.004	-0.44	.006	11	B0 V	567,1586,1649
		G 138 - 21		16 16 52	+07 47.7	1	15.00		1.11		0.86		1		3060
		CS 22872 # 26		16 16 52	−02 01.5	1	14.38		0.14		0.12		1		1736
146918	−33 11103			16 16 52	−33 36.1	1	10.62		0.23		0.04		1	B9 III	1770
		NGC 6087 - 87		16 16 52	−57 46.8	1	12.47		0.30				2		1385
328113	−46 10655			16 16 54	−46 33.3	1	10.18		0.31		-0.51		2	B2.5II	540
146805	−52 09604			16 16 54	−52 15.2	4	9.43	.010	0.00	.006	-0.67	.014	7	B2 III/IV	115,181,540,976
		CpD -51 09323	⋆	16 16 56	−51 16.0	1	11.20		0.32		-0.80		2		1649
		HP Nor		16 16 56	−54 46.2	1	13.08		0.11		-0.66		1		1471
146973	−20 04464			16 16 58	−20 39.8	1	8.15		0.28		0.12		2	A5 II/III	1770
	−52 09606			16 16 58	−53 04.9	1	11.27		0.24		-0.09		1		115
	−50 10412	LSS 3561		16 16 59	−50 16.3	1	11.28		0.83		-0.04		3	B0 V	1649
	+22 02955			16 17 03	+21 54.5	1	8.80		1.22		1.13		3	K0	1733
146998	−25 11477	V953 Sco		16 17 07	−25 44.3	2	9.56	.007	0.63	.004	0.34	.015	4	A8/9 V	40,1770
147009	−19 04358	IDS16142S1949	B	16 17 08	−19 55.5	3	8.05	.011	0.29	.015	0.13	.019	10	A0 V	26,206,1054
147011	−22 04153			16 17 09	−22 24.0	1	10.18		0.61		0.07		1	F3/5 V	1770
146976	−31 12786			16 17 09	−31 19.4	1	10.18		0.25		0.16		1	A0 IV	1770
146954	−39 10412	HR 6080	⋆ AB	16 17 09	−39 18.7	3	6.12	.005	-0.08	.005	-0.22	.012	8	B9 V	404,1770,2032
147010	−19 04359	V933 Sco	⋆ A	16 17 10	−19 56.2	4	7.40	.010	0.15	.011	-0.27	.007	12	B9 II/III	26,206,1054,1776
	−52 09615			16 17 10	−52 28.3	1	10.48		0.15		-0.37		1		115
		NGC 6101 sq1 8		16 17 11	−72 00.3	1	9.96		1.57		1.97		2		89
147013	−25 11478			16 17 12	−25 31.6	2	9.09	.009	0.44	.008	0.25	.013	4	A0 V	40,1770
146821	−57 07911	NGC 6087 - 2		16 17 12	−57 44.7	1	10.03		0.02				1	A0	1385
		NGC 6101 sq1 13		16 17 12	−72 09.5	1	12.02		0.84		0.36		3		89
		G 17 - 10		16 17 13	+03 08.3	1	11.25		0.90		0.51		1	K1	333,1620
147012	−25 11479			16 17 16	−25 29.4	2	9.76	.017	0.51	.009	0.20	.005	4	B9 IV	40,1770
		CS 22872 # 8		16 17 18	−04 42.0	1	14.90		0.43		0.06		1		1736
		Nor sq 2 # 75		16 17 18	−52 04.	1	11.42		0.22		-0.52		2		181
	−71 01987	NGC 6101 sq1 10		16 17 18	−72 02.7	1	10.24		1.21		1.14		2		89
146557	−71 01988	NGC 6101 sq1 14		16 17 19	−72 09.9	1	10.30		0.15		0.10		2	A0	89
	−52 09623			16 17 24	−52 48.7	1	10.78		0.24		-0.33		1		115
147046	−29 12459	IDS16143S2916	A	16 17 25	−29 23.7	1	7.75		0.22		0.15		1	A1 V	1770
		CS 22872 # 1		16 17 26	−05 57.7	1	15.30		0.19		0.12		1		1736
147000	−38 10960	LSS 3566		16 17 26	−38 33.0	3	10.42	.003	0.61	.008	-0.27	.021	4	B8	400,540,1770
		RV CrB		16 17 27	+29 50.0	1	11.14		0.17		0.16		1	A1	668
	−49 10562	LSS 3564		16 17 27	−49 26.4	2	10.49	.025	0.76	.025	-0.07	.010	5	B2 III:	1649,1737
147265	+39 02977			16 17 28	+38 54.8	1	8.70		0.21		0.06		3	A3 V	833
147407	+62 01470			16 17 28	+62 32.6	1	7.32		1.10		0.96		2	K0	1733
146919	−52 09626	LSS 3562		16 17 30	−52 55.0	6	8.54	.010	0.55	.012	-0.46	.013	9	B0.5Ia	115,138,181,403,1737,8100
147083	−21 04338			16 17 31	−21 23.4	1	8.59		0.35		0.11		2	A7 III/IV	1770
147117	−12 04479			16 17 33	−12 41.4	1	10.02		0.41		0.26		2	A1mA4-F2	1003
147103	−19 04361	IDS16146S1953	AB	16 17 34	−19 59.9	2	7.56	.005	0.48	.036	0.08	.014	5	B9/A0 V	206,1771
		LP 917 - 26		16 17 34	−31 51.0	1	13.73		0.64		0.13		4		3062
147104	−19 04361	IDS16146S1953	C	16 17 35	−19 59.7	2	8.41	.007	0.40	.005	0.11	.035	5	Ap	206,1770

HD	DM	Other Id	N	Rem	α_{1950}	δ_{1950}	S	V	σ_V	B−V	σ_{B-V}	U−B	σ_{U-B}	n	Spectrum	References
147045	−28 12059	IDS16145S2830		AB	16 17 36	−28 37.5	1	9.32		0.58		0.22		1	F3 V	1770
		CS 22872 # 2			16 17 37	−05 11.9	1	15.38		0.38		0.25		1		1736
147084	−23 12849	HR 6081			16 17 37	−24 03.0	9	4.55	.017	0.84	.006	0.64	.046	27	A4 II/III	15,26,40,1075,1770,2012*
	−52 09635				16 17 37	−52 57.6	2	10.21	.010	0.11	.005	-0.55	.000	2		115,181
		CS 22872 # 42			16 17 38	−03 20.4	1	14.66		0.19		0.17		1		1736
146955	−52 09643				16 17 42	−52 27.4	5	9.66	.004	0.16	.013	-0.44	.010	11	B5 Ib	115,181,540,976,1021
		G 168 - 42			16 17 43	+22 45.8	2	11.50	.005	0.70	.005	0.08	.000	2		1064,3077
147147	−11 04115				16 17 43	−11 24.7	1	10.28		1.01		0.85		2	K5 V	3072
		CS 22872 # 7			16 17 46	−04 58.9	1	13.50		0.25		0.27		1		1736
147001	−47 10677				16 17 46	−48 04.2	1	6.52		-0.05		-0.31		1	B7 V	567
	−51 10053				16 17 46	−52 00.7	1	10.39		0.21		-0.18		1	B8	115
147352	+49 02491	HR 6090			16 17 47	+49 09.4	1	5.91		1.37				2	gK6	71
		G 17 - 11			16 17 47	−04 08.9	7	10.69	.018	1.39	.020	1.28	.016	16	K7-M0	158,679,1017,1620*
147105	−25 11483	V961 Sco			16 17 47	−25 16.5	2	8.81	.006	0.49	.007	0.29	.003	4	Ap SrCrEu	40,1770
		G 202 - 38			16 17 50	+51 52.4	1	10.35		1.10				2	K4	1619
147137	−22 04156				16 17 51	−22 28.5	1	8.98		0.44		0.17		2	A9 V	1770
146755	−69 02533	IDS16127S6907		A	16 17 51	−69 14.1	1	9.50		-0.03		-0.49		3	B3/7 Vne	1586
147266	+21 02902	HR 6087			16 17 54	+21 15.1	3	6.03	.011	0.94	.010	0.67	.044	8	G7 IIIb	71,3077,8100
		CS 22872 # 44			16 17 55	−03 37.0	1	14.84		0.21		0.27		1		1736
147179	−12 04481				16 17 58	−12 34.8	1	9.33		0.38		0.26		1	A4 V	1770
147035	−49 10569				16 17 58	−49 21.6	1	9.10		0.42		-0.07		1	F2 V	1649
	−49 10571	LSS 3569			16 17 59	−49 26.1	1	11.18		0.55		-0.31		1	B0:n	1649
	−49 10570	LSS 3570			16 18 00	−49 51.5	1	10.79		0.62		-0.14		2	B2 V	1649
	−51 10061	LSS 3567			16 18 00	−51 25.1	2	10.75	.015	0.49	.015	-0.38	.019	3	B1:V:	115,1649
147148	−29 12468				16 18 06	−29 59.0	1	8.25		0.30		0.07		1	A5/7 + (F)	1771
147165	−25 11485	HR 6084, σ Sco		⋆ A	16 18 09	−25 28.5	10	2.87	.038	0.14	.006	-0.69	.012	42	B2III+O9.5V	15,542,1054,1075,1088*
147165	−25 11485	IDS16151S2521		B	16 18 09	−25 28.5	3	8.44	.034	0.32	.031	0.05	.110	11	B1 III	321,542,3024
147662	+68 00868	HR 6101			16 18 10	+68 40.4	1	6.36		1.34		1.54		2	K0	1733
	−50 10435	LSS 3572			16 18 10	−51 05.8	2	11.10	.005	0.58	.010	-0.13	.070	4		33,1649
147365	+40 03005	HR 6091		⋆ AB	16 18 12	+39 49.6	2	5.47	.010	0.40	.002	-0.05	.030	5	F3 V	1733,3016
147069	−48 10745				16 18 13	−49 02.3	1	8.62		0.68		-0.30		1	B7 V	1649
147394	+46 02169	HR 6092		⋆ AB	16 18 14	+46 25.9	24	3.90	.004	-0.15	.004	-0.56	.008	282	B5 IV	1,15,61,71,154,667*
147049	−52 09668	LSS 3573		⋆ AB	16 18 16	−52 36.5	7	7.69	.025	0.35	.015	-0.55	.026	11	B1 Ia/ab	115,181,403,540,976*
328053	−44 10800				16 18 18	−44 40.4	1	10.39		0.40		0.29		2	F0	1730
		CS 22872 # 47			16 18 19	−04 00.9	1	15.21		0.21		0.28		1		1736
147196	−23 12852				16 18 19	−23 35.4	3	7.04	.012	0.17	.008	-0.15	.015	10	B5/8 Vn	26,1054,1770
147124	−44 10802				16 18 21	−44 41.9	1	8.59		0.50		-0.08		2	F6 V	1730
147676	+67 00937				16 18 22	+67 22.7	1	9.22		0.58		0.01		5	G5	3026
147207	−20 04472				16 18 22	−21 08.9	1	10.41		0.60		0.16		1	F2/3 (IV)	1770
328355	−45 10718				16 18 22	−43 52.2	1	11.05		0.28		-0.19		3	B8	1586
147074	−54 07563				16 18 22	−54 12.3	1	8.08		1.26		1.42		3	K2 III	8100
147220	−21 04342				16 18 23	−21 59.5	1	8.63		0.44		0.23		2	A9 V	1770
	−54 07564	LSS 3574			16 18 24	−54 27.4	1	11.56		0.22		-0.59		2		181
		CS 22872 # 35			16 18 25	−02 05.6	1	14.99		0.22		0.27		1		1736
147090	−52 09685				16 18 27	−52 18.5	4	9.62	.035	0.07	.011	-0.58	.011	7	B3 III	115,181,540,976
		CS 22872 # 43			16 18 28	−03 24.2	1	12.32		0.43		0.22		1		1736
		Steph 1331			16 18 29	+54 34.7	1	11.68		1.46		1.84		1	M0	1746
		LP 685 - 52			16 18 29	−06 31.9	1	13.20		0.81		0.07		2		1696
147075	−54 07566				16 18 29	−54 31.0	1	7.90		1.20				4	K1 III	2012
147395	+37 02741				16 18 30	+37 05.6	1	6.60		1.52		1.84		6	M2 III	3016
		Nor sq 1 # 75			16 18 30	−51 25.	1	11.39		0.43		-0.02		1		115
		HRC 256, V895 Sco			16 18 31	−26 05.4	1	13.34		1.25		-0.06		1		649
147208	−29 12474				16 18 31	−29 14.8	1	9.90		0.35		0.24		1	A8/F0	1770
	−48 10750	LSS 3575			16 18 31	−49 03.0	1	11.25		0.57		-0.19		1	B0 V	1649
147221	−29 12475				16 18 33	−30 01.9	1	9.86		0.25		0.14		1	B9 V	1770
		NGC 6101 sq1 12			16 18 33	−72 03.8	2	13.03	.030	0.54	.005	0.02	.000	6		89,371
		GD 199			16 18 34	−04 13.8	1	14.71		0.68		-0.04		1		3060
147197	−34 10954				16 18 34	−34 46.7	1	10.04		0.41		0.29		1	B9 III	1770
		NGC 6101 sq2 7			16 18 35	−71 58.9	1	13.79		0.66		0.19		5		371
	−71 01993	NGC 6101 sq1 7			16 18 37	−72 00.1	2	10.43	.018	0.48	.009	0.07	.000	9		89,371
147211	−33 11125				16 18 38	−33 28.2	1	10.33		0.39		0.30		1	A2/4 (IV)	1770
147127	−53 07865				16 18 39	−53 47.2	1	8.32		0.69				4	G5 V	2033
147198	−40 10334				16 18 42	−40 32.	1	10.53		0.84		0.39		2	B7/9 Ib/II	1770
328026	−43 10721				16 18 42	−43 37.6	1	10.89		0.35		0.00		5	B3	567
	−44 10809	LSS 3579			16 18 42	−44 49.1	1	10.37		1.02		-0.07		2		1737
		NGC 6101 sq1 11			16 18 42	−72 03.1	2	12.89	.034	0.87	.005	0.49	.025	8		89,371
		WLS 1620-10 # 7			16 18 43	−12 05.7	1	12.83		0.85		0.36		2		1375
147152	−49 10591	HR 6083			16 18 43	−49 27.3	4	5.33	.032	-0.05	.008	-0.43	.004	12	B6 IV	26,567,1649,2035
147181	−42 11227				16 18 44	−43 05.0	1	7.62		0.16		-0.03		4	B9 V(n)	567
147295	−13 04412				16 18 46	−13 39.6	1	9.20		0.22		0.13		1	A0 V	1770
		NGC 6101 sq2 3			16 18 48	−71 58.3	1	12.71		1.05		0.76		3		371
	−52 09696	LSS 3576			16 18 49	−53 03.4	2	10.73	.010	0.39	.005	-0.46	.020	2		115,181
	−49 10593	LSS 3577			16 18 50	−49 41.0	1	10.94		0.50		-0.11		1	B3 Ve	1649
147284	−24 12671				16 18 54	−24 52.3	1	8.80		0.68				4	G3 V	2033
	−51 10081				16 18 54	−52 09.8	1	10.98		0.22		-0.17		1		115
148048	+76 00596	HR 6116		⋆ A	16 18 56	+75 52.3	4	4.96	.030	0.35	.012	0.04	.020	7	F3 V	254,1118,1193,3026
147296	−21 04343				16 18 56	−21 21.7	1	10.12		0.47		0.20		1	F2 IV/V	1770
147283	−24 12672				16 18 56	−24 22.7	2	10.27	.000	0.82	.018	0.63	.006	6	A1 IV	40,1770
330695	−49 10594				16 18 56	−49 59.0	1	10.42		0.60				4	B3 V	1021
147225	−43 10724	HR 6085		⋆ A	16 18 57	−43 47.7	2	5.89	.005	1.15	.005	0.83		6	G3 II	13,2007

Table 1   883

HD	DM	Other Id	N Rem	α_{1950}	δ_{1950}	S	V	σ_V	B–V	σ_{B-V}	U–B	σ_{U-B}	n	Spectrum	References
147225	−43 10723	IDS16154S4340	B	16 18 57	−43 47.7	1	9.74		1.49		1.64		2		13
		CS 22872 # 48		16 18 58	−04 08.0	1	14.88		0.24		0.31		1		1736
147269	−33 11130			16 19 01	−34 10.1	1	9.52		0.36		0.22		1	A2/3 IV	1770
147430	+22 02958			16 19 02	+22 21.4	2	6.91	.005	1.26	.000	1.12	.005	5	K1 III	1501,1648
		G 202 - 43		16 19 03	+51 17.5	1	12.18		0.59		-0.07		2		1658
		L 338 - 152		16 19 04	−48 32.6	1	11.84		1.49				1	M3	1705
	−54 07590	LSS 3578		16 19 05	−54 52.9	1	12.17		0.31		-0.42		1		181
	−50 10459	LSS 3580		16 19 06	−50 41.9	2	11.14	.004	0.85	.009	0.04	.008	6	B2:V:	1586,1649
		Steph 1330		16 19 07	+23 44.3	1	10.44		1.52		1.86		2	M2	1746
		CS 22872 # 40		16 19 07	−03 01.0	1	14.21		0.38		0.22		1		1736
	−52 09708			16 19 08	−52 52.9	1	10.24		0.14		-0.40		1		115
	+41 02695			16 19 12	+41 04.7	1	8.98		1.30		1.17		2	M0	3072
147342	−21 04345			16 19 12	−21 32.2	1	8.60		0.34		0.16		2	A5 IV/V	1770
147487	+27 02629			16 19 13	+27 29.4	1	8.54		0.60		0.03		2	G0 V	3026
147226	−48 10762			16 19 13	−48 45.6	1	10.15		0.11		-0.35		5	B8 V	567
	−51 10088			16 19 14	−51 53.1	2	10.16	.010	0.04	.015	-0.56	.040	2		115,181
147325	−30 13082			16 19 17	−31 03.9	1	8.02		0.23		0.16		1	A2 III	1770
	−51 10090			16 19 18	−52 09.1	1	10.98		0.11		-0.17		1		115
147343	−24 12674			16 19 19	−24 14.8	2	9.35	.014	0.66	.003	0.41	.017	5	A1 III/IV	40,1770
147313	−32 11669			16 19 19	−32 22.8	1	10.26		0.35		0.27		1	A0 IV	1770
147356	−22 04158	IDS16164S2224	AB	16 19 20	−22 31.2	1	9.53		0.47		0.09		2	B8 II	1770
		LP 685 - 6		16 19 21	−06 22.5	1	12.60		0.89		0.27		2		1696
	−48 10766	LSS 3585		16 19 21	−48 24.3	2	10.95	.015	0.68	.020	-0.29	.020	4	O9e	1649,1737
	−51 10091			16 19 22	−51 55.4	2	10.74	.005	0.20	.020	-0.66	.015	2	B8	115,181
147468	+13 03124			16 19 23	+13 00.0	1	8.22		1.06				1	K0	882
147444	+7 03152	G 17 - 12		16 19 24	+07 17.7	1	9.85		0.79		0.33		1	G5	333,1620
		CS 22872 # 34		16 19 25	−01 57.4	1	14.49		0.19		0.24		1		1736
	−52 09720	LSS 3584		16 19 26	−52 35.8	2	10.66	.030	0.25	.015	-0.63	.025	2		115,181
		CS 22872 # 33		16 19 28	−01 47.4	1	16.11		-0.10		-0.91		1		1736
328050	−44 10817			16 19 31	−44 40.9	2	10.39	.010	0.31	.010	-0.41		6	B3 III	540,1021
		NGC 6101 sq1 3		16 19 31	−71 54.2	1	12.28		0.37		0.13		3		89
147449	+1 03215	HR 6093		16 19 32	+01 08.7	10	4.82	.015	0.33	.009	0.03	.008	36	F0 V	15,253,254,1075,1088*
147508	+13 03126			16 19 35	+13 34.4	1	7.44		1.33				1	K2	882
147328	−40 10336	NGC 6124 - 1		16 19 35	−40 29.9	2	9.09	.015	1.78	.025			4	G6 (III)	1653,2026
147384	−24 12675			16 19 36	−24 16.0	2	8.60	.019	0.41	.001	0.21	.004	5	A0 V	40,1770
		CS 22872 # 46		16 19 37	−03 50.8	1	13.66		0.39		0.26		1		1736
147229	−58 06775			16 19 37	−58 11.8	1	10.29		0.19		0.18		2	A0	540
		WLS 1620-10 # 5		16 19 41	−07 34.8	1	12.46		0.83		0.42		2		1375
		Nor sq 1 # 83		16 19 42	−52 41.	1	11.76		0.53		0.07		1		115
147547	+19 03086	HR 6095, Gam Her	★ A	16 19 43	+19 16.2	10	3.75	.011	0.27	.006	0.20	.025	70	A9 III	1,3,15,254,1077,1084*
147547	+19 03085	IDS16175N1923	B	16 19 43	+19 16.2	1	9.59	.015	1.11	.019	0.95	.019	7		1084,3072
147664	+41 02697			16 19 43	+41 33.1	1	7.91		1.05		0.84		3	K0	196
325504	−40 10337	NGC 6124 - 2		16 19 43	−40 40.1	1	10.93		0.71				2	A0	2026
		CS 22872 # 39		16 19 46	−02 44.6	1	13.89		0.21		0.01		1		1736
147274	−52 09731			16 19 46	−52 55.7	3	8.27	.004	-0.01	.011	-0.33	.012	7	B8 III/IV	540,976,1586
147471	−6 04416			16 19 47	−06 44.9	1	8.93		0.59		0.21		1	F0	289
		CS 22872 # 50		16 19 48	−04 12.7	1	13.17		0.37		0.26		1		1736
	−49 10618	LSS 3587		16 19 49	−49 13.2	1	10.38		0.42		0.06		1	B0:	1649
147765	+55 01836			16 19 52	+55 22.2	1	9.61		0.56		0.06		3	K0	196
147432	−22 04159	IDS16169S2253	A	16 19 52	−23 00.1	1	7.52		0.27		0.12		2	A1 III/IV	1770
330637	−48 10776			16 19 53	−48 23.0	1	11.34		0.56		-0.31		2		1649
		L 339 - 96		16 19 54	−48 33.	1	13.26		0.96		0.29		2		3062
147400	−33 11140			16 19 55	−33 54.2	1	9.49		0.79		0.47		1	A1/2mF2-F2	1770
147387	−34 10966			16 19 55	−34 48.5	1	7.17		0.34		0.01		8	F2 V	588
147372	−40 10338	NGC 6124 - 3		16 19 56	−40 18.6	1	9.82		0.71				2	G2 V	2026
325558	−40 10339	NGC 6124 - 4		16 19 56	−40 30.4	1	11.34		0.63				1	A0	2026
147318	−51 10109	LSS 3588		16 19 58	−52 01.6	7	9.40	.021	0.44	.013	-0.53	.029	13	B0 Ia/b	115,181,336,403,432*
147302	−55 07498			16 19 59	−55 20.3	5	7.72	.017	0.00	.011	-0.71	.010	13	B2 IIIn	164,181,540,976,1586
147550	−1 03174	HR 6096		16 20 03	−01 57.8	4	6.24	.018	0.06	.011	-0.02	.025	13	B9 V	252,1079,1088,3050
		CS 22872 # 41		16 20 03	−03 17.7	1	12.18		0.22		0.22		1		1736
		CS 22872 # 32		16 20 04	−01 37.5	1	15.09		0.43		-0.01		1		1736
147331	−52 09741	LSS 3590		16 20 04	−52 10.9	8	8.74	.013	0.32	.015	-0.64	.022	17	O9.5/B0 Ia	115,125,181,336,403*
	−59 06719	RT Nor		16 20 04	−59 13.9	2	10.15	.092	1.09	.028	0.88	.090	2		1589,1699
	−24 12677			16 20 06	−24 36.0	2	10.23	.009	1.48	.011	1.19		5	M2	158,1705
147677	+31 02845	HR 6103	★ A	16 20 09	+31 00.4	7	4.86	.006	0.97	.006	0.80	.015	22	K0 III	15,667,1080,1363,3051*
	−38 10980			16 20 09	−39 06.8	6	11.01	.025	-0.14	.011	-0.96	.018	14	DA	7,158,232,832,1698,1765
147665	+25 03071	IDS16181N2459	A	16 20 11	+24 51.5	2	8.75	.040	0.56	.012	0.08		5	F8 V	20,3026
147665	+25 03071	IDS16181N2459	B	16 20 11	+24 51.5	1	12.27		1.12		0.88		2		3016
		Steph 1332		16 20 14	+08 09.4	1	12.27		1.49		1.18		1	K7-M0	1746
		CS 22872 # 30		16 20 14	−01 14.8	1	15.13		0.43				1		1736
147359	−51 10114			16 20 14	−51 52.9	4	9.63	.006	0.13	.009	-0.46	.013	9	B3/5 Ib/II	115,125,540,976
147347	−54 07624			16 20 15	−55 06.3	2	7.79	.027	0.03	.007	-0.45	.010	4	B5 III	540,976
147492	−28 12081			16 20 16	−28 56.0	1	10.05		0.70		0.26		1	F2 V	1770
		NGC 6101 sq1 5		16 20 16	−71 58.1	2	11.61	.010	1.04	.015	0.83	.015	5		89,371
		NGC 6101 sq2 6		16 20 16	−71 58.1	1	13.10		1.21		1.21		3		371
	−52 09747			16 20 18	−52 58.8	1	11.22		0.63		0.00		1		115
147491	−26 11310	V972 Sco		16 20 19	−26 15.3	3	9.68	.012	0.63	.005	0.06	.000	108	G0 V	521,806,816
330587	−46 10698			16 20 19	−46 59.2	1	9.67		1.17				4	G8 III	1021
147434	−40 10343	NGC 6124 - 5		16 20 20	−40 39.6	3	9.44	.011	0.62	.015	0.27	.004	30	B6 II/III	1653,1770,2026
147456	−40 10344	NGC 6124 - 6		16 20 23	−40 31.2	2	10.43	.010	0.62	.010	0.08		4	G5	1653,2026

HD	DM	Other Id	N Rem	α_{1950}	δ_{1950}	S	V	σ_V	B–V	σ_{B-V}	U–B	σ_{U-B}	n	Spectrum	References
147362	−54 07632			16 20 23	−54 36.6	2	7.99	.036	-0.04	.002	-0.62	.005	4	B3 III/IV	540,976
330643	−48 10788			16 20 25	−48 35.6	1	10.71		0.39				4	B6 III	1021
147749	+34 02773	HR 6107	⋆ A	16 20 28	+33 54.9	1	5.20		1.60		1.94		5	M2 IIIab	3055
	−51 10117	Ru 118 - 1		16 20 28	−51 50.4	1	10.90		0.64		0.48		4		251
		NGC 6101 sq1 4		16 20 30	−71 57.6	1	11.66		0.46		0.02		2		89
		GD 200		16 20 31	+26 02.2	1	15.53		-0.17		-1.08		1	DAwk	3060
	+40 03009			16 20 31	+40 00.0	1	9.89		0.54		0.07		122	F8	1655
		GD 508		16 20 33	+58 13.6	1	13.62		0.57		-0.21		3		308
		Ru 118 - 2		16 20 33	−51 50.0	1	12.25		0.33		0.28		2		251
	+40 03010			16 20 34	+40 08.8	1	9.56		1.03		0.82		122		1655
		CS 22872 # 56		16 20 34	−05 35.8	1	15.10		0.35		0.35		1		1736
147644	−0 03106			16 20 35	−00 35.4	1	8.19		0.57		0.02		2	F9 V	3016
147767	+34 02774	HR 6108	⋆ B	16 20 36	+33 49.1	3	5.40	.009	1.52	.004	1.88	.024	10	K5 III	1080,3016,4001
		NGC 6121 - 2		16 20 36	−26 24.	1	11.67		1.12		0.74		1		521
		NGC 6121 - 1103		16 20 36	−26 24.	2	13.82	.000	0.41	.000	0.36	.000	10		806,816
		NGC 6121 - 1203		16 20 36	−26 24.	2	14.89	.010	1.12	.010	0.50		5		521,1515
		NGC 6121 - 1207		16 20 36	−26 24.	2	13.40	.000	0.45	.000	0.39	.000	12		806,816
		NGC 6121 - 1401		16 20 36	−26 24.	2	14.34	.005	1.07	.000			7		420,1515
		NGC 6121 - 1402		16 20 36	−26 24.	2	13.72	.010	1.28	.000			7		420,1515
		NGC 6121 - 1403		16 20 36	−26 24.	3	12.11	.008	1.51	.029	1.25		9		420,521,1515
		NGC 6121 - 1408		16 20 36	−26 24.	2	11.80	.000	1.46	.000			10		420,1515
		NGC 6121 - 1411		16 20 36	−26 24.	5	11.07	.005	1.75	.007	1.84	.011	73		521,806,816,1515,3036
		NGC 6121 - 1412		16 20 36	−26 24.	1	10.09		1.65		1.61		5		521
		NGC 6121 - 1514		16 20 36	−26 24.	1	10.76		1.87		1.68		2		3036
		NGC 6121 - 1520		16 20 36	−26 24.	2	13.58	.000	0.48	.000	0.41	.000	16		806,816
		NGC 6121 - 1605		16 20 36	−26 24.	1	12.50		1.45		1.07		1		3036
		NGC 6121 - 1610		16 20 36	−26 24.	1	13.63		0.86		0.32		1		3036
		NGC 6121 - 1617		16 20 36	−26 24.	1	12.17		1.39		1.02		1		3036
		NGC 6121 - 1620		16 20 36	−26 24.	1	13.65		0.86		0.36		1		3036
		NGC 6121 - 1623		16 20 36	−26 24.	1	13.62		0.87		0.36		1		3036
		NGC 6121 - 2104		16 20 36	−26 24.	2	13.21	.000	0.42	.000	0.38	.000	12		806,816
		NGC 6121 - 2107		16 20 36	−26 24.	3	14.76	.007	1.32	.000	0.97	.005	11		521,828,1515
		NGC 6121 - 2108		16 20 36	−26 24.	1	13.40		0.92		0.31		2		521
		NGC 6121 - 2109		16 20 36	−26 24.	3	13.91	.007	0.92	.008	0.36	.025	11		521,828,1515
		NGC 6121 - 2113		16 20 36	−26 24.	2	13.72	.020	0.96	.005	0.26		5		420,1515
		NGC 6121 - 2114		16 20 36	−26 24.	3	13.66	.005	0.48	.000	0.41	.000	27		420,553,1515
		NGC 6121 - 2206		16 20 36	−26 24.	3	11.89	.002	1.46	.003	1.29	.003	100		521,806,816
		NGC 6121 - 2207		16 20 36	−26 24.	2	10.27	.010	1.74	.005	2.01	.010	9		521,1515
		NGC 6121 - 2208		16 20 36	−26 24.	2	12.30	.015	1.54	.019	1.24		6		420,1515
		NGC 6121 - 2209		16 20 36	−26 24.	2	15.02	.019	1.12	.015			6		420,1515
		NGC 6121 - 2307		16 20 36	−26 24.	2	11.56	.032	1.68	.023	1.64	.061	5		521,3036
		NGC 6121 - 2312		16 20 36	−26 24.	2	13.59	.005	0.92	.000			6		420,1515
		NGC 6121 - 2314		16 20 36	−26 24.	2	13.06	.005	1.47	.000			8		420,1515
		NGC 6121 - 2315		16 20 36	−26 24.	2	12.84	.005	1.30	.015	0.72		8		420,1515
		NGC 6121 - 2320		16 20 36	−26 24.	2	15.16	.035	0.81	.000			7		420,1515
		NGC 6121 - 2414		16 20 36	−26 24.	2	13.27	.000	0.70	.000	0.19	.000	8		806,816
		NGC 6121 - 2415		16 20 36	−26 24.	2	13.50	.000	0.45	.000	0.41	.000	14		806,816
		NGC 6121 - 2426		16 20 36	−26 24.	2	13.51	.010	0.96	.000			7		420,1515
		NGC 6121 - 2602		16 20 36	−26 24.	1	13.37		0.49		0.44		1		3036
		NGC 6121 - 2603		16 20 36	−26 24.	1	13.83		1.17		0.43		1		3036
		NGC 6121 - 2604		16 20 36	−26 24.	1	13.97		1.02		0.49		1		3036
		NGC 6121 - 2608		16 20 36	−26 24.	1	12.30		1.41		1.14		1		3036
		NGC 6121 - 2626		16 20 36	−26 24.	1	12.54		1.41		1.01		1		3036
		NGC 6121 - 3105		16 20 36	−26 24.	1	11.29		0.62		0.17		1		521
		NGC 6121 - 3108		16 20 36	−26 24.	1	13.13		0.44		0.36		3		521
		NGC 6121 - 3111		16 20 36	−26 24.	1	13.37		0.54		0.33		4		521
		NGC 6121 - 3114		16 20 36	−26 24.	1	13.42		0.96		0.37		2		521
		NGC 6121 - 3207		16 20 36	−26 24.	1	11.91		1.19		0.85		2		521
		NGC 6121 - 3209		16 20 36	−26 24.	2	10.93	.014	1.68	.005	1.79	.014	4		521,1515
		NGC 6121 - 3215		16 20 36	−26 24.	1	12.22		1.20		0.73		3		521
		NGC 6121 - 3216		16 20 36	−26 24.	2	12.68	.000	0.59	.000	0.30	.000	4		521,828
		NGC 6121 - 3225		16 20 36	−26 24.	3	15.10	.000	1.08	.008	0.40	.021	10		521,828,1515
		NGC 6121 - 3303		16 20 36	−26 24.	1	11.84		1.60		1.67		2		521
		NGC 6121 - 3316		16 20 36	−26 24.	2	13.04	.000	0.71	.000	0.28	.000	14		806,816
		NGC 6121 - 3408		16 20 36	−26 24.	2	13.29	.000	0.40	.000	0.37	.000	10		806,816
		NGC 6121 - 3413		16 20 36	−26 24.	2	11.33	.005	1.51	.014	1.40	.009	5		521,1515
		NGC 6121 - 3506		16 20 36	−26 24.	1	13.66		1.26		0.74		1		3036
		NGC 6121 - 3601		16 20 36	−26 24.	1	12.72		1.12		0.54		2		3036
		NGC 6121 - 3612		16 20 36	−26 24.	1	11.82		1.50		1.34		2		3036
		NGC 6121 - 3621		16 20 36	−26 24.	1	12.67		1.34		0.83		1		3036
		NGC 6121 - 3622		16 20 36	−26 24.	1	13.44		0.90		0.28		1		3036
		NGC 6121 - 3624		16 20 36	−26 24.	1	11.82		1.55		1.40		1		3036
		NGC 6121 - 3625		16 20 36	−26 24.	1	14.66		1.06		0.31		1		3036
		NGC 6121 - 3629		16 20 36	−26 24.	1	13.38		0.78		0.33		1		3036
		NGC 6121 - 4117		16 20 36	−26 24.	2	13.44	.000	0.74	.000	0.24	.000	10		806,816
		NGC 6121 - 4119		16 20 36	−26 24.	2	12.51	.014	1.41	.019	1.39	.005	4		521,1515
		NGC 6121 - 4208		16 20 36	−26 24.	1	12.03		1.42		1.25		3		521
		NGC 6121 - 4302		16 20 36	−26 24.	1	12.28		1.19		0.79		1		521
		NGC 6121 - 4304		16 20 36	−26 24.	1	14.41		1.54		1.66		2		521
		NGC 6121 - 4305		16 20 36	−26 24.	1	13.46		1.23		0.74		3		521

Table 1 885

HD	DM	Other Id	N Rem	α_{1950}	δ_{1950}	S	V	σ_V	B–V	σ_{B-V}	U–B	σ_{U-B}	n	Spectrum	References
		NGC 6121 - 4306		16 20 36	−26 24.	1	14.37		0.73		0.20		2		521
		NGC 6121 - 4315		16 20 36	−26 24.	2	13.49	.000	0.43	.000	0.38	.000	12		806,816
		NGC 6121 - 4316		16 20 36	−26 24.	4	13.56	.000	0.37	.005	0.30	.005	16		521,806,816,828
		NGC 6121 - 4404		16 20 36	−26 24.	1	12.96		1.28		0.94		2		521
		NGC 6121 - 4408		16 20 36	−26 24.	1	13.03		0.44		0.30		1		521
		NGC 6121 - 4501		16 20 36	−26 24.	6	10.39	.004	0.84	.012	0.31	.009	105		420,521,553,806,816,1515
		NGC 6121 - 4508		16 20 36	−26 24.	1	12.99		1.35		0.83		1		3036
		NGC 6121 - 4509		16 20 36	−26 24.	1	13.04		1.36		0.89		1		3036
		NGC 6121 - 4611		16 20 36	−26 24.	1	10.97		1.97		2.23		2		3036
		NGC 6121 - 4613		16 20 36	−26 24.	1	10.80		1.93		2.04		2		3036
		NGC 6121 - 4630		16 20 36	−26 24.	1	12.16		1.44		1.13		2		3036
		NGC 6121 - 4633		16 20 36	−26 24.	1	11.80		1.41		1.01		2		3036
		CS 22872 # 45		16 20 37	−03 40.8	1	15.09		0.16		0.03		1		1736
	−51 10121	Ru 118 - 5		16 20 37	−51 49.7	1	12.29		0.25		-0.08		2		251
	−51 10120	Ru 118 - 3		16 20 37	−51 50.3	1	11.82		0.38		0.06		3		251
147513	−38 10983	HR 6094		16 20 38	−39 04.7	7	5.39	.012	0.63	.011	0.11	.048	19	G5 V	15,158,232,1075,2024*
		L 339 - 42		16 20 39	−46 36.3	1	12.26		1.41				1		1705
		Ru 118 - 7		16 20 40	−51 50.1	1	12.60		0.73		0.26		2		251
	−54 07638	LSS 3593		16 20 40	−54 31.2	2	9.51	.015	0.18	.025	-0.15	.035	4	A0 Ib	181,403
	+40 03012			16 20 41	+40 17.3	1	8.27		0.89		0.56		122	K0	1655
		CS 22872 # 51		16 20 41	−04 14.8	1	14.50		0.12		-0.02		1		1736
		Ru 118 - 6		16 20 42	−51 51.	1	13.54		0.40		0.33		2		251
		Nor sq 1 # 88		16 20 42	−52 45.	1	11.43		0.19		-0.37		1		115
		BPM 24107		16 20 42	−53 37.	1	15.64		-0.03		-0.69		1		3065
	−55 07515	LSS 3592		16 20 42	−55 54.7	1	11.81		0.17		-0.67		2		181
	−11 04126			16 20 43	−11 28.8	1	10.42		1.00		0.81		1	K3 V	3072
147592	−25 11495			16 20 43	−26 09.3	5	8.92	.005	0.31	.003	0.21	.010	103	A0 V	40,521,806,816,1770
147553	−32 11687	HR 6097	⋆ A	16 20 43	−33 05.0	1	6.99		-0.01		-0.08		1	A0 V	1771
147553	−32 11687	HR 6097	⋆ AB	16 20 43	−33 05.0	2	6.45	.005	0.02	.005	-0.02		7	A0 V +A0 V	404,2035
		Ru 118 - 4		16 20 43	−51 49.8	1	11.82		0.47		0.22		2		251
	−59 06723	LSS 3591		16 20 43	−59 56.7	4	9.75	.023	0.31	.006	-0.20	.032	14	A3 Iab	400,403,540,976
		Nor sq 3 # 408		16 20 45	−51 43.0	1	13.37		0.72		0.18		2		336
330652	−48 10797	LSS 3597	⋆ AB	16 20 46	−48 42.3	2	9.86	.005	0.46	.030	-0.34		6	B0 V	1021,1649
		CS 22872 # 52		16 20 47	−04 20.4	1	15.80		0.16		0.13		1		1736
147593	−27 10923			16 20 48	−27 14.2	1	10.04		0.25		-0.03		1	B9 III/IV	1770
325559	−40 10347	NGC 6124 - 7		16 20 48	−40 36.9	1	10.25		0.65				2	A0	2026
147349	−62 05325	HR 6089		16 20 48	−63 00.7	2	6.14	.005	0.03	.000			7	A1 V	15,2030
147421	−53 07928	LSS 3596		16 20 50	−53 21.0	3	9.02	.022	0.09	.008	-0.81	.046	4	B0 II	115,181,403
147750	+17 03012	G 138 - 24		16 20 52	+17 34.8	2	8.44	.005	0.72	.010	0.26	.015	2	G0	1658,3060
146164	−83 00604			16 20 56	−83 28.9	1	7.91		0.39				4	F3 IVp	2012
		LS IV -12 001		16 20 57	−12 05.6	2	11.14	.000	-0.09	.000	-0.99	.000	3	sdO	405,1650
325560	−40 10348	NGC 6124 - 8		16 20 58	−40 37.0	2	10.12	.020	0.54	.030	0.20		4	A0	1653,2026
	−71 02002	NGC 6101 sq1 2		16 20 58	−71 56.0	1	10.88		0.54		0.08		2		89
325531	−40 10349	NGC 6124 - 9		16 20 59	−40 14.5	1	10.94		1.01				2	G5	2026
		NGC 6101 sq1 6		16 20 59	−71 59.0	1	12.88		0.53		0.10		3		89
147648	−25 11498			16 21 00	−25 18.0	2	9.43	.011	0.78	.003	0.21	.020	7	B8 II	40,1770
		NGC 6124 - 10		16 21 00	−40 40.2	1	12.17		0.80				1		2026
147835	+32 02716	HR 6110	⋆ A	16 21 01	+32 26.9	3	6.39	.005	0.08	.016	0.12	.016	8	A4 Vn	542,1022,3024
147835	+32 02716	IDS16191N3234	B	16 21 01	+32 26.9	2	9.75	.010	0.87	.000	0.42	.005	4	G7 IV	542,3024
		CS 22872 # 31		16 21 01	−01 28.1	1	15.28		0.17		0.19		1		1736
		CS 22872 # 55		16 21 01	−05 03.1	1	15.42		0.23		0.20		1		1736
147649	−25 11497			16 21 02	−26 01.0	4	9.62	.009	0.44	.001	0.28	.018	25	A5 III	40,521,806,816
147612	−33 11152			16 21 02	−33 54.6	1	9.80		0.50		0.34		1	A1/2 III	1770
147613	−34 10978			16 21 03	−35 04.4	1	9.88		0.60		0.29		1	A7/9 (III)	1770
325557	−40 10352	NGC 6124 - 11		16 21 03	−40 29.8	1	10.51		0.66				2	A0	2026
330642	−48 10802	LSS 3598		16 21 03	−48 35.6	2	10.36	.005	0.52	.020	-0.33		6	O9	1021,1649
	−49 10637	LSS 3599		16 21 07	−49 35.1	1	11.50		0.62		-0.17		2	O7	1649
	−51 10132			16 21 07	−51 52.8	1	10.98		0.59		0.04		5		125
147650	−30 13112			16 21 09	−30 55.4	1	9.79		0.48		0.25		1	A8/F0 V	1770
325556	−40 10356	NGC 6124 - 12		16 21 09	−40 29.8	1	11.02		0.63				2	A0	2026
147700	−19 04365	HR 6104		16 21 10	−19 55.3	5	4.49	.008	1.01	.009	0.83	.010	20	K0 III	15,1075,1088,2012,8015
147628	−37 10778	HR 6100	⋆ AB	16 21 11	−37 27.1	3	5.42	.007	-0.11	.004	-0.40	.003	7	B8 V	567,1770,2006
		BPM 24123		16 21 12	−51 09.	1	11.06		0.62		-0.08		1		3065
147752	−2 04176			16 21 13	−02 51.6	1	9.36		0.90		0.56		2	K2 V	3072
		NGC 6124 - 221		16 21 13	−40 33.9	1	12.77		0.86		0.59		3		1653
147666	−30 13113			16 21 14	−30 55.4	1	9.95		0.27		0.20		1	A2/3 III	1770
147559	−50 10504			16 21 14	−50 38.0	1	7.88		0.07		-0.09		1	B9/9.5IV	96
325555	−40 10358	NGC 6124 - 13		16 21 15	−40 26.8	2	10.83	.543	0.63	.248	0.10		6	A0	1653,2026
	−51 10137	LSS 3600		16 21 15	−51 39.2	2	10.82	.078	0.37	.000	-0.34	.034	3		33,115
148016	+53 01863			16 21 17	+53 39.4	1	8.64		0.98		0.73		2	K0	1566
147667	−32 11690			16 21 17	−33 00.7	1	10.48		0.33		0.02		1	B8 Ib/II	1770
		Ton 257		16 21 18	+24 51.	1	15.94		-0.48		-1.14		1	sdO	1036
147701	−24 12682	IDS16183S2448	AB	16 21 19	−24 54.6	3	8.36	.004	0.56	.012	-0.07	.012	7	B5 III	26,1054,1770
147614	−45 10633	HR 6099	⋆ AB	16 21 19	−45 14.1	1	6.33		0.19				4	A2/3 V	2009
147539	−53 07947			16 21 20	−53 16.1	1	10.02		0.10		-0.40		1	B2/5 II	115
147702	−25 11499			16 21 23	−25 35.5	2	9.14	.006	0.46	.002	0.29	.009	4	A1/2 V	40,1770
147703	−26 11327			16 21 25	−27 02.2	2	7.47	.000	0.20	.020	0.00	.010	5	A0 IV	40,1771
147653	−40 10359	NGC 6124 - 15		16 21 25	−40 28.5	3	9.26	.008	0.68	.028	0.34	.001	23	B7 II/III	1653,1770,2026
147654	−40 10360	NGC 6124 - 14		16 21 25	−40 39.1	2	9.16	.015	1.78	.025			16	K0	1653,2026
147836	+11 02977			16 21 26	+11 42.4	1	7.76		1.35		1.59		2	K0	1648

HD	DM	Other Id	N Rem	α₁₉₅₀	δ₁₉₅₀	S	V	σ_V	B–V	σ_B–V	U–B	σ_U–B	n	Spectrum	References
	−10 04311			16 21 28	−11 06.2	1	9.89		0.63		0.04		1	G0	565
147683	−34 10981	V760 Sco		16 21 28	−34 46.8	3	6.99	.005	0.18	.018	-0.38	.008	6	B4 V	540,976,1770
	−40 10361	NGC 6124 - 16		16 21 28	−40 33.3	2	11.61	.050	0.64	.005	0.40		4		1653,2026
147601	−50 10508			16 21 28	−50 37.9	1	8.28		0.07		-0.18		1	B9.5V	96
147723	−29 12513	HR 6106	⋆ AB	16 21 31	−29 35.4	3	5.41	.005	0.60	.005	0.12		11	G0 IV	15,404,2012
		NGC 6124 - 111		16 21 31	−40 26.1	1	12.66		0.84		0.56		2		1653
147776	−13 04418	G 153 - 49	⋆ AB	16 21 32	−13 31.5	4	8.41	.007	0.95	.006	0.69	.025	10	K2 V	196,1705,2012,3077
148017	+51 02088			16 21 33	+50 52.4	1	8.57		1.11		1.05		2	K0	1566
		AJ77,733 T28# 12		16 21 34	−52 03.7	1	12.31		0.23		0.15		3		125
		NGC 6124 - 17		16 21 35	−40 38.8	1	12.02		0.78				1		2026
147603	−51 10147	X Nor		16 21 35	−51 48.8	1	9.54		2.98		3.61		2	Nb	864
147743	−26 11329			16 21 36	−26 41.5	3	8.40	.003	0.62	.000	0.19	.002	29	G0 V	521,806,816
		LP 918 - 19		16 21 36	−30 12.0	1	14.74		0.86		0.15		4		3062
147686	−40 10367	NGC 6124 - 205		16 21 38	−40 31.7	2	8.75	.013	0.64	.014	0.36	.008	2	B8	567,1770
147656	−46 10724			16 21 38	−46 17.0	1	9.81		0.24		-0.22		1	B8 Ib/II	567
147636	−50 10513			16 21 39	−50 35.1	1	8.97		1.03		0.76		1	G5 III	96
		MN229,227 # 9		16 21 40	−48 50.9	1	11.51		1.83		1.82		1		1649
325548	−40 10370	NGC 6124 - 18		16 21 41	−40 29.2	2	10.47	.005	0.68	.005	0.27		4		1653,2026
147617	−51 10148	LSS 3601		16 21 41	−51 54.7	8	9.81	.012	0.25	.013	-0.71	.016	25	O8	33,115,125,181,336*
		MN229,227 # 16		16 21 42	−49 03.5	1	12.49		0.43		0.18		1		1649
		NGC 6124 - 216		16 21 43	−40 38.8	1	12.14		0.73		0.50		2		1653
147869	+7 03164	HR 6111		16 21 45	+07 03.7	2	5.84	.005	-0.01	.000	0.01	.000	7	A2 p(Sr)	15,1256
147754	−31 12855			16 21 45	−31 27.1	1	8.67		0.30		0.05		1	A9 V	1770
325562	−40 10372	NGC 6124 - 19		16 21 46	−40 40.2	1	10.03		0.59				3	B9	2026
		LP 685 - 53		16 21 48	−03 52.0	1	12.28		0.76		0.10		2		1696
325545		NGC 6124 - 20		16 21 48	−40 29.7	2	10.83	.015	0.66	.045	0.20		4	A0	1653,2026
		G 17 - 13		16 21 49	+04 23.0	1	13.24		1.25		0.88		1		1696
		NGC 6124 - 403		16 21 49	−40 32.7	1	11.69		0.78		0.58		4		1653
		NGC 6124 - 215		16 21 49	−40 39.1	1	12.53		0.72		0.56		2		1653
		CS 22872 # 72		16 21 50	−02 42.5	1	14.25		0.09		-0.28		1		1736
325553		NGC 6124 - 22		16 21 51	−40 35.8	2	11.23	.030	0.54	.020	0.20		4		1653,2026
325563	−40 10377	NGC 6124 - 24		16 21 51	−40 40.1	2	10.86	.015	0.61	.015	0.22		4	A0	1653,2026
325564	−40 10375	NGC 6124 - 21		16 21 51	−40 41.9	2	11.33	.010	0.66	.025	0.37		4	A0	1653,2026
330684	−49 10651			16 21 51	−49 22.8	2	9.78	.015	0.75	.025	-0.32	.000	5	B1 Ibk	1649,1737
		NGC 6124 - 401		16 21 52	−40 30.8	1	11.99		1.46		1.03		1		1653
	−50 10516	LSS 3602		16 21 52	−50 54.0	1	10.02		0.97		0.13		2	B5:II:	1649
147927	+16 02938			16 21 53	+16 16.2	1	8.58		1.01				1	K0	882
147779	−31 12856			16 21 53	−31 18.5	1	7.14		0.00		-0.09		1	A0 V	1770
325544	−40 10380	NGC 6124 - 23		16 21 53	−40 29.2	2	9.24	.020	0.66	.040	0.39		5	A0	1653,2026
325552	−40 10382	NGC 6124 - 25		16 21 54	−40 34.3	1	10.02		0.66				1		2026
148293	+69 00845	HR 6126		16 21 55	+69 13.5	1	5.25		1.12		1.11		2	K2 III	3016
		MN229,227 # 13		16 21 55	−48 44.4	1	13.07		0.66		0.19		1		1649
147707	−49 10653			16 21 55	−49 18.6	1	7.35		1.01		0.80		1	K0 III	1649
147809	−25 11501			16 21 57	−25 14.4	2	8.60	.006	0.42	.002	0.23	.017	4	A1 V	40,1770
		NGC 6101 sq1 16		16 21 57	−72 13.6	1	11.46		0.48		0.04		2		89
		MN229,227 # 21		16 21 58	−48 57.5	1	12.41		1.06		0.35		1		1649
		AJ77,733 T28# 13		16 21 58	−51 58.8	1	12.54		0.42		0.05		2		125
147791	−31 12857			16 21 59	−31 35.6	1	10.01		0.23		0.06		1	B9 IV/V	1770
325534	−40 10384	NGC 6124 - 27		16 21 59	−40 20.5	1	10.20		0.60				2	A0	2026
		NGC 6124 - 431		16 21 59	−40 31.6	1	13.08		0.80		0.61		2		1653
		V445 Oph		16 22 00	−06 25.2	2	10.59	.064	0.44	.031	0.34	.029	2	F2.5	668,699
325565		NGC 6124 - 26		16 22 00	−40 40.6	2	11.64	.010	0.85	.045	0.66		4	A0	1653,2026
		AJ77,733 T28# 15		16 22 01	−52 04.9	1	12.92		0.62		0.16		3		125
	−40 10385	NGC 6124 - 29		16 22 02	−40 32.5	1	9.29		1.80		1.56		2		1653,2026
		MN229,227 # 20		16 22 02	−48 59.4	1	13.05		0.58		0.34		1		1649
	−40 10386	NGC 6124 - 28		16 22 03	−40 31.5	2	10.09	.050	0.64	.010	0.16		5		1653,2026
147780	−40 10387	NGC 6124 - 30		16 22 03	−40 37.0	1	9.72		0.50				2	B8 II	2026
		AJ77,733 T28# 11		16 22 03	−51 53.9	1	12.08		0.50		0.06		2		125
		NGC 6124 - 335		16 22 04	−40 27.9	1	12.82		0.82		0.63		1		1653
		MN229,227 # 12		16 22 04	−48 44.7	1	13.00		0.86		0.48		1		1649
	−50 10518			16 22 04	−50 27.3	1	9.54		0.42		0.08		1		96
	−51 10155	LSS 3604		16 22 04	−51 41.7	4	10.33	.025	0.31	.019	-0.50	.016	9	B8	33,115,125,336
	−51 10156			16 22 04	−51 50.4	1	11.20		0.48		0.00		2		125
		AJ77,733 T28# 9		16 22 06	−51 50.8	1	11.50		0.48				2		125
		MN229,227 # 19		16 22 08	−48 59.9	1	12.49		1.13		0.52		1		1649
147728	−50 10520			16 22 08	−50 26.9	1	10.12		0.13		0.03		1	A0 V	96
		CS 22872 # 57		16 22 10	−05 15.4	1	14.90		0.28		0.28		1		1736
147756	−45 10640	LSS 3607		16 22 10	−45 25.8	3	8.60	.020	0.25	.003	-0.69	.012	7	B2 (V)ne	540,976,1586
		HRC 257	V	16 22 11	−23 12.4	1	13.41		1.80		0.38		2		649
147793	−40 10388	NGC 6124 - 31		16 22 11	−40 40.2	1	8.96		0.52				19	F8 V	2026
		LSS 3605		16 22 12	−49 54.8	1	11.87		0.45		0.17		2	B3 Ib	1649
147690	−55 07540			16 22 13	−55 42.5	2	8.01	.045	0.11	.009	-0.35	.006	4	B5 III	540,976
		CS 22872 # 66		16 22 14	−03 56.2	1	15.76		0.40		0.19		1		1736
147792	−40 10389	NGC 6124 - 32		16 22 14	−40 32.5	2	9.96	.045	0.63	.055	0.29		5	A0	1653,2026
		AJ77,733 T28# 14		16 22 16	−51 54.4	1	12.85		1.29		1.02		2		125
	+41 02701			16 22 17	+41 21.6	1	10.19		0.50		0.04		3		196
	−21 04352			16 22 17	−21 49.1	3	10.40	.005	1.47	.009	1.18		6	K5	158,1705,1746
	−24 12683	ROC # 28		16 22 18	−24 20.0	1	10.07		0.90		0.41		1		7003
		CS 22872 # 68		16 22 19	−03 23.9	1	14.33		0.32		0.23		1		1736
325543	−40 10390	NGC 6124 - 33		16 22 19	−40 33.9	2	8.68	.015	1.89	.025			12	M0	1653,2026

Table 1 887

HD	DM	Other Id	N	Rem	α_{1950}	δ_{1950}	S	V	σ_V	B–V	σ_{B-V}	U–B	σ_{U-B}	n	Spectrum	References
		CS 22872 # 82			16 22 20	−01 15.0	1	14.82		0.28		0.22		1		1736
		LP 625 - 10			16 22 21	−00 49.1	1	15.70		0.74				1		1759
		MN229,227 # 11			16 22 21	−48 49.7	1	12.08		0.98		0.85		1		1649
148086	+37 02746	IDS16206N3716	AB		16 22 23	+37 09.0	1	7.82		0.86		0.47		2	K0	938
147889	−24 12684	ROC # 32			16 22 23	−24 21.1	6	7.90	.012	0.83	.014	-0.16	.016	17	B2 III/IV	26,285,1012,1054,1770,7003
		ROC # 33		⋆	16 22 23	−24 22.9	1	15.19		1.04		-0.92		1		7003
		AJ77,733 T28# 10			16 22 23	−51 52.5	1	11.56		0.68		0.06		5		125
	−51 10164				16 22 23	−51 59.0	1	10.81		0.17		-0.26		2		125
147888	−23 12860	IDS16196S2313	DE		16 22 24	−23 20.8	3	6.74	.006	0.30	.012	-0.35	.006	12	B3/4 V	26,1054,1770
147823	−40 10391				16 22 24	−40 30.4	1	10.27		0.59		0.10		1	B9	1770
147930	−15 04320				16 22 25	−16 04.6	1	9.18		0.54		0.41		1	A1 IV	1770
147911	−21 04353				16 22 25	−21 34.4	1	9.18		0.48		0.30		2	A0 V	1770
		CS 22872 # 64			16 22 27	−04 18.9	1	12.90		0.43		0.22		1		1736
148238	+59 01721				16 22 28	+58 49.7	1	8.95		0.60		0.12		2	G5	1733
325540	−40 10393	NGC 6124 - 34			16 22 28	−40 35.5	1	10.64		0.62				2	A0	2026
147872	−31 12864	IDS16193S3128	A		16 22 29	−31 35.4	1	8.17		0.33		0.10		1	A8 V	1771
147931	−19 04368				16 22 30	−19 43.4	1	9.02		0.50		0.28		1	A0 V	1770
	−51 10169	LSS 3608			16 22 30	−51 41.2	7	9.64	.017	0.22	.014	-0.57	.014	18	B2 III	33,115,125,336,540*
147890	−29 12529	V936 Sco	⋆ AB		16 22 31	−29 17.2	2	7.66	.009	0.23	.020	-0.12	.034	6	Ap Si	26,1054
147839	−40 10397	NGC 6124 - 35			16 22 34	−40 26.7	2	8.99	.015	1.74	.025			4	K0	1653,2026
147840	−40 10395	NGC 6124 - 36	⋆ A		16 22 34	−40 44.2	2	9.16	.015	1.88	.025			4	G2 (III)	1653,2026
147932	−23 12862	IDS16196S2313	C		16 22 35	−23 17.5	2	7.27	.000	0.31	.011	-0.35	.004	6		40,1770
147933	−23 12861	HR 6112	⋆ AB		16 22 35	−23 20.0	6	4.60	.017	0.23	.006	-0.57	.011	24	B2 IV	15,938,1054,1770,2007,8015
		MN229,227 # 14			16 22 36	−48 52.9	1	12.40		1.20		1.04		1		1649
		CS 22872 # 58			16 22 37	−05 07.0	1	15.39		0.27		0.33		1		1736
325541	−40 10398	NGC 6124 - 37			16 22 37	−40 32.9	2	11.69	.015	0.69	.025	0.45		4		1653,2026
		WLS 1620-10 # 9			16 22 39	−09 54.2	1	10.33		1.85		2.14		2		1375
		G 202 - 45			16 22 40	+48 28.3	2	10.27	.000	1.49	.002	1.10		3		694,3078
147912	−31 12866				16 22 40	−32 03.0	1	9.73		0.31		0.21		1	A0 V	1770
325536	−40 10400	NGC 6124 - 38			16 22 41	−40 24.0	2	10.67	.010	0.70	.030	0.30		4	A0	1653,2026
325537	−40 10402	NGC 6124 - 39			16 22 42	−40 29.3	2	10.31	.000	0.65	.020	0.17		4	A0	1653,2026
		Nor sq 2 # 94			16 22 42	−53 31.	1	11.27		0.55		-0.43		2		181
148097	+28 02564				16 22 43	+27 52.0	1	8.57		0.33		0.14		3	A5	1776
		NGC 6124 - 40			16 22 44	−40 30.8	2	12.09	.050	0.76	.020	0.53		5		1653,2026
	−29 13323	NGC 6144 sq1 1			16 22 47	−26 01.9	1	10.58		1.59		1.71		3		828
147955	−26 11336				16 22 47	−26 27.3	3	8.09	.015	0.25	.002	0.01	.006	9	B9 V	40,521,1770
		NGC 6124 - 409			16 22 49	−40 31.5	1	11.92		0.77		0.47		2		1653
147893	−40 10403	NGC 6124 - 41			16 22 49	−40 34.8	2	8.91	.015	1.87	.025			6	G8 (III)	1653,2026
		CS 22872 # 81			16 22 53	−01 20.1	1	14.49		0.21		0.24		1		1736
		CS 22872 # 74			16 22 53	−02 17.1	1	15.19		0.36		0.18		1		1736
		MN229,227 # 15			16 22 54	−48 55.4	1	13.18		0.48		0.13		1		1649
148280	+55 01843				16 22 55	+55 16.4	1	8.74		1.55		1.91		3	K5	196
		HRC 259, V2058 Oph			16 22 55	−24 14.0	2	12.82	.070	1.48	.065	0.12	.085	2	K5e	649,7003
	−39 10433				16 22 56	−39 25.7	2	10.93	.003	0.50	.017	-0.32	.013	5	B1.5V	540,567
		G 138 - 25			16 22 57	+15 48.4	3	13.49	.013	1.41	.016	1.15	.055	7		203,1764,3078
325538	−40 10405	NGC 6124 - 42			16 22 57	−40 33.1	2	11.51	.045	0.69	.025	0.38		4	A0	1653,2026
148127	+24 03003				16 23 00	+24 10.2	1	7.79		1.60		1.95		2	K5	1625
		NGC 6144 sq1 8			16 23 00	−25 52.9	1	14.29		0.91		0.37		5		828
		BPM 24150			16 23 00	−54 05.	1	15.74		0.10		-0.80		2		3065
		NGC 6144 sq1 2			16 23 01	−25 49.8	1	11.77		2.02		2.28		3		828
		ROC # 53			16 23 02	−24 16.8	1	13.95		2.13		1.68		1		7003
		NGC 6144 sq1 14			16 23 03	−25 53.4	1	15.56		1.53				5		828
		CS 22872 # 71			16 23 04	−02 50.5	1	14.69		0.34		0.32		1		1736
		CS 22872 # 63			16 23 04	−04 26.8	1	12.15		0.38		0.29		1		1736
147584	−69 02558	HR 6098			16 23 04	−69 58.5	6	4.90	.008	0.55	.000	0.03	.007	21	G0 V	15,1075,2024,2038*
148112	+14 03049	HR 6117, ω Her	⋆ AB		16 23 06	+14 08.8	8	4.57	.014	0.00	.006	-0.04	.009	29	B9 p Cr	15,1007,1013,1022*
148374	+62 01478	HR 6130	⋆ AB		16 23 07	+61 48.6	1	5.67		0.96				2	G8 III	71
		CS 22872 # 69			16 23 07	−03 21.7	1	15.01		0.41		0.26		1		1736
		ROC # 57			16 23 08	−24 27.4	1	12.00		1.41		0.65		1		7003
147894	−47 10752	IDS16194S4749	AB		16 23 08	−47 55.9	3	7.19	.010	-0.01	.012	-0.49	.009	6	B5 III	540,567,976
	−53 07985				16 23 10	−53 42.5	1	9.46		1.03		0.73		2	G7 II	8100
		LP 505 - 15			16 23 12	+14 38.7	1	13.28		1.12		1.06		1		1696
148147	+17 03022	IDS16210N1732	AB		16 23 12	+17 25.0	1	7.98		0.58		0.13		3	G0	3016
148329	+55 01844	G 225 - 61			16 23 12	+55 27.0	1	6.94		0.55		0.03		2	F8	3016
148054	−13 04425	IDS16204S1354	AB		16 23 13	−14 00.0	1	9.84		0.94		0.62		1	K1 V	3072
		CS 22872 # 67			16 23 14	−03 32.2	1	14.66		0.24		0.26		1		1736
		ROC # 63			16 23 16	−24 13.6	1	14.33		1.74		2.09		1		7003
		ROC # 62			16 23 16	−24 15.6	1	16.68		0.63				1		7003
148387	+61 01591	HR 6132	⋆ AB		16 23 18	+61 37.6	7	2.73	.014	0.91	.007	0.68	.025	17	G8 IIIab	15,542,1080,1363,3016*
		CS 22872 # 77			16 23 18	−01 56.0	1	15.02		0.11		0.11		1		1736
147880	−52 09838				16 23 18	−53 07.7	2	8.63	.001	0.04	.005	-0.38	.008	5	B8 Ib/II	540,976
147787	−63 03923	HR 6109	⋆ A		16 23 18	−63 56.8	2	5.31	.015	0.37	.005	-0.03	.005	4	F0 IV	13,3053
147787	−63 03923	IDS16187S6350	B		16 23 18	−63 56.8	1	9.42		0.61		0.33		2		13
148330	+55 01845	HR 6127			16 23 19	+55 19.1	1	5.74		0.00		0.01		2	A2p Si	253
		HRC 260	V		16 23 19	−23 36.6	1	12.48		1.02		0.52		1	F5:	649
	−71 02010				16 23 21	−71 45.6	1	8.25		0.04		0.03		2		89
		ROC # 67			16 23 22	−24 14.3	1	14.82		2.02		1.07		1		7003
147999	−35 10949	IDS16201S3525	AB		16 23 22	−35 32.0	1	10.00		0.38		0.10		1	B9 III	1770
		NGC 6144 sq1 6			16 23 23	−26 01.1	1	12.60		1.25		0.79		4		828
		NGC 6144 sq1 3			16 23 23	−26 02.7	1	12.06		1.32		1.10		3		828

HD	DM	Other Id	N Rem	α_{1950}	δ_{1950}	S	V	σ_V	B–V	σ_{B-V}	U–B	σ_{U-B}	n	Spectrum	References
147985	−43 10792	V348 Nor		16 23 24	−43 41.2	1	7.95		0.14		−0.64		3	B1/2 II/III	567
328280	−46 10754			16 23 25	−46 18.5	1	10.98		0.21		−0.24		3	B8	1586
		LSS 3610		16 23 27	−49 50.9	1	11.96		0.46		−0.33		1	B3 Ib	1649
		Steph 1340		16 23 28	+47 29.2	1	12.03		1.00		0.85		1	K4	1746
	−51 10182			16 23 28	−51 18.6	1	11.26		0.54		−0.02		3		1586
		G 169 - 6		16 23 29	+27 05.4	2	14.78	.014	1.56	.047			4		419,1759
147958	−48 10836	IDS16198S4834	A	16 23 29	−48 40.5	1	7.68		1.54		1.86		1	K4 II	1649
148087	−21 04356			16 23 31	−21 40.9	1	9.96		0.73		0.20		1	F3/5 V	1770
147971	−47 10765	HR 6115	⋆A	16 23 31	−47 26.6	4	4.46	.005	−0.07	.000	−0.53	.004	14	B2 V	15,1075,2012,8015
148041	−32 11714			16 23 33	−32 57.7	1	9.16		0.33		0.19		3	B9 IV	1770
147942	−53 07991			16 23 33	−54 00.1	3	9.25	.017	0.02	.006	−0.65	.015	8	B3 III	181,540,976
		G 169 - 7		16 23 34	+27 04.6	2	12.49	.000	1.42	.005			5		419,1759
		LP 137 - 21		16 23 34	+52 23.9	1	13.60		1.44		1.17		1		1773
148206	+19 03098	HR 6119, U Her		16 23 35	+19 00.3	1	8.50		1.53		0.62		1	M4	3001
148072	−27 10943			16 23 35	−27 16.1	1	10.13		0.34		0.17		2	A0 III/IV	1770
		CS 22872 # 79		16 23 36	−01 40.0	1	15.00		0.49		−0.16		1		1736
148001	−43 10795			16 23 36	−43 16.6	2	9.08	.001	0.25	.014	−0.36	.016	3	B5 II/III	567,1586
	−50 10553			16 23 36	−50 52.3	1	10.27		0.01		0.07		1		96
147972	−50 10552			16 23 36	−50 53.2	1	9.19		1.07		0.75		1	G8 (IV)	96
		Nor sq 1 # 95		16 23 36	−51 33.	1	11.45		0.53		0.22		1		115
		Nor sq 2 # 96		16 23 36	−53 29.	1	11.86		0.21		−0.50		2		181
148283	+37 02750	HR 6123		16 23 37	+37 30.4	3	5.54	.005	0.17	.000	0.09	.008	7	A5 V	985,1501,1733
		LP 137 - 22		16 23 37	+52 24.6	1	11.77		1.08		0.98		1		1773
		NGC 6144 sq1 7		16 23 37	−26 06.2	1	13.57		0.83		0.22		4		828
148073	−28 12114			16 23 37	−28 56.3	1	10.28		0.72		0.23		2	F5 V	1770
148023	−40 10410	NGC 6124 - 43		16 23 38	−40 26.6	2	9.52	.030	0.66	.000	0.04		5	B3 III/IV	567,2026
147987	−50 10554			16 23 40	−50 57.9	3	8.78	.013	−0.03	.013	−0.52	.012	6	B3 V	96,540,976
148088	−28 12115	IDS16206S2904	AB	16 23 43	−29 10.0	1	11.06		0.31		0.17		1	B8/9 (III)	1770
		CS 22872 # 70		16 23 45	−03 12.7	1	15.52		0.26		0.29		1		1736
	+2 03101	G 17 - 14		16 23 46	+02 17.8	1	10.07		0.96		0.71		1	K2	333,1620
	+47 02339			16 23 46	+46 58.8	1	9.22		0.54		0.03		2	F8	1601
330822	−48 10838	MN229,227 # 8		16 23 46	−48 35.2	1	9.88		1.34		1.49		2	K5	1649
		HRC 261, V2251 Oph		16 23 47	−23 08.1	1	13.84		1.66		0.65		1		649
148346	+45 02404			16 23 48	+45 29.5	1	7.72		0.85		0.40		2	G5	1601
		NGC 6134 - 176		16 23 48	−49 00.1	1	12.14		1.29		1.01		3		1781
		NGC 6134 - 114		16 23 48	−49 02.4	1	12.08		1.31		1.08		3		1781
148119	−28 12117			16 23 49	−28 59.6	1	9.27		0.34		0.23		2	A2 IV/V	1770
148228	+11 02984	HR 6121		16 23 50	+11 31.2	1	6.11		1.03				1	G8 III	71
148056	−40 10414	NGC 6124 - 45		16 23 50	−40 30.7	1	9.80		0.29				3	A0	2026
148057	−40 10412	NGC 6124 - 44		16 23 50	−40 38.3	1	10.21		0.37				2	B8 II	2026
148010	−48 10840			16 23 50	−48 38.9	1	9.19		0.55		0.12		2	F4 III	1649
		NGC 6134 - 67		16 23 50	−49 02.2	1	11.70		1.44		1.06		4		1781
		NGC 6134 - 205		16 23 50	−49 06.9	1	11.89		1.45		1.29		3		1781
148265	+26 02840			16 23 51	+26 15.2	1	9.68		−0.11		−0.66		3		963
148118	−26 11344			16 23 52	−27 01.6	1	9.45		0.45		0.27		3	A5 IV	40
		NGC 6134 - 62		16 23 52	−49 03.3	1	11.89		1.31		1.05		4		1781
		NGC 6134 - 204		16 23 52	−49 06.4	1	11.67		1.42		1.12		3		1781
	−53 07997			16 23 52	−53 25.4	5	9.73	.036	0.21	.015	−0.51	.015	10	B6 Iab	115,181,540,976,1586
		NGC 6134 - 203		16 23 53	−48 57.0	1	11.69		1.32		1.15		3		1781
		NGC 6134 - 70		16 23 53	−49 01.5	1	10.76		0.76		0.24		3		185
		NGC 6134 - 109		16 23 53	−49 04.0	1	11.70		0.40		0.22		3		185
148117	−26 11345			16 23 56	−26 37.5	1	10.53		0.27		0.08		3	Ap Si	40
		NGC 6134 - 107		16 23 56	−49 05.0	1	12.34		1.22		0.89		3		1781
		NGC 6134 - 157		16 23 56	−49 06.0	1	12.27		1.17		0.96		3		1781
148182	−12 04510	V Oph		16 23 57	−12 18.9	2	8.03	.662	3.62	.341	3.30		3	C	817,3038
325616	−40 10417	NGC 6124 - 46		16 23 57	−40 37.4	1	10.21		0.71				2	B8	2026
		CS 22872 # 62		16 23 58	−04 26.2	1	14.05		0.41		0.19		1		1736
		NGC 6134 - 75		16 23 58	−48 59.9	1	12.39		1.28		1.02		3		1781
		NGC 6134 - 27		16 23 58	−49 02.0	1	12.31		0.94		0.36		3		185
		NGC 6134 - 29		16 23 59	−49 00.7	1	12.19		1.27		0.90		4		185
		NGC 6134 - 28		16 23 59	−49 01.6	2	12.27	.051	1.27	.016	0.93	.015	7		185,1781
		NGC 6134 - 7		16 23 59	−49 03.3	1	11.07		0.36		0.15		2	B8	185
		NGC 6134 - 8		16 24 00	−49 03.2	1	11.77		1.41		1.20		1		1781
		NGC 6134 - 105		16 24 00	−49 04.7	1	14.51		0.85		0.15		4		185
148183	−13 04429	IDS16212S1337	A	16 24 01	−13 44.2	1	9.21		0.55		0.28		1	A2/3	1770
		NGC 6134 - 30		16 24 01	−49 00.9	1	11.84		1.27		1.01		4		1781
147977	−58 06800	HR 6114		16 24 01	−58 29.3	4	5.67	.009	0.00	.000			15	B9 II/III	15,2013,2020,2030
		NGC 6134 - 129		16 24 02	−48 58.8	1	12.27		1.31		1.04		4		1781
		NGC 6134 - 102		16 24 03	−49 04.9	1	14.85		0.88		0.08		5		185
		NGC 6134 - 154		16 24 03	−49 05.5	1	13.88		0.73		0.23		4		185
148152	−25 11508			16 24 04	−26 08.8	1	8.26		1.22		1.07		4	K0 III	521
		NGC 6134 - 152		16 24 04	−49 06.2	1	13.96		0.62		0.36		3		185
		Ru 119 - 9		16 24 05	−51 23.5	1	11.98		0.45		0.01		2		251
302745				16 24 05	−57 05.	1	10.74		0.09		−0.02		2		761
		NGC 6139 sq2 1		16 24 06	−38 40.0	1	13.66		1.24		0.82		7		1638
		NGC 6134 - 101		16 24 06	−49 04.8	1	13.44		0.66		0.01		3		185
		NGC 6134 - 151		16 24 06	−49 05.9	2	12.33	.039	1.28	.025	1.02	.077	7		185,1781
148184	−18 04282	HR 6118, khi Oph		16 24 07	−18 20.7	8	4.48	.083	0.25	.032	−0.78	.014	42	B2 Vne	15,26,681,815,1054*
		NGC 6134 - 79		16 24 07	−49 00.3	1	11.99		1.24		0.91		4		1781
		NGC 6134 - 47		16 24 08	−49 02.9	1	12.28		1.29		0.96		3		1781

Table 1   889

HD	DM	Other Id	N Rem	α_{1950}	δ_{1950}	S	V	σ_V	B–V	σ_{B-V}	U–B	σ_{U-B}	n	Spectrum	References
		ROC # 107		16 24 09	−24 12.5	1	13.76		1.63		0.80		1		7003
148076	−48 10848	NGC 6134 - 150		16 24 09	−49 05.6	1	9.33		0.00		-0.52		3	B4 V	185
148167	−29 12548			16 24 10	−29 21.1	1	9.04		0.29		0.21		1	A3 III	1770
148185	−21 04359			16 24 11	−22 04.2	1	10.15		0.38		0.26		2	B9/A0 IV/V	1770
	−48 10847	NGC 6134 - 34		16 24 12	−49 01.5	1	11.14		1.45		1.36		5		1781
		NGC 6134 - 99		16 24 12	−49 03.9	1	11.63		1.36		1.16		3		1781
148135	−35 10958			16 24 13	−35 20.7	1	9.74		0.21		-0.25		1	B8/9 II	1770
		NGC 6139 sq1 7		16 24 13	−38 41.5	1	14.95		1.96				6		1638
		NGC 6134 - 39		16 24 13	−49 02.0	1	12.20		1.27		1.00		4		1781
		LSS 3611		16 24 13	−49 25.7	1	11.66		0.56		-0.06		3	B3 Ib	1649
	−51 10198	Ru 119 - 1		16 24 13	−51 20.8	1	10.94		0.49		0.04		2		251
148433	+51 02097			16 24 14	+51 15.1	1	7.26		0.32		-0.07		3	F0	3016
		G 202 - 48		16 24 14	+54 25.0	3	10.11	.018	1.60	.005	1.28	.000	9	M2	679,1663,8105
		Ru 119 - 2		16 24 14	−51 22.2	2	11.66	.012	0.46	.014	-0.11	.010	5		251,1586
148211	−21 04360			16 24 15	−22 00.7	5	7.69	.009	0.55	.008	-0.06	.017	9	F8/G0 V	742,1770,2012,3037,6006
	−51 10199	Ru 119 - 11		16 24 15	−51 23.7	1	12.06		0.35		-0.17		2		251
	−51 10200	Ru 119 - 8		16 24 15	−51 25.3	1	11.79		0.43		-0.02		2		251
	−48 10843	NGC 6134 - 91		16 24 16	−49 02.5	1	11.64		1.14		0.74		3		1781
	−51 10200	He3 1196		16 24 16	−51 25.0	1	11.03		0.57		-0.40		3		1586
	−53 08005	Nor sq 1 # 97		16 24 16	−53 12.7	1	11.01		0.21		-0.13		1		115
148169	−32 11723			16 24 17	−32 10.6	1	10.10		0.40		0.09		1	A9 V	1770
		NGC 6139 sq1 6		16 24 17	−38 40.9	1	14.02		1.76		1.71		6		1638
148122	−43 10807			16 24 17	−43 55.2	1	8.09		0.17		-0.35		3	B3/5 II/III	567
		HRC 263, V2247 Oph		16 24 18	−24 35.0	1	13.27		1.55		1.02		1		649
148077	−50 10564			16 24 18	−51 09.8	1	8.39		0.05		-0.18		1	B9 IV/V	96
		Ru 119 - 3		16 24 18	−51 23.0	1	10.90		0.38		-0.21		2		251
147884	−67 03119			16 24 18	−67 39.4	1	8.68		0.99		0.68		1	G8 II	565
		NGC 6101 sq1 17		16 24 18	−72 07.3	1	12.76		0.37		0.18		2		89
148287	+2 03106	HR 6124		16 24 19	+02 27.6	2	6.06	.000	0.92	.000	0.61	.000	9	G8 III	770,1088
		Ru 119 - 6		16 24 19	−51 24.5	2	11.05	.020	0.55	.024	-0.39	.002	5		251,1586,1649
		Ru 119 - 7		16 24 20	−51 24.2	1	12.14		0.42		-0.01		2		251
148232	−13 04430	IDS16216S1325	A	16 24 21	−13 31.4	1	9.64		0.38		0.24		1	A1 V	1770
330839	−48 10841	NGC 6134 - 136		16 24 21	−49 01.5	2	11.11	.048	2.06	.029	1.82	.002	5	G5	185,1781
148028	−56 07706			16 24 21	−56 17.2	1	8.63		0.10		-0.35		3	B5 IV/V	1586
148296	+11 02987	V746 Her		16 24 22	+11 06.2	1	6.70		1.61		1.81		1	M1	3040
148199	−29 12551			16 24 22	−29 10.6	2	7.02	.019	0.09	.009	-0.19	.020	7	Ap Si	26,1054
		NGC 6101 sq1 18		16 24 22	−72 05.9	1	12.40		1.08		0.75		2		89
	−51 10205	Ru 119 - 4		16 24 23	−51 21.1	1	11.58		0.35		-0.36		2		251
		Ru 119 - 5		16 24 23	−51 21.4	1	11.97		0.42		0.22		2		251
		Ru 119 - 10		16 24 25	−51 25.3	1	11.92		0.44		0.00		2		251
148105	−50 10568			16 24 27	−50 36.6	1	8.42		0.22		0.11		1	A3 V	96
		NGC 6134 - 201		16 24 28	−49 02.6	1	12.06		1.30		1.02		3		1781
		Ru 119 - 12		16 24 28	−51 24.5	1	12.46		0.51		0.04		2		251
148104	−48 10854			16 24 29	−48 38.1	1	8.69		1.04		0.69		2	G5 II-III	1649
		Ru 119 - 13		16 24 30	−51 23.9	1	12.05		0.54		0.00		2		251
		NGC 6134 - 202		16 24 31	−49 00.7	1	11.62		1.46		1.35		2		1781
148063	−57 08026			16 24 32	−57 47.7	2	8.76	.029	0.00	.000	-0.76	.009	4	B2 II	540,976
		LSS 3613		16 24 33	−49 31.5	1	12.20		0.60		-0.16		2	B3 Ib	1649
		CS 22872 # 60		16 24 34	−04 44.5	1	15.48		0.23		0.26		1		1736
	−50 10570			16 24 35	−50 35.0	1	9.85		0.51		0.02		1	F8	96
148188	−39 10444			16 24 36	−39 50.4	1	9.46		0.75		0.37		2	B9 III	1770
148106	−53 08012			16 24 38	−54 05.5	3	9.04	.027	0.03	.009	-0.62	.005	6	B2/3 II/III	181,540,976
	−24 12689	HRC 264, V2129 Oph		16 24 39	−24 15.4	1	11.40		1.26		0.63		1	K3.5e	649,1763,7003
		ROC # 123		16 24 39	−24 18.5	1	16.13		1.50				1		7003
148434	+41 02707			16 24 41	+40 55.3	1	6.91		1.11		1.05		3	K1 III	196
302744				16 24 41	−57 04.9	1	10.51		0.00		-0.15		2		761
148270	−18 04283	IDS16207S1816	AB	16 24 42	−18 21.9	1	10.29		0.57		0.43		2	A0 III/IV	1771
148141	−50 10572			16 24 43	−50 45.7	1	9.64		0.10		0.02		1	A0 IV/V	96
		LSS 3616		16 24 44	−48 35.7	1	11.28		1.17		0.32		2	B3 Ib	1649
		NGC 6101 sq1 19		16 24 44	−72 06.9	1	10.83		1.65		1.34		2		89
		ROC # 126		16 24 48	−24 19.0	1	15.28		1.37				1		7003
148256	−27 10950			16 24 48	−27 11.0	1	10.04		0.39		0.34		1	A1/2 III	1770
148405	+24 03008			16 24 50	+24 20.8	1	8.96		0.91		0.52		3	K0	1502
148288	−22 04168			16 24 52	−22 29.0	1	9.90		0.65		0.15		1	F2/3 V	1770
	−47 10779			16 24 52	−47 55.3	1	11.18		0.49		-0.14		3		1586
		HRC 265, ROC 128	V	16 24 54	−24 19.6	2	14.23	.265	0.98	.110	-0.53	.005	2		649,7003
		ROC # 127		16 24 54	−24 20.2	1	12.98		1.29		1.57		1		7003
148247	−36 10783	HR 6122		16 24 54	−37 04.1	1	5.79		1.10				4	K1 III	2035
		MN229,227 # 26		16 24 56	−49 14.5	1	10.91		1.27		1.31		1		1649
		LSS 3617		16 24 57	−49 18.1	1	11.06		0.75		-0.26		3	B3 Ib	1649
		CS 22872 # 75		16 24 58	−02 07.1	1	13.91		0.54		0.04		1		1736
148320	−17 04574			16 24 60	−17 25.2	1	9.54		0.73		0.54		1	A0 III/IV	1770
	−55 07561	LSS 3614		16 25 01	−55 44.1	2	10.70	.009	0.21	.013	-0.58	.004	4	B2 II-III	181,540
148109	−59 06747			16 25 01	−59 28.4	1	9.15		0.27		0.21		24	A2mA5-A8	1699
148349	−7 04292	HR 6128, V2105 Oph		16 25 02	−07 29.1	4	5.21	.035	1.73	.011	2.06	.032	12	M2.5 III	31,1088,1311,3055
		ROC # 129		16 25 02	−24 19.9	1	15.27		2.56				1		7003
148491	+41 02708			16 25 03	+41 01.5	1	9.09		0.52		0.01		2	G0	1733
148302	−26 11350			16 25 04	−26 20.6	1	10.01		0.49		0.24		3	A9/F0 V	40
		LP 745 - 76		16 25 05	−12 32.7	1	10.60		0.79		0.38		2		1696
		MCC 765	AB	16 25 06	−03 28.	1	10.81		1.15		1.00		1	M0	497

HD	DM	Other Id	N Rem	α_{1950}	δ_{1950}	S	V	σ_V	B–V	σ_{B-V}	U–B	σ_{U-B}	n	Spectrum	References
148367	−8 04243	HR 6129	⋆ AB	16 25 06	−08 15.7	9	4.63	.007	0.17	.010	0.09	.020	39	A3 m	3,15,355,1075,1088*
		ROC # 132		16 25 08	−24 06.6	1	13.66		1.43		0.88		1		7003
		CS 22872 # 88		16 25 11	−01 25.9	1	14.02		0.16		0.21		1		1736
302743				16 25 11	−57 00.	1	10.81		0.11		-0.49		1		761
148321	−25 11513			16 25 12	−25 20.6	4	6.99	.016	0.19	.007	0.08	.024	11	A1mA4-A9	40,355,1770,8071
		G 17 - 15		16 25 13	−00 57.6	1	14.68		1.32		0.82		2		1620
148260	−44 10888	LSS 3621		16 25 13	−44 55.4	4	7.91	.009	0.42	.012	-0.49	.021	10	B1 Ib	403,540,567,976
		G 17 - 16		16 25 14	−00 57.5	1	9.63		0.72		0.13		2		1620
148408	−0 03119	G 17 - 16	⋆ A	16 25 14	−00 57.5	2	9.63	.004	0.71	.004	0.14	.004	6	G5	516,1064
148259	−44 10889	OZ Nor		16 25 15	−44 42.2	5	7.38	.040	0.13	.011	-0.60	.010	15	B2 II	540,567,976,1586,2012
		HA 156 # 622		16 25 16	−29 51.5	1	11.58		0.81		0.28		8		1499
330806	−48 10868			16 25 16	−48 17.7	1	9.89		0.33				4	B9:III:	1021
148528	+39 02996			16 25 17	+38 57.2	1	8.12		0.35		-0.02		3	F3 V	833
		CS 22872 # 84		16 25 18	−01 06.7	1	13.79		0.15		0.20		1		1736
328274	−45 10684			16 25 20	−45 59.6	1	10.20		0.16		-0.26		3	B5	1586
148334	−26 11352			16 25 21	−26 31.7	1	9.96		0.22		-0.11		3	B9 III/IV	40
		G 138 - 30		16 25 22	+18 24.5	1	13.00		0.93		0.48		1		3060
		HA 156 # 639		16 25 22	−29 53.1	1	11.60		0.54		0.37		8		1499
		LSS 3619		16 25 22	−49 10.1	1	12.09		0.78		0.14		3	B3 Ib	1649
148352	−24 12690			16 25 23	−24 38.4	2	7.51	.004	0.41	.008	-0.02	.007	5	F2 V	40,1770
148335	−32 11733			16 25 23	−32 32.2	1	9.51		0.68		0.33		2	A9 V	1771
		CS 22872 # 65		16 25 24	−04 02.8	1	14.58		0.30		0.30		1		1736
	−51 10226	LSS 3620		16 25 27	−51 13.8	4	9.18	.016	0.54	.014	-0.45	.014	8	B0 Ib	115,403,1649,1737
		CS 22872 # 61		16 25 28	−04 26.9	1	14.87		0.20		0.19		1		1736
	−52 09877	LSS 3618		16 25 28	−52 19.3	3	10.28	.033	0.70	.020	-0.33	.031	6	O7	115,403,1737
		G 138 - 31		16 25 30	+09 19.2	3	16.12	.007	0.37	.007	-0.44	.015	7	DC	538,1663,3060
		ApJ144,259 # 39		16 25 30	+28 01.	1	15.60		-0.33		-1.23		7		1360
148369	−26 11354			16 25 31	−26 34.3	1	9.44		1.65		1.93		5	K3/4 (III)	521
148467	+7 03180	G 17 - 18		16 25 32	+07 25.2	3	8.86	.022	1.24	.019	1.21	.017	4	K5	333,1017,1620,3072
		ROC # 137		16 25 32	−24 12.8	1	14.35		1.87		0.47		1		7003
148493	+21 02925			16 25 33	+20 57.2	1	8.40		0.42		0.13		2		3016
148218	−57 08035	HR 6120		16 25 34	−57 38.8	1	6.06		1.48				4	G8 Ib	2007
148492	+21 02926	IDS16234N2107	AB	16 25 36	+21 00.4	1	8.24		0.60		0.07		3		3026
147675	−78 01103	HR 6102		16 25 43	−78 47.3	2	3.88	.005	0.91	.000	0.62	.000	7	G8/K0 III	15,1075
		HRC 266, V853 Oph		16 25 44	−24 21.7	1	12.83		0.89		-0.75				649
328209	−44 10894	LSS 3624		16 25 45	−44 21.7	3	9.73	.052	0.79	.035	-0.21	.012	9	O9.5Ia:	540,954,1021
148416	−21 04364			16 25 46	−22 06.6	1	9.96		0.56		0.41		2	A3/5 II/II	1770
	−50 10585	LSS 3623		16 25 49	−50 31.6	1	9.26		1.11		0.85		2	F2:II	1649
	−9 04395	V2205 Oph		16 25 51	−09 13.0	2	10.57	.017	0.06	.009	-0.83	.009	6		166,272
148438	−17 04580			16 25 51	−17 52.5	1	7.17		0.30		0.18		1	A0 IV/V	1770
148308	−52 09884			16 25 53	−52 17.8	1	8.08		1.48		1.68		8	K2/3 III	1673
	+17 03030			16 25 54	+17 35.5	1	9.20		0.41		0.00		2	F0	1648
148496	−0 03122			16 25 57	−00 41.6	1	9.02		0.16		0.09		2	A0	1371
		LP 505 - 25		16 25 58	+10 42.4	1	15.00		1.36		1.06		1		1696
148428	−24 12691			16 25 58	−24 25.5	1	7.54		0.72		0.35		8	K0 III	1763
		CS 22872 # 105		16 25 59	−03 25.9	1	15.61		-0.13		-1.02		1		1736
	−52 09886			16 26 00	−53 05.3	1	10.89		0.20		-0.29		1		115
148513	+0 03529	HR 6136		16 26 01	+00 46.5	4	5.38	.005	1.46	.004	1.79	.009	10	K4 IIIp	15,37,1417,3077
148469	−16 04288			16 26 03	−17 03.6	1	9.84		0.59		0.42		1	A1 III	1770
148177	−66 02968			16 26 03	−66 36.3	1	8.40		1.31		1.31		1	K1 Ib/II	565
148379	−45 10697	HR 6131, QU Nor		16 26 04	−46 08.1	14	5.34	.019	0.55	.018	-0.46	.025	53	B1.5 Iape	26,164,540,681,815,976*
148530	+3 03203	G 17 - 19		16 26 06	+03 22.2	3	8.82	.005	0.77	.004	0.38	.009	5	G9	333,1003,1620,3008
148556	+15 03008			16 26 07	+15 32.4	1	7.32		0.99		0.79		2	K0	1648
148515	−7 04299	HR 6137	⋆ AB	16 26 07	−08 01.1	4	6.49	.020	0.41	.012	-0.01	.004	15	F2 V	253,254,938,1088
	−51 10235			16 26 11	−51 24.0	1	10.46		0.43		0.00		1	B9	115
	−52 09888			16 26 11	−53 09.7	1	10.79		0.18		-0.35		1		115
		CS 22872 # 85		16 26 13	−01 06.9	1	15.56		0.12		0.14		1		1736
148455	−30 13185			16 26 13	−30 41.6	1	9.62		0.55		0.26		1	A8 V	1770
148442	−33 11201			16 26 14	−34 07.5	1	10.12		0.19		0.10		1	B9.5 V	1770
328222	−44 10899			16 26 14	−44 47.5	1	10.27		0.69		-0.21		2	B1.5III	540
330842	−49 10720			16 26 16	−49 12.1	1	10.35		0.14		-0.37		2	B3	1081
148456	−32 11745			16 26 18	−32 18.2	1	9.65		0.18		0.09		1	A0 IV	1770
148478	−26 11359	HR 6134, α Sco	⋆ A	16 26 20	−26 19.4	9	0.99	.072	1.83	.014	1.26	.058	87	M1.5Iab-Ib+	1,15,26,1024,1088*
		CS 22872 # 95		16 26 22	−02 11.5	1	15.15		0.23		0.26		1		1736
148516	−21 04366			16 26 22	−21 27.6	1	8.31		0.49		0.00		1	F5 V	1770
148419	−44 10903			16 26 22	−44 54.2	2	9.50	.008	0.51	.017	-0.39	.012	5	F8	540,567
148311	−59 06756			16 26 22	−59 50.5	1	7.61		0.40		0.13		26	F0 III/IV	1699
148291	−61 05701	HR 6125		16 26 22	−61 31.5	1	5.20		1.23				4	K0 II/III	2006
148457	−37 10813			16 26 24	−37 11.9	1	10.28		0.50		0.36		3	B9 III	1770
148499	−27 10962			16 26 31	−27 27.7	2	9.84	.016	0.38	.005	0.12	.060	3	B9 III	1054,1770
148667	+29 02834			16 26 32	+29 11.1	1	7.26		0.09		0.08		3	A2	1625
		G 180 - 56		16 26 33	+32 04.1	1	14.08		1.05				1		1759
	−51 10241			16 26 33	−51 36.7	1	10.57		0.32		-0.32		1	B9	115
		CS 22872 # 87		16 26 34	−01 22.7	1	14.04		0.25		0.23		1		1736
		Steph 1343	AB	16 26 39	+05 20.9	1	11.26		1.41		1.18		1	M0	1746
		G 180 - 57		16 26 39	+36 52.4	4	13.84	.032	0.17	.026	-0.66	.016	16		419,974,1663,3028
148653	+18 03182	G 138 - 34	⋆ AB	16 26 41	+18 31.1	5	7.01	.019	0.85	.009	0.45	.006	16	K3 V +K3 V	938,1080,1197,1381,3060
148534	−29 12576			16 26 41	−29 57.6	1	9.01		0.23		0.17		1	A1 V	1770
148533	−28 12146			16 26 42	−28 57.4	1	9.96		0.54		0.02		1	F5 V	1770
		CS 22872 # 102		16 26 44	−02 58.4	1	13.65		0.59		-0.10		1		1736

Table 1   891

HD	DM	Other Id	N Rem	α1950	δ1950	S	V	σV	B-V	σB-V	U-B	σU-B	n	Spectrum	References
		G 180 - 58		16 26 45	+44 47.6	3	11.32	.010	0.69	.010	-0.06	.014	3		1064,1658,1759
148801	+51 02105	IDS16255N5149	AB	16 26 45	+51 42.1	1	7.30		1.02		0.82		2	K0	1566
148576	-17 04585			16 26 46	-18 07.9	1	8.93		0.38		0.28		1	A1 V	1770
		CS 22872 # 96		16 26 48	-02 25.0	1	13.99		0.26		0.29		1		1736
148593	-14 04431			16 26 49	-14 28.6	1	9.12		0.55		0.32		1	Ap SrEuCr	1770
	+40 03020			16 26 50	+39 53.1	1	8.79		0.84		0.53		2	F8	1601
		CS 22872 # 108		16 26 51	-04 53.3	1	13.20		0.07		-0.42		1		1736
148563	-26 11363			16 26 51	-26 28.9	2	8.62	.005	0.24	.002	0.14	.015	4	A1 V	40,1770
148562	-24 12693			16 26 52	-24 52.2	2	7.81	.08	0.18	.006	0.09	.005	4	A2 V	40,1770
148422	-56 07729	LSS 3628		16 26 53	-56 23.2	4	8.64	.016	0.11	.018	-0.77	.019	9	B1 Ia	403,540,976,8100
148537	-38 11011			16 26 56	-38 17.2	3	9.43	.010	0.56	.007	-0.05	.022	8	B9 Ib	540,976,1770
148604	-14 04433	HR 6140		16 26 57	-14 26.6	1	5.68		0.82				4	G5 III/IV	2007
148579	-24 12694			16 26 57	-25 02.4	3	7.33	.010	0.27	.008	-0.02	.008	11	B9 V	26,1054,1770
		L 411 - 46		16 26 59	-41 58.9	1	14.70		1.25		0.90		2		3062
148783	+42 02714	HR 6146, G Her		16 27 00	+41 59.4	8	5.00	.099	1.54	.021	1.19	.040	22	M6 III	15,814,1363,3053,8015*
		CS 22872 # 113		16 27 01	-05 33.1	1	12.31		0.50		0.41		1		1736
	-52 09912			16 27 01	-52 22.2	1	11.31		0.19		-0.15		1		115
148546	-37 10817	LSS 3631		16 27 02	-37 51.9	7	7.71	.011	0.29	.012	-0.70	.017	21	B0 Ia	138,400,403,540,567*
148683	+10 03012	IDS16247N1049	A	16 27 03	+10 42.0	1	7.60		0.93		0.61		3	G5 III	3016
148683	+10 03012	IDS16247N1049	B	16 27 03	+10 42.0	1	9.60		0.46		0.06		3		3016
148484	-50 10608			16 27 04	-50 51.8	1	7.42		0.38		0.00		1	F0/2 IV/V	96
		CS 22872 # 100		16 27 05	-02 45.1	1	13.83		0.35		0.26		1		1736
		Ara sq 1 # 2		16 27 06	-49 17.	1	11.37		0.54		0.02		2		1081
		BPM 24203		16 27 06	-53 31.	1	15.39		0.60		-0.08		1		3065
148538	-44 10912			16 27 07	-44 42.5	1	9.29		0.36		-0.40		3	B1 V	567
148594	-27 10967			16 27 09	-27 48.5	3	6.89	.009	0.10	.002	-0.25	.006	10	B8 V	26,1054,1770
148605	-24 12695	HR 6141		16 27 10	-25 00.4	8	4.78	.007	-0.13	.018	-0.73	.022	24	B3 V	15,26,1054,1075,1586,1770*
		CS 22872 # 101		16 27 11	-02 49.5	1	14.11		0.54		0.26		1		1736
	+49 02510			16 27 13	+49 31.5	1	9.12		0.98		0.72		2	K0	1733
	+19 03109			16 27 14	+19 36.0	2	10.34	.052	1.56	.080	1.28		2	R2	1238,8005
148506	-51 10255			16 27 16	-51 33.4	1	8.08		0.42		0.24		2	A2 II:	1649
148624	-26 11368			16 27 17	-26 20.2	1	10.38		0.40		0.23		3	A7/8 III	40
148473	-55 07569			16 27 18	-55 50.0	2	7.92	.038	0.09	.012	-0.32	.011	4	B8 II	540,976
		CS 22872 # 86		16 27 20	-01 08.4	1	13.53		0.31		0.16		1		1736
148711	+6 03236			16 27 24	+06 04.7	1	6.95		0.42		0.11		4	F3 II	8100
	+60 01681			16 27 24	+59 53.4	1	9.90		0.47		0.02		2	F8	1502
		CS 22872 # 99		16 27 24	-02 45.4	1	13.14		0.25		0.27		1		1736
		G 153 - 57		16 27 24	-14 33.3	2	12.36	.023	1.50	.029	1.43		5		3062,7009
148567	-46 10799	LSS 3632	★ABC	16 27 24	-46 22.3	3	7.79	.064	0.29	.037	-0.56	.025	7	B2 (II)ne	540,976,1586
148880	+51 02106	HR 6150	★AB	16 27 27	+51 31.0	1	6.29		1.05				2	G9 III	71
	-12 04523			16 27 31	-12 32.3	7	10.10	.016	1.59	.016	1.18	.017	31	M4	158,1006,1197,1494,1764*
		CS 22872 # 106		16 27 33	-04 13.6	1	13.77		0.50		0.23		1		1736
148655	-29 12590			16 27 33	-29 22.5	1	7.78		0.31		0.09		1	A2/3mA7-F2	1770
	+2 03114			16 27 36	+02 22.2	2	9.38	.046	1.27	.005	1.40	.005	2	K2	1697,1748
		L 339 - 85		16 27 36	-48 18.	1	13.32		0.82		0.14		2		3062
	-49 10743			16 27 37	-49 17.9	1	11.38		0.59		0.42		2		1081
		CS 22872 # 112		16 27 38	-05 31.7	1	13.29		0.31		0.25		1		1736
	+2 03115			16 27 41	+02 04.5	2	9.40	.015	1.12	.006	1.03	.000	2	K0	1697,1748
148701	-21 04377			16 27 42	-21 21.0	1	10.29		0.55		0.23		2	A3/7 II/II	1770
148671	-28 12152	IDS16246S2807	A	16 27 42	-28 13.7	1	10.19		1.02		0.90		1	K4 V	3072
148671	-28 12152	IDS16246S2807	B	16 27 42	-28 13.7	1	10.23		1.04		0.93		1	K3 V	3072
148549	-54 07736	IDS16237S5451	AB	16 27 44	-54 57.5	2	8.79	.036	0.14	.013	-0.39	.012	4	B5/6 III	540,976
		CS 22872 # 98		16 27 47	-02 28.5	1	15.76		0.24		0.23		1		1736
148610	-48 10900			16 27 47	-49 01.4	2	8.22	.000	-0.03	.010	-0.59	.005	4	B2/3 IV	96,1081
		AO 740		16 27 48	+02 08.7	2	11.59	.007	1.05	.007	0.97	.025	2		1697,1748
148743	-7 04305	HR 6144	★	16 27 48	-07 24.4	3	6.49	.012	0.37	.013	0.22	.106	12	A7 Ib	1088,3016,8100
148657	-38 11017			16 27 49	-38 16.5	1	8.56		0.28		0.17		3	A0 V	434
	-46 10801			16 27 49	-46 40.1	1	11.36		0.78		0.04		3		1586
	+31 02856	G 169 - 13		16 27 50	+30 48.4	1	10.24		0.61		0.03		3	G2	1658
330905	-47 10825			16 28 00	-47 41.6	1	10.77		0.98		0.12		2	G0	1649
148816	+4 03195	G 17 - 21		16 28 01	+04 18.3	8	7.28	.010	0.54	.006	-0.07	.012	18	F8 V	22,792,908,1003,1509*
		HRC 267		16 28 03	-23 58.1	1	14.09		1.38		0.19		1	M0	1763
		HRC 267, V2252 Oph		16 28 03	-23 58.1	1	12.29		1.46		1.22		1		649
148856	+21 02934	HR 6148	★A	16 28 04	+21 35.8	8	2.78	.015	0.93	.010	0.69	.017	20	G7 IIIa	15,1080,1194,1363*
149212	+69 00850	HR 6161		16 28 04	+68 52.6	6	4.96	.014	-0.05	.004	-0.11	.012	26	A0 III	15,1079,1363,3023*
		HBC 644		16 28 04	-23 58.2	1	12.28		1.40		0.78		1	K3	1763
148817	+0 03536			16 28 06	-00 02.7	3	8.34	.005	1.49	.002	1.79	.020	34	K2	989,1371,1729
	-3 03952			16 28 06	-03 52.9	1	9.58		1.25		1.22		1	K8	1746
148804	-3 03951			16 28 06	-03 52.9	1	9.31		0.68		0.17		3	F8	1064
148703	-34 11044	HR 6143		16 28 07	-34 35.8	8	4.23	.008	-0.17	.011	-0.80	.021	28	B2 III-IV	15,26,1075,1586,1770,2012*
		WLS 1620-10 # 8		16 28 08	-10 14.0	1	13.12		0.81		0.16		2		1375
148717	-31 12947			16 28 08	-31 36.5	1	10.32		0.31		0.20		1	A0/1 V	1770
148704	-38 11019	V716 Oph		16 28 08	-38 54.1	3	7.25	.016	0.86	.005	0.38	.117	10	K0 V	285,2012,3008
				16 28 09	-05 23.7	1	11.28		0.39		0.31		1	F2	597
	-18 04289			16 28 09	-18 30.8	1	10.81		1.16		1.06		1	K5 V	3072
	-29 12595			16 28 09	-29 50.5	1	10.17		0.58		0.04		2	F8	1730
	-49 10752	LSS 3634		16 28 10	-49 20.2	1	10.90		0.50		-0.38		3		1081
148731	-33 11223			16 28 11	-33 20.4	1	8.85		0.41		0.25		1	A5 V	1770
	-3 03954	G 153 - 60		16 28 12	-03 56.9	1	10.51		0.74		0.20		3		1696
149198	+67 00942			16 28 13	+67 09.1	1	6.57		1.58		1.69		1	M1	3025

HD	DM	Other Id	N	Rem	α_{1950}	δ_{1950}	S	V	σ_V	B–V	σ_{B-V}	U–B	σ_{U-B}	n	Spectrum	References
148688	−41 10695	HR 6142, LSS 3636		★AB	16 28 13	−41 42.6	12	5.32	.015	0.35	.015	-0.65	.020	45	B1 Iae	15,567,681,815,954,1018*
330938	−48 10907	LSS 3635			16 28 14	−48 52.4	2	10.44	.005	0.52	.000	-0.32		6	B3 V	1021,1081
	+10 03017				16 28 15	+09 50.3	1	8.82		1.47		1.72		3	K5	1733
		CS 22872 # 111			16 28 15	−04 56.1	1	11.38		0.38		0.33		1		1736
148786	−16 04298	HR 6147		★A	16 28 16	−16 30.3	8	4.27	.009	0.92	.006	0.71	.006	29	G8 IIIa	3,15,444,1075,1088*
		CS 22872 # 94			16 28 17	−02 10.9	1	14.38		0.48		0.15		1		1736
148760	−26 11379	HR 6145			16 28 18	−26 25.8	1	6.10		1.08				4	K1 III	2006
		AO 741			16 28 20	+01 48.7	2	10.68	.009	0.40	.007	0.01	.016	2	F0 V	1697,1748
		LP 565 - 23			16 28 20	+07 09.0	1	13.11		1.00		0.46		1		1696
		CS 22872 # 92			16 28 20	−01 34.2	1	16.07		0.07		0.01		1		1736
148614	−58 06825				16 28 21	−58 58.7	1	10.30		-0.04		-0.73		4	B1/2 Vn	400
148857	+2 03118	HR 6149		★AB	16 28 23	+02 05.5	6	3.82	.012	0.01	.009	0.03	.021	20	A0 V +A4 V	15,938,1256,1363,3023,8015
	+2 03117				16 28 23	+02 17.6	2	9.42	.019	1.55	.003	2.00	.011	2	K2	1697,1748
	+12 03027				16 28 23	+12 14.0	1	10.28		0.39				1		1213
148897	+20 03283	HR 6152			16 28 23	+20 35.2	3	5.24	.005	1.28	.024	1.19	.021	7	G8 IIp	1080,3077,8100
		G 153 - 61			16 28 23	−11 13.2	2	11.59	.000	1.11	.019	0.99	.019	4		1773,3062
	−46 10808				16 28 23	−46 18.3	1	11.65		0.63		0.01		1		1586
148833	−7 04308				16 28 24	−07 26.0	1	7.88		0.14		0.01		2	A0	1375
		AAS35,161 # 135			16 28 24	−44 24.9	1	14.56		1.61		0.69		1		5005
		CS 22872 # 97			16 28 26	−02 18.8	1	15.31		0.22		0.21		1		1736
148768	−29 12599				16 28 27	−29 56.2	1	10.66		0.43		0.26		1	A8/9 V	1770
	+18 03186	VX Her			16 28 28	+18 28.1	2	9.91	.023	0.10	.011	0.09	.012	2	A3	668,699
148911	+18 03187				16 28 28	+18 44.5	1	7.93		0.41		-0.06		1	F2	289
149681	+79 00498	HR 6173			16 28 28	+79 04.3	2	5.56	.000	0.25	.010	0.06	.010	5	F0 V	1733,3016
		G 153 - 62			16 28 28	−11 11.6	2	12.05	.018	1.23	.018	1.17	.045	5		1773,3062
		NGC 6171 sq2 35B			16 28 29	−12 57.6	1	12.35		0.72		0.49		1		519
		NGC 6171 sq2 35C			16 28 30	−12 56.0	1	12.70		1.93		1.83		1		519
148587	−63 03955				16 28 30	−63 44.0	2	7.39	.005	0.58	.000	0.01		5	G1 V	2012,3037
		HRC 268		V	16 28 32	−24 21.2	2	12.57	.025	1.35	.020	0.48	.015	2	K3	649,1763
	+2 03119				16 28 33	+01 52.6	2	9.69	.007	0.50	.009	0.07	.018	2	F2	1697,1748
		CS 22872 # 91			16 28 33	−01 28.0	1	14.39		0.19		0.25		1		1736
148788	−29 12601				16 28 33	−29 46.9	1	8.24		1.11		0.92		2	K1 III	1730
		NGC 6171 sq2 35A			16 28 38	−12 57.5	1	13.49		1.47		1.20		1		519
149222	+65 01122				16 28 40	+64 53.6	1	7.70		0.58		0.09		6	G0	3026
		NGC 6171 sq2 35			16 28 40	−12 55.1	1	12.20		0.86		0.27		1		519
	−39 10467	LSS 3639			16 28 41	−39 24.4	1	10.93		0.84		-0.18		3	B1.5 Ia	1737
148807	−31 12958				16 28 42	−31 32.4	1	9.02		0.19		-0.13		1	B8/9 III	1770
		He3 1205			16 28 42	−50 48.	1	12.35		0.36		0.13		3		1586
149081	+49 02514	HR 6156			16 28 43	+49 04.1	1	6.45		0.03		0.00		3	A1 V	3050
		LSS 3637			16 28 43	−50 48.4	1	11.57		0.74		-0.15		2	B3 Ib	1649
		CS 22872 # 103			16 28 44	−03 22.9	1	14.87		0.24		0.25		1		1736
148860	−17 04591				16 28 45	−17 36.4	3	8.04	.007	0.16	.003	-0.05	.012	7	B9.5III	26,1054,1770
148822	−26 11383				16 28 45	−26 30.4	1	9.65		0.48		0.14		3	F3 V	40
		RZ Nor			16 28 45	−53 10.8	1	11.10		1.24		0.77		1		1589
		NGC 6171 sq2 38			16 28 46	−13 08.6	1	10.80		0.87		0.58		2		519
148488	−70 02256	HR 6135			16 28 46	−70 53.0	3	5.49	.004	1.23	.010	1.24	.000	13	K1 III	15,1075,2038
		G 169 - 15	A		16 28 48	+27 25.5	1	12.00		1.49		1.24		8	M1	7010
		G 169 - 15	B		16 28 48	+27 25.5	1	13.86		0.90		0.80		4		7010
		BPM 10197			16 28 48	−63 21.	1	15.00		0.42		-0.27		1		3065
148912	+1 03246				16 28 49	+01 24.9	1	6.78		1.39		1.52		1	K0	116
149105	+48 02407				16 28 50	+48 03.8	3	7.00	.020	0.57	.005	0.07	.009	7	G0 V	1003,1733,3016
148720	−52 09965	LSS 3638			16 28 51	−52 16.0	2	7.97	.006	0.15	.002	-0.65	.004	5	B2 Ib/II	540,976
		NGC 6171 sq2 14			16 28 52	−12 58.4	1	13.71		1.77				2		519
		NGC 6171 sq2 15			16 28 52	−13 00.6	1	12.81		0.97		0.45		3		519
		NGC 6171 sq2 37			16 28 52	−13 07.4	1	11.88		1.51		1.49		2		519
148861	−21 04380				16 28 52	−21 25.4	1	9.99		0.39		0.26		2	B9/A0 IV	1770
148842	−26 11386				16 28 53	−26 33.3	1	10.62		0.26		0.14		2	F0 IV/V	40
		NGC 6171 sq2 23			16 28 55	−13 05.1	1	12.93		0.77		0.22		3		519
		CS 22878 # 17			16 28 56	+11 07.2	1	14.87		0.47		-0.08		1		1736
		NGC 6171 sq2 10			16 28 56	−12 53.1	1	13.35		1.39		1.23		4		519
		NGC 6171 sq2 36			16 28 56	−13 08.4	1	11.23		0.76		0.24		2		519
	+12 03028	DY Her			16 28 57	+12 06.8	1	10.15		0.26		0.15		1	A7 III	668
	−52 09969				16 28 57	−52 28.6	1	10.65		0.23		-0.06		2	B8 V	540
		NGC 6171 sq2 21			16 29 01	−13 04.2	1	12.93		0.81		0.20		1		519
148771	−49 10761				16 29 02	−49 29.5	1	9.32		0.12		-0.14		1	B8/9 V	96
	+2 03122				16 29 03	+02 05.5	2	10.51	.014	1.11	.005	1.01	.012	2	G8 III	1697,1748
148872	−28 12174				16 29 03	−28 59.8	1	10.31		0.24		0.16		2	A0 III/IV	1770
		CS 22872 # 107			16 29 04	−04 21.6	1	13.82		0.37		0.30		1		1736
149009	+22 02983	HR 6154			16 29 05	+22 18.1	2	5.77	.004	1.61	.006	1.95		5	K7 III	71,1501
148979	+8 03215	IDS16267N0830	A		16 29 06	+08 24.0	1	6.98		0.91		0.61		2	G5	3016
148931	−6 04446	IDS16264S0650	AB		16 29 07	−06 55.0	1	7.41		0.18				5	A0	173
148943	−1 03206	IDS16266S0203	AB		16 29 09	−02 09.2	2	8.39	.019	0.63	.002	0.14		8	F5	173,3016
		NGC 6171 sq2 11			16 29 09	−12 53.4	1	14.63		1.03		0.58		2		519
148998	+8 03216	IDS16267N0830	B		16 29 10	+08 24.3	1	8.28		0.59		0.08		2	G5	3016
		NGC 6171 sq2 24			16 29 10	−13 07.1	1	11.57		1.36		1.26		4		519
148898	−21 04381	HR 6153, ω Oph			16 29 10	−21 21.7	9	4.45	.006	0.13	.007	0.16	.036	213	Ap SrEuCr	3,15,26,1075,1202,1770*
		CS 22872 # 139			16 29 12	−01 26.1	1	14.58		0.11		0.10		1		1736
149084	+35 02828	HR 6157			16 29 13	+35 19.9	2	6.25	.005	1.64	.005	1.87	.020	5	K5	985,1733
	−49 10764	LSS 3640			16 29 15	−49 26.5	5	9.86	.025	0.81	.032	-0.05	.027	8	B3 IV	96,954,1081,1649,1737
148899	−28 12178				16 29 16	−28 56.2	1	8.12		0.21		0.18		2	A0 V	1770

Table 1   893

HD	DM	Other Id	N Rem	α_{1950}	δ_{1950}	S	V	σ_V	B−V	σ_{B-V}	U−B	σ_{U-B}	n	Spectrum	References
	+12 03030			16 29 17	+12 26.3	3	9.37	.005	0.61	.006	0.11	.001	18	G0	830,1230,1783
149028	+12 03031			16 29 17	+12 31.6	1	8.52		0.72		0.33		1		1230
		NGC 6171 sq2 25		16 29 17	−13 06.6	1	11.90		0.80		0.28		4		519
		NGC 6171 sq2 39		16 29 20	−13 09.3	1	10.12		1.20		0.98		2		519
		G 138 - 37		16 29 22	+17 40.7	2	12.78	.019	1.57	.010	1.16		6		203,1759
	+19 03113	IDS16272N1930	A	16 29 23	+19 23.8	1	10.31		0.76		0.25		3	K0	3016
	+19 03113	IDS16272N1930	B	16 29 23	+19 23.8	1	10.45		0.78		0.33		3		3016
		AO 746		16 29 25	+01 48.3	2	10.75	.008	0.99	.005	0.69	.002	2	K1 V	1697,1748
		NGC 6171 sq2 20		16 29 25	−13 02.6	1	13.41		1.30		1.07		2		519
148914	−25 11530			16 29 25	−26 02.0	1	10.42		0.58				10	K0/2	1655
		NGC 6171 sq1 10		16 29 26	−12 54.8	1	13.97		1.55		1.43		1		449
148837	−49 10765			16 29 26	−49 29.0	1	7.81		1.82		1.95		1	M1/2 III	96
		NGC 6171 sq1 1		16 29 27	−12 55.7	1	10.63		1.48		1.54		1		449
		NGC 6171 sq1 13		16 29 27	−12 59.1	2	14.23	.005	1.07	.005	0.70	.005	4		449,519
		NGC 6171 sq1 4		16 29 28	−12 58.5	2	12.90	.014	1.23	.005	0.75	.000	4		449,519
		NGC 6171 sq2 26		16 29 29	−13 05.7	1	11.83		0.76		0.27		2		519
		AAS35,161 # 136	V	16 29 29	−44 51.0	1	14.77		1.22		-0.01		1		5005
148969	−17 04594			16 29 30	−17 53.8	1	9.78		0.62		0.09		1	A8/9 (IV)	1770
	+2 03124			16 29 31	+02 06.4	2	10.07	.008	1.20	.009	1.27	.003	2	K2 III	1697,1748
	+19 03115	IDS16273N1931	AB	16 29 31	+19 24.2	1	9.58		0.77		0.30		3		3032
	+19 03115	IDS16273N1931	C	16 29 31	+19 24.2	1	12.30		0.58		0.02		2	K0	3016
		WLS 1612 60 # 8		16 29 32	+59 50.0	1	13.52		0.81		0.55		2		1375
		CS 22878 # 13		16 29 33	+09 15.2	1	13.27		0.47		-0.13		1		1736
148851	−48 10932			16 29 33	−48 19.9	1	8.72		-0.03		-0.58		4	B5 IV	1081
	−50 10640	LSS 3641		16 29 33	−50 10.7	1	10.76		0.75		0.01		2	B2 V	1649
148947	−28 12182			16 29 34	−29 08.9	1	10.12		0.47		0.31		1	A5/7 II	1770
		LP 745 - 25		16 29 36	−11 34.1	1	11.91		1.05		0.93		1		1696
149141	+33 02742	IDS16278N3344	AB	16 29 38	+33 37.2	1	7.11		0.08		0.08		2	A0	3016
148917	−38 11026			16 29 38	−38 22.7	1	10.27		0.48		0.37		1	B9/A0 V	1770
148887	−41 10709			16 29 38	−41 53.5	3	9.40	.018	0.30	.004	-0.37	.003	8	B4 III	540,976,976
		G 180 - 60		16 29 39	+40 58.0	1	14.84		1.72				1		906
148949	−30 13232			16 29 40	−30 46.2	1	8.93		0.23		0.17		1	A1 IV	1770
149132	+29 02844			16 29 41	+29 42.7	1	8.03		1.13		1.27		5	K2 II	8100
		NGC 6171 sq1 2		16 29 41	−12 59.5	2	11.70	.010	0.76	.015	0.18	.000	11		449,519
		NGC 6171 - 1		16 29 42	−12 57.	1	15.44		0.49		0.32		1		519
		NGC 6171 - 4		16 29 42	−12 57.	1	15.50		0.77		0.30		2		519
		NGC 6171 - 5		16 29 42	−12 57.	1	15.19		1.30		1.01		1		519
		NGC 6171 - 14		16 29 42	−12 57.	1	15.66		0.94		0.44		1		519
		NGC 6171 - 18		16 29 42	−12 57.	1	16.20		1.20				1		519
		NGC 6171 - 43		16 29 42	−12 57.	1	15.44		1.32		1.03		2		519
		NGC 6171 - 45		16 29 42	−12 57.	1	15.88		0.82				1		519
		NGC 6171 - 48		16 29 42	−12 57.	1	15.54		0.97		0.36		1		519
		NGC 6171 - 49		16 29 42	−12 57.	1	16.33		1.16		0.68		1		519
		NGC 6171 - 73		16 29 42	−12 57.	1	15.92		1.19		0.26		1		519
		NGC 6171 - 98		16 29 42	−12 57.	1	15.55		0.78		0.30		1		519
		NGC 6171 - 99		16 29 42	−12 57.	1	15.85		1.02		0.44		1		519
		NGC 6171 - 102		16 29 42	−12 57.	1	16.41		1.02				1		519
		NGC 6171 - 105		16 29 42	−12 57.	1	16.61		0.92		0.35		1		519
		NGC 6171 - 106		16 29 42	−12 57.	1	15.57		0.83		0.26		1		519
		NGC 6171 - 130		16 29 42	−12 57.	1	16.22		1.00		0.32		1		519
		NGC 6171 - 200		16 29 42	−12 57.	1	14.45		0.92		0.35		1		519
		NGC 6171 - 202		16 29 42	−12 57.	1	15.67		0.57		0.41		1		519
		NGC 6171 - 204		16 29 42	−12 57.	1	15.65		1.06		0.40		1		519
		NGC 6171 - 205		16 29 42	−12 57.	1	14.58		1.46		1.28		1		519
		NGC 6171 - 206		16 29 42	−12 57.	1	15.67		1.04		0.48		1		519
		NGC 6171 - 217	V	16 29 42	−12 57.	1	13.93		1.79		1.56		1		1468
		NGC 6171 - 273		16 29 42	−12 57.	1	13.18		1.83		1.98		1		1468
		NGC 6171 - 281		16 29 42	−12 57.	1	15.71		1.01		0.46		1		519
		NGC 6171 - 283		16 29 42	−12 57.	1	16.98		1.22		0.70		1		519
		NGC 6171 - 284		16 29 42	−12 57.	1	18.37		1.05				1		519
		NGC 6171 - 285		16 29 42	−12 57.	1	18.50		1.31				1		519
		NGC 6171 - 286		16 29 42	−12 57.	1	18.84		1.00				1		519
		NGC 6171 - 287		16 29 42	−12 57.	1	18.28		1.17				1		519
		NGC 6167 - 1151		16 29 43	−49 44.9	1	11.39		0.35		0.24		1		302
		G 153 - 63		16 29 44	−08 57.2	1	13.34		1.01		0.22		2		3062
		CS 22872 # 109		16 29 47	−04 47.5	1	15.15		0.39		0.36		1		1736
		G 153 - 64		16 29 48	−08 27.1	1	11.39		0.69		0.02		2		1696
148877	−50 10644			16 29 48	−51 09.5	3	8.50	.018	0.25	.007	-0.49	.018	9	B5/7 V	540,976,1586
		L 266 - 129		16 29 48	−52 53.	1	13.69		1.63				1		3062
		LP 445 - 30		16 29 49	+19 57.3	1	15.72		1.32				1		1759
		NGC 6171 sq2 2		16 29 49	−12 50.6	1	11.22		1.44		1.42		3		519
148740	−65 03324			16 29 49	−65 54.6	2	7.34	.000	-0.09	.005	-0.53		8	B5 III	400,2012
		NGC 6171 sq2 1		16 29 50	−12 47.4	1	10.91		1.87		1.98		3		519
148878	−52 09998	LSS 3642		16 29 52	−53 06.3	4	7.70	.015	0.39	.019	-0.49	.027	8	B1/2 Ia	115,403,540,976
		CS 22872 # 127		16 29 53	−02 45.3	1	15.38		0.19		0.09		1		1736
		NGC 6171 sq2 28		16 29 53	−13 03.0	1	13.20		1.06		0.83		3		519
148919	−45 10738			16 29 54	−45 40.8	1	8.80		1.19				4	G0/1 V	2012
	−62 05377	RT Tra		16 29 54	−63 02.0	1	9.62		0.74				1	G2 I	688
330944	−48 10940			16 29 55	−49 09.7	1	9.47		0.46		-0.03		1		96
		NGC 6171 sq1 3		16 29 56	−12 57.7	1	11.78		1.18		0.74		2		449

HD	DM	Other Id	N Rem	α_{1950}	δ_{1950}	S	V	σ_V	B–V	σ_{B-V}	U–B	σ_{U-B}	n	Spectrum	References
		NGC 6171 sq2 29		16 29 56	−13 04.0	1	14.48		1.00		0.43		1		519
		LSS 3644		16 29 56	−48 51.7	1	11.81		0.62		-0.27		2	B3 Ib	1649
149032	−20 04509			16 29 59	−20 13.3	1	9.75		0.46		0.29		2	A1 III/IV	1770
148922	−48 10941	LSS 3645		16 30 00	−49 04.1	4	7.68	.012	0.66	.012	0.45	.091	7	A4 Ib	96,954,1649,8100
		BSD 108 # 409		16 30 02	+00 46.7	1	10.76		1.41		1.44		1	dK3	116
149034	−28 12194			16 30 02	−28 36.7	1	10.53		0.53		0.34		2	A7/9 (II)	1770
149015	−31 12984			16 30 07	−31 59.7	1	9.29		0.08		-0.34		1	B3/5 V	1770
149121	+5 03223	HR 6158		16 30 08	+05 37.6	3	5.64	.012	-0.05	.004	-0.15	.023	16	B9.5III	3,1079,1088
		G 138 - 38		16 30 08	+08 57.8	1	14.29		0.97		0.73		1		3060
		NGC 6167 - 1133		16 30 09	−49 35.8	1	13.07		0.93		0.37		1		302
148937	−47 10855	LSS 3646	⋆ AB	16 30 10	−48 00.4	9	6.73	.010	0.34	.015	-0.66	.015	20	O6	26,403,540,914,976,1081*
	−12 04542	G 153 - 65		16 30 11	−12 29.0	2	10.60	.005	1.36	.017	1.30		5	M0	158,1705
		NGC 6171 sq2 31		16 30 11	−12 57.0	1	13.97		1.39		1.22		2		519
		NGC 6167 - 1137		16 30 12	−49 38.3	1	12.75		0.59		0.06		1		302
325729	−42 11365			16 30 13	−42 19.1	2	10.25	.004	0.48	.017	-0.20	.011	7	B3 IV	540,567
	−1 03209			16 30 15	−01 40.4	1	10.28		1.63		2.03		3	K5 V	3072
		NGC 6167 - 1138		16 30 15	−49 38.6	1	11.79		1.38		0.30		1		302
149161	+11 03008	HR 6159		16 30 16	+11 35.6	10	4.83	.011	1.49	.004	1.83	.013	40	K4 III	3,15,30,369,1003,1080*
	−49 10772	LSS 3647		16 30 16	−49 19.8	1	10.52		0.86		-0.11		2	O9	1649
		NGC 6167 - 1139		16 30 16	−49 39.7	1	12.84		0.85		0.28		1		302
149303	+45 02422	HR 6162	⋆ A	16 30 17	+45 42.2	2	5.64	.025	0.11	.005	0.08		4	A4 Vn	1733,3024
149303	+45 02422	IDS16288N4549	B	16 30 17	+45 42.2		8.89		0.61				4		3024
		NGC 6167 - 1141		16 30 17	−49 40.7	1	12.57		0.42		0.19		1		302
	+12 03035			16 30 19	+11 59.5	1	8.88		0.44		-0.06		2	F5	1375
		CS 22872 # 114		16 30 19	−05 30.3	1	15.09		0.32		0.19		1		1736
	−49 10773	NGC 6167 - 499		16 30 21	−49 29.8	3	10.55	.052	0.72	.026	-0.32	.013	9	O8:e	1081,1586,1649
149240	+29 02847			16 30 22	+29 12.2	1	8.93		0.73		0.34		2	K0	1502
148989	−48 10946	LSS 3650		16 30 22	−48 46.6	7	8.81	.006	0.61	.011	-0.41	.014	19	O9 I	540,954,976,1081,1649*
330954	−49 10774	NGC 6167 - 1131		16 30 22	−49 35.6	1	10.64		0.61		0.11		1	F5	302
149162	+3 03215	G 17 - 22		16 30 23	+03 21.2	3	8.85	.004	0.87	.010	0.54	.008	6	K0	196,333,1620,3008
149412	+58 01645			16 30 25	+57 56.7	1	7.72		1.40		1.67		2	K5	1375
149069	−26 11396			16 30 26	−26 59.6	2	10.49	.011	0.22	.011	0.11	.009	3	A0 IV/V	40,1770
		NGC 6169 - 4		16 30 26	−43 54.7	1	12.73		1.73		1.52		2		251
		G 138 - 39		16 30 27	+12 43.1	1	12.13		1.47				3		538
		G 138 - 40		16 30 29	+09 56.6	1	13.07		1.66				1		906
		CS 22872 # 118		16 30 29	−05 07.4	1	14.39		0.42		0.30		1		1736
149087	−28 12201			16 30 29	−28 30.0	1	10.83		0.25		0.10		2	B9 IV/V	1770
149070	−31 12991			16 30 29	−31 11.1	1	7.98		0.09		0.04		1	A0 V	1770
	+0 03544			16 30 31	+00 11.2	1	9.84		1.05		0.71		1		116
149144	−6 04455			16 30 31	−06 39.0	1	10.18		0.80		0.22		2	G5	1064
149038	−43 10900	NGC 6169 - 1	⋆ V	16 30 31	−43 56.6	6	4.90	.019	0.09	.022	-0.83	.046	15	O9.7Iab	251,1018,1034,2001*
330947	−49 10772	NGC 6167 - 628		16 30 31	−49 20.4	1	10.52		0.90		-0.10		2	A5	1081
		NGC 6171 sq2 30		16 30 32	−12 58.6	1	13.39		1.45		1.36		2		519
		NGC 6169 - 3		16 30 32	−43 54.2	1	13.21		0.57		0.44		2		251
149062	−35 11022			16 30 34	−35 26.3	1	7.71		1.00				4	F8 IV	2012
	+0 03545			16 30 35	+00 05.7	1	10.68		0.61		0.00		1		116
149134	−21 04386			16 30 35	−21 15.0	1	9.43		0.33		0.20		2	A0 V	1770
	−48 10949	LSS 3652		16 30 35	−49 01.8	2	11.31	.009	0.57	.019	-0.28	.000	4	B0 IV	1081,1649
		He3 1210		16 30 36	−49 21.	1	12.29		0.68		-0.37		3		1586
		NGC 6167 - 1132		16 30 36	−49 30.	1	14.06		0.96		0.49		1		302
		NGC 6167 - 1134		16 30 36	−49 30.	1	13.71		0.88		0.41		1		302
		NGC 6167 - 1135		16 30 36	−49 30.	1	13.66		0.64		0.11		1		302
		NGC 6167 - 1136		16 30 36	−49 30.	1	14.66		0.96		0.70		1		302
		NGC 6167 - 1140		16 30 36	−49 30.	1	13.34		0.90		0.52		1		302
		NGC 6167 - 1142		16 30 36	−49 30.	1	13.80		1.03		0.45		1		302
		NGC 6167 - 1143		16 30 36	−49 30.	1	14.47		0.75		0.19		1		302
		NGC 6167 - 1144		16 30 36	−49 30.	1	15.58		0.94				1		302
		NGC 6167 - 1145		16 30 36	−49 30.	1	15.22		1.31				1		302
		NGC 6167 - 1146		16 30 36	−49 30.	1	13.66		3.12				1		302
		NGC 6167 - 1147		16 30 36	−49 30.	1	15.09		1.03				1		302
		NGC 6167 - 1148		16 30 36	−49 30.	1	13.86		1.19		0.49		1		302
		NGC 6167 - 1149		16 30 36	−49 30.	1	14.96		1.06				1		302
		NGC 6167 - 1150		16 30 36	−49 30.	1	13.45		1.84				1		302
		NGC 6167 - 1152		16 30 36	−49 30.	1	13.51		0.88		0.38		1		302
		NGC 6167 - 1153		16 30 36	−49 30.	1	14.54		1.04		0.76		1		302
		NGC 6167 - 1154		16 30 36	−49 30.	1	13.22		0.64		0.20		1		302
		NGC 6167 - 1165		16 30 36	−49 30.	1	12.52		0.79		0.33		2		412
149019	−49 10778	NGC 6167 - 1155		16 30 36	−49 40.0	8	7.40	.020	0.88	.015	0.17	.029	22	A0 Iab	96,164,302,412,954*
		NGC 6169 - 2		16 30 38	−43 55.0	1	13.18		0.48		0.41		2		251
	−49 10780	NGC 6167 - 207		16 30 39	−49 32.2	4	10.89	.024	0.70	.008	-0.41	.027	6	O7 V	412,1081,1649,1737
330953	−49 10779	NGC 6167 - 1161		16 30 39	−49 38.3	1	10.81		0.26		-0.10		1		412
		CS 22878 # 20		16 30 40	+12 00.7	1	14.22		0.03		0.13		1		1736
149089	−36 10829			16 30 41	−36 11.6	1	10.04		0.38		0.22		1	B9.5 V	1770
		NGC 6167 - 871		16 30 43	−49 40.3	1	12.94		0.78				2		412
149165	−12 04546			16 30 44	−12 27.4	2	7.37	.000	1.98	.005	2.23	.029	5	M1 III	1375,1657
148907	−61 05734			16 30 44	−61 47.5	1	9.30		-0.08		-0.56		3	B5/7 V	1586
148839	−66 02978	LV Tra		16 30 45	−67 01.4	2	8.33	.016	0.94	.008	0.67	.032	5	C1,0Hd	158,1589
		He2 171		16 30 47	−34 59.2	1	14.74		1.13		-0.76		1		1753
		NGC 6167 - 870		16 30 47	−49 40.4	1	12.22		0.79		0.18		2		412
		CS 22878 # 19		16 30 48	+11 36.8	1	15.04		0.08				1		1736

Table 1 895

HD	DM	Other Id	N Rem	α_{1950}	δ_{1950}	S	V	σ_V	B–V	σ_{B-V}	U–B	σ_{U-B}	n	Spectrum	References
149148	−20 04511			16 30 48	−20 26.1	1	10.41		0.37		0.26		2	A2/5 II/II	1770
325691	−40 10510			16 30 48	−40 34.8	2	10.58	.003	0.63	.012	−0.30	.010	6	B1 II-III	540,567,954
		NGC 6167 - 1162		16 30 48	−49 38.8	1	12.35		0.76		0.36		2		412
		NGC 6167 - 949		16 30 48	−49 39.9	1	11.65		0.74		0.13		2		412
		NGC 6167 - 1163		16 30 48	−49 40.1	1	12.51		0.82		0.28		2		412
149149	−24 12704			16 30 50	−24 11.2	1	8.60		0.71		0.19		15	G6 V	1763
149054	−49 10784			16 30 50	−49 13.2	1	7.91		1.04		0.70		1	G6/8 III	96
		NGC 6167 - 953		16 30 50	−49 39.5	1	11.83		0.63		0.24		2		412
		NGC 6167 - 868		16 30 50	−49 40.7	1	12.35		0.81		0.35		2		412
		NGC 6167 - 867		16 30 51	−49 40.4	1	12.80		0.80		0.50		2		412
	+0 03546			16 30 52	+00 39.7	1	9.55		1.18		1.00		1	K2	116
		NGC 6167 - 1164		16 30 52	−49 40.8	1	12.03		0.88		0.37		2		412
		NGC 6167 - 936		16 30 54	−49 38.8	1	11.56		0.82		0.35		2		412
		NGC 6167 - 1158		16 30 55	−49 41.0	1	11.17		0.77		0.11		2		412
	−49 10786	NGC 6167 - 1157		16 30 55	−49 41.0	2	10.66	.054	1.21	.355	0.72	.429	3		412,1649
		NGC 6167 - 861		16 30 55	−49 41.4	1	11.74		0.84		0.38		2		412
149065	−49 10785	NGC 6167 - 1156		16 30 55	−49 45.0	4	8.44	.015	-0.02	.007	−0.70	.028	8	B2 IV	96,164,302,1081
148890	−65 03331	HR 6151		16 30 55	−65 23.5	3	5.51	.007	0.94	.005	0.71	.018	10	G8/K0 III	15,1075,1637
	−0 03132			16 30 57	−00 25.0	1	11.18		1.13		0.88		1		116
330950	−49 10789	LSS 3657		16 30 57	−49 27.0	6	9.51	.059	0.48	.031	−0.50	.041	17	B1 Ve	96,954,1021,1081,1586,1737
		NGC 6167 - 860		16 30 57	−49 41.1	1	11.86		0.78		0.36		2		412
149076	−46 10843			16 30 58	−46 54.0	6	7.37	.018	0.48	.009	−0.12	.021	17	B9/A0 Iab/b	164,403,540,976,1081,8100
149168	−26 11400			16 30 59	−26 24.1	2	9.91	.016	0.08	.002	−0.31	.003	4	B9	1054,1770
149151	−31 13000			16 30 59	−31 10.9	1	8.16		0.13		-0.03		1	Ap Si(Cr)	1770
		WLS 1640 10 # 6		16 31 00	+10 33.9	1	11.73		0.64		0.13		2		1375
149077	−49 10790	LSS 3658		16 31 00	−49 17.5	5	7.43	.082	0.47	.021	0.10	.049	9	A1 Ib	96,403,954,1649,8100
	−37 10839	LSS 3664		16 31 02	−37 44.4	1	10.84		0.48		-0.42		2		954
	−49 10791	NGC 6167 - 1160		16 31 02	−49 37.7	1	11.21		0.58		0.19		2		412
149125	−41 10724			16 31 03	−41 23.8	3	9.68	.008	0.46	.008	−0.31	.019	9	B7 Ib	540,567,976
		CS 22872 # 130		16 31 04	−02 15.3	1	15.02		0.62		0.30		1		1736
330946	−49 10792	LSS 3659		16 31 05	−49 13.7	5	8.94	.013	0.42	.021	−0.52	.017	11	B5 IV	96,540,954,976,1081
		NGC 6167 - 1159		16 31 07	−49 39.7	1	11.05		0.74		0.40		2		412
		BSD 108 # 435		16 31 08	+00 46.5	1	11.38		0.57		-0.07		1	F3	116
	−0 03135			16 31 12	−00 49.2	1	10.15		0.53		0.09		1	F0	116
149202	−29 12641			16 31 12	−29 36.6	1	9.32		0.39		0.10		1	A9 IV	1770
		Cr 307 - 11		16 31 13	−50 53.2	1	12.48		1.78		1.38		2		412
149153	−40 10514			16 31 14	−40 14.6	2	9.24	.010	0.59	.010	0.04		3	G1 V	742,1594
149126	−45 10752			16 31 15	−45 19.4	1	9.65		0.93				4	K1 III	2012
149380	+25 03097			16 31 16	+25 47.7	2	8.34	.000	0.56	.005	-0.07	.020	5	F8	833,1625
		CS 22872 # 133		16 31 16	−01 51.2	1	14.16		0.25		0.28		1		1736
149203	−32 11813			16 31 17	−32 44.7	1	9.22		0.46		0.29		1	A8/9 V	1770
		Cr 307 - 10		16 31 17	−50 53.5	1	12.89		0.76		0.58		2		412
		Cr 307 - 9		16 31 17	−50 54.2	1	11.84		1.83		1.68		2		412
		CS 22878 # 3		16 31 20	+08 01.6	1	14.42		0.46		-0.15		1		1736
		Cr 307 - 12		16 31 20	−50 56.5	1	12.43		0.85		0.36		2		412
149228	−25 11541			16 31 21	−25 26.8	2	9.97	.004	0.33	.001	-0.18	.005	5	Ap Si	40,1770
		Cr 307 - 8		16 31 21	−50 54.0	1	11.58		0.67		0.24		2		412
	−49 10799	NGC 6167 - 1122		16 31 22	−49 40.6	2	11.04	.033	0.64	.005	-0.06	.042	4		412,954,1081
149100	−53 08064			16 31 22	−53 32.7	8	7.20	.026	-0.07	.015	−0.76	.009	8	B2 III	540,976,2012
		LSS 3662		16 31 23	−51 36.2	1	12.17		1.75		0.22		2	B3 Ib	1649
		CS 22878 # 2		16 31 24	+07 59.9	1	14.36		0.45		-0.22		1		1736
		HRC 269	V	16 31 25	−24 07.5	1	12.11		1.52		1.01		1		649
149229	−31 13008			16 31 26	−31 11.9	1	9.59		0.42		0.23		1	A9 V	1770
149431	+38 02791	UY Her		16 31 27	+38 10.5	1	8.74		0.20		0.17		7	A2	1222
149217	−31 13006			16 31 27	−31 50.3	1	10.17		0.12		-0.08		1	B9.5 III	1770
	−0 03136			16 31 28	−00 59.3	1	11.14		0.64		0.00		1		116
331042	−49 10801	NGC 6167 - 253		16 31 28	−49 32.4	1	10.63		0.51		-0.41		2	B8	954
149348	+1 03255			16 31 29	+01 32.1	1	7.70		0.26		0.06		1	A2	116
		CS 22872 # 120		16 31 29	−04 19.3	1	15.37		0.37		0.33		1		1736
149128	−51 10319			16 31 29	−52 09.0	1	9.29		0.05		-0.35		3	B7 V	1586
		CS 22872 # 116		16 31 30	−05 11.9	1	13.40		0.40		-0.07		1		1736
		LP 862 - 143		16 31 30	−22 03.2	1	11.54		0.85		0.45		2		1696
		Cr 307 - 7		16 31 30	−50 56.7	1	11.10		1.22		1.08		2		412
149420	+30 02834	IDS16296N3043	AB	16 31 32	+30 36.2	1	6.87		0.23		0.13		3	F0 III	8071
149174	−44 10964	HR 6160		16 31 32	−45 08.5	1	6.46		1.34				4	K2/3 III	2007
328389	−44 10965			16 31 35	−44 16.7	1	9.64		0.39		-0.50		2	B3	540
		CS 22872 # 119		16 31 36	−04 22.4	1	15.01		0.24		0.26		1		1736
149254	−28 12218	IDS16284S2809	AB	16 31 36	−28 15.4	1	10.18		0.67		0.17		1	F3 V	1770
	+0 03549			16 31 37	−00 08.5	1	10.60		0.45		-0.01		1		116
		G 225 - 65		16 31 39	+61 55.6	1	15.22		1.29				1		1759
		Cr 307 - 6		16 31 39	−50 53.3	1	11.92		2.05		2.02		2		412
		LSS 3665		16 31 42	−48 45.8	2	12.28	.030	0.75	.025	0.17	.000	4	B3 Ib	954,1649
		Cr 307 - 5		16 31 42	−50 54.0	1	11.75		0.67		0.14		2		412
149650	+61 01598	HR 6170		16 31 43	+60 55.7	2	5.93	.015	0.03	.010	0.05	.005	5	A2 V	985,1733
149329	−16 04308			16 31 44	−16 48.9	1	10.09		0.42		0.27		1	A5/7 II	1770
149382	−3 03967	LS IV -03 001		16 31 45	−03 54.7	5	8.94	.009	-0.28	.003	-1.11	.034	36	sdOB	166,272,989,1729,1732
149290	−28 12223			16 31 45	−28 14.8	1	10.27		0.39		0.26		2	A0/1 III	1770
149255	−33 11268			16 31 45	−34 07.1	4	8.79	.004	0.10	.005	−0.29	.004	7	B8 II	540,567,976,1770
149190	−48 10961			16 31 45	−49 07.6	1	9.14		0.07		-0.33		1	B8 V	96
		Cr 307 - 3		16 31 45	−50 52.2	1	13.22		0.82		0.58		2		412

HD	DM	Other Id	N Rem	α_{1950}	δ_{1950}	S	V	σ_V	B–V	σ_{B-V}	U–B	σ_{U-B}	n	Spectrum	References
		Cr 307 - 4		16 31 45	−50 55.2	1	12.50		0.53		0.15		2		412
149231	−43 10916			16 31 46	−43 33.0	1	8.34		0.16		−0.52			B3 V	567
		CS 22872 # 115		16 31 47	−05 46.7	1	11.86		0.59		0.41		1		1736
149363	−5 04318	LS IV -06 001		16 31 48	−06 02.0	5	7.78	.010	0.03	.007	−0.84	.011	18	B0.5III	158,1003,1423,1509,1775
149289	−25 11546	IDS16287S2559	AB	16 31 48	−26 05.3	1	10.04		0.68		0.12		1	F2/3 V + F	1770
		Ara sq 1 # 22		16 31 48	−47 51.	1	11.06		0.95		−0.16		2		1081
		Cr 307 - 2		16 31 49	−50 53.2	1	11.65		1.10		0.75		2		412
		Cr 307 - 1		16 31 50	−50 52.9	1	11.22		0.85		0.53		4		412
149383	−7 04324			16 31 51	−08 02.8	1	7.62		1.69		1.95		3	K2	1657
149291	−31 13017			16 31 52	−31 47.7	1	8.79		0.28		0.24		1	A4 IV	1770
149273	−34 11099			16 31 52	−34 18.2	5	9.26	.023	0.01	.007	−0.82	.014	11	B2 Ib/II	400,540,567,976,1311
149208	−48 10964			16 31 52	−49 07.8	1	10.17		0.26		0.16		1	B9 V	96
149308	−24 12712			16 31 53	−25 07.7	1	10.33		0.23		0.15		2	A1 IV/V	1770
149274	−35 11037			16 31 55	−35 37.3	1	6.64		−0.07		−0.21		2	B9 V	1770
		LP 685 - 55		16 31 58	−08 55.6	1	11.51		0.79		0.13		2		1696
149292	−38 11051			16 32 01	−38 27.9	1	9.81		0.09		−0.20		1	B9 III	1770
149351	−28 12226			16 32 02	−28 09.8	1	10.71		0.20		0.08		2	B9 IV/V	1770
	−60 06579			16 32 04	−60 50.8	2	10.07	.011	0.08	.000	−0.29	.032	5		400,1586
		CS 22872 # 136		16 32 05	−01 48.4	1	14.41		0.56		−0.09		1		1736
149414	−3 03968	G 17 - 25	★ A	16 32 05	−04 06.7	10	9.62	.011	0.74	.007	0.10	.014	30	G5 V	158,196,516,742,1064,1620*
149414	−3 03968	IDS16295S0400	B	16 32 05	−04 06.7	1	13.88		1.43		1.46		2		3059
149130	−62 05389			16 32 05	−62 32.8	2	8.49	.023	0.34	.000	0.07		4	A8W	1594,6006
		NGC 6178 - 17		16 32 08	−45 33.0	1	13.69		0.77		0.38		2		251
149257	−45 10768	NGC 6178 - 1		16 32 08	−45 31.2	3	8.47	.009	−0.02	.014	−0.78	.013	8	B2 Vp	251,540,976
	−45 10766	NGC 6178 - 3		16 32 08	−45 31.5	1	9.95		0.06		−0.46		2	B5 V	251
	−45 10767	NGC 6178 - 8		16 32 08	−45 32.5	1	10.14		0.52		0.05		2		251
	−62 05390			16 32 08	−62 31.7	1	8.70		1.55		1.84		1		742
		G 17 - 26		16 32 10	+03 24.2	1	11.57		1.48		1.24		1	M0	333,1620
149748	+63 01281			16 32 10	+62 57.5	3	7.36	.013	0.33	.013	0.13	.018	8	Am	1375,3016,8071
149367	−26 11412	IDS16291S2616	AB	16 32 10	−26 22.6	3	8.52	.007	0.15	.002	−0.12	.013	9	B8/9 IV/V	173,1054,1770
149276	−41 10740			16 32 10	−42 02.7	3	10.00	.015	0.35	.014	−0.42	.014	8	B6 Ib	540,567,976
		NGC 6178 - 10		16 32 11	−45 30.4	1	12.03		0.21		0.16		2		251
		NGC 6178 - 2		16 32 11	−45 31.2	1	10.58		0.09		−0.31		2	B7 V	251
		NGC 6178 - 5		16 32 11	−45 32.0	1	10.93		0.15		−0.10		2		251
149277	−45 10769	NGC 6178 - 9		16 32 11	−45 34.6	4	8.38	.015	0.04	.013	−0.66	.009	8	B2 IVn	251,540,976,1081
331013	−48 10966			16 32 11	−48 36.9	1	9.67		1.44		1.13		2	F5 Ia	1649
		NGC 6178 - 4		16 32 12	−45 31.8	1	12.99		0.47		0.13		2		251
		NGC 6178 - 11		16 32 12	−45 32.4	1	13.80		0.68		0.16		2		251
		NGC 6178 - 7		16 32 12	−45 33.0	1	11.30		0.13		−0.11		2	B7 V	251
150010	+72 00734	HR 6180		16 32 13	+72 42.9	2	6.28	.048	1.31	.010	1.29		15	K2 III	252,1258
		NGC 6178 - 12		16 32 13	−45 32.3	1	12.84		0.48		0.15		2		251
		NGC 6178 - 6		16 32 14	−45 31.8	1	11.80		0.26		0.20		2	B8 V	251
		NGC 6178 - 16		16 32 14	−45 31.8	1	13.04		0.56		0.08		2		251
		NGC 6178 - 14		16 32 15	−45 32.3	1	14.16		0.44		0.21		2		251
		NGC 6178 - 13		16 32 15	−45 32.5	1	12.59		0.28		0.16		2		251
		NGC 6178 - 15		16 32 16	−45 32.0	1	13.01		1.85				2		251
149385	−26 11413			16 32 17	−26 26.0	1	10.15		0.38		0.20		2	B9 IV	1770
		CS 22872 # 129		16 32 19	−02 13.9	1	15.30		0.30		0.33		1		1736
149387	−30 13268			16 32 19	−30 09.9	2	9.20	.010	0.17	.007	−0.33	.022	3	B7 II/III	1054,1770
149313	−41 10743	LSS 3671		16 32 19	−42 01.3	5	9.22	.021	0.50	.012	−0.54	.037	18	B2 Ia/be	540,567,954,976,1586
		NGC 6178 - 19		16 32 21	−45 31.5	1	12.24		0.36		0.21		2		251
149295	−45 10771	NGC 6178 - 18	★ AB	16 32 21	−45 31.5	1	9.85		0.21		−0.12		2	A1 III	251
		CS 22872 # 128		16 32 22	−02 16.8	1	14.98		0.40		0.40		1		1736
		CS 22878 # 9		16 32 23	+08 46.1	1	13.23		−0.05		−0.46		1		1736
		Steph 1362		16 32 23	+40 21.4	1	11.22		1.17		1.08		1	K7	1746
		BSD 108 # 814		16 32 26	+00 53.1	1	11.56		0.18		0.02		1	B9p	116
149298	−49 10812	LSS 3669		16 32 26	−49 09.6	7	9.94	.023	0.41	.023	−0.44	.024	15	B2 Vne	96,540,954,1021,1081*
	+0 03550			16 32 26	+00 24.6	1	9.65		0.49		0.11		2	F2	116
149630	+42 02724	HR 6168	★	16 32 29	+42 32.3	4	4.20	.010	−0.02	.009	−0.12	.019	10	B9 V	15,1363,3023,8015
149434	−15 04346			16 32 29	−16 09.9	1	8.73		0.44		0.34		1	A1 V	1770
		L 339 - 106		16 32 30	−49 11.	2	12.42	.024	1.48	.015			3		1705,3062
149880	+67 00950	R Dra		16 32 31	+66 51.5	4	9.62	.467	1.26	.054	0.69	.111	4	M5-M9eIII	817,8005,8022,8027
		CS 22872 # 124		16 32 31	−02 56.6	1	14.38		0.53		0.30		1		1736
		CS 22872 # 125		16 32 32	−02 54.6	1	15.63		0.39		0.41		1		1736
		LP 918 - 1		16 32 35	−30 44.5	1	12.68		1.54		1.02		2		3078
328402	−44 10975	AAS35,161 # 139		16 32 36	−44 40.0	1	10.45		0.27		−0.36		1	B8	1586,5005
	−0 03140			16 32 36	−01 01.2	1	9.44		1.40		1.33		1	K0	116
149463	−9 04413			16 32 36	−09 34.2	1	8.71		0.62		0.09		2	F8	3026
	−0 03141			16 32 37	−00 02.9	1	10.22		1.27		1.22		1		116
		CS 22872 # 137		16 32 38	−01 41.1	1	14.10		0.22		0.25		1		1736
149801	+61 01599			16 32 39	+61 20.2	1	7.74		−0.01		−0.08		1	B9 V	1776
	+0 03551			16 32 44	+00 22.7	1	9.68		0.59		0.00		1	F8	116
149422	−31 13032	IDS16296S3123	A	16 32 45	−31 29.5	1	9.02		0.15		−0.03		1	B9 III	1770
150275	+77 00627	HR 6191	★	16 32 46	+77 32.8	3	6.35	.010	0.99	.005	0.70	.005	6	K1 III	15,1003,3016
149437	−25 11557			16 32 46	−25 11.7	1	10.42		0.42		0.21		2	B9 (III)	1770
149438	−27 11015	HR 6165		16 32 46	−28 06.8	10	2.82	.008	−0.25	.017	−1.03	.019	41	B0 V	3,15,26,1034,1054,1075*
149506	+0 03552	IDS16302N0008	A	16 32 48	+00 02.0	4	9.20	.002	0.39	.004	0.09	.004	48	F5	281,989,1729,6004
		LP 745 - 31		16 32 48	−13 35.0	1	15.55		1.48				1		1759
	−72 01965			16 32 48	−72 43.4	1	9.63		0.34		0.17		3		400
149659	+38 02798	TZ Her		16 32 50	+38 05.9	1	8.87		0.42		−0.06		18	F5	1222

Table 1   897

HD	DM	Other Id	N	Rem	α_{1950}	δ_{1950}	S	V	σ_V	B–V	σ_{B-V}	U–B	σ_{U-B}	n	Spectrum	References
		G 17 - 27			16 32 51	−03 51.4	1	13.86		1.45		1.41		1		1620
149404	−42 11399	HR 6164, V918 Sco			16 32 51	−42 45.5	9	5.47	.012	0.39	.009	−0.64	.019	31	O9 Iabe	138,401,403,567,681,815*
		CS 22872 # 126			16 32 54	−02 49.8	1	15.15		0.56		0.42		1		1736
149464	−28 12233				16 32 55	−28 15.4	2	8.58	.011	0.19	.024	−0.04	.007	4	A0 V(n)	1054,1770
149342	−54 07765				16 32 55	−54 53.0	1	7.19		1.48				4	K2 III	2012
		CCS 2348			16 32 56	−46 15.7	1	10.99		3.08				2		864
331044	−49 10818				16 32 56	−49 44.6	1	10.13		0.79		−0.08		2	B2 IV	1081
		CS 22872 # 131			16 32 57	−02 01.8	1	14.90		0.38		0.20		1		1736
	+9 03230	G 138 - 42			16 32 59	+08 55.3	1	9.15		0.61		0.03		2	G5	3060
		LP 862 - 185			16 32 59	−25 56.8	1	11.64		1.32		1.22		1		1696
		He2 173		V	16 32 59	−39 45.6	1	13.83		1.47		0.05		1		1753
		PG1633+099			16 33 01	+09 53.9	1	14.40		−0.19		−0.97		6	sdB	1764
		PG1633+099A			16 33 03	+09 53.9	1	15.26		0.87		0.32		6		1764
149465	−31 13037	IDS16299S3119		AB	16 33 03	−31 25.4	1	9.09		0.11		−0.09		2	B9 IV	1770
149237	−63 03978			AB	16 33 03	−64 08.9	1	7.54		0.97				4	K1 III	2012
149447	−34 11112	HR 6166			16 33 05	−35 09.3	4	4.15	.005	1.56	.005	1.93	.005	14	K5 III	15,1075,2012,8015
149416	−44 10982				16 33 05	−45 03.4	1	7.44		0.35		0.25		2	A5 II/III	1732
		LSS 3673			16 33 07	−48 46.4	1	12.33		0.87		0.29		2	B3 Ib	1649
		G 202 - 61			16 33 08	+54 29.5	1	11.84		1.49				1		1759
		HA 108 # 1361			16 33 09	+00 03.2	1	11.39		1.10		0.95		5	G7	281
149479	−30 13279				16 33 09	−30 59.1	1	8.90		0.28		0.23		1	A1/2 III	1770
		PG1633+099B			16 33 11	+09 52.5	1	12.97		1.08		1.01		6		1764
149632	+17 03053			⋆ A	16 33 12	+17 09.5	3	6.43	.013	0.06	.015	0.01	.004	6	A2 V	985,1022,3016
149631	+17 03054	IDS16309N1715		B	16 33 12	+17 12.1	1	7.25		0.10		0.04		1	A5	3016
		LSS 3674			16 33 13	−49 13.5	1	11.55		0.30		−0.42		2	B3 Ib	954,1649
		CS 22872 # 122			16 33 14	−03 29.7	1	13.57		0.69		0.19		1		1736
		HA 108 # 1749			16 33 15	+00 13.1	1	11.24		1.29		1.44		1	K3	281
		PG1633+099C			16 33 15	+09 52.4	1	13.23		1.13		1.14		6		1764
149426	−48 10980	LSS 3675			16 33 15	−48 33.8	3	9.49	.021	0.60	.024	−0.39	.008	5	B0:Ib:	954,1081,1649
		HA 108 # 1372			16 33 16	+00 03.2	1	11.69		0.67		0.12		1	G2	281
		G 180 - 62			16 33 16	+33 24.2	2	11.04	.009	1.37	.024	1.27		2	K7	801,1017
		PG1633+099D			16 33 17	+09 52.8	1	13.69		0.54		−0.03		5		1764
	−49 10823	LSS 3676			16 33 18	−49 42.0	2	10.80	.134	0.63	.098	−0.25	.085	5		954,1081
149615	+1 03262				16 33 19	+01 18.8	1	9.18		0.05		−0.01		1	A0	116
149482	−37 10854				16 33 20	−38 06.7	1	9.36		0.15		−0.06		1	B8/9 IV/V	1770
149509	−26 11421				16 33 21	−26 20.9	1	10.07		0.67		0.24		1	F2 V	1770
149510	−27 11018				16 33 22	−27 17.8	1	10.67		0.33		0.29		1	A1 III	1770
149512	−31 13042				16 33 22	−31 15.5	1	9.64		0.14		0.03		1	A0	1770
149616	−0 03143				16 33 23	−00 18.7	6	8.21	.004	0.56	.006	0.03	.013	55	G2 V	116,281,989,1371,1729,6004
149450	−45 10787				16 33 23	−45 13.1	1	8.19		0.01		−0.70		10	B3 III	1732
		CS 22878 # 21			16 33 24	+11 31.0	1	15.30		−0.02		−0.65		1		1736
		G 180 - 63			16 33 25	+43 23.8	1	14.84		0.45		−0.40		1		1658
	−14 04454				16 33 25	−15 04.1	2	10.35	.005	0.95	.002	0.69	.035	5	K3	803,3062
149528	−30 13285				16 33 26	−30 28.9	1	9.02		0.41		0.31		1	A3 III/IV	1770
		G 225 - 67		⋆ V	16 33 27	+57 15.1	1	12.90		1.60		1.05		1	M6e	3078
		G 225 - 68			16 33 28	+57 15.6	1	15.00		0.49		−0.36		1	DC	3078
149545	−25 11567				16 33 30	−25 39.0	1	9.82		0.63		0.32		1	F0 V	1770
149452	−46 10884	LSS 3679			16 33 30	−47 01.8	4	9.06	.007	0.58	.017	−0.42	.021	8	O8/9	540,954,976,1081
149652	+1 03263				16 33 31	+01 29.0	1	7.24		0.52		−0.04		1	F5	116
149546	−29 12678				16 33 32	−29 26.8	1	9.04		0.35		0.24		1	A2 V	1770
149515	−36 10851				16 33 32	−37 02.4	1	7.95		0.22		−0.07		2	B8/9 III	1770
149563	−28 12242				16 33 33	−28 53.0	1	10.20		0.29		−0.09		2	B9/A0 III	1770
		HA 108 # 1775			16 33 34	+00 11.2	1	11.20		0.61		0.02		1	G1	281
		HA 108 # 719			16 33 36	−00 19.4	2	12.71	.007	1.00	.011	0.67	.008	29		281,1764
149595	−18 04297				16 33 36	−19 10.4	1	9.36		0.50		0.33		1	A0 V	1770
149547	−31 13048				16 33 37	−31 49.4	1	8.60		0.33		0.24		2	A4 III	1770
149580	−19 04390				16 33 38	−20 02.9	1	9.81		0.64		0.26		2	A9 V	1770
		HA 108 # 1061			16 33 39	−00 14.0	1	11.43		1.36		1.52		6	K0	281
		HA 108 # 727			16 33 39	−00 19.5	1	12.73		0.70		0.23		26	G0	281
		HA 108 # 728			16 33 40	−00 19.7	1	13.65		0.96		0.56		11		281
149634	−15 04351				16 33 41	−15 36.1	1	10.51		0.54		0.30		1	A2 III	1770
		BPM 890			16 33 42	−87 19.	1	14.58		0.22		−0.63		1		3065
		G 138 - 43			16 33 43	+08 54.9	1	13.79		1.65				3		538
149619	−19 04392				16 33 43	−19 30.0	1	9.49		0.50		0.20		1	A9 V	1770
149661	−1 03220	HR 6171, V2133 Oph		⋆ A	16 33 44	−02 13.2	12	5.76	.017	0.83	.012	0.49	.027	43	K2 V	15,22,101,252,621,770*
149635	−20 04526				16 33 46	−20 26.8	1	9.37		0.23		0.12		2	A0 IV/V	1770
149455	−54 07769				16 33 46	−54 45.5	1	7.69		0.03		−0.30		5	B7 III/IV	1732
149597	−26 11427	IDS16308S2652		AB	16 33 51	−26 58.4	1	9.56		0.24		0.13		2	A0 V	1770
149754	+17 03057				16 33 54	+17 34.4	1	7.85		0.36		0.16		2	F3 III?	1648
149803	+30 02843				16 33 56	+29 50.8	1	8.54		0.49		0.03		2	F7 V	3026
		G 138 - 44			16 33 59	+10 06.5	1	12.89		1.39		1.05		3		7010
149604	−37 10864				16 33 59	−37 18.8	1	10.62		0.42		0.26		1	B9 IV/V	1770
	−0 03144				16 34 05	−01 11.4	1	10.54		1.29		1.20		1	K2	116
	+39 03017				16 34 08	+39 38.7	1	9.02		1.34		1.58		2	K5	1601
149606	−40 10550				16 34 10	−40 47.0	2	8.95	.015	0.98	.007	0.75		6	K3 V	2033,3008
		AAS35,161 # 140			16 34 10	−44 48.3	1	12.75		1.24		0.15		1		5005
	+38 02803	UU Her			16 34 12	+38 04.1	1	8.19		0.09		−0.19		1	F2 Ib	793
149654	−31 13057				16 34 12	−31 25.8	1	10.26		0.39		0.17		1	A7/8 III	1770
		BPM 24290			16 34 12	−57 43.	1	13.30		0.72		0.12		2		3065
149498	−56 07793				16 34 14	−56 09.6	1	6.55		1.57		1.77		3	K0 V	3016

HD	DM	Other Id	N Rem	α_{1950}	δ_{1950}	S	V	σ_V	B–V	σ_{B-V}	U–B	σ_{U-B}	n	Spectrum	References
331003	−47 10900	LSS 3683		16 34 16	−48 06.5	2	10.01	.005	0.53	.000	−0.37		6	B3 IV	1021,1081
331019	−48 10994	LSS 3681		16 34 16	−48 45.0	2	10.89	.127	0.36	.059	−0.46	.044	5		954,1081
150077	+60 01688	TX Dra		16 34 17	+60 34.2	1	7.14		1.61		1.55		1	M2	3001
		HA 108 # 468		16 34 19	−00 26.5	1	12.07		0.69		0.16		5	F9	281
149589	−48 10995			16 34 19	−48 53.4	1	9.39		0.26		−0.55		2	B2/3 III	1081
		HA 108 # 1123		16 34 22	−00 10.3	1	10.89		0.59		0.10		1	G0	281
	−0 03146			16 34 22	−00 46.7	1	11.01		0.53		0.08		1		116
		HA 108 # 1848		16 34 24	+00 11.9	3	11.71	.009	0.57	.003	0.07	.005	12	G0	281,1499,1764
149757	−10 04350	HR 6175, ζ Oph		16 34 24	−10 28.0	16	2.57	.005	0.02	.007	−0.86	.006	119	O9.5Vn	3,15,26,198,1006,1011*
149673	−31 13060			16 34 24	−31 18.0	1	10.40		0.11		−0.03		1	B9 III	1770
		HA 108 # 475		16 34 25	−00 28.7	2	11.31	.005	1.38	.004	1.47	.012	67	K0	281,1764
		LP 745 − 34		16 34 25	−10 36.7	1	12.62		1.50				1		3062
149822	+15 03029	HR 6176, V773 Her		16 34 27	+15 35.9	2	6.38		−0.09	.013	−0.19	.004	9	B9p SiCr	1022,3033
149485	−60 06594	HR 6167, DQ Dra		16 34 27	−60 53.4	2	6.18	.005	−0.07	.005	−0.38		8	B7 Vn	158,2007
150706	+80 00519			16 34 28	+79 53.7	2	7.03	.020	0.60	.005	0.08	.010	12	G0	1067,3026
149640	−44 11003			16 34 28	−44 12.0	3	7.95	.007	1.17	.009	1.05		15	G8/K0 III	1075,1673,2033
149610	−47 10905			16 34 28	−48 07.2	2	9.06	.003	0.03	.008	−0.57	.009	5	B2/3 Vnne	1081,1586
		MtW 108 # 70		16 34 29	−00 20.9	1	13.55		0.72		0.14		5		397
		CS 22878 # 24		16 34 30	+10 53.1	1	14.18		0.03				1		1736
149890	+31 02873	G 169 − 18		16 34 30	+31 02.9	4	7.11	.008	0.54	.011	−0.01	.005	8	F8 V	308,1003,1658,3026
		G 202 − 65		16 34 30	+45 57.9	1	11.22		0.36		−0.16		2	sdF3	1658
		MtW 108 # 77		16 34 31	−00 21.4	1	16.73		0.68		0.10		3		397
		G 17 − 28		16 34 33	−01 25.5	1	14.21		1.42		1.07		2		3078
149687	−33 11300			16 34 33	−34 01.6	1	10.20		0.10		−0.03		1	B9 V	1770
148527	−82 00687	HR 6138		16 34 33	−83 08.6	2	6.57	.004	1.57	.002			10	K4 III	1486,2038
		MtW 108 # 92		16 34 34	−00 22.6	1	11.54		0.60		0.18		5		397
149806	+0 03553	IDS16320N0027	AB	16 34 35	+00 21.2	1	7.07		0.87		0.52		1	K0	116
		MtW 108 # 101		16 34 36	−00 18.9	1	16.55		0.65		0.04		3		397
		RW Dra		16 34 37	+57 56.5	2	11.44	.393	0.19	.169	0.10	.037	2	A5	668,699
		MtW 108 # 110		16 34 37	−00 19.3	1	14.74		0.64		0.04		5		397
	−0 03147	IDS16320S0105	AB	16 34 37	−01 11.8	1	9.90		0.64		0.05		1		116
		HA 108 # 1863		16 34 39	+00 08.5	3	12.23	.008	0.81	.004	0.41	.021	12	G5	281,1499,1764
		BSD 108 # 869		16 34 39	+00 17.8	1	10.92		0.25		0.10		1	A0p	116
149775	−18 04298			16 34 39	−18 16.3	1	9.03		0.43		0.28		1	A0 V	1770
328507	−44 11006	LSS 3686		16 34 39	−44 48.6	1	10.22		0.37		−0.46		1	B3	954
149825	+0 03554			16 34 40	+00 03.3	4	9.06	.003	0.97	.006	0.62	.008	56	G9 III	281,989,1729,6004
149881	+14 03086	V600 Her		16 34 41	+14 34.5	1	7.03		−0.19		−0.97		5	B0.5III	399
		LSS 3685		16 34 41	−50 56.5	1	10.86		0.92		−0.03		1	B3 Ib	954
149956	+36 02756			16 34 44	+36 08.4	1	7.07		1.59		1.88		2	M1	3040
150030	+46 02194	HR 6183		16 34 44	+46 42.8	3	5.83	.019	1.04	.017	0.86	.020	8	G8 II	71,1501,8100
149761	−27 11035			16 34 44	−27 46.9	1	10.02		0.45		0.20		1	A9/F2 V	1770
149705	−34 11142			16 34 44	−34 16.7	1	9.08		0.59		0.10		1	A1 III	1770
149658	−48 11004			16 34 45	−48 42.9	1	9.63		0.39		−0.50		2	B2 IV	1081
149907	+23 02965			16 34 46	+22 58.2	1	6.88		0.99		0.85		2	K0 III	1625
		HA 108 # 512		16 34 46	−00 33.7	1	10.72		1.24		1.17		1	G8	281
	+8 03241			16 34 47	+08 12.6	1	9.65		1.56		1.90		2	M0	1375
149845	−0 03148			16 34 47	−00 18.8	6	7.96	.007	1.31	.009	1.42	.004	49	K0 III	116,281,397,989,1729,6004
149612	−57 08091			16 34 51	−58 09.3	3	7.02	.003	0.61	.005	0.03	.002	9	G3 V	285,688,2012
149957	+31 02875	G 169 − 20		16 34 52	+31 12.2	1	9.49		1.20		1.16		2	K5	3072
		HA 108 # 523		16 34 53	−00 27.9	2	11.31	.003	1.38	.005	1.46		27		989,1729
	−0 03149			16 34 53	−00 39.6	2	10.03	.015	0.55	.010	0.06	.005	2	G0	116,281
	−5 04325	IDS16322S0518	A	16 34 53	−05 24.6	1	10.60		1.18		1.08		4	K7 V	3072
	−5 04325	IDS16322S0518	B	16 34 53	−05 24.6	1	10.64		1.15		0.97		4	K7 V	3072
149711	−43 10959	HR 6174	⋆ AB	16 34 54	−43 18.0	2	5.82	.005	−0.02	.002	−0.61		7	B2 V	26,2007
331038	−49 10844			16 34 54	−49 30.8	1	10.84		0.57		−0.02		3	B8	1586
149736	−37 10879			16 34 57	−37 18.1	1	10.75		0.23		0.06		2	B9 III/IV	1770
		HA 108 # 1535		16 34 59	+00 01.5	1	12.34		0.52		0.04		1	G0	281
150100	+53 01875	HR 6184	⋆ C	16 35 00	+53 00.0	2	5.52	.005	−0.05	.005	−0.15	.005	6	B9.5Vn	1084,1733
		vdB 72 # b		16 35 00	−48 46.9	1	10.79		0.41		−0.43		4	B2 V	434
328530	−45 10804	LSS 3688		16 35 01	−45 30.1	2	10.87	.165	0.46	.057	−0.50	.080	2	B3	954,1081
150117	+53 01876	HR 6185	⋆ AB	16 35 02	+53 01.5	2	5.08	.000	−0.04	.010	−0.16	.000	6	B9 V	1084,1733
149678	−52 10080			16 35 07	−52 22.3	1	7.48		0.09		0.12		3	A0/1 V	1586
		G 138 − 46		16 35 10	+13 47.0	1	13.97		1.52		1.24		1		3060
		G 169 − 21		16 35 10	+31 25.2	1	12.11		0.67		−0.14		2		1658
		HA 108 # 870		16 35 11	−00 22.4	1	11.86		1.22		1.14		13	G2	281
149847	−22 04181			16 35 12	−23 09.7	1	9.83		0.26		0.12		1	B9 IV	1770
149827	−24 12737			16 35 12	−24 47.3	2	9.64	.007	0.21	.008	0.05	.092	4	A0 V	1054,1770
149933	+0 03555	G 17 − 29		16 35 13	+00 08.6	6	8.05	.009	0.77	.009	0.40	.009	34	G9 V	116,281,333,989,1620*
		CS 22878 # 27		16 35 13	+10 28.1	1	14.41		0.44		−0.20		1		1736
	−0 03152			16 35 13	−00 27.1	4	10.71	.002	0.17	.003	0.17	.010	62	A0	116,281,989,1729
		HA 108 # 551		16 35 14	−00 27.1	1	10.70		0.18		0.18		76		1764
		HA 108 # 872		16 35 15	−00 16.4	1	11.92		0.84		0.37		10	G3	281
		HA 108 # 1918		16 35 16	+00 05.3	2	11.38	.002	1.43	.001	1.74	.046	10	Ma	281,1764
149828	−31 13078			16 35 16	−31 54.2	1	8.80		0.08		−0.40		1	B3 V	1770
150254	+62 01492			16 35 17	+62 29.0	1	9.91		0.57		−0.03		1	F2	1776
		LSS 3689		16 35 18	−50 48.3	1	11.08		1.06		−0.11		3	B3 Ib	1737
		LP 745 − 79		16 35 19	−13 25.4	1	10.13		0.77		0.21		2		1696
149866	−22 04182			16 35 19	−22 47.6	1	7.56		0.42		−0.01		2	F3 V	1770
149779	−43 10964	V954 Sco		16 35 20	−44 03.5	2	7.51	.013	0.23	.001	−0.61	.001	5	B2 IV	540,976
149911	−6 04467	HR 6179		16 35 21	−06 26.3	2	6.09	.014	0.16	.005	0.11	.014	9	A0 p CrEu	220,1088

Table 1

HD	DM	Other Id	N Rem	α_{1950}	δ_{1950}	S	V	σ_V	B–V	σ_{B-V}	U–B	σ_{U-B}	n	Spectrum	References
149867 −24 12739	IDS16323S2502		A	16 35 21	−25 07.7	1	10.13		0.47		0.28		2	A3	1770
149867 −24 12739	IDS16323S2502		B	16 35 21	−25 07.7	1	10.31		0.62		0.17		1	A3	1770
		G 138 - 47		16 35 22	+13 46.7	1	16.90		0.41		-0.50		1	DC	3060
+51 02121				16 35 22	+51 39.6	1	9.42		0.68		0.21		3	G3	3016
−50 10716				16 35 22	−50 20.3	1	9.48		0.77		0.04		3		1586
		Ara OB1 # 557		16 35 24	−48 49.7	1	12.42		0.49		0.12		3		574
		vdB 72 # c		16 35 24	−48 49.9	1	11.50		0.66		-0.29		3	B1 Vp?	434
149849 −34 11154				16 35 28	−34 39.4	1	10.07		0.19		0.04		1	B8/A0 III	1770
149729 −52 10092	LSS 3690			16 35 28	−52 26.3	2	9.07	.016	0.01	.004	-0.69	.002	9	B2 V(n)	540,976
		HA 108 # 570		16 35 29	−00 29.1	1	12.12		0.52		0.11		9	F6	281
150012 +13 03177	HR 6181			16 35 30	+13 47.2	2	6.31	.005	0.41	.005	0.01	.005	4	F5 III-IV	985,1733
149871 −31 13083				16 35 30	−31 14.5	1	9.22		0.42		0.05		1	A9 V	1770
		Ara sq 1 # 35		16 35 30	−49 31.	1	11.76		0.74		-0.21		1		1081
149883 −26 11439				16 35 31	−26 53.4	2	8.43	.010	0.14	.006	-0.05	.010	4	B9 IV	1054,1770
149893 −23 12880	IDS16325S2356		A	16 35 32	−24 02.0	1	8.63		0.49		0.11		1	F2 V	1770
331021 −48 11021	Ara OB1 # 566			16 35 33	−48 48.8	1	10.22		0.25		-0.25		3		574
149914 −17 04606				16 35 34	−18 07.3	1	6.73		0.28		0.11		2	B9.5 IV	1770
331051 −49 10854				16 35 34	−49 44.8	2	9.69	.090	0.45	.066	-0.33	.038	6	B3 V	954,1081
		Ara OB1 # 554		16 35 35	−48 51.4	1	12.23		0.51		-0.11		2		574
149730 −56 07804	R Ara		⋆ AB	16 35 35	−56 53.8	1	6.66		0.09		-0.35		1	B9 IV/V	1588
		HA 108 # 1959		16 35 36	+00 09.8	1	12.50		0.54		0.03		1	G0	281
−48 11026	Ara OB1 # 572			16 35 37	−48 47.2	1	10.40		1.51		1.57		2		574
+0 03557				16 35 38	+00 39.5	1	11.05		1.22		1.13		1	K2	116
149885 −31 13085				16 35 39	−31 47.6	1	9.46		0.22		-0.27		1	B5 III/V	1770
−67 03184				16 35 40	−67 46.0	1	10.75		-0.13		-0.85		2	B5 III	400
−4 04138				16 35 41	−04 55.4	1	10.47		1.19		1.21		1	K5 V	3072
		Ara OB1 # 581		16 35 41	−48 45.9	1	11.59		0.57		0.15		2		574
149996 −2 04219	G 17 - 30			16 35 42	−02 20.4	4	8.49	.017	0.61	.009	0.03	.034	8	G0	516,742,803,1620
+76 00614				16 35 44	+76 04.6	1	9.92		1.12		0.90		1	K5	1746
149886 −36 10879	HR 6178		⋆ AB	16 35 44	−37 07.1	3	5.93	.005	-0.03	.013	-0.11	.011	9	B9/A0 V +B	13,1770,2006
149886 −36 10879	IDS16324S3701		C	16 35 44	−37 07.1	1	11.34		0.34		0.13		2		13
149876 −43 10974	NGC 6192 - 254			16 35 45	−43 12.1	1	9.69		0.55		0.12		1	F2 V	609
149834 −48 11033	Ara OB1 # 696			16 35 45	−48 45.2	2	9.15	.019	0.21	.000	-0.57	.005	7	B2/3 III	574,1081
331023 −48 11032	LSS 3691			16 35 45	−48 57.2	1	9.67		0.54		-0.36		4	B2 V	1081
150102 +27 02661			V	16 35 47	+27 08.6	1	6.98		1.71		0.19		3	M2 III	985
		WD1635+608		16 35 49	+60 49.1	1	15.82		-0.14		-1.06		1	DA3	1727
		Steph 1369		16 35 51	+28 37.7	1	11.51		1.45		1.21		1	M0	1746
		Ara OB1 # 593		16 35 51	−48 42.1	1	12.52		0.60		0.06		3		574
149978 −17 04608				16 35 53	−17 48.0	1	9.86		0.73		0.40		1	A8/9 IV/V	1770
149324 −77 01221	HR 6163		⋆ A	16 35 53	−77 25.0	3	4.23	.008	1.06	.000	0.96	.004	9	K0 III	15,1075,3077
		HA 108 # 1986		16 35 56	+00 09.7	1	12.16		0.40		0.12		1	F9	281
150203 +43 02624				16 35 56	+43 39.8	1	7.28		0.09		0.02		2	A2	1601
149814 −51 10393				16 35 56	−52 01.4	1	9.05		0.11		-0.20		3	B5/7 III/V	1586
149855 −49 10860	IDS16322S4909		A	16 35 57	−49 15.2	1	8.98		0.33		-0.47		3	B1/2 III	1081
−48 11036	Ara OB1 # 590			16 35 58	−48 43.9	1	9.48		1.80		2.08		3		574
149999 −21 04394				16 36 01	−21 56.8	1	9.25		0.52		0.01		1	F5/6 V	1770
		Ara OB1 # 601		16 36 01	−48 41.4	1	11.61		0.79		0.22		2		574
150018 −18 04304				16 36 06	−18 25.6	1	9.15		0.72		0.38		1	A9 IV(m)	1770
149980 −29 12708				16 36 06	−29 49.6	1	7.15		0.28		0.17		1	A5 III	1770
149979 −26 11445				16 36 08	−26 40.6	1	10.71		0.37		0.25		2	A1/2 III	1770
		CS 22878 # 74		16 36 10	+11 48.4	1	14.81		0.06		0.05		1		1736
		CS 22878 # 73		16 36 10	+11 49.7	1	14.62		0.44		-0.22		1		1736
149770 −62 05406	IDS16316S6221		AB	16 36 10	−62 27.4	2	8.04	.004	-0.11	.005	-0.62	.004	9	B4/5 Vn	400,1732
		CS 22878 # 34		16 36 12	+09 14.1	1	14.35		0.11		0.18		1		1736
		Ara OB1 # 543		16 36 12	−48 54.0	1	12.49		0.50		0.13		3		574
150000 −26 11447				16 36 13	−26 21.9	1	9.35		0.58		0.14		1	F3 IV/V	1770
149671 −68 02789	HR 6172			16 36 13	−68 12.0	4	5.90	.009	-0.08	.002	-0.41	.004	16	B7 IVe	15,1075,1637,2038
		HA 108 # 941		16 36 14	−00 19.3	1	11.77		1.00		0.64		11	G5	281
		NGC 6192 - 255		16 36 14	−43 14.1	1	11.90		1.61		1.40		1		609
149902 −47 10923				16 36 15	−47 57.1	1	7.93		0.95		0.68		1	G8 III/IV	96
		Wool 9571		16 36 16	+05 31.9	1	10.22		0.99				1		1705
150034 −23 12881				16 36 16	−23 43.7	1	9.82		0.44		0.32		2	A1 V	1770
150088 −0 03155				16 36 17	−00 44.2	1	8.31		1.12		0.84		3	G5	1657
150205 +29 02860				16 36 18	+29 46.4	1	7.51		0.73		0.28		4	G5 V	3026
−48 11039	NGC 6193 - 15		⋆	16 36 18	−48 41.0	2	11.00	.041	0.38	.008	-0.31	.012	7		574,1081
		vdB 72 # a		16 36 18	−48 42.	1	11.03		0.38		-0.30		5	B3 V	434
−48 11039	Ara OB1 # 430			16 36 19	−48 40.8	1	11.72		0.52		0.45		2	B3	574
		BSD 108 # 918		16 36 21	+00 31.5	1	11.09		0.41		0.15		1	B8	116
150255 +39 03021				16 36 22	+39 40.6	1	6.74		1.13		1.10		2	K0	1601
		HA 108 # 949		16 36 22	−00 18.2	1	12.85		0.78		0.35		19		281
		NGC 6192 - 256		16 36 22	−43 18.4	1	13.01		0.74		0.41		1		609
150429 +63 01289	HR 6198			16 36 24	+63 10.4	1	6.16		1.53				2	K5	71
		GD 202		16 36 25	+16 00.1	1	15.60		0.14		-0.63		1		3060
325833 −42 11440				16 36 25	−43 08.7	1	10.17		0.39		-0.37		2	B8	540
150035 −27 11054	V955 Sco			16 36 26	−27 11.3	2	8.67	.014	0.24	.000	0.14	.019	3	Ap CrEuSr	1054,1770
		NGC 6192 - 158		16 36 26	−43 16.4	1	13.05		0.59		0.34		1		609
149837 −60 06603	HR 6177		⋆ AB	16 36 26	−60 20.9	3	6.17	.007	0.48	.000			11	F6 V	15,2013,2030
331053 −49 10868				16 36 27	−49 57.0	1	10.24		0.44		-0.27		2	B2 IV	1081
		G 138 - 49		16 36 28	+05 46.7	1	15.76		0.77		0.09		1	DAwk	1658,3060
−48 11043	Ara OB1 # 432			16 36 28	−48 41.3	1	10.23		0.46		0.21		2	F0	574

HD	DM	Other Id	N	Rem	α_{1950}	δ_{1950}	S	V	σ_V	B–V	σ_{B-V}	U–B	σ_{U-B}	n	Spectrum	References
	−0 03156				16 36 29	−01 10.6	1	10.24		0.68		0.13		1		116
		HA 108 # 646			16 36 30	−00 29.8	1	10.81		0.58		0.08		1	F5	281
		He3 1233			16 36 30	−44 51.	1	11.88		0.29		-0.26		1		1586
		NGC 6192 - 157			16 36 32	−43 16.4	1	13.13		0.54		0.17		1		609
		NGC 6192 - 155			16 36 33	−43 16.8	1	13.33		0.65		0.31		1		609
	−48 11046	NGC 6193 - 16			16 36 35	−48 48.8	2	10.90	.019	0.39	.005	-0.41	.000	4		574,1081
149922	−53 08122				16 36 36	−54 02.4	3	7.88	.025	-0.02	.011	-0.68	.008	12	B2 III	540,976,1732
		KUV 16366+3506			16 36 37	+35 06.1	1	14.91		-0.26		-1.16		1	DA	1708
		KUV 433 - 3			16 36 37	+35 06.1	1	14.18		-0.15		-1.12		1		974
150080	−24 12751				16 36 37	−24 54.6	1	9.34		0.40		0.28		2	A2 IV	1770
325860	−43 10983				16 36 38	−43 47.1	1	9.78		0.47		0.03		3		1586
		NGC 6192 - 99			16 36 39	−43 17.4	1	12.21		0.54		0.21		1	A5	609
		NGC 6192 - 131			16 36 41	−43 18.8	1	12.52		0.65		0.26		1	F0	609
		HA 108 # 981			16 36 42	−00 19.3	2	12.06	.009	0.48	.010	0.23	.003	13	F4	281,1764
		Ara OB1 # 472			16 36 42	−48 45.6	1	12.87		2.13		1.72		3		574
		NGC 6192 - 70			16 36 43	−43 16.1	1	12.87		0.65		0.33		1		609
	−0 03157				16 36 44	−00 20.8	3	10.44	.011	0.55	.013	0.02	.019	12	F9	116,281,6004
		HA 108 # 985			16 36 44	−00 21.6	1	13.02		1.03		0.59		2		281
		Ara OB1 # 327			16 36 45	−48 38.5	1	11.53		1.84		1.94		3		574
	−48 11050	Ara OB1 # 477			16 36 45	−48 46.5	1	10.25		0.24		0.21		3	A2	574
		NGC 6192 - 102			16 36 46	−43 17.4	1	12.67		0.59		0.28		1		609
150069	−36 10893				16 36 47	−36 23.1	3	8.28	.016	0.18	.008	-0.35	.008	6	B5 III	540,976,1770
150055	−36 10892				16 36 47	−37 05.8	1	10.99		0.25		0.16		1	B8/9 Ib/II	1770
150006	−47 10933				16 36 48	−47 54.4	1	8.73		0.18		0.14		1	A0 V	96
	−48 11051	NGC 6193 - 17			16 36 48	−48 47.4	3	10.34	.040	0.31	.029	-0.54	.019	8	B1p	574,954,1081
150059	−39 10562				16 36 49	−39 22.3	3	8.85	.015	0.28	.005	-0.20	.005	9	B5 Ib/II	540,976,1586
325802	−41 10785				16 36 49	−42 00.5	1	10.98		0.62		-0.08		3	B5	567
	−43 10986	NGC 6192 - 109			16 36 51	−43 16.7	1	10.87		0.77		0.60		1	F0	609
		NGC 6192 - 93			16 36 52	−43 16.4	1	12.83		0.51		0.14		1		609
321841	−38 11108	LSS 3697			16 36 53	−38 22.0	1	9.75		0.37		0.26		2		954
		NGC 6192 - 39			16 36 53	−43 13.5	1	13.40		0.94		0.52		1		609
	−43 10987	NGC 6192 - 45			16 36 53	−43 14.9	1	11.69		1.56		1.24		1		609
		Steph 1374			16 36 54	+48 59.3	1	11.97		1.41		1.21		1	K7-M0	1746
		He3 1235			16 36 54	−47 39.	1	11.19		0.71		-0.43		1		1586
150177	−9 04430	HR 6189			16 36 55	−09 27.3	2	6.39	.071	0.49	.014	-0.10	.014	9	F3 V	254,1088
325832		NGC 6192 - 92			16 36 55	−43 16.0	1	11.26		0.56		0.26		1	A0	609
150631	+70 00887	IDS16372N6958	AB		16 36 56	+69 53.5	1	8.00		0.43		0.00		4	F5	3030
150631	+70 00887	IDS16372N6958	C		16 36 56	+69 53.5	1	11.01		0.57		0.05		4		3030
		Ara OB1 # 226			16 36 56	−48 45.4	2	12.24	.005	0.89	.000	0.42	.010	3		574,1743
		NGC 6193 - 18			16 36 58	−48 43.0	3	10.71	.008	0.18	.000	-0.50	.004	12	B3 V	574,1081,1743
	−48 11055	Ara OB1 # 608			16 36 58	−48 43.1	2	9.81	.000	0.38	.010	0.19	.000	6		574,1743
		Ara OB1 # 251			16 36 58	−48 43.7	1	12.12		0.25		-0.11		2		574
		NGC 6192 - 22			16 36 59	−43 15.4	1	12.76		0.63		0.36		1		609
150041	−48 11056	NGC 6193 - 19			16 36 59	−48 39.6	9	7.06	.005	0.09	.009	-0.81	.021	52	B0 III	96,540,574,954,976,1081*
150449	+56 01907	HR 6199			16 37 00	+56 06.8	2	5.29	.004	1.06	.009	0.90	.000	3	K1 III	1080,3016
		NGC 6192 - 30			16 37 00	−43 13.6	1	13.75		0.59		0.43		1		609
150295	+24 03038				16 37 01	+24 47.7	1	7.85		0.41		-0.03		2	F5	1625
150125	−25 11600				16 37 01	−25 57.8	1	7.69		0.32		0.24		2	A2/3 III/I	1770
150127	−30 13353				16 37 02	−30 42.3	1	9.88		0.22		0.16		1	A1 IV	1770
325830	−43 10989	NGC 6192 - 28			16 37 03	−43 13.1	1	10.63		2.05		2.30		1	gK4	609
		NGC 6192 - 11			16 37 03	−43 17.9	1	12.14		0.52		0.17		1	A5	609
		LP 805 - 25			16 37 04	−15 43.8	2	12.46	.005	0.87	.007	0.09	.013	4		1696,3062
		NGC 6192 - 14			16 37 04	−43 17.2	1	13.47		0.52		0.35		1		609
		NGC 6192 - 24			16 37 05	−43 14.9	1	13.01		0.56		0.22		1		609
	−43 10991	NGC 6192 - 9			16 37 06	−43 16.6	1	11.40		1.53		1.17		1	gK0	609
	−47 10939	LSS 3696			16 37 06	−47 38.7	1	11.13		0.70		-0.41		2	B0 IV	1081
	−48 11059	Ara OB1 # 245			16 37 06	−48 43.6	2	11.24	.000	0.26	.010	-0.23	.000	5		574,1743
150147	−26 11459				16 37 07	−26 21.4	1	8.57		0.32		0.06		1	F0 V	1770
149840	−67 03190				16 37 07	−67 23.4	1	7.98		1.62		1.94		5	K3/4 III/V	1704
150149	−28 12278				16 37 08	−28 47.8	1	9.92		0.31		0.17		1	A5/7 V	1770
	−48 11061	NGC 6193 - 20			16 37 08	−48 39.7	3	11.21	.000	0.22	.004	-0.42	.004	16		574,1081,1743
		L 333 - 19			16 37 09	−45 54.2	1	12.65		1.60				1		1705
	−47 10941	LOD 41+ # 15			16 37 12	−47 39.8	3	10.06	.049	0.53	.036	-0.42	.025	4	F5	96,96,1081
		Ara sq 1 # 49			16 37 12	−47 40.9	1	10.09		0.50		-0.42		2		1081
150304	+22 03000				16 37 14	+22 06.3	1	6.82		1.31		1.32		2	K0	3016
150093	−40 10594				16 37 14	−41 01.8	1	7.92		0.13		-0.26		3	B3/5 II/III	1586
		Ara OB1 # 342			16 37 14	−48 39.0	1	13.40		0.84		0.45		7		574
150152	−35 11066				16 37 15	−35 58.9	1	10.06		0.33		0.07		1	B8 IV/V	1770
150693	+70 00888	IDS16376N7000	AB		16 37 16	+69 55.2	1	7.94		0.33		0.06		3	F0	3030
150193	−23 12887				16 37 16	−23 47.9	2	8.81	.018	0.54	.007	0.36	.011	3	A1 V	1588,1770
325894	−41 10788	LSS 3699			16 37 16	−41 21.8	1	10.37		0.41		-0.27		2	B8 II	954
		Ara OB1 # 240			16 37 17	−48 43.5	2	13.14	.052	0.64	.028	0.15	.052	4		574,1743
	−48 11062	NGC 6193 - 21			16 37 17	−48 47.8	3	11.40	.012	0.28	.006	-0.31	.004	8		574,1081,1743
		Ton 259			16 37 18	+25 27.	2	15.21	.005	-0.14	.020	-0.06	.005	3		98,1036
150151	−35 11067				16 37 19	−35 35.4	3	6.64	.008	0.20	.017	-0.28	.004	7	B5 III	540,976,1770
	−48 11063	Ara OB1 # 291			16 37 19	−48 40.8	2	10.85	.010	0.27	.000	-0.44	.010	6	B3 V	574,1743
		NGC 6192 - 251			16 37 21	−43 19.3	1	10.84		0.46		0.24		1		609
		Ara OB1 # 274			16 37 21	−48 41.3	1	12.21		0.58		0.07		3		574
		NGC 6193 - 1274			16 37 21	−48 41.3	1	12.21		0.55		0.09		4		1743
	−48 11065	NGC 6193 - 22			16 37 21	−48 46.5	1	11.03		0.25		-0.41		2	B3	1081

Table 1 901

HD	DM	Other Id	N	Rem	α₁₉₅₀	δ₁₉₅₀	S	V	σ_V	B−V	σ_B−V	U−B	σ_U−B	n	Spectrum	References
150281	+5 03246	G 17 - 31			16 37 22	+05 36.5	1	8.65		0.88		0.60		1	K0	333,1620
150450	+49 02531	HR 6200		⋆ A	16 37 23	+49 01.5	2	4.90	.000	1.56	.011	1.76		8	M2 III	1363,3055
150207	−23 12888				16 37 23	−23 33.2	1	10.19		0.30		0.08		2	A0 III/IV	1770
150083	−47 10942	IDS16337S4733		AB	16 37 23	−47 39.0	4	7.23	.009	0.02	.012	-0.34	.007	8	B5 III	96,540,976,1081
		NGC 6193 - 28			16 37 23	−48 46.6	1	11.06		0.26		-0.45		2		1743
		CS 22878 # 60			16 37 24	+09 19.7	1	13.26		0.05		0.09		1		1736
		LSS 3706			16 37 24	−31 22.0	1	12.47		0.60		0.35		2	B3 Ib	954
	−32 11900	SU Sco			16 37 25	−32 17.0	1	8.64		4.42				2	C5,5	864
		Ara OB1 # 272			16 37 26	−48 41.2	1	12.00		0.51		-0.25		2		574
		NGC 6193 - 1272			16 37 26	−48 41.2	1	12.00		0.50		-0.31		4		1743
		Ara OB1 # 184			16 37 26	−48 45.5	2	13.10	.033	0.59	.005	0.35	.019	4		574,1743
		Ara OB1 # 181			16 37 27	−48 44.4	1	12.89		0.52		0.08		3		574
		CS 22878 # 76			16 37 28	+11 49.5	1	14.74		0.55		0.04		1		1736
		Ara OB1 # 180			16 37 28	−48 44.4	1	12.25		0.41		0.29		1		574
		NGC 6193 - 14			16 37 30	−48 41.4	1	13.97		0.91		0.42		2		251
150223	−27 11063				16 37 31	−27 59.7	1	10.17		0.28		0.17		1	A1 IV/V	1770
		LSS 3708			16 37 32	−30 59.1	1	13.07		0.63		0.30		2	B3 Ib	954
		NGC 6193 - 11		⋆	16 37 34	−48 38.8	3	12.54	.045	0.52	.023	0.41	.024	5		251,574,1743
150135	−48 11070	NGC 6193 - 2		⋆ C	16 37 34	−48 40.0	5	6.88	.012	0.17	.019	-0.79	.016	14	O7 V	251,574,954,1081,1743
150136	−48 11070	NGC 6193 - 1		⋆ AB	16 37 35	−48 40.0	9	5.55	.093	0.16	.016	-0.79	.021	27	O5 V + O6:	251,540,574,954,976,1081*
		G 180 - 65			16 37 36	+33 31.3	3	14.63	.083	0.22	.083	-0.58	.021	5		419,974,3028
	−48 11069	NGC 6193 - 8		⋆ AC	16 37 36	−48 39.3	4	9.54	.026	0.20	.018	-0.65	.014	9	B1 V	251,574,1081,1743
150259	−20 04537	HR 6190			16 37 37	−20 18.8	1	6.26		1.08				4	K0 III	2007
		NGC 6193 - 1175			16 37 37	−48 45.0	2	12.09	.000	0.31	.015	-0.15	.000	7		574,1743
150341	+17 03067				16 37 39	+16 55.3	1	8.09		1.54		1.86		2	K5	1648
		NGC 6193 - 9			16 37 39	−48 40.1	2	12.30	.025	0.29	.020	-0.24	.005	4		251,1743
		NGC 6193 - 10			16 37 39	−48 40.5	2	12.26	.035	0.31	.025	-0.16	.025	4		251,1743
150134	−47 10944				16 37 41	−47 52.4	1	9.84		0.38		-0.24		1	B3 II/III	96
	−48 11071	NGC 6193 - 7			16 37 41	−48 39.5	6	8.45	.015	0.17	.013	-0.72	.017	17	B1 V	251,574,954,1081,1737,1743
	+36 02762				16 37 42	+36 21.7	1	9.98		0.61				3	G0	209
	−48 11072	Ara OB1 # 173			16 37 42	−48 45.0	2	10.72	.005	0.13	.005	-0.26	.000	7	B9.5V-A0 III	574,1743
		Ara OB1 # 187			16 37 42	−48 46.3	1	12.08		0.62		0.47		2		574
		NGC 6193 - 13			16 37 43	−48 40.6	1	14.75		1.25		1.16		2		251
150224	−32 11902	IDS16344S3256		AB	16 37 44	−33 03.0	1	9.82		0.36		0.21		1	B9 IV	1770
	−48 11074	NGC 6193 - 3		⋆	16 37 44	−48 42.1	3	9.05	.008	1.00	.011	0.64	.012	10		251,574,1743
		Ara OB1 # 171			16 37 44	−48 44.4	2	11.52	.005	0.55	.010	0.34	.000	10		574,1743
150509	+51 02125				16 37 45	+50 55.6	1	7.04		1.04		0.85		2	K0	1566
150243	−31 13119				16 37 45	−31 09.7	1	10.24		0.10		0.02		1	A0 IV	1770
		NGC 6193 - 1058			16 37 45	−48 36.8	1	13.25		0.47		0.18		1		1743
		Ara OB1 # 65			16 37 46	−48 41.0	1	11.50		0.74		0.20		3		574
		G 180 - 66			16 37 47	+34 22.5	1	11.00		0.60		-0.01		3		1658
150242	−30 13364				16 37 47	−30 17.6	1	9.95		0.16		-0.02		1	B9 IV	1770
328686	−46 10931	LSS 3705			16 37 47	−46 49.0	3	10.45	.027	0.45	.021	-0.33	.015	8	B3 Vnn	954,1021,1081
		NGC 6193 - 12			16 37 47	−48 41.0	1	12.62		0.88		0.50		2		251
		Ara OB1 # 144			16 37 47	−48 42.2	2	11.94	.010	0.46	.000	0.12	.010	7		574,1743
	+36 02764				16 37 48	+36 35.3	1	9.20		0.92				3	K0	209
	−48 11076	NGC 6193 - 23			16 37 48	−48 28.2	1	10.10		0.39		-0.50		2	B2 V	1081
	−48 11075	NGC 6193 - 6			16 37 48	−48 39.3	3	10.05	.010	0.25	.013	-0.57	.005	11	B3 IV-V	251,574,1743
	−48 11078				16 37 50	−48 37.5	1	10.02		0.26		-0.57		2	B2 V	1081
		Ara OB1 # 75			16 37 50	−48 37.6	2	11.75	.015	0.32	.029	-0.08	.019	3		574
	−48 11077	NGC 6193 - 5			16 37 50	−48 40.6	4	10.43	.045	0.28	.019	-0.33	.010	9	B3 V	251,574,954,1743
		BSD 108 # 952			16 37 51	+00 21.0	1	11.67		0.07		0.12		1	B8	116
	−48 11080	NGC 6193 - 4			16 37 51	−48 41.5	4	10.32	.010	0.25	.018	-0.48	.008	14	B4 V	251,574,1081,1743
		Ara OB1 # 191			16 37 52	−48 46.5	2	12.03	.019	0.52	.000	0.11	.015	6		574,1743
150168	−49 10890	HR 6188			16 37 53	−49 33.3	6	5.65	.014	-0.02	.009	-0.86	.017	28	B1 Iab/Ib	15,444,1081,1637,2012,8100
152643	+85 00278				16 37 55	+85 43.2	1	8.62		0.44		-0.05		3	F8	1733
150462	+35 02849				16 37 56	+35 36.7	1	7.64		0.10		0.10		2	A3 V	1601
		LSS 3709			16 37 56	−34 02.5	1	12.61		0.58		0.08		2	B3 Ib	954
	−43 11010				16 37 56	−43 53.0	1	12.73		1.64		1.20		1	M3	1705,3073
150431	+25 03113				16 37 57	+25 37.9	1	8.23		0.97				3	G8 III	20
		Ara OB1 # 11			16 37 57	−48 32.9	1	13.19		1.47				3		574
		AF Her			16 38 00	+41 12.5	1	12.23		0.13		0.08		1		668
	+52 01986	IDS16368N5249		AB	16 38 00	+52 42.7	2	9.99	.083	1.14	.024	0.91		4	K8	1017,3072
150285	−27 11070				16 38 00	−27 49.6	1	10.43		0.41		0.28		1	A8/9 V	1770
150198	−47 10950				16 38 00	−47 58.1	2	8.92	.000	0.05	.005	-0.32	.010	3	B9 Ib/II	96,1081
150197	−47 10951	LSS 3707			16 38 01	−47 27.4	1	9.52	.010	0.41	.005	-0.58	.005	3	B0 I	1081,1737
	−48 11082	NGC 6193 - 24			16 38 01	−48 41.9	3	10.37	.009	0.25	.003	-0.55	.009	9	B2 V	574,1081,1743
	−48 11083	Ara OB1 # 194			16 38 01	−48 45.9	2	10.36	.005	1.04	.000	0.71	.018	5		574,1743
		UKST 873 # 12			16 38 02	+00 05.3	1	14.52		1.12				2		1584
		Ara OB1 # 30			16 38 03	−48 34.8	1	12.49		0.42		0.20		2		574
		G 138 - 51			16 38 04	+15 46.2	2	11.79	.000	1.00	.005	0.82	.015	2		1696,3060
		Ara OB1 # 41			16 38 04	−48 36.2	1	12.18		0.29		0.18		2		574
		Ara OB1 # 43			16 38 04	−48 36.4	1	9.91		1.11		0.91		2		574
	−48 11085	NGC 6193 - 26			16 38 04	−48 36.4	3	10.04	.011	0.31	.008	-0.39	.006	7	B2.5V	574,1081,1743
	−48 11085	NGC 6193 - 1043			16 38 04	−48 36.4	1	9.90		1.10		0.87		5		1743
150379	+4 03234	HR 6194		⋆ B	16 38 06	+04 18.2	3	6.93	.012	0.13	.005	0.10	.014	14	A3 IV	13,1084,1088
		UKST 873 # 6			16 38 06	−00 11.5	1	11.30		1.24				2		1584
150085	−61 05766				16 38 08	−61 41.4	3	6.58	.005	0.79	.005	0.52	.000	9	F8 II	285,688,2012
		CS 22878 # 51			16 38 09	+08 29.3	1	13.63		0.17		0.14		1		1736
	−48 11086	NGC 6193 - 25			16 38 09	−48 39.6	3	10.38	.011	0.38	.000	-0.36	.000	14	B2.5V	574,1081,1743

HD	DM	Other Id	N	Rem	α_{1950}	δ_{1950}	S	V	σ_V	B–V	σ_{B-V}	U–B	σ_{U-B}	n	Spectrum	References
150378	+4 03235	HR 6195		⋆A	16 38 10	+04 18.9	3	5.78	.013	-0.01	.005	-0.05	.019	14	A1 V	13,1084,1088
		Ara OB1 # 110			16 38 10	-48 40.7	2	11.55	.025	1.88	.015	1.99	.080	6		574,1743
		Ara OB1 # 130			16 38 10	-48 41.4	2	12.39	.005	0.41	.005	-0.03	.000	4		574,1743
		UKST 873 # 17			16 38 12	+00 37.5	1	16.10		0.71				1		1584
		G 202 - 68			16 38 12	+50 40.0	2	11.81	.014	1.58	.020	1.35		6	M2.5	1723,1746
	-0 03162				16 38 12	-01 06.9	1	9.90		0.67		0.12		1	F8	116
150248	-45 10847				16 38 12	-45 16.3	2	7.03	.005	0.65	.019	0.17		8	G3 V	1311,2012
		S044 # 3			16 38 12	-75 23.9	1	10.71		1.07		0.86		2		1730
		UKST 873 # 13			16 38 15	-00 18.0	1	14.62		0.74				2		1584
	-48 11088	NGC 6193 - 27			16 38 15	-48 36.8	2	10.59	.000	0.39	.015	-0.37	.000	7	B2.5V	574,1743
		Ara OB1 # 96			16 38 15	-48 39.7	2	11.54	.010	0.30	.025	0.19	.025	5		574,1743
		CS 22878 # 52			16 38 17	+08 31.8	1	14.54		0.60				1		1736
		NGC 6193 - 1065			16 38 17	-48 37.1	1	11.52		0.72		0.22		4		1743
150026	-67 03196	HR 6182			16 38 17	-67 20.3	4	6.02	.006	0.02	.002	0.03	.008	16	A0 Vn	15,1075,1637,2038
		UKST 873 # 7			16 38 19	+00 24.2	1	11.72		0.89				2		1584
		Ara OB1 # 135			16 38 19	-48 42.2	2	11.96	.005	0.61	.005	-0.12	.000	5		574,1743
		Ara OB1 # 66			16 38 20	-48 37.0	1	13.11		0.67		0.26		3		574
	-48 11090				16 38 20	-48 37.2	1	10.55		0.39		-0.37		2	A3	1081
150708	+60 01691	WW Dra		⋆A	16 38 22	+60 47.8	1	8.60		0.72				2	G0 V	3024
150708	+60 01691	IDS16377N6054		B	16 38 22	+60 47.8	1	9.67		0.57		0.05		4	G1 IV/V	3024
150322	-30 13371				16 38 22	-30 29.6	1	9.49		0.41		0.34		1	A1 IV	1770
150365	-17 04616				16 38 24	-17 57.8	1	6.70		0.16		0.11		1	A3 III	1770
150345	-24 12762				16 38 24	-24 18.4	1	8.26		0.48		0.01		1	F5 V	1770
		NGC 6193 - 1611			16 38 25	-48 35.5	1	11.17		0.40		-0.39		1		1743
	-48 11094	Ara OB1 # 106			16 38 26	-48 40.2	2	11.09	.000	0.69	.010	0.19	.015	6		574,1743
		BSD 108 # 624			16 38 27	+00 15.1	1	10.81		1.51		1.42		1	K0	116
		Ara OB1 # 108			16 38 27	-48 40.6	1	12.64		0.59		0.10		2		574
		UKST 873 # 11			16 38 28	+00 00.4	1	14.41		0.74				2		1584
150347	-28 12293				16 38 28	-28 29.5	2	8.96	.000	0.10	.009	-0.11	.012	3	Ap Si	1054,1770
150323	-32 11911				16 38 29	-32 43.6	1	7.60		0.06		-0.29		4	Ap Si	400
150433	-2 04230	IDS16359S0239		A	16 38 31	-02 45.3	2	7.24	.005	0.63	.010	0.07		7	G0	2033,3026
150331	-32 11913	HR 6192			16 38 31	-33 03.0	4	5.86	.008	0.65	.005			12	G3 III	15,444,2020,2028
150483	+12 03063	HR 6203			16 38 32	+12 29.4	1			0.05		0.03		3	A3 Vn	1022
150366	-24 12765	HR 6193			16 38 34	-24 22.4	6	6.07	.008	0.21	.008	0.07	.018	21	A7 III	15,1770,2013,2020,2028,8071
	-0 03166				16 38 35	-00 29.8	1	9.67		1.71		1.78		1		116
150288	-46 10939	LSS 3710		⋆A	16 38 35	-46 55.3	5	8.68	.015	0.15	.015	-0.62	.010	13	B2 (I)N	540,954,976,1081,1586
150579	+37 02788				16 38 37	+37 08.0	1	8.93		0.60				3	G5	209
150451	-0 03168	HR 6201			16 38 37	-00 54.3	3	6.25	.013	0.30	.004	0.06	.012	10	A7 III	15,1417,8071
	+0 03561				16 38 40	+00 30.4	1	10.19		0.40		0.08		1		116
		G 138 - 53			16 38 41	+18 12.3	1	14.64		0.93		0.40		1		3060
150416	-17 04618	HR 6196			16 38 41	-17 38.8	3	4.94	.014	1.11	.000	0.85	.018	9	G7.5 II	206,3016,8100
150567	+29 02867				16 38 43	+29 00.6	1	7.66		1.21		1.34		1	K3 III	3077
150493	+1 03286				16 38 44	+01 20.5	1	6.40		1.04		0.84		1	K0	116
150383	-26 11477	AX Sco			16 38 44	-27 00.6	1			1.71		1.39		5	M5 III	897
	+37 02789				16 38 45	+36 52.9	1	10.66		0.96				3		209
150312	-48 11102			AB	16 38 49	-48 55.7	1	8.92		0.42		-0.42		2	B3 II/III	1081
		BSD 108 # 634			16 38 51	+00 06.4	1	11.07		0.51		0.06		1	A8	116
		Steph 1376			16 38 51	-19 20.9	1	11.03		1.25		1.26		1	K7	1746
328584	-43 11020	LSS 3711			16 38 51	-44 03.2	2	10.27	.020	0.63	.010	-0.38	.045	6	O9.5 Ia	954,1737
	-88 00138				16 38 51	-89 09.3	1	10.62		0.52		0.01		4		826
	-0 03169				16 38 53	-00 58.7	1	9.90		0.10		-0.06		1		116
		AG Her			16 38 54	+40 42.8	1	11.99		0.07		0.12		1		668
150417	-30 13380				16 38 55	-30 26.1	1	7.70		1.36		1.38		11	K2 III	1673
150580	+25 03115	HR 6208			16 38 56	+24 57.2	1	6.07		1.31		1.34		2	K2	71
150434	-22 04187				16 38 56	-22 58.4	1	10.74		0.41		0.26		2	A1/2 III	1770
150453	-19 04406	HR 6202		⋆A	16 38 57	-19 49.8	2	5.57	.000	0.44	.018	-0.05		5	F3 V	2007,3053
150437	-28 12301				16 38 58	-29 01.7	2	7.86	.005	0.68	.005	0.28		8	G5 V	164,1075
150525	+5 03254				16 39 02	+04 58.0	1	6.98		0.16		0.15		3	A0	3016
		Ross 812			16 39 02	+36 24.6	1	11.50		1.50				1		1746
150419	-32 11923				16 39 03	-32 41.7	2	9.06	.029	0.63	.005	0.12		3	G0/2 V (+F)	742,1594
		AJ74,1000 M13# 4			16 39 04	+36 32.9	1	10.94		1.10				3		209
		G 17 - 33			16 39 06	-05 43.6	1	12.06		1.50		1.25		1		1620
		AJ74,1000 M13# 3			16 39 07	+36 25.1	1	10.67		1.37				3	M2	209
150557	+1 03290	HR 6205			16 39 10	+01 16.5	4	5.75	.035	0.33	.011	0.06	.013	15	F4 IV	253,254,1088,3016
		AJ74,1000 M13# 5			16 39 12	+36 33.8	1	10.85		0.55				3		209
328678	-46 10946				16 39 12	-46 44.4	1	10.37		0.35		-0.35		3	B3 IV	1081
150373	-47 10971				16 39 13	-47 26.4	2	9.92	.000	0.33	.005	-0.54		6	B3 V	1021,1081
150495	-22 04190				16 39 14	-22 36.5	1	9.86		0.64		0.12		1	F3 V	1770
		G 138 - 54			16 39 16	+11 45.3	1	11.15		1.37		1.29		3	M2	7010
		LP 685 - 44			16 39 16	-07 46.8	1	11.77		0.63		-0.05		2		1696
		WLS 1640 10 # 5			16 39 17	+12 29.6	1	11.96		0.65		0.27		2		1375
150679	+36 02767				16 39 17	+36 17.8	1	7.32		0.24		0.06		2	A8 V	1601
		UKST 873 # 16			16 39 17	-00 08.4	1	15.58		0.69				4		1584
150472	-32 11927				16 39 17	-32 12.3	1	8.50		0.07		-0.09		2	B9 V	1770
		CS 22878 # 58			16 39 18	+09 10.8	1	14.51		0.00		0.31		1		1736
150513	-19 04407				16 39 19	-19 58.1	1	8.77		0.29		0.25		1	A2/3 III	1770
		G 138 - 55			16 39 20	+09 29.8	1	14.58		1.01		0.51		1		1696
		G 138 - 56			16 39 20	+15 18.6	3	15.70	.018	0.36	.043	-0.61	.028	5	DAs	419,1759,3060
150632	+19 03146				16 39 21	+19 01.0	1	7.32		1.47		1.64		2	K2	1648
	-0 03170				16 39 21	-00 36.1	1	8.51		1.65		1.85		1	K5	116

Table 1   903

HD	DM	Other Id	N	Rem	α_{1950}	δ_{1950}	S	V	σ_V	B–V	σ_{B-V}	U–B	σ_{U-B}	n	Spectrum	References
150514	−19 04408				16 39 21	−20 05.6	2	8.62	.018	0.12	.008	-0.21	.013	5	B8 III	1054,1770
	+30 02860	IDS16374N3018		AB	16 39 22	+30 11.6	1	9.96		1.00		0.87		2	K6	3030
150401	−47 10973	LSS 3712			16 39 22	−48 06.0	3	9.25	.008	0.49	.021	-0.30	.020	5	B3 II/III	96,954,1081
	+37 02790				16 39 23	+37 04.3	1	9.73		1.17				3	K0	209
150680	+31 02884	HR 6212		★ AB	16 39 24	+31 41.5	12	2.81	.010	0.64	.010	0.21	.008	52	F9 IV +G7 V	1,15,22,254,938,1077*
150421	−45 10858	HR 6197			16 39 24	−45 58.6	3	6.26	.011	0.83	.007	0.52	.052	10	F5 Iab	1081,2035,8100
150475	−37 10930	LSS 3715			16 39 25	−37 46.3	3	8.72	.011	0.27	.005	-0.68	.013	8	B0/0.5Ib	540,954,976
150558	−6 04485				16 39 26	−06 50.5	1	8.40		1.66		1.97		1	K5	1746
150422	−46 10952	LSS 3714			16 39 28	−46 24.8	5	8.93	.008	0.26	.011	-0.63	.012	14	B1/2ne	540,954,976,1081,1586
	−47 10977	NGC 6200 - 10			16 39 29	−47 25.0	2	10.65	.015	0.32	.005	-0.56	.044	3		537,954
150423	−48 11115				16 39 29	−48 10.9	2	9.27	.015	0.19	.010	-0.56	.020	3	B2 III	96,1081
		WLS 1640 10 # 7			16 39 30	+07 49.6	1	11.88		1.22		1.02		2		1375
150633	+13 03196				16 39 31	+13 14.8	1	7.96		0.69		0.22		1	K0	3060
150497	−32 11931				16 39 32	−32 35.6	1	8.78		0.11		-0.17		2	B9 V	1770
150528	−28 12310				16 39 34	−28 49.6	1	10.50		0.38		0.29		1	A3/5 II/II	1770
150618	+1 03293				16 39 35	+01 20.2	1	7.59		0.41		-0.07		1	F2	116
150682	+27 02668	HR 6213			16 39 35	+27 00.7	2	5.93	.005	0.39	.000	-0.08	.010	3	F2 III	254,3016
150542	−22 04192				16 39 35	−23 01.3	1	10.38		0.44		0.33		2	A8/9 V	1770
150543	−26 11484				16 39 41	−27 05.6	1	7.84		1.74		2.19		5	M0/1 III	897
		G 138 - 59			16 39 43	+10 32.3	2	15.08	.050	1.62	.014	0.82	.009	5		538,3078
	−49 10917				16 39 44	−49 15.0	1	10.00		0.58		0.08		3	B5 III	1081
		GD 356			16 39 49	+53 46.9	1	15.04		0.33		-0.52		2	DP	308
		CS 22878 # 78			16 39 50	+11 43.0	1	14.40		0.58		0.05		1		1736
150560	−29 12770				16 39 51	−29 21.5	1	9.12		0.22		0.17		1	A1/2 IV	1770
		NGC 6205 - 1			16 39 54	+36 33.	1	14.35		0.67		0.22		2		350
		NGC 6205 - 28			16 39 54	+36 33.	1	15.14		0.77		0.27		2		350
		NGC 6205 - 56			16 39 54	+36 33.	1	15.00		0.74		0.21		2		350
		NGC 6205 - 66			16 39 54	+36 33.	1	14.10		0.63		0.18		3		350
		NGC 6205 - 70			16 39 54	+36 33.	1	12.12		1.59		1.69		3		350
		NGC 6205 - 72			16 39 54	+36 33.	1	12.32		1.30		1.16		3		350
		NGC 6205 - 77			16 39 54	+36 33.	1	12.77		1.20		1.03		3		350
		NGC 6205 - 82			16 39 54	+36 33.	1	15.37		0.66		0.22		2		350
		NGC 6205 - 96			16 39 54	+36 33.	1	12.52		1.27		1.09		3		350
		NGC 6205 - 109			16 39 54	+36 33.	1	13.32		0.95		0.53		3		350
		NGC 6205 - 114			16 39 54	+36 33.	1	13.45		1.00		0.60		3		350
		NGC 6205 - 139			16 39 54	+36 33.	1	14.00		0.75		0.24		3		350
		NGC 6205 - 140			16 39 54	+36 33.	1	14.24		0.68		0.17		3		350
		NGC 6205 - 144			16 39 54	+36 33.	1	14.20		0.69		0.15		2		350
		NGC 6205 - 158			16 39 54	+36 33.	1	12.70		1.19		0.95		3		350
		NGC 6205 - 168			16 39 54	+36 33.	1	13.62		0.90		0.40		3		350
		NGC 6205 - 172			16 39 54	+36 33.	1	13.56		1.00		0.62		2		350
		NGC 6205 - 193			16 39 54	+36 33.	1	13.63		0.97		0.54		3		350
		NGC 6205 - 199			16 39 54	+36 33.	1	12.20		1.39		1.22		2		350
		NGC 6205 - 201			16 39 54	+36 33.	1	13.18		1.06		0.76		3		350
		NGC 6205 - 240			16 39 54	+36 33.	1	12.34		1.37		1.30		3		350
		NGC 6205 - 244			16 39 54	+36 33.	1	12.67		1.24		1.11		2		350
		NGC 6205 - 252			16 39 54	+36 33.	1	12.67		1.22		1.03		3		350
		NGC 6205 - 268			16 39 54	+36 33.	1	13.52		1.01		0.56		2		350
		NGC 6205 - 271			16 39 54	+36 33.	1	14.87		0.81		0.25		3		350
		NGC 6205 - 316			16 39 54	+36 33.	1	12.58		1.23		1.08		2		350
		NGC 6205 - 353			16 39 54	+36 33.	1	12.83		1.09		0.64		3		350
		NGC 6205 - 530			16 39 54	+36 33.	1	14.59		0.78		0.27		2		350
		NGC 6205 - 588			16 39 54	+36 33.	1	14.39		0.81		0.32		2		350
		NGC 6205 - 687			16 39 54	+36 33.	1	12.96		1.04		0.76		2		350
		NGC 6205 - 691			16 39 54	+36 33.	1	14.88		0.74		0.32		1		350
		NGC 6205 - 719			16 39 54	+36 33.	1	14.47		0.74		0.06		2		350
		NGC 6205 - 738			16 39 54	+36 33.	1	13.95		0.92		0.54		3		350
		NGC 6205 - 745			16 39 54	+36 33.	1	12.54		1.28		1.14		3		350
		NGC 6205 - 766			16 39 54	+36 33.	1	13.54		0.99		0.65		std		350
		NGC 6205 - 773			16 39 54	+36 33.	1	13.21		1.00		0.63		3		350
		NGC 6205 - 800			16 39 54	+36 33.	1	13.80		0.93		0.51		std		350
		NGC 6205 - 828			16 39 54	+36 33.	1	14.68		0.71		0.03		2		350
		NGC 6205 - 833			16 39 54	+36 33.	1	14.08		0.76		0.33		2		350
		NGC 6205 - 834			16 39 54	+36 33.	1	14.35		0.75		0.23		2		350
		NGC 6205 - 848			16 39 54	+36 33.	1	13.15		1.10		0.86		2		350
		NGC 6205 - 919			16 39 54	+36 33.	1	13.03		1.11		0.88		2		350
		NGC 6205 - 920			16 39 54	+36 33.	1	13.72		0.86		0.40		2		350
		NGC 6205 - 921			16 39 54	+36 33.	1	13.39		1.01		0.72		2		350
		NGC 6205 - 932			16 39 54	+36 33.	1	13.98		0.77		0.23		2		350
		NGC 6205 - 946			16 39 54	+36 33.	1	14.28		0.75		0.20		2		350
		NGC 6205 - 954			16 39 54	+36 33.	1	12.09		1.54		1.64		2		350
		NGC 6205 - 984			16 39 54	+36 33.	1	13.69		0.75		0.24		2		350
		NGC 6205 - 1019			16 39 54	+36 33.	1	12.70		0.67		0.15		2		350
		NGC 6205 - 1027			16 39 54	+36 33.	1	14.18		0.67		0.15		2		350
		NGC 6205 - 1051			16 39 54	+36 33.	1	14.12		0.88		0.53		2		350
		NGC 6205 - 1057			16 39 54	+36 33.	1	15.19		0.77		0.27		1		350
		NGC 6205 - 1060			16 39 54	+36 33.	1	14.05		0.85		0.42		3		350
150586	−23 12898				16 39 55	−23 37.3	1	9.96		0.65		0.37		2	A7 V	1770
150585	−22 04193				16 39 57	−23 08.1	1	10.57		0.59		0.09		1	F2 V	1770
150588	−24 12780				16 39 57	−25 07.9	1	10.28		0.18		0.06		2	B9 III/IV	1770

HD	DM	Other Id	N	Rem	α_{1950}	δ_{1950}	S	V	σ_V	B–V	σ_{B-V}	U–B	σ_{U-B}	n	Spectrum	References
150601	−21 04401				16 39 59	−22 07.7	1	10.12		0.59		0.42		2	A1 IV	1770
150569	−30 13396				16 39 60	−30 18.2	1	9.24		0.19		0.18		1	A2 III	1770
150589	−26 11488				16 40 00	−26 21.9	1	8.44		0.38		0.05		2	F2 V	1770
		LSS 3720			16 40 01	−33 59.9	1	12.69		0.65		0.17		2	B3 Ib	954
150502	−48 11122				16 40 02	−48 24.5	1	9.82		0.07		-0.13		1	B8/9 IV/V	96
150711	+13 03198				16 40 03	+13 08.1	1	8.56		0.58		0.09		2	G5	1648
150500	−46 10958				16 40 03	−47 00.7	3	7.05	.004	-0.02	.019	-0.50	.009	10	Ap Si	540,976,1081
150684	+0 03565				16 40 04	+00 07.7	2	9.81	.005	0.13	.010	0.05		2	A0	116,1584
		UKST 873 # 14			16 40 04	−00 06.1	1	15.08		0.75				3		1584
150533	−45 10869	LSS 3716			16 40 05	−45 22.2	6	9.46	.013	0.62	.017	-0.40	.024	13	B0	540,954,976,1081,1586,1737
150696	+0 03566				16 40 07	+00 37.2	2	8.46	.000	0.35	.010	0.09		2	A3	116,1584
	+25 03118				16 40 07	+25 31.2	1	9.20		0.48				3	F7 IV-V	20
150603	−26 11490	IDS16370S2632		A	16 40 07	−26 37.4	1	9.30		0.40		0.29		2	A8/9 V	1770
150799	+25 03119				16 40 10	+25 31.5	2	8.87	.025	0.54	.010	-0.05		5	F8 IV	20,105
150634	−22 04194				16 40 10	−22 15.1	1	9.97		0.64		0.08		1	F3 V	1770
328867	−47 10993	NGC 6200 - 11		⋆A	16 40 10	−47 25.1	1	9.69		0.31		-0.64		2	B0.5p	537
328867	−47 10993	NGC 6200 - 11		⋆AB	16 40 10	−47 25.1	3	9.45	.154	0.32	.012	-0.56	.027	6		540,1081,1737
328693	−47 10991	NGC 6200 - 9			16 40 10	−47 32.4	5	10.28	.013	0.35	.013	-0.55	.035	5	B0.5III	537,954,1081
	−48 11123				16 40 10	−48 23.2	1	10.65		0.49		0.07		1		96
150535	−48 11124				16 40 10	−48 25.1	1	8.12		0.10		0.02		1	B9.5IV/V	96
		UKST 873 # 19			16 40 12	+00 31.1	1	16.59		0.78				2		1584
		Ara sq 1 # 81			16 40 12	−47 46.	1	10.53		0.67		0.36		2		1081
150336	−65 03353				16 40 13	−65 46.9	1	8.89		-0.02		-0.33		1	B8 Vn	1586
150636	−28 12326				16 40 16	−28 34.3	1	10.41		0.14		-0.06		1	B9 III/IV	1770
150573	−40 10649	HR 6206		⋆B	16 40 17	−41 01.5	3	6.21	.005	0.14	.003	0.15	.008	9	A4 V	401,486,2007
150732	+0 03569	IDS16378N0016	AB		16 40 19	+00 10.1	1	8.21		1.08		0.63		1	G0	116
		BSD 108 # 1025			16 40 21	+00 13.2	1	11.10		1.15		0.78		1	dK0	116
150608	−37 10942	HR 6210		⋆A	16 40 24	−38 03.8	1	6.05		-0.06				4	B9 II/III	2006
	−47 10999	NGC 6200 - 8			16 40 24	−47 21.9	2	10.22	.025	0.32	.000	-0.47	.010	4		537,976
150638	−31 13161	HR 6211			16 40 25	−32 00.8	2	6.47	.007	-0.08	.003	-0.34		6	B8 V	1770,2006
150591	−40 10653	HR 6209		⋆A	16 40 25	−41 01.2	3	6.13	.008	-0.08	.004	-0.53	.000	10	B6/7 V	401,486,2007
150627	−47 11001	NGC 6200 - 6		⋆A	16 40 26	−47 22.6	1	9.60		0.35		-0.49		2	B0 IV	540
150627	−47 11001	NGC 6200 - 6		⋆ABC	16 40 26	−47 22.6	2	9.18	.015	0.34	.020	-0.60	.025	12		537,1081
		NGC 6200 - 7		⋆BC	16 40 26	−47 22.6	1	10.64		0.47		-0.36		1		537
328862	−47 11000	NGC 6200 - 4			16 40 27	−47 13.5	3	10.12	.014	0.28	.013	-0.54	.031	6	B0.5III	537,954,1081
150669	−24 12787				16 40 28	−24 57.0	1	9.61		0.34		0.25		2	A6 III	1770
150574	−45 10875	LSS 3723			16 40 28	−46 02.9	4	8.50	.021	0.23	.007	-0.72	.011	11	O8/9	540,954,976,1081
		LP 685 - 47			16 40 30	−05 51.2	1	12.55		0.76		0.05		2		1696
328864	−47 11002	NGC 6200 - 5		⋆AB	16 40 30	−47 19.1	5	9.21	.013	0.32	.012	-0.52	.015	11	B1 IV	537,540,954,976,1081
		NGC 6200 - 12			16 40 30	−47 23.	1	11.06		0.32		-0.46		2		537
		NGC 6200 - 14			16 40 30	−47 23.	1	12.89		0.52		0.30		2		537
		NGC 6200 - 15			16 40 30	−47 23.	1	13.78		0.62		-0.21		1		537
		NGC 6200 - 13		⋆B	16 40 30	−47 23.	1	11.78		0.30		-0.47		2		537
151067	+62 01501				16 40 32	+62 24.1	1	7.07		0.24		-0.25		1	B8	1776
		BSD 108 # 1029			16 40 33	+00 10.2	1	10.13		1.51		1.67		1	dK2	116
		WD1640+113			16 40 33	+11 22.2	1	16.03		-0.02		-0.91		1	DA2	1727
	−48 11128	LSS 3724			16 40 33	−48 13.2	1	10.38		0.47		-0.38		2		954
151101	+64 01145	HR 6223			16 40 34	+64 41.0	2	4.86	.017	1.21	.003	1.26		6	K1p	1363,3052
		AJ74,1000 M13# 9			16 40 36	+36 26.3	1	10.83		1.31				3		209
		Ara sq 1 # 84			16 40 36	−47 49.	1	11.29		0.41		0.20		2		1081
		LP 625 - 44			16 40 38	−01 49.7	1	11.85		0.69		-0.05		2		1696
325916	−41 10824	LSS 3726			16 40 38	−41 51.6	3	9.46	.018	0.36	.011	-0.60	.013	7	O9 V	540,954,976
150714	−22 04196				16 40 39	−22 38.6	1	7.61		0.14		-0.06		2	Ap Si	1770
150713	−20 04547				16 40 40	−20 36.2	1	7.82		0.31		0.23		2	A1/2 IV	1770
	+0 03570				16 40 41	+00 01.0	2	9.34	.005	1.66	.019	2.04		3	M0	116,1584
151623	+79 00511	HR 6238			16 40 41	+79 00.8	3	6.32	.006	1.13	.015	1.22	.025	11	G9 III	71,196,7008
151625	−44 11081				16 40 41	−44 40.8	1	9.45		0.38		-0.25		3	B8	1586
150576	−52 10161	HR 6207		⋆A	16 40 42	−53 03.6	1	5.96		1.26				4	G8 III	2007
150699	−26 11503				16 40 43	−27 01.5	1	10.34		0.32		0.26		1	A8/9 V	1770
		UKST 873 # 5			16 40 44	+00 00.8	1	11.13		0.71				1		1584
150715	−24 12792				16 40 44	−25 07.5	1	10.11		0.17		0.11		1	A(p SrCrEu)	1770
150700	−29 12788				16 40 46	−29 29.7	1	8.29		0.05		-0.10		1	B9 III/IV	1770
150716	−25 11637				16 40 47	−25 41.1	1	9.71		0.20		-0.11		2	Ap Si	1770
		G 169 - 24			16 40 48	+19 27.7	2	10.84	.036	0.60	.000	-0.08	.034	8		1696,5010
		Ara sq 1 # 86			16 40 48	−47 36.	1	10.94		0.49		-0.54		2		1081
150734	−22 04197				16 40 49	−23 02.3	1	8.25		0.52		0.07		1	F5 V	1770
150717	−28 12336				16 40 49	−28 34.4	1	8.91		0.29		0.17		1	A8 V	1770
150735	−24 12793				16 40 51	−24 59.1	1	10.77		0.19		0.09		2	A1/3	1770
		AJ74,1000 M13# 10			16 40 52	+36 01.7	1	10.52		0.95				3		209
150736	−25 11638				16 40 52	−25 11.4	1	10.18		0.50		0.30		3	A1(mA9-F0)	1770
		CS 22878 # 57			16 40 54	+09 09.7	1	13.20		0.03		-0.01		1		1736
		Ara sq 1 # 89			16 40 54	−47 43.	1	11.28		0.30		-0.45		3		1081
	+0 03571				16 40 56	+00 01.5	1	10.03		0.64				2	G0	1584
150754	−29 12791				16 40 57	−29 12.3	1	9.21		0.05		-0.13		1	B9 III/IV	1770
150658	−46 10979				16 40 59	−46 56.8	1	8.87		0.13		-0.40		9	B7 Ib/II	1081
150786	−17 04624				16 41 00	−18 04.9	1	9.35		0.41		0.28		1	A1 III/IV	1770
		UKST 873 # 9			16 41 01	−00 17.0	1	13.55		0.89				1		1584
		Steph 1384			16 41 02	−08 13.2	1	11.02		1.27		1.21		1	K7	1746
150767	−23 12903				16 41 02	−23 55.2	1	8.94		0.96		0.57		2	B9 III	1770
150548	−60 06629				16 41 02	−60 28.3	1	7.55		-0.08		-0.56		8	B3 V	1732

Table 1

HD	DM	Other Id	N Rem	α_{1950}	δ_{1950}	S	V	σ_V	B–V	σ_{B-V}	U–B	σ_{U-B}	n	Spectrum	References
	−47 11015	NGC 6200 - 3		16 41 04	−47 17.8	2	11.03	.035	0.52	.010	−0.38	.065	4	B1 IV	537,954
328869				16 41 06	−47 22.	1	10.04		0.22		−0.53		3	B2 V	1081
		Ara sq 1 # 93		16 41 06	−48 02.	1	11.14		0.36		−0.53		2		1081
151044	+50 02319			16 41 08	+50 01.9	3	6.48	.032	0.53	.018	0.02	.000	5	F8 V	1007,1013,8023
150675	−47 11017			16 41 10	−47 33.2	2	7.09	.013	1.19	.004	1.04	.022	5	K1 II	1081,8100
150997	+39 03029	HR 6220	★ A	16 41 11	+39 01.0	6	3.49	.026	0.92	.007	0.59	.008	16	G8 IIIb	15,1080,1363,3016*
150768	−27 11103	HR 6216	★ AB	16 41 11	−27 21.8	3	6.58	.004	0.10	.010	0.07	.001	7	A2 V	1770,1771,2007
328712	−44 11088			16 41 11	−44 19.0	2	10.38	.020	1.42	.025	0.35	.030	6	F8	954,1737
150742	−40 10661	HR 6214	★ AB	16 41 14	−40 44.9	3	5.66	.028	−0.11	.008	−0.64	.004	12	B3 V	141,486,2009
150660	−53 08160			16 41 14	−53 51.2	1	9.78		0.10		−0.39		3	B3e	1586
150595	−58 06885			16 41 14	−58 48.7	1	9.41		−0.03		−0.41		1	B5/7 III	400
328870	−47 11018	NGC 6200 - 1		16 41 16	−47 26.5	2	10.66	.019	0.34	.000	−0.58	.024	4	B8	537,1081
150998	+36 02772			16 41 17	+36 36.1	1	6.89		1.50				3	K2	209
150814	−22 04199			16 41 17	−22 25.9	1	7.57		0.15		0.00		2	B9.5 V	1770
		UKST 873 # 8		16 41 18	+00 10.9	1	12.23		0.84				2		1584
		3C 345 # 1		16 41 18	+39 54.	2	13.84	.042	0.67	.004	0.12	.013	7		157,1595
		3C 345 # 2		16 41 18	+39 54.	2	14.23	.051	0.66	.025	0.18	.021	7		157,1595
		3C 345 # 3		16 41 18	+39 54.	1	15.29		0.58		0.04		1		157
		3C 345 # 4		16 41 18	+39 54.	2	15.21	.017	0.89	.009	0.49	.092	6		157,1595
		3C 345 # 5		16 41 18	+39 54.	2	15.18	.022	1.26	.004	1.20	.052	6		157,1595
		3C 345 # 6		16 41 18	+39 54.	1	16.52		0.65		0.00		1		157
		GD 357		16 41 19	+38 46.7	1	14.41		0.28		−0.69		3	DA	308
150875	−2 04235			16 41 20	−02 32.3	2	8.31	.020	0.58	.005	0.13		5	F5	742,1594
150816	−29 12795			16 41 24	−29 30.2	1	9.39		0.12		0.04		1	B9/A0 V	1770
151698	+78 00565			16 41 28	+77 57.8	1	8.06		1.08		0.89		2	K0	3025
150819	−31 13179			16 41 28	−31 31.0	1	9.22		0.06		−0.15		1	B9 V	1770
150772	−41 10836			16 41 29	−41 09.9	1	9.33		0.12		−0.32		4	B9 III	141
151023	+37 02792			16 41 30	+37 08.6	1	8.51		0.99				3	K0	209
		Ara sq 1 # 96		16 41 30	−48 57.	1	10.68		0.88		−0.09		3		1081
		V340 Ara		16 41 30	−51 14.5	1	9.63		1.25		1.02		1		689
234340	+53 01887			16 41 31	+53 40.7	1	9.13		0.31		−0.02		2	A7	1566
		UKST 873 # 10		16 41 31	−00 05.8	1	14.06		0.79				3		1584
		UKST 873 # 18		16 41 32	−00 15.0	1	16.54		0.90				1		1584
150846	−25 11645			16 41 33	−25 29.1	1	9.96		0.55		0.23		2	F0 V	1770
328865	−47 11020	NGC 6200 - 2		16 41 33	−47 17.4	3	10.83	.014	0.38	.008	−0.50	.026	6	B0.5II	537,954,1081
150834	−30 13426			16 41 35	−30 48.2	1	10.13		0.18		−0.07		1	B8/9 V	1770
328806		LSS 3733		16 41 36	−45 59.3	2	11.49	.269	0.35	.005	−0.41	.010	5		954,1081
		Ara sq 1 # 97		16 41 36	−47 28.	1	11.30		0.38		−0.47		2		1081
150549	−66 03009	HR 6204, LP TrA	★ A	16 41 37	−67 01.1	4	5.12	.004	−0.08	.005	−0.44	.005	12	Ap Si	15,1075,2016,2038
		UKST 873 # 15		16 41 44	+00 08.1	1	15.26		0.71				1		1584
150836	−34 11239			16 41 44	−34 12.5	1	9.65		0.21		−0.22		1	B7 II	1770
	+37 02793			16 41 45	+36 51.1	1	10.20		0.64				3	G0	209
148542	−86 00333	HR 6139		16 41 50	−86 16.9	3	6.03	.004	0.04	.005	0.02	.000	15	A1/2 IV/V	15,1075,2038
150894	−28 12358	HR 6218	★ AB	16 41 52	−28 25.1	2	6.01	.014	0.09	.003	0.13		5	A3 IV	1770,2035
151086	+36 02773			16 41 53	+36 28.3	1	8.77		1.10				3	K2	209
		AJ74,1000 M13# 8		16 41 55	+36 26.5	1	11.43		1.05				3		209
151199	+55 01872	HR 6226		16 41 57	+55 46.9	3	6.16	.005	0.08	.005	0.11	.005	6	A2 p Sr	15,1003,3050
	−15 04383			16 41 58	−15 24.8	1	10.40		0.96		0.73		2	K2 V	3072
150908	−25 11649			16 41 58	−25 45.1	1	10.10		0.48		0.32		2	A9/F0 III	1770
150893	−27 11116			16 41 58	−27 14.5	1	10.28		0.26		0.18		1	A2 IV	1770
151087	+34 02830	HR 6222		16 42 01	+34 07.8	1	6.01		0.28		0.04		2	F3 III	254
328833	−46 10996			16 42 03	−46 39.0	1	10.68		0.29		−0.35		3	B3	1081
	−47 11027	LSS 3735		16 42 03	−47 50.5	3	9.80	.013	0.63	.025	−0.30	.040	8	B0 II	954,1021,1081
150745	−58 06889	HR 6215		16 42 04	−58 24.8	1	5.74		−0.09				4	B2 III/IV	2007
151025	+9 03262			16 42 05	+09 02.2	1	8.56		0.40		0.06		2	F5	1375
		AH Her		16 42 06	+25 20.5	1	13.77		0.29		−0.77		1	K2	1471
150937	−22 04205			16 42 07	−23 05.5	1	6.98		0.44		0.07		1	F3 V	1770
		vdB 73 # e		16 42 18	−41 08.	1	14.45		0.80		−0.07		3		434
		L 74 - 113		16 42 18	−72 54.	2	11.35	.005	1.58	.025	1.25		2		1705,3073
328709	−44 11104	LSS 3736		16 42 19	−44 17.1	1	10.27		0.47		−0.38		1	B5	954
328847	−46 11002	NGC 6204 - 3		16 42 19	−46 55.9	1	9.61		0.27		−0.07		2	B9 III	251
150951	−28 12366			16 42 20	−28 59.5	1	8.90		0.18		0.13		2	A1 V	1770
150927	−37 10964			16 42 21	−38 04.6	2	9.44	.010	0.32	.003	−0.52	.021	5	B2/3 Ib	540,976
150839	−51 10482			16 42 21	−52 07.6	2	8.76	.003	0.21	.005	−0.42	.002	5	B3 III	540,976
151188	+43 02639	IDS16408N4340	AB	16 42 23	+43 34.0	3	8.36	.018	1.05	.014	0.92	.000	8	K5	938,1381,3072
328848	−46 11003	NGC 6204 - 5		16 42 23	−46 54.6	1	10.32		0.28		0.05		2	A2 III	251
150884	−46 11005			16 42 24	−46 23.7	3	7.02	.007	0.85	.008	0.47	.027	8	F8 II	403,1081,8100
150970	−28 12368			16 42 25	−28 45.0	1	7.63		0.14		0.14		2	A1 IV/V	1770
234342	+52 01993			16 42 26	+52 35.7	1	8.74		1.39		1.49		2	K7	1566
150897	−46 11008			16 42 26	−46 26.5	1	6.47		0.10		0.04		3	A2 V	1081
		NGC 6204 - 13		16 42 26	−46 55.3	1	12.73		0.57		0.46		2		251
		NGC 6204 - 9		16 42 26	−46 55.6	1	12.52		0.36		0.20		2		251
		NGC 6204 - 11		16 42 26	−46 56.8	1	12.92		0.48		0.46		2		251
	+21 02978	G 169 - 25		16 42 27	+20 49.9	1	9.20		0.71		0.27		2	G8	1064
		NGC 6204 - 7		16 42 27	−46 55.2	1	12.29		0.63		0.55		2		251
		NGC 6204 - 8		16 42 27	−46 55.4	1	11.88		0.31		0.07		2		251
	−46 11007	NGC 6204 - 2		16 42 27	−46 56.1	1	10.70		0.25		−0.14		2		251
	−46 11006	NGC 6204 - 1	★ AB	16 42 27	−46 56.1	1	10.82		0.27		−0.11		2		251
		NGC 6204 - 12		16 42 29	−46 56.7	1	13.22		0.52		0.46		2		251
		vdB 73 # c		16 42 30	−41 07.	1	11.36		0.43		−0.40		3		434

HD	DM	Other Id	N	Rem	α_{1950}	δ_{1950}	S	V	σ_V	B–V	σ_{B-V}	U–B	σ_{U-B}	n	Spectrum	References
328855	−46 11009	NGC 6204 - 4			16 42 31	−46 54.8	1	9.76		0.32		−0.12		2	B7 IV	251
		NGC 6204 - 10			16 42 31	−46 55.2	1	13.00		0.45		0.47				251
151090	+6 03288	G 17 - 34		⋆B	16 42 32	+06 08.2	2	10.31	.014	1.04	.000	0.92	.011	5		333,1620,3016
151010	−18 04321				16 42 33	−18 23.9	1	9.46		0.46		0.30		1	A9 V	1770
150985	−26 11529				16 42 33	−26 28.7	1	9.94		0.32		0.13		2	A9 V	1770
151090	+6 03288	G 17 - 35		⋆A	16 42 34	+06 11.0	2	6.55	.005	0.89	.014	0.51	.005	9	K0 V	333,1620,3077
151061	−2 04242	V2111 Oph			16 42 34	−02 59.6	1	6.90		1.79		1.64		1	M6 III	3040
151001	−25 11655				16 42 35	−25 36.4	1	9.31		0.39		0.12		1	F0 V	1770
328872	−47 11035	LSS 3738			16 42 36	−47 34.2	2	10.07	.025	0.35	.025	−0.41	.030	5	B2 V	954,1081
	−40 10692				16 42 41	−40 43.4	1	10.15		1.75		2.02		3		486
151012	−26 11533				16 42 42	−26 33.5	1	7.03		0.04		0.02		2	B9.5 V	1770
150954	−40 10691				16 42 42	−41 07.8	1	8.66		1.16		0.91		4	G8 III	486
	−44 11110				16 42 42	−44 22.0	1	11.93		0.63		−0.28		1		1586
		Ara sq 1 # 104			16 42 42	−46 55.	1	10.91		0.36		−0.52		3		1081
151541	+68 00883	G 240 - 25			16 42 44	+68 11.3	4	7.56	.016	0.77	.014	0.34	.017	9	K1 V	22,308,1658,3026
	+36 02775				16 42 45	+36 20.9	1	8.80		1.42				3	K5	209
		Steph 1392			16 42 47	+47 52.3	1	11.02		1.30		1.27		1	K7	1746
151033	−25 11657				16 42 47	−25 35.9	1	9.83		0.55		0.05		1	F3 V	1770
	−46 11014				16 42 47	−46 56.6	1	10.49		0.59		0.10		1		1081
		LSS 3744			16 42 48	−32 29.3	1	12.56		0.35		−0.30		2	B3 Ib	954
150974	−40 10696				16 42 48	−40 44.9	1	9.07		0.33		0.19		4	A1mA8-A9	486
		vdB 73 # d			16 42 48	−41 06.	1	14.28		0.92		−0.08		3		434
		Ho 22 - 10			16 42 48	−46 59.9	1	12.64		2.05		2.04		2		251
		Ho 22 - 5			16 42 49	−46 58.6	1	10.47		0.58		0.07		2		251
		Ho 22 - 9			16 42 49	−46 59.0	1	13.00		0.92		0.54		2		251
328871	−47 11037				16 42 50	−47 27.6	1	10.38		0.44		−0.40		2	A2	1081
328856	−46 11016	Ho 22 - 2			16 42 52	−46 59.3	3	8.50	.004	0.43	.017	−0.55	.015	20	O9 II	251,954,1081
		AJ74,1000 M13# 12			16 42 53	+36 37.9	1	10.79		0.65				3		209
		MCC 773			16 42 54	+21 12.8	1	10.43		1.17				1	K5	1017
	−46 11017	Ho 22 - 3			16 42 54	−46 59.8	4	9.21	.014	0.39	.012	−0.58	.018	6	B0 IV	251,954,1081,1737
		Ho 22 - 6			16 42 56	−47 00.2	1	12.45		0.44		−0.11		2		251
151133	+1 03298	HR 6224			16 42 57	+01 06.6	3	6.03	.014	−0.01	.005	−0.15	.014	12	B9.5III	1079,1088,3016
151202	+23 02990				16 42 57	+23 48.5	1	7.35		0.96		0.83		2	K0	1625
150991	−41 10855				16 42 57	−41 09.6	1	9.36		0.18		−0.42		4	B3 III	486
150958	−46 11019	Ho 22 - 1		⋆AB	16 42 58	−46 59.8	7	7.29	.012	0.35	.013	−0.63	.014	27	O7e	251,540,954,976,1081*
		vdB 73 # b			16 43 00	−41 08.	1	13.57		0.87		0.30		1		434
		Ho 22 - 4			16 43 00	−47 01.	2	10.25	.015	0.33	.020	−0.58	.005	5	B0 III	251,1081
		Ho 22 - 7			16 43 00	−47 01.	1	12.58		0.60		−0.31		2		251
		Ho 22 - 8			16 43 00	−47 01.	1	14.31		0.40		−0.02		2		251
		L 110 - 59			16 43 00	−68 55.	1	12.29		0.65		−0.20		2		3062
150898	−58 06893	HR 6219			16 43 03	−58 15.1	5	5.57	.007	−0.07	.012	−0.93	.048	17	B0.5 Ia	26,1018,2001,2007,8100
151003	−41 10856	LSS 3745			16 43 04	−41 31.2	6	7.07	.014	0.18	.010	−0.77	.012	22	O9.5Iab/b	141,158,486,540,954,976
151237	+28 02607	IDS16411N2832	AB		16 43 06	+28 26.9	1	7.20		0.50		0.14		2	F8 II	3016
		CS 22878 # 101			16 43 07	+08 20.2	1	13.79		0.80		0.21		1		1736
151203	+16 03013	HR 6227			16 43 07	+15 50.2	1	5.57		1.66				1	M3 IIIab	71
151074	−28 12384				16 43 07	−29 04.6	1	9.44		0.27		0.12		1	A8/9 V	1770
		LSS 3749			16 43 10	−30 13.1	1	12.54		0.63		0.53		2	B3 Ib	954
	−41 10857	LSS 3748			16 43 10	−41 09.5	2	11.41	.017	0.40	.012	−0.38	.011	8		486,954
150930	−56 07854				16 43 10	−56 56.7	1	8.86		0.08		−0.26		4	B7 III	1586
151168	−0 03178	IDS16406S0035	AB		16 43 12	−00 40.4	1	8.36		0.55		−0.04		3	F8	3030
151039	−40 10707				16 43 14	−41 03.6	2	8.69	.036	0.19	.011	−0.40	.005	7	B3 III	141,486
151288	+33 02777	G 181 - 6			16 43 15	+33 35.7	4	8.10	.005	1.37	.003	1.29	.007	16	K7 V	1,1067,1197,3072
151426	+55 01873	G 202 - 70			16 43 17	+55 00.7	1	8.08		0.63		0.11		2	G0	3016
151018	−45 10927	LSS 3746			16 43 17	−45 47.9	6	8.72	.021	0.62	.009	−0.41	.022	15	O9.5Iab	540,954,976,1081,1737,8100
		L 21 - 38			16 43 18	−83 05.	1	11.12		0.63		0.10		2		1696
151063	−40 10712				16 43 21	−40 47.1	1	9.06		0.30		0.18		4	A1mA7-A7	486
326003	−40 10710				16 43 21	−41 05.0	1	10.01		0.20		−0.41		4	B9	486
151051	−41 10863				16 43 21	−41 23.5	3	7.35	.014	1.79	.007	2.06	.027	9	M2/3 II	486,2012,3040
150798	−68 02822	HR 6217			16 43 21	−68 56.3	4	1.91	.012	1.45	.007	1.55	.024	12	K2 IIb-IIIa	15,1034,1075,2038
		WD1643+143			16 43 22	+14 23.1	1	15.38		0.38		−0.82		1	DA1	1727
151078	−39 10677	HR 6221			16 43 22	−39 17.2	1	5.48		0.98				1	K0 III	2007
328857	−47 11047	LSS 3747			16 43 22	−47 13.9	3	9.95	.005	0.54	.023	−0.41	.025	8	B0.5Ia:	954,1021,1081
151136	−23 12913				16 43 23	−23 22.9	1	9.60		0.52		0.28		2	A2 III	1770
151127	−28 12387				16 43 23	−28 26.5	1	9.67		0.36		0.21		1	A9 V	1770
151217	+8 03271	HR 6228		⋆A	16 43 26	+08 40.3	5	5.14	.006	1.54	.004	1.92	.011	20	K5 III	3,15,1080,1256,3016
151125	−26 11540				16 43 26	−26 13.9	1	9.95		0.19		0.19		2	A1 III/IV	1770
151040	−45 10931				16 43 26	−46 04.8	1	9.98		0.23		−0.29		3	B7/9 (III)	1081
151149	−22 04207				16 43 28	−22 13.6	1	10.06		0.58		0.08		1	F2 V	1770
151150	−22 04208	IDS16405S2227	A		16 43 29	−22 32.3	1	8.36		0.58		0.10		1	F5 V	1770
151126	−27 11140				16 43 29	−27 23.5	1	8.28		0.26		0.21		2	A5 IV	1770
		NGC 6204 - 6			16 43 29	−46 54.8	1	11.86		0.45		0.25		2		251
326004	−41 10870				16 43 33	−41 08.9	1	10.50		0.22		−0.20		5	B8	486
	−30 13458				16 43 34	−30 32.0	1	10.58		1.41		1.25		4	K5	158
326005	−41 10869				16 43 35	−41 13.3	1	10.11		0.66		0.17		4	K0	486
151109	−39 10684				16 43 36	−39 26.7	1	7.00		−0.06		−0.25		3	B9 IV/V	141
151554	+61 01609				16 43 37	+61 03.6	1	7.89		0.11		−0.02		1	A0	1776
		CS 22878 # 88			16 43 38	+10 05.4	1	13.67		0.09		0.14		1		1736
151388	+43 02642	HR 6230			16 43 38	+43 18.5	1	6.05		1.40				1	K4 III	71
151110	−39 10687				16 43 39	−39 29.6	1	8.48		0.13		0.09		2	B9/9.5V	141
	+34 02835				16 43 40	+33 55.3	1	9.45		0.96		0.66		2	K5	3072

Table 1   907

HD	DM	Other Id	N	Rem	α_{1950}	δ_{1950}	S	V	σ_V	B–V	σ_{B-V}	U–B	σ_{U-B}	n	Spectrum	References
150869	−66 03017				16 43 40	−66 59.7	1	6.87		1.56				4	K5 III	2012
	+21 02982				16 43 41	+21 42.9	1	9.26		0.39		0.02		2	F5	1733
326006	−41 10872	LSS 3751			16 43 41	−41 11.2	4	10.13	.022	0.17	.010	−0.69	.010	15	B2 V	141,434,486,954
151137	−33 11444				16 43 42	−34 01.8	1	10.63		0.35		0.22		1	B9/A0 V	1770
151193	−17 04632	IDS16408S1732		AB	16 43 43	−17 37.7	1	9.74		0.65		0.39		1	A8/9 V	1770
151170	−28 12394				16 43 43	−28 09.7	1	9.93		0.08		0.04		1	B9/A0 IV	1770
151111	−41 10873	LSS 3752			16 43 45	−41 31.9	2	7.96	.008	0.49	.009	0.57	.025	6	A5 II	486,954
151367	+30 02871	IDS16418N3011		A	16 43 46	+30 05.7	1	8.77		0.35		0.02		3	G5	3032
151367	+30 02871	IDS16418N3011		B	16 43 46	+30 05.7	1	9.63		0.63		0.15		3		3032
151368	+30 02872	IDS16418N3011		C	16 43 46	+30 05.7	1	8.80		0.38		0.00		3	G5	3032
151368	+30 02872	IDS16418N3011		D	16 43 46	+30 05.7	1	10.96		0.54		0.05		1	G5	3032
151179	−25 11667	HR 6225			16 43 47	−25 26.4	1	6.71		1.18				4	K0 II	2007
326013	−41 10874				16 43 48	−41 15.1	1	11.52		0.24		−0.09		5	A3	486
151097	−47 11052				16 43 48	−47 30.1	1	7.24		1.13		0.78		2	F6 Iab	403
	−50 10813				16 43 49	−50 30.9	1	10.53		0.43		−0.12		3	B3	1586
151367	+30 02871	IDS16418N3011		S	16 43 50	+30 05.0	1	13.70		1.17				1		3032
151157	−34 11259				16 43 50	−35 02.6	1	9.57		0.44		0.27		1	B9 IV	1770
151113	−45 10941	IDS16402S4517		AB	16 43 52	−45 22.3	1	7.49		0.02		−0.50		3	B5 II/III	1586
151115	−48 11174	IDS16400S4808		B	16 43 52	−48 14.8	1	8.13		0.41		0.00		1	F3/8 (III)	1081
		HBC 651		V	16 43 53	−15 09.3	1	13.10		1.06		−0.70		1	K7-M0	1763
151139	−41 10875	LSS 3753			16 43 53	−41 43.2	5	7.57	.011	0.28	.012	−0.57	.013	14	B2 Ib/II	141,486,540,954,976
		He3 1254			16 43 54	−15 08.	1	13.02		1.10		−0.91		1		1586
326012	−41 10878				16 43 55	−41 15.4	1	10.84		1.15		0.87		3	K0	486
329034	−46 11028	IDS16402S4657		AB	16 43 55	−47 02.3	1	10.31		0.43		−0.36		2	B3 V	1081
151083	−51 10503	LSS 3750			16 43 57	−51 40.8	4	8.99	.029	0.18	.007	−0.70	.017	10	B2 Vn	540,954,976,1586
		G 17 - 36			16 43 58	−04 18.7	1	14.09		1.47		1.24		1		3062
		AA48,187 # 14			16 43 58	−41 16.0	1	11.70		1.70		1.84		2		486
		G 138 - 64			16 44 00	+16 34.3	2	11.63	.009	1.49	.004	1.21		6		1723,1759
151428	+35 02867	IDS16422N3549		AB	16 44 00	+35 43.2	1	7.34		0.48		0.03		2	G0	1601
151005	−61 05784				16 44 00	−61 24.1	1	7.19		1.15		1.06		5	K0 III	1704
151369	+26 02896				16 44 01	+26 07.6	1	8.42		0.77		0.31		3	G2 IV	3016
151158	−42 11540	LSS 3754		⋆ AB	16 44 02	−42 47.3	1	8.22		0.22		−0.49		2	B2 Ib/II	954
		AA48,187 # 16			16 44 03	−41 18.7	1	11.36		1.92		2.29		2		486
326014	−41 10882				16 44 05	−41 18.9	1	11.78		0.42		0.31		2	A0	486
		AA48,187 # 17			16 44 08	−41 19.3	1	12.13		1.45		1.60		2		486
326015	−41 10884	AA48,187 # 18			16 44 08	−41 20.4	1	10.77		1.95		2.70		2	K7	486
		AA48,187 # 23			16 44 09	−41 30.3	1	12.78		0.44		0.33		2		486
326037	−41 10885				16 44 11	−41 36.5	1	9.74		0.44		0.05		2	F5	486
326019					16 44 12	−41 28.6	1	11.94		0.48		0.29		2	A5	486
		AA48,187 # 22			16 44 12	−41 29.6	1	12.99		0.56		−0.09		2		486
		AA48,187 # 27			16 44 13	−41 38.9	1	10.78		1.42		1.34		2		486
		AA48,187 # 20		AB	16 44 15	−41 27.2	1	12.80		0.50		0.30		3		486
		CpD -41 07625			16 44 15	−41 30.4	1	11.88		0.52		0.09		2		486
326018	−41 10890				16 44 17	−41 30.7	1	11.30		0.26		−0.01		2	A0	486
326017	−41 10892				16 44 18	−41 25.1	1	10.96		0.27		0.25		2	A2	486
329033	−46 11032	LSS 3755			16 44 18	−47 07.9	3	10.01	.024	0.50	.015	−0.38	.025	8	B1 III	954,1021,1081
151482	+34 02839				16 44 19	+33 53.0	1	7.90		0.29		0.09		2	A9 III	1625
151271	−24 12826				16 44 19	−24 37.1	1	9.51		0.53		0.06		1	F5 V	1770
151261	−27 11155				16 44 19	−27 44.8	1	10.10		0.39		0.17		1	A8/9 (V)	1770
151613	+57 01702	HR 6237		⋆	16 44 21	+56 52.2	7	4.84	.016	0.37	.017	−0.06	.006	15	F2 V	15,1118,1193,1363*
326040	−41 10891	LSS 3757			16 44 21	−41 53.4	2	10.14	.069	0.30	.000	−0.50	.000	5	B3	486,954
		G 226 - 24			16 44 22	+58 54.1	1	13.69		1.36		1.12		1		1658
151212	−40 10727	LSS 3758			16 44 22	−41 03.2	5	9.19	.023	0.11	.008	−0.76	.017	18	B1/2 Ib	141,486,540,954,976
329032	−47 11059				16 44 23	−47 09.2	2	10.80	.005	0.59	.009	−0.27	.019	4	B2 V	954,1081
151309	−21 04412				16 44 24	−21 41.3	1	9.82		0.64		0.36		2	A0 III	1770
151211	−40 10729	IDS16409S4047		A	16 44 25	−40 53.0	1	8.20		0.47		−0.05		4	F5 V	486
151196	−45 10951				16 44 26	−45 51.4	1	6.67		0.37		0.04		5	F2 IV	1081
	−74 01569				16 44 27	−74 27.1	1	10.15		−0.17		−1.01		3	O9.5V	400
		Ton 261			16 44 30	+26 43.	2	16.17	.025	−0.27	.010	−0.97	.005	3	DOp	1036,3028
151310	−21 04414				16 44 30	−22 05.1	3	9.38	.013	0.03	.003	−0.59	.007	7	B3 IV	400,1054,1770
151241	−39 10703				16 44 31	−39 52.6	1	7.88		0.18		0.14		3	A0 III/IV	141
151274	−33 11452				16 44 34	−34 00.0	1	10.04		0.19		−0.17		1	B9 III/IV	1770
151650	+56 01917				16 44 36	+55 51.8	1	8.28		0.95		0.68		2	G5	1502
		NGC 6218 - 1			16 44 36	−01 52.	1	10.49		0.64		0.33		3		90
		NGC 6218 - 2			16 44 36	−01 52.	1	10.63		1.11		1.03		2		90
		NGC 6218 - 3			16 44 36	−01 52.	1	10.60		1.21		1.24		5		90
		NGC 6218 - 4			16 44 36	−01 52.	1	10.86		0.15		−0.19		4		90
		NGC 6218 - 5			16 44 36	−01 52.	1	10.79		1.36		1.58		4		90
		NGC 6218 - 6			16 44 36	−01 52.	1	12.12		1.68		2.08		4		90
		NGC 6218 - 7			16 44 36	−01 52.	1	12.80		1.40		1.37		2		90
		NGC 6218 - 8			16 44 36	−01 52.	1	13.81		0.76				1		90
		NGC 6218 - 9			16 44 36	−01 52.	1	13.80		0.65		0.07		2		90
		NGC 6218 - 10			16 44 36	−01 52.	1	14.03		1.05				2		90
		NGC 6218 - 11			16 44 36	−01 52.	1	14.38		0.74		0.24		4		90
		NGC 6218 - 12			16 44 36	−01 52.	1	15.26		1.23				1		90
		NGC 6218 - 13			16 44 36	−01 52.	1	15.73		0.16				1		90
		NGC 6218 - 14			16 44 36	−01 52.	1	16.23		0.13				1		90
		NGC 6218 - 15			16 44 36	−01 52.	1	16.74		0.76				1		90
		NGC 6218 - 17			16 44 36	−01 52.	1	16.87		0.99				3		90
		NGC 6218 - 50			16 44 36	−01 52.	1	12.29		1.48		1.63		1		90

HD	DM	Other Id	N	Rem	α_{1950}	δ_{1950}	S	V	σ_V	B–V	σ_{B-V}	U–B	σ_{U-B}	n	Spectrum	References
		NGC 6218 - 51			16 44 36	−01 52.	1	12.41		1.00				1		90
		NGC 6218 - 52			16 44 36	−01 52.	1	13.95		0.96		0.48		1		90
		NGC 6218 - 53			16 44 36	−01 52.	1	14.63		1.22				1		90
		NGC 6218 - 54			16 44 36	−01 52.	1	15.84		0.87				1		90
		NGC 6218 - 55			16 44 36	−01 52.	1	17.33		1.02				1		90
		Wes 1 - 1			16 44 36	−45 45.	1	11.71		1.99		1.90		2		33
151213	−47 11061	LSS 3759			16 44 36	−47 11.6	5	7.65	.008	0.29	.010	-0.63	.011	26	B1 Iab	403,540,954,976,1081
151431	+2 03175	HR 6232		⋆ A	16 44 38	+02 09.2	2	6.10	.019	0.14	.005	0.14	.009	9	A3 V	1088,3032
151431	+2 03175	IDS16421N0215		B	16 44 38	+02 09.2	1	9.69		1.12		1.01		2		3032
		CS 22878 # 97			16 44 39	+08 44.3	1	15.01		0.20				1		1736
		LP 806 - 4			16 44 40	−19 39.1	1	12.02		1.50		1.22		1		1773
151358	−14 04484				16 44 41	−14 32.3	1	9.88		0.49		0.31		1	B9.5 IV	1770
151311	−29 12850				16 44 41	−29 38.4	1	10.27		0.10		-0.09		1	B9 IV/V	1770
	−0 03182	G 19 - 2			16 44 42	−01 05.9	2	10.77	.044	1.32	.019	1.19	.010	3	K6	1620,3062
151346	−23 12923				16 44 45	−23 53.2	3	7.91	.014	0.42	.009	-0.16	.014	9	B8 II	26,1054,1770
		Smethells 1			16 44 48	−64 58.	1	10.33		1.33				1		1494
151347	−28 12413				16 44 49	−28 17.5	1	8.14		0.41		-0.01		1	F2 V	1770
		LSS 3761			16 44 55	−31 16.8	1	13.38		0.57		0.14		2	B3 Ib	954
151651	+47 02381				16 44 59	+47 37.9	1	8.01		0.44		-0.07		2	F5	1733
151377	−24 12833				16 44 60	−24 37.7	1	10.52		0.30		0.23		1	A1 IV	1770
		vdB 74 # a			16 45 00	−48 00.0	1	11.21		0.15		-0.12		3	B8 V	434
151614	+39 03043				16 45 08	+39 11.6	1	8.72		0.45		-0.02		1	F4 V	1501
151300	−46 11044	LSS 3760		⋆ AB	16 45 09	−47 04.8	4	9.29	.006	0.46	.015	-0.53	.015	12	O7f	540,954,976,1081
151432	−19 04429				16 45 10	−19 17.1	1	10.12		0.47		0.32		1	A1 III	1770
151318	−48 11194				16 45 10	−48 12.6	1	8.29		0.03		-0.52		6	B3 IV	1081
151415	−24 12834	IDS16422S2421		A	16 45 11	−24 26.3	2	7.08	.020	1.69	.005	2.04		7	K4 III	2012,3040
151433	−21 04417				16 45 12	−21 26.1	1	10.42		0.40		0.29		1	A0 IV	1770
151395	−30 13483				16 45 13	−31 07.0	3	6.82	.010	-0.04	.013	-0.50	.015	7	B4 V	540,976,1770
151418	−30 13487				16 45 16	−30 16.6	1	8.26		-0.04		-0.31		2	B9 IV	1770
151337	−47 11068				16 45 16	−47 37.9	3	7.38	.000	0.91	.005	0.63	.021	10	K0 IV/V	158,1075,3008
		NGC 6208 - 110			16 45 16	−53 34.1	1	15.06		0.86		0.30		4		185
		NGC 6208 - 111			16 45 16	−53 34.4	1	13.82		1.57				3		185
		NGC 6208 - 27			16 45 17	−53 38.0	1	11.09		1.20		0.96		4		185
	+23 02998				16 45 18	+23 18.2	2	9.53	.377	1.20	.005	1.13		2	R0	1238,8005
152303	+77 00634	HR 6267		⋆ AB	16 45 18	+77 36.1	2	5.99	.004	0.41	.002	-0.04	.005	4	F4(v)V	1733,3016
		Ara sq 1 # 123			16 45 18	−47 17.	1	11.01		0.38		-0.26		3		1081
	−51 10517				16 45 18	−51 47.0	1	10.74		0.12				4		1021
151525	+5 03272	HR 6234, V776 Her		⋆ A	16 45 19	+05 20.1	3	5.23	.010	-0.02	.006	-0.01	.014	16	B9 p (Cr)	1088,1202,3032
151525	+5 03272	IDS16428N0526		B	16 45 19	+05 20.1	1	11.16		1.00		0.69		2		3032
		NGC 6208 - 113			16 45 20	−53 35.2	1	13.80		0.62		0.31		3		185
151420	−32 12038				16 45 21	−32 50.0	1	9.91		0.21		0.06		1	B9 IV	1770
		NGC 6208 - 31			16 45 21	−53 37.1	1	11.60		1.17		0.89		4		185
151468	−20 04562				16 45 22	−21 04.6	1	10.48		0.54		0.35		1	B8/9 IV/V	1770
		NGC 6208 - 21			16 45 22	−53 37.7	1	11.66		0.18		0.07		3	B7	185
	−53 08183	NGC 6208 - 78			16 45 22	−53 39.5	1	9.96		1.09		0.81		3	K0	185
151397	−39 10720	LSS 3762			16 45 23	−39 41.0	3	9.77	.008	0.19	.006	-0.70	.013	9	B1 Ib	540,954,976
		NGC 6208 - 84			16 45 23	−53 40.1	1	12.93		0.69		0.15		2		185
151495	−15 04393				16 45 24	−15 16.4	1	7.75		1.28		0.97		13	G8 III	1763
		NGC 6208 - 115			16 45 24	−53 34.3	1	14.26		0.63		0.19		4		185
		CS 22878 # 122			16 45 25	+11 49.8	1	15.78		0.10				1		1736
151625	+28 02613				16 45 25	+28 29.2	1	8.03		0.60		0.15		3	G0 IV	3016
		GD 358, V777 Her			16 45 25	+32 33.7	2	13.68	.047	-0.18	.103	-1.03	.019	4	DB	308,974
		NGC 6208 - 19			16 45 25	−53 37.6	1	10.88		1.29		1.13		2	K0	185
		NGC 6208 - 1			16 45 27	−53 36.8	1	12.79		0.67		0.31		2		185
151249	−58 06906	HR 6229		⋆ A	16 45 27	−58 57.3	3	3.75	.004	1.56	.004	1.93	.000	11	K5 III	15,1075,2012
		CS 22878 # 121			16 45 30	+11 44.5	1	13.99		0.54		0.01		1		1736
	+86 00250	G 256 - 36			16 45 30	+85 59.0	1	10.41		0.82		0.36		3	K0	308
		NGC 6208 - 121			16 45 31	−53 34.3	1	13.00		0.68		0.13		2		185
151809	+56 01919				16 45 34	+56 27.9	1	9.50		-0.16		-0.71		3		963
		NGC 6208 - 48			16 45 34	−53 35.2	1	11.65		1.31		1.05		3		185
150995	−73 01759				16 45 34	−73 38.2	1	6.71		-0.01		-0.16		8	B8/9 V	1628
		NGC 6208 - 49			16 45 35	−53 35.8	1	13.41		0.56		0.11		2		185
		NGC 6208 - 68			16 45 35	−53 40.3	1	11.49		0.75		0.08		2	A7	185
		CS 22878 # 104			16 45 36	+09 28.0	1	14.75		0.04				1		1736
151527	−14 04486	HR 6235			16 45 36	−14 49.3	1	6.03		0.20				4	A0 Vn	2007
151496	−21 04420				16 45 36	−21 51.2	1	8.99		0.23		0.07		2	B9 V	1770
151504	−19 04431				16 45 37	−19 11.8	1	8.09		0.78				4	G8 V	2012
151485	−28 12431				16 45 39	−29 04.1	1	9.10		0.22		0.10		1	A1 V	1770
		NGC 6208 - 9			16 45 39	−53 37.8	1	12.18		1.16		0.84		2		185
		Ton 262			16 45 42	+26 45.	1	15.42		0.07		-0.33		1		1036
151732	+42 02749	HR 6242, V636 Her			16 45 44	+42 19.6	2	5.85	.018	1.61	.005	1.60		3	M4+III-IIIa	71,3001
151616	+11 03045				16 45 45	+11 13.2	1	7.42		0.94		0.60		2	K0	1648
	+13 03224	V652 Her			16 45 46	+13 20.9	4	10.55	.009	-0.18	.012	-0.97	.007	40	B2	308,843,1775,3016
		CS 22878 # 116			16 45 48	+10 36.3	1	14.65		0.49		-0.17		1		1736
151458	−40 10757				16 45 48	−40 43.3	3	9.63	.039	0.14	.018	-0.44	.006	9	B3/5 II	141,1021,1586
		NGC 6208 - 54			16 45 48	−53 38.2	1	11.48		0.54		0.08		2		185
151507	−25 11695				16 45 49	−25 25.3	1	8.68		0.19		0.17		1	A2/3 V	1770
151627	+13 03225	HR 6239			16 45 50	+13 40.7	1	6.35		0.88				1	G5 III	71
	−47 11070				16 45 50	−47 19.5	1	11.06		1.42		0.28		1		1586
		LP 806 - 8			16 45 54	−15 38.9	1	10.94		1.50				1		1746

Table 1 909

HD	DM	Other Id	N	Rem	α_{1950}	δ_{1950}	S	V	σ_V	B–V	σ_{B-V}	U–B	σ_{U-B}	n	Spectrum	References
151473	−43 11139	IDS16424S4346	AB		16 45 56	−43 51.5	1	7.47		0.19		0.03		2	A0 IV/V	401
151510	−32 12048				16 45 57	−32 40.8	1	9.06		0.13		-0.12		1	B9 III/IV	1770
151532	−30 13491				16 45 58	−30 24.7	1	10.35		0.14		0.03		1	B9 V	1770
151699	+25 03137				16 46 01	+25 38.0	1	8.14		0.40		-0.04		2	F8	105
		IBVS1668# 1			16 46 04	−40 50.8	1	10.42		0.14		-0.56		1		799
		CS 22878 # 111			16 46 05	+09 21.8	1	14.25		0.04		0.12		1		1736
151558	−24 12847				16 46 05	−24 49.2	1	10.50		0.67		0.27		1	B9/A0 III	1770
238610	+59 01757				16 46 06	+59 45.3	1	9.44		0.33		-0.06		1		1776
151576	−19 04434	IDS16432S1926	A		16 46 06	−19 31.2	1	9.51		0.50		0.09		1	A3/5 V	1770
151576	−19 04434	IDS16432S1926	B		16 46 06	−19 31.2	1	9.72		0.53		0.00		1		1770
151560	−28 12441				16 46 06	−28 37.4	1	9.78		0.42		0.27		1	Ap Sr(EuCr)	1770
151561	−28 12442				16 46 06	−29 02.1	1	8.13		0.20		0.15		5	A4 V	1770
151475	−46 11054				16 46 07	−47 02.6	3	8.09	.082	0.17	.001	-0.53	.000	22	B3 II/III	540,976,1081
		IBVS1668# 3			16 46 13	−40 50.9	1	12.10		0.41		0.30		1		799
		IBVS1668# 2			16 46 14	−40 50.3	1	11.36		0.41		0.25		1		799
151595	−25 11699				16 46 15	−25 09.5	1	9.47		0.31		0.17		1	A7 IV	1770
		IBVS1668# 6			16 46 16	−40 52.8	1	12.34		0.28		-0.05		1		799
		IBVS1668# 7			16 46 16	−40 52.9	1	13.38		0.48		0.37		1		799
		IBVS1668# 4			16 46 17	−40 51.3	1	12.17		0.98		0.62		1		799
151515	−41 10925	LSS 3765			16 46 17	−41 55.0	11	7.17	.015	0.16	.017	-0.77	.016	31	O6(f)	32,158,390,486,540,954*
152011	+63 01303				16 46 18	+62 53.2	1	8.24		0.10		-0.06		1	A2	1776
		IBVS1668# 8			16 46 18	−40 53.1	1	12.72		1.29		1.06		1		799
		IBVS1668# 9			16 46 19	−40 52.7	1	14.84		0.75		0.21		1		799
		IBVS1668# 10			16 46 20	−40 51.1	1	12.17		0.99		0.61		4		799
326243	−43 11145	LSS 3766			16 46 20	−43 15.3	1	10.19		0.45		-0.05		3	A0 II	954
		IBVS1668# 5			16 46 21	−40 51.6	1	14.56		0.51		0.44		1		799
		V567 Sco			16 46 22	−40 50.0	1	12.21		2.17		2.21		1		799
		IBVS1668# 13			16 46 22	−40 50.1	1	13.01		1.59		1.80		1		799
326189	−41 10926				16 46 23	−42 00.7	1	10.11		1.32		1.22		2	K7	1583
		IBVS1668# 11			16 46 25	−40 49.7	1	13.90		0.52		0.41		1		799
		IBVS1668# 12			16 46 25	−40 49.7	1	13.58		0.57		0.33		1		799
		HRC 270, V1121 Oph			16 46 26	−14 18.4	1	11.41		1.23		0.06		1	K5	1763
326188	−41 10927				16 46 26	−41 57.7	1	9.80		0.51		0.11		2	F0	1583
326138	−40 10778	LSS 3767			16 46 27	−41 05.2	2	10.09	.008	0.12	.017	-0.67	.012	5	B3	486,954
326190					16 46 27	−42 00.8	1	11.01		0.42		0.18		2		1583
		AAS57,205 # 305			16 46 29	−41 49.7	1	11.23		1.40		1.80		2		1583
151582	−38 11234	LSS 3770			16 46 30	−38 27.5	2	9.44	.003	0.34	.003	-0.49	.017	4	B3II/III(e)	540,954
151564	−41 10930	LSS 3768			16 46 31	−41 32.1	7	8.00	.015	0.12	.009	-0.73	.015	17	B0 III	32,486,540,954,976*
	−41 10929				16 46 33	−41 50.6	1	11.48		0.53		0.38		2		1583
329028	−46 11061	IDS16429S4657	AB		16 46 33	−47 01.0	1	10.68		0.53		0.05		3	B9	1586
151676	−15 04395	HR 6240, V1010 Oph			16 46 36	−15 34.9	1	6.20		0.23				4	A3 V	2007
151658	−21 04422	V2106 Oph			16 46 36	−21 46.0	3	7.40	.009	1.94	.009	2.12	.004	27	M2/3 III	1657,2007,3040
151637	−28 12451				16 46 36	−28 54.6	1	10.26		0.37		0.13		1	A8/9 III	1770
		LSS 3773			16 46 38	−31 17.8	1	12.59		0.47		-0.03		2	B3 Ib	954
151636	−26 11581				16 46 40	−26 52.8	1	9.12		0.20		0.14		1	A2 V	1770
		LSS 3772			16 46 40	−41 58.9	1	11.49		0.92		-0.09		2	B3 Ib	1737
151659	−24 12857				16 46 41	−24 33.2	1	6.74		0.14		0.09		1	A1mA3/5-A7	1770
326178	−41 10933				16 46 44	−41 53.0	1	11.12		0.36		0.03		2	B5	1583
329027	−46 11063	LSS 3769			16 46 44	−47 04.2	5	9.86	.019	0.82	.011	-0.23	.027	15	B0 Ia	403,954,1021,1081,1737
151641	−32 12061				16 46 45	−32 22.9	1	9.19		0.18		-0.07		1	B9.5 IV/V	1770
328990	−45 10984	LSS 3771			16 46 45	−46 04.2	1	10.61		0.38		-0.31		2	A0	954
151678	−24 12858				16 46 46	−24 13.6	1	9.90		0.39		0.26		1	A8 V	1770
151566	−49 10998	HR 6236	★	AB	16 46 46	−49 57.5	4	6.45	.007	0.32	.006	0.07		15	A5 III +	15,404,2013,2030
326191	−41 10937				16 46 49	−42 00.6	1	10.43		0.24		-0.25		2	B5	1583
	−41 10938				16 46 49	−42 02.3	1	10.63		1.19		0.87		2		1583
151877	+37 02804	G 181 - 9			16 46 50	+37 06.3	2	8.38	.025	0.82	.005	0.44		4	K7 V	1013,1197
151692	−24 12859				16 46 52	−24 21.6	1	9.63		1.10		1.02		1	K3 V	3072
329038	−47 11089				16 46 53	−47 33.5	1	10.01		0.55		-0.33		2	B2 IV	1081
151680	−34 11285	HR 6241			16 46 55	−34 12.3	8	2.29	.009	1.15	.003	1.15	.005	157	K2 IIIb	3,15,26,1020,1034*
	−31 13275	LSS 3775			16 46 56	−31 45.2	1	12.94		0.45		-0.17		2		954
322040	−37 11013	LSS 3774			16 46 56	−37 28.1	2	9.75	.024	0.46	.009	-0.52	.012	3	O9 V	540,954
151706	−28 12459				16 46 58	−28 39.2	1	10.10		0.17		0.18		1	A1/2 IV	1770
		L 154 - 163			16 47 00	−64 22.	3	10.98	.038	0.71	.011	0.04	.005	8		1696,1705,3062
151441	−65 03365	HR 6233			16 47 01	−65 17.5	4	6.12	.004	-0.02	.006	-0.28	.000	18	B8 II/III	15,1075,1637,2038
151769	−10 04394	HR 6243			16 47 04	−10 41.8	9	4.65	.019	0.47	.005	0.07	.018	35	F7 IV	3,15,254,1075,1088*
151721	−26 11585				16 47 04	−26 39.4	1	7.49		0.11		0.06		1	A3 V	1770
151722	−28 12461				16 47 04	−28 38.2	1	10.14		0.19		0.18		1	A2 IV	1770
		Ton 264			16 47 05	+25 15.2	2	14.11	.015	-0.08	.025	-1.01	.025	3	sdB	98,1036
	+3 03281	TT Oph			16 47 06	+03 43.0	1	9.52		1.12		1.07		1	F5p e	793
		LP 686 - 15			16 47 09	−06 27.7	1	11.99		0.91		0.41		2		1696
151736	−23 12933				16 47 09	−23 49.6	1	8.72		0.57		0.39		1	A0 V	1770
	−63 04022				16 47 09	−64 03.3	1	9.62		0.49		0.04		2	F5	1696
151404	−67 03232	HR 6231			16 47 09	−67 35.8	3	6.31	.004	1.29	.005	1.51	.000	14	K2 III	15,1075,2038
151786	−15 04397				16 47 12	−15 20.5	1	8.81		0.57		0.03		2	G3 V	1763
		LQ Ara			16 47 12	−60 56.7	1	11.60		2.96				1	SC	864
151737	−26 11587				16 47 14	−26 44.5	1	9.94		0.06		-0.14		1	B9 IV/V	1770
326176	−41 10943	LSS 3776			16 47 14	−41 39.7	8	9.14	.021	0.68	.016	-0.34	.022	19	B1 IV	32,141,486,540,954,1021*
326267	−43 11160				16 47 14	−43 41.4	1	10.06		1.10		1.05		4	K2	158
151935	+35 02870				16 47 15	+35 02.5	1	6.92		1.23		1.34		3	K4 III	1501
151862	+13 03233	HR 6246	★	AB	16 47 16	+13 20.8	2	5.88		0.04	.005	-0.01	.010	5	A1 V	1022,3050

HD	DM	Other Id	N	Rem	α_{1950}	δ_{1950}	S	V	σ_V	B–V	σ_{B-V}	U–B	σ_{U-B}	n	Spectrum	References
326175	−41 10944				16 47 17	−41 48.0	1	10.32		1.80		2.43		2	M0	1583
151937	+30 02880				16 47 18	+30 03.0	5	6.59	.007	1.25	.005	1.17	.010	13	K1 II-III	1003,1080,1733,3040,8100
151758	−30 13518				16 47 22	−31 08.5	1	9.94		0.10		-0.15		1	B9 III	1770
151707	−41 10946				16 47 24	−41 44.7	1	7.24		1.27		1.18		2	K0 IIICNII	1583
151787	−24 12862				16 47 25	−24 25.9	1	8.84		0.11		-0.02		1	A0 V	1770
	−41 10947				16 47 26	−41 46.0	1	11.14		0.52		0.27		2		1583
151686	−46 11069	IDS16438S4638	AB		16 47 26	−46 43.3	1	9.39		0.41		-0.02		3	B9	321
326167	−41 10948				16 47 27	−41 29.9	2	10.13	.004	0.20	.009	-0.55	.022	5	B3	1406,1707
152125	+55 01880	G 226 - 28	★	A	16 47 33	+55 13.1	2	8.92	.005	0.70	.017	0.22		3	G5	906,3016
		G 226 - 27			16 47 33	+55 13.5	1	11.22		1.16				1		906
151879	+9 03282				16 47 35	+09 30.4	1	6.91		0.91		0.51		2	G5	1375
151801	−26 11596				16 47 35	−26 32.3	1	8.98		0.07		-0.02		1	B9.5 V	1770
		KUV 16476+3733			16 47 36	+37 33.5	1	14.98		-0.10		-0.94		1	DA	1708
		KUV 433 - 9			16 47 36	+37 33.5	1	15.01		-0.15		-0.96		1		974
151771	−37 11023	HR 6244	★	AB	16 47 37	−37 25.8	2	6.10	.000	0.13	.005	-0.15		7	B8p Si	404,2007
		G 226 - 29	★	V	16 47 38	+59 08.7	2	12.24	.000	0.16	.002	-0.62	.005	6	DA	538,1658
		CS 22878 # 105			16 47 40	+08 16.4	1	13.56		0.10		0.11		1		1736
151939	+15 03058				16 47 41	+15 27.7	1	6.99		0.47		0.04		2	F6 V	1648
	−41 10952				16 47 42	−41 59.0	1	9.75		1.62		1.84		1		1583
151741	−41 10951				16 47 42	−42 01.8	1	8.85		0.59		0.12		2	G5 IV	1583
326196	−41 10950				16 47 42	−42 06.2	1	9.96		0.40		-0.22		2		1583
151953	+14 03125				16 47 44	+14 07.2	1	7.73		1.02		0.87		2	K0	1648
152107	+46 02220	HR 6254, V637 Her	★	A	16 47 46	+46 04.2	9	4.82	.011	0.09	.005	0.04	.011	32	A2 p SrCrEu	15,667,1007,1013,1202*
151900	−2 04259	HR 6248			16 47 46	−02 34.1	2	6.31	.005	0.42	.000	-0.06	.010	7	F1 III-IV	15,1417
	+30 02881				16 47 47	+29 51.7	1	10.33		0.90		0.57		2		1375
		G 17 - 37			16 47 47	−01 41.2	2	13.56	.010	0.85	.010	0.11	.010	3		1620,3029
		PG1647+056			16 47 51	+05 38.0	1	14.77		-0.17		-1.06		3	sdB	1764
151831	−26 11602				16 47 53	−26 41.4	2	10.55	.008	0.08	.001	-0.28	.041	2	B8/9 Ib	1054,1770
151696	−56 07880				16 47 53	−56 34.3	1	9.00		0.01		-0.41		1	B3/5 III	400
151956	+7 03256	HR 6250			16 47 54	+07 20.0	6	5.48	.020	0.11	.009	0.14	.026	29	A3 m	3,15,1256,3023,8071,8100
151995	+19 03174				16 47 54	+18 59.2	1	8.90		1.02		0.89		2	K0	3072
328898	−44 11193	LSS 3779			16 47 57	−44 20.6	1	9.67		0.44		-0.34		1	A5	954
151856	−27 11207				16 47 58	−27 53.7	1	7.88		1.30				4	K2 III	2012
322089	−39 10764	IDS16446S3934	AB		16 47 58	−39 38.7	3	8.82	.011	0.75	.016	-0.27	.027	5	B0	540,954,1737
151865	−19 04443	IDS16450S1940	AB		16 47 59	−19 45.2	2	8.84	.004	0.17	.017	-0.17	.010	4	B9 III	1054,1770
151857	−28 12479				16 48 00	−28 56.3	1	9.05		0.01		-0.10		1	B9 IV/V	1770
152032	+26 02907				16 48 02	+26 17.6	2	7.22	.035	0.95	.015	0.80		5	G8 II-III	20,8100
151884	−16 04360				16 48 02	−16 27.7	1	7.08		0.24		-0.32		2	B5 V	1770
328953		KQ Sco			16 48 02	−45 20.7	1	9.31		1.68		1.38		1	K7	1587
		G 169 - 28			16 48 04	+22 24.2	1	11.26		0.57		-0.14		2		1658
151804	−41 10957	HR 6245, V973 Sco			16 48 04	−41 08.8	7	5.23	.010	0.07	.008	-0.84	.011	59	O8 f	32,131,486,1406,1586*
151805	−41 10956	LSS 3781			16 48 05	−41 41.6	4	8.87	.019	0.14	.010	-0.70	.015	8	B1 Ib	32,954,1406,1707
152153	+43 02654	HR 6256			16 48 07	+43 30.9	1	6.13		1.26		1.31		2	K0 IV	252
151866	−25 11727				16 48 08	−25 31.8	1	9.96		0.37		0.14		1	A9 V	1770
326174					16 48 10	−41 51.2	1	11.64		0.59		0.20		2		1583
151886	−27 11209				16 48 11	−27 52.6	1	9.72		0.36		0.20		1	A1/2 III	1770
		WLS 1640 10 # 8			16 48 12	+10 05.0	1	13.00		0.66		0.14		2		1375
152223	+53 01897				16 48 14	+53 00.1	2	7.08	.005	1.09	.005	0.92	.005	6	K0	985,1733
		LP 686 - 18			16 48 14	−08 47.1	1	13.09		1.04		0.59		1		1696
152014	+4 03268				16 48 16	+04 32.1	1	8.74		1.48		1.80		1	K5	1746
151905	−28 12487				16 48 18	−28 22.8	1	9.67		0.04		-0.13		1	B9 IV	1770
	+32 02794				16 48 20	+32 04.7	1	10.55		1.12		1.12		2	K2	1375
326173					16 48 20	−41 50.0	1	11.45		0.56		0.13		2		1583
		AAS57,205 # 318			16 48 21	−41 46.7	1	11.59		0.40		0.39		2		1583
		KUV 16484+3706			16 48 23	+37 06.2	1	15.83		-0.20		-1.20		1	DA	1708
151903	−26 11612				16 48 23	−27 01.7	1	8.94		0.15		0.11		1	A0 V	1770
151835	−47 11100	LSS 3782			16 48 25	−48 06.5	4	8.67	.015	0.78	.015	-0.26	.025	7	B0.5Ia	540,954,976,1737
151927	−27 11214				16 48 26	−27 08.2	1	10.08		0.11		0.00		1	A0 IV	1770
151869	−39 10780				16 48 26	−39 18.0	1	10.04		0.62		0.24		1	A1 Iab	1586
326171	−41 10963				16 48 26	−41 46.9	1	11.42		0.21		-0.37		2	A3	1583
151849	−45 11011				16 48 26	−45 22.9	1	8.44		0.52		0.16		4	F2 III	164
151940	−25 11733				16 48 28	−25 39.5	1	10.31		0.46		0.22		1	A9/F0 V	1770
151775	−55 07722				16 48 28	−55 16.5	1	9.65		0.13		0.00		1	B9 II/III	400
151890	−37 11033	HR 6247, μ1 Sco	★		16 48 29	−37 57.8	5	3.04	.023	-0.21	.007	-0.85	.005	16	B1.5IV +	15,26,1088,8015,8023
151941	−26 11615				16 48 32	−26 36.7	1	9.90		0.36		0.21		1	A9p SrEuCr	1770
152360	+60 01706				16 48 33	+60 13.2	1	8.05		0.49		-0.04		1	F2	1776
		AAS57,205 # 320			16 48 34	−41 49.6	1	11.87		0.43		0.13		2		1583
148451	−87 00259	HR 6133			16 48 34	−87 29.7	5	6.56	.005	0.90	.008	0.56	.000	24	G5 III	15,1075,2012,2038,3040
326172		CpD -41 07669			16 48 35	−41 48.5	1	10.57		0.28		0.21		2	A2	1583
151908	−39 10785	LSS 3783			16 48 37	−39 09.7	1	8.14		0.67		0.68		2	A3 II	954
152112	+10 03083				16 48 40	+09 57.8	1	7.13		1.67		1.86		8	M3 III	3040
152173	+30 02884	HR 6258			16 48 42	+29 53.4	2	5.73	.011	1.61	.012	1.92		3	M1 IIIa	71,3001
152113	+9 03287	IDS16464N0935	AB		16 48 44	+09 29.4	1	6.66		0.49				5	F6 IV	173
151912	−41 10969				16 48 44	−41 43.3	1	9.49		0.25		0.25		3	B9/A0 IV/V	131
151981	−26 11617				16 48 45	−26 37.4	1	10.16		0.28		0.17		1	A8/9 V	1770
151964	−32 12092				16 48 46	−32 59.2	1	9.94		0.07		-0.36		1	B8 IV	1770
151911	−41 10971	IDS16453S4126	ABC		16 48 47	−41 31.3	1	9.09		0.19		-0.55		2	B2 II/III	1707
	+12 03097				16 48 47	+12 35.0	1	8.85		0.93		0.61		2	K0	1648
151932	−41 10972	HR 6249, V919 Sco			16 48 48	−41 46.3	6	6.49	.025	0.24	.016	-0.65	.018	19	WN7a	32,131,954,1583,1707,2007
		G 169 - 29			16 48 50	+22 31.8	1	14.11		1.75				1		906

Table 1   911

HD	DM	Other Id	N Rem	α_{1950}	δ_{1950}	S	V	σ_V	B–V	σ_{B-V}	U–B	σ_{U-B}	n	Spectrum	References
152224 +32 02795		HR 6259		16 48 50	+32 38.2	1	6.13		1.01		0.78		2	K0 III	252
		G 138 - 65		16 48 51	+15 57.0	2	13.77	.005	0.88	.005	0.46	.166	3		1658,3060
152127 +1 03323		HR 6255	⋆ AB	16 48 53	+01 18.0	3	5.51	.008	0.05	.002	0.03	.008	8	A2 Vs	15,1417,3023
326365 −42 11608				16 48 53	−42 09.4	1	9.80		0.30		-0.26		2	B3	141
152017		CpD -27 05503		16 48 54	−27 56.	1	10.94		0.14		0.08		1	B9 V	1770
151985 −37 11037		HR 6252		16 48 57	−37 56.1	5	3.56	.005	-0.21	.005	-0.84	.014	23	B2 IV	15,26,1075,2012,8015
152055 −21 04430				16 48 58	−21 14.8	1	9.69		0.41		0.27		1	A9 IV/V	1770
152038 −24 12889				16 48 59	−24 42.6	1	10.35		0.29		0.20		1	A0 III	1770
151965 −40 10841		V911 Sco		16 48 59	−40 38.4	2	6.36	.005	-0.14	.000	-0.57	.010	5	Ap Si	401,1707
152262 +42 02753		HR 6264		16 49 00	+41 58.8	1	6.29		1.08				2	K0	71
+47 02391				16 49 03	+47 49.4	1	9.42		1.18		1.10		2	K5	3072
		He3 1268		16 49 06	−41 30.	1	11.63		0.36		-0.26		3		1586
326364 −42 11614		LSS 3787		16 49 06	−42 11.7	5	9.60	.015	0.35	.005	-0.60	.009	12	B0 IV	141,540,1021,1406,1707
		KUV 433 - 11		16 49 07	+35 38.7	1	15.11		-0.25		-1.15		1	sdOB	974
151986 −41 10977		IDS16456S4112	A	16 49 07	−41 16.8	1	9.57		0.10		-0.52		2	B2/3 V	1707
152002 −38 11271				16 49 08	−38 21.2	2	7.42	.032	0.19	.006	-0.55	.012	4	B2 III	540,976
326324 −41 10976		LSS 3789		16 49 09	−41 44.5	5	10.72	.034	0.50	.012	-0.31	.037	11	B0	141,954,1583,1586,1707
151873 −56 07887				16 49 09	−56 56.6	1	9.17		0.06		0.00		5	Apec Shell	1586
322172 −38 11272				16 49 10	−38 33.7	1	10.32		0.23		-0.21		2	B8	1586
+38 02847				16 49 11	+38 14.0	1	10.56		1.60		1.86		3	K8	3072
152071 −25 11743		IDS16461S2526	A	16 49 11	−25 31.0	1	6.91		0.06		0.01		7	A0 V	1770
151933 −51 10564				16 49 14	−51 38.0	1	8.93		0.60				4	G2 V	2012
152264 +29 02889				16 49 15	+29 39.3	1	7.78		0.55		0.07		3	G0 V	3026
322290 −40 10849				16 49 16	−40 45.0	1	11.50		0.22		-0.42		2	A7	1539
326323 −41 10979				16 49 16	−41 37.5	1	10.30		0.43		0.38		2	A0	1539
152003 −41 10980		LSS 3790		16 49 17	−41 42.2	8	7.03	.031	0.37	.016	-0.60	.019	22	O9.5 Ia	32,131,954,1406,1539,1707*
152093 −25 11744				16 49 19	−25 55.2	1	10.48		0.16		0.11		1	B9.5/A0 IV	1770
326347 −41 10981				16 49 19	−41 58.0	2	11.30	.005	0.36	.000	-0.19	.000	4	B3	141,1583
		AAS57,205 # 324		16 49 20	−41 59.4	1	11.70		0.43		-0.24				1583
−41 10983				16 49 21	−42 03.6	1	10.98		1.44		1.61		2		1583
−49 11019				16 49 21	−49 53.2	1	10.47		0.35		-0.30		1	B8	1586
		LP 506 - 19		16 49 22	+11 02.9	1	14.06		1.01		0.84		1		1696
326325 −41 10984				16 49 22	−41 45.5	2	8.83	.014	1.93	.010	2.33	.033	7	M0	131,1583
		AAS57,205 # 3		16 49 23	−41 24.5	1	12.11		0.51		0.43		2		1539
		LP 506 - 20		16 49 24	+12 14.2	1	12.67		1.20		1.11		1		1696
152130 −23 12947				16 49 24	−23 28.6	1	10.53		0.44		0.20		1	B9 IV/V	1770
326305 −41 10985		LSS 3791		16 49 25	−41 30.2	4	9.96	.025	0.17	.009	-0.64	.004	9	B5	32,131,1539,1707
322291 −40 10852				16 49 26	−40 45.6	1	11.21		0.21		-0.51		2	A0	1539
152042 −41 10986				16 49 27	−41 29.0	7	8.18	.014	0.12	.011	-0.74	.010	17	B1 Ib/II	32,131,540,976,1406*
152306 +28 02623				16 49 28	+28 12.3	2	7.06	.024	0.90	.022	0.56		5	G2 V	20,1375
		AAS57,205 # 6		16 49 28	−41 23.8	1	11.41		0.16		-0.38		2		1539
+7 03260				16 49 29	+07 41.7	1	10.74		0.35		0.14		2	A5	1375
322292				16 49 29	−40 45.3	1	9.06		0.16		-0.61		2	F2	1539
		AAS57,205 # 7		16 49 29	−41 24.5	1	11.61		0.22		-0.26		2		1539
322286 −40 10854				16 49 30	−40 30.6	1	12.83		0.23		-0.61		2	K2	1539
152060 −41 10988				16 49 30	−41 19.5	4	9.58	.012	0.06	.006	-0.73	.012	11	B1 III	540,1021,1406,1586,1707
151990 −52 10300		LSS 3788		16 49 31	−52 32.9	3	9.44	.015	0.09	.008	-0.85	.019	7	O9/9.5Ib/II	540,954,976
−41 10989				16 49 32	−41 48.6	1	11.08		0.47		-0.43		2		1583
		LSS 3792		16 49 32	−44 35.4	2	11.33	.000	0.61	.003	-0.27	.000	6	B3 Ib	1586,1737
		AAS57,205 # 8		16 49 35	−41 36.8	1	10.92		0.38		-0.20		2		1539
		LSS 3794		16 49 36	−41 38.8	1	10.85		0.35		-0.22		2	B3 Ib	954
152076 −41 10991		LSS 3794		16 49 37	−41 38.8	5	8.48	.014	0.21	.010	-0.68	.013	15	B0/1 III	32,131,1406,1539,1707
		AAS57,205 # 329		16 49 37	−41 48.6	1	11.37		0.51		-0.24		2		1583
		AAS57,205 # 11		16 49 38	−41 40.1	1	11.45		0.59		0.54		2		1539
−41 10993				16 49 38	−41 40.8	1	10.85		0.34		-0.43		2		1539
152077 −43 11198		LSS 3793		16 49 39	−43 40.0	3	9.06	.033	0.28	.007	-0.59	.014	5	B2 Iab/b	540,954,976
152326 +24 03069		HR 6270		16 49 41	+24 44.4	4	5.03	.036	1.25	.007	1.30	.016	8	K0.5 IIIa	1080,1363,3016,8100
		HA 180 # 1618		16 49 41	−45 02.9	1	12.05		0.54		0.19		10		1499
152342 +25 03150				16 49 43	+25 29.0	1	7.11		0.33		-0.04		3	F4 III	833
		AAS57,205 # 12		16 49 44	−41 25.9	1	11.79		0.25		-0.19		2		1539
		AAS57,205 # 13		16 49 44	−41 43.4	1	10.81		1.67		1.92		2		1539
152096 −41 10994				16 49 45	−41 17.3	2	9.40	.005	0.06	.000	-0.66	.005	4	B2 Ib/II	1539,1707
		AAS57,205 # 14		16 49 45	−41 22.8	1	12.25		0.40		0.37		2		1539
152078 −45 11028				16 49 45	−45 22.7	1	8.16		0.03		-0.40		2	B3 III	401
151967 −57 08157		HR 6251		16 49 45	−57 49.6	3	5.93	.007	1.59	.005	1.92		9	M1 III	15,2012,3005
		AAS57,205 # 17		16 49 46	−41 22.5	1	12.24		0.19		-0.23		2		1539
326306 −41 10997				16 49 46	−41 24.5	3	9.79	.014	0.10	.004	-0.68	.019	8	B1 V	1021,1539,1707
326326 −41 10996				16 49 46	−41 42.6	1	10.92		0.30		-0.41		2	B2	1539
326348 −41 10995		LSS 3796		16 49 47	−41 58.0	5	9.92	.015	0.35	.020	-0.54	.007	14	B1 IV	32,141,1021,1583,1707
152308 +15 03066		HR 6268, V823 Her		16 49 48	+15 03.4	3	6.52	.015	-0.03	.012	-0.08	.006	13	B9.5 p(Cr)	985,1022,3016
322293 −40 10861				16 49 48	−40 44.9	1	10.61		0.21		-0.53		2	B8	1539
−45 11029		LSS 3795		16 49 48	−45 19.5	2	11.27	.056	1.05	.007	-0.12	.022	4		1586,1737
		AAS57,205 # 18		16 49 49	−41 34.0	1	12.26		0.41		-0.16		2		1539
326342 −41 10998				16 49 49	−41 50.8	1	10.77		0.81		0.74		2	A0	1583
		HA 180 # 1644		16 49 49	−45 03.1	1	11.82		0.93		-0.13		10		1499
326299		CpD -41 07692		16 49 50	−41 14.3	1	10.76		0.18		-0.20		2	B9	1539
326307 −41 11000				16 49 50	−41 31.5	3	10.49	.005	0.25	.009	-0.44	.013	6	B5	32,1539,1707
152380 +28 02624		IDS16479N2850	AB	16 49 51	+28 45.7	1	6.66		0.47		-0.02		3	F4 V +G3 V	938
152158 −33 11525				16 49 52	−33 12.0	1	7.65		0.10				4	A1 IV	1075
152180 −31 13317				16 49 53	−31 47.9	3	7.55	.012	0.11	.006	-0.36	.004	5	B3/5 III	540,976,1770

HD	DM	Other Id	N Rem	α_{1950}	δ_{1950}	S	V	σ_V	B–V	σ_{B-V}	U–B	σ_{U-B}	n	Spectrum	References
326341	−41 11001			16 49 53	−41 55.1	1	10.93		0.29		0.17		2	A0	1583
		G 169 - 30		16 49 54	+23 56.5	1	11.64		1.42		1.16		5		1723
		AAS57,205 # 333		16 49 54	−41 06.9	1	11.16		0.20		-0.48		2		1583
152179	−31 13318			16 49 55	−31 20.8	4	8.87	.009	0.23	.005	-0.56	.010	10	B2 IV	400,540,976,1770
152192	−30 13560			16 49 56	−30 20.9	2	7.03	.000	0.16	.005	0.16	.000	5	A3 III	210,244
326308	−41 11004			16 49 57	−41 25.5	3	10.26	.005	0.20		-0.18	.000	6	B8	32,1539,1707
152147	−41 11003	LSS 3798		16 49 57	−42 02.4	8	7.27	.028	0.39	.022	-0.58	.016	22	B0 Ia	32,131,138,540,976,1406*
152241	−19 04455			16 49 59	−20 06.0	1	8.98		0.68		0.53		2	A2/3 III	1770
		AAS57,205 # 335		16 49 59	−41 16.6	1	12.55		0.23		-0.16		2		1583
	−41 11005			16 49 59	−41 30.7	1	10.91		0.35		-0.39		2		1539
326298	−41 11006			16 50 01	−41 15.7	1	10.26		1.97		2.19		2		1583
		AAS57,205 # 337		16 50 01	−41 51.9	1	11.81		0.75		0.18		2		1583
	+15 03068	G 138 - 66		16 50 02	+15 44.0	1	9.13		0.70		0.18		1	G0	3060
		AAS57,205 # 24		16 50 04	−41 36.1	1	11.74		0.29		-0.23		2		1539
		AAS57,205 # 25		16 50 07	−41 08.1	1	11.58		0.43		0.24		2		1539
		NGC 6231 - 26		16 50 07	−41 40.8	1	13.79		1.51		1.19		4		113
152551	+51 02142			16 50 08	+50 51.3	1	7.47		1.08		0.93		2	K0	1566
326327	−41 11007	NGC 6231 - 28	⋆V	16 50 08	−41 42.9	6	9.76	.022	0.26	.011	-0.59	.011	20	B1.5IVe+Sh	32,113,141,1021,1539,1707
152673	+63 01308			16 50 09	+63 13.3	2	8.82	.005	0.26	.015	0.02	.054	3	F0	1733,1776
		AAS57,205 # 27		16 50 09	−41 21.5	1	12.23		0.39		0.27		2		1539
		AAS57,205 # 28		16 50 09	−41 34.1	1	11.64		1.27		1.14		2		1539
		NGC 6231 - 27		16 50 09	−41 42.5	1	13.96		0.63		0.06		4		113
152161	−42 11627	HR 6257		16 50 09	−42 58.1	1	5.96		1.73				4	M3 II/III	2007
		AAS57,205 # 338		16 50 10	−41 23.7	1	13.13		0.48		0.32		2		1583
		AAS57,205 # 30		16 50 11	−41 34.6	1	12.46		0.29		-0.05		2		1539
		NGC 6231 - 30		16 50 11	−41 43.3	1	11.61		0.32		-0.18		2		1539
152578	+54 01841	IDS16491N5357	AB	16 50 12	+53 52.1	1	9.17		0.23		0.16		2		1566
152182	−41 11009			16 50 12	−41 14.8	4	9.12	.013	0.08	.007	-0.75	.009	8	B1 II	540,976,1539,1707
		NGC 6231 - 25		16 50 12	−41 41.3	1	14.41		1.03		0.32		5		113
326349	−41 11008			16 50 12	−42 00.1	2	10.91	.020	0.37	.010	-0.25	.015	4	B8	141,1583
326350	−41 11008			16 50 12	−42 00.1	2	10.97	.005	0.32	.005	-0.41	.010	4	B5	141,1583
		AAS57,205 # 340		16 50 12	−42 00.3	1	11.25		1.71		1.53		2		1583
		LSS 3799		16 50 13	−45 13.5	1	11.04		0.89		-0.23		2	B3 Ib	1737
326328	−41 11010	NGC 6231 - 34		16 50 14	−41 44.1	4	10.22	.018	0.21	.010	-0.60	.017	10	B1 V	32,141,1539,1707
152148	−46 11089	IDS16466S4645	B	16 50 14	−46 49.8	1	8.12		0.26		0.24		1	A1 V	401
		AAS57,205 # 33		16 50 15	−41 00.9	1	12.88		0.32		-0.11		2		1539
		AAS57,205 # 342		16 50 15	−41 04.6	1	12.64		0.26		-0.25		2		1583
	−41 11011			16 50 15	−41 31.8	1	10.58		1.19		0.82		2		1539
152197	−40 10877			16 50 16	−40 14.1	2	8.31	.003	0.16	.005	-0.73	.003	5	B3 III	540,976
152149	−46 11090	IDS16466S4645	A	16 50 16	−46 50.2	1	7.92		0.12		-0.40		2	B2/5 (III)	401
		AAS57,205 # 343		16 50 17	−41 05.9	1	12.67		0.69		0.27		2		1583
		NGC 6231 - 297		16 50 17	−41 39.7	1	12.47		0.45		-0.22		2		1539
		AAS57,205 # 37		16 50 18	−40 58.8	1	12.02		0.34		0.20		2		1539
		AAS57,205 # 38		16 50 18	−41 27.6	1	11.93		0.40		-0.04		2		1539
		AAS57,205 # 39		16 50 18	−41 33.0	1	11.92		0.33		-0.16		2		1539
		NGC 6231 - 14		16 50 18	−41 40.5	1	11.82		0.34		-0.29		2		1539
	+12 03100			16 50 19	+12 29.2	1	9.59		0.07		0.00		2	A0	1375
		AAS57,205 # 344		16 50 19	−41 18.1	1	13.03		0.32		-0.03		2		1583
152198	−41 11015	LSS 3804		16 50 21	−41 13.7	3	8.30	.018	0.15	.005	-0.72	.009	6	B3 II	32,1406,1539,1707
152199	−41 11016	LSS 3803		16 50 21	−41 27.8	3	8.63	.018	0.23	.017	-0.69	.005	9	B0/1 (Ib)	32,131,1707
		NGC 6231 - 15		16 50 21	−41 40.8	1	13.60		0.55		0.22		4		113
152200	−41 11014	NGC 6231 - 266		16 50 21	−41 45.7	5	8.41	.014	0.13	.009	-0.76	.016	15	O9.5V(n)	32,113,131,1406,1707
152243	−32 12125			16 50 22	−32 40.7	1	9.11		0.15		-0.56		4	G8 V	976
		AAS57,205 # 345		16 50 22	−41 03.1	1	11.75		1.55		1.44		2		1583
		NGC 6231 - 16		16 50 22	−41 39.8	2	11.17	.000	1.78	.005	1.82	.131	6		113,1539
152495	+35 02878			16 50 23	+35 34.3	1	7.52		1.55		1.84		3	K5	1601
		NGC 6235 sq1 1		16 50 23	−22 01.7	1	13.14		0.82				3		869
152257	−26 11642			16 50 23	−27 03.8	1	10.10		0.10		0.01		1	B9 V	1770
326351	−41 11017	LSS 3805		16 50 23	−42 01.3	5	9.21	.017	0.31	.018	-0.61	.007	14	B3	32,131,141,1583,1707
152217	−41 11019	LSS 3806	⋆AB	16 50 24	−41 10.4	6	8.46	.016	0.16	.010	-0.71	.019	10	B2 II	32,540,954,976,1539,1707
		NGC 6231 - 20		16 50 24	−41 41.2	1	13.12		0.29		0.19		4		113
		NGC 6231 - 18		16 50 25	−41 40.9	1	12.39		0.95		0.49		4		113
152219	−41 11018	NGC 6231 - 254		16 50 25	−41 48.0	5	7.66	.037	0.15	.015	-0.76	.020	14	O9.5III	32,113,131,1406,1707
326309	−41 11020			16 50 26	−41 23.1	2	10.00	.015	0.26	.005	-0.56	.015	5	B5	1539,1707
		NGC 6231 - 282		16 50 26	−41 43.7	2	9.78	.023	0.17	.011	-0.60	.003	7	B2 V:+B2V:	772,1707
152391	+0 03593	G 19 - 4		16 50 27	+00 04.5	8	6.64	.011	0.76	.009	0.33	.017	16	G8 V	22,101,245,333,1003*
322303	−40 10879	AsApS57,205 # 347		16 50 27	−40 52.9	1	11.09		0.19		-0.42		2	A2	1583
326320		CpD -41 07710		16 50 27	−41 33.8	3	9.85	.052	0.17	.000	-0.60	.016	7	A0	32,1539,1707
152446	+18 03261			16 50 28	+18 08.6	1	6.82		0.51		0.01		3	F8 IV	985
152311	−20 04572	HR 6269		16 50 28	−20 20.0	2	5.88	.003	0.67	.011			6	G5 IV	1619,2007
		AAS57,205 # 46		16 50 28	−41 31.3	1	12.21		0.22		-0.11		2		1539
		NGC 6231 - 248		16 50 28	−41 47.7	3	9.17	.019	0.17	.012	-0.71	.017	6	B1 V	32,131,1707
152235	−41 11021	HR 6261, V900 Sco		16 50 28	−41 54.8	11	6.33	.020	0.51	.015	-0.47	.017	44	B1 Ia	32,131,681,815,954,1583*
152236	−42 11633	HR 6262, κ1 Sco		16 50 28	−42 16.9	11	4.73	.016	0.48	.016	-0.55	.022	53	B1 Iae	32,131,681,815,1018,1034*
152218	−41 11022	NGC 6231 - 2		16 50 29	−41 38.0	5	7.60	.017	0.17	.012	-0.75	.012	18	O9 IV	32,113,131,1539,1707
		NGC 6231 - 276		16 50 29	−41 44.5	1	12.80		0.31		-0.09		7	B8 Vp:	772
	−41 11028	NGC 6231 - 261	⋆A	16 50 29	−41 46.3	2	10.54	.385	0.16	.029	-0.56	.068	3	B2 IV	131,1707
152286	−28 12539	LSS 3814		16 50 30	−28 19.3	4	9.65	.010	0.01	.004	-0.68	.010	9	B1 Ib	400,540,954,976
		AAS57,205 # 49		16 50 30	−41 34.1	1	10.21		0.19		-0.57		2		1539
152283	−24 12909			16 50 31	−25 02.3	1	10.20		0.27		0.19		1	A0 III/IV	1770

Table 1 913

HD	DM	Other Id	N	Rem	α_{1950}	δ_{1950}	S	V	σ_V	B–V	σ_{B-V}	U–B	σ_{U-B}	n	Spectrum	References
152285	−26 11643				16 50 31	−26 09.1	1	10.55		0.24		0.15		1	A3 III/IV	1770
		CpD -41 07717			16 50 31	−41 34.2	2	10.21	.000	0.19	.000	-0.59	.015	5		32,1539
152234	−41 11024	NGC 6231 - 290		⋆ AB	16 50 31	−41 43.5	6	5.45	.016	0.20	.016	-0.73	.015	23	B0 Iab	32,113,131,1406,1707,8100
		NGC 6231 - 259			16 50 31	−41 46.9	3	10.93	.005	0.24	.007	-0.46	.013	9		32,113,1406
152245	−40 10883				16 50 32	−40 27.1	4	8.39	.012	0.09	.015	-0.80	.011	8	B0 Ib	32,540,976,1707
	−41 11030	NGC 6231 - 272			16 50 32	−41 45.2	5	9.47	.015	0.13	.009	-0.67	.015	18	B1 V	32,113,772,1406,1707
152233	−41 11025	NGC 6231 - 306		⋆ F	16 50 33	−41 42.6	3	6.57	.040	0.13	.010	-0.80	.005	11	O6 III(f)	32,131,1707
		AAS57,205 # 52			16 50 34	−40 52.5	1	11.83		0.24		-0.19		2		1539
326288	−41 11026				16 50 34	−41 09.1	1	9.92		0.26		0.15		2	A0	1539
		AAS57,205 # 51			16 50 34	−41 31.8	1	12.22		0.24		-0.25		2		1539
	−41 11032	NGC 6231 - 289			16 50 34	−41 44.5	4	9.50	.027	0.14	.015	-0.69	.023	13	B0.5V	113,772,1406,1707
		NGC 6231 - 274			16 50 34	−41 44.9	1	11.73		0.22		-0.38		5	B3 V	772
	−41 11048	NGC 6231 - 295			16 50 34	−41 45.2	3	9.43	.027	0.14	.008	-0.65	.003	9	B1 V	772,1406,1707
	−41 11029	NGC 6231 - 309			16 50 34	−41 46.2	1	8.42		0.16		-0.74		2		1707
		NGC 6235 sq1 2			16 50 35	−22 03.5	1	13.50		0.65				2		869
152246	−40 10884	LSS 3808			16 50 36	−40 59.9	6	7.31	.015	0.17	.011	-0.77	.011	17	O9 Ib	32,164,540,954,976*
		NGC 6231 - 312			16 50 36	−41 39.4	2	9.87	.095	0.20	.020	-0.56	.010	4	B1 V	1539,1707
		Smethells 2			16 50 36	−65 10.	1	11.38		1.35				1		1494
	−41 11031	NGC 6231 - 6			16 50 37	−41 39.6	2	9.91	.030	0.18	.000	-0.62	.005	4	B0 V	1539,1707
		NGC 6231 - 273			16 50 37	−41 45.4	1	12.75		0.26		-0.23		6	B9 IV:p	772
		NGC 6231 - 286			16 50 37	−41 46.0	6	9.43	.017	0.15	.016	-0.63	.004	48	B0.5V	32,113,141,772,1406,1707
		AAS57,205 # 57			16 50 38	−41 03.0	1	12.85		0.48		0.18		2		1539
326289		CpD -41 07729			16 50 38	−41 09.6	1	10.63		0.21		-0.56		2	B8	1539
326340	−41 11039	NGC 6231 - 70			16 50 38	−41 53.0	4	9.94	.008	0.25	.012	-0.60	.000	12	B0.5V	32,141,1583,1707
326362	−42 11636				16 50 38	−42 09.1	1	11.19		0.46		-0.12		2	A0	1583
		AAS57,205 # 58			16 50 39	−41 18.4	1	12.02		0.59		0.51		2		1539
		AAS57,205 # 59			16 50 39	−41 35.2	1	11.49		0.27		-0.05		2		1539
		NGC 6231 - 209			16 50 39	−41 43.0	4	10.63	.010	0.23	.012	-0.49	.014	13	B2 IV-V	32,113,772,1406
		NGC 6231 - 287			16 50 39	−41 44.1	5	9.27	.029	0.17	.015	-0.59	.020	42	B1 V	113,131,772,1406,1707
152248	−41 11033	NGC 6231 - 291		⋆ A	16 50 39	−41 44.7	6	6.13	.035	0.15	.009	-0.80	.008	25	O7 Ib:(f)	32,131,772,1406,1586,1707
		AAS57,205 # 61			16 50 40	−40 45.1	1	12.22		0.15		-0.31		2		1539
		AAS57,205 # 60			16 50 40	−41 02.0	1	12.60		0.32		0.01		2		1539
		AAS57,205 # 62			16 50 41	−41 15.5	1	12.03		0.31		-0.21		2		1539
152247	−41 11035	LSS 3811			16 50 41	−41 33.7	4	7.17	.012	0.19	.013	-0.73	.011	12	O9/B0 Iab/b	32,131,954,1707
152249	−41 11036	NGC 6231 - 293		⋆	16 50 41	−41 46.1	7	6.46	.019	0.20	.015	-0.74	.017	27	O9 Iabp	32,131,772,1018,1406*
		G 19 - 5			16 50 42	−05 23.8	1	15.16		1.41		1.15		1		3062
		NGC 6235 sq1 3			16 50 42	−22 08.8	1	13.66		0.88				2		869
		AAS57,205 # 63			16 50 42	−40 56.5	1	10.84		1.62		1.65		2		1539
		AAS57,205 # 64			16 50 42	−41 01.0	1	12.27		0.79		0.26		2		1539
326329	−41 11038	NGC 6231 - 292		⋆ AB	16 50 42	−41 45.3	5	8.60	.111	0.17	.018	-0.72	.022	13	O9.5V	32,131,772,1406,1707
	−41 11037	NGC 6231 - 323			16 50 42	−41 45.7	3	7.86	.053	0.17	.012	-0.74	.019	7	O9 III	32,131,1707
		NGC 6231 - 236			16 50 42	−41 46.3	1	11.40		0.23		-0.44		7	B2.5V	772
		NGC 6231 - 80		V	16 50 42	−41 50.2	1	10.29		0.32		-0.52		2	B0 Vn	1707
152364	−18 04347				16 50 43	−19 06.0	1	10.33		0.33		0.19		1	A1/2 IV	1770
322304	−40 10888				16 50 43	−40 42.7	1	11.18		0.12		-0.47		2	F0	1539
		AAS57,205 # 350			16 50 43	−40 49.3	1	12.70		0.21		-0.09		2		1583
		AAS57,205 # 65			16 50 43	−41 22.2	1	11.18		0.34		-0.46		2		1539
		NGC 6231 - 235			16 50 43	−41 46.1	1	13.14		0.40		-0.03		4	B8.5V	772
		NGC 6235 sq1 4			16 50 44	−21 58.5	1	11.60		0.90				1		869
		AAS57,205 # 66			16 50 44	−40 58.0	1	12.35		0.51		0.28		2		1539
		AAS57,205 # 351			16 50 44	−41 05.5	1	11.86		0.59		-0.09		2		1583
		NGC 6231 - 294		AB	16 50 44	−41 45.6	3	10.41	.030	0.23	.012	-0.46	.013	9	B1.5V	772,1406,1707
152082	−63 04032	HR 6253		⋆ AB	16 50 44	−63 11.4	1	6.02		0.05				4	A0 III	2007
	−41 11040				16 50 45	−41 14.1	1	10.38		1.39		1.26		2		1539
		AAS57,205 # 67			16 50 45	−41 16.9	1	12.70		0.72		0.44		2		1539
		NGC 6231 - 19			16 50 45	−41 40.9	1	13.04		0.32		0.06		4		113
		NGC 6231 - 234			16 50 45	−41 46.2	1	12.80		0.37		-0.05		4	B8 Vn	772
152365	−19 04459				16 50 46	−19 19.6	1	9.87		0.11		-0.38		1	B7 II	1770
	−40 10890				16 50 46	−40 47.3	1	10.77		1.15		0.85		2		1539
		NGC 6231 - 189			16 50 46	−41 39.8	1	11.55		0.32		-0.18		2		1539
		NGC 6231 - 213			16 50 46	−41 43.3	1	11.08		0.20		-0.46		5	B2 IVn	772
		NGC 6231 - 225			16 50 46	−41 45.5	1	12.97		0.36		-0.12		3	B9 V	772
326330		NGC 6231 - 238		⋆ V	16 50 46	−41 46.8	4	9.62	.043	0.18	.009	-0.67	.013	11	B1 V(n)	113,772,1406,1707
		NGC 6231 - 253		⋆ V	16 50 46	−41 47.3	2	9.58	.010	0.16	.010	-0.68	.039	3	B1 V:+B1V:	131,1707
152250	−44 11252				16 50 46	−44 48.7	2	7.41	.005	0.38	.005	0.25		8	A7/8 III	244,2012
152268	−40 10891	LSS 3813			16 50 47	−40 54.2	5	8.11	.014	0.12	.010	-0.75	.008	12	B1/2 Ib/II	32,540,976,1539,1707
		AAS57,205 # 70			16 50 47	−40 57.9	1	12.98		0.25		-0.01		2		1539
		He3 1275			16 50 48	−40 21.	1	12.68		1.13		-0.17		3		1586
152269	−41 11043				16 50 48	−41 27.6	4	8.49	.013	0.13	.008	-0.45	.010	8	B9 Ib/II	32,131,1539,1707
	−41 11042			⋆ AB	16 50 48	−41 45.4	4	8.29	.036	0.21	.007	-0.69	.011	14	O9 IV	113,131,772,1707
		NGC 6231 - 227			16 50 48	−41 46.1	1	12.50		0.33		-0.19		5	B7 Vn	772
326339	−41 11047	NGC 6231 - 73			16 50 48	−41 52.8	4	10.12	.009	0.42	.015	-0.36	.014	9	B0.5III	32,141,1583,1707
		AAS57,205 # 75			16 50 49	−41 19.2	1	12.28		0.78		0.42		2		1539
152270	−41 11041	NGC 6231 - 220		⋆ AB	16 50 49	−41 44.3	4	6.61	.041	0.18	.033	-0.54	.027	16	WC7	32,113,131,1707
		NGC 6231 - 334			16 50 49	−41 44.7	1	10.13		0.23		-0.54		2	B1 V	1707
		NGC 6231 - 232			16 50 49	−41 46.9	2	9.67	.017	0.23	.002	-0.61	.006	6	B0.5V	772,1707
		L 203 - 139			16 50 49	−57 46.5	1	12.85		1.30		0.98		1		3062
		AAS57,205 # 76			16 50 50	−40 40.0	1	11.54		0.25		-0.08		2		1539
		AAS57,205 # 354			16 50 50	−41 04.7	1	12.33		0.19		-0.39		2		1583
		AAS57,205 # 77			16 50 50	−41 15.0	1	12.00		0.40		-0.18		2		1539

HD	DM	Other Id	N Rem	α_{1950}	δ_{1950}	S	V	σ_V	B–V	σ_{B-V}	U–B	σ_{U-B}	n	Spectrum	References
		NGC 6231 - 217		16 50 50	−41 44.0	1	12.23		0.30		-0.24		8	B6 Vn	772
		NGC 6231 - 223		16 50 50	−41 45.2	1	11.70		0.29		-0.22		4	B6 Vn	772
		AAS57,205 # 78		16 50 51	−41 27.2	1	10.24		1.16		0.86		2		1539
		AAS57,205 # 79		16 50 51	−41 33.3	1	11.94		0.72		0.28		2		1539
		AAS57,205 # 80		16 50 52	−40 36.5	1	11.66		0.35		0.25		2		1539
152366	−24 12916			16 50 53	−24 59.3	1	8.12		0.04		-0.27		1	Ap Si	1770
		AAS57,205 # 82		16 50 53	−41 21.9	1	11.95		0.45		-0.04		2		1539
		AAS57,205 # 81		16 50 53	−41 29.0	1	11.28		0.22		-0.32		2		1539
		NGC 6231 - 108		16 50 53	−41 48.0	1	11.44		0.30		-0.42		6	B2 V	772
322268	−40 10895			16 50 54	−40 26.	1	11.03		0.39		0.16		2	B9	1707
		AAS57,205 # 83		16 50 54	−41 47.1	1	11.94		0.45		0.40		2		1539
152292	−40 10896			16 50 54	−40 56.9	5	8.53	.013	0.12	.006	-0.71	.010	10	B2 II/III	401,540,976,1539,1707
152348	−30 13581			16 50 55	−30 40.3	1	8.59		0.18		0.15		1	A2 IV	210
		AAS57,205 # 86		16 50 55	−40 59.7	1	11.86		0.20		-0.38		2		1539
		AAS57,205 # 85		16 50 55	−41 31.3	1	12.28		0.34		0.30		2		1539
326331	−41 11044	NGC 6231 - 338	⋆ AB	16 50 55	−41 45.1	3	7.52	.070	0.18	.017	-0.74	.010	8	O7 III: f:	32,131,1707
		NGC 6231 - 92	⋆ V	16 50 55	−41 51.2	2	12.54	.102	3.51	.010	3.23		3	F5 Ia	643,901
152293	−42 11642	HR 6266		16 50 55	−42 23.9	2	5.85	.011	0.63	.004	0.22		5	F5 Ib-II	1034,2007
152291	−40 10897	LSS 3816		16 50 56	−40 34.3	2	8.76	.066	0.13	.030	-0.80	.021	5	B2 (II)	954,1586
322306	−40 10898			16 50 56	−40 43.5	1	10.28		0.85		0.53		2	F2	1539
		AAS57,205 # 88		16 50 56	−40 58.6	1	11.40		1.84		2.03		2		1539
		AAS57,205 # 87		16 50 56	−41 24.8	1	11.99		0.30		-0.32		2		1539
		NGC 6231 - 222		16 50 56	−41 45.5	1	11.84		0.28		-0.30		6	B4 Vn	772
326352	−41 11045			16 50 56	−42 04.1	1	10.62		0.71		0.20		2	F5	1583
		AAS57,205 # 91		16 50 57	−40 49.8	1	12.19		0.32		0.13		2		1539
326310	−41 11046			16 50 57	−41 23.4	2	9.90	.005	0.25	.000	-0.52	.010	4	B3	1539,1707
		AAS57,205 # 92		16 50 57	−41 34.5	1	12.32		0.49		0.24		2		1539
152395	−19 04460	IDS16480S1909	AB	16 50 58	−19 13.9	1	8.45		0.53		0.08		1	F6 V	1770
		AAS57,205 # 95		16 50 58	−40 34.4	1	11.89		0.35		0.33		2		1539
		AAS57,205 # 94		16 50 58	−40 55.5	1	12.00		0.21		-0.38		2		1539
		AAS57,205 # 96		16 50 58	−40 57.8	1	13.45		0.32		0.25		2		1539
		AAS57,205 # 97		16 50 59	−41 01.6	1	11.87		0.67		0.07		2		1539
		CpD -41 07746		16 50 59	−41 34.4	1	9.24		0.23		-0.68		2		32
322282	−40 10900	LSS 3818		16 51 00	−40 36.5	3	8.91	.020	0.21	.049	-0.71	.052	8		1539,1586,1707
	−40 10901			16 51 00	−40 38.7	1	11.17		0.20		-0.38		2		1539
		AAS57,205 # 99		16 51 00	−40 41.2	1	10.84		0.16		-0.48		2		1539
	−41 11054	NGC 6231 - 102		16 51 00	−41 50.6	1	9.98		0.47		0.33		5		32
153751	+82 00498	HR 6322, ε UMi	⋆ A	16 51 01	+82 07.4	6	4.23	.014	0.90	.013	0.54	.012	17	G5 III	15,667,1118,1363,3016,8015
152419	−18 04350	IDS16481S1844	ABC	16 51 01	−18 49.5	1	8.80		0.30		0.20		1	A2 IV/V	1770
		AAS57,205 # 103		16 51 01	−40 47.3	1	12.69		0.42		-0.11		2		1539,1750
152314	−41 11050	NGC 6231 - 161		16 51 01	−41 43.5	8	7.86	.050	0.19	.012	-0.70	.014	17	O8.5III	32,113,131,954,1406,1539*
		NGC 6231 - 137		16 51 01	−41 46.7	1	11.79		0.36		-0.19		4		113
		AAS57,205 # 105		16 51 02	−40 58.5	1	11.67		0.23		-0.34		2		1539
326311	−41 11052			16 51 02	−41 24.6	1	10.50		0.26		-0.51		2	A7	1539
326319	−41 11051			16 51 02	−41 34.4	3	10.59	.013	0.35	.013	0.27	.005	6	A0	32,1539,1707
		AAS57,205 # 104		16 51 02	−41 35.5	1	11.88		0.30		-0.28		2		1539
		AAS57,205 # 356		16 51 02	−42 04.0	1	12.01		0.63		0.09		2		1583
		NGC 6231 - 109		16 51 03	−41 47.9	1	13.81		0.46		0.15		4		113
152598	+31 02925	HR 6279	⋆ A	16 51 04	+31 47.0	2	5.34	.000	0.29	.010	-0.01	.030	4	F0 V	254,3026
		AAS57,205 # 357		16 51 04	−40 23.2	1	11.27		0.29		0.22		2		1583
326332	−41 11053	NGC 6231 - 343		16 51 04	−41 45.5	3	9.66	.000	0.24	.004	-0.63	.000	14	B1 III	32,141,1707
152334	−42 11646	HR 6271		16 51 04	−42 16.7	5	3.62	.015	1.38	.009	1.64	.027	11	K3 III	15,1034,1075,2012,8015
	−45 11034	Ly 14 - 5		16 51 04	−45 10.4	3	11.08	.015	0.92	.018	-0.16	.036	6		92,412,1737
	−45 11051	LSS 3815	⋆ A	16 51 04	−45 10.6	1	10.96		0.96		-0.10		2	O8.5	92
322270	−40 10903			16 51 05	−40 26.8	1	10.34		0.21		-0.42		2	B5	1583
	−41 11056	NGC 6231 - 110	⋆ V	16 51 05	−41 48.8	3	9.83	.011	0.25	.011	-0.63	.000	10	B1 V	32,141,1707
326450	−43 11220			16 51 05	−43 54.9	1	10.13		0.59				4	B3 V	1021
152333	−41 11055	LSS 3821		16 51 06	−41 20.6	5	8.03	.027	0.22	.006	-0.68	.011	9	B1/2 Ib/II	32,540,976,1539,1707
		Steph 1409		16 51 07	+13 21.0	1	12.10		1.23		1.18		1	K5	1746
152420	−24 12924			16 51 07	−24 23.8	1	9.12		0.29		0.10		1	A3 II/III	1770
		AAS57,205 # 108		16 51 07	−40 55.4	1	12.95		0.39		0.08		2		1539
		NGC 6231 - 131		16 51 07	−41 47.9	1	12.56		0.46		0.03		4		113
		AAS57,205 # 109		16 51 08	−40 35.9	1	11.17		0.26		-0.28		2		1539
326318	−41 11057			16 51 08	−41 28.4	1	10.81		0.71		0.16		2	F0	1539
		NGC 6231 - 112		16 51 08	−41 48.9	3	10.12	.013	0.24	.010	-0.61	.004	12	B1 V	32,141,1707
322280	−40 10904			16 51 09	−40 40.8	1	9.77		0.05		0.09		2	B9	1539
		NGC 6231 - 130		16 51 09	−41 47.9	1	12.93		0.60		0.16		4		113
		AAS57,205 # 359		16 51 09	−41 59.3	1	12.11		0.44		0.17		2		1583
	−40 10906			16 51 10	−40 35.8	2	10.42	.025	0.19	.002	-0.51	.013	5		1539,1586
		AAS57,205 # 114		16 51 10	−40 38.5	1	10.25		0.17		-0.51		2		1539
		AAS57,205 # 116		16 51 10	−40 43.7	1	11.32		0.11		-0.47		2		1539
		AAS57,205 # 113		16 51 10	−41 18.0	1	11.99		0.30		-0.24		2		1539
326317	−41 11058			16 51 10	−41 34.4	3	10.12	.005	0.21	.000	-0.60	.012	7	B2	32,1539,1707
		NGC 6231 - 140		16 51 10	−41 46.8	1	11.52		0.43		-0.13		4		113
		AAS57,205 # 360		16 51 10	−42 01.8	1	12.32		0.31		-0.34		2		1583
		AAS57,205 # 380		16 51 10	−42 01.8	1	11.59		0.12		-0.43		2		1583
152426	−18 04352			16 51 11	−18 36.4	1	9.56		0.47		0.21		1	A7 II	1770
152384	−33 11546			16 51 11	−33 23.7	1	7.02		0.04				4	A0 V	1075
		AAS57,205 # 117		16 51 11	−41 04.0	1	11.56		0.40		-0.29		2		1539
		AAS57,205 # 120		16 51 11	−41 15.2	1	11.73		0.38		0.27		2		1539

Table 1   915

HD	DM	Other Id	N	Rem	α_{1950}	δ_{1950}	S	V	σ_V	B–V	σ_{B-V}	U–B	σ_{U-B}	n	Spectrum	References
		AAS57,205 # 119			16 51 11	−41 16.7	1	10.71		0.28		-0.20		2		1539
		AAS57,205 # 361			16 51 11	−41 59.8	1	11.98		0.40		-0.30		2		1583
		AAS57,205 # 122			16 51 12	−40 56.3	1	12.55		0.76		0.22		2		1539,1750
326296	−41 11060				16 51 12	−41 16.1	2	9.81	.005	0.23	.015	-0.58		6	B2 V	1021,1539
326333	−41 11059	NGC 6231 - 150		⋆ V	16 51 12	−41 45.0	2	9.62	.019	0.20	.005	-0.63	.014	7	B1 V(n)	32,1707
		AAS57,205 # 362			16 51 12	−41 59.4	1	11.70		0.37		-0.34		2		1583
		AAS57,205 # 124			16 51 13	−40 44.6	1	11.85		0.14		-0.35		2		1539
		AAS57,205 # 123			16 51 13	−40 50.1	1	12.40		0.69		0.21		2		1539,1750
		AAS57,205 # 126			16 51 13	−40 51.3	1	12.93		0.68		0.56		2		1539,1750
		AAS57,205 # 125			16 51 13	−40 59.1	1	11.67		0.25		-0.31		2		1539,1750
		NGC 6231 - 121			16 51 13	−41 51.6	1	11.22		0.33		-0.42		2		1583
152354	−42 11648	IDS16477S4219	AB		16 51 13	−42 23.6	1	9.23		0.12		-0.13		4	A0 IV	141
		AAS57,205 # 127			16 51 14	−40 40.9	1	12.10		0.45		0.50		2		1539
		NGC 6231 - 128			16 51 14	−41 48.1	1	13.68		0.60		0.27		4		113
329083	−44 11260				16 51 14	−44 56.4	1	10.39		0.43		0.23		4	A3	244
322279	−40 10910				16 51 15	−40 40.6	1	9.54		0.04		-0.75		2	B9	1539
152353	−40 10909	IDS16478S4043	A		16 51 15	−40 48.0	1	8.73		0.07		-0.72		3	B2 Ib/II	1707
152429	−25 11770				16 51 16	−25 44.9	1	7.23		0.55		0.13		1	G0 IV	1770
	−40 10905				16 51 16	−40 35.4	1	9.86		0.52		-0.11		2		1539
322307	−40 10911				16 51 17	−40 49.2	2	9.63	.015	0.10	.000	-0.66	.010	5	B5	1539,1707
		AAS57,205 # 129			16 51 17	−41 25.6	1	11.77		0.13		-0.11		2		1539
322278	−40 10912				16 51 18	−40 41.2	1	9.65		0.17		0.08		2	B8	1539
		AAS57,205 # 132			16 51 18	−41 26.8	1	12.55		0.30		-0.20		2		1539
152454	−27 11258	IDS16482S2724	A		16 51 19	−27 29.3	1	8.38		0.14		0.03		1	A0 V	1770
		AAS57,205 # 134			16 51 19	−40 40.4	1	12.25		0.29		-0.05		2		1539
	−41 11062	LSS 3824		⋆ AB	16 51 19	−41 56.6	1	9.52		0.44		-0.49		2	B5	1583
152355	−44 11261				16 51 19	−44 59.7	1	8.74		0.45		-0.03		4	F5 V	244
152761	+53 01902				16 51 20	+53 00.7	1	8.47		0.27		0.09		27	A3	1603
322274	−40 10913				16 51 20	−40 33.9	2	9.48	.001	0.16	.003	-0.44	.001	5	B5	1539,1586
		AAS57,205 # 135			16 51 20	−40 52.8	1	12.21		0.62		0.54		2		1539
		AAS57,205 # 137			16 51 21	−40 46.9	1	12.78		0.78		0.26		2		1539
		Ly 14 - 14			16 51 21	−45 10.6	1	13.89		1.50		1.01		1		412
329172	−46 11100				16 51 21	−46 40.2	1	9.72		0.53		0.02		1	G0	96
152010	−70 02326	IDS16457S7057	A		16 51 21	−71 01.9	1	6.46		0.26		0.15		10	A5 IV/V	1628
322277	−40 10914				16 51 22	−40 40.7	3	9.51	.040	0.08	.007	-0.67	.010	5	B1.5IV	540,976,1539
		AAS57,205 # 139			16 51 22	−40 42.6	1	13.04		0.30		0.16		2		1539
		AAS57,205 # 365			16 51 22	−41 04.1	1	10.97		0.21		-0.48		2		1583
		AAS57,205 # 142			16 51 22	−41 04.3	1	10.98		0.19		-0.50		2		1539
		AAS57,205 # 141			16 51 22	−41 08.3	1	11.94		0.25		-0.29		2		1539
326440	−43 11225				16 51 22	−43 51.6	2	10.30	.024	0.71	.014	-0.20	.035	5	B0.5V	540,954
	−62 05464				16 51 22	−62 22.7	1	9.35		0.96		0.78		1	K3	1696
152404	−36 11056	AK Sco			16 51 23	−36 48.5	2	8.95	.093	0.62	.025	0.17	.001	2	F5 V	776,1588
		AAS57,205 # 379			16 51 23	−40 20.6	1	11.52		0.65		0.06		2		1583
		AAS57,205 # 144			16 51 23	−40 40.8	1	11.84		0.22		-0.14		2		1539
		AAS57,205 # 143			16 51 23	−40 52.4	1	12.44		0.31		-0.08		2		1539
	−78 01119				16 51 23	−78 39.7	1	10.26		0.68		0.00		2		1696
152431	−30 13594	HR 6273			16 51 24	−30 30.4	4	6.34	.000	0.21	.008	0.14	.009	12	A5 III(m)	210,244,2006,8071
	−40 10917				16 51 24	−40 48.3	1	10.47		1.43		1.52		2		1539
152385	−40 10916	IDS16479S4054	A		16 51 24	−40 59.1	1	9.04		0.16		-0.62		3	B1 Ib	1707
		AAS57,205 # 145			16 51 24	−41 05.7	1	12.83		0.79		0.25		2		1539
		AAS57,205 # 147			16 51 25	−41 25.2	1	11.89		0.56		0.32		2		1539
		Ly 14 - 3			16 51 25	−45 09.7	1	11.66		1.22		0.07		3		412
152356	−46 11101				16 51 25	−46 42.1	1	9.73		0.37		0.19		1	A9 V	96
		AAS57,205 # 149			16 51 26	−40 49.6	1	12.81		0.51		0.52		2		1539,1750
		AAS57,205 # 148			16 51 26	−40 54.6	1	13.09		0.69		0.51		2		1539,1750
		Ly 14 - 7			16 51 26	−45 08.9	1	13.14		1.55		1.28		1		412
152453	−25 11773				16 51 27	−25 44.0	1	8.29		0.27		0.14		1	A6 IV	1770
152405	−40 10918	LSS 3827			16 51 27	−40 26.7	9	7.20	.014	0.12	.012	-0.78	.018	24	B0 Ia	32,540,954,976,1539,1583*
		AAS57,205 # 151			16 51 27	−40 46.2	1	11.34		1.14		0.95		2		1539
		AAS57,205 # 154			16 51 27	−40 46.2	1	12.71		0.57		0.62		2		1539
		AAS57,205 # 152			16 51 27	−41 05.7	1	11.10		0.26		-0.32		2		1539
		AAS57,205 # 366			16 51 27	−42 02.1	1	11.11		1.46		1.13		2		1583
		Ly 14 - 9			16 51 27	−45 09.6	1	13.74		1.24		0.31		2		412
		Ly 14 - 15			16 51 27	−45 10.1	1	14.81		1.39		0.23		1		412
		AAS57,205 # 157			16 51 28	−40 30.2	1	11.44		1.15		0.93		2		1539
		AAS57,205 # 155			16 51 28	−40 43.4	1	12.35		0.16		-0.21		2		1539
		AAS57,205 # 156			16 51 28	−40 58.8	1	12.31		0.24		-0.15		2		1539
		Ly 14 - 6			16 51 28	−45 09.2	1	13.08		1.24		0.23		2		412
		Ly 14 - 8			16 51 28	−45 09.7	1	12.64		0.54		0.15		2		412
		Ly 14 - 10			16 51 28	−45 09.7	1	13.33		1.24		0.39		2		412
		Ly 14 - 2			16 51 28	−45 10.4	1	12.00		1.11		-0.03		2		412
		AAS57,205 # 158			16 51 29	−40 51.2	1	12.95		0.65		0.22		2		1539,1750
152407	−40 10922				16 51 29	−41 00.8	2	8.77	.010	0.29	.000	0.15	.010	5	B5 Ib	1539,1707
152408	−40 10919	HR 6272, LSS 3828		⋆ AB	16 51 29	−41 04.3	9	5.79	.013	0.16	.009	-0.77	.015	36	O8 If	32,131,486,954,1539,1586*
152386	−44 11266	LSS 3825			16 51 29	−44 54.6	6	8.13	.008	0.51	.011	-0.49	.012	19	O5/6Fe	164,244,432,540,976,1586
		Ly 14 - 11			16 51 29	−45 09.8	1	13.73		1.24		0.31		2		412
		Ly 14 - 1			16 51 29	−45 10.8	1	11.16		1.19		1.01		4		412
		AS 212, HK Sco			16 51 30	−30 18.3	1	13.48		1.34		-0.31		1		1753
152406	−40 10920	IDS16480S4025	A		16 51 30	−40 30.0	1	7.80		1.80		2.08		2	K3/4 III	1539
		AAS57,205 # 162			16 51 30	−40 53.0	1	12.80		0.59		0.45		2		1539

HD	DM	Other Id	N Rem	α_{1950}	δ_{1950}	S	V	σ_V	B–V	σ_{B-V}	U–B	σ_{U-B}	n	Spectrum	References
326294	−41 11067			16 51 30	−41 13.0	1	10.35		0.35		0.07		2	A3	1539
		AAS57,205 # 163		16 51 30	−41 16.9	1	11.20		0.35		0.29		2		1539
		Ly 14 - 13		16 51 30	−45 07.6	1	13.08		0.96		0.94		1		412
		Ly 14 - 12		16 51 30	−45 12.8	1	13.07		2.18		1.82		1		412
152372	−48 11288			16 51 30	−48 43.5	2	8.85	.001	0.20	.006	-0.60	.009	5	B2 III/IV	540,976
152485	−24 12930			16 51 31	−24 15.4	1	10.85		0.30		0.18		2	A0 III	1770
322275	−40 10921			16 51 31	−40 38.0	1	9.72		0.09		-0.69		2	B5	1539
		AAS57,205 # 165		16 51 31	−40 54.1	1	13.93		0.66		0.46		2		1539,1750
		AAS57,205 # 168		16 51 31	−40 54.8	1	13.15		0.41		0.40		2		1539
		AAS57,205 # 167		16 51 31	−40 59.4	1	11.36		0.16		-0.42		2		1539
	−41 11066			16 51 31	−41 56.9	1	10.84		0.48		-0.36		2		1583
		Ly 14 - 16		16 51 31	−45 09.9	1	15.31		1.24		0.31		1		412
		AAS57,205 # 169		16 51 32	−40 54.8	1	13.42		0.49		0.10		2		1539
		AAS57,205 # 170		16 51 32	−41 07.3	1	12.78		0.47		0.36		2		1539
		AAS57,205 # 171		16 51 32	−41 26.1	1	11.77		0.64		0.25		2		1539
		AAS57,205 # 370		16 51 32	−41 55.3	1	11.56		0.58		0.10		2		1583
152424	−41 11068 LSS 3829		⋆ A	16 51 32	−42 00.6	8	6.31	.020	0.40	.021	-0.58	.012	18	B0 Ib/II	32,131,138,540,954,976*
		Ly 14 - 4		16 51 32	−45 11.5	1	12.38		0.93		-0.18		2		412
		AAS57,205 # 173		16 51 33	−40 51.1	1	13.12		0.67		0.51		2		1539,1750
		AAS57,205 # 174		16 51 33	−40 52.4	1	13.09		0.64		0.54		2		1539
		AAS57,205 # 175		16 51 33	−40 52.4	1	14.16		0.30		0.00		2		1539
		AAS57,205 # 172		16 51 33	−40 54.1	1	13.32		0.69		0.54		2		1539,1750
326295	−41 11070			16 51 33	−41 20.1	1	10.47		0.45		-0.36		2	B5	1539
152473	−30 13598			16 51 34	−30 54.0	2	7.49	.005	1.15	.005	1.10	.005	5	K1 III	210,244
		AAS57,205 # 177		16 51 34	−40 46.1	1	12.38		0.21		-0.13		2		1539
		AAS57,205 # 179		16 51 34	−40 52.1	1	12.88		0.60		0.48		2		1539
		AAS57,205 # 178		16 51 34	−40 52.4	1	13.01		0.89		0.52		2		1539
152569	−1 03268 HR 6277		⋆ A	16 51 35	−01 31.8	2	6.25	.015	0.28	.003	0.09	.002	10	F0 V	253,1088
		AAS57,205 # 181		16 51 35	−40 53.6	1	12.68		0.62		0.58		2		1539
		AAS57,205 # 182		16 51 36	−40 54.1	1	13.13		0.72		0.45		2		1539
		AAS57,205 # 183		16 51 36	−41 15.6	1	12.17		0.30		0.10		2		1539
152387	−46 11102			16 51 36	−46 39.1	1	8.93		0.24		0.16		1	Ap CrEuSr	96
		AAS57,205 # 184		16 51 37	−40 52.3	1	13.20		0.59		0.37		2		1539
322309	−40 10924			16 51 37	−40 52.3	1	10.44		0.18		-0.59		2	B8	1539
		AAS57,205 # 186		16 51 37	−40 54.8	1	12.93		0.87		0.70		2		1539
152614	+10 03092 HR 6281			16 51 38	+10 14.8	6	4.37	.008	-0.09	.013	-0.34	.016	13	B8 V	15,1022,1079,1363*
322276	−40 10923			16 51 38	−40 40.4	1	9.75		0.09		-0.57		2	B3	1539
		AAS57,205 # 189		16 51 38	−40 52.3	2	13.21	.000	0.54	.000	0.32	.000	4		1539,1539
		AAS57,205 # 187		16 51 38	−41 16.4	1	12.25		0.38		0.29		2		1539
152516	−21 04443			16 51 39	−21 48.0	4	8.03	.009	0.08	.007	-0.56	.011	7	B2 III	400,540,976,1770
		AAS57,205 # 190		16 51 39	−40 52.0	1	12.93		0.82		0.58		2		1539
	−30 13603 CL Sco			16 51 40	−30 32.5	1	13.25		0.94		-0.62		1		1753
152436	−40 10926			16 51 40	−40 14.7	2	9.12	.025	0.20	.005	-0.38	.015	4	B3 II	1583,1707
		AAS57,205 # 193		16 51 40	−40 50.9	1	13.35		0.74		0.52		2		1539
		AAS57,205 # 192		16 51 40	−40 51.6	1	12.51		0.64		0.36		2		1539
		AAS57,205 # 191		16 51 40	−40 52.6	1	12.93		0.58		0.43		2		1539
152533	−14 04499			16 51 41	−14 30.1	1	9.73		1.02		0.93		2	K3 V	3072
152517	−21 04442 IDS16487S2153		AB	16 51 41	−21 58.1	1	8.44		0.39		0.28		1	A5 V	1770
		AAS57,205 # 373		16 51 41	−40 30.4	1	12.07		0.26		0.28		2		1583
		AAS57,205 # 195		16 51 41	−40 50.9	1	12.93		1.42		0.73		2		1539
		AAS57,205 # 194		16 51 41	−40 51.5	1	13.51		0.38		0.26		2		1539
		AAS57,205 # 197		16 51 41	−40 51.9	1	13.84		0.80		0.57		2		1539
		AAS57,205 # 202		16 51 41	−40 52.1	1	13.17		0.62		0.52		2		1539
		AAS57,205 # 198		16 51 41	−40 52.5	1	13.22		0.76		0.57		2		1539
		AAS57,205 # 201		16 51 41	−40 52.6	1	13.28		0.71		0.50		2		1539
		AAS57,205 # 196		16 51 41	−40 52.8	1	13.82		0.72		0.58		2		1539
		AAS57,205 # 199		16 51 41	−40 54.6	1	12.80		0.75		0.51		2		1539
152437	−41 11071			16 51 41	−41 51.9	3	9.15	.009	0.14	.012	0.06	.086	5	B9 III/IV	32,131,1583
152456	−38 11316			16 51 42	−38 20.0	2	7.21	.026	0.01	.014	-0.30	.003	4	B8/9 III(p)	540,976
322308	−40 10925			16 51 42	−40 50.0	1	9.08		1.28		1.30		2	K0	1539
		AAS57,205 # 203		16 51 42	−40 51.1	1	13.62		0.68		0.51		2		1539,1750
		AAS57,205 # 204		16 51 42	−40 53.3	1	13.50		0.66		0.42		2		1539
		AAS57,205 # 205		16 51 42	−40 54.3	1	13.79		0.74		0.93		2		1539,1750
		AAS57,205 # 207		16 51 43	−40 49.5	1	13.46		0.81		0.19		2		1539,1750
		AAS57,205 # 206		16 51 43	−40 55.2	1	12.49		0.57		0.56		2		1539,1750
322272	−40 10928			16 51 44	−40 26.9	1	10.45		0.57		0.02		2		1583
	−40 10929			16 51 44	−40 29.0	1	11.10		0.21		-0.03		2		1583
		AAS57,205 # 210		16 51 44	−40 49.2	1	13.28		0.48		0.34		2		1539,1750
		AAS57,205 # 209		16 51 44	−40 52.4	1	14.57		0.80		0.34		2		1539
		AAS57,205 # 208		16 51 44	−40 52.9	1	13.60		0.71		0.32		2		1539
		AAS57,205 # 214		16 51 46	−40 46.4	1	12.29		0.50		0.07		2		1539
		AAS57,205 # 211		16 51 46	−40 51.4	1	13.29		1.64		1.33		2		1539
		AAS57,205 # 213		16 51 46	−40 54.8	1	12.85		0.64		0.58		2		1539,1750
		AAS57,205 # 212		16 51 46	−41 00.5	1	11.89		1.19		0.91		2		1539
		AAS57,205 # 216		16 51 47	−40 35.7	1	11.50		0.66		0.17		2		1539
		AAS57,205 # 217		16 51 47	−40 55.5	1	12.35		0.71		0.61		2		1539
		AAS57,205 # 215		16 51 47	−41 07.9	1	12.26		0.44		0.25		2		1539
		AAS25,287 # 2		16 51 48	+39 51.8	1	13.23		0.83		0.55		3		511
		AAS57,205 # 376		16 51 48	−40 48.8	1	12.50		0.40		0.10		2		1583
		AAS57,205 # 218		16 51 48	−40 51.3	1	12.73		1.54		1.08		2		1539

Table 1   917

HD	DM	Other Id	N Rem	α_{1950}	δ_{1950}	S	V	σ_V	B–V	σ_{B-V}	U–B	σ_{U-B}	n	Spectrum	References
152600	−3 04023			16 51 49	−04 05.0	1	7.58		1.22		1.00		3	G5	1657
322273	−40 10931			16 51 49	−40 32.5	1	9.49		0.30		0.18		2	A2	1539
326353				16 51 49	−41 58.7	1	11.14		0.16		0.12		2		1583
		AAS57,205 # 221		16 51 50	−40 40.7	1	11.51		0.12		-0.42		2		1539
		AAS57,205 # 220		16 51 50	−40 52.3	1	13.39		0.64		0.76		2		1539,1750
		AAS57,205 # 222		16 51 51	−40 56.3	1	11.70		0.62		0.37		2		1539,1750
152459	−41 11073			16 51 51	−41 28.7	2	8.53	.015	0.19	.005	-0.25	.000	4	B5 II	32,1707
152457	−40 10933			16 51 52	−40 51.4	2	9.55	.010	0.16	.005	-0.50	.005	4	A3 III(m)	1539,1707
		AAS57,205 # 223		16 51 52	−40 53.3	1	10.85		1.31		1.04		2		1539
		G 138 - 67		16 51 53	+11 59.5	3	10.76	.013	1.44	.009			3	K7	1705,1746,1759
152585	−11 04231	HR 6278		16 51 53	−11 42.7	1	6.57		0.14				4	A2 IV	2007
		AAS57,205 # 378		16 51 53	−40 13.0	1	11.72		0.33		-0.41		2		1583
	−40 10935			16 51 53	−40 25.0	1	11.12		0.15		-0.32		2		1583
		Tr 24 - 34		16 51 53	−40 36.1	1	11.34		0.26		-0.33		2		1539
322452	−40 10938	IDS16484S4037	AB	16 51 53	−40 42.2	1	9.71		0.14		-0.53		2	B3	1539
		AAS57,205 # 228		16 51 53	−40 46.6	1	11.58		0.49		0.46		2		1539
152458	−40 10933	LSS 3832		16 51 53	−40 51.4	2	10.01	.015	0.16	.010	-0.62	.005	4	B8	1539,1707
152438	−45 11062	IDS16483S4510	AB	16 51 53	−45 15.2	1	8.21		0.04		-0.35		4	B6 III	244
152812	+47 02400	HR 6286		16 51 54	+47 29.8	1	6.00		1.32		1.44		2	K2 III	1080
	−40 10936			16 51 54	−40 23.4	1	11.25		0.16		-0.46		2		1583
		AAS57,205 # 381		16 51 54	−40 33.0	1	11.17		1.51		1.57		2		1583
		AAS57,205 # 229		16 51 54	−40 52.7	1	11.58		1.31		1.34		2		1539
		He3 1281		16 51 54	−41 54.	1	11.22		0.59		0.02		3		1586
152601	−5 04374	HR 6280		16 51 55	−06 04.4	2	5.27	.025	1.09	.005	1.05	.010	12	K2 III	1088,3016
		AAS57,205 # 230		16 51 55	−40 38.3	1	11.62		0.56		0.12		2		1539
152520	−31 13349	IDS16487S3104	AB	16 51 56	−31 08.6	1	8.17		1.02		0.76		1	G6 (III)	210
152521	−31 13348	IDS16487S3114	AB	16 51 56	−31 19.3	2	6.72	.006	0.06	.005	0.04	.005	5	A0/1 V	210,244
		NGC 6242 - 60		16 51 56	−39 24.5	1	12.21		0.31		0.10		1		251
152792	+43 02659			16 51 57	+42 54.6	2	6.83	.013	0.62	.018	0.09	.009	5	G0 V	1003,3026
		NGC 6242 - 61		16 51 57	−39 25.0	1	11.59		0.26		-0.16		1		251
		AAS57,205 # 231		16 51 57	−41 05.0	1	12.16		0.66		0.28		2		1539
329094	−44 11272			16 51 57	−44 59.3	1	10.33		0.53		-0.05		4	B8	244
322211	−39 10857	NGC 6242 - 59	★ AB	16 51 58	−39 24.2	2	9.08	.000	0.23	.015	-0.31	.003	5	B5 III	251,1730
	−39 10855	NGC 6242 - 129		16 51 59	−39 22.4	1	10.82		0.18		-0.27		1		251
		AAS57,205 # 232		16 51 59	−40 42.4	1	11.13		1.11		0.90		2		1539
		AAS57,205 # 233		16 51 59	−40 43.1	1	11.99		0.22		-0.15		2		1539
	−39 10858	NGC 6242 - 62		16 52 00	−39 25.4	2	9.88	.018	0.24	.024	-0.33	.006	8	B2 III	251,1586
152490	−40 10939			16 52 00	−40 54.0	1	7.74		0.22		0.18		2	A5 m	1539
		NGC 6242 - 82		16 52 01	−39 24.5	1	9.89		0.25		-0.37		1	B5 Vn	251
322408	−40 10940			16 52 01	−40 09.4	1	10.22		0.20		-0.45		2	F0	1583
152491	−43 11233			16 52 01	−43 14.2	1	6.77		0.03		0.06		2	A1 V	401
	−39 10861	NGC 6242 - 49		16 52 02	−39 21.1	2	9.74	.041	0.22	.012	-0.44	.020	5	B3 Ve	251,1586
		NGC 6242 - 50		16 52 02	−39 21.7	1	11.83		0.24		-0.10		1		251
	−39 10859	NGC 6242 - 52	AB	16 52 02	−39 22.4	1	10.47		0.18		-0.36		1		251
		Tr 24 - 32		16 52 02	−40 36.1	1	11.94		0.79		0.16		2		1539
		NGC 6242 - 63		16 52 03	−39 25.3	1	13.13		0.42		0.34		1		251
322429	−40 10941			16 52 03	−40 26.1	1	11.53		0.16		-0.38		2	B9	1583
		AAS57,205 # 236		16 52 03	−40 48.8	1	12.80		0.35		0.22		2		1539
152748	+27 02706			16 52 05	+27 40.3	1	7.91		0.97		0.85		2	G8 II	8100
		AAS25,287 # 3		16 52 05	+39 49.0	1	12.65		0.88		0.70		3		511
	−39 10863	NGC 6242 - 51		16 52 05	−39 21.7	1	10.35		0.20		-0.29		1	B6 V	251
152504	−39 10862	NGC 6242 - 77	★ AB	16 52 05	−39 23.2	1	8.61		0.23		-0.29		3	B5 IV	251
		AAS57,205 # 237		16 52 05	−40 48.6	1	12.18		0.20		-0.31		2		1539
		NGC 6242 - 80		16 52 06	−39 23.8	1	12.13		0.34		0.00		1		251
		NGC 6242 - 83		16 52 06	−39 24.7	1	12.54		0.33		0.08		1		251
		NGC 6242 - 64		16 52 06	−39 25.6	1	13.13		0.40		0.40		1		251
		Tr 24 - 29		16 52 06	−40 37.8	1	11.99		0.42		0.20		2		1539
		AAS25,287 # 4		16 52 07	+39 57.5	1	14.22		0.58		0.18		3		511
		NGC 6242 - 78		16 52 08	−39 23.5	1	12.16		0.33		0.09		1		251
		Tr 24 - 40		16 52 08	−40 31.4	1	11.97		0.28		-0.12		2		1539
322451	−40 10943	Tr 24 - 31		16 52 08	−40 35.1	1	11.00		0.31		0.19		2	B8	1539
322459	−40 10942			16 52 08	−40 49.5	1	9.39		1.82		2.23		2	K5	1539
		AAS57,205 # 242		16 52 08	−40 52.9	1	12.13		0.35		0.26		2		1539
		AAS57,205 # 240		16 52 08	−41 00.1	1	12.56		0.27		0.15		2		1539
152477	−47 11137			16 52 08	−48 04.1	2	9.03	.035	0.66	.012	-0.27	.032	5	B1 Ib	540,976
152602	−18 04357			16 52 09	−18 41.3	1	10.19		0.25		0.15		1	B9 V	1770
	−39 10864	NGC 6242 - 79		16 52 09	−39 23.9	1	10.03		0.24		-0.27		1	B5 V:	251
	−39 10866			16 52 09	−39 24.0	1	10.03		0.26		-0.30		2		1730
152603	−18 04358			16 52 10	−18 48.3	1	7.77		0.17		0.03		2	B9.5 II/II	1770
		NGC 6242 - 66		16 52 10	−39 24.6	1	12.70		0.38		0.24		1		251
152524	−39 10870	NGC 6242 - 65	★ A	16 52 11	−39 25.6	1	7.28		1.81		1.91		1	K0 II	251
152574	−32 12151			16 52 13	−32 25.5	3	7.17	.005	0.57	.005	0.32	.005	6	F0 III	219,657,1075
322407	−39 10868			16 52 13	−40 04.4	1	10.73		0.26		-0.48		2	A2	1583
322409	−40 10945			16 52 13	−40 10.4	1	9.00		1.50		1.77		2	K5	1583
		AAS57,205 # 244		16 52 14	−40 51.5	1	12.70		0.64		0.24		2		1539
		AAS57,205 # 245		16 52 15	−40 58.4	1	12.00		0.53		0.41		2		1539
		Tr 24 - 30		16 52 16	−40 36.0	1	11.72		0.27		-0.17		2		1539
		AAS57,205 # 247		16 52 16	−40 47.1	1	12.75		0.19		-0.19		2		1539
		AAS57,205 # 246		16 52 16	−40 48.1	1	11.95		0.20		-0.31		2		1539
		AAS57,205 # 249		16 52 16	−40 48.1	1	13.25		0.28		0.07		2		1539

HD	DM	Other Id	N	Rem	α_{1950}	δ_{1950}	S	V	σ_V	B–V	σ_{B-V}	U–B	σ_{U-B}	n	Spectrum	References
152604	−26 11674	IDS16493S2619		AB	16 52 17	−26 24.1	1	9.91		0.31		0.16		1	A0 III	1770
		NGC 6242 - 33			16 52 17	−39 25.4	1	10.86		0.24		-0.26		1		251
	−39 10874	NGC 6242 - 28			16 52 17	−39 25.6	2	11.06	.015	0.24	.010	-0.28	.019	3	B4 V	251,1730
		Tr 24 - 39			16 52 17	−40 31.6	1	12.35		0.31		0.20		2		1539
152478	−50 10905	HR 6274			16 52 17	−50 35.8	5	6.30	.019	0.01	.017	-0.68	.007	18	B2 (III)ne	15,1075,1586,1637,2007
	−39 10876	NGC 6242 - 31			16 52 18	−39 25.9	1	11.07		0.24		-0.25		1	B6 V	251
		L 203 - 138			16 52 18	−57 46.	3	11.64	.036	0.48	.008	-0.21	.020	4		1097,1696,3077
		Tr 24 - 46			16 52 20	−40 30.3	1	12.01		0.29		-0.23		2		1539
		NGC 6242 - 34			16 52 21	−39 25.3	1	11.53		0.26		-0.08		1	B7 V:	251
		AAS57,205 # 252			16 52 21	−40 49.8	1	11.75		0.23		-0.34		2		1539
152631	−19 04464				16 52 22	−19 21.7	1	9.26		0.42		0.23		1	A7 II	1770
152505	−48 11295				16 52 22	−48 57.3	1	8.88		0.12		-0.36		3	B2/3 Vne	1586
	−40 10947				16 52 23	−40 22.7	1	9.86		1.26		0.94		2		1583
		Tr 24 - 93			16 52 23	−40 42.6	1	12.48		0.59		0.52		2		1539
		AAS57,205 # 388			16 52 25	−40 23.6	1	11.56		0.28		0.01		2		1583
		Tr 24 - 38			16 52 25	−40 33.0	1	11.70		0.41		0.32		2		1539
		AAS57,205 # 255			16 52 25	−40 50.8	1	11.61		0.55		0.39		2		1539
152619	−30 13612				16 52 26	−30 34.3	1	9.69		0.17		0.05		1	A0 IV	210
322428	−40 10950				16 52 26	−40 19.6	1	10.86		0.36		0.35		2	A5	1583
152559	−40 10949	Tr 24 - 92	★		16 52 26	−40 42.0	5	8.44	.020	0.12	.010	-0.76	.010	9	O9.5III	32,540,976,1539,1707
152560	−40 10948	LSS 3839			16 52 26	−40 56.8	4	8.28	.012	0.12	.008	-0.74	.009	8	B2 (V)n	32,540,976,1707
152645	−19 04465				16 52 27	−19 26.5	1	8.99		0.25		0.17		1	A1 V	1770
326473	−41 11079	LSS 3841			16 52 28	−41 26.8	1	10.21		0.43		-0.50		2	B0.5V	540
		Tr 24 - 2			16 52 30	−40 34.2	1	12.24		0.35		-0.09		2		1539
		Tr 24 - 24			16 52 30	−40 38.2	1	11.08		0.30		0.19		2		1539
322450	−40 10951	Tr 24 - 23			16 52 31	−40 38.0	1	10.02		0.12		-0.61		2	B5	1539
		AAS57,205 # 260			16 52 31	−40 46.2	1	12.00		0.46		0.19		2		1539
		LHS 3253			16 52 32	−08 02.6	1	14.64		1.41				1		1759
322453		Tr 24 - 90			16 52 32	−40 42.3	1	10.32		0.10		-0.19		2	B8	1539
152541	−46 11115	IDS16489S4641		AB	16 52 32	−46 46.2	2	7.45	.001	0.06	.001	-0.25	.005	4	B7 V	401,540
152527	−52 10333	HR 6275			16 52 32	−52 12.3	1	5.94		-0.08				4	B9 V	2032
152655	−21 04449	IDS16496S2124		A	16 52 33	−21 29.4	1	6.62		0.14		-0.15		2	B9 III	1770
152656	−21 04448				16 52 33	−21 37.6	1	10.18		0.23		0.07		1	B9 IV/V	1770
		AAS57,205 # 391			16 52 33	−40 43.1	1	12.91		0.44		0.06		2		1583
		AAS57,205 # 390			16 52 33	−40 44.7	1	13.18		0.28		0.41		2		1583
		Tr 24 - 1			16 52 34	−40 34.4	1	11.84		0.29		-0.36		2		1539
152591	−40 10952	Tr 24 - 4	★		16 52 34	−40 35.6	5	8.40	.019	0.12	.009	-0.74	.008	11	B0 IV	32,540,976,1539,1707
	−15 04408				16 52 35	−15 57.6	1	10.00		0.90		0.56		3	K2 V	3072
322458	−40 10953				16 52 35	−40 50.1	1	10.66		0.15		-0.60		2	B8	1539
152561	−45 11067	NGC 6250 - 33			16 52 35	−45 58.2	2	9.11	.000	0.24	.010	0.10	.040	2	B9.5V	96,5002
322449	−40 10964	Tr 24 - 5			16 52 36	−40 36.0	1	9.22		1.58		1.88		2	K5	1539
322454	−40 10955	Tr 24 - 89			16 52 36	−40 42.4	2	9.29	.015	0.16	.010	-0.66	.005	4	B5	1539,1707
		AAS25,287 # 1			16 52 37	+39 48.2	1	11.90		0.92		0.68		3		511
152590	−40 10956	LSS 3843			16 52 37	−40 16.2	6	8.43	.019	0.14	.011	-0.78	.028	56	O8	32,540,976,1583,1707,1750
		Tr 24 - 52			16 52 38	−40 31.2	1	12.69		0.15		0.00		2		1583
329089	−44 11287				16 52 38	−44 47.9	3	9.32	.005	0.40	.007	-0.44	.008	10	B1.5IV	244,540,976
		AAS25,287 # 5			16 52 39	+39 52.6	1	13.65		0.79		0.51		3		511
152657	−25 11789				16 52 39	−25 27.3	1	7.59		0.02		-0.26		2	B8 II	1770
152635	−31 13366				16 52 39	−31 23.5	3	7.70	.013	-0.04	.012	-0.42	.005	5	B7 II	210,540,1770
		Tr 24 - 98			16 52 39	−40 42.1	1	11.84		0.36		0.25		2		1539
		Tr 24 - 99			16 52 39	−40 42.2	1	12.23		0.30		0.19		2		1539
		Tr 24 - 94			16 52 41	−40 34.7	1	10.40		0.14		-0.48		2		1539
		Tr 24 - 9			16 52 41	−40 34.9	1	12.56		0.21		-0.22		2		1583
		Tr 24 - 7			16 52 41	−40 35.5	1	11.53		0.16		-0.42		2		1539
152636	−33 11570	HR 6282			16 52 42	−33 25.7	4	6.35	.011	1.72	.011	2.06		14	K5 III	15,1075,1311,2007
		Tr 24 - 53			16 52 42	−40 31.8	1	12.74		0.61		0.35		2		1583
		L 74 - 120			16 52 42	−72 54.	1	14.34		1.02		0.78		1		3062
152658	−26 11679	IDS16496S2700		A	16 52 43	−27 04.9	1	9.96		0.28		0.24		1	A2 III/IV	1770
		Tr 24 - 61			16 52 43	−40 27.2	1	10.91		0.38		-0.14		2		1583
322430	−40 10959	Tr 24 - 50			16 52 43	−40 30.7	1	10.21		0.17		-0.52		2	B5	1539
		Tr 24 - 97			16 52 43	−40 38.1	1	11.95		0.30		-0.30		2		1539
		Wolf 629			16 52 44	−08 14.1	1	11.76		1.68		1.26		3	sdM4	1764
322448	−40 10960	Tr 24 - 96			16 52 44	−40 38.1	1	9.80		0.13		-0.67		2	B5	1539
		AAS25,287 # 6			16 52 45	+39 51.9	1	14.43		0.57		0.09		3		511
		LTT 6750			16 52 45	−08 13.8	1	11.70		1.70				1		8105
152815	+21 03002	HR 6287			16 52 46	+21 02.3	3	5.40	.005	0.97	.005	0.70	.009	8	G8 III	374,1080,1355
152905	+43 02661				16 52 46	+43 29.2	2	7.03	.020	1.01	.000	0.77	.005	5	K0 III	1501,1601
152623	−40 10961	Tr 24 - 15	★		16 52 46	−40 34.9	4	6.68	.007	0.09	.012	-0.81	.013	8	(O8)	32,219,954,1707
		Tr 24 - 19			16 52 46	−40 36.3	1	12.02		0.20		-0.35		2		1539
	−40 10962	Tr 24 - 14			16 52 47	−40 34.1	1	9.05		0.11		-0.73		2		1539
152751	−8 04352	V1054 Oph	★	AB	16 52 48	−08 14.7	6	9.01	.024	1.59	.013	1.08	.015	22	M3 Ve	158,1017,1197,1705,3072,8006
152751	−8 04352	IDS16501S0809		C	16 52 48	−08 14.7	2	11.73	.044	1.70	.006	1.35		6		1705,3078
152751	−8 04352	IDS16501S0809		D	16 52 48	−08 14.7	1	16.66		2.05		0.00		2		3003
		AAS57,205 # 398			16 52 48	−40 19.1	1	11.86		0.71		-0.36		2		1583
		Tr 24 - 20			16 52 48	−40 37.7	1	11.62		0.22		-0.36		2		1539
152622	−40 10964	LSS 3846			16 52 49	−40 25.0	6	8.14	.015	0.18	.013	-0.74	.019	15	B1/2 Ib/II	32,540,954,976,1583,1707
322447	−40 10963	Tr 24 - 17	★		16 52 49	−40 36.1	4	8.89	.013	0.12	.011	-0.70	.008	14	B1 IV	32,1539,1586,1707
		Tr 24 - 55			16 52 50	−40 30.9	1	12.11		0.29		-0.18		2		1539
329223	−44 11290				16 52 50	−44 46.6	1	9.80		0.57		0.10		4	F2	244
		Tr 24 - 12			16 52 51	−40 33.6	1	11.32		0.22		-0.25		2		1539

Table 1

919

HD	DM	Other Id	N Rem	α_{1950}	δ_{1950}	S	V	σ_V	B–V	σ_{B-V}	U–B	σ_{U-B}	n	Spectrum	References
326514	−42 11678			16 52 51	−42 09.2	2	10.14	.019	0.11	.000	-0.20	.015	6	B8 V	141,1707
322330				16 52 53	−38 09.5	1	10.20		0.23		-0.21		1	B5	1586
322412	−40 10967			16 52 55	−40 12.0	1	10.39		0.48		0.28		2	F0	1583
322431	−40 10970	Tr 24 - 64		16 52 56	−40 28.5	1	10.11		0.12		-0.66		2	B3	1583
152780	−8 04354			16 52 57	−08 51.9	1	10.05		0.90		0.61		1	K3 V	3072
152624	−46 11120			16 52 57	−46 12.2	1	8.34		0.19		0.12		1	A1/2 V	96
152830	+13 03258	HR 6290, V644 Her		16 52 58	+13 42.0	1	6.35		0.34		0.06		2	F3 Vs	3058
152863	+25 03156	HR 6292	⋆ A	16 52 59	+25 48.6	5	6.08	.008	0.92	.006	0.62	.005	12	G5 III	15,1007,1013,3024,8015
152863	+25 03156	IDS16510N2554	B	16 52 59	+25 48.6	1	10.80		0.70		0.15		3		3024
		vdB 75 # a		16 53 06	−40 10.9	1	11.43		0.35		-0.55		5	B1 Vp?	434
152667	−40 10975	HR 6283, V861 Sco		16 53 07	−40 44.7	12	6.18	.055	0.26	.026	-0.70	.020	42	B0.5 Iae	15,32,540,681,815,954*
152781	−16 04371	HR 6284	⋆ A	16 53 08	−16 43.7	2	6.35	.015	0.94	.015	0.68		6	K0 IV	2007,3005
		Tr 24 - 68		16 53 08	−40 29.0	1	11.48		0.27		-0.52		2		1583
322432	−40 10977	Tr 24 - 69		16 53 08	−40 29.4	2	9.07	.025	0.15	.010	-0.69	.005	4	B5	1583,1707
152879	+18 03266	HR 6293	⋆ AB	16 53 10	+18 30.7	3	5.35	.008	1.41	.004	1.66	.004	5	K4 III	15,1003,1080
152686	−42 11686			16 53 12	−42 17.1	3	8.95	.018	0.25	.014	-0.66	.003	10	B1/2 Ib	141,540,976
152685	−40 10980	LSS 3852		16 53 15	−41 04.6	5	7.46	.017	0.19	.013	-0.60	.012	15	B1 Ib	32,540,976,1707,2012
153204	+63 01312			16 53 19	+63 10.7	1	8.58		0.13		-0.04		1	A2	1776
152782	−21 04454			16 53 19	−21 57.9	1	10.21		0.12		-0.79		1	B9 III/IV	1770
326598	−43 11249			16 53 20	−43 46.5	1	10.17		0.66		0.21		1	B8	1586
153166	+60 01713			16 53 21	+60 26.5	1	7.04		1.31		1.38		3	K0	985
		L 484 - 31		16 53 24	−36 58.7	1	11.44		1.51				1	M1	1705
152687	−45 11076	NGC 6250 - 32		16 53 24	−46 04.0	2	8.69	.000	0.15	.010	0.02	.035	2	B9 V	96,5002
152723	−40 10986	LSS 3854	⋆ AB	16 53 26	−40 26.1	7	7.21	.063	0.13	.017	-0.81	.014	17	O7/8	32,540,954,976,1583*
152640	−55 07751			16 53 26	−55 46.5	3	8.21	.031	0.00	.013	-0.45	.004	10	B5 III	400,540,976
		NGC 6250 - 31		16 53 28	−45 51.8	1	10.63		1.48		1.49		1		5002
		LP 686 - 26		16 53 29	−03 45.3	1	13.63		1.37		1.29		1		1696
152741	−42 11691			16 53 29	−42 11.7	1	9.66		0.10		-0.33		4	B7 II/III	141
322426	−40 10987			16 53 31	−40 18.5	1	10.02		1.23		1.02		2		1583
152742	−42 11692			16 53 31	−42 52.3	1	9.13		1.24		1.03		2	B4 II/III	401
152706	−45 11081	NGC 6250 - 22		16 53 32	−45 47.9	2	10.14	.000	0.22	.015	0.12	.010	2	B9.5V	412,5002
322423	−40 10989			16 53 36	−40 14.6	1	9.83		0.46		-0.05		2	G0	1583
322424	−40 07654			16 53 36	−40 20.2	1	10.64		0.24		-0.54		2		1583
152755	−40 10990			16 53 36	−40 52.7	2	8.04	.010	0.04	.005	-0.45	.005	4	B4 II/III	401,1707
152769	−39 10907			16 53 38	−40 03.7	1	9.18		0.22		-0.40		2	B2/3 II/III	141
	−40 10995			16 53 40	−40 26.1	1	10.60		0.35		-0.50		2		1583
152756	−43 11256	LSS 3857		16 53 40	−43 38.6	2	9.03	.001	0.63	.007	-0.34	.017	6	B3/5 Iab/b	540,976
153093	+47 02404			16 53 41	+47 00.4	1	9.08		0.33		0.03		3	F0	1601
152804	−31 13392			16 53 41	−31 54.7	1	8.68		0.14		-0.28		1	B7 III/IV	1770
152743	−45 11082	NGC 6250 - 21		16 53 41	−45 47.3	3	9.10	.004	0.16	.004	-0.54	.008	4	B2 V	412,540,5002
153032	+40 03074			16 53 42	+39 56.1	1	8.38		1.49		1.82		2	K5	1601
152785	−42 11695			16 53 45	−42 19.6	1	8.89		0.24		0.15		4	B8 II	141
		Steph 1412		16 53 46	+05 22.2	1	11.48		1.39		1.25		1	K7	1746
152849	−22 04249	HR 6291	⋆ AB	16 53 47	−23 04.4	2	5.58	.006	-0.02	.003	-0.09		5	A0 V	1770,2007
		LP 919 - 21		16 53 48	−30 27.0	1	11.06		1.02		1.00		1		1696
322435	−40 10998	IDS16503S4021	AB	16 53 48	−40 26.6	1	10.33		0.36		-0.35		2	B8	1583
		NGC 6250 - 24		16 53 50	−45 53.0	2	12.40	.005	0.69	.015	0.12	.025	2		412,5002
152834	−26 11696			16 53 51	−26 52.4	1	8.86		0.03		-0.18		1	Ap Si	1770
		NGC 6250 - 23		16 53 51	−45 52.8	2	11.57	.000	0.41	.005	0.28	.000	2		412,5002
	−3 04029	NGC 6254 sq1 1		16 53 52	−04 05.3	1	9.93		1.31				3		485
329269	−45 11085	NGC 6250 - 19		16 53 54	−45 49.4	2	11.09	.000	0.51	.000	0.38	.005	2		412,5002
152868	−23 12992			16 53 55	−23 29.9	1	10.22		0.40		0.25		1	A8/F0 V	1770
152820	−33 11590	HR 6288		16 53 55	−33 10.9	3	5.49	.013	1.60	.010			11	K5 III	15,1075,2035
322422	−40 11001	LSS 3859		16 53 55	−40 17.0	4	9.75	.011	0.21	.008	-0.42	.018	12	B0.5IIIe	92,1583,1586,1707
		G 19 - 6		16 53 56	−00 17.9	3	13.68	.024	0.94	.020	0.20	.008	8		316,1620,3062
329271	−45 11087	NGC 6250 - 17		16 53 57	−45 53.6	3	10.65	.000	0.23	.009	-0.32	.022	3		96,412,5002
		NGC 6249 - 6		16 53 58	−44 43.2	1	12.38		0.36		0.11		1		251
329268		NGC 6250 - 20		16 53 58	−45 50.0	2	11.59	.005	0.40	.000	0.32	.005	2	A3	412,5002
		NGC 6254 sq1 10		16 54 00	−04 04.3	1	12.19		0.50				4		485
152260	−76 01186			16 54 00	−76 08.5	1	6.91		0.53				4	G0 IV/V	2035
329215	−44 11303	NGC 6249 - 7		16 54 01	−44 45.1	1	10.46		0.34		-0.29		1	B4 V	251
329211	−45 11089	NGC 6250 - 18		16 54 01	−45 54.2	2	11.09	.000	0.34	.010	0.09	.015	2		412,5002
	−45 11088			16 54 01	−45 56.0	1	11.10		0.30		0.00		1		96
		NGC 6249 - 5		16 54 02	−44 42.9	1	11.32		0.28		-0.25		1	B3 V	251
329216	−44 11304	NGC 6249 - 2	⋆ AB	16 54 02	−44 43.4	2	9.91	.030	0.95	.004	0.63	.007	5	G8 II	251,1685
152798	−45 11092			16 54 02	−45 16.2	1	8.77		0.59				4	G2 V	2012
152799	−45 11091	NGC 6250 - 2		16 54 03	−45 53.1	3	8.74	.000	0.16	.009	-0.46	.015	4	B2 III	96,412,5002
		NGC 6249 - 4		16 54 04	−44 42.6	2	10.30	.021	1.18	.003	1.03	.006	4	G7 II	251,1685
		NGC 6250 - 25		16 54 04	−45 54.7	2	12.41	.000	0.36	.000	0.25	.000	2		412,5002
329217	−44 11305	NGC 6249 - 3		16 54 05	−44 44.2	1	10.83		0.33		0.06		1	A0	251
		NGC 6249 - 9		16 54 05	−44 42.5	1	13.03		0.57		0.42		1		251
		NGC 6249 - 10		16 54 06	−44 42.2	1	12.89		0.40		0.35		1		251
		NGC 6250 - 26		16 54 06	−45 54.3	2	12.85	.010	0.41	.000	0.34	.005	2		412,5002
152909	−19 04471	HR 6294	⋆ A	16 54 07	−19 27.7	1	6.62		0.06		-0.39		1	B6 V	1770
152909	−19 04471	HR 6294	⋆ AB	16 54 07	−19 27.7	3	6.27	.006	0.07	.002	-0.30	.005	8	B6 V + B7 V	540,1770,2007
		NGC 6249 - 15		16 54 07	−44 43.9	1	13.51		0.72		0.39		1		251
322421	−40 11004			16 54 10	−40 19.8	2	9.82	.020	0.14	.015	-0.62	.015	5	B2 IV	141,1583
		NGC 6249 - 11		16 54 10	−44 42.4	1	13.04		0.48		0.40		1		251
329213	−44 11307	NGC 6249 - 1		16 54 10	−44 46.5	1	9.78		0.21		-0.49		2	B1.5V	251
		NGC 6249 - 12		16 54 11	−44 43.0	1	12.84		1.39		1.15		1		251

HD	DM	Other Id	N Rem	α_{1950}	δ_{1950}	S	V	σ_V	B–V	σ_{B-V}	U–B	σ_{U-B}	n	Spectrum	References
		NGC 6249 - 14		16 54 11	−44 43.7	1	12.34		0.56		0.46		1		251
	−44 11306	NGC 6249 - 8		16 54 11	−44 44.6	1	11.10		0.32		−0.21		1		251
152564	−69 02666	HR 6276		16 54 11	−69 11.5	2	5.78	.005	−0.10	.000	−0.43	.005	7	Ap Si	15,1075
322436	−40 11005			16 54 12	−40 27.2	1	10.09		0.48		0.24		2	F0	1583
		NGC 6250 - 27		16 54 12	−45 54.4	1	12.48		0.75		0.23		1		412
		NGC 6249 - 13		16 54 13	−44 43.3	1	14.48		0.59		0.53		1		251
		NGC 6250 - 7		16 54 13	−45 51.4	2	11.86	.010	0.23	.000	0.07	.010	2		412,5002
	−40 11008			16 54 14	−40 25.0	1	10.52		1.46		1.57		2		1583
		NGC 6250 - 6		16 54 15	−45 52.0	2	11.98	.025	0.42	.010	0.28	.005	2		412,5002
153344	+62 01520	G 226 - 36		16 54 16	+62 10.7	4	7.07	.019	0.67	.012	0.22	.010	8	G5 IV	308,1003,1658,3010
		AAS57,205 # 416		16 54 16	−40 19.1	1	11.69		0.33		0.24		2		1583
152822	−45 11095	NGC 6250 - 3	⋆ AB	16 54 16	−45 51.6	3	9.08	.009	0.18	.008	−0.34	.020	3	B5 IV	96,412,5002
	−45 11096	NGC 6250 - 5		16 54 17	−45 52.0	2	11.09	.010	0.19	.005	−0.27	.010	2		412,5002
152912	−25 11815	UU Oph		16 54 18	−25 43.3	1	10.29		0.24		0.14		1	A1 IV	1770
326476	−41 11114			16 54 18	−41 25.	1	10.38		0.32		0.09		2	B9	1707
		NGC 6250 - 36		16 54 18	−45 43.	1	13.92		1.02		0.05		1	B1	5002
		NGC 6250 - 9		16 54 19	−45 50.5	2	12.15	.005	0.26	.010	0.05	.005	2		412,5002
152884	−37 11115			16 54 20	−37 41.2	1	8.28		1.12		0.87		4	G8 IV	119
		NGC 6250 - 14		16 54 20	−45 50.9	1	13.59		0.49		0.34		1		412
		NGC 6250 - 30		16 54 21	−45 49.2	1	12.66		0.33		0.18		1		5002
		NGC 6250 - 10		16 54 21	−45 50.7	2	12.18	.005	0.36	.000	0.16	.035	2		412,5002
		NGC 6250 - 13		16 54 22	−45 51.0	2	13.07	.005	0.46	.015	0.36	.005	2		412,5002
		NGC 6250 - 11		16 54 22	−45 51.2	2	12.76	.030	0.62	.015	0.10	.000	2		412,5002
		NGC 6250 - 8		16 54 22	−45 51.9	2	12.59	.005	0.39	.010	0.23	.000	2		412,5002
		NGC 6250 - 12		16 54 23	−45 51.1	2	12.88	.025	0.33	.030	0.29	.010	2		412,5002
		NGC 6250 - 28		16 54 23	−45 55.0	2	12.43	.005	0.36	.005	0.25	.005	2		412,5002
152871	−40 11011			16 54 24	−40 23.2	3	9.81	.027	0.18	.014	0.14	.014	7	B5 II	141,1583,1707
153033	+6 03318			16 54 25	+06 34.7	1	7.33		1.60		1.89		1	K5 III	3040
		NGC 6250 - 15		16 54 25	−45 49.0	2	11.99	.010	0.25	.005	0.09	.015	2		412,5002
		G 19 - 7		16 54 26	−04 16.1	2	12.31	.020	1.62	.025	1.13	.010	4		1620,3062
152824	−50 10924	HR 6289		16 54 26	−50 33.9	1	5.54		0.02				4	B9 V	2007
153145	+31 02932			16 54 27	+31 05.3	1	8.16		0.17		0.13		2	A2	1733
153720	+75 00608			16 54 27	+75 28.3	1	6.84		0.31		0.06		2	F0	1502
152886	−40 11014			16 54 28	−40 16.5	1	9.73		0.42		0.09		2	F2/3 V	1583
152853	−45 11098	NGC 6250 - 1		16 54 28	−45 54.4	5	7.94	.006	0.13	.008	−0.66	.012	12	B2 II-III	96,412,540,976,5002
152786	−55 07766	HR 6285		16 54 28	−55 54.8	3	3.12	.004	1.60	.000	1.96	.000	11	K3 III	15,1075,2012
		NGC 6254 sq1 5		16 54 29	−04 10.8	1	11.87		0.63				5		485
152901	−37 11118	V883 Sco	⋆ AB	16 54 29	−37 55.2	2	7.46	.005	0.12	.006	−0.55	.007	5	B2 III	540,976
		NGC 6254 - 1002		16 54 30	−04 02.	1	12.00		1.80		1.98		2		3074
		NGC 6254 - 1060		16 54 30	−04 02.	1	13.17		1.30		1.02		2		3074
		NGC 6254 - 2024		16 54 30	−04 02.	1	12.05		1.63		1.52		1		3074
		NGC 6254 - 2049		16 54 30	−04 02.	1	14.22		1.04		0.57		2		3074
		NGC 6254 - 2050		16 54 30	−04 02.	1	12.54		1.42		1.16		1		3074
		NGC 6254 - 2072		16 54 30	−04 02.	1	11.85		0.79		0.25		5		3074
		NGC 6254 - 2098		16 54 30	−04 02.	1	13.87		1.14		0.65		3		3074
		NGC 6254 - 2104		16 54 30	−04 02.	1	13.54		1.21		0.81		2		3074
		NGC 6254 - 2105		16 54 30	−04 02.	1	12.94		1.33		0.98		2		3074
		NGC 6254 - 3021		16 54 30	−04 02.	1	12.10		1.65		1.44		1		3074
		NGC 6254 - 3028		16 54 30	−04 02.	1	13.48		1.22		0.74		2		3074
		NGC 6254 - 3029		16 54 30	−04 02.	1	13.25		1.26		0.82		2		3074
		NGC 6254 - 3053		16 54 30	−04 02.	1	13.83		1.19		0.65		1		3074
		NGC 6254 - 3055		16 54 30	−04 02.	1	12.95		1.26				1		3074
		NGC 6254 - 3056		16 54 30	−04 02.	1	13.88		1.19		0.73		1		3074
		NGC 6254 - 3085		16 54 30	−04 02.	1	13.73		1.38		1.00		1		3074
		NGC 6254 - 4044		16 54 30	−04 02.	1	11.60		1.16		0.80		4		3074
152958	−24 12969			16 54 32	−24 27.5	1	8.43		0.29		0.16		1	A5 V	1770
322437	−40 11015			16 54 32	−40 24.8	1	10.15		1.78		1.99		2	M0	1583
		NGC 6250 - 29		16 54 33	−45 50.4	1	12.12		1.23		1.17		1		5002
		NGC 6254 sq1 4		16 54 34	−04 06.8	1	11.75		1.17				3		485
152902	−40 11017			16 54 38	−40 16.6	2	7.55	.032	1.62	.004	1.93	.015	12	K4 III	1583,1673
		NGC 6254 sq1 11		16 54 39	−04 08.8	1	12.50		1.48				3		485
329267		NGC 6250 - 16		16 54 39	−45 51.0	2	11.02	.005	1.22	.015	1.00	.025	2	K5	412,5002
153021	−10 04417	HR 6296		16 54 40	−10 53.1	4	6.17	.013	1.00	.000			15	G8 III-IV	15,2013,2020,2028
322396	−39 10925	LSS 3866		16 54 40	−39 54.2	1	9.83		0.50		−0.41		1	B0.5V	540
153286	+47 02407			16 54 41	+47 26.9	2	7.02	.010	0.30	.005	0.16	.010	9	Am	3016,8071
		vdB 76 # a		16 54 42	−45 39.	1	13.95		1.04		−0.04		6		434
322414	−40 11020			16 54 45	−40 10.5	1	10.15		0.23		−0.35		2	B5	1583
		AAS57,205 # 421		16 54 45	−40 10.9	1	10.26		1.72		2.20		2		1583
	−40 11023	LSS 3868		16 54 48	−40 14.3	2	10.25	.005	0.79	.005	−0.38	.010	5	O6 III	1583,1737
153299	+50 02345	HR 6306		16 54 49	+50 07.0	3	5.66	.023	1.62	.005	1.97	.015	6	M2 III	1566,1733,3001
		NGC 6254 sq1 6		16 54 49	−03 56.8	1	12.03		1.75				1		485
		G 19 - 8		16 54 50	−00 05.6	2	12.03	.015	1.12	.019	1.00	.024	3		1620,3062
153036	−15 04417			16 54 50	−15 44.1	1	10.06		0.68		0.59		1	B9/A0 II/I	1770
153224	+29 02905			16 54 52	+29 40.0	1	8.29		0.55		0.09		3	F8 V	3026
152988	−26 11719	IDS16518S2615	AB	16 54 52	−26 20.0	1	9.98		0.32		0.18		1	A5 V	1770
152917	−45 11102	NGC 6250 - 4		16 54 53	−45 50.6	3	7.63	.009	0.26	.005	0.04	.036	3	A6 III/IV	96,412,5002
		LP 506 - 36		16 54 55	+11 04.6	1	10.61		0.76		0.33		2		1696
153596	+70 00906			16 54 56	+70 32.5	1	7.01		0.07		0.11		2	A2	985
		NGC 6254 sq1 13		16 54 57	−04 08.6	1	12.69		0.71				3		485
		NGC 6254 sq1 7		16 54 59	−03 52.2	1	12.09		0.64				2		485

Table 1   921

HD	DM	Other Id	N Rem	α_{1950}	δ_{1950}	S	V	σ_V	B–V	σ_{B-V}	U–B	σ_{U-B}	n	Spectrum	References
		GD 203		16 55 00	+21 00.8	1	16.60		-0.21		-1.18		1	DA	3060
		G 169 - 34	⋆ V	16 55 01	+21 31.7	2	14.06	.020	0.25	.010	-0.56	.050	5	DAss	316,3060
152963	−40 11029			16 55 02	−40 19.6	1	9.67		0.20		-0.35		2	B5 Ib/II	1583
153004	−33 11607	RV Sco	⋆ AB	16 55 03	−33 32.0	4	6.65	.041	0.78	.034	0.54	.016	4	G0 Ib	657,688,1490,6011
153068	−22 04254			16 55 05	−22 57.1	1	9.42		0.28		0.10		1	A0 V	1770
153003	−32 12206			16 55 05	−32 58.3	1	9.12		0.46				15	F3 V	6011
322420	−40 11030			16 55 05	−40 17.0	1	10.61		0.19		-0.16		2	B8	1583
		NGC 6254 sq1 14		16 55 06	−03 59.0	1	12.85		0.89				3		485
		AAS57,205 # 425		16 55 06	−40 12.8	1	13.36		0.27		1.10		2		1583
153024	−30 13652			16 55 07	−30 21.3	1	9.64		0.51		0.34		1	A5/8	210
153494	+65 01156			16 55 08	+65 00.4	1	8.81		0.27		-0.08		1	F2	1776
322415	−40 11031			16 55 08	−40 09.8	1	10.02		0.32		-0.29		2	B9	1583
153226	+14 03155	HR 6301		16 55 14	+13 57.6	2	6.38	.009	0.93	.009	0.68	.014	7	K0 V	1080,3051
322416	−40 11032			16 55 15	−40 08.5	1	10.76		0.94		0.64		2	K2	1583
		NGC 6254 sq1 12		16 55 16	−04 04.8	1	12.62		1.83				2		485
153081	−24 12977			16 55 16	−24 16.7	1	9.18		0.47		-0.04		1	A0	1770
152979	−45 11106	NGC 6250 - 35		16 55 16	−46 03.2	6	8.19	.012	0.16	.009	-0.61	.011	16	B2 IV	96,434,540,976,1586,5002
	−18 04365			16 55 17	−19 09.1	2	9.68	.003	0.06	.001	-0.81	.018	5	B0 IV	540,976
153210	+9 03298	HR 6299, κ Oph		16 55 18	+09 27.1	6	3.20	.007	1.16	.015	1.16	.016	15	K2 III	15,1256,1363,3016*
153084	−29 13057	LSS 3875		16 55 18	−29 19.8	3	8.51	.021	-0.06	.012	-0.79	.009	7	B2 II	400,540,976
322439	−40 11033			16 55 18	−40 28.4	1	11.75		0.30		0.10		2	A0	1583
326527	−42 11718	V590 Sco		16 55 18	−42 17.6	2	10.25	.132	0.58	.028	-0.07	.009	4	B5	1406,1707
		WLS 1712 50 # 6		16 55 20	+50 14.6	1	12.53		0.79		0.27		2		1375
	−44 11324			16 55 21	−44 19.4	1	9.38		2.30				1		8051
		NGC 6254 sq1 2		16 55 22	−03 56.7	1	10.78		1.28				5		485
153083	−26 11731			16 55 22	−26 16.8	1	8.92		0.06		-0.10		1	B9 IV/V	1770
153071	−33 11611			16 55 24	−33 43.8	1	9.34		0.45				13	F5 V	6011
153212	−0 03206			16 55 26	−01 04.1	1	9.12		0.56		0.20		1	A3	1776
153026	−39 10940	IDS16520S3925	AB	16 55 26	−39 29.2	1	8.33		1.16				4	K3/4 V	2012
322417	−40 11035	LSS 3873		16 55 27	−40 10.0	5	10.16	.071	0.84	.036	-0.34	.032	11	O5 III	92,1583,1583,1707,1737
153287	+25 03166	HR 6305		16 55 28	+25 25.7	3	6.29	.005	0.91	.008	0.57	.020	8	K0	833,1625,1733
153072	−37 11131	HR 6298	⋆ AB	16 55 29	−37 32.7	4	6.08	.011	0.19	.004	0.09	.002	9	A3 III	15,219,688,2012
		vdB 77 # a		16 55 30	−40 31.9	1	10.19		0.68		0.19		3	B5 III:	434
322419	−40 11036			16 55 32	−40 19.8	1	10.14		0.97		-0.14		2	B2	1583
	−42 11721	V921 Sco		16 55 32	−42 37.4	1	11.43		1.28		-0.12		6	Bep	434
152923	−59 06877	IDS16512S5910	B	16 55 33	−59 14.6	1	10.22		0.34		0.24		2		1279
152923	−59 06876	IDS16512S5910	A	16 55 33	−59 15.2	1	7.12	.030	0.47	.004	-0.02		6	F5/6 V	1279,2012
	+27 02725	G 169 - 37		16 55 35	+26 58.7	3	9.63	.007	0.80	.004	0.40	.014	11	K0	308,1658,1775
		AAS57,205 # 432		16 55 35	−40 21.8	1	11.72		0.65		0.37		2		1583
152980	−52 10372	HR 6295		16 55 35	−53 05.2	3	4.06	.008	1.45	.004	1.71	.000	14	K4 IIIab	15,1075,2012
153312	+24 03095	HR 6307		16 55 38	+24 27.4	2	6.35	.024	1.09	.010	0.95		5	K0 III	71,3016
	+47 02408			16 55 41	+47 37.5	1	10.90		0.58		0.07		2		1375
153495	+57 01716	IDS16548N5720	AB	16 55 41	+57 15.9	1	8.87		0.60		0.07		4	G0	3016
153087	−40 11039			16 55 41	−40 13.4	1	9.07		1.68		2.02		2	K(2) (III)	1583
		LP 919 - 13		16 55 42	−27 58.0	1	12.94		0.50		0.10		1		1696
153102	−38 11396			16 55 42	−38 29.5	1	7.59		0.04				4	B3 II/III	2012
153172	−23 13012			16 55 44	−23 54.7	1	8.23		0.27		0.06		1	A8/9 V	1770
153088	−40 11040			16 55 44	−40 22.1	1	9.39		0.41		-0.28		1	B2/5(e)	1583
153597	+65 01157	HR 6315		16 55 45	+65 12.7	8	4.89	.013	0.48	.009	-0.03	.005	41	F6 V	15,1013,1118,1197*
322418	−40 11041			16 55 45	−40 14.2	1	11.17		0.81		0.30		2	F8	1583
153136	−33 11617			16 55 46	−33 27.1	4	7.80	.005	0.58	.002	0.29	.002	10	A1/2 +G/K	285,657,1075,1490
		RW TrA		16 55 46	−66 35.7	1	10.89		0.34		0.20		1	A7	700
153073	−45 11111	NGC 6250 - 34		16 55 47	−45 48.6	2	9.09	.010	0.22	.020	-0.53	.010	2	B2 III	96,5002
		vdB 78 # a		16 55 48	−40 05.	1	13.13		1.20		0.02		3		434
153598	+62 01521			16 55 50	+62 26.8	1	6.97		0.05		-0.02		1	A0	1776
153229	−14 04509	HR 6302		16 55 51	−14 47.6	2	6.51	.063	0.38	.005	-0.01		5	F3 IV/V	254,2007
152924	−64 03588	IDS16512S6424	AB	16 55 58	−64 28.6	3	8.01	.041	0.45	.005	-0.10	.000	4	F3 V	742,1594,6006
153214	−24 12985			16 55 59	−24 18.6	1	9.12		0.22		0.16		1	A2 IV	1770
153106	−46 11147			16 55 59	−46 24.5	4	7.75	.006	-0.01	.011	-0.76	.007	7	B2 IV/V	96,219,540,976
153053	−54 07947	HR 6297	⋆ A	16 56 02	−54 31.3	3	5.64	.004	0.19	.005			11	A5 IV/V	15,2020,2028
	+25 03173			16 56 07	+25 49.8	2	9.70	.043	1.52	.038	1.20		3	M2	1017,3072
		LP 746 - 37		16 56 07	−14 35.6	1	13.52		1.01		0.56		1		1696
153697	+65 01159	HR 6319	⋆ AB	16 56 10	+65 06.9	2	6.40	.004	0.37	.013	0.02		5	F1 V	938,1733
153177	−41 11147			16 56 12	−41 54.2	1	8.53		0.19		0.00		4	B9 III	141
153140	−46 11150	LSS 3877		16 56 13	−46 14.6	4	7.51	.012	0.37	.007	-0.53	.007	40	B1 II	96,540,976,1732
153472	+42 02774	HR 6313		16 56 15	+42 35.3	2	6.34	.000	1.28	.000	1.41	.000	4	K3 III	15,1003
153254	−25 11842			16 56 15	−25 31.2	1	10.54		0.22		0.11		1	B9/A0 III	1770
153253	−24 12988			16 56 16	−25 00.9	1	10.39		0.32		0.27		1	A0 III/IV	1770
153178	−43 11302	LSS 3878		16 56 16	−43 54.8	2	8.94	.008	0.65	.002	-0.29	.013	6		540,976
		SW Her		16 56 19	+21 37.2	1	13.52		0.12		0.09		1		668
153525	+47 02411	G 203 - 36	⋆ C	16 56 19	+47 26.2	3	7.89	.009	1.00	.018	0.82	.008	7	K0	497,679,3072
153075	−57 08215			16 56 19	−57 13.1	3	7.01	.009	0.57	.015	-0.02	.010	18	G0/2 V	1311,2012,3077
153159	−46 11152			16 56 20	−46 59.5	2	9.65	.003	0.81	.016	-0.08		6	B5 Ib	540,1021
153376	+15 03089			16 56 22	+15 31.6	1	6.90		0.63		0.17		2	F8 V	1648
153199	−43 11304	LSS 3880		16 56 23	−43 08.8	1	8.14	.011	0.20	.006	-0.51	.014	9	B3 II/III	401,540,976,1586
153306	−18 04370			16 56 28	−18 41.8	1	9.31		0.17		0.05		1	B9.5 V	1770
153557	+47 02415	G 203 - 37	⋆ A	16 56 30	+47 26.3	1	7.83		0.98		0.80		1	K0	679
153557	+47 02415	G 203 - 37	⋆ AB	16 56 30	+47 26.3	3	7.74	.005	1.00	.016	0.81	.004	7	K0	497,679,3072
153557	+47 02415	IDS16551N4731	B	16 56 30	+47 26.3	1	11.19		1.47		1.04		1		679
153681	+59 01772			16 56 31	+59 35.0	1	8.51		0.48		-0.07		1	F5	1776

HD	DM	Other Id	N Rem	α_{1950}	δ_{1950}	S	V	σ_V	B–V	σ_{B-V}	U–B	σ_{U-B}	n	Spectrum	References
	+68 00901	G 240 - 35		16 56 31	+68 05.8	3	8.70	.013	0.66	.009	0.11	.028	8	G0	308,1658,3026
153316	−15 04421			16 56 31	−15 59.4	1	7.79		1.66		2.00		3	K5 III	1657
153234	−44 11339			16 56 36	−44 54.8	1	6.50		0.38		-0.04		4	F3 V	244
153294	−31 13442			16 56 38	−31 38.1	2	8.28	.001	0.08	.009	-0.43	.001	5	B7 Ib/II	540,976
153221	−48 11360	HR 6300	⋆ AB	16 56 40	−48 34.4	2	5.99	.005	0.88	.000			7	G8/K0 III+G	15,2027
153222	−49 11105			16 56 40	−49 10.8	3	8.90	.030	0.26	.015	-0.65	.006	9	B1 Ib/IIIe	540,976,1586
		V841 Oph		16 56 42	−12 49.0	1	13.36		0.40		-0.59		21		866
	−48 11361	LSS 3882		16 56 43	−48 55.8	2	9.61	.005	0.27	.008	-0.62	.016	3	B0.5IV	540,976
153377	−1 03278			16 56 45	−01 36.8	1	7.51		0.47		-0.05		1	F2	1776
		NGC 6259 - 1008		16 56 46	−44 35.3	1	13.11		0.62		0.38		2		385
153327	−24 12996	IDS16537S2501	AB	16 56 48	−25 05.9	1	9.73		0.21		0.11		1	A0 V	1770
		NGC 6259 - 4031		16 56 48	−44 37.8	1	14.10		0.60		0.42		2		385
		NGC 6259 - 1013		16 56 51	−44 34.5	1	12.47		1.76		1.71		3		385
		NGC 6259 - 4038		16 56 52	−44 38.5	1	14.14		0.62		0.44		2		385
153335	−24 12998	IDS16539S2403	AB	16 56 53	−24 07.7	1	9.11		0.34		0.09		1	A4/5 V	1770
153295	−42 11737	LSS 3885		16 56 53	−42 14.6	5	9.04	.026	0.55	.014	-0.58	.013	14	O/Be	141,540,976,1586,1707
153258	−45 11123	HR 6303		16 56 53	−45 22.7	2	6.64	.005	1.80	.000			7	K4 III	15,2030
153336	−24 12997	HR 6308		16 56 54	−25 01.1	2	5.86	.000	1.59	.005	1.94	.040	9	M3 III	1024,3055
		NGC 6259 - 4028		16 56 54	−44 37.7	1	14.37		0.74		0.50		2		385
		NGC 6259 - 4029		16 56 56	−44 37.8	1	14.00		0.56		0.35		1		385
153667	+50 02352			16 56 57	+50 46.2	1	7.94		0.96		0.63		2	G5	1566
		NGC 6259 - 4039		16 56 57	−44 38.6	1	11.76		2.08		2.22		2		385
		NGC 6259 - 1		16 56 57	−44 41.7	1	11.89		1.89		1.98		4		385
		NGC 6259 - 1001		16 57 00	−44 35.5	1	13.71		0.52		0.32		3		385
		NGC 6266 sq3 7		16 57 01	−30 12.4	1	11.82		0.79		0.18		3		718
		NGC 6259 - 1038		16 57 01	−44 32.3	1	12.92		0.66		0.32		1		385
		NGC 6259 - 4014		16 57 01	−44 37.2	1	12.80		1.81		1.79		1		385
		NGC 6259 - 1027		16 57 02	−44 33.5	1	13.77		0.87		0.43		2		385
		NGC 6259 - 4040		16 57 03	−44 38.7	1	13.49		0.72		0.51		2		385
		GD 516		16 57 05	+64 37.2	1	14.64		0.00		-0.05		3		308
153363	−24 13002	HR 6310		16 57 06	−24 54.9	1	5.75		0.41				4	F3 V	2007
	−29 13096	IDS16539S2954	AB	16 57 06	−29 58.6	2	9.92	.010	1.15	.035	0.74		6		156,438
		PG1657+078B		16 57 07	+07 46.6	1	14.72		0.71		0.07		1		1764
		PG1657+078		16 57 07	+07 47.9	1	15.01		-0.15		-0.94		4	sdB	1764
		NGC 6266 sq3 14		16 57 07	−30 10.5	1	13.92		1.52		1.29		3		718
		NGC 6259 - 2030		16 57 07	−44 33.8	1	13.91		0.60		0.41		1		385
		NGC 6259 - 2007		16 57 07	−44 34.9	1	12.45		0.60		0.37		1		385
		PG1657+078A		16 57 08	+07 46.8	1	14.03		1.07		0.73		1		1764
		PG1657+078C		16 57 10	+07 46.9	1	15.23		0.84		0.39		1		1764
		NGC 6259 - 3044		16 57 11	−44 39.3	1	13.63		0.74		0.39		1		385
153379	−25 11856			16 57 12	−25 15.7	1	9.58		0.15		0.10		1	A0 V	1770
153364	−30 13687			16 57 12	−30 10.0	2	8.51	.005	0.33	.005	0.01	.015	7	F0 V	156,718
		NGC 6259 - 2032		16 57 13	−44 34.0	1	13.58		0.59		0.40		1		385
		NGC 6266 sq2 30		16 57 14	−30 02.0	1	11.16		0.62				3		438
		NGC 6259 - 3017		16 57 14	−44 37.1	1	12.66		1.64		1.51		2		385
		NGC 6259 - 3025		16 57 14	−44 37.4	1	13.77		0.62		0.35		2		385
154099	+73 00751	HR 6335		16 57 15	+73 12.2	2	6.29	.005	0.24	.000	0.11	.005	4	F0 V	985,1733
		NGC 6266 sq2 36		16 57 16	−30 05.6	2	12.63	.010	1.48	.010	1.50		3		438,718
153367	−35 11237	IDS16540S3536	AB	16 57 16	−35 40.9	1	8.89		0.29		-0.46		3	B8	1586
		NGC 6266 sq2 37		16 57 17	−30 05.9	2	13.09	.005	0.55	.000	0.43		5		438,718
153368	−35 11236	HR 6311		16 57 17	−35 51.6	1	5.97		1.16				4	K2 III	2009
		NGC 6259 - 3037		16 57 17	−44 38.1	1	13.26		0.58		0.32		2		385
		NGC 6259 - 3035		16 57 18	−44 38.9	1	14.92		0.75				4		385
		G 181 - 19		16 57 21	+34 56.4	2	12.63	.000	0.81	.002	0.22	.090	3		1658,3060
		NGC 6259 - 3018		16 57 21	−44 36.9	1	14.98		0.69				4		385
		NGC 6259 - 2037		16 57 22	−44 33.5	1	15.29		0.86				4		385
	−29 13100			16 57 24	−29 57.2	2	10.04	.050	0.88	.000	0.43	.010	5		156,718
		L 203 - 178		16 57 24	−58 52.	1	11.65		0.56		-0.15		2		1696
		NGC 6259 - 2023		16 57 26	−44 35.0	1	11.64		0.34		0.12		1		385
329379	−45 11128	NGC 6250 - 37	⋆	16 57 26	−45 37.6	2	9.64	.010	1.05	.005	0.14		5	B1:V	1021,5002
153261	−58 06964	HR 6304, V828 Ara		16 57 27	−58 53.1	4	6.14	.028	-0.06	.020	-0.93	.017	16	B1 Vne	15,1586,1637,2012
		NGC 6266 sq3 11		16 57 28	−30 09.2	1	13.60		1.69		1.79		3		718
		He3 1310		16 57 30	−41 12.	1	11.67		0.61		-0.31		1		1586
153262	−58 06965			16 57 33	−58 58.5	1	7.58		-0.04		-0.54		3	B0/3 Vnne	1586
153382	−41 11171	IDS16540S4155	AB	16 57 34	−41 59.9	2	7.32	.018	0.13	.009	-0.27	.005	5	B8 II	55,141
	−29 13102			16 57 35	−30 05.7	2	10.13	.010	1.30	.010	0.84		7		438,718
153480	−20 04600			16 57 36	−20 30.4	1	8.92		0.10		-0.14		1	B9 III	1770
153460	−23 13039			16 57 36	−23 26.5	1	8.48		0.28		0.17		1	A7 V	1770
153562	+10 03114			16 57 37	+09 52.3	1	7.55		0.01		-0.08		1	A0	1776
153459	−22 04261			16 57 38	−23 04.0	1	9.94		0.43		0.14		1	A9 V	1770
		NGC 6266 sq2 40		16 57 41	−30 05.4	1	13.55		1.92				2		438
		NGC 6266 sq2 44		16 57 43	−30 05.4	1	14.57		2.36				2		438
153284	−61 05825			16 57 44	−61 29.6	1	8.84		1.03		0.92		3	K3/4 V	3072
153426	−38 11431	LSS 3890		16 57 48	−38 07.9	1	7.47		0.14		-0.82		2	B9 II/III	92
		L 269 - 41		16 57 48	−52 43.	1	15.85		-0.16		-1.18		2		3065
		WLS 1700 30 # 10		16 57 52	+29 43.9	1	12.46		0.60		0.03		2		1375
153698	+27 02733			16 57 53	+27 23.3	1	7.54		1.55		1.67		2	M III	8100
153370	−50 10955	HR 6312	⋆ AB	16 57 53	−51 03.5	2	6.44	.005	0.28	.005			7	A9 III	15,2027
154159	+71 00817	G 240 - 37		16 57 54	+71 32.2	2	8.17	.010	0.62	.010	0.08	.015	8	G5	308,3026
153605	−0 03209			16 57 58	−00 40.2	1	8.53		0.48		0.23		1	A3	1776

Table 1   923

HD	DM	Other Id	N Rem	α_{1950}	δ_{1950}	S	V	σ_V	B–V	σ_{B-V}	U–B	σ_{U-B}	n	Spectrum	References
153653	+6 03332	HR 6317		16 58 03	+06 39.4	1	6.58		0.23		0.07		8	A7 V	1088
	−29 13310	NGC 6266 sq2 3		16 58 07	−30 05.0	1	11.08		0.55				2		438
153608	−17 04685			16 58 17	−17 16.4	1	7.97		0.26		0.00		1	B9 III/IV	1770
153609	−20 04606	IDS16553S2017	A	16 58 17	−20 21.8	1	7.37		0.11		−0.09		1	B7/8 V	1770
		WLS 1700 30 # 7		16 58 18	+28 13.0	1	13.39		0.66		0.20		2		1375
		NGC 6266 sq2 31		16 58 18	−29 53.4	1	11.22		0.63				3		438
		NGC 6266 sq2 34		16 58 18	−29 54.1	1	12.32		0.41				3		438
153565	−25 11872			16 58 19	−25 34.0	1	10.07		0.13		−0.03		1	A0	1770
153808	+31 02947	HR 6324		16 58 22	+30 59.9	5	3.92	.007	−0.02	.009	−0.11	.010	13	A0 V	15,1022,1363,3023,8015
153631	−13 04528			16 58 22	−13 29.4	3	7.13	.010	0.60	.015	0.10	.029	11	G2 V	158,196,1067
153832	+39 03069			16 58 24	+39 10.3	1	7.26		1.04		0.84		5	K0 III	833
153687	−4 04215	HR 6318	⋆ A	16 58 25	−04 09.0	5	4.82	.006	1.48	.004	1.82	.017	24	K4 III	15,1075,1088,1509,2012
		NGC 6266 sq2 32		16 58 25	−30 01.0	1	11.81		0.73				3		438
153956	+56 01934	HR 6330		16 58 26	+56 45.7	1	6.03		1.16				2	K0	71
153574	−33 11657	IDS16552S3313	AB	16 58 27	−33 17.7	1	7.17		0.28				4	A3 III	1075
329313	−44 11359			16 58 29	−44 11.6	1	10.23		0.33		−0.51		2	B1.5II-III	540
		NGC 6266 sq2 39		16 58 30	−30 03.9	1	13.54		1.08				2		438
		NGC 6266 sq2 47		16 58 32	−30 02.4	1	15.84		1.17				3		438
153575	−37 11175			16 58 32	−37 13.8	1	7.74		0.13		−0.09		3	B0 III/IV	1586
153654	−20 04609			16 58 33	−20 52.1	1	9.84		0.49		0.14		1	A9/F0 V	1770
		NGC 6266 sq2 33		16 58 33	−30 03.1	1	12.30		0.81				2		438
	−29 13122			16 58 37	−29 14.2	1	10.65		0.69		0.17		3		718
		NGC 6266 sq2 43		16 58 38	−30 02.0	1	13.71		2.10				4		438
153613	−31 13473	HR 6316	⋆ A	16 58 38	−32 04.3	3	5.03	.006	−0.10	.003	−0.37	.003	9	B8 V	26,1770,2006
		NGC 6273 sq1 2		16 58 42	−26 11.2	1	10.18		1.73				3		485
322548	−39 11024			16 58 44	−39 32.9	1	9.95		0.34		−0.40		5	B8	1586
		NGC 6273 sq1 6		16 58 45	−26 11.3	1	12.68		1.37				2		485
153809	+16 03083	IDS16566N1644	ABC	16 58 48	+16 39.9	2	7.28	.021	0.11	.001	0.08	.012	3	A0	695,1733
153896	+39 03071			16 58 50	+38 58.9	1	7.34		1.02		0.74		2	K0 III	105
153834	+22 03045	HR 6325		16 58 51	+22 42.3	1	5.65		1.33				2	gK3	71
		NGC 6266 sq3 6		16 58 55	−30 01.0	1	11.55		1.97		2.24		3		718
153727	−18 04381	HR 6321		16 58 56	−18 48.8	2	6.28	.019	1.36	.019	1.26		6	K1 III	2007,3005
153707	−24 13027			16 58 58	−24 59.2	1	8.47		0.04		−0.14		1	Ap Si	1770
		NGC 6266 sq3 10		16 58 58	−30 00.4	1	13.17		1.34		0.81		3		718
		BPM 24601		16 59 00	−53 11.	1	13.47		0.10		−0.70		1	DA	3065
153742	−21 04489			16 59 01	−21 13.4	1	10.16		0.47		0.25		1	A(p EuCrSr)	1770
153708	−26 11809	NGC 6273 sq1 1		16 59 01	−26 15.6	2	9.94	.001	0.09	.019	−0.32		6	B8 II/III	485,1586
154081	+58 01689			16 59 02	+58 32.1	1	6.88		0.11		0.11		3	A3 V	1501
153711	−29 13131			16 59 02	−29 56.6	2	9.87	.020	0.19	.055	−0.01	.065	6	B9 III/IV	156,718
153864	+14 03171	G 139 - 6		16 59 05	+13 53.8	1	10.07		0.71		0.25		2	G0	1658
	+25 03179			16 59 06	+25 02.3	1	8.38		1.54		1.95		1	M0	1746
	+29 02920			16 59 06	+29 32.2	1	9.28		0.48		−0.04		3	F5	196
		GD 204		16 59 07	+21 00.1	1	14.10		0.54		−0.17		1		3060
153897	+27 02738	HR 6328		16 59 09	+27 16.1	3	6.57	.023	0.40	.008	−0.11	.005	5	F5 V	71,254,3016
153580	−53 08316	HR 6314	⋆ A	16 59 09	−53 09.9	2	5.29	.000	0.49	.011	0.01		5	F5 V	2007,3053
153639	−44 11371			16 59 10	−44 47.0	1	7.01		1.35		1.22		4	G8 III	244
153865	+12 03125			16 59 14	+12 38.9	1	8.86		0.98		0.71		2	K0	1648
153641	−46 11183	IDS16555S4636	AB	16 59 14	−46 40.8	1	8.21		0.46				4	A0 IV	2012
153882	+15 03095	HR 6326, V451 Her	⋆ A	16 59 17	+15 01.3	4	6.29	.009	0.04	.007	0.04	.010	14	B9 p CrEu	695,1022,1202,3024
153882	+15 03095	IDS16570N1505	B	16 59 17	+15 01.3	1	10.70		0.63		0.05		2	F8 V	3050
153677	−40 11088			16 59 17	−40 50.5	3	8.14	.021	0.12	.005	−0.46	.011	8	B2/3 II	540,976,1707
153762	−27 11399			16 59 18	−27 47.4	1	10.00		0.20		−0.21		1	B8 II/III	1770
154319	+69 00884	HR 6345		16 59 19	+69 15.6	1	6.41		0.71		0.31		2	K0	1733
153867	+8 03335			16 59 20	+07 53.8	1	9.15		0.79		0.28		5	G5	1186
153786	−29 13137			16 59 26	−29 14.4	1	8.85		0.06		−0.11		1	B9 IV/V	1770
153618	−54 07992			16 59 27	−54 17.2	1	8.36		0.07		−0.19		2	B8/9 II	1730
153800	−28 12738			16 59 28	−28 20.2	1	9.19		0.21		0.05		1	B9 III	1770
		G 139 - 8		16 59 30	+16 13.5	2	11.50	.029	0.47	.005	−0.21	.010	3		1696,3029
153801	−28 12739			16 59 30	−28 24.8	1	10.17		0.23		0.06		1	B8/9 IV/V	1770
		CCS 2388		16 59 30	−32 39.2	1	10.50		4.48				2	C6-7,2-4	864
	−54 07993			16 59 30	−54 24.0	1	10.60		1.12		0.94		2		1730
153825	−20 04614			16 59 31	−21 08.3	1	9.92		0.50		0.25		1	A9 V	1770
153766	−36 11168	IDS16562S3643	AB	16 59 32	−36 47.4	1	7.94		0.12		−0.67		1	B2 III	219
153914	+8 03337	HR 6329	⋆ AB	16 59 35	+08 31.3	2	6.32	.005	0.09	.000	0.05	.005	7	A4 V	15,1256
153851	−11 04280	IDS16568S1145	A	16 59 36	−11 49.4	1	8.98		0.89		0.57		2	K2 V	3072
153814	−27 11406			16 59 37	−28 03.4	1	10.58		0.18		0.11		1	B9 III/IV	1770
153803	−30 13730			16 59 37	−30 59.3	1	8.44		0.09		−0.14		2	B8/9 V	1770
153767	−37 11188			16 59 37	−37 38.9	1	7.43		0.02		−0.07		25	A0 V	453
153804	−35 11268	LSS 3895		16 59 39	−35 43.4	1	8.54		−0.06		−0.81		1	B5 Vn	954
		WLS 1700 30 # 5		16 59 41	+32 29.2	1	13.02		0.60		−0.04		2		1375
		NGC 6281 - 77		16 59 41	−37 42.0	1	11.12		0.21		−0.37		2	B8	453
	−37 11191	NGC 6281 - 79		16 59 42	−37 53.1	1	11.33		0.32		0.25		1		453
		NGC 6281 - 78		16 59 43	−38 04.9	1	11.26		0.33		0.19		1	A5	453
	−49 11137	LSS 3893		16 59 43	−49 58.5	1	10.27		0.39				4		1021
154029	+33 02817	HR 6332		16 59 45	+33 38.4	3	5.27	.036	0.02	.026	0.03	.013	8	A3 IV	985,1022,3050
153838	−28 12744			16 59 50	−28 36.9	1	10.33		0.18		−0.03		1	B9 IV	1770
153884	−20 04616			16 59 54	−21 05.2	1	10.48		0.20		0.04		1	B9 III/IV	1770
153839	−32 11285			16 59 54	−32 47.6	2	8.25	.012	0.07	.004	−0.41	.007	3	B7 II	540,976
	+30 02921			16 59 57	+29 54.1	1	9.80		0.44		−0.07		3	F2	196
153854	−28 12749			16 59 57	−28 23.3	1	10.01		0.24		0.06		1	B8/9 IV/V	1770

HD	DM	Other Id	N	Rem	α_{1950}	δ_{1950}	S	V	σ_V	B–V	σ_{B-V}	U–B	σ_{U-B}	n	Spectrum	References
153873	−26 11830				16 59 58	−27 01.7	1	7.37		1.32				33	K2 III	6011
322652	−37 11193	NGC 6281 - 80			16 59 58	−37 58.3	1	9.80		1.28		1.21		1	K7	453
322479	−37 11192	NGC 6281 - 81			16 59 58	−38 02.9	1	10.85		0.34		0.25		1	A0	453
153791	−46 11191	HR 6323	⋆	AB	16 59 58	−47 05.4	1	6.08		0.07				1	A2/3 V	2008
		NGC 6273 sq1 5			17 00 01	−26 14.9	1	12.66		1.72				1		485
322651	−37 11195	NGC 6281 - 82			17 00 02	−37 56.5	1	10.74		0.22		0.18		1	A3	453
153855	−31 13502	LSS 3897			17 00 04	−31 32.7	5	6.98	.008	-0.08	.008	-0.91	.006	14	B2 II	158,540,976,1732,1770
154199	+52 02018				17 00 05	+52 40.5	2	6.88	.029	0.08	.006	0.05		30	A0	225,1603
		NGC 6281 - 76			17 00 06	−37 43.2	1	11.75		0.44		0.22		1		453
153771	−50 10977	IDS16562S5056	B		17 00 06	−51 00.8	1	8.83		0.07		-0.56		3	B8	434
153886	−26 11832				17 00 07	−26 52.8	1	8.51		0.53				33	F2 V	6011
153772	−50 10978	IDS16562S5056	A		17 00 07	−51 00.8	1	8.32		0.06		-0.61		4	B2 V	434
153875	−30 13736				17 00 08	−30 34.3	1	9.24		0.08		-0.31		1	B5 IV/V	1770
	+29 02923				17 00 09	+29 40.5	1	9.76		0.68		0.17		3	G5	196
	−5 04394				17 00 09	−05 59.9	1	10.85		1.50		1.22		2	M1	1746
153716	−57 08265	HR 6320			17 00 09	−57 38.5	3	5.74	.007	-0.10	.002	-0.56	.015	12	B4 IV	26,1637,2035
		NGC 6273 sq1 4			17 00 15	−26 13.4	1	12.45		1.76				2		485
322650	−37 11197	NGC 6281 - 83			17 00 15	−37 55.7	1	10.99		0.29		0.24		1	A5	453
154084	+25 03183	HR 6333			17 00 16	+25 34.5	3	5.76	.011	1.02	.005	0.80	.005	6	K0	71,105,3016
		NGC 6273 sq1 3			17 00 16	−26 11.2	1	11.90		1.25				2		485
322648	−37 11199	NGC 6281 - 28			17 00 18	−37 50.0	2	10.38	.140	0.37	.135	0.18	.025	4	A0	383,453
153827	−47 11221				17 00 19	−47 22.2	2	8.64	.005	0.07	.001	-0.52	.017	3	B3 III/IV	540,976
322649	−37 11200	NGC 6281 - 37			17 00 20	−37 51.5	2	10.82	.015	0.31	.035	0.20	.020	4	A5	383,453
		HA 61 # 225			17 00 24	+29 49.9	1	11.38		1.11		0.98		8		1499
154126	+32 02835	HR 6336			17 00 24	+31 57.3	1	6.36		1.13				2	K0	71,1637
	+30 02923				17 00 25	+30 03.7	1	10.34		0.54		0.04		3	F8	196
153890	−37 11201	NGC 6281 - 1	⋆	V	17 00 26	−38 04.9	4	5.90	.009	0.38	.010	-0.03	.005	12	F3 V	15,383,453,2027
153960	−18 04390				17 00 27	−18 42.6	1	10.20		0.10		-0.13		1	B9 III	1770
		NGC 6281 - 84			17 00 28	−37 44.2	1	11.38		0.32		0.19		2	A7	453
		HA 61 # 228			17 00 29	+29 50.1	1	12.56		0.66		0.15		8		1499
153976	−19 04500				17 00 29	−20 00.9	1	8.77		0.34		0.16		2	A6 IV	1771
155153	+81 00568				17 00 32	+80 56.0	1	6.62		1.01		0.80		2	G5	1733
	−37 11205	NGC 6281 - 45			17 00 32	−37 49.7	2	11.16	.015	0.46	.010	0.25	.010	2	A3	383,453
153919	−37 11206	NGC 6281 - 2	⋆	V	17 00 33	−37 46.5	6	6.55	.029	0.26	.009	-0.73	.019	92	O6.5Iaf+	92,219,358,383,453,1586
322676	−37 11203	NGC 6281 - 24			17 00 34	−38 04.4	2	10.18	.015	0.16	.000	0.14	.000	5	A0	383,453
	+30 02924				17 00 37	+30 03.1	1	10.91		0.36		-0.06		3		196
		NGC 6281 - 75			17 00 37	−37 47.4	1	12.12		0.49		0.13		2		453
153892	−45 11176				17 00 37	−45 10.3	2	7.64	.009	1.21	.001	1.15		10	K0 III	1673,2012
154391	+60 01728	HR 6348			17 00 38	+60 43.2	2	6.12	.015	1.01	.005	0.82		6	K1 III	252,1379
153998	−21 04493				17 00 38	−21 19.1	1	9.79		0.29		0.25		1	A2 III	1770
	−37 11209	NGC 6281 - 42			17 00 39	−37 49.4	2	11.05	.020	0.47	.010	0.18	.020	4		383,453
154034	−14 04527				17 00 40	−14 32.7	1	9.65		0.72		0.22		1	B9 Ib/II	1770
153977	−24 13055				17 00 40	−24 46.0	4	9.43	.027	0.03	.009	-0.46	.020	9	B5 II/III	400,540,1586,1770
		NGC 6281 - 59			17 00 42	−37 50.0	1	11.78		0.69		0.21		1		383
322675	−37 11211	NGC 6281 - 10	⋆	AB	17 00 42	−38 03.9	3	8.94	.020	1.22	.008	1.20	.038	9	K2	383,453,1685
322677	−38 11492	NGC 6281 - 85			17 00 42	−38 11.4	1	11.05		0.26		0.21		2	A0	453
153947	−37 11215	NGC 6281 - 9	⋆	V	17 00 45	−37 41.9	2	8.80	.000	0.09	.005	-0.34	.019	6	A0p	383,453
	−37 11213	NGC 6281 - 54			17 00 45	−37 50.2	2	11.54	.039	0.38	.005	0.20	.000	3		383,453
322673	−37 11212	NGC 6281 - 25			17 00 45	−38 01.4	2	10.21	.023	0.16	.005	0.13	.000	4	A0	383,453
		NGC 6281 - 65			17 00 46	−37 58.6	1	12.08		0.44		0.12		1		383
322674	−37 11214	NGC 6281 - 12			17 00 46	−38 05.4	2	9.23	.005	0.18	.005	-0.46	.020	5	B5	383,453
	+30 02925				17 00 47	+29 53.4	1	8.98		0.44		-0.08		3	F5	196
153879	−51 10676				17 00 49	−51 20.4	3	8.94	.029	0.10	.006	-0.67	.014	6	B1/2 Vne	540,976,1586
154143	+14 03179	HR 6337			17 00 50	+14 09.7	8	4.98	.006	1.60	.009	1.93	.011	59	M3 III	3,15,667,1003,1363*
	−37 11217	NGC 6281 - 61			17 00 50	−37 47.4	2	11.82	.000	0.44	.034	0.13	.058	3		383,453
		NGC 6281 - 64			17 00 51	−37 48.1	2	11.83	.151	0.45	.029	0.18	.024	3		383,453
		NGC 6281 - 74			17 00 51	−37 48.1	1	11.95		0.51		0.15		2		453
153948	−37 11216	NGC 6281 - 15	⋆	V	17 00 51	−37 59.0	2	9.34	.009	0.07	.005	-0.11	.000	5	Ap	383,453
	+30 02926				17 00 52	+30 05.5	1	9.86		0.56		0.04		3	F8	196
154183	+25 03187				17 00 53	+25 42.6	1	8.65		0.66		0.14		3	G0 V	3026
		NGC 6281 - 55			17 00 53	−37 43.8	1	11.61		0.39		0.24				383
154160	+14 03180	HR 6339			17 00 54	+14 35.0	2	6.54	.000	0.76	.002	0.45	.000	5	G5 IV:	1733,3016
	+17 03148	G 170 - 10	A		17 00 54	+17 48.4	1	10.51		0.99		0.76		4	K3	7010
	+17 03148		B		17 00 54	+17 48.4	1	12.77		0.42		-0.11		2		7010
	−40 11121	LSS 3900			17 00 54	−40 6.9	1	11.24		1.06		-0.08		3		1737
154021	−25 11915				17 00 55	−25 37.7	1	6.65		-0.09		-0.30		2	B9 IV/V	1770
322647	−37 11220	NGC 6281 - 34			17 00 56	−37 42.9	2	10.72	.010	0.24	.005	0.18	.010	5	A5	383,453
153950	−43 11380				17 00 56	−43 14.4	2	7.41	.010	0.56	.000	-0.05		6	F8 V	2012,3037
154344	+52 02019				17 00 57	+52 40.7	1	7.83		0.24		0.11		2	A2	1375
154066	−12 04651				17 00 57	−12 47.6	1	8.73		0.28		-0.17		1	B8 V	1770
322653	−37 11221	NGC 6281 - 31			17 00 58	−37 50.5	2	10.54	.000	0.31	.010	0.19	.000	5	A7	383,453
154068	−20 04623				17 00 59	−20 24.3	1	7.87		0.25		0.20		2	A2/3 III	1770
		NGC 6281 - 73			17 01 00	−37 49.5	1	13.54		0.79		0.31		1		383
		NGC 6281 - 70			17 01 00	−37 50.4	1	12.70		0.58		0.08		2		383
322655	−37 11222	NGC 6281 - 27			17 01 00	−37 51.5	2	10.25	.015	0.16	.005	0.14	.010	4	A3	383,453
		NGC 6281 - 52			17 01 03	−37 53.3	1	11.39		0.63		-0.03		1		383
322657	−37 11223	NGC 6281 - 26			17 01 03	−37 58.6	2	10.20	.010	0.17	.005	0.14	.010	7		383,453
		NGC 6281 - 50			17 01 04	−37 57.7	2	11.27	.054	0.40	.029	0.15	.039	3		383,453
322656	−37 11224	NGC 6281 - 13			17 01 05	−37 53.1	2	9.30	.005	0.16	.027	0.02	.014	5	A0	383,453
		NGC 6281 - 51			17 01 05	−37 58.2	2	11.30	.044	0.37	.073	0.15	.058	3		383,453

Table 1   925

HD	DM	Other Id	N Rem	α_{1950}	δ_{1950}	S	V	σ_V	B–V	σ_{B-V}	U–B	σ_{U-B}	n	Spectrum	References
154227	+29 02927			17 01 06	+29 32.9	1	7.72		1.25		1.34		3	K2	196
322654	−37 11225	NGC 6281 - 21		17 01 07	−37 50.2	2	9.93	.005	0.16	.010	0.09	.005	5	A5	383,453
		G 181 - 24		17 01 08	+34 50.7	2	14.99	.009	1.65	.023			4		419,906
		NGC 6281 - 67		17 01 08	−37 56.4	1	12.59		0.68		0.13		1		383
	−37 11226	NGC 6281 - 23		17 01 10	−37 54.2	2	10.05	.014	0.53	.019	0.05	.014	4	F3	383,453
		NGC 6281 - 56		17 01 10	−38 00.9	1	11.72		0.39		0.17		1		383
154345	+47 02420			17 01 12	+47 08.4	5	6.76	.012	0.73	.005	0.28	.016	24	G8 V	22,1013,1197,1355,3026
322659	−37 11230	NGC 6281 - 11		17 01 12	−37 50.6	2	9.01	.005	0.07	.005	-0.09	.009	9	A0	383,453
322658	−37 11228	NGC 6281 - 4		17 01 12	−37 55.4	3	8.10	.015	1.15	.007	0.86	.016	7		383,453,1685
	−37 11229	NGC 6281 - 19	★ AB	17 01 12	−37 56.6	1	9.67		0.24		0.13		2		383
326799	−41 11230	LSS 3902		17 01 12	−41 51.6	1	11.62		0.91		-0.06		3	G0	1737
		NGC 6281 - 41		17 01 13	−37 51.9	2	10.98	.056	0.23	.042	0.17	.028	4		383,453
		NGC 6281 - 60		17 01 13	−37 52.0	2	11.84	.020	0.41	.020	0.14	.020	2		383,453
		NGC 6281 - 43		17 01 15	−37 54.6	2	11.07	.037	0.27	.009	0.19	.019	4		383,453
154103	−20 04624			17 01 16	−20 58.4	1	8.79		0.08		-0.26		1	B8 II/III	1770
153924	−54 08008			17 01 16	−54 32.2	1	9.54		0.40		0.09		2	F2 V	1730
		NGC 6281 - 72		17 01 17	−37 52.9	1	13.08		0.68		0.14		1		383
		NGC 6281 - 57		17 01 17	−37 54.5	1	11.72		2.06		1.30		1		383
		NGC 6281 - 40		17 01 18	−37 47.0	1	10.91		1.29		1.12		1		383
		NGC 6281 - 36		17 01 18	−37 53.9	2	10.78	.023	0.32	.014	0.17	.023	4	A5	383,453
		NGC 6281 - 48		17 01 18	−37 55.7	2	11.27	.060	0.44	.030	0.11	.015	4		383,453
154088	−28 12769			17 01 19	−28 30.6	3	6.58	.005	0.83	.005	0.54	.000	9	K1 V	285,657,2012
		NGC 6281 - 44		17 01 19	−37 54.8	2	11.16	.029	0.28	.010	0.14	.029	3		383,453
	−37 11234	NGC 6281 - 38		17 01 19	−37 56.1	2	10.85	.000	0.24	.010	0.17	.030	4	A5	383,453
153980	−46 11200			17 01 19	−46 54.2	1	7.55		-0.01		-0.21		2	B7/8 V	401
	+5 03312			17 01 20	+04 55.4	1	10.25		0.37		0.19		7	A5	1222
		NGC 6281 - 32		17 01 20	−37 52.6	2	10.61	.015	0.18	.020	0.18	.005	4	A7	383,453
		NGC 6281 - 53		17 01 20	−37 52.6	2	11.36	.084	0.35	.010	0.19	.030	5		383,453
154228	+13 03292	HR 6341	★ A	17 01 21	+13 40.5	3	5.92	.008	0.01	.004	-0.02	.008	7	A1 V	695,1022,3050
	−37 11232	NGC 6281 - 22		17 01 21	−38 04.0	2	10.02	.010	0.61	.010	0.08	.015	5		383,453
		NGC 6281 - 71		17 01 22	−37 50.8	1	13.07		0.59		0.13		1		383
154040	−39 11086			17 01 22	−39 14.8	2	10.05	.022	0.28	.010	-0.41	.038	5	B2	540,1586
322660	−37 11239	NGC 6281 - 3	★ A	17 01 23	−37 49.1	3	7.94	.016	1.12	.015	0.89	.006	10	G9 II-III	383,453,1685
		NGC 6281 - 58		17 01 23	−38 00.5	2	11.70	.045	1.65	.015	1.51	.225	2		383,453
		NGC 6281 - 33		17 01 24	−37 50.	1	10.64		0.21		0.17		1		383
		NGC 6281 - 63		17 01 24	−37 50.	1	11.90		0.42		0.15		1		383
		NGC 6281 - 35	★ B	17 01 24	−37 50.	2	10.63	.061	0.26	.005	0.21	.009	4		383,453
	−37 11237	NGC 6281 - 6	★ AB	17 01 24	−37 53.3	2	8.60	.000	0.09	.000	-0.06	.004	6	A2	383,453
	−37 11238	NGC 6281 - 14		17 01 24	−37 54.1	2	9.26	.031	0.30	.013	-0.05	.026	6	A0	383,453
	−37 11236	NGC 6281 - 8		17 01 24	−38 01.0	2	8.78	.022	0.13	.000	0.01	.004	6	B8	383,453
	−37 11241	NGC 6281 - 5		17 01 24	−38 02.3	2	8.33	.019	0.10	.005	-0.02	.033	4	B9	383,453
	−37 11240	NGC 6281 - 17	★ AB	17 01 24	−38 02.3	2	9.55	.015	0.14	.010	0.06	.005	4	A0	383,453
322661	−37 11243	NGC 6281 - 18		17 01 25	−37 46.6	2	9.63	.005	0.09	.009	-0.02	.014	5	A2	383,453
		NGC 6281 - 62		17 01 25	−37 50.9	1	11.84		0.47		0.20		1		383
154025	−45 11188	HR 6331		17 01 26	−45 26.0	4	6.28	.013	0.07	.008	0.09	.010	14	A2 V	244,401,1637,2009
		NGC 6281 - 30		17 01 27	−37 51.4	2	10.41	.005	0.18	.023	0.13	.005	4	A2	383,453
		NGC 6281 - 49		17 01 28	−37 53.4	1	11.33		0.48		0.12		1		383
322662	−37 11244	NGC 6281 - 16		17 01 28	−37 55.3	2	9.50	.004	0.16	.013	0.06	.000	6	A2	383,453
		TX Oph		17 01 32	+05 03.1	1	9.83		1.07		0.70		1	G0	793
154090	−33 11706	HR 6334	★ A	17 01 32	−34 03.3	8	4.86	.006	0.26	.007	-0.69	.008	37	B2 Iab	15,26,681,815,1075*
154276	+17 03154	G 170 - 11		17 01 36	+17 15.6	3	9.13	.007	0.65	.017	0.10	.000	4	G2 V	1003,1064,3077
154043	−46 11203	LSS 3903		17 01 36	−47 00.0	5	7.09	.009	0.63	.005	-0.37	.014	12	B2 Iab	285,540,976,1737,2012
154042	−46 11189			17 01 38	−46 05.0	2	9.21	.003	0.33	.011	-0.47	.026	3	B2 II	540,976
154277	+16 03091			17 01 39	+16 05.5	1	7.87		1.11		1.07		2	K1 III	37
154188	−16 04411			17 01 39	−16 28.4	1	10.03		0.27		0.13		1	B8 (IV)ne:	1770
154278	+13 03295	HR 6342	★ B	17 01 40	+13 38.3	5	6.07	.029	1.05	.016	0.83	.040	9	K1 III	15,252,1003,1080,4001
154300	+20 03386			17 01 40	+19 52.1	1	7.80		0.20				3	G5	225
154150	−27 11432	LSS 3908		17 01 40	−27 25.6	3	8.79	.007	0.10	.009	-0.79	.008	4	B0/1 (III)	540,976,1737
154301	+19 03218	HR 6343	★ AB	17 01 42	+19 45.5	1	6.35		1.52				2	K4 p	71
154356	+35 02911	HR 6346		17 01 42	+35 29.0	3	6.25	.017	1.56	.012	1.66	.031	9	M4 III	1501,1733,3001
322663	−37 11247	NGC 6281 - 20		17 01 44	−37 52.9	2	9.69	.015	0.18	.005	0.09	.005	6	A0	383,453
		NGC 6281 - 46		17 01 46	−37 46.2	1	11.24		0.34		0.14		2		383
		NGC 6281 - 69		17 01 46	−37 46.2	1	12.69		0.62		0.15		1		383
		NGC 6281 - 66		17 01 46	−37 49.9	1	12.32		0.32		0.34		1		383
154204	−20 04627	HR 6340		17 01 47	−20 25.6	2	6.30	.004	-0.02	.005	-0.42		5	B7 IV/V	1770,2007
		BPM 40663		17 01 48	−45 10.	1	13.55		0.59		-0.13		1		3065
		NGC 6281 - 47		17 01 50	−37 42.3	2	11.23	.020	0.32	.015	0.17	.015	2	A0	383,453
154136	−37 11250	NGC 6281 - 7		17 01 50	−38 04.8	2	8.70	.005	0.50	.005	0.03	.010	3	F3 V	383,453
322665		NGC 6281 - 39		17 01 52	−37 50.1	2	10.92	.020	0.28	.005	0.16	.030	4	A5	383,453
154165	−26 11859			17 01 53	−26 30.3	1	9.33		0.06		-0.14		3	B9 IV/V	1586
		LTT 6819		17 01 54	−14 39.	1	11.80		0.64		-0.06		4		3062
		NGC 6281 - 68		17 01 55	−37 50.7	1	12.61		0.44		0.33		1		383
154633	+64 01170	HR 6360		17 01 58	+64 40.2	2	6.11	.005	0.96	.000	0.71		5	G5 V	71,3016
		G 19 - 10		17 02 00	+03 47.9	2	11.91	.015	1.18	.015	1.02	.029	3		1620,3062
		G 19 - 11		17 02 00	+03 48.6	2	11.13	.015	0.97	.010	0.75		3	K2	1620,3062
326775	−41 11249	LSS 3906		17 02 00	−41 27.2	1	10.60		0.99		-0.17		2	O7 V	1737
154431	+34 02890	HR 6351		17 02 04	+34 51.5	1	6.07		0.19		0.08		2	A5 V	1733
154231	−20 04628			17 02 04	−20 45.4	1	9.90		0.30		0.21		1	A2 III/IV	1770
322664	−37 11253	NGC 6281 - 29		17 02 05	−37 51.3	1	10.40		0.18		0.12		2	A0	383
		G 170 - 12		17 02 08	+17 00.6	2	12.25	.010	1.55	.005	1.13	.063	3		316,3078

HD	DM	Other Id	N Rem	α_{1950}	δ_{1950}	S	V	σ_V	B–V	σ_{B-V}	U–B	σ_{U-B}	n	Spectrum	References
154357	+20 03389			17 02 09	+20 07.0	1	8.34		1.42		1.71		1	K2	1746
		G 203 - 42		17 02 11	+51 28.2	1	13.61		1.68				1		1759
		vdB 82 # a		17 02 12	−39 54.9	1	11.60		0.62		−0.16		3	B3 V	434
154153	−43 11396	HR 6338		17 02 12	−44 02.3	2	6.18	.005	0.29	.005	0.01		6	A4 III	401,2006
154279	−14 04538			17 02 14	−14 39.0	1	8.62		0.30		0.20		1	B9.5 V	1770
329530	−46 11208	LSS 3907		17 02 14	−46 23.2	1	10.87		0.77		−0.27		2		1737
154217	−35 11293	LSS 3912		17 02 16	−35 51.7	2	8.54	.017	0.45	.001	−0.46	.008	3	B3(n)	540,976
154111	−52 10438			17 02 17	−52 24.9	3	8.27	.007	0.05	.002	−0.38	.009	6	B5 IV/V	540,976,1586
		G 19 - 12		17 02 18	+01 34.8	1	13.36		1.48		1.31		1		3062
154241	−25 11935			17 02 18	−25 24.6	2	8.23	.000	0.40	.000	0.16		5	F2 III/IV	657,2012
154154	−48 11424	LSS 3909		17 02 20	−48 21.1	3	8.62	.012	0.12	.004	−0.76	.012	7	B2 Vnne	540,976,1586
154218	−36 11217			17 02 21	−36 40.4	3	7.47	.109	0.08	.007	−0.46	.005	6	B3 Vne	540,976,1586
154711	+64 01171			17 02 22	+64 05.7	1	8.76		0.24		−0.04		1	A3	1776
		BPM 10552		17 02 24	−67 04.	1	12.02		0.53		−0.11		1		3065
154363	−4 04225	G 19 - 13	⋆ A	17 02 27	−04 59.0	12	7.73	.015	1.16	.011	1.05	.013	45	K5 V	1,116,158,1006,1020*
154508	+36 02823			17 02 30	+36 49.3	1	7.71		1.21		1.19		2	K2	1601
154293	−21 04505			17 02 30	−22 00.3	3	7.05	.005	0.11	.003	−0.51	.012	5	B5 III	540,976,1770
154441	+19 03220	HR 6352	⋆ AB	17 02 31	+19 40.2	2	6.20		−0.01	.000	−0.11	.014	6	B9.5V	1022,1733
154243	−36 11221	LSS 3914		17 02 31	−36 31.3	3	7.99	.027	0.25	.003	−0.57	.009	6	B3 Vnne	540,976,1586
154306	−21 04506			17 02 32	−21 38.2	1	10.46		0.26		0.10		1	A0 IV	1770
154712	+59 01783	G 226 - 43	⋆ A	17 02 37	+59 39.0	1	8.65		1.04		0.94		1	K4 V	3072
154712	+59 01783	G 226 - 43	⋆ AB	17 02 37	+59 39.0	1	8.43		1.10		0.92		2	K4 V	1003
154363	−4 04226	G 19 - 14	⋆ B	17 02 37	−05 00.7	13	10.08	.010	1.43	.012	1.08	.018	59	M0 V	1,158,694,830,1006,1197*
154712	+59 01783	G 226 - 44	⋆ B	17 02 38	+59 39.2	1	10.31		1.40		1.33		1		3016
154578	+46 02258			17 02 39	+46 09.7	1	7.85		0.47		−0.09		1	F7 V	1620
154509	+31 02956			17 02 41	+31 28.8	1	7.77		1.29		1.37		1	K2	3040
153987	−67 03284			17 02 42	−67 07.4	1	7.39		0.07		0.06		31	A0/1 IV	978
154417	+0 03629	HR 6349, V2213 Oph		17 02 44	+00 46.5	3	6.01	.008	0.57	.005	0.04	.012	11	F9 V	15,1417,3037
154333	−28 12800			17 02 47	−28 48.3	1	7.54		0.01		−0.07		1	B9.5 V	1770
154247	−44 11436			17 02 48	−44 30.3	1	8.21		0.03		−0.34		2	B6 IV	401
	+29 02933			17 02 51	+29 44.6	2	7.99	.022	1.36	.027	1.61	.000	5	K5	196,1375
154383	−17 04705			17 02 53	−18 02.9	1	7.87		0.13		−0.18		1	B9 II	1770
154350	−29 13190			17 02 53	−29 14.4	1	8.90		0.13		0.03		1	B9.5 V	1770
	−41 11259	LSS 3915		17 02 55	−41 18.2	1	11.23		0.87		−0.21		2	O7 II	1737
154445	−0 03224	HR 6353	⋆	17 02 57	−00 49.5	2	5.62	.010	0.14	.020	−0.64	.015	11	B1 V	1088,8031
154310	−37 11274	HR 6344	⋆ AB	17 02 57	−37 09.6	3	5.99	.015	0.08	.005	0.09		6	A2 IV	404,2009
154365	−26 11880	BF Oph		17 02 59	−26 30.8	3	6.98	.011	0.69	.011	0.45		3	G5 IV	1484,1490,6011
154494	+12 03142	HR 6355	⋆ A	17 03 03	+12 48.5	6	4.90	.005	0.12	.006	0.13	.031	27	A4 IV	3,15,1363,3023,6001,8015
	−33 11723			17 03 03	−33 47.1	1	10.21		0.31		−0.39		2	B2.5V	540
154313	−42 11832	LSS 3916		17 03 05	−42 16.8	2	8.86	.009	0.76	.004	−0.28	.018	4	B0 Iab	540,976
		G 169 - 44		17 03 06	+28 06.3	2	11.75	.002	0.75	.002	0.27	.041	3		1658,3060
154175	−59 06903			17 03 06	−59 36.1	1	7.44		1.14		1.12		7	K0 IIICNII	1673
154368	−35 11306	HR 6347, LSS 3919	⋆ AB	17 03 08	−35 23.1	8	6.12	.010	0.51	.015	−0.49	.039	17	O9.5Iab	15,26,540,954,976*
	+67 00984			17 03 12	+67 07.4	1	8.73		1.07		0.95		2	G5	1733
154418	−21 04512	HR 6350		17 03 12	−21 29.8	2	6.30	.001	0.12	.004	0.10		13	A1mA2-A5/7	1770,2006
155154	+75 00613	HR 6379		17 03 13	+75 22.0	1	6.18		0.30		0.02		2	F0 IVn	1733
154314	−44 11443			17 03 17	−44 40.0	1	10.05		0.13		−0.36		3	B9 II/III	1586
154339	−46 11218	V616 Ara		17 03 19	−46 56.2	2	8.27	.029	0.35	.005	−0.42	.015	4	B3 II/III	540,976
	+27 02754	G 169 - 45	A	17 03 20	+26 59.8	2	10.02	.013	0.58	.016	0.03	.004	5	G1	1658,7010
	+27 02754		B	17 03 20	+26 59.8	1	13.34		0.78				3		7010
154385	−35 11310	LSS 3922		17 03 20	−36 00.7	3	7.38	.005	0.34	.012	−0.54	.015	8	B1 Ib	540,954,2012
326823	−42 11834	LSS 3918		17 03 20	−42 32.7	4	9.02	.039	0.85	.022	−0.23	.033	9	B2:III:ne	540,1021,1586,1737
154419	−28 12815			17 03 21	−28 31.3	1	9.31		0.07		0.01		2	B9 V	1770
	−30 13795	LSS 3926		17 03 25	−30 08.6	1	10.24		0.14		−0.08		3		1586,1737
154407	−35 11315	LSS 3924	⋆ ABC	17 03 27	−35 49.2	1	8.14		0.36		−0.53		2	B2 Vn	540
154518	−5 04401			17 03 28	−06 06.0	1	8.91		1.01		0.90		1	K2 V	3072
154732	+49 02583	HR 6363		17 03 30	+48 52.3	2	6.11	.014	1.09	.001	1.00		6	K1 III	71,3016
154713	+44 02652	HR 6362		17 03 33	+43 52.8	2	6.41	.024	0.10	.015	0.16	.005	3	A3 IV	252,3050
	−35 11313			17 03 35	−35 34.2	1	10.57		0.50		−0.06		3		1586
154498	−20 04635			17 03 36	−20 38.6	1	10.32		0.09		−0.08		1	B9 III	1770
154635	+25 03197			17 03 39	+25 34.4	2	8.01	.015	1.00	.010	0.80	.010	6	K0 II	833,8100
154759	+47 02426	IDS17023N4706	AB	17 03 41	+47 02.1	1	8.15		1.20		1.19		2	K3 III	1003
154450	−35 11320	56 Sco, LSS 3928		17 03 45	−35 41.6	5	7.90	.055	0.52	.032	−0.55	.020	9	Be	540,954,976,1586,1737
154610	+9 03322	HR 6358		17 03 47	+09 48.0	2	6.36	.005	1.45	.000	1.73	.000	7	K5 III	15,1256
		LP 506 - 55		17 03 47	+12 14.4	1	14.60		0.95		0.68		1		1696
154481	−26 11896	HR 6354		17 03 47	−26 26.8	4	6.28	.012	−0.04	.003	−0.26	.006	10	B8/9 II	26,1490,1770,2007
154388	−48 11444			17 03 47	−48 30.6	2	8.68	.004	0.18	.005	−0.56	.017	3	B2 IV	540,976
155089	+71 00823			17 03 49	+70 54.1	1	8.51		0.99		0.74		2		1733
154499	−24 13102			17 03 49	−24 29.6	2	8.63	.005	0.02	.006	−0.38	.003	8	B5 IV/V	1732,1770
154619	+10 03142	HR 6359		17 03 51	+10 31.2	2	6.40	.016	0.88	.009	0.57	.009	9	G8 III-IV	1733,3016,4001
154426	−46 11223			17 03 53	−46 36.9	2	6.88	.014	0.26	.005	0.06		8	A7 III	158,1075
154656	+12 03148	G 139 - 11		17 03 59	+12 40.3	1	8.53		0.75		0.38		1	G5	1658
154653	+15 03108			17 04 00	+15 17.8	2	7.10	.005	0.96	.036	0.67		7	K0 V	1619,7008
154582	−16 04416			17 04 02	−16 22.6	1	9.44		0.32		0.22		1	B9 V	1770
154621	+0 03633	IDS17015N0047	AB	17 04 03	+00 43.1	2	8.31	.023	0.72	.003	0.24		6	G0	173,938
	−7 04396	IDS17013S0742	AB	17 04 03	−07 45.5	1	10.27		1.09		0.95		1	K5 V	3072
154500	−34 11489			17 04 03	−34 25.3	1	9.94		0.22		−0.56		2	B2 Iab/b	540
		V 455 Sco		17 04 04	−34 01.3	1	13.73		1.11		−0.51		1		1753
		AS 218		17 04 06	−27 09.	1	12.17		1.13		0.46		3	B II	1586
154485	−42 11852			17 04 09	−42 41.3	1	7.95		0.05		−0.31		2	B7 III	401

Table 1 927

HD DM	Other Id	N	Rem	α_{1950}	δ_{1950}	S	V	σ_V	B−V	σ_{B-V}	U−B	σ_{U-B}	n	Spectrum	References
154733 +22 03073	HR 6364	A	⋆	17 04 11	+22 09.0	4	5.56	.020	1.30	.008	1.52	.015	6	K4 III	15,1003,1080,1355
	Ton 266			17 04 12	+25 47.	1	14.93		0.05		-0.83		3	sdB	1036
154535 −34 11493	LSS 3930			17 04 14	−34 26.5	2	8.34	.016	0.06	.015	-0.74	.004	3	B2 IV	540,976
154760 +26 02952	IDS17023N2639	AB		17 04 17	+26 35.0	1	8.74		0.62		0.06		3	G2 V	3026
154660 −1 03292	HR 6361	A	⋆	17 04 17	−01 35.4	3	6.37	.008	0.21	.008	0.09	.008	12	A9 V	13,1088,3024
154660 −1 03292	IDS17017S0131	B		17 04 17	−01 35.4	2	9.73		0.69	.030	0.17	.025	7		13,3024
154906 +54 01857	HR 6370	AB	⋆	17 04 18	+54 32.1	3	4.93	.008	0.47	.004	0.01	.019	10	F7 V	938,1381,3016
154602 −23 13124	IDS17013S2312	AB		17 04 19	−23 16.2	1	10.40		0.28		0.14		1	A5 II/III	1770
154486 −48 11450				17 04 20	−48 49.1	1	6.92		1.48		1.60		9	K2/3 III	1628
−40 11169	LSS 3932			17 04 27	−40 36.6	1	10.65		1.22		0.16		2		954
	He3 1337			17 04 30	−41 57.	1	12.15		0.81		-0.16		1		1586
326993 −42 11858				17 04 30	−42 10.0	1	10.67		0.55		-0.38		2	B0.5III	540
154430 −59 06905				17 04 31	−59 13.4	1	8.74		1.56		1.69		1	K1 III	565
154590 −41 11285				17 04 36	−41 39.3	3	8.31	.014	1.05	.004	0.87	.034	10	K3 V	158,2012,3008
154956 +53 01918				17 04 42	+53 48.9	1	8.29		0.46		-0.04		2	F5	1566
154538 −49 11187	IDS17009S4920	AB		17 04 42	−49 24.2	1	8.63		0.14		-0.42		4	B3 V	1586
−40 11175	LSS 3937			17 04 45	−40 56.5	1	10.90		1.16		0.15		2		954,1737
154589 −40 11176	LSS 3938			17 04 47	−40 57.6	1	6.88		0.59		0.38		2	F2 III	954
	Oph sq 1 # 9			17 04 48	−25 44.8	1	12.27		0.88				1		1473
	He3 1339			17 04 48	−37 20.	1	12.88		1.57		0.38		3		1586
163545 +88 00105				17 04 50	+88 41.8	1	8.32		0.87		0.54		3	K0	196
	FQ Sco			17 04 51	−32 37.7	1	12.96		0.17		-0.55		1		1471
154721 −15 04455	R Oph			17 04 53	−16 01.7	1	7.15		1.72		1.30		1	M3/5e	817
154888 +35 02917				17 04 54	+35 23.3	1	7.39		0.01		-0.01		2	A0	1601
154643 −34 11503	LSS 3941			17 04 55	−34 56.4	4	7.17	.020	0.28	.004	-0.66	.016	7	B0/0.5Ib	219,540,976,8100
	G 226 - 45			17 04 56	+59 29.4	1	13.70		1.49				1		1759
	G 170 - 17			17 05 01	+21 37.1	1	11.61		1.56				1		694
154664 −33 11753				17 05 04	−33 48.1	2	8.74	.012	0.05	.013	-0.75	.006	3	B2 V	540,976
	LP 447 - 63			17 05 05	+19 29.8	1	13.00		1.38		0.87		1		1696
−20 04645				17 05 07	−20 19.4	1	10.14		1.48		1.76		1	K6	1746
154739 −27 11474				17 05 11	−28 02.3	1	8.53		1.55		1.80		3	K2/3 III/IV	1657
323062 −39 11258				17 05 13	−36 17.9	2	10.61	.010	0.98	.035	-0.07	.000	4	K0	954,1737
	V2051 Oph			17 05 14	−25 44.6	1	14.98		-0.13		-0.82		1		1471
	G 139 - 12			17 05 17	+07 26.2	1	14.02		1.88		1.44		3		203
	Oph sq 1 # 3			17 05 17	−25 48.0	1	11.17		0.49				1		1473
	G 181 - 28			17 05 18	+34 25.5	2	12.02	.005	0.44	.000	-0.25	.015	2		1658,3060
154779 −17 04717	HR 6365			17 05 20	−17 32.7	1	5.99		1.01				4	K0 III	2007
	Oph sq 1 # 7			17 05 24	−25 39.8	1	11.67		0.34				1		1473
	Ross 815			17 05 29	−18 24.0	1	12.32		1.16		1.08		1		1696
154804 −21 04521				17 05 29	−21 16.4	1	10.43		0.33		0.16		1	A0 IV	1770
154818 −17 04718				17 05 31	−18 06.3	1	8.34		0.06		-0.07		1	B9 V	1770
154555 −61 05842	HR 6356			17 05 32	−61 36.7	1	6.39		-0.04				4	B8 II/III	2007
	LP 687 - 2			17 05 34	−05 07.1	1	11.72		0.90		0.61		1		1696
154831 −20 04646	IDS17026S2005	A		17 05 34	−20 09.2	1	7.78		0.16		0.09		2	A0 IV	1771
154783 −30 13840	HR 6366			17 05 36	−30 20.4	4	5.94	.016	0.27	.017	0.16		13	Fm δ Del	15,2018,2029,3013
	G 139 - 13			17 05 37	+03 01.6	2	15.19	.019	0.47	.015	-0.24	.034	3		419,3060
154895 −0 03230	HR 6367	AB	⋆	17 05 39	−01 00.9	2	6.05	.000	0.08	.000	0.03		11	A1 V +F3 V	176,1088
154577 −60 06718				17 05 40	−60 40.4	3	7.40	.018	0.89	.004	0.52	.013	10	K2 V	285,1075,3008
	He3 1341			17 05 42	−17 22.7	1	12.94		0.79		-1.01		1		1753
326947 −41 11303				17 05 42	−41 07.8	1	8.62		2.25				1	M3	8051
154744 −47 11276				17 05 46	−47 11.9	1	7.74		0.11		-0.23		2	B8 V	401
154931 +4 03336				17 05 53	+04 29.5	1	7.28		0.59		0.07		4	G0 V	3077
	He3 1342			17 05 53	−23 19.8	1	13.33		2.14		0.46		1		1753
+18 03309	G 170 - 18			17 05 54	+18 01.6	1	10.46		0.51		-0.18		2	F6	1696
154868 −27 11484				17 05 58	−27 20.1	1	9.21		1.46		1.60		5	K2 (III)	1763
154974 +16 03102				17 06 01	+16 09.4	1	6.66		0.50		0.05		2	F8 IV	1648,3077
154834 −37 11319				17 06 01	−37 12.0	2	9.79	.012	0.17	.010	-0.58	.008	3	B3 II	540,976
155102 +40 03103	HR 6376			17 06 09	+40 34.8	2	6.35	.010	0.03	.000	0.04	.005	5	A2 IV	985,1733
154811 −46 11250	LSS 3943			17 06 09	−46 58.1	7	6.92	.009	0.40	.010	-0.60	.012	14	B0 II/III	158,219,401,540,976*
154810 −45 11251				17 06 10	−45 34.4	1	8.14		0.50				4	F5/6 V	2012
	LP 863 - 30			17 06 13	−22 52.4	1	12.77		0.87		0.08		2		1696
155103 +36 02827	HR 6377	AB	⋆	17 06 16	+35 59.9	1	5.39	.025	0.31	.009	0.04	.005	9	A9 III-IV +	1381,3058,8071
155103 +36 02827	IDS17045N3604	C		17 06 16	+35 59.9	1	12.10		1.25		1.05		2		3030
154962 −3 04063	HR 6372			17 06 17	−03 49.1	2	6.36	.005	0.70	.005	0.34	.005	12	G8 IV-V	1088,3077
155092 +28 02677				17 06 23	+28 17.9	2	7.07	.005	0.41	.005	0.01	.005	5	F2 IV	1501,1625
155104 +24 03127				17 06 29	+24 33.0	1	6.85		0.12		0.08		2	A0	1625
154965 −20 04653				17 06 30	−20 42.6	2	8.58	.015	0.42	.005	0.32	.025	4	A7 III	262,1771
155072 +15 03117				17 06 33	+15 44.7	1	9.69		1.23				2	K2	882
154873 −46 11258	LSS 3945	AB	⋆	17 06 38	−46 40.7	5	6.75	.105	0.29	.006	-0.59	.012	12	B1 Ib	158,401,540,1737,8100
	HRC 272, IX Oph			17 06 40	−27 13.2	1	10.86		1.30		0.91		1		776
	HRC 272, IX Oph			17 06 40	−27 13.2	1	11.46		1.34		0.82		1	Fpe	1763
154899 −40 11205				17 06 40	−40 59.4	1	8.56		0.11		-0.30		2	B5 III	401
	HA 196 # 1795			17 06 40	−60 03.6	1	12.00		1.33		1.47		10		1499
154556 −70 02361	HR 6357			17 06 40	−70 39.5	2	6.21	.005	1.06	.000	1.05	.005	7	K1 IV	15,1075
154911 −38 11593	HR 3947			17 06 41	−38 33.7	3	9.05	.012	0.43	.013	-0.59	.020	8	B(2)ne	540,976,1586
	HA 196 # 1801			17 06 41	−60 02.8	1	12.76		0.65		0.10		10		1499
	LP 627 - 4			17 06 42	−01 52.5	1	10.95		0.79		0.32		2		1696
	Steph 1439			17 06 43	+11 30.3	1	11.59		1.18		1.15		2	K7	1746
155093 +15 03118				17 06 46	+15 01.4	1	7.07		1.39				2	K0	882
	G 181 -B5		B	17 06 48	+33 16.8	1	15.92		0.17		-0.55		1	DA	3060

HD	DM	Other Id	N	Rem	α₁₉₅₀	δ₁₉₅₀	S	V	σ_V	B−V	σ_B−V	U−B	σ_U−B	n	Spectrum	References
155343	+55 01912				17 06 50	+55 49.8	1	6.85		1.19		1.22		2	K0	1502
155048	−20 04655				17 06 59	−20 37.1	1	7.54		1.26		1.07		2	K0 II/III	262
	−69 02698				17 06 59	−70 01.5	1	9.37		−0.09		−0.88		2	B2p	400
	−27 11501	HRC 273, KK Oph			17 07 00	−27 11.6	2	11.01	.690	0.64	.025	0.33	.115	2	Ae	776,1763
155078	−10 04445	HR 6375			17 07 02	−10 27.6	3	5.47	.052	0.47	.020	0.00	.002	7	F5 IV	254,2007,3053
155328	+51 02178	HR 6383		⋆ A	17 07 03	+50 54.3	1			0.00		−0.14		1	A1 V	1079
154948	−44 11502	HR 6371		⋆ A	17 07 05	−44 29.7	6	5.07	.010	0.87	.007	0.57	.007	21	G8/K0 III	15,278,1075,2011,2024,2038
155064	−20 04656				17 07 09	−20 43.3	2	9.38	.022	0.22	.005	0.08	.027	3	B9 V	262,1770
		G 170 - 21			17 07 10	+22 48.0	1	12.58		0.60		−0.11		2		1658
155479	+63 01326				17 07 10	+63 30.2	1	7.88		0.55		0.00		1	F5	1776
155095	−19 04547	IDS17044S1919			17 07 18	−19 22.5	2	6.98	.046	0.05	.029	−0.31	.068	4	B8 Ib/II	262,1770
154949	−48 11489				17 07 19	−48 42.1	1	9.21		0.20		−0.21		3	B8 II	1586
		G 139 - 16			17 07 24	+08 08.4	1	12.60		0.68		−0.11		2		1696
		LP 687 - 4			17 07 25	−06 06.6	1	11.89		1.36		1.18		1		1696
154343	−78 01126				17 07 27	−78 20.4	1	7.47		0.45		0.08		9	F3/5 IV/V	1628
155109	−19 04549				17 07 28	−19 27.0	2	9.76	.003	0.18	.027	0.01	.008	3	B9.5IV	262,1770
155125	−15 04467	HR 6378		⋆ AB	17 07 30	−15 39.9	7	2.43	.010	0.06	.004	0.09	.005	25	A2.5Va	15,1020,1075,1088*
155125	−15 04467	IDS17046S1536		C	17 07 30	−15 39.9	1	12.34		0.83		0.25		3		3016
		vdB 88 # c			17 07 30	−45 51.	1	13.97		0.93		0.14		3		434
		vdB 88 # b			17 07 30	−45 56.	1	13.41		0.71		0.11		3		434
		vdB 88 # a			17 07 30	−46 04.	1	12.75		0.85		0.06		3		434
		Steph 1446			17 07 31	+73 33.4	1	11.13		1.27		1.07		1	M0	1746
155031	−41 11331				17 07 31	−41 22.7	1	7.80		0.11		0.04		2	A0 V	401
155513	+61 01640				17 07 32	+61 13.3	1	6.70		0.60		0.06		1	F5	1776
		WLS 1712 50 # 10			17 07 34	+49 38.7	1	12.93		0.95		0.63		2		1375
155051	−41 11333	LSS 3949			17 07 34	−41 38.2	2	7.91	.022	0.45	.003	−0.49	.007	3	B1 Ib	540,976
319511					17 07 37	−34 09.2	1	10.96		0.23		−0.13		2	B8 IV	1085
155032	−43 11473				17 07 37	−44 02.2	1	9.31		0.12		−0.62		2	B2 II/III	401
155020	−46 11267	IDS17039S4603		AB	17 07 37	−46 07.2	3	7.52	.001	0.11	.004	−0.50	.009	5	B2 IV	401,540,976
	−32 12419				17 07 38	−32 27.1	1	10.75		0.23		0.03		2		1085
155231	+0 03649				17 07 42	+00 32.6	1	6.90		0.97		0.68		4	G8 II-III	8100
		HRC 274		V	17 07 42	−27 36.	1	13.04		1.43		0.48		1		776
155142	−21 04534				17 07 43	−21 56.6	1	8.92		0.39		0.26		1	A2/3 IV	1770
154970	−55 07931				17 07 43	−55 34.2	2	8.18	.041	−0.01	.020	−0.41	.011	4	B4/5 III	540,976
		G 181 - 30			17 07 45	+39 13.7	1	13.83		1.65				1		1759
155160	−18 04437				17 07 45	−18 41.6	1	9.67		0.11		0.03		1	B9 II/III	1770
155035	−48 11492	HR 6374			17 07 50	−48 48.8	3	5.86	.037	1.81	.031	1.81	.033	10	M1/2 III	1637,2007,3005
155394	+38 02891				17 07 53	+38 21.3	1	7.39		0.46		0.03		2	F5	1601
	−32 12426				17 07 54	−32 15.1	1	10.76		0.24		0.02		2		1085
155131	−33 11796				17 07 55	−33 12.0	1	10.38		0.30		0.14		1	B8/9 III/V	121
155410	+40 03109	HR 6388		⋆	17 07 56	+40 50.3	3	5.08	.008	1.28	.005	1.40	.010	6	K3 III	1080,1363,3016
	−19 04553				17 07 57	−19 54.4	1	10.45		0.80		0.50		2	F6	262
	−19 04552				17 07 57	−20 05.4	1	9.84		1.20		0.67		2	K0 III	262
		G 203 - 47			17 07 58	+43 45.1	2	11.81	.024	1.47	.010	0.91	.024	10	M3	1663,1723
155146	−33 11798	IDS17047S3316		AB	17 07 58	−33 19.0	1	9.59		0.14		−0.23		2	B9 V	1085
155161	−32 12429	AH Sco			17 08 02	−32 15.9	1	8.20		2.33				1	M4 (III)	8051
155233	−20 04659				17 08 06	−20 35.5	1	6.81		1.04		0.91		2	K1 III	262
155134	−41 11340	LSS 3952			17 08 06	−41 42.8	2	8.82	.004	0.45	.000	−0.34	.010	3	B1/2 II	540,976
155234	−20 04660				17 08 07	−20 45.7	1	7.99		1.15		0.84		2	K0 III	262
154903	−67 03296	HR 6368		⋆ A	17 08 10	−67 08.2	4	5.89	.011	1.07	.010	0.88	.022	11	K0/1 III	15,1075,1642,3005
155066	−53 08445				17 08 13	−53 18.9	1	7.37		1.60		1.64		4	M4 III	1673
319536	−35 11389	LSS 3954			17 08 14	−35 46.2	1	10.64		0.59		−0.34		2	B5	1085
319524					17 08 21	−34 53.2	1	10.52		0.43		−0.16		2	B8	1085
155217	−32 12434				17 08 22	−32 09.4	3	8.65	.007	0.00	.004	−0.77	.015	5	B2 II/III	540,976,1085
155291	−20 04661				17 08 24	−20 21.9	2	7.48	.042	0.05	.025	−0.01	.102	4	B9 IV	262,1770
155269	−21 04540				17 08 25	−21 50.2	1	9.04		0.50		0.37		1	A5 IV	1770
155375	+12 03161	HR 6385			17 08 26	+12 31.7	5	6.58	.012	0.08	.004	0.09	.005	16	A1 m	355,1022,1199,3058,8071
	−32 12436				17 08 26	−32 25.5	1	10.69		0.23		0.04		2		1085
319493					17 08 26	−33 23.7	1	10.92		0.15		−0.06		2	A3	1085
155480	+38 02892				17 08 29	+38 47.8	1	7.81		1.24		1.21		2	K2 III	105
		Sco sq 2 # 108			17 08 29	−33 28.3	1	12.13		0.33		0.19		2		1085
		LSS 3955			17 08 29	−35 28.8	1	11.42		0.85		0.04		2	B3 Ib	1085
		JL 1			17 08 30	−87 07.	2	14.71	.029	−0.13	.058	−0.96	.015	3		132,832
	−32 12438				17 08 31	−32 24.8	2	11.66	.144	0.24	.020	−0.09	.055	5		1085,1101
155456	+24 03137	G 170 - 22			17 08 34	+24 35.4	1	8.35		0.87		0.62		3	K2 V	196
155203	−43 11485	HR 6380			17 08 34	−43 10.5	10	3.33	.005	0.40	.004	0.09	.008	38	F2 V	15,125,278,1020,1075,1770*
		LP 747 - 111			17 08 36	−14 44.2	1	14.27		0.04		−0.77		5		3062
155255	−33 11807				17 08 36	−33 38.9	1	10.29		0.09		−0.29		2	A0/2 III/IV	1085
155763	+65 01170	HR 6396		⋆	17 08 38	+65 46.6	7	3.17	.010	−0.12	.011	−0.43	.004	26	B6 III	15,667,1203,1363,3016*
	−34 11549				17 08 38	−34 13.2	1	11.17		0.75		−0.18		2		1085
155274	−34 11548				17 08 38	−34 56.9	2	9.68	.013	0.56	.002	−0.26	.001	5	B2/5(n)	540,1085
319521	−34 11550	LSS 3956			17 08 39	−34 39.3	2	10.39	.010	0.70	.018	−0.33	.001	4	B9	540,1085
155099	−58 07014	IDS17043S5828		AB	17 08 39	−58 32.2	2	6.83	.005	0.39	.009			8	F0/2 V	1075,2033
155185	−46 11288				17 08 40	−46 29.1	2	9.19	.005	0.85	.005	0.44		6	K0 V	2011,3077
155421	+15 03122				17 08 41	+14 58.9	1	8.41		0.43		−0.02		2	F5	1648
		Sco sq 2 # 112			17 08 41	−33 28.9	1	11.67		0.20		0.02		2		1085
		Sco sq 2 # 113			17 08 43	−32 33.8	1	11.88		0.30		0.06		2		1085
		Sco sq 1 # 27			17 08 44	−33 13.5	1	13.31		0.96		0.38		1		121
	−33 11811	Sco sq 1 # 28			17 08 47	−33 14.5	1	10.56		1.20		1.00		1	K2	121
155259	−39 11182	HR 6381			17 08 49	−39 26.8	3	5.66	.007	0.04	.000			11	A0/1 V	15,2013,2030

Table 1

HD	DM	Other Id	N Rem	α₁₉₅₀	δ₁₉₅₀	S	V	σ_V	B–V	σ_B-V	U–B	σ_U-B	n	Spectrum	References
		Sco sq 2 # 114		17 08 50	−33 32.7	1	11.29		0.19		0.05		2		1085
319507				17 08 50	−34 02.2	1	10.53		0.25		0.12		2	A0	1085
		LSS 3958		17 08 50	−36 50.6	1	11.12		1.08		0.05		2	B3 Ib	1737
155276	−38 11632	HR 6382		17 08 50	−38 45.7	1	6.30		1.05				4	K1 III	2007
155313	−33 11816			17 08 54	−33 50.0	1	10.23		0.10		−0.33		2	Ap Si	1085
	+27 02773			17 08 55	+27 46.5	1	10.66		1.18		0.99		2	M0	1375
155524	+32 02862			17 08 55	+32 14.4	1	7.21		0.98		0.63		2	K2	1375
155312	−33 11815			17 08 55	−33 33.2	1	10.17		0.16		0.09		2	A1 III/IV	1085
	−38 11636	LSS 3959		17 08 55	−38 25.9	1	10.79		0.14		0.18		2	O9	432,1737
		Sco sq 2 # 117		17 08 56	−33 32.3	1	11.69		0.15		−0.18		2		1085
		Sco sq 2 # 6		17 08 56	−36 08.8	1	12.05		0.47		0.22		2		1085
	+47 02437			17 08 57	+46 57.7	1	9.63		0.48		−0.04		2	F8	1601
155514	+24 03140	HR 6391, V620 Her		17 08 58	+24 17.8	1	6.20		0.21		0.10		3	A8 V	1733
155337		CpD -33 04244		17 09 00	−33 20.8	1	10.22		0.03		−0.48		2	B9	1085
	−35 11396	LSS 3960		17 09 00	−35 29.3	1	11.31		0.80		−0.09		2		1085
156648	+81 00574			17 09 02	+81 26.3	1	8.94		0.70		0.25		4	K0	3026
155336	−32 12444	LSS 3961		17 09 02	−33 02.5	2	9.47	.045	0.25	.010	−0.58	.002	4	B1/2 Ib	540,1085
		Sco sq 2 # 119		17 09 02	−33 14.4	1	11.32		0.13		−0.26		2		1085
155320	−34 11554			17 09 02	−34 18.6	3	8.35	.004	0.09	.009	−0.74	.013	5	B2 Ib/II	540,976,1085
155377	−18 04443			17 09 03	−18 24.5	1	10.58		0.27		0.21		1	B9 IV/V	1770
		Sco sq 2 # 122		17 09 05	−33 31.1	1	11.00		0.30		−0.44		2		1085
	−33 11818			17 09 05	−33 49.9	1	10.84		0.30		0.12		2		1085
155068	−65 03409			17 09 07	−65 54.2	1	7.13		1.06		0.88		25	K0 III	978
155674	+54 01861	IDS17081N5437	A	17 09 08	+54 33.4	3	8.85	.023	1.15	.016	1.06	.027	5	K0	497,1028,3072
155674	+54 01862	IDS17081N5437	B	17 09 08	+54 33.4	3	9.34	.004	1.26	.004	1.18	.032	5		497,1028,3072
146516	−22 04139			17 09 08	−22 29.0	1	10.14		0.82				4	G0/2 V	370
155349	−31 13694			17 09 08	−32 05.2	2	9.47	.030	0.15	.004	−0.37	.014	5	B5/7 Vne	1085,1586
319506				17 09 08	−34 59.6	1	11.13		0.16		−0.35		2	B8	1085
155280	−46 11292	LSS 3957		17 09 08	−46 22.8	2	8.71	.015	0.35	.007	−0.35	.020	3	B2/3 II	540,976
155379	−25 12018	HR 6386		17 09 09	−25 11.7	1	6.54		−0.04				4	A0 pHg	2007
155543	+24 03141			17 09 12	+24 18.7	1	6.93		0.37		−0.03		2	F2	1733
		LP 747 - 1		17 09 12	−13 41.0	1	14.00		1.42		1.10		1		3062
	−32 12448			17 09 12	−32 13.8	1	10.62		0.21		−0.05		2		1085
319484				17 09 14	−32 49.5	1	10.88		0.21		0.08		2	A0	1085
		Sco sq 1 # 29		17 09 14	−33 15.4	1	12.01		0.26		0.16		1	A0	121
	−23 13201	LSS 3967		17 09 15	−23 44.8	1	12.45		0.38		−0.11		2		1737
155350	−33 11823			17 09 15	−33 21.8	1	9.51		−0.02		−0.60		2	B2 V	1085
		Sco sq 2 # 129		17 09 15	−33 30.1	1	11.19		0.19		0.03		2		1085
155401	−27 11516	HR 6387		17 09 17	−27 42.1	2	6.13	.009	−0.05	.009	−0.21		6	B9 V(n)	1770,2007
		G 19 - 15		17 09 18	−01 47.3	1	11.45		1.60		1.23		2	M3	3062
		vdB 83 # a		17 09 18	−42 39.9	1	10.92		0.33		−0.05		3	B8 V	434
155526	+16 03120	V463 Her		17 09 19	+16 28.2	1	8.46		0.99				1	K0	882
155500	+8 03367	HR 6390		17 09 20	+07 57.3	4	6.34	.013	1.04	.003	0.85	.012	10	K0 III	15,1256,3016,4001
		Sco sq 2 # 130		17 09 20	−33 29.9	1	11.60		0.26		0.00		2		1085
155414	−22 04299			17 09 21	−22 51.9	1	7.85		0.37		0.27		3	A5 V	1771
327086	−40 11231	LSS 3963		17 09 21	−40 34.7	2	9.92	.030	1.49	.094	0.50	.129	5	B3 Ia	954,1737
155711	+52 02032	HR 6395		17 09 22	+52 28.2	2	6.31		−0.02	.012	−0.15	.020	4	B9 V	985,1079
155384	−32 12453			17 09 23	−32 53.6	1	8.31		0.98		0.61		1	K0 III +G	121
		Sco sq 1 # 30		17 09 23	−33 09.1	1	12.92		1.06		0.50		1		121
155385	−33 11826			17 09 23	−33 13.2	1	9.46		−0.03		−0.64		2	B3 II/III	1085
		Sco sq 1 # 31		17 09 25	−33 09.9	1	12.24		0.45		0.24		1	A2	121
155387	−34 11556			17 09 26	−34 12.4	1	10.29		0.24		−0.08		2	B8/9 III	1085
155352	−42 11933			17 09 27	−42 34.5	2	8.31	.098	0.08	.014	−0.54	.012	5	B2 V	401,1586
	−32 12454			17 09 28	−32 58.1	1	11.44		0.18		−0.08		2		1085
		Sco sq 1 # 32		17 09 28	−33 09.7	1	11.47		1.33		1.37		1	K2	121
		Sco sq 2 # 133		17 09 28	−33 55.8	1	11.53		0.20		−0.21		2		1085
155461	−19 04558			17 09 30	−19 43.3	1	9.58		1.26		0.93		2	G8/K0 (III)	262
155402	−33 11830	IDS17062S3314	A	17 09 30	−33 18.2	1	7.77		−0.04		−0.79		2	B2 II	540
155403	−33 11828			17 09 31	−33 37.2	3	8.21	.012	−0.04	.004	−0.66	.012	5	B2 III	540,976,1085
155459	−17 04743			17 09 34	−18 04.6	1	9.12		0.01		−0.42		1	B8 II	1770
	−18 04448			17 09 34	−18 43.3	1	10.33		1.82		1.88		2	M0 III	262
	−39 11193	LSS 3966		17 09 35	−39 50.5	1	11.18		0.72		−0.26		3		1586
319491				17 09 36	−33 25.7	1	10.82		0.16		−0.08		2	A0	1085
319546				17 09 36	−36 16.3	1	10.63		0.38		−0.25		2	B3	1085
		Sco sq 1 # 33	AB	17 09 38	−33 06.2	1	12.05		0.57		0.04		1	F2	121
	−33 11833			17 09 39	−33 23.3	1	11.14		0.15		−0.12		2		1085
	−32 12459			17 09 40	−32 38.9	1	11.19		0.13		−0.32		2		1085
155299	−56 08091			17 09 40	−56 44.5	1	7.02		0.43		0.12		1	F3/5 III	1592
		vdB 84 # a		17 09 42	−38 44.	1	12.44		0.99		0.34		4	B2 V:	434
155503	−14 04568			17 09 43	−15 06.3	1	7.94		0.15		0.09		1	B9 V	1770
155450	−32 12460	HR 6389, LSS 3975	⋆ A	17 09 43	−32 22.8	6	6.00	.014	0.07	.009	−0.74	.019	19	B1 Ib	15,26,1085,2013,2028,8100
	−39 11197	LSS 3968	⋆ A	17 09 43	−39 51.8	1	10.49		0.91		−0.09		2	B1.5 Iab	1737
	−39 11197	LSS 3968	⋆ AB	17 09 43	−39 51.8	1	10.31		0.77		−0.05		2	B1.5 Iab	954
155449	−32 12464	IDS17066S3209	AB	17 09 44	−32 11.8	1	9.52		0.09		−0.46		2	B9	1085
322941	−39 11196	LSS 3969		17 09 44	−39 38.1	1	9.97		0.78		−0.20		2	B1 II-Ib	540
155416	−37 11383	LSS 3971		17 09 45	−37 53.6	3	6.60	.013	0.37	.015	−0.17	.017	5	B8 Iab/b	540,954,976
319486				17 09 46	−33 11.4	1	10.59		−0.04		−0.51		2	A0p	1085
		Sco sq 2 # 10		17 09 46	−34 41.6	1	11.72		0.63		0.02		2		1085
319514				17 09 47	−34 27.6	1	11.16		0.34		−0.30		2	F0	1085
155389	−47 11340			17 09 48	−47 15.6	1	7.11		0.03		−0.04		2	A0 V	401

HD	DM	Other Id	N Rem	α_{1950}	δ_{1950}	S	V	σ_V	B–V	σ_{B-V}	U–B	σ_{U-B}	n	Spectrum	References
		L 203 - 131		17 09 48	−57 34.	1	15.10		0.03		-0.76		3		3065
	−32 12462	Sco sq 1 # 34		17 09 49	−33 05.5	1	10.75		1.20		0.99		1	K2	121
322956	−40 11234	LSS 3970		17 09 49	−40 13.6	1	9.96		1.67		0.55		2	G0	954
		GD 205		17 09 50	+23 04.7	1	14.90		-0.09		-0.98		1	DBn	3060
	−32 12466			17 09 50	−32 16.5	1	10.06		0.09		-0.51		2		1085
	−39 11198	LSS 3972		17 09 50	−39 51.6	2	10.69	.035	0.89	.050	-0.11	.010	4	B1 Ib	954,1737
322942	−39 11200	LSS 3973		17 09 51	−39 37.3	1	10.79		0.56		-0.03			B5 V	540
155727	+39 03086	IDS17082N3923	AB	17 09 55	+39 19.1	1	8.39		0.48		-0.05		3	F2 V	833
		Sco sq 1 # 35	AB	17 09 55	−33 08.3	1	11.88		0.56		0.27		1	A5	121
		Sco sq 1 # 36		17 09 56	−33 13.4	1	11.73		1.13		0.84		1	K0	121
	−39 11201	LSS 3976W		17 09 56	−39 44.2	2	10.89	.075	0.75	.030	-0.22	.015	4	B3 V	954,1737
		Sco sq 1 # 37		17 09 57	−33 12.7	1	12.05		0.66		0.07		1	F2	121
155642	+21 03063	IDS17078N2121	AB	17 09 58	+21 17.3	1	7.26		1.24		1.24		3	K0	938
		LSS 3978		17 09 58	−24 08.8	1	12.46		0.36		-0.50		2	B3 Ib	1737
	−33 11839			17 09 58	−33 07.2	1	11.53		0.09		-0.22		2		1085
319504				17 09 58	−34 13.5	1	10.99		0.32		0.08		2	A2	1085
	−39 11201	LSS 3976E		17 09 58	−39 44.2	1	10.90		0.58		-0.03		2	B1 IIIn?	1737
155341	−56 08098	HR 6384, V829 Ara		17 09 59	−56 49.8	2	6.13	.029	1.78	.003	1.34		17	M1/2 II/III	1592,2007
155472	−33 11840			17 10 00	−33 07.8	1	9.67		0.11		-0.06		1	B9.5V	121
319540	−35 11404	LSS 3977		17 10 02	−36 00.1	1	10.33		0.70		-0.28		2	B5	1085
155675	+25 03212	IDS17080N2522	AB	17 10 04	+25 18.4	1	8.50		0.53		-0.09		3	F8 V	3026
		Sco sq 1 # 38		17 10 04	−33 17.0	1	12.74		0.43		0.08		1	F2	121
155436	−44 11526	LSS 3974		17 10 05	−44 40.3	3	9.18	.011	0.78	.008	-0.23	.017	6	B2/3	540,976,1586
155644	+10 03165	HR 6393	★ A	17 10 06	+10 38.7	2	5.36	.020	1.60	.005	1.67	.030	5	M2 IIIa	1733,3001
	−32 12471			17 10 06	−32 25.3	1	10.84		0.24		-0.08		2		1085
154972	−74 01610	HR 6373		17 10 06	−74 28.6	3	6.24	.007	0.00	.005	0.00	.000	14	A0 V	15,1075,2038
155838	+51 02187			17 10 07	+51 48.8	1	7.14		0.10		0.08		2	A0	1566
		Sco sq 2 # 146		17 10 08	−33 41.8	1	11.23		0.18		-0.03		1		1085
	+29 02958			17 10 11	+29 35.2	1	8.58		0.97		0.77		3	G8 II-III	8100
		G 19 - 16		17 10 11	−05 03.6	4	11.64	.013	1.47	.014	1.11	.040	5		1620,1658,1705,3062
155506	−33 11844			17 10 11	−33 16.9	4	7.75	.006	-0.07	.005	-0.81	.012	10	B2 V(n)	540,976,1085,1732
155418	−51 10761			17 10 11	−52 02.5	3	9.55	.004	0.01	.018	-0.61	.009	6	B2 V	400,540,976
155902	+56 01954			17 10 14	+56 43.2	1	6.98		0.70		0.22		2	G5	3016
155784	+41 02804			17 10 15	+41 47.3	1	6.75		1.37		1.53		2	K2	1601
		G 19 - 17		17 10 15	−02 05.6	1	14.67		1.42		0.46		1		3062
155088	−72 02047	IDS17043S7226	AB	17 10 16	−72 30.0	1	8.06		1.25		1.28		5	K2 III	1704
155646	+0 03654	HR 6394		17 10 21	+00 24.7	2	6.64	.005	0.49	.010	0.04	.015	10	F6 III	1088,3016
155519	−32 12475			17 10 21	−33 03.1	1	9.35		-0.03		-0.61		2	B3/5 Vn	1085
155409	−55 07965			17 10 21	−55 57.0	2	7.89	.034	-0.06	.013	-0.67	.006	4	B2 III	540,976
319544				17 10 23	−36 06.6	1	10.75		0.58		-0.02		2	B8	1085
155860	+49 02604	HR 6399	★ AB	17 10 24	+49 48.3	2	6.13	.005	0.16	.000	0.11	.000	5	A5 III	1501,1733
155536	−38 11663	IDS17070S3810	AB	17 10 28	−38 14.2	1	6.75		0.46		0.12		4	F5 V	938
155438	−54 08115			17 10 28	−54 18.8	3	8.44	.022	0.02	.007	-0.43	.008	8	B5/7 III/Vn	540,976,1586
155550	−32 12479	FV Sco		17 10 29	−32 47.7	2	7.98	.021	0.07	.003	-0.55	.028	5	B4 IV	1085,1768
	−19 04560			17 10 31	−19 52.2	1	9.80		1.41		1.34		2	K2 III	262
		GD 206		17 10 32	+19 46.7	1	15.08		0.65		-0.04		1		3060
	−39 11208	LSS 3979		17 10 35	−39 20.2	1	11.27		1.15		0.13		2		1737
		WLS 1700 30 # 8		17 10 36	+30 04.4	1	12.16		0.96		0.64		2		1375
155509	−45 11303			17 10 36	−45 23.4	2	9.28	.016	0.37	.003	-0.48	.022		B2 Ib	540,976
155569	−33 11848			17 10 39	−33 20.0	1	9.84		0.17		0.09		2	B9.5III/IV	1085
155876	+45 02505	G 203 - 51	★ AB	17 10 40	+45 44.8	4	9.37	.015	1.50	.013	1.01		10	K5	906,1197,1381,3078
155600	−32 12485			17 10 42	−32 11.0	1	7.93		0.15		-0.25		2	B8 V	1085
		WLS 1712 50 # 7		17 10 43	+47 37.9	1	13.87		0.89		0.33		2		1375
155648	−18 04457			17 10 45	−18 55.2	1	8.57		0.34		0.15		2	A9 V	262
	−32 12488			17 10 47	−32 13.1	1	11.32		0.35		-0.04		2		1085
319529				17 10 47	−35 41.2	1	10.53		0.53		0.16		2	B8	1085
327202	−43 11512	LSS 3980		17 10 50	−43 29.1	1	10.23		0.67		-0.22		3	B3	1586
		Sco sq 2 # 153		17 10 51	−33 14.8	1	12.30		0.33		0.18		2		1085
155921	+43 02702			17 10 52	+43 47.6	1	7.24		0.66		0.15		2	G5	1601
319517				17 10 52	−34 29.3	1	10.29		0.42		0.02		2	A2	1085
322944	−39 11210			17 10 53	−39 46.2	3	10.18	.024	0.89	.032	-0.17	.023	6	B3	524,954,1737
155614	−33 11853			17 10 54	−33 09.7	1	9.06		-0.04		-0.68		2	B8	1085
155615	−33 11855	IDS17077S3309	AB	17 10 55	−33 12.5	1	9.13		0.05		-0.38		2	B9/A0 V	1085
319489				17 10 56	−33 27.2	1	10.43		0.22		0.11		2	A0	1085
		NGC 6304 sq1 19		17 10 57	−29 31.5	1	12.16		1.62		1.67		1		474
		LP 687 - 12		17 10 59	−08 21.4	1	12.05		1.61		1.18		1		3073
319499				17 10 59	−34 06.3	1	10.75		0.13		-0.42		2	B8	1085
155603	−39 11212	HR 6392, V915 Sco	★ A	17 11 00	−39 42.6	7	6.48	.058	2.25	.029	2.23	.471	23	K0 Ia	524,584,1018,1088*
155603	−39 11212	IDS17075S3939	B	17 11 00	−39 42.6	2	10.42	.015	0.54	.019	-0.31	.005	6		524,584
155603	−39 11212	IDS17075S3939	C	17 11 00	−39 42.6	2	11.36	.015	1.57	.015	1.36	.075	2		524,584
155603	−39 11212	IDS17075S3939	D	17 11 00	−39 42.6	2	14.02	.065	0.84	.015	0.37	.030	2		524,584
		NGC 6304 sq1 7		17 11 01	−29 28.5	2	10.74	.028	1.00	.005	0.46	.009	8		474,512
	−36 11349	LSS 3983		17 11 01	−36 21.7	1	11.38		0.42		-0.25		2		1085
		NGC 6304 sq1 20		17 11 02	−29 29.7	2	11.46	.015	1.39	.010	1.12	.073	3		474,512
		LP 687 - 13		17 11 03	−04 57.6	1	15.10		0.99				1		1759
156051	+58 01707			17 11 06	+58 01.5	2	7.14	.005	0.44	.000	-0.05	.005	4	F5	1733,3016
		He3 1358		17 11 06	−31 19.	1	12.35		0.24		-0.07		3		1586
319483				17 11 06	−32 43.4	1	9.95		0.26		-0.24		2	B5 IV	1085
155685	−26 12012			17 11 07	−26 55.6	1	6.64		0.43		0.01		27	F2/3 V	978
		NGC 6304 sq1 9		17 11 07	−29 27.7	1	12.46		0.56		0.35		6		474

Table 1

HD	DM	Other Id	N Rem	α_{1950}	δ_{1950}	S	V	σ_V	B–V	σ_{B-V}	U–B	σ_{U-B}	n	Spectrum	References
319518				17 11 08	−34 40.9	1	11.18		0.72		0.34		2		1085
319542	−35 11417	LSS 3985		17 11 09	−36 00.2	1	10.82		0.68		−0.20		2	F8	1085
155878	+28 02694			17 11 10	+27 59.4	1	8.08		1.04		0.84		3	G8 II	8100
		Sco sq 2 # 159		17 11 11	−32 19.1	1	12.12		0.37		−0.04		2		1085
155665	−33 11859			17 11 11	−33 46.7	1	9.88		0.02		−0.56		2	B3 III	1085
155605	−44 11540			17 11 11	−44 24.1	2	9.55	.013	0.34	.010	−0.37	.011	3	B1/2 II	540,976
		LP 627 - 13		17 11 12	+01 42.8	1	15.33		1.50				1		1759
155721	−19 04565			17 11 12	−19 12.5	1	9.47		0.79		0.33		2	G5 V	262
155666	−34 11575			17 11 12	−34 45.0	1	8.44		2.17				1	K3/4	8051
	−26 12015	LSS 3989		17 11 13	−26 33.4	1	10.00		0.51		−0.34		2	B1 V	540
		Sco sq 2 # 161		17 11 14	−32 44.5	1	11.80		0.21		−0.03		2		1085
319527				17 11 14	−35 38.5	1	10.85		0.72		0.01		2	F2	1085
319487				17 11 17	−33 04.7	1	11.17		0.16		−0.08		2		1085
319528	−35 11419	LSS 3987		17 11 17	−35 44.0	1	10.12		0.81		−0.09		2	B5	1085
		VZ Her		17 11 18	+36 02.0	2	10.72	.007	0.05	.004	0.06	.007	2	A9.2	668,699
	−32 12499			17 11 18	−33 04.4	1	11.86		0.22		−0.02		2		1085
155688	−33 11860			17 11 18	−33 48.0	1	9.73		0.08		−0.36		2	B7 V	1085
	−39 11217	LSS 3986		17 11 19	−39 43.4	1	11.07		0.88		−0.19		1		524
155842	+15 03129			17 11 20	+15 33.8	1	9.45		0.95				2	G5	882
319558				17 11 21	−32 53.7	1	10.95		0.22		0.09		2		1085
		NGC 6304 sq1 5		17 11 22	−29 19.9	1	13.29		1.48		1.58		5		474
		NGC 6304 sq1 4		17 11 24	−29 16.3	1	11.71		0.47		0.30		5		474
		NGC 6304 sq1 3		17 11 24	−29 18.9	1	11.76		1.62		1.71		6		474
		NGC 6304 - 5		17 11 24	−29 24.	1	14.88		0.93				3		512
		NGC 6304 - 7		17 11 24	−29 24.	1	13.95		1.65				2		512
		NGC 6304 - 9		17 11 24	−29 24.	1	18.95		1.63				1		512
		NGC 6304 - 10		17 11 24	−29 24.	1	16.28		1.49				2		512
		NGC 6304 - 11		17 11 24	−29 24.	1	18.73		1.43				1		512
		NGC 6304 - 12		17 11 24	−29 24.	1	17.47		1.81				1		512
		NGC 6304 - 13		17 11 24	−29 24.	1	15.35		1.51				3		512
		NGC 6304 - 14		17 11 24	−29 24.	1	17.15		1.17				1		512
		NGC 6304 - 15		17 11 24	−29 24.	1	16.61		1.61				2		512
		NGC 6304 - 16		17 11 24	−29 24.	1	13.57		1.01				3		512
		NGC 6304 - 17		17 11 24	−29 24.	1	13.88		2.01				2		512
		NGC 6304 - 18		17 11 24	−29 24.	1	18.87		1.47				2		512
		NGC 6304 - 21		17 11 24	−29 24.	1	16.04		1.02				1		512
		NGC 6304 - 24		17 11 24	−29 24.	1	15.46		2.00				1		512
		NGC 6304 - 29		17 11 24	−29 24.	1	15.48		0.92				2		512
		NGC 6304 - 30		17 11 24	−29 24.	1	16.55		1.04				1		512
		NGC 6304 - 31		17 11 24	−29 24.	1	15.16		2.09				2		512
		NGC 6304 - 33		17 11 24	−29 24.	1	14.21		1.66				2		512
		NGC 6304 - 34		17 11 24	−29 24.	1	15.27		0.81		0.43		3		512
		NGC 6304 - 35		17 11 24	−29 24.	1	17.07		1.30				1		512
		NGC 6304 - 36		17 11 24	−29 24.	1	13.31		0.80				2		512
		NGC 6304 - 38		17 11 24	−29 24.	1	18.28		2.22				1		512
		NGC 6304 - 40		17 11 24	−29 24.	1	19.15		1.65				2		512
		NGC 6304 - 41		17 11 24	−29 24.	1	15.35		2.09				3		512
		NGC 6304 - 42		17 11 24	−29 24.	1	17.71		1.73				1		512
		NGC 6304 - 43		17 11 24	−29 24.	1	15.21		1.56				2		512
		NGC 6304 - 44		17 11 24	−29 24.	1	13.83		1.15				2		512
		NGC 6304 - 45		17 11 24	−29 24.	1	17.14		1.14				1		512
		NGC 6304 - 51		17 11 24	−29 24.	1	16.04		1.12				2		512
		NGC 6304 - 52		17 11 24	−29 24.	1	16.55		1.87				2		512
		NGC 6304 - 53		17 11 24	−29 24.	1	15.50		1.74				1		512
		NGC 6304 - 54		17 11 24	−29 24.	1	14.09		2.17				1		512
		NGC 6304 - 55		17 11 24	−29 24.	1	16.07		1.93				1		512
		NGC 6304 - 56		17 11 24	−29 24.	1	15.54		0.95		0.40		1		512
		NGC 6304 - 57		17 11 24	−29 24.	1	16.48		1.69				1		512
		NGC 6304 - 58		17 11 24	−29 24.	1	15.09		2.12				1		512
		NGC 6304 - 4238		17 11 24	−29 24.	1	14.72		1.92				2		512
	+42 02810	G 203 - 52	★ AB	17 11 25	+42 23.6	1	10.07		1.28		1.14		1	K7	3078
155802	−8 04400	IDS17087S0817	AB	17 11 25	−08 20.9	1	8.49		0.93		0.56		1	K3 V	3072
155790	−17 04753			17 11 27	−17 24.5	1	8.13		0.09		−0.10		1	B9 III/IV	1770
155703	−33 11865			17 11 27	−33 20.8	1	9.29		0.09		−0.39		2	B7 III	1085
	+14 03203			17 11 31	+14 37.9	1	8.10		1.49		1.75		3	K5	196
155923	+21 03070			17 11 31	+21 29.3	1	6.82		1.01		0.82		2	K0	1648
156192	+63 01334			17 11 31	+63 11.1	1	8.02		0.01		−0.06		1	A2	1776
	−39 11223	LSS 3988		17 11 31	−39 58.8	2	10.82	.005	0.97	.024	−0.08	.019	3	B0 Ia	954,1737
		NGC 6304 sq1 2		17 11 32	−29 19.8	1	12.38		1.42		1.31		7		474
155574	−57 08394			17 11 32	−57 27.3	1	8.18		0.10		−0.75		9	G5 II/III	1732
	+15 03130			17 11 33	+14 56.9	1	9.63		0.39		0.07		3	F0	196
		HA 85 # 356		17 11 34	+15 01.7	1	11.50		1.07		0.96		3		196
	−19 04567			17 11 41	−19 25.5	1	11.00		2.05		2.47		2	M1 III	262
319583				17 11 41	−33 43.3	1	9.88		0.07		−0.41		2	B5 V	1085
319604				17 11 41	−34 30.0	1	10.51		0.32		−0.27		2	B8	1085
155754	−33 11867			17 11 42	−33 41.4	3	7.94	.012	−0.02	.002	−0.77	.018	5	B2 II	540,976,1085
		G 226 - 49		17 11 44	+58 36.2	1	13.28		0.64		−0.05		1		1658
		NGC 6304 sq1 1		17 11 45	−29 21.7	2	10.25	.000	1.21	.010	0.91	.025	9		474,512
155737	−39 11228	IDS17084S3932	AB	17 11 45	−39 35.7	1	8.85		2.06				1	M0/2	8051
156012	+39 03091			17 11 46	+39 02.5	1	6.99		1.18		1.09		2	K1 III	105

HD	DM	Other Id	N	Rem	α_{1950}	δ_{1950}	S	V	σ_V	B–V	σ_{B-V}	U–B	σ_{U-B}	n	Spectrum	References
327083	−40 11253				17 11 46	−40 16.7	1	9.84		1.66		0.57		3		1586
155774	−32 12508				17 11 47	−32 45.1	1	9.64		0.03		-0.57		2	B3/5 V	1085
		Sco sq 2 # 170			17 11 49	−32 36.2	1	11.76		0.31		-0.01		2		1085
155793	−31 13747				17 11 50	−31 16.3	1	8.82		0.47		0.28		2	A(5)mA8-F3	918
319601	−34 11585				17 11 50	−34 11.0	1	11.53		0.30		0.28		1		121
155670	−50 11118				17 11 50	−51 01.6	2	7.01	.073	0.02	.002	-0.41	.006	3	B7/8 II	540,976
		NGC 6304 sq1 21			17 11 51	−29 24.8	1	12.27		1.25		0.97		1		474
155846	−20 04685				17 11 54	−20 54.8	1	7.67		1.64		1.94		3	K4 III	1657
	+15 03132				17 11 56	+15 20.7	1	9.41		0.97		0.67		3	G5	196
155845	−19 04569				17 11 56	−19 48.4	1	8.41		1.87		2.02		2	M1 III	262
155847	−20 04686				17 11 56	−21 02.5	1	9.87		0.19		-0.28		1	B5 (III)	1770
	−59 06926				17 11 56	−59 26.1	2	10.84	.101	-0.03	.011	-0.89	.003	7	B3e	400,1586
156074	+42 02811				17 11 57	+42 09.8	3	7.60	.006	1.15	.005	0.90	.009	4	C0	1238,3035,8005
155775	−38 11680	LSS 3995			17 11 57	−38 09.4	4	6.70	.020	-0.01	.007	-0.83	.008	11	B1 V	540,954,976,2012
155805	−33 11874				17 11 59	−33 08.2	1	8.32		0.05		-0.28		2	B7 Ib/II	1085
156110	+45 02509				17 12 00	+45 25.8	1	7.56		-0.17		-0.74		3	B3 Vn	399
		LSS 3999			17 12 00	−31 31.0	1	11.81		0.55		-0.35		2	B3 Ib	918
155806	−33 11875	HR 6397			17 12 02	−33 29.5	6	5.53	.008	0.01	.008	-0.97	.014	17	O8 Ve	540,681,815,976,1085,2007
155967	+14 03206				17 12 03	+14 36.5	2	7.43	.000	0.44	.005	-0.06	.005	2	F6 V	196,271
156162	+54 01869	IDS17109N5414		AB	17 12 03	+54 11.8	1	6.90		0.29		0.12		2	F0	3016
155883	−14 04583				17 12 03	−14 40.3	1	10.06		0.36		0.09		1	B8/9 V	1770
155823	−30 13953				17 12 03	−30 28.9	1	8.27		1.79		1.72		2	M3 III	918
155824	−31 13753				17 12 03	−31 28.7	1	9.81		0.22		-0.17		2	B8 II	918
155865	−19 04570				17 12 04	−19 29.2	2	9.52	.017	0.19	.009	0.11	.016	3	B9 IV	262,1770
		HA 85 # 483			17 12 06	+15 06.6	1	10.34		1.67		1.92		3		196
156295	+63 01336	HR 6421			17 12 06	+62 55.9	1	5.56		0.21		0.05		2	F0 IV	3023
	+59 01796	G 226 - 50			17 12 07	+59 05.2	1	9.95		1.12		0.96		3	K4	7010
155849	−26 12025	IDS17092S2627		D	17 12 07	−26 28.5	1	7.64		1.28				1	G8 III/IV	1642
	−48 11542				17 12 08	−48 48.5	1	10.89		0.67		0.06		2		540
319585					17 12 09	−33 42.9	1	10.41		0.17		-0.06		2	A0	1085
155756	−45 11330	LSS 3993			17 12 09	−45 51.3	3	9.29	.012	0.53	.008	-0.46	.011	4	B0 Ia	540,976,1737
		NGC 6316 sq1 1			17 12 11	−28 04.0	1	11.52		0.53		0.37		2		156
155826	−38 11686	HR 6398		★ AB	17 12 11	−38 32.0	7	5.96	.004	0.58	.003	0.07	.009	23	F7 V + G2 V	15,219,1020,2012,2012*
	−39 11236	LSS 3997			17 12 12	−39 30.5	2	10.21	.055	1.24	.095	0.21	.040	4	B0 Ib	954,1737
155850	−31 13758				17 12 13	−31 53.8	2	9.49	.000	0.06	.005	-0.59	.005	3	B2 V	121,918
319598					17 12 13	−33 57.6	1	11.01		0.10		-0.33		2	A0	1085
155867	−27 11538				17 12 14	−27 54.9	1	8.54		0.86		0.55		2	G8 IV	156
155832	−33 11881				17 12 14	−33 12.9	1	9.84		0.09		-0.15		2	B8/9 V	1085
	+29 02963				17 12 15	+29 07.6	1	8.46		0.62		0.08		5	G0 V	3026
323059	−39 11238	LSS 4000			17 12 15	−39 44.5	1	9.87		0.80		-0.19		2	B0.5II-III	540
155886	−26 12026	HR 6402		★ AB	17 12 16	−26 31.8	8	4.32	.015	0.86	.007	0.49	.017	27	K0 V + K1 V	15,150,938,1013,1075*
155855	−34 11589				17 12 16	−34 26.6	1	9.39		0.23		-0.30		2	B5 III	1085
155868	−30 13959				17 12 17	−30 12.6	1	8.36		0.20		0.14		2	A5 V	918
155869	−31 13761				17 12 18	−31 41.8	1	10.18		0.18		-0.17		2	B5 V	918
155851	−32 12518	LSS 4003			17 12 18	−32 38.0	5	8.16	.019	0.09	.042	-0.77	.018	10	B(3/5)nne	540,918,976,1085,1586
155555	−66 03080	V824 Ara		★ A	17 12 18	−66 53.7	4	6.72	.020	0.78	.015	0.30	.009	31	K1 Vp	1621,1641,1642,2033
		Sco sq 2 # 177			17 12 19	−33 20.5	1	11.21		0.12		-0.35		2		1085
155927	−20 04689				17 12 20	−20 13.0	1	9.36		0.37		0.17		2	A7/9 (V)	262
156013	+15 03133				17 12 21	+15 22.0	1	9.39		0.35		0.20		3	F5	196
155928	−20 04690				17 12 21	−20 14.3	1	9.14		0.24		0.19		2	A3 (V)	262
156014	+14 03207	HR 6406, α Her		★ AB	17 12 22	+14 26.8	6	3.06	.022	1.45	.011	1.01	.005	20	M5 Ib-II	15,667,1363,8003,8015,8032
		G 170 - 27			17 12 23	+21 30.7	1	16.51		0.32		-0.52		1		3060
319607					17 12 25	−34 17.3	1	10.48		0.24		0.08		2	A0	1085
155887	−30 13964				17 12 26	−31 03.9	1	10.09		0.44		0.31		2	A5/7 IV/V	918
155888	−33 11885				17 12 28	−33 37.1	1	8.69		0.06		-0.49		2	B4 V	1085
156223	+52 02037				17 12 29	+52 21.5	1	8.26		0.32		0.07		2	A5	1375
155970	−14 04585	HR 6404		★ AB	17 12 30	−14 31.7	1	5.99		1.10				4	K1 III	2007
		Sco sq 2 # 179			17 12 30	−32 36.8	1	11.11		0.29		0.07		2		1085
155910	−31 13763	IDS17093S3156		AB	17 12 31	−31 59.4	2	8.23	.005	0.15	.005	-0.32	.005	4	B7 III/IV	918,1085
		Sco sq 2 # 181			17 12 31	−33 47.4	1	12.16		0.23		0.13		2		1085
		Sco sq 2 # 180			17 12 31	−33 59.3	1	10.48		0.09		-0.46		2		1085
156280	+55 01921				17 12 32	+55 31.9	1	7.61		0.41		0.00		2	F5	3016
155889	−33 11887	LSS 4006		★ AB	17 12 33	−33 40.9	5	6.55	.007	-0.02	.007	-0.88	.010	11	B1/2 Ib/II	219,540,976,1085,1732
155890	−33 11886	IDS17093S3356		AB	17 12 33	−33 59.5	1	9.30		0.05		-0.54		2	B5 V	1085
	−32 12527				17 12 35	−32 39.0	1	10.95		0.22		-0.11		2		1085
		Sco sq 2 # 185			17 12 35	−33 59.1	1	12.18		0.20		0.05		2		1085
	−32 12528				17 12 36	−32 40.0	1	10.42		0.21		-0.21		2		1085
155873	−42 11979	LSS 4002			17 12 36	−42 44.3	3	8.24	.007	0.65	.014	-0.34	.012	7	B0 Ib	540,976,1586
		Sco sq 2 # 187			17 12 37	−32 48.2	1	11.93		0.31		0.18		2		1085
155929	−32 12530				17 12 38	−32 31.3	3	9.66	.015	0.17	.005	-0.55	.010	6	B5	540,918,1085
155940	−30 13968	HR 6403			17 12 40	−30 09.3	2	6.21	.000	-0.02	.010	-0.09		6	B9/9.5V	918,2007
		L 421 - 36			17 12 42	+26 58.7	1	12.37		1.52		1.24		4		1663
319581					17 12 42	−33 30.9	1	10.69		0.17		-0.02		2	A0	1085
155896	−42 11983	IDS17092S4214		AB	17 12 44	−42 17.0	4	6.75	.030	0.13	.015	-0.57	.004	10	B2/3 (V)nne	219,401,1586,2012
155930	−33 11890				17 12 45	−33 08.0	1	10.02		0.22		-0.23		2	B5 IV/V	1085
155912	−38 11696				17 12 45	−38 16.9	1	8.26		0.06		-0.28		8	B7 V	1732
156094	+15 03137				17 12 46	+15 27.1	1	8.39		1.59				3	K5	882
	+15 03136				17 12 47	+14 57.2	2	10.54	.000	0.70	.005	0.20	.005	6		196,271
		HA 85 # 270			17 12 48	+14 53.7	1	11.67		1.70				5	M3	271
155941	−32 12533				17 12 48	−32 15.8	2	9.89	.009	0.26	.009	0.16	.024	4	A0 V	918,1085

Table 1 933

HD	DM	Other Id	N Rem	α_{1950}	δ_{1950}	S	V	σ_V	B–V	σ_{B-V}	U–B	σ_{U-B}	n	Spectrum	References
		G 139 - 21		17 12 49	+05 01.6	2	14.66	.014	1.72	.009	1.19		5		419,1759
156558	+69 00900			17 12 49	+69 21.8	1	8.45		0.67		0.15		3	K0	3026
156023	−10 04462	IDS17101S1011	AB	17 12 49	−10 14.6	2	7.33	.025	0.46	.000	0.04		9	F5	173,3016
		LSS 4010		17 12 49	−35 40.5	2	11.71	.000	0.86	.013	-0.05	.018	5	B3 Ib	754,1085
155942	−33 11891			17 12 50	−33 25.3	1	9.52		0.10		-0.18		2	B8/9 V	1085
319579				17 12 51	−33 24.1	1	10.25		0.10		-0.16		2	A0	1085
155913	−42 11986	LSS 4007		17 12 52	−42 36.8	3	8.26	.010	0.48	.003	-0.55	.008	5	O5/6	401,540,976
		HA 85 # 382		17 12 54	+14 57.2	1	11.49		1.20		1.25		3	G9	271
		Sco sq 2 # 195		17 12 54	−32 34.6	1	11.24		0.26		-0.31		2		1085
		LSS 4009		17 12 55	−39 39.0	2	10.54	.020	0.88	.015	-0.10	.030	4	B3 Ib	954,1737
156034	−9 04525	IDS17102S0942	AB	17 12 56	−09 45.2	1	6.93		0.83				4	F5	173
156164	+25 03221	HR 6410	⋆ A	17 12 59	+24 53.8	11	3.12	.010	0.08	.004	0.07	.007	61	A3 IV	1,15,374,667,1022*
155974	−35 11426	HR 6405	⋆ A	17 13 01	−35 41.4	5	6.11	.006	0.48	.007	-0.02	.003	14	F6 V	15,219,688,1311,2012
	+15 03139			17 13 03	+15 22.8	1	10.38		0.85		0.42		2	G5	271
	−26 12035			17 13 04	−26 12.4	1	9.71		1.29				1	K2	1642
319597	−33 11899			17 13 04	−33 56.2	1	11.66		0.44		0.34		1	A2	121
156057	−13 04577			17 13 06	−13 26.3	1	9.61		0.36		0.25		1	B9.5 V	1770
	+15 03140			17 13 07	+15 13.4	1	11.46		1.40		1.67		5		271
155959	−40 11270	LSS 4012		17 13 08	−40 48.7	2	8.73	.024	0.64	.000	-0.31	.020	5	B1 II	540,976
156144	+15 03141			17 13 09	+14 58.0	3	8.28	.007	1.05	.012	0.86	.005	8	K0	196,271,882
156026	−26 12036	V2215 Oph	⋆ C	17 13 09	−26 28.6	8	6.33	.011	1.14	.015	1.04	.030	15	K5 V	219,678,1013,1075*
		Sco sq 2 # 196		17 13 09	−32 24.2	1	11.82		0.46		0.29		2		1085
327297				17 13 09	−42 15.1	1	10.47		0.22		-0.56		2	A0	1539
		HA 85 # 282		17 13 10	+14 55.4	1	12.32		0.78		0.44		2	G0	271
	−36 11378	LSS 4014		17 13 10	−36 21.0	2	11.31	.005	1.02	.005	0.02	.005	4		754,1085
156389	+56 01959	IDS17122N5615	AB	17 13 11	+56 11.4	2	7.81	.035	0.45	.000	-0.09		6	F2	1381,3030
319618				17 13 12	−34 27.8	1	10.96		0.27		-0.06		2	B8	1085
	+14 03210			17 13 16	+14 47.1	1	8.89		1.37		1.50		3	K5	196
		LSS 4015		17 13 16	−38 19.1	2	10.37	.040	0.98	.035	0.00	.035	4	B3 Ib	954,1737
156283	+36 02844	HR 6418		17 13 18	+36 51.9	7	3.16	.018	1.44	.010	1.67	.011	19	K3 IIab	15,1080,1363,3034*
		G 19 - 18		17 13 18	−02 46.7	2	11.52	.005	1.41	.005	1.24	.019	3	K5	1620,3062
156004	−32 12538	LSS 4018		17 13 18	−32 17.1	5	7.81	.007	0.01	.009	-0.82	.012	8	B2 II	219,540,918,976,1085
156077	−16 04461			17 13 19	−16 12.1	1	9.77		0.18		0.00		1	B9 V	1770
319565				17 13 20	−33 01.1	1	11.67		0.21		-0.01		2	K2	1085
319587				17 13 21	−33 40.6	1	10.23		0.15		-0.24		2	B8	1085
156039	−29 13344			17 13 22	−29 40.6	1	9.04		0.59		0.09		2	F5 V	918
319586				17 13 22	−33 43.0	1	10.21		0.19		-0.37		2	B8	1085
323075	−40 11273			17 13 22	−40 08.8	2	10.52	.078	0.84	.078	-0.13	.039	3	F8	954,1737
	−10 04463			17 13 24	−10 50.0	1	10.60		1.03		0.71		2	K7 V	3072
156115	−15 04502			17 13 24	−15 10.2	1	6.59		1.79		2.48		6	K5 III	1024
	−35 11431			17 13 24	−35 54.2	2	11.70	.004	0.74	.018	-0.16	.004	5		754,1085
	−19 04577			17 13 25	−19 21.0	1	10.19		1.66		1.77		2	K4 III	262
156040	−31 13784			17 13 25	−31 46.5	1	10.03		0.18		-0.22		3	Ap Si	918
		Sco sq 2 # 201		17 13 25	−32 37.0	1	11.76		0.42		0.00		2		1085
156095	−19 04578	IDS17106S1913	AB	17 13 26	−19 16.5	1	8.53		0.44		0.17		2	A9 V	262
		Sco sq 2 # 202		17 13 26	−33 52.3	1	11.84		0.40		0.12		2		1085
		Sco sq 2 # 27		17 13 26	−34 24.3	1	12.51		0.44		0.11		2		1085
		HA 85 # 287		17 13 27	+14 47.4	1	11.25		0.66		0.17		3		196
		Sco sq 2 # 203		17 13 27	−33 42.8	1	12.04		0.30		0.03		2		1085
155985	−44 11564	LSS 4013		17 13 27	−44 43.5	8	6.46	.007	0.25	.006	-0.68	.007	37	B0.5Ib	116,278,540,976,1075,1586*
319620				17 13 28	−34 43.6	1	11.17		0.55		0.29		2	A0	1085
		Sco sq 2 # 204		17 13 31	−33 12.3	1	12.18		0.21		0.06		2		1085
156041	−35 11432	LSS 4021		17 13 32	−35 25.6	4	9.34	.004	0.57	.013	-0.39	.011	10	B0 Ia/ab	540,754,976,1085
	−28 12986			17 13 33	−28 12.8	1	9.67		0.36		0.15		2		156
156064	−31 13789			17 13 33	−31 33.2	1	9.88		0.28		0.27		3	A1 III	918
156065	−32 12542			17 13 33	−32 19.3	1	8.95		1.23		1.09		2	G8/K0 III	918
		Ross 817		17 13 34	−23 09.5	1	11.41		0.90		0.59		2		1696
155556	−74 01618			17 13 35	−74 16.9	1	8.15		1.55		1.90		2	K3 III	1730
156066	−33 11913			17 13 36	−33 37.1	1	10.46		0.24		0.04		2	B8/9 III	1085
		Sco sq 2 # 31		17 13 36	−35 54.1	1	11.77		0.65		0.22		2		1085
156225	+11 03148			17 13 37	+10 56.8	1	8.97		0.26		0.18		3	A2	1776
156284	+23 03070	HR 6419		17 13 37	+23 47.8	1	5.96		1.31				2	K2 III	71
	−32 12543			17 13 37	−32 11.0	1	11.67		0.26		-0.15		2		1085
319631				17 13 37	−35 26.6	1	10.82		0.45		-0.01		2	B9	1085
	−28 12989			17 13 38	−28 19.5	1	10.44		0.30		0.16		2		156
	+19 03268			17 13 39	+19 03.3	1	10.36		1.55		1.20		2	M2	1746
	−38 11711	LSS 4022		17 13 39	−38 32.9	2	10.43	.049	0.86	.015	-0.17	.010	3	O9.5 III	954,1737
	+11 03149	G 139 - 23		17 13 41	+11 07.7	1	10.82		1.38		1.27		2	M2.5	7010
156096	−30 13986			17 13 42	−30 17.8	1	7.28		0.30		0.01		2	F0 V	918
156083	−33 11914			17 13 42	−33 53.8	1	10.37		0.15		-0.28		2	B8 II/III	1085
156208	+2 03283	HR 6412		17 13 43	+02 14.4	3	6.17	.016	0.23	.004	0.14		29	A2 V	175,1088,8097
		GD 360		17 13 44	+33 16.4	1	14.54		-0.11		-0.89		1	DA	1727
	−32 12544			17 13 44	−32 16.0	1	11.34		0.29		0.10		2		1085
156097	−31 13792			17 13 45	−31 05.8	1	9.51		0.69		0.18		2	G3/5 III	918
		Sco sq 2 # 208		17 13 45	−32 50.7	1	12.34		0.44		0.27		2		1085
		Sco sq 2 # 210		17 13 45	−33 16.4	1	11.24		0.19		-0.28		2		1085
319628	−35 11435	LSS 4024		17 13 45	−35 11.3	2	10.82	.004	0.78	.004	-0.16	.004	5		754,1085
		Sco sq 2 # 211		17 13 46	−33 27.2	1	11.77		0.15		-0.08		2		1085
156117	−30 13987			17 13 48	−30 17.1	1	9.01		0.48		0.22		2	F2 V	918
156098	−32 12545	HR 6409		17 13 48	−32 36.5	3	5.54	.022	0.51	.010	0.04	.005	8	F6 V	918,2007,3053

HD	DM	Other Id	N	Rem	α_{1950}	δ_{1950}	S	V	σ_V	B−V	σ_{B-V}	U−B	σ_{U-B}	n	Spectrum	References
156099	−33 11919				17 13 48	−33 13.4	1	8.50		0.09		−0.21		2	B8/9 III	1085
156081	−33 11917				17 13 48	−33 30.3	1	10.00		0.04		−0.56		2	B9	1085
156100	−33 11918				17 13 50	−34 03.9	1	9.67		0.19		0.02		2	B9 V	1085
		G 181 - 33			17 13 52	+30 41.6	1	14.66		1.40				1		1759
156118	−31 13794				17 13 52	−31 24.8	1	9.66		0.13		−0.42		2	B3 III	918
156044	−44 11572				17 13 52	−44 43.4	2	8.34	.002	0.13	.000	0.10	.000	12	A1 V	977,1075
	+15 03144				17 13 53	+15 05.6	2	10.59	.005	1.04	.000	0.82	.015	5		196,271
156133	−30 13989				17 13 54	−30 42.7	1	8.66		1.79		2.18		2	M1 III	918
156119	−32 12547				17 13 54	−32 43.8	1	9.75		0.21		−0.40		2	B3 III	1085
		LP 747 - 18			17 13 55	−12 58.0	1	11.66		0.74		0.08		2		1696
329807	−44 11573				17 13 56	−44 33.4	2	9.54	.006	1.29	.002	1.14		9	K2	863,1075
156559	+58 01713				17 13 57	+58 26.3	1	7.90		0.38		−0.04		2	F2	3016
		Bo 13 - 9			17 13 57	−35 29.3	1	13.05		0.97		0.48		1		412
		Bo 13 - 10			17 13 57	−35 29.6	1	14.22		0.71		0.44		1		412
327341	−43 11560				17 13 57	−43 21.1	1	9.81		0.53		−0.26		3	B3	1586
319571					17 13 58	−33 06.0	1	10.33		0.21		−0.17		2	B9 V	1085
319630	−35 11438	LSS 4029			17 13 58	−35 21.1	2	10.68	.000	0.72	.022	−0.17	.004	5	A0	754,1085
156247	+1 03408	HR 6414, U Oph		* A	17 13 59	+01 15.9	4	5.91	.005	0.05	.007	−0.46	.003	17	B5 Vnn+B5 V	588,1079,1088,3024
156247	+1 03408	IDS17114N0119		B	17 13 59	+01 15.9	1	12.14		0.83				5		3024
156362	+27 02780				17 13 59	+27 11.4	1	6.56		1.18		1.16		2	K2 III	3077
		Bo 13 - 12			17 13 59	−35 29.4	1	13.38		0.68		0.34		1		412
		LSS 4030			17 13 59	−35 55.4	2	11.10	.004	0.79	.009	−0.05	.004	5	B3 Ib	754,1085
319588					17 14 00	−33 45.0	1	11.25		0.30		−0.26		2	B8	1085
156134	−35 11441	Bo 13 - 2			17 14 00	−35 29.7	5	8.06	.008	0.65	.012	−0.37	.025	12	B0 Ib	412,540,754,976,1085
		Bo 13 - 5			17 14 00	−35 30.	1	12.35		0.66		−0.15		2		412
		Bo 13 - 8			17 14 00	−35 30.	1	12.54		1.19		1.06		1		412
		Bo 13 - 11			17 14 00	−35 30.	1	13.92		0.85		0.61		1		412
		Bo 13 - 4			17 14 01	−35 28.3	1	12.69		0.81		0.69		2		412
322998	−37 11439	LSS 4027			17 14 01	−37 44.5	3	10.25	.054	0.50	.016	−0.49	.018	7	B3	432,540,1586
156227	−6 04575	HR 6413			17 14 02	−06 11.5	1	6.08		1.10		0.97		8	K0	1088
156266	−0 03255	HR 6415		* AB	17 14 03	−00 23.4	5	4.72	.007	1.14	.010	1.13	.015	20	K2 III	15,1075,1088,1363,8015
156152	−31 13799				17 14 04	−32 00.7	1	8.13		0.64		0.16		2	G3 V	918
		LSS 4028			17 14 05	−39 33.0	1	10.55		0.87		−0.02		1	B3 Ib	954
156341	+16 03139				17 14 06	+16 42.9	1	7.81		0.00		−0.04		3	A0	3016
156154	−35 11445	Bo 13 - 1			17 14 06	−35 29.0	7	8.05	.011	0.61	.010	−0.45	.022	17	O7/8	412,540,754,976,1085*
		LSS 4035			17 14 07	−35 23.4	2	10.73	.004	0.63	.022	−0.37	.004		B3 Ib	754,1085
156085	−45 11362				17 14 07	−45 13.4	2	9.17	.014	0.22	.002	−0.49	.011	3	B2 II	540,976
319564	−32 12554				17 14 08	−32 52.8	1	10.31		0.10		−0.49		2	A0	1085
		Bo 13 - 3			17 14 08	−35 28.1	2	11.77	.010	0.58	.005	−0.24	.005	5	B3	412,754
156184	−30 13996	IDS17110S3003		AB	17 14 09	−30 06.7	2	6.96	.010	0.82	.000	0.44		6	K1 III +A7	918,2012
319572					17 14 09	−33 07.7	1	10.61		0.12		−0.28		2	A0	1085
322961	−36 11385	LSS 4034			17 14 09	−36 41.4	1	10.06		0.76		−0.20		2	B8	754
156070	−49 11306				17 14 09	−49 24.0	1	7.53		−0.01		−0.59		1	B2 II/III	219
		Sco sq 2 # 39			17 14 10	−34 22.4	1	11.92		0.50		0.03		2		1085
156103	−44 11576				17 14 10	−44 19.4	2	8.67	.020	0.57	.005	−0.29	.018	3	B2 II	540,976
	−32 12556				17 14 11	−32 09.1	1	9.54		1.60		2.03		2		918
		Bo 13 - 6			17 14 11	−35 30.3	1	12.19		0.84		0.48		2		412
323078	−37 11657				17 14 13	−40 01.2	2	10.89	.040	0.83	.035	−0.10	.005	4	F5	954,1737
	−32 12558				17 14 14	−32 05.3	1	10.73		0.13		−0.33		2		1085
156171	−33 11926				17 14 14	−33 40.7	1	10.26		0.21		−0.33		2	B5 II/III	1085
		Bo 13 - 7			17 14 14	−35 30.8	1	12.29		1.61		2.00		1		412
156185	−30 13998				17 14 15	−30 22.3	1	8.98		1.66		2.08		2	K4/5 (III)	918
156186	−31 13802				17 14 15	−31 26.9	1	10.15		0.49		0.34		2	A9 V	918
		Sco sq 2 # 220			17 14 15	−32 45.0	1	11.57		0.35		0.04		2		1085
319626	−34 11612	LSS 4038			17 14 15	−35 02.1	2	11.21	.013	0.75	.009	−0.06	.004	5	B5	754,1085
	+15 03145				17 14 16	+15 07.9	1	10.95		0.58		0.12		5		271
156187	−32 12559				17 14 16	−32 06.7	2	9.23	.014	0.09	.005	−0.62	.019	4	B3/5 V	918,1085
319592					17 14 16	−33 53.5	1	11.49		0.28		−0.18		2		1085
		G 139 - 27			17 14 17	+08 06.8	3	11.48	.019	1.55	.032	1.14	.123	7	M3.5	265,906,7009
156342	+14 03213	IDS17120N1447		AB	17 14 17	+14 43.7	2	8.28	.010	0.67	.005	0.26	.000	5	G5	196,271
156198	−30 13999				17 14 17	−30 39.8	1	9.37		0.17		0.05		2	B9 V	918
156377	+18 03336				17 14 18	+18 04.4	1	7.12		−0.04		−0.30		2	B9	1733
	−32 12560				17 14 18	−33 01.5	1	11.62		0.28		0.19		2		1085
		Sco sq 2 # 41			17 14 18	−36 26.1	1	11.92		0.84		0.40		2		1085
156212	−27 11554	LSS 4044			17 14 19	−27 42.8	6	7.91	.011	0.51	.016	−0.53	.024	9	B0 Iab	219,540,918,954,976,1737
319617	−34 11613	LSS 4040			17 14 19	−34 31.5	1	9.67		0.97		−0.07		2	F2	1085
	+14 03214				17 14 22	+14 38.5	1	10.48		0.40		0.14		3	F0	196
		HA 85 # 540			17 14 22	+15 09.5	1	10.93		1.17		1.06		3		196
		Sco sq 2 # 226			17 14 22	−32 56.1	1	12.19		0.52		0.38		2		1085
319591					17 14 24	−33 54.0	1	10.29		0.46		−0.13		2	B5 IV	1085
156201	−35 11448	LSS 4043			17 14 25	−35 10.2	5	7.89	.007	0.69	.014	−0.36	.012	12	B0.5Ia/ab	540,754,976,1085,8100
156157	−43 11567				17 14 25	−43 31.5	1	7.02		−0.04		−0.29		2	B8 V	401
156137	−45 11364	IDS17107S4559		A	17 14 25	−46 02.3	1	8.88		0.11		−0.33		2	B5/7 II	401
		Sco sq 2 # 228			17 14 26	−32 39.8	1	11.97		0.37		0.10		2		1085
155694	−74 01619				17 14 26	−74 18.3	1	8.90		1.08		0.88		2	G8/K0 III	1730
319562					17 14 27	−32 50.6	1	10.91		0.27		−0.22		2	F0	1085
156172	−41 11445	LSS 4037			17 14 27	−42 00.5	2	8.18	.005	0.39	.004	−0.56	.013	3	O9.5Ia/ab	540,976
	−23 13290	NGC 6325 sq1 2			17 14 28	−23 47.8	1	10.63		1.91				3		438
	−32 12563				17 14 28	−32 25.0	1	10.79		0.28		0.10		2		1085
319632					17 14 28	−35 26.7	1	10.91		0.74		0.52		2	B9	1085

Table 1

HD	DM	Other Id	N	Rem	α_{1950}	δ_{1950}	S	V	σ_V	B–V	σ_{B-V}	U–B	σ_{U-B}	n	Spectrum	References
		LSS 4041			17 14 29	−38 31.6	1	10.85		1.60		0.87		2	B3 Ib	954
156229	−30 14004				17 14 30	−30 43.0	1	8.96		0.66		0.16		2	G2/3 V	918
156213	−33 11935				17 14 31	−33 33.6	1	9.64		0.18		-0.42		2	B3 III	1085
156189	−42 12006	NGC 6322 - 17			17 14 31	−42 50.7	2	7.64	.000	0.14	.004	0.05	.016	8	A3 III/IV	310,412
156230	−31 13806				17 14 32	−31 16.4	1	10.17		0.27		-0.10		2	B9 IV	918
156252	−26 12048	IDS17114S2631		A	17 14 33	−26 34.6	1	6.96		-0.03		-0.14		2	B9.5V	1770
156252	−26 12048	IDS17114S2631		AB	17 14 33	−26 34.6	2	6.82	.002	0.02	.000	-0.11	.000	20	B9.5V	978,1770
156252	−26 12048	IDS17114S2631		B	17 14 33	−26 34.6	1	9.07		0.45		-0.02		1		1770
319616					17 14 33	−34 42.3	1	11.06		0.42		0.27		2	F0	1085
156289	−18 04483				17 14 34	−19 06.0	1	10.73		0.28		0.22		1	B9/A0 (II)	1770
319560					17 14 35	−32 36.5	2	10.86	.005	0.20	.009	-0.26	.014	4		918,1085
156454	+26 02990	IDS17126N2641		ABC	17 14 36	+26 38.0	1	8.81		0.72		0.16		1	G2 V	3026
319561	−32 12568				17 14 37	−32 42.4	1	10.34		0.70		0.17		1	K5	121
319635	−35 11451	LSS 4046			17 14 37	−35 46.8	2	10.68	.000	0.71	.013	-0.23	.009	5	G0	754,1085
156299	−17 04763				17 14 38	−17 49.0	1	9.46		0.04		-0.11		1	B9 V	1770
156254	−31 13808				17 14 38	−31 54.7	2	9.27	.019	0.09	.005	-0.50	.005	3	B9 V	121,918
		NGC 6322 - 11			17 14 39	−42 48.1	1	12.61		0.57		0.22		1		412
155875	−69 02715	HR 6400		★ AB	17 14 39	−69 59.5	5	6.53	.008	0.60	.000	0.14	.005	22	F8/G2 IV	15,1075,2012,2038,3077
		NGC 6325 sq1 7			17 14 42	−23 49.4	1	11.77		0.82				3		438
156256	−34 11617				17 14 42	−34 09.7	3	8.25	.007	0.05	.005	-0.60	.015	5	B3 V	540,976,1085
		Sco sq 2 # 47			17 14 42	−36 21.7	2	11.85	.015	0.96	.020	0.03	.035	3		754,1085
		NGC 6322 - 18			17 14 42	−42 50.7	2	12.01	.012	0.64	.033	0.23	.033	2		310,412
323016	−38 11731	LSS 4049			17 14 43	−38 22.9	4	9.49	.021	0.75	.013	-0.26	.014	11	B0.5II-Ib	432,540,914,976
156430	+15 03147				17 14 44	+15 52.5	1	7.43		1.05				1	K0	882
		NGC 6322 - 48			17 14 44	−42 52.7	2	13.16	.040	0.62	.010	0.33	.160	2		310,412
156300	−27 11560				17 14 45	−27 43.1	1	8.65		0.30		0.07		2	Ap Si	918
		NGC 6322 - 44			17 14 45	−42 51.7	1	13.40		0.65		0.58		1		412
	−42 12008	NGC 6322 - 50			17 14 45	−42 53.3	2	10.26	.033	0.41	.038	-0.46	.005	7		310,412
156319	−17 04764				17 14 46	−17 46.6	1	9.21		0.10		0.07		2	B9 IV/V	1770
		NGC 6322 - 87			17 14 46	−42 55.2	1	13.70		1.22				1		310
156091	−59 06954	HR 6408		★ A	17 14 46	−59 38.5	1	5.91		1.37				4	K2 III	2035
	+15 03148				17 14 47	+15 01.6	1	9.54		1.19		1.17		3		196
156431	+15 03149				17 14 47	+15 19.5	1	8.64		0.21		0.12		2	A2	1648
156582	+44 02681				17 14 47	+44 48.8	1	8.56		0.34		0.00		2	A0	1601
156269	−33 11937				17 14 47	−33 09.8	3	9.44	.010	0.05	.004	-0.65	.017	5	B2 III	540,976,1085
		NGC 6322 - 49			17 14 47	−42 52.9	1	13.59		0.64		0.50		1		412
156432	+14 03215				17 14 48	+14 44.0	1	8.27		0.50		0.04		3	F8	196
		HA 85 # 436			17 14 48	+14 57.2	1	11.18		1.11		0.84		3		196
		NGC 6322 - 53			17 14 48	−42 54.0	1	12.86		0.66		0.15		1		412
		NGC 6322 - 83			17 14 48	−42 55.5	2	11.92	.009	0.46	.044	-0.13	.048	6		310,412
		BPM 24754			17 14 48	−54 43.	1	15.55		0.27		-0.59		1		3065
	−23 13292	NGC 6325 sq1 3			17 14 49	−23 35.0	1	11.08		0.40				3		438
		NGC 6322 - 7			17 14 49	−42 47.8	1	13.10		0.49		0.14		1		412
		NGC 6322 - 81			17 14 49	−42 54.5	1	12.27		0.60		0.03		1		412
327327	−42 12009	NGC 6322 - 51			17 14 50	−42 53.3	2	9.54	.024	0.41	.038	-0.46	.005	7	B8	310,412
		NGC 6322 - 85			17 14 50	−42 56.2	1	11.84		0.45		-0.14		1		412
	−50 11162				17 14 50	−50 28.9	2	10.57	.014	0.00	.009	-0.84		6	B0.5II-III	540,1021
		Ross 818			17 14 51	−20 40.5	1	11.83		0.69		0.10		3		1696
		Sco sq 2 # 235			17 14 51	−32 32.8	1	12.40		0.47		0.23		2		1085
156233	−42 12011	NGC 6322 - 52			17 14 51	−42 52.8	2	9.09	.048	0.41	.038	-0.53	.019	7	B0	310,412
319590					17 14 52	−33 48.2	1	10.63		0.27		-0.41		2	B8	1085
		NGC 6322 - 22			17 14 52	−42 51.5	2	10.97	.010	0.40	.029	-0.32	.005	7		310,412
156234	−42 12012	NGC 6322 - 82			17 14 52	−42 55.7	3	7.74	.020	0.35	.015	-0.60	.015	9	O9.5II	310,401,412
		NGC 6322 - 86			17 14 52	−42 56.1	1	12.16		1.65		1.52		1		412
	+15 03150				17 14 53	+14 59.1	1	9.98		0.42		0.04		3		196
		NGC 6322 - 20			17 14 53	−42 49.9	2	12.05	.016	0.36	.008	-0.06	.016	8		310,412
		NGC 6322 - 21			17 14 53	−42 51.2	2	11.00	.010	0.39	.020	-0.32	.025	8		310,412
		NGC 6322 - 55			17 14 53	−42 53.6	1	12.95		0.71		0.54		1		412
		Sco sq 2 # 237			17 14 54	−33 43.9	1	11.79		0.67		0.14		2		1085
		Sco sq 2 # 48			17 14 54	−35 50.2	1	12.04		0.66		0.14		2		1085
		NGC 6325 sq1 6			17 14 55	−23 38.6	1	11.55		0.46				3		438
319634					17 14 55	−35 43.0	1	11.13		0.50		-0.26		2	A2	1085
		NGC 6322 - 57			17 14 55	−42 53.1	1	12.91		0.61		0.37		1		412
		LP 447 - 34			17 14 56	+18 10.3	1	11.15		0.93		0.71		1		1696
		LP 747 - 19			17 14 56	−11 46.1	1	12.89		1.46		1.25		1		3062
319573					17 14 56	−33 08.3	1	11.53		0.37		0.23		1		121
319575					17 14 56	−33 27.1	1	11.18		0.28		-0.34		2	G5	1085
		LP 747 - 20			17 14 57	−11 46.6	2	12.82	.030	1.48	.010	1.14	.105	2		1773,3062
156350	−24 13255	HR 6425		★ A	17 14 57	−24 13.9	1	6.80		0.51		0.05		2	G8/K0 II	3051
		NGC 6322 - 60			17 14 57	−42 52.6	1	13.50		1.75		1.82		1		412
156349	−24 13255	HR 6424		★ A	17 14 58	−24 14.1	1	5.26		0.97		0.98		1	K0 II/III	3051
156349	−24 13255	IDS17119S2411		★ AB	17 14 58	−24 14.1	1	4.92		0.90				4	K0 II/III ⊢	2007
156236	−46 11367				17 14 58	−46 44.8	7	7.21	.005	0.62	.007	0.56	.029	21	F0 III	219,460,657,727,977*
156839	+64 01184				17 14 59	+64 29.5	1	8.74		0.19		-0.08		1	A2	1776
		NGC 6322 - 76			17 15 00	−42 54.8	1	13.36		0.59		0.31		1		412
156483	+15 03153				17 15 01	+15 14.1	1	8.38		-0.02		-0.14		3	A0	196
		Sco sq 2 # 239			17 15 01	−32 14.9	1	11.36		0.24		0.05		2		1085
		NGC 6322 - 78			17 15 01	−42 53.7	2	12.89	.013	0.49	.022	0.51	.048	6		310,412
156320	−30 14015				17 15 02	−30 23.1	1	9.48		0.48		0.21		2	A9 V	918
319659					17 15 02	−33 38.0	1	10.96		0.23		0.03		2	A2	1085

HD	DM	Other Id	N Rem	α_{1950}	δ_{1950}	S	V	σ_V	B–V	σ_{B-V}	U–B	σ_{U-B}	n	Spectrum	References
		NGC 6322 - 25		17 15 02	−42 50.2	2	11.53	.021	1.20	.135	1.11	.017	7		310,412
		NGC 6322 - 64		17 15 02	−42 52.2	1	13.23		1.99		2.51		1		412
		NGC 6322 - 77		17 15 02	−42 54.2	2	11.99	.004	0.53	.031	−0.09	.013	6		310,412
		Sco sq 2 # 241		17 15 03	−32 31.1	1	10.36		0.15		−0.32		2		1085
		Sco sq 2 # 242		17 15 03	−34 10.9	1	12.04		0.29		0.12		2		1085
156365	−23 13297			17 15 04	−24 01.2	1	6.61		0.62				2	G3 V	1619
156327	−34 11622	LSS 4057	⋆ AB	17 15 04	−34 21.4	1	9.32		0.63		−0.07		2	WC	1085
		NGC 6322 - 63		17 15 04	−42 52.4	1	13.34		0.58		0.31		1		412
		L 413 - 156		17 15 04	−43 22.7	1	13.05		1.57		1.23		2		3078
	+49 02613			17 15 05	+49 37.5	1	9.16		0.71		0.20		2	G5	1375
156351	−28 13006	LSS 4061		17 15 05	−29 03.4	2	9.62	.002	0.36	.006	−0.62	.018	3	B1/2 Ib	540,976
156321	−32 12574			17 15 05	−32 16.4	4	8.15	.007	0.01	.004	−0.76	.011	7	B2 V	540,918,976,1085
156325	−32 12573	HR 6422	⋆ A	17 15 05	−32 30.1	7	6.36	.014	0.15	.011	−0.37	.013	20	B5 Vne	26,540,976,1085,1586*
		G 19 - 19		17 15 06	+04 07.9	2	10.90	.010	1.17	.005	1.15	.015	3	K5	1620,3062
319673				17 15 06	−33 48.7	1	10.27		0.39		−0.01		2	A0	1085
156323	−32 12575	IDS17119S3216	B	17 15 07	−32 19.7	1	9.32		0.08		−0.68		2		918
156322	−32 12575	LSS 4058	⋆ A	17 15 07	−32 19.9	1	9.01		0.06		−0.68		2		918
323015	−38 11742		A	17 15 07	−38 24.6	3	9.99	.012	0.78	.018	−0.28	.035	6		432,954,1737
323015	−38 11742		B	17 15 07	−38 24.6	1	11.44		0.81		−0.16		2		432
156271	−42 12015	NGC 6322 - 99		17 15 07	−42 56.4	4	9.28	.031	0.28	.022	−0.57	.015	10	B3 II/III	310,412,540,976
		NGC 6235 sq1 5		17 15 08	−23 37.9	1	14.36		1.30				2		869
		NGC 6325 sq1 5		17 15 08	−23 38.0	1	11.17		0.71				3		438
		NGC 6235 sq1 11		17 15 08	−23 38.3	1	14.33		1.64				1		869
156324	−32 12576	LSS 4060	⋆ AB	17 15 08	−32 21.1	2	8.75	.005	0.10	.009	−0.57	.009	4	B2 V	918,1085
		Sco sq 2 # 248		17 15 08	−32 57.4	1	11.65		0.19		−0.18		2		1085
156326	−33 11943			17 15 08	−34 03.4	1	9.94		0.13		−0.53		2	B2/3 III	1085
		NGC 6322 - 65		17 15 08	−42 52.5	1	13.03		1.46		0.99		1		412
		NGC 6322 - 200		17 15 10	−42 55.0	1	11.79		0.60		0.30		1		412
		Sco sq 2 # 250		17 15 11	−32 17.9	1	12.09		0.25		−0.04		2		1085
319658				17 15 11	−33 31.7	1	11.51		0.30		0.08		2	B8 V	1085
156292	−42 12018	NGC 6322 - 27		17 15 11	−42 50.4	5	7.51	.013	0.26	.010	−0.67	.020	14	O9.5II-III	219,310,412,540,976
156293	−43 11572	HR 6420		17 15 11	−44 04.7	11	5.76	.003	−0.05	.004	−0.13	.004	76	B9 V	15,116,125,278,863,977*
156367	−29 13363	IDS17120S2946	AB	17 15 12	−29 49.3	1	7.85		0.22		0.14		2	A3 V	918
		NGC 6341 sq1 5		17 15 13	+43 19.5	1	14.22		0.58		0.03		1		448
156353	−32 12578			17 15 13	−32 40.9	1	9.09		0.29		0.10		2	B9 V	1085
319646	−32 12577			17 15 13	−33 00.9	1	10.32		0.52		0.27		1	A3	121
		NGC 6322 - 40		17 15 13	−42 51.6	2	11.98	.009	0.55	.061	0.09	.078	6		310,412
		NGC 6322 - 75		17 15 13	−42 54.6	2	11.51	.004	0.23	.013	0.19	.004	6		310,412
156368	−30 14020			17 15 14	−30 31.4	1	9.24		1.23		1.12		2	G8/K0 III	918
	−42 12019	NGC 6322 - 39		17 15 14	−42 50.3	2	11.10	.017	1.01	.100	0.71	.126	6		310,412
		NGC 6322 - 28		17 15 14	−42 50.5	1	12.56		0.51		0.31		1		412
156275	−48 11588			17 15 14	−48 38.7	1	7.87		0.01		−0.24		2	B9 III/IV	401
322987	−37 11455	LSS 4059		17 15 15	−37 17.1	3	9.75	.007	0.89	.013	−0.17	.042	7	O5 V	432,954,1737
156274	−46 11370	HR 6416	⋆ A	17 15 15	−46 35.1	2	5.56	.033	0.78	.005	0.36	.005	2	G8/K0 V	1279,3078
156274	−46 11370	HR 6416	⋆ AB	17 15 15	−46 35.1	12	5.47	.007	0.81	.009	0.35	.011	59	G8/K0 V	15,116,278,863,977,977*
156274	−46 11370	IDS17115S4632	B	17 15 15	−46 35.1	2	8.70	.004	1.40	.004	0.89	.013	3		1279,3078
156274	−46 11370	IDS17115S4632	C	17 15 15	−46 35.1	1	13.34		1.03				2		1279
156380	−29 13364			17 15 16	−29 54.8	1	8.31		1.36		1.45		2	K2 IV	918
		HM 1 - 12		17 15 16	−38 46.7	1	12.57		1.52		0.37		1		570
	−23 13299	NGC 6325 sq1 1		17 15 17	−23 35.0	1	10.35		1.36				3	K0	438
156294	−45 11383			17 15 17	−45 54.8	1	9.86		2.57		4.18		2	N3	864
156563	+25 03231			17 15 18	+25 04.6	1	8.98		0.71		0.26		4	G8 V	3031
156562	+25 03232			17 15 18	+25 26.9	1	8.26		0.39		0.05		3	F5	833
		HM 1 - 20		17 15 18	−38 46.9	1	13.45		1.35		0.17		1		570
		NGC 6322 - 29		17 15 19	−42 50.2	1	12.64		0.59		0.15		1		412
		NGC 6341 sq1 21		17 15 20	+43 19.0	1	13.64		0.65		0.16		4		448
	−32 12580			17 15 21	−32 11.6	1	10.72		0.16		−0.23		2		1085
319657				17 15 21	−33 30.2	1	10.60		0.28		−0.21		2	A2	1085
		HM 1 - 13		17 15 21	−38 46.9	1	12.77		1.50		0.38		1	O8	570
		NGC 6322 - 30		17 15 21	−42 49.3	1	12.99		0.58		0.12		1		412
		NGC 6235 sq1 10		17 15 22	−23 41.4	1	15.90		1.10				2		869
		NGC 6325 sq1 10		17 15 22	−23 41.5	1	12.05		0.89				2		438
156381	−32 12579	IDS17121S3215	AB	17 15 22	−32 18.8	2	9.71	.014	0.17	.009	−0.34	.000	4	B9/A0 III	918,1085
319660				17 15 23	−33 39.5	1	11.52		0.32		−0.04		2	B5	1085
156383	−33 11950			17 15 24	−34 05.0	1	8.97		0.23		−0.01		2	B9 II/III	1085
		G 139 - 29		17 15 25	+11 43.6	3	15.12	.007	1.81	.038			7		419,906,1663
		NGC 6341 sq1 19		17 15 25	+43 19.8	1	14.24		0.57		0.03		2		448
156462	−16 04470	HR 6428		17 15 26	−16 15.6	3	6.33	.040	1.68	.023	1.98	.065	8	M2 III	252,2007,3005
		HM 1 - 3	⋆	17 15 26	−38 46.9	2	11.98	.021	1.64	.057	0.47	.096	2	WN8+OB	570,1586
		HM 1 - 11		17 15 26	−38 47.0	1	13.30		0.68		0.20		1		570
		HM 1 - 16		17 15 27	−38 48.1	1	13.20		1.78		0.60		1		570
327335	−43 11577			17 15 28	−43 26.9	1	11.11		0.37				4	A0	2012
156633	+33 02864	HR 6431, U Her	⋆ AB	17 15 29	+33 09.2	4	4.79	.056	−0.17	.015	−0.76	.000	17	B1.5Vp +	15,154,1363,8015
		HM 1 - 5		17 15 29	−38 44.7	1	11.47		0.49		0.21		1		570
156593	+23 03074	HR 6430		17 15 30	+23 08.6	1	6.22		1.55		1.84		2	K2	1733
156517	−1 03309		A	17 15 30	−01 13.9	1	9.47		0.94		0.75		1	K3 V	3072
156517	−1 03309		B	17 15 30	−01 13.9	1	11.83		1.43		1.42		1		3072
		Sco sq 2 # 257		17 15 30	−32 14.0	1	12.36		0.30		0.13		2		1085
		NGC 6325 sq1 4		17 15 31	−23 46.1	1	11.08		1.43				2		438
156395	−32 12587			17 15 31	−32 10.8	2	9.71	.005	0.07	.005	−0.56	.019	4	B9	918,1085

Table 1

HD	DM	Other Id	N	Rem	α_{1950}	δ_{1950}	S	V	σ_V	B–V	σ_{B-V}	U–B	σ_{U-B}	n	Spectrum	References
319693					17 15 31	−35 31.1	1	10.25		0.42		0.13		2	A2	1085
		HM 1 - 7			17 15 31	−38 47.0	1	12.01		0.34		0.02		1		570
		HM 1 - 17			17 15 31	−38 49.3	1	12.91		0.62		0.18		1		570
		NGC 6341 sq1 18			17 15 32	+43 18.7	1	13.35		0.69		0.25		6		448
156753 +49 02614		HR 6437			17 15 32	+49 44.6	1	7.50		1.03		0.79		2	K2	1733
156407 −32 12588					17 15 33	−32 54.9	1	9.72		0.54		0.08		1	F3/5 V	121
156384 −34 11626		HR 6426		★ AB	17 15 33	−34 56.2	8	5.91	.010	1.04	.009	0.83	.015	25	K3 V + K5 V	15,938,1013,1075,1586*
156384 −34 11626		IDS17121S3453		C	17 15 33	−34 56.2	1	10.22		1.57		1.17		2		3073
319648					17 15 34	−33 13.6	1	10.61		0.15		−0.42		2	B5	1085
	−38 11746				17 15 34	−38 45.7	1	11.03		1.47		0.51		1		1586
		HM 1 - 1		★	17 15 34	−38 45.8	1	11.02		1.54		0.41		1	WN8+OB	570
		HM 1 - 18			17 15 34	−38 46.3	1	13.41		1.46		0.30		1		570
156490 −19 04589					17 15 35	−19 55.9	1	10.02		0.23		0.05		1	B9 III	1770
		Sco sq 2 # 260			17 15 35	−32 11.9	1	12.27		0.33		0.12		2		1085
		HM 1 - 10			17 15 35	−38 45.9	1	13.64		1.52		0.30		1		570
		HM 1 - 19			17 15 35	−38 46.1	1	14.15		1.58		0.39		1		570
		NGC 6341 - 2			17 15 36	+43 12.	1	15.11		0.82		0.37		1		448
		NGC 6341 - 10			17 15 36	+43 12.	1	15.47		0.02		0.03		1		448
		NGC 6341 - 13			17 15 36	+43 12.	2	14.63	.019	0.53	.005	-0.08	.005	3		350,448
		NGC 6341 - 22			17 15 36	+43 12.	1	15.41		0.72		0.02		2		350
		NGC 6341 - 502			17 15 36	+43 12.	2	14.66	.000	0.58	.005	0.01	.019	3		350,448
		NGC 6341 - 506			17 15 36	+43 12.	1	15.14		0.75		0.08		2		350
		NGC 6341 - 512			17 15 36	+43 12.	1	14.68		0.80		0.19		2		350
		NGC 6341 - 518			17 15 36	+43 12.	1	14.04		0.92		0.68		1		448
		NGC 6341 - 524			17 15 36	+43 12.	1	14.54		0.78		0.16		1		448
		NGC 6341 - 525			17 15 36	+43 12.	1	15.25		0.66		0.19		2		350
		NGC 6341 - 528			17 15 36	+43 12.	2	15.07	.030	0.44	.005	-0.07	.010	5		350,448
		NGC 6341 - 539			17 15 36	+43 12.	2	14.36	.015	0.76	.060	0.16	.175	2		350,448
		NGC 6341 - 540			17 15 36	+43 12.	1	15.82		-0.08		-0.30		4		448
		NGC 6341 - 570			17 15 36	+43 12.	1	13.10		0.99		0.49		2		350
		NGC 6341 - 589			17 15 36	+43 12.	1	14.03		0.78		0.21		2		350
		NGC 6341 - 604			17 15 36	+43 12.	1	15.38		0.66		0.06		1		350
		NGC 6341 - 1004			17 15 36	+43 12.	2	14.15	.030	0.72	.010	0.14	.005	4		350,448
		NGC 6341 - 1011			17 15 36	+43 12.	1	15.23		0.69		0.07		1		448
		NGC 6341 - 1012			17 15 36	+43 12.	1	13.29		0.77				1		448
		NGC 6341 - 1013			17 15 36	+43 12.	2	12.13	.078	1.35	.019	1.18	.054	3		448,3036
		NGC 6341 - 1027			17 15 36	+43 12.	1	16.15		-0.12		-0.39		4		448
		NGC 6341 - 1065			17 15 36	+43 12.	1	12.49		1.19		0.81		1		350
		NGC 6341 - 1081			17 15 36	+43 12.	1	14.37		0.65		0.01		2		350
		NGC 6341 - 1082			17 15 36	+43 12.	1	13.33		0.97		0.43		1		350
		NGC 6341 - 1502			17 15 36	+43 12.	2	13.55	.030	0.91	.005	0.44	.010	5		350,448
		NGC 6341 - 1510			17 15 36	+43 12.	1	13.42		0.95		0.45		20		448
		NGC 6341 - 1513			17 15 36	+43 12.	1	15.38		0.69		0.09		1		448
		NGC 6341 - 1517			17 15 36	+43 12.	1	15.51		0.03		0.02		7		448
		NGC 6341 - 1527			17 15 36	+43 12.	1	15.20		0.18		0.11		3		448
		NGC 6341 - 1540			17 15 36	+43 12.	1	13.94		0.80		0.28		2		350
		NGC 6341 - 1579			17 15 36	+43 12.	1	13.47		0.92		0.41		2		350
		NGC 6341 - 1587			17 15 36	+43 12.	1	15.10		0.70		0.08		2		350
		NGC 6341 - 1614			17 15 36	+43 12.	2	13.84	.018	0.87	.000	0.36	.000	56		350,448
		NGC 6341 - 2002			17 15 36	+43 12.	1	13.50		1.05		0.95		1		448
		NGC 6341 - 2023			17 15 36	+43 12.	1	15.79		0.02		-0.21		1		448
		NGC 6341 - 2036			17 15 36	+43 12.	1	15.58		-0.02		-0.15		1		448
		NGC 6341 - 2044			17 15 36	+43 12.	1	15.30		0.70		0.05		2		350
		NGC 6341 - 2045			17 15 36	+43 12.	1	12.83		1.04		0.60		2		350
		NGC 6341 - 2055			17 15 36	+43 12.	1	15.30		0.16		0.11		1		448
		NGC 6341 - 2502			17 15 36	+43 12.	1	15.45		0.68		0.21		2		448
		NGC 6341 - 2503			17 15 36	+43 12.	1	15.33		0.09		0.09		1		448
		NGC 6341 - 2506			17 15 36	+43 12.	1	13.16		0.84		0.35		3		448
		NGC 6341 - 2507			17 15 36	+43 12.	2	13.56	.020	0.70	.005	0.13	.025	5		350,448
		NGC 6341 - 2510			17 15 36	+43 12.	1	15.24		0.13		0.10		4		448
		NGC 6341 - 2518			17 15 36	+43 12.	2	13.75	.019	0.84	.019	0.29	.005	3		350,448
		NGC 6341 - 2555			17 15 36	+43 12.	1	15.25		-0.04		-0.03		1		448
		NGC 6341 - 2561			17 15 36	+43 12.	1	15.31		0.14		0.12		1		448
		NGC 6341 - 2574			17 15 36	+43 12.	1	14.25		0.56		0.07		1		448
		NGC 6341 - 3001			17 15 36	+43 12.	1	14.58		0.89		0.49		1		448
		NGC 6341 - 3005			17 15 36	+43 12.	1	15.08		0.66		0.10		2		350
		NGC 6341 - 3010			17 15 36	+43 12.	1	13.70		0.84		0.26		1		350
		NGC 6341 - 3012			17 15 36	+43 12.	2	14.68	.015	0.75	.025	0.27	.095	4		350,448
		NGC 6341 - 3018			17 15 36	+43 12.	2	12.15	.034	1.31	.005	1.04	.019	6		350,448
		NGC 6341 - 3524			17 15 36	+43 12.	1	14.13		0.74		0.20		2		350
		NGC 6341 - 3544			17 15 36	+43 12.	1	14.06		0.83		0.26		2		350
		NGC 6341 - 4005			17 15 36	+43 12.	2	13.12	.024	0.61	.005	0.08	.024	3		350,448
		NGC 6341 - 4006			17 15 36	+43 12.	1	14.65		0.82		0.24		2		350
		NGC 6341 - 4012			17 15 36	+43 12.	2	14.53	.010	0.56	.025	0.01	.065	4		350,448
		NGC 6341 - 4013			17 15 36	+43 12.	1	14.03		0.85		0.33		std		350
		NGC 6341 - 4030			17 15 36	+43 12.	1	14.07		0.80		0.38		2		350
		NGC 6341 - 4049			17 15 36	+43 12.	1	13.89		0.84		0.34		2		350
		NGC 6341 - 4504			17 15 36	+43 12.	2	12.83	.024	1.16	.000	1.06	.024	7		350,448
		NGC 6341 - 4510			17 15 36	+43 12.	1	14.36		0.61		0.11		2		350
		NGC 6341 - 4549			17 15 36	+43 12.	1	12.22		1.28		0.84		2		3036

HD	DM	Other Id	N Rem	α_{1950}	δ_{1950}	S	V	σ_V	B–V	σ_{B-V}	U–B	σ_{U-B}	n	Spectrum	References
		NGC 6341 - 4565		17 15 36	+43 12.	1	14.42		0.72		0.16		2		350
		NGC 6341 - 5001		17 15 36	+43 12.	1	16.04		-0.04		-0.34		1		448
		NGC 6341 - 5002		17 15 36	+43 12.	1	14.26		0.73		0.25		1		448
		NGC 6341 - 5003		17 15 36	+43 12.	1	13.55		0.89		0.45		2		350
		NGC 6341 - 5004		17 15 36	+43 12.	1	14.28		0.72		0.12		2		350
		NGC 6341 - 5008		17 15 36	+43 12.	1	14.07		0.59		0.02		1		448
		NGC 6341 - 5011		17 15 36	+43 12.	1	15.90		0.02		-0.39		1		448
		NGC 6341 - 5013		17 15 36	+43 12.	2	14.77	.068	0.53	.015	-0.01	.029	3		350,448
		NGC 6341 - 5014		17 15 36	+43 12.	1	13.81		0.87		0.35		2		350
		NGC 6341 - 5501		17 15 36	+43 12.	1	15.11		0.18		0.08		2		448
		NGC 6341 - 5502		17 15 36	+43 12.	1	12.89		0.67		0.09		5		448
		NGC 6341 - 5505		17 15 36	+43 12.	1	14.88		0.76		0.17		2		350
		NGC 6341 - 5506		17 15 36	+43 12.	1	15.84		-0.01		-0.15		1		448
		NGC 6341 - 5508		17 15 36	+43 12.	1	12.76		1.06		0.57		5		448
		NGC 6341 - 5524		17 15 36	+43 12.	1	15.94		-0.07		-0.19		2		448
		NGC 6341 - 5531		17 15 36	+43 12.	1	13.95		0.85		0.27		2		350
		NGC 6341 - 5534		17 15 36	+43 12.	1	13.45		0.89		0.39		2		350
	-32 12590			17 15 36	-32 10.5	1	10.72		0.13		-0.40		4		1586
		vdB 85 # a		17 15 36	-35 37.9	1	11.86		0.90		-0.12		3	B0V	434
	-35 11463			17 15 36	-35 39.3	1	11.59		1.05		0.07		2		1085
		Sco sq 2 # 52		17 15 36	-36 17.4	1	12.51		0.62		0.25		2		1085
		HM 1 - 4		17 15 37	-38 47.6	1	11.39		0.40		0.29		1		570
		HM 1 - 8		17 15 38	-38 46.0	1	12.52		1.56		0.37		1	O8	570
156331	-49 11324	HR 6423	⋆ AB	17 15 38	-50 00.7	1	6.27		0.41				4	F8 III+B9 V	2009
		V452 Oph		17 15 39	+11 07.5	2	11.69	.002	0.31	.012	0.28	.029	2	F5	597,699
	-32 12591			17 15 39	-32 12.9	1	10.79		0.14		-0.38		2		1085
156424	-32 12593	IDS17124S3214	AB	17 15 39	-32 17.6	2	8.72	.009	0.06	.005	-0.60	.005	4	B2 Ib/II	918,1085
	-38 11748	HM 1 - 2	⋆	17 15 39	-38 45.6	2	11.17	.008	1.52	.017	0.36	.009	2	O4f	570,1586
		HM 1 - 9		17 15 39	-38 46.0	1	12.99		1.59		0.41		1		570
		HM 1 - 6		17 15 40	-38 43.7	1	11.64		1.54		0.35		1	O5f	570
155918	-75 01368			17 15 40	-75 17.7	4	7.00	.009	0.60	.008	-0.01	.011	11	G2 V	158,742,2012,3077
156330	-49 11327			17 15 41	-49 23.4	1	8.55		0.02		-0.36		3	B5/7	1586
		HM 1 - 15		17 15 43	-38 45.5	1	13.20		0.68		0.18		1		570
156437	-30 14027			17 15 44	-30 08.8	1	9.64		0.40		0.29		2	A5 V	918
	-32 12597			17 15 45	-32 14.4	1	10.89		0.11		-0.37		2		1085
		HM 1 - 14		17 15 46	-38 45.0	1	12.60		0.62		0.52		1		570
	-29 13730			17 15 47	-29 46.9	1	9.87		1.82				1	K5	1702
156465	-30 14028	LSS 4072		17 15 47	-31 03.6	3	8.25	.008	0.04	.007	-0.75	.010	5	B3/6 (V)	540,918,976
156466	-31 13832			17 15 47	-31 08.1	1	8.41		0.23		-0.09		2	B9 (II)	918
156438	-32 12598			17 15 47	-32 37.3	2	8.99	.005	0.50	.015	0.03	.000	3	F5/6 V	121,918
	-35 11465	LSS 4071		17 15 47	-35 34.7	1	11.44		0.76		-0.04		4		754
156398	-44 11595	HR 6427	⋆ AB	17 15 47	-44 10.3	6	6.64	.010	0.20	.001	0.05	.024	42	B9.5III	15,278,1075,1770,2011,3021
156439	-32 12599			17 15 48	-32 52.4	1	10.01		0.19		-0.22		2	B8 V	1085
156409	-39 11312			17 15 48	-39 45.4	2	8.82	.068	0.38	.014	-0.37	.012	6	B2 (II)ne	540,976
		NGC 6341 sq1 1		17 15 49	+43 21.2	1	14.94		0.70		0.27		1		448
156385	-45 11392	LSS 4066		17 15 49	-45 35.3	2	6.95	.016	0.02	.019	-0.42	.005	12	WC	401,1732
	-32 12600			17 15 50	-32 07.3	2	11.29	.024	0.54	.005	-0.18	.000	4		918,1085
156653	+17 03216	HR 6432		17 15 52	+17 22.2	1			0.01		0.02		3	A1 V	1022
		NGC 6341 sq1 4		17 15 52	+43 22.3	1	13.54		0.96		0.66		1		448
156947	+60 01743	HR 6448, VW Dra		17 15 53	+60 43.4	1	6.32		1.09				2	K1.5IIIb	71
319655				17 15 53	-33 29.0	1	11.25		0.32		0.14		2	B9	1085
		HM 1 - 21		17 15 54	-38 48.0	1	12.09		0.60		0.32		1		570
156467	-32 12601			17 15 55	-32 38.4	1	10.18		0.34		0.19		2	B9/A0 V	1085
319701	-35 11468	LSS 4073		17 15 55	-35 51.0	2	10.09	.009	1.07	.009	0.03	.013	5		754,1085
156729	+37 02864	HR 6436		17 15 57	+37 20.6	5	4.63	.030	0.04	.008	-0.01	.016	13	A2 V	15,1022,1363,3023,8015
		NGC 6341 sq1 2		17 15 57	+43 20.6	1	12.27		1.07		0.98		1		448
		LSS 4079		17 15 59	-21 49.3	1	12.01		0.52		0.01		2	B3 Ib	1737
319647				17 15 59	-33 07.1	1	10.79		0.23		0.06		2	A0	1085
		Sco sq 2 # 54		17 15 59	-35 30.4	1	12.00		0.72		0.48		2		1085
156468	-37 11463	LSS 4074		17 15 59	-37 57.2	5	7.78	.023	0.20	.011	-0.58	.011	11	B2 (V)ne	92,540,914,976,1586
156597	-4 04261			17 16 00	-04 13.7	1	9.88		0.65		0.53		2	A0	1776
	-30 14030			17 16 00	-30 18.6	1	9.61		1.75		2.19		2		918
		L 341 - 182		17 16 00	-48 07.	1	12.34		0.63		-0.15		2		1696
156507	-32 12605			17 16 02	-32 05.6	2	9.60	.005	0.20	.005	0.04	.028	4	B9 V	918,1085
156506	-31 13838			17 16 03	-31 18.6	1	7.01		0.12		0.08		3	A2 V	918
323117				17 16 04	-38 07.8	1	11.05		0.38		-0.21		2	B5	432
156411	-48 11605			17 16 04	-48 29.7	1	6.67		0.62				4	G1 V(w)	2033
		G 19 - 20		17 16 05	+02 00.2	3	14.31	.046	0.12	.012	-0.58	.010	9		1,538,3060
	-32 12606			17 16 05	-32 44.2	1	11.26		0.24		-0.06		2		1085
		G 19 - 21		17 16 06	+02 00.0	3	14.00	.009	1.54	.007	1.32	.013	9		1,538,3028
156508	-32 12607			17 16 06	-32 20.4	2	10.55	.000	0.21	.014	0.12	.028	4	B9/A0 V	918,1085
		Sco sq 2 # 55		17 16 06	-35 36.6	1	11.99		0.51		-0.03		2		1085
		NGC 6341 sq1 6		17 16 07	+43 22.4	1	14.96		0.67		0.18		1		448
		NGC 6341 sq1 3		17 16 08	+43 20.4	1	13.32		0.92		0.71		1		448
	-33 11966			17 16 09	-33 35.7	1	11.39		0.44		-0.22		2		1085
319699	-35 11470	LSS 4078		17 16 09	-35 39.5	4	9.62	.009	0.80	.009	-0.25	.019	13	O7	390,540,754,1085
156543	-30 14037			17 16 11	-30 11.1	1	9.51		0.55		0.10		2	A9/F0 V	918
156525	-31 13840			17 16 11	-31 41.4	1	9.41		0.63		0.31		2	F0 V	918
156526	-32 12609			17 16 12	-32 10.8	1	9.46		1.15		0.88		2	G8	918
156681	+11 03156	HR 6433		17 16 16	+10 55.0	5	5.02	.011	1.54	.007	1.84	.020	13	K4 II-III	15,1003,1080,3053,8100

Table 1   939

HD	DM	Other Id	N Rem	α₁₉₅₀	δ₁₉₅₀	S	V	σ_V	B–V	σ_B-V	U–B	σ_U-B	n	Spectrum	References
	−15 04515			17 16 20	−15 51.6	1	9.84		0.61		0.23		1	A5	979
	−35 11472			17 16 22	−35 55.8	1	11.57		0.80		0.09		2		1085
	−32 12613			17 16 23	−32 41.9	1	11.59		0.27		0.09		2		1085
319703	−35 11473	LSS 4080	B	17 16 23	−36 02.8	4	11.17	.067	1.25	.009	0.09	.047	7	O6	390,434,954,1537
329804	−44 11599	LSS 4076		17 16 23	−44 40.8	2	10.36	.009	0.50	.006	-0.35		6	B2 II	540,1021
156774	+27 02787			17 16 24	+26 59.3	1	7.51		1.26		1.30		2	K2 III	37
319703	−35 11473	LSS 4081	⋆ A	17 16 24	−36 02.8	5	10.68	.035	1.16	.023	0.02	.025	11	O7	390,434,954,1085,1537
156731	+16 03146			17 16 25	+16 20.1	1	8.04		1.45				1	K2	882
	−10 04471	IDS17136S1101	AB	17 16 25	−11 04.1	1	10.45		1.41		1.30		4	M0	158
		Steph 1457	B	17 16 25	−11 04.2	1	11.98		1.40				1	K4	1746
		Steph 1457		17 16 25	−11 04.3	1	10.77		1.36				1		1746
156697	+6 03386	HR 6434, V2112 Oph		17 16 26	+06 08.2	4	6.50	.007	0.41	.021	0.16	.030	12	F0/2 IV/Vn	15,206,1256,1776
319688				17 16 27	−35 05.8	1	10.58		0.54		0.10		2	A0	1085
		LP 447 - 59		17 16 28	+16 27.0	1	12.35		0.51		-0.10		2		1696
		NGC 6341 sq1 11		17 16 28	+43 10.6	1	14.43		0.55		0.02		1		448
156572	−31 13849			17 16 29	−31 38.4	1	8.76		1.14		0.83		2	G5 V	918
156190	−69 02719	HR 6411	⋆ AB	17 16 30	−70 04.4	3	5.39	.012	-0.05	.005	-0.22	.005	12	B9 V +B9.5V	15,1075,2038
		NGC 6341 sq1 9		17 16 32	+43 25.0	1	12.48		0.51		-0.02		1		448
		NGC 6341 sq1 12		17 16 33	+43 10.1	1	14.22		0.59		0.00		1		448
	−32 12615	LSS 4085		17 16 33	−32 09.5	1	11.12		0.46		-0.27		2		918
		NGC 6341 sq1 13		17 16 34	+43 08.2	1	10.99		0.77		0.33		1		448
		NGC 6341 sq1 8		17 16 34	+43 23.1	1	11.96		0.98		0.61		1		448
		NGC 6341 sq1 17		17 16 35	+43 06.0	1	14.30		0.51		-0.08		1		448
156359	−62 05531			17 16 37	−62 52.1	1	9.67		-0.14		-0.98		1	B0 Ia/ab	55
		NGC 6341 sq1 10		17 16 38	+43 25.0	1	14.05		0.73		0.29		1		448
319678	−34 11636	LSS 4086		17 16 38	−34 10.9	2	10.07	.005	0.63	.015	-0.15	.015	3	B2	754,1085
	−35 11476	LSS 4084		17 16 38	−35 12.0	2	10.82	.000	1.02	.009	0.05	.024	4		754,1085
		NGC 6341 sq1 16		17 16 40	+43 09.1	1	14.40		0.73		0.22		1		448
	−32 12617			17 16 40	−32 29.2	1	9.58		1.77		1.92		2	K0	918
156891	+38 02910	HR 6444		17 16 42	+38 51.7	2	5.95	.003	1.00	.007	0.81		4	K0 III	71,105
		NGC 6341 sq1 15		17 16 42	+43 09.9	1	13.46		0.50		-0.09		1		448
		AA69,51 # 25		17 16 42	−35 55.	1	12.09		1.06		-0.03		4		754
		vdB 85 # b		17 16 42	−35 57.	1	13.23		1.06		-0.01		3		434
		AA69,51 # 24		17 16 42	−36 03.	1	12.35		2.01		1.22		2		754
	−40 11328	LSS 4082		17 16 43	−40 25.8	1	11.03		1.00		-0.02		3		1586
156925	+42 02822			17 16 44	+42 09.7	1	8.32		0.45				1	F8	652
		NGC 6341 sq1 14		17 16 44	+43 08.5	1	10.94		0.40		0.00		1		448
	−2 04333	IDS17141S0215	AB	17 16 44	−02 18.2	1	10.21		0.98		0.76		1	K3 V	3072
156715	−2 04332			17 16 44	−02 41.9	1	7.28		1.16		0.94		4	K0	1657
	−35 11477	LSS 4087		17 16 44	−35 53.6	2	11.11	.009	0.90	.004	-0.03	.004	6		754,1085
156277	−67 03310	HR 6417		17 16 45	−67 43.0	3	4.77	.003	1.22	.005	1.27	.000	19	K1 III	15,1075,2038
	−34 11637	LSS 4088		17 16 47	−34 50.8	2	11.32	.009	1.17	.009	0.17	.014	4		754,1085
156639	−30 14045			17 16 48	−30 22.7	1	8.80		0.49		-0.01		2	F5 V	918
156874	+28 02719	HR 6443		17 16 51	+28 52.4	1	5.65		0.98				2	K0 III	71
319698	−35 11478	LSS 4089		17 16 51	−35 40.7	3	10.78	.015	0.64	.015	0.26	.017	7	A0 II	754,954,1085
		G 181 - 36		17 16 53	+34 47.6	1	10.69		0.76		0.24		3	K1	7010
156822	+11 03157			17 16 56	+11 13.4	1	7.70		0.04		-0.07		2	A0	1776
156575	−45 11403	LSS 4083		17 16 56	−45 59.8	6	7.34	.017	0.21	.011	-0.64	.008	9	B1 Ib/II	219,401,540,954,976,1737
156717	−17 04773	HR 6435	⋆ AB	17 16 59	−17 42.4	1	6.02		0.04				4	A2 Vnn	2007
156684	−27 11584			17 16 59	−27 43.4	1	8.62		0.19		0.15		2	A1 V	918
		OT 129 # 5		17 17 01	+17 48.	1	15.65		0.95				1		930
		OT 129 # 6		17 17 01	+17 48.	1	17.19		0.75				1		930
323242	−40 11332			17 17 01	−40 57.7	1	10.43		0.73		-0.21		2	B2 II-Ib	540
156965	+42 02823	TX Her		17 17 02	+41 56.3	1	8.14		0.29				1	A9 V	652
156672	−29 13403			17 17 02	−30 00.8	1	9.81		0.31		0.32		2	A2/3 III	918
319697	−35 11480			17 17 05	−35 39.6	1	10.33		0.66		-0.24		3	A5	754
	−35 11482	LSS 4093		17 17 05	−35 41.0	3	10.64	.044	0.73	.014	-0.27	.019	7		754,1085,1586
	−31 13864			17 17 06	−31 47.6	1	9.95		1.23		0.91		2	G5	918
		vdB 85 # c		17 17 06	−35 47.	1	12.25		1.34		0.05		3		434
		OT 129 # 4		17 17 07	+17 48.5	1	14.81		0.58		0.33		1		930
156760	−16 04478			17 17 08	−16 37.9	1	8.63		0.12		0.03		1	B9 IV	1770
156623	−45 11408			17 17 10	−45 22.2	6	7.25	.006	0.09	.003	0.06	.009	40	A0 V	116,278,977,1075,1770,2011
156825	−3 04087			17 17 11	−03 59.9	1	7.86		0.46		-0.01		2	F5	1776
157103	+53 01932	IDS17161N5344	AB	17 17 12	+53 41.1	1	8.94		0.42		0.00		2	F5	1566
	−35 11483	LSS 4095		17 17 12	−36 02.5	1	11.65		1.03		-0.10		4		754
156876	+13 03349			17 17 13	+13 14.3	1	8.41		1.20				2	K5	882
		OT 129 # 3		17 17 14	+17 46.0	1	12.91		0.90		1.00		1		930
156860	+2 03296	V2113 Oph		17 17 15	+02 11.4	1	6.70		1.68		1.44		1	M2	3040
		OT 129 # 2		17 17 17	+17 46.3	1	14.32		0.80		0.69		1		930
156802	−7 04427	IDS17146S0754	A	17 17 17	−07 58.2	3	7.97	.012	0.66	.010	0.07	.011	5	G2 V	742,1003,3026
		Sco sq 2 # 66		17 17 18	−35 58.9	1	11.76		0.89		0.03		2		1085
156688	−37 11487			17 17 18	−37 58.3	4	7.20	.007	0.08	.007	-0.64	.008	9	B2 II	540,914,976,2012
156826	−5 04426	HR 6439		17 17 19	−05 51.9	5	6.31	.008	0.85	.005	0.47	.004	19	G9 V	15,1256,2018,2028,3077
156674	−42 12052			17 17 19	−42 36.2	1	8.50		0.50				4	F6 V	2012
156912	+12 03195			17 17 20	+12 33.0	1	9.55		0.20		0.18		2		1776
156779	−18 04494	LSS 4102		17 17 20	−18 46.4	4	9.29	.010	0.17	.011	-0.53	.006	8	B2 II	400,540,1225,1770
156911	+19 03281			17 17 23	+19 24.8	1	7.54		1.37		1.56		2	K5	1648
		OT 129 # 1		17 17 24	+17 48.0	1	12.10		0.83		0.81		1		930
156966	+27 02790			17 17 24	+27 20.0	2	6.80	.063	1.66	.004	1.98	.022	5	M1 III	1003,3040
156702	−38 11773	LSS 4097		17 17 24	−38 36.2	4	8.74	.023	0.16	.006	-0.57	.012	8	B2ne	432,540,976,1586

HD	DM	Other Id	N	Rem	α_{1950}	δ_{1950}	S	V	σ_V	B–V	σ_{B-V}	U–B	σ_{U-B}	n	Spectrum	References
156662	−45 11411	V831 Ara		★	17 17 24	−45 56.0	3	7.83	.006	0.17	.004	-0.65	.012	5	B2 III	401,540,976
156736	−29 13413	IDS17142S2918		AB	17 17 25	−29 20.9	1	9.03		0.52		-0.03		1	F5 V	918
156427	−66 03095				17 17 26	−66 13.2	2	7.41	.003	1.49	.002	1.66		26	K3/4 III	1621,1642
319702	−35 11486	LSS 4100			17 17 29	−35 48.8	4	10.13	.038	0.93	.014	-0.13	.020	12	O9	390,540,754,1085
	−35 11484	LSS 4099			17 17 29	−35 49.4	2	11.33	.004	0.95	.004	0.05	.009	5		754,1085
156738	−35 11487	LSS 4101			17 17 31	−36 01.4	4	9.36	.006	0.90	.012	-0.15	.020	11	B7/8 Iab/b	390,540,754,1085
156830	−17 04777				17 17 32	−17 10.7	1	9.61		0.46		0.02		1	B8/9 III	1770
	−36 11424				17 17 35	−36 06.3	1	11.19		0.83		-0.04		2		1085
156724	−41 11508				17 17 35	−41 25.5	1	9.92		0.22		-0.44		3	B2/3	1586
		L 269 - 45			17 17 36	−53 00.	1	10.95		1.12		1.10		1		1696
156846	−19 04605	HR 6441		★ AB	17 17 38	−19 16.9	4	6.50	.010	0.58	.000	0.10		13	G1 V	15,2020,2028,3053
156781	−32 12642				17 17 38	−32 06.6	1	8.79		0.21		0.14		2	A1/2 V	918
156831	−24 13276				17 17 39	−24 13.3	2	8.80	.014	0.27	.007	-0.20	.044	4	B3 Vnne	1586,1770
	−30 14064				17 17 39	−30 14.3	1	9.39		1.11		0.81		2	G5	918
	−58 07076				17 17 39	−58 31.0	1	9.85		1.49				1		1705
319666	−34 11646				17 17 40	−34 06.2	1	11.68		0.52		0.38		1		121
		Sco sq 2 # 71			17 17 41	−35 32.4	1	12.30		0.57		0.43		2		1085
156832	−29 13419				17 17 43	−29 11.3	1	9.46		0.57		0.30		2	F3 V	918
156989	+16 03151				17 17 46	+15 56.1	1	7.41		1.06				3	K0	882
	−31 13867	LSS 4105			17 17 46	−31 06.3	1	11.19		0.78		-0.19		2		918
156968	+9 03366	G 139 - 31			17 17 49	+09 30.9	2	7.98	.005	0.59	.007	0.06	.002	4	G0 V	1003,3026
323110	−37 11490	LSS 4103			17 17 51	−37 56.2	5	9.73	.061	1.07	.018	0.07	.035	8	B2.5 Iab	432,540,914,954,1737
		G 170 - 34		★ V	17 17 53	+26 32.8	2	12.98	.049	1.63	.037	0.97	.188	3	M3	8088,3003
319696					17 17 53	−35 33.7	1	11.77		0.42		0.11		2	B8	1085
		G 170 - 35		★ V	17 17 54	+26 32.8	2	11.39	.033	1.56	.007	1.17	.017	2		8088,3003
	−31 13885				17 17 54	−31 37.4	1	9.45		1.85		1.99		2		918
319695					17 17 55	−35 31.8	1	11.31		0.48		0.32		2	A0	1085
156808	−39 11357	LSS 4104			17 17 56	−39 27.1	1	8.63		0.76		0.76		2	Ap EuCr	954
156895	−18 04499				17 17 57	−18 51.5	1	10.18		0.18		0.02		1	B8 III	1770
	−27 11597				17 17 57	−27 54.5	1	10.42		0.36		-0.39		1	B8	918
		AA69,51 # 73			17 17 58	−35 30.6	1	13.40		2.12		2.01		4	K3 III	1537
156897	−20 04731	HR 6445		★ AB	17 18 00	−21 03.7	7	4.39	.013	0.38	.010	-0.05	.009	19	F2/3 V	15,1075,1586,2012*
156865	−30 14071				17 18 00	−31 04.4	1	9.11		1.70		2.09		2	K(4/5)	918
156952	−6 04581	IDS17153S0700		AB	17 18 01	−07 03.4	2	8.77	.025	0.75	.005	0.29		11	G5	173,3016
156928	−12 04722	HR 6446		★ A	17 18 01	−12 47.9	6	4.33	.012	0.03	.005	0.05	.008	23	A2 V	15,1075,1088,2012*
319771	−37 11494				17 18 04	−37 56.6	2	11.01	.025	0.85	.005	-0.06	.045	4		954,1737
156834	−39 11361				17 18 04	−39 58.7	1	9.48		0.18		-0.56		2	B2/3 III	540
156971	−10 04477	HR 6449		★ AB	17 18 06	−10 38.8	3	6.45	.037	0.34	.004	-0.01		8	F1 III	15,254,2028
156867	−35 11495				17 18 06	−35 33.1	1	8.81		0.31		-0.14		2	B8 IV	1085
157049	+18 03351	HR 6452, V656 Her			17 18 07	+18 06.4	2	4.99	.009	1.62	.001	2.06		4	M2 IIIab	1363,3055
157087	+25 03246	HR 6455			17 18 07	+25 35.2	2	5.38		0.05	.015	0.11	.005	5	A3 III	1022,3023
157050	+16 03153				17 18 10	+16 08.9	1	7.78		1.04				3	K0	882
		G 181 - 37		AB	17 18 11	+36 42.9	1	13.74		1.42				1		1759
156954	−12 04726				17 18 11	−13 02.2	2	7.68	.005	0.29	.000	-0.02		4	A9 V	742,1594
156883	−31 13896				17 18 12	−31 32.0	1	7.60		0.18		0.10		4	A5 IV	918
		vdB 87 # a			17 18 12	−44 05.	1	15.16		0.99		-0.14		2		434
156900	−30 14074				17 18 13	−30 53.7	1	9.81		0.60		0.09		2	G5	918
156902	−31 13895				17 18 13	−31 43.4	1	10.25		0.38		0.33		2	A1/3 IV/V	918
156901	−30 14075				17 18 14	−30 59.8	1	9.28		1.22		0.99		2	G8/K1 (III)	918
	+26 03002				17 18 17	+26 15.9	1	9.55		1.03		0.87		2	K2	8088
		G 181 - 39			17 18 17	+41 46.2	3	11.38	.015	1.56	.008			5	M2	538,1746,1759
	−31 13897				17 18 19	−32 03.4	1	9.37		1.85		1.94		2	G5	918
	−34 11654				17 18 19	−34 34.3	1	11.60		1.14		0.26		2		1085
156974	−17 04783				17 18 21	−18 00.0	2	9.41	.007	0.12	.012	0.03	.012	2	B9.5V	1225,1770
		AA69,51 # 72			17 18 23	−35 32.1	1	14.33		1.55		0.48		4	B0.5	1537
157224	+45 02521				17 18 24	+45 21.4	1	6.67		0.31		0.03		2	F0	1601
		L 341 - 45			17 18 24	−46 01.	1	15.00		0.68		0.00		4		3062
156934	−31 13901				17 18 25	−31 49.5	1	10.48		0.16		-0.03		2	B8 II	918
156751	−58 07086	IDS17142S5822		AB	17 18 27	−58 25.4	1	6.76		0.24				4	A7 II/III	2012
157150	+26 03003				17 18 29	+26 36.1	1	8.87		0.98		0.72		2	G8 III	8088
	−32 12653	LSS 4113			17 18 31	−33 01.9	1	11.02		0.88		0.04		3		1586
157121	+17 03225				17 18 32	+17 05.3	1	7.77		1.37		1.42		2	K2	1648
157089	+1 03421				17 18 36	+01 29.3	7	6.97	.021	0.58	.016	-0.02	.013	14	F9 V	792,908,1003,1509*
156768	−57 08478	HR 6438		★ AB	17 18 36	−57 57.8	6	5.86	.006	1.08	.004	0.86	.001	31	G8 Ib/II	15,404,1637,1637,2013,2030
157225	+42 02830				17 18 38	+41 55.1	1	7.91		0.33				1	A2	652
323099	−37 11501	LSS 4112			17 18 38	−37 51.4	2	9.59	.012	0.78	.018	-0.21	.025	4	B1 II-Ib	432,540
		LSS 4111			17 18 38	−38 06.1	2	11.01	.065	0.77	.010	-0.13	.020	4	B3 Ib	954,1737
		LP 447 - 44			17 18 39	+15 35.8	1	11.81		0.61		-0.05		2		1696
157105	+1 03422	IDS17160N0137	A		17 18 42	+01 36.3	1	8.04		0.53		0.07		1	F8	3016
157105	+1 03422	IDS17160N0137	B		17 18 42	+01 36.3	1	10.80		0.71		0.15		1		3032
157105	+1 03422	IDS17160N0137	C		17 18 42	+01 36.3	1	11.15		1.35		1.39		1		3032
	−16 04484				17 18 43	−16 46.4	1	10.51		1.16				2		1726
	−32 12660	LSS 4116			17 18 43	−32 30.4	1	10.84		0.66		-0.09		2		918
	−31 13908	LSS 4117			17 18 44	−31 52.0	1	10.77		0.54		-0.37		2		918
		AA69,51 # 34			17 18 45	−35 37.0	3	13.15	.022	1.77	.017	0.68	.038	6	O9.5	754,914,1537
157214	+32 02896	HR 6458		★ A	17 18 47	+32 31.9	13	5.39	.011	0.62	.003	0.07	.013	38	G0 V	1,15,22,254,792,908*
		G 181 - 40			17 18 48	+28 27.8	1	11.22		1.05		1.00		3	K3	7010
	−39 11373	LSS 4114			17 18 48	−39 19.3	2	11.53	.040	0.88	.025	0.00	.040	4		954,1737
157198	+24 03167	HR 6457		★ A	17 18 51	+24 32.9	3	5.14	.024	-0.01	.022	0.02	.005	6	A2 V	1022,1363,3023
319665	−34 11657				17 18 52	−34 08.6	1	11.40		0.50		0.37		2	K5	1085

Table 1 941

HD	DM	Other Id	N Rem	α_{1950}	δ_{1950}	S	V	σ_V	B–V	σ_{B-V}	U–B	σ_{U-B}	n	Spectrum	References
319747	−36 11437			17 18 52	−36 40.4	1	10.22		0.32		−0.06		3	B9	1586
323203	−39 11375	LSS 4115		17 18 52	−39 48.3	1	9.98		0.51		−0.35		2	B1.5III	540
		AA69,51 # 76		17 18 53	−34 35.5	1	13.06		1.99		1.87		3	K2 III	1537
		AA69,51 # 74		17 18 53	−35 35.0	1	14.07		2.11		1.60		3	G8 III	1537
	−43 11630			17 18 53	−43 06.3	2	10.44	.006	0.63	.003	0.03		10	G0	863,2011
157255	+32 02898			17 18 54	+32 43.4	1	7.03		0.03		0.05		2	A0	252
157323	+47 02459			17 18 54	+46 59.8	1	9.76		0.32		0.08		2	F2	1601
156854	−56 08191	HR 6442		17 18 54	−56 28.7	4	5.79	.009	0.99	.007	0.77		12	G8/K0 III	15,1592,2013,2030
	−42 12079			17 18 55	−42 54.6	2	9.81	.010	0.64	.006	0.18		10	Bep	863,2012
157325	+46 02293	HR 6464		17 18 56	+46 17.3	1	5.58		1.57		1.89		3	M0 III	3025
157056	−24 13292	HR 6453, θ Oph		17 18 56	−24 57.1	8	3.26	.009	−0.21	.011	−0.86	.013	30	B2 IV	15,26,1020,1075,2012*
157003	−32 12662			17 18 56	−32 29.0	1	9.66		0.30		−0.33		2	B3 II/III	918
157016	−30 14090			17 18 57	−30 27.2	1	7.18		0.10		−0.20		2	B8 V	918
157017	−31 13912			17 18 59	−31 36.6	1	9.03		0.59		0.11		2	G2 V	918
	−32 12664	LSS 4119		17 19 00	−32 31.1	1	10.85		0.41		−0.26		2		918
		G 226 - 55		17 19 01	+53 52.8	1	11.24		1.14		1.15		4	K7	7010
157090	−17 04787			17 19 02	−17 43.6	1	9.36		1.56				2	K2 (III)	1726
157035	−31 13913			17 19 02	−31 32.3	1	7.75		0.20		0.15		2	A5 V	918
157199	+13 03357			17 19 03	+13 40.7	1	7.81		1.12				2	K2	882
		AA69,51 # 77		17 19 03	−34 37.3	1	11.63		2.88		3.52		3	M3 III	1537
		G 226 - 56		17 19 04	+57 21.6	1	14.05		1.11		0.75		1		1658
157124	−14 04619			17 19 04	−15 00.2	1	8.43		0.22		−0.29		1	B3 II	1770
157327	+39 03108			17 19 05	+39 08.7	1	7.96	.000	1.08	.005	1.00	.010	8	K1 III	833,1601
156979	−45 11441	V636 Sco		17 19 05	−45 34.0	2	6.40	.025	0.81	.015	0.54			F7/8 Ib/II	657,1484
157074	−27 11604			17 19 08	−27 57.3	1	8.66		0.36		0.32		3	A2 III	918
157058	−31 13918			17 19 08	−31 15.0	1	8.28		0.35		0.01		2	F0 V	918
		AA69,51 # 75		17 19 09	−35 45.3	1	12.93		2.50		2.25		3	K2 III	1537
157075	−30 14096	IDS17160S3006	A	17 19 12	−30 09.2	1	8.02		0.51		0.00		2	G0 V	918
157075	−30 14096	IDS17160S3006	B	17 19 12	−30 09.3	1	11.23		1.27		1.02		2		918
157373	+48 02506	HR 6467		17 19 13	+48 14.2	2	6.41	.045	0.41	.015	−0.10	.005	5	F6 Va	254,3037
157038	−37 11507	HR 6450, V975 Sco	★AB	17 19 15	−37 45.5	7	6.33	.015	0.76	.014	−0.17	.017	13	B4 Ia	540,914,954,976,1018,1034*
157060	−35 11505	HR 6454		17 19 16	−35 51.9	3	6.45	.016	0.54	.005			11	F8 V	15,2013,2028
157226	+14 03232			17 19 17	+14 30.7	1	8.19		1.73				2	K5	882
157257	+16 03163	HR 6463		17 19 19	+16 46.7	2	6.35	.005	1.61	.005	1.97	.010	5	M2.5 IIIab	1733,3001
156838	−62 05558	HR 6440		17 19 19	−62 49.1	3	5.70	.007	−0.15	.009	−0.77	.012	8	B2 III	26,400,2007
157095	−31 13923			17 19 20	−31 23.4	1	10.09		0.14		−0.06		2	B9/A0III/IV	918
		G 203 - 63		17 19 27	+49 18.9	3	14.41	.030	1.68	.034			8		538,1705,1759
157170	−17 04789			17 19 28	−17 17.6	1	7.97		0.14		0.07		1	A0 V	1225
		AA69,51 # 79		17 19 28	−34 52.1	1	13.64		2.75		2.39		3	K1 III	1537
157111	−32 12673			17 19 29	−32 31.5	1	9.65		0.68		0.20		2	F3/5 V	918
157097	−37 11512	HR 6456	★A	17 19 31	−37 10.4	3	5.92	.004	1.07	.005			11	K1 III	15,1075,2006
157042	−47 11484	HR 6451, ι Ara	★A	17 19 31	−47 25.3	10	5.23	.015	−0.11	.010	−0.83	.014	36	B2 IIIne	15,278,681,815,1075*
157774	+70 00925			17 19 33	+70 50.2	1	7.14		0.03		0.04		4	A0	985
	+7 03348	UZ Oph		17 19 34	+06 57.6	1	9.93		0.95		0.72		1	G2	793
157358	+28 02728	HR 6466	★AB	17 19 34	+28 48.3	1	6.35		0.70				2	G0 III	71
157098	−38 11805			17 19 34	−38 30.0	2	9.60	.011	0.09	.003	−0.56	.002	4	B3 IV/V	540,976
157062	−44 11644			17 19 34	−44 40.2	4	7.88	.005	1.19	.003	0.91	.005	20	G8/K0II/III	863,977,1586,2011
157063	−44 11643			17 19 34	−44 44.7	5	8.49	.005	0.11	.005	−0.24	.008	27	Ap Si	460,977,1075,1586,2011
323154	−38 11806	LSS 4121		17 19 35	−38 00.8	3	9.25	.008	1.06	.028	0.05	.015	7	B3 lane2+	954,1586,1737
157184	−17 04790			17 19 36	−17 17.8	1	9.44		0.23		0.12		1	A0 V	1225
317543	−32 12676			17 19 36	−32 26.5	1	9.81		1.19		0.89		2	G0	918
157297	+11 03166			17 19 37	+11 43.2	2	7.07	.010	0.08	.010	0.05	.040	4	A0	1648,1776
157330	+23 03090	RS Her		17 19 37	+22 58.1	1	9.61		1.61		0.84		1	M4	817
319729	−35 11512	LSS 4123		17 19 38	−35 45.3	1	10.00		0.87		−0.09		2	B2 II-Ib	540
157099	−42 12090			17 19 39	−42 47.0	2	8.80	.025	0.12	.014	−0.53	.001	5	B3 Vne	540,1586
157141	−30 14109			17 19 40	−30 52.6	1	9.29		0.52		0.01		2	F6 V	918
157201	−17 04791			17 19 41	−17 39.4	1	8.22		0.12		0.21		1	A1 IV/V	1225
157579	+59 01804			17 19 44	+59 14.8	1	8.35		0.17		0.16		4	A2	1371
157280	−4 04269			17 19 46	−04 41.5	1	9.08		0.69		0.56		2	A2	1776
157127	−38 11810			17 19 47	−38 30.1	2	9.22	.017	0.23	.008	−0.41	.008	4	B3 (V)nn(e)	540,976
156942	−60 06800	HR 6447		17 19 48	−60 37.7	2	5.76	.005	−0.08	.000			7	B8 Ib/II	15,2030
157375	+21 03103			17 19 53	+21 12.1	1	7.03		1.19		1.26		2	K2	1648
157115	−45 11459			17 19 56	−45 23.5	2	8.59	.005	0.05	.006	−0.42	.004	10	B5 IV	401,1732
157190	−32 12683			17 19 58	−32 07.4	1	8.98		1.81		1.90		2	M0 (III)	918
157117	−48 11653			17 19 59	−48 39.7	1	7.56		1.03		0.80		8	K0 III	288
157189	−31 13935			17 20 00	−31 26.6	1	9.04		0.08		−0.15		2	B9 V	918
157540	+50 02391			17 20 01	+50 53.8	1	8.90		0.45		−0.05		2	K0	1566
157131	−45 11462	IDS17164S4504	AB	17 20 01	−45 07.3	4	9.08	.009	0.43	.006	−0.02	.012	22	F0/2 V	116,863,1770,2011
323097	−37 11520	LSS 4125		17 20 04	−37 57.3	1	11.30		1.23		1.14		1		1586
157482	+40 03136	HR 6469, V819 Her	★	17 20 05	+40 00.8	2	5.55	.019	0.69	.012	0.19	.000	3	G0 Vn:	254,3016
157376	+14 03236			17 20 07	+14 49.8	1	8.50		0.98				2	G5	882
157333	−3 04092			17 20 07	−03 41.0	1	8.60		0.74		0.49		2	A0	1776
323093	−37 11523			17 20 08	−37 46.6	1	10.46		0.34		0.30		2	F0	540
	+87 00166			17 20 09	+87 47.9	1	8.96		1.14		1.06		2	K0	1733
157235	−27 11617			17 20 09	−27 55.3	1	9.42		0.82		0.36		2	G8/K0 V	918
157282	−22 04336			17 20 11	−22 57.7	1	7.41		0.06		−0.14		2	B8/9 V	1770
		AA69,51 # 78		17 20 11	−34 41.5	1	13.65		2.21		1.69		3	G8 III	1537
319734	−36 11457	LSS 4126	AB	17 20 11	−36 11.8	2	11.02	.010	0.99	.035	−0.01		4	F8	725,954
157237	−30 14122			17 20 12	−30 47.9	1	9.99		0.15		−0.07		2	B9 V	918
	−48 11655			17 20 12	−48 34.3	1	10.14		0.43		0.02		9	F0	288

HD	DM	Other Id	N Rem	α_{1950}	δ_{1950}	S	V	σ_V	B–V	σ_{B-V}	U–B	σ_{U-B}	n	Spectrum	References
157236	−28 13081	HR 6459		17 20 13	−28 05.8	2	5.33	.020	1.55	.005	1.80		9	K5 III	2007,3016
157347	−2 04343	HR 6465		17 20 15	−02 20.4	3	6.28	.004	0.68	.000	0.23	.005	11	G5 IV	15,1256,2020
157238	−31 13941			17 20 16	−31 57.1	1	9.60		0.45		−0.30		2	B6/7 Ib	918
	−48 11659	V819 Ara		17 20 17	−48 39.4	2	9.34	.031	1.83	.020	1.96	.062	10		288,868
157264	−31 13945			17 20 18	−31 39.2	1	7.75		0.38		−0.02		2	F0 V	918
157177	−48 11661			17 20 21	−48 27.5	1	9.22		0.96		0.67		8	G8/K0 III	288
157466	+25 03252			17 20 24	+24 55.7	3	6.87	.000	0.51	.012	−0.06	.013	7	F8 V	516,1620,3016
157284	−30 14127			17 20 26	−31 01.0	1	8.93		1.31		1.16		2	G8/K0 III	918
157285	−31 13951			17 20 26	−31 27.4	1	9.74		1.16		0.80		2	G8	918
157283	−30 14128			17 20 27	−30 21.2	1	9.36		0.27		0.24		2	A0 IV	918
		LSS 4128		17 20 29	−35 50.4	1	11.06		0.97		−0.11		2	B3 Ib	1737
		Ross 855		17 20 33	−14 10.3	1	11.81		0.74		0.10		2		1696
		LP 920 - 1		17 20 35	−32 11.8	1	11.64		1.51				1	M2	1705
157243	−44 11669			17 20 35	−44 07.0	9	5.11	.004	−0.06	.006	−0.40	.009	57	B7 III	15,26,116,278,977,977*
		IC 4651 - 197		17 20 35	−49 57.6	1	12.99		0.54		0.11		2		3011
319755	−37 11530	LSS 4129		17 20 36	−37 05.6	2	9.83	.015	1.40	.015	0.45	.035	4	B2.5 Ia	954,1737
		IC 4651 - 188		17 20 36	−49 53.6	1			0.57		0.05		4		3011
		IC 4651 - 198		17 20 36	−49 57.8	1	13.52		0.54		0.08		3		3011
		IC 4651 - 180		17 20 37	−49 50.4	1	12.07		0.58		0.10		1		3011
		IC 4651 - 194		17 20 37	−49 56.2	1	12.34		0.56		0.10		2		3011
		IC 4651 - 93		17 20 38	−49 53.3	1	12.52		0.55		0.06		5		3011
157162	−54 08267			17 20 38	−54 21.7	1	8.57		0.42				4	F0/2 IV/V	2012
		IC 4651 - 92		17 20 39	−49 52.4	1	12.13		0.31		0.24		3	A3	3011
		IC 4651 - 199		17 20 39	−49 58.2	1	12.06		0.61		0.12		2		3011
157681	+53 01937	HR 6479		17 20 40	+53 28.0	1	5.67		1.47				2	K4 III	71
		IC 4651 - 177		17 20 40	−49 49.6	1	12.80		0.52		0.10		2		3011
	−49 11399	IC 4651 - 90		17 20 40	−49 52.6	1	10.46		1.25		1.23		1		3011
		IC 4651 - 201		17 20 40	−49 56.6	1	13.15		0.51		0.03		2		3011
		IC 4651 - 200		17 20 40	−49 57.0	1	13.55		0.57		0.03		2		3011
		IC 4651 - 94		17 20 41	−49 53.6	1	12.76		0.56		0.05		2		3011
157336	−29 13474			17 20 42	−29 46.5	1	7.58		0.92		0.61		3	K0 IV/V	918
		IC 4651 - 91		17 20 42	−49 52.5	1	12.18		0.59		0.10		3		3011
		IC 4651 - 102		17 20 42	−49 54.2	1	12.34		0.52		0.14		1		3011
	−49 11400	IC 4651 - 419		17 20 43	−50 01.1	1	10.74		1.44		1.61		2		185
		IC 4651 - 88		17 20 44	−49 52.0	1	13.05		0.51		0.06		1		4002
		IC 4651 - 89		17 20 44	−49 52.3	1	13.28		0.57		0.08		2		3011
		IC 4651 - 97		17 20 44	−49 53.7	1	10.88		1.27		1.18		2		3011
		IC 4651 - 107		17 20 44	−49 54.9	1	12.58		0.54		0.04		2		3011
		IC 4651 - 206		17 20 44	−49 58.2	1	13.78		0.63		0.06		2		3011
157337	−31 13957			17 20 45	−31 25.2	1	9.60		0.45		0.26		2	A8/9 V	918
		IC 4651 - 109		17 20 45	−49 55.1	1	13.52		0.58		0.01		3		3011
		IC 4651 - 202		17 20 46	−49 57.1	1	12.68		0.50		0.10		2		3011
323484	−42 12104	LSS 4130		17 20 47	−42 08.6	2	10.08	.002	0.47	.009	−0.40		6	B1 III	540,1021
		IC 4651 - 203		17 20 47	−49 57.4	1	12.46		0.55		0.11		2		3011
	−34 11664	LSS 4133		17 20 48	−34 50.7	1	10.46		1.21		0.25		2		754
		IC 4651 - 175		17 20 48	−49 50.3	1	12.72		0.54		0.09		2		3011
		IC 4651 - 98		17 20 48	−49 53.9	1	13.22		0.56		0.08		2		3011
		IC 4651 - 801		17 20 48	−49 54.	1	15.62		0.56		0.10		2		3011
		IC 4651 - 802		17 20 48	−49 54.	1	14.93		0.93		0.63		2		3011
		IC 4651 - 803		17 20 48	−49 54.	1	15.30		0.70		0.30		2		3011
		IC 4651 - 804		17 20 48	−49 54.	1	14.40		0.66		0.09		2		3011
		IC 4651 - 805		17 20 48	−49 54.	1	14.55		0.74		0.41		2		3011
		IC 4651 - 806		17 20 48	−49 54.	1	14.96		0.78		0.25		2		3011
		IC 4651 - 807		17 20 48	−49 54.	1	15.39		0.81		0.42		1		3011
		IC 4651 - 808		17 20 48	−49 54.	1	14.63		0.77		0.23		1		3011
		IC 4651 - 99		17 20 49	−49 54.4	1	13.60		0.62		0.14		2		3011
	−49 11401	IC 4651 - 113		17 20 49	−49 56.4	1	10.40		1.11		0.90		2	K0	3011
		IC 4651 - 174		17 20 50	−49 50.3	1	12.46		0.53		0.09		2		3011
		IC 4651 - 15		17 20 51	−49 53.9	1	13.48		1.43				1		3011
		IC 4651 - 78		17 20 52	−49 51.9	1	12.20		0.60		0.07		2		3011
		IC 4651 - 14		17 20 53	−49 53.2	1	12.12		0.62		0.09		3		3011
		IC 4651 - 208		17 20 53	−49 58.2	1	12.92		0.53		0.08		2		3011
		IC 4651 - 77		17 20 54	−49 52.1	1	13.53		0.56		0.04		2		3011
		IC 4651 - 31		17 20 54	−49 55.4	1	12.40		0.59		0.11		2		3011
	+28 02730	G 181 - 41		17 20 55	+28 52.1	1	9.63		0.80		0.62		2	K2	7010
317524	−31 13961			17 20 55	−31 06.1	1	9.79		0.51		−0.24		2	B5	918
	−49 11402	IC 4651 - 76	⋆ AB	17 20 55	−49 51.4	2	10.93	.015	1.17	.012	1.03	.002	3		3011,4002
		IC 4651 - 24		17 20 55	−49 53.9	1	12.68		0.55		0.09		2		3011
		IC 4651 - 116		17 20 55	−49 56.1	1	13.64		0.60		0.08		2		3011
		IC 4651 - 26		17 20 56	−49 53.9	1	13.78		0.58		0.06		2		3011
	−49 11404	IC 4651 - 27		17 20 56	−49 54.0	2	10.88	.023	1.22	.013	1.17	.027	4		3011,4002
		IC 4651 - 29		17 20 57	−49 54.7	2	10.89	.009	1.13	.003	1.02	.034	4		3011,4002
157269	−49 11406	IC 4651 - 293		17 20 57	−50 03.0	1	9.67		0.16		0.07		3	A1 V	185
		IC 4651 - 18		17 20 58	−49 53.1	1	12.92		0.59		0.05		2		3011
		IC 4651 - 75		17 20 59	−49 51.5	1	11.68		1.30		1.33		1		3011
		IC 4651 - 28		17 20 59	−49 54.6	1	12.88		0.55		0.05		2		3011
		IC 4651 - 117		17 20 59	−49 56.1	1	11.75		0.54		0.17		2		185
		IC 4651 - 118		17 20 59	−49 57.0	2	11.81	.070	0.59	.018	0.13	.000	4		185,3011
157363	−31 13964			17 21 00	−32 04.0	1	10.29		0.13		−0.07		2	Ap Si	918
		IC 4651 - 211		17 21 00	−49 58.4	1	14.28		0.65		0.11		2		3011

Table 1   943

HD	DM	Other Id	N Rem	α_{1950}	δ_{1950}	S	V	σ_V	B−V	σ_{B-V}	U−B	σ_{U-B}	n	Spectrum	References
		IC 4651 - 22		17 21 01	−49 53.8	1	12.12		0.61		0.13		3		3011
		IC 4651 - 33		17 21 01	−49 55.7	2	13.14	.015	0.55	.010	0.15	.054	6		185,3011
157382	−31 13966			17 21 02	−31 56.4	1	8.50		1.11		0.69		2	G3/5 III	918
	−49 11415	IC 4651 - 72		17 21 02	−49 50.4	2	10.43	.015	1.32	.005	1.27	.017	4	K2	185,3011
		IC 4651 - 7		17 21 02	−49 53.1	1	13.50		0.62		0.12		2		3011
		IC 4651 - 1		17 21 02	−49 53.6	1	12.24		0.69		0.14		3		3011
157316	−44 11680			17 21 03	−44 57.8	8	6.65	.007	0.38	.003	-0.01	.004	54	F3 V	116,278,863,977,1075,1586*
		IC 4651 - 2		17 21 04	−49 53.7	1	12.30		0.58		0.10		2		3011
		IC 4651 - 34		17 21 04	−49 55.1	2	12.44	.010	0.30	.015	0.14	.015	6		185,3011
		IC 4651 - 119		17 21 05	−49 56.4	2	14.04	.010	0.55	.017	0.07	.029	6		185,3011
		IC 4651 - 67		17 21 06	−49 51.5	1	13.68		1.26		1.17		2		3011
		IC 4651 - 4		17 21 06	−49 54.3	1	13.98		1.57				1		3011
	−49 11410	IC 4651 - 122		17 21 06	−49 59.3	1	10.78		1.11		0.82		2		3011
157416	−29 13484			17 21 08	−29 49.5	1	7.63		0.05		-0.22		3	B9 III	918
		IC 4651 - 61		17 21 08	−49 51.8	1	13.18		0.55		0.05		2		3011
157244	−55 08100	HR 6461		17 21 08	−55 29.1	3	2.84	.004	1.46	.000	1.56	.000	11	K3 Ib-IIa	15,1075,2012
157543	+14 03243			17 21 09	+14 48.6	1	8.39		1.04				2	K0	882
		AA69,51 # 62		17 21 09	−34 08.4	1	15.07		1.09		0.36		1	B8	1537
		IC 4651 - 64		17 21 09	−49 51.2	1	12.04		0.77		0.28		3		3011
		IC 4651 - 35		17 21 09	−49 55.0	1	12.20		0.55		0.10		3		185
		IC 4651 - 36		17 21 09	−49 55.5	1	12.24		0.57		0.08		2		3011
		Pis 24 - 15		17 21 10	−34 12.1	2	12.31	.005	1.30	.025	0.21	.059	5		251,754
157317	−46 11478			17 21 10	−46 09.4	2	6.76	.005	-0.06	.000	-0.46	.005	6	B7 III	158,401
		IC 4651 - 60		17 21 10	−49 51.9	1	13.84		0.58		0.05		2		3011
		IC 4651 - 123		17 21 10	−49 58.4	1	11.49		0.41		0.15		2		3011
157246	−56 08225	HR 6462	⋆ A	17 21 11	−56 20.0	6	3.33	.006	-0.14	.011	-0.95	.011	21	B1 Ib	15,26,1034,1075,1637,2012
		Pis 24 - 14		17 21 12	−34 08.5	2	13.40	.020	0.83	.005	0.44	.060	4		251,754
		IC 4651 - 62		17 21 12	−49 51.2	1	11.64		0.61		0.12		3	A2	3011
	−49 11413	IC 4651 - 37		17 21 12	−49 55.7	3	10.20	.013	1.04	.011	0.77	.025	9	G8	185,1730,3011
		IC 4651 - 126		17 21 12	−49 57.8	1	12.56		0.47		0.08		2		3011
157498	−9 04546	IDS17185S0916	AB	17 21 13	−09 18.8	2	7.86	.030	0.62	.000	0.06		6	G0	173,3026
319711	−33 12025	LSS 4138		17 21 13	−33 57.2	2	9.50	.011	1.23	.034	1.11	.058	3	A2 Ib	754,954
	−49 11414	IC 4651 - 44	⋆ AB	17 21 13	−49 54.0	2	10.54	.000	0.28	.007	0.18	.020	4	A5 IV	185,3011
		IC 4651 - 38		17 21 13	−49 55.0	1	13.17		0.56		0.03		2		3011
157434	−28 13101			17 21 14	−28 36.5	1	8.07		0.28		0.07		2	B9.5IV/V	918
315761				17 21 14	−30 00.0	1	9.98		0.43		0.30		3		918
		IC 4651 - 48		17 21 14	−49 53.3	1	12.46		0.55		0.08		2		3011
		IC 4651 - 131		17 21 14	−49 57.2	1	12.35		0.62		0.11		2		3011
157400	−35 11531			17 21 15	−35 47.6	1	9.57		0.50		-0.38		2	B3/5	540
		IC 4651 - 49		17 21 15	−49 52.7	1	12.89		0.47		0.11		2		3011
		IC 4651 - 45		17 21 15	−49 53.9	1	12.61		0.55		0.03		2		3011
		IC 4651 - 127		17 21 15	−49 58.3	1	11.58		0.56		0.10		2		3011
315773				17 21 16	−30 05.3	1	10.74		0.26		-0.23		2		918
		Pis 24 - 11		17 21 16	−34 10.6	1	14.35		1.57		0.30		2		251
		IC 4651 - 47		17 21 16	−49 53.5	1	13.50		0.52		0.04		2		3011
		IC 4651 - 132		17 21 16	−49 57.3	1	12.56		0.53		0.11		2		3011
		Pis 24 - 10		17 21 17	−34 11.3	2	13.05	.025	1.41	.005	0.38	.035	4		251,754
319725	−35 11532			17 21 17	−35 20.4	1	9.32		1.34		0.90		4	M0	119
	−49 11415	IC 4651 - 56		17 21 17	−49 51.3	1	8.93		1.67		2.07		3	K3 II	1730
	−49 11417	IC 4651 - 56		17 21 17	−49 51.3	2	8.96	.010	1.65	.030	1.96	.045	5	K3 II	185,3011
	−49 11420	IC 4651 - 153		17 21 17	−49 51.3	2	12.08	.020	0.61	.000	0.13	.002	5	A1	185,3011
		IC 4651 - 39		17 21 17	−49 55.7	1	13.07		0.56		0.04		2		3011
	−38 11837	LSS 4137		17 21 18	−38 41.2	3	10.58	.025	0.75	.021	-0.32	.010	7		954,1586,1737
157354	−45 11488			17 21 18	−45 51.0	2	8.02	.002	1.61	.013	1.85		12	K3 III	977,2011
		IC 4651 - 46		17 21 18	−49 53.6	1	12.83		1.27				2		3011
		Pis 24 - 8		17 21 20	−34 12.3	3	13.00	.011	1.47	.015	0.50	.051	8	B1	251,754,1537
		Pis 24 - 9	⋆	17 21 21	−34 12.8	2	14.30	.049	1.39	.010	0.31	.112	3	B3	251,1537
		IC 4651 - 52		17 21 21	−49 52.2	1	13.80		0.63		0.09		1		3011
		IC 4651 - 137		17 21 21	−49 56.8	1	11.00		0.27		0.16		2	A0	185
		Pis 24 - 4	⋆	17 21 22	−34 09.4	2	13.95	.029	1.46	.039	0.43	.122	3	B2	251,1537
		IC 4651 - 43		17 21 22	−49 54.1	1	14.27		0.62		0.05		1		3011
		IC 4651 - 136		17 21 22	−49 57.3	1	12.06		0.62		0.12		2	B9	3011
		AA69,51 # 60		17 21 23	−34 09.3	1	15.91		1.64		2.37		1	M3 III?	1537
		IC 4651 - 149		17 21 23	−49 52.1	1	13.85		1.15		0.84		1		3011
		IC 4651 - 42		17 21 23	−49 54.5	1	12.63		0.56		0.07		2		3011
		Pis 24 - 12		17 21 24	−34 09.0	1	13.88		1.47		0.38		2		251
		Pis 24 - 3		17 21 24	−34 10.7	2	12.77	.020	1.41	.000	0.32	.069	5		251,754
		IC 4651 - 148		17 21 24	−49 52.2	1	14.60		0.71		0.34		1		3011
		IC 4651 - 50		17 21 24	−49 52.9	1	13.68		0.54		0.04		2		3011
319718	−34 11671	Pis 24 - 1	⋆ A	17 21 25	−34 09.2	7	10.34	.051	1.47	.022	0.41	.024	24	F8	251,390,754,914,954*
		Pis 24 - 2		17 21 25	−34 10.0	2	11.96	.005	1.39	.020	0.39	.064	5		251,754
		G 258 - 4		17 21 26	+71 08.9	1	12.91		0.57		-0.11		2		1658
		AA69,51 # 57		17 21 26	−34 09.3	1	11.58		1.56		0.47		3	O6	1537
		AA69,51 # 58		17 21 26	−34 09.3	1	14.15		1.50		1.22		1	K0 III	1537
		Pis 24 - 6		17 21 26	−34 12.3	2	13.77	.030	1.62	.020	1.52	.094	5		251,754
		IC 4651 - 143		17 21 26	−49 55.1	1	14.35		0.74		0.15		2		3011
		IC 4651 - 142		17 21 26	−49 55.5	1	12.70		0.52		0.09		2		3011
		AA69,51 # 65		17 21 27	−34 02.1	1	14.21		0.78		0.27		1		1537
		Pis 24 - 13		17 21 27	−34 07.0	5	12.74	.034	1.47	.021	0.28	.060	13	B1	251,434,754,914,1537
157485	−26 12112	LSS 4146		17 21 28	−26 52.8	1	9.07		0.56		-0.30		2	B1/2 Ib	540

HD	DM	Other Id	N Rem	α_{1950}	δ_{1950}	S	V	σ_V	B–V	σ_{B-V}	U–B	σ_{U-B}	n	Spectrum	References
157447	−30 14149			17 21 28	−31 03.6	1	9.59		0.15		0.08		2	B9 V	918
		AA69,51 # 64		17 21 28	−34 04.1	1	13.05		0.86		0.18		1	B4	1537
		AA69,51 # 63		17 21 28	−34 05.8	1	15.23		1.08		0.33		1	B9	1537
		Pis 24 - 5		17 21 28	−34 11.3	2	13.62	.025	0.87	.010	0.82	.055	5		251,754
		IC 4651 - 151		17 21 28	−49 51.4	1	13.85		1.57				1		3011
		Pis 24 - 7		17 21 29	−34 12.6	2	13.42	.040	1.70	.020	0.66	.080	4		251,754
		IC 4651 - 236		17 21 30	−49 50.3	1	10.61		1.01		0.69		2		3011
		IC 4651 - 141		17 21 30	−49 56.0	1	12.80		0.53		0.07		2		3011
		AA69,51 # 66		17 21 31	−34 01.4	1	12.59		0.75		0.24		1		1537
		NGC 6352 sq1 10		17 21 32	−48 25.0	1	12.88		0.57		0.19		3		288
		IC 4651 - 147		17 21 32	−49 52.8	1	13.12		0.56		0.08		2		3011
		IC 4651 - 146		17 21 32	−49 53.1	1	10.94		1.14		0.94		2		3011
	−49 11421	IC 4651 - 139		17 21 32	−49 56.2	1	10.68		1.12		0.89		2		3011
157617	+8 03405	HR 6476		17 21 34	+08 53.9	2	5.76	.005	1.25	.000	1.27	.000	7	K2	15,1256
		AA69,51 # 67		17 21 35	−34 01.2	1	13.10		0.57		0.23		1		1537
323285	−37 11549	LSS 4143		17 21 35	−37 42.9	4	9.21	.012	0.84	.004	-0.17	.031	6	B1 II-Ib	540,954,976,1737
		Steph 1467		17 21 36	+25 05.1	1	11.87		1.48		1.20		2	M2	1746
		NGC 6352 - 1001		17 21 36	−48 25.	1	11.54		1.66		1.47		5		143
		NGC 6352 - 1011		17 21 36	−48 25.	1	15.22		1.10		0.81		3		143
		NGC 6352 - 1012		17 21 36	−48 25.	1	13.97		0.73		0.17		3		143
		NGC 6352 - 1013		17 21 36	−48 25.	1	14.14		0.68				3		143
		NGC 6352 - 1014		17 21 36	−48 25.	1	15.02		1.13				2		143
		NGC 6352 - 1015		17 21 36	−48 25.	1	14.85		1.18				3		143
		NGC 6352 - 1016		17 21 36	−48 25.	1	15.03		1.31				3		143
		NGC 6352 - 1027		17 21 36	−48 25.	1	12.05		1.56		1.94		4		288
		NGC 6352 - 1029		17 21 36	−48 25.	1	12.20		0.61		0.19		4		288
		NGC 6352 - 1081		17 21 36	−48 25.	1	13.54		1.87		2.44		2		288
		NGC 6352 - 1094		17 21 36	−48 25.	1	12.92		1.69		1.84		5		288
		NGC 6352 - 111		17 21 36	−48 25.	1	14.40		1.38				3		143
		NGC 6352 - 113		17 21 36	−48 25.	1	13.18		1.86				3		143
		NGC 6352 - 118		17 21 36	−48 25.	1	15.03		1.30		0.98		3		143
		NGC 6352 - 125		17 21 36	−48 25.	1	15.16		1.03		0.61		3		143
		NGC 6352 - 128		17 21 36	−48 25.	1	12.70		0.54		0.03		4		143
		NGC 6352 - 135		17 21 36	−48 25.	1	12.67		0.69		0.11		5		143
		NGC 6352 - 138		17 21 36	−48 25.	1	13.97		1.42				3		143
		NGC 6352 - 142		17 21 36	−48 25.	1	14.38		1.47		1.39		2		143
		NGC 6352 - 161		17 21 36	−48 25.	1	13.76		1.65		1.53		2		143
		NGC 6352 - 162		17 21 36	−48 25.	1	14.50		1.51				2		143
		NGC 6352 - 163		17 21 36	−48 25.	1	14.79		1.38				2		143
		NGC 6352 - 167		17 21 36	−48 25.	1	15.18		1.05				2		143
		NGC 6352 - 195		17 21 36	−48 25.	2	12.06	.005	1.20	.015	1.05	.030	7		143,288
		NGC 6352 - 901		17 21 36	−48 25.	1	13.25		1.62		2.14		5		288
		NGC 6352 - 903		17 21 36	−48 25.	1	15.80		1.22		1.27		1		288
		NGC 6352 - 904		17 21 36	−48 25.	1	14.14		0.73		0.27		3		288
		NGC 6352 - 906		17 21 36	−48 25.	1	15.95		0.61		0.12		2		288
		NGC 6352 - 907		17 21 36	−48 25.	1	12.80		1.49		1.82		5		288
		NGC 6352 - 908		17 21 36	−48 25.	1	11.66		1.51		1.82		5		288
		NGC 6352 - 911		17 21 36	−48 25.	1	10.98		0.42		0.07		5		288
		NGC 6352 - 912		17 21 36	−48 25.	1	16.78		0.89		1.07		1		288
		NGC 6352 - 913		17 21 36	−48 25.	1	13.86		1.11		0.83		2		288
		NGC 6352 - 914		17 21 36	−48 25.	1	15.28		1.30		1.63		1		288
		NGC 6352 - 915		17 21 36	−48 25.	1	15.78		1.23		1.19		1		288
		NGC 6352 - 916		17 21 36	−48 25.	1	16.36		1.33		0.95		1		288
		NGC 6352 - 919		17 21 36	−48 25.	1	15.77		0.70		0.32		1		288
		NGC 6352 - 920		17 21 36	−48 25.	1	19.00		1.27				1		288
		NGC 6352 - 921		17 21 36	−48 25.	1	14.52		1.41		1.42		3		288
		NGC 6352 - 922		17 21 36	−48 25.	1	14.09		1.62		1.93		2		288
		NGC 6352 - 923		17 21 36	−48 25.	1	17.59		0.96				1		288
		NGC 6352 - 924		17 21 36	−48 25.	1	12.63		0.58				1		288
		NGC 6352 - 925		17 21 36	−48 25.	1	17.99		0.98				1		288
		NGC 6352 - 926		17 21 36	−48 25.	1	16.84		0.79				2		288
		NGC 6352 - 927		17 21 36	−48 25.	1	15.18		1.13		1.07		3		288
		NGC 6352 - 928		17 21 36	−48 25.	1	16.69		0.76				2		288
		NGC 6352 - 929		17 21 36	−48 25.	1	15.86		1.41				2		288
		NGC 6352 - 931		17 21 36	−48 25.	1	17.89		0.82				1		288
		NGC 6352 - 932		17 21 36	−48 25.	1	17.37		0.87				1		288
		NGC 6352 - 933		17 21 36	−48 25.	1	14.92		1.74		2.36		2		288
		NGC 6352 - 934		17 21 36	−48 25.	1	17.57		0.99				1		288
		NGC 6352 - 935		17 21 36	−48 25.	1	16.21		1.05				1		288
		NGC 6352 - 936		17 21 36	−48 25.	1	17.05		0.84				2		288
		NGC 6352 - 937		17 21 36	−48 25.	1	14.65		0.56		0.19		3		288
		NGC 6352 - 938		17 21 36	−48 25.	1	14.20		0.73		0.21		1		288
		NGC 6352 - 939		17 21 36	−48 25.	1	15.76		0.95				1		288
		NGC 6352 - 940		17 21 36	−48 25.	1	13.61		1.53		1.76		2		288
		NGC 6352 - 941		17 21 36	−48 25.	1	17.19		1.24				1		288
		NGC 6352 - 942		17 21 36	−48 25.	1	13.24		1.68		2.15		2		288
		NGC 6352 - 943		17 21 36	−48 25.	1	15.90		0.99				1		288
		NGC 6352 - 944		17 21 36	−48 25.	1	15.07		1.07		0.73		1		288
		NGC 6352 - 945		17 21 36	−48 25.	1	17.97		0.96				1		288
		NGC 6352 - 946		17 21 36	−48 25.	1	16.35		1.29		1.72		2		288

Table 1 945

HD	DM	Other Id	N Rem	α_{1950}	δ_{1950}	S	V	σ_V	B–V	σ_{B-V}	U–B	σ_{U-B}	n	Spectrum	References
		NGC 6352 - 947		17 21 36	−48 25.	1	14.45		1.64		2.02		2		288
		NGC 6352 - 948		17 21 36	−48 25.	1	16.81		0.89		0.87		2		288
		NGC 6352 - 949		17 21 36	−48 25.	1	17.61		0.79				1		288
		NGC 6352 - 950		17 21 36	−48 25.	1	18.20		0.85				1		288
		NGC 6352 - 951		17 21 36	−48 25.	1	17.31		1.04				1		288
		NGC 6352 - 952		17 21 36	−48 25.	1	18.10		0.60				1		288
		NGC 6352 - 953		17 21 36	−48 25.	1	19.22		0.75				1		288
		NGC 6352 - 954		17 21 36	−48 25.	1	15.16		0.77		0.20		1		288
		NGC 6352 - 955		17 21 36	−48 25.	1	18.42		1.24				1		288
		NGC 6352 - 956		17 21 36	−48 25.	1	18.23		0.87				2		288
		NGC 6352 - 957		17 21 36	−48 25.	1	13.53		1.43				1		288
		NGC 6352 - 958		17 21 36	−48 25.	1	14.07		1.36		1.41		1		288
		NGC 6352 - 959		17 21 36	−48 25.	1	16.32		1.37				1		288
		NGC 6352 - 960		17 21 36	−48 25.	1	13.43		1.76				1		288
		NGC 6352 - 961		17 21 36	−48 25.	1	13.20		1.37				1		288
		NGC 6352 - 962		17 21 36	−48 25.	1	15.82		0.86		0.66		1		288
		NGC 6352 - 963		17 21 36	−48 25.	1	13.66		0.93		0.62		1		288
		NGC 6352 - 964		17 21 36	−48 25.	1	14.94		1.32		0.80		1		288
		NGC 6352 - 1112		17 21 36	−48 25.	1	15.46		1.44				3		143
		NGC 6352 - 1213		17 21 36	−48 25.	2	12.38	.000	0.67	.015	0.10	.020	8		143,288
		NGC 6352 - 1215		17 21 36	−48 25.	1	15.49		0.80				2		143
157473	−32 12711			17 21 37	−32 27.1	1	9.61		0.11		−0.32		2	B8/9 II/III	918
		AA69,51 # 68		17 21 37	−34 01.4	1	11.64		0.51		0.33		1		1537
		L 485 - 8		17 21 37	−35 26.0	1	13.25		1.56		1.21		1		1773
	−44 11695		AB	17 21 37	−44 48.7	2	9.94	.008	0.54	.001	0.05		11	G0	863,2011
		L 989 - 38	AB	17 21 38	+00 04.0	1	12.19		1.51				1		1746
317525	−31 13982			17 21 38	−31 19.7	1	9.73		1.47		1.40		2	K5	918
	−37 11550	LSS 4145		17 21 38	−37 41.0	2	10.55	.030	0.87	.000	−0.14	.035	4	B0.5 V	954,1737
		LP 687 - 31		17 21 39	−04 19.1	1	12.14		1.51		1.09		1	M2	1696
157546	−18 04516	HR 6473		17 21 41	−18 24.1	3	6.36	.009	0.04	.017	−0.09	.003	7	B9 Vn	252,1225,1770,2009
157527	−21 04597	HR 6472	⋆ AB	17 21 42	−21 23.8	2	5.85	.005	0.93	.005	0.67		6	K0 III	2035,3077
157486	−34 11674	HR 6470		17 21 43	−34 39.1	2	6.15	.012	0.03	.009	0.03		8	A0p SiCr	1637,2009
	−32 12714			17 21 46	−32 28.8	1	10.65		0.63		0.12		2		918
	−48 11683			17 21 46	−48 27.5	1	10.48		1.56		1.62		7		288
		G 19 - 23		17 21 47	−01 10.4	1	14.94		1.36		0.85		1		3062
157504	−34 11675	LSS 4148		17 21 50	−34 08.6	1	10.68		1.42		0.70		3	WC	754
157273	−63 04103			17 21 50	−63 46.3	1	8.91		0.28		−0.30		3	B8/9 Vn	1586
157666	+14 03246			17 21 51	+14 28.0	1	8.89		0.23		0.16		2	A3	1776
	−48 11685			17 21 53	−48 10.2	1	9.98		1.58		1.98		9		288
157806	+43 02724			17 21 55	+43 01.3	1	8.80		1.42		1.67		2	K2	1733
157779	+37 02878	HR 6485	⋆ APB	17 21 57	+37 11.4	5	4.17	.009	−0.03	.025	−0.03	.033	14	B9.5III	15,938,1022,1363,8015
157728	+23 03100	HR 6480		17 22 01	+23 00.3	2	5.73	.010	0.21	.000	0.05	.005	5	F0 IV	985,1733
157477	−45 11506	IDS17184S4510	AB	17 22 01	−45 13.3	17	8.09	.006	0.24	.005	0.20	.009	328	A5 V	116,278,460,657,863,977*
157487	−44 11703			17 22 02	−44 44.1	17	7.64	.012	1.25	.008	1.14	.011	432	G8 III	116,125,278,863,977,977*
157588	−24 13325	HR 6474		17 22 03	−24 12.0	1	6.19		1.10				4	K0 III	2006
156513	−80 00828	NO Aps, HR 6429		17 22 04	−80 49.1	3	5.89	.013	1.65	.015	1.80	.000	12	M3 III	15,1075,2038
157550	−31 13994			17 22 05	−31 31.2	1	9.71		0.32		0.18		2	A0/1 IV	918
157519	−41 11599			17 22 05	−41 56.1	2	7.85	.005	0.05	.003	−0.38	.008	4	B5 IV	401,540
		Pis 24 - 17		17 22 06	−34 23.	1	14.50		0.84		0.10		1		754
		Pis 24 - 16	⋆ BC	17 22 06	−34 23.	1	13.44		1.01		0.46		2		754
157457	−50 11269	HR 6468	⋆ A	17 22 06	−50 35.4	3	5.22	.008	1.05	.005			11	G8 III	15,1075,2007
157455	−49 11431			17 22 07	−49 41.5	1	6.94		−0.06		−0.61		2	B3 III	3011
157506	−44 11704			17 22 09	−44 48.4	1	9.99		0.18				4	A0 V	2011
157488	−48 11687			17 22 09	−48 08.9	1	9.80		0.35		0.13		8	A9 IV	288
		AA69,51 # 51	A	17 22 10	−34 22.6	1	11.91		1.45		0.44		2	O9	1537
		AA69,51 # 51	AB	17 22 10	−34 22.6	1	11.58		1.27		0.57		3		754
		AA69,51 # 71		17 22 10	−34 22.8	1	12.47		1.04		0.61		1	G2 III	1537
	−16 04505			17 22 12	−16 40.8	1	10.35		1.47				2		1726
	+37 02879	G 181 - 43		17 22 14	+37 19.8	2	8.88	.015	0.81	.005	0.45	.025	2	K0	1658,3060
		AA69,51 # 50		17 22 14	−34 21.4	2	12.27	.010	1.41	.079	0.51	.020	5	O9.5	754,1537
		AA69,51 # 70		17 22 14	−34 21.9	1	13.86		1.37		0.22		2	B2	1537
		AA69,51 # 49		17 22 15	−34 20.6	2	11.85	.040	1.72	.015	0.77	.059	5	O5	754,1537
		AA69,51 # 69		17 22 15	−34 21.5	1	13.59		0.94		0.35		2	G0 V	1537
157740	+16 03174	HR 6481		17 22 17	+16 20.7	2	5.71		0.07	.010	0.12	.020	5	A3 V	252,1022
157741	+15 03179	HR 6482	⋆ AB	17 22 18	+15 39.0	2			−0.02	.014	−0.18	.019	4	B9 V	1022,1079
317520	−30 14172			17 22 19	−30 56.6	1	9.95		0.58		0.06		2	F8	918
157520	−45 11512			17 22 19	−46 02.8	1	9.74		1.16				4	G8/K0 (III)	2011
	−34 11680	LSS 4151		17 22 20	−34 06.0	1	10.92		1.20		0.19		3		754
157853	+38 02928	HR 6488	⋆ AB	17 22 21	+38 37.6	1	6.51		0.72		0.19		1	F8 IV	254
157623	−30 14173			17 22 23	−30 58.5	1	9.59		0.28		0.27		2	A1 IV	918
157522	−48 11693			17 22 24	−48 32.8	1	8.82		0.81		0.40		8	K1 V	288
157572	−42 12126			17 22 25	−42 30.9	1	8.60		0.33		−0.35		2	B2 IV	540
157907	+47 02470			17 22 26	+47 03.9	1	8.45		1.12		1.09		2	K0	1601
157622	−30 14174			17 22 28	−30 40.7	1	9.21		0.22		0.06		2	A0 V	918
319858	−37 11566			17 22 29	−37 26.0	2	10.40	.015	1.06	.000	0.07		4	B5	954,1737
157523	−50 11275			17 22 29	−50 05.0	1	8.96		−0.02		−0.44		2	B5 II/III	3011
		G 181 - 45		17 22 30	+41 02.4	2	12.32	.017	0.71	.027	0.03	.054	3	G7	1658,3060
157641	−32 12721	IDS17193S3222	AB	17 22 34	−32 25.3	1	9.45		0.63		0.35		2	F2 V	918
157573	−44 11716			17 22 34	−44 30.7	2	9.10	.000	−0.01	.000	−0.43		8	B7 II	1075,2011
	−44 11715			17 22 34	−44 49.5	1	10.13		0.40				4	F0	2011

HD	DM	Other Id	N	Rem	α_{1950}	δ_{1950}	S	V	σ_V	B–V	σ_{B-V}	U–B	σ_{U-B}	n	Spectrum	References
157555	−48 11697				17 22 34	−48 34.3	1	7.04		0.48		0.01		8	F6/7 V	288
		LSS 4154			17 22 35	−35 58.5	1	11.91		1.07		0.15		2	B3 Ib	954
157533	−51 10877				17 22 35	−51 15.0	1	9.83		0.00		−0.26		3	B9 V	3011
	−45 11520				17 22 36	−45 07.9	4	10.54	.005	0.17	.006	0.04	.011	21	B8	116,863,1460,2011
157640	−32 12722				17 22 37	−32 14.2	1	10.03		0.18		−0.06		2	B8/9 III/IV	918
		V478 Sco			17 22 38	−35 29.6	1	15.69		0.37		−0.54		1		1471
157692	−28 13126				17 22 41	−28 36.6	1	9.02		0.76		0.36		1	G3/6 V	918
317531	−31 14010				17 22 41	−31 29.1	1	9.76		1.76		1.85		2		918
	−44 11719				17 22 41	−44 26.9	1	9.35		1.09				4	G5	2011
	−47 11538				17 22 41	−47 59.7	1	10.99		0.39		0.12		2		540
157910	+37 02882	HR 6491	⋆	A	17 22 43	+36 59.7	2	6.22	.050	0.88	.005	0.52	.010	6	G5 III	1084,3024
157910	+37 02882	IDS17210N3702		B	17 22 43	+36 59.7	2	9.92	.039	0.45	.010	−0.03	.010	7		1084,3024
157556	−49 11438				17 22 44	−49 51.0	1	7.92		−0.03		−0.48		2	B4 IV	3011
157672	−31 14011				17 22 46	−32 00.1	1	9.88		0.15		0.00		2	B9 V	918
157646	−41 11615				17 22 48	−42 03.5	1	9.70		0.13		−0.61		2	B3 II	540
	−44 11722			AB	17 22 50	−44 40.5	1	10.91		0.26				4		2011
157693	−34 11685				17 22 51	−34 30.9	2	8.77	.007	0.47	.010	−0.14	.011	3	B4 II	540,976
	−44 11723				17 22 52	−44 41.1	1	10.92		0.24		0.20		6		863
157624	−44 11725				17 22 53	−44 35.2	5	7.87	.004	−0.05	.004	−0.57	.003	114	B3 III	474,977,1075,1586,2011
157598	−50 11280				17 22 53	−50 14.3	1	9.64		−0.01		−0.43		2	B8 IV	3011
	−44 11724				17 22 55	−45 00.4	3	9.78	.006	1.13	.001	0.96	.002	12	K0	116,863,2011
157947	+39 03124	IDS17214N3918		B	17 22 57	+39 13.8	2	8.42	.005	0.40	.000	−0.02	.010	5	F2 V	833,1601
157746	−21 04604				17 22 58	−21 22.4	1	9.29		0.38		0.02		1	Ap Si	400
157736	−27 11652	IDS17198S2731		AB	17 22 58	−27 33.4	1	7.50		0.09		0.03		1	B9/9.5V	918
157719	−28 13132				17 22 58	−28 29.9	1	7.78		0.87		0.49		2	G8 IV/V	918
157599	−51 10881	HR 6475			17 22 58	−51 54.4	2	6.19	.005	−0.03	.005	−0.25		7	B8/9 V	2035,3011
157948	+38 02932	G 182 - 7	⋆	AB	17 23 01	+38 05.0	1	8.10		0.76		0.31		2	G5	1658
317604	−31 14017				17 23 01	−31 36.0	1	9.29		1.46		1.24		2	K0	918
	−44 11728				17 23 03	−44 37.1	1	11.24		0.47				4		2011
157694	−39 11456				17 23 04	−39 43.1	1	9.16		−0.01		−0.52		7	B3 III/IV	1732
	+20 03471				17 23 06	+19 58.0	1	8.59		1.64		1.80		2	M0	1648
157749	−28 13136				17 23 06	−28 22.4	1	7.22		0.38		0.12		2	F3 III	918
		L 341 - 114			17 23 06	−47 09.	1	13.39		1.53		1.10		1		3073
319788	−34 11687				17 23 07	−34 14.5	2	9.62	.003	0.50	.001	−0.35	.017	4	B1.5V	92,540
		G 20 - 1			17 23 08	+01 01.9	2	13.87	.034	1.50	.015	0.96		3		202,3062
157661	−45 11531	HR 6477	⋆	AB	17 23 10	−45 48.0	8	5.28	.004	−0.06	.004	−0.34	.002	53	B7 V+B9.5 V	15,278,977,977,1075,1770*
	−44 11730				17 23 11	−45 00.1	1	10.93		0.48				4		2011
158064	+50 02400				17 23 12	+50 28.2	1	7.63		0.40		0.00		2	F2	1566
	+59 01809				17 23 13	+59 49.8	1	9.71		0.51		0.00		4	F7	1371
157697	−44 11731				17 23 14	−44 39.6	9	10.00	.008	0.05	.011	−0.26	.008	87	B9 II/III	116,863,1460,1628,1657,1673*
319826	−36 11499				17 23 15	−36 22.8	1	12.97		0.65		0.50		1	B8	1586
157881	+2 03312	G 19 - 24			17 23 16	+02 10.2	17	7.54	.016	1.36	.010	1.26	.012	80	K7 V	1,116,202,265,989,1006*
323374	−39 11461	LSS 4158			17 23 16	−39 40.1	1	10.80		0.44		−0.31		3		1586
		G 182 - 8			17 23 18	+35 37.1	1	12.73		1.27		1.09		1		1658
157750	−32 12737				17 23 18	−32 55.5	2	8.03	.005	0.67	.000	0.18		5	G8 IV/V+(F)	657,1075
		E7 - h		AB	17 23 18	−45 03.	1	11.85		0.80				4		2011
157662	−50 11283	HR 6478	⋆	A	17 23 18	−50 35.3	2	5.91	.005	0.09	.012	0.05		6	B9 II	2007,3011
157792	−24 13337	HR 6486			17 23 19	−24 07.9	8	4.17	.012	0.28	.004	0.11	.013	24	A3mF0 (IV)	15,614,1020,1075,1586*
	−40 11454	LSS 4159			17 23 19	−40 12.0	1	9.64		0.05		−0.31		1		1586
157698	−47 11549				17 23 19	−47 05.6	2	7.14	.010	−0.05	.000	−0.46		6	B7 III	401,2012
		G 170 - 37			17 23 20	+19 26.2	1	15.26		1.53				1		1759
157711	−46 11513				17 23 21	−46 33.3	1	8.63		−0.04		−0.46		2	B9 II/III	401
157856	−1 03329	HR 6489			17 23 22	−01 36.6	2	6.43	.005	0.46	.000	−0.03	.010	7	F3 V	15,1256
		G 19 - 25			17 23 22	−02 41.7	4	11.68	.019	0.77	.007	0.10	.013	8		202,1620,1696,3062
158996	+80 00544	HR 6529			17 23 23	+80 11.0	2	5.74	.018	1.50	.000	1.80		5	K5 III	71,3016
158096	+50 02402				17 23 24	+50 43.9	1	7.57		0.70		0.26		2	G5	1566
	+60 01751				17 23 24	+60 46.2	1	8.70		1.17		1.20		4	K0	1371
		vdB 90 # a			17 23 24	−35 50.9	1	10.62		0.41		−0.22		3	B3 V	434
		E7 - d			17 23 24	−44 58.	1	11.14		0.46				4		2011
157524	−62 05590	HR 6471			17 23 24	−62 59.7	1	6.22		−0.09				4	B7/8 V	2007
	−29 13541				17 23 25	−29 37.2	1	11.27		0.70				1		1702
157781	−31 14029				17 23 26	−31 58.8	1	7.57		0.31		0.11		2	A1mF0-F0	918
	−45 11536				17 23 26	−45 04.6	1	10.39		1.12				4		2011
	+47 02477				17 23 30	+47 48.9	1	9.62		0.19		0.10		2	A5	1375
		E7 - i			17 23 30	−45 01.	1	12.55		0.45				4		2011
		E7 - l			17 23 30	−45 04.	1	12.44		1.23				4		2011
		L 156 - 46			17 23 30	−62 24.	1	12.73		1.51		1.10		1		3073
157857	−10 04493	LS IV -10 002			17 23 31	−10 57.0	1	7.78		0.17		−0.79		6	O7 e	400
157793	−32 12739	LSS 4162			17 23 32	−32 32.4	2	8.31	.010	0.75	.005	0.53	.025	4	A1 II/III	918,954
		E7 - u			17 23 35	−45 00.3	2	15.03	.023	0.57	.001	0.12		7		1075,1460
157860	−17 04805				17 23 36	−18 02.8	1	10.27		0.21		0.13		1	B9 III	1225
		E7 - p			17 23 36	−45 00.	1	13.19		1.12				4		1075
		E7 - k			17 23 36	−45 01.	1	12.90		0.34				4		1075
		E7 - t			17 23 36	−45 01.	1	13.88		1.70				4		1075
157967	+17 03241	HR 6495, V640 Her	⋆	A	17 23 41	+16 57.6	2	6.03	.070	1.62	.004	1.66		3	M4 IIIa	71,3001
317582	−30 14203				17 23 42	−31 03.0	1	8.89		2.35		1.27		2	K0	918
		E7 - m			17 23 42	−44 59.0	2	12.50	.001	1.31	.003	1.32		7		1460,2011
		E7 - o			17 23 42	−45 00.	1	12.82		1.51				4		1075
		E7 - v			17 23 42	−45 00.	1	14.02		1.75				4		1075
315791					17 23 43	−28 34.3	1	11.55		0.35		0.13		2		918

Table 1 947

HD	DM	Other Id	N	Rem	α_1950	δ_1950	S	V	σ_V	B−V	σ_B−V	U−B	σ_U−B	n	Spectrum	References
		G 19 - 26			17 23 44	+03 32.0	3	12.57	.044	0.96	.010	0.70	.005	5		202,1620,3062
		E7 - s			17 23 44	−44 59.5	1	14.23		0.66		0.22		4		1075,1460
	−44 11736				17 23 44	−45 03.0	2	10.96	.002	0.62	.004	0.39		7		1460,2011
158038	+27 02808				17 23 46	+27 20.8	1	7.47		1.04		1.00		3	K2 II	8100
	−44 11739				17 23 46	−44 52.8	4	10.77	.015	0.03	.003	-0.39	.014	22	A0	116,863,1075,1460
157783	−40 11461				17 23 48	−40 26.1	2	9.27	.001	0.15	.004	-0.42	.005	3	B3 II/III	540,976
		E7 - n			17 23 48	−45 00.	1	12.99		1.16				4		2011
		E7 - r			17 23 48	−45 00.	1	13.35		1.60				4		1075
		E7 - g			17 23 48	−45 01.	1	12.02		0.47				4		2011
		E7 - q			17 23 48	−45 01.	1	13.03		1.63				4		1075
157897	−17 04808				17 23 49	−17 50.4	1	9.84		0.23		0.19		1	A0 IV/V	1225
157864	−25 12160	HR 6490			17 23 49	−25 54.1	1	6.44		-0.06				4	A0 V	2035
157866	−28 13156				17 23 51	−28 45.0	1	9.73		0.18		-0.05		4	B8 IV	918
157812	−38 11885	LSS 4165			17 23 51	−38 26.5	1	8.78		0.72		0.35		2	B9 Ib	954
157927	−14 04644				17 23 53	−15 05.2	1	8.26		0.17		0.00		2	B9 V	1770
157978	+7 03368	HR 6497			17 23 54	+07 38.3	2	6.04	.005	0.59	.010	0.27	.005	11	B9.5 V+G0 V	1088,8032
317573					17 23 55	−30 34.0	1	11.29		0.22		-0.27		2		918
	−44 11742			AB	17 23 55	−45 00.6	1	11.99		0.24				4		2011
157795	−43 11711				17 23 56	−43 11.0	3	7.25	.007	0.43	.006	-0.03	.011	16	F3/5 V	977,1075,2011
157753	−52 10662	HR 6483		★ A	17 23 58	−52 15.3	1	5.75		1.17				4	K2 III	2007
157950	−4 04275	HR 6493			17 23 59	−05 02.6	9	4.53	.011	0.39	.011	-0.04	.019	38	F3 V	15,254,1075,1088,1586*
315790					17 24 00	−28 34.4	1	11.21		0.42		0.35		2		918
158067	+27 02809	HR 6499		★ A	17 24 01	+26 55.2	2	6.40	.010	0.10	.005	0.12	.000	5	A5 IV	985,1733
157815	−44 11743				17 24 01	−44 45.9	1	9.05		0.96				4	G6 III	2012
157999	+4 03422	HR 6498			17 24 02	+04 10.9	10	4.33	.014	1.50	.010	1.60	.019	38	K3 II	15,1075,1080,1088*
157798	−46 11527				17 24 02	−46 54.5	4	8.15	.005	0.70	.005	0.31	.005	23	G3 V	863,977,1770,2011
	+10 03209				17 24 03	+10 41.3	1	10.36		0.28		0.32		2	A2	1776
157846	−41 11645	IDS17205S4113		AB	17 24 04	−41 46.1	2	8.15	.005	0.06	.007	-0.55	.004	4	B2/3 III	401,540
157884	−32 12748				17 24 05	−32 30.3	1	8.37		0.45		-0.04		2	F5 V	918
157918	−28 13161				17 24 06	−28 43.7	1	7.78		0.07		0.05		4	A0 V	918
	−44 11744				17 24 06	−45 03.5	1	10.81		0.14				4	A0	2011
315863	−30 14215	LSS 4169			17 24 07	−30 23.5	1	10.55		0.39		-0.34		2	B5	918
	−50 11294				17 24 07	−50 26.3	1	10.53		-0.06		-0.86		2	B0.5IV	540
315858	−30 14216				17 24 08	−30 12.6	1	11.20		0.37		-0.29		1		1586
159831	+83 00511				17 24 09	+83 44.9	1	8.30		0.13		0.08		8	A2	1219
315780					17 24 10	−27 49.8	1	10.72		0.50		0.43		2		918
315782					17 24 10	−28 00.5	1	11.23		0.29		-0.04		2		918
157917	−28 13162				17 24 10	−28 12.0	1	9.99		0.34		0.28		2	A3 III	918
157919	−29 13557	HR 6492			17 24 10	−29 49.4	7	4.29	.011	0.40	.002	0.11	.037	31	F3 III	3,15,614,1075,2012*
157832	−46 11530	V750 Ara		★ AB	17 24 10	−46 59.1	6	6.65	.009	0.03	.009	-0.86	.009	19	B2ne	219,278,1075,1586*
		LP 864 - 1			17 24 11	−25 06.7	1	13.75		1.60		1.12		1		3062
		LSS 4170			17 24 12	−30 11.5	1	11.11		0.41		-0.20		2	B3 Ib	918
157901	−32 12751				17 24 12	−32 16.0	1	9.65		0.25		-0.17		2	B9 III	918
323325	−38 11898				17 24 12	−38 23.7	1	9.43		0.81		-0.16		3		1586
	−2 04354	LS IV -02 003		★ V	17 24 13	−02 21.8	1	10.69		0.92		0.68		1	A5	793
157968	−12 04750	HR 6496			17 24 14	−12 28.2	2	6.22	.009	0.51	.000	0.03		5	F7 V	2007,3053
	−44 11746				17 24 14	−45 02.1	2	11.96	.020	0.10	.025	-0.08		7		116,2011
315792					17 24 16	−28 36.6	1	11.00		0.28		-0.17		2		918
158116	+29 03012	IDS17224N2933	A		17 24 17	+29 29.9	1	7.68		0.29		0.19		2	Am	3016
158116	+29 03012	IDS17224N2933	AB		17 24 17	+29 29.9	1	7.54		0.35		0.23		3	Am	8071
158116	+29 03012	IDS17224N2933	B		17 24 17	+29 29.9	1	9.60		0.94		0.61		2		3016
158116	+29 03012	IDS17224N2933	C		17 24 17	+29 29.9	1	10.82		0.98		0.78		1		3016
157870	−44 11748				17 24 19	−44 40.8	6	8.65	.003	0.32	.003	0.19	.006	46	A3 II(m)	977,1460,1628,1657,1770,2011
319816	−35 11578	LSS 4171			17 24 24	−35 47.9	2	9.94	.015	1.20	.025	0.13	.060	4	O6 Ib(f)	954,1737
	−45 11561				17 24 25	−45 05.1	1	10.20		0.87				4		2011
157955	−29 13563	HR 6494			17 24 26	−29 41.0	1	6.00		0.00				4	B9.5IV	2009
157938	−32 12755				17 24 27	−32 35.0	1	9.93		0.24		0.14		2	B8/9 III	918
157819	−55 08144	HR 6487		★ A	17 24 30	−55 07.7	1	5.94		1.11				4	G8 II/III	2007
		WLS 1712 50 # 13			17 24 31	+52 17.6	1	12.94		0.68		0.07		2		1375
157970	−28 13171				17 24 31	−28 08.3	1	8.63		0.23		0.12		2	A1 IV/V	918
	−30 14225				17 24 33	−30 24.7	1	10.67		1.10				1	K0	1702
319798	−34 11704				17 24 34	−34 55.3	1	9.80		0.65		0.13		4	F8	119
		LP 627 - 28			17 24 36	−00 49.9	1	11.90		1.29		1.15		1		1696
157988	−28 13173				17 24 36	−28 19.0	1	8.10		0.16		-0.03		2	B8/9 III	918
158148	+20 03481	HR 6502			17 24 40	+20 07.3	2	5.53	.005	-0.13	.005	-0.58	.010	6	B5 V	154,3016
315783	−28 13175				17 24 40	−28 07.7	1	8.99		1.15		0.68		2	K0	918
157957	−36 11531				17 24 40	−36 50.5	1	7.78		0.18		-0.80		2	B7 III	1011
157990	−32 12762				17 24 46	−32 26.5	1	9.63		0.45		0.00		1	F2 V	918
315810					17 24 47	−29 07.2	1	11.47		0.29		0.20		3		918
157931	−46 11539				17 24 47	−46 49.9	5	8.52	.010	0.84	.005	0.50	.007	24	G8 IV	116,863,977,1770,2011
		WLS 1712 50 # 8			17 24 48	+49 38.3	1	11.35		1.09		0.99		2		1375
315811	−29 13572				17 24 48	− 29 04.2	1	9.73		0.71		0.28		2	G0	918
	−30 14234	LSS 4176			17 24 48	−30 32.1	1	11.22		0.41		-0.36		4		918
158004	−32 12765				17 24 49	−32 08.5	1	9.23		1.21		0.89		1	G8 III	918
158226	+31 03027	G 181 - 47		★ A	17 24 50	+31 06.0	4	8.50	.008	0.59	.006	-0.01	.010	11	G1 V	1003,1620,1775,3026
158003	−31 14059				17 24 50	−31 30.0	1	9.50		0.19		0.06		1	B9 III	918
323599	−39 11495	LSS 4173			17 24 50	−39 46.9	1	10.34		0.54		-0.43		2	B0.5II-Ib	540
158226	+31 03025	G 181 - 46		★ B	17 24 51	+31 07.1	2	9.65	.042	0.77	.052	0.17	.009	4		1003,3026
	−2 04358				17 24 52	−02 29.6	1	10.54		0.66		0.22		18		1222
157942	−44 11759				17 24 52	−44 36.6	2	9.81	.006	0.02	.007	-0.48	.016	3	B3 II/III	540,976

HD	DM	Other Id	N	Rem	α_{1950}	δ_{1950}	S	V	σ_V	B–V	σ_{B-V}	U–B	σ_{U-B}	n	Spectrum	References
157943	−45 11576				17 24 53	−45 16.3	2	9.22	.008	0.15	.000	0.11		13	A0 V	1770,2011
	−46 11540				17 24 53	−46 50.6	2	9.36	.005	1.53	.005	1.21		6	K5	2011,3078
157973	−40 11483	LSS 4174			17 24 54	−40 17.9	2	7.96	.036	0.59	.007	-0.34	.029	4	B1/2 Iab	540,976
158261	+34 02971	HR 6506			17 24 58	+34 44.2	3	5.94		-0.02	.005	-0.05	.018	6	A0 V	985,1022,1079
158020	−31 14063	IDS17217S3108	A		17 24 58	−31 10.0	2	7.74	.006	1.31	.005	1.17	.009	11	K0 III	918,1673
	−29 13573	Tr 26 - 19			17 24 59	−29 30.2	1	10.91		1.25		1.30		2		918
	−29 13574	Tr 26 - 22			17 24 59	−29 31.6	1	11.80		0.48		0.03		2		918
		Tr 26 - 39			17 25 01	−29 28.5	1	12.73		0.54		0.56		1		918
158460	+60 01754	HR 6511			17 25 02	+60 05.4	2	5.64	.010	0.02	.005	0.01		6	A1 Vn	252,1379
315857	−30 14241				17 25 02	−30 12.1	1	10.31		1.31				1	K0	1702
319881	−34 11711	LSS 4177			17 25 02	−34 30.0	2	10.06	.015	1.10	.020	0.00	.055	4		954,1737
315813	−29 13575				17 25 03	−29 11.4	1	9.91		0.17		0.06		2	A0	918
		Tr 26 - 28			17 25 03	−29 23.9	1	13.06		0.68		0.29		1		918
		Tr 26 - 27			17 25 05	−29 23.7	1	12.57		0.56		0.53		1		918
		Tr 26 - 55			17 25 05	−29 32.4	1	13.04		0.58		0.47		2		918
		Tr 26 - 42			17 25 06	−29 26.8	1	12.17		0.62		0.62		2		918
		Tr 26 - 54			17 25 06	−29 32.2	1	13.04		0.58		0.40		2		918
158054	−30 14243				17 25 06	−30 46.2	1	9.29		0.11		-0.44		3	B9 Ib/II	918
315784					17 25 07	−28 06.1	1	10.71		0.62		0.39		2		918
		LSS 4180	A		17 25 08	−16 36.9	1	12.50		0.67		0.19		3	B3 Ib	1737
		LSS 4180	B		17 25 08	−16 36.9	1	12.65		0.62		0.18		3	B3 Ib	1737
		Tr 26 - 73			17 25 08	−29 26.9	1	13.92		0.51		0.61		1		918
		Tr 26 - 90			17 25 08	−29 32.2	1	11.64		1.23		1.16		2		918
158633	+67 01014	HR 6518			17 25 09	+67 20.9	4	6.43	.011	0.76	.008	0.30	.010	28	K0 V	15,22,1013,1197
	−29 13577	Tr 26 - 89			17 25 10	−29 32.5	1	12.12		0.15		0.00		2		918
317597	−31 14069				17 25 10	−31 36.0	1	9.81		1.72				1	K1	1702
		Tr 26 - 76			17 25 11	−29 27.9	1	13.14		0.50		0.47		2		918
		Tr 26 - 59			17 25 12	−29 22.4	1	12.35		1.22		1.07		3		918
315827	−29 13579	Tr 26 - 105			17 25 13	−29 24.6	1	9.80		1.34		1.23		3		918
		Tr 26 - 74			17 25 13	−29 27.2	1	12.70		0.52		0.46		2		918
		Tr 26 - 109			17 25 14	−29 26.0	1	11.92		1.47		1.51		2		918
158072	−32 12773	IDS17220S3211	AB		17 25 14	−32 13.5	1	9.89		0.44		0.26		1	A7 IV	918
		LP 507 - 25			17 25 15	+12 32.9	1	13.41		1.34		0.99		1		1696
		Tr 26 - 123			17 25 15	−29 28.7	1	13.54		0.59		0.45		3		918
		Tr 26 - 98			17 25 16	−29 22.5	1	12.83		0.53		0.42		2		918
		Tr 26 - 101			17 25 16	−29 23.2	1	13.32		0.57		0.36		2		918
		Tr 26 - 104			17 25 16	−29 24.8	1	15.06		0.96		0.12		2		918
		Tr 26 - 107			17 25 16	−29 25.7	1	14.15		0.60		0.49		2		918
		Tr 26 - 114			17 25 16	−29 27.2	1	12.34		0.52		0.41		3		918
158122	−20 04775	IDS17223S2053	AB		17 25 17	−20 55.4	1	7.96		0.45		-0.05		3	F5 V	3030
158088	−28 13185				17 25 17	−29 01.1	1	7.29		1.86		1.89		2	M0 III	918
		Tr 26 - 103			17 25 17	−29 24.6	1	14.28		0.52		0.46		2		918
	−29 13580	Tr 26 - 122			17 25 17	−29 29.0	1	9.94		1.54		1.15		2		918
158006	−45 11580				17 25 17	−45 11.7	1	9.12		0.44				4	F0 V	2011
		Tr 26 - 120			17 25 18	−29 28.7	1	12.82		1.25		1.05		2		918
		Tr 26 - 121			17 25 18	−29 29.0	1	13.07		0.54		0.40		2		918
158170	−8 04444	HR 6504			17 25 19	−08 10.0	4	6.37	.013	0.58	.007	0.15	.005	12	F5 IV	15,254,1256,2020
		Tr 26 - 115			17 25 19	−29 27.3	1	14.07		0.65		0.42		3		918
		Tr 26 - 116			17 25 19	−29 27.7	1	13.44		0.74				3		918
		Tr 26 - 124			17 25 19	−29 29.6	1	14.41		0.77		0.56		2		918
158228	+8 03418	V2114 Oph			17 25 20	+08 29.0	1	6.45		1.66		1.81		1	M1	3040
158485	+58 01731	HR 6514			17 25 20	+58 41.6	2	6.49	.005	0.14	.000	0.09	.005	6	A4 V	985,1501
315794					17 25 20	−28 38.6	1	11.67		0.37		0.32		2		918
		Tr 26 - 143			17 25 20	−29 26.2	1	12.80		0.47		0.32		3		918
		Tr 26 - 144			17 25 21	−29 26.3	1	11.47		0.53		0.46		2		918
		Tr 26 - 169			17 25 22	−29 29.8	1	15.02		2.42		1.25		1		918
158073	−33 12096	IDS17221S3330	AB		17 25 22	−33 32.6	2	8.97	.038	0.21	.005	-0.50	.021	3	B2 II(p)	540,976
315824		Tr 26 - 135			17 25 23	−29 22.2	1	11.12		0.37		0.27		3	B9	918
		Tr 26 - 142			17 25 23	−29 25.4	1	14.24		0.62		0.52		2		918
158102	−30 14252				17 25 23	−30 52.8	1	9.14		0.12		-0.22		3	B9 IV/V	918
158042	−43 11741	IDS17218S4353	AB		17 25 23	−43 56.0	4	6.19	.018	-0.02	.009	-0.35	.015	14	B5 III	116,278,977,1075,2011
158414	+48 02517	HR 6509			17 25 25	+48 18.1	3	5.85	.004	0.12	.002	0.15	.006	57	A4 V	753,879,1733
315799	−28 13189				17 25 25	−28 53.8	1	9.97		1.52		1.59		3	K0	918
315815	−29 13585				17 25 25	−29 05.2	1	8.63		1.88		1.88		4	K2	918
		Tr 26 - 146			17 25 25	−29 26.0	1	14.53		1.18		0.94		2		918
		G 139 - 39			17 25 26	+14 31.7	2	13.69	.009	1.76	.009	1.22	.014	5		203,3078
		Tr 26 - 148			17 25 26	−29 27.0	1	11.20		1.77		2.20		2		918
315796					17 25 27	−28 35.4	1	11.62		0.33		0.29		2	A2	918
		Tr 26 - 199			17 25 27	−29 26.5	1	11.92		1.38		1.84		2		918
158103	−31 14077				17 25 27	−31 20.6	1	7.65		0.36		-0.01		2	F0 V	918
		Tr 26 - 197			17 25 28	−29 25.8	1	13.47		0.58		0.48		2		918
		Tr 26 - 200			17 25 28	−29 26.8	1	12.08		0.61		0.49		1		918
		Tr 26 - 190			17 25 29	−29 23.7	1	12.10		1.52		1.45		2		918
158262	+11 03183	IDS17232N1128	B		17 25 30	+11 25.9	1	8.62		0.41		-0.02		4		3024
315798					17 25 31	−28 52.5	1	11.37		0.16		0.03		2		918
315816					17 25 31	−29 05.8	1	11.47		0.36		0.11		2		918
		Tr 26 - 185			17 25 31	−29 21.7	1	11.75		0.37		0.27		2		918
	−29 13586	Tr 26 - 201			17 25 31	−29 26.7	1	11.08		1.16		1.02		3		918
158263	+11 03184	IDS17232N1128	A		17 25 32	+11 25.8	1	7.12		0.13		0.12		3	A3	3024
315795					17 25 32	−28 34.0	1	11.22		0.47		0.38		2		918

Table 1

HD	DM	Other Id	N Rem	α1950	δ1950	S	V	σV	B–V	σB–V	U–B	σU–B	n	Spectrum	References
158123	−30 14256			17 25 32	−30 53.5	1	7.77		0.20		0.12		2	A0 III	918
	−29 13589	Tr 26 - 223		17 25 33	−29 23.5	1	11.87		0.57		0.45		2		918
158105	−36 11546	HR 6501		17 25 33	−36 44.3	1	6.02		1.11				4	K0 III	2009
158058	−44 11772			17 25 34	−44 45.2	2	9.23	.000	0.25	.000	0.08		8	A1 III/IV	1075,2011
158332	+26 03023	G 170 - 38		17 25 35	+26 49.9	2	7.72	.014	0.81	.005	0.51	.005	5	K1 IV	1658,3016
158331	+27 02817	G 170 - 39		17 25 35	+27 03.7	1	8.69		0.73		0.33		2	G8 IV-V	1658
315787				17 25 35	−28 21.3	1	10.25		0.40		0.27		2		918
317599				17 25 35	−31 34.3	1	10.54		0.37		-0.36		3		918
		Tr 26 - 225		17 25 37	−29 24.1	1	11.79		1.26		1.04		2		918
		Tr 26 - 231		17 25 38	−29 25.3	1	13.17		0.48		0.25		1		918
158141	−31 14081			17 25 39	−31 32.2	1	9.74		0.11		-0.11		1	B9 III	918
	−43 11743			17 25 39	−43 28.3	2	9.73	.008	0.55	.002	0.16		10	G0	863,2011
158154	−29 13590	IDS17225S2931	AB	17 25 40	−29 33.2	1	8.67		0.10		-0.05		5	B8/9 V	918
158078	−46 11557			17 25 40	−46 39.4	2	7.62	.000	-0.06	.005	-0.52	.005	3	B4 IV	219,401
158155	−32 12786	V499 Sco		17 25 46	−32 57.9	2	8.38	.161	0.40	.001	-0.53	.004	3	B1 III	540,976
317602	−31 14087			17 25 47	−31 43.6	1	9.67		0.57		-0.01		1	G0	918
158185	−28 13195			17 25 48	−28 48.3	1	9.28		0.10		-0.30		3	B6 III	918
		WD1725+586		17 25 57	+58 40.0	1	15.45		-0.09		-1.01		1	DA1	1727
315820	−29 13600			17 25 58	−29 26.7	1	9.70		1.30		1.19		2	K2	918
315856	−30 14266			17 25 58	−30 09.4	1	10.75		1.18				1	K0	1702
158186	−31 14091	LSS 4182		17 25 58	−31 29.7	7	7.00	.012	0.03	.009	-0.85	.012	17	B2/3 II	158,285,432,540,918*
158156	−38 11927	HR 6503	★ AB	17 25 59	−38 28.7	2	6.38	.005	0.09	.000			7	A1 V	15,2030
238698	+59 01814			17 26 01	+59 47.4	1	9.71		-0.14				3	B8	1379
	−29 13601	LSS 4183		17 26 01	−29 51.4	1	11.13		0.82		-0.24		2		918
	+17 03248			17 26 02	+17 33.0	1	9.37		0.66		0.08		1	G0	979
158214	−28 13200			17 26 02	−28 56.0	1	8.87		0.33		0.15		3	A9 V	918
		LP 920 - 24		17 26 05	−30 44.9	1	11.50		0.63		-0.02		3		1696
		BPM 24866		17 26 06	−57 53.	1	15.27		-0.04		-1.02		1	DB	3065
158198	−33 12107			17 26 07	−33 43.2	1	8.52		0.68				13	G5 V	6011
158159	−46 11564			17 26 09	−46 58.1	3	8.38	.004	0.44	.007	-0.03	.003	15	F3 V	116,977,2011
	−29 13605	LSS 4184		17 26 10	−29 54.4	1	10.84		0.57		-0.35		1		918,1586
158158	−44 11783			17 26 10	−44 55.9	3	9.39	.007	0.06	.000	-0.07	.001	14	B9 V	863,1075,2011
158144	−45 11593			17 26 10	−45 40.8	2	9.41	.022	-0.03	.000	-0.50	.020	3	B9 Ib/II	540,976
157512	−78 01133	IDS17184S7803	AB	17 26 10	−78 06.0	1	8.39		0.22		0.22		2	A4 V	1776
315818				17 26 11	−29 06.1	1	10.96		0.39		0.27		2		918
		ApJ144,259 # 41		17 26 12	−15 11.	1	15.95		0.11		-0.87		3		1360
315817				17 26 12	−29 03.3	1	11.51		0.29		0.21		2		918
158175	−44 11784			17 26 13	−44 42.2	4	7.54	.007	-0.05	.004	-0.35	.010	19	Ap Si	116,977,1075,2011
	−51 10924	IDS17223S5133	A	17 26 15	−51 35.8	1	9.58		1.46		1.30		4	M0	158
315797	−28 13204			17 26 16	−28 47.7	1	10.46		1.11				1	K0	1702
158352	+0 03697	HR 6507		17 26 17	+00 22.2	3	5.42	.013	0.23	.006	0.09	.005	8	A8 V	1149,1363,3023
158256	−31 14098			17 26 19	−31 15.8	1	9.44		0.43		0.09		2	F2 V	918
158111	−54 08349			17 26 20	−54 12.1	2	7.77	.009	-0.02	.000	-0.44	.000	4	B8 Ib	219,400
317707	−31 14099	LSS 4185	★ AB	17 26 22	−31 43.3	3	9.74	.027	0.59	.047	-0.41	.019	5	B3	432,725,918
317710	−31 14100			17 26 23	−31 53.0	1	10.01		0.49		0.04		1	F2	918
158319	−16 04526			17 26 25	−16 33.1	3	8.87	.047	0.08	.015	-0.50	.018	4	B8 Vne	400,1586,1770
315786	−28 13206			17 26 25	−28 07.1	1	9.05		1.26				1	K0	918,1702
158202	−46 11571			17 26 26	−46 09.7	4	9.40	.010	1.55	.010	1.22	.005	23	Ap Si(Fe)	863,977,977,1460
317669	−30 14281			17 26 28	−30 48.4	1	10.18		0.51		-0.25		1		1586
158270	−32 12805			17 26 29	−32 30.5	1	9.73		0.25		-0.41		2	B5 Ib/II	918
158304	−28 13210			17 26 33	−28 35.8	1	9.02		0.14		-0.55		2	B3 III	918
315915	−29 13614			17 26 34	−29 37.0	1	9.64		1.32		1.19		2	K5	918
158094	−60 06842	HR 6500	★ A	17 26 35	−60 38.7	5	3.60	.012	-0.10	.000	-0.30	.006	17	B8 Vn	15,1034,1075,1637,2012
		NGC 6362 - 1		17 26 36	−67 01.	1	11.45		0.68		0.21		3		38
		NGC 6362 - 2		17 26 36	−67 01.	1	15.07		1.07				2		38
		NGC 6362 - 3		17 26 36	−67 01.	1	15.24		1.01		0.62		2		38
		NGC 6362 - 4		17 26 36	−67 01.	1	13.32		1.25		1.47		2		38
		NGC 6362 - 5		17 26 36	−67 01.	1	14.30		1.01				1		38
		NGC 6362 - 6		17 26 36	−67 01.	1	13.23		1.35				2		38
		NGC 6362 - 7		17 26 36	−67 01.	1	14.83		1.19		0.86		2		38
		NGC 6362 - 8		17 26 36	−67 01.	1	15.60		0.81				2		38
		NGC 6362 - 9		17 26 36	−67 01.	1	15.39		1.04				2		38
		NGC 6362 - 10		17 26 36	−67 01.	1	14.89		0.80		0.48		2		38
		NGC 6362 - 11		17 26 36	−67 01.	1	14.76		1.17		1.04		2		38
		NGC 6362 - 12		17 26 36	−67 01.	1	13.13		1.26				1		38
		NGC 6362 - 13		17 26 36	−67 01.	1	14.42		1.02				2		38
		NGC 6362 - 14		17 26 36	−67 01.	1	12.77		1.65				2		38
		NGC 6362 - 15		17 26 36	−67 01.	1	13.99		1.06				2		38
		NGC 6362 - 16		17 26 36	−67 01.	1	12.57		1.19				1		38
		NGC 6362 - 17		17 26 36	−67 01.	1	12.71		1.60				1		38
		NGC 6362 - 18		17 26 36	−67 01.	1	13.60		1.23		1.15		2		38
		NGC 6362 - 19		17 26 36	−67 01.	1	12.54		1.49				2		38
		NGC 6362 - 20		17 26 36	−67 01.	1	13.71		1.17				2		38
		NGC 6362 - 21		17 26 36	−67 01.	1	13.29		0.86		0.84		2		38
		NGC 6362 - 22		17 26 36	−67 01.	1	13.03		1.44		1.57		2		38
		NGC 6362 - 23		17 26 36	−67 01.	1	13.71		1.23		0.87		2		38
		NGC 6362 - 24		17 26 36	−67 01.	1	13.71		1.06		0.78		2		38
		NGC 6362 - 25		17 26 36	−67 01.	1	13.32		1.46				2		38
		NGC 6362 - 26		17 26 36	−67 01.	1	15.73		0.29				1		38
		NGC 6362 - 27		17 26 36	−67 01.	1	14.48		1.16				1		38

HD	DM	Other Id	N	Rem	α_{1950}	δ_{1950}	S	V	σ_V	B−V	σ_{B-V}	U−B	σ_{U-B}	n	Spectrum	References
		NGC 6362 - 28			17 26 36	−67 01.	1	14.73		1.03				1		38
		NGC 6362 - 29			17 26 36	−67 01.	1	14.45		0.20				1		38
		NGC 6362 - 30			17 26 36	−67 01.	1	14.34		1.01				1		38
		NGC 6362 - 31			17 26 36	−67 01.	1	13.43		0.70		0.37		3		38
		NGC 6362 - 32			17 26 36	−67 01.	1	13.37		1.41				3		38
		NGC 6362 - 33			17 26 36	−67 01.	1	13.27		1.27		1.02		3		38
		NGC 6362 - 34			17 26 36	−67 01.	1	12.49		1.64				3		38
		NGC 6362 - 35			17 26 36	−67 01.	1	13.14		1.38		1.34		3		38
		NGC 6362 - 36			17 26 36	−67 01.	1	12.73		1.53		1.62		3		38
		NGC 6362 - 37			17 26 36	−67 01.	1	14.03		1.10		1.03		3		38
		NGC 6362 - 38			17 26 36	−67 01.	1	14.60		0.59		0.37		3		38
		NGC 6362 - 39			17 26 36	−67 01.	1	13.57		1.12		1.05		3		38
		NGC 6362 - 40			17 26 36	−67 01.	1	13.73		1.25				2		38
		NGC 6362 - 41			17 26 36	−67 01.	1	14.55		0.78		0.29		2		38
		NGC 6362 - 42			17 26 36	−67 01.	1	14.48		1.28				2		38
		NGC 6362 - 43			17 26 36	−67 01.	1	14.01		1.25		1.19		2		38
		NGC 6362 - 44			17 26 36	−67 01.	1	13.49		1.18				2		38
		NGC 6362 - 45			17 26 36	−67 01.	1	14.30		0.95				2		38
		NGC 6362 - 46			17 26 36	−67 01.	1	14.28		1.09				1		38
		NGC 6362 - 47			17 26 36	−67 01.	1	14.40		1.00				1		38
		NGC 6362 - 48			17 26 36	−67 01.	1	13.77		0.91				1		38
		NGC 6362 - 49			17 26 36	−67 01.	1	13.84		1.30		0.98		2		38
		NGC 6362 - 50			17 26 36	−67 01.	1	14.47		0.90		0.67		2		38
		NGC 6362 - 51			17 26 36	−67 01.	1	12.73		0.61		0.17		2		38
		NGC 6362 - 52			17 26 36	−67 01.	1	13.40		1.19		0.88		3		38
		NGC 6362 - 53			17 26 36	−67 01.	1	14.89		1.02		0.68		2		38
		NGC 6362 - 54			17 26 36	−67 01.	1	15.14		0.26		0.40		2		38
		NGC 6362 - 55			17 26 36	−67 01.	1	14.13		1.08		0.93		2		38
		NGC 6362 - 56			17 26 36	−67 01.	1	15.40		0.12				1		38
		NGC 6362 - 57			17 26 36	−67 01.	1	15.35		0.92				2		38
		NGC 6362 - 58			17 26 36	−67 01.	1	14.23		1.01				1		38
		NGC 6362 - 59			17 26 36	−67 01.	1	15.03		0.51				1		38
		NGC 6362 - 60			17 26 36	−67 01.	1	15.09		0.99				1		38
		NGC 6362 - 61			17 26 36	−67 01.	1	15.26		0.56				1		38
234439	+50 02409				17 26 37	+50 36.9	1	9.40		0.64		0.14		3	G0	196
315917	−29 13616				17 26 38	−29 44.5	2	9.69	.005	1.64	.000	1.41		3	K2	918,1702
		G 203 - 68			17 26 40	+43 03.5	1	14.33		0.85		0.32		1		1658
	+59 01817				17 26 42	+59 49.5	1	10.16		0.70				1	G5	1379
158338	−28 13211				17 26 42	−28 40.0	1	9.51		0.31		−0.05		2	B9 II/III	918
315918					17 26 42	−29 44.4	1	10.88		0.17		−0.17		2		918
158521	+26 03027				17 26 46	+26 45.9	1	8.14		0.52		−0.01		1	F6 V	3026
158377	−19 04644	TW Oph			17 26 47	−19 25.6	2	7.99	.170	4.58	.229			3	C5,5	109,864
158320	−33 12117	LSS 4187		⋆ AB	17 26 48	−33 40.7	7	6.67	.006	0.14	.006	−0.77	.011	101	B1/2 Ib/II	158,285,358,540,658*
158320	−33 12117	IDS17235S3337		C	17 26 48	−33 40.7	1	11.99		0.31		−0.07		1		658
158320	−33 12117	IDS17235S3337		D	17 26 48	−33 40.7	1	9.99		0.19		−0.45		1		658
	−30 14290				17 26 49	−30 40.3	1	10.39		1.48				2	K3	1702
		LS IV -04 002			17 26 50	−04 29.6	1	11.82		0.48		−0.27		2		405
		AT Ara			17 26 51	−46 03.6	1	12.61		0.08		−0.69		1		1471
315878	−28 13214				17 26 52	−28 14.7	1	9.28		1.89		2.29		2	K5	918
315947	−29 13622				17 26 52	−30 03.8	1	9.89		1.63				1	K4	1702
158378	−30 14296				17 26 59	−30 13.2	1	8.94		0.08		−0.21		2	B9 III	918
158357	−31 14121				17 27 00	−32 00.1	1	9.95		0.10		−0.30		1	A0	918
158358	−32 12820				17 27 00	−32 05.0	1	7.49		1.61		1.85		1	K4/5 III	918
	−33 12119	V924 Sco			17 27 03	−33 43.4	2	10.10	.070	2.09	.004	1.05	.133	5		358,1586
158323	−42 12197	IDS17235S4259		AB	17 27 04	−43 01.4	1	8.49		0.02		−0.44		2	B5 III	540
158243	−53 08660				17 27 04	−53 26.5	1	8.15		0.00		−0.82		1	B1 Ib	55
315877	−28 13218				17 27 05	−28 08.1	1	9.45		1.33		1.47		2	K5 V	918
158379	−31 14124				17 27 05	−31 24.9	1	9.02		0.58		0.04		2	F5 V	918
158463	−5 04450	HR 6512		⋆ AB	17 27 07	−05 52.8	2	6.36	.005	0.93	.000	0.67	.000	7	K0 III	15,1256
158392	−31 14128				17 27 08	−31 05.9	1	9.90		0.35		0.31		2	A2 (IIM)	918
158220	−56 08304	HR 6505			17 27 08	−56 53.0	3	5.94	.004	-0.08	.000			11	B7 II/III	15,2013,2030
315879					17 27 10	−28 34.5	1	11.12		0.27		−0.07		2		918
315898	−29 13629				17 27 11	−29 10.1	1	9.66		1.33		0.93		2	K0	918
		MM Sco			17 27 12	−42 08.9	1	16.20		1.34		0.18		1		1471
		G 226 - 62			17 27 14	+53 39.4	2	14.72	.015	1.67	.015			3		1663,1759
158421	−29 13632				17 27 15	−29 23.5	1	9.77		0.11		−0.23		2	B5/8 II/III	918
315896					17 27 16	−28 59.4	1	11.06		0.41		0.31		2		918
158393	−33 12122	V965 Sco			17 27 16	−33 37.0	2	8.62	.010	1.08	.000	0.62	.000	2	G8 III +(F)	576,1641
	−30 14306				17 27 17	−30 07.5	1	10.13		1.59				2	K0	1702
	+47 02487				17 27 18	+47 24.8	1	10.40		0.95		0.58		2		1375
	−30 14309				17 27 18	−30 10.1	1	10.93		1.10				1	K0	1702
	−28 13224				17 27 22	−28 59.8	1	11.37		0.67				1	K3	1702
158408	−37 11638	HR 6508			17 27 22	−37 15.5	6	2.70	.007	-0.23	.006	−0.82	.015	21	B2 IV	15,26,1075,1637,2012,8015
	+29 03029	G 181 - 49		⋆ AB	17 27 24	+29 26.0	3	9.01	.017	1.14	.021	0.99		9	K4	1017,1381,3031
158422	−31 14133				17 27 25	−32 01.5	1	7.73		0.15		0.08		1	A0 V	918
315895	−28 13226				17 27 26	−28 54.7	1	10.09		1.41				1	K2	1702
315880	−28 13227				17 27 28	−28 30.9	1	10.36		1.23				1	K0	1702
317666	−30 14310				17 27 28	−30 54.6	1	11.03		0.67				1	K0	1702
		ApJ218,617 # 7			17 27 30	−21 24.4	1	12.84		0.94		0.40		2		596
158454	−31 14134				17 27 30	−31 18.4	1	9.94		0.17		0.01		2	A8/9 V	918

Table 1   951

HD	DM	Other Id	N	Rem	α_{1950}	δ_{1950}	S	V	σ_V	B–V	σ_{B-V}	U–B	σ_{U-B}	n	Spectrum	References
158443	−33 12126	V482 Sco			17 27 31	−33 34.3	4	7.71	.258	0.84	.073	0.58	.039	4	F8/G0 II	657,1490,1490,6011
317701	−31 14136				17 27 32	−31 32.9	1	9.85		0.54		0.07		1	G0	918
	−31 14139				17 27 36	−31 46.1	1	10.86		1.20				1	K0	1702
158495	−29 13641				17 27 43	−29 36.0	1	9.62		0.12		0.00		2	B9 V	918
315942					17 27 43	−29 51.1	1	11.20		0.35		-0.13		2		918
234442	+50 02411				17 27 44	+50 18.9	1	9.64		0.90		0.55		3	G5	196
		G 20 - 5			17 27 44	−06 44.6	3	12.68	.030	1.35	.013	1.17	.010	5		202,1620,3062
158577	−9 04562	IDS17250S1001	AB		17 27 44	−10 03.5	1	8.11		0.71		0.22		3	G5	1657
158514	−28 13231				17 27 44	−29 00.5	1	9.40		0.59		0.19		2	F3 V	918
		ApJ218,617 # 6			17 27 45	−21 28.5	1	12.94		0.71		0.40		6		596
		ApJ218,617 # 5			17 27 46	−21 28.6	1	12.58		1.69		1.57		6		596
158513	−28 13234				17 27 47	−28 22.3	1	9.31		0.43		-0.27		2	B3 III	918
158614	−0 03300	HR 6516		⋆ AB	17 27 49	−01 01.4	5	5.32	.017	0.72	.005	0.30	.005	18	G8 IV-V	15,1067,1197,1256,3077
315944	−29 13643	LSS 4192			17 27 51	−30 01.7	1	10.85		0.33		-0.31		2	B5	918
315876					17 27 54	−28 08.6	1	10.58		0.29		0.13		2		918
	+5 03409	G 139 - 43		⋆ A V	17 27 55	+05 35.4	4	9.32	.013	1.47	.013	1.19	.026	5	M1 V	265,497,1705,3072
159023	+59 01819				17 27 56	+59 04.4	1	8.05		0.67		0.15		4	G0	1371
315954	−30 14322				17 27 56	−30 17.1	1	9.57		1.79		2.02		2	M0	918
158427	−49 11511	HR 6510, α Ara		⋆ A	17 27 58	−49 50.3	8	2.94	.021	-0.17	.011	-0.71	.019	35	B2 Vne	15,26,681,815,1075*
317690	−31 14152	V700 Sco		⋆ AB	17 27 59	−31 20.5	1	9.59		0.60		-0.21		2		918
158528	−33 12134				17 27 59	−33 18.5	1	8.35		0.25				15	A9 V	6011
315940					17 28 01	−29 54.9	1	10.74		0.10		-0.01		2		918
158456	−45 11625				17 28 01	−45 30.3	3	8.50	.005	0.33	.004	0.10	.005	21	F0 V	977,1770,2011
158716	+12 03234	HR 6521			17 28 03	+11 57.7	2	6.48		0.05	.010	0.02	.010	5	A1 V	1022,1733
		GD 208			17 28 03	+13 48.6	1	15.10		0.23		-0.56		1		3060
158562	−29 13651				17 28 03	−30 03.3	1	8.92		1.15		1.00		2	K1 III	918
158596	−21 04626	IDS17251S2124	AB		17 28 04	−21 26.9	1	8.94		0.28		-0.02		4	Ap Si	596
315956	−30 14324				17 28 04	−30 17.0	1	9.58		0.53		0.05		3	F2	918
158311	−62 05621				17 28 04	−62 12.0	2	7.51	.005	0.89	.005			8	G8 III/IV	1075,2033
158544	−32 12846				17 28 06	−32 18.0	1	9.38		0.29		0.19		2	A1/2 IV	918
158515	−40 11551				17 28 06	−41 00.5	1	8.53		0.23		0.23		1	A1 V	401
158476	−45 11626	HR 6513		⋆ AB	17 28 07	−46 00.0	1	6.03		0.83				4	F8/G0 Ib	2007
317689	−31 14153				17 28 08	−30 57.9	1	10.19		1.32				1	G5	1702
158582	−27 11700	IDS17250S2758	AB		17 28 09	−28 01.5	1	9.47		0.45		0.18		6	A9 V	918
158563	−31 14155				17 28 09	−31 56.0	3	8.09	.013	0.06	.016	-0.50	.008	5	B3 II	540,918,976
158659	−11 04393	LS IV -11 002			17 28 11	−11 08.5	2	10.26	.010	0.22	.005	-0.72	.005	6	B0 V	400,405
158531	−40 11553	IDS17247S4058	AB		17 28 11	−41 00.2	1	7.19		0.06		-0.23		2	B7/8 (IV)	401
317691	−31 14159				17 28 12	−31 24.7	1	10.94		0.95				1	G6	1702
		Steph 1486			17 28 15	+09 04.1	1	11.48		1.22		1.14		1	K7	1746
234444	+50 02415				17 28 18	+50 18.6	1	9.58		0.97		0.69		3	K0	196
	−18 04551				17 28 18	−18 42.9	1	9.58		1.47		1.28		2		262
158661	−17 04834	LSS 4198		⋆ A	17 28 19	−17 06.3	2	8.20	.011	0.19	.015	-0.73	.015	5	B0 II	400,1225
315938	−29 13656	LSS 4196			17 28 19	−29 58.7	1	11.76		0.34		-0.08		2	K0	918
315939	−29 13658	IDS17252S2953	AB		17 28 21	−29 55.7	1	9.42		0.58		0.18		2	F5	918
158598	−31 14163				17 28 21	−31 27.4	1	9.69		0.21		0.17		2	A0 III/IV	918
158643	−23 13412	HR 6519			17 28 22	−23 55.5	8	4.80	.009	0.00	.004	-0.06	.012	24	B9.5 Ve	15,614,1075,1075,1637*
	−31 14166				17 28 22	−31 11.7	1	10.90		0.43		0.13		2	B4	918
317715	−31 14165				17 28 22	−31 50.2	1	8.73		1.86		1.33		2	K5	918
158618	−30 14333	LSS 4197			17 28 23	−30 12.6	3	7.97	.010	0.24	.008	-0.63	.010	5	B0/2(Ib)(n)	540,918,976
315957	−29 14334	IDS17252S3013	AB		17 28 23	−30 15.4	1	8.30		1.30		1.11		3	K0	918
317658	−30 14332				17 28 23	−30 42.2	2	10.19	.005	1.36	.010	1.41		3	G5	918,1702
315882	−28 13248				17 28 24	−28 30.5	1	10.12		1.27				1	K0	1702
158599	−32 12854				17 28 25	−32 30.2	1	9.66		0.26		0.01		2	B8 II	918
		ApJ218,617 # 4			17 28 28	−21 34.7	1	11.22		0.74		0.20		6		596
315902	−29 13659				17 28 29	−29 10.6	1	10.62		1.08				1	G9	1702
158619	−33 12149	HR 6517			17 28 29	−33 40.0	1	6.44		1.19				4	K2 III	2009
158680	−21 04628	IDS17255S2129	AB		17 28 30	−21 31.5	1	9.40		0.68		0.21		6	G5 V	596
317656	−30 14336				17 28 30	−30 37.0	1	10.53		1.07				1	K0	1702
	−21 04630				17 28 31	−21 29.2	1	9.92		1.31		1.10		6		596
158644	−32 12855				17 28 33	−32 51.3	2	8.85	.005	0.26	.007	-0.54	.021	3	B3 II	540,976
	+60 01759				17 28 34	+59 57.6	1	9.96		0.62				2	G0	1379
238707	+59 01821				17 28 38	+59 53.0	2	9.31	.010	0.37	.015	0.05		6	A7	1371,1379
158704	−26 12152	HR 6520			17 28 38	−26 14.0	3	6.06	.016	-0.06	.005	-0.37	.002	11	B9 II/III	240,1637,2009
315926	−29 13662	LSS 4199			17 28 38	−29 36.7	1	9.72		0.52		-0.12		3	B3	918
158663	−31 14170				17 28 38	−31 28.7	1	9.55		0.56		-0.04		2	G2/3 V	918
158835	+15 03201				17 28 39	+15 13.8	1	9.12		0.18		0.10		2	A3	1648
158585	−46 11604	IDS17249S4633	AB		17 28 39	−46 35.6	2	8.63	.008	0.62	.000	0.19		11	G6 (III) +G	1770,2011
158899	+26 03034	HR 6526			17 28 43	+26 08.8	5	4.41	.008	1.43	.008	1.69	.012	12	K3 III	15,1080,1363,3016,8015
315886	−28 13256				17 28 44	−29 00.0	1	9.84		1.13		0.80		2	K0	918
158681	−30 14346				17 28 44	−30 21.9	1	8.20		0.07		-0.43		2	B6 V	918
315927	−29 13666	LSS 4200	A		17 28 46	−29 36.5	1	10.88		1.27				1	O5 III	954
315927	−29 13666	LSS 4200	AB		17 28 46	−29 36.5	3	10.55	.012	0.85	.011	-0.23	.030	7	O5 III	918,954,1737
158722	−28 13259				17 28 47	−28 04.9	1	9.40		0.55		0.01		2	F8 V	918
158682	−31 14174				17 28 47	−31 47.5	1	8.36		0.13		-0.03		2	A1 II	918
	−31 14176				17 28 49	−31 33.8	1	10.10		1.72				1	K2	1702
158503	−58 07170				17 28 49	−58 31.3	1	7.03		0.17		-0.51		3	B7 Vnnep	1586
158809	−2 04381	G 19 - 27			17 28 50	−02 30.1	7	8.14	.009	0.66	.011	0.07	.017	15	G0	202,516,742,803,1620*
158837	+2 03337	HR 6524		⋆ AB	17 28 51	+02 45.6	1	5.58		0.84		0.57		8	G8 III	1088
	+2 03336				17 28 52	+02 00.7	2	9.35	.066	1.89	.028	2.03		2	C4,1	1238,8005
158705	−31 14178	LSS 4201			17 28 53	−31 30.8	3	7.95	.019	0.78	.018	-0.20	.016	5	B1/2 Iab	540,918,976

HD	DM	Other Id	N Rem	α_{1950}	δ_{1950}	S	V	σ_V	B–V	σ_{B-V}	U–B	σ_{U-B}	n	Spectrum	References
	−29 13670			17 28 54	−29 11.0	1	11.32		0.47		-0.09		3		918
158740	−28 13260			17 28 55	−28 48.0	1	8.79		1.08		0.68		2	G8 III/IV	918
315961	−30 14352			17 28 55	−30 25.9	1	10.47		1.35				1	G9	1702
		PSD 157 # 564		17 28 57	−28 05.	1	11.51		0.87				1	K2	1702
315874	−28 13262			17 28 58	−28 04.8	1	9.10		2.09		2.57		2	K7	918
158723	−32 12866			17 28 59	−32 14.8	1	8.88		0.52		-0.02		2	F6 V	918
158855	+1 03450			17 28 59	+01 42.5	3	7.19	.004	1.35	.001	1.43	.010	14	K3 III	1003,1509,4001
159026	+39 03147	HR 6531		17 29 00	+38 55.1	2	6.44	.034	0.49	.010	0.22		3	F5 II	105,254
158974	+31 03047	HR 6528		17 29 02	+31 11.7	1	5.61		0.95				2	G8 III	71
317757	−33 12155	LSS 4203	⋆ AB	17 29 03	−33 18.3	2	9.05	.027	0.78	.010	-0.29	.007	4	B0 III	540,1012
158759	−30 14353	IDS17259S3023	AB	17 29 05	−30 25.5	1	8.29		0.30		0.08		3	A6/7 IV/V	918
		Steph 1489		17 29 06	+11 16.2	1	11.15		1.14		1.11		1	K7	1746
158741	−34 11757	HR 6522, V949 Sco		17 29 06	−34 14.6	1	6.17		0.35				4	F2 IV	2009
158667	−46 11609			17 29 06	−46 13.2	3	8.12	.008	1.14	.001	1.06	.011	15	K0 III/IV	116,977,2011
315873				17 29 16	−27 54.2	1	10.51		0.38		-0.06		2		918
159181	+52 02065	HR 6536	⋆ AB	17 29 18	+52 20.3	7	2.80	.027	0.97	.023	0.63	.013	18	G2 Ib-IIa	15,1080,1118,3016*
315872				17 29 19	−27 44.5	1	11.63		0.49		0.40		2		918
159266	+59 01823			17 29 21	+59 43.5	1	8.00		1.00				2	K0	1379
158816	−30 14361			17 29 22	−30 05.0	1	8.79		1.36		1.24		2	K1/2 III	918
323703	−37 11663	LSS 4204		17 29 22	−37 34.4	1	10.10		1.22		1.04		2	F0 II	954
158551	−61 05969			17 29 23	−61 29.8	1	7.45		1.60		1.97		5	M0 III	1704
	−79 00923			17 29 25	−79 18.6	1	10.38		0.34		0.26		2	B8 V	400
159329	+63 01358			17 29 26	+63 53.5	2	7.66	.024	0.58	.009	0.01	.005	4	F9 V	1003,3026
317770	−30 14362			17 29 27	−30 39.9	1	9.78		1.74				1	K2	1702
158746	−44 11846			17 29 27	−44 27.8	4	8.23	.006	0.01	.009	-0.16	.006	19	B9 IV	116,977,1075,2011
317800	−31 14193			17 29 31	−31 35.8	1	10.48		1.14				1	K2	1702
	−31 14195			17 29 32	−31 43.2	1	11.41		0.59				1	K4	1702
	−37 11667			17 29 33	−37 13.1	1	11.25		0.68		0.13		2	B7 III	540
158747	−45 11652	IDS17259S4526	AB	17 29 33	−45 28.6	1	7.08		0.01		-0.18		2	B9.5III	401
158842	−28 13274			17 29 35	−28 56.0	1	9.67		0.97		0.58		2	K0 III +(F)	918
	−29 13689			17 29 35	−29 45.3	1	10.60		1.21				1	K0	1702
320067	−37 11669	LSS 4205		17 29 35	−37 08.1	1	8.96		1.11		0.42		2	B6	954
158799	−41 11742	HR 6523		17 29 36	−41 08.3	2	5.85	.010	0.05	.005	-0.18		6	B9 Ib/II	401,2009
158630	−59 07063			17 29 38	−59 44.2	1	7.63		0.60				4	G2 V	2033
317782	−31 14200			17 29 44	−31 04.6	1	9.25		1.75		2.09		2	M0	918
158883	−30 14369			17 29 47	−30 20.6	4	7.47	.008	0.97	.011	0.69	.007	9	G8/K0 IV	657,918,2012,3008
317794	−31 14201			17 29 47	−31 21.2	1	10.08		0.42		0.26		2	A5	918
317889	−33 12167	LSS 4207		17 29 48	−33 21.3	2	10.11	.005	0.90	.020	-0.17	.055	4	O4 III(f)	954,1737
158859	−32 12883	LSS 4209		17 29 49	−33 01.4	1	7.26		0.17		-0.59		1	B2 II	219
158884	−31 14202			17 29 51	−31 56.7	1	8.81		0.52		-0.02		2	F5 V	918
159139	+28 02767	HR 6533		17 29 52	+28 26.5	1	5.66		0.00	.000	-0.03	.015	5	A1 V	1022,1733
158902	−29 13694	LSS 4210		17 29 52	−29 37.0	5	7.21	.008	0.36	.008	-0.26	.014	8	B3 II	540,918,954,976,1737
320059	−36 11632	LSS 4206		17 29 52	−36 50.5	1	9.45		0.98		0.84		2	A5 II	954
158901	−29 13695			17 29 53	−29 05.2	1	8.97		1.42		1.31		2	K1 III	918
159051	+2 03344			17 29 55	+02 26.8	1	8.56		1.08		0.80		7	K0	4001
159082	+12 03241	HR 6532		17 29 55	+11 57.9	3	6.45		-0.02	.006	-0.19	.010	6	B9.5V	1022,1079,1733
159330	+57 01774	HR 6540		17 29 55	+57 54.8	1	6.21		1.43		1.57		2	K2 III	1733
315883	−28 13281			17 29 55	−28 24.4	1	10.05		0.73		0.25		2	G0	918
315884				17 29 55	−28 57.4	1	11.19		0.23		0.20		2		918
158846	−42 12245	IDS17263S4220	AB	17 29 55	−42 21.9	1	7.34		-0.04				4	B3 IV/V	2012
316005	−29 13696			17 29 56	−29 44.2	1	9.62		0.29		-0.29		2	B3	918
158903	−31 14211			17 29 59	−31 07.3	1	9.78		0.21		0.13		2	A0 III	918
316006				17 30 03	−29 47.5	1	10.38		0.19		0.11		2		918
316023				17 30 06	−30 00.4	1	11.44		0.28		0.22		2		918
159119	+14 03270			17 30 07	+14 25.2	2	6.99	.015	1.51	.005	1.88	.032	4	K5 III	1648,3040
159239	+40 03162			17 30 08	+40 00.6	1	8.75		0.25		0.09		2	A2	1601
158942	−31 14215			17 30 08	−31 03.6	1	8.83		1.20		0.92		2	G8/K0 III	918
	−28 13284			17 30 09	−28 19.4	1	11.91		0.75		0.22		1		1586
315984				17 30 09	−29 15.1	1	11.11		0.48		-0.15		2		918
315983				17 30 10	−29 00.1	1	10.61		0.23		0.13		2		918
317802	−31 14216			17 30 11	−31 39.5	2	9.53	.095	1.00	.015	0.61	.005	7	G0	119,918
158981	−27 11713			17 30 12	−27 26.1	1	9.48		0.59		0.14		2	G2/3 V	918
317803				17 30 12	−31 31.5	1	10.32		0.18		0.16		2		918
		L 156 - 34		17 30 12	−62 07.	1	12.16		0.91		0.53		3		3062
159222	+34 02989	HR 6538		17 30 13	+34 18.3	5	6.53	.020	0.64	.004	0.16	.029	19	G5 V	101,1067,1355,1758,3026
158926	−37 11673	HR 6527, λ Sco	⋆ A	17 30 13	−37 04.2	6	1.62	.008	-0.23	.008	-0.90	.013	20	B1.5IV +B	15,26,1034,1075,2012,8015
317793	−31 14219			17 30 14	−31 19.1	1	10.39		1.36				1	K0	1702
158888	−43 11821	IDS17266S4400	A	17 30 14	−44 02.5	1	8.14		-0.04		-0.48		2	B5 V	401
158864	−45 11660	V830 Ara		17 30 15	−45 35.6	5	8.18	.027	-0.02	.036	-0.91	.004	12	B2 Ib/IIep	401,540,976,1586,2012
315973	−28 13286			17 30 16	−28 18.7	1	10.24		1.35				1	K0	1702
		AS 233		17 30 18	−22 49.	1	10.93		0.68		-0.01		3	Be	1586
	−27 11715	LSS 4215		17 30 18	−27 38.0	1	11.00		0.66		-0.16		2		918
158982	−30 14385	IDS17271S3028	AB	17 30 18	−30 30.4	1	9.36		0.27		0.18		3	A2 IV/V	918
317771	−30 14384	LSS 4213		17 30 19	−30 36.4	2	9.09	.000	1.50	.045	1.23	.075	4	F2 II	918,954
158906	−46 11632			17 30 20	−46 34.0	4	7.65	.004	-0.13	.004	-0.77	.006	13	B2/3 II	401,540,976,1732
316022				17 30 21	−29 55.8	1	11.06		0.15		0.05		2		918
		Ross 858		17 30 22	−15 46.6	1	13.64		1.31		1.04		2	K4	3073
316004				17 30 22	−29 40.3	1	10.61		0.32		0.26		2		918
	−29 13708			17 30 23	−29 17.9	1	11.08		0.34		-0.21		3		918
159240	+28 02771	IDS17285N2853	AB	17 30 26	+28 50.3	1	8.06		0.35		-0.02		2	F2	3016

Table 1   953

HD	DM	Other Id	N Rem	α_{1950}	δ_{1950}	S	V	σ_V	B–V	σ_{B-V}	U–B	σ_{U-B}	n	Spectrum	References
315982	−28 13288			17 30 26	−28 53.2	1	10.13		1.16		0.68		2		918
317816	−31 14222			17 30 26	−31 55.4	2	9.75	.005	1.69	.015	1.91		3	M0	918,1702
323815	−40 11588	LSS 4211		17 30 26	−40 41.4	2	10.23	.015	0.55	.002	-0.26		6	B3 II	540,1021
159224	+25 03283			17 30 27	+25 45.7	1	8.54		0.12		0.09		3	A2	833
315975				17 30 27	−28 32.7	1	11.10		0.24		0.03		2		918
316002				17 30 28	−29 36.5	1	11.40		0.41		-0.15		2		918
316007				17 30 30	−29 43.2	1	11.75		0.26		0.20		2		918
317784	−31 14224			17 30 30	−31 06.0	1	9.83		1.69				1	K1	1702
315974				17 30 31	−28 25.6	1	11.40		0.41		0.33		2		918
	−31 14223			17 30 31	−31 13.0	1	11.21		0.73				1	K1	1702
158928	−44 11859			17 30 31	−44 58.4	2	7.01	.010	-0.02	.010	-0.29	.025	5	B9 III	156,401
159088	−22 04366			17 30 34	−22 08.3	1	7.75		1.73		2.02		3	K4/5 III	1657
316000	−29 13713			17 30 35	−29 28.1	1	10.57		1.13				1	K0	1702
323771	−39 11602	LSS 4214	⋆	17 30 36	−39 21.6	3	11.07	.054	0.30	.016	-0.35	.032	5	Bep	434,1586,1588
315999				17 30 38	−29 25.3	1	10.59		0.10		-0.29		2		918
	+23 03130	G 170 - 47		17 30 40	+23 46.8	1	8.95		0.64		-0.30		1	G0	1658
		GD 209		17 30 42	+24 11.0	1	15.40		0.51		-0.18		1		3060
159089	−29 13717			17 30 43	−29 11.4	1	8.39		1.75		2.02		2	K5 III	918
317861	−32 12908	NGC 6383 - 76		17 30 45	−32 38.6	2	9.82	.015	0.21	.030	-0.42	.021	2		737,1586
317785	−31 14229			17 30 46	−31 08.2	1	10.76		1.13				1	K0	1702
		NGC 6383 - 8		17 30 46	−32 34.4	1	13.47		0.81		0.29		1		3021
317859	−32 12910	NGC 6383 - 85		17 30 46	−32 35.8	1	9.48		0.45		0.05		1	F0	737
159070	−31 14230			17 30 47	−31 35.0	1	10.03		0.20		0.23		2	A1/2 IV	918
		G 182 - 15		17 30 48	+33 36.1	1	10.21		0.96		0.93		2	M1	7010
317860	−32 12912	NGC 6383 - 9		17 30 48	−32 37.2	2	9.79	.023	1.03	.000	0.63	.050	5	K7	737,3021
		LP 628 - 18		17 30 49	−03 04.8	1	11.61		0.84		0.28		2		1696
159090	−30 14399	LSS 4220		17 30 49	−30 22.7	5	7.40	.006	0.15	.007	-0.78	.017	11	B0.5Ia/ab	158,219,540,918,976
159170	−5 04461	HR 6534		17 30 50	−05 42.6	1	5.61		0.18		0.07		8	A5 V	1088
316021				17 30 50	−29 54.1	1	11.35		0.15		-0.10		2		918
		PSD 157 # 682		17 30 50	−31 35.	1	10.58		1.26				2	K0	1702
159071	−33 12191	LSS 4218		17 30 50	−33 35.3	2	9.77	.034	0.27	.018	-0.35	.013	3	B3/5 Ib(e)	540,1586
	−30 14402			17 30 51	−30 20.6	1	9.48		1.28		1.13		2		918
	−32 12914	NGC 6383 - 91		17 30 52	−32 31.5	1	11.09		1.21		0.99		2		737
159035	−40 11595			17 30 52	−40 29.9	1	7.13		0.03		-0.19		2	B8 IV	401
		G 203 - 71		17 30 53	+50 27.0	1	12.73		1.47		1.02		3		1663
159108	−30 14403			17 30 54	−31 02.0	1	8.59		1.12		0.88		3	K0 III	918
159037	−43 11832			17 30 54	−43 07.6	2	8.83	.001	0.01	.010	-0.39		12	B5 II/III	977,2011
317843	−32 12915	NGC 6383 - 11		17 30 55	−32 26.2	2	9.86	.050	1.18	.010	1.20	.165	2	K0	737,3021
		NGC 6383 - 4		17 30 55	−32 35.1	2	13.34	.117	0.93	.039	0.47	.141	3		737,3021
317804				17 30 56	−31 41.0	1	9.91		0.53		0.08		2		918
		NGC 6383 - 12		17 30 56	−32 26.8	1	13.02		0.73		0.15		1		3021
		NGC 6383 - 13		17 30 58	−32 26.3	1	12.03		0.23		0.13		1		3021
		NGC 6383 - 132		17 30 58	−32 32.3	1	16.61		2.40				1		737
316008				17 30 59	−29 37.	1	10.48		0.22		0.16		2		918
317775	−30 14406			17 30 59	−30 51.6	1	9.96		1.67				1	K2	1702
		NGC 6383 - 15		17 30 59	−32 26.3	3	11.89	.023	0.17	.007	-0.17	.031	5		708,737,3021
159039	−45 11669			17 30 59	−45 06.0	2	8.75	.004	0.94	.000	0.62		12	G8 III	977,2011
317805	−31 14233			17 31 00	−31 43.6	1	9.44		1.75		1.84		2	M2	918
		NGC 6383 - 136		17 31 00	−32 30.5	1	17.70		2.49				1		737
159073	−38 12010			17 31 00	−39 02.9	2	9.75	.021	0.31	.010	-0.32	.006	3	B3/5 II/III	540,976
317858	−32 12919	NGC 6383 - 83		17 31 01	−32 34.5	2	9.53	.005	0.15	.015	-0.51	.010	5	B0 V:	737,918
317825				17 31 02	−32 02.9	1	10.54		0.03		-0.20		2		918
159125	−32 12923	IDS17279S3205	AB	17 31 03	−32 06.3	1	9.24		0.03		-0.62		2	B2 III	918
	−30 14410	316043		17 31 05	−30 31.7	1	10.52		1.32				1		1702
		NGC 6383 - 96		17 31 05	−32 29.6	1	11.38		0.68		0.08		3		737
317845	−32 12921	NGC 6383 - 6		17 31 06	−32 31.7	5	9.02	.007	0.09	.018	-0.70	.011	15	B5	540,737,918,976,3021
		NGC 6383 - 7		17 31 06	−32 31.9	1	12.73		0.32		0.10		1		3021
159057	−45 11670			17 31 06	−45 05.2	2	8.76	.007	1.16	.002	1.05		12	G8 III	977,2011
		NGC 6383 - 81		17 31 07	−32 35.8	1	12.68		1.91		2.16		4		737
317813				17 31 07	−34 49.3	1	11.19		0.38		0.23		2		918
158895	−59 07071	HR 6525		17 31 07	−59 48.8	3	6.27	.007	-0.08	.000			11	B5 II/III	15,2013,2030
317844	−32 12924	NGC 6383 - 100	⋆ V	17 31 08	−32 28.3	3	8.53	.461	0.11	.009	-0.62	.015	7	B5	540,976,3021
159110	−41 11776			17 31 08	−41 17.5	2	7.58	.009	-0.02	.010	-0.69	.012	3	B4 Ib	540,976
		NGC 6383 - 612		17 31 09	−32 25.8	1	12.60		0.37		0.11		2		708
		NGC 6383 - 702	⋆ V	17 31 09	−32 25.8	1	11.87		0.18		-0.36		2	A0 Ve	3021
		NGC 6383 - 19		17 31 10	−32 26.4	1	11.57		1.21		0.90		2		3021
		NGC 6383 - 29		17 31 10	−32 27.8	1	13.57		0.74		0.18		3		737
		NGC 6383 - 128		17 31 10	−32 36.0	1	15.66		1.16		0.82		3		737
159541	+55 01944	HR 6554	⋆ B	17 31 11	+55 13.1	10	4.89	.008	0.25	.006	0.03	.006	144	A6 V	1,15,1028,1084,1118*
159332	+19 03354	HR 6541		17 31 12	+19 17.5	3	5.66	.014	0.49	.015	0.30	.030	12	F6 V	3,254,3037
316020				17 31 12	−29 52.	1	10.41		0.20		-0.44		2		918
		NGC 6383 - 107		17 31 12	−32 29.5	1	14.85		0.92		0.50		3		737
		NGC 6383 - 130		17 31 12	−32 33.4	1	17.77		1.47				1		737
		NGC 6383 - 80		17 31 12	−32 36.6	1	13.30		0.72		0.16		2		737
320074	−37 11687	LSS 4222		17 31 12	−37 13.7	2	8.95	.017	1.03	.005	0.42	.033	4	B8 II-Ib	540,954
		NGC 6383 - 28		17 31 13	−32 29.9	1	12.53		0.24		0.00		2		737
316019				17 31 14	−29 47.	1	10.78		0.55		0.27		2		918
		NGC 6383 - 139		17 31 14	−32 26.6	1	17.81		1.71				1		737
		NGC 6383 - 700		17 31 15	−32 31.6	1	12.71		0.49		0.39		1		3021
		NGC 6383 - 108		17 31 16	−32 29.4	1	15.75		1.21		0.57		2		737

HD	DM	Other Id	N	Rem	α_{1950}	δ_{1950}	S	V	σ_V	B–V	σ_{B-V}	U–B	σ_{U-B}	n	Spectrum	References
		NGC 6383 - 110			17 31 16	−32 30.1	1	16.87		0.92				1		737
159560	+55 01945	HR 6555		⋆A	17 31 17	+55 12.4	11	4.87	.010	0.28	.005	0.07	.010	145	A4 m	1,15,1028,1084,1118*
		NGC 6383 - 109			17 31 17	−32 28.8	1	16.76		1.07				1		737
159018	−53 08682	HR 6530			17 31 17	−53 19.2	1	6.10		0.02				4	B9 III	2035
		NGC 6383 - 26			17 31 18	−32 32.6	2	12.89	.025	0.47	.035	0.22	.020	4	A2 Vep	708,737
317857	−32 12926	NGC 6383 - 3			17 31 18	−32 34.2	3	10.28	.018	0.30	.010	0.09	.017	9	A1 IV:p	708,737,3021
		NGC 6383 - 133			17 31 18	−32 38.4	1	16.42		2.54				1		737
	−49 11554				17 31 18	−49 24.3	2	10.70	.030	0.41	.005	-0.57	.007	5		540,1586
317847		NGC 6383 - 2			17 31 19	−32 31.8	3	10.35	.038	0.12	.021	-0.54	.019	7	B2 V	708,737,3021
		NGC 6383 - 71			17 31 19	−32 38.0	1	12.75		0.80		0.37		2		737
159111	−42 12271				17 31 19	−42 53.1	1	8.04		-0.01		-0.34		2	B5/7 IV	401
		M1-21			17 31 20	−19 07.4	1	13.52		1.09		-0.56		1		1753
317774	−30 14415				17 31 20	−30 49.9	1	10.94		0.53		-0.04		3		1586
		G 204 - 15			17 31 21	+43 47.3	1	14.02		1.16		1.02		1		1658
		NGC 6383 - 31			17 31 21	−32 27.2	1	14.00		0.69		0.20		4		737
		NGC 6383 - 30			17 31 21	−32 27.7	1	13.39		0.85		0.29		4		737
		NGC 6383 - 18			17 31 21	−32 33.4	3	13.39	.023	0.82	.005	0.29	.030	7		708,737,3021
317807					17 31 22	−31 36.8	1	11.30		0.48		0.25		2		918
159175	−32 12932				17 31 22	−32 08.7	1	9.16		0.52		-0.05		2	F7/8 V	918
	−32 12931	NGC 6383 - 261			17 31 22	−32 31.7	1	10.84		0.14		-0.35		3		708
		NGC 6383 - 20			17 31 22	−32 31.8	3	11.41	.058	0.18	.032	-0.14	.011	6	B9 IV	708,737,3021
317846	−32 12929	NGC 6383 - 14		⋆AB	17 31 22	−32 33.0	3	9.86	.024	0.11	.008	-0.47	.045	4	B5	708,737,3021
		NGC 6383 - 27		⋆V	17 31 22	−32 34.4	2	12.61	.005	0.59	.015	0.33	.044	3	A5 IIIp	708,737
315972					17 31 23	−28 15.5	1	10.59		0.43		0.43		1		918
315987					17 31 23	−29 17.2	1	11.51		0.47		0.24		2		918
317848	−32 12933	NGC 6383 - 34			17 31 23	−32 24.0	1	10.34		1.24		1.24		1	G5	737
		NGC 6383 - 111			17 31 23	−32 29.7	1	15.66		0.90		0.35		2		737
159174	−31 14242				17 31 24	−31 50.4	1	7.15		1.35		1.35		2	K1/2 III	918
317828					17 31 24	−32 13.1	1	10.31		0.04		-0.50		2		918
		NGC 6383 - 105			17 31 24	−32 32.1	1	15.37		1.21		0.59		3		737
		NGC 6383 - 263			17 31 24	−32 32.1	1	12.62		0.28		0.11		2		708
		NGC 6383 - 106			17 31 24	−32 32.2	1	15.15		0.87		0.41		2		737
		NGC 6383 - 262			17 31 24	−32 32.2	1	12.82		0.32		0.19		2		708
		NGC 6383 - 102			17 31 24	−32 32.4	1	14.62		0.83		0.16		2	B8	737
	−48 11837				17 31 24	−48 39.6	2	10.14	.018	1.56	.006	1.21		5	K5	158,1705
159353	+16 03218	HR 6542			17 31 25	+16 21.1	1	5.69		1.01		0.82		2	gK0	1355
		ST Oph			17 31 25	−01 02.8	2	11.36	.031	0.28	.003	0.23	.003	2	A6.7	668,699
		NGC 6383 - 104			17 31 25	−32 32.9	1	15.15		0.93		0.47		2		737
159192	−31 14243				17 31 26	−31 16.7	1	8.78		0.13		-0.01		2	A0 IV	918
		NGC 6383 - 112			17 31 26	−32 30.3	1	17.18		1.10				2		737
		NGC 6383 - 103			17 31 26	−32 32.6	1	16.26		1.02		0.45		2		737
		NGC 6383 - 613			17 31 26	−32 32.6	1	12.30		0.26		0.20		2		708
159176	−32 12935	NGC 6383 - 1		⋆AB	17 31 26	−32 32.9	7	5.69	.023	0.04	.011	-0.86	.011	14	O7 V +O7 V	540,708,918,954,976*
		NGC 6383 - 619		⋆B	17 31 26	−32 33.0	1	10.89		0.12		-0.48		2		708
		NGC 6383 - 617		⋆C	17 31 26	−32 33.0	1	10.17		0.13		-0.56		1		708
159354	+14 03279	HR 6543, V642 Her			17 31 27	+14 52.6	1	6.45		1.60		1.62		22	M4 IIIa	3042
		NGC 6383 - 101			17 31 27	−32 32.9	1	14.57		0.97		0.33		3		737
		NGC 6383 - 16			17 31 27	−32 33.3	3	10.81	.013	0.11	.029	-0.50	.029	6	B8	708,737,3021
		NGC 6383 - 21			17 31 28	−32 31.7	3	11.92	.026	0.71	.004	0.18	.023	11	F2 IV:p	708,737,3021
		NGC 6383 - 70			17 31 28	−32 38.7	1	13.05		0.51		0.26		2		737
		LP 688 - 39			17 31 29	−04 18.6	1	11.72		0.80		0.19		2		1696
315976	−28 13305				17 31 29	−28 27.1	1	10.13		1.52		1.48		2	K0	918
159212	−29 13735				17 31 29	−29 13.4	1	8.73		1.65		1.75		2	K3 III	918
316009					17 31 29	−29 41.0	1	11.67		0.26		0.23		3		918
316035	−30 14418				17 31 29	−30 25.4	1	10.16		1.34				1	K0	1702
		NGC 6383 - 25			17 31 29	−32 33.2	3	12.51	.030	0.26	.019	0.08	.031	7		708,737,3021
	−34 11794				17 31 29	−34 14.9	1	9.98		2.82				1	M3 Ia	8051
159213	−30 14420				17 31 30	−30 11.8	1	9.29		0.14		-0.32		2	B7 IV	918
		NGC 6383 - 23			17 31 30	−32 32.	3	13.84	.033	0.99	.022	0.46	.136	5		708,737,3021
		NGC 6383 - 118			17 31 30	−32 32.	1	15.02		0.85		0.32		3		737
		NGC 6383 - 131			17 31 30	−32 32.	1	16.92		1.21				1		737
		NGC 6383 - 264			17 31 30	−32 32.	1	13.77		0.52		0.28		3		708
		NGC 6383 - 388			17 31 30	−32 32.	1	15.10		1.06		0.72		3		708
		NGC 6383 - 266		V	17 31 30	−32 32.	1	15.30		0.90		0.60		3		708
		NGC 6383 - 127			17 31 31	−32 35.7	1	14.65		0.83		0.38		2		737
		NGC 6383 - 69			17 31 31	−32 39.2	1	12.76		0.83		0.36		2		737
159501	+41 02850	HR 6550			17 31 32	+41 16.7	1	5.74		1.09		0.96		3	K1 III	1355
		NGC 6383 - 24			17 31 32	−32 33.3	3	11.37	.012	0.19	.014	-0.22	.030	7	B9 Ve	708,737,3021
		NGC 6383 - 55			17 31 32	−32 35.4	1	12.80		0.65		0.44		2		737
		NGC 6383 - 56			17 31 32	−32 35.5	1	13.79		0.83		0.31		3		737
		G 170 - 50			17 31 33	+18 46.2	1	13.26		1.50		1.10		3		419
		NGC 6383 - 22			17 31 33	−32 33.4	3	12.34	.022	0.53	.012	0.29	.021	8	A3 V:e	708,737,3021
		NGC 6383 - 137			17 31 33	−32 36.9	1	17.69		2.05				1		737
315988					17 31 34	−29 17.9	1	11.36		0.37		0.31		4		918
		NGC 6383 - 126			17 31 34	−32 35.9	1	15.28		0.86		0.36		3		737
		NGC 6383 - 125			17 31 34	−32 37.5	1	15.77		1.07		0.66		1		737
	−35 11714	LSS 4227			17 31 34	−35 15.8	2	11.44	.020	1.35	.030	0.33	.015	4		954,1737
		NGC 6383 - 140			17 31 35	−32 37.5	1	14.82		1.57		1.47		2		737
	−32 12939	NGC 6383 - 41			17 31 36	−32 27.5	1	10.88		1.11		0.81		3		737
		NGC 6383 - 119			17 31 36	−32 34.3	1	14.89		0.90		0.30		3		737

Table 1 955

HD	DM	Other Id	N Rem	α_{1950}	δ_{1950}	S	V	σ_V	B−V	σ_{B-V}	U−B	σ_{U-B}	n	Spectrum	References
		NGC 6383 - 124		17 31 37	−32 37.9	1	15.41		1.08		0.72		2		737
		NGC 6383 - 42		17 31 38	−32 32.0	1	13.63		0.76		0.12		2		737
		NGC 6383 - 123		17 31 38	−32 37.9	1	14.41		1.02		0.79		2		737
	−29 13741			17 31 39	−29 15.3	1	11.75		0.60		-0.09		3		918
		NGC 6383 - 54		17 31 39	−32 33.6	1	12.29		0.59		0.10		2	F0 Ve	737
		NGC 6383 - 120		17 31 39	−32 34.4	1	15.98		0.99		0.64		1		737
316017				17 31 40	−29 57.6	1	10.82		0.53		-0.13		2		918
317812				17 31 40	−31 48.8	1	11.26		0.17		0.13		2		918
		NGC 6383 - 115		17 31 40	−32 30.7	1	16.11		1.02		0.39		1		737
		NGC 6383 - 117		17 31 40	−32 33.1	1	16.15		3.37				2		737
316016				17 31 41	−29 50.0	1	10.98		0.24		0.20		2		918
316044	−30 14421			17 31 41	−30 32.4	1	9.92		0.53		0.03		3	F8	918
317849	−32 12954	NGC 6383 - 47		17 31 41	−32 25.3	1	9.99		0.35		0.21		2	A5	918
		NGC 6383 - 116		17 31 41	−32 33.3	1	15.65		1.05		0.61		3		737
		NGC 6383 - 121		17 31 41	−32 34.5	1	15.95		0.99		0.79		1		737
317856	−32 12943	NGC 6383 - 10		17 31 41	−32 38.4	3	10.02	.011	0.10	.023	-0.47	.005	4	B8	737,918,3021
		NGC 6383 - 138		17 31 42	−32 30.3	1	17.91		1.56				2		737
		NGC 6383 - 122		17 31 42	−32 37.8	1	13.99		0.74		0.23		3		737
317853	−32 12944	NGC 6383 - 5		17 31 43	−32 34.1	2	11.26	.000	0.15	.033	-0.27	.005	4	A0	737,3021
		NGC 6383 - 134		17 31 43	−32 34.9	1	17.51		1.45				1		737
159255	−29 13743			17 31 44	−29 06.9	1	9.19		0.51		-0.05		1	F5/6 V	918
		NGC 6383 - 129		17 31 44	−32 29.4	1	18.12		1.27				1		737
		NGC 6383 - 114		17 31 44	−32 30.1	1	15.96		0.93		0.32		2		737
		NGC 6383 - 135		17 31 45	−32 27.8	1	17.98		1.13				2		737
	−30 14423	LSS 4232		17 31 46	−30 34.2	1	11.41		0.64		-0.24		2		918
		NGC 6383 - 43		17 31 46	−32 31.0	1	13.10		0.47		0.21		3		737
317854	−32 12946	NGC 6383 - 57		17 31 46	−32 35.7	1	10.64		0.26		0.16		3	A5	737
315989				17 31 47	−29 09.7	1	10.58		0.49		0.23		2		918
		NGC 6383 - 53		17 31 48	−32 33.0	1	12.32		0.78		0.22		2		737
320126		V487 Sco		17 31 48	−34 21.7	1	11.19		0.48		0.34		1	F8	700
		NGC 6383 - 113		17 31 49	−32 29.9	1	15.33		1.31		1.02		1		737
		NGC 6383 - 58		17 31 49	−32 34.5	1	12.38		0.37		0.22		2		737
317904	−33 12216	LSS 4231		17 31 49	−33 39.3	2	9.16	.010	1.38	.015	0.58	.055	4	B9	954,1737
		NGC 6383 - 45		17 31 50	−32 29.5	1	13.22		2.46		2.26		3		737
317791	−31 14248			17 31 52	−31 26.8	1	10.49		0.16		0.14		2	A3	918
		G 226 - 64		17 31 53	+54 22.0	1	12.26		1.45				1		1759
315977				17 31 53	−28 28.8	1	11.12		0.40		0.30		1		918
159370	−1 03360			17 31 54	−01 07.2	2	9.57	.010	0.53	.000	0.10		4	F0	742,1594
		NGC 6383 - 50		17 31 54	−32 31.3	1	13.50		0.90		0.45		2		737
		NGC 6383 - 52		17 31 54	−32 33.0	1	12.30		0.68		0.14		2		737
		L 558 - 60		17 31 54	−32 34.	1	14.28		1.19		0.90		3		3062
317835	−32 12955			17 31 55	−32 13.0	1	9.99		0.62		0.13		2	G0	918
159479	+26 03043			17 31 56	+26 42.3	1	8.15		1.10				5	K2 III	20
159217	−46 11661	HR 6537		17 31 56	−46 28.4	9	4.58	.008	-0.03	.004	-0.08	.017	35	A0 V	15,278,1020,1075,1637,1770*
315978				17 31 58	−28 36.8	1	10.74		0.29		0.14		2		918
159291	−30 14429	IDS17287S3004	A	17 31 58	−30 05.8	1	9.87		0.24		-0.01		3	B9 (IV)	918
159358	−11 04411	HR 6544	⋆ A	17 31 59	−11 12.6	3	5.55	.005	0.01	.005	-0.17	.012	7	B8 Vn	26,1079,2007
		GD 210		17 31 59	−13 57.4	2	13.09	.065	0.38	.050	0.11	.015	7	sdA-F	272,3060
159292	−30 14431			17 31 59	−30 10.1	1	8.68		1.16		0.79		3	K0 III	918
159293	−31 14254			17 32 00	−31 22.1	1	10.02		0.16		0.13		2	A0 V	918
159309	−28 13313			17 32 01	−28 10.3	1	8.45		1.30		1.19		2	K1 III	918
159519	+29 03049			17 32 02	+28 56.5	1	8.06		0.44		-0.06		2	F5	3016
		Steph 1494		17 32 04	−06 52.2	1	11.31		1.46		1.23		1	M0	1746
159234	−44 11888	NGC 6388 sq2 2		17 32 04	−44 42.3	1	10.21		0.06		0.01		4	B9	895
	−31 14256			17 32 06	−31 12.5	1	11.48		0.73				1	K1	1702
159278	−37 11700			17 32 06	−37 15.1	2	8.47	.006	0.18	.004	-0.49	.013	3	B2 II/III	540,976
		NGC 6388 sq2 6		17 32 08	−44 45.8	2	12.31	.023	0.77	.009	0.34	.014	4		156,895
159310	−31 14257	IDS17289S3145	A	17 32 09	−31 47.1	1	9.28		0.08		-0.20		2	B9 V +B(9)	918
159310	−31 14257	IDS17289S3145	B	17 32 09	−31 47.1	1	9.73		0.08		-0.31		2		918
159966	+68 00938	HR 6566	⋆ A	17 32 10	+68 10.0	3	5.06	.009	1.08	.004	0.92	.000	6	K0 III	15,1003,1363
159503	+16 03220	HR 6551		17 32 12	+16 32.2	1	6.60		0.20		0.12		2	A8 Vn	1733
	−30 14438			17 32 12	−30 51.3	1	11.15		0.61				1	K4	1702
159480	+9 03424	HR 6548	⋆ A	17 32 14	+09 37.1	2	5.82	.005	0.04	.005	0.03	.040	9	A2 V	1084,3050
159480	+9 03424	HR 6548	⋆ AB	17 32 14	+09 37.1	2	5.48	.005	0.02	.000	0.01	.000	7	A2 V	15,1256
159480	+9 03423	IDS17299N0939	B	17 32 14	+09 37.1	2	7.81	.000	0.29	.010	0.07	.020	8		1084,3016
	−28 13318	LSS 4243		17 32 14	−28 17.2	2	11.54	.058	0.76	.019	0.07	.005	3		918,954
159311	−33 12228	LSS 4237		17 32 14	−33 42.5	2	8.45	.014	0.96	.005	-0.09	.040	4	B1 Ia	954,1737
		NGC 6388 sq2 7		17 32 14	−44 38.3	1	12.49		1.27		1.18		3		895
317898	−33 12227	Tr 27 - 103		17 32 15	−33 25.8	3	10.69	.012	0.98	.016	-0.06	.031	6	B1 II	544,954,1737
	−34 11811	LSS 4235		17 32 15	−34 44.0	1	11.61		0.89		-0.08		3		1586
320086	−33 12229	LSS 4239		17 32 16	−33 53.7	2	9.95	.015	1.41	.015	0.52	.050	4	B5 Ia	954,1737
159482	+6 03455	G 139 - 48	⋆ A	17 32 18	+06 02.5	8	8.39	.014	0.57	.005	-0.06	.014	18	G0 V	22,265,979,1003,1509*
159376	−21 04659	HR 6545, V2125 Oph		17 32 18	−22 00.7	2	6.53	.027	0.02	.005	-0.18		5	Ap Si	220,2009
315967	−27 11732			17 32 18	−27 19.9	1	10.08		0.64		0.16		2	F5	918
320130	−34 11813	LSS 4241		17 32 18	−34 42.6	2	11.42	.110	0.92	.025	-0.11	.015	4		954,1737
159312	−37 11702	HR 6539		17 32 19	−37 24.5	1	6.48		0.01				4	A0 V	2006
159481	+6 03456	IDS17299N0606	A	17 32 22	+06 03.4	4	7.40	.025	0.52	.009	0.08	.010	9	F8	742,979,1028,3024
159481	+6 03456	IDS17299N0606	B	17 32 22	+06 03.4	3	10.50	.017	0.90	.023	0.62	.034	7		979,1028,3016
320105	−34 11814	LSS 4242		17 32 22	−34 02.3	2	10.40	.005	0.82	.005	-0.20	.030	4		954,1737
		NGC 6388 sq2 4		17 32 22	−44 48.3	2	11.68	.042	0.20	.028	0.21	.009	4		156,895

HD	DM	Other Id	N	Rem	α₁₉₅₀	δ₁₉₅₀	S	V	σV	B–V	σB-V	U–B	σU-B	n	Spectrum	References
	+25 03297	IDS17303N2524		A	17 32 23	+25 22.5	1	10.38		0.57		0.04		1		8084
	+25 03297	IDS17303N2524		BC	17 32 23	+25 22.5	1	11.23		0.97		0.80		1		8084
		NGC 6388 sq2 9			17 32 25	−44 47.2	1	12.85		0.80		0.28		3		895
159608	+29 03054				17 32 26	+29 47.7	1	7.72		1.55		1.75		2	M2 III	1733
159285	−47 11660				17 32 27	−47 19.5	2	9.12	.005	1.19	.005	1.04	.005	9	K1 III	156,895
316032	−30 14447				17 32 28	−30 18.4	1	8.98		2.15		2.17		3	M0	918
		NGC 6388 sq1 6			17 32 29	−44 39.3	1	11.77		1.77		1.91		1		156
		NGC 6388 sq2 5			17 32 30	−44 47.3	1	12.05		0.87		0.47		3		895
159341	−38 12035				17 32 31	−38 26.5	1	7.73		1.51		1.67		9	K3 IV	1673
	−28 13322	LSS 4248			17 32 32	−28 59.7	2	11.13	.010	1.15	.024	0.05	.019	3		918,954
315971					17 32 36	−28 08.8	1	10.97		0.34		-0.02		2		918
159561	+12 03252	HR 6556			17 32 37	+12 35.7	9	2.08	.013	0.16	.006	0.10	.011	71	A5 III	15,1006,1007,1013*
315968					17 32 38	−27 28.7	1	10.84		0.48		0.26		2		918
315980	−28 13325				17 32 39	−28 54.7	1	9.35		2.09		2.62		1	K5	918
159378	−33 12237	Tr 27 - 102		⋆ V	17 32 39	−33 24.1	1	8.39		1.94		1.71		2	G0 Ia	544
161948	−37 11914	NGC 6388 sq2 3			17 32 39	−44 47.3	1	11.01		1.23		1.19		3	B9	895
		NGC 6388 sq2 8			17 32 40	−44 38.2	1	12.58		0.59		0.07		3		895
159870	+57 01780	HR 6560		⋆	17 32 42	+57 35.5	2	6.17	.010	0.58	.023	0.33	.081	5	G5 III+A5 V	253,254
	−29 13762				17 32 42	−29 12.2	1	9.94		1.74				1	K4	1702
	−30 14454				17 32 42	−30 58.0	1	11.28		0.64				1	G8	1702
315990	−29 13763				17 32 46	−29 17.7	1	9.99		0.43		-0.19		3	B3	918
159418	−31 14273				17 32 46	−31 34.5	1	10.00		0.07		-0.22		2	B9 II/III	918
		Tr 27 - 6			17 32 46	−33 28.6	1	11.68		1.30		1.24		2	K5 III	544
		Tr 27 - 81			17 32 48	−33 28.2	1	16.32		1.19						544
317952					17 32 51	−31 47.3	1	10.52		0.42		0.25		2		918
		Tr 27 - 80			17 32 51	−33 29.0	1	14.80		1.40				1		544
		Tr 27 - 108			17 32 52	−33 27.7	1	12.70		1.55		0.28		1		544
		Tr 27 - 8			17 32 52	−33 29.1	1	11.88		1.66		0.56		2		544
	−33 12241	Tr 27 - 1		⋆ AB	17 32 53	−33 26.8	1	8.79		3.12		3.32		3	M0 Ia	544
		Tr 27 - 2			17 32 53	−33 27.0	2	10.60	.058	1.28	.005	0.22	.049	3	O9 Ia:	190,544
		Tr 27 - 5			17 32 53	−33 28.2	1	12.16		1.23		0.08		1		544
		Tr 27 - 82			17 32 53	−33 28.5	1	17.29		1.22				2		544
		Tr 27 - 110			17 32 55	−33 27.3	1	12.98		1.34		0.30		2		544
		Tr 27 - 111			17 32 55	−33 27.3	1	13.28		1.45		0.33		2		544
		Tr 27 - 47			17 32 55	−33 27.5	1	14.34		1.86		0.95		1		544
318014	−33 12244	Tr 27 - 46			17 32 55	−33 27.5	3	8.80	.013	1.58	.027	0.67	.033	8	B9 Ia(+?)	544,954,1737
320089	−33 12242	LSS 4254			17 32 55	−33 51.9	4	9.28	.018	1.10	.021	0.01	.037	9	B1 Ia	954,1012,1737,8100
		Tr 27 - 3			17 32 56	−33 27.0	1	13.16		1.17		0.19		1		544
		Tr 27 - 109			17 32 56	−33 27.3	1	11.54		1.73		0.77		2		544
		Tr 27 - 4			17 32 56	−33 28.2	1	11.93		0.59		0.16		2		544
		Tr 27 - 9			17 32 56	−33 29.6	2	12.44	.065	0.72	.020	0.13	.010	2		190,544
		Tr 27 - 43			17 32 57	−33 27.4	1	10.48		1.99		1.06		2	B9 Ia	544
317951					17 32 58	−31 46.0	1	9.79		0.49		0.32		2		918
159401	−43 11862	IDS17294S4334		AB	17 32 58	−43 36.6	1	9.29		0.49				4	F3/5 (V)	2011
159455	−33 12246	LSS 4258			17 32 59	−33 16.5	2	8.39	.020	0.79	.010	-0.25	.040	4	B0.5Ia	954,1737
		Tr 27 - 107			17 32 59	−33 36.3	1	11.46		0.94		-0.16		2	B0 V	544
159472	−30 14458				17 33 00	−30 30.7	1	8.93		1.20		0.98		3	G8 III	918
		Tr 27 - 49			17 33 00	−33 28.2	1	14.32		1.54		0.69		1		544
		Tr 27 - 106			17 33 00	−33 34.8	1	11.43		1.02		0.07		2	B2 III:	544
		V784 Oph			17 33 01	+07 47.2	1	11.85		0.26		0.23		1	B8Ib-II	597
159402	−44 11900				17 33 01	−44 32.0	7	8.11	.006	0.00	.007	-0.57	.005	24	B3 III/IV	156,540,976,977,1075*
		Tr 27 - 28			17 33 02	−33 24.3	1	13.38		1.77		0.86		2	WN5	544
159384	−44 11899				17 33 02	−44 50.9	5	7.38	.009	1.43	.005	1.69	.014	25	K4 III	156,863,977,2011,2024
317950					17 33 04	−31 42.9	1	10.41		0.40		0.12		2		918
		Tr 27 - 52			17 33 04	−33 26.8	1	14.14		1.24		0.27		1		544
	−33 12247	Tr 27 - 14			17 33 06	−33 29.9	1	11.12		1.18		0.15		2	B0 Ib	544
159433	−38 12044	HR 6546			17 33 06	−38 36.1	5	4.28	.008	1.09	.003	0.90	.003	32	K0 IIIb	15,1075,1311,2012,8015
	−33 12248	Tr 27 - 16			17 33 07	−33 31.3	2	10.76	.029	1.09	.000	0.00	.010	3	O9.5II:	190,544
		Tr 27 - 13			17 33 08	−33 29.2	1	11.78		0.96		-0.01		2	B2 V	544
	+18 03407	G 170 - 52			17 33 09	+18 55.1	2	10.06	.028	0.80	.019	0.41	.009	4	K0	1064,1696
159509	−28 13334	IDS17300S2830		AB	17 33 09	−28 32.2	1	9.43		0.65		0.41		2	A3/5 III	918
317956	−31 14286				17 33 09	−31 52.3	1	10.05		1.41				1	K1	1702
318015	−33 12250	Tr 27 - 23			17 33 10	−33 27.8	4	10.07	.028	1.41	.023	0.38	.035	7	B0.5Ib	190,544,954,1737
		L 270 - 137			17 33 12	−54 24.	2	15.81	.009	0.46	.000	-0.45	.000	6		782,3065
	−33 12255	Tr 27 - 24			17 33 15	−33 26.0	2	10.78	.025	1.02	.020	0.69	.005	2		190,544
		Tr 27 - 19			17 33 15	−33 30.0	2	12.82	.080	1.03	.005	0.09	.025	2		190,544
	−44 11904	NGC 6388 sq1 10			17 33 16	−44 48.2	1	10.62		0.51		-0.01		4		156
		Tr 27 - 44			17 33 17	−33 28.5	1	12.11		0.92		-0.10		2	B1 II::	544
316031	−30 14467				17 33 18	−30 21.9	1	10.42		1.28				1	K0	1702
		Tr 27 - 21			17 33 18	−33 28.4	1	12.59		0.90		-0.06		1		544
		Steph 1497			17 33 19	+38 54.5	1	12.73		1.30		0.57		2	M0	1746
		Tr 27 - 25			17 33 20	−33 25.5	2	11.38	.049	1.41	.005	0.34	.005	3		190,544
		Tr 27 - 22			17 33 20	−33 28.5	1	11.48		1.17		0.74		1		544
317954	−31 14289				17 33 21	−31 46.8	1	10.29		1.45				1	G2	1702
	−33 12257	Tr 27 - 104			17 33 22	−33 19.5	1	10.69		0.79		-0.29		2	O9 III	544
		NGC 6388 sq1 11			17 33 22	−44 52.2	1	12.75		1.57		1.67		1		156
		Tr 28 - 69			17 33 25	−32 28.6	1	13.24		0.68		0.11		1		190
159439	−49 11575	IDS17296S4911		A	17 33 25	−49 12.9	1	6.92		-0.03		-0.24		7	B8 III	1732
		IRC +20 328		AB	17 33 26	+15 36.9	1	12.75		2.23				2		8019
159571	−28 13337	IDS17303S2856		AB	17 33 28	−28 57.1	1	9.11		0.15		-0.11		2	B8 V	918

Table 1

HD	DM	Other Id	N	Rem	α₁₉₅₀	δ₁₉₅₀	S	V	σ_V	B–V	σ_B-V	U–B	σ_U-B	n	Spectrum	References
	-31 14291				17 33 28	-31 34.6	1	11.50		0.56				1	K0	1702
		Tr 27 - 34			17 33 28	-33 30.1	1	12.94		1.03		0.03		2	B1 V:	544
	-44 11909				17 33 28	-44 16.5	2	10.95	.005	1.65	.002	1.20		2	M5	1705,3078
	-33 12258	Tr 27 - 26			17 33 29	-33 24.8	1	11.34		0.72		0.20		2	F7 V	544
160078	+60 01764				17 33 30	+60 07.4	1	7.31		0.48				2	F5	1379
		Tr 28 - 60			17 33 30	-32 28.4	1	11.60		0.34		0.25		1		190
159474	-45 11709				17 33 30	-45 52.2	2	8.58	.006	0.16	.002	0.11		12	A2 IV	977,2011
159736	+12 03256				17 33 31	+12 04.6	1	6.77		1.63		1.94		2	K5	1648
317970	-32 12997	Tr 28 - 4			17 33 32	-32 24.3	1	9.99		1.38		1.34		3	K5	190
159489	-45 11712				17 33 33	-45 07.7	4	8.24	.012	-0.02	.001	-0.61	.003	9	B3 IV	400,401,540,976
317971	-32 12998	Tr 28 - 3			17 33 34	-32 25.4	1	9.84		0.47		-0.29		1	B3	190
		Tr 28 - 61			17 33 34	-32 29.6	1	14.51		0.71		0.29		1		190
159463	-49 11577	HR 6547			17 33 34	-50 01.7	4	5.91	.009	1.11	.008			15	K0 III	15,2013,2020,2028
316104					17 33 36	-30 03.5	1	10.56		0.24		0.05		2		918
159462	-49 11579	IDS17298S4920		AB	17 33 36	-49 22.6	1	8.98		0.63				4	G3 V	2012
159616	-28 13340				17 33 39	-28 09.6	1	10.07		0.40		0.31		2	B9.5/A0 IV	918
316103	-30 14478	LSS 4279			17 33 41	-30 05.6	1	10.46		0.65		-0.12		2	A3	918
159594	-31 14295				17 33 42	-31 32.0	1	9.62		0.26		0.23		2	A2 IV	918
159532	-42 12312	HR 6553			17 33 43	-42 58.1	8	1.86	.013	0.40	.007	0.21	.021	19	F0 II	15,278,1034,1075,2001*
		Tr 28 - 1			17 33 44	-32 27.3	1	12.35		0.58		-0.06		1		190
159515	-45 11714				17 33 44	-45 47.0	1	9.38		0.16				4	A3 V(p)	2011
316102					17 33 45	-30 05.6	1	11.55		0.29		0.16		3		918
159573	-38 12054				17 33 46	-38 45.3	2	8.40	.035	0.29	.006	-0.37	.007	3	B3 III	540,976
317975	-32 13007	Tr 28 - 44			17 33 47	-32 31.9	2	9.97	.034	0.41	.019	0.20	.019	3	F0	190,918
159554	-46 11688				17 33 50	-46 05.0	1	8.53		-0.05		-0.47		2	B7 II	401
159834	+21 03157	HR 6559		⋆A	17 33 51	+21 01.6	3	6.11	.010	0.18	.008	0.16	.008	8	A7 IV	542,1022,3016
159834	+21 03157	IDS17317N2104		B	17 33 51	+21 01.6	2	9.71	.113	0.46	.063	-0.02	.002	5	F6 IV-V	542,3016
159631	-32 13008				17 33 53	-32 11.1	1	8.91		0.06		-0.31		2	Ap Si	918
		Tr 28 - 25			17 33 53	-32 27.7	1	11.92		1.34		0.81		1		190
	-44 11918	NGC 6388 sq1 12			17 33 55	-44 40.3	1	9.73		1.05		0.84		4	G5	156
		Tr 27 - 105			17 33 56	-33 26.2	1	11.85		1.37		0.49		2		544
317936					17 33 57	-31 22.7	1	9.89		0.14		-0.44		2		918
159925	+37 02908	HR 6563			17 33 59	+37 19.9	2	6.13	.030	0.98	.002	0.82		5	G9 III	71,3016
317935	-31 14303				17 33 59	-31 21.4	1	9.51		0.63		0.18		2	F5	918
159492	-54 08403	HR 6549			17 33 59	-54 28.2	5	5.24	.007	0.20	.003	0.08	.003	15	A5 IV/V	15,1637,2012,2024,3052
159664	-29 13787				17 34 00	-29 50.0	1	9.23		1.28		1.15		2	K1 III	918
159652	-35 11748				17 34 00	-35 41.8	1	8.30		0.17		-0.31		22	B3 IV	1699
159633	-37 11723	HR 6557			17 34 01	-38 02.2	1	6.26		1.25				4	G2 Ib	2032
159651	-34 11851	LSS 4283			17 34 02	-35 00.7	1	9.16		1.41		1.30		2	F2 Iab/b	954,8100
	+26 03048				17 34 03	+25 59.5	1	9.07		1.57		1.95		1	M0	1746
160538	+74 00717	DR Dra			17 34 03	+74 15.6	1	6.55		1.05		0.81		2	K0 III	1003
316093	-29 13788				17 34 03	-29 42.8	1	9.49		1.32		1.02		3	K0	918
	-31 14307				17 34 03	-31 50.8	1	11.55		0.56				1		1702
159926	+28 02787	HR 6564			17 34 10	+28 12.9	1	6.40		1.37		1.67		2	K5	1733
316060					17 34 10	-28 38.9	1	10.63		0.41		0.32		2		918
159684	-35 11750	LSS 4286			17 34 10	-35 18.2	3	8.23	.048	0.42	.013	-0.45	.006	7	B2 Vne	540,976,1586
159682	-32 13015				17 34 11	-32 09.3	1	9.31		0.60		0.08		2	G1/2 V	918
		NGC 6396 - 19			17 34 11	-35 00.4	1	11.44		0.85		0.02		2		412
		He3 1442			17 34 12	-33 48.	1	12.19		0.91		0.11		2		1586
		Steph 1498			17 34 13	+05 50.3	1	11.59		1.65		2.04		1	M0	1746
		NGC 6396 - 18			17 34 13	-35 01.5	1	14.16		0.83		0.40		2		412
	-31 14313	317933			17 34 14	-31 20.3	1	10.07		1.42				1		1702
		NGC 6396 - 2		⋆AB	17 34 14	-34 58.9	1	10.80		0.80		-0.06		2	B1 V:	412
		NGC 6396 - 12			17 34 14	-35 00.4	1	13.93		1.00		0.32		2		412
	-34 11859	NGC 6396 - 1		⋆AB	17 34 15	-34 58.9	1	9.79		0.79		-0.06		8	B1 II/III	412
		NGC 6396 - 15			17 34 15	-35 01.0	1	13.23		1.68				1		412
		NGC 6396 - 17			17 34 15	-35 01.5	1	13.23		0.86		0.51		2		412
		V973 Oph			17 34 16	-27 11.0	1	10.66		2.18		2.28		1		842
		NGC 6396 - 11			17 34 16	-35 00.2	1	13.89		0.85		0.35		2		412
		NGC 6396 - 16			17 34 16	-35 01.5	1	13.81		0.76		0.21		2		412
	+2 03363				17 34 17	+02 52.6	1	10.68		0.68		0.03		2	G3	1696
317959	-31 14314				17 34 17	-31 57.1	1	10.57		1.18				1	K0	1702
		NGC 6396 - 13			17 34 17	-34 59.6	1	14.29		0.74		0.07		2		412
		NGC 6396 - 3			17 34 17	-34 59.8	1	11.69		0.40		0.26		2		412
		NGC 6396 - 14			17 34 18	-34 59.1	1	13.77		0.82		0.04		2		412
	-34 11862	NGC 6396 - 5			17 34 18	-35 00.4	1	11.42		0.66		-0.12		2		412
		NGC 6396 - 4			17 34 18	-35 00.5	1	11.57		0.74		0.00		2		412
159656	-42 12320				17 34 18	-42 32.0	6	7.16	.006	0.65	.003	0.16	.017	52	G2/3 V	278,977,1075,1499,1770,2011
		NGC 6396 - 6			17 34 19	-35 00.8	1	12.34		0.67		-0.11		2		412
159723	-30 14497				17 34 20	-30 08.6	1	9.81		0.40		0.26		3	A3 III	918
		NGC 6396 - 10			17 34 20	-35 00.4	1	13.73		0.84		0.30		2		412
159747	-28 13354				17 34 21	-28 13.0	1	9.03		1.37		1.22		2	K2 III	918
		NGC 6396 - 21			17 34 21	-34 59.1	1	13.67		0.89		0.16		2		412
	-34 11864	NGC 6396 - 20			17 34 21	-34 59.4	1	10.60		2.44		2.63		2		412
		NGC 6396 - 9			17 34 21	-35 00.4	1	14.07		0.82		0.33		2		412
		Ru 127 - 13			17 34 21	-36 17.4	1	12.11		0.77		-0.12		2		412
159968	+27 02849				17 34 22	+27 35.8	3	6.41	.010	1.59	.018	1.74	.024	10	M1 II	20,3040,8100
		AO 729			17 34 22	+68 22.6	1	11.49		1.27		1.46		1		1697
		NGC 6396 - 22			17 34 22	-34 59.0	1	13.99		2.73				1		412
		NGC 6396 - 8			17 34 22	-35 00.4	1	13.50		0.85		0.42		2		412

HD	DM	Other Id	N Rem	α_{1950}	δ_{1950}	S	V	σ_V	B–V	σ_{B-V}	U–B	σ_{U-B}	n	Spectrum	References
		NGC 6396 - 7		17 34 22	−35 00.7	1	13.54		1.58				1		412
159948	+25 03308			17 34 23	+25 38.8	2	7.37	.015	1.20	.019	1.17		6	K2 III	20,105
		Ru 127 - 3		17 34 23	−36 15.9	3	11.38	.023	0.81	.027	−0.12	.028	6		412,954,1737
	−36 11717	Ru 127 - 2		17 34 23	−36 16.3	1	10.45		0.85		−0.02		6	B2 II	412
		PSD 157 # 847		17 34 24	−29 44.	1	11.43		0.62				1	G7	1702
		Ru 127 - 11		17 34 24	−36 17.1	1	13.76		0.68		0.30		2		412
		Ru 127 - 14		17 34 25	−36 16.0	1	13.77		0.94		0.34		2		412
		Ru 127 - 10		17 34 25	−36 17.0	1	14.29		1.00		0.10		2		412
	−36 11718	Ru 127 - 1	★ AB	17 34 26	−36 17.2	1	10.35		1.95		0.71		2		412
		Ru 127 - 12		17 34 26	−36 17.9	1	12.47		0.87		0.02		2		412
159704	−37 11734	IDS17310S3748	AB	17 34 26	−37 49.8	1	6.68		0.77				4	G8 IV/V	2012
		Ru 127 - 9		17 34 27	−36 17.3	1	13.47		0.97		0.34		2		412
160269	+61 01678	HR 6573	★ AB	17 34 28	+61 54.8	16	5.23	.012	0.60	.009	0.10	.006	55	F9 V +K3 V	1,15,22,680,938,985*
160269	+61 01678	IDS17340N6157	C	17 34 28	+61 54.8	1	10.00		1.45		1.19		1		680
		Ru 127 - 16		17 34 28	−36 15.2	1	12.98		0.75		−0.10		2		412
		Ru 127 - 15		17 34 28	−36 15.4	1	14.23		0.76		0.20		2		412
		Ru 127 - 17		17 34 29	−36 14.6	1	13.04		0.78		0.06		2		412
		LSS 4301		17 34 30	−29 04.9	1	12.00		0.75		−0.15		2	B3 Ib	918
		Ru 127 - 4		17 34 30	−36 17.3	1	12.05		0.86		−0.03		2		412
		Ru 127 - 8		17 34 31	−36 16.8	1	14.04		0.81		0.23		2		412
		LSS 4303		17 34 32	−28 25.1	1	11.05		0.88		0.18		2	B3 Ib	918
		Ru 127 - 18		17 34 32	−36 15.7	1	12.66		0.72		−0.08		2		412
159783	−28 13358			17 34 33	−28 37.1	1	9.05		0.19		−0.20		2	B5/7 III	918
	−36 11719	Ru 127 - 5		17 34 33	−36 16.4	3	11.05	.027	0.80	.025	−0.11	.031	5	B3 III	412,954,1737
		Ru 127 - 7		17 34 33	−36 17.1	1	13.79		0.95		0.25		2		412
159707	−42 12327	HR 6558		17 34 33	−42 51.1	10	6.09	.011	−0.06	.006	−0.25	.005	48	B8 V	15,116,278,285,977,1075*
		Ru 127 - 6		17 34 34	−36 16.6	1	13.26		0.84		0.10		2		412
159782	−28 13359			17 34 35	−28 24.5	1	7.98		0.23		−0.04		2	B9 III	918
		LSS 4299		17 34 36	−35 16.4	2	11.62	.025	0.75	.010	−0.18	.010	5	B3 Ib	954,1737
159784	−30 14505			17 34 37	−30 52.8	1	8.31		0.58		0.19		4	G0 V	918
320156	−35 11760	LSS 4300	V	17 34 38	−35 21.4	5	9.76	.016	0.84	.018	−0.11	.014	17		540,1514,1586,1699,1737
	−9 04582			17 34 40	−09 28.1	1	10.64		2.29		2.52		2	M2	1746
160054	+30 03033	HR 6570		17 34 42	+30 48.9	3	6.03	.013	0.16	.000	0.09	.000	8	A5 V	985,1022,1733
159876	−15 04621	HR 6561	★ A	17 34 43	−15 22.1	8	3.54	.007	0.26	.009	0.12	.006	28	F0 IIIp	15,1020,1075,1088*
159877	−15 04622	HR 6562		17 34 44	−15 32.5	1	5.94		0.37				4	F0 IV	2007
159864	−17 04864	LSS 4308		17 34 44	−17 47.9	2	8.57	.013	0.24	.000	−0.71	.004	5	B1 Ib	400,1012
159804	−29 13805			17 34 44	−30 01.0	1	9.40		0.19		−0.07		3	B8/9 IV	918
160361	+62 01559			17 34 45	+62 29.5	1	7.04		1.31		1.47		2	K2	985
159805	−31 14326			17 34 45	−31 49.1	1	10.01		0.33		0.17		2	A2	918
159994	+14 03289			17 34 46	+14 52.8	1	7.59		0.99		0.74		2	K0	1648
159821	−28 13363			17 34 46	−28 07.4	1	8.76		1.44		1.41		4	K1 III	918
159822	−29 13806			17 34 46	−29 26.8	1	10.07		0.13		−0.14		3	B9 III/IV	918
159845	−24 13405			17 34 50	−24 56.2	1	8.62		0.68		0.14		3	B3 III	1586
159749	−44 11933			17 34 50	−44 50.8	4	7.92	.017	1.52	.004	1.70	.010	19	K3 III	156,460,977,2011
	−29 13812	LSS 4306N		17 34 53	−29 04.9	2	9.53	.005	0.76	.019	−0.29	.024	3	O9 V	918,1737
	−29 13809	LSS 4306S	★ AB	17 34 53	−29 05.4	6	9.72	.013	0.67	.017	−0.33	.025	11	O9.5Ib	540,918,954,1011,1012,1737
159823	−32 13033			17 34 53	−32 21.6	1	10.03		0.18		0.06		2	B9 V	918
318032	−33 12293			17 34 56	−33 43.6	2	10.38	.015	0.83	.005	−0.22	.045	4		954,1737
316055				17 34 59	−28 03.7	1	11.03		0.29		0.17		3		918
159807	−42 12334			17 35 02	−42 42.3	1	7.33		0.03		0.01		2	A1 IV	401
		G 226 - 66		17 35 03	+61 43.0	1	9.97		1.48		1.16		3	M1	196
159881	−27 11764			17 35 03	−28 01.1	3	6.83	.027	1.91	.016	2.35	.010	12	K5 III	897,918,2012
	−30 14512			17 35 04	−31 00.3	1	12.16		0.61				2	K0	1702
159792	−46 11709			17 35 07	−46 17.3	3	9.43	.010	0.14	.005	−0.63	.009	7	B2 II	400,540,976
159975	−8 04472	HR 6567		17 35 08	−08 05.4	8	4.62	.006	0.12	.005	−0.18	.023	32	B8II-III(p)	3,15,1075,1079,1088*
159882	−31 14332			17 35 08	−31 58.8	1	9.47		0.76		0.30		2	G6 IV/V	918
159848	−37 11755			17 35 08	−37 52.9	1	9.35		0.12		−0.68		3	B3 (V)nne	1586
316058				17 35 11	−28 29.5	1	11.41		0.32		0.30		2		918
159809	−45 11742	IDS17315S4542	AB	17 35 11	−45 43.8	2	7.44	.000	1.01	.005	0.86		5	K1 IV	219,1075
	+27 02853	G 170 - 53	★ AB	17 35 13	+27 55.2	1	9.25		1.16		1.14		2	M0 V:p	3072
	+27 02853	IDS17331N2757	C	17 35 13	+27 55.2	1	11.85		1.44		1.29		1		3072
316054				17 35 13	−27 57.1	1	10.20		0.55		0.44		3		918
159916	−28 13376			17 35 13	−28 14.8	1	9.56		0.45		0.27		2	A0 IV	918
316050	−27 11768			17 35 14	−27 20.8	2	10.08	.016	0.38	.002	−0.40	.003	5	B2 IV	540,918
159896	−32 13043			17 35 17	−32 23.9	1	9.81		0.14		−0.13		2	A0 IV	918
160290	+48 02542	HR 6574		17 35 19	+48 36.8	1	5.36		1.15		1.06		2	K0	1355
159897	−32 13044			17 35 20	−33 01.4	1	8.34		0.04		−0.20		6	B9 III/IV	1732
317929	−31 14336			17 35 21	−31 17.8	1	10.77		0.80				1	K1	1702
320138	−34 11889	LSS 4309		17 35 22	−34 53.4	2	10.44	.025	0.94	.005	−0.09	.035	4	O9 III	954,1737
160018	−10 04528	HR 6568		17 35 23	−10 53.9	1	5.75		1.23				4	gK0	2007
316057	−28 13378			17 35 24	−28 34.6	1	9.80		1.65				1	K0	1702
159917	−32 13047			17 35 25	−32 27.4	1	9.38		0.47		−0.01		2	F5 V	918
159868	−43 11901			17 35 25	−43 06.9	10	7.24	.009	0.72	.005	0.25	.010	51	G5 V	278,474,863,977,1075,1499*
160181	+24 03218	HR 6571		17 35 27	+24 20.3	4	5.76	.005	0.11	.014	0.04	.012	10	A2 Vn	985,1022,1501,1733
159558	−67 03343			17 35 28	−67 49.6	1	6.45		1.03		0.73		10	G8 III	1628
159952	−31 14340			17 35 30	−31 03.7	1	9.98		0.14		−0.05		2	B9 IV/V	918
234462	+52 02079			17 35 32	+52 36.2	1	9.38		0.33		0.02		3	F0	1566
160245	+35 03016			17 35 35	+35 49.6	1	7.57		0.54		0.11		2	G0	1601
159979	−30 14531			17 35 35	−30 06.2	1	8.70		0.59		0.14		3	F8/G0 V	918
	+18 03421			17 35 38	+18 36.3	1	9.62		1.53		1.08		2	M1	3078

Table 1 959

HD	DM	Other Id	N	Rem	α_{1950}	δ_{1950}	S	V	σ_V	B–V	σ_{B-V}	U–B	σ_{U-B}	n	Spectrum	References
159980	−31 14342				17 35 38	−31 29.0	1	9.87		0.17		0.20		2	A1 IV/V	918
		G 170 - 55			17 35 40	+18 36.6	2	9.61	.004	1.57	.074			8	M0:	1017,1197
	−53 08705				17 35 40	−53 41.8	1	11.20		0.18		0.16		2		558
		NGC 6405 - 414			17 35 41	−32 07.1	1	14.56		1.18				1		936
		NGC 6405 - 415			17 35 41	−32 07.7	1	12.70		0.24		0.04		1		936
		NGC 6397 sq2 15			17 35 41	−53 41.7	1	13.37		0.16		0.12		3		1662
		LP 688 - 9			17 35 43	−05 04.5	1	11.87		0.87		0.14		2		1696
	+23 03151	IDS17336N2301	A		17 35 44	+22 59.1	1	10.03		1.35		1.24		2	M0	3072
	+23 03151	IDS17336N2301	AB		17 35 44	+22 59.1	1	9.89		1.57				1		1017
	+23 03151	IDS17336N2301	B		17 35 44	+22 59.1	1	10.23		1.37		1.30		2		3016
317945	−31 14347				17 35 45	−31 40.5	1	10.47		1.22				1	K0	1702
320325	−35 11772	LSS 4311			17 35 45	−35 26.4	3	9.85	.007	0.56	.004	−0.41	.008	20	B0.5II	540,976,1699
		GD 211			17 35 46	+09 14.8	1	14.22		0.64		−0.06		2		3060
		G 227 - 10			17 35 48	+63 35.6	1	12.44		0.70		0.07		2		1658
		NGC 6405 - 412			17 35 48	−32 06.5	1	11.91		0.36		0.17		1		936
		NGC 6405 - 413			17 35 48	−32 07.0	1	12.22		0.94		0.49		1		936
160043	−28 13387				17 35 51	−28 23.0	4	7.69	.015	0.44	.013	−0.06	.008	9	F5 V	219,918,1075,3037
159958	−43 11905				17 35 56	−43 49.9	1	7.77		−0.04		−0.40		2	B8 II	401
	−31 14352				17 35 59	−31 36.1	1	11.28		0.76				1	G5	1702
318071	−31 14353				17 36 00	−31 41.9	1	9.44		1.81		1.99		2	K5	918
160044	−32 13059	NGC 6405 - 15			17 36 01	−32 06.1	2	9.86	.015	0.11	.005	−0.01	.010	3	B9 V	145,918
160090	−27 11783				17 36 02	−27 24.9	1	8.93		0.34		0.20		3	A0 III/IV	918
	−31 14354	318072			17 36 03	−31 46.6	1	11.21		0.60				1		1702
160046	−37 11765	LSS 4312			17 36 04	−37 20.2	1	8.40		0.98		0.83		2	A7 II	954
159959	−45 11755				17 36 04	−45 54.1	1	8.18		−0.06		−0.44		2	B5 III	401
	+18 03423	G 170 - 56			17 36 05	+18 35.3	4	9.78	.011	0.48	.008	−0.10	.008	5	F6 V	792,1003,1658,3077
	−32 13062	NGC 6405 - 106			17 36 05	−32 18.9	1	11.37		0.35				1	A5	145
		AO 730			17 36 06	+68 16.1	1	10.43		0.30		0.04		1		1697
160092	−31 14357				17 36 07	−31 04.3	1	9.93		0.14		0.15		2	A0 IV	918
		NGC 6405 - 105			17 36 08	−32 18.3	1	13.32		0.70				2		145
	+68 00942				17 36 09	+68 33.6	1	9.12		0.82		0.41		1	K2	1697
160108	−29 13829				17 36 09	−29 30.2	2	6.71	.004	0.95	.007	0.68	.019	2	K1 III	918,1754
160091	−30 14542				17 36 09	−30 47.5	1	9.51		0.68		0.17		2	G5 V	918
160068	−35 11782	LSS 4313			17 36 09	−35 11.6	2	9.59	.015	0.54	.005	−0.42	.025	3	B1 Ia/ab	540,976
160065	−33 12318	LSS 4314			17 36 10	−33 36.3	4	8.59	.014	0.92	.025	−0.10	.030	6		540,954,976,1737
159495	−73 01852				17 36 10	−73 22.0	1	7.28		0.05		0.00		10	A0 V	1628
		NGC 6405 - 104			17 36 11	−32 19.1	1	13.54		0.67				1		145
160233	+4 03467	LS IV +04 001			17 36 12	+04 21.8	4	9.10	.011	−0.06	.002	−0.80	.031	32	B1 V	963,989,1729,1732
160095	−33 12319				17 36 12	−33 31.6	4	8.67	.069	0.01	.008	−0.66	.037	13	B3 V	540,976,1586,1732
	−32 13069	NGC 6405 - 33			17 36 13	−32 10.8	1	11.71		0.42		0.24		1		145
160093	−32 13066	NGC 6405 - 405			17 36 13	−32 10.8	1	9.42		0.19		−0.07		3	B9 III	918
		G 140 - 2			17 36 14	+05 17.9	2	15.87	.014	0.38	.117	−0.72	.117	4	DA	316,3060
		NGC 6405 - 80			17 36 14	−32 15.7	1	12.62		0.68		0.50		1		145
160109	−29 13830				17 36 15	−29 56.0	3	7.47	.005	0.04	.003	−0.46	.014	4	B5 IV	540,918,976
		NGC 6405 - 102			17 36 15	−32 17.4	1	13.63		0.81				1		145
		NGC 6405 - 101			17 36 15	−32 17.9	1	13.80		1.08				1		145
		NGC 6405 - 79			17 36 16	−32 15.8	1	13.97		0.78				1		145
160094	−32 13068	NGC 6405 - 103		⋆ AB	17 36 16	−32 19.4	1	9.20		0.13		−0.02		3	B9 V	145
160028	−45 11763				17 36 17	−45 06.8	3	8.66	.006	1.45	.008	1.66	.002	19	K2/3 III/IV	863,977,2011
		NGC 6397 sq2 2			17 36 17	−53 47.4	1	13.74		0.84		0.35		4		1662
	−29 13831	LSS 4317			17 36 18	−29 46.1	3	10.98	.020	0.71	.023	−0.25	.035	5		954,1702,1737
160420	+41 02869				17 36 19	+41 02.3	1	8.08		0.56		0.03		2	G5	1601
		LP 448 - 18			17 36 20	+18 07.4	1	13.06		0.94		0.65		1		1696
	+29 03070				17 36 22	+29 10.5	1	10.42		0.18		−0.57		3		963
160186	−18 04598	LSS 4323			17 36 22	−18 23.1	4	9.07	.004	0.18	.007	−0.70	.018	21	B1/2 Ib/II	830,1012,1775,1783
		NGC 6405 - 315			17 36 22	−32 03.1	1	13.28		1.06		0.83		1		936
160124	−32 13072	NGC 6405 - 100		⋆ V	17 36 22	−32 17.6	3	7.18	.015	0.01	.035	−0.47	.014	3	B3 IV/V	145,219,3021
	−65 03477				17 36 22	−65 16.2	1	9.66		1.20		1.25		4	F8	119
	−28 13401	LSS 4319			17 36 23	−28 34.5	2	11.17	.005	0.99	.019	0.04	.078	6		358,954
		NGC 6405 - 129			17 36 23	−32 16.4	1	10.17		0.24		0.19		1		145
		NGC 6405 - 78			17 36 23	−32 16.5	1	13.25		0.81				1		145
		NGC 6405 - 76			17 36 23	−32 16.9	1	10.97		0.59		0.06		1		145
		NGC 6405 - 120			17 36 23	−32 21.5	1	11.14		0.30		0.29		1	A7	145
316053	−27 11786				17 36 25	−28 00.6	1	10.09		1.22				1	G8	1702
	+2 03371				17 36 26	+02 48.1	1	9.20		0.63		−0.38		4		358
318035	−30 14546	IDS17332S3050	AB		17 36 26	−30 52.0	1	9.49		1.21		0.98		2	K0	918
318107	−32 13074	NGC 6405 - 77		⋆ V	17 36 26	−32 17.3	1	9.34		0.04		−0.28		2	B8	145
		NGC 6405 - 119			17 36 26	−32 20.5	1	12.08		0.61		0.15		1		145
		Steph 1505			17 36 27	−12 18.6	1	12.05		2.07		2.28		2	M4	1746
	−12 04796				17 36 27	−12 19.1	1	10.92		1.86		2.15		2	M0p	1746
318102	−32 13076	NGC 6405 - 8			17 36 27	−32 04.3	1	8.77		0.10		0.12		1	A2	145
	−29 13835	316073			17 36 28	−29 17.8	1	10.27		2.38				1		1702
		NGC 6397 sq2 3			17 36 28	−53 44.7	1	14.44		0.79				3		1662
		Ross 132			17 36 29	−22 39.7	1	11.38		1.40		1.19		1	K5	1696
		NGC 6397 sq2 4			17 36 29	−53 44.7	1	14.40		0.90		0.12		3		1662
		NGC 6397 sq2 26			17 36 29	−53 45.5	1	12.75		0.59		0.03		2		1662
160097	−43 11915				17 36 31	−43 42.0	1	9.40		0.05				4	B9 III	2011
160239	−15 04635	IDS17337S1542	AB		17 36 32	−15 44.0	1	9.08		0.76				5	G6/8 V	173
160032	−49 11616	HR 6569			17 36 32	−49 23.2	6	4.76	.006	0.40	.002	−0.05	.009	16	F2 V	15,1075,1586,2012*
160314	+2 03372	IDS17341N0205	B		17 36 33	+02 04.9	2	7.74	.000	0.41	.010	0.05	.005	8	F2 IV	1084,3024

HD	DM	Other Id	N	Rem	α_{1950}	δ_{1950}	S	V	σ_V	B–V	σ_{B-V}	U–B	σ_{U-B}	n	Spectrum	References
		NGC 6405 - 118			17 36 34	−32 20.4	1	10.94		0.40		0.16		1	F0	145
160451	+36 02912				17 36 35	+36 46.5	1	6.77		1.19		1.28		2	K0	1601
		NGC 6405 - 116			17 36 36	−32 21.6	1	12.41		0.47		0.12		1		145
		NGC 6397 sq2 21			17 36 36	−53 29.4	1	11.55		1.06		0.55		2		1662
160315	+2 03373	HR 6575		⋆ A	17 36 37	+02 03.3	5	6.26	.011	1.03	.011	0.83	.006	20	K0 III	15,1084,1417,1509,3024
	−32 13080	NGC 6405 - 74			17 36 37	−32 16.0	1	9.66		0.07		-0.13		2	B9	145
	−32 13079	NGC 6405 - 117			17 36 37	−32 22.3	1	10.51		0.14		0.20		1		145
	−36 11752	LSS 4320			17 36 37	−36 5.0	1	10.84		0.90		-0.17		2	B0.5 Ia	1737
	−53 08716				17 36 37	−53 48.5	1	10.50		1.28		0.94		3		558
160166	−32 13084	NGC 6405 - 21			17 36 38	−32 08.8	2	9.08	.024	0.09	.000	-0.12	.005	3	A0	59,145
	−32 13083	NGC 6405 - 75			17 36 38	−32 15.4	1	9.88		0.01		0.08		1	A0	145
160167	−32 13081	NGC 6405 - 115			17 36 38	−32 20.9	1	8.62		0.10		-0.30		2	B8 V	145
		NGC 6397 sq2 22			17 36 38	−53 29.3	1	13.92		0.62		0.20		2		1662
160365	+13 03421	HR 6577			17 36 40	+13 21.3	1	6.12		0.56				2	F6 III	71
		NGC 6405 - 210			17 36 41	−32 10.7	2	12.71	.024	0.61	.019	0.10	.000	3		936,3021
		NGC 6397 sq2 7			17 36 41	−53 46.2	1	12.40		0.94		0.38		4		1662
160861	+68 00946	G 240 - 63		⋆ AB	17 36 42	+68 23.1	3	9.16	.027	1.50	.013	1.07	.016	11	F5	1,1197,3078
		NGC 6405 - 209			17 36 42	−32 10.5	1	13.40		0.64		0.09		2		3021
		NGC 6397 sq2 9			17 36 42	−53 46.3	1	14.29		0.78		0.12		3		1662
		NGC 6405 - 208			17 36 43	−32 10.5	1	12.16		1.74		1.00		2		3021
160189	−32 13085	NGC 6405 - 128			17 36 43	−32 25.0	1	8.08		0.03		-0.25		2	B8 V	145
160202	−32 13086	NGC 6405 - 32		⋆ V	17 36 45	−32 10.5	5	6.73	.028	0.01	.020	-0.46	.013	20	B3 V:	59,145,1586,1637,3021
		NGC 6405 - 54			17 36 45	−32 11.4	3	10.41	.021	0.23	.016	0.15	.040	3	A2	59,145,3021
		NGC 6405 - 211			17 36 45	−32 11.5	1	14.28		0.99		0.44		2		3021
318126	−32 13087	NGC 6405 - 99			17 36 46	−32 19.1	1	9.41		0.10		-0.23		1	A2	145
		G 20 - 7			17 36 47	+01 04.9	3	11.47	.008	1.28	.013	1.16	.024	5	M2	202,333,1620,3062
		G 259 - 15			17 36 47	+82 07.0	1	14.21		1.73				1		906
318101	−32 13090	NGC 6405 - 20			17 36 47	−32 08.0	2	8.27	.015	0.03	.029	-0.52	.044	3	B9	59,145
160346	+3 03465				17 36 48	+03 35.0	7	6.53	.014	0.95	.006	0.78	.009	23	K3 V	101,770,1013,1067*
	−30 14555	316167			17 36 48	−30 19.5	1	10.22		1.40				1		1702
		NGC 6405 - 212			17 36 48	−32 11.	1	14.51		0.90				1		3021
		NGC 6405 - 416			17 36 48	−32 11.	1	13.20		0.50		0.17		1		936
		NGC 6405 - 417			17 36 48	−32 11.	1	11.82		0.38		0.32		1		936
	−32 13089	NGC 6405 - 53			17 36 48	−32 11.7	3	9.86	.023	0.26	.023	0.15	.079	3	A0	59,145,3021
		NGC 6397 - 2			17 36 48	−53 39.	1	13.89		0.81		0.35		1		366
		NGC 6397 - 4			17 36 48	−53 39.	1	13.43		0.18		0.20		3		205
		NGC 6397 - 12			17 36 48	−53 39.	1	13.58		0.83		0.26		1		366
		NGC 6397 - 25			17 36 48	−53 39.	1	12.22		0.96		0.42		2		366
		NGC 6397 - 26			17 36 48	−53 39.	1	13.46		0.17		0.13		1		205
		NGC 6397 - 28			17 36 48	−53 39.	1	11.81		0.94		0.40		1		366
		NGC 6397 - 33			17 36 48	−53 39.	2	12.80	.020	0.91	.020	0.29	.060	4		366,558
		NGC 6397 - 43			17 36 48	−53 39.	1	10.94		1.12		0.74		2		366
		NGC 6397 - 44			17 36 48	−53 39.	1	13.27		0.18		0.21		2		205
		NGC 6397 - 48			17 36 48	−53 39.	1	13.08		0.21		0.21		3		205
		NGC 6397 - 56			17 36 48	−53 39.	1	13.94		0.12		-0.10		3		205
		NGC 6397 - 59			17 36 48	−53 39.	1	13.89		0.86		0.26		1		366
		NGC 6397 - 70			17 36 48	−53 39.	2	13.03	.024	0.26	.000	0.31	.010	3		205,366
		NGC 6397 - 74			17 36 48	−53 39.	1	12.44		0.67		0.17		1		366
		NGC 6397 - 75			17 36 48	−53 39.	1	12.12		0.87		0.29		1		366
		NGC 6397 - 77			17 36 48	−53 39.	2	13.39	.044	0.18	.005	0.19	.000	3		205,366
		NGC 6397 - 128			17 36 48	−53 39.	1	12.24		0.97		0.44		2		366
		NGC 6397 - 132			17 36 48	−53 39.	1	12.68		0.94		0.35		2		366
		NGC 6397 - 211			17 36 48	−53 39.	1	10.16		1.46		1.30		2		366
		NGC 6397 - 219			17 36 48	−53 39.	1	12.31		1.11		0.63		1		366
		NGC 6397 - 220			17 36 48	−53 39.	1	13.84		0.85		0.08		1		366
		NGC 6397 - 221			17 36 48	−53 39.	1	13.60		0.15		0.07		2		205
		NGC 6397 - 229			17 36 48	−53 39.	1	13.79		0.81		0.17		1		366
		NGC 6397 - 428			17 36 48	−53 39.	1	11.50		1.05		0.56		2		366
		NGC 6397 - 438			17 36 48	−53 39.	2	13.76	.010	0.11	.005	-0.05	.010	2		205,366
		NGC 6397 - 447			17 36 48	−53 39.	2	11.25	.005	0.15	.010	0.14	.010	5		366,1662
		NGC 6397 - 451			17 36 48	−53 39.	2	12.98	.019	0.71	.009	0.21	.019	4		366,1662
		NGC 6397 - 468			17 36 48	−53 39.	1	11.50		1.05		0.60		2		366
		NGC 6397 - 484			17 36 48	−53 39.	1	12.66		0.72		0.16		2		366
		NGC 6397 - 531			17 36 48	−53 39.	1	12.73		0.44		0.05		1		366
		NGC 6397 - 545			17 36 48	−53 39.	1	12.82		0.89		0.28		1		366
		NGC 6397 - 548			17 36 48	−53 39.	1	13.02		0.86		0.28		2		366
		NGC 6397 - 555			17 36 48	−53 39.	1	12.25		0.80		0.25		1		366
		NGC 6397 - 557			17 36 48	−53 39.	1	13.37		0.57		0.09		1		366
		NGC 6397 - 584			17 36 48	−53 39.	1	13.22		0.20		0.21		2		205
		NGC 6397 - 585			17 36 48	−53 39.	1	13.80		0.83		0.27		1		366
		NGC 6397 - 591			17 36 48	−53 39.	2	12.65	.025	0.94	.020	0.30	.059	5		366,558
		NGC 6397 - 603			17 36 48	−53 39.	1	10.35		1.33		1.00		1		366
		NGC 6397 - 608			17 36 48	−53 39.	1	12.92		0.86		0.25		1		366
		NGC 6397 - 616			17 36 48	−53 39.	1	13.18		0.84		0.27		1		366
		NGC 6397 - 620			17 36 48	−53 39.	1	13.19		0.84		0.23		1		366
		NGC 6397 - 632			17 36 48	−53 39.	2	13.06	.044	0.29	.000	0.25	.039	3		205,366
		NGC 6397 - 644			17 36 48	−53 39.	2	9.47	.010	0.53	.005	0.12	.025	9		366,1662
		NGC 6397 - 657			17 36 48	−53 39.	2	13.93	.044	0.07	.000	-0.12	.005	3		205,366
		NGC 6397 - 660			17 36 48	−53 39.	1	12.82		1.31		1.22		3		558
		NGC 6397 - 661			17 36 48	−53 39.	2	13.35	.014	0.17	.019	0.18	.009	4		205,366

Table 1   961

HD	DM	Other Id	N Rem	α_{1950}	δ_{1950}	S	V	σ_V	B–V	σ_{B-V}	U–B	σ_{U-B}	n	Spectrum	References
		NGC 6397 - 662		17 36 48	−53 39.	1	12.64		0.59		0.12		1		366
		NGC 6397 - 669		17 36 48	−53 39.	1	10.50		1.28		0.96		2		366
		NGC 6397 - 670		17 36 48	−53 39.	3	13.54	.043	0.12	.026	0.05	.020	5		205,366,558
		NGC 6397 - 679		17 36 48	−53 39.	1	13.07		0.16		0.20		2		205
		NGC 6397 - 683		17 36 48	−53 39.	1	12.46		0.66		0.18		1		366
		NGC 6397 - 685		17 36 48	−53 39.	2	12.00	.015	0.99	.020	0.43	.020	5		366,558
		NGC 6397 - 686		17 36 48	−53 39.	1	13.50		0.80		0.26		1		366
		NGC 6397 - 689		17 36 48	−53 39.	1	12.99		0.46		0.13		1		366
		NGC 6397 - 691		17 36 48	−53 39.	1	12.55		0.54		0.15		1		366
		NGC 6397 - 692		17 36 48	−53 39.	1	11.17		1.17		1.06		1		366
		NGC 6397 - 696		17 36 48	−53 39.	1	13.88		0.75		0.36		1		366
		NGC 6397 - 697		17 36 48	−53 39.	1	13.46		0.70		0.07		1		366
		NGC 6397 - 699		17 36 48	−53 39.	1	12.28		0.60		0.12		1		366
		NGC 6397 - 709		17 36 48	−53 39.	1	13.24		0.18		0.21		1		205
		NGC 6397 - 717		17 36 48	−53 39.	1	12.51		0.92		0.37		2		366
		NGC 6397 - 718		17 36 48	−53 39.	1	12.24		0.77		0.39		1		366
		NGC 6397 - 719		17 36 48	−53 39.	1	13.81		0.11		-0.01		2		205
		NGC 6397 - 720		17 36 48	−53 39.	1	13.40		0.52		0.11		1		366
		NGC 6397 - 721		17 36 48	−53 39.	1	12.65		0.93		0.34		2		366
		NGC 6397 - 744		17 36 48	−53 39.	1	14.43		1.26		1.39		1		366
		NGC 6397 - 746		17 36 48	−53 39.	2	14.85	.020	0.69	.045	0.06	.064	5		366,558
		NGC 6397 - 749		17 36 48	−53 39.	2	14.72	.029	1.26	.000	1.06		3		366,558
		NGC 6397 - 752		17 36 48	−53 39.	1	13.89		0.80		0.23		1		366
		NGC 6397 - 798		17 36 48	−53 39.	2	16.09	.024	0.58	.005	-0.14	.015	3		366,558
		NGC 6397 - 873		17 36 48	−53 39.	1	13.84		0.11		-0.01		2		205
		NGC 6397 - 1006		17 36 48	−53 39.	1	15.56		0.74				1		366
		NGC 6397 - 1012		17 36 48	−53 39.	1	15.85		0.73				1		366
		NGC 6397 - 1016		17 36 48	−53 39.	1	15.34		0.70		0.21		1		366
		NGC 6397 - 1021		17 36 48	−53 39.	1	14.48		0.71		0.17		1		366
		NGC 6397 - 1024		17 36 48	−53 39.	1	15.67		0.86				1		366
		NGC 6397 - 1029		17 36 48	−53 39.	1	15.09		0.87		0.31		1		366
		NGC 6397 - 1037		17 36 48	−53 39.	1	15.84		0.69				1		366
		NGC 6397 - 1044		17 36 48	−53 39.	1	15.64		0.69				1		366
		NGC 6397 - 1045		17 36 48	−53 39.	1	15.83		0.60				1		366
		NGC 6397 - 1051		17 36 48	−53 39.	1	16.17		0.89		0.11		1		366
		NGC 6397 - 1052		17 36 48	−53 39.	1	16.40		0.75				2		366
		NGC 6397 - 1062		17 36 48	−53 39.	1	16.01		0.62		0.05		1		366
		NGC 6397 - 1068		17 36 48	−53 39.	1	14.52		0.80		0.14		2		366
		NGC 6397 - 1100		17 36 48	−53 39.	1	14.15		0.75		0.21		2		366
		NGC 6397 - 1101		17 36 48	−53 39.	1	16.26		0.58		-0.17		1		366
		NGC 6397 - 1115		17 36 48	−53 39.	1	16.59		0.63				1		366
		NGC 6397 - 4006		17 36 48	−53 39.	1	15.50		0.78		0.07		2		558
		NGC 6397 - 4012		17 36 48	−53 39.	1	15.83		0.76		0.17		2		558
		NGC 6397 - 4016		17 36 48	−53 39.	1	15.28		0.74		0.12		3		558
		NGC 6397 - 4100		17 36 48	−53 39.	1	14.14		0.84		0.13		2		558
	−32 13092	NGC 6405 - 73		17 36 49	−32 14.1	2	10.01	.070	0.10	.020	0.07	.010	2	B8	59,145
		NGC 6405 - 98		17 36 50	−32 18.3	2	12.24	.045	0.58	.010	0.08	.015	2		145,936
		G 204 - 20		17 36 51	+40 15.2	1	13.52		0.71		0.08		3		1658
		NGC 6405 - 207		17 36 51	−32 10.0	1	14.45		0.93		0.39		3		3021
	−32 13093	NGC 6405 - 52		17 36 51	−32 12.3	3	10.23	.022	0.12	.010	0.07	.017	5	B9	59,145,3021
	−32 13097	NGC 6405 - 51		17 36 53	−32 12.7	3	10.00	.032	0.11	.020	0.03	.027	7	B8	59,145,3021
		NGC 6405 - 368		17 36 53	−32 17.7	1	14.14		1.07		0.43		1		936
	−30 14558			17 36 54	−30 15.	1	11.16		0.72				1	G2	1702
		NGC 6405 - 206		17 36 54	−32 09.4	1	15.42		1.02				2		3021
		NGC 6405 - 30		17 36 54	−32 10.1	1	11.51		0.57		0.06		1		59,145
		NGC 6405 - 204		17 36 54	−32 10.1	1	13.66		1.78				3		3021
		NGC 6405 - 31	⋆ V	17 36 54	−32 10.6	2	11.59	.035	0.34	.048	0.23	.052	6		145,3021
		NGC 6405 - 203		17 36 54	−32 10.8	1	14.08		0.85		0.33		3		3021
318129	−32 13094	NGC 6405 - 127		17 36 54	−32 25.1	2	10.31	.025	0.14	.050	0.19	.010	2	B9	145,936
		NGC 6397 sq2 10		17 36 54	−53 46.7	1	12.65		1.04		0.45		3		1662
		NGC 6405 - 205		17 36 55	−32 09.9	1	14.21		1.05		0.66		2		3021
	−32 13099	NGC 6405 - 49		17 36 55	−32 11.4	2	9.84	.010	1.72	.030	1.93	.100	2		59,145
		NGC 6405 - 48		17 36 55	−32 11.5	2	10.62	.025	0.25	.015	0.19	.005	2		59,145
160221	−32 13102	NGC 6405 - 72		17 36 55	−32 13.8	3	7.26	.031	0.00	.027	-0.48	.018	8	B5 V	59,145,3021
160222	−32 13100	NGC 6405 - 97		17 36 55	−32 18.8	1	8.53		0.03		-0.33		2	B9	145
		NGC 6405 - 50		17 36 56	−32 12.5	3	13.89	.053	0.72	.017	0.09	.025	6		59,145,3021
318110	−32 13101	NGC 6405 - 96		17 36 56	−32 17.9	1	8.96		0.04		-0.22		2	B8	145
318100	−32 13103	NGC 6405 - 19	⋆ V	17 36 57	−32 08.0	2	9.82	.019	0.07	.005	-0.13	.010	3	B9	59,145
160171	−43 11920			17 36 57	−43 37.7	1	9.08		0.72				4	G0/2 V	2011
160205	−41 11893	TT Sco		17 36 58	−41 36.3	1	8.18		3.76				1	C	109
160173	−44 11963			17 36 58	−45 01.9	1	8.70		-0.05		-0.38		2	B5/7 III	401
160507	+32 02964	HR 6579		17 36 59	+32 46.0	1	6.56		0.99		0.71		2	gG5	1733
		G 154 - 4		17 36 59	−14 18.3	1	14.86		0.89		0.13		3		3062
		NGC 6405 - 397		17 37 00	−32 28.9	1	11.37		0.78		0.22		1		936
		NGC 6397 sq2 11		17 37 00	−53 46.9	1	11.64		1.04		0.54		4		1662
		GD 363		17 37 01	+41 54.0	1	15.51		-0.08		-1.08		2	DA	308
	−32 13106	NGC 6405 - 71		17 37 01	−32 14.4	2	10.20	.040	0.12	.035	0.15	.050	2		59,145
318127	−32 13104	NGC 6405 - 114		17 37 01	−32 22.1	1	9.83		0.07		0.01		1	B9	145
		NGC 6397 sq2 13		17 37 01	−53 45.5	1	11.67		0.56		0.19		4		1662
160964	+71 00850			17 37 02	+71 54.4	1	8.62		1.10		1.04		2	K4 V	3072

HD	DM	Other Id	N	Rem	α_{1950}	δ_{1950}	S	V	σ_V	B–V	σ_{B-V}	U–B	σ_{U-B}	n	Spectrum	References
318099	−32 13111	NGC 6405 - 29			17 37 02	−32 09.4	2	9.86	.005	0.07	.019	-0.05	.019	3	A0	59,145
	−32 13109	NGC 6405 - 47			17 37 02	−32 11.4	2	10.46	.049	0.16	.019	0.14	.010	3		59,145
		MWC 265, RT Ser			17 37 04	−11 55.1	1	14.95		1.73		-0.59		1	A8	1753
160259	−32 13113	NGC 6405 - 28		⋆ AB	17 37 04	−32 09.2	3	8.78	.020	0.05	.012	-0.14	.019	4	B9 V	59,145,3021
160933	+69 00933	HR 6598			17 37 05	+69 36.1	3	6.36	.033	0.59	.011	0.04	.019	4	G0 V	254,1620,3026
		NGC 6405 - 14			17 37 05	−32 06.6	1	11.71		0.40		0.23		1		145
		NGC 6397 sq2 12			17 37 05	−53 46.8	1	13.19		0.85		0.19		3		1662
160387	−4 04324	IDS17345S0455	B		17 37 06	−04 56.5	1	8.46		0.46		0.01		3	F2	3016
160260	−32 13115	NGC 6405 - 70			17 37 06	−32 14.3	3	8.36	.020	0.04	.014	-0.36	.004	5	B9	59,145,3021
160388	−4 04324	IDS17345S0455	A		17 37 08	−04 56.5	1	7.76		0.47		0.05		3	F2	3016
160388	−4 04325	IDS17345S0455	C		17 37 08	−04 56.5	1	12.71		0.89		0.81		1		3016
		NGC 6405 - 278			17 37 08	−32 19.7	1	12.27		0.63		0.12		1		936
	−32 13112	NGC 6405 - 113			17 37 08	−32 21.8	1	10.82		0.20		0.25		1	A4	145
318135	−32 13117	NGC 6405 - 126			17 37 08	−32 30.1	1	9.41		0.58		0.03		1	F5	145
160207	−44 11967				17 37 09	−44 56.2	5	8.40	.004	-0.02	.005	-0.42	.003	21	B7/8 Ib	158,400,401,977,2011
	−32 13119	NGC 6405 - 7			17 37 10	−32 04.6	1	10.94		0.24		0.38		1	A2	145
318095	−32 13120	NGC 6405 - 13			17 37 11	−32 05.9	1	10.55		0.13		0.19		1	A0	145
160508	+26 03054				17 37 12	+26 47.1	2	8.13	.010	0.54	.013	0.00		6	F8 V	20,3026
		AO 735			17 37 12	+68 05.2	1	10.00		0.00		1.83		1		1697
318094		NGC 6405 - 42			17 37 12	−32 10.6	2	10.60	.029	0.20	.019	0.13	.015	3	A2	59,145
158787	−82 00710				17 37 12	−82 45.2	1	8.67		0.47				4	F5/6 IV/V	2012
160319	−28 13418	LSS 4324			17 37 13	−28 53.8	2	7.19	.011	0.21	.012	-0.40	.006	3	B3 Vne	540,976
		NGC 6405 - 69			17 37 13	−32 15.0	1	11.99		0.53		0.04		1		145
160922	+68 00949	HR 6596		⋆ A	17 37 14	+68 46.9	5	4.79	.011	0.43	.008	-0.02	.007	12	F5 V	15,1118,1620,3026,8015
160133	−53 08727	NGC 6397 - 681			17 37 14	−53 47.8	3	9.07	.010	1.17	.004	1.05	.026	11	K0 III	366,558,1662
		NGC 6405 - 27			17 37 15	−32 08.8	1	12.51		0.67		0.09		2	A0	145
		NGC 6405 - 66			17 37 15	−32 13.8	2	12.28	.010	0.47	.010	0.11	.005	2		145,3021
		NGC 6405 - 68			17 37 15	−32 15.2	1	11.97		0.45		0.11		1		145
	+2 03375	G 20 - 8			17 37 16	+02 26.5	9	9.94	.021	0.45	.015	-0.24	.019	15	A5	202,516,927,979,1003*
160297	−32 13123	NGC 6405 - 41			17 37 16	−32 10.4	3	8.91	.031	0.08	.005	-0.10	.012	4	A0 V	59,145,3021
		NGC 6405 - 46			17 37 16	−32 13.2	1	14.87		0.93				1		145
		NGC 6405 - 45			17 37 17	−32 12.7	2	14.02	.040	0.89	.030	0.43		2		59,145
318113	−32 13124	NGC 6405 - 95			17 37 17	−32 17.4	1	10.17		0.08		0.14		2	A0	145
		NGC 6405 - 18			17 37 18	−32 08.3	1	12.27		0.54		0.02		2		145
		NGC 6405 - 26			17 37 18	−32 09.2	1	11.80		0.46		0.04		2	A2	145
		NGC 6405 - 43			17 37 18	−32 12.3	1	11.55		0.48		0.12		1		145
		NGC 6405 - 67			17 37 18	−32 16.2	1	11.26		0.32		0.22		2		145
		LP 448 - 19			17 37 20	+18 53.9	1	11.57		0.56		-0.11		2		1696
		NGC 6405 - 6			17 37 20	−32 05.0	1	12.81		0.62				1		145
		AE Ara			17 37 20	−47 01.8	1	12.14		0.52		-0.89		1		1753
		G 154 - 5			17 37 21	−19 35.1	1	14.24		0.95		0.10		6		3062
		NGC 6405 - 12			17 37 21	−32 06.2	1	11.90		0.51		-0.14		1		145
		NGC 6405 - 65			17 37 21	−32 14.0	1	12.62		0.59		0.02		1		145
		NGC 6405 - 44			17 37 22	−32 12.7	1	10.62		0.19		0.20		2	A0	145
160298	−32 13125	NGC 6405 - 94			17 37 22	−32 18.1	1	8.76		0.03		-0.26		2	B9 V	145
		NGC 6405 - 273			17 37 22	−32 19.3	1	15.47		0.89				1		936
	−29 13845				17 37 23	−29 35.9	1	11.06		0.87				2	K0	1702
160177	−53 08730	NGC 6397 - 627			17 37 23	−53 46.7	3	9.41	.016	0.87	.014	0.55	.015	11	K1 IV/V	366,558,1662
318093		NGC 6405 - 25			17 37 24	−32 09.8	3	9.12	.036	0.10	.015	-0.04	.054	4	A0	59,145,3021
		L 270 - 140			17 37 24	−54 30.	1	13.26		0.95		0.44		2		3062
	+2 03377				17 37 25	+02 26.7	3	10.03	.000	0.58	.000	0.14	.000	3	F2	742,979,8112
		NGC 6405 - 40			17 37 27	−32 12.5	2	12.86	.049	0.63	.063	0.08	.054	3		145,3021
318114	−32 13128	NGC 6405 - 64			17 37 27	−32 15.3	1	10.23		0.10		0.09		2	A0	145
160335	−32 13129	NGC 6405 - 2			17 37 28	−32 07.8	3	7.27	.030	-0.01	.022	-0.57	.027	21	B3 V	59,145,3021
		NGC 6405 - 39			17 37 28	−32 11.3	1	13.85		0.70				2		145
		NGC 6405 - 23			17 37 31	−32 08.3	1	13.55		0.86				1		145
		NGC 6405 - 24			17 37 31	−32 09.8	1	12.56		0.97		0.10		1		145
	−32 13131	NGC 6405 - 62			17 37 31	−32 16.0	1	10.66		1.50				2	A0	145
160263	−46 11747	HR 6572			17 37 31	−46 53.8	7	5.79	.003	-0.01	.002	-0.03	.006	58	A0 V	15,278,977,977,1075*
		NGC 6405 - 38			17 37 32	−32 10.1	1	13.21		0.68				1		145
	−53 08732				17 37 33	−53 39.6	2	10.77	.005	1.02	.000	0.66	.020	5		366,558
318118	−32 13136	NGC 6405 - 63			17 37 34	−32 15.9	2	10.21	.029	0.12	.019	0.06	.024	3	A0	59,145
	+25 03316				17 37 35	+25 00.9	1	8.49		1.40		1.70		2	K2	1625
318117	−32 13134	NGC 6405 - 37			17 37 35	−32 12.0	2	8.76	.015	0.03	.020	-0.30	.010	5	B9	145,3021
160471	−2 04425	HR 6578			17 37 36	−02 07.6	2	6.19	.010	1.67	.000	1.86	.010	6	K2.5 Ib	1149,3005
318125	−32 13132	NGC 6405 - 112			17 37 36	−32 22.1	1	10.76		0.32		0.28		1	A0	145
		L 158 - 53		V	17 37 36	−61 56.	2	14.90	.000	0.11	.000	-0.69	.000	8		782,3065
		NGC 6405 - 131			17 37 38	−32 13.4	1	11.96		0.55		0.00		2		145
318119	−32 13137	NGC 6405 - 92			17 37 38	−32 17.2	1	10.40		0.15		0.15		2	A0	145
318092	−32 13138	NGC 6405 - 22			17 37 39	−32 09.0	1	9.48		0.08		-0.10		2	A0	145
160337	−37 11789	LSS 4326		⋆ AB	17 37 39	−37 37.7	2	7.98	.014	0.44	.006	-0.51	.013	3	B1/2 Ia/ab	540,976
318120	−32 13139	NGC 6405 - 93			17 37 40	−32 19.3	1	10.01		0.17		0.10		2	A0	145
160371	−32 13142	NGC 6405 - 1		⋆ V	17 37 43	−32 11.4	3	5.95	.285	1.75	.107	1.57	.006	15	K2.5Ib	59,145,3049
		NGC 6405 - 91			17 37 43	−32 17.4	1	11.98		0.48		0.02		1		145
160408	−27 11801	V551 Oph			17 37 44	−27 22.2	1	8.27		1.82				5	M2 III	897
318091	−32 13144	NGC 6405 - 5			17 37 44	−32 04.6	1	10.51		0.14		0.18		2	A0	145
160392	−32 13146	NGC 6405 - 17			17 37 44	−32 08.1	1	9.06		0.06		-0.21		2	A0	145
		NGC 6405 - 88			17 37 46	−32 17.5	1	13.85		0.79				1		145
		NGC 6397 sq2 24			17 37 46	−53 32.8	1	13.46		0.70		0.17		2		1662
160430	−23 13501	LSS 4329			17 37 47	−23 48.7	3	7.90	.003	0.36	.008	-0.50	.010	5	B2 II	240,540,976

Table 1   963

HD	DM	Other Id	N Rem	α_{1950}	δ_{1950}	S	V	σ_V	B-V	σ_{B-V}	U-B	σ_{U-B}	n	Spectrum	References
318121	-32 13147	NGC 6405 - 61		17 37 47	-32 15.7	1	9.67		0.69		0.14		1	G0	145
		NGC 6405 - 90		17 37 48	-32 19.0	1	13.73		0.69				1		145
		NGC 6405 - 125		17 37 48	-32 25.8	1	11.83		0.66				1		145
		NGC 6405 - 85		17 37 49	-32 17.9	1	12.98		0.62		0.17		1		145
		NGC 6405 - 89		17 37 49	-32 18.0	1	13.30		0.39		0.15		1		145
	-32 13149	NGC 6405 - 123		17 37 49	-32 23.2	1	11.40		0.38		0.12		2		145
	-32 13148	NGC 6405 - 130		17 37 50	-32 13.3	3	10.92	.022	0.28	.007	0.20	.031	6	A5	145,936,3021
		NGC 6405 - 87		17 37 51	-32 16.8	1	12.77		0.84				1		145
		NGC 6405 - 111		17 37 51	-32 22.3	1	12.16		0.62		0.18		2		145
		NGC 6405 - 122		17 37 52	-32 24.7	1	14.54		0.84				1		145
		Steph 1508	B	17 37 53	+05 45.6	1	12.93		1.44				2	K7	1746
318090	-32 13155	NGC 6405 - 11		17 37 53	-32 05.4	1	10.70		0.32		0.16		2	A2	145
160409	-32 13152	NGC 6405 - 60		17 37 53	-32 15.2	1	9.19		0.33		0.12		1	A5 IV	145
		Steph 1508	A	17 37 54	+05 45.0	1	11.24		1.28		1.20		1		1746
		NGC 6405 - 36		17 37 54	-32 12.5	1	12.57		0.72				2		145
		NGC 6405 - 110		17 37 54	-32 21.3	2	14.26	.025	0.87	.005	0.41		2		145,936
	-32 13153	NGC 6405 - 124		17 37 54	-32 26.0	1	11.19		0.28		0.23		2	A5	145
		NGC 6405 - 4		17 37 55	-32 04.7	1	10.58		0.68		0.13		2	F0	145
		NGC 6405 - 132		17 37 55	-32 13.6	1	13.74		0.70				1		145
160693	+37 02926	G 182 - 19		17 37 56	+37 13.2	5	8.37	.018	0.58	.016	-0.03	.033	8	G0 V	22,792,1003,1620,1658
	-32 13156	NGC 6405 - 109		17 37 57	-32 20.1	1	11.74		0.28		0.22		1	A4	145
		NGC 6405 - 86		17 37 58	-32 18.9	1	13.88		0.68		0.22		1		145
	-32 13157	NGC 6405 - 121		17 37 58	-32 22.7	1	11.18		0.66		0.06		3		145
	-32 13158	NGC 6405 - 16		17 37 59	-32 06.3	1	9.88		1.83		1.27		2	G8	145
		NGC 6405 - 108		17 37 59	-32 20.9	2	13.03	.020	0.38	.015	0.27		2		145,936
		AO 767		17 37 60	+68 22.4	1	10.91		0.60		0.02		1		1697
		NGC 6405 - 107		17 38 00	-32 20.4	1	14.10		0.71				1		145
	-7 04478		A	17 38 01	-07 44.9	1	10.26		1.07		0.77		2	K7 V	3072
	-7 04478		B	17 38 01	-07 44.9	1	13.68		1.58		1.18		2	K7 V	3072
		NGC 6405 - 84		17 38 02	-32 18.4	1	14.22		0.70				1		145
160762	+46 02349	HR 6588, ι Her	⋆ A	17 38 03	+46 01.9	9	3.80	.000	-0.18	.003	-0.70	.010	74	B3 V	1,15,154,1077,1118*
		NGC 6405 - 10		17 38 03	-32 05.3	2	12.96	.005	1.55	.029	1.16		3		145,936
318122	-32 13160	NGC 6405 - 59		17 38 03	-32 15.5	1	11.06		0.13		0.24		2	A0	145
160677	+31 03075	HR 6584		17 38 04	+31 13.7	2	6.05	.028	1.58	.005	1.98		3	M2 IIIab	71,3001
161178	+72 00800	HR 6606		17 38 05	+72 29.0	1	5.86		1.01				2	G9 III	71
		NGC 6405 - 58		17 38 05	-32 15.5	1	11.60		0.54		0.15		2	A7	145
		ApJS33,459 # 96		17 38 05	-36 46.3	1	15.24		2.27		-0.61		1		1753
160342	-50 11474	HR 6576, V626 Ara		17 38 09	-50 29.2	1	6.35		1.73		1.96		5	M3 III	3042
		NGC 6405 - 57		17 38 14	-32 13.9	1	11.80		0.42		0.14		2	A7	145
		NGC 6405 - 83		17 38 16	-32 19.1	1	13.08		0.67		0.30		1		145
		NGC 6405 - 56		17 38 18	-32 13.6	1	13.37		0.51				1		145
159964	-72 02086	HR 6565	⋆ AB	17 38 18	-72 12.0	4	6.49	.004	0.47	.010	-0.05	.010	24	F7 IV+F5 V	15,1075,2012,2038
160608	+5 03447		V	17 38 19	+05 03.2	1	5.47		0.70		0.43		1	K0	1754
160397	-48 11945			17 38 19	-48 48.3	1	9.77		-0.01		-0.44		3	B5 II	400
		NGC 6405 - 82		17 38 20	-32 17.0	1	13.54		0.59				1		145
	-32 13168	NGC 6405 - 35		17 38 21	-32 12.8	1	11.38		0.36		0.19		2	A7	145
160491	-32 13169	NGC 6405 - 9		17 38 22	-32 05.3	1	7.88		0.05		-0.23		1	B9 IV	145
		NGC 6405 - 81		17 38 22	-32 19.2	1	13.23		0.85				1		145
318089	-32 13170	NGC 6405 - 34		17 38 24	-32 11.9	1	10.71		0.18		0.24		2	A2	145
		NGC 6405 - 55		17 38 24	-32 12.9	1	13.14		0.59				1		145
	+44 02748	G 204 - 23		17 38 30	+44 06.4	1	9.02		0.63		0.07		1	G0	1658
158651	-84 00560			17 38 31	-84 05.6	1	7.64		0.59		0.14		6	G0 V	1704
		G 204 - 24		17 38 32	+37 42.2	1	15.50		1.56				2		1663
318084	-32 13172	NGC 6405 - 3		17 38 32	-32 04.1	1	9.51		0.34		-0.15		1	B8	145
160883	+49 02678			17 38 33	+49 48.3	1	6.62		0.00		0.00		4	A0	753
	-15 04644			17 38 33	-15 46.8	1	10.00		1.74				1	M3	369
160613	-12 04808	HR 6581, o Ser		17 38 36	-12 51.0	6	4.25	.008	0.07	.005	0.09	.010	21	A2 Va	15,1075,1088,2012*
160529	-33 12361	V905 Sco		17 38 41	-33 28.8	3	6.57	.036	1.22	.015	0.36	.098	9	A9 Ia	954,1012,8100
	+26 03059	G 170 - 59		17 38 47	+26 29.0	1	9.74		0.86		0.72		2	K1	7010
160822	+31 03076	HR 6591	⋆ A	17 38 48	+31 18.8	2	6.31	.031	1.05	.000	0.88		4	K0 III	71,3016
160625	-17 04889			17 38 48	-17 47.0	1	9.28		0.26		0.16		37	G2/3 IV	843
160482	-49 11642			17 38 54	-49 58.5	1	7.75		0.51				4	F5 V	2012
160765	+15 03246	HR 6589		17 38 55	+15 12.1	3	6.27	.112	0.04	.008	0.02	.008	7	A1 V	192,253,1022
160641	-17 04890	V2076 Oph		17 38 55	-17 52.7	4	9.86	.007	0.15	.007	-0.85	.007	28	O9.5 Ia(p)	55,400,843,1737
160575	-35 11827	LSS 4333		17 38 56	-35 46.7	2	7.61	.000	0.36	.008	-0.52	.012	3	B1/2 II	540,976
		AO 736		17 38 58	+68 22.7	1	10.50		1.21		1.28		1		1697
160590	-34 11971	LSS 4335		17 38 58	-34 06.2	3	9.17	.013	0.80	.014	-0.23	.028	7	O9.5Ia(e)	954,1586,1737
160435	-57 08687	V Pav	⋆ A	17 39 00	-57 42.1	2	6.85	.253	4.01	.248	3.36		3	C6,4	109,864
160589	-32 13185	V703 Sco		17 39 01	-32 30.0	2	7.64	.107	0.34	.060	0.15		2	A9 V	700,1497
160835	+24 03225	HR 6592	⋆ A	17 39 02	+24 32.2	4	6.37	.014	1.20	.005	1.18	.000	9	K1 III+F4 V	15,1028,3024,8015
160835	+24 03225	IDS17370N2434	B	17 39 02	+24 32.2	2	9.41	.009	0.46	.004	-0.06	.009	5		1028,3024
160578	-38 12137	HR 6580, κ Sgr		17 39 02	-39 00.4	6	2.41	.004	-0.23	.012	-0.89	.015	16	B1.5III	15,26,1034,1075,2012,8015
	+68 00951			17 39 05	+68 14.8	1	9.82		1.08		0.97		1	K0	1697
160781	+6 03498	HR 6590		17 39 06	+06 20.2	3	5.95	.008	1.27	.009	1.18	.004	8	G7 III	15,1256,4001
160950	+43 02781	HR 6599		17 39 07	+43 29.7	2	6.35	.015	1.19	.000	1.22	.005	4	K0	1601,1733
160702	-21 04706			17 39 16	-21 42.2	2	9.30	.010	0.69	.005	-0.29	.020	3	A7 III/IV	540,976
160483	-56 08405	IDS17350S5658	AB	17 39 17	-57 00.0	1	6.64		-0.02				3	B9 III	233
	+71 00851	G 240 - 65	⋆ A	17 39 18	+71 21.2	2	9.17	.051	1.12	.030	1.01		3	M0	1017,3072
160617	-40 11755			17 39 19	-40 17.5	6	8.73	.012	0.46	.010	-0.18	.008	11	FW	158,742,1594,1696*
160704	-23 13515	LSS 4341		17 39 22	-23 42.8	1	9.29		0.72		-0.30		2	B0 (II)	1012

HD	DM	Other Id	N	Rem	α₁₉₅₀	δ₁₉₅₀	S	V	σ_V	B–V	σ_B–V	U–B	σ_U–B	n	Spectrum	References
		AO 768			17 39 24	+68 08.0	1	10.89		0.58		0.05		1		1697
160823	+4 03482				17 39 27	+04 23.4	1	6.97		0.94		0.48		2	G2 II	8100
160730	−24 13435	LSS 4342			17 39 27	−24 16.1	5	9.88	.011	0.65	.011	-0.40	.011	9	O9 III	540,976,1011,1012,1737
160668	−36 11804	HR 6583			17 39 27	−36 55.4	1	5.54		1.55				4	K5 III	2009
160647	−40 11758				17 39 27	−40 58.4	1	8.69		0.02		-0.33		1	B5/7 II	401
		G 20 - 9			17 39 29	−08 47.4	3	13.52	.022	1.60	.017	1.28	.055	6		419,1759,3073
		CE Her			17 39 39	+15 06.0	2	11.59	.064	0.13	.023	0.16	.011	2	F5	597,699
		G 154 - 8			17 39 39	−16 36.7	2	13.05	.033	1.56	.019			4		940,1705,3062
160648	−46 11789				17 39 40	−46 33.9	2	7.55	.007	-0.04	.000	-0.39	.002	4	B5/7 III	401,1586
160836	−9 04592				17 39 41	−09 03.1	1	9.74		1.05		0.94		1	K3 V	3072
160910	+16 03256	HR 6594		★ AB	17 39 44	+15 58.4	2	5.55	.005	0.38	.000	-0.05	.005	5	F4 V	1733,3016
320483	−35 11843				17 39 44	−35 15.0	1	10.82		0.34		-0.31		2	B8	1586
161284	+65 01203				17 39 45	+65 01.5	1	8.37		0.93		0.67		3	K0	196
160885	+5 03454	G 140 - 6			17 39 46	+04 57.0	1	9.67		0.95		0.78		1	K4	265
161162	+57 01791	HR 6605			17 39 46	+57 20.0	2	6.77	.005	0.92	.005	0.58	.005	5	K0	985,1733
318242	−32 13206	NGC 6416 - 377		V	17 39 47	−32 20.9	1	11.56		0.43		0.00		1	F5	1768
		LP 44 - 37			17 39 48	+69 29.0	1	15.78		1.46				1		1759
160748	−32 13208	HR 6587			17 39 50	−33 01.7	3	6.42	.023	1.78	.027	1.99	.005	10	M1 III	1637,2007,3005
160715	−45 11825			V	17 39 54	−45 56.9	4	6.92	.001	-0.03	.006	-0.16	.002	39	B9 IV/V	278,1075,1770,2011
		POSS 509 # 2			17 39 59	+13 16.6	1	15.46		1.03				2		1739
160810	−35 11852				17 40 05	−35 16.6	1	6.96		2.00				1	M0 III	8051
161002	+25 03327				17 40 07	+25 26.9	1	8.23		0.22		0.10		3	A2	833
160839	−27 11850	HR 6593			17 40 09	−27 51.7	2	6.37	.023	0.47	.007			5	F0 III/IV	2007,6007
		V816 Oph			17 40 10	+04 58.8	1	11.71		0.25		0.18		1		597
160691	−51 11094	HR 6585			17 40 10	−51 48.6	7	5.13	.012	0.70	.000	0.24	.000	21	G3 IV/V	15,678,1075,2024,2032*
	−28 13479	LSS 4345			17 40 11	−28 40.3	1	10.56		0.81		-0.20		2	B1II-III:nn	1012
161193	+51 02243	HR 6607			17 40 13	+51 50.5	1	5.99		1.05				2	gK0	71
160826	−32 13221	NGC 6416 - 367		★ AB	17 40 13	−32 12.6	1	8.43		0.17		-0.22		1	B9 V	190
		G 20 - 10			17 40 15	+01 38.0	2	14.96	.070	1.47	.013	1.05		6		538,3062
160812	−38 12147	IDS17368S3809		AB	17 40 17	−38 10.4	2	8.03	.024	0.16	.001	-0.46	.001	6	B3 IV	540,976
160797	−44 12008				17 40 23	−44 36.8	1	8.24		-0.05		-0.42		2	B5 III	401
161074	+24 03231	HR 6602		★ A	17 40 25	+24 35.3	3	5.53	.029	1.45	.013	1.69	.021	6	K4 III	15,1003,1080
		GD 212			17 40 26	+24 33.5	1	14.02		0.58		-0.06		1		3060
160915	−21 04712	HR 6595			17 40 26	−21 39.6	7	4.87	.006	0.46	.005	-0.04	.006	24	F6 V	15,1075,1586,1637*
318231	−32 13230	NGC 6416 - 357			17 40 30	−32 18.2	1	10.81		0.26		0.06		1	A0	190
160872	−35 11860	LSS 4346			17 40 30	−35 28.7	2	8.21	.004	0.20	.002	-0.57	.017	3	B1/2 III	540,976
318251	−32 13232	NGC 6416 - 430			17 40 31	−32 31.2	1	9.84		1.12		0.84		1	G9 III	190
	+13 03435				17 40 35	+13 26.7	1	8.78		1.27		1.28		2	K0	1648
160720	−57 08703	HR 6586		★ AP	17 40 37	−57 31.5	1	6.01		0.90				4	G8 III	2009
161112	+26 03066	IDS17386N2636		AB	17 40 38	+26 34.4	1	7.51		0.99				3	K0 III	20
	−29 13906	LSS 4348			17 40 38	−29 57.8	2	10.35	.005	1.36	.020	0.32	.050	4	B0 Ia	954,1737
160876	−41 11965				17 40 39	−41 38.4	1	7.86		0.03		-0.32		2	B7 III	540,976
		G 170 - 60			17 40 41	+19 14.3	1	14.14		0.93		0.54		2		1658
		POSS 509 # 1			17 40 46	+14 08.2	1	14.73		1.46				2		1739
160635	−64 03662	HR 6582			17 40 49	−64 42.2	5	3.61	.012	1.19	.008	1.17	.011	22	K2 II	15,1034,1075,1075,2038
		NGC 6416 - 406			17 40 50	−32 26.5	1	11.95		0.30		0.12		1		190
318239	−32 13245	NGC 6416 - 405			17 40 51	−32 24.7	1	11.16		0.14		-0.05		1	B8	190
161318	+53 01974				17 40 54	+53 28.1	1	9.17		0.11		0.12		2	A2	1566
	−22 04400	LSS 4349			17 40 55	−22 04.4	2	9.57	.010	0.41	.010	-0.54	.016	3	B3	540,976
160878	−44 12018				17 40 55	−44 11.3	4	8.66	.005	-0.05	.004	-0.69	.010	9	B2 III/IV	400,401,540,976
318238	−32 13249	NGC 6416 - 101			17 40 58	−32 23.2	1	10.14		0.22		0.02		1	B9.5V	190
		V439 Oph			17 40 58	+03 37.5	1	11.85		0.67		0.45		1		1462
161023	−13 04732	HR 6600			17 40 59	−13 29.1	4	6.37	.033	0.39	.011	-0.09		12	F0 V	15,254,2020,2028
161096	+4 03489	HR 6603			17 41 00	+04 35.2	29	2.77	.009	1.17	.007	1.24	.009	254	K2 III	1,3,15,26,30,198,985*
160917	−45 11850				17 41 01	−46 01.3	1	6.72		-0.01		-0.08		2	B9 V	401
160974	−34 12008	IDS17377S3434		AB	17 41 03	−34 35.7	2	8.74	.019	0.39	.006	-0.47	.014	3	B2/3 II	540,976
161149	+14 03321	HR 6604			17 41 05	+14 19.0	5	6.24	.019	0.42	.012	0.17	.014	10	F5 II	192,253,254,3016,8100
161056	−7 04487	HR 6601		★	17 41 05	−07 03.5	3	6.31	.016	0.37	.008	-0.48	.026	15	B1.5V	154,1088,8031
161004	−27 11866	IDS17380S2725		AB	17 41 06	−27 26.2	1	8.76		0.21		-0.17		3	B9 IV(e)	1586
160973	−32 13256	NGC 6416 - 106			17 41 07	−32 21.2	1	8.71		1.28		1.36		1	K2 (III)	190
160928	−42 12431	HR 6597		★ AB	17 41 07	−42 42.5	2	5.88	.005	0.16	.000	0.08		6	A1 V+F0 IV	401,2035
161197	+24 03235	IDS17391N2450		AB	17 41 08	+24 48.5	1	7.92		0.73				3	G2 IV	20
234475	+50 02451				17 41 08	+50 57.0	1	9.33		0.64		0.10		2	G0	1566
161198	+21 03198	G 170 - 61			17 41 09	+21 38.4	5	7.51	.017	0.75	.011	0.30	.014	17	K0 V	22,1003,1080,1499,3026
		NGC 6416 - 139			17 41 14	−32 20.9	1	12.80		0.58		0.14		1		190
161114	−6 04638	XX Oph			17 41 15	−06 14.9	2	9.29	.291	1.03	.040	-0.14	.091	3	A pec	379,1588
		NGC 6416 - 141			17 41 15	−32 22.7	1	13.93		1.89		1.72		1		190
161239	+24 03237	HR 6608			17 41 18	+24 20.9	2	5.71	.006	0.65	.006	0.27		5	G2 IIIb	71,272
		BPM 41238			17 41 18	−46 41.	1	13.34		0.63		-0.10		2		3065
		POSS 509 # 7			17 41 19	+13 54.6	1	18.59		1.19				5		1739
		GD 213			17 41 22	+21 12.3	1	12.87		0.50		-0.17		1		3060
	−16 04602			V	17 41 22	−16 18.8	1	9.84		1.91				3	M2	369
161165	+5 03465	IC 4665 - 22			17 41 23	+05 26.5	3	8.73	.012	0.09	.006	-0.21	.005	47	B8.5V	1160,1327,1654
161199	+15 03259				17 41 23	+15 02.1	1	8.23		1.29		1.48		3	K2	1648
	−47 11775				17 41 23	−47 22.1	1	11.32		0.35		-0.18		3		1081
161184	+5 03466	IC 4665 - 23		★ AB	17 41 25	+05 52.0	6	8.04	.013	0.07	.005	-0.17	.010	52	B8 V	369,379,1159,1160*
		NGC 6416 - 143			17 41 25	−32 21.9	1	13.08		0.50		0.19		1		190
	+5 03467	IC 4665 - 24			17 41 27	+05 34.6	1	9.34		1.25		1.26		3	gG7	1160
318264	−33 12414				17 41 27	−33 08.9	1	9.90		0.40		-0.36		2	B6 Iab	540
		IC 4665 - 27			17 41 28	+05 26.1	2	10.31	.010	0.27	.002	0.18	.010	6	A3 V	1160,1327

Table 1

965

HD	DM	Other Id	N Rem	α_{1950}	δ_{1950}	S	V	σ_V	B–V	σ_{B-V}	U–B	σ_{U-B}	n	Spectrum	References
161185	+5 03468	IC 4665 - 26		17 41 28	+05 28.6	1	8.07		1.26		1.23		3	K0	1160
161061	−28 13519	LSS 4352		17 41 28	−28 09.4	1	8.47		0.77		-0.31		2	B0 Ib/II	1601
316256	−30 14647	LSS 4351		17 41 28	−30 26.9	2	10.82	.015	0.96	.015	0.09	.040	4	B1 III	954,1737
		HA 109 # 71		17 41 31	−00 23.8	5	11.49	.006	0.32	.002	0.15	.003	79	A0	281,989,1729,1764,5006
		HA 109 # 375		17 41 34	−00 16.9	1	11.33		0.83		0.26		2	F8	281
	−0 03351			17 41 36	−00 21.6	4	11.73	.008	0.71	.006	0.23	.010	82	G0	281,989,1729,1764
	−0 03350			17 41 36	−00 32.3	1	10.38		0.69		0.15		1	G0	281
160993	−45 11860			17 41 36	−45 37.0	2	7.72	.009	0.02	.000	-0.77	.033	4	B1 Iab	55,8100
161223	+6 03514	IC 4665 - 28		17 41 37	+06 04.9	3	7.44	.006	0.33	.004	0.16	.015	13	A2	369,1159,1499
161369	+44 02757	HR 6612		17 41 37	+44 06.3	1	6.34		1.54		1.84		2	K4 III	985
161103	−27 11872	V3892 Sgr		17 41 38	−27 12.5	6	8.45	.051	0.44	.006	-0.66	.013	10	Be	445,540,976,1012,1586,1737
		HA 109 # 949		17 41 39	−00 01.3	1	12.83		0.81		0.36		1		1764
	+22 03200			17 41 40	+22 19.2	1	9.60		0.33		-0.53		2	G5	666
		HA 109 # 954		17 41 41	−00 01.1	2	12.44	.002	1.30	.002	0.94	.007	30		281,1764
320459	−34 12019	LSS 4354	★ AB	17 41 41	−34 57.4	1	10.80		0.42		-0.25		1	G0	954
160995	−48 11996			17 41 42	−48 18.2	2	10.38	.000	-0.04	.010	-0.56	.010	3	B3 II/III	55,400
		HA 109 # 956		17 41 43	−00 01.3	2	14.63	.003	1.29	.003	0.79	.030	11		281,1764
		HA 109 # 959		17 41 44	−00 00.9	1	12.79		0.92		0.42		21		281
161242	+5 03469	IC 4665 - 30		17 41 45	+05 16.3	5	7.80	.005	1.28	.003	1.09	.006	113	K2	147,369,1159,1509,6005
		HA 109 # 396		17 41 46	−00 15.0	1	11.33		0.84		0.28		2	dG4	281
161261	+5 03471	IC 4665 - 32		17 41 49	+05 44.1	9	8.28	.011	0.05	.006	-0.15	.011	157	B8 V Shell	147,369,379,1159,1160*
161262	+3 03483	IDS17394N0301	AB	17 41 54	+02 59.3	1	8.31		1.22		0.94		2	A2 V	8100
		POSS 509 # 4		17 41 54	+13 16.9	1	17.41		1.54				5		1739
161088	−41 11992			17 41 55	−41 13.0	1	9.28		0.01		-0.55		2	B3 II	401
	−29 13925	LSS 4359		17 41 56	−29 56.2	4	10.73	.016	1.08	.042	0.04	.036	7		725,954,1012,1737
161208	−18 04634	SZ Sgr	★ AB	17 42 00	−18 38.2	2	8.62	.146	2.31	.000	1.57	.117	3	C7,3	414,864
161321	+14 03329	HR 6611, V624 Her	★ A	17 42 01	+14 25.8	4	6.20	.006	0.21	.006	0.18	.013	20	A3 m	355,588,3058,8071
161321	+14 03329	IDS17397N1427	B	17 42 01	+14 25.8	1	12.04		1.24		1.25		2		3024
320491	−35 11892	LSS 4358		17 42 02	−35 23.4	1	10.16		0.34		-0.48		1	B2	954
161270	+2 03390	HR 6609		17 42 03	+02 35.9	1	6.18		0.07		-0.07		2	A1 IV-V	1733
161270	+2 03390	HR 6609	★ AB	17 42 03	+02 36.0	1	5.58		0.07		-0.02		4	A1 IV-V	1149
161897	+72 00803	G 258 - 16	★ A	17 42 04	+72 26.3	1	7.61		0.72		0.22		2	K0	3026
161289	+2 03391	HR 6610	★ B	17 42 05	+02 35.9	2	6.54	.005	0.06	.005	0.00	.032	3	A0 V	1733,3050
161141	−40 11805			17 42 07	−40 46.8	2	9.19	.013	0.00	.004	-0.60	.007	5	B2 II/III	540,976
	+70 00950	G 240 - 67		17 42 10	+70 28.5	1	9.72		0.76		0.32		2	G5	1658
316204	−28 13537	LSS 4360		17 42 11	−28 24.2	1	9.29		0.32		-0.52		2	B2 III	1012
161263	−14 04748			17 42 13	−14 52.1	1	9.19		1.91				3	K5/M0 (III)	369
		LSS 4362		17 42 15	−28 55.3	1	10.86		0.96				2	B3 Ib	725
	+6 03516	IC 4665 - 35		17 42 16	+06 15.3	1	10.56		0.21				2	A2 V	1160
161304	−0 03352			17 42 17	−00 06.9	6	8.48	.002	0.30	.005	0.23	.012	57	A0	281,989,1509,1729,5006,6004
	+5 03472	IC 4665 - 36		17 42 20	+05 13.4	3	9.64	.017	1.27	.006	1.04		6	G8 III	369,1159,1160
		IC 4665 - 37		17 42 21	+05 26.7	2	11.39	.025	0.53	.016	0.11	.005	6	A8	1160,1327
161306	−9 04598	LS IV -09 002		17 42 22	−09 47.7	3	8.18	.008	0.58	.018	-0.43	.008	7	B0: en	399,400,1586
161229	−27 11888	IDS17392S2801	A	17 42 22	−28 02.7	1	8.69		0.10		-0.36		4	B7 IV/V	1732
161229	−27 11888	IDS17392S2801	B	17 42 22	−28 02.7	1	10.24		0.23		-0.01		3	B7 IV/V	1732
		L 205 - 128		17 42 24	−57 16.9	4	10.75	.013	1.67	.030	1.21		8		158,1494,1705,3073
	+43 02796	G 204 - 27		17 42 25	+43 24.7	2	10.50	.000	1.54	.010	1.23		3	M3	694,3045
	+52 02092			17 42 25	+52 06.6	1	9.98		1.50		1.85		1	K5	1746
		G 20 - 11		17 42 25	−07 59.4	2	11.44	.029	1.43	.019	1.24		3	K6	202,1620
		HA 109 # 197		17 42 26	−00 25.3	1	10.59		1.14		0.80		4	dG7	281
		IC 4665 - 38		17 42 27	+05 26.4	2	10.76	.030	0.53	.005	0.08	.025	6	F0	1160,1327
		HA 109 # 199		17 42 27	−00 28.3	3	10.97	.008	1.74	.004	1.92	.031	10	dG5	281,1764,5006
		G 20 - 12		17 42 28	−00 59.8	2	14.66	.029	1.50	.024	0.94		3		202,1620
316232	−29 13940	LSS 4366		17 42 33	−29 12.1	5	10.28	.015	0.66	.024	-0.36	.027	11	O9 IV	954,1011,1012,1021,1737
161370	+5 03473	IC 4665 - 39		17 42 34	+05 32.5	5	9.38	.004	0.29	.005	0.19	.020	60	Am	369,1159,1160,1327,1654
	+5 03474	IC 4665 - 40		17 42 35	+05 51.2	2	10.20	.014	1.29	.003	1.27		3	K2	369,1159
161291	−27 11899	LSS 4369		17 42 42	−27 11.9	3	8.87	.017	0.73	.010	-0.25	.015	5	B0.5Iab	540,976,1012
161249	−36 11867			17 42 42	−36 38.4	3	8.46	.007	-0.08	.003	-0.63	.008	11	B3 V	540,976,1732
160580	−76 01220			17 42 42	−76 11.1	1	7.48		1.09		0.99		8	K1 III	1704
	−0 03353			17 42 46	−00 24.7	6	9.33	.004	1.46	.005	1.60	.005	132	K3	281,989,1729,1764,5006,6004
161252	−40 11815			17 42 46	−40 42.5	1	8.54		0.64				4	G2/3 V	2012
162003	+72 00804	HR 6636	★ A	17 42 49	+72 10.4	5	4.59	.015	0.43	.020	0.01	.009	14	F5 IV-V	15,1084,1118,3026,8015
161426	+5 03476	IC 4665 - 43		17 42 50	+05 41.2	5	9.08	.005	0.16	.006	0.02	.014	48	A1 V	369,1159,1160,1327,1654
162004	+72 00805	HR 6637	★ B	17 42 51	+72 10.9	3	5.81	.011	0.52	.013	0.03	.010	7	G0 V	1084,1758,3016
		HA 109 # 243		17 42 52	−00 24.8	1	11.96		0.60		0.32		8	F0	281
161427	+0 03766			17 42 56	+00 05.4	4	9.02	.001	0.80	.002	0.35	.005	38	G5 IV	281,989,1729,6004
161445	+5 03477	IC 4665 - 44		17 42 56	+05 34.9	1	10.10		0.12		0.01		4	A0	1159
161693	+53 01978	HR 6618		17 42 56	+53 49.3	2	5.75	.005	0.01	.000	0.02	.010	5	A2 V	1501,1566
		HA 109 # 255		17 42 58	−00 22.0	1	11.16		0.92		0.59		2	dG8	281
161372	−13 04741			17 42 58	−14 00.2	1	7.91		1.88				3	M3 III	369
		Cr 347 - 20		17 42 58	−29 18.8	1	12.80		0.78		0.27		1		412
	−29 13952	Cr 347 - 16		17 42 58	−29 19.8	3	11.76	.012	0.69	.037	0.14	.035	6		412,725,954
	+6 03520	IC 4665 - 47		17 42 59	+06 00.5	2	9.78	.015	0.56	.003			3	F9	369,1159
	−24 13464	LSS 4376		17 43 6	−24 5.0	1	10.73		0.82		-0.25		2	O8 III	1737
		IC 4665 - 48		17 43 02	+06 02.9	2	11.58	.000	0.51	.017	0.00		6	F7	1160,1327
	−29 13953	Cr 347 - 19		17 43 02	−29 18.5	2	11.24	.005	0.53	.020	0.05	.005	4		412,954
161583	+35 03059			17 43 03	+35 14.2	1	6.62		1.12		1.09		2	K0	1601
		G 154 -B5	A	17 43 04	−13 17.3	1	11.91		1.44		1.20		2	dM3	3016
		G 154 -B5	B	17 43 04	−13 17.3	1	14.29		0.30		-0.45		5	DA	3060
	−0 03356			17 43 06	−00 21.4	5	10.36	.007	0.61	.005	0.23	.003	111	F3p	281,989,1729,1764,6004

HD	DM	Other Id	N Rem	α_{1950}	δ_{1950}	S	V	σ_V	B–V	σ_{B-V}	U–B	σ_{U-B}	n	Spectrum	References
161481	+5 03479	IC 4665 - 50		17 43 07	+05 26.7	6	9.09	.006	0.25	.004	0.10	.009	63	A1 V +A2 V	369,1159,1160,1327*
161480	+5 03478	IC 4665 - 49		17 43 07	+05 44.1	5	7.72	.016	0.04	.010	-0.43	.019	52	B6 Vp	1159,1160,1327,1654,1732
		Cr 347 - 8		17 43 08	-29 17.7	1	12.91		0.73		0.14		2		412
		Cr 347 - 9		17 43 09	-29 18.9	1	12.74		0.95		0.10		2		412
	-29 13955	Cr 347 - 1		17 43 10	-29 19.1	1	10.65		0.20		0.02		8		412
	-29 13958	Cr 347 - 2		17 43 10	-29 19.1	2	10.72	.010	0.99	.025	-0.01	.045	4		412,954
		Cr 347 - 10		17 43 10	-29 19.5	1	13.08		0.65		0.42		2		412
		Cr 347 - 11		17 43 11	-29 20.0	1	13.13		1.06		0.08		2		412
161482	+5 03480	IC 4665 - 51		17 43 12	+05 13.0	4	9.86	.023	0.34	.014	0.18	.005	9	A2 V	369,1159,1160,1327
		Cr 347 - 15		17 43 12	-29 18.0	1	11.46		0.62		0.56		2		412
	-25 12274	LSS 4379		17 43 13	-25 54.5	1	10.69		0.84		-0.23		2	B0.5 Ib(n)	1737
		IC 4665 - 53		17 43 14	+05 34.9	2	11.45	.040	0.50	.008	0.05		6	F0	1160,1327
		Cr 347 - 14		17 43 14	-29 18.6	1	14.26		0.67		0.24		2		412
		Cr 347 - 3		17 43 14	-29 19.9	2	11.34	.005	0.87	.005	-0.11	.055	4		412,954
		Cr 347 - 7		17 43 15	-29 18.2	1	13.17		0.55		0.23		2		412
		Cr 347 - 12		17 43 15	-29 20.5	1	14.67		0.95				1		412
		Cr 347 - 17		17 43 15	-29 21.5	1	12.67		0.60		0.12		2		412
		Cr 347 - 6		17 43 16	-29 18.5	1	13.03		0.91		0.18		2		412
		Cr 347 - 5		17 43 16	-29 18.8	1	12.56		0.44		0.24		2		412
		Cr 347 - 4	AB	17 43 17	-29 19.0	1	11.72		0.84		0.05		2		412
		Cr 347 - 13		17 43 17	-29 19.8	1	13.88		0.87		0.03		2		412
	-47 11798			17 43 17	-47 14.9	1	10.63		0.53		-0.39		2		1081
161387	-26 12327	V777 Sgr		17 43 19	-26 10.9	1	8.60		1.81		1.14		2	K2/3	588
160931	-71 02173			17 43 19	-71 04.7	1	7.44		0.19		0.14		8	A3 V	1628
161312	-43 12014	IDS17397S4327	A	17 43 21	-43 28.5	1	7.64		-0.06		-0.45		2	B6 II/III	401
	-43 12015			17 43 21	-43 29.6	1	10.03		0.37		-0.09		1		401
		Cr 347 - 18		17 43 25	-29 22.1	1	12.48		0.64		0.22		2		412
	-31 14532	NGC 6425 - 61		17 43 27	-31 26.0	1	10.69		1.16		0.93		1		190
161542	+5 03481	IC 4665 - 56		17 43 28	+05 55.6	3	7.50	.005	0.13	.004	0.08	.009	8	A0	369,1159,1160
161377	-36 11879			17 43 28	-36 31.7	1	8.16		0.03		-0.01		2	A0 V	1732
		LSS 4381		17 43 29	-29 17.0	2	11.01	.005	0.92	.075	-0.10		4	B3 Ib	725,954
318347	-32 13309	LSS 4380		17 43 29	-32 36.7	2	10.04	.100	0.92	.025	-0.01	.000	4		954,1737
161572	+5 03482	IC 4665 - 58		17 43 30	+05 42.8	6	7.60	.006	0.01	.012	-0.49	.018	54	B6 V	369,1159,1160,1327,1654,1732
		IC 4665 - 57		17 43 30	+05 49.9	2	11.14	.015	0.47	.007	0.06	.010	6		1160,1327
	+17 03325			17 43 30	+17 14.0	2	8.71	.019	1.18	.005	1.13		2		1238,8005
		IC 4665 - 59		17 43 31	+05 08.2	1	11.03		1.38		1.42		3	K5	1327
161552	+4 03502	IC 4665 - 61		17 43 34	+04 41.4	1	9.47		0.61				3	F0 V	1160
161378	-40 11828			17 43 35	-40 50.5	2	7.91	.018	0.00	.013	-0.36	.002	5	B8 II	540,976
318300	-31 14539	NGC 6425 - 49		17 43 36	-31 34.6	1	11.27		0.33		0.17		1	A0	190
318301	-31 14540	NGC 6425 - 50		17 43 38	-31 33.9	1	10.82		0.22		0.07		1	A0	190
161573	+5 03483	IC 4665 - 62		17 43 40	+05 32.9	7	6.87	.015	0.04	.019	-0.55	.018	81	B3 IV	369,1066,1159,1160*
161543	-1 03388			17 43 40	-01 07.9	1	8.81		0.13		-0.28		2	B8	1586
161796	+50 02457	V814 Her		17 43 41	+50 03.8	1	7.12		0.47		0.32		2	F3 Ib	8100
		LSS 4382		17 43 41	-29 17.2	1	11.50		0.56		-0.28		4	B3 Ib	954
161390	-38 12189	HR 6613	★ AB	17 43 41	-38 05.6	3	6.42	.004	-0.02	.000			11	A0 V	15,1075,2007
		IC 4665 - 63		17 43 43	+06 00.0	2	10.59	.025	0.34	.018	0.25	.005	6	A3 V	1160,1327
161603	+5 03484	IC 4665 - 64		17 43 44	+06 40.6	7	7.36	.008	0.03	.014	-0.46	.015	57	B5 IV	369,1159,1160,1327,1654,1732
	+6 03523			17 43 45	+06 06.7	2	11.32	.010	0.52	.002	-0.05		3		369,1159
		NGC 6425 - 47		17 43 45	-31 32.7	1	11.71		0.32		0.25		1		190
		POSS 509 # 5		17 43 46	+13 38.9	1	17.93		1.40				4		1739
		G 20 - 13		17 43 46	-08 41.3	1	12.71		1.67				2		202
161695	+31 03090	HR 6619		17 43 47	+31 31.4	4	6.23	.005	0.00	.007	-0.26	.049	10	A0 Ib	252,1022,1079,8100
161658	+21 03208	IDS17417N2156	A	17 43 48	+21 54.4	1	8.24		0.18		0.08		2	A0	3016
		NGC 6425 - 34		17 43 48	-31 23.7	1	13.47		2.56		1.24		1		190
		BPM 25114	★ V	17 43 48	-52 05.	2	15.68	.060	0.00	.095	-0.88	.115	2		832,3065
161658	+21 03209	IDS17417N2156	B	17 43 49	+21 54.1	1	9.21		0.29		0.02		2		3016
		G 154 - 19		17 43 49	-12 57.0	1	14.54		1.26		0.90		1		3062
	+5 03485	IC 4665 - 65		17 43 52	+05 57.2	3	10.62	.020	0.43	.015	0.14	.009	8	A5 V	1159,1160,1327
324327	-40 11834	IDS17404S4051	AB	17 43 53	-40 52.3	1	9.08		-0.04		-0.64		2	A0	540,976
	+5 03486	IC 4665 - 66		17 43 56	+05 48.7	2	10.40	.010	0.30	.000	0.17	.010	6	A3 V	1160,1327
		NGC 6425 - 20		17 43 59	-31 33.0	1	13.44		0.68		0.37		1		190
161621	+5 03487	IC 4665 - 67	★ AB	17 44 00	+05 43.1	5	8.81	.006	0.22	.008	0.14	.013	47	A2 V	369,1159,1160,1327,1654
318311	-31 14560	NGC 6425 - 21		17 44 00	-31 32.2	1	10.76		0.47		0.28		1	B9	190
		TY Pav		17 44 00	-62 35.0	1	12.09		0.26		0.19		1	F1	700
161622	+5 03488	IC 4665 - 68		17 44 01	+05 24.9	6	7.94	.006	0.45	.005	0.00	.013	32	F0	369,1066,1159,1160*
	-15 04677			17 44 01	-15 40.8	1	10.01		1.97				3	M3	369
161638	+6 03524	IC 4665 - 69		17 44 02	+06 15.2	5	7.88	.004	1.03	.006	0.79	.026	18	K0	369,1066,1159,1160,1509
		NGC 6425 - 19		17 44 03	-31 32.7	1	12.67		0.82		0.26		1		190
161623	-1 03389	IDS17415S0111	ABC	17 44 04	-01 11.8	1	8.85		0.66		0.18		1	G0	219
161471	-40 11838	HR 6615	★ A	17 44 05	-40 06.6	7	3.02	.006	0.51	.009	0.26	.045	19	F2 Iae	15,1018,1020,1034*
	+5 03489	IC 4665 - 70		17 44 06	+05 18.7	2	10.27	.005	1.27	.000	1.04	.030	5	G9	1159,1160
161508	-31 14563	NGC 6425 - 2		17 44 06	-31 22.3	1	9.63		0.75		0.34		1	G6 V	190
161511	-35 11923	SX Sco		17 44 06	-35 41.2	2	7.88	.054	2.84	.010	4.58		3	C5,4	109,864
161660	+6 03525	IC 4665 - 72		17 44 10	+06 08.3	5	7.76	.005	0.01	.007	-0.56	.007	51	B7 V	369,1159,1160,1327,1654
	+46 02361			17 44 11	+46 52.1	1	10.76		1.40		1.24		3	M1	7008
		POSS 509 # 3		17 44 12	+13 59.0	1	16.46		1.32				3		1739
316332	-29 13979	LSS 4386		17 44 12	-29 37.3	5	9.46	.023	1.30	.053	0.31	.004	5	B3 Ia	954,1012,1737
161677	+5 03490	IC 4665 - 73		17 44 14	+05 47.5	4	7.15	.016	0.04	.009	-0.50	.019	47	B5 IV	1159,1160,1327,1654
	-71 02179			17 44 16	-71 17.6	2	8.53	.042	0.42	.005	-0.07		4	F5	742,1594
161698	+5 03491	IC 4665 - 76	★ AB	17 44 19	+05 34.9	5	8.22	.004	0.12	.006	-0.31	.012	50	B8.5Vp	369,1159,1160,1327,1654

Table 1 967

HD	DM	Other Id	N Rem	α_{1950}	δ_{1950}	S	V	σ_V	B–V	σ_{B-V}	U–B	σ_{U-B}	n	Spectrum	References
161832	+39 03219	HR 6626, V826 Her	⋆ A	17 44 19	+39 20.4	1	6.68		1.39		1.46		2	K3 III	542
161832	+39 03219	IDS17427N3922	B	17 44 19	+39 20.4	1	9.70		0.75		0.19		1	F5 V	542
316311	−28 13588	LSS 4389		17 44 20	−28 56.7	4	10.23	.010	0.89	.024	-0.11	.028	9	B1 Ib	954,1012,1021,1737
	+5 03492	IC 4665 - 78		17 44 21	+05 12.9	3	10.66	.005	0.56	.008	0.38	.015	6	A2	369,1159,1160
	−29 13984	LSS 4388		17 44 21	−29 24.0	3	10.97	.091	0.64	.033	-0.34	.017	5		954,1012,1737
	−29 13987	LSS 4391		17 44 24	−29 34.6	2	10.45	.055	0.85	.035	-0.09	.010	4	B1 II	954,1737
161592	−27 11930	HR 6616, X Sgr		17 44 25	−27 48.8	3	4.25	.020	0.60	.003	0.39		3	F7 II	1484,1754,6007
161531	−40 11845			17 44 26	−40 58.0	3	9.06	.007	-0.03	.007	-0.66	.008	5	B2/3 Ib/II	401,540,976
		G 183 - 2		17 44 28	+18 42.5	1	12.61		0.70				2		202
161420	−55 08312	HR 6614		17 44 28	−55 23.2	1	6.11		0.28				4	F0 IV	2009
161731	+10 03295	G 140 - 10		17 44 29	+10 08.2	1	8.53		0.54		-0.06		1	F8	1620
		POSS 449 # 2		17 44 29	+18 42.6	1	14.39		1.49				1		1739
161734	+5 03493	IC 4665 - 81		17 44 30	+05 26.6	4	8.87	.007	0.12	.005	-0.27	.008	48	B8 V	1159,1160,1327,1654
161797	+27 02888	HR 6623	⋆ A	17 44 30	+27 44.9	13	3.42	.011	0.75	.008	0.39	.013	66	G5 IV	1,15,22,1084,1118*
161797	+27 02888	IDS17425N2747	B	17 44 30	+27 44.9	3	9.77	.008	1.50	.011	1.01	.018	9		1,1084,3072
	−13 04747			17 44 30	−13 07.0	1	10.21		1.96				3	M2	369
		L 342 - 60		17 44 30	−46 14.	1	13.90		0.53		-0.11		3		3065
161643	−18 04645	IDS17416S1804	AB	17 44 32	−18 05.4	1	7.44		0.28		0.09		1	A7 V	569
324369	−42 12486			17 44 33	−42 16.2	2	10.00	.008	0.03	.003	-0.79	.014	3	B0.5V	540,976
161733	+5 03494	IC 4665 - 82	⋆ A	17 44 35	+05 42.5	4	8.00	.005	0.07	.007	-0.45	.014	46	B6 Vp	1159,1160,1327,1654
		G 20 - 14		17 44 35	−01 28.8	1	13.76		1.57		0.91		2		3062
		IC 4665 - 83	⋆ B	17 44 36	+05 42.9	3	10.23	.013	0.30	.014	0.19	.017	7	A2 V	1159,1160,1327
161817	+25 03344			17 44 39	+25 46.1	4	6.98	.010	0.15	.010	0.14	.009	12	A2 VI	1003,1064,1499,3077
161612	−33 12476			17 44 39	−33 59.7	3	7.18	.021	0.72	.017	0.30	.015	7	G6/8 V	219,2012,3077
	+5 03496	IC 4665 - 84		17 44 40	+05 31.5	2	9.81	.005	0.67	.001	0.23		4	F7	369,1159
	+67 01034	G 240 - 68		17 44 40	+67 19.0	1	9.69		0.70		0.16		3	K0	3026
	−29 13995	LSS 4394		17 44 40	−29 30.9	2	10.78	.061	0.81	.057	-0.13	.014	4	B1 Ib-II	954,1012
	+5 03495	IC 4665 - 85		17 44 41	+05 14.5	2	10.75	.000	0.36	.007	0.22		3	A0	369,1159
	−8 04501	G 20 - 15		17 44 43	−08 45.5	5	10.61	.020	0.60	.013	-0.11	.023	11	F8	202,1064,1620,1658,3077
161664	−22 04423	HR 6617		17 44 44	−22 27.7	1	6.18		1.49				4	G6 Ib	2007
161701	−14 04770	HR 6620		17 44 46	−14 42.5	2	5.95		0.02	.015	-0.33		5	B9 p HgMn	1079,2007
	+6 03529	IC 4665 - 86		17 44 48	+06 11.2	2	10.37	.025	0.54	.015	0.12		6	F8	1160,1327
		Steph 1528		17 44 50	+22 48.9	1	11.74		1.49		1.19		2	M1	1746
316326	−29 14001	LSS 4398		17 44 50	−29 18.8	2	10.04	.034	0.62	.000	-0.32		6	B1 II	1012,1021
	−12 04837			17 44 52	−12 56.5	1	9.94		1.86				2	M3	369
	+39 03226			17 44 53	+39 20.2	1	10.21		-0.29		-1.21		3		963
	−32 13345	LSS 4396		17 44 53	−32 52.0	1	10.33		0.99		0.12		2		954
		IC 4665 - 88		17 44 54	+05 37.0	2	10.90	.030	0.42	.012	0.16	.020	6	A3	1160,1327
	+5 03497	IC 4665 - 89		17 44 54	+05 46.6	3	9.86	.021	0.24	.014	0.16	.020	9	A2 V	1159,1160,1327
316343	−30 14717			17 44 54	−30 11.8	1	9.19		1.11				17	K0	6011
161833	+17 03334	HR 6627	⋆ AB	17 44 56	+17 42.8	2	5.61		0.02	.010	0.01	.005	5	A1 V	1022,1733
	−29 14004	LSS 4400		17 44 57	−29 06.9	2	11.05	.000	0.81	.035	-0.21		5		725,954
	+11 03259			17 45 01	+11 26.5	1	9.11		0.39		0.02		2	F0	1648
161820	+5 03498	IC 4665 - 90		17 45 02	+05 37.2	5	8.33	.008	1.72	.009	2.04	.020	12	K5	369,1066,1159,1160,1389
161770	−9 04604	G 154 - 21		17 45 02	−09 35.1	7	9.68	.034	0.67	.009	-0.04	.023	12	G0	516,742,979,1064,1594*
161653	−38 12201	LSS 4395		17 45 02	−38 07.0	6	7.19	.018	0.01	.011	-0.79	.015	17	B2 II	240,540,954,976,1732,2012
		POSS 449 # 4		17 45 04	+19 25.3	1	17.23		1.46				2		1739
	+4 03508	IC 4665 - 91		17 45 05	+04 51.3	3	9.33	.001	1.74	.010	2.06	.039	24	K3	989,1160,1729
316285	−27 11944	LSS 4405		17 45 05	−27 59.9	2	9.10	.019	1.68	.015	0.56	.072	4	Be	954,1588
316325	−29 14009	LSS 4404		17 45 05	−29 19.7	1	10.60		0.62		-0.29		2	B1 IV	1012
	−30 14724			17 45 07	−30 29.6	1	10.88		0.44		-0.27		2	B5e	1586
320603	−34 12090	LSS 4402		17 45 07	−34 02.0	1	10.68		0.38		-0.43		3	B5	954
161958	+36 02937			17 45 08	+36 06.3	1	6.52		1.51		1.82		2	K5	985
	−29 14011	LSS 4406		17 45 08	−29 23.2	1	11.00		0.72		-0.20		2		954
316274	−27 11946	LSS 4407		17 45 09	−27 44.2	2	10.39	.019	0.71	.005	-0.14		6	B2 II	1012,1021
		POSS 509 # 6		17 45 10	+12 59.5	1	18.03		1.67				3		1739
161834	+5 03499	IC 4665 - 94		17 45 11	+05 08.0	1	10.20		0.27		0.16		2	A2	1159
	+6 03531	IC 4665 - 95		17 45 13	+06 16.4	2	9.89	.010	1.67	.005	1.71		6	K5	1160,1327
	−37 11893	LSS 4403		17 45 13	−37 25.2	1	11.28		0.54		0.32		2		954
161667	−39 11862			17 45 13	−39 53.0	1	6.99		-0.03		-0.29		2	B8 Ib/II	401
161718	−30 14727			17 45 14	−30 17.0	1	7.96		0.53				15	F7/8 V	6011
161633	−46 11860			17 45 15	−46 55.5	1	9.86		0.12		-0.95		1	B1 Iab/b	55
161848	+4 03509	IC 4665 - 96	⋆	17 45 16	+04 57.6	4	8.91	.005	0.83	.020	0.46	.023	7	K1 V	265,1003,1160,3077
		POSS 449 # 3		17 45 16	+16 38.4	1	15.92		1.48				2		1739
	−29 14013	LSS 4410		17 45 18	−29 35.1	1	11.19		0.67		-0.19		3		954
161756	−26 12367	HR 6621, V3894 Sgr		17 45 20	−26 57.5	4	6.31	.014	0.12	.004	-0.44	.002	13	B4 IVe	26,1586,1637,2007
161867	+5 03500	IC 4665 - 98		17 45 21	+05 22.7	2	8.40	.000	1.24	.003	1.04		4	K2	369,1159
161868	+2 03403	Cr 359 - 13	⋆	17 45 23	+02 43.5	24	3.74	.009	0.04	.007	0.04	.021	196	A0 V	1,3,15,30,125,418,667*
316341	−29 14016	LSS 4412		17 45 23	−29 56.5	5	9.31	.166	0.54	.055	-0.60	.032	9	B0.5Vpe :	445,540,1012,1586,1737
161739	−34 12100	LSS 4408		17 45 23	−34 02.3	1	8.99		0.33		-0.47		2	B2 Ib	954
316354	−30 14730	V500 Sco		17 45 24	−30 27.6	1	8.39		1.14				1	K0	6011
161885	+6 03532	IC 4665 - 99		17 45 27	+06 25.2	2	7.54	.010	1.62	.010	1.74	.040	6	M1	1160,1327
161705	−41 12056			17 45 28	−41 41.9	2	8.68	.007	-0.02	.001	-0.32	.005	5	B8 III	401,1586
		MCC 175		17 45 29	+50 04.2	1	10.58		1.36				1	K4	1017
161743	−38 12209			17 45 31	−38 06.2	1	7.68		0.00		-0.08		5	B9 IV	1732
161774	−33 12490			17 45 33	−33 50.8	3	8.69	.024	0.22	.010	-0.50	.023	5	B(5 V)nne	540,976,1586
		POSS 449 # 5		17 45 36	+17 43.5	1	17.93		1.36				3		1739
		IC 4665 - 100a		17 45 37	+04 53.	1	10.97		0.64		0.18		3	G2	1160
		IC 4665 - 101		17 45 40	+06 26.3	1	10.76		1.14				3	G3	1160
	+27 02891	G 182 - 26	⋆ A	17 45 41	+27 48.8	1	9.47		0.64		0.12		4	G0	3077

HD	DM	Other Id	N Rem	α₁₉₅₀	δ₁₉₅₀	S	V	σ_V	B–V	σ_B–V	U–B	σ_U–B	n	Spectrum	References
159517	−85 00469	HR 6552		17 45 43	−85 12.3	3	6.44	.008	0.43	.005	0.01	.010	12	F5 IV	15,1075,2038
161940	+6 03533	IC 4665 - 102	⋆A	17 45 44	+06 17.3	3	9.27	.005	0.15	.007	0.22		43	A2 V	1160,1327,1654
161849	−19 04708			17 45 44	−19 45.9	1	8.40		2.04				3	M2 III	369
		LSS 4414		17 45 44	−37 51.8	1	12.26		1.08		0.97		2	B3 Ib	954
161789	−32 13365	LSS 4416		17 45 45	−32 52.5	3	9.20	.015	0.44	.011	−0.46	.006	5	B2/3 II	540,954,976
161791	−36 11908	IDS17424S3629	AB	17 45 45	−36 30.5	2	8.44	.021	−0.02	.003	−0.50	.003	7	B3 V	503,1732
162132	+47 02537	HR 6641		17 45 47	+47 37.7	2	6.49	.031	0.10	.001	0.10	.023	25	A2 Vs	379,773
		G 154 - 23		17 45 48	−16 21.4	1	16.09		0.40		−0.54		1		3060
	+5 03502			17 45 49	+05 30.4	2	10.69	.015	0.52	.004	0.25		3		369,1159
161941	+3 03493	HR 6633		17 45 51	+03 49.2	3	6.20	.017	0.15	.019	0.00	.054	9	B9.5V	15,252,1256
161777	−38 12212			17 45 51	−38 46.6	1	9.84		0.07		−0.12		1	B8/B9 V	503
		LSS 4420		17 45 54	−29 45.7	1	11.18		0.54		−0.34		3	B3 Ib	954
161839	−31 14608			17 45 54	−31 16.4	3	9.37	.023	0.34	.018	−0.46	.017	5	B5/7 II/III	432,540,976
161980	+5 03503	IC 4665 - 104		17 45 55	+05 19.5	2	8.97	.010	1.24	.001	1.07		3	K2	369,1159
161840	−31 14609	HR 6628		17 45 55	−31 41.3	6	4.82	.005	−0.04	.000	−0.29	.007	20	B8 Ib/II	15,614,1075,1075,2012,3023
161807	−38 12215	LSS 4417		17 45 57	−38 58.1	6	7.01	.011	−0.08	.006	−0.93	.006	17	O9.5 (V)(n)	503,540,954,976,1586,2012
162093	+34 03049			17 45 58	+34 56.3	1	8.01		0.11		0.11		2	A2	1733
	−29 14032	LSS 4421		17 45 58	−29 13.6	1	11.18		0.59		−0.34		2	B1 III	1737
		vdB 92 # a		17 46 00	−31 22.7	1	12.58		0.71		0.02		4		434
161961	−2 04458	LS IV -02 005		17 46 01	−02 10.8	4	7.78	.008	0.22	.005	−0.73	.021	35	B0.5III	400,989,1423,1729
161853	−31 14612	LSS 4422		17 46 02	−31 11.4	4	7.93	.020	0.23	.010	−0.68	.025	7	B0/1 II	92,540,954,976
	−24 13510	LSS 4425		17 46 06	−24 13.4	2	9.89	.013	0.69	.003	−0.44	.019	5	B0 Ia:e2+	1586,1737
		AS 244		17 46 06	−29 45.	1	12.17		0.48		0.09		3		1586
162317	+58 01755			17 46 07	+58 39.3	1	8.60		1.19		1.24		2	K2	1733
	−29 14034	LSS 4424		17 46 07	−29 41.4	2	11.11	.050	0.55	.025	−0.28	.010	5	B2.5 IIIn	954,1737
324407	−38 12216	LSS 4419		17 46 07	−38 58.9	1	11.08		0.08		−0.65		2	B8	954
163240	+80 00555			17 46 08	+80 18.1	1	7.09		1.64		1.98		3	M1	1502
		G 20 - 16	A	17 46 08	−06 48.8	1	14.91		1.54				3		538
		G 20 - 16	AB	17 46 08	−06 48.8	1	14.03		1.54				2		202
		G 20 - 16	B	17 46 08	−06 48.8	1	14.76		1.54				3		538
161925	−16 04622	IDS17432S1613	AB	17 46 08	−16 14.1	1	9.02		0.88		0.54		1	K1 V	3072
161855	−35 11954			17 46 08	−35 21.8	2	7.36	.070	0.03	.023	−0.09	.014	3	B9.5V	503,1232
161841	−41 12074			17 46 14	−41 25.6	2	7.54	.001	−0.08	.010	−0.49	.002	4	Ap Si	401,540,976
158413	−86 00346	S Oct		17 46 14	−86 47.6	1	8.06		1.70		1.29		1	M4/6e	975
162028	+5 03504	IC 4665 - 105		17 46 16	+05 43.0	5	7.51	.009	0.03	.003	−0.42	.007	50	B6 V	369,1159,1160,1327,1654
162076	+20 03570	HR 6638		17 46 16	+20 34.8	3	5.71	.012	0.94	.004	0.68	.014	8	G5 IV	1355,3016,4001
		H1-36		17 46 24	−37 00.6	1	12.00		1.48		−0.16		1		1753
161877	−41 12077			17 46 26	−41 21.9	1	7.93		−0.06		−0.47		2	B6 II/III	401
161892	−37 11907	HR 6630	⋆A	17 46 27	−37 01.8	3	3.20	.004	1.16	.004	1.19	.000	11	K2 III	15,1075,2012
162056	+5 03505	IC 4665 - 108	⋆A	17 46 29	+04 58.7	2	7.47	.015	0.45	.002	0.18	.002	5	F0 IV	1160,3016
162056	+5 03505	IC 4665 - 109	⋆B	17 46 30	+04 59.0	2	9.81	.015	0.70	.009	0.13	.007	5	F0 IV	1160,3016
		LSS 4432		17 46 31	−21 41.5	1	11.91		0.66		−0.28		2	B3 Ib	1737
161858	−47 11839			17 46 33	−47 13.0	1	7.57		0.23		−0.37		2	B5/7 Vne	1586
320650	−35 11962			17 46 34	−35 18.3	1	10.52		0.22		−0.23		2		1232
318406				17 46 36	−31 20.6	1	10.57		0.36		−0.41		2	B0	432
324410	−39 11895	LSS 4428		17 46 36	−39 11.1	1	11.91		0.22		−0.47		2	A0	954
162161	+19 03435	HR 6642		17 46 37	+19 16.2	1			0.02		−0.09		3	A1 V	1022
320651	−35 11963			17 46 39	−35 18.8	1	9.76		1.48		1.65		2		1232
161912	−40 11886	HR 6631, LSS 4430	⋆A	17 46 41	−40 04.6	5	4.81	.008	0.26	.008	0.04	.083	17	A2 Ib	15,954,1075,2012,2016
161929	−38 12225			17 46 42	−38 07.4	1	8.73		0.08		−0.58		1	B2 V(n)	503
		NGC 6441 sq2 2		17 46 43	−37 03.0	1	10.16		0.08		−0.10		5		503
		L 205 - 83		17 46 45	−56 33.4	1	12.13		1.46		1.15		1		3078
162211	+25 03353	HR 6644		17 46 47	+25 38.3	3	5.11	.019	1.15	.007	1.09	.024	5	K2 III	1080,3016,8083
162113	+2 03406	HR 6639		17 46 48	+01 58.5	2	6.46	.005	1.24	.005	1.25	.010	10	K0 III	252,1088
162039	−13 04759			17 46 48	−13 27.1	1	9.75		1.91				2	M0/1	369
		NGC 6441 - 1		17 46 48	−37 02.	1	11.70		1.07		0.77		5		503
		NGC 6441 - 2		17 46 48	−37 02.	1	12.91		1.89				3		503
		NGC 6441 - 3		17 46 48	−37 02.	1	12.44		0.71		0.26		9		503
		NGC 6441 - 4		17 46 48	−37 02.	1	14.76		0.79		0.29		2		503
		NGC 6441 - 5		17 46 48	−37 02.	1	13.89		2.21		2.44		6		503
		NGC 6441 - 6		17 46 48	−37 02.	1	14.04		0.70		0.64		2		503
		NGC 6441 - 7		17 46 48	−37 02.	1	15.91		2.17				3		503
		NGC 6441 - 8		17 46 48	−37 02.	1	15.69		1.65				2		503
		NGC 6441 - 9		17 46 48	−37 02.	1	14.30		0.80		0.34		2		503
		NGC 6441 - 10		17 46 48	−37 02.	1	16.23		1.73				2		503
		NGC 6441 - 11		17 46 48	−37 02.	1	15.76		0.94				1		503
		NGC 6441 - 12		17 46 48	−37 02.	1	16.28		1.97				1		503
		NGC 6441 - 14		17 46 48	−37 02.	1	16.96		1.47				1		503
		NGC 6441 - 17		17 46 48	−37 02.	1	14.27		0.58		0.38		3		503
		NGC 6441 - 18		17 46 48	−37 02.	1	15.94		1.90				3		503
		NGC 6441 - 20		17 46 48	−37 02.	1	16.74		1.09				3		503
		NGC 6441 - 21		17 46 48	−37 02.	1	16.46		1.95				3		503
		NGC 6441 - 30		17 46 48	−37 02.	1	15.77		1.74				1		503
		NGC 6441 - 31		17 46 48	−37 02.	1	15.56		0.85				2		503
		NGC 6441 - 33		17 46 48	−37 02.	1	14.95		1.05				2		503
		NGC 6441 - 35		17 46 48	−37 02.	1	16.05		2.06				1		503
		NGC 6441 - 37		17 46 48	−37 02.	1	16.17		1.79				1		503
		NGC 6441 - 38		17 46 48	−37 02.	1	16.12		1.02				2		503
		NGC 6441 - 39		17 46 48	−37 02.	1	16.90		1.63				1		503
		NGC 6441 - 40		17 46 48	−37 02.	1	16.03		2.14				2		503

Table 1   969

HD	DM	Other Id	N Rem	α_{1950}	δ_{1950}	S	V	σ_V	B–V	σ_{B-V}	U–B	σ_{U-B}	n	Spectrum	References
		NGC 6441 - 41		17 46 48	−37 02.	1	17.50		1.77				1		503
		NGC 6441 - 42		17 46 48	−37 02.	1	16.69		1.61				1		503
		NGC 6441 - 43		17 46 48	−37 02.	1	16.07		0.92				1		503
		NGC 6441 - 44		17 46 48	−37 02.	1	15.40		1.05				2		503
		NGC 6441 - 45		17 46 48	−37 02.	1	17.59		0.95				1		503
		NGC 6441 - 49		17 46 48	−37 02.	1	15.95		1.89				2		503
		NGC 6441 - 51		17 46 48	−37 02.	1	15.58		1.64				2		503
		NGC 6441 - 53		17 46 48	−37 02.	1	15.49		2.10				1		503
		NGC 6441 - 107		17 46 48	−37 02.	1	13.59		1.63				1		503
		NGC 6441 - 111		17 46 48	−37 02.	1	15.20		1.94				1		503
		NGC 6441 - 112		17 46 48	−37 02.	1	15.53		1.11				1		503
		NGC 6441 - 115		17 46 48	−37 02.	1	16.64		1.62				1		503
		NGC 6441 - 3501		17 46 48	−37 02.	1	14.72		1.92				1		503
		NGC 6441 sq2 3		17 46 48	−37 04.3	1	12.54		0.40		0.17				503
161986	−35 11965			17 46 52	−35 20.7	3	8.77	.067	0.47	.018	0.04	.041	7	F3 V	742,1232,1594
162064	−19 04713	LSS 4434		17 46 54	−19 53.5	3	9.26	.004	0.70	.015	-0.35	.013	4	B0/0.5Ia	540,1225,1737
162162	+5 03510	IC 4665 - 115		17 46 55	+05 34.9	1	9.15		0.44		0.17		3	A0	1160
		NGC 6441 sq2 9		17 46 59	−36 58.3	1	11.74		0.20		0.16		3		503
	+5 03511	IC 4665 - 117		17 47 00	+05 22.0	1	8.73		1.60		1.87		3	G9	1160
162247	+29 03122			17 47 00	+29 09.5	1	9.99		0.36		0.23		25	A0	1699
		IC 4665 - 118		17 47 02	+05 15.3	2	10.31	.015	0.32	.016	0.20		6	A3 V	1160,1327
161972	−46 11888			17 47 02	−46 12.9	2	8.34	.000	-0.11	.000	-0.56	.010	6	B3/5 III/V	400,401
162523	+59 01844			17 47 03	+59 50.4	1	8.69		0.61		0.11		2	K0	1502
161915	−50 11573			17 47 04	−50 16.6	1	7.50		1.75		2.05		5	M1 III	1673
162177	+5 03512	IC 4665 - 121		17 47 05	+05 24.2	1	8.61		0.37		0.31		3	A0	1160
162176	+5 03513	IC 4665 - 119		17 47 05	+05 50.3	1	8.90		0.69		0.18		1	G5	1160
	+5 03514	IC 4665 - 120		17 47 05	+05 56.8	2	10.20	.040	1.13	.000	0.91	.020	6	G7	1160,1327
161814	−60 06950	HR 6624	⋆ A	17 47 05	−60 09.1	1	5.78		1.00				4	K0 III	2007
		NGC 6441 sq2 5		17 47 06	−37 08.6	1	10.98		0.65		0.15		5		503
316393	−28 13666			17 47 07	−28 29.4	1	10.71		0.39		-0.38		2	B3 II-III	1012
161917	−53 08812	HR 6632		17 47 08	−53 07.0	2	6.10	.005	0.01	.005			7	B9.5III/IV	233,2007
	−68 02963			17 47 08	−69 01.0	2	9.53	.055	1.05	.005	0.63		5	G5	742,1594
162382	+44 02773			17 47 10	+44 50.8	1	8.55		0.40		-0.02		2	G0	1733
		Hiltner 668		17 47 12	−30 09.	1	11.31		0.56		-0.13		2	B3 IV	1012
		NGC 6441 sq2 6		17 47 12	−37 08.6	1	12.72		0.39		0.24		5		503
	+71 00856			17 47 14	+71 52.4	1	8.54		1.16				1	F5	1017
		NGC 6441 sq2 7		17 47 16	−37 08.3	1	10.96		1.88		2.27		7		503
162021	−42 12536			17 47 17	−42 19.0	1	6.67		1.04				4	K0 III	2012
162067	−35 11973			17 47 18	−35 42.8	1	9.19		0.08		-0.45		1	B5/7 II/III	503
162045	−36 11936			17 47 19	−36 00.0	1	9.75		0.19		-0.28		1	B(7) Ib/II	503
161476	−75 01398	IDS17406S7512	AB	17 47 19	−75 13.4	1	8.14		0.64				4	G2 V	2012
162085	−35 11974			17 47 22	−35 31.1	1	7.79		0.02		-0.28		1	B8 II	503
162047	−38 12235			17 47 22	−38 27.7	3	8.01	.020	0.09	.010	-0.57	.010	6	B2/3 III	503,540,976
162214	−6 04661	RS Oph		17 47 31	−06 41.7	1	11.64		1.15		-0.15		1	O3	1753
162215	−6 04660			17 47 31	−06 46.6	1	9.68		1.27		0.88		9	K0	1343
		RS Oph		17 47 32	−06 41.7	1	11.26		0.92		-0.21		21		866
		NGC 6441 sq2 8		17 47 32	−36 58.7	1	10.25		1.75		1.63		3		503
318470	−33 12529	LSS 4436		17 47 33	−33 29.9	1	10.88		0.52		-0.13		1	B9	954
162102	−33 12533	RY Sco	⋆ AB	17 47 34	−33 41.5	4	7.54	.035	1.28	.017	1.01	.017	4	F6 Ib	657,689,1587,6011
162102	−33 12533	IDS17443S3341	C	17 47 34	−33 41.5	1	10.71		0.57		0.01		1		657
316438	−30 14794	LSS 4437		17 47 36	−30 13.6	1	10.91		0.32		0.25		2	A3	954
162144	−35 11977	NGC 6475 - 131		17 47 41	−35 03.4	1	7.59		0.01		-0.06		1	A0 III/IV	503
		G 154 - 25		17 47 43	−16 58.2	2	13.00	.050	0.72	.020	-0.01	.020	5		1696,3062
162145	−35 11979			17 47 45	−35 20.1	1	7.04		-0.01		-0.10		1	B9 III	503
162168	−32 13411	LSS 4438		17 47 46	−32 58.6	4	8.42	.010	0.60	.017	-0.34	.018	8	B0.5Iab	540,954,976,1012
162319	+9 03485			17 47 47	+09 51.7	1	6.74		1.39		1.53		4	K4 III	1149
162089	−47 11854			17 47 50	−47 48.2	1	9.26		-0.09		-0.63		2	B3 III	400
162365	+15 03285			17 47 52	+15 30.5	2	8.02	.000	-0.13	.005	-0.78	.005	5	B2 IV	399,1733
162579	+50 02468	HR 6656		17 47 53	+50 47.5	6	5.04	.013	0.03	.003	0.01	.007	150	A2 V	753,753,753,879,1363,3050
162283	−6 04663			17 47 53	−06 02.1	4	10.17	.012	1.43	.013	1.26	.016	9	M0 V	158,1705,1775,3072
162428	+24 03264			17 47 58	+24 28.9	1	7.13		-0.08		-0.47		2	A0	1586
162220	−30 14802	HR 6645	⋆ A	17 47 59	−30 32.7	1	6.66		0.04		-0.06		2	B9.5IV/V	1279
162220	−30 14802	HR 6645	⋆ AB	17 47 59	−30 32.7	2	6.47	.005	0.04	.000	-0.04		6	B9.5IV +A0V	404,2007
162220	−30 14802	IDS17448S3032	B	17 47 59	−30 32.7	1	8.06		0.13		0.00		2	A0 V	1279
162189	−40 11905	HR 6643	⋆ A	17 48 02	−40 45.6	2	5.96	.000	1.57	.015	1.83		6	M2 III	2007,3055
162123	−45 11958	HR 6640		17 48 03	−45 30.0	1	6.11		0.95				4	G6 III	2007
316464	−30 14806			17 48 04	−30 37.3	2	10.61	.015	0.49	.005	-0.45	.005	3	B1 V e:	445,1012
316436	−30 14810	LSS 4441		17 48 05	−30 07.4	4	10.22	.015	0.47	.028	-0.17	.014	12	B3:IV:pe	445,954,1012,1586
318456	−33 12544			17 48 10	−33 18.7	1	9.35		1.96				16	K7	6011
		G 183 - 5		17 48 12	+23 46.3	1	13.50		1.61				3		940
		vdB 93 # a		17 48 12	−31 14.9	1	11.80		0.42		-0.14		3	B6 V	434
166205	+86 00269	HR 6789		17 48 18	+86 36.6	4	4.36	.005	0.02	.002	0.03	.000	20	A1 Vn	15,1363,3023,8015
		vdB 93 # b		17 48 18	−31 17.	1	13.68		0.83		0.41		3		434
161955	−65 03507	HR 6634		17 48 18	−65 28.6	3	6.48	.004	1.09	.005	1.01	.005	15	K0/1 III	15,1075,2038
	+37 02952	G 204 - 30		17 48 19	+37 32.2	1	10.22		0.56		0.01		2	F9	7010
316409	−28 05915	LSS 4442		17 48 19	−28 54.8	1	10.56		0.46		-0.30		2	B1 V	1012
162273	−31 14687			17 48 21	−32 01.3	1	9.30		0.42		-0.32		2	B3 II	540
318512	−31 14690	LSS 4443		17 48 23	−31 41.6	2	9.84	.010	0.91	.030	-0.01	.045	4		954,1737
162468	+12 03305	HR 6650		17 48 24	+11 57.6	2	6.21	.033	1.24	.003	1.26		4	K1 III-IV	71,3016
316435	−29 14086	LSS 4444		17 48 24	−29 49.5	2	10.32	.050	0.65	.020	-0.35	.045	4	O8 III	954,1737

HD	DM	Other Id	N	Rem	α1950	δ1950	S	V	σV	B–V	σB–V	U–B	σU–B	n	Spectrum	References
		Hiltner 673			17 48 24	−29 52.9	1	11.60		0.41		−0.02		2	B8 IV?	1012
		Steph 1546			17 48 25	+39 19.2	1	10.89		1.34		1.19		2	K7	1746
162555	+29 03126	HR 6654			17 48 26	+29 20.1	1	5.50		1.05				2	K1 III	71
		LP 808 - 26			17 48 29	−20 30.2	1	11.43		1.18		1.10		1		1696
162287	−35 11992				17 48 31	−35 21.2	1	7.28		0.03		−0.09		1	B9 IV	503
316406	−28 13711				17 48 33	−28 46.6	1	10.51		0.26		−0.46		2	B2 IV	1012
		POSS 449 # 1			17 48 36	+18 00.3	1	13.99		1.25				1		1739
320797	−36 11954	LSS 4447			17 48 38	−36 06.0	1	10.60		0.19		−0.64		2	F2	954
162898	+62 01573	IDS17485N6250		A	17 48 41	+62 48.7	1	6.69		0.37		−0.03		2	F2	1502
162289	−38 12257	LSS 4446			17 48 41	−38 37.2	3	8.88	.008	0.02	.005	−0.65	.007	7	B2 II/III	540,954,976
162570	+22 03227	HR 6655			17 48 42	+22 19.7	3	6.14	.007	0.24	.005	0.11	.008	7	A9 V	1501,1648,1733
	−24 13552				17 48 43	−25 01.2	1	11.19		0.62		−0.05		2		1586
	−29 14092				17 48 43	−29 55.1	1	9.59		2.08				1		8051
162732	+48 02581	HR 6664, V744 Her	7		17 48 45	+48 24.4	7	6.82	.047	−0.10	.017	−0.39	.027	38	B pec Shell	379,753,773,879,911*
		LP 448 - 35			17 48 46	+18 08.6	1	13.91		1.45		1.12		1		1696
	−28 13720	LSS 4452			17 48 47	−28 07.8	1	11.07		0.77		−0.23		2		954,1012,1737
	−28 13719				17 48 50	−28 21.4	1	9.82		1.41				4		1726
162291	−44 12143				17 48 50	−44 22.3	1	8.98		0.02		−0.40		2	B7 V	401
		LS IV -01 002			17 48 51	−01 42.6	2	10.98	.005	0.36	.000	−0.47	.005	4		405,1514
316496	−27 12032	KW Sgr			17 48 51	−28 00.7	2	9.56	.145	2.82	.028	2.16		4		8032,8051
316520	−28 13727				17 48 52	−28 44.8	3	10.71	.026	0.32	.007	−0.28	.009	6	B5:IV:e:nn	445,1012,1586
		G 240 - 72			17 48 53	+70 52.7	3	14.15	.022	0.40	.018	−0.30		7	DXP	538,1705,1759
162374	−34 12165	NGC 6475 - 26		★ V	17 48 53	−34 47.3	6	5.89	.005	−0.11	.009	−0.64	.000	19	B6 V	15,49,1020,2012,2012,2026
162352	−37 11948	LSS 4450			17 48 53	−37 44.3	3	8.29	.026	0.02	.004	−0.58	.012	5	B2/3N(e)	540,954,976
162391	−34 12170	NGC 6475 - 134		★	17 49 00	−34 24.3	1	5.87		1.13				4	G8 III	2007
316515	−28 13730				17 49 01	−28 26.4	1	8.84		1.89				3	M2	1726
316587	−29 14098	LSS 4454			17 49 01	−29 55.2	4	10.53	.016	0.31	.011	−0.48	.016	8	B1:V:e:n	445,954,1012,1586
		G 204 - 31			17 49 02	+45 54.6	1	15.05		1.51				1		1759
162356	−38 12264	LSS 4451			17 49 02	−38 51.0	3	8.26	.018	0.01	.005	−0.82	.006	6	B2 Ib/II	540,954,976
324574	−38 12266				17 49 03	−38 37.7	1	9.80		0.04		−0.46		1	B8	954
162416	−34 12172	NGC 6475 - 30			17 49 07	−34 50.3	1	8.44		0.30				2	A9 IV	2026
162394	−38 12268				17 49 07	−38 21.0	2	8.96	.015	0.04	.003	−0.60	.002	4	B2 V	540,976
162376	−40 11926				17 49 07	−40 04.4	4	9.15	.009	−0.05	.005	−0.66	.003	11	B3 III	401,540,976,1732
316589	−30 14840				17 49 11	−30 02.3	3	10.57	.010	0.28	.014	−0.48	.008	6	B2 IV:enn	445,1012,1586
166926	+86 00272	HR 6811			17 49 12	+86 59.5	1	5.79		0.25		0.07		5	A2 m	15
		G 227 - 16			17 49 13	+64 24.3	1	11.10		0.86		0.55		2		1658
		BI3-14			17 49 14	−29 45.3	1	14.22		0.68		0.46		1		1753
162418	−38 12271	LSS 4455			17 49 16	−38 37.9	3	7.76	.015	0.00	.008	−0.71	.006	5	B2 II	540,954,976
		Steph 1547			17 49 18	+01 40.4	1	10.39		1.87		2.19		1	K4	1746
162899	+54 01917				17 49 18	+54 44.4	1	8.56		0.56		0.13		2	G0	1566
		ADS 10841		★ AB	17 49 18	−15 33.	1	9.28		1.18		1.07		1		8084
162396	−41 12139	HR 6649			17 49 18	−41 59.0	3	6.20	.009	0.54	.002	−0.05		9	F8 IV/V	15,2012,3037
	−20 04896				17 49 19	−20 14.5	1	10.52		0.73		0.41		4		206
320730	−33 12563				17 49 19	−34 00.5	1	11.18		0.52				2	F8	465
320763	−34 12178	NGC 6475 - 33			17 49 20	−34 47.6	1	10.35		0.43				3	F2	2026
162457	−34 12179	NGC 6475 - 34			17 49 20	−34 55.1	1	8.21		0.04				2	A0 IV-Vn	2026
320751	−34 12180	NGC 6475 - 32			17 49 21	−34 31.1	1	10.14		0.31				1	A7	2026
162596	−1 03412	HR 6659			17 49 24	−01 13.5	1	6.34		1.12		0.89		4	K0	1149
162440	−39 11950				17 49 24	−39 35.2	3	8.89	.003	−0.06	.009	−0.72	.005	8	B3 II/III	540,976,1732
162582	−6 04671				17 49 25	−06 23.8	1	9.88		0.39		0.27		2	A0	634
320762	−34 12181	NGC 6475 - 38			17 49 25	−34 45.1	1	9.67		0.27				1	A7 V	2026
313406					17 49 27	−22 52.	1	9.83		2.03				2		369
162495	−33 12567				17 49 27	−33 50.0	1	9.02		1.18				19	K1/2 (III)	6011
162496	−34 12186	HR 6651			17 49 30	−34 06.2	2	6.06	.002	1.24	.010			10	K1 III	465,2007
		L 559 - 195			17 49 30	−34 37.	1	13.68		0.70		0.46		2		3062
162670	+7 03488	IDS17471N0725		AB	17 49 31	+07 24.5	1	8.17		−0.02		−0.42		1	A0	963
162705	+15 03290				17 49 32	+15 01.1	1	7.47		0.35		0.11		3	F0	865
		MN174,213 T2# 4			17 49 33	−33 52.9	1	11.45		0.76				2		465
	−34 12189				17 49 34	−34 04.3	1	10.59		1.01				6		465
162735	+13 03472				17 49 35	+13 33.8	1	8.27		0.72		0.24		2	G5	1648
162598	−7 04513	G 20 - 18			17 49 35	−07 33.3	4	9.96	.024	1.15	.006	1.11	.006	6	K5 V	202,1620,1658,3072
162514	−34 12188	NGC 6475 - 40			17 49 35	−34 34.8	1	8.85		0.25				1	A7 V	2026
162515	−34 12187	NGC 6475 - 42		★	17 49 35	−35 00.5	2	6.54	.041	0.00	.007	−0.07		5	B9 V	15,2007
162826	+40 03225	HR 6669			17 49 36	+40 05.0	2	6.56	.010	0.52	.000	0.03	.005	4	G0 V	1601,1733
316585	−29 14111	LSS 4458			17 49 36	−29 49.3	1	10.88		0.44				4	B1 II:	1021
162541	−30 14855				17 49 36	−30 33.4	2	9.61	.030	0.45	.005	0.04		4	F3 V	1594,6006
162517	−35 12013	HR 6653			17 49 36	−35 36.8	1	6.03		0.34				4	F2 V	2032
	+7 03492	V564 Oph			17 49 37	+07 57.2	1	10.08		1.59		1.51		1	K2	793
162653	−5 04514				17 49 38	−05 25.2	1	8.83		0.38		−0.13		2	B9	634
	−34 12192	NGC 6475 - 43			17 49 38	−34 35.3	1	10.26		0.69				1		2026
162734	+15 03292	HR 6665		★ AB	17 49 43	+15 20.2	1	6.42		0.99		0.78		2	K0 III	1733
320863	−34 12194	NGC 6475 - 47			17 49 44	−34 45.0	2	8.93	.049	0.18	.029	0.11		3	A2 Van	49,2026
		POSS 449 # 6			17 49 45	+18 19.0	1	18.24		1.73				2		1739
162692	−1 03413				17 49 46	−01 25.2	1	7.72		0.33		0.21		21	A2	1699
162672	−5 04516				17 49 47	−05 43.0	1	9.80		0.58		0.41		2	A0	634
320862	−34 12195	NGC 6475 - 48		★ AB	17 49 47	−34 44.2	2	9.14	.049	0.17	.019	0.11		3	A3 V	49,2026
316493	−28 13757	LSS 4462		★ ABC	17 49 49	−28 03.0	1	10.30		0.55		−0.13		2	A2	954
		NGC 6475 - 50			17 49 49	−34 45.2	1	11.09		0.53		0.07		1		49
162673	−7 04514				17 49 50	−07 32.3	1	9.48		1.93		2.08		2	K5	1375
		WLS 1800-20 # 6			17 49 50	−18 57.8	1	10.95		0.74		0.24		2		1375

Table 1   971

HD	DM	Other Id	N Rem	α_{1950}	δ_{1950}	S	V	σ_V	B–V	σ_{B-V}	U–B	σ_{U-B}	n	Spectrum	References
320864	−34 12196	NGC 6475 - 51		17 49 51	−34 49.0	2	9.24	.049	0.20	.010	0.15		3	A3 V	49,2026
		NGC 6475 - 53		17 49 53	−34 46.0	1	11.46		0.57		0.04		1		49
162576	−34 12198	NGC 6475 - 55	⋆ V	17 49 56	−34 36.6	1	6.97		0.04		-0.21		1	B9.5p	49
162714	−6 04672	HR 6661, Y Oph		17 49 58	−06 08.0	6	5.93	.066	1.29	.037	0.98	.038	9	F8 Ib	669,1149,1484,1587,1772,6007
162586	−34 12200	NGC 6475 - 56	⋆ AB	17 49 59	−34 43.2	4	6.14	.016	-0.05	.012	-0.35	.000	20	B8 V	15,49,2012,2026
162715	−7 04515	IDS17473S0756	A	17 50 01	−07 56.2	1	9.83		0.68		0.06		2	G5	3016
162715	−7 04515	IDS17473S0756	B	17 50 01	−07 56.2	1	10.44		0.75		0.15		2		3016
	−28 13765	LSS 4464		17 50 01	−28 56.8	2	10.70	.024	0.31	.008	-0.32	.019	6		954,1586
162774	+1 03528	HR 6667		17 50 03	+01 18.9	2	5.96	.015	1.57	.010	1.90	.005	7	K5 III	1149,3016
162587	−34 12203	NGC 6475 - 58	⋆ AB	17 50 03	−34 53.1	4	5.58	.020	1.09	.017	1.01	.000	15	K3 III	15,49,2007,2026
162568	−42 12580			17 50 04	−42 53.3	3	8.07	.051	-0.07	.015	-0.58	.014	6	B2 III/IVn	540,976,1586
320850	−34 12205	NGC 6475 - 60		17 50 06	−34 35.7	1	9.73		0.31				2	A9 V	2026
		Smethells 4		17 50 06	−44 16.	1	11.15		1.36				1		1494
		JL 2		17 50 06	−87 26.	1	16.71		-0.16		-0.65		2		132
		WLS 1800 15 # 6		17 50 07	+15 22.8	1	12.21		0.72		0.21		2		1375
320865	−34 12204	NGC 6475 - 61	⋆ AB	17 50 08	−34 53.4	2	9.67	.054	0.45	.010	0.12		3	F8	49,2026
162630	−34 12208	NGC 6475 - 63		17 50 09	−34 40.9	2	7.58	.024	0.00	.000	-0.16		3	B9.5p	49,2026
		NGC 6475 - 64		17 50 09	−34 43.9	1	10.15		1.78				1		49
162775	−3 04193			17 50 10	−03 33.2	1	7.50		1.35		1.36		5	K2	897
	−34 12210	NGC 6475 - 65		17 50 10	−34 45.0	1	9.06		0.15		0.11		1	A2 Van	49
		NGC 6475 - 66		17 50 10	−34 48.4	2	10.46	.033	0.47	.009	0.10		4		49,2026
		MN174,213 T2# 5		17 50 11	−34 04.7	1	12.50		0.72				1		465
163859	+78 00612	G 259 - 19		17 50 12	+78 24.1	1	8.58		0.66		0.13		3		3026
	−34 12211	NGC 6475 - 68		17 50 12	−34 45.4	1	8.93		0.36		0.10		1	Am	49
	+7 03495			17 50 15	+07 57.8	1	9.68		1.38		1.51		11	K2	1222
162631	−34 12213	NGC 6475 - 71		17 50 15	−34 51.3	2	7.37	.000	0.06	.034			3	A0 IV	49,2026
		LP 508 - 42		17 50 16	+14 02.6	1	11.83		1.39		1.23		1		1696
162756	−7 04517	G 154 - 32	⋆ AB	17 50 16	−07 54.4	3	7.62	.023	0.62	.004	0.07	.010	5	G0 IV-V V	219,1003,3026
162756	−7 04517	IDS17476S0753	C	17 50 16	−07 54.4	1	13.92		1.49				1		3016
162633	−36 11985	LSS 4465		17 50 16	−36 48.1	3	9.01	.031	-0.05	.004	-0.67	.006	5	B5 (Vn)	540,954,976
162757	−10 04560	HR 6666		17 50 17	−10 53.3	1	6.18		1.11				4	G5	2007
162656	−34 12215	NGC 6475 - 72		17 50 17	−34 43.6	2	8.20	.005	0.02	.019	0.01		3	A0 IVs	49,2026
		NGC 6475 - 74		17 50 17	−34 48.5	2	10.17	.019	0.40	.029	0.06		3		49,2026
162865	+16 03300			17 50 18	+16 54.7	1	6.61		0.41		0.04		3	F3 III-IV	1648
		WLS 1800-5 # 6		17 50 18	−05 08.4	1	11.49		0.78		0.20		2		1375
	−27 12066	LSS 4466		17 50 20	−27 38.8	1	10.53		0.96		0.34		3		954
162740	−17 04941			17 50 22	−17 28.4	1	8.23		1.32		1.01		2	G0 Ib/II	1375
162717	−24 13584	LSS 4468		17 50 22	−24 15.3	2	9.36	.004	0.48	.012	-0.39	.006	5	B2 II	1012,1586
	+17 03364			17 50 23	+17 14.8	1	9.64		0.18		0.13		2	A5	1375
162718	−24 13585	V771 Sgr		17 50 24	−24 45.9	7	8.72	.013	0.64	.019	-0.48	.018	10	B3/5ne	445,540,976,1012,1586*
162679	−34 12217	NGC 6475 - 77		17 50 24	−34 47.1	1	7.16		0.03		-0.20		1	B9 V	49
		NGC 6475 - 78		17 50 24	−34 49.3	1	9.56		0.45		0.11		1	F1 V	49
		NGC 6475 - 79		17 50 25	−34 45.2	1	9.01		0.16		0.08		1	A1 Van	49
162678	−34 12219	NGC 6475 - 141	⋆	17 50 25	−34 46.6	1	6.38		0.00		-0.06		1	A0 IIIs	49
		NGC 6475 - 81		17 50 25	−34 48.3	2	10.27	.025	0.34	.040	0.15		2	F1 V	49,2026
162989	+40 03228	HR 6673		17 50 27	+39 59.5	1	6.03		1.33		1.49		2	gK4	1355
162812	−2 04482	V533 Oph		17 50 27	−02 34.1	2	7.45	.012	1.55	.041	1.25	.012	7	M3	897,1375
162680	−34 12221	NGC 6475 - 82		17 50 27	−34 50.2	1	7.77		0.08				1	A0.5IV	49
320866	−34 12222	NGC 6475 - 85		17 50 31	−34 53.1	1	9.94		0.26				2	F2 V	2026
		G 140 - 16		17 50 32	+14 39.7	2	12.55	.009	0.97	.013	0.50		6		203,1759
162742	−28 13780			17 50 32	−29 00.2	2	8.42	.016	0.28	.004	-0.31	.010	3	B4 Ib/II	540,976
162619	−47 11895			17 50 32	−47 25.0	1	8.75		1.22				4	K2 III	2012
		G 140 -B1	A	17 50 33	+09 49.0	1	9.36		0.95		0.80		2	dK2	3016
		G 140 -B1	B	17 50 33	+09 49.0	1	15.72		0.10		-0.78		2	DC	3060
162950	+27 02905	IDS17485N2713	A	17 50 33	+27 12.3	1	7.31		0.29		0.10		3	Am	8071
	+36 02964	G 182 - 31		17 50 34	+36 24.8	1	10.37		0.42		-0.20		2	sdF8	1658
		NGC 6475 - 152		17 50 34	−34 47.8	1	12.34		0.36				1		49
162834	−5 04523	IDS17479S0554	A	17 50 35	−05 55.1	1	6.86		1.27		1.13		1	K0	669
162724	−34 12226	NGC 6475 - 86	⋆ AB	17 50 35	−34 44.6	5	5.96	.010	-0.01	.008	-0.14	.029	51	B9 V +B9 V	49,588,2007,2007,2026
162722	−32 13472	IDS17473S3227	AB	17 50 36	−32 28.1	1	8.36		0.96				4	G8 III	2012
161988	−76 01226	HR 6635	⋆ A	17 50 37	−76 10.2	3	6.06	.008	1.21	.005	1.28	.000	11	K2 III	15,1075,2038
163075	+46 02379	HR 6674		17 50 38	+46 39.3	2	6.38	.010	1.09	.006	0.79	.037	16	K0 III	753,1733
162725	−34 12228	NGC 6475 - 88	⋆ V	17 50 38	−34 49.3	5	6.43	.010	0.00	.015	-0.07	.007	14	A0p	15,49,657,2012,2026
162723	−34 12227	NGC 6475 - 89		17 50 40	−34 30.0	1	8.56		0.17				2	A0 V	2026
		NGC 6475 - 91		17 50 44	−34 41.7	1	9.61		1.02				3		2026
320861		NGC 6475 - 92		17 50 44	−34 43.5	2	8.44	.039	0.08	.024	0.06		3	A1 Va	49,2026
		LP 448 - 40		17 50 45	+15 21.8	1	11.86		0.52		-0.10		2		1696
	+30 03075			17 50 45	+30 07.1	1	9.98		1.39		1.48		8	K5	272
320852	−34 12233	NGC 6475 - 94		17 50 46	−34 33.2	1	10.41		0.51				1	F0	2026
320859	−34 12232	NGC 6475 - 96		17 50 46	−34 45.1	2	8.76	.029	0.13	.029	0.12		3	A1 Van	49,2026
		NGC 6475 - 95		17 50 46	−34 47.4	1	11.17		0.48		0.30		1		49
162917	+6 03566	HR 6670		17 50 48	+06 06.6	3	5.77	.004	0.42	.000	-0.03	.010	10	F5 V	15,1256,3016
		L 559 - 194		17 50 48	−34 49.	1	13.52		1.64				1		3062
	−5 04524			17 50 51	−05 36.6	1	10.23		1.48		1.22		4		206
162781	−34 12236	NGC 6475 - 103		17 50 53	−34 48.9	2	7.47	.005	0.08	.024	0.00		3	A0.5IIIn	49,2026
318501	−31 14756	LSS 4473		17 50 54	−31 08.9	1	10.93		0.49		-0.11		2		954
162780	−34 12237	NGC 6475 - 104		17 50 54	−34 43.1	2	6.89	.005	0.04	.019	-0.06		3	B9.5IIIn	49,2026
320867		NGC 6475 - 105		17 50 54	−34 50.2	1	9.47		0.20				2	A7 V	2026
		G 227 - 18		17 50 58	+54 37.4	1	10.59		0.94		0.99		3	K3	7010
316608	−30 14896			17 51 01	−30 12.6	1	9.59		0.17		-0.60		4		1586

HD	DM	Other Id	N Rem	α₁₉₅₀	δ₁₉₅₀	S	V	σ_V	B−V	σ_B−V	U−B	σ_U−B	n	Spectrum	References
163113	+38 03025			17 51 02	+38 49.8	1	7.01		1.59		1.89		2	K7 III	105
162804	−34 12239	NGC 6475 - 108		17 51 02	−34 52.2	1	7.02		−0.01				2	B9.5IIISh	2026
162973	+10 03315	IDS17487N1059	A	17 51 03	+10 58.2	1	8.39		−0.03		−0.48		3	B8	963
162818	−34 12242	NGC 6475 - 109		17 51 05	−34 38.4	1	8.86		0.35				2	Am	2026
162817	−34 12244	NGC 6475 - 110	★	17 51 07	−34 27.5	3	6.11	.046	0.03	.019	−0.10		10	A0 II-III	15,2007,2026
320858	−34 12243	NGC 6475 - 111		17 51 07	−34 46.2	1	9.46		0.20				2	A5 V	2026
162856	−29 14164			17 51 11	−29 50.6	2	9.15	.021	0.12	.005	−0.44	.013	3	B3 Ib/II	540,976
		ApJ144,259 # 43		17 51 12	+10 38.	1	14.71		−0.18		−1.12		8		1360
162786	−42 12602			17 51 12	−42 54.9	1	8.29		0.25		0.16		2	A4mA6-F2	401
162920	−15 04711		V	17 51 14	−15 12.7	1	10.05		1.93				2	M1	369
163077	+25 03368	IDS17492N2501	AB	17 51 16	+25 00.1	2	7.94	.014	0.74	.000	0.26		4	G8 V +G9 V	1381,3030
316569	−29 14168	LSS 4474		17 51 19	−29 46.6	3	9.35	.016	0.29	.008	−0.49	.006	5	B3 II	540,976,1012
316466	−27 12088			17 51 20	−27 17.1	1	9.58		1.58				4	F5	1726
	−28 13809	LSS 4475		17 51 21	−28 16.1	1	11.40		0.69		−0.15		3		954
	+21 03245			17 51 22	+21 20.0	1	8.48		0.95		0.70		2	K0	3072
320855		NGC 6475 - 120		17 51 22	−34 37.9	1	10.08		0.37				2	A9 V	2026
162521	−65 03528			17 51 22	−65 42.6	1	6.36		0.45				4	F5 V	2033
162888	−34 12253	NGC 6475 - 121		17 51 28	−34 32.0	1	6.93		0.02				2	A0 IIIn	2026
		G 20 - 19		17 51 29	+03 04.0	3	13.06	.032	0.71	.007	0.02	.019	5		202,333,1620,1658
316568	−29 14172	LSS 4478		17 51 31	−29 43.3	5	9.69	.031	0.26	.033	−0.63	.020	10	B2 IV-V:pe	445,540,976,1012,1586
162890		NGC 6475 - 123		17 51 32	−34 49.8	1	9.44		0.18				2	A0 IV/V	2026
162889	−34 12254	NGC 6475 - 124		17 51 33	−34 44.6	1	8.37		0.05				3	A0 V	2026
	−24 13609	LSS 4480		17 51 34	−25 00.6	1	10.34		0.71				2		725
162910	−33 12612			17 51 37	−33 11.3	2	9.10	.008	0.15	.000	−0.54	.012	3	B2 II/III	540,976
162049	−77 01277			17 51 37	−77 49.1	1	6.65		1.02		0.66		5	G8 III/IV	1628
163217	+40 03233	HR 6677	★ AB	17 51 40	+40 01.0	3	5.17	.013	1.17	.011	1.11	.010	6	K3 III	1080,1363,3051
163989	+76 00667	HR 6701		17 51 41	+76 58.3	4	5.04	.028	0.49	.012	0.06	.006	6	F6 IV-Vs	254,1118,1620,3026
162892	−39 11995			17 51 44	−39 33.5	3	9.09	.033	0.49	.004	0.00	.009	7	F6/7 V	158,742,1594
162926	−36 12008	HR 6671		17 51 45	−36 28.1	1	6.06		0.08				4	B9.5III	2007
162978	−24 13615	HR 6672, LSS 4481		17 51 49	−24 52.7	8	6.20	.014	0.04	.011	−0.87	.013	19	O8 III	15,240,540,976,1011*
		G 140 - 19		17 51 54	+07 23.4	2	13.13	.023	1.54	.019	1.11		4		316,1759
163151	+11 03283	HR 6676	★ AB	17 51 54	+11 08.5	1	6.39		0.46				2	F5 Vn	71
		LSS 4482		17 51 54	−21 51.5	1	10.94		1.03		−0.06		3	B3 Ib	1737
163787	+72 00814			17 51 56	+72 29.8	1	8.45		1.18		1.20		2	K5	1375
316648	−27 12099			17 51 58	−27 32.0	1	9.51		1.25				4	K2	1726
		G 227 - 20		17 52 00	+64 46.9	1	11.15		1.37		1.29		2	M3	7010
162702	−65 03532			17 52 08	−65 06.4	2	7.30	.005	1.21	.002	1.28		12	K2 III	1499,2012
162913	−48 12154			17 52 09	−48 55.9	1	7.82		1.02		0.77		8	G6/8 III	1673
	+30 03081			17 52 13	+30 31.0	1	10.10		0.39		0.04		1		289
		LSS 4484		17 52 13	−27 15.7	1	11.63		0.87		0.05		2	B3 Ib	954
163153	−7 04523			17 52 15	−07 43.5	3	6.93	.003	0.76	.001	0.42	.007	17	G8 IV	1149,1499,1770
163004	−38 12316			17 52 18	−38 40.2	2	8.07	.028	−0.09	.005	−0.69	.002	4	B3 V	540,976
162806	−61 06079			17 52 19	−61 26.0	1	8.51		1.28		1.12		1	K0 II	565
163467	+52 02110			17 52 20	+52 23.6	1	8.53		0.40		0.14		2	F2	1566
163065	−30 14934	LSS 4485		17 52 20	−30 32.7	5	8.60	.016	0.37	.006	−0.56	.024	9	B1 IabN	240,540,954,976,1012
163417	+44 02789			17 52 28	+44 10.3	1	9.01		0.17		0.15		2	A0	1733
163007	−46 11959			17 52 29	−46 41.7	2	7.48	.012	−0.03	.004	−0.66	.008	3	B3/5 (V)ne	55,1586
		NGC 6494 - 1299		17 52 30	−18 58.8	1	12.61		0.40		0.32		2		1710
		LSS 4488		17 52 33	−22 16.8	1	11.30		1.24		0.19		2	B3 Ib	1737
	−13 04789			17 52 34	−13 59.0	1	9.80		2.05				2	M3	369
	+20 03603	G 183 - 11		17 52 35	+20 17.1	5	9.72	.024	0.44	.000	−0.19	.003	8	sdF8	516,792,1064,1620,3077
		TW Her		17 52 36	+30 25.0	2	10.55	.025	0.09	.012	0.10	.005	2	F6	668,699
163588	+56 02033	HR 6688	★ A	17 52 40	+56 52.8	6	3.75	.007	1.18	.005	1.21	.004	13	K2 III	15,1080,1363,3016*
		NGC 6494 - 1289		17 52 41	−19 06.2	1	12.30		0.53		0.35		2		1710
		LSS 4486		17 52 43	−35 00.9	1	10.61		0.30		−0.17		3	B3 Ib	954
312338		NGC 6494 - 73		17 52 44	−18 58.6	1	10.90		0.38		0.25		2	A0	1710
		NGC 6494 - 104		17 52 44	−18 59.9	1	11.89		0.41		0.31		2		1710
	+47 02555			17 52 45	+47 02.7	1	8.98		0.34		0.09		2	Am A5-F2	1601
163222	−6 04676			17 52 45	−06 41.6	1	10.38		0.63		0.25		2	A3	634
		NGC 6494 - 1287		17 52 45	−19 06.4	1	12.65		0.59		0.30		2		1710
322425	−40 10991			17 52 46	−40 52.8	1	9.09		0.17		−0.71		2	A7	1583
		NGC 6494 - 1284		17 52 47	−19 08.5	1	12.26		0.49		0.46		1		1710
163381	+25 03376			17 52 51	+25 39.1	1	8.78		0.15		0.09		4	A0	833
		NGC 6494 - 101		17 52 52	−18 57.7	1	11.96		0.39		0.32		2		1710
		MCC 324		17 52 55	+03 45.7	2	10.10	.047	1.40	.043			2	K6	1017,1705
		NGC 6494 - 89		17 52 55	−18 57.5	1	11.54		0.41		0.30		3		1710
		NGC 6494 - 1280		17 52 56	−18 47.1	1	12.77		0.73		0.56		1		1710
163208	−18 04685	NGC 6494 - 50		17 52 56	−19 00.3	1	9.73		0.32		0.18		2	B9.5 III/IV	1710
163245	−18 04686	NGC 6494 - 1	★	17 52 58	−18 47.7	3	6.50	.029	0.05	.005	0.05	.000	14	A4 V	15,49,2007
164428	+78 00616	HR 6717		17 52 59	+78 19.0	1	6.24		1.44				2	K5	1705
163181	−32 13517	V453 Sco	★ A	17 53 00	−32 28.1	3	6.52	.087	0.50	.012	−0.49	.009	7	O9.5Ia/ab	540,976,1586
		NGC 6494 - 1263		17 53 01	−18 47.5	1	12.63		0.71		0.34		2		1710
		NGC 6494 - 1273		17 53 01	−19 18.7	1	11.48		0.39		0.37		1		1710
		NGC 6494 - 1267		17 53 03	−19 07.5	1	11.22		0.40		0.34		3		1710
		G 154 - 34		17 53 05	−16 23.8	2	11.30	.005	0.69	.010	−0.13	.002	5		1658,3077
163346	+2 03427	Cr 359 - 12		17 53 06	+02 04.9	1	6.78		0.56		0.36		2	A3	1639
		EP Her		17 53 06	+26 34.	1	12.44		0.12		0.10		1		699
		G 154 - 35		17 53 07	−07 35.6	1	12.13		1.06		0.95		1		3062
		NGC 6494 - 171		17 53 07	−19 02.9	1	11.47		0.55		0.39		2		1710
312349		NGC 6494 - 87		17 53 08	−19 13.5	1	11.42		0.43		0.34		4		1710

Table 1   973

HD	DM	Other Id	N Rem	α_{1950}	δ_{1950}	S	V	σ_V	B–V	σ_{B-V}	U–B	σ_{U-B}	n	Spectrum	References
163145 −44 12201		HR 6675		17 53 08	−44 20.2	3	4.85	.010	1.20	.005	1.22	.000	14	K2 III	15,1075,2012
		NGC 6494 - 1257		17 53 09	−19 09.1	1	13.03		0.62		0.41		1		1710
		NGC 6494 - 169		17 53 11	−18 53.7	1	12.22		0.67		0.44		1		1710
163230 −30 14959				17 53 13	−30 11.5	1	7.31		1.60		1.98		1	K5 III	219
		NGC 6494 - 168		17 53 15	−18 49.8	1	11.78		1.61		1.59		1		1710
		NGC 6494 - 175		17 53 15	−18 59.3	1	12.38		0.52		0.35		1		1710
		NGC 6494 - 38		17 53 15	−19 01.0	1	14.31		0.88		0.42		1		49
163293 −19 04753		NGC 6494 - 3		17 53 16	−19 03.0	2	9.33	.009	0.44	.000	0.33	.023	4	A0 III	49,1710
		NGC 6494 - 1251		17 53 17	−19 07.9	1	13.28		0.80		0.46		1		1710
163608 +45 02621				17 53 18	+45 13.4	1	8.07		-0.01		-0.03		2	A0	1601
312339 −19 04754		NGC 6494 - 63		17 53 18	−19 01.3	1	10.34		0.42		0.29		3	A2	1710
		L 559 - 120		17 53 18	−32 53.	1	13.74		1.52		1.05		1		3062
		NGC 6494 - 1246		17 53 20	−19 01.1	1	12.10		0.56		0.43		2		1710
		NGC 6494 - 1253		17 53 21	−19 13.1	1	12.76		0.63		0.47		1		1710
		NGC 6494 - 1255		17 53 21	−19 16.3	1	13.30		0.70		0.17		1		1710
163296 −21 04779				17 53 21	−21 57.0	2	6.85	.005	0.07	.000	0.07	.005	5	A1 V	379,1598
		NGC 6494 - 179		17 53 22	−18 57.4	1	12.48		0.58		0.41		2		1710
		NGC 6494 - 200		17 53 22	−19 04.7	1	11.91		0.48		0.34		2		1710
		NGC 6494 - 173		17 53 23	−19 01.3	1	11.18		0.46		0.36		3		1710
		NGC 6494 - 32		17 53 23	−19 03.2	2	12.44	.027	0.83	.023	0.25	.014	5		49,1710
163506 +26 03120		HR 6685, V441 Her		17 53 24	+26 03.4	12	5.45	.018	0.34	.011	0.26	.040	38	F2 Ibe	1,15,254,1077,1080*
		NGC 6494 - 1230		17 53 24	−19 02.2	1	13.23		0.70		0.53		3		1710
		NGC 6494 - 42		17 53 24	−19 05.3	1	15.28		1.04		0.44		1		49
		G 182 - 32		17 53 25	+37 45.4	1	11.99		0.56		-0.15		2	F8	1658
		NGC 6494 - 29		17 53 25	−18 52.1	1	11.78		0.74		0.24		1		49
163234 −40 12001		HR 6678		17 53 25	−40 18.0	2	6.45	.015	1.41	.015	1.65		6	K3 III	2007,3005
163336 −15 04722		HR 6681	⋆ AB	17 53 26	−15 48.3	1	5.89		0.05				4	A0 V	2007
		NGC 6494 - 1227		17 53 27	−19 03.5	1	9.73		0.31		0.18		1		1710
163071 −56 08506				17 53 27	−56 53.4	1	6.26		-0.05				3	B4 III	233
		NGC 6494 - 119		17 53 28	−19 00.8	1	12.53		0.67		0.40		2		1710
312341 −19 04756		NGC 6494 - 68		17 53 29	−19 02.3	1	10.84		0.36		0.20		3	B9	1710
163838		CCS 2506		17 53 30	+64 09.	1	10.72		1.12				1	R3	1238
		NGC 6494 - 31		17 53 30	−18 52.6	2	12.26	.009	0.52	.018	0.34	.005	5		49,1710
		NGC 6494 - 106		17 53 30	−18 53.1	1	12.10		0.57		0.57		1		1710
		NGC 6494 - 182		17 53 30	−18 56.8	1	12.97		0.69		0.40		2		1710
		NGC 6494 - 167		17 53 31	−18 54.4	1	12.43		0.69		0.51		1		1710
		NGC 6494 - 35		17 53 32	−18 55.6	1	13.64		0.90		0.55		1		49
312342		NGC 6494 - 70		17 53 32	−19 08.3	1	10.84		0.36		0.25		3	B9	1710
312347		NGC 6494 - 75		17 53 32	−19 14.1	1	11.08		0.50		0.33		3	B8	1710
163318 −28 13878		HR 6680		17 53 32	−28 03.6	2	5.78	.027	0.21	.005	0.07		5	A7 III/IV	1754,2007
		NGC 6494 - 218		17 53 33	−19 11.0	1	12.60		0.59		0.63		1		1710
	+18 03497			17 53 34	+18 30.4	1	9.22		1.18		1.16		2	K5	3072
		NGC 6494 - 28		17 53 34	−18 59.6	2	11.40	.019	0.37	.014	0.26		4		49,1710
		NGC 6494 - 105		17 53 34	−19 04.6	1	12.15		0.47		0.30		2		1710
163302 −34 12293		IDS17503S3444	AV	17 53 35	−34 44.6	1	8.57		0.18		0.04		2	B9 V	1768
	−18 04697	NGC 6494 - 20		17 53 36	−18 56.1	1	10.67		0.57		0.28		1		49
		NGC 6494 - 33		17 53 36	−18 59.6	1	12.88		0.62		0.37		1		49
		NGC 6494 - 177		17 53 36	−18 59.7	1	12.83		0.62		0.40		5		1710
163350 −18 04698		NGC 6494 - 11		17 53 37	−18 54.6	2	9.99	.009	0.40	.005	0.30	.019	4	B9 Ib/II	49,1710
312415 −19 04758		NGC 6494 - 66		17 53 37	−19 04.6	1	10.39		0.32		0.19		3	A0	1710
163254 −41 12221				17 53 37	−41 58.4	5	6.74	.017	-0.09	.013	-0.66	.006	13	B2 IV/V	400,401,540,976,2012
		NGC 6494 - 111		17 53 38	−18 55.9	1	12.27		0.53		0.42		2		1710
		NGC 6494 - 98		17 53 38	−18 57.7	1	11.99		0.49		0.34		5		1710
163351 −18 04699		NGC 6494 - 8		17 53 38	−19 00.7	2	9.73	.010	0.36	.010	0.26	.010	3	A2 V	49,1710
		NGC 6494 - 93		17 53 38	−19 04.3	1	11.72		0.42		0.31		5		1710
		NGC 6494 - 36		17 53 40	−18 59.4	1	13.81		0.78		0.28		1		49
		NGC 6494 - 130		17 53 40	−19 05.1	1	12.81		0.64		0.32		1		1710
		NGC 6494 - 217		17 53 40	−19 09.7	1	12.39		0.61		0.59		1		1710
163338 −30 14970		LSS 4495		17 53 40	−30 16.8	3	8.50	.010	0.26	.010	-0.60	.023	9	B1/2 (II)	540,954,976
		NGC 6494 - 166		17 53 42	−18 54.1	1	12.53		0.59		0.53		1		1710
		Smethells 5		17 53 42	−41 59.	1	11.36		1.51				1		1494
		G 20 - 20		17 53 43	+03 28.8	1	10.48		1.23				2	K7	202
		NGC 6494 - 116		17 53 44	−18 56.9	1	12.41		0.55		0.42		2		1710
312417 −19 04760		NGC 6494 - 49		17 53 44	−19 08.7	3	9.68	.019	1.44	.013	1.20	.018	9	K0	460,1695,1710
163472 +0 03813		HR 6684, V2052 Oph		17 53 45	+00 40.6	3	5.82	.015	0.09	.003	-0.66	.007	12	B2 IV-V	154,1088,1221
163547 +22 03237		HR 6687		17 53 45	+22 28.2	1	5.59		1.24				2	K2	71
163390 −18 04700		NGC 6494 - 52		17 53 45	−18 50.2	1	9.69		0.44		0.36		3	A2	1710
		NGC 6494 - 41		17 53 45	−18 54.7	1	14.79		0.94		0.37		1		49
		NGC 6494 - 97		17 53 45	−19 04.5	1	11.95		0.46		0.35		2		1710
		NGC 6494 - 163		17 53 45	−19 16.8	1	11.99		0.52		0.52		1		1710
312414 −19 04761		NGC 6494 - 60		17 53 46	−19 02.2	1	10.26		0.34		0.26		4	A7	1710
		NGC 6494 - 84		17 53 46	−19 14.2	1	11.43		0.48		0.37		4		1710
		NGC 6494 - 165		17 53 47	−18 53.9	1	12.77		0.56		0.43		1		1710
		NGC 6494 - 110		17 53 47	−18 55.4	1	12.11		0.56		0.21		2		1710
		NGC 6494 - 40		17 53 47	−18 59.1	1	14.56		0.92		0.26		1		49
		NGC 6494 - 1194		17 53 47	−19 12.5	1	13.30		0.76		0.60		1		1710
		NGC 6494 - 201		17 53 48	−19 07.6	1	11.23		0.40		0.25		5		1710
		AS 255		17 53 48	−35 15.3	1	13.01		1.74		0.17		1	K3	1753
		NGC 6494 - 74		17 53 49	−18 58.0	1	11.06		0.37		0.27		5		1710
		NGC 6494 - 79		17 53 49	−19 02.9	1	11.43		0.34		0.23		4		1710

HD	DM	Other Id	N Rem	α₁₉₅₀	δ₁₉₅₀	S	V	σ_V	B–V	σ_B–V	U–B	σ_U–B	n	Spectrum	References
313451	−20 04921			17 53 50	−20 51.3	1	9.49		2.00					2	K7	369
312413	−18 04702	NGC 6494 - 14		17 53 51	−18 56.5	2	10.28	.014	0.32	.009	0.16	.000	5	A0	49,1710
		NGC 6494 - 131		17 53 51	−18 59.6	1	13.23		0.67		0.41		1		1710
		NGC 6494 - 39		17 53 52	−19 00.5	1	14.47		1.15		0.64		1		49
320961	−35 12090	LSS 4496		17 53 52	−35 07.6	1	11.37		0.15		−0.44		2	G5	954
		G 20 - 21		17 53 53	+01 25.1	2	11.50	.019	1.40	.005	1.27		3	M1	202,333,1620
163408	−18 04704	NGC 6494 - 7		17 53 53	−18 58.2	2	9.66	.036	0.32	.000	0.18	.014	5	A0 II/III	49,1710
		NGC 6494 - 139		17 53 54	−18 47.5	1	9.71		0.42		0.36		1		1710
		NGC 6494 - 1175		17 53 54	−18 53.0	1	11.96		0.45		0.36		5		1710
		NGC 6494 - 103		17 53 54	−18 55.9	1	12.09		0.38		0.32		3		1710
		NGC 6494 - 95		17 53 54	−18 56.1	1	11.85		0.38		0.26		5		1710
		NGC 6494 - 34		17 53 54	−18 58.0	1	13.14		2.22		2.33		1		49
163449	−15 04729			17 53 55	−15 40.7	1	7.61		1.27		1.23		3	K0 III	1657
	−18 04705	NGC 6494 - 6		17 53 55	−18 59.7	4	9.64	.025	1.51	.010	1.35	.032	12		49,460,1695,1710
		NGC 6494 - 102		17 53 55	−19 02.5	1	12.10		0.53		0.28		2		1710
	−30 14975	LSS 4497		17 53 55	−30 16.5	1	11.32		0.37		−0.27		3		954
		NGC 6494 - 140		17 53 56	−18 49.0	1	11.75		0.44		0.36		4		1710
		NGC 6494 - 24		17 53 56	−18 56.8	2	10.87	.017	0.36	.004	0.22	.000	7		49,1710
		NGC 6494 - 99		17 53 56	−18 58.5	1	11.99		0.42		0.34		4		1710
		NGC 6494 - 77		17 53 56	−19 00.4	2	11.26	.014	0.36	.019	0.22	.019	8		460,1710
		NGC 6494 - 96		17 53 58	−18 59.5	2	11.90	.005	0.45	.010	0.31	.019	7		460,1710
		NGC 6494 - 1165		17 53 58	−19 00.5	1	13.01		0.58		0.20		1		1710
312418	−18 04765	NGC 6494 - 17		17 53 58	−19 05.0	2	10.57	.014	0.32	.000	0.14	.009	5	B8	49,1710
		NGC 6494 - 123		17 53 59	−19 02.7	1	12.82		0.54		0.21		1		1710
	−19 04766	NGC 6494 - 64		17 53 60	−19 01.8	1	10.47		0.33		0.02		4		1710
	+33 02990	G 182 - 33	⋆ A	17 54 00	+33 26.4	1	10.16		1.18		1.27		2	K7	7010
	+33 02990	IDS17522N3327	B	17 54 00	+33 26.4	1	13.31		1.43		1.00		2		7010
		NGC 6494 - 141		17 54 00	−18 49.5	1	11.98		0.44		0.36		4		1710
		NGC 6494 - 1130		17 54 00	−18 51.4	1	13.14		1.62		2.09		1		1710
		NGC 6494 - 43		17 54 00	−18 55.1	1	15.59		0.98		0.79		1		49
163609	+21 03253	IDS17519N2130	AB	17 54 01	+21 30.0	1	8.21		0.65		0.14		3	G5	3016
	−18 04706	NGC 6494 - 23		17 54 01	−18 54.4	2	10.86	.005	0.42	.000	0.33	.005	5		49,1710
163426	−18 04707	NGC 6494 - 4		17 54 01	−18 56.9	2	9.49	.000	0.35	.009	0.24	.014	5	A3 II/III	49,1710
		NGC 6494 - 19		17 54 01	−19 00.4	3	10.63	.014	0.31	.019	0.13	.020	7		49,460,1710
		NGC 6494 - 100		17 54 01	−19 04.8	1	12.02		0.61		0.44		5		1710
		NGC 6494 - 1152		17 54 01	−19 18.5	1	11.90		0.50		0.38		1		1710
		NGC 6494 - 121		17 54 02	−18 58.4	1	12.59		0.53		0.25		2		1710
		NGC 6494 - 27		17 54 02	−18 59.4	3	11.25	.016	0.32	.011	0.17	.021	10		49,460,1710
		NGC 6494 - 128		17 54 03	−19 02.1	1	13.00		0.56		0.16		1		1710
312419	−19 04767	NGC 6494 - 61		17 54 04	−19 03.2	1	10.34		0.37		0.22		4	B9	1710
163428	−23 13678			17 54 04	−23 56.0	2	6.65	.000	1.99	.040	2.52		5	K5 II	1024,8051
163429	−24 13662	LSS 4500		17 54 04	−24 59.7	1	8.78		−0.01		−0.60		6	B3 V(n)	1732
		NGC 6494 - 25		17 54 05	−18 52.6	2	11.12	.009	0.42	.027	0.30	.014	5		49,1710
163450	−18 04709	NGC 6494 - 9		17 54 05	−18 56.6	2	9.75	.004	0.32	.004	0.16	.008	7	A0 (IV)	49,1710
	−19 04768	NGC 6494 - 22		17 54 05	−19 05.1	2	10.73	.009	0.48	.005	0.35		5		49,1710
		NGC 6494 - 26		17 54 06	−18 54.6	2	11.20	.014	0.39	.000	0.29	.018	5		49,1710
		NGC 6494 - 1150		17 54 06	−19 15.8	1	12.23		0.55		0.44		2		1710
	−18 04710	NGC 6494 - 46		17 54 07	−18 58.9	3	9.42	.006	1.32	.013	1.08	.016	11	K0 III	460,1695,1710
		NGC 6494 - 67		17 54 07	−18 59.5	1	10.77		0.35		0.26		5		1710
		NGC 6494 - 37		17 54 07	−19 01.4	1	14.29		0.98		0.23		1		49
163476	−18 04711	NGC 6494 - 10		17 54 08	−18 51.4	2	9.91	.077	0.34	.005	0.20	.014	5	B9 III/IV	49,1710
163409	−19 04763	NGC 6494 - 5		17 54 08	−19 04.2	2	9.57	.009	0.35	.000	0.23	.009	5	A0 V	49,1710
		NGC 6494 - 1147		17 54 08	−19 13.6	1	12.92		0.61		0.52		1		1710
163532	−4 04376	HR 6686		17 54 09	−04 04.6	1	5.46		1.16		1.06		8	K0	1088
		NGC 6494 - 86		17 54 09	−19 00.1	1	11.54		0.37		0.27		6		1710
163477	−19 04769	NGC 6494 - 47		17 54 09	−19 02.6	1	9.48		0.45		0.37		3		1710
		V690 Sco		17 54 09	−40 33.0	1	10.80		0.22				1	F2.5	700
	−18 04714	NGC 6494 - 21		17 54 10	−18 51.2	2	10.71	.009	0.32	.005	0.18	.005	5		49,1710
	−24 13665			17 54 10	−24 08.3	1	9.95		2.48				1		8051
		NGC 6494 - 1126		17 54 11	−18 50.4	1	11.78		0.42		0.28		5		1710
	−18 04713	NGC 6494 - 12		17 54 11	−19 00.7	2	10.02	.009	0.26	.000	0.12	.026	6	A0 V	49,1710
163478	−19 04770	NGC 6494 - 53		17 54 11	−19 05.1	1	9.84		0.41		0.31		4		1710
		NGC 6494 - 125		17 54 12	−18 58.9	1	12.91		0.55		0.27		1		1710
		NGC 6494 - 109		17 54 12	−19 05.4	1	12.19		0.55		0.30		1		1710
163640	+18 03500	IDS17520N1821	AB	17 54 13	+18 20.0	1	6.66		0.40		0.27		5	A0 III	8100
		NGC 6494 - 162		17 54 14	−18 52.5	1	12.07		0.46		0.35		5		1710
		NGC 6494 - 183		17 54 14	−18 54.7	1	11.86		0.44		0.32		5		1710
		NGC 6494 - 122		17 54 14	−18 59.5	1	12.67		0.58		0.31		5		1710
		NGC 6494 - 71		17 54 14	−19 01.2	1	10.98		0.42		0.29		5		1710
316729	−29 14239	LSS 4501		17 54 14	−29 34.1	1	9.92	.013	0.32	.009	−0.58	.010	8	B1 II:N	540,1012,1021
163430	−32 13544			17 54 14	−32 02.7	2	8.17	.012	−0.04	.003	−0.67	.011	3	B2 II	540,976
163376	−41 12231	HR 6682		17 54 14	−41 42.7	3	4.87	.004	1.65	.000	1.96	.000	11	M0 III	15,1075,2012
		NGC 6494 - 206		17 54 15	−19 06.5	1	12.34		0.56		0.40		4		1710
		NGC 6494 - 85		17 54 15	−19 09.3	1	11.50		0.44		0.31		4		1710
		NGC 6494 - 1109		17 54 16	−19 14.4	1	12.01		0.54		0.39		4		1710
		V2416 Sgr		17 54 16	−21 41.2	1	15.10		2.07				1	M5	1753
163453	−28 13895	LSS 4503		17 54 16	−28 15.0	6	9.66	.238	0.53	.046	−0.52	.023	10	Be	445,540,976,1012,1586,1737
312412	−18 04717	NGC 6494 - 59		17 54 18	−18 58.8	1	10.32		0.32		0.22		4	B9	1710
		NGC 6494 - 207		17 54 18	−19 06.5	1	11.85		0.55		0.32		1		1710
163454	−30 14987	LSS 4502		17 54 18	−31 00.6	7	8.24	.015	0.26	.015	−0.74	.028	12	O/Be	445,540,954,976,1012*

Table 1 975

HD	DM	Other Id	N	Rem	α_{1950}	δ_{1950}	S	V	σ_V	B−V	σ_{B-V}	U−B	σ_{U-B}	n	Spectrum	References
		NGC 6494 - 80			17 54 19	−19 10.9	1	11.35		0.39		0.31		4		1710
163591	+4 03557				17 54 20	+04 50.5		9.03	.006	0.01	.009	−0.41	.024	5	B9	963,1697,1748
312411		NGC 6494 - 62			17 54 20	−18 58.3	1	10.38		0.31		0.17		4	B9	1710
		NGC 6494 - 1106			17 54 20	−19 13.4	1	12.54		0.60		0.46		4		1710
		NGC 6494 - 115			17 54 21	−19 01.6	1	12.42		0.56		0.40		6		1710
		NGC 6494 - 91			17 54 21	−19 02.4	1	11.69		0.50		0.30		1		1710
		NGC 6494 - 1107			17 54 21	−19 17.1	1	12.33		1.44		1.44		1		1710
		NGC 6494 - 208			17 54 22	−19 06.1	1	12.91		0.70		0.47		5		1710
312410	−18 04719	NGC 6494 - 16			17 54 23	−18 53.7	2	10.40	.005	0.37	.000	0.23	.023	4	A0	49,1710
		NGC 6494 - 112			17 54 23	−18 57.9	1	12.27		0.48		0.38		5		1710
163611	+5 03547	V566 Oph			17 54 24	+04 59.5	1	7.73		0.45		−0.07		1	F4 V	1748
		NGC 6494 - 155			17 54 24	−18 52.1	1	12.36		0.70		0.36		5		1710
		NGC 6494 - 209			17 54 25	−19 05.5	1	12.47		0.61		0.47		5		1710
163675	+18 03502				17 54 26	+18 37.1	1	6.57		1.01		0.91		3	K0	1648
	−15 04736				17 54 26	−15 32.8	1	9.82		1.84				2	M2	369
		NGC 6494 - 76			17 54 26	−19 02.6	1	11.20		0.47		0.32		5		1710
318648	−32 13546	LSS 4504			17 54 26	−32 29.6	1	10.90		0.28		−0.28		2		954
		NGC 6494 - 156			17 54 27	−18 50.8	1	11.67		0.40		0.31		4		1710
		NGC 6494 - 82			17 54 27	−18 58.4	1	11.49		0.48		0.33		4		1710
		NGC 6494 - 205			17 54 27	−19 07.4	1	11.76		0.52		0.38		5		1710
	−18 04721	NGC 6494 - 18			17 54 28	−18 53.5	2	10.63	.005	0.32	.005	0.17	.009	5		49,1710
		NGC 6494 - 107			17 54 28	−18 59.1	1	12.15		0.44		0.32		5		1710
		NGC 6494 - 113			17 54 28	−19 00.9	1	12.28		0.50		0.35		5		1710
163929	+55 01995	HR 6699			17 54 29	+55 58.5	3	6.09	.029	0.32	.008	0.06	.015	14	F0 IV	253,254,3016
312406	−18 04724	NGC 6494 - 13			17 54 29	−18 49.8	2	10.27	.005	0.33	.005	0.16	.005	5	B8	49,1710
		NGC 6494 - 30			17 54 29	−18 57.2	2	12.05	.004	0.47	.000	0.34		6		49,1710
		NGC 6494 - 1069			17 54 29	−19 14.0	1	12.89		1.00		0.41		4		1710
163641	+6 03578	HR 6690			17 54 30	+06 29.6	2	6.28	.005	0.00	.000	−0.25	.000	7	B9 III	15,1256
		NGC 6494 - 145			17 54 30	−18 50.2	1	11.41		0.35		0.27		3		1710
163536	−18 04723	NGC 6494 - 2			17 54 30	−18 56.1	1	8.23		0.72		0.30		1	G6 V	49
		NGC 6494 - 197			17 54 30	−19 03.8	1	11.04		0.52		0.32		5		1710
163433	−39 12058	HR 6683	★	AB	17 54 30	−39 07.9	1	6.29		0.01				4	A0 IV/V	2007
		BPM 25260			17 54 30	−55 01.	1	15.58		−0.33		−1.20		4		3065
163624	+0 03816	HR 6689	★	AB	17 54 31	+00 04.3	1	5.96		0.10		0.11		4	A3 V	1149
		NGC 6494 - 1060			17 54 31	−18 50.7	1	11.47		0.44		0.28		1		1710
		NGC 6494 - 196			17 54 31	−19 00.5	1	13.05		0.63		0.32		1		1710
163770	+37 02982	HR 6695			17 54 32	+37 15.4	9	3.86	.022	1.35	.005	1.45	.020	19	K1 IIa	15,369,1080,1355,1363*
		NGC 6494 - 146			17 54 34	−18 49.6	1	12.01		0.49		0.41		4		1710
		NGC 6494 - 1061			17 54 35	−18 51.0	1	10.97		0.40		0.26		3		1710
		NGC 6494 - 1067			17 54 35	−19 11.1	1	11.67		0.46		0.29		3		1710
		NGC 6494 - 1062			17 54 36	−18 51.1	1	12.67		0.52		0.48		1		1710
		NGC 6494 - 195			17 54 37	−19 00.8	1	11.62		0.42		0.26		3		1710
312405	−18 04727	NGC 6494 - 15			17 54 38	−18 49.4	1	10.37		0.44		0.16		1	A0	49
318667	−32 13554	LSS 4505			17 54 38	−32 56.5	1	10.62		0.32		−0.40		2	B5	954
	+16 03319				17 54 39	+16 05.4	1	9.50		1.63				2		1648
		NGC 6494 - 188			17 54 40	−19 00.9	1	11.23		0.42		0.29		4		1710
		NGC 6494 - 1043			17 54 40	−19 01.7	1	12.74		0.63		0.41		2		1710
163573	−21 04791				17 54 40	−21 42.7	1	9.99		1.26		1.19		4	K3/4 V	7009
		NGC 6494 - 186			17 54 41	−18 59.4	1	12.33		0.50		0.36		4		1710
		NGC 6494 - 149			17 54 42	−18 49.7	1	12.03		0.46		0.38		4		1710
		NGC 6494 - 1047			17 54 42	−18 58.9	1	12.96		0.63		0.32		4		1710
		NGC 6494 - 1054			17 54 43	−18 45.1	1	12.14		0.48		0.40		3		1710
		NGC 6494 - 1052			17 54 43	−18 49.0	1	12.03		0.48		0.34		3		1710
		NGC 6494 - 193			17 54 43	−19 04.1	1	12.74		0.95		0.35		5		1710
		NGC 6494 - 1039			17 54 43	−19 05.4	1	11.84		0.41		0.29		5		1710
163594	−18 04730	NGC 6494 - 56			17 54 45	−19 00.4	1	9.91		0.41		0.33		4	A3 II/III	1710
163555	−30 15000				17 54 48	−30 03.7	2	7.63	.004	−0.02	.010	−0.40	.001	3	Ap Si	540,976
		NGC 6494 - 1027			17 54 49	−18 48.4	1	11.66		0.39		0.29		3		1710
		NGC 6494 - 189			17 54 50	−19 02.2	1	11.38		0.39		0.22		4		1710
163697	+4 03558				17 54 52	+04 53.5	2	8.68	.001	0.63	.000	0.14	.002	2	F5	1697,1748
		NGC 6494 - 187			17 54 52	−18 59.6	1	12.77		0.55		0.38		5		1710
		NGC 6494 - 150			17 54 53	−18 55.9	1	12.89		0.58		0.41		2		1710
	+4 03559				17 54 56	+04 36.4	2	10.40	.012	0.64	.009	0.04	.006	2	F8	1697,1748
312408	−18 04735	NGC 6494 - 55			17 54 57	−18 54.1	1	9.80		1.41		1.50		4	K5	460
166798	+85 00294				17 54 58	+85 41.0	1	7.59		0.31		0.04		3	F0	1733
		NGC 6494 - 216			17 54 58	−19 12.8	1	12.17		0.74		0.65		4		1710
163436	−51 11279				17 54 59	−51 36.9	1	9.64		0.92		0.72		4	K3 V	158
		NGC 6494 - 1009			17 54 60	−18 47.1	1	11.70		0.40		0.29		3		1710
	−25 12499	LSS 4510			17 55 05	−25 14.2	1	11.49		0.51		−0.45		2		1737
	−23 13701	LSS 4511			17 55 08	−23 08.3	1	10.66		0.79		−0.27		3	O9 III	1737
163750	+12 03336				17 55 00	+12 38.0	1	7.45		0.51		0.01		2	F8	1375
		NGC 6494 - 1021			17 55 00	−19 03.3	1	12.15		0.57		0.31		4		1710
313486	−22 04470	LSS 4509			17 55 00	−22 10.1	1	9.81		1.55		0.52		2		1375
163613	−28 13921	LSS 4507			17 55 00	−28 08.3	1	8.52		0.33		−0.54		2	B1 I-II	954
163522	−42 12681				17 55 00	−42 28.9	4	8.44	.010	0.00	.004	−0.84	.026	7	B1 Ia	55,540,976,8100
164025	+56 02041				17 55 01	+56 02.3	1	8.85		0.73		0.21		6	K0	3016
	−11 04493				17 55 01	−11 02.3	1	10.06		2.16				2	M3	369
		NGC 6494 - 1012			17 55 02	−19 03.6	1	13.04		0.60		0.27		4		1710
		NGC 6494 - 215			17 55 02	−19 12.9	1	12.38		0.51		0.41		4		1710
		NGC 6494 - 1014			17 55 05	−19 10.4	1	13.01		0.66		0.44		4		1710

HD	DM	Other Id	N	Rem	α_{1950}	δ_{1950}	S	V	σ_V	B–V	σ_{B-V}	U–B	σ_{U-B}	n	Spectrum	References
163772	+11 03299	HR 6696			17 55 06	+11 02.9	2	6.36		0.11	.000	0.10	.012	5	A1 V	1022,3016
	+10 03337	IDS17528N1058	AB		17 55 07	+10 57.4	1	8.49		0.49		-0.01		2		3016
	+10 03337	IDS17528N1058	C		17 55 07	+10 57.4	1	13.05		1.60				1	K5	3016
	+29 03151				17 55 07	+29 32.9	1	8.86		0.51		0.04		6	F8	1371
163645	−18 04736	NGC 6494 - 57			17 55 07	−18 54.7	1	9.83		0.32		0.24		3	A0 III/IV	1710
		NGC 6494 - 1015			17 55 09	−19 11.5	1	12.47		0.55		0.44		4		1710
163840	+24 03283	HR 6697		⋆	17 55 10	+24 00.0	3	6.30	.006	0.64	.013	0.20		4	G2 V	71,6009,8083
		NGC 6494 - 1017			17 55 11	−19 14.4	1	11.92		0.52		0.33		4		1710
163680	−19 04777	NGC 6494 - 45			17 55 14	−19 16.1	1	9.19		1.35		1.22		2	K1 III	1695
163774	+4 03560	IDS17528N0427	A		17 55 15	+04 26.6	2	8.82	.003	0.04	.002	-0.28	.004	2	B9	1697,1748
		NGC 6494 - 1004			17 55 15	−19 12.0	1	12.24		0.52		0.43		4		1710
		AO 771			17 55 18	+04 48.0	2	12.22	.008	1.19	.053	0.95	.043	2		1697,1748
	−29 14271				17 55 21	−29 08.3	1	9.83		1.09				4		1726
163990	+45 02627	HR 6702, OP Her			17 55 22	+45 21.4	5	6.11	.127	1.62	.015	1.54	.019	5	M5 IIb	71,3039,8032
		G 140 - 24		⋆	17 55 23	+04 33.3	12	9.54	.011	1.74	.011	1.28	.026	47	M4 VI	1,538,679,989,1006*
	+30 03089				17 55 25	+30 17.8	1	8.42		1.09		1.09		3	K0	196
		Steph 1563			17 55 26	+41 57.2	1	10.70		1.20		1.18		1	K6	1746
162337	−81 00799	HR 6646			17 55 26	−81 29.2	2	6.34	.005	1.50	.000	1.75	.000	7	K3/4 III	15,1075
164058	+51 02282	HR 6705		⋆ A	17 55 27	+51 29.6	16	2.23	.016	1.52	.006	1.88	.006	141	K5 III	1,15,985,1077,1080,1118*
163683	−27 12187				17 55 27	−27 31.0	1	8.70		-0.04		-0.47		2	B5 III/IV	1732
163667	−31 14893	LSS 4512			17 55 27	−31 47.9	4	8.72	.021	0.40	.016	-0.52	.017	7	B1 Ib/II	540,954,976,1012
163685	−28 13936	HR 6692			17 55 29	−28 45.3	5	6.01	.008	-0.08	.009	-0.56	.015	11	B3 II/III	219,240,540,976,2007
163652	−36 12060	HR 6691		⋆ A	17 55 32	−36 51.3	1	5.74		0.90				4	G8 III	2009
163948	+33 02995				17 55 35	+33 24.3	1	6.88		0.91		0.59		2	K0	1625
163949	+28 02872	IDS17536N2800	AB		17 55 36	+27 59.8	1	8.82		0.47		-0.01		2	F6 V	3026
		UY Dra		⋆ B	17 55 36	+58 13.2	1	10.99		1.16		1.15		1	K2 III-IV	414
		BD +58 01772a		⋆ V	17 55 36	+58 13.2	1	12.48		5.60				1	C6,2e	414
163689	−36 12062				17 55 39	−36 40.1	1	9.69		0.05		-0.44		2	B7 Ib/II	1586
316899	−30 15026				17 55 40	−30 09.6	3	9.35	.016	0.80	.005	0.35	.010	10	K0 V	158,1696,2018
	+29 03154				17 55 44	+29 28.7	1	8.62		1.02		0.81		6	K0	1371
	−30 15030				17 55 45	−30 20.5	1	9.74		1.73				3		1726
163970	+27 02922	IDS17538N2751	A		17 55 47	+27 51.2	1	8.84		0.57		0.00		2	G0 V	3032
163970	+27 02922	IDS17538N2751	B		17 55 47	+27 51.2	1	9.89		0.93		0.56		2		3032
	+30 03090				17 55 48	+30 15.0	1	10.75		1.11		1.01		3		196
163993	+29 03156	HR 6703			17 55 49	+29 15.1	7	3.70	.008	0.94	.007	0.68	.015	14	K0 III	15,1080,1355,1363*
	+10 03340				17 55 50	+10 56.8	1	9.60		1.04		0.83		2	K0	3016
163810	−13 04807	G 154 - 36		⋆ AB	17 55 51	−13 05.0	4	9.63	.007	0.60	.007	-0.07	.018	5	G3 V	742,1003,1658,3077
		AO 16			17 55 53	+04 36.4	2	10.84	.041	1.72	.007	2.01	.078	2	M3 I	1697,1748
163755	−30 15035	HR 6693		⋆ AB	17 55 53	−30 15.0	5	5.01	.045	1.65	.042	1.67	.153	14	M1 Ib	15,404,938,2012,3041
318710	−31 14905				17 55 53	−31 18.5	1	10.85		0.35		-0.24		2		1586
		LSS 4515			17 55 54	−26 42.7	1	11.71		0.76		-0.13		3	B3 Ib	954
	+1 03547	V567 Oph			17 55 55	+01 06.3	2	11.07	.005	0.55	.030	0.35		2	G5	668,700
	+4 03562	IDS17535N0428	AB		17 55 56	+04 27.9	4	9.80	.037	1.19	.004	1.14	.020	4	K4	272,679,1697,1748
164330	+62 01586	IDS17556N6237	A		17 55 56	+62 36.8	1	7.72		0.67		0.27		2	K0	1502
163799	−22 04475				17 55 56	−22 22.8	2	8.81	.005	0.55	.009	-0.06	.005	5	F0/2 V	219,1775
163800	−22 04474	LSS 4518			17 55 56	−22 30.8	9	6.99	.017	0.27	.012	-0.69	.027	21	O7/8	379,401,540,976,1011,1499*
314854	−26 12620				17 55 56	−26 16.3	1	10.68		0.53				4	B3:V	1021
163777	−25 12523	LSS 4517			17 55 57	−25 11.2	3	9.29	.016	0.44	.008	-0.50	.005	5	B0/1 Ia/ab	540,976,1012
314783	−24 13713				17 56 00	−24 43.0	1	11.52		0.35		0.20		3		1586
		G 140 - 26			17 56 01	+11 23.0	1	14.62		1.19		0.97		3		308
163801	−27 12200	IDS17529S2707	AB		17 56 02	−27 07.5	2	9.37	.052	0.46	.006	0.23	.001	18	A8 V +(F/G)	1499,1770
164613	+72 00818	HR 6725			17 56 03	+72 00.6	2	5.44	.010	0.31	.010	0.16	.010	5	F3 II-III	985,1733
164780	+75 00647	HR 6735			17 56 05	+75 10.6	1	6.36		0.98				2	K0	71
		G 154 - 37			17 56 06	−09 23.8	1	13.78		0.93		0.31		2		3056
	−28 13953				17 56 06	−29 00.0	1	9.79		1.24				4		1726
163758	−36 12072	LSS 4513			17 56 06	−36 01.1	7	7.32	.014	0.03	.007	-0.89	.016	11	O5	219,390,540,954,976*
	+30 03091				17 56 07	+30 09.9	1	9.62		1.01		0.70		3	K0	196
163745	−41 12270				17 56 08	−41 29.1	2	6.51	.015	-0.09	.017	-0.45	.003	3	B5 II	540,976
163916	−9 04631				17 56 10	−09 00.9	1	9.26		1.47				2	K0	369
163813	−25 12526				17 56 10	−25 49.0	1	7.99		0.05		-0.34		2	B6 III	401
316786	−27 12204				17 56 10	−27 15.4	1	11.01		0.31		-0.22		2	B5 IV	1012
163816	−29 14296	IDS17530S2928	AB		17 56 11	−29 28.9	1	8.96		0.06		-0.41		1	B9 Ib	742
		AO 18			17 56 12	+04 46.6	2	9.87	.000	0.51	.005	0.05	.014	2	B7 V	1697,1748
163917	−9 04632	HR 6698			17 56 16	−09 46.2	11	3.34	.004	1.00	.007	0.87	.004	43	K0 IIIa	3,15,369,1020,1075*
164079	+27 02925				17 56 22	+27 59.6	1	8.79		0.38		0.02		3	F2 V	3026
		G 259 - 21			17 56 24	+82 45.2	3	14.30	.013	0.34	.015	-0.52	.000	7		203,1369,3078
163976	−5 04543				17 56 25	−05 02.7	1	7.41		0.57				1	F8	6007
163892	−22 04478	LSS 4523			17 56 25	−22 27.9	4	7.44	.004	0.16	.016	-0.79	.010	8	B0.5/1 Ib	390,540,976,1011
		G 204 - 39			17 56 28	+46 35.2	1	11.79		1.56		1.41		1	M1	1510
	+18 03514				17 56 29	+18 54.4	1	9.06		1.30		1.40		2	K5	1733
164043	+14 03378				17 56 31	+14 50.9	2	7.16	.008	0.53	.000	-0.08		4	F8 V	865,1275
341475	+22 03245	MM Her			17 56 32	+22 09.0	1	9.50		0.85		0.40		2	G0	588
		WLS 1800 15 # 10			17 56 35	+13 57.7	1	12.91		0.55		-0.02		2		1375
164136	+30 03093	HR 6707, ν Her			17 56 35	+30 11.5	7	4.41	.010	0.38	.010	0.16	.015	21	F2 II	15,253,254,1118,3026*
164045	+4 03564				17 56 36	+04 57.4	1	8.04		0.21		0.14		1	A2	1748
		Smethells 6			17 56 36	−48 22.	1	12.26		1.45				1		1494
163899	−33 12697	LSS 4522			17 56 38	−33 53.3	4	8.33	.010	0.21	.008	-0.65	.020	10	B2 Ib/II	358,540,954,976
163868	−33 12700	V3984 Sgr			17 56 39	−33 24.4	4	7.32	.057	0.00	.020	-0.69	.019	5	B2/3 (V)ne	219,540,976,1586
		V735 Sgr			17 56 40	−29 33.7	1	14.81		0.89		0.50		1		1471
163869	−35 12140	V540 Sgr			17 56 42	−35 55.5	1	8.61		2.14				1	M3	8051

Table 1 977

HD	DM	Other Id	N Rem	α₁₉₅₀	δ₁₉₅₀	S	V	σ_V	B–V	σ_B–V	U–B	σ_U–B	n	Spectrum	References
163955	−23 13731	HR 6700		17 56 44	−23 48.8	11	4.75	.012	-0.04	.006	-0.04	.040	35	B9 V	3,15,369,401,614,1075*
	+70 00965			17 56 45	+70 06.4	1	10.12		0.64		0.12		2	F8	1375
		WLS 1800-5 # 10		17 56 45	−05 05.7	1	10.65		1.72		1.66		2		1375
163924	−35 12142			17 56 50	−35 38.8	3	8.96	.008	-0.02	.007	-0.76	.006	8	B2 III	540,976,1732
166818	+84 00404			17 56 51	+84 42.7	1	8.29		0.06		0.06		7	A0	1219
		HA 62 # 1340		17 56 52	+30 24.0	1	12.06		0.54		-0.06		3		196
	+29 03158			17 56 53	+29 46.8	2	8.72	.015	0.42	.005	0.11	.000	9	F5	196,1371
	+30 03094			17 56 53	+30 21.1	1	9.88		1.05		0.93		3		196
163981	−25 12544			17 56 55	−25 05.1	1	8.19		0.08		-0.20		3	B8 II/III	240
		HA 62 # 599		17 56 57	+29 51.2	1	10.76		0.25		0.10		3		196
164064	−4 04384	HR 6706		17 56 57	−04 49.1	2	5.86	.000	1.56	.000	1.90	.005	12	K5 III	1088,3077
164097	+2 03447	Cr 359 - 8		17 56 58	+02 20.7	1	8.54		0.17		0.15		2	A0	1639
164280	+36 02986	HR 6711		17 56 58	+36 17.5	1	6.01		0.94				2	K0	71
164394	+52 02119	IDS17558N5214	A	17 56 58	+52 13.3	1	7.52		0.27		0.08		2	B2 V	3016
164394	+52 02119	IDS17558N5214	B	17 56 58	+52 13.3	1	10.04		0.70		0.15		2		3016
163984	−29 14312			17 56 58	−29 50.0	2	8.35	.010	0.08	.004	-0.63	.006	3	B3 II/III	540,976
164002	−22 04484	LSS 4531		17 56 59	−22 32.8	3	7.42	.000	0.01	.004	-0.80	.014	10	B1/2 II	240,401,1732
318699	−30 15073			17 57 00	−30 55.1	1	10.10		0.37		-0.31		2		1586
164028	−20 04940	HR 6704		17 57 02	−20 20.2	1	6.21		1.40				4	K0 III	2007
164018	−23 13741	LSS 4533		17 57 03	−23 07.5	1	9.18		0.65		-0.28		2	B1/2 Ib	1012
316888	−29 14317			17 57 03	−29 53.4	1	8.01		1.74				1	K5	915
164096	+2 03448	Cr 359 - 7		17 57 04	+02 30.4	1	9.70		0.20		0.17		2	A2	1639
316864				17 57 05	−29 30.	1	9.90		1.46				1		915
164253	+30 03096	IDS17552N3003	B	17 57 07	+30 03.1	3	8.54	.015	0.04	.000	-0.01	.012	10	G0	196,1371,8100
164252	+30 03096	IDS17552N3003	A	17 57 08	+30 03.1	3	7.17	.016	0.96	.007	0.69	.024	9	A1 V	196,397,8100
		LP 749 - 16		17 57 08	−13 23.5	1	11.66		1.35		1.40		1		1696
		WLS 1800-20 # 10		17 57 10	−20 02.7	1	10.03		1.98		2.26		2		1375
164019	−28 13994	LSS 4532		17 57 10	−28 37.2	3	9.27	.016	0.23	.012	-0.71	.009	5	B0	540,976,1012
	+30 03097			17 57 12	+30 12.1	1	9.83		0.89		0.47		3		196
314937	−25 12553	LSS 4537		17 57 12	−25 14.1	1	10.85		0.54		-0.47		3	B0.5 Ib	1737
		HA 62 # 620		17 57 13	+29 54.7	1	12.20		0.61		0.17		3		196
	−23 13745	LSS 4538	V	17 57 13	−23 24.5	2	11.30	.105	0.87	.010	0.01		4		725,1737
164032	−29 14322	LSS 4536		17 57 16	−29 49.4	4	7.45	.015	0.12	.009	-0.73	.006	7	B1/2 Ib	240,540,976,1012
316891				17 57 16	−30 05.	1	9.56		1.04				1		915
		LSS 4530		17 57 16	−36 00.4	1	11.90		0.21		-0.70		2	B3 Ib	954
		MtW 62 # 224		17 57 17	+30 00.0	1	12.23		1.19		1.22		5		397
164103	−14 04842			17 57 18	−14 47.3	1	8.12		0.10		-0.45		2	B3 IV	1012
316910				17 57 18	−30 27.	1	10.14		1.28				1		915
		MtW 62 # 237		17 57 20	+29 57.9	1	16.08		1.04		0.58		5		397
		MtW 62 # 236		17 57 20	+29 59.9	1	15.91		0.76		0.27		5		397
	+70 00967			17 57 20	+70 05.1	1	8.99		0.37		-0.04		2	F2	1733
		MtW 62 # 251		17 57 22	+29 59.8	1	15.09		0.74		0.38		5		397
		Near M8 # 4		17 57 23	−24 18.1	1	11.90		0.31		0.22		1	A0 V	5012
164429	+45 02635	HR 6718, V771 Her		17 57 26	+45 28.7	2	6.48		-0.08	.004	-0.21	.026	3	B9 p SiSr	252,1079
		MtW 62 # 287		17 57 28	+30 00.2	1	11.48		1.27		1.34		5		397
164144	−14 04845			17 57 30	−14 12.4	1	7.99		1.69				2	K4 III	369
	+29 03161			17 57 31	+29 54.8	1	10.12		0.26		0.12		3		196
	+30 03099			17 57 34	+30 00.1	1	10.30		1.31		1.47		3		196
		HA 62 # 640		17 57 36	+29 57.8	1	11.01		1.30		1.42		3		196
164165	−12 04890	LS IV -12 002	V	17 57 37	−12 59.2	1	10.96		0.36		-0.47		1	B8/9 III	1699
164105	−24 13743	LSS 4541		17 57 38	−24 40.6	2	8.47	.003	0.09	.002	-0.71	.002	8	B2 III	1586,1732
316873	−29 14333			17 57 38	−29 51.0	1	8.78		1.57				1	K7	915
164258	+0 03832	HR 6709, V2126 Oph		17 57 42	+00 37.8	2	6.37	.005	0.16	.005	0.17	.000	14	A3 p SrCrEu	1088,3016
164283	+5 03568	Cr 359 - 6		17 57 42	+05 32.6	1	9.10		0.26		0.19		2	A0	1639
		HA 62 # 1147		17 57 43	+30 17.0	1	10.56		0.37		0.09		3		196
		LSS 4540		17 57 43	−29 22.1	1	11.64		0.46		-0.39		3	B3 Ib	954
	−15 04766			17 57 45	−15 29.0	1	10.38		1.81				2	M3	369
313599	−23 13762	LSS 4542		17 57 45	−23 03.7	1	9.69		0.83		-0.14		2	B0.5 Ia	1737
164188	−15 04767	LSS 4546		17 57 46	−15 48.1	1	8.70		0.19		-0.71		2	B1 Ib/II	1012
164284	+4 03570	Cr 359 - 5	★ V	17 57 47	+04 22.2	12	4.62	.046	-0.04	.010	-0.83	.024	41	B2 Ve	3,15,30,154,379,1212*
		Near M8 # 3		17 57 48	−23 49.7	1	11.60		0.96		0.41		1		5012
164146	−24 13745	LSS 4543	★ AB	17 57 48	−24 12.5	3	8.17	.004	-0.01	.005	-0.81	.008	5	B2 II/III	401,540,976
164349	+16 03335	HR 6713		17 57 50	+16 45.1	7	4.67	.015	1.26	.011	1.23	.012	18	K0.5 IIb	15,37,1080,1363,3016*
347829	+19 03491			17 57 50	+19 07.0	1	10.27		0.41				1	G0	592
335522	+25 03399			17 57 50	+24 59.8	1	11.65		-0.13		-0.53		1	F0	1298
164259	−3 04217	HR 6710		17 57 50	−03 41.3	15	4.62	.007	0.38	.008	0.00	.013	56	F3 V	1,3,15,116,254,1020*
164169	−22 04491			17 57 52	−22 15.3	1	9.26		0.04		-0.60		2	B3 III	401
163878	−60 06987			17 57 53	−60 08.2	1	6.92		1.09		0.85		7	G8 II/III	1628
		L 157 - 72		17 57 54	−62 58.	1	10.39		1.28				1		1494
		Near M8 # 2		17 57 55	−23 50.4	1	12.46		0.40		-0.22		1	B8 II	5012
		Near M8 # 1		17 57 57	−23 51.0	1	11.20		0.33		-0.16		1	B6 V	5012
164224	−20 04944			17 57 58	−20 59.2	1	8.52		0.19		0.02		3	Ap SiCr	240
164151	−31 14953			17 57 58	−31 58.6	1	9.21		0.00		-0.38		6	B8/9 II	1732
		G 206 - 1		17 57 59	+36 23.7	1	12.44		0.76		0.15		2		1658
		G 204 - 40		17 58 01	+49 02.0	1	11.57		1.50		1.20		3	K7-M0	7010
164175	−29 14342			17 58 01	−29 26.6	1	7.69		1.12				1	K0 III	915
	−42 12745			17 58 02	−42 39.3	1	10.27		-0.12		-0.79		2	B2 V	540
	+29 03162			17 58 03	+29 39.0	1	10.21		0.32		0.07		3		196
	+39 03294			17 58 03	+39 00.2	2	9.43	.010	0.05	.005	0.03	.010	5	A0	1501,1601
313571	−22 04494	LSS 4548		17 58 06	−22 15.0	4	9.91	.016	0.55	.018	-0.47	.013	6		540,976,1586,1737

HD	DM	Other Id	N Rem	α_{1950}	δ_{1950}	S	V	σ_V	B–V	σ_{B-V}	U–B	σ_{U-B}	n	Spectrum	References
164225	−22 04493			17 58 06	−22 35.9	1	8.77		0.05		−0.35		2	B8 II	401
164353	+2 03458	Cr 359 - 3	⋆ AB	17 58 08	+02 55.9	14	3.96	.017	0.02	.013	−0.61	.020	40	B5 Ib +B1 V	15,154,379,450,1075*
164353	+2 03459	Cr 359 - 3	⋆ C	17 58 08	+02 55.9	3	8.12	.000	−0.01	.014	−0.57	.006	9	B2 V	450,1084,1211
234529	+52 02121			17 58 09	+52 01.4	1	7.36		1.42		1.66		5	F8	4001
164264	−20 04946			17 58 09	−20 31.2	1	8.67		2.29				2	K5 (III)	369
164073	−48 12227			17 58 10	−48 48.6	1	8.03		0.03		−0.51		4	B3 III/IV	400
164228	−29 14347	IDS17550S2911	AB	17 58 11	−29 11.3	1	9.42		1.02				1	G8 (III) +F	915
164352	+3 03557	Cr 359 - 4		17 58 12	+03 09.0	1	9.33		−0.01		−0.39		2	B8	1639
	+14 03384	G 140 - 29		17 58 12	+14 44.7	1	9.20		0.68		0.13		2	K0	1696
	−23 13777	LSS 4551		17 58 12	−23 28.0	2	10.43	.020	1.14	.040	0.07		4	O9 Ib	725,1737
	−26 12671	LSS 4549		17 58 14	−26 47.9	1	11.10		0.59		−0.18		2		954
347828	+19 03493			17 58 18	+19 12.2	1	10.53		0.44		0.00		1	G0	1274
164447	+19 03494	HR 6720	4	17 58 18	+19 30.4	4	6.48	.041	−0.06	.004	−0.38	.024	7	B8 Vne	1022,1079,1586,1733
		OS Her		17 58 23	+34 41.	1	13.26		0.12		0.19		1		699
316872	−29 14353			17 58 23	−29 52.2	1	9.61		1.41				1	K5	915
316906	−30 15110			17 58 23	−30 20.9	1	9.38		1.77				1	K5	915
164245	−36 12115	HR 6708		17 58 25	−36 22.7	3	6.31	.013	−0.02	.012	−0.33	.007	7	B7 IV	540,976,2009
164432	+6 03597	Cr 359 - 9	⋆	17 58 26	+06 16.1	5	6.35	.017	−0.08	.006	−0.75	.016	13	B3 IV	15,154,1221,1256,1639
164270	−32 13623	V4072 Sgr		17 58 26	−32 42.9	1	8.74		0.23		−0.45		1	WC	954
164358	−17 04987	HR 6715		17 58 29	−17 09.4	2	6.28	.000	1.80	.002	1.84		9	K2 III	1509,2007
		G 154 - 40		17 58 29	−20 35.0	1	14.55		1.57		1.00		1		3056
164646	+45 02638	HR 6728		17 58 30	+45 30.2	3	5.67	.051	1.56	.004	1.94	.061	6	M0 IIIab	252,1355,3001
	+30 03102			17 58 31	+30 02.0	1	11.06		0.47		0.04		3		196
		LSS 4557		17 58 33	−27 22.0	1	11.89		0.49		−0.24		2	B3 Ib	954
		UV 1758+36		17 58 36	+36 29.	1	11.36		−0.25		−1.07		3		714
164246	−39 12133			17 58 36	−39 39.5	1	9.13		−0.06		−0.34		3	B8/9 IV	1586
	+30 03103			17 58 37	+30 03.0	1	10.30		0.26		0.12		3		196
164359	−22 04500	LSS 4558		17 58 38	−22 07.9	6	7.53	.004	0.00	.007	−0.83	.012	17	B1 II	240,401,540,914,976,1732
		WLS 1800 15 # 5		17 58 41	+17 14.0	1	10.23		1.05		0.83		2		1375
164507	+15 03327	HR 6722		17 58 42	+15 05.7	1	6.26		0.69		0.24		2	G5 IV	252
164384	−23 13789	LSS 4559		17 58 43	−23 10.7	1	8.25	.044	−0.02	.030	−0.78	.017	5	B1/2 Ib/II	540,976,1012
164320	−35 12174			17 58 43	−35 40.2	2	7.57	.001	−0.02	.010	−0.40	.001	3	B7 II	540,976
164595	+29 03165	IDS17569N2934		17 58 44	+29 34.2	2	7.07	.010	0.63	.005	0.12	.005	6	G2 V	196,3026
		NGC 6517 sq1 8		17 58 45	−08 55.0	1	13.56		1.12				1		438
164614	+33 03006	HR 6726		17 58 46	+33 12.9	1	5.99		1.51				2	K5	71
		NGC 6517 sq1 3		17 58 47	−08 55.0	1	12.13		1.04				2		438
313595		NGC 6514 - 24		17 58 47	−22 48.7	1	10.58		0.06		−0.32		1	O9	429
164385	−24 13765			17 58 47	−24 09.4	2	8.05	.004	−0.01	.005	−0.71	.000	9	B2 III	401,1732
164321	−36 12123			17 58 47	−36 21.6	2	7.42	.008	−0.02	.012	−0.39	.003	3	B5 II/III	540,976
		NGC 6517 sq1 4		17 58 49	−09 01.2	1	12.84		1.43				2		438
314904	−24 13768	LSS 4561		17 58 49	−24 19.8	1	9.65		0.99		−0.09		3	B1 Ib	1737
347831	+18 03527			17 58 51	+18 58.0	1	8.27		0.99				70	K0	6011
347830	+19 03496			17 58 51	+19 00.4	1	9.63		0.14				62	A2	6011
		NGC 6514 - 31		17 58 51	−22 57.5	1	13.69		0.61		0.11		2		429
164387	−30 15119			17 58 51	−30 06.4	1	9.66		1.24				1	G8/K0 (III)	915
		NGC 6517 sq1 2		17 58 52	−08 56.8	1	11.27		0.75				2		438
		NGC 6517 sq1 12		17 58 53	−09 03.1	1	14.83		2.06				1		438
164402	−22 04503	NGC 6514 - 55	⋆ AB	17 58 53	−22 46.9	8	5.75	.025	−0.01	.014	−0.85	.037	20	B0 Ib	138,399,401,429,540*
		NGC 6514 - 33		17 58 54	−23 00.2	1	12.93		0.67		0.22		2		429
		Bo 14 - 2		17 58 54	−23 41.6	1	10.67		1.91		0.71		3		412
		Bo 14 - 3		17 58 54	−23 42.	1	12.03		1.31		1.38		2		412
		Bo 14 - 5		17 58 54	−23 42.	1	13.20		1.35		0.47		2		412
		Bo 14 - 6		17 58 54	−23 42.	1	13.64		1.37		0.54		2		412
		Bo 14 - 7		17 58 54	−23 42.	1	13.97		2.02		1.05		2		412
		Bo 14 - 8		17 58 54	−23 42.	1	12.99		1.33		0.51		3		412
		Bo 14 - 9		17 58 54	−23 42.	1	13.50		1.35		0.54		2		412
		Bo 14 - 10		17 58 54	−23 42.	1	13.23		0.98		0.76		2		412
		Bo 14 - 11		17 58 54	−23 42.	1	13.32		0.82		0.48		1		412
164438	−19 04800	LSS 4567	⋆	17 58 55	−19 06.4	5	7.48	.008	0.33	.010	−0.66	.009	8	B0.5Iab	540,976,1011,1012,1737
		NGC 6514 - 32		17 58 55	−23 00.1	1	11.63		1.31		1.13		2		429
317024				17 58 55	−30 17.	1	9.90		1.11				1		915
		G 258 - 26	AB	17 58 56	+71 59.4	1	10.81		1.07		1.11		2		7010
		NGC 6517 sq1 10		17 58 56	−09 03.2	1	14.01		1.23				1		438
	−23 13793	Bo 14 - 1		17 58 56	−23 41.5	2	10.25	.000	1.28	.000	0.19	.005	7		412,1737
		Bo 14 - 4		17 58 56	−23 41.7	1	12.58		1.33		0.41		2		412
164404	−27 12272	IDS17558S2731	AB	17 58 56	−27 30.3	1	8.00	.004	−0.04	.004	−0.63	.007	3	B2 V	540,976
		NGC 6514 - 29		17 58 57	−22 56.9	1	14.27		1.09		0.75		2		429
		NGC 6514 - 65		17 58 58	−22 53.4	1	12.27		1.18		1.07		3		429
347827		BL Her		17 58 59	+19 15.1	2	9.77	.019	0.27	.018			2	F3 II	592,6011
		G 140 - 31		17 59 01	+04 47.9	1	12.90		0.81		0.26		2		1658
164483	−9 04639	NGC 6517 sq1 1		17 59 01	−09 02.9	1	9.96		0.76				2	G2	438
		NGC 6514 - 67		17 59 01	−22 53.9	1	14.11		0.79		0.30		3		429
321131	−34 12383	IDS17557S3433	ABC	17 59 01	−34 33.2	1	9.98		0.04		−0.33		1	B5	1586
		NGC 6517 sq1 7		17 59 02	−08 53.5	1	13.05		1.11				1		438
		NGC 6514 - 36		17 59 02	−23 00.4	1	14.53		0.52		0.12		2		429
		NGC 6514 - 39		17 59 02	−23 03.3	1	14.04		0.78		0.41		3		429
164529	+3 03564	IDS17566N0332	AB	17 59 03	+03 31.4	1	8.36		0.03		−0.30		3	A0	963
164340	−40 12092			17 59 03	−40 05.3	3	9.29	.018	−0.14	.006	−0.95	.008	7	B0/0.5(III)	400,540,976
		NGC 6514 - 61		17 59 04	−22 50.2	1	13.95		0.82		0.19		3		429
	+67 01044			17 59 05	+67 47.7	1	9.27		1.25		1.36		2	K2	1375

Table 1 979

HD	DM	Other Id	N Rem	α_{1950}	δ_{1950}	S	V	σ_V	B–V	σ_{B-V}	U–B	σ_{U-B}	n	Spectrum	References
164557	+3 03565	IDS17566N0332	E	17 59 07	+03 30.3	1	7.99		-0.04		-0.53		3	A0	963
		NGC 6514 - 93		17 59 07	-22 49.1	1	10.84		1.62		2.03		6		429
		NGC 6514 - 94		17 59 07	-22 49.8	1	12.75		0.85		0.37		3		429
324802	-37 12105	LSS 4560		17 59 07	-37 31.3	2	11.25	.034	0.17	.004	-0.53	.002	3		1586,1737
		NGC 6517 sq1 5		17 59 08	-08 53.2	1	12.86		1.09				1		438
164577	+1 03560	Cr 359 - 10	⋆ABC	17 59 13	+01 18.2	8	4.44	.015	0.03	.010	0.01	.017	23	A2 Vn	15,379,1075,1088,1363*
		NGC 6514 - 102		17 59 13	-22 57.1	1	14.72		0.73		0.45		3		429
		NGC 6514 - 44		17 59 14	-23 08.7	1	13.58		0.62		0.13		2		429
	-29 14376			17 59 15	-29 40.1	1	10.69		1.25				3		1726
		L 157 - 110		17 59 18	-64 31.	1	11.54		1.47				1		1494
164492	-23 13804	NGC 6514 - 145	⋆AB	17 59 21	-23 01.9	5	7.29	.097	0.00	.023	-0.86	.019	11	O7 V	540,1011,1012,1732,8084
		NGC 6514 - 146	⋆CD	17 59 21	-23 02.1	1	7.38		0.19		-0.25		1	B0 II	8084
		NGC 6514 - 105		17 59 22	-22 59.2	1	10.40		1.09		0.77		6		429
		G 183 - 16		17 59 23	+20 44.7	1	11.88		0.79		0.23		1		1658
164669	+21 03280	HR 6730	⋆A	17 59 23	+21 35.7	1	4.96		0.12		0.13		2	A5IIIn+G5III	3016
164669	+21 03280	IDS17573N2136	AB	17 59 23	+21 35.7	3	4.28	.005	0.39	.007	0.24	.002	7	A5 IIIn +	253,938,1363
164668	+21 03280	HR 6729	⋆B	17 59 23	+21 35.7	1	5.18		0.95		0.57		2	G8 III	3016
		NGC 6514 - 84		17 59 23	-23 11.5	1	13.72		0.86		0.54		2		429
164455	-33 12759			17 59 24	-33 53.4	3	7.43	.004	-0.09	.006	-0.65	.006	4	B2 III/IV	219,540,976
		NGC 6514 - 115		17 59 26	-22 46.3	1	14.03		0.81		0.22		1		429
		NGC 6514 - 108		17 59 26	-23 01.4	1	13.35		0.41		0.06		1		429
164419	-39 12151			17 59 27	-39 20.3	1	8.59		1.16		1.07		9	K1 III	1673
		NGC 6517 sq1 6		17 59 28	-08 54.6	1	12.96		1.13				1		438
		NGC 6514 - 125		17 59 28	-22 51.9	1	12.77		0.35		0.29		1		429
313597		NGC 6514 - 163		17 59 28	-23 09.1	1	11.64		-0.10		-0.02		1	A5	429
		NGC 6514 - 116		17 59 29	-22 47.3	1	13.46		0.68		0.09		1		429
		NGC 6514 - 122		17 59 29	-22 50.8	1	13.14		0.74		0.27		1		429
		NGC 6514 - 135		17 59 29	-22 59.7	1	12.76		0.53		0.26		3		429
314947	-25 12608	LSS 4572		17 59 29	-25 42.0	1	10.40		0.82		-0.20		2		1737
316995				17 59 29	-29 17.	1	9.54		1.13				1		915
164755	+30 03106			17 59 30	+30 38.7	1	7.08		1.28		1.35		2	K5	1733
		G 182 - 34		17 59 30	+35 35.9	1	13.72		1.54				7		1663
164514	-22 04510	NGC 6514 - 129		17 59 30	-22 54.4	4	7.35	.113	0.98	.028	0.48	.046	10	A5 Ia	138,399,429,8100
		NGC 6530 - 1		17 59 30	-24 14.8	1	12.62		1.47				1		1161
		NGC 6514 - 131		17 59 31	-22 56.4	1	13.15		0.25		0.19		1		429
		NGC 6514 - 132		17 59 31	-22 57.5	1	13.45		1.04		0.74		2		429
		NGC 6514 - 134		17 59 31	-22 58.9	1	14.28		0.60		-0.03		3		429
		NGC 6514 - 149	⋆G	17 59 31	-23 03.6	1	13.17		0.17		-0.03		1		8084
317010				17 59 31	-30 03.	1	10.12		1.62				1		915
164728	+25 03404			17 59 32	+25 29.3	1	7.60		-0.02		-0.10		2	A0	105
164534	-22 04511	NGC 6514 - 167		17 59 32	-22 46.2	1	9.29		0.23		-0.01		7	B8 V	429
		NGC 6514 - 170		17 59 32	-22 50.3	1	12.65		0.39		0.25		1		429
317103				17 59 32	-29 31.8	1	10.19		1.33				1		915
164496	-30 15136			17 59 32	-30 58.2	1	8.60		0.05		-0.02		2	B9 IV	1730
313596		NGC 6514 - 186		17 59 33	-23 00.0	1	10.76		0.09		-0.27		1	B8	429
164808	+35 03128			17 59 34	+35 31.6	1	8.54		0.28		0.09		2	F2	1601
164579	-12 04902			17 59 34	-12 19.2	1	7.67		1.96				2	M2 III	369
		NGC 6514 - 133		17 59 34	-22 58.4	1	9.79		1.25		1.15		3	K0	429
		NGC 6514 - 150		17 59 35	-23 03.5	1	11.67		0.14		0.03		1		429
164536	-24 13783	NGC 6530 - 2	⋆AB	17 59 35	-24 15.4	3	7.13	.011	-0.03	.003	-0.90	.007	9	O7	240,1161,1732
		NGC 6514 - 183		17 59 36	-22 58.9	1	13.26		0.35		-0.22		2		429
		NGC 6514 - 159		17 59 36	-23 09.7	1	13.36		0.65		0.12		1		429
	-24 13785	NGC 6530 - 3	⋆C	17 59 36	-24 14.9	4	8.67	.010	0.00	.005	-0.77	.011	12		720,917,1161,1732
164516	-29 14389			17 59 36	-29 22.1	2	7.86	.043	-0.03	.005	-0.56	.010	4	B3 II	540,976
164581	-20 04952	LSS 4577		17 59 38	-20 44.3	4	6.80	.008	0.12	.010	-0.66	.009	8	B2/3 II	240,401,540,976
		NGC 6514 - 157		17 59 38	-23 07.2	1	13.15		0.56		0.14		1		429
		NGC 6514 - 166		17 59 39	-22 45.8	1	13.03		1.14		0.31		1		429
		NGC 6530 - 211		17 59 39	-24 18.3	2	12.41	.015	0.17	.030	-0.11	.015	4		720,917
		NGC 6514 - 178		17 59 40	-22 56.5	1	14.29		0.81		0.19		1		429
313740	-23 13808	NGC 6514 - 195		17 59 40	-23 06.7	1	10.23		0.15		0.01		6	A0 V	429
		NGC 6514 - 158		17 59 40	-23 09.2	1	14.88		0.67		0.26		1		429
164898	+45 02643	IDS17582N4521	A	17 59 41	+45 21.0	2	7.59	.014	0.09	.004	0.10	.017	14	B9	753,3016
164898	+45 02643	IDS17582N4521	B	17 59 41	+45 21.0	1	10.88		0.71		0.26		2		3016
		NGC 6514 - 169		17 59 41	-22 47.7	1	12.87		0.73		0.25		1		429
		NGC 6514 - 160		17 59 42	-23 09.5	1	12.65		0.64		0.15		1		429
		NGC 6514 - 196		17 59 43	-23 07.9	1	12.29		0.15		0.00		2		429
		NGC 6530 - 210		17 59 43	-24 20.6	2	12.27	.010	0.22	.005	0.09	.015	4		720,917
313710		NGC 6514 - 207		17 59 44	-22 45.4	1	10.44		1.40		1.43		1	G0	429
		NGC 6514 - 222		17 59 45	-22 50.5	1	14.32		0.66		0.27		1		429
164541	-30 15142			17 59 45	-30 58.1	1	8.81		1.32				1	K0/1 (III)	915
164824	+33 03009	HR 6737	⋆A	17 59 46	+33 18.6	1	6.15		1.55				2	K5 III	71
		NGC 6514 - 199		17 59 46	-23 09.4	1	14.18		0.53		-0.07		1		429
		NGC 6514 - 165		17 59 46	-23 14.0	1	15.08		1.08		0.40		1		429
	-30 15144			17 59 46	-30 39.2	1	10.39		0.55		0.04		2		1730
164842	+35 03129			17 59 47	+35 02.7	1	9.62		-0.15		-0.63		3	A5	963
		NGC 6530 - 303		17 59 47	-24 16.4	2	12.58	.540	0.65	.030	0.24	.020	4		720,917
164584	-24 13793	NGC 6530 - 4	⋆	17 59 47	-24 17.0	4	5.39	.010	0.49	.007	0.28	.021	15	F5 II	1161,2035,8071,8100
		NGC 6530 - 320		17 59 47	-24 37.6	2	12.68	.035	0.40	.015	0.22	.040	4		720,917
314900		NGC 6530 - 5		17 59 49	-24 20.4	1	10.42		0.13		-0.50		2	B5	1161
		NGC 6514 - 206		17 59 50	-22 44.5	1	12.54		0.73		0.26		1		429

HD	DM	Other Id	N Rem	α₁₉₅₀	δ₁₉₅₀	S	V	σ_V	B−V	σ_B−V	U−B	σ_U−B	n	Spectrum	References
		NGC 6530 - 321		17 59 50	−24 09.9	2	11.45	.020	1.09	.040	0.01	.030	4		720,917
		WLS 1800-20 # 9		17 59 52	−18 50.8	1	11.85		0.77		0.29		2		1375
164564	−30 15147			17 59 52	−30 47.1	1	9.37		1.01				1	K0 III +(G)	915
		NGC 6514 - 205		17 59 53	−23 12.7	1	13.45		0.59		0.13		1		429
164809	+22 03259			17 59 54	+22 27.3	1	7.40		1.16		1.09		3	K0 II-III	8100
		NGC 6514 - 239		17 59 54	−23 09.7	1	14.23		0.77		0.20		1		429
313714	−22 04514	NGC 6514 - 271		17 59 57	−22 52.4	1	10.11		0.49		0.17		1	F2 V	429
		NGC 6514 - 276		17 59 57	−22 54.2	1	14.71		1.53		0.24		1		429
164606	−29 14399			17 59 57	−29 31.9	2	9.16	.015	−0.02	.015	−0.51	.012	4	B5 II/III	540,976
		Steph 1569		17 59 58	+27 02.3	1	10.57		1.17		1.15		1	K7	1746
		NGC 6514 - 282		17 59 58	−22 57.4	1	13.66		0.94		0.56		1		429
313737	−23 13817	NGC 6514 - 290		17 59 58	−23 05.7	1	9.74		1.38		1.31		1	G2	429
164500	−45 12143			17 59 58	−45 51.4	2	9.63	.015	0.79	.010	0.29		6	G8 V	1696,2012
313734	−23 13818	NGC 6514 - 298		17 59 59	−23 12.2	1	10.74		0.15		−0.03		1	B9	429
164637	−22 04516	HR 6727, LSS 4580		18 00 00	−22 43.2	8	6.73	.007	−0.04	.006	−0.89	.016	17	B0 Ib/II	15,401,540,914,976,1732*
		NGC 6514 - 275		18 00 00	−22 54.0	1	13.07		0.32		0.23		1		429
313713		NGC 6514 - 256		18 00 01	−22 47.4	1	10.02		0.12		−0.24		2	B8 V	429
		NGC 6514 - 259		18 00 02	−22 48.3	1	12.25		0.63		0.20		1		429
		NGC 6530 - 322		18 00 02	−24 07.8	2	13.50	.010	0.50	.065	0.17	.020	4		720,917
		NGC 6530 - 219		18 00 02	−24 15.8	2	12.38	.015	0.57	.075	0.10	.120	4		720,917
	−27 12309	NGC 6520 - 2		18 00 02	−27 53.5	1	10.94		0.21		−0.26		3	B9	1586
		NGC 6530 - 200		18 00 03	−24 24.8	2	12.80	.025	0.30	.015	0.23	.020	4		720,917
164716	−5 04560	HR 6732		18 00 06	−05 21.6	1	6.75		0.16		−0.12		4	B9 V	1149
313736	−23 13823	NGC 6514 - 291		18 00 06	−23 05.9	1	10.23		0.06		−0.41		1	B8 V	429
		NGC 6514 - 316		18 00 09	−22 56.0	1	12.23		0.49		0.13		1		429
		NGC 6514 - 318		18 00 10	−22 58.1	1	13.74		0.62		0.13		1		429
		NGC 6530 - 323		18 00 10	−24 07.6	2	12.73	.070	1.41	.090	0.93	.125	4		720,917
		18H00M11S		18 00 11	+04 28.9	1	10.69		0.71		0.13		2		1696
164760	−1 03435			18 00 12	−01 20.3	2	7.98	.010	0.18	.006	−0.01	.008	22	B9	379,1699
313716	−22 04518	NGC 6514 - 305		18 00 12	−22 49.8	1	9.25		1.23		0.96		1	G8 III	429
164852	+20 03649	HR 6738, V820 Her		18 00 15	+20 49.9	4	5.26	.010	−0.10	.008	−0.62	.013	22	B3 IV	154,1203,1363,3016
		NGC 6514 - 317		18 00 16	−22 57.5	1	14.58		0.99		0.33		1		429
		HIC 88447		18 00 16	−29 36.9	1	10.30		1.80				3		1726
		NGC 6522 sq2 3		18 00 16	−30 03.7	3	12.34	.012	1.60	.012	1.97	.010	21		325,746,1550
164704	−22 04520	NGC 6514 - 309		18 00 17	−22 53.2	5	8.16	.024	−0.03	.011	−0.81	.027	8	B1.5III	401,429,540,914,976
		NGC 6514 - 342		18 00 18	−22 56.1	1	13.29		0.74		0.21		1		429
313732	−23 13828	NGC 6514 - 331		18 00 18	−23 10.9	1	11.02		0.11		−0.34		1	B2 V	429
164427	−59 07218	IDS17559S5913	AB	18 00 18	−59 12.7	4	6.88	.011	0.61	.008	0.13	.015	19	G0 V	285,1499,1770,2033
164765	−8 04549	HR 6734	⋆ AB	18 00 21	−08 10.9	5	4.78	.003	0.39	.012	0.03	.017	22	F2 V	15,938,1075,1088,3026
164765	−8 04549	IDS17576S0811	C	18 00 21	−08 10.9	4	11.28	.012	0.79	.011	0.32	.026	15		158,196,3024,7009
		NGC 6530 - 244		18 00 22	−24 13.7	2	12.04	.010	1.24	.040	1.07	.010	4		720,917
		NGC 6530 - 181		18 00 22	−24 22.7	2	12.78	.120	0.55	.010	−0.01	.100	4		720,917
164717	−22 04522	LSS 4586	⋆ A	18 00 23	−22 37.1	1	8.21		−0.02		−0.73		10	B3nn	1732
		NGC 6530 - 324		18 00 23	−24 06.6	2	11.69	.005	0.35	.035	−0.23	.015	4		720,917
		NGC 6530 - 241		18 00 23	−24 12.9	2	11.88	.025	0.22	.020	−0.26	.005	4		720,917
		NGC 6530 - 178		18 00 24	−24 23.2	2	10.90	.040	1.22	.020	1.06	.265	4		720,917
164900	+22 03260	HR 6741		18 00 25	+22 55.3	2			−0.10	.005	−0.64	.014	4	B3 Vn	1022,1079
	−30 15153			18 00 25	−30 38.7	1	10.80		0.08		0.06		2		1730
		KUV 18004+6836		18 00 26	+68 35.9	1	14.74		−0.20		−1.19		1	sdO	1708
164923	+24 03311			18 00 27	+24 59.5	2	8.18	.005	1.14	.007	1.03	.012	4	K2 III	1733,3040
315124	−27 12324	LSS 4585		18 00 28	−27 10.3	1	9.40		0.71		−0.30		3	B8	1737
164922	+26 03151	G 182 - 35	⋆ A	18 00 29	+26 19.2	5	7.02	.019	0.80	.005	0.47	.007	17	K0 V	22,1003,1080,3026,4001
	−29 14416	ApJS39,135 # 4		18 00 31	−29 56.3	1	10.88		1.10				3		915
		NGC 6522 sq2 2		18 00 31	−30 01.9	3	11.59	.008	0.18	.004	0.17	.000	19		325,746,1550
		ApJS39,135 # 2		18 00 32	−29 51.4	1	12.07		1.19				3		915
		ApJS39,135 # 3		18 00 32	−29 55.3	1	12.69		1.27				2		915
	−30 15157			18 00 32	−30 56.0	1	11.25						2		1730
164719	−26 12728			18 00 33	−26 52.3	1	7.99		0.18		−0.41		3	B3 Ib	240
312602				18 00 36	−21 28.	1	9.85		1.84				2		369
164739	−23 13832			18 00 36	−23 08.4	1	8.46		0.06		−0.23		2	B8/9 III	401
164741	−25 12642	LSS 4590		18 00 39	−25 18.9	2	9.01	.012	0.30	.013	−0.55	.000	4	B2 Ib/II	240,1586
		NGC 6530 - 170		18 00 41	−24 22.3	2	10.03	.020	0.14	.015	−0.52	.015	4		720,917
		WLS 1800-5 # 7		18 00 42	−06 55.6	1	12.11		0.87		0.28		2		1375
		18H00M43S		18 00 43	+10 10.6	1	12.04		0.70		0.13		2		1696
315027		NGC 6530 - 6		18 00 43	−24 16.9	1	10.71		0.12		−0.55		2	A2	1161
164769	−27 12330	LSS 4591		18 00 43	−27 18.4	2	9.24	.010	−0.05	.015	−0.78	.009	4	B2 III	540,976
	−31 15029			18 00 45	−31 04.6	1	11.49		0.49				2		465
164830	−13 04840			18 00 46	−13 15.2	1	9.11		2.30				2	M1 (III)	369
		NGC 6530 - 172		18 00 47	−24 21.2	1	12.15		0.29		−0.46		2		720
		NGC 6530 - 173		18 00 47	−24 21.4	1	12.78		0.37		−0.17		2		720
164794	−24 13814	NGC 6530 - 7	⋆	18 00 48	−24 21.8	13	5.97	.009	0.03	.009	−0.91	.012	83	O4 V((f))	15,540,914,917,976*
317013				18 00 48	−30 02.7	1	9.80		1.31				1		915
321229	−36 12170			18 00 49	−36 58.0	1	11.50		0.00		−0.35		1	B8	675
		NGC 6530 - 168		18 00 50	−24 23.3	2	11.96	.120	0.37	.090	−0.17	.055	4		720,917
		ApJS39,135 # 5		18 00 50	−29 59.4	1	11.79		1.12				4		915
	−12 04908			18 00 52	−12 37.7	1	10.27		2.28				2	M2	369
		NGC 6531 - 10		18 00 52	−22 29.3	1	11.09		0.15		−0.38		1	B5 V	49
		NGC 6530 - 8		18 00 52	−24 26.8	1	12.91		0.28		0.04		1		1161
164816	−24 13816	NGC 6530 - 9		18 00 53	−24 18.9	6	7.09	.010	0.00	.006	−0.87	.014	10	O9.5IVn	540,914,917,976,1161,1737
164833	−22 04533	LSS 4599	⋆ AB	18 00 54	−22 50.4	4	7.14	.008	0.00	.009	−0.85	.016	8	B1/2 II/III	240,401,540,976

Table 1   981

HD	DM	Other Id	N Rem	α_{1950}	δ_{1950}	S	V	σ_V	B–V	σ_{B-V}	U–B	σ_{U-B}	n	Spectrum	References
		NGC 6530 - 11		18 00 54	−24 15.4	1	14.80		0.71				1		1161
		NGC 6530 - 10	★ V	18 00 54	−24 25.8	1	13.63		0.76		0.83		1		1161
		NGC 6530 - 12		18 00 55	−24 23.9	1	11.93		0.60		0.03		2		1161
313692	−22 04534	NGC 6531 - 5		18 00 56	−22 31.6	2	9.86	.133	0.03	.025	-0.56	.014	10	B3 V	49,1655
		NGC 6530 - 13		18 00 56	−24 15.2	1	13.34		0.46		0.15		2		1161
		NGC 6530 - 326		18 00 56	−24 27.1	1	13.67		0.61		0.25		2		720
		NGC 6531 - 28		18 00 57	−22 29.6	1	14.20		0.86		0.32		1		49
		NGC 6531 - 29		18 00 57	−22 30.3	1	14.54		0.96				1		49
		NGC 6530 - 15		18 00 57	−24 24.0	3	11.62	.035	0.14	.034	-0.34	.037	5		720,917,1161
		NGC 6530 - 14		18 00 57	−24 26.7	3	11.43	.077	0.17	.039	-0.31	.006	5		720,917,1161
317011				18 00 58	−29 56.	1	9.86		1.12				1		915
		NGC 6531 - 24		18 00 59	−22 29.8	1	12.74		0.27		0.16		1		49
		NGC 6530 - 16		18 00 59	−24 11.1	3	12.12	.075	0.28	.022	0.09	.032	5		720,917,1161
		NGC 6531 - 14		18 01 00	−22 29.4	1	11.78		0.19		-0.16		1	A0 V	49
		NGC 6530 - 17		18 01 00	−24 21.7	3	11.97	.046	0.15	.045	-0.27	.015	6		720,917,1161
		NGC 6530 - 121		18 01 00	−24 28.7	2	13.81	.060	0.63	.040	-0.14	.185	4		720,917
		ApJS39,135 # 1		18 01 00	−30 04.2	1	10.13		1.72				3		915
164798	−31 15036	LSS 4596		18 01 00	−31 38.9	3	8.07	.010	0.11	.004	-0.64	.030	5	B2/3 II	540,954,976
		NGC 6535 sq1 4		18 01 01	+00 20.4	1	11.33		1.28				1		870
164844	−22 04535	NGC 6531 - 2		18 01 01	−22 34.1	1	8.29		0.01		-0.81		1	B1 V	49
		NGC 6530 - 18		18 01 01	−24 23.2	1	12.72		0.72		0.24		1		1161
164818	−29 14427			18 01 01	−29 17.2	1	9.19		1.51				1	K2/3	915
315101	−26 12742			18 01 02	−26 13.5	1	10.41		0.63				4	B5 III	1021
		WLS 1800 15 # 7		18 01 03	+13 15.9	1	11.77		0.60		0.12		2		1375
		WLS 1800 15 # 9		18 01 03	+15 33.4	1	11.06		1.23		1.16		2		1375
		G 182 - 36		18 01 03	+37 31.6	2	14.89	.000	1.74	.058			6		419,1759
		NGC 6531 - 27		18 01 03	−22 31.6	1	13.09		0.72		0.27		1		49
315025	−24 13821	NGC 6530 - 19		18 01 03	−24 19.6	2	10.74	.049	0.27	.024	-0.50	.029	5	B8	720,1161
		NGC 6530 - 260		18 01 03	−24 22.6	1	13.26		1.14		0.64		2		720
		NGC 6531 - 16		18 01 04	−22 31.2	1	11.87		0.17		-0.14		1	A0 V	49
		NGC 6530 - 163		18 01 04	−24 12.9	1	12.72		0.22		0.11		2		720
		NGC 6530 - 21		18 01 04	−24 17.0	1	14.62		0.81		0.47		1		1161
		NGC 6530 - 20		18 01 04	−24 17.6	1	14.04		0.74		0.33		1		1161
		NGC 6530 - 22		18 01 04	−24 20.8	1	12.87		1.11		0.83		1		1161
		MN174,213 T1# 8		18 01 04	−30 55.7	1	12.73		2.16				3		465
	−30 15174			18 01 04	−30 59.1	1	11.79		0.59				3		465
165029	+19 03508	HR 6744		18 01 05	+19 36.6	2	6.42	.009	0.02	.009	-0.02	.005	6	A0 V	1022,1733
		NGC 6531 - 12		18 01 05	−22 32.1	1	11.49		0.40		0.15		1		49
		NGC 6530 - 23		18 01 05	−24 14.7	1	12.41		1.11		0.49		1		1161
		NGC 6530 - 25		18 01 05	−24 15.9	3	11.46	.026	0.09	.014	-0.31	.021	5		720,917,1161
		NGC 6530 - 28		18 01 06	−24 12.5	1	13.22		0.43		0.26		2		1161
		NGC 6530 - 30		18 01 06	−24 17.1	1	15.70		0.92		0.99		1		1161
		NGC 6530 - 26		18 01 06	−24 20.3	3	11.65	.007	0.21	.010	-0.28	.028	6		720,917,1161
		NGC 6530 - 27		18 01 06	−24 21.0	3	12.95	.016	0.66	.023	0.27	.064	7		720,917,1161
		NGC 6530 - 24		18 01 06	−24 24.4	3	11.94	.020	0.16	.039	-0.23	.021	5		720,917,1161
		NGC 6530 - 33		18 01 07	−24 16.6	3	13.38	.058	0.78	.012	0.46	.028	5		720,917,1161
		NGC 6530 - 32		18 01 07	−24 21.9	3	10.45	.042	0.11	.012	-0.56	.023	6	B3 Ve	720,917,1161
		NGC 6530 - 29		18 01 07	−24 25.0	2	13.16	.019	0.80	.005	-0.09	.146	4		917,1161
		NGC 6530 - 31		18 01 07	−24 25.7	3	11.69	.013	0.25	.035	-0.29	.028	5		720,917,1161
		NGC 6530 - 34		18 01 08	−24 18.0	1	14.83		0.77		0.22		1		1161
	−31 15043			18 01 08	−31 05.3	1	12.18		0.54				3		465
		NGC 6531 - 13		18 01 09	−22 28.5	1	11.53		0.16		-0.22		1	B8 V	49
		NGC 6530 - 36		18 01 09	−24 18.6	1	14.32		0.98		0.22		1		1161
	−24 13822	NGC 6530 - 35		18 01 09	−24 21.9	1	9.84		1.14		0.80		2	K0 III	1161
317015	−30 15177			18 01 09	−30 03.1	1	9.68		1.04				1	K0	915
	−22 04540	NGC 6531 - 4		18 01 10	−22 27.4	2	9.21	.007	0.03	.020	-0.59	.020	10	B2 V	49,1655
		NGC 6531 - 18		18 01 10	−22 27.9	1	12.36		0.21		-0.02		1		49
313693	−22 04538	NGC 6531 - 3	★ AB	18 01 10	−22 29.9	2	8.76	.007	0.02	.092	-0.69	.023	8	B1 Vn	49,1655
		NGC 6531 - 22		18 01 10	−22 30.5	1	12.52		0.28				1		49
		NGC 6530 - 38		18 01 10	−24 13.7	3	12.19	.038	0.51	.016	0.27	.051	6		720,917,1161
		NGC 6530 - 41		18 01 10	−24 16.9	1	12.52		0.37		0.18		1		1161
		NGC 6530 - 39		18 01 10	−24 19.8	1	14.49		1.17		0.77		1		1161
		NGC 6530 - 37		18 01 10	−24 25.1	1	14.60		0.84				1		1161
165042	+19 03509			18 01 11	+19 33.2	2	7.03	.019	1.63	.010	1.54	.015	6	M5 II-III	1648,8100
		NGC 6531 - 23		18 01 11	−22 29.9	1	12.60		0.18		-0.13		1		49
164863	−22 04541	NGC 6531 - 1	★ A	18 01 11	−22 30.3	4	7.26	.007	-0.03	.017	-0.83	.018	12	B0 V	49,1012,1655,1737
164865	−24 13826	NGC 6530 - 45		18 01 11	−24 11.2	6	7.66	.017	0.87	.017	0.15	.010	63	B9 Ia	540,976,1161,1586*
315026	−24 13823	NGC 6530 - 43	★ AB	18 01 11	−24 14.8	2	9.01	.005	0.08	.005	-0.65	.057	4	B5	917,1161
		NGC 6530 - 40		18 01 11	−24 22.1	1	13.24		1.21		0.88		1		1161
315032	−24 13824	NGC 6530 - 42		18 01 11	−24 23.6	3	9.19	.005	0.03	.011	-0.74	.012	5	B2 Vne	720,917,1161
321232	−36 12178			18 01 11	−36 59.0	1	10.56		0.54		0.04		1	G0	675
		NGC 6531 - 9		18 01 12	−22 29.6	1	10.79		0.20		-0.09		1	A2 V	49
		NGC 6530 - 47	★ D	18 01 12	−24 15.1	3	12.25	.021	0.23	.018	0.03	.025	5		720,917,1161
		NGC 6530 - 48		18 01 12	−24 16.8	1	13.60		2.05		0.95		1		1161
		NGC 6530 - 44		18 01 12	−24 22.3	1	14.88		1.33				1		1161
		NGC 6535 sq1 3		18 01 13	+00 22.5	1	14.20		1.63				3		870
		NGC 6530 - 49		18 01 13	−24 17.5	2	11.10	.000	0.15	.017	-0.42	.043	3		917,1161
		NGC 6530 - 50		18 01 13	−24 25.6	1	15.20		0.87		0.35		1		1161
		NGC 6530 - 46		18 01 13	−24 28.8	3	11.62	.013	0.19	.008	-0.20	.021	5		720,917,1161
		NGC 6535 sq1 2		18 01 14	+00 23.1	1	13.88		0.72				1		870

HD	DM	Other Id	N	Rem	α_{1950}	δ_{1950}	S	V	σ_V	B–V	σ_{B-V}	U–B	σ_{U-B}	n	Spectrum	References
		NGC 6531 - 11			18 01 14	−22 32.2	1	11.12		0.54		0.08		1		49
		NGC 6530 - 51			18 01 14	−24 14.7	1	14.15		2.31		1.78		1		1161
164867	−27 12341				18 01 14	−27 23.5	1	7.79		0.03		−0.15		2	B9 II/III	240
321231	−36 12179				18 01 14	−36 57.2	1	10.51		0.03		−0.01		1	A2	675
		NGC 6531 - 25			18 01 15	−22 27.5	1	12.76		0.33		0.28		1		49
		NGC 6530 - 327			18 01 15	−24 35.3	1	13.02		0.56		0.42		2		917
	−30 15179	LSS 4604			18 01 15	−30 46.5	1	11.00		0.48		−0.29		3		954
321230	−36 12180	V2283 Sgr			18 01 15	−36 55.0	1	10.23		0.14		0.08		1	A0	675
		NGC 6531 - 15			18 01 16	−22 28.9	1	11.85		0.20		0.06		1	A0 V	49
164883	−22 04543	NGC 6531 - 30			18 01 16	−22 29.8	1	8.87		0.01		−0.76		2	B1 Vn	1012
		NGC 6530 - 52			18 01 16	−24 14.8	3	12.04	.111	0.30	.003	0.07	.097	5		720,917,1161
		NGC 6530 - 53			18 01 16	−24 25.1	3	13.12	.005	0.64	.065	0.29	.132	5		720,917,1161
		NGC 6530 - 126			18 01 16	−24 30.2	1	11.70		1.35		1.14		2		917
		NGC 6531 - 26			18 01 17	−22 27.9	1	12.80		0.38		0.25		1		49
		NGC 6531 - 19			18 01 17	−22 28.3	1	12.43		0.24		0.12		1		49
315023	−24 13827	NGC 6530 - 55			18 01 17	−24 14.1	2	10.10	.009	0.13	.013	−0.59	.030	3	B2.5Vne	917,1161
	−24 13829	NGC 6530 - 56			18 01 17	−24 21.5	3	9.07	.025	0.10	.010	−0.71	.007	6	B1.5Vne	720,917,1161
		NGC 6530 - 54			18 01 17	−24 23.8	3	11.57	.016	0.26	.012	0.06	.028	5		720,917,1161
	−30 15181				18 01 17	−30 55.1	1	10.18		1.27				4		465
		NGC 6531 - 17			18 01 18	−22 28.8	1	12.13		0.33		0.17		1		49
		NGC 6530 - 57			18 01 18	−24 16.0	1	13.66		0.49		0.23		2		1161
313704	−22 04544	NGC 6531 - 6			18 01 19	−22 29.6	1	9.91		0.05		−0.61		1	B3 V	49
	−24 13830	NGC 6530 - 58			18 01 19	−24 22.4	3	9.80	.040	0.22	.031	−0.65	.014	6	B2 Ve	720,917,1161
315033	−24 13833	NGC 6530 - 59			18 01 19	−24 26.5	3	8.93	.010	0.10	.010	−0.64	.016	6	B2 Vp	720,917,1161
		NGC 6530 - 328			18 01 19	−24 36.5	2	12.50	.005	0.46	.040	0.30	.100	4		720,917
	−1 03438	NO Ser			18 01 20	−01 00.4	3	10.35	.026	0.46	.014	−0.28	.007	5		166,272,1699
		NGC 6530 - 63			18 01 20	−24 15.5	1	12.29		0.84		0.29		2		1161
		NGC 6530 - 62			18 01 20	−24 18.7	1	14.30		1.18		0.36		1		1161
		NGC 6530 - 61			18 01 20	−24 21.2	2	10.29	.000	0.12	.000	−0.61	.000	3	B2 Ve	917,1161
		NGC 6530 - 60			18 01 20	−24 21.6	3	9.65	.010	0.09	.018	−0.65	.019	6	B1 Ve	720,917,1161
164868	−30 15182				18 01 20	−30 52.3	2	9.66	.000	−0.01	.003	−0.41		8	B3 II/III	465,1730
		NGC 6530 - 291			18 01 21	−24 09.5	2	13.12	.055	1.39	.085	0.90	.175	4		720,917
		NGC 6530 - 64			18 01 21	−24 22.9	3	11.67	.014	0.19	.016	−0.29	.014	5		720,917,1161
		NGC 6530 - 263			18 01 21	−24 28.0	2	11.44	.045	0.19	.005	−0.38	.065	4		720,917
		NGC 6530 - 68			18 01 22	−24 18.7	1	15.34		0.95		0.21		1		1161
		NGC 6530 - 67			18 01 22	−24 20.0	2	11.89	.038	0.44	.021	0.02	.017	3		720,1161
	−24 13831	NGC 6530 - 66			18 01 22	−24 20.9	3	10.15	.011	0.11	.000	−0.63	.018	5	B2 Vpe	720,917,1161
164906	−24 13832	NGC 6530 - 65			18 01 22	−24 23.3	6	7.46	.020	0.18	.013	−0.82	.028	12	B0 IVpne	720,917,1161,1586,1737,8027
313696	−22 04545	NGC 6531 - 7			18 01 23	−22 26.5	1	9.96		0.12		−0.63		1	B5 V	49
		NGC 6531 - 21			18 01 23	−22 31.5	1	12.47		0.35		0.16		1		49
		NGC 6530 - 69			18 01 23	−24 19.8	1	12.30		0.75		0.20		2		1161
		NGC 6530 - 71			18 01 23	−24 20.0	1	13.27		1.07				1		1161
		NGC 6530 - 70			18 01 23	−24 23.0	3	10.49	.030	0.14	.021	−0.58	.028	5	B2 Ve	720,917,1161
		NGC 6535 sq1 6			18 01 24	+00 21.3	1	13.98		0.92				2		870
315031		NGC 6530 - 73			18 01 24	−24 21.9	3	8.28	.030	0.08	.007	−0.71	.020	6	B2 IVn	720,917,1161
		NGC 6530 - 72			18 01 24	−24 26.2	1	14.04		0.61				1		1161
	−27 12344	LSS 4614			18 01 24	−27 48.7	1	10.80		0.61		−0.35		2	B1 Ia	1737
		NGC 6528 sq1 1			18 01 24	−30 05.4	1	11.90		1.54		1.79		2		791
321325	−36 12182				18 01 24	−36 34.6	1	10.57		0.15		0.18		1	A0	675
		NGC 6530 - 77			18 01 25	−24 14.3	3	12.00	.010	0.48	.024	0.33	.125	6		720,917,1161
		NGC 6530 - 75			18 01 25	−24 19.1	1	14.30		0.80		−0.62		1		1161
315024	−24 13835	NGC 6530 - 76			18 01 25	−24 19.6	3	9.55	.014	0.05	.005	−0.75	.046	6	B2.5Ve	720,917,1161
		NGC 6530 - 74			18 01 25	−24 25.6	3	10.67	.009	0.12	.012	−0.55	.013	14	B2.5Ve	720,917,1161
		NGC 6530 - 329			18 01 26	−24 06.4	1	13.13		0.65		0.07		2		917
		NGC 6530 - 151			18 01 26	−24 09.8	1	12.19		0.72		0.49		2		917
		NGC 6530 - 78			18 01 27	−24 23.9	1	13.96		0.60		0.09		2		1161
		NGC 6530 - 131			18 01 27	−24 32.5	2	13.10	.020	0.34	.015	0.12	.005	4		720,917
		MN174,213 T1# 5			18 01 27	−30 52.4	1	11.68		1.42				3		465
		LSS 4626			18 01 28	−13 06.3	2	11.10	.000	0.92	.000	−0.25	.000	4	B3 Ib	405,1737
		NGC 6530 - 160			18 01 28	−24 24.1	2	12.51	.180	0.42	.060	0.06	.055	4		720,917
		NGC 6530 - 79			18 01 28	−24 24.4	1	14.66		1.14		0.42		1		1161
		NGC 6528 sq1 5			18 01 28	−30 03.8	2	12.85	.034	1.38	.021	1.43	.063	7		325,791
164870	−35 12229	HR 6739		⋆ A	18 01 28	−35 54.3	1	6.00		1.16				4	K2 III	2009
		NGC 6531 - 20			18 01 29	−22 31.1	1	12.43		1.22		1.07		1		49
164933	−24 13839	NGC 6530 - 85			18 01 29	−24 09.8	4	8.57	.007	0.11	.006	−0.75	.019	6	B0.5V	540,914,976,1161
		NGC 6530 - 83			18 01 29	−24 18.9	3	10.48	.025	0.12	.009	−0.59	.026	6	B2.5Vne	720,917,1161
		NGC 6530 - 82			18 01 29	−24 21.5	2	12.43	.021	0.27	.034	−0.18	.073	3		720,1161
		NGC 6530 - 81			18 01 29	−24 23.1	2	11.71	.017	0.17	.009	−0.44	.047	3		720,1161
	−24 13837	NGC 6530 - 80			18 01 29	−24 23.4	2	9.39	.005	0.07	.000	−0.74	.028	4	B1 Ve	917,1161
		NGC 6530 - 87			18 01 30	−24 19.6	1	12.96		1.02		0.33		1		1161
	−24 13840	NGC 6530 - 86			18 01 30	−24 22.2	3	9.81	.048	0.15	.014	−0.54	.056	6	B2 Vne	720,917,1161
		NGC 6530 - 84			18 01 30	−24 28.3	3	11.90	.011	0.22	.015	−0.32	.062	5		720,917,1161
		NGC 6530 - 90			18 01 31	−24 17.9	1	14.81		1.35				1		1161
		NGC 6530 - 91			18 01 32	−24 18.4	1	13.58		0.64		0.13		2		1161
		NGC 6530 - 92			18 01 32	−24 19.4	1	16.18		1.27				1		1161
315021	−24 13841	NGC 6530 - 93			18 01 32	−24 20.1	4	8.59	.013	0.09	.028	−0.73	.019	25	B2 IVn	720,917,1161,1655
		NGC 6530 - 89			18 01 32	−24 20.4	1	13.35		0.32		−0.54		1		1161
164948	−24 13843	NGC 6530 - 88			18 01 32	−24 27.0	2	8.63	.003	0.58	.001	0.03	.004	18	G0	1161,1655
313703		NGC 6531 - 8			18 01 33	−22 30.5	1	10.67		0.12		−0.45		1	B5 V	49
		NGC 6530 - 94			18 01 34	−24 21.4	1	12.65		0.26		−0.02		1		1161

Table 1 983

HD	DM	Other Id	N Rem	α_{1950}	δ_{1950}	S	V	σ_V	B–V	σ_{B-V}	U–B	σ_{U-B}	n	Spectrum	References
		NGC 6535 sq1 5		18 01 35	+00 19.5	1	11.95		1.22				1		870
		NGC 6530 - 98		18 01 35	−24 19.2	1	13.97		0.69		0.10		2		1161
		NGC 6530 - 97		18 01 35	−24 19.6	3	11.35	.032	0.14	.018	-0.52	.038	6		720,917,1161
		NGC 6530 - 96		18 01 35	−24 23.2	2	12.18	.004	0.22	.013	-0.14	.000	3		720,1161
		NGC 6530 - 95		18 01 35	−24 24.4	1	10.21		1.24		0.91		std	K0 III	1161
	−24 13844	NGC 6530 - 99		18 01 36	−24 23.0	2	10.80	.013	0.10	.017	-0.52	.004	3	B2 Vne	720,1161
164935	−30 15189			18 01 36	−30 34.0	1	8.23		1.55				1	K4 III	915
		NGC 6530 - 101		18 01 37	−24 19.2	1	14.73		1.13		0.40		1		1161
164947	−24 13845	NGC 6530 - 100	⋆ AB	18 01 38	−24 21.2	3	8.88	.004	0.07	.006	-0.59	.032	7	B2 IVe	917,1161,1586
164971	−23 13859	LSS 4625		18 01 39	−23 27.8	3	9.39	.012	0.31	.008	-0.64	.011	7	B0 Ia	432,1586,1737
315022		NGC 6530 - 105		18 01 39	−24 18.8	3	10.54	.005	0.15	.023	-0.41	.093	5	B9	720,917,1161
		NGC 6530 - 106		18 01 39	−24 20.2	1	15.40		0.84		0.21		1		1161
		NGC 6530 - 104		18 01 39	−24 23.1	1	15.63		1.24		0.68		1		1161
		NGC 6530 - 103		18 01 39	−24 23.7	1	14.37		0.90		0.33		2		1161
		OX Her		18 01 40	+38 40.0	1	12.88		0.37		0.04		1	A6.5	699
		NGC 6530 - 107		18 01 40	−24 19.9	1	16.13		1.63		0.19		1		1161
		NGC 6530 - 108		18 01 41	−24 20.8	1	14.19		0.93		-0.26		1		1161
		NGC 6528 sq1 2		18 01 41	−30 06.0	1	12.32		1.06		0.78		2		791
		NGC 6530 - 109		18 01 42	−24 19.6	1	14.61		0.83		0.19		1		1161
		NGC 6530 - 325		18 01 42	−24 20.	2	11.91	.015	1.25	.000	1.19	.005	4		720,917
		NGC 6530 - 134		18 01 42	−24 27.5	2	12.71	.045	0.82	.025	0.24	.045	4		720,917
		Smethells 9		18 01 42	−55 07.	1	11.58		1.33				1		1494
	−13 04843			18 01 43	−13 36.3	1	10.21		2.21				1	M2	369
		NGC 6530 - 111		18 01 43	−24 18.3	3	12.61	.010	0.26	.022	-0.02	.032	6		720,917,1161
313643		LSS 4628		18 01 44	−21 10.1	2	11.93	.061	1.22	.023	0.30	.005	4	WC	1183,1359
		NGC 6530 - 314		18 01 44	−24 26.8	2	13.45	.215	1.12	.060	0.40	.150	4		720,917
		NGC 6530 - 110		18 01 44	−24 28.6	3	11.22	.007	0.16	.007	-0.47	.018	5		720,917,1161
		NGC 6530 - 112		18 01 45	−24 20.5	1	14.83		0.98		0.33		1		1161
164993	−23 13863			18 01 46	−23 36.1	2	9.11	.001	0.11	.005	-0.61	.004	5	B2/3 II	401,1586
		NGC 6530 - 115		18 01 46	−24 17.9	1	15.58		0.92		0.19		1		1161
		NGC 6530 - 113		18 01 46	−24 19.8	1	13.96		1.90		1.78		2		1161
		NGC 6530 - 114		18 01 46	−24 25.9	3	12.00	.020	0.33	.008	-0.24	.086	5		720,917,1161
		AV Sgr		18 01 47	−22 44.1	1	10.67		1.74		1.56		1		748
165145	+11 03326			18 01 49	+11 08.4	1	8.76		-0.02		-0.18		2	A0	1733
315020	−24 13848	NGC 6530 - 116		18 01 49	−24 18.1	3	11.30	.018	0.31	.037	0.12	.075	5	A3	720,917,1161
		NGC 6530 - 158		18 01 49	−24 19.7	1	11.51		0.28		-0.40		2		720
165358	+48 02627	HR 6753	⋆ A	18 01 50	+48 27.7	1	6.23		0.04		0.07		2	A2 V	3024
165358	+48 02627	IDS18005N4828	B	18 01 50	+48 27.7	1	8.91		1.07		0.94		2		3024
164975	−29 14447	HR 6742, W Sgr	⋆ A	18 01 50	−29 35.0	2	4.30	.007	0.54	.007	0.35		2	G0 Ib/II	1754,6007
		NGC 6528 sq1 3		18 01 50	−30 04.0	1	12.53		0.58		0.12		4		791
165049	−15 04803	LSS 4633		18 01 51	−15 22.0	1	8.18		0.37		-0.50		2	B2 Ib/II	1012
315034	−24 13851	NGC 6530 - 137		18 01 52	−24 24.3	2	11.66	.065	0.27	.010	-0.41	.045	4	B5	720,917
165050	−15 04805		V	18 01 54	−15 42.7	1	9.04		1.89				2	M2/3	369
	−20 04978			18 01 54	−20 55.8	1	10.33		1.95				2	M2	369
165016	−24 13853	LSS 4631		18 01 54	−24 41.1	6	7.33	.003	-0.04	.007	-0.87	.011	14	B2 Ib	240,285,401,540,914,976
164725	−60 06997			18 01 54	−60 32.3	1	7.80		1.31		1.33		7	K1 III	1704
315035	−24 13854	NGC 6530 - 117		18 01 55	−24 27.6	3	10.80	.014	0.25	.021	-0.42	.038	6	B2.5Ve	720,917,1161
	−24 13858	NGC 6530 - 141		18 01 58	−24 17.3	2	11.58	.030	0.38	.030	-0.13	.105	4		720,917
		NGC 6530 - 140		18 01 58	−24 18.9	1	13.05		0.36		0.44		2		917
165281	+30 03113			18 01 59	+30 22.7	1	6.78		0.51		-0.04		1	F5 V	1620
		NGC 6530 - 139		18 02 00	−24 25.4	2	12.24	.005	1.41	.065	1.10	.405	4		720,917
		Smethells 10		18 02 00	−48 30.	1	11.31		1.36				1		1494
	−24 13862	NGC 6530 - 142		18 02 04	−24 15.8	2	11.77	.065	0.21	.010	-0.41	.070	4		720,917
165019	−31 15068	LSS 4632		18 02 05	−31 03.4	1	9.09		1.11		1.04		6	F7 Ia/ab	954
165174	+1 03578	Cr 359 - 11	⋆ V	18 02 06	+01 54.9	5	6.14	.011	-0.01	.009	-0.93	.023	13	B0 IIIn	15,154,1221,1256,1639
165052	−24 13864	NGC 6530 - 118		18 02 06	−24 24.2	8	6.87	.008	0.10	.009	-0.83	.010	19	O6.5V	285,540,914,917,976,1161*
		NGC 6530 - 330		18 02 06	−24 32.4	2	12.51	.010	0.69	.015	0.24	.010	4		720,917
165054	−28 14114	IDS17589S2841	AB	18 02 06	−28 41.4	1	8.47		0.67				4	G3/5 V	2012
164806	−58 07333			18 02 07	−58 34.6	1	6.83		-0.10		-0.53		1	B3 III	55
		NGC 6530 - 331		18 02 10	−24 06.3	2	13.02	.005	0.41	.025	0.29	.010	4		720,917
165195	+3 03579			18 02 11	+03 46.6	2	7.30	.000	1.29	.000	0.94	.000	2	K3p	979,8112
		LSS 4636		18 02 13	−26 46.2	1	11.07		0.25		-0.58		3	B3 Ib	954
		WLS 1800-5 # 5		18 02 15	−02 31.4	1	11.43		0.80		0.57		2		1375
164955	−47 12047			18 02 17	−47 41.6	1	7.23		1.37		1.53		25	K2/3 III	978
		NGC 6530 - 102		18 02 18	−24 26.7	1	13.72		0.79		0.16		1		1161
165196	−0 03406			18 02 19	−00 46.7	1	10.54		0.38		0.25		14	A0	1699
		NGC 6530 - 332		18 02 19	−24 20.4	2	13.16	.060	0.68	.005	0.08	.180	4		720,917
165084	−30 15204			18 02 19	−30 50.6	1	8.63		1.01				1	G8 IV	915
	−34 12448	LSS 4634		18 02 19	−34 9.8	1	11.55		0.46		-0.31		2	B9 Ia+e2	1737
165063	−36 12198	IDS17590S3635	ABC	18 02 21	−36 35.1	3	7.46	.003	-0.07	.006	-0.48	.002	4	B3/5ne	540,976,1586
165132	−23 13881	LSS 4639		18 02 26	−23 43.2	6	8.09	.007	0.07	.004	-0.81	.012	17	B5/6 Ib	401,432,540,976,1586,1732
		NGC 6530 - 333		18 02 26	−24 11.2	1	13.39		0.77		-0.03		2		917
165133	−24 13869	LSS 4638		18 02 26	−24 12.4	1	9.35		0.07		-0.66		2	B2 III/IV	1586
165222	−3 04233	G 20 - 22		18 02 28	−03 01.9	8	9.38	.019	1.51	.016	1.21	.012	20	M2 V	158,202,1006,1017,1197,1658*
165179	−17 05011			18 02 31	−17 17.7	1	9.51		0.38		-0.19		2	B7 II	1375
	+35 03145	G 182 - 37		18 02 32	+35 57.3	1	10.83		1.48				1	M2	1746
		AS 270		18 02 35	−20 20.9	1	13.71		1.99		0.01		1	M1	1753
165119	−30 15214			18 02 35	−30 38.7	1	9.17		1.30				1	K0	915
165023	−47 12051			18 02 35	−47 22.1	1	8.27		0.51				4	F5 V	2012
165373	+23 03254	HR 6754		18 02 36	+23 56.3	3	6.36	.014	0.30	.006	0.02	.004	5	F0 IV-V	253,254,8083

HD	DM	Other Id	N	Rem	α_{1950}	δ_{1950}	S	V	σ_V	B–V	σ_{B-V}	U–B	σ_{U-B}	n	Spectrum	References
		NGC 6530 - 334			18 02 36	−24 08.2	2	13.33	.020	0.63	.020	0.10	.010	4		720,917
315057	−25 12696	LSS 4640			18 02 36	−25 19.8	1	11.04		0.38				4	B1 V	1021
165135	−30 15215	HR 6746			18 02 36	−30 25.6	7	2.99	.006	1.00	.010	0.77	.018	20	K0 III	15,465,1034,1075,2012*
321320	−36 12201				18 02 36	−36 35.9	3	10.23	.009	0.58	.021	-0.17	.032	4	F5	1236,1696,3077
		NGC 6530 - 335			18 02 39	−24 29.5	2	12.64	.055	0.43	.000	0.16	.045	4		720,917
165153	−25 12698				18 02 40	−25 14.3	1	6.87		0.76		0.42		1	B8 V	1770
165154	−29 14473				18 02 41	−29 13.6	1	9.22		1.12				1	K0 (III)	915
165204	−23 13889				18 02 44	−23 31.2	1	9.07		0.77		0.31		2	G6/8 V	1064
165205	−26 12789	LSS 4642			18 02 44	−26 34.0	1	9.29		0.18		-0.53		4	B3 Ib/II	954
165024	−50 11720	HR 6743			18 02 44	−50 05.8	7	3.67	.008	-0.09	.009	-0.86	.014	26	B2 Ib	15,26,1034,1075,1075*
	−24 13874	LSS 4643, HRC 281		V	18 02 46	−24 15.6	2	11.15	.025	0.86	.011	-0.25	.027	3		776,1586
		NGC 6530 - 336			18 02 48	−24 34.0	2	11.95	.035	1.35	.020	1.21	.160	4		720,917
165181	−31 15088				18 02 49	−31 29.7	2	8.77	.007	0.01	.006	-0.52	.016	9	B8	240,1732
165223	−22 04555	IDS17598S2213		A	18 02 50	−22 12.6	2	8.87	.074	0.06	.008	-0.55	.053	7	B3 V	401,1732
165225	−26 12792	IDS17596S2659		AB	18 02 50	−26 58.8	2	8.39	.000	0.01	.010	-0.64	.006	4	B2 II/III	540,976
		NGC 6530 - 337			18 02 51	−24 15.3	2	9.90	.015	2.31	.015	2.15	.015	4		720,917
		H2-38			18 02 51	−28 17.4	1	13.76		1.06		-0.26		1		1753
		NGC 6530 - 338			18 02 52	−24 14.2	2	13.38	.055	1.35	.045	0.78	.045	4		720,917
165207	−29 14481				18 02 52	−29 26.2	3	8.25	.004	-0.11	.005	-0.75	.004	10	B2 III	540,976,1732
		NGC 6530 - 339			18 02 54	−24 26.0	2	11.71	.015	0.81	.055	0.46	.030	4		720,917
164871	−64 03796	HR 6740			18 02 54	−64 33.3	3	6.40	.004	1.27	.005	1.41	.000	11	K2/3 III	15,1075,2038
165341	+2 03482	HR 6752		⋆A	18 02 56	+02 30.6	1	4.25		0.78		0.57		1	K0 V	150
165341	+2 03482	IDS17576S0811		AB	18 02 56	+02 30.6	10	4.02	.013	0.86	.010	0.51	.016	43	K0 V +K4 V	15,30,150,938,1013*
165435	+22 03267				18 02 56	+22 54.5	1	7.36		0.43		0.22		2	F3 II	1733
	−58 06970	RS Pav			18 02 57	−58 58.2	1	10.16		0.82				1	Ke	688
165566	+42 02996				18 02 58	+42 51.4	1	7.09		1.17		1.16		2	K0	1601
165266	−19 04835				18 02 58	−19 42.9	1	8.71		0.29		-0.17		3	B9 Ib/II	240
		NGC 6530 - 341			18 02 58	−24 13.9	2	11.96	.025	0.82	.035	0.11	.075	4		720,917
		NGC 6530 - 340			18 02 58	−24 15.0	2	13.11	.015	0.78	.055	0.17	.035	4		720,917
165285	−19 04836	LSS 4647			18 03 01	−19 57.6	6	8.84	.111	0.35	.037	-0.68	.029	11	B1 (I)nn(e)	445,540,976,1012,1586,1737
165246	−24 13880	LSS 4646		⋆AB	18 03 01	−24 12.1	5	7.72	.006	0.09	.005	-0.81	.010	12	O8/9	401,540,976,1732,5012
165185	−36 12214	HR 6748			18 03 01	−36 01.5	3	5.94	.005	0.62	.004	0.07		8	G3 V	15,657,2012
165473	+29 03180	V831 Her			18 03 02	+29 04.6	1	7.06		1.05		0.91		3	K0 II	8100
165504	+33 03019	IDS18012N3316		AB	18 03 02	+33 16.2	1	7.70		0.60		0.10		3	G0	3026
165700	+55 02019				18 03 03	+55 16.2	1	7.78		0.45		-0.10		2	F8	3016
165287	−22 04557	IDS18001S2207		AB	18 03 05	−22 07.3	2	8.93	.106	0.05	.025	-0.58	.037	8	B3/5 V	401,1732
165567	+40 03276	HR 6764			18 03 06	+40 04.7	2	6.54	.005	0.46	.019	0.02	.024	3	F7 V	254,272
		NGC 6530 - 342			18 03 07	−24 24.2	1	12.18		0.53		-0.02		2		917
	−11 04531				18 03 08	−11 42.8	1	9.35		2.05				2	M2	369
165319	−14 04880	LSS 4653			18 03 08	−14 12.2	3	7.92	.009	0.58	.014	-0.50	.025	7	O9.5Iab	1012,1737,8100
		NGC 6530 - 344			18 03 08	−24 19.7	1	12.77		0.65		0.29		2		917
		NGC 6530 - 345			18 03 08	−24 20.7	2	13.47	.000	0.70	.075	0.22	.165	4		720,917
		NGC 6530 - 343			18 03 08	−24 31.4	2	10.79	.035	0.82	.035	0.16	.025	4		720,917
165401	+4 03589				18 03 09	+04 39.4	4	6.81	.006	0.61	.009	0.03	.006	10	G0 V	908,1003,1509,3077
165141	−48 12280	V832 Ara			18 03 12	−48 15.2	1	7.08		1.00		0.61		1	G8 II/IIIp	1641
165189	−43 12272	HR 6749		⋆AB	18 03 13	−43 25.7	4	4.93	.025	0.23	.006	0.16		11	A5 V	15,938,2012,2024
		NGC 6530 - 347			18 03 14	−24 16.3	1	13.26		0.81		0.22		2		917
		NGC 6530 - 346			18 03 14	−24 23.0	2	13.14	.010	0.59	.030	0.18	.135	4		720,917
		G 259 - 22			18 03 16	+80 04.4	1	12.91		1.57				3		940
		NGC 6530 - 348			18 03 16	−24 10.7	2	10.27	.005	2.38	.065	2.43	.015	4		720,917
		NGC 6530 - 349			18 03 19	−24 28.5	1	11.84		1.12		0.93		2		917
313792	−21 04849				18 03 22	−21 40.2	1	9.93		0.08		-0.59		2	B5	401
		NGC 6530 - 350			18 03 22	−24 21.0	1	12.77		0.58		0.13		2		917
165474	+12 03382	IDS18011N1200		B	18 03 23	+11 59.9	1	7.45		0.31		0.17		2	A8p	3032
165524	+21 03300	HR 6763			18 03 23	+21 38.5	1	6.15		1.23				2	gK3	71
165475	+12 03383	HR 6758		⋆A	18 03 24	+11 59.9	1	7.04		0.31		0.12		2	A7 p Cr	3032
165402	−8 04558	HR 6755			18 03 24	−08 19.8	2	5.84		0.21	.005	-0.35	.207	10	B8 III-IV	1079,1088
165508	+13 03514				18 03 27	+13 28.7	1	8.18		1.19				2	K0	882
165645	+41 02968	HR 6767		⋆A	18 03 27	+41 56.4	1	6.36		0.25		0.02		1	F0 V	254
315130	−28 13142	LSS 4161			18 03 27	−23 52.9	2	10.10	.015	0.80	.010	-0.16	.055	5	B0.5 Ib	954,1737
		NGC 6530 - 351			18 03 27	−24 34.2	2	12.15	.130	0.42	.040	0.27	.020	4		720,917
		NGC 6541 sq1 2			18 03 27	−43 42.0	1	11.51		1.62		1.60		6		763
		LSS 4657			18 03 28	−13 43.6	2	11.85	.000	1.18	.000	0.07	.000	4	B3 Ib	405,1737
	−23 13906	AJ87,98 # 6			18 03 28	−23 57.8	1	10.23		0.40		-0.49		1	B0.5Vn	5012
		NGC 6541 sq1 3			18 03 30	−43 44.1	1	12.10		0.56		0.10		5		763
	+4 03593	IDS18001N0433		A	18 03 33	+04 34.6	1	8.33		1.42		1.50		2	K0 III	1732
165462	−0 03414	HR 6757			18 03 33	−00 27.1	2	6.32	.005	1.07	.015	0.79	.044	13	G8 IIp	379,1088
165438	−4 04395	HR 6756			18 03 35	−04 45.4	3	5.77	.009	0.97	.007	0.80	.010	11	K1 IV	1088,3016,4001
		NGC 6530 - 352			18 03 35	−24 38.3	2	12.15	.050	0.78	.035	0.26	.080	4		720,917
165477	+4 03594	IDS18011N0433		A	18 03 36	+04 33.8	2	8.54	.005	-0.02	.001	-0.46	.005	5	A0	963,1732
		NGC 6530 - 353			18 03 38	−24 34.8	2	12.85	.005	0.48	.045	0.14	.140	4		720,917
165069	−62 05792				18 03 38	−62 22.6	1	8.06		0.45		-0.08		1	F5 V	3037
		Ka 102 # 1			18 03 39	+67 35.6	1	12.77		0.68		0.28		1		829
312603	−17 05020	LSS 4659		⋆A	18 03 39	−17 12.4	2	9.32	.024	0.57	.033	-0.46	.019	4	B0 III	432,1012
165271	−46 12117				18 03 41	−46 54.2	2	7.65	.005	0.66	.005	0.19		5	G5 V	219,1075
165590	+21 03302	V772 Her		⋆AB	18 03 42	+21 26.4	1	7.07		0.66		0.13		2	G0 V +G8 V	3030
165590	+21 03302	IDS18016N2126		C	18 03 42	+21 26.4	1	10.62		1.36		1.13		1		3030
		Ka 102 # 3			18 03 46	+67 39.6	1	14.19		1.18		1.09		1		829
		WLS 1800-5 # 9			18 03 46	−04 45.7	1	13.08		1.14		0.84		2		1375
165040	−63 04292	HR 6745			18 03 46	−63 40.4	8	4.34	.004	0.23	.005	0.17	.007	25	A7p Sr	15,285,611,688,1075*

Table 1   985

HD	DM	Other Id	N Rem	α_{1950}	δ_{1950}	S	V	σ_V	B−V	σ_{B-V}	U−B	σ_{U-B}	n	Spectrum	References
166865	+79 00570	HR 6809	⋆ B	18 03 47	+79 59.8	3	6.04	.000	0.51	.000	-0.01	.000	6	F7	1,15,1028
		NGC 6541 sq1 4		18 03 47	−43 41.4	1	12.32		0.53		0.02		4		763
		NGC 6546 - 70	A	18 03 48	−23 18.1	1	12.62		0.44		0.06		3		748
		NGC 6546 - 70	B	18 03 48	−23 18.1	1	13.88		0.48		0.50		1		748
165423	−22 04562			18 03 49	−22 03.2	1	7.72		1.79				3	M1/2 III	369
165424	−26 12812			18 03 49	−26 07.0	1	8.97		0.05		-0.69		6	B3/5 II	1732
		G 20 - 23		18 03 50	+02 03.0	2	11.21	.034	0.65	.000	0.08		3	G2	202,333,1620
		Ka 102 # 4		18 03 50	+67 36.8	1	14.74		0.67		0.27		1		829
165446	−21 04853	LSS 4660		18 03 51	−21 30.7	1	7.96		0.65		0.37		4	F3 II	8100
166866	+79 00571	HR 6810	⋆ A	18 03 53	+80 00.0	3	5.68	.000	0.50	.000	-0.01	.000	6	F7	1,15,1028
	−7 04560			18 03 54	−07 37.0	1	10.79		2.47				2	M3	369
165625	+22 03273	HR 6765		18 03 55	+22 12.8	1	5.06		1.58		2.10		2	M3 III	3055
335623	+25 03427			18 03 57	+25 40.6	1	8.47		0.47		0.01		4	F6 V	3026
165683	+32 03047	HR 6768		18 03 58	+32 13.5	1	5.71		1.16		1.08		2	K0 III	252
		NGC 6546 - 103		18 03 59	−23 21.6	1	11.76		0.35		0.09		2	A5 V	748
		He3 1587		18 04 00	−23 35.	1	11.00		0.85		0.52		2		1586
238812	+59 01875			18 04 01	+59 39.4	1	9.72		0.21		0.08		2	A3	1502
		LSS 4665		18 04 02	−14 57.1	2	10.85	.000	1.11	.000	-0.04	.000	4	B3 Ib	405,1737
		NGC 6546 - 73	A	18 04 02	−23 18.5	1	12.23		1.57		1.29		3		748
		NGC 6546 - 73	B	18 04 02	−23 18.5	1	13.00		0.63		0.28		2		748
		NGC 6546 - 102		18 04 02	−23 20.2	1	11.54		0.30		0.17		2	A0 V	748
		Ka 102 # 2		18 04 03	+67 37.1	1	13.40		0.47		0.12		1		829
	−23 13919	NGC 6546 - 58		18 04 03	−23 14.7	1	11.56		0.74		0.54		3	F0	748
		NGC 6546 - 228		18 04 04	−23 19.1	1	12.65		0.62		0.43		1		748
165626	+15 03358			18 04 05	+15 34.3	1	8.29		0.64				4	K0	882
		NGC 6544 sq1 2		18 04 05	−25 08.1	1	11.27		0.27		0.35		4		984
	−16 04708			18 04 06	−16 53.3	1	9.81		1.94				2	M2	369
		NGC 6546 - 211		18 04 06	−23 17.8	1	12.93		0.49		0.16		1		748
	−23 13922	NGC 6546 - 55		18 04 07	−23 12.6	1	11.14		0.61		-0.25		3	B3	748
		NGC 6546 - 214		18 04 07	−23 16.3	1	12.59		0.36		0.27		1		748
		NGC 6546 - 234		18 04 08	−23 15.9	1	14.77		0.66		-0.04		1		748
		NGC 6546 - 64		18 04 09	−23 16.5	1	11.78		0.29		-0.37		3		748
		NGC 6546 - 230		18 04 09	−23 17.5	1	14.15		2.12		1.96		1		748
		NGC 6546 - 100		18 04 09	−23 20.2	1	10.99		0.46		0.22		3	B8 V	748
315178	−25 12733	LSS 4663		18 04 09	−25 17.2	1	10.61		0.20		0.11		2		1586
165758	+38 03075			18 04 10	+38 59.5	1	8.12		0.30		0.02		2	A8 V	105
165516	−21 04855	HR 6762, LSS 4666		18 04 11	−21 27.0	7	6.25	.015	0.11	.009	-0.78	.017	19	B0.5 Ib	15,26,240,1012,1737*
		NGC 6546 - 221		18 04 11	−23 15.0	1	12.66		0.35		0.23		1		748
		NGC 6546 - 220		18 04 11	−23 15.8	1	13.26		0.59		0.25		1		748
		NGC 6546 - 231		18 04 11	−23 16.0	1	14.32		0.83		0.68		1		748
	−23 13923	NGC 6546 - 99		18 04 11	−23 20.5	1	10.73		1.19		0.86		1	G8	748
		NGC 6546 - 225		18 04 12	−23 20.	1	14.12		1.42		1.50		1		748
		NGC 6544 sq1 3		18 04 12	−25 12.8	1	11.42		1.11		0.93		4		984
		NGC 6546 - 223		18 04 13	−23 14.4	1	14.11		0.72		0.22		1		748
		NGC 6546 - 222		18 04 14	−23 15.8	1	13.49		0.47		0.28		1		748
		NGC 6546 - 229		18 04 14	−23 17.8	1	14.04		1.67		1.91		1		748
		NGC 6546 - 75		18 04 14	−23 18.9	1	11.09		0.38		-0.28		2	B5 V	748
	−19 04845			18 04 15	−19 50.3	1	10.01		1.99				2	M2	369
		NGC 6546 - 224		18 04 15	−23 14.5	1	13.89		0.70		0.66		1		748
		NGC 6546 - 233		18 04 15	−23 16.7	1	15.13		1.50		1.60		1		748
166091	+63 01404			18 04 16	+63 47.5	1	7.48		1.32		1.63		3	K5 II-III	8100
		NGC 6546 - 235		18 04 16	−23 15.4	1	14.44		0.58		0.21		1		748
		NGC 6546 - 210		18 04 16	−23 18.9	1	12.64		1.45		1.60		1		748
		NGC 6546 - 237		18 04 16	−23 19.5	1	13.56		1.43		1.93		1		748
		NGC 6546 - 218		18 04 16	−23 20.0	1	13.90		0.56		0.35		1		748
		NGC 6546 - 215		18 04 16	−23 20.2	1	13.91		0.78		0.36		1		748
		NGC 6546 - 121		18 04 16	−23 24.0	1	11.44		0.43		0.14		2	A0 V	748
164712	−75 01410	HR 6731	⋆ A	18 04 16	−75 53.8	4	5.85	.005	1.25	.010	1.42	.014	14	K2 III	15,1075,2038,3005
		NGC 6546 - 78		18 04 17	−23 17.8	1	12.60		0.42		0.10		3		748
		NGC 6546 - 219		18 04 18	−23 15.7	1	12.33		1.24		1.11		1		748
	−23 13929	NGC 6546 - 61		18 04 18	−23 16.9	1	11.03		0.42		-0.21		3	B5	748
		NGC 6546 - 212		18 04 18	−23 18.8	1	13.13		0.55		0.41		1		748
		NGC 6546 - 241		18 04 20	−23 13.7	1	13.65		0.65		0.31		1		748
165553	−21 04858			18 04 21	−21 12.5	1	7.69		1.03		0.68		4	F8 Ib/II	8100
		NGC 6546 - 217		18 04 21	−23 19.9	1	12.88		0.48		0.42		1		748
		NGC 6546 - 97	A	18 04 22	−23 19.8	1	12.59		1.82		2.04		2		748
		NGC 6546 - 97	B	18 04 22	−23 19.8	1	12.97		1.75		1.96		2		748
165517	−25 12744	LSS 4667		18 04 22	−25 06.7	4	8.44	.007	0.47	.010	-0.54	.017	7	B0 Ia(N)	540,976,1586,1737
165393	−47 12070			18 04 22	−47 57.2	1	8.29		0.17		0.17		26	A3 IV/V	978
347906		DE Her		18 04 23	+20 52.4	1	10.33		1.58		1.24		1	K0	793
		NGC 6546 - 236		18 04 23	−23 16.6	1	14.37		1.60		0.94		1		748
		NGC 6546 - 232		18 04 23	−23 17.2	1	15.30		0.51		0.39		1		748
	−13 04854			18 04 24	−13 30.4	1	10.30		2.08				2	M2	369
	−25 12745	NGC 6544 sq1 19		18 04 24	−25 08.7	1	10.70		1.70				1		984
		NGC 6541 - 1		18 04 24	−43 44.	1	15.67		0.02		0.18		2		110
		NGC 6541 - 2		18 04 24	−43 44.	2	14.07	.028	0.65	.033	0.12		8		110,763
		NGC 6541 - 5		18 04 24	−43 44.	1	13.20		0.97		0.50		2		110
		NGC 6541 - 13		18 04 24	−43 44.	2	12.60	.015	0.45	.005	-0.03	.005	6		110,763
		NGC 6541 - 14		18 04 24	−43 44.	2	14.08	.014	0.94	.014			7		110,763
		NGC 6541 - 15		18 04 24	−43 44.	2	15.11	.034	0.88	.000			7		110,763

HD	DM	Other Id	N Rem	α_{1950}	δ_{1950}	S	V	σ_V	B-V	σ_{B-V}	U-B	σ_{U-B}	n	Spectrum	References
		NGC 6541 - 16		18 04 24	-43 44.	2	11.28	.005	0.60	.005	0.10	.005	6		110,763
		NGC 6541 - 20		18 04 24	-43 44.	1	11.41		1.51		1.50		2		110
		NGC 6541 - 49		18 04 24	-43 44.	1	10.49		0.64		0.10		2		110
		NGC 6541 - 52		18 04 24	-43 44.	1	12.55		1.30		0.79		2		110
		NGC 6541 - 118		18 04 24	-43 44.	1	15.66		0.24				1		110
		NGC 6541 - 119		18 04 24	-43 44.	2	13.82	.015	1.10	.000	0.70		3		110,763
		NGC 6541 - 121		18 04 24	-43 44.	2	12.40	.005	0.79	.009	0.26		4		110,763
		NGC 6541 - 150		18 04 24	-43 44.	1	11.30		0.01				1		110
		NGC 6541 - 245		18 04 24	-43 44.	1	13.37		0.99		0.35		2		110
165554	-23 13932	NGC 6541 - 96		18 04 25	-23 20.3	1	9.14		0.54		0.11		2	F2 V	748
		NGC 6546 - 239		18 04 26	-23 18.3	1	13.39		0.58		0.50		1		748
		NGC 6546 - 238		18 04 26	-23 18.7	1	14.15		1.14		1.01		1		748
		He3 1591		18 04 26	-25 54.2	1	12.96		1.10		0.46		1		1753
165470	-38 12492			18 04 27	-38 34.4	2	7.32	.015	-0.14	.021	-0.66	.002	4	B2 III	540,976
		G 140 - 39		18 04 28	+10 52.2	2	12.50	.023	0.96	.000	0.61	.010	4		203,1658
		NGC 6546 - 240		18 04 28	-23 16.2	1	14.14		1.11		0.46		1		748
313810	-22 04567			18 04 30	-22 06.3	3	9.46	.069	0.07	.004	-0.53	.015	5	B3 IV	401,540,976
		NGC 6546 - 54	A	18 04 30	-23 14.2	1	12.22		0.68		0.15		2		748
		NGC 6546 - 54	B	18 04 30	-23 14.2	1	12.33		1.50		1.60		1		748
		NGC 6546 - 216		18 04 30	-23 21.6	1	13.01		0.72		0.23		1		748
313869	-23 13934	NGC 6546 - 60		18 04 31	-23 16.7	1	9.35		1.81		1.92		3	M2 III	748
		NGC 6546 - 92		18 04 33	-23 20.8	1	11.46		1.88		2.05		2		748
		NGC 6546 - 80		18 04 34	-23 19.1	1	12.62		0.53		0.27		1		748
		NGC 6546 - 93		18 04 34	-23 21.8	1	12.44		0.48		0.26		2		748
	-19 04849			18 04 35	-19 18.3	1	9.70		2.45				1		8051
		NGC 6546 - 91		18 04 38	-23 22.4	1	10.62		0.51		-0.37		3	B3	748
		NGC 6546 - 90		18 04 41	-23 22.9	1	12.22		0.52		0.31		2	A8 V	748
165611	-21 04861		V	18 04 42	-21 52.1	2	8.11	.047	1.76	.027	1.97		6	K4/5 (III)	369,1657
165596	-23 13940	NGC 6546 - 89		18 04 42	-23 21.0	1	8.88		1.55		1.64		2	K3 II-III	748
	-19 04850		V	18 04 44	-19 48.5	1	10.23		2.11				2	M2	369
		NGC 6544 sq1 4		18 04 48	-24 52.2	1	11.57		0.13		0.19		4		984
165582	-34 12498	LSS 4668		18 04 48	-34 35.0	3	9.39	.028	0.00	.000	-0.81	.016	5	B1 Ib	540,954,976
165493	-45 12215	HR 6759	★AB	18 04 48	-45 46.5	3	6.14	.007	-0.08	.000	-0.48		9	B7/8 II	15,404,2012
165687	-17 05028	HR 6769		18 04 54	-17 09.8	4	5.53	.014	1.11	.007	1.09	.009	9	K0 III	15,1003,2007,3005
165760	+8 03582	HR 6770		18 04 55	+08 43.6	9	4.64	.006	0.96	.006	0.74	.007	44	G8 III-IV	3,15,1080,1256,1363*
165634	-28 14174	HR 6766		18 04 55	-28 27.9	8	4.56	.007	0.95	.007	0.74	.013	20	G8 IIIp	15,536,614,1075,1075*
165803	+15 03362			18 04 57	+15 41.1	1	8.55		0.63				3	G5	882
		NGC 6544 sq1 20		18 04 58	-25 04.2	1	12.38		2.09				1		984
		NGC 6544 sq1 1		18 04 58	-25 12.2	1	11.24		0.45		-0.07		4		984
165777	+9 03564	HR 6771	★AB	18 04 59	+09 33.3	5	3.73	.004	0.12	.009	0.10	.013	15	A4 IVs	15,1256,1363,3023,8015
165703	-14 04894			18 05 00	-14 17.9	1	8.55		1.81				2	K5/M0 (III)	369
165688	-19 04854	LSS 4678		18 05 00	-19 24.4	1	9.90		0.43		0.15		3	WC	1359
165655	-25 12769	LSS 4675		18 05 02	-25 21.4	2	8.16	.007	0.31	.001	-0.62	.007	4	B1 (Ia/ab)	540,976
165674	-22 04575	VX Sgr		18 05 03	-22 13.8	2	7.66	.644	2.90	.009	2.31		4	M5/6 (III)	148,8032
	+15 03364	IDS18029N1556	A	18 05 04	+15 56.5	2	8.69	.014	0.65	.005	0.15	.019	4	G0	196,979
165848	+15 03365			18 05 06	+15 54.7	3	6.69	.019	1.12	.005	0.98	.023	6	K0	1355,1733,3040
165908	+30 03128	HR 6775	★AB	18 05 07	+30 33.3	10	5.05	.012	0.52	.009	-0.09	.014	39	F7 V	1,15,254,938,1077*
165908	+30 03128	IDS18032N3033	C	18 05 07	+30 33.3	1	8.45		1.10		1.00		1		3030
	+15 03367			18 05 10	+15 56.9	1	8.21		1.28		1.39		3	K2	196
		G 206 - 8		18 05 11	+29 21.5	1	12.26		0.79		0.34		3		1658
315181	-25 12773	LSS 4679		18 05 13	-25 29.0	1	10.73		0.24				4	B5 V:	1021
166379	+66 01077			18 05 14	+66 56.3	1	6.87		0.68		0.42		4	F5	985
165705	-23 13954	IDS18022S2327	AB	18 05 17	-23 26.4	1	9.59		0.16		-0.56		1	B2 III	5012
165617	-42 12921			18 05 18	-42 39.9	2	7.10	.004	-0.11	.005	-0.50	.007	3	B7 V	540,976
165722	-20 05002			18 05 19	-20 51.3	1	9.20		1.80				2	M1/2	369
		LSS 4683		18 05 20	-21 37.4	1	11.07		1.19		0.22		2	B3 Ib	1737
166012	+36 03019			18 05 21	+36 24.8	1	8.06		0.19		0.07		2	A2	1733
	-13 04858			18 05 21	-13 44.8	1	10.48		2.13				2	M3	369
166356	+65 01241			18 05 22	+65 04.4	1	7.55		0.76		0.37		2	K0	3026
		G 20 - 24		18 05 26	+01 52.2	4	11.09	.015	0.46	.008	-0.14	.005	7	M0	202,927,1620,3062
324924	-37 12227	WX CrA		18 05 27	-37 20.1	1	10.49		1.28		0.78		1	R5	1589
165783	-19 04858	LSS 4686		18 05 29	-19 52.6	2	8.38	.017	0.29	.004	-0.53	.026	5	B3/5 Ib	1012,1586
165763	-21 04864	LSS 4685		18 05 29	-21 15.7	2	7.70	.057	-0.14	.068	-0.47	.043	12	WC	1359,1732
165910	+13 03529	HR 6776	★A	18 05 30	+13 03.8	2	6.63		0.06	.010	0.08	.010	5	A2 Vn	1022,3024
165910	+13 03529	IDS18032N1303	B	18 05 30	+13 03.8	1	10.46		1.32		1.36		2		3024
165782	-18 04799	AX Sgr		18 05 30	-18 33.6	1	7.68		2.14				2	G5 Ia	369
165830	-10 04616			18 05 31	-10 33.5	1	7.81		0.32		0.19		5	Am	355
165764	-21 04865			18 05 33	-21 38.6	1	4.78		0.38				4	G8 IV	2012
165497	-59 07231	HR 6760	★A	18 05 33	-59 03.0	1	6.38		1.55				4	K4 III	2007
166014	+28 02925	HR 6779, o Her	★	18 05 35	+28 45.3	5	3.84	.009	-0.03	.010	-0.05	.017	14	B9.5V	15,1022,1363,3016,8015
165808	-16 04720			18 05 35	-16 25.5	1	7.65		0.19		-0.39		3	B3 V	399
165711	-37 12231			18 05 36	-37 40.5	1	9.64		0.42		0.12		6	F0/2 V	1589
		AS 276		18 05 37	-41 14.0	1	13.46		1.39		-0.14		1	M4	1753
165784	-21 04866	LSS 4689		18 05 39	-21 27.5	5	6.54	.016	0.86	.005	0.28	.056	14	A2/3 Iab	138,206,369,540,976,8100
		G 20 - 25		18 05 40	-00 15.3	2	13.46	.000	1.48	.029	1.05		3		202,1620
165811	-18 04800			18 05 41	-18 23.5	1	10.26		0.38		-0.18		4	B8/9 Ib	206
165785	-26 12851	LSS 4687		18 05 41	-26 52.5	3	9.44	.005	0.22	.000	-0.65	.011	6	B1 (III)	540,954,976
166207	+50 02525	HR 6790		18 05 42	+50 48.8	2	6.31	.013	1.05	.006	0.96		8	K0 III	71,3016
165812	-22 04581	LSS 4691		18 05 44	-22 10.2	4	7.94	.009	0.01	.004	-0.78	.009	11	B1/2 II	401,540,976,1732
165696	-45 12232			18 05 45	-45 56.7	2	7.35	.005	0.50	.000	0.03		5	F6/7 V	219,1075

Table 1   987

HD	DM	Other Id	N	Rem	α_{1950}	δ_{1950}	S	V	σ_V	B–V	σ_{B-V}	U–B	σ_{U-B}	n	Spectrum	References
165499	−62 05797	HR 6761			18 05 47	−62 00.9	5	5.48	.009	0.59	.005	0.06	.010	18	G1 V	15,158,2018,2028,3037
166046	+26 03178	HR 6782		★ B	18 05 48	+26 05.3	2	5.90	.000	0.14	.000	0.08	.000	4	A3 V	15,1028
166045	+26 03178	HR 6781		★ A	18 05 49	+26 05.6	2	5.86	.000	0.12	.000	0.08	.000	4	A3 V	15,1028
165814	−25 12793	HR 6773, V3792 Sgr		★ AB	18 05 49	−25 28.7	3	6.59	.021	0.01	.009	−0.48	.000	11	B3/5 IV	401,588,2007
166228	+49 02732	HR 6792		★ AB	18 05 51	+49 42.1	1	6.50		0.04		0.01		2	A2 V	1733
165852	−17 05034				18 05 54	−17 58.9	1	8.99		1.34		1.41		5	K1/2 III	897
165914	−6 04717				18 05 56	−06 40.3	1	8.32		0.27		0.17		2	A0	1375
165856	−21 04869				18 05 56	−21 24.7	1	8.23		1.87				3	K5 III	369
166093	+29 03186				18 05 57	+29 48.6	1	7.21		1.35		1.61		3	K3 II	8100
165870	−17 05035				18 05 57	−17 31.0	1	9.27		0.19		−0.23		3	B7 III	1586
166208	+43 02892	HR 6791			18 05 58	+43 27.3	5	5.00	.013	0.91	.005	0.68	.007	34	G8 IIIp:	1080,1355,1363,3016,4001
313875					18 05 58	−23 25.0	1	10.31		0.16		−0.54		5		5012
165793	−36 12265	HR 6772, LSS 4688			18 05 59	−36 40.9	10	6.58	.015	−0.04	.012	−0.85	.023	21	B1/2 Ib	26,138,219,240,400*
		3C 371 # 10			18 06 01	+69 47.0	1	13.52		0.67		0.61		2		551
	+32 03052	G 206 - 9			18 06 02	+32 06.9	1	9.58		0.73		0.41		2	K0	7010
313939			V		18 06 04	−21 53.1	1	9.50		1.87				2		369
165892	−21 04871	LSS 4699			18 06 05	−21 25.0	2	9.07	.014	0.41	.002	−0.43	.016	3	B2 II	540,976
		DQ Her			18 06 06	+45 51.0	1	14.72		0.04		−0.78		21	sd:Be	866
165872	−23 13974	IDS18031S2327		A	18 06 06	−23 26.7	1	9.85		0.16		−0.49		1	B3 II/III	5012
165872	−23 13974	IDS18031S2327		AB	18 06 06	−23 26.7	3	9.65	.008	0.19	.031	−0.46	.009	8	B3 II/III	206,963,1732
165872	−23 13974	IDS18031S2327		B	18 06 06	−23 26.7	1	11.76		0.17		−0.32		1		5012
165753	−44 12373				18 06 08	−44 57.3	2	7.03	.005	1.09	.010	0.94		5	K0 III	219,1075
165896	−26 12862	IDS18031S2607		ABC	18 06 09	−26 07.3	1	7.59		0.68		0.07		4	G3 V	1311
165893	−22 04584	IDS18032S2212		AB	18 06 11	−22 11.5	2	8.88	.040	1.08	.005	0.97		8	K3/4 V	176,214
165839	−36 12272				18 06 12	−36 47.2	1	8.75		1.60		1.86		11	M1/2 III	1589
165895	−24 13960				18 06 13	−24 42.6	1	9.95		0.06		−0.40		3	B5 III/V	1586
165259	−73 01888	HR 6751		★ AB	18 06 13	−73 40.8	3	5.86	.007	0.46	.004	0.04	.010	17	F5 V	15,1075,2038
165921	−24 13962	V3903 Sgr			18 06 14	−23 59.9	5	7.33	.028	0.14	.011	−0.80	.013	15	B5/7 (III)	432,540,976,1732,5012
164694	−80 00843				18 06 14	−80 13.8	1	8.81		1.01		0.65		2	G6/8 III/I	1730
		3C 371 # 9			18 06 16	+69 50.8	1	13.24		0.94		0.87		2		551
166095	+14 03427	HR 6784			18 06 17	+14 16.5	4	6.36	.009	0.18	.012	0.20	.019	11	A5 m	355,1022,3058,8071
166253	+41 02988	V566 Her			18 06 17	+41 42.6	1	7.60		1.66		1.89		2	M1	3025
166229	+36 03027	HR 6793			18 06 18	+36 23.7	4	5.50	.030	1.16	.012	1.21	.002	8	K2 III	15,1003,1355,4001
165970	−19 04866				18 06 18	−19 44.0	1	9.00		0.40		−0.29		3	B5 II	1586
166181	+29 03187	V815 Her			18 06 20	+29 41.0	1	7.66		0.72		0.13		3	G5	3026
166276	+39 03327				18 06 22	+39 54.8	1	7.75		0.27		0.04		2	A2	1601
		LkHα 126	★		18 06 22	−23 28.5	1	14.70		0.50		−1.28		1		5012
166097	+9 03576				18 06 23	+09 26.6	2	9.93	.141	1.55	.033	1.77		2	R4	1238,8005
166029	−9 04656	G 154 - 46			18 06 23	−09 27.0	2	9.90	.025	0.90	.023	0.66	.030	4	G5	1696,3062
165971	−21 04873	LSS 4703			18 06 24	−21 31.1	2	9.01	.012	0.52	.005	−0.33	.011	3	B5 Ib	540,976
165953	−27 12482	LSS 4701			18 06 24	−27 52.7	2	10.08	.005	0.28	.002	−0.62	.010	5	O9	540,954
	−45 12240				18 06 25	−45 52.9	1	9.49		1.58				2		1730
165999	−23 13991				18 06 33	−23 34.7	2	7.64	.005	0.26	.015	0.21	.010	3	A2 IV	432,5012
		LkHα 127	★		18 06 34	−23 26.1	1	11.27		0.40		−0.10		1	Be	5012
313959					18 06 36	−22 21.0	2	9.82	.008	0.68	.010	−0.29	.019	3	B0.5IV	540,976
166182	+20 03674	HR 6787		★ A	18 06 37	+20 48.3	6	4.35	.006	−0.16	.013	−0.82	.005	39	B2 IV	15,154,1363,3016,6001,8015
165955	−34 12535	LSS 4702		★ AB	18 06 37	−34 52.7	4	9.19	.013	−0.05	.007	−0.82	.009	7	B2 II/III	400,540,954,976
166230	+20 03675	HR 6794			18 06 43	+20 02.2	2	5.11	.005	0.16	.011	0.18		3	A8 III	1363,3023
165978	−32 13814	HR 6777			18 06 43	−32 43.7	2	6.42	.010	1.03	.010	0.77		6	K0 IV	2006,3077
166033	−23 13998	LSS 4706			18 06 44	−23 39.3	2	9.60	.363	0.15	.064	−0.70	.044	6		1311,5012
	−12 04935	G 154 - 47			18 06 45	−12 02.4	1	10.47		1.38		1.17		1	K7 V	3072
314032	−23 13999				18 06 45	−23 40.0	2	9.94	.015	0.23	.010	−0.29	.005	3	B8	432,5012
166054	−22 04590	IDS18038S2204		A	18 06 47	−22 03.9	1	9.10		0.07		−0.62		10	B3	1732
314031	−23 14002	LSS 4707			18 06 48	−23 37.5	3	9.71	.170	0.15	.047	−0.71	.050	7	B0.5V	432,3060,5012
166055	−23 14003				18 06 50	−23 03.7	1	10.09		0.12		−0.26		2	B8 Ib/II	1586
		WLS 1800-5 # 8			18 06 51	−04 51.3	1	12.81		1.20		0.69		2		1375
166105	−15 04840		V		18 06 51	−15 17.6	1	7.90		2.24				2	M1/2 III	369
166023	−30 15316	HR 6780		★ AB	18 06 52	−30 44.3	9	5.52	.009	0.98	.012	0.74	.031	26	K1 II+F0 IV	15,404,542,614,1075,1075*
166103	−13 04863	HR 6785			18 06 53	−13 56.7	1	6.39		1.42				4	K1 II	2007
166056	−24 13984	IDS18038S2408		AB	18 06 53	−24 07.4	3	9.77	.073	0.37	.026	−0.57	.000	5	B3/5 Ib	432,725,5012
166056	−24 13984	IDS18038S2408		C	18 06 53	−24 07.4	1	11.66		0.52		0.03		1		5012
166579	+58 01789				18 06 54	+58 57.9	1	8.42		0.56		0.04		2	G5	3016
	−24 13985	LSS 4710			18 06 55	−24 07.4	2	10.85	.005	0.43	.005	−0.43		3		725,5012
165933	−46 12157				18 06 55	−46 01.8	1	8.16		1.41		1.61		2	K2 III	1730
166125	−14 04908	LS IV -14 005			18 06 55	−14 12.3	2	9.06	.013	0.53	.004	−0.30	.004	5	B3 II	1012,8100
166256	+13 03540	IDS18047N1328		AB	18 06 57	+13 28.5	1	8.62		0.00		−0.36		1	A0	963
166161	−8 04566				18 06 57	−08 47.2	2	8.12	.035	0.98	.005	0.40		5	G5	742,1594
166184	−0 03426				18 06 58	−00 20.2	1	8.98		1.02		0.85		2	K3 V	3072
166144	−11 04545				18 06 58	−11 44.1	1	7.33		1.79				2	K2	369
166126	−15 04842	W Ser			18 06 58	−15 33.6	1	9.20		0.81				1	F8/G2 Iaep	6009
166079	−23 14004				18 07 00	−23 38.8	2	8.58	.010	0.07	.000	−0.79	.015	3	B9 (II)	432,5012
314030					18 07 02	−23 39.4	2	9.66	.015	0.13	.005	−0.28	.005	3	B8	432,5012
166410	+39 03330				18 07 03	+39 03.3	1	8.39		0.19		0.08		2	A3 V	833
166107	−23 14005	LSS 4712		★ AB	18 07 04	−23 47.0	2	7.96	.006	0.06	.006	−0.75	.006	8	B2 V	219,432,1732,5012
166233	+3 03610	Cr 359 - 1		★ AB	18 07 05	+03 59.0	3	5.72	.004	0.37	.005	0.02	.007	9	F2 V	15,1256,1639
		3C 371 # 7			18 07 05	+69 44.8	1	14.67		0.59		0.26		2		551
		3C 371 # 8			18 07 05	+69 52.6	1	13.60		0.60		0.44		2		551
		3C 371 # 6			18 07 10	+69 45.7	1	15.11		0.86		0.78		2		551
166382	+30 03137	T Her		★ A	18 07 13	+31 00.7	1	8.55		1.60		1.05		1	M4	817
166167	−21 04879	LSS 4716			18 07 14	−21 20.3	2	8.61	.009	0.56	.000	0.04	.047	6	B9.5Iab/b	1012,8100

HD	DM	Other Id	N Rem	α_{1950}	δ_{1950}	S	V	σ_V	B–V	σ_{B-V}	U–B	σ_{U-B}	n	Spectrum	References
166411	+30 03138	HR 6799		18 07 15	+30 27.5	1	6.38		1.20		1.23		2	K2	3025
314029				18 07 16	−23 42.5	1	10.32		0.11		−0.31		1		5012
166284	+3 03612			18 07 17	+03 11.4	1	7.38		1.23		1.30		2	K2 III	37
166303	+6 03639			18 07 17	+06 11.7	1	7.08		1.03		0.82		2	K0	1733
		3C 371 # 3		18 07 18	+69 48.	1	15.42		0.64		0.65		2		551
		3C 371 # 4		18 07 18	+69 48.	1	16.31		0.67		0.65		2		551
		3C 371 # 5		18 07 18	+69 48.	1	15.92		0.92		0.98		2		551
166006	−47 12098	HR 6778	⋆ AB	18 07 18	−47 31.4	3	6.07	.008	1.20	.007	1.15		8	K1 III	15,2012,4001
166188	−18 04815	LSS 4719		18 07 23	−18 12.3	2	9.16	.214	0.38	.048	−0.33		5	B3 (II)e	1586,2023
166285	+3 03613	HR 6797	⋆ AB	18 07 24	+03 06.7	4	5.68	.006	0.44	.019	−0.04	.005	9	F5 V	1149,1620,1758,3016
166435	+29 03190			18 07 26	+29 56.5	1	6.85		0.62		0.09		3	G0	3026
315259	−24 13994			18 07 26	−24 27.9	2	10.21	.007	0.16	.001	−0.60	.020	3	B1.5V	540,976
166191	−23 14016			18 07 27	−23 34.6	1	8.35		0.55				24	F3/5 V	6011
		Steph 1580	AB	18 07 28	+20 48.5	1	11.41		1.30		1.20		1	K7	1746
	−11 04547	G 154 - 49		18 07 28	−11 47.3	1	10.87		0.58		−0.06		1		3062
		G 182 - 41		18 07 29	+27 54.8	1	12.65		0.93		0.48		1		1658
166192	−23 14017	LSS 4718		18 07 29	−23 55.8	3	8.53	.007	0.20	.003	−0.66	.011	7	B2 II	432,1732,5012
166063	−45 12251	HR 6783	⋆ A	18 07 31	−45 57.9	4	4.53	.004	1.01	.003	0.78	.004	20	K0 III	15,1075,2012,8015
166239	−18 04817			18 07 33	−18 11.8	1	9.63		1.81				2	K5	369
166114	−41 12491	HR 6786		18 07 33	−41 22.2	2	5.85	.005	0.30	.005			7	F2 V	15,2030
		AS 281		18 07 35	−27 58.5	1	13.81		1.16		−0.58		1	M5.5	1753
166197	−33 12917	HR 6788		18 07 37	−33 48.7	3	6.15	.007	−0.15	.009	−0.85		11	B2 II/III	15,285,2012
166286	−16 04736	IDS18048S1647	B	18 07 39	−16 45.6	1	8.62		0.19		−0.62		3	B1 II	1732
		S024 # 3		18 07 39	−80 08.6	1	11.24		0.63		0.02		2		1730
		3C 371 # 1		18 07 40	+69 46.6	1	14.21		0.87		0.85		2		551
166286	−16 04736	LSS 4724	⋆ A	18 07 40	−16 45.6	2	8.24	.009	0.25	.004	−0.61	.023	7	B1 II	1012,1732,8100
166291	−19 04882			18 07 41	−19 10.7	1	9.06		0.20		−0.53		2	B3 II	1012
		3C 371 # 2		18 07 42	+69 45.7	1	14.53		0.60		0.27		2		551
166288	−17 05049			18 07 43	−17 44.5	1	9.97		0.37		−0.19		4	B5 Ib	206
166263	−22 04597			18 07 43	−22 15.0	1	7.78		1.23				2	F8 Ib	369
166287	−16 04737	LSS 4726		18 07 44	−16 50.1	2	7.90	.015	0.21	.005	−0.66		7	B1 Ia	399,725
166293	−21 04884	LSS 4725		18 07 47	−21 19.4	1	8.30		0.09		−0.48		2	B3/4 III	240
166304	−16 04739			18 07 49	−16 43.5	1	9.33		0.19		−0.63		2	B2 II	1012
165938	−61 06113			18 07 49	−61 14.8	1	8.21		−0.09		−0.46		1	B4/5 III	55
		G 183 - 28		18 07 50	+16 01.5	2	11.42	.000	0.63	.001	0.00	.030	4	G0	1696,5010
166198	−40 12237			18 07 50	−40 19.1	3	8.16	.010	−0.13	.009	−0.56	.003	9	B7/8 II	540,976,1732
166780	+57 01840			18 07 52	+57 58.2	1	7.33		1.48		1.67		1	K5 III	3025
		LP 629 - 13		18 07 53	+02 48.7	1	11.96		0.84		0.35		1		1696
166479	+16 03390	HR 6803	⋆ AB	18 07 54	+16 27.9	1	6.10		0.58		0.25		2	F7 III+B9V	1733
166620	+38 03095	HR 6806	⋆	18 07 58	+38 27.2	11	6.40	.012	0.88	.007	0.59	.008	45	K2 V	1,15,22,1067,1080*
166323	−18 04822			18 07 58	−18 52.8	1	8.18		2.10				2	M1/2	369
313963	−22 04602	LSS 4729		18 08 01	−22 19.4	2	9.83	.011	0.49	.036	−0.50	.012	3	O8	540,1737
165961	−63 04323	R Pav		18 08 05	−63 37.7	1	12.40		1.46		0.22		1	Me	975
		G 204 - 47		18 08 08	+47 13.0	1	11.98		0.70		0.20		2	G8	1658
166364	−17 05054			18 08 09	−17 29.9	1	9.71		0.37		−0.21		2	B7 Iab/b	1375
166052	−18 04810			18 08 09	−18 50.9	1	7.62		0.12				5	B3 IV	8053
166460	+3 03620	HR 6800		18 08 10	+03 18.8	3	5.50	.007	1.20	.007	1.22	.016	23	K2 III	1088,3016,4001
166640	+36 03039	HR 6807		18 08 14	+36 27.3	2	5.58	.001	0.92	.005	0.67		5	G8 III	71,1501
166393	−19 04886	HR 6798	⋆ AB	18 08 17	−19 51.2	1	6.36		0.16				4	A2 V	2007
166315	−36 12311			18 08 17	−36 02.8	1	9.38		−0.04		−0.17		14	B5/7 III	756
166418	−16 04744	LSS 4735	⋆ AB	18 08 19	−16 42.9	2	8.33	.030	0.47	.025	−0.53	.020	6	B0 Ia/ab	399,8100
166345	−32 13852			18 08 22	−32 25.0	3	8.67	.054	−0.03	.008	−0.42	.017	6	B3 Vn(e)	540,976,1586
166443	−20 05020	LSS 4736		18 08 25	−20 43.5	5	8.70	.018	0.20	.009	−0.78	.013	9	B0/1 (I)ne	445,540,1012,1586,1737
166581	+16 03393			18 08 29	+16 02.7	1	7.74		1.16				3	K2	882
		G 204 - 48		18 08 31	+40 23.4	1	14.37		1.19				1		1759
		UZ Ser		18 08 33	−14 56.3	1	12.81		0.14		−0.57		1		1471
	−36 12314	V659 Sgr	⋆	18 08 33	−36 25.1	1	10.53		2.45				2	N	864
		WLS 1800-20 # 8		18 08 34	−19 53.2	1	12.19		0.21		−0.06		2		1375
		V725 Sgr		18 08 34	−36 07.1	1	12.46		1.28		1.14		1		756
166757	+39 03339			18 08 38	+39 03.9	1	8.14		0.18		0.09		3	A5/6V	833
165861	−70 02507	HR 6774		18 08 39	−70 45.9	5	6.72	.008	−0.03	.002	−0.35	.005	20	B7/8 II/III	15,1075,1499,1770,2038
166464	−23 14047	HR 6801	⋆ A	18 08 40	−23 42.8	5	4.98	.016	1.05	.009	0.90	.020	14	K0 III	285,542,1490,2007,3016
166464	−23 14047	IDS18056S2343	B	18 08 40	−23 42.8	1	11.52		0.66		0.11		2	F8 V	542
166425	−32 13858			18 08 40	−32 39.9	1	8.23		−0.05		−0.35		3	B5/7 III	240
166481	−22 04605	LSS 4740		18 08 42	−22 07.0	2	9.59	.010	0.37	.004	−0.61	.021	3	B0 (III)	540,976
166348	−43 12343			18 08 44	−43 27.1	4	8.38	.028	1.30	.010	1.21	.019	11	M0 V	158,1075,1494,3072
		LP 749 - 24		18 08 46	−09 19.2	1	10.25		1.03		0.89		1		1696
166469	−28 14268	HR 6802, V4045 Sg		18 08 47	−28 54.8	1	6.54		−0.01				4	B9 IV(p)	2007
		Steph 1582		18 08 49	+22 00.7	1	11.29		1.43		1.17		2	K7-M0	1746
312812			V	18 08 50	−20 23.0	1	9.73		1.93				2		369
166539	−15 04854	LSS 4745		18 08 51	−15 35.7	1	8.83		0.25		−0.75		2	B1/2 Ib	1012
166524	−18 04826	IDS18059S1826	AB	18 08 51	−18 23.6	5	9.80	.064	0.46	.025	−0.54	.025	6	B0:Vpe	445,1012,1586,2023,8084
166524	−18 04826	IDS18059S1826	C	18 08 51	−18 23.6	1	11.61		0.35		−0.22		1		8084
166540	−16 04747	V4159 Sgr	⋆	18 08 54	−16 54.4	1	8.14		0.16		−0.73		3	B1 Ib	399
166526	−20 05025			18 08 55	−20 20.8	1	8.84		1.85				2	K5	369
166427	−41 12522			18 08 56	−41 43.7	2	8.14	.005	−0.11	.013	−0.48	.005	4	Ap Si	540,976
166569	−19 04894	LSS 4748		18 08 57	−19 04.2	3	8.90	.012	0.44	.008	−0.48	.010	6	B1 II	540,976,8100
166566	−15 04856	LSS 4750	⋆ AB	18 08 58	−15 41.6	5	7.95	.042	0.24	.024	−0.65	.025	6	B1(III)n(e)	1586,8100
166546	−20 05027	LSS 4747		18 08 58	−20 26.2	3	7.23	.011	0.04	.004	−0.87	.016	5	B1 Ib	540,976,1011
166453	−39 12322			18 09 01	−39 21.3	2	7.25	.013	−0.10	.017	−0.47	.001	4	B5 III	540,976

Table 1   989

HD	DM	Other Id	N Rem	α_{1950}	δ_{1950}	S	V	σ_V	B–V	σ_{B-V}	U–B	σ_{U-B}	n	Spectrum	References
166781	+26 03187			18 09 03	+26 39.5	1	7.05		0.89		0.61		3	G5 III	8100
	+33 03042	IDS18072N3320	AB	18 09 03	+33 20.3	1	9.57		0.63		0.16		19	G0	588
315277	−25 12867	LSS 4744		18 09 03	−25 09.3	2	10.93	.002	0.64	.003	-0.46	.006	5		1586,1737
	−14 04922	LSS 4753		18 09 06	−14 56.9	2	9.73	.000	0.85	.000	-0.30	.000	4	O9.5:II:	1011,1012
166568	−18 04829	LSS 4752		18 09 07	−18 44.3	1	9.24		0.15		-0.52		2	B2 II	540
		He3 1623		18 09 12	−23 20.	1	11.89		0.90		0.21		2		1586
		LSS 4756		18 09 13	−17 44.7	2	11.58	.015	0.79	.040	-0.17		4	B3 Ib	432,725
		G 140 - 44		18 09 14	+14 54.1	1	11.53		0.77		0.23		3		1658
	+32 03065	G 182 - 42		18 09 15	+32 39.7	1	10.48		1.00		0.90		2	K5	7010
166894	+38 03100			18 09 15	+38 56.9	1	7.80		0.12		0.13		4	A2p(SrCrEu)	833
312720				18 09 16	−17 15.	1	9.57		1.77				2		369
166623	−13 04875			18 09 17	−13 49.0	1	8.97		1.96				2	K3/4 (III)	369
166628	−19 04895	LSS 4758		18 09 17	−19 26.8	5	7.18	.021	0.59	.015	-0.30	.015	8	B3 Ia/ab	540,976,1012,2023,8100
		BPM 79402		18 09 18	−01 51.	1	14.18		0.58		-0.14		1		3065
		G 21 - 5		18 09 19	+03 04.1	2	12.38	.007	0.90	.032	0.62	.102	6		316,1658
166759	+12 03419			18 09 19	+12 32.3	1	8.07		-0.05		-0.53		2	A0	1375
	+46 02435			18 09 20	+46 58.1	1	9.09		0.49		0.08		2	F5	1601
		LP 629 - 22		18 09 21	−02 23.1	1	13.09		1.12		0.79		1		1696
166801	+18 03586	NQ Her		18 09 22	+18 18.7	1	8.41		0.02		-0.09		2	A0	1648
	−21 04897	LSS 4760		18 09 22	−21 07.6	1	9.78		2.39		1.21		3		8032
164920	−82 00725			18 09 23	−82 31.4	1	7.64		1.32		1.41		7	K2 III	1704
166611	−26 12929	LSS 4754		18 09 24	−26 44.8	2	10.06	.025	0.20	.023	-0.64	.010	3	B5/7 Ib	540,954
166666	−15 04861	LS IV -15 018		18 09 26	−15 34.8	4	9.15	.029	0.22	.018	-0.63	.012	7	B1/2 II	445,1012,1032,1586
166612	−28 14282	LSS 4757		18 09 28	−28 15.0	3	7.41	.034	0.33	.023	-0.44	.016	6	A8/F2	540,954,976
	−62 05811			18 09 28	−62 57.1	2	9.91	.034	0.27	.005	0.21	.010	6	A2	1696,3077
166251	−63 04334	HR 6796		18 09 28	−63 42.2	1	6.47		1.41				4	K4 III	2007
166842	+25 03453			18 09 30	+25 32.9	1	6.78		1.12		0.99		2	K1 III	105
167042	+54 01950	HR 6817		18 09 30	+54 16.3	4	5.97	.026	0.94	.006	0.72	.004	9	K1 III	15,1003,1080,4001
		LSS 4767		18 09 31	−17 47.6	1	11.67		0.84				2	B3 Ib	725
315335	−26 12934	LSS 4762		18 09 31	−26 49.7	1	10.73		0.14		-0.60		3	A3	954
166895	+30 03144			18 09 32	+30 06.7	1	8.72		0.46		-0.05		3	F6 V	3026
166733	−2 04568			18 09 32	−02 16.3	1	9.51		0.53		0.02		2	F8	1375
166956	+39 03343			18 09 33	+39 05.3	1	7.03		0.02		-0.05		5	A0 V	833
166689	−16 04752	LSS 4768		18 09 34	−16 23.7	1	7.50		0.14		-0.72		3	B1 Ib	399,8100
		AS 289		18 09 35	−11 40.9	1	13.62		2.10		0.67		1	M3	1753
		WLS 1800 15 # 11		18 09 37	+16 58.9	1	11.45		1.09		0.91		2		1375
166734	−10 04625	LS IV -10 003		18 09 38	−10 44.7	3	8.42	.000	1.08	.010	-0.12	.005	5	O8 e	1011,1012,8007
166596	−41 12534	HR 6804, V692 CrA		18 09 40	−41 21.0	2	5.46	.005	-0.18	.005			7	B2 III	15,2012
	+49 02743	G 228 - 20		18 09 42	+49 57.7	1	9.99		1.29				1	K7	906
167103	+53 02042			18 09 42	+53 29.3	1	8.33		0.12		0.08		2	A2	1566
166736	−15 04865			18 09 42	−15 38.3	1	8.78		1.77				2	K5 (III)	369
		GD 375		18 09 43	+28 29.0	1	15.11		0.02		-0.75		3	DA	308
166914	+25 03455			18 09 44	+25 21.9	1	8.90		0.60		0.17		3	F8 IV-V	3016
		G 182 - 43		18 09 47	+32 10.0	1	11.10		0.78		0.38		2	K4	1658
166988	+33 03044	HR 6814	★ AB	18 09 55	+33 26.0	2	5.97		0.03	.015	0.07	.034	5	A3 V	1022,1733
167105	+50 02534			18 09 55	+50 46.8	1	8.95		0.02		0.05		2	A0	1566
166765	−16 04754		V	18 09 55	−16 35.4	1	7.73		1.80				2	K3 III	369
	+5 03640			18 09 56	+05 23.6	1	10.43		0.73		0.21		1	G5	3073
166787	−19 04900	LSS 4773		18 09 58	−19 45.6	3	8.22	.004	0.51	.017	-0.49	.013	4	B0.5Ia	540,976,2023
166767	−23 14072	AP Sgr		18 10 00	−23 07.9	5	6.55	.032	0.62	.019	0.44	.044	5	F7/8 Ib/II	657,1484,1490,1772,6011
166318	−66 03281	IDS18048S6650	AB	18 10 00	−66 49.8	1	5.75		0.20		0.07		1	A0 V	1754
167006	+31 03199	HR 6815, V669 Her		18 10 01	+31 23.5	2	5.00	.024	1.64	.004	1.94		4	M3 III	1363,3055
166803	−15 04868			18 10 04	−15 12.4	1	8.17		0.12		-0.75		3	B1/2 Ib	399
166788	−20 05034			18 10 04	−20 42.3	1	9.43		1.63				2	K2 III	369
166804	−18 04837			18 10 07	−18 19.4	1	8.82		0.04				1	Ap Si	2023
166917	+2 03528			18 10 10	+02 47.9	1	6.70		0.05		-0.62		1	B9	1221
166789	−27 12567	IDS18071S2714	AB	18 10 11	−27 13.8	1	8.06		-0.02		-0.42		3	B6 III	240
166826	−20 05037			18 10 12	−20 24.7	2	9.53	.001	0.21	.000	-0.32	.020	3	B8/9 III	540,976
312890				18 10 15	−18 49.5	1	10.90		1.08		0.99		4		375
166745	−33 12962			18 10 16	−33 30.6	1	8.48		0.77				4	G5 V	2012
312984		LSS 4778		18 10 18	−20 32.2	1	9.05		0.28		0.22		4		1586
166828	−22 04616			18 10 18	−22 50.9	1	8.26		1.47				2	K2/3 III	369
		V3795 Sgr		18 10 18	−25 47.8	1	10.97		1.01		0.24		1		1589
		WLS 1800 15 # 8		18 10 19	+14 59.5	1	10.41		1.49		1.62		2		1375
312891				18 10 20	−18 49.7	1	10.81		1.51		1.23		3		375
		CCS 2551		18 10 24	+14 55.	1	10.71		1.30				1	R5	1238
167470	+65 01247			18 10 24	+65 22.0	1	7.70		1.07		0.73		2	K0	1733
166852	−22 04619	LSS 4781		18 10 24	−22 43.9	3	8.53	.010	0.26	.004	-0.70	.019	4	B0 Ia/ab	540,976,1737
167387	+60 01813	HR 6827		18 10 31	+60 23.8	2	6.49	.005	0.00	.005	-0.05	.010	4	A1 Vnn	985,1733
166960	−4 04415	HR 6813		18 10 31	−04 01.6	4	6.59	.012	0.27	.008	0.15	.014	16	A2 m	15,355,1256,8071
166922	−19 04905	LSS 4788		18 10 37	−19 18.3	3	9.19	.007	0.29	.028	-0.50	.021	6	B2 Ib	413,540,976,2023
166921	−18 04843			18 10 42	−18 24.1	1	7.81		0.04				1	Ap Si	2023
		V1830 Sgr		18 10 42	−27 42.9	1	12.56		0.01		-0.78		1		1471
166832	−36 12348			18 10 42	−36 53.2	3	8.33	.049	-0.07	.010	-0.53	.024	6	B3 III/IV	400,540,976
312985				18 10 43	−20 30.6	1	9.90		1.82				2		369
167191	+35 03177			18 10 44	+35 32.8	1	8.37		0.50		0.02		2	G5	1601
314059	−21 04907	IDS18078S2105	A	18 10 44	−21 03.8	1	9.63		0.11		-0.57		1	B8	8084
	−19 04907			18 10 46	−19 16.0	3	9.36	.026	2.08	.019	2.06	.125	8	M3	369,413,8032
166937	−21 04908	HR 6812, 20 Sgr	★ A	18 10 46	−21 04.4	5	3.86	.008	0.23	.010	-0.50	.030	13	B8 Iape	15,1075,2012,8015,8084
166937	−21 04908	IDS18078S2105	B	18 10 46	−21 04.4	2	8.04	.000	-0.04	.000	-0.11	.000	3		8072,8084

HD	DM	Other Id	N	Rem	α_{1950}	δ_{1950}	S	V	σ_V	B–V	σ_{B-V}	U–B	σ_{U-B}	n	Spectrum	References
166937	−21 04908	IDS18078S2105	C		18 10 46	−21 04.4	2	10.99	.000	0.23	.000	−0.30	.000	3		8072,8084
166937	−21 04907	IDS18078S2105	D		18 10 46	−21 04.4	1	9.63		0.11		−0.57		2		8072
166937	−21 04909	IDS18078S2105	E		18 10 46	−21 04.4	2	9.25	.000	0.04	.000	−0.66	.010	3		8072,8084
	−30 15415	LSS 4784			18 10 46	−30 30.4	1	11.55		0.09		−0.60		3		954
234580	+51 02331	IDS18096N5108	A		18 10 48	+51 09.3	1	9.63		0.68		0.22		2	G5	3032
234580	+51 02331	IDS18096N5108	B		18 10 48	+51 09.3	1	10.07		0.62		0.16		2		3032
234580	+51 02331	IDS18096N5108	C		18 10 48	+51 09.3	1	11.84		0.52		−0.01		1		3032
167348	+52 02158				18 10 48	+52 18.1	1	7.44		0.99		0.76		2	K2	1566
	−19 04908				18 10 48	−19 12.2	1	10.17		0.14		−0.29		3		413
166813		Y CrA			18 10 48	−42 51.6	1	14.35		1.21		−1.08		1	O3	1753
	−12 04949	LSS 4798			18 10 52	−12 53.7	2	10.78	.000	0.39	.000	0.13	.000	3	B9:IV:	1012,1032
166965	−19 04909	LSS 4797			18 10 55	−19 00.3	1	8.71		0.27				1	B2 Ib	2023
166599	−63 04343	HR 6805		⋆A	18 10 56	−63 04.3	1	5.60		0.92				4	K0 III/IV	2007
167045	+3 03633				18 10 57	+03 36.3	2	9.24	.000	0.02	.002	−0.44	.006	8	A2	963,1732
		G 204 - 49			18 10 59	+40 32.7	1	10.85		0.73		0.19		2	G7	1658
166999	−19 04910				18 11 00	−19 07.8	1	9.45		0.09		−0.52		3	B2/5 Ib	413
166968	−27 12588	IDS18079S2732	AB		18 11 02	−27 31.1	1	7.16		−0.02		−0.38		3	B8 II/III	240
166967	−25 12900	LSS 4796			18 11 03	−25 19.6	3	8.38	.018	0.08	.007	−0.64	.009	7	B2 II	540,976,1586
312980	−20 05043	LSS 4800			18 11 05	−20 19.0	4	9.51	.013	0.84	.009	−0.21	.013	6	O6.5 V(f)	1011,1012,1032,1737
234582	+52 02159				18 11 06	+52 02.9	1	8.32		1.68		1.96		2	M0	1566
167304	+41 03011	HR 6824			18 11 07	+41 08.0	2	6.37	.010	1.04	.007	0.98		7	K0 III	1118,3016
		G 21 - 6			18 11 08	−08 39.1	4	12.35	.025	1.03	.019	0.54	.008	12	dK0	202,308,1620,3062
167193	+21 03347	HR 6820			18 11 09	+21 51.9	4	6.11	.011	1.47	.000	1.72	.005	8	K4 III	15,1003,1080,3040
167034	−19 04913	IDS18082S1914	A		18 11 10	−19 12.8	1	9.26		0.49				23	F2/3 V	6011
	−5 04600				18 11 14	−05 21.1	1	10.38		2.12				2	M2	369
167036	−21 04916	HR 6816			18 11 16	−21 43.7	4	5.47	.020	1.53	.018	1.68	.014	10	K2 III	369,1375,2007,3005
167195	+13 03564				18 11 17	+13 37.7	1	7.88		1.16				3	K2	882
167050	−19 04915	LSS 4806		⋆AB	18 11 20	−19 05.0	1	9.36		0.07		−0.62		2	B3 II	540
165737	−80 00847	IDS18023S8016	AB		18 11 21	−80 15.5	1	7.36		1.06		0.85		3	K0 III	1730
		G 140 - 47			18 11 23	+08 12.1	2	13.95	.047	1.25	.009	0.86	.164	4		203,1658
167016	−28 14329				18 11 23	−28 51.3	1	7.95		−0.02		−0.33		2	B8 III/IV	240
167370	+38 03113	HR 6826			18 11 25	+38 45.5	2	6.05		−0.07	.010	−0.22	.010	5	B9 IIIn	105,1022
167003	−33 12990	LSS 4801			18 11 25	−33 09.4	3	8.49	.040	−0.13	.005	−0.95	.007	6	B1 Ib/II	400,540,976
167275	+26 03195				18 11 26	+26 13.8	1	7.26		1.22				3	K1 III	20
167163	+0 03885				18 11 27	+00 11.0	2	9.34	.013	0.35	.009	0.20	.038	6	A0 V	1003,1775
	−17 05078				18 11 29	−17 50.5	1	9.84		2.15				2	M2	369
167067	−18 04851				18 11 29	−18 10.3	1	8.16		0.05				1	B5/7 V	2023
	+42 03025	IDS18100N4249	A		18 11 31	+42 50.7	1	10.17		0.43		0.19		1	F2	8084
314170	−23 14113				18 11 31	−23 11.9	1	9.64		0.95				29	K0	6011
167389	+41 03013				18 11 32	+41 27.7	1	7.41		0.65		0.17		std	F8	1335
	+42 03026	IDS18100N4249	B		18 11 33	+42 50.1	1	10.34		0.40		−0.02		1	F2	8084
167088	−19 04917				18 11 38	−19 04.3	5	8.53	.005	0.01	.015	−0.81	.014	11	B2 Ib/II	413,540,976,1732,2023
167277	+16 03413				18 11 41	+16 24.1	1	7.89		1.01				2	K2	882
		G 206 - 17			18 11 41	+32 47.7	1	16.40		0.28				4	DAwk	940
167143	−17 05080				18 11 43	−17 22.7	1	9.10		0.22		−0.39		4	B3 III	206
		G 206 - 18			18 11 45	+32 47.9	1	17.24		0.50				2	DC	940
167144	−18 04856				18 11 48	−18 47.6	1	8.07		1.96				2	K5 (III)	369
167145	−22 04631				18 11 48	−22 01.3	1	8.18		1.19				2	G6/8 III	369
312929					18 11 50	−19 23.	1	9.33		1.76				2		369
		G 154 - 51			18 11 53	−09 59.8	1	12.22		1.01		0.78		1		1696
167199	−19 04921				18 11 56	−19 14.6	1	9.12		0.31		0.28		4	A2(m)A3-A7	413
317333	−29 14717	VZ Sgr			18 11 57	−29 43.4	1	10.15		0.73		0.10		1	B8	1589
167278	+0 03892	IDS18094N0009	AB		18 12 00	+00 09.6	3	7.60	.017	0.44	.015	−0.05	.025	9	F2	176,938,3016
167279	−2 04578				18 12 00	−02 37.1	1	7.47		2.08				2	K5	369
166913	−59 07243				18 12 01	−59 25.1	5	8.21	.016	0.45	.005	−0.19	.018	11	F5/7 Vw	742,1097,1594,2012,3077
167200	−22 04635				18 12 03	−22 27.7	1	9.10		−0.01		−0.66		4	B5	1732
167203	−24 14088				18 12 03	−24 00.9	1	8.28		0.18		−0.20		3	B7 III/IV	240
167224	−18 04857	LSS 4814		⋆AB	18 12 04	−18 57.6	2	8.07	.001	0.09	.002	−0.62		3	B3 Ib/II	540,2023
167247	−19 04924	IDS18091S1914	AB		18 12 04	−19 13.3	1	8.99		0.17				1	B5 III/V	2023
167175	−25 12920				18 12 06	−25 48.3	1	8.14		1.87		1.96		3	K2 (III)	1657
312989	−20 05052	LSS 4815			18 12 08	−20 30.8	1	9.89		0.71		−0.21		2	B3	432
167472	+28 02960				18 12 10	+28 12.1	3	6.79	.033	1.09	.015	1.06	.045	8	K1 III	20,37,8100
		LSS 4817			18 12 13	−21 25.0	1	10.93		1.14		0.05		3	B3 Ib	1737
167263	−20 05055	HR 6823		⋆AB	18 12 14	−20 24.3	8	5.96	.012	0.02	.013	−0.87	.018	20	O9 II	15,26,540,976,1011*
167264	−20 05054	HR 6822, LSS 4818			18 12 14	−20 44.7	5	5.35	.015	0.05	.012	−0.85	.033	16	B1 Ia	15,26,1012,2007,8100
167502	−22 04644	IDS18104S2217	A		18 12 14	−22 16.2	1	9.40		1.41				2	K1/2 (III)	369
167095	−43 12392				18 12 14	−43 11.1	2	7.05	.003	−0.01	.013	−0.18	.015	3	B9 V	540,976
167096	−44 12456	HR 6818			18 12 14	−44 13.4	1	5.46		0.96				4	G8/K0 III	2006
167285	−17 05087	IDS18094S1740	A		18 12 18	−17 39.3	1	10.52		0.17				1	B9 III	2023
		Ma 38 - 13			18 12 18	−19 00.9	1	12.61		0.25		−0.08		2		413
		Ma 38 - 45			18 12 18	−19 01.3	1	12.87		1.48		1.50		1		413
		Ma 38 - 50			18 12 19	−19 01.0	1	11.77		0.08		−0.43		2		413
		Ma 38 - 11		⋆D	18 12 19	−19 01.5	1	9.92		0.09		−0.63		3		413
		Ma 38 - 2		⋆B	18 12 20	−19 00.2	1	11.32		0.15		−0.43		2	B2 V	413
167287	−19 04928	Ma 38 - 1		⋆A	18 12 20	−19 00.2	2	7.08	.005	0.09	.015	−0.79	.010	8	O9.5III	399,413
		Ma 38 - 8			18 12 20	−19 00.7	1	13.53		0.62		0.31		2		413
		Ma 38 - 9			18 12 20	−19 01.4	1	12.33		0.14		−0.30		2		413
		Ma 38 - 30		⋆E	18 12 20	−19 01.8	1	10.66		0.09		−0.59		3		413
		Ma 38 - 3		⋆R	18 12 20	−19 01.8	1	12.45		0.31		−0.21		2	B1 V	413
		Ma 38 - 4			18 12 21	−19 00.4	1	11.70		0.38		0.23		2		413

Table 1   991

HD	DM	Other Id	N Rem	α_{1950}	δ_{1950}	S	V	σ_V	B–V	σ_{B-V}	U–B	σ_{U-B}	n	Spectrum	References
		Ma 38 - 7		18 12 21	−19 01.2	1	12.64		0.27		-0.11		2		413
		Ma 38 - 31		18 12 21	−19 01.7	1	14.00		0.59		0.30		2		413
		Ma 38 - 5	⋆ C	18 12 22	−19 00.5	1	11.81		0.16		-0.40		2		413
		LSS 4822		18 12 22	−19 59.9	1	11.41		0.54		-0.31		4	B3 Ib	1737
167353	−6 04729			18 12 23	−06 47.9	1	8.88		2.24				2	K5	369
167311	−12 04953	LSS 4826		18 12 23	−12 31.1	4	8.61	.011	0.95	.020	-0.14	.019	8	B0 Ib	138,399,405,1586
		Ma 38 - 27		18 12 24	−19 01.2	1	14.99		0.78		0.18		2		413
167330	−12 04954	LSS 4828		18 12 25	−12 33.2	2	8.23	.000	0.66	.000	-0.45	.000	4	O9.5 Ia/ab	1011,1012
		YY Her		18 12 26	+20 58.2	1	12.84		1.14		-0.27		1	M2 pe	1753
	−5 04606			18 12 26	−05 55.4	1	10.74		2.41				2	M3	369
		Ma 38 - 24		18 12 26	−19 00.9	1	14.37		0.79		0.23		2		413
	−20 05056		A	18 12 26	−20 39.6	1	10.44		1.26				1		369
	−20 05056		B	18 12 26	−20 39.6	2	11.21	.000	2.81	.000			4	M3	369,8032
		V1954 Sgr		18 12 27	−20 39.0	1	10.46		1.28				1		1772
		AS 296, FG Ser		18 12 28	−00 16.2	1	10.55		1.28		-0.10		1		1753
167233	−36 12388			18 12 28	−36 35.5	1	7.00		-0.11		-0.60		1	B3 III	219
167314	−18 04862			18 12 31	−18 56.2	2	7.83	.015	1.99	.010	2.20		5	M1/2 Ib/II	369,413
167336	−18 04863	LSS 4827		18 12 32	−18 22.4	3	8.70	.012	0.51	.011	-0.49	.010	4	B0 II	1012,1737,2023
167356	−18 04864	HR 6825, LSS 4831	⋆ AB	18 12 34	−18 40.7	2	6.09	.005	0.20	.000	-0.34		9	Ap Si	2007,8100
312974				18 12 34	−20 22.	1	9.62		0.19		-0.50		2		432
167235	−41 12597			18 12 36	−41 38.4	2	8.67	.010	-0.04	.013	-0.35	.002	4	B7 III	540,976
312973	−20 05060	LSS 4832		18 12 38	−20 23.0	6	8.81	.032	0.22	.016	-0.78	.016	10	B0:IV:pe	445,540,976,1012,1586,1737
167357	−19 04932			18 12 40	−19 58.3	1	8.83		2.15				3	K3/5 (III)	369
166841	−68 03081	HR 6808	⋆ AB	18 12 40	−68 14.0	4	6.32	.008	-0.03	.005	-0.21	.031	16	B9 V	15,1075,2038,3021
	−16 04776			18 12 43	−16 15.0	1	9.22		1.84				3	M2	369
		HIC 89483		18 12 43	−18 50.8	1	10.85		1.37				2		1726
167374	−17 05091			18 12 45	−17 53.6	1	8.84		0.05				1	A1 IV	2023
167379	−20 05063			18 12 46	−20 11.8	1	8.42		0.12		-0.40		2	B3 II	432
167588	+29 03213	HR 6831		18 12 47	+29 11.6	1	6.55		0.54		0.00		6	F8 V	3026
	−30 15464	LSS 4825		18 12 47	−30 46.8	1	11.99		0.05		-0.73		2		954
167375	−18 04865	LSS 4834		18 12 49	−18 58.1	1	8.24		-0.01		-0.79		3	B2 III	413
		TV Her		18 12 50	+31 48.2	1	10.94		1.50		0.60		1	M4	817
167398	−18 04867			18 12 50	−18 26.7	1	9.87		0.16		-0.30		4	B5 III	413
312903	−18 04866			18 12 50	−18 55.7	1	8.84		0.03		-0.58		3	B3	413
		LP 629 - 23		18 12 51	−02 22.7	1	12.15		0.90		0.49		2		1696
167297	−40 12327	IDS18094S4030	AB	18 12 52	−40 29.4	2	7.34	.018	-0.08	.014	-0.52	.003	4	B7 V	540,976
		LSS 4837		18 12 53	−16 42.7	1	11.15		0.30				2	B3 Ib	725
167128	−56 08706	HR 6819		18 12 55	−56 02.5	6	5.35	.012	-0.06	.007	-0.68	.013	30	B3 IIIep	26,681,815,1637,2006,8028
167363	−31 15328	IDS18097S3111	A	18 12 57	−31 10.6	1	7.65		0.04		-0.35		2	B7 II	240
167321	−38 12634			18 12 57	−38 56.3	1	8.93		-0.04		-0.31		4	B9 II	400
167516	+1 03633		AB	18 12 58	+01 19.9	1	8.46		0.60		0.14		2	G5 III	8100
	−20 05064			18 12 58	−20 50.7	1	10.09		1.79				2	M3	369
167412	−18 04869			18 12 59	−18 27.0	1	8.49		0.08				1	B5 Ib/II	2023
167435	−19 04936	LSS 4838		18 12 59	−19 55.5	2	9.15	.005	0.38	.000	-0.33	.005	3	B5 Ib	540,976
167411	−18 04871	LSS 4841		18 13 00	−18 15.5	1	7.71		0.27				1	B0.5Ia	2023
167413	−18 04870			18 13 00	−18 45.4	1	9.49		0.47		0.17		3	F0 V	413
167436	−20 05065	LSS 4839		18 13 00	−20 02.0	2	9.44	.016	0.03	.011	-0.76	.006	3	B5	540,976
167433	−17 05094			18 13 03	−17 37.5	1	7.51		0.07				1	B7 II/III	2023
167451	−13 04897	LSS 4844		18 13 05	−13 55.5	1	8.23		0.79		-0.27		2	B1 Ib	1012
167257	−51 11460	HR 6821		18 13 05	−51 05.2	1	6.06		-0.06				4	B9 V	2009
348282	+18 03606			18 13 06	+18 28.4	4	10.07	.026	1.34	.001	1.32	.010	6	M1	196,801,1619,7008
167402	−30 15474	LSS 4835		18 13 06	−30 08.6	3	8.99	.036	-0.01	.009	-0.89	.004	6	O9.5 Ib/II	540,954,976
	+36 03064			18 13 07	+36 21.8	1	8.23		1.42		1.60		2	K5	1601
	−12 04957			18 13 07	−12 23.0	1	11.00		2.03				2	M2	369
167740	+36 03066	W Lyr		18 13 12	+36 39.2	1	9.99		1.63		0.96		1	M4	817
314261				18 13 14	−22 33.	1	9.78		1.68				2		369
167626	+12 03436			18 13 16	+13 00.4	1	8.34		1.66				1	K5	882
312875				18 13 16	−18 16.0	1	10.31		0.32		-0.42		2		432
167479	−18 04873			18 13 16	−18 49.0	3	8.13	.009	0.06	.016	-0.63	.009	6	B1/2 Ib/II	413,540,2023
167441	−30 15479			18 13 18	−30 25.4	1	7.78		0.07		-0.25		3	B8 III/IV	240
167478	−18 04875			18 13 19	−18 27.6	1	9.06		0.18		-0.60		2	B2 (III)	432
167564	−3 04259	HR 6830		18 13 20	−03 38.1	1	6.35		0.20		0.15		8	A4 V	1088
167519	−14 04957	LSS 4848		18 13 20	−14 37.7	2	9.81	.000	0.51	.000	-0.49	.000	3	B1/2 Ib	1012,1032
167481	−22 04643			18 13 20	−22 03.0	1	9.02		1.79				2	M1 (III)	369
167543	−14 04959	LS IV -14 026		18 13 24	−14 39.8	1	8.56		0.43		-0.53		2	B1 Ib/II	1012
		G 183 - 32		18 13 25	+20 29.3	1	13.31		1.42		1.20		5		419
315376	−24 14114	LSS 4847		18 13 26	−24 35.0	2	10.10	.008	0.31	.008	-0.65	.007	4	B0 II	540,976
	−13 04901			18 13 28	−12 59.8	1	9.97		1.87				2	M2	369
167544	−14 04960			18 13 28	−14 57.7	1	9.03		1.85				2	M(3)	369
167520	−19 04939	IDS18105S1900	AB	18 13 29	−18 59.4	1	8.80		0.19				1	B7 Ib	2023
		BH Oph		18 13 30	+12 04.8	1	11.63		0.41		0.33		1	F7.1e	1399
		Smethells 12		18 13 30	−49 28.	1	11.28		1.36				1		1494
		G 183 - 33		18 13 33	+18 55.4	1	10.84		1.50				1	M2	906
167654	+2 03547	HR 6834		18 13 35	+02 21.6	1	6.00		1.59		1.55		8	M4 IIIab	1088
168151	+64 01252	HR 6850		18 13 36	+64 22.8	5	5.02	.021	0.41	.018	-0.06	.010	9	F5 V	254,1118,1193,1620,3026
167504	−29 14752			18 13 36	−29 52.6	1	8.98		-0.08		-0.71		7	B3 II/III	1732
167941	+49 02758			18 13 39	+49 08.4	1	6.83		1.25		1.30		2	K2	1601
167568	−18 04880			18 13 41	−18 49.5	1	7.77		1.21		0.99		3	K0 II	413
		AJ84,127 # 2231		18 13 42	+18 39.	1	11.34		0.78		0.24		1		801
	+13 03578	G 140 - 49		18 13 44	+13 54.9	2	10.24	.047	1.41	.012	1.21		9	M0	1619,7008

HD	DM	Other Id	N Rem	α_{1950}	δ_{1950}	S	V	σ_V	B–V	σ_{B-V}	U–B	σ_{U-B}	n	Spectrum	References
167782	+25 03475			18 13 44	+25 46.6	2	8.29	.030	1.08	.010	0.82		5	G8 II	20,8100
		G 154 - 53		18 13 46	−12 47.6	1	13.79		1.38		0.92		1		3062
		G 21 - 7		18 13 48	+01 30.7	4	12.51	.013	1.62	.037	1.34		5		202,333,1510,1620,1705
168092	+56 02080	HR 6849	★ A	18 13 48	+56 34.2	1	6.39		0.33		0.00		2	F1 V	254
167576	−27 12658			18 13 49	−27 44.0	4	6.68	.000	1.26	.005	1.37	.012	8	K1 III	657,2012,2017,3077
313076				18 13 52	−18 59.1	2	9.40	.012	0.36	.002	-0.58	.017	3	B1.5Iab	540,976
167611	−18 04882	LSS 4857		18 13 53	−18 34.1	2	9.48	.035	0.18	.015			5	B1 Iab/Ib	1021,2023
		LSS 4862		18 13 55	−17 00.0	1	11.04		1.08		-0.07		2	B3 Ib	1737
167638	−19 04940	LSS 4858		18 13 55	−19 47.8	4	9.85	.009	0.24	.018	-0.58	.016	9	B2 II	206,432,540,976
167633	−16 04786	LSS 4864		18 13 56	−16 32.2	1	8.14		0.27		-0.72		2	O6	1011
167636	−18 04883			18 13 56	−18 41.9	1	9.56		-0.04		-0.65		2	B5 (III)	413
167656	−15 04906			18 13 58	−15 40.8	1	9.46		0.36		-0.52		2	B7/8 (III)	1012
167635	−17 05107			18 13 58	−17 57.7	1	8.65		0.15				1	B8 Ib/II	2023
167856	+30 03162			18 13 59	+30 22.9	1	6.78		1.03		0.83		2	K0	1625
167807	+13 03581			18 14 02	+13 09.7	1	8.51		1.08				1	K2	882
167659	−19 04944	LSS 4865		18 14 02	−18 59.2	5	7.38	.006	0.21	.009	-0.74	.010	8	B8 Ib	540,976,1011,1012,2023
167599	−31 15344			18 14 02	−31 19.0	1	7.48		0.03		-0.38		1	B7/8 II	540,976
	−11 04579	LS IV -11 007		18 14 03	−11 49.4	1	10.44		1.08		-0.05		2	B0:V:	1012
167660	−19 04945	WZ Sgr		18 14 03	−19 05.7	4	7.50	.035	1.11	.020	0.82	.020	4	F7 (III)	689,1484,1587,6011
167965	+42 03035	HR 6845		18 14 05	+42 08.5	4	5.58	.013	-0.11	.023	-0.47	.018	20	B7 IV	154,1118,1335,3023
		LSS 4867		18 14 05	−17 00.8	1	10.49		0.98		-0.11		4	B3 Ib	1737
168009	+45 02684	HR 6847		18 14 06	+45 11.6	2	6.29	.006	0.63	.015	0.12		6	G2 V	71,3026
313094	−19 04946	IDS18112S1954	A	18 14 06	−19 54.1	1	11.29		0.31		-0.42		2	B5	432
		AS 299, V2905 Sgr		18 14 06	−28 11.	1	12.17		0.69		-0.68		1		1586
		LP 922 - 18		18 14 06	−30 52.0	1	10.04		0.63		0.03		2		1696
167300	−61 06125			18 14 06	−61 57.3	2	9.17	.027	0.59	.009	0.02		5	F8 V	742,1594
168093	+50 02549			18 14 08	+50 57.2	1	7.66		0.53		0.02		2	F8	1566
313095	−19 04946	LSS 4866		18 14 08	−19 53.0	2	11.08	.034	0.39	.019	-0.43	.034	6	B5	206,432
313077	−19 04947			18 14 11	−19 02.0	3	9.83	.003	0.27	.005	-0.39	.004	7	B5 II	540,976,1021
167665	−28 14408	HR 6836		18 14 14	−28 18.4	4	6.39	.004	0.53	.007	-0.02	.013	10	F8 V	15,657,2012,3077
167666	−28 14407	HR 6835		18 14 14	−28 40.3	1	6.19		0.18				4	A5 V	2009
	−52 11059	NGC 6584 sq1 1		18 14 14	−52 22.9	1	10.15		1.41		1.59		3		1585
165338	−84 00573			18 14 14	−84 24.7	1	6.41		0.08		-0.17		8	B8/9 V	1628
167618	−36 12423	HR 6832, η Sgr	★ AB	18 14 15	−36 46.7	6	3.11	.010	1.56	.003	1.70	.011	19	M2 III	15,1024,1075,2012*
167768	−3 04263	HR 6840		18 14 16	−03 01.3	5	6.02	.008	0.89	.000	0.52	.013	27	G3 III	15,1003,1417,3077,4001
167719	−15 04907	LS IV -15 039		18 14 16	−15 25.1	1	9.46		0.31		-0.61		2	B2 II	405
167720	−17 05112	HR 6838		18 14 17	−17 23.6	2	5.79	.029	1.57	.010	1.70		6	K4 II-III	2007,3005
167647	−34 12673	HR 6833, RS Sgr	★ A	18 14 17	−34 07.6	2	6.11	.068	-0.10	.015	-0.61	.013	4	B3 V	540,976
167722	−19 04948			18 14 19	−19 45.5	3	9.23	.155	0.10	.040	-0.69	.041	5	B2 III	432,540,976
		NGC 6584 sq1 4		18 14 20	−52 22.0	1	11.84		0.66		0.14		2		1585
167746	−16 04790			18 14 23	−16 56.9	1	8.63		1.35		0.98		4	F7 II	206
		Sgr sq 1 # 17		18 14 23	−18 30.4	1	11.75		1.18		0.89		2		413
		Sgr sq 1 # 18		18 14 23	−18 30.4	1	13.20		0.25		0.03		2		413
		G 206 - 20		18 14 27	+29 11.3	1	12.86		0.60		-0.11		3		1658
313084	−19 04951	LSS 4870		18 14 27	−19 12.1	1	9.56		0.54				4	B5	1021
		NGC 6584 sq1 3		18 14 27	−52 12.1	1	11.12		0.58		0.03		6		1585
		Sgr sq 1 # 16		18 14 28	−18 30.2	1	10.76		0.20		-0.47		3		413
313079	−19 04950	LSS 4871		18 14 28	−19 06.3	2	9.57	.012	0.53	.004	-0.35	.015	2	B3	540,976
		LSS 4872		18 14 28	−19 12.1	1	10.13		0.58		-0.34		3	B3 Ib	1737
313098				18 14 28	−19 44.0	1	10.55		0.12		-0.50		2	B5	432
		G 154 - 55		18 14 29	−09 17.2	2	11.74	.039	0.90	.039	0.43	.019	3		1696,3062
		RS Ser		18 14 29	−13 04.7	1	11.21		0.78				1		1777
		LSS 4875		18 14 30	−16 43.4	1	10.28		0.52				1	B3 Ib	725
167772	−19 04952			18 14 30	−19 30.4	1	7.66		0.48				30	F5/6 V	6011
		AA65,65 # 105		18 14 30	−31 12.9	1	12.42		1.45		1.49		1		713
		NGC 6584 sq1 6		18 14 31	−52 23.3	1	12.81		1.25		1.30		4		1585
167858	+0 03907	HR 6844		18 14 32	+00 59.2	1	6.62		0.31		0.02		8	F2 V	1088
167771	−18 04886	HR 6841, LSS 4874	★ AB	18 14 32	−18 29.0	5	6.54	.011	0.09	.018	-0.86	.020	13	O8	15,413,1011,2023,2035
167686	−33 13052			18 14 33	−33 24.9	2	7.01	.011	-0.07	.015	-0.37	.004	3	B8 II	540,976
	−31 15361	AA65,65 # 119		18 14 35	−31 10.6	1	10.63		0.69		0.18		1		713
	−11 04581			18 14 36	−11 46.4	1	11.32		0.93		-0.10		2	B1:II:	1012
	−12 04964	LSS 4876		18 14 36	−12 19.6	2	9.82	.000	0.91	.000	-0.22	.000	4	O7 V	1011,1012
		NGC 6584 sq1 7		18 14 38	−52 22.6	1	13.54		1.15		0.98		3		1585
167833	−9 04678	HR 6843		18 14 39	−09 46.6	1	6.30		0.38		0.24		8	A8 V	1088
167810	−12 04965			18 14 40	−12 44.6	1	7.74		1.76				3	K3 III	369
167815	−19 04953	IDS18117S1942	AB	18 14 42	−19 41.5	4	7.56	.012	0.16	.010	-0.72	.014	8	B1/2 III	540,976,1732,2023
167834	−12 04969	LSS 4879		18 14 43	−12 07.5	2	9.61	.000	0.92	.000	-0.20	.000	4		405,1737
168038	+27 02991			18 14 45	+27 04.6	1	8.31		0.52				3	F7 IV	20
167795	−23 14187			18 14 45	−23 17.8	1	7.67		0.05		-0.33		3	B8 V	240
315487				18 14 45	−25 13.6	2	10.10	.010	0.10	.019	-0.50	.002	4	B6 II-Ib	540,976
167775	−29 14775			18 14 45	−29 16.6	3	8.60		-0.03	.016	-0.43	.009	4	B5 IV/Vn(e)	540,976,1586
		LSS 4880		18 14 46	−12 07.1	2	10.48	.000	0.87	.000	-0.23	.000	4	B3 Ib	405,1737
167838	−15 04911	LSS 4878		18 14 46	−15 27.0	1	6.73		0.46		-0.36		2	B3 Ia/ab	1012
		AA65,65 # 118		18 14 47	−31 15.1	1	11.38		1.77		2.05		1		713
238839	+59 01885			18 14 49	+59 53.4	1	9.29		0.37		0.06		2	F0	1502
	−18 04887			18 14 49	−18 24.0	1	10.47		0.14		-0.35		3		413
		NGC 6611 - 559		18 14 50	−13 30.6	2	11.57	.015	0.61	.005	-0.09	.005	3	G0	771,1703
167425	−63 04370	HR 6828	★ AB	18 14 50	−63 54.2	7	6.17	.008	0.58	.006	0.07	.005	32	F9 V	15,1075,1311,2013*
		LP 629 - 26		18 14 52	−00 24.1	1	12.17		0.87		0.28		2		1696
167863	−18 04888	IDS18120S1851	A	18 14 55	−18 49.1	1	6.71		0.04				1	B5/7 II/III	2023

Table 1

HD	DM	Other Id	N Rem	α_{1950}	δ_{1950}	S	V	σ_V	B–V	σ_{B-V}	U–B	σ_{U-B}	n	Spectrum	References
167818	−27 12684	HR 6842		18 14 55	−27 03.8	4	4.65	.015	1.66	.000	1.80	.004	14	K3 III	15,1075,2012,8015
167996	+11 03411			18 14 57	+11 50.8	1	9.74		-0.01		-0.28		1	B9	963
	−12 04970	LSS 4883		18 14 58	−12 31.2	1	8.78		1.02		-0.11		2	B0.5Ia	1012
		AA65,65 # 120		18 14 58	−31 00.2	1	12.63		1.27		1.07		1		713
	−46 12278			18 14 58	−46 33.5	1	9.46		1.39		1.59		6	K0	1589
	−12 04971	NGC 6604 - 7		18 14 59	−12 17.0	1	9.15		1.28		1.13		1	K0	412
167861	−17 05117			18 14 59	−17 50.9	1	8.69		2.03				3	M1	369
167862	−18 04890			18 15 00	−18 23.8	1	10.08		0.10		0.00		3	B9/A0 III	413
	−13 04912	NGC 6611 - 556	⋆ V	18 15 01	−13 52.1	3	9.97	.045	1.17	.014	0.13	.021	7	B0	771,1021,1703
	+45 02688	G 204 - 55		18 15 05	+45 32.0	2	10.30	.061	1.44	.052	1.20		4	M0	196,1510
313080	−19 04955	LSS 4882		18 15 05	−19 08.4	7	8.92	.023	0.95	.015	-0.07	.023	16	B2 II-Ib	413,540,976,1012,1021*
		AJ84,127 # 2255		18 15 06	+45 25.	1	12.46		0.70				1		1510
168269	+48 02668			18 15 06	+48 21.0	1	7.45		1.65		1.92		2	K5	1733
	−12 04973	NGC 6604 - 6		18 15 06	−12 14.1	3	10.81	.016	0.88	.020	-0.31	.055	6	B1	412,712,1021
		NGC 6604 - 17		18 15 06	−12 16.9	1	13.82		0.52		0.19		1		712
167756	−42 13101	HR 6839		18 15 06	−42 18.5	4	6.30	.012	-0.13	.008	-0.98	.006	9	B0.5 Ia	400,540,976,2009
	−18 04892			18 15 07	−18 38.3	1	10.02		0.38				4	B2	1021
	−46 12279	RS Tel		18 15 07	−46 34.1	1	9.96		0.84		0.36		1		1589
314217				18 15 08	−21 15.1	1	9.86		1.84				2		369
	−12 04975	NGC 6604 - 8		18 15 09	−12 17.6	3	10.47	.012	0.76	.018	-0.40	.061	10	O7	412,712,1021
		NGC 6604 - 19		18 15 10	−12 15.7	1	13.83		1.67		-1.14		1		712
		AA65,65 # 117		18 15 11	−31 15.3	1	11.52		0.33		0.22		1		713
167846	−34 12687			18 15 11	−34 42.4	2	6.91	.010	-0.06	.016	-0.39	.003	3	B7/8 III	540,976
		NGC 6604 - 20		18 15 12	−12 13.9	1	13.20		2.26		3.41		1		712
		NGC 6604 - 9		18 15 12	−12 15.3	2	11.62	.049	0.68	.015	-0.39	.068	3		412,712
		LSS 4892		18 15 12	−13 20.7	2	10.58	.000	1.64	.005	0.57	.000	4	B3 Ib	405,1737
	−12 04978	NGC 6604 - 4	⋆ AB	18 15 13	−12 15.3	5	10.40	.045	0.67	.024	-0.38	.061	8		412,712,1011,1012,1032
	−12 04979	NGC 6604 - 3		18 15 15	−12 15.8	2	9.72	.034	0.74	.024	-0.48	.005	3	B1 IV	405,412
		NGC 6604 - 16		18 15 15	−12 16.8	1	12.21		0.69		-0.27		4		712
		NGC 6604 - 12		18 15 15	−12 17.6	1	12.51		1.90		1.76		4		712
167820	−37 12418			18 15 15	−37 14.7	1	7.53		1.09		0.84		8	G8 III	1673
	−11 04586	LS IV -11 010		18 15 16	−11 18.8	2	9.40	.000	1.00	.000	-0.14	.000	4	O8 I:	1011,1012
		NGC 6604 - 10		18 15 17	−12 15.8	1	12.85		0.79		-0.09		1		412
		NGC 6604 - 11		18 15 17	−12 16.0	1	12.95		0.81		-0.06		1		412
		NGC 6584 sq1 2		18 15 17	−52 07.8	1	10.60		1.66		2.12		2		1585
		NGC 6584 sq1 5		18 15 17	−52 16.7	1	12.10		1.29		1.42		3		1585
		NGC 6604 - 15		18 15 18	−12 15.	1	7.94		0.18		-0.31		2		712
	−12 04981	NGC 6604 - 2		18 15 18	−12 15.3	6	10.15	.007	0.62	.015	-0.44	.032	24	B0 V	405,412,712,1547,1586,1737
167971	−12 04980	NGC 6604 - 1	⋆ V	18 15 18	−12 15.7	7	7.47	.021	0.76	.008	-0.35	.016	31	O8 Ib(f)p	412,712,1011,1012*
315482	−25 12991			18 15 18	−25 08.0	1	10.54		0.07		-0.54		3	B5	1586
		BPM 11316		18 15 18	−62 35.	1	14.86		0.56		-0.31		2		3065
		NGC 6604 - 18		18 15 21	−12 16.1	1	12.16		0.69		-0.44		2		712
	−12 04982	NGC 6604 - 5		18 15 22	−12 12.0	2	9.22	.034	0.62	.024	-0.56	.063	3	B1 Ib	412,712
	−15 04913	LSS 4895		18 15 22	−15 36.1	1	10.29		0.43				4	B0.5III	725
		NGC 6604 - 14		18 15 23	−12 15.1	1	12.10		0.70		-0.38		6		712
		NGC 6604 - 13		18 15 23	−12 17.1	1	15.43		0.73		0.83		1		712
		G 21 - 8		18 15 24	−05 40.8	1	13.51		1.46				2		202
		NGC 6604 - 21		18 15 24	−12 13.4	1	13.61		0.76		-0.09		2		712
		AA45,405 S47 # 1		18 15 24	−15 37.	1	10.58		1.46				3		725
	−15 04914			18 15 24	−15 38.2	2	9.85	.033	1.23	.023	1.12		4		432,725
	−22 12818	LSS 4891		18 15 24	−22 46.2	1	11.06		0.46		-0.56		3		1737
		L 158 - 59		18 15 24	−61 55.	1	11.35		0.49		-0.21		2		1696
	−66 03307	AR Pav		18 15 24	−66 06.1	1	11.14		0.71		-0.59		1	B1	1753
		G 140 - 53		18 15 25	+05 25.9	1	10.93		0.77		0.31		2	K1	1658
		NGC 6584 sq1 8		18 15 25	−52 17.2	1	13.63		1.05		0.70		10		1585
167976	−18 04895			18 15 27	−18 29.8	1	7.39		2.01				4	K2/3 III	369
		NGC 6611 - 79		18 15 28	−13 44.8	1	12.41		0.67		0.41		3		1703
		NGC 6611 - 90		18 15 30	−13 47.4	2	11.73	.019	0.38	.014	-0.21	.005	7	B0	914,1703
168131	+11 03415	LS IV +11 001		18 15 32	+11 51.5	1	7.08		-0.07		-0.64		1	B8	963
168322	+40 03332	HR 6853		18 15 32	+40 55.0	5	6.12	.016	0.98	.018	0.70	.004	10	G9 III	15,1003,1080,1118,4001
		NGC 6611 - 103		18 15 33	−13 52.7	1	12.49		0.39		0.22		3	B8	1703
	−14 04982			18 15 33	−14 41.9	1	9.81		1.89				1	M3	369
168653	+68 00984	HR 6865		18 15 34	+68 44.3	1	5.95		1.06				2	K0	71
		NGC 6611 - 578		18 15 34	−13 46.8	1	11.74		0.31		0.16		2	A0	771
		NGC 6611 - 112		18 15 35	−13 54.6	1	13.90		0.51		0.11		3		1187
	−13 04920	NGC 6611 - 125		18 15 36	−13 51.3	6	10.06	.016	0.47	.010	-0.51	.023	17	B1.5V	49,73,771,914,1187,1703
167979	−25 12995	HR 6846		18 15 36	−25 37.5	2	6.52	.020	1.35	.005	1.39		8	K0 III	285,2009
		NGC 6611 - 135		18 15 37	−13 56.3	1	12.84		0.75		0.51		1		1703
168046	−14 04986	NGC 6611 - 130		18 15 37	−14 00.6	1	9.36		1.37		1.36		2	K2 III	1187
	−8 04584			18 15 38	−08 00.3	2	10.32	.000	0.73	.000	0.23	.000	17	K0	830,1783
		NGC 6611 - 140		18 15 38	−13 48.2	1	14.69		1.03		0.60		1		49
168428	+51 02344			18 15 39	+51 12.6	1	7.93		1.57		1.89		2	K5	1566
	−13 04921	NGC 6611 - 150		18 15 40	−13 51.2	7	9.87	.007	0.48	.004	-0.54	.031	45	B0.5V	49,73,771,914,1187*
		NGC 6611 - 163		18 15 41	−13 38.3	1	13.09		0.90		0.28		1		1703
		NGC 6611 - 161		18 15 41	−13 44.4	5	11.28	.021	1.03	.015	-0.22	.039	17	F9	49,771,914,1187,1703
		NGC 6611 - 171		18 15 42	−13 48.4	1	14.24		0.75		0.36		1		49
		NGC 6611 - 166		18 15 42	−13 50.0	5	10.38	.014	0.57	.010	-0.55	.026	16	O9 V	73,771,914,1187,1703
168198	+17 03520	IQ Her		18 15 43	+17 57.6	2	7.14	.115	1.61	.000	1.21	.028	19	M2	1648,3042
		NGC 6611 - 181		18 15 43	−13 46.3	1	11.15		1.02		0.17		1		1703
	−13 04923	NGC 6611 - 175		18 15 43	−13 46.4	6	10.07	.016	0.82	.013	-0.37	.034	17	O3:III(f*)	73,771,914,1011,1187,1703

HD	DM	Other Id	N Rem	α_{1950}	δ_{1950}	S	V	σ_V	B–V	σ_{B-V}	U–B	σ_{U-B}	n	Spectrum	References
		NGC 6611 - 174		18 15 43	−13 52.9	1	13.00		0.93		0.32		3		1703
		NGC 6611 - 188		18 15 44	−13 42.2	1	13.09		1.35		0.02		2		1703
	−13 04924	NGC 6611 - 184		18 15 44	−13 55.6	1	9.76		1.36		1.25		2	K2 III	1187
168199	+13 03593	HR 6851		18 15 45	+13 45.4	2	6.29	.015	−0.03	.005	−0.53	.035	5	B5 V	154,252
		NGC 6611 - 196		18 15 45	−13 44.2	1	16.56		0.99		−0.46		1		1187
	−15 04917	LSS 4904		18 15 45	−15 47.6	1	9.94		0.94		−0.09		4	B0.5III	1737
	−31 15388	AA65,65 # 121		18 15 45	−31 11.9	1	11.09		1.68		2.03		1		713
		NGC 6611 - 202		18 15 46	−13 46.4	1	14.35		0.95		0.23		2	A0	1701
168075	−13 04925	NGC 6611 - 197		18 15 46	−13 48.9	7	8.75	.016	0.45	.006	−0.63	.049	21	O6 V((f))	73,771,914,1187,1703*
168076	−13 04926	NGC 6611 - 205		18 15 46	−13 49.3	9	8.19	.022	0.46	.008	−0.64	.042	26	O4 V((f))	49,73,771,914,1187,1586*
		NGC 6611 - 203		18 15 46	−13 52.1	1	14.03		0.58		0.36		2		1187
		NGC 6611 - 194		18 15 46	−13 52.8	1	13.80		0.58		0.20		2		1703
168200	+11 03416			18 15 47	+11 52.8	1	9.51		0.02		−0.28		1	B9	963
168480	+55 02047			18 15 47	+55 40.2	1	8.75		0.31		−0.04		2	F2	1502
		NGC 6611 - 222		18 15 47	−13 44.9	3	13.03	.051	1.28	.027	−0.03	.058	11		914,1187,1703
		NGC 6611 - 213		18 15 47	−13 46.8	2	14.07	.005	1.56	.010	0.86	.097	3		1701,1703
		NGC 6611 - 207		18 15 47	−13 48.8	3	11.96	.092	0.53	.023	−0.39	.026	7		771,1187,1703
		NGC 6611 - 210		18 15 47	−13 48.9	5	11.42	.013	0.48	.013	−0.56	.026	12	B1.5V	73,771,1187,1586,1703
		NGC 6611 - 211		18 15 47	−13 51.1	1	12.93		0.76		0.30		2		1703
		NGC 6611 - 216		18 15 47	−13 51.7	1	14.47		1.11				2		1187
168097	−14 04988	NGC 6611 - 206		18 15 47	−13 59.9	3	8.25	.017	1.18	.009	0.96	.015	13	K0 II	49,1187,8100
	−15 04918			18 15 47	−15 12.8	1	9.66		2.60				1	M2	369
168021	−18 04896	HR 6848, LSS 4901	★ AB	18 15 47	−18 38.4	5	6.60	.174	0.26	.040	−0.51	.085	9	B0 Ia/ab	1423,2007,2023,6007,8100
335828	+26 03215	G 183 - 37		18 15 48	+26 39.0	2	9.59	.000	1.01	.010	0.82	.019	3	K8	1510,3072
		NGC 6611 - 228		18 15 48	−13 45.7	2	13.44	.040	0.93	.000	−0.11	.087	4		914,1187
		NGC 6611 - 231		18 15 48	−13 47.2	5	12.74	.015	0.75	.011	−0.22	.042	10		49,914,1187,1701,1703
		NGC 6611 - 223		18 15 48	−13 47.8	6	11.19	.010	0.85	.019	−0.38	.031	16	B1 V	73,771,914,1187,1701,1703
		NGC 6611 - 227		18 15 48	−13 48.4	4	12.87	.035	0.61	.035	−0.23	.046	9	B5	49,771,1701,1703
168021	−18 04897	LSS 4902	★ C	18 15 48	−18 38.2	1	7.87		0.24				1		138
		NGC 6611 - 236		18 15 49	−13 45.9	1	14.84		1.00		0.74		2		1187
		NGC 6611 - 235		18 15 49	−13 48.0	4	11.04	.030	0.81	.010	−0.46	.030	14		771,1187,1701,1703
		NGC 6611 - 234		18 15 49	−13 49.1	1	10.82		0.47		−0.54		1		1703
		NGC 6611 - 232		18 15 49	−13 49.5	1	14.81		0.90		1.32		2	A0	1701
		NGC 6611 - 240		18 15 49	−13 50.5	1	14.52		0.58				2		1701
		NGC 6611 - 256		18 15 50	−13 43.8	1	14.18		0.86		0.53		2		1187
	−13 04927	NGC 6611 - 246		18 15 50	−13 46.6	8	9.53	.024	0.85	.012	−0.29	.048	38	O7 Ib(f)	49,73,771,914,1187,1655*
		NGC 6611 - 251		18 15 50	−13 47.6	1	13.11		0.71		−0.28		1		1701
		NGC 6611 - 245		18 15 50	−13 48.3	1	13.11		0.87		0.23		2	A0	1701
		NGC 6611 - 239		18 15 50	−13 55.8	3	11.45	.020	0.39	.000	−0.47	.007	11	B3	771,1187,1703
		NGC 6611 - 259		18 15 51	−13 46.7	4	11.62	.024	0.76	.016	−0.34	.041	14		771,914,1187,1703
		NGC 6611 - 254		18 15 51	−13 48.1	4	10.82	.021	0.47	.014	−0.46	.029	14	B2	771,914,1187,1703
		NGC 6611 - 260		18 15 51	−13 50.1	1	14.37		0.70		0.36		1		49
168080	−18 04900	LSS 4906	A	18 15 51	−18 11.6	1	7.61		0.12		−0.67		1	B1 Ib/II	8100
168080	−18 04900		B	18 15 51	−18 11.6	1	9.41		0.04		−0.66		2		8100
		NGC 6611 - 266		18 15 52	−13 47.8	2	14.25	.131	1.17	.039	0.41		3		1701,1703
		NGC 6611 - 275		18 15 52	−13 48.1	2	12.01	.063	0.51	.049	−0.47	.019	3	B5	771,1703
		NGC 6611 - 273		18 15 52	−13 48.8	1	14.55		0.40		0.37		2	A0	1701
		NGC 6611 - 276		18 15 52	−13 50.0	1	13.58		0.81		0.06		1		49
		NGC 6611 - 265		18 15 52	−13 53.4	1	11.86		0.40		0.31		4		1703
168112	−12 04988	LSS 4912		18 15 53	−12 07.6	2	8.52	.000	0.69	.000	−0.40	.000	4	O5	1011,1012
	−13 04928	NGC 6611 - 280		18 15 53	−13 48.1	8	10.08	.019	0.46	.027	−0.56	.030	23	B0.5Vne	445,771,914,1012,1032,1187*
		NGC 6611 - 294		18 15 54	−13 46.3	1	16.45		1.18				1		1187
		NGC 6611 - 297		18 15 54	−13 47.1	2	12.86	.005	0.68	.000	−0.19	.045	8		1701,1703
		NGC 6611 - 289		18 15 54	−13 50.2	3	12.57	.030	0.51	.039	−0.26	.054	7	B5	771,914,1703
168056	−28 14441			18 15 54	−28 14.2	1	8.90		0.09		−0.30		1	B8 IV	1586
		NGC 6611 - 305		18 15 55	−13 46.7	4	13.57	.040	1.10	.015	−0.08	.030	8		49,914,1187,1703
		NGC 6611 - 306		18 15 55	−13 46.9	3	12.73	.026	0.66	.018	−0.26	.024	8		771,1701,1703
		NGC 6611 - 301		18 15 55	−13 47.7	4	12.16	.093	0.64	.050	−0.32	.033	11	B5	771,1187,1701,1703
		NGC 6611 - 307		18 15 55	−13 48.7	2	14.23	.102	0.89	.136	0.20	.166	3		49,1701
		NGC 6611 - 300		18 15 55	−13 49.1	1	12.62		0.58		−0.19		1		1703
		NGC 6611 - 311		18 15 55	−13 49.1	3	12.73	.148	0.59	.008	−0.21	.099	6		1187,1701,1703
		NGC 6611 - 296		18 15 55	−13 49.2	4	11.80	.005	0.49	.009	−0.48	.019	14	B2 V	73,771,1187,1703
		NGC 6611 - 299		18 15 55	−13 50.6	2	14.14	.039	0.70	.039	0.50	.015	3	A3	1701,1703
		NGC 6611 - 290		18 15 55	−13 57.6	3	12.14	.000	0.43	.025	−0.36	.009	6	F5	771,1187,1703
	−18 04902			18 15 55	−18 01.2	1	10.27		0.16		−0.56		4		413
168270	+18 03623	HR 6852		18 15 56	+18 06.7	2	6.23		0.01	.020	−0.12	.015	5	B9 V	1022,1733
	−13 04929	NGC 6611 - 314		18 15 56	−13 47.8	9	9.85	.022	0.62	.021	−0.48	.045	25	O9.5V	49,73,771,914,1012,1032*
		NGC 6611 - 313		18 15 56	−13 50.7	2	13.08	.166	0.49	.107	−0.17	.024	3		1701,1703
		NGC 6611 - 323		18 15 57	−13 49.1	1	13.45		0.57		−0.18		2		1187
	−21 04944			18 15 57	−21 03.6	1	10.94		1.32				2		1726
		NGC 6611 - 333		18 15 58	−13 47.4	1	16.75		1.09		−0.33		1		1187
		NGC 6611 - 327		18 15 58	−13 54.2	1	15.07		0.61		0.83		1		1703
167935	−46 12292	IDS18122S4631	AB	18 15 58	−46 29.7	1	8.56		0.88		0.55		12	G5 III	1589
168159	−6 04738			18 15 59	−06 43.3	1	9.28		1.06		0.91		3	K3 V	3072
		NGC 6611 - 341		18 15 59	−13 47.3	1	15.79		1.14		−0.13		2		1187
		NGC 6611 - 335		18 15 59	−13 47.6	1	16.58		0.87		−0.08		2		1187
		NGC 6611 - 339		18 15 59	−13 48.1	1	13.75		0.89		0.48		2		1701
		NGC 6611 - 343		18 15 59	−13 48.1	3	11.79	.044	0.85	.013	−0.14	.020	11	F3	771,1701,1703
		NGC 6611 - 336		18 15 59	−13 49.3	2	13.29	.005	0.51	.024	−0.16	.170	2		1187,1703
		NGC 6611 - 349		18 16 00	−13 48.8	4	11.63	.015	1.44	.024	1.03	.030	16	K0 II	49,1187,1701,1703

Table 1   995

HD	DM	Other Id	N	Rem	α_{1950}	δ_{1950}	S	V	σ_V	B–V	σ_{B-V}	U–B	σ_{U-B}	n	Spectrum	References
		NGC 6611 - 345			18 16 00	−13 54.3	1	14.12		1.02		0.45		2		1703
167954	−45 12390				18 16 00	−45 43.1	2	6.85	.005	0.53	.005	0.05		8	F7 V	158,2012
		LP 629 - 27			18 16 01	−01 23.7	1	12.73		1.19		1.07		1		1696
		NGC 6611 - 351			18 16 01	−13 49.5	4	11.28	.022	0.46	.009	-0.55	.026	14	B1 Vne	73,771,1187,1703
168101	−21 04946				18 16 01	−21 34.1	1	9.12		1.80				2	K4	369
168135	−12 04991				18 16 02	−12 28.5	1	7.93		0.17		-0.27		2	B8 IV/V	1586
	−13 04930	NGC 6611 - 367			18 16 02	−13 51.0	10	9.43	.015	0.27	.013	-0.73	.049	26	O9.5V	49,73,771,914,1011,1012*
168323	+23 03299	HR 6854	★	A	18 16 03	+23 16.6	1	6.47		1.74		2.06		2	K5	1733
		NGC 6611 - 374			18 16 03	−13 47.4	2	13.16	.175	0.95	.005	0.43	.195	3		1701,1703
		NGC 6611 - 371			18 16 03	−13 48.0	1	13.55		0.61		0.10		2		1701
168114	−16 04804				18 16 03	−16 41.0	1	9.39		0.09		-0.51		1	B4 II	963
168138	−19 04958				18 16 03	−19 28.7	3	9.28	.005	0.03	.004	-0.64	.007	6	B2 III	963,1732,2023
		LSS 4914			18 16 04	−12 06.5	2	11.52	.000	0.69	.000	-0.31	.000	4	B3 Ib	405,1737
		NGC 6611 - 407			18 16 04	−13 34.5	1	11.45		0.56		0.51		2	B9	1703
		NGC 6611 - 375			18 16 04	−13 50.2	1	15.02		0.80				2		1187
		NGC 6611 - 587			18 16 04	−13 56.9	1	11.95		0.43		-0.32		2		771
168136	−13 04931			V	18 16 05	−13 33.7	1	7.94		1.76				3	K4 III	369
		NGC 6611 - 388			18 16 05	−13 50.0	1	13.16		0.41				2		1701
		NGC 6611 - 396			18 16 05	−13 50.0	2	13.97	.044	1.22	.015	0.56	.010	3		1701,1703
		NGC 6611 - 389			18 16 05	−13 56.0	1	14.38		0.56				3		1187
		NGC 6611 - 406			18 16 06	−13 46.1	1	11.53		1.95		2.46		1		1703
		NGC 6611 - 402			18 16 06	−13 47.7	2	11.50	.000	1.48	.018	1.35		10	K0 II	1701,1703
		NGC 6611 - 400			18 16 06	−13 48.2	2	12.79	.063	0.68	.093	0.24		6	A2	1701,1703
168137	−13 04932	NGC 6611 - 401			18 16 06	−13 49.8	7	8.94	.009	0.40	.010	-0.67	.046	22	O8 V	49,73,771,914,1187,1703,1732
		NGC 6611 - 411			18 16 07	−13 45.4	2	12.08	.122	1.63	.024	1.72	.088	3		771,1703
		NGC 6611 - 409	★	V	18 16 07	−13 53.5	2	12.82	.000	0.41	.020	-0.31	.055	4		914,1703
168183	−14 04991	NGC 6611 - 412			18 16 08	−14 00.8	6	8.24	.023	0.32	.009	-0.70	.041	16	O8 V	49,771,963,1187,1703,1732
168162	−15 04921				18 16 09	−15 29.7	1	9.57		0.50		0.12		3	B2 Ib	3063
		NGC 6611 - 444			18 16 10	−13 44.0	1	12.66		0.79		-0.13		1		1703
		NGC 6611 - 440			18 16 10	−13 47.9	1	12.29		1.39		1.15		1		1703
		NGC 6611 - 432			18 16 10	−13 56.8	1	11.23		0.52		0.37		1	A0	1703
168165	−16 04805				18 16 10	−16 40.7	1	8.17		1.08		0.75		1	G8 III	963
	+37 03070				18 16 12	+37 57.6	1	8.65		0.87		0.54		2	K0	1601
		NGC 6611 - 455			18 16 13	−13 48.6	3	12.05	.004	0.61	.004	0.36	.020	7	A5	1187,1701,1703
		NGC 6611 - 462			18 16 14	−13 44.0	1	9.47		0.27		-0.78		1		1703
	−12 04994	LSS 4918			18 16 15	−12 07.9	2	9.81	.000	0.70	.000	-0.39	.000	2	O9:II:	1011,1012
		NGC 6611 - 472			18 16 15	−13 46.0	2	12.80	.136	0.49	.015	-0.31	.015	3		914,1703
	−13 04933	NGC 6611 - 469			18 16 15	−13 49.6	7	10.71	.014	0.41	.010	-0.59	.024	19	B1.5V	49,73,771,914,1187,1586,1703
		NGC 6611 - 473			18 16 15	−13 54.8	2	12.48	.045	0.34	.025	-0.12	.075	4		771,1703
	−13 04934	NGC 6611 - 468			18 16 15	−13 56.1	4	9.47	.013	0.28	.007	-0.72	.027	21	B1 V	49,771,1187,1703
	−13 04935	NGC 6611 - 483			18 16 16	−13 44.8	2	10.97	.010	0.44	.005	-0.20	.024	6	F8	771,1703
		NGC 6611 - 599			18 16 16	−13 59.1	3	10.16	.017	1.34	.027	1.14	.030	8		771,1586,1703
		NGC 6611 - 489			18 16 17	−13 44.4	3	11.56	.012	0.44	.017	0.02	.024	6	F2	49,771,1703
		NGC 6611 - 484			18 16 17	−13 46.4	1	12.21		0.52		0.03		1		1703
		POSS 391 # 5			18 16 18	+24 58.7	1	17.75		1.60				3		1739
	+38 03134		A		18 16 19	+38 19.2	1	10.14		0.31		0.12		2	F0	3032
	+38 03134		B		18 16 19	+38 19.2	1	11.00		0.39		0.05		1	Am	3032
	+68 00986	G 258 - 31	★	A	18 16 19	+68 37.4	1	10.13		0.69		0.11		3	G8	1064
		NGC 6611 - 500			18 16 19	−13 44.5	2	11.24	.010	0.45	.010	-0.09	.005	6	B9	771,1703
		NGC 6611 - 494			18 16 19	−13 52.0	1	11.24		1.73		1.91		2		1703
168206	−11 04593	CV Ser			18 16 20	−11 39.3	2	9.11	.019	0.76	.029	-0.24	.014	7	C8	1012,1095
		NGC 6611 - 504			18 16 20	−13 50.4	1	12.84		0.45		-0.01		3		1703
168344	+14 03482				18 16 21	+14 21.5	1	7.63		1.06				2	K2	882
	−13 04936	NGC 6611 - 503			18 16 21	−13 58.0	7	9.76	.087	0.55	.043	-0.62	.052	14	B1 Vne	49,445,1012,1032,1187*
		NGC 6611 - 525			18 16 24	−13 41.0	1	13.29		1.74		1.53		2		771
		NGC 6611 - 520			18 16 24	−13 53.7	1	11.68		0.46		-0.09		4	F8	1703
		G 204 - 58			18 16 25	+38 45.8	1	11.86		1.58				1		940
		G 204 - 57			18 16 25	+38 45.9	1	13.52		1.77				1		940
		WLS 1824 70 # 10			18 16 27	+69 54.8	1	12.53		1.03		0.67		2		1375
		NGC 6611 - 534			18 16 27	−13 52.8	2	12.97	.124	1.37	.030	1.14	.059	5		771,1703
		NGC 6611 - 536			18 16 28	−13 57.0	3	11.43	.008	0.35	.010	-0.55	.013	11	F8	771,1187,1703
168385	+19 03586				18 16 29	+19 56.0	1	9.04		0.03		-0.18		1		963
	−13 04937	NGC 6611 - 551			18 16 29	−13 55.7	1	10.65		0.38		-0.48		1	B2 V	49
168227	−15 04923	FO Ser			18 16 29	−15 38.1	3	8.52	.065	1.83	.033	2.00	.000	4	C4,5i	1238,3039,8005
168060	−45 12402				18 16 29	−45 56.5	1	7.32		0.76				4	G5 V	2033
		NGC 6611 - 601			18 16 30	−13 55.7	2	10.66	.000	0.35	.015	-0.56	.040	4	B2	771,1703
168229	−18 04909	LSS 4922			18 16 31	−18 14.4	1	8.61		0.19		-0.78		2	B2/3 Ib/II	1737
167918	−56 08740				18 16 31	−56 57.3	1	7.78		-0.06		-0.47		36	B5 III	1699
168637	+55 02049				18 16 32	+55 26.1	1	8.16		0.16		0.11		3	A2	963
168230	−18 04908	LSS 4921			18 16 32	−18 52.4	1	9.54		0.51		-0.26		2	B2 Ib/II	540
		NGC 6611 - 617			18 16 34	−13 11.2	1	12.29		1.57		1.85		2		771
168601	+53 02064				18 16 36	+53 43.4	1	8.63		0.47		0.07		2	F8	1566
		LSS 4923			18 16 36	−15 20.1	1	10.45		0.84		-0.30		3	B3 Ib	1737
		NGC 6611 - 605			18 16 38	−14 03.3	1	12.26		0.80		0.23		2		771
		G 21 - 9			18 16 40	−05 47.6	3	12.62	.036	1.44	.015	1.13	.010	4		202,1620,1658
		NGC 6611 - 611			18 16 41	−13 35.7	1	11.98		1.63		1.34		2		771
		NGC 6618 - 79			18 16 42	−16 10.0	1	12.54		1.54		0.79		2	K0	456
168144	−38 12692				18 16 42	−38 50.0	1	9.54		-0.08		-0.41		3	B7 V	1586
		Steph 1597			18 16 43	+16 42.4	1	12.32		1.33		1.37		1	K7	1746
		NGC 6611 - 606			18 16 43	−14 03.1	1	10.50		1.24		1.00		2		1703

HD	DM	Other Id	N	Rem	α_{1950}	δ_{1950}	S	V	σ_V	B–V	σ_{B-V}	U–B	σ_{U-B}	n	Spectrum	References
168276	−16 04811	NGC 6618 - 60		⋆ AB	18 16 43	−16 01.9	1	9.34		0.56		0.03		2	F7 V	456
168387	+7 03629	HR 6857		⋆ A	18 16 44	+07 14.3	3	5.41	.027	1.08	.015	1.04	.027	10	K2 III-IV	15,1256,3024
168387	+7 03629	IDS18143N0713		B	18 16 44	+07 14.3	1	12.54		0.43		0.19		4		3024
168564	+45 02690				18 16 45	+45 08.1	1	7.85		0.09		0.10		2	A0	1733
168302	−16 04812	NGC 6618 - 62			18 16 46	−16 02.7	2	9.34	.057	0.23	.019	-0.35	.009	4	B3 III	456,1012
		NGC 6618 - 80			18 16 46	−16 11.3	1	14.56		0.53		0.01		1		456
168430	+14 03486				18 16 48	+14 33.2	1	8.04		1.27				3	K5	882
313171	−18 04914				18 16 50	−18 29.2	1	9.45		1.06		0.77		3	K0	413
168147	−44 12514				18 16 50	−44 06.0	1	7.18		1.62				4	M3 III	2012
		NGC 6613 - 21			18 16 51	−17 08.7	1	12.56		1.38		1.50		3		161
		NGC 6613 - 47			18 16 51	−17 09.7	1	10.60		0.70		0.32		2	F8	161
		KUV 18169+6643			18 16 52	+66 43.0	1	16.71		-0.28		-1.32		1	DB	1708
		NGC 6613 - 20			18 16 52	−17 08.3	1	12.55		0.31		0.01		3	B9	161
168413	+0 03918				18 16 53	+00 49.2	1	7.93		1.95				2	K5	369
		NGC 6611 - 614			18 16 53	−13 10.2	1	12.33		0.66		0.68		2	F8	771
167468	−75 01417	HR 6829			18 16 53	−75 04.1	3	5.47	.004	0.02	.000	0.04	.000	14	A0 V	15,1075,2038
168214	−40 12394				18 16 57	−40 17.7	1	7.96		0.87		0.40		1	G6p Ba	565
		NGC 6613 - 16			18 16 59	−17 07.3	1	10.33		0.24		-0.40		3	B6	161
168352	−17 05136	NGC 6613 - 15			18 17 00	−17 05.7	2	8.67	.025	0.32	.000	-0.43	.020	10	B2 II	161,206
		NGC 6613 - 3			18 17 00	−17 09.	1	13.46		2.30				3		161
168251	−32 14024				18 17 01	−32 22.0	1	9.04		0.05		-0.21		3	B8 II/III	1586
		NGC 6618 - 134			18 17 03	−16 10.0	1	15.22		0.90		0.60		2		456
		NGC 6613 - 28			18 17 03	−17 09.8	1	10.47		0.27		-0.44		2	B6	161
168235	−32 14027				18 17 03	−32 42.1	1	9.44		0.02		-0.19		3	B9	1586
		NGC 6618 - 136			18 17 04	−16 10.3	1	12.69		0.38		-0.08		2		456
		NGC 6618 - 137			18 17 04	−16 10.7	1	12.85		1.83		1.48		2		456
		NGC 6613 - 1			18 17 04	−17 06.4	1	11.93		0.33		-0.18		3	B7	161
		NGC 6613 - 2			18 17 04	−17 06.8	1	13.47		0.44		0.18		4		161
168393	−11 04596				18 17 05	−11 18.6	2	7.41	.000	0.56	.003	0.12	.021	8	F5 II	253,8100
	−12 05004	LSS 4925			18 17 05	−12 31.9	2	10.47	.000	0.89	.000	-0.20	.000	4	O9 IV	405,1737
	−17 05137	NGC 6613 - 13			18 17 05	−17 04.9	2	9.25	.000	0.47	.005	0.40	.025	7	A6 V	161,206
168331	−24 14189				18 17 05	−24 58.0	1	9.27		0.16		-0.35		3	B6/7 III	1586
168532	+24 03381	HR 6860			18 17 07	+24 25.4	2	5.26	.005	1.51	.020	1.76	.015	6	K4 II	1080,8100
		NGC 6613 - 12			18 17 08	−17 06.4	1	10.39		0.22		-0.22		3	B5 V	161
168368	−17 05139	NGC 6613 - 38			18 17 09	−17 04.9	2	9.37	.010	0.30	.015	-0.43	.040	7	B2 Ib/II	161,206
		NGC 6613 - 11			18 17 09	−17 06.8	1	12.02		1.26		1.07		3		161
168442	−1 03474				18 17 15	−01 57.7	5	9.66	.010	1.36	.014	1.22	.009	13	K7 V	158,196,1705,1775,3072
	−13 04941	LSS 4926			18 17 15	−13 04.5	3	9.75	.005	1.06	.029	-0.07	.005	4	O9.5IV	1012,1032,1737
		NGC 6618 - 195			18 17 15	−16 09.1	1	12.96		0.78		0.14		2		456
169027	+68 00989				18 17 16	+68 43.3	1	6.79		-0.07		-0.23		4	A0	3016
168415	−15 04927	HR 6858			18 17 16	−15 51.2	3	5.38	.018	1.47	.014	1.64		9	K3 III	369,2007,3005
		NGC 6618 - 197			18 17 16	−16 09.5	1	14.65		0.75		0.23		2		456
	−17 05140	NGC 6613 - 31			18 17 17	−17 08.4	1	9.97		0.25		-0.41		4	B3	206
314372	−22 04677				18 17 17	−22 31.6	1	9.44		1.74				2	M0	369
168499	+10 03473	IDS18150N1014		AB	18 17 19	+10 15.1	2	8.24	.020	0.61	.010	0.06		5	G1 V +G6 V	176,3016
168443	−9 04692	G 155 - 11			18 17 19	−09 36.9	1	6.92		0.70		0.31		1	G8 V	3062
168638	+39 03380				18 17 20	+39 14.7	1	7.86		1.33		1.34		4	K2 III	833
		NGC 6618 - 255			18 17 20	−16 09.3	1	15.41		0.56		0.17		2		456
234610	+51 02350				18 17 21	+51 31.5	1	8.76		1.27		1.47		1		1746
	−12 05008	LSS 4928			18 17 21	−12 39.7	3	9.82	.006	1.00	.006	-0.08	.002	9	B0.5III	854,1012,1032
168421	−20 05095				18 17 21	−20 32.1	2	9.56	.002	0.22	.009	-0.63	.007	5	B2/3 Iab	540,1732
		G 154 - 57			18 17 22	−13 51.9	1	13.52		1.39		1.15		1		3062
168418	−17 05141	NGC 6613 - 82			18 17 22	−17 01.6	2	9.43	.000	0.31	.009	-0.45	.009	6	B2 III	206,1012
168419	−18 04917				18 17 22	−18 36.9	1	8.88		0.27		0.04		2	A3/5 II	634
		NGC 6618 - 275			18 17 23	−16 15.7	1	13.02		0.55		0.32		1		456
168444	−14 05002	LSS 4929			18 17 24	−14 52.2	2	8.76	.005	0.60	.029	-0.45	.005	3	O9.5/B0 Iab	405,432
		NGC 6618 - 256			18 17 24	−16 09.4	1	12.75		1.47		1.37		2	K2	456
238846	+55 02053				18 17 25	+55 25.5	1	9.65		-0.17		-0.73		3	A5	963
		NGC 6618 - 273			18 17 25	−16 15.2	1	14.22		0.67		-0.67		1		456
	−15 04930	LSS 4930			18 17 28	−15 05.3	2	9.41	.010	0.73	.005	-0.39	.015	3	O6.5V((f))	1011,1737
		NGC 6618 - 247			18 17 28	−16 06.2	1	12.64		0.48		0.01		2		456
		NGC 6618 - 269			18 17 28	−16 13.0	1	14.96		1.06		0.25		1		7002
168461	−12 05009	LSS 4931			18 17 29	−12 11.6	3	9.54	.000	0.68	.000	-0.42	.000	5	(O9) Ia	1011,1012,1032
		NGC 6618 - 259			18 17 29	−16 10.1	1	11.68		1.72		2.10		1		7002
		NGC 6618 - 266			18 17 29	−16 11.7	1	12.65		0.74		0.18		1		7002
		NGC 6618 - 276			18 17 29	−16 16.2	1	14.81		0.91		0.73		1		456
		NGC 6618 - 258			18 17 30	−16 09.9	1	13.69		1.30		0.04		1		7002
		NGC 6618 - 704			18 17 30	−16 11.3	1	14.81		0.97		0.29		1		7002
		NGC 6618 - 737			18 17 30	−16 11.5	1	15.80		0.95		0.23		1		7002
168357	−37 12457	HR 6856			18 17 30	−37 30.6	1	6.45		1.31				4	K2 II	2009
		NGC 6618 - 325			18 17 31	−16 09.6	2	11.14	.090	0.73	.105	0.07	.060	2	F8	456,7002
		NGC 6618 - 712			18 17 31	−16 10.0	1	15.55		1.42		0.72		1		7002
		NGC 6618 - 331			18 17 31	−16 10.5	2	13.15	.015	0.58	.005	0.41		3		725,7002
		NGC 6618 - 336			18 17 31	−16 11.3	1	14.93		1.02		0.16		2		456
		NGC 6618 - 339			18 17 31	−16 11.9	1	15.24		0.90		0.90		1		7002
168220	−54 08835				18 17 31	−54 12.8	1	8.85		1.09				3	K0 III	1594
		NGC 6618 - 306			18 17 32	−16 03.1	1	14.17		0.73		0.20		2		456
		NGC 6618 - 334			18 17 32	−16 11.1	2	14.45	.093	0.80	.000	0.22	.107	3		456,7002
		NGC 6618 - 705			18 17 32	−16 11.7	1	15.55		1.47		0.48		1		7002
		NGC 6618 - 355			18 17 32	−16 18.1	1	14.68		1.12		0.97		2		456

Table 1

997

HD	DM	Other Id	N	Rem	α_{1950}	δ_{1950}	S	V	σ_V	B–V	σ_{B-V}	U–B	σ_{U-B}	n	Spectrum	References
		G 260 - 1			18 17 33	+68 32.8	1	15.40		1.77				1		1759
	−5 04630	LS IV -05 004			18 17 33	−05 05.0	1	10.31		1.10		0.17		2		1586
		NGC 6618 - 323			18 17 33	−16 09.5	1	12.97		0.34		0.33		1		7002
		NGC 6618 - 326			18 17 33	−16 09.9	1	14.20		1.67		0.52		1		7002
		NGC 6618 - 728			18 17 33	−16 11.6	1	15.44		1.46		0.53		1		7002
		NGC 6618 - 703			18 17 33	−16 12.2	1	14.53		1.04		0.55		1		7002
		NGC 6618 - 707			18 17 33	−16 12.4	1	15.74		1.37		0.51		1		7002
		NGC 6618 - 362			18 17 33	−16 20.1	1	13.35		0.32		0.21		1		456
168448	−17 05142	NGC 6613 - 73			18 17 33	−17 08.2	1	9.18		0.23		-0.45		4	B2 II	206
		NGC 6618 - 305			18 17 34	−16 03.0	1	14.80		0.93		0.40		2		456
		NGC 6618 - 327			18 17 34	−16 10.0	1	14.52		0.44		-0.05		1	O5	7002
		NGC 6618 - 345			18 17 34	−16 14.9	1	11.06		1.23				2	O8	725
		NGC 6618 - 356			18 17 34	−16 18.3	1	11.54		0.29		0.18		5		456
		NGC 6618 - 335			18 17 35	−16 11.3	2	12.98	.073	0.94	.029	0.55	.024	3	G6	456,7002
		NGC 6618 - 338			18 17 35	−16 11.7	1	14.43		0.74		0.58		1		7002
		NGC 6618 - 727			18 17 35	−16 12.5	1	16.68		1.44				1		7002
	−16 04816	NGC 6618 - 343			18 17 35	−16 13.4	3	9.68	.048	1.62	.009	1.87	.083	8	K2	456,725,7002
	+60 01820	IDS18170N6048		AB	18 17 36	+60 48.9	1	11.15		0.60		0.07		1		8084
	+60 01820	IDS18170N6048		C	18 17 36	+60 48.9	1	10.72		0.52		0.06		1		8084
	+60 01820	IDS18170N6048		D	18 17 36	+60 48.9	1	10.73		0.60		0.02		1		8084
		NGC 6618 - 298			18 17 36	−16 00.8	1	13.83		0.87		0.29		1		456
		NGC 6618 - 324			18 17 36	−16 09.6	1	15.18		1.05		0.69		1		7002
		NGC 6618 - 330			18 17 36	−16 10.5	1	15.80		1.02		0.50		1		7002
		NGC 6618 - 723			18 17 36	−16 12.5	1	16.12		1.77				1		7002
168463	−17 05143	NGC 6613 - 74			18 17 36	−17 07.5	1	9.70		0.13		-0.40		4	B3 Ib	206
314334					18 17 36	−21 23.	1	9.81		1.91				1		369
		NGC 6618 - 341			18 17 37	−16 12.2	1	14.13		2.15		0.76		1		7002
		NGC 6618 - 363			18 17 37	−16 20.1	1	15.05		0.91		1.14		1		456
		NGC 6618 - 332			18 17 38	−16 10.7	2	12.88	.175	0.32	.010	0.04	.049	3		456,7002
		NGC 6618 - 706			18 17 38	−16 11.5	1	15.41		1.44		0.59		1		7002
		NGC 6618 - 725			18 17 38	−16 11.5	1	15.28		1.61		0.74		1		7002
	−14 05003	G 154 - 58			18 17 39	−14 22.8	1	10.62		1.14		1.07		1		1696
		NGC 6618 - 329			18 17 39	−16 10.4	1	14.91		0.73		0.25		1		7002
		NGC 6618 - 709			18 17 39	−16 13.0	1	15.56		1.02		-0.02		1		7002
		NGC 6618 - 361			18 17 39	−16 20.0	1	13.87		1.31		1.25		1		456
168486	−16 04817	NGC 6618 - 365			18 17 39	−16 20.7	1	10.69		0.13		-0.30		1	B9 V	456
		NGC 6618 - 302			18 17 41	−16 02.6	1	12.07		0.60		0.07		1	F5	456
		NGC 6618 - 299			18 17 42	−16 00.9	1	12.97		0.70		0.31		1		456
		NGC 6618 - 328			18 17 42	−16 09.9	1	13.14		1.07		1.12		1		7002
		NGC 6618 - 333			18 17 42	−16 11.0	1	13.55		0.94		0.64		1		7002
		NGC 6618 - 337			18 17 42	−16 11.5	2	11.21	.122	1.12	.014	-0.08	.066	4		456,7002
	−16 04818	NGC 6618 - 342			18 17 42	−16 12.2	3	9.94	.061	0.79	.004	-0.28	.005	4	B1 Vp	456,725,7002
		NGC 6618 - 414			18 17 43	−16 12.4	1	13.22		0.78		-0.17		1		7002
		NGC 6618 - 402			18 17 44	−16 08.1	1	15.02		0.69				1		456
		NGC 6618 - 425			18 17 48	−16 14.2	1	14.01		1.41		1.38		1		456
168454	−29 14834	HR 6859		★ A	18 17 48	−29 51.1	5	2.70	.000	1.38	.005	1.55	.005	22	K3 IIIa	15,1075,2012,3005,8015
		NGC 6618 - 401			18 17 49	−16 07.2	1	11.64		0.44		0.30		2		456
		NGC 6618 - 432			18 17 49	−16 17.1	1	12.41		1.28		0.99		1	G8	456
		NGC 6618 - 435			18 17 50	−16 18.5	1	15.19		0.72		0.09		1		456
		NGC 6618 - 390			18 17 52	−16 03.9	1	11.35		0.34		-0.28		6	A6	456
		LSS 4936			18 17 53	−14 47.8	2	11.08	.000	0.71	.000	-0.21	.000	4	B3 Ib	405,1737
		NGC 6618 - 400			18 17 54	−16 07.0	1	15.23		1.14		0.76		2		456
	−16 04822	NGC 6618 - 458			18 17 55	−16 03.0	1	9.75		0.53		-0.32		3	B0 V	456
		NGC 6618 - 389			18 17 55	−16 03.2	1	13.52		0.43		-0.27		1		456
168694	+29 03236	HR 6867			18 17 56	+29 38.6	1	5.99		1.29				2	gK4	71
		NGC 6618 - 492			18 17 56	−16 12.7	1	13.52		0.35		0.13		1		456
		NGC 6618 - 497			18 17 56	−16 13.3	1	14.08		0.49		0.50		1		456
168606	+6 03701				18 17 57	+06 31.3	1	8.27		0.06		-0.15		2	A0	212
	−13 04946	LSS 4939			18 17 57	−13 34.5	2	10.54	.000	1.13	.000	0.02	.000	4	B0.5 III	405,1737
		NGC 6618 - 467			18 17 57	−16 04.2	1	13.33		0.34		0.23		2		456
	−5 04631				18 17 58	−05 31.3	1	10.38		0.92		0.64		1	K3 V	3072
		NGC 6618 - 518			18 17 58	−16 18.1	1	11.55		0.27		0.05		4	A0	456
		NGC 6618 - 499			18 18 01	−16 13.9	1	13.85		0.73		0.08		1		456
	−4 04445				18 18 03	−04 57.8	1	10.85		2.38				2	M2	369
		NGC 6618 - 513			18 18 03	−16 17.9	1	14.46		0.55		0.13		2		456
168469	−37 12468	IDS18147S3717		AB	18 18 03	−37 16.3	2	8.18	.003	-0.04	.001	-0.41	.005	3	B7 III	540,976
	−14 05007				18 18 04	−14 40.9	1	9.83		1.96				2	M3	369
		NGC 6618 - 481			18 18 04	−16 09.8	1	12.78		0.39		0.24		2		456
168695	+25 03493	IDS18160N2529		AB	18 18 05	+25 30.5	2	8.83	.015	0.22	.015	0.13	.005	5	A0	379,833
168775	+36 03094	HR 6872			18 18 06	+36 02.5	5	4.34	.012	1.16	.013	1.18	.012	10	K2 IIIab	15,1080,1363,3016,8015
168794	+39 03383				18 18 06	+39 04.1	1	8.08		-0.02		-0.09		5	A0 V	833
168552	−17 05149				18 18 06	−17 10.5	4	8.09	.010	0.31	.020	-0.48	.018	10	B2/3 Ib/II	138,206,399,8100
335916	+26 03219				18 18 07	+26 29.1	1	8.68		1.56				2	K3 III	20
		NGC 6618 - 530			18 18 07	−16 18.3	1	13.79		0.61		0.30		1		456
168568	−15 04937				18 18 08	−15 19.4	1	9.64		1.84				2	B8 III/IV	369
	−16 04826	NGC 6618 - 569			18 18 09	−16 02.5	1	9.89		0.76		-0.37		2	O3 V	1011
168555	−19 04972				18 18 09	−18 59.6	1	10.31		0.03		-0.08		2	B9.5III	634
168811	+39 03385				18 18 10	+39 34.3	1	8.45		1.50		1.76		2	K5	1601
168567	−15 04938				18 18 10	−15 15.3	1	6.93		1.76				2	K3 III	369
		NGC 6618 - 597			18 18 10	−16 09.9	1	14.72		0.85		0.39		3		456

HD	DM	Other Id	N Rem	α_{1950}	δ_{1950}	S	V	σ_V	B–V	σ_{B-V}	U–B	σ_{U-B}	n	Spectrum	References
168720	+21 03390	HR 6868		18 18 11	+21 56.3	6	4.94	.009	1.59	.011	1.98	.010	26	M1 IIIb	15,1355,1363,3016*
		NGC 6618 - 623		18 18 11	−16 17.0	1	13.81		0.58		0.31		2		456
		NGC 6618 - 631		18 18 11	−16 19.2	1	14.23		0.57		0.26		1		456
168912	+52 02184			18 18 12	+52 37.8	1	7.33		1.22		1.22		2	K5	1566
168583	−15 04939			18 18 12	−15 00.1	1	8.10		1.86				2	M0 III	369
		G 141 - 2		18 18 13	+12 37.4	2	15.91	.020	0.53	.005	-0.19	.020	2		1658,3060
		NGC 6618 - 576		18 18 13	−16 05.2	1	11.70		0.26		0.08		2		456
168569	−16 04827	NGC 6618 - 578		18 18 13	−16 08.0	1	10.00		0.23		-0.05		3	A1 II/III	456
		NGC 6618 - 602		18 18 14	−16 11.2	1	16.16		1.26		-0.64		1		456
		NGC 6618 - 604		18 18 14	−16 11.5	1	15.90		0.48				1		456
168571	−17 05151	LSS 4947		18 18 14	−17 24.3	2	7.84	.038	0.52	.005	-0.41	.009	4	B1 Iab/b	1012,8100
		NGC 6618 - 577		18 18 16	−16 05.9	1	14.10		0.79		0.22		2		456
168585	−16 04828	NGC 6618 - 565		18 18 17	−16 01.8	1	9.28		0.19		-0.26		1	B8 V	456
		NGC 6618 - 605		18 18 17	−16 11.9	1	11.72		0.23		0.04		5	A0	456
168572	−21 04961			18 18 18	−21 23.8	1	9.22		1.73				2	K4 (III)	369
	+2 03569			18 18 19	+02 47.9	1	9.21		1.73				1	K2	246
168657	+3 03679			18 18 19	+03 07.0	1	8.34		1.51		1.57		1	K5	246
		NGC 6618 - 613		18 18 19	−16 13.6	1	16.26		0.97		0.92		1		456
		NGC 6618 - 621		18 18 19	−16 16.0	1	16.48		0.70		-0.20		1		456
		NGC 6618 - 635		18 18 19	−16 20.7	1	15.48		0.79		-0.09		1		456
		G 184 - 4		18 18 20	+16 06.9	1	13.56		0.78		0.22		2		1658
		LSS 4953		18 18 20	−15 51.8	1	11.87		0.67		-0.44		2	B3 Ib	1737
		NGC 6618 - 679		18 18 20	−16 12.3	1	13.47		1.13		0.71		4		456
		NGC 6618 - 675		18 18 21	−16 11.2	1	13.25		0.74		0.36		3		456
		V1290 Sgr		18 18 21	−32 39.6	1	11.82		0.55		0.31		1		689
168656	+3 03680	HR 6866	⋆ A	18 18 22	+03 21.2	10	4.85	.008	0.91	.008	0.61	.005	56	G8 III	15,1075,1080,1088*
168656	+3 03680	IDS18159N0320	B	18 18 22	+03 21.2	1	12.05		1.09		0.50		3		3024
		NGC 6618 - 680		18 18 22	−16 13.1	1	14.80		0.92		0.47		1		456
168607	−16 04829	V4029 Sgr		18 18 22	−16 24.0	3	8.25	.052	1.55	.015	0.41	.005	7	(B1/7) Iae	1012,1586,1737
		NGC 6618 - 701		18 18 23	−16 21.8	1	14.15		0.85		0.22		1		456
		G 21 - 10		18 18 24	−01 03.9	2	12.66	.039	1.62	.034	1.22		3		202,3078
		NGC 6618 - 672		18 18 24	−16 09.9	1	11.56		0.37		0.20		2	B9	456
		NGC 6618 - 702		18 18 24	−16 22.7	1	15.60		0.79		0.31		1		456
168625	−16 04830	V4030 Sgr		18 18 26	−16 23.9	3	8.39	.024	1.41	.016	0.37	.004	6	B2/5 Ia(e)	1012,1586,1737
168608	−18 04926	HR 6863, Y Sgr		18 18 26	−18 53.0	2	5.39	.002	0.67	.005			2	F8 II	1484,6007
168574	−24 14219	HR 6861, V4028 Sg		18 18 27	−24 56.4	3	6.24	.015	1.90	.036	1.98	.031	26	M3 III	1024,3042,6002
238850	+56 02086			18 18 28	+56 10.6	1	9.66		1.34		1.58		2		1746
	−16 04831	NGC 6618 - 652		18 18 28	−16 00.6	1	10.69		0.90		-0.21		3	B0	1737
	−8 04591			18 18 29	−08 19.5	1	10.90		2.60				2	M2	369
		LSS 4955		18 18 29	−18 29.8	1	10.12		0.85		-0.21		3	B3 Ib	1737
		L 158 - 28		18 18 30	−61 14.	1	12.49		0.63		0.00		2		1696
	−13 04948			18 18 31	−13 36.9	1	10.13		1.91				2	M2	369
168627	−18 04927			18 18 32	−18 33.5	1	9.95		0.14		-0.03		2	A0 III/IV	634
	+2 03572			18 18 33	+02 36.6	1	9.22		1.88				1	K5	246
		LSS 4950		18 18 35	−29 46.6	2	12.15	.020	0.24	.005	-0.64	.030	5	B3 Ib	954,1737
168641	−16 04833		V	18 18 36	−16 54.7	1	9.63		1.74				2	K5/M1	369
168339	−61 06140	HR 6855	⋆ AB	18 18 37	−61 31.2	4	4.36	.004	1.47	.014	1.55	.000	18	K4 III	15,1075,1075,2012
		LSS 4961		18 18 38	−14 49.6	2	11.27	.000	1.00	.000	-0.06	.000	4	B3 Ib	405,1737
168723	−2 04599	HR 6869	⋆ A	18 18 43	−02 54.8	16	3.25	.009	0.94	.006	0.65	.006	114	K2 IIIab	1,3,15,22,26,125,1007*
		WLS 1824 70 # 7		18 18 45	+68 12.2	1	11.36		0.99		0.76		2		1375
168663	−18 04931			18 18 45	−18 32.6	1	8.94		0.38		-0.02		2	F0 V	634
169028	+51 02357	HR 6880		18 18 46	+51 19.5	1	6.29		1.10		1.08		2	K2	1355
168560	−42 13165			18 18 46	−42 57.6	1	8.09		1.30		1.28		1	K1p Ba	565
168762	+2 03575			18 18 48	+02 57.0	1	9.21		0.24		0.14		2	A2	246
168874	+27 03003	IDS18168N2729	AB	18 18 50	+27 30.3	1	7.05		0.61		0.09		6	G2 IV	3026
168646	−28 14495	HR 6864		18 18 50	−28 27.3	2	6.16	.009	0.26	.000	0.23		5	A3 III	285,2008
168779	+2 03576			18 18 51	+02 50.2	1	9.73		0.37		0.34		2	A3	246
		G 258 - 33		18 18 52	+66 10.4	2	13.48	.033	1.82	.014			4		940,1759
	−12 05018	LSS 4967		18 18 52	−12 25.7	2	10.01	.000	1.37	.000	0.17	.000	6	B0.5 Ia	405,1737
168592	−38 12729	HR 6862		18 18 52	−38 40.9	1	5.10		1.49				4	K4/5 III	2009
168701	−16 04836			18 18 56	−16 21.0	1	7.65		1.22		0.25		4	K1 III	1657
168725	−15 04943			18 18 59	−15 14.6	1	9.14		1.88				2	B5/7 Ib	369
168476	−56 08755	PV Tel		18 19 00	−56 39.3	6	9.29	.020	-0.01	.012	-0.68	.011	25	Bpec	55,400,843,1112,1699,2033
168797	+5 03704	Cr 359 - 2	⋆	18 19 01	+05 24.7	7	6.13	.012	-0.03	.008	-0.63	.018	19	B3 Ve	154,212,1149,1212*
168913	+29 03241	HR 6876		18 19 01	+29 50.0	1	5.61	.015	0.22	.009	0.07	.026	12	A5 m	1022,1355,8071,8100
		LSS 4971		18 19 03	−13 57.0	2	11.03	.000	1.53	.000	0.40	.000	4	B3 Ib	405,1737
168914	+28 02981	HR 6877		18 19 04	+28 50.7	2	5.11	.005	0.21	.007	0.15		4	A7 V	1363,3023
168729	−18 04934			18 19 05	−18 52.4	1	10.16		0.13		0.02		2	A0 III	634
168709	−24 14244			18 19 09	−24 55.8	1	8.80		0.17		-0.08		3	B9.5II	1586
315593	−26 13080	LSS 4968		18 19 13	−26 36.0	1	10.18		0.17		-0.53		3	B5	954
168750	−26 13081	LSS 4969		18 19 14	−26 26.5	1	8.27		0.08		-0.74		4	B1 Ib	954
	−17 05160			18 19 15	−17 23.0	1	9.92		1.82				2	M3	369
168956	+26 03222			18 19 16	+26 40.9	2	8.40	.049	0.49	.022	-0.06		4	F6 V	20,3026
168749	−24 14247		V	18 19 16	−24 10.0	2	9.73	.289	1.15	.000	0.79	.069	5	G8 (IV)	842,1589
168853	+5 03705			18 19 17	+05 19.1	1	9.53		0.13		-0.10		4	A0	212
	−14 05012			18 19 19	−13 59.3	1	9.69		1.95				2	M3	369
	−12 05019	LSS 4979		18 19 22	−12 38.7	2	9.99	.005	1.64	.000	0.45	.000	4	B1 Ia	405,1737
168815	−15 04946	IDS18165S1508	ABC	18 19 23	−15 06.8	1	7.18		0.99		0.24		3	F0 V +(K)	938
		LP 630 - 39		18 19 23	+00 44.5	1	12.44		0.58		-0.42		1		1696
		POSS 391 # 4		18 19 23	+25 04.5	1	17.01		1.44				2		1739

Table 1

HD	DM	Other Id	N	Rem	α₁₉₅₀	δ₁₉₅₀	S	V	σ_V	B–V	σ_B–V	U–B	σ_U–B	n	Spectrum	References
313321	−20 05108	LSS 4977			18 19 23	−20 05.5	4	9.15	.007	0.52	.004	-0.47	.018	8	B0 IV	540,976,1012,8100
		AO 774			18 19 24	+72 52.7	1	11.21		1.12		1.11		1		1697
168957	+24 03395				18 19 25	+25 01.9	2	7.01	.015	-0.10	.020	-0.57	.005	5	B3 V	379,1212
168814	−14 05013	LSS 4980			18 19 25	−14 24.8	1	7.17		0.65		0.08		2	A1 Ib	8100
		AO 775			18 19 28	+72 24.6	1	10.79		0.47		-0.03		1		1697
168733	−36 12524	HR 6870, V4050 Sg			18 19 30	−36 41.7	1	5.37		-0.14				4	B7 Ib/II	2007
	−14 05014	LSS 4981		★ AB	18 19 31	−14 38.6	3	10.48	.014	0.74	.019	-0.34	.000	5	O9.5 V	405,432,725,1737
168785	−30 15604	LSS 4973			18 19 33	−30 10.0	5	8.49	.016	0.03	.012	-0.74	.004	18	B2 II	400,540,954,976,1732
168890	+0 03923				18 19 35	+00 08.0	1	7.69		1.72				2	K5	369
168959	+16 03472				18 19 35	+16 06.9	1	7.71		1.03				2	K0	882
168817	−17 05162				18 19 35	−17 33.2	1	8.80		2.09				2	K5 (III)	369
168915	+2 03577				18 19 36	+02 14.1	1	9.75		0.38		0.29		2	A0	246
		G 141 - 4			18 19 43	+06 19.0	3	12.56	.054	1.48	.026	1.00	.059	6		203,1705,3078
	−16 04846				18 19 45	−16 10.1	1	9.55		1.85				2	M3	369
168962	+3 03687				18 19 49	+03 51.9	1	8.97		0.18		0.11		2	A0	246
		G 206 - 23			18 19 49	+34 47.6	1	12.09		0.46		-0.20		2		1658
168894	−14 05016	LSS 4983			18 19 50	−14 40.1	3	9.39	.010	0.63	.010	-0.38	.010	5	B0 Ia	1012,1032,8100
169221	+49 02776	HR 6886			18 19 52	+49 42.0	1	6.40		1.07				2	K1 III	71
169666	+71 00884				18 19 54	+71 29.6	1	6.67		0.42		-0.01		4	F5	3016
168917	−14 05017	LSS 4985			18 19 54	−14 23.4	1	8.44		0.43		-0.58		2	B0.5II/III	1011
	+6 03716	LS IV +06 003			18 19 58	+06 24.2	1	10.25		0.11		-0.66		4		212
169110	+23 03316	HR 6882			18 20 03	+23 15.5	1	5.41		1.60				2	M0 IIIab	71
		V 443 Her			18 20 03	+23 25.8	1	11.36		0.90		-0.72		1	M3 pe	1753
168936	−17 05171				18 20 06	−17 42.2	1	7.59		0.14		-0.09		3	A1 II(p)	8100
168838	−36 12537	HR 6874			18 20 06	−36 15.9	1	5.55		1.02				4	K0 III	2007
169354	+53 02072				18 20 11	+53 23.0	1	8.41		0.20		0.11		2	A2	1566
	+2 03580				18 20 13	+02 29.2	1	9.53		1.01		0.81		2		246
		LSS 4992			18 20 13	−13 22.2	2	11.19	.005	1.44	.000	0.29	.000	4	B3 Ib	405,1737
168987	−16 04850	LSS 4991			18 20 15	−16 37.2	2	8.01	.015	0.87	.015	-0.19	.015	6	B1 Ia/ab	399,1737
169111	+11 03442	HR 6883			18 20 16	+12 00.2	3	5.96	.015	0.05	.014	0.06	.012	6	A2 V	1022,1733,3050
169305	+49 02782	HR 6891			18 20 16	+49 05.7	2	5.00	.033	1.62	.026	1.94		4	M2 III	1363,3055
169009	−10 04673	HR 6878			18 20 16	−10 14.7	1	6.34		0.14				4	B9.5V	2035
168940	−24 14268				18 20 17	−24 02.8	1	9.10		1.46		1.42		2	G8/K0 (III)	1589
168941	−27 12783	LSS 4987			18 20 18	−26 58.8	4	9.36	.015	0.04	.013	-0.84	.021	9	B0 III/IV	540,954,976,8100
		NGC 6624 sq1 9			18 20 20	−30 23.7	3	11.46	.005	0.61	.005	0.07		6		539,745,1465
167714	−80 00849	HR 6837			18 20 20	−80 15.7	3	5.95	.004	1.17	.010	1.27	.000	11	K2 III	15,1075,2038
168988	−20 05118				18 20 21	−20 40.8	1	7.63		2.10				2	K5/M0 (III)	369
	−12 05023				18 20 23	−12 40.3	2	10.30	.159	1.66	.122	0.42		4		867,8032
169113	+7 03661				18 20 24	+07 10.8	1	7.04		1.36		1.36		2	K1 III	37
169033	−12 05024	HR 6881			18 20 24	−12 02.5	3	5.72	.015	0.01	.002	-0.22	.020	7	B8 IV-Ve	1079,1586,2007
	−16 04853				18 20 24	−16 02.0	1	9.49		1.79				1	M3	369
		AO 776			18 20 25	+72 39.4	1	13.53		1.82		0.17		1		1697
169010		LSS 4994			18 20 26	−13 45.0	3	12.03	.022	0.97	.021	0.54	.004	4	WC	1012,1183,1359
169034	−13 04958	LSS 4995			18 20 28	−13 37.2	2	8.13	.009	1.20	.031	0.16	.027	4	B2 Ia	1012,8100
169114	+2 03583				18 20 30	+02 11.4	1	9.38		0.56		0.53		2	A0	246
169154	+15 03443				18 20 30	+15 45.7	1	8.19		0.72				2	G8 IV	882
168791	−55 08658				18 20 30	−54 59.1	2	7.69	.004	1.52	.000	1.70	.001	8	K3 II/III	565,1673
		POSS 391 # 6			18 20 32	+24 48.2	1	18.05		1.46				5		1739
		FR Sct			18 20 34	−12 42.5	1	10.21		2.14				2	M3 Ta ep	867
169191	+17 03555	HR 6885			18 20 36	+17 48.0	2	5.24	.005	1.27	.000	1.31	.009	4	K3 III	1080,3016
		NGC 6624 sq1 8			18 20 38	−30 23.8	3	12.07	.014	1.26	.008	1.08		13		539,745,1465
168905	−44 12569	HR 6875		★ A	18 20 40	−44 08.2	4	5.24	.006	-0.18	.013	-0.71	.006	14	B2.5 Vn	15,26,1732,2012
169245	+26 03230				18 20 41	+26 11.9	1	8.67		0.54		0.01		2	F8 V	3031
168871	−49 12105				18 20 41	−49 40.7	2	6.45	.009	0.59	.009	0.05		5	G1/2 V	219,2033
	−11 04620	LS IV -11 017			18 20 42	−11 57.3	3	10.17	.000	0.80	.000	-0.31	.000	5	O4 III(f)	1011,1012,1032
169040	−26 13108				18 20 42	−26 23.8	1	8.82		0.38		0.31		1	A2/3 III	795
		G 21 - 11			18 20 43	+00 56.7	1	14.25		1.49		1.12		1		1658
238860	+58 01806	G 227 - 27			18 20 45	+58 49.6	1	8.69		0.65		0.14		2		1733
		G 227 - 28			18 20 45	+60 59.9	1	15.65		0.78				1		1759
		WLS 1824 70 # 5			18 20 47	+72 17.3	1	12.23		1.01		0.66		2		1375
	−12 05028				18 20 47	−12 41.3	1	10.00		0.98				1	G0	867
168740	−63 04406	HR 6871			18 20 47	−63 02.9	1	6.14		0.20				4	A2 III	2032
		IO Lyr			18 20 48	+32 55.3	1	11.83		0.43		0.15		1	F3.5	699
	−21 04977				18 20 48	−21 19.1	1	10.11		1.71				2		1726
169223	+16 03478	HR 6887			18 20 49	+16 39.7	1	6.22		1.21				2	K0	71
169022	−34 12784	HR 6879		★ A	18 20 51	−34 24.6	7	1.84	.012	-0.03	.004	-0.07	.064	23	B9.5III	3,15,1034,1075,2012*
169156	−9 04712	HR 6884			18 20 55	−08 57.7	5	4.68	.004	0.95	.004	0.72	.006	20	K0 III	15,1075,1088,2012,8015
		NGC 6626 sq1 10			18 20 56	−24 50.5	1	12.87		1.21		1.00		3		896
		POSS 391 # 2			18 20 57	+25 08.6	1	15.41		1.54				2		1739
		NGC 6626 sq1 8			18 20 57	−24 54.5	1	12.64		1.84		2.34		4		896
169357	+35 03228				18 20 58	+35 04.5	1	9.20		0.33		0.14		2		1601
315643	−26 13111	LSS 4996			18 20 58	−26 09.4	2	10.01	.002	0.09	.010	-0.79	.016	4	B0.5III	540,976
169204	+4 03727				18 20 59	+04 04.9	1	9.15		0.46		0.35		2	A0	246
	−14 05029	LSS 5001			18 21 00	−14 10.3	2	9.61	.000	1.15	.000	-0.02	.000	3	B5 Ia	1012,1032
169205	+3 03695				18 21 01	+03 03.8	1	8.39		0.80		0.49		1	K0	246
169413	+39 03402				18 21 01	+39 13.9	2	8.15	.010	1.01	.010	0.74	.020	5	K0 III	833,1601
		G 205 - 19			18 21 02	+37 56.5	1	11.61		1.43		1.28		5	M2	1723
169118	−26 13114				18 21 03	−26 28.5	1	7.57		1.50				4	K2/3 III	2012
	+2 03586				18 21 04	+02 42.8	1	9.86		0.59		0.44		2	A2	246
		NGC 6626 sq1 6			18 21 04	−24 56.5	1	12.23		1.75		2.12		3		896

HD	DM	Other Id	N	Rem	α_{1950}	δ_{1950}	S	V	σ_V	B–V	σ_{B-V}	U–B	σ_{U-B}	n	Spectrum	References
168241	−77 01298				18 21 04	−77 23.7	1	6.75		1.19		0.98		5	K0 III	1628
169224	+5 03720				18 21 05	+05 18.3	1	7.50		−0.01		−0.16		2	B9	212
		LSS 5003			18 21 06	−13 35.1	1	10.51		0.67				4	B3 Ib	1021
169159	−19 04987				18 21 06	−19 25.3	1	7.69		1.98				2	K5 III	369
168910	−54 08867				18 21 06	−54 40.0	1	9.12		0.35		0.21		28	A8 V	1699
	+5 03721				18 21 07	+05 23.0	1	9.78		0.17		0.08		3	A0	212
		G 21 - 12			18 21 07	−05 10.5	4	11.15	.027	1.40	.018	1.12	.015	6		202,1620,1658,3062
	−9 04713	LS IV -09 006			18 21 07	−09 55.5	4	9.74	.030	0.46	.010	−0.45	.029	8	B2:V:pen	445,1012,1212,1586
169100	−30 15647				18 21 09	−30 09.9	2	9.17	.003	0.28	.005	0.20	.006	3	A1 III	540,976
	−24 14284	GU Sgr			18 21 12	−24 16.8	1	10.22		1.19		0.75		1		1589
		NGC 6626 sq1 4			18 21 14	−24 48.1	1	12.13		1.59		1.71		2		896
		G 155 - 15			18 21 16	−13 10.0	1	15.60		0.42		−0.60		1	DAwk	3060
		Steph 1607			18 21 17	+25 25.6	1	10.53		1.58		1.95		2	M0	1746
		NGC 6626 sq1 2			18 21 17	−24 58.7	1	10.84		1.79		1.81		3		896
169266	+2 03587				18 21 18	+02 26.3	1	9.23		0.46		0.06		3	F5	246
		LSS 5005			18 21 20	−14 59.4	2	11.36	.000	1.02	.000	−0.09	.000	4	B3 Ib	405,1737
		NGC 6626 sq1 1			18 21 20	−24 59.8	1	10.28		0.58		0.27		5		896
	−45 12460				18 21 20	−45 30.1	3	10.95	.009	0.51	.010	−0.13	.015	7	F8	158,1236,3077
	+46 02473				18 21 21	+47 01.6	1	9.00		0.48		0.08		2	F5	1601
		NGC 6626 sq1 7			18 21 21	−24 43.8	1	12.34		0.40		0.31		4		896
183030	+88 00112	HR 7394		⋆ A	18 21 22	+89 03.1	2	6.39	.010	1.58	.005	1.80	.005	7	M1 III	15,3001
169227	−12 05033	LSS 5006			18 21 23	−12 15.5	2	8.59	.030	1.25	.025	0.11	.005	4	B1.5Ia	379,405
169226	−12 05034	LSS 5007		⋆ AB	18 21 24	−12 14.5	4	8.54	.015	1.37	.022	0.22	.017	8	B0.5 Ia(n)	379,405,1586,1737
		NGC 6626 sq1 9			18 21 24	−24 50.5	1	12.82		1.72		1.93		3		896
169268	−3 04277	HR 6890			18 21 26	−03 36.6	1	6.37		0.34		−0.04		8	Am?	1088
		LP 810 - 17			18 21 27	−20 38.8	1	11.74		1.13		0.98		1		1696
	−14 05030	LSS 5008			18 21 28	−14 11.4	3	9.52	.000	1.00	.019	0.02	.009	6	B5 Ia	1012,1032,8100
170000	+71 00889	HR 6920, ϕ Dra		⋆ AB	18 21 29	+71 18.7	4	4.22	.009	−0.10	.003	−0.35	.024	10	A0 p (Si)	15,1363,3033,8015
		NGC 6626 sq1 5			18 21 29	−24 44.7	1	12.20		0.60		0.39		4		896
		NGC 6626 - 9003		V	18 21 30	−24 54.	1	12.90		2.11		2.24		1		1468
		NGC 6626 - 9017		V	18 21 30	−24 54.	1	11.94		1.48		1.33		1		1468
		G 21 - 13			18 21 33	+01 39.6	2	11.83	.005	1.55	.005	1.29		3		202,333,1620
169414	+21 03411	HR 6895		⋆ AB	18 21 34	+21 44.7	6	3.83	.010	1.18	.007	1.17	.005	24	K2.5 IIIab	3,15,1003,1080,1363,8015
	+32 03115				18 21 34	+32 49.9	1	9.16		0.39		−0.03		1	F5	289
	−10 04682	LS IV -10 009			18 21 34	−10 50.2	3	9.63	.000	0.56	.000	−0.55	.000	5	O7 V	1011,1012,1032
		LP 866 - 18			18 21 36	−23 19.0	1	11.97		0.89		0.35		2		1696
		KUV 18216+6420			18 21 37	+64 20.3	1	15.06		−0.41		−1.24		1		1708
	−14 05032				18 21 41	−14 46.9	1	8.29		1.59		1.65		3		1762
169274	−19 04992	Tr 33 - 75			18 21 41	−19 45.7	1	7.51		2.08				2	K5 III	369
169363	+3 03697				18 21 42	+03 53.4	1	9.50		0.26		0.10		2	A2	246
	−13 04965	W Sct			18 21 43	−13 41.3	1	10.01		0.74		−0.05		4	B3n +B0	206
		NGC 6626 sq1 3			18 21 44	−24 48.0	1	12.10		1.81		2.25		6		896
169365	+3 03698				18 21 46	+03 36.8	1	8.90		0.15		−0.30		2	A2	246
		AO 777			18 21 46	+72 36.1	1	10.87		1.35		1.61		1		1697
	+72 00840				18 21 46	+72 38.7	1	9.11		0.95		0.65		1	F8	1697
169273	−18 04952	LSS 5010			18 21 47	−18 45.9	2	9.24	.017	0.23	.004	−0.69	.011	3		540,1586
		NGC 6626 sq1 11			18 21 48	−24 54.3	1	13.42		1.44				4		896
169233	−30 15661	HR 6888			18 21 49	−30 47.0	2	5.59	.005	1.14	.000			7	K0 III	15,2012
169315	−16 04859	XX Sgr			18 21 51	−16 49.5	1	8.41		0.93				1	F7/8 II	6011
		POSS 391 # 3			18 21 52	+25 13.2	1	16.44		1.56				5		1739
169333	−12 05038				18 21 53	−12 30.2	1	8.74		1.99				2	M0/1 (III)	369
	−17 05185				18 21 54	−17 16.4	1	9.37		2.39				2	M2	369
170153	+72 00839	HR 6927		⋆ A	18 21 57	+72 42.7	18	3.57	.009	0.49	.005	−0.06	.005	249	F7 V	1,15,22,61,71,985*
	−12 05039	LS IV -12 050			18 21 58	−12 06.1	2	10.79	.015	1.21	.024	0.05		7	O6	390,1021
	+6 03734	NGC 6633 - 2			18 21 59	+06 32.4	1	9.68		0.23		0.17		4	A3	212
169370	−7 04598	HR 6892			18 22 00	−07 06.2	3	6.30	.007	1.16	.004	1.09	.006	8	K0	15,1067,1256
169334	−16 04861				18 22 00	−16 18.3	1	8.01		1.99				2	M0 III	369
169236	−36 12589	HR 6889			18 22 00	−36 01.2	1	6.15		1.00				4	K0 III	2007
		GD 378			18 22 01	+41 02.2	1	14.39		−0.19		−0.96		3	DB	308
169292	−24 14296				18 22 01	−24 26.6	1	8.50		0.22		−0.07		2	B9 III/IV	240
169275	−30 15665				18 22 02	−30 57.4	1	8.52		0.02		−0.34		6	B6 III	1732
		G 227 - 29			18 22 03	+62 01.9	2	13.45	.005	1.72	.010			3		1705,1759
169435	−1 03484				18 22 03	−00 59.8	1	9.43		1.36				2	K5	369
169335	−18 04953				18 22 04	−18 34.0	1	7.27		1.74				2	M0 III	369
	−6 04756			A	18 22 08	−06 23.1	1	10.52		1.18		1.11		3	K7 V	3072
	−6 04756			B	18 22 08	−06 23.1	1	13.65		1.56		1.15		3		3072
169418	−14 05035				18 22 09	−14 08.8	1	9.17		0.30		−0.34		3	B9.5 III	1762
169373	−16 04863				18 22 09	−16 47.5	1	9.11		2.04				25	K2/3 (III)	6011
169492	+5 03727	NGC 6633 - 7			18 22 13	+05 40.9	1	8.87		0.17		0.18		2	A1 Va	212
169533	+13 03632				18 22 14	+13 40.7	1	7.82		1.30				3	K2	882
		G 184 - 7			18 22 14	+27 15.6	1	14.43		0.83		0.22		2		1658
		LSS 5016			18 22 14	−13 53.1	2	11.68	.000	0.93	.000	−0.02	.000	4	B3 Ib	405,1737
	−14 05037	V433 Sct			18 22 15	−14 40.7	4	8.24	.024	1.35	.023	0.16	.017	9	B1.5Ia	1012,1586,1737,8100
169646	+38 03160	HR 6901		⋆ A	18 22 17	+38 42.7	1	6.37		1.44		1.68		2	K2	1733
	−12 05042	LSS 5019			18 22 17	−12 41.5	2	10.63	.000	1.06	.000	−0.11	.005	4	O9 Ib	405,1737
		AS 304			18 22 17	−28 37.7	1	11.19		0.50		−0.80		1	M4	1753
169512	+7 03676	NGC 6633 - 10		⋆ AB	18 22 20	+07 12.1	1	7.91		0.07		−0.19		2	B9	212
		UX Lyr			18 22 21	+39 04.	1	14.47		0.11		0.17		1		597
169493	−1 03486	HR 6898		⋆ AB	18 22 21	−01 36.5	4	6.15	.023	0.38	.010	0.13	.020	16	A9 III +	176,938,1088,3016
169419	−17 05187	LSS 5018			18 22 21	−17 32.8	1	9.35		0.51		−0.47		4	B2 Ia/ab	8100

Table 1   1001

HD	DM	Other Id	N Rem	α_{1950}	δ_{1950}	S	V	σ_V	B–V	σ_{B-V}	U–B	σ_{U-B}	n	Spectrum	References
		RZ Dra		18 22 22	+58 52.5	1	10.05		0.27				1	A5	859
169420	−20 05134	HR 6896	⋆ AB	18 22 22	−20 34.2	7	4.80	.009	1.31	.008	0.92	.015	21	K2 II	15,285,369,1075,1754*
		G 141 - 6		18 22 24	+07 20.7	1	15.25		0.87		0.38		1		1658,3060
		AA45,405 S53 # 1		18 22 24	−13 15.	1	12.35		0.64				2		725
		AA45,405 S53 # 2		18 22 24	−13 15.	1	12.80		0.66				2		725
	−1 03487			18 22 25	−01 34.5	1	9.10		1.22		1.01		2	G5	3016
169454	−14 05039	V430 Sco		18 22 25	−14 00.4	4	6.62	.010	0.91	.011	-0.21	.014	9	B1 Ia	1012,1586,1737,8100
169513	−1 03488			18 22 32	−01 45.7	1	9.59		0.99		0.76		1	K3 V	3072
169668	+32 03118			18 22 33	+32 06.1	1	8.14		0.02		-0.01		2	A0	1733
169702	+39 03410	HR 6903		18 22 35	+39 28.7	3	5.11	.010	0.03	.007	0.08	.005	7	A3 IVn	1022,1363,3023
169398	−34 12802	HR 6893		18 22 36	−33 58.5	3	6.31	.012	-0.08	.008	-0.51	.012	7	B7 III	540,976,2007
169494	−7 04600			18 22 37	−07 49.5	1	9.45		0.72		0.35		1	G0	963
169778	+42 03074	IDS18211N4225		18 22 38	+42 26.3	1	6.65		0.08		0.07		3	A1 V	1501
		LSS 5022		18 22 38	−15 39.8	1	11.05		1.13		-0.06		3	B3 Ib	1737
168951	−69 02889			18 22 38	−69 20.7	1	8.19		1.11		0.99		7	K0 III	1704
		LSS 5023		18 22 39	−14 40.4	1	11.84		1.18		0.36		1	B3 Ib	1359
		G 155 - 16		18 22 39	−18 04.4	1	12.88		1.45				1		3062
169459	−22 04718			18 22 39	−22 02.3	1	10.83		0.43		-0.52		3	A0 V	1737
169597	+6 03741	NGC 6633 - 16		18 22 40	+06 45.3	1	9.13		0.20		0.22		2	A2	212
169425	−31 15510			18 22 40	−31 47.0	3	7.37	.010	0.02	.009	-0.42	.017	6	B5 III	240,540,976
169578	+5 03730	HR 6900		18 22 41	+05 03.3	3	6.73	.009	0.02	.008	-0.20	.013	9	B9 V	212,1079,1149
		Case *M # 49		18 22 42	−13 47.	1	11.12		3.14				1	M2 Iab	148
169885	+53 02079	HR 6911		18 22 43	+53 16.4	1	6.32		0.16		0.14		3	A3 m	8071
169515	−12 05045	RY Sct		18 22 43	−12 43.2	3	9.22	.035	1.09	.012	-0.08	.009	5	O(8)Fe	405,1588,1737
	−13 04973			18 22 44	−13 47.4	1	11.03		3.06		3.21		2		8032
	+4 03735			18 22 46	+04 38.7	1	9.19		0.18		0.13		3	A0	212
169628	+14 03524			18 22 46	+14 07.0	1	7.96		0.02		-0.44		1	B9	963
	−14 05040	LSS 5025	V	18 22 47	−14 46.9	1	10.80		1.10				2		725
		AO 780		18 22 48	+72 28.6	1	11.23		1.10		1.00		1		1697
		AO 781		18 22 52	+72 47.3	1	11.24		0.89		0.54		1		1697
	−17 05190	LSS 5026		18 22 52	−17 06.3	1	9.31		0.83		-0.21		3	B1 Ia	1737
169618	+4 03736			18 22 55	+04 10.9	1	9.75		0.24		0.07		2	A0	246
169629	+7 03679	NGC 6633 - 19		18 22 56	+07 05.4	1	9.75		0.17		0.06		3	A0	212
169582	−9 04729	LS IV -09 011		18 22 58	−09 47.0	3	8.70	.001	0.55	.007	-0.50	.003	6	O5 e	390,1011,1586
		POSS 391 # 1		18 22 59	+24 23.9	1	14.13		1.36				1		1739
169718	+27 03016	HR 6904	⋆ AB	18 22 59	+27 22.0	3	6.27	.005	0.06	.011	0.02	.001	7	A0 V +A4 V	938,1022,1733
170507	+75 00667			18 22 59	+75 31.3	1	8.56		0.05		0.03		2	A0	1733
	−20 05138			18 23 04	−20 46.8	1	10.35		1.91				2	M2	369
		LSS 5034		18 23 06	−12 51.3	2	11.53	.000	1.08	.005	-0.05	.000	4	B3 Ib	405,1737
169405	−48 12505	HR 6894		18 23 06	−48 08.8	1	5.46		0.84				4	K0 III +F	2007
170073	+58 01809	HR 6923	⋆ AB	18 23 11	+58 46.3	5	4.99	.013	0.08	.008	0.05	.012	19	A0 V +F6 V	15,1084,1363,3024,8015
169669	+4 03737			18 23 12	+04 35.4	2	8.40	.000	0.18	.000	-0.05	.000	4	A0	212,246
		MN174,213 T3# 7		18 23 12	−27 32.3	1	11.90		1.18				2		465
169689	+7 03682	HR 6902	⋆ V	18 23 14	+08 00.1	3	5.64	.005	0.91	.010	0.45	.000	10	G8 III-IV +	15,1256,1363
169383	−51 11566			18 23 14	−51 54.8	1	7.97		0.60		0.03		1	G3 V	219
		G 206 - 30		18 23 15	+30 15.1	1	11.70		1.33		1.39		3		1723
238865	+58 01810	IDS18225N5845	C	18 23 15	+58 47.7	2	7.90	.020	0.50	.002	-0.04	.000	5	F8	1084,3024
	−14 05043	LSS 5037	⋆	18 23 15	−14 39.1	1	10.30		0.99		-0.05		3		1762
169720	+14 03531			18 23 16	+14 43.3	1	7.53		1.19				3	K0	882
169633	−14 05044			18 23 16	−14 49.4	1	8.19		0.48		0.03		3	F7/8 V	1762
169467	−46 12379	HR 6897		18 23 16	−45 59.9	4	3.51	.003	-0.17	.005	-0.64	.007	20	B3 IV	15,1075,2012,8015
169603	−20 05139			18 23 17	−20 39.6	1	9.20		0.24		0.18		3	B8/9 Ib/II	212
		Smethells 13		18 23 18	−57 45.	1	11.35		1.33				1		1494
	+1 03669			18 23 19	+01 59.5	1	9.11		1.92				2	M2	369
169797	+26 03247	IDS18213N2602	A	18 23 19	+26 03.3	2	7.67	.019	0.98	.005	0.68		6	G8 III	20,3024
169797	+26 03247	IDS18213N2602	D	18 23 19	+26 03.3	1	10.95		0.63		0.09		3		3024
169797	+26 03247	IDS18213N2602	E	18 23 19	+26 03.3	1	11.94		0.70		0.19		3		3024
169634	−16 04872			18 23 21	−16 51.2	1	9.20		0.74				19	G8 V	6011
169798	+22 03358			18 23 22	+22 40.6	1	6.79		-0.10		-0.62		2	B2.5IV-V	555
		IC 4725 - 31		18 23 23	−19 05.4	2	10.72	.005	0.34	.005	0.11	.015	2	A1 V	1163,1226
		LSS 5039		18 23 24	−14 52.7	3	11.22	.014	0.95	.005	-0.17	.010	7	B3 Ib	405,1737,1762
		MN174,213 T3# 5		18 23 24	−27 25.3	1	11.69		1.38				2		465
		LSS 5040		18 23 25	−15 17.5	1	10.77		0.48		-0.47		3	B3 Ib	1762
169691	−1 03490	EG Ser		18 23 26	−01 42.6	1	8.22		0.22		0.15		10	A0	588
	−27 12846			18 23 26	−27 31.1	1	9.64		1.73				6		465
	−15 04969	LSS 5041	⋆	18 23 28	−15 19.0	1	10.82		0.43		-0.57		3		1762
169723	+4 03738			18 23 29	+04 36.5	1	9.14		0.70				2	F8	246
169587	−34 12814			18 23 30	−34 01.7	1	9.68		-0.03		-0.41		3	B8 V	1586
	+1 03671			18 23 31	+01 33.5	1	9.07		2.09				2	M2	369
	+6 03747	NGC 6633 - 28		18 23 31	+06 33.6	1	9.93		0.26		0.19		2	A5	104
169673	−15 04970	LSS 5042		18 23 31	−15 39.6	2	7.30	.040	0.08	.010	-0.78	.030	6	B1 Ib	1762,8100
		LS IV -15 049		18 23 33	−15 09.3	1	11.30		0.59		-0.05		3		1762
169820	+14 03533	HR 6906		18 23 39	+14 56.2	2			0.01	.056	-0.30	.047	4	B9 V	1022,1079
	−27 12848			18 23 40	−27 25.6	1	11.32		1.15				3		465
	+6 03748	NGC 6633 - 31		18 23 44	+06 40.1	1	9.97		0.29		0.20		2	B9	104
		CS 20321 # 476		18 23 44	−33 22.8	1	11.88		0.15		0.11		1		1736
		NGC 6633 - 33		18 23 46	+06 26.3	1	11.12		0.75		0.27		4	G4	104
	+6 03749	NGC 6633 - 32		18 23 46	+06 34.5	1	10.22		0.35		0.19		2	A5	104
		MN174,213 T3# 6		18 23 46	−27 27.0	1	11.76		0.95				2		465
	+4 03740			18 23 47	+04 26.9	1	10.04		0.45		-0.04		2		246

HD	DM	Other Id	N Rem	α_{1950}	δ_{1950}	S	V	σ_V	B–V	σ_{B-V}	U–B	σ_{U-B}	n	Spectrum	References
169822	+8 03689	G 141 - 8	⋆ A	18 23 47	+08 45.2	4	7.84	.024	0.70	.005	0.18	.018	13	G7 V	22,265,1003,3060
	−14 05047	LSS 5045		18 23 48	−14 40.3	4	10.43	.093	0.71	.037	-0.48	.028	10		405,1586,1737,1762
	+3 03710			18 23 49	+03 34.3	1	9.35		1.89				3	M2	369
169753	−9 04736	RZ Sct		18 23 49	−09 13.9	1	7.53		0.71		-0.15		6	B3 Ib	399
	+29 03257			18 23 51	+29 22.6	1	7.89		1.10		1.05		4	K2 II	8100
		G 205 - 20		18 23 51	+38 19.9	3	11.28	.004	1.46	.026	1.19		6	M1	1746,1759,7010
169727	−13 04979	LSS 5046		18 23 51	−13 40.9	4	9.28	.007	0.79	.017	-0.37	.005	5	O6.5 III(f)	1011,1012,1032,1737
		CS 20321 # 13		18 23 51	−36 19.5	1	14.14		0.18		0.15		1		1736
169528	−50 11938			18 23 51	−50 01.8	1	8.79		0.51				4	F5 V	2012
	−14 05048	LSS 5047		18 23 54	−14 57.0	1	11.59		0.50		-0.39		3		1762
169842	+6 03752	NGC 6633 - 39		18 23 56	+06 49.6	1	9.12		0.21		0.17		2	A1 Vp	104
169754	−11 04631	LS IV -11 021		18 23 56	−11 23.2	2	8.44	.036	1.05	.009	-0.03	.013	5	B0.5Ia	1012,8100
	BD -14 05050a		⋆	18 23 56	−14 32.7	1	10.30		0.45		-0.54		3	OB	1762
	+6 03750	NGC 6633 - 40		18 23 57	+06 23.3	1	9.81		0.24		0.17		2	A2	104
169755	−14 05050	LSS 5049		18 23 58	−14 31.8	5	9.25	.014	0.51	.009	-0.51	.005	10	B3/4 Ib:	1011,1021,1032,1737,1762
	+4 03742			18 23 59	+04 34.9	1	10.36		0.35				2		246
169889	+8 03692	G 141 - 9		18 23 59	+08 35.5	3	8.29	.017	0.76	.015	0.32	.028	9	G9 V	22,265,3026
		MN174,213 T3# 11		18 23 59	−27 26.6	1	12.56		1.59				2		465
		MN174,213 T3# 9		18 23 59	−27 31.2	1	12.41		1.53				6		465
170109	+48 02692	IDS18227N4842	AB	18 24 00	+48 43.9	1	7.15		0.84		0.46		1	G5	8084
		NGC 6633 - 44		18 24 01	+06 29.8	1	10.73		0.39		0.18		3	A5	104
169782	−12 05049			18 24 01	−12 42.0	1	10.08		0.36		-0.35		6	B5 II/III	854
169757	−17 05196	IDS18211S1715	A	18 24 01	−17 13.7	1	7.97		0.19		0.15		3	A3 III/IV	1762
169679	−36 12632			18 24 01	−36 02.8	2	7.11	.011	-0.03	.010	-0.27	.006	3	B8/9 III	540,976
169784	−16 04877			18 24 02	−16 02.7	1	9.95		0.61		0.14		3	G3/5 V	1762
		AA65,65 # 222		18 24 02	−34 22.4	1	12.55		1.28		1.18		1		713
	+5 03738			18 24 03	+05 12.8	1	9.90		0.13		-0.28		1		212
169981	+29 03259	HR 6917		18 24 03	+29 47.9	6	5.84	.024	0.06	.010	0.09	.007	16	A2 IV	15,1007,1013,1022*
169802	−14 05051			18 24 03	−14 13.3	1	9.93		0.41		0.17		3	B9 II	1762
		MN174,213 T3# 12		18 24 03	−27 34.7	1	12.67		0.61				3		465
169928	+18 03684			18 24 04	+18 50.3	2	8.44	.000	-0.07	.000	-0.55	.000	6	B9	963,963
		CS 20321 # 21		18 24 04	−35 57.5	1	14.29		0.13		0.15		1		1736
		NGC 6633 - 47		18 24 06	+06 21.3	1	10.92		0.47		0.22		3	A9	104
		NGC 6633 - 202		18 24 06	+06 21.3	1	11.08		0.83		0.34		3		104
		NGC 6633 - 203	AB	18 24 06	+06 25.0	1	10.95		0.71				2		104
		NGC 6633 - 46		18 24 06	+06 29.2	1	10.46		0.33		0.18		3	A6	104
		MN174,213 T3# 8		18 24 06	−27 29.3	1	12.40		1.68				3		465
	+6 03755	NGC 6633 - 48	⋆ B	18 24 07	+06 30.6	1	10.41		0.21		0.11		2	A0	104
		NGC 6633 - 49		18 24 08	+06 28.6	1	10.53		0.33		0.17		3		104
	+6 03756	NGC 6633 - 50	⋆ A	18 24 08	+06 29.7	2	8.31	.010	1.07	.010	0.75	.005	12	G8 IIIab	104,460
169785	−18 04968			18 24 08	−18 16.0	1	8.91		1.78				2	K4/5 (III)	369
	+3 03712			18 24 11	+03 43.1	1	10.44		0.42		0.09		1		246
		NGC 6633 - 205		18 24 11	+06 33.1	1	15.10		1.27		1.15		2		104
		NGC 6633 - 204		18 24 11	+06 33.4	1	12.91		0.52		0.15		2		104
169955	+19 03643			18 24 11	+19 09.0	1	8.09		1.68		2.03		2	K5	1648
169912	+6 03757	NGC 6633 - 52		18 24 12	+06 30.6	1	9.00		0.25		0.18		3	A3 V	104
	+4 03744			18 24 13	+04 12.8	1	9.79		1.40		1.39		2	K0	246
		LSS 5051		18 24 13	−14 18.9	2	11.06	.010	0.37	.019	-0.46		7	B3 Ib	1021,1762
	−16 04878			18 24 13	−16 20.4	1	9.84		0.58		0.12		7	G0	830
		NGC 6633 - 206	AB	18 24 14	+06 17.0	1	12.64		0.86		0.28		2		104
		NGC 6633 - 208		18 24 14	+06 25.9	1	14.54		1.17		0.86		2		104
		NGC 6633 - 207		18 24 14	+06 28.0	1	13.28		0.85		0.31		2		104
		NGC 6633 - 210		18 24 14	+06 47.2	1	11.07		0.46		0.16		2		104
	+72 00841			18 24 14	+72 52.0	1	9.27		0.37		0.04		1	F5	1697
		CS 20321 # 96		18 24 14	−32 58.9	1	12.55		0.32		0.14		1		1736
169706	−33 13225			18 24 14	−33 26.5	2	7.04	.015	1.02	.000	0.83		6	G5 III	262,2033
		NGC 6633 - 209		18 24 15	+06 26.4	1	13.87		1.30		0.94		1		104
169805	−19 05007	LSS 5050		18 24 15	−18 59.1	3	8.10	.017	0.15	.004	-0.68	.009	6	B2 Vne	540,976,1586
	−27 12853			18 24 15	−27 31.5	1	11.53		0.41				7		465
		NGC 6633 - 211		18 24 16	+06 26.0	1	11.93		0.97		0.52		2		104
169827	−17 05198			18 24 16	−17 18.6	1	8.12		0.24		-0.12		3	B9/A0 II	1762
		NGC 6633 - 212		18 24 17	+06 30.7	1	13.69		0.81		0.19		2		104
169825	−16 04880			18 24 17	−16 02.4	1	10.32		0.19		-0.11		3	B9 II	1762
		AA65,65 # 220		18 24 17	−34 35.1	1	11.04		0.38		0.17		1		713
	+6 03758	NGC 6633 - 54		18 24 19	+06 22.4	1	10.10		0.29		0.18		4	A4	104
170028	+26 03255			18 24 19	+26 11.8	1	6.90		-0.13		-0.60		2	B3 V	555
	+6 03759	NGC 6633 - 55		18 24 20	+06 19.1	1	10.39		0.38		0.19		2	A9	104
169847	−17 05199			18 24 20	−16 57.9	1	8.99		0.13		-0.28		3	B8/9 II	1762
		NGC 6633 - 213		18 24 21	+06 28.3	1	13.01		0.73		0.17		2		104
		NGC 6633 - 214		18 24 21	+06 28.4	1	14.58		0.72		0.23		2		104
		Steph 1612		18 24 22	+19 56.4	1	10.94		1.63		1.98		2	K7-M0	1746
169869	−14 05053	LSS 5052	⋆ A	18 24 22	−14 43.7	1	8.27		0.82		0.53		3	A1 Iab/b	1762
	−27 12856	IDS18212S2728	AB	18 24 22	−27 27.0	1	9.80		1.36				1		465
169957	+8 03696			18 24 23	+08 04.3	1	8.82		0.66		0.24		2	G2 IV	1003
		NGC 6633 - 215		18 24 24	+06 25.1	1	13.87		0.85		0.37		2		104
	−14 05054	LS IV -14 066		18 24 24	−14 45.6	1	10.05		0.12		-0.50		3		1762
		CS 20321 # 35		18 24 24	−35 31.3	1	14.88		0.12		0.17		1		1736
		NGC 6633 - 216		18 24 25	+06 17.0	1	11.77		1.07		0.66		2		104
	+6 03761	NGC 6633 - 57		18 24 26	+06 21.4	1	8.77		0.22		0.20		3	A2 V	104
		AA65,65 # 221		18 24 26	−34 26.6	1	12.45		0.52		0.04		1		713

Table 1   1003

HD	DM	Other Id	N	Rem	α_{1950}	δ_{1950}	S	V	σ_V	B–V	σ_{B-V}	U–B	σ_{U-B}	n	Spectrum	References
169959	+6 03762	NGC 6633 - 58		⋆AB	18 24 27	+06 23.5	1	7.57		0.09		-0.13		5	A0 III	104
169958	+6 03763	NGC 6633 - 61			18 24 28	+06 27.2	1	8.26		0.24		0.21		2	A2.5V	104
		NGC 6633 - 217			18 24 28	+06 30.7	1	14.61		0.71		0.17		2		104
170051	+26 03257				18 24 28	+26 25.8	2	7.10	.015	-0.10	.000	-0.59	.005	5	B2 V	399,555
169870	-15 04971				18 24 28	-15 43.7	1	9.69		0.55		0.06		3	F7 V	1762
	+6 03764	NGC 6633 - 62			18 24 29	+06 19.1	1	10.00		0.31		0.21		3	A7	104
		NGC 6633 - 63			18 24 29	+06 26.1	1	11.07		0.45		0.14		3		104
		NGC 6633 - 218			18 24 29	+06 31.6	1	12.22		0.73		0.17		2		104
		NGC 6633 - 219			18 24 29	+06 32.0	1	13.93		0.76		0.26		3		104
		NGC 6633 - 220			18 24 29	+06 32.6	1	13.95		0.84		0.34		1		104
		NGC 6633 - 221			18 24 30	+06 34.0	1	13.50		0.95		0.49		2		104
		NGC 6633 - 222			18 24 31	+06 29.6	1	12.88		1.63				1		104
	-14 05055	LSS 5055			18 24 32	-14 40.1	2	10.31	.024	0.52	.022	-0.43	.028	6		1586,1762
	+6 03765	NGC 6633 - 178			18 24 34	+06 23.3	1	9.50		0.32		0.22		3	A2 V Shell	104
		NGC 6633 - 223			18 24 34	+06 28.9	1	12.19		0.62		0.08		3		104
		NGC 6633 - 224			18 24 34	+06 36.0	1	13.64		0.80		0.27		2		104
	+72 00842				18 24 35	+72 35.8	1	9.28		0.26		0.00		1	F0	1697
	-14 05057	LS IV -14 068			18 24 35	-14 43.9	1	9.97		0.51		-0.31		3		1762
		MN174,213 T3# 10			18 24 35	-27 34.4	1	12.54		1.83				4		465
	+6 03767	NGC 6633 - 65			18 24 36	+06 14.6	1	9.95		0.66		0.13		2	G0	104
170011	+6 03766	NGC 6633 - 179			18 24 36	+06 23.0	1	9.00		0.25		0.22		4	A2.5V	104
		NGC 6633 - 225			18 24 36	+06 28.9	1	15.11		0.35		0.33		2		104
		NGC 6633 - 226			18 24 36	+06 35.3	1	14.67		1.27		0.87		1		104
		CS 20321 # 65			18 24 36	-34 13.2	1	13.44		0.29		0.30		1		1736
		NGC 6633 - 67			18 24 37	+06 24.0	1	9.51		0.29		0.18		3	Am	104
		NGC 6633 - 227			18 24 37	+06 25.5	1	14.44		0.86		0.41		2		104
170009	+7 03694				18 24 37	+07 42.1	1	8.00		0.09		-0.14		2	B8	212
169851	-26 13184	HR 6909		⋆AB	18 24 37	-26 40.0	1	6.32		0.26				4	A8 V + F2 V	2007
	+6 03768	NGC 6633 - 68			18 24 38	+06 09.9	1	10.05		0.47		0.24		3	A6	104
		NGC 6633 - 228			18 24 38	+06 29.4	1	13.30		0.80		0.18		2		104
		NGC 6633 - 229			18 24 38	+06 35.8	1	14.42		0.85		0.34		2		104
169830	-29 14965	HR 6907			18 24 38	-29 50.9	1	5.92		0.52				4	F8 V	2007
169985	+0 03936	HR 6918, D Ser		⋆AP	18 24 39	+00 09.9	7	5.21	.010	0.49	.008	0.22	.010	26	G0 III+A6 V	15,938,1088,1363,1477*
		NGC 6633 - 231			18 24 39	+06 26.0	1	14.59		1.18		1.00		2		104
		NGC 6633 - 232			18 24 39	+06 29.6	1	13.99		0.82		0.23		2		104
170010	+6 03769	NGC 6633 - 70			18 24 39	+06 30.3	2	8.21	.000	0.80	.020	0.45	.010	14	G0 IIIa+A3	104,460
		NGC 6633 - 230			18 24 39	+06 32.7	1	10.64		1.27		1.21		2		104
170052	+15 03467				18 24 40	+15 07.2	2	8.74	.010	1.12	.005	0.86		5	K0	882,3077
170111	+26 03259	HR 6924		⋆AB	18 24 40	+26 25.1	2	6.51	.010	-0.11	.005	-0.59	.000	6	B3 V	154,555
		NGC 6633 - 233			18 24 41	+06 26.0	1	13.48		0.79		0.24		2		104
		NGC 6633 - 235			18 24 42	+06 23.0	1	13.37		0.58		0.32		2		104
		NGC 6633 - 234		AB	18 24 42	+06 30.4	1	13.72		1.29		1.04		2		104
	-13 04985	LSS 5057			18 24 42	-13 28.7	1	10.19		0.66		-0.41		2	B0 III	405
	+3 03714				18 24 43	+03 40.4	1	10.11		0.23		0.13		4		246
		NGC 6633 - 236			18 24 43	+06 30.4	1	12.12		0.92		0.50		2		104
170030	+6 03770	NGC 6633 - 72			18 24 44	+06 59.4	1	9.48		0.31		0.19		2	A4 Vn	104
		G 21 - 15			18 24 45	+04 02.0	7	13.91	.015	0.07	.019	-0.56	.016	49		1,202,203,281,1620*
		NGC 6633 - 237			18 24 45	+06 35.8	1	12.34		0.63		0.08		3		104
		NGC 6633 - 74		AB	18 24 47	+06 23.4	1	11.20		0.62		0.10		3		104
	+6 03771	NGC 6633 - 75			18 24 47	+06 26.7	2	9.68	.005	0.30	.010	0.19	.005	6	Am	104,460
170054	+6 03772	NGC 6633 - 77			18 24 48	+06 29.2	1	8.18		0.03		-0.41		4	B6 IV	104
	-12 05055	UY Sct			18 24 48	-12 29.7	2	9.31	.892	2.91	.079	1.96		5	M3 Iab	148,8032
		He3 1700			18 24 48	-18 08.	1	12.32		0.45		0.34		1		1586
170053	+6 03773	NGC 6633 - 78			18 24 49	+06 58.6	1	7.31		1.43		1.54		2	K2 II	104
169934	-16 04882				18 24 49	-15 58.5	1	9.69		0.33		0.27		3	A3 II/III	1762
		NGC 6633 - 238			18 24 51	+06 31.9	1	12.37		0.44		0.33		2		104
		NGC 6633 - 79			18 24 52	+06 11.2	1	10.16		0.35		0.16		3	A2	104
		NGC 6633 - 239			18 24 52	+06 24.3	1	13.14		0.81		0.40		2		104
		NGC 6633 - 240			18 24 53	+06 24.2	1	13.24		1.44		1.49		2		104
		NGC 6633 - 241			18 24 53	+06 34.1	1	12.68		0.61		0.14		2		104
		LS IV -14 070			18 24 53	-14 48.7	1	11.36		0.37		-0.26		3		1762
169916	-25 13149	HR 6913			18 24 53	-25 27.1	9	2.82	.015	1.04	.005	0.90	.008	44	K1 IIIb	3,15,1020,1075,1340*
	+6 03774	NGC 6633 - 81			18 24 54	+06 27.8	1	9.70		0.32		0.17		3	A2	104
		NGC 6633 - 80			18 24 54	+06 37.0	1	10.66		0.52		0.20		2	A3	104
		LS IV -14 071			18 24 54	-14 54.7	1	11.42		0.36		-0.34		3		1762
		NGC 6633 - 82			18 24 55	+06 27.6	1	10.92		0.42		0.15		4		104
		NGC 6633 - 243			18 24 55	+06 35.5	1	13.49		0.62		0.21		2		104
169872	-32 14171				18 24 55	-32 30.1	2	7.77	.005	0.01	.011	-0.38	.002	3	B8 II	540,976
169873	-33 13239				18 24 55	-33 22.9	1	8.74		0.02		-0.23		2	B8/9 III	262
		CS 20321 # 33			18 24 55	-35 36.9	1	12.15		0.15		0.11		1		1736
		NGC 6633 - 242			18 24 56	+06 25.4	1	14.66		0.68		0.22		2		104
170079	+6 03775	NGC 6633 - 83			18 24 57	+06 06.7	1	9.01		0.26		0.22		2	A2.5V	104
		NGC 6633 - 244			18 24 57	+06 31.8	1	13.13		0.75		0.22		2		104
321760	-34 12834	AA65,65 # 219			18 24 57	-34 31.5	1	10.35		0.32		0.15		1	A5	713
		NGC 6633 - 245			18 24 58	+06 26.0	1	14.72		0.78		0.14		2		104
	-8 04612			V	18 24 58	-08 42.9	1	10.27		2.31				2	M3	369
		NGC 6633 - 246			18 24 59	+06 30.1	1	13.12		0.26		0.10		2		104
169989	-14 05060				18 24 59	-14 14.8	1	10.51		0.22		0.10		3	B9	1762
169938	-26 13192	HR 6914			18 24 59	-26 47.3	1	6.27		0.16				4	A3/4 V	2007
169767	-49 12153	HR 6905			18 24 59	-49 06.0	4	4.13	.011	1.01	.006	0.82	.005	15	G8/K0 III	15,1075,2012,3077

HD	DM	Other Id	N	Rem	α_{1950}	δ_{1950}	S	V	σ_V	B–V	σ_{B-V}	U–B	σ_{U-B}	n	Spectrum	References
170095	+6 03776	NGC 6633 - 84			18 25 00	+06 12.5	1	9.51		0.25		0.19		2	A2.5V + Shl	104
	+6 03777	NGC 6633 - 85			18 25 00	+06 22.3	1	10.08		0.35		0.22		2	A4	104
		NGC 6633 - 247			18 25 00	+06 32.4	1	12.46		0.65		0.05		2		104
169853	−39 12626	HR 6910			18 25 00	−39 01.6	1	5.64		0.13				4	A2mA2-F0	2007
234650	+52 02209				18 25 01	+52 13.7	1	9.47		0.20		0.12		3	A2	1566
169990	−17 05203	HR 6919			18 25 01	−17 49.9	2	6.20		0.00	.004	-0.31		5	B8 III/IV	1079,2035
		MWC 297			18 25 02	−03 53.2	1	12.31		2.03		0.95		1		1586
		NGC 6633 - 248			18 25 03	+06 30.9	1	11.79		0.51		0.13		2		104
170094	+6 03778	NGC 6633 - 87			18 25 03	+06 33.4	2	9.29	.000	0.23	.005	0.19	.005	8	Am	104,460
		LS IV -14 072			18 25 04	−14 45.8	1	11.13		0.38		-0.35		3		1762
	+6 03779	NGC 6633 - 88			18 25 05	+06 30.9	1	9.62		0.35		0.20		2	Am	104
		AA65,65 # 218			18 25 05	−34 29.8	1	12.74		1.69		1.57		1		713
170096	+2 03598				18 25 06	+02 21.0	1	8.48		0.57		0.05		1	G0	246
		NGC 6633 - 249			18 25 06	+06 31.2	1	12.08		0.61		0.07		2		104
		NGC 6633 - 250		AB	18 25 07	+06 38.8	1	13.52		0.65		0.36		2		104
169877	−38 12814				18 25 07	−38 53.8	1	7.86		1.32		1.13		1	B8/9 IV	3045
		NGC 6633 - 89			18 25 08	+06 54.1	1	10.69		0.46		0.19		2	A8	104
		NGC 6633 - 251			18 25 09	+06 27.7	1	12.98		0.54		0.32		2		104
170114	+6 03780	NGC 6633 - 90		⋆ AB	18 25 10	+06 31.0	1	8.52		0.34		0.22		2	A4 V	104
170061	−14 05062	LSS 5059			18 25 10	−14 44.3	5	9.42	.055	0.41	.035	-0.58	.022	10	B0	445,1012,1032,1586,1762
		NGC 6633 - 252			18 25 11	+06 38.7	1	13.79		0.78		0.32		2		104
		NGC 6633 - 253			18 25 12	+06 31.1	1	11.24		0.51		0.12		2		104
		NGC 6633 - 254			18 25 12	+06 35.4	1	15.01		1.02		0.51		2		104
170134	+6 03781	NGC 6633 - 91			18 25 12	+06 37.0	1	9.47		0.32		0.11		5	A8	104
	−30 15732				18 25 12	−30 07.9	1	11.57		0.35		0.04		4		1099
		L 206 - 117			18 25 12	−56 37.	1	13.71		0.73		0.13		2		1696
170135	+6 03782	NGC 6633 - 92			18 25 13	+06 30.0	1	8.44		0.26		0.24		2	A5 V	104
		NGC 6633 - 255			18 25 13	+06 35.8	1	14.54		1.25		0.94		2		104
170060	−14 05064				18 25 13	−14 26.6	1	10.38		0.23		0.10		3	A0 Iab/b	1762
170136	+6 03783	NGC 6633 - 93			18 25 15	+06 05.3	1	9.71		0.28		0.18		2	A0	104
		NGC 6633 - 256			18 25 15	+06 28.4	1	12.13		1.34		1.17		2		104
170357	+45 02716	G 205 - 22			18 25 15	+46 03.0	2	8.31	.000	0.61	.004	0.03	.018	5	G1 V	1003,3026
170082	−14 05065				18 25 15	−14 52.2	1	9.38		0.55		0.38		3	A1 II/III	1762
		LS IV -15 050			18 25 15	−15 05.6	1	10.88		0.30		0.14		3		1762
		NGC 6633 - 257			18 25 17	+06 35.2	1	14.67		1.71				2		104
		NGC 6633 - 94			18 25 18	+06 29.1	1	10.95		0.50		0.16		2	F0	104
		NGC 6633 - 258			18 25 18	+06 39.3	1	11.31		0.59		0.07		2		104
	+40 03374	G 205 - 23			18 25 18	+40 40.8	2	9.90	.025	0.87	.005	0.48	.015	3	K1	1064,1510
		LS IV -14 075			18 25 18	−14 39.1	1	11.48		0.40		-0.25		3		1762
		LS IV -15 051			18 25 18	−15 04.2	1	11.55		0.47		0.27		3		1762
		NGC 6633 - 95			18 25 19	+06 27.4	1	10.88		0.42		0.18		2	F0	104
		NGC 6633 - 259			18 25 19	+06 34.6	1	13.70		0.88		0.33		2		104
170157	+6 03784	NGC 6633 - 96			18 25 19	+06 41.3	1	9.82		0.26		0.22		3	A0	104
170083	−16 04887	IDS18225S1606		A	18 25 19	−16 03.9	1	9.21		0.29		-0.24		3	B5 Ib	1762
		NGC 6633 - 260			18 25 20	+06 37.0	1	13.90		0.69		0.23		2		104
170263	+28 03002	IDS18234N2852		AB	18 25 20	+28 53.8	1	8.61		-0.01		-0.13		3	B8	555
	+0 03942				18 25 21	+00 13.8	1	10.73		0.53		0.20		20		1655
170137	+3 03716	HR 6925			18 25 21	+03 43.0	4	6.07	.010	1.63	.010	1.84	.013	10	K3 III	15,1256,1355,4001
170158	+6 03786	NGC 6633 - 97			18 25 21	+06 25.4	1	9.08		0.28		0.22		2	A2 V	104
		NGC 6633 - 261			18 25 21	+06 33.5	1	13.39		0.83		0.28		2		104
169942	−38 12818				18 25 21	−38 03.7	1	7.65		1.58				4	K4 III	2012
		NGC 6633 - 262			18 25 23	+06 32.7	1	13.50		0.83		0.32		3		104
	+72 00844				18 25 23	+72 43.9	1	9.92		0.44		-0.14		1	F5	1697
	−11 04642				18 25 24	−11 06.9	1	9.55		2.40				2	M2	369
	−30 15735				18 25 24	−29 59.9	1	10.76		0.78		0.36		5		1099
		NGC 6633 - 263			18 25 25	+06 36.1	1	13.76		0.68		0.10		2		104
	+6 03787	NGC 6633 - 98			18 25 25	+06 56.4	1	10.09		0.40		0.04		1	F0	104
170020	−33 13248				18 25 25	−33 39.7	1	8.64		1.27		1.36		2	K1 III	262
		NGC 6633 - 99			18 25 26	+06 47.9	1	10.18		0.45		0.22		2	A0	104
	−29 14981				18 25 26	−29 33.1	1	11.00		0.80		0.22		3		1099
		CS 20321 # 94			18 25 26	−33 08.4	1	14.75		0.22		0.24		1		1736
		NGC 6633 - 264			18 25 27	+06 35.3	1	14.09		0.84		0.30		2		104
	−13 04990	LSS 5062			18 25 28	−13 07.9	1	10.62		0.46				4	B0	1021
	+4 03752				18 25 29	+04 30.5	1	10.74		0.26				2		246
	+6 03789	NGC 6633 - 101			18 25 29	+06 13.2	1	10.78		0.31		0.26		2	A7	104
170174	+6 03788	NGC 6633 - 100			18 25 29	+06 34.0	2	8.30	.005	1.12	.010	0.85	.005	14	G8 IIIab	104,460
		HA 158 # 3149			18 25 30	−29 53.9	1	12.78		2.01		1.44		4		1099
170097	−16 04888	V2349 Sgr			18 25 31	−16 44.0	1	8.63		0.13		-0.74		3	B1/2 Ib	399
	−18 04974				18 25 31	−18 51.1	1	10.29		1.96				2	M3	369
170200	+6 03790	NGC 6633 - 102		⋆	18 25 32	+06 09.7	5	5.73	.006	-0.03	.008	-0.36	.008	22	B8 III-IV	15,104,460,1079,1256
		HA 158 # 3163			18 25 32	−29 57.1	1	11.78		1.82		1.15		3		1099
170019	−33 13249				18 25 32	−33 31.3	1	8.54		1.55		1.84		2	K3 (III)	262
		NGC 6633 - 265			18 25 33	+06 23.8	1	11.47		0.52		0.07		2		104
170341	+35 03255				18 25 33	+35 29.3	1	8.30		-0.07		-0.24		1	A0	963
		HA 158 # 3178			18 25 33	−29 53.8	1	13.18		0.98		0.40		4		1099
		NGC 6633 - 103			18 25 34	+06 16.2	1	10.96		0.58		0.02		2		104
		NGC 6633 - 104			18 25 34	+06 18.5	1	10.50		0.42		0.24		2	A7	104
		NGC 6633 - 105		AB	18 25 34	+06 29.8	1	10.91		0.55		0.11		2	F4	104
170231	+6 03791	NGC 6633 - 106			18 25 34	+06 52.9	2	8.66	.005	1.09	.010	0.77	.024	6	G8 IIb	104,460
	−9 04742	LS IV -09 013			18 25 35	−09 37.1	2	10.43	.005	0.74	.010	-0.24		6	B2:V:	1012,1021

Table 1   1005

HD	DM	Other Id	N Rem	α_{1950}	δ_{1950}	S	V	σ_V	B–V	σ_{B-V}	U–B	σ_{U-B}	n	Spectrum	References
	−29 14985			18 25 35	−29 55.7	1	11.70		0.76		0.11		3		1099
169943	−43 12564	HR 6915		18 25 35	−43 52.8	1	6.36		0.90				4	G8 III	2007
		CS 20321 # 5		18 25 36	−36 33.6	1	14.94		0.27		0.18		1		1736
		NGC 6633 - 266		18 25 37	+06 36.4	1	12.86		0.80		0.14		3		104
		NGC 6633 - 267		18 25 37	+06 37.9	1	13.32		0.63		0.12		2		104
		NGC 6633 - 268		18 25 37	+06 39.5	1	11.88		0.79		0.36		2		104
170139	−15 04977			18 25 37	−15 50.2	1	8.87		1.36		1.29		3	K0 III	1762
170138	−14 05067			18 25 38	−14 11.6	1	9.72		0.47		0.04		3	F2/3 V	1762
170119	−17 05207			18 25 38	−17 00.6	1	8.57		0.18		-0.32		3	B9 II	1762
170120	−19 05015			18 25 38	−19 48.2	1	8.29		2.35				2	M0/1 (III)	369
		HA 158 # 3206		18 25 38	−29 54.4	1	13.72		0.86		0.20		6		1099
169836	−57 09063	HR 6908	⋆ AB	18 25 38	−57 33.4	1	5.76		0.98				4	K0 III	2007
312343	−19 04755	NGC 6494 - 48		18 25 39	−18 24.2	3	9.56	.007	1.28	.005	0.97	.017	9	K0	460,1695,1710
	+1 03683			18 25 40	+01 58.5	1	8.88		1.97				2	M2	369
	+6 03793	NGC 6633 - 110		18 25 40	+06 40.4	2	10.16	.000	0.35	.025	0.17	.000	6	A3	104,460
	−17 05208	LSS 5064		18 25 40	−16 58.6	1	9.77		0.26		-0.71		2		1737
		NGC 6633 - 269		18 25 41	+06 36.1	1	12.33		0.57		0.32		2		104
170064	−33 13252			18 25 42	−33 34.6	1	9.56		0.65		0.24		2	F3 V	262
169977	−47 12315			18 25 42	−47 07.2	1	7.60		1.14		1.10		5	K0 III	1673
		NGC 6633 - 270		18 25 43	+06 13.3	1	12.81		0.39		0.25		2		104
170327	+25 03535			18 25 43	+25 33.7	2	8.97	.005	0.19	.005	0.10	.010	5	B9	833,1625
		HA 158 # 3234		18 25 44	−29 54.6	1	14.12		0.72		0.02		5		1099
	−4 04474			18 25 46	−04 31.1	1	10.13		2.61				1	M2	369
170122	−27 12876			18 25 46	−27 17.7	1	8.86		0.15		-0.13		2	B9 III	240
		NGC 6633 - 272		18 25 47	+06 38.5	1	11.99		1.17		0.79		1		104
170271	+7 03701	NGC 6633 - 114		18 25 47	+07 19.3	1	8.93		0.24		0.22		2	A1.5Va	212
170232	−3 04288		A	18 25 47	−03 54.3	1	8.31		0.76		0.29		2	G0	1064
170232	−3 04288		B	18 25 47	−03 54.3	1	8.40		0.63		0.14		1		1064
	−29 14991			18 25 47	−29 37.5	1	10.30		0.26		0.03		std		1099
164461	−87 00274	HR 6721		18 25 47	−87 39.2	5	5.27	.004	1.28	.005	1.60	.000	16	K3 III	15,1075,1075,2017,2038
		NGC 6633 - 271	AB	18 25 48	+06 12.5	1	12.39		0.74		0.18		2		104
	+6 03794	NGC 6633 - 116		18 25 48	+06 28.8	1	10.01		0.31		0.22		2	A5	104
170159	−13 04992	LSS 5065		18 25 48	−13 01.8	2	8.36	.000	0.59	.000	-0.42	.005	4	B1 Iab	1012,8100
	+6 03795	NGC 6633 - 118		18 25 49	+06 32.3	1	10.12		1.28		1.21		2	F5	104
		NGC 6633 - 273		18 25 49	+06 36.8	1	11.51		0.56		0.10		2		104
170177	−13 04993	LSS 5066		18 25 49	−13 32.1	2	9.45	.004	0.78	.012	-0.34	.000	4	O8/B0	1032,8100
	−16 04892			18 25 49	−16 52.0	1	9.39		1.80				2	M3	369
		CS 20321 # 56		18 25 49	−34 32.5	1	14.79		0.17		-0.16		1		1736
170693	+65 01271	HR 6945		18 25 50	+65 31.9	4	4.83	.016	1.18	.013	1.10	.010	10	K2 III	15,1363,3016,8015
	−20 05153		V	18 25 50	−20 11.9	1	9.71		2.03				2	M3	369
170141	−26 13206	HR 6926	⋆ A	18 25 50	−26 36.9	1	6.50		0.11				4	A3 III	2007
170040	−38 12824	HR 6921		18 25 50	−38 53.1	2	6.62	.005	-0.02	.000			7	B9 III	15,2012
170247	+3 03719			18 25 51	+03 56.0	1	7.92		0.12		-0.21		2	A0	246
	+6 03796	NGC 6633 - 119		18 25 52	+06 44.0	2	8.95	.005	1.02	.015	0.66	.025	8	G8 III	104,460
	−58 07400			18 25 52	−58 17.6	2	9.85	.005	1.47	.014	1.20		5		158,1494
170123	−29 14992			18 25 53	−29 56.9	1	8.41		0.72		0.14		9	G6 V	1099
		NGC 6633 - 274		18 25 54	+06 29.6	1	12.24		0.65		0.04		2		104
		NGC 6633 - 121		18 25 54	+06 33.4	1	10.72		1.23		0.95		2	dK0	104
170274	+3 03720			18 25 55	+03 44.8	1	7.89		0.28		0.26		2	F0	246
170202	−14 05069			18 25 55	−14 13.3	1	9.15		0.57		0.26		3	F0 V	1762
170203	−15 04979	LS IV -15 052		18 25 55	−15 35.5	1	9.55		0.11		-0.49		3	B5 Ib	1762
170291	+6 03797	NGC 6633 - 123		18 25 56	+06 48.1	1	7.63		0.49		0.02		4	F5	104
170293	+6 03799	NGC 6633 - 125		18 25 57	+06 22.8	1	8.60		0.33		0.27		3	A6 IV	104
170292	+6 03798	NGC 6633 - 126		18 25 57	+06 40.5	2	8.78	.010	1.05	.010	0.73	.020	7	G7 IIIab	104,460
		LS IV -14 079		18 25 57	−14 46.2	1	10.93		0.24		-0.54		3		1762
		NGC 6633 - 275		18 25 58	+06 11.6	1	11.34		0.54		0.14		2		104
	+6 03800	NGC 6633 - 127		18 26 00	+06 54.4	1	9.93		0.37		0.23		2	A3	104
	−30 15749			18 26 00	−30 02.5	1	11.26		1.32		0.92		6		1099
		CS 20321 # 76		18 26 03	−33 55.1	1	13.51		0.36		0.18		1		1736
170316	+4 03758	IDS18236N0447	A	18 26 05	+04 48.9	1	7.69		0.02		-0.11		2	A0	212
169883	−59 07277			18 26 05	−59 19.6	1	7.85		1.68		2.03		2	M1/2 III	1730
	+3 03721			18 26 06	+03 40.8	1	9.44		0.40		0.04		3	F5	246
170316	+4 03759	IDS18236N0447	B	18 26 06	+04 48.2	1	8.80		0.25		0.06		4		212
		L 206 - 121		18 26 06	−56 42.	1	13.44		0.76		0.20		2		3062
		NGC 6633 - 276		18 26 08	+06 36.1	1	12.40		0.80		0.20		2		104
	+6 03803	NGC 6633 - 131		18 26 08	+06 51.2	1	9.96		0.37		0.24		2	A3	104
		NGC 6633 - 277		18 26 09	+06 09.3	1	11.81		0.64		0.00		4		104
	+6 03802	NGC 6633 - 132		18 26 09	+06 32.5	1	10.22		0.32		0.24		2	A0	104
		NGC 6633 - 278		18 26 09	+06 37.3	1	12.30		0.56		0.17		2		104
		AE Dra		18 26 09	+55 27.5	1	12.42		0.29		0.14		1	F1	699
170249	−16 04893			18 26 09	−16 26.9	1	8.58		1.37		1.17		3	G8 III	1762
		CS 20321 # 169		18 26 10	−36 09.7	1	14.47		0.06		0.05		1		1736
170069	−47 12319	HR 6922		18 26 10	−47 15.3	1	5.70		1.26				4	K2 III	2007
	−12 05063			18 26 12	−12 48.5	1	9.74		1.97				1	M2	369
170346	+6 03804	NGC 6633 - 134		18 26 13	+06 45.4	1	8.74		0.27		0.22		3	A3 V	104
		HRC 282, VV Ser		18 26 14	+00 06.7	1	12.14		1.07		0.51		1	A2e	776
		LS IV -14 080		18 26 14	−14 45.0	1	10.91		0.33		-0.44		3		1762
		V1304 Sgr		18 26 14	−32 24.5	1	12.30		0.78		0.52		1		689
		NGC 6633 - 279	AB	18 26 17	+06 46.7	1	11.53		0.65		0.09		2		104
		NGC 6633 - 135		18 26 17	+06 49.4	1	10.36		0.40		0.23		2	A5	104

HD	DM	Other Id	N	Rem	α_{1950}	δ_{1950}	S	V	σ_V	B–V	σ_{B-V}	U–B	σ_{U-B}	n	Spectrum	References
170235	−25 13170	HR 6929, V4031 Sg			18 26 17	−25 17.4	8	6.57	.032	0.07	.020	−0.74	.035	24	B2 Vnne	26,240,540,976,1212*
	+6 03805	NGC 6633 - 136		⋆ V	18 26 18	+06 07.7	1	10.04		0.43		0.05		2	F4	104
	−4 04475				18 26 18	−04 18.2	1	10.68		1.19		1.16		2	K5 V	3072
170182	−32 14203				18 26 19	−32 15.3	1	9.51		−0.08		−0.64		6	B2 III	1732
169884	−63 04422				18 26 19	−63 04.5	1	7.98		0.14		0.10		2	A4 V	861
		CS 20321 # 168			18 26 20	−36 06.0	1	13.85		0.09		0.12		1		1736
170296	−14 05071	HR 6930			18 26 21	−14 36.0	7	4.70	.007	0.06	.004	0.05	.010	24	A3 Vn	15,369,1075,1088,2012*
		LS IV -15 053			18 26 24	−15 03.9	1	11.07		0.30		−0.40		3		1762
170318	−15 04982	IDS18236S1521	A		18 26 25	−15 19.0	1	9.40		0.17		0.03		3	B9 II/III	1762
170212	−33 13262				18 26 25	−33 32.5	1	8.75		0.07		−0.03		2	B9 IV	262
170467	+29 03264				18 26 26	+29 59.6	1	8.31		−0.03		−0.17		1	A0	963
		Steph 1618			18 26 26	+50 14.1	1	10.61		1.29		1.22		2	K7	1746
	−14 05074	LS IV -14 081			18 26 28	−14 19.7	1	10.00		0.42		−0.40		3		1762
	+16 03516				18 26 29	+16 44.4	1	8.44		0.64		0.11		2	G5	1648
		G 21 - 16			18 26 30	−04 31.6	2	14.53	.010	0.24	.005	−0.56		5		202,3060
	−8 04617	LS IV -08 007			18 26 31	−08 35.7	2	9.36	.000	0.91	.000	−0.21	.000	4	O8.5V:	1011,1012
	−19 04977				18 26 31	−18 42.3	1	9.31		1.20				4		1726
169570	−74 01682	HR 6899			18 26 31	−74 00.0	3	5.89	.004	0.99	.005	0.84	.005	16	K0 III	15,1075,2038
		CS 20321 # 102			18 26 32	−32 37.5	1	15.00		0.23		0.21		1		1736
		CpD -19 06854			18 26 33	−19 21.8	2	10.18	.017	0.11	.002	−0.63	.006	3		540,976
170278	−29 15006				18 26 33	−29 51.7	1	8.91		1.02		0.54		7	G8 V	1099
		CS 20321 # 151			18 26 33	−35 21.2	1	14.81		0.11		0.17		1		1736
171299	+77 00696				18 26 37	+77 31.8	1	7.07		1.01		0.77		4	K0	3016
170330	−17 05213				18 26 37	−17 23.2	1	9.63		0.38		−0.18		3	B9 II	1762
		NGC 6633 - 142			18 26 38	+06 45.8	1	11.02		0.45		0.19		2	A9	104
170251	−33 13267				18 26 38	−33 30.8	1	9.26		0.89		0.66		2	K1 III	262
	+6 03806	NGC 6633 - 143			18 26 39	+06 57.1	1	10.12		0.60		0.02		2	dG5	104
170426	+7 03709	NGC 6633 - 144			18 26 40	+07 10.3	1	9.02		0.26		0.20		2	A0	212
		CS 20321 # 103			18 26 40	−32 40.6	1	14.37		0.21		0.20		1		1736
		PASP75,194 # 38			18 26 40	−33 01.1	1	13.78		1.33				3		1190
		CS 20321 # 178			18 26 40	−37 08.6	1	12.17		0.26		0.15		1		1736
170448	+11 03478				18 26 42	+11 37.0	1	7.49		0.57		0.11		2	G0	1733
170281	−33 13268				18 26 42	−33 34.9	1	8.27		1.68		1.94		2	G6/K0 (III)	262
169978	−62 05879	HR 6916			18 26 42	−62 18.8	5	4.64	.006	−0.12	.005	−0.39	.005	19	B7/8 III	15,1075,1075,2012,3023
	+4 03761				18 26 43	+04 51.9	1	9.93		2.05				2	M2	369
170348	−16 04899				18 26 43	−16 07.7	1	8.22		0.41		0.17		3	A9/F0 V	1762
		LS IV -13 055			18 26 44	−13 04.3	1	11.35		0.58				4		1021
170349	−16 04900	LSS 5068			18 26 44	−16 40.5	1	8.42		0.67		0.55		3	A9 V	1762
170280	−33 13269				18 26 44	−33 31.8	1	8.02		0.01		−0.33		2	B8 V	262
170363	−14 05076				18 26 45	−14 46.5	1	9.73		0.61		0.08		3	G5 V	1762
	−24 14393				18 26 45	−24 33.7	1	10.57		0.19		−0.20		3		1586
170279	−33 13270				18 26 45	−33 05.0	2	7.14	.029	0.22	.000	0.11	.092	14	A3 III	262,1190
170451	+6 03809	NGC 6633 - 147			18 26 47	+06 45.2	1	9.50		0.60		0.07		3	G2	104
	−9 04748	LS IV -09 015			18 26 48	−09 13.8	1	10.83		0.94				4		1021
170321	−33 13271				18 26 48	−33 41.8	1	9.34		1.06		0.91		2	G5	262
		NGC 6633 - 148			18 26 49	+06 50.1	1	10.44		0.39		0.25		2	A4	104
		PASP75,194 # 27			18 26 50	−33 00.7	1	12.87		1.04				2		1190
170171	−49 12173				18 26 50	−49 50.7	1	9.10		0.46				4	F5 V	2012
170470	+10 03528	V451 Oph			18 26 53	+10 51.5	1	7.87		0.08		−0.07		5	B9 V	588
170472	+6 03811	NGC 6633 - 151			18 26 54	+06 49.0	1	9.10		0.08		−0.26		3	A0	104
	−29 15011				18 26 55	−29 55.2	1	10.63		0.56		0.02		8		1099
170397	−14 05077	HR 6932, V432 Sco			18 26 56	−14 37.0	3	6.04	.028	−0.03	.004	−0.13	.010	10	Ap CrEu	252,2007,3016
170378	−16 04902				18 26 56	−16 44.9	1	9.51		0.25		−0.20		3	B8 Ib	1762
		NGC 6633 - 280			18 26 57	+06 33.0	1	11.86		0.33		0.20		2		104
	+6 03813	NGC 6633 - 155			18 26 57	+06 34.1	1	10.55		0.88		0.62		2	F9	104
	+6 03812	NGC 6633 - 154	AB		18 26 57	+06 36.3	1	10.43		1.23		1.06		2	G7	104
170320	−33 13272				18 26 58	−33 09.0	2	8.37	.035	0.05	.000	−0.02	.054	11	B9 IV/V	262,1190
		CS 20321 # 144			18 26 59	−34 57.7	1	13.83		0.28		0.26		1		1736
	−35 12664				18 26 59	−35 50.0	1	10.91		1.26		1.17		1		1696
170811	+59 01899	HR 6949			18 27 01	+59 30.9	1	6.42		0.99		0.72		2	K0 IV	1733
	−29 15015				18 27 01	−29 31.6	1	11.12		0.18		−0.01		5		1099
	+7 03711				18 27 02	+07 48.4	1	9.75		0.14		0.00		2		212
238877	+55 02077				18 27 02	+55 45.3	1	9.14		0.34		0.04		2	F0	1502
	−29 15016				18 27 04	−29 28.3	1	10.32		0.72		0.15		9		1099
		CS 20321 # 136			18 27 04	−34 41.1	1	13.85		0.24		0.18		1		1736
170474	−2 04641	HR 6935			18 27 05	−02 01.2	4	5.41	.013	0.96	.004	0.76	.010	15	K0 III	37,369,1149,3016
		CS 20321 # 167			18 27 05	−36 03.8	1	12.26		0.21		0.11		1		1736
		PASP75,194 # 3			18 27 08	−33 05.0	1	10.42		0.52		0.26		9		1190
		PASP75,194 # 29			18 27 10	−33 10.2	1	12.03		1.58				8		1190
170453	−14 05081	LSS 5071			18 27 11	−14 14.6	2	9.07	.005	0.44	.005	−0.49	.010	5	B1 Iab/b	1762,8100
170454	−14 05080				18 27 11	−14 50.4	1	9.36		0.26		0.03		3	B9 II	1762
170452	−13 05003	LS IV -12 066		⋆ AB	18 27 12	−12 59.0	1	8.75		0.53		−0.48		1	B9.5II/III	1011
170455	−14 05082				18 27 15	−14 53.2	1	9.81		0.38		0.27		3	A1 IV	1762
170456	−16 04903				18 27 15	−16 13.5	1	7.86		0.25		0.16		3	A8 V	1762
170433	−18 04982	HR 6933			18 27 15	−18 45.7	2	5.65	.023	1.05	.009	0.93		5	K0 III	1754,2007
170493	−1 03500	G 21 - 17			18 27 16	−01 51.0	8	8.04	.011	1.09	.013	1.06	.004	37	K3 V	202,989,1013,1197,1620,1658*
	−12 05074	LS IV -12 067			18 27 17	−12 52.4	1	10.33		0.57				4	B1	1021
		CS 20321 # 166			18 27 21	−35 60.0	1	14.90		0.15		0.22		1		1736
170563	+6 03816	NGC 6633 - 161			18 27 23	+06 43.7	1	8.15		0.15		0.08		2	Am	104
170850	+55 02078				18 27 23	+55 38.4	1	7.78		0.36		−0.04		2	F5	1502

Table 1   1007

HD	DM	Other Id	N Rem	α_{1950}	δ_{1950}	S	V	σ_V	B–V	σ_{B-V}	U–B	σ_{U-B}	n	Spectrum	References
171606	+79 00587			18 27 23	+79 11.4	1	6.53		1.09		0.95		2	K0	985
170475	−19 05030	IC 4725 - 1		18 27 25	−19 24.1	1	8.97		0.23		-0.27		4	B8 III/IV	50
170564	+6 03817	NGC 6633 - 163		18 27 26	+06 11.8	1	9.64		0.20		0.09		2	A0	212
		HIC 90710		18 27 27	−20 59.2	1	10.98		1.44				4		1726
170195	−59 07285	IDS18230S5938	A	18 27 29	−59 35.9	1	7.73		0.43		0.01		2	F3 IV/V	1730
	+6 03818	NGC 6633 - 165		18 27 30	+06 42.3	1	9.46		1.25		1.11		2	gG7	104
170579	+9 03737			18 27 30	+09 10.0	1	9.25		0.54		0.01		1	F5	1620
170384	−41 12871	HR 6931		18 27 30	−41 56.9	1	6.04		0.14				4	A3 V	2007
170650	+23 03347	HR 6943		18 27 31	+23 49.9	2	5.90	.010	-0.09	.005	-0.51	.005	6	B6 IV	154,555
		NGC 6633 - 281		18 27 32	+06 44.4	1	11.16		0.29		0.20		2		104
170517	−14 05083			18 27 33	−14 37.6	1	9.77		0.24		0.08		3	B9 IV	1762
170778	+43 02989			18 27 34	+43 54.2	1	7.52		0.59		0.09		2	G5	1601
170547	−5 04675	HR 6940		18 27 34	−05 45.5	1	6.27		0.95		0.65		8	G8 II-III	1088
	+48 02699			18 27 35	+48 10.7	1	10.81		0.89		0.55		3	G0	196
170580	+3 03727	HR 6941	⋆ A	18 27 36	+04 01.8	4	6.69	.017	0.11	.009	-0.52	.011	11	B2 V	154,246,1149,3016
170580	+3 03727	IDS18251N0400	B	18 27 36	+04 01.8	1	11.26		0.56		0.30		2		3016
		CS 20321 # 152		18 27 37	−35 34.6	1	15.16		0.12		0.20		1		1736
170385	−43 12589	IDS18240S4348	AB	18 27 37	−43 45.9	1	7.90		-0.14		-0.63		1	B5 III/V	55
		NGC 6638 sq1 14		18 27 38	−25 31.7	1	13.39		1.44		1.41		3		474
	−33 13278	PASP75,194 # 30		18 27 39	−33 11.7	1	10.89		0.37				3		1190
	−59 07287			18 27 42	−59 29.9	1	9.85		0.55		0.07		2	F9	1730
	+48 02700			18 27 43	+48 10.6	1	11.02		1.03		0.70		3		196
	−19 05032	IC 4725 - 6		18 27 44	−19 04.5	1	9.54		0.23		-0.26		2	B6	50
		NGC 6638 sq1 8		18 27 44	−25 33.3	1	13.07		1.66		2.05		4		474
170651	+15 03483			18 27 45	+15 54.6	1	7.17		1.03		0.81		2	K0	1648
170498	−25 13196	NGC 6638 sq1 5		18 27 46	−25 35.3	1	9.84		0.57		0.28		4	F9	474
170549	−15 04991			18 27 48	−15 46.6	1	8.32		0.40		0.38		3	A2 II	1762
	−19 05033	IC 4725 - 4		18 27 48	−19 25.4	1	10.40		0.36		-0.06		3		50
170479	−33 13281	HR 6936	⋆ AB	18 27 48	−33 01.5	3	5.38	.019	0.18	.009	0.11	.070	12	A5 V	262,1190,2007
170418	−43 12591	IDS18242S4326	A	18 27 49	−43 23.9	2	8.59	.104	0.67	.000	0.17	.032	5	G3 V	219,3062
170418	−43 12591	IDS18242S4326	B	18 27 49	−43 23.9	1	10.62		0.84		0.49		2		3062
	−8 04623	LS IV -08 008		18 27 51	−08 40.0	2	10.51	.015	1.07	.019	-0.04	.019	3	B0.5III	405,432
		NGC 6638 sq1 4		18 27 51	−25 36.7	1	11.63		1.50		1.51		4		474
170634	+1 03694			18 27 52	+01 11.3	1	9.85		0.58		0.10		4	B7 V	206
	+4 03770			18 27 52	+04 44.4	1	9.75		0.41		0.31		1	A0	212
170737	+26 03279			18 27 53	+26 37.4	2	8.09	.005	0.82	.000	0.35	.000	4	G8 III-IV	1003,3077
		G 155 - 19		18 27 53	−10 39.0	1	14.25		0.14		-0.63		1	DA	3060
		NGC 6638 sq1 17		18 27 53	−25 27.4	1	13.07		0.77		0.29		5		474
		NGC 6638 sq1 3		18 27 54	−25 35.9	1	12.48		1.42		1.37		4		474
		NGC 6638 sq1 7		18 27 55	−25 33.4	1	13.25		1.74				3		474
170584	−16 04906			18 27 56	−15 57.9	1	9.48		0.08		-0.35		3	B8 Ib/II	1762
170582	−14 05085			18 27 57	−14 49.7	1	9.62		0.44		-0.27		3	A9 V	1762
		CS 20321 # 124		18 27 57	−34 04.2	1	15.40		0.19		0.24		1		1736
		LS IV -12 071		18 27 58	−12 14.4	1	11.06		0.65				4		1021
170583	−15 04993			18 27 58	−15 19.0	1	10.04		0.19		-0.01		3	B9 II	1762
		IC 4725 - 8		18 27 58	−19 11.1	1	11.48		0.41		0.21		2		50
170581	−13 05011	LSS 5075		18 27 59	−13 40.2	1	9.44		0.44		-0.46		1	B2 Ia	1032
170676	+7 03719			18 28 00	+07 29.9	1	9.41		0.20		0.19		3	A0	212
		IC 4725 - 7		18 28 01	−19 17.0	1	10.41		0.38		0.00		4		50
170603	−15 04995	IDS18252S1500	AB	18 28 03	−14 57.7	1	9.45		0.19		-0.40		1	B(5) V +(A)	1032
170465	−45 12550	HR 6934		18 28 03	−45 57.0	3	4.95	.007	-0.12	.004	-0.44		10	B6 IV	15,26,2012
	+4 03773	IDS18256N0412	AB	18 28 04	+04 14.3	1	8.82		0.26		-0.12		2	A0	246
170604	−16 04907	LSS 5076		18 28 05	−16 36.6	3	8.43	.068	0.16	.012	-0.73	.017	9	B1 Ib	399,1762,8100
170519	−33 13285	IDS18248S3319	AB	18 28 05	−33 17.0	1	9.31		0.59		0.15		2	K0/1 III +F	262
		IC 4725 - 10		18 28 06	−19 10.5	1	10.91		0.33		0.00		2	A1 V	50
		NGC 6638 sq1 18		18 28 06	−25 28.2	1	12.62		1.34		1.22		3		474
		NGC 6637 - 2018		18 28 06	−32 23.	1	13.39		1.13				1		579
		NGC 6637 - 3043		18 28 06	−32 23.	2	13.54	.010	1.64	.000			2		579,610
		NGC 6637 - 4012		18 28 06	−32 23.	2	13.16	.015	1.11	.030			2		579,610
		NGC 6637 - 4020		18 28 06	−32 23.	1	15.75		0.44				1		579
		NGC 6637 - 4047		18 28 06	−32 23.	1	14.32		1.36				1		610
170756	+21 03459	AC Her		18 28 09	+21 49.9	2	7.24	.205	2.23	.605	0.58	.240	2	F4 Ibp	793,3001
		NGC 6638 sq1 6		18 28 09	−25 32.7	1	12.22		1.53		1.65		5		474
	−9 04754	LS IV -09 016		18 28 11	−09 04.1	1	10.40		0.83				4	B2	1021
170620	−15 04996			18 28 11	−15 24.8	1	9.83		0.03		0.28		3	A0 III/IV	1762
		IC 4725 - 22		18 28 11	−19 02.0	1	11.60		1.25		1.09		1		50
	−19 05034	IC 4725 - 18		18 28 11	−19 08.3	2	10.21	.013	0.32	.004	-0.10	.009	3	B9 V	50,1226
	+2 03604			18 28 12	+02 30.5	1	10.23		0.32		0.28		2	A0	246
		IC 4725 - 24		18 28 12	−19 16.7	1	10.89		0.35		-0.01		1		50
		IC 4725 - 17		18 28 13	−19 09.7	2	10.60	.009	0.28	.004	-0.02	.004	3	A0 V	50,1226
		IC 4725 - 23		18 28 14	−19 02.2	1	12.15		0.46		0.24		1		50
		IC 4725 - 3		18 28 14	−19 22.1	1	10.09		0.40		-0.07		3		50
	−19 05035			18 28 15	−19 36.5	1	10.10		0.29				1		1297
170712	+3 03729			18 28 19	+03 07.1	1	7.83		0.12		-0.03		5	A0	246
	+2 03605			18 28 20	+02 08.1	1	10.17		0.38		0.44		2	A0	246
		CS 20321 # 118		18 28 20	−33 46.4	1	13.76		-0.02		-0.40		1		1736
170521	−43 12600	HR 6937		18 28 20	−43 32.6	1	5.72		1.34				4	K2 III	2009
170523	−45 12589	HR 6938		18 28 20	−45 47.6	3	5.07	.012	-0.13	.008	-0.55	.029	10	B3 IV/V	26,2009,8100
170657	−18 04986	IC 4725 - 26		18 28 23	−18 56.5	7	6.81	.018	0.86	.025	0.56	.030	17	K2 V	50,158,196,669,1067*
		IC 4725 - 40		18 28 23	−19 20.3	2	12.57	.026	0.43	.013	0.22	.017	3		50,1226

HD	DM	Other Id	N Rem	α_{1950}	δ_{1950}	S	V	σ_V	B–V	σ_{B-V}	U–B	σ_{U-B}	n	Spectrum	References
		KUV 18284+6650		18 28 24	+66 50.4	1	16.65		0.11		-0.56		1	DA	1708
		IC 4725 - 30		18 28 24	-19 04.3	3	12.39	.012	0.53	.014	0.28	.009	4		50,1163,1226
		IC 4725 - 41		18 28 24	-19 20.9	2	11.80	.004	0.76	.009	0.14	.004	3		50,1226
170781	+7 03721			18 28 25	+07 53.0	1	8.70		0.24		0.16		3	A0	212
		IC 4725 - 201		18 28 25	-19 09.4	1	14.85		0.90		0.48		1		1163
170679	-14 05089			18 28 26	-14 42.6	1	9.35		0.68		0.21		3	G5 V	1762
		IC 4725 - 29		18 28 26	-19 03.7	3	12.07	.017	0.46	.017	0.30	.015	3		50,1163,1226
	+4 03776			18 28 27	+04 35.1	1	10.55		0.40				2	A0	246
170700	-14 05090	LSS 5078		18 28 27	-14 09.0	2	8.64	.099	0.24	.010	-0.63	.005	5	B1/2 Ib/II	1762,8100
		IC 4725 - 67		18 28 27	-19 06.6	3	10.44	.009	0.28	.004	-0.10	.019	5	B7 V	50,1163,1226
		IC 4725 - 65		18 28 27	-19 07.4	1	11.66		0.40		0.21		2		50
170682	-19 05036	IC 4725 - 50	⋆ G	18 28 27	-19 11.7	5	7.92	.023	0.34	.013	-0.20	.023	20	B6 V	50,916,1163,1226,1586
170714	-5 04678	LS IV -05 005		18 28 28	-05 49.5	1	7.37		0.36		-0.35		3	B1 Vne	399
	-19 05040	IC 4725 - 73		18 28 29	-18 58.8	1	9.71		0.21		-0.25		3	B6 V	50
		IC 4725 - 43		18 28 29	-19 19.4	1	12.55		0.54		0.35		2		50
170680	-18 04988	HR 6944	⋆ A	18 28 30	-18 26.3	1	5.14		0.00				4	A0 Vn	2007
		IC 4725 - 68		18 28 30	-19 06.4	2	12.73	.024	1.38	.004	1.08	.004	4		50,1163
		IC 4725 - 76		18 28 31	-18 57.8	1	10.95		0.32		0.08		3		50
	-19 05037	IC 4725 - 51	⋆ T	18 28 31	-19 13.2	3	9.22	.021	0.37	.003	-0.15	.010	7	A0 V	50,1163,1226
	-19 05038	IC 4725 - 49		18 28 31	-19 16.6	3	9.06	.015	1.93	.044	2.05	.051	14	M1 III	50,1226,8032
	-18 04989	IC 4725 - 75		18 28 32	-18 53.0	1	9.80		0.29		-0.16		2		50
		IC 4725 - 71		18 28 32	-19 00.9	1	11.54		0.35		0.33		2		50
170681	-19 05041	IC 4725 - 70		18 28 32	-19 01.1	2	8.97	.016	0.43	.008	0.05	.000	4	A9 V	50,1226
		IC 4725 - 66		18 28 32	-19 07.3	1	12.33		0.52		0.31		2		50
	-19 05039	IC 4725 - 42		18 28 32	-19 20.5	2	9.68	.009	0.36	.009	-0.37	.009	3	B2 V	50,1226
170829	+20 03821	HR 6950		18 28 33	+20 47.0	2	6.47	.025	0.80	.005	0.40	.000	6	G8 IV	1080,3077
		UZ Sct		18 28 33	-12 57.8	1	10.79		1.71		1.27		1	G0	689
		IC 4725 - 54		18 28 33	-19 11.9	3	12.53	.009	0.67	.020	0.36	.008	6		50,1163,1226
		IC 4725 - 45		18 28 33	-19 18.1	2	10.42	.013	0.38	.000	-0.04	.018	5	B9 V	50,1226
		CS 20321 # 105		18 28 33	-32 18.7	1	15.24		0.15		0.11		1		1736
170782	+5 03772			18 28 34	+05 43.0	1	7.81		0.14		0.11		4	A2	212
	-14 05092			18 28 34	-14 23.5	1	9.36		1.61				1	M3	369
170701	-14 05091			18 28 34	-14 48.9	1	10.12		0.17		-0.42		3	B3 Ib	1762
		IC 4725 - 69	AB	18 28 34	-19 05.9	1	10.70		0.43		0.20		2		50
		IC 4725 - 64	AB	18 28 34	-19 07.9	2	11.74	.000	0.56	.013	0.34	.017	3		50,1163
		IC 4725 - 61		18 28 34	-19 10.5	2	13.10	.009	1.39	.009	1.09	.017	3		50,1163
		IC 4725 - 46		18 28 34	-19 17.2	3	11.78	.032	0.52	.017	0.32	.015	5		50,1163,1226
170716	-12 05083	LSS 5081		18 28 35	-12 22.5	3	9.48	.010	0.41	.010	-0.56	.009	6	B1/2 Ia	1012,1032,8100
		LS IV -14 086		18 28 35	-14 32.2	1	10.65		0.32		-0.22		3		1762
		IC 4725 - 72		18 28 35	-19 00.5	1	12.73		0.52		0.26		2		50
		IC 4725 - 63		18 28 35	-19 08.3	2	11.89	.026	0.44	.004	0.27	.017	3		50,1163
		IC 4725 - 47		18 28 35	-19 16.7	3	11.31	.050	0.44	.011	0.18	.014	4		50,1163,1226
170783	+4 03778			18 28 36	+04 35.5	2	7.73	.000	0.19	.000	-0.30	.000	4	B5	212,246
		IC 4725 - 60		18 28 36	-19 10.5	2	14.10	.026	0.82	.017	0.56	.009	3		50,1163
		IC 4725 - 58		18 28 36	-19 11.3	2	12.67	.024	0.65	.015	0.34	.020	5		50,1226
		IC 4725 - 55		18 28 36	-19 15.2	2	12.18	.021	0.70	.017	0.37	.004	3		50,1226
		IC 4725 - 48		18 28 36	-19 16.3	2	11.57	.016	0.49	.008	0.23	.016	4		50,1226
		IC 4725 - 74		18 28 37	-18 59.5	1	11.47		0.36		0.17		2		50
		IC 4725 - 53		18 28 37	-19 11.9	1	14.02		1.91				1		1163
		IC 4725 - 44		18 28 37	-19 18.2	3	10.22	.041	0.41	.016	-0.15	.044	9	B5 V	50,1226 ,1586
170718	-16 04910			18 28 38	-16 00.8	1	8.23		0.10		-0.33		3	B8 II	1762
		IC 4725 - 62		18 28 38	-19 08.9	3	13.71	.033	0.83	.025	0.22	.020	4		50,1163,1226
170638	-30 15795			18 28 38	-30 06.4	4	8.50	.041	-0.03	.014	-0.57	.022	9	B3 II/III	400,540,976,1586
170740	-10 04713	HR 6946	⋆ AB	18 28 39	-10 49.9	2	5.72	.000	0.24	.000	-0.45		8	B2 V +B9 V	206,2007
170740	-10 04713	IDS18259S1052	B	18 28 39	-10 49.9	1	9.42		0.47		0.21		5	B9 V	321
		IC 4725 - 198		18 28 39	-19 11.7	1	14.28		1.13		0.77		1		1163
170931	+32 03141			18 28 40	+32 36.3	1	7.62		0.14		0.11		2	A0	1375
		IC 4725 - 92		18 28 40	-19 08.4	3	11.21	.008	0.39	.008	0.13	.011	6		50,1163,1226
		IC 4725 - 59		18 28 40	-19 10.8	3	11.65	.019	0.47	.009	0.23	.018	6		50,1163,1226
		IC 4725 - 57		18 28 40	-19 12.4	3	10.52	.009	0.42	.009	0.04	.007	5	B8 V	50,1163,1226
170719	-19 05042	IC 4725 - 91	⋆ S	18 28 41	-19 07.5	3	8.08	.010	0.29	.016	-0.15	.028	5	B9 V	50,1163,1226
		IC 4725 - 56		18 28 41	-19 14.3	3	10.55	.010	0.39	.010	0.01	.017	4	B8 V	50,1163,1226
		AA45,405 S58 # 1		18 28 42	-08 31.0	1	12.77		1.45		0.60		1		432
	-13 05015	LSS 5083		18 28 42	-13 34.9	3	10.01	.010	0.52	.013	-0.53	.010	4	O7 III	1011,1032,1737
		IC 4725 - 86		18 28 42	-19 05.2	3	12.81	.013	0.89	.015	0.40	.016	5		50,1163,1226
		IC 4725 - 124		18 28 42	-19 16.7	2	12.41	.000	0.46	.000	0.29	.021	3		50,1226
		IC 4725 - 185		18 28 42	-19 17.	1	13.82		0.62		0.44		1		1163
		IC 4725 -1001		18 28 42	-19 17.	1	10.91		0.25				1		1297
		IC 4725 -1002		18 28 42	-19 17.	1	10.93		0.38				1		1297
		IC 4725 -1003		18 28 42	-19 17.	1	11.01		0.36				1		1297
170816	+6 03820	NGC 6633 - 171		18 28 43	+06 23.0	1	9.38		0.16		-0.09		2	A0	212
		IC 4725 - 90		18 28 43	-19 07.0	3	12.06	.013	0.97	.025	0.45	.022	5		50,1163,1226
		IC 4725 - 107		18 28 43	-19 12.4	2	12.03	.004	0.51	.017	0.31	.004	3		50,1226
		IC 4725 - 125		18 28 43	-19 17.4	2	11.51	.000	0.43	.004	0.23	.017	3		50,1226
		IC 4725 - 85		18 28 44	-19 04.3	3	12.05	.023	0.47	.022	0.27	.017	5		50,1163,1226
		IC 4725 - 95		18 28 44	-19 08.4	3	12.17	.013	0.53	.010	0.32	.012	5		50,1163,1226
		IC 4725 - 121		18 28 44	-19 15.0	1	14.67		0.98		0.44		1		1163
		IC 4725 - 123		18 28 44	-19 15.9	1	13.73		0.72		0.39		1		50
	-19 05043	IC 4725 - 126		18 28 44	-19 17.7	2	9.60	.028	0.39	.000	0.06	.028	4	A1 I	50,1226
		IC 4725 - 96	⋆ F	18 28 45	-19 09.1	3	9.68	.020	0.32	.009	-0.17	.011	7	B8 V	50,1163,1226

Table 1

1009

HD	DM	Other Id	N Rem	α_{1950}	δ_{1950}	S	V	σ_V	B–V	σ_{B-V}	U–B	σ_{U-B}	n	Spectrum	References
		IC 4725 - 122		18 28 45	−19 15.4	2	11.80	.038	1.57	.005	1.49		2		50,1163
		CS 20321 # 195		18 28 45	−36 31.2	1	15.00		0.17		0.23		1		1736
		IC 4725 - 89		18 28 46	−19 06.4	3	12.17	.005	0.46	.009	0.32	.011	5		50,1163,1226
		IC 4725 - 94		18 28 46	−19 07.9	3	12.23	.009	1.38	.024	1.11	.007	4		50,1163,1226
	−19 05044	IC 4725 - 97	⋆ E	18 28 46	−19 09.2	3	8.76	.019	0.39	.011	-0.20	.019	6	B7 V	50,1163,1226
		IC 4725 - 119		18 28 46	−19 14.4	3	12.31	.029	0.74	.010	0.34	.043	4		50,1163,1226
		IC 4725 - 127		18 28 46	−19 22.3	1	10.82		0.46		0.15		2		50
	−19 05045	IC 4725 - 110	⋆ C	18 28 47	−19 12.6	1	9.09		0.36		-0.17		2	B8 V	50
		IC 4725 - 197	⋆ D	18 28 47	−19 12.6	1	10.05		0.38		-0.09		2	B6 V	50
		IC 4725 - 203		18 28 47	−19 13.6	1	13.90		0.86		0.32		1		1163
		IC 4725 - 77		18 28 48	−18 59.9	1	11.95		0.51		0.26		2		50
		IC 4725 - 88		18 28 48	−19 05.8	3	13.14	.030	0.62	.006	0.46	.043	5		50,1163,1226
		IC 4725 - 93	⋆ M	18 28 48	−19 07.6	3	10.39	.010	0.38	.011	-0.04	.015	6	B8 V	50,1163,1226
		IC 4725 - 99	⋆ K	18 28 48	−19 08.8	3	9.87	.017	0.38	.009	-0.13	.015	6	B8 V	50,1163,1226
		IC 4725 - 98	⋆ L	18 28 48	−19 09.3	3	10.19	.012	0.41	.007	-0.08	.022	5	B9 V	50,1163,1226
		IC 4725 - 100		18 28 49	−19 08.7	3	10.06	.023	0.41	.011	-0.10	.012	6	B8 V	50,1163,1226
		IC 4725 - 109	⋆ Q	18 28 49	−19 11.8	3	10.67	.019	0.45	.015	-0.01	.015	5		50,1163,1226
	−17 05224			18 28 50	−17 26.8	1	9.18		1.90				2	M2	369
170742	−18 04991	IC 4725 - 273		18 28 50	−18 47.1	1	10.30		0.31				1	B6 IV	1297
		IC 4725 - 280		18 28 50	−19 10.8	1	15.33		0.73				1		736
170878	+16 03529	HR 6955		18 28 51	+16 53.5	3	5.78	.005	0.05	.004	0.06	.021	8	A2 V	985,1022,3016
170897	+21 03465			18 28 51	+21 37.1	1	7.41		1.07		0.83		26	K0	1222
	+48 02704			18 28 51	+48 34.1	1	9.18		1.05		0.92		3	K0	196
		IC 4725 - 78		18 28 51	−19 01.5	1	10.26		0.71		0.20		2		50
		IC 4725 - 79		18 28 51	−19 02.1	2	12.35	.009	0.46	.004	0.21	.004	3		50,1226
170625	−42 13358	V668 CrA		18 28 51	−42 21.4	1	8.74		0.21		0.15		2	A4/5 V	3014
		IC 4725 - 82		18 28 52	−19 03.6	2	13.53	.040	0.84	.008	0.30	.036	4		50,1163
		IC 4725 - 87		18 28 52	−19 03.6	2	13.78	.034	0.86	.030	0.54	.026	3		50,1163
		IC 4725 - 101	⋆ H	18 28 52	−19 09.2	3	10.27	.018	0.43	.015	0.04	.045	11	B8 V	50,1163,1226
		IC 4725 - 102	⋆ BU	18 28 52	−19 10.1	3	9.26	.023	0.46	.011	-0.07	.022	11	B6 V	50,1163,1226
170642	−39 12696	HR 6942		18 28 52	−39 44.4	1	5.16		0.08				4	A3 Vn	2032
		IC 4725 - 284		18 28 53	−19 06.7	1	15.70		2.09				1		736
		IC 4725 - 81		18 28 54	−19 03.0	3	12.62	.032	0.63	.013	0.35	.022	5		50,1163,1226
		IC 4725 - 83		18 28 54	−19 04.2	2	13.05	.012	0.71	.012	0.36	.020	4		50,1163
		IC 4725 - 104		18 28 54	−19 07.2	3	13.28	.008	0.76	.011	0.45	.020	5		50,1163,1226
		IC 4725 - 274		18 28 54	−19 08.2	1	14.77		2.22				1		736
		IC 4725 - 282		18 28 54	−19 11.0	1	15.48		1.46				1		736
	−19 05046	IC 4725 - 111	⋆ P	18 28 54	−19 12.5	3	9.00	.019	0.38	.004	0.10	.011	5	A1 V	50,1163,1226
		IC 4725 - 117		18 28 54	−19 14.4	3	11.57	.004	0.51	.016	0.28	.012	5		50,1163,1226
		IC 4725 - 118		18 28 54	−19 14.7	3	11.03	.010	0.44	.015	0.11	.015	5	A5	50,1163,1226
		CS 20321 # 196		18 28 54	−36 29.8	1	15.15		0.11		0.16		1		1736
170762	−16 04911			18 28 55	−16 09.7	1	9.78		0.44		0.32		3	A8/9 V	1762
		IC 4725 - 278		18 28 55	−19 06.4	1	15.23		2.01				1		736
		IC 4725 - 279		18 28 55	−19 10.1	1	15.31		0.88				1		736
		IC 4725 - 112		18 28 55	−19 12.9	3	12.27	.000	0.51	.005	0.33	.012	4		50,1163,1226
		IC 4725 - 84		18 28 56	−19 05.6	3	11.68	.015	0.57	.018	0.32	.024	5		50,1163,1226
		IC 4725 - 113		18 28 56	−19 12.7	1	12.50		1.54				1		1163
		IC 4725 - 116		18 28 56	−19 14.1	3	12.71	.015	0.68	.013	0.40	.020	5		50,1163,1226
		IC 4725 - 128		18 28 56	−19 20.4	2	9.81	.009	0.30	.000	-0.21	.004	5	B8 V	50,1226
		IC 4725 - 276		18 28 57	−19 06.5	1	14.98		2.43				1		736
		IC 4725 - 275		18 28 57	−19 07.6	1	14.88		2.18				1		736
170764	−19 05047	IC 4725 - 103	⋆ A	18 28 57	−19 09.7	5	6.38	.030	0.94	.036	0.60	.034	6	F7 II	669,1490,1754,6007,8100
		IC 4725 - 281		18 28 57	−19 10.6	1	15.36		1.01				1		736
		IC 4725 - 130		18 28 57	−19 17.6	3	11.05	.037	0.40	.010	0.10	.024	6		50,1163,1226
		IC 4725 - 129		18 28 57	−19 19.2	1	11.34		1.65		0.00		2		50
170787	−15 05001			18 28 58	−15 13.1	1	7.28		0.31		0.21		3	A7/8 IV	1762
		IC 4725 - 277		18 28 58	−19 07.1	1	15.13		0.72				1		736
		IC 4725 - 285		18 28 58	−19 07.6	1	16.32		1.36				1		736
		IC 4725 - 114		18 28 58	−19 12.8	3	11.48	.021	0.43	.008	0.19	.005	4		50,1163,1226
		IC 4725 - 115		18 28 59	−19 13.2	2	13.22	.009	0.48	.017	0.28	.009	3		50,1163
170881	+6 03824	NGC 6633 - 177	⋆ AB	18 29 00	+06 25.4	1	7.89		0.12		-0.14		3	A0 V +B9Vn	212
		CS 20321 # 217		18 29 00	−35 08.1	1	14.86		0.01		-0.19		1		1736
		IC 4725 - 283		18 29 01	−19 09.6	1	15.56		1.17				1		736
		CS 20321 # 189		18 29 01	−36 47.7	1	14.05		0.12		0.16		1		1736
170899	+7 03729			18 29 02	+07 59.4	2	7.42	.000	1.02	.009	0.51	.000	4	G0	979,1733
		IC 4725 - 108		18 29 02	−19 08.4	1	14.14		0.90		0.50		1		1163
		IC 4725 - 106	⋆ R	18 29 02	−19 09.7	3	11.22	.022	0.48	.013	0.17	.027	12		50,1163,1226
		CS 20321 # 106		18 29 02	−32 31.6	1	14.62		0.14		0.10		1		1736
		IC 4725 - 199		18 29 03	−19 09.0	1	14.60		1.62		1.76		1		1163
170796	−15 05002			18 29 04	−15 42.5	1	9.63		0.11		-0.36		3	B7 Ib	1762
		IC 4725 - 145	⋆ AB	18 29 04	−19 08.0	2	13.65	.017	0.83	.009	0.31	.030	3		50,1163
		IC 4725 - 134		18 29 04	−19 14.0	2	13.71	.033	0.86	.024	0.35	.090	2		50,1163
		CS 20321 # 181		18 29 04	−37 11.7	1	13.60		0.16		0.10		1		1736
	+34 03222	IDS18273N3412	A	18 29 05	+34 14.2	1	9.22		0.61		0.10		1	F8	8084
	+34 03222	IDS18273N3412	B	18 29 05	+34 14.2	1	11.53		0.49		-0.06		1		8084
	+34 03222	IDS18273N3412	C	18 29 05	+34 14.2	1	10.90		1.14		0.96		1		8084
	−19 05049	IC 4725 - 200		18 29 05	−19 07.8	2	10.10	.028	0.40	.023	0.08	.098	4	A1	1163,1226
		IC 4725 - 133		18 29 05	−19 14.8	2	13.68	.043	0.58	.004	0.41	.090	3		50,1163
		IC 4725 - 132		18 29 05	−19 16.4	2	13.55	.102	1.40	.000	1.03	.055	3		50,1163
		CS 20321 # 243		18 29 06	−33 56.9	1	14.34		0.20		0.19		1		1736

HD	DM	Other Id	N Rem	α₁₉₅₀	δ₁₉₅₀	S	V	σ_V	B–V	σ_B–V	U–B	σ_U–B	n	Spectrum	References
		IC 4725 - 202		18 29 07	−19 10.3	1	14.30		1.72		1.89		1		1163
		IC 4725 - 165		18 29 07	−19 17.4	2	13.59	.034	0.98	.009	0.39	.013	3		50,1163
170724	−33 13306			18 29 07	−32 58.6	1	7.64		0.25		0.24		2	A3mA7-F0	262
170525	−58 07418	HR 6939		18 29 07	−58 44.7	5	6.43	.007	0.69	.007	0.17	.006	15	G5 IV	15,861,2020,2028,3031
		IC 4725 - 144		18 29 08	−19 06.0	3	13.01	.010	0.73	.014	0.41	.033	3		50,1163,1226
		IC 4725 - 149		18 29 08	−19 12.3	3	10.41	.005	0.41	.013	-0.02	.012	7	A0 V	50,1163,1226
		IC 4725 - 166		18 29 08	−19 17.8	2	11.83	.103	1.91	.008	2.09	.024	4		50,1226
170770	−27 12936			18 29 08	−27 15.5	1	7.76		0.12		-0.36		3	B5 IV	240
		IC 4725 - 148		18 29 09	−19 11.1	3	13.23	.020	0.84	.035	0.30	.033	6		50,1163,1226
	−10 04715	G 155 - 21		18 29 10	−10 26.5	1	10.78		0.68		0.06		2		3062
		IC 4725 - 147		18 29 10	−19 10.3	3	12.89	.009	0.64	.016	0.34	.030	5		50,1163,1226
		ApJ144,259 # 46		18 29 12	+26 53.	1	15.07		-0.19		-1.11		8		1360
170919	+5 03775			18 29 13	+05 53.4	1	9.68		0.18		0.04		2	A0	212
		IC 4725 - 158		18 29 13	−19 13.1	2	13.72	.033	0.67	.047	0.33	.047	2		50,1226
		G 155 - 22		18 29 14	−08 01.9	1	14.88		1.07		0.40		2		3062
		IC 4725 - 157		18 29 14	−19 12.9	3	11.94	.015	0.54	.011	0.32	.018	6		50,1163,1226
	−19 05050	IC 4725 - 159		18 29 14	−19 13.4	3	10.19	.017	0.35	.007	-0.13	.011	8	B6 V	50,1163,1226
	−19 05051	IC 4725 - 168		18 29 14	−19 22.7	1	10.35		0.31		-0.08		2	B5 V	50
170836	−19 05052	IC 4725 - 167		18 29 15	−19 18.8	2	8.95	.003	0.31	.000	-0.23	.003	11	B8 II	50,1226
		CS 20321 # 256		18 29 15	−33 14.3	1	13.27		0.27		0.22		1		1736
170820	−19 05053	IC 4725 - 150		18 29 16	−19 09.7	5	7.37	.009	1.57	.018	1.33	.018	18	K0+IIa:	50,219,1163,1226,1490
		IC 4725 - 164		18 29 16	−19 17.0	3	12.12	.023	0.57	.013	0.30	.015	5		50,1163,1226
		IC 4725 - 160		18 29 17	−19 14.0	2	13.46	.019	0.83	.004	0.37	.026	5		50,1163
170935	+7 03730			18 29 18	+07 11.7	1	7.38		0.15		-0.04		2	B8	212
		WLS 1840 35 # 6		18 29 19	+35 18.2	1	13.51		0.53		-0.02		2		1375
		IC 4725 - 182		18 29 20	−19 20.2	1	13.99		0.88		0.28		1		50
		G 227 - 35		18 29 21	+54 45.2	1	15.50		0.49		-0.44		5	DXP	940
		IC 4725 - 151		18 29 21	−19 06.3	2	11.83	.005	0.47	.014	0.27	.005	2		50,1226
		IC 4725 - 161		18 29 21	−19 13.1	3	11.82	.006	0.44	.017	0.32	.013	5		50,1163,1226
170835	−19 05055	IC 4725 - 163		18 29 21	−19 15.0	4	8.83	.008	0.24	.006	-0.51	.007	23	B2.5Vne	50,540,976,1226
171010	+22 03394			18 29 22	+22 56.9	1	7.39		0.98		0.73		2	G5	1733
170920	−1 03504	HR 6957		18 29 22	−01 02.4	3	5.94	.004	0.16	.000	0.18	.027	10	A4 III	15,1256,8071
		IC 4725 - 204		18 29 23	−19 13.7	1	14.24		1.61		1.90		1		1163
	−12 05091			18 29 24	−12 01.7	1	10.02		2.04				2	M2	369
		LS IV -14 087		18 29 24	−14 41.4	1	10.94		0.22		-0.55		3		1762
		IC 4725 - 152		18 29 24	−19 07.1	2	10.50	.004	0.32	.013	-0.01	.021	3		50,1226
		IC 4725 - 15		18 29 24	−19 11.9	1	11.97		0.73		0.26		2		50
	−11 04665		A	18 29 25	−11 47.9	1	10.20		1.88				1		725
	−11 04665		B	18 29 25	−11 47.9	2	11.20	.025	0.82	.005	0.36		2		432,725
170884	−16 04915			18 29 25	−16 08.4	1	9.47		0.21		0.02		3	A0 III	1762
		IC 4725 - 175		18 29 25	−19 14.6	2	12.52	.020	0.42	.004	0.19	.020	4		50,1163
		IC 4725 - 172		18 29 25	−19 21.8	1	12.57		0.60		0.30		2		50
	−62 05888	Smethells 15		18 29 25	−62 46.9	1	9.54		1.43		1.73		1		1494
		G 141 - 15		18 29 27	+08 33.9	2	13.08	.015	0.54	.020	-0.19	.005	2		1658,3060
170860	−19 05058	IC 4725 - 153	★ AB	18 29 27	−19 08.0	2	9.40	.012	0.32	.000	-0.22	.008	4	B9 IV/V	50,1226
170838	−22 04761			18 29 27	−22 25.6	2	10.02	.010	0.37	.005	-0.71	.010	5	A1 III/IV	540,1737
170885	−16 04916			18 29 28	−16 15.6	1	8.40		1.28		1.02		3	G8 III	1762
	−19 05056	IC 4725 - 248		18 29 29	−19 00.8	1	9.62		0.42		0.10		5	A0 I	1586
		IC 4725 - 173		18 29 29	−19 20.6	1	11.60		1.60		1.69		2		50
170902	−14 05098	HR 6956		18 29 30	−14 40.9	1	6.37		0.22				4	A4 V	2007
170773	−39 12704	HR 6948		18 29 31	−39 55.7	3	6.23	.010	0.42	.000	-0.06		10	F5 V	2006,2018,3037
	−11 04667	LS IV -11 025		18 29 32	−11 19.6	2	9.80	.000	0.87	.005	-0.13	.010	3	B1:V:pe	445,1012
170886	−19 05059	IC 4725 - 251		18 29 33	−19 00.6	3	6.95	.006	1.39	.012	1.09	.011	6	G2 III	657,1754,2033
		IC 4725 - 191		18 29 33	−19 04.5	1	10.36		0.50		0.35		3	A0 V	50
170903	−15 05003			18 29 34	−15 34.3	1	8.58		0.27		-0.01		3	B9 V	1762
170904	−16 04919			18 29 34	−15 58.1	1	8.34		0.15		-0.25		3	B8 II	1762
	−19 05060	IC 4725 - 174		18 29 34	−19 19.8	1	8.94		2.04				1	M3 II-III	50
170973	+3 03737	HR 6958, MV Ser		18 29 37	+03 37.3	3	6.42	.005	-0.04	.007	-0.25	.007	8	A0 p SiCr	15,1079,1256
	+6 03832			18 29 39	+06 47.6	1	9.78		0.23		0.19		2		212
		IC 4725 - 192		18 29 41	−19 04.9	1	10.06		0.29		-0.15		2	B8 V	50
		IC 4725 - 196		18 29 41	−19 08.3	1	10.88		0.29		0.00		2	A0 V	50
170938	−15 05004	LSS 5084		18 29 45	−15 44.4	4	7.88	.021	0.84	.005	-0.26	.011	12	B1 Ia	1012,1737,1762,8100
	+1 03705			18 29 46	+01 50.2	1	10.62		0.56		-0.06		1		246
	+3 03738			18 29 48	+03 44.6	1	10.17		0.44		-0.05		2		246
	+37 03148			18 29 50	+37 12.3	1	8.96		1.53		1.87		2	M0	1375
171067	+13 03677			18 29 52	+13 41.8	2	7.20	.013	0.68	.018	0.24		7	G8 V	1619,7008
170975	−14 05099	HR 6959		18 29 52	−14 54.2	3	5.50	.043	2.00	.015	2.17		8	K3 I-II	369,2007,3005
	−6 04787	LS IV -06 009		18 29 54	−06 15.7	3	10.21	.005	0.95	.038	-0.13	.000	7	O5.5III(f)	390,405,1021
	−12 05095	LSS 5085		18 29 55	−12 50.1	1	10.05		0.46		-0.51		1	B0 IV	1032
170890	−33 13323			18 29 55	−33 01.2	1	8.64		0.08		0.14		2	A1 V	262
171050	+2 03611			18 29 56	+02 53.1	1	8.90		1.14		0.77		1	K0	246
170868	−38 12895	HR 6952	★ AB	18 29 56	−38 45.5	1	5.19		-0.06				4	A0 III	2007
170845	−42 13378	HR 6951		18 29 56	−42 21.0	5	4.64	.003	1.01	.007	0.76	.004	24	G8 III	15,1075,1075,2012,8015
170976	−15 05006			18 29 58	−15 52.4	1	8.73		1.74		1.87		3	K1/2 (III)	1762
170990	−14 05101			18 29 59	−14 40.0	1	9.92		0.31		0.13		3	B9 III/IV	1762
		CS 20321 # 232		18 30 00	−34 30.4	1	11.03		0.14		0.06		1		1736
170575	−66 03382			18 30 01	−66 12.9	1	9.03		0.29		0.02		4	F0 V	1731
171242	+44 02919	G 205 - 27		18 30 03	+44 58.7	1	8.03		0.52		-0.01		3	G0	3026
		AA45,405 S61 # 1	AB	18 30 06	−05 03.	1	12.34		1.16				1		725
		BPM 11655		18 30 06	−65 57.	1	14.73		0.55		-0.11		1		3065

Table 1 1011

HD	DM	Other Id	N Rem	α_{1950}	δ_{1950}	S	V	σ_V	B–V	σ_{B-V}	U–B	σ_{U-B}	n	Spectrum	References
171088	+7 03734			18 30 08	+07 37.5	1	9.85		0.22		0.18		2	A0	212
	−23 14483	V4066 Sgr		18 30 08	−23 06.9	1	10.02		1.36		1.27		2	M5	1768
170991	−16 04922			18 30 09	−16 35.9	1	6.94		1.46				2	K0/1 II/III	369
171089	+4 03785			18 30 10	+04 15.5	1	7.36		1.84				2	K5	369
170978	−24 14462			18 30 11	−24 08.9	3	6.81	.008	0.05	.017	-0.48	.009	6	B3 III/IV	219,540,976
171090	+2 03612			18 30 12	+02 53.1	1	9.57		0.32		0.23		2	A2	246
		CS 20321 # 203		18 30 12	−36 11.4	1	13.88		0.33		0.18		1		1736
171313	+52 02225			18 30 13	+52 24.5	1	8.40		1.11		1.01		2	K0	1566
171012	−18 04994	LSS 5086		18 30 14	−18 24.4	5	6.87	.042	0.47	.023	-0.61	.010	11	B0 Ia/ab	138,1423,1586,1737,8100
		CS 20321 # 233		18 30 14	−34 28.9	1	13.88		0.26		0.20		1		1736
	−19 05062			18 30 15	−19 00.6	1	10.47		0.32				1		1297
		NGC 6649 - 113		18 30 16	−10 25.2	1	12.41		1.02		0.14		2		323
171052	−10 04717			18 30 17	−10 44.5	1	8.96		0.74				36	G5	6011
171054	−14 05102	LS IV -13 069		18 30 18	−13 57.0	4	9.07	.030	0.16	.019	-0.65	.025	7	B2 II	445,1012,1032,1586
	+4 03786			18 30 19	+04 38.5	1	10.46		0.48				2		246
171126	+4 03787			18 30 24	+04 10.5	1	8.26		0.28		0.18		3	F0	246
		NGC 6649 - 71		18 30 29	−10 27.8	1	14.63		1.19				1		323
		NGC 6649 - 97		18 30 29	−10 28.8	1	14.81		1.10				1		323
171032	−27 12960			18 30 29	−27 31.6	1	8.91		0.15		-0.26		4	B8 III	1586
		NGC 6649 - 103		18 30 30	−10 23.9	1	14.40		1.21				1		323
		NGC 6649 - 74		18 30 30	−10 26.2	1	14.23		1.20				1		323
		NGC 6649 - 96		18 30 30	−10 29.6	1	13.32		1.04		0.43		2		323
171074	−16 04928			18 30 30	−16 04.9	1	7.50		1.36		1.14		3	G8 III/IV	1762
170873	−52 11158	HR 6954		18 30 30	−52 55.8	1	6.22		1.25				4	K2 III	2007
	+4 03788			18 30 32	+04 21.1	1	10.33		0.54				2		246
		NGC 6649 - 104		18 30 32	−10 24.1	1	14.20		1.21				1		323
		NGC 6649 - 114		18 30 33	−10 21.5	1	14.88		1.62				2		323
		NGC 6649 - 78		18 30 33	−10 23.9	1	15.11		1.10				1		323
		NGC 6649 - 392		18 30 33	−10 28.0	1	16.35		1.20		0.89		1		406
171094	−14 05105			18 30 33	−14 08.8	2	8.01	.183	2.27	.015	2.24		5	M2 Iab	148,8032
170994	−32 14255	IDS18273S3253	AB	18 30 33	−32 50.5	1	8.91		0.47		0.09		2	F3 V	262
171232	+25 03564			18 30 34	+25 27.1	2	7.44	.005	0.89	.010	0.60	.015	5	G5	833,1625
		NGC 6649 - 391		18 30 34	−10 28.2	1	18.00		1.53				1		406
		NGC 6649 - 70		18 30 34	−10 28.6	2	13.89	.039	1.13	.005	0.42	.015	6		323,406
		NGC 6649 - 49		18 30 35	−10 25.1	1	12.57		2.58		2.79		4		406
		NGC 6649 - 46		18 30 35	−10 26.2	1	15.73		1.28				2		406
		NGC 6649 - 44		18 30 35	−10 27.8	1	13.81		1.18		0.51		4		406
		NGC 6649 - 532		18 30 35	−10 28.8	1	16.50		2.78				1		406
	+36 03168	T Lyr		18 30 36	+36 57.6	4	8.19	.045	5.51	.014	8.67	.222	3	C8	1238,8005,8008,8022
		NGC 6649 - 79		18 30 36	−10 23.2	1	15.35		1.27				1		323
		NGC 6649 - 48		18 30 36	−10 25.3	1	13.44		1.33		0.62		1		406
		BTT13,19 # 27		18 30 36	−10 27.	1	12.34		1.34		0.69		2		1586
		NGC 6649 - 43		18 30 36	−10 27.9	1	14.97		1.20		0.60		2		406
171110	−15 05010			18 30 36	−15 01.3	1	9.35		0.18		-0.13		3	B9 Ib/II	1762
		LP 867 - 11		18 30 36	−24 18.0	1	11.89		0.85		0.22		2		1696
		NGC 6649 - 115		18 30 37	−10 22.1	1	14.41		1.28				2		323
		NGC 6649 - 107		18 30 37	−10 22.6	1	14.31		1.34				1		323
		NGC 6649 - 24		18 30 37	−10 26.3	1	14.46		1.20		0.53		3		406
		NGC 6649 - 21		18 30 37	−10 27.0	1	14.28		1.15				2		323
		NGC 6649 - 80		18 30 38	−10 23.7	1	15.54		1.15				1		323
		NGC 6649 - 53		18 30 38	−10 25.2	1	15.33		1.35		0.85		3		406
		NGC 6649 - 299		18 30 38	−10 25.5	1	17.08		1.41				1		406
		NGC 6649 - 289		18 30 38	−10 26.2	1	17.55		1.44				1		406
		NGC 6649 - 20		18 30 38	−10 27.0	1	14.98		1.26		0.62		1		406
		NGC 6649 - 264		18 30 38	−10 27.8	1	16.85		1.66				1		406
	−10 04718	NGC 6649 - 42	⋆ AB	18 30 38	−10 28.7	1	9.52		1.91		1.32		1		323
		NGC 6649 - 95		18 30 38	−10 30.3	1	15.05		1.05				2		323
		NGC 6649 - 94		18 30 38	−10 30.8	1	14.54		1.10				2		323
		NGC 6649 - 81		18 30 39	−10 23.2	1	14.08		1.50				2		323
		NGC 6649 - 600		18 30 39	−10 23.7	1	16.20		1.37		1.39		2		406
		NGC 6649 - 52		18 30 39	−10 24.4	1	13.07		1.29		0.58		2		323
		NGC 6649 - 300		18 30 39	−10 25.5	1	15.99		1.41				1		406
		NGC 6649 - 17		18 30 39	−10 26.0	2	13.99	.110	1.25	.015	0.51		4		323,406
		NGC 6649 - 290		18 30 39	−10 26.2	1	18.75		1.75				1		406
		NGC 6649 - 18		18 30 39	−10 26.4	1	15.10		1.32		0.72		3		406
		NGC 6649 - 19		18 30 39	−10 26.7	2	12.10	.010	1.73	.060	1.36	.075	24		323,406
		CS 20321 # 190		18 30 39	−36 45.7	1	15.18		0.08		0.06		1		1736
		NGC 6649 - 212		18 30 40	−10 26.2	1	16.76		1.44				1		406
171245	+23 03363	HR 6966	⋆ A	18 30 41	+23 34.7	1	5.84		1.47				2	gK5	71
		NGC 6649 - 116		18 30 41	−10 21.8	1	14.40		1.16				1		323
		NGC 6649 - 602		18 30 41	−10 23.8	1	16.10		1.34		1.10		1		406
		NGC 6649 - 8		18 30 41	−10 26.5	1	14.08		1.26		0.53		4		406
		NGC 6649 - 206		18 30 41	−10 26.9	1	18.75		1.81				1		406
		NGC 6649 - 204		18 30 41	−10 27.1	1	17.43		1.46				2		406
		NGC 6649 - 119		18 30 41	−10 27.7	1	15.53		1.30		0.57		2		406
		NGC 6649 - 117	⋆ C	18 30 41	−10 27.8	2	12.05	.019	2.65	.122	3.28		6		323,406
171034	−33 13338	HR 6960		18 30 41	−33 03.3	3	5.29	.005	-0.11	.008	-0.67	.048	9	B2 III/IV	26,262,2007
		HRC 284	⋆ AB	18 30 42	−05 00.4	1	12.49		0.93		0.46		1		776
171149	−6 04791	HR 6963		18 30 42	−05 57.0	1	6.35		0.02		-0.04		8	A0 Vn	1088
		NGC 6649 - 108		18 30 42	−10 22.8	1	14.78		1.27				1		323

HD	DM	Other Id	N Rem	α_{1950}	δ_{1950}	S	V	σ_V	B–V	σ_{B-V}	U–B	σ_{U-B}	n	Spectrum	References
		NGC 6649 - 14		18 30 42	−10 25.7	1	13.45		1.34		0.77		1		406
		NGC 6649 - 15		18 30 42	−10 25.9	1	14.95		1.33		0.86		2		406
		NGC 6649 - 199		18 30 42	−10 26.	1	15.56		1.64				2		323
		NGC 6649 - 601		18 30 42	−10 26.	1	16.80		1.36				2		406
		NGC 6649 - 215		18 30 42	−10 26.3	1	17.86		1.39				1		406
		NGC 6649 - 9		18 30 42	−10 26.5	2	11.76	.005	1.25	.045	0.14	.060	19		323,406
		NGC 6649 - 7		18 30 42	−10 26.8	1	14.55		1.23		0.53		3		406
		NGC 6649 - 6		18 30 42	−10 26.9	1	15.10		1.19		0.46		2		406
		MCC 181		18 30 42	−11 40.	1	14.69		0.16		-0.69		1	M0	1036
	−11 04672	G 155 - 23		18 30 42	−11 40.3	4	10.01	.016	1.27	.008	1.21	.013	12	K7 V	1013,1017,1663,1705,3072
171111	−16 04931			18 30 42	−16 51.4	1	9.50		0.31		0.24		3	A0 V	1762
		NGC 6649 - 54		18 30 43	−10 24.5	2	14.26	.020	1.32	.020	0.71		4		323,406
		NGC 6649 - 13		18 30 43	−10 25.7	1	14.00		1.33		0.59		2		406
		NGC 6649 - 5		18 30 43	−10 27.2	1	14.03		1.27		0.96		3		406
		NGC 6649 - 4		18 30 43	−10 27.4	1	15.08		1.31		0.71		3		406
		NGC 6649 - 3		18 30 43	−10 27.6	1	14.58		1.24		0.66		4		406
		NGC 6649 - 369		18 30 43	−10 28.0	1	16.11		1.24		0.88		2		406
		G 21 - 18		18 30 44	−06 55.9	3	12.19	.014	1.32	.027	1.11		4		202,1620,1759
		NGC 6649 - 28		18 30 44	−10 25.0	2	12.37	.019	1.31	.034	0.64	.019	6		323,406
		NGC 6649 - 245		18 30 44	−10 26.2	1	18.18		1.71				1		406
		NGC 6649 - 222		18 30 44	−10 26.5	1	17.24		1.49				1		406
		NGC 6649 - 368		18 30 44	−10 28.3	1	16.15		1.23				1		406
		NGC 6649 - 41		18 30 44	−10 28.7	1	15.01		1.22		0.77		3		406
		NGC 6649 - 34		18 30 45	−10 26.4	2	13.83	.054	1.19	.034	0.50	.005	6		323,406
		NGC 6649 - 36		18 30 45	−10 27.3	1	15.14		1.19		0.73		2		406
		NGC 6649 - 38		18 30 45	−10 27.6	1	15.94		1.32		0.65		2		406
	−13 05028			18 30 45	−13 10.3	1	10.21		1.92				2	M3	369
171194	+6 03840			18 30 46	+06 41.9	1	8.82		0.25		0.18		4	B9	212
		NGC 6649 - 55		18 30 46	−10 24.5	1	16.07		1.35		0.85		1		406
		NGC 6649 - 30		18 30 46	−10 25.0	1	14.22		1.27		0.70		1		406
		NGC 6649 - 33		18 30 46	−10 26.1	2	12.74	.019	1.20	.054	0.56	.005	6		323,406
		NGC 6649 - 243		18 30 46	−10 26.4	1	18.42		1.79				2		406
		NGC 6649 - 35		18 30 46	−10 26.6	2	13.09	.039	1.12	.010	0.49	.039	6		323,406
		NGC 6649 - 31		18 30 47	−10 25.3	1	14.12		1.18		0.48		3		406
		NGC 6649 - 323		18 30 47	−10 25.8	1	17.22		1.46				1		406
		NGC 6649 - 502		18 30 47	−10 28.2	1	17.56		1.94				1		406
		CS 20321 # 185		18 30 47	−37 01.6	1	14.72		0.09		0.15		1		1736
	+3 03740			18 30 48	+03 54.1	1	10.39		0.41				2		246
172864	+83 00536	HR 7025		18 30 48	+83 08.5	2	6.16	.005	0.05	.005	0.07	.010	4	A2 V	985,1733
		NGC 6649 - 56		18 30 48	−10 24.3	2	13.90	.005	1.29	.034	0.62		6		323,406
		NGC 6649 - 57		18 30 48	−10 25.2	1	14.76		1.28		0.49		4		406
		NGC 6649 - 40		18 30 48	−10 26.9	2	14.53	.117	1.15	.034	0.60		6		323,406
		NGC 6649 - 353		18 30 48	−10 27.8	1	16.48		1.26				2		406
		NGC 6649 - 496		18 30 48	−10 28.2	1	17.70		1.57				1		406
171130	−14 05106	HR 6962		18 30 48	−14 53.6	1	5.76		0.04				4	A2 V	2007
171060	−33 13341			18 30 48	−33 00.0	1	7.17		0.22		-0.08		2	B9 IV/V	262
		NGC 6649 - 348		18 30 49	−10 27.5	1	19.12		1.82				1		406
		NGC 6649 - 64	⋆ V	18 30 49	−10 28.0	3	11.48	.095	1.74	.028	1.29		3		682,1484,1772
171219	+5 03784			18 30 50	+05 24.4	2	7.66	.010	0.19	.001	-0.16	.027	4	B8	212,1586
		NGC 6649 - 58		18 30 50	−10 25.0	1	12.20		1.32		0.82		4		406
		NGC 6649 - 59		18 30 50	−10 25.2	1	14.07		2.10				1		323
		NGC 6649 - 63		18 30 50	−10 27.2	1	15.59		1.27		0.72		2		406
		V367 Sct		18 30 50	−10 27.7	1	11.29		1.64				1		6011
171115	−24 14472	HR 6961		18 30 50	−24 04.3	1	5.49		1.79				4	K4 Ib	2007
		NGC 6649 - 84		18 30 51	−10 24.7	1	15.68		1.39				1		406
		NGC 6649 - 61		18 30 51	−10 26.7	2	13.49	.005	1.17	.024	0.52	.005	6		323,406
		NGC 6649 - 109		18 30 52	−10 22.9	1	14.63		2.58				2		323
		NGC 6649 - 60		18 30 52	−10 25.6	2	15.34	.102	1.20	.010	0.64		6		323,406
		CS 20321 # 244		18 30 54	−33 56.7	1	14.82		0.18		0.23		1		1736
170948	−52 11161			18 30 54	−52 25.6	1	6.83		1.13		1.02		7	K0 III	1628
171301	+30 03223	HR 6968	⋆ AB	18 30 55	+30 30.9	3	5.48		-0.10	.002	-0.35	.005	8	B8 IV	1022,1079,3023
171247	+8 03741	HR 6967	⋆ A	18 30 59	+08 13.8	3	6.41	.005	-0.03	.007	-0.34	.006	8	B8 IIIpSi	15,1079,1256
	+13 03683	G 141 - 19		18 30 59	+13 07.3	6	10.57	.007	0.65	.017	-0.08	.008	12	F5	516,1064,1620,1658*
		NGC 6649 - 110		18 30 59	−10 28.2	1	14.04		2.17				2		323
		NGC 6649 - 111		18 30 59	−10 28.9	1	13.30		2.25				2		323
171184	−14 05110			18 30 59	−14 27.9	1	7.94		0.21		-0.18		3	Ap Si	1762
171234	+1 03712			18 31 00	+01 51.9	1	7.95		0.20		0.12		3	A3	246
	−11 04674	LS IV -11 029		18 31 00	−11 13.7	1	10.18		1.11				4	O9	1021
		G 155 - 24		18 31 01	−12 22.9	2	11.41	.010	0.71	.020	0.04	.000	4		1696,3062
171132	−25 13252	LSS 5090		18 31 02	−25 25.1	1	8.98		0.11		-0.66		4	B2 II/III	1732
171461	+52 02232	HR 6974		18 31 03	+52 04.6	3	6.56	.005	-0.07	.008	-0.18	.013	6	B9.5V	985,1079,1501
171286	+18 03728			18 31 04	+18 52.6	1	6.84		1.06		0.90		2	K0	1648
171653	+65 01276	HR 6979	⋆ A	18 31 05	+65 23.8	1	6.60		0.33		0.13		2	A8m	1733
171198	−12 05104	LSS 5093		18 31 05	−12 18.1	3	9.55	.004	0.58	.004	-0.48	.000	4	O7(f)	1011,1012,1032
		CS 20321 # 261		18 31 06	−33 07.2	1	14.88		0.27		0.17		1		1736
171263	+5 03786			18 31 09	+05 34.1	1	7.76		0.19		-0.16		2	A0p	212
171200	−17 05237			18 31 10	−16 57.4	1	8.79		0.20		-0.21		3	B9.5 III	1762
171384	+38 03213	IDS18295N3846	A	18 31 11	+38 47.7	1	6.95		0.63		0.15		2	G0 IV	105
171264	+4 03791			18 31 12	+04 33.2	1	8.09		0.84		0.46		1	K0	246
171314	+22 03406	V774 Her	⋆ AB	18 31 12	+22 16.9	3	8.90	.038	1.13	.010	1.06	.004	11	K4 V	22,1197,3072

Table 1 1013

HD	DM	Other Id	N Rem	α_{1950}	δ_{1950}	S	V	σ_V	B−V	σ_{B-V}	U−B	σ_{U-B}	n	Spectrum	References
	+3 03742			18 31 13	+03 15.2	1	9.58		0.48		-0.19		2	A0	246
171224	-15 05016			18 31 13	-15 48.4	1	8.01		1.40		1.35		3	K1 III	1762
	-17 05238			18 31 18	-17 44.1	1	9.09		2.03				2	M3	369
170915	-61 06193			18 31 18	-61 15.3	1	9.32		1.16				1	K3 V	1494
		CS 20321 # 253		18 31 19	-33 24.8	1	13.10		0.06		0.02		1		1736
171201	-21 05052	LSS 5095		18 31 21	-21 35.6	3	9.59	.007	0.20	.009	-0.77	.013	5	B0 Ia	540,976,1737
171202	-22 04781			18 31 21	-22 15.0	1	9.15		0.02		-0.52		6	B6 II	1732
		G 155 - 25		18 31 22	-08 18.7	1	13.50		1.50		1.10		2		3032
171235	-13 05031			18 31 22	-13 08.0	1	8.14		2.03				2	K5/M0 (III)	369
		CS 20321 # 279		18 31 26	-32 20.8	1	14.23		0.12		0.13		1		1736
		CS 20321 # 328		18 31 26	-34 50.1	1	13.41		0.05		-0.01		1		1736
		CS 20321 # 362		18 31 26	-36 18.6	1	13.64		0.40		0.13		1		1736
	+4 03792			18 31 27	+05 02.7	1	9.27		0.32		-0.01		3	A0	212
171406	+30 03227	HR 6971		18 31 29	+30 51.2	2	6.57	.010	-0.12	.000	-0.53	.010	6	B4 V	154,3016
171251	-17 05240			18 31 29	-16 58.7	1	8.81		0.22		-0.33		3	B3/5 IV/V	1762
171237	-24 14479	HR 6965		18 31 29	-24 15.7	4	6.51	.004	0.53	.003	0.48	.015	20	F2 II	3,219,418,2007
171290	-2 04655			18 31 33	-02 05.1	1	8.97		1.90				2	K5	369
171141	-46 12489			18 31 34	-45 58.9	2	8.37	.007	-0.22	.000	-0.96	.002	6	B2 II/III	55,1732
171485	+40 03411	IDS18300N4005	A	18 31 37	+40 07.2	1	7.26		0.51		-0.05		1	F5	289
	+1 03716			18 31 39	+01 45.3	1	10.54		0.56		0.04		1		246
	-8 04634	LS IV -08 009		18 31 42	-08 08.3	2	9.44	.000	0.91	.000	-0.18	.000	3	O9:V:p	1011,1032
171635	+56 02113	HR 6978		18 31 43	+57 00.4	8	4.79	.032	0.61	.010	0.43	.011	25	F7 Ib	15,254,1080,1118,3026*
		CS 20321 # 353		18 31 43	-35 58.9	1	15.46		0.11		0.11		1		1736
		KX Lyr		18 31 44	+40 08.2	2	10.43	.055	0.13	.029	0.21	.038	2	F0	668,699
174878	+86 00282			18 31 48	+86 37.7	1	6.53		1.64		1.90		3	M1	985
		CS 20321 # 352		18 31 48	-35 58.0	1	15.65		0.19		0.23		1		1736
		G 184 - 12		18 31 49	+19 43.7	1	16.45		0.26		-0.58		3		538
171316	-10 04723			18 31 50	-10 34.1	1	8.32		0.64				38	G0	6011
		CS 20321 # 322		18 31 52	-34 33.1	1	14.58		0.19		0.26		1		1736
	+3 03745			18 31 55	+04 02.3	1	10.12		0.53				2		246
171367	+3 03746			18 31 56	+03 47.5	1	7.68		1.12		0.90		1	K0	246
171318	-16 04936			18 31 56	-16 10.4	1	9.78		0.31		-0.11		3	B8/9 II	1762
169904	-81 00813	HR 6912		18 32 01	-81 51.2	3	6.26	.008	-0.12	.013	-0.36	.005	11	B8 V	15,1075,2038
171410	+1 03718			18 32 03	+02 02.3	1	8.49		0.52		0.16		3	G0	246
	-14 05120			18 32 04	-14 08.6	1	9.87		1.93				2	M3	369
		CC Lyr		18 32 05	+31 34.2	1	11.81		0.39		0.20		1		1399
171344	-13 05039	LS IV -13 076		18 32 08	-13 54.8	1	9.52		0.29				4	B2 Ia	1021
	-8 04635			18 32 09	-08 39.1	1	10.18		2.53				2	M2	369
171487	+20 03847	HR 6975		18 32 10	+20 25.6	3	6.57	.000	0.11	.007	0.08	.007	6	A3 V	985,1063,1733
172340	+77 00699	HR 7006		18 32 10	+77 30.6	1	5.64		1.18				2	K0	71
	+3 03749			18 32 11	+04 03.5	1	10.21		0.41		0.32		1		246
171391	-11 04681	HR 6970		18 32 16	-11 01.1	4	5.13	.012	0.92	.012	0.60		13	G8 III	15,1075,2007,3016
	-6 04798	LS IV -06 011		18 32 17	-06 32.9	1	10.65		0.71		-0.09		3	B3	1195
		NGC 6656 sq3 2		18 32 17	-23 39.4	1	9.98		1.53		1.61		4		1677
238901	+59 01906			18 32 20	+59 50.9	1	9.00		0.96		0.73		2	K5	1502
171348	-22 04790	LSS 5096	⋆ A	18 32 20	-22 07.9	4	8.00	.037	0.12	.015	-0.78	.017	11	B2 Vnne	540,976,1212,1586
	-3 04312			18 32 22	-03 24.6	1	10.05		2.34				2	M2	369
171369	-20 05189	HR 6969	⋆ V	18 32 22	-20 52.9	2	6.48	.002	0.28	.001			5	F0 IV/V	2035,6007
171654	+46 02508			18 32 23	+46 10.7	1	6.74		-0.07		-0.18		2	A0	985
	-16 04939			18 32 24	-16 15.5	1	9.51		1.46		1.20		3	K5	1762
		CS 20321 # 282		18 32 25	-32 27.1	1	12.91		0.13		0.06		1		1736
171505	+10 03573	HR 6976		18 32 26	+10 51.1	2	6.46		0.04	.029	0.01	.029	5	A1 V	1063,1733
171394	-19 05077			18 32 27	-19 18.6	1	6.80		1.77		1.71		5	M3/4 III	3040
171413	-16 04941			18 32 28	-16 10.9	1	7.80		0.43		-0.01		3	F3/5 V	1762
171443	-8 04638	HR 6973		18 32 29	-08 16.8	11	3.85	.013	1.33	.006	1.53	.010	39	K2 III	15,37,1003,1020,1075*
171431	-14 05124			18 32 29	-14 43.6	1	8.36		0.15		-0.11		3	Ap Si	1762
171490	+4 03796			18 32 30	+04 12.0	1	8.06		0.71				1	K0	246
	-6 04799			18 32 36	-06 34.0	1	10.72		0.78		0.56		2	A1	1195
171353	-33 13377			18 32 36	-33 13.0	1	8.46		1.64		2.14		2	K5 III	262
		CS 20321 # 347		18 32 36	-35 50.7	1	13.76		0.26		0.23		1		1736
	+40 03416			18 32 40	+40 52.8	1	9.23		1.54		1.85		1	K5	1746
171620	+34 03239			18 32 41	+34 22.3	3	7.57	.010	0.50	.000	-0.11	.034	7	F6p	516,1620,3026
171432	-18 05008	LSS 5099		18 32 41	-18 35.5	5	7.09	.019	0.22	.013	-0.70	.014	11	B0.5 Ib	138,399,540,1737,8100
171726	+47 02653			18 32 43	+47 04.4	1	8.82		-0.10		-0.48		2	A0	1601
234677	+51 02402	BY Dra	⋆	18 32 45	+51 41.0	2	8.11	.056	1.21	.023	0.99		4	K3 V +MVe	196,1746
234678	+52 02237	IDS18317N5216		18 32 46	+52 18.8	1	8.81		0.26		0.07		3	F0	3024
	-23 14523	NGC 6656 sq3 6		18 32 47	-23 38.7	1	10.60		1.15		0.77		4		1677
		CS 20321 # 335		18 32 47	-35 08.6	1	15.31		0.12		0.14		1		1736
171416	-29 15123	HR 6972		18 32 48	-29 44.4	1	6.37		1.27				4	K1 III	2007
		AQ Lyr		18 32 49	+26 31.1	1	12.49		0.25		0.18		1		699
171779	+52 02238	HR 6983	⋆ AB	18 32 49	+52 18.8	3	5.38	.023	1.07	.014	0.91	.033	10	K1III+G9III	1080,1381,3024
	-44 12736			18 32 49	-44 20.9	4	10.24	.076	0.61	.024	-0.02	.015	16	G0	158,1698,1765,3062
	+48 02721			18 32 50	+48 24.9	1	10.70		-0.20		-0.81		3		963
171469	-15 05024			18 32 50	-15 45.9	2	9.42	.018	0.33	.005	-0.28	.001	6	B5 V(e)	1586,1762
	+4 03797	IC 4756 - 4		18 32 51	+04 53.9	1	8.56		1.68		2.04		18	gK5	1327
		NGC 6656 sq3 3		18 32 52	-23 41.4	1	10.16		1.72		2.07		5		1677
	+1 03721		V	18 32 53	+01 57.2	1	10.16		2.33				2	M2	369
	-4 04503	LS IV -04 007		18 32 53	-04 50.3	1	10.83		0.75		-0.31		2	O7	1011
	-8 04640	LS IV -08 011		18 32 53	-08 42.2	1	10.20		0.86		-0.13		2		405
171679	+35 03294			18 32 54	+35 45.8	1	7.78		0.12		0.06		2	A0	1601

HD	DM	Other Id	N Rem	α_{1950}	δ_{1950}	S	V	σ_V	B–V	σ_{B-V}	U–B	σ_{U-B}	n	Spectrum	References
	−8 04641			18 32 55	−08 19.3	1	10.12		0.98		0.80		std		1195
		NGC 6656 sq2 6		18 32 55	−23 48.4	2	13.08	.005	1.21	.000	0.80	.005	13		559,1677
	+5 03802	IC 4756 - 5		18 32 57	+05 15.1	1	10.00		0.28		0.11		2	A0	1327
	−11 04683		V	18 32 57	−11 42.8	1	10.79		2.14				2	M2	369
		NGC 6656 - 3093		18 32 57	−23 58.6	1	12.72		1.22				2		546
171418	−33 13379	IDS18297S3316	AB	18 32 57	−33 13.6	1	9.31		1.09		1.31		2	K1 III	262
		LP 229 - 17		18 32 58	+40 05.1	1	11.42		1.42				1		1746
		NGC 6656 sq2 9		18 32 59	−23 48.5	2	13.32	.005	0.75	.010	0.14	.000	7		559,1677
171623	+18 03740	HR 6977		18 33 00	+18 09.7	2	5.79		-0.01	.005	-0.11	.005	6	A0 Vn	1063,1733
	+4 03799	IC 4756 - 6		18 33 01	+04 57.6	1	9.11		0.64		0.14		3	dG5	1327
		NGC 6656 - 2031		18 33 01	−23 55.4	2	11.95	.058	1.50	.024	1.12		3		546,3057
		NGC 6656 - 3106		18 33 01	−24 00.4	1	12.19		1.49				2		546
		CS 20321 # 298		18 33 01	−33 25.1	1	14.73		0.15		0.22		1		1736
		CS 20321 # 318		18 33 01	−34 20.1	1	13.80		0.32		0.17		1		1736
		G 155 - 27		18 33 02	−08 18.2	2	13.70	.000	1.34	.024	1.25		3		419,3078
		CS 20321 # 375		18 33 03	−37 03.3	1	13.71		0.26		0.16		1		1736
		CS 20321 # 344		18 33 04	−35 39.5	1	13.92		0.21		0.16		1		1736
171496	−24 14502			18 33 05	−24 29.3	2	8.50	.069	1.10	.000	0.65		5	G8 IV	742,1594
171451	−35 12756	V3877 Sgr		18 33 05	−35 29.3	2	7.08	.027	1.69	.022	1.80		21	M4 III	2012,3042
171042	−70 02568			18 33 05	−70 16.7	1	7.50		1.58		1.90		5	K5 III	1704
		NGC 6656 - 2042		18 33 06	−23 56.7	1	12.70		1.31				1		546
		NGC 6656 sq2 3		18 33 07	−23 51.2	2	11.23	.024	1.90	.005	1.92	.034	9		559,1677
		NGC 6656 - 3052		18 33 07	−23 56.9	1	11.56		1.66		1.55		2		3057
		NGC 6656 - 3087		18 33 07	−23 59.1	1	13.13		1.28		0.71		1		3057
171586	+4 03801	IC 4756 - 8	⋆ AB	18 33 08	+04 53.7	3	6.46	.015	0.07	.005	0.06	.011	30	A2p	603,1202,1327
		NGC 6656 sq2 2		18 33 08	−23 49.0	1	10.55		0.38		0.24		13		1677
		CS 20321 # 345		18 33 08	−35 40.2	1	14.60		0.18		0.25		1		1736
		NGC 6656 sq3 12		18 33 09	−23 50.3	1	12.53		1.32		0.75		5		1677
		NGC 6656 - 3086		18 33 09	−23 59.4	1	12.54		1.29		0.97		1		3057
	+5 03804	IC 4756 - 10		18 33 10	+05 05.5	1	9.50		0.31		0.21		3	A3	1327
	−23 14532			18 33 10	−23 49.0	2	10.54	.010	0.39	.005	0.25		9		559,560
		NGC 6656 sq3 22		18 33 10	−23 50.8	1	14.47		1.01		0.37		3		1677
		NGC 6656 - 3079		18 33 10	−24 00.8	2	13.08	.050	1.20	.002	0.67		4		546,3057
		NGC 6656 - 3012		18 33 11	−23 57.0	1	11.56		1.63		1.50		2		3057
171871	+50 02618			18 33 12	+51 04.3	3	7.79	.007	-0.17	.015	-0.80	.018	11	B2 IIp	399,1733,8040
		NGC 6656 - 3014		18 33 12	−23 57.4	1	11.16		1.79		1.58		2		3057
		NGC 6656 sq3 10		18 33 12	−24 03.4	1	12.01		1.14				2		1677
	+2 03621			18 33 13	+02 36.6	1	10.14		0.67		0.25		2	B8	246
	+3 03751			18 33 13	+04 01.4	1	8.77		1.67				2	M2	369
		NGC 6656 - 3003		18 33 14	−23 56.9	1	11.01		1.82		1.76		2		3057
	−6 04803			18 33 15	−06 21.0	1	11.00		2.39				2	M2	369
		NGC 6656 sq3 4		18 33 15	−23 44.3	1	10.23		0.39				3		1677
		NGC 6656 - 3075		18 33 15	−24 00.8	1	13.10		1.14		0.44		1		3057
		CS 20321 # 301		18 33 15	−33 37.2	1	12.53		0.28		0.19		1		1736
		NGC 6656 - 3035		18 33 17	−23 58.9	1	12.44		1.34		0.86		1		3057
		CS 20321 # 294		18 33 17	−33 05.8	1	13.23		0.09		-0.04		1		1736
	−5 04697			18 33 18	−05 26.8	1	9.82		2.53				2	M2	369
		NGC 6656 - 2009		18 33 18	−23 53.1	1	12.72		1.29				1		546
	+5 03805	IC 4756 - 12		18 33 20	+05 17.8	1	9.54		1.03		0.85		2	dG8	1327
		NGC 6656 sq3 13		18 33 20	−23 49.1	1	12.96		1.24				2		1677
		G 206 - 34		18 33 21	+28 39.6	3	11.39	.007	0.42	.020	-0.24	.006	4		1064,1658,3077
171911	+51 02404			18 33 21	+51 44.5	1	6.67		1.58		1.63		2	M1	3016
171587	−10 04727	G 155 - 28		18 33 22	−10 55.8	1	8.48		0.67		0.10		1	G5	3062
171589	−14 05131	LSS 5100		18 33 22	−14 09.5	3	8.29	.046	0.28	.013	-0.70	.003	7	O7	1011,1586,1762
		NGC 6656 - 1098		18 33 23	−23 54.3	1	12.40		1.27		0.87		1		3057
171827	+39 03463			18 33 24	+39 29.3	1	7.70		-0.01		-0.05		2	A0	401
	−4 04506			18 33 24	−04 15.5	1	10.50		2.27				2	M3	369
		He3 1715		18 33 24	−22 40.	1	10.70		0.12		-0.31		3		1586
		NGC 6656 - 1011		18 33 24	−23 53.9	1	12.85		1.32		0.89		2		3057
		NGC 6656 - 1012		18 33 24	−23 53.9	1	11.76		1.43		1.11		2		3057
		NGC 6656 - 1010		18 33 24	−23 58.	1	13.38		1.15				1		546
		NGC 6656 - 1052		18 33 24	−23 58.	1	13.94		1.15		0.45		3		3057
		NGC 6656 - 1062		18 33 24	−23 58.	1	12.12		0.77		0.07		2		3057
		NGC 6656 - 1075		18 33 24	−23 58.	1	11.95		1.22		0.91		1		3057
		NGC 6656 - 1081		18 33 24	−23 58.	2	13.66	.010	0.68	.005			3		546,3057
		NGC 6656 - 2015		18 33 24	−23 58.	1	13.37		1.57				1		546
		NGC 6656 - 2026		18 33 24	−23 58.	1	11.70		1.42		1.39		2		3057
		NGC 6656 - 2029		18 33 24	−23 58.	1	11.05		0.78		0.20		4		3057
		NGC 6656 - 2092		18 33 24	−23 58.	1	12.55		1.43				1		546
		NGC 6656 - 3091		18 33 24	−23 58.	1	13.61		0.88				1		546
		NGC 6656 - 3097		18 33 24	−23 58.	1	13.21		1.28				2		546
		NGC 6656 - 3108		18 33 24	−23 58.	1	11.85		1.13				2		546
		NGC 6656 - 3109		18 33 24	−23 58.	1	11.63		1.26				2		546
		NGC 6656 - 4095		18 33 24	−23 58.	1	13.51		1.71				1		546
171745	+23 03385	HR 6980	⋆ AB	18 33 25	+23 33.8	5	5.63	.026	1.00	.017	0.81	.021	9	G9III+G7III	37,71,938,1381,3030
171780	+34 03245	HR 6984		18 33 25	+34 25.0	2	6.10	.009	-0.12	.009	-0.56	.005	6	B5 Vne	154,1212
	−8 04644	NGC 6664 - 29		18 33 25	−08 14.1	1	10.63		1.71		1.26		2	K0	1195
		NGC 6656 - 1008		18 33 26	−23 53.6	1	12.06		1.45		1.00		2		3057
		BP CrA		18 33 26	−37 28.3	1	14.01		0.11		-0.67		1		1471
171106	−70 02570			18 33 26	−70 50.9	1	7.12		1.26				4	K2 III	2012

Table 1   1015

HD	DM	Other Id	N	Rem	α_{1950}	δ_{1950}	S	V	σ_V	B–V	σ_{B-V}	U–B	σ_{U-B}	n	Spectrum	References
		NGC 6656 - 4003			18 33 27	−23 57.2	1	14.39		1.09				2		546
171624	−2 04664				18 33 28	−02 25.0	1	8.99		1.17				53	K0	6005
		NGC 6656 - 1027			18 33 28	−23 55.4	1	12.37		1.17		0.60		2		3057
		NGC 6656 - 1086			18 33 29	−23 54.1	1	12.36		1.47		0.85		3		3057
		NGC 6656 - 1085			18 33 30	−23 53.7	1	12.54		1.24		0.73		1		3057
		NGC 6656 - 1092			18 33 30	−23 55.0	1	11.61		1.48		1.26		1		3057
171658	+5 03808	IC 4756 - 14			18 33 31	+05 22.5	1	8.86		0.86		1.22		2	K5	1327
		NGC 6656 sq3 9			18 33 31	−24 04.8	1	11.30		1.73				2		1677
		NGC 6656 - 1080			18 33 33	−23 52.8	2	12.55	.010	1.38	.002	0.94		2		546,3057
		NGC 6656 - 1054			18 33 33	−23 55.9	1	13.21		1.29		0.84		2		3057
		NGC 6656 - 1051			18 33 33	−23 56.6	1	12.70		1.26		0.60		3		3057
		NGC 6656 - 4102			18 33 33	−24 00.2	2	11.10	.050	1.83	.030	1.70		4		546,3057
		NGC 6656 sq3 8			18 33 33	−24 02.2	1	11.17		1.84				2		1677
		LP 923 - 2			18 33 33	−32 33.8	1	13.95		0.12		1.10		1		3062
		NGC 6656 sq2 4			18 33 34	−23 50.0	3	12.17	.014	1.40	.005	1.00	.014	11		559,560,1677
		NGC 6656 - 1082			18 33 34	−23 53.7	2	13.74	.047	1.17	.000	0.41		4		546,3057
171685	+4 03803	IC 4756 - 17		⋆ AB	18 33 35	+05 03.9	1	8.90		0.53		0.04		3	G0	1327
		NGC 6656 - 1057			18 33 35	−23 55.8	2	11.99	.065	1.56	.040	1.16		2		546,3057
		NGC 6656 - 1053			18 33 35	−23 56.6	1	12.71		1.29		0.84		1		3057
	+5 03811	IC 4756 - 20			18 33 36	+05 09.2	1	10.46		0.68		0.28		2	F0	1327
		He3 1716			18 33 36	−19 22.	1	11.73		0.48		0.39		3		1586
		NGC 6664 - 60			18 33 37	−08 11.6	1	11.89		0.62		0.29		1	B6 IV	1165
	−8 04645				18 33 37	−08 55.0	2	10.44	.060	2.53	.015	2.69		4	M3	369,8032
		NGC 6656 - 4097			18 33 38	−24 00.9	2	11.13	.130	1.88	.090	1.65		4		546,3057
	−8 04646	NGC 6664 - 55			18 33 39	−08 13.4	1	10.97		0.53		-0.26		2	B3 IV	1165
171746	+16 03560	HR 6981		⋆ ABC	18 33 40	+16 56.0	3	6.22	.003	0.54	.014	0.00	.005	7	G2 V + G2 V	71,938,3016
		NGC 6656 sq2 7			18 33 40	−23 49.8	2	13.17	.000	1.18	.005	0.50	.000	11		559,1677
		NGC 6656 - 1068			18 33 40	−23 54.4	1	12.56		1.24		0.66		1		3057
		NGC 6656 - 4096			18 33 40	−24 00.7	1	13.00		1.26				2		546
171872	+38 03229				18 33 41	+38 51.2	1	6.97		0.03		0.03		2	A0 V	105
	−8 04647	NGC 6664 - 52			18 33 42	−08 11.0	1	10.33		1.96		2.05		3	G8 II	1165
		NGC 6656 - 4099			18 33 42	−24 01.2	1	13.20		1.26				2		546
		JL 3			18 33 42	−83 16.	1	16.15		-0.07		-0.68		1		832
	+6 03853				18 33 44	+06 57.8	1	10.27		0.63		0.08		2	F8	1696
		WLS 1824 70 # 9			18 33 44	+70 00.1	1	11.62		0.67		0.12		2		1375
171612	−23 14543	NGC 6656 sq2 1			18 33 44	−23 48.4	3	8.61	.009	1.86	.007	2.02	.014	18	M2/3 III	559,560,1677
	+83 00537				18 33 45	+83 54.6	1	10.86		-0.03		-0.45		5	A0	1219
171830	+27 03053				18 33 46	+27 10.2	1	8.04		1.01		0.77		2	G8 III	1733
		NGC 6664 - 75			18 33 46	−08 11.6	1	15.09		0.75		0.58		2		1165
		NGC 6656 - 4085			18 33 46	−23 57.4	1	13.37		1.22				1		546
	+4 03804	BQ Ser			18 33 47	+04 21.4	2	9.22	.003	1.32	.040			2	F5 III	1772,6011
171640	−17 05259				18 33 47	−17 15.0	1	8.01		0.56		0.49		3	F0 IV	1762
		AA45,405 S60 # 1			18 33 48	−06 45.3	1	11.49		1.20		0.14		4		432
	−8 04649	NGC 6664 - 51			18 33 48	−08 12.9	1	10.93		1.98		2.15		7	K3 II	1165
		NGC 6664 - 61			18 33 48	−08 15.4	1	11.94		0.52		-0.21		2	B3 V	1165
	+45 02743	G 205 - 30			18 33 49	+45 41.7	2	9.83	.010	1.42	.000	1.15		6	M2	1197,3016
	+2 03624				18 33 50	+02 26.8	1	10.30		0.45				2	F8	246
		AA45,405 S60 # 2			18 33 50	−06 45.5	1	12.45		0.75		-0.14		4		432
		NGC 6664 - 79			18 33 50	−08 10.6	1	17.70		3.00				1		1165
		NGC 6664 - 64			18 33 50	−08 13.3	1	12.44		0.67		0.07		1		1165
		NGC 6664 - 65			18 33 50	−08 15.8	1	12.65		0.62		0.31		1		1165
		NGC 6664 - 68			18 33 50	−08 15.9	1	13.05		0.70		0.42		1		1165
	−8 04650	NGC 6664 - 54			18 33 50	−08 18.1	1	10.81		2.10		2.39		1		1165
		NGC 6664 - 78			18 33 51	−08 12.3	1	17.69		1.27				2		1165
171749	+5 03815	IC 4756 - 23		⋆ AB	18 33 52	+05 13.1	1	9.12		0.36		0.27		2	A0	1327
		NGC 6664 - 70			18 33 52	−08 10.1	1	13.81		2.09		2.45		1		1165
		NGC 6664 - 67			18 33 53	−08 15.2	1	13.05		0.56		0.20		1		1165
171731	−2 04667				18 33 54	−02 31.9	3	9.06	.008	1.13	.002	1.04	.007	50	K2	147,1509,6005
		NGC 6664 - 63			18 33 54	−08 13.3	1	12.42		0.58		0.05		2		1165
171627	−28 14765				18 33 54	−28 33.3	2	6.79	.000	0.98	.002	0.70		7	K0 IV	2012,3077
		NGC 6664 - 73			18 33 55	−08 14.1	1	14.28		0.65		0.50		2		1165
171662	−16 04952				18 33 55	−15 59.3	2	7.02	.006	1.95	.028	2.06	.025	6	K3/4 III	1657,1762
171849	+25 03581	IDS18341N2538	A		18 33 56	+25 39.8	1	8.66		0.23		0.10		2	A5	105
	+38 03231	G 205 - 31			18 33 56	+39 01.0	1	9.25		0.58		-0.06		1	F8 V	1748
171732	−3 04316				18 33 56	−03 07.2	3	9.12	.005	0.28	.002	-0.11	.016	99	B9	147,1509,6005
	−8 04652	NGC 6664 - 80		⋆ V	18 33 56	−08 13.6	2	9.98	.006	1.11	.013			2	G0 II	1484,6011
		CS 20321 # 280			18 33 56	−32 23.3	1	8.66		0.23		0.16		1		1736
172041	+54 02013				18 33 57	+54 10.0	1	7.97		1.12		1.04		2	K2	1566
		NGC 6664 - 71			18 33 57	−08 13.3	1	13.82		0.50		0.18		2		1165
		NGC 6664 - 77			18 33 57	−08 14.3	1	17.26		1.35				1		1165
171689	−16 04953				18 33 57	−16 45.3	1	8.93		0.35		0.07		3	B9 III/IV	1762
		NGC 6664 - 72			18 33 58	−08 13.5	1	14.14		0.66		0.60		2		1165
	−8 04653	NGC 6664 - 53			18 33 58	−08 21.5	1	10.75		2.13		2.08		1		1165
171767	+4 03806	IC 4756 - 25		⋆ A	18 34 01	+04 54.8	2	6.75	.020	1.10	.049	1.15	.005	5	K1 III	37,1327
171733	−6 04809				18 34 01	−06 27.9	1	8.44		1.94				2	K5	369
171782	+5 03816	IC 4756 - 26			18 34 02	+05 14.8	1	7.84		0.02		-0.24		2	A0p	1327
	−8 04654	NGC 6664 - 50			18 34 02	−08 12.6	2	10.61	.000	0.47	.002	0.17	.025	10	A0 IV	1165,1195
		NGC 6664 - 74			18 34 03	−08 11.5	1	14.64		0.52		0.46		1		1165
		NGC 6664 - 76			18 34 03	−08 13.8	1	15.41		0.61		0.67		2		1165
		NGC 6664 - 66			18 34 03	−08 18.1	1	13.02		0.50		0.20		1		1165

HD	DM	Other Id	N	Rem	α_{1950}	δ_{1950}	S	V	σ_V	B–V	σ_{B-V}	U–B	σ_{U-B}	n	Spectrum	References
171707	−16 04955	IDS18312S1622		A	18 34 03	−16 20.0	1	8.41		1.18		0.88		3	G8/K0 III	1762
		NGC 6664 - 56			18 34 04	−08 15.8	1	11.02		0.44		0.05		2	A0 III	1165
		NGC 6664 - 62			18 34 04	−08 19.5	1	11.99		0.50		0.09		1	B9 V	1165
		NGC 6664 - 59			18 34 04	−08 20.1	1	11.84		0.50		−0.05		1		1165
	+5 03818	IC 4756 - 28			18 34 05	+05 10.1	2	8.98	.017	1.35	.017	1.26	.004	6	dG9	460,1327
171802	+9 03783	HR 6985			18 34 05	+09 04.9	4	5.39	.023	0.37	.010	−0.03	.006	10	F3 V	15,603,1256,3026
		NGC 6664 - 69			18 34 07	−08 12.8	1	13.22		0.43		0.13		1		1165
		NGC 6664 - 58			18 34 07	−08 18.9	1	11.79		0.75		0.32		1		1165
	−11 04687			V	18 34 07	−11 52.5	1	10.27		2.32				2	M2	369
171665	−25 13291				18 34 07	−25 42.6	2	7.44	.015	0.69	.010	0.19		8	G5 V	158,2033
		CS 20321 # 424			18 34 07	−34 36.0	1	13.84		0.05		0.01		1		1736
	+5 03819	IC 4756 - 31			18 34 08	+05 30.3	1	9.96		0.28		0.16		2	A0	1327
		NGC 6664 - 57			18 34 09	−08 14.0	1	11.75		1.67		1.41		1		1165
	+5 03820	IC 4756 - 32			18 34 10	+05 05.6	2	9.51	.000	1.13	.067	0.85	.022	5	G8	460,1327
171360	−64 03935				18 34 10	−64 49.9	1	9.70		0.28		0.08		4	A8 V	1731
171834	+6 03855	HR 6987		★ AB	18 34 13	+06 37.9	4	5.45	.004	0.37	.011	−0.05	.015	11	F3 V	15,254,1256,3037
	−4 04512				18 34 16	−04 39.3	1	10.47		2.15				2	M2	369
171753	−16 04956				18 34 16	−16 18.5	1	8.53		1.31		0.97		3	G3/5 Ib/II	1762
171770	−15 05030				18 34 17	−15 04.2	1	7.46		0.39		0.01		3	F5 IV/V	1762
171381	−64 03936				18 34 17	−64 54.4	1	7.53		1.30		1.40		4	K2/3 III	1731
171874	+12 03598				18 34 18	+12 56.6	1	7.36		0.50				1	F6 II	6009
171754	−19 05099				18 34 18	−19 27.0	1	8.30		0.26		−0.21		3	B5 V	1586
	+4 03808	IC 4756 - 34			18 34 19	+04 59.7	1	9.39		0.44		0.04		3	F3	1327
	−16 04957				18 34 19	−16 04.7	1	10.13		0.61		0.17		3	F0	1762
	+17 03648				18 34 21	+17 29.4	1	9.35		1.26		1.24		2	K0	1648
		CS 20321 # 384			18 34 22	−36 55.0	1	15.44		0.20		0.20		1		1736
	+5 03824	IC 4756 - 35			18 34 23	+05 55.3	1	10.36		0.28		0.19		2	A8	1327
	−7 04634				18 34 23	−07 49.8	1	10.75		0.59				1		725
171804	−7 04633	RX Sct			18 34 24	−07 38.7	5	8.93	.041	2.85	.041	4.75	.235	5	C5 II	109,1238,8005,8022,8027
		CS 20321 # 385			18 34 24	−36 56.3	1	15.28		0.17		0.19		1		1736
	+5 03825	IC 4756 - 36			18 34 26	+05 33.1	1	9.72		0.39		0.31		2	A6	1327
171757	−28 14778				18 34 31	−28 01.7	4	8.98	.042	0.11	.022	−0.82	.009	10	B2nne	400,540,976,1586
	+4 03809	IC 4756 - 37			18 34 34	+04 46.3	1	10.37		2.06				2		369
172068	+41 03100	IDS18330N4112		AB	18 34 36	+41 14.1	1	6.94		0.07		0.03		2	A0	401
	+5 03829	IC 4756 - 38			18 34 37	+05 14.9	2	9.79	.027	1.10	.000	0.75	.000	5	dK0	460,1327
		G 21 - 19			18 34 37	−00 56.0	3	11.88	.020	0.63	.004	−0.05	.000	5		202,1620,1696
		L 44 - 95			18 34 42	−78 08.	1	15.45		−0.06		−0.80		1		3065
	+4 03810				18 34 46	+04 12.1	1	9.71		0.69				71	G5	6011
171459	−65 03746				18 34 46	−65 56.1	1	8.35		0.37		0.21		4	A7mF0-F0	1731
171930	+5 03831	IC 4756 - 39			18 34 47	+05 45.5	1	8.08		0.96		0.58		2	K0	1327
172044	+33 03154	HR 6997		★ AB	18 34 47	+33 25.5	4	5.41	.012	−0.11	.012	−0.50	.006	16	B8 II-IIIp	1063,1079,1224,3033
171931	+5 03830	IC 4756 - 40			18 34 48	+05 16.4	1	9.19		0.15		−0.23		2	B9p	1327
171951	+7 03773	IDS18324N0727		B	18 34 48	+07 29.3	1	9.79		1.73				2	G0	369
	+4 03812				18 34 49	+04 13.0	1	10.00		0.44				74	F2	6011
		AO 1160			18 34 49	+39 01.5	1	9.84		1.09		0.99		1	K1 III	1748
172323	+63 01439	G 227 - 37		★ A	18 34 49	+63 39.5	3	8.07	.005	0.57	.010	0.02	.020	5	F9 V	1003,1658,3024
172323	+63 01439	G 227 - 38		★ B	18 34 50	+63 39.5	2	10.69	.009	1.08	.028	0.95	.023	4		1658,3024
	+38 03233				18 34 51	+38 27.6	1	9.92		1.58		2.01		1	M2 III	1748
171975	+11 03530	HR 6992			18 34 52	+11 22.7	2	6.56		−0.01	.005	−0.23	.005	5	B9 V	1063,1733
		LP 630 - 151			18 34 52	−00 27.8	1	14.33		0.71		−0.03		1		1696
	+5 03832	IC 4756 - 41			18 34 54	+05 32.1	1	10.20		0.49		0.25		2	A4	1327
	+5 03833	IC 4756 - 42			18 34 54	+05 51.1	1	9.46		0.97		0.62		2	dG4	1327
171994	+16 03563	HR 6995			18 34 54	+16 09.2	2	6.31	.020	0.90	.002	0.56	.000	5	G8 IV	1080,3016
171858	−23 14565				18 34 54	−23 14.2	1	9.84		−0.21		−0.95		6	B7/8 II	1732
		LP 867 - 13			18 34 54	−26 49.0	1	10.67		0.65		0.05		3		1696
		IC 4756 - 206			18 34 55	+05 25.7	1	10.59		0.43		0.24		2		1327
	−8 04658				18 34 55	−08 28.4	1	9.93		1.10				40	G0	6011
171856	−21 05076	HR 6988			18 34 55	−21 26.5	1	5.94		0.19				4	A8 IIIn	2007
		CS 20321 # 408			18 34 56	−35 36.4	1	13.21		0.20		0.14		1		1736
	+5 03834	IC 4756 - 43			18 34 57	+05 38.7	1	9.53		0.47		0.29		2	F2	1327
	−8 04660				18 34 57	−08 44.7	1	10.07		0.70				36	G0	6011
171893	−17 05271				18 34 57	−17 16.5	1	6.83		0.61		0.47		7	F2 V	1628
		AO 1162			18 34 58	+38 50.8	1	10.72		0.62		0.11		1	G0 V	1748
	+38 03235	IDS18333N3236		A	18 34 60	+38 38.9	1	10.00		1.12		1.00		1	K0 III	1748
		IC 4756 - 244			18 35 01	+05 18.1	1	12.37		0.70		0.12		2		460
		IC 4756 - 242			18 35 01	+05 20.1	1	11.94		0.53		0.16		2		460
	+5 03835	IC 4756 - 44			18 35 02	+05 09.6	2	9.78	.009	1.09	.004	0.73	.018	5	G5	460,1327
171978	−0 03521	HR 6993			18 35 02	−00 21.2	7	5.76	.022	0.06	.007	0.07	.018	27	A2 V + A1 V	15,252,1007,1013,1149*
		CS 20321 # 436			18 35 02	−34 12.1	1	14.63		0.13		0.15		1		1736
171977	+0 03978				18 35 03	+00 47.8	1	8.98		0.31		0.18		4	A5	1371
	+4 03813	IC 4756 - 45			18 35 03	+05 00.3	1	9.73		0.45		0.27		2	pF4	1327
171921	−17 05273				18 35 04	−17 00.1	1	9.16		0.18		−0.25		3	B8 III	1762
	+5 03836	IC 4756 - 47			18 35 05	+05 05.5	1	10.02		0.41		0.30		2	F0	1327
	+5 03837	IC 4756 - 48			18 35 05	+05 14.7	2	9.28	.022	0.81	.004	0.45	.004	11	G0	460,1327
	−10 04735			V	18 35 05	−10 21.5	1	11.28		2.37				2	M3	369
	+5 03839	IC 4756 - 49			18 35 07	+05 25.9	2	9.44	.013	1.12	.043	0.82	.009	6	G8	460,1327
172149	+38 03237				18 35 07	+38 28.6	1	8.01		0.39		−0.02		1	F5	1748
	+5 03840	IC 4756 - 52			18 35 08	+05 13.0	2	7.98	.049	1.38	.009	1.39	.009	5	gG9	460,1327
	+5 03841	IC 4756 - 50			18 35 08	+05 39.7	1	9.78		0.13		−0.15		2	B7	1327
171955	−6 04816	EW Sct			18 35 09	−06 50.5	2	7.97	.077	1.76	.035	1.40		25	K0	1772,3049

Table 1 1017

HD	DM	Other Id	N Rem	α_{1950}	δ_{1950}	S	V	σ_V	B–V	σ_{B-V}	U–B	σ_{U-B}	n	Spectrum	References
	+4 03814	IC 4756 - 55		18 35 10	+04 55.3	1	10.47		0.45		0.23		2	G0	1327
		IC 4756 - 56		18 35 11	+05 19.7	2	10.45	.013	0.69	.004	0.23	.022	5	G0	460,1327
172012	+5 03843	IC 4756 - 58		18 35 12	+05 29.3	1	9.19		0.15		-0.11		2	A0p	1327
	+5 03844	IC 4756 - 59		18 35 13	+05 27.1	1	9.98		0.47		0.32		2	A3	1327
		IC 4756 - 250		18 35 14	+05 16.5	1	11.15		0.53		0.16		4		460
		IC 4756 - 251		18 35 14	+05 18.8	1	12.33		0.69		0.13		2		460
172187	+43 03027	HR 7003		18 35 14	+43 10.7	1	6.20		0.24		0.09		3	F0 V	985
171957	−14 05139	HR 6989	★ AB	18 35 14	−14 02.9	2	6.51		0.20	.009	-0.20		2	B9 IV	1079,2008
172167	+38 03238	HR 7001, α Lyr	★ A	18 35 15	+38 44.2	14	0.03	.012	0.00	.005	-0.01	.006	68	A0 Va	1,15,1006,1063,1077,1107*
		CS 20321 # 435		18 35 17	−34 10.1	1	15.18		0.16		0.23		1		1736
171999	−6 04817	G 21 - 20		18 35 18	−06 50.6	3	8.33	.017	0.84	.005	0.53	.005	7	G5	202,1064,1620
	−8 04663	Y Sct		18 35 19	−08 24.8	2	9.26	.035	1.42	.096	1.07		2	F7	689,6011
	+38 03239			18 35 20	+38 49.8	1	9.62		1.12		0.93		1		1748
		IC 4756 - 254		18 35 21	+05 15.5	1	11.59		1.07		0.69		2		460
	+5 03845	IC 4756 - 61		18 35 21	+05 16.9	1	10.08		0.39		0.25		2	F2	1327
172046	+5 03846	IC 4756 - 62		18 35 21	+05 50.3	1	6.63		-0.04		-0.43		2	B8	1327
172268	+51 02408			18 35 21	+51 50.0	1	7.90		1.27		1.16		1	K5	1769
	+5 03847	IC 4756 - 63		18 35 22	+05 27.5	1	10.11		0.18		0.04		2	A1	1327
	+13 03698			18 35 22	+13 46.3	1	9.39		0.58		-0.08		2	A0	516
172028	−0 03523	IDS18328S0028	AB	18 35 22	−00 25.8	1	7.83		0.55		-0.26		1	B2 V	1423
		IC 4756 - 255		18 35 23	+05 16.0	1	11.63		0.47		0.23		2		460
		CS 20321 # 415		18 35 23	−35 10.5	1	14.80		0.09		0.05		1		1736
171960	−23 14570			18 35 24	−23 08.8	1	7.28		1.67				7	K3 III	8053
	+8 03773			18 35 25	+08 23.8	1	9.40		1.96				1	M2	369
		IC 4756 - 258		18 35 26	+05 12.0	1	11.70		0.35		0.18		2		460
171880	−37 12754			18 35 27	−37 42.0	1	8.74		-0.06		-0.49		7	B5 III	1732
171819	−48 12644	HR 6986		18 35 27	−47 57.3	1	5.86		0.23				4	A7 IV/V	2035
171961	−23 14572	HR 6990		18 35 28	−23 33.0	2	5.81		-0.01	.011	-0.42		5	B8 III	1079,2007
		IC 4756 - 66		18 35 30	+05 35.7	1	10.13		0.43		0.34		2	A5	1327
		CS 20321 # 464		18 35 30	−32 40.8	1	15.46		0.16		0.21		1		1736
	+11 03535			18 35 31	+11 48.5	1	9.94		0.00		-0.52		3	B8	963
	+5 03848	IC 4756 - 67		18 35 32	+05 11.5	2	9.95	.021	0.41	.004	0.29	.009	6	A8	460,1327
		Steph 1634		18 35 32	+34 24.1	1	11.30		1.53		1.84		2	M0	1746
	+5 03849	IC 4756 - 68		18 35 33	+05 32.6	1	9.90		0.51		0.40		2	F0	1327
172169	+29 03302			18 35 33	+29 32.4	2	6.70	.005	1.29	.009	1.47	.002	4	K3 III	1501,3040
		IC 4756 - 262		18 35 35	+05 20.3	1	12.83		0.85		0.37		2		460
	+5 03850	IC 4756 - 69		18 35 38	+05 21.9	2	9.20	.021	1.06	.004	0.77	.004	6	dG3	460,1327
	−13 05056			18 35 38	−13 50.8	2	9.69	.000	0.78	.000	-0.19	.000	3		1012,1032
		CS 20321 # 423		18 35 38	−34 38.0	1	14.07		0.12		0.13		1		1736
171652	−62 05911			18 35 38	−62 10.5	1	8.63		0.05		-0.15		2	B9 Vn	861
		IC 4756 - 207		18 35 39	+05 08.1	1	10.92		0.39		0.21		2		1327
		IC 4756 - 70		18 35 39	+05 17.2	1	10.63		0.41		0.18		2	F0	1327
		CS 20321 # 463		18 35 39	−32 43.8	1	14.79		0.09		0.03		1		1736
	+2 03632			18 35 41	+02 46.6	1	9.80		2.23				2	M3	369
	+5 03852	IC 4756 - 72		18 35 41	+05 18.3	1	9.70		0.51		0.22		2	A5	1327
		IC 4756 - 73		18 35 41	+05 25.1	1	10.56		0.41		0.22		2	A0	1327
	+5 03851	IC 4756 - 71		18 35 41	+05 29.2	1	10.40		0.44		0.30		2	F0	1327
		CS 20321 # 472		18 35 41	−32 18.8	1	15.21		0.09		0.14		1		1736
172102	+5 03853	IC 4756 - 74		18 35 42	+05 19.2	1	9.38		0.22		0.19		2	A0	1327
		POSS 281 # 1		18 35 43	+35 42.5	1	15.61		1.61				1		1739
	+43 03030	Y Lyr		18 35 43	+43 54.6	1	12.58		0.05		0.11		1		668
172073	−8 04665			18 35 43	−08 39.0	1	9.38		0.56				41	F0	6011
		CS 20321 # 471		18 35 43	−32 20.1	1	14.80		0.15		0.20		1		1736
		IC 4756 - 75		18 35 44	+05 11.1	2	10.42	.004	0.45	.013	0.24	.009	5	F2	460,1327
172103	−1 03529	HR 7000		18 35 44	−01 09.5	1	6.65		0.42		-0.02		4	F1 IV-V	1149
172032	−16 04963			18 35 44	−16 21.3	2	7.76	.064	0.47	.005	0.22	.000	5	Fm δ Del	379,1762
	+5 03854	IC 4756 - 77		18 35 45	+05 18.9	1	10.31		0.33		0.36		2	A1	1327
	+35 03310			18 35 45	+35 36.9	1	8.73		1.05		0.92		2	K0	1733
	−10 04738		V	18 35 45	−10 03.3	2	9.57	.020	2.34	.030	2.19		4	M3	369,8032
		AO 1166		18 35 46	+38 49.4	1	10.14		0.78		0.33		1	G8 V	1748
172088	−3 04331	HR 6999	★ AB	18 35 46	−03 14.3	2	6.49	.010	0.55	.000	0.04	.000	10	F9 V +F9 V	1088,3077
		CS 20321 # 392		18 35 47	−36 27.5	1	15.31		0.16		0.18		1		1736
	+5 03856	IC 4756 - 78		18 35 48	+05 37.0	1	9.97		0.54		0.38		2	F0	1327
172309	+45 02747			18 35 48	+45 37.7	1	8.36		0.23		0.08		2	F0	1601
	+4 03818	IC 4756 - 79		18 35 49	+04 55.2	1	8.87		1.59		1.85		2	K8	1327
		IC 4756 - 270		18 35 49	+05 19.0	1	12.93		1.26		1.06		2		460
	+5 03857	IC 4756 - 80		18 35 50	+05 35.6	2	9.52	.009	1.04	.004	0.65	.027	5	dG7	460,1327
		WLS 1840 35 # 10		18 35 51	+34 50.1	1	12.13		0.51		0.03		2		1375
		IC 4756 - 273		18 35 52	+05 20.1	1	12.79		0.69		0.11		2		460
	+5 03858	IC 4756 - 81		18 35 53	+05 23.3	2	9.40	.036	1.01	.004	0.59	.000	5	gG6	460,1327
	+34 03266	G 206 - 35		18 35 53	+34 39.2	1	10.78		1.34		1.22		2	K7	7010
	−4 04525	IDS18332S0423	A	18 35 53	−04 20.9	1	9.24		2.21				2	M2	369
		G 155 - 29		18 35 54	−14 31.7	1	11.28		1.56		1.23		6	M1 V:	7009
172051	−21 05081	HR 6998		18 35 54	−21 05.7	3	5.86	.012	0.67	.011	0.14	.005	7	G5 V	1754,2007,3077
	+5 03859	IC 4756 - 82		18 35 55	+05 31.8	1	10.44		0.93		1.48		2	G5	1327
171982	−37 12762			18 35 55	−37 44.7	1	8.44		0.90		0.59		2	K0 V	1732
	+5 03861	IC 4756 - 83		18 35 58	+05 51.1	1	9.67		0.29		0.22		2	A1	1327
172171	+8 03780	HR 7002, X Oph	★ AB	18 35 58	+08 47.3	1	6.37		1.29		0.91		1	K1 III +	3001
	+5 03860	IC 4756 - 84		18 35 59	+05 07.1	1	10.10		0.39		0.19		2	A9	1327
172052	−23 14580			18 36 00	−23 13.6	2	6.73	.035	0.66	.005	0.34		12	F5 II	8059,8100

HD	DM	Other Id	N Rem	α_{1950}	δ_{1950}	S	V	σ_V	B–V	σ_{B-V}	U–B	σ_{U-B}	n	Spectrum	References
171967	−43 12699	HR 6991		18 36 00	−43 13.9	2	5.41	.034	1.65	.024	1.95		6	M2 III	2007,3055
172155	−0 03525			18 36 01	+00 04.2	1	8.01		0.46		0.03		6	F5	1371
172136	−3 04334			18 36 01	−03 44.1	1	9.10		2.02				2	K5	369
171940	−45 12654			18 36 01	−45 16.9	1	9.39		−0.14		−0.69		3	B3 III/IV	1732
172569	+65 01283	HR 7013		18 36 04	+65 26.6	2	6.06	.000	0.28	.000	0.09	.005	5	F0 V	985,1733
		IC 4756 – 283		18 36 07	+05 24.9	1	12.91		0.63		0.05		2		460
	+5 03863	IC 4756 – 91		18 36 08	+05 32.1	1	9.94		0.38		0.31		2	A8	1327
		CS 20321 # 417		18 36 08	−35 10.6	1	14.51		0.27		0.20		1		1736
171968	−45 12655			18 36 08	−45 16.8	1	8.90		1.25		1.24		2	K2 III	1732
	+4 03822	IC 4756 – 92		18 36 09	+04 58.9	1	9.90		0.38		0.24		2	A3	1327
	+7 03784			18 36 09	+07 48.2	1	9.16		1.82				2	M2	369
172189	+5 03864	IC 4756 – 93		18 36 10	+05 25.2	1	8.73		0.36		0.25		2	A2	1327
172190	+4 03823	IC 4756 – 95	★ AB	18 36 12	+04 48.6	1	6.71		1.09		1.01		3	K0	1327
		IC 4756 – 204		18 36 12	+04 53.2	1	10.11		0.90		0.53		2		1327
		IC 4756 – 94		18 36 12	+05 14.4	1	10.55		0.39		0.19		2	F6	1327
	+5 03866	IC 4756 – 97		18 36 13	+05 38.3	1	10.62		0.33		0.26		2	A3	1327
	+5 03865	IC 4756 – 98		18 36 13	+05 45.9	1	10.21		0.30		0.24		2	A3	1327
	+5 03867	IC 4756 – 100		18 36 14	+05 39.9	1	9.82		0.30		0.26		2	A4	1327
172122	−20 05220			18 36 14	−20 22.9	2	8.73	.011	0.08	.001	−0.28	.073	4	B3 II	540,976
172324	+37 03183			18 36 15	+37 23.4	1	8.58		0.02		−0.40		1	B9 Ib	8100
	+5 03868	IC 4756 – 101		18 36 16	+05 11.6	2	9.37	.009	1.04	.014	0.70	.009	4	G5	460,1327
	+5 03869	IC 4756 – 102		18 36 16	+05 15.2	1	9.53		0.34		0.22		2	A9	1327
	+5 03870	IC 4756 – 103		18 36 17	+05 24.0	1	9.69		0.31		0.21		2	A1	1327
	+5 03871	IC 4756 – 104		18 36 18	+05 17.3	1	9.88		0.32		0.26		2	A8	1327
172228	+14 03603	IDS18340N1500	A	18 36 18	+15 02.3	1	6.78		0.01		−0.04		2	A0	985
172310	+28 03039	G 206 – 36		18 36 19	+28 53.2	2	8.45	.019	0.70	.002	0.16	.038	4	G5 V	1003,3026
		LP 570 – 22		18 36 20	+04 43.4	1	11.21		1.46		1.21		1		1696
	−6 04825	LS IV −06 013		18 36 20	−06 19.4	1	10.23		0.81		0.09		2	A0	1318
		IC 4756 – 302		18 36 21	+05 18.2	1	11.57		1.33		1.06		2		460
172175	−7 04642	V442 Sct	★	18 36 21	−07 54.3	2	9.44	.003	0.64	.007	−0.44	.005	4	O6 e	1011,1586
172246	+16 03572	IDS18342N1627	AB	18 36 23	+16 29.8	1	7.74		0.60		0.19		2	G5	3077
172669	+66 01117			18 36 24	+66 52.2	1	7.59		0.62		0.10		4	G3 V	3026
		IC 4756 – 110		18 36 25	+04 59.9	1	10.38		0.45		0.24		2	A5	1327
	+5 03873	IC 4756 – 109		18 36 25	+05 17.5	2	9.03	.013	1.12	.038	0.87	.004	6	dG6	460,1327
	+5 03874	IC 4756 – 107		18 36 25	+05 37.6	1	9.62		0.38		0.32		2	A0	1327
349063	+20 03876			18 36 26	+20 35.1	3	9.33	.010	0.65	.005	0.08	.000	8	G0	979,3026,8112
		IC 4756 – 305		18 36 27	+05 25.7	1	12.91		0.65		0.08		2		460
172380	+39 03476	HR 7009, XY Lyr		18 36 27	+39 37.4	4	6.05	.009	1.63	.011	1.43	.121	11	M4-5 II	71,1501,3001,8032
172393	+42 03123	G 205 – 33		18 36 28	+42 37.2	1	8.34		0.82		0.39		4	G5	3016
		IC 4756 – 208		18 36 30	+05 24.	1	10.92		0.50		0.00		1		1327
		IC 4756 – 209		18 36 30	+05 24.	1	10.97		0.53		0.04		2		1327
		IC 4756 – 210		18 36 30	+05 24.	1	10.98		0.39		0.21		1		1327
		IC 4756 – 360		18 36 30	+05 24.	1	13.22		1.20		0.87		1		460
		IC 4756 – 365		18 36 30	+05 24.	1	13.57		0.49		0.20		1		460
		IC 4756 – 368		18 36 30	+05 24.	1	13.53		0.75		0.16		1		460
172158	−22 04817			18 36 30	−22 00.7	2	7.90	.001	0.19	.002	−0.21	.013	4	B8 II	540,976
	+5 03877	IC 4756 – 116		18 36 35	+05 29.4	1	9.57		0.39		0.24		2	F4	1327
172233	−0 03526			18 36 36	+00 03.6	1	8.03		0.61		0.13		6	G0	1371
		G 21 – 22		18 36 36	+00 04.5	6	10.74	.024	0.53	.009	−0.11	.006	10		202,516,1064,1620*
172248	+5 03878	IC 4756 – 117		18 36 36	+05 24.8	1	8.97		0.07		−0.42		2	A0p	1327
172271	+5 03879	IC 4756 – 118		18 36 36	+05 32.6	1	9.07		0.05		−0.13		2	A0p	1327
	−13 05061	LSS 5107		18 36 36	−13 53.4	2	9.86	.002	0.90	.011	0.12	.006	4	B9 Ia	1032,1586
172140	−29 15193			18 36 37	−29 23.1	2	9.95	.011	−0.05	.011	−0.89	.010	3	B1 II	55,540
172094	−42 13493			18 36 38	−41 58.9	1	8.28		−0.15		−0.88		1	B2 Ib/II	55
	+38 03242			18 36 41	+38 46.8	1	10.25		1.33		1.58		1		1748
	+5 03881	IC 4756 – 123		18 36 43	+05 40.9	1	10.91		0.47		0.14		2	F8	1327
171909	−62 05922			18 36 47	−62 03.4	1	8.83		1.12		0.90		4	K0 III	1731
		Smethells 17		18 36 48	−60 28.	1	11.24		1.42				1		1494
	+5 03882	IC 4756 – 125		18 36 50	+05 11.0	2	9.31	.030	1.05	.021	0.71	.009	6	G5	460,1327
		IC 4756 – 127		18 36 51	+05 20.6	1	10.25		0.34		0.22		2	A8	1327
	+5 03884	IC 4756 – 128		18 36 51	+05 37.1	1	10.03		0.86		0.60		2	G5	1327
172127	−39 12802			18 36 51	−39 47.6	1	10.48		−0.12		−0.76		3	B3/5 II/III	400
172252	−12 05132	LS IV −11 035		18 36 52	−11 55.5	3	9.47	.037	0.66	.009	−0.38	.022	4	B0 V:e	445,1012,1032
	+4 03827	IC 4756 – 134	V	18 36 57	+04 47.3	1	9.73		2.11				2	Ma	369
		IC 4756 – 133		18 36 57	+05 21.2	1	10.38		0.33		0.22		2	A9	1327
	−4 04534	LS IV −04 011		18 36 57	−04 38.5	1	10.86		0.57		−0.10		4		845
172275	−7 04645	LS IV −07 010		18 36 57	−07 24.0	3	9.36	.017	0.77	.005	−0.37	.015	6	O7 III	1011,1012,1195
172237	−20 05224			18 36 57	−20 43.3	2	8.56	.005	0.50	.000	0.03		5	F6 V	657,2033
	−56 08902	Smethells 18		18 36 57	−56 55.3	1	10.49		1.21				1		1494
		G 141 – 24		18 36 59	+04 09.0	1	15.15		1.60				4		538
		IC 4756 – 137		18 36 59	+05 17.1	2	10.17	.014	0.34	.009	0.23	.000	4	F3	460,1327
		IC 4756 – 139		18 36 59	+05 18.8	2	9.63	.028	1.04	.005	0.70	.038	4	G2	460,1327
	−7 04646	LS IV −07 011		18 37 00	−07 07.1	1	10.85		0.46		−0.15		2	B3	1195
	+5 03889	IC 4756 – 142		18 37 03	+05 26.3	1	9.54		0.41		0.22		2	A3	1327
	+39 03480			18 37 03	+39 45.0	1	9.19		1.04				1	K0	8097
	+5 03890	IC 4756 – 143		18 37 04	+05 28.7	1	10.43		0.38		0.17		2		1327
		GD 214		18 37 04	+15 04.8	1	13.52		0.81		0.07		2		3060
172144	−44 12797			18 37 04	−44 13.2	2	7.39	.005	0.55	.000	0.07		6	F8 IV	2012,3037
		LP 570 – 23		18 37 05	+03 41.8	1	12.64		0.67		−0.01		2		1696
172397	+20 03879			18 37 06	+20 27.1	1	7.66		−0.02		−0.33		3	B8	555

Table 1   1019

HD	DM	Other Id	N	Rem	α_{1950}	δ_{1950}	S	V	σ_V	B−V	σ_{B-V}	U−B	σ_{U-B}	n	Spectrum	References
172728	+62 01637	HR 7018			18 37 06	+62 28.8	3	5.74	.000	-0.05	.005	-0.16	.004	7	A0 V	985,1501,1733
172327	+0 03989				18 37 07	+00 06.1	1	8.16		1.07		0.91		6	K0	1371
172365	+5 03891	IC 4756 - 145		★	18 37 09	+05 13.1	4	6.35	.048	0.79	.016	0.50	.030	10	F8 Ib-II	1080,1149,1327,8100
	+9 03805				18 37 09	+09 59.9	1	10.93		1.70				2	M2	369
172256	-22 04820	V4131 Sgr		★	18 37 10	-22 42.6	3	8.71	.015	0.06	.008	-0.67	.009	7	B5 II(n)	540,976,1586
172421	+20 03880	IDS18350N2051	AB		18 37 11	+20 53.2	1	7.67		-0.02		-0.41		3	B4 III+B7 V	555
171759	-71 02353	HR 6982		★ A	18 37 12	-71 28.5	5	4.00	.005	1.13	.008	1.02	.000	24	K0 III	15,1075,1075,2038,3077
	-12 05133	LSS 5109			18 37 13	-12 26.8	2	10.78	.000	0.70	.000	-0.23	.000	3	B1 V	1012,1032
	+5 03892	IC 4756 - 146			18 37 16	+05 15.4	1	9.70		0.48		0.06		2	F9	1327
172401	+8 03791				18 37 16	+08 41.2	1	6.98		1.07				1	K0 III	1214
	+5 03894	IC 4756 - 148			18 37 17	+05 09.5	1	8.97		1.08		0.97		2	gG6	1327
		Sct sq 1 # 113			18 37 17	-07 12.0	1	12.72		0.54		0.52		2		1195
		IC 4756 - 149			18 37 18	+05 25.4	1	10.78		0.27		0.17		2	A5	1327
172347	-7 04649				18 37 18	-07 12.4	1	9.74		0.53		0.31		2	A0	1195
172348	-7 04648	HR 7007		★ A	18 37 18	-07 50.2	1	5.83		1.55		1.82		8	K0	1088
		V348 Sgr			18 37 18	-22 57.4	1	12.26		0.45		-0.46		1		1599
171971	-64 03941				18 37 18	-64 32.8	1	8.21		1.42		1.64		5	K3/4 III	1704
		Sct sq 1 # 114			18 37 20	-07 16.6	1	10.88		0.66		0.01		2		1195
	+3 03767			V	18 37 21	+03 44.7	1	9.54		2.11				2	M3	369
	+5 03895	IC 4756 - 151			18 37 21	+05 33.4	1	9.66		0.37		0.26		2	F2	1327
	-4 04540	LS IV -04 012			18 37 23	-04 38.1	1	11.26		0.66		0.22		4		845
	+5 03896	IC 4756 - 152			18 37 25	+05 31.7	1	10.90		0.33		0.20		2	A5	1327
172424	+7 03798	HR 7010		★ A	18 37 26	+07 18.7	3	6.28	.008	0.97	.010	0.69	.008	10	G8 III	15,1256,3016
172424	+7 03798	IDS18350N0716	B		18 37 26	+07 18.7	1	11.68		1.14		0.85		3		3016
172386	-5 04717				18 37 26	-05 49.2	1	8.84		0.74				54	G5	6011
	-6 04827				18 37 28	-06 40.0	1	10.46		1.89		0.58		3		8032
172367	-7 04650	LS IV -07 014			18 37 28	-07 17.8	2	9.51	.009	0.45	.000	-0.51	.004	12	B0 V	1012,1195
172021	-64 03942	HR 6996			18 37 30	-64 41.5	4	6.37	.009	0.15	.012	0.14	.008	17	A5 V	15,1075,2038,3021
		G 155 - 30			18 37 32	-10 30.4	2	11.49	.014	1.55	.002			5		158,1705
172500	+25 03594				18 37 34	+25 50.2	2	8.78	.000	0.15	.000	0.09	.010	5	A0	833,1625
336601	+26 03320				18 37 36	+26 08.5	1	8.92		0.58		0.07		3	F8 V	3026
	-5 04718	LS IV -05 010			18 37 36	-05 32.6	1	9.92		0.81		-0.10		1	B2:II:p	1032
172403	-9 04790	LS IV -09 018			18 37 37	-09 11.3	1	8.48		0.80		0.12		1	B9:Ib:	1032
	-17 05283	G 155 - 31			18 37 39	-17 02.7	1	10.64		1.00		0.91		1		3062
172370	-21 05096				18 37 39	-21 32.5	1	8.57		0.18		-0.40		4	B5 III	1732
	+5 03900	IC 4756 - 158			18 37 41	+05 20.0	1	9.64		0.39		0.22		2	F0	1327
	+5 03901	IC 4756 - 159			18 37 41	+05 46.6	1	10.00		0.81		0.30		2	F7	1327
		IC 4756 - 160			18 37 42	+05 24.7	1	10.32		0.42		0.18		2	F0	1327
	+9 03811				18 37 44	+09 46.4	1	9.40		1.73				2	M2	369
172223	-48 12668	HR 7005			18 37 44	-48 08.5	4	6.49	.005	1.22	.000			18	K2 III/IV	15,1075,2018,2030
172427	-10 04749	LS IV -10 014			18 37 46	-10 46.0	1	9.46		0.48		-0.44		1	B1 IV	1032
		IC 4756 - 162			18 37 47	+05 26.2	1	10.40		0.34		0.25		2	A4	1327
172541	+24 03489				18 37 47	+24 57.7	1	8.47		1.06		0.83		2	K0	1625
		BPM 11668			18 37 48	-61 55.	1	14.67		0.11		-0.73		1		832
173127	+72 00858				18 37 50	+72 13.3	1	8.58		0.56		0.00		2	G5	1375
172098	-62 05931				18 37 50	-62 22.4	1	9.13		1.14		0.90		4	G8 III/IV	119
	+5 03906	IC 4756 - 164			18 37 51	+05 16.0	1	9.27		1.08		0.81		2	dG6	1327
172453	-5 04719				18 37 51	-05 45.3	1	8.05		1.50				52	K0	6011
		LSS 5112			18 37 55	-17 07.5	1	11.93		0.45		-0.44		2	B3 Ib	1737
172671	+40 03446	HR 7017		★	18 37 56	+40 53.3	1	6.25		-0.06		-0.18		2	B9 V	401
172522	+8 03797				18 37 59	+08 49.3	1	6.91		0.22				1	A2 III	1214
172692	+39 03485				18 38 04	+40 04.3	1	7.79		0.94				2	K0	8097
172488	-8 04675	LS IV -08 019			18 38 04	-08 46.0	1	7.62		0.54		-0.40		2	B0.5V	1012
172475	-14 05152	IDS18352S1436	AB		18 38 04	-14 33.2	1	7.97		1.60				2	K0 III +B	369
	-17 05287	G 155 - 32			18 38 04	-17 12.6	1	10.44		0.73		0.10		2		3062
172507	-2 04696				18 38 06	-02 44.7	1	10.05		0.88		0.54		1	K3 V	3072
172631	+30 03262	HR 7016			18 38 08	+30 48.1	1	6.40		0.91		0.61		2	K0	1733
172508	-4 04547				18 38 08	-04 32.6	1	7.61		1.81				2	K0 II-III	369
	-13 05069				18 38 08	-13 25.1	3	10.63	.004	1.51	.004	1.20	.005	6	M0	158,801,1705
		HA 110 # 229			18 38 11	-00 01.0	1	13.65		1.91		1.39		16		1764
		LS IV +05 005			18 38 16	+05 57.3	1	10.97		0.56		-0.46		2		405
		HA 110 # 230			18 38 17	-00 00.5	1	14.28		1.08		0.73		11		1764
172527	-4 04548	RR Sct			18 38 17	-04 07.9	1	8.94		0.86		0.34		4	A0 Ib	845
		HA 110 # 232			18 38 18	-00 00.9	1	12.52		0.73		0.15		16		1764
		HA 110 # 233			18 38 18	-00 02.0	1	12.77		1.28		0.81		10		1764
		G 184 - 17			18 38 19	+19 33.4	1	14.10		0.87		0.09		2		1658
172184	-62 05938				18 38 19	-62 12.4	1	9.92		0.24		0.11		4	A1 III/IV	1731
172650	+26 03324				18 38 20	+26 05.0	1	6.85		0.01		-0.02		3	B9 V	1501
172588	+8 03799				18 38 22	+08 44.7	1	7.22		0.35				1	F0 II-III	1214
		LS IV -04 014			18 38 22	-04 29.1	1	11.94		1.05		0.11		1		1183
173084	+67 01087	G 260 - 6		★ AB	18 38 27	+67 04.7	1	7.74		0.62		0.04		2	G5	3030
	-6 04834	LS IV -06 014			18 38 28	-06 49.8	1	10.97		0.50		-0.18		2	B3	1318
172741	+38 03254	HR 7019			18 38 30	+38 19.2	1	6.45		0.21		0.14		3	A6 m	8071
	-13 05073	LSS 5115			18 38 32	-13 53.3	4	10.35	.087	0.74	.053	-0.22	.025	7	B1:IV:pe	445,1012,1032,1586
	-4 04549	LS IV -04 015	A		18 38 34	-04 25.2	2	10.35	.015	0.89	.010	-0.24	.015	6		405,845
	-4 04549	LS IV -04 015	B		18 38 34	-04 25.2	1	11.23		0.99		-0.17		4		845
		LS IV -04 016			18 38 38	-04 23.5	1	10.80		1.01		-0.05		4		845
172533	-27 13097				18 38 40	-27 29.6	3	8.30	.005	-0.02	.011	-0.49	.005	6	B2 III/IV	400,540,976
	-13 05074	LS IV -13 085			18 38 42	-13 26.5	1	10.51		0.51		-0.23		3		1586
	-4 04551	LS IV -04 017			18 38 43	-04 05.1	1	11.22		1.05		0.23		4		845

HD	DM	Other Id	N	Rem	α_{1950}	δ_{1950}	S	V	σ_V	B–V	σ_{B-V}	U–B	σ_{U-B}	n	Spectrum	References
172883	+52 02263	HR 7028			18 38 44	+52 08.9	3	5.99	.002	-0.07	.001	-0.24	.005	67	A0 p:Hg:	879,1079,1769
		HA 110 # 239			18 38 45	−00 02.7	1	13.86		0.90		0.58		1		1764
172211	−64 03943	HR 7004			18 38 45	−64 36.0	4	5.77	.012	0.96	.007	0.77	.009	17	K0 III	15,1075,1637,2038
		LS II +15 001			18 38 48	+15 21.0	1	11.27		0.41		-0.39		2		405
172546	−23 14625	HR 7011			18 38 49	−23 52.9	1	6.23		0.23				4	A3mA7-A9	2007
172594	−14 05156	LSS 5118			18 38 52	−14 36.8	1	6.42		0.81				4	F2 Ib	2007
172462	−44 12817				18 38 52	−44 17.4	1	8.75		0.35				4	F0 IV	2012
		HA 110 # 339			18 38 53	+00 05.5	1	13.61		0.99		0.78		1		1764
170848	−83 00662				18 38 53	−83 37.0	1	7.66		1.10		0.99		5	K0 III	1704
172651	+0 03993				18 38 54	+00 31.0	4	7.47	.003	1.45	.007	1.71	.012	36	M0 III	281,989,1729,6004
172652	+0 03992				18 38 55	+00 12.5	5	10.03	.004	0.30	.004	0.13	.005	106	A5	281,989,1729,1764,6004
172535	−32 14416	IDS18357S3228		AB	18 38 55	−32 25.0	4	7.77	.011	-0.08	.006	-0.58	.007	10	B3 IV/V	438,540,976,1732
172655	−4 04552	IDS18363S0358		AB	18 38 56	−03 55.5	1	9.61		0.62				24	G0	6011
		NGC 6681 sq1 10			18 38 56	−32 25.6	1	10.96		0.24				2		438
172656	−7 04661				18 38 60	−07 19.2	1	8.35		1.50				3	K2	1726
	+31 03330	G 206 - 38		⋆ A	18 39 01	+31 29.9	1	8.54		0.94		0.63		2	K3 V	3016
	+31 03330	G 206 - 38		⋆ AB	18 39 01	+31 29.9	1	8.50		0.94		0.65		2		1003
	+31 03330	IDS18371N3128		B	18 39 01	+31 29.9	1	11.54		1.58		1.18		4		3016
		LP 630 - 178			18 39 01	−00 18.5	1	11.28		0.49		-0.16		2		1696
		G 21 - 23			18 39 03	+00 53.4	3	12.24	.035	1.41	.008	1.22	.014	7		202,333,1620,3078
		HA 110 # 135			18 39 03	−00 08.0	1	11.07		0.74		0.30		2	G5	281
172677	−6 04837	LS IV -06 016			18 39 05	−06 39.3	2	9.77	.049	0.35	.010	-0.56	.034	3	B8	1195,1318
		HA 110 # 477			18 39 10	+00 23.8	1	13.99		1.35		0.71		11		1764
		POSS 281 # 6			18 39 10	+35 45.2	1	18.05		1.22				6		1739
		HA 110 # 246			18 39 17	+00 02.4	2	12.71	.007	0.58	.008	-0.07	.043	3		1696,1764
172730	−4 04553	RU Sct			18 39 17	−04 09.6	1	8.82		1.35				1	G5	6011
		CN Lyr			18 39 19	+28 40.3	1	11.11		0.45		0.28		1	F5	1399
		HA 110 # 346			18 39 21	+00 07.1	1	14.76		1.00		0.75		1		1764
172865	+30 03271	IDS18374N3012		AB	18 39 21	+30 14.8	1	6.94		0.82		0.41		5	G5 IV+G5 IV	3030
172865	+30 03271	IDS18374N3012		C	18 39 21	+30 14.8	1	8.83		0.54		0.04		5		3030
	−2 04706				18 39 22	−02 27.0	1	10.44		2.35				2	M3	369
		LSS 5120			18 39 22	−13 30.3	2	11.78	.000	0.53	.000	-0.44	.000	4	B3 Ib	405,1737
172694	−15 05063	HR 7014, LSS 5119		⋆	18 39 22	−15 54.3	3	8.14	.039	0.19	.010	-0.72	.008	7	B2 Ibe	379,1212,1586
172945	+41 03116				18 39 24	+41 52.1	1	7.73		0.01		0.01		2	A0	1733
		HA 110 # 248			18 39 26	+00 00.3	1	10.86		0.68		0.50		2	A2	281
		LS IV -03 005		A	18 39 26	−03 56.9	1	11.46		0.85		0.08		7		845
		LS IV -03 005		B	18 39 26	−03 56.9	1	11.89		0.72		0.23		5		845
172579	−39 12836	IDS18360S3923		AB	18 39 26	−39 20.1	1	7.02		-0.06		-0.49		3	B5 III/IV	1586
		LS II +16 001			18 39 27	+16 03.7	1	11.17		0.92		0.36		1	F6/8II-III	1215
172696	−20 05240				18 39 27	−20 21.7	1	7.10		0.19				std	Ap Si	1323
		TY Sct			18 39 29	−04 20.5	1	10.64		1.64		1.23		1	F8	689
	−4 04555	IDS18369S0403		AB	18 39 30	−04 00.4	1	10.55		2.40				2	M3	369
	−6 04840				18 39 30	−05 58.6	1	9.87		1.35				3		1726
172804	+6 03898	V679 Oph			18 39 31	+06 46.2	1	8.97		2.25				2	S4.8	369
	−1 03542	LS IV -01 004			18 39 31	−01 21.1	2	9.12	.055	1.55	.026	0.41	.021	3	B8 Ia:	1032,1586,8100
	−3 04356				18 39 31	−03 46.0	1	10.57		2.68				1	M2	369
172748	−9 04796	HR 7020, δ Sct		⋆ A	18 39 32	−09 06.1	6	4.72	.020	0.36	.010	0.14	.024	14	F2 IIIp δ Del	15,1075,1462,2012*
		HA 110 # 249			18 39 33	−00 06.6	1	12.36		0.72		0.15		6	G5	281
172976	+44 02963				18 39 34	+44 13.3	2	7.32	.025	0.29	.007	0.17	.000	8	F0 III	3016,8071
		MtW 110 # 38			18 39 36	+00 10.7	1	15.53		1.36		0.84		4		397
	−6 04842	LS IV -06 021			18 39 38	−06 39.6	2	10.63	.029	0.21	.019	-0.44	.019	3	B3	1195,1318
172721	−22 04833				18 39 38	−22 13.2	1	8.87	.010	0.12	.004	-0.47	.002	5	B5 II/III	540,976
		HA 110 # 349			18 39 39	+00 07.3	1	15.10		1.09		0.67		1		1764
172641	−40 12732				18 39 39	−40 45.2	1	8.59		0.38		0.09		2	F2 V	1730
		MtW 110 # 49			18 39 42	+00 10.4	1	14.77		1.93		1.03		6		397
		HA 110 # 352			18 39 43	+00 08.1	2	11.35	.002	0.57	.000	0.10	.010	15		281,397
172829	+0 03995	HK Aql		⋆	18 39 44	+00 06.3	5	8.45	.010	1.99	.013	2.26	.047	65	K5 III	281,397,989,1729,6004
		MtW 110 # 54			18 39 44	+00 10.0	1	14.91		1.34		0.64		8		397
		HA 110 # 355			18 39 45	+00 05.4	1	11.94		1.02		0.50		8		1764
		NGC 6681 sq1 12			18 39 46	−32 11.8	1	11.21		1.52				2		438
	−6 04844				18 39 47	−06 03.4	1	10.34		0.26		-0.37		2	B3	1195
172958	+31 03332	HR 7030			18 39 48	+31 34.1	2	6.41	.000	-0.06	.000	-0.21	.008	7	B8 V	253,401,1063
		CM Sct			18 39 48	−05 23.3	1	10.81		1.22				1		6011
172582	−50 12100				18 39 48	−50 12.8	2	9.33	.014	0.68	.009			5	G6 V	1705,2012
172977	+33 03180				18 39 51	+34 03.0	1	7.44		-0.16		-0.60		2	B9	401
	+7 03813				18 39 52	+07 33.1	1	10.05		1.88				2	M2	369
172809	−7 04669				18 39 52	−07 22.2	1	9.12		2.02				2	K5	369
172793	−15 05068	IDS18370S1553		AB	18 39 52	−15 50.4	1	9.28		0.12		-0.46		1	B5 Ib	963
		NGC 6681 sq1 17			18 39 53	−32 17.0	1	13.46		0.99				2		438
172831	−7 04670	HR 7024			18 39 54	−07 07.4	1	6.14		1.00		0.89		8	K1 III	1088
172722	−32 14427				18 39 55	−32 16.5	1	9.89		1.06				3	K0	438
	−33 13497				18 39 55	−33 24.8	1	10.61		1.05		0.82		1	M2	1696
		HA 110 # 144			18 39 56	−00 07.8	1	11.95		0.72		0.27		6	dG5	281
		FAIRALL 51# 4			18 39 56	−62 24.3	1	14.10		0.75		0.15		4		1687
		FAIRALL 51# 1			18 39 57	−62 27.1	1	12.83		0.80		0.34		3		1687
172816	−19 05134	HR 7023, V3879 Sgr			18 39 58	−19 20.0	2	6.32	.039	1.73	.024	1.56		16	M4 III	3042,6002
172737	−32 14428				18 39 58	−32 20.2	1	11.15		0.59		0.01		2		540
		NGC 6681 sq1 11			18 40 02	−32 16.8	1	11.13		0.55				3		438
171161	−83 00663	HR 6964			18 40 02	−83 22.3	2	7.15	.005	1.26	.000	1.30	.000	7	K1 III	15,1075
		Tr 35 - 67			18 40 03	−04 12.0	1	12.96		0.99		0.27		1		49

Table 1   1021

HD	DM	Other Id	N Rem	α_{1950}	δ_{1950}	S	V	σ_V	B–V	σ_{B-V}	U–B	σ_{U-B}	n	Spectrum	References
		FAIRALL 51# 5		18 40 03	−62 26.8	1	14.36		0.70		0.20		3		1687
		HA 110 # 358		18 40 04	+00 12.0	1	14.43		1.04		0.42		1		1764
		HA 110 # 360		18 40 06	+00 06.2	1	14.62		1.20		0.54		2		1764
		Tr 35 - 107		18 40 09	−04 17.9	1	14.56		1.11		0.28		1		49
172776	−32 14432			18 40 09	−32 13.8	1	10.07		0.03				2	B9 V	438
		Tr 35 - 91		18 40 10	−04 17.2	1	14.36		1.16		0.35		1		49
		Tr 35 - 304		18 40 10	−04 18.0	1	16.66		1.23				1		49
172266	−72 02300			18 40 10	−72 51.9	1	6.80		0.49		0.04		8	F5 V	1628
171990	−77 01314	HR 6994		18 40 10	−77 55.4	2	6.38	.005	0.60	.000	0.12	.005	7	F8 V	15,1075
		HA 110 # 361		18 40 11	+00 05.1	1	12.43		0.63		0.04		17		1764
		Tr 35 - 302		18 40 11	−04 17.8	1	15.80		1.12				1		49
172889	−6 04849	LS IV -06 025		18 40 11	−06 32.3	2	10.15	.015	0.25	.020	-0.59	.030	2	B1 V	1032,1195
		G 155 - 34		18 40 12	−11 11.7	1	14.18		0.15		-0.61		1	DAs	3060
		HA 110 # 362		18 40 14	+00 03.4	1	15.69		1.33		3.92		1		1764
		Tr 35 - 88		18 40 14	−04 15.9	1	14.47		1.21		0.44		1		49
		Tr 35 - 303		18 40 14	−04 17.0	1	15.85		1.09				1		49
		HA 110 # 266		18 40 15	+00 02.1	1	12.02		0.89		0.41		3		1764
		WLS 1840 35 # 5		18 40 15	+37 19.8	1	11.46		0.67		0.19		2		1375
		Tr 35 - 90		18 40 15	−04 17.3	1	15.25		1.13		0.78		1		49
		NGC 6681 sq1 14		18 40 15	−32 12.8	1	12.39		1.36				2		438
		FAIRALL 51# 6		18 40 15	−62 24.	1	15.61		0.84				2		1687
		110 L1		18 40 16	+00 04.1	1	16.25		1.75		2.95		1		1764
		Tr 35 - 60		18 40 16	−04 13.7	1	13.41		1.15		0.79		1		49
		Tr 35 - 87		18 40 16	−04 16.1	1	13.73		0.99		0.19		1		49
		LS IV -03 006		18 40 17	−03 56.0	1	11.24		1.00		-0.03		4		845
		Tr 35 - 61		18 40 17	−04 14.0	1	13.08		1.01		0.50		1		49
		Tr 35 - 86		18 40 17	−04 16.2	1	15.06		1.04		0.33		1		49
172902	−5 04734	Z Sct		18 40 17	−05 52.3	2	9.37	.326	1.35	.316	1.75		2	G0	689,6011
		HA 110 # 364		18 40 18	+00 04.9	1	13.61		1.13		1.10		7		1764
		Tr 35 - 59		18 40 18	−04 14.7	1	13.19		1.05		0.35		1		49
		Tr 35 - 63		18 40 19	−04 14.4	1	12.96		1.04		0.31		1		49
		Tr 35 - 85		18 40 19	−04 16.0	1	15.39		1.50				1		49
		Tr 35 - 100		18 40 19	−04 16.3	1	13.27		1.11		0.39		1		49
172903	−9 04802			18 40 19	−09 13.2	1	8.52		1.18				3	K0	1726
172854	−22 04835	IDS18373S2230	AB	18 40 19	−22 27.7	2	7.70	.010	0.23	.002	-0.30	.002	4	B3 III	540,976
173087	+34 03285	HR 7033	⋆AP	18 40 20	+34 41.8	2	6.47	.005	-0.13	.005	-0.58	.014	6	B5 V	154,1084
173000	+15 03537			18 40 21	+15 09.0	1	6.92		0.04		0.02		2	A0	1648
		NGC 6681 sq1 8		18 40 21	−32 30.5	1	10.94		1.29				2		438
172777	−38 13036	HR 7021	⋆A	18 40 21	−38 22.4	1	5.13		0.09				4	A2 Vn	2032
	+34 03286			18 40 22	+34 41.8	1	8.05		-0.05		-0.25		3	A0	1084
		Tr 35 - 80		18 40 22	−04 14.5	1	13.80		1.00		0.32		1		49
		HA 110 # 365		18 40 23	+00 04.3	1	13.47		2.26		1.89		20		1764
		WLS 1840 35 # 7		18 40 23	+32 46.5	1	12.66		0.73		0.33		2		1375
		HA 110 # 157		18 40 23	−00 11.9	1	13.49		2.12		1.68		3		1764
		FAIRALL 51# 3		18 40 24	−62 26.5	1	13.51		0.71		0.16		3		1687
		HA 110 # 273		18 40 25	−00 06.7	1	14.69		2.53		1.00		2		1764
		Tr 35 - 301		18 40 25	−04 14.5	1	15.58		0.62				1		49
	−32 14436			18 40 25	−32 11.2	1	9.89		1.17				2		438
		HA 110 # 496		18 40 26	+00 28.1	1	13.00		1.04		0.74		4		1764
		NGC 6681 sq1 16		18 40 26	−32 12.8	1	13.28		1.69				2		438
		FAIRALL 51# 7		18 40 27	−62 27.3	1	14.38		0.76		0.19		2		1687
		Tr 35 - 73		18 40 28	−04 13.6	1	12.48		2.30		2.50		1		49
172834	−32 14435			18 40 28	−32 32.4	1	9.30		1.26				1	K1/2 III	438
		HA 110 # 497		18 40 29	+00 27.9	1	14.20		1.05		0.38		1		1764
173398	+62 01641	HR 7042		18 40 30	+62 42.0	1	6.09		0.98				2	K0 III	71
		FAIRALL 51# 2		18 40 30	−62 25.4	1	13.37		1.07		0.80		3		1687
		HA 110 # 280		18 40 33	−00 06.6	1	13.00		2.15		2.13		12		1764
172555	−64 03948	HR 7012		18 40 33	−64 55.3	9	4.78	.007	0.20	.004	0.08	.007	36	A5 IV/V	15,285,611,688,1075*
		HA 110 # 499		18 40 34	+00 25.0	2	11.74	.006	1.00	.011	0.62	.019	35	F5	281,1764
173169	+36 03239			18 40 35	+36 30.1	1	7.42		-0.08		-0.22		2	B9	401
173054	+13 03713			18 40 36	+13 31.0	1	7.68		0.11		0.06		2	A0	1648
172855	−32 14439			18 40 36	−32 31.9	1	9.13		0.53				2	F6 V	438
		HA 110 # 502		18 40 37	+00 24.6	2	12.35	.015	2.29	.027	2.25	.061	59		281,1764
		HA 110 # 503		18 40 38	+00 26.7	2	11.78	.008	0.68	.004	0.49	.018	32	A0	281,1764
		HA 110 # 504		18 40 38	+00 27.0	1	14.02		1.25		1.32		2		1764
173027	+6 03905			18 40 39	+06 43.4	1	8.70		2.16				2	K2	369
172630	−61 06229	HR 7015		18 40 39	−61 08.8	1	6.04		1.46				4	K3 III	2007
172584	−62 05951	FAIRALL 51# 8		18 40 40	−62 28.3	2	9.05	.000	0.37	.000	0.01	.010	7	F0	1687,1731
	+46 02536			18 40 41	+46 58.0	1	8.95		1.07		0.80		2	K0	1601
172933	−20 05243	IDS18377S2019	AB	18 40 41	−20 16.0	1	9.55		0.59				19	F2 V	6011
		HA 110 # 289		18 40 45	+00 02.2	1	11.30		0.77		0.30		2		281
173010	−9 04805	LS IV -09 023		18 40 45	−09 22.3	3	9.16	.057	0.82	.010	-0.26	.014	5	B0 Ia e:	1032,1318,1586
172875	−36 12946	HR 7026		18 40 45	−36 46.2	1	6.32		0.98				4	K0 III	2035
		HA 110 # 507		18 40 46	+00 26.4	2	12.44	.000	1.15	.004	0.90	.054	3		281,1764
		HA 110 # 506		18 40 46	+00 27.4	2	11.34	.015	0.56	.003	0.05	.003	11	F5	281,1764
173006	−5 04737	LS IV -05 014		18 40 46	−05 50.2	2	10.01	.044	0.26	.024	-0.54	.029	3	B0.5IV	1032,1195
	−6 04854	LS IV -06 028		18 40 46	−06 32.3	1	10.35		0.24		-0.28		1	B5	1195
	−3 04361	LS IV -03 007		18 40 47	−03 40.8	1	10.21		0.99		-0.18		4		845
	−4 04562	Tr 35 - 156		18 40 47	−04 12.2	1	10.86		0.74		0.16		1		49
		HA 110 # 290		18 40 48	−00 04.3	1	11.90		0.71		0.20		2		1764

HD	DM	Other Id	N Rem	α_{1950}	δ_{1950}	S	V	σ_V	B–V	σ_{B-V}	U–B	σ_{U-B}	n	Spectrum	References
	−3 04362			18 40 48	−03 34.9	1	10.34		2.66				2	M3	369
173009	−8 04686	HR 7032	⋆ A	18 40 48	−08 19.6	6	4.88	.013	1.13	.011	0.88	.014	20	G8 II	15,1075,1088,3016*
		NGC 6681 sq1 13		18 40 48	−32 21.3	1	11.59		0.96				2		438
		LS IV -03 008		18 40 50	−03 49.3	3	10.85	.024	1.06	.016	-0.03	.048	8		725,845,1586
173034	−8 04687			18 40 51	−08 25.2	1	7.05		1.36		1.14		1	K0	1490
173032	−6 04855	LS IV -06 029		18 40 53	−06 31.2	2	10.11	.058	0.10	.044	-0.61	.024	3	B8	1195,1318
		Sct sq 1 # 278		18 40 53	−08 12.5	1	12.01		0.37		0.17		2		1195
		Sct sq 1 # 279		18 40 53	−08 16.0	1	12.09		0.30		-0.16		3		1195
172909	−32 14445			18 40 53	−32 30.7	1	9.22		0.91				1	K1 III (+G)	438
		GD 215		18 40 57	+04 17.3	2	14.87	.086	0.17	.050	-0.60	.014	5	DAss	940,3060
		G 227 - 44		18 40 57	+58 31.3	1	12.33		0.61		-0.06		1		1658
172910	−35 12876	HR 7029	⋆ AB	18 40 58	−35 41.6	8	4.86	.013	-0.18	.008	-0.73	.020	29	B2 V	15,26,1075,1586,1586*
	−40 12758			18 40 58	−40 47.4	1	10.43		1.29				2		1730
	−6 04856			18 40 59	−06 18.4	1	10.60		0.20		-0.08		2	B9	1195
		HA 110 # 441		18 41 00	+00 16.6	4	11.12	.006	0.55	.005	0.12	.006	72	G1	281,989,1729,1764
173058	−7 04683	SS Sct		18 41 01	−07 46.9	5	7.96	.086	0.83	.039	0.54	.050	5	F8	1462,1484,1490,1772,6011
		CS 22936 # 13		18 41 01	−35 44.7	1	11.46		1.05		0.88		1		1736
173074	−1 03551			18 41 02	−01 36.6	1	7.18		1.93				2	M1	369
172912	−40 12760			18 41 02	−40 52.0	1	9.72		0.01		-0.12		2	B9.5 V	1730
173190	+25 03608			18 41 03	+25 25.0	1	8.76		0.05		-0.22		4	A2	833
	−4 04564	Tr 35 - 169		18 41 04	−04 15.7	2	9.92	.013	0.64	.006	0.12		25	G2 V	49,6011
	−40 12762			18 41 04	−40 49.4	1	10.09		1.21				2		1730
		G 155 - 35		18 41 05	−20 42.4	1	12.86		1.00		0.56		2		3059
173291	+36 03243	HK Lyr		18 41 06	+36 54.5	1	7.65		3.08				2	C5 II	1601
		G 155 - 36		18 41 06	−08 12.7	1	13.86		0.72		0.08		2		3056
173239	+30 03279			18 41 07	+31 05.1	1	7.91		-0.08		-0.36		2	B9	401
173093	−6 04859	HR 7034		18 41 10	−06 52.1	3	6.31	.005	0.48	.000	0.01	.009	8	F7 V	15,657,1256
172781	−57 09180	HR 7022		18 41 10	−56 56.0	1	6.22		1.38				4	K3 III	2007
		HA 110 # 311		18 41 14	−00 03.5	1	15.51		1.80		1.18		2		1764
		HA 110 # 312		18 41 16	−00 03.0	1	16.09		1.32		-0.79		1		1764
	−9 04812			18 41 16	−09 22.0	1	10.78		1.38				2		1726
		HA 110 # 450		18 41 18	+00 19.9	4	11.59	.008	0.95	.006	0.68	.016	78		281,989,1729,1764
		POSS 281 # 4		18 41 18	+36 35.8	1	17.28		1.62				3		1739
		HA 110 # 316		18 41 18	−00 02.0	1	14.82		1.73		4.36		1		1764
		HA 110 # 315		18 41 18	−00 02.3	1	13.64		2.07		2.26		4		1764
172969	−37 12817			18 41 18	−36 56.3	1	8.63		1.66		2.00		4	M2 III	1673
173077	−20 05246			18 41 20	−20 15.9	1	9.21		1.12				16	G6 V	6011
		HA 110 # 319		18 41 21	−00 01.1	1	11.86		1.31		1.08		2		1764
	−10 04774			18 41 25	−10 45.6	1	11.08		2.10				2	M2	369
		HA 110 # 529		18 41 26	+00 23.8	1	11.41		0.76		0.22		14	G0	281
		LS IV -05 016		18 41 26	−05 25.8	1	11.10		0.44		-0.51		1		1318
173216	+8 03819			18 41 27	+08 34.4	1	7.14		0.52		0.12		10	F8	8040
342919	+22 03466	LS II +22 001		18 41 27	+22 34.4	1	10.57		0.54		0.19		2	F8 Ib	8100
173399	+44 02973	IDS18400N4450	A	18 41 27	+44 52.5	1	7.21		0.90		0.50		2	G5 IV	3024
173399	+44 02974	IDS18400N4450	B	18 41 27	+44 52.5	1	8.92		0.42		-0.03		2		3024
	−5 04741			18 41 29	−05 03.4	1	9.57		2.16				2	M3	369
172991	−39 12864	HR 7031		18 41 29	−39 44.3	1	5.43		0.87				4	K3 II + B7	2009
		Smethells 19		18 41 30	−51 28.	1	11.54		1.39				1		1494
		MV Sgr		18 41 33	−21 00.4	1	13.09		0.32		-0.62		1		1589
		LS IV -03 009		18 41 35	−03 51.1	1	11.94		1.40		0.36		1		1183
		WLS 1840 35 # 9		18 41 36	+35 16.8	1	13.07		0.55		0.04		2		1375
173383	+39 03505	HR 7041	⋆ A	18 41 37	+39 14.9	2	6.45	.005	1.59	.000	1.97	.044	2	K5 III	833,1733
173198	−1 03553	V1331 Aql		18 41 37	−01 36.4	2	7.77	.014	0.30	.019	-0.53	.014	8	B1 V	399,401
		G 184 - 20		18 41 38	+15 57.7	3	12.61	.016	0.67	.005	-0.03	.030	6		1658,1696,5010
		G 155 - 37		18 41 38	−16 16.3	1	13.85		1.32		1.14		1		3062
	−7 04686	LS IV -06 030		18 41 39	−06 55.0	2	10.27	.049	0.21	.015	-0.53	.024	3	B3	1195,1318
173203	−8 04690			18 41 39	−08 23.9	1	9.98		0.22		-0.15		2	B9	1195
173140	−19 05148	YY Sgr		18 41 39	−19 26.5	1	10.03		0.19		-0.31		12	B8/9 II	1768
173524	+55 02107	HR 7049	⋆ A	18 41 40	+55 29.3	2	5.05	.005	-0.11	.015	-0.30		4	B9.5 p(Hg)	1363,3033
172782	−64 03953			18 41 43	−64 05.7	1	7.96		1.17		1.12		3	K0 III	861
173202	−7 04687	IDS18390S0743	AB	18 41 45	−07 40.2	1	8.38		0.51				20	F0	6011
173117	−25 13394	HR 7035		18 41 45	−25 03.8	2	5.83		0.04	.009	-0.35	.019	4	B8 III	26,1079
		RZ Lyr		18 41 46	+32 44.7	2	11.14	.223	0.20	.027	0.13	.030	2	A2	668,699
	−8 04691			18 41 47	−07 58.1	1	10.27		0.31		0.17		2	A0	1195
173292	+11 03581	LS IV +11 003		18 41 48	+11 09.8	1	8.61		0.03		-0.60		2	B8	1586
173416	+36 03246	HR 7043		18 41 51	+36 30.3	1	6.01		1.04				2	K0	71
173219	−7 04689	LS IV -07 016		18 41 51	−07 09.8	5	7.81	.016	0.21	.012	-0.78	.015	9	B1:V:pen	445,1012,1195,1212,1586
173220	−7 04688			18 41 52	−07 38.2	1	8.97		0.47				24	F0	6011
		AA45,405 S69 # 4		18 41 54	+00 20.1	1	12.00		1.52		0.55		1		432
		POSS 281 # 2		18 41 54	+36 46.8	1	16.16		1.49				4		1739
	−5 04743			18 41 57	−05 07.4	1	10.20		2.28		2.32		3		8032
		Ross 714		18 41 57	−07 14.5	1	12.42		1.01		0.86		1		1696
173417	+31 03348	HR 7044		18 41 59	+31 52.6	1	5.72		0.33		0.03		1	F1 III-IV	254
173278	−6 04869			18 42 00	−06 35.3	1	6.63		1.90		2.16		1	K5	332
		LS IV -07 018		18 42 00	−07 08.5	1	10.23		0.30		-0.37		1		1318
173251	−14 05172	LSS 5125		18 42 02	−14 24.3	1	9.09		0.67		-0.28		1	B1/2 Ib	1032
		G 141 - 34		18 42 03	+06 24.6	1	13.50		0.98		0.54		1		3060
173329	+1 03763			18 42 06	+01 49.3	1	7.72		1.81				2	M1	369
173208	−28 14952			18 42 10	−28 09.9	1	9.35		-0.02		-0.36		3	B8 III	1586
173164	−37 12830			18 42 10	−37 41.5	1	9.25		0.40		0.15		19	F3 IV/V	1699

Table 1

1023

HD	DM	Other Id	N	Rem	α₁₉₅₀	δ₁₉₅₀	S	V	σ_V	B-V	σ_B-V	U-B	σ_U-B	n	Spectrum	References
		Smethells 20			18 42 12	−62 12.	1	12.00		1.42				1		1494
173739	+59 01915	G 227 - 46		★ A	18 42 13	+59 33.2	4	8.90	.013	1.54	.009	1.10	.025	15	dM3	1,497,1017,3072
173739	+59 01915	G 227 - 46		★ AB	18 42 13	+59 33.2	3	8.49	.018	1.54	.021	1.10	.002	8	dM3 + dM5	938,1197,8039
173740	+59 01915	G 227 - 47		★ B	18 42 14	+59 33.0	3	9.69	.006	1.59	.006	1.14	.010	14	dM5	1,1476,3072
173319	−7 04692				18 42 15	−07 37.6	1	9.44		0.15		−0.29		1	B8	1195
173370	+1 03766	HR 7040			18 42 18	+02 00.5	6	5.02	.006	-0.06	.005	-0.23	.044	24	B9 V	3,154,1149,1203,1363,3016
173297	−20 05253	V350 Sgr			18 42 19	−20 42.0	4	7.14	.059	0.74	.025	0.50	.010	4	F8 Ib/II	657,1484,1490,6011
173282	−21 05131	HR 7038			18 42 20	−21 03.2	2	6.36	.007	0.45	.007	0.05		5	F5 V	1490,2009
173371	−0 03543				18 42 22	−00 25.6	1	6.88		-0.01		-0.31		2	B9 III	1586
		NGC 6694 - 139			18 42 22	−09 25.3	1	13.46		1.32		0.90		1		49
		NGC 6694 - 255			18 42 22	−09 30.2	1	14.05		0.56		0.31		1		49
		NGC 6694 - 183			18 42 24	−09 29.9	1	11.63		2.06				1		49
173182	−42 13592				18 42 24	−42 35.1	1	7.50		1.05				4	G8 II	2012
173664	+53 02126	HR 7060			18 42 25	+53 49.2	2	6.20	.015	0.08	.005	0.11	.005	4	A2 IV	1566,1733
173348	−9 04819	NGC 6694 - 29			18 42 27	−09 27.1	1	9.17		0.48		0.27		1	A0 II-III	49
		NGC 6694 - 125			18 42 27	−09 28.8	1	14.70		0.51		0.42		1		49
173400	+6 03917				18 42 28	+06 58.1	1	9.31		0.67		0.28		1	G0	1535
		NGC 6694 - 91			18 42 28	−09 24.5	1	13.29		0.51		0.20		1		49
		NGC 6694 - 90			18 42 28	−09 24.8	1	12.36		0.40		-0.05		1	A0 V	49
		NGC 6694 - 39			18 42 28	−09 25.3	1	13.05		0.48		0.17		1		49
172941	−62 05959	IDS18378S6250		AB	18 42 28	−62 47.1	2	8.40	.002	0.61	.001	0.17	.005	6	F8 V	861,1731
		NGC 6694 - 40			18 42 29	−09 25.1	1	13.84		0.69		0.38		1		49
		NGC 6694 - 76			18 42 29	−09 28.0	1	12.92		0.52		0.06		1		49
		NGC 6694 - 175			18 42 29	−09 30.0	1	12.33		0.41		-0.04		1	B9 V	49
173183	−42 13595				18 42 29	−42 37.2	1	6.92		0.42				4	F3 V	2012
173420	+6 03918				18 42 30	+07 05.0	1	9.13		1.22		1.21		1	K0	1535
		NGC 6694 - 7			18 42 30	−09 25.8	1	12.99		0.46		0.04		1		49
		NGC 6694 - 249			18 42 30	−09 27.	1	13.80		0.55		0.18		1		49
		NGC 6694 - 266			18 42 30	−09 27.	1	12.76		0.46		-0.02		1		49
		LS IV -04 018			18 42 31	−04 49.5	1	10.88		0.93		-0.11		4		845
173300	−27 13170	HR 7039			18 42 32	−27 02.6	6	3.17	.010	-0.11	.006	-0.36	.007	23	B8.5III	15,1075,1637,2012*
		NGC 6694 - 98			18 42 33	−09 23.8	1	14.65		0.51		0.42		1		49
		NGC 6694 - 173			18 42 33	−09 29.5	1	15.40		0.79		0.40		1		49
173494	+23 03439	HR 7047			18 42 35	+23 32.3	2	6.32	.027	0.40	.006	-0.02		4	F6 V	71,254
	−9 04822	NGC 6694 - 214			18 42 35	−09 21.8	1	10.93		0.45		-0.08		1	B6 V	49
		NGC 6694 - 19			18 42 36	−09 26.9	1	13.77		0.54		0.36		1		49
		NGC 6694 - 65			18 42 36	−09 27.3	1	12.27		0.45		-0.05		1	B6 V	49
	−4 04573	LS IV -04 019			18 42 37	−04 50.5	2	9.68	.015	1.07	.004	0.21	.004	5	B8 Iab:	845,1032
		NGC 6694 - 58			18 42 38	−09 26.3	1	12.60		0.52		0.10		1		49
		NGC 6694 - 117			18 42 38	−09 28.6	1	14.21		0.60		0.42		1		49
		NGC 6694 - 167			18 42 38	−09 29.1	1	15.14		0.93		0.40		1		49
		NGC 6694 - 57			18 42 39	−09 25.9	1	14.63		0.78		0.56		1		49
		NGC 6694 - 60			18 42 40	−09 26.8	1	14.34		0.64		0.50		1		49
173582	+39 03509	HR 7051		★ CD	18 42 41	+39 37.0	3	4.67	.010	0.17	.013	0.11	.033	6	A4 V	938,1363,3023
		NGC 6694 - 115			18 42 42	−09 27.8	1	13.24		1.02		0.43		1		49
173375	−17 05310				18 42 42	−17 35.9	2	7.14	.005	0.18	.001	-0.32	.007	6	B5 III	26,399
173607	+39 03510	HR 7053		★ AB	18 42 43	+39 33.6	2	4.59	.005	0.18	.004	0.08		4	A8 Vn	1363,3023
		NGC 6694 - 104			18 42 43	−09 25.9	1	13.49		0.58		0.18		1		49
		NGC 6694 - 106			18 42 44	−09 26.3	1	13.10		0.47		0.03		1		49
		NGC 6694 - 109			18 42 45	−09 26.9	1	14.34		0.60		0.40		1		49
		NGC 6694 - 161			18 42 46	−09 26.2	1	12.88		1.55				1		49
		LS IV -04 020			18 42 47	−04 04.2	1	11.08		1.22		0.10		4		845
		NGC 6694 - 160			18 42 47	−09 25.6	1	13.70		0.64		0.25		1		49
		LP 923 - 15			18 42 48	−28 59.0	1	12.65		1.53		1.12		1		3062
173526	+22 03472				18 42 50	+22 30.6	1	7.47		1.00		0.68		4	G4 II	8100
173438	−4 04575	LS IV -04 021			18 42 50	−04 39.2	2	8.21	.013	0.78	.013	-0.30	.021	6	B0.5Ia	845,1012
173631	+38 03278				18 42 55	+39 01.1	1	7.99		0.12		0.08		3	A1 V	833
		LS IV -02 016			18 42 56	−02 04.3	3	11.14	.012	0.90	.027	-0.23	.025	4		432,725,914
173797	+57 01898				18 42 57	+57 53.7	1	7.42		0.43		0.06		2	F2	3016
		G 155 - 38			18 43 00	−19 50.1	2	12.23	.034	0.91	.015	0.50	.039	3		1696,3062
173495	+5 03941	HR 7048		★ AB	18 43 01	+05 26.8	3	5.83	.012	0.04	.005	0.03	.008	8	A1 V + A1 V	15,603,1071
229377	+12 03640				18 43 01	+12 11.4	1	10.22		0.23				1	A2	851
173648	+37 03222	HR 7056		★ A	18 43 03	+37 33.1	8	4.36	.014	0.19	.010	0.16	.011	25	Am	1,15,401,1028,1084*
173476	−5 04749				18 43 04	−05 13.0	1	10.21		0.84		0.50		2	K0	1696
173425	−19 05154	HR 7045			18 43 04	−19 39.6	1	6.42		1.65				4	M4 III	2007
173649	+37 03223	HR 7057		★ D	18 43 05	+37 32.5	6	5.73	.011	0.28	.009	0.06	.000	14	F0 IV	1,15,401,1028,1084,1355
173701	+43 03058				18 43 05	+43 46.8	1	7.54		0.84		0.57		3	K0	196
173263	−50 12135	HR 7037			18 43 07	−50 08.9	1	6.54		0.28				4	A9 IV	2007
		G 184 - 23			18 43 09	+16 57.3	1	13.90		0.95		0.18		3		1658
173339	−43 12805				18 43 11	−43 50.5	1	7.41		0.27				4	F0/2 V	1075
173409	−31 15954				18 43 13	−31 23.8	1	9.54		0.89				1	C	1238
173530	+4 03870				18 43 14	+04 31.5	1	8.82		0.44		-0.04		2	B9	1586
173689	+34 03302				18 43 16	+34 57.3	1	7.18		-0.02		-0.22		2	B9	401
	−8 04697				18 43 17	−08 38.6	1	9.60		2.12				2	M3	369
173460	−22 04854	HR 7046		★ A	18 43 20	−22 26.8	2	5.38	.010	1.62	.019	1.90		6	K5 III	2007,3005
173431	−32 14495				18 43 20	−31 57.7	1	7.28		1.01				4	K1 III	2012
	+7 03832	RZ Oph			18 43 21	+07 10.0	1	9.85		1.02		0.45		1	F3 Ibe	1535
173532	−6 04885				18 43 21	−06 18.4	1	7.86		1.30		1.15		5	K0	1657
	−23 14703				18 43 23	−23 17.9	1	10.28		0.34		-0.51		2		540
173650	+21 03550	HR 7058, V535 Her			18 43 28	+21 55.9	2	6.50		0.02	.008	-0.10	.009	8	B9 p Si(Cr)	1063,1202

HD	DM	Other Id	N Rem	α_{1950}	δ_{1950}	S	V	σ_V	B–V	σ_{B-V}	U–B	σ_{U-B}	n	Spectrum	References
		Steph 1644		18 43 28	+33 34.1	1	11.26		1.54		1.96		1	K7	1746
173667	+20 03926	HR 7061	⋆ A	18 43 31	+20 29.8	13	4.20	.009	0.46	.006	0.01	.011	105	F6 V	1,3,15,254,369,1077*
173667	+20 03926	IDS18414N2027	B	18 43 31	+20 29.8	1	12.69		0.53				1		3024
173667	+20 03926	IDS18414N2027	D	18 43 31	+20 29.8	1	13.92		0.69				1		3024
229413	+13 03737			18 43 34	+14 03.7	1	8.41		0.09		0.07		1	A0	401
	−35 12911			18 43 34	−35 27.7	1	11.56		0.59		0.14		3		1586
229414	+12 03643	BB Her		18 43 35	+12 17.0	1	9.79		0.89				1	G5	851
		POSS 281 # 5		18 43 35	+35 53.6	1	18.09		1.29				4		1739
172881	−73 01939	HR 7027	⋆ AB	18 43 35	−73 03.1	3	6.05	.004	0.01	.009	−0.09	.000	15	B9.5IV/V	15,1075,2038
173949	+60 01845	HR 7075	⋆ AB	18 43 43	+60 59.7	2	6.02	.024	0.97	.003	0.73		6	G7 IV	71,3016
173168	−65 03754	HR 7036		18 43 43	−65 07.9	5	5.72	.007	0.25	.008	0.10	.006	22	A8 V	15,1075,1637,1754,2038
174205	+70 01023	HR 7082		18 43 44	+70 44.4	1	6.27		1.26		1.24		2	K2	1733
	−5 04752	LS IV −04 022	A	18 43 44	−04 54.4	1	11.39		0.65		−0.11		3		845
	−5 04752	LS IV −04 022	B	18 43 44	−04 54.4	1	12.02		0.95		0.14		1		845
173502	−30 16198			18 43 44	−30 00.9	3	9.73	.023	−0.10	.007	−0.91	.011	6	B1/2 Ib	400,540,976
	−6 04891	LS IV −06 036		18 43 53	−06 06.6	2	10.51	.019	0.39	.005	−0.52		6	B1	1021,1318
173654	−1 03559	HR 7059	⋆ A	18 43 54	−01 01.0	3	5.89	.010	0.14	.008	0.11	.004	11	A2 m	355,1007,3058
173654	−1 03559	HR 7059	⋆ AB	18 43 54	−01 01.0	5	5.64	.011	0.14	.010	0.08	.000	21	A2 m	15,1013,1256,8015,8052
173654	−1 03559	IDS18413S0104	B	18 43 54	−01 01.0	1	7.51		0.31		0.08		4		3024
173654	−1 03559	IDS18413S0104	C	18 43 54	−01 01.0	1	11.31		1.72		1.20		1		3024
173920	+54 02034	HR 7071		18 43 55	+54 50.6	1	6.23		0.82				2	G5 III	71
173637	−8 04702	LS IV −07 021		18 43 55	−07 59.2	4	9.36	.012	0.24	.006	−0.69	.037	26	B1 IV	989,1032,1586,1729
173638	−10 04797	HR 7055	⋆ AB	18 43 58	−10 10.8	4	5.71	.009	0.60	.011	0.51	.057	13	F2 Ib-II	253,2007,3016,8100
	−15 05090			18 43 58	−15 02.5	1	10.19		0.35		−0.28		2		1586
173780	+26 03349	HR 7064		18 44 03	+26 36.4	7	4.83	.009	1.20	.004	1.23	.010	35	K3 III	15,369,1080,1363,3016*
	−5 04754		V	18 44 03	−04 57.7	1	10.68		1.92				2	M3	369
173673	−6 04893			18 44 05	−06 44.7	1	7.65		0.01		−0.36		2	B8	401
173656	−12 05163	LSS 5126		18 44 07	−12 20.1	2	9.38	.000	0.57	.000	0.25	.000	3	B2/3 II(n)	1012,1032
173539	−38 13089	V CrA		18 44 07	−38 12.8	2	10.66	.821	0.92	.188	0.39	.217	2	R0	1589,1699
173540	−40 12807	HR 7050		18 44 15	−40 27.7	3	5.23	.008	0.78	.000			11	G5/6 III	15,1075,2009
	−82 00743			18 44 17	−82 28.8	1	10.78		0.50		−0.15		2		1696
173694	−12 05166	LSS 5128		18 44 18	−12 11.8	3	9.82	.010	0.68	.020	−0.37	.005	4	O9.5 II	1012,1032,1737
174156	+64 01289			18 44 20	+64 45.3	1	7.28		1.02		0.82		2	K0	1733
173344	−63 04451			18 44 20	−63 19.6	1	7.39		0.18		0.14		3	A2mA5-A9	861
173725	−3 04378			18 44 21	−03 50.1	1	11.93		0.75		0.35		2	A2	1682
		G 155 - 40		18 44 21	−17 29.5	1	14.42		1.32		1.05		1		3062
	−1 03562			18 44 28	−01 53.8	1	9.58		2.25				1	M3	369
173744	−6 04897			18 44 28	−05 57.2	1	7.05		0.09		0.12		21	A0	1222
	−10 04800			18 44 29	−10 28.7	1	10.45		1.88				2	M2	369
173833	+18 03817	HR 7067		18 44 30	+18 39.1	1	6.18		1.59				2	K5	71
	+40 03481			18 44 30	+40 13.5	1	8.64		1.17		1.16		2	K2	1733
173764	−4 04582	HR 7063		18 44 31	−04 48.2	10	4.22	.007	1.10	.005	0.83	.026	39	G5 II	3,15,369,418,1075*
	+3 03799			18 44 32	+03 41.7	1	8.58		1.80				2	M2	369
173799	+9 03866			18 44 33	+09 21.8	1	7.66		1.85				2	M1	369
173869	+25 03623			18 44 34	+25 29.0	1	7.94		0.17		0.11		2	A0	105
	−37 12867	V413 CrA		18 44 34	−37 47.6	1	10.29		0.32		0.14		1		700
173936	+41 03137	HR 7073		18 44 37	+41 23.2	2	6.07		−0.13	.004	−0.47	.034	3	B6 V	985,1079
		Steph 1643		18 44 38	+24 21.5	1	11.13		1.69		2.05		2	M1	1746
229463	+12 03652			18 44 39	+12 14.3	1	9.66		0.65				2	G5	851
173560	−50 12149			18 44 39	−50 45.4	3	8.70	.004	0.64	.010	0.11	.003	27	G3/5 V	158,1621,2012
173834	+12 03653			18 44 40	+12 27.7	1	8.14		0.07		−0.13		2	A0	963
173783	−9 04835	LS IV −09 028		18 44 40	−09 21.8	2	9.32	.005	0.52	.005	−0.52		2	O9 I	138,1011
		Smethells 21		18 44 42	−62 05.	1	10.70		1.44				1		1494
		G 155 - 41		18 44 43	−17 06.6	1	13.63		1.38				1		3062
173817	+2 03683			18 44 44	+02 32.8	1	8.63		0.15		−0.26		2	B8	1586
229473	+16 03610			18 44 45	+16 58.5	1	9.64		0.72		0.28		13	K5	1308
	−3 04379			18 44 45	−03 49.5	1	10.95		0.62		0.01		2		1682
173659	−37 12872			18 44 46	−37 38.4	1	9.23		0.31		0.09		25	F0 V	1699
173880	+18 03823	HR 7069	⋆ A	18 44 49	+18 07.5	5	4.36	.006	0.13	.012	0.07	.006	15	A5 III	15,355,1363,3023,8015
173819	−5 04760	HR 7066, R Sct		18 44 49	−05 45.6	4	5.16	.035	1.45	.019	1.53	.078	4	G0 Iae	15,793,3001,8015
173818	−3 04380	G 21 - 24		18 44 50	−03 41.5	4	8.80	.015	1.29	.011	1.21	.017	9	K5	158,202,1620,3072
	+5 03950	DR Ser	A	18 44 53	+05 24.1	1	9.60		3.79				1	C6 II	414
	+5 03950		B	18 44 53	+05 24.1	1	11.81		0.73		0.49		2		414
173950	+38 03292	IDS18432N3815	AB	18 44 53	+38 17.8	1	8.10		0.82		0.38		2	G9 V +K1 V	3030
173820	−6 04903	LS IV −06 038		18 44 53	−06 21.9	4	9.93	.027	0.29	.016	−0.73	.009	9	O8	1011,1021,1032,1318
		MWC 960		18 44 58	−20 09.2	1	12.09		1.50		−0.30		1	M0	1753
		3C 390 3 # 4		18 45 00	+79 44.9	1	14.65		0.75		0.28		2		327
173822	−16 05033	IDS18422S1653	A	18 45 06	−16 50.0	1	6.95		1.73				14	K2/3 III	6011
173789	−24 14715			18 45 06	−24 49.7	1	8.90		1.54		1.10		5	A3 III/IV	3078
	−17 05319			18 45 07	−16 56.5	1	9.68		0.90				13	G0	6011
173770	−32 14530	IDS18419S3158	AB	18 45 08	−31 55.3	2	7.58	.005	−0.04	.008	−0.46	.006	4	B6 V	540,976
173697	−45 12772			18 45 10	−45 19.1	1	7.26		0.99				4	K0 III	2012
		G 155 - 42		18 45 11	−14 38.1	1	12.13		1.51		1.18		1		3062
	+44 02989			18 45 11	+44 37.8	1	9.45		0.90		0.53		2	G5	1733
173715	−43 12841	HR 7062		18 45 14	−43 44.2	3	5.48	.004	0.13	.005			11	A3 V	15,2018,2030
	+28 03078	IDS18433N2819	A	18 45 16	+28 22.3	1	9.10		0.63		0.12		3	G0 V	3026
	+28 03078	IDS18433N2819	B	18 45 16	+28 22.3	1	9.87		0.62		0.04		2		3032
173854	−19 05168			18 45 17	−19 15.4	1	6.85		1.68		1.49		1	K4/5 III	3040
	+28 03079			18 45 18	+28 20.9	1	9.69		1.19		1.10		2	K0	3032
173790	−32 14532			18 45 18	−32 13.9	1	9.69		1.60		1.14		5	A2 IV	3078

Table 1 1025

HD	DM	Other Id	N	Rem	α_{1950}	δ_{1950}	S	V	σ_V	B–V	σ_{B-V}	U–B	σ_{U-B}	n	Spectrum	References
174022	+31 03365	IDS18434N3118	AB		18 45 19	+31 21.0	1	7.22		0.89		0.55		3	G8 II	8100
174063	+35 03361				18 45 21	+35 49.2	1	7.48		0.09		0.05		2	A0	1601
		G 229 - 16			18 45 25	+51 44.5	1	13.75		1.29				1		1759
174064	+33 03205				18 45 26	+34 02.5	1	7.03		1.16				1	K0	8097
		3C 390 3 # 2			18 45 31	+79 44.4	1	14.28		0.76		0.13		3		327
173954	+4 03884	HR 7076			18 45 34	+04 11.1	1	6.20		1.51		1.78		4	K5	1149
174177	+46 02551	HR 7080			18 45 34	+46 15.6	1	6.52		0.07		0.17		2	A2 IV	252
174237	+52 02280	HR 7084, CX Dra			18 45 36	+52 55.9	2	5.88	.009	-0.09	.005	-0.74	.000	6	B2.5Ve	154,1212
	−8 04711				18 45 39	−08 28.3	1	10.25		1.84				2	M3	369
	−5 04767	G 21 - 25			18 45 42	−05 08.5	4	10.34	.011	0.71	.021	0.14	.023	8		202,1620,1658,1696
174104	+28 03085				18 45 46	+28 39.8	1	8.30		0.73		0.30		2	G0 Ib	8100
173791	−45 12779	HR 7065			18 45 46	−45 52.1	2	5.80	.005	0.90	.000			7	G8 III	15,2012
174126	+28 03086				18 45 47	+28 35.2	1	7.59		1.29		1.51		1	K2 II	8100
173938	−12 05173	LS IV -12 104			18 45 48	−12 21.8	1	10.14		0.36		-0.21		3	A1 Ia/b	1586
		Smethells 22			18 45 48	−62 07.	1	10.74		1.48				1		1494
	−5 04769	LS IV -05 028			18 45 50	−05 34.5	2	10.40	.000	0.75	.000	-0.35	.000	3	O8 I:	1011,1032
174043	+10 03663	IDS18435N1032	AB		18 45 52	+10 35.2	1	8.09		1.07		0.80		2	G5	1733
		3C 390 3 # 1			18 45 52	+79 42.5	1	11.71		1.03		0.84		4		327
173987	−6 04910	LS IV -06 039			18 45 52	−06 31.3	1	8.92		0.36		-0.64		1	B0.5Iab	1032
173988	−7 04721				18 45 54	−07 11.4	1	8.31		1.74				2	K5	369
		MCC 189			18 45 58	+75 56.1	1	10.71		1.38				1	M0	1017
173902	−34 13128	HR 7070			18 45 58	−34 48.3	1	6.62		1.08				4	K2 IV	2009
174005	−6 04913	IDS18433S0607	A		18 45 59	−06 03.7	2	6.52	.000	0.25	.000	0.15	.002	3	A2	683,3016
174005	−6 04912	IDS18433S0607	B		18 45 59	−06 03.7	1	9.61		0.51		0.03		2		3016
173861	−43 12854	HR 7068			18 45 59	−43 29.5	1	5.61		-0.08				4	B9 IV	2009
174179	+31 03369	HR 7081			18 46 04	+31 42.0	1	6.05		-0.13		-0.67		3	B3 IVp	154
174080	+10 03665	G 141 - 37		⋆ A	18 46 07	+10 41.7	3	7.96	.020	1.07	.010	0.98	.005	10	K0	1197,1355,3077
174066	+0 04023				18 46 08	+00 28.0	1	8.55		1.12		1.03		3	K2	3016
		G 22 - 1			18 46 10	−02 36.8	3	13.55	.011	1.57	.008	1.12	.005	10		202,316,3078
174160	+23 03461	HR 7079		⋆	18 46 11	+23 27.4	2	6.12	.030	0.50	.009	0.02		4	F8 V	71,8100
174069	−8 04714	LS IV -08 026		⋆ A	18 46 15	−08 31.0	1	7.79		0.12		-0.54		3	B1.5V	399
	−8 04715			V	18 46 15	−08 31.0	1	10.24		1.88				2	M2	369
	+2 03691				18 46 19	+02 21.9	1	9.45		2.17				2	M3	369
		G 22 - 2			18 46 19	+03 24.3	2	13.03	.005	1.51	.005	1.17		3		202,333,1620
174260	+36 03270				18 46 19	+37 01.0	1	7.34		-0.09		-0.47		2	B8	401
		LS IV +00 003		V	18 46 21	+00 31.1	2	11.45	.040	-0.07	.016	-0.99	.014	219		405,866
174366	+48 02767	HR 7090		⋆	18 46 21	+49 01.1	1	6.67		0.06		0.01		2	A1 V	1733
174413	+55 02115				18 46 23	+55 43.7	1	8.40		0.00		-0.01		2	A0	1502
174223	+19 03796				18 46 29	+19 45.1	2	8.08	.010	0.00	.000	-0.55	.005	3	B9	963,1733
174073	−19 05179	LSS 5129			18 46 33	−19 04.0	2	9.31	.004	0.08	.001	-0.63	.019	3	B2/3 III	540,976
174089	−16 05041	YZ Sgr			18 46 35	−16 46.9	4	7.02	.012	0.88	.008	0.57	.030	4	G2 Ib	657,1484,1772,6011
174296	+33 03209				18 46 39	+33 42.9	1	7.05		1.37				2	K0	8097
174115	−19 05182	HR 7077			18 46 39	−19 12.0	7	6.76	.015	0.19	.009	0.13	.011	20	A1mA5-F2	15,355,657,1075,2007*
229590	+17 03729	G 184 - 29			18 46 40	+17 23.2	3	9.20	.019	1.29	.025	1.22	.005	9	M0	1013,1197,3072
	−23 14742				18 46 41	−23 52.6	1	10.44		1.75				1		1705
	−62 05977				18 46 41	−62 53.2	1	9.79		0.22		0.18		2	A3	861
174262	+19 03798	HR 7086			18 46 42	+19 16.3	2	5.88		0.03	.005	0.02	.005	4	A1 V	1063,3050
174116	−20 05277	HR 7078		⋆ A	18 46 42	−20 23.0	1	5.24		1.41				4	K2 III	2035
		Ross 154		⋆ V	18 46 45	−23 53.2	3	10.46	.018	1.74	.027			7	M3	158,1619,1746,3045
172882	−80 00858				18 46 45	−80 47.4	1	6.50		0.13		0.01		8	A0 V	1628
174261	+21 03560	IDS18447N2103	AB		18 46 49	+21 06.6	1	7.13		0.01		-0.48		3	B3 III+B5 V	555
174134	−20 05278				18 46 49	−20 04.1	1	8.48		1.14				22	G8 III	6011
	−2 04752				18 46 50	−02 24.6	2	10.41	.029	1.60	.019	0.60		3		725,914
174298	+23 03465				18 46 51	+23 59.9	2	6.55	.010	-0.07	.010	-0.80	.010	6	B1.5IV	399,555
174347	+32 03221				18 46 51	+32 26.3	1	8.55		0.11		0.20		2	A2	1375
173813	−62 05979				18 46 51	−62 42.3	1	7.61		0.12		0.09		2	A2 V	861
174732	+67 01096				18 46 52	+67 42.9	1	7.15		-0.03		-0.10		2	A0	1375
	−10 04821	RT Sct			18 46 52	−10 26.8	1	9.80		1.92				3	M2	369
174058	−33 13611				18 46 52	−33 39.3	1	8.11		0.44		-0.03		2	F5 V	1730
174074	−33 13612				18 46 55	−33 34.9	1	8.74		0.47		0.03		2	F3 V	1730
173994	−47 12565				18 46 55	−47 50.3	2	7.06	.006	-0.15	.001	-0.72	.005	6	B2 IV	55,1732
174481	+48 02770	HR 7096			18 46 57	+48 42.6	3	6.12	.005	0.21	.005	0.08	.000	6	A7 III	985,1501,1733
	+10 03674			V	18 47 00	+10 12.3	1	9.92		1.83				2	M2	369
174208	−6 04922	HR 7083		⋆ A	18 47 00	−05 58.3	3	5.96	.023	1.63	.013	1.69	.031	37	K2 Ib	542,1088,1222
174208	−6 04923	IDS18443S0602	C		18 47 00	−05 58.3	1	8.14		1.05		0.75		1	K0 II-III	542
173830	−63 04453				18 47 01	−63 53.6	1	8.41		0.18		0.14		2	A3 V	861
174980	+73 00835	HR 7117			18 47 02	+74 01.7	2	5.34	.100	0.95	.030	0.85	.050	4	K0 II-III	3016,8100
		LP 751 - 16			18 47 03	−10 01.2	1	14.20		0.23				1		940
174240	+0 04027	HR 7085			18 47 04	+00 46.7	1	6.24		0.04		0.01		4	A1 V	1149
		S396 # 4			18 47 07	−33 28.3	1	11.09		0.55		0.09		2		1730
174504	+45 02777				18 47 08	+45 12.2	1	6.91		0.34		-0.03		3	F0	196
174225	−8 04721				18 47 08	−08 15.1	1	8.54		1.70				2	K5	369
174369	+24 03545	HR 7091			18 47 10	+24 59.3	1	6.74		0.08		0.07		2	A1 V	1733
	−4 04593	LS IV -04 025			18 47 11	−04 23.1	1	10.51		0.95		-0.15		2		405
174433	+34 03326				18 47 12	+34 28.8	1	8.90		0.15		0.09		1	F5	1620
		S396 # 3			18 47 13	−33 28.0	1	10.14		1.16		1.09		2		1730
	+78 00654				18 47 14	+78 19.7	1	9.04		0.53		0.04		2	F8	1733
229635	+13 03769	G 141 - 39			18 47 21	+13 09.8	1	8.59		0.73		0.21		1	G5	3060
	+45 02778				18 47 21	+45 17.1	1	10.45		0.96		0.81		3		196
		HA 38 # 379			18 47 22	+45 14.4	1	11.41		0.43		0.01		3		196

HD	DM	Other Id	N	Rem	α_{1950}	δ_{1950}	S	V	σ_V	B–V	σ_{B-V}	U–B	σ_{U-B}	n	Spectrum	References
174283	−8 04723				18 47 22	−08 26.7	1	8.69		1.91				2	K5	369
174350	+6 03951				18 47 24	+06 48.3	1	7.89		1.16		1.04		2	K1 III	37
174281	−6 04925				18 47 24	−06 28.5	1	8.74		1.28				2	K5	1726
174391	+15 03583				18 47 27	+15 52.6	1	6.67		−0.02		−0.60		3	B3 V	399
174435	+23 03469				18 47 28	+23 50.3	1	8.43		0.55				35	F8	8097
174152	−41 13159	IDS18440S4111	AB		18 47 28	−41 07.3	1	6.43		−0.09				4	B5/7 III	2012
174076	−51 11792				18 47 28	−51 34.8	1	7.42		0.94		0.59		5	G8/K0 III	1673
		G 22 - 3			18 47 31	+03 02.0	4	10.73	.011	1.44	.027			6		202,1017,1619,1705
174323	−3 04388				18 47 31	−03 40.8	1	6.95		1.51				1	K2	2017
		G 141 - 41			18 47 33	+11 15.3	2	13.71	.010	1.39	.005	1.12		3		203,1759
173948	−62 05983	HR 7074, λ Pav	★ A		18 47 35	−62 14.9	9	4.21	.016	−0.15	.009	−0.88	.007	47	B2 II-IIIe	15,26,681,815,1075*
174325	−8 04726	HR 7089, S Sct	★ A		18 47 37	−07 58.0	3	6.82	.070	3.13	.105	4.04		7	C6,4	109,864,1149
174415	+10 03678				18 47 39	+10 33.4	1	8.40		0.02		−0.44		2	A0	963
174600	+45 02779				18 47 39	+45 15.8	1	8.56		1.28		1.38		3	K0	196
		G 141 - 42			18 47 40	+04 04.8	2	15.79	.009	1.65	.072	1.03		5		538,1759
174328	−13 05119				18 47 42	−13 37.9	1	6.48		1.71		1.73		5	K1 II/III	1628
174153	−44 12917				18 47 42	−44 32.0	2	7.56	.015	0.53	.000	−0.03		6	F7 V	2012,3037
		Cr 394 - 75			18 47 45	−20 19.5	1	11.11		0.59		0.05		3		1572
174307	−20 05283				18 47 45	−20 20.0	1	10.36		0.26		−0.06		3	B8/9 II	1572
174621	+43 03085				18 47 49	+43 40.5	1	6.58		0.81				8	G8 IV	8097
174567	+31 03373	HR 7098			18 47 50	+31 34.2	2			0.03	.024	−0.09	.024	3	A0 Vs	1063,1079
174309	−22 04881	HR 7088	★ A		18 47 50	−22 13.3	3	6.31	.078	0.39	.009	0.14	.107	7	F2 IV	254,2007,3013
	+0 04030				18 47 51	+00 43.7	2	10.42	.020	2.48	.000	2.42		5	M3	369,8032
349425	+20 03950	AD Her	★ A		18 47 51	+20 39.7	1	9.70		0.40		0.31		2	G5	3032
349425	+20 03950	IDS18457N2036	B		18 47 51	+20 39.7	1	11.97		1.18		1.13		3		3032
		G 155 - 43			18 47 53	−16 25.6	1	13.72		0.72		0.01		2		3056
174585	+32 03227	HR 7100	★ A		18 47 54	+32 45.2	3	5.92	.022	−0.17	.013	−0.72	.005	5	B3 IV	154,542,1224
174585	+32 03227	IDS18460N3242	B		18 47 54	+32 45.2	1	10.89		1.05		0.76		1		542
		G 184 - 32			18 47 57	+28 02.3	1	12.06		0.50		−0.12		2		1658
174419	−5 04779				18 47 57	−05 49.8	1	10.17		0.28		0.10		2	A0	1586
		NGC 6704 - 5			18 47 58	−05 16.2	1	13.24		0.64		0.15		4		712
		NGC 6705 - 1816			18 47 58	−06 15.7	1	14.06		0.82		0.44		1		133
174395	−7 04739				18 47 58	−07 53.9	2	8.15	.003	0.04	.005	−0.31	.005	6	B9	963,1732
		NGC 6705 - 1583			18 47 59	−06 21.1	1	13.44		1.45				1		1782
174487	+7 03862				18 48 00	+07 24.0	1	6.86		1.32		1.54		2	K4 III-IVp	37
175286	+75 00682	HR 7124			18 48 00	+75 22.6	1	5.35		0.05		0.04		2	A1 Vn	3023
173997	−63 04456				18 48 00	−63 47.0	1	9.68		0.27		0.15		2	A5/9	861
174602	+32 03228	HR 7102	★ A		18 48 01	+32 29.5	5	5.25	.012	0.09	.011	0.11	.022	9	A3 V	401,1063,1224,1363,3052
		NGC 6704 - 81			18 48 01	−05 18.4	1	13.66		2.64				1		712
		NGC 6705 - 1748			18 48 01	−06 20.3	2	12.70	.060	0.41	.005	0.17		2		133,1782
	+35 03370				18 48 02	+35 42.1	1	9.51		−0.16		−0.85		4		963
174383	−20 05286	BB Sgr			18 48 02	−20 21.3	6	6.66	.017	0.85	.018	0.57	.027	6	G0 Ib	245,657,688,1490,1572*
		NGC 6705 - 1717			18 48 03	−06 16.6	1	15.11		0.50		0.29		1		133
		NGC 6704 - 44			18 48 05	−05 17.5	1	13.18		0.83		0.80		4		712
		NGC 6705 - 1664			18 48 05	−06 15.6	2	11.64	.049	1.50	.005	1.40	.007	3		133,3070
		NGC 6705 - 1627			18 48 06	−06 17.5	1	13.43		1.33		1.02		1		133
		NGC 6705 - 1625			18 48 06	−06 20.9	2	11.68	.068	1.69	.000	1.93	.027	3		133,3070
174402	−20 05287				18 48 06	−20 12.6	1	9.27		1.25		1.09		1	K0/1 (III)	1572
174586	+29 03361	IDS18462N2942	A		18 48 07	+29 45.4	1	7.63		−0.02		−0.10		2	B8	555
		NGC 6705 - 1634			18 48 07	−06 13.0	1	14.36		1.65				1		133
		NGC 6705 - 1617			18 48 07	−06 19.2	2	14.07	.035	0.43	.030	0.32		2		133,1782
		NGC 6705 - 1610			18 48 09	−06 19.5	1	13.43		1.40		1.36		1		133
174403	−20 05288	V4088 Sgr			18 48 09	−20 21.6	3	7.50	.025	0.15	.019	−0.27	.020	35	B8 II/III	245,634,1490,1572
	+33 03222	IDS18464N3315	E		18 48 10	+33 19.0	1	9.81		0.33		0.20		4	G5	1246
		NGC 6704 - 80			18 48 10	−05 16.8	1	14.40		1.84		1.33		2		712
		NGC 6705 - 1549			18 48 10	−06 17.6	1	13.11		0.28		0.08		1		133
		NGC 6705 - 1556			18 48 10	−06 19.1	2	11.95	.019	0.33	.000	0.18		3	A3	133,1782
		NGC 6705 - 1554			18 48 10	−06 19.6	1	14.17		0.39				1		1782
		NGC 6705 - 1566			18 48 10	−06 20.3	1	12.69		0.39				1	A3	1782
		NGC 6704 - 39			18 48 11	−05 17.7	1	11.35		2.04		2.15		4		712
		NGC 6705 - 1538			18 48 11	−06 19.4	1	15.00		0.56				1		1782
		NGC 6704 - 79			18 48 12	−05 16.5	1	14.40		1.14		0.65		3		712
		NGC 6705 - 1523			18 48 12	−06 19.2	1	13.98		0.42				1		1782
174360	−30 16323				18 48 12	−30 10.7	1	8.10		1.09				4	K1 III/IV	2012
229680		AP Her			18 48 13	+15 52.7	1	10.46		0.63				1	F2Ib-II	934
		NGC 6705 - 1487			18 48 13	−06 15.5	1	15.18		0.44		0.35		1		133
		NGC 6705 - 1466			18 48 13	−06 18.5	1	11.81		0.32				1	A0	1782
		NGC 6705 - 1474			18 48 13	−06 20.4	1	11.45		0.36				1	A1	1782
174464	−9 04859	HR 7094	★		18 48 13	−09 50.0	2	5.83	.010	0.60	.007	0.38	.066	11	F2 Ib	253,1088
	−9 04858	LS IV -09 030			18 48 13	−09 53.2	1	8.95		0.20				2		1586
174638	+33 03223	HR 7106, β Lyr	★ A		18 48 14	+33 18.2	4	3.42	.045	0.01	.006	−0.56	.010	13	B7 Ve +	15,154,1363,8015
174639	+33 03223	IDS18464N3315	C		18 48 14	+33 18.2	1	13.43		1.22		1.05		2	B2	1246
174638	+33 03223	IDS18464N3315	D		18 48 14	+33 18.2	1	15.15		0.79		0.18		2		1246
		NGC 6705 - 1461			18 48 14	−06 16.1	1	11.64		0.38		0.23		3	A2	133
		NGC 6705 - 1439			18 48 14	−06 19.4	1	11.78		0.28				1	B9 V	1782
		NGC 6705 - 1446			18 48 14	−06 19.9	2	11.84	.037	1.52	.079	1.35		2		1782,3070
174439	−13 05123				18 48 14	−13 30.4	1	8.35		1.73		1.95		4	K5 III	7009
174233	−49 12378				18 48 14	−49 43.5	1	8.61		0.40		−0.04		27	F2 V	978
		NGC 6705 - 1429			18 48 15	−06 15.2	1	14.58		0.52		0.32		1		133
		NGC 6705 - 1423			18 48 15	−06 21.8	1	11.44		1.61		1.74		4		3070

Table 1

HD	DM	Other Id	N	Rem	α_{1950}	δ_{1950}	S	V	σ_V	B–V	σ_{B-V}	U–B	σ_{U-B}	n	Spectrum	References
		NGC 6705 - 1434			18 48 15	−06 22.1	1	11.12		1.96		2.42		2	M3 II-III	3070
174664	+33 03224	IDS18464N3315		B	18 48 16	+33 17.6	1	7.22		-0.08		-0.49		3	B7 V	1246
	+33 03225	IDS18464N3315		F	18 48 16	+33 19.6	1	10.16		0.32		0.07		4	G5	1246
		NGC 6705 - 1339			18 48 17	−06 14.8	1	15.76		1.59		1.79		1		133
		NGC 6705 - 1364			18 48 17	−06 17.4	1	11.90		1.24		0.95		2		3070
		BD Her			18 48 18	+16 28.2	1	12.09		0.44		0.19		1	F2	699
		Cr 394 - 73			18 48 18	−20 22.9	1	12.47		1.25		0.99		1		1572
229684	+10 03682	NGC 6709 - 413			18 48 19	+10 22.5	1	9.85		0.12		-0.23		1	B6 III	49
229687	+15 03586				18 48 19	+15 52.5	2	9.30	.018	1.02	.023	0.64		13	K0	934,1308
		NGC 6705 - 1286			18 48 19	−06 16.3	2	11.86	.045	1.40	.000	1.15	.005	4		133,3070
		NGC 6705 - 1256			18 48 19	−06 20.6	1	11.63		1.73		1.68		1		3070
		NGC 6705 - 1184			18 48 20	−06 13.0	1	11.46		1.56		1.48		1		3070
		NGC 6705 - 1187			18 48 21	−06 14.4	1	14.25		0.42		0.32		1		133
		NGC 6705 - 1143			18 48 22	−06 16.0	1	12.36		0.42		0.25		2	A2	133
		NGC 6705 - 1158			18 48 22	−06 23.0	1	12.84		0.38				1		1782
		LP 751 - 19			18 48 22	−11 51.8	1	10.42		0.79		0.13		2		1696
174571	+8 03866	IDS18460N0835		AB	18 48 23	+08 38.6	3	8.87	.065	0.57	.028	-0.29	.018	6	B3 V:pe	445,1012,1212
		NGC 6705 - 1090			18 48 23	−06 24.3	1	11.94		1.56		1.54		2		3070
		Cr 394 - 72			18 48 23	−20 22.9	1	11.59		1.37		1.30		2		1572
174569	+10 03685	HR 7099		⋆ A	18 48 24	+10 55.0	1	6.36		1.44		1.64		2	K5 III	1733
		NGC 6705 - 2065			18 48 24	−06 20.	1	15.81		0.83		0.24		2		133
		NGC 6705 - 2066			18 48 24	−06 20.	1	15.90		2.00				1		133
		NGC 6705 - 2067			18 48 24	−06 20.	1	16.01		1.65		1.84		1		133
		NGC 6705 - 2068			18 48 24	−06 20.	1	15.45		0.75		0.33		1		133
		Cr 394 - 71			18 48 24	−20 22.5	1	12.56		0.60		0.31		3		1572
		NGC 6705 - 963			18 48 26	−06 16.3	2	11.77	.040	1.51	.005	1.43	.000	5		133,3070
	−20 05290				18 48 26	−20 16.8	1	10.87		0.50		0.33		3		1572
175305	+74 00792	G 259 - 35			18 48 27	+74 40.0	6	7.18	.008	0.75	.009	0.15	.009	10	G5 III	908,979,1003,1064*
		NGC 6705 - 898			18 48 27	−06 21.4	1	11.67		1.73				1		1782
174513	−7 04740	LS IV -07 027			18 48 27	−07 51.5	2	8.74	.009	0.00	.002	-0.76	.008	3	B1 V:pen	963,1586
		NGC 6705 - 887			18 48 28	−06 16.2	1	15.21		1.60		1.70		1		133
		NGC 6705 - 878			18 48 28	−06 22.6	1	12.84		0.38				1		1782
		NGC 6705 - 827			18 48 29	−06 19.4	1	11.45		1.50		1.43		2	K2 II-III	3070
		NGC 6705 - 779			18 48 30	−06 18.2	1	11.52		1.74		2.02		1		3070
		NGC 6705 - 800			18 48 30	−06 21.3	1	11.85		0.40				1		1782
		NGC 6705 - 788			18 48 30	−06 23.0	1	11.97		0.46				1		1782
		NGC 6705 - 793			18 48 30	−06 23.1	2	10.84	.030	0.62	.010	0.12		2		133,1782
174492	−20 05291				18 48 30	−20 30.0	1	9.93		0.17		-0.28		5	B5/7 III	1572
		L 207 - 33			18 48 30	−57 10.	1	12.17		1.46		1.13		1		3073
	+37 03250				18 48 31	+37 57.2	1	10.36		0.64		0.12		2	F8	1375
		NGC 6705 - 745			18 48 31	−06 22.6	1	13.92		1.34				1		1782
		NGC 6705 - 761			18 48 31	−06 22.6	1	13.88		0.39				1		1782
		NGC 6705 - 756			18 48 31	−06 23.0	1	12.86		0.38				1		1782
		Cr 394 - 70			18 48 31	−20 17.6	1	12.25		0.33		0.25		3		1572
		NGC 6705 - 732			18 48 32	−06 16.3	2	15.06	.005	0.64	.005	0.43		3		133,1782
		NGC 6705 - 720			18 48 32	−06 22.3	2	11.95	.034	0.33	.010	0.16		3	A3	133,1782
		NGC 6705 - 689			18 48 33	−06 16.6	2	12.88	.015	0.40	.000	0.19		3		133,1782
174603	+10 03687				18 48 34	+10 50.0	1	8.73		1.71				3	M1	369
		NGC 6705 - 675			18 48 34	−06 17.2	2	14.94	.015	1.58	.015	1.81		2		133,1782
		NGC 6705 - 686			18 48 34	−06 20.5	1	11.90		1.47		1.34		2		3070
		NGC 6705 - 672			18 48 34	−06 21.2	2	12.00	.034	0.34	.000	0.04		3	A0	133,1782
		NGC 6705 - 669			18 48 34	−06 22.2	2	11.99	.033	1.54	.033	1.50		4		1782,3070
174491		CpD -20 0723			18 48 34	−20 08.3	1	10.36		0.11		-0.23		3	B8/9 III	1572
		Cr 394 - 68			18 48 34	−20 25.5	1	11.17		1.48		1.41		1		1572
		NGC 6705 - 663			18 48 35	−06 18.7	1	12.04		0.42		0.12		1		133
		NGC 6705 - 660			18 48 35	−06 21.9	2	11.58	.151	1.50	.002	1.47		3		1782,3070
		NGC 6705 - 628			18 48 36	−06 15.7	2	11.68	.039	0.44	.015	0.25		3	A0	133,1782
		NGC 6705 - 617			18 48 36	−06 17.3	2	14.80	.058	2.40	.034	1.96		3		133,1782
		NGC 6705 - 615			18 48 36	−06 17.5	2	14.56	.015	0.62	.035	0.41		2		133,1782
174553	−6 04932	NGC 6705 - 624		⋆ V	18 48 36	−06 24.8	2	9.35	.023	0.50	.008	0.20	.004	11	F8	133,683
229700	+10 03689	NGC 6709 - 372			18 48 37	+10 14.5	2	10.31	.010	0.18	.004	-0.19	.004	6	A0 V	49,1655
		NGC 6705 - 593			18 48 37	−06 18.0	1	13.07		0.38				1		1782
	−6 04933	NGC 6705 - 609			18 48 37	−06 24.1	1	9.21		1.25		1.10		1	K0	133
		NGC 6705 - 574			18 48 38	−06 16.6	1	11.95		0.43				1	A3	1782
		NGC 6705 - 565			18 48 38	−06 17.8	1	13.06		1.60				1		1782
	−5 04783				18 48 39	−05 01.9	1	9.85		2.21				2	M2	369
		NGC 6709 - 347			18 48 41	+10 23.5	1	9.82		0.32		0.15		1		49
		NGC 6705 - 503			18 48 41	−06 17.9	1	12.02		1.46				1		1782
		NGC 6709 - 339			18 48 42	+10 17.8	1	11.61		0.16		-0.10		1	B9.5V	49
		NGC 6705 - 476			18 48 42	−06 17.5	2	14.58		0.56	.030	0.41		2		133,1782
		NGC 6705 - 475			18 48 42	−06 22.1	1	14.38		0.55		0.33		1		133
	−20 05293				18 48 42	−20 35.0	1	9.86		1.35		1.19		1		1572
174295	−52 11268	HR 7087			18 48 42	−52 10.0	1	5.20		0.96				4	G8/K0 III	2007
		Smethells 23			18 48 42	−57 12.	1	12.11		1.48				1		1494
	+9 03894	LS IV +10 002			18 48 43	+10 03.5	1	10.64		0.26		-0.67		2		405
		NGC 6705 - 460			18 48 43	−06 19.0	1	14.75		0.45		0.41		1		133
		NGC 6705 - 467			18 48 43	−06 21.2	1	14.18		0.76		0.43		1		133
174386	−44 12931				18 48 43	−44 23.8	1	8.17		0.42				4	F2 V	2012
		AA65,65 # 323			18 48 44	−36 12.8	1	11.91		0.68		0.07		1		713
174387	−46 12669	HR 7092			18 48 44	−46 39.4	1	5.54		1.63				4	M0 III	2007

HD	DM	Other Id	N	Rem	α_{1950}	δ_{1950}	S	V	σ_V	B–V	σ_{B-V}	U–B	σ_{U-B}	n	Spectrum	References
229715	+10 03691	NGC 6709 - 337			18 48 45	+10 15.4	1	10.66		0.23		-0.16		1	B9 V	49
229710	+14 03664				18 48 45	+14 57.6	1	9.09		0.39		0.19		2	A3	1648
174714	+24 03552	HS Her			18 48 45	+24 39.6	1	8.56		0.03		-0.42		2	B5 III	1733
174589	−3 04392	HR 7101			18 48 45	−03 22.7	1	6.09		0.30		0.06		4	F2 III	1149
		Cr 394 - 64			18 48 45	−20 27.2	1	12.33		0.69		0.39		3		1572
		NGC 6705 - 411			18 48 46	−06 18.5	1	11.64		1.67		1.80		2		3070
174538	−20 05294	Cr 394 - 63			18 48 46	−20 15.7	1	10.37		0.18		-0.08		3	A2	1572
		Cr 394 - 62			18 48 47	−20 27.4	1	10.93		0.23		0.09		5		1572
		NGC 6709 - 293			18 48 48	+10 19.8	1	10.88		0.18		-0.08		1	A0 V	49
		NGC 6709 - 291			18 48 48	+10 21.4	1	10.42		0.24		0.17		1		49
229716	+10 03694	NGC 6709 - 303			18 48 49	+10 14.5	2	9.02	.016	1.31	.001	1.09	.006	38	K0	49,1655
		NGC 6709 - 299			18 48 49	+10 17.2	1	11.67		0.20		-0.10		1		49
174524	−27 13296				18 48 49	−27 11.9	1	9.77		0.03		-0.51		4	B3/5 II/III	400
		AA65,65 # 322			18 48 50	−36 17.3	1	11.28		0.35		0.13		1		713
174559	−21 05164				18 48 51	−20 55.3	1	8.82		0.59				18	F6 V	6011
	+15 03592				18 48 52	+15 54.7	1	9.60		1.17				1	G5	934
174666	+7 03867				18 48 53	+08 01.3	1	8.27		1.80				2	K5	369
		NGC 6709 - 513			18 48 55	+10 17.5	1	14.67		0.78		0.17		1		49
		NGC 6709 - 266			18 48 56	+10 13.7	1	11.06		0.24		-0.04		1	B9.5V	49
		NGC 6709 - 265			18 48 57	+10 12.1	1	12.80		0.86		0.42		1		49
		NGC 6709 - 514			18 48 57	+10 17.8	1	15.01		0.69		0.19		1		49
	+10 03692	NGC 6709 - 275			18 48 57	+10 18.6	1	12.72		0.47		0.26		1		49
		Cr 394 - 36			18 48 57	−20 14.8	1	12.04		0.40		0.28		3		1572
		Cr 394 - 35			18 48 57	−20 15.8	1	11.25		0.22		0.17		3		1572
174591	−16 05051				18 48 58	−16 34.3	1	8.49		1.08		0.74		5	G8 III +A	897
		NGC 6709 - 267			18 48 59	+10 13.8	1	11.71		0.28		0.03		1		49
		NGC 6709 - 272			18 48 59	+10 16.3	1	13.23		0.47		0.32		1		49
174809	+39 03551				18 49 00	+39 16.8	2	7.30	.000	0.70	.010	0.41	.010	5	G4 III	833,1601
		NGC 6709 - 246			18 49 02	+10 15.0	1	13.18		0.52		0.31		1		49
		NGC 6709 - 244			18 49 02	+10 16.5	1	14.19		0.72		0.21		1		49
		NGC 6709 - 245			18 49 03	+10 16.2	1	13.23		0.46		0.31		1		49
174594	−20 05295	Cr 394 - 61			18 49 03	−20 26.2	1	8.20		0.11		-0.38		5	A0	1572
		NGC 6709 - 515			18 49 04	+10 14.9	1	15.37		1.15		0.25		1		49
		NGC 6709 - 238			18 49 04	+10 17.4	1	13.19		0.70		0.21		1		49
		NGC 6709 - 235			18 49 04	+10 18.0	1	11.05		0.25		-0.12		1	B9 V	49
		Cr 394 - 34			18 49 04	−20 15.9	1	12.62		0.50		0.25		3		1572
	−20 05296	Cr 394 - 33			18 49 04	−20 19.1	1	10.52		0.21		-0.07		3		1572
		Cr 394 - 32			18 49 05	−20 15.1	1	12.30		0.57		0.16		3		1572
		G 22 - 4			18 49 06	+02 42.9	2	11.30	.024	1.37	.015	1.27		3		202,333,1620
174610	−20 05299				18 49 06	−20 08.1	1	9.10		0.07		-0.30		5	B8 V	1572
	−20 05298	Cr 394 - 31			18 49 06	−20 24.0	1	10.03		0.16		-0.15		5		1572
174595	−20 05297				18 49 06	−20 39.7	1	9.08		0.08		-0.42		2	B9 II	1572
		NGC 6709 - 250			18 49 08	+10 11.3	1	11.43		0.25		-0.06		1	B9.5V	49
		Cr 394 - 30			18 49 09	−20 13.6	1	11.56		0.32		0.20		3		1572
174578	−25 13495				18 49 09	−25 24.7	1	9.47		1.24				3	K0w(F)	1594
		AA65,65 # 324			18 49 09	−36 01.6	1	12.44		1.34		1.33		1		713
	+10 03697	NGC 6709 - 208		⋆ C	18 49 10	+10 15.2	2	9.09	.033	1.54	.003	1.58		6		49,1655
		NGC 6709 - 516			18 49 11	+10 17.1	1	16.78		0.76				1		49
		Cr 394 - 29			18 49 11	−20 20.3	1	12.16		0.45		0.31		3		1572
		AA65,65 # 321			18 49 11	−36 20.2	1	11.51		0.81		0.30		1		713
		NGC 6709 - 205			18 49 12	+10 13.4	1	13.96		0.73		0.31		1		49
174880	+43 03094				18 49 12	+43 53.8	1	7.04		1.16				6	K0	8097
175938	+79 00604	HR 7160			18 49 12	+79 53.1	2	6.40	.005	0.28	.005	0.04	.000	4	A8 V	985,1733
		Cr 394 - 28			18 49 12	−20 21.3	1	10.76		0.26		0.04		5		1572
		LP 867 - 19			18 49 12	−24 49.0	1	10.32		1.11		1.03		1		1696
		NGC 6709 - 201			18 49 13	+10 11.4	1	10.60		0.20		-0.26		1	B9 V	49
	−20 05300	Cr 394 - 27			18 49 14	−20 12.9	1	10.34		0.14		-0.17		3		1572
174429	−50 12190	PZ Tel			18 49 14	−50 14.4	2	8.46	.014	0.78	.008	0.31	.011	20	K0 Vp	1621,1641
174474	−48 12769	HR 7095			18 49 15	−48 25.3	2	6.18	.005	0.13	.009	0.08		8	A2 V	1311,2007
174430	−52 11273	HR 7093			18 49 15	−51 59.6	1	6.31		-0.09				4	B4 III	2007
174612	−24 14778				18 49 16	−24 04.0	1	9.69		0.14		-0.25		3	B8 Vn	1586
		Cr 394 - 25			18 49 17	−20 22.9	1	11.84		0.79		0.17		5		1572
		Cr 394 - 26			18 49 17	−20 23.5	1	11.22		0.23		0.13		5		1572
174500	−46 12676	HR 7097			18 49 17	−46 38.9	2	6.19	.005	0.03	.005	0.05		5	A1 IV/V	663,2007
174651	−20 05302	Cr 394 - 22			18 49 18	−20 15.7	1	8.47		0.11		-0.31		4	B8 IV	1572
174652	−20 05301	Cr 394 - 24			18 49 18	−20 22.7	2	9.06	.004	0.15	.009	-0.30	.000	8	B9	1572,1586
		WLS 2136 85 # 11			18 49 19	+86 20.1	1	11.66		0.56		-0.01		2		1375
	−6 04938				18 49 19	−05 58.7	1	9.43		2.24				2	M2	369
		Cr 394 - 21			18 49 19	−20 16.3	1	10.45		1.16		0.88		3		1572
		Cr 394 - 23			18 49 19	−20 19.7	1	10.86		1.76		2.05		5		1572
229762		NGC 6709 - 175			18 49 20	+10 18.6	1	10.74		0.22		-0.06		1	B9 V	49
		CX Lyr			18 49 20	+28 43.7	1	12.15		0.33		0.25		1	F4	699
	−1 03584				18 49 22	−01 46.2	1	9.89		2.55				2	M2	369
174704	−9 04868				18 49 22	−09 08.6	1	7.74		0.49		0.31		3	Fp	8071
		Cr 394 - 20			18 49 22	−20 18.6	1	12.44		0.47		0.28		3		1572
		Cr 394 - 19			18 49 22	−20 24.0	1	11.93		0.47		0.22		3		1572
174630	−26 13562	HR 7103			18 49 22	−26 42.7	1	6.29		0.94				4	G8/K0 III	2009
		NGC 6709 - 145			18 49 25	+10 10.9	1	11.14		1.17		0.92		1		49
		BS Sct			18 49 25	−06 18.3	1	11.00		0.33		-0.24		1	A7	683
		Cr 394 - 17			18 49 25	−20 19.5	1	14.26		0.45		0.24		2		1572

Table 1 1029

HD	DM	Other Id	N	Rem	α_{1950}	δ_{1950}	S	V	σ_V	B–V	σ_{B-V}	U–B	σ_{U-B}	n	Spectrum	References
		Cr 394 - 18			18 49 25	−20 24.2	1	12.58		0.64		0.15		3		1572
174701	−7 04746	IDS18467S0723		A	18 49 26	−07 19.2	1	8.00		0.03		-0.33		1	B9	963
		Cr 394 - 14			18 49 26	−20 16.9	1	11.54		0.25		0.18		3		1572
		Cr 394 - 16			18 49 26	−20 18.1	1	12.97		0.82		0.27		3		1572
		Cr 394 - 15			18 49 26	−20 20.5	1	13.15		0.63		0.24		3		1572
		Cr 394 - 13			18 49 27	−20 27.2	1	10.70		0.54		0.07		2		1572
174631	−29 15449	HR 7104			18 49 27	−29 26.4	1	6.13		1.35				4	K1 III	2009
174705	−11 04786	LS IV -11 047			18 49 29	−11 41.6	2	8.36	.040	0.26	.005	-0.52	.040	6	B2 Vne	399,1586
		Cr 394 - 11			18 49 29	−20 19.1	1	11.52		0.63		0.18		3		1572
174632	−30 16356	HR 7105			18 49 29	−30 47.7	3	6.65	.013	-0.03	.016	-0.28	.003	8	B7/8 IV	540,976,2009
		Cr 394 - 10			18 49 30	−20 24.2	1	11.11		0.22		0.13		5		
		CS 22936 # 145			18 49 30	−34 45.8	1	14.24		0.45		-0.12		1		1736
		NGC 6709 - 124			18 49 32	+10 17.5	1	11.40		0.46		0.33		1		49
		AA65,65 # 320			18 49 32	−36 17.1	1	12.13		1.13		0.89		1		713
174684	−20 05306				18 49 33	−20 02.3	1	8.30		0.07		-0.41		5	B3/5 III	1572
		Cr 394 - 7			18 49 33	−20 12.2	1	12.86		0.48		0.21		3		1572
		Cr 394 - 9			18 49 33	−20 18.5	1	12.94		1.36		1.39		2		1572
		Cr 394 - 8			18 49 34	−20 12.5	1	12.76		1.79		2.74		1		1572
174685	−20 05304	Cr 394 - 12			18 49 34	−20 21.6	1	9.52		0.10		-0.32		3	B7 III	1572
		Cr 394 - 5			18 49 35	−20 13.3	1	12.82		0.53		0.22		3		1572
174706		Cr 394 - 6			18 49 35	−20 14.1	1	10.30		0.21		0.03		3	B7 II	1572
229793					18 49 38	+16 31.7	2	10.13	.024	1.51	.046			2		1017,1705
174881	+28 03104	HR 7112			18 49 38	+28 43.4	1	6.18		1.18				2	K1 II-III	71
174912	+38 03327	G 207 - 5			18 49 42	+38 33.9	2	7.16	.010	0.54	.010	-0.04	.075	2	F8	1620,1658
		Cr 394 - 4			18 49 43	−20 17.2	1	12.39		0.49		0.29		3		1572
174853	+13 03787	HR 7109, V822 Her			18 49 44	+13 54.3	3	6.16		-0.01	.019	-0.32	.022	5	B8 Vnn	985,1063,1079
		Cr 394 - 3			18 49 44	−20 18.2	1	11.19		0.22		0.09		3		1572
		Cr 394 - 1			18 49 45	−20 17.1	1	10.26		0.21		0.10		3		1572
174723	−20 05308	Cr 394 - 2			18 49 45	−20 18.0	1	8.88		0.11		-0.39		3	B8	1572
174958	+39 03553				18 49 46	+39 15.6	1	7.27		-0.01		-0.16		2	A0	401
174796	−3 04397				18 49 48	−03 47.2	1	7.17		1.91				2	K5	369
	+8 03876				18 49 49	+08 49.6	1	10.30		2.18				2	M2	369
174709	−28 15107				18 49 50	−28 12.4	1	8.17		1.12		0.89		5	K0 III	1657
174959	+36 03295	HR 7115			18 49 51	+36 28.7	1	6.08		-0.11		-0.49		3	B6 IV	154
174756	−20 05310				18 49 55	−20 34.7	1	9.25		0.10		-0.26		2	B5/7 III	1572
		G 22 - 5			18 49 56	+03 52.4	4	13.09	.023	1.12	.017	0.83	.011	11		202,316,333,1620,3060
		NGC 6712 sq1 14			18 49 57	−08 41.7	1	12.22		0.21		0.03		1		641
174961	+31 03382				18 49 59	+31 19.9	1	8.14		0.04		0.02		2	B9	1733
174775	−19 05201				18 50 00	−19 43.8	1	8.74		0.09		-0.34		4	B5 Ve	1586
174564	−54 09142	IDS18459S5429		A	18 50 01	−54 25.1	1	9.18		1.13		1.01		2	K3 V	3072
174564	−54 09142	IDS18459S5429		B	18 50 01	−54 25.1	1	10.58		1.27		1.12		2		3072
174897	+14 03680	IDS18478N1425		A	18 50 02	+14 28.4	1	6.55		1.05		0.85		4	F7 V	3024
174897	+14 03680	IDS18478N1425		B	18 50 02	+14 28.4	1	9.80		0.55		0.04		5		3024
234727	+52 02290				18 50 03	+52 14.7	1	9.25		0.18		0.07		2	A7	1566
		NGC 6712 sq1 16			18 50 03	−08 42.0	1	12.81		0.35		0.21		1		641
		NGC 6712 sq1 15			18 50 03	−08 43.9	1	13.01		0.64		0.05		1		641
		NGC 6712 sq1 1			18 50 04	−08 46.2	1	12.44		0.50		0.11		6		641
		WLS 1840 35 # 8			18 50 05	+34 55.6	1	12.23		1.19		1.05		2		1375
		NGC 6712 sq1 2			18 50 07	−08 50.2	1	11.10		1.09		0.75		1		641
174933	+21 03582	HR 7113			18 50 08	+21 21.8	4	5.39	.090	-0.06	.025	-0.44	.018	8	B9 p Hg	1063,1079,3023,8100
		NGC 6712 sq1 13			18 50 11	−08 37.7	1	11.92		0.44		0.20		1		641
	−9 04875				18 50 12	−08 58.2	1	9.82		1.89				2	M3	369
		G 184 - 38			18 50 13	+25 15.4	1	12.38		0.79		0.32		2		1658
		AS 327			18 50 14	−24 26.6	1	12.89		1.54		-0.29		1		1753
		L 489 - 58			18 50 15	−38 39.8	1	12.70		1.53		1.12		2		3078
174866	−9 04876	HR 7110			18 50 17	−09 38.3	3	6.33	.004	0.21	.008	0.11	.009	11	A7 Vn	15,253,1071
175081	+37 03262	IDS18486N3724		A	18 50 18	+37 27.3	4	7.33	.010	-0.08	.008	-0.37	.008	11	B5 n	399,401,555,3016
		NGC 6712 sq1 4			18 50 19	−08 40.7	1	11.42		0.64		0.23		2		641
		NGC 6712 sq1 6			18 50 23	−08 38.8	1	10.58		0.56		0.01		1		641
		NGC 6712 sq1 19			18 50 24	−08 41.7	1	13.14		0.78		0.26		2		641
175036	+26 03379				18 50 26	+26 28.2	1	8.28		0.57		0.02		3	G0 V	3026
	+37 03263				18 50 26	+37 08.7	1	9.68		0.13		0.12		1	A2	3066
		NGC 6712 sq1 5			18 50 26	−08 43.6	1	11.81		0.67		0.11		1		641
		Steph 1650			18 50 27	+38 23.7	1	11.73		1.59		1.96		2	K7-M0	1746
175225	+52 02294	HR 7123			18 50 28	+52 54.6	3	5.51	.011	0.83	.008	0.51	.002	7	G9 IVa	1080,1118,4001
175306	+59 01925	HR 7125, o Dra	★	A	18 50 28	+59 19.6	5	4.65	.016	1.18	.013	1.04	.006	15	G9 IIIb	15,1080,1363,3016,8015
174886	−10 04848				18 50 29	−10 17.0	2	7.76	.020	0.13	.027	-0.40	.000	6	B8	1212,1586
		Cr 394 - 56			18 50 29	−20 02.3	1	10.63		1.33		1.52		1		1572
175132	+41 03167	HR 7118			18 50 31	+41 19.3	1			-0.09		-0.32		1	B9p Si	1079
174916	−4 04607				18 50 32	−04 47.7	1	7.40		0.39		0.17		4	Am	355
174869	−20 05314				18 50 33	−20 30.9	1	10.27		0.13		-0.07		2	B8 III/IV	1572
	+12 03697				18 50 36	+12 26.9	1	10.05		1.98				3	M3	369
		NGC 6712 sq1 3			18 50 36	−08 41.8	1	11.24		0.64		0.10		1		641
		Ste 1 - 1			18 50 39	+36 50.2	1	11.93		1.12		1.35		1		604
234731	+50 02681				18 50 39	+50 54.4	1	9.26		0.10		0.08		2	A5	1566
174730	−50 12206	HR 7108			18 50 41	−49 56.5	1	6.60		0.08				4	A2 V	2007
		LB 423?			18 50 48	+36 31.	1	10.87		-0.18		0.90		1		3066
234734	+54 02047				18 50 50	+54 50.6	1	8.81		1.30		1.33		2	K2	1733
174919	−19 05211				18 50 51	−19 51.6	1	6.95		1.45		1.51		5	K2 III	1572
175018	+3 03825	IDS18484N0316		A	18 50 54	+03 19.4	1	9.10		0.46		-0.01		4	F5	3016

HD	DM	Other Id	N Rem	α₁₉₅₀	δ₁₉₅₀	S	V	σ_V	B–V	σ_B-V	U–B	σ_U-B	n	Spectrum	References
175018	+3 03825	IDS18484N0316	B	18 50 54	+03 19.4	1	9.75		0.48		-0.02		4		3016
174945	-20 05317			18 50 54	-20 01.2	1	9.92		0.01		-0.26		2	A0	1572
	+10 03711	G 141 - 47		18 50 55	+10 33.8	2	10.54	.000	0.54	.000	-0.09	.005	2	F8	1658,3060
	-20 05318			18 50 55	-20 21.6	1	10.11		0.13		-0.20		2		1572
		FN Sgr		18 50 57	-19 03.5	1	12.94		1.08		-0.47		1		1753
174946		CpD -20 07304		18 50 59	-20 11.3	1	10.19		0.19		0.14		1	B9/A0 V	1572
	-0 03584	LS IV -00 007		18 51 00	-00 37.4	2	9.98	.000	0.62	.005	-0.45	.000	4	O8 V	1011,1012
		G 141 - 48		18 51 01	+07 29.9	1	16.13		1.43		1.13		1		3060
		NGC 6716 - 98		18 51 01	-19 58.3	1	11.38		0.71		0.24		1		1713
174947	-21 05176	HR 7114		18 51 01	-21 25.4	2	5.69	.000	1.23	.005	1.05		6	K1 Ib	2018,8100
175039	-5 04798	IDS18484S0540	AB	18 51 02	-05 36.4	1	8.78		0.72				4	G5	176
		NGC 6715 sq1 3		18 51 06	-30 33.8	1	11.29		0.46				2		438
175100	+9 03911			18 51 07	+09 35.7	1	7.32		1.87				2	M1	369
	+48 02781			18 51 08	+48 20.5	1	9.84		-0.17		-0.79		2	B8	963
		Steph 1651		18 51 09	+29 19.8	1	11.44		1.28		1.23		1	K7	1746
		NGC 6716 - 63		18 51 09	-19 55.1	1	13.65		1.12		0.72		1		1713
174974	-22 04907	HR 7116	★ AB	18 51 09	-22 48.5	7	4.83	.008	1.41	.010	1.25	.035	28	K1 II	15,285,542,1075,1088*
174974	-22 04907	IDS18481S2252	C	18 51 09	-22 48.5	1	11.21		1.25		1.20		2		542
		NGC 6716 - 65		18 51 11	-19 56.6	2	10.76	.029	0.13	.024	0.08	.015	3	B8	1572,1713
174995	-23 14813			18 51 12	-22 58.0	2	8.48	.005	0.65	.000	0.12		5	G3/6	657,2033
175290	+37 03267			18 51 15	+37 15.3	1	8.00		0.44		-0.05		1	F5 V	604
229922	+12 03701			18 51 17	+12 57.9	1	8.10		1.73				1	K5	1205
175204	+25 03654	IDS18492N2515	AB	18 51 17	+25 18.8	1	7.60		0.98		0.63		3	G5 III	833
229923	+12 03703		V	18 51 20	+12 46.6	2	9.10	.005	1.79	.040	2.01		5	M2	369,8032
181518	+87 00181			18 51 20	+87 45.8	1	8.58		0.19		0.10		6	A3	1219
		NGC 6716 - 242		18 51 20	-19 58.1	1	15.01		0.97		0.93		1		1713
175043	-20 05321	NGC 6716 - 46		18 51 21	-19 59.8	5	8.30	.032	0.03	.004	-0.40	.013	10	B7 III	161,540,976,1572,1713
175227	+24 03568	DI Her	AB	18 51 22	+24 12.9	1	8.42		0.02		-0.49		6	B5 III	588
		NGC 6716 - 71		18 51 22	-20 03.7	1	13.23		0.49		0.17		1		1713
	+69 01006	BF Dra		18 51 23	+69 49.2	1	9.82		0.49		-0.02		2	F8	1768
		WLS 1900 0 # 6		18 51 24	+00 55.8	1	10.42		0.91		0.36		2		1375
		NGC 6716 - 20		18 51 24	-19 59.2	2	13.11	.014	0.59	.059	0.03	.054	5		161,1713
		NGC 6716 - 21		18 51 24	-19 59.6	2	12.28	.042	0.66	.005	0.10	.028	4		161,1713
		NGC 6716 - 224		18 51 25	-19 57.4	1	14.48		0.59		0.24		1		1713
175174	+12 03707			18 51 26	+12 56.1	1	8.29		0.25				1	A3	1205
		NGC 6716 - 18		18 51 26	-19 56.3	1	12.23		1.58		0.10		1		1713
		NGC 6716 - 22		18 51 27	-19 59.2	2	12.66	.005	1.10	.015	0.75	.097	3		161,1713
		HR Lyr		18 51 28	+29 09.8	1	15.00		0.30		-0.60		21		866
	+36 03305			18 51 28	+36 44.1	2	10.01	.005	0.51	.018	0.06	.003	4	F7 V	604,3066
		NGC 6716 - 24		18 51 28	-20 00.3	1	11.38		1.56		1.89		1		1713
		CS 22936 # 213		18 51 28	-36 18.3	2	14.90	.000	0.19	.000	0.17	.008	2		966,1736
174874	-50 12214			18 51 28	-50 07.6	1	8.02		0.34		0.17		28	F2 III	978
		NGC 6716 - 25		18 51 29	-20 01.0	1	12.95		0.61		0.16		1		1713
		CoD -22 13408		18 51 31	-22 38.5	1	10.05		1.08		1.12		4	K4	7009
175331	+38 03336			18 51 32	+38 52.1	2	7.79	.010	-0.08	.000	-0.36	.020	4	B9 V	105,401
		NGC 6716 - 14		18 51 32	-19 56.0	1	11.21		0.12		0.07		3	A1	1572
175091	-20 05322	NGC 6716 - 3		18 51 33	-19 59.6	3	9.17	.004	0.05	.013	-0.22	.023	8	B8 V	161,1572,1713
	-20 05323	NGC 6716 - 27		18 51 33	-20 01.2	5	9.83	.014	0.02	.030	-0.27	.028	10	B9	161,540,976,1572,1713
		NGC 6716 - 13		18 51 34	-19 55.1	1	12.51		0.43		0.15		2		1572
174851	-56 09026			18 51 34	-56 29.3	1	6.79		1.61		1.93		5	K5 III	1628
	-20 05324	NGC 6716 - 12		18 51 35	-19 55.3	1	10.00		0.06		-0.10		2	B8	1572
		WLS 1824 70 # 8		18 51 36	+69 45.9	1	12.19		0.55		0.02		2		1375
		NGC 6716 - 2		18 51 36	-19 57.3	2	11.73	.024	0.28	.019	0.16	.010	3		1572,1713
		NGC 6716 - 4		18 51 36	-19 58.8	2	11.10	.019	0.20	.010	0.10	.068	3	A2	161,1713
		Ste 1 - 7		18 51 37	+36 46.1	1	12.15		1.46		0.40		2		3066
		NGC 6716 - 1		18 51 37	-19 58.0	2	12.95	.014	1.07	.014	0.89	.009	4		161,1713
		Ste 1 - 8		18 51 38	+36 48.8	1	12.51		1.12		1.04		3		3066
		WLS 1840 35 # 11		18 51 38	+37 13.9	1	10.96		1.12		1.06		2		1375
		NGC 6716 - 256		18 51 38	-19 53.9	1	15.54		0.67		-0.19		1		1713
		NGC 6716 - 8		18 51 39	-19 56.8	1	13.79		1.89				2		161
	-30 16409			18 51 40	-30 38.6	1	10.06		0.12				2		438
		Ste 1 - 9		18 51 41	+36 47.2	1	12.59		0.69		0.16		3		3066
		Ste 1 - 1004		18 51 41	+36 51.0	1	15.24		0.93		0.54		2		3066
		Ste 1 - 1006		18 51 41	+36 51.5	1	14.37		0.67		0.12		2		3066
		Ste 1 - 1005		18 51 41	+36 51.7	1	13.10		0.87		0.42		2		3066
		NGC 6716 - 9		18 51 41	-19 55.5	3	10.69	.021	0.24	.012	0.12	.047	5	B9	161,1572,1713
		NGC 6716 - 51		18 51 42	-20 02.8	1	10.56		1.66		1.69		2	K5	161
174978	-44 12967			18 51 42	-43 57.0	1	9.18		0.48				4	F5 III/IV	2012
		Ste 1 - 1003		18 51 43	+36 50.8	1	15.54		0.88		0.48		2		3066
		NGC 6716 - 36		18 51 43	-19 55.7	1	13.04		0.57		0.25		4		1572
		Ste 1 - 11		18 51 44	+36 57.9	1	12.17		0.62		0.13		2		3066
175179	-4 04617	G 22 - 6		18 51 44	-04 39.8	7	9.07	.013	0.58	.005	-0.05	.012	14	G0	202,516,742,1064,1620*
		CS 22936 # 169		18 51 44	-32 27.8	2	11.52	.000	0.23	.000	0.15	.008	2		966,1736
		Ste 1 - 1012		18 51 45	+36 51.0	1	16.00		0.68		0.13		1		3066
		NGC 6716 - 38		18 51 45	-19 55.1	1	11.79		0.34		0.16		2	A2	1572
		Ste 1 - 1013		18 51 46	+36 50.9	1	16.43		0.82		0.31		1		3066
		NGC 6716 - 37		18 51 46	-19 55.5	1	11.83		0.29		0.16		2	A1	1572
		Ste 1 - 12		18 51 47	+36 58.7	1	12.00		1.60		1.93		1		3066
		NGC 6716 - 30		18 51 47	-19 58.6	1	13.64		0.66		-0.02		1		1713
		Ste 1 - 1002		18 51 48	+36 50.6	1	14.08		0.69		0.18		2		3066

Table 1 1031

HD	DM	Other Id	N	Rem	α_{1950}	δ_{1950}	S	V	σ_V	B–V	σ_{B-V}	U–B	σ_{U-B}	n	Spectrum	References
		NGC 6716 - 35			18 51 48	−19 56.2	1	13.35		0.49		0.00		1		1713
		NGC 6716 - 31			18 51 48	−19 58.2	1	13.01		1.61		1.69		1		1713
174694	−67 03603	HR 7107, κ Pav			18 51 48	−67 18.0	4	3.94	.017	0.46	.020	0.34	.009	4	G5 Iab/b	611,688,1484,1754
		Ste 1 - 1027			18 51 49	+36 40.5	1	13.61		0.56		0.03		3		3066
		Ste 1 - 1001			18 51 49	+36 50.3	1	13.92		1.10		0.87		3		3066
		Ste 1 - 13			18 51 50	+36 41.4	1	13.26		0.76		0.23		4		3066
		Ste 1 - 14			18 51 50	+37 02.7	1	11.55		0.50		-0.01		2		3066
175141	−20 05326	NGC 6716 - 55			18 51 50	−19 55.7	4	9.24	.011	0.04	.023	-0.31	.031	7	B8 IV	161,400,1572,1713
		Ste 1 - 1028			18 51 51	+36 39.9	1	14.13		0.95		0.52		3		3066
	+36 03306				18 51 51	+36 44.9	2	9.77	.009	0.22	.005	0.09	.017	4	A5	604,3066
175156	−15 05143	HR 7119			18 51 51	−15 40.0	2	5.09	.010	0.17	.000	-0.39		6	B3 II	2007,3016
	−20 05327	NGC 6716 - 33			18 51 51	−19 57.0	1	9.92		0.05		-0.27		3	B8	540
		CS 22936 # 192			18 51 51	−36 13.9	2	11.98	.000	0.14	.000	0.31	.008	2		966,1736
229970	+12 03711				18 51 52	+12 59.8	1	8.47		1.51				1	K2	1205
		Ste 1 - 16			18 51 52	+36 48.8	1	12.86		0.67		0.17		4		3066
		Ste 1 - 15			18 51 52	+36 50.5	1	11.69		0.51		0.00		4		3066
		NGC 6716 - 329			18 51 52	−19 52.6	1	14.67		0.69		0.52		1		1713
		Ste 1 - 1010			18 51 53	+36 50.1	1	16.36		1.46		1.18		4		3066
		Ste 1 - 1030			18 51 53	+36 53.1	1	14.00		0.54		0.08		2		3066
		NGC 6716 - 275			18 51 53	−20 01.7	1	14.49		0.96		0.96		1		1713
		Ste 1 - 1014			18 51 54	+36 41.4	1	15.20		0.86		0.49		1		3066
		Ste 1 - 1007		A	18 51 54	+36 49.7	1	14.46		1.08		0.88		2		3066
		Ste 1 - 1007		B	18 51 54	+36 49.7	1	17.34		0.78		0.30		3		3066
175073	−37 12969				18 51 54	−37 33.5	2	8.00	.072	0.86	.014	0.47		5	K1 V	2012,3072
		Ste 1 - 1011			18 51 58	+36 52.0	1	13.54		1.18		1.12		2		3066
		Ste 1 - 1026			18 51 59	+36 41.9	1	13.15		0.62		0.09		4		3066
175426	+36 03307	HR 7131		⋆ A	18 51 59	+36 54.5	4	5.60	.024	-0.15	.006	-0.66	.008	21	B2.5V	154,399,604,3066
175535	+50 02686	HR 7137			18 51 59	+50 38.7	5	4.92	.008	0.90	.002	0.57	.005	28	G8 III	15,1080,1363,6001,8015
175293	+10 03720				18 52 00	+10 44.7	1	6.52		1.29		1.46		2	K2	985
		Ste 1 - 1009			18 52 00	+36 50.5	1	15.93		1.83		0.94		1		3066
	−5 04806				18 52 00	−05 06.2	1	9.34		2.38				2	M2	369
		Ste 1 - 22			18 52 01	+36 49.4	1	10.90		0.62		0.12		2		3066
		Ste 1 - 1008			18 52 01	+36 50.3	1	14.88		0.86		0.44		1		3066
		Ste 1 - 23			18 52 01	+36 59.0	1	11.70		0.54		0.03		2		3066
175188	−16 05074	UX Sgr			18 52 01	−16 35.4	2	8.32	.745	1.75	.030	1.71	.235	10	M5/6 III	897,897
		NGC 6715 sq1 2			18 52 01	−30 27.6	1	11.23		0.55				2		438
175334	+13 03807				18 52 03	+13 19.6	1	7.21		0.04				1	B9	1205
		NGC 6716 - 83			18 52 03	−19 51.2	1	11.59		0.26		0.23		1	A8	1713
175426	+36 03308	IDS18502N3650		B	18 52 04	+36 57.2	2	9.68	.011	1.28	.034	1.31	.014	5		604,3066
		LS II +18 002			18 52 06	+18 04.8	1	11.21		0.69		0.30		1	F6 II	1215
175190	−22 04915	HR 7120			18 52 06	−22 44.1	6	4.98	.010	1.33	.009	1.50	.008	18	K3 II-III	15,1256,3077,4001*
175309	+10 03721	V913 Aql			18 52 07	+10 34.1	2	7.50	.122	1.75	.005	1.51		6	M4 III	148,8032
		Ste 1 - 25			18 52 08	+36 50.0	1	12.36		1.38		1.50		2		3066
175191	−26 13595	HR 7121		⋆ A	18 52 10	−26 21.6	9	2.08	.034	-0.21	.017	-0.75	.011	46	B2.5V	3,15,26,1034,1075*
		Ste 1 - 28			18 52 11	+36 56.3	1	11.64		0.53		-0.01		2		3066
	−4 04625	LS IV -04 029			18 52 12	−04 25.3	1	10.20		0.76		-0.28		1	B1 II	1032
		NGC 6715 sq1 6			18 52 12	−30 39.0	1	11.95		1.27				2		438
175443	+27 03150	HR 7132			18 52 14	+27 50.8	1	5.62		1.35				2	K4 III	71
174877	−62 06002	HR 7111			18 52 14	−62 52.0	1	6.48		1.53				4	K3 III	2007
	+36 03310				18 52 16	+36 46.5	1	10.29		0.32		0.12		4	F2 III	3066
	+36 03311				18 52 17	+36 44.7	1	11.04		0.39		0.01		4	F2 III	3066
	+36 03312				18 52 17	+36 47.5	2	8.73	.025	1.03	.028	0.78	.012	6	K0 III	604,3066
173661	−82 00747				18 52 17	−82 37.4	1	9.00		0.47				4	F3 V	2012
175405	+20 03982				18 52 18	+20 17.7	1	7.17		1.11		0.87		2	G8 III	1648
	+36 03313				18 52 18	+36 49.2	2	9.80	.006	0.09	.029	0.13	.019	12	A0	604,3066
		Ste 1 - 34			18 52 19	+37 03.5	1	12.67		0.99		0.58		1		604
	+9 03920				18 52 20	+10 03.4	2	9.36	.000	1.85	.010	2.17		4	M3	369,8032
		NGC 6715 sq1 7			18 52 21	−30 31.1	1	14.15		1.29				2		438
		AJ81,106 # 3			18 52 23	−07 47.2	1	10.84		0.24		-0.30		2		505
175253	−19 05226				18 52 23	−19 02.5	2	8.53	.002	0.01	.007	-0.45	.001	4	B8 II	540,976
		NGC 6715 sq1 8			18 52 23	−30 32.4	1	14.65		1.34				2		438
		NGC 6715 sq1 12			18 52 23	−30 32.9	1	15.66		0.96				1		438
		Ste 1 - 40			18 52 24	+36 47.3	1	11.83		0.53		0.02		4		3066
		NGC 6715 sq1 5			18 52 24	−30 32.3	1	11.68		1.70				4		438
		Ste 1 - 1022			18 52 25	+36 59.9	1	15.48		0.74		0.20		2		3066
		Ste 1 - 42			18 52 25	+37 02.0	2	11.84	.019	0.34	.010	0.10	.023	5		604,3066
	+36 03314				18 52 27	+36 51.0	2	8.65	.005	0.08	.008	0.04	.029	3	A0	604,3066
		NGC 6715 sq1 4			18 52 27	−30 26.4	1	11.50		1.08				2		438
		CS 22936 # 201			18 52 27	−35 21.3	2	12.59	.000	-0.04	.000	-0.31	.012	2		966,1736
175538	+36 03315				18 52 28	+36 40.2	2	7.63	.001	-0.11	.010	-0.45	.007	12	B9	604,3066
		Ste 1 - 1021			18 52 28	+37 00.0	1	17.30		0.97		0.50		1		3066
175468	+19 03836				18 52 31	+19 46.7	1	7.74		0.10		0.06		2	A0	401
230017	+10 03724	IDS18502N1050		AB	18 52 32	+10 54.6	1	9.42		1.37		1.30		2	M0	3072
		Ste 1 - 1020			18 52 33	+37 01.0	1	17.35		0.72		0.03		2		3066
		AJ81,106 # 9			18 52 34	−07 45.5	1	11.97		1.32		1.13		1		505
		AJ81,106 # 5			18 52 34	−07 47.3	1	11.50		1.47		1.40		2		505
		Ste 1 - 49			18 52 35	+36 42.2	1	13.24		0.44		0.16		3		3066
		AJ81,106 # 6			18 52 35	−07 47.3	1	14.06		1.09				1		505
		Ste 1 - 1015			18 52 36	+37 00.7	1	16.42		0.70		0.14		2		3066
		Ste 1 - 50			18 52 37	+36 42.0	1	12.04		0.44		0.01		4		3066

HD	DM	Other Id	N Rem	α_{1950}	δ_{1950}	S	V	σ_V	B–V	σ_{B-V}	U–B	σ_{U-B}	n	Spectrum	References
	+36 03317			18 52 37	+36 47.3	2	8.80	.000	0.03	.009	-0.07	.020	5	A0	604,3066
		Ste 1 - 1016		18 52 37	+37 00.5	1	16.30		0.76		0.13		2		3066
175576	+41 03174	HR 7138		18 52 37	+41 09.6	1	7.33		0.46		0.03		2	F5	1733
174928	-63 04461			18 52 37	-63 26.7	1	9.08		0.00		-0.04		2	B9 V	861
175492	+22 03524	HR 7133	⋆ A	18 52 38	+22 34.8	4	4.59	.012	0.78	.004	0.50	.005	7	G4 III+A6 V	15,1363,1769,8015
		Ste 1 - 52		18 52 38	+36 43.6	1	12.47		1.11		1.00		1		3066
		Ste 1 - 1017		18 52 38	+37 00.8	1	17.18		0.08		0.12		2		3066
175317	-16 05078	HR 7126		18 52 38	-16 26.4	2	5.57	.010	0.44	.010	-0.01		7	F4 V	254,2007,3053
		Ste 1 - 1018		18 52 39	+37 00.6	1	16.42		0.73		0.16		2		3066
		AJ81,106 # 4		18 52 39	-07 49.1	1	15.36		1.22				1		505
		Ste 1 - 1019		18 52 40	+37 00.8	1	16.66		0.69		0.12		2		3066
	+42 03187	G 205 - 42		18 52 40	+42 54.4	1	9.95		0.61		-0.06		2	G2	1658
		AJ81,106 # 1		18 52 41	-07 47.5	1	12.29		0.66		0.23		2		505
		MN120,163 # 10		18 52 41	-07 47.5	1	11.26		1.74		0.82		2		375
		Ste 1 - 54		18 52 42	+36 57.0	1	10.90		0.38		0.02		2		3066
175377	-8 04764	T Sct		18 52 43	-08 15.4	1	9.29		2.59				1	C5 II	109
175219	-42 13761	HR 7122		18 52 43	-42 46.5	2	5.35	.005	1.00	.000			7	K0 III	15,2012
		AJ81,106 # 2		18 52 44	-07 47.9	1	14.75		1.18		0.66		2		505
175588	+36 03319	HR 7139, δ Lyr	⋆ A	18 52 45	+36 50.0	8	4.28	.021	1.67	.014	1.65	.016	21	M4 II	15,604,1363,3066,4001*
		AJ81,106 # 7		18 52 46	-07 48.7	1	13.03		0.67		0.41		1		505
		AJ81,106 # 8		18 52 46	-07 48.7	1	10.51		1.23		1.00		1		505
	+36 03320	IDS18515N3654	AB	18 52 49	+36 56.0	1	9.68		0.38		0.08		2		3066
		Ste 1 - 1029		18 52 52	+36 46.5	1	13.14		0.74		0.25		2		3066
175823	+57 01915	HR 7153		18 52 54	+57 25.4	1	6.22		1.23				2	K0	71
175445	-2 04784			18 52 57	-02 22.3	1	7.79		0.11		0.11		2	A0	1375
175360	-23 14844	HR 7128		18 52 59	-23 14.3	2	5.93		0.00	.007	-0.42		5	B5/7 III	1079,2007
175514	+9 03928	V1182 Aql		18 53 00	+09 16.9	2	8.59	.000	0.59	.000	-0.45	.000	4	O8:Vnn	1011,1012
		Ste 1 - 71		18 53 00	+36 37.7	1	11.53		0.60		0.08		2		3066
175515	+6 03978	HR 7135		18 53 01	+06 33.1	5	5.58	.020	1.04	.004	0.87	.009	20	K0 III	15,1118,1355,1417,3016
		Ste 1 - 1024		18 53 01	+36 44.5	1	13.76		0.54		0.04		4		3066
		LS II +17 002		18 53 02	+17 08.1	1	11.08		0.64		0.15		1	F4 III:	1215
175635	+33 03257	HR 7140	⋆ AB	18 53 02	+33 54.2	4	5.99	.021	0.92	.008	0.63	.012	8	G8 III +A2	150,401,542,3016
175634	+33 03256	IDS18512N3350	C	18 53 02	+33 55.0	4	7.68	.017	-0.06	.012	-0.29	.018	9		150,401,542,3032
175447	-6 04965	CW Sct		18 53 02	-06 02.0	1	9.87		0.23		-0.23		3	B9	1768
		CS 22936 # 218		18 53 02	-36 56.6	2	15.49	.000	0.02	.000	-0.02	.011	2		966,1736
175222	-50 12226			18 53 02	-49 54.7	1	8.52		1.38				4	K2 III	1075
		MCC 188	⋆ V	18 53 03	+08 20.3	1	10.12		1.53				1	M2	1017,1705
		Ste 1 - 1025		18 53 03	+36 44.4	1	14.46		0.98		0.62		2		3066
175863	+59 01929			18 53 03	+59 57.2	1	7.06		-0.14		-0.52		3	B4 Ve	1212
	+36 03321			18 53 04	+36 45.5	1	10.72		0.36		0.06		2	F1 V	3066
175590	+23 03497			18 53 07	+23 09.3	1	7.91		1.69		1.98		3	K5	8088
		Ste 1 - 78		18 53 08	+36 58.6	1	12.91		0.55		0.08		5		3066
175579	+13 03813			18 53 10	+13 09.4	2	7.04	.003	0.06	.007	-0.04		2	A0	695,1205
	+19 03839			18 53 10	+19 09.4	1	10.05		1.85				2	M2	369
		Ste 1 - 82		18 53 10	+36 37.9	1	11.10		0.45		0.03		4		3066
		Ste 1 - 84		18 53 11	+36 45.6	1	12.35		0.45		0.04		3		3066
		Ste 1 - 83		18 53 11	+36 56.6	1	11.70		0.62		0.13		2		3066
175541	+4 03911			18 53 12	+04 12.1	1	8.05		0.90		0.56		3	dK0	3016
230080	+17 03773			18 53 12	+17 10.3	1	9.56		1.79				2	K5	369
175544	+0 04055	LS IV +00 006		18 53 13	+00 12.0	4	7.40	.008	0.10	.003	-0.65	.029	31	B2 V	914,989,1012,1729
175518	-5 04811	G 22 - 7		18 53 13	-05 48.3	6	7.46	.011	0.76	.007	0.40	.007	13	K0 IV-V	202,219,1003,1509*
		Ste 1 - 1023		18 53 15	+36 42.8	1	13.43		1.17		0.96		2		3066
		Ste 1 - 88		18 53 15	+36 50.5	2	10.57	.027	0.77	.021	0.31	.031	6		604,3066
		Ste 1 - 91		18 53 16	+36 57.9	1	11.72		0.76		0.22		2		3066
175740	+41 03177	HR 7146	⋆ AB	18 53 16	+41 32.3	2	5.44	.003	1.04	.007	0.91		4	K0	71,1733
175545	-0 03595			18 53 17	-00 48.2	1	7.40		1.20		1.20		2	K2 III	3077
175362	-37 12982	HR 7129, V686 CrA		18 53 17	-37 24.5	3	5.37	.005	-0.15	.005	-0.70	.001	13	B3 V	26,285,2007
	-3 04416	LS IV -02 017		18 53 18	-02 53.0	1	10.44		1.03		0.11		2		1586
		CS 22936 # 240		18 53 18	-33 56.2	2	13.95	.000	0.09	.000	0.14	.010	2		966,1736
175543	+3 03836	IDS18508N0319	AB	18 53 19	+03 23.0	1	7.07		0.15		0.15		4	A5 V +A5 V	3030
		G 141 - 49		18 53 19	+10 37.7	2	12.51	.010	0.92	.005	0.57	.015	2		1696,3060
175558	+3 03837	IDS18508N0315	AB	18 53 21	+03 18.6	1	8.24		0.34		0.03		1	F0	3030
		Ste 1 - 96		18 53 21	+36 49.2	2	11.34	.027	0.46	.012	-0.01	.008	14		604,3066
		CS 22936 # 238		18 53 21	-34 04.2	2	14.23	.000	0.43	.000	0.13	.012	2		966,1736
175224	-56 09037	IDS18492S5606	AB	18 53 21	-56 03.1	3	8.85	.033	1.44	.011	1.17		6	K7 V +K5 V	1484,2012,3072
175701	+32 03254			18 53 22	+32 20.1	1	7.66		-0.08		-0.26		2	A0	401
175741	+36 03324			18 53 24	+36 40.6	2	7.95	.001	0.03	.009	0.05	.022	4	A0	604,3066
175326	-47 12624			18 53 25	-47 27.4	1	7.30		1.10				4	K0 III	2012
	-2 04786	LS IV -02 018		18 53 27	-02 41.5	1	11.85		2.55				1	M3	369
175824	+48 02793	HR 7154	⋆ AB	18 53 28	+48 47.8	1	5.79		0.42		0.00		1	F3 III	254
		CS 22936 # 211		18 53 29	-36 13.4	2	15.11	.000	0.12	.000	0.11	.009	2		966,1736
		Ste 1 - 99		18 53 30	+37 02.2	1	10.42		0.23		0.16		2		3066
		CS 22936 # 209		18 53 34	-36 12.0	2	14.49	.000	0.10	.000	0.11	.010	2		966,1736
175718	+25 03663			18 53 39	+25 37.8	1	7.94		0.06		-0.02		2	A0	105
	+36 03327			18 53 40	+36 42.8	1	9.58		0.06		0.10		3	A0	604
	+36 03328			18 53 40	+37 02.3	1	9.60		0.12		0.13		2	A2	3066
175396	-44 12991			18 53 41	-44 22.1	1	7.54		0.97		0.66		5	G8 III	1673
175617	-4 04636	G 22 - 9		18 53 42	-04 28.5	3	10.10	.005	0.71	.006	0.16	.010	9	G5	202,1620,3059
		G 22 - 8		18 53 42	-04 29.1	4	13.52	.040	0.96	.007	0.33	.038	9		202,419,1620,3060
175638	+4 03916	HR 7141	⋆ A	18 53 44	+04 08.2	8	4.61	.015	0.16	.007	0.11	.025	28	A5 V	1,3,116,1028,1149*

Table 1   1033

HD	DM	Other Id	N Rem	α_{1950}	δ_{1950}	S	V	σ_V	B−V	σ_{B-V}	U−B	σ_{U-B}	n	Spectrum	References
175638	+4 03916	HR 7141	★ AB	18 53 44	+04 08.2	5	4.05	.013	0.17	.008	0.10	.021	13	A5 V +A5 Vn	3,15,369,1509,8015
175987	+58 01844			18 53 44	+58 40.3	1	7.70		0.37		-0.05		2	F2	3016
175639	+4 03917	HR 7142	★ B	18 53 45	+04 08.1	7	4.98	.019	0.20	.008	0.07	.008	25	A5 Vn	1,116,1028,1149,1327*
175677	+7 03898			18 53 45	+07 52.2	1	8.05		0.16		-0.20		1	B9	1682
		CS 22936 # 237		18 53 45	−34 04.0	2	15.31	.000	0.08	.000	0.18	.010	2		966,1736
175742	+23 03500	V775 Her		18 53 47	+23 29.7	2	8.08	.017	0.91	.005	0.55	.010	4	K0	196,8088
175640	−1 03602	HR 7143		18 53 47	−01 51.9	1	6.21		-0.05		-0.30		4	B9 III	1149
175865	+43 03117	HR 7157, R Lyr		18 53 49	+43 52.8	8	4.03	.045	1.59	.016	1.43	.027	13	M5 III	15,1363,3052,4001*
175744	+17 03778	HR 7147, V828 Her		18 53 51	+17 55.7	4	6.67	.024	-0.05	.008	-0.40	.046	7	B9p Si	592,1063,1079,6007
		AA45,405 S76 # 1		18 53 52	+07 44.5	1	11.66		0.50		0.39		2		1682
		CS 22936 # 249		18 53 52	−33 14.2	2	15.25	.000	0.03	.000	0.19	.010	2		966,1736
175785	+30 03351			18 53 53	+30 14.8	1	7.62		-0.01		-0.26		2	A0	401
175743	+17 03779	HR 7148		18 53 54	+18 02.5	5	5.71	.021	1.10	.015	1.03	.018	8	K1 III	592,1355,1462,3016,6007
175679	+2 03730	HR 7144	★ A	18 53 55	+02 24.3	4	6.15	.019	0.96	.008	0.70	.010	11	G8 III	15,252,1417,3016
175623	−15 05154	IDS18510S1500	A	18 53 55	−14 55.5	1	7.15		0.22		-0.16		8	B8 II/III	1088
		AA45,405 S76 # 2		18 53 56	+07 43.6	1	12.77		1.45		0.84		1		1682
175644	−14 05231			18 53 57	−14 05.7	1	7.81		0.08		-0.10		8	B9 V	1088
		AA45,405 S76 # 3		18 54 00	+07 44.1	1	14.25		1.14		0.50		1		1682
175884	+41 03182			18 54 00	+41 59.6	1	6.50		1.31		1.37		2	K0	252
175703	+4 03918			18 54 01	+04 19.3	1	8.43		2.29				2	K5	369
		CS 22936 # 242		18 54 05	−33 45.8	1	14.10		0.47		-0.22		1		1736
175866	+35 03411			18 54 07	+35 44.7	1	6.67		1.20		1.24		2	K0	1601
175529	−39 13012	HR 7136		18 54 07	−39 53.1	1	6.31		0.20				4	A5 IV/V	2009
175803	+19 03848			18 54 08	+19 46.9	3	8.02	.008	0.01	.011	-0.58	.005	8	B3 V	399,401,555
175329	−60 07213	HR 7127		18 54 10	−60 16.1	7	5.13	.009	1.36	.012	1.42	.023	28	K2 III	15,1075,1075,1311*
		AA45,405 S76 # 5		18 54 14	+07 44.3	1	10.50		1.76		1.82		1		1682
175687	−20 05339	HR 7145		18 54 22	−20 43.4	2	5.07	.010	0.13	.005	-0.14			B9/A0 Ib	2009,8100
175751	−6 04976	HR 7149		18 54 23	−05 54.8	6	4.83	.012	1.08	.004	1.01	.017	21	K1 III	15,1003,1075,1088*
175786	+4 03923			18 54 24	+04 37.0	1	7.82		1.78				2	K5	369
337275	+24 03586			18 54 24	+24 23.4	1	8.55		1.25		1.25		1	K0	1769
		CS 22936 # 239		18 54 25	−33 57.6	2	13.40	.000	-0.04	.000	-0.03	.012	2		966,1736
175805	+2 03737	IDS18519N0220	A	18 54 27	+02 23.7	1	7.63		0.54		0.13		2	F8	1375
175510	−53 09402	HR 7134		18 54 28	−53 00.4	1	4.87		-0.05				4	B9.5IV/V	2007
		CS 22936 # 250		18 54 30	−33 10.8	2	14.66	.000	0.06	.000	0.25	.010	2		966,1736
	−5 04819	LS IV -05 033		18 54 33	−05 35.5	1	10.30		0.40		-0.61		1	B2 IV:pe	1032
	+12 03729			18 54 36	+12 05.1	1	10.39		1.92				2	M2	369
	−3 04423			18 54 36	−03 27.6	1	10.00		2.21				2	M2	369
175754	−19 05242			18 54 39	−19 13.2	6	7.01	.009	-0.08	.005	-0.97	.008	15	B2 Ib/II	219,400,540,976,1011,1586
		CS 22936 # 227		18 54 39	−35 46.6	2	15.24	.000	0.06	.000	0.07	.010	2		966,1736
175775	−21 05201	HR 7150		18 54 45	−21 10.4	4	3.51	.005	1.18	.012	1.13	.004	20	K1 III	15,1075,2012,8015
175869	+2 03738	HR 7158		18 54 46	+02 28.1	3	5.56	.003	0.00	.003	-0.23	.040	7	B9 III(p)	1079,1149,1586
		CS 22936 # 225		18 54 47	−36 05.2	2	14.93	.000	0.26	.000	0.15	.009	2		966,1736
		Ste 1 - 4		18 54 48	+36 51.	1	12.13		0.47		-0.06		2		604
		CS 22936 # 236		18 54 48	−34 13.2	2	15.50	.000	0.11	.000	0.17	.009	2		966,1736
		JL 6		18 54 48	−70 38.	1	13.80		-0.19		-0.73		2		132
		CS 22936 # 224		18 54 49	−36 07.7	2	15.49	.000	0.04	.000	0.01	.010	2		966,1736
		CS 22936 # 247		18 54 51	−33 24.9	2	14.71	.000	0.01	.000	0.03	.011	2		966,1736
	+6 03984	V840 Aql		18 54 52	+06 37.9	1	9.45		1.98				2	M3	369
		CS 22936 # 233		18 54 52	−34 43.3	2	15.29	.000	0.05	.000	0.26	.010	2		966,1736
175922	+13 03826			18 54 54	+13 18.3	4	7.20	.015	0.32	.015	0.23	.014	18	Am	355,1205,1346,8071
		CS 22936 # 252		18 54 55	−32 32.3	2	14.92	.000	0.04	.000	0.03	.011	2		966,1736
175606	−51 11851			18 54 56	−51 31.9	3	9.72	.147	0.44	.008	-0.22	.005	10	F5	1097,1594,3077
175923	+6 03985			18 54 59	+06 54.7	1	8.79							B9	1118
175401	−66 03404	HR 7130		18 54 59	−66 43.3	4	5.99	.007	0.98	.009	0.80	.007	18	K0 III	15,1075,1754,2038
	+0 04064	UW Aql		18 55 00	+00 23.3	3	8.84	.040	2.58	.016	2.12		8	M0 Iab:	148,369,8032
		CS 22936 # 243		18 55 00	−33 47.8	2	14.85	.000	0.10	.000	0.19	.009	2		966,1736
176524	+71 00915	HR 7180		18 55 01	+71 13.8	4	4.83	.029	1.15	.005	1.08	.010	13	K0 III	15,1363,3016,8015
175959	+13 03827			18 55 02	+13 11.9	2	8.89	.011	1.54	.016	1.35	.038	11	K2	1205,1346
175905	−0 03607			18 55 02	−00 35.6	1	7.66		1.15		1.09		2	K1 III	3040
	+15 03631			18 55 05	+15 52.7	1	10.27		2.12				2	M3	369
175674	−48 12816	IDS18513S4838	AB	18 55 06	−48 34.5	3	6.63	.004	1.32	.010	1.36	.006	8	K3 III	219,565,993
176051	+32 03267	HR 7162	★ AB	18 55 09	+32 50.2	4	5.22	.009	0.59	.010	0.02	.004	12	F9 V +K1 V	938,1197,1363,3030
175794	−31 16189	HR 7151		18 55 09	−31 06.2	1	6.12		1.35				4	K3 III	2009
176052	+31 03412			18 55 10	+32 04.3	1	8.43		0.00		-0.18		2		401
176228	+56 02164			18 55 10	+56 35.7	1	7.81		1.54		1.82		3	K5	1502
176008	+17 03790			18 55 12	+17 08.8	1	10.18		0.38		0.18		2	A2	634
		CS 22936 # 246		18 55 12	−33 28.2	2	14.94	.000	-0.02	.000	-0.04	.011	2		966,1736
		CS 22936 # 229		18 55 12	−35 18.0	2	15.18	.000	0.02	.000	-0.22	.011	2		966,1736
		G 229 - 18		18 55 13	+54 28.0	1	10.41		1.34		1.39		2	M0	7010
176795	+75 00683	HR 7199	★ AB	18 55 13	+75 43.2	1	6.33		0.01				2	A1 V	1733
175876	−20 05344	IDS18522S2033	A	18 55 13	−20 29.5	4	6.93	.010	-0.12	.008	-1.00	.015	13	O7/8	26,400,1011,8040
	−61 06290			18 55 13	−61 37.0	1	10.54		1.19		1.16		1	G5	861
181203	+87 00180			18 55 14	+87 14.3	1	8.49		0.19		0.13		5	A5	1219
		CS 22936 # 235		18 55 15	−34 15.0	2	15.73	.000	0.02	.000	0.01	.011	2		966,1736
230220	+17 03791			18 55 16	+17 14.1	1	10.69		0.59		0.19		2	K0	634
175852	−25 13574	HR 7155		18 55 17	−24 56.7	1	6.62		0.08				4	B8 III	2007
	−48 12818			18 55 17	−48 19.2	1	11.14		1.43				1	M4	1494
		CS 22936 # 244		18 55 20	−33 35.0	2	13.62	.000	0.08	.000	0.08	.010	2		966,1736
175813	−37 13001	HR 7152, ε Cra		18 55 21	−37 10.5	1	4.75		0.39		0.01		6	F3 IV/V	3053
175892	−22 04928	HR 7159		18 55 24	−22 35.9	1	6.14		0.09				4	A1 V	2007

HD	DM	Other Id	N	Rem	α_{1950}	δ_{1950}	S	V	σ_V	B–V	σ_{B-V}	U–B	σ_{U-B}	n	Spectrum	References
		R CrA T # 110			18 55 29	−37 23.5	1	14.89		1.25				1		5009
		BTT13,19 # 43			18 55 30	+07 59.	1	12.94		0.78		0.22		2		1586
176029	+5 03993				18 55 34	+05 51.4	5	9.22	.016	1.45	.019	1.25	.073	11	M2 V	202,1003,1017,1197,1705,3078
		CS 22936 # 256			18 55 35	−33 07.2	2	13.15	.000	0.29	.000	0.12	.009	2		966,1736
175893	−29 15574	V4152 Srg			18 55 37	−29 34.4	3	9.26	.034	1.14	.020	0.79	.007	3	C0-3,OHd	1238,1589,1699
		CS 22936 # 258			18 55 38	−33 18.2	2	15.11	.000	0.07	.000	0.13	.010	2		966,1736
		G 207 - 9		⋆ V	18 55 40	+33 53.1	2	14.63	.009	0.16	.004	-0.61	.004	6		419,3028
176054	+6 03987				18 55 41	+06 46.2	1	9.15		0.32				4	B9	1118
175855	−39 13032	HR 7156		⋆ AB	18 55 44	−39 36.2	1	6.49		-0.04				4	B9.5V	2007
	−13 05166	LSS 5132			18 55 45	−13 26.1	1	10.34		0.24		-0.70		1	BOII-III	1032
176316	+51 02463				18 55 49	+51 26.2	1	8.31		-0.06		-0.16		2	A0	401
175856	−44 13018				18 55 49	−44 12.4	2	8.61	.000	0.52	.015	-0.05		5	F7 V	219,1075
		GJ 2143			18 55 54	+06 06.	1	10.57		0.57		0.45		4		7009
176408	+57 01922	HR 7175			18 55 54	+57 44.9	2	5.66	.000	1.15	.000	1.19	.004	5	K1 III	37,1355
		LP 924 - 34			18 55 54	−27 33.0	1	13.60		1.48		0.97		1		3062
		NGC 6723 sq1 3			18 55 56	−36 40.8	1	12.46		0.52		0.03		3		349
176095	+6 03989	HR 7163			18 55 57	+06 10.4	2	6.20	.005	0.47	.005	0.00	.010	7	F8 IV-V	15,1417
176156	+16 03684				18 55 59	+17 00.1	1	9.54		1.86				2	K5	369
176155	+17 03799	HR 7165, FF Aql		⋆ AB	18 56 01	+17 17.5	4	5.23	.076	0.71	.048	0.45	.025	5	F8 Ib	71,592,1462,6007
175607	−66 03406				18 56 04	−66 15.4	1	8.61		0.70				4	G6 V	2012
		NGC 6723 sq1 1			18 56 07	−36 40.5	1	10.42		0.58		-0.02		std		349
176097	−1 03610	LS IV -01 011			18 56 08	−01 40.7	1	9.41		0.64		0.37		2	A0	180
		CS 22936 # 275			18 56 09	−34 48.3	2	14.85	.000	0.21	.000	0.21	.008	2		966,1736
176176	+15 03639				18 56 10	+15 31.3	1	9.09		1.89				2	M1	369
176077	−7 04794	LS IV -07 029			18 56 11	−07 24.8	1	9.55		0.33		-0.67		1	B1 Ia	1032
		NGC 6723 - 171			18 56 12	+36 42.	1	12.96		1.58		1.84		1		1468
		NGC 6723 - 456			18 56 12	+36 42.	1	12.94		1.57		1.88		1		1468
		NGC 6723 - 9025	V		18 56 12	+36 42.	1	12.82		1.79		2.26		1		1468
		NGC 6723 - 9026	V		18 56 12	+36 42.	1	12.93		1.80		2.22		1		1468
176598	+65 01309	HR 7187			18 56 12	+65 11.5	1	5.62		0.95		0.69		1	G8 III	1355
230283	+13 03835				18 56 14	+13 31.5	1	9.94		1.90				2	M7	369
176200	+14 03729	UV Aql	A		18 56 14	+14 17.5	1	8.39		3.55				1	C6 II	414
176200	+14 03729		B		18 56 14	+14 17.5	1	12.05		1.53		1.50		1		414
		CS 22936 # 267			18 56 14	−34 00.3	2	12.11	.000	1.15	.000	1.04	.025	2		966,1736
176319	+36 03345				18 56 19	+36 32.0	1	8.28		-0.07		-0.20		1	B9	604
176318	+38 03373	HR 7174			18 56 19	+38 11.8	1			-0.17		-0.52		2	B7 IV	1063
		CS 22936 # 263			18 56 19	−33 37.2	2	15.27	.000	0.06	.000	0.09	.010	2		966,1736
	+13 03837		V		18 56 20	+13 51.8	1	9.91		1.74				2	M2	369
176018	−36 13204				18 56 21	−36 51.5	1	8.88		0.60		0.10		21	F7 IV	1763
176047	−34 13301				18 56 26	−34 32.4	1	8.10		0.96				4	K0 III	2012
176124	−19 05255				18 56 27	−19 20.9	1	6.57		1.62				5	M3 III	6002
176232	+13 03838	HR 7167, V1286 Aql			18 56 29	+13 50.3	4	5.89	.017	0.25	.008	0.09	.008	9	F0 p SrEu	695,1118,1202,1205
		AA45,405 S75 # 1			18 56 30	+07 01.5	1	12.74		0.86		0.53		2		1682
176158	−7 04798				18 56 31	−06 54.7	1	7.52		0.07		-0.40		1	B9	963
176123	−18 05155	HR 7164			18 56 31	−18 38.2	1	6.37		0.99				4	G3 II	2007
176254	+20 04007				18 56 32	+20 33.2	3	6.75	.012	0.04	.008	-0.56	.005	8	B2 IV	399,401,555
176159	−7 04799				18 56 33	−07 12.2	1	9.01		0.15		-0.30		2	B9	1586
		CS 22936 # 281			18 56 33	−35 53.6	2	14.37	.000	0.25	.000	0.17	.009	2		966,1736
176301	+19 03858	HR 7171			18 56 35	+19 43.5	2			-0.04	.010	-0.43	.010	3	B7 III-IV	1063,1079
176391	+42 03206				18 56 35	+42 37.9	1	8.32		0.38		0.12		3	F5 III	253
176162	−13 05172	HR 7166		⋆ AB	18 56 35	−12 54.6	1	5.53		-0.04				4	B5 IV	2007
		AA45,405 S75 # 2			18 56 36	+07 03.8	1	14.14		0.91		0.26		1		1682
		Smethells 27			18 56 36	−52 26.	1	12.10		1.37				1		1494
		CS 22936 # 282			18 56 37	−36 03.2	2	14.87	.000	0.02	.000	-0.03	.011	2		966,1736
176234	+7 03919				18 56 38	+07 11.5	1	8.26		0.56		-0.02		1	F8	1682
176560	+58 01849	HR 7184		⋆ AB	18 56 39	+58 09.4	3	6.46	.011	0.08	.008	0.04	.016	5	A2 V +A3 V	938,1733,3016
	−2 04811				18 56 39	−02 50.5	1	10.96		2.26				2	M2	369
		AA45,405 S75 # 3			18 56 40	+07 07.1	1	13.68		1.43		0.77		2		1682
176375	+36 03348	IDS18549N3617	AB		18 56 40	+36 21.4	1	8.16		0.30		0.16		1	F0 V	604
		CS 22936 # 262			18 56 40	−33 24.1	2	12.50	.000	0.17	.000	0.35	.008	2		966,1736
		R CrA T # 106			18 56 40	−36 41.3	1	13.92		1.45				1		5009
		R CrA T # 105			18 56 40	−36 41.9	1	15.36		1.29				1		5009
176320	+25 03687				18 56 41	+25 23.9	1	7.43		1.09		1.00		3		8100
		CS 22936 # 268			18 56 43	−34 00.0	2	15.03	.000	0.35	.000	0.11	.010	2		966,1736
		CS 22936 # 285			18 56 44	−36 10.9	2	14.58	.000	0.11	.000	0.06	.000	2		966,1736
176668	+62 01669	HR 7191		⋆ A	18 56 47	+62 19.7	2	6.45	.014	0.96	.019	0.63	.019	4	G5 IV+G8 V	150,3016
176668	+62 01669	IDS18563N6216	B		18 56 47	+62 19.7	2	9.75	.178	0.74	.009	0.23	.012	4		150,3016
176303	+13 03841	HR 7172		⋆ AB	18 56 48	+13 33.3	7	5.24	.018	0.53	.001	0.07	.010	25	F8 V	15,1007,1013,1118*
		LS IV -14 109			18 56 49	−14 30.4	2	11.17	.014	0.32	.005	-0.30	.009	8		405,1514
	+0 04073				18 56 50	+00 50.6	1	9.90		1.21		1.30		48		1764
230304	+17 03802				18 56 50	+17 10.8	1	10.24		0.56		0.37		2	F0	634
		CS 22936 # 260			18 56 51	−33 21.8	2	12.86	.000	0.39	.000	0.17	.011	2		966,1736
	−37 13012				18 56 53	−37 19.9	1	9.62		0.63		0.10		26	G2	1763
		CS 22936 # 255			18 56 54	−32 53.8	2	13.85	.000	0.40	.000	0.11	.012	2		966,1736
176304	+9 03951	HR 7173		⋆	18 56 55	+10 04.3	4	6.74	.014	0.25	.008	-0.44	.015	11	B2 Vp	15,154,1118,1417
176377	+29 03423				18 56 55	+30 06.5	1	6.79		0.58		0.00		4	G2 V	3026
	−3 04438				18 56 55	−03 23.0	1	10.51		2.18				2	M2	369
	+1 03845				18 56 56	+01 07.6	1	10.42		0.21		0.11		2		1375
176305	+5 04002				18 56 58	+05 10.7	1	10.57		3.19		2.14		3	A2	8032
		Case *M # 55			18 57 00	+05 18.	1	10.31		3.23				1	M0 Ia	148

Table 1   1035

HD	DM	Other Id	N Rem	α_{1950}	δ_{1950}	S	V	σ_V	B–V	σ_{B-V}	U–B	σ_{U-B}	n	Spectrum	References
		L 562 - 14		18 57 00	−31 24.	1	13.66		0.73		0.14		1		3062
		CCS 2687, UX Aql		18 57 03	+01 46.	1	12.28		2.95				1	CS	864
		BSD 63 # 110		18 57 03	+29 02.8	1	11.41		-0.15		-0.63		3	G0	1026
176203	−23 14928			18 57 03	−23 07.9	2	8.78	.014	0.74	.000	0.22		4	F5 V	742,1594
176437	+32 03286	HR 7178	⋆ A	18 57 04	+32 37.2	18	3.25	.009	-0.05	.005	-0.10	.022	138	B9 III	1,15,369,667,985,1006*
349606	+20 04010	LS II +20 002		18 57 05	+20 47.5	1	9.64		0.70		0.22		2	F8 II	8100
176502	+40 03544	HR 7179	⋆ A	18 57 08	+40 36.6	2	6.22	.009	-0.15	.014	-0.66	.005	4	B3 V	154,542
176502	+40 03544	IDS18555N4033	B	18 57 08	+40 36.6	2	9.85	.050	0.24	.030	0.20	.135	2	A5 V	542,542
	+7 03922	G 141 - 52		18 57 09	+07 55.7	1	10.86		1.43		1.26		2	M2	7010
		LP 691 - 18		18 57 10	−07 50.7	1	10.14		0.80		0.40		2		1696
176393	+16 03690			18 57 12	+17 01.6	1	8.98		0.06		-0.17		2	B9	634
	+5 04006			18 57 16	+06 00.8	1	9.58		2.04				2	M3	369
		CS 22936 # 264		18 57 16	−33 50.1	2	14.80	.000	0.15	.000	0.22	.009	2		966,1736
		CS 22891 # 23		18 57 18	−60 27.7	1	14.41		-0.27		-1.05		1		1736
230327	+15 03645			18 57 19	+15 36.2	1	9.77		1.89				2	M7	369
176411	+14 03736	HR 7176	⋆ A	18 57 21	+14 59.9	4	4.02	.004	1.08	.005	1.04	.005	11	K2 III	15,1080,1363,8015
176246	−25 13614	HR 7168		18 57 21	−25 00.8	1	6.36		1.25				4	K0 III	2009
230334	+11 03687	LS II +11 001		18 57 23	+11 20.8	1	9.61		0.65		0.09		1	B8 II	8100
176562	+41 03198			18 57 23	+41 11.7	1	9.17		-0.11	.000	-0.49	.010	4	B9	401,963
176105	−50 12274			18 57 24	−50 15.2	1	7.96		1.25		1.19		1	K0p Ba	565
176438	+19 03865			18 57 26	+19 25.0	1	7.51		-0.01		-0.25		2	B9	401
176443	+13 03850			18 57 27	+13 11.6	1	9.49		0.00				1	A0	1205
		CS 22936 # 283		18 57 27	−36 06.3	2	14.50	.000	0.11	.000	0.15	.009	2		966,1736
		G 141 - 54		18 57 29	+11 54.3	1	15.52		0.20		-0.57		3		3060
		CS 22936 # 270		18 57 30	−34 04.0	2	14.49	.000	0.37	.000	0.06	.011	2		966,1736
		CS 22936 # 287		18 57 30	−36 21.6	2	14.72	.000	0.09	.000	0.05	.010	2		966,1736
176582	+39 03602	HR 7185		18 57 31	+39 08.8	3	6.41	.017	-0.17	.000	-0.70	.013	9	B5 IV	154,555,833
		CS 22936 # 274		18 57 36	−34 36.6	2	15.13	.000	0.15	.000	0.23	.009	2		966,1736
		CS 22936 # 278		18 57 38	−35 24.9	2	14.16	.000	0.11	.000	0.22	.009	2		966,1736
176270	−37 13018	HR 7170	⋆ AB	18 57 42	−37 08.0	1	5.82		-0.03				4	B9 V	2009
176527	+26 03418	HR 7181		18 57 43	+26 09.6	1	5.27		1.24		1.27		4	K2 III	1080
		CS 22936 # 284		18 57 43	−36 06.6	2	14.66	.000	0.02	.000	0.04	.011	2		966,1736
		HBC 675		18 57 44	−37 02.3	1	14.02		1.99		1.53		1		1763
176707	+50 02705	HR 7196		18 57 45	+50 44.3	1	6.30		0.98				2	G8 III	71
		CS 22936 # 286		18 57 45	−36 12.9	2	14.66	.000	0.16	.000	0.20	.009	2		966,1736
		R CrA T # 14		18 57 45	−37 02.3	1	13.94		1.93				1		5009
		HRC 286, S CrA	AB	18 57 47	−37 01.3	4	11.30	.086	0.85	.055	-0.16	.058	4	G5:e	776,1588,1763,5009
176021	−65 03763			18 57 47	−64 59.6	2	7.62	.014	0.59	.011	-0.03	.042	4	G0/1 III	462,3077
176541	+22 03549	HR 7183		18 57 51	+22 44.6	2	6.33	.035	1.70	.051	2.00		4	M3.5 IIIab	71,3001
176669	+42 03212			18 57 53	+42 56.6	1	7.53		-0.06		-0.20		2	B8	401
176367	−28 15269			18 57 57	−28 47.1	1	8.50		0.56		0.04		23	G2 V	1699
		R CrA T # 101		18 57 58	−36 52.3	1	15.52		1.57				1		5009
	+68 01037			18 58 05	+68 26.2	1	8.36		1.10		0.98		2	K2	1733
		ApJ144,259 # 51		18 58 06	−18 17.	1	15.42		-0.12		-1.04		4		1360
176670	+31 03424	HR 7192		18 58 08	+32 04.5	5	4.94	.015	1.46	.006	1.65	.018	10	K3 III	1080,1363,3041,4001,8100
		CS 22936 # 306		18 58 10	−33 28.0	2	12.66	.000	0.14	.000	0.09	.009	2		966,1736
175986	−68 03180	HR 7161	⋆ AB	18 58 10	−68 49.7	6	5.88	.007	0.55	.013	0.12	.009	26	G0 IV	15,173,1075,1075,2038,3077
	+0 04078	LS IV +00 008		18 58 11	+00 19.8	2	10.02	.005	0.76	.015	-0.06		3		180,851
		R CrA T # 13	V	18 58 12	−37 05.2	2	11.16	.055	1.12	.015	0.55		2		776,5009
	−37 13022	HBC 676	V	18 58 12	−37 05.2	1	11.30		1.15		0.67		1	K0-2 IV	1763
176386	−37 13023	IDS18549S3702	AB	18 58 17	−36 57.7	1	7.22		0.12		-0.04		1	B9 IV	1588
		CS 22936 # 294		18 58 19	−35 14.7	2	15.36	.000	0.17	.000	0.15	.008	2		966,1736
		HBC 678		18 58 19	−36 56.3	1	10.54		0.81		0.31		1	G8 IV	1763
	−37 13024	TY CrA		18 58 19	−36 56.9	2	9.37	.167	0.50	.026	0.10	.011	2	B2	776,1588
176473	−23 14953			18 58 22	−23 18.0	1	7.80		1.09		0.79		5	G8 III	1657
	+9 03958			18 58 23	+09 25.5	1	9.45		2.12				2	M2	369
	+2 03752			18 58 24	+02 27.1	1	10.01		0.73		0.24		2	G0	1375
		Aql sq 1 # 26		18 58 27	+04 38.6	1	12.51		0.58		0.20		1		97
176603	+4 03945			18 58 29	+05 06.6	1	8.77		0.20		0.17		2	A3	97
176535	−13 05186			18 58 29	−13 45.8	1	9.78		1.05		0.95		1	K3 (V)	3072
		CS 22936 # 299		18 58 29	−34 29.4	2	14.65	.000	0.19	.000	0.24	.008	2		966,1736
		Aql sq 1 # 31		18 58 31	+05 08.8	1	13.62		1.04		0.40		2		97
		HRC 288, R CrA		18 58 31	−37 01.5	1	13.53		0.83		0.31		1	F0pe	5009
		Aql sq 1 # 30		18 58 32	+04 43.8	1	13.51		0.84		0.34		5		97
		HRC 289, DG CrA		18 58 32	−37 27.9	2	13.94	.255	1.46	.075	0.90		2		776,5009
	+5 04011	G 22 - 12		18 58 33	+05 23.6	2	10.62	.035	1.08	.010	1.02		4	K3	202,333,1620
230409	+18 03911	G 184 - 42		18 58 33	+19 00.6	2	10.09	.015	0.70	.005	0.09	.005	5	G0	1696,3077
		Aql sq 1 # 15		18 58 34	+04 50.4	1	11.38		0.51		0.08		2		97
		G 260 - 13		18 58 34	+72 46.1	1	11.30		0.78		0.35		2		1658
	−37 13027	R CrA		18 58 34	−37 01.4	1	11.51		0.88		-0.20		1	A5 IIe	1588
		Aql sq 1 # 19		18 58 35	+04 44.0	1	11.63		0.69		0.21		3		97
		Aql sq 1 # 23		18 58 35	+04 45.7	1	12.05		0.64		0.15		3		97
		HRC 290, T CrA		18 58 36	−37 02.3	2	13.27	.094	0.90	.189	0.48	.280	2	F0e	1588,5009
		Aql sq 1 # 22		18 58 37	+04 50.6	1	11.92		0.61		0.10		2		97
176649	+9 03960			18 58 37	+09 21.6	1	8.02		0.10				2	A0	1118
176537	−22 04946	HR 7182		18 58 37	−22 46.1	1	6.24		1.66				4	K3 III	2007
176425	−42 13839	HR 7177		18 58 37	−41 59.0	3	6.22	.007	0.00	.000			11	A0 V	15,2013,2030
176650	+2 03753			18 58 40	+02 24.8	1	7.07		1.03		0.83		2	K0	3008
176844	+40 03555	HR 7201		18 58 41	+40 36.8	1	6.58		1.60		1.63		2	M4 IIIa	1733
176593	−15 05185	HR 7186		18 58 42	−15 21.3	2	6.31	.005	1.00	.000	0.79	.005	7	K0 III	15,1256

HD	DM	Other Id	N Rem	α_{1950}	δ_{1950}	S	V	σ_V	B–V	σ_{B-V}	U–B	σ_{U-B}	n	Spectrum	References
		Aql sq 1 # 24		18 58 43	+04 45.6	1	12.15		0.83		0.43		5		97
		Aql sq 1 # 17		18 58 43	+04 49.9	1	11.49		0.53		0.09		4		97
	−0 03627	V336 Aql		18 58 45	+00 03.4	1	9.51		1.15				1	G0	851
		Aql sq 1 # 25		18 58 45	+05 00.8	1	12.43		0.61		0.18		2		97
176675	+4 03947			18 58 45	+05 06.1	1	9.56		0.46		0.13		1	B9	97
176845	+39 03605			18 58 45	+39 15.9	1	7.87		0.51		0.12		3	F8 V	1501
176427	−44 13057			18 58 46	−44 12.8	1	8.34		0.33				4	F0 IV	2012
		Aql sq 1 # 27		18 58 52	+05 07.9	1	13.01		0.89		0.33		1		97
176776	+19 03879	HR 7198		18 58 54	+19 14.3	1	6.46		1.14		1.10		3	K1 III	1733
176869	+39 03606	IDS18572N3942		18 58 54	+39 46.4	1	7.86		-0.06		-0.28		2	B9	401
176895	+39 03607			18 58 56	+39 47.7	1	7.65		0.85		0.49		1	G5	401
		Aql sq 1 # 20		18 58 57	+05 00.2	1	11.66		1.17		0.57		2		97
176609	−17 05419			18 58 57	−17 25.4	1	8.04		0.45				2	F5 V	1594
337459	+26 03578			18 58 58	+25 59.5	1	9.36		0.37		-0.23		2	F8	1003
177003	+50 02708	HR 7210		18 58 58	+50 27.7	2	5.37	.004	-0.18	.011	-0.76		5	B2.5IV	154,1363
		HBC 679		18 58 58	−36 59.7	1	13.88		1.53		0.57		1	K4-5	1763
	+4 03949			18 58 59	+04 48.2	1	9.63		0.61		0.09		std		97
176735	+7 03931			18 58 59	+07 12.4	1	8.91		0.33		0.02		2	F5	1733
176803	+19 03880	IDS18568N2002	AB	18 58 59	+20 05.1	2	7.34	.000	0.06	.000	-0.32	.005	4	B8	401,555
		G 205 - 53		18 58 59	+45 37.4	1	11.22		0.90		0.53		2	K3	7010
176354	−57 09295			18 58 59	−57 04.1	1	7.06		0.89				4	K0 III	2012
176678	−5 04840	HR 7193		18 59 01	−05 48.7	6	4.02	.000	1.09	.004	1.04	.008	28	K1 III	15,1075,1088,2012*
176737	+2 03756			18 59 03	+02 30.9	1	7.02		1.70		2.01		2	K4 II-III	8100
176896	+33 03287	HR 7204		18 59 05	+33 43.8	1	6.01		0.97				2	gK0	71
176736	+5 04013			18 59 08	+05 09.1	1	9.05		0.51		0.01		2	G0	97
176818	+21 03634			18 59 09	+21 26.5	1	7.04		0.12		-0.60		3	B1 V	555
176613	−29 15629			18 59 09	−29 24.8	1	8.30		0.37		0.03		18	F0 V	1699
		R CrA T # 114		18 59 09	−37 02.5	1	14.00		1.76				1		5009
	+0 04081			18 59 10	+00 24.1	1	10.57		0.71				1		851
176913	+33 03289			18 59 11	+33 28.5	1	8.14		-0.07		-0.44		1	A0	401
177410	+69 01018	HR 7224		18 59 12	+69 27.6	2	6.49		-0.17	.014	-0.55	.009	4	A0p Si	985,1079
		MCC 191	AB	18 59 12	+75 14.	1	10.04		1.21		1.09		1	M0	679
176819	+20 04022	HR 7200		18 59 13	+20 45.7	1	6.68		0.02		-0.70		2	B2 IV-V	154
176821	+16 03704			18 59 15	+16 19.0	1	8.83		1.88				2	K5	369
176871	+26 03429	HR 7202		18 59 15	+26 13.1	1	5.68		-0.08		-0.55		3	B5 V	154
		BPM 42512		18 59 16	−42 50.5	1	11.19		1.24				1	K7	1494
		Aql sq 1 # 13		18 59 18	+04 35.7	1	11.24		0.87		0.62		1		97
		Aql sq 1 # 16		18 59 19	+04 52.5	1	11.39		0.73		0.21		3		97
176914	+28 03153			18 59 19	+28 20.4	1	7.04		-0.08		-0.70		3	B5 I	555
	+19 03881		V	18 59 21	+20 02.6	1	10.59		1.98				3	M3	369
178738	+82 00572			18 59 21	+82 18.1	2	6.90	.010	0.00	.000	-0.02	.010	4	A0	985,1733
176805	+4 03953			18 59 22	+05 05.9	1	8.21		0.47		0.26		std	A5	97
		Aql sq 1 # 11		18 59 23	+04 45.0	1	11.04		0.60		0.02		3		97
176806	+4 03954			18 59 23	+04 45.8	1	9.18		0.28		0.09		2	A5	97
176826	+7 03934			18 59 23	+07 44.5	1	8.35		1.97				2	K5	369
		Aql sq 1 # 14		18 59 24	+04 48.0	1	11.36		0.65		0.07		4		97
176704	−25 13655	HR 7195		18 59 24	−24 55.1	3	5.65	.005	1.24	.009	1.38		9	K2 III	15,2033,3077
176687	−30 16575	HR 7194	⋆ AB	18 59 26	−29 57.2	6	2.60	.005	0.08	.000	0.06	.012	25	A2 III+A4IV	15,1020,1075,2012*
176744	−13 05195			18 59 27	−12 58.0	1	9.64		0.35		-0.18		3	B(8)(Ib/II)	1586
176578	−47 12692			18 59 27	−47 07.4	1	6.86		0.96				4	K0/1 III	2012
		Aql sq 1 # 29		18 59 28	+04 41.6	1	13.47		0.69		0.86		2		97
		Aql sq 1 # 18		18 59 28	+05 12.9	1	11.56		0.63		0.06		1		97
		Aql sq 1 # 12		18 59 30	+04 50.0	1	11.12		0.41		0.23		3		97
176939	+24 03608	HR 7206		18 59 31	+24 57.2	1	6.73		1.49		1.54		2	K2	1733
177195	+56 02177			18 59 31	+56 27.2	1	8.79		-0.20		-0.84		1	B8	963
	+17 03825			18 59 32	+17 14.3	1	10.29		2.07				2	M2	369
177061	+41 03208			18 59 33	+41 20.5	1	8.07		0.05		-0.01		2	A0	401
		Aql sq 1 # 21		18 59 35	+04 41.8	1	11.76		2.06		1.98		1		97
176638	−42 13855	HR 7188		18 59 35	−42 10.1	5	4.74	.004	-0.03	.005	-0.07	.004	19	B9.5 V	15,1075,2012,2020,3023
		Aql sq 1 # 10		18 59 36	+04 44.8	1	10.76		0.55		0.32		3		97
		G 142 - 2		18 59 36	+15 59.6	1	10.38		0.65		0.05		2		1658
176557	−50 12292			18 59 36	−50 23.4	2	7.18	.000	1.46	.005	1.71		5	K4 III	613,1075
177006	+32 03299			18 59 37	+32 19.2	2	7.25	.000	-0.14	.005	-0.62	.005	5	B5	401,555
176875	+7 03935			18 59 38	+07 10.6	1	8.37		0.22				2	B9	1118
176940	+22 03560			18 59 39	+22 48.9	1	8.31		0.13		-0.20		2	B9 V	555
	+63 01477	G 260 - 14		18 59 40	+63 59.6	1	9.17		0.69		0.23		1	G5	1658
176971	+22 03561	HR 7207		18 59 41	+22 11.4	1	6.45		0.12		0.12		2	A4 V	1733
177249	+55 02137	HR 7218		18 59 44	+55 35.2	1	5.48		0.86				2	G5 IIb	71
		HRC 291, VV CrA		18 59 44	−37 17.0	2	13.07	.045	1.26	.060	-0.12	.015	2		776,5009
176754	−29 15641	V523 Sgr		18 59 45	−29 13.0	1	9.57		0.42		0.13		10	A3/5 II/II	1768
176853	−10 04926	V599 Aql		18 59 48	−10 47.7	2	6.65	.020	0.21	.005	-0.06	.439	8	B2 V	399,3016
	+4 03955			18 59 51	+04 42.9	1	10.37		0.85		0.39		2		97
	+4 03956			18 59 52	+04 42.2	1	10.08		0.66		0.14		1	F8	97
176723	−38 13300	HR 7197, V701 CrA	⋆	18 59 53	−38 19.6	3	5.73	.004	0.33	.007			11	F2 III/IV	15,2013,2030
176981	+8 03951	HR 7208		18 59 57	+08 18.0	4	6.30	.004	1.65	.025	1.76	.045	18	K2 III	15,1355,1417,8040
177109	+33 03295	HR 7212	⋆ A	18 59 57	+33 32.9	1	6.38		-0.12		-0.63		3	B5 IV	154
	+7 03939			19 00 00	+07 39.0	1	10.07		1.98				2	M2	369
177196	+46 02602	HR 7215	⋆ A	19 00 01	+46 51.8	6	5.01	.014	0.19	.005	0.08	.010	16	A7 V	15,1007,1013,1363*
178089	+76 00712	HR 7247		19 00 04	+76 58.9	2	6.55	.015	0.38	.000	-0.04	.027	4	F2 V	1733,3016
176664	−51 11893	HR 7190	⋆ A	19 00 04	−51 05.5	5	5.93	.008	1.24	.009	1.38		20	K0/1 III	15,1075,1075,2007,3005

Table 1   1037

HD	DM	Other Id	N Rem	α_{1950}	δ_{1950}	S	V	σ_V	B–V	σ_{B-V}	U–B	σ_{U-B}	n	Spectrum	References
	+0 04086	LS IV +00 010		19 00 05	+00 57.3	1	10.84		0.68		0.04		1		180
		CS 22947 # 36		19 00 05	−49 47.2	1	13.30		0.32		0.07		1		1736
177198	+43 03143			19 00 07	+43 46.4	1	8.57		1.20		1.38		3	K2 III	37
176884	−19 05273	HR 7203	⋆ A	19 00 07	−19 19.2	1	6.05		1.29		1.09		2	K0 II/III	542
176884	−19 05273	HR 7203	⋆ AB	19 00 07	−19 19.2	1	5.97		1.19				4	K0 II/III	2007
176884	−19 05273	IDS18572S1923	B	19 00 07	−19 19.2	1	9.17		0.47		0.09		3		542
176886	−20 05379			19 00 09	−20 48.0	1	8.49		0.53		0.10		1	F5 V	219
176982	−0 03631	IDS18576S0051	A	19 00 10	−00 47.0	3	8.38	.015	0.73	.003	0.27	.005	8	G5	158,196,1705
176982	−0 03632	IDS18576S0051	BC	19 00 10	−00 47.0	1	8.80		0.74				1	G5	1705
176983	−0 03632	IDS18576S0051	BC	19 00 11	−00 47.1	1	8.81		0.72		0.29		4	dG6	158
176903	−19 05275	HR 7205		19 00 11	−19 10.6	1	6.37		0.48				4	F5 V	2007
230499	+18 03923			19 00 12	+18 59.1	1	9.02		1.69				3	M0	369
	−3 04459			19 00 12	−03 07.2	1	9.28		2.24				2	M2	369
	+2 03762	LS IV +02 002		19 00 13	+02 15.9	1	10.32		0.53				2	B0.5IV	725
177044	+15 03667	KP Aql		19 00 14	+15 43.6	1	9.42		0.42		0.15		4	A3	1768
	+11 03707	LS IV +11 005		19 00 15	+11 10.5	2	10.57	.000	1.04	.000	−0.10	.000	3	B1 Ia:	1012,1032
176984	−3 04460	HR 7209	⋆ AB	19 00 16	−03 46.4	3	5.42	.008	0.00	.002	−0.06	.017	9	A1 V	15,1071,3016
177082	+14 03755			19 00 20	+14 29.6	1	6.92		0.58		0.09		2	G2 V	3016
		CS 22936 # 291		19 00 22	−35 44.1	2	14.95	.000	0.21	.000	0.24	.008	2		966,1736
176756	−51 11899			19 00 26	−51 15.0	1	8.79		1.31				4	K3 III	1075
		L 751 - 1		19 00 29	−13 38.1	1	14.78		1.54				1	M5	1705
		CCS 2691		19 00 31	−41 20.4	1	11.43		4.66				3		864
		AA45,405 S78 # 5		19 00 34	+13 58.4	1	11.27		0.36		0.32		1		1682
176891	−38 13306	IDS18572S3813	AB	19 00 34	−38 08.5	1	8.91		0.66		0.08		6	G5 V	173
177015	−20 05381			19 00 36	−20 12.2	2	7.77	.021	−0.01	.027	−0.42	.013	5	B3/5 Vn	1212,1586
177014	−19 05280			19 00 38	−19 30.5	1	9.27		0.19		−0.03		3	B9 III	400
		AA45,405 S78 # 7		19 00 39	+13 59.4	1	13.35		0.44		0.06		1		1682
		G 260 - 15		19 00 40	+70 35.2	4	13.19	.007	0.05	.005	−0.86	.005	9		1,538,1281,3028
177088	−2 04835	LS IV -02 022		19 00 41	−02 36.6	2	9.85	.000	0.81	.000	0.41		4	B9	1594,6006
177199	+19 03888	HR 7216		19 00 42	+19 35.2	1	6.09		1.34				2	K1 III	71
		Steph 1669		19 00 42	+22 08.3	1	11.13		1.27		1.18		1	K7	1746
177017	−22 04958	SU Sgr	⋆ AB	19 00 43	−22 47.2	2	8.31	.105	1.50	.057	0.89	.039	6	M7 III	897,3076
		CS 22936 # 341		19 00 45	−35 12.5	2	11.48	.000	0.12	.000	0.07	.009	2		966,1736
177115	+0 04088			19 00 46	+00 30.1	1	6.76		1.51		1.75		1	K2	934
		AA45,405 S78 # 1		19 00 46	+14 09.4	1	10.75		0.49		−0.02		2		1682
		AA45,405 S78 # 4		19 00 49	+14 03.7	1	12.65		1.24		0.86		1		1682
177175	+12 03780		V	19 00 50	+12 10.7	1	8.44		2.05				2	S7.2	369
177202	+13 03879			19 00 51	+14 06.2	1	9.34		0.51		−0.01		2	F2	1682
176961	−37 13038			19 00 51	−36 58.8	1	9.51		1.20		0.93		33	G6/8 (III)	1763
		V 919 Sgr		19 00 52	−17 04.4	1	11.76		0.77		−0.51		1		1753
		LB 3114		19 00 54	−62 37.	1	12.48		0.12		0.14		10		45
177483	+52 02326	HR 7229	⋆ ABC	19 00 56	+52 11.3	1	6.31		1.00				2	K2	71
		CM Aql		19 00 58	−03 07.7	1	13.56		1.57		−0.21		1		1753
176863	−50 12308			19 00 58	−50 24.5	1	8.36		1.00				4	G8 III	1075
177178	+1 03865	HR 7214		19 01 00	+01 44.7	1	5.82		0.08		0.08		4	A4 V	1149
177095	−20 05385			19 01 00	−20 31.7	3	9.61	.015	0.65	.015	0.05	.015	4	G6 V	1696,2017,3077
177227	+13 03881			19 01 01	+13 59.6	1	9.07		1.22		1.18		2	K0	1682
		CS 22891 # 5		19 01 04	−62 11.8	1	13.32		0.08		0.04		1		1736
349695	+20 04028		V	19 01 06	+20 12.0	1	9.29		1.99				3	M0	369
		G 22 - 13		19 01 06	−02 20.5	2	11.10	.015	1.25	.015	1.23		3		202,1620
		LS IV -00 013		19 01 09	−00 07.8	1	11.31		0.27		−0.68		2		180
		LP 751 - 22		19 01 09	−14 24.7	1	12.17		0.61		−0.03		2		1696
177254	+10 03769			19 01 11	+10 27.5	1	8.48		0.38				3	A0	1118
177229	+0 04090			19 01 12	+00 40.8	1	8.94		1.42		1.42		1	K2	934
		AA45,405 S72 # 2	A	19 01 12	+02 09.6	1	13.07		0.64		0.23		1		1682
		AA45,405 S72 # 2	AB	19 01 12	+02 09.6	1	11.87		0.69				2		725
		L 45 - 4		19 01 12	−74 48.	1	12.99		0.60		−0.12		2		1696
		AA45,405 S72 # 3		19 01 13	+02 08.2	1	12.89		0.74				1		725
177074	−31 16306	HR 7211		19 01 13	−31 07.3	1	5.50		0.03				4	A0 V	2032
177207	−2 04839	LS IV -01 013		19 01 14	−01 55.8	1	9.14		0.44		−0.06		2	B9	180
177137	−17 05446			19 01 14	−17 50.2	2	8.69	.000	0.04	.000	−0.36	.000	7	B9 II/III	963,1732
177347	+25 03710			19 01 15	+25 45.1	2	6.98	.005	−0.01	.015	−0.53	.000	4	B8	105,1733
177230	−4 04678	V1420 Aql	⋆ A	19 01 20	−04 23.5	1	11.12		0.48		−0.28		2	WN8	1359
	+14 03763	LS II +14 005		19 01 21	+14 51.8	1	10.67		0.56		−0.45		2	B1	1318
	+28 03167			19 01 21	+28 34.4	1	9.24		0.69		0.17		2	G0	3016
177166	−21 05233	IDS18584S2141	AB	19 01 21	−21 36.4	2	7.18	.004	0.55	.035	0.04		8	F8 V	176,938
		GD 217		19 01 22	+13 27.6	1	13.59		0.70		−0.14		1		3060
		AA45,405 S72 # 1		19 01 24	+02 13.4	1	10.74		0.59				2		725
177351	+12 03784			19 01 28	+12 40.3	1	8.67		0.20		−0.28		2	A2	963
230566	+14 03764			19 01 28	+14 35.5	1	9.52		1.89				2	M7	369
	+0 04092			19 01 29	+00 44.1	1	10.41		0.42		0.33		2		1375
		R CrA T # 109		19 01 29	−37 32.0	1	15.68		0.82				1		5009
		LS IV -08 028		19 01 30	−08 04.5	1	11.46		0.13		−0.78		2		405
		AS 338		19 01 32	+16 21.8	1	12.93		0.98		−0.28		1	M3	1753
177392	+21 03648	HR 7222, LT Vul		19 01 33	+21 11.6	1	6.63		0.31		0.13		3	F2 III	1733
177284	−2 04840	V337 Aql		19 01 34	−02 06.3	2	8.63	.018	0.49	.000	−0.52	.004	5	B0.5p	588,1012
	+2 03771	LS IV +03 002		19 01 39	+03 01.3	2	9.21	.000	0.63	.000	−0.39	.005	3	B0 III	1012,1032
177332	+3 03882	HR 7219		19 01 41	+03 15.3	3	6.73	.008	0.14	.015	0.13	.004	10	A5 m	15,1071,8071
177241	−21 05237	HR 7217	⋆ A	19 01 41	−21 49.0	6	3.76	.005	1.01	.008	0.85	.011	18	K0 III	15,58,1075,2012,4001,8015
177334	−0 03639			19 01 43	−00 22.5	1	8.03		1.84				2	K5	369

HD	DM	Other Id	N Rem	α_{1950}	δ_{1950}	S	V	σ_V	B–V	σ_{B-V}	U–B	σ_{U-B}	n	Spectrum	References
177336	−5 04858	HR 7220, V Aql		19 01 44	−05 45.6	4	6.89	.084	4.17	.113			8	C5,4	109,864,1149,3001
177060	−51 11910			19 01 44	−51 07.1	1	9.22		0.22				4	A5 V	1075
	+0 04094			19 01 46	+00 13.1	1	10.54		2.04				2	M3	369
230579	+10 03774	LS II +11 002	★ AB	19 01 47	+11 01.9	3	9.23	.083	0.56	.022	−0.38	.027	5	B1.5:IV:en	1012,1032,1586
		R CrA T # 107		19 01 48	−36 59.0	1	12.29		0.76				1		5009
337446	+26 03437	IDS18598N2633	AB	19 01 50	+26 37.2	1	8.98		0.51		−0.05		2	F8	3016
349755	+19 03892			19 01 51	+20 06.0	1	9.62		1.82				2	G5	369
177433	+14 03771			19 01 52	+15 01.5	1	7.28		1.10		0.86		3	K0 II-III	8100
177290	−18 05193			19 01 52	−17 52.7	1	8.33		0.18		0.13		2	A3 V	379
177291	−18 05191			19 01 52	−18 47.0	1	8.65		0.40		−0.26		2	B8 Vnn(e)	379
177459	+17 03842			19 01 56	+17 28.8	1	6.95		0.44		−0.04		2	F5	3016
		LS IV -00 014		19 02 01	−00 11.1	1	11.58		0.20		−0.52		2		180
		LS IV +00 011		19 02 04	+00 07.6	1	11.16		0.44		−0.58		2		180
177490	+12 03790	LS II +12 001		19 02 04	+12 55.3	1	7.77		0.37		−0.08		1	A0 II	8100
177593	+33 03309	IDS19002N3400	A	19 02 05	+34 04.5	2	7.30	.000	−0.10	.010	−0.51	.000	5	B5	401,555
	+29 03460			19 02 06	+30 05.5	1	9.35		0.01		−0.05		2	A0	1733
177441	+1 03877	SZ Aql		19 02 07	+01 13.8	4	8.01	.033	1.07	.021	0.86	.182	4	K2	934,1484,1772,6011
177440	+1 03878			19 02 12	+01 43.5	1	9.28		0.52				20	G0	6011
177462	+1 03879			19 02 14	+01 16.9	1	8.55		0.28				17	A2	6011
177494	+2 03776		V	19 02 16	+02 54.5	1	8.66		1.96				2	M2	369
177380	−22 04968			19 02 16	−21 55.4	1	9.79		−0.08		−0.81		5	B2 II	1732
		WLS 1900 0 # 7		19 02 18	−02 44.2	1	11.48		0.70		0.36		2		1375
177444	−7 04844	LS IV -07 030	★ A	19 02 18	−07 13.1	1	8.56		0.06		−0.83		1	B8	963
177463	−4 04684	HR 7225	★ A	19 02 19	−04 06.4	1	5.41		1.12		1.01		8	K1 III	1088
177697	+39 03630			19 02 22	+40 02.5	1	7.76		1.64		1.90		2	K5	1601
177319	−33 13903			19 02 22	−33 51.6	1	7.52		1.11		1.03		5	K0 III	1673
177171	−52 11356	HR 7213		19 02 23	−52 25.0	7	5.16	.005	0.53	.006	0.04	.000	21	F6 V	15,611,613,1075,2013*
		JL 9		19 02 24	−72 35.	2	13.28	.068	−0.28	.005	−1.19	.015	3	sdO	132,832
177548	+8 03960			19 02 29	+08 24.0	1	8.51		0.35				2	A0	1118
177648	+23 03549	IDS19004N2311	A	19 02 32	+23 15.2	1	7.24	.034	0.12	.024	−0.52	.015	5	B2 Ve	379,1212
177550	+1 03880			19 02 33	+01 31.9	1	7.94		2.21				2	M1	369
177427	−29 15692			19 02 37	−29 09.5	1	7.08		−0.01		−0.25		3	B9 III(e)	1586
177674	+25 03715			19 02 40	+25 42.6	2	8.17	.010	0.04	.010	−0.17	.005	5	A0	833,1625
		R CrA T # 108		19 02 40	−37 13.6	1	15.79		0.62				1		5009
177552	−1 03642	HR 7231		19 02 43	−01 35.4	1	6.52		0.35		−0.07		4	F2 V	1149
177300	−51 11917	BL Tel		19 02 44	−51 29.7	2	7.11	.035	0.47	.005	0.28	.030	2	F5 Iab/b	611,613
177624	+9 03979			19 02 46	+09 33.9	2	6.89	.014	0.19	.000	−0.36		11	B3 V	1118,8040
178001	+57 01942	BH Dra	★ A	19 02 46	+57 22.9	1	8.38		0.05		−0.04		1	A0	627
178001	+57 01942	BH Dra	★ AB	19 02 46	+57 22.9	1	7.98		0.09		−0.01		3	A0	627
178001	+57 01942	IDS19019N5719	B	19 02 46	+57 22.9	1	9.25		0.20		0.07		2		627
		LP 751 - 23		19 02 47	−13 41.8	1	12.33		0.69		0.02		2		1696
177517	−15 05223	HR 7230	★ A	19 02 49	−15 44.2	2	5.98	.010	−0.01	.005	−0.25		8	B8 III(pSi)	2007,3033
177518	−18 05199			19 02 49	−18 05.8	1	9.64		0.60		0.03		2	G2 V	1696
177623	+10 03782			19 02 50	+10 20.4	1	9.09		0.38				2	B9	1118
		Steph 1676	AB	19 02 55	+63 55.0	1	10.55		1.42		1.24		2	M0	1746
177745	+22 03579	G 184 - 47		19 03 00	+22 59.8	1	8.55		0.81		0.42		3	K0	196
177744	+25 03717			19 03 00	+25 45.7	1	7.58		0.11		0.06		3	A0	833
177559	−19 05292	V4197 Sgr		19 03 01	−19 33.5	1	8.10		0.01		−0.66		3	B2/3 V(n)	400
177474	−37 13048	HR 7226/7	★ AB	19 03 02	−37 08.2	5	4.21	.012	0.52	.009	0.03	.025	20	F8 V + F8 V	15,938,1075,2012,3026
177809	+30 03409	HR 7238		19 03 03	+30 39.4	2	6.06	.002	1.53	.015	1.88		4	M2.5 IIIab	71,3001
177808	+31 03453	HR 7237		19 03 03	+31 40.1	5	5.56	.039	1.55	.005	1.92	.025	11	M0 III	15,252,1003,1355,3001
177365	−50 12326			19 03 03	−50 24.0	1	6.30		−0.10				4	B9 V	1075
177931	+45 02825			19 03 05	+45 50.7	1	7.18		0.02		0.00		2	B9	1601
177724	+13 03899	HR 7235	★ AB	19 03 07	+13 47.3	14	2.99	.004	0.01	.005	0.01	.024	190	A0 Vn	3,15,61,71,369,374*
177877	+38 03424			19 03 07	+38 40.0	2	8.24	.008	−0.08	.011	−0.31	.020	8	A0	401,833,1775
177725	+11 03727			19 03 09	+11 11.6	1	7.61		0.10				3	B9	1118
177406	−48 12901	HR 7223		19 03 10	−48 22.6	2	5.97	.005	−0.02	.005	−0.02		5	A0 V	663,2007
177747	+11 03728			19 03 14	+11 33.8	1	8.73		0.16				2	A0	1118
		LS II +12 003		19 03 15	+12 46.4	1	10.72		1.00		−0.10		2		405
178326	+65 01319			19 03 15	+65 30.6	1	7.25		0.92		0.60		3	K0	3016
177272	−60 07259			19 03 15	−59 58.9	1	9.64		0.37		−0.02		3	F0 V	786
177830	+25 03719			19 03 18	+25 50.7	1	7.22		1.03		1.10		3	K0	3026
177609	−22 04972			19 03 21	−22 40.7	1	8.49		1.10		0.95		5	K0 III/IV	897
177706	−9 04987			19 03 26	−09 42.6	1	6.68		1.07		0.63		3	G2 Ib	8100
		CS 22947 # 47		19 03 26	−48 52.6	1	12.75		0.31		0.16		1		1736
177810	+11 03732			19 03 28	+12 07.8	1	7.50		1.94				2	K0	369
177752	−1 03645	LS IV -00 015		19 03 29	−00 55.0	1	8.40		0.30		−0.39		2	B3	180
177565	−37 13049	HR 7232		19 03 30	−37 53.0	7	6.16	.005	0.71	.008	0.28	.008	20	G5 IV	15,219,1020,2012,2012*
177832	+10 03784			19 03 33	+11 07.4	1	8.58		0.50				2	B8	1118
		G 22 - 14		19 03 35	+00 26.2	2	13.20	.029	1.47	.024	1.18		3		202,333,1620
177812	+3 03893	LS IV +03 006		19 03 35	+03 11.3	3	8.60	.000	0.63	.000	−0.37	.005	6	B1 Ib	1012,1032,8100
177756	−5 04876	HR 7236		19 03 36	−05 47.5	8	3.43	.010	−0.10	.005	−0.26	.007	29	B9 Vn	15,1020,1034,1075*
177566	−41 13325			19 03 37	−41 47.9	2	10.19	.005	−0.21	.003	−1.00	.001	8	B5/7 Ib	55,1732
177758	−12 05278	IDS19009S1203	AB	19 03 38	−11 58.2	4	7.25	.006	0.57	.004	−0.03	.005	54	G1 V	158,588,1003,3077
178002	+36 03395			19 03 39	+36 51.8	1	8.33		0.00		−0.13		2	A0	401
		LS IV +03 007		19 03 46	+03 12.7	1	11.11		0.68		−0.38		2		405
		G 205 - 55		19 03 48	+45 02.9	1	13.88		1.61				1		1759
178207	+53 02178	HR 7251		19 03 48	+53 19.2	2	5.40	.007	−0.01	.002	−0.09	.003	49	A0 Vn	879,3023
177716	−27 13564	HR 7234		19 03 49	−27 44.7	6	3.32	.006	1.18	.010	1.16	.014	32	K1 III	3,15,418,1075,2012,8015
178003	+29 03472	HR 7244		19 03 50	+29 50.7	2	6.31	.000	1.66	.015	2.00	.020	5	M0 III	1733,3001

Table 1 1039

HD	DM	Other Id	N	Rem	α_{1950}	δ_{1950}	S	V	σ_V	B–V	σ_{B-V}	U–B	σ_{U-B}	n	Spectrum	References
177879	+6 04020				19 03 51	+06 34.0	1	8.88		0.30				3	A0	1118
178208	+49 02929	HR 7252		⋆ A	19 03 52	+49 50.8	1	6.43		1.37		1.60		2	K3 III	1733
178030	+28 03186				19 03 54	+29 02.1	1	7.60		0.00		-0.18		2	A0	401
177940	+8 03970	HR 7243, R Aql		⋆ A	19 03 58	+08 09.2	2	6.12	.050	1.35	.160	0.68	.040	2	M7 III ev	817,3001
177880	−1 03649	IDS19014S0130	AB		19 04 00	−01 25.5	1	6.76		0.10		-0.41		4	B5 V +A0 V	1149
177817	−16 05153	HR 7239		⋆ AB	19 04 00	−16 18.4	2	6.03	.007	-0.02		-0.36		5	B7 V	1079,2007
		CS 22891 # 10			19 04 01	−61 39.5	1	14.83		0.04				1		1736
177983	+15 03690				19 04 04	+15 47.0	2	7.29	.015	0.38	.010	0.21	.010	7	AmA5-F2	8071,8100
177688	−43 13123				19 04 06	−42 59.8	1	9.01		0.53				4	F6 III	2012
349733	+20 04046				19 04 07	+20 34.2	1	8.41		1.41		1.61		1	K2	1748
177962	+1 03890				19 04 10	+01 39.8	1	10.64		2.01				2	F5	369
177192	−71 02365				19 04 10	−71 17.0	1	8.98		1.14		0.96		1	G8p Ba	565
177863	−18 05206	HR 7241		⋆ V	19 04 13	−18 49.0	2	6.29		-0.05	.004	-0.42		5	B8 V	1079,2007
177843	−22 04980				19 04 14	−22 48.7	1	9.06		1.00		0.61		2	K0 III	1746
178032	+20 04047				19 04 15	+21 05.0	1	8.44		1.19		1.13		1	K2	1748
230681	+11 03735				19 04 17	+11 28.5	1	9.17		1.95				2	K5	369
178036	+8 03974				19 04 18	+09 04.2	1	8.37		0.29				2	A0	1118
	+0 04103				19 04 19	+00 38.3	1	9.24		2.06				2	M2	369
178010	+6 04024				19 04 22	+06 32.3	1	9.19		0.38				4	A0	1118
177846	−28 15403	HR 7240			19 04 22	−28 42.9	1	6.04		1.61				4	K3 III	2009
177222	−71 02366				19 04 22	−71 37.8	1	6.58		1.05		0.75		5	G8/K0 III	1628
		AO 1170			19 04 23	+20 54.3	1	9.69		1.79		1.97		1	K4 III	1748
	+28 03192				19 04 29	+28 38.4	1	9.20		0.55		0.04		3	F8 V	3026
	+0 04105				19 04 30	+00 15.5	1	9.79		2.03				2	M2	369
		EV Aql			19 04 30	+14 51.1	1	11.61		1.43				1		1772
178123	+15 03693				19 04 33	+15 52.6	1	8.18		1.69				2	M1	369
349728	+20 04050				19 04 33	+20 47.1	1	10.37		0.78		0.27		1	G5	1748
178187	+24 03640	HR 7250			19 04 33	+24 10.3	1	5.79		0.10		0.11		2	A4 III	1733
178014	−5 04877				19 04 35	−04 58.6	1	7.99		1.52		1.39		5	K2	1657
	−28 15407	LP 924 - 2			19 04 35	−28 13.2	1	11.04		0.90		0.65		2		1696
178065	+0 04106	HR 7245			19 04 36	+00 33.8	3	6.55	.003	0.06	.007	-0.29	.004	10	B9 III	1079,1088,6007
177389	−68 03185	HR 7221			19 04 36	−68 30.3	6	5.32	.011	0.90	.004	0.60	.009	25	G8/K0III/IV	15,285,611,1075,1075,2038
		L 78 - 9			19 04 36	−70 19.	1	11.95		0.60		-0.10		2		1696
178125	+10 03787	HR 7248, Y Aql		⋆ AB	19 04 37	+10 59.6	4	5.08	.005	-0.06	.010	-0.41	.018	8	B8 III	1063,1079,1118,3023
178126	+7 03967	G 22 - 15			19 04 38	+07 32.8	4	9.22	.025	1.06	.010	0.93	.013	7	K5 V	202,801,1003,3008
		LP 811 - 24			19 04 38	−19 41.6	2	11.61	.005	0.56	.024	-0.12	.029	3		1696,3062
178124	+11 03738	LS II +11 003			19 04 39	+11 21.5	2	9.45	.000	0.41	.000	-0.36	.000	3	B1 V nn	1012,1032
178233	+28 03193	HR 7253			19 04 39	+28 32.9	2	5.53	.020	0.29	.001	0.04		2	F0 III	3023,6009
178329	+41 03232	HR 7258			19 04 40	+41 20.1	2	6.47	.015	-0.16	.005	-0.65	.000	6	B3 V	154,399
177989	−18 05211				19 04 41	−18 48.3	1	9.33		-0.05		-0.89		2	B2 II	400
		AO 1172			19 04 42	+20 36.9	1	11.71		1.86		2.40		1	K2 I	1748
178020	−15 05239				19 04 42	−15 14.4	1	9.84		0.50		0.00		2	F8/G0wF0	6006
177693	−55 09001	HR 7233			19 04 44	−55 47.9	3	6.48	.007	1.10	.000			11	K1 III	15,2018,2030
		AO 1173			19 04 45	+20 59.6	1	10.65		1.99		2.34		1	M3 III	1748
178162	+8 03977				19 04 47	+08 16.2	1	8.84		0.13				2	B9	1118
178129	+3 03902	V1403 Aql		⋆	19 04 48	+03 21.8	3	7.46	.022	0.53	.016	-0.37	.011	9	B3 Ia	1012,1032,8100
		NGC 6752 sq1 12			19 04 50	−60 11.5	1	14.59		1.03				1		368
177873	−40 13061	HR 7242			19 04 52	−40 34.6	3	4.58	.005	1.10	.008	1.03	.005	14	K1 III	15,1075,2012
		NGC 6752 sq1 11			19 04 53	−60 11.5	1	12.57		1.07				1		368
		CS 22947 # 49			19 04 55	−48 37.6	1	14.47		0.61		-0.01		1		1736
		NGC 6755 - 8			19 04 56	+04 09.9	1	12.89		0.67		0.18		1		49
178070	−15 05242				19 04 56	−15 12.1	2	8.68	.005	0.02	.010	-0.35	.005	6	B8/9 III	963,1732
		G 184 - 48			19 04 57	+20 48.8	2	10.77	.002	1.52	.035	1.31		4		8105,7009
		NGC 6755 - 3			19 04 58	+04 08.9	1	11.37		1.32		1.06		1		49
		NGC 6752 sq1 1			19 04 59	−60 05.7	1	10.12		1.50		1.74		1		368
	+28 03198				19 05 00	+29 07.4	1	8.70		0.64		0.24		3	G2 V	3026
		JL 11			19 05 00	−72 27.	1	16.73		0.00		-1.17		1		132
		AO 1175			19 05 01	+20 54.3	1	9.84		0.65		0.12		1	G4 V	1748
349723	+20 04053				19 05 01	+20 59.6	1	9.69		0.26		0.16		1	A0	1748
178277	+22 03594	IDS19029N2226	A		19 05 01	+22 30.3	2	7.46	.010	0.22	.007	0.08	.005	6	F0	379,3016
178277	+22 03594	IDS19029N2226	B		19 05 01	+22 30.3	1	9.61		0.58		0.04		4		3016
177994	−31 16374	V526 Sgr		⋆ AB	19 05 02	−31 25.7	2	9.72	.030	0.14	.021	0.05		7	A0 V	661,1768
178263	+12 03811				19 05 05	+12 32.7	1	8.36		0.35				2		1118
		G 184 - 49			19 05 05	+20 48.1	3	10.77	.007	1.52	.045	1.30	.009	5		1748,7009,8105
234799	+53 02182				19 05 05	+53 43.2	1	9.18		0.20		0.14		2	A5	1566
177694	−60 07268				19 05 05	−59 58.9	1	9.36		1.16		1.09		3	K0 III	786
		LS IV +03 014			19 05 06	+03 30.6	1	10.84		0.65				4		1021
		NGC 6755 - 22			19 05 06	+04 06.4	1	15.38		0.88		0.54		1		49
		NGC 6755 - 13			19 05 07	+04 06.5	1	14.66		0.85		0.55		1		49
		NGC 6755 - 11			19 05 07	+04 13.7	1	13.65		0.98		0.46		1		49
178217	−2 04865				19 05 07	−02 32.0	1	9.02		1.91				1	M1	369
178140	−15 05243				19 05 07	−15 18.6	2	9.56	.025	0.62	.005	-0.03	.005	4	F8/G0 V(w)	1003,1064
	+3 03907	NGC 6755 - 1			19 05 08	+04 08.5	1	10.35		0.66		-0.16		1	B2 III	49
		NGC 6755 - 18			19 05 08	+04 09.5	1	14.90		0.92		0.57		1		49
		NGC 6755 - 5			19 05 08	+04 14.2	1	11.53		0.54		0.12		1	F0 III	49
		NGC 6752 sq1 3			19 05 09	−59 58.2	1	11.32		1.36		1.24		1		368
		NGC 6755 - 19			19 05 11	+04 09.0	1	14.94		0.98		0.40		1		49
		NGC 6755 - 6			19 05 11	+04 09.8	1	12.50		0.85		0.34		1	A0 V	49
178075	−24 15041	HR 7246			19 05 11	−24 44.2	2	6.30		0.02	.007	-0.25		5	B9.5V	1079,2007
		NGC 6755 - 23			19 05 12	+04 08.8	1	15.70		1.06		0.50		1		49

HD	DM	Other Id	N Rem	α_{1950}	δ_{1950}	S	V	σ_V	B–V	σ_{B-V}	U–B	σ_{U-B}	n	Spectrum	References
		NGC 6755 - 15		19 05 14	+04 07.6	1	14.61		0.94		0.54		1		49
		NGC 6755 - 14		19 05 14	+04 13.0	1	14.60		0.89		0.40		1		49
178330	+20 04055			19 05 14	+20 20.9	1	7.40		1.06		0.94		2	K0	1648
		NGC 6755 - 10		19 05 15	+04 07.2	1	13.65		0.78		0.27		1		49
		NGC 6755 - 21		19 05 15	+04 12.7	1	15.82		0.91		0.59		1		49
		NGC 6755 - 17		19 05 16	+04 06.9	1	14.90		0.92		0.56		1		49
		NGC 6755 - 7		19 05 18	+04 15.5	1	12.77		2.31				1		49
		NGC 6755 - 16		19 05 18	+04 15.7	1	14.68		0.93		0.39		1		49
		NGC 6755 - 2		19 05 18	+04 18.0	1	10.92		1.47		1.18		1		49
177996	−42 13922			19 05 18	−42 30.4	2	7.87	.009	0.86	.000	0.52		8	K1 V	158,1075
		CS 22891 # 64		19 05 18	−58 43.9	1	13.78		-0.09		-0.55		1		1736
		NGC 6755 - 12		19 05 19	+04 15.4	1	14.04		0.63		0.32		1		49
178175	−19 05312	HR 7249, V4024 Sg		19 05 20	−19 22.2	8	5.52	.037	-0.06	.011	-0.67	.012	34	B2 Ve	681,815,1212,1586*
		NGC 6755 - 9		19 05 21	+04 16.1	1	12.92		0.80		0.26		1		49
		NGC 6755 - 20		19 05 23	+04 13.0	1	15.06		0.91		0.50		1		49
178332	+15 03696			19 05 24	+15 25.7	1	8.10		0.57		0.21		1	G0 III	8100
178110	−31 16384			19 05 24	−31 23.7	1	8.69		-0.03				1	B8/9 V	661
		JL 12		19 05 30	−72 22.	1	15.86		0.16		-0.71		1		832
		NGC 6755 - 4		19 05 31	+04 14.4	1	11.52		0.74		0.27		1	G8 IV	49
178475	+35 03485	HR 7262	⋆	19 05 31	+36 01.2	3	5.27	.009	-0.11	.005	-0.52	.015	10	B6 IV	154,1203,1363
178449	+32 03326	HR 7261	⋆ AB	19 05 32	+32 25.3	3	5.21	.045	0.36	.016	0.05	.005	6	F0 V	254,1118,3026
178450	+30 03425	V478 Lyr		19 05 35	+30 10.4	1	7.63		0.73		0.21		1	G8 V	3026
		AO 1177		19 05 36	+20 41.4	1	9.46		0.16		-0.11		1	B9 V	1748
177078	−78 01217			19 05 38	−77 56.8	1	6.98		1.86		2.14		5	M0 III	1628
178287	−7 04861	V496 Aql		19 05 39	−07 31.1	3	7.59	.029	1.07	.027	0.77	.013	3	G5	657,1490,6011
178359	+1 03899	TT Aql		19 05 41	+01 13.1	4	6.55	.034	0.95	.018	0.72	.040	4	F5 I	934,1484,1593,1772,6007
178591	+40 03596			19 05 41	+40 58.5	2	7.16	.015	-0.04	.000	-0.23	.010	4	B5	401,555
178428	+16 03752	HR 7260	⋆ AB	19 05 43	+16 46.7	5	6.08	.013	0.71	.008	0.27	.000	14	G5 V	1118,1197,1355,1758,3026
178146	−41 13349			19 05 43	−41 20.7	1	7.89		1.66		1.57		2	M2/3 III	1732
		CS 22891 # 86		19 05 43	−61 23.0	1	14.91		0.05				1		1736
		MV Lyr		19 05 44	+43 56.4	1	13.15		0.01		-0.96		10	Op	949
		G 207 - 16	AB	19 05 47	+32 27.2	1	11.25		1.70				4	M3	538
178403	+10 03793			19 05 49	+10 21.6	1	9.56		0.61				27	G0	6011
177901	−59 07453			19 05 49	−59 37.8	1	8.24		1.51		1.83		3	K4 III	786
178338	−7 04862			19 05 51	−07 41.0	1	9.21		0.69				24	G5	6011
177925	−59 07454			19 05 52	−59 48.0	1	9.89		0.47		0.06		2	F5 IV	786
178476	+21 03672	HR 7263		19 05 55	+21 37.1	2	6.24	.009	0.41	.014	0.00	.009	4	F3 V	254,3037
178540	+24 03650			19 05 58	+24 38.7	2	6.60	.010	-0.02	.005	-0.49	.010	5	B5 V	555,1501
		NGC 6752 - 64		19 06 00	−60 03.	1	14.15		0.16				1		38
		NGC 6752 - 65		19 06 00	−60 03.	1	14.37		0.07				1		38
		NGC 6752 - 70		19 06 00	−60 05.	2	12.81	.054	1.36	.049	1.59		3		38,318
		NGC 6752 - 71		19 06 00	−60 05.	1	13.12		1.02				1		38
178454	+10 03794			19 06 02	+10 18.0	1	8.29		0.35				29	F0	6011
	+42 03250			19 06 03	+42 13.3	2	10.64	.005	-0.20	.000	-0.95	.010	4		401,963
		CS 22891 # 51		19 06 03	−58 11.5	1	14.49		0.10		0.11		1		1736
178253	−38 13350	HR 7254		19 06 04	−37 59.1	5	4.10	.005	0.04	.005	0.08	.009	15	A2 V	15,1075,2012,3023,8015
178661	+38 03441	IDS19044N3846	AB	19 06 05	+38 50.9	1	7.61		0.26		0.07		2	A3 V	105
178512	+12 03819			19 06 06	+13 01.0	2	7.27	.005	0.01	.000	-0.44		4	B8	1118,1733
180499	+82 00578			19 06 06	+82 35.7	1	8.03		0.02		-0.19		5	B9	1219
178408	−7 04864			19 06 06	−07 33.0	1	8.59		0.35				23	A5	6011
		NGC 6752 - 63		19 06 06	−60 03.	2	12.72	.030	0.79	.020	0.14		2		38,318
		NGC 6752 - 66		19 06 06	−60 04.	1	13.78		0.77				1		38
		NGC 6752 - 67		19 06 06	−60 05.	1	13.55		0.82				1		38
		NGC 6752 - 68		19 06 06	−60 05.	2	11.99	.025	1.11	.000	0.86	.070	4		38,318
		NGC 6752 - 69		19 06 06	−60 05.	2	12.61	.030	0.51	.010	0.04	.055	4		38,318
178149	−48 12933			19 06 07	−48 33.7	2	9.07	.005	0.66	.010	0.03	.015	5	G5/6 V	1696,3077
349779	+20 04059			19 06 08	+20 36.3	1	9.81		0.16		0.09		1	A0	1748
178409	−7 04865			19 06 11	−07 41.9	1	7.91		0.24		0.17		1	A0	1490
178151	−50 12355			19 06 12	−50 02.4	1	8.22		0.29				4	A9 V	1075
		NGC 6752 - 57		19 06 12	−60 02.	1	13.96		0.65				1		38
		NGC 6752 - 59		19 06 12	−60 02.	2	10.92	.020	1.59	.005	1.76	.015	4		38,318
		NGC 6752 - 60		19 06 12	−60 03.	1	12.50		0.86				1		38
		NGC 6752 - 1		19 06 12	−60 07.	2	12.27	.109	0.92	.035	0.54	.065	7		38,318
178593	+25 03735			19 06 13	+25 17.7	1	7.48		0.51		-0.03		3	F8	833
178254	−40 13074	HR 7255		19 06 13	−39 54.5	1	6.46		1.06				4	K0 III	2007
177999	−60 07269	NGC 6752 - 1013	⋆ AB	19 06 13	−60 07.6	3	7.41	.002	-0.06	.008	-0.17	.001	22	B9 II/III	318,786,1732
178484	−2 04872			19 06 16	−02 22.1	1	6.56		1.35		1.25		2	K0	1375
178299	−36 13355	HR 7256		19 06 16	−36 14.8	1	6.54		-0.02				4	A0 III/IV	2009
		NGC 6752 - 56		19 06 18	−60 02.	1	13.88		0.71				1		38
		NGC 6752 - 58		19 06 18	−60 02.	1	12.54		0.79		0.42		2		38
		NGC 6752 - 61		19 06 18	−60 03.	2	11.66	.055	1.14	.015	0.97		2		38,318
		NGC 6752 - 62		19 06 18	−60 03.	1	13.28		1.06				1		38
		NGC 6752 - 3		19 06 18	−60 06.	2	11.96	.044	1.10	.010	0.82	.063	6		38,318
		NGC 6752 - 2		19 06 18	−60 07.	2	12.52	.044	0.88	.019	0.45	.054	6		38,318
178515	−3 04494			19 06 21	−03 25.3	1	9.52		0.47		-0.07		2	B9	1586
	−60 07270			19 06 23	−60 03.4	1	9.39		1.19		1.16		1		368
		NGC 6752 - 51		19 06 24	−60 01.	1	14.69		0.92				1		38
		NGC 6752 - 52		19 06 24	−60 01.	1	14.42		0.02				1		38
		NGC 6752 - 53		19 06 24	−60 01.	2	13.52	.005	0.72	.005	0.09		2		38,318
		NGC 6752 - 54		19 06 24	−60 02.	1	13.60		0.57				1		38

Table 1   1041

HD	DM	Other Id	N	Rem	α_{1950}	δ_{1950}	S	V	σ_V	B–V	σ_{B-V}	U–B	σ_{U-B}	n	Spectrum	References
		NGC 6752 - 55			19 06 24	−60 02.	2	12.28	.019	1.09	.029	0.66		3		38,318
		NGC 6752 - 18			19 06 24	−60 04.	1	10.86		1.45				1		38
		NGC 6752 - 1002			19 06 24	−60 04.	1	14.37		0.00		0.03		2		318
		NGC 6752 - 1003			19 06 24	−60 04.	3	11.50	.004	1.26	.011	1.09	.023	12		318,786,3036
		NGC 6752 - 1004			19 06 24	−60 04.	1	14.58		-0.07		-0.09		2		318
		NGC 6752 - 1005			19 06 24	−60 04.	1	14.26		0.01		0.05		2		318
		NGC 6752 - 1006			19 06 24	−60 04.	1	15.84		0.74		-0.01		3		318
		NGC 6752 - 1007			19 06 24	−60 04.	1	14.72		0.79		0.18		2		318
		NGC 6752 - 1008			19 06 24	−60 04.	2	14.05	.010	0.79	.005	0.24	.034	3		318,786
		NGC 6752 - 1009			19 06 24	−60 04.	2	12.37	.000	1.04	.010	0.70	.005	12		318,786
		NGC 6752 - 1010			19 06 24	−60 04.	4	13.42	.025	0.63	.022	0.03	.010	8		318,368,786,3036
		NGC 6752 - 1011			19 06 24	−60 04.	1	16.02		0.69		0.11		3		318
		NGC 6752 - 1012			19 06 24	−60 04.	1	13.73		0.67		0.23		2		318
		NGC 6752 - 1021			19 06 24	−60 04.	1	15.68		-0.17		-0.75		1		318
		NGC 6752 - 1022			19 06 24	−60 04.	1	15.04		-0.12		-0.47		1		318
		NGC 6752 - 1023			19 06 24	−60 04.	1	15.74		-0.13		-0.62		1		318
		NGC 6752 - 1024			19 06 24	−60 04.	1	14.79		-0.04		-0.34		2		318
		NGC 6752 - 1031		V	19 06 24	−60 04.	1	13.00		0.40		0.13		1		318
		NGC 6752 - 1041			19 06 24	−60 04.	1	14.22		0.65		0.09		1		318
		NGC 6752 - 1042			19 06 24	−60 04.	1	13.97		0.56		0.02		1		318
		NGC 6752 - 1043			19 06 24	−60 04.	1	12.63		0.65		0.16		1		318
		NGC 6752 - 1045			19 06 24	−60 04.	1	13.81		0.08		0.15		1		318
		NGC 6752 - 1046			19 06 24	−60 04.	1	14.23		0.67		0.19		1		318
		NGC 6752 - 1047			19 06 24	−60 04.	1	14.16		0.58		0.02		1		318
		NGC 6752 - 1048			19 06 24	−60 04.	1	12.61		0.83		0.34		1		318
		NGC 6752 - 1049			19 06 24	−60 04.	2	13.73	.030	0.60	.025	0.06		2		318,368
		NGC 6752 - 1101			19 06 24	−60 04.	2	14.45	.000	-0.02	.030	-0.17	.020	8		318,786
		NGC 6752 - 1102			19 06 24	−60 04.	1	14.11		0.67		0.10		3		318
		NGC 6752 - 1103			19 06 24	−60 04.	1	15.40		0.74				1		318
		NGC 6752 - 1104			19 06 24	−60 04.	2	12.11	.015	0.94	.030	0.60	.000	2		318,3036
		NGC 6752 - 1105			19 06 24	−60 04.	1	13.76		0.83		0.27		2		318
		NGC 6752 - 1106			19 06 24	−60 04.	1	13.60		0.86		0.34		2		318
		NGC 6752 - 1107			19 06 24	−60 04.	1	12.34		1.09		0.64		3		318
		NGC 6752 - 1108			19 06 24	−60 04.	1	14.67		0.73		0.13		2		318
		NGC 6752 - 1109			19 06 24	−60 04.	1	15.44		0.75		0.02		2		318
		NGC 6752 - 1110			19 06 24	−60 04.	1	14.02		0.98		0.83		2		318
		NGC 6752 - 1111			19 06 24	−60 04.	1	12.78		0.90		0.56		1		318
		NGC 6752 - 1112			19 06 24	−60 04.	1	12.70		0.78		0.29		2		318
		NGC 6752 - 1113			19 06 24	−60 04.	1	14.72		-0.04		-0.32		1		318
		NGC 6752 - 1114			19 06 24	−60 04.	1	13.61		0.83		0.28		2		318
		NGC 6752 - 1115			19 06 24	−60 04.	1	13.55		0.87		0.44		2		318
		NGC 6752 - 1116			19 06 24	−60 04.	1	14.72		0.72		0.14		1		318
		NGC 6752 - 1117			19 06 24	−60 04.	1	14.10		0.82		0.49		2		318
		NGC 6752 - 1118			19 06 24	−60 04.	2	12.19	.005	1.08	.010	0.81	.010	6		318,786
		NGC 6752 - 1119			19 06 24	−60 04.	1	13.00		0.93		0.51		2		318
		NGC 6752 - 1120			19 06 24	−60 04.	1	15.15		0.73		0.06		1		318
		NGC 6752 - 1121			19 06 24	−60 04.	1	13.47		0.85		0.45		2		318
		NGC 6752 - 1122			19 06 24	−60 04.	1	12.73		1.00		0.53		2		318
		NGC 6752 - 1123			19 06 24	−60 04.	1	12.20		1.07		0.68		2		318
		NGC 6752 - 1124			19 06 24	−60 04.	1	13.22		0.86		0.42		2		318
		NGC 6752 - 1125			19 06 24	−60 04.	1	12.16		1.08		0.72		3		318
		NGC 6752 - 1126			19 06 24	−60 04.	1	11.30		1.30		1.13		7		318
		NGC 6752 - 1127			19 06 24	−60 04.	1	12.13		0.75		0.32		2		318
		NGC 6752 - 1128			19 06 24	−60 04.	1	13.73		0.85		0.42		3		318
		NGC 6752 - 1129			19 06 24	−60 04.	1	15.59		0.66		0.02		3		318
		NGC 6752 - 1130			19 06 24	−60 04.	1	15.59		0.76		0.04		3		318
		NGC 6752 - 1131			19 06 24	−60 04.	1	12.12		0.53		0.04		3		318
		NGC 6752 - 1132			19 06 24	−60 04.	1	14.74		0.72		0.20		2		318
		NGC 6752 - 1133			19 06 24	−60 04.	1	14.66		0.61		0.00		3		318
		NGC 6752 - 1135			19 06 24	−60 04.	1	11.44		1.15		0.86		6		318
		NGC 6752 - 1138			19 06 24	−60 04.	1	15.21		0.75		0.04		2		318
		NGC 6752 - 1139			19 06 24	−60 04.	3	13.34	.019	0.61	.019	0.14	.013	10		318,786,3036
		NGC 6752 - 1140			19 06 24	−60 04.	1	13.75		0.65		0.05		5		318
		NGC 6752 - 1141			19 06 24	−60 04.	1	13.95		0.00		0.10		1		318
		NGC 6752 - 1142			19 06 24	−60 04.	1	14.27		-0.06		0.00		1		318
		NGC 6752 - 1143			19 06 24	−60 04.	1	14.77		-0.07		-0.33		2		318
		NGC 6752 - 1144			19 06 24	−60 04.	1	14.35		-0.01		-0.03		1		318
		NGC 6752 - 1145			19 06 24	−60 04.	1	14.47		-0.06		-0.04		1		318
		NGC 6752 - 2001			19 06 24	−60 04.	1	18.46		0.57				1		368
		NGC 6752 - 2003			19 06 24	−60 04.	1	15.20		0.64				1		368
		NGC 6752 - 2004			19 06 24	−60 04.	1	17.15		0.22				1		368
		NGC 6752 - 2005			19 06 24	−60 04.	1	15.97		0.45				1		368
		NGC 6752 - 2006			19 06 24	−60 04.	1	16.42		0.64				1		368
		NGC 6752 - 2015			19 06 24	−60 04.	1	18.01		0.80				1		368
		NGC 6752 - 2016			19 06 24	−60 04.	1	13.68		0.70				1		368
		NGC 6752 - 2018			19 06 24	−60 04.	1	14.68		0.58				1		368
		NGC 6752 - 2020			19 06 24	−60 04.	1	13.61		1.03		0.92		1		368
		NGC 6752 - 2031			19 06 24	−60 04.	1	13.03		1.12		1.10		1		368
		NGC 6752 - 2032			19 06 24	−60 04.	1	11.42		1.05		0.79		1		368
		NGC 6752 - 2040			19 06 24	−60 04.	1	10.02		0.51		-0.07		1		368

HD	DM	Other Id	N Rem	α_{1950}	δ_{1950}	S	V	σ_V	B–V	σ_{B-V}	U–B	σ_{U-B}	n	Spectrum	References
		NGC 6752 - 2041		19 06 24	−60 04.	1	11.72		0.25				1		368
		NGC 6752 - 2042		19 06 24	−60 04.	1	11.53		0.64		0.11		1		368
		NGC 6752 - 2109		19 06 24	−60 04.	1	13.52		0.54		0.09		1		368
		NGC 6752 - 2111		19 06 24	−60 04.	1	15.52		0.70		0.78		1		368
		NGC 6752 - 2133		19 06 24	−60 04.	1	14.24		1.14				1		368
		NGC 6752 - 2201		19 06 24	−60 04.	1	10.69		0.36				1		368
		NGC 6752 - 3001		19 06 24	−60 04.	1	16.43		0.91		0.65		1		786
		NGC 6752 - 3002		19 06 24	−60 04.	1	16.82		0.63		0.08		1		786
		NGC 6752 - 3003		19 06 24	−60 04.	1	16.97		0.73		0.19		1		786
		NGC 6752 - 3004		19 06 24	−60 04.	1	17.19		0.49		-0.09		2		786
		NGC 6752 - 3005		19 06 24	−60 04.	1	17.38		0.59		-0.21		1		786
		NGC 6752 - 3006		19 06 24	−60 04.	1	17.60		0.61		0.00		1		786
		NGC 6752 - 3007		19 06 24	−60 04.	1	17.70		0.54		0.01		1		786
		NGC 6752 - 3008		19 06 24	−60 04.	1	17.80		0.42		-0.19		1		786
		NGC 6752 - 3009		19 06 24	−60 04.	1	17.98		0.73		-0.08		1		786
		NGC 6752 - 3010		19 06 24	−60 04.	1	18.37		0.61		0.14		1		786
		NGC 6752 - 3011		19 06 24	−60 04.	1	18.43		0.53		-0.17		1		786
		NGC 6752 - 4		19 06 24	−60 07.	1	13.70		0.85		0.38		2		38
		NGC 6752 - 5		19 06 24	−60 07.	1	13.53		0.89				2		38
		NGC 6752 - 6		19 06 24	−60 07.	1	14.34		-0.04		0.04		2		38
		NGC 6752 - 7		19 06 24	−60 07.	1	13.48		0.85		0.46		2		38
178619	+16 03758	HR 7267		19 06 26	+16 46.3	2	6.67	.017	0.44	.009	-0.01		6	F5 IV-V	254,1118,3016
234806	+50 02731	IDS19052N5047	AB	19 06 26	+50 52.2	1	9.70		0.22		0.15		2	A5	1566
178322	−42 13933	HR 7257		19 06 27	−41 58.4	3	5.87	.004	-0.08	.005	-0.48		10	B6 V+F1 IV	15,26,2012
178574	+1 03905	IDS19040N0112	AB	19 06 29	+01 16.3	2	7.56	.003	0.47	.012	-0.02		2	F5	934,1593,6007
178487	−10 04972	LS IV -10 028		19 06 29	−10 17.9	1	8.66		0.16		-0.78		3	B0 Ia	400
178370	−32 14907			19 06 29	−32 00.1	1	9.51		-0.12		-0.69		1	B2/3 II	55
		NGC 6752 - 48		19 06 30	−60 01.	2	12.99	.055	0.92	.020	0.51		2		38,318
		NGC 6752 - 46		19 06 30	−60 02.	2	12.88	.035	0.87	.030	0.25		2		38,318
		NGC 6752 - 47		19 06 30	−60 02.	1	14.39		0.97				1		38
		NGC 6752 - 49		19 06 30	−60 02.	1	14.93		0.87				1		38
		NGC 6752 - 50		19 06 30	−60 02.	1	14.44		0.09				1		38
		NGC 6752 - 44		19 06 30	−60 03.	1	11.68		1.25				1		38
		NGC 6752 - 45		19 06 30	−60 03.	2	11.56	.015	1.23	.005	1.12	.020	4		38,318
		NGC 6752 - 8		19 06 30	−60 07.	2	11.99	.035	1.08	.035	0.91	.104	5		38,318
		NGC 6752 - 9		19 06 30	−60 07.	2	11.23	.047	1.21	.009	0.92	.028	4		38,318
		NGC 6752 - 10		19 06 30	−60 07.	2	12.72	.033	0.78	.014	0.31	.052	4		38,318
		Ross 727		19 06 32	−14 48.9	2	12.02	.005	1.49	.000	1.08		3		1705,3062
178085	−60 07272			19 06 32	−60 21.2	1	8.31		0.58		0.06		3	G0 V	786
178596	+5 04040	HR 7266		19 06 33	+05 59.6	15	5.22	.008	0.34	.007	0.01	.004	145	F5 IV	61,71,985,1088,1118,1363*
178345	−39 13146	HR 7259		19 06 35	−39 25.3	4	4.10	.005	1.20	.000	1.07	.000	14	K0 II/III	15,1075,2012,8015
		NGC 6752 - 42		19 06 36	−60 03.	1	11.67		1.20				2		38
		NGC 6752 - 43		19 06 36	−60 03.	1	12.41		1.09				2		38
		NGC 6752 - 11		19 06 36	−60 07.	2	12.24	.005	0.65	.014	0.32	.066	4		38,318
		NGC 6752 - 12		19 06 36	−60 07.	2	11.22	.023	1.36	.009	1.33	.052	4		38,318
178637	+11 03749			19 06 37	+11 12.8	1	6.64		1.16		1.17		2	K1 III	37
178496	−21 05273	IDS19037S2137	AB	19 06 41	−21 32.8	1	8.21		0.63		0.13		3	G3 V	3026
		NGC 6752 - 40		19 06 42	−60 03.	1	14.04		0.77				1		38
		NGC 6752 - 41		19 06 42	−60 03.	1	13.88		0.10				1		38
		NGC 6752 - 36		19 06 42	−60 04.	2	11.58	.010	1.16	.005	0.79	.065	4		38,318
		NGC 6752 - 31		19 06 42	−60 05.	2	10.85	.050	1.64	.035	1.62	.035	4		38,318
		NGC 6752 - 33		19 06 42	−60 05.	2	12.24	.040	1.03	.010	0.67	.020	4		38,318
		NGC 6752 - 13		19 06 42	−60 06.	1	14.05		0.36				1		38
		NGC 6752 - 28		19 06 42	−60 06.	1	13.17		0.89				1		38
		NGC 6752 - 14		19 06 42	−60 07.	1	13.72		0.90				2		38
		NGC 6752 - 15		19 06 42	−60 07.	1	14.31		1.10				2		38
		NGC 6752 - 16		19 06 42	−60 07.	2	11.49	.028	1.56	.000	1.78		4		38,318
178153	−59 07457			19 06 44	−59 37.6	1	10.13		0.54		0.07		3	F8 V	786
	+3 03918	IDS19042N0315	ABC	19 06 45	+03 19.4	1	9.55		0.50		0.22		1		8084
		LS II +21 003		19 06 46	+21 14.3	1	11.26		0.79		0.30		1	F6 III	1215
178524	−21 05275	HR 7264	★ AB	19 06 47	−21 06.3	8	2.89	.009	0.36	.012	0.18	.049	21	F2 II/III	15,369,1020,1075,2001*
		BTT13,19 # 44		19 06 48	+02 25.	1	11.24		0.99		0.70		2		1586
		NGC 6752 - 39		19 06 48	−60 03.	1	14.46		-0.03				1		38
		NGC 6752 - 35		19 06 48	−60 04.	1	12.31		1.03				1		38
		NGC 6752 - 37		19 06 48	−60 04.	1	14.04		0.84				1		38
		NGC 6752 - 38		19 06 48	−60 04.	1	14.67		0.85				1		38
		NGC 6752 - 32		19 06 48	−60 05.	1	13.65		0.92				1		38
		NGC 6752 - 34		19 06 48	−60 05.	2	12.38	.040	0.84	.010	0.40	.080	4		38,318
		NGC 6752 - 26		19 06 48	−60 06.	1	14.37		0.81				2		38
		NGC 6752 - 29		19 06 48	−60 06.	2	11.81	.029	1.16	.015	0.92	.024	3		38,318
		NGC 6752 - 21		19 06 48	−60 07.	1	14.18		0.89				2		38
		NGC 6752 - 24		19 06 48	−60 07.	1	14.50		0.93				2		38
		NGC 6752 - 25		19 06 48	−60 07.	1	15.21		0.79				1		38
		NGC 6752 - 17		19 06 48	−60 08.	1	13.94		0.90				2		38
		JL 13		19 06 48	−71 15.	1	16.01		-0.32		-1.23		2		132
178555	−20 05415	HR 7265	★ AB	19 06 51	−19 53.1	3	6.13	.018	1.16	.005	1.13		13	K1 III	176,214,2007
178395	−42 13943			19 06 51	−42 48.7	1	9.29		1.08				4	G8 III	2012
178694	+11 03751	IDS19046N1142	AB	19 06 53	+11 47.2	1	8.23		0.22				2	A0	1118
178695	+10 03800	FM Aql		19 06 54	+10 28.3	3	7.93	.074	1.17	.045	0.74	.032	3	F2 I	851,1772,6011
		NGC 6752 - 27		19 06 54	−60 06.	1	15.50		0.16				2		38

Table 1

HD	DM	Other Id	N Rem	α_{1950}	δ_{1950}	S	V	σ_V	B–V	σ_{B-V}	U–B	σ_{U-B}	n	Spectrum	References
		NGC 6752 - 30		19 06 54	−60 06.	2	12.16	.014	1.10	.028	0.78	.089	4		38,318
		NGC 6752 - 19		19 06 54	−60 07.	2	12.75	.034	0.81	.005	0.34	.024	6		38,318
		NGC 6752 - 20		19 06 54	−60 07.	2	14.13	.144	0.70	.114	0.22	.119	5		38,318
		NGC 6752 - 22		19 06 54	−60 07.	1	14.73		0.89		1.01		2		38
178849	+34 03437	IDS19051N3436	A	19 06 56	+34 40.7	2	7.03	.025	-0.13	.010	-0.63	.005	6	B3 V	555,3016
178717	+10 03801		3	19 07 00	+10 09.6	3	7.14	.011	1.86	.021	2.06	.038	11	Kp	851,993,3016
		NGC 6752 - 23		19 07 00	−60 07.	1	15.00		0.72				2		38
178443	−43 13162	IDS19035S4326	AB	19 07 03	−43 21.5	2	9.99	.013	0.70	.023	0.12		4	F8	1594,6006
	+33 03339			19 07 04	+33 58.8	1	9.45		1.25		1.16		1	K8	3072
	−76 01313			19 07 08	−76 43.3	1	10.19		-0.15		-0.82		1	B2 V	400
178719	+1 03911			19 07 10	+01 09.3	1	8.89		0.74		0.47		1	G5	934
178445	−47 12773		5	19 07 10	−47 13.4	5	9.36	.007	1.32	.016	1.24	.019	12	K5/M0 V	158,1494,1705,2033,3077
		ADS 12100	⋆ AB	19 07 12	+37 19.	1	11.61		1.01		0.91		1		8084
178349	−56 09109			19 07 12	−56 23.4	1	7.40		1.61		1.90		5	K4 III	1673
178912	+33 03340			19 07 14	+33 09.4	2	8.33	.010	-0.17	.010	-0.81	.000	4	B5	401,555
178911	+34 03439	HR 7272	⋆ A	19 07 14	+34 31.0	1	6.74		0.63		0.16		7	G5 V	3032
178911	+34 03438	IDS19054N3426	B	19 07 14	+34 31.0	1	8.12		0.75		0.33		2		3016
179094	+52 02350	HR 7275, V1762 Cyg		19 07 15	+52 20.7	1	5.81		1.09		0.87		3	K1 IV	1080
179093	+53 02188			19 07 15	+53 32.8	1	8.30		0.04		0.06		2	A0	1566
179095	+50 02734			19 07 17	+50 16.9	1	6.94		-0.04		-0.27		2	A0	401
178744	−0 03662	HR 7269		19 07 17	−00 30.6	3	6.33	.005	-0.04	.002	-0.48	.010	6	B5 Vn	1079,1149,1586
178853	+14 03807			19 07 18	+14 39.2	1	9.23		1.14				1	K0	882
		CS 22891 # 93		19 07 18	−62 01.9	1	14.27		0.05		0.10		1		1736
	+9 04000			19 07 24	+09 42.1	1	10.25		2.15				2	M3	369
178890	+20 04062			19 07 24	+20 10.5	1	9.26		0.48		0.08		2	G5	1648
178973	+32 03335			19 07 25	+32 47.2	1	7.57		1.27		1.31		2	K0	1733
178947	+30 03442	IDS19055N3024	A	19 07 27	+30 29.1	2	6.92	.020	-0.06	.015	-0.24	.139	5	B9	401,3016
		CS 22947 # 99		19 07 27	−48 54.5	1	14.03		-0.09		-0.69		1		1736
178972	+34 03442			19 07 29	+35 06.2	1	8.68		1.68		2.01		2	M1	1733
178991	+29 03496			19 07 35	+29 16.6	1	8.17		-0.05		-0.51		2	A0	401
178628	−39 13156	HR 7268		19 07 36	−39 05.3	1	6.36		0.01				4	B7 II/III	2007
		Steph 1680		19 07 38	+39 07.4	1	11.33		1.53		1.21		2	M0	1746
178880	+2 03804			19 07 39	+02 16.5	1	7.86		0.39		0.18		2	A0	1375
		G 142 - 11		19 07 39	+17 35.4	2	13.55	.019	1.78	.098	1.42		4		203,1759
		V 916 Aql		19 07 41	+12 27.1	1	10.33		1.61				1		1772
178913	+12 03832			19 07 43	+12 23.5	1	9.03		1.47				2	K2	1655
178861	−12 05298			19 07 54	−12 33.6	3	8.48	.009	0.13	.005	-0.50	.004	9	B2 III	400,963,1732
		G 22 - 16		19 08 01	+02 43.7	1	13.85		1.42				3		202
179143	+37 03357			19 08 02	+37 42.7	1	6.77		0.35		0.17		3	Am	8071
178952	+3 03928			19 08 03	+03 22.2	1	8.42		1.81				2	K5	369
179305	+52 02354			19 08 05	+52 30.5	1	8.41		0.50		0.60		2	F5	1566
		G 22 - 17		19 08 07	+01 27.4	2	12.17	.010	1.51	.005	1.22		3		202,333,1620
178840	−29 15804	HR 7270		19 08 09	−29 35.1	1	6.30		-0.04				4	B8/9 V	2007
180427	+79 00614			19 08 10	+79 34.1	1	7.93		0.97		0.73		3	K0	3016
	+18 03972			19 08 11	+18 38.1	1	10.61		2.16				2	M3	369
	+79 00615			19 08 14	+79 40.4	2	9.72	.038	1.10	.038	0.96		3	K5	1017,3072
	−52 11381			19 08 16	−52 19.6	1	9.96		1.49				1	K3 III:	1494
179018	+3 03931			19 08 18	+03 26.9	1	8.05		0.15		-0.42		1	B9	757
179002	−7 04876	IDS19056S0735	AB	19 08 19	−07 30.6	1	6.76		0.80		0.44		2	G0 IV+G0 IV	3030
		CoD -56 07638		19 08 19	−55 57.7	3	11.31	.004	1.45	.014	1.10		6		158,1494,1705
		CS 22891 # 57		19 08 20	−57 39.2	1	15.03		0.16				1		1736
	+28 03223	IDS19064N2810	AB	19 08 21	+28 15.0	1	9.60		0.36		0.12		2	F0	3016
178710	−52 11382	IDS19046S5158	B	19 08 22	−51 53.8	1	8.42		0.48				4	F6 IV	1075
178977	−14 05313			19 08 24	−14 07.1	1	8.28		0.00		-0.45		4	B5 III	1149
179100	+10 03813			19 08 28	+10 15.8	1	7.09		1.42		1.09		1	K0	851
178534	−64 03979			19 08 28	−63 57.7	1	7.98		0.20		0.10		2	A7 V	1730
178734	−52 11383	IDS19046S5158	A	19 08 29	−51 53.4	1	7.04		1.48				4	K5 III	1075
178611	−59 07461			19 08 31	−59 46.4	1	8.64		0.48		-0.01		4	F5 V	786
179003	−15 05265			19 08 32	−15 17.6	2	9.00	.000	0.02	.000	-0.25	.000	7	B8 V	963,1732
	+28 03225			19 08 37	+28 15.9	1	8.75		0.47		-0.05		2	F6 V	3026
343978	+21 03683			19 08 39	+21 11.4	1	8.66		0.33		0.07		1	F0	963
179367	+44 03073			19 08 41	+44 28.6	1	7.35		0.22		0.10		2	A5 p:	1601
		LP 632 - 16		19 08 42	−01 51.4	1	11.23		0.85		0.32		3		1696
		FN Lyr		19 08 47	+42 22.5	1	11.91		0.15		0.12		1		597
178937	−37 13090	HR 7273		19 08 47	−37 40.0	3	6.56	.004	1.01	.005			11	F8/G0 Ib/II	15,1075,2035
	−64 03980			19 08 48	−63 59.4	1	9.93		1.01		0.69		2		1730
179394	+42 03258			19 08 51	+42 31.1	1	7.51		-0.10		-0.36		2	B8	401
179281	+31 03485			19 08 52	+31 14.3	1	7.80		-0.02		-0.21		2	B9	401
178845	−50 12377	HR 7271	⋆ AB	19 08 55	−50 34.2	3	6.12	.004	0.95	.000			11	G8 III	15,1075,2007
179395	+41 03253			19 08 57	+41 56.3	1	7.09		0.01		-0.08		2	B9	401
		LP 692 - 12		19 08 58	−06 48.2	2	12.22	.075	0.59	.025	-0.12	.015	4		1696,3062
		Be 82 - 6		19 08 59	+13 01.3	1	13.36		0.86		0.32		4		1608
		Be 82 - 1		19 08 59	+13 01.7	1	11.23		0.82		0.12		4		1608
179007	−36 13397			19 08 59	−36 01.6	1	10.00		-0.13		-0.68		1	B7/8 Ib	55
		Be 82 - 2		19 09 00	+13 02.0	1	10.46		2.12		2.00		4		1608
		Be 82 - 5		19 09 01	+13 00.5	1	14.10		0.95		0.53		4		1608
		Be 82 - 22		19 09 01	+13 01.2	1	13.80		0.99		0.47		2		1608
		Be 82 - 23		19 09 01	+13 01.3	1	15.05		1.18		0.66		2		1608
		Be 82 - 16		19 09 02	+13 01.0	1	14.20		1.01		0.61		3		1608
		Be 82 - 3		19 09 02	+13 02.0	1	10.20		1.78		1.35		4		1608

HD	DM	Other Id	N Rem	α_{1950}	δ_{1950}	S	V	σ_V	B–V	σ_{B-V}	U–B	σ_{U-B}	n	Spectrum	References
		Be 82 - 25		19 09 02	+13 02.6	1	15.48		2.59				2		1608
		Be 82 - 13	A	19 09 03	+13 00.4	1	13.94		0.95		0.53		2		1608
		Be 82 - 13	B	19 09 03	+13 00.4	1	13.69		0.98		0.53		2		1608
		Be 82 - 20		19 09 03	+13 01.4	1	13.40		2.45		1.79		3		1608
		Be 82 - 9		19 09 03	+13 02.0	1	14.15		1.14		0.67		4		1608
		Be 82 - 7		19 09 03	+13 02.3	1	13.41		0.91		0.48		4		1608
		Be 82 - 15		19 09 04	+13 01.6	1	14.48		0.97		0.56		3		1608
179338	+28 03227			19 09 04	+28 23.1	1	9.34		0.44		0.01		2		3016
		Be 82 - 14		19 09 05	+13 01.4	1	14.41		1.07		0.82		3		1608
		Be 82 - 12		19 09 06	+13 00.6	1	13.70		0.94		0.48		4		1608
		Be 82 - 4		19 09 06	+13 01.9	1	13.24		0.94		0.52		4		1608
		Be 82 - 24		19 09 06	+13 03.0	1	14.52		1.25		0.71		2		1608
		EG Lyr		19 09 06	+38 29.4	2	10.81	.330	1.64	.100	1.23	.100	2	M5III:	693,793
179421	+38 03464			19 09 09	+39 05.3	1	8.14		1.23		1.32		3	K2 III	833
		Be 82 - 10		19 09 10	+13 01.9	1	14.07		1.20		0.84		4		1608
		PASP94,977 # 2		19 09 19	+04 54.6	1	13.50		1.07		0.47		4		972
		PASP94,977 # 3		19 09 19	+04 55.0	1	12.99		0.77		0.25		4		972
	+0 04129			19 09 20	+00 41.6	1	10.39		0.53		-0.01		2		1375
		PASP94,977 # 4		19 09 20	+04 55.0	1	12.79		0.81		0.33		4		972
		PASP94,977 # 1		19 09 20	+04 55.5	1	11.48		1.48		1.23		4		972
		PASP94,977 # 26		19 09 20	+04 57.7	1	12.92		0.76		0.19		4		972
179506	+41 03257 IDS19077N4137		A	19 09 21	+41 41.5	2	7.91	.010	-0.04	.000	-0.38	.005	4	B8 V	401,555
	+64 01332	XZ Dra		19 09 24	+64 46.6	2	9.95	.364	0.27	.146	0.14	.014	2	F0	668,699
179484	+38 03466	G 207 - 21	⋆AB	19 09 26	+38 41.9	1	8.44		0.73		0.22		4	G6 V	3030
179484	+38 03466	G 207 - 21	⋆ABC	19 09 26	+38 41.9	3	8.09	.338	0.82	.039	0.41	.058	8	G7 V	1003,1381,3030
179116	-34 13476 IDS19062S3401		BC	19 09 27	-33 56.2	1	6.78		0.12		-0.33		4	B3/5	244
179422	+26 03474	HR 7280		19 09 29	+26 39.1	6	6.36	.013	0.41	.009	-0.03	.004	20	F5 V	71,253,254,3016,6009,8095
179201	-21 05292	HR 7276		19 09 29	-21 44.6	1	6.41		1.13				4	K0 III	2009
179202	-23 15152			19 09 29	-23 07.3	1	8.35		0.00		-0.44		1	B5 III	55
179315	+4 04009		V	19 09 30	+04 16.2	1	7.65		1.31		0.89		1	K2	1772
		S104 # 4		19 09 31	-63 55.5	1	10.95		1.08		0.81		2		1730
179343	+2 03815 IDS19070N0227		AB	19 09 32	+02 32.3	3	6.94	.028	0.10	.008	-0.18	.014	4	B9	379,695,3032
179343	+2 03815 IDS19070N0227		C	19 09 32	+02 32.3	1	11.24		1.42		1.35		1		3032
179617	+50 02738			19 09 33	+50 38.2	1	8.65		0.02		0.04		2	A2	1566
		PASP94,977 # 9		19 09 35	+04 55.9	1	13.69		0.84		0.31		4		972
179933	+65 01326	HR 7290		19 09 35	+65 53.7	2	6.26	.005	0.00	.000	-0.01	.005	4	A0 V	985,1733
179424	+14 03822			19 09 36	+14 29.9	1	8.01		0.00				3	B9	1118
178274	-76 01319			19 09 37	-75 53.2	1	6.62		0.16				4	A3 IV	2009
		G 22 - 18		19 09 39	+02 48.5	4	11.09	.021	1.51	.006	1.04	.009	9		333,940,1620,1705,3078
		PASP94,977 # 10		19 09 39	+04 56.0	1	12.67		1.55		1.54		4		972
		PASP94,977 # 11		19 09 41	+04 55.7	1	13.29		0.96		0.60		4		972
179034	-51 11983			19 09 41	-50 53.5	4	8.06	.011	0.46	.011	0.23	.000	10	F2/3 III	611,613,1075,2012
		LP 868 - 16		19 09 42	-25 42.9	1	12.67		0.58		-0.10		2		1696
179461	+26 03477 IDS19077N2605		A	19 09 43	+26 09.9	2	7.77	.005	-0.02	.009	-0.27	.033	4	F0 IV	8071,8080
179583	+40 03620	HR 7284		19 09 43	+40 20.7	1	6.18		0.09		0.09		2	A3 V	401
	+3 03938			19 09 44	+03 25.4	1	9.31		1.26		1.21		1	K2	851
178905	-60 07277			19 09 44	-60 01.0	1	8.87		0.35		0.09		3	F2 III	786
179527	+31 03497	HR 7283, V471 Lyr		19 09 51	+31 11.9	4	5.96	.023	-0.07	.009	-0.29	.012	13	B9p Si	220,252,1063,3033
179406	-8 04887	HR 7279		19 09 58	-08 01.5	3	5.34	.005	0.13	.004	-0.44	.024	12	B3 V	154,1149,2032
179405	-6 05063	LS IV -06 049		19 09 59	-06 32.6	2	8.92	.014	0.39	.007	-0.58	.009	5	B5	405,1586
		Ir-Ch # 32		19 10 00	+06 32.	1	11.17		0.71		0.14		2		1586
		G 142 - 12		19 10 01	+11 33.2	1	15.27		1.90				1		1759
		G 185 - 10		19 10 03	+24 02.8	1	11.22		1.41		1.31		6	M0	7010
179407	-12 05308	LS IV -12 108		19 10 05	-12 40.1	1	9.41		0.09		-0.81		3	B0 (II)	400
		CS 22947 # 101		19 10 08	-48 53.8	1	13.21		0.45		-0.15		1		1736
179706	+39 03676			19 10 09	+39 56.2	1	8.18		-0.07		-0.36		2	A0	401
179323	-26 13936	HR 7277		19 10 09	-25 59.5	2	5.80	.005	1.38	.005	1.42		8	K2 III	2032,8100
337843	+26 03479			19 10 10	+26 20.6	1	8.61		1.16				1	K5	8097
179510	+6 04051		V	19 10 13	+06 47.9	1	9.48		1.86				2	M1	369
179558	+16 03774 IDS19081N1641		CD	19 10 13	+16 46.0	2	7.99	.038	0.76	.000	0.32	.014	4	G5 V	1003,3026
		G 22 - 19		19 10 14	+06 38.5	3	12.64	.005	0.68	.005	-0.01	.011	6	G0	1064,1620,1658
179586	+17 03887			19 10 15	+17 55.0	1	7.48		0.36		0.12		2	F0	3016
		WLS 1920 20 # 13		19 10 16	+17 45.1	1	10.83		0.40		0.13		2		1375
179494	+3 03941	FN Aql		19 10 17	+03 28.3	5	8.13	.034	1.12	.026	0.80	.059	5	G5	757,851,1484,1772,6011
		CS 22947 # 100		19 10 18	-48 45.2	1	13.14		0.22		0.20		1		1736
179733	+39 03677			19 10 19	+39 20.1	2	7.52	.010	0.07	.010	0.06	.000	5	A1 V	833,1601
179588	+16 03775	HR 7285	⋆AB	19 10 20	+16 45.7	2	6.73		-0.01	.005	-0.29		3	B9 IV	1079,1118
179708	+32 03354 IDS19085N3205		AB	19 10 21	+32 10.1	1	7.37		0.05		-0.06		2	A0	401
	+30 03463			19 10 22	+30 15.1	1	9.58		1.30		1.18		1		401
337909	+24 03674			19 10 24	+24 43.4	1	9.95		2.00				2	M0	369
	-9 05039			19 10 24	-09 38.1	1	10.98		0.75		0.14		2		1696
179709	+30 03464 IDS19085N3011		AB	19 10 27	+30 15.7	1	7.65		0.01		-0.24		2	A0	401
179648	+21 03690	HR 7286		19 10 28	+21 28.1	2	6.04		0.06	.019	0.06	.028	4	A2 Vn	1063,1733
179497	-12 05311	HR 7282		19 10 28	-12 22.1	1	5.51		1.44				4	K3 III	2035
179735	+29 03514	LS II +29 001		19 10 30	+29 55.3	1	7.93		0.37		0.20		2	F2 II	8100
179140	-58 07594 IDS19063S5810		AB	19 10 32	-58 05.4	2	7.23	.010	0.64	.005	0.17		5	G3 V	219,1075
		G 142 - 15		19 10 33	+18 43.4	1	10.20		0.81		0.47		1		1658
179518	-14 05333 IDS19078S1437		AB	19 10 36	-14 31.9	1	7.64		0.47		-0.05		4	F5 V	1149
		JL 15		19 10 36	-72 49.	1	16.94		-0.22		-0.98		1		132
179782	+35 03523			19 10 37	+36 05.8	1	6.93		-0.03		-0.08		2	A0	401

Table 1 1045

HD	DM	Other Id	N Rem	α_{1950}	δ_{1950}	S	V	σ_V	B–V	σ_{B-V}	U–B	σ_{U-B}	n	Spectrum	References
		LP 924 - 38		19 10 37	−29 47.6	1	11.94		0.69		0.12		2		1696
179890	+46 02641	FL Lyr		19 10 38	+46 14.3	1	9.34		0.56		0.00		4	G0 V	1768
179065	−64 03982			19 10 38	−63 59.4	1	9.39		0.37		0.00		2	F0/2 V	1730
180006	+56 02209	HR 7295		19 10 44	+56 46.4	4	5.13	.009	1.01	.010	0.84	.015	9	G8 III	37,1080,1363,3016
179454	−33 14035			19 10 45	−33 31.4	1	8.58		1.26		1.43		4	K3 III	244
179957	+49 02959	HR 7293	⋆ B	19 10 47	+49 45.6	3	6.75	.000	0.64	.000	0.17	.000	6	G4 V	1,15,22,1028
179626	−0 03676	G 22 - 20		19 10 47	−00 40.4	6	9.18	.022	0.52	.011	-0.08	.013	11	F7 V-VI	22,202,1003,1620,2017,3077
179958	+49 02959	HR 7294	⋆ A	19 10 48	+49 45.7	4	6.57	.005	0.64	.010	0.21	.000	9	G5 V	1,15,22,1028
179958	+49 02959	IDS19080N0338	AB	19 10 48	+49 45.7	1	5.85		0.67				7	G5 V + G4 V	1197
		Smethells 32		19 10 48	−47 00.	1	12.91		1.50				1		1494
179818	+30 03470			19 10 54	+30 31.0	1	8.84		-0.02		-0.26		2		963
		LP 752 - 16		19 10 54	−10 28.8	1	13.72		1.00		0.74		1		1696
337861	+25 03752	S Lyr		19 10 55	+26 01.3	1	11.96		2.80				1		864
337825	+26 03485			19 10 55	+26 33.5	1	8.10		0.99				2	K0 Ib	8097
179784	+14 03829			19 10 59	+14 57.0	2	6.67	.014	1.40	.009	1.19	.024	7	G5 Ib	1355,8100
179785	+14 03830			19 11 01	+14 51.4	3	7.31	.060	1.53	.007	1.75	.004	8	K3 II-III	882,1355,8100
179909	+39 03682			19 11 01	+39 24.7	2	8.29	.015	-0.07	.010	-0.33	.010	4	A0	401,963
180777	+76 00717	HR 7312		19 11 01	+76 28.7	6	5.14	.014	0.31	.007	0.00	.004	23	A9 V	15,1007,1013,1363*
179433	−45 13054	HR 7281		19 11 02	−45 16.8	1	5.92		0.90				4	G8 III	2007
179737	+9 04024			19 11 03	+09 36.4	1	8.07		1.15		0.96		4	G8 II-III	8100
179742	+4 04022			19 11 04	+04 10.7	1	7.66		0.32				24	F0	6011
179786	+14 03831			19 11 05	+14 31.9	2	7.58	.010	1.66	.005	2.00		4	M2	369,3040
179009	−69 02962	HR 7274		19 11 08	−69 16.7	3	6.26	.004	0.16	.005	0.13	.010	12	A6 IV/V	15,1075,2038
179761	+2 03824	HR 7287, V1288 Aql	⋆ A	19 11 11	+02 12.4	7	5.14	.013	-0.07	.008	-0.41	.009	21	B8 II-III	15,695,1079,1202,1363*
179791	+5 04081	HR 7288		19 11 16	+05 25.8	1	6.48		0.09		0.10		2	A3 V	1149
180161	+57 01961	G 229 - 26		19 11 16	+57 34.8	3	7.03	.011	0.78	.021	0.44	.025	9	G8 V	1355,3016,7008
179576	−32 14979			19 11 16	−32 30.8	2	8.21	.010	0.99	.005	0.67		8	G8 III/IV	244,2012
349853	+19 03948			19 11 17	+19 53.2	1	10.40		1.16		0.57		1	B8 Ib	1586
343986	+24 03681			19 11 17	+24 19.6	1	9.74		1.91				2	K5	369
		G 142 -B2	A	19 11 20	+13 31.3	1	12.68		1.54		1.30		2	sdM2	3016
		G 142 -B2	B	19 11 20	+13 31.3	1	14.00		0.12		-0.60		2		3060
179669	−30 16800			19 11 26	−29 55.1	1	6.88		1.62				4	K5 III	2012
179722	−18 05262			19 11 28	−17 59.3	1	7.96		1.01		0.66		5	G8/K0III/IV	1657
179892	+7 04000			19 11 34	+07 25.8	1	7.81		0.25		0.11		3	Am	8071
230938	+11 03777	G 142 - 17		19 11 46	+11 43.5	1	10.84		0.69		0.20		3	G2	7010
180008	+25 03757			19 11 50	+25 40.3	1	7.01		0.05		-0.19		2	B9 V	105
180077	+38 03489	IDS19101N3852	B	19 11 50	+38 57.4	1	8.65		0.98		0.64		3		833
		CS 22891 # 61		19 11 50	−58 29.9	1	14.26		0.22		-0.72		1		1736
230949	+18 03995			19 11 54	+18 40.0	1	9.91		1.96				2	M2	369
		CS 22947 # 322		19 11 56	−48 27.1	1	13.98		0.79		0.28		1		1736
		CS 22891 # 94		19 11 59	−62 16.0	1	13.48		0.16		0.20		1		1736
180163	+38 03490	HR 7298	⋆ A	19 12 03	+39 03.5	8	4.39	.011	-0.15	.007	-0.65	.006	32	B2.5IV	15,154,555,1084,1363*
180163	+38 03491	IDS19104N3858	B	19 12 06	+39 03.6	1	8.58		0.08		-0.06		3		1084
		L 419 - 114		19 12 09	−42 28.0	1	13.60		1.57				1		3062
179419	−65 03783			19 12 09	−65 19.0	1	6.37		-0.03		-0.27		7	B8/9 V	1628
179989	+4 04031			19 12 10	+04 10.1	1	8.88		1.29				23	K2	6011
180026	+10 03838	IDS19098N1028	A	19 12 10	+10 28.1	1	8.23		0.41		0.23		2	F0	1733
179366	−66 03417	HR 7278	⋆ AB	19 12 11	−66 45.0	3	5.52	.004	0.17	.010	0.12	.010	14	A5 V + A8 V	15,1075,2038
180124	+24 03687			19 12 15	+24 55.5	1	7.20		0.00		-0.52		2	B2 V	555
180028	+5 04087	LS IV +05 011		19 12 18	+05 57.6	3	6.97	.035	0.83	.008	0.52	.030	13	F6 Ib	1355,1509,8100
		G 142 - 20		19 12 19	+14 23.4	1	14.25		0.74		0.19		1		1658,3060
180214	+33 03378	IDS19105N3403	A	19 12 23	+34 07.7	1	7.68		-0.07		-0.40		2	B9	401
180347	+50 02744			19 12 23	+50 49.3	1	8.40		0.25		0.11		2		1566
	+1 03942	G 22 - 21	⋆ A	19 12 24	+02 04.1	1	10.21		1.21				3	K7	1619
	+1 03942	G 22 - 21	⋆ B	19 12 24	+02 04.1	1	11.16		1.30				1		1619
179950	−25 13866	HR 7292	⋆ AB	19 12 29	−25 20.7	6	4.86	.014	0.57	.010	0.32	.006	16	K0 III +A	15,58,1075,2012,3077,8015
179949	−24 15161	HR 7291		19 12 30	−24 16.0	3	6.25	.010	0.54	.000	0.07		6	F8 V	219,1705,2035
179904	−33 14068			19 12 31	−33 37.1	1	7.53		0.01		0.00		4	A0 V	244
180711	+67 01129	HR 7310	⋆ A	19 12 33	+67 34.4	5	3.07	.006	0.99	.018	0.78	.006	13	G9 III	15,1355,1363,3016,8015
179926	−33 14069			19 12 34	−33 40.1	2	8.92	.020	-0.14	.005	-0.78	.010	7	B2/3 II/III	240,244
	+4 04036		V	19 12 35	+04 48.6	1	9.70		2.00				2	M2	369
180126	+9 04037			19 12 39	+09 43.2	1	8.01		0.17		-0.45		3	B3p	399
	+30 03482	LS II +30 002		19 12 39	+30 58.1	1	10.33		0.55		0.07		1	F6 II	1215
179775	−52 11405			19 12 40	−52 31.6	3	6.76	.000	1.02	.000	0.83	.005	6	K0 III	611,613,1075
230984	+14 03844		V	19 12 42	+14 35.0	1	8.27		2.01				2	K5	369
179886	−45 13072	HR 7289		19 12 43	−45 33.3	1	5.40		1.35				4	K3 III	2007
180164	+10 03840	G 142 - 21		19 12 47	+10 29.2	1	9.41		0.52		-0.15		1	F8	1620
180242	+19 03956	HR 7299		19 12 52	+20 06.9	1	6.00		0.88				4	G8 III	1118
180110	−15 05290			19 12 52	−14 55.6	1	7.79		0.02		-0.40		8	B8 II	1088
180243	+14 03845	IDS19107N1455	B	19 12 57	+14 59.6	4	7.64	.027	0.08	.010	0.04	.013	10	A0	150,196,1084,3016
179887	−50 12407	IDS19091S5010	AB	19 12 57	−50 04.5	1	7.93		0.37				4	F0/2 V	1075
180316	+27 03307			19 12 59	+27 52.4	9	6.91	.003	-0.07	.005	-0.50	.004	253	B8	985,1502,1566,1569,1569*
180610	+57 01968	HR 7309		19 13 02	+57 37.1	1	4.98		1.16		1.17		2	K2 III	1080
180068	−30 16828			19 13 02	−30 32.9	1	7.29		0.19		0.10		7	A3 IV	1628
179930	−46 12902	IDS19094S4603	AB	19 13 02	−45 58.0	2	9.37	.009	1.42	.005			5	K5/M0 V	1494,2012
180262	+14 03846	HR 7300	⋆ A	19 13 03	+14 59.7	7	5.56	.021	1.08	.017	0.86	.014	15	G8 II-III	150,196,1084,1118*
180501	+48 02858			19 13 05	+48 37.9	1	7.44		0.03		0.04		2	A0	401
180317	+20 04088	HR 7301		19 13 08	+21 08.6	1	5.64		0.11				4	A4 V	1118
230997	+15 03746			19 13 09	+15 13.2	1	9.57		1.40		1.33		3	K2	196
		HA 87 # 311		19 13 11	+14 58.1	1	12.14		0.59		0.32		3		196

HD	DM	Other Id	N Rem	α_{1950}	δ_{1950}	S	V	σ_V	B–V	σ_{B-V}	U–B	σ_{U-B}	n	Spectrum	References
239025	+59 01973			19 13 12	+59 30.9	1	9.53		0.36				1	G0	70
180070	−34 13532			19 13 12	−33 53.9	1	7.54		0.12		0.11		4	A3 IV	244
338030	+24 03692	G 185 - 14		19 13 14	+24 48.4	1	9.71		1.29		1.28		3	M0	7008
180712	+59 01974			19 13 15	+59 27.7	2	7.97	.005	0.60	.008	0.06		4	F8	70,3016
180352	+19 03959			19 13 16	+19 19.9	1	7.19		1.07		0.90		3	K0 III	37
180093	−33 14076	HR 7296, RY Sgr		19 13 17	−33 36.7	1	11.12		1.12		0.30		1	G0 Iab:pe	817
180002	−46 12905			19 13 17	−45 59.0	1	9.77		0.49		0.05		1	F3/5 V	742
180194	−15 05293			19 13 19	−15 41.1	1	9.02		0.00		-0.42		1	B6/7 IV	963
180377	+18 04011	IDS19111N1820	A	19 13 21	+18 25.6	1	6.43		1.75				2	M2 III	369
		T Sgr		19 13 21	−17 03.6	1	12.24		0.70		0.13		1	Se	440
179863	−58 07600	IDS19091S5822	AB	19 13 21	−58 17.3	1	9.55		0.34		0.05		1	F0 V	1700
180450	+30 03491	HR 7302		19 13 28	+30 26.3	2	5.86	.008	1.66	.016	2.03		4	M0 III	71,3001
231014	+14 03849			19 13 32	+15 08.0	2	8.17	.024	1.04	.012	0.74		3	G5	882,3016
180778	+59 01976			19 13 32	+59 36.1	3	7.69	.011	0.16	.009	0.19	.005	7	A2 p:	70,3016,8071
	−33 14081			19 13 34	−33 34.6	1	10.24		0.50		0.10		4		244
180398	+12 03861			19 13 39	+13 01.1	2	7.92	.005	0.07	.005	-0.43	.005	5	B9	379,1212
	+59 01977			19 13 40	+59 18.9	1	10.34		0.76				1		70
180401	+9 04044			19 13 43	+09 14.3	2	7.69	.005	0.19	.032	-0.26	.018	18	B8	627,1603
239030	+59 01978			19 13 45	+59 43.4	1	9.45		0.13				1	A2	70
180451	+15 03747	IDS19115N1559	AB	19 13 46	+16 04.3	1	7.16		0.31		0.15		4	F0	3016
180275	−19 05367	R Sgr		19 13 46	−19 23.8	1	6.10		1.47		0.53		1	M2e	814
179682	−70 02606			19 13 47	−70 13.6	1	9.67		0.14		0.11		2	A2 V	1730
		ApJ86,509 R4# 2		19 13 48	+22 51.9	1	12.95		2.30				1		8008
180553	+27 03313	HR 7305	★AB	19 13 56	+27 22.0	2			-0.07	.015	-0.59	.010	3	B5 V	1063,1079
239032	+59 01979	IDS19132N5936	AB	19 13 56	+59 40.8	1	9.53		0.61				1	G0	70
180583	+27 03314	HR 7308, V473 Lyr		19 13 59	+27 50.2	5	6.16	.032	0.61	.025	0.33	.028	17	F6 Ib-II	253,1080,6009,8095,8100
180756	+49 02968	HR 7311	★ABV	19 14 00	+49 58.9	1	6.31		0.93		0.61		2	G8 III	1733
	−33 14089	IDS19108S3342	AB	19 14 01	−33 36.9	1	10.80		0.54		0.05		1		244
180022	−59 07469			19 14 01	−58 57.9	1	7.91		0.20		0.17		5	A6 V	1700
180482	+4 04045	HR 7303		19 14 02	+04 44.7	3	5.59	.009	0.10	.013	0.09	.010	15	A3 IV	1149,3023,8097
180613	+30 03494			19 14 02	+31 09.1	1	6.79		-0.08		-0.62		2	B9	401
180554	+21 03713	HR 7306	★A	19 14 04	+21 18.1	7	4.76	.013	-0.06	.012	-0.55	.010	22	B4 IV	15,154,1118,1203,3016*
180612	+31 03517			19 14 04	+31 32.0	1	8.28		-0.02		-0.37		2	A0	401
180682	+40 03645			19 14 04	+40 16.3	1	6.96		1.32		1.33		3	K0	1601
		FO Aql		19 14 05	+00 02.2	1	14.30		0.38		-0.57		1		1471
		LB 3116		19 14 06	−64 41.	1	12.59		-0.12		-0.71		9		45
180555	+14 03852	HR 7307	★A	19 14 09	+14 27.3	1	5.71		-0.02		-0.12		1	B9.5V	542
180555	+14 03852	HR 7307	★AB	19 14 09	+14 27.3	4	5.66	.014	-0.01	.018	-0.13	.013	9	B9.5V +F8 V	938,1063,1118,3023
180555	+14 03852	IDS19119N1422	B	19 14 09	+14 27.3	1	9.07		0.14		-0.05		1	F8 V	542
231034	+19 03966			19 14 11	+19 19.5	1	10.18		0.25		0.17		2	A0	1375
180134	−53 09513	HR 7297		19 14 11	−53 28.6	4	6.37	.007	0.49	.000	-0.02		12	F7 V	15,2018,2030,3037
180638	+28 03268	IDS19122N2806	AB	19 14 12	+28 11.5	1	8.24		0.19		0.17		3	Am	8071
	−33 14092			19 14 15	−33 34.9	1	10.44		1.14		1.06		1		244
		CS 22891 # 122		19 14 22	−59 52.0	1	14.39		-0.05		-0.75		1		1736
		CS 22896 # 21		19 14 23	−54 48.5	1	15.14		0.08		0.19		1		1736
239033	+59 01981	IDS19136N5936	AB	19 14 24	+59 41.1	1	9.02		0.39				1	F0	70
180587	+10 03849	LS II +10 006		19 14 25	+10 53.4	1	8.28		0.27		-0.11		4	B9 II	8100
		CS 22896 # 11		19 14 27	−56 05.0	1	13.83		0.09		0.14		1		1736
		G 207 - 23		19 14 28	+36 58.8	1	11.09		0.66		0.02		5	K2	7010
180617	+4 04048	G 22 - 22	★A	19 14 29	+05 05.8	11	9.12	.010	1.50	.008	1.15	.014	42	M3.5V	1,116,202,333,1006,1075*
180617	+4 04048	V1298 Aql	★B	19 14 29	+05 05.8	1	17.30		2.12				4		3078
180617	+4 04048	IDS19120N0500	C	19 14 29	+05 05.8	1	11.05		1.80		1.40		2		3072
	+3 03962			19 14 30	+04 04.0	1	10.37		0.47				2		246
		LS II +14 008		19 14 31	+14 28.4	1	10.68		0.68		-0.28		2		405
239035	+59 01982			19 14 34	+59 36.7	1	9.06		0.44				1	F5	70
180183	−56 09141	IDS19104S5619	A	19 14 34	−56 14.1	3	6.80	.009	-0.17	.003	-0.76	.006	6	B2 V	55,1700,1732
180809	+37 03398	HR 7314	★A	19 14 38	+38 02.6	8	4.37	.012	1.26	.010	1.23	.015	15	K0 II	15,1080,1363,3016*
180686	+12 03867			19 14 39	+13 01.7	1	8.69		0.20		0.13		3	A0	379
181066	+59 01983			19 14 39	+59 20.0	1	9.93		1.61				1	M2	70
180660	+9 04047	IDS19123N0911	A	19 14 40	+09 14.9	2	8.23	.010	1.36	.005	1.35	.026	7	K2 II	627,8100
180639	+9 04048	V342 Aql	★B	19 14 40	+09 15.2	1	8.72		0.34		-0.05		1		627
		CS 22896 # 15		19 14 40	−55 30.2	2	14.25	.010	0.61	.014	-0.08	.000	3		1580,1736
180642	+0 04159	LS IV +00 016		19 14 42	+00 58.1	1	8.27		0.22		-0.66		6	B1.5II-III	399
180685	+15 03754			19 14 42	+15 54.1	1	8.84		0.09				2		1118
180540	−19 05379	HR 7304		19 14 43	−19 02.6	3	4.92	.025	1.02	.011	0.80	.005	13	K0 III	2007,3016,8100
		G 125 - 1		19 14 44	+41 32.4	2	14.12	.005	1.02	.005	0.61		4		538,1658
181044	+54 02113	IDS19137N5447	AB	19 14 47	+54 52.4	1	8.08		0.53		-0.07		2	K0	3016
180715	+15 03755			19 14 48	+15 59.9	1	8.19		0.16				2	B9	1118
		CS 22947 # 187		19 14 49	−47 19.0	1	12.95		0.65		0.03		1		1736
	+33 03395			19 14 50	+33 18.6	1	9.23		0.92		0.62		2	K0	401
180844	+32 03379	IDS19130N3257	A	19 14 51	+33 02.2	2	7.22	.005	-0.11	.005	-0.55	.010	4	B5	401,555
		CS 22891 # 134		19 14 51	−58 31.6	1	14.46		0.42		-0.17		1		1736
180717	+7 04013	LS IV +07 007		19 14 54	+07 52.3	1	8.67		1.33				1	G5	6009
180629	−17 05564	IDS19121S1706	AB	19 14 58	−17 00.7	1	8.10		-0.05		-0.50		1	B3 III	963
180257	−58 07603			19 14 60	−58 48.8	1	8.39		0.63		0.20		6	G3 V	1700
		ApJ144,259 # 57		19 15 00	+25 32.	1	17.66		0.20		-0.89		2		1360
180915	+38 03514			19 15 01	+38 47.5	1	8.21		0.24		0.06		3	A2	833
	+1 03959			19 15 03	+01 20.2	1	10.02		1.93				2	M3	369
180811	+22 03644			19 15 03	+22 20.1	1	7.86		0.12		-0.26		4	B9	865
344098	+24 03704			19 15 03	+24 14.6	1	8.62		1.91				2	M0	369

Table 1 1047

HD	DM	Other Id	N Rem	α_{1950}	δ_{1950}	S	V	σ_V	B–V	σ_{B-V}	U–B	σ_{U-B}	n	Spectrum	References
		CS 22896 # 16		19 15 04	−55 28.1	1	15.36		0.06		0.14		1		1736
180914	+38 03515			19 15 05	+39 02.6	1	8.42		0.09		0.04		3	A1 V	833
180543	−33 14106			19 15 06	−33 14.6	1	7.65		−0.06		−0.43		4	B5 IV	244
	−70 02611			19 15 07	−70 01.9	1	10.68		0.48		−0.36		2		1730
180544	−34 13566			19 15 08	−34 03.3	1	8.81		0.22		0.19		4	A3 IV	244
		CS 22891 # 143		19 15 08	−57 50.2	1	12.67		0.05		0.07		1		1736
180782	+1 03960	HR 7313		19 15 17	+01 56.4	3	6.18	.005	0.02	.007	0.00	.005	7	A1 Vn	379,695,1149
180889	+21 03719			19 15 23	+21 43.4	1	6.83		0.19		0.14		2	A3	865
	+71 00943	G 260 - 18		19 15 24	+71 26.3	1	9.55		0.77		0.35		2	K0	1658
181096	+46 02658	HR 7322		19 15 25	+46 54.3	4	6.02	.005	0.46	.013	−0.03	.016	6	F6 IV:	254,272,1620,3037
	+87 00183	IDS19453N8807	A	19 15 27	+88 13.8	1	9.96		1.06		0.92		2	K8	3072
180653	−33 14109			19 15 27	−33 35.7	1	8.31		1.26		1.34		4	K1/2 III	244
180868	+11 03790	HR 7315		19 15 28	+11 30.2	3	5.28	.003	0.21	.006	0.23	.032	10	F0 IV	3,1118,3023
180818	−1 03702			19 15 28	−01 06.0	1	8.35		1.75		2.12		4	K2	1657
180968	+22 03648	HR 7318, ES Vul	⋆ AB	19 15 37	+22 56.1	1	5.42		0.02		−0.80		3	B0.5IV	154
181069	+38 03520			19 15 37	+39 01.8	1	6.57		1.13		1.06		2	K1 III	105
180734	−26 14043			19 15 37	−26 10.0	1	8.37		0.05		−0.03		5	B9.5 V	1732
180990	+25 03779			19 15 41	+25 45.6	2	8.45	.005	0.17	.005	0.12	.000	5	A0	833,1625
180702	−33 14114	IDS19125S3327	AB	19 15 44	−33 22.1	3	6.94	.000	0.58	.001	0.07	.001	25	G1 V	173,244,1589
180575	−51 12029	IDS19118S5145	AB	19 15 44	−51 39.9	1	6.58		0.04				4	A1 V	1075
180473	−58 07604			19 15 44	−58 04.3	1	8.64		0.30		0.13		3	A1mA5-F3	1700
		Smethells 36		19 15 48	−53 48.	1	10.82		1.58				1		1494
181047	+25 03780	IDS19137N2511	A	19 15 49	+25 16.5	1	8.34		0.68		0.19		3	G8 V	3026
180748	−33 14116	IDS19126S3320	AB	19 15 52	−33 15.1	2	7.78	.010	0.70	.010	0.26		8	G6/8 V +F/G	244,2012
180576	−54 09321			19 15 56	−54 33.1	1	8.30		1.06		0.88		2	K0 III	1730
181276	+53 02216	HR 7328		19 15 57	+53 16.5	8	3.79	.023	0.96	.011	0.74	.013	22	K0 III	15,1080,1363,3016*
180972	+0 04168	HR 7319	⋆ AB	19 16 00	+00 59.6	6	5.15	.031	1.14	.015	1.03	.043	15	K2 II-III V	542,1149,1355,3016*
181048	+14 03862			19 16 05	+14 22.9	1	7.55		1.08				1	K0	882
181119	+30 03502	HR 7324		19 16 05	+30 55.8	1			0.08		0.13		2	A3 V	1063
181098	+24 03708			19 16 06	+24 19.3	2	7.21	.010	1.09	.005	1.06	.005	6	K1 III-IV	37,3016
181163	+34 03490			19 16 09	+34 40.4	1	8.35		−0.04		−0.18		2	A0	401
180928	−15 05310	HR 7317	⋆ A	19 16 09	−15 37.5	6	6.09	.020	1.42	.012	1.57	.010	21	K4 III	15,252,2035,3077,8011,8015
239041	+58 01888			19 16 13	+59 03.9	1	9.92		0.39				1	G0	70
180996	−4 04768			19 16 15	−03 52.1	1	7.95		1.16		0.96		1	G8 IV	565
181099	+16 03809			19 16 16	+16 36.3	2	7.46	.000	0.23	.009	0.21		5	Am	1118,8071
181053	+0 04170	HR 7321	⋆ A	19 16 17	+00 14.8	2	6.42	.010	1.05	.000	0.78	.009	20	K0 IIIa	1088,4001
		EP Lyr		19 16 17	+27 45.1	1	10.12		0.57		0.23		1	A4Ib:e	793
		CS 22939 # 28		19 16 17	−30 00.3	1	14.08		0.49		0.11		1		1736
180953	−16 05272	V1942 Sgr		19 16 18	−16 00.1	5	6.95	.055	2.46	.057	4.36	.109	8	C6,4	109,864,1238,3038,8005
		CS 22896 # 19		19 16 19	−55 24.6	1	15.65		0.21		0.18		1		1736
180885	−35 13393	HR 7316		19 16 21	−35 30.9	2	5.58	.005	−0.13	.001	−0.58		7	B4 III	26,2035
181120	+15 03762			19 16 23	+15 35.3	1	7.92		0.07				2	A0	1118
	+58 01889			19 16 24	+59 09.0	1	10.03		1.17				1		70
		CS 22896 # 22		19 16 24	−54 44.1	1	14.48		0.30		0.09		1		1736
181164	+25 03786			19 16 25	+25 57.9	1	7.53		−0.02		−0.70		2	B5	555
	−54 09323			19 16 25	−54 42.3	1	9.91		0.62		0.13		2		1730
338046	+27 03336			19 16 28	+27 45.1	1	9.55		0.44		−0.01		24	F5	1222
		L 491 - 30		19 16 28	−37 06.9	1	13.88		1.45				1		3062
239044	+59 01993			19 16 29	+59 23.1	1	8.69		1.59				1	M0	70
181122	+9 04057	HR 7325		19 16 30	+09 31.6	8	6.31	.013	1.06	.009	0.88	.013	23	G9 III	15,627,1256,1355,1509*
181226	+29 03550			19 16 30	+29 51.9	1	7.86		−0.03		−0.35		2	A0	401
181984	+73 00857	HR 7352		19 16 31	+73 15.8	6	4.46	.007	1.26	.011	1.45	.004	15	K3 III	15,1363,3016,6001*
181007	−20 05493			19 16 31	−20 31.1	2	9.59	.008	0.80	.015	0.18		4	G0/3(w)	1594,6006
181166	+15 03765	V889 Aql		19 16 34	+16 09.4	1	8.57		0.21		0.13		2		1768
		CS 22896 # 7		19 16 34	−56 34.3	1	14.01		0.13		−0.71		1		1736
231124	+14 03863	LS II +14 009		19 16 35	+14 14.1	1	10.87		0.62		−0.32		1	B2 III:	1032
181567	+59 01994			19 16 42	+59 16.4	1	8.86		0.46				1	G5	70
	−54 09326			19 16 44	−54 43.7	1	10.30		1.38				4		1730
239047	+59 01995			19 16 46	+59 26.3	1	9.91		0.21				1	A5	70
		CS 22891 # 115		19 16 47	−60 19.5	2	15.15	.010	0.40	.007	0.14	.010	2		1580,1736
		MCC 194	A	19 16 50	−01 42.	2	10.31	.000	1.26	.044			2	M0	1017,1619
		MCC 194	B	19 16 50	−01 42.	1	13.76		1.68				2		1619
		CS 22896 # 23		19 16 50	−54 47.9	1	14.78		0.02		0.07		1		1736
		G 22 - 23		19 16 54	+01 46.4	2	11.77	.044	1.10	.005	0.98		3	K3	202,333,1620
		CS 22891 # 133		19 16 57	−58 45.9	1	14.19		0.37		−0.75		1		1736
	+60 01922			19 17 03	+60 17.7	1	9.53		0.98				1	G5	70
181131	−19 05398			19 17 03	−19 27.6	1	7.47		1.04		0.79		23	G8 III	978
		CS 22896 # 13		19 17 03	−55 50.5	1	14.27		0.09		0.16		1		1736
181675	+59 01996			19 17 04	+59 44.6	1	7.63		0.05				7	A0	1379
		L 347 - 14		19 17 06	−45 37.	1	12.23		1.70		1.22		1		3078
		CS 22896 # 4		19 17 06	−56 54.5	1	15.06		0.01		0.05		1		1736
181593	+53 02219			19 17 07	+53 21.9	1	8.41		0.04		−0.04		2	A0	1566
181231	−0 03705			19 17 08	−00 08.6	1	8.69		0.26		−0.18		2	B9	1586
181360	+23 03625			19 17 09	+23 12.7	2	7.60	.005	0.06	.010	−0.48	.005	5	B3 V	399,1733
181409	+33 03409	HR 7335	⋆	19 17 11	+33 17.8	3	6.59	.008	−0.19	.004	−0.91	.004	8	B2 IV	154,401,555
181596	+49 02976			19 17 12	+50 08.1	1	7.51		1.60		1.93		2	K5	1601
239049	+59 01997			19 17 12	+60 04.8	1	10.25		0.52				1	F8	70
181109	−32 15071	HR 7323		19 17 13	−31 54.7	4	6.57	.011	1.67	.007	2.00		15	K5/M0 III	15,1075,2012,3005
181361	+19 03976	IDS19151N1932	A	19 17 15	+19 37.5	1	8.18		0.06		−0.11		6	A0	1603
181470	+37 03413	HR 7338		19 17 15	+37 21.1	2	6.26		−0.01	.014	−0.08	.009	4	A0 III	1063,1733

HD	DM	Other Id	N	Rem	α_{1950}	δ_{1950}	S	V	σ_V	B–V	σ_{B-V}	U–B	σ_{U-B}	n	Spectrum	References
		G 125 - 3			19 17 15	+38 38.0	2	14.57	.019	0.44	.005	-0.37	.014	4	DC	316,3060
181469	+39 03719	IDS19156N3905		AB	19 17 16	+39 10.4	1	7.92		0.33		0.01		3	A5 V	833
		CS 22896 # 36			19 17 16	−53 24.4	1	14.89		0.09		0.03		1		1736
		CS 22896 # 6			19 17 17	−56 38.2	1	14.81		0.20		0.07		1		1736
181521	+40 03665				19 17 18	+40 16.0	1	6.85		0.02		-0.01		2	A0	401
181597	+49 02977	HR 7341			19 17 18	+49 28.6	1	6.31		1.12				2	K1 III	71
	+60 01924				19 17 18	+60 19.4	1	9.53		0.26				1	A5	70
181333	+12 03879	HR 7331, V1208 Aql	⋆	AC	19 17 19	+12 16.9	4	5.52	.015	0.26	.018	0.17	.011	9	F0 III	253,254,1118,3058
	+59 01998				19 17 19	+59 29.9	1	10.33		1.07				1		70
	+12 03880	IDS19150N1211		B	19 17 20	+12 15.9	1	9.02		0.50				1	G5	3032
	+41 03306	G 125 - 4			19 17 22	+41 33.0	5	8.85	.017	0.81	.011	0.35	.019	9	K0 V	22,792,908,1003,1658
181110	−38 13422				19 17 23	−38 50.4	1	7.28		1.53		1.68		6	K3 III	1673
		CS 22896 # 25			19 17 26	−54 42.2	1	15.64		-0.08		-0.55		1		1736
181677	+51 02552				19 17 28	+52 09.3	1	7.88		1.12		1.06		2	K0	1566
181308	−1 03711				19 17 30	−01 41.4	1	8.66		0.29		-0.25		2	B8	1586
338113	+25 03793				19 17 31	+26 01.1	1	9.83		1.85				2	K7	369
181799	+60 01926				19 17 31	+60 52.0	1	7.12		-0.08				7	B9	1379
181383	+11 03802	HR 7332			19 17 32	+11 26.5	2	6.02		0.08	.000	0.07		5	A2 V	1063,1118
181411	+14 03872				19 17 32	+14 35.8	1	8.67		0.09				2	A0	1118
181367	+2 03852				19 17 33	+02 14.2	1	9.34		0.53		-0.06		2	B8	1586
181492	+31 03544				19 17 33	+31 52.4	2	6.82	.005	-0.09	.005	-0.59	.000	6	B3 V	399,555
	−3 04564				19 17 33	−03 08.2	1	9.90		0.88		0.55		3	K3 V	3072
181240	−22 05063	HR 7327			19 17 38	−22 29.8	2	5.59	.009	0.28	.014			5	A5/7mA5-F2	2035,6009
181312	−10 05035	IDS19149S1045		AB	19 17 39	−10 39.3	1	7.05		5.75		1.36		1	M5 III	3040
181386	+3 03978	IDS19152N0352		A	19 17 41	+03 57.1	1	8.22		1.08		0.77		2	G5 II	8100
181386	+3 03978	IDS19152N0352		B	19 17 41	+03 57.1	1	8.98		1.05		0.77		1		8100
		ApJ144,259 # 61			19 17 42	+46 09.	1	17.39		-0.34		-1.23		2		1360
181414	+4 04071				19 17 50	+04 41.1	1	7.06		0.12		0.13		2	A2	1733
181523	+19 03978				19 17 50	+19 17.7	1	8.52		0.14		0.11		2	A2	1603
		CS 22896 # 2			19 17 50	−57 15.4	1	14.41		0.00		-0.08		1		1736
	−4 04778	G 22 - 24			19 17 51	−04 35.7	2	10.54	.034	0.90	.005	0.58		3	K0	202,1620
		LP 692 - 15		A	19 17 51	−07 45.3	2	12.14	.020	1.64	.010	1.37	.050	6	M2	1,272,3028
		LP 692 - 15		AB	19 17 51	−07 45.3	2	11.66	.015	1.50	.038	1.33		4	M2	1705,7009
		CS 22896 # 32			19 17 51	−53 54.0	1	15.17		-0.11		-0.63		1		1736
181655	+37 03417	HR 7345			19 17 53	+37 14.4	2	6.29	.005	0.67	.005	0.19	.015	7	G8 V	1080,3016
181391	−5 04936	HR 7333	⋆	A	19 17 53	−05 30.6	3	5.01	.009	0.92	.012	0.63	.002	11	G8 III-IV	1088,3016,4001
		LP 692 - 16			19 17 53	−07 45.6	6	12.28	.024	0.06	.015	-0.82	.020	13	DAwk	1,203,832,1705,3060,7009
181494	+13 03985				19 17 54	+14 02.7	1	7.88		0.41				3	B9	1118
		CS 22896 # 35			19 17 54	−53 41.9	1	15.29		0.09		0.17		1		1736
181496	+8 04049				19 17 55	+08 45.1	1	10.60		0.72				1	F8	851
		CS 22896 # 30			19 17 56	−54 07.9	1	15.07		0.09		0.17		1		1736
181497	+8 04050				19 17 59	+08 34.7	1	10.41		0.71				1	G0	851
181495	+8 04051	V600 Aql			19 17 59	+09 07.3	3	9.72	.007	1.32	.023			3	M1	851,1772,6011
181440	−1 03716	HR 7336			19 18 01	−00 59.2	4	5.48	.008	-0.05	.006	-0.22	.005	11	B9 III	15,1079,1256,3023
		G 22 - 25			19 18 02	−02 03.5	1	13.93		1.40				2		202
181657	+35 03573	IDS19162N3521		A	19 18 03	+35 26.8	1	8.20		0.94		0.67		2	A0	1601
		BTT13,19 # 47			19 18 06	+08 25.	1	11.35		1.78		1.92		1		1586
181475	−4 04781				19 18 09	−04 35.8	1	6.96		2.07		2.14		2	K5 II	8100
181603	+20 04108				19 18 11	+20 46.2	1	8.19		1.23		1.24		2	K1 III	37
181321	−35 13422	HR 7330			19 18 12	−35 04.6	3	6.48	.008	0.63	.004	0.09		8	G5 V	15,219,2012
231187	+16 03819				19 18 13	+16 44.8	1	8.54		0.51				3	F8	1118
		GD 218			19 18 14	+11 05.0	2	16.15	.080	-0.04	.070	-0.86	.000	2	DA	1727,3060
		Smethells 37			19 18 18	−46 15.	1	11.70		1.38				1		1494
231193	+17 03930				19 18 21	+17 10.7	1	9.05		0.41				2	B8	1118
		CS 22896 # 28			19 18 21	−54 14.4	2	14.12	.010	0.43	.008	-0.17	.008	3		1580,1736
181605	+9 04071				19 18 23	+09 11.5	1	9.14		1.97				2	K5	369
231195	+14 03881	LS II +14 010	⋆	A	19 18 23	+14 19.5	3	7.71	.018	1.48	.084	1.21	.139	5	F5 Ia	369,1355,8100
		CS 22896 # 10			19 18 24	−56 09.9	1	14.64		0.56		0.02		1		1736
		CS 22896 # 14			19 18 25	−55 36.0	1	14.16		0.01		0.06		1		1736
		CS 22896 # 9			19 18 27	−56 09.9	1	14.76		0.08		0.16		1		1736
181606	+7 04034				19 18 28	+07 15.6	1	9.65		0.38		-0.20		2	B8	1586
181428	−29 16026				19 18 29	−29 42.0	1	7.10		0.57				4	G1 V	2012
		CS 22891 # 144			19 18 29	−57 46.8	1	13.69		0.04		0.07		1		1736
181960	+54 02123	HR 7351			19 18 30	+54 16.9	1	6.26		0.03		0.03		2	A1 V	401
181750	+31 03550				19 18 31	+32 00.3	1	6.78		-0.10		-0.52		2	B9	401
181683	+13 03988				19 18 35	+13 28.9	1	6.90		1.11				1	K0	369
181401	−42 14133	HR 7334			19 18 39	−42 06.7	3	6.33	.004	1.13	.010			11	K1 III	15,1075,2032
181480	−26 14096				19 18 40	−26 15.7	1	7.27		1.44				4	K4 III	2012
181558	−19 05412	HR 7339	⋆	A V	19 18 41	−19 19.8	3	6.25	.005	-0.09	.002	-0.51	.003	28	B5 III	978,1079,2035
	−45 13161				19 18 41	−45 02.0	2	10.76	.021	0.51	.004	-0.24	.021	3		1236,1696
		LS II +28 002			19 18 43	+28 09.6	1	11.88		0.58		0.04		2		405
181828	+34 03503	HR 7346	⋆	A	19 18 43	+35 05.5	3	6.31		-0.10	.013	-0.42	.041	6	B9 V	985,1063,1079
		CS 22896 # 8			19 18 43	−56 26.4	1	14.46		0.04		0.08		1		1736
181577	−18 05322	HR 7340, ρ1 Sgr			19 18 46	−17 56.6	10	3.93	.007	0.22	.008	0.16	.039	45	F0 III/IV	3,15,253,418,1020*
181613	−15 05324				19 18 47	−14 52.2	1	10.08		-0.03		-0.50		1	B3 III	963
181707	+9 04075	IDS19164N0938		A	19 18 48	+09 43.3	1	7.74		2.09				2	K5	369
	+50 02781				19 18 49	+50 50.2	1	7.23		1.65		2.04		1		692
181296	−54 09339	HR 7329			19 18 49	−54 31.1	2	5.04	.005	0.02	.000			7	A0 Vn	15,2027
181709	+5 04118				19 18 50	+05 19.6	1	8.77		0.47		-0.04		2	B8	1586
181615	−16 05283	HR 7342, υ Sgr	⋆	AB	19 18 52	−16 03.0	2	4.61	.000	0.10	.000	-0.53	.000	6	B2 Ia Shell +	15,8015

Table 1

HD	DM	Other Id	N Rem	α₁₉₅₀	δ₁₉₅₀	S	V	σ_V	B–V	σ_B–V	U–B	σ_U–B	n	Spectrum	References
231213	+17 03935	IDS19167N1740	AB	19 18 53	+17 45.7	1	9.32		0.41				2	B8	1118
181544	−29 16041			19 18 54	−29 37.1	1	7.10		0.58				4	G1 V	2012
181019	−68 03218	HR 7320		19 18 54	−68 28.1	3	6.33	.004	1.24	.005	1.25	.000	13	K2 III	15,1075,2038
181961	+45 02877			19 18 56	+45 30.0	1	8.62		1.07		0.96		2		1733
181645	−18 05325	HR 7344		19 18 56	−18 24.2	1	5.87		1.06				4	K0 III	2007
181801	+14 03886			19 19 00	+14 32.3	2	7.70	.024	−0.01	.010	−0.32		5	A0	963,1118
231217	+14 03885			19 19 00	+14 48.8	1	9.73		0.54		0.16		2	A0	379
		NGC 6791 - 4		19 19 00	+37 45.	1	13.23		0.51				1		87
		NGC 6791 - 9		19 19 00	+37 45.	1	15.03		0.54				1		87
		NGC 6791 - 17		19 19 00	+37 45.	1	15.95		0.60				1		87
		NGC 6791 - 24		19 19 00	+37 45.	1	15.11		0.54				1		87
		NGC 6791 - 31		19 19 00	+37 45.	1	16.55		0.62				1		87
		NGC 6791 - 58		19 19 00	+37 45.	1	16.92		0.06				4		87
		NGC 6791 - 120		19 19 00	+37 45.	1	10.76		0.96		0.65		1		87
		NGC 6791 - 121		19 19 00	+37 45.	1	11.19		0.30		0.08		1		87
		NGC 6791 - 122		19 19 00	+37 45.	1	11.31		0.48		0.10		1		87
		NGC 6791 - 124		19 19 00	+37 45.	1	14.32		0.80				2		87
		NGC 6791 - 125		19 19 00	+37 45.	1	14.66		0.82				5		87
		NGC 6791 - 126		19 19 00	+37 45.	1	14.66		1.00				4		87
		NGC 6791 - 127		19 19 00	+37 45.	1	15.95		0.81				4		87
		NGC 6791 - 128		19 19 00	+37 45.	1	16.52		0.87				4		87
		NGC 6791 - 129		19 19 00	+37 45.	1	16.78		1.09				1		87
		NGC 6791 - 130		19 19 00	+37 45.	1	17.02		0.70				2		87
		NGC 6791 - 131		19 19 00	+37 45.	1	17.30		0.71				1		87
		NGC 6791 - 132		19 19 00	+37 45.	1	17.41		0.88				2		87
		NGC 6791 - 133		19 19 00	+37 45.	1	17.59		1.05				5		87
		NGC 6791 - 134		19 19 00	+37 45.	1	18.40		1.11				1		87
		NGC 6791 - 135		19 19 00	+37 45.	1	19.38		1.17				4		87
		NGC 6791 - 136		19 19 00	+37 45.	1	19.60		0.69				4		87
		NGC 6791 - 137		19 19 00	+37 45.	1	19.93		1.02				2		87
		NGC 6791 - 138		19 19 00	+37 45.	1	19.96		1.54				1		87
		NGC 6791 - 139		19 19 00	+37 45.	1	19.97		0.58				3		87
		NGC 6791 - 140		19 19 00	+37 45.	1	14.71		0.70				1		87
		NGC 6791 - 141		19 19 00	+37 45.	1	15.62		0.54				1		87
		NGC 6791 - 142		19 19 00	+37 45.	1	15.63		0.58				1		87
		NGC 6791 - 143		19 19 00	+37 45.	1	10.63		0.13		0.02		2		907
		NGC 6791 - 144		19 19 00	+37 45.	1	11.51		1.09		0.85		4		907
		NGC 6791 - 145		19 19 00	+37 45.	1	10.07		0.40		0.11		4		907
		NGC 6791 - 1401		19 19 00	+37 45.	2	15.31	.045	1.38	.015	1.50		2		907,1503
		NGC 6791 - 1414		19 19 00	+37 45.	1	15.85		1.18		1.37		1		907
		NGC 6791 - 1425		19 19 00	+37 45.	1	18.03		0.85		0.19		2		907
		NGC 6791 - 1459		19 19 00	+37 45.	1	17.25		1.21				1		907
		NGC 6791 - 2001		19 19 00	+37 45.	1	13.71		1.60		1.83		4	K5 III	907
		NGC 6791 - 2002		19 19 00	+37 45.	1	17.00		1.10		0.71		4	K4 III	907
		NGC 6791 - 2008		19 19 00	+37 45.	1	13.84		1.64		1.96		3		907
		NGC 6791 - 2014		19 19 00	+37 45.	1	14.64		1.39				1		1503
		NGC 6791 - 2015		19 19 00	+37 45.	1	16.11		0.48		0.10		1	F2 IV	907
		NGC 6791 - 2017		19 19 00	+37 45.	1	15.04		0.42		0.20		1	F2 IV	907
		NGC 6791 - 2019		19 19 00	+37 45.	1	15.52		0.76		0.30		1	G2 V	907
		NGC 6791 - 2023		19 19 00	+37 45.	1	15.07		0.76		0.26		1	G0 V	907
		NGC 6791 - 2027		19 19 00	+37 45.	1	16.21		0.90		0.49		1	G0 V	907
		NGC 6791 - 2028		19 19 00	+37 45.	1	15.41		0.74		0.25		1	G0 V	907
		NGC 6791 - 2031		19 19 00	+37 45.	1	15.04		1.12		1.00		1	G8 IV	907
		NGC 6791 - 2035		19 19 00	+37 45.	1	15.78		0.81		0.17		1	G5 IV	907
		NGC 6791 - 2038		19 19 00	+37 45.	1	14.12		1.62		1.86		3	K3 III	907
		NGC 6791 - 2044		19 19 00	+37 45.	1	15.01		1.14		0.99		1	K1 IV	907
		NGC 6791 - 2048		19 19 00	+37 45.	1	16.25		1.27		1.44		2		907
		NGC 6791 - 2051		19 19 00	+37 45.	2	14.71	.005	1.46	.010	1.52		3		907,1503
		NGC 6791 - 3004		19 19 00	+37 45.	1	12.76		0.61		0.01		1		1503
		NGC 6791 - 3009		19 19 00	+37 45.	1	14.73		1.39				1		1503
		NGC 6791 - 3010		19 19 00	+37 45.	1	14.19		1.67				1		1503
		NGC 6791 - 3012		19 19 00	+37 45.	1	14.03		1.65		1.47		1		1503
		NGC 6791 - 3016		19 19 00	+37 45.	1	14.57		1.37		1.46		1		1503
		NGC 6791 - 3019		19 19 00	+37 45.	1	14.69		1.34		1.56		1		1503
		NGC 6791 - 572		19 19 00	+37 45.	1	14.09		1.50				2		87
		NGC 6791 - 576		19 19 00	+37 45.	1	14.66		1.36				1		87
		NGC 6791 - 579		19 19 00	+37 45.	1	14.74		1.31				1		87
		NGC 6791 - 584		19 19 00	+37 45.	1	15.27		1.29				2		87
	+60 01930			19 19 01	+60 45.6	1	9.63		0.79				1	K0	70
181454	−44 13277	HR 7337	⋆ A	19 19 03	−44 33.3	1	4.01		−0.10		−0.38		4	B9 V	1279
181454	−44 13277	HR 7337	⋆ AB	19 19 03	−44 33.3	4	3.92	.004	−0.09	.004	−0.32	.008	12	B8 V	15,1075,2012,3023
		CS 22896 # 33		19 19 03	−53 48.9	1	14.46		0.05		0.11		1		1736
231219	+17 03937			19 19 05	+17 18.3	1	10.03		0.53				3	F5	1118
181484	−44 13278	IDS19154S4439	B	19 19 05	−44 33.2	1	7.21		0.33		−0.07		4	F0 V	1279
		CS 22896 # 38		19 19 06	−52 48.2	1	14.81		0.03		0.06		1		1736
181829	+20 04114			19 19 08	+20 52.0	2	7.43	.005	0.00	.005	−0.57	.010	3	A0	401,963
181803	+8 04053			19 19 11	+08 46.6	1	9.03		0.44		−0.03		2	B9	1586
	+22 03670		V	19 19 11	+22 32.6	1	10.06		2.17				2	M2	369
231223	+16 03824			19 19 12	+17 02.8	1	10.62		0.55				4	G0	1118
231225	+17 03938			19 19 13	+17 38.6	1	9.31		0.36				2	B9	1118

HD	DM	Other Id	N	Rem	α_{1950}	δ_{1950}	S	V	σ_V	B–V	σ_{B-V}	U–B	σ_{U-B}	n	Spectrum	References
181620	−32 15110				19 19 13	−32 02.0	1	6.91		1.69				4	M2 III	2012
		ArkAstr3,439 # 436			19 19 14	+17 20.6	1	11.08		0.70				3		1118
344237	+21 03743				19 19 14	+22 06.8	1	9.63		0.53		0.07		2	G2	1375
181986	+38 03550				19 19 16	+38 41.6	1	8.38		0.43		0.05		3	F2 II	833
	+14 03887	LS II +14 011			19 19 17	+14 47.1	3	9.87	.019	1.62	.013	0.36	.038	5		379,445,1012
		CS 22947 # 208			19 19 17	−48 41.6	1	14.89		1.30		0.69		1		1736
	+10 03872	LS IV +10 006			19 19 18	+10 39.9	3	10.44	.006	0.91	.013	0.17	.011	18	B5 Ib	830,1318,1783
231226					19 19 18	+16 51.	1	10.70		0.39				3	A7	1118
182190	+57 01986	HR 7356			19 19 21	+57 33.0	3	5.91	.005	1.63	.025	1.99	.015	6	M1 IIIab	71,1733,3001
		CS 22939 # 58			19 19 21	−28 14.9	1	14.80		0.32		0.16		1		1736
		CS 22891 # 131			19 19 21	−59 12.4	1	14.17		0.22		−0.46		1		1736
		GD 219			19 19 23	+14 34.9	1	13.01		0.06		−0.66		1	DA	3060
231229	+16 03825				19 19 23	+16 58.3	1	9.55		0.39				3	A5	1118
		Vul R2 # 16			19 19 23	+21 09.3	1	10.96		0.42		0.22		1	A0 V	5012
182308	+64 01344	HR 7361		⋆ A	19 19 23	+64 17.8	1			−0.04		−0.48		3	B9 p HgMn	1079
181963	+25 03802				19 19 24	+25 30.5	3	7.43	.005	−0.03	.007	−0.76	.000	7	B2 V	399,833,963
	+37 03424	UZ Lyr			19 19 24	+37 50.5	1	10.27		0.08		−0.09		1	A2	87
		ArkAstr3,439 # 439			19 19 30	+17 30.0	1	10.74		0.65				4		1118
181935	+14 03890				19 19 31	+14 26.8	1	7.36		1.38				1	K0	882
	+17 03941				19 19 31	+17 28.5	1	10.90		0.53				4		1118
		G 185 - 18			19 19 31	+20 47.1	2	13.41	.025	1.71	.000			7		419,3078
181783	−14 05407				19 19 34	−13 49.7	1	8.65		0.13		0.09		22	A0 V	978
181987	+25 03803	Z Vul		⋆ A	19 19 35	+25 28.7	1	7.28		0.09		−0.41		1	B5 V	963
181987	+25 03803	IDS19175N2523	B		19 19 35	+25 28.7	1	12.00		1.22		0.39		20		1352
181786	−20 05514				19 19 35	−20 24.8	1	8.42		1.19				1	K0 III	1642
	+60 01932				19 19 37	+60 45.6	1	9.10		1.20				1	K2	70
181623	−45 13171	HR 7343			19 19 37	−44 53.7	6	4.28	.004	0.34	.005	0.05	.015	23	F2/3 V	15,1075,1075,2012*
		G 22 - 26			19 19 38	+06 57.2	2	12.32	.015	1.68	.020	1.31		5		202,203
181858	−8 04950	HR 7347			19 19 38	−08 17.8	4	6.66	.006	−0.03	.004	−0.54	.008	18	B3 IVp	15,154,1088,2030
181720	−33 14164				19 19 38	−33 00.6	2	7.85	.000	0.58	.005	−0.02	.044	5	G1 V	1311,3077
182056	+30 03524				19 19 41	+30 16.1	1	7.70		1.20		1.14		2	K2 II	8100
		CS 22896 # 34			19 19 41	−53 40.5	1	15.45		−0.01		−0.05		1		1736
231240	+17 03942				19 19 42	+17 51.9	1	10.02		0.42		0.28		2	A3	1375
181462	−59 07478				19 19 42	−59 35.4	1	8.90		1.46		1.75		1	K4 III	1700
181809	−20 05516	V4138 Sgr			19 19 43	−20 44.3	2	6.76	.049	1.03	.001	0.75		2	K2 III	1641,1642
	+30 03526	LS II +31 001			19 19 44	+31 04.0	1	9.93				0.06		2	A0 Ib-IIe	379
	−24 15272				19 19 46	−24 14.3	1	9.87		0.29		−0.07		1		1696
181907	−0 03725	HR 7349			19 19 47	−00 20.9	3	5.82	.005	1.09	.005	0.97	.008	8	K0	15,657,1256
181671	−45 13173				19 19 47	−45 07.8	1	9.52		0.17		0.08		1	A3 V	742
182032	+22 03674	IDS19177N2219	AB		19 19 49	+22 24.6	1	7.43		−0.01		−0.56		4	B2.5V	399
182010	+17 03943				19 19 50	+17 39.4	2	7.05	.015	0.06	.005	−0.38		7	A0	1118,1733
		G 185 - 19			19 19 50	+28 34.0	2	11.54	.029	1.42	.005	1.09		7		419,1759
231243	+16 03826	LS II +16 006			19 19 51	+16 26.6	1	9.72		0.72				3	A2 II	1118
181863	−14 05409				19 19 51	−14 13.3	1	8.18		1.27		1.20		17	K1/2 III	978
	+30 03527				19 19 52	+30 57.6	1	9.18		0.15		0.08		2	A0	379
	+29 03570			V	19 19 53	+29 40.2	1	9.82		1.64				2	M2	369
181862	−14 05411				19 19 53	−14 08.0	2	8.68	.002	1.32	.004	1.46	.002	36	K1/2 III	978,1621
		BPM 42769			19 19 54	−46 52.	1	13.70		0.08		0.25		3		3065
231247	+16 03829	IDS19177N1628	AB		19 19 58	+16 34.0	1	8.70		0.40				4	B9	1118
		CS 22896 # 26			19 19 58	−54 32.9	1	14.31		0.08		0.13		1		1736
		G 125 - 5			19 20 00	+35 56.8	1	11.22		0.56		−0.08		2		1658
181989	+4 04080			V	19 20 02	+04 30.1	1	9.46		2.00				2	M1	369
	+37 03426				19 20 03	+37 51.8	1	9.69		0.94		0.60		1		87
		NGC 6791 - 123			19 20 04	+37 42.2	1	13.19		0.55				1		87
182033	+7 04043				19 20 05	+07 34.2	1	9.45		0.06		−0.46		1	B8	963
182078	+22 03675				19 20 05	+22 26.4	2	7.81	.009	−0.05	.000	−0.66	.000	4	B2 V	399,963
	−12 05373				19 20 05	−11 59.0	1	10.30		1.62		1.93		2	M0	1746
		LS II +34 001			19 20 06	+34 57.1	1	11.22		0.14		−0.62		2		405
	+6 04112				19 20 07	+06 40.5	1	10.05		2.02				2	M2 III	369
181943	−14 05413			V	19 20 07	−14 21.3	2	9.51	.197	0.81	.012	0.33	.009	14	G8 V	1621,1641
181743	−45 13178				19 20 07	−45 09.5	7	9.66	.020	0.46	.010	−0.25	.016	20	F3/5w	742,1236,1311,1594*
	+37 03427				19 20 09	+37 28.7	1	8.76		1.48		1.80		1	K5	87
231255	+16 03831				19 20 12	+16 14.3	1	9.09		0.18				2	A2	1118
182100	+14 03892				19 20 13	+14 57.7	1	7.65		1.70				1	K2	882
181433	−66 03431				19 20 14	−66 34.3	2	8.39	.015	1.04	.000	0.96		6	K2 IV/V	2012,3077
	+4 04082				19 20 19	+04 29.1	1	9.84		0.45		−0.15		2	B5	1586
182038	−7 04942	HR 7353			19 20 23	−07 29.8	2	6.31	.000	1.45	.004	1.68		9	K0	1088,6007
182040	−10 05057				19 20 24	−10 48.0	5	7.00	.018	1.08	.011	0.63	.013	6	C1,2 Hd	109,864,1238,1589,8005
182101	+9 04081	HR 7354			19 20 25	+09 48.9	3	6.33	.017	0.45	.008	−0.03	.015	10	F6 V	15,1256,8100
182564	+65 01345	HR 7371			19 20 25	+65 37.1	4	4.59	.005	0.03	.012	0.06	.000	10	A2 IIIs	15,1363,3023,8015
181869	−40 13245	HR 7348			19 20 25	−40 42.7	4	3.96	.004	−0.10	.005	−0.33	.005	15	B8 V	15,1075,2012,2016
349969	+19 03995				19 20 29	+20 05.1	1	9.44		0.57		0.15		1	G5	401
239065	+59 02008	IDS19198N6003	AB		19 20 29	+60 08.5	1	9.95		0.49				2	F5	1379
	+37 03429				19 20 30	+37 52.0	2	9.85	.010	0.11	.005	0.08	.019	3	A2	87,907
182440	+57 01993	HR 7365			19 20 31	+57 40.2	1	6.58		1.34		1.42		2	K2	1733
182271	+37 03430				19 20 32	+37 47.3	1	8.68		0.17		0.11		1	A0	87
		CS 22896 # 70			19 20 32	−55 30.9	1	14.75		0.13		0.19		1		1736
182085	−6 05125				19 20 36	−06 40.9	1	9.72		1.10		0.95		1	K5 V	3072
182272	+33 03434	HR 7359			19 20 41	+33 25.3	1	6.06		1.03				2	K0 III	71
		G 207 - 30			19 20 42	+29 20.6	1	15.35		1.42				1		1759

Table 1   1051

HD	DM	Other Id	N Rem	α1950	δ1950	S	V	σV	B–V	σB–V	U–B	σU–B	n	Spectrum	References
182256	+25 03810			19 20 43	+25 13.7	2	8.65	.034	0.42	.013	0.03	.009	4	F5 IV	1603,3016
231267	+15 03787	LS II +15 011		19 20 44	+15 57.6	1	9.84		0.44				3	B7 II	1118
182471	+55 02190			19 20 47	+55 38.5	1	8.28		0.19		0.10		2	A3	1502
182255	+25 03811	HR 7358		19 20 48	+26 09.9	7	5.19	.011	-0.12	.003	-0.51	.012	38	B6 III	15,154,1203,1363,3016*
181925	-43 13352	HR 7350		19 20 48	-43 49.2	1	6.17		1.60				4	M1/2 III	2009
182196	+7 04052			19 20 50	+07 25.7	2	8.21	.043	1.64	.047	2.01		8	K5 V	1619,7008
182239	+14 03896	HR 7357		19 20 51	+14 49.4	1	6.66		0.26		0.09		2	F1 V	1733
		LS II +13 004		19 20 52	+13 45.4	1	10.87		1.59				3		725
182511	+55 02191			19 20 53	+55 46.9	1	9.07		0.11		0.09		2	A0	1502
		LP 692 - 20		19 20 54	-03 23.3	1	11.55		0.69		0.03		2		1696
		CS 22896 # 81		19 20 58	-57 17.1	1	15.29		0.12		0.17		1		1736
182274	+19 03996			19 21 01	+19 16.7	1	7.81		0.48				2	F6 V	1025
182293	+19 03997			19 21 01	+20 10.7	4	7.10	.021	1.14	.013	1.20	.004	9	K3 IVp	1003,1025,1080,3077
		CS 22891 # 162		19 21 02	-58 44.6	1	15.62		0.02				1		1736
		ArkAstr3,439 # 447		19 21 03	+16 41.6	1	11.52		0.71				3		1118
231285	+14 03898	LS II +15 012		19 21 07	+15 07.1	1	9.50		0.75		-0.32		2	B0 III	1012
	+15 03793			19 21 08	+16 08.2	1	11.27		0.65				3		1118
		ArkAstr3,439 # 448		19 21 08	+17 18.4	1	12.16		0.84				3		1118
		CS 22896 # 71		19 21 08	-55 50.2	1	14.73		-0.07		-0.47		1		1736
		LS II +13 005		19 21 09	+13 29.0	1	11.93		0.54				3		725
	+28 03312	IDS19192N2830	AB	19 21 11	+28 36.0	1	9.16		1.67				2	M2	369
		CS 22896 # 42		19 21 11	-53 12.2	1	14.05		0.10		0.20		1		1736
181773	-62 06081	IDS19166S6222	AB	19 21 11	-62 17.0	2	7.60	.009	0.46	.005	0.00		8	F5/6 V	158,1075
182296	+8 04072			19 21 14	+08 33.7	2	7.01	.036	1.32	.000	1.14	.018	5	G3 Ib	1355,8100
182335	+20 04123			19 21 14	+20 28.4	2	7.96	.045	0.57	.010	0.15		7	G2 IV wls	81,1025
		ArkAstr3,439 # 449		19 21 15	+17 01.6	1	12.27		0.79				4		1118
182442	+39 03748	IDS19196N3943	AB	19 21 18	+39 48.8	1	9.00		-0.01		-0.14		2	A0	401
231292	+16 03835			19 21 21	+17 05.2	1	9.40		0.41				2	A2	1118
182180	-28 15767	HR 7355		19 21 23	-27 57.8	2	6.04	.000	-0.12	.004	-0.69		7	B2 Vnn	26,2007
182156	-31 16655			19 21 23	-30 54.0	1	7.65		1.15				4	K1 III	1075
	+29 03581			19 21 24	+29 42.0	1	10.25		1.70				2	M2	369
182487	+42 03325			19 21 25	+42 52.6	3	7.01	.000	0.02	.002	-0.12	.000	33	A0	28,1276,1303
239068	+59 02013			19 21 29	+60 02.0	1	9.39		0.47				2	F8	1379
		L 114 - 31		19 21 30	-65 41.	1	13.84		0.84		0.06		2		3062
		Vul R2 # 5		19 21 31	+23 04.0	1	12.89		0.90		0.66		1		5012
		CS 22896 # 49		19 21 31	-53 52.4	1	14.53		0.00		-0.01		1		1736
182357	+8 04074			19 21 34	+08 18.8	1	7.85		1.24				31	K0	6011
		CS 22896 # 48		19 21 34	-53 52.2	1	15.25		0.08		0.19		1		1736
182381	+15 03798			19 21 35	+15 55.0	1	7.61		0.15				2	B9p	1118
182549	+46 02675			19 21 35	+46 11.5	1	7.99		0.90		0.70		2	G6 II	8100
182422	+19 04000	HR 7364		19 21 36	+20 10.0	5	6.41	.019	0.02	.020	-0.08	.019	16	B9.5V	81,401,1025,1063,1118
	-22 13916			19 21 36	-22 08.3	2	10.93	.015	1.44	.015	1.16		5	K4	1705,7008
		CoD -22 13916		19 21 36	-22 09.3	1	10.89		1.41		1.15		4	K5	158
	+6 04120			19 21 38	+06 43.1	1	10.21		2.01				2	M5 III	369
231303	+16 03837			19 21 39	+16 22.7	1	9.47		1.96				2	M2	369
184146	+83 00552	HR 7425	⋆A	19 21 39	+83 22.2	2	6.52	.010	0.10	.000	0.11	.005	4	A3 V	985,1733
182298	-9 05123			19 21 39	-09 26.2	1	7.22		1.57		1.85		3	K5	1657
182488	+32 03411	HR 7368		19 21 41	+33 07.3	5	6.38	.015	0.80	.008	0.47	.010	21	G8 V	101,253,1067,1080,1197
		JL 21		19 21 42	-70 54.	1	15.73		0.02		-0.62		1		832
231309	+17 03946			19 21 44	+17 19.7	1	9.26		0.69				3	G5	1118
182613	+50 02789			19 21 44	+50 56.2	1	8.59		0.17		0.12		2	A2	1566
		CS 22891 # 184		19 21 48	-60 40.1	2	13.82	.010	0.51	.011	-0.03	.005	3		1580,1736
182444	+17 03949			19 21 51	+17 55.2	1	7.84		0.14				4	A2	1118
182301	-20 05531			19 21 51	-20 41.5	1	8.69		1.10				1	K0 III	1642
182300	-19 05438			19 21 54	-19 59.6	1	9.37		1.09		0.74		1	Kp Ba	565
		BF Cyg		19 21 55	+29 34.5	1	11.12		0.80				2		867
		MWC 315, BF Cyg		19 21 55	+29 34.5	1	12.00		0.88		-0.64		1		1753
182286	-29 16104	HR 7360	⋆AB	19 21 55	-29 24.5	1	5.93		1.26				4	K3 III	2009
		CS 22896 # 47		19 21 57	-53 52.6	1	14.77		0.23		0.09		1		1736
182489	+18 04055			19 22 00	+18 38.6	2	7.91	.009	0.01	.005	-0.25		4	B8 V	401,1025
		CS 22896 # 82		19 22 02	-57 07.8	1	14.85		0.21		0.14		1		1736
		CS 22896 # 79		19 22 04	-56 41.0	1	15.74		0.17		0.15		1		1736
182615	+40 03700			19 22 05	+40 41.3	1	7.98		-0.05		-0.20		2	B9 V	555
182691	+49 02994	HR 7381		19 22 05	+50 10.4	3	6.51	.004	-0.08	.004	-0.30	.047	5	B9 III	985,1079,6003
350014	+18 04056			19 22 06	+18 37.3	1	10.29		1.45		1.33		1	K5	401
	+24 03734		V	19 22 06	+24 54.3	1	10.36		1.93				2	M2	369
182490	+16 03839	HR 7369	⋆A	19 22 07	+16 50.3	5	6.25	.002	0.07	.009	0.04	.009	60	A2 III-IV	61,71,1063,1118,8071
349993	+20 04126			19 22 08	+20 25.9	2	8.67	.019	0.52	.019	0.04		7	F6 V	1025,1375
344341	+21 03767			19 22 08	+21 31.1	1	8.37		1.06				2	G8 III	1025
		CS 22891 # 161		19 22 08	-58 36.3	1	14.87		0.16		0.17		1		1736
182568	+29 03584	HR 7372	⋆A	19 22 09	+29 31.3	6	4.98	.008	-0.11	.015	-0.71	.008	26	B3 IV	15,154,1363,3016,6001,8015
182411	-8 04966			19 22 10	-08 12.1	1	9.46		0.08		-0.44		1	B8	963
181673	-73 02016			19 22 12	-73 11.5	1	7.00		0.31		0.10		6	A7/8 III/IV	1628
	-53 09556			19 22 13	-52 57.9	1	9.17		1.13		0.90		1	K0	565
182369	-24 15303	HR 7362	⋆AB	19 22 14	-24 36.4	2	5.02	.005	0.24	.005			7	Am	15,2012
182635	+36 03539	HR 7376		19 22 18	+36 21.1	1	6.43		1.08		0.92		2	K1 III	1733
		ArkAstr3,439 # 456		19 22 19	+16 15.9	1	11.36		0.73				4		1118
182160	-55 09080			19 22 19	-54 57.8	1	7.55		1.06				4	K0 III	2012
182618	+27 03379	HR 7374		19 22 22	+27 59.3	3	6.53		-0.08	.003	-0.56	.009	7	B5 V	253,1063,1079
182694	+43 03229	HR 7382		19 22 22	+43 17.4	3	5.86	.023	0.92	.006	0.63	.001	6	K0 III	71,3016,4001

HD	DM	Other Id	N Rem	α_{1950}	δ_{1950}	S	V	σ_V	B–V	σ_{B-V}	U–B	σ_{U-B}	n	Spectrum	References
182475	−5 04964	HR 7366		19 22 22	−04 59.0	1	6.51		0.33		0.09		8	A9 V	1088
182570	+21 03768	IDS19202N2119	AB	19 22 24	+21 25.0	1	8.17		0.36				2	F2 V	1025
234884	+54 02136			19 22 24	+54 38.0	1	9.06		1.41		1.61		2		1733
182671	+38 03575			19 22 27	+39 06.8	1	7.47		1.38		1.60		2	K4 III	105
182416	−24 15307	HR 7363		19 22 28	−24 03.7	5	5.45	.011	1.44	.007	1.69	.018	28	K3 III	3,418,1311,2035,3005
		ArkAstr3,439 # 457		19 22 29	+16 18.2	1	12.10		0.42				4		1118
182571	+16 03842	IDS19199N1644	B	19 22 30	+16 51.5	3	6.82	.002	0.15	.004			53	A0	61,71,1118
		CCS 2730		19 22 30	+30 32.	1	10.67		1.39				1	R5	1238
182477	−14 05428	HR 7367		19 22 32	−13 59.9	2	5.72	.025	1.40	.025	1.35		7	K2 III	2007,3005
182620	+19 04004			19 22 33	+19 50.4	3	7.18	.022	0.16	.028	0.10		9	A2 V	81,1025,1118
	−73 02017			19 22 33	−72 53.8	1	8.69		0.74		0.29		2		1696
		AO 785		19 22 34	+42 41.4	2	10.89	.010	1.28	.005	1.52	.038	2	K3 III	1697,1748
182572	+11 03833	HR 7373	★ A	19 22 35	+11 50.2	9	5.16	.017	0.76	.018	0.43	.010	28	G8 IV	15,22,37,1003,1013*
344312	+22 03686	LS II +22 005		19 22 36	+22 38.0	1	9.38		0.62		−0.33		2	F2 II	8100
		CS 22896 # 62		19 22 36	−54 57.7	1	13.03		0.26		0.14		1		1736
231348	+12 03900			19 22 37	+12 53.8	1	9.61		1.99				2	M0	369
182754	+46 02681	IDS19212N4615	AB	19 22 37	+46 20.4	1	7.56		−0.06		−0.35		2	A0	401
		CS 22896 # 50		19 22 37	−54 03.0	1	14.47		0.14		0.18		1		1736
344313	+22 03687	LS II +22 006	★ A	19 22 38	+22 40.5	3	9.39	.018	0.64	.012	−0.36	.023	4	B2 V:pe	445,1012 ,7007
		Vul R2 # 3		19 22 39	+22 40.6	1	11.11		0.51		0.30		1	B9 V	5012
		LP 752 - 17		19 22 39	−12 02.8	1	11.88		0.54		−0.12		2		1696
		CS 22939 # 87		19 22 39	−29 32.0	1	13.52		0.08		0.05		1		1736
	+9 04094			19 22 40	+09 19.2	1	9.59		1.98				2	M3	369
		ArkAstr3,439 # 461		19 22 40	+16 12.8	1	11.97		0.53				3		1118
231351	+16 03843			19 22 41	+17 01.9	1	9.18		0.18				2	A2	1118
182755	+42 03331			19 22 42	+42 49.4	2	8.63	.005	0.44	.000	0.13	.014	2	F6 II-III	1276,1697
182737	+36 03543			19 22 44	+37 05.3	1	7.90		0.15		−0.17		2	A0	401
350000	+19 04005			19 22 48	+19 42.0	1	8.87		0.09				3	B9 V	1025
		CS 22896 # 46		19 22 48	−53 50.1	1	15.26		0.09		0.20		1		1736
		LS II +31 002		19 22 50	+31 10.4	1	10.33		0.48		0.16		1	F3 II	1215
231358	+17 03956			19 22 52	+17 13.5	1	9.40		1.26				3	K0	1118
350009	+18 04064			19 22 53	+19 06.2	1	10.09		0.47		0.02		2	F2	1375
231360	+16 03844			19 22 54	+16 38.5	1	9.84		0.47				2	B9	1118
231363	+17 03957			19 22 56	+17 17.4	1	8.92		0.19				2	A0	1118
182806	+38 03578			19 22 57	+38 39.9	1	8.30		0.17		0.13		3	A2	833
344338	+21 03772	IDS19208N2138	AB	19 22 58	+21 44.1	1	9.24		0.17				2	A0	1025
182640	+2 03879	HR 7377	★ A	19 22 59	+03 00.8	12	3.36	.006	0.32	.004	0.03	.009	64	F0 IV	15,851,1013,1197,1256*
182988	+60 01939			19 23 01	+61 00.0	1	8.55		1.09				4	K2	70
		LP 752 - 18		19 23 01	−11 15.9	1	11.50		0.71		0.06		2		1696
182718	+19 04007			19 23 03	+19 33.7	2	8.36	.049	0.25	.005	0.11		5	Ap	81,1118
		CS 22896 # 75		19 23 03	−56 23.9	1	14.94		0.21		0.12		1		1736
		CS 22896 # 52		19 23 04	−54 13.8	1	15.39		0.29		−0.06		1		1736
	+42 03333			19 23 05	+42 30.8	1	9.88		1.02		0.92		1	K0 III	1697,1748
		CS 22896 # 60		19 23 06	−54 43.8	1	14.87		0.20		0.19		1		1736
349999	+19 04008			19 23 10	+20 08.3	2	9.71	.029	0.50	.005	−0.01	.000	3	F9	81,401
		Cr 399- 109		19 23 12	+20 05.	1	12.28		0.59		0.34		1		81
		Cr 399- 110		19 23 12	+20 05.	1	12.07		0.96		0.35		1		81
		Cr 399- 115		19 23 12	+20 05.	1	11.45		0.60		0.35		5		81
		Cr 399- 121		19 23 12	+20 05.	1	11.38		0.62		0.09		4		81
		Cr 399- 122		19 23 12	+20 05.	1	12.63		0.62		0.07		5		81
		Cr 399- 123		19 23 12	+20 05.	1	13.35		0.72		0.21		4		81
		Cr 399- 249		19 23 12	+20 05.	1	13.58		0.92		0.30		2		81
		Cr 399- 250		19 23 12	+20 05.	1	14.16		2.53						81
182761	+19 04009	HR 7384		19 23 12	+20 10.3	4	6.32	.020	−0.01	.014	−0.08	.005	13	A0 V	81,401,1025,1118
182781	+25 03823			19 23 12	+25 36.5	2	8.70	.010	0.14	.010	0.07	.010	5	B9	833,1625
182917	+49 02999	CH Cyg		19 23 14	+50 08.5	4	6.17	.479	0.75	.308	−0.63	.051	4	M7 III	692,1591,1753,6003
231376	+16 03846			19 23 16	+16 58.4	1	9.19		0.42				2	F2	1118
231373	+17 03960			19 23 16	+17 35.1	1	9.69		0.45				2	B5	1118
182762	+19 04010	HR 7385	★ AB	19 23 17	+19 42.0	5	5.18	.026	0.98	.020	0.81	.010	14	K0 III	81,1025,1118,1355,3016
182760	+21 03775			19 23 18	+21 11.4	2	8.60	.039	0.04	.010	−0.46	.015	3		401,963
182645	−15 05348	HR 7378		19 23 20	−15 09.2	3	5.72	.005	0.02	.006	−0.36	.009	7	B7 IV	252,1079,2007
182629	−22 05105	HR 7375		19 23 20	−21 52.6	5	5.58	.013	1.23	.008	1.23	.031	20	K3 III	15,978,1075,2007,3005
		LP 752 - 19		19 23 21	−14 22.3	1	13.20		0.80		0.20		2		1696
		CS 22939 # 115		19 23 21	−31 04.1	1	14.08		0.06		−0.14		1		1736
182807	+24 03737	HR 7386	★ A	19 23 22	+24 49.3	6	6.18	.017	0.51	.022	−0.03	.013	11	F7 V	15,22,254,1003,1620,3077
	+42 03334			19 23 24	+42 36.	2	10.19	.003	0.24	.006	0.12	.019	2	A7 IV	1697,1748
183052	+59 02025			19 23 26	+59 45.9	1	8.99		0.20				5	F0	1379
183077	+60 01943			19 23 26	+60 15.0	1	7.65		1.07				7	K0	1379
183098	+60 01944	IDS19228N6100	AB	19 23 27	+61 06.2	1	7.68		1.04				4	K2	70
		DH Aql		19 23 27	−10 21.4	1	18.25		−0.09		−0.82		1		1471
231382	+15 03808			19 23 28	+16 08.3	1	9.27		0.32				2	B9	1118
		CS 22939 # 106		19 23 28	−30 39.5	1	13.44		0.23		0.16		1		1736
	+42 03335			19 23 29	+42 47.1	2	10.36	.005	1.10	.000	1.07	.001	2	K0	1697,1748
231385	+19 04012			19 23 31	+19 21.6	1	9.45		0.17				3	B9 V	1025
338268	+26 03550			19 23 34	+26 59.6	1	9.56		1.90				2	M0	369
182678	−14 05435	HR 7379		19 23 34	−14 39.1	1	6.70		0.05				4	A0 V	2007
350035	+19 04013			19 23 36	+20 07.2	1	9.79		0.46		0.02		2	A5	81
		CS 22896 # 83		19 23 36	−57 11.3	1	15.26		0.05		0.06		1		1736
	−0 03741			19 23 41	−00 21.3	2	9.27	.004	0.09	.004	−0.41	.003	6	B8	963,1732
182163	−70 02639			19 23 42	−70 25.0	2	8.76	.005	0.41	.000	−0.07		3	F2 V	742,1594

Table 1   1053

HD	DM	Other Id	N	Rem	α_{1950}	δ_{1950}	S	V	σ_V	B–V	σ_{B-V}	U–B	σ_{U-B}	n	Spectrum	References
182830	+15 03811	IDS19215N1521	AB		19 23 44	+15 27.4	1	8.05		0.82		0.50		4	F0	206
239080	+59 02027				19 23 44	+60 08.1	1	9.80		0.97				2	K0	70
182786	−0 03742				19 23 44	−00 12.0	1	9.06		0.74		0.25		3	B8	1732
		CS 22896 # 43			19 23 44	−53 27.2	1	14.19		0.09		0.17		1		1736
		CS 22896 # 78			19 23 44	−56 38.9	1	15.05		0.12		0.20		1		1736
		CS 22891 # 171			19 23 45	−59 30.5	1	14.29		0.86		0.20		1		1736
		CS 22896 # 40			19 23 46	−52 49.0	1	14.70		0.25		0.15		1		1736
182681	−29 16140	HR 7380	2		19 23 47	−29 50.6	2	5.66		-0.02	.007	-0.11		5	B9 V	1079,2009
	+42 03337				19 23 48	+42 51.	2	10.28	.006	0.70	.002	0.25	.010	2	G5 V	1697,1748
182509	−54 09371	HR 7370		⋆ A	19 23 48	−54 25.6	3	5.70	.005	1.41	.005	1.68	.002	7	K4 III	58,2035,4001
185778	+85 00330				19 23 49	+85 59.2	1	8.67		0.84		0.40		2	G5	1733
182989	+42 03338	G 125 - 7, RR Lyr	3		19 23 52	+42 41.2	3	7.23	.203	0.19	.064	0.15	.032	7	F5	668,764,1748
231395	+15 03812				19 23 53	+16 00.8	1	8.63		1.03				3	G5	1118
182899	+24 03742				19 23 53	+24 57.1	1	7.95		0.16		-0.03		6	B9	1352
182835	+0 04206	HR 7387		⋆ A	19 23 58	+00 14.2	13	4.66	.018	0.59	.007	0.45	.081	46	F2 Ib	3,15,369,450,1075*
182835	+0 04204	IDS19214N0008	B		19 23 58	+00 14.2	1	9.42		0.31		0.16		3		450
		CS 22939 # 68			19 23 58	−27 33.0	1	14.36		0.56		0.03		1		1736
		CS 22896 # 57			19 23 58	−54 20.8	1	15.39		0.03		0.09		1		1736
		CS 22896 # 76			19 23 58	−56 05.0	1	15.18		0.02		-0.05		1		1736
		ArkAstr3,439 # 473			19 24 01	+16 28.5	1	10.65		0.68				3		1118
		ArkAstr3,439 # 474			19 24 02	+16 12.0	1	12.45		0.69				2		1118
182919	+19 04015	HR 7390	5		19 24 02	+19 59.8	5	5.64	.017	-0.01	.020	-0.04	.005	24	A0 V	81,401,1025,1063,1118
		CS 22896 # 39			19 24 02	−52 33.2	1	15.44		0.11		0.16		1		1736
183122	+53 02238				19 24 03	+53 40.1	1	8.50		0.02		-0.08		2	A0	1566
182918	+22 03693		3		19 24 04	+22 39.4	3	8.61	.005	0.16	.007	-0.34	.004	8	B6 V	206,963,7007
182900	+12 03907	HR 7389	4		19 24 05	+12 55.3	4	5.77	.011	0.45	.011	0.03	.013	10	F6 III	254,1118,3053,8100
		CS 22896 # 53			19 24 08	−53 49.7	1	13.24		0.00		-0.08		1		1736
231410	+17 03970				19 24 10	+17 14.6	1	8.84		0.17				2	A2	1118
231409	+17 03969				19 24 10	+17 16.6	1	9.42		0.27				3	A3	1118
182990	+35 03614				19 24 10	+36 05.1	1	7.48		1.62		1.54		2	K5	1601
182901	+11 03840				19 24 11	+11 44.9	1	6.92		0.41		0.01		2	F5 III	3016
182937	+18 04075				19 24 13	+18 53.6	1	8.14		0.04				3	A1 V	1025
231416	+16 03850				19 24 15	+16 56.3	1	9.18		0.21				5	B8	1118
231415	+17 03971				19 24 16	+17 13.7	1	9.27		0.16				2	A2	1118
		AO 793			19 24 16	+42 26.1	2	11.23	.003	1.52	.004	1.95	.026	2	M0 III	1697,1748
182955	+19 04017	HR 7391	6	⋆ A	19 24 17	+19 47.4	6	5.83	.024	1.55	.020	1.94	.032	18	M0 III	81,542,1025,1118,3016,8100
182955	+19 04016	IDS19221N1942	C		19 24 17	+19 47.4	2	10.48	.020	0.58	.030	0.12	.035	2		81,542
182708	−40 13288				19 24 17	−40 11.9	1	7.16		0.55		0.16		11	F8 V	1621
	+42 03339				19 24 18	+42 37.2	2	10.89	.002	0.61	.002	0.09	.014	2	G1 IV-V	1697,1748
234892	+52 02425				19 24 18	+52 18.1	1	9.15		0.05		0.04		2	A0	1566
183239	+59 02030				19 24 18	+59 40.2	1	8.65		1.65				1	M1	70
		LS II +16 007			19 24 20	+16 49.3	1	11.69		0.86		-0.05		2	B1 III	946
182972	+19 04019		4		19 24 20	+20 09.5	4	6.66	.017	0.01	.018	-0.05	.005	16	A1 V	81,401,1025,1118
183279	+61 01859				19 24 20	+61 20.0	1	9.12		0.39				1	G5	70
183056	+36 03557	HR 7395, V1741 Cyg	4		19 24 21	+36 13.0	4	5.15	.005	-0.12	.008	-0.43	.015	11	B9p Si	196,1063,1363,3033
231422	+15 03814				19 24 23	+15 54.1	1	9.50		0.58				2	F8	1118
183124	+44 03133				19 24 25	+44 49.8	1	6.64		1.04		0.85		2	K1 III	985
		ArkAstr3,439 # 484			19 24 28	+16 41.5	1	12.55		0.61				2		1118
182746	−40 13292				19 24 28	−40 50.6	1	9.39		0.55		0.06		26	F6 V	978
183032	+27 03391	IDS19225N2707	AB		19 24 32	+27 13.2	3	7.47	.012	0.57	.030	-0.02	.021	10	F8	938,1381,3030
183203	+50 02791				19 24 32	+50 45.1	1	7.36		1.63		1.93		1	K5	401
350036	+19 04021				19 24 33	+20 05.8	1	10.17		0.91		0.66		2	K2	81
		Vul R2 # 4	V		19 24 34	+23 48.0	1	11.67		0.89		0.41		1	F0 V:e	5012
	+27 03392				19 24 35	+27 40.6	1	9.98		1.83				2	M3	369
183013	+21 03782				19 24 36	+21 33.1	2	7.38	.024	0.09	.000	-0.65		5	B2 IV	401,1025
231437	+14 03917				19 24 38	+15 00.2	1	8.64		2.00				2	M0	369
231438	+19 04022				19 24 38	+19 39.0	1	9.01		0.40				2	F5 V	1025
350037	+19 04023				19 24 38	+20 03.7	1	9.35		0.73		0.27		2	F8 V	81
182776	−41 13525	V4139 Sgr	2		19 24 38	−40 56.2	2	8.58	.166	1.22	.001	1.04	.007	12	K2/3 III	1621,1641
183014	+20 04139	IDS19225N2058	A		19 24 39	+21 03.7	1	8.04		0.02		-0.52		1	B7 V	7007
183014	+20 04139	IDS19225N2058	AB		19 24 39	+21 03.7	3	7.34	.019	0.02	.012	-0.40	.040	6	B7 V	81,401,1025
183014	+20 04139	IDS19225N2058	B		19 24 39	+21 03.7	1	8.18		0.06		-0.32		1		7007
234893	+50 02792				19 24 40	+50 48.0	2	9.41	.020	-0.11	.015	-0.61	.005	4	B9	401,963
231435	+16 03852				19 24 42	+16 39.4	1	10.01		0.47				2	F2	1118
182485	−67 03646	IDS19197S6731	A		19 24 43	−67 24.7	1	7.69		0.94		0.69		7	G8/K0 III	1704
	+61 01860				19 24 44	+61 15.5	1	9.25		1.26				1	K0	70
182800	−40 13294				19 24 44	−40 33.8	1	8.70		1.10		0.94		26	K1 III	978
183035	+16 03853				19 24 45	+16 26.2	1	8.07		0.08				2	A0	1118
184102	+79 00628	HR 7423	3		19 24 45	+79 30.3	3	6.06	.005	0.07	.000	0.07	.009	6	A3 V	985,1502,1733
231442	+15 03816	IDS19225N1554	AB		19 24 46	+15 59.8	1	8.94		0.18				1	A2	1118
183016	+6 04136				19 24 47	+06 51.9	1	9.52		1.94				2	M2 III	369
182975	−2 04998	LS IV -02 032	3		19 24 47	−02 07.4	3	8.37	.008	0.12	.008	-0.51	.007	12	B2 IV	399,400,627
344434	+20 04141	IDS19226N2057	AB		19 24 48	+21 02.9	2	9.26	.020	0.31	.005	-0.18	.120	2	A2	401,7007
183058	+20 04142	IDS19226N2054	A		19 24 49	+21 00.3	4	7.13	.038	0.16	.006	-0.63	.019	9	B5	81,401,401,1025
		VES 3			19 24 49	+25 31.9	1	11.84		1.82		2.12		2		363
	+42 03342				19 24 50	+42 58.4	2	9.78	.002	0.26	.002	0.06	.019	2	A9 V	1697,1748
231446	+17 03972				19 24 51	+17 19.1	1	9.74		0.19				4	A2	1118
183339	+57 01999	HR 7401	4		19 24 52	+57 55.6	4	6.59	.012	-0.14	.010	-0.56	.007	14	B8 IV He wk	252,253,1079,1379
231448	+16 03854				19 24 54	+16 20.3	1	9.00		0.13				2	A2	1118
231447	+17 03973				19 24 54	+17 11.5	1	9.51		0.41				2	F0	1118

HD	DM	Other Id	Rem	α_{1950}	δ_{1950}	S	V	σ_V	B–V	σ_{B-V}	U–B	σ_{U-B}	n	Spectrum	References
350042	+19 04024			19 24 54	+19 56.6	2	8.66	.019	0.41	.028	-0.04		4	F3 V	81,1025
		G 185 - 24		19 25 00	+22 31.3	1	14.44		0.86		0.32		2		1658
231456	+16 03855			19 25 01	+17 06.2	1	9.60		0.32				3	A3	1118
183204	+39 03767			19 25 02	+39 51.6	1	7.39		-0.01		-0.07		2	A0	401
183255	+49 03009	G 208 - 22		19 25 02	+49 21.2	4	8.02	.020	0.92	.019	0.66	.039	14	K3 V	692,1197,3016,7008
	+14 03920			19 25 04	+14 36.9	1	9.86		2.11				2	M3	369
		VES 4		19 25 04	+28 08.6	1	12.27		1.22		1.26		1		363
	+31 03597		V	19 25 04	+31 29.0	1	9.98		1.73				3	M3	369
183184	+35 03623			19 25 04	+35 48.3	1	7.72		-0.02		-0.08		2	A0	401
183298	+50 02794			19 25 05	+50 53.1	1	8.11		0.58		0.09		1	G0	692
		CS 22939 # 148		19 25 05	-30 23.3	1	13.74		0.16		0.17		1		1736
231458	+15 03818			19 25 06	+15 56.3	1	10.16		0.28				2	A0	1118
		CS 22896 # 77		19 25 07	-56 45.7	1	14.74		0.26		0.17		1		1736
231465	+16 03858			19 25 09	+16 59.3	1	9.76		1.86				2	M0	369
231464	+16 03857			19 25 09	+17 06.4	1	8.91		0.40				2	F2	1118
183161	+20 04146	IDS19230N2028	AB	19 25 10	+20 34.0	1	8.32		0.28		-0.43		1	B9	7007
183161	+20 04146	IDS19230N2028	ABC	19 25 10	+20 34.0	4	8.16	.028	0.30	.018	-0.39	.026	9	B9	81,401,963,1025
		CS 22896 # 61		19 25 10	-54 52.7	1	15.03		0.01		-0.03		1		1736
		Vul R2 # 10		19 25 11	+20 34.3	1	10.23		0.45		-0.02		1	B7 V	5012
183143	+18 04085	HT Sge		19 25 13	+18 11.6	4	6.84	.017	1.18	.018	0.17	.031	12	B7 Ia	1012,8007,8040,8100
231475	+19 04026			19 25 15	+19 46.4	1	9.01		0.97				3	K0	1025
		CS 22939 # 153		19 25 15	-29 53.5	1	14.52		0.39		0.16		1		1736
183144	+13 04020	HR 7396		19 25 16	+14 10.8	3	6.32	.000	-0.07	.007	-0.51	.005	6	B4 III	154,252,1118
181466	-82 00770			19 25 17	-81 51.6	1	6.53		1.16		0.94		5	K0 III	1628
350040	+19 04027	IDS19232N1957	AB	19 25 22	+20 03.1	4	8.83	.023	0.17	.018	-0.08	.007	6	B8 V	81,401,1025,7007
183063	-12 05409	IDS19226S1221	AB	19 25 23	-12 15.1	3	7.57	.020	0.74	.002	0.30	.010	13	G8 V	176,214,3026
183240	+30 03578			19 25 25	+30 19.3	1	8.40		-0.08		-0.49		2	A2	401
		CS 22896 # 41		19 25 26	-53 00.3	1	13.23		0.13		0.14		1		1736
		CS 22896 # 54		19 25 27	-53 46.4	1	14.92		0.45		0.01		1		1736
183222	+24 03750			19 25 29	+24 26.9	1	7.97		1.71				2	K5	369
	+31 03603			19 25 32	+31 42.1	1	9.15		1.68				3	M2	369
183471	+59 02032			19 25 32	+59 19.6	1	7.99		0.96				1	K0	70
183129	-1 03749			19 25 32	-01 11.6	1	8.13		0.11		-0.29		2	B5	400
		CS 22896 # 59		19 25 32	-54 39.3	1	13.34		0.03		0.05		1		1736
		CS 22939 # 167		19 25 36	-28 56.6	1	13.53		0.22		0.22		1		1736
		G 261 - 6		19 25 42	+75 26.9	1	15.38		1.73				2		1759
231494	+16 03862			19 25 43	+16 36.0	1	8.54		0.54				2	G0	1118
183261	+19 04028	LS II +20 004		19 25 43	+20 08.7	4	6.89	.029	-0.03	.015	-0.67	.015	14	B3 II	81,401,1025,8100
183133	-15 05362			19 25 45	-15 12.3	4	6.77	.014	-0.03	.007	-0.56	.005	10	B2 IV	26,219,399,400
		CS 22896 # 55		19 25 45	-54 05.7	1	13.53		0.43		0.01		1		1736
231496	+15 03825			19 25 46	+15 53.5	1	9.82		0.59				3	G0	1118
182893	-55 09096	HR 7388		19 25 49	-55 32.7	2	6.14	.010	0.97	.003	0.81		6	K0/1 III	58,2035
183227	+2 03892	HR 7397		19 25 50	+02 49.6	3	5.84	.000	0.00	.004	-0.39	.007	13	B6 III	379,1079,1149
344365	+24 03754	BN Vul		19 25 50	+24 14.7	1	10.63		0.68		0.39		1	A9.2	699
231509	+17 03975	IDS19236N1727	AB	19 25 51	+17 32.7	1	8.27		1.06				2	K0	1118
183362	+37 03465	HR 7403		19 25 51	+37 50.3	2	6.34	.033	-0.15	.009	-0.75	.005	6	B3 Ve	154,1212
183007	-43 13395	HR 7392		19 25 51	-43 32.9	3	5.71	.011	0.22	.000	-0.01		15	A0 III	15,1311,2012
183262	+17 03976			19 25 52	+17 44.6	2	6.84	.009	0.31	.014	0.09		6	Am	1118,8071
183282	+21 03790			19 25 52	+21 39.1	1	8.64		0.06				3	B8 V	1025
183611	+62 01716	HR 7413		19 25 52	+62 27.3	2	6.37	.010	1.40	.015	1.71		11	K5 III	252,1379
183363	+36 03566	IDS19241N3620	AB	19 25 53	+36 25.6	1	7.57		-0.03		-0.30		2	A0	401
231510	+13 04024	G 23 - 1	★ A	19 25 55	+13 53.9	5	9.01	.008	0.78	.007	0.25	.004	8	K0 V	516,979,1620,1658,8112
		G 92 - 1		19 25 56	+07 01.7	1	13.54		1.43		1.15		1		333,1620
231512	+12 03917	IDS19236N1220	AB	19 25 56	+12 26.0	1	9.19		1.10		0.90		2	K2	3072
		CS 22896 # 64		19 25 56	-55 02.2	1	15.05		0.22		0.11		1		1736
	+41 03353	TT Lyr		19 25 58	+41 35.9	1	10.88		0.22		-0.35		1	B2 V	963
	-13 05367			19 25 58	-13 05.5	1	8.40		1.68		2.02		2	M0	1746
182709	-68 03251	HR 7383		19 26 00	-68 32.3	4	5.96	.009	1.65	.010	1.92	.000	18	K4/5 III	15,1075,2038,3005
	+29 03606			19 26 01	+29 14.1	1	9.30		1.40				2	M3	369
	+68 01065	UZ Dra		19 26 02	+68 50.0	1	9.58		0.48		-0.04		2	F8	1768
183416	+41 03354			19 26 03	+41 33.0	1	8.73		-0.02		-0.19		2	A0	963
183302	+9 04114		V	19 26 05	+09 31.6	1	8.19		1.75				2	M1	369
231518	+15 03827	IDS19238N1605	AB	19 26 07	+16 11.1	1	9.09		0.49				3	F7 IV	1118
	-38 13504	UV Sgr		19 26 07	-38 11.7	1	6.31		0.95		0.53		1	M4	1772
		CS 22891 # 194		19 26 09	-62 26.8	2	14.48	.010	0.43	.008	-0.18	.008	3		1580,1736
		AJ67,79 # 93		19 26 12	+19 29.	1	12.78		1.60		0.59		3		1359
183534	+52 02434	HR 7408		19 26 12	+52 13.1	1	5.75		0.00		-0.02		2	A1 V	401
		Smethells 38		19 26 12	-55 03.	1	11.33		1.37				1		1494
183267	-6 05158			19 26 18	-06 16.7	2	8.59	.039	0.19	.003	-0.35	.026	3	B8	634,963
		WLS 1920 20 # 8		19 26 19	+20 19.3	1	11.74		0.74		0.12		2		1375
231533	+17 03978	G 142 - 32		19 26 23	+18 09.6	1	9.92		0.76		0.39		2	K0	7010
183558	+47 02838			19 26 25	+48 04.4	1	8.31		-0.13		-0.82		2	A5	401
183286	-7 04967			19 26 26	-07 17.0	1	9.58		0.25		-0.27		2	B9	634
183324	+1 04010	HR 7400		19 26 29	+01 50.8	2	5.79	.005	0.08	.000	0.06	.005	7	A0 V	15,1071
231539	+19 04034			19 26 29	+19 21.4	1	9.05		0.44				2	F2 V	1025
183216	-31 16750			19 26 30	-30 53.9	1	7.13		0.60				4	G2 V	1075
344441	+20 04155			19 26 31	+20 56.6	1	9.35		0.33				2	A5 V	1025
183365	+7 04087			19 26 32	+07 56.2	1	8.41		1.72				2	M0 III	369
183418	+21 03794			19 26 32	+21 52.6	2	7.74	.005	0.96	.005	0.67		5	G8 III	1025,8100
183028	-55 09100	HR 7393		19 26 32	-55 12.9	1	6.30		0.45				4	F5 V	2035

Table 1   1055

HD	DM	Other Id	N Rem	α_{1950}	δ_{1950}	S	V	σ_V	B–V	σ_{B-V}	U–B	σ_{U-B}	n	Spectrum	References
183419	+18 04094			19 26 35	+18 22.9	2	8.22	.042	0.07	.014	-0.42		4	B6 IV	401,1025
	+27 03401			19 26 35	+27 28.1	1	9.91		2.58				1	M3	369
183439	+24 03759	HR 7405	⋆ A	19 26 37	+24 33.7	9	4.44	.020	1.50	.004	1.81	.013	36	M0 III	15,369,1003,1355,1363*
		G 23 - 2		19 26 38	+07 03.2	1	10.71		1.41		1.21		1	K7	333,1620
	+32 03443			19 26 38	+32 15.6	1	9.42		0.18		0.14		2	A2	401
183401	+12 03923	IDS19243N1215	A	19 26 40	+12 22.0	1	7.94		0.30		0.10		2	A5	3016
183401	+12 03923	IDS19243N1215	B	19 26 40	+12 22.0	1	11.06		0.70		0.15		2		3016
183344	−7 04968	HR 7402, U Aql	⋆ AB	19 26 40	−07 08.9	5	6.28	.198	0.94	.104	0.68	.068	8	F8	1149,1484,1772,3032,6007
183344	−7 04968	IDS19240S0715	C	19 26 40	−07 08.9	1	14.00		0.67				3		3032
		CS 22896 # 65		19 26 41	−55 09.4	1	15.45		0.05		0.14		1		1736
183402	+10 03916			19 26 43	+10 31.6	1	8.48		1.73				3	M1	369
		CS 22896 # 56		19 26 43	−54 14.9	1	14.66		0.24		0.12		1		1736
183387	−0 03760	HR 7404		19 26 44	+00 08.5	3	6.25	.011	1.32	.004	1.38	.004	9	K2	15,1256,1355
		CS 22896 # 113		19 26 44	−52 44.8	1	15.26		0.14		0.11		1		1736
183459	+21 03795			19 26 46	+21 31.2	1	8.37		0.03				3	B8 V	1025
183612	+46 02693			19 26 46	+46 55.3	1	7.75		1.01		0.85		2	G5	1601
182687	−73 02027			19 26 46	−73 36.2	1	7.50		0.98		0.82		6	K0 II	1704
183275	−27 14004	HR 7398	⋆ A	19 26 47	−27 05.4	1	5.52		1.12		1.35		2	K3 III	542
183275	−27 14004	HR 7398	⋆ AB	19 26 47	−27 05.4	5	5.47	.022	1.12	.015	1.12	.038	13	K3 III	15,219,542,2012,3051
183275	−27 14004	IDS19237S2711	B	19 26 47	−27 05.4	1	8.69		0.86		0.32		1		542
		CS 22896 # 201		19 26 49	−56 39.3	1	15.55		0.20		0.16		1		1736
183536	+34 03566			19 26 50	+34 30.2	3	8.20	.015	0.53	.007	-0.07	.016	6	G5	979,3026,8112
		CS 22896 # 58		19 26 50	−54 35.4	1	15.61		0.12		0.17		1		1736
183535	+36 03572			19 26 51	+36 40.5	3	8.63	.005	-0.16	.007	-0.92	.007	5	B5	401,555,963
183491	+24 03761	HR 7406	⋆ B	19 26 52	+24 39.9	2	5.80	.044	1.02	.010	0.87		6	K0 III	1118,1355
	+0 04221	G 92 - 3		19 26 53	+00 25.6	1	10.73		1.28		1.28		1	K5	333,1620
		G 92 - 4		19 26 54	+00 24.4	1	11.80		1.42		1.28		1	M0	333,1620
		IRC +20 408		19 26 56	+23 09.6	1	10.98		2.33		1.60		2		363
183331	−27 14006			19 26 56	−27 27.2	1	8.32		1.60		2.01		3	K4 III	1657
183613	+40 03745			19 26 58	+41 05.6	1	8.84		0.21		0.13		2		1733
		CS 22896 # 108		19 27 01	−53 59.1	1	14.28		0.23		0.18		1		1736
183312	−32 15233	HR 7399		19 27 02	−32 11.8	2	6.59	.005	0.40	.005			7	F3 V	15,2012
		CS 22896 # 68		19 27 02	−55 34.7	1	15.31		0.15		0.20		1		1736
		CS 22896 # 66		19 27 03	−55 24.0	1	15.39		0.24		0.13		1		1736
231564	+12 03927	LS II +12 007		19 27 04	+12 28.0	2	10.24	.005	0.79	.000	0.25	.000	3	B1 Ib	1012,1032
183492	+14 03936	HR 7407		19 27 04	+14 29.5	2	5.57	.014	1.04	.014	0.91		4	K0 III	252,1118
		CS 22896 # 67		19 27 06	−55 32.0	1	15.52		0.07		0.14		1		1736
183586	+28 03356			19 27 08	+28 41.5	1	9.25		1.70				2	M1	369
		G 92 - 5		19 27 09	−02 02.7	2	13.68	.014	1.38	.009	1.02	.103	4		203,1620
183537	+19 04039	HR 7409		19 27 10	+20 10.5	2	6.34	.028	-0.11	.012	-0.52	.005	16	B5 Vn	81,154,1025,1118
183614	+30 03589			19 27 11	+30 11.6	1	8.30		-0.08		-0.47		2	B9 Siλ3955	401
		CS 22891 # 181		19 27 11	−60 31.1	1	13.79		0.22		0.16		1		1736
		G 92 - 6		19 27 12	+00 55.3	3	11.77	.021	0.60	.018	-0.07	.010	7		308,333,1620,1696
183561	+26 03566			19 27 12	+26 36.7	1	8.05		0.17		-0.54		2	B2 III	1012
183649	+34 03572			19 27 15	+35 05.5	2	8.83	.023	-0.11	.009	-0.62	.005	4	B2 V	555,963
		CS 22896 # 84		19 27 16	−56 55.4	1	15.38		0.04		0.00		1		1736
		CS 22947 # 287		19 27 18	−50 20.8	1	13.87		0.23		0.10		1		1736
183679	+39 03781			19 27 19	+39 49.9	1	8.40		-0.05		-0.15		1	A0	963
183562	+21 03800			19 27 20	+21 13.7	1	8.51		-0.01				3	B9 Siλ3955	1025
184657	+79 00629			19 27 22	+79 40.8	1	8.13		0.41		0.02		2	F5	1502
183650	+31 03618	G 207 - 33		19 27 26	+31 30.6	2	6.97	.009	0.72	.005	0.31	.018	7	G7 IV	1355,3026
183480	−7 04969	IDS19247S0747	AB	19 27 27	−07 41.3	1	10.07		0.45		0.25		2	A0	634
231588	+14 03938			19 27 29	+14 46.0	1	9.31		1.95				2	K7	369
183702	+35 03648			19 27 29	+35 12.9	1	8.10		0.09		0.04		2	A0	401
		JL 22		19 27 30	−74 39.	1	12.78		0.14		-0.70		1		832
	+30 03594			19 27 34	+30 17.5	2	9.53	.015	1.25	.107	1.70		11	K2 III	1344,1413
		CS 22939 # 142		19 27 35	−30 56.2	1	14.49		-0.22		-1.11		1		1736
		VES 6		19 27 37	+23 15.8	1	13.09		1.66		1.39		1		363
338466	+25 03852	LS II +25 002		19 27 37	+25 54.5	1	10.72		0.79		0.66		1	F2 II	1215
		CS 22939 # 179		19 27 38	−27 40.5	1	15.41		-0.27		-1.05		1		1736
183563	+3 04042			19 27 39	+03 25.2	1	9.21		0.45				26	A0	6011
	−7 04970			19 27 39	−07 07.4	1	9.90		0.60		0.08		2	F5	634
183589	+2 03904	HR 7412	⋆ A	19 27 40	+02 47.9	1	6.08		1.82		2.01		4	K5 Ib	1149
		CS 22896 # 112		19 27 41	−53 09.7	1	15.60		0.13		0.21		1		1736
	+7 04097			19 27 44	+07 41.6	1	10.38		1.85				3	M3	369
		CS 22896 # 99		19 27 44	−55 18.2	1	13.10		0.28		0.14		1		1736
183681	+22 03712			19 27 45	+22 36.1	1	7.31		1.81				2	M0 III	369
183566	−7 04971			19 27 47	−07 22.7	1	10.03		0.28		-0.11		2	A0	634
231600	+18 04099	IDS19256N1813	AB	19 27 48	+18 19.6	1	10.53		0.28		0.22		2	K0	684
		CS 22896 # 114		19 27 48	−52 45.1	1	15.59		0.05		0.08		1		1736
		CS 22896 # 104		19 27 48	−54 57.2	1	15.36		0.14		0.21		1		1736
		LP 692 - 21		19 27 55	−08 24.0	1	11.26		0.84		0.46		1		1696
183570	−16 05337			19 27 55	−16 16.3	2	7.42	.005	-0.01	.000	-0.44	.005	4	B5 III	399,400
183753	+28 03363			19 27 56	+28 37.2	1	7.68		1.35		1.51		1	K3 II	8100
	+60 01956			19 27 56	+60 24.5	1	10.88		0.17				1		70
	+60 01957			19 27 56	+60 39.3	1	10.66		0.69				1		70
183545	−21 05410	HR 7410		19 27 56	−21 25.1	1	6.13		0.12				4	A2 V	2007
344502	+21 03804	G 185 - 27		19 27 57	+21 34.4	1	9.98		1.16		1.02		2	K5	3072
350080		NGC 6802 - 28		19 27 58	+20 02.8	1	8.08		0.06				2	A2p	1118
		CS 22896 # 95		19 27 58	−55 39.7	1	15.05		0.21		0.17		1		1736

HD	DM	Other Id	N	Rem	α_{1950}	δ_{1950}	S	V	σ_V	B–V	σ_{B-V}	U–B	σ_{U-B}	n	Spectrum	References
350073	+20 04163	NGC 6802 - 2			19 28 01	+20 13.4	2	9.91	.002	0.24	.003	0.15	.018	5	A7 V	49,1655
		CS 22896 # 105			19 28 01	−54 32.5	1	14.75		0.31		0.01		1		1736
231606	+18 04100				19 28 02	+18 36.0	1	8.25		1.43				2	K3 III	1025
183656	+3 04043	HR 7415, V923 Aql			19 28 03	+03 20.3	2	6.05	.010	0.00	.020	-0.34	.035	6	A0 Shell	379,1149
183630	−3 04612	HR 7414			19 28 03	−02 53.7	2	5.04	.020	1.74	.015	2.03	.005	10	M1 III	1088,3005
350074	+20 04164	NGC 6802 - 1			19 28 04	+20 13.5	2	9.08	.002	1.07	.001	0.77	.006	36	K1 III-IV	49,1655
350079		NGC 6802 - 4			19 28 07	+20 06.1	1	11.10		0.26		0.07		1	F2	49
231621	+18 04103				19 28 09	+19 03.8	1	9.51		0.43				3	B5 III	1025
231616	+17 03986				19 28 10	+18 09.3	2	10.02	.039	0.74	.005	-0.28	.010	6	B0.5III	342,684
		NGC 6802 - 6			19 28 10	+20 06.7	1	13.04		0.84		0.35		1	A0 V	49
183574	−24 15387				19 28 10	−24 12.0	1	8.19		0.46				4	F2 V	2012
		NGC 6802 - 10			19 28 12	+20 07.8	1	14.47		0.91		0.59		1		49
183658	−6 05170	IDS19255S0643	A		19 28 12	−06 37.1	3	7.28	.005	0.64	.005	0.17	.004	10	G0	158,196,1775
		NGC 6802 - 9			19 28 13	+20 06.9	1	14.15		1.10		0.68		1		49
		NGC 6802 - 12			19 28 15	+20 10.4	1	14.53		0.92		0.61		1		49
		NGC 6802 - 24			19 28 15	+20 11.9	1	16.19		1.14				1		49
		NGC 6802 - 22			19 28 15	+20 12.3	1	15.66		1.22				1		49
		NGC 6802 - 15			19 28 17	+20 08.3	1	14.83		1.02				1		49
183735	+0 04234	G 26 - 9			19 28 18	+00 56.3	1	9.90		0.97		0.50		2	A2	1620
		JL 23			19 28 18	−77 52.	1	16.25		-0.32		-1.20		1		132
350062	+20 04167				19 28 20	+20 38.6	1	8.53		0.96				3	G5 III	1025
		NGC 6802 - 23			19 28 22	+20 05.9	1	16.08		1.19				1		49
	+60 01958				19 28 22	+60 34.2	1	10.49		0.46				1	F5	70
		NGC 6802 - 19			19 28 23	+20 12.9	1	15.25		1.11				1		49
		NGC 6802 - 11			19 28 23	+20 13.4	1	14.52		1.40		0.85		1		49
		NGC 6802 - 7			19 28 24	+20 11.1	1	13.88		0.72		0.48		1		49
		NGC 6802 - 21			19 28 25	+20 13.0	1	15.34		1.25				1		49
183886	+33 03480				19 28 25	+33 37.4	2	6.56		1.14		1.08		2	K0	1625
		NGC 6802 - 18			19 28 27	+20 05.6	1	15.13		1.17		0.55		1		49
		NGC 6802 - 5			19 28 27	+20 13.9	1	12.04		0.62		0.33		1	F3 II-III	49
184006	+51 02605	HR 7420			19 28 27	+51 37.3	5	3.78	.009	0.15	.009	0.14	.021	19	A5 Vn	15,1363,3023,8015,8040
		NGC 6802 - 8			19 28 28	+20 13.5	1	14.09		2.14				1		49
		LS II +16 008			19 28 30	+16 49.1	1	10.63		1.09		0.00		2	B1.5Iab	405
183791	+6 04172				19 28 32	+06 16.7	1	8.04		1.15		0.78		4	G2 II	8100
183813	+11 03868	IDS19262N1125	AB		19 28 32	+11 31.1	1	8.97		1.74				3	M2	369
183864	+24 03768	IDS19265N2455	AB		19 28 32	+25 00.9	1	7.32		1.27		0.86		2	G2 Ib	8100
		WLS 1944 55 # 6			19 28 32	+55 04.1	1	11.89		1.51		0.37		2		1375
		NGC 6802 - 20			19 28 33	+20 08.6	1	15.33		1.02		0.76		1		49
		NGC 6802 - 25			19 28 33	+20 08.8	1	16.33		0.94				1		49
		NGC 6802 - 13			19 28 33	+20 10.0	1	14.64		1.06		0.75		1		49
		NGC 6802 - 17			19 28 34	+20 12.8	1	14.86		1.25				1		49
		NGC 6802 - 16			19 28 35	+20 11.0	1	14.84		0.98		0.74		1		49
		Ross 739			19 28 36	−05 15.7	1	12.12		1.06		0.90		1	K5	1696
		CS 22939 # 173			19 28 36	−28 22.0	1	17.01		1.18		0.65		1		1736
350077	+19 04044	NGC 6802 - 3			19 28 37	+20 05.8	2	10.58	.000	0.60	.005	0.06	.007	5	G5 V	49,1655
	+8 04122	LS IV +08 005			19 28 38	+08 23.6	1	9.86		0.25		-0.67		1	B1 Ia	1032
		NGC 6802 - 14			19 28 39	+20 21.1	1	14.68		1.90				1		49
183910	+31 03630				19 28 39	+31 46.6	1	8.68		-0.01		-0.16		2	B9	401
183794	−2 05024	V822 Aql	★ AB		19 28 40	−02 13.0	1	7.03		0.18		-0.39		1	B8	627
183794	−2 05024	IDS19260S0220	C		19 28 40	−02 13.0	1	11.10		1.16		0.76		6		627
183930	+30 03603		V		19 28 41	+30 28.9	1	7.93		1.63				3	M2 III	369
183912	+27 03410	HR 7417	★ A		19 28 42	+27 51.2	10	3.09	.010	1.13	.011	0.63	.013	40	K3 II+B0.5V	1,15,150,245,667,1080*
184055	+52 02440				19 28 42	+52 28.4	1	8.69		0.37		0.10		2	F2	1375
183606	−42 14274				19 28 42	−42 23.1	1	9.73		0.08		-0.04		1	A0 V	742
183914	+27 03411	HR 7418	★ B		19 28 44	+27 51.5	6	5.12	.012	-0.10	.009	-0.31	.010	15	B8 Ve	1,15,150,667,1079,3023
184168	+60 01959	IDS19280N6028	A		19 28 44	+60 33.9	1	7.61		1.18				2	K2	70
350061	+20 04172				19 28 45	+20 44.1	2	9.19	.000	0.34	.035	-0.04		4	A3 II	1025,8100
183887	+19 04046				19 28 48	+19 19.7	2	7.59	.035	1.57	.015	1.77		3	K2 III	1025,8100
183761	−17 05658				19 28 49	−17 20.2	1	8.92		-0.10		-0.98		1	B2 III	963
	−42 14278				19 28 53	−42 18.6	1	10.22		0.46		-0.25		2		1696
		CS 22947 # 299			19 28 54	−49 33.5	1	14.52		-0.08		-0.41		1		1736
183552	−53 09585	HR 7411			19 28 57	−53 17.6	2	5.76	.010	0.30	.000	0.19		10	Fm δ Del	355,2007
183316	−68 03260				19 28 57	−68 03.1	1	7.97		0.74				4	F7 V	2012
183986	+35 03658	HR 7419	★ A		19 28 58	+36 07.3	2	6.26	.005	-0.03	.005	-0.12	.010	5	B9.5III	196,401
	+60 01960				19 29 03	+60 28.9	1	10.34		1.21				1		70
183915	+11 03871				19 29 04	+11 31.3	1	7.31		1.31		1.12		8	Kp	993
		CS 22896 # 85			19 29 07	−57 24.2	1	15.89		0.05		0.03		1		1736
		CS 22947 # 281			19 29 08	−51 08.0	1	13.55		0.16		-0.68		1		1736
350125	+19 04048	LS II +20 006			19 29 10	+20 05.5	1	9.39		0.40		-0.03		2		946
183987	+27 03414				19 29 11	+27 39.4	1	8.53		1.66				2	M1	369
		LS II +19 007			19 29 12	+19 51.3	1	12.33		0.64		0.21		2	A1 Ib	946
183783	−35 13554				19 29 14	−35 33.7	1	8.71		1.11		0.98		2	K3/4 V	3072
		CS 22896 # 103			19 29 14	−55 05.3	1	14.54		0.33		0.00		1		1736
		AO 1179			19 29 15	+69 51.8	1	10.55		1.05		0.69		1	G5 III	1748
	−24 15398	V440 Sgr			19 29 18	−23 58.0	2	9.65	.045	0.18	.020	0.11	.010	2	F0	668,700
		CS 22896 # 89			19 29 18	−56 53.8	1	14.26		0.47		0.07		1		1736
		G 92 - 7			19 29 19	+07 59.3	1	12.61		1.27		1.19		1		333,1620
184010	+26 03573	HR 7421			19 29 19	+26 30.6	2	5.86	.020	0.93	.015	0.63		4	K0 III-IV	1080,1118
183870	−11 05030				19 29 20	−11 22.9	1	7.57		0.93		0.64		1	K2 V	3072
		LS II +17 007			19 29 21	+17 25.9	1	12.38		0.65		0.00		2	B4 III	946

Table 1   1057

HD	DM	Other Id	N	Rem	α_{1950}	δ_{1950}	S	V	σ_V	B–V	σ_{B-V}	U–B	σ_{U-B}	n	Spectrum	References
183989	+17 03991				19 29 21	+17 51.1	1	7.53		0.11		0.13		2	A0	1375
		CS 22896 # 98			19 29 21	−55 23.1	1	15.30		0.11		0.17		1		1736
338443	+27 03417				19 29 22	+27 18.2	1	8.87		1.75				2	M0	369
184057	+33 03487				19 29 23	+33 21.8	1	6.71		0.08		0.10		2	A2	401
231683	+17 03992	IDS19272N1735	AB		19 29 24	+17 40.6	2	8.27	.009	0.64	.002	0.05	.005	4	G2 V	1003,3016
		VES 7			19 29 24	+24 20.8	1	12.95		1.06		0.20		1		363
	+35 03659	G 125 - 13			19 29 24	+36 03.5	2	10.26	.050	0.62	.124	0.02	.129	3		1003,1620
		CS 22939 # 164			19 29 24	−28 47.8	1	13.35		0.16		0.14		1		1736
183877	−28 15936				19 29 33	−28 07.0	3	7.15	.009	0.67	.007	0.15	.044	7	K0 +F/G	1075,1311,3077
		G 142 - 36			19 29 36	+11 50.7	1	14.99		0.96		0.66		1		1658
231693	+11 03874				19 29 41	+11 47.0	1	9.08		1.73				3	K5	369
183994	−1 03766				19 29 41	−01 32.6	1	8.60		0.06		-0.49		1	B8	963
183899	−26 14280				19 29 41	−26 16.3	1	9.80		-0.08		-0.79		1	B1/2 II	55
344510	+21 03810				19 29 46	+21 12.4	1	8.68		0.31				3	F0 V	1025
183806	−45 13296	HR 7416, PW Tel			19 29 46	−45 22.8	1	5.61		-0.04				4	Ap CrEuSr	2007
184216	+46 02708				19 29 47	+46 57.2	1	8.86		0.29		0.08		2		1601
183414	−70 02664				19 29 47	−70 04.9	1	7.92		0.66				4	G3 V	2012
184108	+20 04175				19 29 52	+20 49.3	2	6.96	.000	-0.02	.015			7	B9 III	1025,1118
184171	+34 03590	HR 7426			19 29 55	+34 20.7	6	4.74	.010	-0.14	.011	-0.66	.014	19	B3 IV	15,154,555,1363,3016,8015
		CS 22891 # 217			19 29 58	−60 08.4	1	14.28		0.47		0.10		1		1736
184293	+49 03034	HR 7427			19 30 00	+50 11.9	1	5.53		1.25				2	K0	71
		BPM 12527			19 30 00	−67 16.	1	13.90		0.51		-0.10		2		3065
344495	+21 03813				19 30 01	+21 56.0	1	9.66		0.18		0.18		2	A2	1375
184151	+25 03864				19 30 01	+25 29.3	1	6.92		0.42		-0.03		2	F5 V	105
		VES 8			19 30 02	+27 01.7	1	11.52		0.32		-0.25		1		363
	+0 04241	G 92 - 8			19 30 04	+00 28.2	4	10.41	.012	1.43	.015	1.25	.009	4	M1	1017,1620,1705,3008
183784	−55 09121				19 30 07	−55 08.5	1	9.74		0.59		0.02		1	G5 V	1696
184398	+55 02215	HR 7428, V1817 Cyg ★			19 30 10	+55 37.5	1	6.36		1.16		0.91		2	K2 II-IIIe+	1355
		L 160 - 102			19 30 10	−62 56.9	1	12.17		1.48				1		1705
183861	−51 12147				19 30 11	−50 51.7	1	8.47		0.07		0.07		2	A1 V	1730
		Smethells 39			19 30 12	−52 33.	1	12.80		1.55				1		1494
183296	−75 01525				19 30 12	−74 56.0	1	8.48		1.27				4	K2 III	2012
		L 45 - 66			19 30 12	−76 34.	1	10.98		0.73		0.17		3		1696
184467	+58 01929	G 229 - 33		★	19 30 18	+58 29.1	6	6.60	.016	0.85	.012	0.53	.012	14	K1 V	22,70,1003,1355,1758,3016
		CS 22896 # 111			19 30 18	−53 18.4	1	15.41		0.07		0.14		1		1736
183906	−50 12554				19 30 25	−50 40.5	1	7.89		1.16		1.18		5	K2 III	1673
183997	−34 13760				19 30 26	−34 18.4	1	6.96		0.03		-0.03		5	A0 IV/V	1628
184152	+7 04124	G 23 - 3		★ AB	19 30 28	+07 18.0	1	9.38		0.69		0.14		1	G5 V	333,1620
338529	+26 03578				19 30 29	+26 17.1	4	9.36	.005	0.39	.011	-0.22	.004	8	B5	792,1064,1658,3077
		LS II +16 009			19 30 31	+16 55.3	1	11.34		0.74		-0.14		2	B2 IV	946
184078	−17 05665				19 30 31	−17 09.9	1	8.98		0.70		0.29		2	G3 V	1375
		VES 9			19 30 32	+25 25.8	1	12.17		0.62		-0.18		1		363
		VES 10			19 30 32	+26 05.7	1	13.69		2.13		1.08		1		363
184400	+48 02909	G 208 - 29			19 30 32	+48 28.8	2	8.52	.005	0.46	.000	0.00	.020	2	F6	1620,1658
		LS II +19 008			19 30 33	+19 52.2	1	11.09		0.68		-0.30		2		946
	+29 03633				19 30 34	+29 44.6	1	9.00		1.63				3	M2	369
		AO 1180			19 30 36	+69 16.3	1	10.34		1.40		1.42		1	K2 III	1748
184334	+39 03803	IDS19289N3907	A		19 30 40	+39 12.1	2	8.55	.005	0.42	.010	0.05	.005	5	F3 V	833,1601
184203	+3 04061	IDS19282N0328	A		19 30 43	+03 33.6	1	9.16		0.33		0.31		2	B9	1586
		LS II +19 009			19 30 43	+19 24.8	1	11.02		1.11		0.83		2	A2 V	946
184448	+49 03038	G 229 - 34			19 30 43	+50 04.2	3	8.06	.011	0.63	.011	0.08	.020	4	F8	1620,1658,3016
184035	−40 13356	HR 7422, V4089 Sg			19 30 43	−40 08.6	1	5.90		0.09				4	A2 V	2006
184275	+21 03819				19 30 47	+21 20.9	1	7.83		1.23				4	K1 III	1025
		CS 22896 # 102			19 30 47	−55 07.6	1	15.29		0.15		0.17		1		1736
		CS 22896 # 90			19 30 48	−56 38.6	1	14.73		0.35		-0.03		1		1736
	+10 03951				19 30 49	+10 31.5	1	10.13		1.73				3	M2	369
		LS II +19 010			19 30 50	+19 24.1	1	11.89		0.94		-0.19		2	B0 IV	946
	+32 03463				19 30 50	+32 32.9	1	9.26		1.71				2	M2	369
184160	−13 05395				19 30 50	−13 15.4	1	9.47		1.29		1.16		2	K1 III	1375
		CS 22896 # 101			19 30 50	−55 07.3	1	15.98		0.11		0.17		1		1736
184296	+20 04178				19 30 51	+20 53.7	2	7.74	.019	0.08	.019			6	A2 V	1025,1118
		V 885 Cyg			19 30 51	+29 54.8	1	9.97		0.19		0.19		3		1768
		CS 22891 # 200			19 30 51	−61 49.0	2	13.91	.014	0.87	.005	0.22	.005	4		1580,1736
184677	+63 01534				19 30 54	+64 10.8	1	7.94		1.30		1.30		2	K2	1733
184381	+30 03622				19 30 58	+31 07.6	1	6.69		0.39		0.20		2	F5 V	8100
184004	−51 12152				19 30 59	−50 54.6	1	8.94		0.13		0.13		2	A3 V	1730
		VES 11			19 31 00	+23 35.4	1	14.09		1.56		1.70		1		363
		G 185 - 30			19 31 01	+23 53.5	2	12.52	.030	1.03	.005	0.63	.030	5		316,1658
184601	+60 01963				19 31 01	+60 45.4	1	8.24		0.54				1	G0	70
	−50 12560				19 31 04	−50 50.3	1	10.23		1.24				2		1730
184360	+20 04179	IDS19289N2012	A		19 31 06	+20 18.3	1	7.53		0.24		0.15		3	A5p	3016
184360	+20 04179	IDS19289N2012	AB		19 31 06	+20 18.3	5	7.29	.011	0.26	.019	0.11	.012	14	A5 p	938,1003,1025,1118,8071
184360	+20 04179	IDS19289N2012	B		19 31 06	+20 18.3	1	9.22		0.41		0.03		2		3016
184279	+3 04065	V1294 Aql			19 31 07	+03 39.1	10	7.00	.124	0.03	.011	-0.74	.064	43	B0.5IV	116,197,219,237,379*
184384	+27 03428				19 31 09	+27 16.6	1	8.04		0.47		-0.02		3	F7 V wls	3016
	−23 15543				19 31 10	−23 07.2	1	10.11		1.01		0.79		3	K3 V	3072
		CS 22896 # 115			19 31 10	−52 53.8	2	14.12	.005	0.43	.008	-0.23	.008	3		1580,1736
184251	−15 05393				19 31 12	−15 19.4	1	10.10		0.54		0.05		2	F3/6 V	1375
184469	+39 03809				19 31 13	+39 33.2	1	7.75		-0.08		-0.39		2	B9	401
		CS 22896 # 94			19 31 14	−55 44.4	1	14.28		0.76		0.34		1		1736

HD DM Other Id	N Rem	α_{1950}	δ_{1950}	S	V	σ_V	B–V	σ_{B-V}	U–B	σ_{U-B}	n	Spectrum	References
184498 +46 02716		19 31 15	+46 12.1	1	7.94		1.02				1	K0	8097
184385 +21 03822		19 31 16	+21 44.0	3	6.89	.006	0.72	.020	0.35	.004	8	G5 V	1025,1733,7008
184873 +69 01052		19 31 16	+69 25.1	1	7.63		0.95		0.63		1	K0	1748
184313 +5 04190 V450 Aql		19 31 18	+05 21.4	2	6.41	.058	1.64	.053	1.48	.024	16	M8 V	897,3042
231799 +18 04123		19 31 19	+18 35.8	1	8.61		0.57				3	G0 V	1025
184266 −16 05359		19 31 23	−16 25.4	3	7.59	.012	0.57	.012	0.11	.041	9	F2 V	462,742,1594
184958 +70 01073 HR 7450		19 31 25	+70 52.9	1	6.07		1.41				2	K2	71
184484 +33 03500		19 31 27	+33 58.1	1	8.23		−0.06		−0.17		2	A0	401
239124 +56 02257 XZ Cyg		19 31 27	+56 16.8	2	9.26	.109	0.17	.045	0.17	.036	2	A2	668,699
184283 −16 05360 AQ Sgr		19 31 27	−16 29.0	3	6.98	.081	3.24	.185	4.27	.127	8	C7,4	109,864,897
CS 22896 # 106		19 31 29	−54 13.0	1	14.40		0.18		0.18				1736
184127 −48 13161 HR 7424		19 31 31	−48 12.5	3	4.89	.007	1.09	.005			11	K0 III	15,2013,2030
184268 −24 15421		19 31 33	−23 58.1	1	6.43		1.65		2.04		5	K4/5 III	1024
184426 +16 03893		19 31 35	+16 45.1	1	8.70		2.00				2	M1	369
184499 +32 03474		19 31 35	+33 05.3	6	6.63	.008	0.58	.007	0.01	.012	10	G0 V	22,792,908,1003,1658,3026
LS II +17 008		19 31 36	+17 04.8	1	12.15		0.77		−0.12		2	B0 V:	946
184926 +67 01162		19 31 38	+67 35.5	1	7.75		0.97		0.76		3	K0	3016
184406 +7 04132 HR 7429	⋆ A	19 31 39	+07 16.3	8	4.45	.008	1.18	.005	1.25	.010	29	K3 IIIb	15,1003,1118,1355*
231827 +18 04125		19 31 39	+18 44.5	1	9.12		0.14				3	A1 V	1025
184757 +59 02055		19 31 39	+60 11.0	1	8.78		0.46				2	F8	70
231825 +12 03964 G 23 - 4		19 31 42	+12 31.8	1	8.83		0.81		0.37		1	G5	333,1620
CS 22939 # 230		19 31 51	−31 00.7	1	14.27		0.60		0.00		1		1736
CS 22896 # 110		19 31 51	−53 32.9	2	13.55	.015	0.54	.012	−0.02	.003	3		1580,1736
184603 +38 03650 HR 7436		19 31 52	+38 39.1	2	6.61		0.10	.045	0.09	.009	5	A3 Vn	833,1063
184501 +21 03827		19 31 53	+21 54.6	1	8.08		0.15				3	A7 V	1025
+32 03477	V	19 31 53	+32 36.2	1	10.06		1.75				2	M2	369
CS 22896 # 88		19 31 54	−56 58.5	1	15.73		0.06		0.14		1		1736
184537 +25 03875		19 31 57	+25 57.1	1	7.01		0.26		0.17		3	Am	8071
184538 +25 03876		19 31 58	+25 41.8	1	7.42		1.32		1.41		2	K2 III	105
184486 +8 04145 IDS19296N0811	A	19 32 00	+08 17.5	1	9.29		1.19		1.43		1	F8	679
184486 +8 04145 IDS19296N0811	B	19 32 00	+08 17.5	1	12.42		0.77		0.39		1		679
AO 1182		19 32 01	+69 40.9	1	11.54		0.62		0.05		1	F6 V	1748
231848 +19 04061		19 32 02	+19 16.6	1	8.55		1.05				3	K0 IV	1025
184502 +15 03866		19 32 03	+16 09.3	1	7.01		−0.02		−0.63		3	B3 III	399
184604 +32 03480		19 32 07	+32 14.7	1	8.88		0.36				27		6011
184489 +4 04157 G 23 - 5		19 32 09	+04 28.1	5	9.35	.005	1.39	.009	1.22	.010	13	K5	333,1013,1197,1620,1705,3008
184590 +25 03877		19 32 09	+25 14.6	2	7.09	.010	1.79	.010	2.09	.020	4	M1 III	1733,8100
344590 +21 03829		19 32 12	+21 36.0	1	9.22		0.10				3	B8 Si	1025
184902 +59 02059 IDS19316N5957	B	19 32 13	+60 03.3	2	8.11	.005	0.95	.020	0.76		2	K0 III	70,3032
LS II +20 007		19 32 15	+20 43.4	1	9.41		0.41		0.00		2	B8 V	946
184786 +48 02914 HR 7442, V1743 Cyg		19 32 19	+49 09.2	2	5.93	.030	1.65	.011	1.77		4	M4.5 IIIas	71,3001
184903 +59 02058		19 32 19	+59 17.2	1	7.81		−0.02				2	A0p	70
184631 +25 03879 IDS19303N2522	AB	19 32 20	+25 28.4	1	8.79		0.15		−0.26		1		963
338506 +27 03435		19 32 21	+27 27.4	1	8.03		1.65				2	M0	369
+29 03647		19 32 21	+29 15.4	1	9.35		1.76				2	M2 III	369
184492 −10 05122 HR 7430		19 32 22	−10 40.3	3	5.12	.001	1.12	.004	0.90	.006	14	G9 IIIa	3,37,2007
184606 +19 04063 HR 7437	⋆ AB	19 32 23	+19 39.8	8	5.00	.019	−0.09	.015	−0.43	.006	29	B8 IIIn	15,369,1025,1063,1079*
184936 +59 02060 HR 7448	⋆ A	19 32 23	+60 02.9	2	6.30	.014	1.57	.011	1.98		4	K4 III	70,3016
185327 +74 00828		19 32 23	+74 39.6	1	8.42		0.24		0.13		2	A5	1733
231870 +18 04132		19 32 24	+19 11.3	1	8.20		1.34				3	K2 III	1025
185144 +69 01053 HR 7462	⋆ A	19 32 27	+69 34.6	26	4.68	.009	0.79	.007	0.39	.011	247	K0 V	1,15,22,61,71,101*
−44 13426		19 32 28	−44 10.8	1	10.54		1.19				4		2011
184592 +11 03895 G 23 - 6		19 32 33	+11 18.8	1	7.91		0.68		0.13		1	G5	333,1620
CS 22896 # 86		19 32 36	−57 25.7	1	13.58		−0.03		−0.23		1		1736
VES 12		19 32 39	+27 26.8	1	11.78		0.28		−0.36		2		363
184788 +41 03397		19 32 39	+41 19.1	1	7.14		−0.06		−0.18		2	B9	401
184787 +41 03398		19 32 39	+41 49.0	1	6.68		−0.01		−0.08		2	A0	401
CS 22896 # 92		19 32 39	−56 12.2	1	14.71		0.44		0.13		1		1736
350165 +19 04066		19 32 40	+19 56.6	1	8.52		1.31				2	K3 III	1025
CS 22896 # 97		19 32 40	−55 25.9	1	13.88		0.20		0.16		1		1736
CS 22896 # 91		19 32 41	−56 19.0	1	12.90		0.11		0.17		1		1736
+34 03612		19 32 44	+34 58.6	1	10.19		1.73				2	M3	369
184574 −12 05461 HR 7433		19 32 46	−12 21.8	1	6.27		1.09				4	K0 III	2007
MCC 811	A	19 32 48	+08 18.	1	10.38		1.46		1.22		1	M0	679
MCC 811	B	19 32 48	+08 18.	1	12.52		1.52				1		679
+34 03615	V	19 32 48	+35 11.3	1	9.60		1.74				2	M3	369
184573 −7 04998 HR 7432		19 32 48	−07 34.3	2	6.33	.001	1.12	.007	0.92		9	K0	1088,6007
IRC +30 375		19 32 49	+30 39.7	1	12.42		1.48		0.59		2		363
350151 +20 04189		19 32 50	+20 29.0	1	9.61		0.57				27	G0	6011
338582 +24 03786 LS II +24 006		19 32 52	+24 28.1	1	8.66		0.96		0.84		1	A6 Ib	8100
184759 +29 03651 HR 7441	⋆ AB	19 32 52	+29 21.1	3	5.39	.015	0.55	.009	0.30	.020	7	F8 III+A0 V	71,401,3026
+38 03659		19 32 53	+39 07.3	2	8.87	.000	0.46	.005	0.08	.000	2	F2	1501,1601
184663 +2 03932 HR 7438		19 32 54	+02 48.1	1	6.37		0.41		−0.05		4	F6 IV	1149
LDS 6831		19 32 54	−13 36.	1	13.60		1.49		1.40		3		3028
LDS 6832		19 32 54	−13 36.	1	15.95		0.05		−0.76		2	DA	3028
184722 +18 04137		19 32 55	+18 53.6	3	6.83	.013	1.68	.032	1.96		6	K7 III	369,1025,8100
Ross 740		19 32 58	−04 16.7	1	12.71		0.71		−0.06		2		1696
−55 09130 LSS 3520		19 32 58	−55 24.6	2	11.47	.025	0.39	.045	−0.47	.040	2		115,181
184740 +21 03836		19 32 59	+22 02.8	1	7.39		0.08				4	B9	1118
+28 03402		19 33 00	+28 58.6	1	8.88		0.44		−0.03		3	F5 V	3026

Table 1   1059

HD DM Other Id	N Rem	α_{1950}	δ_{1950}	S	V	σ_V	B−V	σ_{B-V}	U−B	σ_{U-B}	n	Spectrum	References
184552 −24 15442 HR 7431		19 33 00	−24 49.8	4	5.64	.016	0.19	.003	0.17	.018	17	A1mA2-F0	15,216,355,2027
184553 −26 14337		19 33 00	−26 00.7	1	8.02		0.50				4	F3 III	2012
JL 25		19 33 00	−76 08.	2	13.29	.010	-0.19	.005	-1.10	.015	3	sdOB	132,832
184724 +11 03901 IDS19307N1144	A	19 33 02	+11 49.6	1	7.92		1.74				3	K5	369
184960 +50 02815 HR 7451		19 33 02	+51 07.7	6	5.73	.019	0.47	.016	-0.01	.009	14	F7 V	15,254,1007,1013,3037,8015
231915 +18 04139		19 33 03	+19 10.0	2	10.37	.018	0.57	.004	0.01	.025	10	G5	946,1655
184827 +33 03507 V1919 Cyg		19 33 03	+33 41.1	1	6.69		1.60		1.84		6	M3 III	1501
184875 +42 03386 HR 7444		19 33 03	+42 18.1	1	5.35		0.05		0.09		1	A2 V	3023
AO 1183		19 33 05	+69 45.1	1	9.98		0.22		0.21		1	A2 V	1748
CS 22896 # 93		19 33 06	−55 46.2	1	15.55		0.06		0.06		1		1736
GY Sge		19 33 08	+19 05.7	2	9.90	.003	2.04	.043	1.78		2		946,1772
184905 +43 03290 V1264 Cyg		19 33 09	+43 50.1	1	6.61		-0.03		-0.23		8	A0p	1202
CS 22891 # 206		19 33 09	−61 12.8	2	14.87	.010	0.47	.010	-0.15	.006	3		1580,1736
184700 −0 03786 G 92 - 11	A	19 33 11	−00 20.9	3	8.83	.000	0.66	.006	0.10	.003	6	G2 V	1003,1620,3077
+1 04042 G 23 - 7		19 33 12	+01 37.2	3	10.01	.019	0.96	.005	0.39	.040	5	G3	333,927,1620,1696
350156 +20 04190		19 33 12	+20 14.6	1	9.94		0.13		0.06		2	A2	634
CS 22896 # 107		19 33 13	−54 13.3	1	14.18		0.13		0.20		1		1736
184977 +47 02870 HR 7453		19 33 14	+48 03.3	2	6.78	.005	0.25	.005	0.06	.005	4	A9 V	1601,1733
350174 +19 04071		19 33 16	+19 15.2	1	10.13		1.84		1.99		3	K5	946
184938 +46 02727		19 33 16	+46 21.4	1	7.17		1.06				1	G5	8097
+49 03050		19 33 17	+49 51.7	1	9.82		1.55		1.97		3		1746
−23 15564		19 33 18	−23 18.0	1	10.03		1.10		0.99		2	K3 V	3072
344669 +20 04191		19 33 19	+20 46.6	1	8.95		0.10				1	B8 V	1025
VES 14		19 33 19	+30 59.8	1	12.41		1.72		1.65		2		363
184848 +27 03437	V	19 33 20	+27 45.2	1	8.24		1.66				3	K2	369
+35 03693		19 33 20	+35 31.0	1	8.57		-0.08		-0.31		2	K5	1601
G 92 - 13		19 33 20	−01 12.5	1	12.04		1.40		1.26		1		1620
184829 +20 04193 IDS19312N2036	AB	19 33 21	+20 42.6	2	8.25	.005	-0.01	.002	-0.20		6	B8 V	634,1025
184906 +35 03694		19 33 21	+35 35.7	1	8.09		-0.09		-0.33		2		1601
183966 −75 01532		19 33 21	−75 21.3	1	9.93		0.34				4	A8 III/IV	2012
344556 +22 03736		19 33 23	+22 56.8	1	9.34		0.11		0.06		2	A0	106
184939 +39 03826		19 33 25	+39 54.8	1	8.88		-0.05		-0.46		1	A2	963
184768 −0 03788 IDS19309S0007	AB	19 33 27	−00 00.9	3	7.56	.010	0.68	.007	0.15	.010	5	G5	265,1620,3077
184849 +22 03737		19 33 29	+22 23.4	1	7.80		0.06				4	A0	1118
184940 +34 03620		19 33 30	+34 34.6	2	7.12	.025	-0.10	.010	-0.47	.005	4	B8	401,3016
AO 1184		19 33 34	+69 33.1	1	11.61		0.90		0.51		1	K0 V	1748
184927 +30 03645 V1671 Cyg		19 33 35	+31 09.9	3	7.45	.013	-0.17	.004	-0.82	.009	12	B2 V	267,399,401
G 125 - 19		19 33 35	+31 41.9	1	12.74		1.21		1.15		1		1658
CS 22896 # 134		19 33 35	−53 54.7	1	15.30		0.07		0.14		1		1736
231934 +17 04020		19 33 36	+17 15.7	1	9.46		2.02				2	K7	369
ApJ86,509 R4# 1		19 33 36	+22 13.8	1	14.76		3.00				1		8008
184790 −3 04645		19 33 36	−02 55.1	4	8.12	.002	0.17	.003	-0.33	.010	58	B8	147,963,1509,6005
184941 +32 03486		19 33 38	+32 15.9	1	8.46		-0.14		-0.73		2	B8	401
CS 22896 # 117		19 33 38	−52 50.1	1	14.75		0.11		0.19		1		1736
350149 +20 04195		19 33 40	+20 24.0	1	9.57		0.48				27	F5	6011
344619 +23 03705 IDS19316N2321	AB	19 33 40	+23 27.1	1	9.68		0.41		-0.04		2	A0	106
184707 −25 14184 HR 7440	★ AB	19 33 40	−24 59.7	6	4.60	.007	-0.07	.009	-0.15	.009	16	B8/9 V	15,369,1075,1079,2012,801
G 229 - 35		19 33 42	+53 08.4	1	12.21		1.58				1		1759
184884 +10 03981 HR 7445	★ A	19 33 46	+11 02.3	1	6.71		0.08		0.07		2	A3 V	1733
184855 +4 04165 G 92 - 15		19 33 47	+04 39.4	1	9.20		0.66		0.08		2	G0	333,1620
+30 03647		19 33 47	+31 00.8	1	9.27		1.82				2	M0 III	369
184711 −40 13389		19 33 47	−39 51.3	4	7.98	.012	1.34	.013	0.90	.023	12	G5wA/F	119,742,1594,3077
184943 +23 03706 LS II +23 010		19 33 49	+23 44.7	4	8.18	.016	0.73	.009	-0.07	.011	7	B8 Ia	106,1012,1032,8100
+27 03441 V909 Cyg		19 33 52	+28 10.0	1	9.51		0.14		0.13		2	A0	1768
CS 22891 # 216		19 33 53	−60 14.1	1	14.71		0.26		0.16		1		1736
185239 +57 02029		19 33 55	+57 39.7	1	8.16		0.44		-0.04		2	F8	1375
CS 22896 # 122		19 33 55	−53 16.1	1	12.62		0.13		0.18		1		1736
184978 +25 03891		19 33 57	+25 26.1	2	8.40	.005	0.12	.015	0.08	.005	5	A0	833,1625
184944 +14 03974 HR 7449		19 33 58	+14 16.8	2	6.38	.005	1.04	.005	0.89		5	K0 II-III	37,71
184961 +22 03741 HR 7452		19 34 00	+22 28.4	4	6.33	.005	-0.06	.011	-0.27	.009	14	B9 p (SiCr)	1063,1079,1118,3033
185037 +36 03619 HR 7457		19 34 00	+36 50.0	2			-0.11	.005	-0.35	.045	5	B8 Vne	1063,1079
VES 15		19 34 01	+28 49.7	1	12.22		1.65		1.24		1		363
185015 +35 03700		19 34 01	+35 37.3	1	9.13		-0.07		-0.19		2		401
CS 22896 # 96		19 34 01	−55 27.4	1	14.22		0.27		0.11		1		1736
HA 111 # 669		19 34 02	+00 03.0	1	12.11		0.88		0.44		2		281
184860 −10 05130 IDS19313S1039	AB	19 34 02	−10 33.1	2	8.40	.000	1.00	.005	0.81	.009	4	K2 V	1003,3072
LS II +22 007		19 34 03	+22 38.8	1	11.43		0.53		0.04		2		405
231961 +11 03908 LS II +11 007		19 34 04	+11 55.3	1	9.25		0.58		0.34		1	F3 Ib-II	8100
184914 −4 04855		19 34 07	−04 24.7	6	8.17	.005	1.19	.005	0.94	.006	85	K5	147,989,1509,1657,1729,6005
231970 +12 03981		19 34 08	+12 51.6	3	9.83	.005	-0.03	.004	-0.66	.004	17	B2 V	830,963,1032,1783
184930 −1 03782 HR 7447	★ A	19 34 08	−01 23.9	8	4.35	.008	-0.08	.008	-0.44	.008	37	B5 III	15,154,1075,1088,1203*
CS 22896 # 137		19 34 08	−54 26.6	1	15.20		0.13		0.20		1		1736
184358 −69 03034		19 34 08	−69 27.1	1	6.95		1.30				4	K3 III	2012
184835 −18 05432 HR 7443		19 34 09	−18 20.6	2	5.66	.015	1.24	.015	1.08		6	K3 III	2007,3005
231977 +19 04077 IDS19320N1925	AB	19 34 10	+19 31.6	1	9.51		0.08		-0.04		2	A0	634
AO 1185		19 34 10	+69 28.0	1	10.58		1.47		1.61		1	K3 III	1748
184999 +21 03843		19 34 12	+21 57.0	1	8.82		0.05				2		1118
185055 +33 03516		19 34 12	+33 38.1	1	7.42		1.18		1.23		2	K2 III-IV	37
184915 −7 05006 HR 7446	★	19 34 12	−07 08.4	14	4.96	.007	0.00	.010	-0.87	.010	131	B0.5III	3,15,154,247,1006*
185016 +23 03707		19 34 13	+23 34.6	1	7.95		0.27				3	A3	1118

HD	DM	Other Id	N Rem	α_{1950}	δ_{1950}	S	V	σ_V	B–V	σ_{B-V}	U–B	σ_{U-B}	n	Spectrum	References
		VES 16		19 34 14	+23 32.5	1	13.73		0.79		0.61		1		363
184585	−58 07627	HR 7434		19 34 14	−58 05.8	2	6.19	.008	0.98	.003	0.81		6	K0 III	58,2035
	+0 04257			19 34 15	+00 23.6	2	10.71	.025	0.62	.000	0.37	.020	5		196,281
		VES 17		19 34 15	+29 08.8	1	12.23		0.84		0.23		1		363
	+23 03709			19 34 16	+23 31.4	1	9.56		2.53				2	M2	369
185118	+34 03625			19 34 17	+35 06.1	1	6.75		1.52		1.82		1	K2	963
		HA 111 # 710		19 34 19	+00 00.9	1	11.86		0.74		0.18		8	G0	281
184982	+7 04151			19 34 19	+07 26.3	1	7.59		0.17				2	A2	1025
		VES 18		19 34 19	+26 58.8	1	12.06		0.17		−0.43		2		363
187216	+85 00332			19 34 19	+85 15.6	1	9.57		1.69				1	R3	1238
		VES 20		19 34 20	+29 26.1	1	11.67		0.71		−0.25		1		363
184965	−0 03796			19 34 21	+00 00.8	5	8.54	.006	0.42	.005	0.22	.008	41	F0 p:	281,989,1509,1729,6004
		HA 111 # 3264		19 34 21	+00 37.5	1	10.74		1.65		1.89		3		196
		VES 21		19 34 21	+29 12.6	1	11.21		0.14		0.08		2		363
344609	+23 03710	IDS19322N2344	A	19 34 23	+23 50.5	1	10.46		0.73		0.06		1	B9	363
185058	+23 03711	IDS19323N2347	AB	19 34 23	+23 53.4	1	8.44		0.09				3	A0	1118
185119	+33 03518			19 34 23	+33 20.9	1	7.27		−0.10		−0.47		2	B8	401
185561	+69 01056			19 34 23	+69 41.0	1	8.58		0.53		0.02		1	G0	1748
185002	+6 04217			19 34 24	+06 56.9	1	9.35		0.16				2	A0	1025
185059	+20 04200	HR 7458, U Vul		19 34 26	+20 13.2	3	6.79	.010	1.14	.019	0.80	.072	3	F2 I	851,1772,6011
185171	+39 03837			19 34 26	+39 13.4	2	8.21	.005	0.29	.010	0.18	.010	5	A2	833,1601
184984	−0 03797			19 34 27	+00 01.5	4	9.18	.005	0.30	.005	0.19	.010	16	A0	196,281,5006,6004
		LS II +12 009		19 34 27	+12 58.8	1	11.28		0.19		−0.75		2		405
		CS 22896 # 123		19 34 27	−53 17.6	1	15.37		0.26		0.16		1		1736
185003	+0 04259			19 34 31	+00 30.8	5	9.70	.003	0.16	.003	−0.39	.022	41	B9	281,989,1729,5006,6004
185018	+10 03984	HR 7456	★ A	19 34 31	+11 09.6	1	5.98		0.88				2	G0 Ib	71
185173	+35 03705	IDS19327N3527	A	19 34 32	+35 33.9	1	8.13		−0.09		−0.33		2		401
185172	+35 03706			19 34 34	+35 34.5	1	8.47		−0.05		−0.38		1		401
185264	+49 03059	HR 7465		19 34 35	+50 07.5	2	6.50	.029	1.06	.005	0.84	.005	7	G9 III	814,1733
	+31 03681			19 34 36	+31 38.1	1	9.11		1.63				2	M0 III	369
185061	+9 04174			19 34 40	+09 26.8	1	8.02		1.26				2	K0	1025
185024	−0 03798			19 34 41	+00 15.0	3	9.24	.017	1.68	.010	1.78		13	K3 III	281,5006,6004
185063	+8 04163	IDS19323N0805	AB	19 34 41	+08 12.1	1	8.31		0.28				2	A3	1025
185025	−0 03800			19 34 42	+00 04.2	7	8.97	.005	0.20	.005	−0.21	.007	114	A0	196,281,989,1509,1729*
		HA 111 # 775		19 34 42	+00 05.3	4	10.75	.008	1.74	.006	2.03	.018	84		281,989,1729,1764
185026	−0 03799			19 34 43	−00 36.3	1	7.93		0.11		−0.44		1	B9	963
344611	+23 03713			19 34 45	+23 55.5	1	10.15		0.14		0.10		2	A0	106
184985	−14 05479	HR 7454		19 34 45	−14 24.8	4	5.47	.009	0.50	.003	0.04	.014	18	F7 V	3,158,253,2035
185175	+23 03714			19 34 46	+23 25.8	1	8.18		0.08		−0.65		1	B9	963
		VES 22, EH Cyg		19 34 47	+28 00.9	1	10.54		1.81		1.12		1		363
		CS 22896 # 124		19 34 53	−53 36.3	1	14.60		0.07		0.15		1		1736
185224	+30 03658	V1744 Cyg	★ AB	19 34 54	+30 12.7	1	7.59		−0.08		−0.47		2	B9	401
		CS 22896 # 152		19 34 54	−56 49.4	1	13.32		0.64		0.12		1		1736
		HA 111 # 1925		19 34 55	+00 18.2	2	12.40	.006	0.40	.002	0.24	.020	39	A3	281,1764
	+34 03631	LS II +35 002	★ A	19 34 56	+35 07.9	1	10.16		−0.11		−0.88		2	B2 V	1012
185414	+56 02272			19 34 56	+56 52.4	1	6.74		0.61		0.08		2	G0	3016
184586	−66 03445	HR 7435	★ A	19 34 57	−66 48.0	3	6.38	.007	0.02	.000	0.04	.000	13	A1 V	15,1075,2038
185091	−1 03790			19 34 59	−01 08.9	1	8.21		0.16		−0.06		1	B9	963
185194	+16 03918	HR 7463	★ A	19 35 02	+16 21.0	2	5.66	.015	1.02	.025	0.81	.025	7	G8 III	1084,1355
185155	+6 04223			19 35 03	+06 32.0	1	9.27		0.00				2	B8	1025
		CS 22896 # 143		19 35 04	−55 16.3	1	15.65		0.28		0.20		1		1736
185351	+44 03185	HR 7468		19 35 05	+44 35.0	5	5.17	.010	0.93	.007	0.69	.007	8	K0 III	1080,1355,1363,3016,4001
350219	+19 04086	LS II +19 011		19 35 06	+19 19.4	1	9.76		1.34		0.94		2	A0 Iab	946
185395	+49 03062	HR 7469	★ AB	19 35 06	+50 06.3	9	4.48	.015	0.38	.008	−0.03	.007	30	F4 V	15,254,1013,1118,1197,1620*
		CS 22896 # 142		19 35 06	−55 13.8	1	15.55		0.05		0.07		1		1736
	+31 03685			19 35 07	+31 56.6	2	9.05	.010	−0.12	.019	−0.57	.005	3		401,963
		CS 22896 # 121		19 35 07	−52 57.0	1	14.93		0.08		0.06		1		1736
		HA 111 # 1965		19 35 08	+00 20.0	2	11.43	.009	1.70	.005	1.85	.011	40	dK2	281,1764
232029	+16 03919	LS II +16 011	★ B	19 35 08	+16 21.2	2	8.28	.010	0.50	.010	−0.19	.010	4	B9 Ib	1084,8100
185124	−4 04861	HR 7460		19 35 08	−04 45.6	2	5.45	.000	0.43	.000	−0.01	.009	8	F3 IV	254,1149,3016
	+0 04260			19 35 09	+00 19.0	5	10.39	.012	1.95	.018	2.30	.011	90	M5 III	196,281,989,1729,1764
185176	+7 04157			19 35 09	+07 29.4	1	9.38		0.46				2		1025
		Vul R2 # 17		19 35 09	+25 43.2	1	12.49		0.68		0.20		1		5012
		NGC 6811 - 73		19 35 09	+46 13.4	1	9.89		1.64		2.01		5	K5	151
185268	+29 03670	HR 7466	★ AP	19 35 10	+29 13.2	3	6.42	.021	−0.09	.004	−0.51	.000	8	B5 V	154,401,1118
185125	−6 05213			19 35 10	−06 15.4	1	8.26		1.14				1	K0	1642
185126	−6 05212			19 35 10	−06 39.5	1	9.87		1.30				1	K0	1642
185207	+9 04177			19 35 11	+09 36.6	1	9.17		0.25				2	A2	1025
		G 185 - 32	★ V	19 35 11	+27 36.5	2	12.98	.013	0.17	.000	−0.56	.004	6	DA	538,3060
185269	+28 03412			19 35 11	+28 23.3	1	6.70		0.58		0.17		2	G0 IV	3016
	+33 03522			19 35 11	+33 45.3	1	9.54		1.60				2	M3	369
185330	+38 03677	HR 7467		19 35 11	+38 16.2	3	6.50		−0.15	.007	−0.62	.008	5	B5 II-III	401,1063,1079
		CS 22896 # 145		19 35 11	−55 31.2	1	13.62		0.01		−0.11		1		1736
185270	+25 03895			19 35 12	+26 02.5	1	8.48		0.53		0.00		3	F8 V	3026
185290	+25 03896			19 35 18	+25 26.2	1	8.19		0.23		0.17		3	A2	833
	+4 04170	G 92 - 16		19 35 19	+04 12.0	1	10.18		0.70		0.17		1	G0	333,1620
		CS 22896 # 120		19 35 19	−52 49.4	1	13.15		0.09		0.14		1		1736
	+0 04262			19 35 20	+00 19.2	5	10.61	.004	0.89	.012	0.52	.010	39	K2 V	196,281,989,1729,6004
		NGC 6811 - 26		19 35 22	+46 16.5	1	11.35		0.29		0.23		4	A4	151
		NGC 6811 - 79		19 35 22	+46 17.7	1	10.30		1.53		1.96		4		151

Table 1   1061

HD	DM	Other Id	N Rem	α_{1950}	δ_{1950}	S	V	σ_V	B–V	σ_{B-V}	U–B	σ_{U-B}	n	Spectrum	References
185331 +30 03660		IDS19334N3012	AB	19 35 24	+30 18.6	1	8.09		-0.03		-0.16		2	A0	401
344629 +22 03747				19 35 25	+22 39.2	1	9.64		0.18		0.03		3	A0	106
185291 +21 03852				19 35 26	+22 06.1	1	8.50		0.15				2	A0	1118
+0 04264		IDS19329N0033	A	19 35 27	+00 40.1	1	9.96		0.62		0.12		3		196
185244 +9 04179				19 35 27	+09 30.7	1	9.42		0.03				2	A0	1025
185271 +12 03995				19 35 27	+12 49.5	1	8.04		0.97		0.57		2	K0	1733
		NGC 6811 - 24		19 35 27	+46 15.9	1	11.16		0.96		0.63		4	G8	151
		CS 22896 # 119		19 35 27	-52 39.7	1	14.55		0.06		0.10		1		1736
		NGC 6811 - 18		19 35 28	+46 13.6	1	12.03		0.30		0.30		3	A4	151
		G 260 - 27		19 35 28	+69 49.1	3	9.80	.008	0.61	.010	0.05	.025	4	G0 V	1658,1748,5010
		HA 111 # 2036		19 35 29	+00 19.6	1	12.45		0.79		0.22		20	G2	281
185456 +49 03064		R Cyg	⋆ A	19 35 29	+50 05.2	3	8.13	.174	2.00	.076	1.87	.705	3	S3.9 e	765,814,817
		NGC 6811 - 17		19 35 30	+46 13.7	1	13.84		0.70				7		151
		NGC 6811 - 23		19 35 30	+46 15.6	1	14.02		0.75				5		151
		CS 22896 # 151		19 35 30	-56 49.9	1	14.84		0.12		0.19		1		1736
		HA 111 # 2039		19 35 31	+00 25.4	2	12.40	.002	1.37	.000	1.23	.008	6	K2	281,1764
		NGC 6811 - 31		19 35 31	+46 16.2	1	13.29		0.35		0.24		8		151
185456 +49 03065		IDS19341N4959	B	19 35 31	+50 06.7	1	9.91		-0.09				1		765
-30 17239				19 35 31	-30 40.1	1	10.02		1.43		1.64		1		552
-0 03803				19 35 32	-00 02.3	1	10.31		0.79		0.22		3		196
		CS 22896 # 133		19 35 32	-53 47.8	1	15.44		0.23		0.10		1		1736
+31 03688				19 35 33	+31 48.5	1	8.15		1.62				2	M0 III	369
		NGC 6811 - 70		19 35 33	+46 12.6	1	10.84		0.37		0.22		5	A4	151
		VES 23, V917 Cyg		19 35 36	+29 49.2	1	12.71		2.08		0.71		1		363
		NGC 6811 - 14		19 35 36	+46 14.2	1	13.61		1.15				5		151
185799 +69 01058				19 35 36	+69 41.6	1	6.92		1.56		1.49		1	M2	1748
185334 +23 03717				19 35 37	+24 08.8	1	7.67		0.14				3	A0	1118
-31 16911				19 35 37	-31 20.7	1	11.54		1.12		0.98		2		552
		HA 111 # 326		19 35 38	-00 08.3	1	11.08		0.65		0.25		3		196
+28 03417				19 35 39	+28 56.2	1	9.42		-0.01		-0.56		1		963
		LP 633 - 11		19 35 41	+01 35.2	1	11.95		1.59		1.55		1		1696
185131 -31 16912				19 35 41	-31 03.2	2	10.32	.000	0.54	.000	-0.01		5	F6/8 V	438,552
185337 +18 04168				19 35 42	+18 28.7	1	7.14		0.23		0.07		2	A5	1733
185352 +24 03814				19 35 42	+24 54.2	1	8.20		0.14				2	B9	1118
		NGC 6809 sq2 12		19 35 42	-31 03.7	1	14.11		1.01				1		438
185308 +9 04181				19 35 43	+09 33.0	1	9.61		0.31				2	B8	1025
		NGC 6811 - 4		19 35 43	+46 16.7	1	12.57		0.33		0.11		4	A7	151
185353 +22 03749		IDS19336N2234	AB	19 35 44	+22 41.0	1	7.56		0.89		0.58		1	G5 III	8100
		NGC 6811 - 3		19 35 44	+46 16.5	1	15.12		0.85				7		151
+27 03448				19 35 45	+28 12.9	1	9.37		1.72				2	M0 III	369
185309 +8 04168				19 35 46	+08 48.3	1	8.69		0.23				2	A2	1025
185310 +8 04169				19 35 47	+08 38.0	1	7.79		0.05				2	A0	1025
		NGC 6811 - 5		19 35 47	+46 16.4	1	11.73		0.25		0.21		4	A2	151
185297 +0 04265		IDS19332N0007	AB	19 35 48	+00 13.8	5	7.22	.004	0.28	.004	0.12	.005	38	A2	196,281,989,1729,6004
		HA 111 # 2088		19 35 48	+00 24.2	2	13.18	.011	1.61	.005	1.67	.004	6		281,1764
185375 +24 03815				19 35 48	+24 35.9	1	7.73		0.13				3	A2	1118
		NGC 6811 - 57		19 35 48	+46 12.3	1	13.82		0.44		0.15		3		151
		NGC 6811 - 35		19 35 48	+46 16.9	1	13.80		0.57		0.17		9		151
-31 16915				19 35 48	-31 08.6	2	10.90	.035	1.65	.020	1.55		7		438,552
185273 -1 03795				19 35 49	-01 02.3	2	8.63	.010	0.11	.007	-0.41	.031	6	B9	963,1732
		HA 111 # 2093		19 35 50	+00 24.6	2	12.55	.010	0.63	.003	0.27	.006	18	G0	281,1764
185435 +34 03637				19 35 50	+34 54.5	1	6.42		1.57		1.93		3	K5	1625
		NGC 6809 sq2 4		19 35 51	-30 54.7	1	11.36		1.09				1		438
		VES 24		19 35 52	+21 05.8	1	12.73		1.67		0.48		2		363
+32 03498				19 35 52	+32 20.7	1	9.77		0.31				1	A5	851
185416 +25 03900				19 35 53	+25 54.1	1	7.80		0.09				4	B9	1118
		NGC 6811 - 43		19 35 53	+46 16.1	1	12.65		0.40		0.14		5		151
+32 03499				19 35 54	+32 19.2	1	10.46		0.20				1		851
		NGC 6811 - 42		19 35 54	+46 16.6	1	12.45		0.40		0.15		5		151
		HA 111 # 960		19 35 54	-00 01.5	1	11.21		0.72		0.30		3		196
185338 +9 04183				19 35 55	+09 52.0	1	8.75		0.02				3	A0	1025
		VES 25		19 35 55	+28 27.3	1	12.26		1.72		1.14		1		363
		G 92 - 17		19 35 55	-02 58.1	2	10.67	.020	1.12	.030	1.07	.010	2	K4	1620,1658
		CS 22896 # 136		19 35 55	-54 17.3	2	14.71	.000	0.41	.007	-0.17	.000	3		1580,1736
		NGC 6811 - 38		19 35 56	+46 17.7	1	13.19		1.08				5		151
232078 +16 03924			V	19 35 57	+16 41.6	3	8.66	.087	2.01	.062	2.15		5	K3 IIp	369,1391,3025
		CS 22896 # 118		19 35 57	-52 27.9	1	14.79		0.18		0.18		1		1736
		NGC 6811 - 47		19 35 58	+46 15.6	1	13.53		0.43				6		151
		NGC 6811 - 46		19 35 59	+46 15.8	1	12.95		0.40		0.03		5		151
185355 +9 04184				19 36 01	+09 58.2	1	9.04		-0.02				3	A0	1025
+32 03502				19 36 01	+33 08.9	1	9.71		1.69				2	M3	369
		NGC 6809 sq2 5		19 36 01	-31 00.4	1	11.52		1.59				3		438
		CS 22896 # 146		19 36 04	-55 42.2	1	15.37		0.12		0.19		1		1736
		VES 26		19 36 05	+25 08.8	1	11.17		0.61		0.15		1		363
		NGC 6809 sq1 9		19 36 05	-31 23.9	1	14.54		0.78				2		422
		NGC 6809 sq2 8		19 36 06	-31 07.2	1	12.85		0.97				1		438
185436 +20 04210		HR 7472		19 36 07	+20 40.1	1	6.48		1.08		0.98		1	K0 III	851
		NGC 6809 sq1 7		19 36 07	-31 22.8	1	13.09		1.28		0.90		2		422
185139 -45 13354		HR 7461, QQ Tel		19 36 07	-45 23.6	2	6.24	.005	0.29	.005			7	F2 IV	15,2012
		HA 111 # 2135		19 36 08	+00 22.3	1	10.64		1.52		1.40		3		196

HD	DM	Other Id	N Rem	α_{1950}	δ_{1950}	S	V	σ_V	B−V	σ_{B-V}	U−B	σ_{U-B}	n	Spectrum	References
		VES 27		19 36 08	+31 42.4	1	12.22		1.62		1.26		1		363
		NGC 6809 sq2 13		19 36 08	−31 08.1	1	14.77		0.12				1		438
	−31 16920			19 36 08	−31 27.2	1	10.83		0.53		0.04		2		552
185376	+9 04185			19 36 09	+09 55.8	1	8.33		1.10				3	K0	1025
		LS II +27 007		19 36 09	+27 04.3	1	10.27		0.39		-0.54		2		405
		VES 28		19 36 09	+29 34.0	1	14.91		2.73				1		363
	−30 17251			19 36 10	−30 49.3	1	10.84		0.54		0.05		1		552
185398	+7 04164			19 36 11	+07 30.0	1	8.23		1.11				3	K0	1025
		NGC 6809 sq2 11		19 36 11	−31 02.5	1	13.69		0.93				1		438
185342	−10 05143	IDS19334S1034	A	19 36 12	−10 27.2	3	8.07	.023	0.21	.018	0.15	.021	9	A0	1028,1731,3024
185342	−10 05142	IDS19334S1034	B	19 36 12	−10 27.2	3	9.49	.013	0.47	.024	0.08	.022	9		1028,1731,3024
185418	+16 03928	IDS19340N1702	A	19 36 13	+17 08.6	1	7.45		0.22		-0.67		2	B0.5V	1012
		IRC +20 420		19 36 13	+21 15.7	1	14.04		2.77		0.44		2		363
185378	+0 04266			19 36 14	+00 29.5	5	8.30	.007	1.70	.010	2.02	.030	43	M1 III	196,281,989,1729,6004
186063	+74 00831	IDS19373N7409	AB	19 36 14	+74 15.9	1	6.96		1.06		0.82		2	K0	985
		NGC 6809 sq2 6		19 36 14	−30 57.8	1	11.90		1.02				1		438
		NGC 6809 sq1 4		19 36 15	−31 22.3	1	12.25		0.66		0.15		2		422
	−31 16922			19 36 16	−31 27.9	1	9.72		1.47		1.55		6		552
		G 92 - 18		19 36 17	+06 22.7	1	12.55		1.34		1.22		1		333,1620
		V924 Cyg		19 36 17	+32 22.5	2	10.57	.013	0.81	.021			2	F6	851,6011
185423	+3 04097	HR 7471		19 36 19	+03 16.0	2	6.35	.015	0.04	.000	-0.55	.019	7	B3 III	154,1149
185075	−54 09438	HR 7459		19 36 19	−54 32.0	3	6.26	.004	1.00	.000			14	K0 III	15,1075,2030
		CS 22891 # 211		19 36 19	−60 53.6	1	13.64		0.27		0.14		1		1736
		NGC 6809 sq1 13		19 36 22	−31 21.2	1	16.12		0.88				2		422
185438	+8 04174			19 36 23	+09 02.3	1	9.35		0.10				2	A2	1025
	+44 03197	G 208 - 32		19 36 23	+44 52.6	2	9.62	.019	0.50	.013	-0.09	.016	4	F9	1658,7010
		NGC 6809 sq1 5		19 36 25	−31 18.3	1	12.77		0.99		0.66		4		422
		LS II +17 010		19 36 26	+17 55.5	1	11.22		0.73		-0.28		2	B0.5IV	405
185528	+25 03903			19 36 26	+25 51.8	1	7.92		0.23				4	A3	1118
		NGC 6809 sq1 6		19 36 27	−31 19.9	1	12.79		1.18		1.00		5		422
	−31 16924			19 36 28	−31 20.0	3	11.48	.020	0.68	.047	0.24	.025	11		422,552,1730
		HA 111 # 2182		19 36 29	+00 18.8	1	11.48		0.76		0.26		3		196
350202	+19 04095	LS II +20 009		19 36 30	+20 00.9	1	10.31		0.72		-0.24		2	B5	946
	−31 16925			19 36 30	−31 18.7	2	10.83	.020	0.92	.010	0.63	.010	8		422,552
		NGC 6809 sq1 11		19 36 30	−31 19.2	1	14.67		0.72		0.05		4		422
		HA 111 # 2188		19 36 31	+00 26.6	2	11.61	.000	1.35	.000	1.15		5	dG5	281,5006
		LS II +19 012		19 36 31	+19 49.2	1	10.96		0.86		-0.10		2	B2 V	946
185477	+9 04187			19 36 32	+09 14.4	1	8.73		0.27				2	A2	1025
338664	+24 03822			19 36 32	+24 38.3	1	10.14		0.09		-0.08		2	B8	106
185257	−39 13371	HR 7464, V4090 Sg		19 36 32	−39 32.9	3	6.60	.004	0.24	.007			11	A1mA6-F0	15,2013,2030
185657	+48 02922	HR 7477	★ A	19 36 33	+49 10.1	2	6.46	.005	1.00	.015	0.75	.032	4	G6 V	1733,3016
	−31 16926			19 36 33	−31 16.6	1	11.27		1.11		0.91		3		552
		GD 222		19 36 34	+32 46.5	2	13.59	.010	-0.11	.010	-0.89	.010	2	DA	1727,3060
		HA 111 # 1058		19 36 34	−00 00.2	1	12.18		0.63		0.22		3		196
184588	−75 01537			19 36 34	−75 31.0	2	8.00	.003	0.55	.005	0.05		6	G0 V	278,2012
	−31 16929			19 36 36	−31 22.7	2	10.54	.010	0.63	.010	0.08	.029	7		422,552
	+35 03720			19 36 37	+35 52.3	1	9.75		1.63				2	M2	369
		G 142 - 44		19 36 38	+16 18.8	2	11.15	.003	0.66	.005	-0.01	.005	4	G5	1658,1696
	−2 05072	G 92 - 19		19 36 38	−02 43.5	2	10.31	.010	0.66	.005	0.08	.024	3		1620,1658
		LP 693 - 19		19 36 39	−03 20.6	1	12.52		0.89		0.44		2		1696
185505	+7 04170			19 36 41	+07 15.7	1	9.41		0.67				2	F5	1025
		LS II +19 013		19 36 41	+19 46.9	1	10.63		0.89		-0.09		2	B2 Ib	946
185507	+5 04225	HR 7474, σ Aql	★ A	19 36 44	+05 16.9	4	5.18	.017	0.02	.012	-0.61	.004	15	B3 V +B3 V	154,1149,1363,3024
185507	+5 04225	IDS19342N0510	B	19 36 44	+05 16.9	1	12.31		1.62		1.70		2	B3 V	3024
		VES 30		19 36 44	+23 23.3	1	11.98		0.83		-0.27		1		363
		ApJS18,429 # 2		19 36 44	+30 24.5	1	12.76		0.38		0.20		11		646
		VES 31		19 36 45	+20 00.1	1	12.34		0.96		0.12		1		363
	−31 16933			19 36 45	−30 56.8	1	11.36		0.76		0.30		4		552
184996	−66 03447	HR 7455		19 36 48	−65 58.2	3	6.08	.004	1.56	.005	1.95	.000	12	M0 III	15,1075,2038
185404	−23 15618	HR 7470	★ AB	19 36 49	−23 32.6	1	6.34		0.03				4	B9.5V+A3 IV	2007
		G 92 - 20		19 36 52	−02 23.1	1	11.37		1.28		1.19		1	K5	1620
		ApJS18,429 # 1		19 36 53	+30 24.0	1	11.47		0.91		0.60		11		646
		NGC 6809 - 1102		19 36 54	−31 03.	1	13.71		0.97		0.32		2		552
		NGC 6809 - 1210		19 36 54	−31 03.	1	13.58		0.73		0.22		1		552
		NGC 6809 - 1317		19 36 54	−31 03.	1	11.92		1.24		0.94		2		552
		NGC 6809 - 1401		19 36 54	−31 03.	1	12.01		1.25		0.90		2		552
		NGC 6809 - 2102		19 36 54	−31 03.	1	15.14		0.80		0.08		1		552
		NGC 6809 - 2105		19 36 54	−31 03.	1	15.65		0.59		0.07		1		552
		NGC 6809 - 2106		19 36 54	−31 03.	1	15.21		0.66		0.12		1		552
		NGC 6809 - 2110		19 36 54	−31 03.	1	14.17		0.77		0.56		2		552
		NGC 6809 - 2219		19 36 54	−31 03.	1	14.97		0.04		0.16		1		552
		NGC 6809 - 2322		19 36 54	−31 03.	1	11.78		1.15		0.94		2		552
		NGC 6809 - 2420		19 36 54	−31 03.	1	12.50		1.14		0.82		3		552
		NGC 6809 - 2437		19 36 54	−31 03.	1	11.77		1.29		0.94		1		552
		NGC 6809 - 2441		19 36 54	−31 03.	1	11.75		1.34		1.14		3		552
		NGC 6809 - 3202		19 36 54	−31 03.	2	13.69	.005	0.85	.019	0.23	.014	8		552,1677
		NGC 6809 - 3213		19 36 54	−31 03.	2	12.73	.009	1.12	.018	0.63	.009	9		552,1677
		NGC 6809 - 3304		19 36 54	−31 03.	2	13.49	.000	0.95	.010	0.34	.005	9		552,1677
		NGC 6809 - 3314		19 36 54	−31 03.	2	12.76	.005	1.10	.014	0.60	.041	9		552,1677
		NGC 6809 - 3331		19 36 54	−31 03.	1	11.96		1.26		1.00		2		552

Table 1   1063

HD	DM	Other Id	N	Rem	α_{1950}	δ_{1950}	S	V	σ_V	B–V	σ_{B-V}	U–B	σ_{U-B}	n	Spectrum	References
		NGC 6809 - 4101			19 36 54	−31 03.	1	12.31		1.18		0.83		4		552
		NGC 6809 - 4103			19 36 54	−31 03.	1	14.00		0.91		0.30		2		552
		NGC 6809 - 4105			19 36 54	−31 03.	1	14.17		0.59		0.09		3		552
		NGC 6809 - 4107			19 36 54	−31 03.	1	13.18		0.97		0.38		1		552
		NGC 6809 - 4201			19 36 54	−31 03.	1	12.56		1.10		0.65		3		552
		NGC 6809 - 4204			19 36 54	−31 03.	1	13.43		0.88		0.37		3		552
		NGC 6809 - 4209			19 36 54	−31 03.	1	10.90		1.23		1.11		3		552
		NGC 6809 - 4319			19 36 54	−31 03.	1	12.49		1.12		0.67		2		552
		NGC 6809 - 4323			19 36 54	−31 03.	1	12.96		1.03		0.57		2		552
		NGC 6809 - 4406			19 36 54	−31 03.	1	11.63		1.35		1.09		1		552
		NGC 6809 - 4422			19 36 54	−31 03.	1	12.37		1.14		0.71		2		552
		NGC 6809 - 4505			19 36 54	−31 03.	1	11.39		1.37		0.98		1		552
		NGC 6809 sq1 8			19 36 54	−31 19.2	1	13.95		0.61				1		422
185531	−0 03808				19 36 55	+00 04.0	1	9.29		1.35		1.05		3	K2	196
185564	+9 04192				19 36 55	+09 15.9	1	8.25		0.70				2	G5	1025
		LS II +18 005			19 36 55	+18 42.8	1	11.12		0.70		−0.38		2	B0 III	946
185636	+33 03539				19 36 56	+33 13.9	1	7.75		0.37				26	F2	6011
		NGC 6809 sq3 9			19 36 56	−31 16.1	1	11.81		1.28		1.36		3		552
		G 125 - 24		A	19 36 58	+35 05.1	1	15.66		1.73		1.82		5		316
185510	−6 05221	V1379 Aql			19 36 58	−06 10.7	2	8.38	.042	1.16	.013	0.69		2	G5	1641,1642
		CS 22896 # 147			19 36 58	−55 48.8	1	14.80		0.31		0.08		1		1736
		CS 22896 # 129			19 36 60	−53 00.4	1	15.66		0.06		0.16		1		1736
		NGC 6809 sq3 10			19 37 00	−31 14.0	1	13.40		0.88		0.45		2		552
	−31 16937				19 37 03	−31 09.4	1	10.93		1.21				1		438
	+19 04103			V	19 37 04	+20 05.8	1	11.09		2.65				2	M3	369
185585	+9 04195				19 37 05	+09 30.9	1	9.44		0.41				2	F8	1025
185696	+34 03645	IDS19352N3501	AB		19 37 05	+35 07.3	1	8.27		0.54		0.04		2	G0	1625
184735	−75 01540				19 37 06	−74 53.2	2	7.59	.001	1.32	.004	1.43		6	K2/3 III	278,2012
338722	+25 03907				19 37 07	+26 09.8	1	9.30		1.21				2	K0	1118
185467	−23 15625	HR 7473			19 37 07	−23 32.7	1	5.97		1.04				4	K0 III	2007
353339	+16 03935	IDS19349N1621	B		19 37 08	+16 27.6	2	9.45	.015	0.25	.010	−0.34	.005	5	A0	150,1084
185622	+16 03936	HR 7475	⋆ A		19 37 10	+16 27.3	4	6.38	.012	2.03	.030	2.14	.041	11	K4 Ib	150,1080,1084,8100
185449	−31 16940				19 37 10	−31 23.1	1	9.32		1.16		0.99		7	G8 III	552
185606	+12 04012				19 37 11	+12 25.5	1	9.29		−0.02		−0.55		1	A2	963
185516	−20 05644				19 37 13	−20 39.8	1	7.71		1.72		2.03		3	M1 III	1657
185608	+7 04171				19 37 14	+07 43.7	1	7.89		0.31				2	A2	1025
185283	−56 09246				19 37 14	−56 26.3	1	8.99		1.04		0.86		1	K4 V	3072
185607	+7 04172				19 37 15	+07 53.7	1	8.53		0.13				2	A0	1025
185567	−6 05222				19 37 15	−06 16.1	1	8.32		0.62		0.17		23	G0	978
353340	+16 03938	LS II +16 012			19 37 17	+16 35.0	1	10.55		0.63		−0.47		2	B0	405
185733	+34 03648				19 37 17	+34 50.1	1	7.64		1.58				2	K5	369
	+49 03073				19 37 17	+50 05.3	1	9.00		1.50				1	M0	765
338747	+24 03827				19 37 18	+25 06.2	1	9.31		0.10		0.10		2	A2	106
185637	+12 04015				19 37 19	+13 03.8	1	8.80		0.10		−0.42		2	A2	963
185534	−21 05476				19 37 19	−21 25.1	1	7.83		−0.03		−0.32		1	B5/6 IV	400
185663	+18 04182				19 37 20	+19 03.2	1	7.64		1.38		1.35		1	K2 II	8100
185662	+21 03863				19 37 21	+21 13.6	2	7.38	.019	1.48	.019	1.60	.015	3	K2	1080,1355
		He3 1761			19 37 21	−68 14.8	1	10.39		0.48		−0.44		1		1753
		WLS 1940-15 # 10			19 37 22	−14 17.8	1	12.50		0.52		0.01		2		1375
	+13 04091	LS II +13 010			19 37 23	+13 55.9	1	10.55		0.43		−0.55		2	B1	1318
185587	−6 05223				19 37 23	−06 13.2	1	9.09		0.18		−0.02		24	A0	978
185734	+29 03684	HR 7478		⋆	19 37 24	+30 02.2	7	4.70	.016	0.97	.014	0.80	.012	29	G8 III-IV	15,245,1080,1118,1363*
		NGC 6809 sq2 10			19 37 24	−31 12.6	1	12.92		0.87				2		438
185735	+29 03685				19 37 26	+29 48.5	1	8.18		1.69				2	M2 III	369
185780	+40 03824	LS III +40 001			19 37 26	+40 30.7	3	7.74	.008	−0.08	.017	−0.92	.012	8	B0 III	399,401,555
185697	+20 04215				19 37 27	+20 40.0	1	8.20		0.08		0.02		2	B9	634
	−31 16945				19 37 27	−31 21.3	1	11.34		0.63		0.10		1		552
		CS 22896 # 127			19 37 28	−52 44.9	1	13.74		0.02		0.07		1		1736
		G 125 - 25			19 37 30	+37 55.6	1	11.30		0.70		0.17		2	F8	1658
185755	+30 03677				19 37 31	+30 17.5	2	7.15	.035	−0.08	.000	−0.32	.010	5	B9	401,3016
185667	+7 04175				19 37 33	+07 27.0	1	7.22		0.97				3	K0	1025
185912	+54 02193	HR 7484, V1143 Cyg			19 37 34	+54 51.4	2	5.85	.009	0.44	.009	−0.05	.052	4	F6 Va	254,3016
185835	+43 03319				19 37 35	+43 43.4	1	7.97		−0.04		−0.26		2	B8	401
	−30 17273				19 37 35	−30 44.7	1	10.66		0.52		0.06		1		552
185756	+29 03686	IDS19356N2931	A		19 37 36	+29 38.0	1	7.39		−0.04		−0.37		2	A0	401
		NGC 6809 sq2 7			19 37 36	−31 03.3	1	12.62		0.64				2		438
344733	+21 03864				19 37 37	+21 54.7	1	8.60		1.94				4	M3	369
338729	+25 03911				19 37 37	+26 06.7	1	9.67		1.93				2	M2	369
344721	+22 03758				19 37 38	+22 18.5	1	9.92		0.38		0.01		2	A0	106
	+70 01081	G 260 - 28			19 37 38	+70 38.6	2	9.97	.017	0.87	.018	0.51	.013	5	K0	1658,7010
		WLS 1940-15 # 7			19 37 39	−17 00.0	1	11.36		0.84		0.36		2		1375
		CS 22891 # 209			19 37 39	−61 10.8	2	12.16	.005	0.85	.022	0.25	.009	3		1580,1736
185816	+37 03549				19 37 42	+37 34.0	1	8.89		−0.06		−0.39		2	B9	401
338739	+25 03912				19 37 43	+25 17.6	2	10.21	.000	0.70	.000	−0.08	.000	3	B2.5V	1012,1032
185871	+44 03200				19 37 45	+44 19.2	1	7.00		1.10		1.04		3	K1 III	1501
185934	+50 02833				19 37 45	+50 59.3	1	8.02		0.53		0.06		2	F5	1566
185717	+7 04177				19 37 46	+08 00.5	1	9.65		0.19				2	A0	1025
		HA 111 # 1225			19 37 47	+00 09.1	1	11.02		0.42		0.38		3		196
		VES 32, V938 Cyg			19 37 47	+32 05.3	1	11.56		1.78		0.95		1		363
344697	+22 03760				19 37 49	+23 10.6	1	9.99		0.30		0.08		3	A7	1749

HD	DM	Other Id	N Rem	α_{1950}	δ_{1950}	S	V	σ_V	B−V	σ_{B-V}	U−B	σ_{U-B}	n	Spectrum	References
185872	+42 03413	HR 7483		19 37 49	+42 42.1	3	5.42	.009	-0.08	.009	-0.21	.006	4	B9 III	985,1079,3023
353351	+16 03945			19 37 52	+16 12.9	1	9.29		1.94				3	M0	369
185758	+17 04042	HR 7479	⋆ A	19 37 52	+17 53.9	8	4.38	.012	0.78	.014	0.42	.012	19	G0 II	15,1080,1118,1355*
185837	+33 03547	HR 7481	⋆ AB	19 37 52	+33 51.8	3	6.10	.004	0.09	.006	0.13	.015	10	A3 IV	1063,1733,3016
185644	−16 05399	HR 7476	⋆ A	19 37 52	−16 24.6	3	5.30	.017	1.13	.025	1.06	.000	12	K1 III	1084,2035,3024
185644	−16 05399	IDS19350S1631	B	19 37 52	−16 24.6	1	12.64		0.40		0.10		3		3024
185673	−16 05400	IDS19350S1631	C	19 37 54	−16 24.0	2	8.14	.000	0.56	.002	0.08	.005	8	F7/8 V	1084,3024
	−0 03812			19 37 56	+00 01.5	1	9.90		1.34		1.11		3		196
185784	+8 04182			19 38 01	+08 58.6	2	8.67	.009	0.10	.009	-0.38		4	B9	1025,1733
	−10 05151			19 38 02	−10 17.6	1	10.29		1.08		0.80		4	G5	1731
185820	+23 03728	IDS19359N2317	AB	19 38 03	+23 23.7	1	8.62		0.34		0.05		3	F2	1749
185894	+38 03696			19 38 03	+38 46.8	1	8.77		0.10		0.06		3	A1 V	833
185955	+45 02940	HR 7487		19 38 03	+45 50.5	1	6.20		0.90		0.55		2	K0 III	252
185783	+9 04202			19 38 04	+09 39.6	1	8.14		0.27				2	F2	1025
		LS II +22 008		19 38 05	+22 24.6	1	11.78		0.54		0.37		1	B2/5: V:	1215
184997	−74 01829			19 38 06	−74 34.5	2	7.61	.002	0.37	.007	0.04		6	F2 IV	278,2012
		NGC 6819 - 962		19 38 08	+40 12.8	1	9.73		0.40		0.02		1		84
		CS 22896 # 144		19 38 08	−55 29.7	1	14.63		0.07		0.15		1		1736
344684	+23 03730	LS II +23 015		19 38 09	+23 53.7	5	9.30	.011	1.00	.015	0.13	.013	19	B8 Ib	830,1012,1032,1783,8100
		G 260 - 29		19 38 09	+62 30.7	2	10.42	.018	0.80	.029	0.50	.026	4	K0	1658,5010
185762	−0 03813	HR 7480	⋆ A	19 38 09	−00 44.3	3	5.67	.005	0.12	.005	0.08	.012	11	A3 IV	252,1088,3032
185762	−0 03813	IDS19356S0051	B	19 38 09	−00 44.3	1	14.06		1.28		1.31		2		3032
185762	−0 03813	IDS19356S0051	C	19 38 09	−00 44.3	1	13.68		1.29				1		3032
185804	+7 04181	LS IV +08 008		19 38 13	+08 08.1	1	9.19		0.06				2	B8	1025
	+9 04204			19 38 15	+09 37.4	1	10.05		0.07				2	B8	1025
185858	+23 03731			19 38 16	+23 21.7	1	7.10		0.20		0.10		3	A2	1749
185896	+31 03713			19 38 16	+31 56.3	1	8.02		-0.07		-0.19		2	B9	401
185913	+34 03653	IDS19364N3455	A	19 38 16	+35 01.9	1	8.15		-0.03		-0.37		2	A0	401
185859	+20 04218	HR 7482	⋆	19 38 17	+20 21.6	7	6.51	.019	0.41	.016	-0.61	.024	17	B0.5Iae	15,154,851,1012,1118*
		GD 223		19 38 17	+35 06.3	1	16.27		0.35		-0.60		1		3060
185454	−59 07505	IDS19340S5914	AB	19 38 18	−59 07.5	2	7.48	.000	0.72	.005	0.27		5	G5 V	219,2012
185821	+9 04205	RV Aql		19 38 20	+09 48.9	1	9.89		1.63				2	M4	1025
		VES 33		19 38 20	+27 11.6	1	13.02		1.07		0.06		1		363
		G 125 - 26		19 38 20	+42 49.1	1	13.74		1.02		0.45		1		1658
		CS 22896 # 154		19 38 20	−57 05.7	2	13.63	.009	0.64	.015	-0.03	.000	4		1580,1736
225320	+35 01847			19 38 22	+35 07.9	1	8.23		1.45		1.64		2	K2	1625
		VES 34		19 38 23	+29 29.6	1	12.58		1.79		1.85		1		363
344719	+22 03762	LS II +22 009		19 38 27	+22 26.0	1	10.62		0.56		-0.33		3	F0	106
		VES 35		19 38 29	+24 37.8	1	13.26		1.42		0.11		1		363
185915	+23 03733	HR 7485	⋆ AB	19 38 32	+23 36.0	3	6.65	.014	0.00	.005	-0.41	.010	10	B6 IV	154,1118,1749
		CS 22896 # 139		19 38 34	−54 22.9	1	15.37		0.03		-0.01		1		1736
185876	+9 04206			19 38 35	+09 18.3	1	8.85		1.07				2	K0	1025
		G 92 - 21		19 38 35	−03 04.1	1	13.88		1.53		1.05		1		1620
344694	+23 03734			19 38 40	+23 36.8	1	8.19		0.94		0.62		3	K0	1749
		CS 22896 # 130		19 38 41	−53 01.6	1	15.08		-0.10		-0.37		1		1736
185917	+9 04209			19 38 44	+09 45.9	1	9.08		0.41				2	F8	1025
344705	+22 03765	NGC 6823 - 212		19 38 44	+23 07.8	1	9.24		0.06		-0.06		3	A0	1749
185936	+13 04098	HR 7486, QS Aql	⋆ AB	19 38 47	+13 41.9	1	6.00		-0.08		-0.53		5	B5 V	154
185916	+9 04210			19 38 48	+09 55.5	1	9.00		1.14				2	K0	1025
185958	+17 04048	HR 7488		19 38 48	+17 21.5	6	4.37	.010	1.05	.002	0.90	.008	29	G8 IIIa	15,263,1080,1363,8015,8100
185983	+26 03640	IDS19368N3654	A	19 38 49	+27 00.9	1	8.67		0.10		0.08		3	A0	8071
331081	+32 03521			19 38 49	+32 18.0	1	9.76		1.67				2	M1	369
185984	+23 03736			19 38 50	+23 22.3	1	8.17		0.32		0.09		3	F2	1749
	+23 03737			19 38 50	+23 22.7	1	9.60		0.41		-0.02		3	G5	1749
344702	+23 03735			19 38 51	+23 15.8	1	10.10		0.57		0.26		3	F2	1749
331084	+31 03718			19 38 51	+32 05.2	1	8.29		1.56				1	K5	369
355657	+11 03946	LS II +11 009		19 38 52	+11 56.7	2	9.36	.000	0.14	.007	-0.76	.007	5	B0 Ib:n	1032,8100
344681	+23 03738			19 38 53	+24 00.4	1	8.95		0.17		-0.16		2	A0	106
		IRC +30 381	⋆ V	19 38 53	+28 55.4	1	12.62		2.30		0.53		1		363
338691	+26 03641			19 38 54	+27 08.8	1	8.72		1.75				4	M0	369
		CS 22896 # 132		19 38 55	−53 47.3	1	15.08		0.05		0.09		1		1736
185941	+6 04253			19 38 56	+07 03.8	1	9.08		0.48				2	F5	1025
234964	+53 02288			19 38 56	+53 34.1	1	9.54		0.00		-0.02		2	A2	1566
186197	+55 02239	IDS19379N5531	AB	19 38 58	+55 38.2	1	9.12		0.09		0.10		2	A0	1502
185962	+6 04254			19 38 59	+06 54.5	1	9.41		0.35				2	F0	1025
		VES 37		19 38 59	+25 37.7	1	12.77		1.63		1.36		1		363
		CS 22896 # 155		19 38 60	−57 22.8	1	14.76		0.05		0.09		1		1736
185942	+5 04242			19 39 00	+05 19.7	1	8.72		0.02		-0.47		1	A0	963
234963	+52 02481			19 39 01	+52 16.3	1	9.52		0.20		0.12		2	A3	1566
186196	+55 02240			19 39 01	+55 47.4	1	7.28		1.04		0.77		2	K0	1502
	−25 14263			19 39 01	−25 01.6	1	9.94		0.91		0.55		1	K2 V	3072
344704	+22 03766	NGC 6823 - 211		19 39 02	+23 09.1	1	9.09		0.11		-0.22		4	B9	1749
186121	+42 03419	HR 7492		19 39 04	+42 57.6	2	6.13	.024	1.56	.002	1.42		4	M2 III	71,3055
186120	+44 03208			19 39 05	+44 40.3	1	7.18		1.12		1.04		2	K0	1080
353353	+15 03904			19 39 06	+16 07.8	1	9.37		1.79				2	K7	369
186021	+22 03767	HR 7490		19 39 06	+22 20.1	1	6.36		1.53				3	K0 I	1118
186253	+59 02092			19 39 06	+59 43.4	1	7.77		0.36		0.07		2	A5	1733
		CS 22896 # 148		19 39 10	−55 44.1	1	14.98		0.08		0.16		1		1736
344716	+22 03768			19 39 11	+22 25.6	1	10.65		0.18		0.09		2	A3	106
186239	+55 02241			19 39 11	+55 40.6	1	7.37		0.22		0.09		2	A5	1502

Table 1

HD	DM	Other Id	N	Rem	α_{1950}	δ_{1950}	S	V	σ_V	B–V	σ_{B-V}	U–B	σ_{U-B}	n	Spectrum	References
186002	+6 04256				19 39 14	+07 04.2	1	8.57		0.16				2	A0	1025
	+33 03559				19 39 14	+33 44.2	1	10.27		1.85				1	M3	369
185967	−10 05155				19 39 17	−10 12.6	1	9.31		1.70		2.01		4	K5	1731
		CS 22896 # 135			19 39 17	−54 03.4	1	14.58		0.22		0.17		1		1736
		NGC 6819 - 964			19 39 18	+40 06.8	3	10.99	.017	0.55	.007	0.10	.018	9	G2	83,84,284
186155	+45 02949	HR 7495		⋆ A	19 39 18	+45 24.3	5	5.07	.028	0.39	.033	0.15	.003	12	F5 II	253,254,1118,1363,3023
186075	+27 03466				19 39 19	+27 35.2	1	7.90		0.22				1	A3	242
332407	+28 03434	V1818 Cyg		⋆	19 39 20	+29 01.6	1	8.50		0.20		−0.71		2	B1 Ibp	1012
		CS 22896 # 140			19 39 21	−54 28.7	1	15.71		0.04		0.00		1		1736
344703	+22 03769	NGC 6823 - 210			19 39 22	+23 11.9	1	9.06		0.29		0.19		4	A3	1749
		CS 22896 # 126			19 39 23	−52 30.2	1	15.14		0.09		0.17		1		1736
		G 23 - 8			19 39 24	+03 02.4	3	12.88	.019	1.59	.023	1.28	.075	6		203,333,1620,1759
344715	+22 03770				19 39 24	+22 32.6	1	11.14		0.41		0.10		3	A7	1749
		CS 22896 # 131			19 39 24	−53 22.7	1	15.04		0.28		0.12		1		1736
		LB 3121			19 39 24	−61 49.	1	13.44		0.00		0.01		6		45
		NGC 6819 - 486			19 39 25	+40 04.1	1	14.74		0.68				9		284
186340	+60 01991	HR 7500		⋆ A	19 39 25	+60 23.4	1	6.50		0.21		0.09		2	A5 V	1733
	−45 13385				19 39 25	−45 10.1	1	9.90		0.66		0.12		2	G0	1696
		NGC 6819 - 227			19 39 26	+40 06.1	1	14.10		1.17		0.66		1		83
		NGC 6819 - 987			19 39 26	+40 06.4	3	13.11	.027	1.20	.021	1.15	.059	4		83,84,4002
		NGC 6819 - 970			19 39 27	+40 06.5	3	11.54	.018	1.64	.036	1.93	.065	12		83,84,4002
		NGC 6819 - 966			19 39 28	+40 04.1	4	11.65	.017	1.13	.015	0.96	.031	19		83,84,284,4002
		NGC 6819 - 967			19 39 28	+40 04.7	2	11.64	.015	1.27	.027	1.25	.031	11		83,84
		NGC 6819 - 230			19 39 28	+40 06.6	1	15.86		0.61		0.32		1		83
		NGC 6819 - 189			19 39 28	+40 06.8	1	16.44		1.03				1		83
		NGC 6819 - 989			19 39 28	+40 07.7	2	13.16	.005	1.21	.016	1.05	.052	2		84,4002
		NGC 6819 - 991			19 39 28	+40 08.1	2	13.36	.025	1.08	.035	0.73	.051	2		84,4002
		NGC 6819 - 968			19 39 29	+40 02.4	3	11.73	.031	1.56	.014	1.73	.050	16		83,84,284
		NGC 6819 - 670			19 39 29	+40 02.5	1	13.55		0.60		0.08		11		284
		NGC 6819 - 999			19 39 29	+40 03.8	1	11.38		1.63		1.33		1		4002
		NGC 6819 - 550			19 39 30	+40 03.9	2	11.55	.095	1.50	.047	1.15	.290	5		284,4002
		NGC 6819 - 969			19 39 31	+40 01.0	1	11.84		0.56		0.25		1		84
		NGC 6819 - 981			19 39 31	+40 02.7	4	12.90	.023	0.36	.034	0.17	.064	30	B9	84,284,284,284
		NGC 6819 - 973			19 39 31	+40 05.8	3	11.98	.013	1.42	.008	1.31	.027	13		83,84,4002
		NGC 6819 - 965			19 39 31	+40 07.9	3	11.79	.031	1.44	.018	1.33	.110	3		83,84,4002
		NGC 6819 - 983			19 39 31	+40 08.7	2	12.94	.015	1.22	.024	1.12	.023	2		84,4002
		NGC 6819 - 978			19 39 32	+40 04.0	1	12.76		1.06		0.79		1		84
344799	+22 03771				19 39 33	+22 31.5	1	10.77		0.49		−0.01		3	F8	1749
		NGC 6819 - 976			19 39 33	+40 04.3	1	12.44		0.42		0.33		1		84
		G 125 - 27			19 39 34	+32 32.8	1	13.85		1.46		1.02		4		316
		NGC 6819 - 609			19 39 34	+40 03.5	1	13.46		0.53		0.22		7		284
		NGC 6819 - 979			19 39 34	+40 04.1	2	12.89	.083	1.05	.039	0.74	.093	3		84,4002
		NGC 6819 - 396			19 39 34	+40 04.8	1	15.73		0.59		0.01		2		83
		NGC 6819 - 992			19 39 34	+40 08.7	1	13.41		0.60		0.13		1		84
		NGC 6819 - 616			19 39 35	+40 03.1	2	13.05	.118	1.19	.007	1.12		6		284,4002
		NGC 6819 - 982			19 39 35	+40 04.0	2	12.92	.015	0.56	.033	0.32	.007	13		84,284
		NGC 6819 - 182			19 39 35	+40 06.9	1	16.16		0.78		0.33		2		83
186080	+9 04216				19 39 36	+09 29.9	1	8.93		1.25				2	K2	1025
		NGC 6819 - 958			19 39 36	+40 04.	1	11.35		1.49		1.13		6		284
		NGC 6819 - 959			19 39 36	+40 04.	1	12.40		0.49		0.12		7		284
		NGC 6819 - 960			19 39 36	+40 04.	1	11.23		1.71		1.93		2		83
		NGC 6819 - 974			19 39 36	+40 04.7	1	12.11		1.18		0.84		1		84
		NGC 6819 - 395			19 39 36	+40 04.9	2	13.66	.015	1.14	.005	0.97	.098	4		83,4002
186222	+41 03451				19 39 36	+42 01.4	1	8.03		−0.07		−0.24		2	B9	401
		CS 22896 # 138			19 39 36	−54 17.5	1	15.47		0.04		0.07		1		1736
185170	−77 01378				19 39 36	−77 12.5	2	8.70	.004	0.10	.004	0.06		6	A0 V	278,2012
186200	+38 03711				19 39 37	+38 20.4	1	7.52		−0.02		−0.22		2	B9	401
186199	+38 03710				19 39 37	+38 40.1	1	8.64		0.11		0.11		3	A3	833
		NGC 6819 - 994			19 39 37	+40 07.4	1	13.47		0.58		0.17		1		84
		CS 22896 # 156			19 39 37	−57 24.8	1	14.67		0.12		0.18		1		1736
		NGC 6819 - 391			19 39 38	+40 04.9	1	13.00		1.13		0.90		1		4002
		NGC 6819 - 1000			19 39 38	+40 05.2	1	14.97		0.76		0.36		2		4002
		NGC 6819 - 988			19 39 39	+40 06.9	3	13.09	.005	1.23	.096	1.01	.035	4		83,84,4002
		NGC 6819 - 685			19 39 40	+40 02.5	1	14.11		1.21		1.06		4		4002
186005	−16 05413	HR 7489		⋆ AB	19 39 40	−16 14.6	4	5.05	.009	0.34	.005	0.09	.002	11	F3 IV/V	15,253,2012,3026
		HM Sge			19 39 41	+16 37.5	3	11.47	.021	0.15	.031	−0.42	.019	3		1564,1591,1753
		NGC 6819 - 972			19 39 41	+40 05.5	4	12.02	.020	1.51	.060	1.57	.036	17		83,84,284,4002
		CS 22896 # 149			19 39 41	−56 17.3	1	12.02		0.21		0.05		1		1736
		NGC 6819 - 600			19 39 43	+40 03.0	1	13.67		0.58		0.22		9		284
344779	+23 03740	NGC 6823 - 206			19 39 44	+23 15.7	1	8.56		0.64		0.11		3	G2 V	1749
186177	+32 03526	LS II +32 002			19 39 44	+32 57.9	1	7.05		0.24		0.17		3	A5 Ib	8100
		NGC 6819 - 975			19 39 44	+40 04.6	2	12.23	.038	1.35	.004	1.37		7		84,284
186079	+9 04217				19 39 45	+09 52.2	1	8.28		1.42				3	K0	1025
		NGC 6819 - 977			19 39 45	+40 05.4	1	12.52		1.36		1.56		1		84
		NGC 6819 - 265			19 39 45	+40 06.1	1	14.49		0.51				7		284
344758	+24 03843	LS II +24 008			19 39 46	+24 13.8	3	10.35	.010	0.98	.010	0.01	.019	4	O8 V	666,1011,1032
		NGC 6819 - 377			19 39 46	+40 04.9	1	14.30		0.77				10		284
		NGC 6819 - 984			19 39 46	+40 05.3	2	12.98	.004	1.23	.033	1.11		8		84,284
		NGC 6819 - 996			19 39 46	+40 05.8	2	14.05	.057	0.61	.052	0.28		6		84,284
		NGC 6819 - 995			19 39 46	+40 06.2	3	13.84	.019	0.46	.004	0.00	.140	8		84,284,4002

HD	DM	Other Id	N	Rem	α_{1950}	δ_{1950}	S	V	σ_V	B–V	σ_{B-V}	U–B	σ_{U-B}	n	Spectrum	References
		NGC 6819 - 985			19 39 47	+40 06.0	1	13.00		1.15		0.76		1		84
344798	+22 03773				19 39 51	+22 34.3	1	11.08		0.36		-0.08		3		1749
		NGC 6819 - 990			19 39 51	+40 05.5	1	13.25		1.28		1.25		2		84
186122	+11 03954	HR 7493			19 39 52	+12 04.5	2	6.26		-0.02	.036	-0.41	.009	5	B9 III	1063,1079
		NGC 6819 - 715			19 39 52	+40 01.7	2	13.16	.008	1.17	.048	1.08	.157	11		83,4002
		NGC 6819 - 963			19 39 53	+40 02.7	2	10.12	.178	1.57	.033	1.16	.028	4	gM4	83,84
		NGC 6819 - 997			19 39 53	+40 04.6	1	14.57		0.73		0.62		1		84
		ApJ144,259 # 63			19 39 54	+16 58.	1	14.67		0.38		-0.70		6		1360
344797	+22 03774				19 39 54	+22 36.6	1	11.07		0.28		0.01		3	A0	1749
344790	+22 03775				19 39 54	+23 01.3	1	10.32		1.56		1.65		3	K2	1749
		NGC 6819 - 998			19 39 55	+40 05.6	1	16.58		0.72				1		84
186255	+39 03876				19 39 57	+39 54.2	1	6.91		0.16		0.10		std	A3	83
186254	+40 03846				19 39 57	+40 31.5	1	8.60		-0.03		-0.31		2	A0	401
		NGC 6823 - 208		★	19 39 58	+23 11.9	3	10.40	.060	0.79	.017	-0.23	.023	6	B0 IVe	666,886,1749
		NGC 6819 - 986			19 39 58	+40 06.3	1	13.04		1.76		2.33		1		84
186453	+62 01747	IDS19393N6226	AB		19 39 58	+62 32.7	1	7.38		0.50		0.04		2	F5	3016
		NGC 6823 - 245			19 39 59	+23 23.6	2	12.64	.025	0.71	.010	-0.16	.079	5	B3	886,1749
		NGC 6819 - 980			19 40 00	+40 04.6	1	12.82		0.19		0.29		2		84
186142	+7 04188	G 92 - 23			19 40 02	+08 02.0	3	8.97	.012	0.73	.016	0.23	.005	4	G5	333,1025,1620,1658
338801	+26 03645				19 40 02	+26 31.4	1	8.57		1.85				3	M0	369
		NGC 6819 - 961			19 40 02	+40 05.7	1	9.49		0.13		0.19		1		84
		VES 38			19 40 03	+23 03.3	1	12.94		2.08		1.01		1		363
		NGC 6819 - 993			19 40 03	+40 04.1	1	13.45		1.45		1.64		1	A0	84
		NGC 6823 - 248			19 40 04	+23 08.3	2	12.46	.020	0.47	.025	0.18	.089	5		886,1749
344777		NGC 6823 - 205		★	19 40 04	+23 18.9	3	9.46	.015	0.75	.016	-0.37	.041	6	O9.5III	666,886,1749
		G 260 - 30			19 40 05	+71 45.4	3	10.94	.015	1.49	.025	1.25		7	M0	940,1746,7010
186182	+14 04006				19 40 06	+15 08.0	1	8.23		0.07		-0.55		1	B5	963
		NGC 6823 - 252			19 40 07	+23 15.4	1	12.27		0.65		0.03		2	B2	886
332408	+28 03438				19 40 08	+28 52.6	1	8.93		0.01		-0.70		2	B2 IV	1012
185993	-44 13506				19 40 10	-44 15.2	2	7.28	.001	1.22	.010	1.28		9	K1/2 III	1673,2012
186203	+11 03955	HR 7497		★ AB	19 40 13	+11 42.5	2	5.30	.010	0.57	.005	-0.01	.030	5	dF3 + A3	1733,3026
186205	+8 04189	LS IV +09 008			19 40 14	+09 06.5	1	8.53		0.05				2	B5	1025
332427	+28 03440				19 40 14	+28 39.2	1	9.08		1.89				1	M2 III	369
		NGC 6823 - 170			19 40 15	+23 15.9	2	11.99	.000	0.73	.019	-0.14	.049	6		886,1749
186307	+39 03878	HR 7499		★ AB	19 40 15	+40 08.1	1	6.23		0.18		0.05		std	A6 V	83
		LS II +19 015			19 40 17	+19 57.2	1	10.55		0.88		-0.21		2	B1 Ib	946
186042	-37 13322	HR 7491			19 40 17	-37 39.5	1	6.16		0.01				4	B8 IV/V	2035
186012	-43 13541	IDS19368S4341	A		19 40 18	-43 33.6	1	9.15		0.39		-0.01		2	F0 V	1279
186012	-43 13541	IDS19368S4341	AB		19 40 18	-43 33.6	1	9.10		0.38				4	F0 V	2012
186012	-43 13541	IDS19368S4341	B		19 40 18	-43 33.6	1	12.14		0.58		0.07		2		1279
		NGC 6823 - 174		★	19 40 19	+23 13.5	3	11.83	.018	0.65	.029	-0.36	.055	6	B1	666,886,1749
		CS 22896 # 195			19 40 20	-53 34.7	1	15.53		0.10		0.18		1		1736
186226	+8 04190	IDS19379N0809	A		19 40 21	+08 15.8	2	6.88	.005	0.48	.016	0.06		4	F8 IV-V	1025,3016
186226	+8 04190	IDS19379N0809	B		19 40 21	+08 15.8	2	8.79	.009	0.69	.014	0.27		4		1025,3016
		Vul R1 # 15			19 40 21	+22 58.3	1	13.70		1.20		0.19		1		5012
186225	+8 04191				19 40 25	+08 38.7	1	8.93		0.26				2	A0	1025
		CS 22896 # 141			19 40 25	-55 09.8	1	12.55		0.19		0.13		1		1736
344768	+23 03744				19 40 26	+23 40.4	1	10.44		0.55		0.04		3	G0	1749
332431	+27 03478				19 40 26	+28 07.1	1	9.97		1.89				2	M5 III	369
186160	-11 05097				19 40 26	-10 47.8	1	7.89		0.58		0.01		5	F8	1731
		NGC 6823 - 195			19 40 27	+23 00.4	2	12.21	.029	0.58	.024	0.31	.054	6	A2	886,1749
		NGC 6823 - 175			19 40 27	+23 11.8	1	12.61		0.70		-0.19		2	B3	886
	+31 03726				19 40 27	+31 46.8	1	10.13		1.74				2	M2	369
344816	+21 03882				19 40 28	+22 04.2	1	9.63		2.02				2	M0	369
186408	+50 02847	HR 7503		★ A	19 40 29	+50 24.5	8	5.96	.009	0.64	.006	0.19	.009	34	G2 V	1,15,1028,1084,1355*
		LS II +26 004			19 40 30	+26 26.7	1	11.52		0.81		-0.16		1		824
		LP 282 - 3			19 40 30	+37 25.0	1	14.51		-0.09		-0.97		1	DB	3028
186262	+9 04225				19 40 31	+09 38.7	1	9.25		0.24				2	A0	1025
332396		SAO 87604			19 40 31	+29 11.7	1	10.29		0.03		-0.18		2	A0	634
		CS 22896 # 196			19 40 31	-53 28.8	1	14.84		0.08		0.15		1		1736
331102	+30 03704	V541 Cyg			19 40 32	+31 12.3	1	10.35		0.04		-0.01		4	A0	1768
186427	+50 02848	HR 7504		★ B	19 40 32	+50 24.0	8	6.22	.016	0.66	.005	0.20	.008	34	G5 V	1,15,1028,1084,1355*
		CS 22896 # 191			19 40 32	-53 56.0	1	15.84		0.08		0.15		1		1736
186272	+17 04059				19 40 34	+17 50.9	1	7.78		0.10		-0.52		2	B2.5V	1648
		LS II +20 013			19 40 34	+20 51.3	2	10.90	.005	0.99	.000	-0.15	.005	4	B0.5III	405,946
		NGC 6823 - 179			19 40 34	+23 09.8	1	12.50		1.70		1.38		2	K5	886
		LS II +23 018			19 40 35	+23 50.2	1	10.46		0.62		-0.39		1		666
344761	+23 03741	LS II +23 018			19 40 35	+23 50.2	1	10.43		0.62		0.11		3		1749
186309	+28 03445	LS II +28 005			19 40 35	+28 13.7	1	8.29		0.54		0.28		2	F2 II	8100
		LS II +26 005	AB		19 40 36	+26 29.7	1	11.65		0.85		-0.06		2		824
186274	+8 04193				19 40 37	+09 03.6	1	8.80		1.13				2	K2	1025
		LS II +23 021			19 40 37	+23 31.5	2	11.11	.010	0.62	.015	-0.28	.005	2		666,5012
350461	+17 04060				19 40 39	+17 23.4	1	9.05		0.09		-0.38		1	A0	963
		VES 39			19 40 39	+24 17.9	1	12.58		0.84		0.24		1		363
186310	+22 03776				19 40 40	+22 43.9	1	6.54		1.06		0.85		3	K0	1749
186922	+76 00750	IDS19422N7611	AB		19 40 40	+76 18.2	1	8.08		0.88		0.55		2	K0	3016
344795	+22 03777				19 40 42	+22 49.7	1	9.47		1.07		0.79		3	G5	1749
186293	+9 04226				19 40 42	+09 22.2	2	7.82	.018	1.16	.004	1.00		2	K0	1025,1509
344800	+22 03778				19 40 42	+22 26.3	3	10.01	.003	0.38	.003	-0.37	.014	6	B2 V:nne	106,1032,1749
344776	+23 03745	NGC 6823 - 111			19 40 42	+23 20.6	5	8.75	.010	0.66	.016	-0.37	.008	13	B0.5Ib	106,666,886,1012,1749,8100

Table 1 1067

HD	DM	Other Id	N Rem	α₁₉₅₀	δ₁₉₅₀	S	V	σ_V	B–V	σ_B–V	U–B	σ_U–B	n	Spectrum	References
186185	−15 05444	HR 7496	⋆A	19 40 42	−15 35.3	3	5.50	.005	0.46	.006	0.02	.007	8	F5 V	253,2009,3053
186275	+7 04191			19 40 44	+07 27.5	1	9.18		0.41				2	F0	1025
350430	+17 04061	CV Sge		19 40 44	+18 07.2	1	9.74		2.01				2	M4 III:	369
		NGC 6823 - 21		19 40 44	+23 12.3	2	13.66	.005	0.64	.005	0.44	.015	2	A5	49,340
186342	+27 03480			19 40 44	+27 29.7	2	8.11	.025	-0.02	.000	-0.46		4	A0	963,1118
		NGC 6823 - 214		19 40 47	+23 09.2	1	14.29		1.11		0.57		1		49
		NGC 6823 - 20		19 40 47	+23 11.8	3	12.88	.031	0.60	.012	-0.12	.078	6	B3	340,886,1749
185544	−75 01545			19 40 47	−75 01.9	1	8.90		0.93				1	K0 III	2012
344788	+22 03779	NGC 6823 - 189		19 40 49	+23 04.4	2	8.82	.005	0.63	.005	0.07	.009	4	G2 V	49,1749
186357	+28 03447	HR 7501, V1276 Cyg		19 40 49	+29 12.7	5	6.52	.014	0.33	.010	0.14	.051	9	F1 III	592,1118,3016,6007,8075
186377	+32 03531	HR 7502		19 40 49	+32 18.4	2	5.94		0.14	.039	0.17	.034	5	A5 III	985,1063
		NGC 6823 - 19		19 40 50	+23 10.9	2	14.84	.025	0.99	.005	0.31	.015	2		49,340
		NGC 6823 - 74		19 40 50	+23 13.1	5	11.61	.012	0.54	.014	-0.41	.029	8	B1 V	49,340,666,886,1749
186428	+43 03331			19 40 50	+43 54.1	1	8.10		-0.04		-0.40		2	B9	401
		LS II +14 014		19 40 51	+14 43.0	1	11.75		0.31		-0.52		2		405
186358	+27 03482	IDS19388N2749	AB	19 40 51	+27 56.1	1	8.59		0.14				5	A0	1118
186410	+39 03881	IDS19391N3948	A	19 40 51	+39 54.4	1	7.94		0.00		-0.04		2	A0	401
		WLS 1940-15 # 9		19 40 51	−15 29.0	1	11.74		0.77		0.22		2		1375
		NGC 6823 - 18		19 40 52	+23 10.6	2	13.82	.005	0.67	.010	0.09	.020	2	B3	49,340
		LP 813 - 13		19 40 52	−16 02.6	1	12.38		0.62		-0.05		2		1696
186312	+9 04227			19 40 53	+09 22.6	2	7.84	.027	1.46	.005	1.72		7	K2	1025,1509
344781	+23 03746	NGC 6823 - 73		19 40 53	+23 13.4	3	9.43	.018	0.75	.004	0.23	.006	5	K0	49,340,1749
344801	+22 03780			19 40 54	+22 30.5	1	9.81		0.43		-0.02		3	F5	1749
		NGC 6823 - 25		19 40 54	+23 08.6	2	13.69	.052	0.57	.033	-0.10	.033	4	B2	340,1749
		NGC 6823 - 17		19 40 54	+23 10.3	4	12.06	.022	0.53	.014	-0.37	.036	7	B1 V	49,340,886,1749
		VES 40		19 40 55	+25 44.2	1	12.16		0.92		0.28				363
		NGC 6823 - 78		19 40 55	+23 11.2	4	12.10	.023	0.52	.018	-0.32	.022	7	B2 V	49,340,886,1749
186343	+21 03885			19 40 56	+22 10.9	1	8.05		0.20		0.18		3	A2	8071
		NGC 6823 - 79		19 40 56	+23 11.6	1	13.72		0.68		0.16		1		49
		NGC 6823 - 76		19 40 57	+23 12.4	1	13.51		1.34		1.01		1		49
		NGC 6823 - 77		19 40 57	+23 12.4	1	13.72		0.66		0.22		1	B3	340
		NGC 6823 - 75		19 40 57	+23 12.6	3	13.28	.053	1.45	.008	1.24	.015	4	K3	49,340,886
		NGC 6823 - 217		19 40 57	+23 13.2	1	15.60		0.95		0.12		1		49
		NGC 6823 - 16		19 40 58	+23 10.3	2	13.93	.010	0.69	.015	0.20	.010	2		49,340
186532	+55 02245	HR 7509, V1351 Cyg		19 40 58	+55 20.7	2	6.42	.072	1.61	.005	1.56		3	M5 IIIa	71,3001
		CS 22964 # 16		19 40 58	−40 10.9	2	12.48	.000	0.08	.000	0.12	.010	2		966,1736
344783	+22 03781	NGC 6823 - 13		19 40 59	+23 09.0	7	9.76	.020	0.42	.017	-0.56	.015	13	B0 IV	49,340,769,886,1012*
		NGC 6823 - 14		19 40 59	+23 09.4	2	12.02	.015	0.50	.005	-0.17	.010	6	B2 V	886,1749
		CS 22896 # 157		19 40 59	−57 19.5	1	14.08		0.05		0.12		1		1736
		NGC 6823 - 286		19 41 00	+23 11.	1	15.58		0.97		0.15		1		340
186379	+24 03849			19 41 00	+24 28.9	3	6.86	.004	0.58	.022	0.01	.020	4	F8 V	1620,1779,3026
332612	+27 03485			19 41 00	+27 49.4	1	8.56		0.54		0.04		3	F8 V	3026
		NGC 6823 - 12		19 41 01	+23 09.2	2	13.97	.009	0.74	.026	0.10	.166	6		340,1749
344782		NGC 6823 - 1	AB	19 41 01	+23 10.9	2	10.42	.068	0.52	.005	-0.51	.005	6	O7 V((f))	886,1749
		NGC 6823 - 225		19 41 01	+23 11.8	2	13.77	.023	0.62	.027	-0.06	.072	5		340,1749
225482	+35 03762			19 41 01	+35 37.8	1	7.82		1.16		1.04		3	K0	1601
186280	−10 05166			19 41 01	−09 54.4	1	9.98		0.42		0.28		4	A0	1731
186061	−52 11521			19 41 01	−52 27.4	1	8.78		0.99		0.71		2	K2 V	3072
		G 142 - 47		19 41 02	+09 58.1	1	10.00		1.26		1.17		1		3072
		NGC 6823 - 227		19 41 02	+23 10.1	1	14.00		0.80		0.21		1		340
		NGC 6823 - 219		19 41 02	+23 10.6	2	12.50	.097	0.64	.019	-0.21	.068	6		886,1749
		NGC 6823 - 2		19 41 02	+23 10.7	2	11.34	.024	0.57	.010	-0.42	.034	6	B1.5V	886,1749
		NGC 6823 - 223		19 41 02	+23 10.7	1	13.85		0.65		-0.11		4		1749
		NGC 6823 - 71		19 41 02	+23 13.6	2	13.60	.035	0.70	.010	-0.18	.025	2		49,340
344775	+23 03747	NGC 6823 - 109		19 41 02	+23 19.1	3	10.39	.026	0.60	.006	-0.38	.013	7	B1 III	666,1012,1749
		NGC 6823 - 120		19 41 02	+23 22.5	1	10.45		1.97		1.98		2	K8 V	886
186325	+6 04276			19 41 03	+06 21.7	1	8.15		-0.04				3	B9	1025
		NGC 6823 - 298		19 41 03	+23 10.5	1	13.13		0.46		0.07		2		886
344784	+22 03782	NGC 6823 - 3		19 41 03	+23 10.6	7	9.35	.027	0.56	.011	-0.54	.024	13	O7 V((f))	49,340,886,1011,1012*
		NGC 6823 - 218		19 41 03	+23 10.8	2	11.07	.000	0.56	.010	-0.50	.074	5	B1.5V	886,1749
		NGC 6823 - 70		19 41 03	+23 13.1	2	14.15	.005	0.80	.000	0.27	.020	2		49,340
		NGC 6823 - 10		19 41 04	+23 08.9	2	14.51	.010	0.79	.020	0.11	.015	2	B0	49,340
		NGC 6823 - 216		19 41 04	+23 11.3	2	15.18	.010	0.82	.015	0.29	.010	2		49,340
		NGC 6823 - 231		19 41 04	+23 12.0	1	13.52		0.78		0.15		2		340
		NGC 6823 - 11		19 41 05	+23 09.3	4	11.57	.026	0.75	.013	-0.28	.038	7	B0.5V	49,340,886,1749
		NGC 6823 - 4		19 41 06	+23 10.1	1	11.45		0.60		0.07		1	G2	340
		NGC 6823 - 68		19 41 06	+23 11.9	4	11.88	.018	0.73	.018	-0.33	.040	9	O9 V	49,340,886,1749
225488	+35 03763	IDS19393N3512	AB	19 41 06	+35 19.5	1	9.24		0.04		0.01		1	A2	401
186438	+37 03578	LS II +37 001	⋆A	19 41 06	+37 33.5	1	8.01		0.53		0.19		2	F3 Ib	8100
344817	+21 03889			19 41 07	+22 08.6	1	10.13		0.19		0.03		2	A0	106
		NGC 6823 - 9		19 41 07	+23 08.8	4	11.81	.020	0.74	.018	-0.29	.039	7	BO V:pe	49,340,886,1749
		NGC 6823 - 33		19 41 08	+23 06.2	1	12.61		0.65		-0.03		2	F5	886
		NGC 6823 - 8		19 41 08	+23 09.5	2	13.53	.015	0.47	.005	0.12	.010	2	A0	49,340
186439	+34 03675			19 41 08	+34 18.1	1	7.96		-0.01		-0.09		2	A0	401
234979	+54 02209			19 41 08	+55 11.7	1	9.79		1.40		1.57		2	M0	1375
		NGC 6823 - 215		19 41 09	+23 09.1	1	14.69		0.81		0.25		1		49
		NGC 6823 - 213		19 41 09	+23 10.2	1	13.97		0.80		0.20		1		49
		NGC 6823 - 5		19 41 09	+23 10.5	1	13.23		1.36		1.00		1		49
		NGC 6823 - 67		19 41 09	+23 12.0	3	12.62	.045	0.79	.028	-0.25	.018	7	O9.5V	340,886,1749
		NGC 6823 - 220		19 41 09	+23 12.6	1	12.30		0.54		-0.04		3	G0 V:	1749

HD	DM	Other Id	N	Rem	α_{1950}	δ_{1950}	S	V	σ_V	B–V	σ_{B-V}	U–B	σ_{U-B}	n	Spectrum	References
		NGC 6823 - 260			19 41 09	+23 14.9	1	12.65		0.75		0.05		2		886
		NGC 6823 - 82			19 41 09	+23 15.0	1	11.98		0.48		-0.04		2		886
		CS 22964 # 31			19 41 09	-38 35.4	2	15.55	.050	0.13	.000	0.11	.009	2		966,1736
		CS 22896 # 186			19 41 10	-54 34.2	1	14.66		0.11		0.19		1		1736
		NGC 6823 - 7			19 41 11	+23 09.9	2	14.69	.015	0.89	.010	0.28	.020	2		49,340
		NGC 6823 - 65			19 41 11	+23 11.5	2	13.59	.037	0.72	.009	-0.16	.098	4	B2	49,1749
186440	+30 03706	HR 7505			19 41 11	+30 33.5	5	6.07	.019	-0.01	.023	0.00	.020	13	A1 V	252,1063,1079,1118,3016
186454	+35 03764				19 41 11	+35 20.4	1	8.25		-0.07		-0.28		2	A0	401
186485	+42 03425				19 41 12	+42 31.8	1	8.54		-0.01		-0.12		3	B9 V	555
		NGC 6823 - 145			19 41 13	+23 25.6	1	12.99		0.58		0.18		2	A3	886
344791	+22 03783				19 41 14	+22 54.4	2	9.82	.010	0.13	.000	0.04	.000	5	A0	106,1749
		NGC 6823 - 61			19 41 14	+23 10.3	1	13.53		0.75		0.13		1	F0	49
186465	+40 03856	IDS19396N4029		AB	19 41 14	+40 36.1	1	6.71		-0.02		-0.18		2	A0	401
		NGC 6823 - 34		★	19 41 15	+23 05.4	3	11.73	.009	0.83	.026	-0.22	.012	6	B0.5V	666,886,1749
344789		NGC 6823 - 62			19 41 15	+23 11.2	1	12.95		1.38		0.83		2	A3	886
186412	+22 03784				19 41 16	+22 22.5	1	6.82		-0.08		-0.51		3	B5 V	399
186441	+29 03710				19 41 16	+30 06.7	1	7.23		0.07		0.07		2	A0	401
		CS 22964 # 29			19 41 16	-38 46.8	2	15.79	.000	0.16	.000	0.15	.009	2		966,1736
		LS II +18 009			19 41 18	+18 17.4	2	12.13	.000	-0.32	.000	-1.18	.000	3	sdO	405,1650
		LS II +16 013			19 41 19	+16 40.8	1	11.72		0.26		-0.51		2	B1.5V	946
344787	+22 03786	NGC 6823 - 35			19 41 21	+23 03.5	3	9.30	.021	1.18	.007	0.78	.043	9	F8 Ib	769,886,1749
		Vul R1 # 13			19 41 21	+23 36.4	1	12.24		0.63		0.42		1		5012
186505	+39 03885				19 41 21	+39 52.6	1	6.93		0.30		0.01		2	A5	401
		NGC 6823 - 95			19 41 24	+23 15.9	1	12.92		0.68		0.34		2		886
		Smethells 43			19 41 24	-51 32.	1	12.06		1.47				1		1494
		NGC 6823 - 93		★	19 41 25	+23 15.0	4	10.74	.025	0.36	.014	-0.56	.010	9	B1 V	666,769,886,1749
		Vul R1 # 12			19 41 27	+23 38.1	1	12.17		0.95		-0.03		1		5012
225508	+35 03770				19 41 27	+36 07.3	1	8.40		1.70				2	K7	369
		NGC 6823 - 91		★	19 41 29	+23 13.9	3	10.99	.005	0.45	.009	-0.47	.016	6	B1 V	666,886,1749
		G 23 - 9			19 41 30	+05 04.9	1	11.28		1.37		1.42		1	K7	1620
344773	+23 03748				19 41 31	+23 29.3	1	10.51		0.30		0.12		3	A5	1749
		GD 224			19 41 32	+12 27.7	1	14.08		0.88		0.39		1		3060
	-15 05448				19 41 32	-15 07.2	1	9.71		0.53		0.01		3	G0	812
	+22 03788	NGC 6823 - 54			19 41 34	+23 08.9	2	10.23	.005	0.45	.009	0.14	.009	4		49,1749
		CS 22896 # 197			19 41 34	-53 24.2	1	15.11		0.14		0.18		1		1736
		LS II +23 032			19 41 36	+23 56.3	1	10.44		0.79		-0.27		1		666
		JL 26			19 41 36	-78 03.	1	15.35		-0.25		-1.08		3		132
		LS II +27 008			19 41 37	+27 26.3	1	11.46		0.43		-0.42		2		824
186442	+9 04233				19 41 38	+09 23.8	1	6.52		1.21				2	K0	1025
186486	+25 03933	HR 7506			19 41 38	+25 39.1	4	5.49	.035	0.94	.007	0.67	.000	8	G8 III	242,1080,1118,3016
		NGC 6819 - 971			19 41 38	+40 03.2	2	11.87	.000	0.47	.015	0.03	.010	16	F8	84,284
186194	-51 12222				19 41 39	-50 59.2	1	8.63		0.70				4	G3 V	2033
344805	+22 03790	IDS19395N2224		AB	19 41 40	+22 30.7	1	9.28		0.45		0.16		3	F0	1749
		NGC 6823 - 44			19 41 45	+23 04.4	1	9.96		1.80		1.30		3	G5	1749
		WLS 1940-15 # 5			19 41 45	-12 41.6	1	10.10		1.24		1.14		2		1375
185833	-72 02436				19 41 45	-72 32.1	1	9.41		0.33				4	F0 III	2012
344793	+22 03791				19 41 46	+22 47.9	1	10.29		0.46		-0.05		3	G0	1749
344772	+23 03751				19 41 48	+23 33.3	1	8.76		1.21		1.28		3	M0	1749
344767	+23 03750				19 41 48	+23 38.9	1	9.57		0.48		0.05		3	A2	1749
		Vul R1 # 10			19 41 49	+23 20.1	1	13.84		0.83				1		5012
186456	+7 04197				19 41 50	+07 28.0	3	8.15	.013	0.06	.015	-0.54	.040	7	B5	1025,1212,1733
344786	+23 03752	NGC 6823 - 141			19 41 52	+23 13.5	1	9.51		1.20		1.23		3	K5	1749
332454	+30 03714				19 41 52	+31 04.2	1	9.78		1.80				2	K5 III	369
		G 23 - 10			19 41 53	+11 20.4	1	11.16		1.35		1.27		1	K7	333,1620
186507	+18 04213	LS II +18 010			19 41 53	+18 26.6	1	8.68		0.12		-0.11		2	B8 II	8100
		NGC 6823 - 269			19 41 53	+23 01.1	1	12.67		0.54		0.15		2	A3	886
186518	+26 03654	HR 7508, PS Vul		★ AB	19 41 53	+27 00.9	1	6.28		1.10				2	G4 III	1118
		CS 22896 # 159			19 41 53	-56 58.2	1	14.11		0.03		0.03		1		1736
		GD 225			19 41 54	+08 46.6	2	13.80	.040	0.69	.005	-0.04	.020	2		1,3060
186618	+46 02765	IDS19404N4701		A	19 41 54	+47 07.5	4	7.75	.010	-0.19	.014	-0.98	.008	9	A5 V	399,401,555,963
		VES 41			19 41 57	+23 19.6	1	14.82		1.05		0.22		2		363
		CS 22896 # 177			19 41 57	-55 14.8	1	14.69		0.09		0.17		1		1736
		CS 22896 # 158			19 41 57	-57 01.4	1	15.47		0.13		0.18		1		1736
186488	+8 04198				19 41 58	+08 58.2	1	8.24		0.20				2	A2	1025
		VES 42			19 41 58	+26 51.8	1	11.83		0.96		0.56		1		363
186568	+33 03572	HR 7512		★ A	19 41 58	+34 02.5	3	6.07	.004	-0.04	.036	-0.30	.008	16	B8 III	1742,3016,8100
186568	+33 03572	IDS19401N3355		B	19 41 58	+34 02.5	1	13.81		0.48		0.35		4		3016
344804	+22 03793				19 41 60	+22 37.1	1	9.46		0.91		0.58		3	K0	1749
		CS 22964 # 7			19 42 00	-41 18.4	2	14.54	.000	0.19	.000	0.21	.008	2		966,1736
186605	+37 03582	IDS19402N3805		AB	19 42 01	+38 12.1	1	7.12		-0.14		-0.57		2	B8	401
186489	+7 04200				19 42 02	+07 50.4	1	8.32		0.38				2	F8	1025
225556	+35 03773	LS II +35 003			19 42 04	+35 18.4	1	8.60		0.42		0.31		2	F1 II	8100
344833	+21 03894				19 42 05	+21 43.1	1	9.65		1.88				2	K7	369
186619	+41 03469	HR 7514			19 42 05	+41 39.1	2	5.84	.002	1.58	.010	1.99		4	M0 IIIab	71,3001
186814	+63 01556				19 42 05	+63 54.6	1	9.23		0.03		0.02		2	B8 V	555
186402	-29 16494				19 42 06	-29 14.1	1	10.22		0.59		-0.02		2	F2 V	1097
344807	+21 03895				19 42 08	+22 09.8	1	10.42		0.50		-0.27		3	B8	106
		LS II +23 034			19 42 13	+23 09.8	1	11.82		0.48		-0.44		1		666
		LS II +23 033			19 42 13	+23 45.5	1	10.70		0.78		-0.33		1		666
186686	+48 02942	RT Cyg			19 42 13	+48 39.4	2	8.03	.905	1.29	.390	0.91	.365	2	M4	817,3001

Table 1   1069

HD	DM	Other Id	N Rem	α_{1950}	δ_{1950}	S	V	σ_V	B–V	σ_{B-V}	U–B	σ_{U-B}	n	Spectrum	References
186547	+12 04059	HR 7511		19 42 15	+13 10.9	2			-0.03	.005	-0.21	.009	4	B9 III-IV	1063,1079
		Vul R1 # 11		19 42 16	+23 28.1	1	12.02		0.56		-0.19		1		5012
186760	+57 02057	HR 7522		19 42 16	+57 53.8	1	6.24		0.55		0.04		1	F8 IV	254
		CS 22896 # 185		19 42 16	-54 33.9	1	14.54		0.14		0.17		1		1736
186535	+8 04200			19 42 17	+08 36.3	2	6.42	.004	0.95	.013	0.68		7	K0	1025,1509
225573	+37 03585			19 42 17	+37 48.7	1	8.93		-0.03		-0.38		3	B5	555
185712	-77 01383			19 42 17	-77 01.9	2	7.15	.006	0.01	.001	0.00		6	A0 V	278,2012
186569	+18 04216			19 42 18	+18 28.0	1	6.80		1.69				3	M1	369
		LS II +23 036		19 42 18	+23 10.6	1	10.58		0.54		-0.34		1		666
186536	+6 04285			19 42 21	+07 06.3	1	8.10		0.03				2	B9	1025
186478	-17 05736			19 42 21	-17 36.7	1	9.15		0.97				1	K2 III	1594
338808	+25 03941			19 42 22	+26 06.2	2	10.41	.000	0.70	.000	-0.19	.000	3	B1.5V	1012,1032
		CS 22873 # 15		19 42 22	-59 50.5	1	14.08		0.46		-0.21		1		1736
338834	+24 03857			19 42 23	+25 08.1	1	9.68		2.00				2	M0	369
332554	+28 03457			19 42 23	+28 59.9	1	9.21		0.09		-0.43		2	B3	634
		Steph 1717		19 42 25	-04 04.0	1	11.67		1.86		1.93		2	M3	1746
186675	+37 03586	HR 7517		19 42 28	+37 13.9	3	4.90	.020	0.95	.009	0.69	.000	6	G8 III	1080,1363,3016
		CS 22964 # 15		19 42 28	-40 05.2	2	15.47	.000	0.18	.000	0.10	.008	2		966,1736
344764	+23 03756	LS II +24 010		19 42 29	+24 04.6	2	9.10	.034	0.97	.044	0.52	.029	3	A0 II	106,666
		VES 43, V969 Cyg		19 42 29	+33 17.5	1	13.19		1.86		0.64		1		363
		CS 22896 # 168		19 42 29	-55 58.6	1	14.15		0.29		0.09		1		1736
239189	+55 02250			19 42 31	+55 29.5	1	8.69		0.58		0.09		2	G0	3016
		VES 44, V968 Cyg		19 42 33	+28 42.2	1	13.07		1.78		0.86		1		363
186587	+10 04036	IDS19402N1032	AB	19 42 34	+10 39.2	1	7.48		0.10		-0.49		1	B2 V	399
186761	+52 02501			19 42 34	+52 29.4	2	7.63	.015	0.28	.005	0.10	.020	5	AmA5-F0	1566,8071
186701	+38 03739			19 42 38	+39 10.0	1	8.46		0.39		0.02		3	F2 II	833
186815	+56 02291	HR 7526		19 42 38	+56 55.3	1	6.28		0.88				2	K2 III	71
186420	-42 14438			19 42 40	-42 23.3	1	9.08		0.93				1	G6 II	729
		LS II +27 009		19 42 41	+27 43.8	1	11.71		0.42		-0.37		2		824
		CS 22964 # 2		19 42 42	-42 06.4	2	14.61	.000	0.03	.000	0.02	.010	2		966,1736
		CS 22896 # 189		19 42 42	-54 04.7	1	15.23		0.02		0.04		1		1736
186702	+34 03691	HR 7520	V	19 42 45	+34 17.5	2	6.37	.025	1.64	.010	1.66	.035	5	M1 III	1733,3001
186774	+50 02863			19 42 45	+50 46.7	1	8.26		-0.06		-0.16		2	A0	1566
		CS 22896 # 178		19 42 45	-55 11.7	1	13.14		0.10		0.05		1		1736
		CS 22896 # 163		19 42 45	-56 42.5	1	15.16		0.14		0.17		1		1736
		CS 22896 # 164		19 42 48	-56 36.2	1	15.06		0.18		0.15		1		1736
		VES 45		19 42 49	+22 48.3	1	11.51		0.48		0.10		1		363
186688	+28 03460	HR 7518, SU Cyg		19 42 49	+29 08.6	3	6.44	.004	0.42	.012	0.27	.050	3	F2 III	592,6007,8075
186610	-3 04698	LS IV -03 021		19 42 50	-03 16.4	3	9.66	.010	-0.01	.012	-0.76	.011	12	B3	400,963,1732
186500	-32 15443	HR 7507		19 42 50	-32 01.9	1	5.52		0.02				4	B8 III	2007
185996	-74 01838			19 42 53	-74 15.0	2	8.23	.002	0.34	.003	0.04		6	F2 III	278,2012
186657	+7 04207			19 42 57	+07 47.4	1	8.00		0.47				2	F8	1025
186656	+8 04205			19 42 59	+08 52.9	1	8.70		0.49				2	F0	1025
344879	+23 03757			19 42 59	+23 45.9	2	9.57	.000	0.18	.000	-0.10	.025	5	A3	106,1749
344876	+23 03758	LS II +23 042	⋆AB	19 42 59	+23 51.3	1	9.99		0.70		-0.32		1		666
		VES 46		19 42 59	+35 52.3	1	14.02		1.00		0.01		1		363
186452	-48 13255			19 43 00	-48 38.4	1	10.20		0.58		-0.02		2	G5 V	1696
		LS II +18 011		19 43 05	+18 11.7	1	11.81		0.46		-0.51		2	B0.5III	946
186729	+29 03724	LS II +29 011		19 43 05	+29 28.4	1	8.08		0.20		0.28		1	A2 II	592,8100
344873	+23 03759	LS II +23 043		19 43 06	+23 55.7	5	8.78	.025	0.78	.029	-0.35	.012	8	B0 II	106,666,1012,1779,8100
		CS 22896 # 161		19 43 06	-56 49.4	1	14.82		-0.13		-0.73		1		1736
186776	+40 03866	HR 7523, V973 Cyg		19 43 07	+40 35.7	5	6.33	.038	1.63	.014	1.91	.016	7	M3 III	31,71,979,3042,8112
344908	+22 03797			19 43 08	+22 30.0	1	9.86		0.38		-0.23		2	B8	106
225631	+37 03591			19 43 08	+37 49.6	1	9.46		1.72				2	M0	369
		VES 47		19 43 11	+21 12.8	1	11.04		0.45		0.43		1		363
186689	+7 04210	HR 7519		19 43 14	+07 29.4	3	5.90	.013	0.18	.004	0.08	.010	9	A3 IV	15,1025,1256
		Lynds 810 # 2		19 43 14	+27 44.2	1	11.62		0.39		0.24		1		500
		VES 48		19 43 15	+20 09.5	1	11.28		0.29		-0.12		1		363
		Lynds 810 # 4		19 43 15	+27 43.4	1	12.30		1.19		0.98		1		500
186660	-3 04701	HR 7516		19 43 15	-03 00.4	3	6.47	.008	0.05	.008	-0.53	.013	13	B3 III	154,400,1088
		G 142 - 50		19 43 16	+16 20.4	1	14.08		-0.06		-0.86		2		3060
		Lynds 810 # 3		19 43 16	+27 42.5	1	12.18		0.76		0.30		1	K3	500
		Lynds 810 # 1		19 43 16	+27 43.0	1	11.25		0.52		0.04		1	F5 IV	500
		CS 22896 # 199		19 43 16	-52 58.3	1	14.30		0.15		0.20		1		1736
186746	+23 03760	LS II +23 044		19 43 17	+23 49.2	4	7.03	.006	0.93	.004	0.03	.008	9	B8 Ia	106,666,1012,1586
344863	+23 03761	LS II +24 012		19 43 17	+24 12.2	4	8.84	.017	0.68	.026	-0.44	.010	6	B0 II:	106,666,1012,8100
225639	+40 03869			19 43 17	+40 19.8	1	10.21		0.43		0.09		2	F2	401
	+79 00642	G 261 - 10		19 43 17	+79 35.6	1	10.34		0.70		0.14		1	G0	1658
338911	+25 03946			19 43 18	+25 37.3	1	9.30		0.08		0.10		2	A0	106
		Lynds 810 # 5		19 43 20	+27 42.6	1	13.25		0.54		0.14		1		500
186777	+31 03752			19 43 20	+31 18.1	3	7.42	.012	-0.06	.010	-0.43	.008	6	B5	401,1192,3016
		Lynds 810 # 6		19 43 22	+27 44.	1	13.28		1.18		0.80		1		500
		LS II +33 005		19 43 24	+33 51.1	2	10.38	.039	0.14	.005	-0.76	.005	6		405,1514
		CS 22896 # 184		19 43 24	-54 38.9	1	13.57		0.29		0.11		1		1736
186882	+44 03234	HR 7528	⋆AB	19 43 25	+45 00.5	5	2.87	.013	-0.03	.010	-0.09	.015	18	B9 III+F1 V	15,1203,1363,3023,8015
		CS 22896 # 175		19 43 26	-55 38.5	1	15.47		-0.08		-0.55		1		1736
186648	-20 05698	HR 7515		19 43 27	-19 53.0	2	4.89	.015	1.06	.000	0.96		6	K0 III	2009,3016
186117	-73 02061			19 43 27	-73 38.9	3	7.34	.004	0.14	.004	0.13	.020	10	Ap SrCrEu	278,1075,2012
		CS 22964 # 3		19 43 30	-41 52.0	2	14.86	.000	0.18	.000	0.20	.008	2		966,1736
		VES 49, EQ Cyg		19 43 32	+31 19.5	1	13.03		1.83		0.56		1		363

HD	DM	Other Id	N	Rem	α_{1950}	δ_{1950}	S	V	σ_V	B–V	σ_{B-V}	U–B	σ_{U-B}	n	Spectrum	References
		CS 22964 # 21			19 43 32	−39 47.0	2	14.27	.000	0.39	.000	0.10	.011	2		966,1736
	−42 14451				19 43 32	−42 00.9	1	10.51		0.42				1		729
186665	−18 05480	UW Sgr			19 43 33	−18 16.5	1	9.29		2.50		3.62		1	C	109
344880	+23 03762	LS II +23 045	3		19 43 35	+23 51.7	3	9.29	.005	0.78	.008	−0.29	.018	5	B0.5III:nn	106,666,1012
338926	+24 03866	LS II +25 007	6		19 43 35	+25 00.2	6	9.57	.009	1.20	.029	0.05	.015	23	O8.5II(f)	666,830,1011,1012,1032,1783
		CS 22896 # 198			19 43 35	−53 08.3	1	15.02		0.29		0.08		1		1736
		CS 22873 # 6			19 43 35	−61 03.1	1	13.88		0.19		−0.60		1		1736
		AS 360			19 43 36	+18 29.	1	12.43		0.69		0.01		1	M6	1753
		ApJ144,259 # 65			19 43 36	−23 16.	1	15.90		0.09		−0.98		4		1360
338916	+25 03952	LS II +25 008	4		19 43 37	+25 13.9	4	10.16	.009	0.63	.000	−0.42	.000	8	O8	106,1011,1012,1032
	+20 04267			V	19 43 39	+20 27.5	1	10.80		2.55				2	M2	369
344864	+23 03763	LS II +24 013	2		19 43 39	+24 11.9	2	10.03	.009	0.77	.023	−0.23	.052	4	G0	106,666
186858	+33 03582	G 125 - 29	4	⋆ AB	19 43 39	+33 29.1	4	7.67	.015	1.00	.021	0.79	.007	16	K3 V +K3 V	938,1197,1381,3030
		LS II +19 016			19 43 40	+19 39.9	1	12.15		0.90		−0.07		2	B1 IV	946
344894	+22 03800	LS II +23 046	4		19 43 40	+23 04.3	4	9.62	.007	0.32	.007	−0.47	.019	18	B2 III:n	106,830,1012,1783
		LS II +23 047			19 43 40	+23 58.6	1	10.81		0.72		−0.23		1		666
186219	−72 02445	HR 7498	7		19 43 41	−72 37.7	7	5.40	.007	0.23	.006	0.10	.013	29	A3mA3-A8	15,26,278,1075,2012*
344887	+23 03764				19 43 42	+23 31.9	1	10.38		0.61		0.43		3	F0	1749
		G 185 - 37			19 43 42	+27 00.8	1	12.38		1.72				1		906
		CS 22896 # 169			19 43 42	−55 58.1	1	13.72		0.19		0.19		1		1736
186860	+29 03730	V1957 Cyg	2		19 43 44	+30 08.1	2	7.58	.049	1.70	.000	1.63		6	M4 III	148,8032
	+5 04285				19 43 46	+05 51.9	1	8.39		−0.06		−0.52		3	B5	1212,1586
344889	+23 03766				19 43 46	+23 16.7	1	9.56		0.51		0.00		3	F5	1749
	+23 03765				19 43 46	+23 21.7	1	10.82		1.69		1.47		4		1749
186841	+23 03767	LS II +23 048		V	19 43 47	+23 58.4	5	7.87	.017	0.79	.035	−0.29	.015	8	B1 Ia	106,666,1012,1779,8100
186901	+35 03786	HR 7529		⋆ A	19 43 49	+35 58.1	1	6.43		−0.03		−0.13		2	B9.5V	542
186901	+35 03786	HR 7529		⋆ AB	19 43 49	+35 58.1	1	5.98		−0.01		−0.12		2	B9.5V	542
186901	+35 03786	IDS19420N3551		B	19 43 49	+35 58.1	2	7.16	.000	0.02	.000	−0.07	.005	3		542,542
186791	+10 04043	HR 7525		⋆ A	19 43 53	+10 29.4	9	2.72	.012	1.51	.014	1.69	.019	33	K3 II	3,15,1363,1509,3016*
		VES 50			19 43 54	+33 29.6	1	11.88		0.50		0.06		1		363
186651	−43 13580				19 43 55	−43 28.1	1	7.11		0.56				4	G1 V	1075
186543	−56 09290	HR 7510			19 43 57	−56 29.1	2	5.34	.005	0.20	.000			7	A7 III/IV	15,2030
332606	+27 03508				19 43 58	+28 06.6	1	9.13		0.00				1	A2	242
186927	+34 03701	HR 7530		⋆ A	19 43 59	+34 53.4	4	6.11	.035	1.07	.011	1.04	.047	9	K0 II-III +	71,150,542,8100
186927	+34 03702	IDS19421N3446		B	19 43 59	+34 53.4	2	8.53	.024	0.05	.000	0.11	.000	3	A2 V	150,542
		CS 22964 # 8			19 43 59	−41 04.7	2	14.72	.000	−0.06	.000	0.11	.012	2		966,1736
	−42 14460				19 44 01	−42 13.9	1	11.31		0.48				1		729
186820	+7 04216				19 44 02	+07 58.3	1	9.83		0.09				2	A0	1025
		Vul R1 # 4			19 44 02	+24 31.4	1	14.24		0.98		0.24		1		5012
186929	+30 03729				19 44 02	+30 14.5	1	8.66		−0.04		−0.24		2	A0	401
332541	+28 03469				19 44 03	+29 04.5	1	9.55		0.06		−0.12		2		634
186994	+44 03236	LS III +44 001	3		19 44 04	+44 50.5	3	7.50	.004	−0.13	.011	−0.95	.005	8	B0 III	399,401,555
344892	+22 03802				19 44 05	+23 04.4	2	9.70	.010	0.32	.015	0.15	.025	5	A2	106,1749
		CS 22896 # 160			19 44 05	−57 22.8	1	14.80		0.19		0.08		1		1736
186978	+40 03870	IDS19424N4019		AB	19 44 06	+40 25.9	1	7.56		−0.07		−0.50		2	B8	401
		CS 22964 # 19			19 44 07	−39 53.8	2	13.26	.009	0.84	.020	0.24	.008	4		1580,1736
338930	+24 03871				19 44 08	+24 37.3	1	9.82		0.48		0.08		2	A0	106
186843	+6 04300				19 44 09	+06 55.9	1	8.16		0.59				2	G5	1025
		CS 22896 # 167			19 44 09	−56 21.2	1	15.34		0.13		0.19		1		1736
		VES 51			19 44 10	+33 31.0	1	13.58		0.57		0.18		1		363
186842	+7 04218				19 44 11	+07 32.7	1	9.45		0.33				2	A0	1025
		Vul R1 # 3			19 44 11	+24 28.4	1	15.30		1.05		0.72		1		5012
186996	+39 03905	IDS19425N3939		AB	19 44 11	+39 45.9	1	7.48		−0.10		−0.52		2	B9	401
	−42 14462	V3885 Sgr			19 44 12	−42 07.3	2	10.29	.063	−0.01	.001	−0.77	.008	17	DB:p	729,1732
		CS 22896 # 194			19 44 12	−53 31.0	1	15.89		0.44		−0.35		1		1736
		CS 22873 # 30			19 44 12	−57 59.4	1	14.36		−0.30		−1.10		1		1736
186866	+7 04220				19 44 13	+07 23.3	1	8.40		1.18				3	K2	1025
		CS 22964 # 39			19 44 13	−38 06.1	2	15.09	.000	0.08	.000	0.10	.010	2		966,1736
186943		QY Vul			19 44 14	+28 08.9	3	10.26	.034	0.26	.015	−0.65	.010	8	WN5	824,1012,1359
338936	+24 03873	LS II +24 014	3		19 44 16	+24 30.4	3	9.94	.015	0.69	.004	−0.33	.010	6	B0.5V	106,342,666
332706	+29 03734				19 44 16	+29 15.6	1	9.24		0.00		−0.45		2	B8	634
186780	−17 05746	V4026 Sgr			19 44 17	−17 12.1	2	6.89	.010	1.77	.010	1.87		6	M2 III	2012,3040
186682	−46 13217				19 44 17	−45 50.5	1	7.24		0.14				4	A3 V	2012
187035	+44 03237				19 44 18	+44 47.7	1	8.83		−0.04		−0.45		2	B5 V	555
		CS 22964 # 26			19 44 18	−39 21.6	2	15.15	.000	0.08	.000	0.16	.009	2		966,1736
344882	+23 03769				19 44 19	+23 43.9	2	9.16	.000	0.02	.039	−0.22	.015	3	B8 V	106,7007
186888	+6 04301				19 44 20	+06 59.4	1	9.11		0.50				2	G0	1025
		Vul R1 # 2			19 44 20	+24 29.5	1	12.61		0.80		−0.04		1	B1 Vne:	5012
186980	+31 03765	LS II +31 005	5		19 44 20	+31 59.6	5	7.48	.016	0.06	.013	−0.85	.012	11	O7.5	1011,1012,1192,1203,3016
		G 92 - 25			19 44 21	+01 18.9	1	12.99		1.26		0.99		1		333,1620
	+24 03876				19 44 23	+24 55.1	1	9.78		0.41		0.17		3		106
187340	+68 01079	HR 7545			19 44 23	+69 12.9	1	5.92		0.05		0.11		4	A2 III	253
331161	+31 03767	G 125 - 31		⋆ A	19 44 26	+31 53.9	1	10.28		1.48		1.17		2	M1	3072
331161	+31 03767	IDS19425N3147		AB	19 44 26	+31 53.9	3	9.76	.019	1.55	.073	1.13		4	M1	369,1017,3003
186847	−8 05103	IDS19417S0824		AB	19 44 26	−08 16.6	2	8.04	.015	1.17	.000	0.96	.010	7	K0	214,3030
186933	+7 04221				19 44 29	+08 06.6	1	9.80		0.13				2	B9	1025
		G 142 - 52			19 44 29	+11 58.4	2	14.33	.024	1.49	.010	1.15	.033	7		203,3078
186962	+18 04236				19 44 30	+18 41.6	1	7.59		1.07		0.94		2	K0 II-III	8100
		Vul R1 # 7			19 44 30	+25 04.3	1	11.65		0.90		−0.09		1	B1 Vne	5012
		LS II +28 008			19 44 31	+28 40.7	1	11.38		0.34		−0.45		1		824

Table 1 1071

HD	DM	Other Id	N Rem	α_{1950}	δ_{1950}	S	V	σ_V	B–V	σ_{B-V}	U–B	σ_{U-B}	n	Spectrum	References
187013	+33 03587	HR 7534	⋆A	19 44 32	+33 36.6	10	4.99	.016	0.46	.011	0.00	.006	32	F7 V	1,15,254,1028,1084,1118*
		CS 22964 # 32		19 44 32	−38 42.3	2	13.86	.000	0.08	.000	0.15	.010	2		966,1736
186998	+24 03877	HR 7533		19 44 33	+25 00.6	1	6.62		0.27				2	A7 IV	1118
338872	+26 03665			19 44 33	+27 02.1	1	8.98		0.26		0.22		4	F0	650
		CS 22964 # 5		19 44 33	−41 33.3	2	15.24	.000	0.13	.000	0.24	.009	2		966,1736
225732	+33 03589	IDS19426N3330	B	19 44 34	+33 36.8	4	8.55	.012	1.04	.009	0.94	.014	9	K5	1,1028,1084,3072
187037	+34 03706			19 44 35	+34 55.1	2	8.42	.161	-0.12	.000	-0.50	.015	3	A3	401,963
332755	+27 03512	LS II +28 009		19 44 37	+28 07.2	4	8.80	.021	0.19	.010	-0.77	.015	9	O7	342,824,1011,1012
		WLS 1944 55 # 7		19 44 39	+52 51.5	1	13.81		0.80		0.26		2		1375
187038	+32 03558	HR 7535	⋆AB	19 44 40	+32 45.9	1	6.18		1.13				3	K2	1118
		VES 52		19 44 41	+25 04.8	1	13.48		1.72		1.42		1		363
		Vul R1 # 8	A	19 44 41	+25 05.9	1	13.82		1.25		0.18		1		5012
		Vul R1 # 8	B	19 44 41	+25 05.9	1	13.80		1.00		0.18		1		5012
		CS 22896 # 171		19 44 41	−56 03.2	1	14.04		-0.23		-1.07		1		1736
		CS 22964 # 9		19 44 42	−41 00.8	2	15.07	.000	0.24	.000	0.19	.008	2		966,1736
186602	−62 06108	IDS19403S6204	AB	19 44 42	−61 56.3	1	7.26		0.49				3	F7/8 V	173
		CS 22896 # 181		19 44 43	−54 58.4	1	13.23		0.13		0.21		1		1736
332624	+30 03735			19 44 44	+30 54.6	1	8.58		-0.12		-0.48		1	A0	963
		VES 53		19 44 45	+28 49.1	1	12.36		0.44		-0.27		1		363
350584	+18 04256			19 44 47	+17 23.5	1	8.76		0.98		0.78		2	F5	1682
		L 160 - 108		19 44 48	−63 07.	1	10.89		0.42		-0.21		3		3065
225757	+34 03707	LS II +34 003		19 44 49	+34 31.8	1	10.59		-0.04		-0.81		2	B3:II:	1012
		VES 54		19 44 50	+28 49.5	1	12.19		1.75		1.67		1		363
186483	−69 03054			19 44 50	−69 27.9	1	7.25		1.22		1.22		6	K1 III	1704
187074	+29 03740			19 44 51	+29 24.7	1	8.16		-0.05		-0.41		2	A0	401
		CS 22964 # 23		19 44 51	−39 41.0	2	15.58	.000	0.03	.000	0.01	.010	2		966,1736
	+24 03880	LS II +24 015		19 44 52	+24 25.8	2	10.85	.045	0.56	.030	-0.47		2	B0.5V	666,725
186782	−42 14468			19 44 53	−42 12.2	2	7.35	.002	1.04	.009	0.86		3	K0 III	729,1732
	−15 05467			19 44 55	−15 24.3	1	10.08		0.40		0.14		1	A2	1776
		VES 55		19 44 56	+21 36.0	1	13.06		0.93		0.39		2		363
332757	+27 03513	LS II +27 010		19 44 56	+27 55.3	4	8.32	.007	0.39	.019	0.01	.010	6	A2 Ia	242,824,1012,8100
353562	+16 03999			19 44 57	+17 10.4	1	9.37		-0.02		-0.50		1	A0	963
338931	+24 03881	LS II +24 016		19 44 57	+24 43.5	4	9.12	.008	0.69	.004	-0.43	.010	7	O6.5III(f)	106,666,1011,1012
		CS 22964 # 13		19 44 58	−40 26.6	1	11.59		1.41		1.16		1		1736
344898	+22 03807			19 45 00	+22 47.5	1	9.84		0.33		-0.12		2		106
187139	+43 03358			19 45 00	+43 38.3	2	8.18	.005	0.01	.005	-0.16	.005	4	B8	401,555
186584	−67 03680	HR 7513		19 45 00	−66 56.3	1	6.45	.021	1.48	.004	1.67	.045	20	K4 III	15,1075,2038,3021
187003	+0 04314	IDS19425N0051	A	19 45 01	+00 58.0	1	6.79		0.59		0.09		2	G0 IV	1355
		LS II +23 049		19 45 01	+23 40.0	2	11.64	.010	0.74	.020	-0.08		2		363,666
332646	+30 03737			19 45 02	+30 36.4	1	9.52		1.69				3	M0	369
186756	−53 09678	HR 7521		19 45 02	−53 00.8	2	6.26	.005	1.13	.001	1.11		6	K1 III	58,2007
		CS 22964 # 4		19 45 03	−41 31.0	2	15.62	.000	0.05	.000	0.12	.010	2		966,1736
187160	+43 03359			19 45 05	+44 13.5	1	7.08		0.55		0.03		3	G0	3016
187075	+21 03909		V	19 45 08	+21 38.8	1	7.11		1.67				3	K5	369
187076	+18 04240	HR 7536, δ Sge	⋆	19 45 10	+18 24.3	5	3.82	.021	1.41	.004	0.96	.008	15	M2 II +A0 V	15,369,1363,3016,8015
		CS 22964 # 20		19 45 10	−39 49.2	1	15.27		-0.20		-1.07		1		1736
344903	+22 03809			19 45 11	+22 32.2	1	10.03		0.47		-0.05		2		106
338923	+24 03882			19 45 13	+24 51.5	1	9.26		0.17		0.02		2	A0	106
187057	+7 04227			19 45 14	+07 31.5	1	8.35		0.56				2	G0	1025
187078	+9 04260	IDS19428N0940	AB	19 45 14	+09 47.5	1	9.82		0.06				2	A0	1025
345024	+21 03912	LS II +21 012		19 45 14	+21 56.7	1	8.44		0.92		0.88		2	F4 II	8100
	+24 03883	LS II +24 017		19 45 15	+24 26.3	1	10.22		0.71		-0.30		1		666
186984	−14 05555	HR 7532		19 45 15	−13 49.7	3	6.11	.015	0.20	.004	0.12	.010	11	A6m	355,2035,8071
187191	+38 03756			19 45 18	+39 11.4	2	8.36	.015	0.34	.010	0.01	.010	5	F2 V	833,1601
187161	+30 03739			19 45 19	+30 42.9	1	7.62		-0.09		-0.38		1	B9	1192
		CS 22896 # 187		19 45 19	−54 25.1	1	14.97		0.04		0.08		1		1736
187095	+6 04307			19 45 24	+06 48.9	1	9.41		0.17				2	A0	1025
		Vul R1 # 6		19 45 24	+24 54.6	1	11.96		0.67		-0.30		1		5012
187162	+28 03478			19 45 27	+28 21.7	1	8.16		0.98		0.62		3	G8 III	1625
332701	+29 03744			19 45 27	+29 19.5	1	8.95		0.01		-0.34		2	B8	401
186502	−73 02067	IDS19396S7303	A	19 45 27	−72 55.3	4	7.31	.006	0.46	.003	-0.02	.006	13	F6 V	278,861,1075,2012
		LS II +24 018		19 45 29	+24 38.7	1	10.66		0.73		-0.43		1		666
		LS II +27 011		19 45 29	+27 19.7	1	11.87		0.71		-0.06		1		824
187192	+30 03744			19 45 29	+30 53.1	1	8.94		0.02		-0.14		2	A0	401
		VES 57		19 45 30	+20 51.0	1	13.02		1.01		-0.12		1		363
331222	+33 03596			19 45 32	+33 21.6	1	9.72		1.82				2		369
187181	+25 03968			19 45 33	+26 09.4	1	8.92		-0.01		-0.06		1		963,1655
	+25 03970	LS II +25 010		19 45 37	+25 21.3	1	11.19		0.68		-0.18		1	A0 II	666
		LS II +24 019		19 45 39	+24 31.7	1	10.90		0.74		-0.28		1		666
187235	+38 03758	HR 7543	⋆A	19 45 41	+38 17.0	3	5.77		-0.05	.011	-0.38	.019	5	B8 Vn	1063,1079,1192
		CS 22896 # 182		19 45 41	−54 49.9	1	13.94		0.19		0.13		1		1736
		G 92 - 26		19 45 42	−01 55.7	1	14.76		1.22		0.83		1		1620
187193	+25 03972	HR 7540		19 45 43	+25 15.6	2	5.96	.005	1.00	.010			3	K0 II-III	242,1118
187147	+6 04310			19 45 45	+06 49.8	1	9.22		0.35				2	A3	1025
		JL 29		19 45 48	−78 32.	1	16.42		-0.07		-1.06		2		132
187182	+13 04154			19 45 49	+13 19.8	1	6.96		0.41		0.00		2	F5 III	1648
186784	−61 06412			19 45 49	−61 03.0	2	8.01	.005	0.50	.005	0.07		3	F6 IV	742,1594
		CS 22964 # 66		19 45 51	−39 29.5	2	15.32	.000	0.10	.000	0.20	.010	2		966,1736
		CS 22964 # 1		19 45 52	−42 19.2	2	14.68	.000	0.08	.000	0.11	.010	2		966,1736
187111	−12 05540			19 45 53	−12 14.8	2	7.72	.000	1.23	.005	0.93		2	G8wF8III/IV	742,1594

HD	DM	Other Id	N Rem	α_{1950}	δ_{1950}	S	V	σ_V	B–V	σ_{B-V}	U–B	σ_{U-B}	n	Spectrum	References
		CS 22964 # 58		19 45 53	−38 35.3	2	15.27	.000	0.06	.000	0.20	.010	2		966,1736
186975	−46 13237			19 45 53	−45 52.4	1	7.26		1.08				4	K0/1 III	1075
189205	+84 00445			19 45 55	+84 39.1	1	8.24		0.09		0.03		5	A0	1219
187298	+38 03760			19 45 57	+38 47.3	1	8.69		0.11		0.06		3	A2 V	833
		LS II +20 014		19 45 59	+20 25.0	2	11.34	.005	0.83	.000	0.27	.000	4	B8 III	405,946
		C1947+272 # B		19 45 59	+27 07.0	1	11.63		0.33		0.11		3		1617
		C1947+272 # A		19 45 59	+27 13.4	1	11.61		0.64		0.25		3		1617
187237	+27 03516			19 45 59	+27 44.5	3	6.88	.020	0.65	.023	0.13	.007	4	G2 III	934,1355,3026
187372	+47 02916	HR 7547		19 45 59	+47 47.0	3	6.12	.010	1.64	.008	2.01	.020	6	M2 III	71,262,3001
		CS 22896 # 174		19 45 59	−55 41.5	1	13.66		0.26		0.12		1		1736
338933	+24 03887			19 46 00	+24 35.5	1	10.03		0.34		0.06		2	B9	106
186837	−61 06413	HR 7527		19 46 00	−61 11.3	2	6.20	.005	−0.14	.001	−0.57		7	B5 V	26,2035
344977	+23 03778			19 46 01	+23 44.8	1	9.86		0.38		−0.31		3	A2	106
187255	+27 03517			19 46 02	+27 34.7	5	7.57	.014	−0.02	.005	−0.23	.018	12	A0	242,401,650,1118,3016
187279	+31 03779			19 46 02	+31 22.8	1	6.83		0.11		0.11		2	A0	401
187238	+22 03812			19 46 03	+22 38.3	1	7.10		2.04		2.06		1	K3 Iab-Ib	8100
		C1947+272 # C		19 46 03	+27 09.7	1	13.28		0.79		0.16		4		1617
225849	+35 03809			19 46 04	+35 14.4	1	8.70		−0.02		−0.45		2	B8	401
		C1947+272 # D		19 46 05	+27 10.3	1	13.16		0.78		0.15		4		1617
		C1947+272 # E		19 46 05	+27 13.2	1	11.90		1.94		1.94		4		1617
332690	+29 03748			19 46 05	+29 43.2	1	10.38		0.02		−0.08		1	A0	401
332691	+29 03748			19 46 05	+29 44.1	2	8.98	.000	−0.03	.000	−0.26	.010	3	B9	401,963
187098	−29 16546	HR 7538		19 46 05	−28 54.8	3	6.04	.006	0.41	.005	−0.10		8	F3 V	15,2027,3037
		CS 22964 # 85		19 46 05	−41 01.1	2	15.14	.000	−0.01	.000	0.07	.011	2		966,1736
		CS 22964 # 74		19 46 06	−39 50.1	2	14.45	.010	0.39	.007	−0.16	.009	3		1580,1736
338446	+26 03672	C1947+272 # F		19 46 07	+27 13.7	4	10.09	.025	0.68	.015	0.23	.030	23	A2	245,650,1617,1655
187203	+10 04058	HR 7542	⋆	19 46 08	+10 34.1	2	6.44	.006	0.95	.004	0.63	.036	5	F8 Ib-II	1080,8100
187280	+27 03518			19 46 10	+28 12.0	2	7.93	.000	1.17	.023			4	K2 III	20,242
		VES 58		19 46 10	+30 25.6	1	14.45		0.07		0.37		1		363
225860	+34 03720			19 46 12	+34 49.8	1	8.90		1.70				2	M2	369
187258	+18 04248			19 46 14	+18 31.7	1	7.58		0.32		0.09		3	Am	8071
		VES 59		19 46 15	+26 45.8	1	13.06		2.47		0.72		1		363
		C1947+272 # G		19 46 15	+27 07.1	1	13.96		0.83		0.55		4		1617
	+47 02917			19 46 15	+47 25.7	1	9.11		1.33		1.39		2	K2 III	262
186786	−65 03827	HR 7524, NZ Pav		19 46 15	−65 43.8	4	6.04	.007	0.30	.006	0.07	.011	18	F0 II/III	15,1075,2038,3014
187299	+24 03889			19 46 16	+24 53.0	2	7.15	.019	1.61	.009	1.49	.085	4	G5 Iab-Ib	1080,8100
		C1947+272 # H		19 46 16	+27 07.6	1	13.71		0.67		0.43		4		1617
		C1947+272 # M		19 46 17	+27 09.9	1	13.90		1.34		0.77		3		1617
187195	−11 05131	HR 7541		19 46 17	−10 59.8	2	6.02	.015	1.23	.005	1.36		6	K5 III	2007,3016
186503	−76 01368			19 46 17	−76 18.4	1	7.98	.009	0.04	.002	0.05	.000	11	A1 V	278,861,2012
187282		QT Sge		19 46 18	+18 04.6	1	10.50		−0.02		−0.89		4	WN5	1359
		C1947+272 # L		19 46 18	+27 10.4	1	13.16		0.64		0.38		5		1617
		C1947+272 # J		19 46 19	+27 08.0	1	12.59		0.67		0.45		5		1617
		C1947+272 # N		19 46 19	+27 09.6	1	14.09		1.31		0.83		4		1617
187259	+11 03994	HR 7544	⋆ AB	19 46 20	+11 41.4	1	5.74		0.49		0.29		2	F2	1733
		C1947+272 # O		19 46 20	+27 09.3	1	12.43		0.81		0.40		6		1617
		C1947+272 # K		19 46 20	+27 11.4	1	13.81		1.53		1.43		3		1617
338867	+26 03674	S Vul		19 46 21	+27 09.6	4	8.79	.108	1.75	.099	1.35	.008	4	K0	670,793,824,1772
		LS II +27 012		19 46 21	+27 41.0	1	11.46		0.74		−0.13		2		824
187283	+14 04048	IDS19441N1449	AB	19 46 22	+14 56.0	2	7.64	.009	0.43	.005	−0.04		4	F5 V +F6 V	1381,3030
		C1947+272 # P		19 46 23	+27 09.3	1	14.37		0.91		0.21		4		1617
		C1947+272 # Q		19 46 24	+27 09.6	1	14.53		1.24		0.63		3		1617
		C1947+272 # R		19 46 24	+27 10.1	1	14.16		1.91		1.70		4		1617
		C1947+272 # S		19 46 24	+27 11.1	1	14.20		1.78		1.12		3		1617
225882	+38 03763			19 46 24	+38 39.8	1	9.67		0.01		−0.69		3	B2	555
187343	+24 03892	IDS19444N2443	AB	19 46 28	+24 50.2	1	7.41		0.00		−0.27		2	B9	106
		C1947+272 # U		19 46 28	+27 10.1	1	14.20		1.25		0.64		3		1617
		C1947+272 # T		19 46 28	+27 10.2	1	14.19		0.92		0.29		3		1617
	+47 02918			19 46 28	+47 42.2	1	9.19		1.07		1.02		2	K0 III	262
186957	−59 07534	HR 7531	⋆ AB	19 46 31	−59 19.2	1	5.42		0.08				4	A0 IV	2007
187284	+11 03996			19 46 32	+11 33.5	1	6.52		0.91		0.65		2	K0	1733
187323	+18 04253			19 46 32	+18 14.6	1	8.28		0.07		−0.02		3	B5	1682
187320	+19 04162	LS II +19 019		19 46 32	+19 32.1	1	7.39		0.14		−0.64		3	B2 III	399
331313	+31 03786			19 46 35	+31 31.2	1	8.39		1.66				2	M0	369
	+47 02920			19 46 35	+47 29.8	1	10.19		0.18		0.10		2		262
187086	−47 13103	HR 7537		19 46 35	−47 41.0	2	5.95	.005	1.68	.015	1.97		6	M1 III	2007,3005
	−59 07535			19 46 35	−59 18.2	1	9.05		0.82		0.44		1		1696,3060
		G 23 - 11		19 46 37	+08 10.3	1	11.86		1.37		1.09		1		333,1620
	+42 03471	V1154 Cyg		19 46 37	+43 00.1	1	9.01		0.91		0.66		1	G0	592
338998	+25 03978	IDS19445N2533	A	19 46 38	+25 40.9	1	9.18		1.85				1	M0	369
345042	+21 03927			19 46 39	+21 25.0	1	9.97		1.60				3	G5	369
		CS 22896 # 179		19 46 39	−55 04.3	1	14.68		−0.01		−0.08		1		1736
350564	+17 04103			19 46 40	+18 05.3	2	9.42	.000	0.13	.000	0.12	.000	14	A5	830,1783
339039	+24 03893			19 46 40	+24 40.9	2	9.64	.010	0.41	.005	−0.49	.010	3	B1.5V	106,1012
187764	+68 01082	HR 7563, CN Dra		19 46 41	+68 18.8	1	6.34		0.28		0.09		4	F0 III	253
187068	−52 11547			19 46 41	−52 21.7	1	8.72		1.46		1.66		4	K2/3 III	1673
		LS II +19 020		19 46 42	+19 50.6	1	11.43		0.37		−0.50		2	B1 V	946
187399	+29 03754	V1507 Cyg		19 46 42	+29 16.6	5	7.01	.021	0.18	.019	−0.42	.050	14	B7 Ia:e	138,379,401,8040,8100
	+42 03473			19 46 44	+43 01.0	1	9.25		1.23		0.99		1	K0	592
187362	+18 04254	HR 7546	⋆ AB	19 46 45	+19 00.9	7	5.02	.020	0.09	.017	0.06	.024	35	A1 V +A3 V	15,263,1063,1363,3030*

Table 1   1073

HD	DM	Other Id	N	Rem	α_{1950}	δ_{1950}	S	V	σ_V	B–V	σ_{B-V}	U–B	σ_{U-B}	n	Spectrum	References
187362 +18 04254		IDS19445N1853		C	19 46 45	+19 00.9	1	9.03		0.46		-0.03		6		3030
	+42 03472				19 46 45	+42 57.1	1	8.37		0.38				1		592
		LP 869 - 35			19 46 45	-23 11.4	1	11.14		0.68		0.15		1		1696
		AA45,405 S84 # 4			19 46 48	+18 17.5	1	11.69		0.32		0.11		1		1682
339016 +25 03980					19 46 50	+25 21.4	1	10.74		0.16		0.08		4	A0	106
		VES 60			19 46 50	+27 51.9	1	12.03		0.82		0.04		2		363
187458 +34 03727		IDS19450N3504		B	19 46 51	+35 11.1	1	6.66		0.42		-0.06		7	F4 V	1742
187458 +34 03727		HR 7550		★ AB	19 46 52	+35 11.1	2	6.61	.050	0.43	.002	-0.09	.017	5	F4 V +F6 V	254,3030
	+21 03929				19 46 55	+22 04.2	1	11.14		2.12				3	M3	369
332727 +28 03485					19 46 55	+28 41.0	1	9.40		0.10		-0.57		2	B2 V:n	1012
	+32 03574				19 46 55	+32 32.8	1	9.22		1.69		2.68		1		1697
		CS 22896 # 166			19 46 55	-56 18.2	1	15.12		0.49		-0.06		1		1736
345006 +22 03819					19 46 56	+22 37.1	1	9.72		0.37		0.28		2	A2	106
187459 +33 03602		HR 7551, V1765 Cyg			19 46 56	+33 18.7	7	6.46	.020	0.19	.013	-0.75	.022	20	B0.5Ib	15,154,814,1012,1192*
186154 -81 00868		HR 7494			19 46 56	-81 28.8	3	6.38	.007	1.40	.000	1.68	.000	14	K3/4 III	15,1075,2038
187401 +14 04053					19 46 59	+15 04.7	1	7.91		1.18		1.22		2	G5 II	8100
187428 +19 04165					19 46 59	+19 40.1	2	7.91	.010	0.76	.019	0.43	.093	5	F8 Ib-II	253,8100
187350 -1 03834		LS IV -01 023			19 46 59	-01 13.6	2	8.13	.010	0.10	.020	-0.73	.020	6	B1 Vne	399,400
		CS 22964 # 67			19 46 59	-39 35.0	2	15.89	.000	-0.02	.000	0.00	.012	2		966,1736
187460 +29 03756					19 47 03	+29 45.4	1	8.18		0.93		0.65		1	K2 II-III:	8100
187462 +27 03523					19 47 06	+27 36.5	4	6.92	.032	0.61	.019	0.13	.035	9	G0 IV	20,934,3026,8053
		AA45,405 S84 # 2			19 47 07	+18 19.4	1	11.23		0.47		-0.03		2		1682
187503 +32 03578					19 47 08	+32 40.4	1	7.51		1.24				1	K5	765
		LS II +26 006			19 47 11	+26 40.0	2	11.40	.010	0.79	.010	-0.21	.005	5	O9.5III	342,824
	-15 05477				19 47 11	-15 10.3	1	10.16		0.41		0.20		2		1375
		VES 61			19 47 12	+29 23.9	1	12.20		0.42		0.38		1		363
332653 +30 03759					19 47 12	+30 19.3	1	9.67		1.81				2	M0	369
		Steph 1721			19 47 13	+00 17.0	1	11.16		1.06		1.02		1	K7	1746
		G 92 - 28			19 47 13	+08 04.9	1	15.21		1.72				1		333,1620
		IRC +30 393		★ V	19 47 13	+30 17.2	1	12.70		2.25		0.51		1		363
		CS 22873 # 56			19 47 14	-59 48.0	1	14.10		-0.04		-0.29		1		1736
	+30 03761				19 47 16	+31 10.0	1	9.40		1.86				2	M2	369
		DF Cyg			19 47 16	+42 54.7	1	10.35		1.02		0.74		1	G5Iab:	793
		LP 753 - 29			19 47 16	-13 26.8	1	11.60		0.69		-0.03		2		1696
187504 +25 03986					19 47 17	+25 22.8	2	8.19	.010	0.13	.005	0.09	.000	5	A0	833,1625
187524 +28 03486					19 47 18	+28 27.1	1	7.38		0.14		0.11		2	A0	379
187308 -27 14309					19 47 18	-26 50.6	1	7.45		1.06		0.76		1	G8 II	565
		VES 63			19 47 24	+20 58.8	1	11.06		0.53		0.39		1		363
	+28 03487	LS II +28 010			19 47 24	+28 48.6	2	10.11	.005	0.51	.000	-0.41	.005	2	B2 Ib:	824,1032
		VES 64			19 47 24	+30 55.1	1	12.94		1.67		0.93		1		363
187483 +16 04020					19 47 25	+17 02.0	1	8.68		-0.07		-0.48		1		963
187548 +28 03488					19 47 25	+28 28.9	2	7.95	.040	0.52	.002	-0.09		5	G0 V	20,3026
187613 +44 03261		IDS19458N4408		A	19 47 25	+44 15.2	1	7.76		-0.04		-0.31		1	B8	401
187613 +44 03261		IDS19458N4408		ABC	19 47 25	+44 15.2	1	7.18		-0.03		-0.32		2	B8	401
187613 +44 03261		IDS19458N4408		BC	19 47 25	+44 15.2	1	8.12		-0.02		-0.29		1		401
		VES 65			19 47 28	+21 06.2	1	13.40		2.10		0.50		2		363
188013 +72 00911					19 47 28	+72 20.3	1	7.54		1.07		0.88		2	K0	1733
187505 +16 04021					19 47 29	+16 14.9	1	7.70		0.87		0.55		4	G5 II	8100
		VES 66			19 47 29	+25 46.8	1	13.33		1.35		0.37		1		363
187728 +55 02266					19 47 29	+55 35.9	1	7.66		0.41		0.00		2	F8	1502
187565 +29 03760					19 47 31	+29 14.9	2	8.05	.050	0.48	.010	0.01		5	F8 V	20,3026
339010 +25 03988					19 47 33	+25 29.3	1	10.58		0.16		0.14		2	A2	106
		VES 67			19 47 33	+33 23.4	1	11.35		0.38		0.10		1		363
		LS II +26 007			19 47 35	+26 13.6	1	12.31		0.85		-0.10		2		824
187637 +41 03498					19 47 36	+41 27.5	1	7.53		0.51		-0.08		1	F5	1620
		LS II +21 013			19 47 37	+21 02.8	1	10.81		0.40		-0.29		1	B3 II:	1215
		CS 22964 # 84			19 47 37	-40 55.8	2	13.02	.000	0.09	.000	0.17	.010	2		966,1736
339009 +25 03989					19 47 38	+25 28.7	1	10.53		0.33		-0.08		3	A5	106
225985 +32 03583		LS II +32 004			19 47 38	+32 49.8	1	9.08		0.10		-0.77		2	B1 V:pe	1012
187311 -41 13719					19 47 39	-41 09.1	1	10.25		-0.17		-0.67		1	B5 II	55
	+7 04250	G 92 - 29			19 47 40	+07 58.3	1	10.67		1.21		1.20		1	K5	333,1620
	+19 04168				19 47 40	+19 32.4	1	10.48		1.86				3	M3	369
187638 +38 03772		HR 7555		★ A	19 47 41	+38 35.0	1	6.11		0.90				2	G8 III	71
		CS 22873 # 72			19 47 42	-61 45.3	1	14.64		0.41		-0.20		1		1736
225999 +32 03586					19 47 46	+32 45.8	1	9.74		1.22		1.32		1		1697
187639 +32 03587					19 47 46	+33 08.9	1	8.42		0.09		0.12		2	A2	1733
		CS 22964 # 87			19 47 46	-41 17.0	2	15.29	.000	0.05	.000	0.12	.010	2		966,1736
187709 +47 02929					19 47 47	+47 16.9	1	9.42		0.20		0.12		2	A2	262
187767 +53 02303					19 47 51	+53 38.4	1	7.10		-0.11		-0.41		2	B8	401
187567 +7 04252		HR 7554, V1339 Aql			19 47 52	+07 46.5	4	6.51	.018	-0.09	.010	-0.70	.013	13	B2.5IVe	15,154,1212,1417
187614 +26 03678					19 47 52	+26 57.5	2	6.44	.000	0.94	.005			4	G8 III	20,242
187640 +28 03493		HR 7556			19 47 53	+28 18.8	5	6.40	.031	-0.08	.025	-0.50	.007	33	B5 V	242,650,1063,1079,1192
		CS 22964 # 76			19 47 53	-40 07.4	2	14.35	.010	-0.20	.010	-0.89	.040	2		966,1736
		CS 22896 # 172			19 47 53	-55 51.5	1	15.36		0.22		0.01		1		1736
		AA45,405 S89 # 4			19 47 55	+26 19.7	1	11.98		0.72		0.36		1		1682
		AA45,405 S89 # 5			19 47 57	+26 21.9	1	12.13		1.29		0.24		4	O7 Vn	1682
	+38 03773				19 47 57	+38 47.1	2	8.54	.005	-0.09	.000	-0.51	.005	4	A0	833,963
187369 -42 14506					19 48 00	-42 13.3	1	7.85		0.58				4	F8 V	2012
187532 -11 05149		HR 7553		★ A	19 48 02	-10 53.5	1	5.39		0.38				4	F0 V	2007
187569 +0 04329					19 48 04	+00 35.8	1	8.83		0.04		-0.42		2	B9	634

HD	DM	Other Id	N Rem	α_{1950}	δ_{1950}	S	V	σ_V	B–V	σ_{B-V}	U–B	σ_{U-B}	n	Spectrum	References
		AA45,405 S89 # 1		19 48 04	+26 20.0	1	12.11		1.61		1.26		2		1682
		G 125 - 35		19 48 05	+32 27.3	1	12.40		1.62				4		940
350685	+18 04264	LS II +18 015		19 48 06	+18 27.0	1	9.93		0.05		-0.78		1	B5	963
339034	+24 03902	NR Vul		19 48 06	+24 48.2	3	9.30	.071	3.04	.025	2.57		9	M2 Ia	148,369,8032
332907				19 48 06	+28 00.4	1	10.91		0.45		-0.51		1	G0	824
		VES 68		19 48 07	+29 59.6	1	11.43		1.15		1.12		1		363
350686	+18 04267		A	19 48 08	+18 24.8	1	9.85		1.16		0.97		1	A2	963
350686	+18 04267		B	19 48 08	+18 24.8	1	9.82		0.23		0.18		1	A2	963
187688	+32 03589			19 48 08	+32 30.8	1	8.14		-0.01		-0.82		1	B8	1192
		LS II +26 008		19 48 09	+26 50.9	1	12.12		0.76		-0.31		1		824
187876	+57 02074			19 48 11	+57 17.0	1	7.76		0.59		0.06		2	G0	1375
332888	+28 03494			19 48 13	+28 26.8	1	9.62		1.78				2	K7	369
226046	+32 03590			19 48 13	+33 00.5	1	9.21		0.24		0.06		1		1697
		AO 762		19 48 14	+32 50.7	1	10.24		1.57		1.95		1		1697
		CS 22896 # 180		19 48 14	-54 59.5	1	14.82		0.10		0.10		1		1736
187711	+29 03767			19 48 17	+29 53.4	1	8.40		0.03		-0.43		2	A2	401
		AA45,405 S89 # 7		19 48 19	+26 22.6	1	13.22		1.51		1.17		1		1682
187536	-28 16217			19 48 20	-28 21.2	1	9.46		-0.08		-0.86		1	B3 II	55
187442	-44 13579			19 48 20	-44 29.7	2	9.89	.001	0.10	.003	0.12		11	A1 IV/V	863,2011
187642	+8 04236	HR 7557	★ A	19 48 21	+08 44.1	15	0.77	.015	0.22	.010	0.08	.022	91	A7 IV-V	1,3,15,22,1006,1020*
338961				19 48 21	+27 20.3	1	10.84		0.46		-0.49		3	B8	824
332913	+27 03531			19 48 21	+27 51.1	1	8.36		1.12				1	K0	242
		MWC 415, CI Cyg		19 48 21	+35 33.4	3	10.43	.146	1.15	.046	-0.26	.017	3	M6.5	1764,1591,1753
188119	+69 01070	HR 7582	★ AB	19 48 21	+70 08.4	7	3.84	.021	0.89	.005	0.52	.013	15	G7 IIIb	15,1355,1363,3016*
187619	+0 04331			19 48 22	+00 43.0	1	7.58		0.10		0.03		1	A0	634
		VES 69		19 48 22	+25 51.8	1	12.42		0.92		-0.04		1		363
331364	+32 03591			19 48 22	+32 44.0	1	10.53		1.25		1.29		1	F5	1697
226067	+33 03615			19 48 23	+34 13.8	1	8.24		0.01		-0.14		1	A2	963
345093	+23 03801			19 48 24	+23 16.8	1	9.37		1.92				1	K7	369
226070				19 48 24	+38 29.	1	10.19		0.21		0.17		3	B5	555
187645	+3 04168	G 23 - 12		19 48 25	+03 49.2	1	9.55		0.93		0.64		2	G5	333,1620
187474	-40 13514	HR 7552, V3961 Sgr		19 48 27	-40 08.7	5	5.32	.010	-0.06	.008	-0.23	.000	19	Ap Cr(Si)	15,1075,1202,2030,3023
187750	+30 03769			19 48 28	+30 38.3	1	7.34		-0.03		-0.05		2	A0	401
187795	+35 03826			19 48 29	+35 58.2	1	7.08		-0.07		-0.38		2	A0	1601
		He3 1783		19 48 30	+25 45.	1	12.55		0.94		0.05		1		363
		LS II +26 009		19 48 31	+26 03.4	1	12.20		0.95		0.23		1		824
		NGC 6830 - 16		19 48 32	+22 59.8	1	12.52		0.55		0.28		1		49
		NGC 6830 - 161		19 48 33	+22 57.5	1	15.10		0.89		0.31		1		49
187660	-2 05133	HR 7559		19 48 35	-02 35.3	3	6.12	.000	1.57	.005	1.81	.017	11	K5 III	252,1088,3005
		CS 22964 # 83		19 48 35	-40 56.9	2	14.69	.000	0.21	.000	0.18	.008	2		966,1736
		NGC 6830 - 15		19 48 37	+22 59.0	1	12.70		0.46		-0.02		1		49
187691	+10 04073	HR 7560	★ A	19 48 38	+10 17.4	2	5.12	.004	0.55	.002	0.08		7	F8 V	1363,3026
		NGC 6830 - 14		19 48 38	+22 58.6	1	12.02		0.42		-0.10		1	B6 IV	49
187796	+32 03593	HR 7564, khi Cyg	★ A	19 48 38	+32 47.2	8	6.09	.807	1.92	.088	0.74	.301	9	S7.1 e:	15,765,814,817,8003*
187420	-55 09221	HR 7548	★ A	19 48 39	-55 06.0	2	5.73	.018	0.92	.001	0.65	.014	5	G8/K0 III	58,1279
187420	-55 09221	IDS19447S5514	AB	19 48 39	-55 06.0	1	5.31		0.56				4	G8/K0 III	2007
187421	-55 09222	HR 7549	★ B	19 48 40	-55 06.4	1	6.50		0.10		0.13		3	A2 V	1279
353895		TW Aql		19 48 41	+13 50.5	1	10.07		1.14		1.05		1	K7	793
		WLS 1944 55 # 9		19 48 41	+54 52.4	1	12.93		0.69		0.27		2		1375
339008	+25 03996			19 48 42	+25 22.0	1	10.53		0.31		0.21		3	A2	106
		VES 71		19 48 43	+22 27.3	1	12.53		0.40		0.19		1		363
		NGC 6830 - 159		19 48 43	+23 00.4	1	14.14		0.96		0.60		1		49
		NGC 6830 - 22		19 48 43	+23 01.4	1	13.15		0.55		0.10		1		49
		NGC 6830 - 162		19 48 44	+22 56.6	1	15.22		0.84		0.22		1		49
		NGC 6830 - 23		19 48 44	+23 01.0	1	13.18		0.49		0.02		1		49
332848	+29 03772	LS II +29 018		19 48 45	+29 16.1	3	10.10	.005	0.50	.005	-0.53	.005	5	O7e	824,1011,1012
		G 92 - 31		19 48 46	+07 07.0	1	13.91		1.36		1.03		1		333,1620
		NGC 6830 - 46		19 48 46	+22 57.0	1	13.39		0.48		0.11		1		49
		NGC 6830 - 11		19 48 46	+22 58.6	1	13.00		0.75		0.36		1		49
345110	+22 03832			19 48 47	+22 46.0	1	9.79		0.27		-0.23		2	B9	106
		NGC 6830 - 12		19 48 47	+22 58.5	1	12.16		0.49		-0.05		1	B6 IV	49
226107	+35 03827			19 48 48	+35 42.8	1	8.55		-0.04		-0.33		1	B9	1192
187849	+38 03780	HR 7566, V1509 Cyg	★ A	19 48 48	+38 35.6	3	5.15	.066	1.68	.016	2.10		5	M2 IIIa	71,369,3055
		BPM 12772		19 48 48	-71 57.	1	15.10		0.58		-0.21		3		3065
		NGC 6830 - 24	★	19 48 50	+23 00.8	2	11.70	.015	0.43	.005	-0.05	.025	2	B6 IV	49,363
188120	+65 01406	G 260 - 33		19 48 50	+65 49.5	2	8.51	.010	0.52	.005	-0.01	.015	2	F8	1620,1658
332847	+29 03774	LS II +29 019		19 48 51	+29 16.5	3	9.94	.009	0.56	.005	-0.36	.005	5	B2 Ib	824,1012,1318
		NGC 6830 - 7		19 48 52	+22 58.2	1	11.19		0.42		-0.02		1	B7 III	49
		NGC 6830 - 6		19 48 53	+22 56.9	1	13.10		0.57		0.23		1		49
226111	+33 03618	LS II +33 008		19 48 53	+33 32.1	1	9.87		0.55		-0.44		2	B1 Ib	1012
187753	+9 04288	HR 7562		19 48 54	+09 30.1	4	6.25	.015	0.10	.011	0.09	.012	13	A1 m	15,355,1256,8071
187879	+40 03902	HR 7567, V380 Cyg		19 48 54	+40 28.3	2	5.69	.010	-0.05	.010	-0.77	.010	5	B1 III+B8 V	154,588
187811	+22 03833	HR 7565		19 48 55	+22 28.9	8	4.94	.021	-0.15	.017	-0.68	.012	30	B2.5Ve	15,154,374,1118,1212*
345108	+22 03834	NGC 6830 - 5		19 48 55	+22 57.3	2	9.87	.010	0.30	.015	0.06	.214	3	B5 III	49,106
		NGC 6830 - 9		19 48 55	+22 58.8	1	11.00		0.65		0.37		1	A9 III	49
187736	+0 04332			19 48 56	+00 20.2	1	9.78		0.53		0.08		2	F8	634
		NGC 6830 - 30		19 48 56	+23 01.3	1	13.61		0.42		0.14		1		49
187754	+7 04260			19 48 57	+07 16.1	1	8.49		0.17		0.20		2	A2	1375
		CS 22896 # 165		19 48 57	-56 41.5	1	13.74		-0.17		-0.86		1		1736
345111	+22 03835			19 48 58	+22 45.8	1	10.44		0.39		-0.40		3	A3	106

Table 1 1075

HD	DM	Other Id	N	Rem	α_{1950}	δ_{1950}	S	V	σ_V	B–V	σ_{B-V}	U–B	σ_{U-B}	n	Spectrum	References
339003	+25 03998	LS II +25 017			19 48 58	+25 49.5	2	9.94	.012	0.66	.008	-0.40	.008	4	B0.5III	106,1032
		CS 22964 # 60			19 48 58	−38 54.1	1	14.63		-0.30		-1.12		1		1736
345117	+22 03836	LS II +22 010			19 48 59	+22 42.2	3	9.17	.025	0.51	.009	-0.45	.006	6	B1 II	106,1012,8100
187880	+37 03636	HR 7568			19 48 59	+37 41.9	3	6.08	.036	1.69	.016	1.78	.124	5	M4 IIb	369,1375,3001
345105		NGC 6830 - 2			19 49 01	+22 58.8	1	10.44		0.38		-0.15		1	B9 III:e	49
		NGC 6830 - 1			19 49 02	+22 59.3	1	13.05		0.36		-0.04		1	B9.5II	49
187851	+27 03534	LS II +27 017			19 49 02	+27 35.3	1	7.74		0.13		-0.63		2	B2 V:nn	1012
		AO 766			19 49 02	+32 45.1	1	10.58		1.32		1.22		1		1697
187920	+42 03490				19 49 03	+42 15.0	1	8.24		1.47		1.70		2	K0	1375
		CS 22964 # 88			19 49 03	−41 10.2	1	12.78		1.43		1.23		1		1736
		CS 22896 # 173			19 49 03	−55 51.8	1	14.13		0.35		-0.09		1		1736
		G 23 - 13			19 49 05	+02 50.7	1	10.31		1.08		0.96		1	K3	333,1620
		NGC 6830 - 160			19 49 05	+22 56.9	1	14.44		0.46		0.38		1		49
226130	+35 03830				19 49 05	+35 16.3	1	8.92		-0.01		-0.19		1	B9	1192
		CS 22964 # 51			19 49 06	−38 13.7	1	11.66		0.22		0.21		1		1736
		NGC 6830 - 35			19 49 08	+23 00.7	1	13.18		0.49		0.02		1		49
		WLS 2000 40 # 6			19 49 09	+40 04.0	1	12.87		0.33		0.27		2		1375
		NGC 6830 - 34			19 49 10	+23 01.9	1	13.25		0.50		0.20		1		49
		VES 73			19 49 14	+27 27.8	1	12.94		0.68		-0.55		1		363
331372	+32 03598				19 49 16	+32 28.0	1	9.51		0.68		0.20		8	G5	1655
		Steph 1723			19 49 16	−05 09.4	1	12.01		1.26		1.37		1	K4	1746
		VES 74			19 49 17	+27 24.1	1	13.32		1.40		0.10		2		363
226166	+38 03786	LS II +38 001			19 49 18	+38 49.3	1	9.86		0.81		0.52		1	F8 Ib	1215
187739	−19 05631	HR 7561			19 49 18	−19 10.4	2	5.92	.005	0.98	.002	0.64		5	K0 III	2007,4001
345104	+22 03837	NGC 6830 - 72			19 49 20	+22 57.4	3	9.03	.018	0.33	.027	0.05	.023	5	B9.5III	49,106,8100
187758	−12 05563				19 49 20	−12 42.4	1	8.30		1.06		0.81		2	K0 III	1375
	+5 04314	G 23 - 14			19 49 22	+05 29.2	2	10.71	.015	0.75	.015	0.08	.029	3	G5	333,1620,1658
345112		NGC 6830 - 78			19 49 22	+22 51.4	1	10.89		0.62		0.20		1	F9 V:	49
188056	+52 02547	HR 7576			19 49 22	+52 51.6	5	5.02	.007	1.29	.008	1.52	.012	16	K3 III	37,1355,1363,3053,4001
		CS 22964 # 61			19 49 22	−39 05.8	1	14.82		0.47		0.08		1		1736
		NGC 6830 - 74			19 49 25	+22 54.9	1	12.11		0.51		0.30		1		49
		NGC 6830 - 75			19 49 27	+22 54.4	1	13.60		0.70				1		49
187855	+1 04122				19 49 28	+01 28.4	1	7.45		0.10		0.03		2	A2	634
187921	+27 03536	SV Vul			19 49 28	+27 19.9	3	6.84	.158	1.22	.183	1.05	.274	3	F7 Ia	934,1772,6007
		CS 22964 # 48			19 49 28	−37 54.6	1	14.30		0.50		0.02		1		1736
187760	−24 15644				19 49 30	−24 04.7	1	9.50		1.15		1.07		2	K4 V	3072
		CS 22873 # 66			19 49 30	−60 50.4	1	14.81		0.30		0.07		1		1736
345076	+23 03809				19 49 31	+23 45.4	1	10.68		0.88		-0.12		1		1032
		AO 765			19 49 34	+32 44.9	1	10.38		1.19		1.15		1		1697
226178	+32 03600				19 49 34	+32 58.2	1	8.81		0.38		-0.05		1		1697
		CS 22873 # 55			19 49 35	−59 47.8	1	12.65		0.93		0.40		1		1736
350633	+19 04184		V		19 49 36	+19 40.4	1	8.51		1.69				1	M0	369
	+18 04275				19 49 38	+18 21.5	1	10.23		1.85				1	M3	369
187826	−21 05554		V		19 49 41	−21 28.3	1	7.55		0.73		0.39		1	F3 V	1772
226202	+36 03736				19 49 42	+36 47.9	1	9.56		1.84				2	M7	369
187923	+11 04019	HR 7569	★	A	19 49 43	+11 30.2	2	6.15	.015	0.65	.000	0.12	.005	3	G0 V	15,1003,1243,3077
187981	+30 03779	IDS19478N3053		A	19 49 43	+31 00.7	1	7.06		0.31		0.17		2	A5	3016
187981	+30 03779	IDS19478N3053		C	19 49 43	+31 00.7	1	9.70		0.41		-0.06		2		3016
	−76 01372				19 49 43	−76 35.6	1	10.08		1.29				4	K0	2012
187971	+28 03507	NGC 6834 - 206			19 49 44	+29 08.0	1	8.50		0.13		0.16		1	A3 V	49
226206	+34 03757				19 49 46	+34 25.1	1	9.09		0.18		0.11		1	A3	401
		CS 22964 # 77			19 49 48	−40 06.9	2	14.45	.000	0.03	.000	0.06	.011	2		966,1736
		LS II +27 019			19 49 50	+27 17.3	1	11.42		0.59		-0.17		2		824
188074	+47 02937	HR 7577			19 49 50	+47 14.9	2	6.21	.000	0.36	.015	0.03	.010	3	F2 V	254,262
226208	+33 03625	IDS19480N3349		A	19 49 51	+33 56.4	1	7.38		1.07				1	K1	8097
187959	+14 04074				19 49 52	+15 11.6	1	7.74		0.27		0.10		3	AmA3-F0	8071
345071	+23 03811				19 49 52	+23 49.2	1	9.31		0.38		-0.20		2	B5	106
187961	+9 04295	HR 7572	★	A	19 49 53	+10 13.3	1	6.53		-0.01		-0.39		2	B7 V	154
226223	+38 03790	LS II +38 002			19 49 53	+38 37.2	2	9.19	.010	0.67	.020	0.26	.094	5	F6 Ib	247,8100
187869	−15 05492				19 49 53	−15 32.7	1	8.94		0.07		-0.28		1	B9 IV/V	1776
187982	+24 03914	HR 7573			19 49 55	+24 51.8	3	5.59	.026	0.68	.017	0.17	.027	7	A1 Iab	650,1012,1118
		NGC 6834 - 157			19 49 55	+29 17.1	1	12.41		0.60		-0.07		1		49
187929	+0 04337	HR 7570, η Aql			19 49 56	+00 52.6	7	3.68	.142	0.71	.124	0.47	.052	12	F6 Ib	669,851,1149,1363,1484*
		NGC 6834 - 163			19 49 57	+29 15.9	1	14.70		0.66		0.10		1		49
187930	+0 04338	IDS19474N0024		AB	19 49 59	+00 31.1	2	8.02	.025	0.31	.014	0.11	.001	3	A3	634,669
		VES 75			19 49 59	+22 27.5	1	12.46		0.79		-0.14		2		363
332918	+27 03538	LS II +27 020			19 50 00	+27 50.3	1	11.20		0.70		-0.32		2	F6 Ib	824
		NGC 6834 - 73			19 50 01	+29 16.0	1	11.67		0.66		0.15		1	F8 V	49
188036	+31 03813				19 50 01	+31 28.4	1	7.71		0.01		-0.09		2	A0	401
188035	+31 03814	IDS19481N3127		A	19 50 03	+31 34.7	1	7.84		0.00		-0.14		2	A0	401
		NGC 6834 - 87			19 50 05	+29 13.9	1	12.55		0.59		-0.05		1		49
		NGC 6834 - 80			19 50 05	+29 14.3	1	15.03		0.61		0.27		1		49
		NGC 6834 - 79			19 50 05	+29 14.7	1	14.26		0.64		0.17		1		49
		NGC 6834 - 75			19 50 05	+29 15.3	1	14.64		0.64		0.44		1		49
332844		NGC 6834 - 72			19 50 05	+29 16.1	1	11.15		0.42		0.08		1	F2 V	49
		NGC 6834 - 56			19 50 05	+29 16.8	1	11.94		0.61		0.04		1	B5 III	49
		G 125 - 39			19 50 06	+29 10.7	1	14.52		0.74		0.01		3		1658
		NGC 6834 - 77			19 50 06	+29 15.2	1	13.82		0.62		0.04		1		49
		NGC 6834 - 55			19 50 06	+29 16.8	1	13.21		0.59		-0.15		1		49
		CS 22964 # 50			19 50 06	−38 16.1	1	12.22		0.13		0.12		1		1736

HD	DM	Other Id	N	Rem	α_{1950}	δ_{1950}	S	V	σ_V	B–V	σ_{B-V}	U–B	σ_{U-B}	n	Spectrum	References
		WLS 2000 5 # 6			19 50 08	+04 58.3	1	11.65		0.49		0.11		2		1375
188001	+18 04276	QZ Sge, HR 7574		★	19 50 08	+18 32.5	3	6.24	.015	0.01	.000	-0.93	.005	8	O8 f	15,154,1011
332917					19 50 08	+27 55.9	1	9.99		0.82		-0.22		3	G5	824
332877	+28 03508				19 50 08	+28 45.0	1	8.46		0.14				3	A0	1118
188075	+37 03649				19 50 08	+37 14.8	2	7.74	.015	0.34	.010	0.01	.030	4	A5	1733,3016
		NGC 6834 - 67		★	19 50 09	+29 15.7	2	11.72	.004	0.64	.016	0.11	.015	2		363,1586
		NGC 6834 - 42			19 50 09	+29 18.5	1	13.61		0.57		0.06		1		49
	-57 09526				19 50 09	-56 55.3	1	11.25		0.56		-0.19		2		1696
		NGC 6834 - 52			19 50 10	+29 16.8	1	12.73		0.96		-0.18		1		49
	+23 03814				19 50 11	+23 22.2	1	10.22		0.39		-0.16		3		106
		NGC 6834 - 184			19 50 11	+29 12.1	1	13.18		0.70		0.24		1		49
		NGC 6834 - 84			19 50 12	+29 14.4	1	14.91		0.86		0.56		1		49
332843	+29 03779	NGC 6834 - 1			19 50 13	+29 16.7	1	9.65		0.87		0.81		1	F0 Ib	49
		NGC 6834 - 96			19 50 14	+29 13.4	1	14.30		0.66		0.03		1		49
		NGC 6834 - 207			19 50 14	+29 14.6	1	12.99		0.60		0.01		1		49
		NGC 6834 - 11			19 50 14	+29 15.0	1	14.17		0.53		0.08		1		49
		NGC 6834 - 4			19 50 14	+29 16.1	1	12.61		0.56		-0.08		1		49
		VES 78			19 50 15	+28 50.9	1	12.58		1.23		0.92		1		363
188059	+26 03699				19 50 16	+26 36.0	2	8.64	.068	-0.07	.010	-0.40	.030	3	A0	634,963
		LS II +27 023			19 50 16	+27 23.7	1	11.87		0.52		-0.15		2		824
		NGC 6834 - 37			19 50 17	+29 18.0	1	12.79		0.60		0.21		1		49
		AJ89,1229 # 417			19 50 18	+03 44.	1	12.86		1.58				2		1532
		MWC 996			19 50 18	+22 34.	1	11.62		1.28		0.30		1		363
		NGC 6834 - 23			19 50 18	+29 15.7	1	12.57		0.46		-0.16		1	B5 V	49
		L 349 - 68			19 50 18	-47 54.	2	12.53	.000	1.55	.015	1.12		3		1705,3073
		G 92 - 34			19 50 19	+04 31.6	3	13.70	.008	1.38	.020	0.96	.023	5		316,333,1620,1759
187653	-61 06426	HR 7558			19 50 19	-61 18.1	1	6.24		0.16				4	A4 V	2007
		NGC 6834 - 20			19 50 20	+29 15.5	1	13.22		0.48		-0.04		1		49
188037	+22 03840	NS Vul		★ AB	19 50 21	+22 19.4	1	7.76		0.92		0.28		6		8032
		NGC 6834 - 125			19 50 21	+29 17.0	1	10.90		1.40		0.98		1		49
		VES 79			19 50 23	+29 10.1	1	12.80		1.19		0.99		1		363
		NGC 6834 - 131			19 50 23	+29 18.1	1	15.77		0.64		0.36		1		49
		LS II +27 024			19 50 25	+27 20.8	1	11.63		0.55		-0.16		2		824
188149	+36 03744	HR 7583			19 50 26	+36 18.1	2	6.11	.003	1.43	.013	1.62		4	K4 III	71,247
188209	+46 02793	HR 7589		★	19 50 29	+46 53.8	9	5.63	.011	-0.07	.014	-0.96	.016	48	O9.5Ia	15,154,1011,1025*
332845		NGC 6834 - 202			19 50 30	+29 12.1	1	10.35		0.33		-0.15		1	B5 III	49
187987	-7 05090				19 50 30	-07 13.1	1	9.96		0.38		0.24		2	A2	1375
		CS 22964 # 78			19 50 32	-40 22.0	2	15.06	.000	0.15	.000	0.03	.009	2		966,1736
		VES 80			19 50 34	+22 32.6	1	11.37		2.23		1.37		1		363
226280	+34 03766	V480 Cyg			19 50 34	+34 28.9	1	9.25		1.65				2	M0	369
		VES 81			19 50 35	+26 40.3	1	12.40		0.95		0.03		2		363
226273	+33 03630				19 50 35	+33 18.1	1	9.83		1.74				2	M0	369
		CS 22964 # 97			19 50 36	-42 16.3	1	11.64		0.09		0.15		1		1736
187456	-72 02467				19 50 37	-72 29.5	1	8.48		1.07		0.98		2	K4 V	3072
188252	+47 02939	HR 7591		★	19 50 39	+47 48.1	3	5.91	.005	-0.19	.005	-0.85	.004	7	B2 III	262,401,555
188041	-3 04742	HR 7575, V1291 Aql	A		19 50 42	-03 14.7	5	5.64	.000	0.20	.002	0.08	.010	14	A5p	1088,1202
345122	+21 03959				19 50 43	+22 06.8	4	9.84	.013	0.35	.010	-0.36	.021	6	B2 Vpe	106,445,1012,1032
331392	+31 03821				19 50 43	+32 02.2	1	9.56		1.65				2	K7	369
350800	+17 04130	Ha 20 - 1			19 50 46	+18 12.4	1	9.23		0.23		0.20		4	A2	872
		VES 82			19 50 46	+23 17.8	1	12.08		0.48		0.00		2		363
		CS 22964 # 81			19 50 46	-40 35.5	2	14.01	.000	0.33	.000	0.17	.010	2		966,1736
		CS 22964 # 98			19 50 47	-42 14.9	1	13.90		-0.19		-0.92		1		1736
187807	-57 09533				19 50 48	-57 41.9	1	7.97		1.10				4	K1 III	2012
		Ha 20 - 31			19 50 49	+18 11.2	1	13.64		0.52		0.19		2		872
188170	+28 03513				19 50 49	+28 51.7	4	7.33	.010	-0.09	.005	-0.37	.002	80	A0	401,879,1118,3016
		LS II +10 010			19 50 52	+10 10.7	1	11.87		0.56		0.07		1	F6/8 II-V	1215
		Ha 20 - 2			19 50 52	+18 14.1	1	11.72		1.13		0.81		4		872
		CS 22891 # 221			19 50 52	-59 48.5	2	14.42	.009	0.58	.013	-0.22	.002	4		1580,1736
188107	+4 04264	HR 7580			19 50 53	+04 16.2	2	6.52	.005	0.02	.000	-0.11	.000	7	B9.5Vn	15,1071
		Ha 20 - 4			19 50 53	+18 13.7	1	13.37		1.21		0.94		3		872
		G 92 - 35			19 50 54	+07 53.8	1	12.48		1.33		1.23		1		333,1620
		Ha 20 - 6			19 50 54	+18 12.	1	13.59		0.22		0.16		4		872
		Ha 20 - 32			19 50 54	+18 12.	1	13.97		0.68		0.14		1		872
		Ha 20 - 34			19 50 54	+18 12.	1	13.59		1.33		1.46		1		872
		CS 22964 # 126			19 50 54	-39 05.8	2	14.20	.000	0.09	.000	0.16	.009	3		966,1736
		Ha 20 - 33			19 50 55	+18 10.6	1	13.81		1.44		1.16		1		872
		Ha 20 - 5			19 50 55	+18 12.3	1	13.31		0.71		0.24		3		872
		VES 83			19 50 55	+28 10.8	1	12.32		1.56		0.52		1		363
		Ha 20 - 9			19 50 56	+18 11.6	1	11.85		0.14		-0.26		3		872
350801	+17 04132	Ha 20 - 8			19 50 56	+18 13.6	1	8.90		1.32		1.13		4	G0	872
		Ha 20 - 11			19 50 57	+18 12.3	1	13.38		1.22		0.72		1		872
		Ha 20 - 12			19 50 58	+18 15.3	1	12.50		0.19		-0.10		2		872
		WLS 2000-5 # 6			19 50 58	-04 44.8	1	13.03		1.03		0.47		2		1375
		Ha 20 - 14			19 50 59	+18 15.1	1	12.93		0.23		0.07		2		872
		Ha 20 - 13			19 51 00	+18 11.9	1	12.00		0.14		-0.19		3		872
350737	+19 04192	IDS19488N2002	A		19 51 00	+20 09.3	1	11.13		0.24		-0.09		2	B9	363
339143	+24 03918				19 51 00	+24 18.3	1	9.85		0.16		0.22		2	A0	106
		Ha 20 - 15			19 51 02	+18 13.3	1	12.29		0.19		-0.04		2		872
		Ha 20 - 17			19 51 04	+18 14.8	1	13.00		1.56		1.82		2		872
	+47 02942				19 51 04	+47 51.2	1	10.48		1.25		1.35		2	K2 III	262

Table 1 1077

HD	DM	Other Id	N	Rem	α_{1950}	δ_{1950}	S	V	σ_V	B–V	σ_{B-V}	U–B	σ_{U-B}	n	Spectrum	References
		AB Dra			19 51 04	+77 37.0	1	14.54		1.10		0.03		1		698
		Ha 20 - 19			19 51 05	+18 12.9	1	13.47		0.71		0.27		2		872
		Ha 20 - 18			19 51 05	+18 14.0	1	12.73		0.23		-0.06		2		872
		VES 85			19 51 05	+28 59.6	1	13.23		0.75		0.32		1		363
	+40 03924				19 51 05	+41 05.3	1	10.10		1.74				2		369
		Ha 20 - 22			19 51 06	+18 11.0	1	11.90		1.42		1.28		2		872
		Ha 20 - 23			19 51 07	+18 13.3	1	12.62		0.15		-0.14		2		872
188234	+26 03706				19 51 08	+26 27.4	1	8.02		-0.01		-0.05		2	A0	634
		Ha 20 - 25			19 51 09	+18 11.7	1	10.75		0.48		0.05		5		872
188360	+47 02943				19 51 09	+47 58.5	2	8.07	.000	0.02	.020	0.04	.035	4	A0	262,401
		Ha 20 - 35			19 51 11	+18 14.6	1	13.66		0.43		0.35		2		872
188211	+19 04195	IDS19490N2004	A		19 51 11	+20 12.4	1	7.13		0.14		0.03		3	A0	1084
188173	+7 04281				19 51 12	+08 10.3	2	8.77	.028	-0.07	.011	-0.42	.012	7	A0	963,1732
188212	+19 04196	IDS19490N2004	B		19 51 12	+20 11.8	1	7.32		0.31		0.05		3	A0	1084
		LS II +27 025			19 51 13	+27 20.0	1	11.84		0.68		0.02		2		824
	+47 02944				19 51 13	+47 44.2	1	10.29		1.57		1.87		2	K5 III	262
		Ha 20 - 26			19 51 14	+18 11.7	1	11.78		0.13		-0.17		3		872
345168	+23 03818				19 51 14	+23 40.8	1	10.37		0.40		-0.41		3	B8	106
188258	+27 03546				19 51 14	+27 58.1	1	6.94		1.09				3	K2 III	20
		Ha 20 - 27			19 51 15	+18 10.0	1	11.99		1.28		1.12		2		872
		Ha 20 - 28			19 51 15	+18 11.8	1	12.14		1.08		0.74		1		872
188259	+26 03708				19 51 15	+26 22.1	2	7.66	.178	1.00	.136	0.40		4	K1 III	20,3026
188326	+38 03801	G 125 - 41	⋆ A		19 51 15	+38 38.3	3	7.57	.010	0.78	.000	0.36	.010	5	G8 IV	247,1003,3016
188326	+38 03801	IDS19495N3830	B		19 51 15	+38 38.3	1	11.47		1.65		1.38		1		3016
		Ha 20 - 29			19 51 16	+18 11.1	1	12.55		0.70		0.21		2		872
188109	-17 05787				19 51 16	-17 32.6	1	8.72		0.51		0.08		2	F2 V	1375
188257	+28 03515				19 51 17	+28 23.1	2	8.13	.010	0.01	.010	-0.27		5	A3	401,1118
		VES 86, IX Cyg			19 51 17	+32 02.9	1	12.60		2.03		0.77		1		363
		Ha 20 - 30			19 51 18	+18 11.3	1	10.56		1.69		1.95		2		872
345223	+21 03963				19 51 18	+22 03.8	1	10.06		0.28		-0.44		1	B2 V	1032
		VES 87			19 51 18	+32 06.3	1	11.59		0.14		0.19		1		363
188088	-24 15668	HR 7578	⋆ A		19 51 18	-24 04.0	6	6.19	.012	1.02	.008	0.93	.012	23	K2 V	15,158,1075,2013,2030,3072
226356	+34 03772		A		19 51 19	+34 59.1	1	9.46		0.01		-0.34		1	B9	1192
226356	+34 03772		B		19 51 19	+34 59.1	1	8.68		-0.01		-0.26		1		1192
188260	+23 03820	HR 7592	⋆ AB		19 51 20	+23 56.9	5	4.57	.005	-0.05	.014	-0.14	.024	14	B9.5III	15,1063,1118,3023,8015
188111	-19 05640				19 51 20	-19 37.8	1	9.76		0.95		0.70		1	K2 V	3072
188213	+12 04124				19 51 21	+12 37.5	1	9.33		0.41				4	F5	1655
		LS II +14 015			19 51 21	+14 50.2	1	12.21		0.45		0.06		1	F0 Ib	1215
188261	+20 04321				19 51 22	+21 09.5	1	8.48		-0.03		-0.42		1	A0	963
188194	+0 04343				19 51 23	+01 03.8	1	9.00		0.27		0.07		2	A5	634
		LS II +26 010			19 51 23	+26 44.5	1	11.48		0.80		0.16		2		824
188154	-8 05150	HR 7584	⋆ A		19 51 25	-08 42.3	2	5.76	.028	1.63	.014	1.97	.019	9	gK5	1088,3016
188154	-8 05150	IDS19487S0850	B		19 51 25	-08 42.3	1	12.32		0.38		0.21		1		3016
188112	-28 16258				19 51 25	-28 28.3	1	10.18		-0.18		-0.81		8	B9 V	1732
188344	+33 03638	V449 Cyg			19 51 27	+33 49.1	1	7.39		1.66				2	M1	369
		UKST 284 # 8			19 51 28	-45 07.2	1	12.71		0.47		-0.06		3		1584
350763	+19 04198	LS II +19 021			19 51 29	+19 21.0	2	9.97	.005	0.19	.000	-0.77	.010	4	B0	405,946
		NGC 6838 - 151			19 51 30	+18 39.	1	10.13		-0.10		-0.49		1		108
		NGC 6838 - 152			19 51 30	+18 39.	1	10.76		0.50		0.05		1		108
		NGC 6838 - 153			19 51 30	+18 39.	1	11.31		0.33		0.23		1		108
		NGC 6838 - 154			19 51 30	+18 39.	1	12.08		1.84		2.11		1		108
		NGC 6838 - 155			19 51 30	+18 39.	1	12.21		1.92		2.01		1		108
		NGC 6838 - 156			19 51 30	+18 39.	1	12.28		1.72		1.90		1		108
		NGC 6838 - 157			19 51 30	+18 39.	1	12.29		1.20		0.85		1		108
		NGC 6838 - 158			19 51 30	+18 39.	1	12.38		1.53		1.53		1		108
		NGC 6838 - 159			19 51 30	+18 39.	1	12.47		1.78		2.14		1		108
		NGC 6838 - 160			19 51 30	+18 39.	1	12.90		0.29		0.19		1		108
		NGC 6838 - 161			19 51 30	+18 39.	1	13.01		1.55		1.58		1		108
		NGC 6838 - 162			19 51 30	+18 39.	1	13.04		0.58		0.21		1		108
		NGC 6838 - 163			19 51 30	+18 39.	1	13.10		1.56		1.54		1		108
		NGC 6838 - 164			19 51 30	+18 39.	1	13.47		1.25		1.15		1		108
		NGC 6838 - 165			19 51 30	+18 39.	1	13.63		1.43		1.35		1		108
		NGC 6838 - 166			19 51 30	+18 39.	1	13.85		1.42		1.47		1		108
		NGC 6838 - 167			19 51 30	+18 39.	1	13.90		1.20		0.97		1		108
		NGC 6838 - 168			19 51 30	+18 39.	1	14.48		1.03		0.55		1		108
		NGC 6838 - 169			19 51 30	+18 39.	1	14.49		1.05		0.46		1		108
		NGC 6838 - 170			19 51 30	+18 39.	1	14.54		1.00		0.50		1		108
		NGC 6838 - 171			19 51 30	+18 39.	1	14.54		1.13		0.71		1		108
		NGC 6838 - 172			19 51 30	+18 39.	1	14.70		1.21		0.88		1		108
		NGC 6838 - 173			19 51 30	+18 39.	1	14.76		0.55		0.11		1		108
		NGC 6838 - 174			19 51 30	+18 39.	1	14.78		0.62		0.05		1		108
		NGC 6838 - 175			19 51 30	+18 39.	1	14.95		1.17		0.74		1		108
		NGC 6838 - 176			19 51 30	+18 39.	1	15.70		1.02		0.44		1		108
		NGC 6838 - 177			19 51 30	+18 39.	1	15.71		1.07		0.58		1		108
		NGC 6838 - 178			19 51 30	+18 39.	1	15.72		0.72		0.19		1		108
		NGC 6838 - 179			19 51 30	+18 39.	1	15.78		0.65		0.06		1		108
		NGC 6838 - 180			19 51 30	+18 39.	1	15.95		0.60		0.15		1		108
		NGC 6838 - 181			19 51 30	+18 39.	1	15.99		1.07		0.62		1		108
		NGC 6838 - 182			19 51 30	+18 39.	1	16.73		0.75		0.10		1		108
		NGC 6838 - 183			19 51 30	+18 39.	1	16.79		0.58		0.14		1		108

HD	DM	Other Id	N Rem	α₁₉₅₀	δ₁₉₅₀	S	V	σ_V	B–V	σ_B–V	U–B	σ_U–B	n	Spectrum	References
		NGC 6838 - 184		19 51 30	+18 39.	1	16.90		0.98		0.56		1		108
		NGC 6838 - 185		19 51 30	+18 39.	1	16.98		0.82		0.22		1		108
		NGC 6838 - 186		19 51 30	+18 39.	1	17.76		1.25		1.07		1		108
		NGC 6838 - 187		19 51 30	+18 39.	1	17.89		0.78		-0.19		1		108
		NGC 6838 - 188		19 51 30	+18 39.	1	18.20		0.81		0.11		1		108
		NGC 6838 - 189		19 51 30	+18 39.	1	18.29		0.85		0.41		1		108
		NGC 6838 - 190		19 51 30	+18 39.	1	18.31		0.74		0.09		1		108
		NGC 6838 - 191		19 51 30	+18 39.	1	18.36		0.78		0.19		1		108
		NGC 6838 - 192		19 51 30	+18 39.	1	18.42		0.72		0.10		1		108
		NGC 6838 - 193		19 51 30	+18 39.	1	18.44		1.01		0.29		1		108
		NGC 6838 - 194		19 51 30	+18 39.	1	18.69		0.81		0.21		1		108
		NGC 6838 - 195		19 51 30	+18 39.	1	18.74		0.87		0.30		1		108
		NGC 6838 - 196		19 51 30	+18 39.	1	18.77		0.70		-0.05		1		108
		NGC 6838 - 197		19 51 30	+18 39.	1	19.52		0.46		-0.23		1		108
		NGC 6838 - 198		19 51 30	+18 39.	1	19.61		1.47		1.52		1		108
188031	−42 14547			19 51 30	−42 45.9	6	10.14	.019	0.43	.011	-0.21	.023	20	F0 V	863,1236,1594,1696*
188439	+47 02945	HR 7600, V819 Cyg		19 51 32	+47 40.6	4	6.29	.015	-0.12	.012	-0.91	.018	8	B0.5IIIn	154,262,401,1423
		CS 22964 # 138		19 51 32	−38 16.4	1	14.30		0.42		-0.13		1		1736
		LS II +27 027		19 51 34	+27 25.6	1	12.44		0.67		0.02		2		824
188011	−47 13147			19 51 34	−46 55.5	1	7.69		1.08				4	K0 III	2012
188177	−10 05215			19 51 35	−10 45.4	1	9.68		0.43		-0.25		2	A0	3008
345214	+22 03847	IDS19495N2221	AB	19 51 37	+22 28.7	2	9.58	.000	0.23	.000	-0.45	.000	3	B5 III	1012,1032
188366	+26 03711			19 51 41	+26 52.1	1	7.73		0.03				3	A0	1118
188268	+1 04134	G 23 - 15	⋆ A	19 51 43	+01 48.9	2	8.76	.005	0.88	.005	0.62	.005	4	K0	333,1620,3077
		UKST 284 # 28		19 51 43	−44 39.8	1	16.28		0.64				4		1584
		CS 22964 # 121		19 51 44	−39 29.2	2	12.32	.000	0.05	.000	-0.09	.010	2		966,1736
		CS 22964 # 107		19 51 44	−41 23.8	2	14.11	.000	0.19	.000	0.17	.008	2		966,1736
		LS II +27 028		19 51 45	+27 07.6	1	11.41		0.72		-0.13		2		824
		CS 22964 # 110		19 51 45	−40 59.0	1	15.63		0.07		0.10		1		1736
188178		CpD -22 07589		19 51 46	−22 45.1	1	9.83		0.56				2	F5/6 Vw	1594
333015	+28 03518			19 51 47	+28 34.7	1	9.10		-0.07		-0.35		2	A0	401
188383	+26 03712			19 51 48	+26 27.7	2	6.82	.010	0.10	.005			4	A0	242,1118
188310	+8 04261	HR 7595		19 51 49	+08 19.8	6	4.70	.022	1.05	.005	0.89	.016	15	K0 IIIb	15,1355,1363,1417*
188114	−42 14549	HR 7581		19 51 49	−42 00.1	5	4.12	.004	1.08	.002	0.90	.008	15	K0 II/III	15,278,1020,1075,2011
345185	+23 03826	AT Vul		19 51 50	+23 26.0	1	9.10		0.46		-0.38		3	A0	1768
188220	−16 05454			19 51 50	−16 02.7	1	8.04		0.35		0.06		1	A0 V	1776
188158	−33 14560	HR 7585		19 51 53	−33 14.7	1	6.46		1.48				4	K2/3 III	2032
188311	+1 04135	G 23 - 16	⋆ B	19 51 54	+01 48.8	2	8.93	.005	0.90	.000	0.65	.005	4	K0	333,1620,3077
		VES 88		19 51 54	+27 42.2	1	11.45		0.24		0.23		2		363
188418	+34 03778			19 51 54	+34 27.1	1	7.02		-0.13		-0.53		2	A0	401
		VES 89, EW Cyg		19 51 55	+31 24.2	1	11.80		1.97		0.66		1		363
188293	−8 05154	HR 7593	⋆ A	19 51 55	−08 21.5	8	5.71	.005	-0.08	.008	-0.48	.011	32	B7 Vn	1,15,116,154,1075*
		CS 22964 # 111		19 51 55	−40 52.3	1	14.69		0.45		-0.15		1		1736
		CS 22964 # 103		19 51 55	−41 46.1	2	14.67	.000	0.19	.000	0.11	.008	2		966,1736
188461	+40 03931			19 51 56	+41 13.5	3	6.99	.005	-0.16	.005	-0.76	.013	8	B2 IV	399,555,1733
188294	−8 05155	HR 7594	⋆ B	19 51 56	−08 22.1	8	6.49	.006	-0.04	.004	-0.27	.010	32	B8 V	1,15,116,154,1075*
		UKST 284 # 7		19 51 56	−44 23.9	1	12.37		1.08				3		1584
		G 260 - 36		19 51 58	+67 17.6	1	11.22		0.90		0.72		1	K2	1658
188116	−45 13508		V	19 52 01	−45 42.2	1	10.00		0.49				4	F5 V	2011
188367	+16 04056			19 52 02	+16 32.6	1	9.14		0.04		-0.02		2		634
		CS 22873 # 42		19 52 04	−58 28.7	1	16.06		-0.03		-1.10		1		1736
333030	+27 03550	LS II +28 014		19 52 05	+28 12.3	3	8.94	.000	0.98	.000	0.30	.005	5	B9 Ia	824,1012,1318,8100
188501	+38 03809	IDS19503N3851	A	19 52 06	+38 58.5	1	8.07		-0.09		-0.54		2	B8 III	105
226417	+34 03781			19 52 08	+34 19.6	1	8.77		0.00		-0.25		2	B9	401
188350	−0 03871	HR 7596		19 52 11	+00 08.5	4	5.63	.008	0.10	.012	0.06	.046	19	A0 III	1088,1776,6007,8040
		UKST 284 # 18		19 52 11	−44 42.4	1	14.38		1.03				3		1584
188537	+45 03001			19 52 12	+45 20.3	1	7.67		1.23		1.08		2	K0	1601
		UKST 284 # 23		19 52 12	−45 09.6	1	15.34		0.63				1		1584
188385	+6 04351	HR 7598	⋆ AB	19 52 14	+07 00.5	2	6.14	.005	0.03	.000	0.04	.000	7	A2 V	15,1256
226428	+34 03784			19 52 14	+34 24.6	1	9.23		0.05		0.04		1	B9	1192
188484	+33 03642			19 52 15	+33 38.9	1	6.66		1.10				1	K0	8097
		G 208 - 44	⋆ V	19 52 16	+44 17.3	1	13.41		1.90				3		538
188665	+57 02084	HR 7608		19 52 16	+57 23.5	4	5.14	.010	-0.14	.011	-0.55	.000	18	B5 V	154,1203,1363,3016
	−44 13617			19 52 16	−44 45.3	1	11.33		1.19				2		1584
		G 208 - 45		19 52 17	+44 17.3	1	13.99		1.98				3		538
356214	+12 04132			19 52 19	+12 35.3	1	10.24		0.28				1	A2	800
188503	+30 03804			19 52 21	+30 34.3	1	8.54		-0.09		-0.49		2	B8 III	401
188610	+47 02948			19 52 21	+48 09.5	1	8.10		0.10		0.12		2	A0	262
188504	+29 03795	IDS19504N2946	AB	19 52 23	+29 53.6	1	8.21		0.12				3	B7 III	1118
188485	+23 03829	HR 7601		19 52 24	+24 11.2	3	5.53	.019	-0.01	.005	-0.15	.019	7	A0 III	1063,1118,3023
188160	−50 12730			19 52 27	−50 04.9	1	8.82		0.64				4	G3 V	2033
188464	+16 04059			19 52 30	+17 12.8	1	7.90		0.00		-0.38		2	B9	634
356213	+12 04134			19 52 31	+12 38.0	1	10.94		0.35				6	A2	800
		V1165 Aql	A	19 52 31	+12 38.0	1	10.75		1.16				9		800
332996	+28 03524	LS II +28 015		19 52 31	+28 59.0	1	10.17		0.87		0.43		3	A0 II	824
		LS II +24 023		19 52 32	+24 47.4	1	11.07		0.86		0.27		1	B8 III	1215
188427	+3 04191	G 23 - 17		19 52 33	+03 56.0	3	9.34	.075	0.98	.000	0.81	.037	6	K2	196,333,1620,3008
188246	−44 13624			19 52 34	−44 08.7	5	7.17	.004	-0.07	.004	-0.21	.009	30	B8/9 V	278,474,1075,1770,2011
188507	+22 03854			19 52 35	+22 17.8	2	6.73	.015	1.51	.020	1.81	.025	3	K4 II-III	1080,8100
		CS 22964 # 106		19 52 35	−41 23.0	2	15.06	.000	0.37	.000	0.09	.011	2		966,1736

Table 1   1079

HD	DM	Other Id	N Rem	α_1950	δ_1950	S	V	σ_V	B–V	σ_B–V	U–B	σ_U–B	n	Spectrum	References
		UKST 284 # 13		19 52 35	−44 57.8	1	13.63		1.04				1		1584
188525	+26 03717			19 52 36	+27 11.9	1	8.20		0.05				3	A0	1118
		LSE # 21		19 52 36	−23 22.	1	11.78		-0.25		-1.18		1		1650
188405	−7 05102	HR 7599	★ AB	19 52 39	−06 52.0	1	6.50		0.39		0.03		8	F2 V	1088
339174				19 52 41	+26 47.7	1	10.03		0.14		-0.07		1	A0	363
332967	+29 03797			19 52 41	+29 32.5	1	9.25		2.03				2	M2	369
188793	+59 02137	HR 7611	★ A	19 52 42	+59 34.6	1	6.06		0.02		0.02		2	A3 V	401
		CS 22873 # 52		19 52 42	−59 37.4	1	14.03		0.42		0.06		1		1736
	+23 03834	LS II +23 052		19 52 43	+23 20.0	1	9.66		0.86		-0.24		1	B2 V	1032
188510	+10 04091	G 143 - 17		19 52 47	+10 36.2	5	8.83	.012	0.59	.014	-0.12	.017	9	G5 V wle	516,792,1064,1658,3077
226486	+37 03671			19 52 47	+37 54.3	1	10.29		0.72		0.24		2	G8 III	1395
188376	−26 14637	HR 7597		19 52 47	−26 26.0	7	4.70	.007	0.76	.004	0.31	.005	20	G5 V	15,58,1075,2012,3026*
		UKST 284 # 21		19 52 47	−44 23.5	1	15.24		0.65				1		1584
	+20 04332			19 52 48	+21 04.2	1	10.78		1.56				1	M2	369
188527	+11 04048			19 52 50	+11 49.6	1	7.80		-0.20		-0.78		1	B8	963
188543	+15 03985			19 52 50	+15 18.9	1	8.10		1.53		1.63		2	K2	1648
	+23 03835	LS II +23 054		19 52 50	+23 20.6	1	9.70		0.41		-0.32		1	B2 V:	1032
188161	−58 07683	HR 7586		19 52 50	−58 03.5	2	6.54	.005	1.53	.000	1.79		5	K4 II	2007,3005
188512	+6 04357	HR 7602	★ AB	19 52 51	+06 16.8	28	3.72	.011	0.85	.006	0.49	.010	215	G8 IV	1,3,15,26,30,851*
		LP 814 - 1		19 52 51	−20 38.7	1	14.70		0.20		-0.60		6		3060
		VES 91		19 52 54	+31 51.1	1	12.76		0.84		-0.02		1		363
		Smethells 46		19 52 54	−59 25.	1	10.90		1.31				1		1494
188612	+29 03799			19 52 55	+30 00.9	1	8.22		-0.02				3	B6 V	1118
188162	−59 07550	HR 7587		19 52 55	−59 02.1	1	5.26		-0.01				4	B9.5IV	2007
226502	+34 03787			19 52 56	+34 52.4	1	8.57		0.70		0.26		2	F8	394
188650	+36 03766	HR 7606		19 52 59	+36 51.8	2	5.76	.000	0.75	.000	0.35		6	G1 Ib-II	1118,3016
226503				19 53 00	+34 37.	1	11.35		0.17		0.05		2	A0	394
	+47 02949			19 53 00	+47 46.9	2	7.84	.005	1.59	.015	1.85	.060	4	M1 III	262,1733
		Smethells 47		19 53 00	−59 25.	1	10.32		1.20				1		1494
188631	+26 03721			19 53 01	+26 37.3	1	8.51		-0.07		-0.33		2	A0	634
		CS 22964 # 102		19 53 03	−41 58.8	2	14.10	.012	-0.12	.000	-0.63	.002	2		966,1736
188651	+29 03802	HR 7607	★ A	19 53 07	+30 03.7	2	6.59	.038	0.00	.030	-0.54	.064	3	B6 V	1192,8072
188651	+29 03802	HR 7607	★ AB	19 53 07	+30 03.7	4	6.58	.005	-0.10	.014	-0.47	.004	9	B6 V +A5 V	1063,1079,1118,3016
188651	+29 03802	IDS19511N2956	B	19 53 07	+30 03.7	1	9.55		0.03		-0.13		1	A5 V	1192,8072
188651	+29 03802	IDS19511N2956	C	19 53 07	+30 03.7	1	10.22		0.74		0.04		2		8072
		UKST 284 # 12		19 53 12	−44 43.4	1	13.45		0.52				4		1584
		L 80 - 56		19 53 12	−71 31.	3	15.10	.026	-0.04	.015	-0.89	.037	4		132,832,3065
226540	+33 03652	LS II +33 011		19 53 17	+33 21.8	1	9.02		0.86		0.64		3	F2 Ib	247
188753	+41 03535	G 208 - 47	★ AB	19 53 17	+41 44.1	2	7.41	.015	0.80	.005	0.42	.000	5	G8 V +K0 V	1080,3030
	+47 02950			19 53 18	+47 16.8	1	8.90		0.33		0.14		2	K0	262
		CS 22964 # 112		19 53 19	−40 36.1	1	15.40		-0.28		-1.04		1		1736
188318	−54 09538	IDS19494S5430	AB	19 53 20	−54 22.0	1	8.34		0.36				4	F0 IV	2012
	+47 02951			19 53 21	+47 37.2	1	6.70		1.08		0.92		2	K0 III	262
		G 125 - 45		19 53 22	+30 22.2	2	12.92	.037	1.54	.009	1.14		4		203,1759
188819	+46 02804			19 53 23	+47 00.8	1	8.63		0.50		0.04		2	F8	262
		AJ78,401 T4# 35		19 53 23	+47 04.7	1	11.27		1.51		1.70		2	K3 III	262
187915	−74 01854			19 53 23	−74 15.9	2	7.52	.008	0.43	.002	0.09		6	F3 III/IV	278,2012
		Numerova 277		19 53 24	+34 49.8	1	11.57		0.44		0.05		1		394
188392	−46 13302			19 53 24	−45 57.0	3	8.81	.001	0.57	.004	0.13	.027	16	F6/7 IV	977,1075,2011
188474	−31 17179			19 53 26	−31 28.2	1	8.43		0.99				2	K3/4 V	1619
		UKST 284 # 24		19 53 27	−44 40.1	1	15.38		0.87				3		1584
332938	+30 03808	LS II +30 007		19 53 32	+30 22.7	1	9.15		0.50		0.14		2	B9 III	8100
188097	−69 03072	HR 7579		19 53 33	−69 17.9	4	5.73	.011	0.23	.015	0.14	.019	20	A1mA5-F2	15,355,1075,2038
	+25 04025			19 53 35	+25 29.8	1	9.82		1.97				1	M3	369
		UKST 284 # 20		19 53 35	−44 54.5	1	14.68		0.63				3		1584
339212	+25 04026			19 53 36	+25 30.1	1	9.50		0.20				1	A0	369
188303	−59 07552			19 53 37	−59 39.6	1	7.92		0.96		0.72		5	G8/K0 III	1673
		VES 92		19 53 39	+21 54.9	1	11.80		1.88		1.40		1		363
188653	+3 04202	G 92 - 39		19 53 40	+04 10.4	2	9.44	.010	0.71	.010	0.25	.035	2	G5	333,1620,1658
188854	+46 02807			19 53 40	+46 32.0	1	7.59		0.31		0.15		3	A7 p:	8071
188715	+15 03988			19 53 41	+15 49.9	1	9.26		0.01		-0.18		2	A0	634
188727	+16 04067	HR 7609, S Sge		19 53 45	+16 30.1	4	5.32	.023	0.65	.019	0.38	.042	5	F6 Ib	71,851,1772,6007
		AJ78,401 T4# 18		19 53 45	+46 58.6	1	14.16		0.87		0.51		2		262
		AJ78,401 T4# 5		19 53 46	+47 02.7	1	11.80		1.15		0.90		2	G8 III	262
188726	+17 04149	IDS19515N1737	AB	19 53 47	+17 45.2	1	8.35		-0.07		-0.45		1	B9	963
226580				19 53 48	+34 48.	1	10.81		0.08		0.03		2	A0	394
188164	−69 03073	HR 7588		19 53 48	−68 53.8	2	6.38	.005	0.16	.000	0.08	.005	7	A0 V+F5 IV	15,1075
		AJ78,401 T4# 15		19 53 49	+47 00.0	1	15.08		1.29		0.00		2		262
188728	+11 04055	HR 7610		19 53 52	+11 17.4	3	5.28	.000	0.02	.031	-0.02	.020	8	A1 IV	985,1063,3050
188603	−27 14399	HR 7604		19 53 53	−27 18.2	6	4.54	.010	1.47	.010	1.55	.009	79	K3 II	15,58,1075,2012,3047,8015
226594	+32 03630			19 53 54	+33 10.3	1	9.31		1.81				2	K8	369
226593				19 53 54	+34 29.	1	11.17		0.18		0.16		2	A0	394
226591	+34 03794			19 53 54	+34 51.3	1	9.49		0.25		0.07		2	A2	394
		Steph 1735		19 53 54	−10 18.7	1	11.04		1.65		1.96		3	M0	1746
188875	+39 03959			19 53 56	+40 02.3	2	6.57	.005	1.48	.015	1.63	.019	3	K2	1601,1716
		G 92 - 40		19 53 56	−01 10.2	5	13.69	.013	0.27	.017	-0.60	.012	13		316,1281,1620,1705,3060
		LS II +26 011		19 53 57	+26 38.8	1	11.38		0.96		-0.14		2		824
		CS 22964 # 125		19 53 57	−39 09.4	2	14.31	.000	0.27	.000	0.17	.009	2		966,1736
		Numerova 385		19 54 00	+34 44.7	1	11.85		0.26		0.15		1		394
		AJ78,401 T4# 19		19 54 00	+46 59.8	1	12.94		0.70		0.08		2		262

HD	DM	Other Id	N Rem	α_{1950}	δ_{1950}	S	V	σ_V	B–V	σ_{B-V}	U–B	σ_{U-B}	n	Spectrum	References
188891 +40 03948		LS II +40 001	⋆ A	19 54 01	+40 15.5	1	7.30		-0.01		-0.74		1	B1 V	1192
188892 +38 03817		HR 7613		19 54 04	+38 21.2	7	4.94	.008	-0.09	.005	-0.52	.010	44	B5 IV	15,154,1192,1203,1363*
		CS 22964 # 104		19 54 04	-41 29.0	2	15.31	.000	0.71	.000	0.20	.015	2		966,1736
		LS II +28 016		19 54 05	+28 52.4	1	11.76		0.80		-0.14		2		824
188734 -3 04751		IDS19515S0316	AB	19 54 06	-03 08.0	1	8.46		1.32		0.99		2	K0	1375
		CS 22964 # 115		19 54 06	-40 08.1	1	14.94		0.43		-0.17		1		1736
188554 -44 13634				19 54 06	-44 26.4	1	10.01		0.48				2	F3 IV/V	1584
188777 +7 04300				19 54 08	+08 07.3	1	7.47		-0.06		-0.51		1	A0	963
		UKST 284 # 25		19 54 10	-44 34.8	1	15.51		0.62				1		1584
		VES 93		19 54 11	+24 58.3	1	11.65		0.40		0.39		1		363
188555 -46 13306				19 54 11	-45 57.7	1	8.60		0.40				4	F2 IV/V	1075
188876 +32 03634				19 54 12	+32 56.4	2	7.27	.020	-0.07	.010	-0.45	.005	5	A0	401,3016
226630				19 54 12	+39 00.	1	9.45		1.49		0.57		3		1737
		Numerova 436		19 54 13	+34 27.6	1	11.36		0.42		0.16		1		394
		V724 Aql		19 54 15	+00 57.8	1	11.10		0.49		0.16		1	G0	1399
		WLS 2000 5 # 12		19 54 15	+02 32.1	1	11.03		0.31		0.20		2		1375
		LP 814 - 50		19 54 16	-19 51.3	1	12.63		1.22		1.17		1		1696
		UKST 284 # 10		19 54 16	-44 20.1	1	12.91		0.56				5		1584
226642 +34 03796				19 54 18	+35 14.9	1	9.99		0.35		0.20		2	A5	394
190224 +82 00598				19 54 19	+82 19.4	1	7.78		1.07		0.90		2	K0	1733
226643 +34 03797				19 54 20	+35 06.5	2	9.73	.005	0.01	.020	-0.29	.005	4	A0	324,394
189037 +52 02572		HR 7619	⋆ AB	19 54 20	+52 18.3	4	4.92	.004	0.12	.006	0.06	.005	9	A2 IV-V+F4V	15,1363,3023,8015
188605 -44 13639				19 54 20	-44 34.7	1	9.76		0.49				2	F6 V	1584
188642 -38 13776		HR 7605		19 54 21	-38 11.5	2	6.54	.005	0.40	.005			7	F3 V	15,2012
345329				19 54 22	+22 11.5	1	11.24		0.51		-0.40		1	G0	363
189013 +46 02812				19 54 23	+46 57.4	2	6.88	.000	0.14	.000	0.10	.020	4	A2	262,401
189127 +57 02092		HR 7626		19 54 23	+58 07.1	1	6.09		1.02				2	G9 III	71
331549 +30 03813				19 54 25	+31 14.1	1	9.31		-0.08		-0.26		1	B9 V	963
188947 +34 03798		HR 7615	⋆ AB	19 54 26	+34 57.0	19	3.89	.008	1.03	.007	0.88	.007	261	K0 III	15,61,71,247,369,985*
		G 92 - 41		19 54 28	+01 58.9	2	13.18	.000	1.30	.002	0.96	.043	7		316,3062
		Cyg sq 4 # 119		19 54 28	+35 01.7	1	12.52		0.67		0.30		1	G0 V	324
189065 +48 02984		IDS19530N4856	AB	19 54 28	+49 03.7	1	8.09		-0.08		-0.43		2	B9	401
		G 23 - 18		19 54 29	+01 58.8	1	14.41		1.40		1.03		4		316
		Cyg sq 4 # 106		19 54 30	+34 53.8	1	11.95		0.44		0.28		1	B6 V	324
226664				19 54 30	+35 18.	1	9.97		0.39		0.07		2	F0	394
		G 230 - 18		19 54 30	+51 08.2	1	11.87		1.49				1	M2	1759
188877 +16 04073				19 54 31	+16 16.5	1	8.57		-0.03		-0.45		2	A0	634
188928 +28 03539				19 54 32	+28 29.5	1	8.27		0.11				3	A3 V	1118
188948 +32 03637				19 54 32	+32 59.1	1	8.41		-0.02		-0.31		1	B9	1192
188969 +35 03872				19 54 33	+35 39.3	1	7.58		0.83		0.45		2	G5	1601
188807 -12 05594		IDS19518S1250		19 54 33	-12 41.7	4	9.29	.011	1.32	.014	1.20	.009	15	K4 V	158,1197,1705,3072
		UKST 284 # 16		19 54 33	-45 04.8	1	14.21		0.82				2		1584
		UU Aql		19 54 35	-09 27.4	1	16.07		0.22		-0.93		1		1471
188559 -55 09267				19 54 35	-55 04.2	1	8.60		1.08		0.88		2	K3 V	3072
331511 +31 03851		IDS19526N3140	AB	19 54 36	+31 48.3	1	8.64		1.39		1.56		10	A2	1655
226678				19 54 36	+34 59.	2	10.62	.019	0.67	.015	0.06	.005	3	G0	324,394
		ApJ144,259 # 66		19 54 36	-21 45.	1	17.39		0.32		-0.87		4		1360
		UKST 284 # 22		19 54 37	-44 51.0	1	15.32		0.60		-0.02		2		1584
226679 +34 03799				19 54 38	+34 49.9	1	9.87		0.24		0.19		2	A2	394
		UKST 284 # 11		19 54 38	-44 50.6	1	13.30		0.80		0.37		1		1584
226680 +34 03800				19 54 39	+34 44.7	1	10.06		0.33		0.00		2	F0	394
		UKST 284 # 14		19 54 39	-44 49.5	1	13.63		0.72		0.13		1		1584
333125 +29 03814		V1356 Cyg	⋆	19 54 40	+29 51.7	1	10.18		0.45		-0.51		2	B0 V	1012
189208 +59 02145				19 54 40	+59 36.4	1	8.89		0.27		0.10		2	F5	1502
226677				19 54 42	+35 02.	1	10.96		1.05		0.64		1	A2	324
		UKST 284 # 9		19 54 42	-44 50.6	1	12.84		0.71		0.21		7		1584
		Numerova 551		19 54 43	+35 08.7	2	11.20	.005	0.19	.025	0.16	.030	2		324,394
188147 -74 01855				19 54 43	-74 30.1	1	9.79		0.09				4	A1 V	2012
189016 +35 03876				19 54 44	+36 11.1	1	7.80		-0.05		-0.51		1	B9	1192
226687 +35 03874				19 54 46	+35 26.0	1	8.89		0.29		0.14		2	A7	394
188994 +29 03815				19 54 48	+29 23.9	3	8.02	.005	0.43	.011	-0.05	.018	4	F6 V wlm	979,1620,3016
188992 +32 03641				19 54 48	+32 39.3	1	8.29		-0.07		-0.38		2	A0	401
		UKST 284 # 15		19 54 48	-44 18.2	1	13.94		0.63		0.12		6		1584
188971 +20 04351		HR 7616		19 54 49	+20 51.8	2	6.61		0.07	.000	0.03	.068	5	A2 IV	1063,1733
		CS 22964 # 118		19 54 49	-39 57.0	2	11.89	.000	0.09	.000	0.06	.009	2		966,1736
188914 +4 04286		IDS19524N0457	AB	19 54 51	+05 04.7	2	8.53	.005	0.56	.005	0.05		4	F8	173,3016
188228 -73 02086		HR 7590		19 54 51	-73 02.7	8	3.95	.005	-0.03	.004	-0.05	.008	32	A0 V	15,278,1034,1075,1075*
		Cyg sq 4 # 105		19 54 52	+34 52.7	1	12.55		0.70		0.11		1		324
		CS 22964 # 124		19 54 52	-39 05.9	2	14.33	.000	0.95	.000	0.00	.015	2		966,1736
188972 +19 04218				19 54 53	+19 25.8	1	8.51		-0.09		-0.76		1	B5	963
189066 +35 03878		HR 7620		19 54 53	+36 07.0	4	6.02	.006	-0.15	.012	-0.62	.020	7	B5 IV	154,1079,1192,3016
345327 +21 03984				19 54 54	+22 13.5	1	9.75		1.82				1		369
		Cyg sq 4 # 123		19 54 54	+35 05.7	1	12.75		0.53		-0.02		2	F6 V	324
	-21 05582			19 54 54	-21 38.5	1	10.81		1.12		0.94		1	K3 V	3072
		Cyg sq 4 # 107		19 54 55	+34 54.0	1	12.85		0.52		-0.05		1	F8 V	324
226724 +41 03542			V	19 54 58	+41 44.3	1	9.00		1.73				2	M0	369
189276 +58 02013		HR 7633		19 54 58	+58 42.7	4	4.98	.017	1.58	.011	1.90	.046	7	K5 II-III	1080,1363,3016,4001,8100
188229 -74 01859				19 54 58	-74 08.9	3	6.58	.003	1.04	.001	0.93	.005	10	K1 III	278,1075,2012
226709 +34 03802				19 54 59	+34 51.3	1	10.36		0.84		0.39		1	K0	324
		Cyg sq 4 # 126		19 54 59	+35 07.2	1	12.70		0.67		0.03		2	B6 III	324

Table 1   1081

HD	DM	Other Id	N	Rem	α_{1950}	δ_{1950}	S	V	σ_V	B–V	σ_{B-V}	U–B	σ_{U-B}	n	Spectrum	References
339181	+26 03739				19 55 00	+26 25.8	1	8.87		0.30		0.17		1	A2	851
	+30 03817				19 55 00	+30 17.5	1	9.84		1.96				2	M2	369
226722					19 55 00	+34 37.	1	11.03		0.38		0.15		2	F0	394
		Cyg sq 4 # 122			19 55 00	+35 04.4	1	12.94		0.80		0.28		1	G8 V	324
189085	+35 03881				19 55 01	+35 24.1	1	8.20		0.34		0.12		3	Am	394
		Cyg sq 4 # 117			19 55 03	+35 00.6	1	11.31		1.80		2.14				324
188934	−0 03883				19 55 04	+00 06.4	4	9.35	.006	2.02	.027	2.09	.027	32	R4	989,1238,1729,8005
		Cyg sq 4 # 101			19 55 04	+34 46.1	1	12.49		0.60		0.11				324
		G 92 - 43			19 55 04	−03 35.6	1	12.61		1.49		1.08		1		1620
189040	+25 04034				19 55 06	+25 21.2	2	7.75	.000	0.07	.005	0.06	.000	5	A0	833,1625
226721					19 55 06	+34 48.	1	10.79		0.33		0.20		2	A3	394
		Cyg sq 4 # 103			19 55 06	+34 51.7	1	13.26		0.50		0.06		1		324
		Cyg sq 4 # 115			19 55 07	+34 59.7	1	11.39		1.05		0.70		1		324
188899	−15 05516	HR 7614			19 55 07	−15 37.5	2	5.02	.000	0.06	.010	0.07		6	A2 V	2007,3023
189086	+30 03820				19 55 09	+30 38.5	3	6.93	.005	-0.03	.008	-0.12	.000	8	A0	401,1118,3016
226731	+34 03803				19 55 09	+34 51.3	1	10.24		1.34		1.27		1	M0	324
333119	+29 03818				19 55 10	+29 47.6	1	8.69		0.59		-0.02		1	G0 V	3026
356404	+10 04109	V733 Aql			19 55 11	+10 54.5	2	9.73	.012	0.79	.011			2	G0 Ib	851,1772
		Numerova 637			19 55 11	+35 18.6	1	10.88		1.14		0.97		1	F1 V	324
189067	+23 03847				19 55 12	+23 57.3	2	7.21	.005	0.60	.000	0.10	.020	5	G2 V	1733,3016
189107	+30 03826				19 55 12	+30 43.3	1	7.48		-0.06		-0.20		2	B8 V	401
188998	+10 04110				19 55 13	+10 23.3	1	9.55		0.55				1	K0	851
356416	+10 04111				19 55 13	+10 46.4	1	10.41		0.64				1	G0	851
189087	+29 03820	G 125 - 48			19 55 13	+29 41.1	2	7.91	.015	0.80	.005	0.38		7	K1 V	1197,3016
		Cyg sq 4 # 231			19 55 13	+35 04.0	1	12.53		0.87		0.42		1		324
189160	+43 03422	IDS19536N4400		A	19 55 13	+44 08.2	1	7.92		-0.03		-0.42		2	A0 p:	401
189160	+43 03422	IDS19536N4400		B	19 55 13	+44 08.2	1	10.98		0.39		-0.01		1		401
189296	+56 02331	HR 7634			19 55 13	+56 33.1	1	6.16		0.09		0.07		2	A4 Vn	1733
	−74 01860				19 55 14	−74 41.8	1	10.21		0.54				4	F8	2012
226743	+34 03804				19 55 15	+35 02.2	2	9.69	.005	0.21	.029	0.11	.000	3	A2	324,394
		Cyg sq 4 # 214			19 55 16	+34 52.6	1	12.32		0.85		0.25		1		324
188813	−42 14584	RU Sgr			19 55 17	−41 59.1	1	12.10		1.53		0.33		1	M4e	975
189159	+43 03423				19 55 18	+44 12.4	1	9.16		0.12		-0.03		1	B9	401
	−44 13645				19 55 18	−44 25.9	1	11.55		0.49				3		1584
226744	+34 03805				19 55 19	+34 32.8	1	10.37		0.15		0.03		2	A0	394
189148	+34 03806				19 55 20	+34 53.8	3	7.27	.007	0.46	.012	0.00	.021	5	F6 V	247,324,394
		AS 373		⋆ V	19 55 20	+39 41.7	3	11.24	.025	0.01	.045	-0.88	.051	3	M6.5:	1564,1591,1753
188814	−44 13646				19 55 21	−44 41.0	1	9.90		1.30				2	K1 (III)	1584
		Cyg sq 4 # 225			19 55 22	+34 58.1	1	10.80		1.58		1.60		1		324
189253	+50 02930	HR 7632			19 55 23	+50 46.0	1	6.43		-0.01		-0.06		2	A1 V	401
339279	+26 03741	X Vul			19 55 24	+26 25.3	3	8.48	.018	1.28	.035	0.92	.086	3	F8 Ia	851,1772,6011
188815	−46 13313				19 55 24	−46 13.5	6	7.47	.002	0.48	.003	-0.09	.004	54	F5 V	278,863,977,977,1075,2011
		Numerova 684			19 55 26	+34 53.1	1	11.73		0.62		0.01		1		324
351123	+16 04079				19 55 27	+17 14.4	1	8.58		0.07		-0.48		2	A0	379
189089	+16 04080				19 55 28	+17 00.6	1	7.77		-0.03		-0.29		2	B9	634
		UKST 284 # 19			19 55 28	−44 29.2	1	14.46		0.64				2		1584
189090	+16 04081	HR 7622			19 55 29	+16 39.2	4	5.53	.004	-0.05	.007	-0.17	.024	8	B9 III	851,985,1063,3023,6007
189178	+39 03968	HR 7628		⋆ A	19 55 30	+40 13.9	3	5.46	.015	-0.10	.012	-0.53	.006	5	B5 V	154,1192,3016
		Numerova 701			19 55 31	+34 54.0	2	12.14	.020	0.50	.015	0.48	.010	2	A0 V	324,394
188322	−75 01561				19 55 31	−74 54.1	1	7.51		1.67				4	M0 III	2012
		Cyg sq 4 # 230			19 55 32	+35 03.2	1	12.55		0.55		0.00		2	F9 V	324
188584	−67 03695	HR 7603			19 55 32	−67 05.0	4	5.75	.008	1.03	.005	0.80	.005	15	K0 III	15,1075,2038,3077
189044	+4 04292	IDS19531N0440		AB	19 55 34	+04 48.0	2	8.62	.038	0.74	.014	0.33		4	G3 V +G4 V	1381,3030
		CS 22873 # 106			19 55 35	−58 55.2	1	14.19		0.63		0.09		1		1736
		Numerova 710			19 55 36	+32 24.0	1	11.78		0.32		0.27		1		394
189256	+43 03425	AX Cyg		⋆ A	19 55 36	+44 07.6	1	7.85		3.43		1.69		1	C6 II	401
351118	+17 04166				19 55 39	+17 30.1	1	9.23		0.95		0.70		2	A0	379
		Numerova 732			19 55 42	+34 49.5	1	12.12		0.62		0.07		1		324
		Cyg sq 4 # 234			19 55 42	+35 06.0	1	12.46		0.66		0.43		2	F0 V	324
188903	−42 14588				19 55 44	−41 58.1	1	8.27		0.57				4	F8 V	2012
		G 23 - 19			19 55 45	+01 54.6	1	11.95		1.55		1.33		1		333,1620
333162		LS II +28 019			19 55 45	+28 49.0	2	10.44	.008	0.79	.012	-0.24	.008	4		824,1318
		VES 133			19 55 45	+36 09.4	1	10.76		1.08		0.92		1		426
351092	+17 04168				19 55 46	+17 44.0	1	9.57		-0.03		-0.52		1	B8	963
189371	+54 02255	V548 Cyg			19 55 46	+54 39.8	1	8.57		0.09		-0.03		3	A1 V	1768
226782					19 55 48	+34 26.	1	10.64		0.22		0.14		2	A3	394
		Cyg sq 4 # 216			19 55 48	+34 53.2	1	12.92		0.43		0.09		1		324
188981	−30 17525	HR 7617			19 55 48	−30 40.4	1	6.28		1.05				4	K1 III	2032
333172	+27 03570	LS II +28 020			19 55 51	+28 11.6	1	10.42		0.56		-0.40		3		824
189213	+28 03546				19 55 53	+28 44.3	1	7.24		0.16		0.11		2	A7 V	401
226803					19 55 54	+35 04.	1	10.68		0.18		0.06		2	A0	394
226804	+34 03809				19 55 55	+34 53.3	2	9.95	.010	0.23	.015	0.12	.000	4	A3	324,394
189114	−3 04757				19 55 55	−03 41.4	1	6.86		1.92		2.16		3	K5	1657
189005	−26 14682	HR 7618			19 55 55	−26 20.0	6	4.84	.009	0.90	.006	0.55	.012	15	G8 II/III	15,58,369,1075,2012,8015
	−18 05550				19 55 57	−18 20.2	2	9.26	.017	0.91	.013	0.20		6	G0	742,1594
	−46 13316				19 55 57	−46 20.0	2	10.31	.001	0.42	.002	-0.03		11	F8	863,1075
	+21 03992				19 56 01	+22 06.6	1	9.86		2.01				1	M2	369
		Numerova 798			19 56 01	+34 50.2	1	11.41		0.43		0.07		1		324
		WLS 2000 40 # 10			19 56 01	+39 58.7	1	13.15		0.68		0.36		2		1375
189301	+37 03698				19 56 03	+37 59.1	1	7.26		1.55		1.82		2	K3 III	247,8100

HD	DM	Other Id	N Rem	α_{1950}	δ_{1950}	S	V	σ_V	B–V	σ_{B-V}	U–B	σ_{U-B}	n	Spectrum	References
		Cyg sq 4 # 235		19 56 06	+35 06.2	1	12.83		0.68		-0.06		3	B3 V	324
189315	+35 03889			19 56 08	+36 11.5	1	7.31		-0.04		-0.13		1	B9	1192
		Cyg sq 4 # 329		19 56 09	+35 00.9	1	13.58		0.68		0.40		2	A7 V	324
226836	+35 03888			19 56 09	+35 25.4	1	10.96		0.84		0.45		1	K0	324
		Cyg sq 4 # 350		19 56 10	+35 15.0	1	13.54		0.60		0.16		2	F3 V	324
339268	+26 03746			19 56 11	+26 34.6	1	9.55		0.16		0.08		1	A0	851
		Numerova 838		19 56 12	+35 20.2	1	12.01		0.48		0.00		2	F3 V	324
		CS 22964 # 105		19 56 12	-41 20.1	2	14.72	.000	0.07	.000	0.15	.010	3		966,1736
189334	+34 03812			19 56 14	+35 14.6	2	9.08	.005	0.02	.019	-0.08	.005	3	A2	324,394
189377	+41 03549	HR 7638	⋆ ABC	19 56 15	+42 07.5	1	6.46		0.07		0.06		3	A3 V	1733
		Ros 3 - 6		19 56 17	+20 20.1	1	12.81		0.39		0.33		1		1735
		Cyg sq 4 # 332		19 56 17	+35 02.2	1	13.07		0.48		0.06		2	F0 V	324
226847				19 56 18	+34 31.	1	10.90		0.11		-0.02		4	B9	394
		JL 36		19 56 18	-72 06.	1	12.93		-0.16		-0.86		2	sdB	132
189317	+28 03552			19 56 19	+28 28.5	1	7.85		0.44		-0.03		3	F6 V	3026
350989				19 56 20	+20 25.5	1	10.92		0.18		-0.21		1	B9	363
333188				19 56 20	+27 52.9	1	11.27		0.72		-0.08		3	B8	824
189335	+34 03813			19 56 20	+34 17.2	2	7.72	.051	-0.11	.034	-0.44	.030	3	B9	401,1192
		Cyg sq 4 # 361		19 56 20	+35 23.8	1	13.10		0.61		0.27		1	F2 V	324
226850	+33 03669			19 56 21	+33 55.4	1	8.75		1.68				2	M0	369
		UKST 284 # 26		19 56 21	-44 49.3	1	15.61		1.04				2		1584
189316	+28 03553	IDS19543N2843	A	19 56 22	+28 51.1	1	7.67		0.00		-0.18		2	B9 V	401
356388	+11 04068	Ring Aql # 18		19 56 26	+11 28.5	1	9.78		1.21		1.13		2	G5	4
		Cyg sq 4 # 352		19 56 26	+35 15.2	1	12.68		1.08		0.84		2		324
		UKST 284 # 27		19 56 26	-44 39.2	1	16.23		0.71				1		1584
		LS II +27 030		19 56 27	+27 58.6	1	11.82		0.71		-0.04		3		824
189350	+31 03864		V	19 56 27	+31 20.4	1	7.89		1.82				2	M1	369
		Cyg sq 4 # 339		19 56 27	+35 07.2	1	13.23		1.60		1.07		1		324
		UKST 284 # 17		19 56 27	-44 59.6	1	14.24		0.57				2		1584
226867	+34 03814			19 56 28	+35 09.4	2	10.64	.010	0.39	.024	-0.03	.083	3	A7	324,394
		Cyg sq 4 # 353		19 56 28	+35 17.1	1	12.57		1.49		1.55		1		324
189282	+11 04070	Ring Aql # 17		19 56 29	+11 25.7	1	7.36		0.28		0.03		3	A7 V	4
		Ros 3 - 75	AB	19 56 29	+20 23.6	1	11.78		0.31				1	B7 IVnnp	1735
339248	+27 03572			19 56 29	+27 22.4	1	8.09		1.54				2	K7	1655
226868	+34 03815	V1357 Cyg		19 56 29	+35 03.9	3	8.88	.014	0.81	.011	-0.29	.016	8	B0 Ib	180,324,1012
	+34 03816			19 56 29	+35 04.9	2	10.00	.010	0.61	.015	0.06	.005	3		324,394
189103	-35 13831	HR 7623		19 56 29	-35 24.8	5	4.37	.005	-0.15	.000	-0.66	.011	189	B2.5IV	3,15,26,1075,2012
		UKST 284 # 29		19 56 31	-44 21.1	1	17.20		0.79				1		1584
189057	-44 13660			19 56 31	-44 30.5	1	9.88		0.71				2	G3 V	1584
189319	+19 04229	HR 7635		19 56 32	+19 21.3	7	3.48	.028	1.57	.023	1.93	.005	15	K5 III	15,263,369,1080,1363*
189394	+34 03817			19 56 32	+34 22.6	1	8.45		-0.08		-0.41		2	B9p	394
		Ros 3 - 83		19 56 34	+20 23.8	1	14.14		0.40		0.30		1		1735
188480	-75 01564			19 56 34	-75 29.3	2	8.22	.006	0.54	.004	0.08		6	F8/G0 V	278,2012
		Ros 3 - 73		19 56 36	+20 21.2	1	11.38		0.23		-0.21		1	B7 IIInn	1735
		Ros 3 - 82		19 56 36	+20 23.3	1	13.92		0.66		0.78		1		1735
		DG Vul		19 56 36	+27 32.2	1	11.13		1.87				1		1772
331626				19 56 36	+31 19.	1	12.46		-0.08				1	B9	725
		Ros 3 - 81	AB	19 56 37	+20 22.8	1	13.73		0.39		0.33		1		1735
		Ros 3 - 78		19 56 37	+20 23.1	1	13.34		0.42		0.31		1		1735
		Ros 3 - 84		19 56 37	+20 23.4	1	14.26		1.32		1.01		1		1735
189351	+21 03997			19 56 37	+21 15.5	1	8.96		0.01		-0.41		1	A0	963
189378	+32 03651	IDS19547N3300	AB	19 56 37	+33 08.4	1	7.24		0.41		-0.02		2	F5 IV	247,8100
		Numerova 916		19 56 37	+35 01.5	1	11.69		0.60		0.40		1		394
189118	-35 13832	HR 7624	⋆ A	19 56 37	-34 50.1	2	5.30	.000	0.17	.000	0.06		7	A4/5 IV	2009,8071
356395	+10 04123	Ring Aql # 1		19 56 38	+11 12.0	1	10.28		0.45		0.02		2	F0	4
350991		Ros 3 - 72		19 56 38	+20 21.9	1	10.61		0.23		-0.21		1	B6 IIIn	1735
		Ros 3 - 79		19 56 38	+20 22.3	1	13.51		0.44		0.24		1		1735
		Ros 3 - 77	AB	19 56 38	+20 22.3	1	13.16		0.42		0.29		1		1735
350990	+20 04363	Ros 3 - 71		19 56 38	+20 23.3	1	10.24		0.22		-0.18		1	B6 IIIn	1735
		Cyg sq 4 # 324		19 56 38	+34 56.1	1	12.18		0.52		0.00		1		324
189395	+30 03837	HR 7640		19 56 39	+30 50.8	5	5.49	.005	-0.05	.016	-0.19	.032	9	B9 Vn	1063,1079,1118,1192,3023
		Ros 3 - 74		19 56 40	+20 21.9	1	11.52		0.21		-0.21		1	B6 IVn	1735
		Ros 3 - 80		19 56 41	+20 22.7	1	13.62		0.42		0.30		1		1735
	-44 13661			19 56 41	-44 07.0	1	10.50		0.42				4		2011
226889				19 56 42	+34 49.	2	11.07	.019	0.43	.029	0.20	.005	3	F7	324,394
		Cyg sq 4 # 354		19 56 42	+35 19.0	1	12.33		0.53		0.34		1	A4 V	324
		Numerova 926		19 56 42	+36 09.0	1	11.68		0.57		0.38		1	A7 V	324
226891	+34 03818			19 56 44	+34 25.5	1	9.91		0.69		0.07		2	G0	394
		Cyg sq 4 # 321		19 56 44	+34 55.3	1	11.83		1.36		1.00		1		324
		Cyg sq 4 # 331		19 56 44	+35 01.9	1	12.72		0.67		0.10		2	F8	324
		Cyg sq 4 # 360		19 56 44	+35 23.0	1	12.31		1.18		1.06		1		324
189080	-49 12949	HR 7621		19 56 44	-49 29.3	2	6.18	.006	1.05	.006	0.92		6	K0 III	58,2035
339260	+26 03750			19 56 45	+26 44.5	1	9.72		1.91				2	M0	369
189434	+34 03819			19 56 46	+34 55.9	2	8.80	.010	0.03	.029	-0.19	.005	3	A0	394,8070
189432	+37 03703	HR 7642	⋆ AB	19 56 46	+37 58.1	5	6.33	.016	-0.09	.016	-0.52	.010	11	B5 IV	154,938,1118,1192,3016
		Numerova 950		19 56 47	+34 51.6	1	11.55		0.24		0.21		1		324
226901				19 56 48	+34 32.	1	11.26		0.31		0.12		2	A7	394
189121	-45 13546			19 56 49	-44 49.1	3	9.13	.007	0.44	.006	-0.01	.016	14	F3/5 V	863,1075,2011
189322	+0 04375	HR 7636		19 56 50	+01 14.4	4	6.17	.009	1.13	.003	0.90	.012	10	G8 III	15,1256,1355,6007
189337	+10 04126	Ring Aql # 2		19 56 50	+11 10.1	2	6.50	.000	1.12	.010	0.79	.065	4	K0	4,1733

Table 1   1083

HD	DM	Other Id	N	Rem	α_{1950}	δ_{1950}	S	V	σ_V	B–V	σ_{B-V}	U–B	σ_{U-B}	n	Spectrum	References
		Numerova 967			19 56 50	+34 54.0	1	11.07		0.45		0.06		1		324
189242	−20 05781				19 56 51	−20 20.8	1	9.24		0.92		0.56		2	K2 V	3072
		LS II +33 012			19 56 52	+33 30.9	1	11.27		0.46		−0.58		2		405
		AJ89,1229 # 420			19 56 54	+02 08.	1	11.53		1.80				1	M1	1632
345467	+22 03871				19 56 54	+22 29.5	1	9.06		1.79				1	M2	369
226918	+34 03821				19 56 57	+35 13.3	1	9.60		0.73		1.84		1	K5	324
189195	−38 13802	HR 7629		⋆A	19 56 57	−37 50.4	3	5.96	.008	0.99	.007	0.77	.001	7	G8 III	58,2009,4001
		Cyg sq 4 # 338			19 56 58	+35 07.1	1	13.00		0.59		0.41		1	A4 V	324
189140	−43 13735	HR 7627			19 56 58	−43 10.9	4	6.13	.008	1.64	.000			16	M0 II/III	15,1020,1075,2012
		Cyg sq 4 # 325			19 56 59	+34 57.1	1	13.26		0.92		0.34		1		324
226919	+34 03822				19 56 59	+34 57.6	4	9.49	.013	0.28	.038	0.13	.026	5	A3	324,324,394,8070
189474	+35 03895				19 56 59	+35 21.6	1	7.00		0.02		0.00		2	A0	401
226920					19 57 00	+34 42.	1	10.38		0.50		0.04		2	F2	394
		LP 926 - 56			19 57 00	−32 23.3	1	11.16		0.51		−0.20		2		1696
189410	+22 03872	HR 7641			19 57 01	+22 57.8	1	5.66		0.33		0.02		2	F1 IV	1733
		3U 1956+11			19 57 02	+11 34.3	1	18.67		0.31		−0.56		6		715
188887	−67 03698	HR 7612			19 57 02	−67 04.9	4	5.31	.005	1.22	.000	1.29	.000	16	K2 IV	15,1075,2038,3035
		Cyg sq 4 # 343			19 57 03	+35 09.8	1	12.30		0.76		0.18		1	G3 V	324
356393	+10 04129	Ring Aql # 3			19 57 04	+11 07.0	1	9.57		0.94		0.64		2	G5	4
		Cyg sq 4 # 342			19 57 04	+35 09.4	1	12.62		0.37		0.27		2	B9 V	324
		Numerova 1006			19 57 04	+35 13.8	1	11.07		1.44		1.50		1		324
189340	−10 05238	HR 7637		⋆A	19 57 04	−10 05.4	5	5.88	.008	0.58	.005	0.05	.004	13	F8 V	15,1003,1075,2035,3077
		GH Cyg			19 57 07	+29 19.0	2	9.53	.047	1.11	.004	0.76		2	F5	851,1399
		Cyg sq 4 # 424			19 57 07	+34 59.0	1	12.80		0.44		0.37		2	A2 V	324
189245	−34 14082	HR 7631			19 57 07	−33 50.3	4	5.66	.005	0.49	.000	−0.04		18	F7 V	15,158,1075,2012
345514	+21 04002				19 57 08	+21 15.3	1	9.31		0.01		−0.37		1	B3	963
		Cyg sq 4 # 423			19 57 08	+34 57.3	1	12.70		0.71		0.30		2	F7 V	324
189360	−2 05160				19 57 09	−02 38.8	1	9.45		0.38		0.15		1	A0	1462
226938	+34 03824				19 57 11	+35 04.1	3	10.07	.028	0.16	.011	0.15	.019	4	A0	324,394,8070
		G 92 - 45			19 57 12	+01 10.1	1	14.18		1.61		1.13		1		333,1620
		V1020 Cyg			19 57 12	+32 33.5	1	13.40		1.76				1		1772
189411	+11 04075	Ring Aql # 19		⋆A	19 57 13	+11 45.1	1	9.00		0.11		0.02		3	A0 V	4
		Numerova 1042			19 57 13	+35 00.0	1	11.86		0.45		0.36		1	Am	324
		VES 96			19 57 14	+31 18.9	1	12.14		1.54		0.48		1		363
		Numerova 1050			19 57 16	+34 53.7	1	11.98		0.55		0.11		1		324
		Numerova 1054			19 57 16	+34 55.2	1	12.19		0.49		0.14		1		324
	+37 03710	QZ Cyg			19 57 16	+38 07.5	1	10.25		1.75				2	M3	369
189198	−45 13549	HR 7630			19 57 16	−45 15.1	8	5.80	.004	0.29	.005	0.11	.007	45	A8 III	15,278,977,977,1075,1770*
		Cyg sq 4 # 418			19 57 17	+34 54.8	1	12.82		0.74		0.34		1		324
226949	+38 03848	IDS19555N3814		AB	19 57 17	+38 22.6	1	9.34		−0.11		−0.64		1	B5	963
		Ir-Ch # 50			19 57 18	+31 46.	1	12.28		1.40		0.39		3		1359
189529	+33 03676				19 57 18	+33 56.2	1	8.94		−0.01		−0.35		1	B9	1192
189246	−40 13601				19 57 19	−40 20.3	1	8.05		1.61		2.01		1	M2 III	1468
226951	+35 03899				19 57 20	+35 58.7	1	9.12		0.16		−0.74		2	B0.5III	1012
189226	−46 13328				19 57 20	−45 50.3	3	8.74	.003	0.04	.004	0.02	.001	16	A0 V	977,1075,2011
189493	+25 04055				19 57 21	+25 33.0	2	8.28	.005	0.26	.005	0.02	.020	5	A5	833,1625
		VES 134		V	19 57 21	+36 44.2	1	12.95		1.87		1.41		1		426
189247	−44 13671				19 57 21	−44 07.1	4	7.65	.001	0.44	.005	−0.02	.009	11	F5 V	278,1075,1770,2011
189415	+0 04379				19 57 24	+00 30.6	1	8.84		1.25		1.21		1	K2	1462
333261	+29 03835				19 57 25	+29 27.6	1	9.72		1.10				1	F8 V	851
	+32 03655			V	19 57 26	+32 23.9	1	10.09		1.99				2	M3	369
189574	+38 03850				19 57 26	+38 44.6	2	7.77	.005	0.27	.010	0.10	.015	6	A3 m	833,8071
		Cyg sq 4 # 421			19 57 29	+34 55.6	1	12.73		0.49		0.16		1		324
189292	−44 13674				19 57 33	−44 31.4	2	8.88	.001	1.53	.003	1.90		12	K3/4 III	977,1075
189124	−59 07564	HR 7625, NU Pav			19 57 33	−59 30.9	3	5.03	.043	1.52	.020	1.31	.007	7	M6 III	58,2007,3051
189494	+10 04131	Ring Aql # 4			19 57 35	+11 00.2	1	8.72		1.65		1.65		3	K5	4
		Cyg sq 4 # 422			19 57 37	+34 57.0	1	12.62		0.56		0.40		2	A4 V	324
189596	+32 03658				19 57 40	+32 45.4	1	7.54		−0.11		−0.44		1	B7 Vp:	1192
		Cyg sq 4 # 430			19 57 40	+35 04.8	1	11.87		1.37		1.04		1		324
189507	+10 04132	Ring Aql # 5		⋆A	19 57 41	+11 02.8	1	9.35		0.38		−0.06		3		4
333281	+28 03565				19 57 41	+29 11.2	1	10.39		0.58				1	K2	851
		Numerova 1140			19 57 41	+34 53.7	1	11.47		0.37		0.18		1		324
189508	+10 04133	Ring Aql # 6		⋆B	19 57 42	+11 02.6	1	9.24		0.37		−0.03		3		4
226986					19 57 42	+34 36.	1	11.42		0.29		0.16		2	F0	394
		Cyg sq 4 # 426			19 57 42	+35 01.6	1	12.08		0.67		0.40		1		324
		WLS 2000 5 # 10			19 57 43	+05 40.3	1	10.54		0.73		0.27		2		1375
189550	+19 04236				19 57 43	+19 45.1	1	7.65		−0.06		−0.70		3	B2 V	399
	−74 01864				19 57 43	−73 53.4	2	9.56	.015	0.95	.017	0.76		6	K3 V	2012,3072
189684	+45 03025	HR 7646			19 57 46	+45 38.1	1	5.92		0.16		0.15		2	A5 III	252
		Cyg sq 4 # 429			19 57 47	+35 04.5	1	11.66		0.75		0.31		1		324
189577	+17 04183	HR 7645, VZ Sge		⋆A	19 57 48	+17 22.7	2	5.35	.007	1.66	.005	1.75		14	M4 IIIa	71,3042
189307	−47 13199	IDS19542S4740		AB	19 57 48	−47 32.1	5	6.96	.003	0.35	.002	0.09	.007	25	F2 IV	278,977,1075,1770,2011
		VES 97			19 57 49	+19 48.5	1	12.26		1.48		1.05		1		363
		LS II +34 005			19 57 49	+34 51.1	3	11.78	.015	0.38	.018	0.14	.014	5		180,324,394
		CS 22964 # 183			19 57 49	−40 57.5	1	14.48		0.43		−0.12		1		1736
333263					19 57 50	+29 37.5	1	9.51		0.02		−0.19		2	B8.5V	401
189900	+63 01584	HR 7654			19 57 51	+63 23.8	1	6.15		0.04		0.06		2	A3 V	1733
189386	−38 13809	IDS19545S3852		AB	19 57 53	−38 43.5	1	7.70		0.40				3	F3 V	173
227005					19 57 54	+34 54.	2	11.30	.039	0.20	.010	0.14	.015	3	B9 II	324,394
189656	+31 03879				19 57 55	+31 42.5	1	8.14		1.72				2	M1	369

HD	DM	Other Id	N	Rem	α_{1950}	δ_{1950}	S	V	σ_V	B–V	σ_{B-V}	U–B	σ_{U-B}	n	Spectrum	References
	−73 02093				19 57 55	−73 45.9	1	10.58		1.02				4	K0	2012
189551	+11 04079	Ring Aql # 20			19 57 56	+11 49.7	1	8.83		0.27		0.03			A7 IV	4
227018	+34 03828	LS II +35 009			19 57 56	+35 10.3	3	9.06	.117	0.37	.010	−0.63	.014	5	O7	180,1011,1012
189775	+51 02728	HR 7651			19 57 56	+51 55.1	1	6.14		−0.19		−0.66		2	B5 III	154
		LP 514 - 5			19 57 57	+12 40.1	1	11.76		0.84		0.50		1		1696
333313	+38 03568				19 57 58	+28 35.8	1	8.91		0.02		−0.26		3	A2	833
	−30 17560	V1711 Sgr			19 57 59	−30 39.0	1	10.49		0.54				1		688
189818	+57 02106				19 58 00	+57 40.3	2	7.32	.015	−0.18	.000	−0.83	.005	4	B5	401,555
227019	+34 03829				19 58 02	+34 38.4	1	10.26		0.52		0.03		2	F2	394
		VES 98			19 58 03	+25 10.8	1	11.51		2.02		1.86		1		363
189688	+34 03830				19 58 03	+35 09.5	1	7.67		1.75				2	K5	369
189388	−41 13807	HR 7639			19 58 03	−40 57.2	3	6.28	.004	0.11	.005			11	A2/3 V	15,2013,2030
		Cyg sq 4 # 428			19 58 04	+35 04.5	1	12.19		0.35		0.26		1	A6 V	324
189599	+11 04081	Ring Aql # 21			19 58 05	+11 55.2	1	8.77		1.24		0.98		3	K5	4
189687	+36 03806	HR 7647, V1746 Cyg			19 58 05	+36 54.3	4	5.19	.019	−0.17	.008	−0.69	.007	8	B3 IVe	154,1192,1212,1363
		Cyg sq 4 # 511			19 58 08	+35 00.8	1	12.61		0.48		0.32		2	A6 V	324
		G 23 - 20			19 58 10	+09 13.1	3	11.60	.011	0.64	.004	−0.03	.012	6		333,927,1620,1658
189671	+25 04060	IDS19561N2555	AB		19 58 10	+26 03.0	2	6.57	.061	1.08	.047	1.02	.005	4	G8 II	1080,3016,8100
		Cyg sq 4 # 506			19 58 10	+34 55.9	1	12.81		0.89		0.95		1		324
		VES 135			19 58 10	+37 23.3	1	12.59		1.79		1.39		1		426
190061	+68 01095	G 260 - 39			19 58 10	+69 00.1	1	9.26		0.63		0.09		1	G5	1658
189689	+32 03662				19 58 11	+32 39.1	1	7.27		−0.07		−0.34		1	B6/7IV(p)	1192
189404	−44 13680				19 58 11	−44 36.5	2	9.39	.002	0.88	.004	0.45		10	G8 V	863,2011
		Numerova 1252			19 58 12	+34 53.4	1	12.35		0.53		0.15		1		324
		Numerova 1253			19 58 12	+34 54.9	1	12.29		0.53		0.03		1		324
189513	−21 05595				19 58 13	−21 36.1	1	8.29		0.36				4	F2 V	2012
227030	+32 03663				19 58 14	+32 50.7	1	9.48		1.09		0.95		1	K0	363
189690	+29 03838	IDS19562N2939	AB		19 58 15	+29 46.9	1	7.42		0.04		−0.08		1	A0 V	401
189558	−12 05613	IDS19555S1231	AB		19 58 15	−12 23.4	3	7.73	.010	0.57	.012	−0.08	.019	9	F8/G2 V	158,1003,3077
189441	−37 13484				19 58 16	−36 49.1	1	7.61		1.47		1.72		4	K4 III	1673
339373	+26 03757				19 58 19	+26 32.2	1	8.89		0.38				35	F2	6011
189706	+29 03839				19 58 19	+29 41.0	1	7.76		−0.08		−0.40		2	B9 V	401
189751	+36 03807	IDS19565N3608	A		19 58 20	+36 16.5	1	7.01		1.09		1.04		2	K1 III	37
189659	+10 04137	Ring Aql # 7			19 58 22	+10 53.6	1	8.56		−0.05		−0.41		3	B7 V	4
		LS II +33 013			19 58 22	+33 12.5	1	11.19		0.33		0.12		2		180
227046	+34 03831				19 58 22	+34 25.6	1	10.18		0.57		0.03		2	G0	394
		CS 22964 # 157			19 58 22	−38 06.6	1	14.49		0.52		−0.10		1		1736
		Cyg sq 4 # 510			19 58 23	+34 59.9	1	13.11		0.59		0.20		1		324
		G 186 - 7			19 58 24	+27 18.8	1	13.24		1.54				1		1759
		LS II +33 014			19 58 24	+33 07.5	1	12.08		0.97		−0.12		1		1359
		Numerova 1291			19 58 24	+34 23.4	1	11.54		0.52		0.02		1		394
227045					19 58 24	+34 42.	1	10.34		0.27		0.12		2	A3	394
227049	+37 03721				19 58 24	+37 21.1	1	9.19		1.19				3	K2 III	1726
	+20 04375				19 58 25	+20 43.4	1	9.81		1.82				1	M2	369
189777	+34 03832				19 58 26	+35 13.2	1	7.66		1.70		1.74		2	K2	1625
189561	−23 15935	HR 7643			19 58 26	−22 52.6	2	6.00	.003	0.97	.000	0.77		6	G7 V	58,2007
		G 23 - 21			19 58 27	+09 07.2	2	14.37	.039	1.19	.039	0.78	.029	3		1620,3062
189674	+11 04085	Ring Aql # 22			19 58 27	+11 59.1	1	9.55		0.39		0.00		3	F1 V	4
		G 23 - 22			19 58 29	+10 07.2	1	13.04		1.52		1.40		1		333,1620
189753	+26 03758				19 58 29	+27 00.2	1	7.76		1.51		1.82		1	K4 II	8100
		Cyg sq 4 # 514			19 58 29	+35 05.2	1	12.10		1.12		0.73		1		324
		LS II +22 017			19 58 31	+22 06.1	1	11.18		0.46		−0.21		2		405
189778	+34 03833				19 58 31	+34 19.5	1	8.20		−0.03		−0.42		2	A0	401
189695	+8 04300	HR 7648			19 58 34	+08 25.1	2	5.90	.005	1.52	.000	1.89	.000	7	K5 III	15,1071
333226	+29 03842	LS II +30 016			19 58 34	+30 14.6	1	10.11		0.49		−0.42		2	B1 V:e	1012
189732	+23 03868				19 58 35	+23 39.6	1	7.37		−0.16		−0.59		2	B9	401
227060					19 58 36	+35 02.	1	11.13		0.36		0.05		2	F0	394
339376	+26 03760				19 58 37	+26 26.4	1	8.73		1.08				32	K0	6011
189779	+29 03844				19 58 37	+29 45.0	3	8.22	.011	0.09	.010	−0.62	.010	6	B2 III	401,1012,1371,8100
189406	−53 09773				19 58 37	−53 01.6	1	8.43		0.58				4	G0 V	2012
		CS 22964 # 181			19 58 38	−40 50.8	1	13.60		0.92		0.59		1		1736
356528	+10 04138	Ring Aql # 8			19 58 39	+10 51.4	1	10.15		1.23		1.31		2	K7	4
189711	+9 04369				19 58 40	+09 22.5	2	8.41	.052	2.08	.024	2.28		2	N	1238,8005
189502	−44 13686				19 58 40	−44 36.4	8	7.91	.009	−0.04	.006	−0.25	.004	43	Ap Si(Cr)	278,863,977,977,1075,1586*
		Cyg sq 4 # 509			19 58 41	+34 59.4	1	12.48		0.65		0.12		1		324
227073					19 58 42	+34 57.	2	10.28	.010	0.60	.019	0.01	.015	3	G5	324,394
227072	+34 03834	IDS19568N3501	A		19 58 42	+35 09.6	2	10.18	.010	0.31	.024	0.24	.019	3	A3	324,394
		CS 22964 # 176			19 58 44	−40 29.8	1	15.33		0.44		−0.18		1		1736
356486	+11 04089	Ring Aql # 23			19 58 45	+12 01.4	1	9.97		0.84		0.14		2	G0	4
190379	+73 00891	IDS19596N7358	AB		19 58 45	+74 06.4	1	9.18		0.53		0.10		39		588
189796	+29 03845	IDS19568N2933	AB		19 58 46	+29 41.1	1	7.88		0.61		0.11		1	G0 V	3026
189845	+38 03862				19 58 46	+39 04.1	1	7.23		0.03		−0.02		2	A0 V	105
354510	+15 04019		AB		19 58 47	+15 35.2	1	9.72		0.34				1	A3	851
189712	+5 04381				19 58 48	+05 30.6	2	7.45	.000	0.44	.014	−0.04	.010	7	F5	379,1733
		JL 37			19 58 48	−73 07.	1	14.18		−0.12		−0.67		2		132
190252	+69 01084	HR 7666			19 58 50	+70 13.7	1	6.33		0.88		0.54		3	G8 III	3051
189484	−50 12780		A		19 58 50	−50 11.2	1	8.72		1.12		1.07		2	K4 V	3072
189484	−50 12780		AB		19 58 50	−50 11.2	2	8.32	.015	1.12	.005	1.10		8	K4 V	158,2012
339371	+26 03761				19 58 51	+26 30.3	1	9.23		1.07				34	K0	6011
		Steph 1745			19 58 51	−03 33.8	1	10.96		1.27		1.25		1	K7	1746

Table 1   1085

HD	DM	Other Id	N	Rem	α_{1950}	δ_{1950}	S	V	σ_V	B–V	σ_{B-V}	U–B	σ_{U-B}	n	Spectrum	References
354511		KL Aql			19 58 52	+15 39.9	2	9.83	.008	0.80	.027	0.51		2	F6 Iab	851,1399
189864	+36 03816	IDS19571N3619		A	19 58 54	+36 27.0	2	6.68	.025	-0.06	.020	-0.44		5	B9	1118,1192
189846	+32 03673				19 58 58	+32 52.4	1	8.41		-0.08				1	B6 V	1070
		AJ89,1229 # 421			19 59 00	+05 27.	1	10.73		1.64				1	M1	1632
		Case *M # 61			19 59 00	+30 41.	1	9.80		2.73				1	M1.5Iab	148
189847	+30 03853				19 59 00	+31 05.5	1	6.86		-0.08		-0.49		1	B7 V	1192
		AA45,405 S99 # 1			19 59 00	+33 19.	1	13.32		1.11				1		725
189917	+43 03449				19 59 00	+44 04.1	1	10.80		0.37		0.13		2	A2	401
189849	+27 03587	HR 7653, NT Vul			19 59 02	+27 36.9	5	4.63	.013	0.18	.005	0.15	.004	35	A4 III	15,879,1363,3058,8015
		LS IV -12 111			19 59 03	-12 49.6	1	11.33		0.10		-0.81		2		405
		LP 814 - 12			19 59 03	-16 02.7	1	12.54		0.56		-0.10		2		1696
189783	+10 04143	IDS19567N1028		AB	19 59 04	+10 36.5	1	6.98		0.44		0.00		4	F2	3026
189585	-44 13691				19 59 04	-44 21.7	2	8.90	.005	0.97	.005	0.75		12	G8 III	977,2011
189563	-45 13568				19 59 04	-45 20.2	5	6.57	.002	1.22	.002	1.22	.002	19	K1 III	278,977,1075,1586,2011
		LS II +33 015			19 59 05	+33 15.4	1	11.25		0.48		-0.27		3		180
		LS II +33 016			19 59 07	+33 11.8	1	11.76		0.36		-0.31		2		180
189808	+11 04093	Ring Aql # 24			19 59 08	+12 04.4	1	9.01		0.36		-0.02		3	A9 V	4
		VES 100			19 59 09	+31 01.6	1	12.68		0.79		0.19		1		363
189865	+25 04067				19 59 10	+25 32.4	1	8.28		0.23		0.06		2	A3	1625
		MtW 64 # 385			19 59 10	+30 04.0	1	12.58		0.35		0.28		5		397
189901	+37 03723	LS II +37 005		⋆ AB	19 59 10	+37 33.6	1	7.99		-0.02		-0.72		1	B5	1192
189810	+10 04144	Ring Aql # 9			19 59 11	+10 44.1	1	9.65		0.45		-0.07		3	F5	4
		MtW 64 # 395			19 59 11	+30 04.2	1	15.62		1.05		0.69		4		397
		MtW 64 # 414			19 59 11	+30 04.5	1	15.40		0.90		0.26		5		397
		MtW 64 # 394			19 59 11	+30 05.1	1	11.23		0.12		0.01		5		397
189741	-14 05618	HR 7649			19 59 11	-13 46.6	1	5.69		0.08				4	A1 IV	2035
		VES 101			19 59 12	+21 27.0	1	12.51		2.04		1.20		1		363
		MtW 64 # 432			19 59 12	+30 04.0	1	13.32		2.47		2.69		8		397
189762	-7 05147				19 59 12	-07 13.4	1	9.81		0.46		0.09		2	A5	1375
		L 115 - 21			19 59 12	-65 44.	3	11.35	.011	1.48	.012	0.98		4		1494,1705,3073
		L 115 - 22			19 59 12	-65 44.	2	12.82	.005	1.55	.019	1.01		3		1705,3073
333209	+30 03854				19 59 16	+30 43.1	1	10.07		2.65		2.58		3	B9 V	8032
189631	-41 13822				19 59 16	-41 33.5	1	7.56		0.30				4	F0 V	2012
		LS II +22 019			19 59 18	+22 40.3	1	11.63		0.74		0.39		1	F4 II	1215
227130					19 59 18	+37 59.	1	10.81		0.13		0.05		2	A0	1375
189957	+41 03569	LS III +41 001			19 59 18	+41 52.2	3	7.81	.005	0.01	.009	-0.86	.015	8	B0 III	399,405,555
		LS II +35 010			19 59 19	+35 51.0	1	11.21		0.26		-0.45		1		180
		CS 22964 # 158			19 59 20	-38 08.3	1	14.74		0.48		-0.16		1		1736
		VES 136		⋆ V	19 59 21	+36 04.6	1	11.26		1.73		1.24		1		426
227132	+35 03915	LS II +35 011			19 59 23	+35 49.0	1	10.34		0.33		-0.49		1	B2 III	180
		LS II +36 003			19 59 23	+36 11.6	1	9.68		1.44		1.52		1		363
		VES 103			19 59 24	+32 53.7	1	13.32		0.88		-0.05		1		363
		WLS 2000 5 # 7			19 59 25	+03 07.1	1	11.39		1.07		0.87		2		1375
189942	+36 03820	HR 7655			19 59 25	+36 57.5	1	6.20		1.31				2	K0 III	71
333247	+29 03855				19 59 27	+29 53.5	1	8.03		1.67				2	M0	369
		WLS 1944 55 # 8			19 59 27	+54 53.2	1	12.30		1.19		1.10		2		1375
		V765 Aql			19 59 28	-03 10.7	1	12.33		0.82		0.46		1		1462
189667	-45 13571				19 59 28	-44 59.3	2	9.40	.001	1.09	.002	0.98		9	K0 IV	863,1075
189982	+37 03727				19 59 31	+37 50.7	2	7.79	.005	-0.04	.000	-0.14		4	A0	401,1118
		VES 104			19 59 32	+28 16.9	1	13.07		0.97		0.01		1		363
189943	+29 03857				19 59 32	+30 04.7	2	7.72	.006	0.83	.020	0.52	.056	9	G5 III	397,8100
189825	-5 05138				19 59 32	-05 07.8	1	6.77		0.15		0.03		7	A0	1628
190025	+42 03562				19 59 33	+42 54.3	2	7.53	.000	-0.08	.005	-0.55	.000	5	B5	401,555
		Case *M # 62			19 59 35	+30 39.	1	10.29		2.38		1.53		3	M2.5(I) + B	8032
189763	-28 16355	HR 7650, V3872 Sgr			19 59 35	-27 51.0	6	4.53	.045	1.64	.018	1.77	.034	21	M4 III	15,1024,2007,3016*
189944	+24 03975	HR 7656			19 59 37	+24 39.6	3	5.88		-0.14	.015	-0.55	.026	8	B4 V	1063,1079,3016
227150	+34 03839				19 59 37	+35 00.0	1	8.72		0.95				1	G8 III	1070
351236	+17 04195	LS II +17 015			19 59 38	+17 57.2	1	9.30		0.17		-0.28		1	B8 II	8100
189983	+34 03840	IDS19578N3418			19 59 40	+34 26.6	1	8.54		0.01				1	B8 V	1070
333274	+29 03858				19 59 41	+29 23.1	1	9.28		0.47		0.08		1	F6 V	363
190065	+42 03563	V1675 Cyg			19 59 42	+43 05.3	1	7.68		1.67				2	M1	369
190002		LS II +32 012			19 59 43	+32 26.0	1	12.67		0.75		0.41		1	C7	1359
190046	+39 04007	IDS19580N4001		A	19 59 43	+40 09.6	1	7.78		0.02		-0.38		2	A0	401
189920	+11 04100	Ring Aql # 25			19 59 44	+12 10.9	1	9.37		0.11		0.03		3	A0 V	4
189699	-43 13765				19 59 44	-43 24.7	1	9.26		1.61				4	M1 (III)	1075
		GD 226			19 59 46	+05 59.3	1	16.41		0.15		-0.56		1		3060
189921	+10 04147	Ring Aql # 10			19 59 47	+10 35.9	1	6.95		-0.08		-0.57		3	B5. V	4
		G 143 - 23			19 59 47	+14 59.5	1	12.17		0.55		-0.09		2	G0	1658
190001	+32 03679				19 59 47	+32 55.6	3	7.96	.030	-0.08	.007	-0.47	.015	4	B6 V	1070,1192,1733
		VES 106, V823 Cyg			19 59 47	+35 58.1	1	13.03		0.51		0.36		1		363
189853	-13 05557				19 59 47	-12 50.8	1	8.64		0.08		-0.20		12	B8/9 III	1732
	+41 03573				19 59 48	+41 45.5	1	9.44		1.79				2		369
	-77 01404				19 59 48	-77 28.8	1	9.86		1.29				4	M0	2012
189719	-43 13767	IDS19564S4312		AB	19 59 49	-43 03.9	3	7.57	.003	0.37	.004	0.02	.011	16	F3 V	977,1075,2011
		LS II +35 013			19 59 52	+35 24.8	1	10.95		0.48		-0.41		1		180
		VES 107			19 59 53	+24 53.1	1	11.74		1.80		1.98		1		363
190003	+28 03585				19 59 53	+28 24.3	1	8.63		0.00		-0.41		1	A0	963
190004	+24 03977	HR 7657		⋆ AB	19 59 54	+24 47.8	5	5.21	.024	0.37	.009	0.09	.007	16	F2 III	253,1080,3026,8100
333251	+29 03861	LS II +29 034			19 59 54	+29 45.2	2	8.41	.030	0.95	.020	0.89		2	F5 Ib	138,8100
357782	+0 04391				19 59 56	+00 22.1	2	10.49	.015	1.18	.085	0.62		2	F8	592,1462

HD	DM	Other Id	N	Rem	α_{1950}	δ_{1950}	S	V	σ_V	B–V	σ_{B-V}	U–B	σ_{U-B}	n	Spectrum	References
		WLS 2000-5 # 10			19 59 56	−04 35.4	1	12.30		0.52		0.28		2		1375
190047	+30 03862				19 59 57	+30 58.7	1	6.53		-0.05		-0.43		1	B7 V	1192
190147	+49 03158	HR 7660		★ A	19 59 57	+49 57.9	8	5.06	.019	1.12	.011	1.02	.023	44	K1 II-III	37,659,1080,1198,1322*
190088	+38 03869				19 59 58	+38 35.7	1	7.85		-0.06		-0.26		2	A0	401
		LP 814 - 14			19 59 59	−15 12.8	1	11.57		0.67		0.03		2		1696
189767	−44 13702				19 59 59	−44 44.1	4	8.32	.003	0.19	.002	0.13	.011	17	A5 IV/V	977,1075,1586,2011
		V572 Aql			20 00 00	+00 34.0	3	10.99	.017	0.73	.102	0.47	.023	3		592,1399,1462
		LS II +33 017			20 00 00	+33 12.3	1	11.21		0.54		-0.32		1		180
		AA45,405 S100 # 1			20 00 00	+33 21.	1	12.82		0.86				1		725
189923	−3 04771	IDS19574S0337		A	20 00 01	−03 29.0	1	8.21		1.28		0.90		1	G5	1462
339418	+24 03979				20 00 02	+25 00.8	1	8.08		1.66				2	M0	369
190026	+20 04389				20 00 05	+20 47.1	1	8.40		-0.02		-0.10		2	A0	1733
		CS 22964 # 161			20 00 06	−38 47.0	1	14.41		0.49		-0.17		1		1736
190149	+43 03457				20 00 07	+43 58.7	2	6.94	.023	1.62	.014	1.94	.056	4	M0 II-III	247,3016,8100
190114	+34 03847				20 00 09	+35 11.5	2	7.42	.017	-0.07	.009	-0.49		3	B8 V	401,1070
		VES 108			20 00 10	+29 04.8	1	12.66		2.47		0.57		1		363
227196	+34 03846				20 00 10	+34 20.0	1	8.54		1.55				1	K5	1070
190113	+35 03920				20 00 10	+35 30.1	2	7.87	.020	1.50	.020	1.19		4	G8 V	1070,8100
227207	+36 03825				20 00 10	+36 35.6	1	9.93		0.10		-0.55		2	B3	426
190165	+45 03038				20 00 11	+45 20.2	2	7.52	.005	0.31	.000	0.11	.005	5	AmA2-F2	1601,8071
190066	+21 04027	LS II +22 020			20 00 12	+22 00.7	2	6.51	.025	0.18	.000	-0.74	.010	3	B1 Iab	1012,8100
235083	+52 02604				20 00 12	+53 15.5	1	8.02		1.48		1.64		2	G5	1375
		WLS 2000-5 # 5			20 00 13	−02 52.6	1	13.29		1.15		0.64		2		1375
333461	+28 03590				20 00 14	+28 31.4	1	10.10		0.10		-0.09		1		272
189831	−38 13828	HR 7652			20 00 14	−38 04.8	3	4.76	.005	1.41	.000	1.68	.000	14	K4 III	15,1075,2012
190007	+2 04076				20 00 17	+03 11.0	3	7.45	.011	1.14	.011	1.08	.014	8	K4 V	196,285,3072
190067	+15 04026				20 00 17	+15 27.6	4	7.16	.009	0.71	.002	0.22	.017	13	G7 V	22,1355,1758,3026
	+42 03566				20 00 17	+43 01.4	1	9.12		1.77				2		369
		RR Tel			20 00 18	−55 51.7	1	10.84		0.42		-1.36		1	F5 pe	1753
		BPM 26536			20 00 18	−56 12.	1	15.25		-0.30		-1.29		1		3065
357777	+0 04397				20 00 19	+00 37.3	1	10.46		0.82				1	F5	592
190050	+11 04104	Ring Aql # 26			20 00 20	+12 13.1	1	6.79		1.51		1.53		3	K5	4
189834	−44 13705				20 00 22	−43 58.5	4	9.24	.008	0.07	.002	0.07	.016	15	A0 V	861,863,1075,2011
227228	+34 03848				20 00 23	+34 34.0	1	9.22		0.00				1	B7 IV	1070
227242	+36 03827	LS II +36 004			20 00 24	+36 57.2	1	10.59		0.23		-0.69		2	B0 IV:	1012
		HIC 98652			20 00 25	+37 15.2	1	10.85		1.35				2		1726
189987	−6 05347				20 00 25	−06 25.6	1	9.64		1.04		0.98		2	K5 V	3072
189855	−44 13706				20 00 27	−44 46.8	5	8.34	.003	1.08	.004	0.93	.005	19	K0 III	977,1075,1586,2011,2024
227245	+35 03924	LS II +35 014			20 00 29	+35 32.1	3	9.75	.014	0.61	.014	-0.45	.018	5	O7 V	180,1011,1012
227247	+34 03850	LS II +35 016		★ A	20 00 30	+35 10.2	1	9.38		0.32		-0.60		2	B2 II:	180
227247	+34 03850	LS II +35 016		★ AP	20 00 30	+35 10.2	2	9.16	.010	0.34	.025	-0.56	.030	2		401,8084
227247	+34 03850	LS II +35 016		★ APC	20 00 30	+35 10.2	1	9.00		0.32		-0.47		1	B2	8084
227247	+34 03850	IDS19586N3502	BQ		20 00 30	+35 10.2	1	10.58		0.62		-0.41		1		8084
227247	+34 03850	IDS19586N3502	D		20 00 30	+35 10.2	1	12.08		0.44		-0.36		1		8084
		L 349 - 81			20 00 30	−48 50.	2	12.66	.000	0.45	.000	-0.30	.045	4		1696,3062
227250	+33 03697	LS II +33 019			20 00 32	+33 37.6	1	9.28		1.07		0.80		2	F8 Ia	247
		G 23 - 23			20 00 33	+14 07.4	2	11.07	.005	0.85	.010	0.54	.035	2	K3	333,1620,1658
339366	+26 03767				20 00 33	+26 37.7	1	9.01		1.69				2	M0	369
		LS II +35 017			20 00 33	+35 31.8	1	10.61		0.56		-0.39		1		180
190073	+5 04393	V1295 Aql			20 00 34	+05 35.8	2	7.86	.015	0.09	.001	-0.04	.013	3	Ape	379,1588
189567	−67 03703	HR 7644			20 00 34	−67 27.2	6	6.08	.011	0.65	.005	0.08	.008	21	G3 V	15,742,1075,1075,2038,3078
		G 92 - 48			20 00 35	+07 55.4	1	12.02		1.31		1.24		1		333,1620
190254	+44 03318				20 00 36	+44 18.4	1	8.68		-0.02		-0.51		3	B2 III	555
		LS II +33 020			20 00 39	+33 24.5	1	10.77		0.27		-0.52		2		180
333412	+29 03867				20 00 40	+29 30.4	1	9.02		0.35		0.05		1	A7 II	8100
190192	+33 03699				20 00 40	+33 22.6	1	8.67		0.28				1	A5 V	1070
	+41 03583				20 00 40	+42 01.5	1	9.41		1.66				2		369
		L 349 - 18			20 00 41	−45 47.7	1	12.14		1.40				1		1705
190095	+4 04325				20 00 43	+04 35.3	1	6.56		1.54		1.55		2	K5	985
190397	+57 02115				20 00 43	+57 30.7	1	7.68		-0.08		-0.45		2	A0	401
189931	−38 13832				20 00 44	−38 00.8	1	6.93		0.60		0.14		3	G1 V	1311
190167	+28 03595				20 00 45	+28 22.4	1	6.95		-0.01		-0.11		3	A1 V	3016
		G 92 - 49			20 00 45	−03 05.8	2	12.26	.015	1.03	.007	0.41	.045	4		1620,1658
190009	−22 05318	HR 7658			20 00 47	−22 44.3	4	6.44	.008	0.50	.000	-0.01		12	F7 V	15,2018,2030,3053
227290	+34 03851				20 00 48	+34 41.6	1	9.19		0.60				1	G0 V	1070
		MCC 198			20 00 49	+16 58.	1	10.57		1.38				2	M0	1619
227292	+33 03702				20 00 50	+33 29.2	1	9.23		0.02				1	B5 V	1070
		WLS 1944 55 # 11			20 00 50	+53 06.6	1	10.72		0.32		0.08		2		1375
	−75 01570				20 00 50	−74 58.0	1	10.46		0.39				4	K0	2012
190227	+31 03905				20 00 51	+31 49.0	1	6.42		1.20		1.23		2	K1 III	37
	−45 13587				20 00 51	−45 07.1	5	10.71	.005	0.92	.005	0.57	.022	29		861,863,1460,1730,2011
227306	+33 03703				20 00 53	+33 39.3	1	9.69		0.08				1	B8 V	1070
227310	+37 03734				20 00 53	+38 15.2	1	6.92		0.28				1	B5 IV	1070
189933	−45 13588				20 00 53	−45 10.5	9	9.31	.008	0.44	.004	-0.04	.005	62	F3/5 V	861,1460,1628,1657,1673*
190256	+32 03685				20 00 55	+32 26.2	2	8.51	.005	-0.06	.018			5	B7 V	1070,1118
		WLS 2000-5 # 9			20 00 56	−04 31.1	1	12.04		1.52		1.49		2		1375
190228	+27 03593				20 00 57	+28 10.0	1	7.33		0.79		0.31		3	G5 IV	3016
190275	+37 03735				20 00 57	+37 40.3	1	6.93		0.28		0.12		4	Am	8100
190544	+64 01405	HR 7676		★ A	20 00 57	+64 40.8	1	5.27		1.56		1.81		2	M1 III	3001
		WLS 2000 5 # 5			20 00 58	+07 04.3	1	11.34		0.31		0.19		2		1375

Table 1   1087

HD	DM	Other Id	N	Rem	α_{1950}	δ_{1950}	S	V	σ_V	B–V	σ_{B-V}	U–B	σ_{U-B}	n	Spectrum	References
189951	−45 13590				20 00 58	−45 05.3	6	7.83	.003	0.29	.004	0.04	.028	23	A9 IV	143,278,1075,1586,1770,2011
	+63 01588	G 260 - 40			20 01 00	+63 24.3	1	10.03		0.89		0.80		2	K2	7010
333430	+28 03597				20 01 01	+29 07.4	1	8.65		-0.02		-0.45		1	G5	272
227324	+41 03586	V420 Cyg			20 01 01	+42 00.4	1	9.44		1.72				2	K5	369
190426	+55 02312				20 01 01	+55 29.5	1	8.51		0.44		0.06		2	F5	1502
190211	+18 04365	HR 7662		⋆ A	20 01 02	+18 21.6	1	5.96		1.42		1.60		1	K3 II-III	3032,8100
190211	+18 04365	IDS19588N1813	B		20 01 02	+18 21.6	1	10.44		1.03		0.67		2		3032
190960	+76 00771	HR 7686		⋆ C	20 01 02	+76 20.6	3	6.20	.000	1.60	.005	1.89	.005	6	M3 III	15,1003,3055
190155	−3 04773	V1359 Aql			20 01 03	−03 31.1	1	8.83		1.36		1.15		1	G5	1462
239292	+56 02341	G 230 - 23			20 01 07	+56 58.6	1	9.60		0.76		0.28		2	G5	3016
		VES 138			20 01 08	+39 37.3	2	12.57	.000	1.75	.000	2.02	.000	4		426,8103
	−45 13591				20 01 08	−44 53.3	1	10.17		0.57				4	F8	2011
333452	+28 03598	LS II +28 021			20 01 09	+28 34.0	1	9.43		1.02		-0.05		2	B0 III:np	1012
190056	−32 15682	HR 7659		⋆ A	20 01 10	−32 11.9	3	4.99	.003	1.21	.003	1.18	.005	88	K1 III/IV	3,58,2009
		LS II +35 019			20 01 11	+35 32.7	1	12.04		0.52		-0.34				180
190229	+15 04033	HR 7664			20 01 13	+15 53.4	2	5.67		-0.10	.016	-0.50	.005	4	B9 p HgMn	1063,3023,8100
331777	+31 03907	LS II +31 022			20 01 14	+31 46.7	1	8.16		1.54		1.34		2	F8 Ia	1080
239295	+55 02314				20 01 20	+55 44.9	1	9.44		0.57		0.11		2	G0	1502
190172	−7 05159	HR 7661			20 01 20	−07 36.7	1	6.71		0.35		0.04		8	F4 III	1088
190427	+45 03044	LS III +45 002			20 01 21	+45 50.7	3	8.35	.000	0.16	.014	-0.74	.012	8	B0 III	399,401,555
190103	−28 16379				20 01 22	−27 53.3	1	7.47		1.51		1.84		3	K4 III	1657
190336	+33 03708	LS II +33 021			20 01 23	+33 18.5	2	8.61	.017	0.10	.000	-0.76		3	B0.7 II-III	180,1070
		G 23 - 24			20 01 24	+05 51.7	2	12.98	.015	1.54	.015	1.14		2		333,1620,1759
		G 125 - 55			20 01 24	+29 43.7	1	14.38		1.71		1.08		4		316
331808					20 01 24	+31 10.	1	11.02		0.27		-0.30		1	F0	363
190401	+41 03589				20 01 26	+41 19.9	2	7.02	.005	0.35	.003	0.12	.014	4	Am	194,8071
		VES 110			20 01 27	+31 48.9	1	13.31		1.03		0.08		1		363
190508	+53 02343				20 01 27	+53 45.0	1	8.32		0.68		0.24		2	G5	1566
190322	+22 03903				20 01 29	+22 48.0	1	6.42		1.04		0.91		1	K2	1733
190323	+14 04158	LS II +14 016		⋆ A	20 01 31	+14 50.5	2	6.83	.005	0.87	.027	0.72		3	G0 Ia	138,3016,8100
		WLS 2000 40 # 5			20 01 31	+42 21.5	1	11.62		1.04		0.73		2		1375
190382	+33 03709				20 01 32	+33 24.1	2	9.13	.003	-0.01	.002			29	B6 V	1070,6011
227381	+39 04018				20 01 33	+39 30.3	1	8.55		1.26		1.30		7	B3	1655
190360	+29 03872	HR 7670		⋆ A	20 01 34	+29 45.7	7	5.74	.022	0.73	.008	0.36	.026	17	G6 IV	15,22,252,1003,1118*
190360	+29 03872	IDS19594N2936	B		20 01 34	+29 45.7	1	14.40		1.57		1.16		2	M6 V	3032
190057	−46 13374				20 01 34	−46 14.4	7	6.96	.005	1.51	.006	1.85	.004	59	K3/4 III	460,474,863,977,1075*
190296	+10 04165	Ring Aql # 11			20 01 37	+10 32.3	1	8.01		0.12		0.05		3	A2 V	4
		LS II +35 022			20 01 37	+35 22.3	1	11.13		0.56		-0.40		1		180
190429	+35 03930	LS II +35 023		⋆ AP	20 01 37	+35 53.0	5	6.62	.019	0.15	.024	-0.77	.044	8	O5 +O9.5III	1118,1192,1193,1196,8084
190429	+35 03930	IDS19598N3545	C		20 01 37	+35 53.0	1	9.71		0.17		-0.53		1		8084
190429	+35 03930	IDS19598N3545	D		20 01 37	+35 53.0	1	12.90		0.19		-0.57		1		8084
		G 24 - 2			20 01 38	+00 15.1	1	9.61		1.23		0.95		1	K3	333,1620
190324	+11 04111	IDS19593N1204	AB		20 01 38	+12 12.9	1	8.84		0.52		0.01		3	F8 V	4
190446	+39 04020	LS II +40 002			20 01 39	+40 08.3	2	8.21	.005	0.57	.005	0.27	.005	3	F6 Ib	247,8100
	+43 03464				20 01 39	+43 42.4	1	9.67		1.70				2		369
239300	+57 02119				20 01 41	+57 16.8	1	9.62		0.90		0.47		2	K0	1375
190327	+6 04416	HR 7669			20 01 42	+07 08.1	3	5.51	.005	1.06	.004	0.86	.005	8	gK0	15,1256,1355
190403	+29 03873				20 01 43	+29 50.9	2	6.76	.022	1.18	.018	0.94	.013	5	G5 Ib-II	1080,8100
190430	+32 03689				20 01 44	+33 02.1	2	7.81	.000	1.39	.011	1.34		5	K1 II	1070,8100
190178	−28 16387				20 01 45	−28 25.0	2	9.52	.024	0.51	.010	-0.03		3	F3 V	742,1594
		CS 22873 # 139			20 01 45	−59 25.8	1	13.83		0.37		-0.22		1		1736
190431	+31 03915	IDS19598N3128	A		20 01 46	+31 36.2	1	8.29		0.04				3	A1 V	1118
	−45 13597				20 01 46	−44 52.4	4	10.62	.005	0.52	.008	0.07	.006	22		861,863,1460,2011
190404	+22 03908	G 186 - 11			20 01 47	+23 12.7	6	7.27	.015	0.81	.005	0.40	.007	14	K1 V	22,1003,1013,1355,1733,3078
190713	+64 01407	HR 7682		⋆ A	20 01 47	+64 29.6	1	6.25		0.93		0.64		2	G5	1733
		NGC 6866 - 7			20 01 48	+44 01.6	1	11.78		0.35		0.07		1		49
190299	−1 03887	HR 7667			20 01 49	−00 31.3	2	5.67	.005	1.30	.000	1.35	.000	7	K4 III	15,1256
190467	+36 03841	V1362 Cyg		⋆ A	20 01 50	+36 17.0	4	8.15	.030	0.17	.012	-0.47	.012	6	B5 II:N	1012,1070,1192,8100
		NGC 6866 - 20			20 01 50	+44 00.8	1	13.60		0.42		0.10		1		49
190406	+16 04121	HR 7672		⋆ ABC	20 01 51	+16 56.0	7	5.79	.009	0.60	.010	0.11	.010	14	G1 V	15,22,851,1013,1197*
190466	+37 03744	IDS20000N3803	A		20 01 51	+38 11.1	2	7.15	.005	1.75	.019	2.62		3	A2 V	369,8100
190468	+34 03858				20 01 52	+34 43.4	1	8.26		0.24				1	Am	1070
354615	+16 04120	IDS19596N1643	A		20 01 53	+16 51.5	1	8.80		0.46		0.23		2	G5	3054
354615	+16 04120	IDS19596N1643	B		20 01 53	+16 51.5	1	9.74		0.29		0.16		2		3054
354615	+16 04120	IDS19596N1643	C		20 01 53	+16 51.5	1	11.70		1.25		1.00		2		3016
190584	+52 02613				20 01 54	+52 26.0	1	7.78		0.29		0.05		3	A5	1566
339483	+25 04083	LS II +26 014			20 01 55	+26 07.8	1	8.94		0.63		-0.31		2	B1 III	1012
		IRC +30 410			20 01 56	+29 00.9	1	12.85		2.46		1.20		2		363
190266	−22 05328				20 01 56	−22 03.4	1	9.40		0.70				2	F7/8 V	438
190469	+30 03871				20 01 57	+30 58.9	1	7.81		-0.05				3	A0	1118
190405	+17 04206	LS II +17 016			20 01 59	+17 35.5	1	6.86		0.48				1	F2 Ib	138
227415	+35 03932	LS II +35 024			20 02 00	+35 25.4	2	9.43	.012	0.40	.013	-0.58	.024	7	B3	180,1124
	+36 03845	LS II +36 005			20 02 00	+37 00.0	1	10.40		0.44		-0.17		2	B9 Ib	426
		NGC 6866 - 13			20 02 00	+44 03.2	1	12.91		0.45		0.09		1		49
		V1467 Cyg			20 02 01	+32 21.	1	13.50		2.59				1		1772
190283	−21 05609				20 02 02	−21 27.4	2	7.11	.009	1.04	.005	0.66		5	G8/K0 III	219,2018
		Cyg sq 3 # 2			20 02 03	+35 33.0	1	12.30		0.68		0.18		10		1124
190470	+25 04085				20 02 04	+25 38.9	1	7.88		0.91		0.71		3	K3 V	3026
		NGC 6866 - 12			20 02 04	+44 01.6	1	12.64		0.30		0.19		1		49
		NGC 6866 - 25			20 02 05	+44 02.3	1	15.43		1.03		0.06		1		49

HD	DM	Other Id	N	Rem	α_{1950}	δ_{1950}	S	V	σ_V	B–V	σ_{B-V}	U–B	σ_{U-B}	n	Spectrum	References
227423	+33 03713				20 02 06	+33 44.9	1	9.54		0.03				1	B8 V	1070
227421	+35 03933				20 02 06	+35 42.3	1	9.14		0.26				1	A5 III	1070
		NGC 6866 - 23			20 02 06	+43 58.7	1	14.68		0.56		0.08		1		49
		LS II +33 022			20 02 08	+33 25.8	1	10.46		0.23		-0.51		2		180
227433	+35 03934				20 02 08	+35 34.1	1	10.78		0.41		0.21		7	F2 Vp	1124
190331	-19 05697				20 02 09	-18 54.9	1	8.19		0.38		0.20		19	A5 IV	978
190536	+33 03714	IDS20003N3402	A		20 02 10	+34 10.4	2	7.66	.033	1.04	.009	0.80		2	G5 II	1070,8100
		NGC 6866 - 9			20 02 11	+44 02.3	1	12.09		0.21		0.17		1		49
	-74 01872				20 02 11	-74 26.1	1	10.05		0.59				4	F8	2012
		AJ84,127 # 2259			20 02 12	-13 37.	1	11.19		1.26				1		1510
190412	+0 04408				20 02 14	+01 00.8	1	7.69		0.70		0.16		1	G5	1462
190537	+30 03875				20 02 14	+31 06.0	2	6.87	.033	0.27	.014	0.14		6	F0 III	1118,3016
331759	+31 03921	LS II +32 013			20 02 14	+32 00.8	2	8.66	.040	0.49	.030	-0.41	.040	3	B1 Ib	1012,8100
		NGC 6866 - 6			20 02 14	+44 01.8	1	11.67		0.30		0.19		1		49
190549	+37 03748				20 02 15	+37 33.9	2	8.98	.038	-0.04	.009	-0.61		3	B3:III	401,1070
190570	+36 03848				20 02 16	+37 02.4	2	8.12	.017	-0.02	.000	-0.12		3	B9 V	401,1070
		NGC 6866 - 5			20 02 17	+43 59.9	1	11.64		1.00		0.71		1		49
		NGC 6866 - 17			20 02 17	+44 01.3	1	13.41		0.37		0.18		1		49
		XX Cyg			20 02 17	+58 48.7	1	11.28		0.12		0.12		1	A5	668
		MCC 814			20 02 17	-13 35.4	1	10.86		1.26				2	K6	1619
227454					20 02 18	+38 03.	1	10.07		1.11				2	K2	1726
227452	+39 04022	LS II +39 006		★AB	20 02 18	+39 48.9	1	9.84		0.43		0.05		2	F1 II	247
		NGC 6866 - 4			20 02 19	+44 01.0	1	11.30		0.21		0.21		1		49
		NGC 6866 - 8			20 02 19	+44 02.0	1	12.07		0.24		0.17		1		49
190369	-19 05700				20 02 19	-18 51.1	1	9.36		0.56		0.14		19	F2/3 V	978
190390	-12 05641	HR 7671, V1401 Aql		★A	20 02 20	-11 44.5	1	6.34		0.52				4	F1 III	2007
		NGC 6866 - 21			20 02 21	+44 01.9	1	13.89		0.42		0.13		1		49
190306	-33 14700	HR 7668		★AB	20 02 21	-33 08.6	1	6.53		-0.08				4	B4 III	2009
227460	+35 03936				20 02 24	+36 07.2	2	9.50	.017	0.15	.008	-0.69		10	B0.5:V	1118,1124
190287	-35 13902				20 02 24	-35 03.7	2	8.53	.005	0.75	.000	0.12		3	G6 Vw	742,1594
		NGC 6866 - 3			20 02 25	+44 03.4	1	11.18		0.25		0.22		1		49
227450	+32 03692				20 02 26	+32 50.9	1	9.17		0.00				1	B8 V	1070
		CS 22964 # 216			20 02 26	-39 26.9	1	15.76		-0.27		-1.06		1		1736
189723	-75 01572				20 02 27	-74 54.1	3	7.24	.004	1.06	.008	0.95	.019	11	K0 III	278,1117,2012
190571	+33 03716				20 02 28	+33 24.5	2	7.46	.007	0.87	.006			34	G8 V	1070,6011
	+43 03473	NGC 6866 - 1		★A	20 02 28	+43 56.4	1	10.10		0.95		0.70		1	G0	49
		CS 22964 # 203			20 02 28	-41 03.1	1	14.31		0.58		0.04		1		1736
		CS 22964 # 214			20 02 29	-39 36.3	2	13.65	.010	0.40	.007	-0.16	.009	3		1580,1736
227465	+33 03717	LS II +33 023			20 02 31	+33 33.8	3	10.29	.015	0.46	.015	-0.62	.080	6	O7.5 V	180,1011,1012
		VES 140			20 02 31	+36 00.2	2	13.60	.000	2.34	.000	0.97	.000	3		426,8103
		NGC 6866 - 2			20 02 31	+43 55.7	1	10.66		0.36		0.36		1		49
227463	+33 03718	CD Cyg			20 02 32	+33 58.2	5	8.40	.037	0.99	.033	0.68	.046	5	F8 Iab	934,1070,1399,1772,6011
		NGC 6866 - 22			20 02 32	+43 58.3	1	13.99		0.56		0.09		1		49
190269	-45 13603				20 02 32	-45 07.4	5	8.06	.003	0.47	.001	0.00	.014	27	F5 V	143,977,1075,1586,2011
		NGC 6866 - 19			20 02 33	+44 02.0	1	13.49		0.39		0.18		1		49
227464	+33 03719				20 02 34	+33 38.0	1	9.41		1.72				2	M7	369
		Cyg sq 3 # 5			20 02 34	+35 27.2	1	12.26		0.33		0.23		10		1124
		NGC 6866 - 24			20 02 34	+44 00.1	1	14.71		0.67		0.13		1		49
190901	+65 01425				20 02 34	+65 43.2	1	9.65		0.19		0.13		2	B9 V	555
		NGC 6866 - 11			20 02 36	+44 04.0	1	12.24		0.61		0.17		1		49
190940	+67 01222	HR 7685		★A	20 02 36	+67 43.9	6	4.51	.018	1.31	.008	1.51	.013	14	K3 III	15,1355,1363,3016*
		VES 112, EX Vul			20 02 37	+22 10.7	1	11.99		0.50		0.08		1		363
		NGC 6866 - 14			20 02 37	+44 01.8	1	12.98		1.20		0.99		1		49
190498	+1 04196				20 02 38	+01 58.6	1	6.76		0.47		0.05		2	F7 V	3016
190603	+31 03925	HR 7678, V1768 Cyg		★A	20 02 38	+32 04.5	11	5.64	.024	0.54	.010	-0.46	.016	30	B1.5Iae	15,154,450,1012,1025*
190603	+31 03925	IDS20007N3156	B		20 02 38	+32 04.5	1	11.44		0.74		0.34		3		450
		NGC 6866 - 18			20 02 38	+44 03.5	1	13.45		0.38		0.16		1		49
		NGC 6864 sq1 11			20 02 38	-22 05.3	1	12.84		0.67				3		438
227472	+34 03862				20 02 39	+34 51.0	1	8.30		1.58				1	K0 II	1070
		Cyg sq 3 # 7			20 02 40	+35 30.8	1	11.31		0.40		0.15		10		1124
		NGC 6866 - 10			20 02 40	+44 01.3	1	12.14		0.28		0.20		1		49
190454	-13 05569	HR 7675			20 02 40	-12 48.5	2	6.53	.015	0.05	.005	0.00		6	A1 Vn	252,2007
		NGC 6866 - 16			20 02 41	+43 59.4	1	13.40		0.44		0.09		1		49
190309	-44 13725				20 02 41	-44 29.1	5	7.87	.035	1.12	.007	1.06	.033	34	K1 IIICNII	143,705,977,1075,2011
227479					20 02 42	+35 44.	1	10.71		1.24		1.12		7	K5	1124
339479	+26 03780				20 02 43	+26 20.4	1	8.54		0.55		-0.01		3	G0 V	3026
227480	+35 03939				20 02 43	+35 26.4	3	9.30	.011	0.06	.020	-0.02	.000	8	A3 II	1070,1124,8100
190675	+41 03599				20 02 44	+42 03.3	1	8.33		-0.11		-0.64		2	B8	401
		LS II +22 021			20 02 47	+22 12.1	1	12.58		-0.35		-1.24		2	sdO	405
190590	+22 03913	HR 7677			20 02 49	+23 04.0	1	6.53		0.17		0.08		2	A5 Vn	1733
190605	+25 04090				20 02 49	+25 55.0	1	7.72		0.63		0.17		3	G2 V	3026
		NGC 6866 - 15			20 02 49	+44 00.7	1	12.99		1.44		1.41		1		49
190333	-43 13796	IDS19594S4330	AB		20 02 50	-43 21.4	2	9.21	.005	0.67	.005	0.15		6	G3/5 V	2011,3010
190606	+20 04417	X Sge		★A	20 02 53	+20 30.3	3	8.37	.048	3.30	.017			6	C6 II	109,414,3038
190606	+20 04417	IDS20007N2021	B		20 02 53	+20 30.3	1	13.18		0.77		0.16		1	G2	414
190780	+54 02281	IDS20017N5411	A		20 02 53	+54 19.5	1	7.72		0.94		0.65		2	K0	3016
		Ros 4 - 5			20 02 54	+29 04.	1	13.08		0.67		-0.14		3		18
		Ros 4 - 6			20 02 54	+29 04.	1	14.78		0.78		0.19		3		18
		Ros 4 - 8			20 02 54	+29 04.	1	13.61		0.99		-0.06		3		18
		Ros 4 - 9			20 02 54	+29 04.	1	14.98		0.99		0.20		3		18

Table 1 1089

HD	DM	Other Id	N Rem	α_{1950}	δ_{1950}	S	V	σ_V	B–V	σ_{B-V}	U–B	σ_{U-B}	n	Spectrum	References
		Ros 4 - 10		20 02 54	+29 04.	1	12.27		0.73		-0.12		3		18
		Ros 4 - 11		20 02 54	+29 04.	1	15.02		0.92		0.33		3		18
		Ros 4 - 12		20 02 54	+29 04.	1	12.85		0.95		0.00		3		18
		Ros 4 - 13		20 02 54	+29 04.	1	9.70		0.53		0.14		3		18
		Ros 4 - 14		20 02 54	+29 04.	1	11.61		0.73		-0.09		3		18
227496				20 02 54	+35 25.	1	11.26		0.21		0.19		8	A7	1124
190608	+19 04277	HR 7679		20 02 56	+19 50.8	2	5.09	.000	1.06	.000	0.98	.000	4	K2 III	1355,3051
		CS 22873 # 128		20 02 56	−58 43.6	1	13.03		0.69		0.05		1		1736
189794	−75 01573			20 02 56	−75 30.7	1	9.34		1.38				4	K2 III	2012
		VES 141		20 02 58	+39 02.3	2	12.02	.000	1.70	.000	1.07	.000	3		426,8103
190781	+47 03004	HR 7684		20 02 59	+48 05.2	2	6.15	.010	0.04	.000	0.04	.005	4	A2 IV	985,1733
227504	+35 03940			20 03 00	+35 41.8	1	9.96		0.57		-0.04		7	G0	1124
	−45 13607			20 03 00	−45 13.2	1	10.56		1.04				4		2011
227528	+40 04003			20 03 04	+40 53.7	1	9.34		-0.03		-0.64		1	B3	963
		NGC 6864 sq1 4		20 03 04	−22 08.7	1	11.98		1.29				3		438
	+21 04047			20 03 05	+22 07.1	1	10.26		1.84				1	M2	369
		Cyg sq 3 # 11		20 03 05	+35 22.2	1	11.95		0.73		0.30		10		1124
190221	−61 06453			20 03 05	−61 37.3	1	8.50		1.36		1.56		4	K2 III	1731
	+30 03882			20 03 06	+30 31.7	1	10.15		1.93				2	M3	369
227512				20 03 06	+35 47.	1	11.51		0.25		0.15		6	A7	1124
		Cyg sq 3 # 12		20 03 07	+35 28.6	1	12.53		0.43		0.17		10		1124
		G 125 - 59		20 03 07	+35 34.3	2	12.97	.005	1.02	.017	0.70	.002	4		1658,3032
	−45 13609			20 03 07	−44 47.8	2	11.28	.001	0.49	.002	-0.02		5		861,2011
190658	+15 04040	HR 7680	★ AB	20 03 08	+15 21.4	3	6.35	.007	1.61	.029	1.71	.058	5	M2.5 III	31,71,3040
		LS II +34 006		20 03 09	+34 49.5	1	10.78		0.30		-0.68		1		180
190540	−19 05704	V4091 Sgr		20 03 09	−18 50.9	1	8.38		1.20		1.02		1	G8/K0 Vp	1641
189899	−74 01874			20 03 09	−74 21.8	4	7.59	.004	0.51	.006	0.03	.013	17	F7 V	278,1075,1117,2012
227523	+35 03942			20 03 10	+35 50.7	1	10.47		0.20		-0.49		6	F0	1124
		Cyg sq 3 # 15		20 03 12	+36 30.8	1	11.83		0.56		0.05		10		1124
		E8 - o		20 03 12	−44 46.	1	14.68		0.59				4		1075
		G 24 - 3		20 03 15	+03 54.4	4	10.46	.012	0.47	.009	-0.19	.012	5	sdF8	516,1620,1658,3077
356711	+11 04124	Ring Aql # 28		20 03 15	+12 13.6	1	9.07		1.29		1.06		3	K0	4
227535	+36 03859			20 03 15	+36 31.6	1	9.44		1.11		0.77		8	G0:II:	1124
	+43 03477			20 03 15	+43 28.2	1	10.11		2.49		2.21		3		8032
227536	+35 03943	LS II +36 006		20 03 16	+36 00.7	3	9.49	.008	0.21	.008	-0.13	.025	8	B9 V	247,1124,8100
		NGC 6864 sq1 3		20 03 16	−22 07.3	1	11.16		0.70				3		438
		NGC 6864 sq1 14		20 03 18	−22 07.3	1	14.79		0.87				1		438
		E8 - i		20 03 18	−44 46.	1	13.62		0.50				4		1075
190660	+11 04125	Ring Aql # 29		20 03 19	+12 13.5	1	9.39		0.29		0.02		2	A5 IV	4
		VES 142		20 03 19	+38 15.4	1	11.17		0.59		0.18		2		426
		G 143 - 26		20 03 20	+15 15.0	1	12.79		1.44		1.13		3		316
227549	+35 03944			20 03 20	+35 33.5	1	8.54		1.44		1.49		7	G8 III	1124
227546	+36 03860			20 03 20	+36 32.1	1	10.43		0.28		-0.05		7	B9	1124
190771	+38 03896	HR 7683	★ AB	20 03 20	+38 20.0	3	6.17	.017	0.64	.003	0.20	.019	12	G5 IV	1118,3016,7008
227547	+35 03945			20 03 21	+36 09.3	1	9.94		0.65		0.17		6	G5 V	1124
		V1033 Cyg		20 03 23	+32 30.9	1	12.68		1.50				1		1772
	−44 13733			20 03 23	−44 22.2	1	9.78		1.57				4		1075
	−45 13611			20 03 23	−44 56.6	1	11.45		0.49				4		2011
354850	+14 04170			20 03 24	+14 38.0	1	10.42		0.69		0.16		2	G0	1064
333549	+30 03813	LS II +29 041		20 03 24	+29 12.3	3	9.56	.008	1.38	.035	1.11	.023	14	F2 Ib-II	830,1215,1783
227548				20 03 24	+35 57.	1	10.86		0.16		-0.52		7	B9	1124
		G 230 - 25		20 03 24	+51 04.3	1	15.11		1.28				1		1759
		NGC 6864 sq1 10		20 03 24	−22 03.3	1	12.67		0.68				1		438
190480	−44 13734			20 03 26	−44 26.9	10	8.10	.004	1.41	.006	1.59	.018	116	K3/4 III	705,863,977,1460,1628,1657*
190785	+34 03866			20 03 30	+34 50.5	1	8.19		0.06				1	A4 V	1070
		E8 - q		20 03 30	−44 53.	1	14.90		0.72				4		1075
		NGC 6864 sq1 13		20 03 31	−21 59.1	1	14.24		0.70				1		438
		NGC 6864 sq1 12		20 03 32	−21 59.2	1	13.53		0.82				1		438
		NGC 6864 sq1 7		20 03 32	−21 59.7	1	12.60		1.15				1		438
		G 92 - 53		20 03 33	+07 28.2	1	14.96		1.11		0.97		1		333,1620
		G 143 - 27		20 03 33	+14 53.5	2	12.93	.041	0.58	.007	-0.12	.005	3		1658,1773
		LS II +33 024		20 03 33	+33 28.3	1	10.66		0.46		-0.03		3		180
227573	+35 03946			20 03 33	+35 56.5	3	9.60	.021	0.21	.002	-0.20	.004	10	B7 IV	426,1124,8103
189877	−76 01386			20 03 33	−76 01.4	1	9.73		0.56				4	G2/3 V	2012
190664	−4 05013	HR 7681		20 03 34	−04 13.3	2	6.46	.005	1.16	.000	1.00	.000	7	K0	15,1256
		NGC 6864 sq1 8		20 03 34	−22 07.2	1	12.65		1.25				2		438
		E8 - p		20 03 34	−44 51.1	2	14.69	.031	0.77	.012	0.41		8		1075,1460
190421	−53 09794	HR 7673		20 03 34	−53 01.6	5	4.94	.009	1.62	.004	1.85	.022	17	M1 IIab	15,1075,2012,3055,8100
	−45 13613			20 03 35	−45 41.8	1	10.47		0.68				4		2011
191124	+65 01427			20 03 36	+65 53.9	1	9.32		0.12		0.06		3	A0 V	555
		NGC 6864 sq1 6		20 03 36	−22 06.7	1	12.44		0.29				2		438
190222	−66 03473	HR 7663		20 03 36	−66 30.0	3	6.44	.012	1.58	.004	1.98	.000	14	K5 III	15,1075,2012
	+4 04344	G 23 - 25		20 03 37	+05 06.8	1	9.62		1.07		0.95		1	K2	333,1620
227575	+33 03729	LS II +33 025		20 03 38	+33 35.6	2	9.31	.015	1.17	.055	0.95	.005	4	F4 Ia	180,247
227574	+33 03730			20 03 38	+34 03.1	2	9.39	.014	0.29	.047	0.08		2	A3 IV	934,1070
190813	+33 03728	Ros 5 - 1		20 03 39	+33 20.5	3	8.68	.021	0.02	.007	-0.16	.007	7	B7 III	102,1070,1192
190422	−55 09317	HR 7674		20 03 39	−55 09.7	2	6.25	.005	0.53	.000			7	F8 V	15,2012
190558	−39 13615			20 03 40	−39 00.2	1	8.28		1.54		1.91		1	K4 III	1468
190814	+29 03889	IDS20017N3003	AB	20 03 42	+30 11.1	2	8.27	.132	0.47	.069	-0.02	.069	15	F5 V	938,1371
		E8 - k		20 03 42	−44 47.	1	13.68		0.61				4		1075

HD	DM	Other Id	N	Rem	α₁₉₅₀	δ₁₉₅₀	S	V	σ_V	B–V	σ_B-V	U–B	σ_U-B	n	Spectrum	References
		E8 - e			20 03 42	−45 03.	1	12.59		0.50				4		2011
190841	+34 03867	LS II +34 007	4		20 03 43	+34 46.3	4	8.17	.046	0.70	.022	0.64	.028	5	F2 V	180,247,1070,8100
227586	+35 03948	NGC 6871 - 7		⋆ AB	20 03 43	+35 28.8	7	8.82	.033	0.21	.020	-0.66	.008	21	B0.5IVp	49,450,671,1070,1124*
190788	+25 04097	IDS20016N2518	A		20 03 44	+25 27.4	4	7.94	.034	2.29	.012	2.21		14	M2 II	148,369,8032,8053
190772	+19 04285				20 03 45	+19 26.2	1	7.59		1.15		1.18		1	G5	272
190787	+27 03612				20 03 45	+27 59.3	1	7.54		1.72				2	M1	369
190964	+51 02763	HR 7687	3		20 03 45	+51 41.7	3	6.13	.008	1.60	.010	1.96	.009	6	M1 IIIa	71,1501,3001
		NGC 6871 - 14		⋆ V	20 03 46	+35 36.9	1	10.80		0.20		-0.46		1		49
190864	+35 03949	NGC 6871 - 4		⋆ A	20 03 47	+35 27.8	9	7.77	.019	0.19	.016	-0.75	.014	24	O6.5III(f)	49,450,1011,1012,1070*
		E8 - b			20 03 48	−44 46.	1	11.97		0.67				4		2011
		E8 - f			20 03 48	−44 56.	1	12.83		0.40				4		2011
333535	+29 03891	IDS20018N2929	AB		20 03 49	+29 37.4	1	9.18		0.48		0.04		3	G0	3026
190376	−60 07371				20 03 49	−60 29.3	1	9.10		0.48		-0.04		4	F5 V	1731
190248	−66 03474	HR 7665	7		20 03 50	−66 18.7	7	3.56	.010	0.76	.002	0.45	.004	37	G6/8 IV	15,1034,1075,1075*
		E8 - n			20 03 51	−44 48.9	2	14.45	.009	0.59	.005	0.05		9		1075,1460
227613	+33 03734				20 03 52	+34 08.1	2	8.94	.024	1.05	.028	0.84		2	K0	934,1070
		NGC 6871 - 26			20 03 52	+35 41.5	1	11.85		0.30		0.18		1		49
190944	+46 02846	LS III +46 002	4		20 03 52	+46 31.7	4	8.35	.010	0.22	.010	-0.78	.017	9	B1.5Vne	399,401,405,483
		G 230 - 26		⋆ V	20 03 52	+54 17.9	1	11.97		1.52				4		538
190842	+29 03893				20 03 53	+29 20.7	1	8.42		1.16		2.12		1	G8 II	8100
227611	+35 03950	NGC 6871 - 6		⋆ A	20 03 53	+35 45.4	5	8.74	.027	0.35	.009	-0.71	.021	7	B0pe	49,671,1012,1070,1192
227608					20 03 54	+36 11.	1	10.63		0.40		0.06		7	F0	1124
		E8 - g		V	20 03 54	−45 04.	1	13.02		0.40				4		2011
		NGC 6871 - 19			20 03 55	+35 39.4	1	11.52		0.22		-0.38		1		49
190916	+40 04015	LS II +41 001	2		20 03 55	+41 08.1	2	7.64	.015	0.41	.040	-0.13	.025	2	B9 Iab	247,8100
		NGC 6871 - 27			20 03 56	+35 46.5	1	11.92		0.30		-0.35		1		49
		NGC 6871 - 20			20 03 57	+35 36.6	1	11.57		0.29		-0.14		1		49
		NGC 6871 - 16			20 03 57	+35 39.0	1	10.96		0.23		-0.44		1		49
	−22 05336	NGC 6864 sq1 2			20 03 57	−22 10.0	1	10.60		1.19				2		438
	−45 13616				20 03 57	−45 43.8	2	10.44	.006	0.65	.008	0.09		10		863,2011
		G 125 - 60			20 03 58	+29 12.1	1	12.41		0.93		0.71		2		1658
331727	+32 03700				20 03 58	+32 38.5	1	9.23		0.07				1	B6 III	1070
		NGC 6871 - 69			20 03 58	+35 37.0	1	13.98		0.47		0.35		2		305
		NGC 6864 sq1 5			20 03 58	−22 02.5	1	12.17		0.77				2		438
		G 143 - 28			20 03 59	+16 00.9	1	14.50		1.61				4		419
333506	+30 03888				20 03 59	+30 19.0	1	10.77		1.17		2.68		1	K2 II	8100
		WLS 2000 5 # 9			20 04 00	+04 41.2	1	11.44		0.57		0.11		2		1375
		AJ89,1229 # 425			20 04 00	+05 33.	1	12.32		1.70				1		1532
227621		NGC 6871 - 11			20 04 00	+35 33.8	1	10.32		0.17		-0.60		1	B2 V	49
227630	+35 03951	NGC 6871 - 9			20 04 02	+35 28.8	2	9.49	.002	0.62	.002	0.09	.005	8	K5	49,1124
		E8 - h			20 04 02	−44 50.1	2	13.30	.009	0.60	.009	0.00		9		1460,2011
190919	+35 03952	NGC 6871 - 2	4		20 04 03	+35 31.7	4	7.31	.013	0.24	.009	-0.64	.017	4	B0.7Ib	49,672,1070,8100
		NGC 6871 - 29			20 04 03	+35 37.1	2	12.89	.005	0.35	.000	-0.01	.024	3		49,305
227625	+32 03702	Ros 5 - 2			20 04 04	+32 51.4	1	8.58		1.51		1.75		3	A2	102
		NGC 6871 - 24			20 04 04	+35 36.7	2	11.71	.000	0.26	.000	-0.37	.000	3		49,305
		NGC 6871 - 68		⋆ D	20 04 04	+35 38.7	1	9.71		0.21		-0.63		1		305
190886	+25 04099				20 04 05	+25 18.8	1	8.17		0.00		-0.24		3	A0	833
		NGC 6871 - 25			20 04 05	+35 37.0	1	11.83		0.29		-0.34		1		49
190918	+35 03953	NGC 6871 - 1		⋆ AB	20 04 05	+35 38.6	8	6.80	.038	0.13	.015	-0.77	.016	20	WN5 +O9.5III	49,305,1070,1095,1192*
356792	+10 04173	Ring Aql # 13			20 04 06	+10 25.5	1	10.60		0.19		0.11		2	A2	4
333565	+28 03619				20 04 06	+28 46.4	1	8.79		0.61		0.10		3	G0 V	3026
	+35 03955	NGC 6871 - 3		⋆ F	20 04 06	+35 39.2	6	7.36	.025	0.24	.014	-0.65	.018	11	B0.7Iab	49,305,1192,3024,8084,8100
		E8 - a			20 04 06	−44 52.5	2	12.09	.009	0.61	.009	0.10		8		1460,2011
		NGC 6871 - 30			20 04 07	+35 34.7	1	12.91		0.25		-0.05		1		49
	+35 03956	NGC 6871 - 8		⋆ D	20 04 07	+35 37.1	6	8.86	.022	0.20	.007	-0.62	.005	11	B0.5V	49,305,1012,1012,1070,1192
		NGC 6871 - 35		⋆ E	20 04 07	+35 38.5	1	11.42		0.22		-0.33		2		3024
191174	+63 01593	HR 7695		⋆ AB	20 04 07	+63 44.8	1	6.26		0.10		0.04		2	A2 II-III	3024
191174	+63 01593	IDS20035N6336	C		20 04 07	+63 44.8	1	11.38		0.71		0.25		2		3024
		NGC 6871 - 70		⋆ Q	20 04 08	+35 36.6	1	13.18		0.30		-0.20		3		450
227634	+35 03957	NGC 6871 - 5		⋆ AB	20 04 09	+35 37.3	5	7.90	.020	0.25	.009	-0.66	.012	8	B0 II	49,305,1012,1070,1192,8100
		E8 - u			20 04 09	−44 51.4	1	15.91		0.60		0.00		4		1460
190945	+32 03704	Ros 5 - 3			20 04 11	+32 27.6	2	8.75	.004	0.07	.000	0.04		4	A1 V	102,1070
		NGC 6871 - 28			20 04 11	+35 35.5	1	12.68		0.28		-0.14		1		49
		NGC 6871 - 21			20 04 11	+35 37.0	1	11.68		0.25		-0.37		1		49
		NGC 6871 - 23			20 04 11	+35 37.2	1	11.70		0.28				1		49
190850	+7 04366	IDS20018N0716	B		20 04 12	+07 24.8	1	7.59		0.18		0.07		3	A2 V	1084
190849	+7 04367	IDS20018N0716	A		20 04 12	+07 25.9	1	7.14		0.13		0.02		3	A1 V-IV	1084
		E8 - Z			20 04 12	−44 46.	1	11.81		0.59				4		2011
		E8 - r			20 04 12	−44 51.	1	15.38		0.67				4		1075
		E8 - l			20 04 12	−44 53.	1	13.95		0.67				4		1075
		E8 - d			20 04 12	−44 56.	1	12.73		0.29				4		2011
191096	+55 02324	HR 7692	4		20 04 13	+56 11.8	4	6.19	.010	0.41	.007	0.01	.017	5	F4 V	1716,1733,3053,6009
		VES 113			20 04 14	+25 17.8	1	12.89		0.82		0.46		1		363
190649	−41 13871				20 04 14	−41 09.7	1	8.84		0.64		0.10		4	G5 V	158
190991	+39 04033	LS II +39 008			20 04 15	+39 55.2	1	8.22		0.14		-0.78		2	B0 IVp	1012
190967	+34 03871	V448 Cyg	A		20 04 17	+35 14.5	2	8.05	.010	0.39	.024	-0.56		3	B1 Ib-II	1012,1070
		WLS 2000 40 # 9			20 04 21	+40 09.9	1	10.22		1.30		1.29		2		1375
		NGC 6871 - 15			20 04 22	+35 42.4	1	10.81		0.27		-0.50		1		49
190887	+12 04226	IDS20020N1239	AB		20 04 23	+12 47.5	1	7.29		0.43		0.17		3	F2	8100
		NGC 6871 - 17			20 04 23	+35 29.6	1	11.24		0.40		-0.31		1		49

Table 1

HD	DM	Other Id	N Rem	α_{1950}	δ_{1950}	S	V	σ_V	B–V	σ_{B-V}	U–B	σ_{U-B}	n	Spectrum	References
		WLS 2136 85 # 6		20 04 23	+84 33.0	1	11.06		1.67		1.86		2		1375
227659	+35 03958	NGC 6871 - 12		20 04 24	+35 39.4	2	10.37	.004	0.54	.001	0.02	.005	8	G0	49,1124
239319	+57 02135	IDS20034N5743	A	20 04 27	+57 51.2	1	9.16		1.02		0.81		1	K0	3032
239319	+57 02135	IDS20034N5743	BC	20 04 27	+57 51.2	1	12.97		1.69		1.18		1		3032
195252	+87 00187			20 04 28	+87 47.5	1	7.96		1.26		1.38		2	K2	1502
		CS 22873 # 137		20 04 28	−59 10.2	1	13.46		0.39		-0.10		1		1736
		NGC 6871 - 22		20 04 29	+35 43.8	1	11.68		0.25				1		49
227671		NGC 6871 - 13		20 04 30	+35 34.2	1	10.38		0.22		-0.52		1	B8 V	49
191026	+35 03959	HR 7689	★ A	20 04 30	+35 50.0	9	5.36	.013	0.85	.006	0.56	.007	88	K0 IV	15,22,247,1003,1080*
190709	−44 13744			20 04 32	−44 20.3	5	8.48	.003	0.44	.004	-0.05	.018	27	F5 V	143,977,1075,1586,2011
191024	+38 03905			20 04 33	+38 26.7	2	8.29	.049	-0.07	.029	-0.34	.058	3	B8 II	401,8100
190922	+10 04176	IDS20022N1018	A	20 04 34	+10 26.7	1	8.42		0.46		-0.07		3	F5 V	4
227680	+35 03961	LS II +36 008		20 04 34	+36 11.1	1	9.68		0.33		-0.46		2	B3 II	1012
		LP 926 - 51		20 04 35	−31 53.3	1	12.51		1.65				1	M3	1705
		E8 - c		20 04 36	−44 53.	1	12.21		0.65				4		2011
		E8 - m		20 04 36	−44 57.	1	12.67		0.65				4		1075
227682				20 04 37	+35 46.0	1	10.74		0.40		0.19		7	F8	1124
191046	+35 03962			20 04 37	+36 05.0	6	7.03	.024	1.15	.006	1.02	.006	17	G9 III	247,1003,1070,1080*
191045	+38 03906			20 04 37	+38 57.2	2	6.86	.005	1.55	.005	1.84	.005	6	K5 III	247,833
		LS II +34 008		20 04 38	+34 23.3	1	12.43		0.26		0.33		1		180
		VES 144		20 04 38	+38 14.9	2	11.90	.000	1.96	.000	2.16	.000	3		426,8103
		LP 926 - 52		20 04 38	−31 53.6	1	12.22		1.56				1		1705
191010	+25 04103			20 04 39	+25 32.4	2	8.16	.005	1.00	.005	0.66		10	G3 Ib	8059,8100
227695	+35 03963	NGC 6871 - 10	★ V	20 04 41	+35 44.1	1	10.14		0.40		0.18		1	A5p	49
191372	+67 01226	HR 7704		20 04 41	+67 53.0	2	6.30	.017	1.65	.001	2.00		3	M3 IIIa	71,3055
191047	+34 03873			20 04 42	+34 59.2	3	7.89	.009	1.00	.003	0.80	.001	53	G5 II	671,1070,1124,8100
227696	+35 03964	NGC 6871 - 31	★ V	20 04 42	+35 35.8	6	8.28	.045	0.20	.015	-0.67	.012	25	B0.5IV	588,671,672,1070,1124,1192
227693	+35 03965	IDS20028N3552	A	20 04 42	+36 00.3	1	9.19		0.95		0.67		6	G5 IV	1124
190993	+23 03896	HR 7688		20 04 44	+23 28.1	3	5.07	.004	-0.18	.002	-0.69	.010	9	B3 V	154,1363,3016
		NGC 6871 - 18		20 04 44	+35 35.0	1	11.33		0.29		0.21		1		49
191277	+61 01970	HR 7701		20 04 45	+61 51.0	3	5.42	.014	1.19	.006	1.29	.023	7	K3 III	1355,3016,4001
227704	+34 03874	LS II +34 009	★ AB	20 04 47	+34 46.0	3	8.61	.021	0.33	.016	-0.63	.000	4	B0 III	180,1012,1070
227711	+34 03876			20 04 48	+35 12.4	1	9.02		0.09				1	A1 V	1070
190995	+10 04178	Ring Aql # 14	★ A	20 04 49	+10 27.8	1	8.54		0.02		-0.16		3	B9 V	4
191064	+33 03741	Ros 5 - 4	★ A	20 04 50	+33 19.9	2	8.38	.011	-0.04	.011	-0.21		6	B9 III	102,1070
227719	+38 03907			20 04 53	+38 24.3	1	9.67		0.21		0.11		1	A1 IV	401
191082	+32 03707	Ros 5 - 5		20 04 54	+32 27.4	3	8.05	.013	-0.03	.005	-0.28		8	A0 V	102,1070,1118
191195	+52 02623	HR 7697	★ A	20 04 55	+53 01.0	4	5.83	.022	0.42	.018	-0.03	.000	7	F4 V	254,1118,1620,3053
190223	−75 01575			20 04 57	−74 56.0	1	9.07		1.22				4	K2 III	2012
190777	−45 13630			20 04 59	−44 47.8	6	9.52	.005	1.13	.004	1.02	.012	39	K0/1 III	861,863,1460,1628,1730,2011
227722	+34 03877		AB	20 05 00	+35 09.6	1	9.53		0.16				1	B1 III	1070
		GD 228		20 05 01	+05 43.7	1	13.58		0.51		-0.20		3		308
		VES 145		20 05 01	+35 28.8	1	11.72		2.00				2		426
190779	−46 13399	IDS20015S4635	A	20 05 04	−46 26.8	2	8.23	.005	0.44	.015	-0.05		6	F3 IV/V	1075,1279
191139	+35 03966	LS II +36 009		20 05 06	+36 15.1	4	7.97	.011	0.22	.008	-0.67	.002	94	B0.5III	1070,1124,1192,8100
		VES 114		20 05 08	+32 01.0	1	11.24		2.12		1.28		1		363
227733				20 05 08	+35 31.5	1	10.31		0.24		-0.44		2	A5	426
191138	+38 03909			20 05 08	+38 41.9	1	8.92		0.05		-0.09		4		833
356714	+11 04143	Ring Aql # 30		20 05 10	+12 08.8	1	9.07		1.36		1.16		2	K0	4
227740	+36 03878			20 05 10	+36 53.3	1	9.08		0.99		0.60		1	G8 IV	1462
191632	+73 00897			20 05 10	+73 45.9	1	6.88		0.94		0.67		2	K0	985
190803	−46 13400	IDS20015S4635	B	20 05 10	−46 27.0	1	8.94		0.81		0.41		2	K0 V	1279
227741	+36 03879	IDS20033N3634	AB	20 05 11	+36 42.6	1	9.37		0.07				1	B8 V	1070
190827	−45 13631			20 05 13	−45 34.5	2	9.04	.000	1.01	.005	0.76		10	K0 III	863,2011
227749	+37 03771			20 05 15	+37 56.6	2	9.56	.000	0.03	.000	-0.19	.000	3	B8 V	426,8103
		CS 22964 # 231		20 05 16	−38 10.7	1	14.60		0.46		0.03		1		1736
191083	+17 04225			20 05 18	+17 31.3	2	8.76	.000	0.16	.000	0.13	.004	7	A2	1266,1321
		VES 149	V	20 05 19	+35 12.4	1	10.36		0.16		-0.24		1		8103
		WZ Sge		20 05 21	+17 33.5	2	14.66	.300	0.02	.042	-0.99	.148	14	DAep	866,3028
227757	+35 03967	LS II +36 010		20 05 21	+36 12.8	4	9.22	.013	0.19	.013	-0.74	.005	14	B2	1011,1012,1070,1124
190878	−44 13748			20 05 21	−44 33.6	3	9.98	.003	0.89	.002	0.51	.010	13	G8 IV	861,863,1075
191177	+32 03712	Ros 5 - 6		20 05 22	+32 39.1	2	8.80	.011	0.38	.011	0.10		5	F4 V	102,1070
227760	+34 03879			20 05 22	+35 13.8	1	10.36		0.16		-0.24		1	A2	426
191176	+37 03772			20 05 22	+37 34.4	1	8.35		0.02				1	A1 III	1070
		G 92 - 55		20 05 22	−01 41.2	1	13.55		1.62		1.20		1		1620
227758	+35 03968	IDS20035N3517	A	20 05 23	+35 25.7	1	9.18		0.11		0.04		60	B9 V	1124
333617	+27 03621			20 05 24	+27 34.3	1	8.70		1.63				2	K7	369
191104	+8 04344	HR 7693	★ AB	20 05 26	+09 15.2	4	6.44	.014	0.46	.008	-0.04	.022	11	F3 V	15,254,938,1256
227767	+35 03969	IDS20035N3517	APB	20 05 26	+35 26.4	2	8.89	.018	0.03	.015	-0.77		8	B2 III:	1070,1124
191067	−1 03899	HR 7690		20 05 27	−00 49.4	5	5.99	.002	1.02	.004	0.85	.015	57	K1 V	15,37,1071,3047
190879	−47 13262			20 05 29	−47 12.8	2	6.45	.005	1.47	.002	1.78		6	K5 III	2012,3040
		G 23 - 26		20 05 30	+07 17.9	1	11.26		1.38		1.30		1	M0	333,1620
	+35 03971		V	20 05 30	+35 39.8	2	9.98	.000	0.19	.000	-0.62	.000	2		426,8103
		JL 42		20 05 30	−73 45.	1	16.54		0.14		-1.26		1		132
191201	+35 03970	LS II +35 043	★ ABC	20 05 31	+35 34.3	4	7.26	.028	0.13	.010	-0.77	.014	6	B0 III	672,1070,1118,1192
190897	−45 13634	IDS20020S4509	AB	20 05 31	−45 00.5	1	9.46		0.53				4	F7 V	2011
		VES 151		20 05 32	+36 18.7	1	10.97		0.15		-0.26		1		426
	+36 03882	LS II +36 012		20 05 33	+36 24.6	1	9.89		0.38		-0.52		2	B1 III:	1012
190710	−64 04025			20 05 33	−63 48.3	1	7.82		1.69		2.03		6	M1 III	1704
227776	+35 03972			20 05 34	+35 46.0	1	8.49		1.19				1	K0 II	1070

HD	DM	Other Id	N	Rem	α₁₉₅₀	δ₁₉₅₀	S	V	σ_V	B–V	σ_B-V	U–B	σ_U-B	n	Spectrum	References
191226	+36 03883			V	20 05 35	+36 25.3	3	7.29	.035	1.81	.023	1.96	.061	11	K2 II:	369,1124,8100
		VES 152			20 05 36	+35 37.8	1	11.77		0.47		-0.54		2		426
	+42 03594				20 05 36	+42 52.1	1	8.92		0.12		0.01		2	A2	401
191227	+34 03880				20 05 39	+34 35.6	1	8.31		0.30				1	F2 IV	1070
227785	+35 03973				20 05 41	+35 28.4	1	9.44		0.02				1	B9 II	1070
191179	+15 04053				20 05 42	+16 01.2	1	8.34		0.82				1	G5	882
227784	+35 03974				20 05 42	+35 41.6	1	9.58		0.01				1	B8 V	1070
		VES 115			20 05 43	+34 14.3	1	12.24		0.47		-0.20		1		363
191069	-16 05509				20 05 44	-15 52.3	2	8.12	.005	0.67	.000	0.20	.020	3	G5 V	219,1003
351517	+17 04228				20 05 46	+17 31.4	1	9.72		1.03		0.76		2	K0	1321
191243	+34 03881	HR 7699		★	20 05 46	+34 16.6	6	6.11	.022	0.15	.014	-0.45	.034	8	B5 Ib	180,1063,1079,1118*
227787	+33 03746	LS II +33 028			20 05 47	+33 30.5	1	9.63		0.83		0.50		2	F4 Ib	247
239326	+56 02353				20 05 47	+57 13.7	1	9.18		-0.08		-0.63		2	B3	401
191110	-10 05285	HR 7694, AV Cap			20 05 47	-10 12.5	1	6.18		0.06				4	B9.5III	2007
227795					20 05 48	+35 36.	1	9.63		0.07				1	F2	1070
191178	+16 04153	HR 7696			20 05 50	+16 31.1	2	6.41	.010	1.81	.083	1.90	.034	3	M3 III	1733,3001
331821	+32 03714	Ros 5 - 7			20 05 52	+32 31.3	1	8.75		1.09		0.92		3	K0	102
227808					20 05 54	+33 58.	1	9.13		0.21				1	A7	1070
227806					20 05 54	+35 36.	1	10.65		0.21		-0.49		6	F5	1124
191333	+43 03495				20 05 56	+43 51.7	1	8.28		-0.08		-0.36		2	A0	401
	+6 04450	G 23 - 27			20 05 57	+06 31.9	2	9.79	.005	1.18	.019	1.09		2	K8	333,1017,1620
191290	+38 03915				20 05 57	+38 18.4	1	8.28		1.10		1.47		1	K0 II	8100
		LP 754 - 23			20 05 58	-12 17.3	1	12.50		1.03		0.82		1		1696
191007	-44 13755				20 05 58	-44 03.8	5	7.84	.003	1.19	.001	1.24	.007	27	K1 III	116,977,1075,1586,2011
		Steph 1758			20 06 00	+06 21.2	1	10.78		1.72		2.15		2	M0	1746
191292	+32 03715	Ros 5 - 8		★ AB	20 06 00	+32 26.4	3	7.90	.014	0.02	.006	-0.32	.004	8	B9	102,1118,1192
191374	+46 02854				20 06 00	+46 59.9	1	8.76		0.14		0.07		2	A0	1601
	-74 01880				20 06 01	-74 36.4	1	10.02		1.04				4	G5	2012
191291	+34 03882				20 06 02	+34 57.5	3	8.09	.034	-0.07	.014	-0.41	.024	8	B6 III	1070,1124,8100
227828	+35 03978				20 06 05	+35 34.3	1	9.22		1.47				2		1726
		LS II +38 006			20 06 05	+38 04.7	1	11.50		0.39		0.17		1	A0 III	1215
227829					20 06 06	+35 22.	1	11.23		0.29		0.19		6	A7	1124
		VES 116			20 06 08	+33 59.8	1	10.92		0.46		-0.44		2		363
227838					20 06 12	+35 21.	1	11.27		0.22		0.18		6	A7	1124
227836	+35 03981	V425 Cyg			20 06 12	+35 58.6	3	10.23	.043	0.20	.007	0.03	.085	8	B2/5: V:	1124,1192,1215
356879	+11 04149	Ring Aql # 31			20 06 14	+12 06.4	1	8.85		1.67		1.84		2	M0	4
191263	+10 04189	HR 7700			20 06 15	+10 34.7	2	6.31	.012	-0.12	.012	-0.67	.004	21	B3 IV	4,154
191335	+32 03717	Ros 5 - 9			20 06 15	+33 14.6	1	9.37		0.14		0.11		5		102
191334	+32 03718	Ros 5 - 10			20 06 16	+33 15.7	1	8.11		1.06		0.87		6	K0	102,1070
191378	+36 03891				20 06 17	+36 51.6	1	8.96		0.08				27	B9.5V	6011
		VES 153			20 06 17	+38 53.9	2	12.70	.000	2.05	.000	1.90	.000	3		426,8103
		VES 154			20 06 18	+35 16.2	1	13.19		0.69		0.04		2		426
191379	+36 03892				20 06 19	+36 31.5	1	7.54		1.74				2	M1	369
		G 125 - 62			20 06 21	+33 09.1	1	14.88		1.82		1.14		3		203
191395	+39 04049	LS II +39 009			20 06 21	+39 36.3	1	8.34		0.09		-0.75		1	B0.5V:	1192
191356	+29 03909				20 06 24	+30 09.0	1	8.16		1.67		1.96		13	K5	1371
		JL 43			20 06 24	-71 24.	1	13.96		0.64		-0.27		2		132
191423	+42 03599	LS III +42 003			20 06 25	+42 27.6	2	8.03	.000	0.16	.000	-0.76	.005	4	A9 V:	1011,1012
191295	+11 04153	Ring Aql # 32			20 06 28	+12 04.9	1	7.16		-0.04		-0.41		3	B7 III	4
227877	+35 03983	IDS20046N3510		AB	20 06 30	+35 18.7	1	9.04		0.11				1	B1:IV:nn	1070
191396	+37 03783	LS II +37 010		★ A	20 06 30	+37 59.0	3	8.14	.010	0.24	.004	-0.67	.005	6	B0.5II	1012,1070,1775,8100
191116	-42 14714				20 06 34	-41 59.7	3	8.00	.003	1.60	.011	1.94	.001	13	M2 III	977,1468,2011
227886	+33 03756	Ros 5 - 11			20 06 36	+33 29.1	1	9.33		1.01		0.79		5	K2	102
		ADS 13368		★ A	20 06 37	+35 59.2	1	10.88		0.41		0.10		1		8084
		ADS 13368		★ AB	20 06 37	+35 59.2	1	10.16		0.52		0.04		1		8084
191117	-44 13763				20 06 37	-44 02.5	10	6.93	.005	1.02	.004	0.82	.006	140	K0 III	116,278,861,863,977*
227885	+33 03757	Ros 5 - 12			20 06 39	+33 30.9	1	9.39		0.10		0.08		6	A2	102
227885	+33 03757				20 06 40	+33 31.2	1	9.38		0.10				1	A0 V	1070
227900	+40 04032	LS II +41 002			20 06 40	+41 06.2	1	10.44		0.25		-0.55		2	B2 III	1012
		Biu 2 - 129			20 06 41	+35 20.0	1	10.37		0.16		-0.50		2		480
191609	+57 02144				20 06 42	+57 39.2	1	7.18		-0.02				2	A0	401
191456	+36 03896	LS II +36 014			20 06 44	+36 31.5	4	7.44	.009	0.06	.012	-0.78	.021	6	B0.5III	401,1070,1118,1192,8100
227902	+37 03785	LS II +37 011			20 06 44	+37 21.2	2	9.26	.015	0.25	.015	-0.49		3	B1 V	1012,1070
227908	+41 03622				20 06 45	+41 36.2	1	9.23		1.77				2	K7	369
191285	-14 05652				20 06 48	-14 26.0	3	9.75	.004	1.14	.014	1.05	.044	7	K3/4 V	158,1705,3072
		Biu 2 - 125			20 06 49	+35 14.8	1	12.53		1.42		1.10		4		480
227915	+36 03897				20 06 49	+36 22.2	1	9.43		0.00				1	B7 V	1070
191400	+15 04063				20 06 50	+15 55.6	1	8.71		1.31				1	K2	882
191472	+37 03787				20 06 51	+37 47.7	2	8.99	.011	0.25	.002	0.68		28	A5 Ib	6011,8100
		G 23 - 28			20 06 52	+05 13.0	2	12.92	.010	1.02	.005	0.86	.002	3		1620,3062
		VES 155			20 06 52	+35 51.0	1	13.07		0.91		0.18		2		426
		Biu 2 - 121			20 06 53	+35 13.9	1	13.14		0.68		0.17		4		480
227913	+36 03898				20 06 53	+37 01.8	1	9.92		1.20				1	K0 III	592
191528	+45 03068	IDS20053N4605		AB	20 06 53	+46 13.7	1	8.38		0.20		0.18		6	Am	195
193214	+84 00451				20 06 53	+84 31.6	1	6.83		0.15		0.09		2	A2	985
		DO 6498			20 06 54	+06 06.	1	10.75		1.74				2		1532
191493	+35 03985				20 06 54	+36 00.0	3	7.32	.012	1.63	.005	1.91	.003	78	K5 II	369,1124,8100
191473	+36 03900	LS II +37 012			20 06 55	+37 05.4	2	8.62	.026	0.12	.009	-0.71		3	B0 IV	401,1070
227920	+33 03760	Ros 5 - 13			20 06 56	+34 10.3	2	9.17	.004	0.10	.004	-0.42		4	B3 V	102,1070
		VES 117			20 06 57	+25 27.3	1	11.72		0.57		0.04		1		363

Table 1　　　1093

HD	DM	Other Id	N Rem	α_{1950}	δ_{1950}	S	V	σ_V	B–V	σ_{B-V}	U–B	σ_{U-B}	n	Spectrum	References
190984	−64 04030			20 06 57	−64 46.2	2	8.76	.010	0.54	.000	0.02		3	F8 V	742,1594
		Biu 2 - 113		20 06 58	+35 13.4	1	11.41		0.23		-0.39		2		480
	+35 03986	Biu 2 - 117		20 06 59	+35 26.0	2	10.36	.269	0.47	.071	-0.51	.024	4	B0pe	480,1012
191494	+35 03988			20 06 59	+35 59.9	1	8.76		0.03		-0.12		7	B8 V	1124
191401	+10 04192	Ring Aql # 16		20 07 00	+10 38.3	1	7.51		-0.08		-0.52		3	B6 V	4
191495	+35 03987	Biu 2 - 126	★ A	20 07 00	+35 21.9	3	8.40	.029	0.10	.009	-0.83	.016	5	B2	480,1070,1192
		Biu 2 - 124		20 07 00	+35 24.8	1	14.39		0.74		0.27		3		480
		Biu 2 - 115		20 07 01	+35 13.6	1	12.72		0.73		0.33		4		480
227934	+34 03885	Biu 2 - 114		20 07 01	+35 15.2	1	9.37		0.16		0.00		5	A0	480
191546	+43 03505	V395 Cyg		20 07 01	+43 54.8	1	8.62		0.67		0.36		1	F5 Ib	247
191322	−17 05875			20 07 01	−17 15.3	1	9.00		-0.06		-0.56		11	B3/4 IV	1732
		Biu 2 - 118		20 07 04	+35 24.1	1	10.93		0.58		0.02		3		480
191095	−57 09622	HR 7691	★ AB	20 07 04	−57 40.3	2	6.36	.005	0.06	.000			7	B7 V + A3 V	15,2012
		Biu 2 - 116		20 07 05	+35 19.5	1	11.09		0.21		-0.53		2		480
191190	−47 13275	IDS20035S4702	A	20 07 05	−46 53.0	1	6.81		1.17		1.14		2	K1 III	1279
191190	−47 13275	IDS20035S4702	AB	20 07 05	−46 53.0	1	6.81		1.16				4	K1 III	2012
191190	−47 13275	IDS20035S4702	B	20 07 05	−46 53.0	1	11.18		1.13		0.77		2		1279
191496	+32 03720	Ros 5 - 14		20 07 06	+33 02.3	2	7.47	.004	0.26	.011	0.09		6	A9 Ib	102,1070
		Cyg sq 3 # 46		20 07 06	+35 51.4	1	11.31		1.23		1.18		7		1124
227941	+35 03990			20 07 06	+35 53.9	1	9.16		0.23		0.10		51	F0	1124
		VES 156		20 07 06	+38 58.6	1	10.95		0.25		-0.07		1		8103
227950	+38 03920	IDS20053N3847	AB	20 07 07	+38 56.2	1	10.95		0.25		-0.07		2	A7	426
		Biu 2 - 128		20 07 09	+35 18.2	1	13.91		0.76		0.16		2		480
227951	+35 03991			20 07 09	+35 40.8	1	10.16		0.39		-0.25		7	A5:IV:	1124
		VES 157		20 07 10	+35 09.3	1	12.94		1.03		-0.02		2		426
191511	+33 03763	Ros 5 - 15	★ AB	20 07 12	+33 30.5	3	8.15	.019	0.92	.014	0.59	.049	7	G5 II	102,1070,8100
		Biu 2 - 123		20 07 12	+35 18.0	1	14.68		2.17		1.44		3		480
190711	−75 01579			20 07 15	−74 57.8	1	8.49		1.60				4	M4 III	2012
191530	+33 03765	Ros 5 - 16		20 07 16	+33 46.3	3	7.89	.015	-0.01	.008	-0.37	.007	7	B9 II	102,1070,1192,8100
227959	+36 03902	V402 Cyg		20 07 16	+37 00.2	4	9.58	.023	0.89	.036	0.57	.019	4	G0 III	592,1399,1462,6011
	+42 03607	G 125 - 64		20 07 17	+42 42.0	1	10.11		0.51		-0.22		2	F3	1658
191499	+16 04166	IDS20050N1630	AB	20 07 18	+16 39.3	3	7.56	.063	0.81	.018	0.36	.075	4	K0	882,938,1355
		Biu 2 - 119		20 07 18	+35 20.	1	10.28		0.45		-0.06		1		480
		Biu 2 - 120		20 07 18	+35 20.	1	12.94		0.75		0.27		1		480
		EG 137		20 07 18	−21 55.	1	14.46		0.18				1	DA	1705
191120	−61 06465			20 07 18	−61 06.1	1	9.56		0.08		0.08		4	A1 V	1731
	+35 03992	Biu 2 - 112		20 07 19	+35 16.6	1	9.65		0.19		-0.53		2		480
227960	+35 03993	LS II +35 052		20 07 19	+35 54.0	3	9.46	.027	0.25	.006	-0.65	.049	10	B2 II	1124,1192,8100
191567	+35 03994	Biu 2 - 130	★ B	20 07 20	+35 20.2	1	8.75		0.16		-0.59		4	B3	480
		Biu 2 - 127		20 07 20	+35 21.7	1	12.39		1.63		2.30		3		480
191566	+35 03994	Biu 2 - 131	★ A	20 07 21	+35 20.2	4	7.58	.254	0.16	.009	-0.72	.029	9	B3	401,480,1070,1118
		LP 870 - 43		20 07 21	−21 55.4	1	14.40		0.20		-0.70		6	DA	3062
		Biu 2 - 122		20 07 22	+35 18.1	1	14.48		0.88		0.43		4		480
		HIC 99284		20 07 22	+35 48.6	1	10.58		1.26				2		1726
227966	+36 03904			20 07 22	+37 10.3	1	9.51		0.46				1	F2 Ib	592
191273	−43 13836			20 07 22	−43 45.1	7	7.83	.006	0.26	.003	0.13	.015	112	A3mA7-F2	116,278,1066,1075,1586*
		VES 118		20 07 24	+23 44.8	1	11.78		2.11		1.89		1		363
		L 79 - 86		20 07 24	−73 22.	1	11.15		0.64		0.02		3		1696
191391	−20 05833			20 07 26	−20 38.2	2	8.91	.029	1.29	.000	1.21	.002	6	K4 Vp	158,3072
191568	+32 03721	Ros 5 - 17		20 07 27	+33 14.4	2	8.69	.000	-0.01	.003	-0.11		8	B9 V	102,1070
227977	+37 03793	LS II +37 014		20 07 27	+37 21.3	1	9.65		0.18				1	B2 III	1070
	−51 12417			20 07 33	−51 34.2	1	9.71		0.58		0.01		1	K0	1696
191611	+36 03906	LS II +36 020	★ D	20 07 34	+36 20.5	1	8.59		0.37		-0.60		2	B0.5III	1012
191610	+36 03907	HR 7708, V1624 Cyg		20 07 34	+36 41.5	7	4.96	.025	-0.14	.012	-0.77	.024	20	B2.5Ve	15,154,1118,1192,1212*
191612	+35 03995	LS II +35 054		20 07 35	+35 35.1	3	7.80	.027	0.27	.016	-0.68	.020	9	O8	1070,1124,1192
		AJ89,1229 # 431		20 07 36	+02 58.	1	10.34		1.68				3		1532
227995				20 07 36	+35 59.	1	10.82		0.72		0.30		7	G5	1124
191569	+27 03631		V	20 07 37	+28 10.6	1	7.77		1.73				2	M1	369
191589	+33 03766	Ros 5 - 18		20 07 37	+33 32.0	2	7.25	.004	1.54	.007	1.42		5	K5 III	102,1070
227984	+34 03890			20 07 37	+34 33.9	1	8.32		1.17				1	K7	1070
191305	−43 13840			20 07 37	−43 37.4	4	7.69	.003	0.23	.001	0.09	.011	13	A4 V	278,1075,1770,2011
191613	+32 03724	Ros 5 - 19	★ AB	20 07 40	+33 15.8	2	7.83	.007	0.03	.007	-0.02		6	A0	102,1070
		VES 158		20 07 40	+35 17.9	1	12.46		1.22		0.92		2		426
227994				20 07 41	+36 00.	1	10.93		0.38		0.18		7	F7	1124
191533	+8 04358	IDS20053N0809	AB	20 07 43	+08 17.9	1	6.56		0.50		0.06		2	F8	3016
191614	+26 03811			20 07 43	+27 11.0	1	7.95		1.66				2	K5	369
191570	+20 04453	HR 7705	★ A	20 07 44	+20 45.9	1	6.50		0.38		-0.05		2	F5 IV	3016
191570	+20 04453	HR 7705	★ AB	20 07 44	+20 45.9	3	6.48	.000	0.38	.000	-0.04	.000	6	F5 IV	1,15,1028
191570	+20 04453	IDS20055N2037	B	20 07 44	+20 45.9	3	8.89	.005	0.76	.000	0.33	.007	6	F4 V	1,1028,3016
228205	+35 03997			20 07 46	+35 37.0	1	9.87		0.12		-0.53		1	B0.5V:	1192
191349	−44 13775			20 07 48	−43 48.7	7	6.54	.003	0.88	.003	0.57	.004	29	G5/6 III	278,977,977,1075,1586*
191615	+25 04124			20 07 50	+25 23.3	1	7.80		0.94		0.83		2	G8 IV	1003
339659	+26 03812	NGC 6882 - 18		20 07 51	+26 35.4	1	10.18		0.15		-0.14		1	A2	49
		L 565 - 18		20 07 51	−30 21.9	5	12.20	.013	0.06	.011	-0.66	.022	18	DA4	782,1696,1698,1765,3062
191651	+32 03726	Ros 5 - 20		20 07 53	+32 24.6	1	8.47		-0.07		-0.45		3	A0	102
	+47 03032			20 07 54	+48 14.2	1	9.75		1.21		1.24		1	F8	414
191408	−36 13940	HR 7703	★ AB	20 07 55	−36 13.7	7	5.32	.006	0.87	.007	0.44	.010	26	K2 V	15,678,1075,2017,2027*
339666	+26 03813	NGC 6882 - 12		20 07 56	+26 26.6	1	9.57		0.41		0.06		1	F2 II	49
191652	+25 04126	NGC 6882 - 13	★ V	20 07 57	+26 07.8	1	9.58		1.68		1.00		1	M5 III	49
191738	+47 03031	SV Cyg	★ A	20 07 59	+47 43.4	5	8.73	.049	3.21	.039	6.10	.600	5	R3	414,1238,8005,8022,8027

HD	DM	Other Id	N Rem	α_{1950}	δ_{1950}	S	V	σ_V	B–V	σ_{B-V}	U–B	σ_{U-B}	n	Spectrum	References
228033	+33 03768	Ros 5 - 21		20 08 00	+33 45.3	1	9.89		0.14		0.01		6	B9	102
191703	+38 03927			20 08 03	+38 59.3	2	7.12	.005	0.31	.005	0.13	.000	6	F0 V	833,1601
228034	+33 03769	Ros 5 - 22		20 08 04	+33 30.7	1	9.67		0.19		0.12		7	A3	102
228041	+35 03998	LS II +35 057		20 08 05	+35 20.9	1	8.97		0.36		-0.66		2	B0.5V:e	1012
191742	+42 03616			20 08 05	+42 23.6	1	8.14		0.22		0.20		4	A7p	1202
339683		NGC 6882 - 36		20 08 06	+26 05.8	1	10.90		0.46		0.02		1	G0	49
		NGC 6882 - 31		20 08 08	+26 29.4	1	10.72		1.38		1.31		1	A1	49
191690	+30 03922			20 08 08	+31 04.7	1	8.47		-0.07		-0.57		14	B9	1371
228047	+32 03728	Ros 5 - 24		20 08 08	+33 03.1	1	10.37		0.28		0.16		4	A5	102
228052	+36 03915			20 08 08	+36 59.8	1	11.04		0.12		0.93		1	B1 II	8100
331888	+32 03727	Ros 5 - 23		20 08 09	+32 35.0	1	8.90		1.16		0.98		3	G5	102
228053	+36 03914	LS II +36 022		20 08 09	+36 33.1	3	8.79	.052	0.49	.028	-0.41	.093	4	B1 II	1012,1070,8100
191940	+61 01975	IDS20074N6147	AB	20 08 09	+61 55.7	1	6.74		0.03		0.01		2	A0	401
339684	+25 04128	NGC 6882 - 24		20 08 10	+26 07.9	1	10.47		0.53		0.16		1	F8	49
191720	+36 03916			20 08 10	+36 49.8	1	7.79		0.03				1	B9 V	1070
191595	−3 04803			20 08 10	−02 50.4	1	8.85		0.63		0.16		2	G0	1375
191655	+14 04209			20 08 12	+14 24.7	1	7.87		1.58				1	K5	882
339667	+26 03814	NGC 6882 - 10		20 08 12	+26 25.3	1	9.42		0.36		0.18		1	A9 III	49
		VES 159		20 08 12	+32 47.9	2	12.64	.000	1.86	.000	1.88	.000	3		426,8103
228059	+33 03772	Ros 5 - 25		20 08 13	+33 29.7	1	10.09		0.08		0.02		7	A0	102
228056	+33 03773	Ros 5 - 26		20 08 13	+33 52.3	1	9.71		0.35		-0.27		4	B8	102
228062	+35 03999			20 08 14	+36 07.5	2	8.34	.055	2.14	.015	2.30		26	M0	369,1124
191781	+44 03365	LS III +45 003		20 08 14	+45 15.3	2	9.54	.000	0.64	.000	-0.41	.000	5	B0 Ibp	555,1012,8100
228063	+35 04000			20 08 19	+36 02.5	1	8.63		0.00		-0.36		7	B8 V	1124
		VES 119		20 08 20	+34 14.5	1	12.08		0.65		-0.03		1		363
191743	+33 03775	Ros 5 - 27		20 08 21	+33 24.7	2	8.89	.000	-0.01	.003	-0.32		7	B7 III	102,1070
191765	+35 04001	V1769 Cyg		20 08 22	+36 01.7	4	8.07	.017	0.01	.021	-0.47	.018	23	B9	1095,1124,1192,1359
191464	−45 13658			20 08 24	−45 31.3	3	8.89	.001	1.25	.001	1.27	.008	19	K1/2 III	460,977,2011
339665		NGC 6882 - 37		20 08 25	+26 30.3	1	10.91		0.93		0.59		1	K5	49
228079	+36 03918			20 08 25	+36 22.6	1	9.20		-0.02		-0.07		1	B8 II	8100
		NGC 6882 - 34		20 08 27	+26 48.4	1	10.84		0.45		0.02		1		49
191639	−9 05382	HR 7709	★	20 08 27	−08 59.5	7	6.47	.007	-0.16	.004	-0.92	.004	84	B1 V	15,154,989,1071,1509,1729*
191747	+26 03815	NGC 6882 - 2	★	20 08 28	+26 45.3	6	5.50	.022	0.08	.012	0.13	.017	10	A3 V	15,49,379,603,1063,3050
191746	+28 03645			20 08 29	+28 17.1	1	7.17		0.01		-0.67		3	B2 IV	399
333626	+30 03925			20 08 29	+30 40.9	1	9.20		1.61				2	M0	369
228073	+33 03776	Ros 5 - 28		20 08 29	+33 39.2	2	9.23	.014	0.06	.007	-0.02		6	B9	102,1070
		G 92 - 59		20 08 30	−02 39.4	1	13.51		1.57		1.04		1		1620
		HA 205 # 570		20 08 31	−75 07.9	1	12.20		0.48		0.07		std		2
		NGC 6882 - 42		20 08 32	+26 43.4	1	11.33		0.97		0.73		1		49
191872	+44 03368			20 08 32	+44 52.1	1	7.79		-0.16		-0.50		2	B8	401
		VES 120		20 08 33	+23 25.5	1	10.98		0.47		0.29		1		363
228084	+35 04003	NGC 6883 - 5		20 08 33	+35 43.1	1	8.92		1.82		2.00		3	K5	1625
191854	+43 03513	IDS20069N4339	AB	20 08 34	+43 47.7	2	7.44	.015	0.68	.015	0.19	.025	2	G4 V +G8 V	247,3030
351582	+19 04308			20 08 35	+19 37.5	1	9.17		0.00		-0.29		1	B8	963
228081	+32 03730	Ros 5 - 29		20 08 35	+33 14.3	1	10.09		0.12		0.13		5	A0	102
228085	+34 03898			20 08 37	+35 11.6	1	9.67		1.23		1.20		7	K5	1124
191350	−60 07387			20 08 37	−60 24.1	1	8.98		0.35		0.08		4	F0 IV	1731
228087	+32 03731	Ros 5 - 30		20 08 38	+33 14.3	2	8.98	.010	0.04	.003	0.02		8	A0	102,1070
191812	+32 03778	Ros 5 - 31	★ A	20 08 39	+33 29.6	3	7.75	.021	0.08	.004	0.09	.000	14	B9 V	102,401,1070
345957	+23 03912			20 08 40	+23 48.9	2	8.88	.000	0.51	.000	-0.16	.000	2	G0 V we	979, 8112
339660		NGC 6882 - 19		20 08 40	+26 35.8	1	10.23		0.44		0.25		1	A7	49
	+33 03779	Ros 5 - 32		20 08 40	+33 44.5	1	9.14		-0.01		-0.28		9		102
192017	+59 02185			20 08 41	+60 15.3	1	8.04		0.34		0.00		3	F0	3016
		AJ89,1229 # 432		20 08 42	+03 22.	1	12.01		1.68				3		1532
191813	+33 03780	Ros 5 - 33	★ BC	20 08 42	+33 28.8	3	8.18	.014	0.00	.004	-0.35	.000	14	B9 V	102,401,1070
191811	+33 03781	Ros 5 - 34		20 08 42	+33 42.4	3	7.63	.023	-0.06	.009	-0.50	.005	9	B3 V	102,401,1070
191833	+34 03899	IDS20068N3434	AB	20 08 42	+34 42.8	1	8.38		0.58				1	F7 V	1070
191692	−1 03911	HR 7710	★ A	20 08 43	−00 58.3	10	3.24	.014	-0.07	.008	-0.13	.019	46	B9.5III	3,15,1075,1088,1203*
228104	+35 04004	NGC 6883 - 1		20 08 45	+35 43.5	2	8.85	.009	0.27	.005	-0.71	.005	4	B1 IVpe	305,1012
191709	−0 03937	IDS20062S0025	A	20 08 45	−00 16.5	1	7.16		0.44		0.08		2	F0	3016
228101	+37 03804			20 08 46	+37 18.6	1	8.44		0.07		-0.65		2	B1 V	1012
228097	+32 03732	Ros 5 - 35		20 08 48	+33 03.1	2	8.48	.004	1.08	.012	0.96		4	K0	102,1070
228105		NGC 6883 - 169		20 08 48	+35 43.6	1	10.63		0.25		-0.33		1	A7	305
		G 125 - 67		20 08 49	+38 44.7	1	13.00		1.07		1.02		3		1658
191556	−45 13663			20 08 49	−45 00.4	2	9.51	.000	1.22	.008	1.33		9	K2 III	863,1075
191785	+15 04074	G 143 - 34	★ A	20 08 50	+16 02.0	4	7.33	.032	0.83	.020	0.49	.005	8	K1 V	22,882,1003,3077
		NGC 6882 - 29		20 08 50	+26 49.4	1	10.58		1.41		1.52		1	A2	49
191814	+20 04462	HR 7713		20 08 51	+20 59.1	1	6.22		0.90				2	K0	71
228108	+33 03785			20 08 51	+34 12.7	1	9.06		0.44				1	F4 III	1070
346063	+21 04085			20 08 52	+21 23.5	1	10.49		0.58		0.20		2	K7	1375
339702	+25 04137			20 08 52	+25 29.7	1	9.50		1.95				2	M0	369
228110	+33 03783			20 08 52	+33 34.1	1	9.23		0.00				1	B7 V	1070
228110	+33 03783	Ros 5 - 36		20 08 52	+33 34.1	1	10.48		0.10		0.01		4	B9	102
		WLS 2000-5 # 8		20 08 52	−04 50.7	1	11.08		1.37		1.44		2		1375
		NGC 6883 - 33		20 08 53	+35 43.5	1	13.39		0.37		0.25		2		305
339661		NGC 6882 - 30		20 08 55	+26 36.0	1	10.60		0.21		0.14		1	A2	49
		NGC 6882 - 48		20 08 55	+26 44.7	1	11.63		0.59		-0.03		1		49
191722	−6 05397	IDS20062S0627	A	20 08 56	−06 18.0	1	8.33		0.40		-0.01		2	F5	1375
191785	+15 04074	G 143 - 35	★ B	20 08 57	+16 01.9	2	13.99	.058	1.59	.037	1.14		3		1759,3032
191897	+36 03920	IDS20070N3645	AB	20 08 57	+36 53.8	2	7.34	.010	0.96	.005	0.60		3	G3 Ib	247,1118,8100

Table 1   1095

HD	DM	Other Id	N Rem	α₁₉₅₀	δ₁₉₅₀	S	V	σ_V	B–V	σ_B–V	U–B	σ_U–B	n	Spectrum	References
		NGC 6882 - 46		20 08 58	+26 52.9	1	11.52		0.38		0.08		1	F5	49
191584	−43 13855	HR 7706		20 08 58	−42 55.8	4	6.21	.012	1.23	.005	1.33		12	K1 III	15,1075,2006,3077
339649	+26 03817	NGC 6882 - 23		20 09 00	+26 58.2	1	10.47		0.33		0.28		1	F0	49
228123				20 09 00	+32 42.	1	9.32		0.37				1	F5	1070
228130	+33 03786	Ros 5 - 37		20 09 00	+33 34.6	1	9.24		0.00		−0.22		6	K0	102
		NGC 6882 - 64		20 09 01	+26 18.2	1	13.09		0.79		0.51		1		49
191998	+47 03037			20 09 03	+48 04.9	1	6.60		1.14				1	K0	8097
		NGC 6883 - 130		20 09 04	+35 43.1	1	13.36		0.37		0.20		2		305
191917	+35 04006	NGC 6883 - 38		20 09 04	+35 48.2	2	7.79	.008	0.14	.014	−0.68		8	B1 III	1070,1124
		VES 121		20 09 05	+28 44.3	1	11.06		0.43		0.17		1		363
		IRC +40 392		20 09 05	+36 25.5	1	12.18		2.01		0.67		1		426
		AO 795		20 09 05	+38 09.1	1	10.89		0.86		−0.10		1		1697
		LP 454 - 9		20 09 07	+19 10.7	1	12.90		0.76		0.33		1		1696
228152	+38 03935			20 09 07	+38 20.2	1	10.27		0.47		0.32		1		1697
228147	+34 03902			20 09 08	+34 40.2	1	9.29		−0.01				1	B9 V	1070
		NGC 6882 - 56		20 09 09	+26 19.9	1	12.57		0.71		0.19		1		49
191877	+21 04088	HR 7716	⋆	20 09 10	+21 43.5	3	6.27	.014	−0.02	.007	−0.80	.048	8	B1 Ib	138,154,8100
191898	+25 04138	NGC 6882 - 7		20 09 10	+25 59.4	2	8.80	.005	0.58	.000	0.03	.009	4	F7 V	49,3026
339664		NGC 6882 - 21		20 09 10	+26 30.6	1	10.35		0.15		0.10		1	A1 V	49
191964	+40 04048			20 09 10	+40 33.1	1	9.10		1.81				2	M1	369
		HA 205 # 579		20 09 10	−75 03.8	1	13.08		0.61		0.08		std		2
	−75 01583	HA 205 # 580		20 09 10	−75 05.3	1	10.58		1.22		1.14		std		2
		NGC 6883 - 120		20 09 11	+35 43.7	1	14.07		1.34		0.99		1		305
	−73 02115			20 09 11	−73 40.7	1	10.24		0.29				4		2012
228157	+33 03788	Ros 5 - 38		20 09 12	+33 44.6	1	9.78		0.41		−0.01		4	F5	102
191977	+42 03624			20 09 12	+42 56.9	1	7.79		0.00		−0.23		2	B8	401
192034	+51 02787			20 09 12	+52 14.1	1	7.37		1.65		1.78		2	M1	1566
191753	−12 05664	HR 7712		20 09 12	−12 32.5	1	6.34		1.21				4	K0 III	2035
		NGC 6882 - 43		20 09 13	+26 18.2	1	11.47		0.44		−0.08		1		49
339669	+26 03819	NGC 6882 - 25		20 09 13	+26 22.3	1	10.49		0.42		−0.47		1	B8	49
		NGC 6883 - 122		20 09 13	+35 43.2	1	12.14		0.20		−0.28		1		305
		NGC 6883 - 121		20 09 13	+35 43.5	1	11.86		2.03		2.18		2		305
		G 125 - 68		20 09 13	+36 09.7	1	13.28		1.16		1.09		2		1658
		NGC 6882 - 49		20 09 14	+26 44.5	1	11.64		0.45		−0.05		1	G0	49
339648	+26 03820	NGC 6882 - 32		20 09 14	+26 55.7	1	10.77		0.54		0.01		1	G0	49
		NGC 6883 - 119		20 09 14	+35 44.0	1	11.68		0.62		0.15		2		305
191978	+40 04050	LS III +41 003		20 09 14	+41 12.2	2	8.02	.000	0.14	.000	−0.78	.000	4	O8	1011,1012
		NGC 6882 - 63		20 09 15	+26 21.6	1	13.03		0.46		0.22		1		49
		NGC 6882 - 44		20 09 15	+26 46.3	1	11.50		0.71		0.25		1	A2	49
339647	+26 03821	NGC 6882 - 14		20 09 15	+26 50.1	1	9.76		1.13		0.99		1	K2 III	49
		NGC 6882 - 61		20 09 16	+26 21.0	1	13.00		0.53		0.07		1		49
		NGC 6882 - 38		20 09 16	+26 55.1	1	11.00		0.54		0.02		1	A0	49
		NGC 6882 - 76		20 09 17	+26 22.6	1	15.46		1.04				1		49
339675		NGC 6882 - 39		20 09 18	+26 09.3	1	11.06		0.55		0.02		1		49
228167	+33 03790	Ros 5 - 39		20 09 18	+33 25.1	1	9.24		0.02		−0.16		4	A0	102
		VES 161	V	20 09 18	+35 20.3	1	13.03		0.74		−0.28		1		426
192035	+47 03038	RX Cyg		20 09 18	+47 39.8	2	8.18	.005	0.06	.010	−0.81	.015	9	B0III-IV(n)	399,555
192001	+41 03642	LS III +41 004		20 09 19	+41 58.6	1	8.25		0.32		−0.60		1	O9.5IV	1011
		NGC 6882 - 66		20 09 20	+26 19.2	1	13.51		0.66		−0.02		1		49
		NGC 6882 - 74		20 09 20	+26 21.3	1	14.62		1.08		0.56		1		49
		NGC 6882 - 54		20 09 21	+26 21.2	1	12.31		0.49		0.06		1		49
339662		NGC 6882 - 27		20 09 21	+26 38.0	1	10.53		0.35		0.07		1	A5	49
		NGC 6882 - 53		20 09 21	+26 52.2	1	11.85		0.32		0.15		1		49
339663		NGC 6882 - 20		20 09 23	+26 36.9	1	10.33		0.16		0.11		1	A0	49
		NGC 6882 - 26		20 09 23	+26 41.5	1	10.51		1.06		0.88		1	F7	49
		VES 162		20 09 23	+36 33.8	1	12.85		1.42		1.40		2		426
339674	+25 04142	NGC 6882 - 17		20 09 24	+26 10.6	1	9.98		0.60		0.13		1	G5	49
		NGC 6882 - 72		20 09 25	+26 19.0	1	14.42		0.68		0.40		1		49
339670		NGC 6882 - 28		20 09 25	+26 21.8	1	10.56		0.12		0.06		1	A0	49
339676	+25 04144	NGC 6882 - 6		20 09 26	+26 14.2	1	8.68		1.12		1.11		1	K2 III	49
		NGC 6882 - 57		20 09 26	+26 20.1	1	12.63		0.51		0.18		1		49
		NGC 6882 - 41		20 09 26	+26 22.9	1	11.33		0.46		0.03		1		49
192003	+37 03811			20 09 27	+38 04.8	1	8.10		0.12		−0.67		1	B2 IV	1070,1697
		NGC 6882 - 62		20 09 28	+26 20.7	1	13.03		0.55		−0.31		1		49
		NGC 6882 - 73		20 09 28	+26 22.9	1	14.46		0.75				1		49
228187	+36 03925			20 09 28	+37 12.5	1	9.68		0.12				1	B3 II	1070
		NGC 6882 - 69		20 09 29	+26 22.5	1	13.90		1.05		0.62		1		49
		NGC 6882 - 68		20 09 31	+26 25.9	1	13.78		0.62		0.11		1		49
		NGC 6882 - 47		20 09 31	+26 27.7	1	11.62		0.48		0.10		1		49
191946	+19 04312			20 09 32	+19 55.2	1	8.55		−0.04		−0.51		10	B9	8053
		NGC 6882 - 60		20 09 32	+26 21.6	1	12.97		0.56		0.08		1		49
191979	+27 03642			20 09 32	+27 17.6	2	8.78	.005	1.83	.019	2.02		4	M1	369,1375
		VES 250		20 09 32	+42 43.8	1	10.53		0.35		−0.01		1		8056
	+7 04396			20 09 34	+07 30.0	1	9.79		0.49		−0.03		2	F8	1375
192039	+41 03644	LS III +41 005		20 09 34	+41 48.6	1	8.75		0.22		−0.69		2	B0 IV	401
		NGC 6882 - 75		20 09 35	+26 22.4	1	14.99		0.84				1		49
		NGC 6882 - 71		20 09 35	+26 25.7	1	14.22		0.67		0.27		1		49
228199	+36 03927	LS II +36 026	⋆ A	20 09 35	+36 20.9	1	9.25		0.13				1	B0.5:V	1070
		NGC 6882 - 58		20 09 36	+26 19.4	1	12.73		0.69		0.50		1		49
		NGC 6882 - 70		20 09 36	+26 22.1	1	14.04		0.78		−0.25		1		49

HD	DM	Other Id	N	Rem	α_{1950}	δ_{1950}	S	V	σ_V	B–V	σ_{B-V}	U–B	σ_{U-B}	n	Spectrum	References
		NGC 6882 - 65			20 09 36	+26 26.5	1	13.17		0.61		-0.08		1		49
		NGC 6882 - 51			20 09 36	+26 27.5	1	11.82		0.45		0.22		1		49
192020	+37 03812				20 09 36	+38 14.9	4	7.95	.017	0.87	.013	0.55	.008	7	G8 V	1070,247,1118,1193,1775
		WLS 2000 5 # 8			20 09 37	+04 47.2	1	12.85		0.67		0.18		2		1375
		NGC 6882 - 67			20 09 37	+26 24.3	1	13.68		0.60		0.15		1		49
192021	+33 03793	Ros 5 - 40			20 09 37	+33 49.5	2	7.76	.000	0.47	.008	-0.05		4	F6 V	102,1070
192041	+38 03939				20 09 37	+38 40.0	1	7.71		1.45		1.52		1	K2 II	8100
	+26 03824	NGC 6882 - 15			20 09 38	+26 26.3	1	9.87		1.84		1.90		1	K2 III	49
191862	-13 05608	HR 7715		★ A	20 09 39	-12 45.9	2	5.85	.000	0.48	.005	-0.06		6	F5 V	2007,3037
		NGC 6882 - 55			20 09 40	+26 22.9	1	12.48		0.56		0.06		1		49
		NGC 6882 - 52			20 09 42	+26 19.2	1	11.83		0.56		0.14		1		49
		NGC 6882 - 59			20 09 42	+26 23.6	1	12.97		0.54		0.14		1		49
		LS II +20 019		★ V	20 09 43	+20 11.1	3	8.96	.051	1.38	.121	0.80	.063	4	B4:I:	405,787,810
192004	+26 03825	NGC 6882 - 1		★	20 09 43	+26 39.5	4	5.48	.033	1.40	.013	1.53	.032	6	K2 III	15,49,1355,8100
		VES 163			20 09 43	+36 53.8	2	11.90	.000	1.94	.000	1.86	.000	3		426,8103
		NGC 6882 - 40			20 09 44	+26 27.8	1	11.27		0.40		0.01		1		49
192022	+26 03826	NGC 6882 - 4			20 09 45	+26 44.8	1	7.30		-0.10		-0.55		1	B8 III	49
		NGC 6882 - 45			20 09 46	+26 23.4	1	11.52		0.22		0.12		1		49
339671	+25 04145	NGC 6882 - 9			20 09 47	+26 15.8	1	9.26		0.28		0.07		1	A7 III	49
		VES 164			20 09 47	+35 24.3	1	12.08		0.36		-0.06		2		426
		AJ89,1229 # 437			20 09 48	+11 56.	1	11.94		1.71				1		1532
191981	+13 04305				20 09 48	+14 04.6	1	8.31		1.35				1	K2	882
339672	+25 04146	NGC 6882 - 8		★ AB	20 09 48	+26 13.2	1	9.18		0.48		-0.03		1	F6 V	49
192078	+38 03940	IDS20080N3835	AB		20 09 49	+38 43.8	1	7.35		1.53		1.47		1	G5 II	8100
		HA 205 # 586			20 09 49	-75 04.9	1	13.22		1.04		0.73		std		2
		VES 165			20 09 50	+37 28.5	2	11.16	.000	0.19	.000	0.07	.000	3		426,8103
192042	+28 03656				20 09 51	+28 57.8	1	8.44		0.36		-0.04		1	F2 V wlm	979
192043	+26 03827	NGC 6882 - 5			20 09 52	+26 35.6	2	7.59	.006	-0.09	.004	-0.40	.009	2	B5 IV	49,674
		VES 166			20 09 52	+37 11.2	2	11.73	.000	1.08	.000	0.83	.000	3		426,8103
		VES 167			20 09 53	+34 28.5	1	12.95		0.71		-0.25		2		426
339775		NGC 6882 - 22			20 09 54	+26 47.3	1	10.41		0.16		0.14		1	A2	49
192079	+37 03816	LS II +37 018		★ AB	20 09 54	+37 24.0	1	8.76		0.28				1	B0.5IV	1070
192238	+56 02368	IDS20088N5659	AB		20 09 54	+57 08.3	1	8.81		0.63		0.13		3	F8	3030
191902	-13 05610				20 09 54	-13 20.9	1	9.89		1.04		0.95		2	K2/3 V	3072
192044	+26 03828	NGC 6882 - 3		★	20 09 55	+26 19.7	5	5.90	.029	-0.12	.007	-0.43	.020	16	B8 Vn	15,49,379,1063,1079
192122	+40 04056				20 09 56	+41 14.5	1	8.05		-0.08		-0.55		2	B8	401
		NGC 6882 - 35			20 10 00	+26 21.8	1	10.87		0.63		0.11		1	A0	49
228226	+32 03738				20 10 01	+32 51.7	1	9.33		1.14				1	K2	851
192103	+35 04013	V1042 Cyg		★ A	20 10 01	+36 02.8	3	8.12	.027	0.02	.009	-0.39	.021	16	C8	1012,1095,1359
192143	+39 04075				20 10 01	+40 10.8	1	7.05		-0.08		-0.40		2	B9	401
192142	+40 04057			V	20 10 01	+40 41.6	1	8.60		1.59				2	M1	369
191603	-63 04571	HR 7707			20 10 01	-63 34.1	2	6.08	.005	0.32	.005			7	F0 IV	15,2028
191984	+00 04444	HR 7717		★ AB	20 10 02	+00 43.0	3	6.26	.010	0.04	.033	-0.02	.011	9	B9p Cr	15,938,1071
192179	+46 02871				20 10 03	+46 58.1	1	9.00		0.19		0.09		2	F0	1601
191796	-45 13674				20 10 03	-45 44.5	6	7.79	.001	-0.02	.003	-0.03	.008	32	Ap EuCr(Sr)	460,977,1075,1586,1770,2011
331891	+32 03739				20 10 04	+32 38.7	2	9.31	.000	0.29	.061			2	A4 III	851,1070
228229	+34 03907	IDS20082N3418	AB		20 10 04	+34 26.6	1	8.78		1.00				2	G8 IV	6009
339780		NGC 6882 - 33			20 10 07	+26 16.3	1	10.80		0.33		0.01		1	G0	49
192124	+34 03908	IDS20082N3411	AB		20 10 07	+34 19.7	1	7.28		0.22				1	A5 III	1070
228256	+39 04076	LS II +39 011			20 10 08	+39 51.5	1	9.88		0.58		-0.52		2	Bpe	1012
228243	+37 03817	IDS20085N3752	AB		20 10 09	+38 02.5	1	9.73		0.30		0.24		1	A5 IV	1697
		VES 168			20 10 11	+37 26.2	1	12.01		0.73		-0.30		2		426
		WLS 2020 25 # 6			20 10 12	+25 27.9	1	11.15		1.87		1.71		2		1375
228250	+34 03909				20 10 12	+34 32.5	1	9.53		0.06				1	B9 V	1070
228263	+37 03819	LS II +37 019			20 10 12	+37 29.9	2	9.44	.024	0.16	.000	-0.61		3	B1 V	1012,1070
		VES 169		V	20 10 12	+38 06.3	1	15.14		1.12				1		426
		VES 170			20 10 13	+39 14.6	1	11.06		0.40		0.38		1		426
228253	+32 03742				20 10 14	+33 04.8	1	8.82		1.71				2	M0	369
339779		NGC 6882 - 16			20 10 15	+26 19.2	1	9.88		0.63		0.15		1	F5	49
192163	+37 03821	V1770 Cyg			20 10 17	+38 12.2	3	7.50	.014	-0.02	.009	-0.40	.028	12	WN6	245,1095,1359
		HA 205 # 594			20 10 17	-75 04.2	1	14.50		0.74		0.26		std		2
		VES 171			20 10 18	+38 43.8	1	11.57		0.49		-0.35		1		426
	+2 04111				20 10 19	+02 25.1	1	10.42		0.17		0.13		2	A2	1375
192062	+13 04311				20 10 19	+13 35.6	1	8.72		1.25				1	K2	882
191849	-45 13677				20 10 19	-45 18.8	12	7.97	.009	1.43	.009	1.18	.010	133	M1/2 V	158,861,863,977,1460,1494*
		VES 172			20 10 22	+34 57.8	2	11.63	.000	1.96	.000	1.81	.000	3		426,8103
		GD 229			20 10 23	+31 04.6	1	14.85		-0.08		-1.22		5	DP	940
192164	+34 03910	IDS20085N3433	A		20 10 23	+34 42.1	1	7.81		1.50				1	K2 V	1070
331970	+32 03743	MW Cyg			20 10 24	+32 43.4	4	9.14	.019	1.19	.029	0.82		4	F8.5Ib	851,1399,1772,6011
228277	+37 03822				20 10 24	+37 47.3	1	10.22		1.26				3		1726
192182	+38 03946	IDS20086N3809	A		20 10 24	+38 17.5	1	7.21		1.02		0.88		1	G8 III	245
192126	+14 04223				20 10 29	+15 10.7	2	8.34	.034	1.04	.039	0.90		3	K0	882,1648
192239	+45 03091	G 209 - 20			20 10 29	+46 08.7	1	9.06		0.81		0.45		1	K2	1658
228290	+37 03824	LS II +37 020			20 10 31	+37 51.0	1	9.47		0.26				1	B1 II	1070
339781	+26 03829	NGC 6882 - 11			20 10 32	+26 17.4	1	9.55		1.23		1.29		1	K2	49
228286	+34 03913				20 10 32	+34 45.9	1	8.57		1.35				3		1726
192276	+47 03045	HR 7721			20 10 32	+47 35.2	1			-0.12		-0.46		1	B7 V	1079
191829	-52 11643	HR 7714			20 10 32	-52 35.8	1	5.65		1.50				4	K4 III	2006
192031	-15 05584	IDS20078S1544	AB		20 10 34	-15 34.8	2	8.66	.000	0.71	.005	0.14	.014	4	G8 V	1003,1696
		VES 173			20 10 35	+35 35.3	1	13.23		0.74		0.14		2		426

Table 1   1097

HD	DM	Other Id	N Rem	α_{1950}	δ_{1950}	S	V	σ_V	B–V	σ_{B-V}	U–B	σ_{U-B}	n	Spectrum	References
192907	+77 00764	HR 7750	★ AB	20 10 37	+77 33.7	6	4.38	.012	-0.05	.004	-0.12	.017	22	B9 III+A7 V	15,1079,1203,1363*
191935	-44 13819			20 10 38	-44 19.5	1	8.39		0.44				4	F3 V	2012
		Steph 1767	A	20 10 39	+02 47.2	1	10.33		1.27		1.21		1		1746
192107	-1 03920	HR 7720		20 10 39	-01 09.6	2	5.46	.005	1.43	.000	1.29	.005	10	K5 III	1088,3005
		LS II +35 067		20 10 40	+35 47.2	2	11.52	.035	0.49	.010	-0.35	.010	4		405,426
		Steph 1767	B	20 10 41	+02 46.9	1	11.78		1.47		1.22		1	K5	1746
192243	+32 03744	Ros 5 - 41		20 10 41	+32 34.4	3	8.81	.006	0.01	.004	-0.03		30	A0 V	102,1070,6011
		AJ89,1229 # 440		20 10 42	+02 48.	1	11.76		1.47				2		1532
192225	+33 03802	Ros 5 - 42		20 10 43	+33 20.6	2	8.05	.011	-0.03	.004	-0.34		6	B8 III	102,1070
357040	+10 04213			20 10 44	+10 32.8	1	9.40		1.26		1.21		7	K0	1099
192575	+67 01235			20 10 44	+68 07.3	3	6.83	.008	0.17	.012	-0.67	.000	8	B0.5V	399,555,1733
		NGC 6882 - 50		20 10 45	+26 17.2	1	11.64		1.10		0.98		1		49
192302	+45 03094			20 10 45	+45 26.8	1	9.37		1.76				1	M1	369
192455	+61 01983	HR 7727	★ A	20 10 46	+61 55.6	2	5.72	.019	0.50	.019	0.03		4	F7 IV	1118,3053
192281	+39 04082	LS II +40 005		20 10 47	+40 07.0	2	7.55	.000	0.38	.000	-0.61	.000	4	O5 e	1011,1012
192244	+31 03991			20 10 48	+32 09.7	1	8.39		0.98				28	K0	6011
192260	+34 03915			20 10 49	+35 01.6	1	7.54	.028	1.09	.009	1.00		2	K0 IV	247,1070
191453	-75 01587	IDS20048S7507	AB	20 10 51	-74 58.5	3	9.09	.021	0.47	.006	-0.03	.031	17	F6 V	2,1117,2012
		HA 205 # 603		20 10 51	-75 07.4	1	14.68		0.78		0.32		std		2
		VES 175		20 10 53	+35 38.9	1	12.15		1.17		0.86		2		426
228324	+38 03950			20 10 53	+38 25.8	1	10.07		1.00		0.74		1		1697
	+37 03827	LS II +38 013		20 10 55	+38 14.4	1	8.14		0.95		0.75		1	F3 Ib	247,8100
192285	+32 03745	Ros 5 - 43		20 10 56	+33 05.8	2	8.14	.004	0.20	.014	0.11		6	A3 IV	102,1070
191440	-75 01586			20 10 56	-75 27.0	2	9.78	.010	0.30	.019	0.08		9	A9 IV	1117,2012
192303	+37 03828	LS II +38 014		20 10 58	+38 04.6	2	8.91	.012	0.34	.004	-0.57		2	B1 III	1070,1697
192490	+59 02193			20 11 00	+59 32.3	1	7.09		0.43		-0.04		2	F5	1733
192284	+34 03916	IDS20091N3410	AB	20 11 01	+34 19.5	1	8.29	.023	0.18	.014			7	A2 V	1070,1118,1196
	+46 02879			20 11 05	+47 15.2	1	9.96		1.15		0.99		1		1316
192246	+13 04313			20 11 06	+14 15.5	1	7.50		0.96				1	K0	882
192320	+34 03919			20 11 07	+34 40.7	1	8.03		1.55				1	K5	1070
192513	+59 02195			20 11 07	+59 56.2	2	7.86		0.01		-0.05		2	A0	401
192321	+33 03807	Ros 5 - 45	★ A	20 11 08	+34 02.3	3	8.24	.014	0.06	.005	0.05	.010	8	A2 V	102,401,1070
192439	+51 02796	HR 7726	★ AB	20 11 08	+51 18.7	1	6.01		1.14				2	K2	71
192322	+33 03806	Ros 5 - 44		20 11 09	+33 40.8	2	9.06	.010	-0.02	.000	-0.33		7	B7 III	102,1070
192287	+24 04045		V	20 11 10	+25 05.4	1	7.48		1.64				2	M III	369
228346	+36 03935	LS II +36 030		20 11 10	+36 52.1	1	9.98		0.70		0.54		1	A8 Ib	247,8100
228347	+35 04021	LS II +35 068	★ A	20 11 11	+35 53.2	1	9.47		0.64		0.27		1	B2:II	247,8100
192381	+40 04060			20 11 11	+40 34.2	1	7.95		0.03		0.07		2	A0	401
		HA 205 # 607		20 11 11	-75 10.0	1	15.62		1.09		0.63		std		2
228353	+39 04087	LS III +39 013		20 11 12	+39 58.6	1	9.98		1.35		1.04		2	A7 Iab	247,8100
228349	+33 03809	Ros 5 - 46	★ B	20 11 13	+34 03.5	3	9.15	.011	0.03	.005	0.00	.004	7	A0 V	102,401,1070
192361	+38 03952			20 11 15	+38 19.7	1	8.81		-0.01		-0.38		1	B8 III	1697
228365	+40 04061	LS II +40 006		20 11 16	+40 52.6	1	10.02		0.23		-0.68		2	B1 V	1012
	-14 05676			20 11 22	-13 58.9	1	11.18		0.55		0.10		12		1308
		HA 205 # 265		20 11 23	-74 51.0	1	15.16		0.46		0.01		std		2
355058	+16 04192			20 11 24	+16 24.6	1	8.89		0.88		0.40		1	G5	1222
228368	+34 03920	LS II +34 013		20 11 24	+34 52.4	3	8.39	.008	0.49	.004	-0.51	.000	5	O7	1011,1012,1070
192071	-44 13829			20 11 26	-44 12.6	1	8.40		0.63				4	G3 V	2012
192323	+17 04257			20 11 27	+18 01.3	1	8.08		1.39		1.61		2	K2	1648
228380	+34 03922			20 11 27	+34 53.6	1	8.79		0.59		0.10		1		247
192383	+35 04023		A	20 11 27	+35 26.9	1	7.12		1.11				1	G5 III	1070
192383	+35 04023		AB	20 11 27	+35 26.9	1	6.91		1.15		1.34		1	G5 III	8100
		VES 176		20 11 27	+41 17.8	1	13.66		0.64		0.22		2		426
		G 24 - 9	★ V	20 11 29	+06 34.0	2	15.66	.015	0.40	.005	-0.44	.000	5	DC:	1620,1705,3028
192404	+34 03923			20 11 30	+34 45.7	2	7.57	.090	1.27	.033	4.34		2	K0 II	1070,8100
228379	+34 03924			20 11 30	+34 59.6	1	8.78		1.35				2		1726
		G 24 - 10		20 11 32	+06 32.5	5	13.20	.030	1.54	.027	1.22	.045	9	dM5	203,1620,1705,1759,3028
192342	+23 03935	HR 7723	★ AB	20 11 32	+24 05.2	5	6.57	.010	0.30	.031	0.08	.027	15	A1 m	938,1049,1063,3058,8071
192092	-44 13830			20 11 32	-43 49.9	3	8.16	.010	0.22	.004	0.15	.013	16	A7 V	977,1075,2011
192422	+38 03956	LS II +38 017	★ B	20 11 33	+38 36.8	4	7.08	.032	0.50	.010	-0.47	.013	5	B0.5Ib	414,1012,1070,8100
192405	+27 03653			20 11 35	+27 23.0	1	8.16		0.49		0.01		3	F7 V	3026
192443	+38 03957	RS Cyg	★ A	20 11 35	+38 34.6	5	7.71	.279	2.99	.246	3.78	.059	2	C5 II	369,414,8005,8022,8027
		HIC 99664		20 11 36	+36 40.2	1	10.07		1.68				5		1726
192362	+15 04099			20 11 37	+16 16.3	1	8.63		1.33				1	K0	882
192494	+48 03059	IDS20102N4853	A	20 11 38	+49 02.2	1	7.86		0.02		-0.03		2	A0	3016
192494	+48 03059	IDS20102N4853	B	20 11 38	+49 02.2	1	11.07		1.43		1.28		2		3016
		LS II +38 019		20 11 39	+38 05.4	1	11.24		0.81		-0.21		2		405
192445	+35 04026	LS II +36 033		20 11 40	+36 10.6	2	7.23	.030	-0.09	.005	-0.76		5	B0.5III	1118,1212
192444	+38 03958	LS II +38 020		20 11 40	+38 19.6	1	8.40		0.56				1	B1 III	1070
357093	+12 04262			20 11 41	+12 32.3	1	9.96		-0.07		-0.74		1	A2	963
339770	+26 03835	DR Vul		20 11 41	+26 35.9	2	8.96	.306	0.22	.024	-0.06	.576	2	B8	674,774
331993	+31 03996	LS II +31 024		20 11 41	+31 55.8	1	10.49		0.49		-0.13		2	B9 Iab	180
	-14 05679	TW Cap		20 11 41	-13 59.5	2	10.17	.190	0.49	.140	0.36		2	A5 Ib	688,3074
192343	+6 04479	IDS20093N0618	B	20 11 42	+06 25.5	1	7.99		0.68		0.25		2	G4 V	3077
		VES 122		20 11 42	+29 56.9	1	11.29		1.43		1.49		2		363
192344	+6 04480	IDS20093N0618	A	20 11 43	+06 26.2	2	7.71	.004	0.70	.003	0.30	.005	7	G4 IV	1509,3077
331976	+32 03749	LS II +32 019		20 11 43	+32 23.9	2	9.93	.009	0.82	.022	-0.41	.049	5	B0:pe	180,1012
192496	+43 03528			20 11 44	+43 39.5	1	7.76		0.02		-0.02		2	A0	401
192514	+46 02881	HR 7730	★ D	20 11 44	+46 39.8	6	4.82	.008	0.10	.006	0.21	.037	31	A5 IIIn	15,247,659,1363,3023,8015
192385	+18 04431	LS II +18 018		20 11 46	+18 27.1	3	8.42	.023	0.65	.009	0.39	.077	12	F6 Ib	8059,8100,8100

HD	DM	Other Id	N Rem	α_1950	δ_1950	S	V	σ_V	B–V	σ_B–V	U–B	σ_U–B	n	Spectrum	References
191605	−75 01589			20 11 46	−75 09.1	1	8.61		0.93				4	G5/6 III	2012
192388	+16 04197	R Sge		20 11 47	+16 34.4	2	8.97	.125	0.92	.170	0.84	.270	2	G0 Ib	793,3001
228423	+38 03960			20 11 47	+38 33.9	2	9.24	.005	1.93	.005	2.24		2	M0	369,414
	−7 05223			20 11 47	−07 25.7	3	10.19	.025	1.33	.021	1.22		5	K7 V	1017,1619,3072
		HA 205 # 270		20 11 47	−74 48.6	1	15.06		0.76		0.16		std		2
		VES 177		20 11 49	+34 04.5	1	13.22		1.52		1.28		2		426
	−2 05206			20 11 51	−01 53.8	1	10.76		0.57		−0.05		2		1696
		VES 251		20 11 52	+42 03.2	1	12.61		1.87		1.23		1		8056
		VES 178		20 11 53	+36 03.7	1	13.06		1.58				1		426
		LS II +38 021		20 11 53	+38 59.5	1	11.37		0.69		−0.19		2		405
		VES 123		20 11 55	+29 24.5	1	12.67		1.43		1.02		1		363
228437	+36 03945	LS II +37 022		20 11 55	+37 08.9	1	8.85		0.27				1	B0.5V:	1070
192425	+14 04227	HR 7724		20 11 58	+15 02.6	6	4.95	.003	0.07	.012	0.04	.022	19	A2 V	15,1063,1363,3023*
228442	+33 03815			20 12 01	+33 18.3	1	9.11		1.73				2	M0	369
192535	+42 03642	HR 7733	⋆ AB	20 12 02	+43 13.6	2	6.14	.019	1.50	.014	1.83		3	K4 III	71,247
192577	+46 02882	HR 7735, V695 Cyg	⋆ AP	20 12 03	+46 35.3	7	3.80	.012	1.27	.017	0.40	.028	16	K2 II+B3 V	15,659,1084,1363,3041*
192348	−11 05269			20 12 04	−11 02.6	1	7.39		1.27		1.26		3	K0	1657
355267	+14 04231			20 12 05	+14 59.0	1	9.68		0.67		0.17		3	G5	196
228452	+34 03929			20 12 05	+35 14.3	1	9.77		0.66		−0.09		2	B3 V	1012
192579	+46 02883	IDS20105N4626	C	20 12 05	+46 33.6	3	6.99	.005	−0.13	.010	−0.59	.009	5	B9	963,1079,1084
192536	+38 03963			20 12 07	+39 00.4	3	7.09	.023	0.18	.009	0.14	.008	7	A7 III	105,1733,8071
228456	+36 03948			20 12 09	+36 39.0	1	9.86		0.25		−0.50		2	B2 IV	1012
192310	−27 14659	HR 7722		20 12 10	−27 11.0	6	5.73	.005	0.88	.005	0.64	.000	19	K3 V	15,1013,1075,2035*
192518	+28 03675	HR 7731, NU Vul		20 12 11	+28 32.5	3	5.19	.008	0.19	.005	0.16	.005	5	A7 IVn	603,1363,3023
192459	+15 04100			20 12 12	+15 32.7	1	9.17		0.16		0.12		3	A2	196
192538	+36 03949	HR 7734		20 12 12	+36 27.1	5	6.45	.012	−0.01	.005	−0.02		8	A0 V	1063,1070,1118,1193,1196
192516	+33 03817			20 12 13	+33 32.0	1	8.64		−0.01				1	B9 IV	1070
192696	+56 02376	HR 7740		20 12 14	+56 24.8	4	4.30	.010	0.11	.005	0.08	.007	14	A3 IV-Vn	15,1363,3023,8015
331978	+31 04002			20 12 16	+32 11.0	1	9.12		1.80				2	M0	369
228461	+37 03837	LS II +38 022		20 12 16	+38 05.5	2	9.47	.005	0.39	.010	−0.47		3	B2 II:	1012,1070,8100
192517	+29 03948			20 12 17	+30 01.5	1	7.08		−0.14		−0.74		2	B0.5V	401
192539	+31 04001	LS II +31 025		20 12 17	+31 50.7	2	7.29	.000	0.11	.014	−0.64	.009	4	B2 III	180,1012
		LS II +36 037		20 12 17	+36 37.8	1	11.30		0.33		−0.47		2		405
192678	+53 02368	V1372 Cyg		20 12 18	+53 30.4	2	7.36	.012	−0.02	.003	0.00	.011	7	A4p	695,1202
		G 24 - 11		20 12 19	+03 48.7	1	11.77		0.91		0.54		1		333,1620
192557	+34 03930	IDS20104N3504	A	20 12 19	+35 12.6	1	7.59		0.52				1	F6 V	1070
		WLS 2000 40 # 8		20 12 20	+40 14.3	1	12.77		0.57		0.03		2		1375
192679	+52 02657	IDS20110N5249	AB	20 12 21	+52 58.2	1	7.02		0.49		−0.04		1	F5	695
339764	+26 03837			20 12 23	+26 53.7	1	9.58		0.10		0.12		1	A0	674
228471	+32 03752	LS II +33 031		20 12 23	+33 17.0	1	9.88		0.69		−0.23		2	B2 Vn	1012
192558	+34 03931			20 12 23	+34 50.2	2	7.79	.075	1.21	.028	2.07		2	G2 V	1070,8100
228478				20 12 24	+32 52.	1	8.90		0.14				1	A0	1070
193202	+76 00785	G 261 - 19		20 12 24	+77 04.8	2	8.88	.005	1.32	.015	1.28		7	K5	196,1197
192559	+31 04003	IDS20104N3111	AB	20 12 25	+31 20.1	1	8.74		0.24		0.17		3	Ap Sr	8071
355264	+14 04234			20 12 27	+15 02.0	1	10.17		0.32		0.07		3	F5	196
192584	+34 03932			20 12 27	+34 43.9	1	8.88		−0.01				1	B8 V	1070
192604	+35 04031	IDS20106N3512	A	20 12 30	+35 47.1	1	9.09		0.00				1	B8 V	1070
192605	+35 04032			20 12 31	+35 28.2	1	8.70	.011	−0.13	.024			5	B9 V	1070,1118,1196
192603	+35 04033			20 12 31	+36 02.3	1	7.92		1.75		1.50		1	K2 Ib	8100
192606	+34 03934	IDS20106N3506	A	20 12 32	+35 15.2	1	8.49		0.00				1	B7 V	1070
192781	+60 02099	HR 7742		20 12 33	+60 29.2	1	5.79		1.47				2	gK5	71
192461	−3 04825	IDS20100S0348	A	20 12 34	−03 39.4	1	6.92		0.37		0.05		3	F0	3016
192461	−3 04824	IDS20100S0348	B	20 12 34	−03 39.4	1	8.08		0.45		−0.01		3		3016
		G 143 - 41		20 12 35	+18 19.1	1	10.50		0.88				1	K2	1759
		VES 179		20 12 35	+34 57.3	1	13.78		2.48		1.13		1		426
		CS 22885 # 34		20 12 35	−37 42.6	2	13.65	.010	0.41	.008	−0.15	.009	3		1580,1736
	+77 00767			20 12 36	+77 34.0	1	10.28		1.25				1	K8	1017
192659	+41 03668	HR 7737	⋆ AB	20 12 38	+41 57.0	2	6.71		−0.04	.000	−0.34	.028	4	B9 IV-V	1079,3016
192659	+41 03668	IDS20109N4148	C	20 12 38	+41 57.0	1	10.05		0.20		0.11		3		3016
331977				20 12 39	+32 17.6	1	10.42		0.99		0.19		3	B9	180
192641	+36 03956	V1679 Cyg	⋆ B	20 12 39	+36 30.5	3	7.92	.007	0.27	.017	−0.43	.028	17	C7	1012,1095,1359
192639	+36 03958	LS II +37 026		20 12 39	+37 12.0	4	7.11	.008	0.34	.010	−0.62	.007	9	O8 e	1011,1012,1070,1203
192640	+36 03955	HR 7736, V1644 Cyg	⋆ A	20 12 40	+36 39.1	8	4.95	.019	0.15	.018	0.01	.015	19	A2 V	15,247,1063,1118,1196*
192660	+39 04096	LS II +40 007		20 12 40	+40 10.6	2	7.45	.079	0.65	.020	−0.28	.070	3	B0 Ia	1012,8100
192541	+12 04265			20 12 41	+12 56.0	1	7.60		0.24		0.16		3	AmA4-F0	8071
192562	+15 04105			20 12 42	+15 32.9	1	8.64		1.06				1	K0	882
355263	+14 04237			20 12 44	+15 04.1	1	9.24		1.03		0.77		3	G5	196
192433	−30 17773	HR 7725		20 12 44	−30 09.6	1	6.30		1.51				4	K4 III	2035
228506	+32 03755			20 12 45	+32 52.0	1	8.14		0.98				1	G5 V	1070
		LP 350 - 81		20 12 48	−48 35.0	1	15.07		1.38		1.02		2		1773
		LP 350 - 82		20 12 48	−48 35.0	1	13.46		1.13		0.90		2		1773
228519	+38 03965	IDS20110N3837		20 12 49	+38 46.3	1	9.46		0.15				1	B0.5III	1070
239355	+55 02347			20 12 49	+55 48.0	1	8.85		0.31		0.10		2	A7	1502
191937	−73 02121			20 12 51	−73 08.1	2	6.55	.009	1.41	.000	1.67		6	K3 III	278,2012
355244				20 12 54	+15 26.	1	9.96		1.04		0.93		3	K0	196
228530	+41 03670	LS III +41 008		20 12 56	+41 41.3	1	10.20		0.83		−0.22		2	B0.5II	1012
192782	+54 02313	IDS20117N5426	AB	20 12 56	+54 35.5	1	8.93		−0.01		−0.57		2	A2	401
339956	+24 04063			20 12 57	+24 58.0	1	8.82		1.63				2	M0	369
191973	−73 02122			20 12 57	−72 57.8	2	6.94	.004	1.03	.001	0.87		6	K0 II/III	278,2012
		IRC +30 422		20 13 02	+29 36.6	1	12.28		0.81		0.20		1		363

Table 1 1099

HD	DM	Other Id	N Rem	α_{1950}	δ_{1950}	S	V	σ_V	B–V	σ_{B-V}	U–B	σ_{U-B}	n	Spectrum	References
	+57 02161		V	20 13 03	+57 56.7	1	9.60		1.75				1	R0	1238
228534	+36 03961			20 13 04	+37 13.4	1	9.23		0.27				1	O9.5II:	1070
192711	+34 03937	IDS20112N3441	AB	20 13 06	+34 50.4	1	9.16		0.14				1	A2 V	1070
228543	+37 03844	LS II +37 027		20 13 06	+37 59.1	2	8.68	.057	0.39	.024	-0.30		2	B2:II:	1070,8100
192663	+18 04437			20 13 07	+18 18.8	1	7.85		0.51				6	F5	8053
192698	+30 03959			20 13 07	+30 28.0	1	8.72		-0.06		-0.42		2	A0	401
192644	+13 04332			20 13 08	+13 29.7	1	7.85		0.90		0.68		2	G0	1648
192685	+25 04165	HR 7739, QR Vul	★ AB	20 13 09	+25 26.3	6	4.79	.011	-0.18	.004	-0.73	.017	18	B3 V	15,154,263,1363,3016,8015
192472	-36 14011	HR 7728	★ A	20 13 09	-36 36.5	1	6.39		1.54				4	M4 III	2035
228548	+39 04098	LS II +39 015		20 13 10	+39 48.7	1	10.10		0.69		-0.39		2	B0pe	1012
192745	+36 03962			20 13 11	+37 14.2	2	8.16	.000	0.03	.005			4	A0 V	1118,1196
192802	+47 03054			20 13 12	+47 17.2	1	7.24		-0.09		-0.37		2	B9	401
		VES 180		20 13 13	+41 05.2	1	11.20		0.36		0.11		1		426
192486	-35 14020	HR 7729		20 13 13	-35 21.2	3	6.53	.009	0.38	.004			11	F2 V	15,2032,2033
192744	+37 03845	LS II +37 028	★ AB	20 13 14	+37 32.5	2	7.39	.085	0.38	.005	0.01	.070	2	F0 V	247,8100
192686	+14 04240			20 13 15	+15 17.0	1	8.82		-0.02		-0.36		3	A0 p:	196
355243	+15 04108			20 13 16	+15 25.1	1	10.53		0.67		0.24		3	G5	196
		HA 88 # 544		20 13 17	+15 13.7	1	11.39		1.66				3		196
	+36 03963	LS II +36 045		20 13 17	+36 42.3	1	9.94		1.32		0.61		1	B8 II	247
	-74 01887			20 13 17	-74 05.4	1	10.87		0.11				4	A0	2012
228553	+36 03964			20 13 18	+37 06.8	2	8.71	.038	0.08	.017	-0.64		3	B2:IV	401,1070
		JL 48		20 13 18	-72 58.	1	17.30		0.60		0.80		1		132
		VES 126		20 13 19	+29 57.5	1	14.14		2.38				1		363
		LS II +34 017		20 13 19	+34 51.2	1	10.14		0.44		-0.31		2		426
192786	+42 03651			20 13 19	+42 44.3	1	7.95		-0.02		-0.06		2	B9	401
192713	+23 03944	HR 7741, QS Vul		20 13 21	+23 21.3	8	5.17	.015	1.05	.008	0.71	.015	25	G2 Ib	15,263,3016,6001,6009*
192712	+23 03943			20 13 21	+23 35.3	1	7.15		1.00		0.72		2	G5	1375
		VES 182		20 13 22	+35 05.0	1	11.50		0.42		0.00		2		426
192803	+40 04077	IDS20116N4100	AB	20 13 23	+41 08.7	1	7.87		0.03		-0.18		2	A0	401
355256				20 13 24	+15 10.	1	10.69		0.72		0.19		3	G5	196
		VES 252		20 13 24	+43 23.7	1	11.70		1.31		1.05		1		8056
192715	+14 04242			20 13 27	+15 09.8	3	6.86	.017	0.26	.004	0.05	.004	7	F0	196,379,3016
192787	+33 03827	HR 7743	★ AB	20 13 27	+33 34.6	2	5.68	.017	0.93	.015	0.66		3	K0 III	938,1118
192830	+48 03066			20 13 27	+49 15.3	1	8.56		0.76				1	K0	8097
		AJ89,1229 # 447		20 13 30	+01 58.	1	12.28		1.69				1		1532
228592	+39 04104	IDS20112N3919	AB	20 13 32	+39 28.6	1	9.20		0.10		1.06		1	A0 II	8100
192788	+30 03960	IRC +30 423	V	20 13 33	+30 55.0	1	10.09		1.59		0.93		1	M4	363
228587	+36 03966			20 13 33	+36 28.2	1	10.07		0.80		-0.14		2	B1 II	1012
193030	+64 01427			20 13 35	+64 36.7	1	7.36		0.83		0.38		2	G5 IV	3016
332085	+30 03963	LS II +30 027		20 13 36	+30 59.8	1	9.87		0.82		-0.09		2	G0	180
		VES 183		20 13 36	+36 54.7	1	12.23		0.79		0.61		2		426
192699	+4 04395			20 13 37	+04 25.6	1	6.43		0.86		0.49		2	G7 III-IV	985
228599	+37 03851			20 13 37	+37 33.2	1	9.04		1.35		1.42		2	G5 III	1375
192666	-12 05680	HR 7738	★ A	20 13 37	-12 29.5	2	6.32		0.01	.000	-0.18		6	B9 IV-V	1079,2009
192832	+41 03675	LS III +42 006		20 13 40	+42 14.8	1	8.62		0.89		-0.02		2	B5 Ia	1012,8100
192866	+45 03113			20 13 40	+45 35.2	1	7.13		-0.10		-0.44		2	A3	401
192805	+30 03966			20 13 41	+30 35.3	1	8.52		-0.06		-0.51		2	A0	401
192806	+27 03666	HR 7744	★	20 13 42	+27 39.6	5	4.52	.008	1.26	.006	1.11	.011	14	K3 III	15,1080,1363,3016,8015
228602	+36 03968	LS II +36 048		20 13 42	+36 40.7	1	9.84		0.57		-0.30		2	B1 III	1012
		HA 205 # 636		20 13 42	-75 06.5	1	11.90		0.49		0.01		std		2
192867	+43 03541			20 13 47	+43 59.1	2	7.23	.010	1.60	.005	1.93	.005	3	M1 III	247,1733
192981	+59 02206			20 13 51	+59 28.3	1	8.69		0.00		-0.10		2	A2	401
		VES 184		20 13 52	+37 11.6	1	12.94		1.94		1.99		1		426
192869	+41 03678			20 13 53	+42 12.4	2	7.84	.015	0.55	.005	0.07	.010	3	F6 IV	247,1375
228633	+41 03677			20 13 55	+41 40.8	1	9.13		1.66				2	M0	369
192909	+47 03059	HR 7751, o2 Cyg	★ AB	20 13 55	+47 33.6	7	3.98	.009	1.52	.005	1.02	.016	27	K3 Ib+B3 V	15,401,1332,1363*
192718	-7 05235			20 13 56	-07 35.8	4	8.39	.007	0.57	.005	-0.02	.012	7	F8	516,742,1064,1594
192967	+53 02374	IDS20126N5350	AB	20 13 57	+53 59.5	1	7.27		1.55		1.94		1	M1	8084
192967	+53 02374	IDS20126N5350	D	20 13 57	+53 59.5	1	12.45		0.61		0.22		1		8084
355258				20 14 00	+15 15.	1	10.62		0.15		0.13		3	A2	196
		VES 186		20 14 03	+36 17.0	1	12.41		0.81		-0.17		2		426
192653	-37 13646			20 14 03	-36 47.0	1	8.66		0.06		-0.14		10	A0 IV/V	1732
192273	-69 03107			20 14 04	-69 35.5	1	8.85		-0.19		-0.82		11	B2 V	1732
		LP 754 - 52		20 14 06	-10 50.0	1	11.07		1.04		0.85		1		1696
192836	+21 04130	HR 7746		20 14 08	+21 26.6	6	6.14	.009	1.04	.002	0.92	.002	73	K1 III	15,1007,1013,3016*
228654	+40 04082	LS II +40 008		20 14 09	+40 43.0	1	9.49		0.97		-0.11		2	B7	405
192850	+21 04132			20 14 10	+22 07.9	1	6.82		1.23		1.24		2	K2	1648
228647	+36 03973			20 14 10	+36 40.2	1	8.69		1.31		1.29		2	K2 III	1601
192633	-45 13709			20 14 10	-45 23.1	3	8.56	.003	1.09	.002	0.99	.003	18	K0 III	863,977,2011
355285	+14 04246			20 14 11	+14 38.2	1	11.37		0.33		0.10		3	A3	196
192934	+38 03977	HR 7752		20 14 14	+38 44.6	8	6.28	.009	0.01	.010	-0.01	.010	20	A1 V	379,833,1049,1063*
		MCC 818		20 14 14	+42 47.	1	9.98		1.27				2	M0p	1619
192737	-21 05672	RT Cap	★ A	20 14 14	-21 28.9	1	7.41		4.02				1	C	109
192871	+21 04133	LS II +22 023		20 14 15	+22 14.5	2	7.19	.010	0.33	.020	0.13		5	F3 II	6009,8100
228657	+37 03855	IC 4996 - 6		20 14 15	+37 28.6	1	9.34		0.00		-0.05		2	B8 V	49
192983	+49 03236	HR 7755		20 14 15	+50 04.7	1	6.36		0.09		0.06		2	A2 Vn	1733
355259				20 14 16	+15 16.9	1	9.88		1.06		0.86		3	G5	196
332044	+32 03761	LS II +32 021		20 14 17	+32 31.7	2	9.76	.115	0.49	.021	-0.34	.239	3	B3 Ia	180,1215
		IC 4996 - 85		20 14 17	+37 31.3	1	14.54		0.64		0.30		1		49
		IC 4996 - 9		20 14 17	+37 31.6	1	11.25		0.48		0.09		1		49

HD	DM	Other Id	N	Rem	α_{1950}	δ_{1950}	S	V	σ_V	B–V	σ_{B-V}	U–B	σ_{U-B}	n	Spectrum	References
355260	+14 04247				20 14 18	+15 07.1	1	9.65		0.22		0.17		3	A3	196
228658	+37 03856	IC 4996 - 8			20 14 18	+37 24.1	1	10.20		0.38		-0.49		1	B2 II	49
		IC 4996 - 10			20 14 19	+37 25.9	1	12.59		0.47		0.11		1		49
		HA 205 # 466			20 14 20	−74 57.4	1	15.48		0.82		0.06		std		2
192893	+25 04173				20 14 21	+25 43.1	1	7.72		0.30		0.11		3	Am	8071
192913	+27 03668	MW Vul		A	20 14 23	+27 37.2	2	6.64	.034	-0.07	.006	-0.24	.005	7	A0p	603,1202
		VES 187			20 14 23	+34 36.1	1	12.34		0.45		0.09		1		426
192985	+45 03119	HR 7756			20 14 23	+45 25.5	3	5.89	.027	0.40	.021	-0.05	.010	6	F4 V	254,1118,3016
		LS II +30 028			20 14 24	+30 17.9	1	10.10		0.10		-0.32		2		180
		CS 22885 # 52			20 14 25	−39 16.0	1	9.62		1.49		0.73		1		1736
192702	−39 13722	RT Sgr			20 14 25	−39 16.1	1	8.28		1.52		0.68		1	Me	975
		VES 188			20 14 26	+38 27.7	1	13.81		0.64		0.30		2		426
	+6 04489				20 14 27	+06 46.3	1	9.72		1.14		1.08		2	K4	3072
191220	−83 00695	HR 7698			20 14 27	−83 28.2	1	6.16	.005	0.20	.000	0.12	.010	7	A2/3mA8-F0	15,1075
192968	+40 04086	LS II +40 009			20 14 27	+40 48.5	2	7.84	.015	-0.03	.015	-0.83	.010	5	B1 Vn	399,401
		IC 4996 - 90			20 14 29	+37 27.6	1	15.00		0.77		0.29		1		49
		V 396 Cyg			20 14 29	+41 57.2	1	10.98		2.03				1		1772
		IC 4996 - 87			20 14 31	+37 31.6	1	14.60		0.88		0.52		1		49
		IC 4996 - 18			20 14 32	+37 30.7	1	11.74		1.36		1.05		1		49
		HA 205 # 651			20 14 32	−75 01.2	1	11.89		1.35		1.50		std		2
		IC 4996 - 89			20 14 33	+37 29.2	1	14.93		0.92		0.22		1		49
355382	+15 04115				20 14 34	+15 25.7	1	9.49		0.26		0.19		3	A2	196
228684	+34 03952	LS II +34 018			20 14 35	+34 57.2	1	9.96		0.70		-0.13		2	B3 II	1012
		G 186 - 18			20 14 36	+29 22.5	1	11.39		0.82		0.43		2	K0	1658
192987	+36 03978	HR 7757			20 14 36	+36 54.1	2	6.47	.009	-0.09	.004	-0.42		3	B6 III	154,1070
		IC 4996 - 20			20 14 36	+37 25.8	1	11.71		0.44		-0.30		1	B3 V	49
		VES 189			20 14 37	+33 58.2	2	13.49	.000	3.00	.000	1.38	.000	3		426,8103
192938	+16 04208				20 14 39	+16 36.6	1	7.37		1.21				1	K0	882
192944	+24 04075	HR 7753			20 14 39	+24 31.0	1	5.32		0.95		0.67		5	G8 III	3016
228690	+37 03861	LS II +37 036			20 14 39	+37 46.0	1	9.25		0.27		-0.58		2	B0.5V	1012
192531	−63 04576	HR 7732			20 14 40	−63 23.2	1	6.27		1.04				4	K0 III	2035
	+1 04247				20 14 40	+01 34.9	1	10.77		1.67				1	M2 III:	1532
192989	+35 04044				20 14 40	+36 07.3	1	7.09		1.08				2	G5 IV	1118
		IC 4996 - 30			20 14 40	+37 30.7	1	10.46		0.37		-0.48		2		305
193007	+37 03860	IC 4996 - 28		⋆ B	20 14 41	+37 29.2	2	8.00	.015	0.38	.025	-0.52	.005	5	B0.2III	305,399,8084
		IC 4996 - 31			20 14 41	+37 29.6	1	12.60		0.50		-0.12		1		49
	−75 01597	HA 205 # 830			20 14 41	−75 11.6	1	10.85		0.70		0.14		std		2
228693	+35 04045				20 14 42	+35 19.1	1	9.69		0.19		0.12		1	A1 V	401
	+37 03862	IC 4996 - 34		⋆ C	20 14 42	+37 29.4	3	8.97	.045	0.38	.011	-0.48	.005	6	B0 V	305,1012,8084
		IC 4996 - 37			20 14 42	+37 31.1	1	13.43		0.53		0.18		2		305
		L 278 - 36			20 14 42	−51 25.	1	13.98		0.56		-0.21		2		1696
		IC 4996 - 40			20 14 43	+37 28.8	1	12.80		0.49		0.03		2		305
		IC 4996 - 39		⋆ V	20 14 43	+37 29.3	1	11.28		0.39		-0.39		2		305
		VES 190			20 14 43	+38 40.7	1	12.60		0.53		0.35		2		426
		IC 4996 - 84			20 14 44	+37 28.4	1	14.49		0.78		-0.06		1		49
		IC 4996 - 91			20 14 44	+37 28.4	1	12.20		1.89		1.97		1		305
		IC 4996 - 41			20 14 44	+37 30.9	2	11.47	.049	1.28	.000	0.96	.029	3		49,305
		G 143 - 43			20 14 45	+17 06.6	1	10.75		0.75		0.26		2		1658
192990	+35 04047				20 14 45	+35 18.8	2	7.11	.020	-0.03	.000	-0.10		5	B9 IV	401,1118
		IC 4996 - 43			20 14 45	+37 24.9	1	11.92		0.67		0.53		1		49
		IC 4996 - 86			20 14 45	+37 25.9	1	14.56		1.02		0.40		1		49
		IC 4996 - 82			20 14 45	+37 26.4	1	14.18		0.87		0.26		1		49
228699		IC 4996 - 44			20 14 46	+37 31.9	2	9.45	.015	0.38	.005	-0.52	.005	3	B0.5III	49,305
		IC 4996 - 81			20 14 48	+37 28.6	1	13.85		0.54		0.09		1		49
193009	+31 04018	LS II +32 022			20 14 49	+32 13.5	2	7.15	.033	0.09	.000	-0.84	.005	6	B1 V:penn	180,1212
		IC 4996 - 80			20 14 49	+37 25.7	1	13.82		0.67		0.41		1		49
228706	+37 03864	IC 4996 - 47			20 14 50	+37 22.7	1	8.81		0.16		0.18		1	A3	49
		IC 4996 - 83			20 14 50	+37 30.6	1	14.25		0.64		0.44		1		49
228712	+40 04087	LS II +40 011			20 14 50	+40 43.8	2	8.67	.020	1.14	.005	0.05	.010	3	B0.5Ia	401,1012
	+46 02895				20 14 50	+46 24.1	1	10.25		1.88				2		369
192758	−43 13915				20 14 50	−43 00.9	4	7.02	.002	0.32	.007	0.02	.009	23	F0 V	278,1075,1770,2011
193032	+38 03980	LS II +38 026			20 14 51	+38 44.6	2	8.34	.019	0.34	.015	-0.58		3	B0 III	1012,1070
		CS 22885 # 40			20 14 51	−38 15.4	2	15.23	.010	0.41	.008	-0.16	.009	3		1580,1736
192954	+15 04120				20 14 52	+15 43.0	2	7.47	.005	-0.10	.000	-0.35	.025	4	B9	379,1733
192876	−12 05683	HR 7747		⋆ A	20 14 53	−12 39.8	12	4.25	.018	1.07	.010	0.80	.010	41	G3 Ib	15,1075,1088,2001*
		IC 4996 - 51		⋆ V	20 14 54	+37 31.4	1	10.86		0.23		-0.13		1		49
		BPM 26691			20 14 54	−57 31.	1	13.73		-0.17		-1.08		4		3065
193010	+31 04020	IDS20129N3112		A	20 14 55	+31 21.0	1	6.90		-0.01		-0.10		2	A0	401
		IC 4996 - 88			20 14 55	+37 27.0	1	14.76		0.73		0.51		1		49
228710	+32 03767	G 210 - 12			20 14 55	+32 57.4	1	9.26		0.58		0.03		2	G0	1658
		VES 191			20 14 56	+37 32.3	1	12.32		0.68		0.26		2		426
		LS III +41 011		⋆ V	20 14 58	+41 48.4	1	10.88		0.55		-0.35		1		651
193090	+44 03414				20 14 58	+45 11.0	1	7.10		1.50		1.82		2	K5 III	247
		VES 192			20 14 59	+38 03.8	2	16.02	.000	1.69	.000			2		426,8103
192740	−51 12475				20 14 59	−51 39.9	1	7.68		0.42		-0.02		1	F5 IV/V	219
193034	+33 03837				20 15 00	+33 59.0	1	7.81		-0.07		-0.42		2	B9	401
		IC 4996 - 54			20 15 00	+37 26.4	1	13.14		0.58		0.09		1		49
		VES 193			20 15 00	+37 45.8	1	13.08		0.77		-0.11		1		426
		L 494 - 2			20 15 00	−36 20.	1	13.64		0.78		0.23		1		3062
193063	+39 04113	IDS20132N3923		AB	20 15 01	+39 32.5	1	7.73		-0.05		-0.42		2	B9 V	401

Table 1   1101

HD	DM	Other Id	N Rem	α_{1950}	δ_{1950}	S	V	σ_V	B–V	σ_{B-V}	U–B	σ_{U-B}	n	Spectrum	References
228721	+37 03865	IC 4996 - 57		20 15 02	+37 21.8	1	8.60		1.15		1.03		1	G8 III	49
		VES 194	V	20 15 03	+39 46.8	1	12.34		1.17		0.16		1		426
177482	−89 00047	HR 7228, σ Oct		20 15 03	−89 08.3	6	5.45	.011	0.28	.014	0.11	.010	22	F0 IV	15,615,1075,1117,2038,3023
192900	−18 05641	RW Cap		20 15 05	−17 49.8	1	10.22		0.06		0.12		1	A1 III/IV	668
192879	−22 05384	HR 7748		20 15 05	−21 58.0	4	5.86	.009	1.00	.001	0.79		12	G8 III	15,2018,2030,4001
334068	+29 03973	LS II +29 043		20 15 06	+29 57.4	1	9.08		0.38		-0.55		1		180
		CS 22873 # 166		20 15 06	−61 39.8	1	11.82		0.99		0.51		1		1736
193064	+35 04048			20 15 07	+35 48.1	3	7.67	.030	-0.03	.004	-0.28		6	A0	401,1118,1196
193076	+37 03866	IC 4996 - 64		20 15 08	+37 31.6	3	7.64	.015	0.31	.011	-0.56	.005	4	B0.7II	49,1012,1070
193092	+39 04114	HR 7759	★A	20 15 08	+40 12.6	3	5.24	.010	1.65	.007	1.89	.019	6	K5 II	247,1080,1118
193077	+36 03987	LS II +37 043	★A	20 15 09	+37 16.1	2	8.00	.005	0.29	.015	-0.52	.015	11	WN6	1095,1359
193091	+42 03666			20 15 10	+42 42.9	1	8.69		-0.03		-0.13		2	A0	401
332077	+31 04024			20 15 11	+31 23.9	1	9.00		1.96				2	M9	369
346286	+21 04142			20 15 14	+22 14.8	1	10.81		0.40		0.33		2	A3	1375
193117	+40 04090	LS II +40 014		20 15 14	+40 41.3	2	8.70	.000	0.61	.000	-0.42	.000	4	O9.5II	1011,1012
192826	−43 13919			20 15 14	−42 46.5	4	7.45	.002	0.59	.000	0.16	.009	14	F8 IV/V	278,1075,1770,2011
339967	+24 04081			20 15 17	+24 37.0	1	10.53		0.29		0.06		2	A7	1375
192947	−12 05685	HR 7754	★ABC	20 15 17	−12 42.1	10	3.57	.011	0.94	.008	0.68	.006	33	G8 IIIb	15,58,1020,1075,1088*
		VES 195	V	20 15 20	+41 56.8	2	11.92	.060	0.80	.020	-0.34	.020	2		426,8056
		LS III +41 014		20 15 21	+41 48.4	1	11.01	.005	0.60	.000	-0.47	.009	5		483,651
192827	−48 13509	HR 7745		20 15 21	−47 52.1	3	6.29	.007	1.67	.001	1.94	.063	7	M1 III	1770,2035,3005
		VES 196		20 15 24	+36 07.6	1	12.62		2.27				2		426
228759	+41 03689	LS III +41 015		20 15 24	+41 48.0	2	9.47	.014	0.68	.009	-0.44	.009	4	O5.5V	483,651
193159	+40 04093			20 15 26	+40 26.1	1	7.14		0.00		-0.15		2	B8	401
193216	+49 03245	G 230 - 31	★A	20 15 28	+50 07.6	1	8.16		0.75		0.40		2	G5	1625
192844	−44 13869			20 15 28	−44 40.9	4	7.55	.001	0.95	.007	0.63	.009	24	K0 III	460,977,1075,2011
192418	−74 01890			20 15 28	−74 08.1	1	7.83		1.29		1.38		7	K2 III	1704
193094	+28 03695	HR 7760	★AB	20 15 29	+28 59.5	1	6.22		1.01				2	G9 III	71
334041	+30 03977			20 15 31	+30 48.3	1	10.36		0.06		-0.16		1	B8	363
193183	+37 03867	LS II +38 030		20 15 32	+38 04.8	2	6.99	.014	0.45	.005			5	B1.5Ib	1070,1118
		CS 22873 # 165		20 15 32	−61 50.1	1	14.80		0.32		0.01		1		1736
193182	+39 04115	LS II +39 017		20 15 37	+39 26.3	3	6.49	.015	-0.09	.015	-0.23	.033	5	Ape:	379,1012,6009
228766	+36 03991	LS II +37 046		20 15 38	+37 09.2	4	9.14	.008	0.61	.023	-0.38	.008	15	O6 e	1011,1012,1095,1359
193097	+15 04124			20 15 39	+15 51.2	1	8.38		1.61				1	K5	882
192886	−47 13340	HR 7749		20 15 43	−47 44.1	12	6.12	.005	0.47	.006	0.00	.008	65	F5 V	15,208,278,863,977,977*
193217	+42 03670	HR 7762		20 15 46	+42 33.9	2	6.30	.015	1.64	.010	1.91	.000	3	K4 II:	247,1080
		CS 22885 # 54		20 15 49	−39 21.9	2	14.77	.010	0.45	.009	-0.12	.007	3		1580,1736
228791	+38 03991	IDS20141N3818	A	20 15 53	+38 26.8	1	9.03		0.10				1	B6 IV	1070
228789	+38 03993	IDS20141N3902	A	20 15 53	+39 10.6	1	10.21		0.50		-0.31		4	B7	206
		Steph 1769		20 15 53	−02 58.0	1	10.47		1.70		2.08		2	M1	1746
		VES 197		20 15 54	+36 35.8	1	12.03		0.70		0.01		2		426
193237	+37 03871	HR 7763, P Cyg		20 15 57	+37 52.6	7	4.81	.019	0.42	.005	-0.58	.005	18	B2pe	15,154,1012,1118,8003*
193161	+17 04282	IDS20137N1801	AB	20 15 59	+18 10.3	1	7.96		0.12		0.13		3	AmA2-A7	8071
228779	+34 03961	LS II +34 019		20 15 59	+34 39.6	2	8.92	.000	1.31	.000	0.22	.000	4	O9.5Ib	1011,1012
228797	+39 04117	LS II +39 018		20 16 01	+39 52.5	1	9.39		0.87		-0.07		2	B1 II	1012
193238	+32 03773	IDS20141N3253	AB	20 16 02	+33 02.1	1	7.56		-0.02		-0.26		2	B9	401
		CS 22885 # 69		20 16 02	−41 32.1	2	13.71	.010	0.72	.018	0.16	.004	3		1580,1736
193122	+0 04475			20 16 04	+00 29.1	1	6.94		0.19		0.05		2	A3	985
193205	+24 04085	LS II +24 025		20 16 05	+24 20.7	2	8.00	.005	0.85	.005	0.49		11	F6 Iab	8059,8100
193221	+25 04188			20 16 07	+25 21.2	2	7.69	.005	1.11	.000	1.05	.000	5	K2 III	833,1625
228808	+37 03872	LS II +37 053		20 16 07	+37 26.0	1	9.17		1.23		0.94		1	F5 Iab	247
192961	−46 13498			20 16 07	−46 35.1	5	8.72	.018	1.18	.012	1.14	.011	21	M0/1 V	863,977,1494,2011,3072
334060	+29 03982			20 16 09	+30 17.1	1	9.28		0.33		-0.12		2	B8	379
228807	+37 03873			20 16 09	+38 16.2	1	9.00		0.13				1	B5 IV	1070
193220	+25 04189	IDS20140N2520	AB	20 16 10	+25 29.5	3	6.98	.013	-0.15	.005	-0.77	.016	6	B1.5V	105,399,963
		VES 130		20 16 12	+30 33.3	1	12.57		1.08		0.39		1		363
193239	+25 04190			20 16 14	+25 27.7	2	8.19	.000	-0.06	.000	-0.38	.005	5	A0	833,1625
228827	+40 04102	IDS20146N4103	AB	20 16 17	+41 12.1	1	9.44		1.28		0.10		2	B0	405
		CS 22885 # 67		20 16 20	−41 00.7	2	13.98	.010	0.62	.014	0.02	.000	3		1580,1736
193322	+40 04103	HR 7767	★AB	20 16 21	+40 34.5	9	5.83	.010	0.11	.009	-0.77	.008	41	O9 Vn	15,154,245,450,1011*
193322	+40 04103	IDS20146N4025	C	20 16 21	+40 34.5	1	11.11		0.19		-0.29		3		450
193322	+40 04103	IDS20146N4025	D	20 16 21	+40 34.5	1	11.17		0.19		-0.32		3		450
193248	+21 04147			20 16 24	+21 18.4	1	7.45		0.26		0.09		3	Am A5-F0	8071
193270	+29 03984			20 16 24	+30 12.6	1	8.58		0.08		0.06		2	A0	379
334039	+30 03980	LS II +30 029		20 16 25	+30 39.9	2	8.29	.009	0.73	.019	0.12	.019	4	B9 Ib-II	180,1012
193292	+31 04029			20 16 25	+31 57.2	1	7.20		0.25		0.16		3	Am	8071
		LP 754 - 47		20 16 28	−11 38.3	1	11.58		0.77		0.19		2		1696
193150	−19 05776	HR 7761	★A	20 16 31	−19 16.6	3	5.28	.020	1.41	.012	1.55	.015	10	K2 III	2009,3016,8100
193468	+53 02379			20 16 32	+54 09.9	1	6.74		0.43		-0.10		1	F5 V	1716
193344	+35 04059			20 16 34	+36 05.8	3	7.58	.016	-0.04	.005	-0.13		5	B9p	1090,1118,1196
		CS 22885 # 48		20 16 35	−39 06.1	1	14.27		0.34		-0.07		1		1736
193369	+36 03998	HR 7769		20 16 36	+36 50.5	6	5.58	.008	0.06	.010	0.01	.006	19	A2 V	196,245,1063,1118*
228841	+38 04000	LS II +38 036		20 16 40	+38 43.2	3	8.94	.008	0.56	.008	-0.42	.008	7	O7.5p	1011,1012,1209
193002	−55 09365	HR 7758		20 16 40	−55 12.5	3	6.27	.001	1.61	.008	2.00	.015	7	M0/1 III	58,2035,3005
		LP 515 - 3		20 16 43	+12 26.6	1	15.48		1.69				3		1759
193370	+34 03967	HR 7770	★	20 16 44	+34 49.5	9	5.17	.016	0.65	.009	0.48	.012	22	F5 Ib	15,1080,1118,1119*
		LS II +30 030		20 16 49	+30 53.0	1	11.32		0.50		-0.12		2		180
		Dol 42 - 1		20 16 49	+37 56.5	1	11.00		0.52		0.13		4	B5 V	1745
		VES 199	V	20 16 49	+39 46.1	1	11.45		0.62		0.19		1		426
		VES 198	V	20 16 49	+40 00.5	1	11.70		0.64		0.21		1		426

HD	DM	Other Id	N	Rem	α_{1950}	δ_{1950}	S	V	σ_V	B–V	σ_{B-V}	U–B	σ_{U-B}	n	Spectrum	References
193426	+39 04130	LS II +40 016			20 16 52	+40 04.2	1	7.73		1.24		0.37		1	B9 Ia	247
228854	+35 04062	V382 Cyg		⋆ A	20 16 54	+36 11.0	3	8.56	.235	0.71	.004	-0.29	.009	8	O8	588,1011,1012
228859	+36 04000	LS II +37 057			20 16 55	+37 05.9	1	9.63		0.81		-0.16		2	B0.5Ia	1012
193427	+38 04002	LS II +39 022		⋆ A	20 16 57	+39 14.9	1	9.22		0.44		-0.45		2	B1 V	1012
193132	-43 13932				20 16 58	-43 03.6	3	7.76	.002	1.02	.008	0.79	.004	16	K0 III	977,1075,2011
228860	+36 04001	LS II +36 056			20 16 59	+36 48.2	1	9.68		0.60		-0.32		2	B0.5IV	1012
193349	+13 04356	IDS20147N1403	AB		20 17 01	+14 12.8	1	6.75		0.35		0.04		3	A0	3016
		Dol 42 - 3			20 17 01	+38 00.2	1	11.91		1.73		1.92		3		1745
193443	+37 03879	Dol 42 - 2		⋆ AB	20 17 01	+38 07.3	6	7.25	.011	0.39	.009	-0.56	.027	16	O9 III	1011,1012,1070,1118*
193242	-20 05895				20 17 01	-20 04.2	2	9.16	.005	0.77	.014	0.13		4	G6wF2/3	742,1594
193444	+37 03878	LS II +37 059			20 17 02	+37 41.1	1	8.45		0.50				1	B0.5V	1070
		Dol 42 - 4			20 17 02	+38 02.0	1	12.77		0.48		-0.02		2		1745
193664	+66 01281	HR 7783			20 17 02	+66 41.6	5	5.92	.012	0.59	.009	0.06	.004	29	G3 V	15,22,1013,1197,3026
332117	+32 03783				20 17 04	+32 33.5	1	9.50		1.77				2	K7	369
		Dol 42 - 5			20 17 04	+37 55.8	1	12.74		0.50		0.05		2	F5	1745
		Dol 42 - 6			20 17 06	+37 54.0	1	12.76		0.47		0.32		2	A5	1745
		Dol 42 - 7			20 17 06	+37 56.7	1	13.12		0.49		0.35		1	B8	1745
		CS 22950 # 50			20 17 07	-13 34.0	1	13.91		0.69		-0.03		1		1736
193373	+12 04289	HR 7771			20 17 08	+13 03.5	1	6.21		1.63				2	M1 III	71
193469	+38 04003	IDS20153N3842	AB		20 17 08	+38 06.8	3	6.37	.024	1.88	.027	1.84	.021	7	K5 Ib	105,247,1080
193486	+42 03685	IDS20154N4225	AB		20 17 08	+42 34.4	1	8.65		0.01		-0.16		2		401
193329	-1 03951	HR 7768			20 17 08	-01 14.2	1	6.05		1.09		1.02		8	K0	1088
228882	+40 04113	LS II +40 017			20 17 11	+40 32.8	1	9.21		1.15		0.12		2	B0.5Ia	1012
193550	+49 03250				20 17 11	+49 20.4	1	7.85		0.14		-0.31		2	B8	555
193592	+54 02329	HR 7781		⋆ AB	20 17 11	+55 14.4	4	5.76	.006	0.12	.007	0.03	.005	16	A2 Vs	15,1007,1013,8015
		CS 22885 # 105			20 17 12	-38 56.5	2	14.35	.010	0.70	.017	0.01	.003	3		1580,1736
228886	+36 04004	LS II +37 060			20 17 13	+37 14.5	1	9.63		0.90		-0.13		2	B0.5III:	1012
193536	+45 03139	HR 7777, V1773 Cyg			20 17 13	+46 09.9	3	6.45	.007	-0.13	.010	-0.70	.013	5	B2 V	154,247,1423
193390	+14 04259				20 17 14	+14 31.1	1	8.67		-0.07		-0.68		1	A2	963
193487	+36 04005	LS II +36 057		⋆ AB	20 17 14	+36 35.7	1	7.46		0.40		0.12		1	F4 III	247
		Dol 42 - 8			20 17 14	+38 05.8	1	10.56		0.29		0.10		6	A6 V	1745
		Dol 42 - 9			20 17 15	+38 01.0	1	10.92		0.48		-0.24		8	B3 III	1745
193516	+37 03881	LS II +37 061			20 17 16	+37 36.7	1	8.61		0.56				1	B2:III:	1070
		Dol 42 - 10			20 17 16	+38 00.5	1	12.40		2.02				1		1745
193515	+37 03882	Dol 42 - 11			20 17 18	+38 00.2	1	7.59		1.02		0.71		7	K0 IV	1745
		Dol 42 - 12			20 17 19	+37 56.7	1	11.43		0.54		-0.02		7	G5	1745
193514	+38 04006	LS II +39 024			20 17 20	+39 06.9	6	7.41	.015	0.43	.017	-0.53	.029	13	O7 e	253,1011,1012,1070*
		Dol 42 - 13			20 17 21	+38 01.8	1	11.51		0.31		0.20		8	A2 III	1745
		Dol 42 - 14			20 17 22	+37 55.0	1	12.95		0.79		0.36		6		1745
193281	-29 16981	HR 7764		⋆ AB	20 17 23	-29 21.4	2	6.48	.284	0.20	.048	0.24		5	A4/7 V	938,2009
193213	-45 13741	IDS20139S4551	AB		20 17 24	-45 42.2	3	7.32	.002	0.51	.000	0.01	.016	10	G5 III +(F)	278,1075,2011
228899	+36 04006	IDS20156N3616	A		20 17 25	+36 25.7	1	9.32		0.97		0.71		1	G5	3032
228899	+36 04006	IDS20156N3616	B		20 17 25	+36 25.7	1	9.77		0.98		0.69		1		3032
193470	+23 03970				20 17 27	+24 04.5	1	7.79		0.95				7	K0	8053
		Dol 42 - 15			20 17 29	+38 01.0	1	11.34		0.42		0.03		6	F6	1745
		Be 86 - 18			20 17 29	+38 34.9	1	11.29		0.24		0.19		1		1751
		AJ89,1229 # 456			20 17 30	+07 53.	1	11.49		1.75				1		1532
		Dol 42 - 17			20 17 30	+37 59.1	1	12.40		0.35		-0.14		4	B2	1745
		Dol 42 - 16			20 17 30	+38 03.0	1	12.70		0.69		0.50		3		1745
228907	+37 03886	LS II +37 064			20 17 31	+37 29.9	1	9.63		1.13		0.90		1	A0 II	247
		Dol 42 - 18			20 17 31	+38 01.9	1	11.63		0.60		0.11		6	G5	1745
		CS 22885 # 96			20 17 31	-40 03.0	2	13.32	.010	0.71	.017	0.05	.003	3		1580,1736
		Be 86 - 20			20 17 32	+38 32.3	1	12.57		1.32		1.17		1		1751
		Dol 42 - 19			20 17 34	+38 01.2	1	10.20		0.08		0.14		5	A0 V	1745
		Dol 42 - 20			20 17 35	+38 06.2	1	12.42		0.51		0.09		4		1745
193594	+40 04116				20 17 35	+41 14.0	1	7.79		0.00				2	B9	1118
		Dol 42 - 21			20 17 36	+37 58.3	1	12.97		0.66		0.22		4		1745
		Be 86 - 19			20 17 36	+38 32.8	1	13.67		0.93		0.81		1		1751
193302	-36 14057	HR 7765			20 17 38	-35 50.0	1	6.46		1.31				4	K3 III	2009
193472	+13 04360	HR 7774			20 17 40	+13 23.4	3	5.94	.008	0.30	.012	0.12	.011	8	A5 m	252,355,8071
		HIC 100213			20 17 40	+37 02.0	1	10.50		1.58				2		1726
	+38 04008	Be 86 - 17			20 17 41	+38 39.6	1	9.74		0.52		-0.36		1	B0.5 V	1751
228919	+39 04135	LS II +40 019			20 17 41	+40 18.2	1	9.68		0.63		-0.28		2	B1 IV	1012
		Dol 42 - 22			20 17 42	+37 57.7	1	12.29		0.52		0.18		4		1745
193595	+38 04012	LS II +38 041			20 17 42	+38 53.9	2	8.74	.024	0.36	.000	-0.58		3	O6 V	1011,1070
193610	+42 03692	LS III +42 011			20 17 42	+42 50.4	1	8.03		0.86		0.21		5	A0:Ib:	399
		Dol 42 - 24			20 17 43	+37 56.1	1	12.89		0.49		0.38		4	F5	1745
		Dol 42 - 26			20 17 43	+37 56.3	1	10.88		0.30		-0.23		6	B4 V	1745
		Dol 42 - 25			20 17 43	+37 57.9	1	12.91		0.69		0.26		4		1745
193576	+38 04010	V444 Cyg			20 17 43	+38 34.4	2	8.00	.000	0.52	.014	-0.38	.005	9	O6	1012,1095
228913	+35 04067	LS II +36 059			20 17 44	+36 03.5	1	9.96		0.60		-0.35		2	B0.5III:nn	1012
		Dol 42 - 27			20 17 44	+37 57.9	1	11.89		0.47		0.10		4	B3	1745
		Dol 42 - 23			20 17 44	+38 10.8	1	11.42		0.51		0.07		7	G3 V	1745
		Be 86 - 21			20 17 46	+38 33.8	1	13.90		2.04				1		1751
228928	+40 04117	LS II +40 020			20 17 46	+40 29.7	1	9.69		0.89		-0.14		2	B2 Ib:nn	1012
193429	-6 05451	HR 7772			20 17 46	-06 31.2	2	6.62	.005	1.55	.000	1.83	.000	7	K5	15,1256
193379	-25 14710				20 17 46	-24 57.4	1	8.01		1.23		1.32		3	K1 III	1657
193553	+29 03992				20 17 47	+29 34.1	1	6.72		-0.16		-0.67		2	B8	401
		Be 86 - 14			20 17 47	+38 25.8	1	10.70		0.58		-0.25		2	B1 V	1751
193611	+37 03890	V478 Cyg		⋆ A	20 17 48	+38 10.6	1	8.66		0.57		-0.46		5	B0 V	1768

Table 1

1103

HD	DM	Other Id	N Rem	α_{1950}	δ_{1950}	S	V	σ_V	B–V	σ_{B-V}	U–B	σ_{U-B}	n	Spectrum	References
193231 −55 09370		IDS20140S5507	AB	20 17 48	−54 58.2	1	8.39		0.73				4	G5 V	2012
		Dol 42 - 29		20 17 49	+38 07.3	1	12.38		0.64		0.55		4	F0	1745
		Dol 42 - 28		20 17 49	+38 10.2	1	12.05		1.22		1.01		7		1745
228929 +39 04136		LS II +39 027		20 17 49	+39 45.2	1	9.63		1.15		0.07		2	B0.5Ib	1012
193612 +37 03889		IDS20160N3724	AB	20 17 51	+37 33.8	3	8.64	.005	-0.04	.015	-0.44		6	A0 III	401,1118,1196
228932 +37 03891		Dol 42 - 30		20 17 51	+38 01.6	1	9.94		0.24		-0.30		5	B4 V	1745
		CS 22950 # 3		20 17 51	−16 30.4	1	15.06		0.30		0.18		1		1736
		Dol 42 - 31		20 17 52	+37 58.1	1	12.81		0.53		0.39		4		1745
		Be 86 - 12		20 17 52	+38 27.6	1	11.19		0.47		-0.07		2		1751
		Dol 42 - 32		20 17 53	+37 58.7	1	13.07		1.57		1.45		3		1745
193432 −13 05642		HR 7773	⋆ A	20 17 53	−12 55.1	6	4.75	.005	-0.05	.005	-0.10	.005	23	B9 IV	15,1075,1088,2012*
		Dol 42 - 34		20 17 54	+37 54.3	1	13.16		0.27		0.16		1		1745
		Dol 42 - 33		20 17 55	+37 59.0	1	13.05		0.55		0.56		2		1745
	+5 04481	G 24 - 13		20 17 56	+05 52.7	5	10.13	.019	0.61	.013	-0.05	.005	10	G4	516,1064,1620,1696,3077
193621 +36 04008		HR 7782		20 17 56	+36 58.4	3	6.58	.027	0.00	.013	-0.08		5	A0 III	401,1070,1118
193555 +15 04137				20 17 57	+15 23.0	1	6.78		0.53		0.12		2	F8 V	3016
228943 +38 04016		Be 86 - 13	⋆	20 17 57	+38 26.2	2	9.30	.005	0.89	.014	-0.13	.024	4	B0 II	1012,1751
		Be 86 - 35		20 17 58	+38 29.1	1	12.97		0.82		0.33		2		1751
193452 −15 05626		HR 7775	⋆ AB	20 17 58	−14 56.6	2	6.10		-0.02	.000	-0.12		5	A0 III	1079,2007
193634 +37 03892		Dol 42 - 35		20 17 59	+38 11.0	3	7.40	.005	0.31	.004	-0.35	.021	67	B3 III	588,1070,1745
193633 +40 04121		IDS20162N4050	A	20 17 59	+40 58.8	2	7.14	.028	0.03	.005	0.03		4	B9	401,1118
		Be 86 - 34		20 17 60	+38 30.1	1	12.20		0.95		0.01		2		1751
193556 +14 04263		HR 7778		20 18 01	+14 24.6	3	6.16	.012	0.93	.012	0.66	.015	13	G8 III	252,8059,8100
		Dol 42 - 36		20 18 03	+37 59.8	1	12.26		0.69		0.10		4	B3	1745
193680 +47 03077		U Cyg	⋆ A	20 18 03	+47 44.2	4	8.49	.092	3.25	.057	4.22	.600	4	Ce	817,8005,8022,8027
193636 +35 04069		IDS20162N3557	AB	20 18 04	+36 06.6	3	7.97	.026	0.28	.016	0.20		5	A7 III	1090,1118,1196
		Be 86 - 36		20 18 04	+38 27.1	1	13.60		0.75		0.37		1		1751
193307 −50 12929		HR 7766		20 18 04	−50 09.3	5	6.27	.009	0.05	.003	-0.02	.000	23	G0 V	15,1311,2012,2012,3077
193579 +17 04294		HR 7780		20 18 05	+17 38.1	1	5.80		1.49				2	gK5	71
193635 +36 04009		IDSS20162N370	AB	20 18 05	+37 15.1	2	8.41	.000	0.16	.005			7	A3 V	1118,1196
		Dol 42 - 37		20 18 08	+37 56.6	1	12.85		0.44		0.43		5	A3 V	1745
193700 +47 03078		IDS20165N4734	B	20 18 08	+47 44.8	1	7.87		0.80		0.51		1	G0	414
		VES 200		20 18 10	+38 10.3	1	11.92		1.56		1.23		1		426
		Be 86 - 37		20 18 10	+38 25.4	1	12.92		0.53		0.02		1		1751
		Be 86 - 32		20 18 10	+38 31.9	1	13.65		0.87		0.36		2		1751
		Be 86 - 33		20 18 11	+38 30.7	1	12.29		2.37		2.95		1		1751
193701 +44 03429		IDS20166N4503	A	20 18 11	+45 12.3	1	6.67		0.45		0.41		2	F5 V	247
193495 −15 05629		HR 7776	⋆ APB	20 18 12	−14 56.4	7	3.08	.007	0.79	.009	0.27	.004	31	F8 V + A0	15,1075,1088,2012*
		Be 86 - 39		20 18 13	+38 30.4	1	13.08		0.80		0.12		2		1751
		Be 86 - 22		20 18 15	+38 28.0	1	13.94		0.96		0.40		2		1751
		Be 86 - 2		20 18 16	+38 30.7	2	10.67	.005	0.67	.005	-0.26	.014	7		903,1751
		Be 86 - 10		20 18 17	+38 29.5	2	12.65	.000	0.79	.015	0.03	.005	10		903,1751
228963 +35 04071				20 18 18	+35 56.7	1	9.91		0.41		-0.48		2	K2	1012
193682 +37 03894		LS II +37 067		20 18 18	+37 40.3	3	8.41	.007	0.51	.004	-0.48	.000	6	O4: III(f)	1011,1070,1209
		L 210 - 160		20 18 18	−58 31.	1	15.55		0.29		-0.56		1		3065
		Be 86 - 11		20 18 19	+38 31.8	2	11.84	.005	0.70	.024	-0.08	.029	6		903,1751
		Be 86 - 40		20 18 19	+38 33.8	1	13.54		0.60		0.08		1		1751
		Be 86 - 24		20 18 20	+38 29.8	1	13.61		0.70		0.08		1		1751
228969 +38 04018		Be 86 - 1		20 18 20	+38 30.0	3	9.49	.010	0.72	.010	-0.24	.016	13	B7	903,1012,1751
		Be 86 - 9		20 18 21	+38 30.0	2	11.89	.015	1.11	.029	0.22	.024	9		903,1751
		Be 86 - 23		20 18 21	+38 30.0	1	12.48		0.68		-0.04		1		1751
		Be 86 - 31		20 18 21	+38 32.3	1	12.85		0.55		0.07		1		1751
193681 +38 04019		Be 86 - 16		20 18 21	+38 40.5	1	8.26		-0.01		-0.19		1	B9 V	1751
193722 +46 02910		HR 7786, V1584 Cyg		20 18 21	+46 40.7	2	6.47		-0.08	.007	-0.37	.003	8	B9p Si	627,1079
193702 +38 04021		HR 7784	⋆ AB	20 18 26	+39 14.7	9	6.22	.012	0.06	.009	0.03	.010	26	A1 V	15,351,1007,1013,1063*
193462 −34 14309				20 18 26	−34 38.3	1	8.11		0.42				4	F3 V	2012
		Be 86 - 25		20 18 27	+38 29.7	1	14.72		0.84		0.23		1		1751
		Be 86 - 3		20 18 28	+38 29.9	2	10.14	.005	0.66	.010	-0.15	.000	10		903,1751
		Be 86 - 29		20 18 28	+38 31.3	1	14.37		0.95		0.25		1		1751
		Be 86 - 30		20 18 28	+38 31.9	1	12.84		0.75		0.02		1		1751
		Be 86 - 7		20 18 29	+38 30.4	2	10.48	.005	0.63	.005	0.11	.010	8		903,1751
		Be 86 - 8		20 18 30	+38 27.3	2	10.31	.010	1.07	.000	0.97	.005	6		903,1751
		Be 86 - 28		20 18 30	+38 28.1	1	12.69		1.48		2.41		1		1751
		CoD -58 07734		20 18 30	−58 25.4	4	10.59	.019	1.45	.020	1.16	.000	8		158,1494,1705,3073
193683 +31 04042				20 18 31	+31 51.9	1	7.51		-0.09		-0.69		2	B5	401
228989 +38 04025		Be 86 - 4		20 18 31	+38 32.4	4	9.56	.205	0.76	.013	-0.21	.015	12	B7	903,1011,1012,1751
193703 +33 03864				20 18 33	+33 39.5	1	8.42		0.35		-0.04		34	G5	588
		Be 86 - 26		20 18 33	+38 28.7	1	12.62		1.13		0.13		1		1751
		Be 86 - 27		20 18 35	+38 28.1	1	13.90		1.02		0.35		1		1751
		Be 86 - 6		20 18 38	+38 32.4	2	11.83	.010	0.74	.020	-0.05	.020	8		903,1751
		Be 86 - 38		20 18 38	+38 33.8	1	13.29		0.66		0.08		2		1751
229005 +40 04123				20 18 38	+41 13.5	1	10.34		0.08		0.11		10	A0	1655
		LP 695 - 28		20 18 38	−06 35.3	3	11.82	.022	1.28	.017	1.10		4		680,1705,3073
193565 −21 05694				20 18 38	−21 34.6	2	7.78	.015	0.41	.000	-0.08		3	F0 V	742,1594
		Be 86 - 5		20 18 41	+38 32.8	2	11.04	.000	0.70	.000	0.02	.000	7		903,1751
		CS 22950 # 46		20 18 42	−13 26.2	1	14.22		0.91		0.34		1		1736
	+40 04124	V1685 Cyg		20 18 43	+41 12.3	2	10.63	.007	0.76	.019	-0.31	.057	3	B2	351,1283
	−15 05634			20 18 43	−15 07.5	1	10.22		0.78		0.39		2		1064
193736 +34 03978				20 18 44	+35 05.6	1	7.95		-0.07		-0.36		2	B9	401
		CS 22950 # 63		20 18 45	−15 11.6	1	14.22		0.72		0.19		1		1736

Catalogue of mean UBV data

HD	DM	Other Id	N Rem	α1950	δ1950	S	V	σV	B–V	σB–V	U–B	σU–B	n	Spectrum	References
193793	+43 03571	V1687 Cyg	★ AB	20 18 47	+43 41.7	2	6.87	.015	0.40	.025	-0.32	.034	8	O5	1095,1359
193964	+61 02000	HR 7792, DE Dra		20 18 47	+62 05.9	1			-0.05		-0.22		1	B9 V	1079
193407	-54 09673	IDS20150S5435	A	20 18 48	-54 25.4	1	7.95		0.99		0.70		5	G8 III	1673
193706	+21 04167			20 18 49	+21 21.5	1	7.91		1.63		1.95		2	F9 V wls	1648
		Be 86 - 15		20 18 55	+38 33.7	1	11.03		0.51		-0.33		2		1751
		G 230 - 35		20 18 56	+52 04.1	1	9.96		1.08		1.21		2	K4	7010
193814	+37 03897			20 18 58	+38 05.8	3	7.58	.005	-0.07	.008	-0.31		14	B8 V	401,1118,1196
229020	+36 04019	Be 87 - 1		20 18 59	+37 17.3	1	9.05		0.32		0.24		3	F0	964
193838	+42 03702			20 18 59	+42 24.7	1	8.69		-0.03		-0.16		2	A0	401
229033	+37 03898	LS II +37 071		20 19 04	+37 34.4	2	8.90	.161	0.56	.200	-0.23		3	B0 II-III	1012,1070
193571	-42 14836	HR 7779	★ A	20 19 04	-42 12.5	10	5.58	.004	0.00	.003	-0.01	.010	57	A0 V	15,208,278,977,977,1075*
		VES 131		20 19 06	+32 20.3	1	15.30		0.26		1.61		2		363
229042	+36 04021	Be 87 - 2		20 19 11	+37 11.6	1	9.78		0.30		0.06		3	A7	964
193944	+53 02384	HR 7791	★ AB	20 19 11	+53 26.2	1	6.18		1.56				2	K5	71
229049	+38 04031	LS II +38 054		20 19 12	+38 51.3	1	9.62		0.48		-0.44		2	B2 III:p	1012
229043	+36 04022	LS II +36 063		20 19 13	+36 38.9	2	9.94	.000	0.83	.000	-0.20	.000	4	O9.5II	1011,1012
		G 186 - 23		20 19 14	+23 05.8	1	10.78		1.05		0.96		6	K3	1723
332184	+32 03797			20 19 14	+32 36.8	1	9.37		1.77				2	K5	369
	+35 04077	V1749 Cyg	★ A	20 19 16	+35 27.2	3	9.77	.050	2.98	.036	2.07		7	M3 Iab:	148,369,8032
		VES 202		20 19 18	+37 25.0	1	12.85		1.56		0.69		1		426
193855	+38 04032	LS II +38 055		20 19 18	+38 52.3	3	7.75	.019	0.26	.009	-0.45	.016	5	B2 III	833,1070,1423
	-3 04864			20 19 18	-03 22.5	2	10.20	.000	0.83	.000	0.43	.000	4	G5	516,1064
193856	+30 03995			20 19 23	+30 42.3	1	7.64		1.71				2	K5	369
229059	+36 04024	Be 87 - 3		20 19 23	+37 14.9	2	8.71	.004	1.52	.004	0.40	.011	11	B7	964,1012
355547	+13 04370			20 19 25	+13 40.0	1	9.54		0.56		-0.10		1	G0 V	979
		WLS 2020 25 # 5		20 19 27	+27 18.4	1	12.80		0.61		0.00		2		1375
		Be 87 - 5		20 19 27	+37 12.9	1	14.65		1.35		0.34		4		964
		Be 87 - 4		20 19 27	+37 13.8	1	10.92		1.26		0.22		9		964
		BY Del		20 19 28	+12 14.9	1	11.15		0.51		0.11		1	F9	1768
193857	+30 03998			20 19 28	+30 25.8	1	6.73		0.25		0.19		3	Am	8071
		Be 87 - 6		20 19 28	+37 11.1	1	13.40		0.83		0.44		2		964
	+45 03149	LS III +45 004		20 19 29	+45 53.6	1	9.45		0.32		-0.49		2	B1 V	1012
		Be 87 - 7		20 19 31	+37 10.5	1	13.02		1.11		0.28		4		964
	+36 04026	Be 87 - 10		20 19 32	+37 04.3	1	10.63		0.59		0.10		3		964
		Be 87 - 8		20 19 32	+37 10.4	1	11.91		2.24		1.60		2		964
193926	+43 03576	IDS20178N4316	A	20 19 32	+43 25.8	1	7.86		-0.04		-0.12		2	A0	401
		Be 87 - 9	V	20 19 33	+37 13.2	2	12.07	.036	1.34	.028	0.36	.044	9		426,964
193819	+13 04371	IDS20172N1316	AB	20 19 35	+13 25.9	1	7.52		-0.05		-0.38		3	A0	3016
		Be 87 - 11		20 19 36	+37 14.9	1	10.97		0.54		0.06		7		964
		CS 22955 # 28		20 19 37	-24 57.2	1	14.05		-0.16		-0.89		1		1736
	+36 04025	BI Cyg	★ A	20 19 38	+36 45.6	2	9.25	.097	3.04	.034	2.49		6	M4	148,8032
		Be 87 - 12		20 19 38	+37 19.4	1	12.89		0.61		0.10		2		964
		Be 87 - 13		20 19 39	+37 11.1	1	11.32		1.09		0.12		8		964
193928	+36 04028	LS II +36 065	★ B	20 19 40	+36 45.4	2	9.83	.040	0.79	.025	0.08	.005	5	WN6	1012,1359
193945	+40 04132	LS III +41 006		20 19 40	+41 02.0	1	8.53		0.59		-0.41		3	B0 Vnn	399
		Be 87 - 17		20 19 41	+37 09.9	1	12.89		0.54		0.12		3		964
		Be 87 - 16		20 19 41	+37 14.7	1	13.39		1.33		0.55		4		964
		Be 87 - 19		20 19 42	+37 08.8	1	12.33		0.52		0.00		4		964
		Be 87 - 15	★ V	20 19 42	+37 15.2	1	11.84		1.54		0.37		7		964
		Be 87 - 14		20 19 42	+37 15.8	1	14.12		1.33		0.82		1		964
		Be 87 - 18		20 19 43	+37 19.6	1	12.84		1.63		0.66		4		964
		Be 87 - 21		20 19 44	+37 12.0	1	11.13		0.13		0.03		7		964
	+36 04031	Be 87 - 20		20 19 44	+37 13.9	1	9.61		1.12		1.06		9	K0	964
193946	+39 04151	LS II +39 034	★ A	20 19 44	+39 40.1	1	8.92		0.92				1	B2 Ib	138
		Be 87 - 22		20 19 45	+37 14.9	1	13.96		1.46		0.55		3		964
	+36 04032	Be 87 - 25	★ A	20 19 45	+37 15.9	2	10.46	.000	1.28	.013	0.19		7		725,964
		CS 22955 # 27		20 19 45	-25 03.3	1	13.73		0.30		0.14		1		1736
		Be 87 - 24		20 19 46	+37 15.6	1	11.48		1.34		0.37		5		964
		Be 87 - 23		20 19 47	+37 16.9	1	12.94		0.58		0.10		5		964
	+37 03903	Be 87 - 78	★ V	20 19 47	+37 22.4	2	9.56	.466	3.28	.025	3.14		5	M4 Ia	148,8032
		Be 87 - 27		20 19 48	+37 13.0	1	13.51		1.35		0.53		3		964
		Be 87 - 26		20 19 48	+37 15.5	1	11.83		1.42		0.38		5		964
193530	-61 06482			20 19 48	-61 28.2	1	9.11		1.38		1.44		1	K2p Ba	565
		Be 87 - 28		20 19 49	+37 09.2	1	11.87		1.52		1.33		4		964
		Be 87 - 29		20 19 52	+37 12.9	1	12.96		1.43		-0.29		12		964
		Be 87 - 30		20 19 53	+37 09.1	1	12.58		0.66		0.13		4		964
194258	+68 01121	HR 7804, AC Dra		20 19 53	+68 43.2	1	5.55		1.59				2	M5 II-III	401
193911	+23 03986	HR 7789	★	20 19 54	+24 17.1	3	5.52		-0.10	.030	-0.40	.005	6	B8 IIIne	1063,1079,8100
		Be 87 - 33		20 19 54	+37 06.0	1	12.54		2.14		1.70		4		964
		Be 87 - 31		20 19 54	+37 12.8	1	12.32		1.40		0.38		8		964
		Be 87 - 32		20 19 55	+37 16.9	1	11.57		1.34		0.31		5		964
		CS 22955 # 18		20 19 57	-26 10.3	1	14.84		0.15		0.17		1		1736
		LS III +41 019		20 19 58	+41 44.7	1	10.88		1.50		0.39		4		483
		Be 87 - 34		20 19 59	+37 16.5	1	13.32		1.53		0.63		4		964
194009	+38 04036	LS II +38 061		20 20 01	+38 28.1	2	8.68	.019	0.61	.000	-0.14		3	B3 II	1012,1070
194057	+44 03439	LS III +44 010		20 20 01	+44 39.2	2	7.51	.000	0.86	.000	-0.16		5	B1 Ib	1012,8031
		Be 87 - 35		20 20 02	+37 13.9	1	13.85		1.43		0.68		3		964
194175	+60 02117			20 20 02	+60 53.0	1	8.46		0.06		-0.20		2	A2	401
		Be 87 - 36		20 20 03	+37 08.5	1	13.38		0.67		0.28		3		964
229105	+36 04034	Be 87 - 37		20 20 04	+37 11.9	1	9.91		1.28		1.31		6	K2	964

Table 1 1105

HD	DM	Other Id	N	Rem	α_{1950}	δ_{1950}	S	V	σ_V	B–V	σ_{B-V}	U–B	σ_{U-B}	n	Spectrum	References
229108	+38 04038	LS II +39 036			20 20 07	+39 17.9	1	9.48		0.81		-0.19		2	B0.5Ib	1012
		Be 87 - 38		V	20 20 08	+37 16.8	3	12.44	.052	1.59	.015	0.53	.014	6		426,964,8103
229115	+34 03992				20 20 16	+35 04.1	1	9.92		0.67		-0.06		2	B2 IV	1012
194032	+28 03719				20 20 17	+28 55.8	1	8.25		0.39		0.02		3	F3 V	1733
194069	+40 04136	HR 7795			20 20 17	+40 58.3	2	6.39	.010	1.07	.000	0.82	.005	4	G5 III + A	247,1080
193896	−10 05369	HR 7788			20 20 18	−09 49.0	1	6.29		0.91		0.57		8	G5 II	1088
194092	+40 04137	LS II +40 022			20 20 19	+40 49.5	2	8.25	.005	0.13	.000	-0.74	.009	4	B0.5III	401,1012
	+34 03993				20 20 20	+34 30.7	1	9.99		1.69				2	M2	369
		G 210 - 16			20 20 21	+33 16.5	1	12.32		0.66		-0.01		1		1658
194095	+37 03908				20 20 23	+37 56.2	2	7.56	.000	1.14	.000			8	K1 III	1118,1196
235199	+54 02336				20 20 23	+55 04.0	1	9.00		1.29		1.32		3	K5	1733
		LP 815 - 7			20 20 25	−16 28.5	1	11.36		0.58		-0.14		2		1696
194093	+39 04159	HR 7796		⋆ A	20 20 26	+40 05.7	11	2.21	.026	0.67	.012	0.53	.009	33	F8 Ib	1,15,1080,1118,1194*
		NGC 6913 - 217			20 20 27	+38 32.8	1	13.01		0.84		-0.08		2		999
194152	+45 03152	HR 7798			20 20 27	+45 38.0	3	5.58	.017	1.08	.004	1.02	.005	5	K0 III	37,71,247
		NGC 6913 - 219			20 20 29	+38 17.8	1	13.39		0.71		1.01		2		999
194298	+63 01618	HR 7805		⋆ AB	20 20 29	+63 49.2	1	5.69		1.56				2	K5 III	71
		NGC 6913 - 221			20 20 30	+38 14.6	1	12.86		0.64		0.24		2		999
193807	−42 14847	HR 7787		⋆ AB	20 20 30	−42 35.1	9	5.64	.004	0.20	.003	0.10	.005	48	A5 V	15,208,278,977,977,1075*
194094	+38 04043	NGC 6913 - 225		⋆	20 20 31	+38 33.8	9	9.02	.005	0.60	.010	-0.37	.025	6	B0 V	999,1011,1012
194012	+14 04275	HR 7793			20 20 32	+14 23.4	5	6.16	.017	0.50	.008	-0.04	.024	23	F8 V	196,254,1067,8059,8100
		NGC 6913 - 227			20 20 32	+38 23.2	1	13.38		0.57		0.21		2		999
		G 143 - 48			20 20 34	+10 31.3	2	13.97	.015	1.53	.049	1.30		3		203,1759
		NGC 6913 - 228			20 20 34	+38 29.1	1	13.14		0.76		0.23		2		999
229134	+38 04044	LS II +38 066			20 20 34	+38 44.7	1	10.03		0.49		-0.42		2	B1 V	1012
194096	+34 03995				20 20 36	+34 49.8	2	6.83	.005	0.04	.000	0.04		5	A1 V	401,1118
		NGC 6913 - 229			20 20 36	+38 26.3	1	13.53		0.63		0.34		2		999
194097	+30 04005	HR 7797			20 20 37	+31 06.2	1	6.09		1.35				2	K2	71
229138	+33 03878				20 20 37	+33 22.5	1	8.90		0.96		-0.09		1	K7	8100
		NGC 6913 - 236			20 20 37	+38 37.2	1	10.33		0.60		-0.17		2		999
		NGC 6913 - 234			20 20 38	+38 07.6	1	13.07		1.39		0.55		2		999
229141	+36 04036	Be 87 - 39			20 20 39	+37 12.4	1	10.48		0.51		0.09		2	F5	964
193901	−21 05703				20 20 39	−21 31.1	9	8.65	.014	0.55	.013	-0.13	.011	16	F7 V	22,158,742,908,1064*
		NGC 6913 - 238			20 20 41	+38 15.4	1	12.97		1.85		0.33		2		999
194013	+4 04434	HR 7794			20 20 42	+05 10.9	3	5.31	.005	0.97	.010	0.77	.005	12	G8 III-IV	15,1071,1355
		NGC 6913 - 241			20 20 42	+38 32.5	1	12.20		0.41		0.47		2		999
		CS 22955 # 39			20 20 42	−23 19.4	1	14.19		0.38		0.09		1		1736
		VES 206			20 20 43	+32 02.2	2	11.71	.000	1.91	.000	1.81	.000	3		426,8103
194153	+37 03909	LS II +37 080			20 20 44	+37 58.1	1	8.53		1.06				1	B1 Iab	1070
		NGC 6913 - 242			20 20 44	+38 15.0	1	13.28		0.69		0.31		2		999
194204	+50 03067				20 20 44	+50 22.8	1	7.02		0.14		0.12		2	A0	1733
		NGC 6913 - 245			20 20 45	+38 15.7	1	13.41		0.74		0.35		2		999
		NGC 6913 - 246			20 20 45	+38 29.5	1	10.62		1.57		1.74		2		999
		NGC 6913 - 249			20 20 48	+38 07.4	1	13.60		0.80		0.47		2		999
		NGC 6913 - 251			20 20 48	+38 37.8	1	13.23		1.01				2		999
		NGC 6910 - 19			20 20 48	+40 37.5	1	11.12		0.36		0.07		1	F0	749
194074	+10 04264				20 20 52	+10 55.4	1	7.55		0.17		0.08		2	A2	1733
		NGC 6913 - 255			20 20 54	+38 30.0	1	10.91		0.42		0.25		2		999
229153	+37 03910				20 20 56	+37 48.0	1	9.07		1.17		0.15		2	B0 I	1012
		NGC 6913 - 256			20 20 56	+38 14.2	1	13.28		0.83		0.69		2		999
		LP 575 - 39			20 20 58	+05 41.0	1	10.74		0.93		0.64		1		1696
		NGC 6913 - 258			20 20 58	+38 13.0	1	13.59		0.56		0.03		2		999
		NGC 6913 - 259			20 20 58	+38 35.8	1	13.40		0.63		0.21		2		999
194194	+40 04139	NGC 6910 - 1			20 20 58	+40 33.2	1	8.07	.023	0.05	.015	-0.72	.014	4	B1.5IV	49,276,401
194193	+40 04141	HR 7800			20 20 59	+40 51.9	2	5.93	.004	1.61	.014	1.98		3	K7 III	71,3001
		NGC 6913 - 261			20 21 01	+38 20.0	1	13.25		0.78		0.32		2		999
		NGC 6913 - 127			20 21 02	+38 32.2	1	10.16		0.55		-0.14		1	B8	999
		NGC 6913 - 264			20 21 03	+38 30.2	1	13.14		0.51		0.04		2		999
229159	+38 04050	LS II +39 039		⋆ B	20 21 05	+39 02.8	1	8.54		1.00		0.00		2	B1.5Ib	1012
	+45 03158				20 21 05	+46 03.8	1	9.44		1.80				1	M0	369
		NGC 6913 - 116			20 21 06	+38 19.4	1	11.96		0.52		0.11		2	F2	999
		NGC 6913 - 266			20 21 08	+38 34.7	1	13.03		0.83		0.04		2		999
194206	+38 04051	IDS20193N3853		A	20 21 08	+39 03.0	2	6.74	.009	-0.10	.005	-0.49		4	B8 V	401,1118
194240	+42 03718	IDS20194N4224		AB	20 21 08	+42 33.5	1	8.50		-0.03		-0.30		2		401
		VES 207			20 21 09	+35 52.8	1	11.94		0.92		0.71		1		426
		WLS 2020 25 # 9			20 21 11	+24 45.9	1	13.34		0.67		0.32		2		1375
229171	+37 03913	NGC 6913 - 111		⋆	20 21 12	+38 17.7	2	9.33	.000	0.54	.005	-0.31	.024	4	B0.5III:n	999,1012
194205	+38 04053	LS II +39 040			20 21 12	+39 11.0	1	9.04		0.53		-0.32		2	B2 III	1012
		NGC 6910 - 18			20 21 12	+40 36.1	2	10.79	.050	0.75	.010	-0.25	.035	2	B1 V	49,749
		NGC 6910 - 28			20 21 12	+40 36.6	2	12.22	.055	0.75	.020	-0.06	.000	2	F2	49,749
194220	+42 03721	HR 7802		⋆ AB	20 21 12	+42 49.3	4	6.20	.008	0.96	.009	0.74	.012	7	K0 III	71,247,1118,3016
		NGC 6913 - 270			20 21 13	+38 31.7	1	12.77		0.74		0.06		2		999
194163	+23 03999				20 21 14	+23 30.4	1	7.96		-0.01		-0.37		2	B9	1733
		NGC 6913 - 113			20 21 14	+38 17.1	1	12.73		0.91		0.59		2		999
		NGC 6913 - 112			20 21 14	+38 17.3	1	14.95		0.61		0.02		2		999
		NGC 6913 - 118			20 21 14	+38 22.9	1	12.09		0.51		-0.04		2	A3	999
229176		NGC 6910 - 9			20 21 14	+40 43.9	2	10.51	.045	0.22	.015	0.12	.060	2	A2	49,749
334302	+28 03729				20 21 15	+28 38.4	1	8.04		1.20				4	K1 III	20
		NGC 6913 - 271			20 21 15	+38 16.1	1	13.67		0.80		0.25		2		999
		LS II +39 041		⋆ V	20 21 15	+39 20.1	1	10.56		1.25		-0.19		2	B pe	1012

HD	DM	Other Id	N Rem	α_{1950}	δ_{1950}	S	V	σ_V	B–V	σ_{B-V}	U–B	σ_{U-B}	n	Spectrum	References
		NGC 6910 - 174		20 21 15	+40 36.2	1	14.54		0.85		0.41		1		749
		NGC 6910 - 63		20 21 15	+40 36.4	1	14.60		0.79		0.29		1		49
		NGC 6910 - 36		20 21 16	+40 36.9	1	13.02		0.77		-0.03		1		49
	+43 03590			20 21 16	+44 04.7	1	9.97		2.19				2		369
		CS 22955 # 15		20 21 16	-26 26.1	1	14.05		0.25		0.14		1		1736
194300	+48 03107			20 21 17	+48 21.1	1	8.40		1.84				3	M1	369
		NGC 6910 - 172		20 21 18	+40 34.6	1	13.83		0.98		0.31		1		749
		NGC 6910 - 70		20 21 18	+40 34.8	1	13.92		0.94		0.28		1		49
		NGC 6910 - 20		20 21 18	+40 36.2	1	10.95		1.20		1.13		1		49
		NGC 6910 - 175		20 21 19	+40 35.6	1	14.64		1.00		0.78		1		749
		NGC 6910 - 62		20 21 19	+40 35.8	1	14.70		0.93		0.53		1		49
194241	+40 04144	NGC 6910 - 3		20 21 20	+40 38.3	2	7.34	.025	1.17	.000	1.06	.030	2	K1 III-IV	49,749
229189	+40 04145	NGC 6910 - 6	⋆ V	20 21 20	+40 41.4	3	9.99	.032	0.23	.041	0.07	.032	3	A3 V	49,276,749
		NGC 6913 - 126		20 21 21	+38 31.0	1	10.78		0.38		0.09		2	F2	999
229179	+38 04054	V498 Cyg		20 21 21	+39 00.0	2	9.90	.056	1.02	.002	0.04	.027	4	B1:III:	1012,1768
	+40 04146	NGC 6910 - 14	AB	20 21 21	+40 36.6	3	10.38	.013	0.83	.017	-0.18	.009	17	B0.5V	49,276,1655
		NGC 6913 - 125		20 21 22	+38 30.0	1	11.21		1.87		1.56		2		999
		NGC 6910 - 34		20 21 24	+40 36.8	2	12.67	.050	0.87	.000	-0.05	.045	2		49,749
		NGC 6910 - 178		20 21 24	+40 37.5	1	14.81		0.99		0.44		1		749
		NGC 6910 - 56		20 21 24	+40 37.7	1	14.88		0.95		0.49		1	F0	49
229196	+40 04147	NGC 6910 - 4		20 21 24	+40 42.8	6	8.52	.006	0.91	.003	-0.19	.005	331	O5	49,749,1011,1012,1655,8007
		NGC 6910 - 57		20 21 25	+40 38.2	1	14.30		0.82		0.31		1		49
		VES 254		20 21 25	+45 28.7	1	12.03		1.41		1.47		1		8056
		CS 22885 # 106		20 21 25	-38 46.2	2	14.96	.005	0.45	.009	-0.13	.007	3		1580,1736
		NGC 6910 - 55		20 21 26	+40 37.9	1	14.76		1.04		0.49		1		49
		G 210 - 18		20 21 27	+33 19.7	1	11.54		0.95		0.81		2	K3	7010
		CS 22955 # 23		20 21 27	-25 25.5	1	14.50		0.21		0.22		1		1736
	+40 04149	NGC 6910 - 13		20 21 28	+40 35.8	2	10.29	.011	0.91	.002	-0.06	.004	92	O9.5V	49,1655
		CS 22955 # 40		20 21 28	-22 43.5	1	14.23		0.34		0.06		1		1736
		NGC 6910 - 21		20 21 29	+40 38.5	2	11.74	.055	0.75	.010	-0.14	.010	2	B1 V	49,749
		CS 22955 # 12		20 21 29	-26 40.0	1	13.96		0.21		0.17		1		1736
		NGC 6910 - 46		20 21 30	+40 36.5	1	13.39		0.99		0.23		1		749
		NGC 6910 - 7		20 21 31	+40 35.4	2	9.99	.003	0.44	.006	0.00	.008	21	F0 III	49,1655
194279	+40 04150	NGC 6910 - 2		20 21 31	+40 35.8	4	7.03	.015	1.02	.005	-0.03		8	B2 Ia	1012,1025,1234,8031
		NGC 6910 - 24		20 21 31	+40 39.1	1	11.77		0.82		-0.07		1	B1 V	49
		NGC 6913 - 46		20 21 32	+38 23.4	1	11.40		0.55		0.21		2	F3	999
		NGC 6913 - 47		20 21 32	+38 25.1	1	12.98		0.61		0.56		2		999
		NGC 6913 - 44		20 21 33	+38 21.7	1	12.81		1.71				2		999
		NGC 6910 - 31		20 21 33	+40 37.3	1	12.34		1.55				1		49
		NGC 6910 - 38		20 21 33	+40 38.9	1	12.99		0.81		-0.03		1		49
		VES 208		20 21 34	+35 44.8	2	11.96	.000	0.89	.000	0.71	.000	3		426,8103
		NGC 6913 - 107		20 21 34	+38 15.2	1	13.11		1.77		1.51		2		999
		NGC 6913 - 39		20 21 34	+38 16.6	1	13.57		0.76		0.23		2		999
229201		NGC 6910 - 10		20 21 34	+40 34.7	1	10.64		0.29		0.12		1	A0	49
229202	+39 04162	LS III +39 049		20 21 35	+39 59.7	2	9.53	.000	0.87	.000	-0.12	.000	4	B7	1011,1012
		NGC 6913 - 286		20 21 36	+38 07.3	1	13.70		0.77		1.15		2		999
194280	+38 04057	LS II +38 073		20 21 36	+38 46.6	1	8.39		0.76		-0.21		2	B0 Ib	1012
		NGC 6910 - 41		20 21 36	+40 38.6	2	12.80	.075	1.01	.010	0.04	.035	2		49,749
194355	+48 03108			20 21 36	+48 39.0	1	7.93		1.88				3	K5	369
		L 46 - 96		20 21 36	-76 50.	1	13.81		1.72		1.13		1		3078
		NGC 6913 - 40		20 21 37	+38 19.3	1	12.64		0.44		0.20		2		999
		NGC 6910 - 67		20 21 38	+40 38.6	1	14.73		1.06		0.37		1		49
		LS II +39 045		20 21 39	+39 11.2	1	10.62		1.19		0.21		2	B0 V	1012
194282	+31 04061			20 21 41	+31 30.9	1	8.01		-0.07		-0.47		2	B9	401
	+36 04048	LS II +37 086		20 21 41	+37 08.7	1	9.66		1.27		0.21		2	B0 Ib	1012
193924	-57 09674	HR 7790	⋆ A	20 21 42	-56 53.8	5	1.93	.007	-0.20	.006	-0.71	.006	15	B2.5V	15,1034,1075,2012,8015
194303	+36 04049	LS II +36 073		20 21 43	+36 46.0	2	8.63	.019	0.72	.005	-0.10		3	B3 II	1012,1070
		NGC 6913 - 37		20 21 43	+38 17.8	1	11.98		0.67		0.16		2	A0	999
		NGC 6913 - 143		20 21 43	+38 21.0	1	15.23		0.92		0.58		1		49
194334	+38 04059	LS II +38 074		20 21 45	+38 43.2	1	8.77		0.84		-0.16		2	O7.5V	1011
		NGC 6913 - 43		20 21 46	+38 20.7	2	14.10	.035	0.72	.020	0.24	.050	2		49,749
		NGC 6910 - 27		20 21 46	+40 35.9	1	11.77		1.05		0.05		1		49
193607	-74 01906			20 21 47	-74 26.9	3	7.96	.010	1.18	.010	1.21	.004	13	K2 II/III	278,1704,2012
		NGC 6913 - 288		20 21 48	+38 15.6	1	13.19		0.75		0.04		2		999
		NGC 6910 - 43		20 21 48	+40 36.3	1	13.27		0.69		0.09		1		749
		NGC 6913 - 289		20 21 49	+38 11.8	1	13.75		0.70		0.20		2		999
		NGC 6913 - 49		20 21 50	+38 28.0	1	13.50		1.02		0.46		2		999
		NGC 6913 - 290		20 21 51	+38 28.4	1	12.74		1.02		0.50		2		999
194317	+31 04062	HR 7806		20 21 52	+32 01.7	5	4.43	.010	1.33	.009	1.50	.010	20	K3 III	15,1003,1080,1363,8015
194335	+37 03916	HR 7807		20 21 52	+37 18.8	5	5.90	.013	-0.20	.012	-0.89	.027	13	B2 Ven	154,401,1118,1196,1212
		NGC 6913 - 103		20 21 52	+38 10.6	1	13.32		1.36		0.51		2		999
		NGC 6913 - 9		20 21 52	+38 18.8	3	11.79	.010	0.78	.015	-0.03	.008	5	B7	49,140,749
		VES 255		20 21 52	+44 10.9	1	11.83		0.96		0.99		1		8056
		NGC 6913 - 291		20 21 53	+38 35.2	1	12.18		0.44		0.33		2		999
194169	-20 05925			20 21 53	-19 52.4	1	9.40		0.94		0.72		2	K3 V	3072
		NGC 6913 - 8		20 21 54	+38 18.9	4	12.19	.009	0.73	.012	-0.03	.007	7	B7	49,140,749,999
194357	+36 04051	LS II +36 074		20 21 55	+36 51.8	3	6.80	.016	0.12	.004	-0.17	.007	5	B9 III	247,401,1733
229221	+38 04062	NGC 6913 - 6	⋆ C	20 21 55	+38 20.3	4	9.24	.094	0.91	.019	-0.22	.051	8	B0 Ipe	49,140,999,1012,8100
		NGC 6913 - 22		20 21 55	+38 21.6	2	12.22	.000	0.79	.005	0.11	.015	3		49,999
229214	+36 04050	LS II +36 075		20 21 56	+36 21.6	1	9.54		0.70		-0.26		2	B1 Ib	1012

Table 1 1107

HD	DM	Other Id	N	Rem	α_{1950}	δ_{1950}	S	V	σ_V	B–V	σ_{B-V}	U–B	σ_{U-B}	n	Spectrum	References
194356	+37 03918				20 21 56	+37 49.2	2	8.56	.000	0.05	.005			12	B9.5V	1118,1196
346489	+23 04002	LS II +23 059			20 21 57	+23 37.6	1	10.21		0.26		0.12		1	A1 II	1215
		NGC 6913 - 20			20 21 57	+38 20.5	2	14.23	.000	1.00	.010	0.63	.010	3		49,140
		NGC 6913 - 7		⋆ B	20 21 58	+38 20.3	3	12.08	.010	0.89	.008	0.09	.013	5	B7	49,140,999
		NGC 6913 - 106			20 21 59	+38 11.8	1	13.17		0.58		0.05		2		999
		NGC 6913 - 30			20 22 00	+38 18.6	2	12.87	.034	0.81	.019	0.21	.078	3		49,999
194378	+38 04063	NGC 6913 - 10		⋆ A	20 22 00	+38 19.8	3	8.58	.009	0.42	.010	0.09	.010	6	F0 III	49,140,999
		NGC 6913 - 23			20 22 00	+38 22.5	2	13.85	.035	1.02	.015	0.68	.005	2		49,749
194445	+50 03073				20 22 00	+50 37.9	1	8.58		0.12		0.06		2	A0	1566
		LP 755 - 54			20 22 01	−10 33.8	1	10.65		0.77		0.38		1		1696
		VES 209			20 22 02	+33 37.2	1	12.84		0.54		0.33		1		426
		VES 210			20 22 02	+37 25.3	1	13.60		1.85				1		426
		NGC 6913 - 174			20 22 02	+38 34.9	1	13.53		1.08		0.54		2	A5	999
194423	+46 02927	IDS20204N4612	A		20 22 03	+46 21.9	2	8.43	.005	-0.05	.015	-0.52	.000	4	B8	401,1733
229227	+38 04065	NGC 6913 - 5			20 22 04	+38 18.2	5	9.36	.014	0.80	.008	-0.15	.012	9	B0 II	49,140,749,999,1012,8100
		NGC 6913 - 27			20 22 04	+38 19.4	2	11.38	.015	0.87	.005	0.11	.040	4	B2 IV	140,999
194244	+0 04495	HR 7803		⋆ A	20 22 05	+00 54.3	1	6.14		-0.04		-0.14		8	B9 V	1088
		NGC 6913 - 144			20 22 05	+38 21.1	1	15.38		0.94		0.38		1		49
		NGC 6913 - 50			20 22 05	+38 27.2	1	13.15		1.06		0.32		2		999
		NGC 6913 - 38			20 22 06	+38 22.	1	11.62		0.35		0.23		2		999
		NGC 6913 - 42			20 22 06	+38 22.	1	13.95		0.87		0.34		2		140
		NGC 6913 - 190			20 22 06	+38 22.	1	15.09		0.95		0.59		1		749
		NGC 6913 - 198			20 22 06	+38 22.	1	14.60		0.93		0.62		1		749
		NGC 6913 - 211			20 22 06	+38 22.	1	13.31		0.63		0.10		2		999
		NGC 6913 - 212			20 22 06	+38 22.	1	12.04		0.53		0.07		2		999
		NGC 6913 - 220			20 22 06	+38 22.	1	13.46		0.58				2		999
		NGC 6913 - 247			20 22 06	+38 22.	1	13.53		0.66		0.55		2		999
		NGC 6913 - 267			20 22 06	+38 22.	1	12.61		0.88		0.68		2		999
		NGC 6913 - 287			20 22 06	+38 22.	1	13.84		0.76		0.19		2		999
		NGC 6913 - 12			20 22 07	+38 20.8	4	12.89	.016	0.84	.004	0.19	.006	6		49,140,749,999
		NGC 6913 - 32			20 22 08	+38 16.2	1	11.88		1.09		0.41		2		999
		NGC 6913 - 13			20 22 08	+38 20.0	2	13.49	.023	0.91	.005	0.41	.005	4		49,140,999
229233	+38 04068	NGC 6913 - 25			20 22 08	+38 23.9	3	10.56	.008	0.34	.010	0.05	.017	4	F0	49,140,749
229232	+38 04070	LS II +38 079			20 22 08	+38 56.0	2	9.53	.005	0.82	.000	-0.17	.000	4	O5 e	1011,1012
		CS 22950 # 78			20 22 08	−16 39.8	1	14.62		0.41		-0.18		1		1736
		NGC 6913 - 14			20 22 09	+38 20.4	3	11.83	.024	0.46	.004	0.05	.034	5		49,140,999
	+38 04067	NGC 6913 - 4			20 22 09	+38 22.0	4	10.21	.018	0.76	.026	-0.14	.016	7	O9 V	49,140,749,999
		NGC 6913 - 11			20 22 11	+38 19.5	2	13.13	.009	0.92	.019	0.22	.023	4		49,140
229234	+38 04069	NGC 6913 - 2			20 22 11	+38 21.1	6	8.93	.012	0.78	.016	-0.20	.014	12	O7 V	49,140,999,1011,1012,8100
194305	+8 04426				20 22 12	+08 26.1	1	6.79		0.00		-0.10		2	B9	1733
		NGC 6913 - 28			20 22 12	+38 18.5	1	12.54		0.76		0.08		2		999
		NGC 6913 - 172			20 22 12	+38 29.9	1	12.49		1.31		0.36		2		999
		NGC 6913 - 34			20 22 13	+38 14.4	1	12.96		1.32		0.41		2		999
		NGC 6913 - 26			20 22 13	+38 21.4	3	13.25	.011	0.90	.013	0.34	.023	5		49,140,749
		NGC 6913 - 51			20 22 13	+38 26.3	1	8.91		0.78		-0.20		2		999
229238	+38 04071	NGC 6913 - 1			20 22 14	+38 22.5	4	8.86	.016	0.91	.008	-0.06	.005	8	B0.5Ib	49,140,999,1012
194424	+39 04166				20 22 14	+39 22.7	1	7.87		-0.04		-0.38		2	B8 IV	401
194359	+23 04004	IDS20201N2357	A		20 22 16	+24 06.8	1	6.97		0.70		0.39		1	G1 III	1716
229239	+38 04072	NGC 6913 - 3			20 22 16	+38 19.8	6	8.94	.031	0.87	.022	-0.12	.022	10	B1 Iab	49,140,749,999,1012,8100
		G 24 - 14			20 22 16	−00 54.7	2	11.03	.015	1.17	.005	1.16	.005	2		1620,1658
194380	+28 03738	IDS20202N2903	AB		20 22 18	+29 13.1	1	8.39		-0.06		-0.36		1	A0	963
334227	+30 04015				20 22 19	+30 53.2	1	7.90		2.02				2	M0	369
		NGC 6913 - 56			20 22 19	+38 21.1	2	14.07	.000	0.86	.009	0.33	.009	4		49,140
		NGC 6913 - 62			20 22 20	+38 19.5	3	14.37	.016	0.91	.013	0.60	.014	4		49,140,749
		NGC 6913 - 54			20 22 20	+38 23.1	2	14.08	.019	1.20	.019	0.73	.047	4		49,140
		NGC 6913 - 33			20 22 22	+38 16.0	1	12.55		0.71		0.18		1	G2	749
		NGC 6913 - 53			20 22 23	+38 23.8	2	13.22	.000	0.75	.009	0.19	.009	4		49,140
		LS II +39 048			20 22 23	+39 30.9	2	9.86	.000	1.18	.000	0.11	.000	4	O9 V	1011,1012
	+47 03094				20 22 23	+47 39.2	1	10.31		1.85				2		369
194215	−29 17049	HR 7801			20 22 23	−28 49.6	2	5.84	.005	1.10	.000			7	K3 V	15,2012
194479	+44 03457				20 22 25	+44 32.3	1	7.46		1.08		1.01		1	K1 III-IV	247
194467	+37 03922				20 22 26	+37 24.8	3	8.03	.017	-0.01	.012	-0.10		14	B9 V	401,1118,1196
		NGC 6913 - 61			20 22 26	+38 20.1	1	13.67		0.81		0.50		2		999
		NGC 6913 - 57			20 22 27	+38 21.5	2	11.84	.000	1.07	.010	0.85	.005	3		49,999
		NGC 6913 - 60			20 22 28	+38 20.2	1	13.20		1.67		1.35		2		999
194184	−41 14024	HR 7799			20 22 28	−40 57.5	1	6.09		1.36		1.61		2	K3 III	58
		NGC 6913 - 87			20 22 29	+38 08.6	1	10.79		0.49		0.05		2	F3	999
		NGC 6913 - 142			20 22 29	+38 24.3	1	15.10		0.98		0.52		1	B0	49
		NGC 6913 - 141			20 22 30	+38 18.4	1	14.84		0.79		0.17		1		49
		NGC 6913 - 55			20 22 32	+38 23.6	1	12.35		0.52		0.13		2	F6	999
	+41 03731				20 22 32	+42 08.0	2	9.89	.015	0.10	.006	-0.50	.009	5	B3 n	206,351
	+47 03096			V	20 22 32	+47 34.0	1	10.02		1.82				2		369
	+39 04168	LS II +39 049			20 22 33	+39 36.3	2	9.99	.000	1.07	.000	0.03	.000	4	O7	1011,1012
194188	−47 13399				20 22 33	−46 49.4	5	6.73	.002	0.27	.002	0.08	.004	33	A9 III/IV	278,977,1075,1770,2011
		LS II +21 020			20 22 34	+21 11.9	1	11.11		0.26		0.28		1	B8 II	1215
	+34 04004				20 22 34	+35 11.7	1	9.19		1.66				2	M3	369
229253	+38 04073	NGC 6913 - 64			20 22 34	+38 19.3	2	10.12	.010	0.17	.010	-0.29	.015	3	A2	49,999
	+41 03732				20 22 34	+42 06.1	1	9.39		-0.02		-0.14		10	A0	1655
		NGC 6913 - 65			20 22 35	+38 18.8	1	12.80		0.53		0.07		2		999
		NGC 6913 - 58			20 22 35	+38 21.4	1	13.62		1.12		0.33		1		49

HD	DM	Other Id	N Rem	α_{1950}	δ_{1950}	S	V	σ_V	B–V	σ_{B-V}	U–B	σ_{U-B}	n	Spectrum	References
		NGC 6913 - 108		20 22 35	+38 26.2	1	11.72		0.34		0.26		1		49
		NGC 6913 - 170		20 22 35	+38 26.2	1	11.70		0.37		0.29		1	A2	749
194480	+38 04074			20 22 36	+38 48.1	1	8.00		-0.15		-0.63		2	A2	401
		G 186 - 26		20 22 37	+24 53.5	3	10.82	.006	0.40	.006	-0.24	.010	4		1064,1658,3077
		NGC 6913 - 183		20 22 37	+38 19.7	1	13.84		1.01		1.04		2		999
		NGC 6913 - 185		20 22 40	+38 18.0	1	13.68		0.61		0.09		2		999
194406	+12 04322			20 22 42	+13 13.4	1	7.50		0.05		0.06		2	A2	3016
229261		NGC 6913 - 63		20 22 42	+38 20.9	2	10.59	.035	0.21	.020	-0.28	.025	2	B9	49,749,999
		CS 22885 # 92		20 22 42	-40 54.4	1	14.42		0.30		0.13		1		1736
		L 210 - 70		20 22 42	-56 35.	1	12.23		1.45		1.07		1		3078
		NGC 6913 - 80		20 22 45	+38 09.9	1	12.63		1.08		0.24		2		999
		VES 211		20 22 49	+32 09.8	1	11.38		0.67		0.23		2		426
		NGC 6913 - 79		20 22 49	+38 12.0	1	12.70		0.55		0.03		2	A0	999
	-74 01910			20 22 49	-74 24.1	1	9.87		0.94				4	K0	2012
229274	+41 03735			20 22 50	+41 20.4	3	9.08	.019	0.61	.012	0.11	.007	3	G5	979,3025
		NGC 6913 - 84		20 22 51	+38 07.6	1	11.38		0.63		-0.09		2	F5	999
334228	+30 04021	LS II +30 031		20 22 56	+30 52.8	1	10.04		0.38		0.24		2	F2 II	180
		NGC 6913 - 298		20 22 56	+38 09.0	1	13.32		0.56		0.16		2		999
		LP 695 - 90		20 22 57	-06 22.6	1	17.54		1.78				3		1663
		NGC 6913 - 82		20 22 59	+38 08.8	1	11.53		1.43		0.20		2	F0	999
194558	+39 04172	IDS20212N3950	AB	20 23 00	+39 59.4	1	6.74		1.27		1.34		2	K2 III	247
194525	+30 04023			20 23 01	+30 24.3	1	7.94		0.80				5	G2 Ib-II	20
332237	+32 03815	IDS20211N3244	AB	20 23 02	+32 58.1	1	9.33		1.83				2		369
	+41 03737	IDS20213N4203	AB	20 23 02	+42 13.3	1	9.26		0.13		-0.46		4	B3 Vn	206
	+42 03736			20 23 03	+42 47.3	1	9.31		1.74				2		369
194614	+48 03117	IDS20215N4822	A	20 23 03	+48 31.8	1	7.71		-0.04		-0.42		2	B8	401
194510	+25 04231			20 23 06	+25 32.7	1	8.14		0.54				5	F7 IV	20
		VES 212		20 23 06	+34 51.2	1	11.02		1.66		1.56		1		426
194667	+55 02391			20 23 06	+55 39.2	1	8.39		0.03		-0.39		2	A0	1502
194454	-3 04888	HR 7809		20 23 06	-02 57.8	1	6.10		1.19		1.19		8	K0	1088
		Smethells 55		20 23 06	-44 55.	1	11.58		1.26				1		1494
		NGC 6913 - 73		20 23 09	+38 16.2	1	12.77		0.76		0.03		2		999
		LS II +39 051		20 23 09	+39 17.6	1	10.63		1.62		0.65		2	B0.5: III:	1012
194668	+53 02397	HR 7815		20 23 12	+53 23.3	2	6.49		-0.01	.011	-0.08	.041	2	B9.5III	695,1079
	+37 03927	LS II +37 090		20 23 15	+37 18.0	1	10.15		1.25		0.23		2	O8e	1012
		NGC 6913 - 81		20 23 16	+38 09.6	1	12.37		0.45		0.10		2		999
	-41 14032			20 23 16	-41 07.0	1	9.76		1.09		0.96		2		1730
		VES 213		20 23 17	+31 35.2	2	10.57	.000	1.18	.000	1.14	.000	3		426,8103
		NGC 6913 - 70		20 23 18	+38 19.5	1	12.22		0.49		-0.06		2	A5	999
		NGC 6913 - 168		20 23 18	+38 26.6	1	12.08		0.43		0.12		2	A3	999
194526	+9 04526	HR 7810		20 23 20	+09 53.6	4	6.32	.006	1.56	.005	1.93	.013	12	K5 III	15,1071,1355,3012
352179	+20 04557	BP Vul		20 23 20	+20 52.1	1	9.80		0.25		0.06		7	A7	1768
		NGC 6913 - 305		20 23 22	+38 33.1	1	13.34		1.03				2		999
		NGC 6913 - 161		20 23 22	+38 36.1	1	11.14		0.32		0.04		2	A2	999
		NGC 6913 - 166		20 23 23	+38 28.1	1	12.33		0.66		0.14		2		999
194369	-41 14034			20 23 23	-40 52.7	1	8.45		1.07		0.85		2	K0 III/IV	1730
		NGC 6913 - 165		20 23 26	+38 29.7	1	12.93		0.71		0.06		2		999
194370	-41 14036			20 23 27	-41 09.6	1	7.99		0.17		0.11		2	A2/3 IV	1730
194577	+20 04559	HR 7811		20 23 28	+21 14.7	1	5.66		0.93				2	G6 III	71
340201	+26 03896	BE Vul	A	20 23 28	+27 12.3	1	9.78		0.12		0.03		1	A0	627
340201	+26 03896		C	20 23 28	+27 12.3	1	11.94		0.47		0.27		3		627
	+37 03929			20 23 28	+37 32.5	1	10.15		1.25		0.23		2		1011
194595	+26 03897	IDS20214N2623	A	20 23 31	+26 32.4	1	8.28		0.36		0.08		5	F2 IV	627
		NGC 6913 - 306		20 23 32	+38 37.5	1	11.90		1.63		0.46		2		999
194786	+59 02224			20 23 32	+60 07.4	1	8.56		0.06		-0.24		2	A0	401
		NGC 6913 - 71		20 23 34	+38 19.8	1	12.42		0.64		0.19		2		999
194649	+39 04177	LS II +40 030	★ AB	20 23 34	+40 03.2	2	9.00	.000	0.95	.000	-0.09	.000	4	O6.5	1011,1012
		LP 635 - 30		20 23 34	-01 23.1	1	12.46		0.77		0.25		2		1696
		NGC 6913 - 308		20 23 36	+38 37.5	1	12.84		0.45		0.07		2		999
		CS 22885 # 125		20 23 38	-38 07.5	1	14.15		0.27		0.12		1		1736
194433	-37 13741	HR 7808	★ AB	20 23 39	-37 34.0	7	6.24	.005	0.96	.007	0.77	.003	21	K1 IV	15,116,219,938,1020*
194670	+39 04178			20 23 40	+39 37.8	1	7.56		-0.09		-0.42		2	B8 V	401
194684	+41 03742			20 23 40	+41 44.7	1	7.12		0.23		0.07		1	A3 V	1716
		LS II +40 032		20 23 41	+40 03.1	2	11.35	.015	1.05	.020	-0.03		4		405,725
194708	+42 03740			20 23 41	+42 26.4	2	6.90	.010	0.45	.010	0.11		3	F6 III	247,1118
194685	+39 04180	IDS20219N3946	AB	20 23 42	+39 56.0	1	7.79		0.59		0.09		2	F8 V	247
194598	+9 04529	G 24 - 15		20 23 47	+09 17.6	10	8.34	.010	0.48	.006	-0.17	.021	25	F7 V-VI	22,219,333,979,1003*
194616	+19 04408	HR 7813		20 23 47	+19 42.0	1	6.41		1.02				2	K0 III	71
	+36 04063	LS II +37 092		20 23 48	+37 12.6	2	9.71	.000	1.14	.000	0.08	.000	4	O9.5Ib	1011,1012
		CS 22955 # 60		20 23 48	-24 22.1	1	14.69		0.12		0.10		1		1736
194861	+59 02227			20 23 52	+60 11.4	1	8.22		-0.01		-0.15		2	A0	401
229291	+39 04181			20 23 53	+39 50.7	1	8.34		1.60				2	M7	369
194477	-39 13802	IDS20206S3933	AB	20 23 53	-39 22.8	1	9.76		0.73		0.25		2	G8 V	1696
		Case *M # 66		20 24 00	+38 12.	1	10.57		3.64				2	M3.5Ia	148
		VES 214		20 24 03	+33 07.7	1	12.63		0.47		0.60		1		426
194882	+59 02228	HR 7818	★ A	20 24 03	+59 26.2	1	6.44		0.09		0.11		2	A2 IVs	401
194688	+16 04259	HR 7816		20 24 06	+17 09.0	1	6.22		1.01				2	K0	71
194738	+35 04113			20 24 06	+35 38.8	1	7.63		0.41		0.01		2	F5	1601
194779	+40 04164	LS III +41 021		20 24 09	+41 10.3	1	7.77		0.23		-0.52		2	B3 II	1012
		LP 635 - 14		20 24 13	-00 47.0	1	11.33		0.43		-0.20		2		1696

Table 1   1109

HD	DM	Other Id	N Rem	α_{1950}	δ_{1950}	S	V	σ_V	B–V	σ_{B-V}	U–B	σ_{U-B}	n	Spectrum	References
194883	+54 02348	LS III +54 002	⋆A	20 24 15	+54 31.2	2	7.36	.005	-0.04	.000	-0.61	.039	5	B2 Ve	401,1212
194789	+39 04186			20 24 16	+40 14.2	2	6.61	.033	-0.11	.000	-0.45		4	B8 V	401,1118
		ONRS # 2		20 24 18	+39 47.	1	14.16		4.44				1		8008
194790	+38 04081			20 24 20	+38 41.8	3	7.96	.028	0.23	.019	0.17		8	A3 IV	833,1118,1196
		CS 22885 # 143		20 24 24	-39 33.6	2	13.92	.005	0.77	.019	0.19	.006	3		1580,1736
	+49 03280			20 24 25	+49 18.8	1	9.08		0.67				2	G0	1724
		VES 215		20 24 28	+36 45.7	1	12.09		1.35		0.27		1		426
194636	-18 05685	HR 7814	⋆AB	20 24 28	-18 22.6	4	5.12	.057	-0.07	.023	-0.38	.044	7	B3/5 V	938,1079,2007,3023
	+39 04189	V455 Cyg		20 24 32	+39 30.3	1	9.33		1.11		0.09		2	B2p:e:	1012
194552	-43 14001			20 24 33	-42 47.1	1	8.70		0.51		0.02		4	F5/6 V	285
194780	+25 04237			20 24 35	+25 19.4	2	8.18	.005	0.19	.010	0.07	.005	5	A2	833,1625
194839	+40 04165	LS III +41 022		20 24 35	+41 12.9	3	7.49	.005	0.97	.021	-0.09		5	B0.5Ia	1012,1025,1234
194570	-43 14003			20 24 36	-42 47.0	1	8.18		0.51		-0.02		4	F5/6 V	285
332225	+33 03904			20 24 38	+33 51.3	1	8.90		1.62				2	K5	369
194640	-31 17597			20 24 38	-31 01.6	3	6.61	.000	0.73	.009	0.27	.021	10	G6/8 V	158,1075,3077
	-28 16676			20 24 40	-27 54.1	3	11.42	.030	1.51	.007	1.12	.005	7	M3	158,1705,3073
		AJ89,1229 # 468		20 24 42	+11 48.	1	11.68		1.79				1		1532
194840	+36 04072			20 24 44	+36 25.4	1	8.36		-0.05		-0.32		2	A0	401
		CS 22955 # 61		20 24 44	-24 20.7	1	14.96		0.47		0.08		1		1736
		CS 22955 # 66		20 24 44	-25 07.6	1	12.77		1.42		1.18		1		1736
194887	+37 03934			20 24 50	+38 15.5	1	11.14		3.49		2.77		3	A2	8032
194766	-2 05282	IDS20223S0226		20 24 51	-02 17.1	6	7.49	.007	0.52	.013	-0.03	.011	16	F8	1,116,1028,1075,1355,3026
194765	-2 05283	IDS20223S0226	A	20 24 52	-02 16.1	6	6.71	.035	0.51	.012	-0.03	.008	17	F8 V	1,116,1028,1075,1355,3026
194676	-28 16682	T Mic		20 24 53	-28 25.7	1	6.74		1.59		0.89		1	M7/8 III	3076
	+0 04509			20 24 54	+01 11.4	1	9.55		1.66				1	M2	1532
194885	+39 04192			20 24 54	+39 19.8	2	7.09	.000	-0.07	.014	-0.34		4	A0	401,1118
194863	+29 04038			20 24 56	+30 12.3	1	7.33		0.00		-0.10		2	A0	401
194909	+38 04088			20 24 57	+39 17.8	1	7.76		-0.01		-0.08		1	A0	401
194822	+16 04262			20 24 58	+16 32.7	1	7.56		0.32		0.20		3	F0 V?	8071
194711	-26 15036			20 25 02	-25 46.4	1	6.81		0.38		0.10		5	F8	1628
194696	-29 17089			20 25 02	-29 32.4	1	7.32		1.12		1.02		3	K0 III	1657
		Case *M # 68		20 25 06	+36 23.	1	11.38		2.58				1	M4	148
193721	-81 00901	HR 7785	⋆A	20 25 06	-81 08.0	2	5.76	.005	1.14	.000	0.85	.000	7	G6/8 II	15,1075
195984	+82 00613			20 25 07	+82 41.1	1	8.29		0.08		0.07		6	A0	1219
194951	+33 03910	HR 7823	⋆	20 25 10	+34 09.8	3	6.39	.015	0.48	.014	0.45	.016	9	F1 II	247,253,1118
195014	+50 03095			20 25 10	+50 56.4	1	7.71		0.25		0.08		2	A2	1566
195066	+56 02421	HR 7827	⋆A	20 25 11	+56 28.4	4	6.37	.008	0.01	.007	-0.01	.009	9	A2 V	15,401,1028,1084
		VES 216		20 25 14	+34 55.8	1	12.46		2.11				1		426
239413	+56 02422	IDS20240N5619	B	20 25 14	+56 28.2	2	8.40	.000	0.09	.010	0.07	.015	5	A0	1028,1084
194201	-75 01621			20 25 16	-75 25.6	2	8.66	.005	0.32	.001	0.12		6	A9/F0 IV	278,2012
		CS 22950 # 96		20 25 21	-15 52.3	1	13.90		0.46		-0.16		1		1736
195015	+43 03609			20 25 25	+43 54.8	1	8.28		-0.05		-0.41		2	A0	401
		LS II +39 053		20 25 28	+39 34.6	1	10.34		1.01		-0.08		2		405
		CS 22955 # 47		20 25 30	-23 13.9	1	14.30		0.40		0.10		1		1736
195033	+42 03752			20 25 31	+42 58.9	1	7.67		-0.06		-0.34		2	B9	401
195068	+48 03128	HR 7828		20 25 31	+49 13.0	2	5.67	.094	0.27	.035	0.04		3	gF0	247,1118
194783	-36 14166	HR 7817		20 25 34	-35 45.7	1	6.09		-0.11				4	B9p HgMn	2007
		G 210 - 19		20 25 37	+35 49.9	2	14.75	.033	1.64	.028			4		419,1759
		LP 815 - 22		20 25 37	-15 14.2	1	12.35		0.57		-0.15		2		1696
		VES 258		20 25 39	+47 04.3	1	10.72		0.50		0.03		1		8056
194989	+25 04247			20 25 40	+25 58.8	1	7.48		0.18		0.16		3	A5 V	8071
194937	+7 04477	HR 7820		20 25 41	+08 16.2	3	6.24	.005	1.08	.005	0.93	.004	10	G9 III	15,1256,1355
	+38 04098	LS II +38 087		20 25 42	+38 36.4	1	8.96		0.88		0.43		2	B9 Ib	1012
195050	+37 03941	HR 7826		20 25 43	+38 16.5	6	5.62	.016	0.06	.014	0.06	.020	11	A3 V	247,1063,1118,1193*
334452	+29 04041			20 25 45	+29 47.2	1	8.77		1.66				2	M0	369
194953	+2 04175	HR 7824		20 25 46	+02 46.2	1	6.20		0.90		0.56		8	G8 III	1088
194939	-3 04906	HR 7821	⋆AB	20 25 48	-03 31.5	2	6.12		-0.06	.004	-0.18	.020	9	B9 V	1079,1088
194702	-55 09397			20 25 50	-55 17.2	3	9.46	.009	0.38	.007	-0.12	.009	7	F5wF0	158,742,1594
		CS 22940 # 1		20 25 50	-62 18.1	1	13.38		1.48		0.80		2		1736
334426	+30 04040	V442 Cyg		20 25 51	+30 37.5	1	9.70		0.45		0.02		6	F4	1768
195089	+41 03758			20 25 51	+41 52.1	2	7.31	.009	-0.01	.005	-0.62		4	B2 IV	401,1118
195100	+42 03754			20 25 51	+42 53.4	2	7.57	.015	0.88	.025	0.55	.060	4	G5 III	247,1355
194538	-68 03371			20 25 51	-68 11.4	1	8.86		1.16				4	K2 III	2012
194918	-16 05609	HR 7819		20 25 55	-15 54.5	3	6.41	.007	1.00	.005	0.69	.002	7	G8 III	252,2007,4001
	+40 04179	LS II +40 034		20 25 56	+40 25.8	2	9.65	.000	0.57	.000	-0.43	.000	4	O8 V:	1011,1012
194943	-18 05689	HR 7822	⋆AB	20 26 01	-17 58.8	7	4.77	.017	0.38	.008	0.02	.018	22	F3 V	15,1075,1088,2001*
195019	+18 04505	IDS20237N1826	AB	20 26 02	+18 36.2	2	6.89	.015	0.64	.005	0.18	.020	6	G3 IV-V	1648,3016
195020	+11 04273			20 26 04	+11 33.2	1	7.55		0.28		0.14		3	A3	8071
	+37 03943	LS II +37 096		20 26 04	+37 50.3	1	9.39		1.51		1.14		1	F6 Ia	247
195102	+33 03914	IDS20241N3333		20 26 06	+33 43.3	2	7.00	.005	-0.06	.014	-0.17		4	B9	401,1118
		JL 53		20 26 06	-73 51.	1	15.88		-0.27		-1.14		2		132
	-35 14150			20 26 07	-34 47.6	1	10.10		0.48		-0.05		2	F5	1730
194959	-17 05992			20 26 08	-17 36.1	1	6.75		0.65		0.19		3	G2 IV	3026
194960	-18 05691	IDS20232S1809	D	20 26 09	-18 02.5	2	6.55	.010	1.07	.000	0.86	.005	3	K0 III	219,1003
		CS 22940 # 9		20 26 14	-60 00.9	1	14.07		-0.29		-1.08		1		1736
	+37 03945	LS II +38 089		20 26 26	+38 11.3	1	9.53		0.77		-0.17		2	B0 II:nn	1012
	+40 04185	LS III +40 018		20 26 27	+40 28.3	1	9.82		0.84		-0.16		2	B0 V:	1012
195177		LS II +38 090		20 26 32	+38 27.3	1	11.46		1.21		0.75		1	C5	1359
195228	+47 03114	IDS20250N4755	AB	20 26 33	+48 04.9	1	8.34		-0.02		-0.09		2	A0	401
195006	-22 05442	HR 7825		20 26 35	-22 33.5	3	6.16	.005	1.54	.012	1.99	.020	11	M1 III	1024,2007,3005

HD	DM	Other Id	N Rem	α_{1950}	δ_{1950}	S	V	σ_V	B–V	σ_{B-V}	U–B	σ_{U-B}	n	Spectrum	References
195178	+37 03946			20 26 36	+37 37.3	2	7.56	.000	1.60	.005			10	M1	1118,1196
195194	+38 04102			20 26 36	+39 09.7	2	6.98	.000	0.97	.000	0.72	.020	4	G8 III	105,247
195040	−21 05735	AT Cap		20 26 42	−21 17.6	2	9.01	.030	1.15	.012	0.79		2	G6 III/IV	1641,1642
	−35 14159			20 26 42	−34 56.9	1	11.30		0.51		−0.02		2		1730
195213	+40 04187	LS III +40 019		20 26 44	+40 39.0	2	8.74	.000	0.85	.005	−0.26	.000	4	O7	1011,1012
195229	+41 03765	LS III +41 024		20 26 44	+41 50.6	1	7.66		0.16		−0.71		2	B0.2 III	1012
195134	+12 04348			20 26 47	+12 30.8	1	6.99	.010	0.01	.015	−0.18	.025	4	B9	401,3016
195254	+43 03620			20 26 47	+43 21.4	1	7.69		0.01				2	B9	1118
		AJ89,1229 # 474		20 26 48	+04 23.	1	11.12		1.64				1		1532
194521	−75 01624			20 26 48	−74 48.6	1	9.06		1.22				4	K1/3 III	2012
195162	+15 04172	RS Del		20 26 51	+16 06.4	1	7.90		1.50		0.94		15	M3	3042
195026	−35 14161			20 26 53	−34 51.6	1	9.18		0.43		−0.02		2	F3 V	1730
		L 567 - 35		20 26 54	−31 33.	1	12.65		0.94		0.71		1		1696
195216	+27 03755			20 26 55	+27 41.0	1	7.70		1.62				4	K5 III	20
195077	−21 05736			20 26 56	−21 21.7	1	9.47		1.10		0.94		23	K0 III	978
195391	+59 02240			20 26 58	+59 34.0	1	7.96		−0.03		−0.09		2	A0	401
		VES 132		20 27 00	+29 18.6	1	15.37		0.83		0.28		1		363
	−45 13852			20 27 00	−45 41.2	1	10.28		0.51				1		1594
	+35 04138			20 27 01	+35 22.4	2	9.18	.005	1.84	.015	2.11		4	M0	369,8032
	+39 04208	RW Cyg		20 27 01	+39 48.9	3	8.36	.259	2.88	.019	2.42		9	M2 Iab	148,1118,8032
195093	−19 05830	HR 7829	⋆ B	20 27 01	−18 45.2	2	6.74	.000	0.22	.000	0.04	.000	4	A7/8 V	15,1028
195135	−3 04918	HR 7831		20 27 02	−03 03.2	4	4.90	.009	1.16	.006	1.20	.007	21	K2 III	3,418,1088,3016
195094	−19 05831	HR 7830	⋆ A	20 27 02	−18 45.0	2	5.94	.000	0.08	.000	0.30	.000	4	A3 Vn	15,1028
195094	−19 05831	IDS20242S1855	AB	20 27 02	−18 45.0	1	5.51		0.11				4	A3 Vn+A7 V	2007
		CS 22885 # 203		20 27 04	−37 51.4	2	14.22	.010	0.42	.008	−0.22	.008	3		1580,1736
		L 350 - 30		20 27 06	−46 33.	2	13.14	.045	0.54	.007	−0.19	.035	4		1696,3062
195217	+19 04423	HR 7833		20 27 07	+19 55.2	4	6.56	.010	0.23	.008	0.10	.008	10	A3 m	1049,1063,3058,8071
195106	−21 05737			20 27 08	−21 25.5	2	9.20	.003	1.07	.009	0.92		26	K0 III	978,1642
195322	+45 03191			20 27 09	+45 33.1	1	7.40		−0.03		−0.11		2	B9	401
195338	+47 03117			20 27 09	+47 26.6	2	7.48	.000	1.17	.000	0.96	.005	4	G7 II	247,1355
195107	−22 05446			20 27 09	−21 59.7	1	9.62		1.05				1	G8 IV	1642
	−76 01416			20 27 09	−76 25.5	1	10.53		0.92				4	G5	2012
195271	+35 04140	IDS20252N3530	A	20 27 11	+35 39.8	2	7.32	.000	1.45	.005	1.70	.025	5	K2	1601,3016
195271	+35 04140	IDS20252N3530	B	20 27 11	+35 39.8	1	14.29		0.77		0.25		3		3016
		VES 259		20 27 11	+46 57.5	1	11.21		0.36		0.22		1		8056
195353	+48 03135			20 27 17	+48 41.7	1	7.61		0.54				2	G0	1724
195137	−16 05613			20 27 17	−16 24.2	1	9.90		0.65		0.00		2	G3 V	1003
		WLS 2020 25 # 8		20 27 19	+25 10.1	1	11.98		0.49		0.01		2		1375
		VES 260		20 27 20	+44 43.8	1	11.65		0.84		0.14		1		8056
195295	+29 04057	HR 7834	⋆	20 27 21	+30 12.0	11	4.02	.008	0.40	.008	0.25	.024	128	F5 II	15,120,253,1118,1218*
		G 24 - 16	⋆ V	20 27 22	+09 31.3	3	13.05	.013	1.67	.030	1.29	.014	5		203,1705,3029
195273	+26 03912			20 27 22	+26 46.3	1	7.45		1.03				4	K1 III	20
		CS 22955 # 117		20 27 22	−24 38.6	1	14.40		0.44		−0.18		1		1736
195323	+37 03950	IDS20255N3711	AB	20 27 23	+37 20.9	1	7.74		0.66				2	F5	1118
195324	+36 04141	HR 7835	⋆	20 27 26	+36 17.2	4	5.88	.000	0.52	.014	0.10		9	A1 Ib	1063,1118,1193,1196
195356	+39 04210			20 27 34	+39 55.9	2	8.15	.000	−0.03	.005			4	A0	1118,1196
195392	+46 02947			20 27 34	+46 56.2	2	8.73	.004	0.23	.003	0.08		26		1601,6011
		G 262 - 14		20 27 35	+61 50.9	1	11.47		0.77		0.30		2	G8	1658
		CS 22955 # 129		20 27 35	−23 07.5	1	13.81		0.35		0.11		1		1736
332271	+33 03923	LS II +34 022		20 27 37	+34 02.7	1	8.96		1.14		0.78		1	F7 Iab:	247
195377	+38 04111			20 27 37	+38 47.0	2	8.45	.005	0.04	.009	0.01	.005	5	A0 V	401,833
332278	+33 03922		V	20 27 38	+33 40.1	1	8.52		1.63				2	M3	369
		G 24 - 17		20 27 40	+04 24.4	2	11.47	.025	0.87	.000	0.45	.000	2	K2	333,1620,1658
		VES 218		20 27 40	+38 24.1	1	11.96		1.09		0.55		2		426
195237	−5 05288	TZ Aql		20 27 40	−04 55.4	1	8.75		1.66		1.15		15	M3	3042
195275	+1 04304	KN Aql		20 27 43	+01 42.5	1	8.52		1.56		1.34		1	M2	3076
195394	+38 04114	IDS20259N3817	A	20 27 43	+38 26.4	1	8.45		−0.07		−0.26		2	A0	401
195393	+38 04116	IDS20259N3836	AB	20 27 43	+38 46.4	1	8.55		0.01		−0.56		2	B8	401
195405	+41 03770			20 27 43	+42 09.0	1	7.99		0.64		0.27		1	G2 IV	247
	+46 02948	LS III +46 007		20 27 43	+46 29.9	1	10.05		0.28		−0.59		2	B1 V:enn	1012
195406	+37 03952	IDS20260N3747	A	20 27 51	+37 57.4	1	8.19		−0.07		−0.44		1	B6 V	401
195406	+37 03952	IDS20260N3747	AB	20 27 51	+37 57.4	1	7.56		−0.05		−0.45		1	B6 V +B9 IV	401
195406	+37 03952	IDS20260N3747	B	20 27 51	+37 57.4	1	8.46		−0.05		−0.40		1	B9 IV	401
195206	−29 17122	HR 7832		20 27 53	−29 16.9	1	6.39		0.21				4	A5 V	2035
195325	+10 04303	HR 7836	⋆ AB	20 27 54	+10 43.6	3	6.04	.055	−0.01	.030	−0.12	.019	7	A1 Shell	379,1063,1419
346763	+22 04080			20 27 54	+22 48.8	1	9.52		0.00		−0.20		2	B9	1375
195407	+36 04095	LS II +36 081		20 27 54	+36 48.7	4	7.81	.022	0.32	.032	−0.66	.011	8	B0 IV:pe	379,1012,1212,1419
		LTT 8115		20 27 54	−24 58.	1	11.82		0.57		−0.18		1		3062
195277	−5 05291			20 27 55	−05 33.5	1	7.74		1.61		1.96		3	K2	1657
		AJ84,127 # 2234		20 28 00	+26 40.	1	11.14		0.30		0.11		1		801
		Steph 1792		20 28 01	+24 44.1	1	11.38		1.34		1.22		1	K6	1746
195462	+47 03125			20 28 01	+48 15.1	1	7.96		0.68				5	G5	1724
340345	+26 03915	IDS20260N2631	AB	20 28 05	+26 40.7	2	9.69	.020	1.34	.000	1.14	.095	2	A0	801,3016
195463	+45 03195	LS III +45 009		20 28 05	+45 19.6	1	8.90		−0.05		−0.69		2	B5	401
195478	+44 03486			20 28 11	+44 19.1	1	8.28		−0.04		−0.49		2	B9	401
194777	−74 01912			20 28 11	−74 42.9	1	10.07		0.09				4	A1 V	2012
195554	+55 02411	HR 7843	⋆ AB	20 28 12	+55 54.0	2	5.90		−0.06	.009	−0.24	.004	3	B9 Vne	985,1079
195464	+37 03953			20 28 14	+37 26.7	1	8.58		−0.12		−0.39		2	A0	401
195330	−15 05696	HR 7837	⋆ AB	20 28 16	−15 13.5	2	6.11	.008	0.80	.004	0.42		6	K1/2 III +F	58,2007
194778	−75 01632			20 28 17	−74 56.4	1	9.33		0.63				4	G2 IV/V	2012

Table 1   1111

HD	DM	Other Id	N Rem	α_{1950}	δ_{1950}	S	V	σ_V	B–V	σ_{B-V}	U–B	σ_{U-B}	n	Spectrum	References
195409 +18 04519				20 28 19	+18 22.8	1	8.73		0.10		0.08		2	A2	1733
195432 +27 03768		LS II +27 034		20 28 19	+27 39.8	1	7.06		0.64				2	G0 II	6009
195506 +45 03196		HR 7841		20 28 20	+45 45.5	8	6.42	.013	1.14	.014	1.06	.022	35	K2 III	15,247,1003,1080,1322*
		LP 871 - 22		20 28 20	−22 45.9	1	11.90		0.78		0.32		1		1696
		CS 22950 # 147		20 28 24	−14 38.6	1	14.60		0.46		-0.16		1		1736
		VES 261		20 28 28	+41 59.6	1	11.52		0.76		0.29		1		8056
195557 +48 03141				20 28 31	+48 28.4	1	7.61		0.03		0.01		2	A0	401
195556 +48 03142		HR 7844	⋆ A	20 28 31	+48 47.0	5	4.94	.005	-0.09	.004	-0.65	.024	14	B2.5IV	15,154,1363,3016,8015
195508 +37 03955				20 28 32	+37 19.4	1	8.67		0.12		-0.48		2	A5	401
195507 +38 04121		IDS20267N3813	AB	20 28 33	+38 23.3	2	8.66	.000	0.08	.025			4	A0	1118,1196
195590 +53 02421				20 28 33	+53 32.2	1	7.91		0.96		0.59		2	K0	1566
195529 +40 04205				20 28 35	+40 29.7	2	8.00	.000	-0.12	.005			4	B9.5IV	1118,1196
		CS 22955 # 104		20 28 36	−26 01.5	1	15.37		0.08		0.13		1		1736
195434 +4 04470		G 24 - 18	⋆ AB	20 28 44	+05 02.7	1	8.83		0.89		0.44		2	K0	265,1620
195479 +20 04602		HR 7839	⋆ A	20 28 44	+20 26.1	3	6.20	.014	0.14	.012	0.12	.011	10	A1 m	1063,3058,8071
195725 +62 01821		HR 7850		20 28 45	+62 49.5	4	4.21	.009	0.20	.003	0.16	.002	10	A7 III	15,1363,3023,8015
195481 +15 04181		IDS20265N1528	AB	20 28 47	+15 38.3	2	6.85	.000	0.13	.019	0.08		4	A3 V +A3 V	1381,3030
195509 +26 03917				20 28 49	+26 31.0	1	7.40		0.99				4	K0 III	20
195483 +10 04307		HR 7840	⋆ A	20 28 50	+11 05.5	1	7.11		0.03		-0.40		2	B8 V	401
195544 +35 04146				20 28 50	+36 13.5	1	7.75		0.37		-0.02		3	F5	379
195592 +43 03630		LS III +44 017		20 28 53	+44 08.8	3	7.08	.000	0.87	.000	-0.19	.005	5	O9.5Ia	1011,1012,8007
		CS 22955 # 103		20 28 55	−26 17.1	1	16.18		0.40		0.06		1		1736
	+47 03129	LS III +47 005		20 28 56	+47 41.7	2	9.22	.005	0.38	.000	-0.67	.025	4	B0	405,483
195435 −12 05755				20 28 57	−12 03.2	1	8.97		1.46				1	C3,4i	1238
195490 +8 04456				20 29 03	+08 45.1	1	7.80		0.27		0.11		3	AmA5-F0	8071
195593 +36 04105		HR 7847	⋆ AB	20 29 05	+36 46.0	12	6.20	.020	1.00	.012	0.74	.012	33	F5 Iab	1,15,254,1080,1118*
	+40 04209	Cyg OB2 # 1590		20 29 05	+41 09.6	1	8.80		0.04		-0.18		1	A0	443
	−23 16310			20 29 05	−23 42.8	2	10.39	.024	0.85	.000	0.42		3		742,1594
		CS 22955 # 107		20 29 11	−25 54.1	1	14.31		0.45		-0.22		1		1736
195629 +39 04219				20 29 14	+40 15.4	1	7.60		-0.08		-0.41		2	B7/8IV/III	401
	+36 04108	VES 220		20 29 17	+37 00.6	3	10.86	.000	0.50	.000	0.02	.000	3		426,8103
195455 −24 16098				20 29 17	−24 14.3	1	9.20		-0.18		-0.90		1	B1/2 III	55
195710 +48 03148		IDS20278N4853	A	20 29 20	+49 02.6	1	6.62		0.04		0.04		1	A2	3032
195710 +48 03148		IDS20278N4853	B	20 29 20	+49 02.6	1	9.54		0.48		0.00		1		3032
195711 +46 02958				20 29 23	+46 36.4	1	8.14		0.57				4	G5	1724
		Cyg OB2 # 1512	⋆	20 29 24	+41 21.7	3	11.08	.015	1.41	.015	0.31	.000	6	O9 V	1011,1118,1167
	+33 03936			20 29 33	+33 36.3	2	9.20	.051	1.14	.026	1.08		3	K5	1017,3072
		G 209 - 33		20 29 33	+38 23.1	1	13.51		1.67				1		1759
		Cyg OB2 # 1542	⋆ V	20 29 35	+41 21.3	3	10.65	.097	1.18	.058	-0.02	.122	6	B1 Ib	1012,1118,1167
		L 210 - 96		20 29 36	−57 08.	1	12.51		0.47		-0.23		2		1696
195689 +39 04221				20 29 37	+39 46.1	3	7.55	.010	0.04	.017	0.03		6	A1 V	401,1118,1196
195690 +33 03938				20 29 39	+34 09.7	3	6.47	.005	0.40	.024	-0.02		7	F2 V	1118,1196,1501
195564 −10 05423		HR 7845	⋆ AB	20 29 40	−10 01.5	4	5.65	.007	0.69	.004	0.20		12	G3 V	15,1075,2032,3077
195536 −17 06014		IDS20269S1657	AB	20 29 42	−16 46.8	1	7.16		0.32		0.07		3	F0/2 IV/V	3016
	−10 05424			20 29 43	−10 10.8	1	10.15		0.76		0.42		2		1064
		VES 262		20 29 44	+46 09.8	1	10.50		0.53		0.00		1		8056
	+60 02131	NGC 6939 - 27		20 29 44	+60 30.4	1	9.96		0.49		-0.04		2		299
		GD 230		20 29 44	+18 21.0	1	16.31		0.16		-0.59		1		3060
195774 +48 03154		HR 7851	⋆ A	20 29 46	+49 03.0	2	5.43	.008	1.55	.002	1.90		5	M2 IIIab	71,3016
195774 +48 03154		IDS20282N4853	B	20 29 46	+49 03.0	1	10.54		0.41		-0.07		1		3016
		CS 22950 # 153		20 29 46	−15 58.2	1	13.74		0.56		-0.09		1		1736
		TT Ind		20 29 46	−56 44.0	1	12.81		-0.02		-0.80		1		1471
		NGC 6939 - 36		20 29 47	+60 31.2	1	12.31		0.94		0.26		2		299
195746 +41 03789				20 29 48	+42 02.0	3	8.56	.008	-0.04	.011	-0.15		22		401,1118,1196
195692 +25 04272		HR 7849	⋆ AB	20 29 50	+25 38.1	1	6.34		0.26		0.08		3	A3 m	8071
	+40 04212	Cyg OB2 # 920	⋆ AB	20 29 50	+41 03.1	4	10.23	.053	1.63	.012	0.53	.021	8	O9	1011,1012,1118,1167
		G 262 - 15		20 29 51	+65 16.6	1	10.54	.061	1.55	.018	1.23	.057	10	M3	272,1017,3045
195668 +18 04525				20 29 52	+18 27.4	1	7.27		1.62		1.83		3	M4 II-III	8100
195549 −25 14854		HR 7842		20 29 54	−25 06.8	1	6.36		-0.01				4	A0 V	2009
195820 +51 02882		HR 7854		20 29 55	+52 08.3	1	6.18		1.01				2	K0 III	71
195633 +6 04557				20 29 56	+06 20.8	3	8.53	.009	0.52	.010	-0.04	.042	5	G0 V wlm	462,792,3077
		LP 755 - 59		20 29 56	−11 27.8	1	10.60		0.80		0.32		1		1696
		V341 Aql		20 29 58	+00 24.9	2	10.39	.255	0.17	.071	0.17	.030	2	A9.5	668,699
195190 −71 02560				20 29 58	−71 21.7	1	6.34		1.37		1.39		5	K2 III	1628
	+40 04213	Cyg OB2 # 594		20 29 59	+41 07.2	1	9.11		0.41				3	B3 V	1118
195788 +44 03499				20 29 59	+45 00.1	1	7.93		-0.07		-0.22		2	A0	401
195891 +59 02251				20 29 59	+59 42.4	1	8.60		0.49		-0.02		2	G0	1502
		VES 263		20 30 00	+40 27.8	1	12.87		1.88		0.55		1		8056
195872 +56 02440				20 30 00	+56 43.3	1	7.94		0.61		0.05		1	F8	1620
194612 −81 00906		HR 7812		20 30 00	−81 27.7	3	5.90	.008	1.70	.005	2.01	.005	12	K5 III	15,1075,2038
195852 +55 02416				20 30 02	+55 35.9	1	8.37		-0.06		-0.27		2	A0	401
		NGC 6939 - 52		20 30 02	+60 29.9	1	12.13		0.72		0.12		2		299
		NGC 6939 - 54		20 30 02	+60 32.8	1	13.55		1.00		0.18		2		299
		Cyg OB2 # 560	⋆	20 30 03	+41 18.2	1	11.62		1.12		0.03		1	O9.5V	1170
		NGC 6939 - 56		20 30 03	+60 32.2	1	14.68		0.71		-0.03		2		299
	+40 04215	Cyg OB2 # 553	⋆	20 30 04	+41 16.2	1	10.51		1.38		1.28		1		66
196565 +80 00657		IDS20345N8105		20 30 04	+81 16.0	1	6.67		0.99		0.78		1	K0	3032
		NGC 6939 - 65		20 30 06	+60 32.0	1	12.55		1.30		0.75		2		299
195636 −9 05491				20 30 07	−09 32.0	2	9.54	.000	0.64	.005	-0.01	.005	4	B8	1003,3077
195789 +36 04113				20 30 08	+37 12.4	1	8.27		-0.09		-0.40		2	B9	401

HD	DM	Other Id	N Rem	α_{1950}	δ_{1950}	S	V	σ_V	B–V	σ_{B-V}	U–B	σ_{U-B}	n	Spectrum	References
		G 24 - 19		20 30 13	+05 40.4	1	11.93		1.40		1.06		1	M1	333,1620
196142	+72 00957	HR 7868		20 30 14	+72 21.7	2	6.26	.023	1.35	.005	1.54		20	K2	252,1258
	−73 02155			20 30 17	−73 23.1	1	10.21		1.26				4		2012
		Smethells 57		20 30 18	−46 23.	1	11.88		1.40				1		1494
	+47 03131			20 30 19	+48 10.0	1	8.75		0.08		−0.59		2	B5	401
195917	+55 02419			20 30 20	+56 11.1	1	7.66		−0.13		−0.58		2	B9	401
	+43 03642			20 30 23	+43 44.8	1	10.08		1.86				2		369
		NGC 6940 - 1		20 30 25	+27 58.9	1	12.12		0.56		0.08		2		1168
		CS 22885 # 179		20 30 25	−38 52.7	1	14.64		0.32		0.04		1		1736
	+40 04219	Cyg OB2 # 32	★	20 30 26	+41 16.9	3	10.22	.000	1.18	.007	0.10	.004	5	O7 III((f))	1011,1012,1167
195730	+0 04534			20 30 28	+00 27.4	1	8.87		0.19		0.16		1	A0	289
		G 144 - 9		20 30 28	+19 21.5	2	11.64	.014	0.88	.005	0.49	.005	4	K3	203,1658
		Cyg OB2 # 31	★	20 30 29	+41 15.3	3	11.53	.056	1.24	.025	0.23	.023	3	O9 V	66,276,1170
		CS 22955 # 115		20 30 29	−24 35.3	1	14.55		0.30		0.01		1		1736
195569	−44 14020	HR 7846		20 30 29	−44 41.2	2	5.12	.003	1.00	.009	0.80		6	K0 III	58,2009
		CS 22955 # 132		20 30 32	−22 46.0	1	14.37		0.44		0.11		1		1736
	+13 04434	IDS20282N1336	CD	20 30 33	+13 47.8	1	9.82		0.92		0.65		2	K0	3016
195964	+56 02444	HR 7860		20 30 33	+56 36.6	1	6.14		1.43				2	gK5	71
	+13 04435	IDS20282N1336	AB	20 30 34	+13 46.5	1	8.90		0.60		0.09		3	F8	3032
195873	+41 03796	Cyg OB2 # 1384	★ AB	20 30 34	+41 41.8	2	8.73	.000	−0.01	.004			12		1118,1196
	+40 04220	Cyg OB2 # 629	★ AB	20 30 35	+41 08.0	5	9.10	.036	1.68	.022	0.57	.019	9	O7 Ia nfp	443,1011,1012,1118,1167
195834	+28 03786			20 30 37	+28 53.0	1	8.08		1.34		1.60		3	K3 II	8100
	+40 04221	Cyg OB2 # 29	★	20 30 40	+41 14.2	5	10.66	.015	1.23	.008	0.17	.017	9	O8 V	276,1011,1012,1118,1167
		Cyg OB2 # 30	★	20 30 40	+41 16.1	3	11.17	.067	1.20	.025	0.14	.050	3	O8 V	66,276,1170
		Cyg OB2 # 502	★	20 30 40	+41 18.6	1	11.50		1.02		−0.07		1	B0.5V	1170
195791	+12 04372			20 30 43	+12 34.6	1	9.23		0.05		−0.02		2	A0	3016
	+13 04437		V	20 30 44	+13 35.1	1	9.45		1.15		1.16		3	K2	3016
		ApJ144,259 # 71		20 30 48	+47 11.	1	18.95		0.37		−0.66		3		1360
195810	+10 04321	HR 7852		20 30 49	+11 07.9	8	4.03	.012	−0.12	.013	−0.47	.023	30	B6 III	15,154,379,1203,1363*
	+44 03502			20 30 50	+44 23.6	1	10.13		1.95				2		369
		Cyg OB2 # 986		20 30 51	+40 49.8	1	10.26		1.98		0.78		4		483
		Cyg OB2 # 1623	★	20 30 51	+41 14.9	2	10.95	.033	1.20	.024	0.11	.033	2	O7.5V	66,1170
195965	+47 03136	LS III +48 002		20 30 51	+48 02.7	1	6.98		−0.05		−0.90		2	B0 V	1012
195918	+38 04138	IDS20290N3856	A	20 30 53	+39 06.0	4	8.18	.052	−0.07	.027	−0.47	.015	7	B9 V	401,401,833,963
195918	+38 04138	IDS20290N3856	B	20 30 53	+39 06.0	2	10.53	.150	0.91	.135	0.43	.155	2		401,963
	+39 04227			20 30 53	+39 54.3	1	8.86		1.65				2	M3	369
		Cyg OB2 # 41	★ V	20 30 53	+41 04.2	3	11.42	.042	3.07	.126	1.69	.000	5	B5 Ie	618,1011,1012
196017	+53 02433			20 30 56	+53 59.1	1	7.41		−0.10		−0.41		2	B9	401
195402	−70 02792	HR 7838		20 30 56	−69 47.0	3	6.10	.007	1.30	.005	1.41	.000	13	K2 III	15,1075,2038
		VES 264		20 30 58	+39 30.0	1	11.74		0.38		0.14		1		8056
194149	−84 00619			20 31 01	−84 35.0	1	6.87		0.33		0.15		7	A1mA5-F0	1628
		Cyg OB2 # 28	★	20 31 03	+41 13.5	2	11.72	.019	1.39	.019	0.28	.033	2	O8.5V	66,1170
	+6 04570			20 31 04	+07 13.4	1	9.49		1.43				4	K2	339
195907	+31 04126	LS II +31 027		20 31 04	+31 29.1	1	7.82		0.03		−0.74		3	B1.5V	1212
		NGC 6940 - 23		20 31 05	+28 07.5	1	12.52		0.54		0.24		2	F2	1168
195987	+41 03799	Cyg OB2 # 1359	★	20 31 06	+41 43.2	7	7.08	.016	0.80	.014	0.37	.026	10	G9. V	247,276,1003,1080*
195985	+44 03503			20 31 06	+44 48.5	2	7.67	.019	−0.10	.019	−0.55	.010	3	B5	247,401
		CS 22955 # 124		20 31 07	−23 51.5	1	14.85		0.27		0.07		1		1736
195986	+42 03778	HR 7861		20 31 08	+43 01.2	2	6.59	.022	−0.11	.004	−0.57		5	B4 III	154,1118
		NGC 6934 sq1 7		20 31 11	+07 12.9	1	12.27		0.49				1		339
		NGC 6934 sq1 3		20 31 12	+07 13.3	1	10.62		0.32				1		339
195919	+26 03930			20 31 12	+27 17.1	3	8.98	.006	0.05	.001	−0.02	.008	16	A2	830,1375,1783
195876	+9 04578	CZ Del		20 31 13	+09 20.7	1	8.10		1.63		1.75		10	M1	3042
		VES 222		20 31 13	+34 35.0	1	12.11		2.06				1		426
334744	+27 03783	NGC 6940 - 24		20 31 16	+28 18.5	1	9.57		0.31				3	A3	309
196018	+46 02966	SZ Cyg		20 31 16	+46 25.8	3	8.98	.029	1.24	.019	0.94		3	F7.5	934,1399,6011
195990	+39 04232			20 31 17	+39 41.7	1	8.08		−0.04		−0.17		2	A0	401
334745	+27 03784	NGC 6940 - 26		20 31 18	+28 13.3	1	10.44		0.47				3	F2	309
		NGC 6940 - 28		20 31 19	+27 50.5	2	11.56	.005	1.12	.001	0.78	.012	4		1168,4002
		NGC 6940 - 27		20 31 19	+27 58.5	1	11.85		0.54		0.43		2	A5	1168
		Cyg OB2 # 40	★	20 31 21	+41 03.0	1	11.68		2.01		0.70		1	O4 III(f)	1170
		NGC 6940 - 468		20 31 22	+28 11.9	1	12.91		0.54				4		309
		Cyg OB2 # 35	★	20 31 23	+41 04.8	6	10.80	.005	1.93	.019	0.65	.005	13	O5 If+	276,1011,1012,1118*
		NGC 6940 - 30		20 31 24	+28 06.8	2	10.98	.005	1.20	.013	1.02	.048	3	gK0	3046,4002
		NGC 6934 sq1 6		20 31 25	+07 14.8	1	12.24		0.64				5		339
195838	−14 05781	HR 7855		20 31 25	−13 53.7	2	6.13	.005	0.54	.005	−0.02		7	G0 V	2007,3077
		NGC 6934 sq1 5		20 31 26	+07 14.1	1	12.19		1.21				6		339
	+40 04227	Cyg OB2 # 1620	★ A	20 31 27	+41 08.5	1	8.98		1.29		0.14		4	O6 Ib(f)	1209
	+40 04227	Cyg OB2 # 1620	★ AB	20 31 27	+41 08.5	1	8.69		1.25		0.27		4	O5 III(f)	8084
		Cyg OB2 # 1621	★ D	20 31 27	+41 08.5	2	12.06	.094	1.39	.028	0.35	.038	2	O8.5V	1167,8084
		Cyg OB2 # 33	★	20 31 27	+41 10.0	3	10.49	.009	1.44	.005	0.29	.014	6	O3 If	1012,1167,1209
195627	−61 06492	HR 7848		20 31 27	−60 45.1	4	4.76	.005	0.29	.004	0.04	.015	15	F0 V	15,1075,2012,2016
		Cyg OB2 # 683	★	20 31 28	+41 09.9	1	12.61		1.49		0.14		1	O9.5V	1170
196787	+80 00659	HR 7901	★ A	20 31 28	+81 15.2	1	5.42		1.01		0.86		3	G9 III	3032
196787	+80 00659	IDS20345N8105	B	20 31 28	+81 15.2	1	6.67		0.99		0.78		1		3016
196787	+80 00659	IDS20345N8105	C	20 31 28	+81 15.2	1	11.34		0.81		0.43		1		3032
195922	+9 04579	HR 7857		20 31 29	+09 53.3	2	6.55	.005	0.08	.000	0.04	.005	7	A2 Vnn	15,1256
	+40 04228	Cyg OB2 # 689	★ B	20 31 29	+40 08.7	2	10.42	.088	1.41	.044	0.28			O6 f	1011,1167
196006	+32 03862			20 31 30	+32 44.2	3	7.32	.019	−0.15	.010	−0.70	.005	6	B2 V	399,401,963
		Cyg OB2 # 34	★	20 31 30	+41 06.8	2	11.87	.071	1.57	.024	0.46	.080	2	O7.5V	276,1170

Table 1   1113

HD	DM	Other Id	N	Rem	α_{1950}	δ_{1950}	S	V	σ_V	B–V	σ_{B-V}	U–B	σ_{U-B}	n	Spectrum	References
	+40 04227	Cyg OB2 # 692		⋆ C	20 31 30	+41 08.2	1	10.08		1.35		0.10		1	O5 III(f)	1167
		Smethells 58			20 31 30	−51 20.	1	12.14		1.44				1		1494
195909	+4 04486				20 31 31	+04 43.6	1	6.40		1.02		0.79		4	K0-1I-III	1149
195942	+15 04199				20 31 33	+15 36.5	1	8.34		0.21		0.15		3	AmA5-F0	8071
195943	+12 04378	HR 7858			20 31 35	+12 51.3	4	5.38	.008	0.08	.010	0.05	.005	15	A3 IVs	355,1063,1363,3023
		NGC 6940 - 218			20 31 36	+28 00.1	1	13.30		0.80		0.46		1		49
		L 210 - 17			20 31 36	−55 09.	1	12.63		1.39		1.31		1		1696
	+6 04572				20 31 37	+07 13.7	1	9.25		1.12				1	K0	339
196090	+46 02969				20 31 37	+46 59.7	2	7.79	.001	1.41	.008	1.56		26	K3 III	247,6011
		NGC 6940 - 37			20 31 38	+28 06.3	1	12.28		0.44		0.30		2		1168
		Cyg OB2 # 26		⋆	20 31 38	+41 23.1	2	11.61	.094	1.55	.000	0.50	.042	2	O8.5V	276,1170
196022	+27 03788	NGC 6940 - 38			20 31 39	+27 42.0	3	7.70	.007	0.28	.009	0.12	.005	8	Am	309,1168,8071
		NGC 6940 - 39			20 31 40	+28 02.9	1	11.96		1.21		0.89		1	K0 III	49
195814	−38 14108	HR 7853			20 31 40	−38 15.7	1	6.44		0.26				4	A5mA7-A9	2009
		Cyg OB2 # 705		⋆	20 31 42	+41 05.	2	10.94	.104	1.89	.014	0.68	.024	2	O5 Ia	276,1170
		L 495 - 50			20 31 42	−37 16.	1	12.05		0.50		-0.18		2		1696
195759	−50 13038				20 31 42	−49 55.5	1	9.21		0.69		0.16		4	G8 V	158
		NGC 6940 - 41			20 31 43	+28 11.0	1	12.47		0.56		0.20		2	F0	1168
195843	−30 18013	HR 7856			20 31 43	−30 38.8	1	6.40		-0.08				4	B8 V	2007
		NGC 6940 - 43			20 31 44	+28 20.8	1	11.74		0.68		0.22		2		1168
195993	+17 04355				20 31 45	+18 00.8	1	7.01		1.62		1.95		2	K5 III	1648
		NGC 6940 - 47			20 31 45	+28 02.9	1	12.09		0.38		0.34		2	A2	1168
334743	+27 03789	NGC 6940 - 44			20 31 45	+28 18.1	1	10.34		0.72				4	G0	309
		NGC 6940 - 45			20 31 45	+28 21.0	1	12.25		0.47		0.21		2		1168
		NGC 6934 sq1 8			20 31 46	+07 09.9	1	13.77		0.68				3		339
		G 144 - 10			20 31 46	+21 21.2	1	12.91		0.98		0.76		2		1658
		NGC 6940 - 48			20 31 46	+28 05.2	1	12.02		0.51		0.30		3	A3	1168
		NGC 6940 - 49			20 31 47	+28 08.0	1	12.22		0.51		0.22		3	F0	1168
		NGC 6940 - 50			20 31 48	+27 49.1	1	11.55		0.32		0.10		2		1168
		LP 815 - 36			20 31 48	−18 12.7	1	14.84		1.10				1		1759
196034	+25 04284				20 31 49	+25 26.7	1	7.90		1.36		1.48		3	K3 III	833
		NGC 6940 - 52			20 31 49	+28 08.1	1	12.90		0.51		0.28		3	F0	1168
		NGC 6940 - 54			20 31 50	+28 00.5	1	11.87		0.46		0.34		2	A2	1168
195320	−77 01447	IDS20254S7714		A	20 31 50	−77 04.0	1	7.72		0.96				4	G8/K0 III	2012
		NGC 6940 - 55			20 31 51	+28 02.3	1	12.39		0.46		0.30		2	A5	1168
		NGC 6940 - 56			20 31 51	+28 06.5	1	12.12		0.54		0.24		3	F0	1168
196092	+38 04148	IDS20300N3814		AB	20 31 51	+38 24.7	3	8.00	.005	0.03	.005	-0.08		5	A0	401,1118,1196
		Cyg OB2 # 37		⋆	20 31 51	+41 09.1	1	11.06		1.60		0.37		1	O9.5III	1170
	+46 02972	LS III +46 009		⋆ ABC	20 31 53	+46 52.6	1	11.07		0.50				1	O9.5V	725
		NGC 6940 - 60			20 31 54	+27 52.7	2	11.59	.005	1.11	.004	0.79	.051	3	gG5	1168,4002
	+46 02971				20 31 55	+46 34.5	1	9.80		1.29				1	K0	934
196035	+20 04629	HR 7862			20 31 56	+20 48.8	1	6.47		-0.14		-0.67		3	B3 IV	154
		NGC 6940 - 64			20 31 56	+28 14.0	1	12.56		0.39		0.29		2	A3	1168
		NGC 6940 - 65			20 31 57	+28 06.4	1	12.74		0.53		0.30		3	F0	1168
196093	+34 04079	HR 7866		⋆ AB	20 31 57	+35 04.7	6	4.61	.016	1.59	.023	0.76	.034	11	K2 Ib+B3 V	15,247,1355,1363,3016,8015
	+37 03976	LS II +38 092			20 31 57	+38 06.7	1	10.25		0.76		-0.16		2	B1.5Vn	1012
196133	+44 03505				20 31 58	+45 00.2	3	6.70	.012	0.04	.004	0.01	.000	11	A0p	401,1118,3016
		NGC 6940 - 222			20 31 59	+28 01.3	1	15.03		0.80		0.14		1		49
		NGC 6940 - 67			20 31 59	+28 06.4	2	11.18	.005	1.13	.000	0.86	.001	4	gG8	1168,4002
	+41 03804	Cyg OB2 # 27		⋆	20 31 59	+41 22.6	5	9.87	.013	1.53	.007	0.41	.007	9	O9.5Ia	276,1011,1012,1118,1167
		NGC 6940 - 70			20 32 00	+27 43.1	1	9.85		0.46		0.18		2	F5 II-III	1168
		NGC 6940 - 69			20 32 00	+28 00.9	3	11.64	.007	1.12	.014	0.80	.029	8	G8 III	309,1168,4002
		NGC 6940 - 68			20 32 00	+28 05.0	1	12.34		0.56		0.31		2	A5	1168
		NGC 6940 - 577			20 32 01	+27 51.2	1	13.02		1.65				2		309
		NGC 6940 - 71			20 32 01	+28 09.0	1	11.82		0.49		0.23		2	A5	1168
		NGC 6940 - 73			20 32 01	+28 20.2	1	11.87		0.44		0.29		3	A6 III-IV	1168
		VES 223			20 32 01	+34 20.3	1	13.38		2.56				1		426
196119	+38 04150				20 32 01	+38 31.0	2	8.47	.000	-0.01	.005			4	A0	1118,1196
196134	+41 03805	HR 7867			20 32 01	+41 36.1	1	6.49		1.00		0.78		5	K0 III-IV	253
195762	−60 07410				20 32 01	−59 51.1	1	7.56		1.17		1.25		5	K1 III	1673
340613	+24 04166				20 32 02	+24 53.6	1	9.56		0.02		-0.28		1	A0	963
196177	+52 02746				20 32 02	+52 37.9	1	7.61		1.16		1.07		2	K5	1566
		VES 224			20 32 03	+31 16.1	1	11.45		1.94				1		426
		IRC +50 334		V	20 32 03	+46 49.0	1	11.04		1.86		1.33		1		8056
196025	+6 04576	IDS20296N0632		AB	20 32 04	+06 42.3	1	6.97		-0.12		-0.72		3	B2 IV-V	399
334746		NGC 6940 - 77			20 32 04	+28 14.2	2	11.08	.013	0.53	.004	0.26		5	F4	309,1168
196120	+34 04081	IDS20301N3420		AB	20 32 04	+34 30.4	1	6.67		-0.12		-0.48		2	B8	401
196195	+54 02374				20 32 04	+54 37.8	2	7.01	.005	0.36	.005	0.07	.009	4	F2	1566,1716
		NGC 6940 - 78			20 32 05	+27 53.4	1	11.14		1.76				2	gK4	309
		NGC 6940 - 79			20 32 06	+28 04.2	1	12.58		0.48		0.29		2	A5	1168
		NGC 6940 - 81			20 32 07	+28 02.1	1	12.50		0.55		0.19		2		1168
		NGC 6940 - 80			20 32 07	+28 05.8	1	12.44		0.52		0.05		2	F0	1168
		Cyg OB2 # 38		⋆	20 32 07	+41 09.9	2	11.72	.104	1.56	.033	0.48	.052	2	B1 III-Ib	276,1170
196095	+27 03792	NGC 6940 - 83			20 32 08	+27 59.5	2	9.17	.034	0.21	.004	0.10		11	A0 III	49,309
		NGC 6940 - 84			20 32 08	+28 04.1	1	10.89		0.84		0.48		3	F6 III	1168
334749		NGC 6940 - 86			20 32 09	+28 05.8	1	10.93		0.49		0.27		2	F0 III	49
		NGC 6940 - 87			20 32 09	+28 11.9	2	11.34	.010	1.09	.007	0.81	.034	3	gG5	1168,4002
		NGC 6940 - 85			20 32 09	+28 14.8	1	12.76		0.48		0.22		2	F0	1168
		NGC 6940 - 88			20 32 10	+27 57.9	1	11.46		0.59		0.32		2	F0	1168
		NGC 6940 - 89			20 32 10	+28 03.8	1	12.51		0.45		0.27		2	A5	1168

HD	DM	Other Id	N	Rem	α_{1950}	δ_{1950}	S	V	σ_V	B–V	σ_{B-V}	U–B	σ_{U-B}	n	Spectrum	References
		NGC 6940 - 90			20 32 10	+28 06.4	1	11.48		0.54		0.30		2	F0	1168
	+45 03216	LS III +45 010		⋆ AB	20 32 10	+45 29.3	2	9.07	.000	0.39	.000	-0.57	.000	4	O8 V	1011,1012
		NGC 6940 - 217			20 32 11	+28 04.5	1	13.25		0.53		0.20		1	A5	49
196502	+74 00872	HR 7879, AF Dra	4		20 32 11	+74 47.0	4	5.19	.004	0.07	.004	0.10	.019	10	A0 p SrCrEu	1202,1263,1363,3023
		G 24 - 20			20 32 12	+03 10.8	2	11.99	.017	1.43	.008	0.99	.042	7		333,1620,1663
		NGC 6940 - 221			20 32 12	+28 04.3	1	15.00		0.74		0.36		1		49
		Cyg OB2 # 39		⋆	20 32 12	+41 07.2	2	12.08	.123	1.66	.028	0.57	.047	2	O9.5V	276,1170
	+48 03160				20 32 12	+48 43.3	1	9.53		1.80				2		369
		NGC 6940 - 92			20 32 13	+28 17.1	2	11.82	.005	1.15	.005	0.80	.011	3	gG8	1168,4002
		NGC 6940 - 225			20 32 14	+27 59.3	1	15.65		0.80		0.29		1		49
		NGC 6940 - 93			20 32 14	+28 04.3	1	11.92		0.49		0.35		2	A3	1168
340611					20 32 15	+24 54.2	5	11.52	.021	-0.07	.011	-0.87	.006	11	DA	1,316,963,1281,3028
		NGC 6940 - 220			20 32 15	+28 05.5	1	14.03		0.70		0.18		1		49
	+46 02976				20 32 15	+46 45.4	1	9.46		1.13				1	K0	934
340533	+27 03793	NGC 6940 - 96			20 32 16	+27 52.0	1	9.86		0.09				4	A0	309
		NGC 6940 - 95			20 32 16	+27 57.3	2	12.31	.014	0.47	.009	0.27		7	A5	309,1168
		NGC 6940 - 98			20 32 16	+28 02.0	2	11.52	.015	0.62	.000	0.31	.010	3	F4 III	49,1168
		NGC 6940 - 97			20 32 16	+28 05.4	1	12.31		0.49		0.24		2	A5	1168
		VES 266			20 32 16	+44 31.0	1	12.77		0.88		-0.11		1		8056
		NGC 6940 - 99			20 32 17	+28 03.2	1	11.78		0.49		0.18		2	F0	1168
334747		NGC 6940 - 100	3		20 32 17	+28 11.4	3	10.44	.004	0.75	.016	0.43	.013	8	G8 II +A7	309,1168,4002
196178	+46 02977	HR 7870		⋆ A	20 32 17	+46 31.3	3	5.77	.005	-0.15	.006	-0.54	.000	9	B9p Si	1079,1118,3033
		NGC 6940 - 101			20 32 18	+28 14.1	2	11.29	.010	1.09	.008	0.82	.002	3	gG8	1168,4002
		NGC 6940 - 102			20 32 19	+28 05.1	1	11.71		0.60		0.35		2	F0	1168
		NGC 6940 - 103			20 32 19	+28 09.4	1	11.68		0.48		0.25		2	F0	1168
196196	+45 03217				20 32 19	+46 14.6	2	7.64	.024	-0.03	.005	-0.04		4	B9	401,1118
		NGC 6940 - 105			20 32 20	+27 54.8	3	10.67	.008	1.24	.020	1.06	.001	7	K0 II-III	309,1168,4002
		NGC 6940 - 108			20 32 20	+28 03.3	2	11.20	.010	1.05	.006	0.66	.035	5	gG5	1168,4002
		NGC 6940 - 106			20 32 20	+28 05.9	1	12.25		0.42		0.20		2	A5	1168
		NGC 6940 - 107			20 32 20	+28 06.3	1	11.92		0.44		0.30		2	A5	1168
196178	+46 02978	LS III +46 010		⋆ B	20 32 20	+46 33.2	1	9.34		0.44				1		725
		NGC 6940 - 109			20 32 21	+28 03.0	1	12.56		0.42		0.27		2	A5	1168
		NGC 6940 - 111			20 32 21	+28 12.4	2	11.55	.009	1.08	.003	0.81	.006	4	gG5	1168,4002
		NGC 6940 - 110			20 32 21	+28 19.1	1	11.86		0.42		0.25		2		1168
		NGC 6940 - 459			20 32 21	+28 19.1	1	13.31		0.65				3		309
	+41 03807	Cyg OB2 # 324		⋆	20 32 21	+41 26.6	5	10.04	.005	1.43	.013	0.26	.012	11	O5 If+	1011,1012,1118,1167,1209
		NGC 6940 - 112			20 32 22	+28 04.2	1	12.20		0.49		0.32		2	A3	1168
		NGC 6940 - 115			20 32 24	+28 06.2	1	11.59		1.27		1.22		1		4002
		VES 225			20 32 24	+31 28.1	1	12.17		0.38		0.20		1		426
		NGC 6940 - 116			20 32 25	+28 04.2	1	11.64		0.49		0.28		3	A2	1168
		NGC 6940 - 117			20 32 26	+27 55.3	1	11.74		0.44		0.31		3	A5	1168
		NGC 6940 - 118			20 32 26	+28 03.7	2	12.22	.020	0.51	.010	0.27	.000	3	F2 II	49,1168
		Cyg OB2 # 312		⋆	20 32 26	+41 24.7	1	12.04		1.47		0.22		1	O7 V	1170
196925	+80 00660	HR 7908		⋆ A	20 32 26	+80 55.1	2	5.98	.009	0.97	.026	0.65		6	K0 III	1118,3024
196925	+80 00662	IDS20353N8044	B		20 32 26	+80 55.1	2	8.70	.013	0.62	.017	0.01	.026	6		1064,3024
334750	+27 03795	NGC 6940 - 120		⋆ V	20 32 27	+28 06.6	3	9.26	.123	1.77	.005	1.55	.077	58	M5 II	1168,1563,4002
		NGC 6940 - 119			20 32 27	+28 13.9	1	12.87		0.48		0.16		2	F0	1168
		NGC 6934 sq1 4			20 32 29	+07 13.9	1	11.05		1.34				1		339
196145	+27 03796	NGC 6940 - 123			20 32 29	+27 55.5	1	8.54		0.31		0.14		1	A5	49
		NGC 6940 - 219			20 32 29	+28 03.4	1	13.40		0.57		0.19		1		49
		NGC 6940 - 124			20 32 29	+28 05.2	2	12.07	.005	0.52	.000	0.25	.015	4	F2 III	49,1168
		NGC 6940 - 122			20 32 29	+28 11.9	1	12.58		0.40		0.17		2	A5	1168
		NGC 6940 - 125			20 32 29	+28 14.4	1	12.44		0.52		0.22		2	F0	1168
		NGC 6940 - 127			20 32 30	+28 06.2	1	12.83		0.43		0.19		3	A3	1168
		NGC 6940 - 126			20 32 30	+28 06.3	1	12.63		0.39		0.20		3	A5	1168
		G 262 - 16			20 32 30	+65 59.4	1	11.60		0.94		0.80		2	K3	1658
		NGC 6940 - 129			20 32 32	+28 12.1	1	10.70		1.14		1.04		1	gG8	3046
196216	+42 03786				20 32 32	+43 11.4	1	7.28		0.38				2	F2	1118
		NGC 6940 - 130			20 32 33	+28 10.1	2	11.38	.010	1.07	.003	0.77	.008	4	gG5	1168,4002
	-0 04045				20 32 33	-00 02.7	2	9.58	.010	1.63	.005	1.91		2	M0	116,118
		NGC 6940 - 131			20 32 34	+28 02.2	1	12.24		0.47		0.16		2	F0	1168
196281	+52 02750				20 32 34	+52 25.5	1	7.40		1.10		0.88		2	K2	1566
		CS 22955 # 153			20 32 34	-26 51.3	1	14.49		0.42		-0.17		1		1736
195903	-51 12619				20 32 34	-51 04.2	1	7.96		0.98				4	G8 III	2012
		NGC 6940 - 133			20 32 35	+28 11.1	1	11.59		0.52		0.21		3	F0	1168
		NGC 6940 - 132			20 32 35	+28 16.3	1	10.97		1.11		0.90		3	G8 II-III	1168
		VES 226			20 32 35	+32 23.9	1	10.90		2.22				1		426
		NGC 6940 - 135			20 32 36	+28 00.9	1	12.43		0.48		0.23		2	A5	1168
334751		NGC 6940 - 136			20 32 37	+28 05.8	2	11.22	.000	0.54	.024	0.34		5	A3	309,1168
		NGC 6940 - 138			20 32 40	+27 58.7	2	11.37	.010	1.08	.009	0.79	.002	4	gG8	1168,4002
		NGC 6940 - 137			20 32 40	+28 03.0	1	12.48		0.50		0.17		2	F0	1168
		G 230 - 42			20 32 41	+61 34.3	1	12.59		1.50				1		1759
		NGC 6940 - 139			20 32 42	+28 04.4	2	11.37	.015	1.08	.005	0.79	.011	4	gG5	1168,4002
		CS 22950 # 173			20 32 42	-16 03.9	1	14.04		0.41		-0.23		1		1736
		CS 22955 # 155			20 32 42	-27 12.2	1	13.38		0.23		0.12		1		1736
196124	+5 04556	G 24 - 21		⋆ A	20 32 43	+05 57.4	1	8.91		0.96		0.76		1	K2	333,1620
196078	-17 06027	HR 7865			20 32 43	-16 41.9	2	6.18	.005	0.20	.000			7	A7 V	15,2030
		NGC 6940 - 141			20 32 44	+27 48.4	1	12.24		0.48		0.22		3		1168
		NGC 6940 - 140			20 32 44	+28 04.9	1	11.64		0.40		0.25		2	A5	1168
		NGC 6940 - 143			20 32 44	+28 12.1	1	12.56		0.40		0.18		2	A5	1168

Table 1 1115

HD	DM	Other Id	N	Rem	α_{1950}	δ_{1950}	S	V	σ_V	B−V	σ_{B-V}	U−B	σ_{U-B}	n	Spectrum	References
196197 +32 03865		IDS20307N3210	A		20 32 44	+32 20.0	1	6.84		1.12		1.00		2	F2 IV-V	3016
196197 +32 03865		IDS20307N3210	B		20 32 44	+32 20.0	1	9.18		0.35		-0.01		2		3032
		NGC 6940 - 145			20 32 45	+27 55.7	1	11.78		0.58		0.23		3	F0	1168
		NGC 6940 - 144			20 32 45	+28 13.7	1	12.32		0.44		0.20		2		1168
		NGC 6940 - 146			20 32 46	+28 20.0	1	11.61		0.45		0.22		2		1168
196198 +28 03799		NGC 6940 - 147		⋆ AB	20 32 46	+28 22.5	2	8.62	.009	0.04	.005	-0.16	.000	11	B8 III	49,1168
196240 +41 03811					20 32 46	+41 49.4	3	7.95	.005	0.18	.009	0.21		23	Am	1118,1196,8071
334752		NGC 6940 - 149			20 32 47	+28 10.0	1	11.67		0.42		0.23		2	F5	1168
		NGC 6940 - 148			20 32 47	+28 12.0	1	12.49		0.46		0.16		2	F0	1168
		NGC 6940 - 151			20 32 49	+28 05.6	1	12.43		0.48		0.16		2	F2	1168
		NGC 6940 - 153			20 32 51	+28 03.1	1	12.06		0.22		0.18		2	A0	1168
		NGC 6940 - 152			20 32 51	+28 04.0	2	10.83	.004	1.09	.001	0.90	.013	5	gG8	1168,4002
		NGC 6940 - 224			20 32 51	+28 09.3	1	15.15		0.65		0.22		1		49
		NGC 6940 - 154			20 32 52	+28 01.1	1	12.30		0.42		0.28		2	A3	1168
		NGC 6940 - 156			20 32 52	+28 07.7	1	12.23		0.44		0.23		2	A5	1168
		NGC 6940 - 223			20 32 53	+28 05.3	1	15.03		1.30		0.57		1		49
	-0 04048				20 32 54	+01 16.7	2	10.01	.005	1.10	.010	0.92		2		116,118
		NGC 6940 - 159			20 32 54	+27 58.0	1	12.05		0.50		0.24		2	F0	1168
196081 -27 14911					20 32 54	-26 56.8	1	7.20		0.42				4	F5 V	1075
		NGC 6940 - 162			20 32 55	+28 05.6	1	12.51		0.48		0.12		2		1168
		NGC 6940 - 161			20 32 55	+28 06.9	1	11.63		0.38		0.26		3	A3	1168
		NGC 6940 - 160			20 32 55	+28 09.0	2	12.36	.005	0.45	.010	0.20	.035	3	F0 V	49,1168
		Cyg OB2 # 360		⋆	20 32 56	+41 18.7	1	12.36		1.58		0.68		1	B1.5III?	1170
	+44 03511				20 32 56	+45 11.3	1	9.36		0.12		-0.51		1	B5	8056
196180 +14 04353		HR 7871			20 32 58	+14 30.0	5	4.66	.024	0.11	.010	0.11	.015	14	A3 V	15,1063,1363,3023,8015
		GD 231			20 32 58	+18 49.1	2	15.41	.070	-0.05	.015	-0.81	.005	2	DA	1727,3060
334754 +27 03798		NGC 6940 - 163			20 32 58	+28 07.2	1	10.45		0.52		0.04		1	F7 III	49
334753 +27 03797		NGC 6940 - 164			20 32 58	+28 14.5	2	9.97	.015	1.20	.003	1.24	.000	4	K2 II-III	1168,4002
196407 +61 02028					20 32 59	+61 34.2	1	7.04		0.07		0.04		2	A0	401
	+0 04545				20 33 00	+00 33.5	1	9.40		1.64		1.57		1	M5	116
334736		NGC 6940 - 166			20 33 00	+28 20.9	1	11.10		0.46		0.25		2		1168
		LP 695 - 407			20 33 00	-07 24.0	1	11.04		0.62		0.02		2		1696
		NGC 6940 - 168			20 33 01	+27 51.0	1	13.09		0.47		0.21		3		1168
		NGC 6940 - 169			20 33 03	+28 00.5	1	12.04		0.37		0.23		2	A3	1168
334735		NGC 6940 - 172			20 33 03	+28 20.3	1	11.33		0.12		-0.61		2		1168
		NGC 6940 - 174			20 33 04	+28 07.1	1	11.63		0.38		0.27		2	A5	1168
		NGC 6940 - 175			20 33 06	+28 13.7	1	11.49		0.28		0.28		2	A5	1168
	+41 03813				20 33 08	+41 32.2	1	10.20		0.25		0.02		1		401
196305 +41 03814		Cyg OB2 # 1245			20 33 09	+41 32.7	1	7.59		-0.10		-0.47		2	B9	401
	-0 04049				20 33 13	+00 10.1	2	9.75	.025	1.24	.015	1.29		2	K0	116,118
334755		NGC 6940 - 179			20 33 13	+28 05.7	1	11.28		0.30		0.31		2	A0	1168
196203 -0 04050					20 33 13	-00 10.5	2	7.09	.005	0.49	.014	0.04	.005	4	F8	116,118
		NGC 6940 - 180			20 33 14	+28 03.6	1	12.14		0.37		0.24		2	A2	1168
		NGC 6940 - 181			20 33 16	+28 00.2	1	11.78		0.48		0.24		2	A5	1168
		VES 227			20 33 19	+32 09.9	1	10.51		0.25		0.19		1		426
	+33 03960	LS II +33 034			20 33 22	+33 50.3	3	10.29	.122	0.84	.035	-0.24	.012	7		405,426,994
196379 +51 02895		HR 7876		⋆	20 33 23	+51 40.9	1	6.13		0.40		0.40		1	A9 II	254
196421 +57 02219					20 33 23	+57 45.3	2	8.06	.025	0.11	.010	-0.62	.005	5	B2 IV	401,555
		LS III +42 017			20 33 24	+42 10.0	1	10.64		1.42				2		725
		NGC 6940 - 540			20 33 25	+28 04.9	1	12.47		0.52				1		309
		VES 230			20 33 25	+34 24.4	1	11.72		1.49		1.36		1		426
195961 -61 06495		HR 7859, ρ Pav			20 33 25	-61 42.2	2	4.86	.015	0.43	.002	0.19		6	Fm δ Del	58,2007
		NGC 6940 - 493			20 33 26	+28 10.7	1	12.76		0.71				3		309
334757		NGC 6940 - 187			20 33 26	+28 13.8	1	11.63		0.48		0.36		2	K0	1168
334758		NGC 6940 - 188			20 33 27	+28 15.9	1	11.20		0.47		0.26		2	K0 II +A	1168
334756		NGC 6940 - 189			20 33 28	+28 06.3	1	10.58		0.78		0.50		2	K0 II +A	1168
196187 -19 05864					20 33 29	-19 07.2	1	9.87		1.01		0.85		2	K3 V	3072
		Steph 1800			20 33 31	+01 49.5	1	10.69		1.50		1.76		1	K4	1746
196360 +41 03818					20 33 31	+41 43.0	1	6.65		0.93		0.74		1	K0 III	247
340730 +24 04182		G 186 - 32		⋆ A	20 33 33	+25 12.4	1	9.90		0.62		0.24		2	G5	1003
		LS III +46 011			20 33 35	+46 40.8	2	10.90	.025	1.37	.015	0.17	.015	5		483,684
		NGC 6940 - 198			20 33 36	+28 05.7	1	11.72		0.42		0.18		2		1168
		NGC 6940 - 200			20 33 38	+28 08.0	1	12.53		0.43		0.20		2		1168
		LS III +46 012			20 33 40	+46 39.6	3	10.26	.005	0.98	.009	-0.16	.008	7	O6	405,483,684
		He2 467			20 33 43	+20 01.0	1	13.13		1.26		-0.48		1		1753
		NGC 6940 - 204			20 33 43	+28 06.5	1	11.68		0.38		0.23		2		1168
334761		NGC 6940 - 203			20 33 43	+28 08.8	1	11.78		0.47		0.20		2		1168
		CS 22880 # 18			20 33 49	-19 25.4	1	15.60		-0.26		-1.08		1		1736
		LS II +35 077			20 33 50	+35 36.9	1	11.87		0.57		-0.22		5		994
195962 -66 03496					20 33 50	-66 42.6	2	8.32	.014	0.65	.014	0.12		5	G3/5 V	2012,3077
		NGC 6940 - 209			20 33 54	+27 46.2	2	11.43	.005	1.23	.003	1.10	.014	3		1168,4002
		NGC 6940 - 210			20 33 54	+27 51.0	1	12.19		0.54		-0.04		2		1168
		CS 22879 # 12			20 33 55	-41 13.1	2	14.71	.005	0.39	.007	-0.17	.009	3		1580,1736
	+0 04548				20 33 56	+01 02.7	1	10.23		0.26		0.08		1		116
196345 +16 04315					20 33 58	+16 38.4	1	6.57		1.40		1.64		2	K2	1648
196307 -0 04052					20 34 00	+00 06.9	2	8.67	.020	1.15	.010	1.22		2	K2	116,118
196362 +25 04299		HR 7874			20 34 00	+25 42.5	2	6.42		0.19	.038	0.16	.014	4	A5 III	105,1063
196408 +37 03988					20 34 02	+37 27.0	1	8.70		1.78				2	M1	369
196171 -47 13477		HR 7869		⋆ A	20 34 04	-47 28.0	4	3.11	.004	1.00	.004	0.79	.005	16	K0 III	15,1020,1075,2012
		G 24 - 22			20 34 06	-02 51.9	1	12.14		1.48		1.20		1		1620

HD	DM	Other Id	N Rem	α_{1950}	δ_{1950}	S	V	σ_V	B–V	σ_{B-V}	U–B	σ_{U-B}	n	Spectrum	References
196321	−3 04961	HR 7873		20 34 07	−02 43.5	2	4.92	.029	1.62	.015	1.93	.005	13	K5 II	1088,8100
196310	−13 05713	IDS20314S1305	A	20 34 09	−12 54.7	1	7.98		0.38		−0.02		2	F2 IV/V	1028
196310	−13 05713	IDS20314S1305	B	20 34 09	−12 54.7	1	9.11		0.56		0.00		2		1028
196566	+61 02031			20 34 12	+61 51.4	1	8.81		0.12		−0.34		2	B9	401
197508	+83 00588	HR 7930		20 34 13	+83 27.2	1	6.19		0.15		0.13		2	A4m	1733
		VES 231		20 34 17	+31 31.3	1	12.90		1.67		1.54		1		426
195460	−80 00970			20 34 20	−80 42.5	1	8.76		0.23		0.16		2	A2 V	978
196369	+0 04550			20 34 21	+00 46.1	2	8.67	.009	0.47	.004	0.05	.018	13	F5	116,1371
340686	+25 04301			20 34 21	+26 13.3	1	8.58		1.06		0.83		11	K1 III	1222
	+0 04551			20 34 22	+01 09.4	1	9.43		1.37		1.54		1	K5	116
		AO 720		20 34 23	+14 18.9	1	10.43		1.37		1.56		1		1697
340667	+26 03937	V Vul		20 34 24	+26 25.8	1	8.27		0.82		0.61		1	G1 Iab:	793
	+36 04145	LS II +37 099		20 34 24	+37 14.6	3	8.92	.018	0.66	.005	−0.38	.005	9	O9 V	994,1011,1012
352680	+17 04365			20 34 25	+18 06.1	1	8.55		1.10		1.00		5	K0	897
196448	+28 03810			20 34 26	+29 02.0	1	8.87		0.59		0.08		2	G0 V	3026
196382	−0 04055			20 34 27	+01 06.3	1	9.87		0.19		0.18		1	A2	116
	−0 04054			20 34 28	−00 14.2	2	10.70	.010	0.27	.010	0.09		2	A5	116,118
		L 279 - 25		20 34 30	−53 16.	2	14.46	.000	−0.05	.000	−0.94	.000	4	DB	782,3065
		Smethells 59		20 34 30	−55 47.	1	11.36		1.42				1		1494
	+0 04552			20 34 33	+01 10.3	1	10.41		0.58		0.08		1		116
196489	+38 04162			20 34 33	+39 01.2	1	8.48		0.14		−0.45		3	B3 V	833
196395	−1 04014			20 34 33	−00 41.6	5	8.71	.008	1.66	.005	2.04	.020	53	K5	147,989,1509,1729,6005
196348	−15 05732			20 34 33	−15 19.3	2	6.77	.009	1.26	.007	1.25	.028	4	K1 III	1003,3040
		CS 22880 # 13		20 34 35	−20 25.0	1	13.68		0.38		−0.22		1		1736
	+14 04366			20 34 37	+14 21.1	1	10.22		1.25		1.32		1	K1.5III	1697
	+64 01452	G 262 - 20		20 34 37	+64 47.8	1	10.14		1.18		1.07		2	M0	3072
		BSD 112 # 1268		20 34 38	+00 55.3	1	10.60		0.39		0.00		1	A8	116
	+32 03873			20 34 39	+32 49.4	1	8.38		1.62				2	M2	369
		G 262 - 21		20 34 41	+64 43.4	2	13.82	.064	0.78	.000	0.19	.040	5		1658,1773
196373	−21 05776			20 34 42	−21 30.6	1	8.29		0.47				4	F3/5 V	2012
	+31 04152			20 34 45	+32 16.8	1	8.43		1.95		2.09		4	M1 IIIe	206
		G 262 - 22		20 34 45	+64 43.1	2	14.48	.055	1.14	.005	0.92	.025	5		1658,1773
196426	−0 04056	HR 7878		20 34 45	−00 04.7	8	6.21	.015	−0.08	.012	−0.40	.008	46	B8 IIIp	116,118,147,252,1079*
196522	+39 04252			20 34 46	+40 18.3	3	8.08	.037	−0.04	.023	−0.35		4	A0	401,1118,1196
		CS 22880 # 24		20 34 52	−18 29.0	1	14.57		−0.35		−1.18		1		1736
196385	−25 14920	HR 7877		20 34 54	−25 17.1	2	6.36	.005	0.33	.005	−0.02		5	A9 V	2032,3053
	+31 04153			20 34 55	+31 31.1	1	8.73		1.77				2	M2	369
196504	+25 04302	HR 7880		20 34 57	+26 17.2	3	5.60		−0.06	.009	−0.22	.010	8	B9 V	985,1063,1079
	+50 03144			20 34 58	+50 46.3	1	9.11		1.67				2	K7	369
		VES 268		20 34 59	+46 41.6	1	11.85		0.61		−0.16		1		8056
		VES 269		20 35 01	+46 52.8	1	10.74		0.52		0.14		1		8056
		G 144 - 16		20 35 08	+21 46.5	1	11.44		1.50				1	M2	1746
	+14 04368			20 35 09	+14 35.8	1	9.47		0.99		0.82		1	G5	1697
	−0 04057			20 35 09	−00 22.4	1	10.74		0.56		0.00		1	G0	116
	+55 02439	IDS20339N5536	AB	20 35 10	+55 46.4	1	10.40		0.63		0.17		1		8084
	+55 02439	IDS20339N5536	D	20 35 10	+55 46.4	1	12.15		0.53		0.09		1		8084
196524	+14 04369	HR 7882	★ AB	20 35 12	+14 25.2	6	3.62	.022	0.45	.026	0.07	.017	15	F6 III+F6IV	15,254,1118,3030,8003,8015
196589	+36 04150			20 35 13	+36 24.5	1	7.48		0.06				2	A0	1118
196670	+50 03145	IDS20337N5032	AB	20 35 13	+50 42.9	1	7.94		0.12		0.08		2	A0	1566
196627	+38 04168			20 35 18	+38 46.8	1	8.99		−0.04		−0.17		4	A0 V	833
		LP 815 - 43		20 35 21	−20 36.5	1	10.91		0.38		−0.20		2		1696
		G 144 - 19		20 35 22	+18 31.6	1	12.18		1.10		1.11		1	M0	1658
	+37 03998			20 35 23	+38 13.6	1	9.46		1.78				2	M2	369
		L 495 - 42		20 35 24	−37 00.	1	14.88		0.22		−0.68		6		3062
		CS 22940 # 50		20 35 24	−61 51.2	2	14.87	.009	0.51	.010	−0.21	.005	4		1580,1736
	+0 04556			20 35 26	+01 01.4	1	9.93		1.55		1.90		1		116
196544	+10 04339	HR 7883		20 35 26	+11 12.1	3	5.43	.004	0.05	.009	0.04	.002	15	A2 V	355,1063,3058
196607	+30 04105			20 35 26	+30 55.5	1	8.26		−0.06		−0.38		2	A0	401
196605	+31 04158			20 35 27	+31 52.8	1	7.84		−0.04		−0.21		2	A0	401
196805	+61 02037			20 35 28	+61 50.4	1	7.84		−0.08		−0.32		2	A0	401
334802	+29 04110			20 35 29	+30 00.3	1	8.70		0.50		−0.03		3	G0 V	3026
196643	+37 04000			20 35 29	+37 55.4	1	7.09		1.59		1.93		2	K5 III	247
196606	+31 04159	HR 7885	★ A	20 35 30	+31 23.8	1			−0.09		−0.40		4	B8 IIIn	1063
196629	+31 04160	HR 7887	★	20 35 31	+31 20.8	1	6.49		0.34		−0.03		3	F0 V	253
196642	+37 04002	HR 7888		20 35 31	+38 09.2	3	6.22	.026	0.99	.008	0.80	.016	6	K0 III	247,1118,3016
	+48 03176		AB	20 35 31	+49 13.4	1	9.46		0.30		0.41		1	F8	8084
	+45 03230	LS III +45 011		20 35 33	+45 57.8	1	9.40		0.75		−0.27		2	B0 III	1012
196546	−3 04969			20 35 34	−02 57.1	2	8.48	.000	−0.02	.000	−0.09	.000	8	B9	1306,1386
196445	−40 13972			20 35 34	−40 32.7	1	9.08		1.42		1.41		1	K1 (III)	565
196317	−63 04602	HR 7872		20 35 36	−63 05.0	1	6.22		1.10				4	K1 III	2007
196687	+42 03803			20 35 37	+42 48.9	1	7.27		−0.09		−0.38		2	A0	401
		CS 22955 # 147		20 35 37	−25 58.7	1	14.23		0.21		0.19		1		1736
		CS 22879 # 34		20 35 37	−38 24.4	2	14.24	.010	0.29	.004	0.07	.014	3		1580,1736
		CS 22879 # 29		20 35 37	−38 47.4	2	14.44	.010	0.45	.009	−0.20	.007	3		1580,1736
196553	+0 04557			20 35 38	+00 45.8	1	9.58		0.03		0.05		1	A0	116
196610	+17 04370	HR 7886, EU Del		20 35 38	+18 05.5	5	6.06	.094	1.48	.004	1.00	.017	17	M6 III	71,897,979,3042,8112
		AA Aql		20 35 39	−03 04.0	2	11.20	.201	0.13	.083	0.16	.024	2	A8.5	668,699
196573	+0 04558			20 35 43	+00 50.4	5	7.88	.010	1.64	.014	2.01	.021	30	K5	116,118,989,1371,1729
		BSD 112 # 549		20 35 45	+00 41.4	1	10.90		0.55		0.07		1	F7	116
196673	+32 03883	IDS20338N3301	AB	20 35 45	+33 11.5	3	6.97	.011	1.11	.010	0.94	.009	6	K0	471,993,3016

Table 1   1117

HD	DM	Other Id	N Rem	α_{1950}	δ_{1950}	S	V	σ_V	B–V	σ_{B-V}	U–B	σ_{U-B}	n	Spectrum	References
196574	−1 04016	HR 7884	⋆ A	20 35 45	−01 16.9	6	4.32	.012	0.95	.010	0.68	.007	24	G8 III	15,1075,1088,1509*
		VES 232		20 35 46	+36 04.9	1	11.45		1.90				1		426
		VES 233		20 35 48	+35 40.3	1	11.50		0.39		0.05		1		426
		Smethells 60		20 35 48	−43 55.	1	12.26		1.44				1		1494
		Smethells 61		20 35 48	−43 55.	1	12.72		1.48				1		1494
196067	−75 01644	HR 7864	⋆ A	20 35 50	−75 31.6	2	6.51	.035	0.62	.000	0.26	.005	3	G5 III	1279,3077
196067	−75 01644	IDS20298S7542	AB	20 35 50	−75 31.6	4	6.02	.004	0.62	.005	0.19	.005	20	G5 III	15,1075,1075,2038
		LS II +33 035		20 35 52	+33 17.3	1	10.83		0.84		−0.25		4		994
196068	−75 01644	IDS20298S7542	B	20 35 52	−75 31.3	2	7.18	.030	0.64	.000	0.28	.005	3	G5 V	1279,3077
196557	−17 06045			20 35 53	−17 14.5	1	7.85		1.50		1.85		3	K4/5 III	1657
196378	−60 07419	HR 7875		20 35 55	−60 43.1	5	5.11	.007	0.53	.000	-0.01	.005	14	F7 V	15,158,219,2012,3026
196051	−76 01434	HR 7863		20 35 56	−76 21.5	6	5.98	.011	0.45	.004	0.11	.013	28	F6 II/III	15,278,1075,1075,2012,2038
		AJ89,1229 # 486		20 36 00	+03 42.	1	11.56		1.67				1		1532
	−0 04058			20 36 01	+00 03.8	1	10.89		0.57		0.11		1		116
196531	−28 16816			20 36 01	−28 36.2	2	7.95	.010	0.54	.005	-0.01		5	G0 V	219,1075
		VES 270		20 36 03	+39 44.4	1	11.93		0.42		0.35		1		8056
196632	−0 04059			20 36 03	−00 15.5	2	9.15	.020	0.50	.015	-0.03	.010	2	F8	116,118
	−0 04060			20 36 04	+00 07.8	1	10.57		0.51		0.05		1		116
	+0 04560			20 36 08	+01 10.1	1	10.19		0.47		0.00		1		116
196807	+51 02910			20 36 08	+51 24.3	1	8.46		1.79				2	M2	369
		POSS 107 # 5		20 36 09	+59 32.5	1	16.72		1.70				2		1739
		AO 727		20 36 13	+14 40.2	1	10.60		1.14		0.84		1		1697
196789	+42 03807			20 36 13	+42 40.0	1	7.04		0.48				2	F8	1118
	+82 00620			20 36 14	+82 31.8	1	10.17		1.58		1.85		2	F0	1375
		VES 234		20 36 15	+36 18.9	1	11.75		1.36		1.02		1		426
196808	+47 03153	LS III +47 009	⋆ A	20 36 15	+47 46.1	1	7.93		-0.02		-0.08		2	A0	401
196808	+47 03153	IDS20346N4736	B	20 36 15	+47 46.1	1	9.46		0.84		-0.16		1		401
		Ross 763		20 36 16	−06 38.1	2	12.27	.005	0.57	.002	-0.15	.030	5		1696,3062
196724	+20 04658	HR 7891		20 36 17	+21 01.5	6	4.82	.015	-0.03	.017	-0.09	.025	15	A0 V	15,263,1063,1363,3023,8015
235300	+53 02447	IDS20349N5330	AB	20 36 17	+53 40.2	1	8.71		0.61		0.03		2	G5	1566
	+12 04410			20 36 19	+12 23.7	1	9.63		0.37				1	B8	1242
196740	+23 04084	HR 7894		20 36 21	+23 56.4	3	5.04	.009	-0.14	.004	-0.55	.015	10	B5 IV	154,1363,3016
196725	+12 04411	HR 7892		20 36 22	+13 08.3	3	5.69	.034	1.53	.005	1.70	.023	8	K3 Ib	1355,8032,8100
196790	+39 04260			20 36 22	+39 22.2	2	8.06	.015	0.57	.010	0.15	.005	5	F8 IV	247,1775
196810	+39 04261			20 36 22	+39 30.9	1	8.85		1.62				2	M1	369
196809	+44 03524			20 36 22	+44 24.9	1	8.75		-0.04		-0.30		2	A2	401
196970	+63 01640			20 36 22	+64 11.1	1	7.66		1.64		1.95		2	K5	1502
196753	+23 04085	HR 7895	⋆	20 36 24	+23 30.2	2	5.91	.002	0.99	.004	0.39		6	K0 II-III +	71,1501
		G 230 - 44		20 36 24	+51 33.4	1	11.08		0.75		0.33		2	K1	1658
196832	+46 02993			20 36 27	+46 42.6	1	7.97		-0.06		-0.46		2	B9	401
196711	+0 04561			20 36 29	+01 01.3	2	7.75	.005	1.03	.005	0.80	.005	4	K0	116,118
196662	−15 05743	HR 7889	⋆ AB	20 36 29	−15 07.9	2	5.23	.010	-0.13	.010			8	B7 III	176,2007
196819	+41 03836			20 36 30	+41 53.8	2	7.51	.009	1.87	.032	1.99	.014	5	K3 II	206,247
196833	+43 03672			20 36 32	+44 09.4	2	6.65	.028	-0.13	.005	-0.51		4	B7 V	401,1118
196772	+25 04310			20 36 35	+25 22.4	1	7.71		0.17		0.10		3	A0	833
196600	−41 14155			20 36 36	−41 09.0	1	8.84		0.29		0.02		4	A9 V	158
196712	−2 05328	HR 7890		20 36 37	−02 35.4	2	6.21		-0.10	.000	-0.42	.024	9	B7 IIIne	1079,1088
		BSD 112 # 1391		20 36 39	+00 43.7	1	10.81		0.48		-0.03		1	F3	116
196865	+47 03154	IDS20351N4743	A	20 36 40	+47 53.6	1	6.58		0.93		0.63		2	G5	3016
196679	−20 05995			20 36 40	−19 50.9	1	9.34		0.21		-0.50		12	F2 V	1732
196755	+9 04600	HR 7896	⋆ A	20 36 42	+09 54.5	9	5.07	.017	0.70	.013	0.23	.012	25	G5 IV	3,15,1003,1084,1256*
196755	+9 04600	IDS20343N0944	B	20 36 42	+09 54.5	1	8.50		0.92		0.68		3	K2 IV	1084
	+14 04375			20 36 42	+14 23.9	1	9.89		0.69		0.30		1	G0	1697
	−0 04063			20 36 42	−00 04.2	2	8.94	.020	1.35	.025	1.40		2	K2	116,118
		BSD 112 # 1392		20 36 45	+00 12.6	1	10.77		0.38		0.02		1	F0	116
		BSD 112 # 1395		20 36 45	+00 26.7	1	11.37		0.46		-0.03		1	F0	116
196775	+15 04220	HR 7899	⋆ A	20 36 45	+15 39.7	1	5.98		-0.16		-0.71		2	B3 V	1733
		POSS 107 # 7		20 36 45	+59 57.5	1	19.07		1.38				4		1739
340676	+26 03943			20 36 47	+26 31.4	1	8.87		0.68		0.18		3	G8 V	3026
196850	+38 04172	IDS20349N3818	A	20 36 47	+38 27.7	2	6.76	.000	0.61	.007	0.09	.025	5	G2 V	247,3026
		POSS 107 # 4		20 36 48	+60 11.3	1	16.26		1.03				3		1739
		VES 271		20 36 50	+46 35.0	1	12.14		2.00		2.38		1		8056
196758	−0 04064	HR 7897	⋆ A	20 36 51	+00 18.6	2	5.16	.014	1.06	.000	0.93	.008	7	K1 III	1509,3016
196939	+51 02914			20 36 54	+52 12.1	1	7.92		0.06		0.04		2	A0	1566
196794	+9 04602	G 144 - 22	⋆	20 36 56	+09 53.9	1	8.56		0.91		0.62		2	K2 V	3026
196852	+29 04121	HR 7904		20 36 56	+30 09.5	1	5.67		1.09		0.99		2	K2 III	1355
196821	+21 04305	HR 7903		20 36 57	+21 38.4	2	6.09		-0.07	.004	-0.11	.018	5	A0 III	1063,3016
196650	−43 14124			20 36 58	−43 14.5	1	8.02		1.13		1.06		5	K0 III	1673
		BPM 13537		20 37 00	−73 27.	1	15.52		-0.16		-1.09		2		3065
196717	−29 17235	R Mic		20 37 01	−28 58.1	2	10.14	.145	1.34	.195	1.02	.275	2	M4e	817,975
196227	−76 01436			20 37 04	−76 43.5	2	7.67	.005	0.61	.009	0.17		6	G2 V	278,2012
196795	+4 04510	G 24 - 23	⋆ AB	20 37 05	+04 47.6	6	7.88	.015	1.23	.015	1.14	.026	17	K4 V +K8 V	22,333,1003,1197,1619*
		CS 22955 # 176		20 37 07	−24 12.2	1	15.18		0.03		-0.04		1		1736
196866	+25 04312			20 37 12	+25 53.5	2	6.96	.022	1.29	.009	1.34	.004	6	K2 III	979,3077
196777	−18 05738	HR 7900		20 37 12	−18 19.0	4	5.16	.019	1.64	.007	1.99	.015	16	M1 III	1024,2007,3055,6002
196737	−33 15119	HR 7893		20 37 12	−33 36.6	2	5.48	.005	1.12	.002	1.08		6	K1 III	58,2009
196823	+0 04563			20 37 13	+00 53.1	2	8.76	.014	0.62	.009	0.16	.005	4	G0	116,118
196761	−24 16193	HR 7898		20 37 13	−23 57.5	2	6.36	.014	0.72	.004	0.22		12	G8 V	2018,2024,2030,3077
		CS 22940 # 78		20 37 13	−59 56.7	1	13.62		0.47		-0.12		1		1736
196867	+15 04222	HR 7906	⋆ A	20 37 19	+15 44.1	16	3.77	.004	-0.06	.005	-0.20	.012	184	B9 V	1,3,15,26,30,247,1006*

HD	DM	Other Id	N Rem	α₁₉₅₀	δ₁₉₅₀	S	V	σ_V	B–V	σ_B-V	U–B	σ_U-B	n	Spectrum	References
196882	+21 04307	IDS20351N2122	A	20 37 21	+21 32.6	1	8.66		1.32		1.29		3	K3 III	3026
196882	+21 04307	IDS20351N2122	AB	20 37 21	+21 32.6	1	8.49		0.94		0.60		2	K3 III	1003
196882	+21 04307	IDS20351N2122	B	20 37 21	+21 32.6	1	10.53		0.04		-0.14		3		3026
196519	-67 03754	HR 7881		20 37 23	-66 56.3	4	5.15	.005	-0.05	.005	-0.27	.005	14	B9 III	15,1075,2016,2038
196800	-24 16195			20 37 25	-24 17.5	3	7.21	.004	0.60	.004	0.14	.007	8	G1/2 V	219,285,1311
	-30 18090			20 37 27	-30 13.8	1	10.10		1.10		0.95		2	K3 V	3072
196885	+10 04351	HR 7907	⋆A	20 37 28	+11 04.2	2	6.40	.028	0.53	.009	0.04	.019	4	F6 V	254,3053
	-0 04065			20 37 29	+00 04.5	1	10.18		0.62		0.00		1		116
196869	+0 04565	IDS20350N0057	AB	20 37 30	+01 07.7	2	9.17	.000	0.12	.029	0.11	.010	3	A0	116,118
		CS 22880 # 3		20 37 31	-21 42.0	1	14.21		0.39		-0.14		1		1736
		CS 22940 # 77		20 37 31	-60 19.3	2	14.12	.010	0.54	.012	0.01	.003	3		1580,1736
		AE Aqr		20 37 34	-01 03.0	1	11.38		0.84		0.03		1	K5 IV-V	1471
		CS 22955 # 163		20 37 35	-25 51.2	1	13.11		0.39		0.08		1		1736
196815	-27 14959	HR 7902		20 37 36	-26 49.4	1	6.51		0.57				4	F7 V	2009
		G 24 - 25		20 37 42	+00 22.7	2	10.57	.017	0.60	.005	-0.11	.047	5		1620,1658
197036	+45 03233	HR 7912		20 37 42	+45 29.4	3	6.58	.011	-0.06	.004	-0.56	.005	7	B5 IV	154,401,1118
197101	+55 02444	HR 7916		20 37 42	+55 49.7	1	6.45		0.36		0.11		2	F2 Vn	1733
		IRC +40 439	AB	20 37 43	+39 01.5	1	14.58		4.25				2		8019
197018	+40 04266	HR 7911	⋆AB	20 37 43	+40 24.1	2	6.06		-0.16	.024	-0.55		3	B6 IIIp Mn	1079,1118
196857	-16 05663	HR 7905		20 37 44	-16 18.2	2	5.80	.002	1.00	.002	0.74		6	K0 III	58,2007
		LS II +35 078		20 37 45	+35 15.6	1	11.33		0.62		-0.20		5		994
197037	+41 03845			20 37 46	+42 04.4	1	6.82		0.50				2	F7 V	1118
		LP 755 - 45		20 37 46	-09 29.3	1	12.17		0.48		-0.18		2		1696
197054	+34 04111			20 37 58	+35 12.7	1	6.89		0.94		0.67		2	K0	1625
196892	-19 05889			20 37 58	-18 57.9	5	8.23	.015	0.49	.007	-0.13	.008	10	G2wF5	158,742,1003,1594,3037
197637	+78 00716			20 38 02	+79 15.3	2	6.94	.000	-0.13	.005	-0.76	.010	5	B3	399,1733
197433	+75 00752	VW Cep	⋆AB	20 38 03	+75 25.0	1	7.39		0.85		0.44		3	K0 V	22
196829	-42 15034	AV Mic		20 38 04	-42 18.7	3	6.30	.000	1.58	.019	1.72		12	M4 III	1075,2012,3040
197020	+25 04323			20 38 05	+25 51.7	1	8.68		0.53		0.07		3	G0 V	3026
		VES 272		20 38 06	+39 44.8	1	13.57		2.18		1.54		1		8056
197118	+46 03001	IDS20365N4659	AB	20 38 07	+47 10.0	1	7.93		-0.06		-0.24		2	A0	401
		BSD 112 # 671		20 38 09	+00 35.6	1	11.39		0.45		-0.01		1	F3	116
197038	+25 04324			20 38 09	+25 36.7	1	8.21		0.00		-0.34		2	A2	105
		G 144 - 25		20 38 11	+15 18.8	2	13.41	.005	1.71	.010	1.27	.019	3		203,3078
		Steph 1806		20 38 13	-10 17.5	1	11.94		1.51		1.17		2	M0e	1746
197039	+15 04227			20 38 15	+15 27.9	1	6.74		0.44		0.03		2	F7 V	3026
		VES 273		20 38 15	+44 39.6	1	11.71		1.76		2.09		1		8056
197139	+42 03818	HR 7919		20 38 18	+43 16.9	1	5.95		1.19				2	K2 III	1118
196917	-32 16130	HR 7908		20 38 18	-31 46.6	4	5.75	.004	1.54	.005	1.85		13	M1 III	15,1075,2006,3005
		BSD 112 # 683		20 38 20	+01 06.5	1	10.80		0.41		-0.02		1	A8	116
		BSD 40 # 2464		20 38 20	+46 48.3	1	11.82		0.35		0.19		2	B9	117
	+47 03162	LS III +47 010		20 38 24	+47 55.7	1	9.80		0.64		-0.33		2	B1 Ib-II	1012
196947	-26 15192	HR 7910		20 38 25	-26 10.7	1	6.28		1.22				4	K2 III	2009
197076	+19 04484	HR 7914	⋆A	20 38 29	+19 45.2	6	6.44	.010	0.61	.008	0.08	.020	17	G5 V	908,1067,1355,1758,1774,3077
197076	+19 04484	G 144 - 26	⋆B	20 38 29	+19 45.2	3	11.88	.012	1.54	.044	1.25	.043	9		940,1774,7009
		CS 22940 # 70		20 38 30	-61 51.5	1	14.87		0.48		0.17		1		1736
197120	+29 04131	HR 7917	⋆AB	20 38 32	+29 37.6	2	6.02	.009	0.15	.009	0.12	.019	4	A2 V	1063,1733
197161	+44 03538			20 38 33	+44 35.6	1	8.25		-0.04		-0.19		2	A0	401
197119	+30 04134			20 38 34	+31 10.8	1	7.88		0.08		0.06		2	A2	401
196877	-53 09928			20 38 37	-52 51.8	3	8.83	.003	1.32	.006	1.13		15	K5 V	1494,2012,3078
196983	-34 14555			20 38 42	-34 04.0	2	9.07	.005	1.18	.000			8	K1 III	2013,2033
197344	+64 01457			20 38 43	+64 58.0	1	8.58				0.11		2	B8	555
		HA 112 # 595		20 38 44	+00 05.8	4	11.35	.006	1.60	.004	2.00	.019	71	K8	281,989,1729,1764
196982	-32 16135	AT Mic	⋆B	20 38 44	-32 36.6	1	10.24		1.58		0.89		4	Mpe	158
		BSD 40 # 2472		20 38 45	+46 23.4	1	11.30		0.38		0.05		2	B8	117
196998	-22 05504			20 38 45	-22 29.6	3	9.71	.019	1.11	.024	0.89	.013.	7	K5 V	1619,3072,7009
		CS 22955 # 165		20 38 45	-25 28.9	1	14.89		0.43		-0.17		1		1736
	-0 04067			20 38 48	-00 38.6	1	9.66		1.58		2.00		1	K5	116
		VES 235		20 38 49	+37 04.1	1	12.44		2.26				1		426
		BSD 40 # 2476		20 38 51	+46 35.7	1	10.18		0.86		-0.11		2	B5	117
197176	+34 04114			20 38 52	+34 51.7	1	7.23		0.45				1	F2	6007
		CS 22940 # 63		20 38 52	-62 11.7	1	14.62		0.39		0.08		1		1736
	+0 04571			20 38 53	+00 52.2	1	9.81		1.01		0.81		1	K0	116
		POSS 107 # 2		20 38 53	+59 22.5	1	15.51		1.44				3		1739
	+48 03185			20 38 54	+48 44.5	1	9.24		0.59				2	G0	1724
		G 230 - 45		20 38 54	+54 02.3	1	11.43		0.80		0.39		2	K2	1658
197122	+11 04352			20 38 55	+12 18.8	1	8.95		0.32				1	A0	1242
197121	+14 04393	HR 7918		20 38 56	+14 24.2	1	5.99		1.24				2	K2	71
		CS 22940 # 71		20 38 56	-61 08.8	1	14.26		0.45		0.11		1		1736
197105	-0 04068			20 38 59	-00 11.4	2	8.70	.020	0.40	.015	0.15	.020	2	A5	116,118
		VES 274		20 39 00	+46 16.8	1	11.14		0.38		0.27		1		8056
	-0 04069			20 39 01	+00 05.0	3	9.85	.001	0.69	.007	0.13	.007	24	G3	281,989,1729
	-0 04070			20 39 01	+00 09.7	2	10.35	.011	0.50	.000	-0.03	.011	11	F7	116,281
197177	+31 04181	HR 7921	⋆AB	20 39 01	+32 07.7	1	5.51		0.88				2	G8 IIb	71
	-6 05559			20 39 02	-05 41.0	1	10.55		1.30		1.22		3	M1 V:	7009
197204	+40 04276			20 39 04	+41 01.4	3	7.01	.004	-0.06	.010	-0.20	.005	6	B9	401,1118,1733
		HA 112 # 654		20 39 07	+00 03.2	1	10.77		0.45		-0.06		2	F3	281
		LS II +34 023		20 39 08	+34 42.1	1	12.18		0.50		-0.12		5		994
197226	+38 04187	HR 7922	⋆A	20 39 08	+38 54.2	5	6.53	.008	-0.12	.011	-0.48	.021	9	B6 III	105,351,1063,1079,1118
		CS 22955 # 174		20 39 08	-23 60.0	1	14.38		0.50		-0.04		1		1736

Table 1   1119

HD	DM	Other Id	N	Rem	α_{1950}	δ_{1950}	S	V	σ_V	B–V	σ_{B-V}	U–B	σ_{U-B}	n	Spectrum	References
197031	−41 14178				20 39 11	−41 24.4	1	10.21		0.58				2	G5 V	1594
197308	+54 02396				20 39 12	+54 55.7	1	7.32		−0.02		−0.29		2	A0	401
197207	+29 04135				20 39 13	+30 00.8	1	8.21		0.67		0.17		3	G5 V	3026
197373	+59 02272	HR 7925			20 39 14	+60 19.4	2	6.00	.019	0.43	.014	−0.05	.019	4	F5 V	254,3037
	−0 04071				20 39 15	+01 15.7	2	10.31	.035	0.61	.020	0.13		2		116,118
197169	+13 04487				20 39 16	+13 37.8	2	6.81	.037	0.18	.005	0.17	.014	4	A2	8071,8080
		BSD 40 # 1258			20 39 17	+45 42.7	1	11.45		0.30		−0.26		2	B9	117
		Steph 1809			20 39 18	+00 17.8	1	11.73		1.54		1.89		1	M0	1746
	+45 03239				20 39 18	+45 36.2	1	8.71		0.15		−0.30		3	B6 III	117
		HA 112 # 1181			20 39 20	+00 14.8	1	12.03		0.71		0.25		10	F7	281
		POSS 107 # 3			20 39 20	+60 31.8	1	15.97		1.45				2		1739
		POSS 107 # 6			20 39 21	+58 54.8	1	18.20		1.62				3		1739
		CS 22879 # 51			20 39 21	−39 29.3	1	13.88	.010	0.39	.007	−0.21	.009	3		1580,1736
		HA 112 # 704			20 39 28	+00 08.4	2	11.46	.004	1.53	.008	1.73	.011	28		281,1764
		LS III +46 019			20 39 31	+46 24.1	2	11.38	.000	0.61	.000	−0.29	.019	6		405,483
		HA 112 # 712			20 39 32	+00 05.1	1	11.66		0.41		0.03		2	F3	281
	+0 04573				20 39 32	+00 55.1	1	10.31		1.70		1.99		1		116
	+44 03540				20 39 32	+44 56.7	1	9.67		0.14		0.14		2	A2	117
		HA 112 # 1212			20 39 34	+00 12.3	1	13.07		0.62		0.00		18		281
		CS 22940 # 82			20 39 34	−58 52.7	2	14.27	.010	0.65	.015	0.08	.001	3		1580,1736
197292	+35 04218				20 39 35	+35 44.0	1	7.47		1.32		1.53		2	K2	1601
		L 116 - 79			20 39 36	−68 16.	2	13.46	.120	0.07	.073	−0.79	.013	3		1411,3065
197228	+20 04680	IDS20374N2022	AB		20 39 37	+20 32.1	1	7.40		0.33		0.12		2	A3	938
197293	+35 04217				20 39 37	+35 33.1	1	8.05		−0.10		−0.45		2	B9	401
197093	−40 13994	HR 7915			20 39 37	−39 44.3	3	6.28	.004	1.07	.005			11	K1 III	15,1075,2007
197263	+27 03832				20 39 38	+28 05.4	1	8.20		0.58		0.09		3	G0 V	3026
197249	+17 04382	HR 7923, LU Del			20 39 40	+17 20.5	2	6.20	.017	0.94	.002	0.57		7	G8 III	71,897
		PASP82,889 # 7			20 39 41	+18 54.8	1	12.85		1.41		1.68		1		120
		HA 112 # 223			20 39 41	−00 01.8	4	11.43	.005	0.46	.002	0.01	.006	72	A5	281,989,1729,1764
		LP 815 - 51			20 39 41	−20 15.3	1	12.34		−0.07		−0.83		1	DA	3062
		EG 141			20 39 42	−20 16.	1	12.40		−0.07				1	DAw	1705
		L 711 - 10			20 39 42	−20 16.	1	12.33		−0.08		−0.85		1	DA	1727
197310	+35 04219				20 39 43	+35 21.9	2	7.42	.004	0.64	.012	0.12		2	G0	934,6007
197345	+44 03541	HR 7924, α Cyg	⋆ A		20 39 44	+45 06.1	11	1.25	.015	0.09	.007	−0.23	.008	52	A2 Iae	1,15,1077,1118,1196*
		HA 112 # 233			20 39 46	+00 00.6	1	11.86		0.51		0.04		6	F7	281
		PASP82,889 # 6			20 39 46	+18 51.9	1	11.53		0.54		0.01		1		120
		MtW 112 # 159			20 39 47	+00 22.1	1	12.54		0.43		−0.01		5		397
197232	−0 04072				20 39 50	+00 15.9	5	9.09	.008	0.42	.007	0.02	.004	44	F5	281,397,989,1729,6004
		MtW 112 # 178			20 39 51	+00 23.4	1	14.73		0.73		0.09		5		397
		HA 112 # 250			20 39 53	−00 03.1	2	12.09	.003	0.54	.004	−0.01	.013	24	G0	281,1764
		HA 112 # 1253			20 39 54	+00 13.8	1	11.46		0.42		0.09		14	F0	281
		MtW 112 # 195			20 39 54	+00 23.2	1	16.58		0.84		0.15		5		397
197406	+52 02777	V1696 Cyg			20 39 54	+52 24.5	2	10.30	.000	0.56	.010	−0.44	.010	5	WN7	1012,1359
		PASP82,889 # 5			20 39 56	+18 51.9	1	11.39		1.25		1.18		1		120
197391	+44 03542				20 39 56	+44 28.0	1	9.09		0.03		0.02		2	B8 V	117
	−20 06008				20 39 56	−20 11.5	2	9.85	.005	0.87	.000	0.29		4	G0	742,1594
196419	−80 00977				20 39 56	−80 30.4	1	8.47		0.34		0.02		24	F0 V	978
		HA 112 # 1258			20 39 57	+00 13.6	1	11.48		0.42		0.10		14	F3	281
197375	+42 03833				20 39 57	+43 15.8	2	7.86	.000	0.03	.005			4	A0	1118,1196
197374	+43 03687	IDS20381N4329	AB		20 39 57	+43 38.6	1	8.36		−0.09		−0.20		2	B9 V	117
197376	+41 03855				20 39 59	+41 49.7	1	7.96		0.60		0.28		1	F7 IV-V	247
		G 144 - 28			20 40 02	+11 06.9	3	12.92	.013	0.63	.011	−0.13	.018	5		316,1658,1696
197434	+53 02466				20 40 02	+54 01.8	2	7.99	.020	0.12	.015	−0.26	.005	4	A0	379,1733
	−0 04073				20 40 02	−00 03.5	5	9.91	.007	1.21	.005	1.29	.009	59	G8	116,281,989,1729,6004
		HR Del			20 40 04	+18 58.7	1	11.86		0.18		−1.00		377		866
	−19 05899				20 40 04	−19 05.2	2	10.79	.010	1.43	.020			2	M1:	1705,1746
197311	+18 04586				20 40 07	+19 13.0	1	7.85		1.03		0.79		29	K0	120
197392	+41 03856	HR 7926			20 40 08	+41 32.2	2	5.67		−0.12	.005	−0.46		4	B8 II-III	1079,1118
	+45 03242				20 40 09	+45 49.8	1	8.98		0.04		−0.56		2	B5 V	117
		PASP82,889 # 4			20 40 10	+18 52.4	1	12.28		1.64		1.94		1		120
		CS 22880 # 37			20 40 10	−18 24.9	1	14.34		0.46		−0.16		1		1736
		PASP82,889 # 1			20 40 11	+19 00.2	1	13.83		0.99		1.00		1		120
		HA 112 # 805			20 40 12	+00 05.3	4	12.09	.003	0.15	.004	0.15	.008	74	A0	281,989,1729,1764
		VES 275			20 40 12	+45 40.8	2	11.29	.026	0.64	.009	−0.16	.031	6		117,8056
	+56 02471				20 40 13	+57 14.3	1	10.27		1.36		1.30		1	K7	3003
197214	−29 17282				20 40 14	−29 36.1	1	6.94		0.67				4	G3/5 V	1075
		BSD 112 # 797			20 40 15	+00 15.1	1	10.59		1.16		1.07		1	gG8	116
		BSD 112 # 791			20 40 15	+00 16.0	1	12.10		0.11		0.09		1	A0	116
		VES 325			20 40 15	+48 16.7	1	12.64		0.86		0.19		2		8106
		HA 112 # 810			20 40 16	+00 06.3	1	10.58		1.16		1.10		4	G8	281
		PASP82,889 # 3			20 40 16	+18 56.2	1	13.47		0.71		0.45		1		120
		PASP82,889 # 2			20 40 17	+18 58.4	1	14.21		1.16		1.39		1		120
		BSD 40 # 1322			20 40 17	+45 45.2	1	12.25		0.36		0.24		3	B9	117
	+56 02472	G 230 - 47	⋆ A		20 40 17	+57 19.6	1	10.12		0.74		0.23		1	K0	1658
197393	+35 04224				20 40 18	+35 57.9	1	8.75		0.43		−0.06		2	F8	634
197237	−31 17795				20 40 18	−31 43.7	1	9.00		0.98		0.85		1	G8 IV	1738
196520	−80 00978				20 40 18	−80 02.2	2	7.60	.004	1.07	.001	0.83		24	K0 III	978,1642
		HA 112 # 822			20 40 21	+00 04.2	4	11.55	.004	1.03	.003	0.88	.007	57	G7	281,989,1729,1764
	+42 03835	LS III +43 010			20 40 21	+43 00.3	2	9.20	.000	0.91	.000	−0.19	.000	4	O9 III:	1011,1012
		CS 22879 # 66			20 40 23	−41 35.7	2	13.62	.010	0.41	.008	−0.20	.009	3		1580,1736

HD	DM	Other Id	N	Rem	α_{1950}	δ_{1950}	S	V	σ_V	B–V	σ_{B-V}	U–B	σ_{U-B}	n	Spectrum	References
197157	−52 11752	HR 7920			20 40 23	−52 06.1	6	4.50	.007	0.28	.004	0.08	.012	21	A7 III/IV	15,208,1075,1075,2012,3023
197419	+34 04127	HR 7927, V568 Cyg			20 40 25	+35 16.6	2	6.66	.020	-0.16	.000	-0.69	.010	5	B2 IV-Ve	154,401
		BSD 112 # 1670			20 40 27	+00 49.1	1	11.87		0.09		0.11		1	B9	116
		BSD 40 # 1328			20 40 28	+46 09.6	1	9.93		0.52		0.42		2	A0	117
197051	−66 03501	HR 7913			20 40 29	−66 23.1	7	3.42	.009	0.16	.005	0.10	.014	22	A7 III	15,1020,1034,1075*
		L 495 - 82			20 40 30	−39 13.	1	13.80		0.19		-0.57		5		3062
197510	+53 02471				20 40 31	+54 14.4	1	7.85		0.25		0.10		2	F0	379
	+46 03014				20 40 32	+47 00.0	2	8.47	.029	0.51	.005	0.02		3	F7 Vp	247,1724
		VES 326		V	20 40 32	+47 55.5	1	13.05		1.92		1.15		1		8106
197330	−11 05402				20 40 33	−10 55.4	1	8.16		0.98		0.71		4	K0	1731
	+46 03015				20 40 34	+46 26.8	1	10.36		0.39		0.00		2		117
197512	+49 03352	LS III +49 006			20 40 36	+49 33.3	2	8.56	.005	0.06	.005	-0.69	.005	4	B1 V	399,963
		LS IV +10 009			20 40 38	+10 23.3	2	11.98	.005	-0.27	.005	-1.20	.005	3	sdO	405,1650
	−0 04074				20 40 39	+00 15.2	4	9.98	.016	0.63	.006	0.09	.012	11	G0	116,118,281,6004
197488	+45 03245	IDS20390N4528	A		20 40 39	+45 38.7	2	7.61	.020	0.59	.015	0.09		7	G2 V	1724,3026
197511	+49 03353	HR 7929	★	AB	20 40 40	+50 09.6	1	5.38		-0.10		-0.64		3	B2 V	154
197460	+35 04229	LS II +36 083			20 40 45	+36 12.0	2	8.13	.004	0.35	.004	-0.56	.004	7	B0.5Ib	994,1012
		VES 276			20 40 46	+45 43.2	1	11.96		0.91		0.26		1		8056
		VES 327			20 40 46	+49 46.7	1	10.75		0.23		0.02		2		8106
	+45 03246	LS III +45 016			20 40 50	+45 43.3	1	9.67		0.17		-0.64		2	B9 V	1012
		G 187 - 2			20 40 52	+26 37.8	1	10.80		1.08		0.79		3	K3	7010
	+33 03998	LS II +33 036	A		20 40 52	+33 55.6	1	10.52		1.49		1.54		3	F4 II	247
	+33 03998	LS II +33 036	AB		20 40 52	+33 55.6	2	10.30	.000	0.45	.005	0.15	.010	7		247,994
	+46 03018				20 40 57	+46 57.0	1	9.99		0.19		-0.24		2	A0	117
		VES 236			20 40 58	+36 56.3	2	11.47	.005	0.56	.015	-0.20	.015	2		426,8056
197409	−0 04076				20 40 59	+00 17.3	5	8.79	.017	0.14	.012	0.11	.015	24	A0	116,118,281,1371,6004
	+33 03999				20 40 59	+33 42.6	1	9.35		1.87				2	M2	369
197339	−32 16161				20 40 59	−32 06.6	1	7.39		1.29		1.41		5	K2 III	1673
		POSS 107 # 1			20 40 60	+60 57.5	1	15.10		1.46				1		1739
		CS 22880 # 58			20 41 02	−21 47.4	1	14.54		0.37		-0.17		1		1736
		CS 22880 # 52			20 41 03	−20 59.0	1	15.23		-0.07		-0.94		1		1736
	−30 18140				20 41 03	−30 10.7	4	9.94	.018	0.42	.009	-0.21	.014	13	F8	742,1594,1696,2033
		BSD 40 # 2558			20 41 04	+46 22.1	1	11.53		0.33		0.10		2	B8	117
	−30 18142				20 41 05	−30 00.2	1	9.74		1.60		2.06		1		742
		G 24 - 27			20 41 06	−00 21.8	1	11.47		1.48		1.24		1		1620
197461	+14 04403	HR 7928, δ Del			20 41 07	+14 53.6	6	4.43	.016	0.31	.018	0.10	.009	14	A7 IIIp δ Del	15,597,1363,3023,8015,8052
	−21 05811				20 41 07	−21 31.5	3	9.87	.020	1.20	.040	1.08	.036	11	K7 V	1619,3072,7008
197531	+35 04231				20 41 08	+35 35.0	1	8.53		-0.02		-0.16		2	A0	634
197559	+42 03839	IDS20394N4249	AB		20 41 08	+42 59.8	3	8.45	.010	-0.05	.004	-0.26		6		401,1118,1196
		BSD 40 # 2560			20 41 09	+46 35.2	1	12.09		0.56		0.44		3	B2	117
		BSD 40 # 2564			20 41 09	+46 46.2	1	11.59		0.36		0.08		2	B8	117
197489	+25 04347	LS II +25 023	★	A	20 41 11	+25 38.6	2	7.05	.005	0.34	.000	0.25	.045	4	A7 II	105,8100
		MARK A2			20 41 11	−10 56.4	1	14.54		0.67		0.10		10		1764
197549	+33 04003				20 41 15	+33 39.0	1	7.48		-0.05		-0.30		2	A0	401
		Mark 509 # 3			20 41 15	−10 52.5	1	14.02		0.98		0.61		4		899,1687
		MARK A1			20 41 15	−10 58.1	1	15.91		0.61		-0.01		10		1764
		MARK A			20 41 16	−10 58.6	1	13.26		-0.24		-1.16		22		1764
		Mark 509 # 2			20 41 17	−10 54.0	1	13.12		0.55		0.00		4		899,1687
		BSD 40 # 1369			20 41 18	+45 13.1	1	11.48		0.44		-0.18		3	B3	117
		MARK A3			20 41 19	−10 56.5	1	14.82		0.94		0.65		10		1764
197560	+35 04232				20 41 20	+35 33.6	1	7.96		-0.08		-0.35		2	Ap Si	634
		Mark 509 # 8			20 41 20	−10 48.6	1	13.30		0.54		-0.04		4		1687
		Mark 509 # 5			20 41 20	−10 51.3	1	11.38		0.92		0.60		4		899,1687
		BSD 40 # 47			20 41 21	+43 37.6	1	11.60		0.42		0.08		3	B5	117
	+45 03247				20 41 22	+45 39.4	1	9.39		0.05		-0.09		3	B9 V	117
197561	+33 04004				20 41 24	+33 41.3	1	7.24		-0.12		-0.43		2	A0	401
		AJ88,650 # 7			20 41 24	+35 22.4	1	13.46		0.63		0.50		4		994
		LS III +47 012			20 41 24	+47 22.2	1	10.74		0.98		-0.12		2		483
		AJ88,650 # 8			20 41 25	+35 21.6	1	12.82		0.90		0.58		3		994
		Mark 509 # 9			20 41 26	−10 50.1	1	14.43		0.59		-0.03		2		1687
		AJ88,650 # 6			20 41 27	+35 22.0	1	11.17		1.00		0.77		4		994
197572	+35 04234	HR 7932, X Cyg			20 41 27	+35 24.4	3	5.88	.010	0.85	.017	0.55	.023	3	F7 Ib	934,1772,6007
		Mark 509 # 7			20 41 27	−10 50.3	1	12.55		0.58		-0.05		4		1687
197751	+62 01850				20 41 28	+63 10.2	1	7.66		0.07		-0.30		2	B8	555
		Mark 509 # 4			20 41 28	−10 52.8	1	12.41		0.57		0.09		5		899,1687
		BSD 40 # 1385			20 41 29	+45 57.8	1	9.80		0.39		0.30		2	A0	117
197192	−66 03502				20 41 30	−66 21.0	4	9.40	.047	0.92	.008	0.49	.010	14	K2/3 V	119,158,1097,3008
197621	+44 03552	IDS20398N4434	B		20 41 31	+44 45.5	1	9.53		0.04		-0.01		2		117
		Mark 509 # 1			20 41 32	−10 55.6	1	14.54		0.67		0.17		3		899,1687
		AJ88,650 # 14			20 41 33	+35 26.1	1	13.92		1.42		1.23		2		994
		BSD 40 # 2579			20 41 33	+46 37.5	1	10.79		0.36		-0.11		3	B9	117
		AJ88,650 # 4			20 41 34	+35 23.2	1	13.22		0.47		0.31		4		994
197621	+44 03553	IDS20398N4434	A		20 41 34	+44 45.2	1	8.63		-0.08		-0.35		2	B9.5V	401
197666	+51 02932				20 41 34	+52 06.7	1	7.77		0.55				2	G0	1724
197562	+23 04124				20 41 36	+23 36.6	2	6.93	.005	-0.04	.000	-0.25	.000	3	A0	1625,1716
		IRC +40 442	★	V	20 41 36	+43 01.0	2	12.47	.677	2.28	.248	0.40		3		8019,8056
		VES 278			20 41 36	+45 25.9	1	12.32		1.34		0.28		1		8056
197734	+60 02154	HR 7938			20 41 36	+60 25.2	1	6.15		0.01		0.04		2	A2 IV	401
197604	+34 04134				20 41 37	+34 53.2	1	9.17		1.44				1	R4	1238
	+35 04237				20 41 37	+35 24.7	1	9.04		0.62		0.19		8	F8	994

Table 1   1121

HD	DM	Other Id	N	Rem	α₁₉₅₀	δ₁₉₅₀	S	V	σ_V	B–V	σ_B–V	U–B	σ_U–B	n	Spectrum	References
235333	+51 02933				20 41 38	+52 10.8	1	10.12		0.45		0.30		3	A0	1566
	+46 03022				20 41 39	+46 23.7	1	10.32		0.25		0.14		3	Be	117
		G 24 - 28			20 41 40	+04 26.9	1	12.59		1.06		0.93		1		333,1620
		KUV 20417+7604			20 41 40	+76 03.9	1	12.74		-0.11		-0.88		1	sdO	1708
		BSD 40 # 1395			20 41 42	+45 22.3	1	10.40		0.22		0.05		2	B9	117
		LS II +35 080			20 41 47	+35 06.3	1	11.33		0.54		0.20		5		994
197605	+26 03970				20 41 48	+27 16.4	2	8.49	.005	0.57	.005	0.19	.000	4	F5 II	253,8100
		BSD 40 # 1400			20 41 48	+45 19.3	1	11.49		0.27		0.23		2	A0	117
335007	+30 04155			AB	20 41 50	+31 07.8	1	9.95		1.21		1.20		1	K8	680
197678	+44 03556				20 41 51	+44 28.9	1	8.86		0.17		0.15		2	A0 IV	117
		BSD 40 # 1406			20 41 53	+45 54.9	1	12.34		0.49		0.22		3	B9	117
		G 231 - 13			20 41 55	+55 09.2	1	14.74		1.65				1		1759
		VES 280			20 41 56	+38 49.6	1	11.33		0.65		-0.24		1		8056
197770	+56 02477	HR 7940		★	20 41 58	+56 56.0	4	6.31	.005	0.34	.005	-0.47	.004	11	B2 III	15,154,555,1012
		G 144 - 33			20 41 59	+21 43.4	1	10.09		0.76		0.35		2	K2	1658
		BSD 40 # 525			20 42 00	+45 09.9	1	11.75		0.47		0.05		4	B8	117
	+46 03024				20 42 00	+46 57.3	1	10.03		0.25		0.15		3	A0	117
197679	+41 03868				20 42 01	+41 39.6	1	8.50		-0.05		-0.40		2		401
	+46 03025	LS III +46 020			20 42 03	+46 29.3	1	10.70		0.35		0.11		2		117
235338	+51 02937				20 42 03	+52 05.7	1	9.58		0.60				3		1724
197576	-0 04079				20 42 03	-00 17.0	2	8.67	.025	0.27	.010	0.11	.005	2	A5	116,118
197481	-31 17815	AU Mic		★ A	20 42 04	-31 31.1	3	8.77	.055	1.45	.016	1.05	.063	19	M1 Ve	1738,2012,3072
197481	-31 17815	IDS20390S3142		BC	20 42 04	-31 31.1	1	10.25		1.56		0.80		3		3072
		BSD 40 # 1410			20 42 05	+45 35.3	1	12.10		0.46		0.23		2	B7	117
	+46 03028				20 42 05	+46 48.6	1	10.15		0.41		0.27		2		117
	+48 03200				20 42 05	+49 17.3	1	9.19		1.76				2		369
352860	+19 04499	G 144 - 34			20 42 06	+19 34.8	3	10.32	.015	1.44	.021	1.20	.037	12	M2	196,1746,7009
197700	+41 03869				20 42 06	+42 02.3	2	6.99	.000	0.30	.000			4	F0	1118,1196
	-0 04081				20 42 07	+00 20.0	2	10.19	.000	0.51	.010	0.00	.005	2		116,118
	+47 03177				20 42 08	+47 20.5	1	8.86		0.86				2	F8	1724
197795	+54 02408				20 42 08	+55 06.6	1	6.97		-0.08		-0.25		2	B9	401
		BSD 40 # 1420			20 42 11	+45 52.1	1	11.73		0.49		0.45		3	B9	117
197484	-43 14174				20 42 12	-43 07.8	2	8.88	.010	0.64	.015	0.02		6	G3 V	1075,3077
196818	-80 00981			V	20 42 12	-80 19.1	2	8.19	.063	1.14	.027	0.72		2	K0 IIIp	1641,1642
235339	+50 03179				20 42 13	+51 01.2	1	8.68		0.38				2		1724
197540	-27 15014	HR 7931			20 42 13	-27 25.8	1	6.50		0.94				4	G8 III	2009
197701	+35 04240				20 42 15	+35 21.3	1	8.93		0.06		0.04		2	A2	634
197645	+13 04506				20 42 17	+13 22.4	1	7.52		0.05				1	A0	1242
		BSD 40 # 1422			20 42 17	+45 42.9	1	11.24		0.44		0.16		4	B7	117
	-63 04611				20 42 17	-63 34.2	1	10.36		0.47		-0.18		2	F8	1696
		BSD 40 # 1418			20 42 18	+45 18.3	1	11.65		0.45		0.27		2	A0	117
	-0 04083				20 42 20	-00 16.3	1	10.26		0.43		-0.05		1		116
197623	-0 04084				20 42 23	+00 06.6	3	7.55	.020	0.66	.007	0.21	.011	5	G5	116,118,3016
197702	+31 04204	LS II +31 030			20 42 24	+31 30.8	2	7.89	.014	0.18	.005	-0.68	.014	8	B1 III(n)	399,401
		LP 636 - 2			20 42 25	-01 51.7	1	14.08		0.63		-0.09		2		1773
		LP 636 - 3			20 42 26	-01 51.7	2	12.81	.005	0.51	.005	-0.22	.000	6		1696,1773
		VES 281			20 42 27	+47 13.0	1	11.54		2.03		2.17		1		8056
197911	+62 01854				20 42 28	+63 01.7	2	7.65	.010	0.05	.005	-0.77	.015	5	B2 V	401,555
197684	+11 04368	IDS20402N1157		AB	20 42 32	+12 07.8	1	6.65		0.22		0.07		1	A0	1476
197703	+19 04501				20 42 32	+20 18.5	1	6.97		0.25		0.10		1	A5	1648
197950	+66 01318	HR 7945			20 42 34	+66 28.5	2	5.59	.010	0.22	.005	0.06	.010	4	A8 V	985,1733
197416	-60 07424				20 42 34	-60 27.8	2	8.17	.010	0.49	.005	-0.03		5	F6 V	219,1075
		BSD 40 # 1444			20 42 35	+45 53.5	1	11.75		0.36		0.22		3	B9	117
	+12 04451				20 42 36	+12 44.6	1	9.32		0.34				1	A5	1242
		3HLF4 # 1			20 42 36	-45 11.	1	14.43		-0.01		-0.18		1	A0	208
	+34 04138				20 42 37	+34 38.1	1	10.29		1.86				2	M2	369
	+43 03701				20 42 37	+43 46.6	1	8.97		0.01		-0.04		3	B9 V	117
197683	+12 04452	IDS20403N1222		AB	20 42 38	+12 32.7	2	8.05	.009	0.45	.009	-0.06		5	F5 V +F6 V	1381,3032
197683	+12 04452	IDS20403N1222		C	20 42 38	+12 32.7	1	8.12		0.36		0.04		3		3032
197683	+12 04452	IDS20403N1222		E	20 42 38	+12 32.7	1	11.06		1.09		0.91		3		3032
		LP 816 - 58			20 42 38	-17 42.0	1	10.85		0.76		0.29		2		1696
	+0 04581				20 42 40	+01 08.5	1	9.13		0.44		-0.03		1	F8	116
		BSD 40 # 1453			20 42 41	+45 59.4	1	11.22		0.22		0.09		2	A0	117
		BSD 40 # 1447			20 42 42	+45 24.8	1	10.16		0.11		0.06		3	B9	117
197752	+24 04229	HR 7939			20 42 43	+25 05.4	7	4.92	.019	1.18	.010	1.19	.011	19	K2 III	15,263,1003,1080,1363*
239489	+55 02460				20 42 43	+55 38.9	1	8.52		1.15		1.13		2	K2	1502
	+0 04582				20 42 47	+00 28.9	1	10.06		1.03		0.70		1		116
		POSS 397 # 6			20 42 48	+24 49.8	1	15.43		1.44				2		1739
		3HLF4 # 2			20 42 48	-43 49.	1	12.64		0.14		0.15		2	A0	208
	+0 04583				20 42 49	+00 52.3	1	10.29		0.65		0.12		1		116
235346	+50 03183				20 42 51	+51 13.6	1	9.00		0.60				2		1724
197718	+6 04634				20 42 52	+06 51.4	1	8.36		0.12		0.04		3	A0	1733
197753	+17 04397	KP Del			20 42 53	+18 06.6	1	8.28		1.73				5	M1	897
		CS 22880 # 70			20 42 54	-21 47.1	1	12.65		-0.31		-1.10		1		1736
		LP 636 - 4			20 42 56	+02 00.9	1	12.00		1.14		1.03		1		1696
		CS 22880 # 86			20 42 57	-18 44.5	1	14.41		0.63		0.00		1		1736
197690	-10 05493				20 42 58	-10 18.3	1	8.58		0.44		0.00		4	F5	1731
		VES 237			20 42 59	+36 58.5	1	11.55		1.87				1		426
		BSD 40 # 1476			20 42 59	+45 44.2	1	10.90		0.53		0.26		3	B8	117
197739	-0 04087				20 43 01	+00 16.9	2	8.88	.005	1.61	.010	1.98	.010	2	K2	116,118

HD	DM	Other Id	N	Rem	α_{1950}	δ_{1950}	S	V	σ_V	B–V	σ_{B-V}	U–B	σ_{U-B}	n	Spectrum	References
197939	+55 02462	HR 7944			20 43 04	+56 18.4	2	5.87	.106	1.67	.004	1.86		3	M4 III	71,3001
197630	−39 13960	HR 7933			20 43 05	−39 22.9	1	5.49		−0.10				4	B8/9 V	2007
		VES 328			20 43 07	+47 45.8	1	11.62		0.52		−0.13		2		8106
197692	−25 15018	HR 7936			20 43 08	−25 27.1	8	4.14	.004	0.43	.003	0.00	.016	199	F5 V	3,15,26,1075,2012*
197649	−36 14396	HR 7935			20 43 08	−36 18.2	3	6.48	.004	0.41	.011			11	F3 IV	15,2032,2033
		CS 22880 # 74			20 43 10	−21 10.2	1	13.27		0.57		−0.15		1		1736
197812	+17 04401	HR 7941, U Del			20 43 11	+17 54.4	2	6.43	.032	1.59	.061	1.26		7	M5 II-III	71,897
		AJ89,1229 # 491			20 43 12	+09 34.	1	10.96		1.64				1		1532
		KUV 20432+7457			20 43 13	+74 57.1	1	14.46		0.61		−0.61		1		1708
	−0 04088				20 43 13	−00 10.6	1	10.74		0.62		0.08		1		116
197725	−22 05523	HR 7937			20 43 16	−21 41.8	2	5.92	.005	0.05	.018	0.10		5	A1 V	2007,3050
		G 25 - 1			20 43 17	+10 06.4	3	13.76	.004	0.86	.005	0.38	.005	5		316,1658,3062
		G 209 - 41			20 43 17	+44 18.7	1	10.73		1.57		1.17		3		1625
197776	−11 05417				20 43 21	−11 12.2	1	8.86		0.45		−0.04		4	F5	1731
	−23 16484				20 43 21	−23 15.3	1	10.76		1.30		1.28		2	K7 V	3072
	+45 03256				20 43 24	+45 34.3	1	9.02		−0.01		−0.20		3	B8 V	117
		3HLF4 # 4			20 43 24	−43 38.	1	14.23		0.19		0.09		2	A0	208
	+47 03179				20 43 25	+47 31.5	1	8.66		1.93				2	M0	369
197814	−0 04089				20 43 27	−00 31.4	2	6.74	.000	0.99	.040	0.63	.035	2	G8 III	116,118
	+43 03706	IDS20417N4316		AB	20 43 28	+43 26.6	1	9.92		0.13		0.12		2		117
		BSD 40 # 1508			20 43 29	+45 47.3	1	11.27		0.53		0.20		2	B9	117
		VES 282			20 43 30	+39 02.8	1	11.92		0.53		0.07		1		8056
	+43 03708				20 43 32	+44 12.8	1	9.50		0.14		0.08		2	A0 IV	117
		BSD 112 # 1001			20 43 33	+00 11.9	1	11.43		0.38		−0.05		1	A8	116
		G 210 - 33			20 43 33	+40 12.5	1	11.20		0.47		−0.17		3	sdF5	1658
		BSD 40 # 1514			20 43 35	+45 43.0	1	11.68		0.50		0.04		2	B5	117
197912	+30 04167	HR 7942	★	AB	20 43 36	+30 32.2	7	4.22	.011	1.05	.007	0.88	.007	15	K0 III	15,1080,1355,1363*
	+49 03366				20 43 36	+49 57.8	1	9.45		0.47				2	F5	1724
		VES 329			20 43 36	+50 36.2	1	12.12		2.09		1.56		1		8106
		VES 283			20 43 37	+45 33.1	2	11.48	.009	0.40	.000	0.11	.009	4		117,8056
197417	−72 02554	V343 Pav			20 43 38	−72 23.8	1	7.89		0.11		0.11		80	Ap CrEuSr	587
235350	+50 03189	LS III +51 005			20 43 39	+51 01.7	1	9.27		0.29		−0.61		2	B0.5IV	1012
		VES 239			20 43 40	+35 53.3	2	10.94	.000	1.90	.000	2.19	.000	2		426,8103
197961	+45 03258				20 43 40	+46 10.3	2	6.70	.019	−0.03	.005	0.00		4	A1 IV	401,1118
		BSD 40 # 588			20 43 44	+44 11.9	1	10.70		0.31		0.23		2	B9	117
		VES 238			20 43 45	+36 33.2	1	14.85		1.86				1		426
197325	−76 01446				20 43 46	−76 01.1	1	7.89		1.03		0.91		7	K0 III	1704
		BSD 40 # 1526			20 43 48	+45 33.6	1	9.78		0.08		0.08		4	B8	117
197913	+15 04251	IDS20416N1532	A		20 43 53	+15 43.4	1	7.55		0.76		0.34		3	K0	3026
197913	+15 04251	IDS20416N1532	B		20 43 53	+15 43.4	1	8.38		0.78		0.37		3		3026
	+45 03260	LS III +46 022			20 43 54	+46 10.0	3	9.05	.015	0.52	.010	−0.47	.025	6	O9 V	117,1011,1012
197858	−11 05420				20 43 54	−10 53.8	1	8.42		0.48		0.01		4	F8	1731
	+42 03854				20 43 58	+43 15.7	2	8.66	.000	−0.03	.005			4	A0	1118,1196
		CS 22879 # 103			20 43 58	−37 37.9	1	14.30		0.49		0.05		1		1736
		BSD 40 # 2675			20 44 04	+46 30.6	1	11.53		0.47		−0.02		2	B3	117
197788	−42 15087				20 44 04	−42 25.3	1	9.34		0.23		0.10		2	A9 V	208
		VES 284			20 44 07	+45 34.9	1	11.76		0.67		0.21		1		8056
198084	+57 02240	HR 7955	★	A	20 44 07	+57 24.0	5	4.52	.010	0.54	.005	0.09	.010	13	F8 IV-V	15,1118,1363,3026,8015
197925	+12 04457				20 44 08	+12 26.3	2	7.96	.000	0.20	.000	0.12		4	A2	1242,1386
		BSD 40 # 109			20 44 08	+44 09.9	1	11.93		0.29		0.14		3	A0	117
		KUV 20441+7548			20 44 08	+75 48.5	1	17.12		0.30		−0.49		1		1708
		LS II +25 024			20 44 10	+25 53.6	1	10.61		0.26		0.30		1	A5/7 Ib-II	1215
	+45 03264				20 44 10	+45 35.8	1	9.35		0.04		−0.33		2	B8 III	117
198039	+50 03191	IDS20426N5018	A		20 44 10	+50 29.4	1	7.18		0.57				2	F5	1724
197989	+33 04018	HR 7949	★	A	20 44 11	+33 46.9	18	2.47	.011	1.03	.007	0.86	.008	86	K0 III	1,15,985,1077,1080,1355*
197989	+33 04018	IDS20422N3336	B		20 44 11	+33 46.9	1	13.40		1.66				1		3032
		VES 240			20 44 12	+37 27.5	2	10.70	.000	2.01	.000	2.22	.000	2		426,8103
		3HLF4 # 5			20 44 12	−41 00.	1	13.14		0.00		0.04		2	A0	208
	+34 04147				20 44 14	+35 17.4	1	9.18		1.71				2	M2	369
	+35 04258	LS II +35 081	★	AB	20 44 15	+35 21.4	2	9.38	.030	0.03	.015	−0.75	.005	3	B0.5VN	963,1012
198236	+69 01127	HR 7967			20 44 15	+69 34.2	1	6.48		0.98		0.70		2	G8 III	1733
198149	+61 02050	HR 7957	★	AB	20 44 16	+61 38.6	18	3.42	.012	0.92	.006	0.61	.010	120	K0 IV	1,15,22,1077,1080,1118*
		VES 285			20 44 17	+40 15.2	1	13.25		0.97		0.61		1		8056
	+44 03571	BZ Cyg	★	A	20 44 17	+45 07.4	2	10.11	.080	1.58	.022	1.03		2	F7 Ib	934,1399
		BSD 40 # 1551			20 44 18	+45 33.5	1	11.86		0.47		0.19		2	B8	117
197963	+15 04255	HR 7947	★	B	20 44 19	+15 56.6	4	5.15	.012	0.49	.009	0.08	.008	9	F7 V	1,1028,1586,3002
197964	+15 04255	HR 7948	★	A	20 44 20	+15 56.6	5	4.27	.006	1.04	.004	0.97	.011	9	K1 IV	1,1028,1355,1586,3002
197964	+15 04255	IDS20420N1546		AB	20 44 20	+15 56.6	3	3.91	.010	0.85	.007	0.51	.000	8	K1 IV+F7 V	15,1363,8015
		BSD 112 # 1028			20 44 21	+00 12.7	1	11.47		0.34		0.06		1	A5	116
	+44 03572			AB	20 44 21	+45 05.5	1	9.64		0.47				1	A5	934
197928	−12 05835				20 44 25	−12 34.0	1	9.88		0.95		0.71		2	K3 V	3072
	+46 03039				20 44 27	+46 43.8	1	9.20		0.25		0.10		3	F0 V	117
197954	−3 05018	HR 7946			20 44 28	−02 40.3	2	6.26	.005	1.56	.010	1.97	.025	10	K2	252,1088
		BSD 40 # 1569			20 44 29	+46 01.0	1	10.00		0.31		−0.27		4	B8	117
		Ross 189			20 44 30	−10 22.3	1	12.07		0.57		−0.15		2		1696
		L 46 - 63			20 44 30	−79 29.0	1	11.83		1.51				1		1705
		IRC +40 448			20 44 33	+39 56.1	1	16.60		2.04				1		8008
	+37 04048	LS II +37 100			20 44 35	+37 54.8	1	12.01		0.51		0.11		4	B9 II	994
		BSD 40 # 1573			20 44 36	+45 39.2	1	12.50		0.39		−0.23		2	B7	117
197635	−69 03138	HR 7934			20 44 36	−68 57.7	4	5.41	.006	1.12	.000	1.10	.004	13	K0 III	15,1075,2038,3035

Table 1 text.

Table 1   1123

HD	DM	Other Id	N	Rem	α_{1950}	δ_{1950}	S	V	σ_V	B–V	σ_{B-V}	U–B	σ_{U-B}	n	Spectrum	References
235357	+50 03195				20 44 41	+50 53.2	1	9.68		0.21				2		1724
	+39 04313	LS II +39 054			20 44 42	+39 44.8	2	9.51	.043	1.16	.009	0.84	.000	3	F4 II	247,1215
		LP 696 - 4			20 44 42	−04 18.	1	17.02		0.15		-0.48		1	DA5	1727
		Smethells 64			20 44 42	−54 42.	1	11.90		1.36				1		1494
		L 46 - 163			20 44 42	−79 30.	1	11.87		1.49		1.03		1		3073
198056	+31 04217				20 44 45	+32 14.1	1	7.04		-0.04		-0.15		2	B9	401
198068	+33 04022				20 44 47	+34 11.4	2	8.83	.000	0.10	.000	-0.29	.000	2	A5	426,8103
		BSD 40 # 1586			20 44 47	+46 09.7	1	11.24		0.59		0.29		3	B8	117
		BSD 40 # 1591			20 44 48	+45 27.2	1	11.22		0.48		0.02		3	B8	117
	+35 04262				20 44 50	+35 58.7	1	8.44		1.70				2	M2	369
		G 24 - 29			20 44 51	+04 15.1	1	13.88		0.99		0.64		2		3062
	−14 05850				20 44 51	−14 36.2	1	10.96		0.50		-0.24		2	F5	1696
	+10 04379	G 25 - 3			20 44 52	+10 41.0	1	9.72		1.03		0.87		2	K2	333,1620
	+47 03183				20 44 52	+47 54.5	1	9.63		0.53				4	F8	1724
197827	−59 07640				20 44 52	−59 25.1	1	7.53		0.05		-0.01		14	A1 IV	1732
	+11 04380				20 44 53	+11 54.8	3	10.20	.004	0.07	.015	0.10	.000	15	A2 V	830,1242,1783
197947	−30 18186				20 44 53	−30 25.5	1	9.90		0.70				4	G6/8 V	2033
198181	+52 02799	HR 7962			20 44 54	+52 48.8	1	6.33		1.12				2	K0	71
	+47 03184				20 44 55	+48 05.5	1	8.23		0.47				2	F5	1724
197900	−44 14143				20 44 55	−44 22.9	1	6.46		1.16				4	K2 III	1075
197999	−0 04092				20 44 56	−00 36.8	1	9.09		0.24		0.11		1	A2	116
		BSD 112 # 1946			20 44 57	+00 48.8	1	11.10		0.17		0.13		1	A0	116
198151	+45 03270	HR 7958		★ AB	20 44 58	+46 20.9	4	6.30	.035	0.03	.026	0.03	.010	9	A3 V	247,401,1118,1196
198001	−10 05506	HR 7950			20 44 58	−09 40.8	17	3.77	.006	0.00	.008	0.04	.025	118	A1 V	1,3,15,26,30,247,1006*
		VES 286			20 44 59	+39 24.1	1	12.48		0.55		0.02		1		8056
		BSD 40 # 1599			20 45 00	+45 27.2	1	10.36		0.26		0.19		3	B8	117
		IRC +40 449		V	20 45 02	+39 41.5	1	13.20		2.99		1.96		1		8056
	+34 04152	LS II +34 024			20 45 03	+34 24.8	2	9.98	.019	0.27	.014	0.14	.029	7	A0 Ib	247,994
	+11 04381	DX Del			20 45 06	+12 16.7	1	9.52		0.27		0.18		1	A7 III	668
198108	+31 04220				20 45 06	+31 59.2	1	7.42		0.07		0.06		2	A0	401
		BSD 40 # 1604			20 45 06	+45 23.6	1	10.40		0.24		0.17		3	B9	117
198026	−5 05378	HR 7951, EN Aqr			20 45 06	−05 12.7	9	4.43	.013	1.64	.015	1.93	.022	42	M3 III	3,15,30,418,1088,2007*
		3HLF4 # 7			20 45 06	−40 27.	1	12.87		0.31		0.05		2	A5	208
197937	−44 14145	HR 7943		★ AB	20 45 06	−44 10.3	5	5.11	.012	0.35	.004	0.04	.000	14	F2 V	15,158,219,221,2012
	+46 03043				20 45 09	+46 33.8	1	10.35		0.31		-0.16		2		117
	−0 04095				20 45 10	+01 06.9	1	10.68		0.36		0.02		1		116
198134	+33 04028	HR 7956, T Cyg		★ A	20 45 11	+34 11.4	1	4.94		1.31		1.49		2	K3 III	542
198134	+33 04028	HR 7956, T Cyg		★ AB	20 45 11	+34 11.4	5	4.93	.016	1.30	.012	1.46	.009	8	K3 III	542,1080,1355,1363,3016
198134	+33 04028	IDS20432N3400	B		20 45 11	+34 11.4	1	10.03		1.27		0.66		2		542
235361	+51 02950				20 45 16	+51 59.1	1	8.55		0.80				2		1724
		CS 22880 # 100			20 45 16	−17 46.5	1	15.39		0.12		-0.66		1		1736
198070	+2 04250	HR 7954			20 45 17	+03 07.3	1	6.39		-0.02		-0.09		8	A0 Vn	1088
		BSD 40 # 2739			20 45 17	+46 20.2	1	11.59		0.33		-0.18		3	B5	117
		CS 22880 # 73			20 45 18	−21 10.5	1	14.05		0.31		0.04		1		1736
		3HLF4 # 8			20 45 18	−44 33.	1	13.47		0.03		0.05		2	A0	208
198109	+15 04257				20 45 19	+16 03.2	1	7.59		0.48		-0.11		1	F9 V wlm	979
198069	+5 04613	HR 7953		★ AB	20 45 20	+05 49.4	2	5.57	.005	-0.02	.000	-0.09	.000	7	A0 V	15,1071
		BSD 40 # 650			20 45 20	+44 31.9	1	11.13		0.42		0.26		2	B9	117
235364	+54 02417				20 45 20	+54 36.8	1	9.67		0.12		0.05		2	A0	401
198195	+41 03884	IDS20436N4202	A		20 45 22	+42 13.5	1	7.42		-0.08		-0.46		1	B9 III	401
198195	+41 03884	IDS20436N4202	AB		20 45 22	+42 13.5	1	7.18		-0.05		-0.39		2	B9 III+A1Vn	401
198195	+41 03884	IDS20436N4202	B		20 45 22	+42 13.5	1	8.96		0.05		-0.02		1	A1 Vn	401
198212	+45 03273				20 45 22	+46 00.2	1	8.66		0.12		0.08		2	A1 V	117
		BSD 40 # 1624			20 45 23	+46 05.6	1	12.04		0.47		0.35		4	B8	117
	+45 03274				20 45 25	+46 15.0	1	11.01		0.46		0.19		2		117
		CS 22880 # 89			20 45 27	−18 41.4	1	14.77		0.32		0.06		1		1736
198183	+35 04267	HR 7963		★ AB	20 45 28	+36 18.4	7	4.54	.027	-0.12	.007	-0.51	.016	21	B4 V +B7 V	15,154,379,938,1363*
198300	+59 02283	G 230 - 49		★ AB	20 45 28	+59 40.3	1	8.51		0.59		-0.10		1	G0	1658
198196	+35 04268				20 45 30	+36 05.7	1	8.41		0.26		0.10		2	A3	379
	+44 03578				20 45 32	+44 58.8	1	9.90		0.65				1		934
		BSD 40 # 155			20 45 33	+44 00.2	1	10.40		0.41		0.07		2	A8	117
		CS 22879 # 97			20 45 33	−38 42.0	2	14.21	.009	0.47	.009	-0.02	.006	4		1580,1736
		LS II +34 025			20 45 35	+34 53.7	1	12.52		0.05		-0.57		4		994
	+43 03720	IDS20438N4323	AB		20 45 35	+43 34.2	1	10.12		0.17		0.11		3		117
		BSD 40 # 1639			20 45 35	+45 58.4	1	11.35		0.37		-0.05		2	B9	117
		BSD 40 # 2749			20 45 35	+46 14.4	1	11.78		0.36		0.23		3	B5	117
		3HLF4 # 10			20 45 36	−40 50.	1	13.90		0.23		0.00		2	A0	208
		3HLF4 # 9			20 45 36	−41 18.	1	12.92		0.37		-0.17		2	A3	208
		VES 288			20 45 37	+45 26.1	1	11.64		0.53		0.05		1		8056
	+43 03721				20 45 38	+44 16.3	1	10.34		0.24		0.13		2		117
198237	+45 03275	HR 7966		★ A	20 45 38	+45 23.7	3	6.40	.005	1.61	.005	1.99	.023	6	K3 III	369,1501,3001
		BSD 40 # 2761			20 45 40	+46 27.3	1	11.64		0.47		0.10		3	B9	117
	+45 03276				20 45 42	+46 13.5	1	10.92		0.53		0.34		3		117
		3HLF4 # 11			20 45 42	−43 35.	1	14.60		0.04		0.07		2	A0	208
198285	+51 02951	IDS20442N5126	A		20 45 43	+51 36.6	1	8.01		0.32				2	F0	1724
		VES 289			20 45 44	+39 56.8	1	11.64		0.51		0.01		1		8056
		BSD 40 # 162			20 45 45	+43 58.7	1	12.12		0.39		0.18		2	A0	117
198511	+71 01029				20 45 46	+71 35.3	1	8.86		0.49		0.04		2	G0	1733
198263	+48 03210				20 45 47	+48 40.0	1	8.14		0.37				2	F0	1724
198009	−47 13573				20 45 49	−46 50.1	1	7.81		1.21				4	K0 II	2012

HD	DM	Other Id	N Rem	α_{1950}	δ_{1950}	S	V	σ_V	B–V	σ_{B-V}	U–B	σ_{U-B}	n	Spectrum	References
	+44 03579			20 45 50	+45 18.3	1	9.57		0.34		-0.24		3	B5 V	117
198286	+48 03211			20 45 50	+49 12.2	1	8.15		0.46		0.02		2	F6 V	1601
		BSD 40 # 164		20 45 51	+43 36.5	1	10.89		0.66		0.34		2	B9	117
		VES 243		20 45 52	+36 35.4	2	10.57	.000	1.69	.000	1.60	.000	2		426,8103
198253	+41 03889	IDS20440N4142	AB	20 45 52	+41 53.0	2	7.50	.000	0.05	.005			4	A0	1118,1196
	+45 03279			20 45 54	+45 29.4	1	9.53		0.16		-0.09		4	B6 V	117
		BSD 40 # 167		20 45 57	+43 53.6	1	10.92		0.37		0.14		4	B7	117
341196	+24 04245			20 45 59	+24 27.3	1	8.75		1.44		1.57		2	K5	1733
		BSD 40 # 1659		20 46 00	+45 25.2	1	10.98		0.39		-0.24		4	B3	117
		LS II +31 033		20 46 03	+31 03.0	1	12.20		0.14		-0.62		2		405
198048	-46 13718	HR 7952		20 46 03	-46 24.8	4	4.89	.006	1.52	.004	1.86	.009	16	K5 III	15,208,1075,2012
		BSD 40 # 171		20 46 04	+43 36.5	1	11.40		0.35		0.28		2	B9	117
	+37 04057	VES 244		20 46 05	+37 26.4	3	10.37	.053	0.45	.000	-0.22	.017	3		426,8056,8103
	+44 03582			20 46 05	+44 34.9	1	8.51		0.36		0.06		1	F0 IIIn	247
	+45 03280			20 46 05	+45 34.7	1	10.05		0.26		0.13		2		117
		CS 22879 # 98		20 46 05	-38 36.1	2	13.29	.010	0.78	.019	0.22	.006	3		1580,1736
		POSS 397 # 1		20 46 06	+25 57.1	1	13.42		1.15				2		1739
	+34 04157	LS II +35 082		20 46 06	+35 00.4	2	9.96	.010	0.46	.005	0.13	.000	7	F3	247,994
198287	+38 04235	V367 Cyg	⋆ABC	20 46 06	+39 06.1	3	6.98	.015	0.69	.021	0.44	.021	6	A7 Ia	105,247,1601
	-41 14245			20 46 07	-41 09.9	1	10.05		1.00		0.72		2	G8 III	221
		VES 291		20 46 09	+44 36.8	1	11.27		0.62		-0.08		1		8056
198345	+47 03188	HR 7969	⋆A	20 46 10	+47 38.8	2	5.57	.020	1.47	.005	1.78		5	K5 III	247,1118
198378	+55 02468			20 46 11	+55 23.1	1	8.09		0.01		-0.32		2	B9	401
	+43 03724			20 46 14	+43 47.3	1	10.19		0.25		0.15		2		117
		CS 22880 # 67		20 46 14	-21 46.1	1	15.08		0.51		-0.18		1		1736
198142	-30 18202			20 46 14	-30 22.7	1	7.26		1.50				4	M0/1 (III)	2012
		CS 22880 # 78		20 46 16	-20 41.3	1	15.30		-0.07		-0.93		1		1736
		LS II +34 026		20 46 17	+34 16.4	2	11.06	.010	0.18	.005	-0.75	.005	7		405,994
		VES 245		20 46 17	+36 41.5	2	12.62	.000	2.08	.000	1.23	.000	2		426,8103
	+44 03588			20 46 17	+45 10.8	1	9.56		0.08		0.05		4	A0 V	117
		VES 292		20 46 18	+38 42.5	1	12.03		0.51		0.02		1		8056
198269	+17 04421			20 46 19	+17 39.3	1	8.23		1.34				1	R0	1238
	+44 03587			20 46 19	+44 34.6	1	10.61		0.72		0.21		3	B7	117
199095	+81 00718	HR 8002		20 46 20	+82 20.9	2	5.75	.000	0.00	.005	-0.03	.010	4	A0 V	985,1733
198174	-26 15282	HR 7961		20 46 20	-25 58.0	2	5.86		-0.08	.004	-0.47		5	B7 IIIp	1079,2007
	+4 04551	IDS20438N0450	AB	20 46 21	+05 00.8	1	9.61		0.51		-0.17		2	F7 V	979
198188	-21 05840			20 46 22	-20 48.9	2	8.13	.009	0.59	.005	0.08		5	G3 V	158,2017
		LS II +34 027		20 46 23	+34 15.4	1	12.96		0.72		0.25		6		994
		BSD 40 # 2811		20 46 23	+46 12.4	1	11.45		0.55		0.32		4	A0	117
198387	+51 02954	HR 7972	⋆AB	20 46 23	+52 13.4	4	6.27	.013	0.88	.006	0.53	.012	10	G5	1080,1355,1724,3016
	+47 03190			20 46 30	+47 39.1	1	10.19		0.40				2	F5	1724
198208	-18 05783	HR 7964	⋆A	20 46 30	-18 13.3	1	6.21		1.42				4	K3 III	2007
198313	+28 03890			20 46 33	+28 37.0	1	8.31		0.78		0.42		3	K1 IV	3016
		BSD 40 # 191		20 46 34	+43 28.7	1	11.01		0.74		0.22		2	A0	117
		BSD 40 # 1699		20 46 35	+46 03.7	1	11.74		0.45		-0.14		4	B3	117
		G 144 - 39		20 46 36	+19 31.9	1	13.39		1.54				6		1663
		VES 246		20 46 36	+37 36.8	1	10.60		1.85				1		426
	+35 04272			20 46 38	+35 43.5	1	9.02		-0.08		-0.55		2		401
		BSD 40 # 196		20 46 39	+43 50.2	1	11.68		0.52		0.20		3	B9	117
	+45 03287			20 46 41	+45 47.7	1	10.04		0.32		0.22		2		117
		BSD 40 # 1701		20 46 42	+45 47.7	1	11.20		0.30		0.19		3	B9	117
198414	+44 03590			20 46 43	+45 16.0	3	7.63	.045	-0.10	.017	-0.31		6	B7 III	401,1118,1196
	+47 03191			20 46 43	+47 52.6	1	8.92		0.57				2	F8	1724
198272	-1 04057		V	20 46 43	-00 45.0	2	6.36	.050	1.54	.005	1.34	.050	5	M5 III	116,1149
		Feige 113		20 46 48	+30 05.0	1	13.08		0.09		-0.29		3	B8p	308
	+1 04107	G 25 - 5		20 46 49	+01 44.4	2	10.11	.005	0.67	.010	0.14	.015	2	G5	333,1620,1658
	+35 04277	LS II +35 083		20 46 49	+35 54.5	2	9.31	.000	-0.08	.000	-0.88	.014	7		401,994
198273	-9 05590			20 46 49	-08 50.2	1	8.42		0.60		0.00		2	G2 V	1003
198366	+25 04383			20 46 50	+25 29.2	1	7.94		0.04		-0.09		3	B9	833
335167	+29 04193	IDS20448N2910	ABC	20 46 51	+29 21.0	1	8.87		0.22		0.18		1	A3	8084
198232	-34 14660	HR 7965	⋆A	20 46 51	-33 58.0	4	4.92	.029	1.00	.003	0.78	.020	33	G8 III	3,15,1075,2012
		CS 22880 # 124		20 46 52	-22 05.9	1	15.48		-0.16		-0.95		1		1736
		G 144 - 40		20 46 53	+15 54.4	1	13.53		1.36		1.02		3		203
		LP 696 - 51		20 46 53	-07 54.7	1	12.60		0.54		-0.09		2		1696
198379	+25 04384			20 46 55	+25 29.0	1	7.70		-0.02		-0.24		2	B9	105
		CS 22879 # 94		20 46 55	-38 54.0	2	15.19	.010	0.43	.008	-0.19	.008	3		1580,1736
	-42 15112			20 46 57	-42 24.3	1	11.35		0.37		0.16		2	A5	208
	+10 04385	G 144 - 41	⋆AB	20 46 59	+11 13.1	2	9.31	.009	1.07	.024	0.98		6		1381,3072
		BSD 40 # 2848		20 46 59	+46 17.5	1	11.23		0.29		-0.14		4	A0	117
341105	+27 03877			20 47 00	+27 28.9	1	9.72		0.44		0.01		2	A7	634
198424	+38 04239			20 47 07	+38 40.7	2	7.50	.005	-0.11	.015	-0.41	.010	12	B9 V	1377,1733
198512	+53 02495	LS III +53 003		20 47 07	+53 43.2	2	8.28	.005	0.39	.020	-0.54	.010	5	B1 Vpenn	1012,1212
198245	-41 14250			20 47 07	-40 47.4	3	8.95	.018	0.62	.007	0.01	.006	10	G3 V	158,1097,1696
198436	+39 04331	IDS20453N3925	A	20 47 08	+39 36.3	1	7.71		-0.03		-0.28		3	A0	3032
198436	+39 04331	IDS20453N3925	B	20 47 08	+39 36.3	1	9.92		0.09		0.03		1		3032
198436	+39 04331	IDS20453N3925	C	20 47 08	+39 36.3	1	13.64		1.17		0.57		2		3032
198436	+39 04331	IDS20453N3925	D	20 47 08	+39 36.3	1	13.56		0.68		0.08		3		3032
198494	+47 03192			20 47 09	+48 02.8	1	8.50		0.21				2	A1mA7	1119
		G 210 - 36		20 47 10	+37 16.9	1	12.93		0.14		-0.67		3	DA	308
198479	+45 03290	LS III +45 022		20 47 12	+45 26.3	2	8.58	.005	0.41	.019	-0.47	.033	4	B1 III	117,1012

Table 1   1125

HD	DM	Other Id	N Rem	α₁₉₅₀	δ₁₉₅₀	S	V	σ_V	B–V	σ_B–V	U–B	σ_U–B	n	Spectrum	References
198513 +51 02957	HR 7978		⋆ AB	20 47 12	+51 43.5	1		-0.07		-0.57			1	B8 V +A3 V	1079
	3HLF4 # 14			20 47 12	-40 29.	1	14.38		0.03		-0.05		2	A2	208
	BSD 40 # 1737			20 47 13	+45 13.2	1	12.03		0.67		0.35		2	A0	117
+32 03954	IDS20452N3252		A	20 47 14	+33 02.4	2	9.31	.300	3.36	.775	0.68		2	N	414,8085
+32 03954	IDS20452N3252		B	20 47 14	+33 02.4	2	10.76	.035	0.89	.000	0.46	.020	2		414,8085
+32 03954	IDS20452N3252		C	20 47 14	+33 02.4	1	11.94		1.08		0.80		1		414
198437 +35 04282				20 47 14	+35 22.7	3	6.64	.005	1.63	.010	2.01	.029	5	K5	934,1601,1733
198478 +45 03291	HR 7977, V1661 Cyg		⋆ A	20 47 14	+45 55.7	9	4.83	.014	0.40	.009	-0.46	.006	19	B3 Iae	1,15,154,369,450,1012*
198478 +45 03291	IDS20455N4545		B	20 47 14	+45 55.7	1	11.14		0.58		0.09		3		450
198246 -42 15116				20 47 14	-42 40.6	1	8.41		0.57		0.01		2	G2 V	221
198390 +12 04472	HR 7973		⋆ A	20 47 15	+12 21.4	2	6.01	.010	0.42	.000	-0.09	.035	5	F6 V	254,3037
	LS II +36 084			20 47 15	+36 29.4	1	11.67		0.49		-0.13		4		994
235370 +50 03208				20 47 15	+50 55.5	1	9.82		0.28				3		1724
198456 +38 04240	IDS20454N3855		AB	20 47 16	+39 06.2	1	7.89		1.06		0.98		std	K0	1377
	BSD 40 # 2868			20 47 17	+46 23.3	1	11.15		0.35		-0.15		3	B9	117
198373 -5 05390	T Aqr			20 47 18	-05 20.0	2	11.20	.305	1.62	.015	0.76	.335	2	M4	817,975
198480 +42 03873				20 47 20	+42 46.1	1	7.30		0.00		-0.39		2	B8	401
198391 +7 04556	HR 7974			20 47 21	+07 40.6	2	6.32	.005	0.02	.000	-0.02	.000	7	A1 Vs	15,1256
+43 03731				20 47 21	+43 42.7	1	9.32		0.14		0.10		2	A3 V	117
+44 03594	LS III +45 024		⋆ AB	20 47 28	+45 13.5	3	9.76	.020	0.71	.014	-0.32	.050	6	B1:V:pen	117,1012,8056
	G 262 - 27			20 47 28	+70 44.6	2	10.76	.000	1.37	.015	1.27		6	K5	1746,7010
198160 -62 06180	HR 7959		⋆ AB	20 47 29	-62 37.0	3	5.65	.013	0.17	.009	0.07		10	A2/3 IV/V	15,404,2028
198404 +5 04626	HR 7975		⋆ A	20 47 30	+05 21.5	1	6.20		0.98		0.79		8	K0	1088
	CS 22880 # 110			20 47 31	-19 08.1	1	15.01		0.43		-0.16		1		1736
	G 187 - 7			20 47 32	+26 45.0	1	10.53		1.03		0.82		3	K4	7010
	LS III +46 026			20 47 34	+46 00.9	1	10.61		0.61		-0.40		2		405
	BSD 40 # 2893			20 47 34	+46 45.2	1	11.12		0.29		0.18		3	B9	117
+46 03062	LS III +46 027			20 47 35	+46 51.7	1	9.43		0.24		-0.56		3		117
198483 +25 04387				20 47 36	+25 35.0	1	7.70		0.58		0.10		3	G0 V	3026
	BSD 40 # 1772			20 47 36	+45 55.9	1	11.45		0.50		0.23		2	A0	117
+45 03293				20 47 36	+46 51.9	1	10.38		0.10		-0.04		3	B5	117
	BSD 40 # 2901			20 47 40	+46 45.4	1	11.44		0.29		0.05		3	B8	117
+3 04437	TX Del			20 47 42	+03 27.9	3	8.87	.007	0.57	.011	0.50	.015	3	F8	688,851,6011
	ApJ144,259 # 72			20 47 42	+13 22.	1	16.12		-0.33		-1.24		6		1360
198356 -32 16236	HR 7970			20 47 42	-32 14.5	2	6.40	.027	1.51	.018	1.81		5	K5 III	2007,3005
+43 03732				20 47 45	+43 35.2	1	8.90		0.04		0.05		3	A0	117
	VES 294			20 47 46	+45 18.8	1	12.00		2.13		1.65		1		8056
	BSD 40 # 2908			20 47 47	+46 35.6	1	10.85		0.25		0.01		2	B9	117
	VES 330, V750 Cyg			20 47 47	+50 20.7	1	14.55		2.50				1		8106
198357 -38 14250	HR 7971			20 47 48	-38 06.0	2	5.53	.010	1.38	.005	1.63		8	K3 II	2032,8100
	VES 247			20 47 52	+35 36.0	1	12.18		1.66		1.08		1		426
+38 04244	IDS20460N3811		A	20 47 53	+38 22.0	1	8.21		1.97				2	M2	369
198654 +58 02176	IDS20469N5822		B	20 47 53	+58 33.5	2	7.79	.005	0.92	.003	0.61	.001	8	K0	1355,1655
	3HLF4 # 16			20 47 54	-41 43.	1	12.95		-0.08		-0.32		2	A0	208
	3HLF4 # 15			20 47 54	-41 47.	1	14.01		0.31		0.01		2	A2	208
198308 -52 11782	HR 7968			20 47 54	-51 47.8	2	5.07	.013	1.12	.005	1.05		6	K1 II/III	58,2035
+35 04285	LS II +35 084			20 47 55	+35 28.4	2	8.70	.005	0.49	.000	0.14	.014	7	F6 Ib	247,994
198431 -13 05773	HR 7976			20 47 56	-12 43.9	2	5.88	.001	1.08	.005	0.94		6	K1 III	58,2007
	LP 756 - 8			20 47 59	-10 36.4	1	11.88		1.31		1.27		1		1696
198567 +35 04286				20 48 00	+36 08.5	2	8.43	.010	-0.12	.010	-0.72	.010	3	B8	401,963
-41 14259				20 48 00	-40 45.3	1	10.15		0.24		0.12		2	A3	208
198624 +49 03386	IDS20465N4945		A	20 48 03	+49 56.4	2	6.50	.005	1.62	.005	1.63		5	F7 V:	369,1084,8100
198624 +49 03386	IDS20465N4945		B	20 48 03	+49 56.4	1	10.30		0.58		0.01		3		1084
198550 +28 03900				20 48 04	+29 11.9	1	8.41		1.06		0.91		3	K5 V	3026
198638 +50 03211				20 48 04	+51 07.7	1	6.64		0.36				2	F2	1724
	BSD 40 # 2929			20 48 05	+46 37.6	1	11.01		0.43		0.26		3	B9	117
+48 03220				20 48 06	+48 57.0	1	8.98		0.41				2	F5	1724
	VES 331			20 48 07	+50 22.7	1	10.69		0.54		0.00		2		8106
	BSD 40 # 2939			20 48 10	+46 52.5	1	11.32		0.30		-0.10		2	B3	117
+46 03066			V	20 48 11	+46 48.7	1	9.41		1.65				2		369
198517 +2 04256				20 48 13	+02 54.2	1	8.86		0.07		0.15		1	A5	851
198516 +2 04257				20 48 13	+03 04.7	2	9.26	.002	0.21	.002	0.13	.008	25	A2	851,1699
198625 +46 03067	HR 7983		⋆ A	20 48 13	+46 28.4	3	6.33	.015	-0.07	.005	-0.58		5	B4 Ve	154,1118,1196
+46 03068				20 48 16	+46 52.0	1	9.36		0.09		-0.02		2	A0 V	117
198739 +61 02057				20 48 16	+62 20.6	2	7.97	.010	0.03	.005	-0.41	.005	5	B8 II	401,555
198639 +43 03739	HR 7984		⋆ A	20 48 18	+43 52.2	5	5.06	.011	0.19	.010	0.12	.002	11	A4m δ Del	1118,1724,1769,3052,8071
	BSD 40 # 1804			20 48 18	+46 04.0	1	11.20		0.41		0.14		4	B9	117
198552 +17 04431	HR 7981			20 48 19	+17 51.8	2	6.62		0.03	.005	0.03	.019	4	A1 Vs	1063,1733
	VES 332			20 48 23	+48 25.0	1	11.80		0.57		0.08		2		8106
198781 +63 01663	HR 7993			20 48 25	+63 51.3	3	6.45	.005	0.06	.010	-0.77	.000	7	B0.5V	15,154,1012
	POSS 397 # 3			20 48 26	+23 34.3	1	13.74		1.50				2		1739
	VES 295			20 48 28	+38 01.5	1	13.00		0.80		0.37		1		8056
198626 +30 04199	IDS20464N3032		A	20 48 31	+30 43.5	1	6.80		0.26		0.12		2	A2	3016
	BSD 40 # 2970			20 48 35	+46 26.5	1	11.01		0.28		0.17		2	B9	117
	VES 333			20 48 36	+49 45.1	1	11.16		0.60		0.21		2		8106
198501 -28 16955				20 48 36	-28 11.0	1	6.87		0.26		0.16		7	A1mA5-F0	355
+6 04665	G 25 - 7			20 48 38	+06 50.6	2	9.82	.005	0.90	.000	0.68	.019	3	K3	333,1620,1658
	CS 22879 # 82			20 48 44	-40 54.1	1	15.20		-0.35		-1.18		1		1736
198571 -6 05604	HR 7982		⋆ AB	20 48 47	-05 48.9	3	5.99	.013	0.46	.005	0.02	.008	9	F5 V +F8 V	15,938,1256
+1 04381	FQ Aqr		⋆	20 48 49	+02 07.5	3	9.55	.021	0.19	.006	-0.55	.008	5		405,1514,1699

HD	DM	Other Id	N	Rem	α_{1950}	δ_{1950}	S	V	σ_V	B−V	σ_{B-V}	U−B	σ_{U-B}	n	Spectrum	References
198542	−27 15082	HR 7980			20 48 50	−27 06.5	11	4.11	.010	1.64	.015	1.95	.021	73	K5 IIIa	3,15,30,369,1007,1013*
198690	+41 03909				20 48 51	+42 11.6	1	7.28		−0.10		−0.43		2	B8	401
198505	−38 14267	IDS20456S3832		AB	20 48 52	−38 21.2	1	8.27		0.46				4	F3 V	2012
		BSD 40 # 2980			20 48 53	+46 27.3	1	11.78		0.31		−0.06		2	B5	117
198529	−33 15245	HR 7979			20 48 53	−33 22.0	1	6.04		0.03				4	A1 IV	2032
198490	−44 14170				20 48 54	−44 04.3	1	8.72		0.02		0.02		2	A0 V	208
		VES 334			20 48 56	+46 49.2	1	11.27		2.01		1.64		2		8106
		VES 335			20 48 59	+48 31.2	1	11.93		0.64		0.52		2		8106
	+0 04598				20 49 00	+01 19.8	1	10.09		0.29		0.20		21	A2	1699
198793	+56 02495				20 49 00	+56 36.6	1	7.06		−0.08		−0.45		2	B8	401
		V1329 Cyg			20 49 02	+35 23.5	2	12.32	.265	0.82	.205	−0.98	.050	2	M4	426,1753
335263	+28 03905				20 49 07	+28 30.8	1	9.53		0.02		−0.17		2	A0	634
		G 210 - 38			20 49 08	+39 44.1	1	12.73		1.49				3		538
198768	+49 03393				20 49 08	+49 25.7	1	8.90		0.30				2	F0	1724
	+47 03199				20 49 09	+47 45.0	1	9.09		0.41				2	F2	1724
		VES 336			20 49 09	+48 34.5	1	12.00		1.52		1.43		2		8106
	+45 03301				20 49 11	+46 13.8	1	10.29		0.41		0.00		2		117
197908	−80 00985				20 49 11	−80 33.5	1	9.58		0.58				1	G3 V	1642
		TonS 10			20 49 12	−29 18.	1	14.38		0.00		−0.31		1		832
198794	+47 03200				20 49 18	+47 50.5	2	6.90	.055	1.47	.040	2.01		4	K3 Ib	1025,8100
	+45 03303				20 49 19	+45 51.1	1	9.21		0.10		0.01		2	A0 Ib	117,8100
198726	+27 03890	HR 7988, T Vul			20 49 21	+28 03.7	3	5.41	.004	0.48	.010	0.20	.029	3	F5 Ib	71,592,1772,6007
	+46 03073				20 49 23	+47 19.2	1	9.13		0.12				3	A1 V	1025
		CS 22880 # 109			20 49 23	−18 57.8	1	14.62		0.49		−0.14		1		1736
		CS 22880 # 126			20 49 23	−22 17.8	1	15.00		0.17		−0.60		1		1736
	+46 03072				20 49 24	+46 29.4	1	9.56		0.26		−0.05		2	B8	117
		L 47 - 26			20 49 24	−75 50.	1	12.57		0.51		−0.17		4		1696
198667	−6 05606	HR 7985			20 49 30	−05 41.7	6	5.54	.008	−0.08	.008	−0.24	.029	21	B9 III	3,15,1079,1256,1509,3016
	+50 03214				20 49 31	+50 45.7	1	9.32		0.65		0.19		3		1566
		VES 296			20 49 32	+38 02.8	1	12.66		1.97		1.44		1		8056
	+46 03075				20 49 32	+46 27.8	1	9.94		0.24		−0.04		2	A7 V	117
198834	+50 03215	IDS20480N5102		AB	20 49 32	+51 13.7	1	6.90		0.28				2	F0	1724
198797	+38 04254				20 49 34	+39 12.2	1	8.03		0.47		0.09		2	F5 III	247
		VES 337			20 49 34	+52 03.6	1	11.55		1.00		0.99		1		8106
198833	+51 02962				20 49 34	+52 12.4	1	7.10		0.84				4	G5	1724
198784	+37 04076	V1792 Cyg			20 49 35	+37 48.1	2	7.28	.009	0.06	.014	−0.51	.009	8	B2 V	399,401
	+46 03076				20 49 38	+46 26.1	1	9.55		0.26		0.18		2	A7 V	401
		VES 338			20 49 39	+49 00.3	1	13.38		2.26		0.80		2		8106
198708	−5 05402				20 49 41	−04 53.5	1	8.11		1.15		1.17		3	K2	1657
	−60 07431				20 49 41	−59 51.4	1	9.54		0.43		−0.03		2		1730
		3HLF4 # 18			20 49 42	−42 06.	1	14.69		0.09		0.12		2	A0	208
	+3 04445				20 49 45	+03 29.9	1	9.31		1.05				28	K0	6011
	+47 03201	LS III +47 016			20 49 47	+47 35.1	1	9.05		0.74				3	B5 Ib	1119
198895	+54 02429	LS III +55 004			20 49 47	+55 18.0	2	8.32	.015	0.55	.005	−0.30	.005	5	B1 V	5,1012,1212
198858	+47 03202				20 49 51	+47 31.1	2	7.36		1.13				3	K1 III	1119
		BSD 40 # 272			20 49 52	+43 48.8	1	10.79		0.59		−0.13		3	B8	117
		Orsatti 130			20 49 53	+43 24.8	1	13.55		0.77		0.05		1		1766
235381	+51 02964				20 49 55	+51 43.8	1	9.42		0.55				3		1724
199437	+80 00672	HR 8016			20 49 55	+80 22.0	2	5.40	.008	1.14	.014	1.06		5	K1 III	71,3016
198743	−9 05598	HR 7990		*	20 49 57	−09 10.3	10	4.73	.012	0.32	.004	0.10	.020	49	A3 m	3,15,30,216,355,1075*
198820	+32 03974	HR 7996			20 49 58	+32 39.6	3	6.44	.014	−0.14	.005	−0.62	.023	5	B3 III	154,247,252
198809	+26 04017	HR 7995			20 49 59	+26 54.5	9	4.57	.014	0.84	.007	0.47	.015	32	G8 III	15,592,1118,1355,3016*
341209	+27 03897				20 49 59	+27 33.5	1	9.00		0.62		0.34		2	A3	634
198562	−60 07435				20 49 60	−59 52.3	1	8.84		0.39		0.06		2	F2 IV	1730
		LS III +45 030			20 50 01	+45 50.5	1	10.54		0.41		−0.51		2	B0.5 n V	1012
	−43 14241				20 50 02	−42 51.8	1	11.56		0.24		0.09		3	A2	208
	+48 03227				20 50 03	+48 43.9	1	10.16		0.35				3		1119
198846	+34 04184	Y Cyg			20 50 04	+34 28.1	3	7.30	.003	−0.07	.006	−0.96	.011	26	B0 IV	401,600,994
	+47 03203				20 50 05	+47 50.6	1	10.22		0.19				2		1119
198732	−24 16328	HR 7989		* AB	20 50 05	−23 58.3	2	6.31	.005	0.87	.005	0.55		7	K0 III	404,2009
	+46 03079				20 50 06	+46 22.4	1	9.86		0.33				2	A1 V	1119
	+46 03080	RZ Cyg			20 50 06	+46 45.4	1	9.63		0.33		−0.14		2	B8	117
198772	−7 05428				20 50 07	−07 13.0	1	9.68		0.07		−0.15		1	A0	1776
198696	−31 17913				20 50 07	−31 42.2	2	7.33	.015	0.92	.024	0.49	.024	6	G8 III	119,3077
		G 25 - 8			20 50 08	+10 25.5	1	13.98		1.73				6		1663
		LP 816 - 18			20 50 10	−16 48.8	1	12.61		0.61		−0.07		2		1696
		BSD 40 # 3079			20 50 11	+46 44.2	1	11.09		0.35		0.03		2	B7	117
		VES 339		V	20 50 13	+47 10.0	1	11.40		2.59				1		8106
		VES 249			20 50 15	+37 36.2	1	9.81		0.30		0.02		1		426
198915	+46 03081				20 50 15	+46 32.7	2	7.50	.000	−0.07	.014	−0.56		4	B6 V	401,1119
		BSD 40 # 3087			20 50 17	+46 35.8	1	11.97		0.34		0.04		2	A0	117
	+47 03204				20 50 19	+47 24.4	1	9.24		0.14				3	A3 V	1025
198749	−28 16969				20 50 20	−28 12.8	1	8.35		0.58				4	F8/G0 V	2012
		Orsatti 140			20 50 21	+43 52.4	1	12.75		0.50		−0.42		1		1766
		BSD 40 # 279			20 50 22	+43 52.4	1	10.52		0.71		−0.01		2	B8	117
198802	−12 05854	HR 7994			20 50 22	−11 45.8	2	6.37	.005	0.67	.005	0.14		5	G1 V	2007,3077
198751	−31 17917	HR 7991			20 50 22	−30 54.5	1	6.35		1.09				4	K1 III	2009
198698	−44 14181				20 50 23	−44 31.8	1	7.51		1.61		1.99		2	K5 III	1627
198931	+43 03747	LS III +44 024			20 50 24	+44 14.7	2	8.75	.019	0.61	.014	−0.29	.014	4	B1 V e:nn	117,1012
198716	−40 14078	HR 7987			20 50 25	−39 59.9	4	5.34	.007	1.32	.011	1.43		12	K2 III	15,1075,2007,3077

Table 1 1127

HD	DM	Other Id	N Rem	α₁₉₅₀	δ₁₉₅₀	S	V	σ_V	B–V	σ_B–V	U–B	σ_U–B	n	Spectrum	References
198752	−36 14455			20 50 30	−36 23.1	1	7.13		1.44				4	K4 III	2012
		Smethells 65		20 50 30	−53 42.	1	12.07		1.49				1		1494
	−48 13714			20 50 34	−48 24.0	2	10.69	.013	0.51	.009	−0.13		6	G0 V-VI	1097,1594
		3HLF4 # 20		20 50 36	−41 15.	1	14.15		0.15		0.11		2	A0	208
	+42 03890	IDS20488N4251	A	20 50 39	+43 02.2	1	9.32		0.53				2	F8	1724
	+46 03083			20 50 40	+46 43.7	1	9.35		1.30				8	K2	1339
		3HLF4 # 1004		20 50 42	−41 20.	1	13.29		0.43		−0.07		2	F8	221
		VES 297		20 50 43	+39 59.9	1	11.58		1.93		1.78		1		8056
		LP 576 - 14		20 50 45	+06 08.7	1	12.97		0.63		−0.03		2		1696
	−44 14185		AB	20 50 46	−44 17.2	1	11.49		0.19		0.10		3	A2	208
	+43 03750			20 50 47	+43 45.8	1	8.81		1.70				2	M0	369
	+43 03751			20 50 47	+44 00.3	1	9.72		0.53		−0.22		2	B8	117
	+44 03612			20 50 47	+45 19.1	1	10.45		0.75		0.07		3	B4	117
		NGC 6981 - 1		20 50 48	−12 44.	1	9.41		0.47		0.05		1		168
		NGC 6981 - 2		20 50 48	−12 44.	1	11.16		0.52		0.03		2		168
		NGC 6981 - 3		20 50 48	−12 44.	1	11.93		0.69		0.17		4		168
		NGC 6981 - 4		20 50 48	−12 44.	1	13.94		0.65		0.13		5		168
		NGC 6981 - 5		20 50 48	−12 44.	1	12.58		1.10		0.66		5		168
		NGC 6981 - 6		20 50 48	−12 44.	1	14.28		0.93		0.57		4		168
		NGC 6981 - 7		20 50 48	−12 44.	1	14.79		1.14		1.14		4		168
		NGC 6981 - 8		20 50 48	−12 44.	1	12.94		0.63		0.04		3		168
		NGC 6981 - 9		20 50 48	−12 44.	1	12.98		0.58		0.09		5		168
		NGC 6981 - 10		20 50 48	−12 44.	1	12.91		0.98		0.74		6		168
		NGC 6981 - 11		20 50 48	−12 44.	1	12.37		0.54		0.00		3		168
		NGC 6981 - 12		20 50 48	−12 44.	1	13.63		0.98		0.68		1		168
		NGC 6981 - 13		20 50 48	−12 44.	1	11.24		0.85		0.27		1		168
		NGC 6981 - 14		20 50 48	−12 44.	1	13.69		0.94		0.66		2		168
		NGC 6981 - 15		20 50 48	−12 44.	1	13.70		0.71		0.23		1		168
		NGC 6981 - 16		20 50 48	−12 44.	1	14.73		1.28		1.21		2		168
		NGC 6981 - 17		20 50 48	−12 44.	1	15.13		1.12		0.84		2		168
		NGC 6981 - 18		20 50 48	−12 44.	1	14.80		1.17		0.84		1		168
		NGC 6981 - 19		20 50 48	−12 44.	1	16.02		0.88				1		168
		NGC 6981 - 20		20 50 48	−12 44.	1	14.70		0.98		0.77		2		168
		NGC 6981 - 22		20 50 48	−12 44.	1	14.36		0.61		0.12		3		168
		NGC 6981 - 23		20 50 48	−12 44.	1	15.77		1.04		0.63		3		168
		NGC 6981 - 25		20 50 48	−12 44.	1	14.17		0.76		0.34		2		168
		NGC 6981 - 26		20 50 48	−12 44.	1	16.19		0.77				2		168
		NGC 6981 - 27		20 50 48	−12 44.	1	16.47		0.69				1		168
		NGC 6981 - 29		20 50 48	−12 44.	1	16.82		0.93				1		168
		NGC 6981 - 30		20 50 48	−12 44.	1	16.63		0.87				1		168
		3HLF4 # 1005		20 50 48	−41 40.	1	11.51		0.40		0.03		3	F2	221
198918	+14 04461			20 50 50	+15 14.5	1	8.89		0.14		0.07		2	A0	1733
198133	−81 00927			20 50 51	−80 48.8	1	8.36		0.94		0.75		8	K0 III/IV	1704
		VES 298		20 50 52	+40 10.3	1	12.21		1.59		1.44		1		8056
198700	−58 07788	HR 7986	★ A	20 50 55	−58 38.7	3	3.64	.005	1.25	.000	1.23	.000	14	K1 II	15,1075,2012
199067	+54 02433			20 50 57	+54 51.7	1	6.72		0.06		0.08		2	A1 V	401
	−19 05953			20 50 58	−18 58.7	1	9.76		1.09		1.00		2	K3 V	3072
		BSD 40 # 3132		20 50 59	+46 33.3	1	11.43		0.25		−0.03		3	B8	117
	+47 03208			20 51 00	+47 41.8	1	9.23		0.33				2	A5	1724
198976	+29 04221	HR 7999		20 51 01	+29 27.6	1	6.34		1.09				2	K2	71
198766	−51 12748	HR 7992		20 51 01	−50 55.1	1	6.24		−0.12				4	B5 IV	2009
	+45 03308			20 51 02	+45 55.4	1	10.14		0.42		0.11		2	F5	117
199007	+34 04196			20 51 03	+34 33.7	1	7.95		−0.06		−0.30		22	B9	600
		POSS 397 # 5		20 51 04	+22 51.1	1	15.33		0.91				2		1739
199021	+42 03894	LS III +42 019		20 51 04	+42 25.1	1	8.43		0.55		−0.38		3	B0 V	399
	+12 04490			20 51 05	+12 22.0	1	8.78		0.34				1	A5	1242
		BSD 40 # 3141		20 51 05	+46 34.8	1	10.97		0.24		−0.10		2	B8	117
198853	−28 16975	HR 7997		20 51 07	−28 06.9	2	6.41	.000	1.60	.000	1.90		6	M1 III	2035,3005
		VES 340	V	20 51 08	+49 40.7	1	13.71		2.66				1		8106
		3HLF4 # 23		20 51 12	−43 04.	1	13.96		0.29		0.14		2	A2	208
		VES 299		20 51 13	+44 03.7	1	12.06		0.79		0.27		1		8056
	+47 03210	IDS20496N4751	AB	20 51 13	+48 02.2	1	9.99		0.51				3	B0 II-III	1119
199069	+44 03615			20 51 15	+45 16.6	2	8.30	.000	0.17	.015			4	A3 III	1118,1196
		VES 341	V	20 51 19	+52 51.1	1	12.40		2.45				1		8106
198949	−7 05433	HR 7998		20 51 19	−07 04.8	2	6.43	.005	0.37	.000	−0.03	.005	9	F1 IV	1088,1776
198841	−44 14188			20 51 19	−43 54.1	1	9.79		1.27		1.36		1	K2/3 (III)	221
		CS 22940 # 121		20 51 19	−58 12.4	2	14.15	.010	0.56	.012	−0.03	.002	3		1580,1736
		CS 22940 # 99		20 51 20	−61 06.0	1	14.33		0.57		0.00		1		1736
		BSD 40 # 857		20 51 21	+44 25.9	1	10.89		0.30		0.14		2	A0	117
		Orsatti 156		20 51 23	+44 25.9	1	13.47		0.72		−0.17		1		1766
		BSD 40 # 3159		20 51 23	+46 36.5	1	11.51		0.26		−0.14		3	B9	117
	+47 03211	LS III +47 018		20 51 23	+47 44.9	1	9.79		0.77		−0.27		2	B0 II-III	1012
199136	+56 02504			20 51 23	+56 23.7	1	7.44		−0.13		−0.51		2	B8	401
198828	−47 13618	IDS20479S4659	AB	20 51 23	−46 47.4	2	7.39	.020	0.53	.010	0.05		5	F7 V	219,1075
	−3 05059			20 51 24	−02 56.9	1	11.18		1.30		1.11		1	K4:	3073
	+44 03616			20 51 25	+44 30.0	1	9.78		0.19		0.14		1	A0	117
	+48 03236			20 51 28	+49 18.1	1	9.04		0.26				2	A5	1724
341211	+27 03903			20 51 29	+27 33.1	1	9.33		−0.05		−0.39		2	A0	634
199081	+43 03755	HR 8001		20 51 29	+44 11.8	6	4.77	.012	−0.13	.007	−0.59	.015	19	B5 V	15,154,1118,1203,3016,8015
		G 262 - 29		20 51 33	+68 54.7	1	11.70		1.54				1	M3	1746

HD	DM	Other Id	N Rem	α_{1950}	δ_{1950}	S	V	σ_V	B–V	σ_{B-V}	U–B	σ_{U-B}	n	Spectrum	References
199098	+44 03617	HR 8003		20 51 34	+44 59.5	3	5.41	.068	1.11	.008	0.98	.042	6	K0 II	252,1118,1355,8100
	+43 03756			20 51 36	+43 24.0	1	10.18		0.28		0.18		2		117
		BSD 40 # 3179		20 51 36	+46 22.9	1	10.66		0.25		-0.43		2	B9	117
199099	+41 03922	HR 8004		20 51 37	+42 13.2	3	6.66	.020	-0.04	.008	0.00		6	A1 V	401,1118,1196
	+44 03618			20 51 38	+44 28.9	1	10.62		0.34		0.20		2	A0	117
199100	+35 04311			20 51 44	+35 56.5	1	7.76		0.74		0.42		2	G5 IV-V	247
	+37 04092	LS II +37 101		20 51 45	+37 59.8	1	9.66		0.29		-0.57		2	B1 III	1012
	-41 09482			20 51 45	-41 36.1	1	9.55		-0.15		-0.70		15		1732
	-44 14192		A	20 51 45	-44 31.0	1	11.68		0.19		0.12		2	A2	208
	-44 14192		B	20 51 45	-44 31.0	1	11.82		0.71		0.21		2	F5	221
199154	+47 03214			20 51 46	+48 04.6	2	7.38	.005	0.20	.020			5	A5 IV	1025,1724
	-41 14294			20 51 46	-41 36.1	1	9.57		-0.15		-0.74		2	B2	208
199138	+44 03621			20 51 47	+44 57.9	2	8.06	.000	0.01	.005			4	A1 V	1118,1196
199122	+40 04346			20 51 48	+40 51.5	1	7.56		-0.12		-0.48		2	A2	401
199057	+17 04446			20 51 49	+17 50.2	1	8.12		0.93		0.62		2	G5	1648
		BSD 40 # 2004		20 51 49	+45 45.5	1	10.64		0.34		0.27		2	B9	117
199121	+45 03317			20 51 49	+45 50.3	2	8.99	.033	-0.01	.028	-0.24		4	B8 V	117,1119
	+47 03213			20 51 49	+47 35.2	1	9.51		1.82		1.97		3	K2	261
		VES 342		20 51 49	+49 54.1	1	10.72		0.42		-0.33		2		8106
	+48 03238			20 51 50	+49 07.6	1	9.48		0.37				2	F2	1724
		CS 22880 # 134		20 51 50	-20 50.9	1	15.06		-0.06		-0.94		1		1736
199101	+32 03980	HR 8005	★ A	20 51 52	+33 14.8	2	5.47	.006	1.53	.005	1.87		5	K5 III	71,1501
199167	+49 03405			20 51 52	+50 17.2	1	8.62		0.59				3		1724
199191	+53 02511			20 51 53	+54 19.5	2	7.15	.044	0.97	.010	0.63	.029	3	G8 III	1003,1355
		AJ89,1229 # 498		20 51 54	+08 53.	1	11.43		1.68				1		1532
		VES 300		20 51 54	+45 07.8	1	11.48		2.15		2.11		1		8056
		3HLF4 # 25		20 51 54	-45 09.	1	14.73		0.03		0.06		1	A2	208
		VES 343		20 51 55	+45 01.0	1	11.74		0.48		-0.40		1		8106
		Orsatti 167		20 51 56	+44 26.6	1	11.65		0.42		-0.32		3		1766
199012	-18 05805	HR 8000		20 51 58	-18 06.8	3	5.78	.009	1.12	.007	0.99	.016	7	K0 III	58,2007,4001
		AJ89,1229 # 499		20 52 00	+08 00.	1	11.87		1.65				2		1532
	+45 03319			20 52 00	+46 04.7	1	9.99		0.24				2	A0 V	1119
		3HLF4 # 26		20 52 00	-45 26.	1	13.48		0.30		0.02		2	A0	208
200544	+85 00357	G 265 - 39		20 52 01	+85 22.7	1	9.10		0.63		0.08		2	F8	1658
199476	+74 00889	G 261 - 29		20 52 04	+74 35.0	6	7.81	.014	0.68	.013	0.16	.016	16	G8 V	22,908,1003,1658,3026,7008
		VES 301		20 52 05	+41 55.8	1	10.56		0.59		0.14		1		8056
	+46 03092			20 52 05	+46 55.4	1	8.89		1.41		1.42		3	K0	261
199178	+43 03759	V1794 Cyg		20 52 07	+44 11.7	1	7.24		0.79		0.31		8	G2 V	3016
	+36 04308	LS II +36 085		20 52 08	+36 49.8	1	9.84		0.03		-0.76		2		401
	-44 14194	IDS20488S4449	AB	20 52 08	-44 37.6	1	10.67		0.71		0.36		7		1627
		POSS 457 # 4		20 52 09	+18 14.2	1	17.90		0.89				2		1739
	+48 03241			20 52 10	+48 27.5	1	9.82		0.37		-0.03		3		261
		CCS 2940		20 52 12	+07 54.	1	10.33		1.35				1	R0	1238
		3HLF4 # 1008		20 52 12	-43 41.	1	11.70		0.39		-0.06		3	F0	221
	-44 14195			20 52 14	-44 32.3	1	11.40		1.24		1.30		6		1627
199140	+27 03909	HR 8007, BW Vul		20 52 15	+28 19.9	4	6.53	.025	-0.13	.013	-0.89	.026	8	B2 III	154,1278,1319,1423
199216	+48 03242	LS III +49 010		20 52 15	+49 20.6	1	7.03		0.48		-0.46		2	B1 II	1012
198987	-44 14196			20 52 16	-44 39.8	1	9.13		0.35		0.13		45	Fm δ Del	1627
	+45 03323			20 52 17	+45 21.9	1	9.45		0.45		-0.04		3	F5 V	261
235392	+50 03223			20 52 18	+51 04.0	1	9.30		0.35				2		1724
199305	+61 02068	G 231 - 19	★ A	20 52 18	+61 58.5	4	8.55	.037	1.48	.012	1.24	.009	11	M2 V	1,1197,3072,8039
		AAS67,103 # 27		20 52 18	-44 35.6	1	18.85		1.38		1.11		1		1627
198988	-45 14102			20 52 18	-45 02.4	1	9.90		0.54		0.03		3	G0 V	1627
		HA 184 # 915		20 52 19	-44 35.5	1	17.08		0.55		-0.09		1		1627
199206	+44 03624	IDS20506N4444	AB	20 52 21	+44 55.2	1	7.34	.010	-0.03	.008	-0.22		6	B8 II	401,938,1118,1196
	+47 03215			20 52 21	+48 19.7	2	9.39	.010	0.39	.000	-0.01		5	F5	261,1724
199217	+47 03216			20 52 22	+48 14.8	2	7.69	.020	-0.04	.020			5	A0 V	1118,1119
		3HLF4 # 28		20 52 24	-44 28.	1	14.41		0.09		0.15		2	A0	208
199000	-45 14103			20 52 24	-44 46.0	3	7.28	.016	1.64	.008	1.98	.033	23	M2 III	1627,1673,2012
199169	+27 03911	HR 8008		20 52 26	+27 52.0	4	5.03	.025	1.48	.009	1.80	.016	8	K4 III	1080,1355,1363,3016
	+46 03094			20 52 27	+46 51.5	1	9.64		0.07				2	B9 V	1119
		HA 184 # 1187		20 52 27	-44 40.8	1	15.13		0.67		0.25		2		1627
		HA 184 # 919		20 52 28	-44 34.9	1	16.55		1.29		1.30		2		1627
	+42 03899			20 52 29	+42 27.1	1	9.24		1.69				2	M2	369
199306	+58 02187	IDS20513N5856	AB	20 52 29	+59 07.0	1	6.96		0.40		0.01		1	F1 V	3030
		POSS 397 # 2		20 52 30	+24 39.5	1	13.45		1.07				2		1739
199218	+40 04354	HR 8009	★ AB	20 52 30	+40 30.7	3	6.70	.005	-0.07	.008	-0.42	.020	8	B8 V.nn	1079,1118,1212
		BSD 40 # 3226		20 52 30	+46 29.1	1	11.40		0.49		0.17		3	A0	117
		HA 184 # 921		20 52 30	-44 35.8	1	12.89		0.55		0.03		5		1627
		HA 184 # 920		20 52 30	-45 20.4	1	12.29		0.71		0.27		4		1627
199308	+55 02486			20 52 31	+56 10.3	3	7.51	.004	0.14	.009	-0.64	.005	7	B2 IV-V	5,399,555
		AAS67,103 # 24		20 52 31	-44 37.0	1	17.28		0.72		0.19		1		1627
		HA 184 # 1188		20 52 31	-44 41.9	1	14.30		0.43		-0.18		2		1627
		HA 184 # 1190		20 52 32	-44 40.2	1	13.93		0.71		0.25		5		1627
		BSD 40 # 909		20 52 33	+44 51.1	1	12.12		0.52		0.30		2	A0	117
		VES 302		20 52 33	+45 31.2	1	11.45		2.12		1.93		1		8056
199124	-1 04075	HR 8006, EM Aqr		20 52 33	-01 33.9	5	6.55	.005	0.28	.010	0.07	.014	15	A9 Vn	15,253,285,688,1071
		HA 184 # 922		20 52 33	-44 34.9	1	15.86		0.83		0.45		5		1627
		HA 184 # 925		20 52 34	-44 36.1	1	12.67		0.78		0.39		6		1627
		HA 184 # 923		20 52 34	-44 36.9	1	15.08		0.91		0.59		3		1627

Table 1   1129

HD	DM	Other Id	N Rem	α_{1950}	δ_{1950}	S	V	σ_V	B–V	σ_{B-V}	U–B	σ_{U-B}	n	Spectrum	References
		HA 184 # 1191		20 52 34	−44 40.7	1	15.36		0.73		0.22		3		1627
199031	−44 14198			20 52 35	−44 06.1	1	10.67		0.24		0.12		2	A7	208
		AAS67,103 # 26		20 52 35	−44 38.5	1	18.40		1.01		0.97		1		1627
		AAS67,103 # 25		20 52 37	−44 38.0	1	17.95		0.69		−0.03		1		1627
	+46 03097			20 52 38	+46 31.2	1	9.12		0.39				2	B9.5III	1119
		HA 184 # 929		20 52 38	−44 34.7	1	16.45		0.98		0.76		2		1627
		HA 184 # 928		20 52 38	−44 36.8	1	14.49		0.67		0.09		3		1627
	+47 03218			20 52 39	+48 12.1	2	8.74	.024	0.56	.029			5	F9 IV	1025,1724
199331	+56 02509	IDS20514N5648	AB	20 52 40	+56 59.1	1	7.16		0.06		0.07		2	A0	401
	+12 04499	G 144 - 44		20 52 42	+12 58.8	2	8.83	.014	1.05	.000	1.04	.005	4	K3 V	1003,3072
	+45 03325			20 52 42	+45 54.8	1	9.08		1.38		1.41		3	K2	261
		3HLF4 # 1009		20 52 42	−43 16.	1	13.42		1.15		1.24		2		221
		BSD 40 # 2074		20 52 44	+45 29.5	1	11.47		0.70		0.47		3	B9	117
	+46 03098			20 52 44	+46 31.6	1	9.18		1.16		1.02		3	K0	261
199234	+36 04314	IDS20508N3642	AB	20 52 45	+36 52.9	1	7.15		0.02		−0.14		2	A0	401
		HA 184 # 931		20 52 45	−44 37.5	1	15.74		0.44		−0.10		2		1627
		VES 346		20 52 46	+49 14.7	1	11.17		0.56		0.04		2		8106
		LP 756 - 18		20 52 46	−14 15.0	1	14.61		1.73				1	M5:	1705
	+10 04409			20 52 47	+11 04.1	3	9.27	.032	1.69	.009	1.82	.020	3	M2	679,680,3072
		LP 756 - 19		20 52 47	−14 13.3	2	12.46	.000	1.58	.012			2		1705,1759
199290	+48 03243			20 52 48	+48 44.3	2	7.96	.014	0.25	.014	0.09		6	F2 Ib	1119,8071
199236	+25 04415			20 52 53	+25 41.6	2	8.49	.005	0.26	.020	0.22	.005	5	A0	833,1625
199221	+27 03915	IDS20508N2743	AB	20 52 54	+27 54.0	1	7.85		0.63		0.13		2	G1 V +G3 V	3030
		G 25 - 9		20 52 55	+09 39.3	1	13.51		1.50		1.10		3		203
199309	+47 03220			20 52 56	+47 32.9	3	8.67	.013	−0.03	.007	−0.09	.000	7	B8 V	401,1119,1462
199311	+45 03326			20 52 57	+46 02.4	5	6.67	.013	0.06	.016	0.09	.000	13	A2 V	1025,1118,1196,3016,8071
	+46 03101	IDS20510N4711	AB	20 52 58	+47 20.2	1	9.78		1.05		0.75		3	K0	261
		POSS 397 # 4		20 52 60	+25 01.1	1	15.28		0.76				1		1739
199252	+29 04231	UX Cyg		20 53 00	+30 13.4	1	10.76		0.24		0.21		1	M4	699
199312	+44 03628			20 53 00	+44 56.7	1	7.59		−0.07		−0.45		2	A0 V	401
		3HLF4 # 30		20 53 00	−44 47.	1	13.77		−0.07		−0.23		2	A0	208
	+44 03627	LS III +44 025		20 53 01	+44 39.2	2	9.86	.007	0.77	.009	−0.10	.000	5	B3 V	117,1012
235397	+50 03225	G 231 - 20		20 53 01	+51 18.9	1	9.58		0.78				2	G5	1724
199310	+46 03102			20 53 02	+47 05.3	4	8.31	.017	0.88	.000	0.52		9	G8 V	1119,1119,1462,1724
198969	−65 03875			20 53 05	−65 07.8	1	7.71						4	G8 IV	2012
199223	+3 04461	HR 8010	★ AB	20 53 10	+04 20.4	6	6.04	.008	0.82	.016	0.49	.008	16	G6 III-IV	15,285,688,938,1071,3051
	+47 03221			20 53 13	+48 00.9	1	9.41		1.40		0.86		2		261
		BSD 40 # 2104		20 53 14	+45 35.9	1	11.70		0.47		0.07		2	F8	117
199254	+12 04501	HR 8012	★ A	20 53 15	+12 22.6	1	5.54		0.11	.005	0.08	.010	7	A4 V	1063,1733
199253	+13 04572	HR 8011		20 53 15	+13 31.8	3	5.19	.032	1.12	.010	0.96	.018	10	K0 III	3,1355,3016
199266	+19 04564			20 53 17	+20 12.6	1	7.25		1.34		1.38		3	K2	1648
	+44 03629			20 53 17	+44 34.0	1	10.06		0.40		−0.04		2	B8	117
		V520 Cyg		20 53 17	+47 20.5	1	10.53		1.15		0.82		1	G0	1462
	+14 04472			20 53 25	+14 49.6	1	10.89		0.23		0.15		1	A5	1697
	+41 03931	G 212 - 7		20 53 25	+42 07.5	2	10.27	.002	0.61	.007	−0.11	.003	4	G5	1658,7010
199394	+45 03327			20 53 26	+46 09.5	3	7.01	.007	1.01	.012	0.65	.004	8	G5 II	993,1119,3016
199080	−58 07796			20 53 26	−58 21.6	1	10.04		0.94		0.67		2	K2 V	3072
199355	+41 03932	IDS20516N4208	AB	20 53 28	+42 19.3	1	7.00		−0.12		−0.55		2	B9	401
199356	+39 04368	LS II +40 038		20 53 30	+40 06.5	2	7.16	.035	0.16	.010	−0.76	.000	5	B2 IVp:	1012,1212
		BSD 40 # 2113		20 53 32	+45 27.1	1	10.61		0.43		−0.01		3	B8	117
199417	+44 03633			20 53 36	+44 41.3	1	9.01		0.70		−0.02		2	B9 V	117
199415	+46 03108			20 53 37	+47 17.6	1	8.88		0.10				4	B8 V	1119
199395	+42 03907			20 53 38	+43 10.5	1	6.71		1.44		1.75		2	K4 III	247
	+46 03107			20 53 41	+46 25.0	1	10.48		0.17		0.15		2		117
	+47 03222			20 53 41	+47 29.3	2	9.15	.015	0.29	.000			5	F0 III	1119,1724
199280	−4 05307	HR 8014		20 53 41	−03 45.2	5	6.57	.003	−0.08	.008	−0.30	.012	26	B8 Vn	15,989,1079,1256,1729
199397	+38 04282			20 53 42	+38 42.6	1	7.92		0.14		0.10		4	A2 V	833
		BSD 40 # 3281		20 53 42	+46 30.9	1	12.31		0.61		0.45		3	A0	117
	−41 14309			20 53 43	−41 18.0	1	10.80		0.31		0.08		2	A2	208
199396	+39 04371			20 53 45	+39 25.8	1	8.50		0.39		0.13		2	F3 III-IVn	247
		ADS 14438	★ AB	20 53 48	+47 10.	1	10.10		0.46		0.22		1		8084
		ADS 14438	★ C	20 53 48	+47 10.	1	12.34		0.64		0.34		1		8084
199260	−26 15344	HR 8013		20 53 49	−26 29.3	4	5.72	.024	0.50	.001	−0.02	.000	17	F7 V	3,15,2030,3077
199439	+46 03109			20 53 50	+47 17.4	2	8.84	.005	0.29	.005			5	F0 III	1119,1724
199360	+20 04772			20 53 54	+21 14.9	1	7.84		0.31		0.15		3	A5	8071
	−10 05549			20 53 58	−09 50.3	2	10.26	.009	0.52	.014	−0.11	.000	4	F6	1064,1696
	−43 14281			20 53 59	−43 38.6	1	11.79		0.27		0.10		3	A2	208
		3HLF4 # 32		20 54 00	−43 52.	1	14.87		0.02		0.11		2	A2	208
		VES 303		20 54 01	+38 42.1	1	13.01		2.29		2.48		1		8056
		BSD 40 # 2144		20 54 01	+46 09.6	1	11.70		0.48		0.00		2	B3	117
199242	−37 14005			20 54 01	−36 45.5	1	9.22		1.10		1.11		1	K1 III	742
	+44 03636	IDS20523N4425	AB	20 54 03	+44 36.3	1	10.40		0.41		0.13		2	A2	117
		CS 22879 # 149		20 54 03	−38 23.5	1	14.28		−0.23		−0.94		1		1736
	+45 03332			20 54 04	+46 03.6	1	9.80		0.15		0.13		3	A0	117
		Wolf 896		20 54 04	−10 37.0	2	11.47	.035	1.50	.015	1.07		2	M4	1705,3078
	+47 03226	IDS20524N4732	AB	20 54 05	+47 43.1	1	9.84		1.06		0.76		3		261
	+43 03764	IDS20523N4322	AB	20 54 06	+43 33.4	1	10.25		0.16		0.11		4	B9	117
		LP 696 - 55		20 54 07	−08 01.6	1	12.03		0.56		−0.08		2		1696
		G 210 - 40		20 54 08	+38 39.4	1	11.60		1.27		1.29		2	M1	7010
199478	+46 03111	HR 8020	★	20 54 08	+47 13.5	7	5.68	.018	0.46	.017	−0.34	.028	19	B8 Ia	15,252,1012,1079,1118*

HD	DM	Other Id	N Rem	α_{1950}	δ_{1950}	S	V	σ_V	B–V	σ_{B-V}	U–B	σ_{U-B}	n	Spectrum	References
		Ross 193, FR Aqr	A	20 54 08	−05 01.9	2	11.89	.033	1.50	.010	1.06		4	M3enn	1705,3079
		LTT 8292	B	20 54 08	−05 01.9	3	16.64	.057	1.15	.024	0.64	.000	11	DC	940,3016,3079
199479	+43 03766			20 54 12	+44 10.9	3	6.81	.018	−0.05	.010	−0.22	.005	6	B9 V	401,408,1118
199493	+46 03112		AB	20 54 12	+46 54.4	1	7.21		0.82				3	G8 IV	1025
199345	−10 05553	HR 8015	⋆ AB	20 54 12	−09 53.4	2	5.49	.015	1.47	.000	1.71		6	K2	2007,3005
		BSD 40 # 995		20 54 14	+45 10.1	1	11.21		0.45		−0.19		2	B5	117
	+45 03333			20 54 14	+45 25.2	1	9.94		0.42		0.27		3	A0	261
199546	+53 02520			20 54 14	+53 36.8	1	8.14		0.98		0.68		3	K0	1566
	+43 03765			20 54 15	+43 25.0	1	9.12		0.05		0.04		2	A0	117
	−37 14010			20 54 18	−36 44.4	2	9.74	.020	1.01	.015	0.50		5		742,1594
		BSD 40 # 1004		20 54 22	+44 28.2	1	11.41		0.62		0.42		2	B9	117
199288	−44 14214			20 54 22	−44 18.5	6	6.52	.009	0.59	.008	−0.04	.009	20	G0 V	158,742,1075,1627*
335336	+29 04240	UY Cyg		20 54 23	+30 14.1	1	10.59		0.26		0.18		1	F0	668
199511	+42 03911			20 54 23	+43 13.9	1	6.82		−0.12		−0.46		2	B8	401
199496	+35 04332			20 54 25	+36 11.1	1	7.83		−0.01		−0.09		2	A0	401
199287	−44 14213	IDS20511S4425	AB	20 54 25	−44 13.3	1	9.02		0.47		−0.04		2	F5 V	1627
		VES 304		20 54 29	+38 35.9	1	10.67		0.21		−0.23		1		8056
	+34 04210	G 210 - 42		20 54 33	+34 47.1	2	10.81	.021	1.06	.022	1.01	.002	5	F8	1658,7010
	+48 03248			20 54 33	+48 30.4	2	8.73	.019	1.04	.009	0.83		6	K0 III	261,1119
	+42 03914	LS III +42 021	⋆ AB	20 54 35	+42 56.2	1	8.43		1.05		0.08		2	B0 III:	1012
		LS III +47 021		20 54 36	+47 30.0	1	11.45		0.35		−0.22		2		180
	+47 03227			20 54 36	+47 38.5	1	10.13		0.43				2	B5	1119
		3HLF4 # 1010		20 54 36	−40 53.	1	12.48		0.57		0.01		2	F8	221
		3HLF4 # 34		20 54 36	−42 10.	1	13.94		0.18		0.01		2	A0	208
199442	−0 04132	HR 8017	⋆ A	20 54 37	+00 16.3	4	6.05	.007	1.22	.000	1.33	.022	12	K2 III	15,37,1071,3016
199442	−0 04132	IDS20521N0005	B	20 54 37	+00 16.3	1	11.76		1.18		0.99		2		3016
199547	+43 03767			20 54 37	+43 42.5	2	7.05	.015	1.14	.010	1.06	.020	4	K0 III	408,1733
		BSD 40 # 3316		20 54 37	+46 17.8	1	11.64		0.47		0.20		2	A0	117
199578	+50 03232	HR 8022		20 54 38	+50 52.9	1			−0.10		−0.51		1	B5 V	1079
	+43 03768			20 54 39	+44 07.9	1	9.83		0.03		−0.10		2	A0	408
	+35 04333			20 54 40	+36 08.2	1	9.41		0.02		−0.67		2	A2	401
199289	−48 13728			20 54 40	−48 23.6	1	8.29		0.52		−0.13		1	F5 V	219
199713	+65 01518			20 54 42	+65 29.7	1	7.40		−0.07		−0.18		2	B9	401
		3HLF4 # 35		20 54 42	−41 18.	1	15.05		0.02		0.01		2	A2	208
	+36 04330	LS II +37 102		20 54 45	+37 10.4	1	9.10		0.24		0.25		2	B9 II	247,8100
		VES 347		20 54 45	+46 14.6	2	11.14	.007	0.35	.025	−0.04	.000	4		117,8106
199370	−36 14506			20 54 45	−36 38.8	2	9.85	.025	0.45	.005	−0.06		4	F3 V	742,1594
199612	+48 03249	HR 8026		20 54 48	+49 00.2	4	5.90	.020	1.04	.008	0.93	.005	8	G8 II-III	247,1025,1118,1355
199340	−45 14120			20 54 48	−45 06.6	2	9.50	.014	0.21	.005	0.14	.000	4	A0mA3-A7	208,1627
199579	+44 03639	HR 8023	⋆	20 54 49	+44 43.9	6	5.96	.003	0.04	.004	−0.85	.002	19	O6 Ve	15,154,1011,1012,1118,1203
		BSD 40 # 3326		20 54 49	+46 17.7	1	11.51		0.81		0.54		3	A0	117
		PASP87,379 # 11		20 54 50	+44 08.2	1	9.99		0.17		0.07		2		408
199611	+50 03233	HR 8025	⋆ AB	20 54 50	+50 32.1	4	5.81	.017	0.31	.009	0.02		7	F0 III	254,1118,1196,1724
		G 25 - 10		20 54 51	+11 49.3	1	12.33		1.42		1.08		1		333,1620
		BSD 40 # 1038		20 54 52	+44 27.4	1	11.63		0.60		0.41		2	A0	117
199443	−16 05741	HR 8018		20 54 53	−16 13.5	3	5.86	.007	0.19	.009	0.12		10	A2m	15,2030,8071
	+44 03640			20 54 54	+45 20.0	1	10.09		0.41		0.14		2		117
	+45 03338			20 54 54	+45 55.5	1	9.56		0.18		0.11		3	A2 V	261
199596	+46 03118			20 54 55	+47 00.2	1	8.85		0.10				3	A2 V	1119
	+45 03339	LS III +46 030		20 54 56	+46 09.8	2	9.92	.009	0.42	.000	−0.38	.028	4	B1 IV	117,1012
	+47 03231			20 54 56	+47 37.6	1	9.83		0.37				2	A0	1119
	+47 03232			20 54 57	+47 55.3	1	9.84		0.88		0.55		2		261
199661	+56 02515	HR 8029		20 54 57	+56 41.7	3	6.23	.004	−0.17	.008	−0.69	.013	5	B2.5IV	5,154,555
199580	+42 03915	G 212 - 10	⋆ A	20 54 58	+42 42.0	6	7.21	.008	0.96	.008	0.72	.009	63	K0 III-IV	247,1003,1080,3016*
199469	−8 05529			20 54 59	−07 54.4	1	7.51		0.01		−0.20		1	A0	1776
	+37 04115	LS II +37 103		20 55 01	+37 22.2	1	9.12		0.43		0.17		2	F4 II	247,8100
	+42 03917			20 55 02	+42 44.5	1	9.13		0.14		0.13		3	A2	1625
		BSD 40 # 2202		20 55 02	+45 31.0	1	11.37		0.25		−0.15		2	B9	117
199627	+45 03340			20 55 02	+46 07.8	1	8.55		0.18				3	A5 IV	1025
		BSD 40 # 1050		20 55 03	+45 01.1	1	12.05		0.60		0.40		2	A1	117
	+46 03119			20 55 05	+46 35.4	1	9.83		0.68		0.22		3		261
		3HLF4 # 1011		20 55 06	−42 14.	1	13.69		0.57		0.07		2	F8	221
		G 187 - 9		20 55 07	+22 10.4	1	12.00		1.64				1	M2	1746
		BSD 40 # 3337		20 55 07	+46 13.9	1	10.38		0.22		0.11		2	B9	117
235405	+52 02836	IDS20536N5226	AB	20 55 07	+52 37.9	1	9.50		0.58				3	G0	1724
		BT Aqr		20 55 10	−05 52.7	1	11.78		0.14		0.15		1		597
		3HLF4 # 36		20 55 12	−41 35.	1	14.29		0.17		0.08		2	A0	208
	+43 03770			20 55 13	+43 44.8	1	9.14		1.03		0.84		2		408
	+43 03771			20 55 14	+44 06.5	1	10.08		0.47		0.08		2		408
235406	+51 02973			20 55 14	+51 43.3	1	9.40		0.38				3		1724
199662	+47 03234			20 55 15	+47 23.1	1	7.65		0.07				3	A1 V	1119
199629	+40 04364	HR 8028		20 55 18	+40 58.4	5	3.93	.009	0.02	.008	0.00	.014	16	A1 Vn	15,1118,1203,3023,8015
		VES 305		20 55 19	+39 26.2	1	11.41		0.66		0.19		1		8056
	+40 04365		AB	20 55 19	+40 33.6	1	9.63		0.45				1	F0	934
	+45 03341	LS III +46 031		20 55 19	+46 21.1	1	8.73		0.48		−0.48		2	B1 II	1012
199190	−70 02814			20 55 19	−69 46.2	1	6.87		0.62				4	G1 IV/V	2012
199739	+56 02516	IDS20540N5632	AB	20 55 21	+56 44.0	1	7.96		0.47		−0.25		3	B8	555
200039	+75 00764	HR 8043		20 55 21	+75 43.9	1	6.05		0.93				2	G5 III	71
	−41 14319			20 55 21	−41 08.6	1	11.19		−0.01		−0.03		2	A0	208
	+39 04379	VX Cyg		20 55 27	+39 59.0	2	9.57	.020	1.43	.003	1.00		2	F5 Iab:e	934,1399

Table 1   1131

HD	DM	Other Id	N	Rem	α_{1950}	δ_{1950}	S	V	σ_V	B–V	σ_{B-V}	U–B	σ_{U-B}	n	Spectrum	References
199598	+25 04422				20 55 29	+26 12.6	1	6.94		0.58		0.10		3	G0 V	3026
	+39 04380				20 55 30	+40 15.2	1	9.78		1.19				1		934
199679	+41 03943				20 55 34	+41 29.7	1	7.34		0.13				2	A2	1118
199714	+47 03237				20 55 36	+48 06.1	2	8.29	.005	0.29	.009	-0.19		7	B8 Ib	206,1119
		PASP87,379 # 19			20 55 38	+44 08.2	1	11.33		0.47		0.30		1		408
		BSD 40 # 2227			20 55 38	+45 34.6	1	12.01		0.47		0.39		2	A0	117
199693	+41 03944				20 55 39	+42 15.0	1	7.29		1.37				45	K2 III	6011
199715	+45 03344				20 55 39	+46 18.9	1	8.70		0.17				3	A8 V	1025
		POSS 457 # 3			20 55 40	+18 43.8	1	17.25		1.43				2		1739
		PASP87,379 # 17			20 55 40	+44 31.7	1	9.99		0.95		0.71		2		408
199694	+39 04382				20 55 41	+39 28.1	1	7.47		-0.11		-0.51		2	B9	401
		L 352 - 73			20 55 42	-48 18.	1	15.84		0.12		-0.65		1		3065
199761	+46 03121				20 55 47	+46 59.3	3	7.92	.022	0.45	.009	0.11	.015	7	F4 III	247,261,1119
	+48 03252				20 55 47	+48 41.8	1	9.89		0.68		0.29		2		401
		3HLF4 # 38			20 55 48	-41 05.	1	12.81		0.27		0.09		2	A5	208
		3HLF4 # 1012			20 55 48	-44 42.	1	11.73		1.04		0.80		3	G9 III	221
	+47 03239				20 55 50	+48 21.5	1	10.15		0.34		0.24		3		261
		AAS11,365 T9# 18			20 55 54	+47 30.	1	10.69		1.59		1.32		2		261
199603	-15 05848	HR 8024, DV Aqr			20 55 56	-14 40.6	4	6.00	.018	0.23	.005	0.13		12	A9 V	15,1075,1588,2007
199632	+5 04659				20 55 57	+05 23.8	1	8.42		1.48		1.68		3	K2	1733
	+44 03646				20 55 57	+45 16.7	1	8.99		0.39		-0.04		3	A5 Ib	261
	+43 03775				20 55 59	+43 59.0	2	8.65	.013	0.15	.004	0.20	.012	13	A2	408,1655
		PASP87,379 # 16			20 55 59	+44 00.3	1	10.98		0.49		0.03		6		408
199781	+45 03346				20 55 59	+45 49.0	1	8.37		0.43				2	F2 Ib	1119
199665	+10 04425	HR 8030		⋆ A	20 56 01	+10 38.7	1	5.48		0.93				2	K0	71
	+45 03347				20 56 01	+46 10.3	1	10.89		0.29		0.22		2		117
199697	+21 04424	HR 8032			20 56 02	+22 07.9	1	5.31		1.40				2	K5	71
	+38 04298				20 56 04	+38 22.7	1	9.94		1.28		1.32		2		401
199799	+44 03649				20 56 07	+44 35.6	1	7.33		1.67		1.70		2	M1	408
	+49 03423				20 56 07	+49 31.4	1	9.25		1.20				2	G5	1724
199634	-9 05631	IDS20534S0944	AB		20 56 09	-09 32.7	1	8.97		0.46		-0.06		3	F8	3016
199634	-9 05631	IDS20534S0944	P		20 56 09	-09 32.7	1	12.71		0.92		0.61		3	F8	3016
		LP 928 - 41			20 56 09	-29 28.3	1	12.89		1.11		0.97		1		1696
		CS 22897 # 11			20 56 11	-64 49.5	1	14.38		0.42		-0.03		1		1736
199653	-9 05632	IDS20534S0944	C		20 56 12	-09 30.6	1	9.57		0.53		0.02		3	G0	3016
		BSD 40 # 2273			20 56 14	+45 24.3	1	11.42		0.25		-0.17		2	B8	117
	+46 03122				20 56 14	+46 55.4	1	8.69		1.29				3	G8 Ib	1119
199654	-9 05635	IDS20534S0944	D		20 56 15	-09 33.3	1	9.08		0.87		0.48		2	G0	3016
	+44 03651				20 56 16	+45 21.0	1	10.13		0.42		0.10		3	F2	261
	+45 03349	AZ Cyg			20 56 16	+46 16.4	2	8.68	.193	2.59	.035	2.54		5	M2 Ia	369,8032
199821	+44 03652				20 56 17	+44 50.3	1	9.85		0.66		0.19		2	F8	408
	+40 04372				20 56 18	+41 17.9	1	9.61		1.94				2		369
		PASP87,379 # 26			20 56 18	+44 01.2	1	13.31		0.64		0.19		4		408
199668	-13 05810	IDS20536S1336	AB		20 56 20	-13 24.6	1	8.57		0.37		0.05		4	F0 V	3016
		PASP87,379 # 25			20 56 23	+44 00.0	1	12.27		0.87		0.67		4		408
		Steph 1837			20 56 23	-23 39.1	1	12.97		1.50		1.14		1	K7-M0	1746
199620	-34 14779				20 56 23	-33 48.6	1	7.35		0.84		0.49		4	K0 V	119
199908	+54 02452	DQ Cep			20 56 24	+55 17.6	1	7.22		0.33		0.14		1	F2 II	597
	+46 03123				20 56 25	+47 02.7	1	9.75		0.15				2	A0 V	1119
235412	+50 03236				20 56 29	+50 40.6	3	8.84	.004	0.59	.011	0.13	.000	8	G0	1566,1724,1733
		LP 816 - 31			20 56 31	-16 39.3	1	11.61		0.50		-0.17		2		1696
		PASP87,379 # 24			20 56 32	+43 59.3	1	12.06		0.73		0.28		4		408
199870	+43 03777	HR 8035			20 56 32	+44 16.6	4	5.55	.019	0.97	.004	0.83	.011	8	K0 IIIb	247,408,1118,3051
		VES 348			20 56 32	+47 46.3	1	11.31		0.46		-0.17		2		8106
199890	+47 03240				20 56 34	+47 25.4	3	7.50	.008	-0.10	.008	-0.39		7	B5 V	401,1118,1119
199889	+48 03257				20 56 34	+48 34.0	2	8.33	.019	0.11	.009	-0.21		6	B8 V	261,1119
199766	+3 04473	HR 8034		⋆ AB	20 56 35	+04 06.0	5	5.31	.082	0.47	.005	0.01	.009	16	F6 IV+F6 IV	15,938,1071,3030,8015
199766	+3 04473	IDS20541N0355	C		20 56 35	+04 06.0	1	7.35		0.48		-0.03		3	G0 V	3030
199871	+40 04473	IDS20547N4059	A		20 56 35	+41 09.8	1	7.12		1.53		1.68		2	M0 III	247
		G 25 - 11			20 56 37	+03 21.8	2	12.02	.045	1.40	.015	1.10	.015	5		333,1620,3073
	+44 03654				20 56 37	+45 17.0	1	10.32		0.13		0.05		3		261
	-43 14300				20 56 38	-42 43.9	1	11.54		-0.16		-0.83		2	B2	208
199837	+31 04292				20 56 39	+31 27.2	1	7.19		-0.05		-0.21		2	B9	1733
199892	+41 03949	HR 8036			20 56 40	+41 44.7	2	6.16		-0.08	.015	-0.47		3	B7 III	1079,1118
	+44 03655	LS III +44 030			20 56 40	+44 57.3	1	9.24		0.31		-0.51		2	B1 IV	1012
199639	-43 14302				20 56 40	-43 14.0	1	7.28		0.16		0.06		1	A3 V	208
199891	+46 03124				20 56 41	+46 24.5	1	7.42		0.41				3	F6 V	1025
		AJ89,1229 # 503			20 56 42	+04 02.	1	9.91		1.75				3		1532
		3HLF4 # 40			20 56 42	-40 35.	1	13.97		0.06		0.05		1	A0	208
		VES 349			20 56 46	+46 32.2	1	14.37		2.47				1		8106
199728	-19 05982	HR 8033, AO Cap			20 56 46	-19 13.8	1	6.26		-0.13				4	Ap Si	2007
		YZ Mic			20 56 48	-40 46.0	1	13.53		0.32		0.13		2	F0	221
		3HLF4 # 41			20 56 48	-41 24.	1	12.40		0.08		0.19		1	A0	208
	+46 03126			AB	20 56 49	+46 58.1	1	9.19		0.39				3	F3 IV	1025
	-43 14304				20 56 49	-42 50.6	1	11.55		1.57		0.39		1	K3	1753
199623	-51 12778	HR 8027			20 56 49	-51 27.8	3	5.76	.007	0.48	.004	-0.07	.004	7	F5 IV-V	58,2032,3037
		VES 350			20 56 50	+46 20.7	1	11.44		0.55		-0.27		2		8106
199909	+46 03125				20 56 50	+46 37.8	1	8.72		0.19				2	A5 V	1119
199801	+8 04585				20 56 51	+08 26.6	1	8.86		0.57		-0.01		1	K5	908
	+44 03657	LS III +44 031			20 56 51	+44 53.4	1	10.21		0.55		0.06		2		405

HD	DM	Other Id	N	Rem	α_{1950}	δ_{1950}	S	V	σ_V	B–V	σ_{B-V}	U–B	σ_{U-B}	n	Spectrum	References
199684	−36 14530	HR 8031			20 56 51	−36 19.5	3	6.10	.010	0.40	.007	-0.05		9	F2 V	15,2030,3037
199674	−39 14058				20 56 51	−38 55.0	1	9.55		1.07		0.95		2	K3 V	3072
		CS 22879 # 144			20 56 53	−39 15.6	2	15.12	.010	0.41	.008	-0.21	.009	2		1580,1736
199802	+0 04632				20 56 54	+00 54.2	1	8.90		0.57		-0.02		3		3026
199955	+49 03426	HR 8040		★ AB	20 56 54	+50 16.0	2	5.61		-0.15	.014	-0.52		5	B7 Vn +B7 V	1079,1118
		3HLF4 # 42			20 56 54	−43 16.	1	14.84		0.06		0.16		2	A3	208
199475	−68 03398	HR 8019			20 56 55	−68 24.4	3	6.36	.008	0.10	.000	0.10	.005	12	A2 V	15,1075,2038
199939	+43 03779				20 56 56	+44 13.2	2	7.45	.022	1.24	.004	1.08	.028	3	G0 p	408,993
		BSD 40 # 1150			20 56 57	+45 01.0	1	13.06		0.48		0.17		2	A0	117
	+47 03241				20 56 58	+47 22.0	1	8.85		0.97				3	G0	1724
199803	+0 04633				20 56 59	+00 52.1	1	8.63		0.67		0.13		3	K0	3026
199687	−43 14305	3HLF4 # 1014			20 57 00	−43 04.1	1	10.32		0.57		0.05		2	G0	221
		V1057 Cyg			20 57 06	+44 03.7	1	9.47		1.23		0.58		1		8037
		PASP87,379 # 27			20 57 07	+43 59.9	1	13.11		0.68		0.20		8		408
199956	+43 03780				20 57 08	+43 51.9	1	6.66		1.07		0.97		2	K0	408
199986	+45 03352				20 57 10	+46 03.5	5	7.03	.005	0.20	.016	0.09	.000	10	A5 V	379,1025,1118,1196,1769
	+43 03781				20 57 12	+44 06.3	1	9.95		1.32		1.48		6		408
		LP 928 - 43			20 57 12	−31 29.2	1	14.68		0.85		0.18		3		3062
		BSD 40 # 2350			20 57 14	+45 50.6	1	10.98		0.24		0.16		2	B8	117
		BSD 40 # 2337			20 57 14	+46 01.4	1	10.31		0.25		-0.29		3	B3	117
199957	+37 04131				20 57 17	+37 44.1	1	8.00		1.67		2.03		3	K5	1625
199998	+47 03243				20 57 17	+47 27.3	1	8.41		1.38				3	K2 III	1119
		VES 351			20 57 17	+49 19.5	1	11.63		1.09		0.58		2		8106
		3HLF4 # 43			20 57 18	−41 46.	1	13.83		0.16		0.03		2	A0	208
199757	−43 14310	ZZ Mic			20 57 18	−42 51.1	1	9.36		0.22		0.05		1	A5/7 IV	208
		POSS 457 # 1			20 57 20	+18 03.3	1	13.92		0.83				1		1739
200018	+45 03353				20 57 21	+46 14.7	1	7.72		1.11				3	K0 IV	1119
199987	+35 04344	IDS20554N3602		AB	20 57 24	+36 14.0	1	8.03		0.10		-0.21		2	B9	401
	+48 03259				20 57 24	+48 41.1	2	9.27	.010	0.25	.010			5	A1 V	1119,1724
200019	+44 03658				20 57 25	+44 35.7	1	7.16		0.83		0.51		2	G5	408
199854	−15 05858				20 57 27	−15 18.6	2	8.93	.019	0.42	.005	0.07		3	F(7)wA(0)	742,1594
199941	+16 04425	HR 8037		★ A	20 57 31	+16 37.7	3	6.62	.049	0.38	.005	-0.12	.024	3	F4 III	254,3016
199941	+16 04426	IDS20552N1626		B	20 57 31	+16 37.7	1	9.14		0.77		0.25		2		3032
		BSD 40 # 2373			20 57 32	+45 58.2	1	10.54		0.29		0.03		2	B9	117
200030	+41 03956				20 57 34	+42 07.7	2	6.47	.005	-0.08	.000	-0.29		4	B9	401,1118
199942	+6 04718	HR 8038		★ AB	20 57 36	+07 19.2	3	5.99	.004	0.27	.004	0.08	.007	10	A4 V +F3 V	15,253,1256
	+46 03129				20 57 37	+47 08.7	1	9.49		0.06				3	B8 III	1119
200041	+47 03246				20 57 37	+48 06.8	2	9.33	.000	0.06	.014	-0.28		6	B7 V	261,1119
199971	+18 04668				20 57 40	+18 25.4	1	7.12		1.06		0.95		1	K1 III	1716
	+41 03958				20 57 43	+42 00.9	1	10.25		1.20				1		934
		PASP87,379 # 14			20 57 44	+43 46.2	1	10.85		2.53		2.22		1		408
		BSD 40 # 1178			20 57 45	+45 06.8	1	11.05		0.42		-0.07		3	B3	117
		PASP87,379 # 15			20 57 46	+43 57.1	1	11.11		0.57		0.13		3		408
		POSS 457 # 2			20 57 47	+16 56.8	1	16.48		1.34				1		1739
200031	+38 04306				20 57 47	+38 37.6	3	6.76	.004	0.79	.004	0.46	.015	8	G5 III	247,833,6009,8100
200060	+43 03782				20 57 47	+44 01.9	2	7.55	.010	1.17	.015	1.27	.015	6	K2 III	37,408
		VES 306			20 57 49	+38 21.5	1	12.60		1.38		1.10		1		8056
335435	+28 03960				20 57 50	+29 05.0	1	9.01		0.60		0.10		3	F8 V	3026
		PASP87,379 # 32			20 57 51	+44 05.6	1	12.60		0.75		0.31		4		408
		BSD 40 # 1188			20 57 52	+44 46.2	1	10.02		0.20		0.10		2	B9	117
		PASP87,379 # 33			20 57 53	+44 05.3	1	13.79		1.24		1.20		4		408
	+46 03130				20 57 53	+47 19.6	1	9.58		0.03				3	A0 V	1119
199960	−5 05433	HR 8041			20 57 56	−04 55.5	2	6.20	.000	0.63	.000	0.22	.005	9	G1 V	1088,3077
199863	−41 14339				20 57 57	−40 48.5	1	9.32		0.05		0.00		2	A0 V	208
	−42 15193				20 57 57	−42 23.4	1	12.28		0.24		0.11		2	A3	208
	+45 03356				20 57 58	+45 45.9	1	8.67		0.95		0.66		3	K0	261
	+45 03357				20 57 59	+46 04.3	1	9.74		1.94		2.10		3	K0	261
200077	+39 04400	G 210 - 46		★ A	20 58 01	+40 03.6	2	6.58	.010	0.56	.005	0.02	.034	3	F8	247,1620
200089	+46 03131				20 58 01	+46 37.4	4	8.51	.007	0.15	.006	0.07		9	A5 III	261,1118,1119,1196
200102	+44 03661				20 58 05	+44 48.0	2	6.63	.005	1.05	.005	0.72	.000	4	G1 Ib	247,1080
		VES 352			20 58 07	+45 31.1	1	11.45		1.90		1.86		2		8106
200120	+46 03133	HR 8047, V832 Cyg		★ AP	20 58 07	+47 19.5	11	4.71	.109	-0.04	.023	-0.93	.016	32	B1 ne	15,154,180,542,1118*
200120	+46 03133	V832 Cyg		★ APB	20 58 07	+47 19.5	1	4.58		0.00		-0.98		2	B1 ne +A3 V	542
200120	+46 03133	IDS20564N4708		B	20 58 07	+47 19.5	5	9.39	.026	0.21	.016	-0.04	.155	9	A3 V	542,542,3024,8072,8084
200120	+46 03133	IDS20564N4708		C	20 58 07	+47 19.5	3	11.32	.175	0.40	.152	-0.88	.065	5		3016,8072,8084
200120	+46 03133	IDS20564N4708		D	20 58 07	+47 19.5	2	12.62	.083	-0.41	.054	-0.95	.107	3		8072,8084
		VES 353			20 58 08	+45 53.0	1	12.11		0.32		-0.11		1		8106
200044	+18 04675	HR 8044		★ A	20 58 10	+19 08.1	1	5.65		1.61				2	M3 IIIab	71
		G 210 - 48		★ AB	20 58 10	+39 52.6	1	10.12		1.52		1.06		3	M3 Ve	196
		VES 307			20 58 10	+41 54.4	1	12.64		2.22		1.89		1		8056
		G 262 - 32			20 58 10	+64 51.0	1	10.73		0.82		0.38		2	K0	1658
	+44 03664				20 58 11	+45 09.2	1	10.13		0.55		-0.19		3	B9	117
200205	+58 02201	HR 8049			20 58 11	+59 14.6	1	5.51		1.40		1.64		3	K2	1355
		MCC 829			20 58 12	+17 12.	1	9.82		1.13				2	K8	1619
199951	−32 16353	HR 8039		★ A	20 58 14	−32 27.3	6	4.67	.013	0.89	.003	0.54	.007	24	G8 III	3,15,208,1075,2012,4001
		G 231 - 22			20 58 15	+51 42.0	1	14.72		1.11		0.72		1		1658
199435	−78 01370				20 58 15	−77 44.6	1	8.29		1.10		0.68		1	G8 Ib/II	565
	+42 03935	TX Cyg			20 58 16	+42 24.1	4	8.80	.303	1.60	.052	1.19		4	F5.5Ib	934,1399,1772,6011
	+49 03435				20 58 17	+49 52.0	1	9.05		0.41				2	F5	1724
		3HLF4 # 45			20 58 18	−41 47.	1	13.10		0.04		0.16		2	A0	208

Table 1 1133

HD	DM	Other Id	N Rem	α_{1950}	δ_{1950}	S	V	σ_V	B−V	σ_{B-V}	U−B	σ_{U-B}	n	Spectrum	References
+44 03666		LS III +44 033		20 58 19	+44 51.0	1	10.23		0.50		-0.29		2	B2	117
		BSD 40 # 2432		20 58 20	+45 52.7	1	12.34		0.38		0.16		2	A0	117
		BSD 40 # 2433		20 58 26	+46 00.2	1	11.63		0.32		0.18		2	B5	117
200177	+48 03260	IDS20568N4817	AB	20 58 26	+48 29.0	3	7.34	.013	0.01	.011	-0.13		7	B9p	401,1118,1119
	+37 04138			20 58 31	+37 56.1	1	9.11		1.70				2	M2	369
199981	−33 15343			20 58 33	−32 43.1	1	9.40		1.25		1.21		2	K5 V	3072
200007	−28 17057			20 58 34	−28 00.0	1	7.62		0.88		0.55		3	G8/K0 III	1657
199963	−41 14346			20 58 35	−41 07.7	1	9.52		0.22		0.12		2	A1mA7-A8/9	208
		3HLF4 # 46		20 58 36	−41 00.	1	14.31		0.28		0.10		2	A3	208
	+33 04117	G 211 - 5	⋆ A	20 58 41	+33 42.0	1	9.62		0.75		0.32		2	G5	1658
		RV Cap		20 58 42	−15 25.3	1	10.61		-0.01		0.07		1	A8	699
200240	+52 02846			20 58 43	+52 28.7	1	9.30		0.28		0.21		2	A2	1566
		G 25 - 12		20 58 44	+09 35.9	1	12.83		1.20		1.04		1		333,1620
199532	−77 01474	HR 8021		20 58 44	−77 13.0	4	5.14	.009	0.49	.008	0.11	.008	16	G2III+A7III	15,1075,2038,3077
		VES 308		20 58 46	+39 29.0	1	12.12		0.47		0.33		1		8056
	+44 03668	CO Cyg		20 58 47	+44 33.7	1	8.59		1.68		1.99		2	K5	408
	+45 03359			20 58 47	+45 49.2	1	10.21		0.17		0.09		2		117
	+47 03250			20 58 47	+48 19.0	1	9.08		1.04				2	G8 III	1119
200052	−27 15197	HR 8045		20 58 47	−27 04.7	1	6.06		0.06				4	A3 V	2007
	+45 03360	LS III +46 038		20 58 48	+46 03.0	1	10.00		0.20		-0.43		2	B3 V	1012
		3HLF4 # 47		20 58 48	−42 12.	1	14.72		0.09		0.13		2	A2	208
200151	+20 04803			20 58 52	+20 24.7	1	9.25		0.15		0.08		3	A2	1648
	+48 03261			20 58 53	+48 31.2	1	9.13		0.37				2	F2	1724
200011	−43 14325	HR 8042	⋆ A	20 58 54	−43 11.9	5	6.64	.014	0.67	.009	0.16	.010	12	F8/G0 V	15,219,1075,1279,3077
		CS 22897 # 8		20 58 56	−65 17.1	2	13.32	.010	0.71	.017	0.11	.003	3		1580,1736
200241	+44 03669			20 58 57	+44 35.3	1	8.53		0.74		0.33		2		408
		GD 232		20 58 58	+18 09.1	1	15.03		0.01		-0.75		1	DA	3060
	−43 14426			20 58 58	−43 41.9	1	11.81		0.31		0.04		2	A7	208
200083	−23 16657			20 58 59	−22 46.7	1	9.62		1.05		0.89		2	K3 V	3072
200026	−43 14327	IDS20556S4323	B	20 59 00	−43 11.6	3	6.90	.005	0.96	.007	0.72	.000	7	K0 III/IV	219,1075,1279
200138	+0 04642			20 59 01	+01 18.4	1	8.22		-0.06		-0.24		4	A0	379
	+44 03670			20 59 01	+45 18.4	1	8.82		0.72		0.22		3	F8	261
	+48 03263			20 59 02	+48 35.1	1	9.19		0.21				2	A5	1724
200252	+46 03140			20 59 04	+46 53.7	1	9.44		-0.01				4	A0 V	1119
200269	+46 03141			20 59 06	+46 22.9	2	7.28	.014	-0.11	.019	-0.48		4	B5 V	401,1119
		CS 22937 # 17		20 59 06	−38 55.8	1	13.65		0.33		-0.04		1		1736
	+45 03361			20 59 08	+46 11.2	3	8.59	.010	0.13	.005			7	A2 V	1025,1118,1196
200057	−41 14349			20 59 09	−41 01.4	1	9.65		0.34		0.15		2	Fm δ Del	208
		IRC +50 353		20 59 10	+45 11.4	1	13.98		2.44		0.56		1		8056
		CS 22897 # 1		20 59 10	−66 33.9	2	14.03	.010	0.47	.010	-0.16	.006	3		1580,1736
	+46 03142	IDS20575N4706	AB	20 59 12	+47 17.4	1	9.64		0.44		0.08		1		8084
200253	+35 04357	HR 8051		20 59 13	+35 49.8	1	5.97		0.98				2	G5 III	71
	+48 03265			20 59 13	+49 19.8	1	8.41		1.72				2	M0	369
	+47 03253			20 59 14	+47 48.6	1	9.57		0.14				2	A0 V	1119
200073	−39 14079	HR 8046		20 59 15	−38 43.6	2	5.93	.005	1.11	.000			7	K2 III	15,2033
	−45 14155		AB	20 59 16	−45 34.2	1	12.27		0.20		0.10		2	A3	208
		Wolf 906		20 59 20	−06 30.6	1	11.24		1.49				1	M3	1705
	+10 04443			20 59 21	+11 21.6	1	9.44		0.25		0.14		2	A2	1733
		Orsatti 247		20 59 21	+43 50.0	1	12.48		0.68		-0.20		2		1766
		LP 928 - 47		20 59 21	−27 04.2	1	12.63		0.53		-0.19		2		1696
200311	+43 03786			20 59 25	+43 31.5	1	7.70		-0.10		-0.49		2	B9p	401
200310	+45 03364	HR 8053, V1931 Cyg	⋆ AB	20 59 26	+45 57.5	8	5.37	.027	-0.20	.028	-0.94	.017	20	B1 Ve	379,1118,1119,1193*
235425	+50 03245			20 59 27	+51 11.3	1	9.84		0.61				2	A7	1724
		L 211 - 59		20 59 30	−57 09.	2	12.84	.010	1.62	.025	1.36		2	M4	1494,3062
235427	+50 03246			20 59 35	+50 41.1	1	9.93		0.33				3		1724
200289	+29 04282	IDS20575N2914	AB	20 59 36	+29 26.0	1	7.90		-0.11		-0.76		2	B8	1319
199997	−63 04631			20 59 39	−63 40.8	2	7.67	.010	0.84	.005	0.50		6	G1 II	2033,8100
		G 187 - 15		20 59 41	+31 37.0	1	15.07		0.10		-0.79		3	DC	316
	+70 01157			20 59 42	+70 28.4	1	9.38		1.12		1.05		2	K2	3072
	+48 03267			20 59 44	+48 22.9	3	9.24	.010	0.39	.014	-0.05		8	A5 V	261,1025,1724
200369	+46 03144			20 59 45	+47 00.5	2	8.99	.010	0.02	.005	-0.06		7	B8 V	261,1119
	−45 14159			20 59 45	−44 59.0	1	12.25		0.15		0.11		3	A2	208
		G 144 - 51		20 59 46	+19 01.2	1	16.40		0.37		-0.39		4	DAwk	316
200163	−39 14089	HR 8048		20 59 47	−38 49.7	1	5.31		0.41				4	F3 V	2009
		CS 22897 # 7		20 59 47	−65 27.4	1	14.39		0.43		0.08		1		1736
	+49 03439			20 59 50	+49 44.1	1	9.23		0.56				2	F8	1724
200114	−53 09979			20 59 55	−53 05.3	1	7.43		1.56				4	M2 III	2012
200406	+46 03145			20 59 56	+47 18.2	4	7.94	.018	0.52	.007	0.08	.015	10	K2 III	247,261,1025,1724
200405	+47 03256			20 59 56	+47 42.9	2	8.92	.000	0.09	.000	0.04		5	A2p	261,1119
200407	+43 03789	IDS20582N4348	AB	20 59 59	+43 59.4	4	6.74	.019	0.32	.013	0.10	.017	8	Am	408,3016,8071,8080
200449	+46 03146	IDS20584N4610	AB	21 00 05	+46 22.2	2	8.68	.013	0.96	.004	0.69		5	G8 III	261,1119
200448	+47 03257			21 00 05	+47 37.3	1	8.07		1.25				2	K1 II	1119
		3HLF4 # 50		21 00 06	−45 24.	1	13.81		0.12		0.22		2	A0	208
200352	+22 04278			21 00 08	+23 07.6	1	7.38		0.01		-0.10		2	B9	1319
	+46 03147			21 00 08	+46 36.3	1	10.03		0.11		0.05		4		261
200245	−28 17077	HR 8050	⋆ AB	21 00 12	−27 55.8	2	6.25	.006	0.94	.006	0.74		6	K0 III	58,2009
200478	+46 03149			21 00 14	+46 49.3	3	7.79	.009	0.08	.000			7	A1 IV	1118,1119,1196
200391	+27 03952	ER Vul		21 00 16	+27 36.6	1	7.30		0.62		0.07		1	G0 III	588
200477	+47 03264			21 00 23	+47 25.0	2	8.04	.019	0.97	.005			6	G8 IV	1025,1724
200425	+25 04442			21 00 24	+25 57.6	1	7.80		0.55		0.04		4	F8 V	3026

HD	DM	Other Id	N Rem	α1950	δ1950	S	V	σV	B–V	σB-V	U–B	σU-B	n	Spectrum	References
		AS 453, V407 Cyg		21 00 24	+45 34.7	1	11.51		2.26		1.74		1	Mpe	1753
200340	−1 04095	HR 8054		21 00 25	−01 07.4	5	6.50	.003	-0.10	.003	-0.48	.016	24	B6 V	15,989,1071,1079,1729
200465	+38 04321	IDS20585N3907	AB	21 00 26	+39 18.7	1	6.31		1.51		2.39		1	A1 V	8100
	+47 03265		AB	21 00 26	+47 46.7	1	8.72		1.33				2	K0 III	1119
200393	+15 04317			21 00 27	+15 46.1	1	6.84		1.71		2.01		3	M1	1648
200508	+46 03150	IDS20588N4618	AB	21 00 29	+46 29.3	1	8.64		0.87				3	K0 V	1025
		VES 310		21 00 30	+43 45.2	1	10.37		1.16		0.90		1		8056
		L 163 - 69		21 00 30	−61 58.	1	13.32		0.71		0.08		2		3062
200375	+0 04648	HR 8056	⋆ AB	21 00 31	+01 20.1	5	6.22	.015	0.48	.007	0.00	.013	14	F5 V	15,254,938,1256,3016
200261	−41 14359			21 00 32	−40 52.8	1	8.74		0.01		-0.04		2	A0 V	208
200575	+56 02523	IDS20592N5641	AB	21 00 33	+56 52.4	2	6.69	.005	-0.12	.005	-0.52	.005	4	B8	401,1733
200410	+15 04318			21 00 35	+16 01.7	1	8.54		1.02				2	K0	882
200527	+44 03679	HR 8062	⋆ V	21 00 37	+44 35.6	5	6.19	.040	1.69	.012	1.84	.018	10	M3 Ib-II:	71,247,369,3001,8032
200430	+14 04518	HR 8057		21 00 40	+14 31.9	2	6.30	.011	1.67	.004	2.04		3	M1 III	71,3005
200428	+15 04320			21 00 40	+15 33.7	1	7.67		0.93				2	G5	882
200378	−15 05874			21 00 44	−15 21.6	1	7.15		1.28		1.30		4	K1 III	379
	−29 17512			21 00 45	−29 23.6	1	11.01		0.35		0.17		2		1730
200614	+56 02524	HR 8065	⋆ AB	21 00 46	+56 28.3	1			-0.07		-0.35		1	B8 III	1079
	−29 17513			21 00 46	−29 25.7	1	11.36		0.93		0.48		2		1730
		CS 22937 # 6		21 00 47	−41 24.5	2	14.83	.005	0.41	.008	-0.21	.009	3		1580,1736
	+6 04733	G 25 - 13		21 00 48	+06 53.2	1	10.26		0.98		0.82		2	K2	333,1620
200594	+52 02854			21 00 51	+52 22.6	1	8.65		0.02		-0.11		2	A0	401
200530	+36 04379			21 00 52	+36 57.1	1	7.75		0.63		0.13		2	K0	247
200560	+45 03371	G 212 - 17	⋆ CD	21 00 53	+45 41.1	2	7.69	.010	0.97	.005	0.76	.029	12	K2	247,261
200576	+47 03266			21 00 53	+47 49.9	1	6.92		1.69				2	K5 Ib	1119
200433	−2 05434			21 00 54	−01 46.7	1	6.94		0.42		-0.04		6	F5	1628
		VES 354		21 00 56	+51 43.7	1	10.72		0.40		0.17		2		8106
200511	+25 04444			21 00 58	+25 43.3	1	8.27		0.17		0.07		3	A2	833
200531	+34 04252			21 00 58	+34 50.3	1	8.07		0.48		0.00		2	F3 V	247
		CS 22898 # 40		21 00 59	−22 12.9	1	14.84		0.42		-0.15		1		1736
335471	+28 03971	G 187 - 18		21 01 00	+29 17.2	1	10.19		0.61		0.03		2	G0	1658
200775	+67 01283			21 01 00	+67 57.9	4	7.36	.043	0.38	.011	-0.39	.007	10	B2 Ve	206,351,379,1212
	+45 03375			21 01 01	+46 07.4	1	9.13		0.14				3	A0 III	1119
200595	+45 03374	HR 8064	⋆ AB	21 01 03	+45 39.0	3	6.49	.005	-0.15	.007	-0.56		5	B3 Vn	1025,1079,1118
		G 25 - 14		21 01 04	+02 05.6	2	14.20	.005	0.80	.015	0.07	.039	3		1620,1658,3062
		VES 355		21 01 05	+44 37.0	1	12.47		0.84		-0.31		2		8106
200494	+2 04294			21 01 08	+02 43.7	3	7.44	.013	1.23	.022	1.28	.024	18	K0	1509,3016,4001
200577	+38 04325	HR 8063		21 01 08	+38 27.5	2	6.08	.006	1.01	.000	0.79		3	G8 III	71,770
		LS III +47 023		21 01 08	+47 08.1	1	10.96		1.10		0.95		4		180
200615	+47 03267			21 01 08	+48 19.1	1	8.14		0.03				3	B8 V	1119
200334	−46 13822			21 01 09	−46 27.2	1	7.04		0.58				4	G0 V	1075
200361	−44 14270			21 01 11	−44 39.6	2	9.33	.000	0.68	.000			8	G3 V	1075,2012
200596	+39 04418	IDS20593N3936	A	21 01 12	+39 47.6	1	8.63		0.63				33	G0	6011
	+44 03682	G 212 - 18		21 01 15	+44 36.8	2	9.13	.002	0.68	.001	0.16	.014	5	G5 V	1658,1723
200616	+37 04159			21 01 20	+37 57.2	1	8.09		-0.11		-0.50		2	A2	401
200657	+48 03271			21 01 20	+48 50.4	1	8.60		0.28				3	A5 V	1025
		LS III +59 001		21 01 20	+59 14.3	1	11.17		0.85		-0.11		3		483
	+11 04463			21 01 21	+11 49.3	1	9.30		0.48		-0.15		1	F7 V	979
	+46 03155			21 01 21	+46 58.5	1	9.44		0.12				3	B9 V	1119
		VES 356		21 01 21	+49 53.4	1	13.03		2.96				2		8106
200578	+28 03974	IDS20592N2842	AB	21 01 22	+28 53.6	1	6.96		0.98		0.73		2	G8 III	3016
	−6 05663			21 01 22	−06 19.9	1	10.52		0.97		0.71		2		3016
		LS III +47 024		21 01 25	+47 02.3	2	11.55	.090	0.73	.030	-0.19	.100	2		180,8106
200497	−6 05664	HR 8059	⋆ APB	21 01 26	−06 01.3	5	5.53		0.68		0.39		8	G4 III	1088
200516	−11 05511			21 01 27	−11 33.9	1	8.84		0.45		-0.30		4		1732
200304	−60 07445			21 01 30	−60 36.1	1	8.54		1.02		0.80		9	K0 III	1704
200499	−20 06115	HR 8060	⋆ AB	21 01 34	−20 03.2	6	4.83	.005	0.16	.010	0.08	.021	19	A5 V	15,1075,1425,2012*
200365	−55 09509	HR 8055		21 01 34	−54 55.6	2	5.16	.000	1.20	.003	1.22		6	K2 III	58,2007
200565	+3 04493			21 01 35	+03 47.0	1	8.45		0.65		0.13		3	G5	3031
200580	+2 04295	G 25 - 15		21 01 37	+02 48.0	7	7.32	.011	0.54	.007	-0.05	.012	12	F9 V	219,333,979,1003,1620*
		LS III +47 025		21 01 37	+47 16.1	1	11.31		0.52		-0.28		1		180
		CS 22951 # 115		21 01 38	−44 25.2	2	14.02	.010	0.40	.007	-0.22	.010	2		1580,1736
200349	−59 07670			21 01 39	−59 20.3	1	9.48		1.04		0.87		2	K2/3 V	3072
		KUV 21017+7158		21 01 40	+71 58.5	1	12.51		0.54		-0.15		1	NHB	1708
200709	+45 03381			21 01 43	+45 35.6	1	9.21		-0.02				2	B8 V	1119
200739	+50 03256			21 01 43	+50 36.2	2	7.95	.080	0.37	.023	0.22	.000	4	Am	8071,8080
		AJ89,1229 # 507		21 01 48	+02 30.	1	11.13		1.57				1		1532
		AJ89,1229 # 506		21 01 48	+06 05.	1	10.94		1.63				2		1532
200740	+49 03448	HR 8072		21 01 48	+50 09.1	1	6.37		0.98		0.76		5	K0	253
	+46 03157			21 01 49	+46 46.9	1	9.31		0.09				4	B9 V	1119
200722	+44 03685			21 01 52	+45 10.1	1	7.86		-0.06		-0.24		2	A0	401
		VES 311		21 01 53	+42 50.3	1	11.86		0.75		0.30		1		8056
		G 144 - 58		21 01 54	+20 24.9	1	14.56		1.58		1.16		3		203
	+47 03272			21 01 57	+48 16.5	2	9.45	.000	0.09	.009	0.02		6	B8 V	261,1119
201597	+82 00636			21 01 59	+82 47.4	1	8.16		0.15		0.09		2	A5	1733
200723	+41 03987	HR 8071	⋆ A	21 02 00	+41 25.8	1	6.34		0.38		0.13		2	F3 IV	3024
200723	+41 03986	IDS21002N4114	B	21 02 00	+41 25.8	1	8.82		0.48		-0.01		2		3024
		VES 358		21 02 00	+45 09.1	1	11.84		0.39		-0.18		1		8106
200753	+46 03159	HR 8074		21 02 00	+46 39.9	2	6.32	.030	0.24	.005			4	F0 IVn	1118,1119
		VES 359		21 02 03	+44 57.1	1	12.18		0.30		-0.17		1		8106

Table 1   1135

HD	DM	Other Id	N	Rem	α_{1950}	δ_{1950}	S	V	σ_V	B–V	σ_{B-V}	U–B	σ_{U-B}	n	Spectrum	References
	−62 06195				21 02 03	−61 45.6	2	9.83	.019	0.72	.015	0.14		3	G5	742,1594
200644	+4 04606	HR 8066		⋆ A	21 02 05	+05 18.2	4	5.60	.011	1.65	.000	1.98	.039	15	K5 III	15,1071,1355,1509
		Ross 769			21 02 08	−17 08.0	2	11.47	.014	1.46	.022	1.08		4	M3	1705,3078
200776	+45 03384	V1898 Cyg		⋆	21 02 09	+46 07.9	2	7.76	.040	0.02	.005	-0.82		5	B1 IV:p	1012,1119
	−21 05912				21 02 09	−21 03.7	1	10.54		1.25		1.18		3	K3 V	803
200661	+2 04297	HR 8067			21 02 10	+02 44.5	1	6.41		1.05		0.89		8	K0	1088
200663	+1 04418	HR 8068			21 02 13	+02 04.3	1	6.34	.016	0.96	.005	0.62	.007	11	G5	15,1256,1355,3016
200553	−44 14278				21 02 14	−43 43.3	1	7.20		0.94				4	K0 III	2012
200817	+52 02859	HR 8078			21 02 16	+53 05.2	1	5.91		0.99				2	K0 III	71
200324	−70 02830				21 02 16	−69 48.5	1	8.80		0.19		0.14		4	A3/5 IIIs	244
		VES 312			21 02 17	+39 25.9	1	11.69		2.06		2.11		1		8056
200804	+48 03277	LS III +49 015			21 02 17	+49 02.4	1	8.15		0.32		-0.56		2	B3 IV	180
		LP 636 - 38			21 02 17	−00 49.4	1	13.19		0.65		-0.05		1		1696
	+49 03450				21 02 20	+49 54.3	1	9.11		0.47				2	F5	1724
239558	+55 02510				21 02 21	+56 16.9	1	9.50		0.26		-0.54		1	A7	433
	+39 04423	VY Cyg			21 02 22	+39 46.4	4	9.31	.127	1.11	.119	0.87	.077	5	F5.5Ib	247,851,1399,6011
200805	+44 03688	LS III +44 037			21 02 23	+44 57.4	3	8.30	.015	0.81	.008	0.65	.007	9	F5 Ib	247,261,8100
		VES 313			21 02 24	+39 49.8	1	11.85		1.13		1.12		1		8056
200857	+54 02470	LS III +55 005			21 02 26	+55 01.9	2	7.13		0.55	.005	-0.25	.015	3	B3 III	5,1012
200610	−41 14375				21 02 31	−41 17.7	1	10.53		-0.04		-0.05		2	B9 III	208
		LS III +48 015			21 02 32	+48 44.6	1	11.01		0.32		-0.38		2		180
200830	+45 03387				21 02 33	+46 19.9	2	8.60	.005	0.11	.005	-0.31		5	A0	401,1119
		LS III +47 026			21 02 33	+47 39.5	1	10.82		1.17		0.25		2		180
235436	+52 02862				21 02 34	+52 33.4	1	9.88		0.45				2		1724
	+45 03388			AB	21 02 36	+46 19.1	1	9.34		0.07				3	B8	1119
200839	+47 03276				21 02 37	+47 54.1	1	8.26		1.16				3	K0 V	1119
		VES 314			21 02 42	+41 25.4	1	11.76		0.69		0.24		1		8056
200858	+45 03389				21 02 42	+45 50.9	1	7.92		1.10				3	G8 III	1119
200652	−33 15382				21 02 43	−33 40.9	1	8.68		1.08		0.98		4	K2 II	119
200669	−29 17539				21 02 44	−29 19.6	1	9.79		0.42		0.16		2	F2/3 IV	1730
		G 144 - 60			21 02 45	+14 23.4	1	12.81		1.43		1.16		3		203
	+39 04425				21 02 48	+39 47.4	1	10.74		1.01				1		851
200779	+6 04741	G 25 - 16		⋆ AC	21 02 51	+06 52.6	5	8.30	.020	1.22	.012	1.15	.037	10	K5	333,1003,1197,1620,1758,3072
		G 144 - 61			21 02 52	+19 24.5	1	11.61		1.24		1.10		3		7010
	+46 03166				21 02 54	+46 53.1	1	10.06		0.20				3	A0 V	1119
200859	+38 04335				21 02 55	+39 09.3	1	8.83		0.17		0.14		4	A2 V	833
200925	+50 03259			V	21 02 55	+50 35.0	1	8.06		0.42				5	F5 III	1724
	+78 00738	G 261 - 36			21 02 55	+79 05.4	1	8.65		0.68		0.20		1	G8	1658
	−17 06172				21 02 56	−17 06.8	1	10.33		1.42		1.17		2	K7	1746
		CS 22898 # 27			21 02 56	−18 48.9	1	12.76		0.50		-0.16		1		1736
200790	+5 04697	HR 8077		⋆ A	21 02 58	+05 45.6	4	5.94	.004	0.54	.004	0.03	.016	13	F8 V	15,254,1071,3080
	+13 04614	IDS21011N1408		A	21 03 00	+14 20.4	1	10.51		1.15		1.17		1	K5	3032
	+13 04614	IDS21011N1408		B	21 03 00	+14 20.4	1	12.78		1.45		1.08		1		3032
200718	−30 18382	HR 8070			21 03 00	−30 19.5	1	5.68		1.03				4	K0/1 III	2032
200943	+51 02991	IDS21014N5200		AB	21 03 01	+52 12.0	2	7.59	.000	-0.04	.015	-0.21		5	A0	401,1118
200927	+48 03279	IDS21014N4838		AB	21 03 03	+48 49.7	2	8.05		0.13		-0.29		4	B6 V	401,1119
200654	−50 13237				21 03 05	−50 09.7	3	9.09	.012	0.63	.011	-0.11	.010	9	G0	742,1594,1696
		LS III +47 027			21 03 06	+47 45.0	1	10.48		0.48		-0.49		2	B1 V nne	1012
		3HLF4 # 52			21 03 06	−42 26.	1	14.18		0.07		0.16		2	A0	208
200266	−76 01473	HR 8052			21 03 06	−76 24.9	3	6.57	.008	1.23	.000	1.18	.000	11	K1 III	15,1075,2038
200905	+43 03800	HR 8079			21 03 07	+43 43.7	14	3.73	.044	1.65	.011	1.80	.035	37	K5 Ib	1,15,369,1025,1118*
200761	−17 06174	HR 8075			21 03 08	−17 26.0	6	4.06	.005	-0.01	.003	0.01	.005	19	A1 V	15,1075,1425,2012*
		CS 22898 # 47			21 03 08	−21 28.9	1	14.24		0.56		-0.13		1		1736
		CS 22897 # 17			21 03 10	−63 51.1	2	14.32	.010	0.43	.008	-0.18	.008	3		1580,1736
	+45 03393				21 03 11	+46 20.5	1	9.97		0.12				2	A2 Ib	1119
	+49 03457				21 03 11	+49 49.5	1	8.79		0.86				3	G5	1724
200702	−41 14379	HR 8069		⋆ A	21 03 11	−41 35.2	3	5.53	.005	1.35	.000	1.51	.008	8	K3 III	58,221,2009
200945	+46 03167				21 03 14	+46 56.8	2	7.76	.025	1.79	.015	1.96		3	K2 Ib	1119,8100
	+46 03168				21 03 14	+47 18.0	1	9.61		0.34		0.07		2	Am	261
200944	+47 03281				21 03 14	+47 40.2	1	8.54		0.02				2	B9 V	1119
		CS 22898 # 26			21 03 16	−18 45.2	1	13.43		0.23		-0.51		1		1736
200703	−46 13835				21 03 17	−46 28.5	1	8.80		0.60				4	G0 V	2012
200863	+16 04454				21 03 18	+16 41.1	2	8.23	.005	1.48	.010	1.80		4	K5	882,1648
	+45 03394				21 03 18	+45 50.4	2	9.57	.028	0.11	.005	0.17		4	A1 III	1119,8100
		3HLF4 # 53			21 03 18	−41 56.	1	12.96		0.00		0.06		1	A0	208
	+45 03395				21 03 20	+46 20.8	2	8.11	.020	1.88	.025	2.10		3	K2 Ib	1119,8100
	−41 14380	Smethells 71			21 03 21	−41 17.1	1	11.09		1.26				1		1494
201033	+55 02514				21 03 22	+55 23.2	1	8.04		0.26		0.15		3	Am	8071
201063	+59 02313				21 03 22	+60 03.4	1	7.30		0.43		-0.04		1	F5 V	1716
200763	−32 16398	HR 8076			21 03 22	−32 32.6	1	5.18		1.10				4	K2 III	2032
		3HLF4 # 54			21 03 24	−45 08.	1	13.21		0.01		-0.02		2	A0	208
200877	+14 04530				21 03 27	+15 07.6	1	6.64		0.46		-0.03		2	F7 IV	1648
		G 25 - 17			21 03 28	+07 51.2	3	12.77	.034	0.82	.013	0.30	.008	6		1620,1658,3062
200784	−38 14378				21 03 30	−37 46.0	1	8.90		0.94		0.61		4	G8 IV	119
		G 187 - 21			21 03 31	+26 26.5	1	10.74		1.08		0.92		4	K7	1723
	+45 03398				21 03 31	+45 34.1	1	9.51		0.41				2	A4 III	1119
	+46 03170				21 03 31	+47 17.5	1	9.80		0.20				1		1119
200655	−60 07451				21 03 31	−60 11.6	1	6.78		0.64				4	G2 V	2012
201034	+50 03260				21 03 37	+50 57.0	2	8.83	.010	0.46	.005	-0.01		5		1566,1724
200848	−10 05596				21 03 38	−10 35.0	1	10.04		0.20		0.19		2	A0	1700

HD	DM	Other Id	N Rem	α₁₉₅₀	δ₁₉₅₀	S	V	σ_V	B–V	σ_B-V	U–B	σ_U-B	n	Spectrum	References
	+1 04423			21 03 47	+01 59.1	1	8.94		0.95		0.63		11	K0	4001
201049	+48 03282			21 03 47	+48 29.9	2	9.04	.010	0.00	.000	-0.07		7	B8 IV	261,1119
	+47 03283	IDS21021N4734	AB	21 03 50	+47 46.2	1	10.23		0.50		0.05		3		261
201065	+46 03174			21 03 51	+46 45.7	3	7.58	.029	1.79	.027	1.93	.077	6	K5 Ib	247,1119,8100
	-44 14292			21 03 51	-44 04.4	1	11.90		0.18		0.11		2	A0	208
201035	+38 04341			21 03 56	+39 07.5	1	8.15		0.18		0.15		4	A3 V	833
200932	-0 04161		V	21 03 59	-00 18.4	1	6.90		1.51		1.72		4	K5 I:	1149
		3HLF4 # 56		21 04 00	-42 28.	1	14.85		-0.02		0.03		2	A0	208
201076	+47 03284	IDS21023N4724	A	21 04 01	+47 36.2	2	7.45	.000	-0.05	.000	-0.23		5	A0 V	401,1119
		VES 315		21 04 02	+42 20.2	1	11.78		0.71		0.12		1		8056
	+46 03177			21 04 03	+46 47.7	1	9.38		0.46		-0.04		3	A1 V	261
	+47 03285			21 04 03	+48 14.8	1	10.17		1.25		1.03		3		261
	+46 03178			21 04 05	+47 22.3	1	10.25		0.10				2	A1 V	1119
		3HLF4 # 57		21 04 06	-40 13.	1	14.13		0.05		0.13		2	A0	208
201006	+15 04331			21 04 07	+15 47.4	1	8.75		1.66				2	K5 III	882
201089	+45 03403			21 04 10	+45 31.0	1	8.53		0.01				2	B9.5V	1119
		VES 360		21 04 11	+49 05.2	1	10.75		0.58		0.26		1		8106
200525	-73 02192	HR 8061	★ABC	21 04 11	-73 22.3	5	5.67	.007	0.59	.008	0.09	.005	24	F8/G0 V	15,1075,2012,2038,3026
	+48 03284			21 04 12	+48 34.1	1	9.06		0.73		0.31		3	G5	261
200914	-25 15235	HR 8080	★ A	21 04 12	-25 12.4	7	4.49	.008	1.60	.005	1.98	.050	22	M0.5 III	3,15,369,1024,1075*
201051	+26 04073	HR 8082		21 04 13	+26 43.4	1	6.11		1.05		0.62		2	K0 II-III	1080
201114	+47 03286			21 04 14	+47 50.6	1	7.57		0.05				2	B9 V	1119
	+45 03404			21 04 16	+45 50.1	1	9.15		0.58				2	F8	1724
200855	-47 13695			21 04 16	-47 30.8	2	10.16	.064	0.67	.005	0.05	.030	5	G8	1696,3060
		VES 361		21 04 18	+44 11.7	1	13.84		2.59		1.86		1		8106
		3HLF4 # 58		21 04 18	-40 18.	1	13.13		-0.30		-1.13		2	B2	208
200886	-41 14385			21 04 19	-40 42.7	1	10.76		0.13		0.09		2	A2	208
201078	+30 04318	HR 8084, DT Cyg		21 04 24	+30 59.0	5	5.75	.061	0.52	.033	0.31	.045	11	F5.5Ib-II V	71,592,1501,1772,6007
200968	-14 05936	IDS21016S1419	AB	21 04 24	-14 07.4	4	7.10	.015	0.90	.006	0.58	.010	12	K1 V	158,196,2018,3077
		3HLF4 # 59		21 04 24	-44 44.	1	14.31		-0.03		-0.03		2	A0	208
200751	-64 04094	HR 8073		21 04 25	-64 07.9	3	5.74	.007	1.18	.000	1.15	.000	20	K0 III	15,1075,2038
201267	+61 02092	IDS21033N6145	ABC	21 04 26	+61 57.5	1	7.81		0.09		0.04		2	A0	401
		CS 22937 # 4		21 04 26	-41 40.0	2	14.70	.010	0.42	.008	-0.19	.008	3		1580,1736
200856	-49 13335	IDS21010S4920	A	21 04 27	-49 08.5	1	6.67		0.00		-0.06		2	A0 V	1311
		VES 316		21 04 28	+39 41.1	1	11.77		0.53		-0.04		1		8056
200995	-21 05928			21 04 29	-20 42.1	2	8.20	.010	1.27	.010	1.19	.015	3	K0 II	565,3048
		AJ89,1229 # 511		21 04 30	+05 06.	1	11.78		1.69				2		1532
201094	+25 04465			21 04 32	+26 21.4	1	8.27		1.32		1.40		2	K2 II	8100
201174	+44 03701	NGC 7039 - 205		21 04 35	+45 04.1	2	8.78	.024	0.02	.005	-0.02	.039	6	B9p	255,261
	+47 03290			21 04 37	+47 58.1	1	9.45		1.31				3	F8	1724
201092	+38 04344	HR 8086	★ B	21 04 38	+38 29.5	17	6.06	.023	1.35	.024	1.23	.006	88	K7 V	1,15,22,150,770,1028*
201091	+38 04343	HR 8085, V1803 Cyg	★ A	21 04 40	+38 30.0	19	5.23	.021	1.16	.019	1.11	.016	94	K5 V	1,15,22,150,369,770*
201091	+38 04343	HR 8085, V1803 Cyg	★ AB	21 04 40	+38 30.0	1	4.84		1.26				5	K5 V + K7 V	1197
200954	-37 14121			21 04 41	-37 28.0	1	8.43		1.45		1.70		4	K3 III	1673
200973	-31 18080			21 04 42	-31 15.6	1	7.22		0.73				4	F5 V	2034
	+45 03406	LS III +45 050		21 04 46	+45 39.4	2	9.55	.019	0.62	.024	-0.19	.014	4	B5 Iab	1012,8100
	+45 03407	LS III +46 044		21 04 47	+46 04.9	1	9.73		0.52		0.06		1	B9	180
	+42 03972			21 04 49	+42 55.4	1	10.04		2.00				2		369
201251	+47 03292	HR 8089	★ A	21 04 53	+47 26.8	7	4.56	.026	1.57	.012	1.75	.035	17	K4 Ib-IIa	15,1118,1119,1355*
		CCS 2975		21 04 54	-41 48.	1	14.25		1.62		1.44		2		628
201250	+48 03288			21 04 55	+48 25.6	1	8.88		-0.02				2	B8 V	1119
201139	+14 04538			21 04 56	+14 42.2	1	8.68		0.98				2	K0	882
		POSS 108 # 3		21 04 56	+59 13.8	1	15.31		1.51				2		1739
		3HLF4 # 61		21 05 00	-43 06.	1	12.65		-0.02		-0.02		2	A0	208
201269	+47 03294			21 05 01	+47 59.2	3	7.63	.010	0.01	.017			7	B9 V	1118,1119,1196
201194	+30 04322			21 05 02	+30 23.7	2	7.55	.005	-0.12	.019	-0.61	.015	3	B8	592,1733
201099	-6 05683	IDS21025S0558	A	21 05 05	-05 46.3	2	7.59	.010	0.55	.005	0.00	.005	5	G0 V	1311,3037
201270	+45 03410			21 05 06	+45 28.4	1	7.25		0.56		0.34		2	F4 V	1601
		VES 362		21 05 06	+45 33.7	1	11.91		1.98		1.58		1		8106
		NGC 7031 - 12		21 05 06	+50 40.9	1	13.68		1.71				1		49
201503	+72 00976	G 262 - 34		21 05 06	+73 00.0	1	8.71		0.55		-0.02		1	G5	1658
	+40 04413			21 05 08	+41 12.8	1	9.74		0.65				5		1724
	+40 04413			21 05 08	+41 12.8	1	9.80		0.93				5		1724
	+49 03464			21 05 10	+50 13.1	1	9.57		0.12		-0.05		4		261
201196	+15 04340	HR 8088		21 05 12	+15 27.4	2	6.30	.040	1.02	.010	0.78	.000	4	K2 IV	1733,3016
201104	-16 05800			21 05 12	-15 49.6	1	7.70		1.64		2.00		3	M1 III	1657
		L 24 - 52		21 05 12	-82 01.	1	13.62		0.24		-0.61		2		3065
201252	+36 04416	IDS21032N3633	A	21 05 13	+36 44.9	1	8.15		0.00		-0.28		1	A0	963
		NGC 7031 - 15		21 05 13	+50 39.3	1	13.95		0.90		0.49		1		49
201195	+15 04342			21 05 15	+16 10.8	1	8.84		0.85				2	G5	882
		NGC 7031 - 24		21 05 15	+50 40.7	1	14.87		1.18				1		49
		NGC 7031 - 17		21 05 16	+50 43.0	1	14.54		0.98		0.57		1		49
	+45 03414			21 05 19	+46 05.3	1	9.49		0.04				2	B7 V	1119
		NGC 7031 - 21		21 05 20	+50 36.8	1	14.72		1.00		0.47		1		49
		NGC 7031 - 22		21 05 20	+50 38.0	1	14.74		1.08				1		49
		NGC 7031 - 5		21 05 21	+50 38.4	1	12.77		0.70		0.30		1		49
		NGC 7031 - 9		21 05 22	+50 36.6	1	12.97		0.77		0.35		1		49
		NGC 7031 - 23		21 05 22	+50 37.3	1	14.83		0.92				1		49
201177	-5 05474			21 05 22	-04 42.0	1	8.99		0.41		-0.02		1	A5	1776
		3HLF4 # 62		21 05 24	-43 33.	1	12.15		0.35		-0.07		2	A7	208

Table 1 1137

HD	DM	Other Id	N	Rem	α_{1950}	δ_{1950}	S	V	σ_V	B–V	σ_{B-V}	U–B	σ_{U-B}	n	Spectrum	References
201019	−55 09524				21 05 24	−55 29.8	1	8.35		0.34				2	F3 w	1594
		NGC 7031 - 10			21 05 25	+50 39.9	1	13.24		0.81		0.52		1		49
201306	+44 03710	IDS21036N4416	AB		21 05 26	+44 28.3	1	7.38		-0.05		-0.46		2	B9	401
	+43 03809				21 05 27	+43 24.6	1	9.36		1.69				1	M2	369
		NGC 7031 - 27			21 05 29	+50 36.3	1	15.86		1.14				1		49
		NGC 7031 - 7			21 05 29	+50 40.3	1	12.86		0.92		0.48		1		49
	+45 03415				21 05 30	+45 58.5	1	8.75		1.12		0.94		4	K0	261
201320	+47 03297	IDS21038N4720	A		21 05 30	+47 31.9	3	7.33	.005	0.05	.004			8	A0 V	1118,1119,1196
		NGC 7031 - 6			21 05 30	+50 40.4	1	12.78		0.89		0.43		1		49
201179	−10 05609				21 05 30	−10 05.3	1	9.26		0.45		-0.05		3	F8	1700
		3HLF4 # 1016			21 05 30	−43 00.	1	11.34		0.36		-0.03		2	F2	221
		NGC 7031 - 16			21 05 31	+50 40.6	1	14.48		1.04		0.78		1		49
	+46 03188				21 05 32	+47 02.3	1	9.62		0.63				2	F0	1724
		NGC 7031 - 18			21 05 32	+50 39.2	1	14.55		1.04		0.56		1		49
		NGC 7031 - 13			21 05 33	+50 41.2	1	13.72		0.91		0.65		1		49
	+29 04320				21 05 34	+30 10.1	1	8.74		0.57		0.09		2	G0 V	3026
201257	+8 04618				21 05 35	+09 09.6	1	7.37		0.41		0.04		15	F5	8040
201254	+14 04544				21 05 35	+14 28.3	1	7.02		-0.13		-0.67		2	B3 V	1648
		NGC 7031 - 28			21 05 35	+50 36.3	1	16.80		1.27				1		49
		3HLF4 # 63			21 05 36	−44 36.	1	12.82		0.25		0.06		2	A3	208
		NGC 7031 - 20			21 05 37	+50 38.9	1	14.67		1.04		0.72		1		49
		NGC 7031 - 14			21 05 37	+50 42.1	1	13.75		1.02		0.64		1		49
		NGC 7031 - 25			21 05 38	+50 37.3	1	15.26		0.99				1		49
	+45 03417				21 05 39	+45 31.5	1	7.26		0.58				2	A6 III	1119
201359	+46 03191				21 05 39	+47 04.5	5	7.27	.018	-0.05	.010	-0.13		13	B8 V	401,1118,1119,1196,1724
201358	+49 03465				21 05 39	+49 35.3	1	7.56		0.27		0.01		4	A3	261
		NGC 7031 - 2			21 05 39	+50 39.7	1	11.59		0.94		0.36		1	B5 III	49
	+29 04321	IDS21035N2922	AB		21 05 40	+29 34.2	1	9.63		1.04		0.89		2	K8	3072
		NGC 7031 - 26			21 05 40	+50 37.8	1	15.39		1.03				1		49
		NGC 7031 - 29			21 05 41	+50 36.0	1	17.18		0.98				1		49
		NGC 7031 - 11			21 05 41	+50 36.7	1	13.68		0.86		0.45		1		49
		NGC 7031 - 3			21 05 41	+50 40.0	1	11.61		0.90		0.32		1	B8 V	49
		NGC 7031 - 19			21 05 41	+50 42.1	1	14.60		1.19		0.36		1		49
		NGC 7031 - 4			21 05 42	+50 39.8	1	11.79		0.90		0.44		1	B9 V	49
201184	−21 05933	HR 8087	⋆A		21 05 42	−21 23.7	2	5.30	.019	0.00	.007	-0.03		6	A0 V	2032,3023
		L 281 - 9			21 05 42	−50 10.	1	13.36		0.76		0.29		3		3062
		VES 363			21 05 43	+42 54.2	1	11.12		2.04		1.61		1		8106
235444	+51 02997		V		21 05 45	+51 50.8	2	9.23	.010	0.56	.003	0.02		3		1724,1768
201283	+14 04546				21 05 46	+15 03.0	1	8.42		1.08				2	K0	882
	+47 03298		AB		21 05 48	+48 07.0	1	9.01		0.16				2		1119
	+24 04329				21 05 50	+24 58.6	1	9.88		1.16		1.05		1	K8	3072
201151	−43 14391				21 05 50	−43 28.7	1	8.79		0.26		0.12		2	A0mA8-F0	208
201150	−41 14397	IDS21026S4111	A		21 05 51	−40 59.0	1	9.80		0.31		0.04		2	A9 V(m)	208
201345	+32 04060	LS II +33 038			21 05 52	+33 11.7	1	7.66		-0.13		-0.95		1	O9p	1423
201396	+46 03192	IDS21042N4654	AB		21 05 56	+47 06.0	4	8.49	.012	0.22	.011	0.05	.014	9	A3 V	261,401,1118,1196
		NGC 7031 - 8			21 05 57	+50 40.4	1	12.94		0.80		0.27		1		49
		VES 364			21 05 57	+50 53.2	1	12.00		0.88		0.05		1		8106
		VES 317			21 05 59	+42 17.7	1	11.89		0.59		0.06		1		8056
201397	+44 03711				21 05 59	+44 45.7	1	7.52		0.01				2	A0	1118
201416	+48 03289	V1720 Cyg	⋆A		21 05 59	+48 39.2	1	7.78		0.96				3	G5 III	1119
201298	+6 04754	HR 8090			21 06 00	+06 47.2	3	6.14	.005	1.66	.000	1.98	.020	10	K5 III	15,1256,1355
201307	+12 04553				21 06 00	+13 13.2	1	7.96		1.37				2	K0	882
		NGC 7031 - 1			21 06 00	+50 36.8	1	11.27		0.68		0.18		1	G8 III	49
201431	+51 02998				21 06 00	+51 22.9	2	7.37	.005	-0.02	.014	-0.08		4	A0	401,1118
201020	−64 04096				21 06 00	−64 13.7	1	6.90		0.27		0.16		6	A3mA3-A8	1628
201187	−44 14311				21 06 02	−44 00.2	2	9.71	.005	0.43	.010			6	F3 V	1594,2012
201346	+28 03996				21 06 03	+28 25.2	1	8.40		0.85		0.61		2	K1 IV	3016
		L 497 - 56			21 06 04	−37 52.0	1	12.83		0.86		0.45		2		1696
201636	+70 01164	HR 8099			21 06 06	+71 13.9	3	5.87	.010	0.38	.004	0.02	.014	5	F3 IV	254,351,3016
		VES 365			21 06 07	+45 53.2	1	14.83		0.39		0.38		3		8106
	−5 05480		A		21 06 08	−04 37.8	1	9.45		1.15		1.00		4		3072
	−5 05480		B		21 06 08	−04 37.8	1	13.40		1.63				1		3072
	+45 03420				21 06 09	+46 02.3	1	10.08		1.52		1.40		4		261
201399	+34 04294				21 06 12	+35 07.8	1	8.64		-0.01		-0.06		2	A0	1733
	+39 04445				21 06 14	+40 14.5	1	9.36		1.81				2	M2	369
		G 145 - 17			21 06 16	+20 27.0	1	13.64		0.96		0.65		1		1658
		AAS11,365 T10# 2			21 06 18	+46 13.	1	10.12		1.68		1.89		4		261
		3HLF4 # 64			21 06 18	−43 54.	1	14.49		0.11		0.15		2	A0	208
201455	+45 03422				21 06 19	+46 14.2	2	8.35	.015	0.91	.020	0.71		5	K0 III	1119,8100
200924	−73 02195	HR 8081			21 06 21	−72 44.9	3	6.18	.007	1.09	.005	1.04	.005	15	K1 III	15,1075,2038
201244	−41 14403				21 06 22	−41 12.5	1	10.98		0.20		0.14		2	A2	208
		AAS11,365 T10# 3			21 06 24	+46 13.	1	9.92		1.68		1.83		4		261
201651	+69 01148				21 06 25	+69 28.3	2	8.18	.000	0.77	.005	0.25	.010	5	K0	979,3026
201444	+39 04447				21 06 26	+39 28.0	1	8.30		0.51				31	F8	6011
201456	+43 03815				21 06 26	+43 32.6	2	7.84	.023	0.52	.005	0.08		4	F8 V	247,1724
201245	−44 14313				21 06 26	−44 24.8	1	6.51		1.16				4	K1/2 III	2012
		VES 366			21 06 29	+45 42.7	1	11.58		0.62		-0.31		2		8106
201264	−44 14314				21 06 30	−44 39.9	1	10.12		0.36		-0.03		2	F0 V	221
		Smethells 72			21 06 30	−53 16.	1	11.65		1.29				1		1494
		CS 22897 # 42			21 06 30	−66 31.2	1	14.33		0.34		0.10		1		1736

HD	DM	Other Id	N Rem	α_{1950}	δ_{1950}	S	V	σ_V	B–V	σ_{B-V}	U–B	σ_{U-B}	n	Spectrum	References
201433 +29 04324		HR 8094, V389 Cyg	⋆AB	21 06 31	+30 00.2	2	5.61	.020	-0.09	.010	-0.29	.025	3	B9 V	592,1319
		VES 367		21 06 31	+46 11.6	1	11.89		1.82		1.24		1		8106
		Wolf 918		21 06 31	−13 28.5	4	10.87	.001	1.49	.012	1.12	.023	9	M3	1705,1764,1774,3078
201908 +77 00800		HR 8112		21 06 32	+77 55.4	3	5.91	.019	-0.06	.009	-0.23	.026	19	B8 Vn	252,1079,1258
		LS III +45 053		21 06 36	+45 04.3	1	11.52		0.38		-0.39		2		405
		AAS11,365 T10# 4		21 06 36	+46 17.	1	11.58		0.22		-0.09		4		261
		LS III +46 045		21 06 38	+46 55.9	1	10.81		0.76		-0.39		1		180
		G 187 - 24		21 06 39	+24 04.9	1	13.32		0.72		0.12		2		1658
201684 +68 01188				21 06 39	+68 24.8	1	8.17		0.27				1	A3	765
201379 −5 05483				21 06 40	−05 09.4	1	8.39		0.08		-0.10		1	A0	1776
		AAS11,365 T10# 5		21 06 42	+46 24.	1	11.16		0.32		0.16		4		261
235447 +50 03269				21 06 42	+50 32.4	1	9.38		0.66				2		1724
201352 −21 05940		HR 8091		21 06 42	−20 45.5	1	6.25		0.38				4	F2 IV/V	2035
	+47 03302	LS III +47 030		21 06 43	+47 28.3	1	10.49		0.38		-0.51		2	B2 V:pen	1012
		G 231 - 27		21 06 44	+59 32.1	1	13.30		1.55				4		538
201522 +46 03198				21 06 45	+47 03.2	1	7.88		-0.05				3	B0 V	1119
	+41 04015			21 06 46	+42 19.8	1	9.57		0.87				2	G5	1724
201504 +43 03817				21 06 49	+43 37.5	1	8.07		-0.08		-0.35		2	A0	401
201381 −11 05538		HR 8093		21 06 52	−11 34.5	6	4.51	.008	0.94	.005	0.69	.015	22	G8 III	15,58,1075,1088,2012,8015
201458 +16 04465				21 06 53	+16 38.8	1	8.64		1.09				2	K0	882
201317 −43 14402				21 06 53	−43 35.2	1	6.57		-0.08		-0.26		2	B8 V	208
		G 187 - 26		21 06 54	+26 10.1	1	12.69		0.95		0.67		2		1658
	−10 05614			21 06 54	−10 12.5	1	10.57		1.05		0.93		2	K3 V	3072
201403 −10 05615		IDS21042S1005	AB	21 06 55	−09 52.7	1	8.14		1.47		1.74		3	K2	1700
	−45 14227			21 06 59	−44 51.7	1	11.17		0.33		0.06		2	A5	208
		AAS11,365 T10# 6		21 07 00	+46 21.	1	12.28		0.64		-0.16		4		261
		3HLF4 # 67		21 07 00	−42 30.	1	13.33		0.17		0.05		1	A2	208
		G 212 - 24		21 07 01	+46 57.3	1	10.78		0.95		0.69		3		196
239581 +55 02530		LS III +55 007		21 07 01	+55 50.5	1	7.93		0.39		-0.38		1	B2 V	5
201731 +67 01288				21 07 01	+68 03.1	3	6.88	.038	0.11	.014	0.16	.029	8	A2	379,765,814
201490 +29 04329				21 07 03	+30 10.4	1	8.00		0.51		-0.03		2	F7 V	3026
		CS 22937 # 25		21 07 03	−39 06.0	1	14.08		0.52		-0.30		1		1736
	+47 03307			21 07 12	+48 02.5	1	9.52		1.18				2	G0	1724
201599 +46 03201		IDS21054N4652	A	21 07 13	+47 04.0	2	6.95	.005	0.38	.000			4	F2	1118,1724
		VES 318		21 07 17	+40 06.3	1	10.14		0.28		-0.25		1		8056
	−13 05865			21 07 18	−13 23.6	1	10.58		1.11		1.04		2	K5 V	3072
		VES 368		21 07 19	+48 17.2	1	12.39		1.96		2.12		2		8106
	+87 00195			21 07 19	+87 42.0	1	9.47		0.55		-0.01		2	G2	1375
201612 +48 03295				21 07 20	+48 32.5	3	8.58	.005	-0.10	.017	-0.31		6	A0	1118,1196,1733
		LS III +48 018		21 07 22	+48 53.5	2	11.30	.042	0.80	.019	-0.33	.047	4		180,483
201507 +2 04311		HR 8095		21 07 27	+02 44.3	5	6.45	.015	0.37	.005	0.05	.013	14	F5 IV	15,253,254,1071,3016
201545 +18 04719				21 07 28	+19 00.4	1	6.96		0.48		-0.02		2	F8 V	1648
	+45 03426			21 07 28	+46 07.2	1	9.31		0.02		-0.32		3	A0	261
	+19 04635			21 07 30	+20 00.2	1	9.54		-0.13		-0.81		1	B5	963
		VES 319		21 07 30	+43 02.5	1	12.09		1.44		1.17		1		8056
		AAS11,365 T10# 8		21 07 30	+46 21.	1	11.93		0.32		0.11		4		261
		AAS11,365 T10# 7		21 07 30	+46 24.	1	11.39		2.00		1.98		6		261
		Pav sq 1 # 9		21 07 31	−67 56.4	1	10.15		1.55		1.91		2		8098
		LS III +48 019		21 07 34	+48 28.5	1	11.72		0.48		-0.02		1		180
		CP Aqr		21 07 37	−01 55.5	1	11.19		0.14		0.08		1	F0	699
		G 25 - 18		21 07 40	+10 25.6	1	12.49		1.37		1.09		1		333,1620
201210 −70 02834				21 07 40	−70 22.0	1	8.04		1.30		1.43		7	K2 III	1704
201614 +35 04419				21 07 41	+35 36.0	1	8.15		1.01		0.83		3	K0	1601
		AAS11,365 T10# 10		21 07 42	+46 20.	1	11.55		1.09		0.50		5		261
		L 81 - 49		21 07 42	−72 32.	2	11.95	.015	0.46	.022	-0.21	.020	4		1696,3062
		CS 22898 # 43		21 07 45	−21 57.2	1	14.06		0.43		0.02		1		1736
201665 +46 03204				21 07 46	+46 36.5	1	9.14		0.16		0.09		3	Am	261
201666 +45 03427		NGC 7039 - 206		21 07 47	+45 32.1	3	7.64	.013	-0.02	.005	-0.65	.020	6	B2 V	255,401,1118
		LS III +48 020		21 07 47	+48 34.3	1	11.25		0.42		-0.13		2		180
201626 +26 04091				21 07 48	+26 24.6	2	8.14	.020	1.11	.010	0.49	.020	3	G9p	1319,8005
		3HLF4 # 69		21 07 48	−41 27.	1	14.18		0.07		0.14		2	A2	208
		3HLF4 # 1018		21 07 48	−43 22.	1	11.59		1.20		1.28		2	K0 III	221
	+33 04179			21 07 51	+33 43.1	1	9.74		1.13		0.91		1	F5	245
358906 −16 05811		Z Cap		21 07 51	−16 22.7	1	9.23		1.65		1.15		1	M Iab:e	975
201638 +34 04312		LS II +35 085		21 07 52	+35 17.3	1	9.10		-0.14		-0.98		2	B0.5Ib	401
201411 −57 09807				21 07 52	−56 43.3	1	6.91		1.15		1.08		5	K1 III	1628
		AAS11,365 T10# 13		21 07 54	+46 12.	1	12.97		0.82		0.38		6		261
201601 +9 04732		HR 8097, γ Equ	⋆AB	21 07 55	+09 55.7	10	4.70	.025	0.26	.007	0.10	.020	25	F0 IIIp	15,39,220,254,1202*
201701 +45 03429				21 07 56	+45 57.3	1	7.62		0.51		0.04		3	G0	261
201699 +46 03206				21 07 57	+46 23.3	1	8.90		0.20		0.05		3	A5	261
201654 +34 04181				21 07 58	+34 12.1	1	8.83		-0.12		-0.59		2	B8	245
	+2 04314	G 25 - 19		21 07 59	+02 55.0	1	10.84		1.04		0.95		2	K3	333,1620
201655 +33 04180				21 07 59	+33 49.4	1	9.35		-0.01		-0.23		1	A0	245
		AAS11,365 T10# 17		21 08 00	+46 17.	1	13.86		0.92		0.34		6		261
		AAS11,365 T10# 16		21 08 00	+46 18.	1	14.40		1.32		1.15		6		261
		AAS11,365 T10# 15		21 08 00	+46 23.	1	12.99		0.44		0.19		4		261
		3HLF4 # 70		21 08 00	−43 08.	1	14.48		-0.11		-0.56		2	A2	208
201759 +55 02534				21 08 01	+55 29.3	1	8.07		-0.01		-0.04		2	B9	401
201569 −10 05619				21 08 02	−10 24.8	1	8.30		0.98		0.68		4	G5	1700
201616 +9 04735		HR 8098	⋆D	21 08 05	+09 50.6	3	6.07	.008	0.02	.002	0.04	.000	9	A2 Vs	15,1256,3016

Table 1

1139

HD	DM	Other Id	N	Rem	α_{1950}	δ_{1950}	S	V	σ_V	B–V	σ_{B-V}	U–B	σ_{U-B}	n	Spectrum	References
201567	−9 05674	HR 8096			21 08 05	−09 33.5	3	6.26	.004	1.16	.000	1.21	.000	9	K0 III	15,1061,3005
		AAS11,365 T10# 19			21 08 06	+46 18.	1	14.34		1.80		1.27		5		261
		AAS11,365 T10# 18			21 08 06	+46 21.	1	13.70		1.28		0.83		6		261
201640	+14 04550				21 08 07	+15 02.9	1	8.15		1.41				2	K2	882
201668	+33 04182				21 08 07	+34 03.5	1	8.85		0.02		−0.38		2	A0	245
201733	+44 03718	NGC 7039 - 170		⋆	21 08 11	+45 17.9	5	6.62	.025	-0.15	.022	-0.67	.035	12	B3 III	154,255,1118,1196,1212
201484	−45 14237	V Ind			21 08 11	−45 16.7	2	9.71	.380	0.25	.117	0.08	.034	3	F0 V	221,700
		Pav sq 1 # 11			21 08 11	−67 55.4	1	12.15		0.59		0.07		2		8098
		AAS11,365 T10# 20			21 08 12	+46 20.	1	14.34		0.85		0.28		6		261
	+47 03315				21 08 12	+47 57.3	1	9.05		0.32				2	A5	1724
		L 497 - 38			21 08 12	−36 49.	1	12.66		0.68		0.10		2		1696
201670	+21 04485	IDS21060N2203		B	21 08 16	+22 15.2	1	7.70		0.17		0.08		5	A5 V	3032
	+48 03299				21 08 16	+48 57.2	1	9.91		2.09				2		369
201671	+21 04486	HR 8101		⋆ A	21 08 17	+22 15.0	1	6.71		0.01		−0.06		5	A1 V	3032
	+48 03298				21 08 17	+48 48.9	1	8.95		0.47		−0.24		3	B5	261
		3HLF4 # 71			21 08 18	−42 15.	1	13.72		0.42		0.07		2	A7	208
235453	+50 03273				21 08 20	+51 19.8	1	9.29		0.67				2		1724
	−4 05381	RS Aqr			21 08 21	−04 13.9	1	10.59		1.64		1.76		1	M2	817
239587	+57 02289				21 08 22	+57 25.1	1	9.45		0.47		−0.02		2	F2	3016
		VES 369			21 08 23	+42 32.2	1	10.84		1.98		1.41		1		8106
201537	−42 15289				21 08 24	−41 49.9	1	9.94		0.11		0.09		2	A2 IV/V	208
		3HLF4 # 72			21 08 24	−43 27.	1	13.96		-0.02		0.15		1	A0	208
201888	+62 01903	HR 8109			21 08 25	+63 05.5	4	6.53	.015	-0.12	.017	-0.49	.012	8	B7 III	252,351,1079,1379
201554	−42 15290				21 08 27	−42 25.1	1	10.11		0.30		0.02		2	A9 III/IV	208
		V360 Cyg			21 08 29	+30 28.1	1	10.85		0.99		0.77		1	F8	793
		LS III +47 031			21 08 29	+47 40.7	1	12.44		0.26		0.18		1		180
		Smethells 74			21 08 30	−43 48.	1	12.01		1.56				1		1494
		Smethells 73			21 08 30	−45 32.	1	11.95		1.40				1		1494
	+41 04027				21 08 31	+42 17.8	1	9.75		1.78				1		369
201775	+44 03721	NGC 7039 - 1			21 08 31	+45 21.3	1	8.75		-0.06		-0.51		2	B5 V	255
	+44 03720	NGC 7039 - 164			21 08 32	+45 16.1	1	10.14		-0.04		-0.31		2		255
201706	+14 04552				21 08 35	+15 15.1	1	9.04		0.58		0.05		3	G0	196
	+45 03433	NGC 7039 - 25			21 08 35	+45 26.1	1	9.66		0.79		0.22		1		255
	+44 03723	NGC 7039 - 159			21 08 36	+45 12.4	1	9.98		0.09		-0.23		2	B3 III	255
		AAS11,365 T10# 21			21 08 36	+46 28.	1	10.90		1.20		0.81		4		261
		3HLF4 # 73			21 08 36	−44 45.	1	14.43		0.15		0.15		2	A0	208
		G 261 - 37			21 08 39	+80 35.3	1	10.84		1.26		1.27		3		7010
201795	+38 04372	LS II +38 097			21 08 41	+38 45.4	2	7.53	.020	-0.04	.010	-0.85	.025	5	B1 V	105,399
	+45 03434				21 08 41	+46 02.7	1	8.52		1.83		1.89		3	K5	261
201371	−70 02835	HR 8092			21 08 41	−70 19.9	5	5.03	.012	1.58	.005	1.56	.010	20	M1/2 III	15,244,1075,2038,3055
		NGC 7039 - 28			21 08 43	+45 25.0	1	11.32		0.19		-0.14		1		255
201835	+51 03008				21 08 43	+52 13.5	1	8.01		0.46				2	F5	1724
201834	+52 02880	HR 8106			21 08 43	+53 21.5	1			-0.12		-0.46		1	B9 III	1079
201837	+47 03321	IDS21070N4717		C	21 08 45	+47 27.0	1	7.29		1.55		1.38		3	M2	1084
202123	+73 00925	G 261 - 38			21 08 45	+73 29.9	3	8.69	.005	0.84	.010	0.47	.013	6	K1 V	22,1003,1658
		G 187 - 29			21 08 47	+29 13.3	1	11.07		1.34		1.22		3	M1	7010
	+46 03213				21 08 47	+47 00.1	1	9.62		0.91				2	F5	1724
201836	+47 03322	HR 8107		⋆ AB	21 08 47	+47 29.2	3	6.46	.010	0.00	.015	-0.36	.005	7	B6 IV	154,1084,1118
		CS 22937 # 42			21 08 48	−41 38.7	2	13.49	.040	0.64	.012	0.04	.001	3		1580,1736
201751	+15 04362				21 08 49	+15 40.4	3	8.59	.040	1.27	.022	1.35	.005	7	K0	196,271,882
		NGC 7039 - 207			21 08 50	+45 11.1	1	10.94		0.23		-0.14		1	B3 III	255
202012	+67 01291	HR 8113, T Cep			21 08 53	+68 17.2	4	6.90	.854	1.55	.046	0.31	.015	4	M4	207,765,814,817
		LS III +47 032			21 08 54	+47 53.1	1	11.89		0.58		-0.15		2		180
201707	−15 05908	HR 8102, EW Aqr			21 08 56	−14 40.7	3	6.47	.017	0.29	.008	0.15	.010	9	Fm δ Del	252,2007,3016
		VES 370, V579 Cyg			21 08 58	+43 58.4	1	11.12		1.90		0.91		1		8106
	+44 03726	NGC 7039 - 196			21 08 58	+45 22.8	1	10.35		-0.03		-0.18		2	B8 III	255
		VES 371			21 08 59	+45 01.4	1	13.12		0.37		0.28		2		8106
	+45 03435				21 09 00	+45 50.0	1	9.68		0.47		0.08		4	F0	261
201647	−40 14216	HR 8100			21 09 01	−40 28.3	4	5.83	.005	0.45	.009	-0.01		18	F6 V	15,158,1075,2012
201819	+35 04426	HR 8105		⋆ A	21 09 03	+36 05.6	1	6.53		-0.14		-0.93		2	B1 Vp	154
201870	+45 03436	NGC 7039 - 208			21 09 03	+45 53.7	1	8.47		0.35		0.07		2	F2 III	255
	+15 04364				21 09 05	+15 44.6	1	9.84		1.26		1.28		3	K0	196
201695	−31 18118	IDS21061S3100		AB	21 09 07	−30 47.7	1	7.17		0.44		0.00		21	F5 V	978
		CS 22937 # 67			21 09 08	−38 33.2	1	13.91		0.06		-0.83		1		1736
		G 187 - 30		A	21 09 12	+33 19.1	2	11.59	.038	0.93	.008	0.59	.041	3		1658,7010
		G 187 - 30		C	21 09 12	+33 19.1	1	13.42		1.17		1.43		2		7010
		G 187 - 30		D	21 09 12	+33 19.1	1	14.88		0.59				2		7010
		G 187 - 30		E	21 09 12	+33 19.1	1	14.87		0.72		0.18		3		7010
		G 187 - 30		F	21 09 12	+33 19.1	1	14.23		1.05		0.83		3		7010
		G 187 - 30		G	21 09 12	+33 19.1	1	15.63		0.77		0.19		2		7010
	+46 03217				21 09 12	+46 24.6	1	10.03		0.37		-0.02		3		261
		V459 Cyg			21 09 13	+48 56.2	2	10.28	.007	1.28	.002	0.90		2	F8	1399,6011
202030	+65 01552				21 09 13	+65 28.9	1	7.79		-0.03		-0.03		2	B9	401
	+44 03727	NGC 7039 - 209			21 09 14	+45 08.2	1	10.04		0.29		0.24		1	A2 V	255
201859	+26 04097				21 09 16	+27 20.9	1	7.73		0.04		-0.08		2	A0	1319
203836	+86 00319				21 09 17	+86 50.0	2	7.41	.000	0.11	.010	0.05	.040	5	A3	1733,3016
201781	−6 05705				21 09 20	−06 10.1	1	8.40		0.12		-0.13		1	A0	1776
201860	+25 04478				21 09 22	+26 07.7	1	8.68		0.63		0.14		3	G0 V	3026
		VES 372, V581 Cyg			21 09 22	+44 20.2	1	11.75		2.22		1.88		1		8106
201841	+12 04569				21 09 23	+12 44.8	1	7.83		1.46				2	K5	882

HD	DM	Other Id	N	Rem	α₁₉₅₀	δ₁₉₅₀	S	V	σ_V	B–V	σ_B-V	U–B	σ_U-B	n	Spectrum	References
201924 +44 03728		NGC 7039 - 125		★	21 09 24	+45 15.3	2	7.79	.049	0.78	.000	0.40	.019	3	G8 IV	247,255
		LS III +45 054			21 09 24	+45 19.2	1	11.49		0.39		-0.34		2		483
201910 +40 04432		LS III +40 025			21 09 28	+40 58.8	2	7.40	.010	-0.14	.000	-0.65	.000	4	B5 V	247,401
201935 +45 03438		NGC 7039 - 72			21 09 28	+45 28.1	2	6.65	.023	0.35	.005	0.33		4	F0 III	255,6009
		VY Aqr			21 09 28	-09 02.0	1	17.11		0.32		-1.29		1		1471
-45 14247					21 09 29	-45 24.2	1	12.58		0.17		0.15		2	A2	208
		VES 373			21 09 32	+46 16.2	1	11.02		0.29		-0.25		1		8106
202000 +54 02489		V1425 Cyg			21 09 33	+55 07.6	1	7.70		0.08		-0.29		2	B8	588
201842 +8 04627					21 09 34	+08 34.3	1	8.70		0.26		0.12		2	A0	1733
+44 03731		NGC 7039 - 210			21 09 34	+45 17.4	2	9.49	.005	0.06	.005	-0.29	.139	3	B5 III	255,1012
		LS III +55 008			21 09 35	+55 59.8	2	10.36	.005	0.64		-0.37	.010	4		405,483
201824 -8 05603					21 09 37	-08 33.8	2	8.91	.010	1.08	.010	0.67	.032	3	K0	565,3048
201806 -12 05931					21 09 37	-12 07.3	1	8.87		0.37		0.00		2	F3 V	1375
201728 -42 15303					21 09 37	-42 18.3	1	11.07		-0.05		-0.13		2	A0	208
+45 03439		NGC 7039 - 211		★ AB	21 09 38	+45 38.9	1	9.71		0.43		0.01		1		255
		CS 22937 # 50			21 09 38	-41 06.1	1	14.91		0.36		-0.06		1		1736
		HA 89 # 1176			21 09 40	+15 49.6	1	10.61		0.46		0.14		3		196
201891 +17 04519					21 09 40	+17 32.1	9	7.37	.011	0.51	.005	-0.16	.020	16	F8 V-VI	22,792,908,1003,1064,1620*
201890 +22 04331					21 09 40	+22 52.5	1	7.32		1.13		1.05		2	K1 III	1733
201912 +29 04342		IDS21075N2918		A	21 09 40	+29 30.4	1	6.86		-0.13		-0.71		2	B5	1319
		Pav sq 1 # 12			21 09 40	-67 54.2	1	12.95		0.48		-0.07		2		8098
201937 +39 04475				V	21 09 41	+39 49.9	1	8.02		1.60				1	M1	369
201876 +10 04481		IDS21073N1014		A	21 09 42	+10 25.9	1	8.06		0.56		0.04		2	F8	1375
239595 +59 02331					21 09 42	+60 00.9	1	8.87		0.15		-0.60		1	B2	5
		3HLF4 # 76			21 09 42	-42 41.	1	14.06		0.08		0.11		4	A0	208
201875 +14 04554					21 09 43	+14 30.7	1	7.89		1.27				3	K5	882
201874 +15 04368					21 09 43	+15 24.3	1	9.66		0.52		0.03		3	G0	196
201889 +23 04264		IDS21075N2345		AB	21 09 44	+23 57.7	4	8.06	.015	0.59	.003	-0.04	.018	8	G1 V	979,1003,3026,8112
		LP 817 - 61			21 09 45	-16 57.2	1	12.33		0.78		0.36		2		1696
		NGC 7039 - 213			21 09 51	+45 26.4	1	10.93		1.32				1		255
202345 +74 00907					21 09 51	+75 01.4	1	6.90		0.55		0.11		2	F5	1733
201772 -39 14152		HR 8104			21 09 51	-39 37.8	1	5.26		0.44				4	F5 IV+F6 V	2009
		NGC 7039 - 212			21 09 52	+45 17.9	1	10.76		0.08		-0.41		1		255
+45 03440					21 09 52	+46 19.9	1	9.38		0.60		0.13		4	G0	261
+49 03478					21 09 52	+49 42.3	1	9.36		0.06		-0.02		4	A2	261
+49 03479					21 09 53	+49 27.0	1	9.21		0.39				2	F5	1724
		VES 321			21 09 54	+42 11.5	1	10.89		0.54		-0.17		1		8056
201977 +40 04437					21 10 00	+40 34.1	1	7.93		-0.11		-0.62		2	B8	401
		VES 374			21 10 00	+45 09.7	1	11.37		1.78		1.63		3		8106
202107 +57 02295					21 10 01	+57 24.9	2	7.90	.020	0.14	.005	-0.56	.005	5	B8	401,555
+45 03441		NGC 7039 - 214			21 10 05	+45 32.4	1	9.38		0.45		-0.14		1	B5 III	255
		LP 757 - 154			21 10 06	-14 06.7	1	12.82		0.65		-0.19		3		3062
202084 +53 02568					21 10 07	+53 32.4	1	7.05		0.01		-0.01		2	A0	401
201917 -7 05512					21 10 08	-07 21.8	1	8.24		1.00		0.72		3	G5	1657
		CS 22897 # 28			21 10 11	-63 00.1	2	14.15	.010	0.40	.007	-0.16	.009	3		1580,1736
		HA 89 # 559			21 10 12	+15 21.8	1	11.77		0.66		0.14		5	G6	271
201852 -36 14676		HR 8108			21 10 12	-36 37.8	2	5.97	.011	0.97	.008	0.74		6	K0 III	58,2009
201800 -51 12863					21 10 12	-50 48.0	1	8.30		0.33				4	F3 III	2012
		HA 89 # 986			21 10 13	+15 39.3	1	11.61		1.12		0.86		5	K3	271
+44 03734		NGC 7039 - 215			21 10 14	+45 17.2	2	9.10	.015	-0.06	.010	-0.53	.040	4	B5 V	255,401
		LS III +46 050			21 10 14	+46 29.5	2	10.69	.005	0.70	.015	-0.34	.025	4		405,483
		CS 22897 # 46			21 10 15	-67 12.3	1	13.92		0.30		-0.01		1		1736
		959 41 39			21 10 16	+03 07.6	1	11.17		1.03				2		1719
201969 +14 04555					21 10 17	+15 08.0	1	10.34		0.19		0.20		2	A0	271
		NGC 7039 - 216			21 10 17	+45 37.5	1	10.01		1.27		0.98		1		255
		HA 89 # 785			21 10 18	+15 28.0	1	10.41		1.25		1.23		5	K3	271
+15 04370					21 10 18	+15 28.7	2	10.98	.015	0.17	.010	0.12	.010	6		196,271
202068 +45 03442		NGC 7039 - 217		★ A	21 10 19	+45 29.8	2	7.83	.015	-0.11	.005	-0.37	.035	4	B7 V	255,401
		NGC 7039 - 218			21 10 19	+45 32.7	1	10.65		0.43		0.14		1		255
+45 03443		NGC 7039 - 219		★ AB	21 10 20	+45 36.7	1	10.51		0.91		0.51		1		255
201901 -28 17178		HR 8110			21 10 20	-27 49.5	3	5.41	.007	1.42	.000			11	K3 III	15,2018,2030
201853 -40 14223					21 10 20	-40 20.7	1	10.15		0.28		0.05		2	A9 V	208
		HA 89 # 787			21 10 21	+15 24.9	1	11.07		0.49		0.00		3	G0	271
+37 04221					21 10 21	+37 57.4	1	9.93		0.21		0.06		2		401
+44 03735		NGC 7039 - 220			21 10 24	+45 05.8	2	8.64	.072	0.99	.135	0.91	.009	5	G8 III	255,261
201980 +7 04645					21 10 26	+07 29.2	1	7.40		0.43		0.06		2	F2	1375
202088 +37 04222		IDS21086N3809		A	21 10 31	+38 21.6	1	7.34		-0.10		-0.63		2	B9	401
202087 +38 04389					21 10 31	+39 08.5	1	8.57		0.00		-0.13		4	A0 V	833
202017 +14 04556					21 10 32	+15 22.9	3	8.08	.005	0.52	.012	0.12	.024	9	F8	196,271,3026
+45 03444					21 10 32	+46 15.9	1	10.05		0.66				2	G5	1724
202214 +59 02334		HR 8119		★ APB	21 10 32	+59 46.8	7	5.64	.017	0.11	.009	-0.77	.012	16	B0 V	5,15,154,1012,1119*
		959 41 128			21 10 35	+04 13.4	1	12.09		1.10				2		1719
202124 +43 03842		LS III +44 041			21 10 39	+44 19.5	3	7.82	.020	0.23	.015	-0.66	.030	6	O9.5Ib	1011,1012,8100
		VES 375			21 10 41	+43 05.3	1	11.68		1.41		1.40		2		8106
202144 +45 03446		NGC 7039 - 221		★ A	21 10 42	+45 39.2	1	7.70		1.00		0.70		2	G8 III	255
		HA 89 # 801			21 10 46	+15 23.1	1	13.26		0.72				5		271
		VES 376			21 10 46	+45 03.8	1	10.95		0.23		-0.08		2		8106
		Ross 770			21 10 46	-19 31.3	1	11.76		1.19		0.97		4	K4	3078
		HA 89 # 802			21 10 47	+15 29.0	1	11.82		0.72		0.24		4	G6	271
		HA 89 # 592			21 10 48	+15 17.6	1	13.44		0.70				4		271

Table 1

HD	DM	Other Id	N Rem	α_{1950}	δ_{1950}	S	V	σ_V	B–V	σ_{B-V}	U–B	σ_{U-B}	n	Spectrum	References
202109	+29 04348	HR 8115	⋆ A	21 10 48	+30 01.3	10	3.21	.016	0.99	.006	0.76	.010	29	G8 II	15,1218,1355,1363*
202163	+45 03448	NGC 7039 - 222		21 10 49	+45 35.3	2	8.61	.015	-0.02	.020	-0.59	.030	4	B2 V	255,401
202126	+35 04435			21 10 52	+35 35.5	1	6.78		0.14		0.13		3	A2	1601
		V1234 Cyg		21 10 54	+41 09.	2	12.48	.105	2.42	.030	0.00		2		8056,8056
	+45 03449	NGC 7039 - 223		21 10 54	+45 26.1	1	9.42		0.56		0.07		1		255
	−45 14261			21 10 54	−44 47.7	1	11.30		0.04		0.07		2	A0	208
202091	+16 04475	IDS21086N1630	AB	21 10 56	+16 42.7	1	7.31		1.37				2	K2	882
	+47 03334			21 10 57	+47 49.9	1	9.73		0.34		-0.23		3		261
202236	+53 02571			21 10 57	+53 41.5	1	8.41		0.19		0.14		3	Am	8071
		WLS 2120 10 # 12		21 10 58	+12 31.3	1	10.30		0.71		0.19		2		1375
202020	−10 05624			21 10 58	−09 50.1	1	9.28		0.63		0.12		2	G0	1375
202073	+6 04776	IDS21085N0648	AB	21 10 59	+07 00.7	1	7.25		0.32		0.05		4	F0	3016
202073	+6 04776	IDS21085N0648	C	21 10 59	+07 00.7	1	14.01		0.72		0.10		3		3032
		G 25 - 20		21 11 00	+07 13.7	2	16.18	.009	0.55	.150	-0.27	.131	4		1663,3062
		HA 89 # 808		21 11 00	+15 25.3	1	11.53		1.08		0.90		3	G8	271
		Pav sq 1 # 1		21 11 00	−67 46.5	1	11.51		0.47		-0.01		2		8098
202093	+6 04777	IDS21085N0648	D	21 11 01	+06 57.6	1	7.15		0.33		0.04		4	A2	3016
	+14 04557			21 11 02	+15 19.0	2	10.03	.010	0.66	.010	0.31	.015	5	G0	196,271
	+71 01053	G 262 - 35		21 11 02	+71 27.5	2	9.74	.000	0.54	.000	-0.02	.000	4	G0	516,1064
		GD 394		21 11 03	+49 53.7	1	13.09		-0.24		-1.15		4	DAwk	308
	+15 04373			21 11 04	+15 26.2	1	9.94		0.49		-0.02		2	F8	271
201964	−43 14428			21 11 04	−43 00.4	1	8.38		0.32		0.10		2	A2mA7-A8	208
201965	−44 14356			21 11 04	−44 01.6	1	8.93		0.18		0.12		2	A1 IV(m)	208
		Pav sq 1 # 3		21 11 04	−67 53.7	1	12.07		0.47		-0.04		2		8098
202216	+48 03312			21 11 05	+48 23.2	1	8.09		1.07				29	K0	6011
	+45 03451	NGC 7039 - 224		21 11 06	+45 33.8	1	9.43		0.04		-0.20		2	B8 III	255
		G 25 - 21		21 11 07	+00 16.2	1	14.46		1.22		0.86		1		3062
		CS 22898 # 81		21 11 07	−21 50.3	1	12.82		0.90		0.46		1		1736
		AO 1189		21 11 08	+09 57.3	1	11.36		1.35		1.51		1	K3 III	1748
202128	+15 04375	HR 8116	⋆ AB	21 11 08	+15 46.5	3	6.27	.004	0.24	.004	0.06	.017	8	A7 V +A7 V	196,271,3030
		NGC 7039 - 225		21 11 08	+45 26.5	1	10.72		0.59		0.69		1		255
202198	+33 04197			21 11 11	+33 29.2	2	7.28	.010	-0.10	.005	-0.49	.010	3	B8	1625,1716
	+45 03453	NGC 7039 - 226		21 11 11	+45 27.4	1	10.32		0.32		0.29		1		255
202184	+34 04336			21 11 12	+35 17.0	1	7.73		-0.04		-0.39		2	B8	401
		POSS 108 # 1		21 11 12	+58 50.0	1	12.83		1.04				1		1739
		LS III +48 022		21 11 14	+48 11.5	1	12.49		0.45		-0.45		1		180
		VES 323		21 11 16	+41 55.3	1	11.44		2.24		2.07		1		8056
202217	+36 04468			21 11 17	+36 53.7	1	8.23		-0.04		-0.30		2	A0	401
202168	+16 04476			21 11 18	+17 06.5	1	8.83		1.22				3	K0	882
202253	+43 03850	LS III +43 018		21 11 18	+43 40.0	2	7.73	.009	0.16	.009	-0.61	.000	6	B2 III	1012,1775
		3HLF4 # 81		21 11 18	−41 28.	1	14.81		0.02		-0.10		1	A0	208
		VES 377		21 11 19	+51 12.4	1	10.59		0.51		0.07		2		8106
		vdB 141 # 10		21 11 20	+68 18.9	1	9.94		1.67				1		207
		CS 22898 # 62		21 11 21	−18 55.1	1	13.80		0.63		-0.05		1		1736
202169	+13 04643			21 11 22	+13 56.6	1	8.78		1.05				2	K0	882
201831	−68 03411	Pav sq 1 # 2		21 11 22	−67 52.4	1	10.43		0.42		-0.09		3	F8	8098
202237	+39 04483			21 11 24	+40 15.0	1	8.94		0.06		-0.58		2	B8	401
202240	+36 04470	HR 8120	⋆	21 11 26	+36 25.6	4	6.06	.011	0.23	.021	0.21	.015	5	F0 III	247,592,605,1462
		VES 378		21 11 26	+44 42.1	1	11.32		0.22		-0.66		2		8106
	+48 03314	LS III +48 023		21 11 26	+48 25.4	1	10.51		0.46		-0.09		1		180
		959 41 150		21 11 28	+04 32.3	1	11.01		1.01				3		1719
202273	+41 04041			21 11 29	+41 25.6	1	8.38		1.26				29	K5	6011
	+9 04744			21 11 30	+09 28.4	1	9.63		0.99		0.69		1	G5	1748
202200	+16 04478			21 11 31	+17 01.8	1	8.49		1.13				3	K0	882
202380	+59 02342			21 11 31	+59 53.5	1	6.62		2.39				1	M3 Ib	138
202077	−31 18144	BM Mic		21 11 31	−30 57.9	2	8.33	.021	0.84	.001	0.36	.000	27	G6 IV/V	978,1641
		LP 929 - 11		21 11 32	−29 02.0	1	15.05		1.03		0.74		1		1773
		CS 22937 # 72		21 11 33	−37 37.4	2	14.01	.009	0.54	.011	-0.04	.003	4		1580,1736
202149	−11 05553	HR 8118		21 11 34	−10 48.8	2	6.76		-0.07	.004	-0.32		5	B9 p Hg(Mn)	1079,2007
	+42 04008	VES 379		21 11 36	+42 54.7	1	9.79		1.00		0.69		1	G5	8106
		LP 929 - 12		21 11 37	−29 02.1	2	12.94	.005	0.60	.005	-0.12	.005	6		1696,1773
202042	−43 14433			21 11 40	−43 36.2	1	9.48		0.27		0.10		2	A8/9 V	208
		VES 324		21 11 44	+43 17.8	1	11.92		1.86		2.05		1		8056
		VES 380		21 11 44	+48 02.8	1	11.72		2.21		1.21		2		8106
	+14 04562	G 145 - 26		21 11 45	+14 47.0	1	10.33		0.94		0.87		2	K3	7010
	+44 03740	NGC 7039 - 227		21 11 45	+45 22.4	1	9.63		0.17		0.15		2	A5 V	255
202362	+52 02890			21 11 46	+52 27.6	2	8.65	.020	0.17	.010	0.12		4	F2	1566,1724
		AO 1191		21 11 47	+09 56.4	1	11.37		1.13		1.10		1	K1 III	1748
202258	+12 04578			21 11 47	+12 46.8	1	7.91		1.39				1	K5	882
202312	+44 03741			21 11 47	+44 57.7	2	7.36	.025	0.89	.015	0.50	.000	5	G5 II-III	247,8100
	+44 03742			21 11 48	+45 07.1	1	9.57		1.72				1		369
		LS III +48 024		21 11 49	+48 10.9	2	10.51	.114	0.59	.050	-0.22	.045	3	B2 ne (V)	180,1012
		CS 22937 # 68		21 11 49	−38 22.1	1	15.01		0.35		-0.11		1		1736
202134	−31 18145	BN Mic		21 11 52	−31 23.5	2	7.75	.076	1.09	.018	0.89	.011	13	K1 IIIp	1621,1641
202347	+45 03456	NGC 7039 - 228		21 11 53	+45 24.2	3	7.50	.007	-0.10	.018	-0.82	.005	6	B1.5V	255,401,1118
202348	+44 03744	NGC 7039 - 229		21 11 55	+45 19.8	1	7.37		1.50		1.61		2	K3 III	255
202381	+50 03284			21 11 55	+50 38.4	1	9.02		0.12		0.09		2	A2	1566
		HA 89 # 411		21 11 56	+15 11.0	1	11.43		0.55		0.03		1		271
202221	−6 05719			21 11 56	−06 10.2	1	8.16		0.59		-0.12		1	A5	1776
		Pav sq 1 # 20		21 11 56	−68 06.6	1	11.68		1.14		1.12		2		8098

HD	DM	Other Id	N	Rem	α_{1950}	δ_{1950}	S	V	σ_V	B–V	σ_{B-V}	U–B	σ_{U-B}	n	Spectrum	References
		LS III +48 025			21 11 57	+48 34.3	1	11.16		0.62		0.03		1		180
		Pav sq 1 # 4			21 11 58	−67 50.7	1	12.15		0.51		0.02		2		8098
202383	+47 03340				21 11 59	+47 51.2	1	8.40		0.03		−0.02		1	A0	526
202314	+29 04354	HR 8126			21 12 02	+29 41.6	2	6.17	.001	1.10	.016	0.81		3	G2 Ib-IIa	252,6007,8100
202173	−31 18149				21 12 02	−31 11.5	1	7.41		1.18		1.11		12	K1 III	1621
202135	−41 14440	HR 8117			21 12 02	−40 42.9	2	6.20	.005	1.14	.000			7	K2 III	15,2033
202275	+9 04746	HR 8123		⋆ AB	21 12 03	+09 48.2	10	4.50	.016	0.49	.007	−0.02	.009	46	F7 V +G0 V	15,254,1013,1197,1256*
202259	−0 04186	HR 8121			21 12 03	−00 06.9	3	6.38	.008	1.61	.001	1.93	.008	12	M1 III	15,1071,1509
202382	+48 03316				21 12 05	+48 26.1	2	8.74	.002	0.38	.003			30		1724,6011
202349	+37 04235				21 12 06	+37 34.4	1	7.37		−0.20		−0.95		1	B0.5V	1423
		3HLF4 # 82			21 12 06	−44 03.	1	13.39		0.16		0.07		2	A0	208
		L 117 - 72			21 12 06	−66 47.	1	11.89		0.74		0.21		2		1696
202260	−1 04131	IDS21095S0075		AB	21 12 07	−01 02.4	1	7.46		0.11		0.02		5	A1 V +A2 V	3030
202260	−1 04131	IDS21095S0075		C	21 12 07	−01 02.4	1	10.52		0.56		0.05		5		3030
	+9 04747				21 12 10	+09 58.2	1	9.29		1.66		2.06		1	K2	1748
202103	−53 10015	HR 8114		⋆ AB	21 12 12	−53 28.3	2	5.74	.005	0.19	.000			7	A3 Vm	15,2012
		AO 1193			21 12 15	+09 50.2	1	11.08		0.98		0.79		1	K3 V	1748
202304	+13 04647				21 12 15	+13 56.0	2	7.71	.035	0.91	.010	0.61		4	K0	882,1648
202281	−8 05613				21 12 18	−08 33.7	1	7.34		0.42		0.05		2	F5	1375
202261	−17 06216	HR 8122			21 12 19	−17 33.2	2	6.04	.000	0.98	.015	0.68		6	K0 III	252,2007
		POSS 108 # 6			21 12 21	+59 26.1	1	16.64		1.39				3		1739
		CS 22937 # 73			21 12 22	−37 37.4	2	14.21	.005	0.44	.009	−0.15	.007	3		1580,1736
	−45 14276				21 12 22	−44 45.5	1	12.15		0.15		0.15		2	A2	208
		Pav sq 1 # 15			21 12 22	−68 01.8	1	12.05		0.39		−0.11		2		8098
		G 25 - 22			21 12 23	+10 48.3	3	12.33	.023	0.55	.010	−0.19	.017	5	F8	203,333,1620,1658
202404	+38 04409				21 12 26	+39 02.1	1	7.25		1.22		1.26		4	K3 III	833
		VES 381			21 12 32	+47 01.3	1	12.76		0.99		−0.07		2		8106
	+9 04748				21 12 35	+09 37.6	1	10.52		1.14		1.02		1	G0	1748
		VES 382			21 12 35	+47 16.8	1	11.10		1.54		1.37		1		8106
202614	+67 01295				21 12 36	+67 56.9	1	8.48		0.23				1	A0	207
202287	−36 14699	HR 8124			21 12 40	−36 25.2	1	6.12		1.37				4	K3 III	2032
202582	+63 01708	HR 8133		⋆ AB	21 12 41	+64 11.9	2	6.39	.000	0.60	.002	0.06		6	G2 IV+G2 IV	1381,3030
		SW Aqr			21 12 45	−00 07.7	3	10.61	.480	0.14	.048	0.11	.011	3	A3	597,668,700
	+41 04049	V386 Cyg			21 12 46	+41 30.5	3	9.28	.022	1.36	.020	1.01		3	F5.5Ib	851,1399,6011
202320	−21 05974	HR 8127			21 12 47	−20 51.6	2	5.20	.040	1.16	.010	1.11		7	K0 II/III	2007,3016
		HA 89 # 1128			21 12 48	+15 48.9	1	10.56		1.03		0.91		3		196
202444	+37 04240	HR 8130, τ Cyg		⋆ AB	21 12 48	+37 49.9	17	3.73	.008	0.39	.005	0.02	.004	173	F1 IV	15,247,254,605,1118,1197*
202444	+37 04240	IDS21108N3737		C	21 12 48	+37 49.9	1	12.00		1.53		1.14		2		3032
		Pav sq 1 # 14			21 12 50	−68 00.4	1	11.99		0.41		0.05		2		8098
	+9 04750				21 12 51	+09 46.7	1	9.88		1.61		2.04		1	M0	1748
201906	−75 01697	HR 8111			21 12 53	−75 33.4	2	6.62	.005	0.03	.000	0.00	.000	7	A1 V	15,1075
	+9 04752				21 12 58	+09 42.3	1	9.43		0.52		−0.01		1	F8	1748
202269	−49 13403				21 12 58	−49 38.8	1	9.55		0.64		0.12		4	G3/5 V	285
202491	+41 04051				21 12 59	+41 26.2	1	8.56		0.06				29	A2	6011
202369	−15 05935	HR 8128			21 12 59	−15 22.8	3	5.29	.004	1.62	.008	1.89	.030	10	M2 III	1024,2009,3051
		CS 22937 # 54			21 12 59	−40 47.8	2	14.79	.005	0.43	.008	−0.20	.008	3		1580,1736
		G 212 - 32			21 13 03	+48 16.1	1	12.42		0.90		0.60		1	K3	1658
	+49 03485				21 13 06	+49 33.1	3	9.79		0.30		0.16		3	A2	261
202434	+13 04650				21 13 09	+13 42.0	1	8.35		1.17				2	K2	882
202447	+4 04635	HR 8131		⋆ A	21 13 19	+05 02.4	8	3.92	.009	0.53	.004	0.28	.014	28	G0 III+A5 V	15,1020,1075,1088*
	+40 04455				21 13 19	+41 23.3	1	10.23		0.22				1	A0	851
		VES 383			21 13 19	+45 20.7	1	11.12		0.32		−0.26		2		8106
		POSS 108 # 7			21 13 20	+59 52.1	1	17.07		1.47				4		1739
		vdB 141 # 11			21 13 20	+68 10.0	1	11.35		0.62				1	F7 III	207
	−30 18489	Z Mic			21 13 21	−30 29.7	1	11.23		0.35				1	F2.5II	700
202478	+16 04486				21 13 22	+16 31.2	1	7.00		1.08				2	K0	882
	−51 12883				21 13 22	−51 37.1	1	11.90		0.49		−0.25		2	F5	1696
202530	+40 04457				21 13 23	+41 00.1	1	7.92		0.45				3	G0	1724
239618	+59 02344	LS III +59 004			21 13 27	+59 33.2	1	8.45		0.50		−0.42		1	B2 Ve	5
		POSS 108 # 4			21 13 30	+58 42.9	1	15.60		1.47				1		1739
		3HLF4 # 1020			21 13 30	−41 51.	1	11.25		0.36		0.00		2	F0	221
	+41 04055				21 13 31	+41 41.3	1	8.83		0.59				3	G0	1724
202568	+41 04056				21 13 33	+41 49.0	3	6.61	.005	0.90	.005	0.56	.009	7	G8 IV	1601,1716,1724
202569	+36 04492	IDS21116N3650		AB	21 13 34	+37 02.6	1	7.75		−0.06		−0.56		2	B8	401
		G 25 - 23			21 13 40	+06 39.6	1	11.34		1.34		1.38		1	M2	333,1620
		3HLF4 # 84			21 13 42	−40 57.	1	12.42		0.09		0.16		2	A0	208
202573	+24 04357				21 13 45	+25 13.5	2	6.97	.010	0.89	.000	0.49	.005	5	G5 V:	1733,3077
202617	+44 03754				21 13 46	+44 44.7	1	7.44		−0.07				2	A0	1118
	+49 03488				21 13 48	+50 13.0	1	9.80		0.14		0.06		3		261
		3HLF4 # 85			21 13 48	−42 34.	1	13.62		0.14		0.08		2	A3	208
202653	+50 03291				21 13 49	+50 32.2	1	9.07		0.14		0.08		3	A3	261
202654	+47 03348	HR 8136			21 13 52	+47 45.9	3	6.45	.007	−0.16	.008	−0.64	.039	5	B4 IV	154,526,1118
202299	−65 03900	HR 8125			21 13 53	−64 53.5	3	6.30	.003	−0.06	.004	−0.21	.005	20	B8 V	15,1075,2038
	+47 03349				21 13 57	+48 19.3	1	8.78		0.40		−0.05		3	B8	261
	+62 01916	G 262 - 37			21 13 57	+62 37.0	1	9.54	.015	0.75	.005	0.35	.006	7	G5	196,1064,1658
	+46 03245				21 13 58	+46 46.1	1	9.26		0.17		−0.12		3	A0	261
202574	+13 04653				21 14 00	+13 57.2	1	8.19		1.24				2	K5	882
		Smethells 75			21 14 00	−60 02.	1	11.17		1.18				1		1494
	+45 03474	IDS21123N4600		AB	21 14 04	+46 11.8	1	9.49		0.23		0.09		4	A2	261
202554	−2 05495	HR 8132			21 14 04	−01 49.0	1	6.47		0.98		0.81		8	K0	1088

Table 1   1143

HD	DM	Other Id	N	Rem	α_{1950}	δ_{1950}	S	V	σ_V	B–V	σ_{B-V}	U–B	σ_{U-B}	n	Spectrum	References
		LP 697 - 48			21 14 04	−07 02.0	1	11.03		0.62		0.11		2		1696
202575	+8 04638				21 14 05	+09 11.1	4	7.95	.019	1.02	.017	0.84		5	K2	1017,1705,1758,3072
202618	+25 04498				21 14 05	+26 08.5	1	7.19		0.44				1	F2 Ib	6009
		LP 817 - 63			21 14 05	−20 18.8	1	13.35		0.61		−0.14		2		1696
		G 25 - 24			21 14 06	−01 30.6	4	11.63	.004	0.50	.004	−0.20	.000	6		516,1620,1658,3077
202664	+45 03476				21 14 09	+45 31.3	1	7.83		−0.02		−0.38		2	B9	401
	−0 04192				21 14 11	+01 05.4	1	9.68		0.42		−0.02		4		1306
		VES 384			21 14 12	+42 42.2	1	12.77		1.92		1.18		2		8106
		3HLF4 # 86			21 14 12	−40 33.	1	12.45		0.28		0.13		2	A2	208
	+41 04062				21 14 13	+41 25.3	1	8.63		1.63				1	K5	851
202587	+4 04639				21 14 16	+05 02.5	1	7.85		0.01		−0.10		15	A0	8040
	+41 04064	LS III +42 024			21 14 16	+42 19.9	1	9.02		0.29		−0.22		2	B3pnn Shell	1012
		959 41 175			21 14 18	+04 46.9	1	10.83		1.09				2		1719
		AAS11,365 T9# 47			21 14 18	+45 32.	1	9.99		0.53		0.31		4		261
		3HLF4 # 87			21 14 18	−42 28.	1	13.77		0.32		0.02		2	A7	208
	+52 02898				21 14 20	+52 24.2	1	9.62		1.79				2		369
202560	−39 14192	AX Mic			21 14 20	−39 03.7	4	6.69	.030	1.40	.012	1.18	.017	22	M1/2 V	285,1075,2033,3078
202682	+38 04427				21 14 22	+38 53.2	2	8.26	.005	0.02	.010	−0.03	.005	6	A0 V	833,1601
		LP 873 - 27			21 14 23	−21 45.7	1	12.54		0.96		0.76		2		1696
	−44 14320				21 14 24	−43 51.0	1	11.32		0.24		0.09		2	A5	208
202540	−31 18175				21 14 26	−30 57.4	1	6.85		1.17		1.17		25	K1 III	978
202606	−13 05897	HR 8134			21 14 29	−13 29.3	1	6.40		0.04				4	A1 V	2007
202501	−49 13412				21 14 29	−48 55.6	1	6.54		1.17				4	K2 III	2009
235471	+50 03294				21 14 31	+50 30.8	1	9.10		0.54				2		1724
		VES 385			21 14 33	+48 47.2	1	10.68		0.44		−0.02		2		8106
	+49 03489				21 14 35	+49 53.8	1	9.53		0.56				2	F8	1724
		ApJ144,259 # 74			21 14 36	+23 57.	1	17.11		−0.20		−1.19		5		1360
202720	+41 04067	HR 8138			21 14 36	+42 02.5	1	6.21		1.28		1.28		2	K2	1733
	−3 05166				21 14 44	−02 57.5	1	10.84		0.45		−0.15		2		1696
202796	+51 03025				21 14 45	+52 16.3	1	8.25		−0.02		−0.30		1	A0	963
202562	−43 14462				21 14 45	−42 49.2	1	9.51		1.09		0.96		2	G8 III	221
202457	−61 06537				21 14 47	−61 33.4	3	6.60	.013	0.69	.007	0.24	.005	16	G5 V	219,1075,1311
	+42 04040				21 14 50	+42 28.4	1	8.50		0.72				4	G0	1724
202830	+51 03026				21 14 55	+52 19.5	1	8.99		0.04		−0.10		5	A0	848
	+67 01298				21 14 55	+68 08.9	1	9.45		0.65				1	G5	207
		Steph 1876	A		21 14 55	−09 06.8	1	12.11		1.52		1.19		2	M1	1746
202627	−32 16498	HR 8135			21 14 55	−32 23.0	6	4.71	.011	0.06	.004	0.01	.004	21	A0 V	15,208,1075,2012,2020,3023
202699	+7 04658				21 14 58	+08 01.4	1	6.99		1.33		1.24		15	K2	8040
	+7 04657				21 14 59	+07 32.5	1	9.34		0.74		0.36		5	G5	897
202986	+67 01299				21 14 59	+68 08.6	1	7.85		0.42				1	F0	207
235474	+51 03027	LS III +52 003			21 15 00	+52 16.4	1	9.21		0.91		−0.12		6	G5	848
		CS 22897 # 57			21 15 01	−66 03.7	1	14.39		0.34		−0.08		1		1736
202671	−18 05903	HR 8137			21 15 09	−18 11.7	3	5.41	.019	−0.11	.018	−0.50	.000	8	B5 II/III	1079,2009,3023
202831	+46 03256				21 15 11	+47 08.5	1	9.15		0.28		0.09		3	F0	261
202628	−43 14464				21 15 11	−43 32.7	3	6.74	.004	0.64	.009	0.12	.010	10	G2 V	158,221,2012
		Smethells 76			21 15 12	−42 37.	1	11.42		1.18				1		1494
	+44 03764				21 15 13	+44 53.5	1	9.58		0.05		−0.03		4	A0	261
		3HLF4 # 89			21 15 18	−42 15.	1	12.79		0.25		0.04		2	A0	208
202650	−44 14397				21 15 23	−44 32.1	1	9.50		0.17		0.10		2	A1 V	208
		L 24 - 7			21 15 24	−80 17.	1	11.60		1.00		0.83		3		3062
202046	−80 01006	IDS21086S8032	A		21 15 26	−80 19.5	1	7.62		0.81		0.32		2	G8 IV	3062
202850	+38 04431	HR 8143			21 15 27	+39 11.1	11	4.23	.009	0.12	.011	−0.39	.026	39	B9 Iab	15,1049,1063,1119*
	+28 04045	IDS21133N2816	AB		21 15 28	+28 28.1	1	10.24		1.21		1.05		1		8084
202923	+53 02588	HR 8147			21 15 28	+53 47.2	3	6.13	.014	0.05	.014	0.05		10	A1 V	401,1118,1196
		POSS 108 # 5			21 15 28	+60 03.3	1	16.36		1.68				2		1739
202751	−0 04195	IDS21128S0015	A		21 15 28	−00 02.8	3	8.22	.016	0.99	.005	0.84	.009	6	K2	1658,1705,3072
202723	−18 05904	HR 8139			21 15 28	−17 40.4	1	7.05		0.34				4	F0 V	2009
202862	+42 04046	HR 8144			21 15 30	+42 28.4	1			−0.11		−0.44		1	B7 Vn	1079
203501	+80 00690	HR 8174			21 15 31	+81 01.3	2	6.14	.015	0.09	.000	0.11	.020	5	A3 IV	985,1733
	+45 03485	V591 Cyg		★ AB	21 15 34	+45 57.9	1	10.06		2.04				2	M1	369
235477	+49 03491				21 15 34	+50 18.5	2	8.64	.015	0.41	.005	0.06		5	F5	261,1724
	+51 03030				21 15 34	+52 22.7	2	9.93	.051	0.72	.245	0.03		7		369,848
202753	−5 05512	HR 8141			21 15 34	−04 43.8	3	5.81	.005	−0.13	.005	−0.51	.009	8	B5 V	15,1079,1256
202880	+38 04432				21 15 38	+39 00.0	1	7.00		0.34		0.06		2	F2 V	105
	+67 01300				21 15 42	+68 03.3	2	10.66	.117	1.35	.018	1.28		5	G8	206,207
		959 42 134			21 15 44	+04 05.4	1	12.01		1.07				4		1719
202987	+55 02549	HR 8150			21 15 44	+55 35.2	1	5.98	.008	1.45	.000	1.62		4	K3 III	37,71
	+67 01301				21 15 47	+68 14.1	1	9.76		0.32				1	F0	207
202783	−12 05958				21 15 47	−12 28.6	1	7.95		0.52		0.09		3	G1 V	3026
		L 212 - 19			21 15 48	−56 03.	1	14.28		0.26		−0.59		2		3065
202816	+7 04660	RU Equ			21 15 50	+07 33.0	1	8.44		1.53				5	M2	897
		vdB 141 # 7			21 15 51	+68 07.0	1	11.07		0.63				1	F8 V	207
202883	+29 04372				21 15 52	+30 06.8	1	8.23		−0.10		−0.68		2	B5	1319
202904	+34 04371	HR 8146, υ Cyg		★ A	21 15 52	+34 41.2	6	4.39	.094	−0.10	.022	−0.78	.017	13	B2 Vne	15,247,542,542,1212,1363
202904	+34 04371	HR 8146, υ Cyg		★ AB	21 15 52	+34 41.2	1	4.42		−0.10		−0.82		3	B2Vne+F8III	8015
	+67 01302				21 15 54	+68 00.1	1	9.02		1.13				1	G5	207
203025	+57 02309	HR 8153		★ AB	21 15 56	+58 24.1	5	6.42	.016	0.20	.010	−0.50	.017	18	B2 IIIe	5,154,206,1203,1212
202773	−29 17692	HR 8142			21 15 57	−28 58.6	3	6.40	.004	0.97	.005			14	K0 IV	15,1075,2030
	+51 03033				21 15 58	+51 27.5	1	9.79		0.29		0.13		4		848
202819	−8 05626				21 15 58	−08 15.5	1	9.87		1.13		1.06		1	K5 V	3072

HD	DM	Other Id	N Rem	α_{1950}	δ_{1950}	S	V	σ_V	B−V	σ_{B-V}	U−B	σ_{U-B}	n	Spectrum	References
		CS 22937 # 74		21 15 59	−37 32.6	1	14.75		0.51		-0.13		1		1736
		G 93 - 1		21 16 00	+02 23.7	2	14.63	.005	0.60	.030	-0.07		4		202,1658
202704	−49 13423			21 16 00	−49 36.6	1	7.74		0.91		0.59		5	G8 III/IV	1673
203026	+56 02551			21 16 01	+56 33.1	1	7.53		-0.01		-0.42		2	B9	401
202907	+17 04546			21 16 05	+17 30.6	2	7.47	.025	1.12	.015	1.07		4	K5	882,1648
202851	−2 05503			21 16 08	−01 44.7	1	9.57		1.30				1	R2	1238
202884	+8 04648			21 16 09	+08 44.9	1	7.27		0.47		-0.05		2	F5	1733
202908	+10 04514	IDS21138N1109	AB	21 16 10	+11 21.5	1	7.03		0.56		0.07		3	F7 V +G6 V	3016
203027	+51 03034			21 16 11	+51 51.5	1	8.36		0.29		0.12		5	A3	848
		vdB 141 # 8		21 16 11	+68 04.5	1	12.62		0.89				1	G0	207
202926	+17 04548			21 16 12	+17 46.6	1	7.34		0.41		-0.07		3	F6 V	3016
202989	+38 04439			21 16 12	+39 08.6	1	8.67		0.50		-0.04		1	G0	583
203028	+50 03303			21 16 13	+50 59.8	1	8.36		0.23		0.06		3	A2	1566
239626	+59 02350	LS III +59 005		21 16 15	+59 53.4	3	9.27	.015	0.35	.000	-0.55	.015	3	B0 V	5,483
203399	+76 00833	HR 8168		21 16 17	+76 48.1	1	5.95		1.50				2	K5 III	71
		Cyg sq 2 # 1		21 16 18	+48 58.	1	12.66		0.84		0.67		2		293
		CS 22937 # 84		21 16 18	−40 43.8	1	14.80		-0.07		-0.37		1		1736
	−43 14477			21 16 19	−43 38.3	1	12.27		0.28		0.11		2	A2	208
202730	−53 10037	HR 8140	★AB	21 16 19	−53 39.6	5	4.39	.010	0.19	.007	0.10	.010	17	A5 V	15,208,1075,2012,3023
		Cyg sq 2 # 2		21 16 24	+48 56.	1	10.35		0.72		0.31		4		293
		Steph 1877		21 16 25	+30 01.0	1	11.76		1.47		1.17		2	M0	1746
202951	+10 04516	HR 8149		21 16 27	+10 59.5	1	5.96		1.65				2	K5 III	71
202890	−16 05840			21 16 28	−16 23.4	1	6.95		0.93		0.66		2	K0 III	3016
	−44 14410			21 16 29	−44 22.4	1	12.20		0.23		0.12		2	A0	208
	+2 04338			21 16 31	+03 01.8	1	9.78		1.34				1	R1	1238
203014	+36 04510			21 16 31	+36 33.2	1	8.84		0.47		-0.02		1	G0	583
		959 42 99		21 16 34	+03 40.3	1	11.69		1.24				2		1719
202978	+12 04592			21 16 34	+13 21.3	1	7.90		1.05				2	K0	882
203064	+43 03877	HR 8154, V1809 Cyg		21 16 35	+43 44.1	6	5.00	.016	-0.03	.013	-0.97	.025	15	O8	154,1025,1119,1234*
		3HLF4 # 92		21 16 36	−44 18.	1	14.04		-0.08		-0.33		2	A2	208
203047	+38 04440			21 16 37	+38 43.7	1	8.71		0.48		0.02		1	G0	583
203029	+35 04473			21 16 42	+36 23.5	2	8.46	.010	-0.04	.015	-0.58	.005	3	A0	401,963
	+48 03338			21 16 43	+49 18.4	2	9.42	.015	0.07	.000	0.05	.035	7	A2	293,848
203079	+44 03768			21 16 44	+45 14.3	1	8.81		0.22				28	A0	6011
	+49 03496			21 16 44	+49 23.9	1	10.72		0.48				2	F8	1724
203030	+25 04507			21 16 45	+26 01.1	1	8.48		0.75		0.31		2	G8 V	3026
		G 25 - 27		21 16 45	−00 45.9	1	11.55		1.39		1.19		1		1620
		VES 386		21 16 47	+49 56.2	1	12.56		2.38		0.98		2		8106
		YZ Cap		21 16 47	−15 19.7	1	11.06		0.21		0.20		1	F5	668
		VES 387		21 16 48	+47 43.2	1	12.55		2.40		1.20		2		8106
		vdB 141 # 13		21 16 48	+68 12.5	1	11.65		1.41				1	G5	207
		KUV 21168+7338		21 16 49	+73 38.1	1	14.87		-0.31		-1.21		1	DA	1708
203032	+24 04373			21 16 51	+25 16.1	1	8.77		0.01		-0.18		3	A0	833
202940	−26 15541	HR 8148	★AB	21 16 52	−26 33.6	5	6.56	.005	0.73	.008	0.22		22	G5 V	15,219,1075,2030,2033
202874	−45 14302	HR 8145, T Ind		21 16 52	−45 14.1	3	6.04	.014	2.45	.028	3.75		14	C5 II	109,2007,3038
203048	+29 04378			21 16 53	+29 31.9	1	7.28		-0.02		-0.27		2	A0	1319
202875	−45 14303			21 16 53	−45 26.0	1	9.21		0.32				5	A9 V	897
		vdB 141 # 15		21 16 55	+68 03.4	1	12.59		2.05				1	G5	207
202872	−44 14414	IDS21137S4401	AB	21 16 56	−43 48.3	1	8.83		0.15		0.07		2	A2 V	208
		G 231 - 39		21 16 57	+52 11.1	1	12.33		0.94		0.75		2		1658
203096	+40 04485	HR 8155	★A	21 16 59	+40 49.7	1	6.15		0.26		0.21		2	A5 IV	1733
203136	+49 03498			21 16 59	+49 58.3	1	7.76		0.88		0.61		3	K0	293
		AAS42,335 T8# 6		21 17 00	+52 22.	1	11.85		0.60		0.12		5		848
		VES 388		21 17 01	+45 01.9	1	11.51		0.99		0.83		2		8106
203112	+39 04519	IDS21151N3920	A	21 17 02	+39 32.3	1	6.67		-0.03		-0.16		2	A0	401
203137	+49 03499			21 17 03	+49 51.5	3	6.96	.007	1.86	.016	2.17	.040	12	K5	261,293,848
	+49 03500			21 17 10	+49 30.6	1	9.69		2.42		2.25		6		293
203318	+67 01303			21 17 11	+68 21.3	1	8.37		1.04				1		207
203066	+13 04674	IDS21148N1356	AB	21 17 12	+14 08.5	1	8.17		0.88				1	K0	882
		Cyg sq 2 # 7		21 17 12	+49 24.	1	11.11		0.34		0.11		7		293
202946	−44 14417			21 17 17	−43 47.3	1	8.49		0.09		0.11		2	A3 V	208
		Cyg sq 2 # 8		21 17 18	+49 26.	1	11.36		0.45		-0.12		6		293
202917	−53 10042			21 17 18	−53 14.7	1	8.56		0.69				4	G5 V	2012
202993	−30 18522			21 17 19	−30 39.2	1	7.37		1.44				4	K3 III	2012
203156	+37 04271	HR 8157, V1334 Cyg	★AB	21 17 22	+38 01.5	5	5.82	.024	0.49	.015	0.16	.019	6	F5 II	254,401,592,1462,6009
		G 231 - 40		21 17 22	+53 59.9	1	12.35		0.05		-0.63		2	DA	1658
	+50 03310			21 17 23	+50 46.2	1	10.56		0.24		0.16		3		261
203280	+61 02111	HR 8162	★CD	21 17 23	+62 22.4	16	2.46	.011	0.22	.005	0.10	.010	119	A7 IV-V	1,15,985,1197,1355,1363*
	−51 12903			21 17 23	−51 15.2	1	10.63		0.81		0.35		4	G5	158
		Cyg sq 2 # 10		21 17 24	+49 06.	1	12.89		0.75		0.19		5		293
		Cyg sq 2 # 9		21 17 24	+49 06.	2	12.12	.015	0.93	.034	0.26	.078	9		261,293
203040	−20 06185			21 17 25	−20 03.3	2	9.12	.005	1.36	.008	1.26		4	K5 V	1705,3077
203052	−10 05651			21 17 26	−10 25.8	1	8.84		1.04		0.88		2	K0	1375
	+48 03343			21 17 29	+48 51.9	1	8.96		1.20		1.36		6	K0	293
		AAS11,365 T6# 3		21 17 30	+49 05.	1	14.71		1.27		0.54		3		261
		AAS11,365 T6# 2		21 17 30	+49 06.	1	11.98		2.08		2.30		4		261
		Cyg sq 2 # 12		21 17 30	+49 12.	1	13.01		0.65		-0.15		3		293
		Cyg sq 2 # 13		21 17 30	+49 23.	1	12.85		0.79		0.11		6		293
		AAS42,335 T8# 9		21 17 30	+52 05.	1	11.64		0.59		0.08		5		848
	+65 01572	G 262 - 40		21 17 30	+66 00.5	1	10.71		0.93		0.68		1	K2 V	1658

Table 1   1145

HD	DM	Other Id	N Rem	α₁₉₅₀	δ₁₉₅₀	S	V	σ_V	B–V	σ_B-V	U–B	σ_U-B	n	Spectrum	References
		LS III +51 010		21 17 32	+51 30.3	1	11.65		0.47		-0.26		4		848
		959 42 125		21 17 34	+04 00.3	1	11.41		1.30				2		1719
203006	−41 14475	HR 8151, θ 1 Mic		21 17 34	−41 01.3	6	4.82	.008	0.02	.006	-0.08	.010	27	Ap CrEuSr	15,208,1075,1202,2012,8015
		Cyg sq 2 # 15		21 17 36	+49 04.	2	10.72	.010	0.40	.019	-0.01	.010	9		261,293
		AAS11,365 T6# 5		21 17 36	+49 08.	1	14.56		1.09		0.67		3		261
		Cyg sq 2 # 14		21 17 36	+49 08.	1	12.51		0.91		0.40		5		293
		AAS11,365 T6# 4		21 17 36	+49 09.	1	12.53		0.67		0.26		3		261
203171	+26 04130			21 17 37	+27 17.5	1	8.35		0.51		0.01		3	G0 V	3026
		VES 389		21 17 37	+48 08.4	1	11.04		0.54		-0.12		2		8106
203204	+38 04448			21 17 38	+38 35.4	1	8.85		0.46		0.02		1	G0	583
	+48 03344			21 17 38	+48 46.4	2	9.87	.000	0.26	.010	0.17	.040	11		293,848
	−24 16586			21 17 40	−23 49.0	1	10.65		1.02		0.83		1	K7 V	3072
		vdB 141 # 14		21 17 41	+68 10.6	1	12.40		2.19				1	G0	207
203400	+65 01573			21 17 42	+65 57.3	1	8.27		0.83		0.36		2	G5	3016
	+49 03501			21 17 43	+49 24.1	2	8.76	.005	0.31	.005	0.15		7	F0	293,1724
203245	+48 03345	HR 8161		21 17 45	+49 17.9	3	5.75	.012	-0.14	.009	-0.49	.000	7	B6 V	154,848,1118
		VES 390		21 17 46	+50 35.4	1	12.68		2.40				1		8106
203283	+51 03042	IDS21162N5154	A	21 17 46	+52 06.7	2	7.32	.013	-0.02	.004	-0.02		6	A0	848,1118
	+67 01304			21 17 46	+67 49.9	1	9.32		0.28				1	A0	207
		3HLF4 # 1023		21 17 48	−42 21.	1	12.69		0.59		0.02		2	G0 III	221
203010	−50 13325	HR 8152		21 17 51	−50 08.9	4	6.39	.008	1.33	.005	1.56		13	K3 III	15,1075,2009,3077
203284	+48 03346			21 17 52	+49 00.6	2	7.69	.010	0.58	.025	0.08		9	G5	293,1724
203338	+58 02249	HR 8164	⋆ AB	21 17 53	+58 24.7	5	5.67	.036	1.34	.032	0.00	.023	8	M1 Ibep +	5,71,138,3001,8032
203339	+58 02249	IDS21165N5812	B	21 17 53	+58 24.7	1	5.67		0.36		-0.02		2	A0	401
203233	+37 04274			21 17 54	+37 55.7	1	8.92		0.88		0.44		1	G5	583
		AAS11,365 T6# 7		21 17 54	+49 07.	1	13.71		0.95		0.63		4		261
		Cyg sq 2 # 19		21 17 54	+49 14.	1	11.65		0.69		0.27		8		293
203282	+52 02910	IDS21163N5230	AB	21 17 54	+52 42.9	1	8.45		0.50				3	G0	1724
203374	+61 02112	IDS21167N6126	A	21 17 54	+61 38.8	2	6.68	.010	0.31	.005	-0.74	.005	4	B0 IVpe	5,1212
		959 42 186		21 17 55	+04 49.1	1	11.12		1.06				2		1719
		LS III +49 019		21 17 56	+49 55.1	1	11.17		0.64		-0.08		5		848
203206	+21 04521	HR 8158		21 17 58	+21 48.8	2			-0.08	.009	-0.49	.004	6	B6 IV	1063,1079
		Cyg sq 2 # 20		21 18 00	+49 02.	1	12.76		0.81		0.61		5		293
		AAS42,335 T8# 12		21 18 00	+52 11.	1	11.65		0.60		0.09		5		848
	+49 03502			21 18 01	+49 52.8	1	9.12		0.16		0.07		3	A2	261
	+67 01305			21 18 01	+67 55.4	1	9.98		0.89				1	K1	207
		959 42 49		21 18 02	+03 04.3	1	11.16		1.00				2		1719
203186	+11 04541			21 18 02	+11 55.8	1	7.56		0.32		0.03		2	F0	1375
		VES 392		21 18 02	+43 29.1	1	12.49		1.91		1.54		2		8106
		VES 391		21 18 02	+49 29.2	1	12.64		2.33		0.00		2		8106
		959 42 2		21 18 03	+02 35.1	1	12.46		0.92				1		1719
203320	+52 02913	IDS21164N5238	A	21 18 04	+52 50.7	1	6.80		1.05		0.81		2	K2	985
		POSS 108 # 2		21 18 04	+59 50.7	1	13.65		1.45				1		1739
203319	+52 02914			21 18 05	+53 06.0	1	8.93		0.06		0.01		1	A0	401
		VES 393		21 18 06	+48 55.4	1	11.86		2.25				1		8106
		Cyg sq 2 # 21		21 18 06	+49 00.	1	11.04		0.40		-0.06		5		293
		Cyg sq 2 # 22		21 18 06	+49 05.	2	11.43	.030	0.47	.030	0.38	.030	8		261,293
		Cyg sq 2 # 23		21 18 06	+49 10.	1	12.43		0.76		0.48		4		293
		959 42 72		21 18 08	+03 16.2	1	12.42		1.02				1		1719
203321	+46 03275			21 18 10	+47 07.2	1	9.34		0.00				2	A5	1724
203356	+53 02599			21 18 10	+53 44.3	1	8.08		0.26		-0.14		2	B9	1566
235490	+53 02600			21 18 11	+54 06.7	1	9.28		1.86				2	M0	369
		959 42 26		21 18 12	+02 44.8	1	11.59		1.22				2		1719
	+36 04520			21 18 13	+36 35.7	1	8.26		1.74		1.96		3	M0	1733
203340	+48 03347			21 18 13	+48 25.3	1	7.89		0.94				2	G5	1724
203142	−21 05992			21 18 13	−21 02.0	1	7.18		0.38		-0.02		1	F2/3 V	3037
203218	+14 04588			21 18 14	+14 43.4	1	8.40		1.61				1	K5	882
203286	+32 04128			21 18 15	+33 16.3	1	6.91		0.03		-0.03		1	A0 V	1716
203379	+52 02915	IDS21167N5254	AB	21 18 18	+53 06.6	1	8.11		-0.02		-0.13		2	A0	401
	+47 03372			21 18 20	+47 59.1	1	9.58		0.03		-0.13		3	A0	261
203467	+64 01527	HR 8171		21 18 20	+64 39.6	3	5.18	.018	-0.04	.008	-0.59	.009	8	B3 IVe	154,1212,1363
203380	+52 02916	IDS21168N5233	A	21 18 22	+52 45.9	2	7.68	.000	0.57	.002	0.02		7	G5	1724,3032
203380	+52 02916	IDS21168N5233	B	21 18 22	+52 45.9	2	7.87	.010	0.79	.024	0.32		6		1724,3032
		POSS 342 # 2		21 18 24	+31 44.5	1	15.08		1.46				1		1739
	+43 03887			21 18 24	+44 08.2	1	8.95		1.87				1	M0	369
	+45 03496			21 18 25	+45 27.6	1	9.85		0.50				1	F2	592
		LP 873 - 78		21 18 25	−26 09.9	2	12.85	.045	-0.01	.050	-0.86	.115	2		1696,3062
203222	−5 05524	HR 8160		21 18 27	−04 46.4	3	5.87	.008	0.91	.010	0.64	.005	10	K0	15,1256,3016
		Cyg sq 2 # 24		21 18 30	+49 21.	1	12.93		0.55		0.13		4		293
		Steph 1883		21 18 31	+66 24.8	1	11.84		1.39		1.29		2	M0	1746
	+45 03498			21 18 33	+45 37.0	1	9.59		1.10				1	K2	592
203416	+48 03348	IDS21168N4855	A	21 18 35	+49 08.4	1	7.99		1.61		1.22		4	M5 II	293
203291	+6 04802	HR 8163		21 18 36	+07 08.5	4	5.80	.010	1.66	.007	1.97	.000	13	M2 IIIa	15,1071,1363,3001
		Cyg sq 2 # 26		21 18 36	+49 08.	1	10.16		0.44		-0.04		5		293
		Cyg sq 2 # 25		21 18 36	+49 16.	1	11.90		1.60		1.03		5		293
	+48 03349			21 18 40	+49 19.5	1	8.84		0.11		0.15		5	A2	293
203358	+31 04425	HR 8166	⋆ AB	21 18 43	+32 14.4	2	6.44		0.84		0.35		2	G8 IV	938,1355
	+44 03779	V532 Cyg		21 18 43	+45 15.3	3	8.91	.026	1.00	.058	0.67		3	F5	592,1399,6011
203251	−15 05958	AU Cap		21 18 44	−15 22.1	1	8.00		1.22		1.10		1	K2 III	1641
203322	+15 04396			21 18 47	+15 37.0	1	8.06		1.34				1	K5	882

HD	DM	Other Id	Rem	α₁₉₅₀	δ₁₉₅₀	S	V	σ$_V$	B–V	σ$_{B–V}$	U–B	σ$_{U–B}$	n	Spectrum	References
203344	+23 04294	HR 8165		21 18 49	+23 38.7	5	5.57	.011	1.05	.006	0.92	.008	8	K1 III	15,1003,1355,3077,4001
203323	+2 04346	IDS21163N0228	AB	21 18 52	+02 40.4	2	7.45	.068	1.07	.012	0.96	.019	3	K0	1476,3016
203345	+9 04786	IDS21165N0955	AB	21 18 55	+10 07.2	1	6.73		0.53		-0.02		4	F8 V +F8 V	3030
	+47 03376			21 19 01	+48 02.2	1	9.00		0.61				2	G0	1724
203454	+39 04529	HR 8170		21 19 04	+40 08.1	8	6.39	.012	0.53	.006	-0.01	.011	21	F8 V	15,254,1007,1013,1620,1724*
		WLS 2120-10 # 5		21 19 04	-07 39.4	1	11.36		0.70		0.18		2		1375
		DM Cyg		21 19 07	+31 58.7	2	10.95	.017	0.21	.038	0.23	.032	2	F2.5	668,699
203419	+16 04509	.		21 19 11	+16 57.8	1	7.68		1.60				2	K5	882
203550	+60 02226	IDS21179N6042	A	21 19 11	+60 54.4	1	8.65		0.05		-0.09		4	A0	1371
203439	+32 04134	HR 8169		21 19 15	+32 24.0	1			0.03		0.10		3	A1 V	1049
203574	+60 02227	HR 8179		21 19 16	+60 32.6	2	6.10	.010	0.99	.005	0.75		7	G5 III	252,1379
203575	+59 02359			21 19 17	+60 14.2	1	8.67		0.05		-0.08		4		1371
203364	-9 05724			21 19 18	-09 32.5	1	6.86		1.17		1.32		3	K2	3016
		3HLF4 # 1024		21 19 18	-41 19.	1	12.18		0.58		0.01		2	G0 III	221
203471	+27 04049			21 19 24	+28 21.1	1	8.38		0.74		0.28		3	G5 V	3026
203518	+45 03502			21 19 26	+45 30.8	1	9.02		-0.01				29		6011
	-11 05581			21 19 27	-10 43.9	2	10.26	.024	1.24	.018	1.22		4	K7 V	1619,3072
203387	-17 06245	HR 8167		21 19 28	-17 02.9	9	4.28	.008	0.90	.009	0.58	.016	33	G8 III	3,15,30,58,418,1075*
		AAS42,335 T8# 13		21 19 30	+51 45.	1	11.36		0.65		0.13		4		848
		WLS 2120 10 # 10		21 19 31	+10 17.9	1	11.20		0.85		0.35		2		1375
203407	-15 05963	IDS21168S1521	AB	21 19 33	-15 08.3	1	8.27		0.48		0.01		19	F5 V	978
		3HLF4 # 95		21 19 36	-40 22.	1	12.31		0.18		0.11		2	A2	208
203472	+16 04513			21 19 39	+16 59.8	1	7.68		1.48				2	K2	882
		959 42 78		21 19 41	+03 19.6	1	12.44		1.01				2		1719
203503	+25 04519			21 19 42	+25 36.1	1	8.18		0.22		0.16		3	A3	833
203441	-7 05546			21 19 42	-07 03.6	1	10.01		0.05		-0.21		1	A0	1776
	+26 04136	G 187 - 40		21 19 43	+27 14.3	1	10.51		0.53		-0.15		2	sdG0	1658
203577	+47 03381			21 19 43	+47 57.2	2	7.23	.005	1.54	.015	1.81		5	M1	293,369
		LS III +50 015		21 19 45	+50 03.1	1	10.06		0.64		0.20		4		848
203504	+19 04691	HR 8173	★A	21 19 46	+19 35.4	13	4.09	.010	1.11	.005	1.05	.006	54	K1 III	1,15,150,1028,1084,1258*
203504	+19 04690	IDS21175N1923	B	21 19 46	+19 35.4	5	9.15	.016	0.86	.005	0.53	.010	12		1,150,1028,1084,1775
203133	-70 02844	HR 8156, Y Pav		21 19 47	-69 56.9	2	6.37	.127	2.77	.122	3.42	.049	3	C5 II	58,109
	+35 04497			21 19 48	+35 34.8	1	9.06		1.12				1	K0	1253
203627	+58 02255			21 19 48	+58 50.3	1	8.28		1.02		0.80		1	G5	401
203458	-3 05188			21 19 51	-03 20.6	1	6.68		0.38		0.05		3	F5 II-III	8100
		G 187 - 41		21 19 52	+30 53.2	1	10.67		1.21		1.24		2	K5	7010
		LS III +49 020		21 19 56	+49 49.2	1	11.20		0.64		-0.36		4		848
235497	+51 03048			21 19 56	+51 54.6	1	9.52		0.27				3		1724
		959 31 108		21 19 58	+04 01.4	1	12.83		0.99				2		1719
		LS III +48 031		21 19 58	+48 59.8	1	11.13		1.53		0.83		2		848
203609	+50 03317			21 19 58	+50 37.4	1	8.09		-0.01		-0.16		2	A0	401
		AAS42,335 T8# 16		21 20 00	+51 46.	1	11.55		0.47		0.01		5		848
203695	+60 02229	IDS21187N6016	AB	21 20 02	+60 29.2	1	8.77		0.30		0.13		4		1371
		POSS 342 # 1		21 20 03	+31 49.9	1	14.45		1.48				1		1739
	+51 03049			21 20 03	+51 46.9	1	10.11		0.17		0.11		9		848
		WD2120+054		21 20 05	+05 29.7	1	16.38		-0.31		-1.13		1	DA2	1727
204129	+79 00701	IDS21215N7955	AB	21 20 05	+80 08.4	1	7.32		0.50		-0.04		2	F8	1003
203448	-31 18229			21 20 06	-31 01.7	2	7.81	.009	0.58	.009			8	G1/2 V	1075,2033
		VES 394, V597 Cyg		21 20 07	+42 38.0	1	11.89		2.29		1.26		1		8106
		Steph 1881		21 20 07	-15 46.0	1	12.32		1.31		1.48		1	K6	1746
203475	-23 16877	HR 8172		21 20 09	-22 53.0	3	5.62	.015	1.61	.008	2.03	.035	9	M0 III	1024,2007,3005
203537	+7 04673			21 20 11	+07 52.1	1	8.72		1.01		0.66		2	K0	1375
203610	+46 03287			21 20 11	+46 40.7	1	8.37		1.06				3		1724
203525	-9 05728	HR 8175		21 20 15	-09 32.0	3	6.00	.013	1.52	.015	1.89	.010	12	M0 III	15,1061,3016
		POSS 342 # 4		21 20 16	+31 09.4	1	17.11		0.13				2		1739
203644	+48 03357	HR 8185		21 20 16	+49 10.4	3	5.69	.007	1.10	.002	1.01	.019	46	K0 III	261,1118,1355
		VES 395		21 20 17	+43 49.8	1	11.75		1.75		1.83		2		8106
203244	-68 03418			21 20 19	-68 26.7	1	6.97		0.73				4	G5 V	2033
	-42 15420			21 20 24	-42 05.0	1	12.51		0.29		0.09		2	A5	208
203562	+6 04811	HR 8178	★A	21 20 25	+06 35.8	4	5.15	.014	0.05	.011	0.07	.011	12	A3 V	15,1071,1363,3023
203543	-15 05972			21 20 26	-15 02.3	1	9.37		0.53		-0.03		20	F7/8 V	978
203630	+29 04397	HR 8182		21 20 33	+30 05.7	2	6.08	.036	1.08	.001	1.07		4	K1 III	71,3016
203614	+17 04569			21 20 34	+17 50.9	1	7.72		0.06				27	A0	6011
	+41 04116			21 20 34	+42 04.5	1	10.26		1.89				1		369
		VES 396		21 20 35	+45 15.3	1	11.37		1.38		1.01		2		8106
203212	-72 02598	HR 8159	★A	21 20 35	-72 00.9	4	6.08	.003	1.26	.005	1.39	.015	20	K2 III	15,244,1075,2038
		AAS42,335 T8# 18		21 20 36	+51 45.	1	11.75		0.52		-0.23		4		848
	-45 14323			21 20 40	-44 45.1	1	11.34		0.33		0.15		2	A0	208
203746	+48 03360			21 20 45	+49 16.6	3	6.84	.000	0.03	.004	-0.03	.000	8	B9	401,848,1118
203631	+15 04404			21 20 47	+16 16.9	2	7.64	.045	1.52	.010	1.79		4	K5	882,3016
203696	+38 04471	HR 8186		21 20 47	+38 25.2	1	6.64		0.03	.015	0.02		4	A1 V	1049,6009
203764	+51 03050	IDS21191N5152	A	21 20 47	+52 05.0	1	8.69		0.04		0.06		2	A0	848
203423	-61 06549			21 20 47	-61 24.1	1	7.95		1.04		0.87		7	K0 III	1704
		AAS11,365 T2# 3		21 20 48	+51 27.	1	11.93		0.51		0.35		4		261
235502	+51 03051	IDS21191N5152	B	21 20 49	+52 05.4	1	8.92		0.13		0.11		4	A3	848
203605	-12 05980			21 20 50	-12 08.9	1	10.00		1.21		1.14		2	K1 III	1375
239643	+56 02562			21 20 51	+57 19.6	1	8.65		0.54		0.01		5	K0	3032
	+48 03362			21 20 52	+48 55.8	1	9.22		0.45				2	A5	1724
		L 353 - 143		21 20 52	-46 55.2	2	12.46	.010	1.54	.021			2		1494,1705
203802	+56 02564	IDS21194N5708	AB	21 20 53	+57 21.0	1	8.10		0.56		0.08		5	G5	3032

Table 1   1147

HD	DM	Other Id	N Rem	α_{1950}	δ_{1950}	S	V	σ_V	B–V	σ_{B-V}	U–B	σ_{U-B}	n	Spectrum	References
203546	−41 14498			21 20 53	−40 41.3	1	10.12		0.24		0.10		2	A5/7 III/IV	208
203731	+40 04503	LS III +40 027		21 20 54	+40 29.0	1	7.52		0.15		-0.75		3	B1 Vne	1212
203747	+46 03291			21 20 54	+47 24.1	1	8.55		0.17		0.05		3	A0	261
	+45 03513			21 20 55	+46 10.8	1	9.84		2.00				2		369
203554	−40 14329			21 20 57	−40 07.1	1	10.23		0.29		0.18		2	A3/5 II	208
	+45 03515	NGC 7062 - 95		21 20 59	+46 13.3	1	9.23		0.19		0.10		1	A1 V	49
		AAS42,335 T8# 22		21 21 00	+52 03.	1	11.08		0.46		-0.03		4		848
		959 11 36		21 21 01	+06 55.4	1	11.84		0.99				2		1719
203664	+8 04664			21 21 02	+09 43.0	1	8.57		-0.20		-1.00		3	B0.5IIIn	399
203548	−47 13796	HR 8177		21 21 02	−46 49.8	2	6.30	.005	0.20	.000			7	A5mA6-F0	15,2012
235505	+51 03052	IDS21194N5148	AB	21 21 04	+52 00.6	1	8.97		0.31				2		1724
		NGC 7062 - 104		21 21 05	+46 19.8	1	13.51		0.52		0.07		1		49
239644	+55 02569			21 21 05	+56 17.9	1	9.38		0.39		-0.37		1	B3	5
203782	+45 03517	NGC 7062 - 107		21 21 07	+46 16.7	2	8.56	.009	0.44	.005	-0.07		5	A3 V	49,1724
		NGC 7062 - 111		21 21 11	+46 20.8	1	12.49		0.73		0.24		1		49
203699	+13 04692			21 21 12	+13 50.1	1	6.86		-0.11		-0.57		3	B2.5IVne	1212
		NGC 7062 - 114		21 21 12	+46 08.6	1	12.22		0.69		0.25		1		49
		959 11 94		21 21 14	+06 09.2	1	11.24		0.85				2		1719
203585	−41 14503	HR 8180	⋆ AB	21 21 14	−41 13.3	3	5.76	.016	-0.05	.013	-0.21		9	A0 IIIp Si	15,208,2012
		VES 906		21 21 15	+51 04.3	1	10.69		0.31		0.00		1		8113
203639	−23 16889	HR 8184		21 21 16	−22 57.7	1	6.38		1.02				4	K0 III	2009
203783	+38 04476			21 21 18	+39 08.1	2	7.35	.000	-0.02	.005	-0.11	.000	5	B9 V	833,1601
		VES 397		21 21 18	+48 19.7	1	11.36		0.60		-0.05		2		8106
		NGC 7062 - 127		21 21 20	+46 10.3	1	12.87		1.37		1.19		1		49
203638	−21 06007	HR 8183		21 21 20	−21 03.9	2	5.36	.005	1.16	.000	1.12		8	K0 III	2032,3077
203784	+36 04537	HR 8189	⋆ B	21 21 21	+37 11.4	1	6.55		0.49		0.09		2	F6 III	1733
		AAS42,335 T8# 23		21 21 24	+50 55.	1	12.01		0.86		0.00		4		848
		3HLF4 # 1025		21 21 24	−41 13.	1	10.69		0.71		0.18		2	G5 III	221
	+36 04539			21 21 26	+36 48.8	1	9.29		-0.03		-0.41		1	A0	963
		959 31 87		21 21 27	+04 10.9	1	12.80		0.85				3		1719
		NGC 7062 - 142		21 21 28	+46 05.8	1	12.11		1.36		1.05		1		49
	+45 03519	NGC 7062 - 141		21 21 28	+46 09.4	1	10.10		0.18		0.12		1	A5 V	49
203705	−13 05923	HR 8187	⋆ A	21 21 28	−13 05.6	2	5.48	.005	0.30	.005			7	F0 V	15,2030
	+45 03520	NGC 7062 - 144		21 21 29	+46 21.5	1	9.21		0.24				2	A5	1724
203855	+47 03387			21 21 29	+47 51.2	2	8.47	.005	0.97	.025	0.76	.020	7	G5	261,293
		959 11 58		21 21 30	+06 38.9	1	11.92		0.88				5		1719
		AAS42,335 T8# 24		21 21 30	+50 39.	1	13.26		0.63		-0.02		1		848
		NGC 7062 - 336		21 21 33	+46 23.4	1	10.68		0.39		-0.02		1	F0 V	49
		Steph 1888		21 21 34	+25 01.5	1	10.33		1.61		1.99		3	M0	1746
		VES 398		21 21 34	+45 50.4	1	12.28		0.40		-0.16		2		8106
		NGC 7062 - 168		21 21 34	+46 11.3	1	13.05		0.47		0.41		1		49
	+40 04507			21 21 35	+40 38.5	1	8.98		0.49				2	F8	1724
		NGC 7062 - 176		21 21 37	+46 10.2	1	13.26		0.43		0.47		1		49
	+17 04572	AU Peg		21 21 40	+18 03.8	4	9.11	.034	0.77	.042	0.48	.084	4	F8	592,804,1462,6011
	+47 03388			21 21 40	+48 02.4	1	9.68		0.42				2	F8	1724
		NGC 7062 - 188		21 21 41	+46 08.0	1	11.24		1.32		1.15		1		49
202418	−85 00519	HR 8129		21 21 43	−85 01.7	3	6.42	.017	1.41	.005	1.66	.000	13	K3 III	15,1075,2038
203803	+23 04300	HR 8190	⋆ A	21 21 44	+24 03.5	3	5.72	.022	0.31	.004	0.07	.013	9	F1 IV	253,254,3016
		NGC 7062 - 197		21 21 44	+46 13.7	1	13.57		1.49		1.42		1		49
		NGC 7062 - 202		21 21 46	+46 17.4	1	11.93		0.62		0.15		1		49
203857	+36 04543	HR 8193	⋆ A	21 21 47	+37 08.2	1	6.43		1.49		1.74		2	K5	1733
203918	+49 03513	IDS21201N5006	AB	21 21 47	+50 18.6	3	8.09	.005	0.57	.005	0.08	.005	5	F8 II	261,848,1724
		NGC 7062 - 207		21 21 48	+46 12.4	1	14.12		0.74		0.22		1		49
		NGC 7062 - 341		21 21 48	+46 14.3	1	15.65		0.87				1		49
203789	+3 04562	IDS21193N0317	AB	21 21 49	+03 30.1	1	9.27		0.73		0.23		3	G5	3030
		NGC 7062 - 337		21 21 50	+46 05.3	1	14.63		1.16		0.71		1		49
		NGC 7062 - 338		21 21 50	+46 11.3	1	14.72		0.55		0.40		1		49
		NGC 7063 - 15		21 21 52	+36 25.0	1	12.10		0.36		0.06		1		49
203858	+24 04394	HR 8194	⋆ A	21 21 53	+25 05.8	1	6.19		0.02		0.01		3	A2 V	1733
	+45 03521	NGC 7062 - 218		21 21 53	+46 09.0	1	10.39		0.72		0.25		1		49
203885	+34 04403			21 21 54	+34 57.1	1	9.28		-0.04		-0.54		1	A0	963
		AAS42,335 T8# 26		21 21 54	+50 08.	1	11.42		0.65		-0.22		5		848
		VES 399, V604 Cyg		21 21 55	+42 35.2	1	12.72		2.02		1.08		1		8106
	+44 03799			21 21 55	+45 09.5	1	9.47		1.19		1.21		7	K0	1655
		VES 400		21 21 55	+48 02.7	1	11.42		0.49		0.38		2		8106
203676	−44 14457			21 21 55	−44 16.0	1	9.93		-0.06		-0.14		2	B9 III/IV	208
203821	+7 04679			21 21 56	+07 48.1	1	9.35		0.00		-0.12		1	A0	963
		NGC 7062 - 342		21 21 56	+46 13.4	1	15.65		0.91				1		49
		G 25 - 28		21 21 58	+01 09.6	1	14.03		1.42				2		202
203842	+9 04800	HR 8191		21 21 58	+09 57.5	4	6.36	.015	0.47	.004	0.09	.030	9	F5 III	15,254,1003,1071
	+42 04080			21 21 58	+42 24.7	1	9.75		1.87				1		369
		959 31 59		21 21 59	+04 30.4	1	11.15		1.02				2		1719
		NGC 7063 - 28		21 22 00	+36 14.6	1	15.44		0.38				1		49
203919	+42 04081			21 22 00	+42 55.3	1	9.00		-0.01		-0.13		2	A3	1375
		NGC 7063 - 8		21 22 01	+36 11.4	1	10.96		1.03				1		49
		NGC 7062 - 339		21 22 02	+46 10.3	1	14.83		0.72		0.58		1		49
		NGC 7062 - 340		21 22 02	+46 12.2	1	15.32		0.77		0.49		1		49
203938	+46 03294	LS III +46 057	⋆ AB	21 22 02	+46 56.9	1	7.08		0.46		-0.41		2	B0.5IV	1012
	+48 03370			21 22 02	+49 12.9	1	9.41		0.47				2	F5	1724
		NGC 7067 - 8		21 22 03	+47 48.2	1	13.67		0.79		0.59		1		49

HD	DM	Other Id	N Rem	α_{1950}	δ_{1950}	S	V	σ_V	B–V	σ_{B-V}	U–B	σ_{U-B}	n	Spectrum	References
		NGC 7063 - 26		21 22 04	+36 16.3	2	14.93	.070	0.98	.060	0.52	.010	2		49,749
		NGC 7063 - 7		21 22 05	+36 19.5	1	10.85		1.72		2.06		1		49
		V1336 Cyg		21 22 06	+45 15.9	1	14.27		1.17				1		1772
203886	+23 04305	HR 8197		21 22 08	+24 18.8	2	6.36	.045	1.04	.000	0.94		4	K0 III	71,3016
		NGC 7063 - 11		21 22 08	+36 25.6	1	11.83		0.30		0.10		1		49
203771	−23 16895			21 22 09	−23 32.0	1	9.93		1.01		0.89		2	K1 V	3072
	+48 03371			21 22 10	+49 19.1	1	9.39		0.22				2	A5	1724
		959 11 142		21 22 11	+05 33.2	1	12.11		0.91				3		1719
203921	+35 04510	NGC 7063 - 1		21 22 11	+36 12.8	1	8.89		-0.02		-0.36		1	B9 IV	49
		NGC 7063 - 14		21 22 13	+36 20.1	2	12.07	.005	0.36	.005	0.11	.005	2		49,749
		NGC 7067 - 28		21 22 13	+47 46.6	1	15.98		0.94				1		49
239649	+59 02368			21 22 13	+59 38.2	1	9.31		0.19		-0.49		1	B3	5
203988	+49 03514			21 22 14	+50 13.5	1	7.90		1.58				2	K5	369
204089	+64 01535			21 22 14	+64 47.1	1	7.59		-0.04		-0.04		1	B9	401
		G 93 - 8		21 22 15	+02 41.0	2	12.21	.014	1.32	.005	1.26		4		202,333,1620
		NGC 7063 - 16		21 22 15	+36 11.5	1	12.61		1.12				1		49
203843	−4 05444	HR 8192		21 22 15	−03 36.8	3	6.39	.037	0.33	.004	0.12	.012	10	F0 III	15,1071,3016
204001	+51 03055			21 22 17	+51 26.5	2	7.11	.005	-0.11	.000	-0.53	.005	6	B9	401,848
	+35 04511	NGC 7063 - 2		21 22 18	+36 17.5	2	9.60	.005	0.00	.005	-0.25	.005	2	A0 V	49,749
203989	+47 03394			21 22 18	+47 29.5	3	8.01	.010	0.02	.008	-0.02		6	A0	401,1118,1196
		NGC 7063 - 6		21 22 19	+36 10.2	2	10.80	.005	0.12	.005	0.07	.020	2	A0 Vnn	49,749
		NGC 7063 - 24		21 22 19	+36 14.4	1	14.20		0.66				1		49
		NGC 7067 - 14		21 22 19	+47 47.2	1	14.58		0.60		-0.19		1		49
204002	+47 03395			21 22 20	+48 20.5	2	7.56	.000	-0.06	.005			8	A0	1118,1196
		LP 757 - 548		21 22 20	−08 33.0	1	10.70		0.60		0.00		2		1696
203756	−43 14529	3HLF4 # 1026		21 22 20	−43 26.0	1	10.42		0.39		-0.24		2	A3	221
203608	−65 03918	HR 8181		21 22 20	−65 35.6	8	4.21	.008	0.48	.006	-0.12	.020	37	F7 V	15,244,1034,1075,2024*
203925	+25 04531	HR 8198		21 22 21	+25 57.5	2	5.70	.039	0.31	.001	0.06	.018	6	A8 III	253,254
	+35 04512	NGC 7063 - 5		21 22 21	+36 20.3	2	10.23	.005	0.05	.005	-0.05	.005	2	A0 V	49,749
		NGC 7062 - 253		21 22 21	+46 08.7	1	11.55		0.54		0.06		1	G0 V	49
		NGC 7067 - 23		21 22 21	+47 46.9	1	15.09		0.65		-0.06		1		49
	+45 03524	NGC 7062 - 252		21 22 22	+46 19.9	1	8.85		0.94		0.68		1	K0	49
		NGC 7067 - 12		21 22 22	+47 48.8	1	14.29		1.64				1		49
204100	+64 01536			21 22 22	+64 49.2	1	7.34		-0.08		-0.15		2	B9	401
		NGC 7067 - 21		21 22 23	+47 47.4	1	14.91		0.72		0.09		1		49
		NGC 7067 - 25		21 22 23	+47 48.6	1	15.42		0.78		-0.02		1		49
204003	+46 03298			21 22 24	+46 59.4	1	8.91		0.12		-0.01		3		261
		NGC 7067 - 2		21 22 24	+47 46.5	1	11.17		0.69		-0.18		1	B5 III	49
204023	+48 03372			21 22 24	+48 36.2	2	7.43	.000	0.12	.005			8	A2	1118,1196
204037	+51 03057	IDS21208N5201	AB	21 22 24	+52 14.0	2	8.11	.010	0.01	.000	-0.13	.015	6	A0	848,1566
		3HLF4 # 1027		21 22 24	−40 39.	1	14.25		0.04		0.08		1	A2	221
		NGC 7063 - 23		21 22 25	+36 16.7	2	14.01	.020	0.77	.005	0.25	.010	2		49,749
		NGC 7067 - 22		21 22 25	+47 47.1	1	15.07		0.84		-0.01		1		49
		NGC 7067 - 17		21 22 25	+47 48.3	1	14.66		0.54		-0.18		1		49
		NGC 7067 - 19		21 22 25	+47 48.7	1	14.71		0.65		-0.04		1		49
		959 31 167		21 22 26	+03 29.8	1	11.08		1.00				2		1719
	+35 04513	NGC 7063 - 31	★ AB	21 22 26	+36 18.6	1	10.09		0.09		-0.01		1		749
		NGC 7063 - 13		21 22 26	+36 24.8	1	12.06		0.45		0.02		1		49
		NGC 7067 - 15		21 22 26	+47 47.9	1	14.58		0.69		0.06		1		49
		NGC 7067 - 29		21 22 26	+47 48.4	1	16.01		0.79		0.59		1		49
		NGC 7067 - 16		21 22 26	+47 49.2	1	14.61		0.74		-0.11		1		49
		NGC 7067 - 5		21 22 27	+47 47.6	1	12.57		0.68		-0.16		1	B2 V	49
	+35 04515	NGC 7063 - 4		21 22 28	+36 17.0	1	9.73		-0.02		-0.36		1	B7 V	49
		Steph 1891		21 22 29	+08 20.3	1	11.53		1.42		1.28		1	K7	1746
		NGC 7067 - 9		21 22 29	+47 47.2	1	13.84		0.50		-0.22		1		49
204022	+49 03516			21 22 29	+50 13.2	1	7.44		1.48		1.17		4	K0	848
		NGC 7063 - 21		21 22 30	+36 23.3	1	13.72		0.99		0.65		1		49
		NGC 7062 - 262	★	21 22 30	+46 13.8	2	10.78	.005	0.60	.015	0.23	.010	3		49,8106
		NGC 7067 - 24		21 22 30	+47 47.8	1	15.12		0.74				1		49
		NGC 7067 - 13		21 22 30	+47 49.2	1	14.34		1.02		0.62		1		49
		NGC 7063 - 10		21 22 31	+36 16.6	2	11.78	.005	0.52	.005	0.00	.015	2		49,749
		NGC 7067 - 27		21 22 31	+47 48.6	1	15.91		0.56				1		49
204130	+64 01538	IDS21215N6430	AB	21 22 32	+64 42.7	1	8.18		0.14		0.05		2	A1 V +F5 V	401
203875	−10 05668	HR 8195		21 22 32	−09 57.8	3	5.69	.004	0.21	.007			11	F0 IV	15,2018,2030
		NGC 7063 - 19		21 22 33	+36 11.5	2	13.55	.030	1.09	.050	0.86	.020	2		49,749
		NGC 7063 - 33		21 22 33	+36 14.5	1	10.49		0.07		0.00		1		749
		NGC 7063 - 20		21 22 33	+36 16.4	2	13.64	.005	0.62	.015	0.09	.005	2		49,749
		NGC 7063 - 25		21 22 33	+36 17.0	2	14.28	.015	1.00	.010	0.56	.005	2		49,749
		NGC 7067 - 26		21 22 33	+47 46.3	1	15.68		0.91				1		49
		NGC 7067 - 18		21 22 34	+47 46.7	1	14.67		0.96		0.46		1		49
		NGC 7067 - 20		21 22 34	+47 47.6	1	14.71		0.71		-0.15		1		49
203977	+25 04533			21 22 35	+25 39.9	1	7.07		0.03		-0.03		2	A0	105
		NGC 7067 - 6		21 22 36	+47 48.0	1	12.77		0.68		0.18		1		49
		NGC 7067 - 3		21 22 38	+47 48.8	1	11.25		1.87		1.71		1		49
		NGC 7063 - 17		21 22 39	+36 18.3	1	12.98		0.51		0.04		1		49
		959 11 51		21 22 40	+06 43.5	1	11.47		0.98				2		1719
	+17 04575			21 22 41	+17 54.7	1	9.29		0.97		0.70		1	G5	592
203926	−4 05446	HR 8199		21 22 41	−03 46.3	2	5.52	.050	1.44	.020	1.79	.025	10	K4 III	1088,3016
		NGC 7063 - 22		21 22 42	+36 12.4	2	13.75	.030	0.74	.020	0.24	.010	2		49,749
		NGC 7067 - 7		21 22 42	+47 45.9	1	13.22		1.54		1.27		1		49

Table 1   1149

HD	DM	Other Id	N Rem	α_{1950}	δ_{1950}	S	V	σ_V	B–V	σ_{B-V}	U–B	σ_{U-B}	n	Spectrum	References
203760	−55 09586	HR 8188		21 22 42	−54 52.7	2	6.10	.007	0.36	.007	0.04		5	F1 III	1034,2007
		NGC 7063 - 12		21 22 43	+36 12.8	1	11.99		0.49		-0.01		1		49
	+35 04516	NGC 7063 - 29		21 22 43	+36 16.2	1	9.04		1.11		0.96		1	G8 III:	749
		NGC 7063 - 27		21 22 43	+36 20.6	2	15.31	.010	0.61	.040	0.12	.055	2		49,749
		959 11 13		21 22 44	+07 10.9	1	12.78		0.91				3		1719
		NGC 7067 - 11		21 22 44	+47 46.5	1	14.15		0.66		0.42		1		49
204050	+43 03908			21 22 47	+44 04.5	1	8.08		1.20		1.16		2	K0	8100
		3HLF4 # 101		21 22 48	−40 24.	1	14.42		0.13		0.23		1	A0	208
	+47 03396	NGC 7067 - 1	⋆ A	21 22 49	+47 47.9	1	9.46		0.16		0.12		1	A0	49
235508	+50 03329			21 22 50	+51 10.3	1	8.94		0.42				2		1724
		NGC 7063 - 9		21 22 51	+36 25.2	2	11.13	.000	0.26	.010	0.13	.005	2		49,749
		NGC 7067 - 4	⋆ B	21 22 51	+47 47.7	2	11.75	.009	0.91	.023	0.01	.005	5	B0.5V	49,483
		NGC 7067 - 10	⋆ C	21 22 51	+47 48.3	1	13.98		0.78		0.21		1		49
204408	+76 00836			21 22 51	+76 52.5	1	6.58		0.07		0.05		2	A0	985
		959 31 176		21 22 53	+03 21.8	1	12.66		0.97				1		1719
204151	+59 02371			21 22 53	+59 32.7	1	8.30		1.17		1.03		4	K2	1371
		959 11 81		21 22 54	+06 22.2	1	11.54		1.05				2		1719
204038	+33 04252	V1073 Cyg		21 22 55	+33 28.3	1	8.40		0.42		0.14		3	F2 IV	8071
204150	+60 02233			21 22 55	+60 34.9	3	7.69	.011	0.01	.005	-0.74	.015	8	B2 V	5,399,555
203866	−41 14517	IDS21197S4129	AB	21 22 55	−41 16.4	1	10.03		0.07		0.05		2	A0 V	208
204116	+54 02533	LS III +55 009		21 22 57	+55 09.0	2	7.81	.072	0.46	.017	-0.37	.051	7	B1.5IVn	5,399
203993	+1 04477	IDS21204N0137	AB	21 22 58	+01 49.6	1	7.53		0.01		-0.10		3	A0	3016
203993	+1 04477	IDS21204N0137	C	21 22 58	+01 49.6	1	11.53		0.66		0.13		3		3016
		NGC 7063 - 106		21 22 59	+36 13.8	1	12.50		0.55		0.07		1		749
204101	+51 03058			21 23 00	+52 16.7	1	9.17		0.08		-0.05		3		261
		Smethells 79		21 23 00	−42 39.	1	12.67		1.53				1		1494
204091	+45 03529	NGC 7062 - 335		21 23 02	+46 16.4	2	8.05	.014	0.31	.005	0.07		4	F0	49,1724
204102	+48 03375			21 23 02	+49 10.5	4	8.02	.009	0.10	.006	0.05	.012	25	A0	261,848,1118,1196
		POSS 342 # 5		21 23 03	+31 43.0	1	17.17		1.35				2		1739
		WLS 2120-10 # 9		21 23 03	−09 39.1	1	11.91		0.59		0.08		2		1375
204040	+18 04788			21 23 05	+18 26.5	3	8.75	.008	0.39	.014	-0.03	.000	30	F8	592,1462,6011
	+43 03910			21 23 05	+43 37.8	1	9.16		0.50				2	F8	1724
204118	+49 03517	IDS21214N4948	AB	21 23 06	+50 02.0	1	8.68		1.07		1.06		3	G0	293
		JL 76		21 23 06	−82 54.0	1	11.28		0.13		-0.68		1		832
		NGC 7063 - 18		21 23 08	+36 22.4	2	13.23	.010	0.56	.005	0.03	.010	2		49,749
204131	+48 03376	HR 8206, V1934 Cyg	⋆ A	21 23 10	+49 06.4	4	6.57	.008	-0.02	.010	-0.02	.010	11	B9p(SiCrSr)	401,848,1118,1196
	+53 02619			21 23 10	+53 31.0	1	10.18		2.33				2		369
	+43 03913	LS III +44 043		21 23 12	+43 51.5	2	8.89	.025	0.54	.005	-0.34	.010	5	B1.5V:nnpe	1012,1212
		LS III +50 017		21 23 12	+50 33.7	1	11.83		0.73		-0.21		5		848
		959 11 163		21 23 13	+05 16.8	1	12.31		1.37				4		1719
		LS III +47 041		21 23 13	+47 47.4	1	12.44		0.71		0.27		4		483
		VES 402		21 23 14	+45 25.8	1	11.03		0.27		-0.16		1		8106
204119	+40 04518			21 23 16	+40 51.6	1	8.12		0.41				2	F5	1724
203949	−38 14551	HR 8200		21 23 16	−38 02.8	5	5.63	.006	1.18	.005	1.22		17	K2 III	15,1075,2018,2032,3005
204079	+26 04156			21 23 17	+26 59.8	1	8.24		1.12		0.96		2	K1 IV	3077
204041	−0 04215	HR 8203		21 23 18	+00 19.0	2	6.45	.005	0.16	.000	0.04	.005	7	A1 IV	15,1071
		3HLF4 # 1028		21 23 18	−42 20.	1	11.16		1.15		1.11		2	K0 III	221
203850	−56 09645			21 23 18	−56 20.7	4	8.65	.019	0.92	.005	0.61	.022	10	K3 V	158,1696,2012,3072
	+35 04522	NGC 7063 - 3		21 23 19	+36 22.1	1	9.71		0.46		0.05		1		49
	+41 04126			21 23 19	+41 57.1	1	9.34		0.59				2	F8	1724
204170	+49 03519			21 23 19	+49 45.1	1	8.49		0.92		0.71		3	K0	293
235513	+50 03332			21 23 19	+51 09.2	2	8.80	.000	0.46	.000	0.17		6	F5	848,1724
203915	−45 14338			21 23 21	−45 03.8	1	7.87		0.45				4	F5 V	1075
		959 11 2		21 23 22	+07 18.9	1	12.45		1.22				4		1719
		VES 403		21 23 22	+48 08.2	1	12.29		1.28		0.98		2		8106
		959 31 190		21 23 24	+03 14.6	1	12.41		1.43				1		1719
		VES 404		21 23 25	+48 21.3	1	11.26		1.28		0.96		1		8106
	+54 02535			21 23 25	+55 22.8	1	9.74		2.10				2		369
239659	+56 02571	IDS21220N5708	AB	21 23 27	+57 20.8	1	7.16		0.56		0.03		4	A5	3016
	+47 03398			21 23 28	+48 21.5	1	8.48		0.98		0.69		4	K0	261
204153	+46 03305	HR 8208		21 23 29	+46 29.8	6	5.60	.012	0.33	.017	-0.02	.009	12	F0 V	254,401,1118,1196,1620,1724
		UKST 342 # 17		21 23 29	−40 22.0	1	15.50		0.58				2		1584
		VES 405		21 23 30	+48 19.6	1	11.43		0.52		0.09		1		8106
		UKST 342 # 18		21 23 38	−40 58.7	1	15.55		0.67				2		1584
204194	+50 03333			21 23 41	+50 55.0	1	7.80		0.17		0.02		4	A2	848
		G 212 - 35		21 23 43	+42 47.2	1	12.19		1.32		1.35		2	M2	7010
		UKST 342 # 11		21 23 43	−40 59.2	1	13.89		0.98				3		1584
		959 31 77		21 23 44	+04 17.6	1	11.78		1.05				2		1719
204172	+36 04557	HR 8209	⋆ A	21 23 44	+36 27.0	6	5.93	.009	-0.08	.023	-0.94	.024	11	B0 Ib	154,450,542,592,1119,8084
204172	+36 04557	IDS21217N3614	B	21 23 44	+36 27.0	2	12.09	.041	1.01	.194	-0.58		5	F5 V	450,8084
204172	+36 04557	IDS21217N3614	C	21 23 44	+36 27.0	3	10.38	.056	0.47	.005	-0.03	.013	7		450,542,8084
	+37 04305			21 23 44	+38 22.9	1	9.63		-0.02		-0.41		1	A0	963
		VES 406		21 23 45	+49 08.0	1	11.56		1.57		1.23		1		8106
203985	−45 14340	IDS21204S4514		21 23 45	−45 01.7	2	7.48	.005	0.91	.005	0.70		5	K1 IV	219,2012
204133	+17 04582			21 23 47	+17 50.9	1	6.87		0.00		-0.01		2	A0	252
204075		HR 8204	⋆ A	21 23 49	−22 37.7	11	3.74	.008	1.00	.009	0.58	.013	29	G4 Ibp	15,58,565,1020,1075*
204018	−43 14539	HR 8202	⋆ AB	21 23 50	−42 45.9	5	5.50	.012	0.39	.014	0.15	.018	19	A5mF0-F3	216,355,404,938,2035
	+46 03307			21 23 52	+47 17.5	1	10.35		0.29		-0.29		2		117
		UKST 342 # 6		21 23 53	−40 29.8	1	11.95		0.55				2		1584
		959 32 162		21 23 54	+03 32.7	1	11.71		1.23				2		1719

HD	DM	Other Id	N Rem	α₁₉₅₀ δ₁₉₅₀	S	V	σV	B–V	σB–V	U–B	σU–B	n	Spectrum	References
	+40 04520			21 23 54 +40 34.8	1	8.87		0.35				2	F2	1724
204121	+0 04726	HR 8205		21 23 55 +00 53.3	4	6.13	.010	0.45	.008	-0.02	.022	11	F4 V	15,254,1071,2024
	+46 03308			21 23 55 +46 36.5	1	9.21		0.45				2	F5	1724
		POSS 76 # 3		21 23 55 +66 09.5	1	16.16		1.46				1		1739
		3HLF4 # 103		21 24 00 −43 44.	1	13.57		0.09		0.13		1	A2	208
204076	−32 16569			21 24 01 −32 09.4	1	8.79		-0.15		-0.84		1	B2 II	55
		959 11 39		21 24 02 +06 52.8	1	10.82		1.02				2		1719
		POSS 342 # 3		21 24 03 +31 52.5	1	16.20		1.11				2		1739
	+41 04134	IDS21221N4117	A	21 24 03 +41 30.5	1	8.70		0.55				2	G0	1724
204188	+18 04794	HR 8210, IK Peg		21 24 07 +19 09.5	2	6.07	.005	0.23	.010	0.04	.015	6	A8 m	3058,8071
204262	+43 03922			21 24 08 +44 10.7	1	7.88		0.21		0.14		3	Am	8071
	−9 05746	G 26 - 1		21 24 08 −08 37.0	3	11.25	.021	0.48	.013	-0.17	.009	6		202,1620,1696
		LS III +51 012		21 24 10 +51 40.9	1	12.01		0.79		0.10		4		848
		MCC 70		21 24 11 +03 31.2	3	10.52	.037	1.39	.027	0.96		3	K7	1017,1705,3016
204155	+4 04674	G 25 - 29		21 24 12 +05 13.7	6	8.49	.009	0.57	.006	-0.05	.010	11	G5	202,516,979,1620*
		AAS42,335 T8# 36		21 24 12 +52 02.	1	11.27		0.36		-0.18		4		848
204244	+35 04526		V	21 24 14 +35 37.3	1	7.80		1.59		1.51		3	M1	1601
		959 11 3		21 24 17 +07 16.4	1	11.52		0.95				3		1719
204305	+49 03521			21 24 17 +49 37.4	2	8.97	.000	0.12	.009	0.04	.009	13	A2	261,848
		3HLF4 # 104		21 24 18 −44 03.	1	13.96		0.25		0.09		1	A2	208
204214	+22 04394			21 24 19 +22 43.9	1	8.44		1.11		0.94		2	K2	1648
203881	−70 02850	HR 8196, SX Pav		21 24 19 −69 43.4	2	5.64	.135	1.56	.007	1.34	.049	24	M5 III	58,3042
204233	+20 04919			21 24 20 +20 29.7	1	7.69		0.12		0.10		2	A2	1648
	+43 03923			21 24 21 +43 38.3	1	10.39		0.29		-0.14		2		8106
		VES 408		21 24 24 +45 40.6	1	11.30		0.47		0.17		1		8106
204139	−21 06020	HR 8207		21 24 25 −21 24.8	6	5.78	.010	1.44	.008	1.74	.008	22	K3 III	15,1007,1013,2032*
		VES 409		21 24 27 +47 21.6	1	12.79		1.54		1.38		2		8106
	+40 04523			21 24 28 +40 57.4	1	9.63		1.09				2	G5	1724
235515	+50 03337			21 24 29 +50 52.7	1	9.06		0.47				2		1724
	+35 04528		AB	21 24 32 +36 07.8	1	9.38		0.02		-0.76		5		994
204409	+61 02134	SW Cep		21 24 32 +62 21.4	1	8.88		2.36		2.24		2	M3	8032
		CS 22897 # 96		21 24 33 −65 31.6	1	14.99		0.19		0.17		1		1736
		959 32 184		21 24 34 +02 57.8	1	12.98		1.04				1		1719
		WLS 2136 45 # 6		21 24 35 +45 07.7	1	12.64		0.69		0.50		2		1375
		G 26 - 2		21 24 38 −07 03.4	4	11.10	.015	1.55	.028	1.18	.010	6	M0	202,1620,1705,1746
	+66 01401			21 24 41 +66 29.8	1	8.86		1.07		0.71		2	G5	1733
	+49 03523			21 24 42 +49 30.0	1	8.41		2.07				2	M0	369
204374	+48 03386	IDS21230N4850	AB	21 24 43 +49 03.8	2	7.94	.015	-0.12	.005	-0.54	.000	6	A0	401,848
204277	+15 04416			21 24 45 +15 54.5	1	6.72		0.50		0.00		2	F8 V	985
		AO Peg		21 24 46 +18 23.0	1	12.90		0.39		0.12		1	F0	699
		UKST 342 # 16		21 24 47 −40 14.7	1	15.09		1.08		0.68		1		1584
		G 231 - 43		21 24 50 +55 00.4	3	14.70	.044	0.14	.010	-0.68	.005	4		1,419,3028
		POSS 76 # 4		21 24 50 +66 06.3	1	17.16		1.30				2		1739
204128	−53 10073	BH Ind		21 24 51 −53 02.3	1	9.65		1.03		0.53		1	K1 III	1641
204166	−44 14484			21 24 52 −44 39.4	1	8.17		0.96		0.69		2	G8 III	221
	+47 03410	NGC 7092 - 226		21 24 54 +48 21.8	2	9.75	.005	0.21	.019	0.20		6	A5 m	1544,1581
		AAS42,335 T8# 39		21 24 54 +49 15.	1	12.02		0.87		-0.03		5		848
		UKST 342 # 20		21 24 56 −40 26.1	1	15.93		1.31				2		1584
		G 26 - 3		21 24 58 +01 47.9	2	11.36	.024	0.94	.024	0.73		3	K5	202,333,1620
	+7 04692	G 25 - 31		21 24 58 +07 26.8	3	10.64	.015	0.59	.008	-0.01	.005	4	G0	202,1620,1658
		UKST 342 # 10		21 24 58 −40 15.5	1	13.71		0.62		0.07		5		1584
		AAS42,335 T8# 40		21 25 00 +49 50.	1	12.09		0.59		0.10		4		848
		VES 410		21 25 01 +47 44.9	1	12.17		0.61		0.11		1		8106
205083	+82 00648			21 25 04 +82 45.8	1	8.12		0.10		0.09		5	A2	1219
204411	+48 03390	HR 8216		21 25 05 +48 37.0	5	5.31	.010	0.08	.008	0.14	.019	12	A6 p Cr(Eu)	220,848,1118,1263,3033
204428	+52 02939	HR 8218		21 25 06 +52 40.8	2	6.01		-0.14	.007	-0.47	.000	5	B6 V	848,1079
	−49 13484			21 25 09 −49 38.3	1	10.58		0.62		0.10		3		1730
		959 32 78		21 25 12 +04 07.0	1	11.76		1.00				6		1719
	+41 04142			21 25 12 +41 49.3	1	9.55		0.53				2	F8	1724
		959 32 69		21 25 15 +04 17.7	1	12.51		1.08				2		1719
	+46 03317			21 25 16 +46 40.3	1	8.63		0.58				2	G0	1724
	+58 02265	LS III +58 008		21 25 17 +58 53.2	1	9.85		0.45		-0.42		2		401
		959 32 190		21 25 19 +02 54.1	1	13.18		1.01				2		1719
		G 25 - 32		21 25 19 +07 05.2	3	11.60	.007	1.55	.012	1.22		7	M2	202,333,940,1620
204403	+36 04568	HR 8215		21 25 19 +36 53.9	3	5.29	.010	-0.15	.009	-0.66	.014	8	B3 IV	154,1363,3016
		V 609 Cyg		21 25 22 +54 16.5	1	10.48		1.76				1		1772
	+47 03411	NGC 7092 - 227		21 25 24 +48 14.7	2	10.60	.020	0.46	.015	-0.01		5	F6 V	1544,1581
		AAS42,335 T8# 42		21 25 24 +52 20.	1	12.16		0.60		0.14		4		848
		G 25 - 33		21 25 25 −00 52.8	1	11.95		1.42		1.13		1		1620
		959 12 44		21 25 27 +06 33.4	1	12.68		0.79				1		1719
204414	+26 04164	HR 8217		21 25 28 +27 23.4	3	5.39	.013	0.03	.017	0.03	.015	8	A2 Vs	985,1049,3050
		L 714 - 46		21 25 28 −22 31.6	1	12.21		1.56				1		1746
204272	−40 14362			21 25 28 −40 37.0	1	8.66		1.62				2	M0 III	1584
		959 32 12		21 25 30 +04 53.5	1	12.75		1.06				1		1719
		Smethells 81		21 25 30 −47 29.	1	12.20		1.42				1		1494
204228	−54 09872	HR 8211		21 25 30 −53 55.5	1	6.39		1.14				4	K2 III	2007
204363	−12 06005	HR 8212		21 25 32 −11 47.2	1	6.61		0.49				4	F7 V	2035
204481	+47 03412			21 25 34 +47 57.4	1	8.54		-0.03		-0.26		2	A0	401
204258	−49 13487			21 25 35 −49 38.6	1	8.69		1.25		1.33		2	K1/2 (III)	1730
203971	−75 01710			21 25 35 −75 25.3	1	7.65		1.05		0.84		6	K0 III	1704

Table 1 1151

HD	DM	Other Id	N Rem	α_{1950}	δ_{1950}	S	V	σ_V	B–V	σ_{B-V}	U–B	σ_{U-B}	n	Spectrum	References
		WLS 2120 10 # 9		21 25 36	+10 23.3	1	10.29		0.97		0.74		2		1375
		AAS42,335 T8# 43		21 25 36	+51 56.	1	11.13		0.59		0.07		5		848
204303	−41 14527			21 25 37	−40 55.1	1	9.14		0.41				2	F3 V	1584
	+46 03321			21 25 40	+46 31.0	1	9.06		0.34				2	F2	1724
204497	+47 03413			21 25 41	+47 35.1	2	7.51	.004	0.03	.000	-0.02	.028	12	A0	261,1375
		3HLF4 # 1030		21 25 42	−43 01.	1	12.59		0.80		0.36		2	K1	221
		959 12 116		21 25 43	+05 31.3	1	12.14		1.07				1		1719
		NGC 7092 - 228		21 25 44	+48 06.2	2	12.36	.015	0.70	.019	0.07		6		1544,1581
204417	+9 04821			21 25 45	+10 23.5	1	8.72		0.86		0.59		3	K0	196
239674	+59 02380			21 25 47	+60 19.6	1	8.49		1.14				2	K2	1379
		LP 873 - 51		21 25 47	−25 58.2	1	11.11		0.76		0.29		2		1696
		AAS42,335 T8# 44		21 25 48	+52 04.	1	11.04		0.61		0.11		7		848
	−40 14364			21 25 48	−40 11.3	1	11.01		1.11				1		1584
204484	+34 04422			21 25 49	+35 11.4	1	7.13		0.20		0.12		2	A3	1625
		UKST 342 # 8		21 25 49	−40 11.0	1	12.49		0.57				1		1584
		UKST 342 # 7		21 25 49	−40 25.1	1	12.36		0.47				2		1584
204498	+41 04148			21 25 52	+41 46.4	1	8.76		0.25				2	F0	1724
204381	−22 05692 HR 8213			21 25 53	−22 01.6	6	4.50	.008	0.90	.005	0.60	.002	24	G5 III	15,1075,2012,4001*
	−49 13488			21 25 53	−49 36.1	1	11.48		0.44		0.01		3		1730
		NGC 7092 - 229		21 25 54	+48 18.9	1	13.98		0.97		0.35		2		1581
239675	+59 02381 IDS21245N5927	AB	21 25 54	+59 40.4	1	9.16		0.30		-0.25		1	B5	5
		959 32 154		21 25 55	+03 25.3	1	13.18		1.01				2		1719
204445	+7 04696 HR 8219			21 25 57	+07 58.6	3	6.41	.013	1.63	.010	1.90	.000	10	M1	15,1071,3055
203532	−83 00716 HR 8176			21 25 58	−82 54.2	3	6.37	.008	0.12	.010	-0.37	.005	11	B3 IV/V	15,1075,2038
204485	+31 04462 HR 8220			21 26 00	+32 00.3	2	5.80	.020	0.30	.005	0.01	.010	5	F0 V	254,3026
		959 32 87		21 26 01	+04 00.9	1	11.38		1.33				7		1719
204599	+59 02383 HR 8224	★ A	21 26 02	+59 31.9	3	6.09	.011	1.74	.014	1.72		6	M3 IIIa	71,369,3001
204317	−53 10078			21 26 03	−53 07.0	1	8.96		1.06		0.88		33	K1 III	978
204536	+45 03549			21 26 04	+46 20.6	2	6.86	.005	-0.08	.000	-0.59	.000	7	B3 III	399,401
204394	−31 18291 HR 8214			21 26 05	−31 27.5	1	6.50		0.03				4	A1 V	2032
	+40 04535 IDS21242N4044	AB	21 26 07	+40 57.1	1	8.92		0.73				2	G0	1724
239676	+58 02268 LS III +59 008		21 26 08	+59 04.6	1	9.06		0.52		-0.38		1	B2	5
	+45 03550			21 26 14	+45 34.1	1	8.51		1.83				2	M0	369
204570	+44 03825			21 26 15	+45 21.6	1	8.60		0.62		0.11		2	G0	1601
		959 32 98		21 26 16	+03 56.1	1	12.17		0.87				3		1719
		Steph 1899		21 26 16	+23 00.7	1	11.69		1.31		1.31		1	K7	1746
		3HLF4 # 106		21 26 18	−44 37.	1	13.46		0.20		0.11		2	A3	208
		Smethells 82		21 26 18	−55 15.	1	11.56		1.38				1		1494
	+45 03552			21 26 21	+46 01.7	1	9.61		0.70				2	A5	1724
204614	+52 02945			21 26 21	+53 04.2	1	7.18		0.01		-0.11		2	A0	401
	+40 04536			21 26 22	+40 30.4	1	9.00		0.19				2	A5	1724
205072	+79 00707 HR 8239			21 26 22	+80 18.4	1	5.97		0.92				2	K0	71
	+17 04591 IDS21241N1744	AB	21 26 24	+17 56.7	1	9.46		0.57		0.04		1	F8	1462
		UKST 342 # 19		21 26 24	−40 44.1	1	15.61		0.42				1		1584
		3HLF4 # 107		21 26 24	−43 42.	1	14.06		-0.01		-0.20		2	A0	208
		UKST 342 # 9		21 26 25	−40 48.5	1	12.87		0.71		0.18		3		1584
		NGC 7092 - 230		21 26 26	+47 44.6	1	13.47		0.86		0.35		2		1581
	+48 03392			21 26 26	+48 38.1	1	8.08		1.90		1.85		2	K2	562
204509	+10 04554 IDS21240N1039	AB	21 26 27	+10 51.9	1	6.65		0.38				1	F4 II	6009
204541	+24 04409			21 26 27	+24 27.0	1	7.49		0.23		0.16		3	Am	8071
204600	+47 03418			21 26 27	+47 44.1	2	8.03	.005	-0.09	.010	-0.35	.040	4	B9	401,562
	+48 03393 NGC 7092 - 137		21 26 27	+48 45.2	3	9.37	.009	0.23	.005	0.13		6	A7 Vm	562,1544,1724
204540	+25 04542			21 26 28	+25 42.1	2	6.55	.005	1.32	.000	1.44		6	K2 III	20,833
204539	+25 04543			21 26 28	+26 11.6	1	7.58		1.24				5	K2 III	20
204583	+41 04153			21 26 30	+41 39.4	1	8.61		0.01		-0.05		2	A0	1733
		Cyg sq 1 # 2		21 26 30	+50 44.	1	12.66		0.66		0.13		2		293
		Cyg sq 1 # 1		21 26 30	+51 08.	1	12.25		0.57		-0.01		4		293
		UKST 342 # 15		21 26 31	−40 51.2	1	14.87		1.01		-0.22		1		1584
204615	+46 03325			21 26 34	+46 27.4	1	8.86		0.27				5		1724
		AAS11,365 T3# 1		21 26 36	+50 41.	1	11.12		0.63		0.10		4		261
		Cyg sq 1 # 3		21 26 36	+50 55.	1	12.34		0.68		0.22		1		293
		Cyg sq 1 # 4		21 26 36	+51 10.	1	11.74		0.39		0.01		3		293
		3HLF4 # 108		21 26 36	−44 24.	1	13.23		-0.01		-0.02		2	A0	208
204626	+47 03420 NGC 7092 - 138		21 26 37	+48 12.9	1	7.54		0.03		0.04		2	A2	562
204560	+17 04592 HR 8221			21 26 39	+17 41.2	1	6.44		1.39		1.64		2	K5	252
		NGC 7092 - 231		21 26 41	+48 02.3	1	12.63		0.79		0.39		2		1581
	+11 04571 G 126 - 2	★ ABC	21 26 42	+11 58.1	2	11.16	.010	1.08	.020	0.80	.020	6	K3	203,1620
		AAS11,365 T3# 2		21 26 42	+50 41.	1	14.22		0.74		0.71		4		261
204467	−40 14371			21 26 42	−40 37.2	1	8.47		0.99				2	K0 III	1584
204585	+21 04555 HR 8223	★ A	21 26 43	+21 57.6	2	5.86	.064	1.50	.023	1.32		5	M4.5-IIIa	71,3032
204585	+21 04555 IDS21244N2145	B	21 26 43	+21 57.6	1	10.40		0.45		0.04		2		3032
		G 261 - 43		21 26 43	+73 25.8	2	12.81	.047	0.02	.005	-0.65	.005	4		419,3028
		G 126 - 3		21 26 48	+19 16.0	2	10.87	.005	1.30	.026	1.34	.068	3	M0	1620,5010
204673	+49 03531			21 26 48	+50 13.6	1	7.37		-0.01		-0.04		2	A0	401
		AAS11,365 T3# 3		21 26 48	+50 40.	1	14.06		0.69		0.43		2		261
204770	+66 01405 HR 8227			21 26 48	+66 35.4	1	5.43		-0.11		-0.43		3	B7 V	154
	+47 03421 LS III +47 044		21 26 49	+47 49.8	1	9.54		0.58		0.01		2	A0 Iab	562,1012
	+49 03532			21 26 50	+50 10.4	2	8.53	.005	0.29	.010	0.05	.005	9		261,848
		959 12 126		21 26 52	+05 23.9	1	12.06		1.07				1		1719
204734	+63 01737			21 26 52	+64 10.0	1	8.86		0.84		0.45		2	K0	3016

HD	DM	Other Id	N Rem	α_{1950}	δ_{1950}	S	V	σ_V	B–V	σ_{B-V}	U–B	σ_{U-B}	n	Spectrum	References
203955	−80 01017	HR 8201	⋆ A	21 26 52	−80 15.6	2	6.46	.005	0.04	.000	0.05	.000	7	A0 V	15,1075
	+40 04540			21 26 53	+40 55.6	1	9.60		1.09				2	G5	1724
	+47 03423			21 26 54	+48 02.9	2	9.39	.005	0.72	.000	0.47		5	F5	562,1724
		AAS11,365 T3# 4		21 26 54	+50 44.	1	13.72		0.71		0.59		3		261
		Cyg sq 1 # 5		21 26 54	+50 50.	1	12.62		0.50		0.28		4		293
		Cyg sq 1 # 7		21 26 54	+51 13.	1	10.73		0.40		0.10		3		293
	+50 03343			21 26 54	+51 15.9	1	10.33		0.70		0.22		3		293
	−41 14533			21 26 56	−41 21.1	1	11.29		0.20		0.10		2	A3	208
	+46 03330	NGC 7082 - 71		21 26 59	+46 53.8	1	8.59		1.36				3	G5	1724
		959 32 144		21 27 00	+03 32.1	1	12.66		1.14				1		1719
		Cyg sq 1 # 9		21 27 00	+51 05.	1	12.09		0.55		−0.39		4		293
		Cyg sq 1 # 8		21 27 00	+51 10.	1	10.88		0.51		0.02		3		293
		3HLF4 # 111		21 27 00	−40 12.	1	13.22		0.27		0.15		2	A5	208
		3HLF4 # 110		21 27 00	−44 29.	1	13.82		0.18		0.13		2	A2	208
204627	+24 04411			21 27 01	+25 13.7	1	8.71		0.61		0.06		3	G5	3016
235518	+50 03344			21 27 01	+50 36.4	2	8.25	.020	1.25	.010	0.96		6	G5	261,1724
204699	+49 03534			21 27 03	+49 34.6	2	6.79	.005	0.29	.005	0.05	.014	8	A3	261,848
204642	+27 04076			21 27 04	+28 21.9	2	6.75	.015	1.09	.015	1.08		5	K1 III-IV	20,3016
239681	+59 02384	LS III +60 001		21 27 05	+60 01.8	1	9.34		0.27		−0.60		1	B3	5
	+64 01552	IDS21261N6458	AP	21 27 05	+65 11.7	1	9.24		1.50		1.75		1		8084
235519	+50 03345			21 27 07	+50 53.3	2	8.28	.010	1.49	.030	1.68	.035	7	K7	261,293
	+50 03346			21 27 09	+51 03.7	1	10.69		0.44		−0.03		3		293
204658	+28 04105			21 27 11	+28 39.3	2	8.53	.028	0.59	.009	0.05		4	G5	20,3026
204577	−19 06107	HR 8222		21 27 11	−19 22.0	1	6.57		0.41				4	F3 V	2009
		L 164 - 140		21 27 12	−63 43.	1	11.26		0.83		0.49		2		1696
		G 25 - 34		21 27 13	+06 29.0	1	13.73		1.44				2		202
		G 126 - 4		21 27 13	+17 25.2	3	10.32	.020	1.63	.012	1.29	.022	4		1510,1746,3078
204710	+44 03832	LS III +44 044	⋆ AB	21 27 13	+44 42.2	1	6.95		0.26		−0.24		2	B8	1012
		UKST 342 # 14		21 27 13	−40 08.4	1	14.73		0.72		0.13		1		1584
	+39 04573			21 27 16	+40 15.0	1	10.22		0.33				2	A5	1724
	+48 03395	LS III +49 023		21 27 16	+49 07.7	1	9.68		0.33		−0.30		4		848
204587	−13 05945			21 27 16	−12 43.6	6	9.11	.012	1.27	.016	1.26	.014	25	K4/5 V	830,1003,1013,1705,1783,3078
	+50 03347			21 27 17	+51 01.5	1	10.16		0.71		0.33		4		293
		AAS11,365 T3# 8		21 27 18	+50 39.	1	12.65		0.73		0.43		2		261
		AAS11,365 T3# 7		21 27 18	+50 40.	1	12.27		0.74		0.28		6		261
		Cyg sq 1 # 12		21 27 18	+50 58.	1	11.90		1.37		1.25		5		293
		Cyg sq 1 # 13		21 27 18	+51 02.	1	11.08		1.66		1.73		4		293
204754	+54 02544	HR 8226		21 27 18	+55 11.9	1			0.14		−0.29		1	B8 III	1079
		UKST 342 # 12		21 27 18	−40 09.8	1	14.19		0.92		0.63		1		1584
204722	+43 03941	LS III +44 045		21 27 21	+44 07.1	2	7.66	.010	−0.02	.000	−0.83	.015	5	B1.5IV:Np	1012,1212
	−40 14379			21 27 22	−40 22.1	1	10.55		0.54				1	F8	1584
		LP 873 - 56		21 27 23	−25 06.0	1	12.37		1.49		1.20		1		3073
		AAS11,365 T3# 9		21 27 24	+50 41.	2	11.36	.000	0.44	.010	0.06	.010	5		261,848
		Cyg sq 1 # 15		21 27 24	+50 49.	1	12.01		0.32		−0.02		4		293
		Cyg sq 1 # 16		21 27 24	+50 49.	2	12.43	.020	0.57	.045	0.29	.050	6		261,293
		Cyg sq 1 # 17		21 27 24	+51 02.	1	11.35		0.17		−0.25		3		293
	+79 00708			21 27 29	+79 35.2	1	8.61		1.16		1.11		2	K0	1502
		959 32 32		21 27 30	+04 41.7	1	13.08		1.15				2		1719
		VES 411		21 27 31	+44 48.5	1	11.39		0.77		0.38		2		8106
204827	+58 02272	Tr 37 - 1240		21 27 31	+58 31.2	5	7.94	.021	0.81	.015	−0.13	.013	9	O9.5V	5,1012,1119,1252,8031
	+48 03396	NGC 7092 - 139		21 27 32	+48 34.5	1	9.68		0.48		0.20		2	F3 III	562
204609	−25 15459			21 27 33	−25 24.8	2	7.31	.002	1.55	.008	1.92		7	K5/M0 III	1657,2012
	+47 03427			21 27 35	+47 42.2	1	9.10		1.00				2	G5	1724
204771	+45 03558	HR 8228		21 27 36	+46 19.2	2	5.24	.004	0.98	.004	0.77	.018	5	K0 III	1355,3016
		Cyg sq 1 # 18		21 27 36	+50 50.	3	11.35	.010	0.38	.025	0.11	.012	11		261,293,848
		L 425 - 35		21 27 36	−40 55.4	1	13.19		1.62		1.12		1		3078
	−33 15610			21 27 38	−33 06.2	1	10.74		0.41		−0.21		2		1696
204555	−53 10085			21 27 40	−53 22.1	1	9.26		0.46		−0.01		30	F5 V	978
204724	+23 04325	HR 8225	⋆ A	21 27 41	+23 25.1	4	4.55	.021	1.62	.008	1.93	.005	10	M1 III	15,1363,3001,8015
		AAS11,365 T3# 13		21 27 41	+50 45.	1	11.42		0.56		0.36		2		261
		Cyg sq 1 # 19		21 27 42	+50 46.	1	12.58		0.56		0.47		1		293
		Cyg sq 1 # 20		21 27 42	+50 49.	2	10.24	.005	0.14	.014	0.11	.019	8		261,293
		UKST 342 # 13		21 27 43	−40 25.7	1	14.61		0.62				3		1584
204889	+60 02252			21 27 44	+61 12.8	1	7.08		0.50		0.06		2	F5	1733
		NGC 7092 - 232		21 27 45	+48 01.0	3	12.02	.028	0.78	.019	0.32	.000	7		1544,1544,1581
		Cyg sq 1 # 22		21 27 48	+50 58.	1	10.88		0.99		0.70		3		293
		Cyg sq 1 # 21		21 27 48	+51 03.	1	10.99		0.55		0.15		4		293
		3HLF4 # 112		21 27 48	−42 00.	1	14.02		0.22		0.05		1	A3	208
		NGC 7092 - 234		21 27 49	+48 07.9	2	11.25	.010	0.59	.019	0.09		6	G2	1544,1581
		CS 22897 # 102		21 27 50	−66 34.8	2	14.75	.005	0.52	.011	−0.20	.004	3		1580,1736
204811	+47 03341	NGC 7092 - 233	⋆ AB	21 27 51	+47 18.8	1	9.09		0.14		0.13		2	A1 V	1581
	+47 03428			21 27 52	+47 47.6	1	9.99		0.11		0.15		2		562
	+48 03398	V1726 Cyg		21 27 52	+48 45.0	3	8.94	.042	0.86	.018	0.59	.073	5	F5	562,1724,1772
		G 26 - 5	A	21 27 52	−07 20.0	1	14.96		1.30				1		419
		G 26 - 5	AB	21 27 52	−07 20.0	1	14.31		1.41				2		202
		G 26 - 5	B	21 27 52	−07 20.0	1	15.36		1.60				1		419
	−24 16689			21 27 52	−24 23.1	1	10.70		1.22		1.23		2	K7 V	3072
204814	+45 03561	G 212 - 39		21 27 53	+45 40.2	5	7.91	.014	0.76	.008	0.32	.013	11	G5	22,1003,1658,1724,3016
		Cyg sq 1 # 24		21 27 54	+50 38.	1	11.37		0.50		0.05		2		293
		Cyg sq 1 # 23		21 27 54	+51 03.	1	11.63		1.35		0.84		5		293

Table 1 1153

HD	DM	Other Id	N Rem	α_{1950}	δ_{1950}	S	V	σ_V	B–V	σ_{B-V}	U–B	σ_{U-B}	n	Spectrum	References
	+50 03348			21 27 55	+50 55.5	1	9.47		0.21		0.21		3	A5	293
235522	+50 03349			21 27 55	+51 01.5	3	9.02	.008	0.20	.010	0.16	.022	14	B9	261,293,1724
204692	−14 06047			21 27 56	−14 30.7	2	6.64	.027	1.20	.001	1.30	.023	21	K2/3 III	978,3016
204726	+14 04618			21 27 58	+14 38.9	1	9.04		0.45		0.03		2	G0	1648
		VES 413		21 27 58	+47 26.9	1	11.63		0.18		-0.40		2		8106
		G 25 - 35		21 27 59	−01 06.3	1	11.13		1.27		1.17		1	K5	1620
204815	+34 04436			21 28 01	+35 15.1	1	7.32		0.00		-0.18		2	A0 p (Si)	1625
205021	+69 01173	HR 8238, β Cep	★ A	21 28 01	+70 20.5	2	3.19	.034	-0.23	.010	-1.03	.083	6	B2 IV	542,3024
205021	+69 01173	HR 8238, β Cep	★ AB	21 28 01	+70 20.5	7	3.23	.014	-0.22	.006	-0.95	.011	16	B2 IV+A2.5V	15,154,542,1278,1363*
205021	+69 01173	IDS21274N7007	B	21 28 01	+70 20.5	3	7.83	.041	0.12	.044	-0.07	.082	6	A2.5V	542,542,3024
		VES 414	V	21 28 03	+47 59.3	1	11.08		2.02		1.66		1		8106
		959 32 59		21 28 04	+04 23.1	1	12.69		1.09				2		1719
204860	+44 03840			21 28 09	+45 16.5	1	6.90		-0.04		-0.55		2	B5	401
235523	+51 03075			21 28 12	+51 37.8	1	9.25		0.35		-0.16		6	B5	848
		959 32 43		21 28 13	+04 35.0	1	12.83		1.37				2		1719
	+49 03536			21 28 13	+49 27.1	1	9.47		0.21				5	A5	1724
235524	+51 03076			21 28 13	+51 39.4	1	9.38		0.33		-0.19		5	B8	848
	+50 03350			21 28 16	+50 39.4	2	10.12	.009	0.36	.005	0.07	.014	4		261,293
		Cyg sq 1 # 28		21 28 18	+50 36.	1	12.22		0.32		0.01		8		293
		Cyg sq 1 # 29		21 28 18	+50 39.	1	11.70		0.49		0.25		3		293
		CS 22897 # 99		21 28 18	−65 51.1	2	14.40	.010	0.71	.017	0.05	.003	3		1580,1736
204507	−70 02856			21 28 18	−69 42.6	1	7.51		1.22		1.22		4	K2 III	244
		NGC 7086 - 62		21 28 19	+51 22.8	1	10.76		0.40		-0.03		1		49
	+48 03401			21 28 20	+48 36.7	2	9.53	.010	0.73	.005	0.56		4	F2	562,1724
	−43 14579			21 28 20	−43 05.2	1	11.85		0.36		-0.03		2	A7	208
		959 32 22		21 28 21	+04 47.6	1	12.76		1.02				2		1719
		NGC 7086 - 63		21 28 22	+51 22.7	1	15.23		0.93		0.54		1		49
204916	+54 02547			21 28 22	+54 25.9	1	8.03		1.11		1.01		4	K2	848
		AAS11,365 T3# 16		21 28 24	+50 52.	1	13.55		0.65		0.31		5		261
		Cyg sq 1 # 30		21 28 24	+50 54.	1	12.50		0.57		0.40		6		293
		NGC 7086 - 71		21 28 24	+51 25.3	1	14.76		0.80		0.54		1		49
239683	+57 02334	Tr 37 - 677		21 28 24	+57 35.7	2	9.31	.009	0.29	.009	-0.43	.000	6	B3 IV	5,499
204964	+59 02387			21 28 24	+60 09.5	1	7.47		0.17		-0.44		2	B8	401
204706	−39 14289	IDS21253S3900	AB	21 28 25	−38 46.9	1	9.47		1.19		1.09		2	K4/5 V	3072
204818	+9 04834			21 28 27	+09 56.2	1	7.80		0.92		0.63		3	K0	196
		NGC 7086 - 65		21 28 27	+51 16.5	1	10.86		0.32		0.18		1		49
		NGC 7086 - 133		21 28 27	+51 25.4	1	15.66		1.41				1		49
204671	−54 09881			21 28 27	−54 24.8	1	9.29		0.06		0.09		2	A0 V	1730
204906	+46 03346	LS III +46 060		21 28 28	+46 57.1	2	8.46	.005	0.50	.010	0.01		5	G0	1601,1724
		NGC 7086 - 70		21 28 29	+51 25.5	1	15.44		0.86		0.57		1		49
	+47 03431	NGC 7092 - 17		21 28 30	+48 23.5	1	9.76		1.01		0.60		2	gG7	562
		Cyg sq 1 # 31		21 28 30	+50 56.	1	12.89		0.48		0.26		4		293
		G 25 - 36		21 28 31	+03 25.8	1	13.98		1.16		0.73		1		3062
204917	+47 03432	NGC 7092 - 19		21 28 31	+48 10.2	3	7.35	.015	-0.02	.000	-0.02	.030	7	B9.5V	43,401,1581
		NGC 7086 - 64		21 28 31	+51 24.2	1	14.46		0.75		0.44		1		49
		NGC 7086 - 75		21 28 31	+51 27.4	1	11.59		0.45		0.13		1		49
	+11 04582			21 28 32	+12 20.1	1	8.39		1.75		2.06		2	M0	1375
		NGC 7086 - 72		21 28 33	+51 25.4	1	14.23		0.85		0.58		1		49
235527	+50 03351			21 28 34	+50 58.2	2	9.27	.005	0.11	.010	0.11	.020	7	A0	261,293
204843	+15 04435			21 28 35	+16 21.0	1	8.33		0.31		0.00		3	A5	1648
		G 26 - 7		21 28 35	−10 00.7	4	11.95	.035	1.67	.011	1.20	.023	16		202,1705,1764,3078
		AAS11,365 T3# 18		21 28 36	+50 47.	1	13.26		0.76		0.32		6		261
		Cyg sq 1 # 33		21 28 36	+50 58.	1	12.98		0.47		0.18		2		293
		NGC 7086 - 76		21 28 36	+51 27.4	1	13.85		1.15		0.51		1		49
		NGC 7086 - 46		21 28 38	+51 21.4	1	15.02		0.88		0.58		1		49
		NGC 7086 - 131		21 28 39	+51 23.4	1	10.19		1.89		1.78		1		49
		NGC 7086 - 66		21 28 39	+51 24.2	1	10.97		0.65		0.26		1		49
204890	+31 04481			21 28 40	+31 44.2	1	7.02		0.29		0.22		2	A5	1375
	+40 04552			21 28 40	+40 44.9	1	9.21		0.25				2	A5	1724
204965	+52 02957	HR 8237		21 28 40	+52 44.2	2	6.03	.010	0.07	.010	0.11	.005	6	A3 V	401,848
		NGC 7086 - 45		21 28 43	+51 21.5	1	14.71		0.87		0.59		1		49
		NGC 7086 - 58		21 28 43	+51 22.2	1	14.94		1.07		0.33		1		49
205234	+75 00791	EI Cep		21 28 43	+76 11.0	1	7.55		0.34		0.06		9	F2 V	588
204862	+11 04583	HR 8231		21 28 44	+11 55.0	2	6.09		-0.08	.030	-0.20	.010	4	B9.5V	1049,1776
	+36 04593	G 213 - 3		21 28 44	+36 39.9	1	10.66		1.03		1.11		2	K3	7010
		NGC 7092 - 236		21 28 44	+47 46.8	3	11.94	.013	0.71	.017	0.20	.019	7		1544,1544,1581
		NGC 7086 - 44		21 28 44	+51 21.1	1	15.21		0.79				1		49
		NGC 7086 - 68		21 28 44	+51 25.0	1	14.16		0.86		0.51		1		49
	+22 04409	G 145 - 43		21 28 45	+23 07.0	1	9.25		1.05		0.80		2	K8	3072
	+47 03433	NGC 7092 - 25		21 28 45	+48 04.6	2	9.07	.005	0.11	.005	0.12		4	A2 Vn	43,1544
		WD2128+112		21 28 46	+11 16.3	1	16.06		-0.29		-1.08		1	DA2	1727
	+48 03402	NGC 7092 - 29		21 28 46	+48 27.7	1	10.49		0.30		0.25		1	A1	562
		NGC 7086 - 16		21 28 47	+51 23.7	1	14.28		0.82		0.50		1		49
204863	+7 04709			21 28 48	+07 28.8	1	8.46		0.50		0.06		2	F8	1375
		Cyg sq 1 # 34		21 28 48	+50 36.	1	11.76		0.35		-0.34		6		293
		NGC 7086 - 132		21 28 48	+51 22.	1	13.37		0.64		0.26		1		49
		NGC 7086 - 27		21 28 49	+51 21.7	1	13.41		0.68		0.33		1		49
		LP 873 - 63		21 28 49	−22 01.5	1	12.46		0.74		0.15		2		1696
		959 32 108		21 28 50	+03 52.7	1	13.27		1.04				3		1719
		WLS 2140 30 # 6		21 28 51	+30 17.0	1	10.93		1.10		0.69		2		1375

HD	DM	Other Id	N	Rem	α_{1950}	δ_{1950}	S	V	σ_V	B–V	σ_{B-V}	U–B	σ_{U-B}	n	Spectrum	References
		G 232 - 18			21 28 51	+48 38.8	1	10.53		0.60		-0.01		2	G2	1658
204923	+25 04553				21 28 53	+25 49.6	2	8.03	.005	1.36	.010	1.41		5	K3 III	20,3077
		NGC 7086 - 21			21 28 54	+51 23.3	1	11.66		0.72		0.11		1		49
	+47 03435				21 28 55	+47 52.0	1	9.33		1.38		1.26		3		562
		NGC 7086 - 36			21 28 55	+51 21.1	1	13.17		0.90		0.41		1		49
	+47 03438	NGC 7092 - 34			21 28 56	+48 06.7	1	9.15		0.16		0.11		3	A2 V	43
235529	+50 03353				21 28 56	+50 39.4	2	9.30	.010	1.43	.015	1.19	.010	6	M0	261,293
204867	-6 05770	HR 8232		★ A	21 28 56	-05 47.5	15	2.90	.019	0.83	.010	0.57	.017	62	G0 Ib	1,3,15,1075,1088,1509*
204783	-41 14550	HR 8229			21 28 56	-41 24.1	3	5.28	.005	1.11	.010	1.01		9	K0 III	15,58,1075
204921	+29 04435				21 28 58	+30 03.5	1	7.52		1.32				3	K2 III	20
		NGC 7086 - 6			21 28 59	+51 24.8	1	13.74		0.65		0.40		1		49
		NGC 7086 - 5			21 28 59	+51 25.3	1	12.61		0.58		0.21		1		49
	+47 03439	NGC 7092 - 38		★ AB	21 29 00	+48 14.3	1	8.54		0.74		0.26		std	dG1	43
		NGC 7086 - 35			21 29 02	+51 20.7	1	15.31		1.15		0.26		1		49
		LP 518 - 4			21 29 03	+12 27.1	1	11.84		1.26		1.26		1	K5	1696
	+47 03440	NGC 7092 - 39		★ D	21 29 03	+48 15.8	2	10.40	.108	0.39	.019	0.02		4	F0	43,1544
		NGC 7092 - 237			21 29 03	+48 15.9	2	11.93	.039	0.54	.000	0.05	.010	3		1544,1581
		NGC 7086 - 18			21 29 04	+51 23.0	1	13.20		0.79		0.32		1		49
		NGC 7092 - 238			21 29 05	+48 09.4	1	13.79		0.90		0.44		2		1581
	+57 02339	Tr 37 - 682			21 29 05	+57 48.6	1	9.97		0.30		-0.43		5	B5 III	499
	+27 04087				21 29 06	+28 06.1	1	9.78		-0.01		-0.42		2	B8	1375
		NGC 7092 - 42			21 29 06	+48 12.4	2	11.24	.005	0.47	.000	-0.02		6	F5 V	1544,1581
		NGC 7086 - 32			21 29 06	+51 20.4	1	13.57		0.71		0.36		1		49
205039	+55 02597				21 29 06	+55 40.4	1	7.71		1.32		1.39		2	K2	1502
204995	+42 04121				21 29 08	+42 53.1	1	8.66		-0.06		-0.51		2	B9	401
		NGC 7092 - 239			21 29 09	+48 22.9	1	13.06		0.76		0.26		2		1581
	-2 05557	G 26 - 8			21 29 09	-02 08.6	3	10.44	.034	0.84	.011	0.50	.020	4		202,1620,1658
204978	+35 04555				21 29 11	+35 39.2	1	7.54		0.95		0.66		2	K0	1601
	+47 03441	NGC 7092 - 241			21 29 11	+47 34.8	2	9.77	.000	0.22	.005	0.14		6	F0 V	1544,1581
	+47 03441				21 29 11	+47 34.9	1	9.78		0.21				3	A5	1724
		NGC 7092 - 240			21 29 11	+47 35.2	1	13.07		0.81		0.37		2		1581
		Cyg sq 1 # 36			21 29 12	+50 38.	1	11.14		0.38		-0.01		3		293
		Cyg sq 1 # 37			21 29 12	+50 50.	1	12.38		0.71		0.13		6		293
		PHL 25			21 29 12	-17 32.	1	12.01		-0.14		-0.68		3		3060
		NGC 7086 - 19			21 29 13	+51 23.3	1	13.79		0.80		0.45		1		49
204854	-34 15110	HR 8230		★ AB	21 29 13	-34 10.0	1	5.97		0.05				4	A2 V	2035
		G 25 - 37			21 29 14	+08 09.7	2	11.51	.000	1.28	.015	1.14		7	K7	202,333,1620
239689	+56 02584	Tr 37 - 136			21 29 15	+56 58.8	2	8.84	.004	0.18	.004	-0.61	.004	7	B1.5V	5,499
		NGC 7092 - 242			21 29 16	+48 11.4	2	12.79	.005	0.75	.024	0.26	.029	3		1544,1581
	+50 03352	NGC 7086 - 1			21 29 16	+51 29.2	1	12.89		0.95		0.53		1		49
	+49 03539	LS III +49 025			21 29 17	+49 53.2	1	11.33		0.58		-0.34		1	A2	848
204910	-8 05672				21 29 17	-08 07.6	1	9.34		1.06		0.91		2	K0	1375
		NGC 7092 - 243			21 29 20	+48 25.4	2	12.68	.010	0.74	.015	0.25		6		1544,1581
		NGC 7086 - 3			21 29 21	+51 24.7	1	12.00		1.58		1.38		1		49
	+47 03442	NGC 7092 - 140			21 29 22	+47 51.2	2	10.14	.009	0.29	.005	0.12		4	F0 IV	562,1544
205058	+50 03355				21 29 24	+51 01.7	3	8.81	.008	0.07	.008	-0.01	.019	8	A0	261,293,848
	+41 04175				21 29 25	+41 35.1	1	8.96		0.22				2	A5	1724
		NGC 7092 - 244			21 29 25	+47 59.0	3	12.76	.010	0.73	.011	0.28	.015	6		1544,1544,1581
		G 93 - 19			21 29 25	-01 33.3	1	13.74		1.52				2		202
	+47 03443				21 29 26	+47 50.6	1	9.97		0.08		0.13		3		562
		VES 416			21 29 26	+48 26.8	1	10.92		1.40		1.20		2		8106
		NGC 7092 - 54			21 29 30	+48 24.1	2	10.93	.025	0.50	.005	0.00	.005	5	F0	43,1581
	+47 03445	NGC 7092 - 57			21 29 32	+48 13.9	1	9.96		0.27		0.08		3	A7 V	43
	+48 03403				21 29 32	+49 01.0	1	9.62		1.90		1.87		4		261
	+9 04836	II Zw 136			21 29 33	+09 56.2	2	9.95	.025	1.07	.065	1.02	.030	3	K2	899,1375
205011	+23 04329				21 29 34	+23 37.4	1	6.42		1.06		0.78		1	G8 Ib	1716
	+49 03543				21 29 34	+49 59.2	1	9.30		0.02		0.01		2	A0	261
205073	+47 03446	NGC 7092 - 60		★ AB	21 29 35	+48 08.1	2	7.76	.095	0.01	.035	-0.01	.000	4	A1 Vp	43,8071
	-0 04234	G 26 - 9		★ A	21 29 41	+00 01.9	4	9.86	.049	0.96	.011	0.49	.007	10	K2	202,516,1774,3016
	-0 04234	G 26 - 10		★ B	21 29 36	+00 00.0	5	14.71	.048	-0.07	.029	-0.92	.005	15	DA	202,203,1281,1620,3060
204952	-10 05698				21 29 35	-10 30.6	1	10.55		0.60		0.08		2	G0	1375
204941	-21 06035	IDS21240S2083	A		21 29 35	-21 10.6	1	8.48		0.91		0.58		2	K1/2 V	3072
204941	-21 06035	IDS21240S2083	B		21 29 35	-21 10.6	1	11.48		1.46		1.21		2		3072
		Cyg sq 1 # 39			21 29 36	+50 51.	1	11.47		0.43		0.14		5		293
205084	+51 03078				21 29 36	+51 58.4	2	6.70	.005	0.95	.015	0.67	.005	5	K0	261,848
205059	+46 03352				21 29 37	+47 02.5	1	8.36		0.23				2	F0	1724
205139	+59 02395	HR 8243		★	21 29 37	+60 14.3	5	5.53	.010	0.12	.011	-0.74	.004	11	B1 II	5,154,555,1118,1379
		NGC 7092 - 245			21 29 39	+47 55.7	1	13.18		1.01		0.58		2		1581
204971	-12 06026				21 29 39	-12 29.3	2	6.73	.016	1.61	.013	2.01	.013	9	K5 III	978,1375
205051	+38 04519				21 29 40	+38 54.4	2	8.59	.005	0.25	.015	0.12	.010	5	A5 V	833,1601
205060	+42 04123				21 29 40	+42 28.8	2	7.20	.015	-0.02	.024	-0.42	.015	5	B5	379,1212
	+47 03447	NGC 7092 - 63			21 29 41	+48 02.6	2	10.16	.015	0.29	.000	0.07	.015	3	F0	43,1581
204943	-25 15479	HR 8235			21 29 41	-24 48.7	3	6.43	.005	0.20	.000			11	A7 V	15,2020,2028
		NGC 7092 - 246			21 29 42	+47 42.0	2	12.82	.005	0.71	.010	0.28	.000	3		1544,1581
205085	+47 03448	NGC 7092 - 65		★ AB	21 29 42	+48 03.4	1	7.97		0.06		0.08		3	A1 Vn	43
	+48 03405	NGC 7092 - 68			21 29 44	+48 25.2	3	9.79	.009	0.31	.008	0.09	.005	7	F2 V	43,1581,1724
		NGC 7092 - 247			21 29 45	+48 23.6	2	13.11	.010	0.84	.000	0.52		6		1544,1581
205114	+51 03079	HR 8242			21 29 47	+52 23.9	4	6.16	.009	0.90	.011	0.44		9	G2 Ib	71,138,848,6009
		II Zw 136 # 7			21 29 48	+09 54.9	1	15.20		0.79		0.33		1		899
		NGC 7092 - 248			21 29 50	+47 30.1	2	11.00	.005	0.65	.057	-0.01		7	G2 IV	1544,1581

Table 1

HD	DM	Other Id	N Rem	α_{1950}	δ_{1950}	S	V	σ_V	B–V	σ_{B-V}	U–B	σ_{U-B}	n	Spectrum	References
		fld IV # 2		21 29 51	+50 05.3	1	10.74		0.37		0.08		2		1755
204873	−53 10092	HR 8233		21 29 51	−52 57.6	1	6.41		1.48				4	K4 III	2007
		II Zw 136 # 6		21 29 52	+09 54.5	1	14.71		0.69		0.22		1		899
205116	+47 03449	NGC 7092 - 74		21 29 54	+48 21.8	3	6.84	.011	-0.03	.010	-0.08	.031	9	B9.5V	43,401,401
		NGC 7092 - 249		21 29 55	+48 15.2	3	13.43	.014	0.86	.026	0.46	.015	7		1544,1544,1581
205027	+0 04743	G 26 - 11		21 29 56	+00 47.7	2	8.31	.020	0.60	.005	0.05	.010	3	G2 V	333,1003,1620
		II Zw 136 # 5		21 29 56	+09 54.5	1	14.84		0.80		0.53				899
205117	+47 03451	NGC 7092 - 75		21 29 56	+48 15.8	1	7.65		0.01		0.03		3	A1 Vp	43
239693	+57 02343	Tr 37 - 686	★ AB	21 29 57	+57 40.7	2	9.51	.017	0.22	.004	-0.42	.009	6	B3 V	5,499
		DW Peg		21 29 58	+24 15.0	2	10.16	.150	1.64	.045	0.96	.045	2	M6	693,793
		II Zw 136 # 2		21 30 00	+09 56.	1	16.04		0.93		0.48		1		899
		Cyg sq 1 # 40		21 30 00	+50 45.	1	12.92		0.60		0.75		7		293
		Cyg sq 1 # 41		21 30 00	+50 52.	1	12.34		0.52		0.41		5		293
	+47 03452	NGC 7092 - 77		21 30 01	+48 15.6	2	8.88	.005	0.16	.019	0.14	.005	4	Am	43,562
204886	−57 09888			21 30 02	−57 18.2	1	8.15		1.23		1.15		5	K0 Ib/II	1673
	+48 03407			21 30 03	+49 22.5	1	9.18		0.19				2	A5	1724
		fld IV # 3		21 30 04	+50 06.4	1	11.38		0.52		-0.04		2		1755
		fld IV # 1		21 30 04	+50 16.6	1	11.45		0.56		0.07		2		1755
205062	+9 04838			21 30 05	+10 20.6	1	8.40		1.09		0.96		3	K0	196
	+47 03453	NGC 7092 - 78		21 30 06	+48 07.1	3	9.51	.013	0.18	.005	0.12	.010	7	A3 Vn	43,1581,1724
	+53 02640			21 30 08	+53 38.9	1	10.38		2.03				2		369
205196	+56 02589	Tr 37 - 401		21 30 08	+57 16.9	2	7.42	.030	0.58	.010	-0.46	.010	3	B0 Ib	5,1012
		CS 22948 # 6		21 30 10	−39 53.0	2	15.06	.005	0.50	.011	-0.03	.005	3		1580,1736
204960	−45 14367	HR 8236		21 30 10	−45 04.3	5	5.57	.005	1.04	.004	0.85	.002	13	K1 III	15,219,688,1075,2009
205087	+22 04418	HR 8240		21 30 11	+23 10.4	4	6.70		-0.08	.008	-0.27	.011	13	B9 p SiSrCr	1049,1079,1263,3033
		fld IV # 4		21 30 12	+50 08.6	1	10.86		0.21		-0.23		2		1755
		II Zw 136 # 4		21 30 13	+09 55.2	1	13.88		0.13		1.30		1		899
		G 145 - 46		21 30 13	+14 13.3	2	15.27	.009	1.69	.041	1.48		5		316,906
204961	−49 13515			21 30 14	−49 13.2	3	8.67	.004	1.46	.000	1.05		7	M2/3 V	1494,2012,3072
	+47 03454	NGC 7092 - 81		21 30 15	+48 06.8	1	8.92		0.09		0.12		3	A2 V	43
	+49 03547			21 30 15	+49 37.9	1	10.07		0.18				3	F8	1724
	+49 03547			21 30 15	+49 37.9	1	12.39		1.25				3	F8	1724
		NGC 7089 sq1 26		21 30 16	+00 39.7	1	9.92		0.55				3		438
		II Zw 136 # 3		21 30 16	+09 56.0	1	13.12		0.61		0.04		1		899
205172	+47 03455	NGC 7092 - 82		21 30 16	+48 20.7	3	8.09	.022	0.02	.015	-0.07	.017	9	B9 IV	43,401,8071
		NGC 7092 - 251		21 30 17	+47 53.4	2	10.61	.000	0.37	.000	-0.03		6	F2 V	1544,1581
	+46 03355			21 30 18	+46 48.4	1	9.37		0.59				3	G0	1724
		NGC 7092 - 83		21 30 18	+48 15.1	1	10.77		0.60		0.16		3	F9	43
205171	+48 03409	NGC 7092 - 84		21 30 19	+48 29.7	1	8.52		0.01		0.02		3	A0 V	43
		NGC 7089 sq1 25		21 30 23	+00 41.5	1	11.40		1.06				2		438
	+47 03456	NGC 7092 - 85		21 30 23	+48 15.8	2	9.67	.005	0.09	.005	0.09		5	A1.5V + A2:	43,1544
205044	−21 06038			21 30 23	−20 53.9	1	8.54		1.02		0.82		2	K0 III	3072
205197	+51 03082			21 30 24	+51 40.3	1	7.37		0.51		0.17		3	A0	848
		NGC 7089 sq1 24		21 30 25	+00 35.9	1	12.11		0.66				1		438
205208	+49 03548			21 30 25	+50 12.9	1	8.45		0.19				3	F0	1724
205222	+53 02642			21 30 25	+53 47.6	1	8.18		0.05		-0.30		4	B9	848
205198	+48 03411	NGC 7092 - 87		21 30 27	+48 25.1	2	8.25	.025	0.04	.015	-0.01	.025	5	A1.5Vp	43,401
205209	+49 03549			21 30 28	+49 42.6	1	8.67		0.41				3	F8	1724
205210	+47 03457	NGC 7092 - 88		21 30 29	+48 13.3	3	6.56	.011	0.00	.004	0.02	.013	14	B9.5IV	43,401,1581
	+47 03458	NGC 7092 - 90		21 30 30	+48 07.2	2	8.68	.005	0.05	.005	0.08	.005	6	A1 V	43,562
		NGC 7092 - 252		21 30 31	+47 58.7	2	12.41	.010	0.67	.015	0.12	.034	3		1544,1581
		NGC 7092 - 253		21 30 31	+48 57.0	2	11.01	.005	0.62	.005	0.21		6	F6	1544,1581
	−54 09888			21 30 33	−54 34.1	1	10.02		1.13		0.99		3		1730
205067	−28 17347			21 30 35	−28 06.7	2	7.61	.019	0.66	.000			8	G2/3 V	1075,2033
		CS 22897 # 107		21 30 36	−67 07.5	1	13.54		1.06		0.91		1		1736
205128	+9 04840			21 30 38	+09 40.3	1	8.19		1.18		1.07		3	K0	196
	+49 03550			21 30 39	+49 33.1	1	9.94		0.28		-0.25		4	B8	848
	−41 14564			21 30 39	−40 52.0	1	11.24		0.91		0.61		2	G5	1696
	+5 04809			21 30 40	+06 12.1	1	9.88		1.63				1	M0	1532
205142	+23 04334	G 126 - 8		21 30 40	+23 54.3	2	8.55	.005	0.84	.025	0.44	.005	2	K0	1620,1658
	+48 03412			21 30 40	+48 47.2	2	9.08	.019	0.29	.015	-0.01	.049	6	A2	562,848
		NGC 7092 - 254		21 30 40	+48 57.6	1	11.43		0.50		-0.03		3		1581
	−35 14846			21 30 41	−35 37.6	1	9.78		1.21		1.36		1	K2	742
		NGC 7092 - 255		21 30 42	+47 35.4	2	11.79	.005	0.70	.005	0.16		7	G2	1544,1581
		LS III +51 018		21 30 42	+51 01.3	1	11.59		0.49		-0.04		8		848
		CS 22948 # 2		21 30 42	−41 07.0	1	14.84		0.16		0.18		1		1736
205201	+32 04186			21 30 43	+32 33.3	2	7.41	.022	-0.06	.004	-0.47	.018	5	B9	1375,3016
205212	+41 04184			21 30 43	+42 04.5	1	6.92		1.13		1.12		1	K1 III	1716
	+47 03460			21 30 46	+47 45.1	1	9.77		0.15		0.17		3	A2	562
205223	+38 04527	IDS21288N3838	AB	21 30 48	+38 51.0	1	8.81		0.18		0.12		3	A2 V	833
	−35 14849			21 30 48	−35 39.2	4	10.57	.007	0.40	.005	-0.23	.014	8	B8	158,742,1696,3077
205131	−14 06063			21 30 51	−13 40.3	1	7.61		1.08		0.90		21	K0 III	978
	+47 03461			21 30 52	+47 44.2	1	8.85		0.58		0.16		3	G0	562
		NGC 7092 - 256		21 30 52	+47 54.7	1	13.62		0.89		0.53		2		1581
		LS III +50 023		21 30 53	+50 52.6	1	10.45		0.74		-0.31		4		848
205237	+39 04589			21 30 54	+40 11.1	1	7.28		0.91				3	G5	1724
		NGC 7089 - 1009		21 30 54	−01 03.	1	14.72		0.81		0.17		1		3074
		NGC 7089 - 1020		21 30 54	−01 03.	1	14.63		0.81		0.16		2		3074
		NGC 7089 - 1022		21 30 54	−01 03.	1	14.83		0.88		0.43		2		3074
		NGC 7089 - 1023		21 30 54	−01 03.	1	13.67		1.22		0.92		2		3074

HD	DM	Other Id	N	Rem	α_{1950}	δ_{1950}	S	V	σ_V	B–V	σ_{B-V}	U–B	σ_{U-B}	n	Spectrum	References
		NGC 7089 - 1040			21 30 54	−01 03.	1	14.49		0.93		0.49		2		3074
		NGC 7089 - 1044			21 30 54	−01 03.	1	14.29		0.98		0.46		2		3074
		NGC 7089 - 1077			21 30 54	−01 03.	1	14.17		1.13		0.64		2		3074
		NGC 7089 - 2030			21 30 54	−01 03.	1	13.95		1.15		0.90		2		3074
		NGC 7089 - 2031			21 30 54	−01 03.	1	13.52		1.35		1.10		2		3074
		NGC 7089 - 2033			21 30 54	−01 03.	1	14.09		1.10		0.76		3		3074
		NGC 7089 - 2059			21 30 54	−01 03.	1	13.34		1.43		1.25		2		3074
		NGC 7089 - 2069			21 30 54	−01 03.	1	13.20		1.44		1.40		3		3074
		NGC 7089 - 3019			21 30 54	−01 03.	1	13.74		1.18		1.06		2		3074
		NGC 7089 - 3020			21 30 54	−01 03.	1	14.67		0.98		0.50		3		3074
		NGC 7089 - 3045			21 30 54	−01 03.	1	14.02		1.18		0.85		3		3074
		NGC 7089 - 3047			21 30 54	−01 03.	1	15.06		0.82		0.18		1		3074
		NGC 7089 - 3081			21 30 54	−01 03.	1	14.06		1.12		0.80		2		3074
		NGC 7089 - 4011			21 30 54	−01 03.	1	13.80		1.16		0.72		3		3074
		NGC 7089 - 4017			21 30 54	−01 03.	1	13.42		1.30		0.99		1		3074
		NGC 7089 - 4079			21 30 54	−01 03.	1	13.22		1.40		1.40		2		3074
		NGC 7089 - 4080			21 30 54	−01 03.	1	14.63		0.97		0.50		4		3074
		NGC 7089 - 4083			21 30 54	−01 03.	1	13.72		1.17		0.95		2		3074
	−3 05242				21 30 54	−02 43.7	1	10.55		1.28				2	K7	1619
	+47 03462	NGC 7092 - 100		⋆ A	21 30 55	+48 15.7	1	9.03		0.08		0.09		3	A2 V	43
205132	−16 05885	IDS21282S1638		AB	21 30 55	−16 25.3	1	7.21		0.45		-0.02		3	F6 V	3016
		GD 233			21 30 57	−04 45.8	1	14.50		-0.05		-0.73		1	DB	3060
		NGC 7092 - 101			21 30 58	+48 28.9	1	10.60		0.38		0.30		3	A3	43
205299	+50 03360				21 30 58	+51 01.4	1	8.48		0.64				2	G0	1724
		BP Peg			21 31 00	+22 31.2	1	11.69		0.23		0.20		1	A0	597
	+48 03414	NGC 7092 - 104			21 31 00	+48 27.9	2	8.81	.005	1.17	.000	1.12	.000	6	K0	43,1581
	−8 05676				21 31 00	−07 48.4	1	10.83		1.01		0.89		2	K5 V	3072
		G 26 - 12			21 31 02	+00 10.3	4	12.14	.019	0.43	.011	-0.23	.016	8	sdF5	202,203,333,1620,1696
	+46 03357				21 31 04	+47 12.7	1	9.61		0.42				2	F2	1724
	+48 03416				21 31 04	+48 48.0	1	9.80		0.29		-0.18		3		562
		NGC 7089 sq1 23			21 31 05	+00 32.5	1	10.20		0.49				3		438
	+44 03858				21 31 05	+44 45.2	1	9.82		0.50				2	F8	1724
205096	−43 14602	HR 8241			21 31 06	−43 08.9	1	6.32		1.09				4	K1 III+B9 V	2009
	+47 03464				21 31 09	+47 43.7	1	10.08		0.32		0.14		3		562
	+49 03552				21 31 10	+49 34.6	1	8.46		0.46				4	F8	1724
205314	+49 03553	HR 8246			21 31 10	+49 45.3	2	5.76	.005	-0.04	.010	-0.12	.000	6	A0 V	401,848
		G 26 - 13			21 31 10	−07 04.3	1	14.01		1.58				3		202
	+44 03859	IDS21293N4445		AB	21 31 12	+44 57.9	1	9.52		0.32				2	A5	1724
	+47 03465	NGC 7092 - 112			21 31 12	+48 00.8	2	9.67	.000	0.18	.014	0.13		4	A3	562,1544
		fld IV # 17			21 31 12	+50 21.8	1	12.42		0.60		0.28		2		1755
205153	−28 17354				21 31 14	−28 07.4	2	8.20	.000	0.55	.002	-0.19		5	G0/1 V	1075,3037
		Z Gru			21 31 15	−49 21.5	1	12.01		0.23		0.08		1		700
		G 26 - 14			21 31 16	+01 34.2	1	13.35		1.68				2		202
		VES 417			21 31 16	+46 15.2	1	10.13		1.90		1.38		1		8106
205315	+47 03466	NGC 7092 - 141			21 31 17	+47 49.4	1	8.81		0.05		0.08		3		562
		Smethells 84			21 31 18	−49 46.	1	11.13		1.30				1		1494
		fld IV # 5			21 31 19	+50 18.1	1	10.28		0.17		0.16		2		1755
205331	+47 03468	NGC 7092 - 118			21 31 21	+48 04.9	2	6.84	.005	-0.01	.005	-0.01	.005	4	A1 Vp	401,1733
	+48 03418	NGC 7092 - 130			21 31 21	+48 33.3	1	9.94		1.84				2	B2 Ib	369
205287	+26 04197				21 31 23	+27 22.8	1	7.84		1.56		1.89		2	K5 III	1733
		NGC 7089 sq1 22			21 31 24	+00 28.2	1	11.18		0.61				3		438
		NGC 7089 sq1 21			21 31 24	+00 43.8	1	10.61		1.08				2		438
		NGC 7092 - 258			21 31 24	+47 29.2	1	13.91		0.78		0.31		2		1581
		JL 82			21 31 24	−73 02.	1	12.33		-0.16		-0.90		1		832
205349	+45 03584	HR 8248			21 31 25	+45 37.9	3	6.24	.011	1.80	.016	1.97	.076	5	K1 Ib	1218,1355,4001
		fld IV # 16			21 31 31	+50 18.6	1	13.75		0.44		0.20		2		1755
		fld IV # 15			21 31 33	+50 16.3	1	16.20		0.83		0.43		2		1755
205397	+52 02967				21 31 33	+53 22.1	1	7.79		1.01		0.79		2	K0	848
205249	−14 06070	AS Cap			21 31 33	−13 42.4	1	7.61		1.02		0.77		1	K0 IIIp	1641
205156	−50 13411				21 31 33	−50 00.9	4	8.12	.011	0.62	.010	-0.03	.020	8	G3 V	742,1594,2012,3077
		fld IV # 13			21 31 38	+50 16.1	1	10.84		2.06		1.89		2		1755
	+39 04594				21 31 39	+39 43.1	1	9.06		-0.01		-0.49		2	A0	401
		G 93 - 24			21 31 40	+08 08.0	1	13.31		1.17				2		202
	−8 05680				21 31 42	−07 58.3	1	10.74		1.15		1.06		2	K5 V	3072
		fld IV # 11			21 31 44	+50 22.0	1	14.22		0.80		0.32		2		1755
		fld IV # 14			21 31 47	+50 15.2	1	13.45		1.62		1.74		2		1755
		fld IV # 12			21 31 48	+50 19.6	1	15.11		0.77		0.56		2		1755
205352	+23 04337				21 31 52	+23 52.1	1	8.60		0.30		0.12		3	F2	1733
		LS III +51 019			21 31 56	+51 01.7	1	12.20		0.79		-0.26		1		848
205265	−30 18703	HR 8244			21 31 56	−29 55.1	1	6.43		-0.11				4	B8 III	2035
		NGC 7092 - 261			21 31 58	+48 02.0	2	12.69	.019	0.63	.005	0.10		6		1544,1581
		NGC 7092 - 260			21 31 58	+48 17.8	3	11.56	.017	0.51	.020	-0.05	.010	7	F5	1544,1544,1581
		fld IV # 9			21 31 60	+50 19.2	1	14.84		1.07		0.52		2		1755
		NGC 7092 - 262			21 32 00	+48 56.3	2	10.43	.015	0.34	.005	-0.09		6	F2 IV	1544,1581
		fld IV # 8			21 32 00	+50 17.2	1	16.20		1.17		0.64		2		1755
	+48 03421	NGC 7092 - 135			21 32 02	+48 27.1	1	10.33		0.46		0.16		2	F5 V	562
	+47 03471				21 32 03	+47 49.4	2	9.21	.020	0.40	.015	0.06		6	F5	562,1724
		fld IV # 10			21 32 03	+50 22.5	1	14.46		0.46		0.27		2		1755
205321	−9 05770				21 32 03	−09 18.6	1	7.68		1.07		0.96		5	K0	1657
205289	−20 06237	HR 8245			21 32 03	−20 18.5	3	5.71	.022	0.40	.012	-0.06	.035	11	F5 V	379,2007,3037

Table 1 1157

HD	DM	Other Id	N Rem	α_{1950}	δ_{1950}	S	V	σ_V	B–V	σ_{B-V}	U–B	σ_{U-B}	n	Spectrum	References
	+48 03422	NGC 7092 - 263		21 32 04	+48 55.1	2	9.05	.010	0.11	.025	0.08		5	A2 IV	1544,1581
		fld IV # 7		21 32 04	+50 16.3	1	13.45		2.00		1.63		2		1755
	+47 03472	NGC 7092 - 133		21 32 05	+48 20.0	2	9.07	.005	0.21	.005	0.10		6	Am	43,1724
205435	+44 03865	HR 8252		21 32 06	+45 22.2	9	4.01	.012	0.89	.005	0.55	.005	63	G8 III	15,1007,1013,1355*
		fld IV # 6		21 32 06	+50 19.0	1	11.22		0.52		0.34		2		1755
		G 26 - 15		21 32 06	−07 12.9	1	13.56		1.44				3		202
205482	+53 02647			21 32 08	+54 05.6	1	7.56		0.14		0.13		4	A0	848
	+48 03423	NGC 7092 - 264		21 32 09	+48 58.5	2	9.38	.010	0.13	.020	0.15		5	A3 V	1544,1581
	+48 03424			21 32 10	+49 11.4	1	8.71		0.23		-0.54		4	B8	848
205510	+57 02349			21 32 12	+57 58.4	1	8.48		0.31		-0.41		1	A3 p:	5
	+47 03473	NGC 7092 - 134		21 32 13	+48 24.4	1	8.90		1.69		1.69		3	K2 III	562
235544	+52 02970			21 32 13	+52 30.5	1	10.13		0.26		-0.12		4	F7	848
		CS 22948 # 25		21 32 13	−39 32.4	1	13.03		0.29		0.06		1		1736
		LS III +49 028		21 32 15	+49 42.6	1	11.93		0.55		-0.04		4		848
205420	+22 04431	HR 8250		21 32 17	+22 31.9	2	6.47	.039	0.50	.000	0.03	.015	3	F7 V	254,272
	+47 03475			21 32 21	+48 21.9	1	10.56		0.19		-0.19		3		562
		Tr 37 - 3290		21 32 25	+57 27.4	1	11.27		1.74				1		1706
205342	−24 16729	HR 8247		21 32 25	−23 40.7	1	6.40		1.10				4	K1 III	2009
		CS 22948 # 24		21 32 25	−39 25.6	2	14.64	.010	0.61	.014	-0.10	.000	3		1580,1736
205496	+45 03588			21 32 28	+45 46.3	1	7.99		-0.06		-0.41		2	B9	401
205495	+48 03425			21 32 28	+48 31.0	1	8.00		1.10		1.02		3	K2	562
		G 126 - 10		21 32 34	+10 41.9	2	11.83	.000	0.46	.015	-0.17	.015	4	sdF8	1620,1658
	+47 03477	NGC 7092 - 142		21 32 34	+48 00.2	1	9.16		1.08		0.84		2	K0	562
	+45 03591			21 32 35	+45 34.2	1	8.95		0.44				3	F5	1724
	+51 03090			21 32 37	+51 42.3	1	9.71		0.23		-0.14		3		261
		WLS 2136 45 # 10		21 32 39	+45 19.3	1	11.50		0.58		0.20		2		1375
	+44 03866			21 32 39	+45 19.3	1	9.26		1.60		1.57		2	K0	1375
	+52 02972		V	21 32 39	+52 42.4	1	10.19		1.62				2		369
		POSS 76 # 6		21 32 39	+65 57.8	1	18.80		1.68				2		1739
205511	+41 04199			21 32 40	+42 10.9	1	7.59		0.17		0.13		2	A2	379
		G 26 - 16		21 32 40	−03 32.0	3	11.17	.018	1.01	.011	0.84	.010	4		202,1620,1658
	+42 04139	IDS21308N4255	A	21 32 41	+43 08.8	1	9.17		1.00				4	G0	1724
	+42 04139	IDS21308N4255	B	21 32 41	+43 08.8	1	10.17		0.44				3		1724
205423	−4 05489	HR 8251		21 32 41	−04 12.4	2	5.76	.005	1.11	.000	1.05	.005	7	K0	15,1256
205512	+37 04359	HR 8255		21 32 44	+38 18.5	8	4.89	.016	1.08	.009	1.01	.006	44	K1 III	15,37,1003,1355,1363*
205551	+51 03091	HR 8259		21 32 44	+51 28.5	3	6.17	.000	-0.02	.010	-0.27	.008	8	B9 IIIe	848,848,1079
235548	+51 03092			21 32 44	+52 02.8	1	8.97		0.09		0.01		4	B9	848
204904	−79 01158	HR 8234		21 32 48	−79 40.1	2	6.17	.005	0.46	.000	0.01	.010	7	F6 IV	15,1075
205552	+48 03427	LS III +49 029		21 32 51	+49 18.2	1	7.84		0.19		0.06		4	A0	848
		NGC 7092 - 145		21 32 52	+48 26.8	1	10.46		0.24		-0.35		3		562
	+48 03428	NGC 7092 - 143		21 32 52	+48 28.7	3	9.61	.005	0.19	.004	0.13		6	A4 V	562,1544,1724
	+50 03368			21 32 58	+51 16.9	1	9.54		0.12		-0.41		5	B8	848
		CS 22897 # 117		21 32 59	−65 43.6	1	14.85		0.32		-0.16		1		1736
205539	+27 04107	HR 8257		21 33 06	+27 58.4	2	6.28	.033	0.37	.019	-0.04	.023	4	F0 IV	254,3016
	+51 03094	LS III +52 006		21 33 06	+52 06.0	1	9.67		0.18		-0.52		4	B8	848
		L 570 - 14		21 33 06	−31 35.	1	12.62		1.08		0.88		1		1696
		LS III +51 021		21 33 07	+51 26.6	1	11.72		0.48		-0.09		4		848
	+27 04108			21 33 08	+27 47.8	1	8.78		0.58		0.09		2	G0 V	3026
205616	+51 03096			21 33 08	+52 04.6	1	8.37		0.03		0.00		4	A0	848
205572	+34 04461			21 33 09	+34 57.9	1	8.33		0.26		0.11		2	A2	1625
205600	+46 03375			21 33 09	+46 52.0	1	7.84		-0.05		-0.27		2	A0	401
239704	+56 02595	Tr 37 - 421		21 33 09	+57 15.5	1	9.32		0.11		0.06		3	A1 V	499
205541	+23 04346	HR 8258	★ AB	21 33 10	+24 13.7	1	6.22		0.13		0.10		2	A2 V+A4 V	1733
235550	+50 03369			21 33 10	+51 09.5	1	9.04		0.31				2		1724
205601	+43 03975			21 33 13	+43 28.8	1	6.76		-0.12		-0.60		2	B8	1216
	+58 02292	Tr 37 - 1004		21 33 14	+58 25.7	1	9.87		0.41		0.37		4	B2 IV-V	499
205686	+61 02158	LS III +62 001		21 33 14	+62 05.0	1	8.70		0.25		-0.62		2	B0.5V	401
	+45 03596			21 33 16	+45 40.2	1	9.08		0.36				2	F2	1724
205429	−41 14581	IDS21301S4108	AB	21 33 16	−40 54.3	1	8.57		0.52		0.02		1	F7 V	742
205471	−26 15702	HR 8253	★ A	21 33 17	−26 23.7	3	5.72	.007	0.22	.000			11	Am	15,2018,2030
		CS 22948 # 41		21 33 17	−41 48.2	2	14.48	.010	0.68	.016	-0.01	.002	3		1580,1736
205390	−51 12998			21 33 17	−51 04.1	4	7.14	.005	0.88	.005	0.55	.020	11	K2 V	219,1075,2033,3008
		ApJ144,259 # 78		21 33 24	+31 28.	1	13.25		-0.21		-1.18		6		1360
	+47 03480	NGC 7092 - 144		21 33 24	+48 12.8	2	9.54	.015	0.18	.005	0.14		5	A3 IV	562,1544
		NGC 7092 - 267		21 33 25	+48 05.6	2	11.21	.005	0.46	.010	0.00		6	F6	1544,1581
205556	+4 04703			21 33 26	+05 15.1	5	8.31	.005	-0.05	.005	-0.37	.014	105	B9	147,989,1509,1729,6005
		G 212 - 45		21 33 27	+46 20.1	1	16.05		0.70		0.34		2		940
205655	+48 03430			21 33 31	+49 11.9	1	9.01		0.01		-0.22		6	A2	848
	+43 03979			21 33 32	+43 44.8	1	8.97		1.05				2	G5	1724
	+48 03431			21 33 32	+49 08.6	1	9.76		0.24				3		1724
		Ross 203		21 33 32	−13 31.9	1	13.63		0.23		-0.55		3		3062
205618	+29 04453			21 33 33	+29 31.3	1	8.10		0.05		-0.90		3	B2 Vne	1212
		SX Aqr		21 33 37	+03 00.3	3	11.14	.054	0.13	.021	0.10	.022	3	A2	668,699,700
	+43 03980	LS III +43 022		21 33 37	+43 41.2	1	8.96		0.97				2	F1 Ib	1724
235553	+51 03099	IDS21319N5143	AB	21 33 37	+51 56.7	1	9.19		0.11		-0.37		5	B5	848
205626	+25 04575	IDS21315N2555	A	21 33 42	+26 08.7	1	9.24		0.58		0.06		1	F8 V	3016
205626	+25 04575	IDS21315N2555	AB	21 33 42	+26 08.7	1	8.53		0.52		0.06		1	F8 V	3026
205627	+25 04576	IDS21315N2555	B	21 33 43	+26 08.7	1	9.27		0.57		0.06		1		3016
205584	+5 04824			21 33 44	+05 54.8	3	7.72	.008	1.27	.010	1.33	.012	85	K2	147,1509,6005
	+48 03431	NGC 7092 - 268		21 33 45	+48 32.1	2	10.41	.005	0.26	.005	0.06		6	A7	1544,1581

HD	DM	Other Id	N Rem	α_1950	δ_1950	S	V	σ_V	B-V	σ_B-V	U-B	σ_U-B	n	Spectrum	References
	+47 03483			21 33 48	+47 41.7	2	9.06	.005	0.31	.010	0.27		4	A5	379,1724
		Tr 37 - 3318		21 33 48	+57 08.3	1	11.66		0.38		0.16		1		1706
		CS 22897 # 125		21 33 48	-64 17.1	1	13.72		0.42		-0.19		1		1736
205529	-33 15664	HR 8256		21 33 49	-33 16.4	2	6.10	.005	0.22	.000			7	A5 V	15,2012
		G 126 - 11		21 33 50	+18 51.7	1	11.76		1.04		0.93		1	K2	1620
		NGC 7092 - 269		21 33 50	+47 39.6	1	14.24		1.01		0.84		2		1581
205687	+40 04576			21 33 51	+40 49.1	1	8.58		0.57				3	G5	1724
	+56 02596	Tr 37 - 426		21 33 51	+57 18.0	1	10.11		0.29		-0.42		3	B2 Vn	499
	+39 04604			21 33 52	+40 19.9	1	9.70		0.38				2	A5	1724
		AAS42,335 T8# 66		21 33 54	+50 23.	1	10.64		0.00		-0.47		6		848
		NGC 7092 - 270		21 33 55	+48 48.1	1	14.05		0.91		0.80		2		1581
	+58 02294	Tr 37 - 1279		21 33 55	+58 31.9	1	9.95		0.34		-0.37		4	B2.5V:n	499
		G 26 - 17		21 33 56	-10 57.0	1	13.57		1.41				2		202
235554	+52 02979			21 33 57	+52 36.5	1	9.03		0.02		-0.46		4	B8	848
205348	-70 02861			21 33 57	-69 51.3	1	6.74		-0.10		-0.31		4	B8 V	244
	+47 03485	NGC 7092 - 271		21 33 58	+47 53.0	2	9.40	.010	0.15	.005	0.13		6	A3 m	1544,1581
235556	+52 02980			21 33 58	+52 55.5	1	8.84		0.08		-0.30		4	B9	848
205547	-41 14587			21 33 59	-40 54.3	2	8.61	.005	1.23	.030	0.96		4	G5/6 (III)	742,1594
		AAS42,335 T8# 68		21 34 00	+50 28.	1	13.76		0.64		0.11		1		848
		Smethells 85		21 34 00	-63 57.	1	10.63		1.42				1		1494
	+47 03487	V1427 Cyg		21 34 01	+47 41.2	2	9.11	.009	0.68	.000	-0.29	.033	4	Bpe	379,1012
205688	+29 04456	HR 8261	★ AB	21 34 03	+29 49.8	1	6.21		1.02		0.74		2	G8 III-I	1733
205656	+17 04614	G 126 - 12		21 34 05	+18 24.5	1	8.64		0.69		0.17		1	G5	1620
	+43 03982			21 34 05	+43 49.5	1	9.73		0.26				2	A5	1724
235555	+50 03375			21 34 05	+50 33.3	2	8.74	.005	0.02	.015	-0.43	.010	6	B3	401,848
205417	-65 03937	HR 8249		21 34 05	-65 03.0	3	6.20	.004	0.02	.005	0.05	.000	18	A0/1 IV	15,1075,2038
		G 93 - 25		21 34 07	+05 06.2	1	13.36		1.54				2		202
205730	+44 03877	HR 8262, W Cyg		21 34 08	+45 09.0	6	5.50	.105	1.60	.033	1.25	.011	11	M5 IIIae	15,369,765,3051,8015,8022
		NGC 7092 - 272		21 34 08	+48 39.4	1	12.54		0.73		0.20		2		1581
205732	+41 04205			21 34 10	+41 42.5	1	7.80		0.86				3	G5	1724
205700	+28 04137			21 34 11	+29 18.9	1	8.27		0.38		-0.06		5	F5 V	3026
205794	+56 02598	Tr 37 - 427		21 34 12	+57 14.6	2	8.42	.019	0.36	.023	-0.55	.005	4	B0.5V	499,1706
		NGC 7092 - 273		21 34 13	+48 38.6	1	12.38		0.74		0.15		2		1581
	+53 02652			21 34 14	+53 42.1	1	9.32		1.78				2		369
	+44 03878			21 34 15	+45 22.9	1	9.41		0.48				2	F5	1724
205637	-20 06251	HR 8260, ε Cap	★ A	21 34 17	-19 41.5	14	4.64	.035	-0.18	.014	-0.64	.020	53	B2.5 Vpe	3,15,379,681,815,1075*
		GD 234		21 34 18	+21 51.1	1	14.45		-0.04		-0.83		1	DA	3060
205778	+48 03435			21 34 20	+48 26.6	1	8.75		0.04		-0.22		2	A0	401
	+41 04207			21 34 21	+41 45.1	1	9.64		0.24				2	A5	1724
	+44 03880			21 34 21	+45 07.4	1	9.68		1.80				1		369
	+40 04578			21 34 22	+40 30.9	1	9.08		0.21				2	A5	1724
205807	+52 02983			21 34 22	+52 35.5	1	8.09		0.35				2	F2	1724
205795	+49 03562	IDS21326N5003	A	21 34 23	+50 16.6	2	7.18	.005	-0.01	.010	-0.03	.005	6	Ap CrEu	401,848
	+50 03376			21 34 23	+50 56.5	1	10.03		0.09		-0.39		5		848
		Tr 37 - 430		21 34 24	+57 07.2	1	10.39		0.31		-0.31		3	B3 V	499
		Tr 37 - 3312		21 34 24	+57 07.4	1	10.34		0.33		-0.30		1		1706
	+49 03563			21 34 27	+49 34.1	1	9.67		0.31				2	F0	1724
205808	+50 03377			21 34 27	+50 27.3	2	7.27	.010	-0.10	.000	-0.49	.000	6	B9	401,848
	+48 03437	LS III +49 030		21 34 28	+49 07.5	2	8.68	.004	0.17	.004	-0.69	.004	7	B1 Ib	848,1012
	+66 01418			21 34 28	+67 01.2	1	9.29		0.40		-0.04		2	F5	1375
205796	+47 03489			21 34 29	+48 25.0	1	8.63		-0.03		-0.23		1	A0	401
		POSS 76 # 1		21 34 29	+66 10.1	1	14.41		1.30				1		1739
205850	+56 02600	Tr 37 - 431	★ A	21 34 30	+57 14.8	1	9.20		0.43		0.03		3	F3 III	3016
205850	+56 02600	Tr 37 - 432	★ B	21 34 30	+57 14.8	1	9.64		0.41		-0.02		3		3032
		G 26 - 18		21 34 30	-00 20.4	1	12.36		1.33				3		202
205650	-28 17381			21 34 30	-27 51.5	2	9.04	.000	0.53	.019	-0.28		6	F6 V	2012,3077
	+49 03564			21 34 32	+50 01.5	1	10.46		0.01		-0.39		2		261
	+40 04580			21 34 33	+41 05.8	1	9.43		0.37				2	F2	1724
		Tr 37 - 3313		21 34 33	+57 06.6	1	12.80		2.25				1		1706
		AAS42,335 T8# 74		21 34 36	+49 10.	1	10.89		0.15		0.12		4		848
		G 212 - 48	AB	21 34 37	+39 13.9	2	10.15	.099	1.46	.052	1.10		2		497,1017
		NGC 7092 - 275		21 34 37	+47 48.8	1	13.07		0.75		0.42		2		1581
		CS 22948 # 27		21 34 39	-39 40.9	1	12.66		1.13		0.54		1		1736
205746	+11 04613			21 34 40	+11 29.6	1	7.26		0.06		0.03		2	A0	1733
205762	+19 04748			21 34 43	+19 33.5	1	7.23		0.96		0.72		2	K0	1648
		BPM 14184		21 34 48	-64 45.	1	13.40		0.49		-0.28		2		3065
		Tr 37 - 435		21 34 52	+57 24.6	1	11.51		0.49		0.16		1		1706
	+43 03986			21 34 53	+44 20.9	1	9.27		0.34				2	F0	1724
		Tr 37 - 438		21 34 56	+57 26.1	1	12.26		0.63		0.43		1		1706
		Tr 37 - 2095		21 34 56	+57 30.2	1	13.50		0.72		-0.11		1		1706
205835	+39 04612	HR 8266		21 34 57	+40 11.3	3	5.05	.014	0.17	.009	0.10		6	A5 V	1363,1724,3023
	+40 04582			21 34 57	+40 35.5	1	9.61		1.07				3	G5	1724
		G 126 - 14		21 34 59	+15 03.6	1	11.16		0.59		0.00		1		1620
		POSS 76 # 5		21 35 00	+66 39.1	1	17.66		1.34				2		1739
205765	-1 04180	HR 8263	★ A	21 35 00	-00 36.9	6	6.24	.012	0.06	.004	0.05	.004	44	A2 V	15,1007,1013,1071*
205765	-1 04180	IDS21324S0050	B	21 35 00	-00 36.9	1	9.50		0.52		-0.03		5		3024
	+44 03885			21 35 01	+44 44.6	1	9.21		0.36				3	F0	1724
		Tr 37 - 440		21 35 05	+57 20.6	1	11.03		0.48		-0.25		1		1706
205767	-8 05701	HR 8264	★	21 35 05	-08 04.8	8	4.68	.004	0.18	.007	0.16	.029	34	A7 V	3,15,1075,1425,2012*
205896	+46 03389			21 35 09	+47 13.1	1	8.58		0.42				2	G0	1724

Table 1   1159

HD	DM	Other Id	N Rem	α_{1950}	δ_{1950}	S	V	σ_V	B–V	σ_{B-V}	U–B	σ_{U-B}	n	Spectrum	References
		Tr 37 - 2022		21 35 09	+57 05.2	1	14.53		1.60				1		1706
239710	+56 02604	Tr 37 - 441		21 35 09	+57 16.6	2	9.50	.005	0.35	.015	-0.42	.015	7	B2.5IV	206,499
		CS 22951 # 15		21 35 12	-45 49.9	1	14.02		0.17		0.14		1		1736
		CS 22897 # 126		21 35 12	-64 19.1	2	14.86	.010	0.37	.006	-0.18	.010	2		1580,1736
		CS 22897 # 120		21 35 12	-64 55.3	1	15.32		0.34		-0.02		1		1736
205918	+49 03568	IDS21334N5005	AB	21 35 13	+50 18.2	2	8.31	.004	-0.01	.004	-0.57	.000	16	A0	401,848
205811	+5 04830	HR 8265	★ A	21 35 14	+06 23.6	5	6.19	.011	0.03	.005	0.02	.009	15	A2 V	15,1028,1084,1256,3016
205811	+5 04829	IDS21328N0610	B	21 35 14	+06 23.6	3	7.67	.007	0.36	.005	-0.01	.000	9		1028,1084,3016
	+50 03379			21 35 14	+51 09.3	1	9.98		1.74				1		369
		Tr 37 - 2023		21 35 14	+57 05.3	1	14.42		0.88		0.36		1		1706
205948	+56 02605	Tr 37 - 442		21 35 16	+57 21.5	2	8.63	.014	0.28	.018	-0.58	.005	5	B2 IV	5,499
		G 26 - 19		21 35 16	-03 51.3	1	14.41		1.61				2		202
	+49 03570			21 35 17	+50 18.4	2	9.37	.000	1.75	.012	2.06	.020	9	K5	401,848
		Tr 37 - 2070		21 35 17	+57 22.0	1	13.44		0.59		0.17		1		1706
		Tr 37 - 2025		21 35 18	+57 04.7	1	12.42		2.13				1		1706
		Tr 37 - 2060		21 35 19	+57 17.7	1	13.41		1.71		1.08		1		1706
	+47 03498	NGC 7092 - 276		21 35 20	+48 19.9	2	9.71	.005	0.23	.000	0.16		5	A7 V	1544,1581
205852	+18 04827	HR 8267		21 35 25	+19 05.6	3	5.46	.022	0.31	.011	0.13	.008	8	F1 IV	253,254,3026
239712	+57 02354	Tr 37 - 710		21 35 29	+57 54.9	2	8.59	.041	0.43	.014	-0.33	.036	5	B3 IVnpe	5,499
205965	+50 03381			21 35 30	+51 08.3	1	7.68		1.58				1	K5	369
205826	-11 05635	G 26 - 20	★ AB	21 35 30	-11 07.6	2	8.91	.024	0.60	.000	0.00		3	G0	202,1620
		WLS 2140 30 # 10		21 35 32	+29 42.5	1	12.10		0.68		0.24		2		1375
205966	+50 03382			21 35 32	+50 50.3	2	7.17	.020	1.87	.000	1.87	.035	4	M0 III	261,401
205939	+44 03889	HR 8272, CP Cyg		21 35 33	+44 28.3	3	6.24	.024	0.20	.003	0.18		13	A7 III	765,909,1733
235563	+52 02991	IDS21339N5217	ABC	21 35 34	+52 30.8	1	8.64		1.06		0.89		4	A0	848
205855	-2 05588	G 26 - 21		21 35 35	-02 31.5	5	8.62	.015	0.85	.005	0.55	.010	10	K0 V	202,1003,1658,1775,3016
		Tr 37 - 2229		21 35 38	+57 27.0	1	13.09		0.62		0.40		1		1706
		L 570 - 29		21 35 38	-33 52.8	1	12.57		1.62		1.33		1		3078
		Tr 37 - 2108		21 35 39	+56 59.4	1	13.34		0.62		0.46		1		1706
		Tr 37 - 2232		21 35 39	+57 28.7	1	14.25		1.15		0.63		1		1706
205831	-23 17034			21 35 42	-23 22.4	1	9.64		0.72		0.31		1	G6 IV	796
		V538 Cyg		21 35 44	+51 32.2	2	10.20	.008	1.18	.006	0.77		2		851,1399
	+27 04120	G 188 - 9	★ A	21 35 45	+27 30.0	2	9.83	.034	1.49	.009	1.26		3	M0	1017,3016
		Tr 37 - 448		21 35 46	+57 17.8	1	11.84		0.58		0.25		1		1706
205941	+32 04215			21 35 47	+32 58.4	2	7.40	.005	1.06	.005	0.85	.015	5	K0	1355,3016
205952	+38 04558			21 35 47	+39 05.5	2	6.76	.005	-0.04	.015	-0.16	.015	4	B9 V	105,401
		Tr 37 - 450		21 35 47	+57 17.3	1	11.93		0.47		0.03		1		1706
206012	+51 03105			21 35 48	+51 25.9	1	8.19		0.94		0.64		3	K0	848
		Tr 37 - 449		21 35 48	+57 11.7	1	12.44		0.53		0.29		1		1706
205859	-17 06333			21 35 48	-17 05.4	1	8.31		1.13		1.07		5	K1 III	1657
	+40 04587			21 35 49	+40 37.0	1	8.75		0.59				2	F8	1724
205582	-72 02623			21 35 49	-71 43.8	3	7.43	.010	0.53	.005	-0.06	.027	9	F8/G0 V	158,1097,3037
		Tr 37 - 2169		21 35 51	+57 14.7	1	14.02		0.72		0.62		1		1706
		Tr 37 - 712		21 35 51	+57 31.6	1	11.96		0.52		0.03		1		1706
206078	+61 02166			21 35 51	+62 04.6	3	7.16	.021	0.97	.005	0.66	.010	7	G8 III	1003,1355,3016
		Tr 37 - 451		21 35 52	+57 09.6	1	12.19		0.62		0.26		1		1706
		CS 22944 # 22		21 35 53	-13 24.5	1	14.63		0.37		0.11		1		1736
	+45 03611			21 35 55	+45 57.4	1	9.16		0.62				2	G0	1724
205805	-46 14026			21 35 56	-46 19.4	2	10.18	.002	-0.24	.001	-0.94	.006	18	B7 III	711,1732
235565	+50 03384			21 35 57	+51 16.4	1	8.75		0.08		-0.47		6	B2	848
206040	+53 02659	HR 8275		21 35 59	+53 49.0	2	6.16	.008	0.99	.001	0.74		6	K1 III	71,848
205478	-77 01510	HR 8254		21 36 00	-77 36.8	5	3.74	.015	0.99	.004	0.90	.004	16	K0 III	15,1034,1075,2038,3002
	+46 03397			21 36 01	+47 09.9	1	9.24		1.43				2	K5	1724
206135	+67 01332			21 36 01	+67 57.6	1	8.40		0.18		-0.47		4	B3 V	206
205924	+5 04834	HR 8270	★ A	21 36 02	+05 32.7	4	5.67	.004	0.26	.009	0.06	.017	12	A9 IV-Vn	15,254,1071,3016
		Tr 37 - 2234		21 36 02	+57 27.1	1	14.18		1.10		0.55		1		1706
	+39 04621			21 36 06	+40 14.0	1	10.47		0.48				3	F8	1724
		LS III +51 022		21 36 06	+51 21.4	1	11.57		0.63		-0.07		5		848
		LP 874 - 10		21 36 06	-24 22.5	2	13.42	.024	1.56	.000	1.04		3	M5	1705,3078
205872	-34 15163	HR 8268		21 36 06	-33 54.3	1	6.28		0.92				4	G8 IV	2032
		Tr 37 - 454		21 36 08	+57 09.6	1	12.34		0.37		0.20		1		1706
206081	+56 02610	Tr 37 - 452	★ AB	21 36 08	+57 20.1	2	8.28	.068	0.35	.018	-0.50	.005	5	B1 Vn	499,1706
206080	+58 02300	Tr 37 - 1289		21 36 09	+58 32.0	1	7.67		0.11		0.02		4	B9.5V	499
206041	+47 03505	IDS21343N4802	A	21 36 10	+48 15.7	1	7.57		0.07		-0.62		2	B8	401
206041	+47 03505	IDS21343N4802	B	21 36 10	+48 15.7	1	10.76		0.19		0.28		1		401
235568	+51 03107			21 36 10	+51 47.2	2	8.62	.000	0.35	.000	0.27	.005	8	A3	848,848
		Tr 37 - 453		21 36 10	+57 07.5	1	11.38		0.41		0.03		1		1706
		G 126 - 17		21 36 11	+18 48.6	2	10.32	.025	0.73	.006	0.33	.031	3	G5	1620,5010
	+49 03572			21 36 11	+50 08.3	1	8.98		0.15		0.07		4		848
		BD +49 03571a	★	21 36 12	+49 38.3	1	9.57		-0.03		-0.82		11	B2	848
		Tr 37 - 2149		21 36 15	+57 08.2	1	13.90		1.53				1		1706
		Tr 37 - 2203		21 36 15	+57 21.4	1	13.55		1.11		0.61		1		1706
205905	-27 15550			21 36 15	-27 31.9	4	6.74	.009	0.62	.005	0.14		16	G2 V	158,1075,2012,2020
235569	+49 03574	IDS21345N5010	AB	21 36 16	+50 23.5	1	9.24		0.01		-0.51		5	B5	848
		Tr 37 - 2150		21 36 16	+57 08.8	1	13.65		0.70		0.03		1		1706
		WLS 2200 65 # 6		21 36 20	+65 08.8	1	12.61		0.58		0.11		2		1375
		WLS 2136 45 # 7		21 36 21	+43 01.8	1	12.12		2.18		2.18		2		1375
	+53 02661	LS III +53 009		21 36 21	+53 40.1	1	9.70		0.19		-0.43		3		8113
		Tr 37 - 2148		21 36 22	+57 07.5	1	11.42		1.26		1.06		1		1706
		G 261 - 45		21 36 22	+82 49.8	3	13.02	.000	-0.01	.010	-0.72	.000	8		419,1281,3028

HD	DM	Other Id	N	Rem	α₁₉₅₀	δ₁₉₅₀	S	V	σ_V	B–V	σ_B–V	U–B	σ_U–B	n	Spectrum	References
		AAS42,335 T8# 82			21 36 24	+50 31.	1	13.83		0.68		0.14		1		848
		Tr 37 - 2258			21 36 24	+57 28.8	1	13.66		0.61		0.42		1		1706
		Tr 37 - 2257			21 36 24	+57 29.3	1	14.28		0.71		-0.20		1		1706
		L 570 - 17			21 36 24	-31 47.	1	11.79		0.68		0.06		2		1696
		Tr 37 - 455		★	21 36 27	+57 23.4	3	11.12	.016	0.00	.024	0.00	.007	5	B5 Ib-II	483,499,1706
		G 126 - 18			21 36 28	+22 55.7	2	15.22	.050	0.17	.041	-0.51	.009	5		538,1658
		Tr 37 - 2224			21 36 29	+57 24.6	1	13.05		0.53		0.34		1		1706
205726	-69 03187				21 36 29	-69 36.1	1	8.81		0.25		0.11		4	A8 V	244
206027	+24 04445	HR 8274			21 36 30	+25 16.4	2	6.18	.012	1.03	.004	0.82		5	G9 III	71,833
		AAS42,335 T8# 83			21 36 30	+50 08.	1	12.28		0.84		-0.18		6		848
		AAS42,335 T8# 84			21 36 30	+50 32.	1	13.98		0.43		0.19		1		848
235572	+50 03391				21 36 33	+51 13.2	1	8.64		1.40		1.12		4	K5	848
		Tr 37 - 2266			21 36 33	+57 01.1	1	13.65		0.51		0.36		1		1706
206165	+61 02169	HR 8279, V337 Cep			21 36 35	+61 51.4	10	4.74	.013	0.29	.011	-0.54	.011	26	B2 Ib	1,5,15,154,555,1363*
		Tr 37 - 457			21 36 36	+57 13.2	1	11.59		0.34		0.18		1		1706
		Tr 37 - 456			21 36 36	+57 17.9	1	10.64		0.34		-0.10		1		1706
205877	-52 11911	HR 8269			21 36 36	-52 35.2	3	6.20	.014	0.58	.007	0.32		7	G8 III +A	58,1485,2009
206110	+48 03452				21 36 38	+49 12.6	1	8.68		-0.03		-0.12		5	A0	848
206043	+19 04754	HR 8276			21 36 41	+20 02.3	2	5.82	.055	0.30	.008	0.00	.005	4	F2 V	254,3016
206121	+49 03579				21 36 41	+49 34.2	3	7.03	.008	0.82	.006	0.44	.012	10	G5 II	848,1355,1724
		AAS42,335 T8# 88			21 36 42	+50 13.	1	14.18		0.54		0.09		1		848
		AAS42,335 T8# 89			21 36 42	+50 16.	1	13.77		0.44		-0.18		1		848
205933	-43 14647				21 36 42	-43 21.5	1	6.73		1.03		0.81		5	K1 III	1628
		Tr 37 - 2299			21 36 44	+57 06.3	1	12.93		1.45		1.15		1		1706
		Tr 37 - 2469			21 36 45	+57 17.8	1	14.87		1.53				1		1706
		Tr 37 - 460			21 36 45	+57 26.5	1	11.52		0.65		0.47		1		1706
239722	+58 02302	Tr 37 - 1293			21 36 45	+58 56.5	1	9.55		0.67		-0.14		5	B2 IV	499
		Tr 37 - 2221			21 36 46	+57 22.1	1	12.94		1.29				1		1706
		Tr 37 - 2459			21 36 47	+57 16.6	1	14.50		2.29				1		1706
		Tr 37 - 2472			21 36 47	+57 18.4	1	13.69		0.67		0.28		1		1706
206005	-11 05640	HR 8273			21 36 47	-10 48.2	2	6.07	.003	1.03	.002	0.89		6	K0	58,2007
		AAS42,335 T8# 90			21 36 48	+49 59.	1	11.71		0.70		0.08		4		848
		AAS42,335 T8# 91			21 36 48	+50 01.	1	14.18		0.52		-0.25		1		848
	+38 04564				21 36 49	+38 41.9	2	8.95	.010	0.08	.005	-0.38	.015	6	B8	833,1501
239721	+56 02613	Tr 37 - 461			21 36 49	+57 00.9	1	9.08		1.20		1.07		1	K5	1706
235573	+50 03392				21 36 50	+50 53.8	1	8.30		1.83				2	M0	369
206183	+56 02614	Tr 37 - 171		★ AB	21 36 52	+56 44.8	3	7.42	.011	0.13	.012	-0.79	.005	4	O9.5V	5,499,1012
		Tr 37 - 2275			21 36 54	+57 01.3	1	14.68		0.89		0.70		1		1706
		Tr 37 - 2447			21 36 54	+57 15.1	1	11.75		0.69		0.50		1		1706
206166	+52 03003				21 36 55	+53 21.2	1	7.10		0.24		0.18		4	F0	848
		Tr 37 - 2276			21 36 56	+57 01.5	1	13.82		0.98		0.50		1		1706
		Tr 37 - 462			21 36 57	+57 27.8	1	10.71		0.58		0.38		1		1706
206058	-0 04245	IDS21344S0030		AB	21 36 57	-00 16.7	1	6.63		0.52				4	F7 V +F8 V	2012
	+42 04170				21 36 59	+43 21.3	1	8.69		0.42				2	F0	1724
	+49 03582				21 36 59	+49 31.3	1	9.52		-0.08		-0.42		4	B8	848
206067	+1 04517	HR 8277		★ A	21 37 01	+02 01.1	3	5.11	.023	1.03	.005	0.88	.032	15	K0 III	1088,1355,3016
		Tr 37 - 2486			21 37 01	+57 21.4	1	13.83		1.63				1		1706
205935	-56 09700	HR 8271			21 37 03	-55 57.9	3	6.32	.004	1.06	.000			11	K0 II/III	15,1075,2007
206046	-16 05909	AD Cap			21 37 04	-16 14.0	1	9.77		1.00		0.58		17	K0 IV/V	588
	+45 03615				21 37 05	+45 45.6	1	9.61		0.41				2	F2	1724
206184	+51 03109				21 37 05	+52 16.9	1	8.38		0.03		-0.05		2	A0	1566
	+40 04598				21 37 06	+40 35.8	1	8.98		1.05				2	G0	1724
239724	+56 02615	Tr 37 - 463			21 37 06	+57 08.4	2	9.12	.015	0.38	.005	-0.48	.015	2	B1 V	5,1706
		CS 22948 # 57			21 37 11	-39 10.3	2	13.97	.010	0.61	.014	-0.11	.000	3		1580,1736
206185	+43 03996	IDS21353N4344		A	21 37 15	+43 57.4	1	8.97		0.21				2	A5	1724
206125	+17 04619				21 37 19	+17 44.5	1	8.46		0.15		0.05		2	A0	1648
	+22 04454	G 126 - 19			21 37 19	+23 02.1	2	9.50	.000	0.77	.000	0.21	.000	2	G5	516,1620
206196	+41 04222				21 37 19	+41 47.9	1	8.91		0.26				2	F0	1724
		Tr 37 - 465			21 37 19	+57 26.3	1	11.72		1.02		0.63		1		1706
206088	-17 06340	HR 8278			21 37 19	-16 53.3	8	3.67	.013	0.32	.009	0.19	.017	26	A7mF3 (III)	15,216,355,1075,1425*
		POSS 76 # 2			21 37 20	+66 22.9	1	15.45		1.57				1		1739
	+50 03395				21 37 22	+51 18.1	1	9.98		1.02				1		851
	+50 03394				21 37 23	+50 40.4	1	9.59		0.52		-0.02		4		848
		G 93 - 27			21 37 24	+06 09.3	2	11.63	.010	0.88	.005	0.46		3		202,333,1620
206197	+39 04627				21 37 24	+40 24.6	1	8.06		0.01		-0.40		2	A0	401
		AAS42,335 T8# 93			21 37 24	+50 35.	1	11.66		0.64		0.14		4		848
206267	+56 02617	Tr 37 - 466		★ AB	21 37 24	+57 15.7	3	5.74	.036	0.21	.013	-0.73	.037	6	O6 V	5,154,450
206267	+56 02617	Tr 37 - 466		★ ABC	21 37 24	+57 15.7	6	5.64	.036	0.21	.011	-0.73	.008	15	O6 V	15,1011,1203,1209*
206267	+56 02617	Tr 37 - 466		★ C	21 37 24	+57 15.7	3	8.06	.021	0.24	.012	-0.68	.025	7	B0.2 V	5,450,3024
206267	+56 02617	Tr 37 - 466		★ D	21 37 24	+57 15.7	5	8.02	.019	0.19	.017	-0.70	.018	9	B0 V	5,450,1706,3024,8084
		LP 698 - 23			21 37 26	-04 03.8	1	11.40		0.86		0.54		2		1696
		Tr 37 - 2523			21 37 27	+57 23.2	1	12.73		0.90		0.46		1		1706
206224	+41 04224	IDS21355N4116		AB	21 37 29	+41 30.0	1	7.56		1.21		1.17		3	K0	3016
206224	+41 04224	IDS21355N4116		C	21 37 29	+41 30.0	1	12.58		0.80				1		3016
239725	+56 02618	Tr 37 - 174			21 37 30	+56 43.4	3	9.12	.008	0.25	.013	-0.46	.021	3	B2.5IV-V	5,499,1706
		Tr 37 - 2467			21 37 30	+57 16.9	1	14.26		1.20		0.23		1		1706
206225	+41 04225	IDS21355N4116		E	21 37 31	+41 30.3	1	8.53		1.02		0.74		3	K0	3016
235577	+51 03111				21 37 31	+51 34.9	1	9.14		0.07		-0.42		4	B3	848
	+58 02303				21 37 33	+59 18.8	1	9.18		2.14				2		369
206155	+8 04714	EE Peg			21 37 34	+08 57.4	2	6.95	.020	0.12	.015	0.08	.010	7	A7 V	588,8071

Table 1 1161

HD DM	Other Id	N Rem	α_{1950}	δ_{1950}	S	V	σ_V	B–V	σ_{B-V}	U–B	σ_{U-B}	n	Spectrum	References
206259 +51 03112	LS III +52 007		21 37 34	+52 08.1	3	7.53	.005	0.05	.000	-0.57	.000	19	B3 III	261,848,1012
	Tr 37 - 468		21 37 34	+57 06.3	1	11.36		0.41		0.25		1		1706
	Tr 37 - 469		21 37 35	+57 28.5	1	11.58		0.52		0.33		1		1706
206198 +29 04476			21 37 36	+29 43.0	1	8.35		0.26		0.14		2	A3mA9 ?	1733
	Tr 37 - 2302		21 37 36	+57 02.0	1	14.84		1.61				1		1706
	Tr 37 - 2430		21 37 36	+57 11.3	1	12.12		0.48		-0.30		1		1706
	NGC 7099 - 2		21 37 36	-23 25.	2	10.62	.019	0.28	.024	0.08	.010	6		169,1638
	NGC 7099 - 3		21 37 36	-23 25.	3	11.41	.017	0.91	.000	0.43	.015	12		169,738,1638
	NGC 7099 - 4		21 37 36	-23 25.	3	12.79	.022	1.14	.013	0.68	.000	10		169,738,1638
	NGC 7099 - 5		21 37 36	-23 25.	3	12.04	.026	0.77	.008	0.32	.014	12		169,738,1638
	NGC 7099 - 6		21 37 36	-23 25.	2	12.93	.010	0.62	.010	0.02		4		169,738
	NGC 7099 - 7		21 37 36	-23 25.	2	12.32	.024	0.74	.000	0.26	.024	3		169,738
	NGC 7099 - 8		21 37 36	-23 25.	1	11.97		1.18		1.09		1		169
	NGC 7099 - 9		21 37 36	-23 25.	2	13.26	.030	0.42	.005	0.06	.005	5		169,738
	NGC 7099 - 10		21 37 36	-23 25.	3	13.55	.025	1.01	.008	0.41	.011	10		169,738,1638
	NGC 7099 - 11		21 37 36	-23 25.	2	13.95	.056	0.49	.014	-0.21	.000	4		169,738
	NGC 7099 - 12		21 37 36	-23 25.	1	13.65		0.85		0.34		1		169
	NGC 7099 - 13		21 37 36	-23 25.	1	13.67		0.65		0.07		1		169
	NGC 7099 - 14		21 37 36	-23 25.	1	11.92		1.22		0.58		1		169
	NGC 7099 - 15		21 37 36	-23 25.	1	11.83		1.32		0.83		2		169
	NGC 7099 - 17		21 37 36	-23 25.	1	12.61		1.21		0.86		1		169
	NGC 7099 - 18		21 37 36	-23 25.	1	12.59		1.23		0.87		4		169
	NGC 7099 - 19		21 37 36	-23 25.	3	13.07	.045	1.03	.009	0.56	.008	10		169,738,1638
	NGC 7099 - 20		21 37 36	-23 25.	1	13.11		0.57		0.01		2		169
	NGC 7099 - 21		21 37 36	-23 25.	2	13.60	.005	0.78	.019	0.19	.024	3		169,738
	NGC 7099 - 22		21 37 36	-23 25.	1	14.04	.034	0.70	.015	0.06		3		169,738
	NGC 7099 - 23		21 37 36	-23 25.	2	12.12	.063	1.40	.000	1.28		6		169,1638
	NGC 7099 - 24		21 37 36	-23 25.	1	13.06		1.09		0.71		2		169
	NGC 7099 - 26		21 37 36	-23 25.	1	16.84		0.46				1		169
	NGC 7099 - 80		21 37 36	-23 25.	2	14.65	.015	0.80	.000	0.14	.020	5		169,738
	NGC 7099 - 83		21 37 36	-23 25.	1	15.39		0.06		0.09		2		169
	NGC 7099 - 89		21 37 36	-23 25.	2	14.84	.005	0.76	.025	0.09	.020	4		169,738
	NGC 7099 - 91		21 37 36	-23 25.	1	13.00		1.06		0.54		2		169
	Tr 37 - 2294		21 37 37	+56 58.4	1	11.72		2.01				1		1706
	Tr 37 - 2480		21 37 37	+57 17.8	1	13.45		0.68		0.38		1		1706
206327 +60 02276			21 37 38	+61 19.8	2	8.68	.005	0.18	.000	-0.56	.000	5	B2 V	5,401,555
	Tr 37 - 470		21 37 39	+57 16.3	1	11.92		0.50		0.34		1		1706
	VOP2,1 # 1		21 37 39	+57 16.4	1	11.93		0.53		0.35		4		8096
	Tr 37 - 2293		21 37 40	+56 59.3	1	11.69		0.28		-0.05		1		1706
	Tr 37 - 2454		21 37 40	+57 14.9	1	12.34		0.51		0.37		1		1706
	Tr 37 - 2431		21 37 41	+57 09.6	1	13.21		1.20		0.75		1		1706
	VOP2,1 # 2		21 37 41	+57 15.0	1	12.38		0.53		0.48		4		8096
	NGC 7099 sq2 9		21 37 41	-23 27.2	1	14.71		0.66		0.01		9		1638
+50 03397			21 37 42	+51 22.8	1	9.88		1.21				1		851
	AAS42,335 T8# 97		21 37 42	+51 41.	1	11.61		0.46		0.29		4		848
+57 02358	Tr 37 - 721		21 37 42	+57 34.5	1	10.13		0.33		-0.33		3	B3 Vnnpe	499
206246 +43 03999			21 37 43	+44 12.9	1	8.38		0.54				2	F5	1724
	Tr 37 - 467		21 37 44	+57 23.7	1	11.29		0.37		0.05		1		1706
	Tr 37 - 2565		21 37 44	+57 28.6	1	12.51		1.37		1.14		1		1706
	VOP2,1 # 14		21 37 45	+57 10.5	1	15.76		1.66		0.83		3		8096
206094 -43 14655			21 37 45	-43 32.5	2	9.72	.018	0.87	.000	0.55		5	K1 (III)	2034,3077
	Tr 37 - 471		21 37 46	+57 12.2	2	11.59	.000	0.49	.000	-0.06	.032	5		1706,8096
	Tr 37 - 472		21 37 46	+57 28.9	1	10.85		0.48		0.32		1		1706
+32 04224			21 37 47	+33 22.2	1	9.10		1.58		1.91		1	K5	1746
+48 03455			21 37 47	+49 16.0	1	10.06		1.84				2		369
	Tr 37 - 2432		21 37 48	+57 10.4	1	13.81		0.82		0.74		1		1706
	VOP2,1 # 5		21 37 48	+57 10.4	1	13.89		0.86		0.53		3		8096
	Tr 37 - 473		21 37 49	+57 14.1	1	10.58		0.40		0.10		1		1706
	VOP2,1 # 3		21 37 49	+57 14.1	1	10.59		0.46		0.07		4		8096
	Tr 37 - 2497		21 37 49	+57 18.2	2	12.76	.000	0.54	.028	0.18	.075	4		1706,8096
239726 +56 02619	Tr 37 - 474		21 37 49	+57 25.6	1	9.09		1.04		0.79		1	K0	1706
	LS III +50 025		21 37 51	+50 11.7	1	11.36		0.37		-0.51		6		848
	Tr 37 - 2433		21 37 51	+57 10.3	1	14.47		1.12		0.76		1		1706
	VOP2,1 # 8		21 37 51	+57 10.3	1	14.45		1.16		0.74		4		8096
	Tr 37 - 2514		21 37 51	+57 21.5	1	12.46		2.19				1		1706
	LS III +56 010		21 37 52	+56 06.3	1	11.78		0.75		-0.19		3		8113
206172 -2 05600	G 93 - 28		21 37 52	-02 14.6	3	8.53	.027	0.66	.005	0.11	.005	4	G0	202,1620,1658
239727 +56 02621	Tr 37 - 475		21 37 53	+57 05.1	2	9.28	.027	0.86	.014	0.32	.009	5	A2 Ia	499,1706
	Tr 37 - 2498		21 37 53	+57 18.2	1	13.58		1.20		0.81		1		1706
	VOP2,1 # 19		21 37 53	+57 18.2	1	13.58		1.23		0.30		3		8096
+59 02407	G 231 - 52		21 37 53	+60 02.2	1	10.34		0.63		-0.07		2	G1	1658
+49 03586			21 37 54	+50 22.5	1	9.97		1.06		0.78		4	B8	848
239729 +56 02620	Tr 37 - 477	⋆ AB	21 37 54	+57 15.4	3	8.35	.020	0.37	.004	-0.54	.007	5	B0 V	5,499,1706
	Tr 37 - 2575		21 37 54	+57 31.6	1	12.60		1.38		1.08		1		1706
	VOP2,1 # 6		21 37 55	+57 10.6	1	13.72		0.78		0.64		4		8096
	VOP2,1 # 25		21 37 57	+57 16.4	1	14.45		0.75		-0.07		1		8096
	Tr 37 - 2516		21 37 57	+57 20.7	1	14.00		0.69		0.42		1		1706
206280 +43 04002			21 37 58	+44 12.3	2	6.74	.013	-0.05	.004	-0.18	.018	5	B9	1216,3016
	WLS 2136 45 # 5		21 37 58	+47 25.9	1	12.29		0.52		0.04		2		1375
	Tr 37 - 724		21 37 58	+57 31.4	1	12.40		0.37		0.18		1		1706

HD	DM	Other Id	N	Rem	α_{1950}	δ_{1950}	S	V	σ_V	B–V	σ_{B-V}	U–B	σ_{U-B}	n	Spectrum	References
		HA 113 # 440			21 38 01	+00 28.1	2	11.81	.007	0.64	.001	0.18	.011	4	G0	281,1764
	+49 03587	LS III +50 026			21 38 01	+50 12.5	3	9.98	.011	0.07	.015	-0.70	.010	8	B0	725,848,8113
		HA 113 # 221			21 38 02	+00 07.4	1	12.07		1.03		0.87		10		1764
	+82 00656	IDS21405N8232		A	21 38 02	+82 46.2	1	10.08		0.62		0.07		2	G5	1375
		Tr 37 - 2436			21 38 03	+57 09.7	2	14.12	.028	0.77	.042	0.44	.014	4		1706,8096
206178	−23 17050				21 38 03	−23 00.6	1	9.78		0.46				2	F5 V	1594
		Tr 37 - 3510			21 38 04	+56 50.4	1	12.56		0.86		0.49		1		1706
		Tr 37 - 2657			21 38 04	+57 21.5	1	13.07		0.60		0.32		1		1706
	+0 04763				21 38 06	+00 31.1	5	9.65	.011	1.17	.007	1.25	.008	32	K2 III	281,989,1729,5006,6004
206312	+48 03457	IDS21363N4840		AB	21 38 06	+48 54.3	1	7.13		1.22		1.14		2	K1 II	261
	+48 03458				21 38 06	+49 01.3	1	9.11		0.10		-0.36		4	B8	848
		HA 113 # 443			21 38 07	+00 25.9	1	11.96		0.52		0.01		11		281
206363	+52 03005				21 38 07	+53 10.3	1	8.43		0.04		-0.22		2	A0	401
		VOP2,1 # 21			21 38 07	+57 19.1	1	14.72		0.77		0.39		4		8096
		Tr 37 - 2658			21 38 08	+57 21.3	1	10.95		0.18		-0.23		1		1706
		Tr 37 - 2660			21 38 08	+57 21.5	1	13.32		0.45		0.31		1		1706
		G 213 - 9			21 38 09	+36 50.0	1	11.44		1.37		1.25		2	M3	7010
206350	+49 03589				21 38 09	+49 59.3	1	8.70		0.10		-0.29		2	A2	8113
206349	+50 03401				21 38 10	+51 15.0	2	6.72	.009	1.32	.009	1.36	.014	9	K1 II-III	37,848
	+46 03404				21 38 11	+46 35.8	1	9.76		1.98				2		369
		Tr 37 - 2369			21 38 12	+57 09.9	2	11.79	.009	0.55	.005	0.45	.047	4		1706,8096
206330	+42 04177	HR 8284	★	AB	21 38 13	+43 02.8	2	5.11	.009	1.58	.013	1.85	.031	5	M1 IIIab	1355,3001
		Tr 37 - 2320			21 38 13	+57 03.4	1	13.51		0.91				1		1706
		Tr 37 - 2371			21 38 13	+57 10.4	2	13.35	.019	0.71	.023	0.47	.023	4		1706,8096
		VOP2,1 # 11			21 38 13	+57 12.5	1	13.70		0.66		0.43		3		8096
239731	+57 02360	Tr 37 - 1025	★	ABC	21 38 13	+58 01.1	1	9.03		0.30		-0.51		2	B2 IV	499
		113 L1			21 38 14	+00 14.9	1	15.53		1.34		1.18		1		1764
		VOP2,1 # 12			21 38 14	+57 12.2	1	14.94		1.81		2.33		4		8096
		Tr 37 - 2640			21 38 14	+57 19.2	2	10.97	.000	0.42	.037	0.39	.023	4		1706,8096
	+65 01623				21 38 14	+65 37.8	1	8.90		0.61				1	F8	207
206364	+50 03403	IDS21365N5106		AB	21 38 15	+51 19.7	1	8.33		0.13		0.07		4	A2	848
		Tr 37 - 2598			21 38 15	+57 15.1	2	12.38	.014	0.68	.014	0.13	.005	5		1706,8096
		Tr 37 - 481			21 38 15	+57 22.6	1	10.62		0.29		-0.13		1		1706
		HA 113 # 337			21 38 16	+00 14.3	1	14.23		0.52		-0.03		3		1764
		VOP2,1 # 16			21 38 16	+57 14.3	1	14.91		1.01		0.56		4		8096
		LS III +50 027			21 38 18	+50 10.2	1	10.73		0.52		-0.54		2	B0: pe	1012
		Tr 37 - 2321			21 38 18	+57 03.3	1	14.27		0.72				1		1706
		VOP2,1 # 13			21 38 18	+57 12.8	1	15.33		0.47		0.34		3		8096
206383	+53 02680	LS III +53 010			21 38 20	+53 44.5	1	7.57		0.13		-0.50		2	B5	401
		VOP2,1 # 15			21 38 20	+57 13.4	1	13.45		1.49		1.11		4		8096
		HA 113 # 453			21 38 21	+00 24.3	1	10.14		0.95		0.58		9	G5	281
		Tr 37 - 2326			21 38 21	+57 04.1	1	14.38		2.37				1		1706
		Tr 37 - 2641			21 38 21	+57 19.5	2	14.23	.018	0.63	.018	-0.27	.050	5		1706,8096
		HA 113 # 339			21 38 22	+00 14.3	2	12.25	.000	0.57	.006	-0.02	.012	29	G0	281,1764
		Tr 37 - 2312			21 38 22	+57 01.2	1	14.47		0.96				1		1706
206365	+49 03590				21 38 23	+49 27.3	2	7.17	.010	-0.06	.005	-0.09	.010	6	B9	401,848
	−0 04248				21 38 24	+00 15.2	1	10.90		1.01		0.69		3	G2	281
		Tr 37 - 482			21 38 24	+57 22.6	1	9.71		0.12		-0.47		1		1706
		HA 113 # 233			21 38 25	+00 08.3	1	12.40		0.55		0.10		1		1764
		HA 113 # 342			21 38 26	+00 13.9	1	10.88		1.01		0.70		3		1764
		CS 22944 # 14			21 38 28	−14 47.1	1	14.18		0.43		-0.21		1		1736
206331	+33 04323				21 38 29	+33 59.1	1	8.82		0.49		0.03		1	G0	583
206332	+28 04161				21 38 30	+28 31.7	1	7.43		0.60		0.18		4	G0 V	3016
		VOP2,1 # 24			21 38 31	+57 11.8	1	14.13		1.06		0.61		4		8096
		Tr 37 - 2374			21 38 32	+57 09.7	1	12.17		1.48		1.28		1		1706
		G 26 - 22			21 38 32	−07 42.4	3	11.86	.041	0.67	.004	0.10	.010	4		202,1620,1658
		HA 113 # 239			21 38 33	+00 08.9	1	13.04		0.52		0.05		1		1764
		Tr 37 - 2328			21 38 34	+57 04.6	1	15.67		1.76				1		1706
		HA 113 # 241			21 38 35	+00 12.2	1	14.35		1.34		1.45		9		1764
	+49 03591	LS III +50 028	★	AB	21 38 36	+50 18.0	2	9.63	.022	0.51	.005	-0.43	.030	9	O7.5	848,1011,1012
	+65 01625				21 38 36	+65 42.0	1	9.80		0.47				1	G0	207
		Tr 37 - 484			21 38 38	+57 12.2	2	10.83	.014	0.21	.014	-0.24	.023	4		1706,8096
		HA 113 # 245			21 38 39	+00 08.2	1	15.66		0.63		0.11		1		1764
		Tr 37 - 485			21 38 39	+57 28.1	1	11.15		0.34		-0.34		1		1706
		WLS 2140 30 # 5			21 38 40	+32 20.4	1	12.81		0.62		0.09		2		1375
211455	+88 00131				21 38 40	+89 12.0	1	8.93		0.22		0.05		2	A5	1733
		HA 113 # 459			21 38 41	+00 29.4	3	12.13	.002	0.51	.008	0.00	.007	16	G0	281,1764,5006
		G 26 - 23			21 38 41	−01 00.2	1	12.57		1.56		1.19		1		1620
		Tr 37 - 2712			21 38 42	+57 28.8	1	13.02		2.33				1		1706
		CS 22944 # 11			21 38 42	−15 26.5	1	12.00		0.41		-0.19		1		1736
		Tr 37 - 486			21 38 45	+57 17.1	2	10.11	.009	0.22	.019	-0.44	.009	4	B2 Vn	499,1706
	+44 03901				21 38 47	+45 04.9	1	9.38		0.37				2	F2	1724
		Tr 37 - 2634			21 38 47	+57 17.7	1	13.54		0.73		0.36		1		1706
206373	+28 04162	IDS21366N2853		AB	21 38 48	+29 06.8	1	8.30		0.56		0.04		3	G0 V	3026
206482	+56 02623	IDS21372N5707		A	21 38 48	+57 21.3	2	7.42	.025	0.46	.005	0.03	.010	5	F5	379,3016
206482	+56 02623	IDS21372N5707		B	21 38 48	+57 21.3	1	8.76		0.62		0.08		3		3016
		HA 113 # 250			21 38 50	+00 07.0	1	13.16		0.90				1		1764
206374	+26 04237	G 188 - 18			21 38 50	+26 31.4	2	7.47	.004	0.69	.009	0.21	.013	5	G8 V	1355,3026
206301	−14 06102	HR 8283			21 38 50	−14 16.3	3	5.15	.006	0.67	.005	0.17	.015	9	G2 V + G0 V	2007,3077,4001
	+0 04766				21 38 54	+00 26.6	5	10.01	.006	0.46	.007	0.00	.007	97	F5	196,281,989,1764,6004

Table 1 1163

HD	DM	Other Id	N	Rem	α_{1950}	δ_{1950}	S	V	σ_V	B–V	σ_{B-V}	U–B	σ_{U-B}	n	Spectrum	References
206291	−25 15545	HR 8282			21 38 54	−25 19.8	1	6.49		1.19				4	K1 III	2007
235587	+50 03405				21 38 56	+51 16.9	1	9.24		0.38				2		1724
		Tr 37 - 2647			21 38 56	+57 20.0	1	12.83		0.51		0.23		1		1706
		vdB 146 # 26			21 38 56	+65 44.1	1	10.24		1.16				1		207
		Tr 37 - 488			21 38 58	+57 03.0	1	10.03		1.13		0.89		1		1706
206483	+53 02684	RU Cyg	⋆	A	21 38 59	+54 05.8	1	7.29		1.85		1.36		1	M6:e	3024
206483	+53 02684	RU Cyg	⋆	AB	21 38 59	+54 05.8	1	7.53		1.86		1.55		1	M3	8084
206483	+53 02684	IDS21373N5352	B		21 38 59	+54 05.8	1	12.40		0.55		0.18		3		3024
206483	+53 02684	IDS21373N5352	C		21 38 59	+54 05.8	2	11.13	.063	0.33	.019	0.23	.058	3		3024,8084
206484	+50 03406				21 39 01	+51 05.8	2	8.33	.000	0.49	.005	-0.01		5	F8	261,1724
		GD 235			21 39 02	+11 32.8	1	15.81		0.08		-0.68		1	DA	3060
		CG Peg			21 39 02	+24 33.0	2	10.92	.297	0.35	.114	0.19	.001	2	F1	668,699
206276	−41 14616				21 39 02	−41 21.0	3	8.82	.016	1.04	.000	0.87	.039	10	K3 V(p)	158,2033,3077
206509	+54 02595	HR 8290			21 39 04	+54 38.6	4	6.18	.025	1.13	.025	0.96	.005	11	K0 III	71,206,848,3016
206536	+59 02409				21 39 04	+59 31.5	1	6.88		1.02		0.84		2	G8 III-IV	1502
	+15 04475				21 39 06	+16 13.6	1	9.19		1.12		0.92		2	K0	1648
	+41 04231				21 39 06	+41 53.1	1	9.30		0.87				2	G5	1724
	+48 03464				21 39 07	+48 58.3	1	8.71		0.28				3	A5	1724
235589	+50 03407				21 39 08	+51 11.7	1	8.49		1.86		2.17		7	M0	848
	−31 18409				21 39 08	−30 49.4	1	10.52		1.29		1.25		2	K5 V	3072
		HA 113 # 259			21 39 10	+00 04.0	6	11.74	.005	1.20	.005	1.22	.013	73		196,281,989,1729,1764,5006
206356	−23 17057	HR 8285	⋆	AB	21 39 10	−23 29.4	2	5.22	.023	0.99	.012	0.77		7	K0 III	938,2007
206341	−28 17407				21 39 10	−27 55.1	1	7.64		1.06				4	K0 III/IV	1075
206461	+40 04608	IDS21372N4049			21 39 12	+41 03.1	1	8.39		-0.01		-0.17		2	A0	401
		CS 22944 # 2			21 39 12	−16 57.7	1	14.63		0.40		0.07		1		1736
		HA 113 # 260			21 39 14	+00 10.2	1	12.41		0.51		0.07		11		1764
206417	+14 04658				21 39 14	+14 25.8	1	8.26		0.27		0.12		2	A2	1648
	+42 04179				21 39 16	+42 44.3	1	9.40		0.71				3	G5	1724
	+0 04767				21 39 18	+00 25.6	4	10.31	.003	1.06	.005	0.84	.002	69	G9 III	281,989,1729,1764
		HA 113 # 265			21 39 19	+00 04.3	1	14.93		0.64		0.10		4		1764
		HA 113 # 263			21 39 19	+00 11.9	1	15.48		0.28		0.07		1		1764
		G 126 - 24			21 39 19	+17 21.6	1	13.38		0.60		-0.07		2		1658
		HA 113 # 366			21 39 21	+00 15.6	1	13.54		1.10		0.90		1		1764
		VES 419			21 39 21	+45 03.2	1	12.37		1.85		0.67		2		8106
		HA 113 # 268			21 39 23	+00 06.2	1	15.28		0.59		-0.02		2		1764
206404	−0 04249				21 39 23	+00 07.1	5	7.66	.007	0.49	.007	0.02	.005	32	F5 V	196,281,989,1729,6004
		LS III +56 012			21 39 25	+56 50.3	1	11.10		0.99		-0.18		1		483
		Tr 37 - 2337			21 39 25	+57 02.5	1	12.85		2.11				1		1706
		HA 113 # 34			21 39 25	−00 12.6	1	15.17		0.48		-0.05		1		1764
		Ross 206			21 39 25	−12 23.6	2	12.83	.000	1.47	.000			5	M2	1663,1705
206182	−68 03440				21 39 25	−68 17.6	1	7.26		1.64		2.01		4	M3 III	1704
	+41 04234				21 39 26	+41 31.9	1	9.69		0.41				2	F2	1724
		Tr 37 - 2335			21 39 26	+56 58.0	1	12.53		1.44		1.16		1		1706
		HA 113 # 372			21 39 29	+00 14.9	1	13.68		0.67		0.08		1		1764
206524	+43 04011				21 39 29	+44 16.1	1	8.77		0.21				2	A5	1724
	−0 04250	HA 113 # 269			21 39 30	+00 04.3	5	9.48	.007	1.11	.008	1.04	.008	33	gK4	196,281,989,1729,6004
		HA 113 # 149			21 39 32	−00 04.3	1	13.47		0.62		0.04		1		1764
206538	+40 04611	HR 8291	⋆	A	21 39 33	+40 34.7	5	6.11	.007	0.07	.004	0.07	.009	18	A2 V	15,1007,1013,1216,8015
	+48 03467				21 39 34	+49 07.3	1	9.14		0.70				2	G0	1724
		HA 113 # 153			21 39 35	+00 01.4	1	14.48		0.75		0.28		2		1764
206445	+0 04770	HR 8287			21 39 37	+01 03.4	3	5.65	.012	1.44	.000	1.66	.010	9	K2 III	15,252,1071
	+39 04641				21 39 38	+40 19.4	1	10.17		0.54				3	F8	1724
		LS III +51 024			21 39 39	+51 44.7	1	11.53		0.25		-0.26		8		848
	+57 02369	Tr 37 - 743			21 39 39	+57 32.8	1	9.85		0.12		-0.46		1		1706
	+65 01627				21 39 42	+66 05.6	1	8.89		1.42				1	K2	207
		G 26 - 24			21 39 42	−07 48.0	2	12.99	.000	1.44	.039	1.11		3		202,1620
		L 930 - 65			21 39 43	+00 08.0	1	12.32		1.45				1		1746
		G 126 - 25			21 39 43	+13 15.0	1	16.58		0.29		-0.56		5		538
		G 93 - 30			21 39 44	+01 46.0	1	13.62		1.49				2		202
206487	+5 04850	HR 8289			21 39 45	+05 27.1	5	5.29	.026	1.64	.004	1.94	.016	8	M2 IIIab	15,1256,1355
		BD +42 04182a	⋆	V	21 39 45	+42 36.7	1	14.94		0.31		-0.71		21		866
		HA 113 # 272			21 39 47	+00 07.3	1	13.90		0.63		0.07		1		1764
		HA 113 # 156			21 39 47	+00 28.8	2	11.20	.016	0.54	.007	-0.04	.013	4		196,1764
	+43 04012				21 39 47	+43 45.5	1	9.27		0.14				10	A2	909
		HA 113 # 158			21 39 48	+00 00.4	1	13.12		0.72		0.25		2		1764
	+0 04771				21 39 48	+00 26.7	1	10.16		0.70		0.23		5	G8 V	281
		Tr 37 - 2615			21 39 50	+57 11.8	1	12.69		0.81		0.34		1		1706
		Tr 37 - 744			21 39 50	+57 31.6	1	10.52		0.17		-0.23		1		1706
206395	−44 14601				21 39 51	−43 43.5	1	6.66		0.55				4	G8 V	2012
		HA 113 # 491			21 39 52	+00 30.2	1	14.37		0.76		0.31		2		1764
206453	−19 06152	HR 8288			21 39 52	−19 05.7	5	4.72	.005	0.88	.004	0.52	.005	22	G8 III	15,1075,1088,2012,8015
206488	−0 04251	IDS21374N0000	A		21 39 54	+00 12.6	6	8.83	.005	0.49	.008	0.01	.008	41	G5 V	196,281,989,1729,5006,6004
206488	−0 04251	IDS21374N0000	B		21 39 54	+00 12.6	5	9.08	.006	0.65	.009	0.19	.012	36		196,281,989,1729,6004
		HA 113 # 492			21 39 54	+00 24.6	2	12.18	.003	0.56	.003	0.01	.007	17	F8	281,1764
206570	+34 04500	HR 8297, V460 Cyg			21 39 54	+35 16.9	7	6.05	.045	2.53	.045	4.73	.384	9	C6.3	1,15,8003,8005,8015*
	+48 03470				21 39 54	+48 58.2	1	8.60		1.83				2	M0	369
		HA 113 # 493			21 39 55	+00 24.4	2	11.77	.001	0.80	.007	0.38	.006	16	F5	281,1764
		HA 113 # 495			21 39 56	+00 24.3	2	12.44	.001	0.95	.001	0.54	.005	14		281,1764
206601	+48 03471				21 39 57	+49 04.3	1	7.75		1.08		0.99		4	K1 III	848
		WLS 2200 65 # 11			21 39 57	+62 57.4	1	10.99		0.39		0.30		2		1375

HD	DM	Other Id	N	Rem	α_{1950}	δ_{1950}	S	V	σ_V	B–V	σ_{B-V}	U–B	σ_{U-B}	n	Spectrum	References
	+65 01628				21 39 57	+65 40.7	1	10.66		0.31				1	F0	207
		Tr 37 - 2367			21 39 58	+57 04.8	1	12.70		0.73		0.43		1		1706
		G 188 - 20			21 40 00	+30 46.2	1	12.65		0.52		-0.12		2		1658
206586	+41 04237				21 40 00	+42 20.4	1	7.96		0.53				3	G0	1724
		HA 113 # 163			21 40 01	+00 03.0	1	14.54		0.66		0.11		7		1764
		Tr 37 - 2805			21 40 02	+57 08.7	1	11.04		0.26		-0.31		1		1706
		HA 113 # 165			21 40 04	+00 01.8	1	15.64		0.60		0.00		1		1764
		HA 113 # 281			21 40 05	+00 05.3	1	15.25		0.53		-0.03		1		1764
	+56 02626	LS III +56 013			21 40 05	+56 27.9	1	10.47		0.65		-0.35		2	B0.5III	1012
206540	+10 04604	HR 8292			21 40 06	+10 35.7	2			-0.11	.005	-0.51	.009	4	B5 IV	1049,1079
		HA 113 # 167			21 40 07	+00 02.4	1	14.84		0.60		-0.03		1		1764
206499	-23 17068				21 40 09	-23 24.0	1	7.54		1.28				4	K1 III	2012
		G 126 - 26			21 40 11	+17 36.9	1	10.93		1.17		1.20		3	K7	7010
	+47 03528				21 40 11	+47 45.9	1	9.66		0.44				2	G0	1724
206602	+38 04582				21 40 12	+39 17.8	1	7.36		1.26		1.33		1	K2 III	1716
206632	+45 03637	HR 8298, V1339 Cyg			21 40 14	+45 32.2	3	6.08	.045	1.53	.013	1.44	.037	6	M4 III	369,1501,3001
	+36 04654				21 40 19	+36 40.3	1	8.96		-0.12		-0.92		2	B2	963
206672	+50 03410	HR 8301		★	21 40 19	+50 57.7	6	4.66	.011	-0.12	.006	-0.69	.016	17	B3 IV	15,154,369,1363,3016,8015
		G 26 - 25			21 40 19	-01 10.4	1	13.68		1.56				2		202
	+42 04185				21 40 21	+42 38.1	1	10.33		0.46				2	F2	1724
206671	+51 03125				21 40 21	+52 03.8	1	7.04		0.95		0.77		3	K0	848
206561	-15 06046	HR 8295			21 40 21	-14 37.8	1	5.88		0.25				4	A9/F0 V	2007
		HA 113 # 177			21 40 22	+00 01.0	1	13.56		0.79		0.32		13		1764
		G 126 - 27			21 40 22	+20 46.6	2	13.25	.015	0.16	.005	-0.71	.010	7		203,3060
206644	+40 04615	HR 8300		★ AB	21 40 22	+40 50.9	2	5.72	.024	0.04	.024	0.00	.005	4	A0 V	1216,3030
		vdB 146 # 16			21 40 24	+65 50.3	1	11.16		0.13				1		207
	+44 03908				21 40 26	+45 01.5	1	9.15		0.58				2	G0	1724
206546	-20 06270	HR 8293			21 40 26	-19 51.0	6	6.23	.018	0.27	.013	0.15	.015	19	A0mA5-F0	15,216,355,2012,3016,8052
		WLS 2136 45 # 9			21 40 27	+45 32.0	1	13.00		0.18		0.15		2		1375
239738	+57 02372	Tr 37 - 1036			21 40 27	+58 16.3	2	8.58	.009	0.28	.014	-0.46	.019	4	B2 IV	433,499
206429	-57 09940	IDS21369S5747	A		21 40 27	-57 33.2	1	6.49		0.48		-0.02		4	F6/8 IV/V	158
206655	+41 04241				21 40 28	+41 35.5	1	8.06		0.37				2	A5	1724
		Tr 37 - 2842			21 40 28	+57 20.2	1	10.44		0.30		-0.26		1		1706
206428	-57 09941	IDS21369S5747	B		21 40 28	-57 30.7	1	6.86		0.46		-0.03		4	F5/7 IV/V	158
		Tr 37 - 499			21 40 30	+57 25.5	1	11.90		0.33		0.20		1		1706
206656	+40 04617	IDS21385N4035	AB		21 40 32	+40 49.1	2	7.57	.020	0.34	.000	0.00		4	F0	1724,3032
		HA 113 # 182			21 40 34	+00 01.1	1	14.37		0.66		0.06		6		1764
206696	+50 03411				21 40 34	+50 37.7	1	7.24		-0.05		-0.44		2	B9	401
206673	+41 04243	IDS21385N4158	A		21 40 35	+42 12.7	1	7.46		0.14		0.10		2	A0	1601
		Smethells 86			21 40 42	-61 13.	1	11.72		1.40				1		1494
206697	+42 4189a	SS Cyg		★	21 40 44	+43 21.4	2	11.25	.243	0.36	.029	-0.81	.078	3	B8	405,698
206747	+52 03017				21 40 45	+52 46.6	2	8.09	.010	0.57	.010	0.30		6	F2	848,1724
		HA 113 # 187			21 40 46	+00 03.1	1	15.08		1.06		0.97		4		1764
235596	+51 03126				21 40 47	+51 30.8	2	9.40	.000	1.00	.000	0.64		6	G5	848,1724
206731	+48 03480	HR 8304			21 40 49	+49 22.3	2	6.08	.005	0.99	.005	0.72	.028	4	G8 II	252,1355
		Tr 37 - 500			21 40 49	+57 07.0	1	11.48		0.39		0.03		1		1706
		CS 22948 # 61			21 40 49	-38 34.4	1	14.53		0.57		-0.10		1		1736
	+46 03410				21 40 50	+46 54.1	2	8.69	.010	0.48	.005	0.08		4	F2	1601,1724
206773	+57 02374	Tr 37 - 750			21 40 50	+57 30.4	7	6.83	.067	0.23	.036	-0.82	.028	14	B0 Vnnpe	5,379,401,499,1012,1212,1706
		vdB 146 # 15			21 40 50	+65 51.6	1	12.18		0.75				1		207
		HA 113 # 189			21 40 53	+00 03.6	1	15.42		1.12		0.96		4		1764
206674	+24 04455				21 40 53	+25 17.8	1	7.68		0.18		0.09		3	A2	833
	+65 01631				21 40 53	+65 54.5	1	9.21		1.13				1	K0	207
		vdB 146 # 17			21 40 54	+65 45.2	1	11.48		1.70				1	K2 III	207
206636	-0 04254				21 40 55	-00 10.7	1	9.14		0.15		0.08		3	B7 V	196
		HA 113 # 307			21 40 56	+00 04.3	1	14.21		1.13		0.91		1		1764
206763	+52 03019				21 40 56	+53 01.3	1	7.95		0.08		-0.33		2	B8	401
		Tr 37 - 751			21 40 56	+57 37.1	1	10.00		0.15		-0.18		1		1706
		HA 113 # 191			21 40 59	+00 02.2	1	12.34		0.80		0.22		4		1764
	+57 02376	LS III +58 011			21 41 01	+58 00.9	1	9.75		0.34		-0.39		3	B2.5Vpnne	8113
	-0 04255				21 41 02	+00 03.2	1	9.82		0.53				3		196
206399	-71 02632	HR 8286			21 41 05	-71 14.3	3	6.01	.003	-0.10	.000	-0.34	.005	23	B8 IV	15,1075,2038
		HA 113 # 195			21 41 06	+00 03.6	1	13.69		0.73		0.20		5		1764
206749	+40 04623	HR 8306			21 41 06	+40 55.5	2	5.49	.006	1.60	.002	1.96		4	M2 IIIab	71,3001
		AAS11,365 T4# 1			21 41 06	+52 59.	1	11.92		0.32		0.15		3		261
		Tr 37 - 2910			21 41 09	+57 31.6	1	11.67		1.07		0.73		1		1706
206790	+53 02689	NGC 7128 - 1			21 41 11	+53 32.3	2	7.87	.014	1.05	.009	0.86	.009	5	K0	49,848
206952	+70 01193	HR 8317			21 41 12	+71 04.9	6	4.56	.013	1.11	.008	1.11	.014	15	K0 III	15,1355,1363,3016*
206842	+58 02314	HR 8312			21 41 15	+59 02.5	1	6.08		1.34				2	K1 III	71
		vdB 146 # 11			21 41 15	+65 56.8	1	11.89		0.79				1	G8 V	207
	+26 04249	IDS21391N2653	AB		21 41 16	+27 06.5	1	9.30		0.56		0.22		1	G0	8084
239742	+56 02631	Tr 37 - 204			21 41 16	+56 47.3	2	9.41	.000	0.21	.004	-0.55	.004	6	B2 V	5,499
206677	-15 06052	HR 8302			21 41 17	-14 58.8	2	5.97	.015	0.22	.005	0.07		7	A7 IV/V	2007,8071
	+65 01633				21 41 18	+65 53.7	1	10.58		1.08				1	G8	207
206897	+65 01634				21 41 18	+66 08.0	1	8.00		0.17				1	A0	207
239743	+59 02412	LS III +60 005			21 41 19	+60 04.0	1	9.01		0.59		-0.15		1	B2 V	5
206774	+37 04408	HR 8307		★ AP	21 41 21	+38 03.2	2	5.69		-0.01	.009	-0.04	.013	5	A0 V	1049,3016
206791	+43 04024				21 41 22	+44 06.9	1	8.53		0.43				2	F8	1724
		ADS 15262		★ A	21 41 24	+57 25.	1	11.44		1.72		1.93		1		3032
		ADS 15262		★ B	21 41 24	+57 25.	1	11.80		0.51		0.40		1		3032

Table 1   1165

HD	DM	Other Id	N Rem	α_{1950}	δ_{1950}	S	V	σ_V	B–V	σ_{B-V}	U–B	σ_{U-B}	n	Spectrum	References
	+40 04627			21 41 25	+41 16.9	1	9.00		0.61				2	G0	1724
206642	−39 14405	HR 8299		21 41 25	−38 46.8	3	6.29	.004	1.12	.005			11	K1 III	15,2018,2030
		G 261 - 46		21 41 26	+78 35.5	1	13.27		1.44				3		1663
	+47 03536			21 41 27	+47 58.2	1	9.41		0.41				3	F2	1724
	−0 04258			21 41 28	+00 06.2	1	10.03		0.96		0.63		3	K0	196
		AAS11,365 T4# 4		21 41 30	+52 35.	1	11.01		0.81		0.03		2		261
		AAS11,365 T4# 3		21 41 30	+52 40.	1	12.64		1.05		0.79		5		261
		AAS11,365 T4# 2		21 41 30	+52 44.	1	10.59		1.84		2.07		4		261
206807	+37 04410	IDS21393N3750	B	21 41 32	+38 04.5	1	6.98		0.09		0.08		2	A0	3016
		vdB 146 # 10		21 41 32	+65 56.4	1	12.40		0.63		0.37		4	B9.5V	206
206857	+51 03127			21 41 33	+52 10.7	1	7.56		1.45		1.71		2	K2	1566
		Tr 37 - 505		21 41 33	+57 21.0	1	10.82		1.16		0.77		1		1706
		AAS11,365 T4# 6		21 41 36	+52 33.	1	14.16		0.70		0.19		2		261
		AAS11,365 T4# 7		21 41 36	+52 33.	1	12.52		1.30		1.17		2		261
		AAS11,365 T4# 5		21 41 36	+52 45.	1	12.88		0.46		0.24		4		261
	+65 01635			21 41 36	+65 50.8	1	10.71		1.28				1		207
	+65 01636			21 41 37	+65 51.4	1	9.72		0.99				1	G8	207
		vdB 146 # 9		21 41 38	+65 57.1	1	13.25		0.45				1	B9.5V	207
	+57 02377			21 41 39	+58 23.7	1	10.12		0.38		0.18		1		433
206792	+27 04145	IDS21394N2723	AB	21 41 40	+27 37.0	2	7.46	.009	0.44	.000	-0.04		6	F6 V +F6 V	1381,3030
		AAS11,365 T4# 10		21 41 42	+52 33.	1	10.75		0.50		0.02		2		261
		AAS11,365 T4# 9		21 41 42	+52 43.	1	11.92		1.04		0.66		4		261
		AAS11,365 T4# 8		21 41 42	+52 58.	1	13.56		0.70		0.40		3		261
	+65 01637	V361 Cen		21 41 42	+65 52.9	2	10.13	.035	0.41	.001	-0.35	.007	5	B2 nne	206,351
		NGC 7128 - 3		21 41 43	+53 30.5	1	10.65		1.55		1.58		1		49
		G 93 - 33		21 41 44	+06 24.7	3	12.06	.039	1.52	.021	1.21		5		202,333,1532,1620
206778	+9 04891	HR 8308, ϵ Peg	★ A	21 41 44	+09 38.7	12	2.39	.013	1.52	.017	1.66	.040	32	K2 Ib	1,3,15,369,1034,1218*
206778	+9 04891	IDS21393N0925	B	21 41 44	+09 38.7	1	8.65		0.55		0.05		2		3032
206778	+9 04891	IDS21393N0925	C	21 41 44	+09 38.7	1	12.66		0.99		0.80		2		3032
	+26 04251	G 188 - 22		21 41 44	+27 11.2	1	10.05		0.48		-0.14		1	F8	1658
	+48 03484			21 41 44	+49 02.6	1	9.55		0.43				2	F5	1724
206887	+48 03485			21 41 46	+48 39.2	2	8.09	.029	0.58	.005	0.40		6	F2 II-III	206,1724
		vdB 146 # 8		21 41 47	+65 57.9	1	12.46		1.05				1	G8 IV	207
		AAS11,365 T4# 13		21 41 48	+52 43.	1	14.26		0.90		0.35		5		261
		AAS11,365 T4# 12		21 41 48	+52 45.	1	13.91		0.52		-0.03		4		261
		AAS11,365 T4# 11		21 41 48	+52 58.	1	12.97		0.92		0.30		4		261
		CS 22948 # 66		21 41 48	-37 41.8	2	13.46	.005	0.65	.015	0.01	.001	3		1580,1736
239745	+56 02632	Tr 37 - 207		21 41 49	+56 47.6	2	8.91	.005	0.26	.024	-0.63	.005	6	B1 V	401,499
		VES 913		21 41 50	+52 48.5	2	11.22	.015	1.10	.005	-0.03	.010	7		261,8113
	+65 01638			21 41 50	+65 52.4	1	10.18		0.45		-0.36		4		206
239746	+58 02315	Tr 37 - 1318		21 41 51	+58 36.9	2	9.87	.031	0.33	.009	-0.37	.026	6	B2 Vp	433,499
206828	+25 04607	G 188 - 23	★ A	21 41 53	+26 17.3	1	8.45		0.65		0.19		3	G2 V	3026
206826	+28 04169	HR 8309	★ ABD	21 41 54	+28 31.0	6	4.49	.015	0.48	.012	-0.01	.014	18	F6 V	15,938,1118,1197,3026,8015
		AAS11,365 T4# 15		21 41 54	+52 44.	1	14.50		1.18		0.83		4		261
	+24 04460	G 188 - 24		21 41 55	+25 06.5	2	9.48	.045	0.78	.040	0.25	.022	5	K0 V	1003,3026
		NGC 7128 - 22		21 41 56	+53 27.9	1	16.65		1.16				1		49
		NGC 7128 - 23		21 41 57	+53 28.0	1	16.76		1.17				1		49
	+40 04628			21 41 58	+40 47.7	1	8.92		0.67				2	G0	1724
206742	−33 15734	HR 8305	★ A	21 41 58	−33 15.3	7	4.34	.007	-0.05	.005	-0.11	.005	25	B9.5V	15,208,1075,1637,2012*
206936	+58 02316	Tr 37 - 1319	★ A	21 41 59	+58 33.0	10	4.10	.110	2.33	.068	2.40	.044	22	M2 Ia+	1,5,15,369,499,8003*
206690	−50 13463	HR 8303		21 42 01	−49 43.8	5	6.46	.013	1.15	.005	1.12		13	K1 III	15,1075,1485,2035,4001
		NGC 7128 - 2		21 42 02	+53 21.3	1	10.15		0.32		0.15		1		49
		NGC 7128 - 19		21 42 02	+53 28.7	1	15.82		0.93		0.43		1		49
		NGC 7128 - 21		21 42 03	+53 27.7	1	16.52		1.15				1		49
		CoD -60 07777		21 42 03	-60 14.5	1	10.55		0.91		0.64		1		1696
		NGC 7128 - 18		21 42 04	+53 28.8	1	15.54		0.99		0.59		1		49
		WLS 2140 30 # 7		21 42 06	+27 41.4	1	13.26		0.45		-0.05		2		1375
235599	+49 03605	IDS21405N5006	AB	21 42 06	+50 19.9	1	8.72		0.90				3		1724
		NGC 7128 - 25		21 42 06	+53 28.2	1	17.21		0.90				1		49
206860	+14 04668	HR 8314, HN Peg		21 42 07	+14 32.6	3	5.94	.011	0.59	.002	0.03	.003	74	G0 V	71,770,3026
	+43 04028			21 42 07	+43 28.8	1	9.40		0.65				2	G0	1724
239748	+57 02380	Tr 37 - 1055	★	21 42 08	+58 00.6	2	8.75	.000	0.16	.005	-0.68	.014	4	B1 V	5,499
206859	+16 04582	HR 8313		21 42 09	+17 07.2	13	4.34	.018	1.16	.012	0.98	.014	41	G5 Ib	1,3,15,369,1218,1355*
		NGC 7128 - 14		21 42 11	+53 29.5	1	14.75		0.84		0.04		1		49
		NGC 7128 - 10		21 42 12	+53 30.2	1	13.99		1.08		0.49		1		49
206898	+34 04511			21 42 13	+35 01.5	1	8.62		0.03		-0.25		2	A0	1625
		NGC 7128 - 9		21 42 13	+53 29.9	1	13.32		0.84		0.02		1		49
		NGC 7128 - 5	★ V	21 42 14	+53 29.5	1	12.35		0.78		0.00		1	B2 V	49
		NGC 7128 - 17		21 42 14	+53 30.8	1	15.29		1.01		0.56		1		49
		G 26 - 26		21 42 15	+01 43.3	1	13.87		1.48				2		202
		NGC 7128 - 7		21 42 15	+53 29.7	1	12.65		0.77		-0.06		1	B2 V	49
206833	−9 05827			21 42 15	-09 16.0	1	7.08		1.41		1.67		3	K0	3016
		NGC 7128 - 20		21 42 16	+53 30.2	1	16.30		1.01				1		49
		NGC 7128 - 4	★ V	21 42 17	+53 28.3	1	11.50		0.78		-0.01		1	B3 IV	49
		AAS11,365 T4# 16		21 42 18	+52 39.	1	10.24		0.19		-0.07		3		261
206211	−81 00972			21 42 18	-81 28.0	1	9.25		0.53		0.08		1	F7 V	832
206096	−83 00721			21 42 18	-82 39.9	1	7.11		0.57		0.06		7	F6 V	1628
		G 212 - 55		21 42 20	+42 20.3	1	14.15		0.88		0.24		2		1658
		NGC 7128 - 6		21 42 20	+53 29.6	1	12.55		0.72		-0.07		1	B2 V	49
206962	+53 02694			21 42 20	+54 03.2	1	8.14		1.03		0.81		3	K0	848

HD	DM	Other Id	N Rem	α_{1950}	δ_{1950}	S	V	σ_V	B–V	σ_{B-V}	U–B	σ_{U-B}	n	Spectrum	References
		vdB 146 # 14		21 42 20	+65 49.7	1	11.16		0.55				1	F2 IV	207
206834	−9 05829	HR 8311		21 42 20	−09 18.8	4	5.11	.019	1.11	.004	0.97	.013	14	G8 II-III	15,37,1061,8100
	+53 02693			21 42 21	+53 28.4	1	9.46		2.30		1.58		2		401
206652		RR Ind		21 42 22	−65 32.3	1	8.96		2.57		4.39		1	C	109
206901	+24 04463	HR 8315	⋆ AB	21 42 23	+25 24.9	5	4.14	.015	0.42	.015	0.00	.021	14	F3 IV+F3 IV	15,254,1118,3026,8015
	+43 04030			21 42 23	+43 34.5	1	9.06		0.33				2	F0	1724
		NGC 7128 - 13		21 42 23	+53 28.2	1	14.42		0.83		0.00		1		49
		NGC 7128 - 12		21 42 23	+53 28.8	1	14.32		0.76		-0.02		1		49
		NGC 7128 - 16		21 42 23	+53 30.7	1	15.01		0.93		0.33		1		49
206955	+45 03646			21 42 24	+46 13.6	1	8.80		0.28				2	A5	1724
		NGC 7128 - 24		21 42 24	+53 27.8	1	16.92		0.84				1		49
207017	+62 01973			21 42 24	+62 32.5	3	8.58	.005	0.21	.012	-0.50	.017	7	B2 V	5,27,399
		G 26 - 27		21 42 24	−06 00.8	2	12.81	.010	1.62	.044	1.30		3		202,1620
	+48 03488			21 42 25	+48 26.2	1	10.98		2.14				2		369
		NGC 7128 - 11		21 42 25	+53 28.1	1	14.22		1.76				1		49
206954	+48 03489			21 42 27	+49 02.3	1	7.54		-0.09		-0.38		2	A0	401
		NGC 7128 - 15		21 42 27	+53 27.9	1	14.82		0.83		0.16		1		49
		Tr 37 - 509		21 42 28	+57 12.6	1	10.47		0.46		0.16		1		1706
207130	+71 01082	HR 8324		21 42 28	+72 05.4	3	5.19	.020	1.05	.008	0.97	.008	8	K0 III	37,1355,3016
		CS 22944 # 39		21 42 28	−14 55.2	1	14.30		0.55		0.03		1		1736
206744	−54 09940			21 42 28	−54 14.9	1	9.81		0.60				4	G2 V	2033
	+40 04631			21 42 31	+41 21.7	1	9.63		1.34		1.20		2	K0	3016
206963	+46 03422			21 42 31	+46 37.9	1	6.61		0.40				2	F4 V	1724
206691	−63 04703			21 42 34	−62 46.8	1	6.87		1.21		1.18		5	K1 III	1628
		NGC 7128 - 8		21 42 36	+53 29.7	1	12.67		0.60		0.07		1		49
206868	−29 17984			21 42 39	−28 57.3	1	7.64		0.40				4	F3 V	2012
207049	+57 02386	Tr 37 - 763		21 42 47	+57 30.4	1	9.00		0.32		0.01		1	B8 III	1706
	+73 00943			21 42 48	+74 12.8	1	10.18		0.71		0.10		3	G5	3016
		G 26 - 28		21 42 48	−06 07.8	2	13.55	.024	1.59	.049	1.01		3		202,1620
	−21 06087			21 42 48	−21 30.8	1	11.25		0.48		0.01		2		875
		Smethells 87		21 42 48	−54 33.	1	11.25		1.31				1		1494
	+39 04655			21 42 49	+40 20.8	1	10.93		0.59				2	F8	1724
	+42 04196			21 42 52	+42 51.0	1	9.06		0.44				3	F2	1724
235603	+51 03132			21 42 57	+51 34.4	1	9.23		0.40				2		1724
	+51 03133			21 42 57	+52 16.2	1	9.68		0.32		0.04		4		261
	+18 04853			21 42 58	+19 16.2	2	9.85	.028	1.54	.028	1.85		2	K2	694,1364
		WLS 2140 30 # 9		21 42 59	+30 18.8	1	12.06		0.58		0.09		2		1375
	+48 03493			21 42 59	+49 05.9	1	8.68		0.38				2	F0	1724
206855	−47 13920			21 43 00	−47 38.0	1	7.64		1.49		1.78		5	K4 III	1673
	+57 02389			21 43 01	+58 25.6	1	10.24		0.29		-0.08		1		433
239758	+58 02320	Tr 37 - 1331		21 43 02	+58 49.6	4	9.51	.009	0.25	.006	-0.55	.017	11	B2 IV:nnep	5,27,499,1012
206804	−58 07893	IDS21395S5808	AB	21 43 02	−57 55.3	3	8.77	.016	1.32	.004	1.15		7	K5 V G/KIII	1494,2033,3077
206805	−58 07892			21 43 02	−58 13.7	1	9.04		1.63		1.90		2	M1 III	3072
207031	+45 03651			21 43 03	+46 19.2	1	8.39		0.60				3	G5	1724
	+65 01640			21 43 03	+65 38.8	1	10.08		0.63				1	F0	207
		ApJ239,815 # 36		21 43 04	−21 28.8	1	12.97		0.60		0.08		2		875
		ApJ239,815 # 35		21 43 07	−21 25.6	1	11.44		1.06		0.96		2		875
	+65 01641	IDS21421N6553	AB	21 43 14	+66 07.2	1	9.38		0.84				1	F8	207
		CS 22948 # 75		21 43 15	−38 39.6	2	15.04	.010	0.39	.007	-0.17	.009	3		1580,1736
206993	+7 04745			21 43 18	+07 45.4	1	7.24		0.91		0.66		2	G5	1733
	+42 04198			21 43 18	+43 03.3	1	9.12		0.50				2	F8	1724
	+47 03544			21 43 19	+48 05.2	1	9.41		0.49				2	F5	1724
	+45 03653			21 43 21	+45 30.5	1	9.47		0.35				3	F2	1724
	+45 03654			21 43 22	+46 17.5	1	9.42		0.71				2	G0	1724
		G 126 - 34		21 43 24	+20 32.9	1	14.36		1.68				1		906
206983	−15 06060			21 43 25	−15 28.5	2	9.43	.015	1.64	.034	1.61	.007	3	Gp Ba	565,3048
207119	+51 03135			21 43 27	+52 02.2	3	6.42	.000	1.90	.003	1.98	.031	35	K5 Ib	261,848,1355
		ApJ239,815 # 38		21 43 27	−21 20.8	1	13.00		0.58		0.00		2		875
207033	+11 04653	IDS21410N1125	AB	21 43 28	+11 38.8	1	8.81		0.63		0.15		3	G0 V +G2 V	3030
206240	−83 00722	HR 8280	⋆ AB	21 43 28	−82 57.1	3	5.28	.004	0.75	.000	0.46	.005	15	G8/K0 III	15,1075,2038
207111	+48 03497			21 43 29	+49 15.0	1	8.51		0.36				2	A5	1724
		JL 87		21 43 30	−76 35.	1	12.00		-0.10		-0.90		1		832
207198	+61 02193	HR 8327	⋆	21 43 31	+62 13.8	9	5.94	.005	0.31	.007	-0.64	.003	29	O9 IIe	5,15,27,154,555,1011*
		NGC 7142 - 26		21 43 33	+65 32.4	1	16.84		0.60		0.50		1		136
		NGC 7142 - 16		21 43 33	+65 32.8	3	14.12	.027	1.44	.075	1.18	.111	3		49,136,4002
	−64 04166			21 43 33	−63 58.5	2	9.93	.010	0.64	.005	0.01	.000	6	G3	158,1696
		NGC 7142 - 5		21 43 34	+65 35.4	1	12.71		1.71		1.80		1		49
		AAS11,365 T4# 19		21 43 36	+52 43.	1	11.63		0.37		-0.21		3		261
207005	−9 05833	HR 8318, AG Cap		21 43 36	−09 30.5	5	6.02	.025	1.66	.009	1.89	.056	23	M3 III	15,1061,1256,3042,6002
		ApJ239,815 # 39		21 43 36	−21 20.2	1	12.36		1.26		1.14		2		875
207088	+35 04626	HR 8320		21 43 37	+35 37.6	2	6.42	.015	1.00	.005	0.71	.000	4	G8 III	1733,3016
		NGC 7142 - 70		21 43 37	+65 33.8	1	14.87		1.17		0.86		1		4002
		ApJ239,815 # 11		21 43 38	−21 42.0	1	10.87		0.82		0.38		2		875
		ApJ239,815 # 16		21 43 39	−21 35.5	1	12.59		0.58		-0.03		3		875
206948	−46 14065			21 43 39	−46 37.2	2	7.54	.005	1.15	.000	1.16		5	K2 III	219,1075
207071	+24 04471			21 43 40	+25 21.1	1	6.57		-0.06		-0.28		2	B8	105
		ApJ239,815 # 8		21 43 40	−21 27.8	1	14.22		0.85		0.44		5		875
		NGC 7142 - 190		21 43 41	+65 28.5	1	12.82		0.42		0.24		1		49
		AAS11,365 T4# 20		21 43 42	+52 43.	1	10.41		1.35		0.99		2		261
		NGC 7142 - 25		21 43 44	+65 31.7	1	15.46		1.88		1.54		2		136

Table 1   1167

HD	DM	Other Id	N	Rem	α_{1950}	δ_{1950}	S	V	σ_V	B–V	σ_{B-V}	U–B	σ_{U-B}	n	Spectrum	References
		NGC 7142 - 23			21 43 44	+65 34.2	1	15.96		1.00				1		49
		NGC 7142 - 31			21 43 45	+65 31.8	1	16.11		1.49		0.71		1		136
207089	+22 04472	HR 8321			21 43 46	+22 43.1	3	5.29	.000	1.38	.015	1.31	.010	5	K0 Ib	1218,1355,3016
	+39 04663	IDS21417N4001	AB		21 43 46	+40 15.5	1	9.20		0.24				2	A5	1724
239762	+58 02322				21 43 47	+58 46.0	1	9.53		0.37		0.15		1	A7	433
		NGC 7142 - 9			21 43 48	+65 32.0	2	13.42	.004	1.37	.022	1.22	.013	6		49,136
		NGC 7142 - 10			21 43 48	+65 34.6	2	13.49	.010	0.61	.045	0.08	.005	2		49,136
		NGC 7142 - 14			21 43 48	+65 34.8	2	13.68	.020	1.31	.000	0.99	.020	2		49,136
		NGC 7142 - 30			21 43 49	+65 35.5	1	16.72		0.95		0.25		1		136
		NGC 7142 - 52			21 43 50	+65 31.7	1	12.10		1.81		1.94		1		4002
		NGC 7142 - 24			21 43 50	+65 33.9	1	16.30		0.77				1		49
207052	−12 06087	HR 8319			21 43 51	−11 35.8	3	5.57	.012	-0.01	.000	-0.05		7	A1 V	2009,2024,3023
		NGC 7142 - 12			21 43 52	+65 30.8	2	13.63	.023	0.69	.023	0.24	.036	5		49,136
		NGC 7142 - 11			21 43 52	+65 31.2	2	13.57	.037	0.81	.023	0.23	.014	4		49,136
	−21 06090				21 43 52	−21 30.7	1	11.21		0.77		0.37		7		875
		ApJ239,815 # 7			21 43 53	−21 31.0	1	11.68		0.86		0.57		5		875
		G 26 - 29			21 43 54	−03 06.6	2	12.69	.005	1.39	.024	1.14		3		202,1620
		ApJ239,815 # 41			21 43 55	−21 40.4	1	13.05		0.61		0.06		2		875
	+49 03614				21 43 56	+49 52.4	1	9.90		1.99				2	M0	369
		NGC 7142 - 27			21 43 56	+65 32.8	1	15.10		0.90		0.34		2		136
207076	−2 05631	EP Aqr			21 43 56	−02 26.7	2	6.69	.055	1.49	.014	0.77	.002	23	M8 III	1628,3042
		NGC 7142 - 28			21 43 57	+65 32.6	1	15.92		0.85		0.42		1		136
	+42 04201				21 43 58	+42 58.5	1	9.03		0.24				2	F0	1724
	+57 02394				21 43 58	+58 13.9	1	10.46		0.38		0.10		2		433
		NGC 7142 - 6			21 43 58	+65 32.7	2	12.91	.005	0.64	.009	0.13	.019	4		49,136
		ApJ239,815 # 40			21 43 58	−21 41.2	1	13.63		0.71		0.20		2		875
		NGC 7142 - 32			21 43 59	+65 30.7	1	15.94		0.91		0.53		1		136
207260	+60 02288	HR 8334, ν Cep		⋆	21 44 00	+60 53.4	10	4.29	.007	0.52	.011	0.12	.018	28	A2 Ia	5,15,27,1012,1119*
		NGC 7142 - 8			21 44 01	+65 34.6	1	13.41		0.75		0.13		1		49
	+49 03615	LS III +50 031			21 44 03	+50 03.7	1	9.07		0.33		-0.47		2	B1 V:	1012
207182	+41 04261				21 44 04	+42 09.5	1	8.64		0.37				2	F0	1724
		NGC 7142 - 29			21 44 04	+65 32.7	1	17.04		0.98		0.56		1		136
		G 26 - 30			21 44 04	−00 24.0	2	12.71	.058	1.59	.005	1.27		3		202,3073
207134	+24 04473	HR 8325			21 44 07	+25 19.9	5	6.29	.009	1.22	.004	1.26	.011	9	K2 III	15,1003,1355,3040,4001
		NGC 7142 - 19			21 44 07	+65 32.7	2	14.97	.039	0.66	.015	0.49	.088	3		49,136
	+65 01642	NGC 7142 - 2			21 44 07	+65 36.0	3	10.73	.030	1.31	.008	1.07	.027	4		49,136,4002
		NGC 7142 - 22			21 44 08	+65 30.1	1	15.71		0.89		0.34		1		49
		NGC 7142 - 20			21 44 08	+65 30.8	1	15.38		0.90		0.29		1		49
		NGC 7142 - 15			21 44 08	+65 34.6	2	13.74	.005	1.44	.044	1.07	.020	2		49,4002
		NGC 7142 - 21			21 44 09	+65 32.7	1	15.49		0.94		0.35		1		49
		NGC 7142 - 4			21 44 09	+65 33.7	2	12.72	.019	1.54	.005	1.33	.034	3		49,136
		ApJ239,815 # 10			21 44 09	−21 24.6	1	11.03		0.32		0.09		5		875
207219	+42 04203	IDS21423N4226			21 44 12	+42 40.3	1	8.74		0.47				2	G5	1724
235608	+50 03429				21 44 12	+50 40.7	2	9.39	.005	0.80	.005	0.43		4	G0	1566,1724
		NGC 7142 - 18			21 44 12	+65 32.8	1	14.81		1.25		0.31		1		49
		ApJ239,815 # 25			21 44 13	−21 41.3	1	12.26		0.65		0.04		3		875
	+49 03616				21 44 14	+49 41.9	1	8.97		1.88				2	M0	369
207232	+50 03430				21 44 14	+50 26.6	1	7.02		-0.06		-0.40		2	B9	401
		ApJ239,815 # 29			21 44 15	−21 38.3	1	14.18		0.55		-0.07		2		875
		ApJ239,815 # 26			21 44 15	−21 41.1	1	14.30		0.65		0.05		2		875
207218	+42 04204	HR 8329			21 44 17	+42 49.7	2	6.53	.000	0.27	.005	0.19	.000	3	A4 V+(GIII)	351,1733
207098	−16 05943	HR 8322, δ Cas		⋆ A	21 44 17	−16 21.3	7	2.86	.031	0.30	.016	0.09	.015	29	A5mF2 (IV)	15,216,355,1088,1197*
		ApJ239,815 # 13			21 44 18	−21 28.9	1	12.99		0.77		0.45		4		875
207308	+61 02194	LS III +62 003			21 44 19	+62 04.6	4	7.49	.007	0.25	.005	-0.58	.015	10	B0.5V	5,27,399,555
	+51 03138				21 44 20	+52 15.2	1	10.79		1.90				1		369
		ApJ239,815 # 28			21 44 20	−21 39.0	1	13.98		1.35		1.35		2		875
207165	+13 04781				21 44 21	+13 29.3	1	6.67		0.18		0.11		3	A3	1648
		G 93 - 36			21 44 22	+10 00.3	1	13.48		1.47				2		202
		vdB 146 # 23			21 44 23	+66 03.3	1	10.62		1.26				1	K2 III	207
		ApJ239,815 # 27			21 44 23	−21 39.5	1	12.28		0.76		0.28		3		875
		CS 22948 # 79			21 44 25	−39 39.2	2	13.70	.015	0.36	.003	-0.12	.011	2		1580,1736
		NGC 7142 - 7			21 44 26	+65 31.3	3	13.10	.035	1.57	.063	1.61	.114	3		49,136,4002
		NGC 7142 - 17			21 44 26	+65 33.3	1	14.61		1.06		0.23		1		49
		ApJ239,815 # 30			21 44 27	−21 39.7	1	13.41		0.90		0.54		3		875
		NGC 7142 - 3			21 44 28	+65 33.1	2	12.44	.010	0.68	.005	0.07	.010	2		49,136
207201	+19 04784				21 44 32	+19 41.5	1	7.12		1.43		1.55		2	K2	1364
		ApJ239,815 # 32			21 44 32	−21 38.3	1	11.92		1.08		0.99		3		875
208390	+85 00367				21 44 33	+86 12.8	1	8.96		0.16		0.08		7		1219
	−72 02640	AY Ind		⋆ AB	21 44 34	−72 19.8	1	9.80		1.47		1.16		1	M2 Ve	3072
207184	+6 04901				21 44 36	+06 54.6	1	8.61		1.32		1.45		3	K2	196
	+57 02395	Tr 37 - 778		⋆ A	21 44 36	+57 33.7	1	9.53		0.45		-0.27		4	B3 V	499
		Tr 37 - 779		⋆ B	21 44 36	+57 34.2	1	10.08		0.48		-0.15		4	B3 V	499
		ApJ239,815 # 20			21 44 37	−21 25.6	1	11.63		1.17		1.18		3		875
	+47 03549				21 44 38	+48 04.7	1	9.77		0.25				2	A5	1724
	+47 03550				21 44 41	+47 57.8	1	9.73		0.38				2	F2	1724
	−21 06092				21 44 41	−21 25.4	1	11.22		0.59		0.05		3		875
207203	+2 04414	HR 8328			21 44 42	+02 27.2	4	5.64	.008	0.01	.015	0.00	.025	11	A1 V	15,252,1071,3016
207223	+16 04598	HR 8330			21 44 42	+16 57.7	2	6.19	.029	0.33	.000	0.00	.005	3	F3 V	254,3016
207328	+57 02396	Tr 37 - 780			21 44 42	+57 49.9	2	7.39	.030	1.90	.040	1.82		5	M3 Ia	369,499
207107	−45 14450				21 44 42	−45 03.4	1	8.58		0.22		-0.49		3	K1/2	555

HD	DM	Other Id	Rem	α_{1950}	δ_{1950}	S	V	σ_V	B–V	σ_{B-V}	U–B	σ_{U-B}	n	Spectrum	References
239767	+56 02640	Tr 37 - 225	★V	21 44 45	+56 41.1	3	9.21	.022	0.68	.028	-0.35	.021	4	B0.5V:p	5,499,1012
	+65 01645	NGC 7142 - 1		21 44 45	+65 33.9	1	10.13		0.49		-0.01		1	F3 II	49
235613	+51 03143			21 44 47	+52 16.5	1	8.40		1.30		1.34		4	K7	261
		NGC 7142 - 13		21 44 47	+65 33.7	1	13.60		0.68		-0.04		1		49
207329	+51 03144	LS III +51 028		21 44 48	+51 53.5	3	7.60	.000	0.29	.000	-0.55	.004	22	B1.5Ib:e	261,848,1012
	+61 02195			21 44 48	+61 34.9	1	9.67		0.17		0.02		2	F8	1069
207175	-28 17454			21 44 48	-27 59.0	1	7.33		1.21		1.22		7	K2 III	1657
207155	-31 18466	HR 8326	★AB	21 44 48	-31 07.8	3	5.01	.009	0.04	.000			9	A1 V	15,2012,2024
	+44 03939			21 44 50	+44 38.5	1	8.95		0.99				2	G5	1724
	+52 03033			21 44 52	+52 45.3	1	8.96		1.76				2		369
207264	+23 04392			21 44 53	+23 54.8	1	7.42		1.00		0.47		10	K0	865
207416	+65 01647			21 44 53	+65 42.5	1	8.47		0.18				1	A2	207
	-21 06094			21 44 55	-21 25.2	1	10.52		1.14		1.07		3		875
207144	-40 14498			21 44 55	-40 29.0	3	8.63	.018	0.96	.008	0.74	.030	12	K3 V	158,2012,3077
	+5 04868	G 93 - 37		21 44 56	+06 22.2	2	10.16	.010	1.05	.005	0.91		3	K5	202,333,1620
	+43 04046			21 44 56	+43 53.2	1	9.37		0.40				2	F5	1724
		CS 22944 # 35		21 44 56	-14 09.6	1	14.02		0.52		-0.22		1		1736
207330	+48 03504		★	21 44 57	+49 04.7	7	4.24	.014	-0.12	.008	-0.73	.034	23	B3 III	15,154,1203,1216,1363*
		G 26 - 31		21 44 58	-07 58.1	5	14.80	.031	-0.10	.024	-0.95	.010	9		202,316,1620,1705,3060
207235	-6 05827	HR 8332		21 45 00	-06 09.0	2	6.16	.005	0.23	.005	0.04	.010	7	A7 V	15,1061
		CS 22944 # 32		21 45 00	-13 54.3	1	13.28		0.59		-0.09		1		1736
207129	-47 13929	IDS21418S4746	B	21 45 00	-47 30.9	1	8.69		1.31		1.47		3		1279
207267	+18 04861			21 45 01	+18 51.9	1	8.17		1.18		1.09		2	K2	1648
	+45 03669			21 45 01	+46 14.6	1	10.10		1.78				2		369
207129	-47 13928	HR 8323	★A	21 45 01	-47 31.9	6	5.58	.007	0.60	.003	0.07	.006	13	G0 V	15,219,688,1279,2012,3077
207209	-22 05766			21 45 02	-21 34.9	1	9.60		1.46		1.29		2	K0	875
	+41 04267			21 45 05	+42 02.5	1	9.48		0.65				2	G5	1724
207350	+47 03552	IDS21432N4731	A	21 45 07	+47 45.1	1	8.13		0.37				2	F0	1724
235618	+54 02623	LS III +55 014		21 45 10	+55 06.2	1	9.66		0.68		-0.31		2	B1 IV	1012
	+41 04270			21 45 17	+41 34.4	1	9.08		0.54				2	F5	1724
	+19 04787			21 45 19	+20 11.5	2	10.38	.038	1.70	.000	2.03		2		694,1364
207192		R Gru		21 45 19	-47 08.8	1	11.59		1.52		0.55		1	M5e	975
		VES 914		21 45 23	+48 55.3	1	11.47		0.36		0.10		3		8113
	+46 03437			21 45 24	+46 30.3	1	9.55		0.38				3	F0	1724
		G 93 - 39		21 45 26	+05 35.6	2	11.79	.005	1.43	.015	1.15		3		202,333,1620
	+52 03036			21 45 26	+52 27.2	1	9.65		2.26				2		369
		Steph 1939		21 45 33	+12 10.9	1	11.70		1.44		1.27		1	K7	1746
	+50 03436	LS III +50 035		21 45 37	+50 56.4	1	10.31		0.44		-0.01		4		848
	-34 15264			21 45 42	-34 05.6	1	10.08		0.82		0.53		2		1696
207356	+6 04906			21 45 44	+06 47.9	1	8.57		0.94		0.52		2	K0	196
		G 214 - 1		21 45 45	+32 52.5	1	12.08		0.57		-0.16		2		1658
207277	-39 14438			21 45 45	-39 18.7	2	8.11	.010	1.18	.005	1.00	.032	3	K0 Ib/II	565,3048
235623	+53 02705		V	21 45 49	+54 23.2	1	9.71		2.04				2	M0	369
	+54 02629	LS III +55 016		21 45 50	+55 06.2	1	10.54		0.65		-0.31		2	B1 II	1012
	+46 03442			21 45 56	+47 07.0	1	10.44		2.00				2		369
207528	+60 02294	HR 8339		21 45 57	+60 27.6	2	5.49	.023	1.55	.017	1.93		7	M1 IIIb	71,3001
		G 188 - 26		21 46 01	+27 42.1	1	12.04		1.58				6		940
207446	+35 04643	HR 8336		21 46 02	+36 20.9	1	6.26		1.56		1.91		2	K5	1733
	+19 04788	G 126 - 36		21 46 04	+19 44.0	2	9.95	.020	0.61	.000	-0.05	.045	2		1620,1658
		CS 22948 # 68		21 46 05	-37 41.9	1	13.77		0.32		-0.14		1		1736
		UV2146 01		21 46 06	+01 43.0	1	10.75		-0.16		-0.79		3		1732
		Smethells 89		21 46 06	-41 48.	1	11.52		1.57				1		1494
		LP 983 - 4		21 46 07	-34 40.9	1	11.91		0.89		0.58		1		1696
207538	+59 02420	Tr 37 - 1588		21 46 08	+59 28.1	5	7.30	.014	0.32	.008	-0.64	.000	9	O9.5V	5,27,1012,1119,1252
207636	+69 01198	HR 8342		21 46 08	+69 55.1	1	6.45		-0.01		-0.02		2	A0 V	1733
207229	-65 03951	HR 8331		21 46 09	-64 56.7	3	5.61	.004	1.02	.004	0.84	.005	15	K0 III	15,1075,2038
207488	+41 04275			21 46 10	+41 55.6	1	8.59		0.58				4	G5	1724
207515	+50 03438			21 46 10	+51 23.6	1	7.86		0.10		0.07		3	A0	848
207314	-48 14016			21 46 11	-48 12.4	1	8.63		0.62		0.02		4	G1/2 V	158
		BPM 27561		21 46 12	-56 18.	1	13.48		0.46		-0.18		3		3065
207469	+32 04263			21 46 13	+32 33.8	1	6.81		-0.01		-0.15		1	A0 V	1716
207489	+38 04611			21 46 14	+38 43.4	2	7.25	.005	0.68	.010	0.50	.010	5	F5 Ib	833,3016
	-34 15270			21 46 16	-34 09.9	2	10.18	.020	0.41	.005	0.02		4	F2	742,1594
207539	+52 03038			21 46 17	+53 16.2	1	8.65		0.61		0.13		3	O9.5III	848
	+45 03681			21 46 20	+46 13.0	1	9.56		0.51				3	F5	1724
235627	+51 03155			21 46 21	+51 29.6	1	9.79		0.27				2		1724
207516	+37 04427	HR 8338		21 46 25	+38 24.9	2			-0.08	.015	-0.34	.015	3	B8 V	1049,1079
207435	-6 05837			21 46 25	-05 38.2	1	6.75		1.01		0.78		3	K0	3016
207543	+52 03040			21 46 26	+52 51.4	1	7.60		-0.01		-0.18		2	A0	401
207529	+46 03444			21 46 27	+46 35.6	1	8.32		0.53				2	G0	1724
207241	-70 02873	HR 8333		21 46 36	-69 51.8	4	5.51	.009	1.38	.004	1.63	.000	25	K2/3 III	15,1075,2024,2038
207498	+25 04621			21 46 39	+25 28.6	1	7.95		0.35		0.06		2	F0	105
207491	+5 04874	G 93 - 40	★AB	21 46 40	+05 29.4	4	8.66	.025	1.01	.019	0.91	.023	10	K3 V	22,202,3008,7009
		LS III +55 017		21 46 42	+55 55.4	1	10.50		0.75		-0.10		3		8113
	+39 04677			21 46 50	+40 23.2	1	8.51		0.37				2	F0	1724
	+61 02201			21 46 53	+62 18.5	1	9.57		0.25		-0.05		4	A0	27
		L 82 - 34		21 46 54	-72 44.	1	12.60		0.66		0.03		3		1696
	+5 04876			21 46 55	+06 25.7	1	9.76		1.15		1.07		3	K0	196
207503	-13 06027	HR 8337		21 46 59	-12 57.4	1	6.31		0.22				4	A3m	2009
207608	+46 03448			21 47 00	+46 40.2	1	8.06		0.47				2	F8	1724

Table 1 1169

HD	DM	Other Id	N Rem	α_{1950}	δ_{1950}	S	V	σ_V	B–V	σ_{B-V}	U–B	σ_{U-B}	n	Spectrum	References
207480	−27 15639			21 47 01	−27 38.3	1	7.14		0.05		0.09		7	A1 V	1628
207491	+5 04874	G 93 - 41	⋆ C	21 47 04	+05 25.0	2	14.84	.049	1.68	.063			3		202,906
	−12 06102	G 26 - 32		21 47 06	−11 54.9	2	10.84	.005	1.39	.039	1.25		3	K7	202,1620
207563	+19 04793	HR 8341		21 47 07	+20 13.7	4	6.28	.011	-0.10	.004	-0.65	.005	6	B2 V	154,1364,1423,1586
207610	+40 04644			21 47 09	+40 59.3	1	8.04		0.49				3	G5	1724
	+40 04645			21 47 14	+40 51.6	1	9.27		0.54				2	F8	1724
		G 126 - 37		21 47 16	+11 46.1	1	11.71		1.44		1.24		1	M1	1620
208020	+80 00706			21 47 16	+80 28.6	1	8.29		0.27		0.10		2	F2	1502
207662	+47 03565			21 47 19	+48 12.3	1	8.44		0.45				2	F8	1724
207551	−15 06076			21 47 21	−14 57.4	1	10.61		0.60		0.00		2	G6 V	1696
		CS 22948 # 93		21 47 26	−41 21.9	2	15.17	.010	0.38	.017	-0.19	.009	3		1580,1736
207552	−17 06389	HR 8340		21 47 28	−17 04.7	1	6.38		1.42				4	K3 III	2009
207672	+43 04057			21 47 29	+43 58.6	1	8.97		0.13				2	A5	1724
		G 93 - 42		21 47 32	+04 27.7	1	13.21		1.35				2		202
208306	+83 00618			21 47 33	+83 48.4	1	7.14		0.26		-0.04		2	A5	985
		CS 22944 # 61		21 47 36	−13 59.5	1	14.35		0.43		-0.23		1		1736
207650	+29 04525	HR 8343		21 47 38	+29 56.4	3	5.06	.024	-0.02	.017	0.01	.029	5	A1 Vs	1049,1363,3023
207673	+40 04648	HR 8345	⋆	21 47 38	+40 54.9	2	6.47	.005	0.41	.000	0.06		4	A2 Ib	138,399
		G 188 - 27		21 47 39	+28 03.2	1	14.68		-0.01		-0.88		4		1663
239775	+55 02633		B	21 47 42	+56 19.5	1	10.31		1.96				2	K5	369
207585	−24 16860			21 47 45	−24 25.2	2	9.79	.005	0.71	.000	0.07	.010	3	G6wF6 (II)p	565,3048
		G 93 - 43		21 47 46	+09 08.6	1	12.66		1.40				2	K5	202
207652	+16 04612	HR 8344	⋆ AB	21 47 46	+17 03.1	4	5.48	.345	0.37	.023	0.00	.015	19	F2 III-IV	3,254,3026,8053
207180	−79 01176	IDS21421S7903	A	21 47 49	−78 48.9	1	7.53		1.13		1.07		8	K0 III	1704
207780	+60 02300	HR 8347		21 47 51	+61 02.4	2	6.15	.017	1.68	.005	2.03		7	M1 II-III	71,3001
	+42 04225			21 47 52	+42 57.1	1	9.39		1.14		0.95		1	K0	592
207826	+66 01441	IDS21468N6619	A	21 47 59	+66 33.6	4	6.46	.013	0.39	.000	0.05	.006	8	F3 IV	1,1028,1355,3032
207826	+66 01441	IDS21468N6619	B	21 47 59	+66 33.6	3	10.56	.026	0.92	.010	0.58	.010	4		1,1028,3032
		CS 22948 # 91		21 47 59	−41 52.0	2	14.11	.010	0.63	.015	-0.04	.000	3		1580,1736
207739	+43 04060	V1914 Cyg		21 48 00	+43 43.9	1	8.59		0.66				3	G5	6009
		VES 916		21 48 05	+49 00.0	1	11.75		0.21		-0.29		3		8113
	+49 03635			21 48 08	+49 45.5	1	8.50		0.99				2	G5	1724
207755	+42 04226			21 48 09	+42 37.5	2	7.71	.009	0.94	.000	0.65		26	K0	592,6011
		WLS 2136 45 # 11		21 48 10	+47 12.5	1	11.34		0.24		0.18		2		1375
207781	+53 02716			21 48 15	+54 00.4	2	8.31	.020	0.02	.010	-0.10	.000	5	A0	401,848
207793	+52 03043	LS III +52 009		21 48 16	+52 27.8	2	6.58	.009	0.38	.000	-0.56	.000	6	B0.5III	848,1012
207756	+31 04558			21 48 20	+32 25.4	1	6.79		1.14		1.07		1	K1 III	1716
207740	+28 04209			21 48 22	+28 32.0	1	8.00		0.70		0.22		3	G5 V	3026
207782	+38 04618	IDS21463N3857	AB	21 48 22	+39 12.5	1	7.78		0.22		0.14		2	A3 V	105
	+47 03574			21 48 23	+47 58.0	1	10.68		0.24		0.07		3		8113
		LP 518 - 54		21 48 25	+12 29.4	1	10.68		0.85		0.51		2		1696
	+44 03960			21 48 34	+44 40.3	1	9.35		0.51				2	F8	1724
207692	−23 17135			21 48 34	−23 30.3	6	6.87	.036	0.48	.009	-0.04	.019	17	F6 V	742,1149,1594,2034*
207757	+11 04673	AG Peg		21 48 36	+12 23.5	5	8.51	.199	0.97	.135	-0.35	.138	6	WN	867,963,1591,1753,1753
		AAS42,335 T9# 29		21 48 36	+52 53.	1	10.05		0.38		0.05		4		848
	+18 04873			21 48 38	+19 00.8	1	9.50		0.22		0.26		1	A2	1364
207872	+59 02424	LS III +59 012		21 48 38	+59 56.7	1	7.98		0.43		-0.44		3	B5	27
208002	+74 00938			21 48 40	+74 45.9	1	8.99		0.86		0.52		3	K0	196
	+43 04063			21 48 41	+43 36.6	1	9.73		0.85				2	G0	1724
207855	+50 03451			21 48 43	+50 45.9	3	7.93	.004	0.34	.008	-0.02	.005	8	F0	848,1566,1724
		G 26 - 33		21 48 43	−00 43.4	1	14.16		1.50				2		202
		CS 22948 # 104		21 48 44	−38 06.6	2	13.95	.005	0.61	.014	-0.11	.000	3		1580,1736
		RT Gru		21 48 45	−46 12.8	1	12.18		0.07		0.08		1		700
	+7 04755		AB	21 48 46	+07 53.7	1	10.09		1.02		0.74		2	K8	1746
	+7 04755		A	21 48 47	+07 54.9	1	10.22		0.98				2	K5	1619
	+7 04755		B	21 48 47	+07 54.9	1	10.91		1.06		0.74		2		1619
207794	+28 04210	IDS21466N2842	AB	21 48 48	+28 56.3	1	8.82		0.88		0.62		2	K1 V +K5 V	3030
207772	+7 04756	IDS21464N0753	A	21 48 49	+08 06.7	1	8.51		0.67		0.09		1	G5	3016
207772	+7 04756	IDS21464N0753	B	21 48 49	+08 06.7	1	11.48		1.22		1.12		1		3016
		G 18 - 1		21 48 49	+12 36.4	1	13.42		1.50		1.10		4		316
	+29 04530			21 48 53	+30 01.0	1	9.80		1.12		0.92		2	K0	1375
207760	−19 06176	HR 8346		21 48 55	−18 51.4	2	6.18	.025	0.35	.010	0.06		7	F0 V	2009,3016
	+28 04211			21 48 58	+28 37.7	7	10.52	.006	-0.33	.008	-1.25	.016	27		1,830,963,1011,1118*
207816	+22 04493			21 49 00	+22 37.5	1	7.85		1.05		0.92		2	K0	1648
207857	+38 04621	HR 8349, V1619 Cyg		21 49 01	+39 18.1	3	6.17		-0.07	.021	-0.44	.018	7	B9 p HgMn	1049,1079,1216
207898	+54 02634			21 49 02	+55 10.6	1	9.05		0.49		0.08		2		1566
207795	+0 04788	G 26 - 34		21 49 05	+00 36.9	3	8.61	.019	0.83	.000	0.50	.014	6	K2	196,202,333,1620
	+42 04230			21 49 06	+42 36.4	1	8.84		0.93		0.59		2	K0	1375
	+45 03699			21 49 07	+46 25.6	1	9.40		0.73				3	A5	1724
		LP 698 - 55		21 49 07	−05 33.1	1	11.15		0.93		0.59		2		1696
207886	+46 03463	IDS21472N4643	A	21 49 08	+46 56.6	1	8.70		0.30				2	Am	1724
235641	+50 03455			21 49 08	+51 09.4	1	8.99		0.98				3		1724
207840	+19 04797	HR 8348	⋆ A	21 49 13	+19 35.5	4	5.77	.006	-0.10	.008	-0.40	.027	18	B8 III	3,1049,1079,3032
207840	+19 04797	IDS21469N1921	B	21 49 13	+19 35.5	1	11.45		0.58		0.23		3		3032
207840	+19 04797	IDS21469N1921	C	21 49 13	+19 35.5	1	13.51		0.82		0.18		2		3032
207951	+61 02208			21 49 14	+61 34.1	2	8.17	.005	0.14	.000	-0.55	.014	4	B2 V	5,27
		G 126 - 39	A	21 49 15	+17 51.7	1	15.05		1.64		1.24		3		316
		G 126 - 39	B	21 49 15	+17 51.7	1	15.16		1.56		0.84		2		316
207828	+11 04677			21 49 16	+11 48.0	1	7.61		0.17				1	A3	867
207765	−41 14669			21 49 16	−41 38.9	1	6.84		1.30		1.41		4	K2 III	1628

HD	DM	Other Id	N Rem	α_{1950}	δ_{1950}	S	V	σ_V	B–V	σ_{B-V}	U–B	σ_{U-B}	n	Spectrum	References
207859	+18 04874	IDS21470N1850	AB	21 49 19	+19 04.3	1	6.84		0.47		-0.02		3	F6 V	938
	+70 01199	VZ Cep		21 49 23	+71 11.5	2	9.73	.008	0.46	.012	0.01	.028	7	G0	731,1768
	+21 04632			21 49 24	+22 25.3	1	9.34		0.67		0.14		3	G0	1306
	+70 01200			21 49 26	+71 09.6	1	9.82		0.53		0.03		1		731
207860	+12 04704			21 49 27	+12 30.7	2	8.72	.005	0.46	.000	0.01	.005	21	F8	245,8053
207790	+41 14671			21 49 27	-40 48.5	1	8.08		0.65				4	F8/G0 IV	2012
	+7 04757			21 49 29	+07 38.5	1	10.17		1.10				1	K7	1632
207496	-77 01522		AB	21 49 35	-77 34.3	1	8.23		1.00		0.89		4	K3/4 V	158
		Steph 1950		21 49 39	+05 23.4	1	12.11		1.53		1.09		2	M2	1746
	+6 04914	G 93 - 47		21 49 39	+07 24.5	2	10.74	.024	0.64	.005	0.07		3	G5	202,333,1620
	+19 04799			21 49 39	+19 43.9	1	10.67		0.42		0.21		2	F0	1364
	+42 04233	VZ Cyg		21 49 41	+42 53.9	3	8.61	.026	0.72	.021	0.44	.027	3	G0	592,1399,6011
	+31 04564			21 49 43	+32 15.2	1	11.04		1.23		1.16		2	K2	1375
	+41 04290			21 49 45	+42 17.5	1	9.31		0.96				2	G0	1724
	+21 04633	AV Peg		21 49 47	+22 19.3	2	9.87	.003	0.19	.006	0.19	.003	2	F0	668,699
207954	+40 04658			21 49 50	+41 24.8	1	8.34		0.82				2	G5	1724
207966	+41 04291	G 215 - 17	★ A	21 49 52	+42 06.8	2	7.84	.025	0.77	.013	0.31		6	G5	1724,3016
		G 93 - 48		21 49 53	+02 09.4	8	12.74	.010	-0.01	.009	-0.79	.008	97		202,203,281,989,1620,1705*
	+45 03705			21 49 56	+45 48.7	1	9.81		0.55				2	F8	1724
207932	+20 05027	HR 8350, HO Peg		21 49 58	+21 02.2	1	6.33		1.70		1.64		3	M4 III	1733
207933	+11 04681			21 49 59	+12 18.7	1	8.18		1.01				1	K0	867
235642	+50 03460			21 50 01	+50 39.2	1	9.68		1.14				2		1724
207991	+47 03584			21 50 02	+48 12.1	1	6.87		1.60		1.93		4	K5 Ib	1355
	+42 04234			21 50 06	+42 30.3	1	10.06		0.43				2	F5	1724
	+45 03706			21 50 06	+45 59.6	1	9.69		0.35				2	A5	1724
207852	-47 13963	IDS21469S4718	AB	21 50 07	-47 04.1	2	7.42		0.58	.005			8	G0 V	2012,2012
		WLS 2136 45 # 8		21 50 09	+44 40.8	1	11.22		1.39		1.37		2		1375
207920	-4 05564			21 50 09	-04 13.7	2	6.68	.005	0.91	.000	0.60	.055	6	G8 II	3016,8100
208004	+41 04294			21 50 13	+42 24.6	1	8.84		0.01		-0.50		1	A3	963
207992	+39 04694	G 213 - 15		21 50 14	+39 34.0	2	8.25	.025	0.73	.005	0.26	.015	6	G5	1625,3026
207700	-74 02021			21 50 14	-73 40.1	1	7.42		0.69		0.25		1	G5 IV/V	219
207978	+28 04215	HR 8354		21 50 16	+28 33.5	1	5.53	.015	0.41	.002	-0.13	.019	6	F6 IV-V	254,3053
208075	+59 02427			21 50 16	+59 41.7	1	7.75		0.10		0.05		2	A0	1502
208053	+50 03465			21 50 19	+51 11.8	1	8.66		0.02		-0.05		3	A2	848
208063	+55 02638	IDS21486N5519	B	21 50 19	+55 33.4	2	6.64	.026	-0.03	.009	-0.25	.017	6	A1p Si	5,3032
208095	+55 02639	HR 8357	★ A	21 50 19	+55 33.7	3	5.69	.012	-0.12	.007	-0.47	.009	8	B6 IV-V	5,154,3032
208106	+61 02209			21 50 20	+61 42.4	4	7.42	.014	0.14	.008	-0.55	.014	13	B2 V.Np	5,27,399,555
		G 18 - 3		21 50 21	+03 12.8	1	12.98		1.47				2		202
208132	+65 01664	HR 8361	★ AB	21 50 22	+65 31.1	1	6.38		0.26		0.10		2	A1m	1733
208056	+41 04299	IDS21479N4153	D	21 50 30	+42 10.2	1	8.58		0.35				2	F0	1724
208025	+23 04416			21 50 33	+23 28.3	1	8.50		0.96		0.41		22	G5	865
207958	-14 06149	HR 8351		21 50 34	-13 47.3	6	5.08	.012	0.37	.008	0.00	.026	22	F3 IV	3,15,2024,2032,2033,3037
		G 93 - 51		21 50 35	+08 30.3	1	15.64		1.04		0.74		1		3062
		IC 5146 - 2		21 50 36	+47 00.3	1	13.40		0.84		0.22		1		1172
		VES 918		21 50 37	+49 32.8	1	11.95		0.31		-0.30		3		8113
		IC 5146 - 5		21 50 38	+46 59.7	1	16.83		1.51				1		1172
	+46 03471	IC 5146 - 6	★ V	21 50 39	+46 59.6	3	10.15	.020	0.41	.005	0.16	.009	7	B9.5Ve	206,351,1172
		IC 5146 - 7		21 50 41	+47 01.6	1	15.35		1.06		0.73		2		1172
208076	+38 04630			21 50 43	+38 36.0	1	7.08		0.93		0.59		3	G9 III	833
208107	+54 02638			21 50 43	+54 26.8	1	6.84		1.02		0.81		3	K1 III-IV	848
208057	+25 04635	HR 8356		21 50 47	+25 41.3	3	5.07	.008	-0.17	.011	-0.68	.014	9	B3 Ve	154,1212,1363
		IC 5146 - 8		21 50 47	+46 59.4	1	11.61		0.50		0.22		2		1172
208134	+54 02640			21 50 48	+54 48.3	1	7.29		0.12		-0.24		2	B8	401
		PHL 191		21 50 48	-04 02.	1	13.56		-0.28		-1.10		2		622
		IC 5146 - 9		21 50 49	+47 01.5	1	11.66		0.54		0.15		1		1172
		IC 5146 - 11		21 50 51	+46 59.7	1	11.05		0.49		-0.04		2		1172
		Steph 1953		21 50 53	+28 36.0	1	11.47		1.49		1.18		2	M0	1746
	+47 03588	LS III +47 054		21 50 53	+47 47.2	1	9.07		0.09		-0.66		2	B1.5V	1012
	+47 03589			21 50 54	+47 38.8	1	9.43		0.52				3	F8	1724
207971	-37 14536	HR 8353		21 50 54	-37 36.1	7	3.00	.004	-0.12	.002	-0.33	.030	42	B8 III	3,15,1020,1075,1079*
	+0 04793			21 50 55	+00 42.8	1	8.25		0.60				2		202
		IC 5146 - 14		21 50 56	+47 01.6	1	11.28		0.52		0.07		1		1172
		IC 5146 - 13		21 50 56	+47 02.3	1	13.96		0.99		0.69		1		1172
208008	-10 05785	HR 8355	★ AB	21 50 56	-10 32.9	1	6.58		-0.12				4	B9 V	2009
	+49 03653			21 50 57	+49 37.9	1	8.99		0.66				2	G0	1724
208185	+62 01992	IDS21496N6237	AB	21 50 57	+62 51.9	3	7.38	.005	0.12	.010	-0.60	.018	7	B2 V +B3 V	5,401,555
208064	+10 04651			21 50 58	+11 23.9	1	7.79		1.09		0.94		2	K0	1733
		G 26 - 35		21 51 00	-01 52.3	2	12.20	.024	1.37	.029	1.27		3		202,1620
208011	-20 06313	IDS21482S2029	ABC	21 51 00	-20 14.7	1	8.12		0.45		-0.03		3	F8	3030
208011	-20 06313	IDS21482S2029	BC	21 51 00	-20 14.7	1	9.82		0.78				2	F5 V	3030
		IC 5146 - 16		21 51 02	+47 02.2	1	15.50		1.26		0.95		1		1172
		IC 5146 - 17		21 51 04	+47 02.3	1	14.06		0.96		0.31		1		1172
		IC 5146 - 18		21 51 05	+47 02.5	1	11.43		0.47		0.02		1		1172
		JL 92		21 51 06	-70 52.	1	15.24		0.63		-0.50		1		832
	+61 02211			21 51 07	+62 09.8	1	9.07		0.19		-0.33		2	B8	213
		IC 5146 - 20		21 51 10	+47 02.1	1	14.78		0.74		0.14		1		1172
208218	+62 01994	NGC 7160 - 1		21 51 10	+62 28.6	7	6.69	.019	0.24	.007	-0.55	.017	21	B1 II-III	5,49,213,401,555,1203,8096
	+51 03174			21 51 11	+52 05.7	1	9.35		1.76		1.72		5		848
		IC 5146 - 23		21 51 14	+47 01.2	1	12.85		0.76		0.15		2		1172
208108	+18 04879	HR 8358		21 51 16	+19 25.9	4	5.68	.006	0.01	.005	0.03	.034	11	A0 Vs	3,1049,1364,3050

Table 1   1171

HD	DM	Other Id	N	Rem	α_{1950}	δ_{1950}	S	V	σ_V	B–V	σ_{B-V}	U–B	σ_{U-B}	n	Spectrum	References
		NGC 7160 - 68			21 51 22	+62 24.8	1	15.14		0.84		0.24		4		8096
		IC 5146 - 29			21 51 23	+46 58.4	1	15.67		1.17				1		1172
		NGC 7160 - 40			21 51 24	+62 17.3	1	15.84		0.74		0.23		3		8096
		NGC 7160 - 8			21 51 25	+62 16.4	2	10.51	.010	0.44	.000	0.01	.010	3		49,8096
	+35 04669	G 188 - 29			21 51 27	+35 35.7	1	9.13		0.58		0.12		2	G0	1658
208110	+6 04919	HR 8359			21 51 28	+06 37.7	3	6.15	.021	0.80	.004	0.26	.018	8	G0 III	15,1256,1355
		IC 5146 - 32			21 51 28	+47 02.9	1	14.71		1.78		1.50		1		1172
239789	+57 02416	LS III +58 015			21 51 28	+58 07.4	1	9.28		0.42		-0.35		1	B2	5
		NGC 7160 - 100			21 51 28	+62 21.8	1	15.39		0.76		0.16		5		8096
		IC 5146 - 35			21 51 29	+46 59.8	1	13.14		0.69		0.33		1	B9 Ve	1172
208266	+59 02430				21 51 29	+60 24.4	2	8.13	.004	0.26	.000	-0.58	.044	6	B1 V	5,213
		JL 93			21 51 30	−70 49.	1	13.63		0.59		0.02		1		832
		IC 5146 - 37			21 51 31	+47 03.9	1	13.94		1.03		0.55		1		1172
208069	−30 18898				21 51 31	−30 29.5	2	9.22	.019	0.70	.014	0.05		4	G(5/8)wF0	742,1594
207964	−62 06277	HR 8352		⋆ AB	21 51 31	−62 07.3	4	5.90	.006	0.40	.007			18	F3 V	15,1075,2013,2029
		IC 5146 - 39			21 51 32	+47 01.0	1	16.35		1.73				1		1172
		IC 5146 - 40			21 51 32	+47 03.3	1	16.53		0.96				1		1172
		G 93 - 53			21 51 32	−01 31.3	2	14.41	.000	0.27	.005	-0.51		3		202,3060
	+44 03974				21 51 33	+45 13.1	1	9.33		0.63				3	G0	1724
		IC 5146 - 44			21 51 33	+47 00.5	1	14.34		1.21		0.75		1		1172
	+46 03474	IC 5146 - 42		⋆ AB	21 51 33	+47 01.8	2	9.64	.002	0.81	.014	-0.06	.037	11	B2 V	1172,1655
	+46 03475	IC 5146 - 48			21 51 34	+46 57.9	1	9.64		0.59		0.03		2		1172
		IC 5146 - 45			21 51 34	+46 59.7	1	15.49		1.43				1		1172
		IC 5146 - 46			21 51 34	+47 00.3	1	15.18		1.44				1		1172
208253	+53 02727				21 51 34	+53 45.7	3	6.61	.005	0.39	.000	0.23	.005	8	A2	848,1566,6009
208111	−4 05568	HR 8360			21 51 34	−04 30.7	3	5.71	.013	1.18	.005	1.26	.019	9	K2 III	15,1256,3016
		NGC 7160 - 11			21 51 36	+62 16.5	2	11.19	.005	0.28	.015	0.15	.019	3		49,8096
		NGC 7160 - 21			21 51 36	+62 22.0	2	12.57	.026	0.46	.000	0.36	.013	6		49,8096
		L 355 - 62			21 51 36	−47 14.	2	12.01	.045	1.57	.015	1.17		2		1494,3073
		IC 5146 - 50			21 51 38	+47 02.9	1	15.49		1.50		1.50		1		1172
235648	+50 03472				21 51 38	+51 16.0	1	9.04		0.12		-0.55		2	B3	401
		IC 5146 - 53			21 51 39	+46 59.7	1	11.76		0.65		0.10		1	A0 V	1172
		IC 5146 - 52			21 51 39	+47 03.1	1	17.03		0.95				1		1172
	+4 04762	G 18 - 5			21 51 40	+04 44.6	3	9.25	.023	0.63	.007	0.03	.009	5	G3 V	202,333,1003,1620
208174	+27 04191				21 51 40	+28 06.4	2	6.79	.010	0.24	.005	0.09	.002	4	A2	1733,3016
208222	+39 04701				21 51 42	+40 11.1	2	9.08	.005	0.52	.000	-0.03		5	G5	583,1724
	−61 06596				21 51 42	−61 26.4	1	9.90		1.06		0.94		2	K3 V	3072
208220	+42 04241	LS III +43 029			21 51 43	+43 14.8	1	9.49		0.04		-0.75		2	B1 IVe	1012
		IC 5146 - 56			21 51 43	+47 03.0	1	15.61		1.83		0.75		1		1172
208061	−48 14054				21 51 43	−48 00.4	1	9.68		0.48				5	F3 V	1594
208235	+43 04080				21 51 44	+43 36.8	1	9.56		0.18				2	G5	1724
		IC 5146 - 58			21 51 45	+47 04.7	1	14.51		0.85				1		1172
		IC 5146 - 62			21 51 47	+47 01.3	1	12.04		0.32		0.03		1	B9 V	1172
		IC 5146 - 61			21 51 47	+47 01.8	1	16.55		1.43				1		1172
		NGC 7160 - 12			21 51 47	+62 20.8	2	11.43	.014	0.56	.005	0.10	.005	5		49,8096
		CS 22948 # 101			21 51 47	−39 10.2	1	14.43		0.29		0.00		1		1736
		IC 5146 - 64			21 51 48	+47 02.2	1	12.36		0.39		0.09		1	B8 V	1172
		NGC 7160 - 20			21 51 48	+62 22.0	2	12.47	.042	0.45	.004	0.40	.004	7		49,8096
		IC 5146 - 66			21 51 49	+46 58.4	1	13.90		1.12		0.57		1		1172
		IC 5146 - 70			21 51 51	+47 00.2	1	13.13		0.70		0.19		1		1172
		WLS 2200 65 # 10			21 51 51	+64 30.2	1	12.20		0.69		0.00		2		1375
	+44 03978				21 51 54	+45 19.2	1	10.07		0.63				2	F8	1724
208188	+17 04657				21 51 55	+17 47.2	1	7.89		1.57		1.98		3	K2	1364
208202	+19 04814	HR 8364		⋆ A	21 51 56	+19 28.9	1	6.39		0.97		0.68		3	K0 III	3024
208202	+19 04814	IDS21496N1915	B		21 51 56	+19 28.9	1	8.52		0.52		0.00		3	F7 V	3024
		G 215 - 20			21 51 56	+41 32.7	2	10.34	.035	1.37	.014	1.20		6	M0	272,1017,3016
		NGC 7160 - 25			21 51 57	+62 24.4	2	13.04	.008	0.59	.013	0.35	.089	7		49,8096
		NGC 7160 - 10			21 51 58	+62 19.4	2	11.08	.018	1.44	.009	1.43	.027	5		49,8096
208237	+34 04559				21 52 00	+34 35.2	1	9.01		0.46		0.00		1	G5	583
	+61 02213	NGC 7160 - 4		⋆ AB	21 52 00	+62 21.0	3	8.97	.016	0.18	.008	-0.46	.007	7	B3 V	5,49,213
208177	−3 05329	HR 8363		⋆ A	21 52 00	−03 32.3	2	6.19	.005	0.48	.000	0.06	.010	7	F5 IV	15,1071
	+61 02214	NGC 7160 - 6			21 52 04	+62 21.7	3	9.84	.007	0.23	.019	-0.32	.014	5	B5 V	5,49,8096
		LS III +53 022			21 52 06	+53 57.7	1	11.16		0.46		-0.30		2	B2 p (IV,V)	1012
		NGC 7160 - 13			21 52 07	+62 23.1	2	11.55	.052	0.28	.009	0.01	.033	2		49,8096
	+61 02215	NGC 7160 - 5			21 52 08	+62 21.7	4	9.36	.021	0.19	.005	-0.48	.014	9	B3 V	5,49,213,8096
	+45 03715				21 52 09	+46 02.6	1	9.12		0.52				2	F8	1724
		G 18 - 7			21 52 10	+08 40.5	1	14.20		1.50				1		3062
	+47 03599				21 52 10	+48 15.8	1	9.72		0.42				2	F2	1724
		NGC 7160 - 29			21 52 11	+62 23.9	2	13.52	.019	1.03	.005	0.36	.033	4		49,8096
208509	+72 01003				21 52 12	+73 27.9	1	6.64		0.11		0.06		2	A2 V	985
235652	+50 03477				21 52 14	+51 23.9	1	9.11		0.12		-0.46		2	B5	401
		NGC 7160 - 16			21 52 14	+62 21.2	2	11.72	.028	0.28	.014	0.17	.037	4		49,8096
		NGC 7160 - 28			21 52 14	+62 22.4	1	13.47		0.74		0.35		1		49
		NGC 7160 - 22			21 52 16	+62 22.3	1	12.53		0.68		0.40		1		49
		NGC 7160 - 63			21 52 18	+62 22.	1	14.90		0.94		0.25		5		8096
		NGC 7160 - 27			21 52 18	+62 24.1	2	13.24	.074	0.75	.035	0.08	.131	6		49,8096
208267	+18 04884				21 52 20	+19 21.0	1	9.25		0.04		-0.07		2		1364
235654	+51 03182				21 52 21	+51 31.5	2	9.01	.005	0.58	.000	0.04		6	F8	848,1724
208326	+51 03183				21 52 22	+52 05.7	1	8.17		0.41				2	F2	1724
208392	+61 02216	NGC 7160 - 2		⋆ A	21 52 22	+62 22.7	7	7.06	.029	0.26	.010	-0.56	.015	15	B1 IV	5,49,213,1084,1211*

HD	DM	Other Id	N	Rem	α_{1950}	δ_{1950}	S	V	σ_V	B–V	σ_{B-V}	U–B	σ_{U-B}	n	Spectrum	References
		IC 5146 - 76			21 52 23	+46 58.0	1	13.17		0.71		0.43		1		1172
208340	+52 03056				21 52 25	+52 44.9	1	8.96		-0.06		-0.25		2	B9p	171
208440	+61 02217	NGC 7160 - 3		⋆ B	21 52 27	+62 21.8	7	7.94	.017	0.07	.004	-0.70	.019	15	B1 V	5,49,213,555,1084*
		NGC 7160 - 23			21 52 27	+62 23.9	2	12.62	.027	0.52	.036	0.35	.050	5	Am	49,8096
208213	-30 18907				21 52 28	-30 16.3	1	8.42		-0.13		-0.57		1	B3 IV	55
208277	+14 04692				21 52 29	+14 59.3	1	7.74		0.30		-0.02		2	A3	1648
		NGC 7160 - 17			21 52 29	+62 20.2	2	12.08	.005	0.35	.005	0.24	.028	4		49,8096
208341	+43 04084	IDS21505N4335	AB		21 52 31	+43 48.7	1	8.24		0.61				2	G5	1724
208439	+63 01784				21 52 31	+63 30.0	1	7.54		0.03		-0.01		2	A0	401
208313	+31 04574				21 52 32	+32 05.7	1	7.78		0.92		0.64		2	K0	3016
		NGC 7160 - 14			21 52 32	+62 23.3	2	11.61	.054	0.35	.005	0.23	.000	5		49,8096
		NGC 7160 - 19			21 52 34	+62 19.7	2	12.27	.019	0.39	.000	0.25	.019	4		49,8096
	+61 02218	NGC 7160 - 7			21 52 35	+62 22.9	4	10.04	.012	0.15	.007	-0.49	.023	9	B5 V	5,49,213,8096
		NGC 7160 - 46			21 52 35	+62 26.7	1	12.73		0.67		0.12		4		8096
		LP 874 - 65			21 52 35	-22 34.4	1	13.35		0.49		-0.15		2		1696
		L 82 - 11			21 52 36	-71 22.	2	13.03	.005	0.93	.000	0.66	.005	7		1696,1773
		L 82 - 12			21 52 36	-71 22.	2	13.28	.000	1.12	.000	0.95	.015	5		1696,1773
		NGC 7160 - 24			21 52 38	+62 17.4	2	12.86	.037	0.40	.052	0.25	.009	4		49,8096
	+57 02421				21 52 39	+58 18.8	1	9.56		2.12				2		369
208362	+46 03484				21 52 40	+46 55.0	1	7.46		0.35				2	F0	1724
		NGC 7160 - 51			21 52 40	+62 24.8	1	14.29		0.86		0.31		5		8096
		NGC 7160 - 81			21 52 40	+62 25.2	1	14.17		1.23		1.28		4		8096
208196	-45 14506				21 52 40	-45 29.1	1	7.55		1.10		1.02		5	K1 III	1673
	+46 03485	IDS21508N4701	AB		21 52 41	+47 15.1	1	9.75		1.74				2		369
		NGC 7160 - 9			21 52 41	+62 18.1	2	10.84	.018	0.29	.005	-0.15	.023	5	B9 V	49,8096
		CS 22881 # 17			21 52 41	-39 46.0	2	14.23	.005	0.54	.012	0.15	.003	3		1580,1736
	+60 02310				21 52 42	+61 25.8	1	9.64		0.26		-0.04		4	A0	213
		NGC 7160 - 26			21 52 42	+62 23.5	2	13.05	.026	0.92	.004	0.37	.044	6		49,8096
208330	+28 04232				21 52 43	+29 06.7	1	8.38		1.16		1.04		2	K0 III	1375
208345	+31 04577				21 52 43	+32 06.1	1	7.04		1.57		1.83		4	K5	985
208184	-53 10196				21 52 44	-52 42.1	1	6.49		0.93		0.65		5	G6 III	1628
208149	-58 07911	HR 8362			21 52 44	-58 08.2	2	6.17	.005	0.21	.000	0.13		5	A3mA6-A9	219,2035
		NGC 7160 - 15			21 52 45	+62 18.5	2	11.68	.028	0.52	.019	0.42	.028	4		49,8096
208215	-47 13978				21 52 45	-47 09.9	1	6.52		0.46				4	F3 V	2012
	+48 03544				21 52 48	+49 02.4	1	8.67		0.41				2	F2	1724
208742	+78 00768				21 52 48	+79 18.9	1	6.48		1.62		1.95		2	K5	1502
		L 355 - 29			21 52 48	-45 53.	1	14.19		1.26		0.86		2		3073
		NGC 7160 - 30			21 52 50	+62 21.7	1	14.48		1.53		0.74		1		49
		NGC 7160 - 18			21 52 57	+62 21.5	2	12.16	.025	0.37	.025	0.32	.100	2		49,8096
208285	-31 18541	HR 8365			21 53 01	-30 50.6	3	6.41	.011	0.93	.005			14	G8 III	15,1075,2030
208414	+42 04248				21 53 02	+42 39.9	2	8.89	.001	0.22	.003			28	F0	1724,6011
		G 188 - 30			21 53 03	+32 24.4	2	11.06	.054	0.67	.005	-0.04	.027	5		1658,3026
208413	+42 04249				21 53 03	+42 59.7	1	8.99		0.40				2	F0	1724
	+40 04670				21 53 04	+41 04.5	1	9.63		0.43				2	F5	1724
208348	+10 04659	IDS21506N1025	AB		21 53 05	+10 38.7	1	7.72		0.64		0.15		2	G5	3016
		G 26 - 36			21 53 10	-11 44.4	3	12.27	.017	1.24	.029	1.01	.015	4		202,1620,3056
208501	+55 02644	HR 8371		⋆	21 53 12	+56 22.4	6	5.80	.004	0.72	.008	-0.02	.007	20	B8 Ib	5,15,1012,1079,6001,8015
208442	+35 04675				21 53 13	+35 54.5	1	6.72		1.17		1.15		2	K0	1601
208502	+53 02735				21 53 13	+53 41.8	2	6.93	.010	0.47	.007	0.01	.015	5	F5	848,3016
		VES 919			21 53 14	+47 59.5	1	14.10		1.96		0.40		1		8113
		LP 818 - 62			21 53 15	-17 44.5	1	12.29		0.63		0.00		2		1696
208321	-37 14565	HR 8366			21 53 23	-37 29.5	2	5.45	.005	0.08	.000			7	A3 V	15,2030
208512	+49 03673	LW Cyg		⋆ A	21 53 27	+50 16.1	4	8.84	.085	4.11	.070	9.09	.038	4	R3	1238,8005,8022,8027
		G 264 - 15			21 53 27	+61 45.1	1	9.81		0.84		0.54		1	K2	1658
208473	+34 04567	IDS21513N3436	A		21 53 28	+34 50.5	1	8.28		0.04		-0.05		2	A0	1625
	+41 04310				21 53 28	+41 46.2	1	9.67		0.67				2	G0	1724
208513	+44 03985				21 53 31	+44 42.7	2	7.68	.030	-0.03	.005	-0.22	.010	4	B9	401,1733
208525	+50 03486				21 53 34	+50 46.0	1	9.10		0.05		-0.03		2	Ap Sr	1566
208582	+59 02432	LS III +59 013			21 53 34	+59 45.6	1	8.04		0.30		-0.53		5	B9	213
208629	+67 01375				21 53 34	+67 31.0	1	6.96		0.32		0.05		3	F0	985
208323	-47 13982				21 53 37	-46 43.0	1	7.44		0.39				4	F2 V	2012
		G 232 - 40			21 53 43	+55 54.1	2	11.64	.005	0.54	.009	-0.21	.023	4		979,1658
208402	-22 05794				21 53 43	-22 20.4	1	9.30		0.77		0.29		1	G8 V	796
208563	+52 03063				21 53 48	+53 00.3	1	6.68		1.24		1.24		3	K2 II-III	848
208370	-44 14702				21 53 48	-44 33.1	1	8.99		1.01		0.76		2	G8/K0 III	1730
208606	+60 02318	HR 8374		⋆	21 53 51	+61 18.3	2	6.16	.019	1.60	.000	1.58	.028	6	G8 Ib	1355,3016
208360	-50 13533				21 53 53	-49 56.0	3	7.64	.030	0.68	.018	0.12	.018	7	F5/7 +(G)	158,742,1594
208407	-44 14705				21 54 01	-44 29.9	1	9.76		0.76		0.41		2	K1/2 (III)	1730
208435	-38 14801	HR 8367, BZ Gru			21 54 02	-37 59.1	1	6.18		0.33				4	F2/3 V	2009
208527	+20 05046	HR 8372			21 54 04	+21 00.1	3	6.39	.009	1.61	.085	2.09	.056	15	K5 V	985,1733,7009
	+51 03189	IDS21523N5144	AB		21 54 09	+51 59.1	1	10.21		0.39		0.45		1		8084
208617	+50 03491				21 54 10	+50 51.8	1	9.09		0.18		0.14		2	A0	171
208682	+64 01607	HR 8375		⋆ AB	21 54 12	+65 05.0	3	5.85	.053	-0.06	.042	-0.77	.022	8	B2.5Ve	154,379,1212
	+62 02002				21 54 15	+62 36.3	1	9.69		0.16		-0.40		4	B8	213
		LP 639 - 1			21 54 16	-02 08.9	1	14.64		1.76				8		940
		LP 819 - 1			21 54 17	-17 14.9	1	12.12		1.19		1.08		1		1696
208507	-19 06190				21 54 19	-19 25.7	1	7.50		0.93		0.57		6	G8 III	1657
	+42 04255				21 54 20	+42 36.7	1	10.42		0.43				2	F0	1724
235664	+50 03492				21 54 20	+50 48.8	1	8.97		1.02				4		1724
		LP 819 - 2			21 54 20	-18 14.6	1	11.46		0.78		0.28		2		1696

Table 1 1173

HD	DM	Other Id	N Rem	α_{1950}	δ_{1950}	S	V	σ_V	B–V	σ_{B-V}	U–B	σ_{U-B}	n	Spectrum	References
	−34 15340			21 54 20	−34 32.3	1	10.50		1.21		1.13		2	K5 V	3072
	−51 13128	IDS21511S5128	A	21 54 22	−51 13.0	1	10.36		1.51		0.83		4	M0	3060
	−51 13128	IDS21511S5128	B	21 54 22	−51 13.0	1	14.68		0.16		−0.80		4		3060
	+45 03735			21 54 28	+45 46.5	1	9.05		1.12				2	G5	1724
208630	+46 03499			21 54 29	+47 15.8	1	8.47		0.43				2	F8	1724
208565	+11 04696	HR 8373		21 54 30	+11 50.3	2	5.54		0.05	.003	0.11	.061	10	A2 Vnn	3,1049
	+58 02355			21 54 30	+59 16.0	1	10.62		1.98				2		369
208450	−55 09733	HR 8368	⋆ AB	21 54 32	−55 13.9	5	4.39	.004	0.29	.006	0.08	.014	17	F0 IV+F0 IV	15,1075,2012,2024,3023
		LP 459 - 32		21 54 33	+19 13.0	1	12.86		0.97		0.65		1		1696
	+39 04716			21 54 37	+40 25.8	1	9.51		0.53				2	F8	1724
	−66 03602			21 54 37	−65 48.0	1	10.29		0.74		0.24		2		1696
208609	+16 04634			21 54 39	+17 26.6	1	7.12		1.52		1.74		3	K4 III	1364
		G 26 - 37		21 54 39	−09 33.6	1	14.24		1.52				2		202
	+17 04663			21 54 40	+17 57.1	2	9.34	.024	1.08	.028	0.94		2	K2	694,1364
	+61 02222			21 54 40	+62 19.8	1	9.66		0.28		−0.07		4	A0	213
208685	+51 03194			21 54 46	+52 17.4	1	8.61		0.02		−0.40		2	A0	401
		CS 22951 # 77		21 54 47	−43 22.4	2	13.60	.010	0.51	.011	−0.03	.005	3		1580,1736
	+46 03502			21 54 48	+46 51.5	1	9.18		0.44				2	F2	1724
		LDS 7661		21 54 48	−43 42.	2	14.82	.215	−0.10	.030	−0.91	.060	2	DB	832,3065
		LDS 7662		21 54 48	−43 42.	1	14.54		1.54		0.91		1		832
208761	+62 02006			21 54 49	+62 39.4	4	7.64	.018	0.05	.004	−0.75	.004	10	B3 V	27,213,401,555
208744	+59 02436	IDS21533N5918	AB	21 54 52	+59 33.4	1	6.89		0.09		0.07		6	A0	213
235668	+50 03496	LS III +51 030		21 54 54	+51 20.3	1	8.13		0.33		−0.47		3	B2	1212
		AAS11,365 T7# 1		21 54 54	+53 53.	1	10.09		1.35		0.62		3		261
		G 18 - 8		21 54 56	+07 53.8	4	11.02	.006	1.45	.021	1.13	.010	5	dM1	202,333,801,1620,1632
	+30 04565			21 54 57	+30 49.7	1	10.22		0.69		0.21		3		272
		CS 22881 # 7		21 54 57	−41 43.1	1	15.21		−0.31		−1.05		1		1736
239805	+55 02653			21 54 58	+56 25.4	1	10.23		0.56		0.03		3	G7	261
208496	−59 07744	HR 8369, BG Ind		21 54 59	−59 15.1	4	6.15	.039	0.47	.005	−0.04		12	F5 V	15,219,1075,2007
208612	−9 05876			21 55 00	−08 48.2	1	6.85		−0.04		−0.21		5	B9	1628
208686	+34 04571			21 55 03	+35 06.6	1	8.80		0.06		0.03		2	A0	1733
208726	+55 02652	IDS21533N5529	AB	21 55 03	+55 43.6	1	9.07		0.33		0.20		3	A3	261
208728	+45 03741			21 55 07	+46 21.2	1	6.78		1.19		0.86		6	K0 III	245
	+47 03617			21 55 07	+48 08.8	1	9.51		0.84				2	G0	1724
208727	+47 03618	HR 8377		21 55 07	+48 25.8	2	6.44	.029	−0.09	.010	−0.39	.010	3	B8 V	351,401
209111	+79 00721			21 55 13	+80 04.3	1	6.46		1.56		1.57		3	M2	985
208816	+62 02007	HR 8383, VV Cep	⋆ AB	21 55 14	+63 23.2	6	4.94	.060	1.75	.022	0.38	.017	12	M2 Iaep +	5,15,1363,8003,8015,8032
208702	+19 04833			21 55 18	+20 03.8	1	7.25		0.03		0.00		2	A0	1364
		L 213 - 75		21 55 18	−58 12.	1	14.11		1.55		1.18		2		3073
208603	−45 14521			21 55 21	−44 59.7	2	8.28	.005	0.39	.015	−0.04		3	F5 V	742,1594
		CS 22881 # 10		21 55 22	−41 07.8	2	14.79	.010	0.40	.007	−0.23	.009	3		1580,1736
208625	−42 15711			21 55 26	−41 57.6	1	6.76		1.58				4	K5 III	2012
208627	−44 14717			21 55 28	−44 18.1	2	6.54	.000	0.90	.000			8	G8 IV	1075,2012
	+19 04836			21 55 29	+20 16.5	2	9.79	.015	1.74	.020	2.17		3		694,1364
208786	+47 03622			21 55 32	+47 48.0	2	8.33	.000	−0.05	.003	−0.23	.003	3	A0	401,623
		VES 920		21 55 32	+53 59.2	1	12.75		0.79		−0.16		2		8113
	+54 02652			21 55 34	+55 16.8	1	10.30		0.49		0.14		3		261
208750	+26 04316			21 55 35	+26 58.6	1	8.55		0.70		0.32		3	G0 IV	3016
	+46 03506			21 55 35	+47 21.8	1	8.78		1.01				2	G5	1724
208904	+64 01611			21 55 35	+65 21.3	2	7.55	.005	−0.01	.000	−0.60	.000	6	B3 V	399,555
208703	−6 05878	HR 8376		21 55 36	−05 39.8	3	6.33	.008	0.37	.004	−0.03	.015	10	F5 V	15,1256,3016
208787	+42 04256			21 55 37	+42 58.7	2	7.90	.007	0.53	.000			33	G0	1724,6011
	+48 03558			21 55 42	+49 23.3	1	8.99		0.44				2	F5	1724
208704	−13 06060			21 55 43	−12 54.3	1	7.21		0.62		0.12		2	G5/6 V	3026
235672	+50 03502			21 55 45	+50 36.4	1	8.88		0.56				2		1724
208905	+60 02320	LS III +61 005	⋆ AB	21 55 46	+61 03.4	3	6.99	.004	0.09	.000	−0.73	.008	7	B1 Vp:	5,213,1012
235673	+52 03071	LS III +52 016		21 55 49	+52 34.9	2	9.14	.000	0.21	.000	−0.77	.000	4	O7	1011,1012
		PKS 2155-304 # 4		21 55 49	−30 24.8	1	14.28		0.65		0.15		3		1687
		CS 22881 # 6		21 55 49	−41 50.7	1	14.98		0.41		−0.12		1		1736
	+62 02009			21 55 53	+63 00.9	1	9.19		0.10		−0.48		2	B8	27
208835	+46 03512			21 55 54	+46 37.5	1	7.50		−0.04		−0.35		2	A0	401
208947	+65 01691	HR 8384		21 55 54	+65 55.0	2	6.43	.030	−0.05	.005	−0.68	.005	6	B2 V	154,379
	−60 07528			21 55 56	−59 59.4	4	9.74	.023	1.49	.013	1.18	.015	8		158,1494,1705,3072
	−60 07821			21 55 56	−59 59.4	1	9.80		1.49		1.19		2		3073
208735	−21 06131	HR 8378		21 55 57	−21 25.3	3	6.11	.010	1.64	.017	1.83	.028	13	M4 III	1024,2035,3005
208776	+3 04644			21 55 58	+03 32.4	2	6.97	.017	0.58	.011	0.08	.000	6	G0 V	1003,3026
		PKS 2155-304 # 5		21 55 58	−30 27.	1	15.35		0.66		−0.01		4		1687
		CS 22881 # 20		21 55 58	−38 50.9	2	14.09	.000	−0.03	.000	−0.03	.012	2		966,1736
208837	+38 04643			21 56 00	+38 41.3	1	7.67		−0.10		−0.64		3	B5 V	833
208861	+42 04257			21 56 00	+42 54.6	1	7.79		0.12				26	A2	6011
208893	+50 03504			21 56 00	+50 47.9	1	8.91		0.06		0.04		2	A0	1566
		AJ89,1229 # 303		21 56 02	+09 00.8	1	13.53		1.45				2		1532
	+18 04897			21 56 02	+18 56.3	1	10.39		1.36		1.59		1	K0	1364
	+44 03999			21 56 03	+44 44.1	1	9.53		0.42				2	F2	1724
235674	+50 03503	G 215 - 26		21 56 03	+50 36.3	2	8.86	.005	0.74	.005	0.36		3	G5	1658,1724
		LS III +55 019		21 56 06	+55 58.1	2	10.64	.000	0.63	.000	−0.40	.010	4	F5 Ib	405,483
208756	−30 18948	PKS 2155-304 # 1		21 56 08	−30 23.9	1	9.17		0.40		−0.02		3	F0	1687
208817	+14 04705			21 56 09	+15 04.1	1	7.75		1.00		0.73		1	K0	1364
208878	+42 04260			21 56 09	+43 00.0	1	7.43		−0.05		−0.34		2	B9	401
	+47 03630			21 56 09	+47 45.5	1	9.71		0.14		−0.01		1	A0	623

HD	DM	Other Id	N Rem	α_{1950}	δ_{1950}	S	V	σ_V	B–V	σ_{B-V}	U–B	σ_{U-B}	n	Spectrum	References
		PKS 2155-304 # 2		21 56 09	−30 25.1	1	12.05		0.69		0.15		4		1687
208710	−46 14147			21 56 11	−46 35.0	1	7.58		1.24				4	K2 III	2012
	+0 04801	G 26 - 38		21 56 12	+00 34.3	2	10.25	.034	0.80	.005	0.40		3	K0	202,333,1620
		G 26 - 39		21 56 12	−04 00.9	1	14.63		1.67				2		202
		PKS 2155-304 # 3		21 56 12	−30 25.2	1	13.00		0.90		0.58		4		1687
	+44 04003			21 56 13	+45 14.5	1	9.30		0.57				3	F8	1724
208931	+50 03507			21 56 15	+51 22.9	1	8.82		0.08		0.11		4	A0	171
208737	−38 14820	HR 8379		21 56 17	−38 38.1	3	5.50	.004	0.99	.005			11	K0 III	15,1075,2032
208654	−61 06607			21 56 17	−61 00.3	1	7.51		1.50		1.84		8	K3 III	1704
		Smethells 98		21 56 18	−62 19.	1	11.97		1.42				1		1494
208960	+60 02321	IR Cep	★ AB	21 56 19	+60 46.8	2	7.61	.003	0.75	.072	0.45	.032	2	G0	592,1772
208801	−5 05674	HR 8382	★	21 56 19	−04 36.6	4	6.23	.012	1.00	.003	0.85	.025	13	K2 V	15,1071,1509,3077
	+56 02666			21 56 21	+56 58.2	1	10.14		0.62		0.13		3		261
208896	+33 04395			21 56 22	+34 25.2	2	9.07	.034	1.04	.000	0.89		3	K2	1017,3072
208906	+29 04550	IDS21543N2921	A	21 56 28	+29 34.7	4	6.96	.009	0.50	.008	−0.12	.011	8	F8 V-VI	22,792,908,3026
		VES 921		21 56 28	+54 21.2	1	10.81		0.27		−0.54		2		8113
		CS 22881 # 8		21 56 28	−41 38.6	2	14.30	.010	0.82	.004	0.24	.007	3		1580,1736
208808	−23 17199			21 56 29	−23 06.7	1	7.37		0.42		0.00		2	F5 V	3037
		AJ89,1229 # 537		21 56 30	+02 08.	1	10.66		1.57				2		1532
		G 215 - 27		21 56 33	+42 34.1	2	14.26	.014	1.08	.005	0.84	.036	5		419,1658
208940	+43 04098	IDS21546N4413	A	21 56 34	+44 27.2	1	8.55		0.19				2	F0	1724
	+54 02655			21 56 35	+54 43.8	1	10.43		0.20		−0.04		3		261
208961	+55 02658	IDS21549N5547	AB	21 56 35	+56 01.4	1	8.94		0.47		0.38		7	A2	1655
208880	+2 04453	G 18 - 10		21 56 36	+02 57.4	2	8.41	.028	0.75	.005	0.32		4	G0	202,333,1620
208897	+18 04899			21 56 37	+18 46.9	1	6.51		1.01		0.90			K0	1648
		LS III +52 017		21 56 39	+52 41.0	1	12.25		0.43		−0.33		2		483
	+47 03636			21 56 40	+48 21.2	1	9.12		1.24				2	G0	1724
		G 26 - 40		21 56 42	−04 19.6	2	14.16	.083	1.58	.058	1.45		3		202,3062
235679	+53 02749	LS III +54 008		21 56 43	+54 14.7	1	8.86		0.35		−0.55		2	B2 Ia:	1012
208500	−78 01430	HR 8370		21 56 44	−77 54.2	3	6.39	.007	0.22	.000	0.07	.010	15	A5 IV/V	15,1075,2038
	+40 04689			21 56 45	+41 13.6	1	9.56		1.10				2	G5	1724
208812	−44 14727			21 56 45	−43 43.1	2	8.14	.024	0.49	.009	−0.04		8	F7 V	158,1075
		G 18 - 11		21 56 49	+06 05.9	2	11.59	.005	1.39	.005	1.18		3	M2	202,333,1620
208962	+41 04333			21 56 52	+41 27.4	1	7.60		−0.06		−0.48		2	B9	401
		AJ89,1229 # 538		21 56 54	+04 27.	1	12.57		1.51				1		1532
	+16 04641			21 56 56	+16 39.4	1	9.92		0.29		0.22		2	A5	1364
	+27 04219			21 56 57	+28 24.4	1	8.93		0.59		0.10		4	G0 V	3026
		LS III +52 019		21 56 59	+52 40.6	1	12.04		0.44		−0.49		3		483
208973	+32 04310			21 57 00	+33 23.5	2	8.22	.000	−0.10	.005	−0.71	.009	4	B2 V	399,963
	+47 03639	MR Cyg		21 57 00	+47 44.6	1	8.95		0.06		−0.54		2	B3 V	401
208796	−56 09784	HR 8381		21 57 00	−56 07.4	4	6.01	.008	−0.08	.015	−0.12		12	B9 IV/V	15,1034,1075,2035
208963	+14 04710			21 57 06	+15 25.3	1	8.21		−0.06		−0.43		5	A0	1364
		JL 94		21 57 06	−76 07.	1	13.03		−0.23		−1.11		1		832
		G 214 - 5		21 57 07	+40 48.2	1	11.52		0.53		−0.19		2	sdF8	1658
	+13 04818			21 57 10	+14 06.8	1	8.93		1.18		1.13		2	K2	1648
		GD 272		21 57 10	+16 11.2	1	16.19		−0.04		−0.81		1	DA	3060
239817	+58 02366			21 57 10	+58 51.1	1	9.49		0.53		−0.09		2	F8	3016
		CS 22881 # 21		21 57 10	−38 40.4	2	14.82	.000	−0.02	.000	0.00	.012	2		966,1736
	+40 04692			21 57 12	+40 50.8	1	9.33		0.88				2	G0	1724
		CS 22951 # 117		21 57 12	−44 05.8	2	13.85	.010	0.42	.008	−0.12	.008	3		1580,1736
	+43 04101			21 57 13	+44 06.2	1	10.80		0.14		−0.05		3		627
	+16 04644			21 57 18	+17 07.7	1	10.16		0.55		0.15		1	G0	1364
		AAS11,365 T7# 7		21 57 18	+53 34.	1	10.14		1.82		1.95		3		261
	+48 03565			21 57 19	+48 35.3	1	9.31		0.42				2	F2	1724
		G 26 - 41		21 57 21	−00 54.0	3	13.23	.023	0.78	.008	0.17	.020	7		202,1620,1658
208638	−76 01541			21 57 21	−75 55.2	1	6.66		1.43				4	K3 III	2012
209258	+74 00946	HR 8395		21 57 23	+74 45.4	2	6.35	.005	1.57	.010	1.61	.005	5	K5	985,1733
	+41 04337			21 57 24	+42 14.1	1	9.56		0.26				2	A5	1724
	+25 04655	IS Peg		21 57 25	+26 11.6	1	9.69		−0.26		−1.16		1		963
209112	+62 02010	HR 8388		21 57 25	+62 27.5	3	5.90	.028	1.66	.004	1.81		10	M3 IIIab	71,369,3001
209162	+62 02012			21 57 37	+63 01.4	1	8.62		0.15		−0.11		2	B8	27
209008	+6 04940	HR 8385		21 57 38	+06 28.6	4	6.00	.008	−0.12	.014	−0.57	.013	11	B3 III	15,154,1256,1423
209178	+63 01794	IDS21562N6332	AB	21 57 39	+63 45.5	1	8.69		0.24		−0.45		3	B8	555
		G 18 - 12		21 57 42	+08 07.9	1	10.97		0.77		0.33		1		333,1620
209124	+56 02670	HR 8389		21 57 42	+57 25.1	1	6.59		0.03		−0.02		2	A0 III-IV	401
239820	+58 02368			21 57 45	+59 12.5	1	9.71		0.32		−0.47		3	B8	213
209145	+59 02443	LS III +60 010		21 57 45	+60 03.5	2	7.61	.004	0.32	.004	−0.52	.007	12	B1 V	213,1012
	+41 04341			21 57 50	+41 41.9	1	9.23		0.48				2	F8	1724
		G 26 - 42		21 57 51	+01 26.9	3	12.55	.022	0.50	.011	−0.21	.015	4		1620,1658,3029
209113	+44 04012			21 57 53	+44 43.8	1	8.36		0.21		0.18		1	A2	379
	+70 01208			21 57 53	+70 38.7	1	9.17		0.42		−0.05		2	F8	1733
		Smethells 99		21 57 54	−52 00.	1	10.78		1.27				1		1494
	+38 04653			21 57 55	+38 59.5	1	8.94		0.99		0.64		2	G5	1733
209014	−29 18119	HR 8386	★ AB	21 57 58	−28 41.7	11	5.42	.007	−0.10	.009	−0.30	.020	40	B8/9 V+B8/9	15,404,681,815,1079*
		G 263 - 10		21 57 59	+75 20.9	1	10.10		1.47				2	M1	1619
	−3 05357	FF Aqr		21 58 01	−02 58.9	2	9.33	.008	0.87	.003	−0.01	.006	18	G8 III+sdOB	963,1732
	+43 04105			21 58 02	+43 57.1	1	9.46		0.29				2	A5	1724
	+46 03527			21 58 02	+46 43.5	1	9.21		1.06				2	G5	1724
	+47 03649			21 58 03	+47 29.5	1	9.79		1.73				2	M0	369
		LP 819 - 13		21 58 05	−15 22.3	1	11.69		0.47		−0.18		2		1696

Table 1 1175

HD	DM	Other Id	N	Rem	α_1950	δ_1950	S	V	σ_V	B–V	σ_B-V	U–B	σ_U-B	n	Spectrum	References
209163	+49 03711				21 58 06	+50 17.5	1	9.05		0.10				2	A5	1724
	+41 04345				21 58 11	+42 26.8	1	9.88		0.65				2	G0	1724
209215	+60 02324				21 58 11	+60 51.6	1	8.90		0.15		0.12		1	A2	592
		G 26 - 43			21 58 11	−03 51.4	3	13.89	.020	1.61	.032	1.19	.112	6		202,203,1620
	+44 04014	LS III +45 074			21 58 12	+45 21.6	1	9.49		0.06		−0.40		2	B5	379
209149	+32 04316	HR 8391			21 58 14	+32 45.9	1	6.47		0.42		0.00		2	F5 III	1733
		WLS 2200 65 # 5			21 58 14	+66 57.0	1	12.06		0.69		0.17		2		1375
208741	−76 01542	HR 8380		★ A	21 58 14	−76 21.5	4	5.93	.011	0.40	.003	0.10	.010	18	F3 III	15,1075,2020,2038
		ApJ86,509 R1# 1			21 58 18	+56 29.7	1	15.36		3.07				1		8008
208998	−53 10224				21 58 18	−53 19.7	4	7.10	.028	0.57	.011	0.02	.014	11	G0 V	158,462,1311,3077
	+47 03650				21 58 19	+47 41.2	1	9.59		0.45				2	F5	1724
209217	+56 02675				21 58 21	+56 28.5	1	8.24		0.44		0.27		2	A2	379
	+42 04267	BG Lac			21 58 23	+43 12.3	2	8.58	.010	0.82	.039	0.55		2	G0	851,6011
209068	−25 15713				21 58 23	−24 50.7	1	10.01		0.39				2	F6 Vw	1594
208605	−81 00982				21 58 24	−80 52.0	1	9.14		0.42		0.00		1	F5 V	832
	+9 04955	G 18 - 14			21 58 26	+09 42.7	3	10.49	.017	0.98	.005	0.77	.005	4	K4	202,333,1620,3025
209128	−0 04296	HR 8390			21 58 32	+00 21.8	2	5.57	.005	1.28	.000	1.40	.000	7	K0	15,1071
209369	+72 01009	HR 8400		★ A	21 58 33	+72 56.5	2	5.03	.000	0.44	.001	−0.03		5	F5 V	1363,3026
209083	−36 15141				21 58 33	−36 18.0	2	8.87	.000	0.38	.005	−0.05		3	F2 V	742,1594
209166	+12 04737	HR 8392		★ A	21 58 39	+12 52.8	2	5.62	.000	0.34	.005	0.05	.014	7	F4 III	254,3016
	+41 04347				21 58 39	+42 03.3	1	9.40		0.44				2	F5	1724
209220	+43 04109				21 58 39	+43 47.7	1	8.58		0.34				2	F8	1724
209167	+7 04779	HR 8393			21 58 40	+08 01.0	1	5.64		1.44		1.76		8	K2	1088
		CS 22892 # 14			21 58 40	−14 31.3	1	14.32		0.44		−0.17		1		1736
209205	+30 04586				21 58 41	+31 17.6	1	7.78		0.04		0.02		2	A0	1733
239825	+55 02665				21 58 41	+55 47.6	1	9.67		1.72		1.75		3	K7	261
		AAS11,365 T7# 10			21 58 42	+53 17.	1	10.55		0.27		0.12		3		261
		LS III +53 026			21 58 46	+53 23.2	1	11.63		1.01		−0.02		2		483
	+54 02666				21 58 46	+54 59.4	1	9.88		0.25		0.13		3		261
209133	−25 15718				21 58 51	−25 06.4	1	8.34		0.36				4	A9 V	2012
209246	+42 04270	IDS21568N4310	AB		21 58 54	+43 23.7	2	7.65	.005	0.30	.005	0.10		5	A5	627,1724
	+15 04543				21 58 56	+16 00.6	1	8.99		1.69		2.05		13	M0	1364
209296	+56 02676				21 58 57	+56 28.2	3	8.23	.070	0.27	.004	−0.23	.011	7	B6:V:n	379,1012,1212
209287	+46 03531				21 58 58	+47 23.9	1	8.64		0.45				2	G5	1724
209285	+50 03522				21 58 58	+50 59.7	1	8.74		0.03		−0.15		3	A0	1566
	+54 02667				21 58 59	+54 40.6	1	9.50		1.83				2		369
209260	+38 04656	IDS21569N3846	A		21 59 00	+39 00.3	1	7.13		−0.02		−0.29		2	A0	105
		JL 96			21 59 00	−73 53.	1	14.46		0.04		−0.92		2		132
	+47 03656	LS III +47 057			21 59 05	+47 56.6	1	8.78		0.45		−0.39		3	B5	1212
209308	+53 02766				21 59 08	+53 55.1	1	9.11		−0.02		−0.23		4	B9p	171
209339	+61 02233	HR 8399		★ AB	21 59 09	+62 14.8	5	6.65	.009	0.07	.013	−0.82	.007	13	B0 IV	5,154,213,1203,1476
		LP 819 - 17			21 59 12	−19 43.5	1	12.03		1.59				1	M3	1705
	+62 02015				21 59 13	+62 42.0	1	9.60		0.15		0.04		2		27
		AAS11,365 T7# 11			21 59 18	+57 02.	1	11.15		1.38		1.24		1		261
		TZ Aqr			21 59 18	−05 50.5	1	11.98		0.26		0.04		1	F0	699
		G 126 - 49			21 59 24	+16 13.4	2	10.64	.024	1.55	.024	1.21	.052	2		1364,1696
		CS 22881 # 31			21 59 25	−41 29.0	1	15.81		−0.35		−1.17		1		1736
209885	+82 00672				21 59 27	+83 15.1	1	7.79		0.26		0.09		2	F0	1733
209240	−18 06056	HR 8394			21 59 27	−18 08.7	1	6.28		1.01				4	K0 III	2007
209329	+47 03660				21 59 32	+48 03.4	1	8.26		0.36				2	F5 III	1724
	+15 04547				21 59 33	+16 05.5	1	9.55		1.56		1.96		2	K2	1364
209100	−57 10015	HR 8387			21 59 33	−56 59.6	7	4.69	.012	1.06	.007	0.99	.005	26	K4/5 V	15,1075,2012,2024*
209288	+10 04676	HR 8397			21 59 34	+10 43.9	2			−0.10	.014	−0.62	.005	4	B5 IIIn	1049,1079
	+41 04350				21 59 34	+42 26.4	1	9.27		0.44				2	F5	1724
	+56 02679				21 59 37	+56 35.5	1	9.76		0.25				3	A1	1025
	+57 02440				21 59 37	+57 42.9	1	10.14		0.15				5	B9	1025
209342	+40 04704				21 59 38	+41 17.9	1	6.98		1.22		1.16		2	K2	1733
209253	−32 16868				21 59 38	−32 22.5	1	6.66		0.50		−0.04		9	F6/7 V	1628
209290	+0 04810	G 18 - 16			21 59 39	+01 09.7	6	9.17	.011	1.46	.015	1.22	.026	15	M0.5V	202,680,1013,1197,1705,3072
	+48 03576				21 59 39	+49 06.6	1	9.51		0.29				2		1724
	+48 03576	IDS21577N4853	C		21 59 39	+49 06.6	1	10.80		0.56		0.07		1		8084
	+48 03576	IDS21577N4853	D		21 59 39	+49 06.6	1	12.61		0.83		0.34		1		8084
209330	+38 04659				21 59 40	+38 57.8	1	8.38		0.02		−0.06		3	A0 V	833
239828	+58 02373	LS III +59 017			21 59 42	+59 15.6	2	9.36	.013	0.91	.004	−0.02	.004	5	B5 Ia	27,1012
		LS III +53 027			21 59 43	+53 51.8	1	11.84		0.63		−0.22		3		483
209452	+65 01704				21 59 43	+65 50.5	1	8.27		0.15		−0.37		3	B8	555
209278	−17 06422	HR 8396, DX Aqr		★ AB	21 59 43	−17 12.4	1	6.39		0.42				4	K0 III+A2 V	2007
	+54 02670				21 59 44	+54 32.9	1	9.84		1.78				2		369
209277	−16 06003				21 59 45	−16 25.7	1	9.05		0.43		−0.04		24	F3 V	1699
	+41 04353				21 59 46	+41 46.9	1	9.73		0.51				2	F5	1724
235697	+50 03525				21 59 48	+50 51.2	1	9.87		0.39				3		1724
		L 48 - 15			21 59 48	−75 28.	1	15.06		0.16		−0.63		4		3065
	+48 03579				21 59 49	+49 20.8	1	8.47		0.51				2	F8	1724
209391	+53 02768				21 59 49	+53 33.3	1	8.44		0.08		−0.39		2	A0	1566
	+42 04273				21 59 51	+43 06.9	2	8.65	.010	0.66	.015	0.39		3	F5	851,1724
		CS 22881 # 28			21 59 51	−41 08.6	2	14.30	.010	0.39	.007	−0.21	.009	3		1580,1736
209437	+59 02448				21 59 52	+60 20.3	1	9.09		0.24		−0.28		5	F8	213
209454	+60 02329				21 59 53	+61 18.9	3	7.78	.010	0.16	.007	−0.65	.005	7	B2 V	5,213,399
209392	+47 03665				21 59 55	+48 19.9	1	8.17		0.52				2	F8	1724
		L 118 - 273			21 59 55	−70 09.9	1	10.97		1.38				1		1705

HD	DM	Other Id	N Rem	α_{1950}	δ_{1950}	S	V	σ_V	B–V	σ_{B-V}	U–B	σ_{U-B}	n	Spectrum	References
		G 26 - 45		21 59 56	−00 31.5	1	14.29		1.52				2		202
		CS 22881 # 32		21 59 57	−41 28.9	1	15.23		0.37		-0.20		1		1736
	+42 04275			21 59 58	+42 53.4	2	9.15	.005	0.57	.019	0.06		3	F8	851,1724
209346	+18 04913			21 59 59	+18 35.5	1	8.26		0.26		0.09		1	A2	1364
209321	−1 04236			21 59 59	−01 09.6	1	7.80		1.25		1.37		6	K2	1657
209419	+52 03083	HR 8403		22 00 00	+52 38.4	2	5.78	.005	-0.11	.010	-0.50	.015	6	B5 III	154,171
		L 118 - 272		22 00 00	−70 09.8	1	13.70		1.61				1		1705
209393	+43 04116			22 00 04	+44 06.1	1	7.97		0.68				4	G5	1724
		Cep sq 1 # 122		22 00 05	+60 18.9	1	10.55		0.42		-0.01		2		27
209610	+74 00947			22 00 06	+74 49.9	1	8.75		0.10		0.07		5	A0	1733
	+46 03542			22 00 07	+46 49.9	1	9.86		0.39				2	A5	1724
209942	+82 00673	HR 8423	⋆ A	22 00 07	+82 37.8	4	7.00	.033	0.52	.005	-0.01	.009	8	F5	1,15,1028,3016
209357	+17 04677			22 00 10	+18 11.8	1	8.61		0.65		0.08		1	K0	1364
	+45 03767			22 00 10	+45 44.6	1	8.88		0.39				2	F5	1724
		CS 22951 # 108		22 00 10	−46 17.2	2	14.29	.010	0.47	.010	-0.25	.006	3		1580,1736
		Smethells 100		22 00 12	−50 53.	1	12.09		1.57				1		1494
209943	+82 00674	V376 Cep	⋆ B	22 00 13	+82 37.9	3	7.53	.056	0.70	.002	0.18	.010	6	F5	1,1028,3016
209380	+15 04548			22 00 14	+15 44.7	1	6.69		-0.05		-0.11		18	A0	1364
	+42 04277			22 00 18	+42 36.8	1	9.37		0.43				3	F2	1724
		CS 22892 # 11		22 00 20	−15 16.2	1	14.13		0.36		0.11		1		1736
	+40 04705			22 00 22	+40 57.8	1	9.27		0.40				3	F2	1724
		WLS 2200 65 # 7		22 00 23	+63 08.8	1	11.42		0.37		0.32		2		1375
209336	−32 16875	TT PsA		22 00 23	−31 41.2	2	7.29	.072	1.54	.046	1.40		13	M5 III	2012,3042
209481	+57 02441	HR 8406, LZ Cep		22 00 24	+57 45.5	8	5.56	.021	0.06	.011	-0.86	.007	28	O9 V	5,15,154,1011,1025*
209480	+57 02442			22 00 24	+57 48.5	2	7.57	.025	-0.04	.010	-0.48		4		338,1025
209335	−30 18975	HR 8398		22 00 24	−30 08.8	1	7.10		0.32				4	F2 V	2032
		MtW 66 # 125		22 00 26	+30 20.0	1	14.64		0.65		0.08		4		397
		LP 699 - 74		22 00 26	−07 22.6	1	11.02		1.02		0.86		1		1696
209293	−51 13155			22 00 27	−51 26.8	2	9.98	.005	0.63	.000	0.08	.000	6	G6 V	158,1696
	+54 02672		V	22 00 29	+54 28.4	1	8.92		1.85				2	M2	369
	+40 04706			22 00 30	+40 59.9	1	9.73		0.50				2	F5	1724
		AAS11,365 T7# 12		22 00 30	+56 43.	1	12.85		0.46		0.26		1		261
		MtW 66 # 151		22 00 31	+30 18.4	1	11.23		1.22		1.12		6		397
	+29 04566			22 00 35	+30 22.6	1	9.04		1.05		0.83		6	K0	397
		MtW 66 # 183		22 00 36	+30 19.6	1	16.39		0.72		-0.06		4		397
	−17 06424	U Aqr		22 00 36	−16 52.2	2	11.09	.161	0.97	.029	0.31	.047	2		1589,1699
235701	+50 03531			22 00 37	+50 43.4	1	9.63		0.41				2		1724
239833	+57 02376	Tr 37 - 1044		22 00 37	+59 25.1	1	9.74		0.30		-0.38		3	B2.5Vpnne	499
209514	+55 02669			22 00 38	+55 42.3	1	7.24		0.24				2	B8	1025
	+56 02684			22 00 38	+56 36.8	1	9.81		0.41				2	A5	1025
		G 27 - 8		22 00 38	−01 27.7	3	11.39	.007	0.52	.013	-0.19	.012	5		1620,1658,1696
		MtW 66 # 199		22 00 39	+30 19.6	1	15.10		0.88		0.48		5		397
		BL Lac # 1		22 00 39	+42 02.	1	13.31		0.74		0.44		1		334
		BL Lac # 14		22 00 39	+42 02.	1	14.65		0.85		0.23		1		334
		BL Lac # 15		22 00 39	+42 02.	1	15.29		1.02		0.54		1		334
		BL Lac # 2		22 00 39	+42 02.	1	13.72		0.97		0.60		1		334
		BL Lac # 27		22 00 39	+42 02.	1	13.39		0.61		0.33		1		334
		BL Lac # 28		22 00 39	+42 02.	2	12.81	.048	1.72	.031	1.79	.061	6		334,1595
		BL Lac # 29		22 00 39	+42 02.	2	14.21	.031	0.91	.013	0.43	.017	6		334,1595
		BL Lac # 3		22 00 39	+42 02.	1	14.50		1.06		0.71		1		334
		BL Lac # 32		22 00 39	+42 02.	1	13.36		1.04		0.65		1		334
		BL Lac # 34		22 00 39	+42 02.	1	14.31		1.37		0.96		4		1595
		BL Lac # 37		22 00 39	+42 02.	1	15.44		0.82		0.22		4		1595
		BL Lac # 9		22 00 39	+42 02.	1	15.79		1.08		0.40		3		551
209482	+45 03771			22 00 39	+45 58.8	1	8.86		0.17		-0.11		1		379
209396	−7 05688	HR 8401		22 00 39	−06 45.9	2	5.53	.005	0.96	.000	0.74	.005	7	K0 III	15,1061
209469	+42 04280			22 00 41	+42 34.3	1	7.22		-0.05		-0.32		2	B9	401
209557	+60 02330	IDS21592N6037	AB	22 00 44	+60 51.2	2	7.91	.012	-0.01	.012	-0.25	.012	8	A0	213,592
209409	−2 05681	HR 8402, o Aqr		22 00 44	−02 23.9	10	4.71	.017	-0.10	.020	-0.40	.011	41	B7 IVe	15,154,681,815,2024*
	+15 04552			22 00 51	+16 22.1	1	10.71		0.76		0.29		1		1364
209459	+10 04681	HR 8404		22 00 52	+11 08.7	1			-0.07		-0.20		2	B9.5V	1049
		LS III +55 020		22 00 54	+55 00.6	2	11.09	.020	0.75	.020	-0.26	.005	5		405,483
		CS 22892 # 17		22 00 54	−13 41.4	1	15.36		-0.19		-0.94		1		1736
	−35 15116			22 00 54	−34 45.3	1	10.92		0.81		0.42		2		1696
		G 18 - 20		22 00 55	+12 06.3	1	10.83		1.29		1.30		1	K7	333,1620
209515	+43 04119	HR 8407, V1942 Cyg	⋆ AB	22 00 55	+44 24.5	5	5.60	.006	-0.03	.010	-0.10	.004	21	A0 IV	15,1007,1013,3033,8015
		AAS60,495 # 9		22 00 55	−18 34.8	1	13.60		1.10				1		1584
209295	−65 03989			22 00 55	−64 58.2	1	7.32		0.26		0.07		10	A9/F0 V	1628
	+56 02686			22 00 57	+56 35.4	1	9.71		0.24				3	A4	1025
		LP 983 - 80		22 00 57	−36 15.7	1	11.37		1.09		1.00		1		1696
	−35 15117	IDS21580S3502	AB	22 00 59	−34 47.7	1	10.29		0.71		0.17		2	G6	1696
	+62 02020			22 01 00	+62 51.0	1	9.53		0.04		-0.57		2		27
		G 26 - 46		22 01 00	−05 35.2	2	13.93	.044	1.53	.010	1.00		3		202,3062
		L 165 - 99		22 01 00	−63 25.	1	11.75		0.45		-0.19		2		1696
239835	+56 02687			22 01 05	+56 53.7	1	9.39	.010	0.30	.010	0.19		7	A2	261,1025
	+57 02444			22 01 06	+57 58.8	1	10.41		0.49				4		1025
		TOU28,33 T8# 6		22 01 17	+55 54.4	1	10.03		0.33				4		1025
	+16 04655			22 01 18	+16 29.0	1	9.58		0.43		0.01		1	F5	1364
		G 232 - 50	A	22 01 20	+59 23.6	1	13.93		1.45				4		538
		G 232 - 50	AB	22 01 20	+59 23.6	1	13.30		1.46				1		906

Table 1 1177

HD	DM	Other Id	N	Rem	α_{1950}	δ_{1950}	S	V	σ_V	B–V	σ_{B-V}	U–B	σ_{U-B}	n	Spectrum	References
		G 232 - 50	B		22 01 20	+59 23.6	1	14.18		1.45				4		538
209675	+67 01386	IDS22001N6746	A		22 01 21	+68 00.9	1	8.74		0.92		0.57		2	G5	3016
		AAS11,365 T7# 14			22 01 24	+54 22.	1	10.10		0.94		0.44		3		261
209596			A		22 01 27	+45 19.6	1	10.18		2.40		4.05		1	N0	414
209596			B		22 01 27	+45 19.6	1	12.96		0.85		0.39		1		414
	+54 02676	LS III +54 010			22 01 28	+54 50.8	2	10.40	.075	0.46	.000	-0.61	.160	4	B1:V:enn	1012,8113
209691	+65 01712				22 01 29	+65 49.3	1	6.72		0.14		-0.02		2	B8	401
209476	-30 18985	HR 8405			22 01 31	-30 09.6	5	6.46	.016	1.62	.010	1.93	.004	14	K5 III	15,1075,1637,2009,3005
209612	+48 03588				22 01 32	+49 25.3	1	7.57		-0.06		-0.43		2	B9	401
		G 232 - 51			22 01 32	+59 21.6	2	10.93	.005	0.94	.005	0.76		5	K3	538,906
209636	+54 02677				22 01 33	+54 38.3	1	7.01		-0.05		-0.23		2	B9	401
239836	+56 02690				22 01 36	+56 57.4	1	9.32		1.90				2	M0	369
209496	-35 15121				22 01 36	-35 18.9	1	10.03		0.49		0.01		2	F5 V	1730
209654	+55 02670				22 01 39	+55 33.3	2	9.52	.005	0.23	.024	0.14		7	A2	261,1025
239837	+55 02672				22 01 40	+56 01.7	1	9.34		1.33		0.99		3	K5	261
239839	+56 02692				22 01 42	+57 18.9	1	9.40		1.95				2	M7	369
		PKS 2201+04 # 1			22 01 44	+04 26.1	1	13.51		0.74		0.06		2		551
		PKS 2201+04 # 4			22 01 46	+04 22.3	1	15.36		0.61		-0.17		2		551
209522	-27 15757	UU Psa, HR 8408			22 01 46	-27 03.9	9	5.95	.012	-0.17	.008	-0.65	.009	36	B4 IVne	55,681,815,1586,1637*
		PKS 2201+04 # 2			22 01 47	+04 26.1	1	13.40		1.39		1.18		2		551
235706	+52 03086	G 232 - 52			22 01 48	+52 49.6	1	9.36		0.56		-0.18		5	G5	3016
		AAS11,365 T7# 17			22 01 48	+53 52.	1	11.00		0.53		-0.02		3		261
209678	+52 03088	LS III +52 023			22 01 51	+52 58.2	2	8.42	.000	0.37	.013	-0.54	.040	5	B2 I	1012,3016
		PKS 2201+04 # 3			22 01 52	+04 22.7	1	12.39		0.77		0.14		2		551
	+18 04922	G 126 - 52			22 01 53	+19 16.4	1	11.02		0.38		-0.20		2	sdF2	1620
	+46 03558				22 01 58	+47 02.4	1	9.35		0.88				2	G0	1724
		AAS11,365 T7# 18			22 02 00	+56 30.	1	10.87		0.51		0.33		3		261
	+48 03592				22 02 01	+49 09.4	1	9.87		1.80				2		369
209706	+55 02673				22 02 02	+55 36.3	1	7.79		0.49				3	F8	1025
	+62 02023				22 02 02	+63 17.5	1	8.88		0.04		-0.48		2	A0	401
209621	+20 05071	HP Peg			22 02 04	+20 48.6	2	8.86	.033	1.47	.024	1.14		2	C2	1238,3077
209679	+43 04122				22 02 04	+44 06.1	1	6.56		0.10		0.10		1	A1 V	1716
		LS III +59 019			22 02 06	+59 44.2	2	11.03	.010	0.56	.000	-0.38	.000	4		405,483
		G 18 - 21			22 02 07	+11 08.0	2	13.16	.010	0.60	.019	-0.19	.019	3		333,1620,1658
	+46 03562				22 02 07	+47 03.8	1	9.12		0.59				2	G0	1724
209566	-35 15127				22 02 08	-35 25.4	1	8.58		0.76		0.37		2	G8 V	1730
209692	+45 03777	NGC 7209 - 25			22 02 11	+46 06.6	1	8.53		0.34		0.16		1	A0	49
		CS 22892 # 5			22 02 11	-16 04.8	1	14.10		0.43		-0.17		1		1736
209789	+66 01471				22 02 12	+66 51.8	1	8.97		0.20		-0.40		3	B2 V	555
209625	-1 04242	HR 8410			22 02 13	-01 08.9	8	5.30	.016	0.23	.005	0.15	.005	33	A5 m	15,355,1007,1013,1071*
	+15 04559				22 02 14	+16 00.4	1	9.79		1.15		1.17		1	K0	1364
	+14 04727				22 02 15	+15 21.9	1	9.53		0.42		0.06		2	F5	1364
209665	+24 04525				22 02 15	+25 24.9	1	7.22		0.07		0.05		2	A0	105
	+46 03563				22 02 15	+47 09.6	1	9.51		1.03				2	G0	1724
	+51 03239				22 02 15	+51 51.4	1	9.92		0.09		-0.59		2	B2 V	1012
239841	+56 02694				22 02 15	+57 18.7	1	9.40		0.22				4	B3	1025
209744	+59 02456	IDS22006N5920	A		22 02 16	+59 34.3	1	6.66		0.22		-0.58		3	B1 V	3024
209744	+59 02456	IDS22006N5920	AB		22 02 16	+59 34.3	3	6.70	.013	0.23	.004	-0.62	.010	5	B1 V	5,213,401
209744	+59 02456	IDS22006N5920	C		22 02 16	+59 34.3	1	8.90		1.36		2.11		1		3024
	+15 04560				22 02 18	+15 48.1	1	9.07		0.44		0.09		2	F5	1364
		LP 819 - 24			22 02 19	-16 08.2	1	11.69		0.51		-0.10		2		1696
209790	+63 01802	HR 8417		★ AP	22 02 20	+64 23.0	3	4.29	.009	0.34	.007	0.09	.000	10	A3m	15,1363,8015
209680	+29 04573				22 02 21	+29 43.0	1	8.06		1.61		1.92		9	K5 III	1371
239842	+57 02446				22 02 21	+57 45.8	1	9.68		0.31				2	A5	1025
209529	-60 07541	HR 8409			22 02 21	-59 52.7	3	5.62	.016	1.47	.011	1.81	.000	8	K4 III	58,2007,3012
209693	+32 04329	HR 8412		★ A	22 02 22	+32 41.9	2	6.39	.017	1.12	.019	0.80		4	G5 Ia	71,8100
209772	+62 02028	HR 8416, MO Cep			22 02 23	+62 52.6	2	5.30	.007	1.56	.007	1.80	.003	7	M5 IIIab	1355,3051
	-35 15131				22 02 23	-35 31.3	1	10.16		1.06		0.86		2		1730
		AAS11,365 T7# 19			22 02 24	+55 26.	1	10.76		1.72		1.73		1		261
		NGC 7209 - 33			22 02 25	+46 11.3	1	12.53		0.23		0.25		1	A5	49
	+6 04956				22 02 26	+06 37.4	1	9.89		1.62				1	M0	1632
		Cep sq 1 # 137			22 02 26	+59 06.4	1	10.48		0.21		-0.42		2		27
209707	+31 04617				22 02 28	+31 48.5	2	8.85	.000	0.50	.000	0.10	.000	6	G5	196,583
		NGC 7209 - 37			22 02 29	+46 12.7	1	11.48		1.12				1		49
		NGC 7209 - 141			22 02 30	+46 17.0	1	13.19		0.57		0.07		1		49
	+45 03780	NGC 7209 - 38		★ AB	22 02 31	+46 08.5	1	11.00		0.16		0.14		1	B8	49
		G 18 - 24			22 02 32	+08 23.7	2	12.04	.010	0.58	.005	-0.17	.029	3	sdF8	333,1620,1658
		LS III +53 032			22 02 32	+53 38.3	1	10.69		0.40		-0.46		2		483
		G 18 - 25			22 02 34	+04 53.4	1	13.67		1.51		0.98		1		3062
		NGC 7209 - 41			22 02 35	+46 18.4	1	11.59		0.32		0.19		1	A1	49
209809	+59 02459	IDS22006N5920	E		22 02 35	+59 37.4	2	6.98	.009	0.03	.009	-0.18	.099	15	A0	213,3024
209567	-57 10027	IDS21592S5657	AB		22 02 36	-56 42.9	1	8.41		0.60				4	G1 V	2012
	+15 04562				22 02 39	+15 59.7	1	8.42		1.12		1.02		5	K0	1364
209810	+59 02461	IDS22006N5920	D		22 02 39	+59 35.2	2	7.85	.005	0.06	.005	-0.05	.005	5		27,3024
	+45 03782	NGC 7209 - 46		★ V	22 02 42	+46 11.0	2	10.09	.002	0.16	.008	0.17	.005	8	B9 V	49,1768
		NGC 7209 - 144			22 02 44	+46 15.9	1	13.89		0.63		0.26		1		49
		AAS60,495 # 8			22 02 47	-18 44.8	1	13.35		0.67				1		1584
		NGC 7209 - 143			22 02 48	+46 10.5	1	13.82		1.06		0.68		1		49
		NGC 7209 - 142			22 02 48	+46 13.4	1	13.72		1.27		1.12		1		49
	+45 03783	NGC 7209 - 53			22 02 48	+46 17.3	1	10.68		0.17		0.18		1	A0 IV	49

HD	DM	Other Id	N Rem	α_{1950}	δ_{1950}	S	V	σ_V	B–V	σ_{B-V}	U–B	σ_{U-B}	n	Spectrum	References
209725	+13 04842			22 02 49	+14 14.6	1	7.47		1.26				4	K2	882
		NGC 7209 - 145		22 02 49	+46 13.8	1	14.00		0.83		0.31		1		49
209830	+59 02463	IDS22006N5920	F	22 02 49	+59 40.1	2	8.14	.000	0.02	.010	-0.08	.005	6	A0	27,3024
209684	-14 06205			22 02 49	-14 00.8	2	9.84	.009	-0.19	.004	-0.81	.000	18	B2/3 III	963,1732
		NGC 7209 - 147		22 02 50	+46 14.9	1	14.38		0.58		0.10		1		49
209036	-83 00730			22 02 51	-83 36.2	1	7.35		1.23		1.15		7	K1 III	1704
		NGC 7209 - 133		22 02 52	+46 17.5	1	12.95		0.37		0.20		1		49
	+40 04720			22 02 53	+40 44.5	1	9.64		0.58				2	F8	1724
		Cep sq 1 # 138		22 02 53	+58 49.2	1	11.07		0.42		0.17		2		27
209761	+25 04671	HR 8415		22 02 54	+26 25.8	1	5.78		1.25				2	K2 III	71
209745	+29 04578	IDS22006N2923	AB	22 02 54	+29 37.5	2	8.71	.015	0.51	.010	0.04	.015	12	F6 V +G1 V	1371,3026
	+45 03785	NGC 7209 - 61		22 02 54	+46 18.4	1	9.94		0.19		0.20		1	B9 III	49
239843	+56 02695	IDS22012N5620	AB	22 02 54	+56 34.4	1	9.41		0.39				3	A2	1025
209869	+61 02243			22 02 56	+62 09.9	1	8.44		0.15		-0.42		4	A0	213
		NGC 7209 - 134		22 02 57	+46 16.7	1	12.93		0.39		0.21		1		49
209813	+46 03572	HK Lac		22 02 57	+46 59.5	1	6.91		1.04		0.72		2	K0 III	3016
		G 18 - 27		22 02 58	+05 30.7	1	10.57		0.86		0.56		1	K1	333,1620
209661	-44 14763			22 02 59	-44 12.6	1	6.93		1.04				4	K0 III	2012
209792	+38 04677			22 03 00	+39 14.3	2	7.67	.010	0.38	.010	-0.05	.015	5	F3 V	833,1601
	+45 03788	NGC 7209 - 70		22 03 00	+46 09.0	1	10.95		0.38		0.00		1	A5	49
		Cep sq 1 # 139		22 03 00	+59 05.0	1	11.18		0.24		-0.12		2		27
		AAS60,495 # 11		22 03 00	-19 12.6	1	13.96		0.69				1		1584
		NGC 7209 - 71		22 03 01	+46 16.7	1	12.54		0.66		0.20		1	A1	49
209888	+61 02245			22 03 03	+61 43.9	1	9.10		0.21		-0.08		3	A0	213
209717	-19 06217			22 03 03	-18 54.7	1	7.64		0.24		0.11		1	A9 V	1584
209688	-40 14639	HR 8411		22 03 07	-39 47.1	8	4.46	.013	1.37	.000	1.65	.013	26	K3 III	15,208,1034,1075,2024*
209747	+4 04800	HR 8413		22 03 09	+04 48.8	11	4.85	.010	1.45	.008	1.80	.015	51	K4 III	15,1003,1075,1088*
		NGC 7209 - 148		22 03 10	+46 15.9	1	14.69		0.67		0.23		1		49
	-12 06174			22 03 11	-12 09.0	1	10.15		1.45				1	M0	1746
	+45 03791	NGC 7209 - 81		22 03 12	+46 11.3	1	10.61		0.32		0.28		1	A0	49
239845	+56 02698			22 03 12	+56 48.3	1	9.49		0.50				2	G5	1025
209750	-1 04246	HR 8414	★A	22 03 13	-00 33.8	17	2.94	.021	0.97	.011	0.74	.025	59	G2 Ib	1,3,15,1020,1034,1075*
	+11 04725	G 18 - 28		22 03 14	+12 08.3	4	9.55	.004	0.65	.005	0.03	.005	6	G3	516,1620,1658,3077
		NGC 7209 - 84		22 03 16	+46 09.2	1	11.08		0.33		0.30		1	B9	49
		NGC 7209 - 88		22 03 17	+46 11.6	1	11.54		0.22		0.21		1		49
209857	+46 03574	NGC 7209 - 86	★V	22 03 17	+46 30.1	6	6.13	.016	1.61	.015	1.75	.017	22	M4 IIIab	15,71,245,3016,8015,8022
	+45 03792	NGC 7209 - 89		22 03 18	+46 14.4	1	9.44		1.60		1.76		1	K4 III	49
209833	+28 04284	HR 8419		22 03 19	+28 43.2	4	5.70	.003	-0.06	.007	-0.21	.005	9	B9 Vn	985,1049,1716,1733
		TOU28,33 T8# 14		22 03 19	+56 47.3	1	10.59		0.32				2		1025
209733	-28 17592			22 03 19	-28 16.1	1	8.53		0.51		-0.02		5	G0 V	1731
	+45 03793	NGC 7209 - 91		22 03 20	+46 09.1	1	10.57		0.24		0.22		1	A2	49
		NGC 7209 - 146		22 03 20	+46 13.3	1	14.14		1.10		0.85		1		49
		NGC 7209 - 92		22 03 20	+46 18.2	1	12.58		0.28		0.21		1		49
	+39 04748			22 03 22	+40 03.0	1	10.80		0.36				2	F2	1724
209740	-29 18169			22 03 23	-28 56.9	1	7.44		1.29				4	K3 III	2012
209900	+52 03095	LS III +53 033		22 03 24	+53 16.0	1	8.71		0.45		0.23		2	A0 Ib-II	1012
	+56 02699			22 03 24	+57 10.8	1	10.26		0.46				3		1025
	+45 03795	NGC 7209 - 98		22 03 25	+46 10.2	1	10.26		0.18		0.21		1	A0 IV	49
		NGC 7209 - 136		22 03 25	+46 15.7	1	12.53		0.63		0.14		1	K2	49
209796	+0 04820			22 03 28	+01 08.2	6	8.94	.004	1.20	.007	1.17	.009	79		147,989,1509,1729,6005,6009
		AJ89,1229 # 309		22 03 28	+07 58.6	1	11.14		1.39				1	M0	1632
209960	+62 02029	HR 8426		22 03 29	+62 32.5	2	5.29	.015	1.42	.005	1.76	.015	7	K4 III	1355,3016
		G 263 - 11		22 03 29	+78 02.1	1	15.87		1.80		1.64		1		906
	+56 02700			22 03 30	+57 10.3	1	10.08		0.21				3		1025
	+52 03097			22 03 36	+53 09.7	1	10.21		1.77				2		369
209944	+57 02450			22 03 36	+57 59.5	1	7.99		1.15				3	K0	1025
209975	+61 02246	HR 8428	★A	22 03 36	+62 02.2	11	5.10	.013	0.09	.009	-0.84	.014	44	O9.5Ib	5,15,154,213,450,101
209975	+61 02246	IDS22021N6148	C	22 03 36	+62 02.2	1	10.79		1.08		0.75		4		450
209858	+27 04252			22 03 37	+27 43.5	1	7.82		0.53		-0.01		2	F8 V	3026
209918	+50 03551	IDS22017N5039	AB	22 03 37	+50 53.5	1	9.09		0.12		0.07		4	A0	171
	-35 15144			22 03 38	-34 39.7	1	10.26		1.02		0.79		2	K3 V	3072
209742	-45 14576			22 03 38	-45 38.0	3	8.43	.013	0.85	.005	0.54	.021	10	K1 V	158,1075,3008
209890	+32 04335	RZ Peg	A	22 03 39	+33 15.7	1	9.25		3.93		2.66		1	C9 e	414
209890	+32 04335		B	22 03 39	+33 15.7	1	12.31		0.57		0.13		1		414
		NGC 7209 - 104		22 03 42	+46 12.5	1	11.37		0.25		0.20		1	A3	49
		AAS11,365 T7# 20		22 03 42	+56 27.	1	11.32		0.68		0.15		1		261
209819	-14 06209	HR 8418		22 03 44	-14 06.8	9	4.27	.009	-0.07	.006	-0.28	.012	28	B8 V	15,1075,1079,1088*
209919	+41 04378			22 03 46	+41 43.1	1	8.29		0.38				3	F2	1724
209874	+14 04735			22 03 47	+14 36.8	1	8.22		0.96				4	K2	882
	+45 03800	NGC 7209 - 108		22 03 47	+46 27.3	1	9.59		0.54				2	F9	1724
209932	+44 04041	HR 8422		22 03 49	+44 52.1	1	6.44		-0.03		-0.15		2	A0 V	401
	+15 04566			22 03 50	+15 29.6	1	9.02		1.59		2.01		1	K5	1364
209848	-17 06439			22 03 51	-16 53.7	1	9.34		0.11		0.13		24	A1/2 V	1699
	+56 02701	IDS22021N5642	AB	22 03 52	+56 56.9	1	10.55		0.58		0.08		3		261
209961	+47 03692	HR 8427, V365 Lac		22 03 53	+47 59.3	3	6.28	.015	-0.07	.015	-0.71	.000	13	B2 V	15,154,1174
		BPM 44347		22 03 54	-48 33.	1	15.55		-0.05		-1.05		2	DA	3065
209991	+56 02702	IDS22022N5638	A	22 03 55	+56 52.9	1	7.58		0.45				2	F5	1025
239847	+57 02451			22 03 56	+57 42.0	1	9.46		0.21				2	A0	1025
		Cep sq 1 # 141		22 03 56	+58 41.3	1	11.10		0.54		0.11		6		27
		G 18 - 29		22 03 57	+05 40.9	1	10.61		0.75		0.28		1		333,1620

Table 1 1179

HD	DM	Other Id	N	Rem	α_{1950}	δ_{1950}	S	V	σ_V	B–V	σ_{B-V}	U–B	σ_{U-B}	n	Spectrum	References
	+48 03606				22 03 58	+49 00.1	1	9.77		1.91				2	M0	369
209875	+1 04583	G 18 - 30			22 03 59	+01 36.6	3	7.24	.005	0.52	.005	0.02	.007	4	F8 V	219,333,1620,3037
209945	+44 04043	HR 8424			22 04 00	+44 46.2	2	5.08	.050	1.55	.015	1.96	.020	5	K5 III	865,3016
210013	+57 02452	IDS22023N5750	AB		22 04 00	+58 04.0	1	9.18		0.10				2	A0	1025
210040	+61 02248				22 04 00	+62 01.1	1	7.74		0.06		-0.66		3	A3	213
		CS 22892 # 24			22 04 01	-15 23.7	1	14.74		0.49		-0.26		1		1736
		Steph 1971			22 04 02	+09 52.0	1	11.63		1.18		1.02		1	K7	1746
		Cep sq 1 # 142			22 04 02	+58 47.5	1	10.63		0.30		0.19		2		27
		AAS60,495 # 10			22 04 05	-18 53.2	1	13.76		0.70				1		1584
209905	+1 04584				22 04 06	+02 11.7	4	6.51	.003	-0.06	.004	-0.24	.010	50	B9	147,1509,6005,6009
	+55 02678				22 04 06	+55 33.0	2	9.84	.005	0.28	.015	-0.09		3	B7 III	273,1025
		Cep sq 1 # 146			22 04 07	+59 59.3	1	10.41		0.41		-0.31		3		27
209992	+52 03105				22 04 08	+53 22.8	1	6.97		1.01		0.71		3	K0 Ib	3016
209993	+44 04044	HR 8429			22 04 11	+45 00.3	1	6.19		0.09		0.14		2	A3 V	252
		G 18 - 31			22 04 12	+03 10.6	1	13.72		1.60		1.35		1		3062
	+49 03735	LS III +49 040			22 04 13	+49 39.5	1	9.72		0.05		-0.64		2	B1.5V:e:nn	1012
		Cep sq 1 # 143			22 04 14	+59 23.4	1	11.54		0.28		-0.17		3		27
	+56 02705				22 04 16	+57 09.6	1	9.92		0.18				2	B8	1025
235721	+50 03556				22 04 17	+50 50.9	1	10.19		0.57				2		1724
209947	+16 04665				22 04 18	+16 30.3	1	7.75		0.32		0.18		1	A3	1364
209882	-28 17600				22 04 18	-28 17.5	1	9.00		1.11		1.05		5	G8 III	1731
		Cep sq 2 # 4			22 04 19	+55 10.0	1	11.20		0.25		-0.12		1	B7 III	273
		TOU28,33 T8# 20			22 04 22	+55 03.0	2	10.26	.025	0.20	.005	-0.19		3	B7	273,1025
	+57 02454				22 04 22	+58 12.3	1	10.50		0.39				2		1025
		Cep sq 2 # 6			22 04 23	+55 20.7	1	11.92		0.40		0.04		1	B7 V	273
		Cep sq 2 # 9			22 04 28	+54 29.7	1	12.12		0.31		0.10		1		273
210071	+55 02679	HR 8434			22 04 28	+55 05.9	4	6.39	.016	-0.10	.010	-0.45	.006	12	A0 III	401,1025,1063,1079,3033
210072	+54 02683	LS III +55 021			22 04 30	+55 00.2	3	7.66	.009	0.29	.013	-0.36		5	B2 V	273,1012,1025
239850	+55 02680				22 04 30	+55 53.5	1	9.33		0.39		0.02		3	F2	261
	+42 04301				22 04 32	+43 08.0	1	8.54		0.74				3	G0	1724
	+53 02784	LS III +54 012			22 04 34	+54 20.3	1	9.87		0.42		-0.32		1	B3 III	273
	-64 04212				22 04 34	-63 40.1	1	10.60		0.52		-0.01		2		1730
	+7 04795				22 04 35	+08 00.0	2	10.67	.001	-0.12	.002	-0.70	.011	16	B5	963,1732
209977	+11 04730				22 04 35	+11 31.4	1	6.80		1.64		1.97		3	M1 III	3012
		Cep sq 1 # 147			22 04 36	+59 14.4	1	11.53		0.33		0.18		2		27
		Cep sq 2 # 12			22 04 37	+55 07.5	1	11.86		0.32		-0.43		1	B5 II	273
	+40 04723				22 04 40	+40 32.6	1	9.86		1.10				2	G0	1724
239852	+58 02390				22 04 40	+58 38.6	1	9.68		2.20				2	M7	369
		CS 22892 # 20			22 04 40	-14 19.3	1	14.35		0.33		-0.21		1		1736
210027	+24 04533	HR 8430	★	A	22 04 41	+25 06.0	17	3.77	.006	0.43	.007	-0.02	.012	141	F5 V	1,15,254,985,1118,1197*
		AAS11,365 T1# 1			22 04 42	+52 54.	1	11.52		1.52		1.31		2		261
		AAS60,495 # 7			22 04 44	-18 42.6	1	12.02		0.77				1		1584
210100	+51 03248				22 04 46	+51 33.7	2	7.07	.015	-0.09	.010	-0.42	.020	4	B8	171,401
		Cep sq 2 # 13			22 04 47	+55 21.5	1	11.61		0.23		-0.07		1	B7 V	273
	+39 04755				22 04 47	+40 25.2	1	9.76		0.48				2	F5	1724
210128	+56 02706				22 04 49	+56 57.3	1	8.90		0.20				3	A0	1025
	+46 03584				22 04 50	+46 60.0	1	8.98		0.37				2	F2	1724
	+47 03694				22 04 50	+47 37.6	1	9.03		1.07				2	G5	1724
239853	+56 02707				22 04 50	+57 24.4	1	9.18		0.43				2	F8	1025
209969	-28 17604	IDS22020S2833	A		22 04 51	-28 18.2	1	7.96		0.67		0.19		5	G0	1731
210087	+35 04712	IDS22027N3536	AB		22 04 52	+35 51.0	1	7.80		0.23		0.08		2	A3	1601
210144	+52 03112	G 232 - 54			22 04 53	+52 53.5	4	7.80	.016	0.79	.005	0.42	.010	18	G8 V	22,261,1003,3026
		Cep sq 1 # 148			22 04 53	+59 00.9	1	10.40		0.23		-0.09		2		27
	-51 13182		A		22 04 53	-51 27.5	2	10.51	.005	1.20	.003			2	K7	1494,1705
		Steph 1972			22 04 56	+20 42.1	1	10.94		1.33		1.22		1	K7	1746
	+48 03612				22 04 56	+48 34.8	1	9.66		1.88				2		369
		Cep sq 2 # 14			22 04 56	+54 36.0	1	11.18		0.26		-0.10		1	B7 V	273
		CS 22881 # 36			22 04 57	-40 59.0	2	13.95	.010	0.49	.010	-0.25	.005	3		1580,1736
		CS 22892 # 25			22 05 00	-15 29.7	1	14.03		0.39		-0.21		1		1736
210074	+18 04930	HR 8435			22 05 05	+19 13.8	1	5.74		0.31		0.05		3	F1 III	1733
209952	-47 14063	HR 8425	★	A	22 05 05	-47 12.2	8	1.73	.012	-0.13	.005	-0.46	.004	21	B7 IV	15,278,1020,1034,1075*
210090	+17 04693	HR 8436			22 05 06	+17 45.4	1	6.35		1.64				2	M1	71
210030	-10 05837				22 05 07	-10 19.4	1	7.45		1.09		0.91		5	K0	1657
		Cep sq 2 # 17			22 05 14	+54 31.8	1	11.19		0.62		0.21		1	A0 III	273
235726	+53 02788	G 232 - 55			22 05 15	+53 55.8	1	9.15		0.66		0.12		2	G0	3016
		Cep sq 2 # 18			22 05 17	+56 05.2	1	9.95		0.36		-0.31		1	B3 V	273
		Cep sq 2 # 21			22 05 19	+55 19.1	1	11.80		0.47		0.20		1	B8 III	273
		G 232 - 56			22 05 19	+59 15.1	1	13.15		0.73		0.08		2		1658
210180	+51 03251				22 05 20	+52 23.5	1	8.73		0.14		-0.18		2	A0	171
210022	-44 14778				22 05 21	-44 18.2	1	9.22		0.98		0.37		2	G0 V	3072
		Cep sq 2 # 22			22 05 22	+55 12.8	1	11.07		0.27		-0.52		1	B2 III	273
		CS 22892 # 27			22 05 22	-15 50.2	1	12.74		0.32		-0.04		1		1736
		CS 22881 # 34			22 05 22	-42 30.9	2	15.27	.000	-0.01	.000	0.15	.012	2		966,1736
	+57 02456				22 05 23	+57 36.3	1	9.96		0.21				3	B9	1025
		G 127 - 2			22 05 24	+24 23.1	1	11.49		0.87		0.49		2	K3	1620
	+57 02457				22 05 27	+57 52.5	1	10.50		0.33				2	B3	1025
210081	-24 17016				22 05 27	-23 58.7	1	7.84		1.01				4	K0 III	2012
210220	+58 02393	HR 8442			22 05 28	+58 35.8	1	6.32		0.88		0.63		3	G6 III	1355
210049	-33 15922	HR 8431			22 05 28	-33 14.0	7	4.50	.005	0.06	.008	0.05	.010	25	A2 V	15,208,1075,2012,2024*
210129	+21 04695	HR 8438			22 05 29	+21 27.5	4	5.78	.000	-0.10	.008	-0.42	.024	8	B7 Vne	15,1003,1049,1079

HD	DM	Other Id	N	Rem	α_{1950}	δ_{1950}	S	V	σ_V	B–V	σ_{B-V}	U–B	σ_{U-B}	n	Spectrum	References
210130	+12 04760				22 05 31	+12 49.5	1	7.78		0.29		0.11		3	A5	1364
		Cep sq 2 # 23			22 05 31	+54 39.3	1	10.04		0.44		0.35		1	A3 III	273
239856	+55 02682				22 05 31	+55 59.4	1	8.63		1.03				2	F2 Iab	1025
210066	−34 15421	HR 8433			22 05 31	−34 17.3	3	4.98	.004	1.49	.008	1.81		8	K4 III	2007,2024,3053
210221	+52 03114	HR 8443		⋆	22 05 34	+53 03.8	3	6.14	.000	0.41	.017	0.24	.000	7	A3 Ib	15,1012,3016
	+54 02687				22 05 34	+54 28.7	1	9.51		1.48		1.52		3		261
210051	−43 14906				22 05 36	−43 17.2	1	7.01		1.20				4	K1 III	2012
	+41 04390				22 05 37	+42 15.1	1	9.57		0.48				2	F0	1724
235729	+51 03253				22 05 37	+51 28.4	1	8.77		0.05		−0.31		2	B9	401
		NGC 7213 sq1 5			22 05 38	−47 25.3	1	14.09		0.63		0.16		2		1687
	+16 04672				22 05 41	+16 43.0	1	9.89		1.39		1.70		2	K2	1364
		CS 22881 # 40			22 05 41	−39 59.9	2	13.32	.010	0.71	.017	0.12	.003	3		1580,1736
239857	+55 02683				22 05 44	+56 15.8	1	8.65		1.18				3	K0	1025
		NGC 7213 sq1 3			22 05 44	−47 25.1	1	12.52		1.09		1.02		2		1687
210170	+16 04673				22 05 45	+17 18.8	1	7.10		0.05		0.09		3	A0	1364
		Cep sq 2 # 26			22 05 45	+55 15.2	1	10.59		0.35		0.12		1	A0 V	273
		Cep sq 2 # 27			22 05 47	+55 38.6	1	11.02		0.38		0.12		1	B8 V	273
		CS 22892 # 38			22 05 48	−13 55.9	1	15.08		−0.24		−1.04		1		1736
210111	−33 15926	HR 8437			22 05 48	−33 22.3	1	6.37		0.20				4	A2 III/IV	2035
210183	+28 04295				22 05 49	+29 26.3	1	8.06		0.14		0.09		9	A0	1371
210351	+69 01219				22 05 51	+69 59.3	1	8.40		0.20				1	A0	207
		Cep sq 2 # 28			22 05 52	+54 44.3	1	12.64		0.54		0.35		1	A0 II	273
210262	+55 02684				22 05 52	+55 30.0	2	8.59	.018	1.20	.027	0.85		5	K0	261,1025
		AAS6,117 # 9			22 05 54	+52 21.	1	12.31		0.69		0.08		2		171
		Cep sq 2 # 30			22 05 58	+54 50.2	1	11.33		0.53		0.29		1	A2 III	273
210211	+23 04472				22 05 59	+23 54.8	1	6.56		0.85		0.54		2	G2 V	1733
210210	+24 04540	HR 8441			22 05 59	+25 17.9	2	6.10	.009	0.29	.014	0.11	.009	4	F1 IV	39,833
210263	+52 03116				22 05 59	+52 37.4	1	9.14		0.26		0.16		2		261
		Cep sq 2 # 33			22 06 01	+54 39.4	1	11.82		0.46		0.22		1	A2 III	273
		Cep sq 2 # 34			22 06 03	+55 16.1	1	12.30		0.37		−0.14		1	B7 V	273
209855	−76 01547	HR 8420			22 06 03	−76 07.6	3	6.54	.007	1.18	.000	1.30	.000	13	K2 III	15,1075,2038
		LS III +57 023			22 06 05	+57 39.8	1	10.67		0.98		−0.10		3		483
239861	+55 02686				22 06 06	+56 00.9	2	9.29	.000	0.34	.000			3	B9	273,1025
	+51 03256				22 06 07	+52 23.5	1	9.70		0.22		0.24		3		171
	+53 02790	LS III +54 016			22 06 07	+54 16.4	3	9.85	.019	0.27	.023	−0.70	.023	5	B0 III	273,1011,1012
		Cep sq 2 # 38			22 06 08	+55 09.1	1	11.43		0.34		0.25		1	B9.5V	273
		Cep sq 2 # 39			22 06 08	+55 54.9	1	11.30		0.31		0.22		1	B8 III	273
	+60 02340				22 06 08	+60 44.6	1	9.53		0.21		−0.48		2	B5	213
		Cep sq 2 # 40			22 06 09	+54 25.6	1	10.08		0.43		0.12		1	A0 V	273
		Cep sq 2 # 42			22 06 10	+54 14.4	1	11.75		0.59		0.44		1	A1 V	273
239862	+55 02685				22 06 10	+55 39.7	2	9.43	.000	0.31	.000			3	A0	273,1025
210401	+69 01221				22 06 10	+70 27.0	1	7.14		0.16				1	A0	207
		LS III +57 024			22 06 11	+57 15.4	1	10.57		0.91		−0.07		2		483
		AAS11,365 T1# 6			22 06 12	+52 16.	1	14.32		0.56		0.35		6		261
		AAS11,365 T1# 5			22 06 12	+52 17.	1	13.27		0.77		0.27		6		261
		AAS11,365 T1# 4			22 06 12	+52 19.	1	13.51		1.09		0.81		6		261
210322	+55 02688				22 06 12	+56 20.0	2	7.96	.000	0.30	.018	0.02		5	F0	261,1025
	−44 14788				22 06 12	−43 57.0	1	10.36		1.07				4		1075
		Cep sq 2 # 44			22 06 14	+55 11.1	1	11.86		0.22		0.01		2	B8 V	273
		CS 22881 # 43			22 06 14	−38 36.3	2	15.61	.000	0.11	.000	0.09	.009	2		966,1736
210223	+12 04764				22 06 15	+12 44.1	1	8.24		1.51		1.87		4	K5	1364
210191	−19 06227	HR 8439			22 06 15	−18 45.9	2	5.82	.014	−0.15	.032			5	B2 III	1584,2009
	+45 03812				22 06 18	+46 16.5	1	9.04		0.28				3	A5	1724
		AAS11,365 T7# 26			22 06 18	+56 22.	1	11.11		0.62		0.04		1		261
		Cep sq 2 # 45			22 06 19	+56 01.9	1	10.85		0.55		−0.13		1	B3 V	273
210289	+49 03746	HR 8445		⋆ A	22 06 20	+49 33.1	1	6.40		1.60		1.86		2	K5 III	1733
210308	+48 03621	IDS22044N4844		AB	22 06 21	+48 57.6	1	8.00		0.00		−0.22		2	A0	401
		CS 22892 # 41			22 06 21	−13 02.6	1	14.33		0.40		−0.19		1		1736
210198	−19 06228	IDS22036S1858		AB	22 06 21	−18 43.4	1	8.25		0.58				1	G1 V	1584
		LS III +53 040			22 06 22	+53 02.9	1	12.16		0.18		−0.46		2		483
		Cep sq 2 # 48			22 06 22	+54 39.9	1	11.42		0.38		0.28		1	B8 V	273
		AAS11,365 T1# 9			22 06 24	+52 12.	1	14.49		0.87		0.27		6		261
		AAS11,365 T1# 8			22 06 24	+52 16.	1	11.48		1.78		1.95		3		261
		AAS11,365 T1# 7			22 06 24	+52 21.	1	12.88		0.62		0.27		4		261
		AAS6,117 # 11			22 06 24	+52 22.	1	11.29		0.19		0.16		3	A0	171
		AAS6,117 # 12			22 06 24	+52 22.	1	11.32		0.43		0.16		2	B8	171
		Cep sq 2 # 49			22 06 26	+55 03.6	1	12.37		0.30		−0.21		1	B7 III	273
	+53 02792				22 06 29	+54 16.3	1	10.37		0.24		0.09		1	B9.5V	273
		AAS6,117 # 13			22 06 30	+52 14.	1	12.67		0.32		0.21		3		171
		AAS6,117 # 14			22 06 30	+52 16.	1	12.19		0.34		0.40		3	B9	171
		AAS11,365 T1# 10			22 06 30	+52 20.	1	12.99		1.49		1.65		5		261
210386	+63 01807	LS III +63 002			22 06 30	+63 29.4	1	8.01		0.32		−0.54		3	B1.5II-III	555
		G 18 - 33			22 06 32	+02 39.5	2	11.37	.010	1.43	.000	1.27		2	M0	333,1620,1632
210193	−41 14770				22 06 33	−41 28.2	2	7.84	.024	0.66	.000	0.21		8	G3 V	158,1075
		Cep sq 2 # 52			22 06 34	+54 15.8	1	11.86		0.25		−0.02		2	B9 III	273
		Cep sq 2 # 53			22 06 35	+54 45.1	1	11.64		0.39		0.22		1	B9 V	273
		CS 22881 # 39			22 06 35	−40 40.7	2	15.13	.010	0.39	.007	−0.01	.009	3		1580,1736
		AAS11,365 T1# 11			22 06 36	+52 10.	1	11.43		1.23		1.00		2		261
	−47 14075	NGC 7213 sq1 2			22 06 38	−47 28.0	1	10.80		0.46		0.03		2		1687
	+28 04300	IDS22043N2818		AB	22 06 40	+28 29.2	1	10.88		0.67		0.29		1		8084

Table 1   1181

HD	DM	Other Id	N	Rem	α_{1950}	δ_{1950}	S	V	σ_V	B–V	σ_{B-V}	U–B	σ_{U-B}	n	Spectrum	References
210334	+45 03813	HR 8448, AR Lac			22 06 40	+45 29.8	1	6.11		0.74				2	K2 III+G2IV	71
		G 27 - 14			22 06 40	−08 09.8	1	12.94		1.36		1.10		2		203
	+43 04138				22 06 41	+44 18.6	1	8.86		0.68				2	G0	1724
		Cep sq 2 # 56			22 06 41	+55 07.6	1	11.68		0.39		-0.07		1	B5 III	273
	+60 02344	IDS22051N6024	A		22 06 41	+60 39.1	1	9.44		0.24		-0.45		5	B8	213
		AAS6,117 # 15			22 06 42	+52 15.	1	10.67		0.26		0.27		3	A2	171
		AAS11,365 T1# 13			22 06 42	+52 18.	1	12.40		0.18		-0.04		3		261
		AAS11,365 T1# 12			22 06 42	+52 27.	1	10.30		1.65		1.88		1		261
	+54 02691				22 06 42	+54 40.5	1	10.06		0.22		-0.23		1	B7 III	273
210353	+47 03706				22 06 43	+47 41.3	1	6.80		-0.04		-0.19		2	A0	401
210371	+53 02793				22 06 44	+54 02.3	1	8.84		0.46		-0.05		3	F2	261
239864	+57 02461				22 06 44	+57 58.3	1	9.86		0.44				3	F8	1025
210256	−19 06232	IDS22040S1856	A		22 06 44	−18 41.8	1	8.90		0.30				1	A9/F0 (IV)	1584
	−47 14076	NGC 7213 sq1 1			22 06 44	−47 25.0	1	10.99		0.57		0.09		2		1687
210255	−19 06231	IDS22040S1856	B		22 06 45	−18 41.4	1	9.55		0.32				1	A8/F0 V	1584
	+47 03707				22 06 46	+47 47.5	1	9.09		0.30				2	A5	1724
210311	+15 04578				22 06 48	+15 43.0	1	8.48		1.42				4	K2	882
		AAS11,365 T1# 18			22 06 48	+52 19.	1	12.00		0.83		0.29		3		261
		AAS11,365 T1# 17			22 06 48	+52 22.	1	13.92		0.48		0.30		5		261
		AAS11,365 T1# 16			22 06 48	+52 23.	1	10.77		1.48		1.44		3		261
		AAS11,365 T1# 14			22 06 48	+52 42.	1	12.05		0.57		0.05		2		261
210269	−8 05817				22 06 48	−08 25.9	1	7.00		0.97		0.67		3	G8 III	3016
210312	+12 04766				22 06 49	+13 25.6	1	8.63		0.67				2	G5	882
210204	−48 14143	HR 8440			22 06 49	−48 21.2	3	6.42	.004	1.38	.010			11	K3 III	15,1075,2032
		Cep sq 2 # 57			22 06 50	+54 19.9	1	11.05		0.47		0.31		1	A3 III	273
		G 126 - 56			22 06 51	+11 27.1	2	11.73	.010	0.69	.005	0.10	.030	2		1620,1658
	+51 03262		A		22 06 51	+52 11.5	2	9.52	.005	1.76	.010	2.19	.154	5		171,261
	+51 03262		B		22 06 51	+52 11.5	1	10.64		0.27		0.23		3		171
	+51 03263				22 06 51	+52 26.3	1	10.19		0.09		-0.28		2		261
		Cep sq 2 # 58			22 06 51	+55 47.6	1	11.40		0.62		0.45		1	A1 V	273
		Cep sq 2 # 59			22 06 52	+55 19.4	1	11.80		0.32		0.26		1	A1 III	273
210277	−8 05818				22 06 52	−07 47.3	1	6.59		0.75		0.36		1	G8 IV	3077
210432	+58 02395	IDS22052N5848	A		22 06 53	+59 02.9	1	7.28		0.05		-0.02		1	A3	401
	+52 03122	LS III +52 028		★ AB	22 06 55	+52 43.7	3	9.32	.009	0.29	.009	-0.60	.005	8	B1 V:	261,483,1012
		Cep sq 2 # 60			22 06 55	+56 00.5	1	10.92		0.71		0.50		1	A1 III	273
239865	+57 02462				22 06 55	+57 45.1	1	9.31		0.42				3	F5	1025
210433	+58 02395	IDS22052N5848	AB		22 06 55	+59 02.6	1	7.15		-0.04		-0.12		2	A0	401
210443	+62 02039				22 06 55	+62 29.1	1	7.80		0.16		0.03		2	A0	27
210414	+55 02689				22 06 56	+55 47.1	1	8.32		0.48				3	G0	1025
		CS 22892 # 36			22 06 56	−16 04.1	1	14.66		0.36		-0.32		1		1736
	+58 02396	AZ Cep			22 06 58	+59 18.3	1	10.75		2.67		2.33		2	M2 Ia	8032
210549	+72 01016				22 06 59	+72 42.2	1	8.11		1.18				1	K2	207
	+15 04581				22 07 00	+15 53.6	1	9.99		1.63		1.95		2	M0	1364
210354	+32 04349	HR 8449		★ A	22 07 00	+32 55.6	2	5.58	.000	1.01	.010	0.78	.005	5	K0	1355,3016
		AAS11,365 T1# 21			22 07 00	+52 19.	1	13.75		0.31		0.17		5		261
	−5 05715				22 07 00	−04 53.2	2	10.41	.011	1.52	.006	1.12		11	M3	1705,3078
210295	−14 06222				22 07 00	−13 51.0	3	9.57	.005	0.90	.010	0.41	.000	4	G8/K0wF6	742,1594,1658
210271	−34 15430	HR 8444			22 07 01	−34 15.7	2	5.36	.005	0.23	.005			6	A5 IV	2009,2024
210387	+44 04058	IDS22052N4421	D		22 07 02	+44 36.5	2	6.74	.047	-0.03	.009	-0.25	.000	4	A0	3024,8084
210404	+47 03711				22 07 03	+48 16.4	2	7.15	.014	0.81	.010	0.39		7	G5	1724,1733
210533	+69 01224				22 07 03	+69 43.2	1	8.16		0.04				1		207
235738	+50 03580				22 07 04	+50 29.6	2	9.17	.000	1.23	.001			34	G0	1724,6011
210117	−71 02663				22 07 04	−71 06.8	1	8.53		1.31		1.47		8	K2 III	1704
	−43 14916	BPM 44377			22 07 05	−43 09.1	1	10.40		1.27		1.25		2	K0	3072
		AAS11,365 T1# 24			22 07 06	+52 11.	1	13.50		0.68		0.10		6		261
		AAS11,365 T1# 23			22 07 06	+52 20.	1	13.03		0.68		0.12		5		261
		AAS11,365 T1# 22			22 07 06	+52 24.	1	12.12		0.16		-0.03		3		261
235739	+50 03581	Y Lac			22 07 08	+50 48.0	3	8.75	.006	0.53	.015	0.44	.080	3	F8	592,1462,6011
		Cep sq 2 # 63			22 07 08	+54 25.0	1	10.57		0.22		-0.24		1	B5 V	273
210478	+60 02348	LS III +60 012			22 07 08	+60 46.6	2	7.32	.004	0.08	.004	-0.70	.004	12	B1 V	213,1012
210550	+69 01225				22 07 09	+69 44.2	1	7.85		0.04				1	A0	207
210300	−28 17622	HR 8446			22 07 09	−28 32.3	1	6.44		0.15				4	A5 V	2007
		AJ89,1229 # 311			22 07 10	+02 16.9	1	11.58		1.63				1	M1	1632
210388	+34 04610				22 07 10	+34 53.0	1	7.47		0.51		-0.01		2	G0	1625
	+39 04767	IDS22050N4010	AB		22 07 10	+40 24.7	1	9.57		0.22				3	A5	1724
210415	+50 03582	IDS22052N5017	AB		22 07 10	+50 31.8	1	8.71		0.12		0.13		1	A0	592
210405	+44 04059	IDS22052N4421	A		22 07 12	+44 36.0	3	6.73	.017	-0.04	.030	-0.26	.020	5	B9	1476,3024,8084
210405	+44 04059	IDS22052N4421	B		22 07 12	+44 36.0	3	9.95	.032	0.28	.031	0.12	.023	6		3024,8072,8084
210405	+44 04059	IDS22052N4421	C1		22 07 12	+44 36.0	2	13.41	.120	0.69	.040	0.12	.255	2		8072,8084
210405	+44 04059	IDS22052N4421	C2		22 07 12	+44 36.0	1	14.69		1.43				1		3024
		Cep sq 2 # 64			22 07 12	+54 13.5	1	10.75		0.75		0.41		1		273
210325	−18 06077				22 07 12	−18 09.1	1	9.37		0.62				1	F7 V	1584
210206	−64 04221				22 07 12	−63 48.9	1	8.94		1.04		0.83		2	G8 III	1730
210302	−33 15941	HR 8447			22 07 13	−32 47.7	6	4.93	.014	0.49	.004	0.03	.011	25	F6 V	3,15,1075,2024,2032,3053
		Cep sq 2 # 66			22 07 14	+55 05.3	1	10.45		0.19		-0.38		1	B5 V	273
	+60 02349	LS III +61 007			22 07 16	+61 26.9	1	9.27		0.14		-0.50		1	B8	213
210056	−76 01549	HR 8432			22 07 16	−76 21.7	3	6.13	.012	1.00	.000	0.83	.005	14	K0 III	15,1075,2038
	+48 03625				22 07 17	+49 23.6	1	8.71		0.67				5	G0	1724
		AAS11,365 T1# 25			22 07 18	+52 27.	1	12.34		1.23		1.12		2		261
		Steph 1977			22 07 18	−08 40.6	1	11.74		1.58		1.80		1	M1	1746

HD	DM	Other Id	N Rem	α_{1950}	δ_{1950}	S	V	σ_V	B–V	σ_{B-V}	U–B	σ_{U-B}	n	Spectrum	References
210320	−28 17624			22 07 18	−27 55.1	1	8.64		0.73		0.31		5	G6 V	1731
210196	−66 03630			22 07 19	−66 35.2	1	8.45		0.59				4	F8 IV/V	2012
210480	+56 02713			22 07 20	+56 48.1	3	8.71	.012	0.15	.012	0.08	.019	6	A0	261,338,1025
		G 18 - 34		22 07 21	+14 14.9	2	15.64	.023	0.26	.042	-0.55	.019	4	DC	538,3060
	+15 04582			22 07 23	+15 27.7	1	9.77		1.50		1.89		2	K5	1364
210615	+72 01017	DM Cep		22 07 23	+72 31.4	1	6.95		1.93				1	M2	207
239867	+57 02463			22 07 24	+57 46.6	1	9.74		0.36				3	B8	1025
		Cep sq 2 # 67		22 07 26	+54 54.4	1	10.67		0.35		0.25		1	A2 III	273
210457	+46 03600			22 07 27	+46 40.5	1	7.94		0.48				2	F8	1724
	+53 02795			22 07 27	+54 22.7	1	10.41		0.14		-0.14		1	B8 V	273
	+55 02691			22 07 28	+55 29.9	1	11.32		0.25		-0.08		1	B8 IV	273
210272	−56 09825			22 07 29	−55 42.1	1	7.21		0.66		0.13		4	G3 V	158
210481	+47 03714			22 07 30	+47 42.7	1	8.69		0.21				2	A5	1724
		AAS6,117 # 18		22 07 30	+53 52.	1	12.45		0.16		-0.04		6	A0	171
		Cep sq 2 # 69		22 07 30	+54 11.7	1	12.22		0.27		0.03		1	B8 V	273
210494	+55 02692			22 07 30	+55 56.8	3	8.39	.005	0.35	.010	0.26		6	A0	261,273,1025
		JL 99		22 07 30	−76 41.	1	15.06		-0.13		-1.06		2		132
		Cep sq 2 # 72		22 07 31	+56 10.2	1	11.07		0.56		0.20		1	B9 III	273
	+15 04584			22 07 32	+16 15.9	1	10.04		0.20		0.16		3		1364
	+40 04742			22 07 32	+40 32.9	1	9.19		0.70				3	G0	1724
235740	+53 02796			22 07 32	+53 55.8	1	9.04		0.64		0.16		3	G0	171
210329	−41 14777			22 07 34	−40 39.3	2	9.55	.015	0.94	.017	0.69	.009	4	K0 V	1730,3072
		G 214 - 12		22 07 35	+40 47.2	1	12.57		1.49				4		940
	+55 02693	EE Cep		22 07 35	+55 30.6	1	10.72		0.37		-0.18		1	B6 III	273
	+57 02465	LS III +57 025		22 07 36	+57 36.7	1	11.00		0.73		-0.25		2	B0.5V:n	1012
239868	+57 02464			22 07 37	+57 49.0	1	9.19		0.34				3	B8	1025
210418	+5 04961	HR 8450		22 07 41	+05 57.1	8	3.52	.017	0.09	.015	0.08	.014	27	A2 V	15,1020,1075,1088*
210444	+22 04563	IDS22054N2238	AB	22 07 42	+22 52.8	1	7.33		0.38		0.00		2	F6 IV	1648
		Cep sq 1 # 20		22 07 42	+62 29.1	1	10.51		0.20		-0.46		2		27
210419	−4 05623	HR 8451		22 07 45	−04 08.4	2	6.26	.005	-0.01	.000	-0.07	.000	7	A1 Vnn	15,1071
210350	−40 14676			22 07 45	−40 29.9	1	8.89		1.15		1.04		2	K0 III/IV	1730
210459	+32 04352	HR 8454		22 07 46	+32 55.9	8	4.29	.019	0.46	.012	0.18	.011	24	F5 III	15,254,1007,1013,1118*
210640	+69 01226			22 07 48	+69 59.7	1	7.76		0.46				1	F2	207
		LS III +57 026	⋆ V	22 07 49	+57 29.8	1	12.12		0.79		-0.13		1		8113
210534	+46 03601			22 07 50	+47 27.3	1	8.29		0.39				2	F8	1724
239871	+57 02468			22 07 51	+58 01.7	1	9.33		0.40				3	F2	1025
		Cep sq 2 # 80		22 07 52	+55 23.0	1	11.56		0.27		0.13		1	B8 V	273
		vdB 150 # 14		22 07 52	+73 15.4	1	10.61		0.59				1	F0 IV	207
210422	−11 05770			22 07 52	−11 04.0	1	6.70		1.11		0.93		3	K0	3016
		LP 931 - 25		22 07 53	−32 29.3	1	12.52		0.60		-0.09		2		1696
		AAS6,117 # 20		22 07 54	+53 42.	1	13.02		0.28		0.32		4	B8	171
		Cep sq 2 # 81		22 07 54	+55 14.3	1	11.01		0.21		-0.03		1	B8 III	273
210461	+13 04861	HR 8456	⋆ A	22 07 56	+14 23.0	1	6.33		1.08		0.94		3	K0 III	1364
210460	+18 04946	HR 8455		22 07 56	+19 22.3	2	6.19	.003	0.70	.003	0.17		5	G0 V	71,3016
210434	−4 05625	HR 8453		22 07 57	−04 30.8	4	6.01	.008	0.98	.004	0.83	.006	11	K0 III-IV	15,1071,3016,4001
210424	−12 06196	HR 8452		22 07 57	−11 48.7	2	5.43	.018	-0.12	.000			5	B5 III	2007,2024
239872	+56 02720			22 07 59	+57 08.4	1	9.51		0.36				4	B8	1025
	+43 04143			22 08 00	+43 56.1	1	9.52		0.33				2	A5	1724
	+40 04745	DE Lac		22 08 01	+40 40.4	1	10.08		0.41		0.20		1	F6.5	668
	+51 03268	G 232 - 58	A	22 08 01	+52 21.5	1	9.56		0.88		0.75		4	K3	7010
	+51 03268		B	22 08 01	+52 21.5	1	14.13		0.40				2		7010
210483	+18 04947	G 126 - 59		22 08 02	+18 33.0	2	7.59	.000	0.58	.019	0.06	.023	4	G1 V	1620,3026
		WLS 2136 85 # 9		22 08 04	+85 08.4	1	11.68		0.65		0.08		2		1375
		Cep sq 2 # 84		22 08 07	+55 13.3	1	11.67		0.26		0.18		1	B9.5V	273
		Cep sq 2 # 83		22 08 07	+56 09.6	1	11.38		0.43		-0.04		1	B8 III	273
210516	+20 05093	HR 8459		22 08 08	+20 43.9	2	6.46		0.10	.013	0.10	.004	5	A3 III	985,1049
		AAS60,495 # 12		22 08 08	−18 25.1	1	14.26		0.69				1		1584
210502	+10 04701	HR 8458		22 08 10	+11 22.7	2	5.75	.026	1.59	.003	1.94		6	M1 III	71,3001
	+55 02694			22 08 12	+55 32.2	1	9.77		1.26		0.94		3		261
	+22 04567	G 126 - 60		22 08 13	+22 33.1	5	9.19	.043	0.93	.011	0.67	.007	9	K3 V	680,1003,1017,1620,3072
	+72 01018			22 08 14	+72 38.6	2	9.80	.000	0.55	.000	0.14		5	B8 V	206,207
		MCC 205	AB	22 08 15	+07 39.8	1	10.92		1.48				2	K5	1619
		Cep sq 2 # 85		22 08 15	+54 17.7	1	11.25		0.23		0.08		1	B9.5V	273
		Cep sq 2 # 86		22 08 15	+54 28.3	1	10.35		0.20		-0.37		1	B5 III	273
		Cep sq 2 # 87		22 08 16	+54 13.2	1	10.82		0.31		0.21		1	A0 V	273
		Cep sq 2 # 88		22 08 16	+54 54.6	1	11.94		0.40		0.29		1	A0 III	273
210464	−21 06173	HR 8457		22 08 16	−21 28.7	3	6.09	.008	0.50	.000	0.02		8	F6 V	15,2012,3053
235743	+50 03594			22 08 18	+50 45.0	1	8.59		1.04				33	K2	6011
		Cep sq 2 # 89		22 08 19	+54 14.0	1	10.87		0.19		0.02		1	B8 V	273
210627	+56 02721			22 08 19	+57 13.5	2	8.10	.000	1.28	.005	1.34		8	K2	261,1025
	+60 02351			22 08 19	+61 09.0	1	9.62		0.15		-0.55		2	B8	213
	+40 04748			22 08 22	+40 28.8	2	10.05	.010	0.36	.010	0.02		5	A5	1307,1724
210606	+50 03595			22 08 22	+50 36.8	1	9.11		1.82				2	M1	369
210628	+55 02695			22 08 22	+55 50.2	4	6.93	.010	0.09	.009	-0.37	.000	16	B6:V:	261,273,1012,1025
		vdB 150 # 16		22 08 22	+73 13.6	1	12.41		0.79				1		207
210554	+13 04862			22 08 23	+13 42.3	1	8.78		0.30		0.05		3	A3	1364
210441	−44 14805			22 08 23	−44 05.4	4	6.60	.004	0.99	.005	0.72		12	G8/K0 III	116,1020,2012,2012
210643	+57 02471			22 08 26	+57 34.5	1	8.81		0.43				3	F5	1025
	+70 01218	G 263 - 13		22 08 28	+71 05.3	1	9.61		0.78		0.41		1	G5	1658
	+72 01019			22 08 28	+73 10.2	1	9.23		0.28				1	F0	207

Table 1   1183

HD	DM	Other Id	N	Rem	α_{1950}	δ_{1950}	S	V	σ_V	B–V	σ_{B-V}	U–B	σ_{U-B}	n	Spectrum	References
	+54 02695				22 08 29	+54 57.6	1	10.63		0.17		−0.05		1	B9 V	273
	+61 02263				22 08 29	+62 21.9	1	9.82		0.14		−0.47		2	B8	27
	+41 04408				22 08 30	+42 26.9	1	9.34		0.40				2	F0	1724
		vdB 150 # 17			22 08 30	+73 13.0	1	12.47		0.81				1	G1 V	207
210607	+39 04774				22 08 31	+40 01.0	1	8.57		0.34				2	F8	1724
	−13 06123				22 08 31	−13 06.4	1	11.07		1.32		1.25		1	K7 V	3072
		TOU28,33 T8# 45			22 08 32	+55 58.3	1	10.23		0.19				3		1025
		WLS 2200 65 # 9			22 08 32	+65 17.9	1	13.04		0.69		0.12		2		1375
210629	+47 03720	IDS22066N4725	A		22 08 34	+47 40.0	1	7.47		0.49				3	F5	1724
		Cep sq 2 # 95			22 08 34	+55 10.1	1	11.06		0.42		−0.07		1	B6 III	273
		vdB 150 # 18			22 08 35	+73 11.1	1	12.46		1.20				1	G8	207
210594	+29 04604	HR 8460			22 08 36	+30 18.4	1	6.32		0.18		0.16		2	A8 IV	252
210645	+50 03596				22 08 36	+50 33.4	2	8.25	.005	0.00	.010	−0.34	.068	3	B8 IV/V	401,592
	+55 02696		A		22 08 36	+55 59.1	1	9.93		0.30				3	B9 III	1025
	+55 02696		B		22 08 36	+55 59.1	1	10.12		0.51		0.06		1		273
		LS III +56 028			22 08 36	+56 05.6	1	11.89		0.74		−0.30		2		483
	+55 02697				22 08 36	+56 22.8	1	9.65		0.79				4	A4	1025
	+57 02472	LS III +57 027			22 08 36	+57 53.7	2	9.68	.015	1.34	.005	0.19	.025	4		405,483
		G 27 - 17			22 08 37	−02 47.4	1	12.14		1.56		1.25		1		1620
210595	+23 04482	IDS22063N2359	AB		22 08 38	+24 14.5	1	8.60		0.45		−0.05		2	F7 V wle	3054
		Cep sq 2 # 97			22 08 41	+54 52.6	1	11.54		0.29		0.20		1	B9 III	273
		Cep sq 2 # 98			22 08 41	+55 15.4	1	11.70		0.45		−0.05		1	B7 III	273
		Cep sq 2 # 99			22 08 41	+55 47.5	1	10.83		0.28		0.24		1	A0 V	273
	+40 04750				22 08 43	+40 53.7	1	9.94		0.27				2	A5	1724
		vdB 150 # 19			22 08 47	+73 10.6	1	12.48		1.36				1	K0 III	207
		AAS11,365 T7# 33			22 08 48	+55 04.	1	10.75		1.96		2.04		3		261
	+41 04409	IDS22067N4141	AB		22 08 49	+41 56.0	1	9.04		0.59				2	G0	1724
210806	+72 01020				22 08 50	+73 08.7	1	8.38		0.09		−0.37		4	B8 IV	206
210807	+71 01111	HR 8468			22 08 51	+72 05.7	7	4.79	.010	0.92	.005	0.60	.023	26	G8 III	15,1355,1363,3016*
210368	−71 02667				22 08 54	−71 23.1	1	9.65		0.80		0.45		1	G6 III/IV	1696
	−40 14683				22 08 56	−40 24.5	1	11.40		0.70		0.24		3		1730
210743	+63 01818				22 08 57	+64 05.7	1	8.01		0.38		−0.20		2	A0	401
		Cep sq 2 # 101			22 08 59	+55 04.3	1	11.20		0.31		−0.28		1	B5 III	273
210667	+35 04725	G 214 - 13			22 09 01	+36 00.8	3	7.27	.029	0.81	.010	0.46	.005	10	K0	101,196,1067
		Cep sq 2 # 105			22 09 01	+55 40.1	1	11.40		0.63		0.31		1	B9 III	273
	−40 14684				22 09 01	−40 26.1	1	10.61		0.96		0.65		2		1730
210770	+64 01634				22 09 02	+65 16.6	1	7.66		−0.01		0.00		2	A0	401
		VES 924			22 09 03	+54 43.8	1	10.63		0.64		0.23		2		8113
		Steph 1985			22 09 04	−19 37.8	1	11.13		1.38		1.27		1	K7	1746
210647	+16 04684				22 09 05	+17 16.4	1	8.29		0.87				4	K0	882
	+17 04708	G 126 - 62			22 09 06	+17 50.7	11	9.47	.011	0.44	.010	−0.19	.017	18	sdF8	22,308,792,908,927*
		G 126 - 61			22 09 06	+18 10.6	2	10.24	.037	1.49	.012	1.14	.029	8	M2	680,1663
210697	+47 03722				22 09 06	+48 25.8	1	6.57		−0.06		−0.27		2	B9	401
		Cep sq 2 # 106			22 09 07	+54 56.0	1	12.11		0.35		0.09		1	A0 III	273
210745	+57 02475	HR 8465			22 09 07	+57 57.3	10	3.35	.012	1.57	.018	1.73	.016	30	K1 Ib	15,369,851,1025,1194*
210631	+5 04966	G 18 - 35			22 09 08	+05 56.7	1	8.49		0.58		0.00		1	G0	333,1620
		Cep sq 2 # 108			22 09 09	+54 44.1	1	12.29		0.59		0.35		1		273
		Cep sq 2 # 107			22 09 09	+54 50.1	1	11.49		0.35		0.31		1	A2 IV	273
210760	+57 02476	IDS22074N5727	A		22 09 10	+57 41.6	1	7.42		0.18				5	A5	1025
		AJ89,1229 # 312			22 09 11	+02 39.9	1	11.49		1.64				1		1532
		G 214 - 14			22 09 11	+40 46.0	1	11.10		1.46		1.23		3	M2	1723
		G 126 - 63			22 09 12	+17 48.2	3	12.15	.008	0.57	.027	−0.13	.022	4		308,1620,1696
235746	+49 03760				22 09 13	+50 23.7	1	9.24		1.15		1.11		1	M0	1462
210715	+50 03602	HR 8463	★ A		22 09 13	+50 34.5	2	5.39	.005	0.15	.009	0.07	.028	4	A5 V	1462,3023
		Cep sq 2 # 109			22 09 13	+55 13.3	1	12.37		0.33		0.12		1	A1 V	273
		vdB 150 # 15			22 09 15	+73 10.5	1	12.12		1.70				1	G2 V	207
		Cep sq 2 # 111			22 09 16	+54 15.1	1	12.49		0.27		0.06		1	B9.5V	273
210873	+71 01112	HR 8473			22 09 17	+71 51.9	5	6.37	.005	−0.05	.016	−0.18	.010	17	B9 p Hg(Mn)	15,1007,1013,1079,8015
	+51 03275				22 09 18	+51 34.4	1	9.71		1.82				1		369
210731	+46 03611				22 09 20	+46 54.0	1	7.40		0.58				2	F8	1724
235747	+51 03276				22 09 20	+51 54.9	1	9.16		0.56		−0.04		2	F8	3016
		Cep sq 2 # 113			22 09 20	+55 43.6	1	11.01		0.35		0.34		1	A0 V	273
		Cep sq 2 # 115			22 09 22	+54 13.1	1	11.47		0.27		0.17		1	A1 III	273
235748	+53 02804	IDS22075N5317	ABC		22 09 23	+53 31.4	1	8.77		0.98		0.87		1	F5	8084
		vdB 150 # 11			22 09 24	+73 07.0	1	11.51		0.91				1	F8 V	207
		CS 22886 # 13			22 09 24	−09 00.6	1	13.65		0.24		0.14		1		1736
210702	+15 04592	HR 8461			22 09 26	+15 47.6	7	5.95	.008	0.95	.005	0.73	.011	25	K1 III	15,252,1007,1013,3016*
210808	+62 02045	IDS22079N6254	AB		22 09 27	+63 09.2	1	7.32		0.19		−0.63		2	B5	401
210701	+22 04574				22 09 28	+22 52.2	1	8.86		0.49		0.00		2	G0	1375
	+52 03131				22 09 28	+52 54.1	1	9.56		1.83				2	M0 III	369
210573	−57 10066				22 09 28	−56 39.8	1	9.75		1.04		0.91		1	K3 V	3072
210884	+69 01228	HR 8474	★ A		22 09 30	+69 53.1	5	5.50	.008	0.38	.005	−0.04	.003	9	F2 V	1,15,207,1028,3016
210884	+69 01228	IDS22084N6938	B		22 09 30	+69 53.1	3	8.62	.005	0.86	.005	0.49	.000	7		1,1028,3032
	+40 04754	IDS22074N4026	AB		22 09 32	+40 41.1	1	9.21		0.37				3	F0	1724
210787	+52 03132	IDS22077N5230	AB		22 09 33	+52 44.5	1	8.12		0.40		−0.02		3	F0	3016
		Cep sq 2 # 118			22 09 34	+55 13.0	1	11.07		0.35		−0.40		1	B2 V	273
		Cep sq 2 # 119			22 09 34	+55 35.0	1	10.60		0.31		−0.24		1	B5 III	273
		CS 22892 # 45			22 09 36	−14 44.6	1	15.09		0.23		−0.56		1		1736
		Cep sq 2 # 120			22 09 37	+55 27.0	1	11.41		0.40		0.28		1	A0 I:V	273
		CS 22892 # 46			22 09 42	−15 08.3	1	14.55		0.41		−0.04		1		1736

HD	DM	Other Id	N	Rem	α_{1950}	δ_{1950}	S	V	σ_V	B–V	σ_{B-V}	U–B	σ_{U-B}	n	Spectrum	References
210718	+7 04818				22 09 44	+08 12.8	1	8.31		0.49		-0.02		2	F8	1733
210705	-14 06229	HR 8462			22 09 44	-14 26.5	4	6.04	.021	0.37	.009	-0.02	.010	11	F0 V	254,1117,2007,3016
210771	+38 04703				22 09 45	+38 47.5	1	8.64		0.40		-0.01		3	F3 V	833
210809	+51 03281	LS III +52 030			22 09 45	+52 11.0	3	7.56	.011	0.04	.004	-0.87	.008	14	O9 Ib	261,1011,1012
210800	+49 03764	NGC 7243 - 306			22 09 46	+49 41.4	1	7.71		0.81				5	G5	1724
235749	+54 02698	LS III +55 025			22 09 46	+55 01.2	2	8.85	.057	1.67	.052	0.15		4	M0 III	1025,8113
	+54 02699				22 09 46	+55 24.7	1	10.46		0.30		0.20		1	B9.5V	273
210750	+12 04779				22 09 47	+13 14.5	1	8.70		0.66				3	G5	882
210762	+24 04548	HR 8466			22 09 48	+24 42.2	2	5.94	.015	1.50	.012	1.68		5	K1 III	71,1501
		AAS6,117 # 21			22 09 48	+53 57.	1	10.85		0.16		0.10		2	B8	171
	+53 02806				22 09 48	+54 10.1	1	10.33		0.44		0.23		1	A3 III	273
		Cep sq 2 # 123			22 09 48	+54 50.0	1	12.10		0.30		0.18		1	A0 III	273
210839	+58 02402	HR 8469		⋆	22 09 49	+59 10.0	15	5.05	.014	0.24	.012	-0.74	.009	52	O6 If	1,5,15,154,213,1011*
210874	+61 02265	LS III +62 009			22 09 52	+62 17.1	1	8.44		0.07		-0.06		6	A2	213
	+30 04633	G 188 - 46			22 09 53	+31 19.4	3	10.14	.011	1.16	.010	1.16	.052	12	M0	196,1663,1746
		Cep sq 2 # 125			22 09 55	+54 18.1	1	12.04		0.52		0.16		1	A2.3:II	273
		Cep sq 2 # 126			22 09 58	+54 51.7	1	12.16		0.29		0.03		1	B9 III	273
210709	-36 15262				22 09 58	-35 40.7	2	9.23	.009	1.10	.005	0.87	.019	4	G8 II/III	565,3048
210875	+57 02478	IDS22082N5713		AB	22 09 59	+57 28.1	1	8.69		0.35				3	A2	1025
		vdB 150 # 12			22 10 00	+73 05.4	1	10.02		1.67				1	K1 III	207
210819	+49 03767	NGC 7243 - 181		⋆A	22 10 00	+49 56.9	2	7.24	.009	0.43	.000			4	F0	1724,6009
		AAS11,365 T8# 1			22 10 00	+54 38.	1	12.40		0.38		0.23		3		261
210855	+56 02727	HR 8472		⋆A	22 10 00	+56 35.4	5	5.25	.015	0.51	.011	0.06	.028	14	F8 V	254,261,338,1025,3053
210820	+46 03612				22 10 01	+46 50.9	1	6.65		-0.03		-0.17		2	A0	401
		G 126 - 65			22 10 04	+16 49.1	1	11.79		1.38		1.27		1		1620
		Cep sq 2 # 129			22 10 05	+55 55.5	1	11.67		0.37		0.26		1	A0 V	273
210752	-7 05727				22 10 05	-06 43.0	1	7.40		0.52		-0.08		2	G0	1375
		G 18 - 36			22 10 06	+08 18.7	1	12.00		1.56		1.22		2		333,1620
	+15 04594				22 10 06	+15 51.4	1	10.32		1.49		1.96		1		1364
		AA26,31 # 1			22 10 06	+54 43.9	1	12.44		0.25		-0.64		5		483
		AAS11,365 T8# 4			22 10 06	+54 51.	1	14.32		0.83		0.23		2		261
		AAS11,365 T8# 3			22 10 06	+54 53.	1	13.76		0.46		0.32		3		261
		AAS11,365 T8# 2			22 10 06	+54 55.	1	12.82		0.62		0.07		3		261
		Cep sq 2 # 130			22 10 06	+55 00.8	1	11.40		0.23		-0.17		1	A0 V	273
		G 233 - 2			22 10 06	+56 34.9	1	13.75		0.16		-0.62		2		1658
210904	+60 02356				22 10 06	+61 21.1	1	8.43		0.14		-0.63		3	A0	213
	+49 03769				22 10 07	+49 41.3	1	9.63		1.74				2		369
		Cep sq 2 # 131			22 10 07	+55 12.4	1	11.89		0.48		0.27		1	A0 III	273
210857	+48 03640				22 10 08	+49 00.7	1	8.29		0.26				3	F0	1724
		Cep sq 2 # 132			22 10 08	+54 45.1	1	11.20		0.18		-0.23		1	B8 III	273
210763	-5 05732	HR 8467			22 10 08	-04 58.1	3	6.39	.008	0.49	.009	0.03	.036	8	F7 V	15,254,1071
239881	+57 02480				22 10 09	+58 13.1	1	9.74		0.58		-0.03		2	G0	3032
210739	-26 16033	HR 8464			22 10 09	-26 34.5	2	6.16	.005	0.17	.000			7	A3 V	15,2012
		AJ89,1229 # 313			22 10 12	+04 43.9	1	11.31		1.58				1	M1	1632
		AAS11,365 T8# 8			22 10 12	+54 34.	1	11.58		0.16		-0.10		3		261
		AAS11,365 T8# 7			22 10 12	+54 38.	1	13.38		0.49		0.29		3		261
		AAS11,365 T8# 6			22 10 12	+54 39.	1	14.04		0.50		0.27		3		261
		AAS11,365 T8# 5			22 10 12	+54 52.	1	14.45		0.38		0.16		2		261
		Cep sq 2 # 135			22 10 12	+55 15.8	1	11.10		0.48		-0.03		1	B9.5V	273
		vdB 150 # 9			22 10 12	+72 56.3	1	11.13		0.62				1	A2 V	207
		Cep sq 2 # 136			22 10 13	+54 38.1	1	10.92		0.21		0.15		1	A0 V	273
210905	+58 02403	HR 8476			22 10 13	+58 50.2	2	6.28	.020	1.12	.005	1.07	.000	8	K0 III	37,3032
		Cep sq 2 # 138			22 10 16	+55 03.6	1	10.39		0.20		-0.15		1	B7 III	273
210756	-27 15823				22 10 16	-26 44.6	1	7.63		1.01				4	K0 III	2012
	+52 03135	LS III +53 047			22 10 17	+53 27.1	1	9.62		0.33		-0.50		2	B3 II:	1012
210886	+55 02701				22 10 17	+55 30.8	2	9.50	.000	0.18	.000			5		273,1025
	+16 04689				22 10 18	+17 02.7	1	10.37		-0.15		-0.73		1	B5	963
		AAS11,365 T8# 10			22 10 18	+54 45.	1	12.16		1.83		1.98		3		261
		AAS11,365 T8# 9			22 10 18	+54 50.	2	11.32	.014	0.25	.005	0.09	.037	4	B9	261,273
	+56 02729				22 10 19	+56 29.8	1	9.98		0.45				3	A5	1025
211003	+69 01229				22 10 19	+70 08.0	1	8.24		0.34				1	F0	207
		NGC 7235 - 18			22 10 20	+57 02.0	1	13.76		0.72		-0.09		1		49
210510	-75 01744	IDS22058S7529		AB	22 10 20	-75 14.6	1	8.44		0.28		0.16		2	A3 IV[s]	1730
210822	+14 04753				22 10 22	+14 45.6	1	8.16		0.17		0.11		2	A3 V	1364
		NGC 7235 - 97			22 10 22	+57 00.8	1	16.56		0.74		0.15		4		1552
		NGC 7235 - 2			22 10 22	+57 01.5	1	10.56		0.67		-0.32		1	B0 II	49
210939	+60 02358	HR 8479		⋆A	22 10 22	+60 30.7	2	5.35	.005	1.17	.005	1.15		4	K1 III	71,3016
		AAS11,365 T8# 15			22 10 24	+54 44.	1	11.69		1.41		1.33		4		261
		AAS11,365 T8# 14			22 10 24	+54 48.	1	11.51		1.60		1.60		5		261
		AAS11,365 T8# 13			22 10 24	+54 52.	1	15.03		0.94		0.14		1		261
		AAS11,365 T8# 12			22 10 24	+54 58.	1	12.51		0.16		-0.47		3		261
		AAS11,365 T8# 11			22 10 24	+54 59.	1	11.03		0.53		0.01		3		261
		Cep sq 2 # 141			22 10 24	+55 01.5	1	11.38		0.38		0.36		1	A4 III	273
		NGC 7235 - 25			22 10 24	+57 00.1	1	15.37		0.78		0.23		1		49
		Cep sq 2 # 142			22 10 25	+54 19.2	1	11.88		0.31		0.25		1		273
		NGC 7235 - 11			22 10 25	+57 01.3	1	12.82		0.81		0.35		1		49
		Cep sq 2 # 143			22 10 26	+55 19.4	1	11.11		0.46		-0.02		1	B8 III	273
		NGC 7235 - 22			22 10 26	+57 00.0	1	14.26		1.41				1		49
		Cep sq 2 # 144			22 10 27	+54 14.8	1	11.75		0.16		-0.28		1	B5 III	273
		NGC 7235 - 21			22 10 27	+57 01.3	1	14.20		0.75		-0.08		1		49

Table 1   1185

HD	DM	Other Id	N	Rem	α_{1950}	δ_{1950}	S	V	σ_V	B–V	$\sigma_{B–V}$	U–B	$\sigma_{U–B}$	n	Spectrum	References
	+34 04620				22 10 28	+34 45.9	1	8.87		0.41		0.07		2	F5	1625
		NGC 7235 - 10			22 10 28	+57 01.7	1	12.74		0.63		-0.26		1		49
210803	-11 05781	IDS22078S1140		AB	22 10 28	-11 26.2	3	9.43	.010	0.92	.005	0.65	.006	18	G5	830,1064,1783
		G 214 - 15			22 10 29	+33 02.6	1	16.32		1.88				4		538
		Cep sq 2 # 145			22 10 29	+54 33.3	1	11.19		0.27		0.14		1	A0 V	273
		AAS11,365 T8# 18			22 10 30	+54 41.	1	12.42		0.34		0.25		3		261
		AAS11,365 T8# 17			22 10 30	+54 47.	1	13.04		1.14		0.73		5		261
		BPM 44407			22 10 30	-47 03.	1	14.02		0.41		-0.15		1		3065
210922	+54 02702				22 10 31	+54 51.0	2	7.18	.030	1.37	.015	1.39		8	K1 III	261,1025
		Cep sq 2 # 146			22 10 32	+54 14.0	1	11.66		0.88				1	B9 II	273
239883	+56 02730				22 10 32	+57 23.3	1	8.84		1.03				3	K0	1025
		NGC 7235 - 5			22 10 33	+57 00.3	1	11.40		0.65		-0.25		1	B2 III	49
		NGC 7235 - 9			22 10 33	+57 00.8	1	12.26		0.72		-0.25		1		49
210889	+33 04456	HR 8475			22 10 35	+34 21.5	1	5.33		1.13		1.13		3	K2 III	1355
		AAS11,365 T8# 19			22 10 36	+54 44.	1	12.89		1.22		1.04		3		261
		Cep sq 2 # 148			22 10 36	+55 10.0	1	11.32		0.24		0.00		1	B8 V	273
		NGC 7235 - 19			22 10 36	+56 58.3	1	13.92		0.89		0.33		1		49
		NGC 7235 - 20			22 10 37	+56 57.8	1	14.15		0.68				1		49
		NGC 7235 - 24			22 10 37	+56 57.9	1	15.21		0.76		0.30		1		49
		NGC 7235 - 6			22 10 37	+57 01.5	1	11.47		0.66		-0.28		1	B3 V	49
		Cep sq 2 # 149			22 10 38	+54 19.7	1	11.22		0.09		-0.39		1	B2 III	273
		NGC 7235 - 17			22 10 38	+56 57.5	2	13.33	.000	0.64	.010	-0.20	.010	3		49,1552
		NGC 7235 - 135			22 10 38	+56 57.5	1	16.52		0.54		0.50		4		1552
		NGC 7235 - 133			22 10 39	+56 57.2	1	16.40		1.02		0.48		6		1552
		NGC 7235 - 12			22 10 39	+56 59.8	2	12.86	.010	1.42	.005	0.93	.034	3		49,1552
		NGC 7235 - 23			22 10 39	+57 00.3	1	14.52		0.69		0.02		1		49
		VV Peg			22 10 41	+18 12.2	1	11.41		0.16		0.18		1	F1.5	699
		LS III +55 026			22 10 41	+55 17.3	3	11.41	.021	0.40	.015	-0.57	.016	6		273,405,483
		NGC 7235 - 14			22 10 41	+57 01.9	1	12.93		0.64		-0.15		1		49
		LS III +52 031			22 10 42	+52 15.0	2	10.77	.024	0.22	.015	-0.73	.015	3		483,8113
		Cep sq 2 # 152			22 10 42	+54 49.1	1	11.25		0.14		-0.53		1	B3 V	273
		Cep sq 2 # 154			22 10 43	+55 21.8	1	11.95		0.31		0.20		1	B9 V	273
		NGC 7235 - 27			22 10 43	+56 58.6	1	15.72		1.03				1		49
		Cep sq 2 # 155			22 10 45	+54 20.0	1	11.41		0.32		0.22		1	A1 III	273
		NGC 7235 - 26			22 10 45	+56 59.0	1	15.46		0.84		0.28		1		49
		NGC 7235 - 8			22 10 45	+57 00.7	1	11.93		0.64		-0.29		1	B3 V	49
		CS 22886 # 8			22 10 45	-09 49.2	1	14.44		0.47		-0.22		1		1736
210890	+17 04712				22 10 46	+18 01.9	1	6.60		1.29				2	K2	882
		NGC 7235 - 16			22 10 46	+57 01.6	1	13.06		0.87		-0.10		1		49
210981	+55 02702				22 10 47	+55 56.9	2	8.22	.024	1.64	.024	1.80		6	K5	261,1025
239886	+56 02733	NGC 7235 - 1			22 10 47	+57 01.1	3	8.81	.016	0.86	.009	0.12	.000	5	B9 Iab	49,1012,1025
		vdB 150 # 10			22 10 47	+72 56.7	1	11.42		0.62				1	G0 V	207
		CS 22886 # 12			22 10 47	-08 58.6	1	14.52		0.47		-0.16		1		1736
		NGC 7235 - 38			22 10 48	+57 02.	1	16.43		1.03		0.56		4		1552
		NGC 7235 - 140			22 10 48	+57 02.	1	16.43		0.82		0.46		5		1552
		NGC 7235 - 141			22 10 48	+57 02.	1	17.44		0.82		0.35		5		1552
		NGC 7235 - 142			22 10 48	+57 02.	1	17.48		0.72		-0.21		4		1552
		NGC 7235 - 143			22 10 48	+57 02.	1	17.54		0.12		0.07		4		1552
		Cep sq 2 # 156			22 10 49	+54 58.2	1	11.65		0.14		-0.10		1	B8 III	273
211029	+62 02048	HR 8483			22 10 49	+63 02.6	3	5.75	.035	1.62	.052	1.93		8	M3 IIIab	71,369,3001
210942	+44 04071				22 10 50	+44 31.4	1	7.98		0.59				3	G0	1724
		NGC 7235 - 3			22 10 50	+57 00.9	1	10.97		0.69		-0.22		1	B2 II	49
210925	+25 04691				22 10 53	+25 41.7	2	6.59	.010	1.02	.005	0.76	.005	4	K0 IV	105,3016
		NGC 7235 - 15			22 10 54	+57 00.9	1	13.06		0.74		-0.09		1		49
		WLS 2220 25 # 6			22 10 55	+25 21.8	1	10.58		1.14		0.90		2		1375
210944	+26 04379				22 10 55	+27 04.3	1	7.20		0.48		0.00		2	F5 V	1375
		LP 819 - 52			22 10 56	-17 55.8	2	13.58	.036	1.75	.014	1.48		5		940,3073
210848	-25 15815	HR 8470			22 10 56	-25 25.8	2	5.59	.004	0.50	.003	0.22		6	F7 II	58,2007
	+54 02705				22 10 57	+54 53.4	1	9.95		0.07		-0.51		3		261
		G 27 - 19			22 10 58	+00 20.8	1	13.13		1.32		1.10		1		3062
210909	+12 04781				22 10 58	+12 28.4	1	8.36		1.01				2	K2	882
	+51 03289	LS III +52 032			22 10 58	+52 15.1	1	10.43		0.06		-0.61		2	F2 V	483
		Cep sq 2 # 159			22 10 59	+55 55.7	1	11.75		0.31		-0.04		1	B9 III	273
		NGC 7235 - 4			22 10 59	+57 00.3	1	11.39		1.49				1		49
		LP 875 - 32			22 10 59	-24 40.9	1	13.68		0.85		0.51		2		1696
210894	-5 05735				22 11 00	-04 48.5	3	9.17	.005	1.32	.007	1.45	.014	27	K2	989,1375,1729
		AA26,31 # 3			22 11 01	+55 24.9	1	12.37		0.32		-0.44		6		483
		NGC 7235 - 39			22 11 01	+57 01.1	1	16.32		0.81		0.61		4		1552
239890	+57 02481				22 11 01	+58 07.4	1	9.40		1.50		1.32		1	K5	338
211004	+50 03616				22 11 02	+50 56.5	2	8.09	.005	0.95	.010	0.67		4	G5	1566,1724
235756	+50 03615				22 11 03	+50 33.4	1	9.41		0.27				2		1724
		AA26,31 # 4			22 11 03	+55 25.3	1	14.08		0.39		-0.08		4		483
		NGC 7235 - 13			22 11 03	+57 01.7	1	12.86		0.70		-0.21		1		49
	+43 04153				22 11 04	+44 22.6	1	9.39		0.60				2	F8	1724
		Cep sq 2 # 160			22 11 05	+55 00.0	1	11.52		0.24		0.13		1	A0 III	273
239891	+56 02734				22 11 05	+56 29.3	1	9.22		0.43				3	A2	1025
211030	+53 02813				22 11 06	+53 44.2	1	8.54		0.10		-0.34		2	A0	1566
		LS III +55 027			22 11 06	+55 25.5	2	11.74	.054	0.23	.023	-0.28	.018	5	B7	273,483
235758	+53 02814				22 11 09	+54 14.5	1	11.51		0.27		0.23		1	K3 III	273
		LP 699 - 58			22 11 09	-06 41.8	1	14.62		1.45		0.85		1		1773

HD	DM	Other Id	N Rem	α_{1950}	δ_{1950}	S	V	σ_V	B−V	σ_{B-V}	U−B	σ_{U-B}	n	Spectrum	References
		LS III +57 034		22 11 10	+57 24.8	1	10.32		0.65		−0.34		2	B1 III	1012
		LP 699 - 59		22 11 12	−06 41.8	1	14.92		1.54		1.17		1		1773
		Cep sq 2 # 166		22 11 14	+55 00.6	1	10.91		0.32		0.25		1	A1 V	273
		CS 22892 # 48		22 11 14	−15 01.8	1	14.13		0.35		0.01		1		1736
210958	+15 04598			22 11 15	+15 37.4	1	8.46		0.96				4	K0 III	882
211005	+38 04710			22 11 15	+38 58.8	1	8.08		0.75		0.24		3	G0 III	833
211148	+69 01230			22 11 15	+69 38.7	1	8.79		0.19				1	A0	207
		CS 22881 # 60		22 11 16	−40 17.1	2	14.26	.010	0.40	.007	−0.23	.009	3		1580,1736
210946	+0 04838			22 11 17	+01 21.6	2	8.07	.000	1.08	.018	0.78	.032	2	K0:IV	565,3048
235759	+52 03144	V341 Lac		22 11 17	+53 22.6	2	8.72	.109	2.16	.020	1.88		5	M2 III	261,369
		Cep sq 2 # 171		22 11 18	+54 19.5	1	11.31		0.21		−0.43		1		273
		NGC 7235 - 7		22 11 18	+57 00.1	1	11.48		0.67		−0.41		1	B1 V	49
		LTT 16505		22 11 21	+05 01.7	1	11.79		1.49				1	M2 V:	1532
211006	+27 04280	HR 8482		22 11 21	+28 21.6	1	5.89		1.15		1.22		2	K2 III	252
	+19 04885			22 11 22	+19 58.6	1	9.08		0.31		0.04		2	F2	1648
211057	+54 02708			22 11 25	+55 03.9	3	7.60	.028	0.08	.010	−0.17		8	B9	273,401,1025
235760	+53 02817			22 11 27	+53 39.7	1	8.98		1.79		1.65		3	K2 III	261
		Cep sq 2 # 173		22 11 27	+54 51.3	1	11.83		1.02		0.75		1	B7 V	273
		Cep sq 2 # 174		22 11 28	+54 46.5	1	12.32		0.45		0.30		1		273
210934	−28 17653	HR 8478		22 11 29	−28 00.9	5	5.43	.026	−0.13	.013	−0.43	.056	11	B7 V	1034,1079,1637,2007*
211071	+52 03146			22 11 30	+52 54.3	2	7.26	.005	0.21	.000	0.11	.005	4	A5	1566,1733
		AA26,31 # 6		22 11 30	+55 00.7	2	11.77	.049	0.22	.015	−0.33	.024	3		273,483
211070	+54 02709			22 11 31	+55 13.4	1	7.57		1.09				5	K2 III	1025
239895	+56 02735	LS III +57 036		22 11 31	+57 25.1	1	8.63		1.02		0.12		2	B8 Ia	1012
211123	+61 02266			22 11 31	+62 19.5	1	8.57		0.02		−0.16		2	A0	27
210960	−21 06180	HR 8480	⋆ AB	22 11 32	−21 19.4	3	5.33	.013	0.82	.013	0.45	.019	9	K0 III+F2 V	58,938,2007
	+62 02051	LS III +62 010		22 11 33	+62 27.6	1	9.39		0.28		−0.39		2	B8	213
		RU Peg		22 11 36	+12 27.2	1	12.75		0.71		−0.37		1	sd:Be	698
210918	−41 14804	HR 8477		22 11 36	−41 37.2	4	6.23	.013	0.65	.012	0.13	.018	17	G5 V	15,1311,2012,3077
		Cep sq 2 # 179		22 11 37	+54 25.6	1	11.97		0.41		0.35		1	A1 III	273
211094	+48 03649	NGC 7243 - 694		22 11 39	+49 13.4	1	7.08		0.50				3	F5	1724
211124	+56 02736			22 11 40	+57 23.5	1	8.13		0.20				2	A2	1025
	+41 04427	IDS22096N4207	AB	22 11 42	+42 21.8	1	9.05		0.38				2	F2	1724
210975	−32 16963			22 11 42	−32 33.3	1	9.55		1.07		0.93		2	K3/4 V	3072
211031	+11 04757			22 11 44	+12 27.4	1	8.49		1.15				2	K2	882
211073	+38 04711	HR 8485	⋆ A	22 11 44	+39 28.0	6	4.50	.015	1.38	.007	1.43	.023	24	K3 III	15,1355,1363,3016*
211096	+44 04073	HR 8487		22 11 46	+45 11.5	2	5.48	.040	0.02	.020	0.00	.000	5	A0 III	865,1733
		Wolf 1332		22 11 46	−08 59.1	2	11.86	.005	0.87	.014	0.40	.023	4		979,1696
		Cep sq 2 # 181		22 11 48	+54 11.8	1	12.55		0.35		0.28		1	A2 V	273
	+27 04282	G 127 - 11		22 11 49	+27 36.5	2	10.36	.033	1.43	.042	1.24		2		1017,1620
239896	+55 02705			22 11 50	+56 05.4	1	9.68		0.51				2	F2	1025
	+45 03840			22 11 51	+46 11.3	1	10.08		0.36				3	G0	1724
	+54 02711			22 11 51	+54 52.1	1	9.97		0.16		−0.32		1		273
211136	+55 02704	IDS22100N5606	A	22 11 52	+56 21.4	1	9.45		0.21				2		1025
		YZ Aqr		22 11 52	−11 10.5	2	12.13	.037	0.28	.140	0.02	.054	2		699,700
		G 27 - 20		22 11 53	+02 27.5	1	10.40		1.26		1.20		1	K7	333,1620
211076	+16 04694	IDS22095N1642	AB	22 11 54	+16 56.5	2	6.47	.033	1.28	.047	1.39	.170	4	K4 III	1003,1733
		Cep sq 2 # 186		22 11 54	+54 28.1	1	10.96		0.57		0.19		1	B8 III	273
		Cep sq 2 # 187		22 11 54	+55 11.1	1	11.60		0.23		−0.09		1	B6 III	273
211075	+17 04714			22 11 56	+17 46.3	1	8.19		1.22				3	K2	882
211038	−16 06046			22 11 56	−16 03.7	4	6.55	.009	0.89	.005	0.57	.005	14	K0/1 V	1149,2009,2034,3016
	+2 04480			22 11 57	+03 14.1	1	8.29		1.67		1.52		3	G0	1657
	+54 02712	IDS22101N5513	AB	22 11 57	+55 27.9	2	9.80	.000	0.16	.000			4	A0 III	273,1025
	+53 02820	LS III +54 021		22 11 58	+54 09.7	1	9.95		0.10		−0.78		2	B0 IV:n	1012
		Cep sq 2 # 189		22 11 59	+55 20.5	1	11.34		0.22		−0.19		1	B7 II	273
211300	+72 01022	HR 8493	⋆ A	22 11 59	+73 03.5	5	6.05	.039	1.02	.016	0.85	.010	10	K0 II-III	150,207,1084,1355,3032
211300	+72 01022	IDS22111N7249	B	22 11 59	+73 03.5	4	8.48	.024	0.15	.008	0.08	.013	7		150,207,1084,3032
211149	+52 03149			22 12 00	+53 01.2	1	7.92		1.60		1.97		3	K5 III	171
	−3 05408			22 12 00	−02 45.8	1	9.63		1.15		0.99		2	K2	1375
		AA26,31 # 7		22 12 02	+54 38.6	2	11.69	.005	0.16	.019	−0.41	.005	3		273,483
239898	+57 02485			22 12 02	+57 59.5	1	9.04		0.31				4	A7	1025
		AA26,31 # 8		22 12 04	+55 33.4	1	13.27		0.51		−0.26		4		483
211161	+52 03150			22 12 05	+52 31.3	1	8.59		0.04		−0.17		2	A0	171
		NGC 7243 - 338		22 12 07	+49 39.1	1	12.33		0.39		0.27		4	A5	72
		LP 819 - 54		22 12 08	−19 31.5	1	12.06		0.75		0.24		2		1696
211025	−32 16965	IDS22093S3240	AB	22 12 08	−32 24.8	1	8.73		0.41				4	F2/3 V	2012
		Cep sq 2 # 191		22 12 11	+54 23.0	1	12.63		0.30		0.06		1	B8 V:	273
	+56 02737			22 12 11	+56 52.3	1	9.83		0.72		0.23		3	K0	196
	+52 03151	LS III +53 054		22 12 13	+53 25.9	1	9.60		0.22		−0.60		3		483
		NGC 7243 - 471		22 12 14	+49 36.1	1	10.80		0.19		−0.05		4	A0 V	72
	+54 02715			22 12 14	+54 37.5	1	11.43		0.28		0.17		1	B7 V	273
	+69 01231			22 12 14	+70 00.2	2	9.27	.032	0.15	.009	−0.02		5	B9.5V	206,207
211242	+62 02053	HR 8490		22 12 15	+62 54.8	2	6.10		−0.08	.004	−0.43	.004	6	B8 Vn	27,1079
		Cep sq 2 # 193		22 12 19	+55 20.3	1	11.90		0.39		0.26		1	B9 V	273
211243	+61 02267			22 12 21	+62 14.8	2	9.05	.015	0.03	.000	−0.45	.005	2	A2	27,401
		Wolf 1333		22 12 22	−09 04.1	1	11.37		0.62		−0.06		2		1696
211319	+69 01232			22 12 24	+69 53.5	1	8.11		0.12				1	A0	207
	+48 03653			22 12 27	+48 50.6	1	9.33		0.34				2	A5	1724
		AA26,31 # 9		22 12 27	+55 43.7	1	14.33		0.62		−0.19		4		483
211210	+49 03782	NGC 7243 - 475		22 12 29	+49 37.4	1	8.47		0.12		−0.28		11	B5 III	72

Table 1 1187

HD	DM	Other Id	N Rem	α1950	δ1950	S	V	σV	B-V	σB-V	U-B	σU-B	n	Spectrum	References
211227	+51 03298			22 12 29	+52 15.9	1	7.62		1.56		1.52		3	K2 IV	1566
		Cep sq 2 # 197		22 12 29	+54 56.2	1	10.34		0.17		0.10		1	B9 V	273
		NGC 7243 - 476		22 12 30	+49 33.4	1	11.59		0.29		0.17		4	A2 V	72
		LS III +54 022		22 12 30	+54 02.0	1	10.33		0.85		-0.17		2	B1 IV	1012
211226	+54 02716			22 12 30	+55 12.1	3	8.65	.010	0.07	.020	0.03		6	A0	261,273,1025
		Cep sq 2 # 199		22 12 30	+55 27.5	1	11.70		0.30		-0.16		1	B7 III	273
	+6 04987			22 12 31	+06 39.6	1	9.85		0.56		0.12		15	F8	1317
	+41 04434			22 12 31	+41 51.1	1	9.60		0.67				3	G0	1724
		Cep sq 2 # 200		22 12 31	+54 15.3	1	11.25		0.31		0.24		1	A1 V	273
239899	+55 02706	IDS22107N5519	AB	22 12 31	+55 33.8	2	9.23	.000	0.19	.000			6	A2	273,1025
211053	-45 14644	HR 8484		22 12 32	-44 42.1	8	6.10	.013	1.01	.005	0.82	.008	40	G8/K0 III	15,208,278,977,977*
		Cep sq 2 # 204		22 12 35	+54 33.9	1	10.48		0.29		0.26		1	A3 III	273
		LS III +54 023		22 12 36	+54 52.0	2	10.15	.036	1.13	.009	0.19	.023	5		273,483
		Smethells 103		22 12 36	-57 33.	1	11.56		1.41				1		1494
	+69 01233			22 12 37	+69 47.8	1	9.07		0.27				1		207
211088	-41 14810	HR 8486		22 12 37	-41 35.8	4	4.78	.007	0.80	.003	0.47	.000	15	G8 III +G	15,1075,2012,2024
211211	+42 04333	HR 8489		22 12 38	+42 42.3	2	5.71	.005	0.00	.000	-0.04	.000	5	A2 Vnn	985,1733
211054	-52 12000			22 12 38	-52 24.5	1	8.22		1.30		1.36		5	K2/3 III	1673
211181	+14 04761			22 12 39	+15 13.0	1	8.49		1.40				3	K2 III	882
	+43 04160			22 12 39	+44 02.6	1	9.21		0.53				3	F8	1724
	+14 04760			22 12 40	+14 31.8	1	10.05		0.27		0.12		1	A5	1364
		NGC 7243 - 350		22 12 42	+49 38.2	1	10.40		0.45		0.20		5	F5 II	1126
211264	+48 03656	NGC 7243 - 785		22 12 44	+49 06.3	1	7.56		0.57				2	F8	1724
211263	+49 03784	NGC 7243 - 480		22 12 45	+49 36.7	2	8.73	.008	0.11	.003	-0.27	.033	8	B7 IV	72,1126
211386	+69 01234			22 12 46	+69 38.9	1	7.84		1.10				1	K0	207
	+40 04766	IDS22107N4112	AB	22 12 48	+41 26.9	1	9.14		0.51				3	F5	1724
		LS III +54 024		22 12 48	+54 39.6	2	11.27	.023	0.24	.005	-0.57	.023	5		273,483
		AA26,31 # 12		22 12 49	+54 46.0	1	12.42		0.20		-0.37		5		483
	+44 04080			22 12 50	+45 04.9	1	9.06		0.43				3	F5	1724
235766	+50 03625			22 12 50	+50 32.3	1	8.96		1.10				2		1724
	+62 02055			22 12 50	+62 52.4	1	8.62		0.33		-0.33		2	B5	27
		Cep sq 2 # 208		22 12 51	+54 16.4	1	11.03		0.22		-0.22		1	B7 III	273
		NGC 7243 - 355		22 12 52	+49 37.8	1	12.20		0.73		0.29		4	G6	1126
	+60 02366	LS III +60 014		22 12 52	+60 28.8	1	9.74		0.45		-0.46		2		213
		Cep sq 2 # 209		22 12 53	+54 29.6	1	10.75		0.28		0.19		1	A0 V	273
	+6 04990	DH Peg		22 12 55	+06 34.2	1	9.25		0.26		0.21		1	A4	668
	+49 03786	NGC 7243 - 482		22 12 57	+49 32.7	2	9.10	.019	0.13	.005	-0.27	.029	9	B8 V	72,1126
239900	+56 02738			22 12 59	+56 37.0	1	9.01		0.21				3	B9	1025
		AAS6,117 # 24		22 13 00	+52 43.	1	11.40		0.09		0.01		5	B8	171
211320	+56 02739			22 13 00	+57 27.7	1	8.60		0.56				2	G0	1025
211188	-18 06093	IDS22103S1839		22 13 00	-18 23.5	1	9.66		0.91		0.56		2	K2 V	3072
210921	-75 01745			22 13 00	-75 12.0	1	8.82		1.23		1.24		2	K1/2 III	1730
		NGC 7243 - 484		22 13 02	+49 33.8	1	11.15		0.22		0.05		4	A1 V	1126
		Cep sq 2 # 210		22 13 02	+54 24.0	1	11.88		0.50		0.32		1	A0 III	273
		NGC 7243 - 483		22 13 03	+49 36.4	1	12.41		0.48		0.29		4	A6	1126
210853	-78 01442	HR 8471		22 13 03	-77 45.7	3	5.50	.007	0.31	.004	0.10	.015	14	F3 III	15,1075,2038
211229	+5 04982			22 13 04	+06 23.6	2	8.44	.015	0.04	.005	-0.12	.005	6	B9	1307,1317
		NGC 7243 - 485		22 13 04	+49 34.8	1	11.68		0.29		0.22		5	A3 V	1126
	+49 03788	NGC 7243 - 358	★ C	22 13 05	+49 37.7	2	9.39	.021	0.12	.003	-0.27	.032	17	B8 V	72,1126
		Cep sq 2 # 213		22 13 05	+54 38.8	1	10.53		0.22		-0.29		1	B6 III	273
		Cep sq 2 # 214		22 13 05	+55 01.4	1	12.21		0.41		0.21		1	B9.5III	273
211173	-32 16974			22 13 05	-32 06.6	2	8.46	.000	0.98	.002	0.63	.027	3	G8 II/III	565,3048
210728	-81 00992			22 13 06	-81 13.8	1	9.53		0.48		0.07		1	Fm δ Del	832
		NGC 7243 - 488		22 13 07	+49 34.3	1	11.82		0.62		0.18		4	F2	1126
		NGC 7243 - 487		22 13 07	+49 36.5	1	11.20		0.17		0.01		4	Am	1126
235767	+53 02825			22 13 07	+54 15.5	1	9.30		1.15		0.96		3	G5 III	171
	+40 04768			22 13 08	+40 31.2	1	9.28		0.37				2	F0	1724
	+54 02718	LS III +55 028		22 13 09	+55 12.9	2	10.15	.000	0.19	.000	-0.62		3	B1 V	273,1012
		NGC 7243 - 360		22 13 10	+49 38.9	1	10.27		1.75		2.10		4	K2 Ib	1126
		NGC 7243 - 489		22 13 11	+49 36.2	1	11.44		1.44		1.38		4		1126
211336	+56 02741	HR 8494, ε Cep	★ A	22 13 11	+56 47.6	8	4.19	.005	0.28	.005	0.04	.011	40	F0 IV	1,15,39,253,1025,1118*
211337	+49 03790	NGC 7243 - 362	★ A	22 13 12	+49 37.5	1	9.29		0.14		-0.31		4	B9 V	72
211337	+49 03790	NGC 7243 - 361	★ AB	22 13 12	+49 37.7	1	9.64		0.11		-0.34		4	B8 V	72
	+12 04789			22 13 13	+12 32.2	1	10.29		1.41		1.67		1	K2	1364
	+49 03789	NGC 7243 - 490		22 13 13	+49 30.5	1	10.10		0.12		-0.26		4	B7 III	1126
		NGC 7243 - 363		22 13 13	+49 38.3	1	11.79		0.26		0.16		4	A2	1126
		NGC 7243 - 493		22 13 15	+49 32.0	1	10.73		0.21		-0.05		4	A1 V	1126
		vdB 152 # 14		22 13 15	+70 13.0	1	11.11		0.94				1	B9.5V	207
211352	+54 02719			22 13 19	+55 11.1	2	9.15	.000	0.10	.000			5	A2	273,1025
	+58 02412			22 13 19	+58 30.1	1	9.70		0.40		0.26		1		338
	-14 06241			22 13 19	-14 26.2	1	10.40		1.07		0.83		3	K5 V	3072
211234	-15 06174			22 13 19	-14 41.1	2	7.85	.010	1.25	.006	1.40	.005	7	K2 III	861,1117
		NGC 7243 - 364		22 13 20	+49 42.2	1	11.93		0.39		0.23		5	A6	1126
		NGC 7243 - 494		22 13 21	+49 31.2	1	12.70		0.40		0.27		4	A7	72
211402	+58 02413			22 13 21	+58 51.9	1	7.00		0.04		0.04		2	A2	401
		CS 22892 # 51		22 13 21	-17 34.7	1	14.66		-0.36		-1.23		1		1736
211236	-23 17334			22 13 21	-23 15.6	1	7.44		0.24				2	A8/9 IV/V	1594
	+54 02721	IDS22115N5500	AB	22 13 22	+55 14.3	1	9.34		0.65		0.53		1	G0 III:	8084
		Cep sq 2 # 219		22 13 22	+56 01.0	1	10.66		0.32		0.20		1	B9.5V	273
	+60 02369	LS III +61 008		22 13 22	+61 26.6	1	8.18		0.33		-0.51		4	B5	213

HD	DM	Other Id	N	Rem	α_{1950}	δ_{1950}	S	V	σ_V	B–V	σ_{B-V}	U–B	σ_{U-B}	n	Spectrum	References
		NGC 7243 - 366			22 13 23	+49 39.2	1	12.30		0.45		0.36		5	A7	1126
	+55 02707				22 13 24	+55 40.5	2	8.89	.000	0.19	.000			3	B9 V	273,1025
		LS III +60 015			22 13 24	+60 50.9	1	10.68		0.60		-0.17		2	B2 V	1012
	+49 03793	NGC 7243 - 496			22 13 25	+49 33.9	1	10.39		0.13		-0.08		3	A0 V	1126
		LS III +56 036			22 13 26	+56 31.3	1	12.43				-0.16		4		483
211202	-42 15846	HR 8488			22 13 27	-41 52.6	4	5.09	.007	0.92	.005			12	G8 III	15,1075,2024,2032
		NGC 7243 - 497			22 13 29	+49 34.6	1	11.72		0.27		0.19		4	A3 V	1126
211403	+57 02489				22 13 29	+58 01.6	1	8.47		0.50				2	G0	1025
211276	-7 05737	G 27 - 21			22 13 29	-07 20.1	1	8.78		0.68		0.20		1	G5	1620
211287	+7 04834	HR 8491			22 13 30	+08 18.0	2	6.20	.005	0.02	.000	-0.04	.000	7	A1 Vn	15,1256
		NGC 7243 - 501			22 13 31	+49 32.9	1	12.21		0.34		0.24		6	A0	1126
	+5 04984				22 13 32	+06 22.8	2	9.73	.005	0.59	.014	0.06		4	G5	1213,1317
		Steph 1996			22 13 32	+31 30.2	1	11.97		1.28		1.44		1	K4	1746
		NGC 7243 - 503			22 13 32	+49 31.8	1	11.24		0.29		0.21		4	A3 V	1126
		NGC 7243 - 502			22 13 32	+49 33.8	2	12.20	.010	0.41	.009	0.30	.000	7	A3	72,1126
211305	+2 04483				22 13 33	+02 59.8	1	8.79		1.67		2.22		5	K0	897
		Cep sq 2 # 221			22 13 33	+54 25.4	1	12.41		0.43		0.41		1	B7 V	273
		Cep sq 2 # 222			22 13 34	+55 44.8	1	10.56		0.31		0.28		1	A2 V	273
		NGC 7243 - 505			22 13 35	+49 35.2	1	12.87		0.24		-0.06		4	F0	72
	+24 04563	G 127 - 12			22 13 36	+24 41.1	2	8.94	.015	0.66	.010	0.17	.030	2	G0	1620,1658
		NGC 7243 - 368			22 13 36	+49 40.4	1	10.89		0.14		-0.01		5	B9 V	1126
239903	+56 02745				22 13 36	+56 38.3	2	9.76	.005	0.43	.010	0.22		5	A7	338,1025
		NGC 7243 - 369			22 13 37	+49 39.1	1	10.32		0.20		0.11		5	A2 V	1126
		TOU28,33 T8# 66			22 13 37	+55 13.8	2	10.41	.034	0.09	.015	-0.32		6	B7	273,1025
		CS 22881 # 70			22 13 38	-40 59.1	2	14.38	.005	0.37	.007	-0.20	.006	3		1580,1736
		Cep sq 2 # 225			22 13 39	+54 36.9	1	11.86		0.28		0.34		1	B3 III	273
211430	+55 02709				22 13 39	+55 34.2	5	7.46	.006	-0.05	.012	-0.49	.007	34	B9	227,261,273,401,1025
		NGC 7243 - 370			22 13 40	+49 42.3	1	9.97		0.04		-0.35		4	A0 Vp	72
211512	+69 01236				22 13 40	+69 45.1	1	7.83		0.98				1	K0	207
		Cep sq 2 # 226			22 13 41	+55 14.3	1	11.59		0.23		0.17		1	A0 III	273
		CS 22881 # 71			22 13 41	-40 11.3	2	14.24	.010	0.39	.007	0.03	.009	3		1580,1736
211404	+51 03307	IDS22118N5122		AB	22 13 42	+51 37.3	1	8.58		0.07		-0.12		2	B9 IV	171
211405	+43 04165	IDS22116N4324		AB	22 13 43	+43 39.1	1	7.48		0.06		-0.42		1	B9	1476
235771	+53 02829				22 13 43	+54 06.5	1	8.82		0.50		-0.04		3	F5 V	261
211376	+38 04721				22 13 44	+38 36.5	1	7.66		1.12		0.89		3	K0	833
		PG2213-006C			22 13 44	-00 37.2	1	15.11		0.72		0.18		4		1764
		CS 22886 # 22			22 13 45	-08 57.4	1	13.97		0.40		-0.15		1		1736
211388	+37 04526	HR 8498			22 13 47	+37 30.0	7	4.14	.017	1.45	.011	1.65	.028	16	K3 III	15,1363,3016,4001*
		CS 22886 # 32			22 13 47	-11 45.6	1	13.79		-0.14		-0.73		1		1736
		NGC 7243 - 510			22 13 48	+49 37.3	2	10.59	.041	0.19	.002	-0.02	.017	7	A0 V	72,1126
		PG2213-006B			22 13 48	-00 36.8	1	12.71		0.75		0.30		4		1764
211291	-26 16057	HR 8492			22 13 49	-26 08.9	4	6.15	.005	1.11	.006			18	K1 II/III	15,1075,2013,2030
		NGC 7243 - 372			22 13 50	+49 38.8	1	11.80		0.23		0.17		4	A3 V	1126
		Cep sq 2 # 227			22 13 50	+54 57.8	1	12.24		0.40		-0.29		1	B1 V	273
211445	+55 02710				22 13 50	+55 52.6	3	8.85	.009	0.07	.004	0.00		7	A0	273,338,1025
		PG2213-006A			22 13 50	-00 36.4	1	14.18		0.67		0.10		5		1764
		NGC 7243 - 373			22 13 51	+49 39.5	1	10.52		0.62		0.16		4	G4 V	1126
		NGC 7243 - 374			22 13 53	+49 42.3	1	10.66		0.73		0.27		5	G1 IV	1126
		AAS6,117 # 27			22 13 54	+51 51.	1	11.69		0.66		-0.04		2		171
		PG2213-006			22 13 54	-00 36.1	1	14.12		-0.22		-1.12		5	sdB	1764
211295	-36 15292				22 13 54	-36 02.0	1	9.71		0.61		0.01		3	G2/3 V	840
		NGC 7243 - 375			22 13 55	+49 38.0	1	12.54		0.45		0.25		4	F0	72
		NGC 7243 - 376			22 13 55	+49 44.3	1	12.37		0.44		0.28		4	F3	1126
	+49 03796	NGC 7243 - 377			22 13 56	+49 39.4	2	9.25	.029	0.08	.001	-0.35	.032	9	B9 V	72,1126
		vdB 152 # 15			22 13 56	+70 14.1	1	10.48		0.69				1		207
		NGC 7243 - 511			22 13 57	+49 34.8	1	12.42		0.37		0.26		3	A3	1126
		NGC 7243 - 378			22 13 57	+49 41.9	1	12.07		0.27		0.20		4	A2	1126
		LS III +55 029			22 13 59	+55 20.0	1	10.30		0.29		-0.60		5		483
211356	-2 05726	HR 8495			22 13 59	-01 50.8	2	6.14	.005	0.19	.000	0.08	.010	7	A5 Vn	15,1071
		Cep sq 2 # 229			22 14 00	+55 19.7	1	10.23		0.35		-0.56		1	B0 V	273
211261	-54 10051				22 14 00	-54 34.3	1	7.65		0.31		0.09		10	A9/F0 IV/V	1628
		NGC 7243 - 513			22 14 01	+49 35.8	1	11.08		0.70		0.25		3		1126
		NGC 7243 - 512			22 14 01	+49 36.3	1	12.62		0.32		0.17		4	A3 V	1126
	+49 03797	NGC 7243 - 380			22 14 01	+49 41.8	2	9.26	.028	0.09	.002	-0.26	.031	10	B8 III	72,1126
		NGC 7243 - 379			22 14 01	+49 42.8	1	11.74		0.25		0.19		5	A3 V	1126
211472	+53 02831	G 232 - 61		★A	22 14 01	+54 25.3	2	7.52	.000	0.81	.000	0.41	.015	12	K1 V	171,3016
		AA26,31 # 14			22 14 02	+55 25.5	1	13.21		0.28		-0.42		5		483
		Cep sq 2 # 230			22 14 05	+54 26.0	1	10.70		0.26		-0.10		1	B8 III	273
		Cep sq 2 # 231			22 14 06	+54 48.7	1	11.15		0.66		0.23		1	A4 I	273
		NGC 7243 - 382			22 14 08	+49 38.4	1	11.35		0.64		0.16		3	F9	1126
211489	+54 02722				22 14 08	+54 34.6	1	7.98		0.25		0.07		3	F1 V	261
235775	+54 02723				22 14 08	+54 54.4	1	8.63		1.02				2	G5 III	1025
		CS 22886 # 18			22 14 08	-07 36.9	1	14.59		0.38		-0.19		1		1736
211361	-13 06148	HR 8496			22 14 08	-13 04.9	2	5.34	.001	1.13	.004	1.08			K1 III	58,2007
		NGC 7243 - 384			22 14 10	+49 43.2	2	11.82	.024	0.25	.012	0.19	.002	9	A5 V	72,1126
		Cep sq 2 # 233			22 14 10	+54 40.5	1	11.48		0.30		0.14		1	B9.5III	273
211432	+27 04288	HR 8503			22 14 11	+27 33.3	1	6.37		0.99				2	G9 III	71
	+47 03751				22 14 12	+48 25.0	1	8.79		0.24				2	A5	1724
211391	-8 05845	HR 8499			22 14 12	-08 02.0	9	4.17	.013	0.98	.005	0.81	.007	33	G8 III-IV	3,15,369,1075,1425*
211420	+12 04793				22 14 13	+12 40.2	1	7.50		0.99				3	K0	882

Table 1   1189

HD	DM	Other Id	N Rem	α_{1950}	δ_{1950}	S	V	σ_V	B–V	σ_{B-V}	U–B	σ_{U-B}	n	Spectrum	References
		NGC 7243 - 386		22 14 13	+49 42.2	2	10.31	.027	0.14	.006	-0.08	.020	9	B9 V	72,1126
211364	−23 17344	HR 8497		22 14 13	−23 23.4	1	6.17		1.05				4	K0 III	2007
		AA26,31 # 15		22 14 14	+55 17.1	1	12.75		0.89		-0.19		4		483
211392	−9 05948	HR 8500		22 14 14	−09 17.4	2	5.78	.005	1.16	.000	1.14	.000	7	K0	15,1061
211446	+31 04668			22 14 15	+32 22.3	1	7.50		1.35		1.41			K5	1733
		Cep sq 2 # 234		22 14 15	+54 11.0	1	12.24		0.25		-0.45		1	B5 II:	273
211380	−15 06180			22 14 15	−14 54.4	1	7.13		0.47		-0.01		3	F5 V	1117
211433	+22 04601			22 14 19	+22 38.8	1	6.92		0.11		0.12		3	Am	3016
		Cep sq 2 # 236		22 14 19	+55 33.1	1	11.35		0.49		0.14		1	B9 III	273
		CS 22892 # 52		22 14 19	−16 54.4	1	13.18		0.78		0.14		1		1736
235776	+50 03637			22 14 21	+50 50.2	1	8.08		1.24		1.09		3	K0	1566
	+53 02833	LS III +53 068		22 14 21	+53 55.0	2	9.99	.012	0.24	.008	-0.60	.029	7	B1 III	171,1012
		Cep sq 2 # 237		22 14 22	+56 05.6	1	10.65		0.34		0.19		1	A1 V	273
211460	+28 04337			22 14 23	+28 55.4	1	6.69		0.95		0.58		3	G5 IV	3016
235778	+53 02834			22 14 23	+53 47.1	1	9.16		2.10				2	K5 III	369
	+41 04443			22 14 24	+42 10.3	1	9.00		0.89				3	G5	1724
		Cep sq 2 # 238		22 14 24	+55 12.8	1	12.11		0.32		0.05		1	B8 III:	273
	+54 02725			22 14 25	+54 32.3	1	12.20		0.32		0.00		1	B9	273
211496	+42 04343			22 14 29	+42 35.4	1	8.70		0.29				3		1724
211540	+57 02493			22 14 29	+57 43.5	3	7.95	.020	1.96	.024			13	M2	369,1025,1339
		Cep sq 2 # 240		22 14 30	+55 14.6	1	11.64		0.25		-0.31		1	B5 III	273
		AA26,31 # 16		22 14 30	+55 29.8	1	14.09		0.48		-0.29		4		483
211434	−6 05960	HR 8504		22 14 30	−05 38.3	2	5.74	.005	0.88	.000	0.51	.000	7	G6 III	15,1061
		NGC 7243 - 392		22 14 33	+49 42.1	1	10.90		0.12		-0.04		4	A0 V	72
210967	−81 00995	HR 8481, BO Oct		22 14 33	−80 41.4	3	5.10	.004	1.47	.000	1.18	.059	14	M5 III	15,1075,3046
	+48 03662	NGC 7243 - 735		22 14 36	+49 10.1	1	9.26		0.56				2	F8	1724
		AAS6,117 # 30		22 14 36	+51 48.	1	11.56		0.21		0.17		3	B9	171
211475	+12 04796			22 14 38	+13 18.8	1	8.64		1.13				3	K0	882
		Mark 304 # 2		22 14 38	+13 58.4	1	12.91		0.71		0.18		1		829
211554	+56 02746	HR 8506		22 14 38	+56 58.2	2	5.88	.002	0.94	.011			4	G8 III	1025,6007
239907	+55 02711			22 14 41	+56 24.9	2	9.09	.009	1.34	.000	1.40		6	K5	261,1025
		LP 759 - 82	★ B	22 14 42	−09 02.	2	13.68	.632	1.79	.074	1.18		4		203,1705
		Wolf 1561		22 14 42	−09 02.1	1	13.48		1.74				1	M4	1705
		PsA sq 1 # 3		22 14 43	−36 04.8	1	13.54		0.95		0.77		2		840
211476	+12 04797			22 14 45	+12 38.8	3	7.04	.015	0.60	.000	0.08	.005	8	G2 V	22,1003,3077
211451	−24 17099	IDS22120S2413	A	22 14 45	−23 57.5	1	8.91		0.71		0.42		1	K1 V	3062
211451	−24 17099	IDS22120S2413	B	22 14 45	−23 57.5	2	11.00	.030	1.42	.025	1.16	.035	2		1696,3062
	+41 04446			22 14 46	+41 46.8	1	9.69		0.62				2	F8	1724
		CS 22875 # 6		22 14 46	−39 40.5	2	13.45	.000	0.07	.000	0.05	.010	3		966,1736
239908	+57 02496			22 14 48	+58 04.5	1	9.04		0.18				2	A5	1025
		Cep sq 2 # 244		22 14 50	+54 58.5	1	11.93		0.36		-0.19		1	B5 III	273
		PsA sq 1 # 1		22 14 50	−35 55.6	1	12.12		0.94		0.65		2		840
211500	+14 04766			22 14 51	+14 47.8	1	7.10		0.94				2	K0 III	882
	+48 03663			22 14 52	+48 47.8	1	8.63		0.58				3	G5	1724
	+54 02726	LS III +55 031		22 14 53	+55 13.6	3	9.37	.013	0.33	.005	-0.54	.000	5	B1.5II	273,1012,1025
	+52 03164			22 14 55	+53 21.9	1	9.63		0.27		0.15		3	A5	171
		PsA sq 1 # 5		22 14 55	−36 00.7	1	13.86		1.01		0.80		3		840
		Mark 304 # 4		22 14 56	+13 59.	1	14.94		0.94		0.59		1		829
		Mark 304 # 3		22 14 56	+13 59.5	1	14.57		0.73		0.22		1		829
		PsA sq 1 # 2		22 14 58	−35 55.3	1	13.25		0.60		0.06		4		840
211516	+4 04837			22 14 59	+04 53.6	1	7.28		1.78		2.04		3	M1	1733
211415	−54 10055	HR 8501	★ AB	22 15 00	−53 52.1	5	5.36	.013	0.61	.007	0.06	.000	13	G3 V	15,404,2012,2024,3026
239909	+56 02749	IDS22132N5641	AB	22 15 01	+56 55.9	1	9.66		0.40		-0.20		3	A5	261
		Mark 304 # 1		22 15 03	+13 57.7	1	10.89		0.72		0.15		1		829
239911	+57 02498			22 15 04	+57 47.3	1	9.57		1.85				8	K7	1339
	+45 03866			22 15 05	+46 17.5	1	8.99		0.54				3	F8	1724
211604	+55 02713			22 15 05	+55 54.4	3	8.05	.035	0.29	.025	0.06		6	A5	261,273,1025
235781	+52 03167	LS III +53 069		22 15 06	+53 24.0	2	8.59	.000	0.41	.007	-0.27	.004	12	B6 Ib	171,1012
239910	+55 02712	LS III +56 037		22 15 06	+56 17.3	1	9.45		0.85				2	F5	1025
211416	−60 07561	HR 8502	AB	22 15 06	−60 30.6	7	2.85	.009	1.39	.009	1.54	.001	29	K3 III	15,1020,1075,1075*
		PsA sq 1 # 4		22 15 07	−35 55.8	1	13.80		0.82		0.45		3		840
	+29 04630			22 15 08	+29 47.5	1	10.65		1.62		1.96		3	M2	1746
	+67 01424	G 241 - 2		22 15 08	+68 05.4	2	9.23	.000	1.15	.015	1.07		6	K5	196,1663
211517	−4 05647			22 15 08	−03 35.6	1	9.14		0.40		-0.05		3	F2	196
		AA26,31 # 17		22 15 09	+56 00.5	1	13.22		0.80		-0.22		4		483
		Cep sq 2 # 248		22 15 11	+55 10.6	1	11.27		0.27		0.09		1	B8 V	273
211643	+55 02714			22 15 11	+55 55.6	1	7.08		0.11				2	Am:	1025
	+53 02837	LS III +53 070		22 15 13	+53 57.5	1	9.97		0.26		-0.54		2	B2 III:	1012
235783	+53 02838	LS III +54 032		22 15 14	+54 15.4	2	8.68	.000	0.17	.000	-0.71	.000	3	B1 Ib	273,1012
		Cep sq 2 # 251		22 15 14	+54 50.8	1	10.59		0.81				1	B5 Ia	273
		Cep sq 2 # 250		22 15 14	+55 15.6	1	11.54		0.45		0.26		1	A1 V	273
	+49 03800	NGC 7243 - 527		22 15 16	+49 31.3	1	8.83		0.42				3	F0	1724
	−11 05801			22 15 16	−10 56.4	1	10.60		1.18		1.14		1	K4	1746
		CS 22886 # 21		22 15 17	−08 36.1	1	13.73		0.08		-0.80		1		1736
	+54 02728			22 15 19	+54 46.3	1	10.58		0.20		0.04		1	F0 V	273
211659	+55 02715			22 15 21	+55 37.1	2	9.38	.000	0.10	.000			3	A3	273,1025
		Cep sq 2 # 255		22 15 21	+55 56.3	1	10.17		0.26		0.16		1	A0 V	273
		Cep sq 2 # 256		22 15 22	+54 29.2	1	11.43		0.30		0.23		1	A2 V	273
211507	−39 14656			22 15 22	−39 00.5	1	9.98		0.96		0.75		2	K2 (III)	3072
211771	+72 01025			22 15 23	+73 21.2	1	8.12		0.43				1	F2	207

HD	DM	Other Id	N	Rem	α_{1950}	δ_{1950}	S	V	σ_V	B–V	σ_{B-V}	U–B	σ_{U-B}	n	Spectrum	References
		LS III +54 033			22 15 24	+54 56.6	2	10.42	.080	0.25	.015	-0.52	.020	2		273,483
		CS 22875 # 3			22 15 24	-41 33.0	2	14.49	.000	0.07	.000	0.17	.010	2		966,1736
		Cep sq 2 # 258			22 15 25	+54 18.3	1	11.54		0.26		0.18		1	A1 III	273
		Cep sq 2 # 259			22 15 26	+55 27.7	1	11.34		0.49		0.33		1	A2 III	273
	+48 03666				22 15 27	+48 56.7	1	8.30		0.52				3	F8	1724
211606	+26 04399				22 15 28	+26 41.2	3	6.61	.012	1.60	.016	1.99	.054	9	K5 II	985,1501,8100
211557	-3 05423				22 15 28	-03 25.4	1	9.30		1.29		1.42		3	K5	196
	+49 03803				22 15 29	+50 19.1	1	8.71		1.80				2		369
211661	+43 04167				22 15 30	+44 27.1	1	9.31		0.29				3	F2	1724
	+59 02501			V	22 15 30	+60 09.4	1	10.18		2.29				2		369
211575	-0 04333	HR 8507			22 15 30	-00 29.3	2	6.38	.005	0.44	.000	-0.01	.010	7	F3 V	15,1071
		CS 22886 # 28			22 15 30	-10 16.9	1	15.49		0.40		-0.15		1		1736
	+53 02839				22 15 31	+54 19.6	1	11.07		0.21		-0.21		1	B7 IV	273
		Cep sq 2 # 262			22 15 32	+54 32.7	1	10.54		0.21		-0.03		1	B9 V	273
	+41 04449	IDS22134N4202		AB	22 15 33	+42 16.9	1	9.53		0.22				2	A5	1724
		Cep sq 2 # 263			22 15 34	+54 24.2	1	10.04		0.18		0.02		1	A0 V	273
211523	-46 14270				22 15 36	-46 23.3	1	7.53		0.94		0.58		5	G8 III	1673
211607	+13 04887				22 15 37	+13 42.0	1	6.87		0.93				3	K0	882
	+48 03668	NGC 7243 - 649			22 15 37	+49 24.1	1	9.83		0.43				3	F0	1724
	+2 04488	UW Peg			22 15 38	+02 28.8	2	9.44	.105	1.67	.030	1.32	.060	10	M5	897,3042
211594	-6 05964				22 15 38	-06 05.6	4	8.08	.006	1.15	.011	0.78	.016	13	G4 III-Ba4	158,565,993,1775
211645	+38 04727				22 15 39	+38 46.5	1	7.38		1.22		1.17		2	K3 III	105
		LS III +54 034			22 15 39	+54 50.4	1	11.76		0.26		-0.52		3		483
211694	+50 03651				22 15 41	+51 04.0	2	7.61	.020	-0.08	.015	-0.42	.025	4	B7 IV/III	171,401
239914	+57 02501				22 15 41	+57 50.1	1	9.21		0.52				2	G0	1025
		AAS11,365 T7# 45			22 15 42	+53 29.	1	10.21		1.87		2.03		3		261
		Cep sq 2 # 266			22 15 42	+55 22.8	1	10.44		0.24		-0.29		1	B5 V	273
		Cep sq 2 # 267			22 15 43	+54 17.6	1	10.93		0.17		-0.56		1	B3 III	273
211746	+65 01746				22 15 44	+65 52.7	1	7.03		0.12		0.09		2	A0	401
	+54 02729				22 15 45	+54 30.1	1	10.07		0.13		-0.08		1	B9 V	273
	+48 03669	NGC 7243 - 754			22 15 46	+49 16.7	1	9.30		0.32				3	F0	1724
211626	+12 04800				22 15 47	+13 17.1	1	8.75		1.11				2	K2	882
		Cep sq 2 # 270			22 15 49	+54 41.2	1	10.52		0.24		-0.02		1	B8 V	273
235787	+50 03653				22 15 51	+50 42.3	1	8.92		0.52				3		1724
	+13 04888				22 15 53	+13 42.3	1	8.67		1.38		1.75		1	K5	1364
		Cep sq 2 # 271			22 15 53	+54 15.8	1	11.33		0.50		0.23		1	B7 III	273
		Cep sq 2 # 272			22 15 54	+54 25.8	1	11.90		0.32		0.06		1	B8 III	273
	+54 02730				22 15 54	+55 23.0	1	10.51		0.27		0.23		1	A1 V	273
		Cep sq 2 # 273			22 15 55	+54 30.8	1	12.38		0.20		-0.38		1	B8 IIp	273
211583	-30 19094				22 15 55	-30 00.1	1	9.96		1.14		1.13		2	K4 V	3072
		Cep sq 2 # 275			22 15 56	+54 15.5	1	10.87		0.23		0.03		1	B8 V	273
		Cep sq 2 # 276			22 15 59	+54 17.2	1	11.56		0.24		-0.27		1	B6 V	273
		LP 759 - 73			22 16 01	-10 23.3	2	12.50	.035	0.54	.010	-0.24	.045	5		1696,3062
		CS 22881 # 69			22 16 03	-41 38.5	1	13.89		-0.33		-1.22		1		1736
		CS 22886 # 26			22 16 04	-09 57.7	1	14.89		0.43		-0.19		1		1736
	+7 04841	G 18 - 39			22 16 05	+08 12.2	5	10.37	.009	0.46	.000	-0.16	.008	9	F5	516,1064,1620,1696,3077
		LS III +54 039		V	22 16 05	+54 13.0	2	10.64	.206	0.22	.014	-0.68	.033	4		273,483
		Cep sq 2 # 279			22 16 07	+55 31.8	1	11.50		0.38		-0.08		1	B7 V	273
	+47 03767				22 16 08	+48 02.6	1	9.44		0.45				3	F5	1724
		Cep sq 2 # 281			22 16 10	+54 40.6	1	11.40		0.48		0.36		1	A2 III	273
		AA26,31 # 20			22 16 10	+55 51.0	1	13.64		0.56		-0.20		4		483
211774	+54 02732				22 16 12	+55 01.8	3	8.89	.020	0.20	.010	0.11		6	B9 V	261,273,1025
		L 119 - 34			22 16 12	-65 43.	1	14.43		0.14		-0.77		2		3065
	+45 03871				22 16 13	+45 44.5	1	9.81		0.56				2	G0	1724
	+48 03672				22 16 13	+49 03.9	1	8.93		0.33				3	A5	1724
		G 127 - 15			22 16 15	+21 07.2	1	13.40		0.97		0.58		2		1620
		Cep sq 2 # 284			22 16 15	+54 54.0	1	11.20		0.29		0.13		1	B9.5V	273
		LS III +55 032			22 16 15	+55 39.2	3	10.45	.057	0.52	.010	-0.50	.023	6	B0 III:	273,483,1012
	+54 02731	LS III +54 040			22 16 17	+54 45.4	1	10.41		0.24		-0.56		1	B2 V	273
211773	+55 02718	IDS22144N5537		AB	22 16 17	+55 51.9	1	8.13		0.93				2		1025
		Cep sq 2 # 287			22 16 18	+54 23.8	1	10.87		0.23		0.10		1	A2 V	273
211867	+69 01238				22 16 18	+69 48.5	1	7.22		1.02				1	K0	207
		Cep sq 2 # 288			22 16 19	+54 34.8	1	12.05		0.26		-0.25		1	B5 III	273
211676	-14 06255	HR 8508			22 16 20	-13 33.4	2	5.96	.005	1.07	.010	0.94		6	K0 III	252,2009
211718	+19 04897				22 16 23	+19 42.7	1	6.86		1.26		1.42		2	K2	1648
		Cep sq 2 # 290			22 16 23	+54 10.6	1	11.00		0.40		0.28		1	A3 III	273
		Cep sq 2 # 289			22 16 23	+55 15.5	1	11.38		0.36		-0.18		1	B6 III	273
211732	+25 04709				22 16 25	+25 31.4	1	7.72		0.22		0.12		2	A3	105
		Cep sq 2 # 291			22 16 25	+54 35.1	1	12.06		0.33		0.23		1	A0 III	273
239918	+57 02502				22 16 26	+57 56.9	2	8.90	.000	1.32	.041			10	K5	1025,1339
211706	-8 05850	G 27 - 22			22 16 30	-07 33.8	1	8.89		0.61		0.12		1	G0	1620
		CS 22892 # 56			22 16 30	-16 22.3	1	14.11		0.46		-0.19		1		1736
211833	+62 02059	HR 8511			22 16 35	+63 33.2	2	5.75	.012	1.26	.004	1.37		5	K3 III	37,71
	+40 04776				22 16 38	+41 15.9	1	9.61		0.45				3	F2	1724
	+24 04574				22 16 40	+24 43.1	1	9.27		0.93		0.60		2	K0	1375
		Cep sq 2 # 297			22 16 40	+55 03.2	1	10.62		0.07		-0.41		1	B8 III	273
		Cep sq 2 # 299			22 16 43	+54 37.3	1	12.54		0.34		0.27		1		273
211797	+37 04537	HR 8510		★ A	22 16 44	+37 31.1	3	6.19	.015	0.27	.005	0.11	.000	7	A9 IIIp	15,1028,3024
211797	+37 04537	IDS22146N3716		B	22 16 44	+37 31.1	2	8.84	.004	0.60	.000	0.05		5		1028,3024
		AA26,31 # 22			22 16 46	+55 51.2	1	13.26		0.57		-0.32		4		483

Table 1   1191

HD	DM	Other Id	N	Rem	α_{1950}	δ_{1950}	S	V	σ_V	B–V	σ_{B-V}	U–B	σ_{U-B}	n	Spectrum	References
		Cep sq 2 # 300			22 16 47	+54 28.0	1	11.19		0.19		-0.14		1	B7 V	273
211775	+17 04727				22 16 48	+17 49.7	1	7.77		1.56				2	K5	882
211799	+28 04348				22 16 49	+28 35.8	1	7.12		0.52		0.09		2	F8 V	3016
	-49 13807	Smethells 10			22 16 50	-49 18.4	1	10.92		1.30				1	K7	1494
		Cep sq 2 # 302			22 16 51	+54 25.5	1	11.31		0.19		-0.52		1	B2.5V	273
211880	+62 02061	LS III +62 011	⋆	A	22 16 51	+62 58.3	2	7.74	.005	0.31	.005	-0.61	.005	4	B0.5V	27,1012
211880	+62 02060	IDS22152N6243		B	22 16 51	+62 58.3	1	8.57		0.31		-0.49		3		27
	+53 02843	LS III +54 043			22 16 52	+54 01.3	2	9.50	.000	0.21	.000	-0.75	.005	4	O8 III	1011,1012
		Cep sq 2 # 304			22 16 53	+54 28.8	1	11.93		0.38		-0.35		1	B5 V	273
		AA26,31 # 23			22 16 53	+54 56.0	1	12.96		0.28		-0.40		4		483
		PsA sq 1 # 9			22 16 53	-34 55.0	1	12.01		0.98		0.71		1		840
211853	+55 02721	GP Cep			22 16 55	+55 52.5	3	9.02	.008	0.39	.007	-0.55	.000	10	B0:I:	1012,1095,1359
		CS 22875 # 7			22 16 55	-39 17.1	2	14.27	.000	0.23	.000	0.02	.008	2		966,1736
211835	+45 03879				22 16 56	+45 33.1	2	8.48	.005	0.01	.010	-0.80	.005	9	B3:Ve	1174,1212
211723	-37 14738				22 16 56	-37 31.6	1	10.17		1.24		1.18		2	K3 V	3072
211868	+53 02844	LS III +54 046			22 16 57	+54 19.6	1	8.03		0.40		0.41		2	A3	3016
		UW Gru			22 16 57	-54 48.3	1	12.61		0.15		0.10		1		958
211800	+14 04772				22 16 58	+15 17.7	1	6.91		1.66		1.99		2	K2 V	1648
	+49 03810	NGC 7243 - 168			22 16 58	+49 59.2	1	8.61		1.01				2	G5	1724
		Cep sq 2 # 305			22 16 58	+55 02.1	1	11.17		0.35		0.23		1	A0 V	273
235794	+51 03327				22 16 59	+51 46.4	1	9.04		1.06		0.77		3	G8 III	171
		Cep sq 2 # 307			22 17 04	+55 32.6	2	10.82	.054	0.43	.005	-0.28	.029	3	B2 II	273,483
211787	-1 04284				22 17 04	-00 51.0	1	9.00		1.37				1	K2	1238
		Cep sq 2 # 308			22 17 06	+54 33.6	1	11.39		0.42		0.30		1	A2 V	273
	+56 02753				22 17 06	+57 26.7	1	9.67		0.25				3	A0	1025
		LS III +55 036			22 17 08	+55 52.3	2	9.93	.032	0.45	.000	-0.57		5		483,725
211744	-44 14857				22 17 08	-44 36.9	2	9.12	.005	0.76	.000	0.27		3	G3/5 (IV)p	742,1594
		LS III +55 037			22 17 09	+55 52.2	2	10.97	.090	0.36	.086	-0.56		5		483,725
		WW Cep			22 17 09	+69 36.5	1	10.64		0.85		0.39		7	G3	1768
		Cep sq 2 # 309			22 17 10	+54 23.2	1	12.00		0.13		-0.47		1	B5 III	273
235795	+51 03330	LS III +51 035			22 17 11	+51 52.1	1	9.14		0.07		-0.78		2	B1:V:enn	1012
		Cep sq 2 # 310			22 17 11	+54 37.2	1	12.15		0.28		0.19		1	A0 III	273
235796	+51 03331				22 17 12	+51 38.8	1	8.48		1.62		1.96		3	K3 III	171
		TOU28,33 T8# 79			22 17 12	+56 42.7	1	10.19		0.28				3		1025
211903	+55 02723				22 17 15	+56 02.4	2	8.81	.000	0.23	.000			3		273,1025
		Cep sq 2 # 311			22 17 16	+55 02.0	1	12.32		0.26		-0.13		1	B8 V	273
	+55 02722	IDS22154N5537		AB	22 17 16	+55 52.5	2	9.52	.019	0.47	.010	-0.58	.024	3	O8.5V	684,8084
	+55 02722	IDS22154N5537		D	22 17 16	+55 52.5	1	12.56		0.51		-0.31		1		8084
	+55 02722	IDS22154N5537		E	22 17 16	+55 52.5	1	12.98		0.66		-0.15		1		8084
211726	-58 07942	HR 8509			22 17 17	-57 45.7	1	6.34		1.37		1.71		2	K3/4 III	3005
235797	+52 03179				22 17 18	+53 21.2	2	8.68	.020	1.81	.010	2.17		5	M0 III	171,369
		Cep sq 2 # 313			22 17 19	+55 58.0	1	10.56		0.44		-0.12		1	B8 III	273
	+60 02380	LS III +60 017			22 17 20	+60 30.9	3	9.04	.008	0.39	.000	-0.50	.007	14	B2 III	27,213,1012
		G 18 - 40			22 17 21	+05 10.6	2	12.47	.010	0.53	.015	-0.19	.019	3		333,1620,1658
		Cep sq 2 # 315			22 17 21	+54 53.0	1	11.20		0.28		0.02		1	B9 III	273
		AA26,31 # 27			22 17 21	+55 50.3	1	12.70		0.49		-0.41		3		483
		PsA sq 1 # 12			22 17 26	-35 25.0	1	12.45		0.92		0.69		1		840
	+39 04805				22 17 30	+40 27.9	1	9.77		0.38				3	F2	1724
		CZ Lac			22 17 33	+51 13.1	1	10.77		0.46		0.27		1	F1	597
		CS 22886 # 44			22 17 33	-09 56.7	1	14.23		0.51		-0.12		1		1736
211884	+24 04576				22 17 34	+25 28.4	1	7.29		1.59		2.01		3	K5 III	833
212710	+85 00383	HR 8546			22 17 34	+85 51.5	2	5.25	.015	-0.03	.002	-0.17	.057	7	B9.5Vn	15,3023
211838	-8 05855	HR 8512			22 17 34	-08 04.4	7	5.35	.011	-0.06	.003	-0.36	.010	19	B8 IIIp	15,1034,1071,1079*
239923	+58 02421	LS III +58 023			22 17 35	+58 42.1	2	8.88	.004	0.82	.009	-0.13	.004	5	B3 Ib	27,1012
		WLS 2220-5 # 10			22 17 37	-04 12.6	1	11.16		1.75		1.98		2		1375
		MCC 863			22 17 39	+00 52.	1	11.65		1.21				2	M0	1619
		LP 875 - 46			22 17 39	-24 36.1	1	13.76		1.32		1.00		1		3078
	+49 03817				22 17 40	+49 53.4	1	9.09		0.33				3	A5	1724
		AAS6,117 # 37			22 17 42	+52 39.	1	10.44		0.11		-0.16		5	B8	171
		AA26,31 # 28			22 17 42	+56 00.2	1	12.84		0.45		-0.44		5		483
211971	+59 02506	LS III +59 026			22 17 42	+59 53.8	3	6.91	.007	0.98	.032	0.38	.014	7	A2 Ib	27,1012,1733
212150	+75 00820	HR 8525			22 17 45	+76 14.2	2	6.64	.015	0.00	.005	-0.10	.010	6	A1 Vn	985,1733
		CS 22886 # 42			22 17 47	-10 38.5	1	13.26		0.77		0.14		1		1736
211972	+52 03181				22 17 48	+52 37.8	1	7.73		1.57		1.90		3	K5 III	171
		PsA sq 1 # 13			22 17 48	-35 27.3	1	13.50		0.88		0.59		1		840
211982	+55 02724				22 17 49	+55 57.1	1	7.28		1.12				2	K1 III	1025
211999	+60 02383				22 17 51	+60 41.7	1	9.37		0.34		0.15		2	A2	27
211863	-35 15264				22 17 51	-34 46.1	1	6.94		1.16		1.05		8	K1 III	1628
		LS III +54 050			22 17 52	+54 41.8	1	11.90		0.26		-0.53		2		483
		Cep sq 2 # 322			22 17 52	+55 27.2	1	11.45		0.30		0.27		1	A2 V	273
		NGC 7261 - 123			22 17 52	+57 46.1	1	12.30		0.73		0.43		1		49
	+40 04785				22 17 53	+41 26.9	1	8.55		0.70				3	G0	1724
	+53 02847				22 17 53	+54 17.6	1	10.22		0.18		-0.24		1		273
		AA26,31 # 30			22 17 53	+55 46.1	1	12.36		0.45		-0.34		3		483
		NGC 7261 - 148			22 17 53	+57 39.3	1	12.72		0.68		0.42		1		49
		LS III +55 038			22 17 54	+55 09.2	2	10.43	.070	0.19	.005	-0.59	.030	2	B1.5V	273,483
		Cep sq 2 # 324			22 17 54	+55 16.1	1	11.68		0.30		0.01		1	B9 V	273
		NGC 7261 - 119			22 17 54	+57 47.1	1	13.08		1.61		1.43		4		4002
211891	-7 05751				22 17 56	-07 23.3	1	9.31		0.44		0.07		1	F2	301
211924	+5 04998	HR 8513	⋆	AB	22 17 57	+05 32.3	9	5.37	.011	-0.02	.006	-0.49	.011	46	B5 IV	15,154,1256,1601,1625,1648*

HD	DM	Other Id	N Rem	α_{1950}	δ_{1950}	S	V	σ_V	B−V	σ_{B-V}	U−B	σ_{U-B}	n	Spectrum	References
211924	+5 04998	IDS22154N0517	C	22 17 57	+05 32.3	1	11.11		0.36		−0.11		1		8084
	+49 03818			22 17 57	+50 24.2	1	8.96		1.65				2		369
		LS III +55 039		22 17 57	+55 58.6	1	11.74		0.45		−0.48		3		483
		NGC 7261 - 117		22 18 01	+57 47.2	1	14.34		0.94		0.42		1		49
		NGC 7261 - 150		22 18 01	+57 47.8	1	16.18		0.96				1		49
	+45 03885			22 18 02	+46 08.9	1	9.51		0.60				3	G0	1724
212000	+52 03182			22 18 02	+52 37.2	1	8.38		0.02		−0.06		2	B9	171
		LS III +55 040		22 18 02	+55 54.6	1	12.08		0.64		−0.40		4		483
		NGC 7261 - 115		22 18 02	+57 48.2	1	15.85		1.00				1		49
		NGC 7261 - 86		22 18 02	+57 49.1	1	13.59		0.89		0.38		1		49
211907	−4 05658			22 18 03	−03 46.2	1	9.50		0.40		0.05		3	F5	196
		WLS 2220 25 # 7		22 18 05	+22 57.0	1	11.57		0.59		0.11		2		1375
		NGC 7261 - 118		22 18 05	+57 46.9	1	11.93		0.50		0.11		2		49
		CS 22875 # 4		22 18 05	−40 46.5	2	13.84	.005	0.38	.008	−0.20	.000	2		1580,1736
		G 18 - 43		22 18 06	+03 55.9	1	15.12		1.34				1		3062
	+41 04460			22 18 06	+41 35.4	1	8.93		0.46				3	F2	1724
		Cep sq 2 # 326		22 18 06	+55 19.7	1	10.49		0.18		−0.12		1	B8 V	273
239925	+57 02505			22 18 06	+57 39.5	1	9.24		2.16				1	M0	338
		CS 22875 # 1		22 18 06	−42 12.8	1	14.86		0.30		−0.05		1		1736
		LS III +55 042		22 18 09	+55 53.6	1	12.38		0.52		−0.43		3		483
		NGC 7261 - 114		22 18 10	+57 47.9	1	13.89		0.80		0.52		1		49
		NGC 7261 - 87		22 18 11	+57 49.4	1	14.66		0.83		0.53		1		49
212028	+49 03821			22 18 14	+50 02.9	1	8.52		0.05		−0.06		2	A0	401
		NGC 7261 - 143		22 18 14	+57 41.3	2	13.35	.010	1.55	.025	1.18	.036	2		49,4002
		LS III +55 043		22 18 15	+55 07.3	2	11.27	.020	0.18	.025	−0.58	.020	2		273,483
		NGC 7261 - 88		22 18 15	+57 49.6	1	11.52		0.45		0.20		1		49
	+39 04809			22 18 16	+40 07.4	1	9.26		0.45				3	F2	1724
212043	+56 02755			22 18 16	+56 40.0	3	6.55	.015	−0.07	.011	−0.39	.005	6	B6 II	401,1025,1733
		NGC 7261 - 20		22 18 16	+57 55.5	1	12.38		0.53		0.29		1		49
		NGC 7261 - 149		22 18 17	+57 48.8	1	15.07		1.13		0.67		1		49
	−7 05752			22 18 17	−07 24.8	1	10.68		0.59		0.10		1		301
		NGC 7261 - 90		22 18 18	+57 50.5	1	12.81		1.50		1.30		1		4002
212029	+45 03890			22 18 21	+46 10.0	1	8.52		0.51				2	G0	1724
		TOU28,33 T8# 339		22 18 21	+56 10.1	1	9.34		0.19				2		1025
		NGC 7261 - 147		22 18 22	+57 42.4	1	11.57		0.53		0.18		1		49
		NGC 7261 - 110		22 18 22	+57 48.2	1	12.98		1.03		0.46		1		49
		NGC 7261 - 74		22 18 22	+57 51.5	1	13.24		0.86		−0.16		1		49
	+54 02739			22 18 23	+54 34.4	1	10.35		0.99		0.56		1	B9 V	273
		Cep sq 2 # 330		22 18 23	+54 40.3	1	12.27		0.53		0.32		1	A3 III	273
212044	+51 03341	V357 Lac		22 18 25	+51 36.5	2	6.99	.005	0.05	.005	−0.90	.005	5	B1:V:penn	1012,1212
		NGC 7261 - 44		22 18 25	+57 54.6	1	14.66		0.72		0.05		1		49
211965	−7 05753	ST Aqr		22 18 25	−07 13.0	2	9.21	.042	0.43	.006	0.12		3	G8 IV:	1375,6009
211976	+7 04853	HR 8514		22 18 26	+07 56.1	4	6.18	.015	0.45	.006	−0.05	.020	11	F6 V	15,254,1256,3037
		Cep sq 2 # 332		22 18 26	+55 32.7	1	11.47		0.24		0.11		1	A0 V	273
		NGC 7261 - 146		22 18 26	+57 43.5	1	13.79		0.82		0.57		1		49
		LS III +54 051		22 18 27	+54 46.3	1	11.74		0.25		−0.55		2		483
	+55 02725	LS III +56 041		22 18 28	+56 08.2	1	10.08		0.45		−0.53		2	B1pe	1012
	+47 03777			22 18 29	+48 01.6	1	9.09		0.74				2	G5	1724
212070	+55 02726	IDS22166N5537	AB	22 18 29	+55 51.6	1	9.18		0.23				2		1025
	+57 02506	NGC 7261 - 136		22 18 29	+57 44.4	1	10.42		0.55		0.34		1		49
		NGC 7261 - 96		22 18 29	+57 48.9	1	11.49		0.57		0.28		1		49
211954	−27 15871			22 18 29	−27 29.9	2	10.21	.010	1.33	.005	1.13	.005	3	G8/K0 (III)	565,3048
		LS III +55 045		22 18 30	+55 53.8	1	10.45		0.70		−0.36		4		483
	+43 04173			22 18 31	+43 49.1	1	9.86		0.49				3	F8	1724
		NGC 7261 - 71		22 18 31	+57 51.2	1	10.77		0.90		−0.05		1	B2 III	49
		NGC 7261 - 58		22 18 31	+57 52.1	1	13.69		0.78		−0.03		1		49
	+26 04409			22 18 32	+27 18.5	1	10.34		0.54		0.05		2		1375
	+45 03891			22 18 32	+45 36.1	1	9.09		0.41				3	F5	1724
		NGC 7261 - 141		22 18 33	+57 42.2	1	13.57		0.82		−0.09		1		49
		NGC 7261 - 137		22 18 33	+57 44.3	1	12.24		0.57		−0.04		1		49
	+53 02848			22 18 34	+54 21.8	1	9.91		0.06		0.05		1		273
212093	+54 02740			22 18 34	+55 25.9	2	8.25	.033	−0.02	.009	−0.29		4	B9	401,1025
239927	+57 02507	NGC 7261 - 98		22 18 37	+57 50.0	1	9.61		0.84		0.46		1	K0	49
	+41 04463			22 18 38	+41 38.3	1	9.47		0.42				3	F5	1724
		LS III +55 046		22 18 38	+55 33.1	1	12.54		0.35		−0.33		2		483
	−48 14193			22 18 39	−47 40.7	1	10.09		1.37				1		1494
212094	+53 02849	IDS22168N5319	AB	22 18 40	+53 33.8	1	8.30		0.65		0.17		2	A3	1566
212106	+54 02741			22 18 40	+54 35.5	1	7.42		0.49		−0.04		3	F5 V	261
		LS III +55 047		22 18 40	+55 44.4	1	12.48		0.43		−0.37		2		483
		NGC 7261 - 99		22 18 40	+57 50.4	2	13.10	.020	1.62	.021	1.39	.037	2		49,4002
212047	+26 04410	HR 8517		22 18 41	+26 41.0	2	6.47	.005	1.60	.015	1.71	.005	7	M4 III	1733,3001
212071	+50 03673	HR 8519		22 18 41	+50 43.7	1	6.37		1.25		0.99		2	K2	1733
		CS 22875 # 5		22 18 41	−40 21.6	2	15.32	.000	0.09	.000	0.06	.009	2		966,1736
		AAS11,365 T7# 48		22 18 42	+53 29.	1	10.06		1.40		1.43		3		261
		NGC 7261 - 140		22 18 43	+57 41.9	1	10.60		2.46		1.61		1		4002
		NGC 7261 - 138		22 18 44	+57 44.4	1	14.31		0.87		0.52		1		49
212118	+54 02742			22 18 45	+55 02.6	1	8.67		0.19				2	A5	1025
212107	+45 03893			22 18 46	+45 48.6	1	7.43		0.47				3	F5	1724
		VES 927		22 18 49	+55 35.4	1	10.06		0.18		−0.26		2		8113
	+56 02759			22 18 51	+57 04.8	1	10.07		0.38				2	A0	1025

Table 1 1193

HD	DM	Other Id	N Rem	α₁₉₅₀	δ₁₉₅₀	S	V	σV	B-V	σB-V	U-B	σU-B	n	Spectrum	References
212136	+57 02508			22 18 51	+58 09.5	2	6.46	.040	0.97	.013	0.81		5	G8 III	1025,1501
212010	-22 05897	HR 8516		22 18 51	-21 51.0	3	5.15	.031	1.05	.016	0.92		11	K0 III	2009,2024,3016
239928	+55 02727			22 18 54	+55 51.7	1	8.75		0.56				3	G2 V	1025
		NGC 7261 - 106		22 18 54	+57 47.4	1	14.53		0.77		0.31		1		49
212120	+45 03894	HR 8523	⋆ A	22 18 57	+46 17.1	7	4.55	.010	-0.09	.007	-0.53	.006	98	B6 V	15,154,879,1203,1363*
212120	+45 03894	IDS22169N4602	B	22 18 57	+46 17.1	1	11.60		0.25		0.06		3		3024
		AA26,31 # 40		22 18 57	+54 42.9	1	12.83		0.29		-0.52		4		483
	+61 02284			22 18 58	+62 05.2	1	9.52		0.23		-0.33		2	B8	27
	-17 06500			22 18 58	-17 24.5	1	10.71		1.19		1.18		1	K7 V	3072
		LP 931 - 74		22 18 59	-32 05.0	1	14.96		0.10		-0.81		2		3065
212097	+27 04299	HR 8522	⋆ A	22 19 01	+28 04.7	8	4.81	.010	-0.01	.014	-0.21	.013	26	B9 III	15,1049,1079,1203*
		LS III +55 049		22 19 01	+55 53.2	1	11.50		0.57		-0.34		3		483
212074	+14 04782			22 19 02	+14 38.7	1	7.65		1.07				3	K1 IV	882
		TOU28,33 T8# 87		22 19 02	+56 05.7	1	9.19		0.21				2		1025
212076	+11 04784	HR 8520, IN Peg		22 19 03	+11 57.2	7	4.99	.039	-0.13	.034	-0.81	.014	23	B2 IV-Ve	15,154,1212,1363,1586*
235805	+53 02852			22 19 03	+54 11.4	1	8.41		1.40		1.33		3	K0 IV	261
212061	-2 05741	HR 8518	⋆ A	22 19 04	-01 38.4	8	3.85	.011	-0.06	.009	-0.11	.017	32	A0 V	15,1075,1203,1425*
211970	-55 09831			22 19 04	-54 49.0	1	9.06		1.35		1.17		2	M0 V	3072
	+46 03667			22 19 06	+46 48.0	1	9.25		0.46				3	F5	1724
	+45 03895			22 19 07	+45 53.4	1	9.23		0.39				6	F2	1724
	+45 03895			22 19 07	+45 53.4	1	9.37		0.41				6	F2	1724
	+73 00972			22 19 08	+73 55.0	1	10.40		0.72		0.21		1		335
212034	-25 15887			22 19 08	-25 37.1	1	7.53		1.21		1.19		5	K1 III	1657
212081	-6 05974			22 19 09	-06 26.0	1	7.69		1.63		1.84		6	M1	1657
	+45 03897			22 19 10	+46 22.7	1	9.75		0.83				3	F8	1724
212183	+55 02729			22 19 12	+55 44.0	4	8.01	.029	-0.04	.009	-0.39	.010	8	B9 III-IV	261,338,401,1025
235807	+54 02745	LS III +55 050		22 19 17	+55 17.9	3	9.58	.021	0.21	.012	-0.66	.010	2	B0.5IV:N	338,483,1012
		G 18 - 44		22 19 18	+11 43.3	2	11.26	.005	0.98	.005	0.76	.025	2	K3	333,1620,1696
	+47 03782			22 19 21	+47 48.6	1	9.35		0.57				3	F8	1724
		Steph 2004		22 19 23	+13 50.6	1	12.04		1.40		1.31		1	K7	1746
	+43 04177			22 19 23	+44 12.5	1	9.58		0.49				3	F8	1724
	+46 03670			22 19 25	+46 48.3	1	9.86		0.50				3	F5	1724
212109	-2 05744			22 19 25	-02 10.6	1	9.46		0.47		-0.05		3	G0	196
		LS III +55 052		22 19 26	+55 50.5	1	12.39		0.36		-0.47		2		483
	+60 02388			22 19 26	+61 10.2	1	9.69		0.47		0.09		2		27
212110	-3 05433			22 19 26	-02 31.8	1	9.07		1.50		1.89		3	K5	196
212038	-51 13248			22 19 26	-51 02.8	4	8.75	.018	0.84	.012	0.42	.002	10	K2 V	742,1594,2012,3078
		CS 22886 # 54		22 19 28	-08 33.3	1	13.58		0.75		0.17		1		1736
212267	+62 02068			22 19 30	+62 57.7	1	8.04		0.30		0.27		2	A0	1375
239931	+55 02730			22 19 32	+56 12.2	1	9.90		0.28				2	A2	1025
212202	+34 04664			22 19 34	+35 28.4	1	9.26		0.08		-0.06		2	A0	1625
212239	+50 03680		V	22 19 34	+51 12.4	1	8.16		1.73				2	M1	369
	+54 02747			22 19 36	+55 21.0	1	9.90		0.25				2	A2	1025
212069	-46 14291			22 19 38	-46 08.3	1	8.71		1.32		1.50		5	K2/3	897
212186	+14 04786			22 19 39	+15 23.9	1	6.69		-0.01		-0.05		2	A0	252
	+58 02426			22 19 41	+59 24.9	1	10.36		0.57		0.24		3		27
212087	-46 14292	HR 8521, π1 Gru	⋆ AB	22 19 41	-46 12.0	6	6.17	.333	2.05	.097	1.88	.038	28	S5	15,897,2032,3051,8015,8022
212222	+41 04469	HR 8528		22 19 42	+41 49.5	1	6.40		-0.08		-0.52		2	B5 V	154
235810	+54 02749			22 19 43	+55 24.2	1	9.34		0.75				2	G8 IV	1025
	+54 02748	LS III +54 053		22 19 45	+54 50.9	2	9.05	.015	0.11	.005	-0.49	.000	3	B5	338,483
		Cep sq 1 # 35		22 19 45	+60 48.5	1	10.73		0.64		0.51		2		27
	+73 00974			22 19 45	+73 57.0	1	9.64		1.23		0.73		1	B8	335
	+42 04367			22 19 48	+43 06.2	1	8.96		1.03				3	G5	1724
212279	+56 02762			22 19 54	+56 32.0	2	8.47	.018	0.22	.027	0.05		5	A2	261,1025
		CS 22886 # 43		22 19 55	-10 29.4	1	14.72		0.46		0.03		1		1736
	+45 03900			22 19 56	+46 19.5	1	9.88		0.63				3	F8	1724
212146	-30 19122			22 19 56	-29 52.2	1	8.19		0.75				4	G6 V	2034
		G 241 - 4		22 20 00	+68 12.5	1	12.98		0.80		0.25		2		1658
212132	-46 14295	HR 8524	⋆ AB	22 20 04	-46 10.8	8	5.61	.013	0.37	.010	0.02	.017	37	F0 V	15,278,938,977,1075,1770*
212311	+55 02731			22 20 07	+56 16.7	2	8.11	.036	0.08	.000	0.00		5	A0	261,1025
239932	+57 02511			22 20 07	+57 38.0	1	9.73		0.43				2	A2	1025
		LS III +60 018		22 20 07	+60 02.0	2	10.41	.009	0.87	.000	0.01	.005	4	B3 V	27,1012
		CoD -57 08545		22 20 09	-57 27.3	3	10.75	.036	1.47	.018	1.17		3		1494,1705,3073
212312	+54 02750	LS III +55 053		22 20 10	+55 20.1	2	8.44	.005	0.80	.019	0.55		6	F2 Ib	261,1025
		Cep sq 1 # 37		22 20 10	+59 14.5	1	10.69		0.61		0.49		3		27
		LS III +55 054		22 20 13	+55 17.9	1	10.86		0.23		-0.68		2		483
	+61 02288			22 20 14	+62 11.3	1	9.19		0.28		-0.04		2	B8	27
212280	+29 04645	KX Peg	⋆ A	22 20 15	+30 06.3	1	7.51		0.70				3	G0 IV	20
		LS III +55 055		22 20 17	+55 58.3	1	11.98		0.50		-0.46		5		483
	+57 02513	LS III +58 026	⋆ AB	22 20 17	+58 28.1	2	9.53	.010	0.64	.000	-0.37	.005	7	B0 V	342,684
212241	-2 05748			22 20 17	-02 15.3	1	9.68		0.70		0.19		3	G0	196
235813	+54 02751	LS III +54 054		22 20 20	+54 32.8	1	8.84		0.22		-0.74		2	B0 III	1012
212180	-47 14171			22 20 21	-46 55.3	4	6.91	.002	0.06	.004	0.09	.008	17	A1 V	278,1075,1770,2011
211998	-72 02690	HR 8515	⋆ AB	22 20 22	-72 30.0	6	5.28	.007	0.66	.009	-0.06	.012	29	A3 V:+F2 V	15,1075,1075,1265*
206553	-89 00053	HR 8294, CG Oct		22 20 22	-89 04.6	4	6.55	.014	0.29	.003	0.10	.011	20	F0 IV/V	15,826,1075,2038
212289	+30 04695			22 20 23	+30 29.6	2	7.89	.040	1.24	.015	1.36		5	K1 II	20,8100
		Hiltner 1118		22 20 23	+55 23.	2	10.29	.000	0.26	.005	-0.73	.000	4	O5	1011,1012
		LP 820 - 12		22 20 23	-17 51.	1	13.25		1.84				6		940
		Cep sq 1 # 39		22 20 24	+63 06.3	1	10.13		0.34		-0.02		4		27
212250	-3 05437			22 20 24	-02 50.2	1	9.03		1.15		1.09		2	K2	1375

HD	DM	Other Id	N	Rem	α_{1950}	δ_{1950}	S	V	σ_V	B–V	σ_{B-V}	U–B	σ_{U-B}	n	Spectrum	References
212231	−26 16110				22 20 26	−26 05.7	1	7.86		0.61				4	G2 V	2012
		AA26,31 # 47			22 20 27	+55 21.1	1	12.81		0.36		−0.32		3		483
	+56 02763	LS III +56 048	A		22 20 34	+56 59.2	2	9.53	.014	0.65	.009	0.08		6	A1	261,1025
	+56 02763		B V		22 20 34	+56 59.2	1	9.71		2.28				2	M2	369
	−16 06074	BW Aqr			22 20 35	−15 35.1	2	10.33	.006	0.51	.006	0.02	.019	12	F7	588,1768
212303	+12 04817				22 20 37	+12 31.7	1	8.38		0.17		0.16		2	A2	1733
212334	+35 04785				22 20 37	+36 24.3	1	6.43		1.06		0.88		4	K0	985
212291	+8 04856				22 20 39	+09 12.5	2	7.92	.000	0.68	.000	0.16	.009	6	G5	1355,3026
212271	−25 15905	HR 8529			22 20 44	−25 01.0	3	5.55	.026	0.98	.006	0.76	.034	9	K0 III	58,2009,3016
		LS III +55 056			22 20 45	+55 20.0	1	12.33		0.28		−0.57		2		483
		G 156 - 1			22 20 45	−13 28.2	2	12.08	.000	0.59	.002	−0.14	.015	7		1696,3062
		CS 22875 # 12			22 20 46	−39 27.3	2	14.48	.005	0.40	.007	−0.20	.009	3		1580,1736
		WLS 2200 65 # 8			22 20 47	+64 35.1	1	11.62		0.58		0.39				1375
	+54 02752				22 20 50	+54 54.9	1	9.21		1.82				2	M0 III	369
212318	−0 04353				22 20 54	+00 21.3	1	7.01		0.19		0.08		2	A2	379
		Be 94 - 51			22 20 54	+55 37.2	1	14.02		0.50		−0.21		1		724
		Be 94 - 50			22 20 55	+55 37.2	1	12.28		0.31		0.20		1		724
212320	−7 05765	HR 8530			22 20 55	−07 26.9	4	5.94	.022	0.99	.005	0.68	.017	10	G6 III	15,1071,3016,4001
		LS III +55 057			22 20 56	+55 56.3	1	11.90		0.54		−0.33				483
		Be 94 - 56			22 20 57	+55 36.4	1	12.14		0.40		−0.45		2		483
		Be 94 - 40			22 20 58	+55 37.3	1	12.73		0.37		−0.32		1		724
	+48 03702				22 20 59	+48 45.2	1	9.75		0.30				3	A5	1724
		Be 94 - 36			22 20 59	+55 36.9	1	13.30		0.44		−0.40		1		724
		Be 94 - 39			22 20 59	+55 37.3	1	11.59		0.46		−0.44		2		483
	+55 02736	Be 94 - 38			22 20 59	+55 37.3	2	9.66	.010	0.42	.010	−0.56	.015	3		338,483
		Be 94 - 30			22 21 00	+55 37.5	1	12.15		0.46		−0.41		1		724
		Be 94 - 34			22 21 02	+55 37.3	1	14.04		0.56		−0.05		1		724
		Be 94 - 32			22 21 02	+55 37.4	1	13.27		0.53		−0.32		1		724
212378	+16 04724				22 21 03	+17 24.2	1	7.37		1.23				3	K2	882
		LS III +55 061			22 21 03	+55 34.0	1	10.33		0.33		−0.63		2		483
		LS III +55 062			22 21 04	+55 10.5	1	12.23		0.27		−0.41		2		483
		Be 94 - 23			22 21 05	+55 37.7	1	10.16		0.36		−0.64		3		483
212857	+82 00689	G 261 - 49			22 21 05	+83 16.8	1	8.63		0.58		0.06		1	G5	1658
		Be 94 - 24			22 21 06	+55 37.6	1	13.58		0.40		−0.28		1		724
212455	+54 02756	LS III +54 055			22 21 07	+54 58.8	2	7.88	.010	0.38	.000	−0.37	.005	3	B5 Iab	338,1012
212454	+56 02765	HR 8535			22 21 09	+57 01.9	2	6.16		−0.13	.010	−0.56		4	B8 III-IV	1025,1079
		G 189 - 16			22 21 12	+32 12.4	1	10.73		1.57				1	M1	1017
212395	+20 05139	HR 8532		★ A	22 21 14	+20 35.7	1	6.20		0.49		0.00		2	F7 V	1733
		AA26,31 # 56			22 21 14	+55 08.9	1	12.67		0.27		−0.47		3		483
212466	+55 02737	RW Cep			22 21 14	+55 42.6	4	6.63	.052	2.27	.027	2.36	.022	9	G8 Ia	261,338,369,8032
	+48 03703				22 21 15	+49 25.7	1	9.24		0.59				3	G0	1724
212211	−71 02686	HR 8527			22 21 18	−70 41.1	4	5.77	.007	0.38	.004	0.03	.008	23	F5 III	15,1075,1637,2038
212495	+61 02291	HR 8537			22 21 19	+62 10.0	1	6.04		0.05		0.01		2	A1 V	401
	+61 02292				22 21 23	+62 27.7	2	10.03	.044	0.67	.005	−0.10	.010	6	B2 Vn	27,206
		LS III +55 064			22 21 24	+55 04.2	1	10.64		0.42		−0.32				483
		LS III +55 065			22 21 24	+55 22.8	1	10.28		0.26		−0.73		2		483
		G 127 - 26			22 21 28	+24 08.5	2	10.64	.032	0.57	.017	0.01	.007	3	F9	1620,1658
212404	−5 05780	HR 8533		★ AB	22 21 31	−05 05.4	3	5.78	.004	−0.04	.005	−0.12	.013	10	A0 V	15,1071,3016
212404	−5 05780	IDS22189S0521		E	22 21 31	−05 05.4	1	9.78		0.33		0.02		3		3016
	+49 03834			V	22 21 33	+49 33.4	1	9.80		1.68				2		369
212442	+14 04790				22 21 34	+15 01.7	1	6.80		−0.08		−0.26		3	B9 V	985
235825	+54 02758	LS III +54 056			22 21 34	+54 59.9	3	9.29	.005	0.23	.005	−0.72	.010	7	B2	261,1011,1012
212168	−75 01748	HR 8526		★ A	22 21 34	−75 16.2	1	6.10		0.64		0.12		2	G3 IV	3077
212168	−75 01748	IDS22171S7531		AB	22 21 34	−75 16.2	3	6.03	.003	0.63	.004	0.13	.005	26	G3 IV	15,1075,2038
212496	+51 03358	HR 8538			22 21 35	+51 58.7	8	4.44	.012	1.02	.004	0.77	.008	18	G8.5 IIIb	15,1355,1363,3016*
		LB 1502			22 21 36	−46 55.	1	13.02		−0.09		−0.27		3		45
212330	−58 07954	HR 8531		★ A	22 21 38	−58 02.8	7	5.31	.010	0.66	.012	0.13	.003	23	G3 IV	15,678,1075,2024,2030*
		LS III +55 066			22 21 39	+55 12.1	1	11.19		0.29		−0.61				483
212487	+37 04560	HR 8536			22 21 41	+38 19.1	1	6.24		0.48		−0.03		1	F7 IV	254,3050
235827	+53 02866				22 21 42	+53 34.6	1	9.02		0.54		0.04		3	F8 IV	261
		CS 22875 # 14			22 21 42	−40 30.3	2	15.33	.000	0.19	.000	0.01	.007	2		966,1736
		G 27 - 25			22 21 43	−01 52.8	3	14.00	.020	1.48	.022	1.12	.055	5		203,1773,3062
212488	+34 04672				22 21 44	+34 51.3	1	8.11		0.07		0.05		2	A0	1625
212430	−14 06276	HR 8534			22 21 46	−13 47.0	3	5.77	.026	0.97	.004	0.70	.011	9	K0 III	58,2007,3016
		G 27 - 26			22 21 48	−01 53.4	3	13.89	.049	1.45	.007	1.10	.065	4		203,1773,3062
		JL 104			22 21 48	−70 52.	1	15.21		0.81		0.44		1		832
	+54 02761	LS III +55 067			22 21 50	+55 26.8	2	9.98	.000	0.34	.000	−0.68	.000	4	O6 III(f)	1011,1012
235828	+50 03698				22 21 53	+50 49.1	1	10.42		0.49				3		1724
		LS III +55 068			22 21 54	+55 38.2	1	11.80		0.45		−0.44		2		483
	+52 03200			V	22 21 59	+52 49.4	1	10.39		1.71				2	M0 III	369
		AA26,31 # 61			22 21 59	+55 11.9	1	13.37		0.34		−0.33		4		483
	+54 02763	LS III +54 057		★ A	22 22 01	+54 54.1	2	9.91	.009	0.18	.005	−0.46	.005	4		338,483
212566	+57 02520				22 22 01	+58 16.9	1	7.08		0.31		0.14		1	F0	338
	+54 02764	LS III +55 070			22 22 03	+55 08.5	2	9.55	.005	0.28	.000	−0.63	.005	3	B1 Ib	338,1012
		LS III +57 047			22 22 05	+57 59.2	1	10.71		1.65		0.50				483
		Cep sq 1 # 41			22 22 05	+62 40.9	1	10.53		0.35		0.19		4		27
212423	−52 12022				22 22 09	−51 39.1	1	7.59		0.91		0.63		10	G8 III	1628
		LS III +55 071			22 22 11	+55 36.8	1	11.67		0.30		−0.51		2		483
		G 241 - 6			22 22 11	+68 22.0	1	15.65		−0.05		−0.95		3		1663
235829	+51 03361				22 22 12	+51 51.0	1	9.21		0.41		0.16		2	A0	401

Table 1   1195

HD	DM	Other Id	N Rem	α_{1950}	δ_{1950}	S	V	σ_V	B–V	σ_{B-V}	U–B	σ_{U-B}	n	Spectrum	References
212545	+34 04674			22 22 13	+35 10.8	1	7.40		1.57		2.02		3	B5 Iab	8100
	+21 04747	G 127 - 28	⋆ AB	22 22 24	+22 17.9	1	8.82		1.19		1.10		2	K7	3072
212585	+38 04759			22 22 26	+39 10.5	1	8.05		0.64		0.14		3	G2 V	833
	+54 02766	LS III +55 072		22 22 28	+55 24.6	2	10.06	.019	0.35	.000	-0.60	.015	3		338,483
212593	+48 03715	HR 8541	⋆	22 22 29	+49 13.3	6	4.57	.018	0.09	.004	-0.34	.006	11	B9 Iab	15,1079,1119,1363*
	+54 02767			22 22 29	+55 25.2	1	10.03		0.38		-0.03		1	F0 V	338
	+16 04732	G 127 - 29		22 22 31	+17 02.3	1	10.36		0.95		0.89		2	K4	7010
	+75 00826			22 22 31	+75 42.1	1	9.48		1.69		1.88		2	K2	1375
	+54 02768			22 22 32	+54 36.3	1	9.92		1.72				2	M0 III	369
		CS 22886 # 64		22 22 32	-09 27.2	1	14.42		0.62		-0.07		1		1736
		CS 22875 # 13		22 22 35	-39 41.2	2	14.83	.000	0.23	.000	0.07	.008	2		966,1736
		G 127 - 30		22 22 37	+15 02.9	1	11.85		1.18		1.09		1		1620
	-37 14774			22 22 40	-37 22.9	1	10.33		1.10		1.05		2	K5 V	3072
211539	-86 00406	HR 8505		22 22 40	-86 13.4	3	5.75	.007	1.02	.004	0.89	.005	15	K0 III	15,1075,2038
		LP 700 - 10		22 22 42	-05 47.8	1	10.99		0.78		0.34		2		1696
212571	+0 04872	HR 8539, π Aqr		22 22 43	+01 07.4	9	4.63	.046	-0.03	.022	-0.99	.024	27	B1 Ve	15,154,379,815,1075*
212595	+25 04726			22 22 43	+26 21.4	1	7.42		0.22		0.14		3	Am	8071
212651	+53 02868			22 22 44	+54 27.1	1	9.23		0.19		0.13		1	A0	401
	-35 15313			22 22 44	-35 10.1	1	10.55		1.21		1.17		2	K5 V	3072
212537	-38 15044	T Gru		22 22 46	-37 49.4	1	8.47		1.20		0.55		1	M0e	975
212666	+51 03369	LS III +51 038		22 22 54	+51 52.7	1	8.61		-0.01		-0.62		2	B2 III-IV	401
	+60 02392			22 22 54	+61 03.4	1	9.87		1.97				2		369
212563	-25 15916			22 22 54	-24 59.0	1	9.57		0.93		0.65		1	K2 V	3072
212676	+53 02870			22 22 55	+54 28.1	1	8.24		0.10		-0.41		2	B9	401
		LS III +61 010		22 22 56	+61 18.2	3	10.62	.033	0.76	.019	-0.11	.019	6	B3 Vn	180,1012,8113
212586	-11 05829	G 156 - 6		22 22 57	-11 25.8	1	9.67		0.66		0.13		1	G0	1064
		22H22M58S		22 22 58	-04 16.6	1	11.83		0.60		-0.05		2		1696
	-37 14777			22 22 58	-37 26.6	1	10.52		1.34		1.31		2	K7 V	3072
		AA26,31 # 65		22 22 59	+55 21.1	1	12.49		0.37		-0.43		4		483
212539		S Gru		22 23 01	-48 41.2	1	12.09		1.35		0.37		1	M8 (III)e	975
212691	+50 03706			22 23 03	+51 00.1	1	6.52		1.67		2.06		3	M1 III	1501
212522	-63 04789			22 23 09	-62 47.6	1	7.75		1.28		1.39		9	K2/3 III	1704
	+1 04609			22 23 11	+01 36.6	1	10.21		1.63				1	M3 III:	1532
212623	-1 04292			22 23 11	-01 05.3	1	9.26		0.27		0.16		2	A0	355
		3C 446 # 1		22 23 11	-05 12.	2	15.09	.112	0.78	.056	0.32		4		157,1595
		3C 446 # 2		22 23 11	-05 12.	1	15.56		0.61		0.10		3		1595
		3C 446 # 3		22 23 11	-05 12.	2	15.89	.098	0.68	.033	0.23		4		157,1595
		3C 446 # 5		22 23 11	-05 12.	1	13.03		0.65				1		157
		3C 446 # 6		22 23 11	-05 12.	1	14.65		0.96				1		157
	+57 02525	LS III +57 053	⋆ AB	22 23 13	+57 35.2	1	10.35		0.25		-0.59		2	B1.5IV-V:pe	1012
		LS III +55 073		22 23 15	+55 49.0	1	12.45		0.45		-0.29		3		483
212670	+17 04746	HR 8543		22 23 16	+18 11.4	1	6.26		1.22				2	K0	71
212937	+77 00860	HR 8550		22 23 17	+77 59.4	1			-0.06		-0.17		1	A0 III	1079
		LP 875 - 62		22 23 18	-21 05.2	1	12.28		1.00				1		1705
212713	+46 03692			22 23 21	+46 50.9	1	8.42		-0.02		-0.36		2	A0	1601
	+50 03709			22 23 23	+50 29.4	1	10.86		0.52				6	F5	1724
		CS 22875 # 17		22 23 23	-41 14.8	2	14.38	.010	0.36	.006	0.07	.011	3		1580,1736
	+55 02748	LS III +56 052		22 23 25	+56 02.2	1	9.96		0.44		-0.52		2	B0.5V	1012
212643	-24 17171	HR 8542		22 23 25	-23 56.2	1	6.29		-0.03				4	A0 V	2009
212776	+60 02393			22 23 28	+60 34.9	1	8.09		0.10		-0.01		4	A0	27
212658	-19 06274			22 23 28	-19 26.3	1	9.31		1.13		1.03		1	K3/4 V	3072
212672	-5 05790			22 23 30	-05 26.0	1	7.26		0.28		0.04		2	F0	1375
	+47 03791			22 23 33	+47 58.4	1	9.45		0.80				3	G5	1724
212791	+51 03372			22 23 43	+52 11.0	1	8.09		-0.03		-0.54		3	B3 V	1212
		G 18 - 48		22 23 44	+02 45.1	1	13.68		1.64		1.10		3		203
212734	+25 04730			22 23 44	+25 40.4	2	7.05	.010	0.05	.005	0.07	.010	4	A0	105,1733
	+24 04590			22 23 46	+25 26.6	1	9.95		0.39		0.04		2		1375
212581	-65 04044	HR 8540	⋆ AB	22 23 48	-65 13.3	8	4.47	.013	-0.02	.010	-0.06	.010	31	B9.5 V	15,404,1034,1075,1637*
	+20 05152	G 127 - 31		22 23 49	+21 16.3	2	9.93	.020	0.76	.022	0.19	.003	3	K0	1620,7010
212698	-17 06521	HR 8545	⋆ AB	22 23 52	-16 59.8	4	5.56	.014	0.62	.013	0.08	.017	9	G3 V	196,219,938,2007
	+62 02078	LS III +63 006		22 23 53	+63 09.8	3	9.72	.005	1.11	.000	-0.01	.010	6	O7	27,1011,1012
		CS 22875 # 16		22 23 55	-41 16.8	2	13.74	.000	0.01	.000	0.01	.011	2		966,1736
		CS 22875 # 11		22 23 59	-38 17.7	2	14.86	.005	0.31	.015	-0.05	.005	2		966,1736
		JL 106		22 24 00	-82 00.	1	14.83		0.30		-0.60		1		832
212753	+3 04704	G 27 - 28		22 24 01	+04 21.2	1	9.71		0.82		0.39		1	G5	333,1620
212827	+53 02877	LS III +53 077		22 24 01	+53 31.6	1	8.30		0.26		0.12		2	A0 II	1012
212754	+3 04705	HR 8548	⋆ AB	22 24 05	+04 08.3	4	5.75	.009	0.52	.004	0.03	.007	15	F7 V	15,272,1071,1363,3026
212882	+62 02079			22 24 06	+63 04.4	1	6.90		1.56				2	M4 III	369
212725	-37 14788	RU Gru		22 24 06	-37 26.6	1	11.00		0.23		0.15		5	A1/3 (III)	3064
	+54 02775	LS III +55 074		22 24 10	+55 12.8	1	9.57		0.30		-0.49		2	B1 V	1012
212708	-49 13852			22 24 18	-49 37.0	1	7.48		0.73				4	G5 V	2033
		G 241 - 7		22 24 19	+69 16.3	1	10.50		0.62		0.03		1	G2	1658
	+58 02432			22 24 23	+59 18.8	1	10.11		0.49		0.17		6		27
		CS 22875 # 19		22 24 30	-42 14.1	2	13.60	.000	0.29	.000	0.01	.009	2		966,1736
		G 18 - 49		22 24 32	+06 33.9	1	13.22		1.66				2		1532
212883	+36 04835	HR 8549	⋆ AB	22 24 32	+37 11.3	3	6.46	.022	-0.13	.000	-0.75	.000	12	B2 V	15,154,1174
		LS III +56 054		22 24 33	+56 10.5	1	11.09		0.62		-0.42		2		483
		LS III +55 075		22 24 36	+55 17.6	1	12.65		0.28		-0.55		2		483
		LS III +55 076		22 24 36	+55 56.1	1	12.60		0.48		-0.42		3		483
		Smethells 109		22 24 36	-58 21.	1	11.51		1.30				1		1494

HD	DM	Other Id	N	Rem	α_{1950}	δ_{1950}	S	V	σ_V	B–V	σ_{B-V}	U–B	σ_{U-B}	n	Spectrum	References
	−69 03239				22 24 36	−69 26.8	1	9.19		1.05		0.74		2	G5	1730
		CS 22875 # 15			22 24 40	−41 06.6	2	14.12	.000	0.10	.000	0.20	.010	2		966,1736
213022	+70 01240	HR 8557			22 24 43	+70 31.0	1	5.47		1.20				2	K0	71
212822	−34 15593				22 24 48	−33 45.0	1	8.94		0.55				4	G0 V	2013
		S076 # 4			22 24 48	−69 24.2	1	11.05		1.44				2		1730
	−31 18815				22 24 52	−31 35.6	1	11.22		0.70		0.21		2	G8	1696
		G 241 - 8			22 24 53	+63 37.4	1	9.50		1.10		1.17		2	K7	7010
212940	+45 03941				22 24 55	+45 32.0	1	7.44		0.15		0.13		2	A2	1601
		AE Peg			22 24 56	+16 33.	1	12.37		0.26		0.16		1	A9.2	699
	+52 03210	LS III +53 078			22 24 57	+53 23.4	1	10.69		-0.02		-0.78		2	B1 V	1012
		LS III +55 077			22 24 57	+55 22.8	1	10.89		0.35		-0.55		4		483
212728	−68 03493	HR 8547			22 24 58	−67 44.6	4	5.56	.007	0.20	.006	0.08	.007	26	A4 V	15,1075,1637,2038
212986	+55 02750	HR 8554			22 25 05	+56 10.7	1			-0.10		-0.48		2	B5 III	1079
	+76 00860				22 25 05	+77 15.8	1	10.33		0.43		-0.02		2		1375
	+52 03211				22 25 07	+53 13.9	1	9.64		1.88				2	M0 III	369
	+66 01510				22 25 07	+67 17.0	1	9.64		0.25		0.22		2	A2	1375
	−36 15400				22 25 07	−35 39.8	1	10.83		1.11		0.99		2	K5 V	3072
	+53 02885	LS III +53 079			22 25 10	+53 55.6	1	10.46		0.04		-0.76		2	B2 III	1012
		G 18 - 50			22 25 12	+06 42.2	1	11.98		1.19		1.08		1		333,1620
213023	+62 02081	LS III +63 008		★ AB	22 25 12	+63 27.8	1	8.50		0.75		-0.24		3	O9 V:	399
		BN Aqr			22 25 12	−07 44.2	1	11.82		0.10		0.02		1	F3	699
212978	+39 04841	HR 8553			22 25 15	+39 33.3	3	6.15	.009	-0.14	.000	-0.77	.005	12	B2 V	15,154,1174
212895	−39 14719				22 25 18	−39 20.7	1	7.48		1.56		1.87		4	M0 III	1673
212943	+3 04710	HR 8551		★ AB	22 25 20	+04 26.7	12	4.79	.010	1.05	.005	0.88	.008	68	K0 III	3,15,1003,1075,1088*
212977	+41 04504				22 25 20	+41 45.5	1	8.09		0.43		-0.01		2	F8	1733
		VES 930			22 25 20	+55 13.8	1	11.22		0.23		-0.16		1		8113
		G 27 - 29			22 25 20	−05 32.0	3	11.20	.017	0.80	.008	0.26	.014	4		1620,1658,1696
		CS 22875 # 30			22 25 20	−38 46.0	2	15.45	.000	-0.02	.000	-0.01	.011	2		966,1736
		Cep sq 1 # 45			22 25 23	+59 51.2	1	10.43		0.34		0.21		2		27
212988	+31 04701	HR 8555			22 25 29	+31 35.0	1	5.98		1.45				2	K2	71
213087	+64 01664	HR 8561		★	22 25 29	+64 52.6	3	5.48	.026	0.37	.005	-0.59	.000	6	B0.5Ib	15,154,1012
213086	+64 01665				22 25 29	+65 12.7	1	7.97		1.24		1.27		2	G5	1733
		LS III +55 080			22 25 33	+55 55.6	2	10.94	.010	0.42	.010	-0.58	.010	4		405,483
	+57 02536	LS III +58 030			22 25 34	+58 06.7	2	10.31	.011	0.48	.007	-0.44	.010	3	B1 II-III	1012,1722
212806	−70 02932				22 25 34	−69 41.6	1	7.66		1.07		0.86		2	K1 III	1730
		CS 22875 # 18			22 25 37	−41 59.9	2	13.75	.000	0.05	.000	0.17	.010	2		966,1736
212989	+11 04804	IDS22232N1144		AB	22 25 39	+11 59.6	1	7.08		0.90		0.56		1	K0	1355
	+24 04592	G 127 - 33			22 25 39	+24 40.5	2	9.99	.005	0.91	.020	0.62	.032	2	K2	1620,1658
212918	−50 13685				22 25 40	−49 50.4	1	9.71		0.88				4	K2 (V)	2033
212953	−39 14723	HR 8552		★ A	22 25 44	−39 33.4	1	5.47	.005	0.96	.007			13	G8 III	15,1075,2024,2032
213014	+16 04746	IDS22233N1645		AB	22 25 46	+17 00.5	1	7.59		0.90				1	G9 III	882
213089	+50 03726				22 25 54	+51 19.2	1	7.57		0.12		-0.09		2	A0	401
		G 127 - 34			22 25 56	+27 51.4	2	11.01	.009	0.73	.014	0.17	.037	4		308,1620
212748	−77 01544				22 25 59	−76 56.3	1	8.08		0.47				4	F6 V	2012
	+57 02537				22 26 01	+58 09.6	2	10.05	.012	2.08	.001	2.68		3		369,1722
	+59 02522	G 232 - 74			22 26 01	+59 29.3	1	10.65		1.08		1.21		2	K5	7010
		NGC 7293 sq1 7			22 26 04	−21 03.1	1	13.40		0.76		0.20		1		268
	+62 02086				22 26 10	+62 44.8	1	9.25		0.21		-0.27		2	B8	27
239960	+56 02783	DO Cep		★ AB	22 26 12	+57 26.7	5	9.59	.015	1.65	.004	1.27	.008	13	F7 V	1,369,1197,3072,8039
213038	−3 05450				22 26 13	−02 40.6	1	8.93		0.47		-0.04		1	F8	1375
		G 18 - 51			22 26 16	+05 33.9	2	14.17	.019	1.42	.005	1.22		3		1705,3078
213052	−0 04365	HR 8559		★ AB	22 26 16	−00 16.6	9	3.65	.014	0.40	.011	0.00	.024	24	F1 IV	15,938,1020,1075,1311*
213009	−44 14931	HR 8556		★ AB	22 26 17	−43 45.1	8	3.96	.005	1.03	.006	0.81	.003	25	G6/8 III	15,58,278,1020,1075*
		NGC 7293 sq1 8			22 26 19	−20 57.1	1	12.54		0.79		0.38		1		268
213403	+78 00796	HR 8578			22 26 20	+78 31.8	1	5.84		0.16		0.12		2	A2m	1733
		G 127 - 35			22 26 21	+18 40.6	2	10.74	.012	1.42	.065	1.31		4	K7-M0	1619,7010
213159	+50 03730				22 26 24	+51 03.0	1	7.73		0.00		-0.34		2	B9	401
212920	−69 03240				22 26 24	−69 31.6	1	8.59		1.41		1.61		2	K3 III	1730
213042	−30 19175				22 26 25	−30 15.8	3	7.64	.025	1.10	.015	1.00		10	K4 V	1075,2034,3077
		CS 22875 # 27			22 26 28	−40 05.3	2	14.47	.000	0.09	.000	-0.67	.009	3		966,1736
		G 215 - 47			22 26 29	+50 55.0	1	12.18		0.56		-0.12		2	sdF8	1658
		Steph 2011			22 26 29	−13 32.8	1	11.38		1.30		1.28		1	K7	1746
		CS 22875 # 29			22 26 30	−39 13.2	2	13.67	.005	0.41	.008	-0.01	.009	3		1580,1736
		Cep sq 1 # 47			22 26 31	+62 28.6	1	10.69		1.37		0.11		6		27
213072	−25 15945				22 26 31	−24 49.2	1	9.51		1.06		0.89		1	K0 III/IV	855
239964	+57 02541			V	22 26 35	+57 02.5	2	9.72	.016	1.96	.012	2.12		3	M7	369,1722
213224	+57 02542	IDS22248N5755		A	22 26 35	+58 10.0	1	8.49		0.03		-0.02		1	A0	1722
213075	−29 18354				22 26 37	−28 59.0	1	8.56		0.50				4	F5/6 V	1075
213119	+8 04874	HR 8562			22 26 38	+08 52.4	4	5.58	.015	1.56	.008	1.91	.009	12	K5 IIIa	15,1256,1355,3016
213242	+63 01852	HR 8568			22 26 39	+63 49.8	1	6.29		1.08				2	K0	71
		NGC 7293 sq1 6			22 26 39	−21 04.5	1	12.44		0.77		0.37		1		268
213143	+20 05166				22 26 43	+21 08.2	1	7.76		0.37		0.20		3	Fm	8071
213232	+57 02543				22 26 43	+58 17.0	1	7.95		0.15		0.13		1	A5	1722
213177	+29 04674				22 26 44	+29 32.4	2	7.86	.005	1.11	.020	0.95		6	K0 II	20,8100
		NGC 7293 sq1 5			22 26 47	−21 12.0	1	12.41		0.78		0.38		1		268
213080	−44 14935	HR 8560, δ2 Gru		★ A	22 26 47	−44 00.3	9	4.16	.036	1.57	.021	1.70	.004	35	M4.5 IIIa	15,278,1075,2012,2024*
213178	+28 04381				22 26 48	+28 46.4	1	7.16		1.17				2	K1 III	20
213231	+58 02440				22 26 48	+58 35.4	1	8.28		0.38		-0.31		1	B8	338
213179	+26 04439	HR 8564			22 26 49	+26 30.4	1	5.79		1.25				2	K2 II	71
213191	+39 04851	S Lac			22 26 49	+40 03.6	1	10.05		1.57		0.40		1	M4	817

Table 1 1197

HD	DM	Other Id	N	Rem	α_{1950}	δ_{1950}	S	V	σ_V	B–V	σ_{B-V}	U–B	σ_{U-B}	n	Spectrum	References
		NGC 7293 sq1 9			22 26 51	−20 59.5	1	12.02		0.63		0.14		1		268
	+46 03717				22 26 53	+46 54.9	1	8.82		0.29		0.19		2		1601
		CS 22875 # 31			22 26 57	−38 38.5	2	14.13	.000	0.01	.000	0.06	.011	2		966,1736
	+14 04806				22 26 58	+14 38.7	1	8.88		1.07		0.90		3	K0	1648
213135	−27 15932	HR 8563			22 26 58	−27 21.8	3	5.94	.004	0.34	.008			11	F0 V	15,1075,2032
213210	+26 04440				22 27 00	+27 10.3	1	7.68		0.30		0.14		2	F0	1375
		Cep sq 1 # 48			22 27 00	+62 56.2	1	10.09		0.31		-0.40		3		27
239967	+55 02756	LS III +56 059			22 27 02	+56 20.8	1	9.36		0.33		-0.49		2	B3 II	1012
		NGC 7293 sq1 1			22 27 05	−21 03.7	1	13.18		0.50		0.02		1		268
		Cep sq 1 # 49			22 27 07	+63 00.9	1	10.77		0.41		-0.21		4		27
239968	+57 02545				22 27 10	+57 46.7	1	9.92		0.46		0.10		1		1722
213307	+57 02547	IDS22254N5754		C	22 27 17	+58 08.9	2	6.31	.003	-0.03	.006	-0.34	.016	7	A0	542,1358
213155	−42 15947				22 27 18	−42 32.9	9	6.92	.004	-0.04	.004	-0.08	.010	67	B9.5V	278,474,743,863,977,1075*
213306	+57 02548	HR 8571, δ Cep		★ A	22 27 19	+58 09.5	3	3.56	.098	0.48	.061	0.36		5	F5 Ib	851,1363,6007
		NGC 7293 sq1 4			22 27 19	−21 09.0	1	12.70		0.68		0.24		1		268
213198	−13 06204	HR 8565			22 27 21	−13 10.3	3	6.42	.024	0.33	.009	0.01	.037	8	F3 IV	254,2009,3016
213258	+35 04815				22 27 22	+35 40.9	1	7.70		0.32		0.06		2	A3	1601
		LS III +57 055			22 27 23	+57 13.8	1	10.77		0.53		-0.40		2	B1 III	1012
213322	+53 02897				22 27 24	+53 59.4	2	6.77	.000	-0.02	.000	-0.54	.015	5	B2 IV	399,401
213235	+3 04713	HR 8566		★ AB	22 27 26	+04 10.6	6	5.50	.016	0.38	.012	0.10	.016	17	F4 IV +F4 V	15,938,1256,2024,3026,8071
213310	+46 03719	HR 8572		★	22 27 27	+47 27.0	7	4.37	.014	1.68	.011	1.10	.025	20	M0 II +B8 V	15,369,1355,1363,3016*
213272	+34 04700	HR 8569			22 27 28	+35 28.2	1	6.55		0.04		0.04		2	A2 V	1733
239971	+56 02788				22 27 28	+57 27.7	1	8.61		1.91				2	M0	369
	+62 02088				22 27 33	+62 32.0	1	9.91		0.55		-0.24		4		27
		NGC 7293 sq1 3			22 27 33	−21 04.3	1	12.85		0.75		0.25		1		268
		G 215 - 50			22 27 34	+41 13.2	1	13.25		1.65				3		419
213236	−15 06231	HR 8567			22 27 37	−14 50.5	2	6.36	.005	-0.04	.000	-0.37		6	B8 Vs	252,2009
213405	+64 01672	LS III +64 005			22 27 39	+64 51.4	1	7.95		0.47		-0.44		2	B0.5V	1012
		Be 96 - 10			22 27 40	+55 08.9	1	11.67		0.54		0.27		2		1507
	+62 02089				22 27 41	+62 58.9	1	9.72		0.38		-0.30		2	B8	27
213323	+31 04708	HR 8574			22 27 44	+32 19.0	3	5.65	.007	-0.04	.018	-0.12	.031	7	B9.5V	985,1049,1716
213220	−44 14940				22 27 44	−44 21.1	6	6.93	.007	0.07	.002	0.09	.004	33	A0/1 V	278,977,977,1075,1770,2011
		NGC 7293 sq1 2			22 27 45	−21 01.7	1	12.29		0.58		0.08		1		268
		CS 22875 # 20			22 27 48	−42 44.7	2	14.14	.000	0.06	.000	0.16	.010	3		966,1736
213315	+21 04770	IDS22255N2157		AB	22 27 51	+22 12.6	1	8.00		0.34		0.05		2	F0	3030
		Be 96 - 3			22 27 51	+55 08.1	1	11.16		0.34		-0.58		4		1507
		Be 96 - 2			22 27 53	+55 09.2	1	14.30		0.52		-0.14		4		1507
		Be 96 - 1			22 27 53	+55 09.4	1	13.02		0.89		0.38		4		1507
		Be 96 - 4			22 27 55	+55 09.0	1	12.23		0.35		-0.51		5		1507
		Cep sq 1 # 52			22 27 55	+62 06.2	1	10.24		0.71		0.55		3		27
		Be 96 - 8			22 27 56	+55 09.2	1	12.21		0.40		-0.41		3		1507
		LP 820 - 28			22 27 56	−18 58.1	1	13.26		0.70		0.14		2		1696
213240	−50 13701	IDS22249S4956		A	22 27 57	−49 41.3	2	6.80	.005	0.60	.005			8	G0/1 V	1075,2012
	+54 02789	Be 96 - 6			22 27 58	+55 09.1	1	10.31		0.38		-0.60		6		1507
		Be 96 - 9			22 27 58	+55 10.3	1	11.86		1.23		0.92		3		1507
		Be 96 - 5			22 27 59	+55 08.6	1	14.07		0.42		-0.22		3		1507
		Be 96 - 7			22 28 00	+55 09.2	1	12.89		0.43		-0.54		3		1507
213320	−11 05850	HR 8573			22 28 00	−10 56.1	9	4.82	.010	-0.06	.009	-0.10	.020	34	A0 IVs	15,369,1075,1088,2012*
		WLS 2220-5 # 8			22 28 01	−05 04.8	1	10.96		0.98		0.59		2		1375
		LS III +56 061			22 28 07	+56 45.7	1	10.49		0.56		-0.47		2	B0.5 III:	1012
213296	−26 16175	HR 8570			22 28 07	−26 19.8	1	6.43		1.08				4	K1 III	2035
	+21 04772	IDS22257N2202		AB	22 28 08	+22 16.8	3	9.06	.011	0.41	.001	0.09	.012	19	F2	830,1783,3030
	+54 02790	LS III +54 062			22 28 08	+54 32.0	1	9.69		0.20		-0.67		2	B0 IVn	1012
		CS 22875 # 22			22 28 09	−40 55.3	2	13.59	.000	0.19	.000	0.18	.007	2		966,1736
213481	+65 01774	LS III +66 004			22 28 10	+66 13.3	1	8.20		0.55		-0.33		3	B2 II	555
		CS 22875 # 21			22 28 12	−42 29.2	2	14.33	.000	-0.06	.000	-0.19	.013	2		966,1736
213461	+59 02529				22 28 14	+59 43.5	1	8.99		0.54		0.28		2	F5	1502
239978	+56 02793	ST Cep			22 28 16	+56 44.6	2	8.21	.139	2.29	.010	2.32		5	M0 Ib	369,8032
213420	+42 04420	HR 8579		★	22 28 19	+42 52.0	7	4.51	.017	-0.09	.004	-0.74	.007	29	B2 IV	15,154,1119,1174,1363*
213470	+56 02794	IDS22265N5643		AB	22 28 25	+56 58.1	1	6.65		0.56		0.24		2	A3 Ia	1012
		LS III +56 065			22 28 27	+56 19.7	1	9.89		0.60		-0.39		2	B0 IV	1012
213222	−69 03243				22 28 27	−69 21.9	1	7.04		0.49		0.01		10	F5/6 IV/V	1628
213570	+72 01042				22 28 28	+72 45.5	1	9.45		0.52		0.02		2	G5	1375
239981	+57 02549				22 28 29	+57 45.8	1	9.60		0.94		0.12		1		1722
239982	+59 02530	LS III +60 020			22 28 30	+60 23.1	1	8.92		0.71		0.16		2	B3	180
213482	+56 02795	IDS22266N5621		AB	22 28 32	+56 35.9	1	8.09		0.95				1	F9 Ib	6009
213571	+69 01257				22 28 37	+69 54.9	2	7.15	.010	-0.05	.005	-0.62	.005	8	B1 V	399,555
213398	−32 17126	HR 8576		★ AB	22 28 40	−32 36.2	6	4.29	.005	0.01	.004	0.02	.007	17	A1 V	15,1075,1637,2012*
213429	−7 05797	HR 8581			22 28 41	−06 48.6	3	6.15	.017	0.56	.005	0.04	.010	9	F7 V	15,1061,3016
		G 156 - 13			22 28 41	−09 40.5	1	15.62		1.66		1.17		2		3016
		G 156 - 12			22 28 41	−09 40.9	1	11.18		1.23		1.11		2	K5	3016
213428	−3 05460	HR 8580			22 28 43	−03 10.1	2	6.15	.005	1.08	.000	1.01	.005	7	K0	15,1071
	+57 02550				22 28 44	+58 19.3	1	10.52		1.37		1.02		1		1722
213363	−47 14229				22 28 47	−46 50.2	3	7.23	.001	0.46	.000	0.01	.014	10	F5 V	278,1075,2011
		Cep sq 1 # 53			22 28 48	+62 36.3	1	10.28		0.76		-0.12		6		27
		Cep sq 1 # 54			22 28 52	+62 52.5	1	10.51		0.54		-0.13		2		27
	−6 06012				22 28 56	−06 28.7	1	10.45		0.35		0.14		2		1375
213413	−44 14949			V	22 29 00	−43 48.3	1	7.40		1.55				4	M4 III	2011
213464	−11 05855	HR 8583			22 29 02	−11 09.7	2	6.39	.005	0.29	.005	0.07		7	A8 III	2007,8071
		G 18 - 54			22 29 03	+01 54.4	3	10.70	.004	0.48	.008	-0.16	.004	4	sdF7	333,927,1620,1696

HD	DM	Other Id	N Rem	α_{1950}	δ_{1950}	S	V	σ_V	B–V	σ_{B-V}	U–B	σ_{U-B}	n	Spectrum	References
		LP 820 - 33		22 29 07	−16 31.6	1	11.35		0.96		0.79		1		1696
		Steph 2015		22 29 09	+20 29.8	1	11.90		1.42		1.28		2	K7-M0	1746
		AO 986		22 29 10	+58 13.6	1	11.15		1.87		2.26		1		1722
213534	+28 04389	HR 8584, GX Peg		22 29 14	+29 17.1	2	6.33	.000	0.20	.015	0.11	.000	3	A5m	1716,1733
213558	+49 03875	HR 8585	⋆ A	22 29 14	+50 01.5	4	3.76	.016	0.02	.012	0.01	.012	10	A1 V	15,1363,3023,8015
213457	−43 15075			22 29 14	−43 31.2	4	6.90	.002	0.98	.003	0.77	.001	22	K0 III	278,863,1075,2011
		AO 987		22 29 15	+58 06.2	1	11.30		1.92		1.95		1		1722
	+82 00692			22 29 15	+82 30.4	1	10.42		1.17		0.86		2	K2	1375
213415	−59 07808			22 29 16	−59 02.2	3	9.72	.008	0.74	.012	0.19	.037	8	G8 V	1097,1696,3008
213467	−31 18861			22 29 18	−31 25.8	4	8.52	.011	0.72	.005	0.11	.000	9	G5 VwF3	742,1311,1594,2012
213487	−22 05933			22 29 19	−21 51.4	1	9.85		0.84				2	G8 IV	1594
213468	−43 15076			22 29 20	−42 50.2	2	10.93	.005	0.01	.005	0.01	.005	5	B9 IV	1097,3077
	+60 02405	LS III +61 012		22 29 23	+61 21.1	3	9.89	.032	0.52	.024	-0.19	.008	5	B3 V:nn	27,180,1012
213798	+78 00801	HR 8591		22 29 27	+78 34.1	2	5.48	.020	0.07	.010	0.08	.005	4	A3 V	985,1733
		G 27 - 32		22 29 28	−04 57.4	1	12.58		1.30		1.07		1		1620
	+64 01677	LS III +65 007	⋆ AB	22 29 35	+65 12.5	1	8.96		0.64		-0.27		4	B2 IV-III	206
213606	+50 03743			22 29 36	+50 59.0	1	7.96		0.04		-0.05		2	A0	1566
213442	−62 06348	HR 8582, ν Tuc		22 29 38	−62 14.4	4	4.87	.037	1.60	.011	1.72	.005	21	M4 III	1088,2024,2035,3052
	+48 03755	G 215 - 53		22 29 41	+49 26.6	1	9.62		0.80		0.28		3	G8	3016
		CS 22875 # 23		22 29 49	−41 01.2	2	15.50	.000	-0.05	.000	-0.01	.012	2		966,1736
213517	−60 07574			22 30 06	−60 15.0	1	7.90		0.93				4	K0 III	2012
	+14 04813			22 30 07	+14 45.4	1	9.20		1.30		1.31		2	K2	1375
213617	+19 04949	HR 8586		22 30 10	+19 58.3	2	6.43	.010	0.32	.005	0.01	.013	4	F1 V	1733,3026
239989	+59 02535			22 30 10	+60 26.6	1	9.67		0.53		-0.15		2	B7	27
213619	+12 04838			22 30 11	+12 47.2	1	6.52		0.32		0.03		2	F2 III	3016
		G 27 - 33		22 30 11	−06 12.6	2	11.51	.015	0.76	.010	0.20	.019	3		1620,1658
213660	+39 04871	HR 8588	⋆ A	22 30 13	+39 31.3	2	5.88	.000	0.16	.005	0.14	.005	5	A6 V	985,1733
213702	+52 03236	IDS22283N5309	A	22 30 14	+53 24.7	1	8.53		0.01		-0.07		2	A0	1566
		CS 22875 # 25		22 30 16	−40 00.2	2	14.80	.000	0.02	.000	-0.90	.011	2		966,1736
	+22 04660			22 30 18	+23 15.1	1	8.82		0.45		0.00		2	F5	1375
		WLS 2220 25 # 8		22 30 18	+25 17.1	1	12.25		0.33		0.06		2		1375
213720	+53 02910	HR 8589		22 30 19	+53 46.8	2	6.38	.005	1.07	.010	0.90	.005	5	G8 III	1566,1733
213646	+9 05061	G 18 - 55		22 30 20	+10 09.0	1	9.35		0.83		0.37		1	K2	333,1620
213644	+15 04670	HR 8587		22 30 20	+15 36.3	2	6.32	.005	1.19	.000	1.17	.000	6	K0	985,1733
	+8 04887	G 18 - 56		22 30 30	+09 07.1	5	10.35	.025	1.51	.011	1.19	.024	7	dM0	333,680,1017,1620*
213757	+59 02536	LS III +60 021		22 30 32	+60 04.6	3	8.39	.008	0.24	.016	-0.48	.026	8	B1 V	27,180,399
		LP 984 - 59		22 30 35	−35 56.5	1	12.07		0.69		0.00		2		1696
213758	+58 02450			22 30 36	+58 46.1	1	11.74		3.12				3	F3 V	8032
213628	−36 15445			22 30 38	−35 42.1	2	7.79	.000	0.72	.000	0.28		8	G5 V	158,2012
		CS 22875 # 24		22 30 39	−40 30.2	2	14.80	.000	-0.01	.000	0.08	.012	2		966,1736
213677	−9 06001	G 156 - 19		22 30 43	−09 19.2	2	8.74	.009	0.58	.009	0.02	.005	2	G0	1064,1658
		Steph 2018		22 30 44	−09 52.3	1	12.41		1.57		1.16		2	M3	1746
	+53 02911	G 233 - 17	⋆ AP	22 30 46	+53 32.1	1	10.07		1.34		1.14		1	K5 V	3078
213402	−79 01206	HR 8577		22 30 46	−79 01.8	3	6.13	.007	1.37	.009	1.35	.000	15	K1 III	15,1075,2038
213722	+17 04765			22 30 47	+17 31.6	1	7.75		0.56		0.13		2	F8	1375
213776	+48 03763	IDS22287N4852	AB	22 30 49	+49 07.8	1	7.73		0.41		-0.03		5	F4 V +F5 V	3030
		Steph 2019		22 30 51	−05 32.5	1	12.93		1.48		1.13		1	M0	1746
213657	−42 15968			22 30 51	−42 18.5	6	9.66	.015	0.42	.011	-0.23	.016	13	(G0)w(A)	1236,1311,1594,1696*
213656	−45 14732			22 30 56	−44 56.9	4	6.78	.002	0.95	.004	0.66	.011	18	G8 III	977,1075,2011,2024
235874	+50 03748	LS III +50 041		22 30 57	+50 57.5	1	9.64		0.00		-0.70		2	B3 III	1012
	+52 03240			22 31 02	+53 27.6	1	10.21		1.83				2	M2 III	369
	+62 02096			22 31 02	+62 43.7	1	9.91		0.67		-0.13		6		27
		LP 760 - 71		22 31 03	−11 04.9	1	13.41		0.48		-0.20		2		1696
		WLS 2240-20 # 6		22 31 04	−19 22.5	1	11.59		0.73		0.26		2		1375
235876	+54 02798			22 31 08	+54 50.4	1	8.65		0.41		0.03		2	F2	1566
		Cep sq 1 # 59		22 31 14	+61 44.9	1	10.77		0.41		-0.24		5		27
213803	+28 04398			22 31 18	+29 19.9	1	8.05		1.00				2	K0 III	20
213835	+40 04850	IDS22292N4018	AB	22 31 22	+40 33.6	1	8.55		1.07		0.82		6	K0	1655
214035	+75 00836	HR 8599		22 31 24	+75 58.1	3	5.71	.009	0.01	.004	0.02	.007	6	A2 V	695,985,1733
213781	−12 06299			22 31 25	−12 25.0	1	9.04		-0.12		-0.47		3	B8/9 III	1026
213789	−2 05781	HR 8590	⋆ A	22 31 28	−01 49.9	4	5.89	.026	0.98	.010	0.69	.027	12	G6 III	15,252,1071,3016
	+39 04879			22 31 29	+39 29.7	1	10.20		0.36		0.04		2		401
		G 241 - 11		22 31 33	+65 58.2	1	12.29		0.93		0.59		1		1658
213856	+35 04837			22 31 35	+35 36.9	2	8.64	.000	0.46	.000	0.00	.000	6	G5	196,583
213973	+69 01262	HR 8595	⋆ AB	22 31 36	+69 39.3	1	6.01		0.32		0.00		2	A9 III	3030
	−34 15647			22 31 38	−34 09.8	1	11.28		0.74		0.23		2		1696
213857	+28 04401	IDS22293N2913	A	22 31 41	+29 29.1	1	8.24		1.03				2	K0 III	20
		Case *M # 75		22 31 42	+58 38.	1	10.67		3.18				2	M3.5Ia	148
213930	+55 02769	HR 8594		22 31 44	+56 22.0	1	5.71		0.96				2	G8 III-IV	71
239994	+56 02806	V351 Cep		22 31 46	+57 03.6	2	9.27	.025	0.85	.015	0.61		2	F8 Ib	592,1462
213785	−46 14356			22 31 48	−46 02.6	4	7.77	.003	0.36	.003	-0.01	.012	18	F0 V	861,977,1075,2011
	+55 02770	LS III +56 072	A	22 31 49	+56 16.5	1	9.70		0.35		-0.54		2	B1.5III	1012
	+55 02770	LS III +56 072	B	22 31 49	+56 16.5	1	10.10		0.37		-0.54		2	B1.5II	1012
		VES 933		22 31 50	+56 47.8	1	11.26		0.24		-0.01		2		8113
214019	+69 01263	HR 8598	⋆ A	22 31 52	+70 06.9	1	6.34		0.00		-0.06		2	A0 V	3016
214019	+69 01263	IDS22304N6951	B	22 31 52	+70 06.9	1	9.42		0.47		-0.04		2		3016
213845	−21 06251	HR 8592		22 31 58	−20 57.9	5	5.20	.016	0.44	.004	-0.01	.005	11	F7 V	15,254,2012,2024,3026
213893	−0 04383			22 32 02	+00 20.3	3	6.69	.015	1.53	.008	1.84	.014	7	K5	1003,1355,3077
214007	+61 02314			22 32 05	+61 31.2	1	6.56		0.15		0.13		3	A2	985
213990	+53 02918	IDS22302N5334	AB	22 32 10	+53 49.6	1	7.70		0.05		0.02		3	A2	1566

Table 1

1199

HD	DM	Other Id	N Rem	α_{1950}	δ_{1950}	S	V	σ_V	B–V	σ_{B-V}	U–B	σ_{U-B}	n	Spectrum	References
213947	+25 04768			22 32 15	+26 20.4	1	6.86		1.45		1.72		1	K4 III	1716
	+55 02771	LS III +56 073		22 32 15	+56 26.1	1	9.70		0.48		-0.45		2	B1 IV	1012
		LP 640 - 43		22 32 16	-01 20.7	1	14.83		1.70				7		940
		LS III +60 022		22 32 17	+60 43.3	1	10.52		0.66		-0.19		1		180
213830	-55 09862			22 32 17	-55 26.6	1	7.69		1.02				4	G8/K0 III	2012
213976	+40 04854	IDS22301N4015	A	22 32 18	+40 31.0	1	7.02		-0.11		-0.80		6	B1.5V	1174
		LP 760 - 39		22 32 24	-12 44.9	1	11.47		0.76		0.28		2		1696
		Cep sq 1 # 60		22 32 27	+61 22.4	1	10.49		0.44		0.25		6		27
213882	-52 12042			22 32 28	-51 51.8	1	6.77		1.07		0.86		8	K0 III	1628
213797		R Ind		22 32 28	-67 32.8	1	10.12		1.54		0.88		1	M4/6e	975
		G 27 - 34		22 32 32	-04 13.6	1	11.09		1.08		0.99		1		1620
	+21 04785			22 32 34	+21 50.7	1	9.15		0.62		0.19		3	G0	1648
	+59 02541			22 32 34	+60 14.2	2	9.96	.005	2.02	.015	2.19		5	M2	369,8032
	+38 04805			22 32 36	+38 56.1	1	10.22		0.35		0.08		2		401
213996	+11 04831	G 18 - 58		22 32 38	+11 37.6	2	8.65	.005	0.86	.000	0.62	.009	4	K2	1620,1658
		LS III +58 037		22 32 39	+58 04.3	1	10.45		0.42		-0.46		2	B2 III	1012
213884	-58 07971	HR 8593		22 32 39	-58 08.6	2	6.22	.005	0.20	.000			7	A5 V	15,2030
214023	+30 04744	IDS22304N3017	A	22 32 43	+30 32.7	1	7.35		1.34				2	K3 III	20
239996	+56 02808			22 32 43	+57 06.4	1	9.38		0.11				1	A2	592
213985	-18 06151	HM Aqr		22 32 46	-17 31.0	1	8.90		0.18		0.23		2	A0 III	1375
213998	-0 04384	HR 8597	14	22 32 47	-00 22.5	14	4.03	.011	-0.09	.010	-0.27	.027	44	B9 IV-Vn	3,15,1020,1034,1075,1079*
		G 156 - 26		22 32 51	-07 38.4	1	10.27		0.93		0.74		3		7010
		LS III +58 038		22 32 52	+58 02.5	1	9.91		0.42		-0.53		2	O9 III	342
213986	-24 17232	HR 8596		22 32 52	-24 15.0	2	5.96	.006	0.98	.004	0.81		6	K1 III	58,2007
213987	-28 17838			22 32 53	-28 23.4	1	10.26		1.25		1.24		1	K5 V	3072
	-52 12044			22 32 57	-51 53.8	1	9.56		0.60				4	sdG2	2033
213941	-55 09866			22 32 57	-54 52.0	2	7.58	.000	0.67	.005	0.10		8	G5 V	158,2033
214011	-18 06153			22 32 59	-17 52.8	1	7.55		1.42		1.51		5	K1 II/III	1657
214059	+4 04879	G 18 - 60		22 33 05	+05 07.0	4	8.27	.024	0.68	.014	0.16	.030	5	G4 V	333,1003,1620,3026
		LP 700 - 47		22 33 07	-08 07.5	1	12.27		0.80		0.33		2		1696
214028	-18 06154			22 33 07	-17 43.1	1	6.42		1.42		1.75		8	K4 III	1628
		CX Aqr		22 33 10	-00 57.0	1	10.55		0.44		-0.07		4	F2 p	3064
214165	+60 02414			22 33 11	+60 34.0	1	7.08		0.39		-0.09		2	F2 V	3016
214046	-21 06254	IDS22305S2127	A	22 33 11	-21 11.6	3	7.53	.008	1.43	.009	1.79	.025	7	K4 III	1375,1738,2012
214017	-44 14973			22 33 13	-43 50.9	1	9.98		0.49				4	F3/5 V	2011
214112	+38 04807	IDS22314N3907	E	22 33 14	+39 19.6	1	7.24		0.28		0.13		2	F0	401
235885	+51 03419			22 33 14	+51 39.9	1	8.90		0.48		0.04		2	F2	3016
240003	+59 02545			22 33 14	+59 56.9	1	10.31		0.44		0.17		3	F0	27
235887	+53 02922			22 33 18	+54 05.2	1	8.96		0.50		0.03		2	F8 IV	3016
214033	-44 14974			22 33 21	-43 55.2	1	9.60		0.41				4	F2 V	2011
	+29 04705			22 33 23	+29 50.6	1	8.27		0.95				2	G8 III	20
	+5 05039			22 33 24	+06 10.1	1	9.95		1.44				1	K7	1632
214080	-17 06554			22 33 25	-16 38.8	3	6.83	.018	-0.13	.010	-0.87	.033	10	B1/2 Ib	55,285,8100
214114	+4 04880			22 33 28	+05 20.4	1	9.26		0.51		0.13		3	G0	1625
214085	-41 14959	HR 8600		22 33 34	-40 50.5	5	6.26	.011	0.12	.000	0.13		17	A3 Vn	15,1075,2018,2032,3050
214100	-1 04323			22 33 35	-01 05.2	3	9.99	.018	1.44	.015	1.18		4	K5	1017,1705,3072
214167	+38 04808	IDS22314N3907	B	22 33 38	+39 22.1	6	6.46	.022	-0.15	.014	-0.83	.009	15	B2 V	1,154,401,879,1174,8084
214065	-47 14264			22 33 38	-46 43.1	1	9.25		1.20				4	K2 III	2012
214168	+38 04808	HR 8603	⋆ A	22 33 39	+39 22.5	7	5.73	.017	-0.15	.005	-0.90	.006	19	B1 Ve	15,154,401,879,1174*
214168	+38 04808	IDS22314N3907	C	22 33 39	+39 22.5	1	10.77		0.55		-0.10		1		8084
214168	+38 04808	IDS22314N3907	D	22 33 39	+39 22.5	1	9.06		-0.03		-0.26		1		8084
		G 127 - 39		22 33 41	+23 40.1	1	11.90		1.16		1.15		1	K4	1620
214222	+52 03247	IDS22317N5241	AB	22 33 41	+52 56.7	1	8.23		0.61		0.03		1	G1 V +G1 V	3016
214238	+56 02814	IDS22318N5620	A	22 33 43	+56 36.1	5	7.64	.012	0.68	.008	0.17	.010	8	G0	1,1028,1355,1462,3032
214238	+56 02814	IDS22318N5620	B	22 33 43	+56 36.1	3	9.70	.016	0.81	.000	0.42	.000	5		1,1028,3032
214094	-44 14978			22 33 45	-43 43.8	6	6.75	.007	0.53	.003	0.02	.008	31	F8 V	278,977,1075,1770,2011,2024
		DI Lac		22 33 46	+52 27.4	2	14.41	.058	0.15	.067	-0.75	.013	22	sd:Be	698,866
	+60 02415			22 33 46	+61 16.9	1	9.95		0.69		-0.06		5		27
214122	-32 17161	HR 8601	⋆ A	22 33 47	-31 55.4	3	5.82	.013	1.09	.005	0.98	.016	11	K2 III	58,2009,3008
214240	+49 03903	HR 8606		22 33 48	+49 48.7	5	6.29	.006	-0.06	.006	-0.54	.005	17	B3 V	15,154,879,1174,1423
214201	+31 04733			22 33 49	+32 17.2	1	8.73		0.82		0.41		2	K0	1733
214121	-32 17162	IDS22310S3211	B	22 33 50	-31 54.1	2	7.45	.006	1.26	.003	1.24	.017	10	K2 III	1673,3008
214200	+34 04728	HR 8604	⋆ A	22 33 51	+35 19.1	2	6.10		1.00				2	K0	71
214259	+56 02815			22 33 53	+57 08.3	2	8.69	.025	0.15	.000	0.10		2	A0	592,1462
214279	+55 02779	HR 8607		22 33 54	+55 48.7	2	6.38	.002	0.11	.014	0.08		3	A3 V	401,6007
	-14 06313	G 156 - 29	⋆ A	22 33 55	-13 35.9	1	10.97		1.05		0.84		1		1696
214202	+28 04411	IDS22316N2914	A	22 33 56	+29 29.2	1	8.32		1.00				2	G8 III	20
214225	+38 04811			22 33 58	+39 04.7	1	7.24		0.96		0.69		3	G8 III	833
214137	-36 15473			22 34 00	-35 42.9	1	8.93		0.86		0.50		4	K1 V	119
214243	+39 04892			22 34 03	+39 49.8	2	8.31	.004	-0.10	.000	-0.61	.015	4	B6 IV	963,1733
214150	-41 14963	HR 8602	⋆ AB	22 34 04	-40 51.0	5	5.85	.007	0.06	.004	0.05		15	A1 V	15,404,1075,2024,2035
		CS 22875 # 35		22 34 05	-38 12.1	2	14.90	.000	-0.11	.000	-0.43	.014	2		966,1736
214183	-0 04387			22 34 06	+00 06.3	1	7.96		0.35		0.05		5	F0	1509
		JL 111		22 34 06	-77 08.	2	14.03	.034	-0.11	.010	-1.00	.024	3		132,832
214263	+37 04631			22 34 07	+37 35.0	1	6.85		-0.13		-0.78		5	B2 V	1174
214203	+10 04781	HR 8605		22 34 08	+11 26.2	1			0.00		0.06		3	A1 III	1049
		CS 22875 # 34		22 34 09	-38 07.1	2	13.97	.000	0.01	.000	0.10	.011	2		966,1736
		G 242 - 3		22 34 13	+74 25.7	1	12.00		1.45		1.36		2	M1	7010
214161	-41 14965			22 34 13	-40 46.2	1	9.11		0.79				2	G5/8wF(0)	1594
214245	+12 04850			22 34 14	+12 53.8	1	7.00		1.57		1.93		2	K5 III	1375

HD	DM	Other Id	N	Rem	α_{1950}	δ_{1950}	S	V	σ_V	B–V	σ_{B-V}	U–B	σ_{U-B}	n	Spectrum	References
	+59 02547				22 34 16	+60 08.7	1	10.13		0.44		-0.13		2		27
		Smethells 111			22 34 18	-44 25.	1	10.26		1.35				1		1494
214174	-45 14750				22 34 22	-44 55.5	5	7.92	.004	1.25	.003	1.23	.011	26	K1 III	861,863,977,2011,2024
		JL 112			22 34 24	-77 40.	1	16.04		-0.18		-0.98		1		832
214313	+34 04729	HR 8609			22 34 32	+35 23.6	2	6.28	.013	1.39	.012	1.62		5	K5 III	71,1501
214470	+72 01049	HR 8615			22 34 32	+73 23.0	3	5.08	.010	0.39	.004	0.16	.006	9	F3 III-IV	253,1363,3026
214369	+57 02568	W Cep			22 34 33	+58 10.0	1	7.76		1.89				2	K0 Iape	369
214216	-45 14753				22 34 34	-44 50.0	1	9.50		0.59				4	G2/3 V	2011
214164	-60 07580				22 34 35	-60 21.3	3	8.69	.007	0.42	.007	-0.04		7	F2 V	742,1594,2012
214298	+11 04838	HR 8608			22 34 36	+12 19.1	1	6.28		1.44		1.70		2	K5	1733
	-53 10320				22 34 39	-53 08.1	1	9.97		0.96		0.73		1	K5 V	3072
		G 18 - 62			22 34 41	+11 43.6	2	10.58	.014	1.30	.006	1.25		3	K6	333,1619,1620
		CS 22887 # 15			22 34 41	-08 59.1	1	14.08		0.21		-0.50		1		1736
	+59 02548				22 34 44	+59 51.5	1	10.30		1.07		0.61		7		27
	-28 17856	IDS22320S2836		A	22 34 46	-28 20.6	1	10.67		1.29		1.21		1	K5 V	3072
214410	+53 02932				22 34 47	+54 09.8	1	7.42		-0.05		-0.31		2	B9	401
		LS III +60 023			22 34 51	+60 54.3	1	10.89		0.70		-0.42		1		180
214511	+72 01050	IDS22336N7221		AB	22 34 52	+72 37.3	1	7.56		0.48		-0.06		3	F5	3026
214511	+72 01050	IDS22336N7221		C	22 34 52	+72 37.3	1	12.76		0.74		0.21		1		3024
214411	+50 03767				22 34 53	+50 51.1	1	8.09		-0.02		-0.23		2	A0	401
		L 119 - 21			22 34 54	-65 38.	3	11.49	.016	1.60	.015	1.23		3		1494,1705,3073
214419	+56 02818	CQ Cep			22 34 57	+56 38.8	2	8.69	.080	0.40	.009	-0.56	.024	4	WN7	405,1012
214511	+72 01051	IDS22336N7221		D	22 34 58	+72 36.7	1	8.41		0.49		-0.02		2	F5	3016
	+23 04575	IDS22327N2325		AB	22 35 02	+23 40.7	1	9.60		1.05		0.90		3		3072
	+23 04575	IDS22327N2325		C	22 35 02	+23 40.7	1	12.21		1.44		1.29		2	K8	3072
214308	-47 14273				22 35 05	-46 58.1	5	7.68	.001	0.43	.004	-0.01	.011	13	F5 V	278,1075,1770,2012,2024
		CS 22875 # 41			22 35 08	-40 06.9	2	13.80	.000	0.08	.000	0.16	.009	2		966,1736
214398	+23 04576				22 35 10	+23 44.5	1	7.03		0.18		0.07		2	A3	985
	+75 00839	G 242 - 4			22 35 10	+76 13.6	1	9.39		0.54		-0.11		1	G0	1658
214376	-4 05716	HR 8610	★	AB	22 35 10	-04 29.2	4	5.05	.030	1.13	.010	1.15	.025	19	K2 III	15,1071,1075,3016
	+0 04900	CY Aqr			22 35 14	+01 16.5	1	10.44		0.17		0.14		1	B8	668
214432	+38 04817				22 35 14	+39 10.7	2	7.59	.000	-0.10	.005	-0.68	.005	12	B3 V	1174,1775
214373	-0 04394				22 35 14	-00 00.1	1	10.89		1.39		1.24		5	G0	7009
214362	-23 17523				22 35 14	-22 54.2	2	9.09	.005	0.50	.005	0.01		3	GwA/F	742,1594
	+40 04862	IDS22330N4030		AB	22 35 15	+40 44.7	1	9.65		0.08		-0.11		2	B9.5V	401
214401	-0 04395				22 35 17	-00 26.7	1	9.85		0.71		0.23		1	G5	1620
	+62 02101				22 35 18	+62 47.1	1	9.99		0.53		-0.30		4		27
214454	+50 03770	HR 8613			22 35 19	+51 17.2	4	4.63	.006	0.24	.006	0.11	.006	10	A8 IV	15,1363,3023,8015
		AJ89,1229 # 328			22 35 20	+06 45.9	1	10.42		1.72				1		1532
214383	-25 16025				22 35 23	-25 24.0	1	10.18		1.00		0.75		1	K3 IV	3072
214385	-28 17861				22 35 23	-27 42.2	2	7.86	.005	0.64	.009			8	G3 V	1075,2033
214434	+25 04779	IDS22330N2554		AB	22 35 24	+26 09.7	1	7.65		1.27		1.20		3	K2 II	8100
		G 27 - 37			22 35 31	-02 21.7	2	11.97	.030	1.37	.020	1.31	.115	2	K4	1620,1658
214388	-45 14754				22 35 32	-44 51.6	1	9.46		0.41				4	F2 V	2011
214458	+29 04715				22 35 35	+29 39.8	2	7.30	.000	1.23	.000	1.36	.007	3	K2 III	583,3040
214488	+44 04183				22 35 37	+45 08.9	1	7.77		1.06		0.83		2	K0	1601
		LP 580 - 72			22 35 38	+04 42.5	1	11.41		0.58		-0.11		2		1696
214489	+38 04819				22 35 41	+38 41.3	1	7.79		0.53		-0.01		2	F8	401
214448	-8 05912	HR 8612	★	AB	22 35 45	-08 09.5	2	6.22	.005	0.78	.000	0.49	.000	7	G0 III+F0 V	15,1061
		GD 239			22 35 46	+25 45.0	1	13.02		0.62		-0.08		4	sdG	308
		G 156 - 31	★	V	22 35 48	-15 35.2	4	12.36	.010	1.98	.016	1.46	.109	11	M5 eV	621,801,1705,1764,3078
		CS 22875 # 44			22 35 49	-42 40.2	2	14.68	.000	0.05	.000	0.13	.011	2		966,1736
	+9 05074	G 67 - 7			22 35 53	+09 36.0	1	10.59		0.85		0.55		1	K2	333,1620
214524	+40 04866				22 35 56	+40 51.6	1	7.49		-0.04		-0.65		2	B8	401
214462	-29 18414	HR 8614			22 35 58	-29 00.5	3	6.46	.004	1.03	.004			11	K1 III	15,1075,2009
214479	-21 06267	FK Aqr	★	A	22 36 01	-20 52.8	6	9.10	.022	1.50	.007	1.09	.014	21	M0 Vpe	158,801,1294,1705,1738,3072
214479	-21 06267	IDS22333S2108		B	22 36 01	-20 52.8	2	11.44	.010	1.61	.009	1.09	.034	3		801,1294,1705,3072
	+0 04902				22 36 02	+00 55.9	1	9.72		0.76		0.38		std		1402
	+9 05076	G 67 - 8			22 36 02	+10 17.2	2	9.51	.005	0.77	.005	0.31	.010	3	K2	333,1620,3077
214557	+45 04002				22 36 02	+45 34.2	1	7.00		0.56		0.05		2	F8	3016
214484	-33 16160	HR 8616			22 36 02	-33 20.5	3	5.66	.004	0.04	.010			11	A2 Vp	15,1075,2009
214441	-53 10326	HR 8611, CC Gru	★	A	22 36 04	-52 57.2	2	6.64	.005	0.36	.005			7	F3 II/III	15,2012
214558	+44 04185	HR 8617			22 36 07	+44 55.3	1	6.37		0.76		0.44		2	G2 III+A4V	1733
214710	+74 00978	HR 8625			22 36 09	+75 06.7	2	5.75	.034	1.58	.017	1.91		4	M1 IIIab	71,3001
214627	+59 02552				22 36 13	+60 21.8	1	8.66		0.36		0.27		2	A0	27
		G 28 - 10			22 36 15	+08 06.0	1	13.39		1.38				2		1532
		CS 22875 # 40			22 36 16	-39 54.8	2	14.29	.000	0.13	.000	0.18	.009	2		966,1736
214546	+3 04745				22 36 18	+04 16.2	1	6.82		0.20		0.06		2	A3	985
214586	+51 03433	IDS22343N5151		A	22 36 18	+52 06.9	1	9.13		0.05		-0.31		2	A2	401
	+24 04627	G 127 - 43			22 36 22	+25 18.5	2	10.03	.015	0.82	.010	0.48	.015	2	K2	1620,1658
	-47 14285	RV Gru			22 36 25	-47 08.0	1	11.00		0.93		0.67		4		3064
214486	-51 13342				22 36 26	-50 56.5	1	7.84		1.63		1.86		5	M2/3 III	1673
214567	+18 05014	HR 8618	★	AB	22 36 27	+19 15.8	2	5.82	.000	0.93	.005	0.62		4	G8 II	71,1733
214509	-45 14757				22 36 27	-45 00.4	4	8.92	.009	0.34	.001	0.01	.008	22	F0 V	863,977,1075,2011
214520	-43 15121				22 36 31	-43 32.0	2	8.32	.001	1.18	.002	1.22		12	K2 III	977,2011
240010	+55 02783	LS III +55 095			22 36 33	+55 34.5	1	9.48		0.38		-0.60		2	B1:IV:penn	1012
214608	+43 04260	IDS22344N4347		AB	22 36 35	+44 03.2	2	6.84	.005	0.56	.005	0.00		7	F9 V +F9 V	1381,3030
		LS III +61 013			22 36 35	+61 53.8	1	10.70		0.86		-0.23		3		180
214572	-10 05963				22 36 38	-10 17.3	1	7.01		0.59		0.10		3	G0	3026
214665	+56 02821	HR 8621	★	A	22 36 40	+56 32.1	1	5.09		1.58		1.71		3	M4 III	3016

Table 1 1201

HD	DM	Other Id	N	Rem	α_{1950}	δ_{1950}	S	V	σ_V	B–V	σ_{B-V}	U–B	σ_{U-B}	n	Spectrum	References
	+63 01873				22 36 44	+64 16.2	1	9.46		0.41				1	B8	1213
214579	−23 17538				22 36 48	−23 02.4	2	8.21	.005	1.33	.010	1.29	.022	3	Kp Ba	565,3048
214652	+36 04898	IDS22346N3651		A	22 36 49	+37 06.9	2	6.83	.013	-0.12	.017	-0.73	.013	6	B2:V	1174,1423
214734	+62 02102	HR 8627			22 36 52	+63 19.5	2	5.21	.020	0.16	.091	0.00		3	A3 IV	1363,3023
214615	−13 06235	G 156 - 34		⋆ AB	22 36 55	−12 52.4	3	7.74	.019	0.79	.005	0.37	.009	6	G8/K0 V	1705,1775,3032
214599	−28 17873	HR 8619		⋆ A	22 36 57	−28 35.2	3	6.30	.004	1.01	.005			14	K0/1 III	15,2030,2033
		LS III +61 014			22 37 00	+61 03.1	1	11.44		0.76		0.04		3		180
214600	−28 17874	IDS22342S2852		BC	22 37 00	−28 36.5	1	7.11		0.47				4	F5 V +F8 V	2033
214680	+38 04826	HR 8622		⋆ A	22 37 01	+38 47.4	30	4.88	.013	-0.20	.007	-1.03	.011	240	O9 V	1,15,247,450,985,1004*
214680	+38 04826	IDS22348N3832		B	22 37 01	+38 47.4	1	10.02		0.28		0.13		3		450
		G 156 - 35			22 37 02	−12 50.9	1	14.65		1.61		1.45		1		3032
	+10 04791	G 67 - 9			22 37 06	+11 05.5	1	10.44		0.82		0.37		1	K1	333,1620
214764	+58 02465				22 37 10	+58 40.0	1	7.04		0.32		0.11		3	F0	3016
214670	+0 04904				22 37 12	+00 46.7	3	8.55	.009	0.36	.004	0.12	.010	20	A2	281,1371,6004
		LS III +61 015			22 37 14	+61 37.2	1	11.60		0.66		0.08		1		180
214714	+36 04902	HR 8626		⋆ A	22 37 18	+37 19.9	1	6.03		0.86				2	K1 Ib:	71
214712	+37 04651				22 37 18	+38 29.3	2	6.97	.005	0.28	.010	0.06	.015	5	A9 V	401,1501
214539	−68 03518				22 37 18	−67 57.0	1	7.22		0.02				4	B8/9 Ib	2012
	−20 06464				22 37 20	⁻19 53.7	1	10.29		0.36		0.00		2		1375
214645	−30 19248				22 37 20	−30 12.4	1	7.34		0.96				4	K0 III	2012
		CS 22875 # 36			22 37 20	−38 07.0	2	14.61	.000	-0.05	.000	-0.04	.012	2		966,1736
214698	+18 05021	HR 8624			22 37 22	+19 25.2	1			0.00		0.08		3	A2 V	1049
214699	+0 04905	IDS22349N0041		AB	22 37 27	+00 56.8	1	8.54		0.34		0.09		2	G5	3016
	+64 01694	RZ Cep			22 37 28	+64 35.7	1	9.19		0.40		0.30		1	A2	668
		G 233 - 25			22 37 29	+57 03.9	1	10.74		1.28		1.28		2		7010
214686	−10 05966				22 37 30	−09 37.3	1	6.92		0.51		0.02		4	F8 IV	3026
214768	+38 04829				22 37 31	+38 56.0	2	8.30	.010	0.47	.015	-0.05	.005	5	F8 IV	401,833
214781	+46 03777				22 37 32	+46 47.9	1	8.26		1.69		1.74		3	M2	1601
214690	−31 18920	HR 8623			22 37 35	−30 55.0	5	5.88	.009	1.30	.006	1.44	.000	16	K3 III	15,1075,1311,2009,3077
214632	−58 07984	HR 8620			22 37 39	−57 41.0	4	5.97	.007	1.45	.008	1.70		13	K3 III	15,1075,2035,3005
	+52 03261				22 37 45	+53 26.4	1	9.25		0.14		-0.15		2	A0	1375
	+37 04656				22 37 47	+38 22.4	2	8.75	.024	0.46	.000	0.06	.010	6	F2	401,1775
		G 28 - 12			22 37 50	+07 38.5	1	11.77		1.44				1	M2	1632
214847	+55 02787				22 37 50	+55 54.0	2	8.13	.027	1.21	.002			38	G2 Ib	6009,6011
214706	−46 14399				22 37 50	−45 49.6	2	9.39	.011	1.31	.004	1.40		9	K2/3 III	863,2011
		HA 114 # 527			22 37 51	+00 43.4	1	11.14		1.04		0.94		1	K2	281
215743	+86 00335				22 37 51	+87 01.5	1	8.18		0.04		0.04		6	A0	1219
		LP 520 - 69			22 37 52	+13 28.2	1	12.73		1.36		1.28		2		1773
		LP 520 - 70			22 37 53	+13 28.0	1	10.74		0.93		0.70		2		1773
214748	−27 16010	HR 8628			22 37 54	−27 18.3	13	4.17	.012	-0.11	.010	-0.33	.025	58	B8 Ve	15,681,815,1034,1075*
214749	−30 19255				22 37 55	−29 56.1	2	7.84	.019	1.14	.005	1.00		7	K4/5 V	1075,3072
214728	−45 14764				22 37 55	−44 45.9	4	8.89	.006	0.30	.004	0.05	.008	22	A9 V	743,977,1075,2011
214729	−45 14763				22 37 55	−45 27.9	3	8.12	.003	0.47	.004	-0.04	.014	16	F6 V	977,1075,2011
		AJ89,1229 # 332			22 37 56	+07 24.3	1	10.18		1.49				1	K7	1632
		G 27 - 39			22 37 58	−00 23.3	1	13.88		1.35		1.05		1		3062
	−0 04398				22 37 59	+00 14.9	1	10.62		0.57		0.04		2	G0	281
240018	+56 02826				22 37 59	+57 04.1	1	10.06		0.26		-0.17		1	B5	8113
214730	−45 14767				22 38 00	−45 29.2	1	9.41		0.53				4	F6 V	2011
214731	−46 14401				22 38 00	−46 27.8	1	9.20		1.04		0.84		1	K4 V	3072
		HA 114 # 531			22 38 03	+00 36.3	4	12.09	.005	0.73	.004	0.19	.014	69	G2	281,989,1729,1764
214809	+5 05054	IDS22355N0600		AB	22 38 03	+06 16.0	1	9.11		0.44		-0.09		2	F3 V +G8 V	3016
		AJ89,1229 # 333			22 38 05	+10 21.9	1	10.96		1.54				1	M1	1632
		G 233 - 26			22 38 06	+61 27.4	1	11.88		0.67		-0.01		1		1658
214759	−32 17191				22 38 06	−32 15.1	6	7.39	.009	0.80	.009	0.43	.010	21	G8/K0 V	116,1020,1075,1775,2012,3008
		LP 520 - 71			22 38 07	+13 26.9	1	13.64		1.15		0.98		2		1773
	+57 02581	LS III +58 049			22 38 08	+58 10.1	1	10.04		0.61		-0.42		2	B0 III	1012
		HA 114 # 637			22 38 09	+00 47.5	2	12.07	.005	0.80	.000	0.32	.016	15	F8	281,1764
214810	−4 05728	HR 8629		⋆ AB	22 38 13	−03 48.9	3	6.31	.015	0.52	.000	-0.02	.009	8	F6 V	15,1071,3077
		G 127 - 44			22 38 14	+16 43.1	1	11.46		0.83		0.46		1	K2	1620
214837	+23 04589				22 38 15	+23 51.1	1	9.45		-0.10		-0.47		1	A0	963
214878	+53 02950	HR 8633			22 38 16	+53 35.1	1	5.93		0.93				2	K0 III	71
	+42 04471				22 38 17	+42 44.8	2	9.82	.004	1.10	.004	0.98	.025	7		1663,3072
214868	+43 04266	HR 8632			22 38 19	+44 00.9	4	4.48	.021	1.32	.011	1.37	.012	11	K3 III	15,1363,3016,4001,8015
215038	+74 00980	GQ Cep			22 38 19	+75 23.8	2	8.17	.020	-0.04	.005	-0.46	.018	6	A0p	695,1202
214850	+13 04971	HR 8631		⋆ AB	22 38 24	+14 17.2	3	5.73	.014	0.71	.007	0.22		5	G3 V +G8 V	71,1381,3077
214850	+13 04971	IDS22359N1401		C	22 38 24	+14 17.2	1	14.88		1.00				3		3030
		HA 114 # 446			22 38 30	+00 30.4	2	12.07	.002	0.74	.001	0.25	.005	10	dG6	281,1764
		G 67 - 11			22 38 32	+18 40.4	1	10.40		0.92		0.64		1	K2	333,1620
	+0 04909	IDS22361N0032		AB	22 38 39	+00 47.4	1	10.28		0.59		0.04		1	G2 V	281
214885	−0 04401				22 38 45	+00 13.9	1	10.48		0.63		0.10		2	G2 V	281
		WLS 2240 15 # 10			22 38 48	+15 42.1	1	10.95		0.74		0.23		2		1375
		HA 114 # 654			22 38 52	+00 54.5	2	11.83	.001	0.66	.002	0.19	.010	16	G2	281,1764
214975	+56 02829	Z Lac			22 38 53	+56 34.1	3	7.95	.086	0.87	.078	0.62	.002	3	F6 Ib	934,1772,6011
	+58 02468	LS III +59 028			22 38 53	+59 25.6	1	10.32		1.10		0.38		6		27
		Cep sq 1 # 68			22 38 56	+59 24.0	1	10.93		0.73		0.15		3		27
214923	+10 04797	HR 8634		⋆ A	22 38 58	+10 34.2	10	3.41	.009	-0.09	.008	-0.22	.039	37	B8.5V	3,15,1034,1049,1079*
		HA 114 # 151			22 39 00	+00 05.1	1	10.66		0.75		0.29		2	G3	281
		HA 114 # 656			22 39 02	+00 55.5	1	12.64		0.96		0.70		2		1764
214930	+23 04592				22 39 02	+23 35.1	1	7.38		-0.14		-0.68		3	B2 IV	399
		HA 114 # 548			22 39 03	+00 43.5	4	11.60	.001	1.36	.004	1.57	.009	85	dK5	281,989,1729,1764

HD	DM	Other Id	N	Rem	α_{1950}	δ_{1950}	S	V	σ_V	B–V	σ_{B-V}	U–B	σ_{U-B}	n	Spectrum	References
214924	−0 04403				22 39 04	−00 08.8	2	9.08	.003	1.23	.000	1.26		11	K0 III	281,6004
	+64 01699	LS III +64 007			22 39 07	+64 57.2	1	8.92		0.40		−0.56		2	B2	1307
214908	−0 04402				22 39 08	+00 24.1	1	10.11		0.44		−0.09		2	F5	281
214977	+40 04882				22 39 10	+40 46.6	2	9.34	.000	0.03	.005	−0.34	.024	7	B6 V	401,1775
		HA 114 # 750			22 39 11	+00 57.0	4	11.92	.004	−0.04	.002	−0.36	.006	97	A2	281,989,1729,1764
214979	+30 04771	HR 8638			22 39 11	+30 42.2	1	6.36		1.60		1.94		2	K5	1733
215065	+65 01796	G 241 - 18			22 39 11	+66 15.4	2	7.44	.019	0.63	.002	0.06	.012	3	G5	1658,3016
		RW Gru			22 39 11	−44 25.2	1	11.75		0.18				1		700
214993	+39 04912	HR 8640, DD Lac	⋆	A	22 39 14	+39 57.8	6	5.25	.018	−0.14	.008	−0.87	.004	21	B2 III	15,154,1119,1174,1278,1363
214980	+29 04734				22 39 15	+29 31.1	1	10.05		0.42		−0.06		1	G5	583
214925	−21 06283				22 39 17	−21 02.7	1	9.28		1.52				2	K0	1594
215318	+80 00731	IDS22392N8052			22 39 20	+81 07.8	1	6.86		0.75		0.49		3	F8	985
215030	+40 04885	HR 8643			22 39 22	+41 17.2	1	5.92		1.02		0.71		2	G9 III	1733
214994	+28 04436	HR 8641			22 39 24	+29 02.8	7	4.80	.008	−0.01	.004	0.00	.016	28	A1 IV	15,1049,1203,1363*
215013	+20 05208	IDS22370N2054		AB	22 39 27	+21 10.2	1	8.21		0.63		0.10		4	G0	938
240024	+55 02789	RR Lac			22 39 27	+56 10.3	3	8.47	.018	0.74	.026	0.44	.025	3	G5	851,1772,6011
		CS 22875 # 37			22 39 27	−38 59.1	2	14.32	.000	0.02	.000	0.11	.011	2		966,1736
	−0 04404				22 39 28	+00 14.4	1	10.82		0.52		0.01		2	G1	281
214995	+13 04974	HR 8642			22 39 29	+14 15.3	1	5.90		1.11				2	gK0	71
240025	+55 02790				22 39 29	+56 00.2	1	9.75		0.48				36	F8	6011
240026	+55 02791				22 39 30	+56 29.0	3	8.81	.012	1.06	.006	0.79	.018	12	K2	851,934,1655
214943	−44 15015				22 39 30	−44 21.0	2	9.36	.005	1.09	.000	0.97		9	G8 III	863,2011
215066	+53 02958				22 39 32	+54 07.5	1	7.82		0.03		−0.12		2	B9	401
	+0 04910				22 39 34	+01 00.9	3	10.91	.004	0.57	.002	−0.05	.021	25		281,989,1729
214966	−30 19267	HR 8637			22 39 35	−29 37.4	2	6.15	.014	1.55	.028	1.92		8	M3 III	1024,2009
		HA 114 # 670			22 39 36	+00 54.6	4	11.11	.006	1.21	.007	1.22	.008	34	K5	281,989,1729,1764
214967	−31 18933				22 39 38	−30 54.8	1	8.59		0.92		0.58		2	G8 IV	1730
214953	−47 14307	HR 8635	⋆	A	22 39 39	−47 28.1	2	6.02	.000	0.53	.000	0.07	.000	4	G0 V	1279,3077
214953	−47 14307	HR 8635	⋆	AB	22 39 39	−47 28.1	9	5.98	.013	0.57	.009	0.07	.017	45	G0 V	15,278,938,977,977*
214953	−47 14307	IDS22367S4743		B	22 39 39	−47 28.1	2	11.10	.000	1.41	.000			2		1279,3031
214998	−6 06056				22 39 40	−05 59.7	1	9.54		1.04		0.94		1	K2	1696
214952	−47 14308	HR 8636, β Gru			22 39 41	−47 08.8	8	2.11	.014	1.62	.011	1.63	.063	35	M5 III	3,15,678,2011,3052*
		MtW 114 # 88			22 39 43	+00 28.5	1	12.97		1.00		0.71		6		397
215000	−31 18934				22 39 46	−30 52.2	1	10.38		0.74		0.34		2	G6/8	1730
214987	−44 15017	HR 8639			22 39 47	−44 30.6	6	6.06	.007	0.97	.007	0.69	.007	21	K0 III	15,278,1075,2011,2038,4001
	+17 04789	IDS22374N1720		AB	22 39 49	+17 35.9	1	10.12		0.35		0.18		2		1375
215044	+0 04911				22 39 50	+00 30.5	5	8.51	.008	1.01	.007	0.81	.006	38	G8 III	281,397,989,1729,6004
		G 67 - 15			22 40 09	+17 24.1	1	11.83		1.55		1.16		1	M3	3078
215077	−0 04405				22 40 09	−00 11.5	1	7.19		0.37				9	F0	281,6004
		GD 240			22 40 09	−04 30.0	1	15.21		−0.29		−1.15		2	DAwk	1727
		MtW 114 # 160			22 40 11	+00 29.1	1	15.51		0.99		0.68		4		397
		CS 22875 # 42			22 40 11	−40 02.8	2	12.80	.000	0.20	.000	0.13	.008	2		966,1736
		WLS 2240-20 # 5			22 40 12	−17 24.7	1	10.84		0.53		−0.03		2		1375
		MtW 114 # 167			22 40 14	+00 29.7	1	14.15		1.20		1.11		6		397
215176	+59 02561				22 40 14	+59 40.2	1	8.41		0.15		0.10		2	A0	1502
	−22 05981				22 40 14	−22 25.3	1	10.26		0.80		0.44		2		1375
215148	+52 03271				22 40 15	+52 32.5	1	8.06		0.05		−0.28		2	B9	401
215093	−0 04406				22 40 15	−00 01.8	4	6.98	.008	0.31	.003	0.10	.005	35	F0	281,989,1729,6004
	−5 05845	G 27 - 41			22 40 16	−05 15.4	1	10.69		1.14		1.10		1	K4:	1620
		MtW 114 # 176			22 40 17	+00 29.9	1	15.59		0.68		−0.01		6		397
215188	+56 02836				22 40 17	+57 20.8	1	8.15		0.04		−0.05		2	A0	1375
215159	+53 02960	HR 8648			22 40 18	+53 38.8	1	6.12		1.62				2	K2	71
		PHL 382			22 40 18	−15 06.	1	11.34		−0.16		−0.73		1		963
215177	+55 02792	LS III +56 082			22 40 19	+56 02.9	2	8.71	.000	0.24	.000	−0.06	.000	2	A2	851,934
		Cep sq 1 # 69			22 40 19	+60 04.8	1	11.28		0.77		0.28		5		27
	+56 02837			V	22 40 20	+56 32.4	1	8.91		2.18				2		369
		VES 935			22 40 20	+58 05.3	1	13.23		0.62		0.51		1		8113
215110	−0 04407				22 40 24	+00 08.5	6	7.76	.025	0.87	.006	0.48	.021	43	K0 V	281,975,989,1729,3016,6004
		UV2240-15			22 40 24	−15 06.1	1	11.34		−0.17		−0.71		6		1732
215072	−42 16045				22 40 24	−41 48.3	2	8.25	.010	1.56	.005	1.91		12	M0 III	977,2011
215114	−9 06038	HR 8645	⋆	AB	22 40 26	−08 34.4	2	6.44	.005	0.17	.000	0.11	.010	7	A5 V	15,1061
215251	+64 01701				22 40 31	+64 36.6	1	7.23		1.20		0.91		3	F8	196
		G 28 - 13			22 40 31	−01 43.4	3	16.18	.031	0.28	.013	−0.56	.044	6	DA	419,3028,3060
		LS III +59 029			22 40 33	+59 24.6	2	11.16	.015	0.67	.010	−0.27	.015	5		405,483
215144	−7 05837				22 40 36	−07 28.6	1	7.65		0.95		0.71		5	G5	1657
215104	−42 16049	HR 8644	⋆	A	22 40 36	−41 40.5	9	4.84	.007	1.03	.004	0.81	.003	29	K0 III	15,278,977,1020,1075*
215105	−45 14782				22 40 36	−44 49.7	7	9.49	.003	1.22	.005	1.16	.008	106	K2 III	863,1460,1628,1657,1673*
215141	−0 04408				22 40 37	+00 05.5	9	9.24	.004	1.49	.004	1.85	.010	100	M0 III	281,989,1729,1764,6004
215143	−7 05838	HR 8647			22 40 38	−07 13.5	3	6.40	.005	−0.04	.004	−0.11	.013	9	A0 Vn	15,1071,1079
215164	+17 04792	G 127 - 48			22 40 39	+18 13.2	1	8.70		0.60		0.09		1	G5	1620
215182	+29 04741	HR 8650	⋆	A	22 40 39	+29 57.6	8	2.95	.010	0.85	.008	0.54	.011	56	G2 II-III	1,15,1077,1118,1355*
215191	+37 04670	HR 8651			22 40 39	+37 32.4	1	6.43	.000	−0.10	.015	−0.81	.005	13	B1 V	15,154,1174
215224	+50 03797				22 40 40	+50 46.9	1	8.79		0.13		0.08		2	A2	1566
	−46 14417				22 40 41	−45 40.8	2	10.34	.003	0.51	.003	0.04		10	G0	863,2012
		HA 114 # 781			22 40 44	+01 03.6	1	11.64		0.60		0.08		2	G0	281
	+55 02795	LS III +56 083			22 40 45	+56 05.0	1	9.80		0.31		−0.52		2	B1 III	1012
215227	+43 04279	IDS22386N4411		AB	22 40 46	+44 27.6	1	8.75		0.09		−0.77		6	B5: en	1174
215152	−7 05839	G 27 - 42			22 40 46	−06 39.6	1	8.14		0.97		0.79		1	K0	1620
		E9 - f			22 40 48	−44 54.	1	12.82		0.65				4		1075
215268	+55 02796				22 40 54	+56 24.4	1	8.68		0.36				34		6011

Table 1   1203

HD	DM	Other Id	N	Rem	α_{1950}	δ_{1950}	S	V	σ_V	B–V	σ_{B-V}	U–B	σ_{U-B}	n	Spectrum	References
215321	+64 01702				22 40 54	+65 25.1	1	7.91		0.08		0.03		2	A0	401
215167	−19 06324	HR 8649			22 40 54	−19 05.6	9	4.68	.015	1.37	.007	1.55	.007	42	K3 III	15,1007,1013,1075*
215242	+46 03803	HR 8652		★ABC	22 40 55	+46 54.4	2	6.41		0.45	.025	-0.07	.005		A1 V+(G)	963,1079
		Cep sq 1 # 71			22 40 55	+62 51.3	1	10.37		0.49		0.25		5		27
		Cep sq 1 # 70			22 40 58	+60 04.4	1	10.78		0.51		0.24		3		27
		LP 984 - 89			22 40 58	−33 05.4	1	12.75		0.46		-0.27		2		1696
		HA 114 # 785			22 40 59	+01 02.6	1	12.13		0.58		0.05		2	F9	281
	+74 00982				22 41 02	+75 10.5	1	8.71		1.76		2.00		2	K5	1733
215156	−44 15024				22 41 03	−44 22.2	1	9.21		0.40				4	F3 V	2011
215121	−61 06676	HR 8646			22 41 03	−60 45.8	1	6.30		0.56				4	K0 III+A7	2035
214846	−82 00889	HR 8630	6		22 41 04	−81 38.7	6	4.14	.010	0.21	.005	0.09	.012	22	A9 IV/V	15,1034,1075,2038*
		G 27 - 43			22 41 05	+01 43.9	1	13.84		1.48		1.18		1		3062
		E9 - c			22 41 06	−44 53.	1	12.24		0.56				4		1075
		HA 114 # 790			22 41 10	+01 02.9	1	12.13		0.82		0.40		2	dG5	281
215243	+10 04805	HR 8653			22 41 13	+10 40.7	2	6.49	.035	0.47	.005	0.00	.015	5	G8 IV	1733,3016
215322	+55 02797				22 41 14	+56 08.7	1	6.95		0.27		0.07		3	F0	3016
	−45 14785				22 41 14	−45 28.4	2	10.47	.009	0.79	.000	0.30		10		863,2011
240035	+58 02472				22 41 16	+59 14.7	1	9.61		0.47		0.10		4	A7	27
215257	+3 04763	G 27 - 44	6		22 41 18	+03 37.2	6	7.42	.017	0.50	.015	-0.10	.003	10	F9 V	516,979,1620,1658*
215274	+29 04742				22 41 19	+29 49.8	1	8.03		0.67		0.20		3	G5 V	3026
215324	+45 04032	IDS22391N4530		AB	22 41 19	+45 45.8	1	7.28		0.43		0.00		2	F5	3016
215371	+64 01704				22 41 19	+65 04.4	2	6.74	.039	-0.02	.010	-0.67	.010	6	B1.5V	399,401
		G 127 - 51			22 41 20	+24 21.2	1	12.61		1.03		0.78		1		1620
	+53 02964	LS III +53 081			22 41 20	+53 48.5	1	9.19		0.15		-0.71		2	B2 IV:penn	1012
		E9 - d			22 41 24	−44 43.	1	12.40		0.60				4		1075
		E9 - Y			22 41 36	−44 32.	1	11.55		0.62				4		2011
		VES 937			22 41 40	+59 26.1	1	13.23		0.91		0.05		2		8113
	−45 14788				22 41 41	−44 52.3	6	10.37	.002	1.05	.007	0.84	.018	46	G5	863,1460,1628,1657,1673,1704
	+64 01705				22 41 44	+65 18.6	1	9.28		0.26		-0.46		2	B5	401
	−45 14789				22 41 46	−44 41.7	1	11.71		0.73				4		1075
215330	+0 04917				22 41 47	+00 53.8	1	9.06		1.42		1.53		8	K5	1371
		E9 - e			22 41 48	−44 37.	1	12.42		0.66				4		1075
215359	+38 04855	HR 8654		★ AB	22 41 49	+39 12.2	2	5.93	.015	1.49	.012	1.72		5	K5 III	71,1501
215373	+41 04594	HR 8656		★ A	22 41 51	+41 33.4	2	5.12	.027	0.97	.000	0.77		5	K0 III	1363,3016
		LP 521 - 3			22 41 55	+14 04.1	1	11.30		0.54		-0.05		1		1773
		HA 114 # 803			22 41 57	+01 00.5	1	12.49		0.75		0.18		2	F8	281
		LP 521 - 4			22 41 57	+14 03.9	1	12.98		0.80		0.30		1		1773
	+16 04806				22 42 00	+17 17.9	1	10.16		1.16				2	K6	1619
		G 233 - 27			22 42 05	+56 28.2	1	14.84		0.97		0.70		1		1658
215441	+54 02846	GL Lac			22 42 06	+55 19.6	2	8.78	.007	0.03	.003	-0.51	.021	6	A0p	695,1202
	−45 14791				22 42 10	−44 54.6	2	10.68	.007	0.59	.004	0.05		12	G5	1460,2011
		AO 1199			22 42 12	+43 47.8	1	11.76		1.60		1.69		1	K4 III	1748
215442	+50 03809				22 42 12	+51 11.1	1	7.50		0.55		0.06		2	F8	1733
215267	−66 03706				22 42 15	−66 21.8	1	8.01		0.95		0.73		8	G8 III	1704
	+54 02847	LS III +55 099			22 42 18	+55 28.9	1	10.12		0.39		-0.53		2	B1 II:nn	1012
215484	+60 02432	DG Cep			22 42 18	+61 27.9	1	8.84		2.70				1	C6.4	1238
		E9 - h			22 42 18	−44 48.	1	13.00		0.76				4		1075
		G 27 - 45			22 42 20	−02 36.9	5	11.50	.000	0.67	.008	-0.07	.011	12		516,1064,1620,1696,3077
		WLS 2240 15 # 7			22 42 22	+12 52.0	1	10.36		1.23		1.23		2		1375
	−45 14792				22 42 24	−45 06.7	1	11.08		0.89				4		2011
215208	−76 01564				22 42 25	−75 41.1	1	7.29		0.16		0.13		9	A3 V	1628
240038	+58 02474	V1213 Aql			22 42 27	+58 54.2	1	9.16		0.81				32	G5	6011
215472	+43 04295				22 42 30	+44 15.6	1	9.57		1.14		0.97		1	K0	1748
215415	−3 05501	IDS22399S0311		A	22 42 30	−02 55.1	1	9.36		0.56		-0.01		2	F8	3024
215415	−3 05501	IDS22399S0311		BC	22 42 30	−02 55.1	1	9.44		0.66		0.11		3		3024
		E9 - o			22 42 30	−44 50.	1	14.43		0.89				4		1075
		E9 - q			22 42 31	−44 43.5	1	15.20		0.69		0.19		4		1460
215488	+43 04296				22 42 33	+44 06.4	1	9.97		0.86		0.45		1	G5	1748
215501	+55 02800				22 42 34	+55 35.6	1	8.35		0.26		0.14		1	A5 IV	695
215369	−54 10123	HR 8655		★ A	22 42 34	−53 45.8	4	4.84	.008	1.18	.000	1.17	.000	16	K2 III	15,1075,2012,2024
		E9 - n			22 42 36	−44 45.5	2	14.70	.015	0.57	.008	-0.07		9		1075,1460
215473	+38 04858				22 42 37	+38 56.3	1	6.43		1.50		1.80		2	K5 III	105
		E9 - g			22 42 38	−44 33.3	2	12.68	.023	0.92	.027	0.68		7		1075,1460
240041	+58 02476				22 42 41	+59 03.1	1	10.14		1.21				31	G5	6011
		E9 - i			22 42 42	−44 44.	1	13.45		0.69				4		1075
215506	+52 03281			V	22 42 43	+53 02.4	1	8.12		1.78				2	K5	369
215518	+51 03460	HR 8661			22 42 44	+52 15.2	2	6.41	.005	1.56	.019	1.90	.029	5	K5 III	1716,1733
215405	−47 14320	HR 8657	6		22 42 44	−46 48.6	6	5.50	.003	1.32	.002	1.43	.006	33	K2 III	15,278,863,1075,2024,2038
		E9 - s			22 42 47	−44 40.5	1	15.57		0.52		-0.09		4		1460
		E9 - m			22 42 48	−44 46.	1	14.17		0.81				4		1075
		RW Lac			22 42 49	+49 23.6	1	10.64		0.66		0.05		6	F2	1768
		E9 - k			22 42 49	−44 46.1	2	13.95	.009	0.55	.010	-0.07		7		1075,1460
	+48 03827	V363 Lac		★ A	22 42 50	+49 12.9	1	10.29		2.30		3.44		1	N0	414
	+48 03827	IDS22407N4856		B	22 42 50	+49 12.9	1	12.74		0.74		0.08		1		414
		E9 - l			22 42 54	−44 41.	1	13.89		0.88				4		1075
215575	+60 02433				22 42 56	+60 33.1	1	8.02		0.27		-0.59		3	B8	27
	+10 04812	IDS22405N1040		AB	22 42 58	+10 55.9	1	9.82		1.13		1.00		2	K4	3072
	+1 04645				22 42 59	+01 57.8	1	9.92		1.38				1	K7	1632
	+39 04923	IDS22408N3930		A	22 43 00	+39 45.8	1	9.07		0.96		0.58		1		401
	+39 04923	IDS22408N3930		AB	22 43 00	+39 45.8	1	8.86		0.87		0.48		2		401

HD	DM	Other Id	N	Rem	α_{1950}	δ_{1950}	S	V	σ_V	B–V	σ_{B-V}	U–B	σ_{U-B}	n	Spectrum	References
	+39 04923	IDS22408N3930	B		22 43 00	+39 45.8	1	10.78		0.65		0.02		1		401
		E9 - Z			22 43 00	−44 47.	1	11.67		0.54				4		2011
215510	+18 05046	HR 8660			22 43 02	+19 06.1	1	6.26		1.05				2	K0	71
215547	+46 03815				22 43 05	+46 56.4	1	7.92		1.60		1.58		3	M2	1601
215588	+57 02595				22 43 05	+57 53.1	2	6.46	.014	0.42	.005	-0.06	.019	4	F5	985,1716
215467	−43 15148				22 43 05	−42 41.1	3	9.79	.007	0.81	.001	0.38	.022	12	G8 V	863,2011,3008
		G 28 - 15			22 43 08	−01 59.4	1	11.26		0.79		0.26		2		1620
215468	−46 14428				22 43 08	−46 06.3	3	8.01	.000	0.37	.004	-0.04	.019	10	F2 V	278,1075,2011
215456	−49,13955	HR 8658			22 43 08	−49 14.5	1	6.62		0.63				4	G0 V	2006
	+18 05047	G 67 - 17			22 43 09	+18 59.6	1	8.77		0.74		0.25		1	K0	333,1620
		LP 641 - 4			22 43 12	+01 25.6	1	11.95		1.49				1	M0	1632
215549	+29 04753	G 128 - 1		⋆ A	22 43 14	+30 11.0	3	6.38	.014	0.93	.009	0.67	.006	5	K1 III-IV	1003,1355,3016
215606	+56 02851				22 43 19	+56 52.6	1	8.00		0.20		0.19		3	Am:	8071
215605	+57 02597	LS III +57 072			22 43 19	+57 35.3	1	9.43		0.25		0.68		2	B2:IV:enn	1012
240047	+56 02852	LS III +57 073			22 43 20	+57 10.1	1	9.83		0.35		-0.43		2	B2 III	1012
215661	+67 01463	ZZ Cep		⋆ AB	22 43 20	+67 51.9	1	8.60		0.28				3	A2p	3024
215661	+67 01463	IDS22417N6736	C		22 43 20	+67 51.9	1	10.86		0.45				1		3024
		MCC 849			22 43 21	+18 49.1	1	10.65		1.29				2	K6	1619
215576	+43 04299				22 43 21	+44 05.6	1	9.16		1.18		1.06		1	K2	1748
		LS III +55 100			22 43 23	+55 55.3	1	11.03		0.34		-0.56		1		8113
215550	+10 04815				22 43 24	+10 56.2	1	7.20		0.87		0.58		2	G5	1733
		LP 521 - 10			22 43 26	+10 48.2	1	10.76		0.75		0.33		2		1696
		BF Peg			22 43 27	+23 52.7	1	12.33		0.28		0.13		1		699
		G 216 - 15			22 43 27	+45 07.5	1	10.46		1.24		1.20		1		3003
215504	−50 13788	HR 8659			22 43 29	−49 56.7	1	6.48		1.14				4	K1 III	2032
215578	+18 05048	IDS22411N1843	AB		22 43 32	+18 59.2	1	7.47		0.94		0.66		2	K0	3016
215619	+34 04766				22 43 42	+35 24.4	1	7.57		0.43		-0.02		2	F5	1733
		LS III +61 016			22 43 42	+61 34.3	1	10.82		0.71		-0.13		5		180
215544	−44 15037				22 43 42	−44 09.0	2	9.23	.010	1.01	.000	0.78		10	K0 III	863,1075
	−45 14796				22 43 42	−44 42.2	1	11.40		0.84				4		1075
	−45 14797				22 43 45	−45 22.6	1	11.15		0.40				4		2011
215545	−47 14331	HR 8662		⋆ AB	22 43 47	−47 12.2	7	6.56	.006	0.30	.003	0.11	.006	39	A9 (IIIm)	15,278,977,977,1075*
215673		TX Lac			22 43 48	+54 47.8	1	10.16		2.02				1	R6	1238
215559	−44 15038				22 43 48	−44 32.8	1	9.33		0.25				4	A7 V	2011
		LP 701 - 7			22 43 50	−06 55.2	1	15.98		1.83				4		1663
215647	+39 04925				22 43 53	+40 03.7	1	9.09		-0.01		-0.27		2	A0	401
	+39 04926				22 43 55	+39 50.6	2	9.24	.029	0.21	.000	0.12	.000	3	B9 II-III:	1026,1240
215600	−21 06301				22 43 56	−20 31.5	1	8.48		0.80		0.31		2	K3 III	742
215571	−45 14800				22 43 56	−45 18.9	6	8.23	.004	0.36	.007	-0.01	.010	32	F2 V	110,143,395,977,1075,2011
215664	+43 04300	HR 8666			22 43 57	+44 16.9	2	5.85	.010	0.35	.007	0.08	.007	5	F0 III-IV	1733,3037
215561	−55 09896				22 43 59	−54 40.6	1	8.35		0.48				4	F3 V	2011
215601	−32 17234				22 44 01	−32 08.0	1	8.45		0.80				2	G6/8wF3	1594
	+43 04301				22 44 02	+43 46.2	1	10.12		0.52		0.00		1	F4 V	1748
		LS III +58 053			22 44 02	+58 45.8	2	10.33	.020	1.13	.005	0.06	.010	4		405,483
215624	−14 06345				22 44 02	−13 49.1	1	10.58		0.36		0.16		3	A2 Ib	1026
		Steph 2042			22 44 04	−03 47.2	1	12.07		1.43		1.19		1	K7	1746
		WLS 2240 15 # 9			22 44 05	+14 38.7	1	11.61		0.92		0.59		2		1375
215714	+57 02601	NGC 7380 - 1		⋆ A	22 44 06	+57 48.6	1	7.58		0.46		0.02		1	F5 V	49
215665	+22 04709	HR 8667			22 44 07	+23 18.1	6	3.94	.012	1.07	.008	0.93	.020	16	G8 IIIa	15,1118,1355,4001*
240051	+57 02602	NGC 7380 - 3		⋆ B	22 44 09	+57 48.3	2	8.64	.005	0.63	.020	0.14	.025	2	G0 V	49,749
215704	+49 03937	G 233 - 28			22 44 11	+49 56.8	1	7.86		0.80		0.48		3	K0	196
215648	+11 04875	HR 8665		⋆ A	22 44 12	+11 55.0	13	4.20	.007	0.50	.007	-0.01	.017	50	F7 V	1,3,15,22,254,1020*
215629	−44 15041				22 44 14	−44 09.6	3	8.71	.006	1.00	.002	0.84	.004	18	K0 III	863,977,2011
215562	−66 03709	IDS22409S6605	A		22 44 15	−65 49.5	1	6.48		1.24		1.19		8	K1 III	1628
215627	−42 16071				22 44 16	−41 57.4	4	6.83	.001	1.28	.002	1.35	.016	2	K2 III	743,977,1075,2011
	−45 14802				22 44 17	−45 17.3	4	10.36	.004	0.52	.012	0.00	.000	20	G0	110,143,395,2011
	+43 04303				22 44 19	+44 01.7	1	9.84		1.40		1.50		1	K0	1748
		VES 939			22 44 19	+57 03.3	1	12.62		0.48		-0.39		1		1543
		NGC 7380 - 7			22 44 19	+57 54.6	2	10.66	.025	0.56	.005	0.09	.010	2		49,749
215628	−44 15043				22 44 19	−43 52.2	3	8.80	.004	0.55	.005	0.07	.005	16	F8 V	977,1075,2011
	+57 02604	NGC 7380 - 30			22 44 20	+57 55.3	1	9.55		1.67		1.95		1		749
215690	+25 04810				22 44 21	+25 32.4	1	7.96		0.07		0.03		3	A0	833
	+55 02808	LS III +56 093			22 44 25	+56 03.6	1	10.01		0.41		-0.46		2	B2 III	1012
215757	+54 02856				22 44 26	+54 36.5	2	6.85	.001	0.07	.003	-0.02	.003	3	A0	401,695
240059	+58 02478	CR Cep			22 44 27	+59 10.7	3	9.45	.012	1.36	.035	0.98		3	G0	851,1399,6011
		LS III +59 030			22 44 30	+59 17.7	1	10.23		1.22		0.47		3		881
215669	−34 15735	HR 8668			22 44 31	−34 25.6	3	6.28	.004	1.14	.015			11	K1 III	15,1075,2009
215657	−45 14803				22 44 32	−45 13.7	7	7.21	.007	0.60	.007	0.06	.012	37	G3 V	110,143,278,395,863*
		AO 1204			22 44 33	+44 01.8	1	10.94		0.18		0.04		1	B9.5	1748
215733	+16 04814				22 44 35	+16 58.2	2	7.33	.010	-0.13	.005	-0.89	.060	7	B1 II	399,8100
		Per sq 1 # 1			22 44 36	+58 02.	1	12.78		0.51		-0.33		2		881
		LS III +60 024			22 44 38	+60 09.2	1	10.87	.040	0.71	.040	-0.29	.035	7		180,881
215696	−16 06152	IDS22420S1640	AB		22 44 39	−16 24.6	3	7.36	.034	0.68	.000	0.21	.010	6	G5 V	219,1311,3026
	+43 04305	EV Lac		⋆ A	22 44 40	+44 04.6	2	10.05	.180	1.50	.095	1.36	.195	2	M4.5V:e	680,1748,3003
	+43 04305	EV Lac		⋆ AB	22 44 40	+44 04.6	4	10.04	.028	1.38	.009	0.77	.042	11		1017,1197,3016,8039
	+43 04305	IDS22426N4349	B		22 44 40	+44 04.6	1	10.66		0.28		-0.34		1		680,3003
		VES 940			22 44 40	+57 21.7	1	11.27		0.22		-0.31		1		8113
		G 128 - 2			22 44 41	+29 35.3	2	11.60	.071	0.69	.013	0.06	.026	4	G8	1620,1723
215806	+57 02606				22 44 41	+58 01.9	1	9.20		0.42		-0.57		2	B0 Ib	1012
		AO 1206			22 44 43	+43 44.9	1	11.11		0.66		0.12		1	F6 V	1748

Table 1   1205

HD	DM	Other Id	N	Rem	α_{1950}	δ_{1950}	S	V	σ_V	B–V	σ_{B-V}	U–B	σ_{U-B}	n	Spectrum	References
	+17 04808	G 67 - 19			22 44 46	+18 07.2	2	9.00	.005	1.04	.010	0.94	.019	3	G0	333,1620,3072
		NGC 7380 - 25			22 44 47	+57 50.1	1	14.68		0.50		0.44		1		49
215774	+23 04606	IDS22424N2328		A	22 44 48	+23 43.3	1	7.91		0.25		0.07		1	F0	1776
	+53 02986				22 44 48	+54 29.5	1	9.68		0.31		0.20		1		401
215722	-25 16105				22 44 48	-24 49.5	1	10.03		1.17		0.94		1	K2/3 V	3072
215807	+53 02985				22 44 50	+53 41.3	1	8.47		0.52		0.26		2	F5 II	253
215721	-20 06486	HR 8670			22 44 52	-19 52.5	2	5.25	.009	0.94	.000			5	G8 III	2024,2035
215835	+57 02607	NGC 7380 - 2		⋆ A	22 44 54	+57 49.2	4	8.58	.004	0.33	.008	-0.65	.008	8	O5.5	49,1011,1012,1209
215775	+17 04809	G 127 - 55			22 44 55	+17 33.1	1	8.04		0.69		0.16		1	K0	1620
215837	+52 03293				22 44 55	+53 29.6	1	8.30		0.02		-0.20		2	A0	401
215749	-2 05826				22 44 55	-02 03.2	2	7.43	.006	1.15	.004	1.12		9	K2 III	1657,2007
		NGC 7380 - 15			22 44 57	+57 51.3	1	13.12		0.38		-0.22		1		49
		LP 701 - 82			22 44 57	-07 14.5	1	10.43		0.71		0.26		2		1696
215724	-38 15217	HR 8671			22 44 57	-38 29.1	3	6.71	.019	0.48	.000	-0.02		9	F6 V	15,2028,3053
215726	-43 15160				22 44 57	-42 57.4	2	9.06	.004	1.46	.004	1.82		9	K3/4 III	863,1075
215836	+55 02809	LS III +56 097			22 45 00	+56 10.1	2	9.21	.020	0.50	.000	-0.44	.010	3	B1 II	1012,1462
		LS III +60 025			22 45 00	+60 09.2	2	10.95	.005	0.79	.025	-0.23	.040	6		180,881
		NGC 7380 - 12			22 45 04	+57 52.2	1	12.19		0.35		-0.29		1		49
215766	-14 06346	HR 8673		⋆ AB	22 45 04	-14 19.2	2	5.66		-0.02	.011	-0.26		5	B9 V	1079,2009
		G 28 - 16			22 45 06	+06 09.6	1	11.59		0.81		0.30		2	K1	333,1620
		NGC 7380 - 29			22 45 06	+57 46.6	1	15.62		0.90		0.30		1		49
215848	+53 02987				22 45 07	+54 29.3	1	7.78		0.08		-0.46		2	B8	401
215868	+56 02859				22 45 08	+57 19.9	2	8.42	.010	-0.02	.005	-0.56	.005	3	B9	401,963
215682	-62 06369	HR 8669			22 45 08	-61 56.9	4	6.36	.004	1.06	.008			18	K0 III	15,1075,2013,2029
		G 67 - 20			22 45 09	+19 34.9	1	11.63		1.31		1.21		1	M0	333,1620
		NGC 7380 - 27			22 45 10	+57 51.9	1	15.05		0.65		0.07		1		49
		NGC 7380 - 41			22 45 10	+57 54.4	1	12.28		0.36		-0.44		1		749
215782	-26 16324	HR 8674			22 45 12	-26 10.5	1	6.30		0.91				4	G8 III	2032
215756	-44 15049				22 45 12	-44 31.2	8	9.52	.005	1.39	.003	1.61	.005	116	K3 (III)	143,863,1075,1460,1628,1657*
		NGC 7380 - 8			22 45 13	+57 52.9	1	10.67		0.27		-0.58		1	B1 V	49
240065	+55 02811				22 45 15	+56 12.0	1	9.72		0.64				45	G0	6011
		NGC 7380 - 26			22 45 15	+57 46.1	1	14.71		0.65		0.12		1		49
		NGC 7380 - 94			22 45 15	+57 49.5	1	15.13		0.56		0.24		1		749
		NGC 7380 - 22			22 45 15	+57 54.7	2	14.10	.020	0.57	.015	0.19	.010	2		49,749
		NGC 7380 - 11			22 45 15	+57 57.3	1	11.72		0.34		0.24		1		49
215812	-4 05757	G 156 - 48		⋆ AB	22 45 16	-04 29.3	1	6.66		0.65		0.12		2	G5 V	1003
		VES 941			22 45 18	+56 08.2	1	11.87		0.73		-0.39		1		8113
215857	+43 04306				22 45 19	+44 17.4	1	9.57		0.57		-0.02		1	F8	1748
240068	+57 02611	LS III +58 055			22 45 19	+58 13.5	2	9.66	.015	0.48	.005	-0.51	.010	3	B0 III	401,1012
		NGC 7380 - 17			22 45 21	+57 52.0	1	13.21		1.53				1		49
		NGC 7380 - 13			22 45 22	+57 56.4	2	12.71	.000	0.60	.010	0.16	.000	2		49,749
	+61 02346				22 45 22	+62 07.8	1	10.88		0.54		0.36		6		27
		NGC 7380 - 23			22 45 24	+57 55.4	1	14.32		0.55		0.23		1		49
215907	+57 02612	HR 8677			22 45 24	+58 13.1	1	6.36		-0.04		-0.04		2	B9.5IV	401
		LP 985 - 21			22 45 25	-37 03.0	1	11.90		1.51				1	M3	1705
215788	-45 14806	IDS22425S4547		AB	22 45 26	-45 30.8	3	8.68	.006	0.44	.000	-0.05	.014	16	F3/5 V	977,1075,2011
215800	-45 14807				22 45 27	-44 45.9	2	8.86	.009	1.10	.009	1.01		12	K1 III	977,2011
		NGC 7380 - 119			22 45 28	+57 48.7	1	12.26		0.57		0.09		1		749
		NGC 7380 - 20			22 45 28	+57 50.4	1	13.76		0.52		0.34		1		49
		NGC 7380 - 10			22 45 30	+57 44.6	2	11.46	.000	0.55	.010	0.50	.015	2		49,749
		NGC 7380 - 14			22 45 30	+57 50.1	1	13.04		1.53		1.29		1		49
		NGC 7380 - 21			22 45 31	+57 53.3	2	14.00	.080	0.38	.040	-0.11	.040	2		49,749
215922	+57 02613	IDS22435N5801		AB	22 45 31	+58 16.9	1	9.16		0.17		-0.21		1		401
		NGC 7380 - 24			22 45 32	+57 50.0	2	14.62	.015	0.65	.010	0.32	.055	2		49,749
215789	-51 13389	HR 8675			22 45 33	-51 34.8	7	3.48	.007	0.08	.004	0.09	.008	21	A3 V	15,1020,1068,1075*
		NGC 7380 - 18			22 45 34	+57 48.3	1	13.45		0.45		-0.18		1		49
215801	-46 14447				22 45 34	-46 19.5	6	10.04	.016	0.43	.012	-0.23	.022	12	F0w	742,1236,1594,1696*
		NGC 7380 - 34		⋆	22 45 35	+57 51.8	1	11.86		0.36		-0.44		1		749
	+59 02570				22 45 35	+59 38.0	1	10.48		0.45		0.04		2		27
215631	-77 01554	HR 8664			22 45 35	-77 18.9	3	6.71	.007	0.11	.000	0.09	.005	20	A4 V	15,1075,2038
		NGC 7380 - 28			22 45 36	+57 52.8	2	15.25	.050	0.54	.040	0.46		2		49,749
		NGC 7380 - 6			22 45 36	+57 55.4	2	10.32	.005	0.43	.005	0.05	.005	2		49,749
		NGC 7380 - 19			22 45 37	+57 48.0	2	13.65	.005	0.95	.010	0.47	.005	2		49,749
215818	-44 15051				22 45 37	-43 42.6	3	7.06	.002	0.99	.001	0.73	.003	16	G8 III	977,1075,2011
215842	-20 06487				22 45 38	-19 57.8	1	8.30		0.98		0.84		2	K0 III	1375
		NGC 7380 - 9			22 45 39	+57 53.8	1	10.68		0.36		-0.49		1	B1 V	49
215924	+54 02863	U Lac			22 45 40	+54 53.7	2	8.86	.234	2.32	.028	1.32	.206	4	M4 Iab +B	8032,8049
215881	+29 04771				22 45 41	+30 21.7	1	7.88		1.61		1.95		2	K5	1733
		LS III +57 088			22 45 44	+57 01.0	1	10.18		0.45		-0.65		2	B0: III:pe	1012
215729	-71 02726	HR 8672			22 45 45	-70 36.8	3	6.34	.003	0.07	.000	0.09	.005	21	A2 V	15,1075,2038
		NGC 7380 - 16			22 45 49	+57 48.8	2	13.18	.005	0.66	.010	0.09	.005	2		49,749
	+57 02615	NGC 7380 - 4			22 45 49	+57 53.0	1	10.19		0.40		-0.12		1	B6 Vne	49
215573	-80 01055	HR 8663			22 45 49	-80 23.3	3	5.33	.011	-0.14	.007	-0.49	.010	19	B6 IV	15,1075,2038
215910	+38 04867				22 45 50	+39 00.4	1	8.15		0.19		0.09		3	A5 V	833
		NGC 7380 - 118			22 45 52	+57 50.0	1	10.38		0.31		-0.56		1		749
215874	-11 05923	HR 8676, FM Aqr			22 45 52	-10 49.2	1	6.19		0.28				4	A9 III-IV	2007
215943	+36 04934	HR 8678			22 45 53	+37 09.2	2	5.96	.175	1.03	.005	0.84	.039	2	K0	1184,1375
215864	-37 14937				22 45 57	-37 09.1	1	9.76		1.02		0.87		1	K2/3 V	3072
216014	+64 01717	AH Cep		A	22 46 04	+64 47.9	2	6.90	.047	0.28	.016	-0.65	.016	8	B0.5V:nn	399,1012
215877	-45 14810				22 46 04	-44 52.0	5	8.50	.005	0.64	.004	0.13	.006	28	G3 V	143,977,1075,2011,2024

HD	DM	Other Id	N	Rem	α_{1950}	δ_{1950}	S	V	σ_V	B–V	σ_{B-V}	U–B	σ_{U-B}	n	Spectrum	References
215944	+27 04413				22 46 07	+27 51.5	1	8.34		0.51		-0.07		3	F8 V	3026
216001	+58 02483				22 46 07	+58 50.9	1	8.58		0.11		-0.35		2	A0	27
	+62 02114	LS III +62 015			22 46 07	+62 35.9	1	9.60		0.43		-0.42		2	B8	27
	-29 18469				22 46 09	-29 04.6	2	10.70	.039	1.38	.017	1.26	.002	3	K7 V	1773,3072
215956	+27 04414				22 46 10	+28 27.7	1	8.46		0.55		-0.05		3	G0 V	3026
	+57 02617	NGC 7380 - 5			22 46 10	+57 46.5	1	10.20		0.21		0.16		1		49
		LP 761 - 111			22 46 16	-13 38.7	1	12.66		0.96		0.73		1		1696
240072	+55 02814				22 46 22	+55 32.1	1	8.89		0.45		-0.03		2	F5	1502
		LP 932 - 83			22 46 24	-29 07.0	1	13.93		1.58		0.98		2		1773
		CS 22938 # 19			22 46 31	-65 15.4	2	14.25	.010	0.40	.007	0.04	.009	2		1580,1736
240073	+55 02815	V Lac			22 46 35	+56 03.4	5	8.47	.167	0.71	.090	0.49	.043	5	G5	592,1399,1462,1772,6011
240075	+59 02573	LS III +60 026			22 46 36	+60 18.6	3	9.56	.011	0.76	.028	-0.11	.021	11	B2	27,180,881
		G 67 - 83			22 46 38	+22 20.6	1	14.35		0.20		-0.70		2		3028
240074	+58 02485				22 46 38	+58 57.9	1	10.92		2.90		2.68		3	G5	8032
		G 67 - 23			22 46 39	+22 20.5	3	14.34	.027	0.20	.016	-0.67	.010	10		316,1620,3060
216044	+54 02865	LS III +54 076			22 46 39	+54 51.7	1	8.51		0.08		-0.82		2	B0 II	1012
216057	+53 02993	HR 8682			22 46 43	+54 09.0	1			-0.07		-0.50		2	B5 Vne	1079
216039	+25 04820				22 46 48	+26 14.9	1	8.18		0.34		0.14		1	A4mF2	1776
216102	+62 02115	HR 8683			22 46 51	+62 40.5	1	6.06		1.20				2	K0	71
216047	+15 04710				22 46 57	+16 05.3	1	8.27		0.25		0.09		1	A2	1776
	+63 01889	Cep OB3 # 1			22 46 57	+63 31.2	1	10.47		0.56		0.26		2	A0 III	1015
216032	-14 06354	HR 8679		★ A	22 46 57	-13 51.4	9	4.02	.020	1.57	.010	1.94	.008	42	K5 III	3,15,58,1024,1075*
216032	-14 06351	IDS22443S1407		B	22 46 57	-13 51.4	1	9.80		1.24		1.30		2		3016
216009	-45 14819				22 46 57	-44 41.3	10	8.06	.006	0.05	.008	0.09	.013	123	A0 V	116,278,743,977,977,1066*
215985	-55 09907	AX Gru			22 46 57	-55 30.0	1	7.00		1.55				4	M3 III	2012
216105	+55 02817	X Lac			22 47 00	+56 09.8	5	8.26	.085	0.88	.050	0.56	.007	5	G5	851,1399,1462,1772,6011
216048	+9 05111	HR 8681			22 47 02	+10 12.9	4	6.54	.008	0.29	.013	0.04	.009	12	F0 IV-V	15,253,254,1256
216092	+47 03931				22 47 07	+47 39.9	1	7.85		-0.07		-0.70		6	B1 V	1174
216106	+50 03836				22 47 09	+50 42.8	1	6.75		0.54		0.02		2	G0	1566
216042	-33 16244	HR 8680		★ AB	22 47 12	-33 04.2	3	6.32	.004	0.31	.004			11	F0 V+F3 V	15,1075,2007
	-45 14822				22 47 13	-45 23.2	2	10.39	.001	0.49	.000	-0.02		10	F8	863,2011
216172	+67 01468	HR 8687		★ AB	22 47 17	+68 18.2	1	6.19		0.42		0.02		2	F5 V	1733
		LS III +57 093			22 47 19	+57 29.6	1	10.82		0.35		-0.58		3		342
		G 67 - 24			22 47 20	+11 54.3	1	12.50		1.34		1.19		2		333,1620
216054	-42 16092				22 47 20	-41 45.1	6	7.77	.008	0.74	.007	0.27	.025	29	K0 V	743,977,1075,2011*
	-41 15048				22 47 22	-41 13.4	2	10.16	.024	0.53	.019	0.00		3	F8	742,1594
		G 67 - 25			22 47 23	+11 53.9	1	13.69		1.49		1.12		1		1620
		Case *M # 78			22 47 24	+59 02.	1	10.76		2.80				1	M1.5Ib	148
216131	+23 04615	HR 8684			22 47 35	+24 20.2	6	3.49	.011	0.94	.008	0.67	.016	13	G8 III	15,1025,1363,3016*
		Per sq 1 # 2			22 47 36	+61 26.	1	12.35		1.24		0.19		1		881
240078	+55 02819				22 47 38	+56 00.0	2	9.37	.010	1.11	.007	0.50		48	K0	592,6011
216090	-44 15061				22 47 41	-44 18.5	2	10.87	.006	-0.05	.003	-0.20		10	B8/A0III/IV	863,2011
216174	+55 02820	HR 8688			22 47 42	+55 38.2	2	5.38	.056	1.17	.004	1.12		4	K1 III	71,3016
	+61 02350	Cep OB3 # 2		★	22 47 42	+62 04.1	2	9.19	.012	0.71	.008	-0.24	.008	8	B1 V	27,1015
216189	+54 02867	LS III +54 077			22 47 43	+54 34.9	2	7.55	.019	0.00	.019	-0.72	.000	3	B8	401,963
		LS III +57 095			22 47 44	+57 55.2	1	10.60		0.26		-0.61		3		881
216446	+82 00703	HR 8702		★ AB	22 47 44	+82 53.3	5	4.75	.025	1.26	.008	1.34	.013	11	K3 III	15,1363,3016,4001,8015
216133	-7 05871				22 47 44	-07 22.1	3	9.85	.012	1.47	.012	1.23	.041	6	K5	158,680,1705
216141	+12 04894				22 47 47	+12 32.5	1	9.19		0.51		0.04		2	G0	1375
		Per sq 1 # 3			22 47 48	+58 18.	1	14.19		-0.21		-1.18		2		881
216098	-44 15063				22 47 48	-43 52.3	2	9.58	.004	0.56	.002	0.06		10	G2/5 V +A/F	863,2011
216227	+65 01813	IDS22460N6601		AB	22 47 49	+66 17.3	1	7.01		0.06		-0.34		2	B9	401
216135	-14 06357				22 47 50	-13 34.3	6	10.11	.004	-0.12	.006	-0.59	.020	44	B5 V	711,830,989,1026,1729,1783
	+77 00876	G 242 - 9			22 47 51	+77 41.2	1	10.16		0.76		0.28		1	K2	1658
	+61 02352	Cep OB3 # 3		★	22 47 54	+62 04.6	2	10.02	.005	1.43	.000	0.50	.005	4	B8 Ib	1012,1015
216228	+65 01814	HR 8694			22 47 54	+65 56.2	7	3.52	.022	1.05	.009	0.92	.016	15	K0 III	15,1118,1355,3016*
	+58 02486	LS III +58 056			22 47 55	+58 31.5	1	9.72		0.60		-0.24		3		881
		Cep sq 1 # 80			22 47 55	+59 29.0	1	10.96		0.38		0.20		5		27
	+60 02444	IDS22460N6047		AB	22 47 57	+61 02.7	1	9.63		0.56		-0.17		2	B8	342
		L 285 - 14			22 48 00	-50 26.	2	14.93	.019	0.11	.000	-0.72	.000	4	DA	782,3065
		Smethells 117			22 48 00	-59 53.	1	11.77		1.23				1		1494
216206	+49 03954	HR 8692			22 48 01	+50 24.7	3	6.22	.023	1.13	.012	0.87	.023	5	G4 Ib	1218,1355,4001
216177	+7 04935				22 48 02	+08 10.6	2	9.99	.000	0.63	.000	0.20	.000	3	G0	979,8112
		LS III +58 057			22 48 02	+58 33.8	1	12.16		0.62		-0.29		3		881
216200	+41 04623	HR 8690, V360 Lac			22 48 06	+41 41.3	4	5.92	.011	0.08	.007	-0.51	.005	13	B3 IV:e	15,154,1119,1174
		VES 943			22 48 08	+58 23.8	1	11.64		0.37		-0.06		1		8113
216179	+1 04656	G 28 - 17			22 48 12	+01 36.3	5	9.34	.008	0.68	.004	0.08	.013	9		516,979,1620,3077,8112
216201	+18 05059	HR 8691			22 48 12	+18 52.5	2	6.49	.005	1.12	.005	1.00	.005	4	K0	1648,1733
216149	-39 14848	HR 8685			22 48 12	-39 25.3	3	5.41	.005	1.44	.005	1.69		6	K3 III	2024,2035,3005
		Smethells 118			22 48 12	-47 28.	1	11.61		1.39				1		1494
	+61 02353	Cep OB3 # 6		★	22 48 14	+61 41.0	1	9.28		0.46		0.36		3	A2 IV	1015
		G 189 - 38			22 48 15	+34 35.3	2	11.72	.004	1.54	.008			7		419,906
216155	-45 14830				22 48 15	-45 00.1	4	8.43	.001	0.90	.006	0.53	.013	20	G8 IV	743,977,1075,2011
216248	+57 02625	LS III +58 058			22 48 18	+58 23.8	1	9.89		0.44		-0.41		2	B3 II	1012
	-44 15065				22 48 18	-44 05.2	1	10.83		1.07				4		2011
216219	+17 04818				22 48 25	+17 44.2	1	7.45		0.64		0.05		1	G0 IIp	979
		LS III +60 027			22 48 25	+60 10.8	2	11.07	.010	0.87	.030	-0.19	.035	5		180,881
216186	-42 16098				22 48 29	-41 50.8	1	7.45		1.60		1.92		5	M0 III	1673
		Cep OB3 # 4			22 48 32	+62 26.1	1	10.80		0.40		0.18		2	B9.5V	1015
216209	-24 17362				22 48 35	-24 01.8	1	7.54		1.07		0.89		5	K0 III	1657

Table 1   1207

HD	DM	Other Id	N Rem	α_{1950}	δ_{1950}	S	V	σ_V	B–V	σ_{B-V}	U–B	σ_{U-B}	n	Spectrum	References
216210	−30 19324	HR 8693		22 48 36	−29 48.1	2	5.97	.006	0.91	.001	0.59		6	G8/K0 III	58,2007
	+59 02579	IDS22461N5946	A	22 48 37	+60 01.0	1	9.94		0.55		0.15		2		27
216169	−60 07610	HR 8686		22 48 37	−60 08.8	1	6.46		1.13				4	K1 III	2035
		AJ89,1229 # 342		22 48 38	+04 33.5	1	11.10		1.63				1	M1	1632
		WLS 2300 35 # 11		22 48 38	+32 33.8	1	10.66		0.41		0.12		2		1375
		LS III +58 059		22 48 39	+58 30.3	1	11.11		0.65		−0.27		3		881
		Cep OB3 # 5		22 48 41	+62 11.8	1	10.47		0.50		0.39		2	A5 V	1015
240083	+55 02824			22 48 46	+55 55.7	1	9.57		0.49		0.12		1	F2	592
216213	−45 14834			22 48 46	−45 15.8	2	8.70	.001	1.56	.009	1.94		12	K5/M0 III	977,1075
		LS III +58 061		22 48 51	+58 45.8	1	11.26		0.84		−0.22		3		881
		G 28 - 18		22 48 52	−02 33.7	1	11.68		0.97		0.71		1		1620
		G 28 - 19		22 48 54	+05 19.6	1	14.40		1.05		0.80		1		1658
216223	−45 14835			22 48 55	−45 03.0	3	8.34	.006	1.08	.001	0.92	.007	16	K0 III	977,1075,2011
216259	+13 05006	G 67 - 27	⋆ AB	22 48 56	+13 42.1	3	8.30	.016	0.83	.005	0.44	.015	5	K0	333,1620,1733,3016
216224	−47 14371	IDS22460S4718	AB	22 48 56	−47 02.5	1	8.82		0.93		0.58		1	K2 V	3072
216187	−63 04826	HR 8689	⋆ AB	22 48 56	−63 27.2	2	6.13	.002	1.02	.006	0.94		6	K0 III	58,2035
		G 128 - 7		22 48 57	+29 23.7	2	15.53	.014	0.63	.028	−0.12	.028	4	DF	316,3060
		Cep sq 1 # 82		22 48 59	+60 07.1	1	10.85		0.31		−0.15		3		27
216321	+47 03944			22 49 02	+48 28.1	1	6.77		−0.09		−0.47		2	B9	3016
216285	+25 04828	IDS22466N2552	AB	22 49 03	+26 07.5	2	6.94	.009	0.23	.005	0.12		4	A3	1381,3030
216285	+25 04828	IDS22466N2552	C	22 49 03	+26 07.5	1	10.88		1.64		2.00		2	A7 V	3030
216328	+53 03009			22 49 04	+53 39.3	1	8.47		0.04		−0.32		2	A0	401
	+60 02446			22 49 06	+61 10.5	1	9.43		2.04				2		369
216323	+32 04529			22 49 13	+32 33.0	1	7.18		0.17		0.08		3	A2	985
216339	+44 04253			22 49 15	+45 14.7	1	9.02		−0.01		−0.41		1	A0	963
216308	+14 04879			22 49 17	+14 49.2	1	7.00		0.11		0.04		1	A0	1776
216340	+44 04254			22 49 17	+45 13.4	1	8.79		−0.01		−0.35		1	A0	963
		LS III +58 062		22 49 20	+58 39.3	1	11.34		0.43		−0.48		3		881
216331	+29 04786			22 49 22	+29 46.8	1	7.93		0.95		0.53		2	G5 II	8100
216341	+40 04924	IDS22473N4047	E	22 49 23	+41 02.6	1	8.94		0.08		−0.03		1	A0	401
		LS III +58 063		22 49 25	+58 01.9	1	10.71		0.53		−0.42		3		881
216380	+60 02450	IDS22474N6109	⋆ AB	22 49 25	+61 25.8	2	5.61	.010	0.77	.003	0.40	.013	6	G8 III-IV +	3032,4001
216380	+60 02450	HR 8696	C	22 49 25	+61 25.8	1	11.05		1.09		0.98		2		3016
		G 128 - 8		22 49 30	+31 29.4	1	11.65		1.47		1.21		3		1723
216353	+40 04925	IDS22473N4047	C	22 49 30	+41 04.1	1	8.07		0.11		−0.11		1	A0	401
216291	−46 14461			22 49 32	−46 15.4	7	7.25	.003	1.45	.005	1.75	.006	84	K4/5 III	460,474,743,863,977*
216369	+40 04926	IDS22473N4047	A	22 49 33	+41 02.8	1	7.07		0.01		−0.21		2	A0	401
216411	+58 02492	LS III +58 064		22 49 33	+58 44.6	4	7.19	.010	0.61	.005	−0.43	.000	11	B1 Ia	27,1012,1025,1234
	+40 04927			22 49 35	+41 17.3	1	9.49		0.55		0.06		2	F5	401
	+22 04725			22 49 37	+23 09.0	1	9.78		1.20		1.14		2	K8	3072
		G 128 - 10		22 49 39	+31 28.0	2	9.66	.017	1.08	.026	1.03	.053	3	K4:	1620,1723
	+30 04824			22 49 39	+31 28.1	1	9.70		1.06				1	K5	1746
		Cep OB3 # 7		22 49 41	+63 19.9	1	11.36		0.52		0.28		1	B8	1015
240089	+59 02586			22 49 42	+59 31.0	1	9.44		0.14		−0.01		3	B9	27
		LS III +60 028		22 49 42	+60 52.9	3	10.74	.031	1.17	.030	−0.02	.035	7		180,342,483
		WLS 2240-20 # 8		22 49 44	−20 30.7	1	10.87		0.53		0.04		2		1375
	+40 04928			22 49 45	+41 09.6	1	8.77		0.99		0.70		2		401
216336	−33 16270	HR 8695	⋆ AB	22 49 45	−33 08.5	7	4.46	.007	−0.04	.004	−0.13	.031	25	A0 III	15,938,1075,2012,2016*
216397	+42 04521	HR 8699	⋆ A	22 49 46	+43 02.8	3	4.98	.026	1.56	.010	1.93	.009	6	M0 III	1355,1363,3053
		Cep OB3 # 8		22 49 47	+62 08.3	1	11.00		0.69		−0.10		1	B2 V	1015
216333	−21 06315			22 49 47	−21 26.9	1	8.76		0.28		0.05		2	A8 IV/V	1375
216438	+52 03315	LS III +53 083		22 49 51	+53 26.6	1	8.46		0.06		−0.76		2	B1 II	1012
216385	+9 05122	HR 8697	⋆ A	22 49 52	+09 34.1	5	5.16	.022	0.48	.012	−0.02	.014	15	F7 IV	15,22,254,1256,3037
216385	+9 05122	IDS22474N0918	B	22 49 52	+09 34.1	1	13.52		1.66		1.32		3		3024
216383	+15 04719			22 49 54	+15 37.0	1	7.00		1.38		1.59		3	K2	1648
	+14 04881			22 49 56	+14 50.5	1	10.09		1.24		1.36		2		1375
216429	+37 04713		V	22 49 56	+38 28.8	1	8.34		0.19		1.18		7	A3 III	1768
216448	+56 02890	IDS22480N5712	AB	22 49 58	+57 27.1	3	8.03	.018	1.04	.012	0.86	.010	5	K5 III	938,1068,3030
216363	−41 15063			22 49 59	−41 16.2	1	8.31		0.40				4	F0 IV/V	2012
216449	+55 02827			22 50 00	+56 24.2	1	8.45		0.21		0.15		1	A0	851
216386	−8 05968	HR 8698, λ Aqr		22 50 00	−07 50.8	7	3.75	.031	1.64	.006	1.74	.018	33	M2.5 IIIa	3,15,1075,1088,2012*
		LP 877 - 23		22 50 00	−20 51.1	2	12.06	.000	0.62	.000	−0.17	.058	6		1696,3062
		WLS 2300 35 # 6		22 50 06	+35 12.2	1	13.10		0.46		−0.02		2		1375
	+40 04932			22 50 13	+41 06.5	1	9.50		0.12		−0.13		2	A0	401
		G 67 - 30		22 50 18	+17 05.4	1	13.41		1.19		0.94		5		419
216406	−45 14847			22 50 22	−45 24.8	9	6.85	.004	1.12	.005	1.07	.007	119	K0 III	116,278,743,977,977*
216465	+28 04475	IDS22481N2855		22 50 26	+29 10.5	1	8.47		0.43		−0.03		3	F5 V	3026
216479	+41 04626			22 50 26	+42 17.2	1	9.32		0.13		0.02		2	A0	401
240098	+58 02496			22 50 29	+58 54.5	1	9.59		0.12		0.02		2	A0	27
216509	+55 02830			22 50 31	+56 26.6	2	7.39	.025	0.54	.020	0.08	.030	2	G0	851,1462
	+15 04721	BH Peg		22 50 32	+15 30.8	1	9.99		0.27		0.10		1		668
	+61 02355	Cep OB3 # 9		22 50 33	+62 25.2	2	9.62	.010	0.35	.005	−0.13	.010	6	B7 IV	27,1015
216489	+16 04831	HR 8703, IM Peg		22 50 34	+16 34.5	1	5.64		1.12				2	K1 III	71
216532	+61 02356	Cep OB3 # 10	⋆	22 50 34	+62 10.5	4	8.00	.011	0.54	.004	−0.46	.006	12	O8 V	27,1011,1012,1015
	−15 06290			22 50 35	−14 31.2	11	10.17	.021	1.59	.015	1.19	.027	47	dK5	1,116,308,830,1006,1586*
216455	−17 06616	IDS22479S1742	AB	22 50 35	−17 25.9	1	8.74		0.47		0.05		2	F5 V	1375
		G 156 - 58		22 50 36	−14 41.5	1	12.25		0.57		−0.16		1		1696
	+61 02357	Cep OB3 # 11	⋆ AB	22 50 37	+62 02.8	2	9.84	.041	1.20	.000	0.21	.014	7	B1 V	27,1015
216457	−19 06353	DS Aqr		22 50 37	−18 51.5	1	10.54		0.54		0.26		1	F2 II	3064
216533	+58 02497	MX Cep		22 50 40	+58 32.4	2	7.91	.005	0.08	.006	0.07	.022	7	A2p	220,1202

HD	DM	Other Id	N Rem	α_{1950}	δ_{1950}	S	V	σ_V	B–V	σ_{B-V}	U–B	σ_{U-B}	n	Spectrum	References
		Cep OB3 # 12		22 50 40	+63 42.4	1	11.10		0.60		0.22		1	A2	1015
216435	−49 13988	HR 8700		22 50 41	−48 51.8	3	6.03	.007	0.62	.000	0.17		11	G0 V	15,158,2012
216523	+49 03962	HR 8705		22 50 42	+50 08.8	1			-0.04		-0.35		2	B8 V	1079
		VES 944		22 50 42	+57 15.2	1	12.72		0.72		-0.04		1		8113
		GD 242		22 50 45	+23 41.6	2	10.43	.014	0.43	.023	-0.07	.014	4		308,979
		LS III +58 065		22 50 46	+58 12.5	1	11.58		0.22		-0.58		3		881
	+59 02591			22 50 46	+60 09.4	1	10.08		0.27		0.16		4		27
216534	+49 03965			22 50 48	+49 35.9	1	8.50		-0.01		-0.58		5	B3 V	1174
		LP 877 - 25		22 50 48	−24 10.4	2	11.68	.005	0.52	.005	-0.22	.035	4		1696,3062
216494	−12 06371	HR 8704, HI Aqr	★	22 50 51	−11 53.0	11	5.80	.013	-0.09	.006	-0.30	.020	50	B8 IV/V	3,15,116,1006,1075*
	+37 04715			22 50 53	+37 31.3	1	9.96		1.03				1	G5	1184
216538	+39 04957	HR 8706		22 50 54	+39 54.1	3	6.33		-0.08	.005	-0.45	.013	6	B7 III-IV	1049,1079,1716
	+62 02124	Cep OB3 # 13		22 50 55	+62 55.8	1	10.13		0.54		0.44		1	F6 V	1015
	+62 02125	Cep OB3 # 14	★	22 50 55	+63 08.8	1	8.95		0.65		-0.26		3	B1.5V	1015
216606	+67 01475	IDS22490N6727	A	22 50 56	+67 43.4	1	6.98		0.42		0.00		2	F2	1502
	+14 04886			22 50 57	+15 28.6	1	8.60		0.55		0.03		1	G0	289
216560	+44 04263			22 50 58	+45 25.6	1	7.98		1.10		0.87		2	K0	1601
	+59 02594	LS III +59 031		22 50 59	+59 56.8	1	10.13		0.74		-0.02		2		27
		LS III +60 029		22 51 00	+60 55.5	2	11.23	.010	0.96	.030	-0.14	.010	6		180,881
		G 128 - 11		22 51 04	+27 29.5	3	11.52	.053	0.80	.012	0.25	.033	5		1620,1658,1723
216595	+59 02595	HR 8707		22 51 04	+59 50.1	1	6.01		1.75				2	gK2	71
		VES 945		22 51 05	+57 25.4	1	12.08		0.58		-0.31		1		8113
		JL 117		22 51 06	−72 39.	1	14.39		-0.36		-1.23		2	sdO	132
		WLS 2300 75 # 10		22 51 09	+74 12.8	1	12.82		0.76		0.13		2		1375
216562	+29 04797	IDS22488N3013	AB	22 51 13	+30 29.8	1	7.50		0.17		0.12		2	A2	3016
216437	−70 02971	HR 8701		22 51 13	−70 20.5	4	6.04	.009	0.66	.000	0.23	.008	20	G2/3 IV	15,1075,2038,3077
		CS 22893 # 5		22 51 14	−10 15.2	1	14.22		0.53		-0.15		1		1736
		Case *M # 79		22 51 18	+61 01.	2	10.84	.223	2.97	.079	2.47		5	M3	148,8032
216629	+61 02361	Cep OB3 # 15	★ A	22 51 18	+61 52.8	2	9.29	.010	0.72	.005	-0.19	.010	5	B2 IV-Vne	1012,1015
216629	+61 02361	Cep OB3 # 15	★ B	22 51 18	+61 52.8	1	11.91		1.35		0.87		2		1015
		Cep sq 1 # 91		22 51 19	+61 01.5	1	10.67		0.56		0.31		7		27
216531	−46 14470			22 51 22	−46 09.3	4	8.29	.006	0.59	.003	0.03	.004	18	G0 V	460,977,1075,2011
216608	+43 04331	HR 8708	★ AB	22 51 26	+44 29.0	1	5.81		0.26		0.10		4	A3 Vm +F6 V	8071
		G 242 - 14		22 51 27	+79 26.2	1	11.94		0.72		0.08		1		1658
		BO Aqr		22 51 30	−12 37.7	2	11.53	.030	0.22	.045	0.11	.050	2		597,700
216658	+61 02363	Cep OB3 # 16	★ AB	22 51 33	+61 52.1	2	8.90	.009	0.70	.000	-0.28	.005	9	B0.5V	1012,1015
		Cep OB3 # 16s	★ S	22 51 33	+61 52.1	1	13.95		1.02		1.76		6		1015
		G 127 - 62		22 51 36	+26 41.4	1	11.41		1.14		1.11		2	K7	7010
		Cep OB3 # 17	A	22 51 39	+62 20.8	1	10.48		0.82		-0.10		5	B1.5V	1015
		Cep OB3 # 17	B	22 51 39	+62 20.8	1	12.56		1.17		0.50		3		1015
216625	+19 05028			22 51 41	+19 37.7	1	6.97		0.50		0.00		3	F8	3016
		G 28 - 22		22 51 43	+00 12.1	1	13.38		1.56		1.13		1		333,1620
		LP 987 - 52		22 51 43	−36 50.0	1	15.13		0.15		-0.68		1		3060
	+62 02127	Cep OB3 # 18	★	22 51 46	+63 09.1	1	10.16		0.58		-0.29		3	B2 IV-V	1015
		Cep OB3 # 32		22 51 47	+62 20.9	1	10.65		0.44		0.32		1	A1 V	1015
216632	+27 04436	IDS22494N2729	A	22 51 48	+27 45.0	1	7.68		0.50		0.01		3	F8 V	3026
216646	+39 04964	HR 8712		22 51 49	+40 06.6	2	5.81	.006	1.14	.005	1.14		5	K0 III	71,1501
216635	+17 04827			22 51 50	+17 31.7	1	6.56		1.04		0.74		2	K0	1375
		Cep OB3 # 19		22 51 52	+62 55.0	2	10.35	.018	0.40	.005	-0.21	.000	7	B7 V	27,1015
		L 718 - 20		22 51 53	−20 40.2	2	12.06	.001	0.62	.010	-0.14	.014	11		1698,1765
	−18 06214			22 51 56	−18 04.2	1	10.42		1.17		1.12		1	K5 V	3072
216711	+61 02364	Cep OB3 # 20	★	22 51 57	+62 19.8	3	9.06	.015	0.62	.000	-0.33	.005	7	B1 V	27,1012,1015
216637	−7 05886	HR 8710		22 51 58	−07 28.3	3	6.18	.005	1.28	.000	1.41	.000	8	M3 III	15,1061,3005
		G 156 - 61		22 51 58	−13 15.8	1	12.90		0.60		-0.11		2		3062
	+12 04906	G 67 - 31		22 51 59	+12 50.8	1	10.40		0.82		0.37		1	K1	333,1620
216627	−16 06173	HR 8709		22 52 00	−16 05.2	11	3.27	.013	0.06	.014	0.10	.033	69	A3 V	3,15,263,418,1007*
216721	+60 02456			22 52 02	+60 35.2	1	8.62		0.26				1	A3	609
		CS 22893 # 6		22 52 03	−10 19.4	1	14.28		0.53		-0.10		1		1736
		G 67 - 33		22 52 05	+23 15.0	1	11.13		1.40		1.28		1	K7	333,1620
216684	+42 04529			22 52 06	+43 15.7	2	7.76	.000	-0.03	.010	-0.60	.005	5	B8 V	963,1174
216649		CCS 3168		22 52 06	−07 14.	1	10.75		1.21				1	R3	1238
	+60 02457	IDS22502N6058	A	22 52 07	+61 13.4	1	9.33		0.67		0.08		6	B8	27
216640	−17 06619	HR 8711		22 52 07	−16 32.3	5	5.56	.023	1.11	.014	1.08	.015	12	K2.5 IIIb	15,1003,2035,3016,4001
216672	+16 04833	HR 8714, HR Peg		22 52 08	+16 40.5	2	6.32	.029	1.80	.000	1.88	.058	6	S5.1	1733,3039
		Cep OB3 # 21		22 52 08	+63 48.2	1	11.91		0.54		0.35		1	A0	1015
		Cep sq 1 # 95		22 52 09	+59 01.5	1	11.16		0.93		0.58		2		27
216685	+28 04479			22 52 12	+29 05.5	1	8.80		0.50		0.02		3	F8 V	3026
216519	−75 01766			22 52 12	−75 15.7	1	7.74		1.07		1.03		8	K0 III	1704
		NGC 7419 - 1		22 52 18	+60 34.	1	12.97		0.68				1		609
		NGC 7419 - 2		22 52 18	+60 34.	1	15.97		1.37				1		609
		NGC 7419 - 3		22 52 18	+60 34.	1	15.19		1.25				1		609
		NGC 7419 - 4		22 52 18	+60 34.	1	16.02		2.76				1		609
		NGC 7419 - 5		22 52 18	+60 34.	1	16.43		2.15				1		609
		NGC 7419 - 6		22 52 18	+60 34.	1	15.44		1.36				1		609
		NGC 7419 - 7		22 52 18	+60 34.	1	14.50		1.60				1		609
		NGC 7419 - 8		22 52 18	+60 34.	1	16.66		1.43				1		609
		NGC 7419 - 9		22 52 18	+60 34.	1	12.98		3.23				1		609
		NGC 7419 - 10		22 52 18	+60 34.	1	14.96		0.96				1		609
		NGC 7419 - 11		22 52 18	+60 34.	1	15.69		1.67				1		609
		NGC 7419 - 12		22 52 18	+60 34.	1	13.77		3.40				1		609

Table 1   1209

HD	DM	Other Id	N Rem	α_{1950}	δ_{1950}	S	V	σ_V	B–V	σ_{B-V}	U–B	σ_{U-B}	n	Spectrum	References
		NGC 7419 - 13		22 52 18	+60 34.	1	13.97		1.69				1		609
		NGC 7419 - 14		22 52 18	+60 34.	1	10.57		1.25				1		609
		NGC 7419 - 15		22 52 18	+60 34.	1	16.13		1.47				1		609
		NGC 7419 - 16		22 52 18	+60 34.	1	11.82		0.72				1		609
		NGC 7419 - 17		22 52 18	+60 34.	1	13.41		0.86				1		609
		NGC 7419 - 106		22 52 18	+60 34.	1	14.48		3.89				1		609
		NGC 7419 - 115		22 52 18	+60 34.	1	17.24		1.84				1		609
		NGC 7419 - 128		22 52 18	+60 34.	1	18.39		1.64				1		609
		NGC 7419 - 129		22 52 18	+60 34.	1	18.00		1.81				1		609
		NGC 7419 - 130		22 52 18	+60 34.	1	17.55		1.72				1		609
		NGC 7419 - 219		22 52 18	+60 34.	1	17.44		1.16				1		609
		NGC 7419 - 232		22 52 18	+60 34.	1	18.38		1.80				1		609
		NGC 7419 - 237		22 52 18	+60 34.	1	17.94		1.61				1		609
		NGC 7419 - 419		22 52 18	+60 34.	1	17.10		1.89				1		609
		NGC 7419 - 437		22 52 18	+60 34.	1	17.53		1.66				1		609
		L 49 - 19		22 52 18	−75 42.	2	10.41	.018	1.49	.005	1.01		5		1705,3078
	+61 02365	Cep OB3 # 22	★	22 52 21	+62 23.9	2	9.19	.005	0.51	.005	-0.45	.000	7	B1 V	27,1015
		Cep OB3 # 23		22 52 22	+62 58.8	1	10.82		0.47		-0.17		3	B5 V	1015
216655	−49 13996	IDS22494S4900	AB	22 52 22	−48 44.0	2	7.05	.030	0.55	.015	0.01		6	G3/6	1075,3077
216656	−49 13997	IDS22494S4860	AB	22 52 22	−48 45.6	2	6.69	.025	0.46	.000	-0.05		6	F6/8 (V)	1075,3077
216722	+35 04908			22 52 23	+35 43.0	1	7.48		1.01		0.80		2	K0	1601
217157	+84 00517	HR 8736		22 52 24	+85 06.4	1	5.90		1.32				2	K5	71
216701	+0 04939	HR 8715		22 52 26	+00 47.9	3	6.12	.022	0.19	.010	0.15	.015	9	A7 III	15,252,1071
216666	−37 14981	HR 8713		22 52 27	−36 39.3	1	6.40		1.31				4	K2 III	2009
216733	+41 04634			22 52 28	+42 14.6	1	7.83		0.52		0.46		2	A2	401
		Cep sq 1 # 97		22 52 32	+59 27.9	1	11.03		0.45		0.21		3		27
216667	−43 15197			22 52 33	−43 09.1	3	8.46	.003	0.50	.002	-0.01	.006	16	F6 V	977,1075,2011
216718	−5 05885	HR 8716	★ AB	22 52 35	−05 15.3	2	5.71	.005	0.88	.000	0.60	.000	7	K0 III-IV	15,1071
216679	−46 14474			22 52 36	−46 35.7	6	7.53	.007	0.37	.007	0.03	.008	55	F2 III/IV	278,743,863,1075,1770,2011
216754	+50 03868			22 52 38	+50 41.8	1	8.73		0.01		-0.28		2	Ap Si	1566
217158	+84 00516	AR Cep		22 52 38	+84 46.8	1	7.40		1.52		1.26		3	M4 III	1733
	+61 02366	Cep OB3 # 24	★	22 52 39	+62 20.7	2	9.70	.015	0.72	.010	-0.28	.010	5	B1 V	27,1015
216735	+8 04961	HR 8717		22 52 43	+08 32.9	16	4.91	.007	0.00	.008	0.00	.023	78	A1 V	1,3,15,116,1020,1034*
216756	+36 04956	HR 8718		22 52 43	+36 48.6	2	5.88	.037	0.38	.019	-0.06	.028	4	F5 V	254,3037
		Cep OB3 # 25		22 52 43	+62 29.0	1	10.68		0.49		-0.14		3	B5 Vn	1015
	−51 13427			22 52 43	−50 54.2	1	10.15		0.82		0.46		2	G5	1696
	+58 02508	IDS22508N5842	AB	22 52 47	+58 58.4	1	9.65		0.20		0.07		6	A0	27
235994	+50 03872	IDS22507N5100	AB	22 52 49	+51 15.5	1	9.20		0.58		0.09		2	F2	3016
216729	−32 17309	IDS22501S3218	AB	22 52 51	−32 01.7	1	9.68		0.90		0.54		1	K0 V	3072
	+58 02509			22 52 53	+58 58.1	1	10.41		0.65		0.24		3		27
	+40 04938			22 52 55	+40 59.7	1	10.08		0.30		0.15		2		401
216886	+75 00858	IDS22517N7548	AB	22 52 58	+76 04.3	2	7.44	.010	0.25	.006	-0.17	.016	2	A3	695,963
		Per sq 1 # 4		22 53 00	+58 53.	1	12.38		1.03		0.17		2		881
		Cep OB3 # 26		22 53 00	+62 18.3	2	10.31	.025	0.74	.045	0.00	.000	6	B2.5V	27,1015
	+55 02840	LS III +56 108		22 53 03	+56 07.0	2	10.01	.000	0.45	.000	-0.56	.000	4	O7	1011,1012
216743	−43 15198			22 53 05	−42 49.2	4	7.26	.005	0.16	.001	0.07	.009	16	A3 V	278,1075,1770,2011
216761	−32 17312	HR 8719		22 53 06	−31 54.0	2	6.09	.003	1.36	.002	1.52		6	K3 III	58,2009
		Cep OB3 # 27		22 53 07	+62 33.6	1	11.91		0.55		0.40		1	B5	1015
216770	−27 16109			22 53 10	−26 55.4	2	8.11	.010	0.82	.005	0.48		6	K0 V	2012,3008
216763	−33 16303	HR 8720	★ AB	22 53 11	−32 48.4	5	4.20	.012	0.96	.011	0.70	.009	24	G8 III	15,938,1075,2012,8015
216777	−8 05980	G 156 - 65	★ A	22 53 12	−08 05.3	7	8.01	.007	0.64	.007	0.09	.011	16	G6 V	22,742,908,1003,1658*
216777	−8 05980	G 156 - 64	★ B	22 53 12	−08 06.1	1	16.50		0.42		-0.46		3		3060
	+15 04729	IDS22508N1515	AB	22 53 15	+15 30.7	1	8.67		0.67		0.17		3	G5	3016
216828	+51 03503			22 53 18	+52 21.3	1	7.89		0.39		-0.02		2	F5	1566
		G 128 - 18		22 53 23	+25 09.2	1	12.83		1.16		1.08		1		1620
		Cep OB3 # 29		22 53 23	+62 06.3	1	11.56		0.47		0.25		1	A0	1015
		Cep OB3 # 28		22 53 23	+62 25.1	1	10.92		0.31		0.23		3	A0 V	1015
		G 28 - 24		22 53 23	+05 29.5	2	11.23	.035	1.45	.000	1.19		2	dM0	333,1620,1632
216831	+35 04917	HR 8723	★ A	22 53 24	+36 05.1	2			-0.05	.010	-0.39	.030	5	B7 III	1049,1079
		LB 1514		22 53 24	−49 16.	1	13.01		0.01		-0.01		3		45
		L 167 - 14		22 53 24	−60 18.	1	14.08		1.79		1.45		2		3078
		Steph 2065		22 53 31	+17 32.6	1	10.50		1.49		1.19		2	M0	1746
216851	+42 04538			22 53 31	+43 17.5	2	8.02	.005	0.05	.015	-0.52	.025	7	B3 V:n	1174,1212
		AA45,405 S147 # 1		22 53 33	+58 12.0	1	12.89		0.85				1		725
216811	−24 17405			22 53 33	−24 25.1	1	8.37		0.35				4	F0 IV	2012
216852	+42 04539			22 53 34	+42 55.9	1	8.80		0.03		-0.47		2	A2	401
	+61 02369	Cep OB3 # 30		22 53 36	+62 04.2	2	9.32	.010	0.27	.005	0.18	.015	4	A1 V	401,1015
216803	−32 17321	HR 8721, TW PsA		22 53 37	−31 49.8	6	6.49	.012	1.10	.013	1.01	.009	19	K4 Vp	15,1013,1075,2032*
216854	+40 04942			22 53 39	+41 04.1	1	7.32		0.49		0.07		2	F5	401
	+56 02903	LS III +57 101		22 53 41	+57 20.1	1	10.12		0.22		-0.67		2	B0.5:IV:	1012
		Per sq 1 # 5		22 53 42	+58 56.	1	12.76		0.64		-0.23		2		881
216832	+10 04849	G 67 - 35		22 53 43	+10 47.9	2	9.36	.020	0.73	.015	0.23	.010	2	F8	333,1620,1696
216898	+61 02370	Cep OB3 # 31	★	22 53 44	+62 02.4	5	8.01	.013	0.53	.005	-0.48	.004	13	O8.5V	27,401,1011,1012,1015
216888	+53 03032			22 53 46	+53 59.2	1	8.27		1.04		0.71		2	K0	1566
		LP 521 - 45		22 53 47	+14 36.4	1	12.42		0.98		0.68		1		1696
240121	+59 02602			22 53 49	+59 44.5	2	9.59	.034	0.16		0.00	.010	3	B8	27,1502
		Cep OB3 # 33	B	22 53 50	+62 35.0	1	12.28		0.57		0.11		3	B5 V	1015
		Cep OB3 # 33	A	22 53 52	+62 35.2	1	11.36		0.64		0.00		3	B5 V	1015
216823	−48 14364	HR 8722		22 53 53	−48 14.2	3	5.71	.008	0.23	.009	0.18	.019	11	A4mA5-F2	216,355,2007
216863	+16 04838	G 67 - 36	★ AB	22 53 55	+17 11.3	1	8.90		1.00		0.80		1	K0	333,1620

HD	DM	Other Id	N	Rem	α₁₉₅₀	δ₁₉₅₀	S	V	σ_V	B–V	σ_B–V	U–B	σ_U–B	n	Spectrum	References
		Steph 2068			22 53 55	+32 17.3	1	11.86		1.44		1.24		1	K7-M0	1746
	+62 02133	Cep OB3 # 34			22 53 55	+62 34.2	1	10.68		0.35		0.15		1	A1 V	1015
216926	+62 02134	Cep OB3 # 35			22 53 56	+63 12.4	1	8.90		0.26		0.02		4	B9 V	1015
216879	+22 04742	IDS22515N2225		A	22 53 57	+22 41.3	1	7.36		0.39		0.14		1	F0	1776
	+62 02135				22 53 58	+62 37.1	1	10.78		0.47		0.23		2		27
216912	+57 02644				22 53 59	+57 55.7	1	7.07		-0.06		-0.45		2	A0	401
216945	+61 02371	Cep OB3 # 36			22 54 00	+62 09.9	1	6.54		1.82		2.15		1	gK2	1015
	+59 02604				22 54 04	+59 41.6	1	9.73		0.25		0.13		3	A5	27
216928	+55 02850				22 54 05	+56 11.0	2	7.25	.010	0.09	.005	-0.38	.005	4	B9	401,1733
	+59 02603				22 54 05	+59 41.0	1	10.65		0.30		0.19		3		27
		G 189 - 50			22 54 06	+33 37.2	1	12.69		0.67		-0.01		1		1658
216916	+40 04949	HR 8725, EN Lac		★A	22 54 06	+41 20.2	6	5.59	.015	-0.14	.008	-0.84	.007	20	B2 IV	15,154,1119,1174,1278,3024
216916	+40 04949	IDS22518N4104		B	22 54 06	+41 20.2	1	11.32		0.39		0.09		3		3024
216927	+58 02511	LS III +58 066			22 54 06	+58 37.2	1	8.31		0.86		0.05		2	B9 Ia	1012
216992	+67 01482				22 54 09	+67 58.9	1	8.04		0.15		-0.44		3	B3 II	555
		G 28 - 25			22 54 09	-00 29.4	1	13.06		1.51		1.15		1		1620
216899	+15 04733	G 67 - 37			22 54 10	+16 17.4	7	8.67	.013	1.51	.010	1.17	.010	22	M1	1,1017,1197,1705,1775*
216915	+42 04545				22 54 11	+42 44.7	1	8.62		0.16		0.08		2	A0	401
		VES 946			22 54 11	+62 23.8	1	12.35		0.73		0.21		2		8113
		Cep OB3 # 37			22 54 13	+62 53.5	1	10.95		0.56		-0.03		1	B6 Vnnp	1015
217049	+72 01073	IDS22527N7218		AB	22 54 13	+72 34.6	1	8.01		0.08		0.03		5	A0	3032
216946	+48 03887	HR 8726			22 54 14	+49 28.0	7	4.97	.033	1.77	.018	1.95	.019	18	K5 Ib	15,1025,1218,1355*
		AA45,405 S148 # 1		★	22 54 14	+58 15.3	2	12.53	.020	0.88	.055	-0.28		4		725,881
		G 128 - 21			22 54 19	+29 14.2	1	12.35		1.24	.004	1.20	.032	2	K5	1620,5010
216900	+11 04904	HR 8724		★AB	22 54 21	+11 34.9	2	6.54	.014	0.17	.016	0.07	.021	5	A3 Vs	938,1733
		G 28 - 26			22 54 23	+04 49.5	1	13.18		1.00		0.73		1		333,1620
		AA45,405 S149 # 1		★	22 54 23	+58 16.2	2	12.44	.030	0.53	.010	-0.41		4		725,881
		G 241 - 31			22 54 25	+67 59.5	1	14.68		1.70				3		1663
		WLS 2256 55 # 10			22 54 26	+55 18.5	1	12.18		1.16		0.90		2		1375
217200	+80 00739				22 54 26	+81 01.4	1	8.25		0.13		0.08		4	A2	1219
216907	-21 06330	S Aqr			22 54 26	-20 36.7	1	11.95		1.53		0.22		1	M4	975
	+63 01907	Cep OB3 # 38			22 54 31	+64 11.8	2	9.10	.015	0.86	.005	-0.20	.005	4	B0.5Ia	1012,1015
	+25 04848				22 54 32	+26 07.2	1	8.59		1.32		1.32		2	K0	1733
		Cep sq 1 # 107			22 54 32	+59 11.7	1	10.65		0.30		0.15		4		27
216931	-4 05793	IDS22520S0347		ABC	22 54 32	-03 30.8	1	6.59		0.27		0.15		2	A0	938
217035	+62 02136	Cep OB3 # 39		★AB	22 54 33	+62 36.1	2	7.75	.010	0.46	.000	-0.53	.010	6	B0.5Vn	1012,1015
		VES 947			22 54 38	+56 40.4	1	12.39		0.25		-0.27		1		8113
217085	+72 01076	IDS22531N7218		AB	22 54 39	+72 34.2	1	7.38		0.21		0.10		5	A2	3032
217085	+72 01076	IDS22531N7218		C	22 54 39	+72 34.2	1	11.51		0.69		0.19		5		3032
	+27 04445	G 128 - 22		★A	22 54 42	+27 44.2	1	9.93		1.18		1.12		2	M0p	3016
	+27 04445	IDS22523N2728		B	22 54 42	+27 44.2	1	14.29		1.58				2		3016
216953	-5 05894	HR 8727			22 54 42	-05 04.7	3	6.31	.012	0.94	.008	0.71	.016	8	G9 III	15,1071,4001
217061	+61 02372	Cep OB3 # 40		★	22 54 44	+62 21.4	3	8.76	.009	0.68	.007	-0.26	.004	15	B1 Vn	27,1012,1015
		LS III +58 067			22 54 45	+58 30.8	1	10.80		0.60		-0.36		3		881
		LS III +57 105			22 54 46	+57 13.2	1	10.97		0.58		-0.29		1		8113
		LS III +58 068			22 54 47	+58 32.4	1	11.96		0.55		-0.32		3		881
217086	+61 02373	Cep OB3 # 41		★AB	22 54 49	+62 27.6	5	7.65	.008	0.63	.007	-0.43	.007	13	O7n	27,1011,1012,1015,1209
217050	+47 03985	HR 8731, EW Lac			22 54 52	+48 25.0	4	5.42	.025	-0.09	.011	-0.52	.029	9	B4 IIIep	154,879,1212,1363
217382	+83 00640	HR 8748			22 54 53	+84 04.7	5	4.73	.020	1.42	.016	1.70	.003	7	K4 III	15,1363,3016,4001,8015
		LS III +58 069			22 54 54	+58 28.5	1	11.00		0.67		-0.28		3		881
217062	+59 02607				22 54 54	+59 41.7	1	7.17		-0.05		-0.45		2	B9	27
		LP 933 - 101			22 54 54	-27 07.0	1	13.08		1.13		1.10		1		1696
216956	-30 19370	HR 8728			22 54 54	-29 53.3	9	1.16	.005	0.09	.005	0.10	.037	50	A3 V	3,15,1013,1020,1034*
		G 28 - 27			22 54 55	+07 39.8	1	17.20		0.10		-1.00		3		3060
	+46 03884				22 54 55	+46 55.1	1	9.17		0.37		0.14		2	F0p:	1601
217019	+3 04799	HR 8730			22 55 00	+03 32.5	2	6.27	.005	1.12	.000	1.09	.005	7	K1 III	15,1071
217014	+19 05036	HR 8729			22 55 00	+20 30.0	13	5.47	.020	0.66	.012	0.23	.028	64	G2.5 IVa	1,3,15,101,116,1197*
		BPM 44710			22 55 00	-46 33.	1	13.97		0.47		-0.29		3		3065
217004	-26 16395	IDS22523S2638		AB	22 55 05	-26 22.3	1	7.37		0.68				4	G3 V	2034
216989	-45 14879				22 55 06	-45 25.5	5	7.71	.004	0.36	.004	0.05	.008	33	F3 IV/V	278,1075,1770,2011,2024
		LS III +57 106			22 55 09	+57 43.8	1	10.90		0.54		-0.41		3		881
217090	+38 04903				22 55 18	+39 07.3	1	7.43		1.10		0.96		2	K2	1601
217101	+38 04904	HR 8733			22 55 22	+39 02.5	13	6.18	.017	-0.15	.005	-0.81	.017	13	B2 IV-V	15,154,379,1174
	+64 01735				22 55 23	+65 01.8	1	9.28		1.89				2	K5	369
		Cep OB3 # 42			22 55 29	+62 32.1	1	11.44		0.60		0.12		1	A0	1015
217160	+55 02858				22 55 33	+55 43.8	1	8.13		0.37		0.00		3	F8	1502
217174	+61 02374	Cep OB3 # 43		★AB	22 55 34	+62 05.4	1	8.54		0.47		0.39		1	A3 V	1015
		CS 22893 # 10			22 55 34	-12 12.8	1	14.74		0.59		-0.06		1		1736
217047	-51 13438				22 55 34	-51 23.7	2	7.82	.004	0.84	.004	0.43		9	G6/8 IV	1673,2012
	+39 04980				22 55 36	+40 24.5	1	9.99		0.44		-0.03		2		401
240133	+55 02859				22 55 36	+55 34.3	1	8.90		0.52		0.01		3	F5	1502
		VES 949			22 55 37	+58 33.1	1	13.15		2.81		1.02		1		8113
217294	+77 00879	IDS22546N7757		A	22 55 39	+78 13.7	1	7.99		0.92		0.56		5	G8 IV	3026
217294	+77 00879	IDS22546N7757		B	22 55 39	+78 13.7	1	8.96		0.57		0.04		5		3026
217107	-3 05539	HR 8734			22 55 41	-02 39.8	2	6.15	.005	0.74	.000	0.42	.005	7	G8 IV	15,1071
		G 128 - 25			22 55 44	+27 56.1	1			1.37		1.17		1		1620
217084	-46 14497	IDS22528S4603		AB	22 55 44	-45 47.3	2	7.67	.005	0.64	.005	0.06	.058	6	G3 V	173,214
217096	-36 15650	HR 8732			22 55 48	-35 47.4	3	6.12	.009	0.58	.000	0.05		8	F7 V	15,2028,3053
217131	-2 05858	HR 8735			22 55 49	-01 40.7	3	6.37	.013	0.35	.005	0.01	.013	9	F0 V	15,1256,3037
217143	+6 05091				22 55 52	+07 21.8	1	7.96		1.17		0.87		1	K2	3048

Table 1   1211

HD	DM	Other Id	N	Rem	α_{1950}	δ_{1950}	S	V	σ_V	B–V	σ_{B-V}	U–B	σ_{U-B}	n	Spectrum	References
217108	−34 15843				22 55 52	−33 42.6	1	9.15		0.53		0.03		26	G0 V	978
		CS 22893 # 41			22 55 54	−10 05.6	1	14.86		0.01		−1.20		1		1736
217123	−28 18011				22 55 57	−28 25.1	1	8.12		0.50				4	F6 V	2012
217165	+9 05140				22 55 59	+09 33.5	1	7.66		0.59		0.11		3	G0	1733
217224	+67 01485	GT Cep			22 55 59	+68 08.4	2	8.16	.037	0.33	.004	−0.52	.016	7	B5 II	555,1768
217166	+8 04973	HR 8737	★	AB	22 56 03	+09 05.5	5	6.44	.013	0.64	.006	0.14	.023	13	G2 V +G4 V	15,254,1256,3077,8015
		LS III +57 109			22 56 05	+57 44.2	1	11.91		0.57		−0.36		3		881
	+62 02142	Cep OB3 # 44		A	22 56 06	+62 05.7	1	9.04		0.40		−0.36		4	B3 V	1015
	+62 02142	Cep OB3 # 44		B	22 56 08	+63 05.8	1	10.35		0.46		−0.16		2	B5 V	1015
		HA 199 # 828			22 56 08	−60 18.6	1	13.42		0.73		0.21		7		1499
217186	+6 05092	HR 8738			22 56 11	+07 04.4	2	6.32	.005	0.06	.000	0.01	.005	7	A1 V	15,1256
217110	−60 07621				22 56 12	−60 25.5	1	7.38		1.39		1.64		8	K4/5 III	1704
	+62 02143	Cep OB3 # 45			22 56 14	+63 06.7	1	10.56		0.43		−0.04		2	F3 V	1015
217295	+72 01079				22 56 14	+72 52.0	1	6.53		1.07		0.92		2	K0	985
		GD 245			22 56 22	+24 59.7	1	13.68		0.04		−0.91		3	DA	308
217202	+16 04847				22 56 23	+16 47.3	1	9.20		0.54		0.07		3	G0	196
217226	+49 04003				22 56 23	+50 25.8	1	7.42		1.46		1.74		2	K5	1601
217172	−45 14885				22 56 23	−45 27.5	5	7.28	.005	0.14	.003	0.09	.005	15	A3 V	278,1075,1628,1770,2011,2024
217227	+43 04355				22 56 29	+43 34.3	1	7.16		−0.06		−0.61		4	B2:V	1174
		HA 199 # 835			22 56 33	−60 18.9	1	11.92		0.96		0.68		9		1499
	+57 02659				22 56 36	+57 42.4	1	10.50		0.72		0.19		2		1375
217297	+62 02146	Cep OB3 # 47	★		22 56 36	+63 26.3	1	7.41		0.32		−0.56		4	B1.5V	1015
		AA45,405 S152 # 1	★		22 56 37	+58 30.9	2	12.16	.025	0.87	.070	−0.31		4		725,881
		WLS 2256 55 # 5			22 56 38	+57 42.2	1	11.98		0.41		0.29		2		1375
	+42 04558				22 56 40	+42 31.1	1	10.02		0.13		−0.03		2	A0	401
217312	+62 02147	Cep OB3 # 46	★	A	22 56 41	+62 48.5	4	7.41	.010	0.38	.008	−0.54	.010	11	B0.5V	27,401,1012,1015
217312	+62 02147	Cep OB3 # 46	★	B	22 56 41	+62 48.5	1	10.33		1.19		0.82		1		1015
217232	+10 04859	HR 8739	★	AB	22 56 42	+11 27.7	1	5.75		0.32		0.08		3	A8 V +F6 V	176
217231	+11 04913	G 67 - 38			22 56 48	+11 55.6	1	8.38		0.58		0.01		1	G0	333,1620
		Cep OB3 # 48			22 56 49	+63 28.6	1	11.19		0.41		0.34		2	B9 V	1015
	−21 06343				22 56 51	−20 49.6	1	10.95		0.56		−0.06		2		1696
217236	−30 19383	HR 8740			22 56 52	−29 43.8	1	5.50	.004	0.26	.006	0.21		16	F0 V	3,15,2012
217264	+0 04950	HR 8742	★	AB	22 56 54	+00 41.7	2	5.42	.005	0.98	.000	0.84	.005	7	K0	15,1071
217251	−13 06318	HR 8741			22 56 58	−13 20.4	2	6.10	.018	1.46	.005	1.76		5	K5 III	2007,3005
217314	+51 03514	HR 8744			22 57 00	+52 23.1	1	6.30		1.42				2	K2	71
	+68 01345	G 241 - 32	★	A	22 57 01	+68 45.5	3	8.76	.009	0.78	.002	0.40	.014	7	K0 V	22,1003,3026
217315	+51 03515				22 57 02	+52 02.1	1	6.96		−0.03		−0.40		1	B8 V	1716
217284	+15 04740				22 57 03	+16 15.2	1	9.57		0.56		0.14		3	G0	196
217348	+59 02615	HR 8745			22 57 06	+59 32.8	2	6.42		0.04	.014	−0.25	.000	4	B9 III	27,1079
		LB 1515			22 57 06	−46 30.	1	12.79		0.00		−0.02		5		45
	+29 04828	IDS22547N2933		A	22 57 08	+29 49.2	1	9.41		0.72		0.25		3	G5 V	3026
		LS III +58 070			22 57 09	+58 28.6	3	11.30	.018	0.46	.008	−0.51	.015	6	B0 V	342,483,684
	−12 06393				22 57 16	−11 38.9	3	10.56	.004	1.43	.025			7	M2	158,1619,1705
217271	−45 14889				22 57 19	−45 12.4	2	9.32	.000	0.25	.000	0.03	.000	4	A4 V	1097,3014
	+62 02151	Cep OB3 # 49			22 57 20	+62 31.8	1	9.82		0.22		0.11		4	A2p	1015
217337	+24 04694				22 57 21	+25 22.5	1	8.75		0.38		−0.07		3	A5	833
	+40 04962	G 190 - 5			22 57 22	+40 47.8	1	10.99		1.20		1.26		2	K5	7010
	+61 02380	Cep OB3 # 51			22 57 22	+61 42.3	1	9.14		0.17		0.01		2	B9 V	1015
		LS III +58 071			22 57 24	+58 31.8	1	12.12		0.43		−0.52		2		881
	+62 02150	Cep OB3 # 50	★		22 57 24	+62 36.0	1	9.80		0.46		−0.22		4	B3 V	1015
217303	−25 16220	HR 8743			22 57 24	−25 25.9	2	5.66	.004	1.26	.002	1.20		6	K0 II	58,2009
		G 189 - 55			22 57 28	+35 42.0	1	11.32		1.23		1.32		2	M0	7010
217305	−35 15606				22 57 28	−34 55.5	1	9.72		0.39				2	F0 V(w)	1594
217327	−28 18031				22 57 30	−28 11.5	1	8.67		0.49				4	F5 V	2012
	+60 02465				22 57 31	+60 52.3	1	9.24		0.23		−0.20		3	A2	27
		Cep OB3 # 52			22 57 33	+62 25.1	1	11.87		0.40		0.33		1	A0	1015
		WLS 2300 35 # 10			22 57 38	+34 21.4	1	11.30		0.52		0.05		2		1375
		WLS 2300 75 # 5			22 57 38	+77 23.5	1	13.01		0.53		0.00		2		1375
217357	−23 17699				22 57 38	−22 47.6	2	7.90	.005	1.40	.005	1.24	.000	5	K5/M0 V	1013,3072
		VES 950			22 57 40	+62 28.4	1	12.27		0.90		0.45		2		8113
		Per sq 1 # 9			22 57 42	+58 40.	1	12.36		0.57		−0.33		1		881
217358	−26 16419	HR 8746			22 57 42	−25 53.7	1	6.29		1.13				4	K1 III	2009
217344	−34 15853	TZ PsA	★	AB	22 57 42	−34 00.7	1	8.49		0.72		0.23		1	G5 Vp	1641
		LS III +57 112			22 57 43	+57 29.5	2	10.55	.000	0.81	.005	−0.23	.005	5		405,483
217463	+62 02152	Cep OB3 # 54	★		22 57 43	+62 30.5	3	8.99	.016	0.54	.005	−0.34	.007	8	B1.5Vn	27,1012,1015
	+63 01911	Cep OB3 # 53			22 57 43	+63 33.8	1	10.72		0.65		0.53		2	A7 V	1015
		WLS 2256 55 # 7			22 57 47	+52 39.9	1	12.07		0.48		0.03		2		1375
217379	−26 16420	IDS22550S2651		AB	22 57 49	−26 34.9	2	9.62	.030	1.33	.016			7	K5 (V)	1619,1705
		Per sq 1 # 10			22 57 54	+59 07.	1	11.84		0.60		−0.40		2		881
		Per sq 1 # 11			22 57 54	+60 25.	1	11.68		2.06		0.71		2		881
	+62 02153	Cep OB3 # 55			22 57 55	+63 26.2	1	9.98		0.28		0.14		2	B9.5V	1015
		LS III +59 033			22 57 56	+59 03.2	1	10.23		0.55		−0.33		3		881
217364	−53 10382	HR 8747			22 57 56	−53 01.4	6	4.11	.005	0.97	.010	0.71	.008	19	G8/K0 III	15,1075,2012,2024*
	+41 04660				22 57 58	+42 07.0	1	10.13		0.04		−0.17		2		401
217476	+56 02923	HR 8752, V509 Cas			22 57 58	+56 40.6	7	5.03	.091	1.40	.083	1.14	.093	12	G5	1,15,1616,3016,6009*
217394	−36 15666				22 58 02	−35 44.3	1	9.75		1.05		0.99		3	K3/4 V	3072
	+58 02520	LS III +59 034			22 58 03	+59 03.4	1	10.94		0.62		−0.40		3		881
217454	+34 04817				22 58 04	+35 30.2	1	8.00		1.19		1.24		2	K2	1733
217428	−0 04443	HR 8750			22 58 04	−00 05.0	2	6.20	.005	0.89	.000	0.60	.000	7	G4 III	15,1071
217490	+58 02521	LS III +59 035	★	AB	22 58 05	+59 21.1	2	8.73	.000	0.79	.004	−0.28	.004	12	B0.5Ia	27,1012

HD	DM	Other Id	N Rem	α_{1950}	δ_{1950}	S	V	σ_V	B–V	σ_{B-V}	U–B	σ_{U-B}	n	Spectrum	References
		LS III +60 030		22 58 08	+60 39.5	1	10.86		1.01		-0.02		3		881
217459	+2 04594	HR 8751		22 58 10	+02 44.7	4	5.83	.008	1.34	.003	1.55	.013	13	K4 III	15,1256,1355,3077
		G 67 - 39		22 58 10	+22 01.7	1	13.28		1.36		1.09		1		333,1620
217403	-51 13446	HR 8749		22 58 12	-51 13.1	3	5.67	.004	1.41	.010			11	K3 III	15,1075,2032
		CS 22893 # 11		22 58 15	-10 51.5	1	14.54		0.43		-0.17		1		1736
217447	-20 06523			22 58 15	-19 36.9	2	7.49	.005	0.98	.009	0.54	.028	4	G6/K0 IIIp	565,3048
		G 242 - 19		22 58 16	+75 22.9	1	13.20		0.70		-0.01		2		1658
		LS III +57 113		22 58 17	+57 42.5	1	10.87		0.33		-0.50		2		881
217491	+44 04302	HR 8755		22 58 18	+45 06.4	1	6.50		0.29		0.28		2	A3 V	1733
217691	+79 00759	IDS22574N7948	A	22 58 18	+80 04.5	1	7.10		1.22		1.14		2	K2	985
217477	+30 04859	HR 8753	⋆ AB	22 58 19	+30 48.9	1	6.60		-0.04		-0.35		2	B9 p Mn	220
240155	+55 02872			22 58 24	+56 18.1	1	9.38		2.05				3	K7	369
		Cep sq 1 # 116		22 58 32	+59 09.7	1	10.37		0.42		-0.16		3		27
		CoD -55 09220		22 58 32	-54 40.5	1	12.33		0.86		0.25		4		3073
	+62 02154	Cep OB3 # 56	⋆	22 58 33	+63 14.9	1	9.33		0.51		-0.42		2	B1 V	1015
217511	+21 04866			22 58 34	+22 07.3	1	7.66		0.45		0.01		2	F5	1733
217543	+37 04744	HR 8758		22 58 35	+38 26.4	6	6.54	.018	-0.11	.012	-0.69	.022	17	B3 Vp	15,154,379,879,1174,1212
		LS III +58 072		22 58 36	+58 40.7	1	10.88		0.49		-0.46		3		881
217484	-29 18537	HR 8754		22 58 36	-29 07.3	2	5.54	.004	1.35	.002	1.51		6	K3 III	58,2007
		LS III +60 031		22 58 38	+60 20.0	1	11.19		0.50		-0.19		3		881
		Cep OB3 # 57		22 58 38	+62 34.2	1	11.96		0.64		0.83		1	B0	1015
217498	-23 17706	HR 8756		22 58 42	-23 03.6	1	6.28		0.13				4	A2 V	2007
		Cep sq 1 # 117		22 58 44	+58 52.4	1	11.05		0.29		0.17		2		27
217486	-45 14899			22 58 46	-45 15.5	1	9.04		1.01		0.82		3	K3 V	3072
		Cep OB3 # 58		22 58 47	+62 30.6	1	11.25		0.51		0.41		3	A2 V	1015
217531	-7 05910	HR 8757		22 58 48	-07 19.8	3	6.20	.004	1.41	.005	1.76	.015	10	K5 III	15,1071,3005
217515	-19 06384			22 58 48	-18 34.5	2	9.42	.010	0.36	.010	-0.09		3	F2 V	742,1594
	+62 02155	Cep OB3 # 59	⋆	22 58 54	+62 36.8	2	9.80	.010	0.69	.015	-0.14	.015	9	B2 IV	27,1015
217558	+15 04746			22 58 56	+16 07.7	1	9.06		1.01		0.72		3	G5	196
217563	-5 05910	HR 8759		22 58 56	-04 58.8	2	5.93	.005	0.99	.005	0.76	.005	7	K0	15,1071
217505	-60 07623			22 58 59	-59 43.9	1	9.14		-0.21		-0.83		8	B2 III/IV	1732
217617	+58 02526			22 59 01	+59 12.2	1	8.89		0.02		-0.33		2	A2	27
217577	+18 05089			22 59 05	+19 00.1	1	8.67		0.61		0.09		2	G2 V	1648
240157	+58 02527			22 59 06	+58 54.7	1	9.76		0.36		0.18		2	A0	27
		LS III +59 036		22 59 06	+59 40.5	1	11.48		0.72		-0.41		2		881
		Cep OB3 # 60		22 59 08	+62 45.9	1	11.24		0.54		0.39		1	A0	1015
217657	+62 02156	Cep OB3 # 61	⋆	22 59 12	+62 40.4	1	8.14		0.49		0.46		1	B0.5V	1015
		LB 1516		22 59 12	-48 16.	1	12.97		-0.25		-0.93		3	sdB	45
		JL 119		22 59 12	-71 29.	1	13.48		-0.27		-1.14		1	sdOB	132
		Cep OB3 # 62		22 59 13	+62 17.9	1	11.73		0.88		0.63		1		1015
		G 67 - 40		22 59 15	+11 33.1	1	10.66		0.75		0.22		1	K1	333,1620
217580	-4 05804	G 156 - 75		22 59 15	-04 06.9	3	7.46	.023	0.95	.005	0.77	.011	9	K4 V	158,196,3008
		CS 22888 # 10		22 59 15	-33 57.4	2	14.45	.010	0.40	.007	-0.19	.009	3		1580,1736
		Cep OB3 # 63		22 59 22	+62 32.8	1	11.55		0.72		0.83		1	B0	1015
217621	+7 04961			22 59 23	+08 23.3	1	8.63		0.34		0.05		3	F0	196
217673	+56 02927	HR 8761		22 59 23	+56 50.2	2	6.20	.004	1.50	.001	1.53		4	K2 II	1355,6009
		LS III +59 037		22 59 24	+59 27.9	1	11.32		0.77		-0.29		3		881
240160	+56 02928	LS III +56 114		22 59 28	+56 43.5	2	10.02	.000	0.48	.000	-0.52	.000	4	O9	1011,1012
	+61 02382	Cep OB3 # 64		22 59 28	+61 49.2	1	9.61		0.30		0.24		2	A1 V	1015
		CS 22893 # 30		22 59 30	-08 14.9	1	14.21		0.46		-0.16		1		1736
		LP 985 - 74		22 59 30	-38 08.1	1	11.11		1.03		0.90		1		1696
217659	+38 04911			22 59 32	+38 50.3	2	8.16	.010	0.10	.005	0.09	.005	5	A1 V	833,1601
217730	+62 02157			22 59 33	+63 01.4	1	7.25		1.44		1.73		3	K4 III	37
217650	+31 04827			22 59 34	+32 25.4	1	8.21		0.44		0.00		2	F5	1375
	+58 02529			22 59 34	+59 11.6	1	10.50		0.18		-0.10		2		27
217595	-45 14905			22 59 36	-45 34.4	2	7.19	.000	0.45	.005			8	F5 V	1075,2033
217675	+41 04664	HR 8762, o And	⋆ APB	22 59 37	+42 03.4	5	3.62	.008	-0.10	.007	-0.56	.020	17	B6 IIIpe +	15,379,1363,3016,8015
		LS III +60 032		22 59 37	+60 42.0	1	11.59		0.95		-0.16		3		881
217597	-50 13857			22 59 37	-49 57.8	1	7.74		0.86				4	G8 IV	2012
	+56 02929	LS III +56 115		22 59 38	+56 46.3	1	10.32		0.27		0.43		2	B2 V:nn	1012
	+22 04761			22 59 47	+23 27.4	1	8.60		1.42		1.72		1	K5	1746
217642	-37 15047	HR 8760	⋆ AB	22 59 47	-36 41.4	2	6.46	.000	0.93	.005	0.71		6	K1 III	404,2007
217668	-24 17457			22 59 48	-23 51.0	1	8.32		0.45				4	F5 V	2012
217715	+22 04762			22 59 54	+23 04.2	1	6.84		0.05		0.04		1	A0 V	1716
217716	+17 04853	IDS22574N1804	AB	22 59 55	+18 20.4	1	8.90		0.55		0.00		3	F9 V +G1 V	3030
217701	-7 05913	HR 8763		22 59 57	-06 50.6	3	6.13	.012	1.59	.010	1.89	.010	10	M2 III	15,1071,3035
	+56 02930	LS III +57 114		22 59 58	+57 16.6	1	9.68		0.37		-0.51		2	B1 IV	1012
		Per sq 1 # 12		23 00 00	+60 06.	1	12.88		0.69		-0.19		2		881
		MCC 856		23 00 01	+05 26.9	1	11.53		1.37				2	K7	1619
		Steph 2087		23 00 01	+07 29.4	1	13.45		1.30		1.32		1	K7	1746
	+59 02627	AS Cep		23 00 01	+59 32.9	1	9.99		2.66				2	M3	369
217752	+40 04971			23 00 03	+40 48.2	1	9.67	.005	-0.01	.000	-0.09	.032	9		401,1775
	+58 02530			23 00 03	+59 17.2	1	10.10		0.27		0.18		4		27
217796	+61 02384	NN Cep	⋆ A	23 00 04	+62 14.6	1	8.11		0.40		0.10		5	A5	1768
217703	-21 06354	HR 8764		23 00 04	-21 08.3	1	5.97		0.94				4	K0 III	2032
240165	+56 02931	LS III +56 116		23 00 05	+56 55.5	2	10.13	.000	0.47	.000	-0.53	.000	4	B7	1011,1012
	+62 02158	Cep OB3 # 65		23 00 07	+62 33.2	1	10.09		0.26		0.04		6	B9 V	1015
217768	+53 03059			23 00 09	+53 32.4	1	7.66		0.34		0.15		2	A5	1566
	-63 04838			23 00 09	-63 05.1	2	9.40	.000	0.85	.010	0.31		4	G5	742,1594
217754	+31 04829	HR 8765		23 00 10	+31 30.7	1	6.59		0.34		0.06		1	F2 IV	254

Table 1 1213

HD	DM	Other Id	N Rem	α_{1950}	δ_{1950}	S	V	σ_V	B–V	σ_{B-V}	U–B	σ_{U-B}	n	Spectrum	References
	+28 04510			23 00 17	+28 56.1	1	8.76		0.75		0.29		1	G2 V	3026
217782	+41 04665	HR 8766	★ AB	23 00 18	+42 29.3	5	5.10	.005	0.09	.004	0.11	.003	156	A3 Vn	196,379,879,1363,3016
		WLS 2300 35 # 5		23 00 19	+37 12.9	1	12.67		0.66		0.06		2		1375
240168	+55 02878	LS III +56 117		23 00 20	+56 20.6	1	9.17		0.55		-0.39		2	B1 III	1012
217817	+59 02629			23 00 20	+59 35.0	2	6.99	.010	-0.01	.010	-0.59	.010	5	B3 V	27,399
	+62 02159			23 00 21	+62 45.6	1	10.64		0.60		-0.17		2		27
217737	-34 15873	IDS22576S3402	AB	23 00 22	-33 46.2	1	8.85		0.50		0.03		27	F5 V	978
217811	+43 04378	HR 8768, LN And	★ A	23 00 28	+43 47.4	1	6.37		-0.02		-0.60		2	B2 V	542
217811	+43 04378	HR 8768, LN And	★ AB	23 00 28	+43 47.4	4	6.38	.020	-0.01	.009	-0.59	.007	12	B2 V +A1 V	15,154,542,1174
217811	+43 04378	IDS22582N4331	B	23 00 28	+43 47.4	1	9.88		0.08		-0.37		2	A1 V	542
217812	+40 04974			23 00 30	+40 55.8	1	8.74		0.11		0.06		2	A2	401
217833	+54 02900	HR 8770, V638 Cas	★ A	23 00 34	+54 58.0	2	6.50		-0.13	.033	-0.55	.005	2	B9 III Hewk	1079,8084
217833	+54 02900	IDS22584N5441	B	23 00 34	+54 58.0	1	10.37		0.27		-0.12		1		8084
217833	+54 02900	IDS22584N5441	C	23 00 34	+54 58.0	1	10.77		0.44		0.07		1		8084
		LP 701 - 86		23 00 34	-03 43.3	1	12.20		0.87		0.45		2		1696
217903	+71 01181			23 00 36	+72 27.7	1	8.09		0.11		-0.07		4	B9 V	206
240171	+56 02934	LS III +56 118		23 00 37	+56 52.4	1	9.90		0.18		-0.65		2	B1 V	1012
	+8 04982			23 00 38	+08 39.3	1	10.93		1.24		1.13		6	K0	327
217813	+20 05264			23 00 38	+20 39.0	2	6.64	.032	0.60	.003	0.15	.046	7	G5 V	3016,7009
217872	+62 02160			23 00 39	+63 04.3	1	6.82		1.32		1.45		2	K0	1502
217766	-43 15249			23 00 39	-43 20.8	2	7.75	.005	0.55	.005			8	F7 V	1075,2033
217814	+7 04963			23 00 42	+08 23.5	1	9.04		0.54		-0.02		3	G0	196
		NGC 7469 sq1 5		23 00 43	+08 49.2	1	12.33		0.61		0.19		3		327
217777	-39 14921			23 00 43	-38 42.2	1	7.25		1.58		1.86		4	K5 III	1673
217792	-35 15630	HR 8767, π PsA		23 00 44	-35 01.2	4	5.10	.007	0.29	.003	-0.01	.005	10	F0 V + F3 V	158,219,2024,2032
217825	-27 16160			23 00 56	-27 05.0	1	6.78		1.62				4	M1 III	2012
217855	+8 04984			23 00 58	+08 41.6	1	8.17		0.38		0.05		7	F0	327
		NGC 7469 sq1 4		23 01 00	+08 35.8	1	12.67		0.60		0.07		3		327
		LS I +62 060		23 01 00	+62 15.0	1	12.34		0.29		-0.39		2		41
217919	+62 02161	Cep OB3 # 66	★	23 01 01	+63 25.7	2	8.26	.015	0.63	.005	-0.29	.010	5	B0.5IIIn	1012,1015
217808	-45 14913			23 01 02	-44 51.9	1	9.30		0.75				5	G3/5wA/F	1594
217816	-46 14529	IDS22583S4642	AB	23 01 08	-46 26.2	2	8.11	.000	0.51	.005	-0.01		8	G0 V	158,1075
		CS 22893 # 15		23 01 09	-12 03.4	1	14.80		0.44		-0.15		1		1736
		NGC 7469 sq1 3		23 01 10	+08 37.4	1	13.30		0.68		0.32		4		327
217842	-42 16177	HR 8771		23 01 11	-41 45.0	1	5.80	.005	1.07	.003	0.95		6	K0 III	58,2007
217944	+57 02676	HR 8778		23 01 14	+58 17.7	1	6.43	.007	0.90	.003	0.52		5	G8 V	71,3016,6009
240175	+59 02630	IDS22592N5954	B	23 01 15	+60 10.6	5	9.49	.016	0.09	.004	-0.21	.014	10	B9.5V	27,150,542,1084,1211
240174	+56 02938			23 01 16	+57 20.8	1	8.65		0.73		0.30		5	K5	3026
		Per sq 1 # 13		23 01 18	+59 30.	1	11.97		0.77		-0.32		2		881
		LS III +59 038		23 01 18	+59 43.3	1	10.46		0.25		-0.48		3		881
		Mark 315 # 6		23 01 19	+22 25.8	1	12.43		0.64		0.16		1		899
217943	+59 02631	HR 8777	★ A	23 01 19	+60 10.5	6	6.72	.017	-0.02	.005	-0.64	.016	14	B2 V	27,150,154,542,1084,1211
217966	+61 02385	Cep OB3 # 67		23 01 19	+62 22.4	2	8.21	.000	0.04	.000	-0.35	.000	8	B7 V	1015,1015
217979	+62 02162	Cep OB3 # 68	★	23 01 19	+63 16.9	2	8.59	.000	0.35	.000	-0.50	.000	4	B1 V	1015,1015
217891	+3 04818	HR 8773		23 01 20	+03 33.0	8	4.53	.014	-0.12	.005	-0.48	.014	28	B6 Ve	3,15,154,1075,1212*
217906	+27 04480	HR 8775, β Peg	★ A	23 01 21	+27 48.7	7	2.46	.037	1.67	.020	1.95	.026	18	M2 II-III	15,1477,3055,6002*
		WLS 2256 55 # 9		23 01 21	+54 59.9	1	11.76		0.73		0.28		2		1375
217877	-5 05917	HR 8772	★	23 01 21	-05 03.9	5	6.68	.011	0.58	.000	0.05	.008	19	F8 V	15,1071,2018,2029,3077
	+52 03361			23 01 23	+53 15.0	1	9.57		1.82				2		369
		Per sq 1 # 14		23 01 24	+59 18.	1	12.38		0.53		-0.25		2		881
		G 28 - 31		23 01 24	-02 48.7	2	12.62	.020	0.53	.008	-0.20	.023	4		1620,1658
217884	-28 18060			23 01 24	-28 30.5	1	8.24		1.57		1.88		5	M2 III	1657
		G 67 - 41		23 01 25	+10 32.6	1	12.27		1.41		1.27		2	M2	1620
		Mark 315 # 1		23 01 26	+22 20.5	1	10.97		0.84		0.27		1		899
217980	+60 02473			23 01 27	+61 12.4	1	8.07		0.14		-0.34		3	A0	27
217926	+5 05123	HR 8776		23 01 29	+06 20.8	4	6.42	.014	0.39	.010	0.03	.009	10	F6 III	15,253,254,1256
	+57 02678	LS III +58 073		23 01 30	+58 02.4	1	9.79		0.48		-0.49		2	B0.5V	1012
		Mark 315 # 5		23 01 31	+22 25.4	1	13.11		0.79		0.14		1		899
		CS 22893 # 38		23 01 31	-09 12.6	1	16.18		-0.10		-0.37		1		1736
		CS 22888 # 14		23 01 31	-33 45.9	2	14.43	.010	0.40	.007	-0.23	.009	3		1580,1736
		LS III +60 033		23 01 34	+60 01.6	1	11.34		0.93		-0.16		3		483
217831	-69 03301	HR 8769		23 01 35	-69 05.5	3	5.51	.003	0.36	.003	0.11	.010	24	F3 III	15,1075,2038
		Mark 315 # 3		23 01 36	+22 21.	1	16.24		1.18		0.39		1		899
		Mark 315 # 4		23 01 36	+22 21.	1	16.44		0.81		0.13		1		899
		Mark 315 # 2		23 01 36	+22 21.3	1	15.95		0.87		0.40		1		899
	+58 02536	LS III +59 039		23 01 36	+59 26.1	2	9.96	.010	0.32	.000	-0.03	.020	4		27,881
218029	+66 01575	HR 8779		23 01 38	+66 56.4	2	5.25	.015	1.25	.010	1.40	.000	6	K3 III	1355,3016
		AO 988		23 01 42	+75 04.2	1	10.21		1.20		0.91		1		1722
		Smethells 123		23 01 42	-43 45.	1	11.57		1.20				1		1494
217902	-54 10197	HR 8774		23 01 43	-54 14.0	2	5.37	.005	1.45	.000	1.75		6	K5 III	2035,3077
		G 67 - 42		23 01 45	+10 37.5	1	11.96		1.39		1.12		1	K7	333,1620
236031	+53 03066			23 01 47	+53 55.7	1	8.99		0.25		-0.15		2	A0pe	379
		VES 951		23 01 48	+57 40.7	1	11.96		0.29		-0.13		2		8113
	+60 02474	IDS22596N6018	AP	23 01 51	+60 35.4	1	9.13		0.09		-0.12		2	A0	27
		CS 22888 # 2		23 01 51	-36 56.2	2	14.69	.010	0.43	.008	-0.26	.008	3		1580,1736
	+58 02537			23 01 55	+59 11.2	1	10.32		0.56		-0.01		2		27
218031	+49 04028	HR 8780		23 01 56	+49 46.8	6	4.66	.012	1.06	.005	0.90	.009	11	K0 IIIb	15,1355,1363,3016*
217984	+0 04961	G 157 - 9		23 01 58	+01 19.7	1	8.33		0.92		0.46		2	K0	3056
218066	+62 02163	Cep OB3 # 69	★ B	23 01 59	+63 07.9	1	12.29		0.58		0.19		2		1015
218066	+62 02163	Cep OB3 # 69	★ A	23 02 01	+63 07.6	3	7.63	.011	0.40	.008	-0.52	.000	11	B1.5Vp	588,1012,1015

HD	DM	Other Id	N Rem	α_{1950}	δ_{1950}	S	V	σ_V	B–V	σ_{B-V}	U–B	σ_{U-B}	n	Spectrum	References
	+58 02538			23 02 04	+59 11.2	1	10.23		0.20		-0.20		2		27
240181	+59 02637			23 02 04	+60 14.1	1	8.79		0.11		-0.08		2	B8	27
218067	+58 02540			23 02 05	+59 27.4	1	8.60		0.32		0.09		2	Am	27
240183	+59 02636			23 02 05	+59 49.1	1	9.76		0.22		-0.08		4	B5	27
		Smethells 124		23 02 06	-50 16.	1	11.84		1.34				1		1494
		G 128 - 30		23 02 07	+31 06.1	1	10.55		1.16		1.08		1	K5	1620
240182	+57 02682	LS III +58 074		23 02 07	+58 26.2	1	9.99		0.24		-0.29		4	B3	881
	+74 01000			23 02 07	+75 10.4	1	9.24		0.26		0.16		1	A5	1722
218033	+8 04988			23 02 09	+08 35.0	1	7.35		1.24		1.35		3	K5	196
	+41 04671			23 02 11	+41 32.1	1	10.51		0.64		0.13		2		401
218043	+30 04869			23 02 12	+31 02.3	1	6.76		0.37		0.00		2	F4 II	3016
217988	-43 15255	IDS22594S4334	AB	23 02 15	-43 17.8	2	7.70	.000	0.99	.009			8	K0 IV	1075,2033
218045	+14 04926	HR 8781		23 02 16	+14 56.2	15	2.48	.008	-0.04	.005	-0.04	.029	112	B9.5III	3,15,26,198,1006,1020*
218091	+55 02886			23 02 16	+56 11.2	1	7.96		1.90				2	M1	369
		G 233 - 40		23 02 18	+56 58.1	1	11.14		1.19		1.16		2	M0	7010
		JL 122		23 02 18	-71 19.	1	17.50		-0.10		-0.80		1		132
218094	+51 03536			23 02 23	+52 10.4	1	8.86		0.11		-0.04		2		1566
		WLS 2300 75 # 7		23 02 23	+72 54.4	1	12.27		0.74		0.09		2		1375
		G 67 - 44		23 02 24	+12 58.2	1	11.77		1.01		0.85		1	K3	333,1620
218093	+53 03067			23 02 24	+53 45.3	1	8.54		0.75		0.25		2	G5	379
		LS III +58 075		23 02 26	+58 24.4	1	10.72		0.63		-0.34		3		881
		CS 22893 # 32		23 02 27	-08 46.4	1	14.98		0.47		-0.19		1		1736
		VES 953		23 02 31	+58 45.7	1	12.96		1.08		0.06		1		8113
	+65 01846	G 241 - 41		23 02 32	+66 29.7	2	9.89	.004	1.40	.000	1.26		3	M1 V:	1017,3072
218060	-8 06018	HR 8782	★ AB	23 02 33	-07 57.8	4	5.42	.005	0.31	.003	0.04	.010	14	F2 IV +F0 V	15,1061,2024,3026
218061	-17 06661	HR 8783	★ A	23 02 34	-17 20.9	1	6.14		1.37				4	K4 III	2009
218101	+15 04760	HR 8784		23 02 38	+16 17.7	3	6.44	.004	0.83	.011	0.55	.010	5	G8 IV	15,1003,1355
	+58 02541	LS III +58 076		23 02 39	+58 35.9	1	10.72		0.97		0.51		3		881
217987	-36 15693			23 02 39	-36 08.5	4	7.36	.014	1.48	.012	1.17	.005	12	M2/3 V	219,1075,2033,3078
218081	-8 06019	IDS23000S0814		23 02 42	-08 01.5	1	7.06		1.20		1.16		4	G8 III	1657
218103	+0 04963	HR 8785		23 02 44	+01 02.2	4	6.39	.004	0.95	.005	0.71	.010	15	G9 III	15,1256,1355,1509
	+62 02166	Cep OB3 # 70	★	23 02 44	+63 04.9	1	9.36		0.48		-0.41		2	B1 V	1015
218219	+74 01001			23 02 44	+75 14.8	1	7.67		0.05		-0.15		1	A0	1722
		LS III +59 040	★ V	23 02 46	+59 17.8	2	11.48	.115	0.62	.011	-0.31	.035	6		881,1768
218050	-48 14408			23 02 47	-48 08.5	1	8.65		0.71		0.29		4	G5 V	158
	-39 14931			23 02 53	-38 37.8	1	10.61		0.69		0.04		2		1696
240184	+57 02686			23 02 54	+58 11.8	1	8.84		1.00				35	G5	6011
		Per sq 1 # 15		23 02 54	+59 50.	1	13.40		0.71		-0.31		2		881
		JL 123		23 02 54	-77 37.	1	14.34		0.20		-0.51		2		832
	-35 15652	LP 985 - 130		23 02 58	-34 37.9	2	10.78	.005	1.24	.012	1.22	.030	2	K7 V	1696,3072
		CS 22888 # 1		23 02 58	-37 01.8	2	14.89	.010	0.46	.009	-0.18	.007	3		1580,1736
		JL 124		23 03 00	-78 09.	1	13.67		-0.14		-0.77		1		832
218153	+25 04870	KU Peg		23 03 03	+25 44.4	2	7.63	.005	1.13	.015	0.96	.025	7	G8 II	833,8100
218155	+14 04929			23 03 04	+14 41.3	2	6.78	.003	0.01	.012	-0.01	.046	10	A0	252,1499
		LS III +61 019		23 03 04	+61 59.4	1	10.04		0.50		-0.45		2		41,881
218170	+28 04518			23 03 05	+28 43.1	1	7.20		1.57		2.01		1	M2 III	1716
218195	+57 02689	LS III +57 116	★ AB	23 03 05	+57 58.3	3	8.36	.023	0.29	.032	-0.70	.000	5	O8	851,1011,1012
		Per sq 1 # 16		23 03 06	+59 11.	1	12.86		0.82		-0.12		2		881
		Cep OB3 # 71		23 03 07	+61 59.6	1	10.57		0.81		0.37		2	B5 Vep	1015
	+46 03929		V	23 03 08	+47 25.0	1	10.84		0.20		-0.19		3	B9 V	1768
218209	+67 01498	G 241 - 42		23 03 08	+68 08.7	5	7.49	.011	0.65	.005	0.08	.016	14	G6 V	22,908,1355,1658,3026
	+60 02476			23 03 12	+60 31.8	1	10.11		0.27		0.07		3		27
		G 28 - 34		23 03 12	-02 26.7	4	12.96	.012	1.00	.020	0.67	.027	12	dK0	308,419,1620,3056
	+28 04520	G 128 - 31		23 03 13	+28 39.7	2	10.05	.044	0.70	.016	0.20	.057	4	G8	1620,7010
		LS III +60 034		23 03 18	+60 16.8	1	11.87		0.61		-0.35		3		881
218229	+61 02388	Cep OB3 # 72		23 03 18	+62 05.1	1	8.19		0.23		-0.09		3	B8 III	1015
218199	+29 04855			23 03 21	+30 27.2	1	8.17		1.05		0.87		2	K1 II	8100
218183	-15 06345			23 03 23	-15 16.4	2	10.14	.005	0.07	.005	0.17	.044	11	B9 (III)	152,1775
		JL 125		23 03 24	-71 31.	1	17.20		0.00		-1.30		2		132
		Smethells 125		23 03 30	-66 18.	1	10.96		1.25				1		1494
		Cep OB3 # 74		23 03 31	+62 25.2	1	10.43		0.19		0.11		2	A0 V	1015
		LS III +58 077		23 03 33	+58 54.8	1	12.24		0.18		-0.12		3		881
		LS III +58 078		23 03 34	+58 55.1	1	11.53		0.61		-0.15		3		881
218201	+3 04826	G 28 - 35		23 03 36	+04 25.0	1	9.29		0.67		0.09		1	G5	333,1620
		G 190 - 9		23 03 43	+42 03.7	1	11.17		1.38		1.35		2		7010
	+61 02389	Cep OB3 # 73		23 03 46	+63 19.2	1	9.90		0.46		0.45		3	A2 III	1015
218235	+17 04866	HR 8788		23 03 49	+18 14.8	2	6.16	.005	0.44	.007	0.04	.027	5	F6 Vs	254,3026
218216	-35 15659			23 03 49	-35 31.3	1	10.00		0.16		0.09		4	A2 III/IV	152
218259	+36 05003			23 03 50	+36 33.0	1	6.76		1.16		1.11		3	K1 III	1501
218222	-9 06117			23 03 50	-09 01.2	1	9.00		0.56		0.07		3	G5	1700
218107	-75 01771	IDS23002S7534	AB	23 03 52	-75 16.8	1	9.24		0.36		-0.04		1	F5 V	832
		Per sq 1 # 17		23 03 54	+59 59.	1	12.92		0.66		-0.19		1		881
		VES 954		23 03 55	+57 18.1	1	11.22		0.40		-0.01		2		8113
218323	+63 01928	Cep OB3 # 75		23 03 56	+64 01.5	2	7.63	.010	0.60	.005	-0.38	.005	6	B0 III	1012,1015
218236	-9 06118			23 03 58	-08 54.9	1	8.59		0.51		0.00		5	F8	1700
218249	-0 04461	G 28 - 37		23 03 59	-00 28.0	2	9.33	.000	0.86	.000	0.57	.010	2	K0	1620,1658
		MCC 858		23 04 00	+63 39.0	1	10.87		1.44				1	M0	1017
218240	-24 17497	HR 8789	★ AB	23 04 00	-24 00.8	6	4.47	.009	0.90	.002	0.58	.008	15	G8 III	15,58,1075,2012,2024,8015
218261	+19 05058	HR 8792		23 04 03	+19 38.4	2	6.44	.014	0.54	.001	0.02		5	F7 V	71,3026
218227	-44 15149	HR 8787	★ AB	23 04 04	-43 47.5	6	4.28	.007	0.43	.004	0.15	.010	18	F5m δ Del	15,1075,2012,2024*

Table 1 1215

HD	DM	Other Id	N	Rem	α₁₉₅₀	δ₁₉₅₀	S	V	σ_V	B–V	σ_B–V	U–B	σ_U–B	n	Spectrum	References
		LS III +59 041			23 04 07	+59 59.3	1	10.38		0.42		-0.31		3		881
218342	+62 02170	Cep OB3 # 76		★	23 04 07	+62 56.6	4	7.38	.013	0.42	.005	-0.54	.005	11	B0 IV	27,1012,1015,8031
218242	-39 14936	HR 8790	★	AB	23 04 07	-39 09.8	3	5.61	.013	0.01	.000			14	A0 V	15,2012,2012
218292	+9 05158	R Peg			23 04 08	+10 16.4	2	7.93	.135	1.23	.050	0.34	.020	2	M4	8022,8027
218343	+56 02946	IDS23021N5702		AB	23 04 11	+57 16.5	1	9.17		1.88				2	M1	369
		Pal 13 sq # 1			23 04 12	+12 28.	1	18.57		0.85				1		1578
		Pal 13 sq # 103			23 04 12	+12 28.	1	17.07		0.98				1		1578
		Pal 13 sq # 123			23 04 12	+12 28.	1	15.46		0.66				1		1578
		Pal 13 sq # 13			23 04 12	+12 28.	1	16.58		0.95				1		1578
		Pal 13 sq # 18			23 04 12	+12 28.	1	19.65		0.72				1		1578
		Pal 13 sq # 29			23 04 12	+12 28.	1	21.13		0.53				1		1578
		Pal 13 sq # 6			23 04 12	+12 28.	1	17.08		1.16				1		1578
		Pal 13 sq # 7			23 04 12	+12 28.	1	20.51		0.34				1		1578
		WLS 2300 75 # 9			23 04 13	+75 10.1	1	11.94		0.64		0.18		2		1375
		LS III +59 042			23 04 14	+59 23.4	1	11.53		0.28		-0.16		3		881
218279	-22 06064				23 04 14	-21 43.6	1	9.04		0.92		0.69		1	K2 V	3072
	+60 02478				23 04 15	+60 33.6	1	10.50		0.34		-0.24		3		27
218325	+46 03931	IDS23020N4623		A	23 04 16	+46 39.2	1	7.71		0.06		-0.61		4	B3	1174
218255	-50 13885	HR 8791			23 04 17	-49 52.6	1	6.33		1.45				4	K4 III	2007
218344	+50 03946				23 04 18	+50 48.4	1	7.42		-0.11		-0.72		8	B2 V	1174
240189	+56 02947				23 04 18	+57 17.8	1	9.27		0.26		-0.26		2	K0	27
	+32 04584	G 128 - 32			23 04 19	+32 45.5	1	10.15		0.87		0.45		1		1620
		WLS 2300 35 # 9			23 04 22	+35 15.5	1	13.00		1.08		0.90		2		1375
218269	-51 13471	HR 8793	★	AB	23 04 22	-50 57.4	5	5.83	.005	0.48	.000	0.01		18	F6/7 IV/V	15,404,1075,2007,2007
		Steph 2095			23 04 23	+02 58.5	1	11.09		1.26		1.19		1	K7	1746
218108	-80 01064	HR 8786			23 04 24	-79 45.1	3	6.11	.004	0.13	.005	0.09	.005	13	A3/4 V	15,1075,2038
218294	-23 17748			A	23 04 26	-23 25.6	4	9.61	.016	1.28	.011	1.21	.016	10	K5 V	272,1518,1705,3072
218294	-23 17748			B	23 04 26	-23 25.6	1	13.63		1.60		1.22		3		1518
218329	+8 04997	HR 8795			23 04 29	+09 08.3	17	4.53	.013	1.57	.010	1.90	.019	95	M1 IIIab	1,3,15,116,369,1020*
218376	+58 02545	HR 8797		★	23 04 29	+59 09.0	11	4.85	.009	-0.03	.011	-0.86	.007	44	B0.5IV	15,154,369,1012,1119*
218354	+37 04769				23 04 32	+37 58.3	2	8.21	.005	0.65	.000	0.24	.010	5	G5	196,583
		VES 955			23 04 32	+57 50.2	1	12.91		0.87		-0.13		2		8113
	-24 17504	G 275 - 4			23 04 36	-24 08.4	2	12.09	.025	0.42	.025	-0.31	.050	5		1696,3062
218347	+15 04764				23 04 38	+16 00.0	2	8.90	.000	0.49	.012	0.00	.012	7	F7 V	1003,1775
218356	+24 04716	HR 8796			23 04 40	+25 11.9	7	4.75	.028	1.33	.020	1.15	.009	17	G8 Ib	15,1218,1355,1363*
218332	-9 06123				23 04 41	-09 05.1	1	8.04		1.31		1.45		3	K2	1700
218365	+34 04847				23 04 42	+35 21.9	1	6.34		1.19		1.13		2	K0	985
	+11 04935	G 29 - 3			23 04 45	+12 25.3	1	10.34		0.99		0.83		1	K3	333,1620
		LP 933 - 57			23 04 45	-32 36.4	1	12.35		1.01		0.94		1		1696
218350	-32 17395				23 04 46	-32 04.4	1	8.52		0.04		0.05		4	A0 V	152
236044	+53 03076	LS III +54 080			23 04 47	+54 28.6	1	9.60		0.20		-0.66		2	B1 V	1012
		VES 956			23 04 49	+58 41.0	1	11.76		0.27		-0.14		1		8113
	+40 04996				23 04 50	+40 49.4	1	9.82		-0.09		-0.72		1	B8	963
218393	+49 04045	KX And			23 04 51	+49 55.3	1	7.01		0.38		-0.37		2	Bpe	379
	+60 02481			V	23 04 51	+60 38.0	1	9.83		2.05				2		369
		LS III +60 035			23 04 51	+60 55.5	1	11.69		1.04		-0.10		3		881
218416	+52 03371	HR 8801			23 04 57	+52 32.7	1	6.11		1.05				2	K0 III	71
	+55 02899	LS III +55 110			23 04 58	+55 44.1	1	10.21		0.14		-0.67		2	B1 IIIp	1012
218407	+45 04147	HR 8800			23 05 00	+45 47.8	5	6.66	.004	-0.05	.003	-0.68	.004	59	B2 V	15,154,879,1174,1423
218396	+20 05278	HR 8799			23 05 01	+20 51.8	2	5.98	.010	0.25	.005	-0.04	.005	6	A5 V	253,3016
		SW Cas			23 05 02	+58 17.0	2	9.33	.010	0.99	.034	0.63		2	K0	851,1399
		G 28 - 38			23 05 03	+05 23.6	1	12.90		1.53		1.04		1		333,1620
218440	+58 02546	HR 8803			23 05 03	+59 27.4	2	6.39	.010	-0.01	.010	-0.64	.005	5	B2.5IV	27,154
	+74 01004				23 05 03	+74 54.7	1	9.61		0.09		-0.09		1	A0	1722
218395	+32 04587	HR 8798	★	AB	23 05 04	+32 33.3	2	6.14		0.12	.000	0.10	.005	5	A4 Vn	1049,1733
		Cep OB3 # 77		A	23 05 05	+63 03.4	1	10.48		0.63		-0.18		1	B2 V	1015
		Cep OB3 # 77		B	23 05 05	+63 03.4	1	11.48		0.74		0.18		1		1015
		Per sq 1 # 18			23 05 06	+60 50.	1	12.40		0.83		-0.10		2		881
		LS III +59 044			23 05 07	+59 24.4	1	11.09		1.33		0.18		3		881
218450	+61 02394	Cep OB3 # 78			23 05 09	+62 24.4	1	8.54		0.07		-0.14		1	B9 Vn	1015
218288	-74 02054	HR 8794			23 05 09	-73 51.4	3	6.14	.003	1.42	.000	1.68	.000	22	K3 III	15,1075,2038
		LS III +60 036			23 05 12	+60 54.1	1	11.98		0.60		0.05		3		881
		LP 877 - 65			23 05 12	-24 33.1	1	13.47		1.08		0.89		1		1696
218428	+29 04862				23 05 15	+29 47.0	1	7.57		-0.03		-0.10		3	A2 II-III	985
	+60 02484				23 05 17	+60 36.8	1	10.51		0.39		0.17		2		27
		LP 877 - 67			23 05 19	-26 18.8	1	13.27		0.48		-0.21		2		1696
218452	+45 04149	HR 8804	★	A	23 05 22	+46 07.0	5	5.32	.009	1.41	.003	1.70	.013	27	K5 III	15,1007,1013,4001,8015
218454	+29 04863				23 05 23	+30 10.1	1	7.21		1.52		1.77		2	K4 II	8100
218470	+48 03944	HR 8805			23 05 29	+49 01.4	4	5.68	.029	0.42	.024	-0.04	.019	10	F5 V	71,254,379,3053
240191	+57 02694				23 05 29	+58 21.0	2	9.74	.002	0.51	.027			37	F8	851,6011
		G 241 - 45			23 05 29	+68 23.8	1	12.45		1.54				4		940
		G 28 - 39			23 05 33	+03 03.3	4	10.85	.045	1.51	.010	1.21		3	dM1	1619,1620,1632,1705
		LP 581 - 80			23 05 34	+08 53.4	1	14.41		1.33		0.97		1		1773
218422	-25 16291				23 05 34	-25 00.7	1	10.20		1.12		1.02		2	K(4) (V)	3072
		G 28 - 40			23 05 35	+08 53.1	3	11.31	.022	0.74	.017	0.18	.020	4	dG5	308,1620,1658
		Per sq 1 # 19			23 05 36	+60 00.	1	12.35		0.65		-0.25		2		881
		G 190 - 10			23 05 39	+41 35.2	1	11.22		0.61		-0.09		1	G1	1658
218434	-29 18588	HR 8802			23 05 39	-29 05.6	3	5.60	.004	0.88	.002	0.54	.002	7	G8 III	58,2007,4001
	-16 06218	HK Aqr			23 05 41	-15 40.8	4	10.85	.018	1.42	.018	1.01	.010	10	K5	158,1619,1705,1775
		AO 993			23 05 43	+75 15.4	1	10.43		1.49		1.35		1		1722

HD	DM	Other Id	N Rem	α_{1950}	δ_{1950}	S	V	σ_V	B–V	σ_{B-V}	U–B	σ_{U-B}	n	Spectrum	References
218537	+62 02171	Cep OB3 # 79	⋆ AB	23 05 45	+63 21.8	5	6.25	.003	-0.02	.006	-0.59	.005	26	B2.5V	15,154,1015,8015,6001
		Per sq 1 # 20		23 05 48	+59 44.	1	13.37		0.52		-0.13		1		881
		VES 957		23 05 52	+59 10.8	1	13.86		1.39		0.28		1		8113
	+61 02396	Cep OB3 # 80		23 05 52	+62 23.3	1	9.55		0.17		0.12		2	A1 V	1015
218525	+43 04399	HR 8806		23 05 53	+44 17.5	1	6.56		0.17		0.20		2	A2 IV	252
218560	+63 01931	HR 8811		23 05 55	+63 57.1	1	6.21		1.10				2	K0	71
		G 128 - 34		23 05 57	+31 23.9	2	14.66	.005	1.54	.009	1.08	.023	4		203,3078
	+57 02698			23 06 00	+58 00.8	1	10.14		2.20				2		369
		G 275 - 8		23 06 01	-22 01.2	1	13.74		-0.06		-0.72		1		3062
218483	-43 15275			23 06 02	-42 43.2	1	8.82		0.84		0.47		4	K1 V	158
		CS 22893 # 23		23 06 03	-10 03.0	1	14.75		0.40		-0.16		1		1736
		KUV 23061+1229		23 06 05	+12 29.4	1	15.23		-0.09		-0.90		1	DA	1708
	+61 02397	Cep OB3 # 82		23 06 06	+62 19.7	1	9.34		0.08		-0.03		2	A0	1015
218527	+1 04686	HR 8807		23 06 07	+01 51.3	3	5.42	.031	0.91	.005	0.55	.019	9	G8 IV	15,1071,3016
		Cep OB3 # 81		23 06 07	+62 43.3	1	10.92		0.42		0.29		1	A2 V	1015
		G 157 - 20		23 06 09	-08 01.8	2	13.56	.015	0.98	.020	0.52	.119	5		419,3059
	+16 04876			23 06 16	+17 01.3	1	10.94		0.53		0.06		2		1248
218550	+10 04887	IDS23038N1025	AB	23 06 17	+10 41.2	1	7.60		0.45		-0.02		3	F6 V +F7 V	3030
218550	+10 04887	IDS23038N1025	CD	23 06 17	+10 41.2	1	11.16		0.60		0.09		3		3030
218658	+74 01006	HR 8819	⋆ AB	23 06 18	+75 07.0	5	4.41	.007	0.80	.021	0.44	.028	11	G3 III+F3 V	15,1118,3026,4001,8015
		Steph 2100		23 06 19	+06 17.4	1	10.89		1.40		1.21		1	K7	1746
		AJ89,1229 # 355		23 06 20	+06 17.6	1	10.90		1.37				1	dM0	1632
218549	+16 04877	DY Peg		23 06 22	+16 56.6	2	10.21	.180	0.26	.050	0.09	.050	2	F5	668,3026
		G 128 - 36		23 06 23	+26 44.9	1	11.92		0.90		0.57		1	K1	1620
218586	+38 04939			23 06 25	+38 53.8	1	7.38		1.58		1.98		2	K5 III	105
218600	+56 02953	LS III +56 126		23 06 25	+56 37.8	2	8.44	.050	0.90	.050	0.58		6	F2 Ib	6009,8100
		LS III +59 045		23 06 25	+59 46.5	1	11.92		0.46		0.00		2		881
218568	-3 05575	G 157 - 21		23 06 29	-02 49.3	2	9.26	.005	0.66	.002	0.06	.037	3	G0	1658,3056
218511	-68 03561			23 06 33	-68 00.1	1	8.39		1.20		1.14		1	K5 V	3072
218566	-3 05577	G 28 - 41		23 06 34	-02 31.8	4	8.60	.005	1.01	.008	0.95	.011	8	K3 V	22,1620,1658,3008
218609	+38 04940			23 06 35	+38 38.7	1	7.35		0.04		-0.01		2	A0 V	105
218587	+16 04878			23 06 36	+16 52.1	2	9.83	.014	0.58	.003	0.12	.008	12	G0	1248,1783
	+58 02549	LS III +58 079		23 06 36	+58 50.9	1	10.31		0.72		-0.26		2	B0 III	1012
		LP 880 - 521		23 06 36	-30 51.	1	13.37		0.86		0.60		1		3061
236049	+50 03958	FM Cas		23 06 41	+50 47.3	4	8.85	.022	0.85	.027	0.55	.01	0 5	A0	41,851,1399,6011
218594	-21 06368	HR 8812		23 06 47	-21 26.6	7	3.67	.014	1.22	.005	1.24	.009	28	K1 III	3,15,58,1075,1425*
	+64 01760			23 06 48	+64 34.5	1	8.33		0.09		-0.52		2	B8	401
		G 28 - 42		23 06 49	+06 44.9	2	12.24	.005	0.63	.015	-0.08	.015	3		333,1620,1658
	-0 04470	IDS23045N0012	AB	23 07 00	+00 29.7	3	9.94	.018	0.70	.000	0.01	.008	6	G2	1620,1696,3078
218634	+7 04981	HR 8815, GZ Peg	⋆ A	23 07 00	+08 24.4	3	5.03	.053	1.48	.009	1.15	.074	16	M4 IIIs+A2V	15,1071,3039
218674	+48 03950	KY And		23 07 01	+49 22.8	3	6.78	.035	0.00	.018	-0.57	.001	8	B3 IV	379,879,1174
218558	-67 03949	HR 8809		23 07 01	-67 07.8	3	6.45	.007	0.96	.005	0.72	.005	15	K0 III	15,1075,2038
		AO 994		23 07 03	+75 04.1	1	11.27		1.60		1.74		1		1722
218619	-28 18099	HR 8813		23 07 03	-28 21.6	2	5.87	.000	1.31	.002	1.42		6	K2 III	58,2007
		G 28 - 44		23 07 04	-02 14.5	1	12.66		1.54		1.18		1		1620
		Cep OB3 # 83		23 07 05	+62 50.6	1	11.36		0.30		0.20		1		1015
218572	-69 03306			23 07 08	-69 06.9	1	8.78		1.02		0.76		1	K4 V	3072
218630	-43 15281	HR 8814	⋆ AB	23 07 11	-43 07.9	3	5.81	.008	0.48	.004	-0.02		11	F5 V	15,1311,2012
		G 128 - 37		23 07 12	+32 56.4	1	12.75		1.17		1.01		1		1620
218639	-15 06360	HR 8816		23 07 12	-14 46.9	3	6.43	.014	0.01	.007	0.01	.005	10	A0 Vn	152,252,2007
	+74 01008			23 07 13	+74 52.5	1	9.14		0.18		0.12		1	A2	1722
218723	+64 01764			23 07 15	+64 56.4	1	6.68		-0.04		-0.59		2	B5	401
218640	-23 17771	HR 8817	⋆ AB	23 07 15	-22 43.7	7	4.69	.011	0.65	.009	0.39	.015	19	G2 IV + A2	15,58,1075,2012,2024*
		G 275 - 11		23 07 18	-21 13.6	1	11.76		0.46		-0.18		2		1696
218651	-23 17772			23 07 19	-23 22.6	1	8.43		0.38				4	F2 V	2012
	+65 01851			23 07 20	+65 45.0	1	8.62		1.91				2	M0	369
218655	-41 15163	HR 8818		23 07 23	-40 51.7	3	5.90	.027	1.61	.019	1.80	.032	13	M4 III	58,2007,3007
		Per sq 1 # 21		23 07 24	+60 53.	1	12.23		0.59		-0.25		2		881
	+60 02493	LS III +61 020		23 07 28	+61 11.7	1	9.42		0.84		-0.30		2	B0.5pe	1012
218700	+9 05170	HR 8821		23 07 30	+09 33.0	4	5.38	.016	-0.07	.009	-0.29	.013	10	B9 III	15,1079,1256,3023
		Smethells 126		23 07 30	-63 56.	1	11.39		1.36				1		1494
218725	+38 04945			23 07 31	+39 11.9	2	6.67	.010	0.23	.005	0.15	.000	5	A2 V	833,1601
	-26 16501	G 275 - 13		23 07 32	-26 12.2	1	9.75		1.56				1	M2.5	1705
218670	-45 14947	HR 8820		23 07 32	-45 31.1	7	3.90	.008	1.01	.007	0.87	.006	28	K1 III	15,1075,1075,2012*
240207	+59 02655	NGC 7510 - 2	⋆ AB	23 07 34	+60 25.4	1	9.12		0.34		0.13		1	A3	49
218717	+6 05124			23 07 35	+07 05.5	1	7.39		0.02		0.08		12	B9	8040
218753	+58 02552	HR 8822	⋆ A	23 07 35	+59 03.7	3	5.68	.025	0.35	.015	0.29	.014	9	A5 III	401,450,1119
		VES 958		23 07 37	+58 49.5	1	12.54		0.63		1.05		1		8113
218738	+47 04058	KZ And	⋆ B	23 07 39	+47 41.2	1	7.91		0.90		0.57		2	G5	3016
218739	+47 04059	IDS23053N4725	A	23 07 40	+47 41.3	1	7.17		0.64		0.14		2	G5	3016
218739	+47 04059	IDS23053N4725	AB	23 07 40	+47 41.3	1	6.71		0.70		0.25		3	G5	3016
218780	+58 02553	IDS23055N5847	C	23 07 42	+59 01.0	2	8.14	.000	0.14	.014	-0.38	.000	5		401,450
218742	+32 04596		V	23 07 45	+33 29.8	1	6.80		1.48		1.16		3	M2	1625
218766	+49 04059			23 07 49	+49 43.1	1	7.27		1.14		0.94		3	K1 III	1501
218693	-46 14560			23 07 49	-45 43.6	2	9.58	.049	0.92	.005	0.56		6	K1 V	2034,3072
218721	-27 16206			23 07 52	-26 38.2	1	10.36		0.17		0.13		4	A3/5 III	152
218767	+31 04859	IDS23055N3156	AB	23 07 53	+32 12.9	1	7.03		-0.04		-0.18		4	B9	3024
218767	+31 04859	IDS23055N3156	C	23 07 53	+32 12.9	1	9.48		0.20		0.10		4		3024
	+57 02700	LS III +57 118		23 07 53	+57 52.2	1	10.36		0.43		-0.49		2	B2	881
240208	+58 02554	PV Cas		23 07 53	+58 55.8	2	9.72	.000	0.17	.022	0.00	.004	12	B8	401,588

Table 1 1217

HD	DM	Other Id	N Rem	α_{1950}	δ_{1950}	S	V	σ_V	B–V	σ_{B-V}	U–B	σ_{U-B}	n	Spectrum	References
	−18 06261			23 07 56	−17 46.1	1	10.81		0.59		0.00		2		1696
		NGC 7510 - 6		23 08 01	+60 15.3	1	11.27		0.39		0.02		1		49
		NGC 7510 - 15		23 08 03	+60 20.9	1	12.80		1.36		0.80		1		49
		LP 822 - 31		23 08 03	−19 28.7	1	12.47		1.49		1.05		3		3078
		G 216 - 30		23 08 05	+50 21.5	1	10.97		1.20		1.17		3	K7	1723
		LS III +60 037		23 08 05	+60 49.2	1	12.22		0.61		-0.22		3		881
218759	−30 19460	HR 8823		23 08 05	−29 47.8	3	6.50	.004	0.28	.007			11	F0 V	15,2018,2030
218760	−30 19459	G 275 - 16		23 08 05	−30 11.3	1	8.74		0.96		0.74		2	K3 V	3072
		Case *M # 80		23 08 06	+60 58.	1	9.72		2.60				1	M3	138
218559	−81 01024	HR 8810		23 08 06	−81 11.1	3	6.41	.008	1.49	.005	1.84	.000	14	K4 III	15,1075,2038
		Steph 2105		23 08 07	+07 32.2	1	12.81		1.42		1.18		1	K5	1746
218804	+42 04592	HR 8825		23 08 08	+43 16.5	4	5.92	.015	0.44	.012	-0.06	.027	9	F5 IV	15,254,1003,3053
		LS III +59 047		23 08 08	+59 25.1	1	12.14		0.39		-0.40		3		881
	+30 04894			23 08 09	+30 54.2	1	9.00		1.51		1.74		3	K2	1733
		NGC 7510 - 14		23 08 09	+60 21.4	1	12.79		0.63		-0.29		1		49
		NGC 7510 - 9		23 08 10	+60 20.7	1	12.33		0.52		0.28		1		49
218792	+16 04882	HR 8824		23 08 13	+17 19.4	1	5.71		1.34		1.51		2	K4 III	252
218807	+18 05118	G 67 - 49		23 08 15	+18 38.4	2	8.56	.000	0.60	.005	0.05	.005	4	G5	516,1620
		NGC 7510 - 17		23 08 16	+60 18.6	1	13.78		0.91		-0.09		1		49
		LS III +60 038		23 08 16	+60 44.0	1	11.18		0.57		-0.34		3		881
218786	−21 06373			23 08 17	−21 05.3	1	10.03		0.77		0.44		3	G6 V	1696
	−0 04475	G 28 - 45		23 08 18	+00 09.0	1	8.99		0.85		0.51		1	G0	333,1620
		NGC 7510 - 8		23 08 20	+60 14.8	1	12.03		1.50		1.35		1		49
		NGC 7510 - 13		23 08 21	+60 15.4	1	12.76		0.69		0.38		1		49
		NGC 7510 - 20		23 08 21	+60 15.5	1	14.00		0.88		-0.02		1		49
		AO 996		23 08 22	+75 25.1	1	9.85		0.70		0.31		1		1722
		NGC 7510 - 22		23 08 29	+60 19.3	1	15.29		0.93		0.30		1		49
218810	−30 19462			23 08 29	−30 16.4	2	8.43	.005	0.39	.000	-0.12		3	F0 V	742,1594
218851		CCS 3180		23 08 30	+46 02.	1	10.07		1.37				1	R2	1238
218868	+44 04347	IDS23063N4459	A	23 08 31	+45 14.7	1	7.00		0.75		0.40		2	K0	1601
		NGC 7510 - 21		23 08 31	+60 16.5	1	15.26		1.11		0.29		1		49
		NGC 7510 - 19		23 08 32	+60 16.3	1	14.00		0.93				1		49
		CS 22888 # 39		23 08 38	−33 51.6	2	14.55	.010	0.39	.008	-0.23	.009	2		1580,1736
		NGC 7510 - 16		23 08 41	+60 17.6	1	13.70		1.17		0.22		1		49
218853	+4 04975	IDS23061N0428	A	23 08 42	+04 43.9	1	6.53		1.64		1.31		2	M2	1733
218892	+57 02705	IDS23066N5734	A	23 08 44	+57 50.2	1	8.83		0.99				1		1702
240214	+59 02662			23 08 44	+59 52.0	1	9.45		0.17		-0.05		2	A0	401
		NGC 7510 - 4		23 08 46	+60 17.0	1	10.41		0.48		-0.12		1	F2 V	49
		NGC 7510 - 23		23 08 46	+60 19.2	1	15.56		0.95		0.16		1		49
218857	−17 06692			23 08 47	−16 31.3	2	8.96	.009	0.72	.014	0.05		4	G5/8w(A)	742,955,1594
		CS 22888 # 31		23 08 48	−35 43.0	2	14.89	.010	0.42	.008	-0.24	.008	3		1580,1736
240216	+59 02663			23 08 49	+59 41.8	1	9.39		0.26		0.20		2	A2	401
		G 67 - 50		23 08 50	+19 25.1	1	12.44		1.39		1.14		1	M3	333,1620
		NGC 7510 - 7		23 08 51	+60 17.0	1	12.00		0.80		-0.25		1	B3 V	49
		KUV 23089+0942		23 08 52	+09 42.6	1	15.57		-0.23		-1.14		1	sdO	1708
218915	+52 03383	RT And		23 08 52	+52 47.2	2	7.20	.000	0.02	.000	-0.90	.000	4	O9.5Iab	1011,1012
		NGC 7510 - 5		23 08 55	+60 17.2	1	10.70		0.89		-0.25		1	B0.5V	49
218875	−21 06376			23 08 57	−21 16.5	1	9.17		1.15				1	C	1238
	+33 04662			23 08 58	+33 35.1	1	10.39		1.06		0.78		2	G0	1375
		Per sq 1 # 22		23 09 00	+59 21.	1	12.88		0.54		-0.31		2		881
		LS III +59 048		23 09 00	+59 28.8	1	10.48		0.65		-0.29		3		881
		LS III +60 045		23 09 00	+60 30.2	1	11.86		0.53		-0.40		3		881
		TonS 90		23 09 00	−27 28.	1	14.69		0.08		0.07		3		286
218941	+59 02664	NGC 7510 - 3		23 09 04	+60 18.2	1	9.68		0.84		-0.24		1	B1.5II	49
		NGC 7510 - 10		23 09 04	+60 18.5	1	12.68		0.73		-0.27		1		49
		LP 822 - 36		23 09 05	−17 40.3	1	12.44		0.43		-0.19		2		1696
		LP 880 - 594		23 09 06	−27 07.	1	15.69		1.61				1		3061
		NGC 7510 - 18		23 09 09	+60 18.9	1	13.92		0.79		-0.05		1		49
		KUV 23032+1254		23 09 10	+12 53.8	1	15.06		0.39		-0.09		1	NHB	1708
218918	+7 04991	HR 8826		23 09 13	+08 26.9	4	5.15	.006	0.13	.001	0.07	.005	10	A5 Vn	15,1256,1363,3023
240218	+59 02664			23 09 14	+59 58.9	1	9.71		0.83		-0.20		2	B9	1012
218902	−26 16521			23 09 14	−25 47.5	1	10.33		0.05		0.06		4	A1 IV/V	152
		NGC 7510 - 11		23 09 19	+60 18.5	1	12.69		0.81		-0.35		1		49
218949	+46 03964			23 09 22	+46 49.7	1	7.22		0.68		0.23		2	G5	1601
218935	+26 04580	HR 8827	★ A	23 09 23	+26 34.6	3	6.22	.016	0.94	.004	0.69	.006	9	G8 III-IV	1355,3016,4001
		PHL 2115		23 09 23	−05 19.6	2	14.78	.000	-0.05	.000	-0.69	.003	2		966,1736
		WLS 2136 85 # 8		23 09 25	+84 47.6	1	12.43		0.81		0.41		2		1375
218964	+50 03979			23 09 30	+50 55.1	1	9.46		0.14		0.06		2	A0	1566
		LS III +58 080		23 09 30	+58 40.1	1	11.35		0.31		-0.44		3		881
218997	+58 02560	V Cas		23 09 31	+59 25.7	1	8.43		1.73		0.76		1	M4	817
		NGC 7510 - 12		23 09 31	+60 19.4	1	12.73		0.85		0.03		1		49
218981	+56 02962			23 09 32	+56 50.2	1	9.13		0.30		-0.20		1	B8	1697
		LP 985 - 109		23 09 34	−37 38.7	1	12.55		0.47		-0.20		2		1696
	+26 04581	G 128 - 38	A	23 09 37	+26 39.1	2	9.52	.045	1.03	.023	0.78	.023	3	K3	1620,7010
	+26 04581		B	23 09 37	+26 39.1	1	12.84		0.65		0.24		2		7010
		AO 997		23 09 42	+74 54.1	1	9.68		0.57		0.10		2		1722
		CS 22888 # 44		23 09 47	−33 26.8	2	14.56	.009	0.58	.013	-0.09	.002	4		1580,1736
		UV2309+10S		23 09 48	+10 31.0	1	12.95		0.52		0.05		2		1732
		CS 22949 # 4		23 09 49	−06 32.8	2	13.09	.000	0.24	.000	0.17	.009	2		966,1736
		GD 246		23 09 50	+10 30.7	6	13.09	.008	-0.32	.004	-1.19	.008	112	DAwk	281,974,989,1732,1764,3060

HD	DM	Other Id	N Rem	α_{1950}	δ_{1950}	S	V	σ_V	B–V	σ_{B-V}	U–B	σ_{U-B}	n	Spectrum	References
218970	−21 06378			23 09 51	−21 23.1	1	9.75		−0.19		−0.73		3	B2/3 IV/V	1026
		WLS 2256 55 # 8		23 09 57	+54 56.1	1	12.39		0.25		0.12		2		1375
	+51 03557	G 216 - 32		23 09 58	+51 48.3	1	9.49		0.80		0.60		2	K0	7010
219063	+63 01949			23 09 58	+64 26.8	1	7.32		−0.03		−0.54		2	B5	401
		LS III +59 049		23 10 03	+59 19.7	1	10.79		0.34		−0.51		2		881
240221	+59 02668	NGC 7510 - 1	⋆ AB	23 10 03	+60 23.7	2	8.81	.044	0.36	.005	−0.18	.015	3	B5 III	49,881
219018	+1 04694	IDS23075N0209	AB	23 10 05	+02 25.0	2	7.71	.006	0.62	.005	0.11	.008	15	G2 V +G4 V	1499,3030
		LS III +60 050		23 10 10	+60 55.4	1	11.31		0.46		−0.38		3		881
240223	+57 02709			23 10 11	+58 29.7	1	9.61		0.22		−0.07		2	B8	401
		BSD 19 # 899		23 10 13	+61 05.6	1	10.85		0.34		−0.06		3	A0	1069
	+9 05179			23 10 14	+10 30.2	1	10.13		0.58		0.01		4	G0	1732
240224	+56 02963			23 10 14	+56 48.1	1	8.71		1.41		1.23		1	K5	1697
219080	+48 03964	HR 8830		23 10 15	+49 08.0	7	4.53	.013	0.29	.015	0.03	.012	23	F0 V	15,254,1118,1197,1363*
		AO 808		23 10 17	+57 12.8	1	9.90		2.01		2.27		1		1697
	+5 05147	G 28 - 47		23 10 18	+06 15.4	1	10.77		1.18		1.20		1	K5	333,1620
		Smethells 127		23 10 18	−49 04.	1	10.78		1.34				1		1494
		G 29 - 13		23 10 20	+11 51.0	1	11.19		1.10		1.00		1	K3	333,1620
240225	+56 02964			23 10 23	+56 42.0	1	9.44		0.37		0.27		1		1697
		LS III +60 051		23 10 25	+60 44.8	1	11.94		0.56		−0.41		2		881
219023	−50 13915	HR 8828	⋆ AB	23 10 25	−49 53.5	1	6.80		0.92				4	G8/K0 III	2035
		G 28 - 48		23 10 27	+01 31.9	3	11.10	.009	0.85	.019	0.45	.030	5	K1	1620,1658,1696
		VES 960		23 10 30	+58 32.9	1	11.78		0.75		−0.09		1		8113
219126	+64 01773			23 10 30	+64 31.8	1	7.32		0.01		−0.07		2	A0	401
	+60 02499	LS III +60 052		23 10 36	+60 57.5	2	10.11	.024	0.48	.000	−0.41	.033	6	B6 Ib	881,1069
219110	+28 04548	HR 8831		23 10 38	+29 10.2	1	6.35		0.93		0.61		2	G8 III	252
219109	+41 04714			23 10 40	+41 47.4	1	6.92		0.21		0.20		3	A2	1501
219135	+55 02919			23 10 40	+56 15.9	1	7.76		1.10		0.73		3	G0 Ib	8100
219127	+39 05033	IDS23083N3927	A	23 10 44	+39 43.8	1	7.60		0.28		0.18		4	A3	3024
219127	+39 05033	IDS23083N3927	B	23 10 44	+39 43.8	1	9.69		0.45		0.01		3		3024
219074	−49 14089			23 10 44	−49 04.4	1	8.11		1.29		1.27		5	K1/2 III	1673
	+20 05292	G 68 - 3		23 10 45	+20 39.6	2	9.74	.005	0.80	.000	0.32	.024	3	G5	1620,1658
		BSD 19 # 271		23 10 45	+59 24.7	1	11.95		0.43		−0.05		3	B7	1069
219113	+1 04695	SZ Psc		23 10 51	+02 24.2	1	7.40		0.83		0.32		2	K1 III	588
219134	+56 02966	HR 8832	⋆ A	23 10 52	+56 53.5	27	5.57	.009	0.99	.009	0.89	.010	229	K3 V	1,15,22,247,985,1004*
219116	−18 06271			23 10 52	−17 38.5	2	9.24	.024	1.02	.015	0.56	.019	3	G8 III(p)	565,3048
219104	−30 19482			23 10 54	−29 54.7	1	8.53		−0.10		−0.37		4	B8 V	152
219139	+10 04902	HR 8833	⋆ A	23 10 55	+10 47.6	3	5.73	.117	1.00	.009	0.91		3	K0	71,1569
	+34 04870			23 11 01	+35 29.3	1	8.65		1.10		0.97		3	K0	1625
	+60 02501			23 11 01	+61 25.4	1	10.38		0.16		−0.23		3	A0	1069
219209	+66 01596			23 11 01	+66 48.2	1	7.71		1.91				2	M1	369
		CS 22949 # 3		23 11 01	−06 47.1	2	15.00	.000	0.35	.000	0.11	.010	2		966,1736
		LS III +60 053		23 11 02	+60 26.0	1	10.33		1.62		0.74		2		881
219077	−63 04862	HR 8829		23 11 03	−62 58.0	5	6.12	.008	0.80	.011	0.35	.003	19	G8 V	15,1075,1311,2012,4001
	+58 02562			23 11 05	+58 52.6	1	9.77		0.22		−0.17		2		401
		Per sq 1 # 23		23 11 12	+59 20.	1	11.32		2.34		0.82		2		881
		AA45,405 S158 # 1	AB	23 11 12	+61 15.	2	11.86	.024	1.28	.039	0.05		3		342,725
219170	+38 04956			23 11 13	+38 58.8	1	8.33		0.36		0.16		3	A7 V	833
	+38 04955	G 190 - 15		23 11 16	+39 08.9	4	11.02	.016	0.66	.010	−0.14	.013	5	sdG3	979,1064,1658,8112
		G 190 - 15	AB	23 11 16	+39 08.9	1	11.37		1.60				1	F6	1705
240228	+56 02970	IDS23090N5649	A	23 11 16	+57 05.0	1	9.84		0.58		0.19		1		1697
219210	+58 02563			23 11 16	+58 40.6	1	8.15		0.40		−0.15		2	A2	401
219145	−53 10420			23 11 17	−52 57.8	2	9.44	.005	0.46	.010	−0.08		4	F5 V	742,1594
219172	+14 04952			23 11 18	+15 05.8	1	7.37		0.54		0.07		3	F8 V	1648
		BSD 19 # 282		23 11 21	+59 27.9	1	11.50		0.38		−0.09		2	B9	1069
		LP 1034 - 89		23 11 22	−39 55.0	1	13.18		0.58		−0.17		2		1696
240229	+55 02920		V	23 11 23	+56 27.9	1	9.56		0.19		0.17		6		1768
		Smethells 128		23 11 24	−57 06.	1	11.99		1.48				1		1494
		JL 129		23 11 24	−83 10.	1	13.45		0.36		−0.62		1		832
240230	+58 02564			23 11 29	+59 00.9	1	9.02		0.17		−0.39		2	B3	1069
219175	−9 06149	G 157 - 32	⋆ A	23 11 30	−09 11.8	7	7.56	.012	0.55	.012	−0.03	.012	22	F9 V	1,22,116,1028,1084*
219175	−9 06150	G 157 - 33	⋆ B	23 11 30	−09 12.2	5	8.19	.007	0.68	.010	0.10	.006	14		1,1028,1084,1696,3026
219211	+24 04733			23 11 32	+25 11.0	1	8.11		−0.15		−0.72		3	B9	833
	−20 06558		A	23 11 32	−19 55.1	1	10.62		1.46		1.22		2	K7 V	3072
	−20 06558		B	23 11 32	−19 55.1	1	13.80		1.60		1.13		2		3072
		AO 813		23 11 35	+56 51.2	1	10.97		2.19		2.57		1		1697
		Smethells 129		23 11 36	−60 32.	1	11.50		1.25				1		1494
		CS 22888 # 29		23 11 37	−36 01.8	1	14.72		0.43		−0.19		1		1736
219236	+30 04906			23 11 42	+31 03.0	2	8.90	.020	0.42	.020	0.04	.015	5	G0	196,583
219288	+57 02712			23 11 42	+57 33.9	1	8.40		0.25		0.11		2	A0	1375
219215	−6 06170	HR 8834		23 11 44	−06 19.1	10	4.22	.011	1.56	.007	1.90	.011	38	M2 III	3,15,58,369,1075,1425*
219286	+59 02673	LS III +59 050		23 11 48	+59 32.9	3	8.68	.024	0.64	.005	−0.43	.021	9	O8 II(f)	483,684,1069
		G 128 - 40		23 11 51	+33 29.7	1	13.04		1.23		0.86		2		1620
		G 157 - 34	V	23 11 51	−06 49.1	1	15.47		0.21				1	DC	1705
219218	−41 15194			23 11 51	−40 43.0	1	8.58		0.76				4	G6 V	2012
		WLS 2300 35 # 8		23 11 52	+35 14.7	1	11.48		1.01		0.62		2		1375
		CS 22949 # 2		23 11 52	−06 48.8	1	15.35		0.18		−0.65		1		1736
		G 935 - 50		23 11 52	−06 49.4	1	15.42		0.22		−0.63		4		1663
219287	+58 02565	LS III +59 051		23 11 53	+59 06.1	2	9.03	.072	1.06	.022	−0.17	.027	5	B0 Ia	1012,8100
		LS III +59 052		23 11 53	+59 58.2	1	10.66		0.47		−0.43		3		881
219291	+28 04555	HR 8838	⋆ A	23 11 56	+29 30.0	3	6.41	.014	0.45	.005	−0.01	.014	8	F6 IV	253,254,3037

Table 1   1219

HD	DM	Other Id	N Rem	α₁₉₅₀	δ₁₉₅₀	S	V	σ_V	B–V	σ_B–V	U–B	σ_U–B	n	Spectrum	References
219290	+49 04071	HR 8837		23 11 56	+50 20.7	1	6.31		-0.01		-0.04		2	A0 V	1733
		LS III +59 053		23 11 58	+59 13.6	1	11.25		0.44		-0.54		3		881
219221	-49 14097			23 11 58	-49 16.1	2	8.54	.000	0.40	.005	-0.08		3	F0 V	742,1594
	+60 02502			23 12 00	+61 18.3	1	10.13		0.29		0.22		3	A3 II	1069
219160	-75 01776	IDS23085S7504	AB	23 12 01	-74 47.9	1	8.94		0.47		-0.01		1	F7/8 V	832
219279	-11 06032	HR 8836	★ AB	23 12 04	-10 57.6	2	6.12	.000	1.50	.010	1.80		6	K5 III	2007,3005
	+59 02675			23 12 05	+60 11.3	1	10.34		0.70		0.14		2	B6 II	1069
	+71 01190			23 12 05	+71 38.2	1	8.52		1.53		1.58		2	K5	1733
219310	+23 04704	HR 8839		23 12 08	+23 49.8	1	6.34		1.17		1.23		2	K2 III	1733
219293	+18 05128			23 12 10	+18 59.5	1	8.77		0.48		-0.03		1	F8	1776
		Case *M # 81		23 12 12	-59 42.	3	9.94	.060	2.69	.030	2.53		6	M2 Iab	138,148,8032
240233	+59 02676			23 12 13	+60 29.8	1	9.42		0.15				2	B9 V	1119
219263	-41 15197	HR 8835		23 12 13	-41 22.6	4	5.78	.009	1.17	.011	1.22		12	K2 III	15,1075,2007,4001
		LS III +59 054		23 12 14	+59 38.2	2	11.40	.000	0.49	.004	-0.26	.020	5		881,1069
219249	-57 10232			23 12 14	-57 00.2	1	7.96		0.70				4	G6 V	1075
		LS III +59 055		23 12 19	+59 38.6	2	10.93	.010	0.60	.005	-0.42	.030	8		881,1069
240234	+59 02677	LS III +59 056		23 12 24	+59 33.8	4	9.50	.008	0.59	.017	-0.55	.004	8	B0	401,405,483,684
		Per sq 1 # 24		23 12 24	+59 46.	1	12.51		0.45		-0.37		3		881
		LS III +59 057		23 12 25	+59 16.9	3	10.68	.027	0.67	.009	-0.37	.024	6	B0 III	405,1012,1069
		CS 22949 # 1		23 12 27	-07 11.8	2	14.38	.000	0.42	.000	-0.20	.011	2		966,1736
219383	+53 03107			23 12 30	+53 39.6	1	9.05		0.10		-0.30		2		1566
		Per sq 1 # 25		23 12 30	+59 37.	1	12.42		0.59		-0.28		4		881
219361	+27 04517			23 12 31	+27 47.9	1	7.12		0.04		0.02		2	A0	252
	+60 02504		AB	23 12 32	+61 14.8	2	9.87	.015	0.20	.015	0.20		5	B8 V	1069,1119
219341	-37 15160			23 12 36	-36 44.4	1	10.95		0.07		0.09		4	A1 IV	152
		LS III +59 058		23 12 38	+59 38.9	3	10.91	.000	0.46	.005	-0.53	.023	9		405,483,1069
219339	-25 16354			23 12 39	-25 07.5	2	7.27	.011	0.18	.005	0.10		12	A5 IV	1499,2012
	+59 02678			23 12 41	+60 28.6	1	9.43		2.00				2		369
219353	-23 17809			23 12 41	-22 44.2	1	9.51		1.23		1.20		1	K1/2 III	788
		Per sq 1 # 26		23 12 42	+59 39.	1	12.52		0.51		-0.39		4		881
		GD 247		23 12 42	-07 15.7	1	14.27		0.49		-0.09		1		3060
	+59 02679	LS III +60 054		23 12 43	+60 27.9	1	10.44		0.66		-0.42		2		881
		G 157 - 35		23 12 43	-02 26.3	1	16.29		0.51		-0.37		3		538
219364	-14 06429			23 12 44	-14 17.3	1	7.95		1.10		1.00		3	K0 III	3016
		Ma 50 - 3		23 12 46	+60 11.4	1	15.46		0.97		0.70		1		992
		KUV 23128+1157		23 12 48	+11 56.8	1	17.83		-0.37		-0.74		1	DA	1708
		Per sq 1 # 27		23 12 48	+59 38.	1	13.06		0.66		-0.23		4		881
219485	+73 01023	HR 8844		23 12 49	+73 57.5	1	5.89		-0.02		-0.02		2	A0 V	1733
		Ma 50 - 1		23 12 50	+60 10.9	2	12.33	.005	0.45	.005	-0.36	.009	10	B1.5V	881,992
		Ma 50 - 2		23 12 50	+60 11.3	1	13.34		1.45		1.07		2		992
		CS 22949 # 9		23 12 52	-03 04.8	2	15.79	.000	0.18	.000	0.18	.008	2		966,1736
219418	+24 04737			23 12 56	+25 24.0	2	6.80	.005	0.84	.010	0.44	.015	5	G5 III	833,1625
219428	+51 03566			23 12 56	+52 04.7	1	8.25		0.60		0.13		3	G0	1566
		Ma 50 - 29		23 12 57	+60 10.3	1	14.81		0.91		0.48		2		992
		Ma 50 - 36		23 12 57	+60 11.8	1	12.72		0.68		0.19		3	F5 V	992
		Ma 50 - 28		23 12 58	+60 09.7	1	14.10		0.81		0.45		4		992
		LP 642 - 30		23 12 59	+01 22.5	1	12.86		0.57		-0.14		2		1696
	+59 02681	LS III +59 059		23 12 59	+59 37.0	2	10.32	.040	0.48	.000	-0.54	.005	5		401,881
		Per sq 1 # 29		23 13 00	+59 40.	1	12.04		0.67		-0.32		4		881
	+59 02682	Ma 50 - 11		23 13 00	+60 09.3	1	10.95		0.33		0.21		1	A0 V	1315
		Ma 50 - 63		23 13 00	+60 10.	1	11.90		0.58		-0.25		7	B2 V	992
		Ma 50 - 64		23 13 00	+60 10.	1	12.93		0.76		0.03		4		992
219402	-4 05852	HR 8840		23 13 00	-03 46.2	3	5.54	.009	0.06	.005	0.06	.005	14	A3 V	15,1071,1075
		Ma 50 - 31	★ E	23 13 01	+60 10.2	2	11.21	.004	0.64	.007	-0.27	.011	13	B1 III	881,992
		Ma 50 - 27		23 13 02	+60 09.7	1	13.40		0.72		0.28		4		992
219460	+59 02683	Ma 50 - 61	★ AB	23 13 02	+60 10.7	3	9.83	.026	0.61	.005	-0.29	.006	13	WN4.5 +B0	992,1012,1359
219391	-27 16244			23 13 02	-27 27.1	1	8.73		0.21		0.14		4	A(p SrEuCr)	152
		Ma 50 - 26		23 13 03	+60 09.6	1	13.90		0.66		0.06		4		992
		BSD 19 # 338		23 13 03	+60 13.0	1	12.32		0.46		-0.32		3	B7	1069
219409	-30 19491			23 13 03	-30 07.2	2	6.53	.009	1.08	.005	0.90		5	K0 III	2012,3077
		Ma 50 - 30		23 13 04	+60 09.7	1	12.42		0.49		-0.36		7	B2 V	992
		Ma 50 - 25		23 13 05	+60 09.7	1	13.79		0.55		-0.06		1		992
		Ma 50 - 32		23 13 05	+60 11.0	1	12.85		0.76		0.36		6		992
219420	+0 04982			23 13 06	+01 02.2	1	6.81		0.54		0.06		4	G0 II-III	8100
		Ma 50 - 23		23 13 06	+60 09.8	2	10.68	.005	0.46	.005	-0.43	.023	9	B0.5III	881,992
	+56 02972			23 13 07	+57 02.1	1	9.70		0.58		-0.01		1	B8	1697
		Ma 50 - 20		23 13 07	+60 08.6	1	12.71		0.74		0.30		2		992
		Ma 50 - 24		23 13 07	+60 09.4	1	14.04		0.60		0.11		2		992
	+56 02974			23 13 08	+57 02.1	1	10.16		0.29		-0.04		1		1697
		Ma 50 - 16		23 13 08	+60 08.1	1	12.94		0.56		0.27		2		992
		Ma 50 - 42		23 13 09	+60 13.1	2	12.84	.000	0.86	.005	0.03	.015	4		881,992
	+60 02507			23 13 10	+60 57.1	1	11.31		0.28		0.24		3	A0	1069
219410	-50 13936			23 13 13	-50 07.3	1	10.04		0.13		0.15		4	A1 IV	152
219430	-9 06155	G 157 - 37	★ BC	23 13 14	-09 21.1	3	9.17	.013	1.06	.009	0.90	.009	8		308,1084,3077
219449	-9 06156	HR 8841	★ A	23 13 16	-09 21.6	11	4.23	.011	1.11	.005	1.03	.012	31	K0 III	15,58,263,542,1075*
219477	+27 04521	HR 8842		23 13 19	+27 58.5	2	6.49	.005	0.85	.000	0.48	.005	5	G5 III	985,1733
	+61 02408	LS III +61 021		23 13 19	+61 48.4	2	9.72	.015	0.80	.010	-0.18	.024	6	B0 III:p:	1012,1069
		LP 986 - 18		23 13 19	-37 25.4	1	11.10		0.72		0.19		2		1696
219436	-39 14977			23 13 19	-39 32.5	1	8.35		0.10				4	A1/2 V	2012
		BSD 19 # 940		23 13 20	+60 58.9	1	11.22		0.35		-0.02		2	B9	1069

HD	DM	Other Id	N	Rem	α₁₉₅₀	δ₁₉₅₀	S	V	σ_V	B–V	σ_B–V	U–B	σ_U–B	n	Spectrum	References
		BSD 19 # 347			23 13 21	+60 11.6	1	10.55		0.47		-0.35		2	B5	1069
		KUV 23134+1117			23 13 22	+11 16.7	1	16.48		0.21		-0.73		1		1708
219523	+63 01955				23 13 22	+63 59.6	1	7.19		-0.02		-0.51		2	B5	401
		CS 22949 # 7			23 13 22	-04 56.1	2	13.67	.000	0.43	.000	-0.21	.012	2		966,1736
		G 157 - 38			23 13 29	-11 11.3	1	12.57		0.47		-0.26		2		1696
219487	+23 04712	HR 8845			23 13 30	+24 29.9	1	6.60		0.40		-0.05		1	F5 V	253
		Per sq 1 # 32			23 13 30	+59 40.	1	12.70		0.71		-0.16		3		881
	-64 04333				23 13 31	-63 43.5	1	9.60		1.20		0.93		1		565
		G 28 - 50			23 13 36	+06 28.4	1	13.14		1.63		1.26		1		333,1620
		Per sq 1 # 33			23 13 36	+59 08.	1	12.09		0.96		-0.07		3		881
		PHL 433			23 13 36	-02 06.9	6	12.96	.008	-0.23	.006	-1.05	.006	105	DAs	281,711,989,1298,1764,3016
		CS 22949 # 6			23 13 37	-05 02.0	2	13.96	.000	0.05	.000	0.09	.011	2		966,1736
219499	+17 04887				23 13 38	+17 59.1	1	7.58		1.01				2	K2	882
219537	+55 02929	IDS23114N5608	A		23 13 38	+56 24.2	1	7.85		0.02		-0.09		2	A0	401
219586	+70 01311	HR 8851			23 13 41	+70 36.9	2	5.56	.005	0.24	.005	0.12	.000	6	F0 IV	985,1733
		Per sq 1 # 34			23 13 42	+60 03.	1	13.47		0.60		-0.16		2		881
	+61 02411				23 13 46	+61 50.7	1	10.41		0.34		-0.06		4	B8 V	1069
	+5 05152				23 13 47	+06 19.0	1	9.37		1.68				1	M0	1532
	+59 02688	LS III +60 057			23 13 48	+60 02.2	1	11.00		0.63		-0.37		2	G5 IV	881
	+52 03407				23 13 49	+52 43.0	1	9.70		0.11		-0.16		2	A2	1375
219526	+17 04889				23 13 50	+17 34.2	1	8.43		0.54		0.00		1	F8	1776
219538	+29 04890	G 128 - 42			23 13 51	+30 23.8	2	8.09	.005	0.88	.005	0.58	.005	4	K2 V	1620,3026
		BSD 19 # 365			23 13 52	+59 42.1	1	10.66		0.27		0.04		2	A0	1069
		BSD 19 # 363			23 13 52	+60 04.3	1	10.94		0.62		-0.30		3	B3	1069
219507	-45 14982	HR 8846		★ A	23 13 53	-44 45.7	2	5.92	.002	1.05	.003	0.92		7	K1 III	58,2007
		LS III +59 060			23 13 56	+59 16.8	1	11.60		0.29		-0.47		3		881
	-37 15174				23 13 56	-36 51.6	1	10.48		0.94		0.67		2	K3 V	3072
219482	-62 06412	HR 8843			23 13 58	-62 16.4	3	5.66	.008	0.51	.008			14	F8 V	15,1075,2030
219545	-20 06568				23 14 00	-20 14.1	1	9.03		0.16		0.13		4	A0 V	152
		LTT 9447			23 14 00	-41 03.3	1	11.57		0.90				1		1705
		LP 1034 - 92			23 14 00	-41 03.6	1	11.57		0.90		0.52		2		1696
		LS III +59 061			23 14 01	+59 14.2	2	11.11	.018	0.60	.009	-0.32	.031	5		881,1069
219542	-2 05917	IDS23114S0208	A		23 14 01	-01 51.6	1	8.19		0.64		0.18		4	G2 V	3026
219542	-2 05917	IDS23114S0208	AB		23 14 01	-01 51.6	2	7.62	.034	0.66	.005	0.17		6	G2 V +G7 V	2012,3016
219542	-2 05917	IDS23114S0208	B		23 14 01	-01 51.6	1	8.56		0.69		0.20		3	G7 V	3032
		LS III +60 058			23 14 02	+60 11.0	1	10.66		0.50		-0.43		2		881
		LS III +59 062			23 14 04	+59 13.5	2	11.19	.027	0.61	.011	-0.31	.022	5		881,1069
	+60 02510				23 14 05	+61 20.9	1	9.11		0.53				3	F6 V	1119
219531	-41 15205	HR 8847			23 14 05	-41 28.1	3	6.47	.008	1.07	.010			11	K0 III	15,1075,2032
		LS III +59 063			23 14 08	+59 11.0	1	10.41		0.97		-0.16		3		881
219495	-67 03959	IDS23110S6727	B		23 14 10	-67 12.3	2	9.02	.019	1.13	.012	1.06		6	K2/4	2033,3008
		G 67 - 52			23 14 11	+19 21.0	2	11.16	.007	1.50	.031	1.18		5	M0	333,1619,1620
	+61 02412				23 14 12	+62 28.4	1	9.57		1.12		0.11		2	B5 Ib	1012
	+60 02511				23 14 13	+61 15.5	1	9.29		0.17				3	A0 V	1119
219509	-67 03960	IDS23110S6727	A		23 14 14	-67 11.2	2	8.71	.019	1.06	.005	0.96		6	K2/3 V	2012,3008
219576	-8 06076	HR 8850, Khi Aqr			23 14 15	-08 00.0	5	5.00	.064	1.61	.017	1.66	.077	22	M3 III	15,1071,1075,2024,3053
	+4 04988				23 14 16	+05 25.1	2	10.52	.018	1.50	.103			3	dK7	1619,1632
219612	+54 02941				23 14 16	+54 55.1	1	8.11		0.38		-0.08		3	F0	3016
219634	+61 02413	HR 8854, V649 Cas			23 14 17	+61 41.4	3	6.53	.005	0.23	.011	-0.67	.000	7	B0 Vn	401,1118,1733
		Per sq 1 # 35			23 14 18	+59 20.	1	12.55		0.48		-0.43		2		881
219556	-46 14601				23 14 21	-45 56.8	1	9.04		0.76				4	K0 V	2012
		LS III +59 064			23 14 22	+59 17.1	1	11.66		0.35		-0.55		3		881
		LS III +60 059			23 14 22	+60 12.4	2	11.63	.018	0.58	.000	-0.30	.027	5		881,1069
		Per sq 1 # 36			23 14 24	+59 29.	1	11.94		0.69		-0.34		3		881
219623	+52 03410	HR 8853		★ A	23 14 25	+52 56.6	2	5.58	.015	0.52	.012	0.01	.010	5	F8 V	254,3053
	+59 02689				23 14 25	+59 37.6	2	10.21	.010	0.19	.005	0.07		6	A0 V	1069,1119
		CS 22949 # 8			23 14 26	-03 37.0	1	14.15		0.46		-0.19		1		1736
		HA 91 # 516			23 14 27	+15 12.1	1	11.74		0.49		0.01		5	G3	271
		BSD 19 # 958			23 14 27	+60 45.3	1	11.74		0.69		-0.11		2	B5	1069
219617	-14 06437	IDS23119S1422	C		23 14 29	-14 06.4	1	16.46		1.65				1	A3	1705,3078
219583	-39 14983				23 14 29	-38 55.4	1	9.08		0.16		0.11		4	A3 V	152
219617	-14 06437	G 273 - 1		★ AB	23 14 30	-14 06.5	11	8.16	.013	0.47	.014	-0.20	.015	28	F(8)w	22,176,742,908,1003*
219571	-58 08062	HR 8848			23 14 31	-58 30.6	5	3.98	.005	0.40	.006	-0.03	.010	18	F3 IV/V	15,1075,2012,3026,8029
219615	+2 04648	HR 8852			23 14 34	+03 00.5	14	3.70	.011	0.92	.005	0.59	.010	53	G7 III	3,15,22,58,1003,1020*
219635	+50 04000				23 14 34	+50 44.5	1	9.21		0.29		-0.05		3	A2	1566
		LS III +59 065			23 14 34	+59 18.5	1	11.14		0.26		-0.58		3		881
	+61 02414	IDS23125N6115	AB		23 14 37	+61 31.6	1	9.08		0.58				2	F2 V	1119
	-16 06253				23 14 37	-16 18.7	1	10.59		0.94		0.80		1		1696
240244	+58 02571				23 14 38	+59 24.9	1	9.51		0.43				2	A3 V	1119
		LS III +60 060			23 14 38	+60 11.2	1	12.08		0.33		-0.46		3		881
		G 128 - 43			23 14 41	+31 27.3	2	11.30	.025	0.83	.005	0.50	.025	2	K1	1620,1658
240245	+59 02690				23 14 44	+59 40.6	2	9.24	.005	0.09	.005	0.02		5	B8 V	1069,1119
	+60 02512				23 14 44	+60 39.1	1	9.33		1.33		1.21		8	K2 V	1655
219637	+16 04895				23 14 45	+17 10.6	1	8.60		1.13				2	K0	882
219645	+36 05036				23 14 47	+37 29.3	1	8.23		1.24		1.07		2	K5	1375
		Per sq 1 # 37			23 14 48	+60 14.	1	13.61		0.48		-0.23		3		881
219656	+13 05086				23 14 50	+14 25.3	1	8.27		1.11				2	K0 III	882
219630	-42 16263				23 14 50	-42 27.8	4	10.40	.040	1.32	.009	1.21	.010	8	K3/5 V	158,1494,1705,3072
		LP 702 - 111			23 14 52	-03 36.8	1	10.86		0.48		-0.05		2		1696
		LS III +61 023			23 14 54	+61 36.8	1	10.23		1.00		-0.09		2	B0.5 Ib:	1012

Table 1 1221

HD	DM	Other Id	N Rem	α_{1950}	δ_{1950}	S	V	σ_V	B–V	σ_{B-V}	U–B	σ_{U-B}	n	Spectrum	References
		Feige 109		23 14 55	+07 35.7	1	13.77		-0.14		-0.92		1	sdB	974
219668	+44 04368	HR 8857		23 14 55	+44 53.5	3	6.44	.024	1.08	.009	1.01	.024	7	K0 IV	252,1355,3016
	+63 01962	LS III +63 025		23 14 56	+63 31.0	1	8.40		0.32		-0.56		2	B1 III	1012
	+8 05036			23 14 58	+09 24.9	1	9.72		1.13		0.98		2		3072
		G 67 - 53		23 14 58	+19 20.4	1	12.10		1.59		1.20		1	M4	333,1620
		G 190 - 17		23 15 00	+37 56.5	1	11.46		1.54		1.20		4		940
		Per sq 1 # 38		23 15 00	+59 30.	1	13.29		0.60		-0.40		3		881
		G 29 - 19		23 15 01	+06 07.3	1	12.53		1.55		1.28		1		333,1620
		LS III +59 066		23 15 02	+59 44.1	2	10.82	.075	0.47	.013	-0.44	.040	4	B0.5 V	1012,1069
219657	-2 05920	G 157 - 41	★ AB	23 15 03	-01 47.6	3	7.88	.022	0.68	.010	0.17	.014	11	G5	176,938,3016
219659	-12 06461	HR 8856		23 15 03	-11 59.2	2	6.33	.000	0.05	.000	0.00		8	A1 Vn	152,2009
		HA 91 # 371		23 15 04	+14 58.4	1	10.55		0.37		0.06		2		271
		BSD 19 # 420		23 15 04	+59 47.1	1	12.10		0.42		0.11		6	A0	1069
		BSD 19 # 423		23 15 04	+59 51.4	1	11.79		0.99		-0.13		6	F5	1069
		BSD 19 # 426		23 15 04	+59 56.2	1	12.37		0.50		0.22		4	B4	1069
		BSD 19 # 424		23 15 05	+59 26.2	1	11.44		0.66		0.14		1	A0	1069
219670	+15 04794			23 15 06	+16 25.7	1	8.04		0.99				2	K0 V	882
		Per sq 1 # 39		23 15 06	+59 21.	1	13.05		0.64		-0.22		3		881
		Per sq 1 # 40		23 15 06	+59 28.	1	12.11		0.40		-0.44		3		881
		BSD 19 # 59		23 15 11	+58 55.6	1	12.36		0.58		0.43		2	B8	1069
		Per sq 1 # 41		23 15 12	+59 35.	1	12.36		0.67		-0.36		3		881
		MWC 1080		23 15 12	+60 35.	1	11.58		1.20		0.43		1		351
		Per sq 1 # 42		23 15 12	+61 40.	1	12.65		0.67		-0.28		2		881
		HA 91 # 846		23 15 13	+15 35.5	1	11.01		1.03		0.85		5	G8	271
	+63 01964	LS III +63 026		23 15 14	+63 50.9	1	8.46		0.71		-0.38		2	B0 II	1012
		NGC 7582 sq1 5		23 15 14	-42 40.9	1	12.61		0.66		0.02		3		1687
219685	+15 04796			23 15 16	+16 16.5	2	7.71	.015	1.42	.010	1.66		4	K3 III	882,3077
	+60 02513			23 15 16	+61 08.1	1	10.56		1.15		0.99		2	G8	1069
219644	-68 03567	HR 8855		23 15 16	-67 44.7	3	6.13	.003	1.36	.004	1.58	.000	29	K2/3 III	15,1075,2038
		BSD 19 # 427		23 15 17	+59 44.1	1	12.13		0.33		-0.24		2	A1	1069
	+59 02692	LS III +59 067		23 15 17	+59 58.5	3	9.85	.015	0.36	.005	-0.52	.021	8	B7 II	401,881,1069
		LS III +60 061		23 15 18	+60 13.3	2	11.14	.013	0.44	.000	-0.42	.027	5		881,1069
219688	-9 06160	HR 8858, Psi2 Aqr	★	23 15 18	-09 27.3	11	4.39	.011	-0.15	.009	-0.55	.009	31	B5 Vn	15,154,1020,1034,1068*
		NGC 7582 sq1 3		23 15 21	-42 42.8	1	14.05		0.66		0.12		3		1687
		CS 22949 # 11		23 15 22	-02 34.9	1	14.80		0.29		0.06		1		1736
240247	+59 02693			23 15 24	+60 22.6	1	8.36		1.62		1.45		1	K5	1069
		CS 22949 # 18		23 15 24	-06 43.8	1	13.90		0.49		-0.14		1		1736
219734	+48 03991	HR 8860	★ AB	23 15 25	+48 44.5	6	4.83	.023	1.66	.012	1.96	.012	30	M2 III	15,1363,3055,6001*
219693	-41 15211	HR 8859		23 15 25	-41 05.8	4	5.52	.017	0.45	.009	-0.04	.005	11	F5 V	15,1311,2012,3037
219572	-80 01067	HR 8849		23 15 27	-79 44.8	3	6.32	.004	0.91	.000	0.66	.000	14	K0 III	15,1075,2038
219715	+8 05037	G 29 - 20		23 15 28	+08 48.2	2	9.17	.015	0.75	.000	0.15	.019	3	G5	1620,1658
219728	+8 05039			23 15 28	+09 13.8	1	7.62		1.01		0.81		2	K0	1733
	+14 04965			23 15 28	+15 03.7	1	10.20		0.49		0.04		3	K0	271
	+14 04966			23 15 28	+15 11.4	1	8.92		0.98		0.70		4	K0	271
		HA 91 # 383		23 15 29	+15 04.7	1	10.31		1.08		0.84		3		271
		Per sq 1 # 43		23 15 30	+61 40.	1	12.68		0.72		-0.20		2		881
		JL 133		23 15 30	-71 07.	1	17.20		-0.10		-1.20		2		132
219841	+74 01016	HR 8867		23 15 32	+75 01.5	2	6.36	.020	0.03	.005	0.03	.000	5	A2 Vs	985,1733
		NGC 7582 sq1 1		23 15 32	-42 39.2	1	13.39		0.67		0.12		3		1687
		G 128 - 45		23 15 33	+31 15.5	2	11.74	.015	0.75	.000	0.23	.025	2	K2	1620,1658
		HA 91 # 684		23 15 35	+15 24.6	1	12.24		0.58		-0.01		2	F8	271
219749	+44 04373	HR 8861, ET And		23 15 35	+45 12.9	3			-0.03	.011	-0.27	.015	5	B9p Si	1063,1079,1263
	+62 02209			23 15 36	+62 42.9	1	8.37		1.69				2	K5	369
		LS III +61 024		23 15 38	+61 09.9	1	11.62		0.52		-0.41		3		881
219709	-58 08063	IDS23127S5851	B	23 15 38	-58 34.4	1	9.23		1.02		1.05		2	K0 III	1279
219738	+14 04967			23 15 39	+15 19.8	2	8.23	.019	0.56	.005	0.12	.016	12	F8 V	271,3026
219697	-60 07648			23 15 40	-60 16.4	1	7.71		1.18		1.19		9	K1 III	1704
219709	-58 08064	IDS23127S5851	AB	23 15 41	-58 34.7	2	7.40	.093	0.68	.024	0.06		6	G2/3 V	1075,1279
240248	+59 02694			23 15 46	+60 23.0	3	8.84	.028	0.07	.008	0.05	.033	7	B9 V	401,1069,1119
		KUV 813 - 29		23 15 47	+08 55.4	1	14.41		-0.27		-1.12		1		974
		NGC 7654 - 52		23 15 47	+61 42.7	1	11.67		0.82		0.32		4		1069
		HA 91 # 396		23 15 50	+15 05.3	1	12.31		0.51		-0.09		3		271
		NGC 7582 sq1 4		23 15 50	-42 41.5	1	14.93		0.62		-0.01		1		1687
		KUV 23158+1341		23 15 51	+13 40.8	1	16.59		0.30		-0.11		1		1708
		HA 91 # 397		23 15 51	+14 58.3	1	11.70		0.70		0.34		2		271
219768	+16 04901			23 15 52	+17 26.1	1	8.59		1.02				2	K0	882
240250	+59 02695	LS III +59 068		23 15 52	+59 35.2	3	8.78	.018	0.66	.010	0.01	.005	7	B9 Iab	1012,1069,1119,8100
		Steph 2112		23 15 53	-21 15.5	1	11.25		1.21		1.17			K7	1746
		LS III +60 062	AB	23 15 54	+60 00.3	1	10.81		0.36		-0.51		2		881
		HA 91 # 398		23 15 55	+15 01.5	1	12.25		0.57		0.04		9		271
		NGC 7582 sq1 2		23 15 56	-42 40.2	1	13.54		0.58		0.05		3		1687
		G 275 - 31		23 15 58	-22 59.7	1	10.69		1.19		1.15		1	K5	1696
	+45 04188	G 216 - 37		23 15 59	+46 04.3	1	10.88		1.45				3	M1	1625
219761	-48 14472			23 15 59	-47 42.5	1	6.60		-0.06				4	B9 V	2012
240253	+59 02697			23 16 00	+60 22.1	3	9.40	.033	0.24	.011	0.16	.030	7	A1 V	401,1069,1119
		Per sq 1 # 44		23 16 00	+60 47.	1	12.87		0.80		-0.21		2		881
219815	+40 05043	HR 8864, AN And		23 16 01	+41 30.0	3	5.98	.011	0.21	.005	0.17	.010	13	A7 m	1501,1601,1733
240252	+59 02696	LS III +59 069		23 16 01	+59 52.8	2	9.61	.020	0.62	.015	0.51		5	A2 V	401,1119
	-35 15733			23 16 01	-35 12.8	1	10.90		0.99		0.81		3	K3 V	3072
240254	+59 02698	IDS23137N5944	A	23 16 07	+60 00.0	1	10.52		0.37		-0.36		2	B8	1069

HD	DM	Other Id	N Rem	α1950	δ1950	S	V	σV	B–V	σB–V	U–B	σU–B	n	Spectrum	References
219784	−33 16476	HR 8863		23 16 08	−32 48.3	7	4.41	.008	1.13	.005	1.06	.010	30	K1 III	3,15,1075,2012,2024*
		NGC 7582 sq1 6		23 16 09	−42 37.7	1	12.56		0.60		0.05		3		1687
219764	−61 06736			23 16 11	−60 47.6	1	8.97		1.15		1.05		1	K4 V	3072
219854	+58 02572			23 16 12	+59 12.3	1	8.48		0.60				4	A9 III	1119
		L 168 - 33		23 16 12	−61 47.	1	10.78		0.83		0.36		2		1696
219794	−29 18652			23 16 13	−29 00.9	1	7.72		0.24		0.11		4	A0mA4-A9	152
		KUV 23162+1220		23 16 14	+12 19.6	1	15.38		0.16		-0.62		1	DA	1708
		KUV 813 - 14		23 16 14	+12 19.6	1	15.47		0.00		-0.52		1		974
219804	−31 19218			23 16 14	−30 49.6	1	6.94		1.09				4	K0 III	2012
219855	+57 02719			23 16 15	+57 53.6	1	8.60		0.20		0.02		2	B9p	401
		BSD 19 # 457		23 16 17	+60 01.0	1	9.02		2.19		2.22		2	K5	1069
		HA 91 # 875		23 16 19	+15 26.8	1	11.85		1.00		0.69		8		271
219829	+4 04994	G 29 - 21	★ AB	23 16 20	+05 08.0	4	8.02	.012	0.82	.008	0.40	.039	7	K1 V +K4 V	333,938,1003,1620,3030
219832	−10 06094	HR 8865	★ AB	23 16 22	−09 53.1	4	4.98	.011	-0.02	.004	-0.02		10	A0 V	15,2012,2024,3023
	+62 02210	LS III +63 027		23 16 24	+63 13.7	1	8.42		1.15		0.32		2	B9 Ia	1012
	+14 04968			23 16 25	+15 00.0	1	9.71		0.66		0.21		4	F8 V	271
219846	+16 04904			23 16 25	+16 44.8	1	8.53		1.04				2	K1 V	882
219833	−12 06468			23 16 26	−12 26.6	1	7.23		0.00		0.01		4	A0 V	152
		LS III +59 070		23 16 27	+59 35.5	2	11.81	.026	0.43	.004	-0.34	.017	6		881,1069
219856	+39 05058			23 16 28	+40 09.8	2	8.70	.005	0.39	.000	0.02	.005	6	G0	196,583
219834	−14 06448	HR 8866	★ A	23 16 29	−13 43.9	2	5.20	.010	0.79	.000	0.42	.005	3	G8 IV	150,3008
219834	−14 06448	HR 8866	★ AB	23 16 29	−13 43.9	3	5.07	.004	0.79	.010			11	G8 IV +K2 V	15,1075,2035
219834	−14 06448	IDS23139S1400	B	23 16 29	−13 43.9	2	7.56	.040	0.89	.015	0.63	.035		K2 V	150,3008
219916	+67 01514	HR 8872	★ AB	23 16 34	+67 50.3	7	4.75	.014	0.84	.011	0.49	.008	16	K0 III	15,938,1355,1363,3016*
240255	+58 02573			23 16 37	+58 54.0	1	9.64		0.21				3	B9.5V	1119
		PHL 5663		23 16 37	−02 32.5	2	14.14	.000	0.03	.000	0.07	.011	2		966,1736
	−24 17594	DN Aqr		23 16 37	−24 29.5	1	10.73		0.21		0.06		1	F6.2	700
219891	+44 04378	HR 8870		23 16 40	+44 51.8	1	6.52		0.16		0.13		2	A5 Vn	1733
219860	−39 14994			23 16 42	−39 25.9	1	7.05		0.13		0.13		9	A3 V(m)	1628
		CS 22949 # 19		23 16 44	−06 58.0	1	15.23		0.44		-0.18		1		1736
	+69 01326			23 16 46	+70 08.9	1	8.86		1.06		0.72		2	K0	1733
219879	−18 06283	HR 8869		23 16 46	−18 21.0	2	5.95	.015	1.53	.005	2.01		6	K2/3 III	252,2035
219871	−33 16481			23 16 47	−33 18.2	1	8.35		0.26		0.11		4	A3mA5-A9	152
240256	+59 02699	LS III +60 064		23 16 48	+60 09.2	5	8.79	.092	0.49	.032	-0.28	.031	11	B3 Ia	138,401,1069,1119,8100
219877	−5 05966	HR 8868	★ A	23 16 48	−05 23.9	1	5.60		0.38		-0.04		2	F3 IV	3032
219877	−5 05966	HR 8868	★ AB	23 16 48	−05 23.9	7	5.55	.011	0.40	.007	-0.01	.009	26	F3 IV	15,938,1020,1071,2012*
219877	−5 05966	IDS23142S0540	B	23 16 48	−05 23.9	1	10.92		1.37		1.08		2		3032
219861	−56 10049	X Gru		23 16 51	−55 53.1	1	10.60		0.10		0.25		4	A1 V	3064
		BSD 19 # 472		23 16 54	+59 30.3	1	12.34		0.43		-0.21		3	B7 p	1069
		Per sq 1 # 45		23 16 54	+59 58.	1	13.62		0.32		-0.31		3		881
		L 216 - 6		23 16 54	−53 49.	1	12.42		0.79		0.16		2		1696
		G 273 - 13		23 16 57	−17 21.9	3	14.09	.020	0.02	.013	-0.86	.019	16	DC	1698,1765,3062
		CS 22949 # 14		23 16 58	−03 42.9	1	14.55		0.46		-0.15		1		1736
219873	−55 10000	IDS23141S5516	A	23 16 58	−54 59.4	1	8.94		0.86		0.42		2	K1 V	3072
219873	−55 10000	IDS23141S5516	B	23 16 58	−54 59.4	1	11.29		1.13		1.16		1		3072
219926	+35 05007			23 16 59	+35 33.1	2	6.76	.019	-0.02	.009	-0.13	.000	4	A0 V	1601,1716
		CS 22949 # 13		23 17 00	−02 35.0	2	14.80	.000	0.11	.000	0.17	.009	3		966,1736
	+60 02519	NGC 7654 - 133	★ AB	23 17 01	+61 14.7	1	8.57		0.38				2	F4 V	1119
219912	−34 15985	HR 8871	★ A	23 17 01	−33 58.9	2	6.36	.005	1.30	.000			7	K2 III	15,1075
219927	+34 04899	HR 8873		23 17 02	+34 31.2	2			-0.08	.014	-0.39	.009	4	B8 III	1049,1079
		LS III +59 071		23 17 03	+59 42.1	2	12.11	.018	0.36	.011	-0.40	.031	5		881,1069
		CS 22949 # 20		23 17 03	−07 04.0	1	15.07		0.44		0.05		1		1736
219920	−16 06259			23 17 03	−16 03.3	1	7.28		1.08		0.98		1	K0 III	1657
	+2 04651	G 29 - 23		23 17 06	+03 06.0	6	10.20	.013	0.44	.018	-0.20	.017	10	sdF0	308,516,979,1620,1696,3077
		VES 961		23 17 09	+62 42.5	1	11.94		0.72		0.10		2		8113
219945	+47 04110	HR 8874		23 17 10	+48 21.1	2	5.43	.010	1.04	.005	0.80	.015	6	K0 III	1355,3016
219937	+23 04723			23 17 13	+23 33.1	1	7.28		-0.06		-0.33		2	A0	1625
219978	+61 02423			23 17 13	+62 28.0	2	6.74	.030	2.29	.020	2.57		3	K5 Ib	138,1355
		G 29 - 24		23 17 15	+02 18.2	1	13.01		1.34		1.18		1		333,1620
219949	+7 05009			23 17 16	+07 42.5	1	6.92		1.16		1.08		12	K0	8040
219960	+55 02939		V	23 17 17	+55 37.7	1	8.70		1.70				2	M1	369
219853	−76 01576			23 17 17	−75 54.4	1	7.27		0.46		-0.03		9	F5 V	1628
		Per sq 1 # 46		23 17 18	+60 44.	1	12.83		0.73		-0.19		2		881
	−1 04417	G 157 - 47		23 17 18	−00 44.2	1	10.33		0.90		0.60		1	K0	3056
219952	+38 04980			23 17 20	+39 00.9	1	7.61		1.08		0.81		2	K0 III	105
219962	+47 04114	HR 8875	★ A	23 17 20	+48 06.4	4	6.31	.008	1.12	.004	1.06	.008	8	K1 III	15,1003,1355,4001
		G 217 - 2		23 17 20	+58 20.2	1	12.02		0.64		-0.04		2	sdG0	1658
		PG2317+046		23 17 22	+04 36.1	1	12.88		-0.25		-1.14		4	sdB	1764
		PHL 464		23 17 24	−05 26.	4	11.82	.021	-0.31	.009	-1.16	.008	5	DOp	711,1264,1298,1698,1705,3016
219953	+28 04562	G 128 - 48		23 17 28	+28 35.7	4	8.88	.010	0.80	.010	0.38	.019	2	K1 V	22,1003,1620,3077
219981	+41 04752	HR 8876		23 17 29	+41 48.2	2	5.78	.012	1.50	.013	1.90		6	M0 III	71,3001
		LS III +59 072		23 17 29	+59 54.6	1	11.17		0.37		-0.51		3		881
		G 217 - 3		23 17 32	+60 27.0	1	10.52		1.06		0.96		1		1658
		LS III +59 073		23 17 33	+59 52.9	1	11.43		0.34		-0.40		3		881
		Mrk 320		23 17 36	+26 49.9	1	16.43		-0.02		-1.14		1		1527
220016	+58 02577			23 17 36	+59 21.6	2	7.90	.005	0.04	.010	-0.55		5	B3 V	401,1119
		DZ Peg		23 17 37	+15 47.7	1	11.30		0.11		0.09		1	A8	699
	+60 02520	NGC 7654 - 182	AB	23 17 38	+61 04.3	1	9.57		0.29				3	B3 III	1119
		CS 22888 # 47		23 17 38	−34 02.2	2	14.60	.010	0.41	.008	0.00	.009	2		1580,1736
219983	−4 05868			23 17 40	−04 11.5	1	6.65		0.50		0.01		4	F7 V	3037

Table 1   1223

HD	DM	Other Id	N Rem	α_{1950}	δ_{1950}	S	V	σ_V	B–V	σ_{B-V}	U–B	σ_{U-B}	n	Spectrum	References
220008	+6 05143			23 17 46	+06 35.9	1	8.00		0.67		0.12		3	G4 V	3026
220009	+4 04997	HR 8878		23 17 48	+05 06.5	1	5.05	.008	1.20	.004	1.12	.009	12	K2 III	15,1071,1355,1363
220057	+60 02521	NGC 7654 - 197		23 17 48	+60 52.6	3	6.93	.016	0.03	.005	-0.59		7	B2 IV	401,1118,1119
		LS III +61 025		23 17 48	+61 30.3	1	11.83		0.80		-0.05		3		881
		NGC 7654 - 200		23 17 50	+61 30.5	1	10.72		0.81		0.02		3		1069
220140	+78 00826	V368 Cep	★ A	23 17 51	+78 43.7	1	7.54		0.87		0.39		2	G9 V	3016
220056	+61 02424	NGC 7654 - 213		23 17 53	+61 51.7	1	8.47		0.55				1	F7 V	1118
219998	-26 16595			23 17 57	-26 07.0	1	10.60		0.19		0.16		4	A5 IV	152
220073	+61 02426			23 17 59	+62 20.6	1	7.68		0.42				2	F5	1118
		G 128 - 49		23 18 00	+30 21.0	1	10.45		1.30		1.12		1	M0	1620
		G 217 - 4		23 18 00	+54 12.8	1	11.46		0.78		0.32		1	G6	1658
	-7 05988			23 18 00	-06 53.1	1	12.03		0.96		0.47		1		1696
220002	-50 13947	IDS23152S5051	C	23 18 01	-50 35.0	1	8.89		0.61		0.09		3	A7/9	1279
220003	-50 13948	HR 8877	★ AB	23 18 02	-50 34.8	4	6.04	.026	0.42	.011	0.16	.016	15	F5 Fm δ Del	216,355,1279,2007
220058	+55 02942	LS III +55 113		23 18 03	+55 32.0	1	8.58		0.23		-0.76		3	B1 pne	1212
220074	+61 02427	NGC 7654 - 228		23 18 03	+61 41.8	1	6.40		1.67		2.00		4	K1 V	1733
220014	-49 14120			23 18 05	-48 41.5	1	9.67		0.85				4	K0 V	2012
		BSD 19 # 509		23 18 06	+60 12.1	1	12.39		0.42		-0.05		2	B8	1069
220035	-6 06191	HR 8879		23 18 06	-06 10.9	2	6.16	.005	1.07	.000	0.98	.005	7	K0 III	15,1071
220023	-25 16415			23 18 07	-25 25.7	1	9.34	.009	0.02	.009	0.02	.005	5	A1 IV	55,152
220102	+59 02701	LS III +60 066		23 18 08	+60 00.1	3	6.63	.004	0.62	.015	0.46		5	F5 II	401,1119,6009
220061	+22 04810	HR 8880, τ Peg		23 18 10	+23 28.0	6	4.59	.007	0.17	.010	0.13	.022	34	A5 V	15,1363,3023,6001*
		BSD 19 # 1026		23 18 11	+60 46.7	1	9.88		0.19				2	A0	1119
220078	+14 04974			23 18 14	+14 46.5	1	7.60		0.15		0.12		1	A5 II	1776
240264	+59 02702			23 18 16	+60 10.1	2	9.72	.028	0.24	.005	0.06		4	B8 V	401,1119
220116	+57 02724	LS III +58 081		23 18 19	+58 00.2	1	8.69		0.59		-0.43		2	B0.5Vpe	1012
220105	+43 04440	HR 8884	★ A	23 18 21	+43 50.6	1	6.24		0.14		0.10		2	A5 Vn	1733
220088	+29 04908	HR 8882		23 18 22	+30 08.5	2	5.62	.025	1.50	.001	1.90		4	M0 III	71,3016
		LP 702 - 116		23 18 22	-02 45.7	1	9.95		0.70		0.17		2		1696
		CS 22949 # 25		23 18 22	-06 07.7	1	15.22		-0.30		-1.10		1		1736
220130	+61 02428	NGC 7654 - 265	★ A	23 18 23	+61 56.4	2	6.37	.013	1.64	.018	1.68		5	K2 III	1118,1733
	+61 02429	IDS23162N6200	AB	23 18 24	+62 17.1	1	8.61		0.52				2	F5	1118
220107	+13 05096			23 18 27	+14 02.4	2	8.55	.030	1.20	.010	1.22		4	K2	882,1648
220091	+16 04912			23 18 27	+16 58.7	1	6.67		0.32		0.04		3	A9 III	985
220117	+37 04817	HR 8885	★ A	23 18 28	+37 54.5	2	5.80	.005	0.45	.000	0.00	.000	4	F5 IV-V	254,3016
	+60 02522	NGC 7654 - 270		23 18 32	+60 55.2	6	8.66	.020	0.40	.020	-0.61	.016	17	O6.5IIIf	1011,1012,1069,1086*
		CS 22888 # 57		23 18 33	-36 01.2	2	15.01	.010	0.39	.007	-0.23	.009	3		1580,1736
		BSD 19 # 526		23 18 36	+60 00.1	1	12.41		0.42		-0.25		2	B5 p	1069
220109	-9 06173	IDS23160S0913	AB	23 18 36	-08 56.9	1	7.51		0.27		0.06		1	A2	1700
220096	-27 16284	HR 8883		23 18 36	-27 15.6	4	5.65	.010	0.81	.005			18	G5 IV	15,1075,2013,2030
220147	+61 02430	IDS23165N6152	A	23 18 37	+62 08.3	1	8.13		0.08				2	B9p	1118
236102	+51 03586			23 18 40	+52 17.4	1	9.35		0.21		0.10		2	A2	1566
		BSD 19 # 529		23 18 42	+59 55.4	1	11.24		0.42		-0.34		2	B5	1069
		BSD 19 # 535		23 18 48	+59 52.6	1	11.85		0.43		-0.41		2	B5	1069
		PHL 475		23 18 48	-22 37.	1	16.32		-0.27		-1.11		2		286
		JL 134		23 18 48	-70 43.	1	15.32		-0.03		-0.97		2		132
		CS 22949 # 30		23 18 50	-04 02.1	1	13.85		0.42		-0.23		1		1736
220148	+37 04820			23 18 53	+38 18.5	1	7.35		1.10		0.83		2	K0	1601
220167	+59 02703			23 18 54	+60 11.8	1	7.17		1.14				2	K1 IV	1119
220127	-50 13953			23 18 54	-49 45.2	5	10.14	.013	0.69	.010	-0.01	.035	10	G3/5	1236,1594,1696,2034,3077
240267	+58 02581			23 18 55	+59 30.5	2	9.02	.014	0.27	.009	0.16		4	A5 V	401,1119
220168	+35 05012			23 18 58	+36 13.6	1	7.11		1.44		1.71		3	K3 III	1501
	+58 02580	LS III +58 082		23 18 58	+58 40.5	2	10.08	.020	0.80	.005	-0.25	.015	4	B0.5V:	1012,1069
		BSD 19 # 541		23 19 00	+59 50.1	1	12.40		0.50		-0.26		2	B6	1069
220208	+63 01974			23 19 02	+64 28.0	1	7.40		0.02		-0.38		2	B9	401
		G 241 - 52		23 19 04	+67 20.9	1	13.94		0.83		0.28		2		1658
		G 68 - 8		23 19 09	+17 01.8	3	11.70	.014	1.52	.015	1.08	.008	7		333,940,1620,3078
220182	+43 04445	G 190 - 20		23 19 11	+43 49.2	3	7.36	.004	0.80	.005	0.41	.003	5	K1 V	22,1355,1658
		NGC 7654 - 363		23 19 13	+60 44.6	1	11.34		0.34		-0.36		4		1069
	+11 04986	EI Peg		23 19 15	+12 19.3	2	9.57	.215	1.55	.065	1.16	.065	2	M5 III:	693,793
240271	+57 02728			23 19 15	+58 05.7	2	9.49	.029	0.59	.015	-0.07		5	B9 III	401,1119
220172	-10 06098			23 19 15	-10 02.1	3	7.67	.006	-0.21	.004	-0.83	.004	12	B3 Vn	55,399,1732
	+60 02524	NGC 7654 - 372		23 19 16	+60 50.3	3	8.81	.019	1.30	.012	1.19		7	G8 III	1069,1086,1119
220188	-13 06390	IDS23166S1250	AB	23 19 18	-12 32.5	1	9.78		0.74		0.24		4	G6/K1 V	176
		KPS 813 - 202		23 19 19	+09 52.9	1	11.33		1.06		0.81		3		974
220175	-41 15231			23 19 20	-41 28.6	1	9.82		-0.01		-0.05		4	B9.5V	152
		KPS 813 - 204		23 19 25	+10 01.6	1	8.62		1.62		1.63		1		974
220197	+15 04813	G 68 - 10		23 19 26	+16 21.5	1	8.96		0.61		0.04		1	F8 V	333,1620
		KPS 813 - 205		23 19 27	+10 10.4	1	8.60		1.00		0.68		3		974
220253	+62 02218			23 19 27	+62 33.0	1	8.32		1.14		1.06		10	K0	1655
		G 68 - 11		23 19 28	+23 50.7	1	12.75		1.40		1.27		1		333,1620
220222	+31 04897	HR 8887	★ AB	23 19 28	+31 32.3	4	5.34	.015	-0.11	.007	-0.46	.040	6	B6 III	1049,1079,1221,3023
220242	+25 04924	HR 8888		23 19 30	+26 20.1	2	6.62	.029	0.36	.005	0.01	.019	3	F5 V	254,3053
220204	-43 15352			23 19 30	-43 27.7	1	7.26		1.18		1.20		4	K1/2 III	1673
		LS III +61 028		23 19 31	+61 40.7		11.13	.000	0.64	.005	-0.28	.020	5		405,483
220177	-59 07884			23 19 31	-59 13.4	1	10.36		0.98		0.75		1	K3 V	3072
		CS 22949 # 29		23 19 32	-04 07.2	1	14.63		0.47		-0.23		1		1736
		KPS 813 - 206		23 19 35	+09 52.4	1	12.53		0.52		-0.01		3		974
	+60 02525	LS III +60 071		23 19 36	+60 34.3	4	9.63	.016	0.29	.012	-0.58	.044	12	B0 III-IV	1012,1069,1086,1119
		BSD 19 # 554		23 19 37	+59 40.1	1	10.68		1.58		1.69		2	G5	1069

HD	DM	Other Id	N	Rem	α_{1950}	δ_{1950}	S	V	σ_V	B–V	σ_{B-V}	U–B	σ_{U-B}	n	Spectrum	References
		NGC 7654 - 409			23 19 37	+61 26.9	1	10.23		0.51		−0.24		2		1069
		BSD 19 # 556			23 19 43	+59 41.4	1	10.22		0.25				2	A0	1119
	−16 06270	CZ Aqr			23 19 43	−16 12.8	1	10.71		0.27		0.10		7	A5	3064
		G 29 - 25			23 19 46	+11 53.1	1	10.82		0.67		0.02		2	G8	333,1620
		LB 1521			23 19 48	−46 16.	1	12.57		−0.05		−0.05		3		45
		KPS 813 - 208			23 19 49	+09 58.5	1	14.79		1.06		1.07		3		974
		BSD 19 # 561			23 19 49	+59 48.5	1	12.33		0.46		−0.26		3	B6	1069
220256	−1 04420	G 29 - 26			23 19 49	−00 41.1	1	8.58		0.84		0.53		1	K0	1620
	+59 02707	LS III +60 072			23 19 51	+60 05.4	1	9.84		0.08		−0.11		2	B9:V:	1012
		CS 22949 # 35			23 19 52	−02 29.9	1	11.58		1.09		0.76		1		1736
220314	+63 01978				23 19 53	+64 14.7	1	7.95		0.05		−0.23		2	A0	401
220300	+55 02946	LS III +56 131			23 19 54	+56 04.4	2	7.93	.025	0.15	.010	−0.43	.020	4	B8	379,401
220301	+55 02947	IDS23176N5532		AB	23 19 55	+55 48.3	1	8.58		0.14		0.07		2	A0	379
	+59 02708				23 19 57	+60 09.1	2	9.95	.057	0.06	.019	−0.11		4	B9 V	401,1119
220315	+61 02435				23 19 58	+62 16.4	1	8.75		0.08				2		1118
		KPS 813 - 209			23 20 00	+09 59.5	1	14.17		0.93		0.54		3		974
220288	+25 04927	IDS23176N2521	A		23 20 00	+25 38.6	1	6.35		1.54		1.84		2	K3 III	105
220278	−15 06406	HR 8890		★AB	23 20 02	−15 18.8	7	5.19	.011	0.20	.008	0.07	.022	25	A3 V	15,176,1075,2013,2024*
220263	−60 07654	HR 8889			23 20 03	−60 19.8	4	6.08	.004	1.61	.008	1.91		13	M3 III	15,1075,2035,3005
		UZ Scl			23 20 08	−30 23.7	1	11.63		0.14				1		700
220318	+20 05317	HR 8891			23 20 11	+20 33.3	3	6.29	.000	−0.04	.013	−0.09	.013	6	B9.5V	985,1049,1716
		KUV 23202+0718			23 20 13	+07 18.2	1	14.44		0.47		−0.12		1	NHB	1708
		BSD 19 # 568			23 20 13	+59 21.1	1	11.82		0.53		−0.01		2	B8	1069
240275	+58 02583	V398 Cas			23 20 16	+59 02.0	1	8.69		2.89				2	M0	369
220308	−24 17624				23 20 16	−24 18.6	1	10.09		0.95		0.64		2	K1/2 V	3072
220334	+19 05093	IDS23178N2001	A		23 20 18	+20 17.1	1	6.62		0.60		0.11		1	G2 V	3016
220334	+19 05093	IDS23178N2001	AB		23 20 18	+20 17.1	1	6.64		0.63		0.14		3	G2 V	938
220334	+19 05093	IDS23178N2001	B		23 20 18	+20 17.1	1	9.76		1.06		0.74		1		3016
	+21 04923	G 68 - 12		★A	23 20 18	+21 51.8	1	9.68		0.82		0.46		2	F5	1620
220369	+59 02710	HR 8894			23 20 18	+59 51.5	6	5.56	.022	1.68	.019	1.83	.021	17	K3 II	252,1086,1118,1119*
220321	−20 06587	HR 8892			23 20 21	−20 22.4	6	3.96	.013	1.10	.003	0.95	.006	22	K0 III	3,15,1075,2012,4001,8015
		LP 762 - 60			23 20 23	−09 35.6	1	14.06		1.41		1.16		1		1696
220326	−43 15358				23 20 24	−43 07.2	2	9.34	.010	0.25	.005	0.04	.005	6	A9 V	1097,3014
220339	−11 06064	G 157 - 54			23 20 27	−11 02.5	5	7.80	.006	0.88	.005	0.60	.014	19	K2 V	22,861,1013,1197,3008
		CS 22949 # 26			23 20 29	−05 29.4	1	15.23		0.45		−0.15		1		1736
	+27 04541				23 20 30	+27 46.6	1	9.24		1.13		0.89		2	K0	1733
		BSD 19 # 582			23 20 31	+60 09.9	1	12.65		0.67		0.38		2	B9	1069
220363	+11 04993	HR 8893		★AB	23 20 33	+12 02.4	2	5.10	.020	1.32	.005	1.50	.010	6	K3 III	1355,3016
220373	+15 04815				23 20 34	+15 47.7	1	8.64		0.58		0.08		2	F7 V	1733
220350	−25 16435				23 20 34	−24 53.4	1	7.99		0.84				4	G6 III	2012
220364	−17 06742	IDS23180S1706		AB	23 20 36	−16 48.7	1	9.23		0.21		0.09		4	A1 IV/V	152
		NGC 7654 - 568		★	23 20 39	+61 51.8	3	10.78	.023	0.75	.005	−0.33	.023	7		1069,1011,1012
	+63 01982				23 20 39	+63 36.2	1	9.09		0.04				2	A0	1118
		KUV 23207+0840			23 20 42	+08 40.2	1	14.62		0.02		−0.80		1	sdB	1708
		KUV 813 - 21			23 20 42	+08 40.2	1	14.61		0.14		−0.72		1	sdB	974
		G 68 - 13			23 20 54	+15 18.0	1	11.66		1.48		1.05		1	M0	333,1620
		G 157 - 55			23 20 55	−06 16.7	1	11.28		1.06		1.01		1		1696
		BSD 19 # 593			23 20 56	+59 28.8	1	11.72		0.52		−0.32		3	B8	1069
220406	−0 04509	HR 8897		★A	23 20 58	+00 01.0	2	6.30	.005	1.61	.000	1.95	.000	15	K2	15,1071
220401	−43 15360	HR 8896			23 21 01	−43 24.0	3	6.10	.004	1.46	.000			11	K3 III	15,1075,2032
220400	−31 19267				23 21 04	−31 23.1	1	7.43		1.63		1.99		4	M1/2 III	1673
220391	−54 10280	IDS23182S5421	B		23 21 04	−54 05.4	1	7.12		0.26		0.12		2	A6/F0 V	1279
220445	+44 04399	IDS23187N4514	A		23 21 05	+45 31.1	1	8.58		0.95		0.75		2	K0	3072
220445	+44 04400	IDS23187N4514	B		23 21 05	+45 31.1	1	9.33		1.10		1.04		2		3072
220392	−54 10281	HR 8895		★A	23 21 05	−54 05.0	1	6.15		0.26		0.15		2	A4 III	1279
220392	−54 10281	IDS23182S5421		AB	23 21 05	−54 05.0	1	5.79		0.25				4	A4 III	2035
240279	+57 02734				23 21 09	+57 50.9	1	9.78		0.38		0.04		2	F0	401
220436	−9 06183	IDS23186S0900		AB	23 21 10	−08 44.1	1	6.70		0.94		0.58		3	K0	938
	+60 02529	NGC 7654 - 646			23 21 16	+61 27.0	3	9.02	.024	0.33	.018	0.16		6	A7 III	49,1086,1119
		NGC 7654 - 654			23 21 18	+60 31.9	1	12.34		0.41		−0.27		2		1069
220460	+31 04901	HR 8899			23 21 20	+32 15.4	3	6.69	.021	0.44	.007	−0.10	.034	7	F5	254,272,3037
		BSD 19 # 125			23 21 20	+59 09.8	1	11.12		0.57		0.33		2	A0	1069
		BSD 19 # 601			23 21 20	+59 30.7	1	10.88		1.31		1.17		2	K0	1069
		NGC 7654 - 665			23 21 21	+61 18.6	1	12.48		0.58		−0.01		2	A0	1237
219765	−88 00204	HR 8862			23 21 22	−87 45.5	3	5.49	.004	1.28	.005	1.43	.000	15	K2 III	15,1075,2038
	−38 15473				23 21 23	−38 34.8	1	11.16		0.69		0.06		3		1696
240282	+59 02712				23 21 24	+60 12.0	1	8.68		1.75				2	M0	369
		BSD 19 # 1125			23 21 25	+60 28.2	1	11.39		0.44		−0.26		2	B5	1069
220440	−52 12150	HR 8898			23 21 25	−52 09.9	2	5.75	.000	1.59	.000	1.93		6	M0 III	2035,3005
		BSD 19 # 126			23 21 26	+59 13.1	1	10.51		1.27		0.94		2	G3	1069
220466	−22 06119	IDS23188S2219		AB	23 21 26	−22 02.9	1	6.50		0.42		−0.03		4	F3 IV/V	3037
	+60 02530				23 21 30	+60 35.1	1	9.58		1.57		1.42		2	K0	1069
220465	−19 06450	HR 8900			23 21 30	−18 57.8	2	6.18	.005	1.02	.000			7	K0 III	15,2030
220510	+62 02223				23 21 31	+62 32.7	1	8.39		0.50				2	G0	1118
		BSD 19 # 130			23 21 32	+59 10.4	1	11.18		0.52		0.28		2	A0	1069
		NGC 7654 - 694			23 21 33	+61 16.7	1	12.85		0.60		0.01		2	A8	1237
240284	+58 02586				23 21 38	+58 54.0	2	8.91	.010	0.44	.005	0.32		3	B9.5V	1069,1119
		BSD 19 # 607			23 21 38	+59 38.2	1	12.39		0.67		0.41		3	B5	1069
		NGC 7654 - 711		★	23 21 39	+61 05.4	2	11.63	.052	0.50	.009	−0.21	.024	6	B2 Ib	483,1069
236113	+54 02967				23 21 40	+55 23.0	1	9.70		1.85				2	K5	369

Table 1   1225

HD	DM	Other Id	N	Rem	α_{1950}	δ_{1950}	S	V	σ_V	B–V	σ_{B-V}	U–B	σ_{U-B}	n	Spectrum	References
		NGC 7654 - 719			23 21 42	+61 18.2	1	12.39		0.48		-0.07		2	A0	1237
		BSD 19 # 1134			23 21 43	+60 28.5	1	11.45		0.61		-0.27		5	B0	1069
		BSD 19 # 135			23 21 44	+59 06.9	1	11.54		0.76		0.23		1	G0	1069
220506	-37 15237				23 21 45	-37 28.5	1	7.34		0.10		0.14		4	A3 V	152
	+60 02531	NGC 7654 - 731			23 21 46	+60 59.7	1	9.07		0.52				3	F7 III	1119
		G 242 - 33			23 21 46	+72 31.8	1	11.47		1.20		1.10		2		7010
220515	-18 06299	RU Aqr			23 21 47	-17 35.6	1	9.09		1.58		1.15		1	M4/5 III	3076
		BSD 19 # 1798			23 21 48	+61 19.3	1	10.88		0.48				3	A3	1086
	+6 05151	G 29 - 28			23 21 49	+07 29.8	2	11.14	.015	1.14	.000	1.13	.020	2	K2	333,1620,1696
		NGC 7654 - 735			23 21 49	+61 19.4	2	10.60	.000	0.59	.000	-0.06	.020	3	B6 V	49,1237
		BSD 19 # 136			23 21 50	+59 06.1	1	12.28		0.61		0.31		2	B7	1069
220562	+56 02999	IDS23196N5659		AB	23 21 51	+57 15.6	2	6.79	.010	0.25	.005	-0.48	.025	5	B2 V	399,401
		NGC 7654 - 746			23 21 55	+61 15.3	1	12.82		0.49		-0.02		2		1237
		NGC 7654 - 751			23 21 59	+61 15.7	1	14.09		0.52		0.32		2		1237
		NGC 7654 - 753			23 21 59	+61 18.2	1	13.53		0.54		0.18		3		1237
240286	+58 02588	LS III +59 078		B	23 22 00	+59 27.4	2	10.04	.035	0.26	.015	-0.20	.030	3		401,1069
		NGC 7654 - 757			23 22 00	+61 14.9	1	11.00		0.48		-0.07		2	A8	1237
		NGC 7654 - 756			23 22 00	+61 18.6	1	13.67		0.62		0.24		3		1237
		NGC 7654 - 1773			23 22 00	+61 19.	1	12.13		0.45		0.29		2		1069
		NGC 7654 - 759		⋆ C	23 22 00	+61 20.3	2	11.31	.025	0.54	.000	-0.10	.005	4	B8 V	49,1237
240287	+58 02587				23 22 01	+58 56.0	2	9.57	.010	0.36	.005	-0.05		3	B9 V	1069,1119
		NGC 7654 - 770			23 22 01	+61 18.0	1	14.28		0.60		0.34		2		1237
		NGC 7654 - 764			23 22 01	+61 18.7	1	12.88		0.55		0.06		3		1237
	+60 02532	NGC 7654 - 766		⋆ AB	23 22 02	+61 18.8	5	8.31	.017	1.12	.026	0.78	.016	33	F7 Ib	49,1086,1118,1119,1237
		NGC 7654 - 1940			23 22 02	+61 19.5	1	15.35		0.63		0.70		1		49
		NGC 7654 - 765			23 22 02	+61 20.9	2	13.64	.000	0.63	.005	0.16	.015	3		49,1237
220536	-41 15245				23 22 02	-40 41.5	2	9.01	.032	0.58	.014	0.01		7	G1 V	1097,2034
		NGC 7654 - 769		⋆ B	23 22 03	+61 18.3	1	11.09		0.43		-0.19		18	B7	1237
240288	+58 02588	LS III +59 078		A	23 22 04	+59 27.9	2	9.47	.029	0.81	.000	0.14	.015	5	B5	401,1069
		NGC 7654 - 773			23 22 04	+61 22.8	1	12.26		0.54		-0.09		1	B9	1237
		NGC 7654 - 782		⋆ D	23 22 06	+61 20.2	1	10.45		0.57		-0.05		1	B6 V	49
		BSD 19 # 622			23 22 08	+59 42.1	1	10.60		0.80		-0.18		2	B5	1069
		NGC 7654 - 785			23 22 08	+61 18.9	1	13.88		0.61		0.25		1		49
220575	+40 05068	HR 8902			23 22 10	+40 50.3	1	6.72		0.02		-0.28		2	B8 III	1733
		NGC 7654 - 806			23 22 10	+61 19.7	1	13.53		0.53		0.14		2		1237
240289	+58 02589				23 22 12	+58 40.2	1	9.58		0.51				2	F6 V	1119
		NGC 7654 - 814			23 22 12	+61 16.3	1	12.62		0.39		-0.02		2	A0	1237
		NGC 7654 - 817			23 22 13	+60 58.0	1	10.61		0.46		-0.27		4	B7 Ib	1069
		NGC 7654 - 810			23 22 13	+61 15.6	1	13.48		0.65		0.40		2		1237
		NGC 7654 - 815			23 22 13	+61 18.8	1	12.14		0.43		-0.17		3		1237
		NGC 7654 - 813			23 22 13	+61 20.5	2	11.43	.020	0.56	.015	-0.09	.010	3	B6 V	49,1237
	+57 02735	G 217 - 7			23 22 14	+57 35.0	1	10.05		1.51		1.18		1	M2	3016
		NGC 7654 - 816			23 22 14	+61 15.6	1	12.34		0.46		-0.04		2		1237
		NGC 7654 - 821			23 22 14	+61 21.2	1	12.82		0.56		0.14		1		1237
		NGC 7654 - 820			23 22 14	+61 21.6	2	11.33	.020	0.46	.005	-0.12	.010	3	B6 V	49,1237
		NGC 7654 - 828			23 22 15	+61 20.0	1	12.06		0.49		-0.07		1	B9	1237
		NGC 7654 - 824			23 22 15	+61 25.3	2	12.03	.010	0.56	.005	-0.01	.040	4	B7	1069,1237
		NGC 7654 - 832			23 22 16	+61 15.7	1	13.54		0.51		0.21		2		1237
		NGC 7654 - 829			23 22 16	+61 20.7	1	13.10		0.58		0.14		2		1237
		CS 22888 # 49			23 22 16	-34 39.9	2	14.62	.010	0.43	.008	-0.22	.008	3		1580,1736
220583	+22 04829				23 22 17	+23 12.3	1	8.24		1.04		0.81		2	G8 III	1648
220582	+24 04770				23 22 17	+25 13.3	2	7.37	.005	-0.07	.000	-0.47	.010	5	B9	833,1625
220598	+35 05024				23 22 17	+36 05.3	2	7.00	.005	-0.14	.000	-0.73	.014	4	B2 V	399,963
	+60 02533	NGC 7654 - 842			23 22 18	+60 53.0	4	8.97	.011	0.11	.009			10	B9 III	1086,1118,1119,1379
		NGC 7654 - 841			23 22 18	+61 18.4	1	12.87		0.47		0.02		2		1237
		NGC 7654 - 840			23 22 19	+60 20.2	1	13.66		0.52		0.17		1		1237
		NGC 7654 - 839			23 22 19	+61 25.0	1	12.09		0.62		-0.04		2	A0	1237
		BSD 19 # 629			23 22 20	+59 38.4	1	10.84		0.61		-0.31		3	B5	1069
		NGC 7654 - 847			23 22 20	+61 17.0	1	13.28		0.45		0.10		1		1237
		NGC 7654 - 849			23 22 20	+61 21.0	1	13.21		0.57		0.09		1		49
240290	+58 02590				23 22 21	+59 08.0	1	9.18		0.38				2	A0 V	1119
		NGC 7654 - 858			23 22 22	+61 18.6	1	13.38		0.71		0.27		1		1237
		NGC 7654 - 862			23 22 22	+61 18.8	2	14.17	.009	0.64	.075	0.38	.042	2		49,1237
		NGC 7654 - 852			23 22 22	+61 19.9	1	14.07		0.59		0.29		1		1237
		NGC 7654 - 854			23 22 22	+61 25.3	1	11.64		1.38		0.96		2		1237
		NGC 7654 - 859			23 22 23	+61 24.0	1	11.45		0.62		0.02		2	A2	1237
220599	+31 04904	HR 8903			23 22 24	+32 06.6	3	5.57	.005	-0.10	.013	-0.26	.007	5	B9 III	1049,1221,3023
220638	+63 01988				23 22 24	+63 34.8	1	6.74		1.08		0.90		2	K0	1502
		NGC 7654 - 868			23 22 25	+61 18.0	1	12.96		0.48		0.07		1		1237
		NGC 7654 - 867			23 22 25	+61 21.5	2	10.71	.050	0.55	.030	-0.03	.010	3	B8 V	49,1237
		NGC 7654 - 837			23 22 25	+61 22.2	1	14.33		0.68		0.39		1		49
		KUV 23224+1151			23 22 26	+11 51.2	1	16.04		0.09		-0.84		1		1708
		KUV 813 - 24			23 22 26	+11 51.2	1	16.32		0.03		-0.78		1		974
		NGC 7654 - 876			23 22 26	+61 17.9	1	13.30		0.57		0.31		2		1237
		NGC 7654 - 885			23 22 27	+61 15.0	1	12.20		0.52		0.14		2		1237
		NGC 7654 - 879			23 22 27	+61 16.2	1	12.46		0.46		-0.03		2		1237
240293	+59 02715				23 22 29	+59 48.5	3	9.31	.022	0.09	.013	-0.01		8	B9 V	401,1086,1119
		NGC 7654 - 892			23 22 29	+61 16.8	1	14.67		0.71		0.43		2		1237
220572	-57 10268	HR 8901			23 22 29	-57 07.4	3	5.61	.020	1.06	.008	0.97	.004	8	K0 III	58,2007,3077
240294	+59 02716				23 22 30	+59 55.2	2	9.79	.005	0.10	.000	-0.16		6	B9	401,1086

HD	DM	Other Id	N Rem	α_{1950}	δ_{1950}	S	V	σ_V	B–V	σ_{B-V}	U–B	σ_{U-B}	n	Spectrum	References
		NGC 7654 - 902		23 22 30	+61 19.3	1	13.88		0.53		0.32		1		1237
		G 128 - 57		23 22 31	+26 52.2	1	12.07		1.42		1.22		1		1620
		NGC 7654 - 907		23 22 31	+61 10.0	1	11.70		0.56		-0.11		3	B9	1069
		NGC 7654 - 920		23 22 34	+61 15.9	2	13.33	.010	0.45	.035	0.10	.045	3		49,1237
		NGC 7654 - 919		23 22 34	+61 16.7	1	13.99		0.56		0.33		2		1237
		NGC 7654 - 918		23 22 34	+61 18.2	1	11.76		0.41		-0.18		1	B9	1237
240295	+59 02717			23 22 35	+60 02.8	2	9.41	.035	0.46	.010			5	F2 V	1086,1119
		NGC 7654 - 930		23 22 35	+61 17.9	1	11.57		0.51		-0.11		1	B7	1237
	+59 02718			23 22 36	+60 04.0	1	10.02		1.30				2		1086
	+60 02536	NGC 7654 - 931		23 22 36	+60 58.5	1	9.57		0.36		-0.55		2	A2 III	41
		NGC 7654 - 942		23 22 36	+61 00.7	1	11.66		0.57		-0.33		4		1069
		NGC 7654 - 936		23 22 36	+61 19.0	1	12.86		0.52		0.06		1		1237
220652	+61 02444	NGC 7654 - 923	⋆ A	23 22 36	+62 00.5	4	4.97	.011	1.68	.010	2.02	.066	9	M1 III	369,1118,1355,3053
	+62 02225			23 22 37	+62 58.7	1	9.28		0.16				2	A5	1118
		NGC 7654 - 1941		23 22 38	+61 14.4	1	15.44		0.72		0.60		1		49
		NGC 7654 - 941		23 22 38	+61 15.1	2	11.96	.025	0.46	.010	-0.11	.005	3	B6 V	49,1237
		NGC 7654 - 949		23 22 38	+61 19.1	1	13.75		0.58		0.18		1		1237
		NGC 7654 - 889		23 22 39	+61 22.3	2	13.81	.050	0.55	.010	0.23	.020	3		49,1237
		NGC 7654 - 1939		23 22 40	+61 21.8	1	14.86		0.64		0.39		1		49
220609	-49 14140			23 22 40	-49 10.1	1	10.14		0.16		0.13		4	A3/5 (III)	152
		NGC 7654 - 963		23 22 41	+61 17.8	1	13.41		0.61		0.21		1		1237
		G 29 - 30		23 22 42	+00 41.7	2	12.07	.010	1.43	.005	1.16		2	M0	333,1620,1746
		NGC 7654 - 945		23 22 42	+61 14.6	2	14.08	.124	0.59	.040	0.27	.065	3		49,1237
		NGC 7654 - 964		23 22 42	+61 19.4	1	13.90		0.63		0.41		1		1237
		NGC 7654 - 969		23 22 43	+61 17.1	1	14.21		0.62		0.32		1		1237
		NGC 7654 - 981		23 22 43	+61 17.2	1	13.41		0.60		0.26		1		1237
		G 29 - 31		23 22 44	+02 52.2	2	11.44	.015	1.29	.005	1.17	.000	2	M0	333,1620,1696
		NGC 7654 - 988		23 22 46	+61 25.6	1	11.78		0.64		0.12		2	B8	1069
220654	+46 04041			23 22 47	+46 55.2	1	9.28		0.08		0.05		2	A0	1601
		NGC 7654 - 992		23 22 47	+61 17.2	1	13.49		0.54		0.21		1		1237
240296	+57 02736			23 22 50	+58 08.6	1	9.34		0.47				2	A6 III	1119
		BSD 19 # 646		23 22 50	+60 12.2	1	12.14		0.60		0.01		2	B9	1069
		NGC 7654 - 1938		23 22 50	+61 20.8	1	14.51		1.03		0.38		1		49
		NGC 7654 - 980		23 22 51	+61 21.2	1	13.99		0.65		0.28		1		49
		NGC 7654 - 1007		23 22 52	+61 18.0	1	13.48		0.54		0.20		1		49
220657	+22 04833	HR 8905		23 22 53	+23 07.7	7	4.41	.013	0.61	.006	0.14	.005	19	F8 IV	1,15,1004,1118,3077*
	-46 14649			23 22 53	-45 53.0	1	11.28		1.43				1	M0	1494
		JL 135		23 22 54	-71 44.	1	14.48		0.01		-0.77		1		832
220647	-38 15482			23 22 55	-38 02.2	1	9.34		0.29				4	A9 V	2012
		NGC 7654 - 1942		23 22 56	+61 20.9	1	16.07		0.75		0.54		1		49
		NGC 7654 - 1005		23 22 58	+61 20.6	1	13.58		0.56		0.20		1		49
		NGC 7654 - 1035		23 22 59	+61 54.8	1	10.75		0.26		0.00		3	B8 V	1069
		CS 22949 # 49		23 22 59	-06 17.7	1	13.64		0.77		0.21		1		1736
220658	+16 04926			23 23 00	+17 22.6	1	8.46		0.95				2	K2	882
		G 128 - 58		23 23 01	+29 49.9	2	12.09	.005	0.57	.000	-0.09	.035	2		1620,1658
		BSD 19 # 1189		23 23 02	+60 24.6	1	9.67		1.27		1.05		2	G5	1069
	+59 02719	LS III +60 082		23 23 10	+60 22.3	2	10.63	.000	0.63	.020	-0.27	.010	5	B2 III:	1012,1069
		NGC 7654 - 1082		23 23 11	+61 19.1	1	12.83		0.53		-0.02		1		49
		PHL 515		23 23 12	-23 52.	1	15.66		-0.13		-0.59		3		286
		NGC 7654 - 1105		23 23 16	+60 59.8	1	12.32		0.49		-0.20		3		1069
		LS III +59 082		23 23 18	+59 29.2	2	10.95	.009	0.79	.014	-0.25	.014	6		888,1069
	+59 02720			23 23 20	+59 47.7	1	10.11		0.34				2	F0	1119
		G 68 - 16		23 23 23	+24 15.9	1	13.51		1.38		1.17		3		316
		LP 462 - 55		23 23 24	+19 09.3	1	11.00		0.61		0.04		2		1696
		G 128 - 61		23 23 25	+28 55.6	2	11.05	.003	0.96	.018	0.73	.005	4		1625,5010
220704	-21 06420	HR 8906		23 23 25	-20 55.0	11	4.39	.021	1.47	.009	1.80	.017	42	K4 III	15,58,1007,1013,1075*
220732	+52 03441			23 23 26	+53 17.8	1	9.03		0.10		0.04		2	A0	1566
		KUV 813 - 26		23 23 27	+12 04.9	1	14.95		-0.06		-0.96		1		974
		CS 22949 # 50		23 23 31	-06 19.2	1	15.04		0.47		-0.16		1		1736
		BSD 19 # 148		23 23 33	+59 08.1	1	10.08		1.29		0.88		2	G7	1069
220760	+58 02591			23 23 33	+59 23.5	2	8.50	.005	0.32	.000	0.14		4	B9 V	401,1119
		CS 22949 # 48		23 23 34	-06 06.5	1	13.66		0.77		0.19		1		1736
		GD 248		23 23 36	+15 43.8	1	15.11		0.09		-0.78		1	DC:	3060
220750	+38 04999			23 23 36	+39 03.9	1	7.02		-0.02		-0.06		2	A0 V	105
220770	+60 02539	NGC 7654 - 1176		23 23 36	+61 09.6	5	7.83	.020	0.80	.013	0.51	.008	15	A5 Ib	41,49,1012,1086,1119,8100
220734	+24 04774			23 23 37	+25 23.5	1	8.35		0.06		0.01		3	A0	833
		PW Cas		23 23 38	+60 58.5	1	12.85		1.24				1		1772
		G 29 - 33		23 23 39	+08 37.0	8	10.55	.014	1.44	.038	1.23	.032	14	sdM	308,516,1619,1620*
220771	+54 02972			23 23 42	+54 49.9	1	8.93		0.02		-0.10		2	A0	401
	+61 02448	NGC 7654 - 1139		23 23 46	+61 56.9	1	10.82		0.26		0.23		2	A0	1069
220679	-74 02068			23 23 46	-74 06.7	1	7.47		1.10		0.98		9	G8/K0 III	1704
		LS III +63 028		23 23 49	+63 22.0	2	11.05	.005	0.62	.005	-0.43	.010	5		405,483
220729	-53 10461	HR 8907		23 23 49	-52 59.9	1	5.52		0.40				4	F3 V	2032
		BSD 19 # 1233		23 23 50	+60 26.4	1	10.66		0.45		-0.12		2	B4	1069
	+32 04645	G 128 - 64		23 23 52	+32 55.3	2	9.58	.000	0.74	.005	0.29	.015	2	G5	1620,1658
		PHL 5819		23 23 54	-04 32.0	2	13.50	.000	-0.03	.000	-0.04	.012	2		966,1736
		CS 22949 # 43		23 23 54	-05 05.9	1	14.97		0.44		-0.24		1		1736
220773	+7 05030			23 23 55	+08 22.3	1	7.12		0.60		0.13		12	G0	8040
		BSD 19 # 661		23 23 57	+59 32.7	1	10.36		1.22		0.88		2	G5	1069
		CS 22949 # 40		23 23 57	-04 08.8	1	14.64		0.46		-0.17		1		1736

Table 1   1227

HD	DM	Other Id	N	Rem	α_{1950}	δ_{1950}	S	V	σ_V	B–V	σ_{B-V}	U–B	σ_{U-B}	n	Spectrum	References
220782	+16 04928				23 23 58	+17 14.1	1	8.52		1.13				2	K2	882
		CS 22949 # 37			23 23 58	−02 56.3	1	14.36		0.73		0.15		1		1736
220758	−60 07660				23 23 59	−59 55.1	1	7.89		0.46		-0.01		2	F3 V	1730
220819	+60 02540	NGC 7654 - 1267			23 24 05	+60 48.7	4	6.60	.032	0.34	.012	0.35	.034	9	A6 II	41,1118,1119,8100
220832	+58 02593				23 24 09	+59 01.4	3	9.06	.008	0.23	.008	0.12		7	A4 V	401,1118,1119
220787	−11 06076				23 24 09	−11 18.4	2	8.30	.000	-0.17	.004	-0.70	.013	6	B3 III	55,1775
220759	−67 03964	HR 8908			23 24 11	−66 51.4	3	6.45	.004	1.46	.009	1.77	.000	19	K4 III	15,1075,2038
220821	+44 04419	G 216 - 44		⋆ A	23 24 13	+45 03.7	3	7.38	.015	0.64	.004	0.08	.011	6	G0	979,3026,8112
	+59 02723	G 217 - 8			23 24 13	+60 21.0	5	10.49	.017	0.45	.008	-0.22	.012	10	F2	41,516,1064,1658,3026
		LP 702 - 94			23 24 14	−07 40.2	1	11.47		1.01		0.85		1	K4	1696
	+61 02450	LS I +62 001			23 24 15	+62 29.0	1	9.26		1.34		0.94		2	F5	41
220825	+0 04998	HR 8911, κ Psc		⋆ A	23 24 22	+00 58.9	10	4.94	.011	0.04	.006	-0.02	.014	40	A0 p Cr	15,1075,1202,1263*
240299	+57 02739				23 24 22	+58 29.0	1	9.06		0.63				4	F8 V	1119
220802	−50 13976	HR 8910			23 24 23	−50 26.0	3	6.20	.008	-0.07	.005			11	B9 V	15,1075,2007
220790	−59 07890	HR 8909			23 24 25	−58 45.1	3	5.63	.004	0.98	.000			11	G8 III	15,1075,2007
		BSD 19 # 1251			23 24 27	+60 21.3	1	12.47		0.85		-0.05		4		1069
	−2 05958	G 157 - 63			23 24 28	−01 33.8	1	10.35		1.30				1	M0	1017
220826	−12 06491				23 24 29	−11 44.9	1	10.93		0.22		0.07		4	A2/3	152
	+61 02452	NGC 7654 - 1332			23 24 30	+61 33.0	1	9.88		0.51				2	F1 IV	1119
		CS 22949 # 42			23 24 36	−04 40.7	1	13.86		-0.30		-1.12		1		1736
220838	−27 16319				23 24 38	−27 15.5	1	9.30		1.15				3	G6/8 (III)w	955
220816	−60 07661				23 24 39	−59 56.7	1	9.90		0.49		-0.02		2	F7 V	1730
	+58 02594				23 24 40	+59 06.3	1	9.93		0.34				2	B8 V	1119
220858	+0 04999	HR 8912		⋆ A	23 24 41	+00 50.9	2	6.24	.005	1.01	.005	0.81	.005	7	G7 III	15,1071
220906	+67 01525			AB	23 24 41	+67 35.5	1	8.00		0.35		0.12		2	A5	1502
220885	+42 04672	HR 8913			23 24 42	+42 38.2	1			0.00		-0.10		1	B9 III	1079
		NGC 7654 - 1375			23 24 43	+61 09.4	1	11.94		0.50		-0.15		4		1069
	+60 02542	NGC 7654 - 1396			23 24 50	+61 06.2	4	8.83	.034	0.65	.010	0.22	.024	11	A2 Ib	41,1012,1086,1119
220881	−27 16320				23 24 54	−27 33.2	1	7.44		0.28				4	A9 V	2012
220910	+43 04462				23 24 55	+43 35.3	1	8.15		1.62		1.92		2	K5	1601
221021	+80 00766				23 24 55	+81 24.2	1	8.96		0.49		0.01		2	G5	1733
	+49 04124	G 216 - 45			23 24 58	+49 59.2	1	11.08		0.68		0.18		2	K1	1658
		BSD 19 # 684			23 24 58	+59 25.4	1	12.22		0.73		0.16		2	B9	1069
		NGC 7654 - 1407		⋆	23 24 58	+61 44.0	2	11.06	.000	0.74	.000	-0.16	.028	4	B5	41,1069
		LS III +61 040			23 25 02	+61 37.9	1	11.19		0.78		-0.16		1		483
		NGC 7654 - 1418			23 25 05	+62 01.0	1	11.22		0.49		0.06		2		1069
220940	+64 01810	IDS23229N6504		AB	23 25 05	+65 20.8	1	6.84		1.03		0.86		3	K0	985
220896	−51 13570	IDS23223S5054		AB	23 25 05	−50 37.2	2	7.20	.005	0.91	.004	0.58		8	G6 III	1499,2012
		NGC 7654 - 1449		⋆	23 25 08	+61 43.2	2	10.54	.000	0.62	.000	-0.33	.028	4		41,1069
220974	+69 01332	HR 8918			23 25 08	+70 05.1	1	5.60		0.16		0.12		2	A6 IV	1733
220917	−44 15248				23 25 09	−43 58.7	1	9.27		0.20		0.16		4	A2mA5-F2	152
220933	+24 04778	HR 8915, HV Peg			23 25 11	+24 53.5	1			-0.06		-0.17		2	A0 p HgMn	1049
	+59 02727	IDS23228N5953		A	23 25 11	+60 09.2	3	10.03	.043	0.22	.019	-0.30		6	B5 III	1069,1086,1119
220963	+63 01998				23 25 11	+63 39.2	1	7.29		1.36		1.38		2	K0	1502
		CS 22949 # 46			23 25 12	−05 30.7	1	14.95		0.31		-0.20		1		1736
		CS 22949 # 52			23 25 12	−07 03.5	1	13.85		0.39		-0.22		1		1736
	+4 05012	VZ Psc		⋆	23 25 14	+04 34.7	3	10.30	.030	1.12	.020	0.90	.016	5	K5	1620,1696,3062
		CS 22941 # 11			23 25 15	−33 31.4	2	15.36	.005	0.36	.001	-0.19	.007	3		1580,1736
	+60 02544	NGC 7654 - 1470			23 25 16	+61 06.9	1	10.67		0.22		0.18		2	A3 II	1069
240305	+59 02728	IDS23230N6001		AB	23 25 17	+60 17.9	3	9.25	.014	0.20	.000			9	A0 III	1086,1119,1379
		LP 702 - 96			23 25 20	−05 43.5	1	13.52		1.54		1.10		1		1696
220929	−36 15895	HR 8914			23 25 20	−35 49.2	3	6.32	.008	1.20	.000			11	K2 III	15,1075,2035
220950	+42 04676				23 25 22	+42 52.9	1	9.56		0.45		-0.46		2	F0	1649
220954	+5 05173	HR 8916			23 25 26	+06 06.2	10	4.28	.015	1.08	.006	1.00	.009	36	K1 III	3,15,37,1075,1355*
		BSD 19 # 165			23 25 28	+58 44.6	1	10.80		0.88		-0.10		2	A1	1069
220957	−12 06496	HR 8917			23 25 29	−11 43.5	2	6.38	.014	0.89	.009	0.56		5	G6/8 III	963,2009
220945	−45 15036				23 25 32	−44 43.8	1	8.98		0.94				4	K2 V	2013
220999	+58 02595				23 25 37	+59 25.2	3	7.56	.021	0.25	.005	0.04		6	A7 III	401,1118,1119
	+60 02546	NGC 7654 - 1548			23 25 41	+60 34.3	3	8.96	.028	1.05	.004	0.66		7	A0 Ia	41,1086,1119
		PHL 5857			23 25 42	−23 11.	1	15.08		0.12		0.13		3		286
220982	−39 15037				23 25 43	−38 55.3	1	7.34		0.94		0.63		5	G8/K0III/IV	1673
		CS 22941 # 2			23 25 44	−36 12.6	1	14.54		0.16		0.15		1		1736
221008	+58 02596				23 25 45	+59 03.6	1	8.10		1.92				2	M1	369
		BSD 19 # 696			23 25 45	+60 16.2	1	10.88		0.58		0.08		2	F8	1069
220978	−26 16654				23 25 47	−25 41.7	1	6.92		0.16		0.14		10	A3	1628
	+60 02548	NGC 7654 - 1557			23 25 48	+61 04.7	3	9.71	.034	0.14	.017	0.00		8	B9 V	1069,1086,1119
221037	+61 02454	NGC 7654 - 1550			23 25 50	+61 38.7	1	8.12		-0.04				2	B8 V	1118
		CS 22941 # 12			23 25 50	−33 05.2	1	12.45		0.30		-0.15		1		1736
		CS 22941 # 7			23 25 53	−34 47.4	1	14.43		0.38		-0.05		1		1736
221039	+59 02729				23 25 54	+59 46.9	2	7.48	.065	1.07	.020	0.96		3	K0 II	1119,8100
	+60 02549	NGC 7654 - 1575			23 25 54	+60 52.6	2	9.47	.044	0.08	.015			6	B7 V	1086,1119
221038	+60 02550	NGC 7654 - 1568			23 25 54	+61 11.7	3	8.06	.026	0.18	.014			8	A4 III	1086,1118,1119
	+15 04829	IDS23234N1531		A	23 25 55	+15 47.5	1	9.81		1.13		0.98		2	M0	3072
	+15 04829	IDS23234N1531		B	23 25 55	+15 47.5	1	14.00		1.59				1		3054
	+57 02743				23 25 55	+57 34.8	1	9.91		1.96				2		369
221013	+5 05176				23 25 56	+05 31.9	1	8.23		1.36		1.46		2	K0	1733
		LP 582 - 63			23 25 58	+06 39.9	1	12.55		1.36		1.22		1		1696
		CS 22941 # 8			23 26 00	−34 11.9	1	12.98		0.13		0.16		1		1736
		CS 22941 # 1			23 26 01	−36 54.8	1	14.98		0.21		0.04		1		1736
	+58 02597	LS I +59 001			23 26 02	+59 17.6	1	10.08		1.42		0.81		2	A0p	41

HD	DM	Other Id	N Rem	α_{1950}	δ_{1950}	S	V	σ_V	B–V	σ_{B-V}	U–B	σ_{U-B}	n	Spectrum	References
221057	+44 04424			23 26 03	+45 25.1	1	7.67		1.29		1.16		2	K0	1601
	+60 02551	NGC 7654 - 1591		23 26 03	+60 55.1	3	9.84	.028	0.14	.020	-0.04		8	B8 V	1069,1086,1119
221072	+62 02234			23 26 03	+63 06.9	1	8.33		0.47		0.20		10	F5	1655
	+60 02552			23 26 04	+60 31.8	1	9.81		0.18				2	B9 V	1119
221006	-63 04891	HR 8919, CG Tuc		23 26 08	-63 23.2	4	5.67	.010	-0.18	.004			18	Ap Si	15,1075,2013,2029
		NGC 7654 - 1582		23 26 09	+61 57.7	1	11.60		0.83		-0.12		3		1069
221073	+50 04051			23 26 11	+50 47.4	1	9.09		0.74		0.31		2	G5	1566
221059	+10 04939			23 26 14	+11 14.2	1	8.89		1.12		1.03		2	G5	1733
		G 29 - 38	⋆ V	23 26 16	+04 58.5	4	13.03	.026	0.17	.032	-0.63	.008	8	DA	203,1620,1705,3060
		G 190 - 26		23 26 17	+42 25.9	1	13.55		1.39				6		538
221051	-45 15043	HR 8920		23 26 18	-44 46.4	3	6.42	.008	1.17	.005			11	K2 III	15,1075,2006
		NGC 7654 - 1633		23 26 19	+60 45.2	1	10.99		0.49		0.03		3	A5 V	1069
		VY Scl		23 26 21	-30 03.3	1	13.18		-0.11		-1.05		1		760
221081	-10 06120	HR 8921		23 26 25	-09 32.5	2	6.17	.005	1.44	.000	1.70	.000	7	K0	15,1078
	+60 02553	NGC 7654 - 1636		23 26 27	+61 01.8	3	10.09	.025	0.45	.005	-0.43	.021	7	B2 II	41,1012,1069
		NGC 7654 - 1623		23 26 27	+62 03.5	1	10.22		1.30		1.06		2		1069
	-16 06297	G 273 - 65		23 26 29	-15 30.9	1	10.72		0.74		0.26		2		1696
221114	+15 04830			23 26 31	+15 44.2	1	7.11		0.08		0.07		3	A2	985
221124	+52 03454			23 26 31	+53 23.5	2	6.93	.010	1.07	.005	0.89	.010	5	K1 III	1501,1566
	+60 02554	NGC 7654 - 1658		23 26 33	+60 33.2	2	9.79	.030	0.44	.015			4	A5 Ia	1086,1119
221113	+22 04844	HR 8922		23 26 36	+22 46.4	1	6.44		1.10		0.96		2	G9 III	1733
		Smethells 136		23 26 36	-45 54.	1	11.55		1.29				1		1494
221115	+11 05009	HR 8923		23 26 37	+12 29.1	4	4.55	.017	0.94	.002	0.73	.013	8	G8 III	15,1355,1363,8015
221098	-49 14161			23 26 41	-48 40.5	2	8.35	.020	0.36	.005	0.16		7	A1/2mA7-F3	355,2012
221143	+60 02555	NGC 7654 - 1673		23 26 43	+61 09.2	3	8.46	.038	0.11	.017			8	A2 V	1086,1118,1119
	+59 02731			23 26 44	+60 15.0	1	9.63		1.32		1.34		2	K0	1069
		BSD 19 # 1349		23 26 45	+60 52.3	1	10.30		0.16		-0.20		3	A0	1069
221158	+61 02455			23 26 45	+62 18.5	1	8.72		0.24				3	F0	1118
240308	+58 02598	LS III +59 083		23 26 46	+59 28.7	3	8.43	.013	0.55	.019	-0.08	.010	7	B6 III	888,1069,1119
		NGC 7686 - 17		23 26 48	+48 52.3	2	11.91	.006	1.35	.007	1.36		3		49,1655
	-47 14591			23 26 48	-47 18.5	4	10.19	.018	1.32	.014			10	K7 V	158,1494,1705,2034
		NGC 7686 - 13		23 26 49	+48 44.2	2	11.41	.013	0.50	.017	0.22		3		49,1655
	+59 02732	IDS23244N5924	AB	23 26 49	+59 40.3	1	10.79		0.62		0.01		2	A1	1069
221145	+16 04932			23 26 53	+16 32.9	1	8.90		1.42				1	K2	882
		G 190 - 27		23 26 56	+41 11.7	1	12.45		1.61		1.06		6		1663
221146	-1 04443			23 26 56	-01 18.7	1	6.89		0.62		0.22		3	G0	3026
		G 190 - 28		23 26 57	+41 11.9	1	11.85		1.52		1.09		7	M2	1663
221148	-5 05999	HR 8924		23 26 57	-04 48.3	7	6.25	.011	1.10	.010	1.17	.024	16	K3 III	15,37,1003,1071,1075*
		NGC 7686 - 14		23 26 59	+48 52.1	2	11.55	.014	0.95	.004	0.54		3		49,1655
		BSD 19 # 176		23 26 59	+58 56.7	1	12.12		0.59		0.02		2	B7	1069
221170	+29 04940			23 27 00	+30 09.5	3	7.67	.019	1.08	.015	0.61	.042	6	G2 IV	20,1003,3077
	+60 02557	NGC 7654 - 1722		23 27 01	+60 52.3	2	10.28	.025	0.50	.002	-0.33	.035	7	B3	41,1069
	+60 02558	NGC 7654 - 1714		23 27 03	+61 25.7	2	9.79	.033	0.70	.000	-0.18	.005	4	B3	41,1069
	+48 04067	NGC 7686 - 3		23 27 04	+48 57.5	2	8.94	.009	1.22	.005	1.14		3	K0 III	49,1655
		NGC 7654 - 1736	⋆	23 27 04	+60 41.6	2	11.45	.024	0.66	.020	-0.20	.024	5		41,1069
		NGC 7686 - 26		23 27 07	+48 53.9	1	13.63		0.79		0.35		1		49
		NGC 7686 - 30		23 27 08	+48 53.0	1	13.76		1.01		1.08		1		49
240311	+58 02600			23 27 08	+59 19.6	1	9.83		1.35		1.09		1	B7	1069
	+16 04934			23 27 09	+16 50.5	1	10.02		1.07		1.04		1	K2	1746
		G 157 - 67		23 27 10	-06 51.7	2	12.61	.015	1.19	.015	1.14	.050	2		1658,1696
221166	-26 16667			23 27 11	-26 18.4	1	10.49		0.12		0.12		1	A2/3 V	55
		NGC 7686 - 35		23 27 12	+48 51.2	1	14.93		0.78		0.08		1		49
240312	+58 02601	LS III +59 084		23 27 12	+59 19.7	3	8.23	.012	0.46	.005	-0.38	.024	8	B2 V	888,1069,1119
		CS 22941 # 5		23 27 12	-35 29.6	1	14.61		0.27		-0.12				1736
		NGC 7686 - 39		23 27 14	+48 52.9	1	15.74		0.72				1		49
		NGC 7654 - 1752		23 27 14	+60 37.9	1	12.65		0.48		0.13		3		1069
		NGC 7686 - 10		23 27 15	+48 50.6	2	11.08	.005	1.38	.005	1.48		3		49,1655
221203	+48 04066	NGC 7686 - 2		23 27 16	+48 49.9	2	7.74	.002	0.95	.007	0.68		3	G8 III	49,1655
		NGC 7686 - 25		23 27 18	+48 53.6	1	13.48		1.19		1.06		1		49
	+59 02733	IDS23250N5936	AB	23 27 18	+59 51.8	1	10.19		1.43		1.10		2	G8	1069
		NGC 7686 - 37		23 27 19	+48 53.9	1	15.24		0.76				1		49
		NGC 7654 - 1761		23 27 20	+60 55.2	1	11.80		0.32		0.30		2		1069
	+59 02734	LS III +60 084		23 27 21	+60 03.3	1	9.95		0.36		0.00		3	B7	1069
		NGC 7686 - 24		23 27 22	+48 54.6	1	12.91		0.63		0.07		1		49
		NGC 7686 - 36		23 27 24	+48 54.3	1	15.16		0.58		0.09		1		49
	+21 04940	G 68 - 21	⋆ A	23 27 25	+21 55.1	1	10.56		1.16		1.08		1	K3	333,1620
		NGC 7654 - 1747		23 27 25	+61 58.0	1	11.20		0.68		0.01		2		1069
		NGC 7686 - 29		23 27 26	+48 51.4	1	13.73		0.25		0.12		1		49
		NGC 7686 - 32		23 27 29	+48 51.9	1	13.91		0.47		-0.03		1		49
		NGC 7686 - 9		23 27 30	+48 55.7	2	11.07	.036	1.74	.021			3		49,1655
		NGC 7686 - 28		23 27 33	+48 52.7	1	13.68		0.80		0.38		1		49
221237	+57 02747	IDS23254N5800	CD	23 27 33	+58 16.4	4	7.09	.013	0.01	.018	-0.15	.024	18	A1 V	1119,1394,3024,8084
221237	+57 02747	IDS23254N5800	FG	23 27 33	+58 16.4	1	10.26		0.55		0.16		1		8084
221237	+57 02747	IDS23254N5800	H	23 27 33	+58 16.4	1	12.72		0.66		-0.24		1	A0	8084
		NGC 7686 - 23		23 27 34	+48 51.3	1	12.68		1.07		0.58		1		49
221525	+86 03444	HR 8938		23 27 34	+87 01.9	1	5.58		0.23		0.13		5	A7 IV	15
		NGC 7686 - 18		23 27 35	+48 50.9	1	12.48		0.52		-0.02		1		49
		NGC 7686 - 15		23 27 35	+48 54.4	2	11.54	.002	0.66	.002	0.12		3		49,1655
		G 273 - 59		23 27 36	-20 39.8	1	11.16		1.51		1.03		2	M2:	3072
		PHL 5901		23 27 36	-27 00.	1	12.64		-0.02		0.02		2		3060

Table 1 1229

HD	DM	Other Id	N Rem	α_{1950}	δ_{1950}	S	V	σ_V	B–V	σ_{B-V}	U–B	σ_{U-B}	n	Spectrum	References
		NGC 7686 - 16		23 27 40	+48 55.3	2	11.65	.013	0.35	.007	0.10		3		49,1655
	+48 04065	NGC 7686 - 5		23 27 41	+48 46.5	2	10.23	.010	0.44	.008	0.17		3	F1 III	49,1655
		NGC 7686 - 21		23 27 41	+48 50.3	1	12.62		0.79		0.25		1		49
	+60 02562	NGC 7654 - 1801		23 27 42	+60 37.6	1	10.05	.015	0.14	.005	0.13		4	B9 V	1069,1119
221253	+57 02748	HR 8926, AR Cas	⋆ AB	23 27 43	+58 16.4	8	4.91	.034	-0.12	.009	-0.63	.043	21	B3 IV	15,154,1119,1223,1363*
221253	+57 02747	IDS23254N5800	E	23 27 43	+58 16.4	2	11.30	.033	0.64	.159	0.24	.379	4		3024,8084
221253	+57 02747	IDS23254N5800	F	23 27 43	+58 16.4	1	11.06		0.62		0.04		3		3024
221253	+57 02747	IDS23254N5800	G	23 27 43	+58 16.4	1	11.11		0.63		0.05		3	A0	3024
		NGC 7686 - 20		23 27 44	+48 50.3	1	12.59		0.54		-0.01		1		49
221246	+48 04070	NGC 7686 - 1	⋆	23 27 44	+48 51.4	2	6.17	.000	1.46	.000	1.71	.000	2	K5 Ia	15,49
	+48 04068	NGC 7686 - 4		23 27 44	+48 55.9	2	9.53	.000	0.99	.006	0.69		3	G8 III	49,1655
221239	+30 04961	G 128 - 67		23 27 45	+31 25.9	1	8.31		0.93		0.58		1	K0	1620
		LS I +61 007		23 27 45	+61 57.1	2	10.98	.005	0.85	.000	-0.21	.000	4		41,405
		G 68 - 22		23 27 47	+18 11.6	1	11.81		1.15		1.01		1		333,1620
	+59 02735	LS I +59 002		23 27 47	+59 34.3	2	9.88	.004	1.17	.004	-0.04	.000	8	B0 Ib	41,1012
		LP 876 - 88		23 27 48	-20 48.6	1	11.44		0.93		0.68		1		1696
221263	+58 02602		V	23 27 49	+59 08.7	1	8.57		2.04				2	M1	369
		NGC 7686 - 33		23 27 50	+48 54.6	1	14.32		0.79		0.21		1		49
		NGC 7654 - 1815		23 27 50	+61 02.0	1	11.17		0.21		0.13		2		1069
		NGC 7686 - 19		23 27 51	+48 52.1	1	12.49		0.72		0.23		1		49
		BSD 19 # 731		23 27 53	+59 57.0	1	10.46		1.28		0.99		2	G7	1069
		CS 22952 # 4		23 27 53	-04 43.2	1	13.49		0.42		-0.18		1		1736
		NGC 7686 - 22		23 27 54	+48 53.0	1	12.65		1.30		1.00		1		49
		LS I +61 008		23 27 54	+61 14.0	1	12.51		0.73		0.02		2		41
		CS 22949 # 55		23 27 55	-04 43.1	1	13.50		0.42		-0.18		1		1736
		NGC 7686 - 6		23 27 56	+49 00.4	2	10.36	.005	0.19	.008	0.24		3	A1 IV	49,1655
		NGC 7686 - 31		23 27 57	+48 52.9	1	13.82		0.55		0.05		1		49
221264	+30 04963	IDS23255N3017	AB	23 27 58	+30 33.4	2	7.35	.045	0.50	.000	0.02		5	F7 V +F7 V	1381,3024
221264	+30 04963	IDS23255N3017	C	23 27 58	+30 33.4	1	9.88		0.61		0.11		3		3024
		NGC 7686 - 12		23 27 58	+48 48.2	2	11.39	.014	0.73	.008	0.26		3		49,1655
	+48 04071	NGC 7686 - 11		23 27 58	+48 55.1	2	11.19	.028	0.38	.011	0.10		3		49,1655
	+57 02750	V358 Cas		23 28 01	+57 42.7	4	9.50	.049	2.73	.022	2.28		8	M2 Ia-Iab	138,148,369,8032
		NGC 7686 - 34		23 28 02	+48 51.9	1	14.72		0.63		0.16		1		49
		NGC 7686 - 38		23 28 03	+48 49.4	1	15.62		0.49				1		49
	+48 04073	NGC 7686 - 8		23 28 03	+48 52.9	2	10.89	.024	1.25	.019	1.10		3	K0 III	49,1655
		NGC 7686 - 27		23 28 03	+48 55.3	1	13.64		1.20		0.94		1		49
		KUV 23282+1046		23 28 10	+10 45.6	1	15.53		-0.03		-0.90		1	DA	1708
221293	+37 04856	HR 8927		23 28 13	+38 23.2	1	6.05		0.99				2	G9 III	71
	+48 04076	NGC 7686 - 7		23 28 14	+48 52.4	2	10.37	.004	1.26	.003	1.24		3	G7 III	49,1655
221275	-35 15821			23 28 15	-35 22.9	1	8.11		0.80				4	G8/K0 V	2012
		LS III +63 029		23 28 16	+63 24.6	2	10.91	.005	0.91	.005	-0.20	.010	4		405,483
240317	+59 02736			23 28 19	+60 12.7	1	9.44		0.40				3	F5 V	1119
	-7 06034			23 28 20	-06 50.0	1	10.85		0.87		0.55		2		1696
		LP 462 - 52		23 28 23	+15 31.2	1	11.52		1.53		1.18		1	M1	1696
		Steph 2137		23 28 26	+22 30.0	1	11.71		1.38		1.29		1	K7	1746
221308	-7 06036	HR 8928		23 28 26	-06 33.8	2	6.38	.005	1.26	.000	1.29	.000	7	K0	15,1078
		BSD 19 # 1923		23 28 28	+61 45.5	1	12.03		0.57		0.15		3	B7	1069
221335	+57 02752	IDS23262N5751	AB	23 28 32	+58 08.4	1	8.24		0.26				4	Am:	1119
221334	+61 02462	NGC 7654 - 1856		23 28 32	+62 00.8	2	7.67	.014	0.02	.005	-0.37		4	A0	401,1118
	+58 02604			23 28 36	+59 26.6	1	10.09		0.62		0.08		2	B5	1069
	+63 02007			23 28 36	+63 43.4	1	8.83		0.52				3	F8	1118
	-63 04894			23 28 38	-63 12.3	1	9.34		0.78		0.37		1	G5	1696
		CS 22941 # 18		23 28 39	-34 52.2	1	13.19		-0.01		-0.08		1		1736
		BSD 19 # 745		23 28 41	+59 34.8	1	11.83		1.05		0.72		5		1069
	+60 02565			23 28 43	+60 45.4	1	10.04		0.96		0.66		2	G8	1069
		LS I +59 003		23 28 44	+59 58.6	1	11.24		1.29		0.14		2		41
221323	-45 15055	HR 8929		23 28 44	-45 07.2	3	6.01	.008	1.01	.004			11	K1 III	15,1075,2032
		CS 22941 # 16		23 28 48	-34 13.6	1	14.05		-0.03		-0.18		1		1736
221345	+38 05023	HR 8930		23 28 49	+38 57.7	4	5.22	.010	1.03	.006	0.87	.008	9	K0 III	15,1003,1355,4001
	+59 02738		AB	23 28 50	+59 58.9	1	9.91		0.43				1	F5 V	1119
		NGC 7654 - 1932		23 28 50	+61 01.2	1	10.98		1.28		1.07		2		1069
		LP 522 - 61		23 28 53	+12 00.7	1	11.64		0.99		0.84		1		1696
		BSD 19 # 754		23 28 53	+60 16.4	1	10.94		1.29		1.05		2	dG8	1069
221468	+80 00770			23 28 55	+80 43.7	1	8.26		0.21		0.10		7	A0	1219
221354	+58 02605	G 217 - 11		23 28 56	+58 53.3	4	6.74	.009	0.83	.009	0.54	.019	9	K2 V	22,1003,1119,3078
221356	-4 05896	HR 8931		23 28 56	-04 21.6	4	6.50	.009	0.53	.009	-0.03	.020	13	F8 V	15,254,1256,3037
221377	+51 03630	IDS23266N5152	A	23 28 57	+52 08.1	1	7.57		0.39		-0.05		1	F7 V wlm	979
221364	+27 04566			23 29 01	+28 23.4	2	6.53	.010	0.99	.005	0.82	.000	7	K0 III	985,1501
	+61 02465	NGC 7654 - 1917		23 29 02	+62 00.9	1	9.60		0.31		0.11		1	A5	401
		CS 22941 # 19		23 29 02	-34 56.4	1	14.79		-0.04		-0.12		1		1736
221393	+58 02607			23 29 04	+59 11.2	2	7.28	.015	1.71	.015	2.01		3	K5 III	1119,1355
221357	-22 06141	HR 8932		23 29 05	-21 38.7	2	6.25	.036	0.33	.005	0.14		5	F0 V	254,2009
221406	+60 02567	NGC 7654 - 1934		23 29 06	+61 11.1	2	8.82	.025	0.05	.015			2	A1 V	1118,1119
		LS III +59 085		23 29 07	+59 48.8	1	11.42		0.70		0.12		3		888
		G 171 - 2		23 29 09	+40 44.9	1	13.82		0.03		-0.72		1		3060
240319	+58 02608	LS III +58 084		23 29 09	+58 48.9	2	9.99	.004	0.63	.016	-0.17	.013	5	B7	888,1069
		AF Psc		23 29 09	-03 01.7	1	14.43		1.58		0.75		10	M4V:e	1663
		LS I +62 002		23 29 10	+62 14.0	1	11.91		0.89		-0.09		2		41
221411	+56 03018			23 29 11	+57 11.5	1	8.75		0.11		0.04		2	A0	401
		LS I +61 009		23 29 11	+61 06.8	2	11.34	.030	0.52	.015	-0.41	.005	5		41,405

HD	DM	Other Id	N Rem	α_{1950}	δ_{1950}	S	V	σ_V	B–V	σ_{B-V}	U–B	σ_{U-B}	n	Spectrum	References
		PHL 555		23 29 12	−29 09.	1	13.94		-0.27		-0.96		1	DA	1036
	−56 10090			23 29 13	−56 31.5	1	8.96		1.26				4	M7	2018
221394	+27 04568	HR 8933		23 29 14	+28 07.7	3	6.40	.004	0.03	.008	0.04	.004	9	A1 p Sr	1049,1221,3050
	+59 02739			23 29 15	+60 24.1	1	10.25		0.68		0.22		2	G5	1069
221398	−12 06508			23 29 15	−12 13.5	1	7.66		1.01		0.82		5	K0 III	1657
	+60 02568			23 29 16	+60 43.6	1	9.47		1.49		1.66		3	K0	1069
	+19 05116	EQ Peg	⋆ AB	23 29 19	+19 39.7	2	10.36	.028	1.44	.174	1.06		2	M4	1017,3016
		CS 22941 # 14		23 29 22	−33 29.9	2	14.58	.010	0.63	.015	0.02	.001	2		1580,1736
221439	+58 02609			23 29 24	+58 49.0	1	7.36		1.34				2	K1 III:p	1119
221409	−1 04450	HR 8934		23 29 24	−01 21.7	4	6.39	.023	1.18	.014	1.08	.021	12	K1 III	15,252,1071,3016
221438	+59 02740			23 29 25	+60 15.2	2	8.91	.024	0.19	.005			6	A3 V	1118,1119
	+60 02569		A	23 29 29	+60 47.2	1	9.97		0.27				2	B9 V	1119
	+60 02569		B	23 29 29	+60 47.2	1	9.97		0.92				2		1119
		CS 22894 # 1		23 29 29	−02 09.3	2	13.99	.015	0.15	.012	0.14	.035	2		966,1736
240322	+57 02756			23 29 32	+58 29.6	1	9.08		0.63				2	A2 V	1119
		G 128 - 72		23 29 33	+26 42.1	2	15.31	.035	0.20	.013	-0.62	.022	6		316,3060
236159	+51 03632			23 29 33	+52 15.4	1	8.82		1.14		0.97		2	K2	1566
		CS 22894 # 7		23 29 36	−00 10.9	1	13.63		0.19		0.16		1		1736
		LS I +61 010		23 29 37	+61 08.2	1	10.81		0.56		-0.32		3		41
221445	+6 05168	IDS23271N0632	AB	23 29 40	+06 48.6	1	6.81		0.49		-0.02		4	F7 V +F8 V	3030
		LP 462 - 54		23 29 42	+19 34.2	1	10.80		0.97		0.76		1		1696
221537	+77 00909			23 29 44	+77 32.6	1	7.02		0.05		0.03		2	A0	1375
		CS 22894 # 6		23 29 44	−00 21.0	2	15.54	.000	0.11	.014	0.16	.020	2		966,1736
		BSD 19 # 200		23 29 48	+59 16.4	1	12.49		0.57		0.20		3	B8	1069
		BSD 19 # 768		23 29 48	+59 59.5	1	10.38		0.74		0.24		2	F8	1069
		LS I +60 006		23 29 49	+60 28.5	1	11.69		0.74		-0.08		3		41
		G 242 - 35		23 29 49	+81 08.1	1	10.19		0.77		0.46		3	K1	7010
	−22 16355			23 29 49	−22 11.4	1	10.89		0.64		0.03		2		1696
	+55 02973			23 29 50	+56 26.6	1	9.83		1.92				2		369
221472	−11 06098			23 29 50	−11 16.5	1	6.80		0.91		0.59		8	G5	1628
221490	+35 05047			23 29 55	+35 43.9	1	8.59		0.34		0.04		3	F0	1601
221478	+25 04957			23 29 57	+26 14.5	1	7.93		0.99		0.78		3	G8 II-III	8100
221491	+34 04948	HR 8936		23 29 57	+34 40.6	1	6.65		-0.06		-0.24		1	B8 V	1221
		LS I +61 011		23 30 02	+61 53.8	1	11.16		0.90		-0.10		2		41
		WLS 2345 0 # 6		23 30 03	+00 01.1	1	11.45		1.18		1.01		2		1375
		LS I +59 004		23 30 04	+59 48.7	1	11.53		1.05		-0.05		2		41
		LS I +60 007		23 30 04	+60 23.5	1	11.51		0.83		-0.14		3		41
		CS 22894 # 4		23 30 06	−01 12.8	1	14.17		0.41		-0.17		1		1736
	−17 06768	G 273 - 67	⋆ BC	23 30 07	−17 01.3	2	10.39	.014	1.49	.010	0.95		4	M3	196,1705
221420	−78 01473	HR 8935		23 30 08	−77 39.7	3	5.81	.003	0.68	.000	0.30	.005	27	G2 V	15,1075,2038
		CS 22941 # 17		23 30 10	−34 16.0	2	14.99	.010	0.58	.013	-0.10	.002	3		1580,1736
		CS 22941 # 20		23 30 11	−34 58.2	2	14.68	.010	0.40	.007	-0.19	.009	4		1580,1736
221538	+52 03469			23 30 12	+53 24.6	2	6.82	.010	1.07	.005	0.87	.005	5	K0	985,1733
221503	−17 06769	G 273 - 68	⋆ A	23 30 12	−17 07.1	4	8.60	.013	1.29	.019	1.20	.024	10	K5/M0 V	158,196,1619,3072
		UKST 1038 # 7		23 30 14	+09 40.2	1	12.84		0.66				2		1584
		G 68 - 26		23 30 17	+20 03.7	1	11.00		1.37		1.28		1	M1	333,1620
221507	−38 15527	HR 8937		23 30 18	−38 05.7	9	4.37	.008	-0.09	.007	-0.35	.029	34	B9.5IVp	3,15,1034,1075,1202*
221556	+61 02469			23 30 19	+61 59.9	1	9.25		0.09				2	A0	1118
221531	−12 06514			23 30 23	−12 15.8	1	8.30		0.42		-0.01		2	Ap Sr	1375
221586	+57 02758	V436 Cas		23 30 27	+57 37.8	1	7.61		0.11		-0.08		2	A0p	401
221584	+62 02245			23 30 27	+63 00.9	1	8.61		0.49				3	F7 V	1118
		LS I +59 005		23 30 28	+59 40.6	2	11.01	.033	0.85	.005	-0.32	.019	4		41,1069
		CS 22952 # 2		23 30 29	−05 29.4	1	14.38		0.57		0.04		1		1736
		BSD 19 # 1449		23 30 30	+60 22.5	1	10.03		1.22		1.02		2	G7	1069
		CS 22941 # 22		23 30 32	−35 38.6	1	14.87		0.17		0.10		1		1736
	−49 14178			23 30 32	−48 54.4	1	10.81		1.37		1.60		2		1730
221533	−49 14179			23 30 32	−48 58.1	1	9.93		0.54		0.06		2	G0 V	1730
		UKST 1038 # 5		23 30 33	+09 22.2	1	12.00		0.70				3		1584
		LS III +60 085		23 30 33	+60 47.0	1	11.82		0.66		-0.23		3		888
221585	+62 02244	G 241 - 58		23 30 33	+62 52.7	3	7.46	.017	0.74	.005	0.38	.005	6	G8 IV	1003,1355,1658
		LS I +60 008		23 30 38	+60 23.1	2	10.95	.010	0.75	.000	-0.24	.020	5	O9.5V	41,1069
221564	−13 06428			23 30 38	−12 53.1	1	7.07		0.15		0.12		4	A3/5 IV	152
		G 216 - 52		23 30 39	+45 39.9	1	13.69		0.87		0.29		2		1658
221565	−21 06437	HR 8939	⋆ AB	23 30 40	−21 11.5	5	4.70	.018	0.02	.000	0.00	.010	15	A0 V	15,1075,2012,3023,8015
	+61 02472	LS I +61 012		23 30 41	+61 56.4	1	9.65		1.13		0.12		2	B6 Ia	41
		G 68 - 28		23 30 51	+15 25.2	1	11.27		0.92		0.55		1	K2	333,1620
	+9 05236			23 30 53	+09 45.5	1	9.48		0.46				2	F2	1584
		UKST 1038 # 18		23 30 54	+09 13.6	1	15.49		0.94				1		1584
		UKST 1038 # 20		23 30 54	+09 28.0	1	16.12		0.85				1		1584
		CS 22894 # 9		23 30 55	+00 41.0	1	14.14		0.52		-0.11		1		1736
221613	+42 04700	G 171 - 3		23 30 57	+42 34.1	1	7.14		0.58		0.03		3	G0	196
		G 157 - 72		23 30 57	−08 46.9	1	15.95		0.89		0.34		2		3062
221615	+21 04952	HR 8940, HW Peg	⋆	23 30 58	+22 13.4	2	5.33	.008	1.59	.010	1.61		6	M5 IIIa	71,3001
		BSD 19 # 1960		23 30 59	+61 32.2	1	12.33		0.53		-0.27		2	B4	1069
		CS 22894 # 11		23 30 60	+01 11.4	1	15.14		0.31		-0.05		1		1736
240329	+58 02612			23 31 00	+59 01.8	2	8.55	.030	0.51	.010			4	F6 III	1118,1119
221639	+59 02744	IDS23289N5954	B	23 31 00	+60 08.1	2	7.18	.000	0.92	.000	0.71		6	K1 V	1119,1355
		PHL 5950		23 31 00	−21 02.0	1	10.99		0.42		-0.14		1		3060
221627	+17 04938	IDS23285N1715	A	23 31 03	+17 32.5	2	6.81	.010	0.65	.000	0.21	.014	7	G2 IV	1648,3026
221627	+17 04938	IDS23285N1715	B	23 31 03	+17 32.5	1	13.16		0.89		0.38		2		3016

Table 1 1231

HD	DM	Other Id	N	Rem	α_{1950}	δ_{1950}	S	V	σ_V	B–V	σ_{B-V}	U–B	σ_{U-B}	n	Spectrum	References
		G 128 - 77			23 31 07	+32 17.7	1	13.35		0.89		0.41		2		1658
		CS 22894 # 8			23 31 07	−00 27.0	2	13.12	.000	0.33	.000	0.11	.005	2		966,1736
		UKST 1038 # 15			23 31 08	+09 50.2	1	14.72		1.50		1.08		1		1584
		UKST 1038 # 6			23 31 08	+09 52.6	1	12.82		0.88		0.42		2		1584
221617	−12 06517				23 31 09	−12 15.6	1	10.94		0.13		0.12		4	A1/2 (III)	152
		WLS 2345 0 # 5			23 31 11	+02 08.0	1	11.16		1.09		0.87		2		1375
		PG2331+055			23 31 11	+05 30.0	1	15.18		-0.01		-0.49		1		1764
		CS 22894 # 5			23 31 14	−00 48.7	1	14.51		0.47		-0.14		1		1736
221650	+48 04093	Z And			23 31 15	+48 32.5	4	10.52	.021	1.23	.133	-0.37	.046	4	M6.5	1564,1591,1753,8091
		PG2331+055A			23 31 16	+05 30.2	1	13.05		0.74		0.26		1		1764
221671	+59 02745				23 31 16	+59 46.5	2	7.61	.025	0.08	.010			4	A0 II	1118,1119
221670	+59 02746	IDS23289N5954		A	23 31 16	+60 11.5	1	7.34		0.99				2	G9 III	1119
		UKST 1038 # 10			23 31 17	+09 47.9	1	14.10		0.70				3		1584
221661	+44 04441	HR 8941			23 31 17	+44 46.9	1	6.24		0.98				2	G8 II	71
		BSD 19 # 1963			23 31 17	+61 33.8	1	11.13		0.64		-0.13		3	B2	1069
		CS 22941 # 24			23 31 17	−36 20.2	2	15.79	.070	-0.12	.056	-0.21	.023	1	sdO	966,1736
		PG2331+055B			23 31 18	+05 28.5	1	14.74		0.82		0.43		1		1764
		LS I +60 009			23 31 18	+60 16.9	1	11.37		0.90		-0.23		3		41
		TonS 103			23 31 18	−29 08.	1	14.66		-0.23		-1.06		1		1036
		LS I +60 010			23 31 19	+60 48.2	2	10.96	.042	0.51	.000	-0.49	.028	4		41,1069
		CS 22894 # 3			23 31 22	−01 28.8	2	14.64	.000	0.14	.004	0.15	.036	2		966,1736
221662	+20 05352	HR 8942			23 31 25	+20 33.9	2	6.06	.001	1.71	.015	1.88		6	M3 III	71,3001
		CS 22941 # 13			23 31 25	−32 41.3	1	14.04		0.23		0.04		1		1736
221673	+30 04978	HR 8943		⋆ AB	23 31 28	+31 03.0	4	4.97	.013	1.39	.018	1.63	.014	9	K3III+K5III	1355,1363,1381,3030
221692	+56 03028				23 31 29	+56 56.7	1	8.33		0.14		-0.34		2	B9	401
	+60 02575				23 31 30	+61 15.6	1	9.72		2.01				2		369
		CS 22966 # 15			23 31 30	−28 46.8	1	15.54		0.50		-0.08		1		1736
		CS 22941 # 26			23 31 31	−37 16.5	2	15.45	.005	0.10	.001	0.15	.017	2		966,1736
		G 29 - 43			23 31 33	−00 05.3	3	11.17	.015	1.46	.008	1.10	.014	5	M2-3	1620,1705,3078
221675	−2 05986	HR 8944			23 31 35	−01 31.4	4	5.89	.017	0.30	.004	0.15	.011	14	A2 m	15,1071,3016,8071
		LB 1526			23 31 36	−47 30.	1	13.42		-0.31		-1.18		4		45
		BSD 19 # 800			23 31 37	+59 38.7	1	11.66		0.81		0.43		2	B0	1069
240331	+58 02613	IDS23294N5914		AB	23 31 41	+59 30.2	1	9.12		0.52				2	A4 Ib	1119
		UKST 1038 # 19			23 31 44	+09 11.0	1	15.58		0.62				2		1584
	+61 02474				23 31 45	+61 33.5	1	9.35		0.50		0.06		2	F5	401
		UKST 1038 # 17			23 31 46	+09 16.9	1	15.35		0.65		0.10		1		1584
221711	+54 03006				23 31 47	+55 12.8	3	7.53	.007	-0.09	.000	-0.61	.007	7	B2 V	399,401,963
		G 29 - 44			23 31 49	+09 15.7	3	11.24	.023	1.32	.013	1.18	.015	3	M0	333,1620,1632,1696
		UKST 1038 # 13			23 31 52	+09 19.8	1	14.66		0.76		0.24		3		1584
240333	+56 03031				23 31 54	+56 56.4	1	8.89		0.16		-0.21		1	B5	401
		BSD 19 # 218			23 31 55	+58 51.7	1	10.44		1.46		1.17		2	dG5	1069
		BSD 19 # 805			23 31 55	+59 52.7	1	11.26		0.71		0.27		2	dG7	1069
		WLS 2340-10 # 7			23 31 55	−09 33.4	1	11.85		0.59		0.05		2		1375
		UKST 1038 # 14			23 31 58	+09 21.3	1	14.71		0.76		0.21		2		1584
221741	+60 02578				23 32 00	+61 09.7	3	8.88	.016	0.16	.014	0.15		7	A3 V	401,1118,1119
	+33 04737	G 190 - 34			23 32 06	+33 45.7	3	9.04	.000	0.80	.000	0.39	.000	6	K0	516,1064,1620
	+9 05244				23 32 08	+09 32.8	1	9.90		1.15				2	K2	1584
221758	+32 04667	HR 8948			23 32 10	+33 13.2	1	5.63		1.03				2	gK0	71
221756	+39 05114	HR 8947			23 32 10	+39 57.6	5	5.59	.005	0.10	.007	0.08	.004	20	A1 III	15,1007,1013,1049,8015
	−35 15842				23 32 11	−34 56.7	1	10.28		1.18		1.21		2	K5 V	3072
221745	−16 06314	HR 8946			23 32 13	−15 31.3	3	5.95	.009	1.35	.007	1.57	.012	10	K4 III	58,244,2035
221737	−49 14186				23 32 13	−48 54.4	1	9.05		0.72				4	G6 V	2012
		G 29 - 45			23 32 15	+04 37.2	1	11.85		1.14		1.06		1		333,1620
		G 29 - 46			23 32 17	+00 53.0	1	12.60		1.21		1.04		1		333,1620
221738	−57 10297				23 32 17	−57 06.1	1	6.73		0.38		0.08		9	F2 IV	1628
221776	+37 04866	HR 8950		⋆ A	23 32 19	+37 44.8	2	6.17	.012	1.59	.005	1.97		6	K5 III	71,1501
		G 273 - 76			23 32 19	−20 55.0	1	11.91		0.76		0.08		2		1696
		CS 22941 # 27			23 32 19	−37 08.8	2	14.04	.009	0.36	.006	0.01	.011	4		1580,1736
	+60 02580				23 32 21	+61 22.0	1	9.84		0.09				2	A0 V	1119
221739	−62 06428				23 32 21	−61 44.4	1	9.48		0.94		0.66		2	K3 V	3072
		CS 22941 # 21			23 32 22	−35 09.4	1	13.89		0.42		-0.08		1		1736
221760	−43 15420	HR 8949, ι Phe		⋆ AB	23 32 24	−42 53.5	5	4.71	.005	0.08	.006	0.07	.005	21	Ap SrCrEu	15,1075,2012,3023,8015
	+62 02251				23 32 25	+63 22.9	1	9.48		0.17				3	A0	1118
	+0 05017	G 29 - 47			23 32 26	+01 19.7	3	9.59	.008	1.35	.004	1.19	.010	4	M0	333,1620,1705,3072
	+58 02615	LS III +58 085			23 32 28	+58 34.8	1	10.01		0.64		-0.24		3		888
221777	−8 06141				23 32 29	−07 57.2	3	7.37	.012	1.27	.008	1.38	.010	10	K4 III	1375,1509,1657
		G 68 - 29			23 32 31	+25 07.5	1	10.88		0.81		0.38		1	K1	333,1620
		G 128 - 80			23 32 31	+32 44.9	1	11.99		0.82		0.48		1		1658
		G 157 - 77			23 32 35	−02 39.4	1	14.70		1.95				3		3078
	+57 02760				23 32 37	+58 17.7	1	9.69		2.63				2	M2	369
		CS 22941 # 28			23 32 40	−37 05.0	1	14.41		0.05		0.12		1		1736
		G 241 - 59			23 32 42	+68 01.6	1	10.80		1.07		1.00		1		1658
221793	−41 15278				23 32 43	−41 34.2	1	10.21		0.07		0.07		2	A0 V	152
	+86 00345				23 32 44	+87 22.1	1	10.82		0.71		0.21		2	G5	1375
	+9 05246				23 32 47	+09 58.4	1	10.82		0.63				2		1584
221861	+70 01327	HR 8952			23 32 48	+71 21.9	1	5.84		1.80		1.73		1	K0 Ib	1218
221822	+1 04740	G 29 - 48			23 32 49	+01 56.7	1	8.40		0.72		0.19		1	G5	333,1620
		BSD 19 # 818			23 32 49	+59 41.0	1	11.22		0.61		0.12		2	F8	1069
221805	−42 16386				23 32 49	−41 43.8	1	9.68		0.06		0.07		4	A0 V	152
		UKST 1038 # 9			23 32 50	+09 06.7	1	13.13		0.66		0.18		4		1584

HD	DM	Other Id	N	Rem	α_{1950}	δ_{1950}	S	V	σ_V	B–V	σ_{B-V}	U–B	σ_{U-B}	n	Spectrum	References
	+60 02581	LS III +61 045			23 32 50	+61 07.4	2	10.60	.000	0.67	.018	-0.07	.013	5	B3 V	888,1012
221818	-47 14628				23 32 52	-47 13.1	4	8.55	.014	0.79	.007	0.34	.021	14	K0 V	158,1075,2012,3008
		LS I +59 006			23 32 54	+59 43.8	1	11.08		1.02		-0.18		2		41
221862	+66 01619				23 32 54	+67 12.9	1	7.17		1.10		1.00		2	K0	1502
221833	+0 05018				23 32 55	+01 02.2	1	6.48		1.12		1.11		2	K0	3016
		BSD 19 # 822			23 32 55	+59 23.0	1	11.46		0.60		0.12		2	dG3	1069
		BSD 19 # 823			23 32 55	+59 35.1	1	11.15		0.54		0.07		3	dG3	1069
	-2 05993				23 32 55	-02 06.0	1	9.42		0.60		0.20		2	G0	1375
221836	-21 06443				23 32 56	-20 57.9	1	11.14		-0.01		-0.01		4	B8/9 Ib	152
221851	+30 04983	G 128 - 83			23 32 57	+30 53.3	2	7.90	.005	0.85	.005	0.50	.010	2	G5	1620,1658
221835	-8 06142	HR 8951			23 32 57	-07 44.5	3	6.40	.018	0.87	.010	0.57	.005	10	K0	15,1071,3016
		UKST 1038 # 12			23 32 58	+09 06.9	1	14.40		0.81		0.24		2		1584
221830	+30 04982	IDS23305N3027			23 32 58	+30 44.2	3	6.85	.009	0.60	.015	0.06	.009	7	F9 V	22,1003,3077
		CS 22941 # 23			23 32 58	-35 58.0	2	15.69	.015	0.17	.005	-0.02	.013	2		966,1736
		LP 463 - 17			23 33 00	+15 52.9	1	12.47		1.04		0.95		1		1696
221839	-28 18257	IDS23304S2802		AB	23 33 02	-27 45.9	2	6.65	.010	0.56	.000	-0.05		7	F8 V	173,2012
240335	+58 02616				23 33 03	+59 29.4	1	9.36		0.49				2	F7 V	1119
		CS 22966 # 10			23 33 04	-30 38.8	1	15.39		0.52		-0.06		1		1736
221875	+53 03196				23 33 06	+53 44.4	1	8.27		0.30		0.06		2	A2	1566
		CS 22941 # 40			23 33 06	-33 11.1	1	15.00		0.22		0.04		1		1736
	+61 02481	LS I +62 003			23 33 08	+62 01.3	3	9.90	.022	1.13	.009	0.00	.005	4	B0	41,851,1069
221886	+58 02617				23 33 09	+58 39.3	3	8.21	.021	0.16	.005	0.08		6	A2 II	401,1118,1119
	+60 02582	LS I +60 011			23 33 10	+60 38.2	4	8.69	.261	0.77	.016	0.02	.011	20	B8 Iab	41,1012,1069,1119
		UKST 1038 # 16			23 33 11	+09 38.6	1	15.10		0.88				2		1584
221858	-30 19610				23 33 11	-30 12.6	1	9.46		0.92				4	K0 III +F	955
221876	+19 05124	G 68 - 30			23 33 12	+20 18.3	5	9.10	.015	0.56	.000	-0.04	.000	8	G2	516,979,1620,1658,8112
		LP 360 - 51			23 33 12	-46 27.0	1	13.32		1.52		1.17		1		1773
		CS 22894 # 14			23 33 13	+00 36.3	1	14.53		0.12		0.14		1		1736
221871	-46 14707	IDS23306S4642		A	23 33 13	-46 25.9	3	9.71	.015	1.18	.019	1.13		6	K4 V	1494,1773,2033
		G 171 - 5			23 33 15	+41 41.3	1	11.17		1.37		1.27		2		7010
221900	+60 02583				23 33 16	+61 05.5	2	8.70	.025	0.39	.005			4	F3 V	1118,1119
		G 29 - 51			23 33 18	+04 45.0	2	11.96	.040	1.29	.005	1.07		2	K4-5r	333,1620,1746
	-0 04534				23 33 19	+00 11.1	1	10.17		0.90		0.53		1	K0	333,1620
	+60 02584	LS I +60 012			23 33 21	+60 54.7	2	10.30	.005	0.55	.014	-0.59	.005	4	B1 IV:pe	41,1012
		G 30 - 2			23 33 23	+08 13.9	1	12.52		1.02		0.69		2		333,1620
		BPM 45042			23 33 24	-48 13.	1	14.06		0.38		-0.18		4		3065
	+60 02585				23 33 25	+60 51.6	1	9.80		2.10				2		369
221905	+23 04769	HR 8953		V	23 33 26	+24 17.1	3	6.45	.008	1.71	.013	2.03	.038	12	M0 III	985,1501,1733
		CS 22941 # 29			23 33 28	-37 08.1	2	15.06	.010	0.07	.001	0.12	.013	2		966,1736
221935	+60 02586				23 33 30	+60 37.6	3	8.36	.024	0.20	.022	-0.18		7	B7 IV	1069,1118,1119
		CS 22941 # 39			23 33 30	-33 34.8	2	15.72	.040	0.22	.016	0.02	.007	2		966,1736
221914	+17 04946	G 68 - 31			23 33 32	+18 09.8	4	7.65	.011	0.70	.004	0.23	.009	7	G5 V	22,333,1003,1620,3077
	+34 04960				23 33 34	+34 49.5	1	9.46		0.38		-0.03		3	F0	1625
		LS I +61 013			23 33 37	+61 40.4	1	11.76		0.99		-0.08		2		41
		LS I +60 013			23 33 39	+60 51.9	2	12.02	.027	0.54	.004	-0.34	.009	5	B5	41,1069
		CS 22966 # 4			23 33 39	-32 01.4	1	15.53		0.13		0.15		1		1736
		CS 22941 # 34			23 33 39	-36 25.8	1	13.76		0.21		0.01		1		1736
	-33 16613	G 275 - 73		★ A	23 33 40	-33 28.4	2	10.16	.033	1.17	.019	1.14	.014	4	K7 V	1773,3072
		LP 763 - 35			23 33 42	-11 42.2	1	12.87		0.54		-0.18		2		1696
	-33 16613	G 275 - 74		★ B	23 33 43	-33 28.2	1	13.20		1.26		1.21		1		3072
		BSD 19 # 831			23 33 44	+59 23.7	1	11.25		0.58		0.11		3	G0	1069
		CS 29499 # 12			23 33 44	-25 09.7	1	15.37		0.01		0.10		1		1736
		LP 986 - 88			23 33 45	-33 28.1	1	12.48		1.51		1.16		1		1773
		UKST 1038 # 11			23 33 47	+09 54.3	1	14.28		0.88		0.41		1		1584
240338	+57 02763				23 33 48	+58 30.3	2	9.40	.015	0.40	.025	0.06		4	B8 V	1069,1119
	+60 02587				23 33 48	+61 31.1	2	9.49	.009	0.44	.019	-0.09		4	B5 V	401,1119
		Smethells 140			23 33 48	-52 43.	1	11.53		1.09				1		1494
	+8 05085				23 33 49	+09 08.0	1	9.45		0.62				2	F8	1584
221950	+1 04744	HR 8954			23 33 50	+01 49.5	5	5.70	.020	0.44	.012	-0.07	.015	18	F6 Vb	3,15,254,1071,3077
		CS 22894 # 17			23 33 50	-00 12.4	1	14.13		0.48		-0.14		1		1736
		UKST 1038 # 8			23 33 51	+09 53.0	1	12.99		0.71		0.22		3		1584
	+58 02620				23 33 52	+58 56.2	2	9.66	.000	0.52	.007	0.14		4	B6 V	1069,1119
	+32 04670	G 128 - 84			23 33 53	+32 45.4	3	10.12	.007	0.91	.018	0.61	.008	4	K2	1620,1658,7010
221943	-45 15080				23 33 53	-45 10.2	3	6.91	.010	0.20	.004	0.09		16	A3 III/IV	1075,1499,2012
		LS I +60 014			23 33 56	+60 48.0	2	11.70	.024	0.63	.000	-0.26	.009	6		41,1069
		CS 22894 # 13			23 33 56	-01 31.0	2	15.16	.015	0.14	.015	0.22	.003	2		966,1736
		G 157 - 85			23 33 56	-08 42.4	2	11.99	.015	0.52	.010	-0.18	.015	3		1696,3062
		G 217 - 13			23 34 00	+55 13.2	1	11.72		1.46		1.19		3	M1	7010
221970	+32 04671	HR 8955			23 34 01	+32 37.6	2	6.34	.035	0.46	.005	0.00	.015	6	F5 III	254,272
221990	+61 02484				23 34 01	+62 09.2	2	8.07	.000	0.07	.005	-0.19		4	A0	401,1118
		G 29 - 52			23 34 03	+10 06.9	1	12.22		1.10		1.04		1	K5	333,1620
		LP 823 - 22			23 34 06	-17 04.2	1	12.19		0.88		0.58		1		1696
221974	-18 06342	G 273 - 85			23 34 08	-17 30.8	1	9.31		0.92		0.69		2	K1 III	3008
		LP 986 - 92			23 34 08	-36 45.5	1	13.72		1.62		1.36		1		3078
	+60 02589	LS I +61 014			23 34 12	+61 26.5	1	9.83		0.97		0.63		2		41
	+60 02590				23 34 13	+60 56.6	2	9.73	.028	0.18	.005	0.00		4	B9 V	401,1119
		G 29 - 53			23 34 15	+00 53.4	2	13.13	.039	1.55	.034	1.13		3		1705,3078
	+47 04206				23 34 15	+47 57.9	1	9.30		0.37				1	F2	21
		G 241 - 60			23 34 15	+59 41.6	1	12.81		0.65		-0.10		2		1658
		LS I +60 015			23 34 19	+60 00.3	1	11.23		1.18		0.30		3		41

Table 1 1233

HD	DM	Other Id	N	Rem	α_{1950}	δ_{1950}	S	V	σ_V	B–V	σ_{B-V}	U–B	σ_{U-B}	n	Spectrum	References
		LS III +58 086			23 34 20	+58 22.9	1	11.58		0.77		−0.33		3		888
		CS 29499 # 14			23 34 20	−24 01.5	1	15.46		0.13		0.12		1		1736
221984	−42 16401				23 34 22	−42 19.8	1	8.70		0.08		0.08		4	A2 V	152
		G 68 - 32			23 34 23	+23 58.2	1	13.04		1.55		1.20		1		333,1620
222008	+10 04963				23 34 25	+10 44.3	1	7.93		0.23		0.12		2	A5	1733
222017	+47 04209				23 34 26	+47 55.3	1	8.94		0.12				1	A0	21
222004	−32 17593	HR 8956		⋆ AB	23 34 27	−32 08.9	2	6.50	.005	1.25	.005	1.21		6	K1 III	404,2007
	+58 02622				23 34 29	+58 59.1	1	9.94		0.55				2	B8 V	1119
	+47 04210				23 34 30	+48 08.0	1	10.48		0.48				1		21
		PHL 5980			23 34 30	−26 31.	2	12.10	.015	0.14	.019	0.12	.023	2		1736,3060
222012	−26 16713				23 34 33	−26 08.5	1	8.94		0.08		0.07		4	A1 V	152
222013	−46 14713				23 34 34	−45 45.0	2	9.20	.039	0.82	.000	0.45		6	K0 V	2012,3077
222031	+37 04872				23 34 35	+38 27.7	1	8.35		1.48		1.59		2	K2	1601
222033	+29 04971				23 34 36	+30 24.0	2	7.20	.015	0.61	.015	0.13		5	G0 V	20,3026
222046	+47 04211				23 34 36	+48 18.0	1	8.65		−0.06				1	B9	21
	+60 02591				23 34 38	+61 06.7	1	10.04		0.21				2	B8 V	1119
	+60 02592				23 34 38	+61 30.7	1	10.09		0.12				2	A0	1119
222025	−52 12184				23 34 41	−52 00.6	1	8.00		1.10		0.99		7	K0/1 III	1673
222040	−21 06451				23 34 42	−21 12.7	1	10.87		0.10		0.17		4	B9 IV	152
		LS I +60 016			23 34 43	+60 23.5	1	11.79		0.83		−0.21		3		41
	+47 04213				23 34 44	+48 23.6	1	9.38		0.41				1	F2	21
	+60 02593				23 34 44	+61 08.4	2	9.39	.009	0.20	.024	0.13		4	A2 V	401,1119
		CS 22952 # 11			23 34 44	−02 25.1	1	13.75		0.40		−0.19		2		1736
	+24 04803				23 34 47	+25 29.6	1	8.66		0.91				3	G8 III	20
		G 29 - 54			23 34 50	+08 44.6	1	12.30		1.37		1.22		1	M0	333,1620
		G 29 - 55			23 34 52	+12 27.4	1	11.63		1.08		0.89		1	K3	333,1620
222086	+47 04214				23 34 52	+48 12.3	1	8.55		0.00				1	B9	21
		LP 763 - 87			23 34 53	−14 04.2	1	11.99		0.41		−0.24		2		1696
		CS 22952 # 15			23 34 54	−06 04.5	1	13.28		0.75		0.18		1		1736
		JL 141			23 34 54	−80 21.	1	15.51		0.04		0.21		1		832
	+61 02487	RS Cas			23 34 55	+62 09.1	2	9.56	.000	1.27	.012	0.89		2	G0.5Ib	851,1399
	+18 05180				23 34 56	+19 30.0	1	8.98		1.38		1.49		2	K0	1648
		BSD 19 # 1534			23 34 56	+60 28.2	1	11.74		0.53		0.10		2	B8	1069
222089	+35 05066				23 34 59	+35 32.3	1	7.17		0.97		0.73		2	K2	1601
		CS 22966 # 18			23 34 59	−28 16.1	1	14.41		0.09		0.17		1		1736
		CS 22941 # 31			23 34 59	−36 40.9	2	15.06	.025	−0.02	.009	−0.03	.016	2		966,1736
	+47 04215				23 35 01	+48 01.1	1	13.21		0.86				1		21
		BSD 19 # 1535			23 35 01	+61 09.7	1	12.26		0.43		0.03		2	B7	1069
		LS III +61 046			23 35 02	+61 23.8	1	10.89		0.51		−0.14		3		888
	+60 02594				23 35 02	+61 30.3	1	10.15		0.44		−0.06		2		401
222093	−13 06439	HR 8958		⋆ A	23 35 04	−13 20.3	3	5.65	.014	1.02	.014	0.82	.011	9	K0 III	58,2007,3016
222109	+43 04508	HR 8962		⋆ AB	23 35 05	+44 09.1	2	5.80		−0.06	.000	−0.32	.000	4	B8 V	1079,1501
		LTT 9631			23 35 06	−41 27.	1	14.94		−0.01		−1.02		2		3060
222107	+45 04283	HR 8961, λ And		⋆ A	23 35 07	+46 11.2	7	3.82	.025	1.01	.014	0.69	.007	26	G8 III-IV	1,15,1355,1363,4001*
222105	+47 04217				23 35 07	+48 24.0	1	9.47		0.46				1	G0	21
	+59 02749				23 35 07	+60 05.9	1	10.35		0.62		0.20		2	A0	1069
222098	+16 04954	HR 8960		⋆ AB	23 35 08	+16 32.9	2	6.26		0.02	.015	0.03	.037	5	A1 V	1049,3050
		BSD 19 # 243			23 35 09	+59 03.8	1	9.95		0.43				2	A0	1119
222095	−46 14720	HR 8959			23 35 10	−45 46.2	6	4.73	.009	0.08	.005	0.09	.005	22	A2 V	15,1075,2012,2024*
		BM And			23 35 13	+48 07.5	1	12.47		1.06		0.61		1	F8	856
		CS 22966 # 7			23 35 15	−31 22.0	1	15.03		−0.07		−0.20		1		1736
222076	−71 02767				23 35 15	−71 10.8	2	7.47	.000	1.03	.000	0.92		13	K1 III	1499,2012
240344	+59 02751				23 35 18	+60 06.5	2	8.89	.005	0.56	.010	−0.07		5	B5 V	1069,1119
	+47 04218				23 35 20	+48 10.4	1	10.91		0.35				1	A7	21
		CS 22966 # 23			23 35 20	−29 35.4	1	15.28		0.09		0.18		2		1736
222060	−77 01583	HR 8957			23 35 20	−77 08.8	3	5.98	.007	0.91	.005	0.65	.000	15	K0 II/III	15,1075,2038
	+47 04219				23 35 21	+48 22.0	1	10.79		0.47				1	F7	21
		LS I +59 007			23 35 23	+59 25.2	2	11.11	.005	0.91	.020	−0.22	.015	4		41,405
		G 157 - 88			23 35 23	−08 02.1	2	13.55	.019	0.95	.015	0.63	.037	3		1658,3062
		LS I +59 008			23 35 24	+59 14.3	1	11.94		0.74		−0.25		2		41
		WLS 2300 75 # 8			23 35 24	+75 04.9	1	11.39		0.70		0.09		2		1375
		CS 29499 # 9			23 35 24	−25 39.5	1	12.51		0.23		0.08		1		1736
		Smethells 141			23 35 24	−41 50.	1	12.76		1.35				1		1494
		L 120 - 191			23 35 24	−69 22.	1	13.44		1.44		1.22		1		3073
222133	+17 04952	HR 8963, KS Peg		⋆ A	23 35 25	+18 07.4	3	5.50	.051	−0.01	.011	0.01	.037	7	A1 Vn	985,1049,3023
		LS I +61 015			23 35 25	+61 12.3	1	12.17		0.56		−0.26		2		41
222142	+47 04220				23 35 26	+48 13.2	2	9.56	.036	0.16	.009	−0.17		5	B8 V	21,206
240346	+59 02752				23 35 27	+60 25.7	2	9.56	.020	0.48	.020	0.00		5	F2	1119,3016
222125	−15 06464				23 35 27	−15 22.2	1	6.39		1.08		0.92		4	K0 III	244
222126	−24 17776				23 35 28	−24 08.4	1	9.81		0.32		0.16		2	A0mF0/2-F3	3060
222143	+45 04288	HR 8964		⋆	23 35 30	+45 55.4	2	6.58	.000	0.65	.010	0.16	.005	4	G5	1733,3016
		LS III +61 047			23 35 30	+61 03.0	1	11.60		0.55		−0.38		3		888
		CS 22894 # 20			23 35 31	−00 41.9	1	14.86		0.34		−0.05		1		1736
222155	+48 04112				23 35 33	+48 43.3	1	7.11		0.65				1	G0	21
	−17 06785	G 273 - 93			23 35 33	−16 30.8	1	11.34		1.57				1	M2	1746
240348	+59 02753	LS I +59 009			23 35 36	+59 33.3	2	9.26	.024	0.85	.019	0.58		4	B9 V	41,1119
	+58 02624				23 35 38	+59 24.8	1	11.11		0.89		−0.19		2	B3	1069
240349	+58 02625				23 35 40	+59 00.3	1	9.01		0.44				2	F5 V	1119
222173	+42 04720	HR 8965			23 35 41	+42 59.5	7	4.28	.012	−0.10	.010	−0.30	.022	15	B8 V	15,1079,1119,1203*
		VES 964			23 35 41	+61 38.8	1	12.44		0.73		−0.14		1		8113

HD	DM	Other Id	N Rem	α_{1950}	δ_{1950}	S	V	σ_V	B–V	σ_{B-V}	U–B	σ_{U-B}	n	Spectrum	References
	+47 04224			23 35 47	+48 26.8	1	9.63		0.47				1	F5	21
222163	−39 15099			23 35 47	−39 16.8	1	9.02		1.23				4	K2 III	2033
222200	+48 04114			23 35 48	+48 39.3	1	9.06		0.09				1	A2	21
		LS I +59 010		23 35 49	+59 42.3	3	10.82	.014	0.79	.025	-0.34	.015	6	B0.5IV	41,684,1069
		WLS 2345 0 # 10		23 35 50	+00 05.3	1	11.48		0.73		0.23		2		1375
	+47 04225			23 35 54	+48 09.3	1	10.42		0.80				1		21
222167	−63 04913			23 35 57	−63 09.7	1	6.62		0.93				4	G8 III	2033
222218	+57 02770			23 35 59	+58 22.6	1	6.94		1.02				2	K1 III	1119
	−10 06147	BR Aqr		23 35 59	−09 36.2	2	10.77	.001	0.12	.014	0.13	.025	2	F1	597,699
		LS I +60 017		23 36 00	+60 21.6	1	11.37		0.92		-0.11		3		41
		PG2336+004B		23 36 05	+00 26.1	1	12.43		0.51		-0.04		1		1764
		CS 29499 # 22		23 36 08	−22 48.0	1	12.08		0.05		0.15		1		1736
		PG2336+004A		23 36 09	+00 25.8	1	11.28		0.68		0.14		1		1764
		PG2336+004		23 36 10	+00 26.3	1	15.90		-0.17		-0.78		1	HBB	1764
222238	+48 04115			23 36 12	+48 36.1	1	9.29		0.10				1	A2	21
	+60 02596			23 36 12	+61 29.5	1	9.81		0.41				2	F3	1119
	−30 19633	G 275 - 76		23 36 14	−29 45.3	1	11.34		0.59		-0.04		2		1696
222226	−46 14730			23 36 15	−45 53.3	4	7.00	.007	0.30	.007	0.01		12	A7/8 III	116,1020,2012,2012
		CS 22941 # 37		23 36 16	−35 33.2	1	13.32		0.09		0.14		1		1736
	+25 04980			23 36 17	+25 40.2	1	8.21		1.11				2	K1 III	20
	+47 04227			23 36 20	+48 14.9	1	10.82		0.36				1		21
		DW Cas		23 36 22	+59 04.5	1	10.79		1.30		0.76		1	F7	1399
		LS I +61 016	V	23 36 22	+61 46.5	1	11.80		0.81		-0.15		2	OB	41
	+47 04228			23 36 23	+48 15.2	1	11.09		0.39				1		21
		CS 22952 # 18		23 36 25	−04 43.6	1	13.79		0.45		-0.15		1		1736
		CS 29499 # 13		23 36 25	−24 37.4	1	12.77		0.25		0.01		1		1736
	+60 02597		A	23 36 26	+61 19.9	2	10.31	.024	0.10	.014	-0.13		4	A0	401,1119
	+60 02597		B	23 36 26	+61 19.9	1	13.17		1.47				2		1119
		CS 29499 # 19		23 36 26	−23 29.2	1	14.78		0.12		0.16		1		1736
222268	+16 04959			23 36 28	+16 44.2	1	7.97		1.59		1.90		2	K2	1648
222275	+61 02490	LS I +61 017	⋆ A	23 36 28	+61 51.5	2	6.58	.005	0.56	.019	0.49		4	A3 II	41,1118
		CS 29499 # 20		23 36 28	−23 07.9	1	15.58		0.04		0.11		1		1736
222246	−35 15872			23 36 31	−35 11.1	1	9.25		0.47		-0.06		2	F5 V	1730
		G 217 - 14		23 36 41	+59 20.9	1	9.40		0.92		0.71		2		7010
222237	−73 02299			23 36 41	−72 59.3	5	7.09	.012	0.99	.006	0.83	.013	23	K4 V	158,1075,1311,1499,3045
222304	+49 04180	HR 8967		23 36 42	+50 11.7	3	5.35	.030	-0.06	.000	-0.15	.005	5	B9 V	985,1079,3023
		PHL 6002		23 36 42	−27 02.	2	14.62	.037	-0.04	.004	-0.29	.007	4		286,1736
		CS 29499 # 1		23 36 42	−27 41.2	1	14.72		0.01		0.11		1		1736
		CS 22894 # 22		23 36 43	−01 43.3	1	14.84		0.24		0.04		1		1736
	−35 15875			23 36 43	−35 11.2	1	9.39		1.48		1.81		2	K2	1730
		LS I +61 018		23 36 44	+61 41.9	1	10.47		1.15		0.32		2		41
		CS 22894 # 19		23 36 45	−00 12.9	1	13.92		0.44		-0.24		1		1736
222286	−47 14650			23 36 46	−46 43.0	1	9.58		0.08		0.08		4	A0 V	152
222287	−47 14651	HR 8966	⋆ AB	23 36 47	−46 54.9	2	6.07	.005	0.24	.000	0.09		7	A8 V + F0 V	404,2032
	+48 04117			23 36 48	+48 34.5	1	10.49		0.49				1	F8	21
		LS III +59 086		23 36 48	+59 25.4	1	11.68		0.74		-0.06		3		888
222326	+44 04464	IDS23344N4510	AB	23 36 54	+45 26.6	1	7.63		0.43		0.27		2	A2	1601
222340	+61 02492			23 36 58	+61 47.3	2	9.16	.040	0.11	.055			3	A0	851,1118
222317	+27 04588	KT Peg	⋆	23 36 59	+27 58.0	3	7.04	.024	0.65	.013	0.14	.014	6	G2 V	20,333,1620,3026
	−35 15880			23 36 59	−35 14.2	1	11.35		0.68		0.13		2		1730
222386	+74 01032	HR 8971		23 37 01	+75 00.9	2	5.95	.005	0.13	.005	0.12		18	A3 V	252,1258
222332	−23 17984			23 37 05	−22 48.6	1	7.29		0.12				4	A0 V	2012
		LS III +61 049		23 37 07	+61 02.5	2	11.17	.009	0.48	.004	-0.40	.018	5		888,1069
222351	+61 02493			23 37 09	+62 09.6	1	9.03		0.33		0.00		2	F8	401
222387	+73 01047	HR 8972		23 37 09	+73 43.5	1	5.98		0.89				2	G8 III	71
222366	+58 02627			23 37 12	+58 41.6	2	7.50	.030	0.84	.015	0.57		5	K0 V	1119,7008
222345	−15 06471	HR 8968		23 37 12	−14 29.9	2	4.97	.018	0.25	.005			5	A7 IV	2009,2024
222335	−33 16646			23 37 13	−33 01.0	3	7.18	.005	0.81	.000	0.40	.005	14	K1 V	158,1775,2033
		LS I +60 018		23 37 14	+60 46.9	1	11.91		0.78		-0.23		3		41
		BSD 19 # 865		23 37 15	+59 54.6	1	12.12		0.67		0.04		2	F5	1069
		LS I +62 004		23 37 16	+62 23.4	1	11.91		0.88		-0.20		2		41
222404	+76 00928	HR 8974		23 37 17	+77 21.2	13	3.21	.007	1.03	.005	0.95	.004	163	K1 IV	1,15,985,1197,1363*
		LS I +60 019		23 37 21	+60 56.0	1	11.70		0.71		-0.22		3		41
222377	+8 05095	HR 8970		23 37 22	+09 24.0	3	5.99	.023	0.21	.004	0.13	.011	10	A2 m	15,1256,8071
222368	+4 05035	HR 8969	⋆ A	23 37 23	+05 21.3	27	4.13	.006	0.51	.007	0.01	.013	146	F7 V	1,3,15,22,58,418,985*
		LB 1529		23 37 24	−55 48.	1	12.97		0.06		0.13		2		45
		L 121 - 40		23 37 24	−68 19.	1	14.14		0.76		0.20		2		3062
		CS 22894 # 30		23 37 25	−01 05.2	1	14.65		0.41		0.07		1		1736
	+46 04117			23 37 26	+46 55.8	1	9.01		1.14		1.02		2	K0	1601
	+61 02494	LS I +61 019		23 37 26	+61 39.0	2	9.98	.009	0.47	.009	-0.47	.014	10	B0 Vn	342,684
		CS 22894 # 31		23 37 26	−00 47.3	1	14.18		0.07		0.14		1		1736
222369	−20 06629	G 273 - 101		23 37 27	−19 44.0	1	10.58		0.46		-0.17		2	F2	1696
		LS I +59 011		23 37 29	+59 48.8	2	10.63	.000	0.99	.010	-0.19	.022	4		41,405
222390	+26 04671			23 37 30	+27 14.3	1	6.66		1.06				2	K1 III	20
222407	+62 02268	IDS23352N6311	AB	23 37 31	+63 26.9	2	6.85	.000	0.10	.009	0.07		2	A2	401,1118
	+2 04699			23 37 32	+02 35.5	1	9.66		0.46		-0.03		2	F5	1375
222391	+26 04673			23 37 32	+26 31.5	1	7.52		0.57				2	G0 III	20
		CS 22894 # 33		23 37 32	−00 36.0	1	14.99		0.13		0.17		1		1736
222399	+36 05098	HR 8973	⋆ A	23 37 34	+37 22.6	2	6.55	.023	0.35	.000	0.09	.014	4	F2 IV	39,254
	−35 15883			23 37 37	−35 08.7	1	11.10		0.85		0.41		2		1730

Table 1 1235

HD	DM	Other Id	N Rem	α_{1950}	δ_{1950}	S	V	σ_V	B–V	σ_{B-V}	U–B	σ_{U-B}	n	Spectrum	References
	−11 06124			23 37 41	−10 31.1	1	10.82		0.55		0.05		2		1375
222401	−2 06013			23 37 42	−02 01.9	1	8.71		1.07		0.89		2	K0	1375
		PHL 6012		23 37 42	−27 23.	1	14.35		0.02		−0.21		1		1036
		CS 22941 # 43		23 37 42	−33 09.3	2	15.01	.010	0.41	.008	−0.24	.009	3		1580,1736
		L 26 - 23		23 37 42	−76 03.3	1	14.66		0.05		−0.69		2		3065
		G 241 - 62		23 37 44	+60 24.8	1	11.43		1.54		1.20		2	M2	1658
	+62 02269			23 37 44	+63 28.4	1	8.67		1.74				2	K2	369
		LS I +61 020		23 37 46	+61 59.8	1	11.98		0.86		0.02		2		41
222412	−26 16736			23 37 49	−26 28.6	1	7.58		0.44				4	F5 V	2012
		LP 1035 - 48		23 37 49	−43 39.4	1	12.53		0.91		0.57		1		1696
222439	+43 04522	HR 8976	★ A	23 37 56	+44 03.4	6	4.14	.003	−0.07	.006	−0.23	.015	46	B9 IVn	1,15,879,1363,3023,8015
		G 217 - 15		23 37 56	+58 59.0	1	10.47		0.83		0.55		2	K1	1658
	+60 02600	LS III +61 050		23 37 56	+61 04.1	3	9.30	.011	0.28	.034	−0.39	.022	7	B9 V	888,1069,1119
		LP 583 - 27		23 38 01	+09 01.8	1	11.76		0.78		0.33		2		1696
222433	−32 17621	HR 8975		23 38 01	−32 21.0	5	5.31	.005	0.97	.006	0.69		23	K0 III	3,15,1075,2024,2029
	+43 04523			23 38 04	+43 47.9	1	8.96		1.67		2.03		1	M0	1748
222434	−35 15886			23 38 04	−34 58.4	1	8.72		1.08				3	G2/3 V	955
222455	−0 04547	IDS23356S0008	AB	23 38 07	+00 08.3	1	7.40		1.17		1.16		2	K3 III	1003
		LS I +59 012		23 38 07	+59 04.4	1	10.84		0.90		−0.13		2		41
		LS I +59 013		23 38 08	+59 55.1	1	11.33		0.99		−0.10		2		41
		CS 29499 # 8		23 38 08	−25 46.4	1	14.20		0.81		0.48		1		1736
222451	+35 05074	HR 8977		23 38 10	+36 26.6	3	6.24	.011	0.40	.008	−0.03	.029	9	F5 V	71,254,3026
		CS 22966 # 25		23 38 14	−29 55.0	1	14.80		0.13		0.15		1		1736
222474	+19 05135	G 68 - 35	★ AB	23 38 19	+20 05.3	1	8.24		1.12		1.06		3	K2	196,333
	+62 02270	IDS23361N6224	A	23 38 23	+62 40.4	1	9.19		0.50		−0.06		2	F8	401
		LS I +60 020		23 38 24	+60 29.5	1	11.98		0.96		−0.12		4		41
	+60 02604			23 38 25	+61 31.8	1	9.68		0.13				2	A0	1119
	+62 02271			23 38 27	+62 40.8	1	10.06		0.52		−0.05		1	F8	401
222480	−32 17623			23 38 30	−32 20.9	1	7.11	.009	0.67	.005			8	G1 V	1075,2013
222481	−39 15114			23 38 30	−39 17.6	1	10.17		0.07		0.09		4	A0 V	152
222485	−24 17796	HR 8978		23 38 31	−24 26.3	5	6.60	.006	1.57	.003	1.92		19	M1 III	15,1075,2013,2030,3005
		HA 115 # 546		23 38 33	+01 14.5	1	12.12		0.65		0.11		9	dG5	281
222493	−12 06535	HR 8979		23 38 34	−11 57.5	6	5.89	.013	0.98	.008	0.80	.014	26	K0 III	15,1075,2013,2029*
222499	+41 04842			23 38 36	+41 34.4	1	6.86		1.43		1.69		2	K2	1601
222514	+57 02780			23 38 36	+57 33.7	2	7.25	.004	0.18	.004	0.16		5	Am:	1118,8071
222517	+43 04526			23 38 37	+44 07.1	1	8.60		0.08		−0.02		1	A0	1748
222516	+45 04301	IDS23362N4540	AB	23 38 38	+45 56.5	1	6.99		0.44		0.02		2	F5	3016
		CS 22966 # 29		23 38 39	−30 40.6	1	13.69		0.12		0.19		1		1736
240360	+58 02629			23 38 40	+59 07.5	1	8.71		0.65				2	K0	1118
		G 157 - 91		23 38 43	−07 52.8	1	15.04		1.00		0.38		2		3062
		G 273 - 105		23 38 43	−20 04.0	1	13.48		0.85		0.33		2		1696
	+55 02998			23 38 46	+55 38.9	1	11.81		0.49		0.00		1		549
		WLS 2340-10 # 8		23 38 46	−11 51.8	1	11.50		0.35		0.00		2		1375
222508	−42 16433	IDS23362S4208	AB	23 38 49	−41 51.5	2	7.81		0.48	.000			8	F6 V	1075,2033
		AO 1209		23 38 51	+43 57.5	1	11.64		1.09		0.94		1	K2 V	1748
222568	+67 01555	LS I +68 004		23 38 52	+68 05.0	2	7.67	.000	0.42	.015	−0.44	.029	9	B1 IV	399,555
		CS 22952 # 17		23 38 52	−05 49.1	1	14.63		0.27		0.13		1		1736
	+0 05034			23 38 53	+00 53.3	1	10.11		0.54		0.03		1		281
		HA 115 # 552		23 38 54	+01 07.0	1	10.91		0.52		−0.06		1	F8	281
222589	+73 01051	IDS23367N7351	A	23 38 54	+74 07.5	3	8.74	.010	0.78	.011	0.30	.023	6	G5	979,3026,8112
		CS 29499 # 32		23 38 55	−26 31.4	1	14.62		0.06		0.13		1		1736
222554	+51 03684			23 38 56	+52 09.4	1	8.24		0.04		−0.09		2	A0	1566
		HA 115 # 554		23 38 57	+01 09.8	2	11.67	.058	1.02	.005	0.59	.019	14	dK2	281,1764
		LP 986 - 112		23 38 58	−36 19.0	1	13.62		1.07		0.91		1		1696
		HA 115 # 486		23 38 59	+01 00.1	3	12.48	.001	0.49	.002	−0.04	.004	31	G5	281,1499,1764
222570	+48 04127	HR 8981		23 38 59	+49 14.1	1	6.26		0.17		0.16		2	A4 V	1733
222547	−18 06357	HR 8980		23 38 59	−18 18.2	4	5.34	.009	1.58	.016	1.90	.032	14	K5 III	58,2007,3016,8053
		G 241 - 64		23 39 00	+59 08.1	1	12.72		0.75		0.02		2		1658
222556	+39 05143			23 39 01	+40 16.7	1	6.91		0.53		0.13		3	F7 V	1501
		LS I +62 005		23 39 01	+62 07.6	1	11.76		0.58		−0.26		1		41
	+46 04128	WY And		23 39 02	+47 19.1	1	8.86		1.70		1.67		1	K5 III	793
		CS 22894 # 34		23 39 04	−00 15.2	1	14.76		0.19		0.14		1		1736
	+43 04527			23 39 07	+43 36.0	1	9.72		1.12		1.04		1	K1 III	1748
		LP 763 - 50		23 39 09	−09 13.8	1	11.86		0.48		−0.18		2		1696
222574	−18 06358	HR 8982	★ A	23 39 10	−18 05.6	10	4.81	.013	0.82	.006	0.50	.020	43	G2 Ib/II	3,15,58,418,1075,1425*
		CS 22952 # 21		23 39 11	−03 53.2	1	14.59		0.42		−0.22		1		1736
222581	+35 05079			23 39 12	+36 02.8	1	8.27		0.01		−0.14		1	A0	963
	−43 15465	Smethells 14		23 39 13	−43 01.5	1	10.93		1.38				1	K7	1494
222576	−42 16435			23 39 18	−42 32.7	1	7.11		1.02				4	K0 III	2012
		AO 1211		23 39 21	+43 49.9	1	10.67		1.26		1.25		1	K2 III	1748
222591	−0 04553			23 39 21	−00 16.5	2	8.79	.000	0.73	.000	0.27	.000	2	G0	516,1620
		LS I +60 022		23 39 22	+60 21.4	1	11.69		0.96		−0.15		2		41
		VES 965		23 39 23	+61 47.9	1	12.49		0.97		−0.11		1		8113
222592	−16 06341			23 39 23	−15 44.4	2	8.17	.010	1.28	.005	1.26	.015	9	K1 III	244,1775
222602	+6 05183	HR 8983		23 39 24	+06 58.4	2	5.88	.005	0.10	.000	0.07	.005	7	A3 Vn	15,1256
		G 171 - 10	★ V	23 39 26	+43 55.0	3	12.29	.008	1.92	.005	1.48		16		419,694,3078
		HA 115 # 412		23 39 27	+00 52.4	2	12.23	.008	0.55	.009	−0.03	.004	10	dG5	281,1764
222596	−57 10327			23 39 28	−56 39.4	1	7.97		1.14		1.14		7	K2 III	1673
222618	+56 03067	HR 8985		23 39 29	+56 58.9	1	6.22		1.06		0.81		2	G8 III	1733
222603	+0 05037	HR 8984		23 39 30	+01 30.3	11	4.50	.014	0.20	.009	0.08	.007	39	A7 V	15,245,1007,1013,1020*

HD	DM	Other Id	N	Rem	α_{1950}	δ_{1950}	S	V	σ_V	B–V	σ_{B-V}	U–B	σ_{U-B}	n	Spectrum	References
	−67 03975				23 39 31	−66 58.1	1	10.74		0.46		−0.17		2		1696
		CS 22952 # 20			23 39 33	−04 26.8	1	13.89		0.42		−0.22		1		1736
		CS 22894 # 36			23 39 34	+00 16.2	1	14.76		0.28		0.07		1		1736
222629	+67 01557				23 39 34	+68 23.7	1	7.88		0.00		−0.11		2	A0	401
		IT Cas			23 39 35	+51 28.0	1	11.15		0.49		−0.01		6	F6	1768
	−3 05691	G 157 - 92			23 39 37	−02 51.1	3	10.28	.023	1.37	.031			7	M0	158,1017,1619
		CS 22894 # 29			23 39 41	−01 08.2	1	14.37		0.29		−0.06		1		1736
222640	+62 02275				23 39 42	+62 40.7	1	9.16		0.37		−0.23		2		401
		HA 115 # 416			23 39 43	+00 49.0	2	10.92	.000	0.48	.004	0.02		6	F7	281,5006
		LS I +61 021			23 39 44	+61 10.0	1	11.59		0.74		−0.19		2		41
		AO 1212			23 39 46	+43 40.2	1	11.68		1.60		1.85		1	M0 III	1748
222641	+44 04473	HR 8986			23 39 46	+44 42.9	1	6.41		1.51		1.75		2	gK5	1733
		BSD 19 # 1632			23 39 46	+61 00.8	1	11.42		0.48		0.33		3	B8	1069
	+58 02634				23 39 47	+59 11.6	1	10.03		2.07				2		369
		LS I +62 006			23 39 50	+62 08.3	1	12.12		0.65		−0.12		2		41
	+64 01842				23 39 51	+65 22.5	2	9.78	.030	2.46	.005	2.54		5	M2	369,8032
222643	−16 06345	HR 8987			23 39 52	−15 43.5	2	5.27	.005	1.36	.010	1.48		6	K3 III	244,2024
222647	+60 02608				23 39 53	+61 12.8	2	8.78	.040	0.02	.020			5	B7 V	1118,1119
222656	+61 02500				23 39 53	+62 25.2	2	8.30	.000	0.23	.009	0.13		4	A3	401,1118
222644	−17 06802				23 39 55	−17 26.3	1	10.70		0.03		0.07		4	A0/1 V	152
		HA 115 # 268			23 39 56	+00 35.5	3	12.49	.006	0.63	.003	0.07	.003	16	dG5	281,1499,1764
		LS I +62 007			23 39 57	+62 04.3	1	11.43		0.58		−0.31		2		41
222658	+46 04136				23 39 58	+47 28.7	1	8.80		0.47		0.01		2	F8	1733
222670	+63 02038	HR 8989		⋆ AB	23 39 58	+64 14.3	3	6.57	.009	1.88	.015	2.12		6	M2 III:	148,369,8032
222645	−28 18301				23 39 58	−28 27.5	1	8.28		0.52				4	F8 V	2012
		LS I +61 022			23 39 59	+61 24.3	2	11.23	.024	0.70	.028	−0.29	.033	4		41,1069
		VES 966			23 40 00	+63 21.0	1	13.74		1.25		0.42		1		8113
		JL 143			23 40 00	−80 23.	1	14.93		0.53		−0.11		1		832
222653	−31 19399				23 40 01	−31 23.7	1	10.45		0.17		0.07		2	A7/8 III	3060
		HA 115 # 420			23 40 02	+00 49.3	5	11.16	.004	0.47	.004	−0.03	.005	70	F4	281,989,1729,1764,5006
		LS I +61 023			23 40 02	+61 17.2	1	11.31		0.75		−0.30		2		41
222655	−41 15296				23 40 03	−41 31.4	1	9.52		0.76				4	G8 V	2012
		AT And			23 40 05	+42 44.5	2	10.42	.002	0.42	.001	0.16	.014	2	F2.5	668,699
		AO 1213			23 40 05	+43 51.0	1	11.35		0.80		0.12		1	F7 V	1748
	−0 04557				23 40 07	+00 28.4	5	9.70	.004	0.62	.004	0.10	.006	126	dG3	281,989,1729,1764,6004
222682	+60 02609	HR 8990		⋆ AB	23 40 08	+61 24.1	1	6.40		1.24		1.36		3	K2 III	37
222661	−15 06476	HR 8988		⋆ AB	23 40 08	−14 49.3	8	4.48	.018	−0.04	.006	−0.12	.014	25	B9 V	15,244,1075,1425,2012*
222662	−20 06640	G 273 - 116			23 40 09	−20 09.2	1	10.17		1.08		1.02		2	K3/4	3072
222683	+15 04872				23 40 11	+16 03.5	1	6.29		0.96		0.72		3	K0	985
	+12 05016				23 40 15	+13 02.7	1	9.58		0.97		0.59		1	K5	3072
		LS I +62 008			23 40 19	+62 08.2	1	11.98		0.63		−0.13		2		41
		G 29 - 61			23 40 20	+00 39.0	2	14.95	.000	1.61	.009			4		1663,1705,3059
	−0 04559				23 40 24	+00 30.0	2	10.02	.002	0.54	.003	0.04		12	F9	281,6004
	−65 04158	Smethells 14			23 40 25	−65 03.5	1	10.33		1.20				1		1494
222688	−47 14668	IDS23378S4652	AB		23 40 27	−46 35.4	1	6.63		0.91				4	K1 III +(G)	2012
	+31 04958				23 40 28	+31 50.1	1	10.87		−0.13		−0.79		1		963
222708	−8 06175				23 40 29	−07 46.8	1	8.47		1.71		1.87		2	M1	1375
		CS 29499 # 28			23 40 31	−23 49.0	1	12.13		0.16		0.10		1		1736
	−24 17814	G 275 - 90		⋆ A	23 40 33	−24 24.7	1	12.14		1.30		1.05		4	K4	3078
		G 130 - 4			23 40 34	+36 15.7	1	12.67		1.60		1.64		3		203
		LS I +61 024			23 40 34	+61 38.1	1	10.39		0.62		−0.07		3		41
		G 217 - 17			23 40 36	+60 26.6	1	12.31		1.24		1.08		2		1658
		CS 29499 # 35			23 40 36	−26 55.4	1	12.90		0.15		0.13		1		1736
	+58 02636	LS I +58 001			23 40 37	+58 47.6	1	10.17		0.91		−0.17		2		41
	+14 05040				23 40 38	+14 41.7	1	8.24		1.64		2.01		2	M0	1648
222749	+43 04531				23 40 38	+44 07.6	1	8.54		0.08		−0.02		1	A0 V	1748
	−24 17814	G 275 - 92		⋆ B	23 40 38	−24 27.0	1	12.85		1.40		1.16		4		3078
222715	−45 15112				23 40 38	−45 26.1	1	8.74		0.06		0.07		4	A1 V	152
222733	+0 05039				23 40 41	+00 37.6	5	8.60	.004	1.07	.008	0.88	.011	43	G8 III	281,989,1509,1729,6004
222732	+0 05038				23 40 41	+00 50.1	6	8.86	.006	1.17	.005	1.13	.016	43	K0 III	281,989,1509,1729,5006,6004
		LS I +61 025			23 40 41	+61 55.0	1	11.50		0.63		−0.32		2		41
		LP 1035 - 54			23 40 41	−40 15.3	1	10.56		0.75		0.35		3		1696
222750	+42 04742				23 40 42	+43 03.8	1	8.71		0.20		0.20		1	K5	289
	+54 03029	IDS23383N5422	A		23 40 42	+54 38.6	1	10.77		0.20		0.40		1		8084
	+54 03029	IDS23383N5422	B		23 40 42	+54 38.6	1	11.64		0.45		0.54		1		8084
	+54 03029	IDS23383N5422	D		23 40 42	+54 38.6	1	10.68		0.97		0.17		1		8084
222761	+62 02280				23 40 43	+62 40.3	1	9.10		0.17		0.01		2	A0	401
222741	−42 16443				23 40 45	−41 53.0	1	8.40		0.48				4	F5 V	2012
		HA 115 # 350			23 40 46	+00 40.1	1	11.06		0.71		0.24		1	dG5	281
222770	+51 03694				23 40 46	+51 58.2	1	7.63		0.30		0.20		3	Am	8071
240371	+59 02761				23 40 46	+59 40.0	1	9.22		0.47		−0.26		2	B2	1502
240372	+58 02637	LS III +58 002			23 40 48	+58 42.8	1	9.12		0.95		0.31		1	B8 Ib	41
		PHL 2430			23 40 48	−21 08.	1	15.10		0.00		0.01		3		286
222764	+9 05268	HR 8991		V	23 40 50	+10 03.2	3	5.05	.004	1.69	.005	2.03	.000	13	M2 III	15,1256,3016
		G 129 -D1	A		23 40 50	+14 30.3	1	14.14		1.17		1.14		5	sdM0	3029
		G 129 -D1	B		23 40 50	+14 30.3	1	14.37		0.43		−0.23		4	sdF5	3029
		LS I +61 026			23 40 50	+61 11.1	2	9.77	.009	0.54	.009	−0.44	.009	4	B0.5 IV	41,1012
222874	+84 00536				23 40 51	+85 11.5	1	7.68		1.06		0.85		2	K0	1733
222766	−8 06177	G 157 - 93			23 40 52	−08 12.3	4	10.12	.016	0.68	.012	0.05	.016	7	G0	1594,1658,1696,3062
		CS 29499 # 25			23 40 54	−23 07.0	1	13.40		0.05		0.14		1		1736

Table 1 1237

HD	DM	Other Id	N Rem	α_{1950}	δ_{1950}	S	V	σ_V	B–V	σ_{B-V}	U–B	σ_{U-B}	n	Spectrum	References
222786 +50 04130				23 40 56	+50 37.9	1	8.15		0.03		0.03		2	A0	1566
		CS 29499 # 26		23 40 56	−23 14.5	1	14.32		0.00		0.00		1		1736
222794 +57 02787		G 241 - 66	⋆ A	23 40 59	+57 47.8	3	7.14	.019	0.65	.011	0.03	.017	7	G2 V	22,1003,3026
+43 04532				23 40 60	+43 59.1	1	10.69		1.12		0.97		1	K0 III	1748
		CS 22966 # 37		23 41 01	−32 05.0	1	14.00		0.26		-0.05		1		1736
		CS 22894 # 27		23 41 04	−01 42.0	1	15.22		0.09		0.18		1		1736
		HA 115 # 357		23 41 05	+00 44.6	1	11.08		0.84		0.45		1	dG5	281
		LS I +61 027		23 41 07	+61 39.0	1	11.24		0.96		-0.18		3		41
222811 +21 04977				23 41 10	+21 39.8	1	7.87		1.72		1.98		2	M1	1648
		St 17 - 10		23 41 12	+61 53.5	1	12.64		0.90		0.41		2		1624
		CS 22894 # 38		23 41 14	+01 44.9	1	14.60		0.53		0.02		1		1736
222800 −16 06352		HR 8992, R Aqr		23 41 14	−15 33.7	1	11.61		0.98		-0.62		1	M3/5pe	975
+63 02042				23 41 15	+64 00.3	1	9.49		1.77				2		369
222802 −44 15338				23 41 15	−43 52.7	1	9.42		0.13		0.12		4	A2/3 V	152
222804 −46 14749				23 41 15	−45 44.4	1	6.81		1.22				4	K2/3 III	2012
+29 04982				23 41 16	+30 27.1	2	8.95	.030	0.58	.005	0.08		5	G0 V	20,3026
+60 02612				23 41 18	+61 01.8	1	8.71		0.35				2	F0	1118
		CS 22966 # 43		23 41 18	−28 35.2	1	13.56		0.26		-0.05		1		1736
+63 02043				23 41 19	+63 42.4	1	9.10		0.28				3	A5	1118
		St 17 - 8		23 41 20	+61 52.9	1	14.27		1.08		0.32		2		1624
		G 130 - 5		23 41 21	+32 16.2	4	12.93		0.14	.012	-0.60	.037	12		419,1620,3060,7009
		St 17 - 7		23 41 21	+61 53.3	1	12.76		0.52		-0.30		2		1624
222803 −45 15114		HR 8993		23 41 21	−45 21.6	4	6.08	.004	0.98	.005			18	K1 III/IV	15,1075,2013,2029
		HA 115 # 440		23 41 22	+00 49.4	1	11.26		0.84		0.44		5	dG7	281
		LS I +61 028		23 41 22	+61 10.4	1	10.94		0.52		-0.56		2		41
−3 05695				23 41 22	−03 12.2	1	10.54		0.83		0.51		2		1696
		G 130 - 6		23 41 23	+32 19.0	4	11.69	.020	1.55	.017	1.17	.031	13	M2-3	333,419,1620,3028,5009
		St 17 - 6		23 41 23	+61 52.9	1	13.75		0.57		-0.11		2		1624
+61 02509		St 17 - 4	⋆	23 41 23	+61 53.2	3	8.42	.017	0.46	.005	-0.54	.004	7	B0.5Ib	41,1012,1624
		LS I +60 023		23 41 23	+60 10.1	1	11.04		0.94		-0.14		2		41
		St 17 - 5		23 41 24	+61 52.6	1	13.73		0.89		0.56		2		1624
		St 17 - 2		23 41 24	+61 53.	1	12.51		0.73		-0.44		2		1624
		St 17 - 9		23 41 24	+61 53.	1	14.94		0.66		0.22		2		1624
		St 17 - 3		23 41 24	+61 54.0	1	13.09		0.59		-0.25		2		1624
		CS 22941 # 54		23 41 26	−33 18.2	1	14.80		0.16		0.15		1		1736
−45 15115				23 41 26	−45 34.1	1	10.92		0.53		0.03		2		1730
		St 17 - 1		23 41 27	+61 53.1	1	13.22		0.57		-0.18		2		1624
222827 −16 06353				23 41 27	−15 56.4	1	9.11		0.39		-0.05		4	F0 V	244
222842 +28 04627		HR 8997	⋆ AB	23 41 28	+29 05.1	5	4.93	.006	0.93	.013	0.64	.009	17	K0 III	15,1355,1363,6001,8015
		LS I +60 024		23 41 28	+60 53.9	1	12.34		0.74		-0.18		3		41
222820 −65 04159		HR 8996		23 41 28	−64 41.0	4	5.73	.030	1.40	.008	1.51	.097	21	K2 II/III	15,1075,2038,8100
		G 241 - 68		23 41 33	+64 27.8	1	11.30		1.40		1.12		3	M1	1723
222837 −41 15301				23 41 33	−40 57.8	1	7.46		1.45		1.67		7	K3 III	1673
222853 +57 02792				23 41 34	+58 28.1	2	8.30	.000	0.03	.005	0.00	.005	4	A2p	220,401
222805 −71 02771		HR 8994		23 41 35	−70 46.1	3	6.06	.003	0.91	.000	0.62	.000	19	G8/K0 IV	15,1075,2038
		CS 22941 # 46		23 41 36	−35 02.0	1	14.53		0.47		-0.14		1		1736
222847 −19 06500		HR 8998		23 41 37	−18 33.3	5	5.24	.015	-0.09	.007	-0.25	.016	14	B9 Vn	3,1079,2007,2024,3023
222838 −43 15479				23 41 37	−43 18.2	1	7.77		0.21		0.12		4	A2mA3-F0	152
+61 02510				23 41 38	+62 27.1	1	8.69		1.14		1.02		1	K2	401
222849 −55 10080				23 41 38	−54 42.8	1	7.40		1.52		1.37		3	M3 III	1730
+0 05040				23 41 39	+00 57.6	5	10.43	.004	1.03	.003	0.76	.005	84	G8 IV	281,989,1729,1764,5006
		LS I +61 031		23 41 40	+61 50.3	1	12.10		0.43		-0.30		2		41
		LS I +61 032		23 41 41	+61 11.1	1	11.70		0.66		-0.18		2		41
+60 02613		PZ Cas		23 41 41	+61 31.0	2	8.75	.195	2.65	.093	1.32		6	M3.5Ia+(B)	369,8032
222806 −79 01239		HR 8995		23 41 41	−79 04.2	3	5.74	.004	1.11	.000	1.07	.005	14	K1 III	15,1075,2038
		LS I +61 033		23 41 43	+61 53.8	1	11.27		0.46		-0.44		2		41
		LS I +62 009		23 41 43	+62 10.2	1	11.14		0.68		-0.22		2		41
222860 −0 04561				23 41 44	+00 26.1	1	7.97		0.54		0.10		2	F8	1375
		HA 115 # 366		23 41 44	+00 44.5	1	12.11		0.64		0.12		3	dG3	281
−35 15910				23 41 44	−34 43.4	1	10.96		-0.26		-1.00		18		1732
		G 68 - 37		23 41 48	+21 19.4	1	13.31		1.59				4		538
		LS I +62 010		23 41 48	+62 23.0	1	11.25		0.57		-0.14		2		41
222861 −15 06485				23 41 49	−15 15.6	1	9.69		0.41		0.00		4	F2 V	244
222872 −26 16762		HR 8999	⋆ AB	23 41 53	−26 31.4	3	6.17	.009	0.49	.009	0.08		9	F4 V	15,404,2029
		CS 22966 # 42		23 41 53	−28 36.0	1	13.15		0.04		0.11		1		1736
222870 −15 06487				23 41 54	−14 41.8	1	7.99		0.43		-0.01		4	F3 V	244
		CS 22941 # 47		23 41 56	−36 03.5	2	14.69	.010	0.15	.001	0.13	.012	2		966,1736
222885 +57 02793				23 41 57	+58 29.3	1	8.38		0.22		0.08		2	A0	401
		CS 22894 # 46		23 41 57	−00 42.9	1	14.83		0.47		0.05		1		1736
222873 −54 10347				23 41 59	−54 28.5	1	10.27		0.49		-0.03		2	F5/6 IV/V	1730
		HA 115 # 451		23 42 01	+00 55.5	1	12.35		0.50		-0.04		6	dG5	281
222888 +28 04630				23 42 02	+29 29.6	1	8.50		0.98		0.62		2	K0	1733
		LS I +61 034		23 42 07	+61 42.6	1	10.66		0.47		-0.50		2		41
222903 −9 06248				23 42 07	−08 44.4	1	8.22		0.28		0.12		4	A0	152
		LS I +61 035		23 42 08	+61 51.1	1	10.52		0.92		0.74		2		41
222896 −45 15120		IDS23395S4548	AB	23 42 08	−45 31.5	1	8.93		0.48		-0.01		2	F5 IV/V	1730
		CS 22894 # 47		23 42 09	−01 45.5	1	13.97		0.56		0.00		1		1736
222919 +6 05197				23 42 16	+06 54.9	1	6.87		1.09		0.93		2	K0	1733
		LS I +62 011		23 42 16	+62 07.1	1	11.37		0.91		-0.07		2		41
+60 02615		LS I +61 036		23 42 17	+61 23.4	2	9.11	.005	0.59	.009	-0.45	.000	4	B0.5Ib	41,1012

HD	DM	Other Id	N Rem	α_{1950}	δ_{1950}	S	V	σ_V	B–V	σ_{B-V}	U–B	σ_{U-B}	n	Spectrum	References
	−34 16158			23 42 18	−34 03.1	1	10.88		0.62		−0.04		1		788
222932	+55 03010	HR 9000		23 42 21	+55 31.3	2	6.44	.000	1.00	.000	0.70	.005	4	gG4	1502,1733
222931	+55 03011	IDS23400N5555	A	23 42 22	+56 10.8	1	8.17		1.72				2	M1	369
		LS I +60 025		23 42 24	+60 57.7	1	11.05		0.64		−0.38		2		41
		LS I +61 037		23 42 25	+61 13.1	1	10.82		0.53		−0.42		2		41
		G 68 - 38		23 42 26	+17 46.8	1	15.21		1.41				5		538
		LS I +61 038		23 42 27	+61 41.9	1	11.28		0.44		−0.45		2		41
		G 130 - 7		23 42 28	+30 03.6	2	11.74	.015	0.69	.029	−0.06	.034	6		308,1620
		LS I +62 012		23 42 29	+62 02.7	1	10.99		0.52		−0.42		2		41
		G 68 - 39		23 42 32	+14 42.2	1	12.64		1.45		1.12		1		333,1620
		LS I +61 039		23 42 32	+61 23.6	1	11.08		0.55		−0.34		2		41
		G 171 - 15		23 42 33	+44 23.5	1	11.56		0.65		−0.12		2	G0	1658
		IRC +40 545		23 42 34	+43 38.5	1	14.04		2.16				2		8019
		LS I +61 040		23 42 34	+61 37.9	1	11.81		0.58		−0.39		2		41
222925	−62 06441			23 42 34	−62 11.3	2	9.03	.010	0.58	.010	0.07		3	Ap SrEu	742,1594
222935	+28 04634	G 130 - 8		23 42 35	+29 17.1	5	8.39	.005	0.86	.005	0.52	.015	25	K2 V	22,333,1003,1620,1775,3026
222958	+68 01393			23 42 36	+69 28.6	1	7.13		0.06		−0.42		2	B8	401
222946	+14 05045			23 42 39	+14 45.9	1	7.69		0.27		0.08		3	A0	1733
		LS I +61 041		23 42 39	+61 23.6	1	12.22		0.66		−0.15		2		41
		CD Cas		23 42 40	+62 43.5	1	10.39		1.24		0.84		1	G0.5	1399
	−16 06356			23 42 52	−15 37.4	1	9.83		0.58		0.00		4	G0	244
		CS 22894 # 48		23 42 56	−02 14.1	1	15.37		0.02		0.04		1		1736
		CS 22894 # 40		23 42 57	+01 03.2	1	13.48		0.27		0.11		1		1736
	+61 02513			23 42 57	+61 37.7	1	9.06		0.14				2	A0	1118
222993	+56 03080			23 42 58	+57 05.7	1	8.30		−0.01		−0.20		2	B9	401
		LS I +62 013		23 42 58	+62 33.0	1	12.33		0.76		−0.08		2		41
		LS I +59 014		23 43 00	+59 28.2	1	11.63		0.62		−0.34		2		41
224832	−24 17946	IDS23560S2400	A	23 43 00	−23 43.1	2	9.40	.023	1.05	.002	0.96	.009	4	K0	481,3061
	+63 02049			23 43 03	+63 53.4	1	9.66		0.16		−0.04		2	A0	401
		LS I +62 014		23 43 04	+62 59.5	1	10.90		0.88		−0.25		2		41
		G 68 - 40		23 43 07	+20 22.4	1	12.32		1.33		1.10		1		333,1620
		CS 22894 # 41		23 43 08	+00 56.1	1	13.51		0.46		−0.21		1		1736
		LS I +61 042		23 43 09	+61 52.1	1	10.84		0.40		−0.55		2		41
	+61 02514	EH Cas		23 43 10	+62 04.8	1	10.18		2.52				2	M3	369
	+59 02765	QQ Cas		23 43 11	+59 37.7	1	10.23		0.73		−0.36		2	B2	41
223257	+87 00217			23 43 12	+88 03.9	1	8.93		0.14		0.04		4	A0	1219
222987	−41 15313			23 43 13	−41 21.4	1	8.87		0.24				4	A9 IV/V	2012
		LS I +60 026	AB	23 43 14	+60 40.4	1	10.83		0.72		−0.25		2		41
		LS I +62 015		23 43 15	+62 05.2	1	11.61		0.53		−0.29		2		41
223006	−15 06491			23 43 17	−15 02.0	1	7.36		1.52		1.79		4	K3/4 III	244
		G 130 - 9		23 43 18	+35 58.6	2	9.89	.037	1.35	.019	1.20	.014	4	K4	196,333,1620
		LS I +61 043		23 43 18	+61 45.6	1	12.16		0.45		−0.31		2		41
		LS I +60 027		23 43 19	+60 06.4	1	12.14		0.78		−0.26		2		41
		LS I +61 044		23 43 19	+61 15.1	1	12.18		0.52		−0.27		2		41
		LS I +61 045		23 43 19	+61 46.0	1	11.83		0.39		−0.44		2		41
	+61 02515	LS I +62 016		23 43 19	+62 00.9	2	9.97	.014	0.43	.000	−0.50	.005	4	B0.5V	41,1012
		LS I +60 028		23 43 21	+60 04.7	1	11.71		0.95		−0.16		2		41
		LS I +61 046		23 43 23	+61 46.0	1	11.72		0.44		−0.48		2		41
223011	−40 15239	HR 9001		23 43 23	−40 27.6	3	6.30	.004	0.21	.004	0.09		8	A7 III/IV	15,717,2012
223024	−19 06506	HR 9002	⋆ AB	23 43 25	−18 57.4	4	5.28	.012	0.29	.006	0.14		14	F2 III+F2 V	15,938,2013,2029
223029	−0 04563	IDS23409S0018	AB	23 43 26	−04 59.0	1	7.97		0.53		0.03		3	F8	3016
223036	+53 03228	LS III +54 085		23 43 28	+54 27.5	1	8.50		0.03		−0.74		2		555
223023	−15 06494	IDS23409S1541	AB	23 43 28	−15 24.7	1	8.01		1.55		1.73		4	K5 III	244
	+46 04159			23 43 30	+46 59.5	1	9.83		0.11		0.10		2	A2	1601
		LS I +61 047		23 43 30	+61 54.5	1	11.42		0.78		0.16		2		41
	+66 01640			23 43 30	+67 29.7	1	10.14		2.12				2		369
		WLS 2340-10 # 9		23 43 30	−08 59.6	1	11.90		0.49		−0.09		2		1375
		G 29 - 63		23 43 31	+04 39.9	1	12.85		1.32		1.19		1		333,1620
	+61 02517			23 43 32	+62 17.8	1	9.03		0.20		0.16		2	A0	401
223047	+45 04321	HR 9003	⋆ A	23 43 33	+46 08.6	6	4.99	.030	1.11	.015	0.81	.011	15	G5 Ib	1218,1355,1363,3016*
		LS I +61 048		23 43 34	+61 38.5	1	10.70		0.44		−0.49		3		41
		LS I +61 049	A	23 43 35	+61 30.8	1	12.54		0.58		−0.33		2		41
		LS I +61 049	B	23 43 35	+61 30.8	1	13.11		0.80		0.26		1		41
223044	+61 02518			23 43 35	+61 42.8	1	8.43		0.21		−0.42		2	A2	401
223031	−15 06495			23 43 35	−14 46.6	1	9.25		0.68		0.17		4	G3/5 V	244
		LS I +62 017		23 43 36	+62 00.6	1	11.98		0.48		−0.35		3		41
223043	+61 02519			23 43 36	+62 23.5	2	7.72	.014	0.04	.000	−0.06		4	A0	401,1118
223057	+62 02294			23 43 36	+63 02.4	1	7.78		0.10		0.09		2	A0	401
	+34 05002	G 130 - 10		23 43 38	+34 58.2	1	9.13		0.78		0.35		1	G5	333,1620
		LS I +61 050		23 43 39	+61 30.3	1	10.88		0.60		−0.48		2		41
		CS 29499 # 40		23 43 39	−26 01.8	1	14.50		0.07		0.16		1		1736
	+59 02768	IDS23412N5955	DV	23 43 41	+60 11.3	2	9.75	.088	2.11	.010	1.15		3	M3	369,8085
223070	+59 02769	IDS23412N5955	AB	23 43 42	+60 11.7	1	7.02		1.04		0.81		1	K0	8085
223070	+59 02769	IDS23412N5955	C	23 43 42	+60 11.7	1	10.95		0.70		0.15		1		8085
		LS I +61 051		23 43 43	+61 46.3	1	11.36		0.48		−0.43		2		41
		G 30 - 12		23 43 45	+10 00.2	1	11.77		1.40				1	M3	1746
	+61 02520	LS I +62 018	⋆ AB	23 43 45	+62 23.4	1	10.16		0.62		−0.31		1		41
223075	+2 04709	HR 9004, TX Psc		23 43 50	+03 12.6	6	5.04	.041	2.58	.025	3.23	.265	17	C5 II	15,109,3038,6002,8015,8022
		LS I +60 029		23 43 50	+60 44.8	1	12.61		0.65		−0.21		3		41
		CS 22941 # 50		23 43 51	−35 12.7	1	14.27		0.03		0.08		1		1736

Table 1   1239

HD	DM	Other Id	N Rem	α_{1950}	δ_{1950}	S	V	σ_V	B–V	σ_{B-V}	U–B	σ_{U-B}	n	Spectrum	References
223078	−23 18034			23 43 52	−23 18.3	1	9.01		0.88		0.55		2	K1 V	3072
		AJ89,1229 # 375		23 43 54	+08 55.7	1	14.60		1.51				2		1532
223094	+27 04617			23 43 54	+28 25.6	2	6.95	.014	1.63	.011			23	K5 III	20,928
223065	−42 16457	SX Phe		23 43 54	−41 50.9	1	7.20		0.20				4	A3 V	1075
236211	+52 03520			23 43 55	+53 20.1	1	8.80		0.40		−0.03		2	F2	1566
	−14 06541	G 273 - 136		23 44 00	−13 57.1	1	10.95		0.55		−0.10		2		1696
240394	+58 02646			23 44 04	+58 44.1	1	8.76		0.21		−0.01		2	A0	401
		LS I +62 019		23 44 06	+62 38.4	1	10.81		0.79		−0.25		3		41
223110	+54 03043			23 44 07	+54 52.5	1	7.85		0.45		0.01		2	F5 V	1566
		CS 22941 # 49		23 44 08	−36 36.9	1	12.53		−0.05		−0.22		1		1736
223099	−36 16030			23 44 09	−36 16.3	1	9.84		0.20		0.09		4	A2 III/IV	152
		LS I +61 052		23 44 10	+61 51.9	1	11.96		0.48		−0.42		2		41
		CS 22966 # 61		23 44 12	−30 17.1	1	14.01		0.14		0.14		1		1736
223128	+65 01943	HR 9005	⋆	23 44 13	+66 30.3	1	5.94		−0.04		−0.75		2	B2 IV	154
		LS I +62 020		23 44 16	+62 38.9	1	12.47		0.76		−0.22		3		41
223118	−24 17846			23 44 16	−24 34.3	1	8.23		0.02		0.01		4	B9.5V	152
		CS 29499 # 36		23 44 16	−27 04.8	1	14.79		0.03		0.14		1		1736
		LS I +61 053	⋆ A	23 44 20	+61 11.3	1	11.53		0.56		−0.11		3		41
		LS I +61 053	⋆ AB	23 44 20	+61 11.3	1	11.74		0.61		−0.41		2		888
223149	+61 02523			23 44 25	+62 14.0	2	8.99	.005	0.21	.014	0.13		4	A3	401,1118
223042	−84 00658			23 44 25	−84 08.4	1	7.82		0.94		0.69		9	G8/K0 III	1704
		LS I +61 054		23 44 26	+61 11.2	1	11.63		0.61		−0.18		2		41
223154	+26 04685			23 44 27	+26 54.6	1	8.48		0.91		0.63		47	G5	588
		CS 29517 # 2		23 44 29	−16 57.7	1	13.66		0.20		0.15		1		1736
		CS 22966 # 60		23 44 30	−30 19.5	1	14.65		−0.07		−0.30		1		1736
240397	+57 02803			23 44 33	+58 06.1	1	9.40		0.06		−0.48		2	A0	401
		LS I +61 055		23 44 34	+61 41.5	1	11.86		0.37		−0.52		3		41
223173	+56 03085	HR 9010		23 44 35	+57 10.4	2	5.51	.005	1.65	.000	1.77	.045	6	K3 IIb	1355,4001
223165	+57 02804	HR 9008		23 44 36	+58 22.4	3	4.87	.007	1.12	.005	1.03	.020	9	K1 IIIa	1118,1355,3016
		LS I +62 021		23 44 36	+62 07.3	1	10.97		0.50		−0.42		2		41
223145	−50 14047	HR 9006		23 44 37	−50 30.2	3	5.17	.007	−0.20	.004			9	B3 V	15,2012,2024
223158	−31 19445			23 44 38	−30 42.8	1	9.16		0.06				4	A0 V	2012
223161	−47 14693	IDS23420S4716	AB	23 44 38	−46 59.4	1	9.80		0.20		0.09		4	A0 V	152
223148	−69 03335	HR 9007		23 44 39	−68 40.3	3	6.88	.004	0.46	.000	0.05	.010	14	F3 IV	15,1075,2038
223146	−57 10349			23 44 40	−57 17.8	1	9.97		0.76		0.38		2	G5 V	1696
223170	−12 06559	HR 9009		23 44 41	−12 11.3	2	5.74	.003	1.07	.008	0.97		6	K0 III	58,2009
		LS I +62 022		23 44 42	+62 29.7	1	11.43		0.80		−0.23		2		41
		CS 22894 # 43		23 44 44	+00 27.9	1	14.00		0.33		−0.04		1		1736
223171	−48 14610			23 44 44	−48 33.0	1	6.88		0.66				4	G2 V	2012
223200	+59 02773	LS I +60 030		23 44 47	+60 02.2	2	8.47	.005	0.41	.010	−0.40	.000	4		41,401
240400	+59 02648			23 44 48	+59 06.0	1	8.49		0.51				2	G0	1118
223184	−32 17672			23 44 50	−31 44.3	1	10.01		0.17		0.13		4	A2 III/IV	152
		LS I +61 057		23 44 51	+61 10.4	1	11.24		0.58		−0.34		2		41
		LS I +61 056		23 44 51	+61 28.2	1	12.34		0.61		−0.11		2		41
		LS I +60 031		23 44 54	+60 28.7	1	10.08		0.77		−0.26		2		41
	+62 02296	LS I +62 023	⋆ AB	23 44 55	+62 56.5	2	8.64	.004	1.08	.009	0.05	.009	5	B3 Ia:	41,1012
		G 30 - 13		23 44 57	+17 56.8	1	11.83		1.41		1.24		1	M0	333,1620
223211	+24 04834	IDS23424N2501	A	23 44 58	+25 18.1	1	6.78		1.24		1.31		3	K3 III	833
		CS 29499 # 42		23 44 59	−25 11.8	1	13.98		0.23		0.14		1		1736
223209	+63 02054			23 45 00	+63 52.0	2	7.87	.014	0.03	.028	−0.03		4	B9	401,1118
223206	−16 06363			23 45 01	−16 15.6	1	8.32		0.50		−0.04		4	F5 V	244
		LS I +58 003		23 45 03	+58 53.0	1	11.08		0.66		−0.31		2		41
223229	+46 04169	HR 9011	⋆ AB	23 45 04	+46 33.3	1	6.06		−0.14		−0.67		2	B3 IV	154
	+59 02774	LS I +60 032		23 45 04	+60 28.1	1	10.30		0.54		−0.34		2	B2 V	41
		CS 29517 # 6		23 45 04	−15 16.3	1	15.22		0.26		0.04		1		1736
		G 29 - 64		23 45 07	+11 30.7	1	12.30		1.05		0.81		1		333,1620
		Steph 2158		23 45 10	+19 30.8	1	11.88		1.45		1.30		1	K7	1746
223236	−5 06051	IDS23426S0507	AB	23 45 10	−04 50.3	1	8.18		0.18		0.13		4	A0	152
		ApJ144,259 # 84		23 45 12	+51 07.	1	18.49		0.18		−1.00		2		1360
223224	−55 10098			23 45 13	−55 33.6	2	9.58	.010	0.76	.015	0.24		4	G6wF5 IV/V	742,1594
		LS I +61 058		23 45 14	+61 45.7	1	12.06		0.39		−0.47		2		41
	+61 02526	LS I +61 059	⋆ A	23 45 15	+61 46.2	2	8.76	.005	0.38	.005	−0.51	.005	2	B2 Ib	41,1012
223241	−6 06293	G 31 - 3		23 45 17	−05 31.2	3	8.37	.019	0.61	.011	0.07	.010	5	G0	202,1620,1658
223238	+3 04896	G 29 - 65		23 45 18	+03 53.9	4	7.71	.016	0.62	.008	0.13	.013	8	G2 V	202,1003,1620,3060
		CS 29517 # 7		23 45 18	−15 04.7	1	15.78		0.11		0.17		1		1736
		CS 29499 # 45		23 45 20	−24 07.7	1	11.91		0.21		0.12		1		1736
223252	−3 05707	HR 9012	⋆ A	23 45 22	−03 02.4	5	5.49	.014	0.94	.007	0.70	.006	16	G8 III	15,1071,2012,3077,4001
		G 30 - 15		23 45 23	+06 08.2	1	12.25		1.13		0.98		1		333,1620
		CS 22894 # 53		23 45 24	−00 23.7	1	13.82		0.16		0.17		1		1736
		LS I +60 033		23 45 25	+60 49.1	1	10.97		0.61		−0.41		2		41
223258	+62 02298			23 45 25	+62 41.1	1	8.46		0.23				3	A2	1118
		LS I +62 024		23 45 28	+62 54.4	1	11.81		0.90		−0.14		2		41
223274	+67 01562	HR 9013		23 45 30	+67 31.7	2	5.02	.014	0.01	.011	−0.02		4	A1 Vn	1363,3023
		G 171 - 19		23 45 32	+48 44.2	1	12.03		1.64		1.33		5		1723
		CS 22941 # 52		23 45 33	−33 14.9	1	14.71		0.30		0.07		1		1736
		LS I +60 034		23 45 35	+60 57.0	1	11.01		0.70		−0.35		2		41
		LS I +61 060		23 45 35	+61 51.5	1	11.89		0.44		−0.20		2		41
		Smethells 142		23 45 36	−41 47.	1	11.93		1.46				1		1494
		LS I +61 061		23 45 37	+61 03.2	1	12.05		0.77		−0.26		2		41
	+62 02299	LS I +63 007		23 45 38	+63 07.2	2	9.58	.000	0.61	.000	−0.47	.000	4	O8	1011,1012

HD	DM	Other Id	N	Rem	α_{1950}	δ_{1950}	S	V	σ_V	B–V	σ_{B-V}	U–B	σ_{U-B}	n	Spectrum	References
223288	+63 02055	IDS23432N6316	A		23 45 40	+63 32.4	1	7.41		0.08				2	A0	1118
	+58 02651	LS I +59 016			23 45 45	+59 12.3	2	10.03	.055	0.84	.004	0.06	.021	3	B5 Ib	41,1215
	+57 02810	IDS23433N5721	AB		23 45 46	+57 37.0	1	11.06		1.86						369
		LS I +61 062			23 45 50	+61 31.1	1	11.96		0.49		-0.45		2		41
	-13 06464	G 273 - 141	★ AB		23 45 50	-13 15.9	3	9.59	.029	1.26	.000	1.13	.014	6	K6	158,1696,1705
		LS I +58 004			23 45 54	+58 44.8	1	11.61		0.54		-0.24		2		41
		CS 22894 # 54			23 45 55	+00 07.3	1	14.18		0.39		-0.11		1		1736
		LS I +61 063			23 45 56	+61 30.9	1	10.99		0.46		-0.50		2		41
223302	-19 06513				23 45 56	-19 09.7	1	7.37		1.05		0.90		3	K0 III	1657
223311	-7 06086	HR 9014			23 45 58	-06 39.5	5	6.07	.008	1.45	.007	1.71	.000	25	K4 III	15,1071,1075,2013,2030
223312	-9 06264				23 45 58	-09 06.3	1	9.38		0.93				1	G5	1213
223323	+24 04836				23 46 00	+25 22.6	2	7.08	.005	0.41	.005	-0.10	.020	5	F2 IV-V	833,1625
223315	-16 06370	G 273 - 142			23 46 01	-15 29.2	1	8.79		0.75		0.36		4	G5 V	244
223332	+27 04625				23 46 03	+28 05.6	1	7.03		1.46		1.97		1	K5 II	8100
	+61 02529	LS I +61 064			23 46 03	+61 42.6	2	8.66	.002	0.54	.002	-0.46	.002	43	B1 Ib	41,1012
	+61 02531	LS I +61 065			23 46 09	+61 58.1	1	9.42		0.40		-0.54		2	B0 III	41
		LS I +59 017			23 46 11	+59 00.2	1	12.09		0.75		-0.25		2		41
223338	-8 06194	BS Aqr			23 46 12	-08 25.4	1	9.16		0.22		0.09		1	F2	668
		LS I +61 066			23 46 13	+61 42.4	1	11.65		0.43		-0.46		2		41
223358	+64 01861	HR 9017	★ AB		23 46 13	+64 35.9	1	6.41		0.06		0.01		2	A0 p Sr	401
		LS I +61 067			23 46 14	+61 20.3	1	12.30		0.51		-0.40		2		41
223340	-28 18352	IDS23437S2841			23 46 14	-28 23.9	1	9.09		0.82		0.42		1	K1 V	1279
		CS 22876 # 13			23 46 15	-33 33.7	1	13.58		0.78		0.34		1		1736
223346	+1 04773	HR 9015			23 46 16	+01 56.2	3	6.46	.015	0.44	.005	-0.02	.014	8	F5 III	15,254,1256
		LS I +61 068			23 46 17	+61 56.8	1	10.45		0.30		-0.52		2		41
		BPM 28452			23 46 18	-50 43.	1	14.49		0.47		-0.20		3		3065
223352	-28 18353	HR 9016	★ AB		23 46 19	-28 24.4	8	4.57	.008	0.01	.007	-0.01	.029	36	A0 V	3,15,1068,1075,1279*
		LS I +61 069			23 46 20	+61 52.4	1	12.04		0.31		-0.44		2		41
		LP 879 - 94			23 46 20	-21 41.2	1	12.65		0.45		-0.18		2		1696
		CS 22876 # 3			23 46 20	-36 45.9	2	13.83	.005	0.46	.009	-0.18	.007	3		1580,1736
223369	+61 02532				23 46 22	+62 18.7	1	8.72		0.23				2	A0	1118
223385	+61 02533	HR 9018, V566 Cas	★ AB		23 46 23	+61 56.2	3	5.44	.013	0.67	.009	-0.02	.000	12	A3 Iae	15,41,1012
		G 29 - 67			23 46 24	+06 10.3	2	12.30	.010	1.27	.034	1.24		3		202,333,1620
		Steph 2162			23 46 24	+09 32.5	1	12.17		1.52		1.21		1	M0	1746
		LS I +56 001			23 46 24	+56 50.5	1	10.71		0.17		-0.53		2		41
		AJ89,1229 # 377			23 46 25	+09 35.2	1	12.14		1.50				1	dM1	1632
223387	+56 03094	LS I +56 002			23 46 26	+56 55.9	2	8.93	.090	0.20	.005	-0.86	.052	4	Bpe	41,1012
223386	+59 02777	HR 9019			23 46 26	+59 42.0	3	6.33	.005	-0.01	.015	-0.05		7	A0 V	252,1118,1379
		LS I +62 025			23 46 27	+62 43.7	1	11.54		0.90		-0.15		2		41
223392	+5 05223				23 46 32	+06 06.3	2	8.30	.090	1.25	.113	1.01		2	R3	109,1238
	+1 04774	G 29 - 68			23 46 36	+02 08.2	13	8.98	.011	1.46	.024	1.10	.022	62	M2 V	1,116,202,989,1006,1197*
		LP 523 - 52			23 46 40	+14 14.0	1	13.06		0.44		-0.17		2		1696
223404	-10 06172				23 46 40	-09 47.9	2	10.31	.010	0.24	.005	0.12	.020	7	A0	152,1775
		LS I +61 071			23 46 42	+61 15.0	1	11.67		0.67		-0.30		2		41
223421	+58 02653	HR 9020			23 46 44	+58 41.1	3	6.35	.032	0.39	.005	-0.02	.020	6	F4 IV	254,1118,3037
		LS I +60 035			23 46 45	+60 22.5	1	12.41		0.86		-0.04		2		41
	-3 05711				23 46 47	-02 51.1	1	10.72		0.87		0.58		2		1696
		CS 29499 # 37			23 46 47	-26 52.4	1	13.43		0.06		0.13		1		1736
	-39 15165				23 46 47	-39 14.3	1	11.44		0.48		0.02		2		1730
		HA 187 # 460			23 46 48	-44 59.0	1	11.73		0.94		0.65		9		1499
223436	+23 04813				23 46 49	+23 35.9	1	8.15		0.50		0.03		2	F8	1733
		LS I +61 072			23 46 50	+61 41.3	1	11.95		0.38		-0.47		2		41
		NGC 7762 - 125			23 46 51	+67 48.9	1	12.57		1.67		1.24		1		312
		NGC 7762 - 128			23 46 51	+67 50.2	1	10.55		1.36		1.13		1		312
223429	-22 06199				23 46 51	-21 53.5	1	7.03		1.23		1.20		8	K0 III	1628
		NGC 7762 - 36			23 46 52	+67 44.8	1	14.71		1.08				1		312
223438	+0 05054	HR 9022			23 46 54	+00 47.9	3	5.76	.004	0.15	.014	0.14	.054	9	A5 m	15,252,1256
	+34 05013	G 130 - 13			23 46 57	+35 23.2	2	9.38	.024	1.12	.019	1.06	.005	3	M0	333,1620,3072
		Steph 2165			23 46 57	+51 25.6	1	11.37		1.66		1.93		1	M0	1746
		LS I +60 036			23 46 57	+60 50.0	1	10.80		0.58		-0.43		2		41
		NGC 7762 - 35			23 46 57	+67 44.3	1	11.66		1.71		1.36		1		312
		NGC 7762 - 126			23 46 57	+67 49.7	1	14.09		1.14		0.61		1		312
223428	-16 06373	HR 9021			23 46 57	-16 08.3	2	6.22	.010	1.22	.005	1.17		8	K2 IIIb	244,2007
		LS I +62 026			23 46 58	+62 41.6	1	12.41		0.67		-0.30		2		41
		NGC 7762 - 67			23 46 58	+67 47.7	1	12.49		1.29		0.93		1		312
		Steph 2166			23 47 03	+08 04.8	1	11.39		1.45		1.14		2	M1	1746
		NGC 7762 - 33			23 47 03	+67 44.2	1	15.30		1.24		0.81		1		312
		LS I +58 005			23 47 04	+58 18.5	1	11.29		0.39		-0.48		2		41
	+59 02779	LS I +60 037			23 47 04	+60 29.0	1	10.48		0.78		-0.33		2	B2 I	41
	+62 02304	LS I +62 027			23 47 04	+62 49.9	1	9.99		0.69		-0.30		2	B3	41
223441	-39 15166				23 47 04	-39 25.3	1	9.58		0.48		0.00		2	F3 IV/V	1730
223444	-63 04931	HR 9023			23 47 04	-63 07.0	3	6.58	.004	1.47	.010			11	K5 III	15,1075,2035
		CS 29499 # 38			23 47 05	-26 48.3	1	14.31		0.07		0.17		1		1736
		NW # 1			23 47 06	+67 26.	1	11.61		0.72		0.25		3		25
		NGC 7762 - 66			23 47 06	+67 47.3	1	12.68		1.68		1.70		1		312
223461	+28 04649	HR 9025			23 47 07	+28 33.8	5	5.96	.007	0.19	.007	0.13	.018	10	A2 m	985,1501,1733,1769,3058
		LS I +60 038			23 47 08	+60 41.2	1	11.99		0.56		-0.35		2		41
223460	+35 05110	HR 9024, OU And			23 47 10	+36 08.9	1	5.90		0.79				2	G1 IIIe	71
		NGC 7762 - 37			23 47 10	+67 45.4	1	13.65		1.18		0.49		1		312
		CS 22876 # 12			23 47 12	-34 11.5	1	14.79		0.16		0.09		1		1736

Table 1   1241

HD	DM	Other Id	N Rem	α_{1950}	δ_{1950}	S	V	σ_V	B–V	σ_{B-V}	U–B	σ_{U-B}	n	Spectrum	References
		HA 187 # 468		23 47 12	−45 02.6	1	11.83		1.16		1.16		9		1499
223466	−26 16796	HR 9026	⋆ AB	23 47 14	−25 36.5	6	6.42	.009	0.12	.005	0.10	.005	19	A3 V	15,404,1020,2012,2012,3050
		CS 22966 # 54		23 47 16	−29 25.0	1	14.43		0.28		−0.05		1		1736
223479	−39 15167			23 47 16	−39 15.4	1	8.55		0.57		−0.03		2	G0 V	1730
		NGC 7762 - 10		23 47 18	+67 43.6	1	13.73		1.28				1		312
		CS 22894 # 49		23 47 19	−02 14.5	1	14.46		0.45		−0.26		1		1736
		NGC 7762 - 1		23 47 20	+67 45.1	1	13.75		1.17		0.54		1		312
		NGC 7762 - 9		23 47 21	+67 44.0	1	14.25		1.33		0.41		1		312
		G 130 - 15		23 47 23	+29 17.7	1	15.76		0.59		−0.21		3		1658
		NGC 7762 - 2		23 47 23	+67 44.9	1	12.47		1.68		1.58		1		312
		CS 22966 # 59		23 47 23	−30 12.9	1	15.44		0.05		0.14		1		1736
		NW # 2		23 47 24	+67 26.	1	13.46		0.89		0.25		3		25
		LS I +61 074		23 47 25	+61 31.3	1	12.62		0.54		−0.21		2		41
		CS 22876 # 14		23 47 25	−33 05.5	1	14.94		0.10		0.08		1		1736
223501	+61 02537			23 47 26	+61 56.2	1	7.79		0.05		−0.70		3	B2 Vn(e)	1212
	−38 15638			23 47 26	−38 15.3	1	10.00		0.62		0.05		3	G2	1696
		G 29 - 71		23 47 27	+08 26.8	2	11.34	.000	0.53	.005	−0.23	.015	3	sdF8	333,1620,1658
		WLS 0 35 # 6		23 47 27	+35 27.3	1	12.95		0.75		0.27		1		1375
		NGC 7762 - 11		23 47 27	+67 44.8	1	14.86		1.17		0.63		1		312
		LS I +62 028		23 47 28	+62 25.0	1	10.95		0.63		−0.23		2		41
		LS I +62 029	⋆ A	23 47 30	+62 25.4	1	11.47		0.62		−0.23		2		41
		LS I +62 029	⋆ B	23 47 30	+62 25.4	1	12.35		0.61		−0.29		1		41
	+66 01646	LS I +67 003		23 47 30	+67 21.6	1	9.78		0.59		0.00		3	B8 II	25
		NGC 7762 - 7		23 47 30	+67 43.5	1	13.33		1.28		0.56		1		312
		NGC 7762 - 12		23 47 30	+67 44.8	1	15.75		1.22				1		312
223498	+2 04723	G 29 - 72	⋆ A	23 47 31	+02 35.8	6	8.33	.026	0.75	.007	0.35	.015	16	G7 V	22,202,333,1003,1620*
		G 171 - 23		23 47 31	+48 12.0	1	13.24		0.66		0.03		1		1658
		LS I +62 030		23 47 33	+62 23.2	1	11.54		0.55		−0.31		2		41
	−31 19466	G 275 - 111		23 47 34	−30 50.5	1	11.41		0.44		−0.23		2		1696
		CS 22894 # 55		23 47 35	+00 44.9	1	14.96		0.32		0.11		1		1736
	−0 04573			23 47 35	−00 05.0	1	10.94		0.59		0.02		2	G0	1375
		LS I +58 006		23 47 38	+58 18.4	1	10.93		0.57		−0.47		2		41
		LS I +60 039		23 47 38	+60 38.9	1	10.52		0.50		−0.47		2		41
		LS I +61 075		23 47 38	+61 56.0	1	12.15		0.40		−0.34		3		41
223515	−30 19702	G 275 - 113	⋆ AB	23 47 39	−29 40.8	1	7.94		0.84		0.43		2	K1 V	3072
223524	−10 06177	HR 9027		23 47 40	−10 15.2	2	5.94	.000	1.13	.000	1.15		6	K0 IV	2007,3016
223532	+16 05002			23 47 42	+17 22.1	1	7.80		0.41		0.01		2	F2	1648
		LS I +60 040		23 47 43	+60 49.3	1	11.26		0.67		−0.19		2		41
	+62 02308			23 47 44	+63 15.7	1	9.98		0.90		0.24		1	A0	1215
		CS 22894 # 51		23 47 44	−01 33.3	1	14.93		0.18		0.18		1		1736
	+60 02627			23 47 45	+60 50.3	1	9.94		2.55				2		369
		LS I +62 031		23 47 47	+62 14.2	1	12.60		0.71		−0.13		2		41
223552	+50 04165	HR 9028	⋆ A	23 47 52	+51 20.6	2	6.48	.020	0.37	.015	−0.10	.015	5	F4 V	254,3016
223542	−15 06505			23 47 52	−14 51.5	1	6.68		1.37		1.57		4	K3 III	244
223543	−15 06506			23 47 53	−15 15.7	3	7.49	.012	0.20	.008	0.11	.012	11	A2mA3/5-F0	152,244,8071
		CS 22966 # 58		23 47 53	−29 56.8	1	14.13		0.06		0.15		1		1736
	+29 05007	G 130 - 16		23 47 55	+30 04.5	1	9.34		1.26		1.19		1	M0	3072
223567	+66 01648			23 47 56	+67 17.5	1	8.44		1.01		0.70		4	K0	25
223551	−52 12220	IDS23453S5216	AB	23 47 56	−51 58.9	3	7.60	.012	0.77	.005	0.34		17	G8/K0 V	173,214,2024
223559	−15 06507	HR 9029		23 47 59	−14 40.8	3	5.70	.016	1.49	.018	1.83	.034	10	K4 III	244,2007,3016
223558	−2 06047	IDS23454S0238	AB	23 48 00	−02 20.9	1	10.08		0.50		0.03		2	F8	1375
		G 29 - 73		23 48 01	+09 40.0	1	11.51		1.46		1.13		1	M1	333,1620
	+46 04184	IDS23455N4630	AB	23 48 01	+46 46.3	1	8.85		0.58		0.04		4	F8	3016
223579	+61 02538			23 48 02	+61 54.3	1	8.99		0.15				3	A0	1118
		LS I +62 032		23 48 03	+62 09.6	1	10.82		0.49		−0.30		2		41
223580	+60 02628			23 48 04	+60 44.4	1	9.20		0.30				3		1118
		LS I +60 041		23 48 04	+60 52.5	1	10.88		0.74		−0.24		2		41
		LS I +61 076		23 48 04	+61 32.7	1	12.38		0.45		−0.44		2		41
	+46 04185	IDS23456N4632	AB	23 48 05	+46 48.6	1	9.01		0.41		−0.04		2	F5	3016
		NGC 7762 - 91		23 48 05	+67 45.0	1	11.74		1.69		1.29		1		312
	+16 05004	G 30 - 24		23 48 07	+17 03.9	1	10.56		0.88		0.59		1	K3	333,1620
	+60 02629	LS I +60 042	⋆ V	23 48 11	+60 37.9	1	9.59		0.86		−0.26		2	B1Iab:	41
223586	−19 06519			23 48 11	−18 34.1	2	7.05	.015	1.15	.012	0.99	.029	3	G8/K0	565,3048
		NW # 5		23 48 12	+67 16.	1	14.26		1.95				2		25
		CS 29499 # 59		23 48 12	−26 24.7	1	14.28		0.09		0.17		1		1736
		G 68 - 45		23 48 14	+21 17.9	1	10.77		1.07		1.06		1	K5	333,1620
	+61 02543		V	23 48 15	+61 35.9	1	9.03		2.16				2	M2	369
	−53 10535	Smethells 14		23 48 18	−53 01.6	1	10.79		1.19				1		1494
	+61 02541			23 48 20	+62 09.8	1	9.42		1.81				2		369
	+62 02311			23 48 20	+62 59.2	1	8.93		0.41				3	F5	1118
		LP 463 - 62		23 48 22	+19 40.	1	15.35		1.64				5		538
		POSS 192 # 2		23 48 23	+48 44.8	1	11.87		1.20				2		1739
		NW # 6		23 48 24	+67 17.	1	12.41		2.15		2.17		3		25
223600	−49 14270			23 48 24	−48 56.2	1	6.85		1.21		1.25		7	K1 III	1628
		G 29 - 74		23 48 27	+06 27.6	1	12.73		1.43		1.22		1		333,1620
		LS I +62 033		23 48 27	+62 36.0	1	11.87		0.55		−0.43		2		41
		CS 22876 # 11		23 48 27	−34 20.7	1	14.27		0.14		0.17		1		1736
		G 29 - 75		23 48 29	+03 42.8	3	13.40	.029	0.88	.019	0.44	.010	4		202,333,1620,1658
		LS I +60 043		23 48 29	+60 34.6	1	12.27		0.87		−0.15		2		41
223606	−66 03803			23 48 29	−65 35.6	1	8.17		1.04		0.89		2	K1 III	1704

HD	DM	Other Id	N Rem	α_{1950}	δ_{1950}	S	V	σ_V	B–V	σ_{B-V}	U–B	σ_{U-B}	n	Spectrum	References
		WLS 0 35 # 5		23 48 30	+36 54.2	1	11.37		0.75		0.21		2		1375
		LS I +60 044		23 48 31	+60 47.5	1	11.45		0.59		-0.20		2		41
223624	+63 02064			23 48 32	+63 42.4	1	6.82	.005	0.04	.010			4	A0	1025,1118
223617	+1 04786			23 48 33	+01 57.6	2	6.93	.024	1.16	.002	0.96	.029	3	G9 III-Ba2	565,3048
		NGC 7762 - 139		23 48 33	+67 43.9	1	12.76		1.71		1.80		1		312
223630	-24 17880			23 48 41	-23 42.9	1	10.04		0.11		0.06		4	A0 V	152
		CS 22966 # 65		23 48 41	-31 47.2	1	15.36		0.04		0.16		1		1736
		LS I +62 034		23 48 45	+62 08.9	1	11.23		0.24		-0.62		2		41
223649	+61 02547			23 48 45	+62 19.0	3	8.34	.004	0.57	.004			29	F8	1025,1118,6011
223633	-43 15521			23 48 45	-42 38.9	2	7.54	.015	0.46	.005	-0.02		6	F6 V	2012,3037
223640	-19 06522	HR 9031, ET Aqr		23 48 46	-19 11.2	5	5.18	.014	-0.14	.009	-0.41	.021	22	Ap Si	3,15,1202,2012,3016
223636	+39 05174			23 48 47	+39 55.3	2	6.78	.010	0.41	.005	0.00	.005	5	F5 V	985,1501
223637	+8 05127	HR 9030, HH Peg		23 48 48	+09 02.2	4	5.79	.007	1.66	.004	1.73	.004	12	M3 IIIa	15,1256,3055,8015
		LS I +61 077		23 48 49	+61 48.0	1	11.39		0.26		-0.52		2		41
240425	+58 02660			23 48 50	+58 50.4	1	8.76		0.45				2	G0	1118
		CS 29499 # 57		23 48 57	-26 02.5	1	13.85		0.20		-0.01		1		1736
		TU Scl		23 49 02	-29 27.2	2	13.60	.000	0.29	.007	0.09	.000	7		1736,3064
	+7 05093			23 49 03	+08 24.3	1	11.31		0.02		-0.13		3	A0	1026
	-31 19475	G 275 - 117		23 49 03	-30 42.5	1	10.14		0.85		0.54		1	G5	1696
223666	-35 15973			23 49 03	-34 58.2	1	6.71		0.97		0.66		8	K0 III	1628
		LS I +60 045		23 49 04	+60 32.0	1	11.77		0.73		-0.30		2		41
223675	-14 06563			23 49 04	-14 15.8	1	8.51		0.13		0.11		4	A2 IV	152
		POSS 192 # 3		23 49 06	+48 11.8	1	16.75		1.47				3		1739
223676	-37 15410			23 49 07	-37 22.5	1	9.46		0.28		0.09		4	A4-A5-A7	152
		LS I +62 035		23 49 08	+62 56.7	1	11.52		0.49		-0.11		2		41
		NGC 7822 - 8		23 49 10	+67 57.2	1	11.49		0.67		-0.10		2	B5	1285
223688	-7 06095	IDS23466S0710	AB	23 49 12	-06 53.5	1	8.68		0.66		0.21		2	F6 V +F6 V	3030
		G 130 - 18		23 49 14	+31 09.1	1	11.62		0.76		0.29		1		333,1620
223647	-82 00905	HR 9032		23 49 14	-82 17.8	4	5.10	.005	0.93	.015	0.60	.000	21	G7 III	15,1075,1075,2038
		PG2349+002		23 49 19	+00 11.6	1	13.28		-0.19		-0.92		5	sdB	1764
223705	+42 04780			23 49 20	+42 37.9	1	6.80		1.05		0.78		2	K0	1601
223718	+37 04898	IDS23468N3720	AB	23 49 21	+37 36.9	1	6.96		0.45		-0.07		4	F5	938
	+61 02549			23 49 21	+62 20.4	1	10.12		0.32				2	B2 V	1025
		LS I +62 036		23 49 21	+62 44.8	1	11.90		0.56		-0.35		2		41
		LS I +62 037		23 49 22	+62 09.6	1	11.67		0.47		-0.44		2		41
	+59 02781			23 49 23	+60 07.1	1	9.73		1.81				2		369
223719	+2 04725	HR 9033		23 49 24	+02 39.1	6	5.56	.021	1.54	.010	1.86	.013	19	K4 II	15,1256,1355,3016*
		WD2349+286		23 49 24	+28 38.5	1	16.26		-0.25		-1.16		4	DA1	1727
		LS I +62 038		23 49 24	+62 36.3	1	11.46		1.06		0.40		2		41
		LS I +62 039		23 49 30	+62 19.1	1	10.17		0.35		-0.58		2		41
	+62 02313	LS I +63 010		23 49 30	+63 01.5	2	8.83	.010	0.48	.015	-0.25		4	B3 Ib	1012,1025
	+64 01865			23 49 30	+65 21.8	1	9.46		1.95				3	M2	369
223731	+76 00934	HR 9034		23 49 32	+77 19.4	1	6.57		0.43		-0.13		1	F5 V	254
223713	-62 06446			23 49 35	-61 41.5	2	9.43	.029	0.80	.015	0.27		6	K3 Vw	2033,3077
		RY Cas		23 49 38	+58 27.7	2	9.39	.008	1.12	.017	0.85		2	G2	934,1399
	+57 02823	IDS23469N5810	AB	23 49 38	+58 27.8	1	9.58		0.29				1		934
	+48 04176			23 49 40	+49 31.6	1	9.03		1.38		1.40		2	K0	1733
		CS 22876 # 15		23 49 40	-32 53.8	1	12.43		0.05		0.02		1		1736
	+60 02631	LS I +60 046		23 49 44	+60 38.3	1	9.65		0.51		-0.48		2	B0 III	41
		CS 22876 # 6		23 49 44	-36 20.6	1	14.04		0.03		0.10		1		1736
223767	+61 02551	LS I +61 078		23 49 48	+61 36.0	2	7.23	.009	0.64	.022	0.46		5	A5 I	41,1025
223755	+20 05386	HR 9035		23 49 51	+21 23.6	2	6.12	.008	1.60	.002	1.97		6	M2 III	71,3001
	+61 02550	LS I +61 079		23 49 51	+61 50.4	3	9.30	.010	0.32	.010	-0.64	.004	9	O9.5II	41,1012,1025
		CS 22876 # 5		23 49 54	-36 19.9	1	15.20		0.06		0.14		1		1736
	-10 06187	G 273 - 152		23 49 55	-09 41.0	2	10.09	.010	0.76	.005	0.31	.000	3		1658,1696
223768	+18 05231	HR 9036		23 49 56	+18 50.6	3	5.09	.020	1.59	.016	1.88	.001	13	M3 III	3,1363,3001
223774	-15 06515	HR 9037		23 49 56	-14 31.8	3	5.85	.016	1.25	.023	1.37	.018	9	K2 III	244,2007,3005
223778	+74 01047	HR 9038	★ AB	23 49 57	+75 15.9	5	6.39	.020	0.98	.010	0.71	.005	19	K3 V + M2	15,1013,1197,1758,3003
		LS I +62 040		23 49 59	+62 07.4	1	11.12		0.33		-0.31		2		41
223781	+10 05004	HR 9039, HT Peg		23 50 03	+10 40.1	2	5.31	.010	0.18	.000	0.10	.005	4	A4 Vn	1733,3058
		WLS 2340-10 # 5		23 50 03	-08 21.1	1	11.28		0.63		0.10		1		1375
223782	-6 06308	G 31 - 12		23 50 04	-06 16.3	5	9.55	.030	1.08	.012	1.00	.032	9	K0	196,202,1620,1658,1705
		LS I +61 080		23 50 05	+61 59.8	1	11.63		0.33		-0.52		2		41
		LS I +62 041		23 50 05	+62 30.8	1	12.01		0.47		-0.47		2		41
223785	-19 06527			23 50 05	-18 50.4	1	6.81		0.09				4	A1 V	2012
223783	-17 06836			23 50 07	-16 39.2	1	7.59		1.64		1.88		6	M2 III	3007
		Steph 2176		23 50 10	+41 34.5	1	12.00		1.38		1.27		1	K7	1746
	+60 02632	LS I +60 047		23 50 11	+60 42.9	1	9.68		0.50		-0.55		2	B0 V	41
		LS I +60 048		23 50 12	+60 26.5	1	11.86		0.48		-0.35		2		41
		Steph 2177		23 50 13	+18 15.2	1	12.11		1.44		1.27		2	M0	1746
240432	+57 02824			23 50 13	+58 20.1	1	9.79		1.41				1	K5	934,1655
		LS I +61 081		23 50 13	+61 23.2	1	10.97		0.57		-0.29		2		41
		LS I +62 042		23 50 14	+62 10.1	1	11.22		0.43		-0.38		2		41
223818	+60 02633			23 50 15	+61 00.9	1	8.14		1.07				2	K0	1025
223806	-7 06101			23 50 15	-06 48.6	2	9.44	.005	0.23	.000	0.11	.020	10	A0	152,1775
		POSS 192 # 1		23 50 16	+48 51.5	1	14.57		1.58				4		1739
		LS I +60 049		23 50 16	+60 53.2	1	11.49		0.48		-0.37		2		41
223807	-9 06277	HR 9040		23 50 16	-09 16.5	3	5.75	.010	1.17	.003	1.15	.000	8	K0 III	15,688,1078
		WLS 0 35 # 7		23 50 18	+32 38.0	1	11.74		0.69		0.21		2		1375
	+45 04348			23 50 18	+46 11.7	1	9.49		-0.10		-0.74		1	B5	963

Table 1   1243

HD	DM	Other Id	N Rem	α_{1950}	δ_{1950}	S	V	σ_V	B–V	σ_{B-V}	U–B	σ_{U-B}	n	Spectrum	References
		Ki 12 - 1		23 50 18	+61 39.6	1	14.47		0.48		0.14		2		1542
		CS 22966 # 73		23 50 19	−30 27.2	1	15.15		0.34		0.03		1		1736
		Ki 12 - 2		23 50 20	+61 39.9	1	13.23		0.39		-0.17		2		1542
223825	−3 05723	HR 9041	⋆ AB	23 50 21	−03 26.0	4	5.93	.009	1.07	.011	0.96	.012	10	G9 III	15,688,1256,3016
		EQ Cas		23 50 23	+54 44.1	1	11.17		1.02		0.64		1	Fp	793
		LS I +61 082		23 50 23	+61 13.4	1	11.06		0.55		-0.41		2		41
		Ki 12 - 3		23 50 24	+61 40.2	1	12.30		1.53		1.17		2		1542
		Ki 12 - 4		23 50 24	+61 40.5	1	13.25		0.71		0.24		2		1542
		WLS 0 20 # 6		23 50 25	+20 03.9	1	11.05		0.67		0.19		2		1375
	+5 05229			23 50 27	+06 10.0	1	10.12		1.69				1	M0	1532
	+60 02634	TZ Cas		23 50 27	+60 43.5	2	9.17	.005	2.54	.030	2.49		5	M2 Iab:	369,8032
223837	+17 04999			23 50 28	+17 37.3	1	6.56		1.42		1.71		2	K2	1648
		Ki 12 - 5		23 50 28	+61 39.6	1	14.91		0.53		0.21		2		1542
		Ki 12 - 8		23 50 29	+61 39.1	1	14.95		0.45		0.32		2		1542
		Ki 12 - 6		23 50 29	+61 40.5	1	13.62		0.44		-0.25		2		1542
		LS I +60 050		23 50 30	+60 12.2	2	11.58	.020	0.74	.020	-0.33	.015	4	B0 V	41,342
		Ki 12 - 9		23 50 30	+61 39.8	1	15.14		0.58		0.47		2		1542
		Ki 12 - 7		23 50 30	+61 41.1	1	13.88		0.43		-0.07		2		1542
223846	+61 02553			23 50 30	+62 21.8	1	8.46		1.23				2	K5	1025
223855	+1 04792	HR 9042		23 50 31	+01 48.8	2	6.27	.005	-0.01	.000	-0.01	.000	7	A1 V	15,1071
		Ki 12 - 11		23 50 31	+61 39.3	1	15.22		0.57		0.18		2		1542
		Ki 12 - 10		23 50 31	+61 40.5	1	12.77		0.38		-0.27		2		1542
223847	+58 02667			23 50 32	+59 08.6	1	7.86		1.05				2	G7 III	1118
		Ki 12 - 14		23 50 32	+61 40.1	1	13.48		0.43		-0.18		2		1542
		Ki 12 - 12		23 50 32	+61 40.3	1	13.90		0.52		0.08		2		1542
223854	+1 04793			23 50 33	+02 03.1	1	8.02		0.50		-0.04		2	F5	1375
		Ki 12 - 13		23 50 33	+61 41.2	1	14.91		0.50		0.20		2		1542
		Ki 12 - 15		23 50 34	+61 40.3	1	13.52		0.55		0.08		2		1542
	+28 04660			23 50 35	+28 45.1	1	9.74		1.39		1.25		1	M0	3072
223866	+59 02784	IDS23481N6009	AB	23 50 35	+60 25.6	1	6.67		1.56				3	K5	1025
		Ki 12 - 24		23 50 35	+61 38.8	1	14.30		0.55		0.06		2		1542
		Ki 12 - 16		23 50 35	+61 40.7	1	14.76		0.43		0.13		2		1542
		Ki 12 - 18		23 50 35	+61 41.0	1	14.40		0.49		0.01		2		1542
		Ki 12 - 17		23 50 35	+61 41.1	1	13.18		0.36		-0.21		2		1542
		MCC 869		23 50 36	+59 40.0	1	10.71		1.21				1	K8	1017
	+60 02635	LS I +60 051		23 50 36	+60 38.0	1	10.13		0.46		-0.61		2	O9 III	41
		Ki 12 - 22		23 50 36	+61 39.6	1	14.36		0.73		0.27		2		1542
	+61 02555	Ki 12 - 19	⋆ B	23 50 36	+61 40.7	2	10.73	.010	0.36	.000	-0.54	.010	3		41,1542
		Ki 12 - 20		23 50 36	+61 41.5	1	14.47		0.56		0.02		2		1542
	+61 02555	Ki 12 - 21	⋆ A	23 50 37	+61 40.4	2	10.34	.030	0.34	.010	-0.59	.030	4	B0	41,1542
		Ki 12 - 25		23 50 38	+61 41.5	1	13.69		0.73		0.41		2		1542
		Ki 12 - 23		23 50 38	+61 41.6	1	12.62		0.38		-0.32		2		1542
	+61 02554	LS I +61 084		23 50 38	+61 44.1	1	10.77		0.37		-0.53		2	B1.5V	41
		LS I +60 052		23 50 39	+60 19.7	1	11.67		0.83		-0.18		2		41
		Ki 12 - 26		23 50 39	+61 40.7	1	12.50		0.57		-0.35		2		1542
		LS I +62 043		23 50 39	+62 13.3	1	12.34		0.37		-0.41		2		41
		CS 22876 # 7		23 50 39	−36 11.0	1	14.73		0.10		0.12		1		1736
		LS I +59 018		23 50 41	+59 08.4	1	12.15		0.53		-0.36		2		41
		Ki 12 - 27		23 50 41	+61 40.0	1	13.38		0.74		0.27		2		1542
		CS 22945 # 56		23 50 41	−65 46.4	2	14.08	.005	0.40	.007	-0.01	.009	3		1580,1736
		Ki 12 - 28		23 50 43	+61 40.1	1	14.49		0.54		0.00		2		1542
		Ki 12 - 30		23 50 43	+61 40.1	1	14.73		0.67		0.42		2		1542
	+61 02554	Ki 12 - 29		23 50 44	+61 41.1	1	10.90		0.41		-0.48		2		1542
	+61 02556	Ki 12 - 31	⋆	23 50 44	+61 41.1	1	10.95		0.42		-0.51		2	B2	41
223884	−24 17897	HR 9043		23 50 46	−24 30.4	1	6.24		0.19				4	A5 V	2009
		LS I +62 044		23 50 52	+62 11.6	1	12.56		0.47		-0.29		2		41
		LS I +60 053		23 50 53	+60 21.4	1	10.37		0.52		-0.31		2		41
		LS I +61 086		23 50 54	+61 09.2	1	12.54		0.61		-0.27		2		41
240435	+59 02785	LS I +60 054		23 50 55	+60 24.1	1	10.00		0.53		-0.51		2	G7	41
223895	−12 06582			23 50 56	−12 13.7	1	9.40		0.94		0.62		2	G8 III/IV	1375
223906	−19 06531			23 51 00	−18 38.9	1	9.45		0.20		0.13		4	A1 V	152
		LS I +61 087		23 51 01	+61 26.3	1	12.17		0.63		-0.30		2		41
		MCC 870	A	23 51 03	+11 51.2	1	11.12		1.36				2	K4	1619
		MCC 870	B	23 51 03	+11 52.2	1	12.19		1.47				2		1619
		LS I +61 088		23 51 04	+61 20.1	1	10.18		1.03		0.75		2		41
223924	+56 03106	LS I +56 003		23 51 06	+56 32.5	2	8.24	.000	0.02	.000	-0.70	.005	4	B1.5V	41,1012
		LS I +61 089		23 51 07	+61 40.7	1	10.27		0.38		-0.58		2		41
		LS I +62 045		23 51 07	+62 13.9	1	12.28		0.57		-0.29		2		41
	+61 02557			23 51 07	+62 15.6	1	9.45		0.29				2	A5	1025
		SW # 2		23 51 12	+66 16.	1	12.65		2.01		2.15		1		25
		LS I +61 090		23 51 14	+61 37.4	1	11.37		0.55		-0.33		2		41
	+61 02559	LS I +62 046		23 51 14	+62 08.7	3	9.72	.005	0.29	.000	-0.65	.010	6	O9 V	41,1011,1012
	+62 02319			23 51 16	+63 16.4	1	9.32		0.96				2	K0	1025
223959	+75 00897			23 51 16	+76 18.3	1	8.39		0.12		-0.48		4	B5 II	555
		CS 22876 # 10		23 51 17	−34 08.6	1	14.50		0.13		0.16		1		1736
		SW # 3		23 51 18	+66 12.	1	13.68		1.21		1.42		1		25
		SW # 1		23 51 18	+66 17.	1	11.46		2.38		2.78		2		25
		CS 22876 # 9		23 51 18	−34 40.2	1	14.82		0.21		0.08		1		1736
		LS I +60 055		23 51 19	+60 18.8	1	10.67		0.53		-0.35		2		41
223960	+60 02636	LS I +60 056		23 51 20	+60 34.5	4	6.90	.009	0.72	.009	-0.05	.000	16	A0 Ia	41,1012,1025,1119

HD	DM	Other Id	N	Rem	α_{1950}	δ_{1950}	S	V	σ_V	B–V	σ_{B-V}	U–B	σ_{U-B}	n	Spectrum	References
	+62 02320	LS I +62 047			23 51 20	+62 56.8	2	10.11	.024	0.35	.014	-0.47	.014	4	B2 V	41,1012
223948	-38 15668	IDS23487S3758	AB		23 51 20	-37 41.2	1	8.71		0.18				4	A3 V	2012
223954	+17 05001				23 51 21	+17 42.8	1	7.15		1.17		1.16		2	K0	1375
		LS I +61 091			23 51 21	+61 36.6	1	12.11		0.58		-0.33		2		41
		WLS 0 20 # 12			23 51 22	+22 18.8	1	10.57		1.10		0.70		2		1375
		LS I +61 092			23 51 24	+61 26.9	1	11.22		0.49		-0.28		2		41
		G 266 - 1			23 51 24	-19 15.3	1	13.52		0.88		0.34		7		3073
		L 10 - 56			23 51 24	-84 48.	1	12.03		0.67		0.04		2		1696
	+59 02786	LS I +60 057			23 51 26	+60 05.8	2	9.60	.005	0.65	.025	-0.49	.030	4	B0 III	41,342
	+63 02073				23 51 27	+63 59.9	2	10.19	.052	2.34	.066	2.31		4	M0 Ib	138,8032
223963	-10 06192				23 51 29	-09 34.1	2	7.19	.019	1.59	.014	2.00	.005	10	M1 III	1375,1499
	+16 05012	G 30 - 27			23 51 31	+17 16.3	1	10.11		0.96		0.74		1	K2	333,1620
223971	+38 05091	IDS23490N3844	AB		23 51 32	+39 00.2	1	6.63		0.67		0.31		2	G0 III	105
		LP 823 - 90			23 51 33	-19 37.4	1	13.36		0.57		-0.18		2		1696
	+65 01962				23 51 34	+66 00.5	1	10.75		0.33		-0.12		2		25
		LS I +60 058			23 51 38	+60 57.7	1	11.52		0.56		-0.33		2		41
223987	+60 02637	LS I +61 093			23 51 43	+61 19.7	3	7.57	.013	0.50	.009	-0.49	.000	17	B1 Ib	41,1012,1025
		POSS 192 # 6			23 51 45	+48 10.8	1	18.76		1.48				3		1739
223988	+55 03042				23 51 46	+55 42.9	1	8.41		1.16		1.14		2	K2	1502
223991	-27 16479	HR 9044		★AB	23 51 47	-27 19.3	4	6.34	.009	0.20	.008	0.01	.019	19	A2 V + F2 V	15,1068,2012,8071
		LS I +61 094			23 51 51	+61 22.6	1	11.58		0.63		-0.31		3		41
224014	+56 03111	HR 9045, ρ Cas			23 51 52	+57 13.3	6	4.50	.078	1.16	.071	1.01	.081	13	F8 Ia	15,759,1355,1363,3016,8015
		G 30 - 28			23 51 54	+07 53.0	1	13.01		1.50				2		1532
	+61 02561				23 51 54	+61 36.6	1	9.34		0.16				2	B9 V	1025
224013	+62 02323	V375 Cas			23 51 54	+62 54.9	1	8.66		0.19				2	A0	1025
		CS 29499 # 71			23 51 55	-23 57.6	1	15.62		0.06		0.13		1		1736
224016	+28 04665				23 51 57	+29 21.8	1	8.52		0.78		0.49		1	G5	1769
	+47 04313	IDS23495N4742	A		23 52 00	+47 57.3	1	9.67		0.32		0.48		1	A5	8084
	+47 04313	IDS23495N4742	B		23 52 00	+47 57.3	1	12.16		1.01		0.76		1		8084
224022	-40 15285	HR 9046			23 52 02	-40 34.7	6	6.02	.011	0.57	.008	0.11	.005	22	G0 V	15,1075,1311,2013*
		LS I +61 095			23 52 03	+61 05.5	1	12.24		0.54		-0.27		2		41,888
		LS I +59 019			23 52 11	+59 27.0	1	12.05		0.76		0.06		3		41
	+65 01964				23 52 11	+66 10.2	2	8.73	.010	1.98	.024	2.32		6	M2	25,369
		LS I +61 097			23 52 12	+61 03.3	1	12.00		0.46		-0.29		1		41
224055	+61 02562	LS I +61 096			23 52 12	+61 33.6	4	7.18	.017	0.71	.007	-0.21	.005	10	B3 Ia	41,1012,1025,8031
	-22 06219			V	23 52 12	-22 03.2	3	10.82	.047	1.47	.030	1.22		13	M0	1619,1705,7009
		CS 29499 # 67			23 52 12	-25 27.2	1	14.61		0.29		-0.01		1		1736
224062	-0 04585	HR 9047, XZ Psc			23 52 13	-00 10.1	3	5.72	.058	1.60	.009	1.64	.072	17	M5 III	15,1256,3042
	+60 02638				23 52 19	+60 33.1	1	9.44		0.38				2	A0 V	1025
		LS I +60 059			23 52 19	+60 46.2	1	10.80		0.38		-0.56		2		41
224065	-19 06540				23 52 20	-19 24.5	1	8.92		0.68		0.17		2	G6 V	1375
224075	-19 06541				23 52 21	-19 00.9	1	9.52		1.08		1.03		4	K1 (III)	3007
224098	+73 01063	IDS23500N7351	A		23 52 22	+74 07.9	1	6.62		0.02		-0.24		3	B9	985
240439	+59 02788				23 52 23	+60 16.7	1	9.22		0.25				2	A0	1025
		G 171 - 27			23 52 24	+40 11.1	1	14.94		0.18		-0.68		3		940
		CS 22966 # 69			23 52 24	-31 22.6	1	13.83		0.17		0.17		1		1736
	+61 02565				23 52 25	+61 47.7	1	9.92		0.38				2	A5	1025
	+65 01966				23 52 27	+65 56.9	1	11.14		0.72		0.26		2		25
	-14 06575	G 158 - 4			23 52 27	-14 10.5	1	10.55		0.70		0.19		2		1696
224085	+27 04642	II Peg		★	23 52 29	+28 21.3	5	7.45	.054	1.02	.005	0.67	.008	16	K1 V	22,928,1355,1620,3026
224087	+19 05170	G 129 - 42			23 52 30	+20 06.4	2	8.94	.020	0.81	.005	0.36	.040	2	G5	1620,1658
		LS I +61 098			23 52 31	+61 48.8	1	10.26		0.54		-0.25		2		41
224103	+6 05216	HR 9048			23 52 34	+06 47.6	4	6.20	.010	-0.07	.008	-0.18	.004	10	B9 V	15,1079,1256,1319
	+50 00859	IDS03488N5051	AB		23 52 34	+50 59.3	1	9.81		0.34				2	B8	1119
224104	-5 06081				23 52 34	-04 56.8	1	7.84		1.44		1.70		3	K0	1657
224112	-32 17724	HR 9050			23 52 42	-32 09.8	4	6.82	.004	-0.08	.004	-0.34		12		15,960,1075,2032
224113	-32 17723	HR 9049, AL Scl			23 52 42	-32 12.0	5	6.09	.013	-0.09	.007	-0.44	.016	34	B6 V	15,960,1034,1068,2012
		G 158 - 5			23 52 43	-13 36.4	1	14.29		0.86		0.31		2		3062
224116	+57 02829				23 52 44	+57 48.6	1	7.39		1.09		1.10		3	K2 IV	37
		POSS 192 # 5			23 52 45	+48 54.5	1	18.35		1.78				7		1739
		CS 22876 # 8			23 52 46	-35 04.7	1	13.95		0.29		-0.15		1		1736
		LS I +61 099			23 52 47	+61 56.4	1	10.84		0.27		-0.57		2		41
224128	+25 05042	HR 9051			23 52 50	+25 40.6	1	6.50		1.48		1.85		2	K5	1733
224129	+21 04995	G 129 - 43			23 52 51	+21 55.1	1	8.77		1.02		0.81		1	K0	1620
	+35 05133				23 52 53	+35 40.6	1	8.26		1.64		1.91		2	M0	1733
224202	+85 00406				23 52 56	+85 37.7	1	8.78		0.08		0.07		6	A0	1219
		CS 29499 # 66			23 52 56	-25 33.0	1	13.90		-0.01		-0.03		1		1736
		CS 29499 # 64			23 52 58	-25 49.6	1	14.69		0.01		0.09		1		1736
224156	+2 04731	G 31 - 16			23 52 59	+03 13.6	2	7.70	.005	0.75	.015	0.35		3	G5	202,333,1620
		CS 29499 # 69			23 52 59	-24 40.0	1	15.44		0.05		-0.03		1		1736
	+62 02326				23 53 01	+62 51.9	1	9.53		0.37				2	A9	1025
	+62 02325				23 53 01	+63 02.4	1	9.70		0.64				2	F7 V	1025
224143	-21 06498				23 53 01	-21 13.9	1	7.91		0.64		0.11		2	G3/5 V	1375
224165	+46 04214	HR 9053			23 53 02	+47 04.7	2	6.01	.015	1.17	.010	0.94		3	G8 Ib	138,252
		LS I +61 100			23 53 02	+61 27.5	1	11.97		0.47		-0.37		2		41
224151	+56 03115	HR 9052, V373 Cas			23 53 03	+57 08.0	4	6.00	.005	0.21	.015	-0.71	.005	9	B0.5II	15,41,154,1012
224155	+7 05101				23 53 04	+07 56.7	2	6.81	.011	-0.01	.007	-0.03	.018	7	A0	1319,1509
	-31 19511	G 275 - 130			23 53 05	-31 23.2	1	10.58		1.01		0.91		2		1696
	+21 04996	G 129 - 44			23 53 06	+21 32.8	1	10.34		0.63		0.10		1	G0	1658
224166	+45 04363	PV And			23 53 06	+46 04.8	2	6.94	.005	-0.07	.015	-0.37	.002	9	B9	379,1063,3016

Table 1

1245

HD	DM	Other Id	N Rem	α_{1950}	δ_{1950}	S	V	σ_V	B–V	σ_{B-V}	U–B	σ_{U-B}	n	Spectrum	References
	−6 06318	G 31 - 17		23 53 08	−06 24.9	5	11.15	.009	1.47	.015	1.15	.021	11	M2	202,1620,1705,1746,7008
		CS 22966 # 71		23 53 14	−30 52.4	1	15.37		0.02		0.11		1		1736
		NGC 7822 - 16		23 53 15	+68 29.2	1	10.75		0.72		0.10		1	B3	1285
		CS 22957 # 13		23 53 15	−05 39.6	1	14.08		0.73		0.12		1		1736
	+60 02639			23 53 18	+60 58.4	2	9.11	.000	0.23	.010			4	A5 V	1025,1118
	−24 17914			23 53 18	−23 43.6	1	11.60		0.12		0.11		1		1736
224186	+14 05074			23 53 21	+14 57.1	1	6.39		1.56		1.75		3	M2	3035
		LS I +62 048		23 53 21	+62 35.9	1	12.40		0.36		-0.44		2		41
	+62 02327			23 53 27	+62 35.4	2	9.22	.010	0.21	.010			5	B9 V	1025,1118
	+50 00863	IDS03498N5046	AB	23 53 28	+50 55.3	1	9.57		0.26				3	A1 IV	1119
		LS I +61 101		23 53 30	+61 09.2	1	10.64		0.31		-0.53		2		41
224215	+60 02640			23 53 32	+61 23.3	2	8.47	.005	0.22	.005			4	A2 V	1025,1118
	+63 02076			23 53 33	+63 54.9	1	9.51		0.36				2	G5	1025
224225	−22 06225			23 53 33	−22 16.2	2	7.28	.028	1.62	.009	1.98	.042	8	M2 III	1024,3007
224235	+32 04737			23 53 34	+33 12.6	1	7.05		-0.01		-0.15		1	A0 V	1716
		LS I +60 060		23 53 35	+60 38.7	1	11.29		0.48		-0.42		2		41
224233	+58 02674	G 217 - 24		23 53 39	+59 29.1	1	7.67		0.63				3	G0	1118
224234	+55 03047	NGC 7789 - 36		23 53 41	+56 27.2	2	8.44	.010	1.77	.010	1.99		3	M1	369,944
		G 266 - 9		23 53 43	−18 35.9	1	13.75		1.48		0.92		1		3078
	−33 16770			23 53 43	−32 48.8	1	11.51		1.09		0.97		2		937
240450	+59 02790			23 53 44	+60 00.9	1	8.70		0.26				2	B9	1118
		NGC 7789 - 72		23 53 46	+56 23.4	3	11.04	.025	1.86	.078	1.91		8		1175,4002,8092
		NGC 7789 - 80		23 53 48	+56 28.1	1	13.16		1.25		1.13		1		1175
		NGC 7789 - 88		23 53 49	+56 31.4	1	13.10		0.38		0.36		1		944
		NGC 7793 sq1 10		23 53 49	−32 47.3	1	13.71		0.55		0.03		2		937
224257	+55 03051	LS I +55 001		23 53 53	+55 42.7	2	7.99	.009	-0.06	.000	-0.90	.009	4	B0 IV	41,1012
224283	−25 16707	HR 9054		23 53 56	−25 00.9	1	6.31		0.92				4	G8 III/IV	2009
		LS I +59 020		23 53 58	+59 56.3	1	11.16		0.35		-0.44		2		41
224274	+52 03563			23 53 59	+53 26.1	1	7.60		0.97		0.66		2	G5	1566
		NGC 7789 - 155		23 54 01	+56 25.1	1	15.07		0.91				3		1175
		NGC 7789 - 160		23 54 01	+56 27.9	1	13.05		1.22		0.92		4		4002
224279	−18 06403	IDS23514S1749	AB	23 54 01	−17 32.7	1	9.23		0.49		0.02		2	F7 V	1375
236250	+53 03256			23 54 02	+53 40.9	1	7.64		1.42		1.30		3	K2	1566
		NGC 7789 - 169		23 54 03	+56 25.5	1	14.17		0.71				3		1175
		NGC 7789 - 168		23 54 03	+56 27.7	1	13.81		0.35		0.30		1		944
224309	+82 00743	HR 9056, V Cep		23 54 03	+82 54.7	2	6.58	.010	0.06	.010	0.05	.005	5	A3 V	985,1733
		NGC 7789 - 192		23 54 06	+56 25.9	1	13.63		0.33		0.29		1		944
		NGC 7789 - 193		23 54 06	+56 33.0	1	12.59		1.42		1.64		1		1175
		NGC 7793 sq1 8		23 54 06	−32 40.3	1	13.15		0.64		0.13		2		937
224296	−42 16527			23 54 07	−42 28.4	2	7.88	.014	0.44	.014			8	F3 V	1075,2013
224303	+21 04999	HR 9055		23 54 08	+22 22.2	2	6.16	.012	1.61	.004	1.97		6	M2 III	71,3001
		NGC 7789 - 197		23 54 08	+56 25.2	1	13.32		0.43		0.39		4		944
		NGC 7789 - 213		23 54 09	+56 29.9	1	13.62		1.51				1		1175
	+61 02568			23 54 09	+61 54.7	1	9.68		2.32				2		369
	−7 06114	G 158 - 11		23 54 09	−07 07.7	1	10.70		0.64		0.08		2		1696
	+60 02644	NGC 7788 - 1		23 54 10	+61 07.1	1	9.65		0.30		-0.54		2	B2	41
		NGC 7789 - 234		23 54 12	+56 34.3	1	13.50		0.44		0.36		3		944
		LS I +62 049		23 54 12	+62 12.6	1	10.74		0.25		-0.36		2		41
		NGC 7789 - 244		23 54 13	+56 26.2	1	13.16		1.24		0.93		1		4002
		NGC 7789 - 246		23 54 13	+56 27.4	1	10.29		0.21		0.11		10		1175
		NGC 7789 - 253		23 54 14	+56 25.3	1	13.75		0.74		0.25		1		1175
		NGC 7789 - 256		23 54 14	+56 31.5	1	13.92		0.73		0.26		1		1175
	+62 02329	LS I +63 015		23 54 14	+63 13.5	1	10.13		0.40		-0.22		2	B5 V	1012
		NGC 7789 - 282		23 54 17	+56 21.6	2	12.05	.000	0.26	.013	0.26	.054	5		944,1175
		NGC 7789 - 296		23 54 19	+56 24.1	1	12.47		0.62		0.15		1		944
	+60 02645	LS I +60 061	A	23 54 19	+61 22.4	2	9.85	.010	0.42	.005	-0.51	.005	3	B3 V	41,1012
	+60 02645	LS I +60 061	AB	23 54 19	+61 22.4	1	9.41		0.37				2	B3 V	1025
		NGC 7793 sq1 7		23 54 19	−32 42.9	1	12.24		0.65		0.06		2		937
		NGC 7789 - 301		23 54 20	+56 16.9	1	12.29		1.37				2		1175
		NGC 7788 - 95		23 54 20	+61 04.0	1	11.91		0.33		-0.48		2		41
		NGC 7789 - 304		23 54 21	+56 26.8	2	11.08	.030	1.76	.025			10		1175,8092
		NGC 7789 - 316		23 54 22	+56 24.1	1	13.79		0.36		0.33		3		944
		NGC 7789 - 311		23 54 22	+56 25.6	1	14.85		0.60				2		1175
		NGC 7789 - 321		23 54 23	+56 31.6	1	12.92		1.26		0.88		1		1175
	+61 02569			23 54 23	+61 44.3	1	8.82		0.54				2	F7 V	1025
		NGC 7789 - 329		23 54 24	+56 28.5	2	12.31	.014	1.41	.005	1.37	.056	2		1175,4002
		NGC 7788 - 68		23 54 24	+61 06.8	1	11.86		0.26		-0.50		2		41
		NGC 7789 - 342		23 54 25	+56 24.5	2	12.42	.000	0.21	.012	0.11	.028	4		944,1175
		NGC 7789 - 347		23 54 25	+56 25.1	1	13.00		1.23				2		531
		NGC 7789 - 338		23 54 25	+56 27.9	2	11.78	.004	1.19	.009	0.86		3		531,1175
		NGC 7789 - 349		23 54 25	+56 33.2	1	13.39		0.44		0.31		3		944
		NGC 7789 - 351		23 54 26	+56 18.6	2	13.59	.022	0.52	.013	0.17	.022	5		944,1175
		NGC 7789 - 371		23 54 27	+56 30.8	2	12.95	.004	0.36	.004	0.34	.067	4		944,1175
	+6 05221			23 54 29	+06 34.2	1	9.82		1.35				1	M0	1632
		POSS 348 # 5		23 54 30	+30 29.9	1	18.12		1.40				2		1739
		NGC 7788 - 71		23 54 30	+61 08.4	1	10.96		0.27		-0.54		2		41
		LS I +61 105		23 54 30	+61 13.8	1	12.91		0.34		-0.30		2		41
224342	+41 04902	HR 9057		23 54 31	+42 22.8	1	5.99		0.68		0.28		2	F8 III	254
		NGC 7789 - 415		23 54 31	+56 29.3	1	10.66		1.89				2	K4 III	531
		NGC 7789 - 409		23 54 31	+56 30.3	2	12.98		0.31		0.28		3		944

HD	DM	Other Id	N	Rem	α_{1950}	δ_{1950}	S	V	σ_V	B–V	σ_{B-V}	U–B	σ_{U-B}	n	Spectrum	References
		NGC 7789 - 416			23 54 32	+56 26.6	1	13.13		1.23		0.98		3		4002
		NGC 7789 - 430			23 54 33	+56 32.0	2	12.97	.012	0.66	.004	0.16	.012	4		944,1175
		NGC 7789 - 444			23 54 34	+56 28.1	1	12.91		1.24				2		531
	+61 02570				23 54 34	+62 12.8	1	9.50		0.09				2	B9 V	1025
224350	−27 16494	HR 9058			23 54 34	−26 54.1	1	6.25		1.45				4	K3 III	2035
		NGC 7789 - 453			23 54 35	+56 30.1	2	12.65	.020	0.25	.004	0.01	.012	4		944,1175
		NGC 7789 - 466			23 54 36	+56 24.1	2	12.21	.023	1.44	.000	1.46		4		531,4002
		NGC 7789 - 461			23 54 36	+56 25.0	2	11.34	.024	1.67	.016			4		1175,8092
		NGC 7789 - 468			23 54 36	+56 26.7	2	11.16	.020	1.64	.001	1.77		5		531,4002
		NGC 7789 - 462			23 54 36	+56 33.8	2	13.00	.024	0.53	.009	0.32	.028	2		944,1175
224355	+54 03076	HR 9059			23 54 37	+55 25.7	2	5.56	.000	0.48	.023	0.02	.000	8	G8 Ib +F1 V	254,401
		NGC 7789 - 476			23 54 37	+56 32.1	2	13.17	.030	1.25	.013	0.96	.004	3		1175,4002
	+60 02646	IDS23521N6045		A	23 54 37	+61 01.8	1	9.09		0.15				3	B8 V	1025
	+61 02571				23 54 37	+61 48.0	1	9.02		0.53				2	F7 V	1025
		NGC 7789 - 489			23 54 38	+56 24.2	1	11.25		1.49				3		531
	+62 02332	LS I +62 050		⋆V	23 54 38	+62 43.7	1	10.27		0.27		-0.57		2	B3	41
		NGC 7793 sq1 6			23 54 38	−32 43.7	1	13.96	.030	0.61	.010			5		937,1554
		NGC 7789 - 494			23 54 39	+56 26.1	2	10.73	.016	1.77	.016			4	K4 III	1175,8092
		NGC 7789 - 501			23 54 39	+56 27.8	3	11.24	.023	1.78	.066			7		531,1175,8092
240453	+59 02792	IDS23521N6005		AB	23 54 39	+60 21.9	1	9.62		0.26				2	A3	1025
		NGC 7789 - 521			23 54 40	+56 34.5	1	13.52		0.67		0.11		1		1175
		LS I +60 062			23 54 40	+60 12.4	1	11.02		0.56		-0.50		2		41
		LS I +62 051			23 54 40	+62 00.5	1	11.62		0.36		-0.43		2		41
		G 158 - 13		B	23 54 40	−16 47.2	1	14.90		0.98				2	K7	1619
	−17 06852	G 158 - 12		A	23 54 40	−16 47.3	1	10.76		1.30				1	K7	1017
		NGC 7789 - 526			23 54 41	+56 33.3	2	12.85	.033	1.25	.021	1.01	.028	2		1175,4002
		LS I +61 107			23 54 41	+61 23.5	1	10.84		0.31		-0.54		2		41
		NGC 7789 - 549			23 54 42	+56 24.3	1	12.35		1.43		1.44		1		4002
		NGC 7789 - 555			23 54 43	+56 22.7	1	13.10		1.23				2		531
		NGC 7789 - 558			23 54 43	+56 32.4	1	14.65		0.58				1		1175
		NGC 7788 - 78			23 54 43	+61 06.6	1	11.24		0.23		-0.55		2		41
		CS 22957 # 16			23 54 43	−06 11.2	1	14.34		0.45		-0.20		1		1736
	−33 16778				23 54 43	−32 42.0	2	11.35	.019	0.47	.005	-0.07	.009	8		937,1554
		NGC 7793 sq2 8			23 54 43	−32 54.6	1	13.14		0.61		0.06		5		1554
224361	−63 04940	HR 9060			23 54 43	−63 14.1	4	5.96	.006	0.11	.005	0.10		12	A1 IV	15,1075,2007,3050
		NGC 7793 sq2 6			23 54 44	−33 01.0	1	12.55		0.65		0.13		6		1554
224360	−46 14812				23 54 44	−46 23.3	4	7.70	.004	0.45	.004	-0.08		12	F5 V	116,1020,1075,2012
		NGC 7789 - 575			23 54 45	+56 29.6	2	12.13	.015	1.44	.010	1.38		3		531,4002
		AJ89,1229 # 386			23 54 47	+09 27.5	1	11.71		1.66				1		1532
		NGC 7789 - 605			23 54 47	+56 24.2	1	12.83		1.26				2		531
224364	+60 02647	IDS23523N6029		AB	23 54 48	+60 44.9	2	6.59	.005	1.63	.024			5	M1	369,1025
224364	+60 02647	IDS23523N6029		C	23 54 48	+60 44.9	1	9.11		0.53				2		1025
		DD Cas			23 54 49	+62 25.7	4	9.59	.010	1.07	.026	0.74		4	F7	851,1399,1772,6011
		CS 22876 # 1			23 54 49	−37 32.8	1	15.33		0.11		0.16		1		1736
		NGC 7789 - 635			23 54 50	+56 36.4	1	12.97		0.48		0.43		3		944
		LS I +61 109			23 54 50	+61 17.7	1	12.05		0.32		-0.47		2		41
224395	+72 01127	IDS23524N7218		AB	23 54 50	+72 34.8	1	7.84		0.13		-0.50		6	B5	555
224379	−66 03816	R Tuc			23 54 50	−65 39.7	1	9.92		1.16		0.92		1	M4	975
224362	−82 00907	HR 9061			23 54 52	−82 26.9	3	5.72	.004	1.06	.005	0.93	.005	13	K0 III	15,1075,2038
		NGC 7789 - 677			23 54 53	+56 21.3	2	11.15	.020	0.17	.008	0.01	.028	4		944,1175
		NGC 7789 - 676			23 54 53	+56 21.9	1	12.87		1.20				2		531
		NGC 7789 - 669			23 54 53	+56 31.8	4	11.50	.014	1.66	.018	1.73		9		531,1175,4002,8092
		NGC 7789 - 671			23 54 53	+56 33.2	1	13.32		0.68		0.10		1		1175
		NGC 7789 - 684			23 54 54	+56 29.2	1	13.02		1.25		1.00		2		4002
224386	−37 15436				23 54 54	−36 59.0	1	7.38		1.26		1.28		7	K1 III	1673
		NGC 7789 - 696			23 54 55	+56 31.7	1	13.78		0.28		0.19		3		944
		LS I +61 110			23 54 55	+61 20.7	1	10.71		0.36		-0.40		2		41
		LS I +61 111			23 54 56	+61 46.8	1	11.22		0.53		-0.29		2		41
		NGC 7789 - 724			23 54 57	+56 18.7	1	13.23		1.17		0.89		1		4002
		CS 22957 # 19			23 54 57	−07 13.8	1	13.71		0.41		-0.22		1		1736
		NGC 7789 - 732			23 54 58	+56 25.7	1	12.72		1.24				2		531
		NGC 7789 - 737			23 54 58	+56 26.6	2	13.39	.010	1.17	.012	0.92		2		531,4002
		NGC 7789 - 746			23 54 58	+56 27.8	1	12.74		0.39		0.35		3		944
240455	+59 02794	LS I +59 021			23 54 58	+59 40.4	1	9.10		0.92		0.73		2	F6 I	41
		G 31 - 19			23 54 58	−03 47.4	1	15.46		1.61				1		3062
224383	−10 06206	G 158 - 14			23 54 58	−09 55.4	2	7.86	.015	0.64	.010	0.15	.005	3	G2 V	1003,3056
224392	−64 04391	HR 9062			23 54 58	−64 34.6	4	4.99	.008	0.06	.005	0.07	.005	15	A1 V	15,1075,2024,2038
		NGC 7789 - 758			23 54 59	+56 18.6	1	10.46		0.72		0.33		5		1175
		NGC 7789 - 751			23 54 59	+56 34.2	2	10.89	.031	1.85	.009			5	M0 III	1175,8092
224399	−40 15295				23 54 59	−39 57.4	1	9.99		0.03		0.04		4	A0 V	152
		WLS 0 35 # 10			23 55 00	+34 42.5	1	10.65		0.03		-0.13		2		1375
224403	+61 02573				23 55 01	+61 42.8	3	7.68	.009	0.14	.000	0.00		9	B9 IV	206,1025,1118
	+66 01661	NGC 7822 - 6			23 55 01	+67 16.6	4	8.71	.005	0.81	.008	-0.20	.004	29	O9.5V	25,1011,1012,1285
		NGC 7789 - 799			23 55 02	+56 29.3	1	11.78		0.33		0.20		1		1175
224404	+59 02795	HR 9063			23 55 02	+59 44.7	2	6.47		0.01	.010	-0.11		3	B9 III-IV	1079,1118
		BI 1 - 1			23 55 02	−29 58.9	1	9.58		1.07		0.96		6	dK0	1675
224406	+19 05176				23 55 03	+20 03.1	1	7.86		1.27		1.22		2	G5	1648
		LS I +61 112			23 55 03	+61 05.8	1	10.62		0.26		-0.57		2		41
		LS I +61 113			23 55 03	+61 43.8	1	11.96		0.49		-0.17		2		41
	+67 01580	NGC 7822 - 5			23 55 03	+67 34.4	2	10.13	.014	0.58	.038	-0.15	.033	4	B3	25,1285

Table 1   1247

HD	DM	Other Id	N Rem	α_{1950}	δ_{1950}	S	V	σ_V	B–V	σ_{B-V}	U–B	σ_{U-B}	n	Spectrum	References
	−30 19753	BI 1 - 135		23 55 04	−30 06.8	1	10.28		0.41		-0.07		2	F5	3004
240458	+55 03057	LS I +55 002		23 55 05	+55 34.8	1	9.80		0.28		-0.59		2	B7	41
	−33 16781			23 55 05	−32 42.7	2	10.86	.019	1.28	.009	1.50	.023	8		937,1554
		NGC 7789 - 833		23 55 06	+56 33.8	1	15.23		0.83				1		1175
		POSS 348 # 6		23 55 07	+30 53.0	1	18.45		1.05				3		1739
		CS 22876 # 23		23 55 07	−34 17.9	1	14.42		0.21		0.09		1		1736
		G 129 - 47		23 55 08	+23 02.3	2	11.75	.050	1.54	.005	1.05	.002	5	M3	940,3078
		NGC 7789 - 859		23 55 08	+56 28.7	1	12.71		1.39				1		531
	+52 03566	IDS23526N5313	AB	23 55 09	+53 29.1	1	10.06		0.24		0.10		1		8084
		NGC 7789 - 866		23 55 09	+56 25.4	1	12.14		1.22				1		531
	−33 16782			23 55 10	−32 43.7	2	10.93	.019	0.41	.000	-0.04	.000	8		937,1554
		NGC 7789 - 875		23 55 11	+56 22.2	1	12.95		1.22				1		531
	−62 06457			23 55 11	−61 48.9	1	9.85		0.89		0.58		1	K2 V	3072
224427	+24 04865	HR 9064	⋆	23 55 12	+24 51.8	9	4.67	.020	1.58	.008	1.67	.014	60	M3 III	15,369,1007,1013,1363*
		NGC 7789 - 889		23 55 12	+56 23.0	1	11.52		0.21		0.18		1		1175
224421	−50 14088			23 55 12	−49 40.7	1	8.46		0.38				4	Fm δ Del	2012
224429	+10 05013	IDS23526N1055	A	23 55 13	+11 11.8	1	6.62		0.01		-0.04		2	B9 V	1319
		NGC 7789 - 897		23 55 13	+56 29.0	1	12.93		1.23				1		531
		NGC 7793 sq1 3		23 55 13	−32 45.4	2	12.37	.019	0.70	.005	0.20	.028	8		937,1554
		NGC 7789 - 908		23 55 14	+56 27.7	1	13.01		1.21				1		531
224425	+56 03119			23 55 15	+56 51.7	1	7.30		0.22		0.15		2	A2:V:	1012
		LS I +62 052		23 55 15	+62 50.2	1	11.31		0.25		-0.50		2		41
224424	+58 02676	LS I +59 022		23 55 16	+59 26.5	2	8.11	.012	0.77	.020	-0.20	.012	7	B1 Iab	41,1012
		CS 22957 # 15		23 55 16	−06 14.0	1	14.85		0.44		-0.19		1		1736
		G 242 - 44		23 55 18	+78 20.0	2	13.88	.028	1.66	.028			4		538,906
		G 30 - 31		23 55 19	+08 20.3	1	12.25		1.16		0.80		3		333,1620
		NGC 7789 - 971		23 55 19	+56 22.2	2	11.02	.034	1.93	.024	1.87		3		531,4002
		NGC 7793 sq1 4		23 55 19	−32 49.4	1	12.27		0.50		-0.12		2		937
		NGC 7789 - 977		23 55 20	+56 25.7	2	11.00	.004	1.71	.028			4	K4 III	1175,8092
224442	−9 06294	V Cet		23 55 20	−09 14.3	1	13.40		0.95		0.23		1	M4	975
224443	−37 15438			23 55 20	−37 22.7	1	8.46		0.51				4	F7 V	2012
224435	+56 03120	IDS23528N5650	AB	23 55 21	+57 06.6	1	8.34		0.10		-0.25		2	B9	401
	+55 03059			23 55 22	+56 11.2	1	10.20		2.18		1.82		3		8032
224436	+56 03122	LS I +56 004		23 55 23	+56 48.4	1	8.94		0.15		-0.73		2	B1 II	41
224444	−39 15210			23 55 24	−39 14.0	1	7.13		1.57				4	K5 III	2012
		NGC 7789 - 1036		23 55 25	+56 26.5	1	13.03		1.16		0.77		1		4002
	+61 02574			23 55 25	+61 56.0	1	8.81		1.10				2	G8 IIIp	1025
		NGC 7789 - 1047		23 55 26	+56 27.6	1	12.46		0.51		0.40		1		944
		G 242 - 45		23 55 26	+78 19.8	2	13.51	.005	1.60	.014			4		538,906
		AJ89,1229 # 387		23 55 29	+05 45.2	1	13.47		1.51				2		1532
		NGC 7789 - 1071		23 55 29	+56 22.7	1	12.63		1.30		1.16		2		4002
		NGC 7790 - 87		23 55 29	+60 55.5	1	13.90		0.39		-0.24		1		1676
		LS I +61 114		23 55 29	+61 35.5	1	12.23		0.35		-0.39		2		41
		NGC 7790 - 353		23 55 30	+60 54.8	2	12.51	.035	0.55	.000	0.31	.084	11		1176,1676
		CS 22876 # 22		23 55 32	−34 01.9	1	13.81		0.17		0.16		1		1736
224458	+29 05034			23 55 33	+29 41.9	4	8.26	.019	1.03	.009	0.73	.008	15	G8 III	20,979,3016,4001
		NGC 7789 - 1097		23 55 33	+56 30.2	1	14.44		0.68		0.30		1		944
224465	+49 04291			23 55 34	+50 10.0	1	6.64		0.67		0.19		2	G5	3016
		NGC 7790 - 383		23 55 34	+60 57.3	1	16.22		0.83				1		1176
224462	−22 06232			23 55 34	−22 10.8	1	8.82		0.17		0.07		6	A3 V	1068
		NGC 7789 - 1107		23 55 35	+56 24.1	1	12.86		1.24		1.03		1		4002
		NGC 7790 - 351		23 55 35	+60 55.0	3	11.07	.007	0.25	.029	-0.60	.040	44	B2 III-IV	41,1176,1676
		NGC 7790 - 356		23 55 36	+60 56.0	2	12.71	.075	1.83	.108	1.03		2		1176,1676
		NGC 7789 - 1114		23 55 37	+56 23.2	1	13.26		1.20				1		531
		NGC 7790 - 360		23 55 37	+60 54.7	2	13.22	.025	0.52	.015	0.34	.035	4		1176,1676
		NGC 7790 - 385	⋆ ABV	23 55 37	+60 56.1	2	9.86	.021	1.06	.005	0.60		2	F9 Ib	1772,6011
		NGC 7789 - 1119		23 55 38	+56 22.3	1	12.77		0.31		0.23		3		944
	+61 02575	LS I +62 053		23 55 38	+62 19.9	2	8.47	.005	1.09	.014	0.87		4	F8 Ib:	41,1025
		POSS 348 # 3		23 55 39	+30 41.6	1	16.59		1.65				2		1739
		NGC 7790 - 382		23 55 40	+60 58.0	1	15.98		0.80				1		1176
		NGC 7790 - 359		23 55 41	+60 54.7	2	13.13	.030	1.28	.047	0.77	.043	3		1176,1676
		NGC 7790 - 354	AB	23 55 42	+60 53.9	2	12.76	.107	0.46	.013	0.33	.040	5		1176,1676
		BI 1 - 2		23 55 42	−29 53.9	1	10.72		0.52		-0.02		2		1675
		NGC 7790 - 375		23 55 43	+60 57.1	1	14.72		1.35				3		1176
		LS I +62 054		23 55 43	+62 51.4	1	12.16		0.39		-0.40		2		41
		NGC 7790 - 62		23 55 44	+60 55.8	1	13.23		0.50		0.00		1		1676
		NGC 7789 - 1168		23 55 45	+56 33.0	1	13.91		0.44		0.39		3		944
		NGC 7790 - 352		23 55 45	+60 54.9	2	12.25	.061	0.38	.000	-0.10	.005	6		1176,1676
		NGC 7790 - 384	⋆ V	23 55 46	+60 56.6	3	11.00	.147	1.20	.099	0.86		5	G0 Ib	1676,1772,6011
	−30 19758	BI 1 - 132		23 55 46	−30 33.6	2	9.94	.125	0.33	.020	0.00	.042	7	F0	1068,3004
	−54 10402	Smethells 14		23 55 46	−54 05.5	1	10.75		1.35				1		1494
		NGC 7790 - 373		23 55 47	+60 56.0	1	14.55		1.44				3		1176
		NGC 7790 - 368		23 55 47	+60 56.9	2	14.11	.038	0.43	.019	0.04	.052	2		1176,1676
		NGC 7790 - 378		23 55 47	+60 57.2	2	15.23	.030	0.50	.009	0.18	.132	3		1176,1676
		NGC 7790 - 381		23 55 47	+60 58.2	1	15.76		0.51				1		1176
224481	−16 06394	HR 9065		23 55 47	−16 07.5	2	6.25	.005	1.08	.000	1.04	.005	7	K1 III	15,1071
		NGC 7790 - 367		23 55 48	+60 56.2	2	13.71	.005	0.52	.038	-0.13	.108	2		1176,1676
224482	−21 06505			23 55 48	−20 35.8	1	8.86		0.38				4	F3 V	2012
		PHL 603		23 55 48	−32 21.	1	15.24		-0.26		-0.94		3		286
		LB 3126		23 55 48	−74 06.	1	11.93		0.03		0.01		2		45

HD	DM	Other Id	N	Rem	α_{1950}	δ_{1950}	S	V	σ_V	B–V	σ_{B-V}	U–B	σ_{U-B}	n	Spectrum	References
		NGC 7790 - 379			23 55 49	+60 53.6	1	15.29		0.82				1		1176
		NGC 7790 - 374			23 55 49	+60 53.8	2	14.41	.151	0.53	.075	0.03	.019	4		1176,1676
224491	+38 05103				23 55 50	+38 41.1	1	7.91		1.63		2.00		2	K2	1601
		NGC 7790 - 366			23 55 50	+60 57.5	1	13.60		0.79		0.12		2		1176
		NGC 7790 - 355			23 55 51	+60 55.7	2	12.80	.004	0.40	.016	-0.06	.012	4	B5 IV	1176,1676
		NGC 7793 sq2 10			23 55 51	-32 48.6	1	14.06		1.08				3		1554
224490	+50 04202	HR 9066, R Cas	★AB		23 55 52	+51 06.6	1	6.45		1.83		0.08		1	M4	814
224490	+50 04201	IDS23533N5050	C		23 55 52	+51 06.6	1	11.10		0.54		0.01		4	G0	3024
		CS 22876 # 26			23 55 52	-35 24.3	1	14.03		0.24		0.02		1		1736
		NGC 7789 - 1211			23 55 53	+56 25.7	2	11.55	.000	0.18	.015	-0.25	.010	5		877,944
		NGC 7790 - 95			23 55 56	+60 55.3	1	12.46		0.42		-0.25		1	B5 IV	1676
		NGC 7822 - 10			23 55 56	+65 38.1	1	10.53		0.60		-0.40		1	B5	1285
224505	-57 10389				23 55 56	-57 33.5	1	6.69		0.79		0.39		7	G3 III/IV	1628
		G 131 - 6	A		23 55 57	+23 55.5	1	11.68		1.33		1.26		3	K7	7010
		NGC 7790 - 55			23 55 57	+60 56.0	1	13.04		0.46		-0.02		1		1676
		LS I +61 115			23 55 57	+61 50.2	1	12.40		0.48		-0.32		2		41
224514	-24 17934				23 55 57	-24 26.9	2	7.98	.009	0.27	.000	0.09	.014	8	A2mA8-F0	355,1068
224519	+59 02797				23 55 58	+60 24.5	2	8.03	.012	1.11	.003			36	K2	1025,6011
		NGC 7790 - 362			23 55 58	+60 57.0	1	13.25		0.69		0.12		2		1176
		G 30 - 33			23 55 59	+07 23.0	2	11.77	.009	1.51	.005	1.17	.019	4		333,1620,1663
		NGC 7790 - 52			23 55 59	+60 56.1	1	13.00		0.45		-0.01		1		1676
		NGC 7790 - 363			23 55 59	+60 57.5	2	13.36	.020	0.37	.000	-0.11	.016	4		1176,1676
224515	-29 18903	BI 1 - 3	AB		23 55 59	-29 29.3	1	8.42		1.09		0.97		2	K0	1675
		NGC 7790 - 380			23 56 00	+60 53.4	1	15.65		0.91				3		1176
		NGC 7790 - 372			23 56 01	+60 57.4	2	14.51	.000	0.48	.020	0.07	.024	4		1176,1676
		CS 22957 # 24			23 56 01	-04 50.0	1	14.30		0.38		-0.21		1		1736
		CS 22876 # 18			23 56 01	-33 04.5	1	15.12		0.16		0.03		1		1736
		NGC 7790 - 365			23 56 03	+60 55.3	2	13.66	.068	0.38	.013	-0.07	.051	3	B9 IV	1176,1676
	+61 02576				23 56 05	+62 15.3	2	9.82	.005	0.33	.050			3	B9 V	851,1025
		NGC 7790 - 376			23 56 06	+60 57.4	2	14.76	.004	0.43	.030	0.10	.098	3		1176,1676
		NGC 7793 sq2 7			23 56 06	-32 56.1	1	13.05		0.72		0.21		4		1554
	+45 04378	G 171 - 29			23 56 07	+46 27.0	1	9.62		1.44		1.19		3	M0	3072
		NGC 7790 - 36			23 56 07	+60 55.6	1	13.71		0.43		-0.01		2		1676
		NGC 7790 - 16			23 56 07	+60 58.5	1	15.48		0.62				1		1676
		NGC 7790 - 377			23 56 07	+60 58.5	1	14.93		0.73				1		1176
224533	-4 05996	HR 9067	★AB		23 56 07	-03 50.0	3	4.86	.005	0.93	.007	0.70	.000	14	G9 III	15,1071,1075
224536	-30 19761	BI 1 - 4			23 56 07	-29 46.8	1	9.11		0.44		-0.09		2	G0	1675
		LS I +59 023			23 56 08	+59 14.1	1	11.76		0.50		-0.20		2		41
	+61 02577				23 56 08	+62 26.9	1	10.06		1.12				1		851
		NGC 7790 - 357			23 56 09	+60 58.1	1	13.13		0.55		0.26		1		1176
		NGC 7790 - 369			23 56 09	+60 58.4	1	14.32		0.46		-0.29		2		1176
		POSS 348 # 1			23 56 10	+29 36.5	1	14.89		1.65				1		1739
		LS I +60 064	★V		23 56 11	+60 53.0	2	10.22	.000	0.29	.000	-0.59	.000	4	B1 V	41,1012
		NGC 7790 - 364			23 56 11	+60 58.2	1	13.48		0.39		0.09		1		1176
	+62 02337	LS I +63 019			23 56 11	+63 23.0	1	9.57		0.57				2	A2 Ib	1025
224532	+8 05152	G 30 - 34			23 56 13	+08 57.9	1	9.20		0.67		0.12		1	G5	333,1620
224559	+45 04381	HR 9070, LQ And			23 56 13	+46 08.1	5	6.53	.010	-0.08	.013	-0.62	.012	12	B4 Ven	154,252,379,1212,1223
		NGC 7793 sq2 4			23 56 14	-33 02.9	1	12.17		0.60		0.10		3		1554
224544	+31 05012	HR 9068			23 56 16	+32 06.2	2	6.51	.000	-0.11	.024	-0.58	.015	5	B6 IVe	154,1212
224543	+32 04745	G 130 - 29			23 56 16	+33 28.0	1	7.82		0.64		0.15		1	G0	333,1620
224557	+52 03571				23 56 16	+53 23.6	1	8.26		0.11		0.07		3	A0	1566
		NGC 7790 - 358			23 56 16	+60 55.9	1	13.15		0.49		0.29		2		1176
240463	+59 02798				23 56 17	+60 27.8	1	9.95		0.66				2	F5	6011
	-29 18906	BI 1 - 5	★		23 56 17	-29 08.4	2	10.82	.000	0.46	.010	-0.25	.022	4	F2	1675,1696
		WLS 0 20 # 10			23 56 18	+20 39.8	1	11.89		0.57		0.00		2		1375
240464	+59 02799	LS I +59 024			23 56 18	+59 59.2	3	9.60	.015	0.31	.008	-0.62	.015	6	F5	41,1011,1012
224576	+38 05104				23 56 20	+38 50.0	1	7.26		0.33		0.03		2	F2 V	105
		NGC 7790 - 361			23 56 21	+60 54.3	1	13.21		1.52				3		1176
224554	-53 10561	HR 9069			23 56 21	-53 01.5	2	5.13	.000	1.12	.009	1.03		6	K0 III	58,2009
		NGC 7790 - 370			23 56 24	+60 54.7	1	14.30		0.68				1		1176
		NGC 7790 - 371			23 56 24	+60 58.2	1	14.48		0.56		0.34		1		1176
224567	-30 19762	BI 1 - 6			23 56 25	-30 11.6	1	9.45		0.54		-0.13		2	G0	1675
224572	+54 03082	HR 9071	★AB		23 56 28	+55 28.6	6	4.88	.004	-0.07	.013	-0.81	.013	18	B2 III+B3 V	15,41,154,1203,1363,8015
		LS I +61 116			23 56 28	+61 26.7	1	11.70		0.31		-0.35		2		41
224595	-23 19005				23 56 29	-22 39.8	3	10.18	.017	0.30	.005	0.16	.009	19	A1/2mF0-A8	152,1068,1775
		CS 22957 # 28			23 56 32	-03 54.9	1	14.82		0.40		-0.17		1		1736
224600	+55 03064				23 56 33	+55 40.7	1	8.37		0.01		-0.30		2		401
224596	-42 16545				23 56 35	-42 30.9	1	6.77		0.21				4	A4 V	2012
		LB 1538			23 56 36	-50 10.	1	12.11		0.07		0.12		3		45
	-29 18908	BI 1 - 7			23 56 38	-29 09.1	1	10.29		1.12		1.03		2	K1	1675
		CS 22957 # 27			23 56 39	-04 10.5	1	13.59		0.77		0.01		1		1736
	+63 02084	LS I +63 022			23 56 41	+63 32.5	1	9.16		0.31				2	B3 Ia	1025
224599	+59 02801	LS I +59 025			23 56 42	+59 44.7	2	9.58	.018	0.42	.000	-0.52	.009	5	B0.5:V:p,nn	41,1012
224617	+6 05227	HR 9072			23 56 44	+06 35.2	5	4.01	.007	0.42	.003	0.05	.019	14	F4 IV	15,1256,1363,3026,8015
		CS 22876 # 19			23 56 46	-33 33.8	1	14.31		0.18		0.14		1		1736
	+62 02341				23 56 48	+62 54.4	2	9.09	.025	0.19	.005			4	B5	1025,1118
224618	-17 06856	G 266 - 20			23 56 50	-17 13.3	7	8.92	.007	0.77	.007	0.27	.017	21	K0 V	22,742,1003,1097*
	-29 18909	BI 1 - 8			23 56 51	-29 08.1	1	10.10		0.48		-0.04		2	F0	1675
224619	-20 06684	G 266 - 21			23 56 53	-20 18.6	4	7.47	.009	0.74	.005	0.28	.015	9	G8 V	22,1311,2034,3077
224630	-30 19765	HR 9073			23 56 54	-29 45.8	2	5.60	.027	1.59	.018	2.00		5	K5 III	2032,3005

Table 1   1249

HD	DM	Other Id	N	Rem	α_{1950}	δ_{1950}	S	V	σ_V	B–V	σ_{B-V}	U–B	σ_{U-B}	n	Spectrum	References
224624	+56 03127	IDS23544N5707		A	23 56 55	+57 23.6	1	7.20		0.14		0.08		2	A0	401
224635	+32 04747	HR 9074		⋆AB	23 56 56	+33 26.8	3	5.81	.010	0.53	.015	-0.01	.019	8	F8 V + G0 V	938,1733,3026
		LS I +62 055			23 56 56	+62 23.1	1	11.70		0.28		-0.41				41
224648	+50 04208				23 57 00	+50 33.3	1	7.17		-0.04		-0.32		5	B9	814
	-30 19766	BI 1 - 10			23 57 01	-30 08.7	1	10.49		0.95		0.67		6	dG6	1675
		CS 22876 # 27			23 57 01	-35 56.3	2	14.35	.009	0.40	.007	-0.22	.009	4		1580,1736
	+60 02651				23 57 03	+60 32.3	1	9.60		0.15				2	A0 V	1025
		G 31 - 21			23 57 07	+02 16.1	2	12.41	.005	1.44	.029	1.20		3		202,333,1620
		LS I +61 117			23 57 10	+61 11.8	1	10.99		0.24		-0.39		2		41
224662	-25 16732				23 57 10	-24 55.4	1	7.78		1.21		1.17		4	K0 III	1657
	-29 18911	BI 1 - 11			23 57 10	-29 25.6	1	11.26		0.91		0.54		2	G2	1675
224669	+62 02343				23 57 11	+62 42.8	2	7.95	.015	1.66	.004			40	K2	1025,6011
	-33 16796				23 57 11	-32 53.0	1	11.35		0.98		0.71		1		937
		Smethells 148			23 57 12	-44 21.	1	12.82		1.64				1		1494
	+49 04301	G 217 - 25		AB	23 57 13	+49 50.0	1	9.90		0.90		0.78		3	K3	1723
240470	+58 02683	LS I +58 007			23 57 13	+58 55.8	2	9.20	.043	0.77	.004	0.20	.009	3	A0 Ia	41,1215
224677	-1 04514				23 57 13	-00 33.5	1	6.83		1.62		1.95		8	M1	3040
		NGC 7793 sq1 12			23 57 14	-32 52.2	1	13.13		0.85		0.39		1		937
		POSS 348 # 2			23 57 16	+30 04.2	1	15.28		1.48				1		1739
		LP 149 - 14			23 57 17	+47 28.5	1	16.10		1.87				4		1663
224695	-30 19769	BI 1 - 12			23 57 19	-29 36.4	1	9.42		0.61		0.00		2	F8	1675
236265	+54 03090				23 57 20	+55 01.7	1	10.22		0.13		0.09		2	A1 V	1003
224686	-66 03819	HR 9076			23 57 20	-65 51.3	6	4.49	.010	-0.08	.005	-0.27	.014	22	B9 IV	15,1034,1075,2024*
224689	+5 05245				23 57 21	+05 40.7	1	7.57		1.00		0.76		2	K0	1733
		LP 764 - 47			23 57 21	-16 13.7	1	12.68		0.78		0.38		2		1696
		CS 22876 # 25			23 57 22	-35 23.7	1	15.22		0.25		-0.12		1		1736
	+67 01585	NGC 7822 - 4			23 57 25	+67 34.6	2	10.35	.014	0.68	.000	-0.17	.047	4	B2	25,1285
		LOD 10 # 7			23 57 26	-00 16.7	1	12.09		1.10		1.26		2		123
		CS 22957 # 26			23 57 28	-04 23.8	1	13.16		0.44		-0.19		1		1736
		CS 22876 # 20			23 57 30	-33 52.0	1	14.21		-0.02		-0.09		1		1736
	+33 04813	G 130 - 32			23 57 31	+33 54.7	1	8.50		0.65		0.10		1	G0	333,1620
	+19 05185	G 129 - 52			23 57 33	+19 46.3	1	9.74		0.74		0.33		1	G0	1620
	-30 19770	BI 1 - 13			23 57 33	-30 07.3	1	10.57		1.14		1.03		6	K0	1675
	-29 18914	BI 1 - 14			23 57 35	-29 02.3	1	10.42		0.90		0.61		2	F8	1675
	-31 19545	BI 1 - 16			23 57 37	-30 37.0	1	10.96		0.57		0.03		4	F5	1675
	+38 05109	G 171 - 34			23 57 38	+38 46.3	1	10.74		0.82		0.35		2	G8	7010
224739	+57 02841				23 57 38	+57 35.6	1	8.60		0.29		0.22		2	A0	401
		BI 1 - 15			23 57 38	-30 06.0	1	11.69		0.75		0.19		6		1675
224732	-40 15306				23 57 38	-40 28.2	1	8.13		0.43				4	F3 V	2012
	+32 04748				23 57 39	+32 51.7	1	9.26		1.15		1.00		2	K2	1375
224721	+37 04912				23 57 39	+38 01.5	2	6.53	.005	0.95	.005	0.69	.005	6	K0 III	196,583
		LS I +61 118			23 57 39	+61 51.4	1	11.41		0.58		0.13		2		41
	-4 06001	MCC 871			23 57 39	-04 19.9	1	10.95		1.33				2	K5	1619
		TOU28,33 T9# 61			23 57 41	+60 48.1	1	9.81		0.30				2		1025
224750	-44 15420	HR 9077		⋆AB	23 57 45	-44 34.0	2	6.28	.005	0.76	.000			7	G3 IV	15,2012
224758	+26 04727	HR 9078		⋆A	23 57 50	+26 38.4	2	6.47	.015	0.50	.005	0.00	.015	6	F8 V	254,3053
	+61 02579				23 57 50	+62 26.9	1	9.89		0.16				2	B8 V	1025
	-30 19772	BI 1 - 17			23 57 50	-29 58.3	1	10.91		1.07		0.89		3	G0	1675
		LS I +62 056			23 57 53	+62 23.1	1	11.53		0.36		0.07		2		41
	+66 01669	NGC 7822 - 12		⋆AB	23 57 55	+66 56.2	1	11.08		1.35		0.32		1		25
224771	+25 05055				23 57 56	+25 34.0	1	8.20		0.38		0.02		3	F2	833
	+66 01669	NGC 7822 - 11		⋆A	23 57 56	+66 56.3	1	10.76		1.26		0.21		1		25
224784	+58 02685	HR 9079			23 57 58	+59 16.9	3	6.20	.008	1.03	.011	0.81	.011	6	G9 III-IV	1118,3016,4001
		G 30 - 35			23 58 02	+17 41.4	1	10.57		1.03		0.90		1	K3	333,1620
		G 31 - 22			23 58 04	-06 47.3	1	12.74		1.43				2		202
	-29 18916	BI 1 - 18		AB	23 58 04	-29 33.0	1	10.97		0.41		-0.08		2	F2	1675
224792	+61 02580				23 58 08	+61 53.9	2	7.05	.000	0.49	.010			4	G0	1025,1118
	-29 18918	BI 1 - 19			23 58 08	-29 23.6	1	12.32		0.88		0.49		6		1675
224801	+44 04538	HR 9080, CG And			23 58 10	+44 58.5	5	6.34	.027	-0.06	.014	-0.35	.009	15	B9 p SiEu	112,695,1202,1263,3033
	-32 17773				23 58 10	-31 46.4	1	12.54		0.43		-0.25		2		1696
	-29 18919	BI 1 - 20			23 58 12	-29 25.4	1	11.03		0.60		0.05		2	G3	1675
	+60 02654				23 58 13	+60 31.9	1	9.79		0.23				2	B8 III	1025
	-30 19774	BI 1 - 21			23 58 14	-29 38.0	1	11.78		0.51		-0.09		2		1675
224808	+16 05027	G 30 - 36			23 58 15	+16 42.8	1	8.82		0.94		0.72		1	K0	333,1620
		LS I +60 065			23 58 15	+60 46.1	1	11.51		0.45		-0.40		2		41
	-30 19775	BI 1 - 22			23 58 19	-30 07.2	1	9.92		1.51		1.84		2	K4	1675
224821	-51 13741				23 58 19	-50 43.5	1	7.41		1.52		1.82		6	K4 III	1673
224820	-30 19776	BI 1 - 23			23 58 21	-30 20.6	4	8.42	.012	0.03	.013	0.02	.018	12	A0 Va	152,1068,1675,3004
224818	-21 06515	IDS23558S2059		AB	23 58 22	-20 42.1	3	9.85	.020	0.17	.018	0.13	.010	11	A1 IV	152,355,1068
224817	-12 06598	G 158 - 20			23 58 23	-12 06.1	1	8.40		0.55		-0.08		1	G2 V	3056
		POSS 348 # 4			23 58 24	+30 10.0	1	17.63		1.59				2		1739
		LS I +62 057			23 58 25	+62 17.8	1	12.00		0.32		-0.36		2		41
224828	-5 06097	G 158 - 21			23 58 25	-05 12.5	1	8.59		0.64		0.09		1	G5	1658
		CG Cas			23 58 26	+60 41.3	2	10.90	.004	1.06	.004	0.60		2	F5	1399,1772
	+61 02581	LS I +61 119			23 58 26	+61 48.2	1	9.87		0.34		-0.38		2	B3 V	41
	-25 16747				23 58 27	-24 59.4	1	10.62		0.35		-0.06		4	A2	1068
		CS 22876 # 21			23 58 30	-34 04.9	1	14.48		0.31		-0.08		1		1736
224834	-49 14316	HR 9081			23 58 31	-49 05.3	4	5.70	.011	0.91	.006			18	G8 III	15,1075,2013,2028
	-29 18921	BI 1 - 24			23 58 32	-29 27.4	1	10.73		0.79		0.34		2	K1	1675
		LS I +58 008			23 58 35	+58 42.2	1	11.20		0.60		-0.47		2		41

HD	DM	Other Id	N	Rem	α_{1950}	δ_{1950}	S	V	σ_V	B–V	σ_{B-V}	U–B	σ_{U-B}	n	Spectrum	References
		LP 584 - 45			23 58 37	+01 38.9	1	11.53		0.33		-0.15		1		1696
	−29 18923	BI 1 - 25			23 58 37	−29 21.4	1	11.67		0.48		-0.08		2	G2	1675
	+13 05195		A		23 58 39	+13 41.7	2	10.46	.020	1.36	.010	1.27	.005	2	M0	801,3072
	+13 05195		B		23 58 39	+13 41.7	2	11.06	.040	1.45	.005	1.27	.015	2		801,3072
		LS I +58 009			23 58 41	+58 51.9	1	11.25		0.53		-0.33		2		41
224855	+59 02810	WZ Cas		⋆ A	23 58 42	+60 04.6	3	7.14	.013	2.84	.000	4.15	.072	3	C5p	8005,8022,8027
224850	−42 16563				23 58 43	−41 45.9	1	7.03		1.55				4	M1 III	2012
224870	+49 04309	HR 9083			23 58 45	+49 42.2	2	6.21	.004	0.97	.007	0.71		5	G7 II-III	71,1501
224865	−51 13743	HR 9082			23 58 46	−50 37.0	2	5.53	.000	1.60	.000	1.93		6	M2 III	2035,3005
		LP 584 - 47			23 58 48	+03 31.5	1	10.87		0.75		0.31		2		1696
224868	+60 02656				23 58 48	+60 33.7	2	7.26	.010	0.13	.000			5	B0 Ib	1025,1118
224862	−29 18925	BI 1 - 26			23 58 48	−29 00.2	1	7.11		1.10		1.05		2	K0	1675
		G 130 - 34			23 58 49	+34 00.3	1	11.28		1.31		1.33		2	M2	7010
	−17 06862	G 266 - 26			23 58 49	−17 13.3	1	10.76		1.39				2	M1.5V:	1619
		UU Scl			23 58 49	−26 52.1	1	14.16		0.70		0.00		4		3064
224873	+38 05112	IDS23563N3905	AB		23 58 50	+39 20.0	2	8.61	.006	0.79	.003	0.36	.004	7	K0 V +K1 V	938,3030
224869	+59 02812	IDS23561N5948	B		23 58 50	+60 04.6	2	8.30	.005	0.06	.005	-0.70	.000	4		41,401
	+62 02346				23 58 51	+63 13.6	1	9.49		0.43				4	B0 V	1025
	−31 19554	BI 1 - 28			23 58 51	−30 55.7	2	10.51	.005	0.41	.005	-0.07	.008	4	F8 V	1675,3004
224882	+29 05046	IDS23564N3011	AB		23 58 55	+30 27.5	1	7.96		0.65		0.20		3	G0 IV	3016
		WLS 0 20 # 7			23 58 58	+17 45.2	1	11.48		0.74		0.24		2		1375
224895	+27 04664				23 59 01	+28 08.8	1	6.84		1.20				21	K2 III	928
		CS 22957 # 25			23 59 02	−04 33.5	1	14.41		0.61		-0.04		1		1736
224893	+60 02657	HR 9085		⋆	23 59 03	+60 56.7	3	5.55	.015	0.42	.015	0.39		10	F0 III	41,1025,1118
	−30 19780	BI 1 - 29			23 59 03	−30 02.2	1	10.90		1.17		1.13		2	G0	1675
224889	−77 01596	HR 9084			23 59 03	−77 20.5	6	4.78	.005	1.27	.008	1.41	.000	39	K3 III	15,1075,1075,2035*
224890	+72 01135				23 59 04	+73 20.0	1	6.53		0.18		0.11		4	Am:	985
224905	+59 02813	LS I +60 068			23 59 05	+60 10.3	3	8.47	.008	0.15	.010	-0.57	.004	7	B1 V N	41,1012,1025
224892	+61 02582				23 59 07	+61 44.4	2	8.12	.005	0.17	.020			4	A4 V	1025,1118
224913	−31 19557	BI 1 - 30			23 59 07	−30 58.3	2	9.15	.010	1.07	.007	0.93	.021	7	dK0	702,1675
	−29 18928	BI 1 - 31			23 59 08	−29 25.6	1	11.17		0.64		0.04		6	G0	1675
224906	+41 04920	HR 9086			23 59 10	+42 05.3	2	6.25		-0.03	.004	-0.31	.044	3	B9 IIIp	695,1079
	+62 02349	IDS23566N6236	AB		23 59 10	+62 53.0	1	9.02		1.33				2	G8	1025
	−20 06694	G 266 - 28			23 59 11	−20 08.4	1	11.24		0.88		0.49		1		3061
	+60 02658				23 59 12	+60 42.7	2	9.01	.009	0.26	.014	0.12		4	A5 V	401,1025
		CS 22957 # 22			23 59 12	−06 06.5	1	13.34		0.61		-0.02		1		1736
	+66 01673	NGC 7822 - 3			23 59 13	+67 13.7	2	10.09	.005	1.33	.015	0.09	.115	4	B0	25,1285
	−29 18930	BI 1 - 32			23 59 14	−29 18.0	1	11.07		0.99		0.69		2	K3	1675
		G 129 - 57			23 59 15	+21 11.0	2	10.35	.002	1.03	.020	0.78	.004	3	K3	1620,5010
	−30 19783	BI 1 - 33			23 59 15	−29 39.5	1	11.19		0.67		0.09		4		1675
	+63 02093	LS I +64 011		⋆ A	23 59 16	+64 20.8	2	9.99	.009	0.50	.005	-0.50	.014	8	O9.5V	342,684
	+63 02093	LS I +64 011		⋆ AB	23 59 16	+64 20.8	1	9.76		0.76		0.29		2	O9.5V	684
224926	−3 05749	HR 9087			23 59 16	−03 18.3	5	5.11	.007	-0.13	.006	-0.50	.003	15	B7 III-IV	15,1078,1079,1509,2024
224927	−26 16876	G 266 - 29		⋆ AB	23 59 16	−26 04.3	5	8.96	.020	0.33	.028	-0.13	.023	14	A9 V	55,1068,1696,2034,3077
	+62 02350	IDS23568N6246	AB		23 59 18	+63 02.3	1	9.37		0.25				2	A0 V	1025
	−29 18931	BI 1 - 34			23 59 21	−29 23.6	1	11.95		0.80		0.25		4	G1	1675
224936	−37 15469				23 59 22	−37 30.5	1	7.06		1.06		0.90		6	K1 III	1628
224937	−42 16567	IDS23568S4210	AB		23 59 22	−41 53.5	1	7.75		0.42				4	F3/5 III	2012
		G 130 - 36			23 59 23	+25 44.3	1	11.24		0.94		0.53		2		333,1620
224938	+65 01985				23 59 24	+66 09.6	1	7.27		0.04		-0.07		2	B9	401
224935	−6 06345	HR 9089, YY Psc			23 59 24	−06 17.5	8	4.40	.004	1.63	.004	1.85	.010	80	M3 III	15,1075,1075,2012*
224948	−30 19785	BI 1 - 35			23 59 24	−30 26.2	3	9.99	.027	0.32	.006	-0.02	.019	8	F0 V	702,1675,3004
		CS 22876 # 29			23 59 24	−36 57.5	1	14.31		0.32		-0.01		1		1736
224939	+62 02351				23 59 25	+63 04.8	1	8.60		0.10				2	B8 V	1025
		NGC 7822 - 23			23 59 26	+67 08.5	1	12.76		1.19		0.47		4		25
		LP 988 - 88			23 59 26	−43 26.0	2	13.01	.049	0.14	.084	-0.87		3	DAs	1705,3060
224940	+59 02814				23 59 30	+59 39.7	1	7.96		0.97		0.61		2	G9 III	1502
224964	−31 19562	BI 1 - 36			23 59 30	−30 57.8	5	8.99	.004	0.21	.296	0.08	.028	18	A3 Va-	702,1068,1675,2012,3004
224953	−68 03594	IDS23570S6850	AB		23 59 30	−68 33.3	1	9.30		1.39		1.18		1	K5/M0 V	3072
		LS I +62 058			23 59 32	+62 07.6	1	11.24		0.27		-0.57		2		41
224930	+26 04734	HR 9088		⋆ AB	23 59 33	+26 49.0	9	5.74	.011	0.67	.008	0.05	.011	43	G3 V +K7 V	1,15,22,1013,1197*
224957	+38 05113				23 59 33	+38 54.3	1	8.59		0.06		-0.04		2	B9 V	105
224960	−15 06531	HR 9090, W Cet			23 59 34	−14 57.3	2	14.52	.075	1.62	.015	0.65	.105	4	S7,3e	8022,8027
224959	−3 05751				23 59 35	−03 05.8	1	9.50		1.10				1	R0	1238
	+66 01675	NGC 7822 - 14			23 59 36	+67 07.8	4	9.05	.016	1.09	.000	0.02	.007	20	O7	25,1011,1012,1285
	+66 01674	NGC 7822 - 13			23 59 36	+67 09.0	2	9.62	.024	1.10	.005	0.06	.005	17		25,1285
		CS 29517 # 31			23 59 36	−14 25.6	1	15.42		0.09		0.18		1		1736
		NGC 7822 - 22			23 59 38	+67 11.9	1	12.55		1.51		0.48		3		25
		NGC 7822 - 21			23 59 38	+67 13.4	1	12.27		0.84		0.45		3		25
224976	−30 19786	BI 1 - 37			23 59 38	−30 31.7	2	9.89	.000	0.46	.002	-0.05	.046	5	F5 IV	702,1675
		LS I +60 069			23 59 39	+60 09.5	1	11.97		0.40		-0.42		2		41
224973	−12 06600				23 59 41	−12 14.6	1	8.90		0.17		0.12		4	A1mA5/7-A8	152
224980	+59 02816				23 59 43	+60 25.5	2	6.73	.000	1.83	.019			4	M1	369,1025
		LS I +60 070			23 59 43	+60 58.9	1	11.71		0.46		-0.28		2		41
		LS I +61 120			23 59 44	+61 35.9	1	11.81		0.38		-0.38		2		41
		NGC 7822 - 15			23 59 45	+67 08.9	2	11.27	.020	1.07	.035	0.09	.010	3		25,1285
		NGC 7822 - 24			23 59 45	+67 08.9	1	11.65		1.72		0.51		3		25
		NGC 7822 - 2			23 59 46	+67 17.7	2	11.07	.040	1.10	.005	0.20	.005	4	B3	25,1285
224990	−30 19790	BI 1 - 42		⋆ AB	23 59 46	−29 59.9	11	5.02	.014	-0.15	.009	-0.54	.016	40	B5 V	3,15,702,1020,1034*
		BI 1 - 40			23 59 47	−29 46.3	1	12.18		1.07		0.75		2		1675

Table 1

HD	DM	Other Id	N Rem	α_{1950}	δ_{1950}	S	V	σ_V	B–V	σ_{B-V}	U–B	σ_{U-B}	n	Spectrum	References
	−30 19788	BI 1 - 39		23 59 47	−30 03.0	1	11.69		0.62		-0.02		4		1675
	−30 19791	BI 1 - 45		23 59 47	−30 03.0	1	11.17		0.54		0.00		2		1675
	−30 19789	BI 1 - 41		23 59 47	−30 05.6	1	10.73		0.56		0.00		4		1675
224984	+9 05313	IDS23573N1013	C	23 59 48	+10 30.5	1	8.53		0.53		0.03		5	F8	3032
224983	+10 05022			23 59 48	+10 43.7	1	8.47		0.86		0.63		2	K0	1733
	+62 02353	LS I +62 059		23 59 48	+62 37.4	2	9.88	.005	0.24	.000	-0.59	.009	4	B3 II	41,1012
	−30 19778	BI 1 - 27		23 59 48	−30 17.1	1	11.74		0.50		-0.12		2	F1	1675
	−30 19787	BI 1 - 38		23 59 48	−30 25.1	1	10.72		0.48		-0.09		2	F0	1675
224995	+8 05164	HR 9092		23 59 50	+08 40.7	2	6.31	.005	0.18	.000	0.13	.010	7	A6 V	15,1256
224998	−29 18934	BI 1 - 43		23 59 50	−28 50.8	1	8.94		1.10		1.03		3	gK0	1675
	−29 18935	BI 1 - 44		23 59 51	−29 22.0	1	11.09		0.49		-0.06		2	F9	1675
224994	+9 05314	IDS23573N1013	AB	23 59 52	+10 30.0	1	8.74		0.57		0.06		5	G0	3016
		NGC 7822 - 1		23 59 52	+67 25.7	2	10.79	.050	0.94	.015	-0.05	.045	3	B2	25,1285
		Smethells 149		23 59 54	−46 17.	1	12.43		1.49				1		1494
		NGC 7822 - 25		23 59 55	+67 09.0	1	12.82		1.42		0.39		4		25
225003	+7 05121	HR 9093		23 59 56	+08 12.5	4	5.70	.028	0.33	.032	-0.01	.009	13	F0 V	15,39,254,1256,3037
		LS I +60 071		23 59 57	+60 19.0	1	11.90		0.31		-0.33		2		41
		LP 824 - 310		23 59 57	−20 28.0	1	12.00		0.69		0.10		2		1696
		LS I +60 072		23 59 58	+60 50.1	1	11.50		0.36		-0.40		2		41
		LS I +60 073		23 59 59	+60 54.1	1	12.67		0.26		-0.09		2		41

Table 2 1253

Table 2: Stars without coordinates

HD	DM	Other Id	N	Rem	α_{1950}	δ_{1950}	S	V	σ_V	B−V	σ_{B-V}	U−B	σ_{U-B}	n	Spectrum	References
		0159-11 # 1					1	12.90		0.66		0.13		1		157
		0159-11 # 2					1	13.37		0.73		0.28		1		157
		0159-11 # 3					1	13.96		0.68		0.12		1		157
		0159-11 # 4					1	14.52		0.66		0.20		1		157
		0159-11 # 5					1	15.84		0.52		0.07		1		157
		0159-11 # 6					1	16.02		0.80		0.08		1		157
		0237-23 # 1					1	13.38		0.32		-0.04		1		157
		0237-23 # 2					1	14.12		0.57		0.06		1		157
		0237-23 # 3					1	15.25		0.63		0.11		1		157
		0237-23 # 4					1	15.72		0.62		0.25		1		157
		0237-23 # 5					1	16.15		0.46		0.01		1		157
		0350-07 # 1					1	13.08		1.21		1.06		1		157
		0350-07 # 2					1	14.03		0.70		0.42		1		157
		0350-07 # 3					1	14.42		0.79		0.36		1		157
		0350-07 # 4					1	15.09		0.82		0.04		1		157
		0350-07 # 5					1	15.90		0.42		-0.08		1		157
		0350-07 # 7					1	15.45		1.59		0.43		1		157
		0405-12 # 1					1	11.49		1.03		0.73		1		157
		0405-12 # 2					1	11.92		1.00		0.68		1		157
		0405-12 # 3					1	12.54		0.59		0.04		1		157
		0405-12 # 4					1	13.35		0.70		0.22		1		157
		0405-12 # 5					1	14.06		0.63		0.13		1		157
		0405-12 # 6					1	14.34		0.50		-0.04		1		157
		0405-12 # 7					1	15.28		0.60		-0.04		1		157
		0957+00 # 1					1	17.36		0.67		0.08		1		327
		1354+19 # 1					1	13.58		0.74		0.16		1		157
		1354+19 # 2					1	14.25		0.60		0.06		1		157
		1354+19 # 3					1	14.43		0.73		0.12		1		157
		1354+19 # 4					1	14.81		0.79		0.34		1		157
		1354+19 # 5					1	15.12		0.84		0.24		1		157
		1354+19 # 6					1	15.15		0.98		0.81		1		157
		2128-12 # 1					1	13.02		0.69		0.29		1		157
		2128-12 # 2					1	13.51		0.64		0.17		1		157
		2128-12 # 3					1	14.07		0.57		0.05		1		157
		2128-12 # 4					1	13.84		0.79		0.52		1		157
		2128-12 # 5					1	14.61		0.61		0.23		1		157
		2128-12 # 6					1	15.41		0.14		-0.79		1		157
		2128-12 # 7					1	15.49		0.84		0.21		1		157
		2135-14 # 1					1	14.41		0.57		0.23		1		157
		2135-14 # 2					1	14.79		0.95		0.70		1		157
		2135-14 # 3					1	15.51		0.71		0.40		1		157
		2135-14 # 4					1	15.51		0.63		-0.01		1		157
		2135-14 # 5					1	15.12		1.40		1.30		1		157
		2135-14 # 6					1	16.20		0.60		0.04		1		157
		2251+11 # 1					1	14.56		0.27		0.25		1		157
		2251+11 # 2					1	14.29		0.92		0.52		1		157
		2251+11 # 3					1	14.61		0.83		-0.16		1		157
		2251+11 # 5					1	15.11		0.95		0.69		1		157
		2251+11 # 6					1	15.55		0.83		0.06		1		157
		2344+09 # 1					1	15.19		0.51		0.19		1		157
		2344+09 # 2					1	14.47		1.36		1.24		1		157
		2344+09 # 3					1	15.26		0.77		0.35		1		157
		2344+09 # 4					1	15.40		0.61		0.09		1		157
		2344+09 # 5					1	15.56		1.00		0.65		1		157
		1Zw 051+12 # 2					1	14.40		0.62		0.03		1		327
		2Zw 2130+09 # 1					1	14.74		0.87		0.47		2		327
		2Zw 2130+09 # 2					1	16.07		0.94				2		327
		2Zw 2130+09 # 3					1	13.89		0.91		0.53		2		327
		2Zw 2130+09 # 4					1	13.08		0.60		0.12		2		327
		3Zw 008+10 # 1					1	14.36		0.78		0.19		1		327
		3Zw 008+10 # 2					1	13.81		0.92		0.42		1		327
		3C 9 # 3					1	17.05		0.97		1.03		1		327
		3C 47 # 4					1	16.30		0.54		0.21		2		327
		3C 47 # 5					1	16.95		0.97				1		327
		3C 48 # 1 (A)					1	10.91		0.58		0.09		1		157
		3C 48 # 2 (B)					1	11.73		0.59		0.08		1		157
		3C 48 # 3 (C)					1	12.92		0.52		-0.03		1		157
		3C 48 # 5 (D)					1	13.37		0.83		0.40		1		157
		3C 48 # 6 (D')					1	13.06		0.94		0.73		1		157
		3C 48 # 8 (E')					1	14.20		0.54		0.00		1		157
		3C 48 # 11 (G)					1	14.88		0.53		-0.10		1		157
		3C 48 # 12 (G')					1	15.66		0.87		0.44		4		327
		3C 48 # 13 (H)					1	16.46		0.82		0.43		1		327
		3C 57 # 27					1	16.36		0.58		-0.06		1		327
		3C 57 # 28					1	16.62		0.36		-0.21		1		327
		3C 68.2W # 1					1	12.83		1.18		0.96		1		157
		3C 68.2W # 2					1	12.65		1.01		0.75		1		157

HD	DM	Other Id	N Rem	α_{1950}	δ_{1950}	S	V	σ_V	B–V	σ_{B-V}	U–B	σ_{U-B}	n	Spectrum	References
		3C 68.2W # 3				1	13.75		1.22		1.05		1		157
		3C 68.2W # 4				1	15.85		1.03		0.60		1		157
		3C 68.2W # 5				1	14.66		0.64		0.08		1		157
		3C 68.2W # 6				1	13.42		0.61		0.04		1		157
		3C 68.2W # 7				1	15.06		0.57		0.22		1		157
		3C 68.2W # 8				1	13.08		0.33		0.21		1		157
		3C 68.2W # 9				1	15.58		0.82		0.21		1		157
		3C 120 # 1				1	12.72		1.22		0.89		1		157
		3C 120 # 4				1	14.57		0.93		0.62		1		157
		3C 120 # 5				1	14.45		1.23		1.08		1		157
		3C 120 # 7				1	15.41		0.77		0.17		1		157
		3C 147 # 5				1	14.44		0.78		0.16		1		327
		3C 147 # 6				1	13.98		1.31		1.09		1		327
		3C 147 # 7				1	15.83		0.76		0.17		1		327
		3C 147 # 8				1	15.72		0.63		0.25		1		327
		3C 175 # 27				1	16.65		0.64		-0.02		1		327
		3C 190 # 27				1	17.46		-0.20		-0.90		1		327
		3C 190 # 29				1	18.36		0.55		0.09		1		327
		3C 208 # 1				1	14.81		0.52		-0.13		1		327
		3C 217 # 27				1	17.31		0.70		0.10		1		327
		3C 217 # 28				1	18.50		0.25		-0.86		1		327
		3C 217 # 29				1	17.95		0.12		-0.76		1		327
		3C 217 # 30				1	16.47		0.65		0.08		1		327
		3C 217 # 32				1	17.61		0.92		0.56		1		327
		3C 232 # 1				1	13.07		0.70		0.13		1		157
		3C 232 # 2				1	13.75		0.46		-0.06		1		157
		3C 232 # 3				1	13.96		0.89		0.48		1		157
		3C 232 # 4				1	15.20		0.62		0.34		1		157
		3C 232 # 5				1	15.15		0.82		0.47		1		157
		3C 232 # 6				1	15.45		0.96		0.48		1		157
		3C 232 # 7				1	16.18		0.29		0.39		1		157
		3C 232 # A				1	16.40		0.07		0.13		1		327
		3C 263 # 1				1	13.29		0.57		0.08		1		157
		3C 263 # 2				1	13.46		0.79		0.50		1		157
		3C 263 # 3				1	14.18		0.67		0.23		1		157
		3C 263 # 4				1	14.73		0.64		0.12		1		157
		3C 263 # 5				1	16.04		0.80		0.00		1		157
		3C 286 # 1				1	16.05		0.73		0.79		3		327
		3C 286 # 2				1	15.87		0.98		0.82		3		327
		3C 286 # 3				1	15.93		1.50		1.26		2		327
		3C 298 # 3				1	16.89		0.53		-0.13		1		327
		3C 323.1 # 1				1	12.64		0.76		0.37		1		157
		3C 323.1 # 2				1	13.53		0.76		0.33		1		157
		3C 323.1 # 3				1	14.26		0.62		0.02		1		157
		3C 323.1 # 4				1	15.17		0.74		0.32		1		157
		3C 323.1 # 5				1	15.27		0.81		0.30		1		157
		3C 323.1 # 6				1	15.48		0.82		0.19		1		157
		3C 323.1 # 7				1	16.27		0.74		-0.10		1		157
		3C 334 # 1				1	14.17		0.64		-0.01		1		157
		3C 334 # 2				1	14.54		0.75		0.23		1		157
		3C 334 # 3				1	14.68		1.02		0.72		1		157
		3C 334 # 4				1	15.38		0.76		0.25		1		157
		3C 334 # 5				1	15.65		0.83		0.30		1		157
		3C 334 # 6				1	16.41		0.85		0.01		1		157
		3C 351 # 1				1	12.06		0.78		0.35		1		157
		3C 351 # 2				1	13.30		0.60		0.09		1		157
		3C 351 # 3				1	14.15		0.60		0.04		1		157
		3C 351 # 4				1	14.76		0.74		0.74		1		157
		3C 351 # 5				1	14.81		0.78		0.05		1		157
		3C 351 # 6				1	15.71		0.70		0.17		1		157
		3C 454.3 # 1				1	15.85		1.00		0.54		1		157
		3C 454.3 # 2				1	15.16		0.71		0.07		1		157
		3C 454.3 # 3				1	14.43		0.75		0.24		1		157
		3C 454.3 # 4				1	13.85		1.09		1.00		1		157
		3C 454.3 # 5				1	15.88		1.22		1.84		1		157
		3C 454.3 # 6				1	15.21		0.85		0.29		1		157
		3C 454.3 # 7				1	15.42		0.86		0.43		1		157
		3C 454.3 # 8				1	13.65		0.97		0.55		1		157
		3C 454.3 # 9				1	14.94		0.75		0.16		1		157
		A2 57				1	15.67		0.61		0.02		1		98
		A2 269				1	16.56		0.58		-0.11		1		98
		A2 403				1	16.02		0.56		-0.11		1		98
		LkHα 120	⋆ V			1	11.96		1.01		0.40		3		8013
		AA9,95 # 25				1	13.02		1.56		1.00		3	M6 III	35
		AA9,95 # 101				1	12.25		0.50		0.01		2	F5 V	35
		AA9,95 # 103				1	11.83		0.59		0.08		2	F8 V	35
		AA9,95 # 115				1	13.08		0.79		0.46		2	K0 V	35
		AA9,95 # 121				1	13.65		-0.21		-1.01		2	B0.5 Ib	35
		AA9,95 # 122				1	12.24		0.00		-0.58		3	B8.5 Ia	35
		AA9,95 # 201				1	11.12		1.01		0.75		4	G8 III	35

Table 2   1255

HD	DM	Other Id	N	Rem	α_{1950}	δ_{1950}	S	V	σ_V	B–V	σ_{B-V}	U–B	σ_{U-B}	n	Spectrum	References
		AA9,95 # 202					1	12.78		0.04		-0.04		3	B9.5 V	35
		AA9,95 # 210					1	13.03		1.72		1.07		3	M6 III	35
		AA9,95 # 213					1	12.46		0.49		-0.04		3	F5 V	35
		AA9,95 # 214					1	13.36		0.24		0.14		4	A6 V	35
		AA9,95 # 217					1	12.46		0.62		0.10		5	G1 V	35
		AA9,95 # 219					1	12.87		1.61		1.20		2	M4 V	35
		AA9,95 # 227					1	13.28		1.48		0.14		3		35
		AA9,95 # 229					1	13.34		0.72		0.19		3	G2 V	35
		AA9,95 # 233					1	13.71		0.11		0.23		3	A7 Ib	35
		AA9,95 # 235					1	12.54		0.75		0.26		4	G5 V	35
		AA9,95 # 238					1	13.69		0.60		0.05		3	G0 V	35
		AA9,95 # 239					1	14.41		-0.08		-0.70		3	B4 II	35
		AA9,95 # 244					1	13.13		0.16		0.09		3	A5 V	35
		AA9,95 # 249					1	14.22		-0.14		-0.80		3	B3 II	35
		AA9,95 # 251					1	14.31		-0.12		-1.00		3	B0 II	35
		AA9,95 # 301					1	12.94		0.12		0.11		3	A1 V	35
		AA9,95 # 302					1	12.40		0.71		0.29		3	G0 III	35
		AA9,95 # 303					1	12.09		0.57		0.09		3	F5 V	35
		AA9,95 # 306					1	10.88		0.67		0.19		3	G4 V	35
		AA9,95 # 307					1	13.68		1.12		1.05		2	K1 III	35
		AA9,95 # 309					1	13.15		0.11		0.08		2	A3 V	35
		AA9,95 # 311					1	10.92		1.02		0.75		3	G7 III	35
		AA9,95 # 315					1	13.03		0.65		-0.02		3		35
		AA9,95 # 318					1	12.64		0.57		0.02		2	F7 V	35
		AA9,95 # 322					1	10.86		0.42		-0.15		3		35
		AA9,95 # 323					1	11.22		0.52		0.00		4	F7 V	35
		AA45,405 S4 # 1					1	12.22		0.42		0.12		2		432
		AA45,405 S17 # 1					1	11.87		1.18		0.90		2		432
		AA45,405 S65 # 2					1	14.05		0.95		0.19		3		1682
		AA45,405 S65 # 4					1	13.51		1.67		0.41		2	B1 II:	1682
		AA45,405 S65 # 5					1	14.31		1.88		1.08		1		1682
		AA45,405 S68 # 1					1	15.19		1.49		1.09		3		1682
		AA45,405 S68 # 2					1	13.61		0.96		0.67		2		1682
		AA45,405 S68 # 3					1	16.59		-0.01		-0.89		3		1682
		AA45,405 S68 # 4					1	14.04		2.35		1.57		1		1682
		AA45,405 S69 # 1					1	14.05		0.88		0.42		2		1682
		AA45,405 S69 # 2					1	14.47		1.62		0.35		2	B0.5IV	1682
		AA45,405 S69 # 3					1	15.58		0.85		0.28		1		1682
		AA45,405 S72 # 4					1	14.14		1.18		0.93		2		1682
		AA45,405 S72 # 5					1	14.04		0.85		0.31		1		1682
		AA45,405 S72 # 6					1	13.78		0.88		0.16		2		1682
		AA45,405 S72 # 7					1	14.14		0.76		0.06		1		1682
		AA45,405 S74 # 1					1	12.39		1.31		0.08		2	O9 V	1682
		AA45,405 S75 # 5					1	10.53		0.70		0.13		1		1682
		AA45,405 S79 # 1					1	12.16		0.85		0.05		2		1682
		AA45,405 S79 # 2					1	13.20		0.89		0.32		1		1682
		AA45,405 S84 # 5					1	13.53		1.78		2.44		1		1682
		AA45,405 S89 # 3					1	14.41		1.36		0.21		2		1682
		AA45,405 S90 # 2					1	13.63		0.97		0.42		2		1682
		AA45,405 S90 # 3					1	14.38		1.30		0.47		2		1682
		AA45,405 S90 # 4					1	14.01		0.90		0.43		2		1682
		AA45,405 S93 # 1					1	13.64		1.40		0.18		2	O9.5V	1682
		AA45,405 S97 # 1					1	13.06		1.66		1.48		1		1682
		AA45,405 S97 # 2					1	12.66		0.43		-0.39		3		1682
		AA45,405 S97 # 3					1	11.69		0.32		-0.24		2		1682
		AA45,405 S97 # 4					1	11.15		0.15		-0.38		2	B1 V	1682
		AA45,405 S97 # 5					1	12.06		0.46		-0.40		2	B1 V	1682
		AA45,405 S97 # 7					1	12.63		0.95		-0.16		2	O9 V	1682
		AA45,405 S98 # 1					1	13.53		0.72		0.10		2		1682
		AA45,405 S98 # 2					1	13.02		1.68		1.78		1		1682
		AA45,405 S98 # 3					1	11.41		0.55		0.04		1		1682
		AA45,405 S98 # 4					1	12.01		0.48		0.06		2		1682
		AA45,405 S98 # 5					1	13.34		0.69		0.24		1		1682
		AA45,405 S121 # 1					1	14.91		1.86		0.64		4		7006
		AA45,405 S121 # 2					1	15.44		1.76		0.53		4		7006
		AA45,405 S121 # 3					1	15.48		1.76		0.54		4		7006
		AA45,405 S124 # 1					1	12.28		0.95		-0.07		2		7006
		AA45,405 S124 # 2					1	14.46		1.18		0.23		3		7006
		AA45,405 S127 # 1					1	15.80		1.44		0.29		3		7006
		AA45,405 S128 # 1					1	15.25		1.47		0.22		3		7006
		AA45,405 S157 # 1					1	15.30		0.73		-0.10		2		7006
		AA45,405 S157 # 2					1	11.77		0.61		-0.39		3		7006
		AA45,405 S157 # 3					1	13.71		0.64		-0.15		2		7006
		AA45,405 S157 # 4					1	15.91		0.86		0.26		2		7006
		AA45,405 S159 # 1					1	12.02		0.82		-0.20		2		7006
		AA45,405 S207 # 5					1	15.86		0.89		0.14		3		797
		AA45,405 S207 # 6					1	16.56		1.14		0.47		2		797
		AA45,405 S207 # 7					1	15.98		1.21		0.76		1		797
		AA45,405 S208 # 2					1	14.91		1.20		0.52		4		797
		AA45,405 S208 # 4					1	13.52		0.63		-0.31		3		797

HD	DM	Other Id	N Rem	α_{1950}	δ_{1950}	S	V	σ_V	B–V	σ_{B-V}	U–B	σ_{U-B}	n	Spectrum	References
		AA45,405 S208 # 8				1	14.32		0.66		-0.11		2		797
		AA45,405 S208 # 9				1	15.16		0.90		0.51		1		797
		AA45,405 S208 # 10				1	15.44		0.70		0.20		1		797
		AA45,405 S208 # 11				1	15.62		0.83		0.43		1		797
		AA45,405 S208 # 12				1	16.06		0.77		0.10		1		797
		AA45,405 S209 # 1				1	15.18		1.12		0.84		4		7006
		AA45,405 S209 # 2				1	15.67		1.68		0.90		4		7006
		AA45,405 S209 # 3				1	17.53		1.48		0.47		3		7006
		AA45,405 S211 # 1				1	15.78		1.40		0.45		4		7006
		AA45,405 S211 # 2				1	15.23		1.41		0.32		3		7006
		AA45,405 S211 # 3				1	13.54		1.33		0.34		3		7006
		AA45,405 S217 # 11				1	14.49		0.50		-0.22		2		797
		AA45,405 S217 # 14				2	15.01	.010	0.56	.015	-0.14	.035	4		797,7006
		AA45,405 S217 # 15				1	14.96		0.89		0.20		1		797
		AA45,405 S219 # 2				1	14.79		0.67		0.29		1		797
		AA45,405 S219 # 3				1	15.63		0.66		0.30		1		797
		AA45,405 S219 # 4				1	14.45		0.78		-0.11		1		797
		AA45,405 S219 # 5				1	14.47		0.70		0.10		1		797
		AA45,405 S219 # 6				1	15.97		0.83		0.19		1		797
		AA45,405 S219 # 7				1	15.99		1.00		0.01		1		797
		AA45,405 S228 # 1				1	12.62		0.97		0.02		1		7006
		AA45,405 S228 # 2				1	14.74		1.16		0.23		1		7006
		AA45,405 S237 # 4	⋆ C			1	14.11		0.70		-0.24		1		797
		AA45,405 S237 # 5	⋆ E			1	13.99		0.52		-0.19		1		797
		AA45,405 S241 # 3				1	15.36		1.12		0.54		1		797
		AA45,405 S241 # 4				1	14.97		0.86		-0.12		1		797
		AA45,405 S241 # 5				1	15.73		0.94		0.38		1		797
		AA45,405 S241 # 6				1	15.93		0.90		0.21		1		797
		AA45,405 S241 # 7				1	15.72		0.70		0.40		1		797
		AA45,405 S241 # 9				1	16.39		1.16		-0.03		1		797
		AA45,405 S241 # 10				1	14.46		1.31		0.91		1		797
		AA45,405 S241 # 11				1	14.83		0.32		-0.03		1		797
		AA45,405 S241 # 12				1	17.16		0.73		0.70		1		797
		AA45,405 S247 # 3				1	13.46		0.81		0.11		1		797
		AA45,405 S247 # 4				1	14.18		0.62		0.16		1		797
		AA45,405 S247 # 5				1	15.83		1.86		0.70		1		797
		AA45,405 S247 # 9				1	14.35		0.72		0.27		1		797
		AA45,405 S247 # 10				1	14.24		0.64		0.19		1		797
		AA45,405 S254 # 4				1	15.28		1.22		0.40		4		797
		AA45,405 S254 # 5				1	14.37		1.15		-0.03		4		797
		AA45,405 S254 # 6				1	13.84		0.58		0.12		1		797
		AA45,405 S254 # 7				1	12.99		0.44		0.33		1		797
		AA45,405 S254 # 8				1	14.01		0.74		0.13		1		797
		AA45,405 S255 # 1				1	11.77		0.86		-0.17		1		608
		AA45,405 S259 # 1				1	15.43		0.95		-0.01		2		797
		AA45,405 S266 # 1				1	11.95		1.33		-0.39		2		7006
		AA45,405 S271 # 4				1	14.67		0.65		0.35		2		797
		AA45,405 S271 # 5				1	14.74		0.71		0.33		2		797
		AA45,405 S271 # 6				1	14.64		0.92		0.58		2		797
		AA45,405 S283 # 5				1	15.11		0.96		0.78		2		797
		AA45,405 S283 # 6				1	15.90		0.96		0.23		2		797
		AA45,405 S283 # 9				1	13.12		1.30		1.01		1		797
		AA45,405 S285 # 1				1	12.30		0.23		-0.66		3	B0 V	797
		AA45,405 S285 # 3				1	15.39		0.39		-0.03		2		797
		AA45,405 S287 # 8				1	14.13		0.67		0.08		1		797
		AA45,405 S288 # 2				1	14.65		0.55		0.29		2		797
		AA45,405 S289 # 7				1	13.93		0.64		0.28		1		797
		AA45,405 S294 # 6				1	15.69		0.63		0.15		1		797
		AA45,405 S294 # 7				1	17.01		1.06		0.39		1		797
		AA45,405 S294 # 8				1	15.49		0.84		0.20		1		797
		AA45,405 S294 # 14				1	14.72		0.70		0.17		1		797
		AA45,405 S294 # 15				1	15.60		0.68		0.17		1		797
		AA45,405 S294 # 16				1	14.67		0.56		0.26		1		797
		AA45,405 S298 # 1				1	9.92		1.00		0.72		1		1721
		AA45,405 S298 # 3				1	11.87		1.23		1.16		1		797
		AA45,405 S298 # 4				1	10.80		0.56		0.12		1		797
		AA45,405 S298 # 5				1	12.70		0.31		0.31		1		797
		AA45,405 S298 # 6				1	12.94		0.33		-0.01		1		797
		AA45,405 S298 # 7				1	12.16		0.57		0.08		1		797
		AA45,405 S298 # 8				1	13.71		0.26		-0.35		4		797
		AA45,405 S298 # 10				1	14.76		0.74		0.12		1		797
		AA45,405 S298 # 11				1	14.41		0.75		0.24		1		797
		AA45,405 S300 # 3				1	14.86		0.54		0.02		2		797
		AA45,405 S300 # 4				1	15.37		0.46		0.06		2		797
		AA45,405 S301 # 2				1	11.48		0.29		0.16		1		797
		AA45,405 S301 # 7				1	12.67		0.33		0.19		1		797
		AA45,405 S305 # 5				1	13.93		0.44		0.09		2		432
		AA45,405 S305 # 7				1	13.63		0.48		0.09		2		432
		AA45,405 S305 # 8				1	13.64		0.50		-0.02		2		432
		AA45,405 S305 # 9				1	14.08		0.61		0.10		2		432

Table 2 1257

HD	DM	Other Id	N	Rem	α_{1950}	δ_{1950}	S	V	σ_V	B−V	σ_{B-V}	U−B	σ_{U-B}	n	Spectrum	References
		AA45,405 S305 # 10					3	14.38	.013	0.80	.028	-0.11	.011	6		432,797,7006
		AA45,405 S305 # 11					2	14.92	.040	0.84	.040	0.30	.090	4		432,797
		AA45,405 S305 # 12					1	12.68		0.76		0.32		2		797
		AA65,65 # 101					1	13.48		0.59		0.10		1		713
		AA65,65 # 102					1	14.85		0.69		0.19		1		713
		AA65,65 # 103					1	12.54		1.26		1.17		1		713
		AA65,65 # 104					1	15.26		1.21		0.65		1		713
		AA65,65 # 106					1	14.69		1.07		0.79		1		713
		AA65,65 # 107					1	15.22		1.66		2.70		1		713
		AA65,65 # 108					1	15.52		1.43		1.05		1		713
		AA65,65 # 109					1	13.26		1.28		1.17		1		713
		AA65,65 # 110					1	14.93		0.85		0.37		1		713
		AA65,65 # 111					1	13.06		0.49		0.21		1		713
		AA65,65 # 112					1	15.16		0.83		0.36		1		713
		AA65,65 # 113					1	14.29		0.73		0.24		1		713
		AA65,65 # 114					1	13.84		1.72		2.12		1		713
		AA65,65 # 115					1	15.99		0.99		0.72		1		713
		AA65,65 # 116					1	15.90		1.13		0.91		1		713
		AA65,65 # 201					1	14.17		1.27		1.34		1		713
		AA65,65 # 202					1	14.64		1.32		1.20		1		713
		AA65,65 # 203					1	15.72		1.01		0.79		1		713
		AA65,65 # 204					1	15.26		1.49		2.39		1		713
		AA65,65 # 205					1	14.87		0.91		0.42		1		713
		AA65,65 # 206					1	16.34		1.05		0.75		1		713
		AA65,65 # 207					1	14.75		1.25		1.23		1		713
		AA65,65 # 208					1	15.80		0.77		0.24		1		713
		AA65,65 # 209					1	14.58		1.32		1.57		1		713
		AA65,65 # 210					1	15.70		1.09		0.92		1		713
		AA65,65 # 211					1	13.45		1.50		1.65		1		713
		AA65,65 # 212					1	15.14		1.40		1.56		1		713
		AA65,65 # 213					1	14.91		1.14		0.95		1		713
		AA65,65 # 214					1	14.76		0.73		0.20		1		713
		AA65,65 # 215					1	15.87		1.18		0.85		1		713
		AA65,65 # 216					1	15.83		0.86		0.40		1		713
		AA65,65 # 217					1	14.20		1.27		1.60		1		713
		AA65,65 # 301					1	13.62		1.08		0.80		1		713
		AA65,65 # 302					1	15.24		0.99		0.48		1		713
		AA65,65 # 303					1	15.57		1.00		0.61		1		713
		AA65,65 # 304					1	14.87		1.01		0.67		1		713
		AA65,65 # 305					1	14.58		1.04		0.93		1		713
		AA65,65 # 306					1	15.40		0.61		0.07		1		713
		AA65,65 # 307					1	15.74		0.89		0.45		1		713
		AA65,65 # 308					1	15.94		1.16		1.10		1		713
		AA65,65 # 309					1	14.32		1.35		1.52		1		713
		AA65,65 # 310					1	16.47		0.72		0.08		1		713
		AA65,65 # 311					1	15.75		0.79		0.40		1		713
		AA65,65 # 312					1	15.35		0.67		0.20		1		713
		AA65,65 # 313					1	15.74		0.77		0.22		1		713
		AA65,65 # 314					1	14.82		0.86		0.42		1		713
		AA65,65 # 315					1	13.44		0.76		0.29		1		713
		AA65,65 # 316					1	14.92		0.21		0.21		1		713
		AA65,65 # 317					1	14.37		1.03		0.73		1		713
		AA65,65 # 318					1	15.59		1.05		0.89		1		713
		AA65,65 # 319					1	16.50		1.41		0.80		1		713
		AAS38,197 R 20 # 1					1	10.19		0.27		-0.71		2	B0 V	797
		AAS38,197 R 20 # 2					1	10.68		0.41		0.19		1		797
		AAS38,197 R 20 # 3					1	11.15		1.21		1.36		1		797
		AAS38,197 R 20 # 4					1	11.70		0.20		0.22		1		797
		AAS38,197 R 20 # 5					1	11.86		0.35		0.12		1		797
		AAS70,69 T4# 4					1	14.39		0.79		0.54		1		1636
		AAS70,69 T4# 5					1	14.36		0.55		0.50		1		1636
		AAS70,69 T4# 6					1	14.25		0.61		0.03		1		1636
		AAS70,69 T4# 7					1	13.05		0.58		0.30		1		1636
		AAS70,69 T4# 8					1	16.46		1.52				2		1636
		AAS70,69 T4# 9					1	15.43		1.34		1.20		2		1636
		AAS70,69 T4# 10					1	16.43		0.91				2		1636
		AAS70,69 T4# 11					1	13.17		1.48		1.10		1		1636
		AAS70,69 T4# 12					1	15.36		0.96		0.28		2		1636
		AAS70,69 T4# 14					1	14.70		1.46		0.99		1		1636
		AAS70,69 T4# 15					1	16.80		0.76		0.62		2		1636
		AAS70,69 T4# 16					1	17.19		0.95				2		1636
		AAS70,69 T5# 4					1	13.78		0.70		0.26		2		1636
		AAS70,69 T5# 5					1	13.44		0.55		0.14		1		1636
		AAS70,69 T5# 6					1	15.72		1.41				2		1636
		AAS70,69 T5# 7					1	15.30		0.67		0.30		2		1636
		AAS70,69 T5# 8					1	15.37		0.84		0.34		1		1636
		AAS70,69 T5# 9					1	15.44		0.59		0.25		1		1636
		AAS70,69 T5# 10					1	13.64		1.13		0.83		2		1636
		AAS70,69 T5# 11					1	15.92		0.73		0.40		2		1636
		AAS70,69 T5# 15					1	16.34		0.93				2		1636

HD	DM	Other Id	N Rem	α_{1950}	δ_{1950}	S	V	σ_V	B–V	σ_{B-V}	U–B	σ_{U-B}	n	Spectrum	References
		AAS70,69 T5# 16				1	13.93		0.47		-0.09		1		1636
		AAS70,69 T5# 17				1	14.57		0.42		0.34		1		1636
		AJ74,1125 T1# 3				2	11.08	.007	0.95	.010	0.73	.007	4		2,1496
		AJ74,1125 T1# 4				2	12.34	.000	0.70	.018	0.46	.080	4		2,1496
		AJ74,1125 T1# 5				2	12.67	.002	0.78	.008	0.29	.018	4		2,1496
		AJ74,1125 T1# 6				2	13.15	.005	1.11	.005	1.03	.126	4		2,1496
		AJ74,1125 T1# 7				2	14.17	.027	0.10	.005	0.05	.068	5		2,1496
		AJ74,1125 T1# 8				2	14.22	.052	1.20	.022	1.33	.044	6		2,1496
		AJ74,1125 T1# 9				2	14.31	.026	0.11	.022	-0.01	.031	6		2,1496
		AJ74,1125 T1# 10				2	14.50	.028	0.11	.064	-0.30	.041	9		2,1496
		AJ74,1125 T1# 11				1	14.56		0.79		0.22		5		2
		AJ74,1125 T1# 12				2	14.92	.022	-0.11	.013	-0.78	.048	6		2,1496
		AJ74,1125 T1# 13				1	15.20		1.45		1.28		2		2
		AJ74,1125 T1# 14				2	15.31	.014	-0.12	.053	-0.95	.057	7		2,1496
		AJ74,1125 T1# 15				1	15.38		1.32		0.30		2		2
		AJ74,1125 T1# 16				1	16.44		-0.12		-0.54		6		2
		AJ74,1125 T1# 17				1	12.06		-0.03		-0.74		5		1496
		AJ74,1125 T1# 18				1	12.71		2.04		2.25		4		1496
		AJ74,1125 T1# 19				1	14.72		1.72		0.64		4		1496
		AJ74,1125 T3# 5				1	13.09		0.27		-0.03		1		2
		AJ74,1125 T3# 6				1	13.18		0.28		-0.19		1		2
		AJ74,1125 T3# 7				1	13.53		0.36		0.22		1		2
		AJ74,1125 T3# 8				1	13.56		0.32		0.02		1		2
		AJ74,1125 T3# 9				1	14.00		0.31		0.00		1		2
		AJ74,1125 T3# 10				1	14.11		0.37		0.17		1		2
		AJ74,1125 T3# 11				1	14.35		1.47		1.20		1		2
		AJ74,1125 T3# 12				1	14.46		0.42		0.28		1		2
		AJ74,1125 T3# 13				1	14.65		0.38		0.26		1		2
		AJ74,1125 T3# 14				1	14.88		0.52		0.23		1		2
		AJ74,1125 T4# 12				1	12.77		0.37		0.26		4		2
		AJ74,1125 T4# 13				1	13.35		0.26		-0.07		3		2
		AJ74,1125 T4# 14				1	13.44		0.52		-0.28		4		2
		AJ74,1125 T4# 15				1	13.79		0.40		0.39		4		2
		AJ74,1125 T4# 16				1	14.45		0.31		0.23		7		2
		AJ74,1125 T4# 17				1	14.61		0.69		0.23		6		2
		AJ74,1125 T4# 18				1	14.90		0.73		0.30		6		2
		AJ74,1125 T4# 19				1	15.33		0.55		0.24		8		2
		AJ74,1125 T5# 8				1	13.76		0.44		-0.26		2		2
		AJ74,1125 T5# 9				1	13.90		0.52		0.29		1		2
		AJ74,1125 T5# 10				1	14.45		0.45		0.42		7		2
		AJ74,1125 T5# 11				1	15.24		0.76				1		2
		AJ74,1125 T5# 12				1	15.43		1.02		1.28		1		2
		AJ76,1082 # 4				1	13.40		0.48		0.16		2		325
		AJ76,1082 # 30				1	14.54		0.87				2		1702
		AJ76,1082 # 37				1	13.69		0.82		0.32		2		325
		AJ76,1082 # 38				1	12.90		0.31		0.18		2		325
		AJ76,1082 # 41				1	15.07		0.81				2		1702
		AJ76,1082 # 48				1	12.75		1.53		1.58		2		325
		AJ76,1082 # 49				1	13.53		1.30		1.22		2		325
		AJ76,1082 # 59				1	15.15		0.74				2		1702
		AJ76,1082 # 64				1	15.15		0.90				1		1702
		AJ76,1082 # 67				1	15.21		0.76				2		1702
		AJ76,1082 # 70				1	13.22		1.27		1.20		2		325
		AJ76,1082 # 77				1	13.22		1.76		2.07		2		325
		AJ76,1082 # 84				1	13.15		0.48		0.18		2		325
		AJ76,1082 # 88				1	14.28		0.83				2		1702
		AJ76,1082 # 90				1	13.17		1.19		1.08		2		325
		AJ76,1082 # 91				1	15.00		0.89				2		1702
		AJ76,1082 # 93				1	12.51		0.61		0.34		2		325
		AJ76,1082 # 103				1	12.80		0.49		0.34		2		325
		AJ76,1082 # 108				1	14.38		0.82				2		1702
		AJ76,1082 # 104				1	13.84		0.71		0.20		2		325
		AJ76,1082 # 114				1	13.62		1.44		1.51		2		325
		AJ76,1082 # 118				1	13.66		1.42		1.40		2		325
		AJ76,1082 # 126				1	14.00		1.81		2.19		3		325
		AJ76,1082 # 128				1	14.89		1.12				1		1702
		AJ76,1082 # 133				1	11.76		0.21		0.18		2		325
		AJ76,1082 # 137				1	12.17		0.35		0.18		2		325
		AJ76,1082 # 139				1	14.77		1.24				2		1072
		AJ76,1082 # 142				1	15.18		1.53				2		1072
		AJ76,1082 # 156				1	13.61		0.60		0.47		2		325
		AJ76,1082 # 176				1	13.52		2.02		2.56		3		325
		AJ76,1082 # 231				1	13.80		1.96		2.08		1		325
		AJ77,733 T9# 11				1	14.06		0.66		0.29		4		125
		AJ77,733 T9# 12				1	14.46		0.68		0.10		4		125
		AJ77,733 T11# 12				1	13.85		0.60		0.47		4		402
		AJ77,733 T12# 27				1	12.71		0.66		0.10		3		125
		AJ77,733 T12# 33				1	14.07		0.61		0.17		5		125
		AJ77,733 T12# 35				1	14.56		0.70		0.01		5		125
		AJ77,733 T12# 36				1	14.74		0.86		0.16		2		125

Table 2 1259

HD	DM	Other Id	N Rem	α_{1950}	δ_{1950}	S	V	σ_V	B–V	σ_{B-V}	U–B	σ_{U-B}	n	Spectrum	References
		AJ77,733 T12# 37				1	14.81		0.74		0.28		4		125
		AJ77,733 T12# 38				1	14.99		0.75		0.25		4		125
		AJ77,733 T12# 39				1	15.14		0.74		0.20		4		125
		AJ77,733 T12# 40				1	15.84		0.82		0.26		4		125
		AJ77,733 T13# 10				1	13.64		0.38		0.08		5		125
		AJ77,733 T13# 11				1	13.85		0.45		0.20		3		125
		AJ77,733 T13# 12				1	14.23		0.64		0.13		2		125
		AJ77,733 T13# 13				1	14.35		0.42		0.10		4		125
		AJ77,733 T13# 14				1	14.60		0.57		0.06		3		125
		AJ77,733 T13# 15				1	14.68		0.68		0.09		4		125
		AJ77,733 T13# 16				1	14.74		0.57		0.06		3		125
		AJ77,733 T13# 17				1	14.98		0.62		0.11		4		125
		AJ77,733 T13# 18				1	15.32		0.65		0.09		4		125
		AJ77,733 T13# 19				1	15.49		0.67		0.24		4		125
		AJ77,733 T13# 20				1	15.73		0.64		0.21		3		125
		AJ77,733 T17# 9				1	13.11		0.47		0.35		3		125
		AJ77,733 T17# 10				1	13.24		0.60		0.47		4		125
		AJ77,733 T17# 11				1	13.31		0.90		-0.17		3		125
		AJ77,733 T17# 12				1	13.32		0.29		0.10		4		125
		AJ77,733 T17# 13				1	13.37		0.74		0.19		2		125
		AJ77,733 T17# 14				1	13.84		0.49		0.24		3		125
		AJ77,733 T17# 15				1	13.93		0.47		0.12		4		125
		AJ77,733 T17# 16				1	14.14		0.71		0.25		2		125
		AJ77,733 T17# 17				1	14.57		0.71		-0.24		3		125
		AJ77,733 T17# 18				1	14.67		1.05		-0.07		3		125
		AJ77,733 T17# 19				1	14.73		0.76		0.02		2		125
		AJ77,733 T17# 20				1	14.97		0.96		-0.07		3		125
		AJ77,733 T17# 21				1	15.09		0.73		-0.17		3		125
		AJ77,733 T17# 22				1	15.29		0.47		-0.20		5		125
		AJ77,733 T17# 23				1	15.46		1.02		0.45		3		125
		AJ77,733 T17# 24				1	15.71		0.52		-0.04		3		125
		AJ77,733 T18# 17				1	13.26		0.30		-0.18		2		125
		AJ77,733 T18# 18				1	13.62		0.34		0.27		4		125
		AJ77,733 T18# 19				1	14.04		0.26		0.18		3		125
		AJ77,733 T18# 20				1	14.09		0.46		0.31		5		125
		AJ77,733 T18# 21				1	14.12		0.48		0.07		3		125
		AJ77,733 T18# 22				1	14.43		0.37		0.26		3		125
		AJ77,733 T18# 23				1	14.71		0.33		0.24		3		125
		AJ77,733 T18# 24				1	14.85		0.44		0.11		3		125
		AJ77,733 T18# 25				1	14.86		0.63		0.44		3		125
		AJ77,733 T18# 26				1	14.86		0.98		0.63		3		125
		AJ77,733 T18# 27				1	15.05		0.42		0.27		4		125
		AJ77,733 T18# 28				1	15.15		0.42		0.22		5		125
		AJ77,733 T18# 29				1	15.82		0.63		0.22		2		125
		AJ77,733 T19# 13				1	13.19		0.35		0.17		4		125
		AJ77,733 T19# 14				1	13.46		0.45		0.35		4		125
		AJ77,733 T19# 15				1	13.62		0.75		0.21		3		125
		AJ77,733 T19# 16				1	14.06		0.53		0.41		3		125
		AJ77,733 T19# 17				1	14.37		0.75		0.12		3		125
		AJ77,733 T19# 18				1	14.53		1.47		1.19		2		125
		AJ77,733 T19# 19				1	14.62		0.92		0.29		3		125
		AJ77,733 T19# 20				1	16.00		0.73		0.05		3		125
		AJ77,733 T19# 21				1	16.08		0.65		0.13		3		125
		AJ77,733 T20# 6				1	13.34		0.36		0.19		2		125
		AJ77,733 T20# 7				1	14.55		0.92		0.40		3		125
		AJ77,733 T21# 6				1	12.63		1.24		0.89		3		125
		AJ77,733 T21# 7				1	13.65		0.83		0.29		3		125
		AJ77,733 T21# 8				1	14.18		0.65		0.45		3		125
		AJ77,733 T22# 13				1	12.94		0.39		-0.28		3		125
		AJ77,733 T22# 14				1	12.95		0.24		0.44		4		125
		AJ77,733 T22# 15				1	13.01		0.52		0.20		2		125
		AJ77,733 T22# 16				1	13.43		0.58		0.21		2		125
		AJ77,733 T22# 17				1	13.86		0.40		0.23		2		125
		AJ77,733 T22# 18				1	14.15		0.40		0.29		3		125
		AJ77,733 T22# 19				1	14.24		0.27		0.21		2		125
		AJ77,733 T22# 20				1	14.69		0.82		0.43		3		125
		AJ77,733 T23# 3				1	11.83		0.41		0.16		1		125
		AJ77,733 T23# 6				1	13.47		0.63		0.03		1		125
		AJ77,733 T23# 7				1	13.94		0.60		0.27		1		125
		AJ77,733 T23# 8				1	14.14		0.77		0.27		1		125
		AJ77,733 T23# 9				1	14.44		0.63		0.30		1		125
		AJ77,733 T24# 11				1	13.58		0.80		0.35		2		125
		AJ77,733 T24# 12				1	13.68		0.72		0.23		2		125
		AJ77,733 T24# 13				1	14.44		2.01				2		125
		AJ77,733 T24# 14				1	14.70		1.50				2		125
		AJ77,733 T24# 15				1	15.30		0.73		0.40		2		125
		AJ77,733 T25# 9				1	12.80		0.55		0.05		3		125
		AJ77,733 T25# 10				1	12.98		0.63		0.33		3		125
		AJ77,733 T25# 11				1	13.07		0.66		0.07		5		125
		AJ77,733 T25# 12				1	13.37		0.47		0.19		4		125

HD	DM	Other Id	N Rem	α_{1950}	δ_{1950}	S	V	σ_V	B–V	σ_{B-V}	U–B	σ_{U-B}	n	Spectrum	References
		AJ77,733 T25# 13				1	13.55		0.61		0.44		5		125
		AJ77,733 T25# 14				1	13.78		0.62		0.18		4		125
		AJ77,733 T25# 15				1	13.94		0.62		0.05		2		125
		AJ77,733 T25# 16				1	14.13		0.66		0.10		4		125
		AJ77,733 T25# 17				1	15.20		0.92		0.62		3		125
		AJ77,733 T26# 14				1	12.89		0.68		0.19		5		125
		AJ77,733 T26# 15				1	13.00		1.60		1.20		3		125
		AJ77,733 T26# 16				1	13.16		0.68		0.15		5		125
		AJ77,733 T26# 18				1	13.84		0.66		0.35		6		125
		AJ77,733 T26# 19				1	14.28		1.73		1.65		3		125
		AJ77,733 T26# 21				1	14.95		1.07		0.83		4		125
		AJ77,733 T26# 22				1	15.44		0.95		0.48		3		125
		AJ77,733 T26# 23				1	15.69		1.01		0.66		3		125
		AJ77,733 T27# 8				1	12.41		0.70		0.24		2		125
		AJ77,733 T27# 9				1	12.78		0.58		-0.10		5		125
		AJ77,733 T27# 10				1	13.36		0.64		0.02		2		125
		AJ77,733 T27# 11				1	13.73		0.66		0.02		4		125
		AJ77,733 T27# 12				1	14.26		0.74		0.12		5		125
		AJ77,733 T27# 13				1	14.51		0.84		0.43		5		125
		AJ77,733 T27# 14				1	15.21		0.91		0.24		3		125
		AJ77,733 T27# 15				1	15.39		1.00				2		125
		AJ77,733 T28# 16				1	14.05		0.58		0.24		3		125
		AJ77,733 T28# 17				1	14.72		0.88		0.48		3		125
		AJ78,401 T1# 2				1	14.03		0.75		0.16		2		262
		AJ78,401 T1# 3				1	16.90		0.77		0.31		2		262
		AJ78,401 T1# 4				1	17.00		0.74		0.03		2		262
		AJ78,401 T1# 7				1	11.52		1.54		1.67		2		262
		AJ78,401 T1# 8				1	11.39		0.65		0.05		2	G0	262
		AJ78,401 T1# 9				1	16.06		0.71		-0.04		2		262
		AJ78,401 T1# 12				1	14.70		1.03		0.14		2		262
		AJ78,401 T1# 13				1	15.72		0.96		0.36		2		262
		AJ78,401 T1# 16				1	10.88		0.68		0.17		2	G0	262
		AJ78,401 T1# 17				1	12.73		1.26		1.15		2		262
		AJ78,401 T1# 19				1	11.32		1.13		0.74		2	G2 III	262
		AJ78,401 T1# 20				1	12.76		0.76		0.16		2		262
		AJ78,401 T1# 21				1	15.26		0.79		0.08		2		262
		AJ78,401 T1# 26				1	14.00		0.83		0.27		2		262
		AJ78,401 T1# 29				1	16.27		1.29		0.59		2		262
		AJ78,401 T1# 30				1	15.72		1.23				2		262
		AJ78,401 T1# 34				1	15.43		1.06		0.41		2		262
		AJ78,401 T1# 35				1	14.69		0.69		0.25		2		262
		AJ78,401 T1# 36				1	13.89		0.83		0.31		2		262
		AJ78,401 T1# 41				1	12.42		0.63		0.17		2		262
		AJ78,401 T1# 42				1	14.57		1.09		0.58		2		262
		AJ78,401 T1# 43				1	11.78		0.76		0.18		2		262
		AJ78,401 T1# 44				1	10.31		0.47		0.22		2	A7	262
		AJ78,401 T1# 45				1	13.37		0.72		0.18		2		262
		AJ78,401 T1# 46				1	13.80		0.89		0.33		2		262
		AJ78,401 T1# 47				1	13.92		0.36		0.12		2		262
		AJ78,401 T1# 48				1	14.68		0.93		0.50		2		262
		AJ78,401 T1# 49				1	15.18		0.69		0.03		2		262
		AJ78,401 T1# 50				1	15.73		0.72		0.02		2		262
		AJ78,401 T2# 2				1	15.89		0.85		0.15		2		262
		AJ78,401 T2# 3				1	14.05		0.67		0.01		2		262
		AJ78,401 T2# 4				1	17.23		1.20		-0.06		2		262
		AJ78,401 T2# 5				1	16.57		1.13		0.88		2		262
		AJ78,401 T2# 6				1	17.65		1.25		-0.23		2		262
		AJ78,401 T2# 8				1	18.30		0.99		0.42		2		262
		AJ78,401 T2# 10				1	14.92		1.04		0.48		2		262
		AJ78,401 T2# 11				1	14.85		0.89		0.26		2		262
		AJ78,401 T2# 12				1	15.84		0.67		0.30		2		262
		AJ78,401 T2# 17				1	12.17		0.52		0.11		2		262
		AJ78,401 T2# 18				1	13.54		0.85		0.38		2		262
		AJ78,401 T2# 19				1	12.44		0.54		0.09		2		262
		AJ78,401 T2# 20				1	14.19		0.64		0.05		2		262
		AJ78,401 T2# 21				1	14.63		1.81		1.87		2		262
		AJ78,401 T2# 22				1	16.37		0.80		0.68		2		262
		AJ78,401 T2# 25				1	14.47		0.79		0.24		2		262
		AJ78,401 T2# 26				1	16.58		1.23		0.65		2		262
		AJ78,401 T2# 27				1	12.29		0.72		0.28		2		262
		AJ78,401 T2# 28				1	11.23		0.57		0.09		2	F0	262
		AJ78,401 T2# 31				1	16.64		0.82		0.57		2		262
		AJ78,401 T2# 32				1	14.75		1.44		1.46		2		262
		AJ78,401 T2# 33				1	10.70		1.08		0.88		2		262
		AJ78,401 T2# 34				1	13.90		1.26		1.28		2		262
		AJ78,401 T2# 37				1	16.46		1.06		1.38		2		262
		AJ78,401 T2# 38				1	12.14		1.72		0.73		2		262
		AJ78,401 T2# 39				1	14.01		0.69		0.08		2		262
		AJ78,401 T2# 41				1	14.34		1.36		0.54		2		262
		AJ78,401 T2# 42				1	13.18		0.50		0.14		2		262

Table 2 1261

HD	DM	Other Id	N Rem	α_{1950}	δ_{1950}	S	V	σ_V	B–V	σ_{B-V}	U–B	σ_{U-B}	n	Spectrum	References
		AJ78,401 T2# 43				1	12.82		1.34				2		262
		AJ78,401 T3# 3				1	10.43		0.46		0.22		2	F0	262
		AJ78,401 T3# 4				1	13.67		0.58		-0.08		2		262
		AJ78,401 T3# 5				1	13.99		0.76		0.26		2		262
		AJ78,401 T3# 6				1	15.89		0.94		0.39		2		262
		AJ78,401 T3# 16				1	14.83		0.79		0.13		2		262
		AJ78,401 T3# 20				1	10.63		0.19		0.08		2	B9	262
		AJ78,401 T3# 23				1	13.64		0.61		0.14		2		262
		AJ78,401 T3# 24				1	13.61		0.43		0.14		2		262
		AJ78,401 T3# 25				1	14.13		1.13		0.75		2		262
		AJ78,401 T3# 26				1	15.33		0.87		-0.10		2		262
		AJ78,401 T3# 30				1	10.80		0.37				2	F0	262
		AJ78,401 T3# 31				1	12.44		0.26		-0.29		2		262
		AJ78,401 T3# 33				1	15.18		0.73				2		262
		AJ78,401 T3# 35				1	14.80		0.87				2		262
		AJ78,401 T3# 37				1	13.86		1.59		1.59		2		262
		AJ78,401 T3# 38				1	13.82		1.23		1.06		2		262
		AJ78,401 T3# 42				1	10.04		-0.22		-0.86		2	B8	262
		AJ78,401 T3# 43				1	12.66		1.23		1.02		2		262
		AJ78,401 T3# 44				1	12.95		1.20		0.75		2		262
		AJ78,401 T4# 2				1	14.84		0.87		0.45		2		262
		AJ78,401 T4# 4				1	11.47		1.11		0.89		2	K0 III	262
		AJ78,401 T4# 6				1	12.54		1.92		1.97		2		262
		AJ78,401 T4# 7				1	11.57		0.61		0.16		2	G0	262
		AJ78,401 T4# 20				1	15.90		1.03		0.45		2		262
		AJ78,401 T4# 21				1	16.74		0.79		-0.10		2		262
		AJ78,401 T4# 26				1	14.04		0.90		0.44		2		262
		AJ78,401 T4# 27				1	15.52		0.84		0.48		2		262
		AJ78,401 T4# 28				1	16.52		1.11		0.83		2		262
		AJ78,401 T4# 34				1	13.28		0.39		0.22		2		262
		AJ78,401 T4# 37				1	15.95		0.80		0.07		2		262
		AJ78,401 T4# 41				1	12.70		0.62		0.10		2		262
		AJ79,1294 T4# 12				1	14.08		0.75		0.26		5		396
		AJ79,1294 T4# 13				1	14.58		0.50		0.43		2		396
		AJ79,1294 T4# 14				1	14.75		0.65		0.09		3		396
		AJ79,1294 T4# 15				1	15.23		0.72		0.20		4		396
		AJ79,1294 T4# 16				1	15.71		0.57		0.32		2		396
		AJ79,1294 T4# 17				1	15.73		0.70		0.29		2		396
		AJ79,1294 T5# 16				1	14.31		0.48		0.20		4		396
		AJ79,1294 T5# 17				1	14.80		0.38		0.30		4		396
		AJ79,1294 T5# 18				1	15.45		1.19		0.75		3		396
		AJ79,1294 T5# 19				1	15.72		0.76		0.25		6		396
		AJ79,1294 T5# 20				1	15.82		0.58		0.53		6		396
		AJ79,1294 T5# 21				1	16.12		0.86		0.42		6		396
		AJ79,1294 T6# 14				1	13.60		0.66		0.24		3		396
		AJ79,1294 T6# 15				1	14.06		1.38		0.96		3		396
		AJ79,1294 T6# 18				1	14.78		0.92		0.44		4		396
		AJ79,1294 T6# 19				1	15.36		0.71		0.18		4		396
		AJ79,1294 T6# 20				1	15.49		0.66		0.21		3		396
		AJ79,1294 T6# 21				1	15.51		1.23		1.18		2		396
		AJ79,1294 T6# 104				1	11.88		0.18		-0.54		1		354
		AJ79,1294 T6# 112				1	15.47		0.87		0.06		1		354
		AJ79,1294 T6# 119				1	12.12		0.59		-0.29		1		354
		AJ79,1294 T6# 128				1	12.68		0.77		-0.20		1		354
		AJ79,1294 T7# 17				1	14.66		0.78		0.40		4		396
		AJ79,1294 T7# 18				1	15.26		0.74		0.18		3		396
		AJ79,1294 T7# 19				1	15.32		1.06		0.19		4		396
		AJ79,1294 T7# 20				1	15.49		0.79		0.43		3		396
		AJ79,1294 T7# 115				1	14.81		0.98		0.10		1		354
		AJ79,1294 T7# 116				1	14.43		1.01		0.12		1		354
		AJ79,1294 T7# 117				1	14.56		1.00		0.06		1		354
		AJ79,1294 T7# 120				1	14.06		1.42		0.35		1		354
		AJ79,1294 T7# 122				1	14.77		1.02		0.10		1		354
		AJ79,1294 T8# 21				1	12.90		0.40		0.07		3		396
		AJ79,1294 T8# 22				1	13.18		0.39		0.06		4		396
		AJ79,1294 T8# 23				1	13.19		0.42		0.28		3		396
		AJ79,1294 T8# 25				1	13.88		0.37		0.33		3		396
		AJ79,1294 T8# 26				1	14.27		0.66		0.21		4		396
		AJ79,1294 T8# 27				1	14.53		0.70		0.26		4		396
		AJ79,1294 T8# 28				1	14.92		0.68		0.20		4		396
		AJ79,1294 T8# 29				1	15.02		0.58		0.40		7		396
		AJ79,1294 T8# 30				1	15.09		0.74		0.14		3		396
		AJ79,1294 T8# 31				1	15.10		0.53		0.10		4		396
		AJ79,1294 T8# 32				1	15.41		0.51		-0.09		6		396
		AJ79,1294 T8# 33				1	15.66		0.76		0.50		5		396
		AJ79,1294 T8# 34				1	16.11		0.54				3		525
		AJ79,1294 T8# 35				1	16.29		0.74				3		525
		AJ79,1294 T8# 36				1	17.12		0.77				3		525
		AJ79,1294 T8# 130				1	12.34		0.32		-0.54		2		354
		AJ79,1294 T9# 12				1	13.23		0.60		0.13		3		396

HD	DM	Other Id	N	Rem	α_{1950}	δ_{1950}	S	V	σ_V	B–V	σ_{B-V}	U–B	σ_{U-B}	n	Spectrum	References
		AJ79,1294 T9# 13					1	13.94		0.35		0.25		4		396
		AJ79,1294 T9# 14					1	13.97		0.20		-0.18		5		396
		AJ79,1294 T9# 15					1	14.74		0.67		0.35		7		396
		AJ79,1294 T9# 16					1	14.98		0.37		0.21		7		396
		AJ79,1294 T9# 17					1	15.41		0.43		0.32		5		396
		AJ79,1294 T9# 18					1	15.68		0.47		0.30		5		396
		AJ79,1294 T9# 19					1	15.73		0.79		0.21		3		525
		AJ79,1294 T9# 20					1	16.34		0.73				3		525
		AJ79,1294 T9# 21					1	17.12		0.46				2		525
		AJ79,1294 T10# 14					1	14.06		0.66		0.29		4		396
		AJ79,1294 T10# 15					1	14.46		0.68		0.10		4		396
		AJ79,1294 T10# 16					1	14.82		0.65		0.08		3		396
		AJ79,1294 T10# 17					1	15.03		0.65		0.13		3		396
		AJ79,1294 T10# 18					1	15.42		0.93		0.52		3		396
		AJ79,1294 T10# 19					1	15.48		0.98		0.38		3		396
		AJ79,1294 T11# 14					1	14.17		0.84		0.28		3		396
		AJ79,1294 T11# 15					1	14.77		0.82		0.26		3		396
		AJ79,1294 T11# 16					1	15.34		0.96		0.33		3		396
		AJ79,1294 T12# 11					1	14.16		0.64		0.40		3		396
		AJ79,1294 T12# 12					1	14.29		0.64		0.08		4		396
		AJ79,1294 T12# 13					1	14.88		0.71		0.22		3		396
		AJ79,1294 T12# 14					1	15.41		0.92		0.38		3		396
		AJ79,1294 T12# 15					1	15.42		1.15		0.58		3		396
		AJ79,1294 T13# 11					1	13.28		0.73		0.51		4		396
		AJ79,1294 T13# 13					1	14.19		0.87		0.47		5		396
		AJ79,1294 T13# 14					1	14.46		0.72		0.58		3		396
		AJ79,1294 T13# 15					1	14.63		0.80		0.28		2		396
		AJ79,1294 T13# 16					1	14.68		0.85		0.46		3		396
		AJ79,1294 T13# 17					1	14.81		0.96		0.21		2		396
		AJ79,1294 T13# 18					1	14.93		0.94		0.23		2		396
		AJ79,1294 T13# 19					1	14.98		0.84		0.28		5		396
		AJ79,1294 T13# 20					1	15.49		0.94		0.40		6		396
		AJ79,603 # 1					1	13.25		1.66		1.73		4		359
		AJ79,603 # 2					1	14.97		1.01		0.52		4		359
		AJ79,603 # 3					1	16.68		0.87		0.33		4		359
		AJ79,603 # 4					1	16.01		1.26		0.84		4		359
		AJ79,603 # 5					1	16.34		1.46		2.22		4		359
		AJ79,603 # 6					1	11.69		1.30		1.49		4		359
		AJ79,603 # 7					1	14.66		1.14		0.75		4		359
		AJ79,603 # 8					1	17.67		1.13				4		359
		AJ79,603 # 9					1	18.24		0.67				4		359
		AJ79,603 # 10					1	15.13		0.94		0.46		4		359
		AJ79,603 # 11					1	16.84		1.09				4		359
		AJ79,603 # 12					1	15.88		1.52				4		359
		AJ79,603 # 13					1	10.74		0.05		-0.01		4		359
		AJ79,603 # 14					1	13.98		1.19		1.10		4		359
		AJ79,603 # 15					1	14.40		1.47		1.68		4		359
		AJ79,603 # 16					1	15.58		1.14		0.74		4		359
		AJ79,603 # 17					1	15.84		0.80		0.22		4		359
		AJ79,603 # 18					1	15.22		1.08		0.77		4		359
		AJ79,603 # 19					1	14.26		1.16		1.00		4		359
		AJ79,603 # 20					1	17.10		0.88				4		359
		AJ79,603 # 21					1	15.19		1.12		0.89		4		359
		AJ79,603 # 22					1	16.36		1.14				4		359
		AJ79,603 # 23					1	16.14		1.36				4		359
		AJ79,603 # 24					1	16.20		1.16				4		359
		AJ79,603 # 25					1	15.62		1.01				4		359
		AJ79,603 # 26					1	12.57		1.66		2.07		4		359
		AJ79,603 # 27					1	12.62		0.58		0.06		4		359
		AJ79,603 # 28					1	13.77		0.66		0.05		4		359
		AJ79,603 # 29					1	13.33		1.57		1.96		4		359
		AJ82,163 # 6					1	13.61		0.68		0.41		1		533
		AJ82,163 # 9					1	13.83		0.63		0.25		11		533
		AJ82,163 # 11					1	13.53		1.00		0.75		4		533
		AJ82,163 # 12					1	10.76		0.23		0.06		10		533
		AJ82,163 # 13					1	12.78		0.30		0.13		8		533
		AJ82,163 # 14					1	14.22		0.43		0.20		4		533
		AJ82,163 # 15					1	13.14		1.26		1.07		4		533
		AJ82,163 # 16					1	14.07		0.68		-0.01		2		533
		AJ82,163 # 17					1	13.16		1.26		1.38		1		533
		AJ82,163 # 19					1	13.75		0.52		0.01		1		533
		AJ85,38 Anon					1	11.94		0.56		-0.14		1		6006
		AJ88,650 # 2					1	12.77		0.44		0.28		4		994
		AJ88,650 # 3					1	13.87		1.33		1.10		3		994
		AJ88,650 # 5					1	13.81		0.79		0.49		4		994
		AJ88,650 # 9					1	13.73		1.18		0.88		3		994
		AJ88,650 # 10					1	13.73		0.78		0.41		4		994
		AJ88,650 # 11					1	14.46		2.00				2		994
		AJ88,650 # 12					1	14.41		1.05		0.64		2		994
		AJ88,650 # 13					1	14.25		0.74		0.33		2		994

Table 2 1263

HD	DM	Other Id	N Rem	α_{1950}	δ_{1950}	S	V	σ_V	B–V	σ_{B-V}	U–B	σ_{U-B}	n	Spectrum	References
		Anti 3				1	14.58		0.68				1		1702
		Anti 6				1	15.15		0.62				1		1702
		Anti 8				1	15.58		0.66				1		1702
		Anti A				1	10.72		1.01				1		1702
		Anti 24				1	13.81		0.53				1		1702
		Anti 25				1	12.55		1.45				1		1702
		Anti 49				1	13.88		0.42				1		1702
		Anti 54				1	13.20		0.73				1		1702
		Anti 64				1	15.34		0.89				1		1702
		Anti 65				1	15.25		0.64				1		1702
		Anti 66				1	14.02		0.43				1		1702
		Anti 78				1	15.62		0.77				1		1702
		Anti 79				1	16.24		0.49				1		1702
		Anti 84				1	15.04		1.46				1		1702
		Anti 85				1	15.42		0.75				1		1702
		Anti 86				1	15.73		1.15				1		1702
		Anti 96				1	14.22		0.45				1		1702
		Anti 117				1	14.06		0.50				1		1702
		Anti 139				1	15.47		0.70				1		1702
		Anti 142				1	14.67		1.56				1		1702
		Anti 145				1	14.97		0.20				2		1702
		Anti 148				1	14.23		1.51				1		1702
		Anti 152				1	12.84		1.50				1		1702
		AN298,163 # 102				1	15.21		0.67		0.63		1		590
		AN298,163 # 103				1	14.20		0.66		0.06		2		590
		AN298,163 # 104				1	14.96		1.02		0.68		1		590
		AN298,163 # 105				1	15.58		0.50		0.47		2		590
		AN298,163 # 106				1	12.25		0.77		0.25		3		590
		AN298,163 # 109				1	12.91		0.67		0.05		2		590
		AN298,163 # 110				1	13.79		0.64		0.08		3		590
		AN298,163 # 112				1	14.97		-0.10		-0.06		1		590
		AN298,163 # 113				1	13.99		0.71		0.14		1		590
		AN298,163 # 114				1	13.21		0.64		0.01		1		590
		AN298,163 # 204				1	15.33		-0.23		-0.79		1		590
		AN298,163 # 206				1	14.68		1.17		0.36		2		590
		AN298,163 # 209				1	15.47		1.22		0.85		1		590
		AN298,163 # 210				1	15.55		-0.55		-1.05		2		590
		AN298,163 # 211				1	15.18		0.11		-0.18		1		590
		AN298,163 # 214				1	13.81		1.76		1.48		2		590
		AN298,163 # 215				1	15.29		0.83		0.39		1		590
		ASS75,135 # 2				1	14.14		0.69		0.22		2		898
		ASS75,135 # 3				1	15.44		0.69		0.12		2		898
		ASS75,135 # 4				1	15.70		0.61		0.02		2		898
		ASS75,135 # 6				1	17.14		0.67		0.25		2		898
		ASS75,135 # 7				1	14.32		0.96		0.78		2		898
		ASS75,135 # 10				1	12.74		0.72		0.31		2		898
		ASS75,135 # 11				1	17.13		1.64		1.14		2		898
		ASS75,135 # 12				1	15.13		0.73		0.16		2		898
		ASS75,135 # 13				1	16.54		1.33		0.82		2		898
		ASS75,135 # 14				1	17.17		1.00		1.02		2		898
		ASS75,135 # 15				1	14.54		0.82		0.10		2		898
		ASS75,135 # 16				1	14.33		1.05		0.93		2		898
		ASS75,135 # 17				1	13.09		0.84		0.51		2		898
		ASS75,135 # 18				1	15.39		0.57		0.03		2		898
		ASS75,135 # 20				1	16.46		0.72		0.08		2		898
		ASS75,135 # 21				1	14.78		1.12		0.98		2		898
		ASS75,135 # 22				1	12.47		0.70		0.24		2		898
		AcAstr31,363 Anon				1	11.70		0.63		0.14		3		8088
		ApJ110,424R200# 3				1	11.09		0.33		0.10		1		572
		ApJ110,424R200# 4				1	12.48		0.78		0.30		1		572
		ApJ110,424R200# 5				1	13.96		0.78		0.16		1		572
		ApJ110,424R217# 9				1	11.18		0.90		-0.20		3		1359
		ApJ110,424R299# 2				1	12.47		0.80		0.31		1		572
		ApJ110,424R913# 1				1	12.03		1.29		1.07		1		572
		ApJ113,309 # 200				1	11.12		0.21		-0.65		1		211
		ApJ113,309 # 201				1	11.14		0.28		-0.74		1		211
		ApJ146,316 # 1				1	14.97		0.94		0.41		1		515
		ApJ146,316 # 2				1	16.25		1.05		0.43		1		515
		ApJ146,316 # 3				1	14.17		1.16		0.87		1		515
		ApJ146,316 # 4				1	14.47		0.84		0.33		1		515
		ApJ146,316 # 5				1	14.46		0.83		0.26		1		515
		ApJ165,259 # 2				1	11.50		0.64		0.15		5		314
		ApJ165,259 # 3				1	12.61		0.60		0.09		5		314
		ApJ165,259 # 27				1	14.08		1.82		1.46		2		314
		ApJ165,259 # 28				1	13.87		1.05		0.48		3		314
		ApJ165,259 # 29				1	16.26		1.31		0.38		2		314
		ApJ165,259 # 30				1	14.19		1.06		0.42		3		314
		ApJ165,259 # 31				1	14.60		1.06		0.50		2		314
		ApJ165,259 # 32				1	16.48		1.06		0.94		2		314
		ApJ165,259 # 33				1	14.73		1.13		0.71		1		314

HD	DM	Other Id	N	Rem	α_{1950}	δ_{1950}	S	V	σ_V	B–V	σ_{B-V}	U–B	σ_{U-B}	n	Spectrum	References
		ApJ165,259 # 34					1	14.99		0.90		0.70		1		314
		ApJ165,259 # 35					1	17.26		0.86		0.47		1		314
		ApJ165,259 # 36					1	15.74		0.99		0.89		1		314
		ApJ165,259 # 37					1	15.98		1.40		0.42		1		314
		ApJ165,259 # 38					1	15.33		1.66		1.48		1		314
		ApJ165,259 # 39					1	15.85		1.29		0.74		1		314
		ApJ165,259 # 40					1	16.00		1.48		1.14		1		314
		ApJ190,L1 Anon					1	10.76		2.41				1		8051
		ApJ218,617 # 8					1	15.29		2.18				1		596
		ApJ218,617 # 9					1	15.40		1.90				2		596
		ApJ218,617 # 10					1	14.47		1.80				1		596
		ApJ218,617 # 11					1	14.75		1.46				1		596
		ApJ218,617 # 12					1	15.56		1.96				1		596
		ApJ218,617 # 13					1	14.45		1.76				1		596
		ApJ231,384 # 2					1	13.64		2.14				2	M1 Ia	5007
		ApJ231,384 # 5					1	12.73		2.08				2	M0 Ia	5007
		ApJ231,384 # 9					1	12.49		2.01				2	M0 Ia	5007
		ApJ231,384 # 10					1	12.64		1.85				2	K5-M0Iab	5007
		ApJ231,384 # 20					1	12.93		1.87				2	K5-M0I	5007
		ApJ231,384 # 22					1	12.28		1.79				2	M0 Ia	5007
		ApJ231,384 # 25					1	11.34		1.88				2	K5-M0Ia	5007
		ApJ231,384 # 33					1	13.20		1.65				2	K5-M0I	5007
		ApJ231,384 # 44					1	12.96		1.83				2	M0 Iab	5007
		ApJ231,384 # 56					1	12.77		1.99				2	M0 Iab	5007
		ApJ231,384 # 57					1	12.68		1.95				2	K5-M0 I	5007
		ApJ231,384 # 74					1	12.18		0.53				4		8033
		ApJ231,384 # 89					1	11.59		0.15				3		8033
		ApJ231,384 # 98					1	11.53		0.02				2		8033
		ApJ231,384 # 105					1	11.55		0.27				3		8033
		ApJ231,384 # 114					1	11.32		0.00				3		8033
		ApJ231,384 # 117					1	11.19		0.08				3		8033
		ApJ231,384 # 124					1	11.50		-0.01				2		8033
		ApJ231,384 # 136					1	11.32		0.12				3		8033
		ApJ231,384 # 166					1	12.68		2.04				2	M0 Ia-Iab	5007
		ApJ231,384 # 206					1	12.02		1.76				2	K5 Ia	5007
		ApJ231,384 # 234					1	11.96		1.83				2	K5-M0 I	5007
		ApJ231,384 # 235					1	12.87		1.56				2	K5-M0 I	5007
		ApJ231,384 # 243					1	12.95		1.72				2	K5-M0 I	5007
		ApJ231,384 # 252					1	13.11		1.71				2	K5-M0 I	5007
		ApJ231,384 # 257					1	12.69		1.79				2	K5-M0 I	5007
		ApJ231,384 # 260					1	13.29		1.64				2	K5-M0 I	5007
		ApJ231,384 # 275					1	13.24		1.82				2	M0 Ia-Iab	5007
		ApJ231,384 # 287		AB			1	12.24		1.99				2	M0 Ia	5007
		ApJ231,384 # 297					1	12.13		-0.02				2		8033
		ApJ239,112 # 23					2	14.48	.000	0.84	.000	0.26	.000	4		874,1533
		ApJ239,112 # 24					2	14.66	.000	1.09	.000	0.66	.000	4		874,1533
		ApJ239,112 # 25					2	15.00	.000	0.78	.000			2		874,1533
		ApJ239,112 # 26					1	15.34		1.09		0.66		4		1533
		ApJ239,112 # 27					1	15.61		0.88		0.30		2		1533
		ApJ239,112 # 28					1	15.73		1.11		0.43		3		1533
		ApJ239,112 # 29					2	15.81	.000	1.20	.000			2		874,1533
		ApJ239,112 # 30					1	15.88		0.83		0.17		2		1533
		ApJ239,112 # 31					2	16.03	.000	1.01	.000			2		874,1533
		ApJ239,112 # 32					1	16.42		0.98				2		1533
		ApJ239,112 # 33					2	16.61	.000	1.35	.000			2		874,1533
		ApJ239,112 # 34					1	16.65		0.97				2		1533
		ApJ239,112 # 35					1	16.79		0.94		0.46		2		1533
		ApJ239,112 # 36					1	17.26		0.92				2		1533
		ApJ239,112 # 37					1	17.35		0.95				1		1533
		ApJ239,112 # 38					1	17.63		1.18				1		1533
		ApJ239,112 # 39					1	17.78		0.94				1		1533
		ApJ239,112 # 40					1	18.44		1.13				1		1533
		ApJ239,112 # 41					1	18.91		0.86				1		1533
		ApJ239,112 # 42					1	19.91		1.44				1		1533
		ApJ239,112 # 43					1	20.09		0.85				1		1533
		ApJ239,815 # 1					1	14.58		1.58		1.80		4		875
		ApJ239,815 # 2					1	14.83		1.51		1.58		2		875
		ApJ239,815 # 3					1	17.87		0.34				1		875
		ApJ239,815 # 4					1	17.84		0.82		0.29		2		875
		ApJ239,815 # 5					1	17.82		0.75		0.21		2		875
		ApJ239,815 # 6					1	12.36		0.58		0.01		5		875
		ApJ283,254 # 8					1	13.80		0.98		0.59		2		1524
		ApJ239,815 # 12					1	16.59		0.77		0.02		2		875
		ApJ239,815 # 14					1	16.89		0.84		0.16		1		875
		ApJ239,815 # 15					1	13.68		1.66		1.20		4		875
		ApJ239,815 # 17					1	14.67		0.62		0.04		2		875
		ApJ239,815 # 18					1	16.00		0.78		0.09		2		875
		ApJ239,815 # 23					1	15.61		1.58		1.28		1		875
		ApJ239,815 # 24					1	14.62		0.80		0.17		2		875
		ApJ239,815 # 31					1	15.27		0.78		0.35		2		875

Table 2   1265

HD	DM	Other Id	N Rem	α_{1950}	δ_{1950}	S	V	σ_V	B–V	σ_{B-V}	U–B	σ_{U-B}	n	Spectrum	References
		ApJ239,815 # 33				1	17.05		0.60		0.08		1		875
		ApJ239,815 # 37				1	14.88		0.61		0.00		2		875
		ApJ239,815 # 42				1	15.76		0.63		0.03		2		875
		ApJS8,439 T4- 1# 1				1	13.58		0.85		0.30		2		1083
		ApJS8,439 T4- 1# 2				1	14.58		1.05		0.50		2		1083
		ApJS8,439 T4- 2# 1				1	11.25		1.34		1.16		2		1083
		ApJS8,439 T4- 2# 2				1	11.45		0.49		0.15		2		1083
		ApJS8,439 T4- 3# 4				1	12.23		0.55		0.15		2		1083
		ApJS8,439 T4- 4# 1				1	13.67		0.42		0.28		2		1083
		ApJS8,439 T4- 4# 2				1	14.47		1.21		1.19		2		1083
		ApJS8,439 T4- 5# 1				1	14.15		0.84		0.29		2		1083
		ApJS8,439 T4- 5# 2				1	12.53		0.32		-0.23		2		1083
		ApJS8,439 T4- 5# 3				1	14.74		0.72		0.39		2		1083
		ApJS8,439 T4- 5# 4				1	13.86		0.86		0.43		2		1083
		ApJS8,439 T4- 6# 1				1	12.71		0.58		-0.24		2		1083
		ApJS8,439 T4- 6# 2				1	14.35		0.76		0.54		2		1083
		ApJS8,439 T4- 6# 3				1	15.01		0.66		0.32		2		1083
		ApJS8,439 T4- 7# 1				1	10.97		0.84		0.38		2		1083
		ApJS8,439 T4- 7# 3				1	9.64		0.45		0.14		2		1083
		ApJS8,439 T4- 7# 4				1	12.17		0.97		0.28		2		1083
		ApJS8,439 T4- 7# 5				1	13.67		1.35		1.36		2		1083
		ApJS8,439 T4- 8# 1				1	12.32		0.59		0.27		2		1083
		ApJS8,439 T4- 8# 2				1	13.98		0.66		0.11		2		1083
		ApJS8,439 T4- 9# 1				1	14.91		0.34		-0.74		2		1083
		ApJS8,439 T4- 9# 2				1	15.49		1.11		0.83		2		1083
		ApJS8,439 T4- 9# 3				1	11.07		0.19		-0.50		2		1083
		ApJS8,439 T4- 9# 4				1	15.43		1.30		0.48		2		1083
		ApJS8,439 T4- 9# 5				1	12.44		0.48		-0.19		2		1083
		ApJS8,439 T4-10# 1				1	13.98		0.66		0.18		2		1083
		ApJS8,439 T4-10# 2				1	12.31		0.69		0.40		2		1083
		ApJS8,439 T4-10# 3				1	14.82		1.99		2.48		2		1083
		ApJS8,439 T4-10# 4				1	14.28		1.58		1.44		2		1083
		ApJS8,439 T4-11# 2				1	12.38		0.27		-0.37		2		1083
		ApJS8,439 T4-12# 1				1	13.20		0.75		0.16		2		1083
		ApJS8,439 T4-12# 2				1	15.34		0.41		0.16		2		1083
		ApJS8,439 T4-13# 1				1	12.34		0.20		-0.32		2		1083
		ApJS8,439 T4-13# 2				1	11.83		0.14		-0.48		2		1083
		ApJS8,439 T4-14# 0				1	10.30		2.31		2.51		2		1083
		ApJS8,439 T4-14# 1				1	13.97		0.80		0.37		2		1083
		ApJS8,439 T4-14# 2				1	15.12		1.00		0.49		2		1083
		ApJS8,439 T4-14# 3				1	13.56		0.68		0.48		2		1083
		ApJS8,439 T4-14# 4				1	14.62		0.67		0.35		2		1083
		ApJS8,439 T4-16# 3				1	14.56		0.56		0.29		2		1083
		ApJS8,439 T4-17# 1				1	12.98		0.51		0.02		2		1083
		ApJS8,439 T4-17# 2				1	11.14		0.35		-0.41		2		1083
		ApJS8,439 T4-17# 3				1	11.39		0.39		-0.38		2		1083
		ApJS8,439 T4-19# 1				1	12.56		0.57		0.40		2		1083
		ApJS8,439 T4-19# 2				1	12.53		0.43		-0.16		2		1083
		ApJS8,439 T4-20# 1				1	12.99		0.92		0.42		2		1083
		ApJS8,439 T4-20# 2				1	13.55		0.72		0.54		2		1083
		ApJS8,439 T4-21# 1				1	14.03		0.92		0.42		2		1083
		ApJS8,439 T4-22# 1				1	13.91		0.34		0.20		2		1083
		ApJS8,439 T4-22# 2				1	14.74		1.29		0.92		2		1083
		ApJS8,439 T4-22# 3				1	10.42		0.42		0.25		2		1083
		ApJS8,439 T4-24# 1				1	14.26		0.85		0.51		2		1083
		ApJS8,439 T4-24# 2				1	13.54		1.79		1.53		2		1083
		ApJS8,439 T4-26# 1				1	13.28		0.33		-0.16		2		1083
		ApJS8,439 T4-27# 1				1	12.94		0.39		0.25		2		1083
		ApJS8,439 T4-27# 2				1	12.76		0.53		0.22		2		1083
		ApJS8,439 T4-27# 3				1	14.24		0.61		0.40		2		1083
		ApJS8,439 T4-28# 2				1	11.65		0.53		0.10		2		1083
		ApJS8,439 T4-28# 3				1	10.48		1.15		0.79		2		1083
		ApJS8,439 T4-28# 4				1	11.54		0.59		0.31		2		1083
		ApJS8,439 T4-30# 1				1	15.11		0.63		0.43		2		1083
		ApJS8,439 T4-30# 2				1	13.87		0.61		0.31		2		1083
		ApJS8,439 T4-33# 2				1	14.14		1.68		1.68		2		1083
		ApJS8,439 T4-35# 2				1	11.57		0.76		0.22		2		1083
		ApJS8,439 T4-36# 1				1	14.05		0.56		0.15		2		1083
		ApJS8,439 T4-36# 2				1	11.28		0.45		-0.39		2		1083
		ApJS8,439 T4-36# 3				1	12.21		0.57		-0.23		2		1083
		ApJS8,439 T4-36# 4				1	15.77		1.19		0.56		2		1083
		ApJS8,439 T4-37# 1				1	14.27		0.95		0.34		2		1083
		ApJS8,439 T4-37# 2				1	11.95		0.42		0.25		2		1083
		ApJS8,439 T4-38# 1				1	13.17		0.73		0.04		2		1083
		ApJS8,439 T4-38# 2				1	16.24		1.08		0.87		2		1083
		ApJS8,439 T4-38# 3				1	14.10		0.67		-0.03		2		1083
		ApJS8,439 T4-38# 4				1	13.90		0.96		0.42		2		1083
		ApJS8,439 T4-39# 1				1	15.47		1.25		0.91		2		1083
		ApJS8,439 T4-39# 2				1	11.00		0.56		0.09		2		1083
		ApJS8,439 T4-39# 3				1	10.60		0.42		0.19		2		1083

HD	DM	Other Id	N Rem	α_{1950}	δ_{1950}	S	V	σ_V	B–V	σ_{B-V}	U–B	σ_{U-B}	n	Spectrum	References
		ApJS8,439 T4-40# 1				1	13.45		0.54		-0.06		2		1083
		ApJS8,439 T4-40# 2				1	14.70		1.75		1.90		2		1083
		ApJS8,439 T4-40# 3				1	14.60		1.79		2.43		2		1083
		ApJS8,439 T4-40# 4				1	14.94		0.62		0.27		2		1083
		ApJS8,439 T4-40# 5				1	14.68		1.00		0.30		2		1083
		ApJS8,439 T4-41# 1				1	14.60		1.03		0.46		2		1083
		ApJS8,439 T4-41# 2				1	11.76		0.75		0.39		2		1083
		ApJS8,439 T4-42# 1				1	12.84		0.62		-0.06		2		1083
		ApJS8,439 T4-42# 2				1	15.20		0.82		0.47		2		1083
		ApJS8,439 T4-42# 3				1	13.58		0.73		0.27		2		1083
		ApJS8,439 T4-43# 1				1	14.26		0.95		0.22		2		1083
		ApJS8,439 T4-43# 2				1	15.16		0.83		0.47		2		1083
		ApJS8,439 T4-44# 1				1	13.38		0.67		-0.06		2		1083
		ApJS8,439 T4-44# 2				1	13.86		0.77		0.06		2		1083
		ApJS8,439 T4-45# 1				1	13.34		0.79		0.24		2		1083
		ApJS8,439 T4-45# 2				1	12.73		0.84		0.31		2		1083
		ApJS8,439 T4-46# 1				1	12.88		0.83		0.25		2		1083
		ApJS8,439 T4-46# 2				1	14.69		1.02		0.32		2		1083
		ApJS8,439 T4-46# 3				1	11.16		1.18		1.09		2		1083
		ApJS8,439 T4-47# 1				1	14.36		0.83		0.57		2		1083
		ApJS8,439 T4-47# 2				1	13.15		1.11		0.81		2		1083
		ApJS8,439 T4-47# 3				1	14.15		1.20		0.48		2		1083
		ApJS8,439 T4-48# 1				1	14.08		0.75		0.10		2		1083
		ApJS8,439 T4-48# 2				1	11.57		0.62		-0.23		2		1083
		ApJS8,439 T4-49# 1				1	15.07		1.22		0.38		2		1083
		ApJS8,439 T4-49# 2				1	13.41		0.58		-0.08		2		1083
		ApJS8,439 T4-50# 1				1	12.31		0.66		0.09		2		1083
		ApJS8,439 T4-50# 2				1	13.76		1.09		0.63		2		1083
		ApJS8,439 T4-50# 3				1	15.14		0.86		0.23		2		1083
		ApJS8,439 T5- 1# 1				1	13.01		0.93		-0.48		2		1083
		ApJS8,439 T5- 1# 2				1	12.35		1.26		1.01		2		1083
		ApJS8,439 T5- 3# 1				1	11.82		0.79		0.26		2		1083
		ApJS8,439 T5- 4# 1				1	14.80		1.89		1.09		2		1083
		ApJS8,439 T5- 4# 2				1	12.72		0.90		0.56		2		1083
		ApJS8,439 T5- 4# 3				1	14.35		0.82		0.22		2		1083
		ApJS8,439 T5- 4# 4				1	13.87		1.09		0.77		2		1083
		ApJS8,439 T5- 5# 1				1	14.34		1.70		1.21		2		1083
		ApJS8,439 T5- 5# 2				1	12.19		0.47		-0.38		2		1083
		ApJS8,439 T5- 6# 1				1	13.28		0.84		0.42		2		1083
		ApJS8,439 T5- 6# 2				1	12.84		0.74		0.07		2		1083
		ApJS8,439 T5- 6# 3				1	13.30		0.76		0.26		2		1083
		ApJS8,439 T5- 7# 3				1	11.17		0.66		-0.34		2		1083
		ApJS8,439 T5- 7# 4				1	12.00		0.56		-0.33		2		1083
		ApJS8,439 T5- 8# 5				1	12.82		0.72		-0.51		2		1083
		ApJS8,439 T5- 8# 6				1	11.27		0.50		-0.42		2		1083
		ApJS8,439 T5- 9# 1				1	12.22		0.72		-0.21		2		1083
		ApJS8,439 T5- 9# 2				1	15.16		1.25		0.55		2		1083
		ApJS8,439 T5-10# 1				1	12.77		0.60		-0.27		2		1083
		ApJS8,439 T5-10# 2				1	13.67		0.76		0.05		2		1083
		ApJS8,439 T5-11# 1				1	14.67		0.86		0.24		2		1083
		ApJS8,439 T5-11# 2				1	15.03		0.70		-0.30		2		1083
		ApJS8,439 T5-11# 3				1	14.28		0.73		0.33		2		1083
		ApJS8,439 T5-12# 1				1	12.33		0.81		-0.15		2		1083
		ApJS8,439 T5-12# 2				1	11.54		0.55		-0.42		2		1083
		ApJS8,439 T5-13# 1				1	14.04		0.84		0.43		2		1083
		ApJS8,439 T5-13# 2				1	13.83		0.64		0.95		2		1083
		ApJS8,439 T5-13# 3				1	14.84		1.18		0.54		2		1083
		ApJS8,439 T5-14# 1				1	10.93		0.74		0.13		2		1083
		ApJS8,439 T5-14# 2				1	12.35		1.11		0.78		2		1083
		ApJS8,439 T5-15# 1				1	12.14		0.37		0.27		2		1083
		ApJS8,439 T5-15# 2				1	11.87		0.57		0.16		2		1083
		ApJS8,439 T5-16# 1				1	12.36		0.61		0.09		2		1083
		ApJS8,439 T5-16# 2				1	12.42		0.48		0.24		2		1083
		ApJS8,439 T5-17# 3				1	12.53		0.54		-0.25		2		1083
		ApJS8,439 T5-17# 4				1	12.23		0.43		0.17		2		1083
		ApJS8,439 T5-18# 1				1	11.58		0.72		-0.22		2		1083
		ApJS8,439 T5-18# 2				1	11.60		0.64		-0.46		2		1083
		ApJS8,439 T5-18# 3				1	13.19		0.77		0.51		2		1083
		ApJS8,439 T5-19# 1				1	11.60		1.81		1.55		2		1083
		ApJS8,439 T5-19# 2				1	11.57		0.67		0.14		2		1083
		ApJS8,439 T5-20# 1				1	12.82		0.60		-0.08		2		1083
		ApJS8,439 T5-20# 2				1	13.12		0.46		-0.19		2		1083
		ApJS8,439 T5-21# 1				1	10.80		0.37		0.20		2		1083
		ApJS8,439 T5-21# 2				1	12.12		0.61		0.12		2		1083
		ApJS8,439 T6- 1# 1				1	13.78		0.79		0.20		2		1083
		ApJS8,439 T6- 1# 2				1	14.11		0.73		0.57		2		1083
		ApJS8,439 T6- 2# 1				1	15.31		1.07		0.70		2		1083
		ApJS8,439 T6- 2# 2				1	14.42		0.35		0.13		2		1083
		ApJS8,439 T6- 3# 1				1	13.94		0.30		0.05		2		1083
		ApJS8,439 T6- 3# 2				1	14.17		0.65		0.30		2		1083

Table 2 1267

HD	DM	Other Id	N Rem	α_{1950}	δ_{1950}	S	V	σ_V	B–V	σ_{B-V}	U–B	σ_{U-B}	n	Spectrum	References
		ApJS8,439 T6- 4# 1				1	15.07		0.47		-0.27		2		1083
		ApJS8,439 T6- 4# 2				1	13.76		0.91		0.37		2		1083
		ApJS8,439 T6- 4# 3				1	13.21		0.89		0.41		2		1083
		ApJS8,439 T6- 4# 4				1	14.44		1.40		1.07		2		1083
		ApJS8,439 T6- 5# 1				1	15.15		0.66		-0.11		2		1083
		ApJS8,439 T6- 5# 2				1	15.21		0.69		0.29		2		1083
		ApJS8,439 T6- 5# 3				1	14.98		0.81		0.36		2		1083
		ApJS8,439 T6- 6# 1				1	12.08		0.20		-0.48		2		1083
		ApJS8,439 T6- 6# 2				1	12.53		0.29		-0.14		2		1083
		ApJS8,439 T6-11# 1				1	15.14		0.90		0.13		2		1083
		ApJS8,439 T6-11# 2				1	11.89		0.57		0.04		2		1083
		ApJS8,439 T6-14# 2				1	15.80		0.89		0.34		2		1083
		ApJS8,439 T6-16# 1				1	14.22		0.53		0.28		2		1083
		ApJS8,439 T6-16# 2				1	13.19		0.36		0.46		2		1083
		ApJS8,439 T6-16# 3				1	10.86		1.18		0.85		2		1083
		ApJS8,439 T6-17# 1				1	14.33		0.94		0.25		2		1083
		ApJS8,439 T6-17# 2				1	15.70		1.10		0.86		2		1083
		ApJS8,439 T6-17# 3				1	15.59		0.71		-0.08		2		1083
		ApJS13,379 # 201				1	11.21		0.53				1		648
		ApJS13,379 # 202				1	11.26		0.41				1		648
		ApJS13,379 # 203				1	12.58		0.76				1		648
		ApJS13,379 # 204				1	13.06		0.91				1		648
		ApJS13,379 # 205				1	14.02		0.88				1		648
		ApJS13,379 # 206				1	14.05		0.97				1		648
		ApJS13,379 # 207				1	14.24		0.88				1		648
		ApJS13,379 # 208				1	14.51		0.82				1		648
		ApJS13,379 # 209				1	15.16		1.19				1		648
		ApJS13,379 # 210				1	16.27		0.42				1		648
		ApJS13,379 # 211				1	16.84		0.82				1		648
		ApJS13,379 # 212				1	16.84		0.86				1		648
		ApJS13,379 # 213				1	17.14		1.23				1		648
		ApJS13,379 # 301				1	9.72		1.04				1		648
		ApJS13,379 # 302				1	10.42		1.00				1		648
		ApJS13,379 # 303				1	10.68		0.53				1		648
		ApJS13,379 # 304				1	11.20		0.65				1		648
		ApJS13,379 # 305				1	11.28		0.45				1		648
		ApJS13,379 # 306				1	12.92		0.90				1		648
		ApJS13,379 # 307				1	13.94		0.58				1		648
		ApJS13,379 # 308				1	14.73		0.74				1		648
		ApJS13,379 # 309				1	14.81		0.74				1		648
		ApJS13,379 # 310				1	15.24		1.46				1		648
		ApJS13,379 # 311				1	15.97		0.61				1		648
		ApJS13,379 # 312				1	16.11		1.16				1		648
		ApJS13,379 # 313				1	16.48		0.80				1		648
		ApJS13,379 # 314				1	16.77		0.69				1		648
		ApJS13,379 # 315				1	17.24		1.47				1		648
		ApJS13,379 # 316				1	17.67		1.54				1		648
		ApJS13,379 # 401				1	18.07		0.45				1		648
		ApJS13,379 # 402				1	17.09		0.57				1		648
		ApJS17,467 # 104				1	13.68		1.13				1		642
		ApJS17,467 # 105				1	14.06		1.76				1		642
		ApJS17,467 # 106				1	18.17		0.20				1		642
		ApJS17,467 # 107				1	18.24		-0.09				1		642
		ApJS17,467 # 108				1	14.47		0.07				1		642
		ApJS17,467 # 203				1	13.58		0.69				1		642
		ApJS17,467 # 204				1	14.44		0.76				1		642
		ApJS17,467 # 205				1	15.63		0.20				1		642
		ApJS17,467 # 206				1	17.71		-0.45				1		642
		ApJS18,429 # 492				1	10.34		0.94				1		1017
		ApJS39,135 # 33				1	13.64		1.11				1		915
		ApJS53,765 T2# 6				1	13.48		0.41		0.12		1		1480
		ApJS53,765 T2# 7				1	13.58		1.12				1		1480
		ApJS53,765 T2# 8				1	13.76		0.59		0.12		1		1480
		ApJS53,765 T2# 9				1	14.78		0.68				1		1480
		ApJS53,765 T2# 10				1	14.55		0.66		0.18		1		1480
		ApJS53,765 T3# 3				1	13.20		0.59		0.06		1		1480
		ApJS53,765 T3# 4				1	13.37		0.69		0.32		1		1480
		ApJS53,765 T3# 5				1	13.72		0.68		0.25		1		1480
		ApJS53,765 T3# 6				1	13.82		0.59		0.32		1		1480
		ApJS53,765 T3# 7				1	14.56		0.65		0.21		1		1480
		ApJS53,765 T3# 8				1	14.98		0.68		0.13		1		1480
		ApJS53,765 T3# 9				1	15.31		-0.13		0.21		1		1480
		ApJS53,765 T4# 7				1	13.82		2.06				1		1480
		ApJS53,765 T4# 8				1	13.95		0.66		0.16		1		1480
		ApJS53,765 T4# 9				1	14.76		0.78		0.29		1		1480
		ApJS53,765 T4# 10				1	15.58		0.92				1		1480
		ApJS53,765 T5# 5				1	14.10		0.58		0.42		1		1480
		ApJS53,765 T5# 6				1	15.22		0.61		0.45		1		1480
		ApJS53,765 T5# 7				1	15.48		0.94				1		1480
		ApJS53,765 T5# 8				1	13.98		1.16		0.88		1		1480

HD	DM	Other Id	N Rem	α_{1950}	δ_{1950}	S	V	σ_V	B–V	σ_{B-V}	U–B	σ_{U-B}	n	Spectrum	References
		ApJS53,765 T6# 8				1	14.79		0.72		0.69		1		1480
		ApJS57,743 T1# 5				1	13.91		0.52		0.06		1		1577
		ApJS57,743 T1# 6				1	13.53		0.41		0.08		1		1577
		ApJS57,743 T1# 7				1	15.51		0.66		0.34		1		1577
		ApJS57,743 T1# 8				1	16.65		0.56				1		1577
		ApJS57,743 T2# 2				1	13.67		0.46		0.40		1		1577
		ApJS57,743 T2# 5				1	12.86		0.18		0.08		1		1577
		ApJS57,743 T2# 6				1	14.32		0.60		0.16		1		1577
		ApJS57,743 T2# 9				1	13.20		0.50		0.44		1		1577
		ApJS57,743 T2# 11				1	13.48		1.37				1		1577
		ApJS57,743 T3# 8				1	13.89		0.73		0.13		1		1577
		ApJS57,743 T3# 9				1	15.04		0.99				1		1577
		ApJS57,743 T4# 4				1	14.01		0.30		0.22		1		1577
		ApJS57,743 T4# 8				1	13.53		0.28		0.12		1		1577
		ApJS57,743 T4# 9				1	15.78		0.85				1		1577
		ApJS57,743 T4# 10				1	14.91		0.03		-0.25		1		1577
		ApJS57,743 T5# 8				1	14.55		0.48		0.46		1		1577
		ApJS57,743 T5# 9				1	14.21		1.36		1.02		1		1577
		Aql sq 1 # 28				1	13.20		0.53		0.31		1		97
		Aql sq 1 # 32				1	14.18		0.94		0.55		1		97
		Aql sq 1 # 33				1	14.71		0.92		0.21		2		97
		Aql sq 1 # 34				1	14.75		1.12		0.64		3		97
		Aql sq 1 # 35				1	14.78		0.99		0.38		1		97
		Aql sq 1 # 36				1	14.86		1.24		0.55		3		97
		Aql sq 1 # 37				1	15.56		1.03		0.45		2		97
		Aql sq 1 # 38				1	15.68		1.03		0.66		2		97
		Aql sq 1 # 39				1	15.86		1.25				2		97
		Aql sq 1 # 40				1	16.03		1.11				1		97
		Aql sq 1 # 41				1	17.19		1.38		0.21		1		97
		Ara OB1 # 1				1	12.93		0.73		0.59		4		574
		Ara OB1 # 21				1	13.06		0.63		-0.19		7		574
		Ara OB1 # 26				1	13.95		0.77				4		574
		Ara OB1 # 28				1	15.20		1.10				1		574
		Ara OB1 # 29				1	14.54		1.35		0.99		4		574
		Ara OB1 # 38				1	14.43		1.00		0.41		3		574
		Ara OB1 # 39				1	13.86		0.70				1		574
		Ara OB1 # 68				1	13.53		0.70		0.08		3		574
		Ara OB1 # 91				1	15.08		2.21				2		574
		Ara OB1 # 125				1	11.98		0.43		-0.11		2		574
		Ara OB1 # 139				1	15.28		1.02		0.29		3		574
		Ara OB1 # 228				1	13.04		0.38		0.23		2		574
		Ara OB1 # 302				1	13.67		2.13		0.78		3		574
		Ara OB1 # 353				1	13.77		0.61		0.10		5		574
		Ara OB1 # 551				1	13.64		0.95		0.47		2		574
		Ara OB1 # 568				1	13.24		0.78		0.18		2		574
		Ara OB1 # 580				1	12.98		0.56		0.07		3		574
		Ara OB1 # 671				1	12.21		0.30		-0.13		2		574
		Ara OB1 # 672				1	11.98		0.33		-0.17		2		574
		BBB 2				1	12.52		0.58		0.00		1		655
		BBB 5				1	14.34		-0.16		-0.92		1		655
		BBB 6				1	14.39		-0.15		-0.84		1		655
		BBB 7				1	14.90		-0.03		-0.84		1		655
		BBB 8				1	15.00		0.00		-0.80		1		655
		BBB 9				1	15.10		0.03		-0.41		1		655
		BONN82 T6# 14				1	13.15		1.27		0.81		3		322
		BONN82 T6# 16				1	14.22		1.58		0.48		3		322
		BONN82 T6# 18				1	14.74		0.67		0.51		4	K0	322
		BONN82 T6# 19				1	15.48		0.87				4		322
		BONN82 T6# 20				1	15.49		0.70				3		322
		BONN82 T6# 21				1	15.61		0.83				3		322
		BONN82 T7# 17				1	14.02		2.13		2.39		4		322
		BONN82 T7# 18				1	14.94		0.55		0.45		4		322
		BONN82 T7# 19				1	15.18		1.56				2		322
		BONN82 T7# 20				1	15.43		0.86				3		322
		BONN82 T7# 21				1	16.35		1.24		1.05		2	K5	322
		BONN82 T8# 14				1	12.48		1.77		1.81		3		322
		BONN82 T8# 15				1	13.17		1.20		0.93		3	K3	322
		BONN82 T8# 16				1	13.66		1.29		1.02		3	K3	322
		BONN82 T8# 17				1	14.22		0.64		0.07		3	F8	322
		BONN82 T8# 18				1	14.24		0.76		0.17		3	G2	322
		BONN82 T8# 19				1	14.25		1.37		0.97		3	K5	322
		BONN82 T8# 20				1	14.46		0.66		0.50		3	K0	322
		BONN82 T8# 21				1	14.94		0.40		0.15		3	F0	322
		BONN82 T8# 22				1	15.02		0.45		0.23		3	A7	322
		BONN82 T8# 23				1	15.17		0.66		0.47		3	K0	322
		BONN82 T8# 24				1	15.64		0.73		0.37		3	G8	322
		BONN82 T11# 14				1	13.31		1.64		1.40		3	M0	322
		BONN82 T11# 16				1	13.81		0.99		0.51		3	G9	322
		BONN82 T11# 17				1	14.16		1.66		1.11		3	K7	322
		BONN82 T11# 18				1	14.30		0.91		0.81		3	K2	322

Table 2

1269

HD	DM	Other Id	N	Rem	α_{1950}	δ_{1950}	S	V	σ_V	B–V	σ_{B-V}	U–B	σ_{U-B}	n	Spectrum	References
		BONN82 T11# 20					1	14.35	1.12	0.64				3	G9	322
		BONN82 T11# 22					1	14.67	1.25	0.40				3		322
		BONN82 T11# 23					1	14.75	1.29	0.43				3		322
		BONN82 T11# 24					1	14.89	1.07	0.47				3		322
		BONN82 T12# 11					1	12.75	0.62	0.10				3	G0	322
		BONN82 T12# 12					1	13.11	0.40	0.10				3	F0	322
		BONN82 T12# 13					1	13.22	0.80	0.43				3	G9	322
		BONN82 T12# 14					1	13.42	1.89	1.12				4		322
		BONN82 T12# 15					1	13.54	0.77	0.18				3	G0	322
		BONN82 T12# 16					1	13.66	0.49	0.23				3	F0	322
		BONN82 T12# 17					1	14.27	0.83	0.35				3	G7	322
		BONN82 T12# 19					1	14.40	1.52	0.92				3		322
		BONN82 T12# 22					1	15.33	0.75	0.36				3		322
		BONN82 T13# 21					1	14.14	0.51	0.28				3	A9	322
		CMa R1 # 11a					1	11.65	0.18	-0.19				1	B8 V	740
		CMa R1 # 13a					1	12.57	0.63	0.13				1	F5	740
		CMa R1 # 16		V			1	11.07	0.53	-0.13				1	B5 Vn	740
		CMa R1 # 16a					1	13.38	1.01	0.57				1		740
		CMa R1 # 17a					1	11.94	0.51	0.08				1		740
		CMa R1 # 18					1	12.00	0.47	0.27				1	A1 IV-V	740
		CMa R1 # 19					1	12.13	0.24	0.08				1	A0 IV-V	740
		CMa R1 # 20a					1	11.08	0.60	0.08				1		740
		CMa R1 # 20b					1	14.03	0.80	0.34				1		740
		CMa R1 # 23					1	10.77	0.14	-0.17				1	B8 III-IV	740
		CMa R1 # 26					1	12.42	0.64	0.28				1	A1 IV	740
		CMa sq 0 # 26					1	9.99	1.04	0.83				2		388
		CMa sq 0 # 52					1	9.55	0.01	-0.27				2		388
		CMa sq 0 # 70					1	11.15	0.39	-0.36				4		388
		CMa sq 0 # 77					1	10.05	0.08	-0.51				3		388
		CMa sq 0 # 81					1	10.92	0.50	-0.17				3		388
		CMa sq 0 # 91					1	10.90	0.29	-0.33				1		388
		CMa sq 0 # 107					1	10.56	0.03	0.03				3		388
		CMa sq 0 # 119					1	10.53	0.22	-0.32				3		388
		CMa sq 0 # 124					1	10.18	0.08	0.01				2		388
		CMa sq 0 # 242					1	10.21	0.02	-0.09				2		388
		Cen sq 1 # 114					1	13.80	1.61	0.54				3		144
		Cen sq 1 # 115					1	13.96	0.66	0.13				3		144
		Cen sq 1 # 116					1	14.19	1.56	0.44				3		144
		Cen sq 1 # 117					1	14.21	0.73	0.13				3		144
		Cen sq 1 # 118					1	14.69	0.68	0.13				3		144
		Cen sq 1 # 119					1	15.04	0.97	-0.22				3		144
		Cen sq 1 # 120					1	15.14	2.17	0.76				3		144
		Cen sq 1 # 121					1	15.67	0.64	0.33				3		144
		Cen sq 1 # 214					1	13.05	1.52	1.29				3		144
		Cen sq 1 # 215					1	13.25	2.10	1.92				3		144
		Cen sq 1 # 219					1	14.38	0.73	0.20				3		144
		Cen sq 1 # 220					1	14.40	0.76	0.14				3		144
		Cen sq 1 # 221					1	14.69	2.26	1.66				3		144
		Cen sq 1 # 222					1	14.90	1.64	1.94				3		144
		Cen sq 1 # 223					1	15.06	1.66	1.45				3		144
		Cen sq 1 # 224					1	15.45	1.05	0.44				3		144
		Cen sq 1 # 311					1	12.90	1.38	1.27				3		144
		Cen sq 1 # 312					1	13.01	1.30	0.93				3		144
		Cen sq 1 # 315					1	13.87	0.41	0.20				3		144
		Cen sq 1 # 316					1	14.03	0.85	0.24				3		144
		Cen sq 1 # 318					1	14.61	0.84	0.38				3		144
		Cen sq 1 # 319					1	14.71	0.75	0.57				3		144
		Cen sq 1 # 320					1	14.78	0.68	0.42				3		144
		Cen sq 1 # 321					1	15.01	0.91	0.14				3		144
		Cen sq 1 # 322					1	15.53	0.79	0.25				3		144
		Cen sq 2 # 1					1	12.97	0.45	-0.40				2		496
		Cen sq 2 # 3					1	11.43	0.34	-0.55				2		496
		Cen sq 2 # 4					1	13.82	0.76	-0.10				2		496
		Cen sq 2 # 6					1	12.51	0.28	-0.36				2		496
		Cen sq 2 # 7					1	13.65	0.33	-0.46				2		496
		Cen sq 2 # 11					1	13.32	0.76	-0.27				2		496
		Cen sq 2 # 17					1	12.51	0.31	-0.45				2		496
		Cen sq 2 # 19					1	14.39	0.46	-0.40				2		496
		Cet sq 1 # 1					1	12.33	1.04	0.86				2		8098
		Cet sq 1 # 2					1	12.50	0.59	0.04				2		8098
		Cet sq 1 # 3					1	11.81	0.52	-0.07				2		8098
		Cet sq 1 # 4					1	12.28	0.68	0.17				2		8098
		Cet sq 1 # 5					1	14.17	1.02	0.85				2		8098
		Cet sq 1 # 6					1	13.03	1.24	1.15				2		8098
		Cet sq 1 # 7					1	14.28	0.72	0.21				2		8098
		Cet sq 1 # 8					1	10.69	0.99	0.69				2		8098
		Cet sq 1 # 9					1	15.69	0.61	-0.25				2		8098
		Cet sq 1 # 10					1	15.23	0.65	-0.05				2		8098
		Cet sq 1 # 11					1	16.05	0.65	-0.13				2		8098
		Cet sq 1 # 12					1	14.17	0.64	0.00				2		8098

HD	DM	Other Id	N	Rem	α_{1950}	δ_{1950}	S	V	σ_V	B–V	σ_{B-V}	U–B	σ_{U-B}	n	Spectrum	References
		Cet sq 1 # 13					1	15.64		0.67		0.11		2		8098
		Cet sq 1 # 14					1	12.87		0.73		0.15		2		8098
		Cet sq 1 # 15					1	13.63		0.67		0.11		2		8098
		Cet sq 1 # 16					1	11.13		0.87		0.45		2		8098
		Cet sq 1 # 17					1	16.01		1.11		0.60		1		8098
		Cet sq 1 # 18					1	11.65		0.95		0.77		1		8098
		Cet sq 1 # 105					1	9.86		0.47		0.00		7		8098
		Cet sq 1 # 108					1	13.50		1.50		1.21		7		8098
		Cet sq 1 # 114					1	14.11		1.52		1.28		6		8098
		Cet sq 1 # 118					1	15.73		0.56		-0.14		5		8098
		Cyg OB2 # 9					1	10.77		1.90		0.74		1		8007
		Cyg OB2 # 10					1	9.86		1.50		0.39		1		8007
		Cyg OB2 # 12					1	11.48		3.22		2.50		1		8007
		Cyg sq 4 # 116					1	13.88		0.73		0.88		1		324
		Cyg sq 4 # 118					1	12.59		0.56		0.09		1		324
		Cyg sq 4 # 226					1	13.81		0.60		0.63		1	A7 V	324
		Cyg sq 4 # 228					1	13.32		0.64		0.16		1	G0 V	324
		Cyg sq 4 # 233					1	13.62		0.62		0.08		1	F8 V	324
		Cyg sq 4 # 236					1	11.84		1.53		1.56		1		324
		Cyg sq 4 # 237					1	13.18		0.60		0.06		2	F8 V	324
		Cyg sq 4 # 238					1	12.73		0.68		0.35		2	A9 V	324
		Cyg sq 4 # 241					1	12.83		0.89		0.35		1	K0 V	324
		Cyg sq 4 # 243					1	11.96		0.86		0.33		1	G8 V	324
		Cyg sq 4 # 244					1	13.42		0.41		0.32		1	A8 V	324
		Cyg sq 4 # 246					1	11.92		1.31		1.26		1		324
		Cyg sq 4 # 248					1	12.90		0.60		0.13		1		324
		Cyg sq 4 # 250					1	11.91		0.47		0.07		2		324
		Cyg sq 4 # 326					1	13.34		0.83		0.68		2	A5 III	324
		Cyg sq 4 # 328					1	13.16		0.61		0.10		1		324
		Cyg sq 4 # 340					1	12.87		0.82		0.30		1	G5 V	324
		Draco # 24					1	17.07		1.30				2		818
		Draco # 146					1	18.63		0.89				2		818
		Draco # 148					1	18.14		1.33				2		818
		Draco # 169					1	20.02		0.89				2		818
		Draco # 170					1	21.12		0.65				2		818
		Draco # 192					1	20.51		0.01				2		818
		Draco # 194					1	18.28		1.06				2		818
		Draco # 248					1	16.84		0.12				2		818
		Draco # 249					1	17.27		1.44				2		818
		Draco # 263					1	19.48		0.73				2		818
		Draco # 361					1	17.50		1.28				2		818
		Draco # 401					1	18.56		0.96				2		818
		Draco # 419					1	17.07		0.67				2		818
		Draco # 433					1	19.79		0.64				2		818
		Draco # 472					1	19.20		0.57				2		818
		Draco # 473					1	17.57		1.23				2		818
		Draco # 506					1	17.98		1.09				2		818
		Draco # 511					1	19.11		0.86				2		818
		Draco # 517					1	18.55		-0.04				2		818
		Draco # 520					1	17.83		0.74				2		818
		Draco # 522					1	18.09		1.04				2		818
		Draco # 523					1	20.40		0.88				2		818
		Draco # 526					1	18.64		1.57				2		818
		Draco # 562					1	17.21		1.43				2		818
		Draco # 567					1	20.29		0.63				2		818
		Draco # 578					1	18.27		1.51				2		818
		Draco # 581					1	17.66		1.22				2		818
		Draco # 583					1	20.93		-0.06				2		818
		Draco # 585					1	16.65		0.77				2		818
		Draco # 586					1	18.10		0.61				2		818
		Draco # 601					1	15.28		0.78				2		818
		Draco # 603					1	16.12		0.86				2		818
		Draco # 604					1	17.34		0.46				2		818
		Draco # 612					1	18.18		1.16				2		818
		Eggen 13138		★			1	14.36		0.59		-0.02		5	sdG	3029
		Eggen 18672		★			1	11.05		1.15		0.70		3	F8I-II	3024
		Eggen 18673		★			1	11.06		0.61		0.32		3	A7	3024
		Eri sq 1 # 4					1	14.10		0.61		-0.02		2		8098
		Eri sq 1 # 6					1	15.16		0.50		-0.10		6		8098
		Eri sq 1 # 9					1	13.91		0.48		-0.04		2		8098
		Eri sq 1 # 10					1	16.49		0.86		0.40		3		8098
		Eri sq 1 # 11					1	14.17		0.72		0.17		2		8098
		Eri sq 1 # 12					1	14.79		0.87		0.47		2		8098
		Eri sq 1 # 13					1	16.01		1.35		1.97		2		8098
		Eri sq 1 # 14					1	14.13		0.77		0.27		2		8098
		Eri sq 1 # 15					1	16.52		0.64		-0.23		3		8098
		Eri sq 1 # 20					1	16.27		0.59		0.12		1		8098
		Eri sq 1 # 21					1	15.64		0.82		0.10		1		8098
		Fld b # 2					1	11.73		1.06		0.67		3		1712
		Fld d # 1					1	10.42		0.35		0.11		2		1712

Table 2   1271

HD	DM	Other Id	N Rem	α_{1950}	δ_{1950}	S	V	σ_V	B–V	σ_{B-V}	U–B	σ_{U-B}	n	Spectrum	References
		Fld d # 7				1	11.74		1.20		1.17		2		1712
		Fornax sq # 1				1	10.67		0.58				2		819
		Fornax sq # 2				1	11.35		1.14				2		819
		Fornax sq # 3				1	12.15		0.51				2		819
		Fornax sq # 4				1	12.76		0.57				2		819
		Fornax sq # 5				1	12.80		0.61				3		819
		Fornax sq # 6				1	13.58		0.92				2		819
		Fornax sq # 7				1	13.88		0.53				3		819
		Fornax sq # 8				1	14.44		0.97				2		819
		Fornax sq # 11				1	14.97		0.65				2		819
		Fornax sq # 12				1	15.65		0.51				2		819
		Fornax sq # 13				1	16.30		0.86				2		819
		Fornax sq # 14				1	16.30		0.65				2		819
		Fornax sq # 16				1	17.17		1.58				2		819
		Fornax sq # 17				1	17.23		1.34				2		819
		Fornax sq # 18				1	17.02		0.93				1		819
		Fornax sq # 19				1	18.24		1.68				1		819
		Fornax sq # 20				1	18.53		1.63				2		819
		Fornax sq # 21				1	18.85		1.51				1		819
		Fornax sq # 22				1	19.41		1.31				1		819
		Fornax sq # 23				1	19.63		1.07				1		819
		Fornax sq # 24				1	19.70		1.28				1		819
		Fornax sq # 202				1	18.33		1.19				1		819
		Fornax sq # 353				1	18.26		1.26				1		819
		Fornax sq # 673				1	18.93		1.55				1		819
		Fornax sq # 737				1	18.02		1.88				1		819
		Fornax sq # 771				1	18.24		1.55				1		819
		H0538 608				1	16.61		-0.01		-1.23		1		1727
		HM 1 - 22	AB			1	14.58		1.42		1.14		1		570
		HM 1 - 23				1	15.03		1.50		0.39		1		570
		HM 1 - 24				1	15.64		1.21		0.30		1		570
		HV 817				1	13.30		0.38				1		1511
		HV 819				1	13.47		0.49		0.29		1		689
		HV 822				1	13.95		0.44		0.28		1		689
		HV 823				2	13.33	.062	0.59	.014	0.38		2		689,1511
		HV 832				1	13.09		2.06				2	M0 Iab	5007
		HV 837				3	12.83	.036	0.59	.008	0.35	.010	3		689,1511,3075
		HV 838				1	13.35		1.68				2	M0eI	5007
		HV 840				1	13.03		0.47				1		1511
		HV 843				1	14.57		0.58				1		1511
		HV 847				2	13.72	.254	0.75	.241	0.63		2		689,1511
		HV 855				1	14.08		0.91				1		1511
		HV 863				1	12.79		0.39				1		1511
		HV 865				1	12.93		0.71				1		1511
		HV 875				1	12.71		0.65		0.42		1		689
		HV 877				2	13.08	.015	1.06	.045	0.69	.055	2		689,3075
		HV 878				1	12.98		0.42		0.41		1		689
		HV 881				2	12.58	.060	0.59	.030	0.41	.095	2		689,3075
		HV 882				1	12.95		0.53		0.46		1		689
		HV 886				1	12.75		0.55		0.40		1		689
		HV 889				1	13.46		0.72		0.49		1		689
		HV 899				1	12.91		0.59		0.49		1		689
		HV 900				2	12.38	.030	0.74	.015	0.49	.030	2		689,3075
		HV 909				1	12.30		0.59		0.40		1		689
		HV 953				2	11.85	.020	0.53	.015	0.43	.005	2		689,3075
		HV 1002				1	12.36		0.36		0.36		1		689
		HV 1326				1	14.75		0.63				1		1511
		HV 1335				1	14.29		0.39				1		1511
		HV 1338				1	14.78		0.32				1		1511
		HV 1342				2	13.98	.033	0.53	.047	0.12		2		689,1511
		HV 1365				1	14.66		0.49				1		1511
		HV 1384				1	15.63		0.58				1		933
		HV 1431				1	15.52		0.62				1		933
		HV 1445				1	15.41		0.22				1		933
		HV 1451				1	14.17		1.35				1		1511
		HV 1475				1	12.01		1.71				2	K0-K5 Ia	5007
		HV 1610				1	14.16		0.38				1		1511
		HV 1636				1	13.54		0.51				1		1511
		HV 1685				1	12.90		1.66		1.60		3		3074
		HV 1695				1	14.57		0.67				1		1511
		HV 1726				1	15.85		0.29				1		933
		HV 1735				1	15.31		0.57				1		933
		HV 1740				1	15.16		0.32				1		933
		HV 1744				1	14.09		0.43				1		1511
		HV 1773				1	16.37		0.49				2		933
		HV 1778				1	15.76		0.24				1		933
		HV 1787				1	13.99		0.61				1		1511
		HV 1789				1	15.67		0.26				1		933
		HV 1851				1	15.82		0.55				1		933
		HV 1873				1	14.51		0.49				1		1511

HD	DM	Other Id	N Rem	α_{1950}	δ_{1950}	S	V	σ_V	B–V	σ_{B-V}	U–B	σ_{U-B}	n	Spectrum	References
		HV 1877				3	12.99	.056	1.00	.096	0.56	.125	3		689,1511,3075
		HV 1884				1	14.15		0.68				1		1511
		HV 1925				1	13.60		0.47		0.29		1		689
		HV 1933				1	14.13		0.51				1		1511
		HV 1954				2	13.48	.074	0.40	.011	0.35		2		689,1511
		HV 1956				3	11.90	.079	1.26	.055	1.15	.030	3		689,1511,3075
		HV 2017				1	14.27		0.44				1		1511
		HV 2064				2	13.32	.004	0.57	.001	0.34		2		689,1511
		HV 2088				1	14.38		0.68				1		1511
		HV 2195				3	12.55	.092	0.55	.194	0.30	.035	3	G0	689,1511,3075
		HV 2202				1	14.06		0.40				1		1511
		HV 2205				1	13.62		0.56				1		1511
		HV 2209				2	13.51	.007	0.60	.062	0.44		2		689,1511
		HV 2225				1	14.47		0.63				1		1511
		HV 2227				1	14.46		0.60				1		1511
		HV 2230				1	14.35		0.64				1		1511
		HV 2231				1	13.17		0.74				1		1511
		HV 2233				1	13.43		0.37				1		1511
		HV 2257				2	12.50	.000	0.65	.025	0.42	.015	2		689,3075
		HV 2338				2	12.26	.050	0.57	.020	0.40	.020	2		689,3075
		HV 2873				1	15.09		0.51				1		933
		HV 2879				1	15.25		0.58				1		933
		HV 2883				1	12.02		0.91		0.58		1		689
		HV 2919				1	9.36		0.90				1		6011
		HV 3658				1	10.68		1.04				1		6011
		HV 5511				1	15.48		0.46				1		933
		HV 5541				1	15.45		0.32				1		933
		HV 5672				1	14.89		0.39				1		857
		HV 6011				1	15.04		0.49				1		857
		HV 6022				1	14.83		0.49				1		857
		HV 6089				1	15.03		0.42				1		857
		HV 6320				1	14.45		0.42				1		1511
		HV 10366				1	14.12		0.59				1		1511
		HV 10386				1	14.09		0.89				1		1511
		HV 11157				3	12.78	.044	1.01	.033	0.64	.010	3		689,1511,3075
		HV 11182				3	13.46	.061	0.93	.069	0.64	.110	3		689,1511,3075
		HV 11210				1	14.30		0.88				1		1511
		HV 11211				1	13.77		0.59				1		1511
		HV 12406				1	15.23		0.54				1		933
		HV 12408				1	15.44		0.37				1		933
		HV 12409				1	14.83		0.58				1		933
		HV 12417				1	15.74		0.53				1		933
		HV 12645				1	15.12		0.57				1		857
		HV 12648				1	15.83		0.43				1		857
		HV 12686				1	14.81		0.42				1		933
		HV 12694				1	15.61		0.41				1		933
		HV 12832				1	15.24		0.53				1		857
		Hya sq 1 # 26				1	15.13		0.96				1		1473
		IC 5146 sq1 1				1	15.60		0.90				1		5014
		IC 5146 sq1 2				1	15.75		1.49				1		5014
		IC 5146 sq1 3				1	16.96		0.77				1		5014
		IC 5146 sq1 4				1	17.73		0.99				1		5014
		IC 5146 sq1 7				1	18.80		1.07				1		5014
		IC 5146 sq1 8				1	16.14		0.90				1		5014
		IC 5146 sq1 9				1	17.17		1.10				1		5014
		IC 5146 sq1 11				1	15.64		0.75				1		5014
		IC 5146 sq1 12				1	18.22		1.28				1		5014
		IC 5146 sq1 14				1	18.16		1.10				1		5014
		IC 5146 sq1 20				1	17.12		0.88				1		5014
		IC 5146 sq1 21				1	17.04		1.15				1		5014
		K17 sq # 1				1	13.96		0.80		0.38		3		1646
		K17 sq # 2				1	15.68		0.47				3		1646
		K17 sq # 3				1	16.50		1.46				3		1646
		K17 sq # 4				1	16.85		1.18				3		1646
		K17 sq # 5				1	11.62		1.21				3		1646
		K17 sq # 6				1	11.86		0.38		0.02		3		1646
		K17 sq # 7				1	17.12		1.48				3		1646
		K17 sq # 8				1	14.79		0.58				3		1646
		K17 sq # 14				1	16.50		1.11				3		1646
		K17 sq # 24				1	13.13		1.80				3		1646
		K17 sq # 25				1	13.90		0.69				3		1646
		K17 sq # 91				1	13.91		0.68		0.15		3		1646
		K17 sq # 92				1	15.06		0.67		0.19		3		1646
		K17 sq # 93				1	11.18		0.59				3		1646
		K17 sq # 96				1	12.38		0.85				3		1646
		SK X-1 # E				2	13.65	.041	-0.05	.003	-0.87	.019	5		358,430
		LOD 4 # 4				1	13.59		0.74				2		126
		LOD 4 # 11				1	13.79		0.49				2		126
		LOD 4 # 12				1	13.62		0.63				2		126
		LOD 5 # 14				1	13.61		0.57		0.32		2		126

Table 2　　　　　　　　　　　　　　　　　　　　　　　　　　　　　1273

HD	DM	Other Id	N	Rem	α_{1950}	δ_{1950}	S	V	σ_V	B–V	σ_{B-V}	U–B	σ_{U-B}	n	Spectrum	References
		LOD 5 # 15					1	12.60		0.55		0.28		2		126
		LOD 18 # 9					1	12.82		0.41				2		127
		LOD 18 # 16					2	12.21	.020	0.39	.015	0.23		5		125,127
		LOD 18 # 19					2	12.65	.099	0.44	.040	0.27		5		125,127
		LOD 18 # 21					1	14.52		0.54		0.36		2		125
		LOD 19 # 18					1	12.63		0.92				2		127
		LOD 21 # 11					1	12.52		0.43				2		127
		LOD 22- # 20					1	14.14		0.25		0.16		1		125
		LOD 24 # 10					1	13.45		0.29		0.21		2		127
		LOD 24 # 19					1	12.90		0.26		0.28		2		127
		LOD 25 # 11					1	13.01		0.22		-0.50		2		122
		LOD 25 # 12					1	12.90		0.29		-0.27		2		122
		LOD 25+ # 15					1	12.18		0.48		0.03		2		122
		LOD 25+ # 16					1	13.42		0.37		-0.32		2		122
		LOD 26+ # 18					1	13.95		0.10				2		122
		LOD 26+ # 5					1	12.99		0.40				2		122
		LOD 26+ # 7					1	13.65		0.40				2		122
		LOD 26- # 15					1	9.97		1.44		1.39		2		122
		LOD 27 # 12					1	13.65		1.48				2		122
		LOD 35 # 13					1	13.91		0.65				1		96
		LOD 35 # 15					1	13.82		0.58				1		96
		LOD 35 # 20					1	13.64		2.25				1		96
		LOD 35+ # 12					1	12.30		0.43				1		96
		LOD 36 # 13					1	13.71		0.89				1		96
		LOD 36 # 20					1	14.92		0.76				1		96
		LOD 37+ # 2					1	12.44		0.36		0.18		1		96
		LOD 37+ # 9					1	13.04		0.40		0.04		1		96
		LOD 38 # 7					1	12.84		0.41				1		96
		LOD 38+ # 14					1	13.80		0.68		0.00		1		96
		LOD 38+ # 18					1	13.59		0.65		0.41		1		96
		LOD 38+ # 3					1	13.79		0.39		0.26		1		96
		LOD 38+ # 6					1	13.82		0.80		0.28		1		96
		LOD 38+ # 8					1	12.46		0.48		0.07		1		96
		LP 415 - 623					1	14.43		0.94				1		950
		LP 882 - 356					1	15.77		1.40				1		3061
		LP 883 - 312					1	13.68		0.76		0.22		2		3061
		Lambda Ori # 2					1	12.43		0.53		0.34		1		572
		Lambda Ori # 3					1	12.70		0.74		0.20		1		572
		Lambda Ori # 7					1	12.59		0.72		0.17		1		572
		Lambda Ori # 8					1	13.70		0.66		0.31		1		572
		Leo A sq # 1					1	13.07		0.58				5		1530
		Leo A sq # 2					1	13.19		0.74				3		1530
		Leo A sq # 3					1	14.12		0.84				4		1530
		Leo A sq # 4					1	14.85		0.86				2		1530
		Leo A sq # 5					1	14.86		0.83				2		1530
		Leo A sq # 6					1	15.50		0.67				1		1530
		Leo A sq # 7					1	15.73		1.11				3		1530
		Leo A sq # 8					1	15.92		0.68				2		1530
		Leo A sq # 9					1	16.37		0.59				2		1530
		Leo A sq # 11					1	16.44		0.53				1		1530
		Leo A sq # 12					1	16.50		0.75				2		1530
		Leo A sq # 13					1	16.67		0.74				2		1530
		Leo A sq # 14					1	16.92		0.60				3		1530
		Leo A sq # 16					1	17.15		0.98				1		1530
		Leo A sq # 17					1	17.92		0.92				1		1530
		Leo A sq # 18					1	18.37		1.28				1		1530
		Leo II sq # 9					1	15.80		1.37				2		988
		Leo II sq # 12					1	16.37		0.72				3		988
		Leo II sq # 19					1	17.12		0.64				2		988
		Leo II sq # 20					1	16.52		0.83				2		988
		Leo II sq # 21					1	17.31		0.62				2		988
		Leo II sq # 22					1	16.16		0.91				1		988
		Leo II sq # 24					1	16.72		1.19				2		988
		Leo II sq # 25					1	17.13		1.30				2		988
		Leo II sq # 53					1	16.20		1.20				1		988
		Leo II sq # 55					1	17.90		1.01				1		988
		Lod 1 # 20					1	12.09		2.51		1.83		2		837
		Lod 1 # 21					1	12.29		0.44		0.06		2	F4	837
		Lod 1 # 23					1	12.32		0.60		0.04		2	G0	837
		Lod 1 # 25					1	12.54		0.23		0.10		2	A5	837
		Lod 1 # 27					1	12.71		0.63		0.04		2	G0	837
		Lod 1 # 28					1	12.74		0.54		0.08		2	G0	837
		Lod 112 # 28					1	12.62		0.40		0.03		1	F3	549
		Lod 143 # 9					1	9.69		0.91		0.39		1		768
		Lod 172 # 12					1	12.16		0.14		0.12		2		807
		Lod 213 # 19					1	12.44		0.78		0.42		2	K0	837
		Lod 213 # 20					1	12.50		1.01		0.61		2	K1	837
		Lod 213 # 21					1	12.56		0.13		0.05		2	A2	837
		Lod 213 # 22					1	13.03		0.24		0.18		2	A5	837
		Lod 213 # 23					1	13.09		0.18		0.18		2	A5	837

HD	DM	Other Id	N Rem	α_{1950}	δ_{1950}	S	V	σ_V	B–V	σ_{B-V}	U–B	σ_{U-B}	n	Spectrum	References
		Lod 213 # 24				1	13.54	0.17	0.00				2	A0	837
		Lod 246 # 773				1	12.56	0.59	-0.06				1		3061
		Lod 247 # 160				1	12.10	0.89	0.62				3		3061
		Lod 309 # 2				1	11.11	0.17	-0.42				1		287
		Lod 372 # 21				1	12.50	0.28	0.12				1	A0	549
		Lod 372 # 22				1	12.52	0.26	0.17				1	B7	549
		Lod 372 # 23				1	12.84	0.24	0.08				1	B8	549
		Lod 372 # 24				1	12.92	0.34	0.07				1	B8	549
		Lod 372 # 25				1	12.94	0.43	0.08				1		549
		Lod 402 # 33				1	11.59	1.21	1.14				2		837
		Lod 402 # 35				1	11.79	0.55	0.05				2		837
		Lod 565 # 20				1	12.66	0.47	0.32				1	A1	549
		Lod 623 # 37				1	12.78	0.30	-0.10				1	B8	549
		Lod 624 # 42				1	11.89	0.35	0.03				1	B8	768
		Lod 807 # 26				1	11.31	0.70	0.22				1		549
		Lod 995 # 21				1	12.89	0.30	0.24				2	A5	807
		Lod 1002 # 21				1	12.67	0.46	0.34				1		549
		Lod 1002 # 22				1	12.83	0.32	0.23				1		549
		Lod 1171 # 35				1	14.17	0.77	0.09				1		549
		Lod 1296 # 23				1	11.34	1.80	1.58				2		837
		Lod 1296 # 30				1	12.22	1.55					2		837
		Lod 1409 # 102				1	13.07	1.56	1.69				2		1667
		Lod 1409 # 136				1	12.70	1.28	1.09				2		1667
		Lucke 1 - 1				1	13.84	-0.03	-0.86				1		360
		Lucke 1 - 2				1	11.55	0.70	0.14				1		360
		Lucke 1 - 3				1	13.30	-0.16	-0.93				1		360
		Lucke 1 - 4				1	11.80	0.57	0.08				1		360
		Lucke 1 - 5				1	14.37	-0.12	-0.21				1		360
		Lucke 4 - 1				1	11.44	2.02	1.81				1		360
		Lucke 4 - 3				1	14.00	0.13	0.22				1		360
		Lucke 4 - 4				1	12.99	-0.04	-0.70				1		360
		Lucke 4 - 6				1	13.27	-0.03	-0.71				1		360
		Lucke 4 - 7				1	14.62	0.75	0.55				1		360
		Lucke 4 - 8				1	14.74	-0.05	-0.76				1		360
		Lucke 4 - 9				1	15.44	-0.12	-0.97				1		360
		Lucke 4 - 10				1	15.61	0.21	0.36				1		360
		Lucke 4 - 11				1	16.32	-0.12	-0.79				1		360
		Lucke 4 - 12				1	14.84	-0.05	-0.86				1		360
		Lucke 5 - 1				1	13.28	0.51	0.03				1		360
		Lucke 5 - 3				1	15.60	-0.10	-0.74				1		360
		Lucke 5 - 4				1	15.73	0.03	-0.88				1		360
		Lucke 5 - 5				1	16.21	1.29	1.39				1		360
		Lucke 5 - 6				1	15.83	-0.05	-0.84				1		360
		Lucke 6 - 3				1	13.98	-0.20	-1.04				1		360
		Lucke 8 - 1				1	12.60	0.63	0.15				1		360
		Lucke 8 - 2				1	13.29	0.05	-0.41				1		360
		Lucke 8 - 4				1	15.74	-0.05	-0.71				1		360
		Lucke 8 - 5				1	15.28	0.26	0.33				1		360
		Lucke 9 - 3				1	13.27	0.71	0.14				1		360
		Lucke 9 - 4				1	12.30	0.62	0.08				1		360
		Lucke 9 - 5				1	14.76	-0.13	-0.92				1		360
		Lucke 9 - 6				1	15.31	-0.18	-0.94				1		360
		Lucke 9 - 7				1	15.41	-0.15	-1.00				1		360
		Lucke 9 - 8				1	15.99	-0.20	-0.83				1		360
		Lucke 12 - 1				1	12.75	0.71	0.13				1		360
		Lucke 12 - 4				1	13.08	0.12	0.18				1		360
		Lucke 12 - 5				1	14.18	0.32	0.43				1		360
		Lucke 15 - 3				1	12.70	1.89	2.18				1		360
		Lucke 15 - 4				1	11.75	1.10	0.86				1		360
		Lucke 15 - 6				1	11.33	0.29	0.11				1		360
		Lucke 15 - 7				1	11.60	0.37	0.09				1		360
		Lucke 15 - 8				1	11.83	0.57	0.11				1		360
		Lucke 15 - 9				1	12.26	0.55	0.03				1		360
		Lucke 15 - 10				1	13.77	0.80	0.38				1		360
		Lucke 15 - 11				1	13.85	-0.18	-0.88				1		360
		Lucke 15 - 12				1	12.18	0.55	-0.01				1		360
		Lucke 15 - 13				1	13.86	0.80	0.49				1		360
		Lucke 15 - 15				1	15.30	-0.19	-0.88				1		360
		Lucke 15 - 16				1	16.64	1.46	0.00				1		360
		Lucke 15 - 17				1	15.77	0.81	0.57				1		360
		Lucke 15 - 18				1	16.20	-0.19	-0.85				1		360
		Lucke 15 - 19				1	15.93	-0.13	-0.78				1		360
		Lucke 15 - 20				1	14.64	-0.16	-0.75				1		360
		Lucke 18 - 2				1	12.83	0.88	0.42				1		360
		Lucke 18 - 3				1	13.08	0.62	-0.01				1		360
		Lucke 18 - 5				1	12.56	1.85	2.33				1		360
		Lucke 19 - 1				1	10.18	1.00	0.68				1		360
		Lucke 19 - 2				1	12.10	1.21	1.02				1		360
		Lucke 19 - 4				1	12.64	0.67	0.14				1		360
		Lucke 19 - 5				1	14.68	-0.12	-0.96				1		360

Table 2 1275

HD	DM	Other Id	N Rem	α_{1950}	δ_{1950}	S	V	σ_V	B–V	σ_{B-V}	U–B	σ_{U-B}	n	Spectrum	References
		Lucke 26 - 1				1	11.01		0.91		0.71		1		360
		Lucke 26 - 2				1	13.19		1.89		1.74		1		360
		Lucke 26 - 3				1	12.31		0.50		0.02		1		360
		Lucke 26 - 5				1	9.16		1.33		1.36		1		360
		Lucke 26 - 6				1	13.82		0.75		0.23		1		360
		Lucke 26 - 7				1	13.82		-0.08		-0.82		1		360
		Lucke 26 - 8				1	15.24		0.60		0.35		1		360
		Lucke 26 - 9				1	16.17		-0.18		-0.88		1		360
		Lucke 26 - 10				1	16.08		-0.14		-0.86		1		360
		Lucke 26 - 11				1	15.70		-0.10		-0.73		1		360
		Lucke 26 - 12				1	15.99		-0.18		-0.90		1		360
		Lucke 28 - 2				1	11.95		0.49		0.07		1		360
		Lucke 28 - 3				1	12.13		0.94		0.59		1		360
		Lucke 28 - 4				1	12.99		0.70		0.08		1		360
		Lucke 28 - 5				1	13.99		-0.23		-1.03		1		360
		Lucke 29 - 2				1	14.02		1.78		1.77		1		360
		Lucke 29 - 5				1	12.73		1.83		1.94		1		360
		Lucke 31 - 1				1	11.08		0.93		0.73		1		360
		Lucke 31 - 2				1	11.87		0.71		0.24		1		360
		Lucke 31 - 4				1	13.41		0.52		0.04		1		360
		Lucke 32 - 1				1	12.21		1.00		0.60		1		360
		Lucke 32 - 2				1	13.54		1.39		0.22		1		360
		Lucke 36 - 2				1	13.10		1.77		1.74		1		360
		Lucke 36 - 4				1	12.82		0.99		0.55		1		360
		Lucke 36 - 7				1	11.78		0.11		0.07		1		360
		Lucke 43 - 1				1	13.21		1.15		1.12		1		360
		Lucke 43 - 2				1	13.64		0.57		-0.99		1		360
		Lucke 43 - 3				1	15.27		-0.21		-0.97		1		360
		Lucke 43 - 4				1	15.73		-0.07		-0.94		1		360
		Lucke 43 - 5				1	14.32		0.50		0.02		1		360
		Lucke 43 - 6				1	12.02		0.43		-0.22		1		360
		Lucke 43 - 7				1	16.11		-0.12		-0.92		1		360
		Lucke 43 - 8				1	16.04		0.68		0.09		1		360
		Lucke 43 - 9				1	15.10		-0.22		-0.93		1		360
		Lucke 44 - 1				1	12.79		0.63		0.12		1		360
		Lucke 44 - 2				1	12.90		1.00		0.66		1		360
		Lucke 44 - 4				1	14.08		-0.01		-0.01		1		360
		Lucke 44 - 5				1	13.82		0.98		0.61		1		360
		Lucke 44 - 7				1	14.26		0.95		0.58		1		360
		Lucke 44 - 8				1	12.68		0.59		0.05		1		360
		Lucke 44 - 9				1	13.98		1.01		0.55		1		360
		Lucke 44 - 10				1	12.84		0.58		-0.06		1		360
		Lucke 44 - 11				1	15.34		-0.19		-0.89		1		360
		Lucke 44 - 12				1	14.86		0.76		0.26		1		360
		Lucke 44 - 13				1	15.13		-0.02		-0.55		1		360
		Lucke 44 - 14				1	15.63		0.07		-0.56		1		360
		Lucke 44 - 15				1	15.51		0.59		0.07		1		360
		Lucke 44 - 16				1	16.02		1.97		2.46		1		360
		Lucke 45 - 2				1	14.98		-0.11		-0.90		1		360
		Lucke 45 - 3				1	13.77		0.53		0.07		1		360
		Lucke 45 - 5				1	13.06		0.90		0.39		1		360
		Lucke 45 - 7				1	11.89		0.57		-0.03		1		360
		Lucke 45 - 10				1	11.71		1.90		2.09		1		360
		Lucke 45 - 12				1	13.79		-0.07		-0.93		1		360
		Lucke 45 - 13				1	15.78		-0.19		-0.93		1		360
		Lucke 45 - 14				1	15.62		1.40		1.49		1		360
		Lucke 45 - 15				1	15.88		-0.16		-0.92		1		360
		Lucke 45 - 16				1	15.84		-0.20		-0.89		1		360
		Lucke 45 - 17				1	15.94		-0.11		-0.88		1		360
		Lucke 53 - 1				1	13.69		0.08		0.10		1		360
		Lucke 53 - 2				1	11.69		1.03		0.73		1		360
		Lucke 53 - 4				1	13.96		0.01		-0.08		1		360
		Lucke 53 - 5				1	14.07		-0.04		-0.49		1		360
		Lucke 54 - 1				1	11.74		2.15		1.80		1		360
		Lucke 54 - 2				1	12.67		0.48		-0.02		1		360
		Lucke 54 - 3				1	12.82		0.65		0.04		1		360
		Lucke 54 - 4				1	13.58		0.54		0.01		1		360
		Lucke 54 - 5				1	14.30		-0.21		1.08		1		360
		Lucke 54 - 6				1	14.35		-0.25		1.10		1		360
		Lucke 54 - 7				1	14.89		0.64		0.12		1		360
		Lucke 54 - 8				1	14.25		-0.32		1.12		1		360
		Lucke 61 - 1				1	11.65		0.74		0.31		1		360
		Lucke 61 - 2				1	13.29		1.51		1.66		1		360
		Lucke 61 - 3				1	13.28		0.81		0.41		1		360
		Lucke 61 - 4				1	14.62		0.71		0.19		1		360
		Lucke 64 - 2				2	12.30	.010	0.62	.005	0.06	.010	2		10,360
		Lucke 64 - 3				1	12.16		1.93		1.77		1		360
		Lucke 64 - 4				1	13.65		-0.25		-1.03		1		360
		Lucke 64 - 7				1	14.02		-0.18		-0.91		1		10
		Lucke 64 - 8				1	14.85		-0.19		-0.86		2		10

HD	DM	Other Id	N	Rem	α_{1950}	δ_{1950}	S	V	σ_V	B–V	σ_{B-V}	U–B	σ_{U-B}	n	Spectrum	References
		Lucke 64 - 10					1	14.95		0.73		0.05		1		10
		Lucke 64 - 11					1	14.15		-0.20		-1.06		1		10
		Lucke 64 - 12					1	14.57		-0.20		-0.83		2		10
		Lucke 64 - 14					1	15.18		0.04		-0.77		1		10
		Lucke 64 - 15					1	14.57		-0.14		-0.92		3		10
		Lucke 64 - 16					1	13.61		-0.20		-1.10		1		10
		Lucke 64 - 18					1	15.13		-0.10		-0.92		1		10
		Lucke 64 - 19					1	13.93		-0.17		-0.96		1		10
		Lucke 64 - 20					1	15.33		0.65		0.05		1		10
		Lucke 64 - 21					2	15.14	.040	-0.09	.015	-0.87	.025	2		10,10
		Lucke 64 - 23					1	15.19		0.04		-0.89		1		10
		Lucke 64 - 25					1	14.45		-0.12		-0.83		1		10
		Lucke 64 - 27					1	15.14		0.12		-0.70		1		10
		Lucke 64 - 28					1	13.39		0.91		0.61		1		10
		Lucke 64 - 31					1	15.54		-0.15		-0.82		1		10
		Lucke 64 - 32					1	13.39		1.85		1.89		1		10
		Lucke 64 - 33					1	14.02		-0.15		-0.92		4		10
		Lucke 64 - 34					1	13.25		1.88		1.48		4		10
		Lucke 64 - 35					1	15.80		-0.21		-0.93		1		10
		Lucke 64 - 37					1	14.97		-0.16		-0.86		1		10
		Lucke 64 - 38					1	12.92		-0.15		-0.94		2		10
		Lucke 64 - 44					1	14.17		-0.16		-0.91		1		10
		Lucke 64 - 45					1	14.34		-0.21		-1.00		1		10
		Lucke 64 - 46					1	15.37		-0.21		-0.85		1		10
		Lucke 64 - 47					1	14.07		-0.25		-0.97		1		10
		Lucke 64 - 48					1	14.97		-0.10		-0.83		1		10
		Lucke 64 - 49					1	14.34		-0.11		-0.93		1		10
		Lucke 64 - 50					1	14.04		-0.22		-0.96		1		10
		Lucke 64 - 52					1	15.08		0.04		-0.74		1		10
		Lucke 64 - 53					1	13.30		-0.14		-0.92		1		10
		Lucke 64 - 54					1	15.32		0.70		0.18		1		10
		Lucke 64 - 55					1	15.67		-0.09		-0.91		1		10
		Lucke 64 - 56					1	14.70		0.69		0.11		1		10
		Lucke 64 - 57					1	15.67		-0.23		-0.93		1		10
		Lucke 64 - 58					1	15.24		-0.22		-0.93		1		10
		Lucke 64 - 59					1	14.42		-0.15		-0.84		2		10
		Lucke 64 - 60					1	13.76		-0.12		-0.92		1		10
		Lucke 64 - 61					1	14.13		-0.23		-0.93		1		10
		Lucke 64 - 62					1	15.19		-0.05		-0.86		1		10
		Lucke 64 - 63					1	13.76		-0.01		-1.01		1		10
		Lucke 64 - 64					1	14.72		-0.17		-0.75		2		10
		Lucke 64 - 65					1	13.80		-0.06		-1.03		1		10
		Lucke 64 - 67					1	15.55		-0.05		-0.86		1		10
		Lucke 64 - 68					1	14.95		-0.21		-0.93		1		10
		Lucke 64 - 69					1	14.46		0.15		-0.81		1		10
		Lucke 64 - 70					1	12.81		-0.15		-0.98		1		10
		Lucke 64 - 71					1	14.50		-0.22		-0.94		2		10
		Lucke 64 - 72					1	14.70		-0.16		-0.85		1		10
		Lucke 64 - 73					1	13.95		-0.15		-0.97		1		10
		Lucke 64 - 74					1	15.57		-0.10		-0.82		1		10
		Lucke 64 - 76					1	15.41		-0.26		-0.95		1		10
		Lucke 64 - 78					1	16.07		0.02		-0.69		1		10
		Lucke 64 - 79					1	15.53		-0.17		-1.04		1		10
		Lucke 64 - 80					1	16.19		-0.24		-0.86		1		10
		Lucke 64 - 81					1	15.34		-0.16		-0.89		1		10
		Lucke 64 - 82					1	15.61		0.25		-0.75		1		10
		Lucke 64 - 83					1	16.04		-0.03		-0.95		1		10
		Lucke 64 - 84					1	15.41		-0.16		-0.74		1		10
		Lucke 64 - 85					1	15.84		-0.20		-0.89		1		10
		Lucke 64 - 86					1	15.58		-0.13		-0.88		1		10
		Lucke 64 - 87					1	16.18		-0.09		-0.70		1		10
		Lucke 64 - 88					1	16.08		-0.08		-0.81		1		10
		Lucke 64 - 89					1	15.75		-0.12		-0.83		2		10
		Lucke 64 - 90					1	15.90		-0.18		-0.86		1		10
		Lucke 64 - 91					1	15.28		-0.26		-0.92		1		10
		Lucke 65 - 1					1	11.86		0.98		0.74		1		360
		Lucke 65 - 5					1	13.47		0.42		0.30		1		360
		Lucke 69 - 4					1	13.80		0.86		0.64		1		360
		Lucke 69 - 5					1	12.08		0.70		0.11		1		360
		Lucke 72 - 2					1	13.84		0.64		0.05		1		360
		Lucke 72 - 4					1	14.02		0.73		0.17		1		360
		Lucke 72 - 5					1	15.07		0.35		0.39		1		360
		Lucke 79 - 1					1	11.85		0.65		0.11		1		360
		Lucke 79 - 2					1	12.97		0.62		0.14		1		360
		Lucke 79 - 5					1	14.46		-0.16		-0.83		1		360
		Lucke 79 - 6					1	15.79		1.10		0.63		1		360
		Lucke 79 - 7					1	15.12		-0.19		-0.93		1		360
		Lucke 84 - 1					1	13.59		-0.11		-0.91		1		360
		Lucke 84 - 2					1	12.81		0.97		0.02		1		360
		Lucke 84 - 3					1	13.37		-0.08		-0.94		1		360

Table 2   1277

HD	DM	Other Id	N	Rem	α_{1950}	δ_{1950}	S	V	σ_V	B–V	σ_{B-V}	U–B	σ_{U-B}	n	Spectrum	References
		Lucke 84 - 5					1	14.65		-0.18		-0.80		1		360
		Lucke 89 - 1					1	10.56		0.55		0.02		1		360
		Lucke 89 - 2					1	12.17		0.92		0.54		1		360
		Lucke 89 - 5					1	13.74		0.12		-0.76		1		360
		Lucke 93 - 3					1	13.28		0.07		-0.79		1		360
		Lucke 93 - 4					1	13.35		-0.18		-0.99		1		360
		Lucke 93 - 5					1	14.19		-0.13		-0.88		1		360
		Lucke 103 - 1					1	13.45		0.95		0.38		1		360
		Lucke 103 - 3					1	12.70		0.55		0.04		1		360
		Lucke 103 - 5					1	14.22		0.06		-0.84		1		360
		Lucke 103 - 6					1	13.61		0.00		-0.80		1		360
		Lucke 103 - 7					1	14.01		0.97		0.66		1		360
		Lucke 103 - 8					1	14.58		0.09		-0.74		1		360
		Lucke 103 - 9					1	14.68		0.16		0.17		1		360
		Lucke 103 - 10					1	13.32		0.86		0.38		1		360
		Lucke 103 - 11					1	15.47		-0.15		-0.86		1		360
		Lucke 103 - 12					1	15.41		0.62		0.05		1		360
		Lucke 103 - 13					1	15.79		-0.10		-0.86		1		360
		Lucke 103 - 14					1	15.63		0.37		0.26		1		360
		Lucke 103 - 15					1	15.73		1.62		2.01		1		360
		Lucke 103 - 16					1	15.62		-0.01		-0.08		1		360
		Lucke 104 - 1					1	10.31		1.59		1.94		1		360
		Lucke 104 - 2					1	13.38		0.17		-0.62		1		360
		Lucke 104 - 3					1	12.34		0.00		-0.82		1		360
		Lucke 104 - 4					1	12.14		0.40		-0.21		1		360
		Lucke 104 - 5					1	15.03		0.86		0.58		1		360
		Lucke 110 - 1					1	10.41		0.51		0.00		1		360
		Lucke 110 - 2					1	12.50		0.68		0.18		1		360
		Lucke 110 - 3					1	11.69		1.01		0.78		1		360
		Lucke 110 - 4					1	11.68		0.57		0.08		1		360
		Lucke 110 - 5					1	13.18		0.83		0.14		1		360
		Lucke 112 - 1					1	11.73		0.10		-0.09		1		360
		Lucke 112 - 3					1	13.43		-0.20		-0.79		1		360
		Lucke 112 - 4					1	14.32		-0.13		-0.90		1		360
		Lucke 112 - 7					1	12.21		0.70		0.21		1		360
		Lucke 115 - 1					1	11.30		0.40		0.00		1		360
		Lucke 115 - 2					1	12.36		1.29		1.18		1		360
		Lucke 115 - 3					1	13.62		0.62		0.05		1		360
		Lucke 115 - 4					1	14.55		3.98				1		360
		Lucke 115 - 5					1	12.53		1.13		1.11		1		360
		Lucke 115 - 6					1	13.03		0.53		-0.01		1		360
		Lucke 115 - 7					1	14.33		1.28		1.03		1		360
		Lucke 115 - 8					1	12.72		0.72		0.23		1		360
		Lucke 115 - 9					1	14.19		0.64		0.13		1		360
		Lucke 115 - 10					1	14.12		-0.17		-0.85		1		360
		Lucke 115 - 11					1	14.50		-0.15		-0.95		1		360
		Lucke 115 - 12					1	14.90		0.75		0.19		1		360
		Lucke 115 - 13					1	15.35		-0.19		-0.86		1		360
		Lucke 115 - 14					1	15.41		-0.18		-0.86		1		360
		Lucke 115 - 15					1	15.68		-0.10		-0.84		1		360
		Lucke 115 - 16					1	15.74		0.49		0.06		1		360
		Lucke 116 - 2					1	12.66		0.78		0.24		1		360
		Lucke 116 - 3					1	11.99		1.07		1.07		1		360
		Lucke 116 - 5					1	12.71		0.42		-1.01		1		360
		Lucke 116 - 6					1	13.86		0.93		0.66		1		360
		Lucke 116 - 7					1	14.55		-0.19		-0.94		1		360
		Lucke 116 - 8					1	15.44		-0.17		-0.74		1		360
		Lucke 116 - 9					1	15.76		0.72		0.20		1		360
		Lucke 116 - 10					1	16.15		1.03		0.51		1		360
		Lucke 116 - 11					1	16.10		1.36		1.71		1		360
		Lucke 116 - 12					1	16.04		-0.10		-0.41		1		360
		Lucke 117 - 2					1	12.20		0.92		0.64		1		360
		Lucke 117 - 3					1	13.54		-0.05		-0.95		1		360
		Lucke 117 - 4					1	12.81		1.86		1.62		1		360
		Lucke 117 - 5					1	10.98		0.50		0.03		1		360
		Lucke 120 - 3					1	13.01		0.56		0.12		1		360
		Lucke 120 - 4					1	13.57		0.56		0.02		1		360
		Lucke 120 - 5					1	13.56		1.76		0.80		1		360
		Lucke 123 - 3					1	11.07		0.96		0.72		1		360
		Lucke 123 - 4					1	12.34		0.73		0.35		1		360
		Lucke 123 - 5					1	12.67		0.77		0.26		1		360
		Lucke 123 - 6					1	13.14		1.10		0.95		1		360
		Lucke 123 - 8					1	14.19		1.19		1.30		1		360
		Lucke 123 - 9					1	14.29		0.10		-0.03		1		360
		Lucke 123 - 12					1	14.91		-0.12		-0.81		1		360
		Lucke 123 - 13					1	15.20		1.45		1.28		1		360
		Lucke 123 - 15					1	15.38		1.32		0.30		1		360
		LundsII139 # 302					1	10.88		0.24		-0.08		2		1101
		LundsII139 # 305					2	12.87	.034	0.68	.054	0.17		3		1101,1101
		LundsII139 # 315					1	15.38		1.39		0.58		1		1101

HD	DM	Other Id	N Rem	α_{1950}	δ_{1950}	S	V	σ_V	B–V	σ_{B-V}	U–B	σ_{U-B}	n	Spectrum	References
		LundsII139 # 316				1	15.92		1.33				1		1101
		LundsII139 # 317				1	13.92		0.58		0.14		1		1101
		LundsII139 # 320				1	15.82		1.15				2		1101
		LundsII139 # 321				1	15.57		1.34				2		1101
		LundsII139 # 322				1	15.72		1.79				2		1101
		LundsII139 # 402				1	11.63		0.56		0.05		2		1101
		MN83,95 # 102				1	13.33		0.48				1		591
		MN83,95 # 102				1	14.02		0.69				1		591
		MN83,95 # 103				1	14.06		-0.11				1		591
		MN83,95 # 103				1	15.10		1.03				1		591
		MN83,95 # 104				1	14.96		0.29				1		591
		MN83,95 # 104				1	15.44		-0.30				1		591
		MN83,95 # 105				1	15.15		-0.01				1		591
		MN83,95 # 105				1	16.10		0.50				1		591
		MN83,95 # 106				1	15.29		0.22				1		591
		MN83,95 # 106				1	16.26		1.44				1		591
		MN83,95 # 107				1	16.43		0.71				1		591
		MN83,95 # 202				1	13.26		0.55				1		591
		MN83,95 # 203				1	13.50		0.69				1		591
		MN83,95 # 204				1	13.50		0.46				1		591
		MN83,95 # 204				1	14.37		1.45				1		591
		MN83,95 # 205				1	15.42		0.57				1		591
		MN83,95 # 205				1	15.53		0.69				1		591
		MN83,95 # 206				1	15.63		1.39				1		591
		MN83,95 # 206				1	16.01		0.60				1		591
		MN83,95 # 207				1	16.77		1.42				1		591
		MN83,95 # 207				1	16.96		1.55				1		591
		MN83,95 # 301				1	11.84		0.77				1		591
		MN83,95 # 302				1	12.43		0.69				1		591
		MN83,95 # 303				1	13.62		0.70				1		591
		MN83,95 # 303				1	14.98		0.87				1		591
		MN83,95 # 304				1	14.62		0.09				1		591
		MN83,95 # 304				1	15.15		0.16				1		591
		MN83,95 # 305				1	15.01		-0.16				1		591
		MN83,95 # 305				1	15.82		0.93				1		591
		MN83,95 # 306				1	15.42		0.92				1		591
		MN83,95 # 306				1	16.74		0.66				1		591
		MN83,95 # 307				1	16.35		-0.25				1		591
		MN83,95 # 307				1	17.51		0.97				1		591
		MN83,95 # 308				1	17.63		2.00				1		591
		MN83,95 # 401				1	12.51		0.78				1		591
		MN83,95 # 402				1	13.53		0.69				1		591
		MN83,95 # 403				1	14.64		0.77				1		591
		MN83,95 # 404				1	15.46		0.46				1		591
		MN83,95 # 405				1	16.34		0.21				1		591
		MN83,95 # 406				1	16.80		1.52				1		591
		MN83,95 # 409				1	12.85		1.07				1		591
		MN83,95 # 410				1	13.69		0.76				1		591
		MN83,95 # 411				1	14.09		0.03				1		591
		MN83,95 # 412				1	14.17		0.93				1		591
		MN83,95 # 413				1	14.44		0.76				1		591
		MN83,95 # 415				1	14.91		0.13				1		591
		MN83,95 # 416				1	15.20		0.50				1		591
		MN83,95 # 417				1	15.47		0.52				1		591
		MN83,95 # 418				1	15.79		0.83				1		591
		MN83,95 # 419				1	15.88		0.33				1		591
		MN83,95 # 420				1	16.42		0.80				1		591
		MN83,95 # 421				1	16.54		0.88				1		591
		MN83,95 # 422				1	16.62		0.17				1		591
		MN83,95 # 423				1	17.08		1.63				1		591
		MN83,95 # 502				1	13.17		0.84				1		591
		MN83,95 # 503				1	13.72		0.57				1		591
		MN83,95 # 504				1	14.58		-0.20				1		591
		MN83,95 # 505				1	14.87		0.06				1		591
		MN83,95 # 506				1	15.07		-0.06				1		591
		MN83,95 # 507				1	15.59		0.53				1		591
		MN83,95 # 508				1	15.83		-0.13				1		591
		MN83,95 # 509				1	16.02		0.95				1		591
		MN83,95 # 602				1	13.94		0.54				1		591
		MN83,95 # 603				1	14.69		0.79				1		591
		MN83,95 # 604				1	15.01		0.95				1		591
		MN83,95 # 605				1	15.43		2.13				1		591
		MN83,95 # 606				1	16.62		1.36				1		591
		MN83,95 # 607				1	17.01		0.79				1		591
		MN83,95 # 704				1	13.57		1.27				1		591
		MN83,95 # 705				1	13.97		0.78				1		591
		MN83,95 # 706				1	15.05		0.14				1		591
		MN83,95 # 707				1	15.30		-0.20				1		591
		MN83,95 # 708				1	16.36		1.78				1		591
		MN83,95 # 709				1	16.37		1.19				1		591

Table 2 1279

HD	DM	Other Id	N	Rem	α_{1950}	δ_{1950}	S	V	σ_V	B–V	σ_{B-V}	U–B	σ_{U-B}	n	Spectrum	References
		MN83,95 # 807					1	15.41		0.63				1		591
		MN83,95 # 808					1	16.78		0.83				1		591
		MN83,95 # 809					1	16.80		0.85				1		591
		MN83,95 # 906					1	12.75		1.30				1		591
		MN83,95 # 907					1	14.05		0.97				1		591
		MN83,95 # 908					1	14.48		0.78				1		591
		MN83,95 # 909					1	15.57		0.79				1		591
		MN83,95 # 910					1	15.70		0.78				1		591
		MN83,95 # 911					1	16.88		1.40				1		591
		MN83,95 # 912					1	17.59		0.98				1		591
		MN120,163 # 19					1	14.26		0.97		0.83		1		375
		MN137,55 # 280					1	14.88		0.18				1		1513
		MN137,275 # 39					1	16.02		0.75				2		656
		MN137,275 # 57					1	14.53		0.91				3		656
		MN137,275 # 138					1	13.68		1.55		1.59		2		656
		MN137,275 # 139					1	13.38		0.41		0.18		2		656
		MN137,275 # 149					1	13.38		1.33		1.14		3		656
		MN174,213 T1# 9					1	13.21		1.00				3		465
		MN174,213 T1# 10					1	13.33		0.97				3		465
		MN174,213 T1# 11					1	13.38		0.75				3		465
		MN174,213 T1# 12					1	13.68		0.64				3		465
		MN174,213 T1# 13					1	13.85		0.72				3		465
		MN174,213 T1# 14					1	14.26		0.78				3		465
		MN174,213 T1# 15					1	14.37		0.60				4		465
		MN174,213 T1# 16					1	14.61		2.37				1		465
		MN174,213 T1# 17					1	14.96		0.70				3		465
		MN174,213 T1# 18					1	15.15		1.94				1		465
		MN174,213 T1# 19					1	15.22		0.88				3		465
		MN174,213 T1# 20					1	15.40		2.02				3		465
		MN174,213 T1# 21					1	15.60		1.74				2		465
		MN174,213 T1# 22					1	16.60		2.70				2		465
		MN174,213 T2# 6					1	12.84		0.50				2		465
		MN174,213 T2# 7					1	13.10		0.70				2		465
		MN174,213 T2# 8					1	13.47		0.70				1		465
		MN174,213 T2# 9					1	14.19		0.72				2		465
		MN174,213 T2# 10					1	14.37		0.94				2		465
		MN174,213 T2# 11					1	14.58		0.78				3		465
		MN174,213 T2# 12					1	14.64		0.86				2		465
		MN174,213 T2# 13					1	15.35		1.01				3		465
		MN174,213 T2# 14					1	15.73		1.24				2		465
		MN174,213 T2# 15					1	15.76		0.87				4		465
		MN174,213 T2# 16					1	16.09		2.06				2		465
		MN174,213 T2# 17					1	16.12		1.14				2		465
		MN174,213 T2# 18					1	16.32		2.33				1		465
		MN174,213 T3# 13					1	13.42		1.76				2		465
		MN174,213 T3# 14					1	13.80		1.36				2		465
		MN174,213 T3# 15					1	13.81		1.33				2		465
		MN174,213 T3# 16					1	13.98		1.11				3		465
		MN174,213 T3# 17					1	14.35		1.68				1		465
		MN174,213 T3# 18					1	14.39		1.72				3		465
		MN174,213 T3# 19					1	14.43		0.68				3		465
		MN174,213 T3# 20					1	14.49		1.21				2		465
		MN174,213 T3# 21					1	14.97		1.86				2		465
		MN174,213 T3# 22					1	15.04		1.57				4		465
		MN174,213 T3# 23					1	15.30		1.43				1		465
		MN174,213 T3# 24					1	15.49		0.88				1		465
		MN174,213 T3# 25					1	15.52		1.73				3		465
		MN174,213 T3# 26					1	15.57		1.18				2		465
		MN174,213 T3# 27					1	15.99		1.46				4		465
		MN177,59P # 1					2	12.28	.014	1.28	.014	1.01	.009	4		506,509
		MN177,59P # 2					1	13.17		1.20		0.70		1		509
		MN177,59P # 3					2	12.98	.005	0.60	.005	0.08	.020	2		506,509
		MN177,59P # 4					2	12.09	.000	0.47	.015	0.21	.010	3		506,509
		MN177,59P # 5					2	13.81	.030	0.78	.025	0.25	.015	2		506,509
		MN177,59P # 6					2	13.73	.015	0.78	.010	0.21	.060	2		506,509
		MN177,59P # 7					1	13.65		1.26		0.82		1		509
		MN177,59P # 8					1	13.70		1.22		0.78		1		509
		MN177,59P # 9					1	13.12		1.31		0.70		1		509
		MN177,59P # 10					2	14.92	.005	0.81	.010	0.29	.049	3		506,509
		MN177,59P # 13					1	13.09		1.40		1.33		1		509
		MN177,59P # 15					1	13.07		1.02		0.50		1		509
		MN177,59P # 16					2	12.09	.010	1.24	.005	0.97	.015	2		506,509
		MN177,59P # 22					1	13.09		1.36		1.08		1		509
		MN177,59P # 24					1	12.70		1.58		1.75		1		509
		MN177,59P # 25					1	13.43		1.30		0.97		1		509
		MN177,99 # 6					1	13.33		1.17		1.00		1		508
		MN177,99 # 16					1	13.88		1.23		1.64		1		508
		MN177,99 # 17					1	14.47		0.73		0.22		2		508
		MN177,99 # 43					1	14.39		0.74		0.10		1		508
		MN177,99 # 45					1	14.23		0.60		0.03		1		508

HD	DM	Other Id	N Rem	α_{1950}	δ_{1950}	S	V	σ_V	B–V	σ_{B-V}	U–B	σ_{U-B}	n	Spectrum	References
		MN182,283 # 101				1	14.83		1.25		1.13		2		705
		MN182,283 # 102				1	10.54		1.19		1.18		3		705
		MN182,283 # 103				1	12.77		1.37		1.53		2		705
		MN182,283 # 104				1	11.21		1.10		1.01		2		705
		MN182,283 # 105				1	11.03		1.20		1.34		2		705
		MN182,283 # 106				1	15.24		1.37		1.05		2		705
		MN182,283 # 107				1	15.87		1.45		1.32		3		705
		MN182,283 # 108				1	16.90		1.46		0.86		1		705
		MN182,283 # 109				1	14.29		1.27		1.17		1		705
		MN182,283 # 110				1	14.94		1.25		1.32		1		705
		MN182,283 # 111				1	15.08		1.31		1.27		2		705
		MN182,283 # 112				1	12.27		1.57		1.80		2		705
		MN182,283 # 113				1	13.02		1.24		1.26		2		705
		MN182,283 # 114				1	17.13		1.09		0.94		1		705
		MN182,283 # 115				1	17.26		0.92		0.65		1		705
		MN182,283 # 116				1	17.16		1.56		1.26		1		705
		MN182,283 # 117				1	16.74		1.34		1.28		2		705
		MN182,283 # 118				1	16.45		1.46		1.13		3		705
		MN182,283 # 119				1	16.52		1.29		1.18		1		705
		MN182,283 # 120				1	14.30		1.20		1.13		2		705
		MN182,283 # 121				1	15.48		1.53		1.01		2		705
		MN182,283 # 122				1	15.55		0.49		-0.33		2		705
		MN182,283 # 124				1	16.69		1.45		0.41		1		705
		MN182,283 # 125				1	17.08		1.48		1.39		2		705
		MN182,283 # 126				1	17.20		1.28		1.36		1		705
		MN182,283 # 127				1	16.88		1.57		0.79		2		705
		MN182,283 # 128				1	17.45		1.32		0.75		1		705
		MN182,283 # 129				1	17.02		1.28		1.18		1		705
		MN182,283 # 130				1	16.96		1.22		0.95		2		705
		MN182,283 # 131				1	15.65		1.42		1.30		2		705
		MN182,283 # 132				1	16.02		0.83		0.45		3		705
		MN182,283 # 133				1	15.54		1.30		1.08		2		705
		MN182,283 # 134				1	16.39		1.32		1.16		2		705
		MN182,283 # 135				1	16.99		1.15		0.91		2		705
		MN182,283 # 136				1	16.69		1.37		0.98		2		705
		MN182,283 # 137				1	16.73		1.21		1.13		2		705
		MN182,283 # 138				1	15.09		1.45		1.19		2		705
		MN182,283 # 139				1	15.65		1.39		1.27		2		705
		MN182,283 # 140				1	15.94		0.72		0.14		2		705
		MN182,283 # 141				1	16.25		1.39		0.83		1		705
		MN182,283 # 142				1	16.41		1.40		1.30		1		705
		MN182,283 # 143				1	12.39		1.17		1.34		1		705
		MN182,283 # 144				1	16.65		1.34		1.01		1		705
		MN182,283 # 145				1	16.85		1.22		1.14		1		705
		MN195,678 # 1				1	13.38		1.82		1.76		4		1474
		MN195,678 # 2				1	13.89		1.71		0.86		5		1474
		MN195,678 # 14				1	14.03		1.25		0.92		3		1474
		MN229,227 # 10				1	13.33		0.60		0.43		1		1649
		MN229,227 # 17				1	13.14		0.61		0.04		1		1649
		MN229,227 # 18				1	14.41		0.77		0.43		1		1649
		MNASSA33,148 # 1				1	16.10		0.55		-0.11		1		357
		MNASSA33,148 # 2				1	15.88		1.14		1.02		1		357
		MNASSA33,148 # 3				1	16.89		0.64		0.23		1		357
		MNASSA33,148 # 4	V			1	16.90		-0.06		-0.69		1		357
		MNASSA33,148 # 5				1	15.45		0.85		0.58		1		357
		MNASSA33,148 # 8				1	16.86		-0.06		-0.66		16		430
		Mark 106 # 1				1	13.19		0.81		0.39		1		829
		Mark 106 # 2				1	13.91		0.62		0.02		1		829
		Mark 106 # 3				1	14.52		0.70		0.03		1		829
		Mark 106 # 4				1	15.81		0.61		-0.01		1		829
		Mark 478 # 1				1	13.84		0.56		0.00		1		829
		Mark 478 # 2				1	14.64		0.58		0.03		1		829
		Mark 505 # 1				1	9.81		0.67		0.26		2		551
		Mark 505 # 2				1	12.54		0.51		0.05		2		551
		Mark 505 # 3				1	14.59		0.65		0.23		2		551
		Mark 505 # 4				1	14.35		1.08		0.81		2		551
		Mark 505 # 5				1	17.53		0.35		0.39		1		551
		Mark 505 # 6				1	14.34		0.90		0.60		2		551
		NAB 0205+02 # 1				1	11.26		0.46		-0.07		1		829
		NAB 0205+02 # 2				1	12.33		0.56		-0.01		1		829
		NAB 0205+02 # 3				1	13.24		0.95		0.77		1		829
		NAB 0205+02 # 4				1	14.93		0.76		0.24		1		829
		NGC 45 sq1 12				1	14.84		1.13		0.80		4		1554
		NGC 45 sq1 13				1	15.45		1.03				5		1554
		NGC 45 sq1 14				1	15.95		0.46				4		1554
		NGC 45 sq1 15				1	15.99		0.65				4		1554
		NGC 55 sq1 2				2	11.43	.000	0.96	.000	0.65	.010	12		831,1554
		NGC 55 sq1 11				1	14.47		0.86		0.58		4		831
		NGC 55 sq1 12				1	14.60		0.79		0.30		4		831
		NGC 55 sq1 13				1	14.79		0.56		-0.03		4		831

Table 2   1281

HD	DM	Other Id	N Rem	α_{1950}	δ_{1950}	S	V	σ_V	B–V	σ_{B-V}	U–B	σ_{U-B}	n	Spectrum	References
		NGC 55 sq1 14				1	14.90		0.77				4		831
		NGC 55 sq1 15				1	14.91		0.58		0.04		4		831
		NGC 55 sq1 16				1	15.11		0.81				3		831
		NGC 55 sq1 17				1	15.14		1.04				3		831
		NGC 55 sq1 18				1	15.47		0.80				4		831
		NGC 55 sq1 19				1	15.62		0.66		-0.11		5		831
		NGC 55 sq1 20				1	15.79		0.70				4		831
		NGC 55 sq1 21				1	15.86		0.86				3		831
		NGC 55 sq1 22				1	16.05		0.63				4		831
		NGC 55 sq1 23				1	16.20		0.61				6		831
		NGC 55 sq1 24				1	16.95		0.59				8		831
		NGC 104 sq2 1				1	17.77		0.59				1		887
		NGC 104 sq2 2				1	17.88		0.67				1		887
		NGC 104 sq2 3				1	18.83		1.16				1		887
		NGC 104 sq2 4				1	15.94		0.79				1		887
		NGC 104 sq2 5				1	18.09		0.61				1		887
		NGC 104 sq2 6				1	17.00		1.16				1		887
		NGC 104 sq2 7				1	17.13		0.75				1		887
		NGC 104 sq2 8				1	18.80		0.75				1		887
		NGC 104 sq2 9				1	16.28		0.77				1		887
		NGC 104 sq2 10				1	18.08		0.59				1		887
		NGC 104 sq2 11				1	18.29		0.56				1		887
		NGC 104 sq2 12				1	18.53		0.56				1		887
		NGC 104 sq2 13				1	17.86		0.58				1		887
		NGC 104 sq2 14				1	18.17		0.63				1		887
		NGC 104 sq2 15				1	16.17		0.76				1		887
		NGC 104 sq2 16				1	17.50		0.49				1		887
		NGC 104 sq2 17				1	17.48		0.65				1		887
		NGC 104 sq2 18				1	17.99		0.64				1		887
		NGC 104 sq2 19				1	16.86		0.77				1		887
		NGC 104 sq2 20				1	17.17		0.62				1		887
		NGC 104 sq2 21				1	16.87		0.42				1		887
		NGC 104 sq2 22				1	16.34		0.84				1		887
		NGC 121 sq1 17				1	16.03		0.59				4		1409
		NGC 121 sq1 18				1	16.40		0.62				3		1409
		NGC 121 sq1 19				1	16.97		0.53				4		1409
		NGC 121 sq1 20				1	17.29		1.55				5		1409
		NGC 121 sq1 21				1	17.64		0.51				10		1409
		NGC 121 sq1 22				1	17.66		1.33				2		1409
		NGC 121 sq1 23				1	17.88		1.00				6		1409
		NGC 121 sq1 24				1	18.31		1.12				11		1409
		NGC 121 sq1 26				1	19.54		0.59				3		1409
		NGC 121 sq1 53				1	19.80		0.57				4		1409
		NGC 121 sq1 54				1	20.22		0.90				7		1409
		NGC 152 sq1 5				1	16.39		1.96				3		959
		NGC 152 sq1 6				1	15.96		0.74				8		959
		NGC 152 sq1 8				1	17.37		1.39				8		959
		NGC 152 sq1 13				1	19.42		0.90				4		959
		NGC 152 sq1 24				1	16.86		-0.03				5		959
		NGC 176 sq1 1				1	12.12		0.50		-0.02		3		1646
		NGC 176 sq1 13				1	19.42		0.90				3		1646
		NGC 176 sq1 16				1	14.01		0.66				3		1646
		NGC 176 sq1 17				1	13.86		0.08				3		1646
		NGC 176 sq1 18				1	15.62		1.23				3		1646
		NGC 176 sq1 19				1	15.82		0.64				3		1646
		NGC 247 sq2 17				1	15.26		0.97				2		1554
		NGC 253 sq2 9				1	15.07		0.57		0.12		4		1554
		NGC 253 sq2 10				1	15.22		0.84		0.59		3		1554
		NGC 253 sq2 12				1	15.95		1.23				6		1554
		NGC 253 sq2 13				1	16.05		0.60				3		1554
		NGC 253 sq2 14				1	16.53		0.74				3		1554
		NGC 253 sq2 16				1	11.25		0.78		0.39		2		1554
		NGC 288 sq1 12				1	14.36		0.92				3		421
		NGC 288 sq1 14				1	15.42		0.63				5		421
		NGC 288 sq1 15				1	15.62		0.87				3		421
		NGC 288 sq1 16				1	15.98		0.81				3		421
		NGC 288 sq2 4				1	12.98		1.45		1.41		5		1638
		NGC 288 sq2 5				1	13.16		1.32				4		1638
		NGC 288 sq2 6				1	13.92		1.12				6		1638
		NGC 288 sq2 9				1	15.07		0.99				6		1638
		NGC 330 sq1 1				3	13.12	.016	0.71	.005	0.24	.015	8		1464,1571,1669
		NGC 330 sq1 2				1	13.92		0.98		-0.04		3		1464,1571
		NGC 330 sq1 3				1	14.18		0.01		-0.21		2		1571
		NGC 330 sq1 4				1	14.72		1.01				2		1464,1571
		NGC 330 sq1 5				1	14.92		1.36				2		1464,1571
		NGC 330 sq1 6				2	15.02	.139	-0.13	.035	-0.82	.004	6		1464,1571,1669
		NGC 330 sq1 7				1	16.22		-0.05		-0.48		1		1464,1571
		NGC 330 sq1 8				1	16.49		-0.19		-0.67		3		1464,1571
		NGC 330 sq1 10				1	17.01		1.96				1		1464,1571
		NGC 330 sq1 11				1	17.07		-0.03		-0.64		1		1464,1571

HD	DM	Other Id	N Rem	α_{1950}	δ_{1950}	S	V	σ_V	B–V	σ_{B-V}	U–B	σ_{U-B}	n	Spectrum	References
		NGC 330 sq1 12				1	18.41	18.12					1		1464
		NGC 330 sq2 1				1	10.70		0.61				1		1571
		NGC 330 sq2 2				1	11.01		0.74				1		1571
		NGC 330 sq2 3				1	11.09		0.87				1		1571
		NGC 330 sq2 4				1	11.50		0.05				1		1571
		NGC 330 sq2 5				1	11.52		0.35				1		1571
		NGC 330 sq2 6				1	11.54		1.09				1		1571
		NGC 330 sq2 7				1	11.69		0.58				1		1571
		NGC 330 sq2 8				1	11.74		0.48				1		1571
		NGC 330 sq2 10				1	12.64		0.41				1		1571
		NGC 330 sq2 11				1	12.96		0.03				1		1571
		NGC 330 sq2 12				1	13.05		0.03				1		1571
		NGC 330 sq2 13				1	13.32		-0.06				1		1571
		NGC 330 sq2 14				1	13.44		0.21				1		1571
		NGC 330 sq2 16				1	15.11		0.76				1		1571
		NGC 330 sq2 17				1	16.83		0.05				1		1571
		NGC 330 sq2 18				1	16.90		-0.12				1		1571
		NGC 330 sq3 4				1	13.42		0.02		-0.39		5		1669
		NGC 330 sq3 5				1	13.76		1.62		1.25		4		1669
		NGC 330 sq3 7				1	14.30		-0.02		-0.34		4		1669
		NGC 330 sq3 8				1	14.38		0.01		-0.96		5		1669
		NGC 330 sq3 9				1	14.45		1.57				5		1669
		NGC 339 sq1 1				1	15.74		0.66				2		1635
		NGC 339 sq1 2				1	17.27		1.30				2		1635
		NGC 339 sq1 3				1	16.73		0.73				3		1635
		NGC 339 sq1 4				1	16.29		0.53				3		1635
		NGC 339 sq1 5				1	16.15		1.85				2		1635
		NGC 362 sq1 10				3	14.81	.034	1.17	.024	1.06		8		489,1461,1677
		NGC 362 sq1 11				2	15.19	.014	0.95	.010	0.60	.000	10		489,1677
		NGC 362 sq1 12				2	15.31	.005	0.85	.000	0.29	.000	6		489,1677
		NGC 362 sq1 13				3	15.42	.026	0.55	.009	0.12	.009	8		489,1461,1677
		NGC 362 sq1 14				2	15.46	.000	0.88	.000	0.37	.000	6		489,1677
		NGC 362 sq1 15				3	15.71	.009	0.17	.017	0.22	.014	10		489,1461,1677
		NGC 362 sq1 16				2	15.96	.000	0.08	.000	-0.03	.000	6		489,1677
		NGC 362 sq1 17				2	15.97	.000	0.63	.034	0.06	.010	10		489,1677
		NGC 362 sq2 23				1	15.42		0.57		0.08		1		1461
		NGC 362 sq2 27				1	16.18		1.81				1		1461
		NGC 362 sq2 40				1	16.07		0.97		0.41		1		1461
		NGC 362 sq2 41				1	16.61		0.82				1		1461
		NGC 362 sq2 42				1	16.66		0.81		0.15		1		1461
		NGC 362 sq2 43				1	17.19		0.40				1		1461
		NGC 362 sq2 44				1	15.11		0.74		0.15		9		1677
		NGC 362 sq2 45				1	14.23		1.06				7		1677
		NGC 362 sq2 72				1	13.79		1.17				9		1677
		NGC 362 sq2 75				1	13.67		1.18		0.92		9		1677
		NGC 362 sq2 76				1	14.64		1.04		0.58		7		1677
		NGC 362 sq3 5				1	12.63		1.56		1.39		9		1677
		NGC 362 sq3 7				1	12.99		1.44		1.26		7		1677
		NGC 362 sq3 8				1	12.82		1.49		1.35		6		1677
		NGC 362 sq3 18				1	14.74		1.02		0.18		8		1677
		NGC 458 sq1 1				1	16.52		0.38		0.28		2		1635
		NGC 458 sq1 2				1	16.83		0.55		-0.07		2		1635
		NGC 458 sq1 3				1	16.44		0.64		0.21		2		1635
		NGC 458 sq1 4				1	16.14		0.60		0.22		2		1635
		NGC 458 sq1 5				1	17.32		-0.05		-0.13		2		1635
		NGC 458 sq1 6				1	17.15		1.13		0.96		2		1635
		NGC 458 sq1 7				1	17.97		-0.07		-0.43		2		1635
		NGC 458 sq1 8				1	15.05		0.78		0.44		3		1635
		NGC 458 sq1 9				1	16.94		0.94		0.52		3		1635
		NGC 609 sq1 2				1	11.88		0.70		0.12		6		155
		NGC 609 sq1 10				1	16.35		1.98				3		155
		NGC 609 sq1 11				1	15.76		1.08				3		155
		NGC 1068 sq1 5				1	10.79		0.60		0.14		2		327
		NGC 1261 sq2 1				1	13.81		1.42				9		1677
		NGC 1261 sq2 2				1	14.14		0.99		0.69		7		1677
		NGC 1261 sq2 4				1	14.56		1.21				8		1677
		NGC 1261 sq2 5				1	14.66		1.07		0.74		9		1677
		NGC 1261 sq1 6				1	14.15		0.70		0.16		13		1677
		NGC 1261 sq2 7				1	15.08		1.04		0.50		9		1677
		NGC 1261 sq1 10				1	14.97		0.99				14		1677
		NGC 1261 sq1 11				1	15.22		0.58		-0.06		5		789
		NGC 1261 sq1 12				1	15.25		0.71		0.20		5		789
		NGC 1261 sq1 13				1	15.26		0.55		-0.06		5		789
		NGC 1261 sq1 16				1	16.09		1.21				3		789
		NGC 1261 sq1 17				1	16.47		0.89				4		789
		NGC 1261 sq1 18				1	16.64		0.67		0.14		4		789
		NGC 1261 sq1 19				1	16.73		0.59				7		789
		NGC 1777 sq1 7				1	15.51		0.82				2		1574
		NGC 1777 sq1 8				1	15.72		0.58				2		1574
		NGC 1777 sq1 11				1	15.22		0.49				6		1574

Table 2 1283

HD	DM	Other Id	N	Rem	α_{1950}	δ_{1950}	S	V	σ_V	B–V	σ_{B-V}	U–B	σ_{U-B}	n	Spectrum	References
		NGC 1783 sq1 1					1	16.65		0.89				2		1635
		NGC 1783 sq1 2					1	16.13		-0.20		-0.90		2		1635
		NGC 1783 sq1 3					1	16.91		1.31				2		1635
		NGC 1783 sq1 4					1	15.08		0.78				2		1635
		NGC 1783 sq1 5					1	16.41		0.98				2		1635
		NGC 1841 sq1 4					1	15.40		0.79				5		1669
		NGC 1841 sq1 5					1	15.63		1.23				4		1669
		NGC 1851 sq2 1					1	13.43		1.60		1.58		4		1638
		NGC 1851 sq1 6					1	15.53		1.10				2		1638
		NGC 1851 sq1 18					1	15.34		0.63		-0.07		2		473
		NGC 1851 sq1 21					1	15.50		0.62		0.05		2		473
		NGC 1851 sq1 22					1	15.75		0.59		-0.16		2		473
		NGC 1851 sq1 23					1	15.83		0.80		0.34		2		473
		NGC 1851 sq1 24					1	16.18		0.66				2		473
		NGC 1856 sq1 2					1	13.23		-0.02				3		1521
		NGC 1856 sq1 3					1	14.19		0.08				4		1521
		NGC 1856 sq1 4					1	14.18		0.07				4		1521
		NGC 1856 sq1 6					1	14.46		-0.01				3		1521
		NGC 1856 sq1 7					1	14.55		0.42				3		1521
		NGC 1856 sq1 9					1	16.14		0.12				2		1521
		NGC 1856 sq1 10					1	16.83		0.03				2		1521
		NGC 1856 sq1 11					1	16.67		-0.03				5		1521
		NGC 1856 sq1 13					1	15.82		0.47				3		1521
		NGC 1856 sq1 16					1	15.36		0.17				6		1521
		NGC 1856 sq1 18					1	14.63		1.52				2		1521
		NGC 1856 sq1 19					1	16.12		1.26				2		1521
		NGC 1856 sq1 20					1	15.35		1.38				2		1521
		NGC 1868 sq1 6					1	13.46		0.37				2		827
		NGC 1868 sq1 7					1	13.56		0.92				1		827
		NGC 1868 sq1 8					1	13.47		1.09				3		827
		NGC 1868 sq1 9					1	14.89		0.73				3		827
		NGC 1868 sq1 10					1	15.22		0.84				3		827
		NGC 1868 sq1 11					1	15.73		0.60				1		827
		NGC 1868 sq1 12					1	16.83		1.52				2		827
		NGC 1868 sq1 13					1	16.58		0.69				1		827
		NGC 1868 sq1 14					1	17.36		1.67				1		827
		NGC 1868 sq1 17					1	16.08		0.56				1		827
		NGC 1868 sq1 23					1	15.25		0.76				2		827
		NGC 1868 sq1 24					1	14.07		0.99				2		827
		NGC 1904 sq1 16					1	15.20		0.73		0.16		5		488
		NGC 1904 sq1 17					1	15.26		0.62		-0.02		3		488
		NGC 1904 sq1 18					1	15.52		0.82		0.35		5		488
		NGC 1904 sq1 19					1	15.87		0.52		-0.06		3		488
		NGC 1904 sq1 20					1	16.20		0.53		-0.05		3		488
		NGC 1904 sq1 21					1	16.42		0.54		-0.02		3		488
		NGC 1904 sq2 8					1	12.21		0.30		0.24		3		581
		NGC 1904 sq2 11					1	12.92		1.37		1.60		1		581
		NGC 1904 sq2 14					1	14.43		1.01		0.56		1		581
		NGC 1904 sq2 18					1	15.66		0.73		0.37		1		581
		NGC 1904 sq2 19					1	15.68		0.92		0.40		1		581
		NGC 1904 sq2 20					1	16.08		0.80		0.56		2		581
		NGC 1904 sq2 21					1	17.14		-0.03		-0.23		1		581
		NGC 1904 sq2 22					1	17.31		0.67				1		581
		NGC 1904 sq3 6					1	15.06		0.38		-0.03		7		1638
		NGC 1904 sq3 7					1	15.10		0.83				7		1638
		NGC 1978 sq1 5					1	15.87		-0.09				4		1525
		NGC 1978 sq1 11					1	14.18		1.81				5		1525
		NGC 1978 sq1 12					1	16.18		2.08				4		1525
		NGC 1978 sq1 13					1	15.28		0.76				2		1525
		NGC 1978 sq1 14					1	15.66		1.09				3		1525
		NGC 1978 sq1 17					1	16.13		0.40				2		1525
		NGC 1978 sq1 18					1	17.17		-0.21				2		1525
		NGC 1978 sq1 26					1	15.71		-0.05				2		1525
		NGC 1978 sq1 36					1	15.52		-0.08				1		1525
		NGC 1978 sq2 1					1	16.47		-0.09				4		1525
		NGC 1978 sq2 2					1	16.17		0.95				3		1525
		NGC 1978 sq2 3					1	16.41		1.09				2		1525
		NGC 1978 sq2 4					1	16.32		1.48				2		1525
		NGC 1978 sq2 5					1	15.58		-0.09				1		1525
		NGC 1978 sq2 6					1	16.49		-0.10				1		1525
		NGC 1978 sq2 11					1	15.81		-0.14				1		1525
		NGC 2004 sq1 1					1	16.24		1.28				2		1635
		NGC 2004 sq1 2					1	16.61		-0.09		-0.71		2		1635
		NGC 2004 sq1 3					1	15.21		1.34				2		1635
		NGC 2004 sq1 4					1	15.51		-0.18		-0.80		2		1635
		NGC 2004 sq1 5					1	15.79		-0.17		-0.79		2		1635
		NGC 2004 sq1 6					1	16.31		-0.16		-0.74		2		1635
		NGC 2004 sq1 7					1	16.43		-0.14		-0.79		2		1635
		NGC 2004 sq1 8					1	15.63		0.61		0.34		1		1635
		NGC 2058 sq1 1					1	13.77		0.41		0.34		6		970

HD	DM	Other Id	N Rem	α_{1950}	δ_{1950}	S	V	σ_V	B–V	σ_{B-V}	U–B	σ_{U-B}	n	Spectrum	References
		NGC 2058 sq1 2				1	13.43		0.93		0.58		4		970
		NGC 2058 sq1 3				1	13.83		0.50		-0.07		3		970
		NGC 2058 sq1 7				1	14.51		0.23		0.30		6		970
		NGC 2058 sq1 9				1	14.33		1.78		1.59		6		970
		NGC 2058 sq1 10				1	14.16		0.63		0.16		5		970
		NGC 2058 sq1 12				1	13.93		1.57				7		970
		NGC 2058 sq1 13				1	13.25		0.69		0.12		4		970
		NGC 2058 sq1 14				1	14.13		0.46		0.35		4		970
		NGC 2058 sq1 16				1	14.81		0.58		0.33		4		970
		NGC 2058 sq1 17				1	14.56		0.70		0.27		4		970
		NGC 2058 sq1 18				1	15.47		-0.09		-0.52		5		970
		NGC 2058 sq1 19				1	13.67		0.63		0.13		3		970
		NGC 2058 sq1 20				1	14.17		0.58		0.38		4		970
		NGC 2058 sq1 21				1	15.13		0.99				4		970
		NGC 2058 sq1 22				1	13.19		-0.08		-0.68		4		970
		NGC 2058 sq1 24				1	15.25		0.54				3		970
		NGC 2058 sq1 25				1	15.36		0.74				2		970
		NGC 2058 sq1 26				1	16.11		-0.16				2		970
		NGC 2058 sq1 36				1	17.98		-0.06				1		970
		NGC 2058 sq1 37				1	16.30		1.51				1		970
		NGC 2058 sq1 40				1	16.64		1.54				1		970
		NGC 2058 sq1 45				1	17.74		0.77				1		970
		NGC 2164 sq1 2				1	15.02		0.57		0.13		5		347
		NGC 2164 sq1 3				1	15.57		0.81		0.40		5		347
		NGC 2164 sq1 4				1	16.49		-0.11		0.36		8		347
		NGC 2164 sq1 6				1	16.21		1.86				4		347
		NGC 2164 sq1 7				1	16.45		0.81				3		347
		NGC 2164 sq1 8				1	16.91		0.40				4		347
		NGC 2164 sq1 11				1	18.11		-0.04				5		347
		NGC 2164 sq1 12				1	18.06		1.33				5		347
		NGC 2164 sq1 16				1	19.00		-0.14				4		347
		NGC 2164 sq2 1				1	15.15		1.39				2		1635
		NGC 2164 sq2 2				1	15.54		0.80		0.37		2		1635
		NGC 2164 sq2 3				1	17.32		0.74				2		1635
		NGC 2164 sq2 4				1	16.53		-0.12		-0.39		2		1635
		NGC 2164 sq2 5				1	16.36		-0.13		-0.39		2		1635
		NGC 2164 sq2 6				1	16.98		-0.07		-0.36		2		1635
		NGC 2164 sq2 7				1	17.34		1.11				2		1635
		NGC 2203 sq1 2				1	15.78		0.66				3		1646
		NGC 2203 sq1 3				1	15.19		0.82				3		1646
		NGC 2203 sq1 7				1	17.15		0.56				3		1646
		NGC 2203 sq1 8				1	16.83		1.22				3		1646
		NGC 2203 sq1 9				1	16.06		0.64				3		1646
		NGC 2243 sq1 19				1	13.68		0.60		0.18		4		543
		NGC 2243 sq1 23				1	13.80		0.90		0.51		1		543
		NGC 2243 sq1 26				1	14.52		0.83		0.03		2		543
		NGC 2243 sq1 27				1	15.26		0.56		-0.07		3		543
		NGC 2243 sq1 31				1	17.39		0.52				2		543
		NGC 2257 sq2 7				1	14.50		0.22				5		1669
		NGC 2257 sq2 9				1	15.29		0.91				3		1669
		NGC 2298 sq1 8				2	15.19	.003	0.44	.009	0.28		6		365,1665
		NGC 2298 sq1 9				2	15.21	.024	0.74	.036	0.22		6		365,1665
		NGC 2298 sq1 10				2	15.28	.000	0.80	.000	0.29		6		365,1665
		NGC 2298 sq1 11				2	15.92	.052	0.79	.030	0.41		6		365,1665
		NGC 2298 sq1 12				2	16.01	.066	0.72	.026	0.37		6		365,1665
		NGC 2298 sq1 13				2	16.15	.073	0.77	.129	0.01		6		365,1665
		NGC 2298 sq1 14				1	16.22		0.83		0.43		3		365
		NGC 2298 sq1 19				1	15.06		1.06				3		1665
		NGC 2298 sq1 25				1	16.13		0.86				3		1665
		NGC 2298 sq1 26				1	16.65		1.53				3		1665
		NGC 2419 sq1 11				1	15.19		0.50		-0.06		1		439
		NGC 2419 sq1 12				1	15.24		0.78		0.43		4		439
		NGC 2419 sq1 13				1	15.37		0.89		0.67		3		439
		NGC 2419 sq1 14				1	15.55		0.59		-0.05		3		439
		NGC 2419 sq1 16				1	15.55		0.55		0.04		3		439
		NGC 2419 sq1 17				1	15.74		0.71		0.23		1		439
		NGC 2419 sq1 18				1	15.81		0.81		0.31		1		439
		NGC 2419 sq1 19				1	15.82		0.64		-0.08		4		439
		NGC 2419 sq1 20				1	15.87		0.78		0.23		2		439
		NGC 2419 sq1 21				1	16.00		0.78		0.19		2		439
		NGC 2419 sq1 22				1	16.03		1.03		0.76		1		439
		NGC 2419 sq1 24				1	16.69		0.99		0.45		1		439
		NGC 2419 sq1 26				1	17.14		1.35				1		439
		NGC 2808 sq1 23				2	12.92	.005	1.19	.005	1.09		7		438,722
		NGC 2808 sq1 28				1	13.57		1.66		1.54		2		722
		NGC 2808 sq1 33				1	14.09		0.80		0.22		2		722
		NGC 2808 sq1 34				2	14.13	.000	1.27	.000			4		89,438
		NGC 2808 sq1 35				1	14.41		0.86		0.42		2		722
		NGC 2808 sq1 36				1	14.46		0.51		0.13		2		722
		NGC 2808 sq1 37				1	14.48		1.15				4		438

Table 2   1285

HD	DM	Other Id	N Rem	α_{1950}	δ_{1950}	S	V	σ_V	B–V	σ_{B-V}	U–B	σ_{U-B}	n	Spectrum	References
		NGC 2808 sq1 38				1	14.58		0.73		0.16		1		722
		NGC 2808 sq1 39				1	14.72		0.70				4		438
		NGC 2808 sq1 40				1	14.72		0.72		0.25		2		722
		NGC 2808 sq1 41				1	15.11		0.97				2		438
		NGC 2808 sq1 42				1	15.25		1.25				2		438
		NGC 2808 sq1 43				1	15.34		0.82				2		438
		NGC 2808 sq1 44				1	15.82		0.66		0.09		2		722
		NGC 2808 sq1 45				1	16.00		1.02				1		438
		NGC 2808 sq1 46				1	16.18		0.69		0.17		1		722
		NGC 2808 sq1 47				2	16.90	.019	1.18	.023	1.25		4		438,722
		NGC 2808 sq1 48				1	17.29		0.87		0.25		1		722
		NGC 2808 sq1 49				1	17.58		0.82		0.38		1		722
		NGC 2808 sq1 50				1	18.04		1.03		0.30		1		722
		NGC 3201 sq1 10				1	14.08		1.03		0.50		6		484
		NGC 3201 sq1 11				1	14.45		1.25		1.08		6		484
		NGC 3201 sq1 12				1	14.50		1.04		0.83		6		484
		NGC 3201 sq1 13				1	14.89		1.01		0.33		5		484
		NGC 3201 sq1 14				1	15.68		1.01		0.68		3		484
		NGC 3201 sq1 16				1	15.98		0.88		0.22		5		484
		NGC 3201 sq1 17				1	16.30		0.68		0.10		4		484
		NGC 3201 sq1 18				1	16.27		0.78		0.09		4		484
		NGC 3201 sq2 7				2	12.26	.027	1.18	.014	0.92	.005	5		484,547
		NGC 3201 sq2 11				1	14.42		1.27		1.13		3		547
		NGC 3201 sq2 12				1	14.75		0.90		0.55		1		547
		NGC 3201 sq2 16				1	15.23		0.85		0.25		1		547
		NGC 3201 sq2 17				2	15.85	.026	0.64	.004	0.06	.013	6		484,547
		NGC 3227 sq1 1				1	12.83		0.65		0.11		5		327
		NGC 3227 sq1 3				1	14.88		0.92		0.11		1		327
		NGC 3227 sq1 4				1	13.54		0.68		0.10		3		327
		NGC 3311 sq1 4				1	14.56		0.84				1		498
		NGC 3311 sq1 5				1	15.18		0.78				1		498
		NGC 3311 sq1 6				1	16.17		0.51				1		498
		NGC 3311 sq1 7				1	16.31		1.07				1		498
		NGC 3311 sq1 8				1	16.23		0.53				1		498
		NGC 3311 sq1 9				1	14.30		0.63				1		498
		NGC 3311 sq1 10				1	14.10		0.71				1		498
		NGC 3311 sq1 11				1	15.90		0.86				1		498
		NGC 3311 sq1 12				1	17.05		0.41				1		498
		NGC 3516 sq1 6				1	13.42		0.58		-0.04		1		327
		NGC 3783 sq1 8				1	14.90		0.66		0.04		4		1687
		NGC 3783 sq1 9				1	15.42		0.98				4		1687
		NGC 3783 sq1 12				1	14.98		1.03		0.72		5		1687
		NGC 4216 sq1 11				1	16.22		0.46				1		492
		NGC 4216 sq1 12				1	15.90		0.65				1		492
		NGC 4216 sq1 13				1	16.12		0.78				1		492
		NGC 4340 sq1 3				1	15.46		0.31				2		492
		NGC 4340 sq1 8				1	15.21		0.91				1		492
		NGC 4340 sq1 11				1	15.31		1.11				1		492
		NGC 4340 sq1 12				1	15.61		1.23				2		492
		NGC 4372 sq1 12				1	14.97		1.20		0.50		3		372
		NGC 4372 sq1 13				1	15.39		1.08		0.45		3		372
		NGC 4372 sq1 14				1	15.88		0.63		0.45		3		372
		NGC 4372 sq3 2				1	12.37		1.78		1.73		5		518
		NGC 4372 sq3 7				1	14.05		1.31		0.89		2		518
		NGC 4372 sq3 9				1	14.33		0.82		0.26		4		518
		NGC 4372 sq3 10				1	14.48		1.41		1.30		3		518
		NGC 4372 sq3 11				1	14.91		1.85				1		518
		NGC 4372 sq3 12				1	15.15		0.85		0.23		3		518
		NGC 4372 sq3 13				1	15.47		0.95				1		518
		NGC 4372 sq3 14				1	15.57		1.19				1		518
		NGC 4372 sq3 15				1	15.65		1.12		0.64		2		518
		NGC 4372 sq3 16				1	15.68		0.66		0.48		2		518
		NGC 4372 sq3 17				1	15.76		0.53		0.41		1		518
		NGC 4372 sq3 18				1	15.76		0.71				1		518
		NGC 4372 sq3 19				1	15.89		0.76		0.44		3		518
		NGC 4372 sq3 20				1	15.90		0.39		0.27		2		518
		NGC 4372 sq3 21				1	16.07		1.14				1		518
		NGC 4372 sq3 22				1	16.11		0.38		0.24		2		518
		NGC 4372 sq3 23				1	16.16		0.51		0.27		1		518
		NGC 4372 sq3 24				1	16.26		1.20		1.33		1		518
		NGC 4372 sq3 25				1	16.35		0.47		0.41		1		518
		NGC 4372 sq3 26				1	16.35		0.65		0.47		1		518
		NGC 4372 sq3 27				1	16.39		1.16		0.07		1		518
		NGC 4372 sq3 28				1	16.55		1.07		0.41		1		518
		NGC 4372 sq3 29				1	16.99		1.25				1		518
		NGC 4372 sq3 30				1	17.16		1.06				2		518
		NGC 4406 sq1 2				1	15.02		0.74				1		492
		NGC 4406 sq1 3				1	15.12		0.63				1		492
		NGC 4406 sq1 5				1	14.93		1.09				1		492
		NGC 4406 sq1 6				1	15.14		0.97				1		492

HD	DM	Other Id	N Rem	α_{1950}	δ_{1950}	S	V	σ_V	B−V	σ_{B-V}	U−B	σ_{U-B}	n	Spectrum	References
		NGC 4406 sq1 7				1	15.61		0.79				1		492
		NGC 4406 sq1 8				1	15.62		0.64				1		492
		NGC 4458 sq1 13				1	15.94		1.02				2		1531
		NGC 4472 sq1 14				1	10.98		0.68				1		492
		NGC 4564 sq1 9				2	15.13	.019	0.65	.049	0.25	.039	3		492,1531
		NGC 4565 sq1 10				1	15.01		0.62				1		492
		NGC 4565 sq1 11				1	16.02		−0.15				1		492
		NGC 4565 sq1 12				1	14.44		0.65				1		492
		NGC 4590 sq1 19				1	15.01		0.70				2		438
		NGC 4590 sq1 20				1	15.45		0.81				2		438
		NGC 4590 sq1 21				1	17.40		0.79				2		438
		NGC 4590 sq1 22				1	17.45		0.98				2		438
		NGC 4590 sq2 10				1	15.23		0.64		0.10		3		548
		NGC 4590 sq2 11				1	15.26		0.72		0.18		3		548
		NGC 4590 sq2 12				1	15.39		0.86		0.48		3		548
		NGC 4590 sq2 13				1	15.43		0.81		0.29		3		548
		NGC 4590 sq2 14				1	15.75		0.89		0.43		3		548
		NGC 4590 sq2 15				1	15.80		0.17		0.24		3		548
		NGC 4590 sq2 16				1	15.93		0.70		0.07		3		548
		NGC 4590 sq2 17				1	16.38		0.78		0.07		3		548
		NGC 4590 sq2 18				1	16.53		0.83		0.09		3		548
		NGC 4697 sq1 12				1	12.82		0.58				1		492
		NGC 4833 sq1 2				2	12.84	.020	0.49	.000	0.21	.010	9		290,1638
		NGC 4833 sq1 3				2	14.96	.040	1.40	.005	1.23		7		290,1638
		NGC 4833 sq1 4				1	15.01		0.82		0.24		2		290
		NGC 4833 sq1 5				1	13.10		1.50		1.18		3		290
		NGC 4833 sq1 6				1	13.77		1.20		0.63		2		290
		NGC 4833 sq1 7				1	14.03		0.73		0.19		2		290
		NGC 4833 sq1 8				1	14.62		1.15		0.56		1		290
		NGC 4833 sq1 9				1	15.68		0.47		0.30		1		290
		NGC 4833 sq1 10				1	15.84		0.39		0.36		2		290
		NGC 4833 sq1 11				1	15.34		1.06		0.43		2		290
		NGC 4833 sq1 12				1	15.70		0.84		0.28		3		290
		NGC 4833 sq1 13				1	15.88		0.26		0.25		2		290
		NGC 4833 sq1 14				1	15.13		1.06		0.31		2		290
		NGC 4833 sq1 15				1	16.15		0.92		0.41		2		290
		NGC 4833 sq1 16				1	16.09		1.01				2		290
		NGC 4833 sq1 17				1	14.89		1.06		0.39		2		290
		NGC 4833 sq1 18				2	14.10	.005	1.20	.005	0.62	.034	8		290,1638
		NGC 4833 sq1 19				1	15.00		0.31		0.29		1		290
		NGC 4833 sq1 21				1	13.84		1.76		2.01		3		290
		NGC 4833 sq1 22				1	16.56		0.98		−0.05		1		290
		NGC 4833 sq1 23				1	15.51		1.35		1.32		3		290
		NGC 4833 sq1 24				1	16.69		1.19				2		290
		NGC 4833 sq1 25				1	16.21		0.88		0.17		4		290
		NGC 4833 sq1 26				1	15.79		1.15		0.31		1		290
		NGC 4833 sq2 5				1	13.50		1.31		0.80		5		1638
		NGC 4833 sq2 6				1	13.57		1.27		0.74		4		1638
		NGC 4833 sq2 7				1	13.59		1.33		0.85		4		1638
		NGC 4833 sq2 9				1	14.88		0.40		0.29		4		1638
		NGC 5128 sq2 13				1	15.42		0.72		0.15		2		5013
		NGC 5128 sq2 14				1	16.06		0.92		0.47		2		5013
		NGC 5128 sq2 15				1	16.56		1.29		1.14		2		5013
		NGC 5128 sq2 16				1	17.27		0.87		0.24		3		5013
		NGC 5128 sq2 17				1	17.81		1.41				3		5013
		NGC 5139 sq1 81				1	15.09		0.16		0.11		1		517
		NGC 5139 sq1 82				1	14.59		0.74				1		517
		NGC 5139 sq2 1				1	17.52		0.79				1		887
		NGC 5139 sq2 2				1	18.34		0.59				1		887
		NGC 5139 sq2 3				1	18.21		0.60				1		887
		NGC 5139 sq2 4				1	18.11		0.52				1		887
		NGC 5139 sq2 5				1	17.07		0.69				2		887
		NGC 5139 sq2 6				1	18.16		0.71				1		887
		NGC 5139 sq2 7				1	17.83		0.64				1		887
		NGC 5139 sq2 8				1	18.40		0.74				2		887
		NGC 5139 sq2 9				1	17.26		0.84				2		887
		NGC 5139 sq2 10				1	17.42		0.72				1		887
		NGC 5139 sq2 11				1	17.34		0.86				1		887
		NGC 5139 sq2 12				1	18.91		0.39				3		887
		NGC 5139 sq2 13				1	17.64		0.47				1		887
		NGC 5139 sq2 14				1	16.56		0.76				2		887
		NGC 5139 sq2 15				1	18.15		0.83				2		887
		NGC 5139 sq2 16				1	18.78		0.59				2		887
		NGC 5139 sq2 17				1	17.38		0.87				1		887
		NGC 5139 sq2 18				1	19.33		0.41				2		887
		NGC 5139 sq2 19				1	18.10		0.48				2		887
		NGC 5139 sq2 20				1	19.47		1.03				1		887
		NGC 5139 sq2 21				1	18.34		0.86				1		887
		NGC 5139 sq2 22				1	17.11		0.80				1		887
		NGC 5139 sq2 23				1	18.62		0.36				2		887

Table 2   1287

HD	DM	Other Id	N	Rem	α_{1950}	δ_{1950}	S	V	σ_V	B–V	σ_{B-V}	U–B	σ_{U-B}	n	Spectrum	References
		NGC 5139 sq2 24		AB			1	18.18		0.34				2		887
		NGC 5139 sq2 25					1	18.69		0.65				2		887
		NGC 5139 sq2 26					1	19.03		0.51				3		887
		NGC 5139 sq2 27					1	17.49		0.78				1		887
		NGC 5139 sq2 28					1	17.06		0.80				2		887
		NGC 5139 sq2 29					1	18.90		0.50				2		887
		NGC 5139 sq2 30					1	17.38		0.62				2		887
		NGC 5139 sq2 31					1	16.88		0.90				1		887
		NGC 5139 sq2 32					1	16.36		0.76				2		887
		NGC 5139 sq2 33					1	18.56		0.66				2		887
		NGC 5139 sq2 34					1	17.29		0.72				1		887
		NGC 5139 sq2 35					1	17.22		0.75				2		887
		NGC 5139 sq3 6					2	11.21	.005	1.61	.034	1.81		5		2001,3021
		NGC 5139 sq3 7					2	11.55	.041	1.71	.050	1.66		8		2001,3021
		NGC 5139 sq3 8					1	11.89		0.60				1		2001
		NGC 5139 sq3 9					2	11.97	.000	1.63	.061	1.90		4		2001,3021
		NGC 5139 sq3 10					2	11.93	.031	0.43	.020	0.08		6		2001,3021
		NGC 5139 sq3 11					2	12.64	.054	0.45	.021	-0.01		5		2001,3021
		NGC 5139 sq3 16					1	15.16		1.05				1		3021
		NGC 5236 sq1 8					1	15.08		0.65		1.00		2		5013
		NGC 5236 sq1 9					1	15.15		0.57		0.01		2		5013
		NGC 5236 sq1 10					1	15.28		0.67		0.01		2		5013
		NGC 5236 sq1 11					1	16.16		0.55		-0.03		3		5013
		NGC 5236 sq1 12					1	16.75		0.37		-0.07		3		5013
		NGC 5236 sq1 13					1	17.94		0.52		-0.20		2		5013
		NGC 5236 sq1 14					1	18.25		0.94				3		5013
		NGC 5286 sq1 7					1	14.66		1.55		1.23		3		371
		NGC 5286 sq1 8					1	14.91		1.38		0.82		3		371
		NGC 5286 sq1 9					1	15.11		0.69		0.24		3		371
		NGC 5286 sq1 10					1	15.27		1.89				3		371
		NGC 5286 sq1 11					1	15.30		0.85		0.30		3		371
		NGC 5286 sq1 12					1	16.04		0.80		0.09		3		371
		NGC 5286 sq2 8					1	14.18		0.70				1		485
		NGC 5286 sq2 10					1	14.64		0.66				1		485
		NGC 5457 sq1 1					1	12.01		0.65		0.06		5		442
		NGC 5457 sq1 2					1	13.50		0.62		0.04		5		442
		NGC 5457 sq1 3					1	13.99		0.68		0.23		5		442
		NGC 5457 sq1 4					1	15.50		0.59		0.10		4		442
		NGC 5457 sq1 5					1	14.92		1.28				1		442
		NGC 5457 sq1 6					1	15.62		0.65		0.09		3		442
		NGC 5457 sq1 7					1	16.22		0.61		0.25		2		442
		NGC 5457 sq1 8					1	16.25		1.11		1.01		1		442
		NGC 5457 sq1 9					1	16.54		1.45		1.28		2		442
		NGC 5457 sq1 10					1	17.92		0.03		-0.75		1		442
		NGC 5457 sq1 11					1	16.60		1.55		1.27		2		442
		NGC 5457 sq1 12					1	17.94		0.89		0.75		1		442
		NGC 5457 sq1 13					1	18.30		1.14				2		442
		NGC 5457 sq1 14					1	18.76		0.73		0.36		2		442
		NGC 5457 sq1 15					1	19.29		0.54				1		442
		NGC 5457 sq1 16					1	18.80		1.46				1		442
		NGC 5457 sq1 17					1	20.50		0.15				1		442
		NGC 5457 sq1 18					1	20.49		0.32				1		442
		NGC 5457 sq1 19					1	19.92		1.14				1		442
		NGC 5466 sq1 1					1	14.01		0.53				2		1634
		NGC 5466 sq1 2					1	14.16		0.70				2		1634
		NGC 5466 sq1 3					1	16.69		0.03				1		1634
		NGC 5466 sq1 6					1	13.98		0.42				1		1634
		NGC 5466 sq1 8					1	16.62		0.20				1		1634
		NGC 5466 sq1 14					1	13.69		1.17				3		1634
		NGC 5466 sq1 20					1	15.23		0.68				2		1634
		NGC 5466 sq1 21					1	16.66		0.61				1		1634
		NGC 5466 sq1 22					1	15.18		1.10				1		1634
		NGC 5466 sq1 24					1	17.84		0.73				1		1634
		NGC 5466 sq1 25					1	14.15		1.09				2		1634
		NGC 5466 sq1 28					1	16.94		0.73				1		1634
		NGC 5466 sq1 31					1	16.52		0.73				2		1634
		NGC 5466 sq1 32					1	16.32		0.76				2		1634
		NGC 5466 sq1 34					1	15.81		0.70				2		1634
		NGC 5466 sq2 1					1	14.08		0.65				2		1634
		NGC 5466 sq2 2					1	14.06		0.80				2		1634
		NGC 5466 sq2 3					1	13.13		0.93				2		1634
		NGC 5466 sq2 4					1	17.35		0.69				2		1634
		NGC 5466 sq2 5					1	16.11		0.60				2		1634
		NGC 5466 sq2 6					1	15.89		0.80				2		1634
		NGC 5466 sq2 7					1	16.63		0.10				2		1634
		NGC 5466 sq2 8					1	16.57		0.58				2		1634
		NGC 5466 sq2 9					1	17.71		0.68				1		1634
		NGC 5466 sq2 10					1	16.35		0.61				1		1634
		NGC 5466 sq2 11					1	16.66		0.13				1		1634
		NGC 5466 sq2 12					1	16.85		1.62				1		1634

HD	DM	Other Id	N Rem	α_{1950}	δ_{1950}	S	V	σ_V	B–V	σ_{B-V}	U–B	σ_{U-B}	n	Spectrum	References
		NGC 5466 sq2 13				1	14.55		0.66				1		1634
		NGC 5694 sq1 18				1	15.40		0.75				1		438
		NGC 5694 sq1 19				1	16.36		0.85				1		438
		NGC 5823 sq1 1				1	11.21		1.45		1.57		6		534
		NGC 5823 sq1 2				1	11.32		1.97		2.29		2		534
		NGC 5823 sq1 4				1	11.54		0.19		0.17		3		534
		NGC 5823 sq1 5				1	11.72		0.46		0.03		3		534
		NGC 5823 sq1 6				1	12.00		1.63		1.65		2		534
		NGC 5823 sq1 7				1	12.09		2.02		2.33		2		534
		NGC 5823 sq1 8				1	12.26		0.40		0.21		2		534
		NGC 5823 sq1 9				1	12.39		0.83		0.34		2		534
		NGC 5823 sq1 10				1	12.47		0.51		0.09		3		534
		NGC 5823 sq1 11				1	12.65		0.57		0.16		2		534
		NGC 5823 sq1 12				1	12.68		1.60		1.61		1		534
		NGC 5823 sq1 13				1	12.89		0.62		0.13		2		534
		NGC 5823 sq1 14				1	13.03		1.70		1.71		2		534
		NGC 5823 sq1 15				1	13.06		1.58		1.62		2		534
		NGC 5823 sq1 16				1	13.56		0.81		0.17		2		534
		NGC 5823 sq1 17				1	13.76		0.73		0.43		2		534
		NGC 5823 sq1 18				1	13.80		0.75		0.24		2		534
		NGC 5823 sq1 19				1	13.84		0.78		0.22		2		534
		NGC 5823 sq1 21				1	13.88		1.78		1.62		2		534
		NGC 5823 sq1 22				1	13.93		0.90		0.38		2		534
		NGC 5823 sq1 23				1	14.03		0.79		0.21		2		534
		NGC 5823 sq1 24				1	14.11		1.71		1.39		2		534
		NGC 5823 sq1 25				1	14.37		0.66		0.47		2		534
		NGC 5823 sq1 26				1	14.41		0.69		0.52		2		534
		NGC 5823 sq1 27				1	14.63		0.77		0.42		2		534
		NGC 5823 sq1 28				1	15.01		1.18		0.63		2		534
		NGC 5823 sq1 29				1	15.16		0.82		0.55		2		534
		NGC 5823 sq1 30				1	15.27		1.03		0.40		2		534
		NGC 5823 sq1 31				1	15.67		1.10				2		534
		NGC 5823 sq1 32				1	15.71		1.12				2		534
		NGC 5823 sq1 33				1	15.88		0.90		1.21		1		534
		NGC 5823 sq2 17				1	13.46		0.71		0.29		2		534
		NGC 5823 sq2 18				1	13.46		1.56		1.38		2		534
		NGC 5823 sq2 25				1	14.04		1.58		1.36		2		534
		NGC 5823 sq2 27				1	14.79		1.26		0.27		1		534
		NGC 5823 sq2 28				1	14.85		0.85		0.18		1		534
		NGC 5823 sq2 29				1	14.88		1.44		1.48		2		534
		NGC 5823 sq2 30				1	15.13		1.35		0.64		2		534
		NGC 5823 sq2 31				1	15.19		0.74		0.53		1		534
		NGC 5823 sq2 32				1	15.32		1.07		0.67		1		534
		NGC 5823 sq2 33				1	15.40		0.72		0.41		1		534
		NGC 5823 sq2 34				1	15.65		1.01		0.21		1		534
		NGC 5823 sq2 35				1	15.77		0.92				2		534
		NGC 5823 sq2 36				1	15.87		1.12		0.08		2		534
		NGC 5824 sq1 2				1	10.85		1.52				2		438
		NGC 5824 sq1 13				1	13.12		0.72				1		438
		NGC 5824 sq1 18				1	15.38		0.61				2		438
		NGC 5824 sq1 19				1	15.58		0.81				4		438
		NGC 5824 sq1 20				1	16.14		0.87				2		438
		NGC 5824 sq1 21				1	16.70		0.76				1		438
		NGC 5824 sq1 22				1	17.00		1.12				2		438
		NGC 5824 sq1 23				1	17.03		1.07				1		438
		NGC 5897 sq1 5				1	13.86		1.01		0.67		1		514
		NGC 5897 sq1 7				1	14.33		0.99		0.74		1		514
		NGC 5897 sq1 9				1	14.78		0.75		0.43		1		514
		NGC 5897 sq1 10				1	15.04		0.75		0.09		1		514
		NGC 5897 sq1 11				2	15.27	.044	0.94	.054	0.47	.019	3		514,514
		NGC 5897 sq1 12				2	15.68	.029	0.72	.024	0.18		3		514,514
		NGC 5897 sq1 13				1	16.16		0.68		0.20		1		514
		NGC 5897 sq1 14				1	16.38		0.23		0.37		2		514
		NGC 5897 sq1 15				1	16.63		0.75				2		514
		NGC 5897 sq1 16				1	16.80		0.09		0.18		2		514
		NGC 5897 sq1 17				1	16.68		0.13		0.28		1		514
		NGC 5897 sq1 18				2	16.84	.039	0.12	.049	0.26	.093	3		514,514
		NGC 5897 sq1 19				1	17.14		0.73				1		514
		NGC 5897 sq1 20				1	17.15		0.82				1		514
		NGC 5927 sq1 5				1	16.02		0.82		0.34		2		89
		NGC 5927 sq1 6				1	13.39		1.30		1.22		2		89
		NGC 5927 sq1 7				1	13.55		1.86		1.87		2		89
		NGC 5927 sq1 8				1	13.51		0.66		0.27		2		89
		NGC 5927 sq1 10				1	14.62		1.01		0.73		2		89
		NGC 5927 sq3 4				1	12.88		1.64		1.50		4		780
		NGC 5927 sq3 6				1	13.40		0.62		0.33		5		780
		NGC 5927 sq3 7				1	13.53		1.60		1.80		2		780
		NGC 5927 sq3 8				1	13.55		0.69		0.18		5		780
		NGC 5927 sq3 9				1	13.63		0.76		0.24		3		780
		NGC 5927 sq3 10				1	14.10		0.62		0.32		3		780

Table 2   1289

HD	DM	Other Id	N Rem	α_{1950}	δ_{1950}	S	V	σ_V	B–V	σ_{B-V}	U–B	σ_{U-B}	n	Spectrum	References
		NGC 5927 sq3 11				1	14.30		1.40		1.15		1		780
		NGC 5927 sq3 13				1	14.49		1.00		0.63		5		780
		NGC 5927 sq3 14				1	14.91		0.78		0.26		3		780
		NGC 5927 sq3 15				1	14.93		1.21		0.68		3		780
		NGC 5927 sq3 16				1	15.11		1.51		1.14		5		780
		NGC 5927 sq3 17				1	15.20		0.80		0.26		1		780
		NGC 5927 sq3 18				1	15.30		0.82		0.28		5		780
		NGC 5927 sq3 19				1	15.70		1.38		0.91		2		780
		NGC 5927 sq3 20				1	15.76		0.90		0.30		4		780
		NGC 5927 sq3 21				1	15.80		1.68		1.98		2		780
		NGC 5927 sq3 22				1	15.88		1.87				3		780
		NGC 5927 sq3 23				1	15.92		0.81		0.63		1		780
		NGC 5927 sq3 25				1	16.13		0.97				4		780
		NGC 5946 sq1 1				1	15.49		0.81		0.55		1		1465
		NGC 5946 sq1 2				1	14.88		2.01				1		1465
		NGC 5946 sq1 3				1	14.81		1.88				2		1465
		NGC 5946 sq1 4				1	14.85		0.79		-0.08		1		1465
		NGC 5946 sq2 1				1	11.90		0.44		0.11		8		1638
		NGC 5946 sq2 2				1	14.04		1.93		1.71		8		1638
		NGC 5946 sq2 3				1	14.29		1.58		1.31		5		1638
		NGC 5946 sq2 4				1	15.15		0.88		0.34		6		1638
		NGC 5986 sq1 13				1	13.88		1.83				2		485
		NGC 5986 sq1 14				1	14.42		1.43				2		485
		NGC 5986 sq1 16				2	14.38	.060	0.70	.025	0.13		4		485,1538
		NGC 5986 sq1 17				1	14.99		1.61				2		485
		NGC 5986 sq1 18				1	16.01		1.12				2		485
		NGC 5986 sq1 19				1	16.04		0.88				1		485
		NGC 5986 sq1 20				1	16.42		0.95				1		485
		NGC 5986 sq2 9				1	14.17		1.38		1.12		2		1538
		NGC 5986 sq2 10				1	14.32		1.29		0.96		3		1538
		NGC 5986 sq2 12				1	14.56		0.83				3		1538
		NGC 5986 sq2 13				1	14.90		1.25		0.93		3		1538
		NGC 5986 sq2 14				1	15.06		1.01		0.24		2		1538
		NGC 5986 sq2 15				1	15.09		1.19				3		1538
		NGC 5986 sq2 16				1	15.21		0.93		0.18		2		1538
		NGC 5986 sq2 17				1	15.64		1.01		0.45		2		1538
		NGC 5986 sq2 18				1	15.75		1.20		0.79		2		1538
		NGC 5986 sq2 19				1	15.82		1.14				4		1538
		NGC 5986 sq2 20				1	15.90		1.24				3		1538
		NGC 5986 sq2 21				1	16.36		1.09				2		1538
		NGC 5986 sq2 22				1	16.59		0.96		0.45		3		1538
		NGC 6101 sq2 8				1	14.44		1.07		0.73		5		371
		NGC 6101 sq2 9				1	14.87		0.64		0.10		5		371
		NGC 6101 sq2 10				1	14.92		0.73		0.23		2		371
		NGC 6101 sq2 11				1	15.30		0.70		0.07		1		371
		NGC 6101 sq2 12				1	16.12		0.75		-0.06		3		371
		NGC 6121 sq1 4				1	12.29		1.52		1.24		4		828
		NGC 6121 sq1 9				1	13.67		0.48		0.41		2		828
		NGC 6121 sq1 10				1	13.71		0.96		0.26		4		828
		NGC 6121 sq1 13				1	15.01		1.13				3		828
		NGC 6121 sq1 15				2	15.36	.015	1.15	.000			8		420,1515
		NGC 6121 sq1 16				3	15.55	.010	1.69	.004			8		420,828,1515
		NGC 6121 sq1 17				3	15.75	.014	1.11	.025			9		420,828,1515
		NGC 6121 sq1 19				1	16.06		1.05				6		1515
		NGC 6144 sq1 5				1	12.83		1.29		0.72		5		828
		NGC 6144 sq1 11				1	14.66		1.90				6		828
		NGC 6144 sq1 12				1	15.53		1.42				6		828
		NGC 6144 sq1 18				1	15.72		1.22				4		828
		NGC 6144 sq1 19				1	16.34		1.07				7		828
		NGC 6144 sq1 20				1	16.70		1.10				5		828
		NGC 6171 sq1 5				1	13.20		1.85		1.88		1		449
		NGC 6171 sq1 6				2	13.35	.037	1.73	.009	1.80	.009	4		449,519
		NGC 6171 sq1 7				1	13.43		1.70		1.83		1		449
		NGC 6171 sq1 8				1	13.89		1.61		1.58		1		449
		NGC 6171 sq1 9				1	13.93		1.45		1.27		1		449
		NGC 6171 sq1 11				1	13.99		1.43		1.37		1		449
		NGC 6171 sq1 12				2	14.01	.000	1.47	.024	1.05		3		449,519
		NGC 6171 sq1 14				1	14.25		1.50		1.38		1		449
		NGC 6171 sq1 15				1	14.39		1.45		1.48		1		449
		NGC 6171 sq1 16				1	14.41		1.08		0.61		1		449
		NGC 6171 sq1 17				1	14.43		1.62		1.73		1		449
		NGC 6171 sq1 18				1	14.58		1.30		1.02		1		449
		NGC 6171 sq1 19				1	14.77		1.43		1.16		1		449
		NGC 6171 sq1 20				1	14.85		1.26		0.86		1		449
		NGC 6171 sq1 21				1	14.87		1.25		0.86		1		449
		NGC 6171 sq1 22				1	15.34		1.10		0.65		1		449
		NGC 6171 sq1 23				1	15.53		0.99		0.40		1		449
		NGC 6171 sq1 24				1	15.70		1.06		0.53		1		449
		NGC 6171 sq1 25				1	15.71		0.88		0.37		1		449
		NGC 6171 sq1 26				1	15.88		1.11		0.67		1		449

HD	DM	Other Id	N Rem	α_{1950}	δ_{1950}	S	V	σ_V	B−V	σ_{B-V}	U−B	σ_{U-B}	n	Spectrum	References
		NGC 6171 sq1 27				1	15.94		1.22		0.83		1		449
		NGC 6171 sq1 28				1	16.41		1.12				1		449
		NGC 6171 sq2 20A				1	13.46		1.12		0.57		1		519
		NGC 6235 sq1 6				1	15.32		0.93				3		869
		NGC 6235 sq1 7				1	15.50		1.56				3		869
		NGC 6235 sq1 8				1	16.29		0.68				3		869
		NGC 6235 sq1 9				1	16.31		1.41				2		869
		NGC 6235 sq1 12				1	14.54		0.83				1		869
		NGC 6235 sq1 13				1	16.64		0.95				1		869
		NGC 6235 sq1 14				1	16.20		1.22				2		869
		NGC 6235 sq1 15				1	15.37		1.04				1		869
		NGC 6254 sq1 3				1	11.52		1.20				1		485
		NGC 6254 sq1 17				1	14.82		1.04				1		485
		NGC 6254 sq1 18				1	15.15		0.87				2		485
		NGC 6254 sq1 19				1	15.38		1.04				1		485
		NGC 6254 sq1 20				1	15.54		0.93				3		485
		NGC 6254 sq1 21				1	15.72		1.68				3		485
		NGC 6254 sq1 22				1	15.78		1.12				2		485
		NGC 6254 sq1 23				1	15.83		1.02				1		485
		NGC 6254 sq1 24				1	18.05		1.24				1		485
		NGC 6256 sq1 1				1	13.67		0.79		0.53		6		474
		NGC 6256 sq1 2				1	14.29		0.98		0.46		6		474
		NGC 6256 sq1 3				1	14.53		1.18		0.78		6		474
		NGC 6256 sq1 4				1	12.62		0.67		0.43		3		474
		NGC 6256 sq1 5				1	13.90		0.80		0.22		5		474
		NGC 6256 sq1 6				1	14.79		1.22		0.73		4		474
		NGC 6256 sq1 8				1	14.96		2.19				5		474
		NGC 6256 sq1 9				1	16.03		1.00				5		474
		NGC 6256 sq1 10				1	15.40		1.12		0.78		1		474
		NGC 6266 sq1 5				1	10.66		0.64		0.18		3		156
		NGC 6266 sq2 42				1	13.56		1.58				1		438
		NGC 6266 sq2 45				1	15.08		1.06				2		438
		NGC 6266 sq2 46				1	15.57		1.20				2		438
		NGC 6266 sq3 13				1	13.91		1.92		2.32		3		718
		NGC 6266 sq3 15				1	14.08		0.82		0.23		3		718
		NGC 6266 sq3 16				1	14.39		0.91		0.27		3		718
		NGC 6266 sq3 17				1	14.64		1.21		0.53		3		718
		NGC 6266 sq3 18				1	15.01		1.48		1.12		3		718
		NGC 6266 sq3 19				1	15.69		1.68				3		718
		NGC 6266 sq3 20				1	15.83		1.09				3		718
		NGC 6266 sq3 21				1	16.10		1.49				2		718
		NGC 6266 sq3 22				1	16.40		1.40				2		718
		NGC 6266 sq3 23				1	16.72		1.25				3		718
		NGC 6273 sq1 10				1	15.05		1.44				2		485
		NGC 6273 sq1 11				1	15.81		1.13				2		485
		NGC 6273 sq1 12				1	16.00		1.17				2		485
		NGC 6304 sq1 6				1	14.05		1.76		2.39		6		474
		NGC 6304 sq1 8				1	12.82		0.80		0.26		6		474
		NGC 6304 sq1 10				1	12.57		1.05		0.57		6		474
		NGC 6304 sq1 11				1	14.01		2.07				4		474
		NGC 6304 sq1 12				1	14.23		1.68				6		474
		NGC 6304 sq1 13				1	15.33		0.81		0.28		5		474
		NGC 6304 sq1 14				1	13.58		0.81		0.21		4		474
		NGC 6304 sq1 15				1	14.34		1.42		1.24		4		474
		NGC 6304 sq1 16				1	15.06		0.86		0.37		4		474
		NGC 6304 sq1 17				1	15.16		1.43		1.25		5		474
		NGC 6304 sq1 18				1	13.18		0.81		0.33		5		474
		NGC 6304 sq1 22				1	13.60		2.10		2.12		1		474
		NGC 6304 sq1 23				1	15.06		1.12		0.40		2		474
		NGC 6325 sq1 11				1	15.12		1.40				2		438
		NGC 6325 sq1 12				1	15.27		1.14				2		438
		NGC 6341 sq1 7				1	15.06		0.74		0.32		1		448
		NGC 6341 sq1 20				1	15.68		1.16		1.05		2		448
		NGC 6341 sq1 22				1	15.47		1.71				1		448
		NGC 6388 sq1 2				1	12.42		0.38		0.25		1		156
		NGC 6388 sq2 10				1	13.26		0.90		0.40		3		895
		NGC 6388 sq2 11				1	13.75		1.22		0.99		3		895
		NGC 6388 sq2 12				1	13.76		1.04				3		895
		NGC 6388 sq2 13				1	13.85		0.62		0.11		3		895
		NGC 6388 sq2 14				1	14.20		1.69				3		895
		NGC 6388 sq2 15				1	14.74		0.76		0.25		3		895
		NGC 6388 sq2 16				1	14.88		1.06		0.75		3		895
		NGC 6388 sq2 17				1	14.98		1.52				4		895
		NGC 6388 sq2 18				1	15.02		2.04				3		895
		NGC 6388 sq2 19				1	15.34		1.39				3		895
		NGC 6388 sq2 20				1	15.40		1.51				3		895
		NGC 6388 sq2 21				1	15.66		1.83				3		895
		NGC 6388 sq2 22				1	15.90		1.71				1		895
		NGC 6388 sq2 23				1	15.92		1.42				5		895
		NGC 6388 sq2 24				1	16.03		0.94				4		895

Table 2 1291

HD	DM	Other Id	N Rem	α_{1950}	δ_{1950}	S	V	σ_V	B−V	σ_{B-V}	U−B	σ_{U-B}	n	Spectrum	References
		NGC 6388 sq2 25				1	16.20		1.24				5		895
		NGC 6388 sq2 26				1	16.37		0.40				4		895
		NGC 6388 sq2 27				1	16.59		1.34				4		895
		NGC 6388 sq2 28				1	17.49		1.13				2		895
		NGC 6388 sq2 29				1	17.67		0.97				3		895
		NGC 6397 sq1 1				1	18.85		0.57				2		887
		NGC 6397 sq1 2				1	17.73		0.91				1		887
		NGC 6397 sq1 3				1	18.08		0.63				2		887
		NGC 6397 sq1 4				1	17.37		1.03				1		887
		NGC 6397 sq1 5				1	17.74		0.57				1		887
		NGC 6397 sq1 6				1	16.73		1.03				1		887
		NGC 6397 sq1 7				1	16.83		0.73				1		887
		NGC 6397 sq1 8				1	17.96		1.06				2		887
		NGC 6397 sq1 9				1	17.89		0.67				1		887
		NGC 6397 sq1 10				1	17.16		0.46				1		887
		NGC 6397 sq1 11				1	18.48		0.86				2		887
		NGC 6397 sq1 12				1	18.35		1.62				2		887
		NGC 6397 sq1 13				1	17.87		0.52				2		887
		NGC 6397 sq1 14				1	17.36		0.86				1		887
		NGC 6397 sq1 15				1	17.40		0.65				2		887
		NGC 6397 sq1 16				1	16.99		0.94				2		887
		NGC 6397 sq1 17				1	18.15		0.58				1		887
		NGC 6397 sq1 18				1	18.57		0.87				2		887
		NGC 6397 sq1 19				1	17.32		0.58				2		887
		NGC 6397 sq1 20				1	17.59		0.96				2		887
		NGC 6397 sq1 21				1	17.65		0.74				2		887
		NGC 6397 sq1 22				1	17.22		0.81				2		887
		NGC 6397 sq2 8				1	13.70		0.12		-0.04		3		1662
		NGC 6397 sq2 16				1	14.31		0.75		0.24		3		1662
		NGC 6397 sq2 17				1	14.69		1.15		0.82		3		1662
		NGC 6397 sq2 19				1	13.72		0.15		-0.08		3		1662
		NGC 6397 sq2 20				1	16.36		0.48				3		1662
		NGC 6397 sq2 23				1	12.31		0.60		0.12		2		1662
		NGC 6397 sq2 25				1	16.78		0.56				1		1662
		NGC 6402 sq1 1				1	13.21		1.52		1.39		3		297
		NGC 6402 sq1 2				1	14.02		1.66				2		297
		NGC 6402 sq1 3				1	14.43		2.05		2.10		3		297
		NGC 6402 sq1 4				1	14.81		1.17		0.81		2		297
		NGC 6402 sq1 5				1	14.86		1.67		1.72		2		297
		NGC 6402 sq1 6				1	15.02		1.85				3		297
		NGC 6402 sq1 7				1	15.43		1.69		1.29		3		297
		NGC 6402 sq1 8				1	15.69		1.61				2		297
		NGC 6402 sq1 10				1	15.84		1.61				2		297
		NGC 6402 sq1 11				1	15.90		0.50				2		297
		NGC 6402 sq1 12				1	16.18		1.30				1		297
		NGC 6402 sq1 13				1	16.65		1.22		1.08		1		297
		NGC 6440 sq1 1				1	15.84		1.20				1		1465
		NGC 6440 sq1 4				1	13.79		1.78		1.76		4		1465
		NGC 6440 sq1 5				1	13.51		1.68		1.17		3		1465
		NGC 6517 sq1 11				1	14.81		2.11				1		438
		NGC 6517 sq1 13				1	16.20		2.25				1		438
		NGC 6522 sq1 31				1	17.19		2.47				2		325
		NGC 6522 sq1 57				1	15.84		1.10		0.58		3		325
		NGC 6522 sq1 58				1	14.69		1.67		1.88		3		325
		NGC 6522 sq2 1				3	11.35	.009	0.50	.009	0.11	.014	26		325,746,1550
		NGC 6522 sq2 4				3	12.78	.015	0.56	.012	0.09	.015	7		325,746,1550
		NGC 6522 sq2 5				3	13.06	.029	1.31	.005	1.23	.000	20		325,746,1550
		NGC 6522 sq2 6				2	13.26	.005	1.50	.000	1.65	.041	5		325,746
		NGC 6522 sq2 7				3	13.31	.012	0.44	.015	0.36	.042	6		325,746,1550
		NGC 6522 sq2 8				3	13.45	.012	1.87	.012	2.20	.060	12	M1 Var	325,746,1550
		NGC 6522 sq2 9				3	13.53	.012	0.70	.007	0.40	.041	7		325,746,1550
		NGC 6522 sq2 10				3	14.18	.029	0.72	.009	0.15	.055	10		325,746,1550
		NGC 6522 sq2 11				3	14.39	.031	1.31	.010	0.87	.019	5		325,746,1550
		NGC 6522 sq2 12				3	14.27	.054	1.82	.020	1.53	.039	5	M1 Var	325,746,1550
		NGC 6522 sq2 13				3	14.50	.056	1.67	.017	1.53	.156	10		325,746,1550
		NGC 6522 sq2 14				3	14.66	.040	0.70	.032	0.30	.015	7		325,746,1550
		NGC 6522 sq2 15				1	14.90		1.11				1		1702
		NGC 6522 sq2 16				2	15.00	.064	1.47	.023	1.07		9		325,1550
		NGC 6522 sq2 17				3	15.14	.016	1.09	.013	0.53	.058	5		325,746,1550
		NGC 6522 sq2 20				2	15.50	.019	1.31	.037	0.94		4		325,1550
		NGC 6522 sq2 21				1	15.72		1.84				1	M2 Var	1550
		NGC 6522 sq2 22				2	15.74	.004	1.49	.044	1.02		6		325,1550
		NGC 6522 sq2 23				2	15.88	.014	1.91	.019			4	M3	325,1550
		NGC 6522 sq2 25				2	15.95	.014	0.54	.009	0.58		5		325,1550
		NGC 6522 sq2 26				2	15.89	.110	1.43	.010			7		325,1550
		NGC 6522 sq2 27				2	15.95	.063	1.31	.009	0.91		5		325,1550
		NGC 6522 sq2 29				2	15.71	.060	1.32	.010	1.29		2		325,1550
		NGC 6522 sq2 30				2	16.05	.020	1.32	.045	0.62		7		325,1550
		NGC 6522 sq2 31				2	16.08	.000	1.66	.015			8		325,1550
		NGC 6522 sq2 33				2	16.05	.100	1.42	.070			7		325,1550

HD	DM	Other Id	N	Rem	α_{1950}	δ_{1950}	S	V	σ_V	B−V	σ_{B-V}	U−B	σ_{U-B}	n	Spectrum	References
		NGC 6522 sq2 39					1	16.62		0.61				3		1550
		NGC 6522 sq2 43					1	17.06		1.33				2		325
		NGC 6522 sq2 50					1	16.71		1.48				2		325
		NGC 6522 sq2 53					1	17.08		1.69				2	M1	1550
		NGC 6522 sq2 55					2	17.09	.030	1.47	.074			5		325,1550
		NGC 6522 sq2 57					2	17.33	.107	1.10	.044			3		325,1550
		NGC 6522 sq2 58					1	14.96		1.62				2		1702
		NGC 6522 sq2 61					2	17.84	.090	1.12	.140			4		325,1550
		NGC 6528 sq1 4					1	12.70		1.85		2.01		5		791
		NGC 6528 sq1 6					2	12.90	.017	0.42	.000	0.12	.009	6		325,791
		NGC 6528 sq1 7					1	13.84		1.45				1		791
		NGC 6528 sq1 8					1	13.60		0.50		0.18		4		791
		NGC 6528 sq1 9					1	13.87		0.71				1		791
		NGC 6528 sq1 10					1	14.10		1.36				3		791
		NGC 6528 sq1 11					1	14.02		0.76		0.30		3		791
		NGC 6528 sq1 12					1	14.04		0.90		0.50		1		791
		NGC 6528 sq1 13					2	14.15	.035	0.64	.004	0.26	.004	6		325,791
		NGC 6528 sq1 14					1	14.58		0.69				2		791
		NGC 6528 sq1 15					2	14.67	.015	1.74	.049	1.95		3		325,791
		NGC 6528 sq1 17					1	14.79		1.49				1		791
		NGC 6528 sq1 18					1	14.92		1.62				3		791
		NGC 6528 sq1 19					1	16.13		1.75				3		791
		NGC 6528 sq1 20					1	16.70		1.00				2		791
		NGC 6528 sq1 21					1	16.30		1.02				2		791
		NGC 6528 sq1 22					2	15.53	.020	1.62	.160	1.06		2		325,791
		NGC 6528 sq1 23					1	15.24		1.60				1		791
		NGC 6528 sq1 24					2	16.58	.025	1.34	.065	0.83		2		325,791
		NGC 6528 sq1 25					1	16.84		1.50				1		791
		NGC 6535 sq1 1					1	14.48		0.85				3		870
		NGC 6535 sq1 7					1	15.52		0.89				2		870
		NGC 6535 sq1 8					1	16.49		0.96				1		870
		NGC 6535 sq1 9					1	16.36		1.40				1		870
		NGC 6535 sq1 10					1	16.34		1.20				2		870
		NGC 6535 sq1 11					1	16.25		1.29				1		870
		NGC 6535 sq1 12					1	16.23		0.36				1		870
		NGC 6535 sq1 13					1	16.04		1.18				2		870
		NGC 6535 sq1 14					1	16.78		1.06				1		870
		NGC 6535 sq1 15					1	16.85		1.19				1		870
		NGC 6535 sq1 16					1	15.40		1.01				2		870
		NGC 6535 sq1 17					1	17.03		1.10				2		870
		NGC 6541 sq1 7					1	13.10		1.24		1.15		5		763
		NGC 6541 sq1 8					1	13.48		0.58		0.06		5		763
		NGC 6541 sq1 9					1	13.62		0.58		0.00		5		763
		NGC 6541 sq1 13					1	14.14		1.24				5		763
		NGC 6541 sq1 14					1	14.60		0.67		0.00		5		763
		NGC 6541 sq1 15					1	14.93		1.09				7		763
		NGC 6541 sq1 17					1	15.14		0.99				7		763
		NGC 6541 sq1 18					1	15.28		0.69				4		763
		NGC 6541 sq1 19					1	15.85		0.87				5		763
		NGC 6541 sq1 20					1	16.29		0.96				7		763
		NGC 6544 sq1 5					1	12.43		0.42		0.25		3		984
		NGC 6544 sq1 6					1	12.78		0.33		0.25		3		984
		NGC 6544 sq1 7					1	12.86		0.74		0.32		4		984
		NGC 6544 sq1 8					1	13.31		0.78		0.48		4		984
		NGC 6544 sq1 9					1	13.99		0.74		0.38		4		984
		NGC 6544 sq1 10					1	14.11		0.59		0.20		4		984
		NGC 6544 sq1 11					1	14.56		0.81		0.40		4		984
		NGC 6544 sq1 12					1	14.81		1.00		0.55		4		984
		NGC 6544 sq1 13					1	15.29		1.05				4		984
		NGC 6544 sq1 14					1	15.56		1.05				5		984
		NGC 6544 sq1 16					1	16.13		0.97				5		984
		NGC 6544 sq1 17					1	16.33		0.94				4		984
		NGC 6544 sq1 18					1	16.56		0.90				5		984
		NGC 6544 sq1 21					1	14.02		1.56				2		984
		NGC 6584 sq1 10					1	14.78		0.66		0.01		3		1585
		NGC 6584 sq1 9					1	14.04		0.66		0.06		3		1585
		NGC 6584 sq1 11					1	15.12		0.62		-0.04		4		1585
		NGC 6584 sq1 12					1	15.24		0.78		0.31		4		1585
		NGC 6584 sq1 13					1	15.70		0.71		0.26		4		1585
		NGC 6624 sq1 3					2	13.80	.084	0.51	.030	0.28		5		745,1465
		NGC 6624 sq1 5					2	12.99	.010	1.29	.024	1.15		3		539,1465
		NGC 6624 sq1 7					1	14.15		1.91		2.10		2		1465
		NGC 6624 sq1 10					2	12.95	.019	0.65	.008	0.14		3		539,1465
		NGC 6624 sq1 11					1	13.35		1.36		1.43		2		1465
		NGC 6624 sq2 1					1	13.10		1.13				2		745
		NGC 6624 sq2 2					1	16.65		1.09				2		745
		NGC 6624 sq2 9					1	15.99		1.62				2		745
		NGC 6624 sq2 22					1	15.60		0.83				2		745
		NGC 6624 sq2 24					1	14.05		1.89				2		745
		NGC 6624 sq2 30					1	14.13		0.76				2		745

Table 2

1293

HD	DM	Other Id	N	Rem	α_{1950}	δ_{1950}	S	V	σ_V	B−V	σ_{B-V}	U−B	σ_{U-B}	n	Spectrum	References
		NGC 6624 sq2 32					1	15.09		0.89				2		745
		NGC 6626 sq1 12					1	13.46		1.78				3		896
		NGC 6626 sq1 13					1	13.84		1.35		1.09		5		896
		NGC 6626 sq1 14					1	14.12		1.56		1.56		5		896
		NGC 6626 sq1 15					1	14.16		0.66		0.11		5		896
		NGC 6626 sq1 16					1	14.69		0.70		0.32		3		896
		NGC 6626 sq1 17					1	14.71		1.74				5		896
		NGC 6626 sq1 18					1	14.92		1.83				3		896
		NGC 6626 sq1 19					1	15.35		1.82				4		896
		NGC 6626 sq1 20					1	15.53		2.24				5		896
		NGC 6626 sq1 21					1	15.87		1.65				4		896
		NGC 6626 sq1 22					1	15.92		0.77				5		896
		NGC 6626 sq1 23					1	16.47		2.07				4		896
		NGC 6637 sq1 1					1	13.92		0.62				2		579
		NGC 6637 sq1 5					1	15.52		1.07				1		579
		NGC 6637 sq1 6					1	15.92		1.00				1		579
		NGC 6637 sq1 7					1	16.98		0.74				1		579
		NGC 6637 sq1 12					1	14.07		0.99				2		579
		NGC 6637 sq1 13					1	16.34		0.91				1		579
		NGC 6637 sq1 15					1	16.35		0.87				1		579
		NGC 6637 sq1 17					1	14.42		1.43				1		579
		NGC 6637 sq1 41					1	14.63		1.41				1		579
		NGC 6637 sq1 42					1	15.61		1.28				1		579
		NGC 6638 sq1 1					1	14.27		1.48		1.68		4		474
		NGC 6638 sq1 2					1	13.99		1.51		1.86		4		474
		NGC 6638 sq1 9					1	14.20		0.73		0.22		3		474
		NGC 6638 sq1 10					1	15.25		0.84				2		474
		NGC 6638 sq1 11					1	15.71		1.51				2		474
		NGC 6638 sq1 12					1	14.27		1.44		1.42		2		474
		NGC 6638 sq1 13					1	14.42		1.37		1.59		3		474
		NGC 6638 sq1 15					1	14.51		1.65		1.93		4		474
		NGC 6638 sq1 16					1	15.26		1.29		1.32		4		474
		NGC 6656 sq2 8					2	13.32	.014	1.36	.000	1.13	.032	13		559,1677
		NGC 6656 sq2 10					3	13.72	.019	1.10	.008	0.38	.000	11		559,560,1677
		NGC 6656 sq2 11					2	14.09	.000	1.39	.000	1.23	.000	6		559,1677
		NGC 6656 sq2 12					3	14.44	.024	1.76	.046	1.68	.005	9		559,560,1677
		NGC 6656 sq2 13					3	15.07	.028	0.39	.035	-0.05	.000	8		559,560,1677
		NGC 6656 sq2 14					2	15.17	.000	1.65	.000	1.09	.000	6		559,1677
		NGC 6656 sq2 15					3	15.49	.018	1.03	.065	0.23	.000	7		559,560,1677
		NGC 6656 sq2 16					2	15.60	.000	1.49	.000	1.02	.000	6		559,1677
		NGC 6656 sq2 17					2	16.18	.000	1.36	.000			6		559,1677
		NGC 6656 sq2 32					1	12.82		1.28				1		560
		NGC 6656 sq2 34					1	16.17		1.44				1		560
		NGC 6656 sq3 20					1	14.19		1.00				4		1677
		NGC 6681 sq1 18					1	14.18		1.32				3		438
		NGC 6681 sq1 19					1	14.23		0.84				3		438
		NGC 6681 sq1 20					1	15.17		0.87				1		438
		NGC 6681 sq1 21					1	15.78		1.00				1		438
		NGC 6712 sq1 7					1	14.81		0.85		0.28		2		641
		NGC 6712 sq1 9					1	14.44		1.43				2		641
		NGC 6712 sq1 10					1	15.25		1.23		0.88		3		641
		NGC 6715 sq1 11					1	15.59		1.00				1		438
		NGC 6717 sq1 27					1	12.06		1.02				2		811
		NGC 6717 sq1 28					1	13.02		0.57				1		811
		NGC 6717 sq1 29					1	13.54		1.44				2		811
		NGC 6717 sq1 30					1	14.02		0.78				1		811
		NGC 6717 sq1 31					1	14.38		0.91				2		811
		NGC 6717 sq1 32					1	15.22		1.00				4		811
		NGC 6717 sq1 34					1	16.05		1.01				4		811
		NGC 6723 sq1 2					1	11.53		1.02		0.80		4		349
		NGC 6723 sq1 4					1	13.08		0.56		0.08		5		349
		NGC 6723 sq1 5					1	13.30		1.47		1.70		6		349
		NGC 6723 sq1 6					1	13.68		0.63		0.12		2		349
		NGC 6723 sq1 7					1	13.68		1.63		2.08		1		349
		NGC 6723 sq1 8					1	14.68		0.57		0.00		1		349
		NGC 6723 sq1 9					1	15.17		1.05		0.94		1		349
		NGC 6723 sq1 10					1	15.45		0.54		0.29		1		349
		NGC 6723 sq1 11					1	15.68		0.55				1		349
		NGC 6723 sq1 12					1	16.35		1.00				1		349
		NGC 6723 sq1 13					1	15.37		1.02		0.45		1		349
		NGC 6723 sq1 14					1	12.96		1.58		1.97		1		349
		NGC 6723 sq1 15					1	14.40		0.91		0.70		1		349
		NGC 6723 sq1 16					1	16.00		0.71		-0.08		1		349
		NGC 6723 sq1 17					1	12.67		1.04		1.06		1		349
		NGC 6723 sq1 18					1	13.22		0.59		0.07		1		349
		NGC 6723 sq1 19					1	15.31		0.68		0.03		1		349
		NGC 6723 sq1 20					1	15.44		0.97		0.48		1		349
		NGC 6723 sq1 21					1	13.04		1.42		1.79		1		349
		NGC 6723 sq1 22					1	14.18		0.45				1		349
		NGC 6723 sq1 23					1	16.54		1.02				1		349

HD	DM	Other Id	N	Rem	α_{1950}	δ_{1950}	S	V	σ_V	B−V	σ_{B-V}	U−B	σ_{U-B}	n	Spectrum	References
		NGC 6723 sq1 24					1	17.01		0.46				1		349
		NGC 6723 sq1 25					1	17.65		0.44				1		349
		NGC 6723 sq1 26					1	17.64		0.34				2		349
		NGC 6723 sq1 27					1	17.05		0.79				2		349
		NGC 6752 sq1 4					1	11.95		1.57		1.53		1		368
		NGC 6752 sq1 5					1	12.62		0.99		0.63		1		368
		NGC 6752 sq1 6					1	12.78		1.24				1		368
		NGC 6752 sq1 7					1	15.59		0.69		0.28		1		368
		NGC 6752 sq1 8					1	14.20		1.10		0.84		1		368
		NGC 6752 sq1 9					1	14.64		0.81				1		368
		NGC 6752 sq1 10					1	15.00		0.77				1		368
		NGC 6752 sq1 16					1	14.40		0.01				1		368
		NGC 6752 sq2 1					1	17.23		0.54				1		887
		NGC 6752 sq2 2					1	16.60		0.84				2		887
		NGC 6752 sq2 3					1	17.10		0.53				1		887
		NGC 6752 sq2 4					1	16.25		0.61				1		887
		NGC 6752 sq2 5					1	17.94		0.62				1		887
		NGC 6752 sq2 6					1	17.59		0.48				1		887
		NGC 6752 sq2 7					1	17.72		-1.05				1		887
		NGC 6752 sq2 8					1	16.94		0.48				1		887
		NGC 6752 sq2 9					1	17.19		0.31				1		887
		NGC 6752 sq2 10					1	17.89		0.49				1		887
		NGC 6752 sq2 11					1	18.33		0.59				2		887
		NGC 6752 sq2 12					1	18.37		0.48				2		887
		NGC 6752 sq2 13					1	18.18		0.29				2		887
		NGC 6752 sq2 14					1	19.24		0.41				1		887
		NGC 6752 sq2 15					1	18.18		0.61				1		887
		NGC 6752 sq2 16					1	18.51		0.56				2		887
		NGC 6752 sq2 17					1	17.71		0.55				2		887
		NGC 6752 sq2 18					1	18.93		0.52				1		887
		NGC 6752 sq2 19					1	17.53		0.45				1		887
		NGC 6752 sq2 20					1	16.83		0.51				2		887
		NGC 6752 sq2 21					1	17.95		0.26				1		887
		NGC 6752 sq2 22					1	17.45		0.55				1		887
		NGC 6809 sq1 10					1	14.55		1.30				3		422
		NGC 6809 sq1 12					1	14.88		1.01		0.70		3		422
		NGC 6809 sq2 14					1	15.43		0.83				1		438
		NGC 6809 sq2 16					1	16.26		1.16				1		438
		NGC 6809 sq3 11					1	14.59		0.86		0.18		1		552
		NGC 6809 sq3 12					1	14.78		0.15		0.18		2		552
		NGC 6809 sq4 5					1	14.06		0.87				4		1677
		NGC 6809 sq4 6					1	14.26		0.92		0.26		6		1677
		NGC 6809 sq4 7					1	14.51		0.15		0.20		4		1677
		NGC 6809 sq4 8					1	14.61		0.86				7		1677
		NGC 6809 sq4 9					1	15.12		0.76		0.35		4		1677
		NGC 6822 sq1 1					1	12.26		0.74		0.23		12		602
		NGC 6822 sq1 2					1	16.13		1.09		0.57		1		602
		NGC 6822 sq1 3					1	17.50		1.51		1.02		1		602
		NGC 6822 sq1 4					1	19.07		1.06		0.05		1		602
		NGC 6822 sq1 5					2	18.31	.072	0.90	.050	0.94		5		575,602
		NGC 6822 sq1 6					1	20.39		0.71		-0.10		1		602
		NGC 6822 sq1 7					2	15.96	.060	1.06	.025	0.84	.139	7		575,602
		NGC 6822 sq1 8					1	18.22		0.85				1		602
		NGC 6822 sq1 9					1	19.78		1.13				2		602
		NGC 6822 sq1 10					1	17.28		0.88		0.18		1		602
		NGC 6822 sq1 11					3	16.99	.000	0.90	.020	0.31	.000	4		575,602,812
		NGC 6822 sq1 12					1	13.15		0.66		0.12		3		602
		NGC 6822 sq1 13					1	15.82		0.92		0.54		1		602
		NGC 6822 sq1 14					1	16.37		1.02		0.50		1		602
		NGC 6822 sq1 15					1	21.21		0.60				1		602
		NGC 6822 sq1 16					1	20.50		0.35				1		602
		NGC 6822 sq1 17					1	13.40		0.66		0.14		3		602
		NGC 6822 sq1 18					1	17.30		1.11		0.56		2		602
		NGC 6822 sq1 19					1	19.48		0.79				2		602
		NGC 6822 sq1 20					1	19.87		1.79				1		602
		NGC 6822 sq1 21					1	20.82		0.74				1		602
		NGC 6822 sq1 22					2	17.28	.005	0.70	.010	0.11	.000	3		602,812
		NGC 6822 sq2 2					1	16.83		1.30		1.74		3		575
		NGC 6822 sq2 5					1	18.29		0.88				1		575
		NGC 6822 sq2 7					1	16.84		0.99		0.49		1		575
		NGC 6822 sq2 9					1	19.69		0.25				2		575
		NGC 6822 sq2 10					1	19.99		1.14				1		575
		NGC 6822 sq2 11					1	20.69		0.38				2		575
		NGC 6822 sq3 1					1	10.58		1.31		1.20		6		812
		NGC 6822 sq3 2					1	13.41		0.67		0.15		9		812
		NGC 6822 sq3 3					1	15.52		0.97				4		812
		NGC 6822 sq3 4					1	11.49		0.68		0.16		3		812
		NGC 6822 sq3 5					1	13.48		0.72		0.18		2		812
		NGC 6822 sq3 6					1	14.30		0.88		0.38		2		812
		NGC 6822 sq3 7					1	10.98		1.29		1.16		3		812

Table 2 — 1295

HD	DM	Other Id	N Rem	α_{1950}	δ_{1950}	S	V	σ_V	B–V	σ_{B-V}	U–B	σ_{U-B}	n	Spectrum	References
		NGC 6822 sq3 8				1	11.65		0.66		0.16		3		812
		NGC 6822 sq3 10				1	12.75		0.94		0.42		6		812
		NGC 6822 sq3 11				1	14.83		1.12		0.73		1		812
		NGC 6822 sq3 12				1	14.64		0.74		0.12		1		812
		NGC 6822 sq3 13				1	10.03		0.68		0.06		1		812
		NGC 6822 sq3 14				1	15.58		1.19				1		812
		NGC 6822 sq3 15				1	16.68		0.95				1		812
		NGC 6822 sq3 16				1	15.29		0.56		0.22		2		812
		NGC 6822 sq3 17				1	15.83		0.96		0.47		1		812
		NGC 6822 sq3 18				1	13.92		0.99		0.49		1		812
		NGC 6822 sq3 19				1	15.41		1.33				1		812
		NGC 6822 sq3 20				1	15.18		0.99		0.52		3		812
		NGC 6822 sq3 21				1	14.82		0.80		0.24		4		812
		NGC 6822 sq3 22				1	13.44		0.86		0.33		1		812
		NGC 6822 sq3 23				1	13.96		1.58		1.69		5		812
		NGC 6822 sq3 24				1	13.65		1.48		1.40		3		812
		NGC 6822 sq3 25				1	12.42		1.33		1.18		2		812
		NGC 6822 sq3 26				1	13.82		0.76		0.30		2		812
		NGC 6822 sq3 27				1	15.72		0.83		0.44		3		812
		NGC 6822 sq3 28				1	13.33		1.48		1.28		2		812
		NGC 6822 sq3 29				1	16.00		0.95				1		812
		NGC 6822 sq3 30				1	15.45		0.76		0.30		2		812
		NGC 6822 sq3 31				1	16.44		0.32		-0.47		6		812
		NGC 6822 sq3 32				1	15.33		0.30		0.26		4		812
		NGC 6822 sq3 33				1	16.52		0.56		0.42		2		812
		NGC 6822 sq3 34				1	16.47		0.34		-0.42		5		812
		NGC 6822 sq3 35				1	16.80		0.40		0.40		2		812
		NGC 6822 sq3 36				1	16.96		0.21		-0.59		5		812
		NGC 6822 sq3 37				1	16.23		0.99				1		812
		NGC 6822 sq3 38				1	16.46		1.89				5		812
		NGC 6822 sq3 39				1	17.04		0.20		-0.27		4		812
		NGC 6822 sq3 40				1	16.96		2.07				5		812
		NGC 6822 sq3 41				1	17.26		1.45				3		812
		NGC 6822 sq3 42				1	17.06		1.78				3		812
		NGC 6822 sq3 43				1	18.06		1.70				4		812
		NGC 6822 sq3 44				1	17.94		1.46				4		812
		NGC 6822 sq3 45				1	17.90		1.50				3		812
		NGC 6822 sq3 46				1	17.60		2.25				5		812
		NGC 6822 sq3 47				1	16.98		1.70				3		812
		NGC 6822 sq3 48				1	17.70		0.02		-0.80		3		812
		NGC 6822 sq3 49				1	17.80		1.12				2		812
		NGC 6822 sq3 50				1	17.05		0.53		0.32		3		812
		NGC 6822 sq3 51				1	17.13		0.22		-0.56		2		812
		NGC 6822 sq3 52				1	17.38		0.37		-0.23		3		812
		NGC 6822 sq3 53				1	17.33		0.34		-0.25		4		812
		NGC 6822 sq3 54				1	17.49		0.15		-0.71		3		812
		NGC 6822 sq3 55				1	16.97		0.36		0.03		3		812
		NGC 6822 sq3 56				1	17.31		0.12		-0.80		3		812
		NGC 6822 sq3 57				1	16.77		0.34		-0.42		3		812
		NGC 6822 sq3 58				1	16.46		0.24		0.08		2		812
		NGC 6822 sq3 59				1	16.93		0.56		-0.31		2		812
		NGC 6822 sq3 60				1	15.52		0.78		0.18		1		812
		NGC 6822 sq3 61				1	16.09		0.65		0.40		1		812
		NGC 6838 sq1 1				1	10.04		0.08		-0.57		13		263
		NGC 6838 sq1 2				1	12.56		1.19		0.96		8		263
		NGC 6838 sq1 3				1	12.89		0.58		0.02		8		263
		NGC 6838 sq1 4				1	14.30		1.02		0.42		5		263
		NGC 6838 sq1 5				1	10.67		0.49		-0.08		9		263
		NGC 6838 sq1 6				1	11.77		0.24		0.08		5		263
		NGC 6838 sq1 7				1	12.36		1.74		2.20		3		263
		NGC 6838 sq1 8				1	14.22		1.30		1.16		4		263
		NGC 6838 sq1 9				1	12.46		1.32		1.08		4		263
		NGC 6838 sq1 10				1	13.64		0.17		-0.07		3		263
		NGC 6838 sq1 11				1	12.21		1.20		0.65		3		263
		NGC 6838 sq1 12				1	12.76		0.26		-0.16		2		263
		NGC 6838 sq1 13				1	11.99		1.33		1.45		1		263
		NGC 6864 sq1 16				1	15.33		0.88				1		438
		NGC 6864 sq1 17				1	15.51		0.75				1		438
		NGC 6934 sq1 9				1	15.10		0.78				3		339
		NGC 6934 sq1 10				1	15.50		1.22				3		339
		NGC 6934 sq1 11				1	15.82		0.73				1		339
		NGC 6934 sq1 12				1	16.12		0.84				1		339
		NGC 6934 sq1 13				1	16.56		1.00				2		339
		NGC 6934 sq1 14				1	17.31		0.99				1		339
		NGC 6934 sq1 15				1	18.23		0.76				1		339
		NGC 7089 sq1 1				1	13.22		1.41		1.42		3		296
		NGC 7089 sq1 2				1	14.09		1.08		0.78		3		296
		NGC 7089 sq1 3				1	14.68		0.99		0.50		3		296
		NGC 7089 sq1 4				1	16.16		0.20		0.03		3		296
		NGC 7089 sq1 5				1	11.77		0.72		0.31		2		296

HD	DM	Other Id	N Rem	α_{1950}	δ_{1950}	S	V	σ_V	B–V	σ_{B-V}	U–B	σ_{U-B}	n	Spectrum	References
		NGC 7089 sq1 6				1	11.94		0.40		-0.01		4		296
		NGC 7089 sq1 7				1	14.58		0.88		0.59		2		296
		NGC 7089 sq1 8				1	16.08		0.80		0.43		2		296
		NGC 7089 sq1 11				1	15.77		0.78		0.30		2		296
		NGC 7089 sq1 12				1	14.08		1.03		0.73		3		296
		NGC 7089 sq1 13				1	15.01		0.97				2		296
		NGC 7089 sq1 14				1	13.64		1.21		1.05		3		296
		NGC 7089 sq1 16	AB			1	13.05		1.38		1.07		1		296
		NGC 7089 sq1 17				1	13.23		1.36		1.39		3		296
		NGC 7089 sq1 18				1	14.49		0.93		0.49		2		296
		NGC 7089 sq1 19				1	13.74		1.12		0.97		1		296
		NGC 7089 sq1 20				1	13.86		1.22		0.70		1		296
		NGC 7099 sq1 14				1	14.90		0.75		0.29		2		738
		NGC 7099 sq1 15				1	15.32		0.06				2		738
		NGC 7099 sq1 16				1	15.77		0.53		-0.22		3		738
		NGC 7099 sq1 17				1	16.16		0.65				2		738
		NGC 7099 sq1 18				1	16.25		0.75				3		738
		NGC 7099 sq2 8				1	14.22		0.77		0.15		7		1638
		NGC 7099 sq2 10				1	15.15		0.78		0.05		8		1638
		NGC 7099 sq2 11				1	15.18		0.20		0.16		9		1638
		NGC 7099 sq2 12				1	15.29		0.15		0.13		10		1638
		NGC 7213 sq1 4				1	12.78		0.68		0.18		2		1687
		NGC 7293 sq1 *				1	13.54		-0.39		-1.22		2		268
		Nor sq 3 # 33				1	14.74		1.30		0.39		1		336
		Nor sq 3 # 35				1	13.58		1.40		0.55		1		336
		Nor sq 3 # 43				1	14.27		1.31		0.43		1		336
		Nor sq 3 # 47				1	13.88		1.08		0.24		1		336
		Nor sq 3 # 50				1	13.38		0.80		0.08		2		336
		Nor sq 3 # 73				1	12.74		0.68		0.14		1		336
		Nor sq 3 # 87				1	12.87		0.79		0.00		2		336
		Nor sq 3 # 91				1	13.45		0.74		0.21		1		336
		Nor sq 3 # 109				1	12.93		0.73		-0.07		2		336
		Nor sq 3 # 118				1	12.38		0.86		-0.28		2		336
		Nor sq 3 # 201				1	14.72		0.88		0.21		1		336
		Nor sq 3 # 202				1	14.97		0.90		0.27		1		336
		Nor sq 3 # 207				1	14.92		0.81		0.23		1		336
		Nor sq 3 # 211				1	13.95		1.73		1.02		1		336
		Nor sq 3 # 224				1	15.03		1.02		0.31		1		336
		Nor sq 3 # 236				1	13.91		0.72		0.00		1		336
		Nor sq 3 # 239				1	14.66		0.81		0.30		1		336
		Nor sq 3 # 250				1	14.54		0.82		0.21		1		336
		Nor sq 3 # 251				1	15.13		0.75		0.07		1		336
		Nor sq 3 # 302				1	14.59		1.58		0.61		1		336
		Nor sq 3 # 306				1	15.00		1.42		0.46		1		336
		Nor sq 3 # 307				1	13.58		1.30		0.12		3		336
		Nor sq 3 # 311				1	13.75		1.20		0.26		1		336
		Nor sq 3 # 313				1	13.83		1.76		0.78		1		336
		Nor sq 3 # 315				1	14.67		1.64		0.77		1		336
		Nor sq 3 # 317				1	14.48		1.73		0.74		1		336
		Nor sq 3 # 404				1	14.79		0.56		0.08		1		336
		Nor sq 3 # 415				1	14.38		0.67		-0.02		1		336
		Oph sq 1 # 4				1	13.87		1.87				1		1473
		Oph sq 1 # 5				1	16.01		-0.04				1		1473
		Oph sq 1 # 6				1	12.73		1.05				1		1473
		Oph sq 1 # 8				1	14.33		1.02				1		1473
		Orion B 1201				1	16.28		1.74		0.75		1		114
		Orion B 1202				1	15.59		1.48		1.15		1		114
		Orion B 1203				1	16.35		1.39		0.71		1		114
		PASP75,194 # 6				1	15.97		0.81				2		1190
		PASP75,194 # 15				1	13.81		0.40				1		1190
		PASP75,194 # 23				1	13.61		0.70				4		1190
		PASP75,194 # 31				1	12.80		0.48				3		1190
		PASP75,194 # 33				1	16.58		0.72				3		1190
		PASP75,194 # 34				1	15.93		1.40				1		1190
		PASP75,194 # 36				1	16.93		0.95				3		1190
		PASP75,194 # 39				1	17.93		0.55				2		1190
		PASP75,194 # 40				1	19.02		1.05				5		1190
		PASP75,256 # 46				1	14.65		0.78		0.41		1		619
		PASP75,256 # 47				1	15.41		0.66		0.05		1		619
		PASP75,256 # 48				1	15.40		0.86		0.50		1		619
		PASP75,256 # 49				1	15.68		0.84		0.34		1		619
		PASP75,256 # 51				1	15.71		1.30				1		619
		PASP75,256 # 52				1	17.06		0.61				1		619
		PASP75,256 # 53				1	16.88		1.07				1		619
		PASP85,203 # 7				1	13.28		0.48		0.09		7		266
		PASP85,203 # 9				1	15.43		1.25		1.50		14		266
		PASP85,203 # 10				1	14.00		0.95		0.70		6		266
		PASP85,203 # 12				1	13.82		0.69		0.05		6		266
		PASP85,203 # 13				1	15.23		1.26		1.10		6		266
		PASP86,394 # 201				1	14.60		0.89		0.00		2		496

Table 2   1297

HD	DM	Other Id	N Rem	α_{1950}	δ_{1950}	S	V	σ_V	B–V	σ_{B-V}	U–B	σ_{U-B}	n	Spectrum	References
		PASP86,394 # 202				1	13.87		0.83		-0.10		3		496
		PASP86,394 # 203				1	14.11		1.01		0.05		2		496
		PASP86,394 # 205				1	14.76		0.88		0.11		2		496
		PASP86,394 # 206				1	14.35		0.94		0.28		2		496
		PASP87,379 # 28				1	14.80		1.50		1.40		6		408
		PASP87,379 # 29				1	15.51		0.85		0.20		3		408
		PASP87,379 # 30				1	14.99		2.13		1.30		3		408
		PASP87,379 # 31				1	15.42		1.33		2.00		4		408
		PASP87,769 T1# 5				1	12.56		0.63		0.49		1		431
		PASP87,769 T1# 8				1	12.23		-0.06		-0.94		2		431
		PASP87,769 T1# 9				1	14.54		0.23		-0.79		1		431
		PASP87,769 T1# 10				1	14.27		-0.02		-0.86		1		431
		PASP87,769 T1# 11				1	14.03		-0.03		-0.90		1		431
		PASP87,769 T1# 12				1	13.22		0.45		0.01		1		431
		PASP88,699 # 107				1	10.32		0.30		0.10		2		634
		PASP94,905 # 8				1	13.02		1.68		1.91		1		971
		PASP94,977 # 5				1	15.49		1.09		0.59		4		972
		PASP94,977 # 6				1	14.41		0.92		0.62		4		972
		PASP94,977 # 7				1	14.76		0.95		0.32		4		972
		PASP94,977 # 8				1	14.56		1.01		0.38		4		972
		PASP94,977 # 12				1	14.58		0.87		0.29		4		972
		PASP94,977 # 13				1	14.21		0.89		0.37		4		972
		PASP94,977 # 14				1	16.02		1.27		1.01		1		972
		PASP94,977 # 15				1	15.04		0.87		-0.19		1		972
		PASP94,977 # 16				1	14.37		1.19		-0.07		1		972
		PASP94,977 # 17				1	16.00		1.67		1.59		1		972
		PASP94,977 # 18				1	15.74		1.44		0.67		1		972
		PASP94,977 # 19				1	15.13		1.21		1.17		4		972
		PASP94,977 # 20				1	15.39		0.93		0.31		4		972
		PASP94,977 # 21				1	14.78		1.04		0.45		4		972
		PASP94,977 # 22				1	14.20		1.10		0.70		4		972
		PASP94,977 # 23				1	15.26		1.12		0.42		4		972
		PASP94,977 # 24				1	14.43		0.99		0.39		4		972
		PASP94,977 # 25				1	15.26		0.99		0.27		4		972
		PASP96,422# 1				1	13.67		0.51		0.17		6		1534
		PASP96,422# 2				1	14.08		0.51		0.35		6		1534
		PASP96,422# 3				1	13.72		1.68		1.64		3		1534
		PASP96,422# 4				1	12.82		0.40		-0.05		5		1534
		PASP96,422# 5				1	11.77		0.68		0.11		3		1534
		PASP96,422# 6				1	13.81		0.55		0.16		5		1534
		PASP96,422# 7				1	13.42		0.45		0.14		5		1534
		PASP96,422# 8				1	13.02		0.32		-0.09		5		1534
		PASP96,422# 9				1	15.09		0.53		0.24		4		1534
		PASP96,422# 10				1	12.23		0.39		-0.03		4		1534
		PASP96,422# 11				1	14.37		1.52		1.49		2		1534
		PASP96,422# 12				1	14.33		0.70		0.26		1		1534
		PHL 658 # 1				1	13.48		0.73		0.18		1		157
		PHL 658 # 2				1	14.19		0.73		0.24		1		157
		PHL 658 # 3				1	14.30		0.96		0.60		1		157
		PHL 658 # 4				1	15.04		0.74		0.15		1		157
		PHL 658 # 5				1	15.43		0.61		0.02		1		157
		PHL 658 # 6				1	15.59		0.81		-0.23		1		157
		PHL 658 # 7				1	15.89		0.54		-0.07		1		157
		PHL 1377 # 2				1	17.81		0.12		0.03		1		327
		PKS 0846+10 # 1				1	10.12		0.41		-0.03		1		829
		PKS 0846+10 # 2				1	13.29		0.77		0.31		1		829
		PKS 0846+10 # 3				1	13.89		0.54		0.02		1		829
		PKS 0846+10 # 4				1	14.60		0.79		0.38		1		829
		Pav sq 1 # 5				1	13.68		0.82		0.55		2		8098
		Pav sq 1 # 6				1	14.70		0.66		0.24		2		8098
		Pav sq 1 # 7				1	15.30		0.71		0.50		2		8098
		Pav sq 1 # 8				1	13.95		0.66		0.13		7		8098
		Pav sq 1 # 10				1	15.83		0.69		0.12		3		8098
		Pav sq 1 # 13				1	13.91		0.61		0.06		2		8098
		Pav sq 1 # 16				1	13.83		0.84		0.47		2		8098
		Pav sq 1 # 17				1	16.00		1.47		1.41		2		8098
		Pav sq 1 # 18				1	14.28		0.98		0.74		2		8098
		Pav sq 1 # 19				1	14.64		0.83		0.46		2		8098
		Pav sq 1 # 23				1	15.63		0.61		-0.03		2		8098
		Pav sq 1 # 24				1	14.75		0.89		0.57		2		8098
		Pav sq 1 # 25				1	15.25		0.67		0.30		1		8098
		Pav sq 1 # 26				1	14.43		0.99		0.84		1		8098
		Peg sq 1 # 3				1	11.97		0.08		-0.05		1	B5	1364
		Peg sq 1 # 4				1	12.37		0.40		0.23		1	A3	1364
		Peg sq 1 # 6				1	12.98		0.18		-0.04		2	A2	1364
		Peg sq 1 # 8				1	11.88		0.36		0.22		1	A7	1364
		Peg sq 1 # 9				1	11.09		1.59		1.94		1	K3 III	1364
		Peg sq 1 # 12				1	11.64		1.32		1.26		1	K2p	1364
		Peg sq 1 # 17				1	11.36		0.37		0.22		1	A7m	1364
		Peg sq 1 # 19				1	11.65		1.35		1.54		1	K2 III	1364

HD	DM	Other Id	N	Rem	α_{1950}	δ_{1950}	S	V	σ_V	B–V	σ_{B-V}	U–B	σ_{U-B}	n	Spectrum	References
		Peg sq 1 # 22					1	12.98		0.28		0.07		1	A3	1364
		Peg sq 1 # 23					1	13.35		0.71		0.20		1		1364
		Peg sq 1 # 24					1	12.84		0.16		0.14		1	A2	1364
		Peg sq 1 # 26					1	12.25		0.17		0.14		2	A2	1364
		Peg sq 1 # 28					1	14.18		0.76		0.14		1		1364
		Peg sq 1 # 31					1	12.93		0.57		-0.01		1		1364
		Peg sq 1 # 32					1	12.30		0.64		0.07		1	F8	1364
		Peg sq 1 # 33					1	13.51		0.63		0.14		1		1364
		Peg sq 1 # 34					1	11.01		0.29		0.23		1	A2	1364
		Peg sq 1 # 37					1	13.68		0.13		0.24		1	A3	1364
		Peg sq 1 # 38		A			1	11.43		0.59		0.06		1	A0	1364
		Peg sq 1 # 38		B			1	11.97		0.06		-0.38		2	F8	1364
		Peg sq 1 # 41					1	11.52		0.34		0.13		1	A5	1364
		Peg sq 1 # 42					1	12.61		0.33		0.19		2	A5	1364
		Peg sq 1 # 43					1	13.40		0.25		0.28		2	A7	1364
		Peg sq 1 # 44					1	11.18		0.97		0.76		1	K2 III	1364
		Peg sq 1 # 49					1	13.91		0.06		-0.04		1		1364
		Peg sq 1 # 50					1	11.62		0.71		0.18		1	G5 V	1364
		Peg sq 1 # 52					1	12.26		0.30		0.12		1	A5	1364
		Peg sq 1 # 53					1	11.63		0.29		0.14		1	A5	1364
		Peg sq 1 # 59					1	11.99		0.22		0.19		1	A7	1364
		Peg sq 1 # 63					1	12.32		0.32		0.14		1	A5	1364
		Peg sq 1 # 65					1	11.50		1.17		1.00		1	K0 III	1364
		Peg sq 1 # 68					1	10.63		1.18		1.15		1	K0 III	1364
		Peg sq 1 # 69					1	12.09		1.03		0.95		1	K0 III	1364
		Peg sq 1 # 70					1	12.96		0.12		0.24		1	A2	1364
		Peg sq 1 # 71					1	12.39		0.38		0.21		1	F2	1364
		Peg sq 1 # 73					1	11.93		0.33		0.10		1	A7	1364
		Peg sq 1 # 74					1	11.45		0.30		0.17		1	A3	1364
		Peg sq 1 # 76					1	12.45		1.45		1.74		1	K0	1364
		Peg sq 1 # 77					1	11.68		1.40		1.63		1	K2 III	1364
		Peg sq 1 # 83					1	11.70		0.33		0.04		1	F0	1364
		PsA sq 1 # 6					1	15.25		0.97		0.75		4		840
		PsA sq 1 # 7					1	15.98		1.52		1.42		3		840
		PsA sq 1 # 8					1	16.25		0.55				1		840
		PsA sq 1 # 10					1	14.84		0.60		-0.01		1		840
		PsA sq 1 # 11					1	14.89		0.56		-0.01		1		840
		PsA sq 1 # 14					1	15.02		0.95		0.74		1		840
		PsA sq 1 # 15					1	16.06		0.52		-0.03		1		840
		PsA sq 1 # 16					1	16.97		0.62		0.02		1		840
		Psc sq 1 # 5					1	15.08		0.96		1.23		8		625
		Pup sq 4 # 5					1	13.32		0.51		0.38		2		525
		Pup sq 4 # 6					1	14.09		0.84		0.36		3		525
		Pup sq 4 # 7					1	14.15		0.71		0.44		3		525
		Pup sq 4 # 8					1	14.69		0.48		0.10		3		525
		Pup sq 4 # 9					1	14.69		0.49		0.05		3		525
		Pup sq 4 # 10					1	15.40		0.73		0.22		3		525
		Pup sq 4 # 11					1	16.08		0.59		0.10		3		525
		Pup sq 4 # 12					1	16.70		0.80				3		525
		Pup sq 5 # 4					1	13.72		0.30		0.27		1		947
		Pup sq 5 # 5					1	14.47		0.56		0.32		1		947
		Pup sq 5 # 6					1	13.90		0.41		0.33		1		947
		Pup sq 5 # 19					1	13.73		0.26		0.14		1		947
		Pup sq 5 # 23					1	12.48		1.75		2.00		2	C	947
		Pup sq 5 # 25					1	13.70		1.30		1.24		1		947
		Pup sq 5 # 28					1	13.95		0.92		0.50		1		947
		Pup sq 5 # 31					1	13.29		1.05		0.62		1		947
		Pup sq 5 # 37					1	13.89		1.33		0.95		1		947
		Pup sq 6 # 3					1	13.98		1.04				1		1473
		Pup sq 6 # 5					1	14.19		0.61				1		1473
		Pup sq 6 # 7					1	13.76		0.60				1		1473
		Pup sq 6 # 9					1	14.76		0.27				1		1473
		Pup sq 6 # 10					1	14.83		0.49				1		1473
		RMAA1,381 # 17					1	13.26		0.39		-0.34		3		1684
		RMAA1,381 # 141					1	11.19		0.58		-0.25		5		1684
		RMAA1,381 # 167					1	9.92		2.24		0.67		4		1684
		RMAA1,381 # 188					1	12.01		1.19		0.02		5		1684
		RMAA1,381 # 207					1	14.30		1.15		-0.20		5		1684
		Reticulum sq # 3					1	17.18		0.83				1		501
		Reticulum sq # 5					1	19.06		0.75				1		501
		Reticulum sq # 6					1	14.13		1.46				1		501
		Reticulum sq # 11					1	18.46		0.49				1		501
		Reticulum sq # 14					1	20.43		0.36				1		501
		Reticulum sq # 23					1	18.03		0.96				1		501
		Reticulum sq # 24					1	19.25		0.84				1		501
		Reticulum sq # 34					1	16.79		0.66				1		501
		Reticulum sq # 43					1	17.41		0.30				1		501
		Reticulum sq # 45					1	19.71		0.89				1		501
		Reticulum sq # 46					1	16.24		1.30				1		501
		Reticulum sq # 50					1	17.35		1.02				1		501

Table 2 1299

HD	DM	Other Id	N Rem	α_{1950}	δ_{1950}	S	V	σ_V	B–V	σ_{B-V}	U–B	σ_{U-B}	n	Spectrum	References
		Reticulum sq # 123				1	17.95		1.53				1		501
		Reticulum sq # 156				1	19.14		-0.01				1		501
		SL 4 sq # 11				1	14.66		1.36				3		1646
		SL 4 sq # 12				1	15.20		0.72				3		1646
		SL 4 sq # 93				1	17.38		0.83				3		1646
		SL 4 sq # 94				1	16.38		1.08				3		1646
		San +23.008				1	14.19		1.56				1	dM5	685
		San +23.066				1	13.97		1.46		1.25		1		685
		San +23.085				1	14.57		1.40		1.65		1	dM2	685
		San +24.082				1	12.14		1.49		1.90		3	gM0	685
		San +25.035				1	14.53		1.43				1	dM3	685
		San +26.024				1	15.24		1.59				1	dM3	685
		San +26.095				1	15.11		1.71				1	dM5	685
		San +26.118				1	13.44		1.47		1.27		1	dM1	685
		San +27.007				1	15.30		1.66				1	dM3	685
		San +27.083				1	12.66		1.28		1.33		3	dM0	685
		San +27.094				1	13.10		1.46		1.13		3	dM1	685
		San +27.128				1	10.86		1.52		1.87		3	gM4	685
		San +27.142				1	14.73		1.48				1	dM3	685
		San +28.076				1	15.78		1.50				2		694
		San +28.078				1	15.79		1.51				2		694
		San +28.079				1	16.18		1.58				2		694
		San +28.080				1	15.99		1.48				2		694
		San +28.081				1	16.50		1.74				2		694
		San +28.082				1	16.88		1.62				2		694
		San +28.083				1	15.84		1.56				2		694
		San +28.084				1	13.60		1.42				2		694
		San +28.085				1	15.58		1.50				2		694
		San +28.086				2	13.26	.000	1.48	.000			3		694,784
		San +28.087				1	14.36		1.58				2		694
		San +28.088				1	16.11		1.44				2		694
		San +28.089				1	14.27		1.54				2		694
		San +28.091				1	13.72		1.27				2		694
		San +28.093				1	15.40		1.58				2		694
		San +29.116				2	16.60	.000	1.56	.000			3		694,784
		San +29.117				2	15.21	.000	1.54	.000			3		694,784
		San +29.120				2	17.27	.000	1.58	.000			3		694,784
		San +29.122				1	16.61		1.59				2		694
		San +30.036				1	14.49		1.52				1	dM2	685
		San +30.098				1	15.48		1.36				2		694
		San +31.017				1	13.21		1.41		1.29		3	dM1	685
		San +31.054				1	14.10		1.48				1	dM3	685
		Sculptor sq # 5				1	14.78		0.83				1		527
		Sculptor sq # 6				1	14.93		0.80				3		527
		Sculptor sq # 7				1	15.06		1.32				1		527
		Sculptor sq # 8				1	17.47		1.36				1		527
		Sculptor sq # 9				1	17.55		0.36				3		527
		Sculptor sq # 10				1	17.92		1.55				3		527
		Sculptor sq # 11				1	18.09		1.05				1		527
		Sculptor sq # 12				1	18.42		1.11				1		527
		Sculptor sq # 13				1	18.83		0.79				1		527
		Sculptor sq # 14				1	19.12		1.01				1		527
		Sculptor sq # 15				1	19.82		0.81				1		527
		Sculptor sq # 16				1	20.24		0.80				1		527
		Sculptor sq # 33				1	20.41		-0.08				1		527
		Sculptor sq # 173				1	20.62		0.08				1		527
		Sextans B sq # 2				1	15.59		0.92				1		1579
		Sextans B sq # 3				1	13.48		0.59				2		1579
		Sextans B sq # 4				1	15.95		0.94				1		1579
		Sextans B sq # 6				1	16.31		0.53				1		1579
		Sextans B sq # 7				1	13.44		0.47				1		1579
		Sextans B sq # 8				1	16.27		0.68				1		1579
		Sextans B sq # 9				1	14.59		0.66				2		1579
		Sextans B sq # 10				1	13.30		0.81				1		1579
		Sextans B sq # 23				1	15.63		0.55		0.02		1		1579
		Sgr sq 1 # 1				1	11.25		1.07		1.00		3		413
		Sgr sq 1 # 3				1	11.32		1.23		1.16		3		413
		Sgr sq 1 # 4				1	13.33		0.32		0.28		3		413
		Sgr sq 1 # 5				1	12.30		0.38		-0.05		3		413
		Sgr sq 1 # 6				1	10.76		0.44		0.24		3		413
		Sgr sq 1 # 7				1	10.33		0.20		-0.42		3		413
		Sgr sq 1 # 8				1	10.25		0.38		0.06		3		413
		Sgr sq 1 # 9				1	10.88		0.16		-0.24		3		413
		Sgr sq 1 # 12				1	12.63		0.88		0.53		2		413
		Sgr sq 1 # 13				1	11.30		0.60		0.05		3		413
		Sgr sq 1 # 15				1	11.60		0.21		-0.11		2		413
		Sgr sq 1 # 20				1	11.98		0.26		0.18		2		413
		Sgr sq 1 # 21				1	14.02		0.37		0.38		2		413
		Sgr sq 1 # 22				1	12.16		0.32		0.32		2		413
		Sgr sq 1 # 26				1	10.79		0.04		-0.33		3		413

HD	DM	Other Id	N Rem	α_{1950}	δ_{1950}	S	V	σ_V	B–V	σ_{B-V}	U–B	σ_{U-B}	n	Spectrum	References
		Sgr sq 1 # 28				1	10.14		0.06		-0.48		3		413
		Sgr sq 1 # 32				1	12.36		0.32		-0.04		4		413
		Sgr sq 1 # 34				1	11.46		0.17		-0.35		3		413
		Sgr sq 1 # 35				1	10.99		0.00		-0.54		3		413
		Sgr sq 1 # 36				1	11.92		0.26		0.22		3		413
		Sgr sq 1 # 37				1	11.08		0.24		-0.13		3		413
		Sgr sq 1 # 39				1	13.94		0.40		0.10		2		413
		Sgr sq 1 # 41				1	10.95		0.37		0.24		3		413
		Sgr sq 1 # 102				1	15.02		0.86		0.26		2		413
		Sgr sq 1 # 201				1	13.50		0.61		0.16		1		413
		Sgr sq 1 # 202				1	13.05		0.50		0.11		1		413
		Sgr sq 1 # 203				1	14.66		1.03		0.47		1		413
		Sgr sq 1 # 204				1	15.62		1.55		1.02		1		413
		Sgr sq 1 # 302				1	15.66		0.60		0.60		1		413
		Sgr sq 1 # 401				1	12.45		0.32		0.20		1		413
		Sgr sq 1 # 402				1	15.51		1.12		1.50		1		413
		Sgr sq 1 # 502				1	14.99		0.86		0.24		1		413
		Sgr sq 1 # 602				1	14.79		0.87		0.28		1		413
		Sgr sq 1 # 701				1	10.33		1.06		0.78		4		413
		Sgr sq 1 # 702				1	14.70		0.61		0.62		1		413
		Sgr sq 1 # 801				1	10.50		0.08		-0.48		4		413
		Sgr sq 1 # 802				1	15.11		0.82		0.02		1		413
		Sgr sq 1 # 901				1	14.66		0.73		-0.01		1		413
		Sgr sq 1 # 902				1	14.69		0.66		0.12		1		413
		Sgr sq 1 # 903				1	14.63		0.66		0.07		2		413
		Sgr sq 1 # 904				1	15.36		0.64		-0.11		1		413
		Sgr sq 1 # 905				1	14.54		1.52		1.63		1		413
		Sgr sq 1 # 906				1	14.89		0.57		-0.10		1		413
		Sgr sq 1 # 907				1	13.84		0.37		0.15		1		413
		Sgr sq 1 # 908				1	15.34		0.74		0.39		2		413
		Sgr sq 1 # 909				1	15.06		0.62		0.20		1		413
		Sgr sq 1 # 910				1	14.68		0.77		0.03		1		413
		Sgr sq 1 # 911				1	15.10		0.65		0.22		1		413
		Sgr sq 1 # 912				1	14.81		1.02		0.08		1		413
		Sgr sq 1 # 913				1	14.72		0.66		0.27		2		413
		Sgr sq 1 # 914				1	13.88		0.69		0.34		1		413
		Sgr sq 1 # 915				1	14.80		0.60		0.33		1		413
		Sgr sq 1 # 916				1	14.18		0.59		0.18		2		413
		Sgr sq 1 # 917				1	14.79		0.60		0.04		2		413
		Sgr sq 1 # 918				1	15.02		0.70		0.16		1		413
		Sgr sq 1 # 919				1	14.60		0.60		0.16		2		413
		Sgr sq 1 # 920				1	14.40		0.66		0.29		1		413
		Sgr sq 1 # 921				1	14.87		0.77		0.25		1		413
		Sgr sq 1 # 922				1	14.30		0.82		0.21		1		413
		Sgr sq 1 # 923				1	14.80		0.92		0.42		1		413
		Sgr sq 1 # 924				1	14.81		0.73		0.17		2		413
		Sgr sq 1 # 925				1	14.74		0.85		0.13		1		413
		Sgr sq 1 # 926				1	14.86		0.76		0.22		2		413
		Sgr sq 1 # 927				1	14.95		0.70		0.20		1		413
		Sgr sq 1 # 928				1	15.11		0.59		0.08		2		413
		Sgr sq 1 # 931				1	15.04		0.75		0.20		2		413
		Sgr sq 2 # 1				1	12.70		1.40				2		915
		Sgr sq 2 # 2				1	13.34		1.34				2		915
		Sgr sq 2 # 3				1	13.65		1.51				2		915
		Sgr sq 2 # 4				1	13.80		1.50				2		915
		Sgr sq 2 # 5				1	13.03		1.47				2		915
		Sgr sq 2 # 6				1	13.57		1.52				2		915
		S244 # 1				1	9.86		0.74		0.29		2		1730
		S244 # 2				1	10.90		0.63		0.16		2		1730
		Tau sq 1 # 1				1	13.79		0.91		0.45		5		23
		Tau sq 1 # 2				1	12.99		0.63		0.11		26		23
		Tau sq 1 # 3				1	13.89		1.16		1.03		5		23
		Tau sq 1 # 4				1	12.27		1.08		0.74		45		23
		Tau sq 1 # 5				1	16.09		0.92		0.53		7		23
		Tau sq 1 # 6				1	14.31		0.57		0.12		4		23
		Tau sq 1 # 7				1	12.29		0.53		0.01		6		23
		Tau sq 1 # 8				1	15.99		0.64		0.15		2		23
		Tau sq 1 # 9				1	13.62		0.82		0.33		5		23
		Tau sq 1 # 10				1	13.43		0.99		0.66		5		23
		Tau sq 1 # 11				1	14.59		0.86		0.44		6		23
		Tau sq 1 # 12				1	13.12		0.57		0.07		6		23
		Tau sq 1 # 13				1	14.71		0.88		0.46		4		23
		Tau sq 1 # 14				1	14.26		0.83		0.36		4		23
		Tau sq 1 # 15				1	12.71		0.57		0.06		5		23
		Tau sq 1 # 16				1	13.69		0.57		0.05		5		23
		Tau sq 1 # 17				1	13.23		1.08		0.75		5		23
		Tau sq 1 # 18				1	14.76		0.61		0.08		7		23
		Tau sq 1 # 19				1	15.50		0.76		0.26		6		23
		Tau sq 1 # 20				1	13.44		0.60		0.02		6		23
		Tau sq 1 # 21				1	14.18		1.06		0.68		7		23

Table 2

1301

HD	DM	Other Id	N Rem	α_{1950}	δ_{1950}	S	V	σ_V	B–V	σ_{B-V}	U–B	σ_{U-B}	n	Spectrum	References
		Tau sq 1 # 22				1	11.86		0.49		0.03		5		23
		Tau sq 1 # 23				1	13.86		0.68		0.10		6		23
		Tau sq 1 # 31				1	14.40		0.66		0.05		7		23
		Tau sq 1 # 32				1	16.46		0.72		0.14		6		23
		Tau sq 1 # 33				1	14.10		0.74		0.27		5		23
		Tau sq 1 # 34				1	14.71		0.70		0.09		9		23
		Tau sq 1 # 35				1	14.00		0.85		0.46		13		23
		Tau sq 1 # 36				1	15.90		0.52		0.07		8		23
		Tau sq 1 # 38				1	14.00		1.11		0.79		6		23
		Tau sq 1 # 40				1	13.44		1.75		1.88		5		23
		Tau sq 1 # 42				1	13.67		0.49		0.10		5		23
		Tau sq 1 # 46				1	14.87		0.35		0.25		2		23
		Tau sq 1 # 49				1	15.33		1.05		0.54		6		23
		Tau sq 1 # 50				1	15.80		1.18		0.96		6		23
		Tau sq 1 # 51				1	13.51		0.66		0.14		5		23
		Tau sq 1 # 52				1	14.58		0.63		0.17		7		23
		Tau sq 1 # 53				1	12.42		0.76		0.26		6		23
		Tau sq 1 # 54				1	16.32		1.02		0.79		6		23
		Tau sq 1 # 55				1	14.54		0.92		0.46		6		23
		Tau sq 1 # 57				1	14.38		0.42		0.14		11		23
		Tau sq 1 # 58				1	15.90		0.55		0.34		6		23
		Tau sq 1 # 59				1	15.47		0.74		0.36		3		23
		Tau sq 1 # 60				1	16.80		0.81		0.14		8		23
		Tau sq 1 # 61				1	15.99		1.00		0.68		8		23
		Tau sq 1 # 104				1	15.66		1.06		0.36		6		23
		Tau sq 1 # 120				1	15.37		0.86		0.27		4		23
		Tau sq 1 # 129				1	16.23		1.15		1.07		4		23
		Tau sq 1 # 141				1	14.56		1.90		2.17		2		23
		Tau sq 1 # 150				1	15.67		0.91		0.33		6		23
		Tau sq 1 # 152				1	15.02		0.89		0.57		6		23
		Tau sq 1 # 153				1	16.73		1.07		0.75		6		23
		Tau sq 1 # 154				1	15.78		1.05		0.58		6		23
		Tau sq 1 # 155				1	15.41		1.00		0.51		6		23
		Tau sq 1 # 156				1	16.65		0.86		0.50		4		23
		Ton 256 # 1				1	13.96		0.80		0.38		1		157
		Ton 256 # 2				1	14.64		0.55		-0.06		1		157
		Ton 256 # 3				1	14.72		0.66		0.11		1		157
		Ton 256 # 4				1	14.82		0.77		0.39		1		157
		Ton 256 # 5				1	15.31		0.54		-0.19		1		157
		Ton 256 # 6				1	15.36		0.63		0.06		1		157
		Ton 256 # 7				1	15.32		1.03		0.64		1		157
		Ton 730 # 4				1	14.52		0.76		0.35		1		157
		Ton 730 # 6				1	15.37		0.74		0.40		1		157
		Ton 730 # 7				1	15.87		0.52		-0.33		1		157
		Ton 730 # 8				1	15.92		0.51		0.06		1		157
		Ton 882 # 1				1	11.86		0.67				1		157
		Ton 882 # 2				1	13.74		0.55				1		157
		Ton 882 # 3				1	13.99		0.70		0.20		1		157
		Ton 882 # 4				1	14.85		0.54		-0.01		1		157
		Ton 882 # 5				1	15.11		0.68		0.24		1		157
		Ton 882 # 6				1	15.40		0.93		0.70		1		157
		Ton 882 # 7				1	15.48		0.86		0.34		1		157
		Ton 882 # 8				1	15.80		0.55				1		157
		Ton 1542 # 1				1	13.27		0.56		-0.05		1		157
		Ton 1542 # 3				1	13.89		0.58		-0.05		1		157
		Ton 1542 # 4				1	14.09		0.85		0.37		1		157
		Ton 1542 # 5				1	14.26		0.97		0.65		1		157
		Ton 1542 # 6				1	15.00		0.70		0.16		1		157
		Ton 1542 # 7				1	15.24		0.61		0.17		1		157
		UMi sq # 9				1	15.59		0.57				1		1618
		UMi sq # 12				1	15.38		0.55				1		1618
		UMi sq # 13				1	17.43		0.86				1		1618
		UMi sq # 14				1	15.09		0.70				1		1618
		UMi sq # 23				1	18.17		1.49				1		1618
		UMi sq # 41				1	15.57		0.77				1		1618
		UMi sq # 44				1	15.59		1.28				1		1618
		UMi sq # 48				1	16.15		0.72				1		1618
		UMi sq # 60				1	17.35		0.63				1		1618
		UMi sq # 196				1	19.90		0.06				1		1618
		UMi sq # 197				1	18.64		0.79				1		1618
		UMi sq # 229				1	18.39		0.63				1		1618
		UMi sq # A				1	15.67		1.09				1		1618
		UMi sq # B				1	15.97		0.70				1		1618
		UMi sq # C				1	17.81		1.05				1		1618
		UMi sq # D				1	17.12		0.53				1		1618
		UMi sq # E				1	16.72		1.28				1		1618
		UMi sq # K				1	16.95		1.33				1		1618
		UMi sq # L				1	16.52		0.71				1		1618
		UMi sq # M				1	16.78		1.77				1		1618
		UMi sq # N				1	18.02		0.99				1		1618

HD	DM	Other Id	N Rem	α_{1950}	δ_{1950}	S	V	σ_V	B–V	σ_{B-V}	U–B	σ_{U-B}	n	Spectrum	References
		UMi sq # P				1	17.94		0.87				1		1618
		UMi sq # T				1	17.64		0.94				1		1618
		UMi sq # X				1	17.23		1.12				1		1618
		VLM sq # A				1	15.81		0.88				1		1600
		VLM sq # AA				1	12.49		0.56				1		1600
		VLM sq # AB				1	13.64		0.66				1		1600
		VLM sq # AC				1	15.55		0.72				1		1600
		VLM sq # B				1	14.93		0.42				2		1600
		VLM sq # C				1	14.53		0.83				3		1600
		VLM sq # F				1	18.39		0.52				1		1600
		VLM sq # G				1	17.67		0.84				1		1600
		VLM sq # H				1	16.49		0.75				1		1600
		VLM sq # I				1	12.75		0.57				1		1600
		VLM sq # K				1	18.39		0.67				1		1600
		VLM sq # L				1	18.63		0.84				1		1600
		VLM sq # P				1	17.89		0.97				1		1600
		VLM sq # V				1	14.91		0.69				1		1600
		VLM sq # Z				1	14.84		0.71				1		1600
		Vir sq 1 # 18				1	13.91		0.69		0.20		1		900
		Vir sq 1 # 27				1	15.18		1.43		1.17		2		900
		Vir sq 1 # 28				1	15.29		1.02				1		900
		Vir sq 1 # 29				1	15.32		0.87		0.66		1		900
		Vir sq 1 # 30				1	15.56		0.80		0.16		1		900
		Vir sq 1 # 31				1	15.58		0.55		0.04		1		900
		Vir sq 1 # 32				1	15.60		0.64		0.08		1		900
		Vir sq 1 # 33				1	15.67		0.87		0.44		1		900
		Vir sq 1 # 34				1	15.77		0.77		0.39		1		900
		Vir sq 1 # 35				1	15.89		0.58		0.02		1		900
		Vir sq 1 # 36				1	15.95		1.37		1.37		3		900
		Vir sq 1 # 37				1	15.96		0.81		0.22		1		900
		Vir sq 1 # 38				1	16.00		0.54		-0.13		1		900
		Vir sq 1 # 39				1	16.20		0.66		0.05		1		900
		Vir sq 1 # 40				1	16.21		0.58				1		900
		Vir sq 1 # 41				1	16.22		0.65		0.24		1		900
		Vir sq 1 # 42				1	16.62		0.55		-0.12		1		900
		Vir sq 1 # 43				1	16.75		0.51		-0.19		3		900
		Vir sq 1 # 44				1	16.80		0.73				1		900
		Vir sq 1 # 45				1	16.99		0.71		0.04		1		900
		Vir sq 1 # 46				1	17.31		1.50				1		900
		Vir sq 1 # 47				1	17.65		0.70				1		900
		Vir sq 1 # 48				1	17.65		1.39				1		900
		Wat 3 - 1				1	12.87		0.16		-0.50		2	B1 III	797
		Wat 3 - 2				1	13.48		0.24		-0.22		1		797
		Wat 3 - 3				1	12.85		0.21		0.14		1		797
		Wat 3 - 4				1	14.62		0.29		0.15		2		797
		Wat 3 - 5				1	15.00		0.21		-0.17		2		797
		Wat 3 - 6				1	14.67		0.16		-0.28		2		797
		Wat 3 - 7				1	14.85		1.11		0.48		2		797
		Wray 1470				1	13.82		1.11		-0.52		1	G8 V	1753
		Zwl 0120+34 # 1				1	12.46		0.47		0.01		1		157
		Zwl 0120+34 # 6				1	14.73		0.64		0.11		1		157
		van Wijk 14				1	13.42		0.26		0.02		3		1099
		van Wijk 15				1	13.65		0.59		0.00		2		1099
		van Wijk 16				1	14.05		0.78		0.26		3		1099
		van Wijk 22				1	13.66		0.84		0.70		2		1099
		vdB K				1	14.10		1.56				1		1702
		vdB L				1	14.60		1.24				2		1702
		vdB O				1	14.90		1.11				1		1702
		vdB P				1	15.31		1.47				1		1702
		vdB 2 # b				1	13.40		0.51		0.00		4		434
		vdB 5 # h				1	11.03		0.02		-0.36		3		434
		vdB 7 # 1				1	11.01		1.41		1.09		4		206
		vdB 15 # e				1	13.72		0.66		0.16		1		434
		vdB 19 # 1				1	11.35		0.97		0.32		4		206
		vdB 30				1	14.54		0.87				2		1702
		vdB 33 # e				1	13.65		0.18		-0.11		2		434
		vdB 33 # f				1	14.31		0.99		0.68		1		434
		vdB 39 # b				1	13.30		0.58		0.14		3		434
		vdB 39 # d				1	14.17		0.93		0.69		2		434
		vdB 39 # e				1	11.70		0.35		0.21		3		434
		vdB 41				1	15.07		0.81				2		1702
		vdB 44 # c				1	13.42		0.46		-0.02		4		434
		vdB 44 # d				1	13.04		0.32		-0.33		4		434
		vdB 44 # g				1	13.50		0.63		-0.13		2		434
		vdB 46 # c				1	11.34		0.21		-0.58		3	B2 Ve	434
		vdB 46 # d				1	11.38		0.18		-0.61		3	B2 Ve	434
		vdB 48 # e				1	13.10		1.39		1.52		1		434
		vdB 53 # c				1	12.77		0.48		0.26		1		434
		vdB 53 # d				1	12.59		1.37		1.14		1		434
		vdB 55 # b				1	11.62		0.21		-0.27		3		434

Table 2 1303

HD	DM	Other Id	N Rem	α_{1950}	δ_{1950}	S	V	σ_V	B–V	σ_{B-V}	U–B	σ_{U-B}	n	Spectrum	References
		vdB 57 # b				1	13.76		0.73		-0.19		4		434
		vdB 59				1	15.15		0.74				2		1702
		vdB 64				1	15.15		0.90				1		1702
		vdB 64 # 1				1	10.56		0.07		-0.15		4		206
		vdB 67				1	15.21		0.76				2		1702
		vdB 70				1	13.21		1.26				1		1702
		vdB 80 # 2				1	11.33		0.40		-0.24		4	B3 n	206
		vdB 80 # 3				1	11.15		0.69				1	A2 III	207
		vdB 80 # 5				1	10.27		0.30				1	B2 IV	207
		vdB 81 # c				1	11.41		0.23		0.09		3	A0 V	434
		vdB 81 # d				1	11.43		0.59		0.30		4		434
		vdB 81 # e				1	11.23		0.12		-0.23		4	B8 V	434
		vdB 87 # 3				1	11.15		0.15		-0.62		4		206
		vdB 88				1	14.28		0.83				2		1702
		vdB 90 # 2				1	11.35		0.22		-0.58		4		206
		vdB 91				1	15.00		0.89				2		1702
		vdB 92 # 3				1	10.05		0.08		-0.50		4		206
		vdB 108				1	14.38		0.82				2		1702
		vdB 114				1	13.47		1.58				1		1702
		vdB 114 # 2				1	11.13		0.32		-0.16		4		206
		vdB 118				1	13.78		1.50				1		1702
		vdB 119 # 2				1	11.03		0.52		-0.50		4		206
		vdB 128				1	14.89		1.12				1		1702
		vdB 130 # 2				1	11.13		0.47		-0.20		4		206
		vdB 139				1	14.77		1.24				2		1702
		vdB 142				1	15.18		1.53				1		1702
		vdB 146 # 2				1	11.90		0.80		0.03		4	A7 III	206
		vdB 146 # 3				1	13.26		0.72				1	F2 IV	207
		vdB 146 # 7				1	12.35		0.61				1	B7 V	207
		vdB 152 # 2				1	12.65		0.79				1	G8 V	207
		vdB ZETA				1	14.96		1.62				2		1702

Table 3: References for the UBV data sources

No	Reference
1	**Johnson H.L., Morgan W.W.** 1953, ApJ 117, 313 *Fundamental stellar photometry for standards of spectral type on the revised system of the Yerkes spectral atlas*
2	**Bok B.J., Bok P.F.** 1969, AJ 74, 1125 *Photometric standards for the southern hemisphere*
3	**Gutierrez-Moreno A., Moreno H., Stock J., Torres C., Wroblewski H.** 1966, Publ. Univ. Chile No 1 *A system of photometric standards*
4	**Isserstedt J.** 1969, A&A 3, 210 *Lichtelektrische Untersuchung eines nahen Sternringes in Aquila*
5	**Simonson S.Ch. III** 1968, ApJ 154, 923 *A spectroscopic and photometric investigation of the association Cep OB2*
6	**Feinstein A.** 1969, MNRAS 143, 273 *The OB stars in Carina-Centaurus*
7	**Hiltner W.A., Stephenson C.B., Sanduleak N.** 1968, ApJ Lett. 2, 153 *Spectroscopic and photometric observations of peculiar stars noted on southern objective prism plates*
8	**Sanduleak N., Philip A.G.D.** 1968, AJ 73, 566 *A stellar group in line of sight with the Large Magellanic Cloud*
9	**Alcaino G.** 1969, ApJ 156, 853 *The globular clusters NGC 2808 and NGC 1851*
10	**Walker G.A.H., Morris S.C.** 1968, AJ 73, 772 *UBV photometry of the associations anon b4 and NGC 2081 in the LMC*
11	**Krzeminski W., Serkowski K.** 1967, ApJ 147, 988 *Photometric and polarimetric observations of the nearby strongly reddened open cluster Stock 2*
12	**Landolt A.U.** 1968, PASP 80, 749 *UBV observations of selected double systems*
13	**Landolt A.U.** 1969, PASP 81, 443 *UBV observations of selected double systems. II*
14	**van den Bergh S., Hagen G.L.** 1968, AJ 73, 569 *UBV photometry of star clusters in the Magellanic Clouds*
15	**Johnson H.L., Mitchell R.I., Iriarte B., Wisniewski W.Z.** 1966, Comm. Lunar Plan. Lab. IV No 63 *UBVRIJKL photometry of the bright stars*
16	**FitzGerald M.P., Wilson W., Stegman J.E.** 1969, PASP 81, 804 *Four southern photoelectric sequences*
17	**Hill G., Perry C.L.** 1969, AJ 74, 1011 *Photometric studies of southern galactic clusters. II. IC 2602*
18	**Racine R.** 1969, AJ 74, 816 *Ros 4, a distant open cluster associated with nebulosities*
19	**van Altena W.F.** 1969, AJ 74, 2 *Low luminosity members of the Hyades cluster. II.*
20	**Bakos G.A.** 1969, AJ 73, 187 *Photoelectric photometry of selected AG stars in the 25° to 30° zone*
21	**Aveni A.F., Hunter J.H.** 1969, AJ 74, 1021 *Observational studies relating to stars formation. II.*
22	**Cowley A.P., Hiltner W.A., Witt A.N.** 1967, AJ 72, 1334 *Spectral classification and photometry of high proper motion stars*
23	**Landolt A.U.** 1967, AJ 72, 1012 *Photoelectric UBV sequences in Taurus*
24	**Racine R.** 1968, AJ 73, 588 *Distance of the cepheid SU Cassiopeiae*
25	**McConnell D.J.** 1968, ApJS 16, 275 *A study of the Cepheus IV association*
26	**Gutierrez-Moreno A., Moreno H.** 1968, ApJS 15, 459 *A photometric investigation of the Scorpio-Centaurus association*
27	**Sarg K., Wramdemark S.** 1970, A&AS 2, 251 *Photoelectric photometry of early-type stars in a Milky Way field in Cepheus*
28	**Preston G.W., Smak J., Paczynski B.** 1965, ApJS 12, 99 *Atmospheric phenomena in the RR Lyrae stars. II. The 41-day cycle of RR Lyrae*
29	**Dickens R.J., Kraft R.P., Krzeminski W.** 1968, AJ 73, 6 *Effect of rotation on the colors and magnitudes of stars in Praesepe*
30	**van den Bergh S.** 1967, AJ 72, 70 *UBV photometry of globular clusters*
31	**Smak J.** 1964, ApJS 9, 141 *Photometry and spectrophotometry of long-period variables*
32	**Schild R.E., Hiltner W.A., Sanduleak N.** 1969, ApJ 156, 609 *A spectroscopic study of the association Scorpius OB1*
33	**Westerlund B.E.** 1969, AJ 74, 882 *OB stars near the supernova remnant RCW 103 and the galactic structure in Norma*
34	**Westerlund B.E.** 1969, AJ 74, 879 *OB stars near the supernova remnant RCW 86*
35	**Dachs J.** 1970, A&A 9, 95 *Photometry of bright stars in the SMC*
36	**Dachs J.** 1970, A&A 5, 312 *Photoelectric UBV photometry of the open cluster NGC 2516*
37	**McClure R.D.** 1970, AJ 75, 41 *A photometric investigation of strong-cyanogen stars*
38	**Alcaino G.** 1970, ApJ 159, 325 *The globular clusters NGC 6752 and NGC 6362*
39	**Danziger I.J., Dickens R.J.** 1967, ApJ 149, 55 *Spectrophotometry of new short-period variable stars*
40	**Garrison R.F.** 1967, ApJ 147, 1003 *Some characteristics of the B and A stars in the Upper Scorpius complex*
41	**Haug U.** 1970, A&AS 1, 35 *UBV observations of luminous stars in three Milky Way fields*
42	**Perry C.L.** 1969, AJ 74, 899 *Photometric studies of southern galactic clusters. I. IC 2391*
43	**Johnson H.L.** 1953, ApJ 117, 353 *Magnitudes, colors, and spectral types in M39*
44	**McClure R., Crawford D.L.** 1971, AJ 76, 31 *The density distributions, color excess, and metallicity of K giants at the north galactic pole*
45	**Hill P.W., Hill S.R.** 1966, MNRAS 133, 205 *Faint blue stars in the far southern hemisphere*
46	**Evans L.T.** 1969, MNRAS 146, 101 *The open cluster IC 2581*
47	**Ishida K.** 1969, MNRAS 144, 55 *UBV observations of IC 1805*
48	**Wesselink A.J.** 1969, MNRAS 146, 329 *Photometry of NGC 4103*
49	**Hoag A.A., Johnson H.L., Iriarte B., Mitchell R.I., Hallam K.L., Sharpless S.** 1961, Publ. US. Nav. Obs. XVII Part VII, 347 *Photometry of stars in galactic cluster fields*
50	**Johnson H.L.** 1960, ApJ 131, 620 *The galactic cluster M25 = IC 4725*
51	**Ahmed F.** 1962, Publ. Roy. Obs. Edinburgh 3, 60 *Three-color photometry of southern galactic clusters. II. NGC 3766*
52	**Przybylski A.** 1968, MNRAS 139, 313 *The analysis of the spectrum of the Large Magellanic Cloud supergiant HD 33579*
53	**Shobbrook R.R., Herbison-Evans D., Johnston I.D., Lomb N.R.** 1969, MNRAS 145, 131 *Light variation in Spica*
54	**Evans L.T.** 1970, MNRAS 149, 311 *K-type stars in a field near the south galactic pole*
55	**Hill P.W.** 1970, MNRAS 150, 23 *Photometry and spectral classification of early-type stars away from the galactic plane*
56	**Sher D.** 1965, MNRAS 129, 237 *Distance of five open clusters near eta Carinae*
57	**West F.R.** 1967, ApJS 14, 384 *Photographic photometry of the galactic cluster NGC 2420*
58	**Oja T.** 1970, Private communication *UBV-fotometri Danska Telescopet*
59	**Talbert F.D.** 1965, PASP 77, 19 *UBV photometry of NGC 6405*
60	**Fernie J.D.** 1961, ApJ 133, 64 *Cepheids in galactic clusters. VII. S Nor and NGC 6087*
61	**Häggkvist L., Oja T.** 1969, Ark. Astr. 5, 303 *UBV photometry of standard stars*

No	Reference
62	**Feinstein A.** 1967, ApJ 149, 107 *Collinder 121, a young southern open cluster similar to h and χ Persei*
63	**Jones D.O.M., Hoag A.A.** 1968, PASP 80, 531 *Photographic photometry of stars in NGC 457*
64	**Johnson H.L., Hiltner W.A.** 1956, ApJ 123, 267 *Observational confirmation of a theory of stellar evolution*
65	**West F.R.** 1967, ApJS 14, 359 *Photographic photometry of the galactic cluster M37*
66	**Schulte D.H.** 1956, ApJ 124, 530 *New members of the association VI Cygni*
67	**Walker M.F.** 1956, ApJS 2, 365 *Studies of extremely young clusters. I. NGC 2264*
68	**Pismis P.** 1970, Tonantzintla Tacubaya 5, 293 *Studies on star clusters. The open cluster Tr 9*
69	**Breger M.** 1966, PASP 78, 293 *Photoelectric observation of the galactic cluster NGC 6087*
70	**Häggkvist L., Oja T.** 1968, Ark. Astr. 5, 125 *Photoelectric BV photometry of 368 northern stars*
71	**Häggkvist L., Oja T.** 1970, Private communication *Results of BV photometry 1969-1970 (Uppsala refractor)*
72	**Hill G., Barnes J.V.** 1971, AJ 76, 110 *A spectroscopic and photometric investigation of NGC 7243*
73	**Hiltner W.A., Morgan W.W.** 1969, AJ 74, 1152 *UBV photometry and spectral types in NGC 6611*
74	**Williams P.M.** 1967, MNASSA 26, 126 *The open cluster Cr 140*
75	**Pesch P.** 1961, ApJ 134, 602 *Photometric and objective prism observations in three galactic clusters*
76	**Alcaino G.** 1967, ApJ 147, 112 *The galactic cluster NGC 2281*
77	**Ljunggren B.** 1965, Ark. Astr. 3, 535 *A photoelectric and spectroscopic investigation of a region near the north galactic pole*
78	**Wild P.A.T.** 1969, MNASSA 28, 48 *UBV photometry of 152 stars in a region near the galactic plane*
79	**Wild P.A.T.** 1969, MNASSA 28, 123 *UBV photometry of 137 stars in an area near the south galactic pole*
80	**Williams P.M.** 1967, MNASSA 26, 30 *Photoelectric photometry of NGC 2451*
81	**Hall D.S., van Landingham F.G.** 1970, PASP 82, 640 *The nearby poor cluster Collinder 399*
82	**Sandage A., Walker M.F.** 1955, AJ 60, 230 *The globular cluster NGC 4147*
83	**Burkhead M.S.** 1971, AJ 76, 251 *Photometric observations of the star cluster NGC 6819*
84	**Purgathofer A.** 1966, Mitt. Univ. Sternw. Wien 13 No 2 *Der galaktische Sternhaufen NGC 6819*
85	**Burkhead M.S.** 1969, AJ 74, 1171 *Photometric observations of the star cluster IC 166*
86	**Purgathofer A.** 1966, Mitt. Univ. Sternw. Wien 13 No 3 *Der Kugelhaufen NGC 5053*
87	**Kinman T.D.** 1965, ApJ 142, 655 *The star cluster NGC 6791*
88	**Arp H. C., Sandage A., Stephens C.** 1959, ApJ 130, 80 *Cepheids in galactic clusters. IV. DL Cas in NGC 129*
89	**White R.E.** 1970, AJ 75, 167 *Secondary UBV comparison stars near some southern globular clusters. I. Stars brighter than 16th magnitude*
90	**Racine R.** 1971, AJ 76, 331 *The C-M diagram of M12 = NGC 6218*
91	**Schild R.E.** 1970, ApJ 161, 855 *Red supergiants in open clusters*
92	**Crampton D.** 1971, AJ 76, 260 *Observations of stars in HII regions. Spectral classification and UBV photometry*
93	**Brosterhus E.** 1963, Astr. Abh. Hamburg. Sternw. VII No 2 *Zur Automatisierung photographischer Dreifarben-Photometrie mit Anwendung auf M37*
94	**Lyngå G.** 1970, IAU Symp. 38, 270 *Photometric and spectroscopic data for some distant O and B stars*
95	**Bruck M.T., Smyth M.J., McLachlan A.** 1968, Publ. R. Obs. Edinburgh 6, 208 *Three-color photometry of southern galactic clusters. IV. NGC 5822, 5823*
96	**Lodén L.O., Nordström B.** 1969, Ark. Astr. 5, 231 *Photometric standard sequences in Norma*
97	**Purgathofer A.T.** 1969, Lowell Obs. Bull. 7, 98 *UBV sequences in selected star fields*
98	**Barbieri C.** 1970, Publ. Oss. Padova No 159 *A study of faint blue stars in high galactic latitudes*
99	**Roslund C.** 1969, Ark. Astr. 5, 249 *A photoelectric sequence near NGC 5128*
100	**Lindoff U.** 1969, Ark. Astr. 5, 223 *The open cluster NGC 559*
101	**Fernie J.D., Hagen J.P., Hagen G.L., McClure L.** 1971, PASP 83, 79 *The color index of the Sun from the Mgb triplet*
102	**Lee P.D., Perry C.L.** 1971, AJ 76, 464 *Photometric studies of northern galactic clusters. I. Roslund 5*
103	**Lyngå G.** 1968, Observatory 88, 20 *A study of the cluster Pismis 20*
104	**Hiltner W.A., Iriarte B., Johnson H.L.** 1958, ApJ 127, 539 *The galactic cluster NGC 6633*
105	**Walker R.J. Jr** 1971, PASP 83, 177 *UBV photometry of 173 PZT stars*
106	**Thé P.S., van Paradijs J.A.** 1971, A&A 13, 274 *UBV photoelectric photometry of early type stars in the direction of the associations Vul OB1 and Vul OB2*
107	**Alcaino G.** 1971, A&A 13, 287 *The globular cluster NGC 4833*
108	**Arp H.C., Hartwick F.D.A.** 1971, ApJ 167, 499 *A photometric study of the metal-rich globular cluster M71*
109	**Richer H.B.** 1971, ApJ 167, 521 *Some intrinsic properties of carbon stars*
110	**Alcaino G.** 1971, A&A 13, 399 *The globular cluster NGC 6541*
111	**Clariá J.J.** 1971, AJ 76, 639 *Photometric study of the galactic cluster NGC 5460*
112	**Blanco C., Catalano F.A.** 1971, AJ 76, 630 *Photoelectric observations of magnetic stars. III. HD 124224 - HD 140160 - HD 224801*
113	**Seggewiss W.** 1968, Veröff. Astr. Inst. Bonn No 79 *Photometrische Untersuchung des offenen Sternhaufens NGC 6231*
114	**Walker M.F.** 1969, ApJ 155, 447 *Studies of extremely young clusters. V. Stars in the vicinity of the Orion nebula*
115	**Haug U., Pfleiderer J., Dachs J.** 1966, Z. Astrophys. 64, 140 *Sterne frühen Spektraltyps in Norma und Circinus*
116	**Pfleiderer J., Dachs J., Haug U.** 1966, Z. Astrophys. 64, 116 *Lichtelektrische UBV-Photometrie von Standardsternen und in vier Sternenfelder am Aequator*
117	**Bigay J.H., Garnier R.** 1970, A&AS 1, 15 *Photométrie photoélectrique UBV d'étoiles chaudes des Selected Areas 40 et 49 situées au voisinage du plan galactique*
118	**Haug H., Walter K.** 1970, A&AS 1, 29 *UBV Messungen in den Selected Areas 100 und 112*
119	**Alexander J.B.** 1971, MNASSA 30, 139 *A search for metal-deficient stars in the southern hemisphere*
120	**Barnes T.G., Evans N.R.** 1970, PASP 82, 889 *UBV photometry of Nova HR Delphini 1967*
121	**Roslund C.** 1967, Ark. Astr. 4, 341 *Investigation of a Milky Way field in Scorpius. IV. A photoelectric sequence of A-K stars*
122	**Lodén L.O.** 1967, Ark. Astr. 4, 375 *Photometric standard sequences in the Crux region*
123	**Lodén L.O.** 1968, Ark. Astr. 4, 425 *Photometric standard sequences in Vela*
124	**Lindoff U.** 1968, Ark. Astr. 4, 433 *Photoelectric sequences in Puppis*

No	Reference

125 **Bok B.J., Bok P.F., Miller E.W.** 1972, AJ 77, 733 *Photometric standards for the southern hemisphere. II*

126 **Lodén L.O.** 1969, Ark. Astr. 5, 149 *Photometric standard sequences in Puppis*

127 **Lodén L.O.** 1969, Ark. Astr. 5, 161 *Photometric standard sequences in Carina*

128 **Hogg A.R.** 1961, Observatory 81, 69 *A southern galactic cluster*

129 **van den Bergh S.** 1968, Astrophys. Letters 2, 71 *Associations of reflection nebulae and local galactic structure*

130 **Jerzykiewicz M., Serkowski K.** 1966, PASP 78, 546 *Fifty standard stars for yellow and blue magnitudes in the UBV system*

131 **Feinstein A., Ferrer O.E.** 1968, PASP 80, 410 *The open cluster NGC 6231*

132 **Jaidee S., Lyngå G.** 1969, Ark. Astr. 5, 345 *A search for faint violet stars in southern galactic latitudes*

133 **Johnson H.L., Sandage A.R., Wahlquist H.D.** 1956, ApJ 124, 81 *The galactic cluster M11*

134 **Johnson H.L., Sandage A.R.** 1955, ApJ 121, 616 *The galactic cluster M67 and its significance for stellar evolution*

135 **Eggen O.J., Sandage A.R.** 1964, ApJ 140, 130 *New photoelectric observations of stars in the old galactic cluster M67*

136 **van den Bergh S., Heeringa R.** 1970, A&A 9, 209 *The old open cluster NGC 7142*

137 **Wooden W.H. II** 1970, AJ 75, 324 *UBV photoelectric photometry in four southern Milky Way fields*

138 **Humphreys R.M.** 1970, AJ 75, 602 *The space distribution and kinematics of supergiants*

139 **Feinstein A., Marraco H.G.** 1971, PASP 83, 218 *The open cluster NGC 4609 behind the Coalsack*

140 **Morgan W.W., Harris D.L.** 1956, Vistas in Astr. 2, 1124 *The galactic cluster M29 (NGC 6913)*

141 **Schild R.E., Neugebauer G., Westphal J.A.** 1971, AJ 76, 237 *Interstellar absorption and color excess in Sco OB1*

142 **Alcaino G., Contreras C.** 1971, A&A 11, 14 *The globular cluster NGC 1261*

143 **Alcaino G.** 1971, A&A 11, 7 *The globular cluster NGC 6352*

144 **McGruder C.H., Schnur G.** 1971, A&A 14, 164 *Three UBV sequences in Centaurus*

145 **Rohlfs K., Schrick K.-W., Stock J.** 1959, Z. Astrophys. 47, 15 *Lichtelektrische Dreifarben-Photometrie von NGC 6405 (M6)*

146 **Landolt A.U.** 1971, PASP 83, 650 *UBV observations of selected double systems. III*

147 **Crawford D.L., Golson J.C., Landolt A.U.** 1971, PASP 83, 652 *A UBV equatorial-extinction star network*

148 **Sharpless S.** 1966, IAU Symp. 24, 345 *Spectroscopic parallaxes of M-type supergiants*

149 **Ardeberg A., Brunet J.P., Maurice E., Prevot L.** 1972, A&AS 6, 249 *Spectrographic and photometric observations of supergiants and foreground stars in the direction of the Large Magellanic Cloud*

150 **Lutz T.E.** 1971, PASP 83, 488 *UBV observations of visual double stars*

151 **Lindoff U.** 1972, A&A 16, 315 *The open cluster NGC 6811*

152 **Alexander J.B., Carter B.S.** 1971, R. Obs. Bull. No 169 *Motion of A0 stars perpendicular to the galactic plane. V. UBV photometry of A stars in the south galactic cap*

153 **Morgan W.W., Lodén K.** 1964, Vistas in Astr. 8, 83 *Some characteristics of the Orion association*

154 **Crawford D.L., Barnes J.V., Golson J.C.** 1971, AJ 76, 1058 *Four-color, Hβ, and UBV photometry for bright B-type stars in the northern hemisphere*

155 **Lee V.L., Burkhead M.S.** 1971, AJ 76, 467 *Preliminary photometry of the open cluster NGC 609*

156 **White R.E.** 1971, AJ 76, 419 *Secondary UBV comparison stars near some southern globular clusters. II. Stars brighter than 13th magnitude*

157 **Angione R.J.** 1971, AJ 76, 412 *Photoelectric sequences for the brighter QSOs*

158 **Corben P.M., Carter B.S., Banfield R.M., Harvey G.M.** 1972, MNASSA 31, 8 *UBV photometry of 500 southern stars*

159 **Karlsson B.** 1966, Medd. Lunds Obs. Ser. 2, No 148 *Investigation of a region in Monoceros*

160 **Ciatti F., Bernacca P.L.** 1971, A&A 11, 485 *New helium-weak stars in the Orion association*

161 **Lindoff U.** 1971, A&A 15, 439 *The open clusters NGC 6613 and NGC 6716*

162 **Feinstein A.** 1971, PASP 83, 800 *The open cluster NGC 6025*

163 **Clariá J.J.** 1972, A&A 19, 303 *Photometric study of the open cluster NGC 2232*

164 **Cousins A.W.J.** 1972, MNASSA 31, 75 *UBV photometry of some southern stars*

165 **Grubissich C., Purgathofer A.** 1962, Z. Astrophys. 54, 41 *Der galaktische Sternhaufen NGC 2301*

166 **McConnell D.J., Frye R.L.** 1972, PASP 84, 388 *Discoveries on southern objective-prism plates. III. Three hydrogen-deficient stars and a bright B-type subdwarf*

167 **Landolt A.U., Blondeau K.L.** 1972, PASP 84, 394 *UBV observations of the eclipsing binary LY Aurigae*

168 **Dickens R.J.** 1972, MNRAS 157, 281 *The colour-magnitude diagram of the globular cluster NGC 6981*

169 **Dickens R.J.** 1972, MNRAS 157, 299 *The colour-magnitude diagram of the globular cluster NGC 7099*

170 **Tifft W.G., Conolly L.P., Webb D.F.** 1972, MNRAS 158, 47 *NGC 2818, an open cluster containing a planetary nebula*

171 **Bern K., Virdefors B.** 1972, A&AS 6, 117 *Photoelectric UBV photometry in Cygnus and Cassiopeia*

172 **Schild R.E.** 1973, AJ 78, 37 *Spectral types and UBV photometry of G-K giants at the north galactic pole*

173 **Knipe G.F.G.** 1969, Republic Obs. Circ. 7, 197 *Photometric observations of binary stars*

174 **Knipe G.F.G.** 1969, Republic Obs. Circ. 7, 198 *Photometric observations of RV Pictoris*

175 **Knipe G.F.G.** 1969, Republic Obs. Circ. 7, 198 *A minimum of U Ophiuci*

176 **Knipe G.F.G.** 1969, Republic Obs. Circ. 7, 260 *Photometric observations of binary stars*

177 **Knipe G.F.G.** 1971, Republic Obs. Circ. 8, 7 *The light curve and minima of BU Velorum*

178 **Knipe G.F.G.** 1971, Republic Obs. Circ. 8, 8 *The light curve and minima of TU Muscae*

180 **Lutz J.H., Lutz T.E.** 1972, AJ 77, 376 *UBV and Hγ observations of early-type field stars*

181 **Drilling J.S.** 1972, AJ 77, 463 *UBV and Hβ photometry of OB stars in a Milky Way field in Norma*

182 **Clariá J.J.** 1972, AJ 77, 868 *UBV and Hβ photometry of the galactic cluster NGC 2343*

183 **Vogt N., Moffat A.F.J.** 1972, A&AS 7, 133 *Southern open clusters I. UBV and Hβ photometry of 15 clusters between galactic longitudes 231° and 256°*

184 **Vogt N., Moffat A.F.J.** 1973, A&AS 9, 97 *Southern open clusters II. UBV and Hβ photometry of 11 clusters between galactic longitudes 259° and 280°*

185 **Lindoff U.** 1972, A&AS 7, 231 *An investigation of four southern open clusters*

186 **Ardeberg A., Sarg K., Wramdemark S.** 1973, A&AS 9, 163 *Photoelectric UBV photometry in four areas of intermediate-high galactic latitude*

187 **Lindoff U.** 1973, A&AS 9, 229 *The open cluster NGC 2527*

No	Reference
188	**Feinstein A., Marraco H.G., Mirabel I.** 1973, A&AS 9, 233 *Photometric study of the open cluster NGC 2516*
189	**Clariá J.J.** 1973, A&AS 9, 251 *Photometric study of the galactic cluster NGC 2335*
190	**Thé P.S., Stokes N.** 1970, A&A 5, 298 *A study of the southern open clusters Tr 27, Tr 28, NGC 6416 and NGC 6425*
191	**Humphreys R.M.** 1973, A&AS 9, 85 *Spectroscopic and photometric observations of luminous stars in Carina-Centaurus (l=282° - 305°)*
192	**Zissell R.** 1972, AJ 77, 610 *Eclipsing binary HR 6611*
193	**Feltz K.A. Jr** 1972, PASP 84, 497 *Interstellar reddening near the north galactic pole*
194	**Hall D.S., Wawrukiewicz A.S.** 1972, PASP 84, 541 *The eclipsing binary WW Cygni, an unlikely candidate for pre-main-sequence contraction*
195	**Hall D.S., Garrison L.M. Jr** 1972, PASP 84, 552 *UBV photometry of the ring in SW Cygni*
196	**Epps E.A.** 1972, R. Obs. Bull. No 176 *UBV photoelectric observations*
197	**van der Wal P.B., Nagel C., Voordes H.R., Boer K.S.** 1972, A&AS 6, 131 *Photoelectric observations of six eclipsing binaries*
198	**Pfleiderer J., Mayer U.** 1971, AJ 76, 691 *Near-ultraviolet surface photometry of the southern Milky Way*
199	**Graham J.A.** 1971, AJ 76, 1079 *The distances of two faint OB star groups in Monoceros*
200	**Cousins A.W.J.** 1972, MNASS 31, 69 *UBV photometry of some very bright stars*
201	**Jones B.F.** 1973, A&AS 9, 313 *Relative proper motions of faint stars in the Pleiades*
202	**Wanner J.F.** 1972, MNRAS 155, 463 *On the stellar luminosity function from the method of mean absolute magnitude*
203	**Priser J.B.** 1961, Publ. US. Nav. Obs. XX Part III, 27 *First catalogue of trigonometric parallaxes of faint stars. Photometric results.*
204	**Brunet J.P., Prevot L., Maurice E., Muratorio G.** 1973, A&AS 9, 447 *Additional observations of supergiants and foreground stars in the direction of the Large Magellanic Cloud*
205	**Newell E.B., Rodgers A.W., Searle L.** 1969, ApJ 156, 597 *The blue horizontal-branch stars of NGC 6397*
206	**Racine R.** 1968, AJ 73, 233 *Stars in reflection nebulae*
207	**Aveni A.F., Hunter J.H.** 1972, AJ 77, 17 *Observational studies relating to stars formation. III.*
208	**Drilling J.S.** 1971, AJ 76, 1072 *UBV photometry of early-type stars in two regions at high galactic latitudes*
209	**McClure R.D., Racine R.** 1969, AJ 74, 1000 *The reddening of M3, M13, M31, M33 from photometry of late-type field stars*
210	**Cousins A.W.J., Lagerwey H.C.** 1971, MNASSA 30, 149 *Comparison stars for RR Scorpii*
211	**Havlen R.J.** 1972, A&A 17, 413 *OB stars distribution in Puppis*
212	**Ardeberg A., Wramdemark S.** 1970, Ark. Astr. 5, 387 *Photoelectric photometry of early-type stars in a field in Ophiucus*
213	**Ardeberg A., Sarg K.** 1970, Ark. Astr. 5, 391 *Photoelectric photometry of luminous B type stars in a Milky Way field in Cepheus*
214	**Alexander J.B.** 1970, MNASSA 29, 44 *U, B, V photometry of some visual binaries*
216	**Feinstein A.** 1967, PASP 79, 184 *Multicolor UBVRI observations of metallic-line stars*
217	**Eggen O.J., Sandage A.R.** 1969, ApJ 158, 669 *New photometric data for the old galactic cluster NGC 188. The presence of a gap, chemical composition, and distance modulus*
218	**Pesch P.** 1972, ApJ 178, 203 *UBV photometry of some selected stars in the Hyades*
219	**Cousins A.W.J.** 1973, MNASSA 32, 11 *UBV photometry of some southern stars*
220	**Huchra J., Willner S.P.** 1973, PASP 85, 85 *UBV photometry of selected Ap stars*
221	**Drilling J.S.** 1973, AJ 78, 44 *Photoelectric UBV photometry of late-type stars in two regions at high galactic latitude*
222	**Garrison R.F., Hiltner W.A.** 1973, ApJ 179, L117 *CPD -31°1701, an extremely helium-rich subluminous, O-type star*
223	**Drilling J.S.** 1973, ApJ 179, L31 *A new helium-rich B-type star*
224	**Sanwal N.B., Parthasarathy M., Abhyankar K.D.** 1973, Obs. 93, 30 *UBV photometry of zeta Aurigae during the 1971-72 eclipse*
225	**Winiarski M.** 1971, Acta Astron. 21, 517 *Photoelectric observations of the eclipsing variable AI Draconis*
226	**Jerzykiewicz M.** 1971, Acta Astron. 21, 501 *UBV photometry of the β Cephei type variable stars. III. KP Persei. (HD 21803)*
227	**Stepien K.** 1970, Acta Astron. 20, 117 *Photometry of HD 211853 - a Wolf-Rayet eclipsing binary*
228	**Jerzykiewicz M.** 1970, Acta Astron. 20, 93 *UBV photometry of the β Cephei type variable stars. I. γ Pegasi*
229	**Jones D.H.P.** 1969, Acta Astron. 19, 53 *Three colour photometry of the eclipsing binary AG Pegasi*
230	**Stepien K.** 1968, Acta Astron. 18, 537 *RW Camelopardalis - a cepheid with a compagnon*
231	**Jerzykiewicz M.** 1971, Lowell Obs. Bull. 7, 199 No 155 *UBV photometry of the β Cephei type variable stars. II. δ Ceti*
232	**Alexander J.B., Lourens J.V.B.** 1969, MNASSA 28, 95 *The white dwarf CoD -38°10980*
233	**Knipe G.F.G.** 1968, MNASSA 27, 29 *The eclipsing and spectroscopic binary HD 161783*
234	**Heiser A.M.** 1968, PASP 80, 334 *UBV photometry of the large proper-motion star L1159-16*
235	**Landolt A.U.** 1968, PASP 80, 331 *UBV magnitudes and colors for parallax stars*
236	**Cousins A.W.J., Harvey G.M.** 1973, MNASSA 32, 27 *UBV photometry of stars in F2*
237	**Dahn C.C., Guetter H.H.** 1973, ApJ 179, 551 *Recent photometric variability of HD 184279*
238	**Walborn N.R.** 1973, ApJ 179, 517 *Some characteristics of the eta Carinae complex*
239	**Lindoff U.** 1968, Ark. Astr. 4, 587 *The open clusters NGC 2533, NGC 2567, NGC 2571 and Ru 58*
240	**Dixon M.E.** 1970, MNRAS 151, 87 *B-type stars within ten degrees of the galactic centre. I.*
241	**Garnier R., Lortet-Zuckermann M.C.** 1971, A&A 14, 408 *The bright condensation 3C153.1 inside the HII region NGC 2175*
242	**Dworak T.Z., Winiarski M.** 1972, Acta Astron. 22, 33 *Photoelectric observations of Nova Vulpeculae 1968(1)=LV Vul*
243	**Cousins A.W.J., Lagerwey H.C., Shillington F.A.** 1969, MNASSA 28, 63 *Comparison stars for the long period variables*
244	**Cousins A.W.J., Lagerwey H.C.** 1970, MNASSA 29, 7 *Comparison stars for the long period variables and RY Sagittarii*
245	**Fernie J.D.** 1969, J. Roy. Astron. Soc. Canada 63, 133 *UBV observations of miscellanous stars*
246	**Schalen C.** 1972, Reports Obs. Lund No 3 *A study of interstellar reddening in a region in Ophiucus*
247	**Appenzeller I.** 1966, Z. Astrophys. 64, 269 *Polarimetrische, photometrische und spektroskopische Beobachtungen von Sternen im Cygnus und Orion*
248	**Seggewiss W.** 1971, Veröff. Astr. Inst. Bonn No 83 *Photoelectric observations of stars in the southern open clusters. NGC 2335, NGC 2343, NGC 2453, NGC 4439 and H5*

No	Reference
249	**Newell E.B., Rodgers A.W., Searle L.** 1969, ApJ 158, 699 *The blue horizontal-branch stars of ω Centauri*
250	**Schild R.E.** 1965, ApJ 141, 979 *Spectral classification in h and χ Persei*
251	**Moffat A.F.J., Vogt N.** 1973, A&AS 10, 135 *Southern open stars clusters. III. UBV-Hβ photometry of 28 clusters between galactic longitudes 297° and 353°*
252	**Rybka E.** 1969, Acta Astron. 19, 229 *The corrected magnitudes and colours of 278 stars near S.A. 1-139 in the UBV system*
253	**Breger M.** 1968, AJ 73, 84 *UBV and narrow-band uvby photometry of bright stars*
254	**Imagawa F.** 1967, Mem. Coll. Sci. Univ. Kyoto A 31 No 2 *Observational results of three color photometry for F-type stars. - II.*
255	**Zelwanowa E., Schöneich W.** 1971, Astron. Nachr. 293, 155 *Pekuliare A-Sternen in offenen Sternhaufen. I. Photoelektrische UBV- Photometrie und MK-Klassifikation von Sternen im Gebiet von NGC 7039*
256	**Zelwanowa E.** 1971, Astron. Nachr. 293, 163 *Pekuliare A-Sternen in offenen Sternhaufen. II. Resultate der Suche nach Pekuliare A-Sternen im Gebiet des Haufens Tr 2*
257	**Iriarte B.** 1967, Tonantzintla Tacubaya 4, 79 *UBV photoelectric photometry for faint stars in the Pleiades cluster*
258	**Cousins A.W.J.** 1973, MNASSA 32, 43 *UBV photometry of some southern stars. (Third List)*
259	**Hartwick F.D.A., Hesser J.E., McClure R.D.** 1972, ApJ 174, 557 *A photometric study of the open cluster NGC 2477*
260	**Hardorp J.** 1960, Astr. Abh. Hamburg. Sternw. V No 7 *Dreifarben-Photometrie einer Sternhaufengruppe*
261	**Wramdemark S.** 1973, A&AS 11, 365 *Photoelectric UBV photometry in some Milky Way fields*
262	**Wehinger P.A., Hidajat B.** 1973, AJ 78, 401 *UBV photometry of the Groningen-Palomar variable star fields*
263	**Cuffey J.** 1973, AJ 78, 408 *NGC 1893 and NGC 6838, standard stars*
264	**Sistero R.F., Sistero M.E.C.** 1973, AJ 78, 413 *UBV light variation and orbital elements of HD 101799*
265	**Muzzio J.C.** 1973, PASP 85, 358 *UBV photometry of large proper motion stars*
266	**Landolt A.U.** 1973, PASP 85, 203 *A magnitude sequence for the variable star ST Monocerotis*
267	**Higginbotham N.A., Lee P.** 1973, PASP 85, 215 *The helium-rich star HD 184927*
268	**Dahn C.C., Behall A.L., Christy J.W.** 1973, PASP 85, 224 *Trigonometric parallax determination for the central star in the planetary nebula NGC 7293*
269	**Häggkvist L.** 1973, A&AS 12, 85 *UBV sequences in six of Kapteyn's Selected Areas*
270	**Perry C.L.** 1969, AJ 74, 139 *The galactic force law K(z)*
271	**Penston M.J.** 1973, MNRAS 164, 121 *Photoelectric UBV photometry of stars in ten Selected Areas*
272	**Penston M.J.** 1973, MNRAS 164, 133 *Photoelectric UBV observations made on the Palomar 20-inch telescope*
273	**Barbier M., Bernard A., Bigay J.H., Garnier R.** 1973, A&A 27, 421 *Galactic structure in the direction of Cepheus*
274	**Degewij J.** 1973, A&A 27, 473 *UBV photoelectric photometry of the magnetic variable star HD 133029*
275	**Philip A.G.D.** 1968, AJ 73, 1000 *Photoelectric photometry of A-type stars near the north galactic pole*
276	**Lawrence L.C., Reddish V.C.** 1965, Publ. Roy. Obs. Edinburgh 5, 111 *The Cygnus II association*
277	**Millis R.L.** 1973, PASP 85, 410 *UBV photometry of 1 Monocerotis*
278	**Cousins A.W.J.** 1973, Mem. R. Astr. Soc. 77, 223 *Revised zero points and UBV photometry in the Harvard E and F regions*
279	**Lyngå G., Wramdemark S.** 1973, A&AS 12, 365 *Carina arm studies I. Deep photoelectric UBV sequences in Vela, Carina and Crux*
280	**Häggkvist L., Oja T.** 1973, A&AS 12, 381 *Photoelectric photometry of stars near the north galactic pole*
281	**Landolt A.U.** 1973, AJ 78, 959 *UBV photoelectric sequences in the celestial equatorial Selected Areas 92-115*
282	**Perry C.L.** 1973, IAU Symp. 50, 192 *Multicolor photometry of the galactic cluster NGC 2362*
283	**Hartwick F.D.A., Hesser J.E.** 1973, ApJ 183, 883 *NGC 2660 and its nearby carbon star*
284	**Lindoff U.** 1972, A&AS 7, 497 *The old open cluster NGC 6819*
285	**Cousins A.W.J.** 1973, MNASSA 32, 117 *UBV photometry of some southern stars (Fourth List)*
286	**Iriarte B.** 1970, Tonantzintla Tacubaya 5, 213 *Photoelectric photometry of faint blue stars at high galactic latitudes*
287	**Lodén L.O.** 1973, A&AS 10, 125 *A list of suspected clusters in the southern Milky Way*
288	**Hartwick F.D.A., Hesser J.E.** 1972, ApJ 175, 77 *The abnormally metal-rich globular NGC 6352*
289	**Stepien K.** 1972, Acta Astron. 22, 175 *Photometric behaviour of field RR Lyrae stars*
290	**Menzies J.** 1972, MNRAS 156, 207 *The globular cluster NGC 4833*
291	**Penston M.J.** 1972, MNRAS 156, 103 *Observations of three short period RR Lyrae variables*
292	**Smak J.** 1967, Acta Astron. 17, 213 *UBV photometry of visual binaries*
293	**Stenholm B., Soderhjelm S.** 1972, A&AS 7, 385 *Two UBV photoelectric sequences in Cygnus*
294	**Lodén K.** 1968, Stockholm Obs. Ann. 23 No 1 *Magnitude, colours and spectral classes for 446 stars in Selected Area 193*
295	**Rodgers A.W.** 1971, ApJ 165, 581 *Solar composition A-stars in the galactic halo*
296	**Demers S.** 1969, AJ 74, 925 *Photometry of population II cepheids in globular clusters. I. M2*
297	**Demers S., Wehlau A.** 1971, AJ 76, 916 *Photometry of variables in globular clusters. III. M14 and the period-luminosity relation of population II cepheids*
298	**Pismis P.** 1970, Tonantzintla Tacubaya 5, 219 *Studies on star clusters. A new small cluster near NGC 2175 and some remarks on the latter*
299	**Chincarini G.** 1963, Contr. Asiago No 138 *The open cluster NGC 6939*
300	**Winzer J.E.** 1973, AJ 78, 618 *A detailed photometric study of the cepheid RT Aur*
301	**Gleim J.K.** 1973, AJ 78, 622 *Photoelectric photometry of ST Aquarii*
302	**Bruck M.T., Smyth M.J.** 1966, Publ. Roy. Obs. Edinburgh 5, 195 *Three-colour photometry of southern galactic clusters. III NGC 6167*
303	**Sturch C.** 1973, PASP 85, 724 *A new UBV comparison of M67 and Hyades stars*
304	**McClure R.D.** 1972, ApJ 172, 615 *The color excesses and metallicities of the open clusters NGC 2360 and NGC 3680*
305	**Purgathofer A.** 1964, Ann. Univ. Sternw. Wien 26 No 2 *Dreifarben Photometrie in offenen Sternhaufen sowie in zwei Sternfeldern im Cygnus*
306	**Hiltner W.A.** 1966, IAU Symp. 24, 373 *The HR diagram of NGC 1893*
307	**Burkhead M.S., Burgess R.D., Haish B.M.** 1973, AJ 77, 661 *Photometric observations of the star cluster NGC 2141*
308	**Bern K., Wramdemark S.** 1973, Lowell Obs. Bull. 8, 1 no 161 *Photoelectric UBV magnitudes of 119 proper motion stars and suspected white dwarfs*
309	**Larsson-Leander G.** 1960, Stockholm Ann. Obs. 20 No 9 *A study of the galactic cluster NGC 6940*

No	Reference
310	**Lohman W., Schnur G.** 1962, Astron. Nachr. 287, 17 *Photometrie des offenen Sternhaufens NGC 6322*
311	**Rahim M.A.** 1966, Astron. Nachr. 289, 75 *Dreifarben-photometrie von NGC 5316*
312	**Chincarini G.** 1966, Mem. Soc. Astr. Italiana 37, 243 *The galactic cluster NGC 7762*
313	**Iriarte B.E.** 1973, Bol. Inst. Tonantzintla 1, 23 *Narrow band photometry for faint blue stars at high galactic latitudes*
314	**van den Bergh S.** 1971, ApJ 165, 259 *Optical studies of Cassiopeia A. II. UBV photometry of field stars*
315	**Feinstein A., Marraco H.G., Muzzio J.C.** 1973, A&AS 12, 331 *A single young open cluster comprising Tr 14 and Tr 16*
316	**Routly P.M.** 1971, Publ. US. Nav. Obs. XX Part VI *Second catalog of trigonometric parallaxes of faint stars*
317	**Ardeberg A.** 1972, A&A 19, 384 *On the extended van Wijk sequence in the Large Magellanic Cloud*
318	**Cannon R.D., Stobie R.S.** 1973, MNRAS 162, 227 *Photometry of southern globular clusters II. Bright stars in NGC 6752*
319	**Cannon R.D., Stobie R.S.** 1973, MNRAS 162, 207 *Photometry of southern globular clusters I. Bright stars in ω Centauri*
320	**Brandi E., Clariá J.J.** 1973, A&AS 12, 79 *Southern peculiar stars with abnormal spectra*
321	**Andrews P.J., Thackeray A.D.** 1973, MNRAS 165, 1 *Absolute magnitude of B-type primaries in visual binaries*
322	**Seggewiss W.** 1971, Veröff. Astr. Inst. Bonn No 82 *Six photoelectric standard sequences in the southern Milky Way*
323	**The P.S., Roslund C.** 1963, Contr. Bosscha Obs. No 19 *Photometry of the galactic cluster NGC 6649*
324	**Bregman J., Butler D., Kemper E., Koski A., Kraft R.P., Stone R.P.S.** 1973, Lick Obs. Bull. No 647 *Colors, magnitudes, spectral types and distances for stars in the field of the X-ray source Cyg X-1 (HDE 226868)*
325	**van den Bergh S.** 1971, AJ 76, 1082 *UBV photometry in the nuclear bulge of the Galaxy*
326	**Wing R.F.** 1973, AJ 78, 684 *UBV sequences in the field of radio sources*
327	**Penston M.J., Penston M.V., Sandage A.R.** 1971, PASP 83, 783 *Stars observed photoelectrically near quasars and related objects*
328	**Appenzeller I.** 1972, PASJ 24, 483 *Light variations of high luminosities O and B stars in the Large Magellanic Cloud*
329	**Stobie R.S., Alexander J.B.** 1970, Observatory 90, 66 *HR 4768 a cepheid variable of small amplitude*
330	**Stobie R.S.** 1972, Observatory 92, 12 *HR 2957 a cepheid variable of small amplitude*
331	**Penston M.V., Wing R.F.** 1972, Observatory 93, 149 *A photoelectric sequence in the field of OJ 287*
332	**Wisse P.N.J., Wisse M.** 1973, A&A 23, 463 *Photoelectric observations of the bright RV Tauri stars R Scuti and U Monocerotis*
333	**Giclas H.L., Burnham R. Jr, Thomas N.G.** 1971, Lowell Obs. *Lowell proper motion survey. The G Numbered Stars*
334	**Bertaud C., Wlerick G., Véron P., Dumortier B., Bigay J., Paturel G., Duruy M., Saevsky P.** 1973, A&A 24, 357 *Etude photométrique de BL Lacertae*
335	**Frölich H.-E., Rössiger S.** 1971, Mitt. Ueber Veränbliche Sterne, Sonneberg 6, 1 *Eine Untersuchung des Entfernungverlaufes der interstellaren Extinktion in der Umgebung von SV Cephei*
336	**Muzzio J.C., Forte J.C.** 1975, AJ 80, 1037 *UBV and Hβ photometry of faint early-type stars in Norma*
337	**Penston M.V.** 1973, ApJ 183, 505 *Multicolor observations of stars in the vicinity of the Orion nebula*
338	**Kubinec W.R.** 1973, Publ. Warner Swasey Obs. 1 No 3 *The galactic structure in Cepheus near NGC 7235*
339	**Harris W.E., Racine R.** 1973, AJ 78, 242 *The globular cluster NGC 6934*
340	**Bigay J.H., Lunel M.** 1966, J. Obs. 49, 411 *Etude photométrique de l'amas NGC 6823*
341	**Sato F.** 1970, Ann. Tokyo Astron. Obs. 2nd Ser. 12, 34 *A study of interstellar matter in the Cassiopeia-Perseus regions. III*
342	**Georgelin Y.M., Georgelin Y.P., Roux S.** 1973, A&A 25, 337 *Observations de nouvelles régions HII galactiques et d'étoiles excitatrices*
343	**Menzies J.** 1973, MNRAS 163, 323 *Photoelectric photometry of 47 Tuc*
344	**Kitamura M., Yamasaki A.** 1971, Tokyo Astron. Bull. 2nd Ser. No 209 *Photoelectric observations of the close binary system TX Cancri*
345	**Whelan J.A.J., Worden S.P., Mochnacki S.W.** 1973, ApJ 183, 133 *TX Cancri, the golden wonder*
346	**Preston G.W., Stepien K., Wolff S.C.** 1969, ApJ 156, 653 *The magnetic field and light variations of 17 Comae Berenices and κ Cancri*
347	**Hodge P.W., Flower P.J.** 1973, ApJ 185, 829 *A color-magnitude diagram for the rich cluster NGC 2164 in the Large Magellanic Cloud*
348	**Smak J.** 1967, Acta Astron. 17, 255 *Light variability of HZ 29*
349	**Menzies J.** 1974, MNRAS 168, 177 *UBV photometry of bright stars and variables in the globular cluster NGC 6723*
350	**Cathey L.R.** 1974, AJ 79, 1370 *UBVR photometry of stars in the globular clusters M92, M13 and 47 Tuc*
351	**Breger M.** 1974, ApJ 188, 53 *Pre-main-sequence stars. III. Herbig Be/Ae stars and other selected objects*
352	**Robinson E.L., Kraft R.P.** 1974, AJ 79, 698 *On the variability of dwarf K- and M-type stars in the Pleiades and Hyades*
353	**Moffat A.F.J., FitzGerald M.P.** 1974, A&A 34, 291 *Distant galactic structure in Puppis. The Ruprecht 44 aggregate*
354	**McCarthy C.C., Miller E.W.** 1974, AJ 79, 1396 *Faint O-B3 stars in Puppis*
355	**Feinstein A.** 1974, AJ 79, 1290 *Photoelectric UBVRI observations of Am stars*
356	**Moffat A.F.J., Vogt N.** 1974, A&A 30, 381 *HD 52942, a new early-type eclipsing binary with a double-line spectrum*
357	**Warren P.R., Penfold J.E.** 1974, MNASSA 33, 148 *UBV observations of stars in the LMC-X3 error box*
358	**Jones C.A., Chetin T., Liller W.** 1974, ApJ 190, L1 *Optical studies of Uhuru sources. VIII. Observations of 92 possible counterparts of X-ray sources*
359	**van den Bergh S., Herbst E.** 1974, AJ 79, 603 *Photometry of stars in the nuclear bulge of the Galaxy through a low absorption window at l=0°, b=-8°*
360	**Lucke P.B.** 1973, Thesis *The OB stellar associations in the Large Magellanic Cloud*
361	**Lucke P.B.** 1974, ApJS 28, 73 *The OB stellar associations in the Large Magellanic Cloud*
362	**Denoyelle J.** 1977, A&AS 27, 343 *The spatial distribution of young stars in Vela (l=257° to 284°)*
363	**Coyne G.V., Lee T.A., de Graeve E.** 1974, Vatican Obs. Pub. 1 No 5 *A survey of Hα emission objects in the Milky Way I. Vulpecula - Cygnus*
364	**Iriarte B.E.** 1974, Bol. Inst. Tonantzintla 1, 73 *A note on the UBV photometry in the Pleiades cluster*
365	**Alcaino G.** 1974, A&AS 13, 55 *The globular cluster NGC 2298*
366	**Cannon R.D.** 1974, MNRAS 167, 551 *Photometry of southern globular clusters. III. Bright stars in 47 Tucanae, NGC 6397 and NGC 288*
367	**McCuskey S.W., Pesch P., Snyder G.A.** 1974, AJ 79, 597 *The space distribution and radial velocities of some early-type stars in the Perseus spiral arm*
368	**Wesselink A.J.** 1974, MNRAS 168, 345 *A colour-magnitude diagram of NGC 6752*

No	Reference
369	**Neckel H.** 1974, A&AS 18, 169 *Photoelectric catalogue of 1030 BD M-type stars located along the galactic equator*
370	**Harris W.E., Racine R.** 1974, AJ 79, 472 *A C-M diagram for M80 obtained with a new photographic calibration technique*
371	**Alcaino G.** 1974, A&AS 18, 9 *Photometry on the bright stars of the globular clusters NGC 5286 and NGC 6101*
372	**Alcaino G.** 1974, A&AS 13, 345 *The globular cluster NGC 4372*
373	**Muzzio J.C., Rydgren A.E.** 1974, AJ 79, 864 *Faint blue stars beyond the Perseus arm*
374	**Elliott J.E.** 1974, AJ 79, 1082 *Study of delta Scuti variables*
375	**Muzzio J.C., Marraco H.G., Feinstein A.** 1974, PASP 86, 394 *Faint OB stars behind the Coal Sack*
376	**Sanwal N.B., Pathasarathy M.** 1974, A&AS 13, 91 *UBV photometry of EU Tau*
378	**Philip A.G.D.** 1974, ApJ 190, 573 *Four-color observations of early-type stars. IV. South galactic pole*
379	**Haupt H.F., Schroll A.** 1974, A&AS 15, 311 *Photoelektrische Photometrie von Shell-Sternen*
380	**Moffat A.F.J., FitzGerald M.P.** 1974, A&AS 16, 25 *NGC 3105, a young very distant cluster in Vela*
381	**Moffat A.F.J.** 1974, A&AS 16, 33 *NGC 2345, a moderately young open cluster in Canis Major*
382	**Moffat A.F.J., FitzGerald M.P.** 1974, A&AS 18, 19 *NGC 2453, a moderately young open cluster in Puppis*
383	**Feinstein A., Forte J.C.** 1974, PASP 86, 284 *The open cluster NGC 6281*
384	**McClure R.D., Forrester W.T., Gibson J.** 1974, ApJ 189, 409 *The old open cluster NGC 2420*
385	**Hawarden T.G.** 1974, MNRAS 169, 539 *NGC 6259, southern image of M11*
386	**Strom S.E., Strom K.A., Carrasco L.** 1974, PASP 86, 798 *A study of the young cluster IC 348*
387	**McClure R.D.** 1974, ApJ 194, 355 *Possible abundance difference among giant stars in NGC 188*
388	**Clariá J.J.** 1974, AJ 79, 1022 *Investigation of a Milky Way field in Canis Major*
389	**Grenier S.** 1974, A&AS 16, 269 *Photoelectric photometry of late type stars in the direction opposed to galactic rotation*
390	**Thackeray A.D., Andrews P.J.** 1974, A&AS 16, 323 *A supplementary list of southern O stars*
391	**Kilkenny D.** 1974, Observatory 94, 4 *CPD -72°1184, a high velocity blue star*
393	**Moffat A.F.J.** 1974, A&A 34, 29 *The Wolf-Rayet star HDE 311884 in the open cluster Hogg 15*
394	**Margon B.** 1974, A&A 30, 467 *Photometry in the Cygnus X-1 field*
395	**Alcaino G.** 1974, A&AS 13, 305 *UBV photometry of thirty-eight southern galaxies*
396	**Miller E.W., McCarthy C.C.** 1974, AJ 79, 1294 *Photometric standards for the southern hemisphere. III.*
397	**Priser J.B.** 1961, Publ. US. Nav. Obs. XX Part VII *UBV sequences in Selected Areas*
398	**FitzGerald M.P., Moffat A.F.J.** 1974, AJ 79, 873 *Puppis clusters Haffner 19 and 18ab and the 15 kiloparsec arm*
399	**Guetter H.H.** 1974, PASP 86, 795 *UBV photometry of 180 early-type stars*
400	**Hill P.W., Kilkenny D., van Breda I.G.** 1974, MNRAS 168, 451 *UBV photometry of southern early-type stars*
401	**Deutschman W.A., Davis R.J., Schild R.E.** 1976, ApJS 30, 97 *The galactic distribution of interstellar absorption as determined from the Celescope catalog of ultraviolet stellar observations and a new catalog of UBV, Hβ photoelectric observations*
402	**Denoyelle J.** 1975, A&AS 19, 45 *Six UBV photoelectric sequences in Vela (l=257° to 281°)*
403	**Humphreys R.M.** 1975, A&AS 19, 243 *Spectroscopic and photometric observations of luminous stars in the Centaurus-Norma (l=305° - 340°) section of the Milky Way*
404	**Hurly P.R.** 1975, MNASSA 34, 7 *Combined-light UBV photometry of 103 bright southern visual doubles*
405	**Drilling J.S.** 1975, AJ 80, 128 *UBV photometry of OB+ stars north of 1950.0 declination −15°*
406	**Madore B.F.** 1975, ApJ 197, 55 *UBV photometry of the cepheid V367 Scuti in the open cluster NGC 6649*
407	**White S.D.M.** 1975, ApJ 197, 67 *UBV photometry of NGC 2439*
408	**Landolt A.U.** 1975, PASP 87, 379 *Photometry of V1057 Cygni and neighbouring stars*
409	**FitzGerald M.P., Moffat A.F.J.** 1975, A&AS 20, 289 *On the nature of the Puppis cluster NGC 2483*
410	**Moffat A.F.J., Vogt N.** 1975, A&AS 20, 85 *Southern open star cluster. IV. UBV-Hβ photometry of 26 clusters from Monocerotis to Vela*
411	**Moffat A.F.J., Vogt N.** 1975, A&AS 20, 125 *Southern open star clusters. V. UBV-Hβ photometry of 20 clusters in Carina*
412	**Moffat A.F.J., Vogt N.** 1975, A&AS 20, 155 *Southern open star clusters. VI. UBV-Hβ photometry of 18 clusters from Centaurus to Sagittarius*
413	**Moffat A.F.J., Vogt N.** 1975, A&A 44, 413 *The bright star cloud in Sagittarius at l=12°, b=−1°*
414	**Olson B.I., Richer H.B.** 1975, ApJ 200, 88 *The absolute magnitude of carbon stars: carbon stars in binary systems*
415	**Isserstedt J.** 1975, A&AS 19, 259 *Lichtelektrische Photometrie von Ueberriesen in der Grossen Magellanschen Wolke*
416	**Azzopardi M., Vigneau J.** 1975, A&AS 19, 271 *Small Magellanic Cloud, first list of probable members*
417	**Brunet J.P., Imbert N., Martin N., Mianes P., Prevot L., Rebeirot E., Rousseau J.** 1975, A&AS 21, 109 *Recherches dans le Grand Nuage de Magellan. I. Un catalogue de 272 étoiles O-B2 nouvelles*
418	**Celis L.S.** 1975, A&AS 22, 9 *Photoelectric photometry of late-type variable stars*
419	**Harrington R.S., Dahn C.C., Behall A.L., Priser J.B., Christy J.W., Riepe B.Y., Ables H.D., Guetter H.H., Hewitt A.V., Walker R.L.** 1961, Publ. US. Nav. Obs. XXIV Part I *Third catalog of trigonometric parallaxes of faint stars*
420	**Alcaino G.** 1975, A&AS 21, 5 *M4 the closest globular cluster*
421	**Alcaino G.** 1975, A&AS 21, 15 *The globular cluster NGC 288*
422	**Alcaino G.** 1975, A&AS 22, 193 *The globular cluster NGC 6809*
423	**Grasdalen G.L., Carrasco L.** 1975, A&A 43, 259 *NGC 2175, the cluster age and the nature of the nebulosity surrounding S252a*
424	**Bond H.E.** 1975, ApJ 202, L47 *A new CH star in the globular cluster omega Centauri*
425	**Azzopardi M., Vigneau J.** 1975, A&AS 22, 285 *List of 506 stars, probable Small Magellanic Cloud members*
426	**Coyne G.V., Wisniewski W., Corbally C.** 1975, Vatican Obs. Pub. 1 No 6 *A survey for Hα emission objects in the Milky Way II. Cygnus*
427	**Hawarden T.G.** 1975, MNRAS 173, 801 *The old, metal-poor open cluster NGC 2243*
428	**Townsed R.E.** 1975, PASP 87, 753 *UBV photometry of Collinder 463*
429	**Ogura K., Ishida K.** 1975, PASJ 27, 119 *UBV photometry of the stars in the fields of emission nebulae. I. M20*
430	**Warren P.R., Penfold J.E.** 1975, MNRAS 172, 41p *Photometry of LMC X-ray candidates*
431	**Dufour R.J., Duval J.E.** 1975, PASP 87, 769 *UBV photometry of stars in the direction of three LMC X-ray sources*
432	**Vogt N., Moffat A.F.J.** 1975, A&A 45, 405 *A comparison of photometric and kinematic distances for southern HII regions*

No	Reference
433	**Polyakova T.A.** 1975, Astrophysics 10, 36 *Intrinsic polarisation in the radiation of μ Cephei*
434	**Herbst W.** 1975, AJ 80, 212 *R associations I. UBV photometry and MK spectroscopy of stars in southern reflection nebulae*
435	**Madore B.F.** 1975, A&A 38, 471 *Tentative membership of the 11-day cepheid TW Normae in the open cluster Lyngå 6*
436	**Joshi U.C., Sagar R., Pandey P.** 1975, Pramana 4, 160 *Photoelectric photometry of the open cluster NGC 1778*
437	**Grasdalen G., Joyce R., Knacke R.F., Strom S.E., Strom K.M.** 1975, AJ 80, 117 *Photometric study of the Chamaeleon T-association*
438	**Harris W.E.** 1975, ApJS 29, 397 *New color-magnitude data for twelve globular clusters*
439	**Racine R., Harris W.E.** 1975, ApJ 196, 413 *A photometric study of NGC 2419*
440	**Culver R.B., Ianna P.A.** 1975, ApJ 195, L37 *The distance and absolute magnitude of the super-lithium S star T Sagittarii*
441	**Norris J., Bessell M.S.** 1975, ApJ 201, L75 *Abundance variations on the lower giant branch of omega Centauri*
442	**Sandage A., Tammann G.A.** 1974, ApJ 194, 223 *Steps toward the Hubble constant. III. The distance and stellar content of the M101 group of galaxies*
443	**Hall D.S.** 1974, Acta Astron. 24, 69 *The O9+Of eclipsing binary V729 Cygni in Cygnus OB2*
444	**Blaauw A., Danziger I.J., Schuster H.E.** 1975, A&A 44, 469 *A new southern planetary nebula*
445	**Haug U.** 1970, A&A 9, 453 *Luminosity determination of OB stars by Hβ indices*
446	**Krzeminski W.** 1972, Acta Astron. 22, 391 *Photometry and intrinsic period of HZ 29 (=AM CVn)*
447	**Wildey R.L.** 1961, ApJ 133, 430 *The color-magnitude diagram of 47 Tuc*
448	**Sandage A., Walker M.F.** 1966, ApJ 143, 313 *Three-color photometry of the bright stars in the globular cluster M92*
449	**Sandage A., Katem B.** 1964, ApJ 139, 1088 *Three-color photometry of the metal-rich globular cluster NGC 6171*
450	**Burnichon M.L., Garnier R.** 1976, A&AS 24, 89 *Photométrie photoélectrique UBV des composantes de 20 systèmes multiples stellaires jeunes*
451	**Breger M.** 1976, ApJ 204, 789 *A polarization survey of stars near the Orion nebula*
452	**Havlen R.J.** 1976, A&A 49, 307 *NGC 2533 and the cepheid BN Puppis*
453	**Bolton C.T., Herbst W.** 1976, AJ 81, 339 *Photometry of stars near 3U1700-37=HD153919 and 3U0933-40=HD77581*
454	**Warren P.R., Hesser J.E.** 1977, ApJS 34, 115 *A photometric study of the Orion OB1 association I. Observational data*
455	**Havlen R.J.** 1976, A&A 50, 227 *Photoelectric photometry and interpretation of the Puppis cluster NGC 2483*
456	**Ogura K. Ishida K.** 1976, PASJ 28, 35 *UBV photometry of the stars in the fields of emission nebulae. II. M17*
457	**Clariá J.J.** 1976, PASP 88, 225 *Basic parameters of the open cluster NGC 2571*
458	**Perry C.L., Franklin Jr C.B., Landolt A.U., Crawford D.L.** 1976, AJ 81, 632 *Multicolor photometry of the open cluster NGC 4755*
459	**Sagar R.** 1976, Ap&SS 40, 447 *Photometry of the galactic cluster NGC 2169*
460	**Oja T.** 1976, Private communication
461	**McNamara B.J.** 1976, AJ 81, 845 *Pre-main-sequence masses and the age spread in the Orion cluster*
462	**Lee P., Perry C.L.** 1976, PASP 88, 135 *Photometry of some southern metal-poor stars*
463	**Clariá J.J.** 1976, AJ 81, 155 *Relation of NGC 3590, Hogg 10, and Collinder 240 to the structure of the Carina spiral feature*
464	**Hawarden T.G.** 1976, MNRAS 174, 225 *NGC 2204, an old open cluster in the halo*
465	**Warren P.R., Penfold J.E., Hawarden T.G.** 1976, MNRAS 174, 213 *Galactic centre sequences in B and V*
466	**Wramdemark S.** 1976, A&AS 23, 231 *Carina arm studies II. Photometry of faint early-type stars in Carina*
467	**FitzGerald M.P., Hurkens R., Moffat A.F.J.** 1976, A&A 46, 287 *Bochum 15, a new young stellar aggregate in Puppis*
468	**Hawarden T.G.** 1976, MNRAS 174, 471 *The old open cluster Melotte 66*
469	**Havlen R.J.** 1976, A&A 47, 193 *UBV, Hβ photometry of OB stars in group: Pup OB2*
470	**Dahn C.C., Harrington R.S.** 1976, ApJ 204, L91 *LP 380-5/6, a binary system containing a late-type degenerate star*
471	**Culver R.B., Ianna P.A.** 1976, PASP 88, 41 *HD 196673, a binary system containing a barium star*
472	**Feinstein A., Marraco H.G., Forte J.C.** 1976, A&AS 24, 389 *Collinder 228 and the η Carinae complex*
473	**Alcaino G.** 1976, A&A 50, 299 *The X-ray globular cluster NGC 1851*
474	**Bernard A.** 1976, A&AS 25, 281 *Séquences standards photoélectriques UBV au voisinage des amas NGC 6256, 6304, 6638 et photométrie globale de NGC 4590, 6256, 6304, 6401, 6638*
475	**Moffat A.F.J.** 1976, A&A 50, 429 *Mass loss from the M3 supergiant HD 143183 in a young compact star cluster in Norma*
476	**Forte J.C.** 1976, A&AS 25, 271 *UBV photometry of luminous stars in the field of NGC 3372*
477	**van den Bergh S., Harris G.L.H.** 1976, ApJ 208, 765 *The 10.8 day cepheid TW Normae and the cluster Lyngå number 6*
478	**van den Bergh S., Herbst E., Harris G.L.H., Herbst W.** 1976, ApJ 208, 770 *The cepheid SV Crucis and the cluster Ruprecht 97*
479	**Herbst W.** 1976, ApJ 208, 923 *On the extinction law in the Carina nebula*
480	**Dupuy D.L., Zukauskas W.** 1976, RASC. Jour. 70, 169 *The galactic cluster Byurakan 2*
481	**Warren P.R.** 1976, MNRAS 176, 667 *The 'very red stars' at the south galactic pole*
482	**Harris G.L.H., van den Bergh S.** 1976, ApJ 209, 130 *The cepheid CS Velorum and the cluster Ruprecht 79*
483	**Wramdemark S.** 1976, A&AS 26, 31 *Distant early-type stars in some northern Milky Way fields*
484	**Alcaino G.** 1976, A&AS 26, 251 *The globular cluster NGC 3201*
485	**Harris W.E., Racine R., de Roux J.** 1976, ApJS 31, 13 *New color-magnitude diagrams for four southern globular clusters*
486	**Sherwood W.A., Dachs J.** 1976, A&A 48, 187 *The distance and mass of the large elephant trunk, a CO cloud pointing towards NGC 6231*
487	**McGimsey B.Q., Miller H.R., Williamson R.M.** 1976, AJ 81, 750 *Photoelectric UBV sequences in the fields of extra-galactic sources*
488	**Alcaino G.** 1976, A&AS 26, 353 *The globular cluster NGC 1904*
489	**Alcaino G.** 1976, A&AS 26, 359 *The globular cluster NGC 362*
490	**Vogt N.** 1976, A&A 53, 9 *Galactic structure in Monoceros, a photometric study of luminous stars*
491	**Butler C.J.** 1976, MNRAS 178, 159 *Photoelectric sequences near the south galactic cluster NGC 3532*
492	**Hanes D.A., Harris W.E., Madore B.F.** 1976, MNRAS 177, 653 *UBV sequences for selected Virgo galaxies*
493	**Turner D.G.** 1976, ApJ 210, 65 *The value of R in Monoceros*
494	**Havlen R.J.** 1976, PASP 88, 685 *A new subdwarf O star, LS 630*
495	**Rieke G.H., Grasdalen G.L., Kinman T.D., Kintzen P., Wills B.J., Wills D.** 1976, Nature 260, 754 *Photometric and spectroscopic observations of the BL Lacertae object AO 0235+164*

No	Reference
496	**Muzzio J.C., Feinstein A., Orsatti A.M.** 1976, PASP 88, 526 *UBV and Hβ photometry of faint early-type stars in Crux and Centaurus*
497	**Kamper K.W.** 1976, PASP 88, 444 *Astrometric and photometric observations of nearby binary stars*
498	**Smith M.G., Weedmann D.W.** 1976, ApJ 205, 709 *Globular clusters in the Hydra I cluster of galaxies*
499	**Garrison R.F., Kormendy J.** 1976, PASP 88, 865 *Some characteristics of the young open cluster Trumpler 37*
500	**Herbst W., Turner D.G.** 1976, PASP 88, 308 *Lynds 810 - an interesting globule in northern Vulpecula*
501	**Demers S., Kunkel W.E.** 1976, ApJ 208, 932 *The Reticulum system, an analog to the Draco dwarf galaxy near the Large Magellanic Cloud*
502	**Herbst W., Racine R.** 1976, AJ 81, 840 *R associations V. Monoceros R2*
503	**Hesser J.E., Hartwick F.D.A.** 1976, ApJ 203, 97 *Metal-rich globular clusters in the Galaxy. III. The 'X-ray' globular cluster NGC 6441*
504	**Ogura K., Ishida K.** 1976, PASJ 28, 651 *UBV photometry of the stars in the fields of emission nebulae. III. IC 1795*
505	**van den Bergh S.** 1976, AJ 81, 106 *UBV observations of Nova Scuti 1975*
506	**van den Bergh S.** 1976, AJ 81, 104 *UBV observations of the X-ray Nova in Monoceros*
507	**Cousins A.W.J., Lagerwey H.C.** 1967, MNASSA 26,83 *Magnitudes of some very bright stars*
508	**Wegner G.** 1976, MNRAS 177, 99 *On the reddening and the effective temperature of HD 101065*
509	**Wickramasinghe D.T., Warren P.R.** 1976, MNRAS 177, 59p *The distance to A0623-00 from reddening*
510	**Shinohara M., Ishida K.** 1976, PASJ 28, 437 *On the HII region NGC 1499 and ξ Persei*
511	**Véron P., Véron M.P.** 1976, A&AS 25, 287 *UBV sequences in the fields of Mark 421 and Mark 501*
512	**Hesser J.E., Hartwick F.D.A.** 1976, ApJ 203, 113 *Metal-rich globular clusters in the Galaxy. IV. A color-magnitude diagram for NGC 6304*
513	**Williamon R.M.** 1976, AJ 81, 1134 *Photometric study of CW Canis Majoris*
514	**Sandage A., Katem B.** 1968, ApJ 153, 569 *The color-magnitude diagram for the globular cluster NGC 5897*
515	**Sandage A., Westphal J.A., Strittmatter P.A.** 1966, ApJ 146, 316 *On the optical identification of Sco X-1*
516	**Sandage A.** 1969, ApJ 158, 1115 *New subdwarfs. II. Radial velocities, photometry, and preliminary space motions for 112 stars with large proper motion*
517	**Geyer E.H.** 1967, Z. Astrophys. 66, 16 *Eine Dreifarbenphotometrie von omega Centauri (NGC 5139)*
518	**Hartwick F.D.A., Hesser J.E.** 1973, ApJ 186, 1171 *The color-magnitude diagram of the metal-poor southern globular cluster NGC 4372*
519	**Dickens R.J.** 1970, ApJS 22, 249 *Photometry of RR Lyrae variables in the globular cluster NGC 6171*
520	**Clariá J.J.** 1977, A&AS 27, 145 *NGC 3324. A very young open cluster in the Carina spiral feature*
521	**Lee S.W.** 1977, A&AS 27, 367 *The C-M diagram of the globular cluster M4*
522	**Lee S.W.** 1977, A&AS 27, 381 *UBV photometry of bright stars in 47 Tuc*
523	**Sarg K., Wramdemark S.** 1977, A&AS 27, 403 *Photoelectric UBV photometry in a Milky Way field in Cassiopeia*
524	**Moffat A.F.J., FitzGerald M.P.** 1977, A&A 54, 263 *Some very luminous supergiants associated with compact groups of luminous OB stars*
525	**Hilditch R.W., Dodd R.J.** 1977, MNRAS 178, 467 *New photoelectric sequences in Puppis (l= 250°, b=0°)*
526	**Williamon R.M.** 1977, PASP 89, 44 *UBV observations of V1500 Cygni*
527	**Kunkel W.E., Demers S.** 1977, ApJ 214, 21 *The Sculptor dwarf galaxy. Photoelectric sequence and a preliminary color-magnitude diagram*
528	**Demarque P., McClure R.D.** 1977, ApJ 213, 716 *The significance of the star clusters NGC 2420 and 47 Tucanae for galactic evolution*
529	**Hesser J.E., Hartwick F.D.A.** 1977, ApJS 33, 361 *Metal-rich globular clusters in the galaxy. V. A photometric study of 47 Tucanae*
530	**Ardeberg A., Maurice E.** 1977, A&AS 28, 153 *A spectrographic and photometric study of stars and interstellar medium surrounding HD 101205*
531	**Janes K.A.** 1977, AJ 82, 35 *DDO and UBV photometry of NGC 7789*
532	**Guinan E.F.** 1977, AJ 82, 51 *Multiband photoelectric photometry of the eclipsing binary R Canis Majoris*
533	**Turner D.G.** 1977, AJ 82, 163 *Absolute magnitude of the cepheid VY Carinae*
534	**Schnur G.F.O.** 1977, Astron. Nachr. 298, 167 *Two UBV sequences, north and south of the galactic cluster NGC 5823*
535	**FitzGerald M.P., Jackson P.D., Moffat A.F.J.** 1977, A&A 59, 141 *NGC 3105. Further photoelectric UBV observations*
536	**Dean C.A., Lee P., O'Brien A.** 1977, PASP 89, 222 *Some observations of weak-G band stars*
537	**FitzGerald M.P., Jackson P.D., Moffat A.F.J.** 1977, Observatory 97, 129 *NGC 6200, a loose young open cluster in the Sagittarius-I arm extension*
538	**Dahn C.C., Harrington R.S., Riepe B.Y., Christy J.W., Guetter H.H., Behall A.L., Walker R.L., Hewitt A.V., Ables H.D.** 1976, Publ. US. Nav. Obs. XXIV Part III *Fourth catalog of trigonometric parallaxes of faint stars*
539	**Harvel CH.A., Martins D.H.** 1977, ApJ 213, L52 *Preliminary photometry in the nuclear region of NGC 6624*
540	**Klare G., Neckel Th.** 1977, A&AS 27, 215 *UBV, Hβ and polarization measurements of 1660 southern OB stars*
541	**Joshi U.C., Sagar R.** 1977, Ap&SS 48, 225 *Photoelectric photometry of the open cluster Tr-1*
542	**Lutz T.E., Lutz H.** 1977, AJ 82, 431 *Spectral classification and UBV photometry of bright visual double stars*
543	**van den Bergh S.** 1977, ApJ 215, 89 *The old open cluster NGC 2243*
544	**Moffat A.F.J., FitzGerald M.P., Jackson P.D.** 1977, ApJ 215, 106 *Trumpler 27. A heavily reddened young open cluster with blue and red supergiants*
545	**McMillan R.S.** 1977, ApJ 216, L41 *Walker No. 67 in NGC 2264. A candidate for strong interstellar circular polarization*
546	**Hesser J.E., Hartwick F.D.A., McClure R.D.** 1977, ApJS 33, 471 *Cyanogen strengths and ultraviolet excesses of evolved stars in 17 globular clusters from DDO photometry*
547	**Lee S.W.** 1977, A&AS 28, 409 *UBV photometry of bright stars in NGC 3201*
548	**Alcaino G.** 1977, A&AS 29, 9 *The metal-poor globular cluster NGC 4590*
549	**Lodén L.O.** 1977, A&AS 29, 31 *A study of some loose clusterings in the southern Milky Way*
551	**McGimsey B.Q., Miller H.R.** 1977, AJ 82, 453 *Photoelectric comparison sequences in the fields of optically active extragalactic objects*
552	**Lee S.W.** 1977, A&AS 29, 1 *The C-M diagram of the globular cluster M55*
553	**Sturch C.R.** 1977, PASP 89, 349 *UBV photometry of RR Lyrae stars in M4*
554	**Culver R.B., Ianna P.A., Franz O.G.** 1977, PASP 89, 397 *Visual binaries among the barium stars. II. HD 105902*

No	Reference
555	**Hill P.W., Lynas-Gray A.E.** 1977, MNRAS 180, 691 *UBV photometry and MK spectral classification of northern early-type stars at intermediate galactic latitudes*
556	**Williams P.M., Brand P.W.J.L., Longmore A.J., Hawarden T.G.** 1977, MNRAS 180, 709 *The young cluster NGC 5367 and A1353-40*
557	**Harris G.L.H., Harris W.E.** 1977, AJ 82, 612 *The Hyades-age cluster NGC 1817*
558	**Alcaino G.** 1977, A&AS 29, 397 *The globular cluster NGC 6397*
559	**Alcaino G.** 1977, A&AS 29, 383 *The metal-poor globular cluster NGC 6656*
560	**Arp H.C., Melbourne W.G.** 1959, AJ 64, 28 *Color-magnitude diagram for the globular cluster M22*
561	**Drilling J.S.** 1977, AJ 82, 714 *UBV photometry of stars in a region near the south galactic pole*
562	**McNamara B.J., Sanders W.L.** 1977, A&AS 30, 45 *Membership of M39*
563	**Rakos K.D., Havlen R.J.** 1977, A&A 61, 185 *Photoelectric observations of Sirius B*
564	**Turner D.G.** 1977, AJ 82, 805 *Study of the NGC 2439 association*
565	**Catchpole R.M., Robertson B.S.C., Warren P.R.** 1977, MNRAS 181, 391 *Spectroscopic radial velocity and photometric observations of barium stars*
566	**Cester B., Giuricin G., Mardirossian F., Pucillo M.** 1977, A&AS 30, 227 *Photoelectric photometry of the open cluster M34*
567	**Haug U., Bredow K.** 1977, A&AS 30, 235 *A contribution to the study of the Norma dark cloud*
568	**Jackson P.D., FitzGerald M.P., Moffat A.F.J.** 1977, A&A 60, 417 *Photoelectric and spectroscopic observations of Bochum 15*
569	**Vogt N., Maitzen H.M.** 1977, A&A 61, 601 *On the decline stage of Nova V3888 Sgr (1974)*
570	**Havlen R.J., Moffat A.F.J.** 1977, A&A 58, 351 *A new cluster containing 2 Wolf-Rayet-stars and 2 Of-stars*
571	**Hutchings J.B.** 1977, MNRAS 181, 619 *The 580-day and other periodicities in X Persei*
572	**Murdin P., Penston M.V.** 1977, MNRAS 181, 657 *The lambda Orionis association*
573	**Ardeberg A., Maurice E.** 1977, A&AS 30, 261 *Observations of supergiant stars in the Small Magellanic Cloud*
574	**Herbst W., Havlen R.J.** 1977, A&AS 30, 279 *Ara OB1, NGC 6193 and Ara R1, an optical study of a very young southern complex*
575	**Hodge P.W.** 1977, ApJS 33, 69 *The structure and content of NGC 6822*
576	**Bord D.E., Mook D.E., Petro L., Thomas J., Hiltner W.A.** 1977, PASP 89, 720 *UBV observations of CoD −33°12119*
577	**Herbst W., Racine R., Richer H.B.** 1977, PASP 89, 663 *A luminous carbon star in Canis Major OB1*
578	**Landolt A.U.** 1977, PASP 89, 403 *A photoelectric magnitude sequence for AA Aurigae*
579	**Harris W.E.** 1977, PASP 89, 482 *Photometric calibration in the metal-rich globular cluster M69*
580	**Connolly L.P.** 1977, PASP 89, 528 *The NGC 2818 photoelectric sequence*
581	**Stetson P.B., Harris W.E.** 1977, AJ 82, 954 *A photometric study of NGC 1904*
582	**Upgren A.R., Weis E.W.** 1977, AJ 82, 978 *Photometry of new possible members of the Hyades cluster*
583	**Woolley R., Martin W.L., Penston M.J., Sinclair J.E., Aslan S.** 1977, MNRAS 179, 81 *Space motions of galactic G- and K-type stars*
584	**Andrews P.J.** 1977, MNRAS 178, 131 *HR 6392 - a double star with very high luminosities*
585	**Broglia P., Conconi P.** 1977, A&AS 27, 285 *Two-colour photometry and elements of VV UMa*
586	**Surdej A., Surdej J.** 1977, A&AS 30, 121 *Rotation period and photoelectric lightcurves of asteroid 471 Papagena*
587	**Lapasset E.** 1977, Ap&SS 46, 155 *UBV light variation and orbital elements of MW Pavonis*
588	**Popper D.M., Dumont P.J.** 1977, AJ 82, 216 *UBV photometric program on eclipsing binaries at Palomar and Kitt Peak*
589	**Sandage A., Katem B., Johnson H.L.** 1977, AJ 82, 389 *Halo globular cluster NGC 5053*
590	**Hering R., Klare G., Schnur G.F.O.** 1977, Astron. Nachr. 298, 163 *Photoelectric UBV and RI sequences in the Small Magellanic Cloud*
591	**Martin W.L.** 1977, Mem. R. Astr. Soc. 83, 95 *Photoelectric (B,V) magnitude sequences in the Magellanic Clouds*
592	**Szabados L.** 1977, Mitt. Sternw. Ungar. Akad. Budapest No 70 *Photoelectric UBV photometry of northern cepheids I.*
593	**Heiser A.M.** 1977, AJ 82, 973 *uvbyβ photometry of NGC 2244*
594	**Clariá J.J.** 1977, PASP 89, 803 *Two sparse open clusters in the region of Collinder 132*
595	**Markkanen T.** 1977, Obs. Astrophys. Lab. Univ. Helsinki Report 1 *The magnetic pocket, an observational study of the structure of the galactic magnetic field and of interstellar dust in the direction of the α Persei cluster*
596	**van den Bergh S., Kamper K.W.** 1977, ApJ 218, 617 *The remnant of Kepler's supernova*
597	**Bookmeyer B.B., Fitch W.S., Lee T.A., Wisniewski W.Z., Johnson H.L.** 1977, Rev. Mex. Astron. Astrofis. 2, 235 *Photoelectric UBV observations of RR Lyrae variable stars, Second List*
598	**Grauer A.D., McCal J., Reaves L.C., Tribble T.L.** 1977, AJ 82, 740 *Period study and UBV light curves of TV Cassiopeiae*
599	**Herbst W.** 1977, PASP 89, 795 *Stars in reflection nebulae near the Herbig - Haro objects in the Gum nebula*
600	**O'Connell D.J.K.** 1977, Ric. Astron. 8 No 29 *Photoelectric observations and apsidal motion of Y Cygni*
601	**Rossiger S., Wenzel W.** 1974, Astron. Nachr. 295, 47 *Lichtwechsel und Extinktion des jungen Sterns RR Tauri*
602	**Kayser S.E.** 1967, AJ 72, 134 *Photometry of the nearby irregular galaxy, NGC 6822*
603	**Winzer J.E.** 1974, AJ 79, 124 *UBV photometry of four magnetic stars*
604	**von Bronkalla W.** 1963, Astron. Nachr. 287, 249 *UBV-Photometrie des Sternfeldes um δ Lyrae*
605	**Millis R.L.** 1969, Lowell Obs. Bull. 7, 113 *Photoelectric observations of HR 8157*
607	**Moffat A.F.J.** 1974, A&A 35, 315 *On the distance and reddening of the massive HII region NGC 3603*
608	**Pismis P., Hasse I.** 1976, Ap&SS 45, 79 *Study of a triple emission nebula in Orion*
609	**Handschel G.** 1972, Thesis (Univ. Münster) *UBV - photometrie der offenen Sternhaufen NGC 7419, K10 und NGC 6192*
610	**Evans T.L., Menzies J.W.** 1971, Observatory 91, 35 *The metal rich globular cluster NGC 6637 (M69)*
611	**Cousins A.W.J., Lagerwey H.C.** 1971, MNASSA 30, 12 *UBV observations of variable stars*
612	**Cousins A.W.J., Lagerwey H.C., Shillington F.A., Stobie R.S.** 1969, MNASSA 28, 25 *Photometry of three δ Scuti variables*
613	**Cousins A.W.J., Lagerwey H.C.** 1969, MNASSA 28, 17 *BL Telescopii (1967-68)*
614	**Cousins A.W.J., Lagerwey H.C.** 1968, MNASSA 27, 138 *UBV photometry of cepheid variables*
615	**Lagerwey H.C.** 1968, MNASSA 27, 17 *Three colour observations of 31 Mensae (HR 2059)*
616	**Nariai K.** 1963, PASJ 15, 7 *Line blanketing effect on the peculiar star HD 30353*

No	Reference
617	**Kiyokawa M.** 1967, PASJ 19, 209 *Photoelectric observations of zeta Aurigae during the 1963-64 eclipse*
618	**Morgan W.W., Johnson H.L., Roman N.G.** 1954, PASP 66, 85 *A very red star of early type in Cygnus*
619	**Mihalas D.** 1963, PASP 75, 256 *Decay times of type I supernovae light curves*
620	**Faulkner D.J.** 1963, PASP 75, 269 *On the distance of the η Carinae nebula*
621	**Bessel M.S.** 1981, Proc ASA 4,212 *Alpha Centauri*
622	**Sanduleak N., Philip A.G.D.** 1968, PASP 80, 437 *A spectroscopic survey of early stars from the Tonantzintla list, near the south galactic pole*
623	**Hall D.S., Hardie R.H.** 1969, PASP 81, 754 *A UBV photometric study of MR Cygni*
624	**Demers S.** 1969, PASP 81, 861 *Two new southern δ Scuti variables*
625	**Landolt A.U.** 1970, PASP 82, 1364 *Magnitude sequences for the variable stars UZ Bootis and W Piscium*
626	**Hall D.S., Gertken R.H., Burke E.W.** 1970, PASP 82, 1077 *A UBV photometric study of IQ Persei*
627	**Hall D.S., Weedman S.L.** 1971, PASP 83, 69 *UBV photometry of eclipsing binaries with visual companions*
628	**Drilling J.S.** 1972, PASP 84, 35 *Two new faint carbon stars at high galactic latitudes*
629	**Guinan E.F.** 1972, PASP 84, 56 *UBVβ photometry of EU Tauri*
630	**Drilling J.S., Landolt A.U.** 1972, PASP 84, 810 *Minima for the eclipsing binary RT Persei*
631	**Honeycutt R.K.** 1972, PASP 84, 823 *On the ratio of total to selective absorption for carbon stars*
632	**Landolt A.U.** 1973, PASP 85, 117 *A minimum each for GL Carinae and SV Centauri*
633	**Vidal N.V.** 1974, PASP 86, 317 *Photoelectric observations of HD 77581*
634	**Feltz K.A., McNamara D.H.** 1976, PASP 88, 699 *Color excesses of classical cepheids I.*
635	**Landolt A.U.** 1968, PASP 80, 450 *UBV observations of long-period variable stars, IV*
637	**de Vaucouleurs G.** 1958, ApJ 128, 465 *Photoelectric photometry of the Andromeda nebula in the UBV system*
639	**Arp H., Cuffey J.** 1962, ApJ 136, 51 *The star cluster NGC 2158*
640	**Smak J.** 1964, ApJ 139, 1095 *On the colors of T Tauri stars and related objects*
641	**Sandage A., Smith L.L.** 1966, ApJ 144, 886 *The color-magnitude diagram of the metal-rich globular cluster NGC 6712*
642	**Hodge P., Wright F.W.** 1969, ApJS 17, 467 *Studies of the Large Magellanic Cloud X. Photometry of variable stars*
643	**Seggewiss W.** 1969, ApJ 155, L1 *An extremely red star in the region of the young open cluster NGC 6231*
645	**Sandage A.** 1970, ApJ 162, 841 *Main-sequence photometry, color-magnitude diagrams, and ages for the globular clusters M3, M13, M15, and M92*
646	**Mumford G.S., Krzeminski W.** 1969, ApJS 18, 429 *Binary stars among cataclysmic variables. X. Photoelectric observations of EM Cygni*
648	**Kinman T.D., Wirtanen C.A., Janes K.A.** 1966, ApJS 13, 379 *An RR Lyrae star survey with the Lick 20-inch astrograph. IV. A survey of three fields near the north galactic pole*
649	**Rydgren A.E., Strom S.E., Strom K.M.** 1976, ApJS 30, 307 *The nature of the objects of Joy. A study of the T Tauri phenomenon*
650	**Grygar J., Kohoutek L.** 1969, BAC 20, 226 *Nova Vulpeculae 1968 No 1 I. Photoelectric observation at Ondrejov in 1968*
651	**Mayer P., Macak P.** 1971, BAC 22, 46 *Spectral classification and photometry of stars in two HII regions*
652	**Vetesnik M., Papousek J.** 1973, BAC 24, 57 *Photoelectric photometry of TX Herculis*
653	**Papousek J.** 1974, BAC 25, 152 *Photoelectric photometry of TV Cassiopeiae*
654	**Haupt W., Moffat A.F.J.** 1973, ApJ Lett. 13, 77 *ADS 2859B as an alternative candidate to X Per for the X-ray source 2U 0352+30*
655	**Basinski J.M., Bok B.J., Bok P.F.** 1967, MNRAS 137, 55 *A colour-magnitude array for a region in the core of the Small Magellanic Cloud*
656	**Wesselink A.J.** 1967, MNRAS 137, 275 *Magnitudes and colours for 276 stars in the neighbourhood of NGC 6067*
657	**Stobie R.S.** 1970, MNRAS 148, 1 *Photometry of bright southern cepheids*
658	**Penny A.J., Penfold J.E., Balona L.A.** 1975, MNRAS 171, 387 *The spectroscopic binary system HD 158320 (=3U 1727-33)*
659	**O'Connell D.J.K.** 1964, Ric. Astron. 6 No 23 *The 1961-1962 eclipse of 31 Cygni*
660	**O'Connell D.J.K.** 1964, Ric. Astron. 6 No 25 *The 1963-1964 eclipse of zeta Aurigae*
661	**O'Connell D.J.K.** 1967, Ric. Astron. 7 No 11 *Photometric orbit and apsidal motion of the eclipsing binary V526 Sagittarii*
662	**O'Connell D.J.K.** 1968, Ric. Astron. 7 No 13 *Orbital elements and apsidal motion of the eclipsing binary HH Carinae*
663	**Alexander J.B.** 1972, R. Obs. Bull. No 174 *UBV observations of the RR Lyrae variable HD 176387 (MT Telescopii)*
664	**Lyutyi V.M.** 1976, Sov. Astron. Lett. 2, 43 *UBV photometry of the object identified with the X-ray source A0620-00*
665	**Walker R.L.** 1976, Inf. Bull. Var. Stars No 1148 *On the secondary minimum of θ¹ Ori A*
666	**Mayer P.** 1964, Acta Univ. Carol. Math. Phys. 1, 25 *Photoelectric measurements of OB stars*
667	**Piirola V.** 1976, Obs. Lab. Univ. Helsinki Report 1 *UBV and uvby photometric systems and extinction conditions at Metsahovi Observatory*
668	**Fitch W.S., Wisniewski W.Z., Johnson H.L.** 1966, Commun. Lunar Plan. Lab. V No 71 *Photoelectric UBV observations of RR Lyrae variable stars*
669	**Schmidt E.G.** 1971, ApJ 165, 335 *A photometric study of four classical cepheids*
670	**Fernie J.D.** 1970, AJ 75, 244 *On the nature of S Vulpeculae*
671	**Cohen H.L.** 1974, A&AS 15, 181 *V453 Cyg. An early type Algol binary*
672	**Wachmann A.A.** 1974, A&A 34, 317 *UBV photometry and orbital elements of V453 Cygni*
673	**Arp H.C.** 1960, ApJ 131, 322 *Cepheids in galactic clusters. V. CV Mon*
674	**O'Connell D.J.K.** 1972, Ric. Astron. 8 No 14 *Photometric orbit and apsidal motion of DR Vulpeculae*
675	**Swope H.H.** 1974, Ric. Astron. 8 No 25 *V2283 Sgr, an eclipsing star with a rotating apse*
676	**Menzies J.** 1974, MNRAS 169, 79 *Photographic UBV photometry of the globular cluster NGC 5927*
677	**Kohoutek L., Wehmeyer R.** 1975, A&A 41, 451 *Concerning the nature of the nebula V-V 1-7 surrounding BD −18° 1967*
678	**Feinstein A.** 1966, Contr. CTIO No 8 *Photoelectric observations of southern late-type stars*
679	**Upgren A.R., Kerridge S.J.** 1973, PASP 85, 721 *Five-color photometry of nearby red dwarfs*
680	**Upgren A.R.** 1974, PASP 86, 294 *Five-color photometry of nearby red dwarfs. II*
681	**Feinstein A.** 1968, Z. Astrophys. 68, 29 *A survey of southern Be stars*
682	**Talbert F.D.** 1975, PASP 87, 341 *UBV photometry of NGC 6649*

No	Reference
683	**Hall D.S., Mallama A.D.** 1974, Acta Astron. 24, 359 *BS Scuti, possibly a blue straggler in M11*
684	**Mayer P., Macak P.** 1973, BAC 24, 50 *Photometry and classification of stars in several HII regions*
685	**Pesch P.** 1972, ApJ 177, 519 *Study of a sample of faint M stars in the direction of the north galactic pole*
686	**Pesch P.** 1976, AJ 81, 1117 *Further observations of faint M stars in the direction of the north galactic pole*
687	**Evans T.L.** 1977, MNRAS 181, 591 *Near infrared photometry of globular clusters - VIII. The anomalous giant branch of ω Centauri*
688	**Dean J.F., Cousins A.W.J., Bywater R.A., Warren P.R.** 1977, Mem. R. Astr. Soc. 83, 69 *VBI photometry of some southern cepheid and RR Lyrae variables*
689	**Madore B.F.** 1975, ApJS 29, 219 *Photoelectric UBV photometry of cepheids in the Magellanic Clouds and in the southern Milky Way*
690	**Coyne G.V., Wisniewski W., Otten L.B.** 1978, Vatican Obs. Pub. 1 No 12 *A survey for Hα emission objects in the Milky Way V. Perseus*
691	**Hall D.S.** 1969, in Mass loss from stars. Ed.M.Hack p. 175 *A gross secular expansion of the primary in RW Persei*
692	**Cester B.** 1969, in Mass loss from stars. Ed.M.Hack p. 329 *Photoelectric observations of CH Cygni during the explosive phases of 1967-68*
693	**Dawson D.W.** 1977, PASP 89, 919 *Observations of variable M stars in the DDO photoelectric system*
694	**Weistrop D.** 1977, ApJ 215, 845 *The velocity dispersion of faint red dwarf stars*
695	**Stepien K.** 1968, ApJ 154, 945 *Photometric behavior of magnetic stars*
696	**Preston G.W., Stepien K.** 1968, ApJ 154, 971 *The light, magnetic, and radial-velocity variations of HD 10783*
697	**Seggewiss W.** 1970, AJ 75, 345 *DI Car: a cepheid*
698	**Mumford G.S.** 1966, ApJ 146, 411 *Binary stars among cataclysmic variables. VIII. Photoelectric observations of twelve old and dwarf novae*
699	**Sturch C.** 1966, ApJ 143, 774 *Intrinsic UBV colors of RR Lyrae stars*
700	**Clube S.V.M., Evans D.S., Jones D.H.P.** 1969, Mem. R. Astr. Soc. 72, 101 *Observations of southern RR Lyrae stars*
701	**Feinstein A., Clariá J.J., Cabrera A.L.** 1978, A&AS 34, 241 *Multicolor photometry of the open cluster NGC 2287*
702	**Perry C.L., Walter D., Crawford D.L.** 1978, PASP 90, 81 *Multicolor photometry of the ζ Sculptoris open cluster*
703	**Clariá J.J., Rosenzweig P.** 1978, AJ 83, 278 *The nearby open cluster Collinder 140*
704	**Chromey F.R.** 1978, AJ 83, 162 *UBV photometry of faint blue stars near the galactic anticenter*
705	**Penfold J.E.** 1978, MNRAS 182, 283 *Observations of faint red stars at intermediate galactic latitude*
706	**Hawarden T.G.** 1978, MNRAS 182, 31p *The broad giant branch of Melotte 66*
707	**van den Bergh S.** 1978, A&A 63, 275 *UBV photometry of the luminous young cluster NGC 3603*
708	**FitzGerald M.P., Jackson P.D., Luiken M., Grayzeck E.J., Moffat A.F.J.** 1978, MNRAS 182, 607 *NGC 6383 - I. The central core*
709	**Dachs J., Issertedt J., Rahe J.** 1978, A&A 63, 353 *On the photometric variations of the red giant HD 65750 and of the surrounding reflection nebula IC 2220*
710	**Havlen R.J.** 1978, A&A 64, 295 *The OB stars in the vicinity of the long period cepheid VZ Puppis*
711	**Newell B., Sadler E.M.** 1978, ApJ 221, 825 *The blue horizontal-branch stars of NGC 6752*
712	**Forbes D., Dupuy D.L.** 1978, AJ 83, 266 *UBV photometry of the open clusters NGC 6604 and NGC 6704*
713	**Loibl B., Schröder R., Haug U.** 1978, A&A 65, 65 *UBV photometry in the nuclear bulge of the Galaxy*
714	**Giddings J.R., Dworetsky M.M.** 1978, MNRAS 183, 265 *UV1758+36, a hot subluminous B star*
715	**Margon B., Thorstensen J.R., Bowyer S.** 1978, ApJ 221, 907 *The optical counterpart of 3U1956+11*
716	**Williams P.M.** 1978, MNRAS 183, 49 *The open cluster Cr 140 revisited*
717	**Elst E.W.** 1978, A&AS 32, 161 *UBV photometric observations of SX Phe*
718	**Alcaino G.** 1978, A&AS 32, 379 *The globular cluster NGC 6266*
719	**Sagar R., Joshi U.C.** 1978, Bull. Astron. Soc. India 6, 12 *Study of the galactic cluster NGC 581*
720	**Sagar R., Joshi U.C.** 1978, MNRAS 184, 467 *Study of the extremely young open cluster NGC 6530*
721	**Sagar R.** 1978, Private communication *UBV photometry of two stars in NGC 1778*
722	**Harris W.E.** 1978, PASP 90, 45 *Photoelectric UBV and DDO photometry in the anomalous globular cluster NGC 2808*
723	**Heiser A.M., Uckotter C.L., Uckotter D.G.** 1978, PASP 90, 105 *Photometry of the ultraviolet stellar group in Auriga*
724	**Wramdemark S.** 1978, A&A 66, 137 *Be 94, a young distant cluster in Cepheus*
725	**Crampton D., Georgelin Y.M., Georgelin Y.P.** 1978, A&A 66, 1 *First optical detection of W51 and observations of new HII regions and exciting stars*
726	**Warren P.R., Bywater R.A.** 1978, Observatory 98, 120 *On the numbers of yellow stars in the Large Magellanic Cloud (paper II)*
727	**Johansson K.L.V.** 1978, A&AS 33, 107 *A photoelectric UBV sequence in the region of the wing of the Small Magellanic Cloud*
728	**Turner D.G., Lyons R.W., Bolton C.T.** 1978, PASP 90, 285 *HD 118246 - a new Be star at high galactic latitude*
729	**Bond H.E.** 1978, PASP 90, 216 *CoD $-42°14462$, a dwarf nova in permanent outburst*
730	**Turner D.G.** 1978, AJ 83, 1081 *The value of R in IC 2581*
731	**Rössiger S.** 1978, Inf. Bull. Var. Stars No 1474 *VZ Cephei is an eclipsing binary*
732	**Przybylski A.** 1978, PASP 90, 451 *HD 27507 - a star escaping from our galaxy*
733	**Isserstedt J.** 1978, A&AS 33, 193 *Lichtelektrische Photometrie in der kleinen Magellanschen Wolke*
734	**Marraco H.G., Forte J.C.** 1978, ApJ 224, 473 *Further observations of stars in the field of the cometary globule NGC 5367*
735	**Piccirillo J., Stein W.L.** 1978, AJ 83, 971 *Preliminary photometry of the open cluster IC 361*
736	**van den Bergh S.** 1978, AJ 83, 1174 *UBV observations of the open cluster M25*
737	**Evans T.L.** 1978, MNRAS 184, 661 *The open cluster NGC 6383*
738	**Alcaino G.** 1978, A&AS 33, 185 *The metal-poor globular cluster NGC 7099*
739	**Walker M.F.** 1978, ApJ 224, 546 *Spectroscopic and photometric observations of YY Orionis*
740	**Herbst W., Racine R., Warner J.W.** 1978, ApJ 223, 471 *Optical and infrared properties of the newly formed stars in Canis Major R1*
741	**Yoshizawa M.** 1978, PASJ 30, 123 *Study of the intermediate - age galactic cluster NGC 2281. I. UBV photoelectric observations, binary frequency, and the luminosity function of bright members*
742	**Carney B.W.** 1978, AJ 83, 1088 *Southern subdwarf photometry*

No	Reference

743 **Alcaino G.** 1978, A&AS 34, 431 *UBV photometry for star clusters in the Small Magellanic Cloud*

744 **Williamon R.M., Collins T.F., Kwan-Yu Chen** 1978, A&AS 34, 207 *The eclipsing variable AQ Tucanae*

745 **Liller M.H., Carney B.W.** 1978, ApJ 224, 383 *A photometric study of the X - ray globular cluster NGC 6624*

746 **Haug U.** 1978, A&AS 34, 417 *Photoelectric and photographic photometry in the open clusters NGC 5617, Tr 22 and NGC 5662*

747 **Feinstein A.** 1978, Rev. Mex. Astron. Astrofis. 2, 331 *Photoelectric measures of hydrogen lines in helium-weak stars*

748 **Grayzeck E.J., FitzGerald M.P., Luiken M.** 1978, PASP 90, 742 *Study of the open cluster NGC 6546 and cepheid AV Sagittarii*

749 **Heiser A.** 1978, Private communication

750 **O'Connell D.J.K.** 1978, Ric. Astron. 8 No 30 *The eclipsing binary CQ Aurigae*

751 **Gallagher J.S., van den Bergh S.** 1978, PASP 90, 665 *A peculiar nebula surrounding the K giant HD 20722*

752 **Mikulasek Z., Harmanec P., Grygar J., Zdarsky F.** 1978, BAC 29, 44 *Photoelectric photometry at the Hvar Observatory III. The Ap star CQ UMa*

753 **Harmanec P., Horn J., Koubsky P., Kriz S., Zdarsky F., Papousek J., Doazan V., Bourdonneau D., Daldinelli L., Ghedini S., Pavlovski K.** 1978, BAC 29, 278 *Properties and nature of Be and shell stars*

754 **Neckel T.** 1978, A&A 69, 51 *UBV, VRI and Hβ observations of stars in the HII regions NGC 6334 and NGC 6357*

755 **Fernie J.D.** 1968, J. Roy. Astron. Soc. Canada 62, 214 *Observations of RU Camelopardalis*

756 **Demers S.** 1973, J. Roy. Astron. Soc. Canada 67,19 *The unique variable V725 Sagittarii*

757 **Vasiljanovskaja O.P.** 1977, Var. Stars 20, 467 *Photoelectric observations of two classical cepheids*

758 **Kitamura M., Yamasaki A.** 1972, Tokyo Astron. Bull. 2nd Ser. No 220 *Photoelectric observations of the close binary system SZ Camelopardalis*

759 **Landolt A.U.** 1973, PASP 85, 606 *UBV observations of XX Camelopardalis and rho Cassiopeiae*

760 **Burrell J.F., Mould J.R.** 1973, PASP 85, 627 *VY Sculptoris, a new rapid blue variable*

761 **Sundman A.** 1979, A&AS 35, 327 *An investigation of the interstellar extinction in 11 selected directions in the Carina-Crux-Centaurus region of the Milky Way*

762 **Yamasaki A.** 1979, Ap&SS 60, 173 *W UMa-type stars BV Draconis and BW Draconis. Photoelectric observations*

763 **Alcaino G.** 1979, A&AS 35, 233 *The globular cluster NGC 6541*

764 **Schöneich W., Lange D.** 1979, Inf. Bull. Var. Stars No 1557 *UBV observations of RR Lyr*

765 **Hopp U.** 1979, Inf. Bull. Var. Stars No 1553 *BV observations of six red variable stars*

766 **Mezzetti M., Cester B., Giuricin G., Mardirossian F.** 1979, Inf. Bull. Var. Stars No 1545 *Photoelectric observations of 53 Cam*

767 **Azzopardi M., Vigneau J.** 1979, A&AS 35, 353 *Small Magellanic Cloud, additional lists of probable members and foreground stars*

768 **Lodén L.O.** 1979, A&AS 36, 83 *Photometry of loose clusterings in the southern Milky Way*

769 **Turner D.G.** 1979, J. Roy. Astron. Soc. Canada 73, 2 *A study of the extinction in the young open cluster NGC 6823*

770 **Blanco C., Catalano S., Marilli E.** 1979, A&AS 36, 297 *Photoelectric observations of stars with variable H and K emission components. III*

771 **Sagar R., Joshi U.C.** 1979, Ap&SS 66, 3 *Study of the young open cluster NGC 6611*

772 **Garrison R.F., Schild R.E.** 1979, AJ 84, 1020 *The main sequence of NGC 6231*

773 **Nakagiri M., Hirata R.** 1979, Inf. Bull. Var. Stars No 1565 *UBV observations of 88 Her*

774 **Rovithis P., Rovithis-Livaniou H.** 1979, Inf. Bull. Var. Stars No 1564 *Photoelectric observations of DR Vul*

775 **Christian C.A., Janes K.A.** 1979, AJ 84, 204 *UBV photometry of the anticenter cluster Berkeley 21*

776 **Bastian U., Mundt R.** 1979, A&AS 36, 57 *UBV photometry of T Tauri stars and related objects*

777 **O'Connell D.J.K.** 1979, Ric. Astron. 8 No 31 *The eclipsing binary AR Aurigae*

778 **FitzGerald M.P., Boudreault R., Fich M., Luiken M., Witt A.N.** 1979, A&AS 37, 351 *The open clusters Pismis 6 and 8, and Wat 6*

779 **Muzzio J.C.** 1979, AJ 84, 639 *Photometric study of faint early-type stars in the southern Milky Way*

780 **Alcaino G.** 1979, Acta Astron. 29, 281 *The metal rich globular cluster NGC 5927*

781 **Feinstein A., Marraco H.G.** 1980, PASP 92, 266 *The open cluster NGC 3293 and the OB complex in Carina*

782 **Hintzen P., Jensen E.** 1979, PASP 91, 492 *Observations of degenerate star candidates*

783 **Chambliss C.R., Leung K.C.** 1979, ApJ 228, 828 *The early-type semidetached system SX Aurigae*

784 **Weistrop D.** 1979, PASP 91, 193 *Photoelectric photometry of a complete sample of faint red stars*

785 **Clariá J.J.** 1979, Observatory 99, 202 *Pismis 13, a small, very compact open cluster in Vela*

786 **Carney B.W.** 1979, AJ 84, 515 *The main sequence of the globular cluster NGC 6752*

787 **Stone R.P.S.** 1979, PASP 91, 389 *Spectro- and UBV photometry of the anomalous planetary nebula FG Sagittae, 1968-77*

788 **Schober H.J.** 1979, A&AS 38, 91 *387 Aquitania and 776 Berbericia, two slow spinning asteroids with rotation periods of nearly one day*

789 **Alcaino G.** 1979, A&AS 38, 61 *The globular cluster NGC 1261*

790 **Anthony-Twarog B.J.** 1979, ApJ 233, 188 *The old(est) open cluster Melotte 66*

791 **van den Bergh S., Younger F.** 1979, AJ 84, 1305 *The metal-rich globular cluster NGC 6528*

792 **Carney B.W.** 1979, ApJ 233, 211 *Subdwarf ultraviolet excesses and metal abundances*

793 **Dawson D.W.** 1979, ApJS 41, 97 *A photometric investigation of RV Tauri and yellow semiregular variables*

794 **de Sistero M.E.C., Sistero R.F., Candellero B.** 1979, A&AS 38, 171 *UZ Octantis UBV light curves*

795 **Kappelmann N., Walter K.** 1979, A&AS 38, 161 *Photometric investigation of the Algol system XZ Sagittarii*

796 **Schober H.J., Surdej J.** 1979, A&AS 38, 269 *UBV photometry of the asteroids 9 Metis, 87 Sylvia and 247 Eukrate during their oppositions in 1978 with respect to lightcurves*

797 **Moffat A.F.J., FitzGerald M.P., Jackson P.D.** 1979, A&AS 38, 197 *The rotation and structure of the Galaxy beyond the solar circle I. Photometry and spectroscopy of 276 stars in 45 HII regions and other young stellar groups toward the galactic anticentre*

798 **Isserstedt J.** 1979, A&AS 38, 239 *Photoelectric photometry in the Large Magellanic Cloud*

799 **van den Bergh S., Younger F.** 1979, Inf. Bull. Var. Stars No 1668 *A search for V567 Scorpii*

800 **Szabados L.** 1979, Inf. Bull. Var. Stars No 1667 *Variability of V1165 Aql is questioned*

801 **Ianna P.A.** 1979, AJ 84, 127 *Parallaxes of 44 stars determined from plates taken with the McCormick 67-cm refractor*

No	Reference
802	**Upgren A.R., Weis E.W., DeLuca E.E.** 1979, AJ 84, 1586 *Photometry of Praesepe in BVRI colors*
803	**Augensen H.J.** 1979, AJ 84, 1557 *UBV colors for southern high-velocity stars*
804	**Harris H., Olszewski E.W., Wallerstein G.** 1979, AJ 84, 1598 *The binary cepheid AU Peg*
805	**Whelan J.A.J., Worden S.P., Rucinski S.M., Romanishin W.** 1979, MNRAS 186, 729 *AH Cancri: a contact binary in M67*
806	**Cacciari C.** 1979, AJ 84, 1542 *UBV photometry of RR Lyrae variables in the globular cluster NGC 6121*
807	**Lodén L.O.** 1979, A&AS 38, 355 *Continued studies of loose clusterings in the southern Milky Way*
808	**Guetter H.H.** 1979, AJ 84, 1846 *Photometric studies of stars in Ori OB1 (Belt)*
809	**Jones B.F., Herbig G.H.** 1979, AJ 84, 1872 *Proper motions of T Tauri variables and other stars associated with the Taurus-Auriga dark clouds*
810	**Jurcsik J., Szabados L.** 1979, Inf. Bull. Var. Stars No 1722 *Period increase in FG Sge*
811	**Goranskii V.P.** 1979, Sov. Astron. 23, 284 *The globular cluster NGC 6717*
812	**van den Bergh S., Humphreys R.M.** 1979, AJ 84, 604 *UBV photometry in the dwarf irregular galaxy NGC 6822*
813	**Turner D.G.** 1978, J. Roy. Astron. Soc. Canada 72, 248 *CV Monocerotis - The forgotten cluster cepheid*
814	**Nakagiri M., Yamashita Y.** 1979, Ann. Tokyo Astron. Obs. 2nd Ser. 17, 221 *UBV observations of Mira variables of types Me and Se*
815	**Feinstein A., Marraco H.G.** 1979, AJ 84, 1713 *The photometric behavior of Be stars*
816	**Cacciari C.** 1979, IAU Coll. No 46, 347 *UBV photometry of RR Lyrae variables in the globular cluster M4*
817	**Landolt A.U.** 1973, PASP 85, 625 *UBV observations of long-period variable stars, VIII*
818	**Stetson P.B.** 1979, AJ 84, 1149 *The dwarf spheroidal galaxy in Draco. I. New BV photometry*
819	**Demers S., Kunkel W.E., Hardy E.** 1979, ApJ 232, 84 *The giant branch of Fornax*
820	**Johansson K.L.V.** 1980, A&AS 41, 43 *Colour excess and stellar distribution in five selected directions of the Milky Way in Carina, Crux, Centaurus and Norma*
821	**Lunel M., Garnier R.** 1980, A&AS 39, 7 *UBV photoelectric measurements of O-B stars in SA 98*
822	**Fernandez J.A., Salgado C.W.** 1980, A&AS 39, 11 *Photometric study of the southern open cluster NGC 3532*
823	**Tempesti P.** 1980, A&AS 39, 115 *Photoelectric observations of the long-period eclipsing system AZ Cassiopeiae*
824	**Turner D.G.** 1980, ApJ 235, 146 *The case for membership of the 67 Day cepheid S Vulpecula in Vulpecula OB2*
825	**Mundt R., Bastian U.** 1980, A&AS 39, 245 *UBV photometry of young emission-line objects*
826	**Soonthornthum B., Tritton K.P.** 1980, Observatory 100, 4 *A UBV photoelectric sequence at the south celestial pole*
827	**Flower P.J., Geisler D., Hodge P., Olszewski E.W.** 1980, ApJ 235, 769 *NGC 1868: a metal-poor intermediate-age cluster in the Large Magellanic Cloud*
828	**Alcaino G.** 1980, A&AS 39, 315 *The globular cluster NGC 6144 and its neighbouring region*
829	**Vanderriest CH., Herpe G.** 1980, A&AS 39, 395 *Morphologie des quasars de faibles décalages spectraux et objets apparentés. II. 8 séquences photoélectriques UBV*
830	**Neckel Th., Chini R.** 1980, A&AS 39, 411 *Sixty faint UBVRI standards*
831	**Alcaino G.** 1980, A&A 84, 354 *A photoelectric sequence in the region of the Sculptor galaxy NGC 55*
832	**Wegner G.** 1980, AJ 85, 538 *UBV photometry of faint southern blue stars*
833	**Guetter H.H.** 1980, PASP 92, 215 *UBV photoelectric photometry of 259 PZT stars*
834	**Feinstein A., FitzGerald M.P., Moffat A.F.J.** 1980, AJ 85, 708 *The young open cluster Trumpler 15 in the eta Carinae complex*
835	**Ardeberg A.** 1980, A&AS 42, 1 *Photoelectric photometry of stars in the Small Magellanic Cloud*
836	**Turner D.G., Grieve G.R., Herbst W., Harris W.E.** 1980, AJ 85, 1193 *The young open cluster NGC 3293 and its relation to Car OB1 and the Carina nebula complex*
837	**Lodén L.O.** 1980, A&AS 41, 173 *Concluding observations of loose stellar clusterings in the southern Milky Way*
838	**Jackson P.D., FitzGerald M.P., Moffat A.F.J.** 1980, A&AS 41, 211 *Photoelectric UBV observations of the open cluster Berkeley 11*
839	**Christian C.A.** 1980, AJ 85, 700 *Preliminary BV photometry of the anticenter cluster Be 19*
840	**Bunclark P.S., Fraser C.W., Dodd R.J.** 1980, A&AS 40, 81 *A UBVR photo-electric sequence in Piscis Austrinus*
841	**Abt H.A., Perry C.L., Olsen E.H., Grauer A.D.** 1980, PASP 92, 60 *A small cluster near IC 1805*
842	**Rao N.K.** 1980, Ap&SS 70, 489 *UBV observations of R CrB-type variables and related objects*
843	**Landolt A.U.** 1973, PASP 85, 661 *UBV observations of helium stars*
844	**Wramdemark S.** 1980, A&A 86, 64 *A study of early-type stars in an area in Puppis*
845	**Turner D.G.** 1980, ApJ 240, 137 *Association membership for the 20 day cepheid RU Scuti*
846	**Wramdemark S.** 1980, A&AS 41, 33 *A study of early-type stars in direction close to the Carina arm*
847	**Clariá J.J.** 1980, Ap&SS 72, 347 *Photoelectric UBV and DDO photometry of NGC 5138*
848	**Lindgren H., Bern K.** 1980, A&AS 42, 335 *UBV sequences in three northern Milky Way regions and a comment on the interstellar extinction around l = 90°*
849	**Turner D.G.** 1980, J. Roy. Astron. Soc. Canada 74, 64 *The reddening of beta Doradus*
850	**Turner D.G., Moffat A.F.J.** 1980, MNRAS 192, 283 *Anomalous extinction in the Carina nebula?*
851	**Szabados L.** 1980, Comm. Konkoly Obs. No 76 *Photoelectric UBV photometry of northern cepheids, II*
852	**Clariá J.J., Kepler S.O.** 1980, PASP 92, 501 *Wide- and narrow-band photometry of stars in a field around Collinder 135*
853	**Milone E.F., Chia T.T., Castle K.G., Robb R.M.** 1980, ApJS 43,339 *RW Comae Berenices . I. Early photometry and UBV light curves*
854	**Ciatti F., Mammano A., Margoni R., Milano L., Strazzulla G., Vittone A.** 1980, A&AS 41, 143 *Photoelectric UBV light curves of the eclipsing binary RY Scuti*
855	**Surdej J., Schober H.J.** 1980, A&AS 41, 335 *Rotation period and photoelectric light curves of asteroids 68 Leto and 563 Suleika*
856	**Walker M.F.** 1980, PASP 92, 66 *Simultaneous spectroscopic and photometric observations of BM Andromedae*
857	**Connolly L.P.** 1980, PASP 92, 165 *Sinusoidal cepheids as first-overtone pulsators*
858	**Morris S., Dupuy D.L.** 1980, PASP 92, 303 *A photoelectric study of three southern δ Scuti stars*
859	**Mallama A.D.** 1980, PASP 92, 463 *Photoelectric light curve and period study for RZ Draconis*
860	**Vogt N., Breysacher J.** 1980, ApJ 235, 945 *The dwarf Nova BV Centauri: a spectroscopic binary*
861	**Griersmith D.** 1980, AJ 85, 789 *Photometric properties of bright early-type spiral galaxies. I. The data: multiaperture UBV photometry for 251 galaxies*

No	Reference
862	**FitzGerald M.P., Moffat A.F.J.** 1980, MNRAS 193, 761 *Luminous stars beyond the solar circle: investigation of a galactic field at l = 231°*
863	**Menzies J.W., Laing J.D.** 1980, SAAO Circ. No 1, 175 *Photoelectric standards of intermediate brightness in the E-regions I. UBV observations*
864	**Walker A.R.** 1979, SAAO Circ. No 1, 112 *Photometry and radial velocities of southern carbon stars*
865	**Wesselink T., van Paradijs J., Staller R.F.A., Meurs E.J.A., Kester D.** 1980, Inf. Bull. Var. Stars No 1800 *UBV - observations of the eclipsing binaries Z Her, AR Lac, AW Peg and RS Vul*
866	**Bruch A.** 1980, Inf. Bull. Var. Stars No 1805 *UBV - Observations of some cataclysmic variables*
867	**Burchi R.** 1980, Inf. Bull. Var. Stars No 1813 *Photoelectric observations of symbiotic stars*
868	**Hesser J.E.** 1980, Inf. Bull. Var. Stars No 1758 *A low-amplitude red variable star near the globular cluster NGC 6352*
869	**Liller M.H.** 1980, AJ 85, 673 *A color-magnitude diagram for the globular cluster NGC 6235*
870	**Liller M.H.** 1980, AJ 85, 1480 *A color-magnitude diagram for the globular cluster NGC 6535*
871	**Culver R.B., Ianna P.A.** 1980, PASP 92, 829 *Visual binaries among the barium stars. III. HD 126313 and ζ Capricorni*
872	**Turner D.G.** 1980, PASP 92, 840 *Harvard 20, a cluster containing a yellow supergiant*
873	**Stone R.C.** 1980, PASP 92, 426 *A UBV photometric study of the open cluster NGC 654*
874	**van den Bergh S., Demers S., Kunkel W.E.** 1980, ApJ 239, 112 *The dying globular cluster E3*
875	**Harris W.E., Canterna R.** 1980, ApJ 239, 815 *Color-magnitude photometry to the main sequence for the "anomalous" globular cluster Palomar 12*
876	**Dürbeck W.** 1960, Z. Astrophys. 49, 214 *Dreifarbenphotometrie des offenen Sternhaufens NGC 2354*
877	**Breger M., Wheeler J.C.** 1980, PASP 92, 514 *Photometry and polarimetry of the extreme blue straggler K1211 in NGC 7789*
878	**Chevalier C., Janot-Pacheco E., Mauder H., Ilovaisky S.A.** 1980, A&A 81, 368 *UBV photometry of V616 Mon (A0620-00)*
879	**Harmanec P., Horn J., Koubsky P., Zdarsky F., Kriz S., Pavlovski K.** 1980, BAC 31, 144 *Photoelectric photometry at the Hvar observatory*
880	**Kriz S., Arsenijevic J., Grygar J., Harmanec P., Horn J., Koubsky P., Pavlovski K., Zverko J., Zdarsky F.** 1980, BAC 31, 284 *Strongly interacting binary RX Cas*
881	**Wramdemark S.** 1981, A&AS 43, 103 *A study of early-type stars in a Perseus arm area*
882	**Yoss K.M., Karman R.A., Hartkopf W.I.** 1981, AJ 86, 36 *The space velocities of 393 G and K stars in the +15° Selected Area zones*
883	**Turner D.G.** 1981, AJ 86, 222 *Comments on the cluster main-sequence fitting method. I. The distance of Ruprecht 44*
884	**McClure R.D., Twarog B.A., Forrester W.T.** 1981, ApJ 243, 841 *The old open cluster NGC 2506 and its similarity to NGC 2420*
885	**Johansson K.L.V.** 1981, A&AS 43, 421 *A study of some stars in the region of the open cluster NGC 3532 and the regions of five Lodén cluster candidates in the southern Milky Way*
886	**Sagar R., Joshi U.C.** 1981, Ap&SS 75,465 *Study of the galactic cluster NGC 6823*
887	**Cannon R.D.** 1981, MNRAS 195, 1 *Photometry of southern globular cluster-IV. Faint photoelectric sequences in ω Centauri, NGC 6397, NGC 6752 and 47 Tucanae*
888	**Wramdemark S.** 1981, A&AS 44, 115 *The distribution of early-type stars and dust around l = 114°*
889	**Johansson K.L.V.** 1981, A&AS 44, 127 *A study of a small cluster candidate in Norma*
890	**Lodén L.O.** 1981, A&AS 44, 155 *A photometric study of two stellar clusterings in the southern Milky Way (and a general consideration on previous and present data concerning galactic clusterings)*
891	**Anthony-Twarog B.J.** 1981, ApJ 245,247 *A photometric and spectroscopic search for white dwarfs in the young open cluster IC 2602*
892	**Ogura K., Ishida K.** 1981, PASJ 33, 149 *UBV photometry of NGC 2244*
893	**Christian C.A.** 1981, ApJ 246, 827 *King 8: a metal-poor disk cluster*
894	**de Vaucouleurs G., Buta R.** 1981, PASP 93, 294 *Bigourdan's forgotten observations of SN1885 (S Andromedae) in M31*
895	**Alcaino G.** 1981, A&AS 44, 33 *The tightly-bound globular cluster NGC 6388*
896	**Alcaino G.** 1981, A&AS 44, 191 *The globular cluster NGC 6626*
897	**Wisse P.N.J.** 1981, A&AS 44, 273 *Three colour observations of southern red variable giant stars*
898	**Fishman G.J., Duthie J.G., Dufour R.J.** 1981, Ap&SS 75, 135 *Optical studies of the regions of the gamma-ray bursts of 19 November, 1978 and 5 March, 1979*
899	**Miller H.R.** 1981, AJ 86, 87 *Photoelectric comparison sequences in the fields of five Seyfert galaxies*
900	**Harris W.E., Smith M.G.** 1981, AJ 86, 90 *A deep BV photometric analysis of background field in Virgo*
901	**Schild R., Oke J.B., Searle L.** 1974, ApJ 188, 71 *The energy distribution of the very red star in NGC 6231*
902	**Weis E.W.** 1981, PASP 93, 437 *Photometry of Praesepe in BVRI colors. II.*
903	**Forbes D.** 1981, PASP 93, 441 *The young open clusters Berkeley 62 and Berkeley 86*
904	**Hube D.P.** 1981, PASP 93, 490 *A spectroscopic and photometric study of the triple system HD 7215*
905	**Janes K.A.** 1981, AJ 86, 1210 *Photometry of the open clusters NGC 3960 and NGC 5823*
906	**Weistrop D.** 1981, AJ 86, 1220 *The nature of the Giclas +4 stars*
907	**Harris W.E., Canterna R.** 1981, AJ 86, 1332 *Photometry in the ancient open cluster NGC 6791*
908	**Carney B.W., Aaronson M.A.** 1979, AJ 84, 867 *Subdwarf bolometric corrections*
909	**Mallama A.D., Skillman D.R.** 1979, PASP 91, 99 *Photometry of Nova Cygni 1978*
910	**Graham J.A.** 1981, PASP 93, 29 *A photoelectric UBVRI sequence near NGC 300 for the calibration of faint stellar magnitudes*
911	**Baldinelli L., Ferri A., Ghedini S.** 1981, Inf. Bull. Var. Stars No 1993 *UBV observations of 88 Her (V744 Her)*
912	**Hartwick F.D.A.** 1977, ApJ 214, 778 *Studies of late-type dwarfs. II. A photometric metal abundance index and application to galactic evolution*
913	**DeLuca E.E., Weis E.W.** 1981, PASP 93, 32 *A search for red-dwarf members of the Coma star cluster*
914	**Neckel Th., Chini R.** 1981, A&AS 45, 451 *The interstellar extinction law in some dusty HII regions*
915	**Janes K.A.** 1979, ApJS 39, 135 *Evidence for an abundance gradient in the galactic disk*
916	**Feast M.W., Evans T.L.** 1967, Observatory 91, 286 *The variable star HD 173682 in M25*
917	**Chini R., Neckel Th.** 1981, A&A 102, 171 *UBV and Hβ observations of stars towards M8*

No	Reference
918	**Terzan A., Bernard A.** 1981, A&AS 46, 49 *A photometric catalogue of stars in the direction of the bright cloud B in Sagittarius*
919	**Penny A.J.** 1981, MNRAS 197, 693 *Three UBV sequences in the LMC*
920	**Weis E.W., DeLuca E.E., Upgren A.R.** 1979, PASP 91, 766 *Photometry of possible members of the Hyades cluster. III*
921	**Stauffer J.R.** 1980, AJ 85, 1341 *Observations of pre-main-sequence stars in the Pleiades*
922	**Ando H.** 1980, Ap&SS 71, 249 *Preliminary study of light variations of the eclipsing binary AB Cassiopeiae*
923	**Hartwick F.D.A., Hesser J.E.** 1974, ApJ 192, 391 *Further observations of stars in the intermediate-age open cluster NGC 2477*
924	**Forte J.C., Orsatti A.M.** 1981, AJ 86, 209 *OB stars in the field of the Carina nebula*
925	**Weis E.W., Upgren A.R., Dawson D.W.** 1981, AJ 86, 246 *Radial velocities from microdensitometer scans of objective-prism spectra*
926	**Stetson P.B.** 1981, AJ 86, 687 *A photometric study of the X-ray globular cluster NGC 1851*
927	**Sandage A.** 1981, AJ 86, 1643 *New subdwarfs. III. On obtaining the vertical galactic metallicity gradient from the kinematics of nearby stars*
928	**Raveendran A.V., Mohin S., Mekkaden M.V.** 1981, MNRAS 196, 289 *Photometric study of HD 224085*
929	**Walker A.R.** 1981, MNRAS 197, 241 *Spectroscopy of four subdwarf O stars*
930	**Miller H.R.** 1981, PASP 93, 564 *Photoelectric comparison sequences in the fields of four radio sources*
931	**Martin W.L.** 1981, SAAO Circ. No 6, 28 *A (B,V) photoelectric sequence of stars in omega Centauri*
932	**Drilling J.S.** 1981, ApJ 250, 701 *The spectra of two new intermediate helium stars*
933	**Martin W.L.** 1981, SAAO Circ. No 6, 96 *Multicolour photoelectric photometry of Magellanic Cloud cepheids. IV. B,V observations of 20 short period cepheids*
934	**Szabados L.** 1981, Comm. Konkoly Obs. No 77 *Photoelectric UBV photometry of northern cepheids, III*
935	**Clariá J.J.** 1982, A&AS 47, 323 *Membership, basic parameters and luminosity function of the southern open cluster NGC 2547*
936	**Anthony-Twarog B.J.** 1982, ApJ 255, 245 *Masses of white dwarf progenitors from open cluster studies*
937	**Hanes D.A., Grieve G.R.** 1982, MNRAS 198, 193 *UBV sequences for south polar cap galaxies*
938	**Rakos K.D., Albrecht R., Jenkner H., Kreidl T., Michalke R, Oberlerchner D., Santos E., Schermann A., Schnell A., Weiss W.** 1982, A&AS 47, 221 *Photometric and astrometric observations of close visual binaries*
939	**Norris J., Freeman K.C.** 1982, ApJ 254, 143 *The anticorrelation of carbon and nitrogen on the horizontal branch of 47 Tucanae*
940	**Dahn C.C., Harrington R.S., Riepe B.Y., Christy J.W., Guetter H.H., Kallarakal V.V., Miranian M., Walker R.L., Vrba F.J., Hewitt A.V., Durham W.S., Ables H.D.** 1982, AJ 87, 419 *U.S. Naval Observatory parallaxes of faint stars. List VI*
941	**Cerruti M.A., Sistero R.F.** 1982, PASP 94, 189 *The close binary V757 Centauri*
942	**Hopp U., Witzigmann S., Geyer E.H.** 1982, Inf. Bull. Var. Stars No 2148 *Photometry of the shell star BU Tau (Pleione) 1980 - 1982*
943	**Bartolini C., Custodi P., Dell'atti F., Guarnieri A., Piccioni A.** 1982, Inf. Bull. Var. Stars No 2139 *The variability of BD +60° 562*
944	**Breger M.** 1982, ApJ 263, 199 *Polarization in NGC 7789 and the membership of blue stragglers*
945	**Feinstein A.** 1982, AJ 87, 1012 *Faint stars in the open cluster Trumpler 16*
946	**Forbes D.** 1982, AJ 87, 1022 *Possible membership for the 51-day cepheid GY Sagittae in a new OB association*
947	**Reed B.C., FitzGerald M.P.** 1982, A&AS 49, 521 *A photoelectric UBV sequence in a low extinction Puppis field*
948	**Upgren A.R., Philip A.G.D., Beavers W.I.** 1982, PASP 94, 229 *Upgren 1, an old cluster near the end of its life*
949	**Szkody P., Downes R.A.** 1982, PASP 94, 328 *IUE and optical observations of MV Lyrae at intermediate and low states*
950	**Weis E.W., Upgren A.R.** 1982, PASP 94, 475 *Photometry of possible members of the Hyades cluster. IV*
951	**Stauffer J.** 1982, AJ 87, 899 *The faint end of the Hyades main sequence*
952	**Isserstedt J.** 1982, A&AS 50, 7 *Photoelektrische UBV-Photometrie in der Grossen Magellanschen Wolke*
953	**Clariá J.J., Lapasset E.** 1982, A&AS 50, 13 *BR Muscae: a new early-type contact binary*
954	**Dachs J., Kaiser D., Nikolov A., Sherwood W.A.** 1982, A&AS 50, 261 *UBV-Hβ photometry of luminous stars between (l=335° and l=6°)*
955	**Pastoriza M.G., Storchi T.B., Livi S.H.B.** 1982, PASP 94, 347 *DDO photometry of G and K stars*
956	**Turner D.G.** 1982, PASP 94, 655 *New UBV photometry for the open cluster Stock 14 and its cepheid-like variable V810 Centauri (= HR 4511)*
957	**Stauffer J.** 1982, PASP 94, 678 *Photometry of low-mass stars in Praesepe*
958	**Bernard A.** 1982, PASP 94, 700 *UW Gruis: a normal AB-type RR Lyrae star*
959	**Hodge P.W.** 1982, ApJ 247, 894 *The intermediate age globular cluster NGC 152 in the Small Magellanic Cloud*
960	**Dean J.F., Laing J.D.** 1981, MNASSA 40, 48 *The eclipsing binary HR 9049*
961	**Impey C.D., Brand P.W.J.L.** 1982, MNRAS 201, 849 *The calibration of a radio-independent search for BL Lac objects*
962	**Bassino L.P., Dessaunet V.H., Muzzio J.C., Waldhausen S.** 1982, MNRAS 201, 885 *A photometric and spectroscopic study of faint OB stars in the southern Milky Way*
963	**Dworetsky M.M., Whitelock P.A., Carnochan D.J.** 1982, MNRAS 201, 901 *Optical observations of ultraviolet objects - II. Classification and photometry (l=0 to 145°)*
964	**Turner D.G.** 1982, PASP 94, 789 *Berkeley 87, a heavily-obscured young open cluster associated with the ON2 star-formation complex and containing the WO star Stephenson 3*
965	**Stauffer J.R.** 1982, AJ 87, 1507 *Observations of low-mass stars in the Pleiades: has a pre-main sequence been detected?*
966	**Pier J.R.** 1982, AJ 87, 1515 *Stars of spectral types A and B in the southern galactic halo. I. UBV photometry*
967	**Vigneau J., Azzopardi M.** 1982, A&AS 50, 119 *Two photoelectric UBVRI sequences in the bar of the Small Magellanic Cloud*
968	**Livi S.H.B., Bergmann T.S.** 1982, AJ 87, 1783 *Rapid optical variation of semiregular variable R Crt*
969	**Rucinski S.M., Kaluzny J.** 1982, Ap&SS 88, 433 *BV Dra and BW Dra: two contact systems in one visual binary*
970	**Flower P.J.** 1982, PASP 94, 894 *Color-magnitude diagrams and ages of two young Magellanic Cloud clusters*
971	**Adler D.S., Janes K.A.** 1982, PASP 94, 905 *The distant open cluster Tombaugh 2*
972	**Leibowitz E.M., Mendelson H.** 1982, PASP 94, 977 *Photoelectric standard stars near V1343 Aquilae (SS 433)*
973	**Turner D.G.** 1982, PASP 94, 1003 *On the possibility of cluster membership for the cepheid V Centauri*

No	Reference
974	**Kondo M., Watanabe E., Yutani M., Noguchi T.** 1982, PASJ 34, 541 *UBV photometric observations of the ultraviolet-excess objects and the Kiso photometric standard stars. I.*
975	**Celis L.** 1982, AJ 87, 1791 *Red variable stars. I. UBVRI photometry and photometric properties*
976	**Schild R.E., Garrison R.F., Hiltner W.A.** 1983, ApJS 51, 321 *UBV photometry for southern OB stars*
977	**Cousins A.W.J.** 1983, SAAO Circ. No 7, 36 *UBV photometry of E region standard stars of intermediate brightness*
978	**Evans T.L., Koen M.C.J., Hultzer A.A.** 1983, SAAO No 7, 82 *Local photometric standards for CaII emission stars*
979	**Carney B.W.** 1983, AJ 88, 5 *A photometric search for halo binaries. I. New observational data*
980	**Feitzinger J.V., Isserstedt J.** 1983, A&AS 51, 505 *Photoelectric UBV-Photometry of Wolf-Rayet stars in the Large Magellanic Cloud*
981	**Joshi U.C., Sagar R.** 1983, MNRAS 202, 961 *Photometry of the open cluster NGC 654*
982	**Turner D.G.** 1983, J. Roy. Astron. Soc. Canada 77, 31 *The open cluster Tombaugh 1 and its neighbouring cepheid XZ Canis Majoris*
983	**Joshi U.C., Sagar R.** 1983, J. Roy. Astron. Soc. Canada 77, 40 *Study of the open cluster IC 1805*
984	**Alcaino G.** 1983, A&AS 52, 105 *The globular cluster NGC 6544*
985	**Oja T.** 1983, A&AS 52, 131 *UBV photometry of FK4 and FK4 supplement stars*
986	**Rucinski S.M.** 1983, Inf. Bull. Var. Stars No 2277 *Observations of V711 Tauri in October 1981 / January 1982*
987	**Weis E.W.** 1983, PASP 95, 29 *Photometry of possible members of the Hyades cluster. V.*
988	**Demers S., Harris W.E.** 1983, AJ 88, 329 *A color-magnitude diagram for Leo II*
989	**Landolt A.U.** 1983, AJ 88, 439 *UBVRI photometric standard stars around the celestial equator*
990	**Rucinski S.M.** 1983, Ap&SS 89, 395 *Photometry of the fast-rotating late-type star W92 in NGC 2264*
991	**Sagar R., Joshi U.C.** 1983, MNRAS 205, 747 *Study of the open cluster NGC 2264*
992	**Turner D.G., Moffat A.F.J., Lamontagne R.** 1983, AJ 88, 1199 *The WN4.5 component of HD 219460 in the open cluster Markarian 50*
993	**Landolt A.U.** 1983, PASP 95, 644 *A search for light variations in barium stars*
994	**Turner D.G.** 1983, AJ 88, 650 *A search for stars physically associated with the 16-day cepheid X Cygni. I. Luminous stars in the field*
995	**Bouchet R., The P.S.** 1983, PASP 95, 474 *Notes on the open cluster NGC 1252 with the variable carbon star TW Horologii as probable member*
996	**Sistero R.F., Grieco A., Candellero B.** 1983, Ap&SS 91, 427 *The early-contact system BH Centauri: UBV photometry*
997	**Walker M.** 1983, ApJ 271, 642 *Studies of extremely young cluster. VII. Spectroscopic observations of faint stars in the Orion nebula*
998	**Feinstein A.** 1983, Ap&SS 96, 293 *The open cluster Tr 14*
999	**Joshi U.C., Sanwal B.B., Sagar R.** 1983, PASJ 35, 405 *Photometry of the open cluster NGC 6913*
1003	**Roman N.G.** 1955, ApJS 2, 195 *A catalogue of high-velocity stars*
1004	**Sharpless S.** 1952, ApJ 116, 251 *A study of the Orion aggregate of early-type stars*
1006	**Johnson H.L., Harris III D.L.** 1954, ApJ 120, 196 *Three-color observations of 108 stars intended for use as photometric standards*
1007	**Miczaika G.R.** 1954, AJ 59, 233 *The color-luminosity diagram of the Ursa Major group*
1008	**Naur P.** 1955, ApJ 122, 182 *Magnitudes and colors of bright F stars*
1009	**Harris III D.L.** 1956, ApJ 123, 371 *Photometry of the Perseus aggregates*
1011	**Hiltner W.A., Johnson H.L.** 1956, ApJ 124, 367 *The law of interstellar reddening and absorption*
1012	**Hiltner W.A.** 1956, ApJS 2, 389 *Photometric, polarization, and spectrographic observations of O and B stars*
1013	**Johnson H.L., Knuckles C.F.** 1957, ApJ 126, 113 *Three-color photometry of nearby stars*
1014	**Johnson H.L.** 1957, ApJ 126, 134 *The color-magnitude diagram for I Orionis*
1015	**Blaauw A., Hiltner W.A., Johnson H.L.** 1960, ApJ 130, 69 *Photoelectric photometry of the association III Cephei*
1017	**Mumford III G.S.** 1956, AJ 61, 213 *Photoelectric observations of red dwarf stars*
1018	**Westerlund B.** 1959, PASP 71, 156 *Three-color photometry of bright southern supergiants*
1019	**Harris III D.L., Upgren A.R.** 1964, ApJ 140, 151 *Photoelectric magnitudes and colors of stars near the north galactic pole*
1020	1961, IAU Trans. IIa, 251 *Supplementary standard stars for V and B-V*
1021	**Feast M.W., Stoy R.H., Thackeray A.D., Wesselink A.J.** 1961, MNRAS 122, 239 *Spectral classification and photometry of southern B stars*
1022	**Osawa K.** 1959, ApJ 130, 159 *Spectral classification of 533 B8-A2 stars and the mean absolute magnitude of A0 V stars*
1024	**Westerlund B.** 1961, Ark. Astr. 3, 21 *Three-colour photometry of bright M giants*
1025	**Bouigue R., Boulon J., Pedoussaut A.** 1961, Ann. Obs. Toulouse 28, 33 *Contribution aux recherches de photométrie photoélectrique dans la galaxie, VII-VIII*
1026	**Klemola A.R.** 1962, AJ 67, 740 *Mean absolute magnitude of the blue stars at high galactic latitude*
1028	**Johnson H.L.** 1953, ApJ 117, 361 *Photoelectric observations of visual double stars*
1029	**Johnson H.L., Mitchell R.I.** 1958, ApJ 128, 31 *The color-magnitude diagram of the Pleiades cluster II*
1030	**Hogg A.R.** 1960, PASP 72, 85 *The galactic cluster IC 2391*
1032	**Hiltner W.A., Iriarte B.** 1955, ApJ 122, 185 *Photometric and spectroscopic studies of early-type stars between galactic longitude $l = 338^o$ and $l = 33^o$*
1034	**Hogg A.R.** 1958, Mount Stromlo Obs. Mimeo. No 2 *Photometric observations of 244 bright stars*
1036	**Iriarte B.** 1959, Lowell Obs. Bull. 4, 130 No 101 *Photoelectric photometry of faint blue stars*
1040	**Lyngå G.** 1959, Ark. Astr. 2, 379 *UBV sequences in five southern galactic clusters*
1041	**Seyfert C.K., Hardie R.H., Grenchik R.T.** 1960, ApJ 132, 58 *A study of the II Persei association*
1046	**Pesch P.** 1960, ApJ 132, 696 *The galactic cluster NGC 654*
1048	**Mitchell R.I.** 1960, ApJ 132, 68 *Photometry of the α Persei cluster*
1049	**Osawa K., Hata S.** 1960, Ann. Tokyo Astron. Obs. 2nd Ser. 6, 148 *Three colour photometry of B8-A2 stars*
1053	**Whiteoak J.B.** 1961, MNRAS 123, 245 *A study of the galactic cluster IC 2602, I. A photoelectric and spectroscopic investigation*
1054	**Hardie R.H., Crawford D.L.** 1961, ApJ 133, 843 *A study of the II Scorpii association*
1058	**Johnson H.L., Mitchell R.I., Iriarte B.** 1962, ApJ 136, 75 *The color-magnitude diagram of the Hyades cluster*

No	Reference
1059	**Johnson H.L.** 1962, ApJ 136, 1135 *The galactic cluster NGC 2244*
1060	**Sharpless S.** 1962, ApJ 136, 767 *Evolutionary effects in the Orion association*
1061	**Cousins A.W.J.** 1962, MNASSA 21, 20 *Photometric data for stars in the equatorial zone (First List)*
1063	**Osawa K., Hata S.** 1962, Ann. Tokyo Astron. Obs. 2nd Ser. 7, 209 *Three-color photometry of B8-A2 stars (II)*
1064	**Sandage A.R.** 1964, ApJ 139, 442 *Results of a pilot program to discover new subdwarfs in the solar neighborhood*
1066	**Westerlund B.E.** 1963, MNRAS 127, 71 *An OB association in the region of RS Puppis*
1067	**Marlborough J.M.** 1964, AJ 69, 215 *Frequency of ultraviolet excesses among late-type dwarfs in the solar neighborhood*
1068	**Westerlund B.E.** 1963, MNRAS 127, 83 *Three-colour photometry of early-type stars near the galactic poles*
1069	**Bigay J.H.** 1963, J. Obs. 46, 319 *Mesures photoélectriques en 3 couleurs (U.B.V.) d'étoiles O-B et A0 dans les Selected Areas du plan galactique, SA 19, 9 et 24*
1070	**Bouigue R., Boulon J., Pédoussaut A.** 1963, Ann. Obs. Toulouse 29, 17 *Contribution aux recherches de photométrie photoélectrique dans la galaxie. IX. Mesures dans des champs galactiques (3)*
1071	**Cousins A.W.J.** 1964, MNASSA 23, 10 *Photometric data for stars in the equatorial zone (Sixth list)*
1075	**Cousins A.W.J., Stoy R.H.** 1962, R. Obs. Bull. No 64 *Photoelectric magnitudes and colours of southern stars*
1077	**Oja T.** 1963, Ark. Astr. 3, 273 *Reduction to the UBV system of photoelectric measurements made with a 1P21 photomultiplier at the Uppsala observatory*
1078	**Cousins A.W.J.** 1962, MNASSA 21, 61 *Photometric data for stars in the equatorial zone (Second List)*
1079	**Crawford D.L.** 1963, ApJ 137, 530 *UBV and Hβ photometry for the bright B8- and B9-type stars*
1080	**Argue A.N.** 1963, MNRAS 125, 557 *UBV photometry of 300 G and K type stars*
1081	**Whiteoak J.B.** 1963, MNRAS 125, 105 *An association of O and B stars in Ara*
1082	**Hogg A.R.** 1963, MNRAS 125, 307 *The galactic cluster NGC 3228*
1083	**Wildey R.L.** 1964, ApJS 8, 439 *The stellar content of h and χ Persei-cluster and association*
1084	**Tolbert C.R.** 1965, ApJ 139, 1105 *A UBV study of 94 wide visual binaries*
1085	**Roslund C.** 1964, Ark. Astr. 3, 357 *Investigations of a Milky Way field in Scorpius. I. Magnitudes and colours of O and B stars*
1086	**Elvius T., Eriksson P.I., Linden H., Ljunggren B.** 1963, Ark. Astr. 3, 317 *The new photoelectric photometer of the Uppsala observatory. Preliminary results for a field in Cassiopeia (SA 19)*
1087	**Feinstein A.** 1964, Observatory 84, 111 *A group of stars around the helium star HD 96446*
1088	**Cousins A.W.J.** 1964, MNASSA 23, 175 *Photometric data for stars in the equatorial zone (Seventh list)*
1089	**Hardie R.H., Heiser A.M., Tolbert C.R.** 1964, ApJ 140, 1472 *A study of the B stars in the Orion Belt region*
1090	**Landolt A.U.** 1964, ApJ 140, 1494 *Preliminary UBV light-curves for the eclipsing binary V382 Cygni*
1095	**Demers S., Fernie J.D.** 1964, PASP 76, 350 *Photometry of Wolf-Rayet stars*
1096	**Feinstein A.** 1964, PASP 76, 399 *Intrinsic colors of Wolf-Rayet stars*
1097	**Przybylski A., Kennedy P.M.** 1965, MNRAS 129, 63 *Radial velocities and three-colour photometry of 52 stars with large proper motions*
1099	**Bok B.J., Bok P.F.** 1960, MNRAS 121, 531 *Four standard sequences in the southern hemisphere*
1100	**Johnson H.L.** 1953, ApJ 117, 356 *Magnitudes and colors in NGC 752*
1101	**Lyngå G.** 1964, Medd. Lunds Obs. Ser. 2, No 139 *Studies of the Milky Way from Centaurus to Norma. I. UBV photometry*
1103	**Sjoegren U.** 1964, Ark. Astr. 3, 339 *Photoelectric and spectrophotometric observations with a discussion of the interstellar absorption in the region of Kapteyn's Selected Area 8*
1105	**Gehrels T., Coffeen T., Owings D.** 1964, AJ 69, 826 *Wavelength dependence of polarization. III. The lunar surface*
1107	**Oke J.B.** 1964, ApJ 140, 689 *Photoelectric spectrophotometry of stars suitable for standards*
1110	**Johnson H.L., Morgan W.W.** 1955, ApJ 122, 429 *Photometric and spectroscopic observations of the double cluster in Perseus*
1111	**Johnson H.L.** 1954, ApJ 119, 185 *Magnitudes and colors in M34*
1112	**Hill P.W.** 1964, MNRAS 127, 113 *The spectra of helium stars. I. Wavelengths and equivalent widths for HD 168476 and HD 124448*
1117	**Elvius T., Lyngå G.** 1965, Ark. Astr. 3, 467 *Three-colour photometry of 43 stars in Kapteyn's Selected Areas at southern galactic latitudes*
1118	**Ljunggren B., Oja T.** 1964, Ark. Astr. 3, 439 *Photoelectric measurements of magnitudes and colours for 849 stars*
1119	**Bouigue R.** 1959, Ann. Obs. Toulouse 27, 47 *Contribution aux recherches de photométrie photoélectrique dans la galaxie. I, II, III, IV, V*
1124	**Lodén L.O., Lodén K.** 1963, Ark. Astr. 3, 299 *A photometric standard region in Cygnus*
1126	**Mianes P., Daguillon J.** 1957, J. Obs. 40, 65 *Photometrie en 3 couleurs de l'amas N.G.C. 7243*
1127	**Johnson H.L., Knuckles C.F.** 1955, ApJ 122, 209 *The Hyades and Coma Berenices star clusters*
1137	**Sarma M.B.K., Walker M.F.** 1962, ApJ 135, 11 *The color-magnitude diagram of NGC 2420*
1140	**Johnson H.L.** 1952, ApJ 116, 640 *Praesepe, magnitudes and colors*
1141	**Upgren A.R., Rubin V.C.** 1965, PASP 77, 355 *An old open cluster near the north galactic pole*
1146	**Lohmann W.** 1961, Astr. Nachr. 286, 105 *Photometrie des offenen Sternhaufens NGC 4349*
1148	**Argue A.N.** 1963, MNRAS 127, 97 *Photographic photometry with the Cambridge Schmidt telescope. II. Photoelectric UBV calibrations in the Coma region*
1149	**Cousins A.W.J.** 1965, MNASSA 24, 120 *Photometric data for stars in the equatorial zone (Eighth list)*
1159	**Johnson H.L.** 1954, ApJ 119, 181 *The standard region near IC 4665*
1160	**Hogg A.R., Kron G.E.** 1955, AJ 60, 365 *The galactic cluster IC 4665*
1161	**Walker M.F.** 1957, ApJ 125, 636 *Studies of extremely young clusters. II. NGC 6530*
1163	**Sandage A.R.** 1960, ApJ 131, 610 *Cepheids in galactic clusters. VI. U Sgr in M25*
1165	**Arp H.C.** 1958, ApJ 128, 166 *Cepheids in galactic clusters. III. EV Sct in NGC 6664*
1167	**Johnson H.L., Morgan W.W.** 1954, ApJ 119, 344 *A heavily obscured O-association in Cygnus*
1168	**Walker M.F.** 1958, ApJ 128, 562 *Photoelectric observations of the galactic cluster NGC 6940*
1170	**Schulte D.H.** 1958, ApJ 128, 41 *New members of the association VI Cygni. II*
1172	**Walker M.F.** 1960, ApJ 130, 57 *Studies of extremely young clusters. III. IC 5146*
1174	**Harris III D.L.** 1955, ApJ 121, 554 *Photometry of the Lacerta aggregate*
1175	**Burbidge E.M., Sandage A.R.** 1958, ApJ 128, 174 *The color-magnitude diagram for the galactic cluster NGC 7789*

No	Reference
1176	**Sandage A.R.** 1958, ApJ 128, 150 *Cepheids in galactic clusters. I. CF Cas in NGC 7790*
1182	**Pesch P.** 1959, ApJ 130, 764 *The galactic cluster NGC 457*
1183	**Hiltner W.A., Iriarte B.** 1955, ApJ 121, 556 *Four faint Wolf-Rayet stars*
1184	**Brownlee R.R.** 1957, ApJ 125, 372 *Photoelectric observations of SW Lacertae*
1185	**Bigay J.H., Lunel M.** 1965, J. Obs. 48, 171 *Photométrie photoélectrique U.B.V. de 263 étoiles B et A de la S.A. 8*
1186	**Popper D.M.** 1965, ApJ 141, 126 *Rediscussion of eclipsing binaries. VII. WZ Ophiuchi and other solar-type stars*
1187	**Walker M.F.** 1961, ApJ 133, 438 *Studies of extremely young clusters. IV. NGC 6611*
1190	**Hogg A.R.** 1963, PASP 75, 194 *A southern photoelectric magnitude sequence*
1192	**Golay M.** 1958, Publ. Obs. Geneve Serie A, FasC. 57 *Etude de l'absorption interstellaire dans une région obscurcie du Cygne*
1193	**Ljunggren B., Oja T.** 1961, Ann. Uppsala Astron. Obs. 4, No 10 *The Uppsala spectral classification. Intrinsic colours and absolute magnitudes*
1194	**Popper D.M.** 1959, ApJ 129, 647 *Remarks on bolometric corrections and effective temperatures*
1196	**Malmquist K.G., Ljunggren B., Oja T.** 1960, Ann. Uppsala Astr. Obs. 4, No 8 *A spectrophotometric survey of stars along the Milky Way. Part II. Galactic longitudes 40-60°*
1197	**Nikonov V.B., Nekrasova S.V., Polosuhina N.S., Rachkovsky D.N., Chuvajev K.K.** 1957, Izv. Krymsk. Astrofiz. Obs. 17, 42 *The colour-luminosity diagram for stars in the vicinity of the Sun*
1198	**Lindblad P.O., Pipping G.** 1963, Ark. Astr. 3, 307 *Photoelectric observations of 31 Cygni and 32 Cygni in 1961-62*
1199	**Osawa K.** 1958, PASJ 10, 102 *On colors of metallic-line stars*
1200	**Grigorian K.A.** 1959, Comm. Byurakan Obs. 27, 69 *Electrocalorimetry and electropolarimetry of stars near lambda Orionis*
1201	**Osawa K.** 1958, PASJ 10, 104 *On the ultraviolet excess of a high-velocity star HR 3018*
1202	**Abt H.A., Golson J.C.** 1962, ApJ 136, 35 *Colors and variability of magnetic stars*
1203	**Belyakina T.S., Chugainov P.F.** 1960, Izv. Krymsk. Astrofiz. Obs. 22, 257 *On the precision of the determination of spectral classes and colour excesses of O-A2 stars by the method of two-colour diagrams*
1205	**Bronkalla W., Notni P.** 1961, Astr. Nachr. 286, 179 *Light curve and color of Nova Her 1960-V446 Her*
1206	**Greenstein J.L., Wallerstein G.** 1958, ApJ 127, 237 *The helium-rich stars, σ Orionis E*
1207	**Hardie R.H., Seyfert C.K., Gulledge I.S.** 1960, ApJ 132, 361 *A study of the I Geminorum association*
1209	**Johnson H.L., Morgan W.W.** 1955, ApJ 122, 142 *Some evidence for a regional variation in the law of interstellar reddening*
1210	**Koch R.H.** 1962, AJ 67, 130 *A three-color photoelectric investigation of δ Librae*
1211	**Lenouvel F., Fiogére C.** 1957, J. Obs. 40, 37 *Observations photoélectriques*
1212	**Mendoza V E.E.** 1958, ApJ 128, 207 *Spectroscopic and photometric study of the Be stars*
1213	**Spinrad H.** 1959, ApJ 130, 539 *Photoelectric observations of RR Lyrae stars*
1214	**Fernie J.D.** 1959, ApJ 130, 611 *The binary system X Ophiuchi*
1215	**Pesch P.** 1963, ApJ 137, 547 *Spectroscopic and photometric observations of some stars from the luminous stars in the northern Milky Way I and II catalogues*
1216	**Wallerstein G.** 1963, ApJ 137, 991 *On the absolute magnitude and reddening of SS Cygni*
1217	**Giclas H.L.** 1954, AJ 59, 128 *Photoelectric magnitudes and colors of Uranus*
1218	**Kraft R.P., Hiltner W.A.** 1961, ApJ 134, 850 *Color excesses for supergiants and classical cepheids. VI. On the intrinsic colors and the Hess diagram of late-type supergiants*
1219	**Abt H., Golson J.C.** 1962, ApJ 136, 363 *Interstellar absorption in the north equatorial polar region*
1220	**O'Dell C.R.** 1962, ApJ 136, 809 *Photoelectric observations of the Crab nebula*
1221	**Wallerstein G., Greenstein J.L., Parker R., Helfer H.L., Aller L.H.** 1963, ApJ 137, 280 *Red giants with extreme metal deficiencies*
1222	**Preston G.W., Krzeminski W., Smak J., Williams J.A.** 1963, ApJ 137, 401 *A spectroscopic and photoelectric survey of the RV Tauri stars*
1223	**Crawford D.L.** 1963, ApJ 137, 523 *Photometry of the stars in the Cassiopeia-Taurus group*
1224	**Wood D.B., Walker M.F.** 1960, ApJ 131, 363 *Photoelectric observations of β Lyrae*
1225	**Sandage A., Wallerstein G.** 1960, ApJ 131, 598 *Color-magnitude diagram for the disk globular cluster NGC 6356 compared with halo clusters*
1226	**Wampler E.J., Pesch P., Hiltner W.A., Kraft R.P.** 1961, ApJ 133, 895 *Cepheids in galactic clusters. VIII. A reinvestigation of U Sgr in M25 (=IC 4725)*
1227	**Varsavsky C.M.** 1960, ApJ 132, 354 *The gravitational contraction times of stars in very young clusters*
1228	**Spinrad H.** 1961, ApJ 133, 479 *SU Draconis and line blanketing in the RR Lyrae stars*
1229	**Popper D.M.** 1961, ApJ 133, 148 *Rediscussion of eclipsing binaries. V. RS Canum Venaticorum*
1230	**Hardie R.H., Lott S.H.** 1961, ApJ 133, 71 *Three-color photometry of DY Herculis*
1231	**Fitch W.S.** 1960, ApJ 132, 701 *The light-variation of CC Andromedae*
1232	**de Vaucouleurs G.** 1960, ApJ 132, 681 *Nova V723 Scorpii 1952*
1233	**Wallerstein G., Carlson M.** 1960, ApJ 132, 276 *On the ultraviolet excess in G dwarfs*
1234	**Bouigue R.** 1958, C.R. 248, 2956 *Sur la mesure photoélectrique de l'absorption interstellaire*
1236	**Deeming T.J.** 1961, MNRAS 123, 273 *Some new southern subdwarfs*
1237	**Pesch P.** 1960, ApJ 132, 689 *The galactic cluster NGC 7654 (M52)*
1238	**Vandervort G.L.** 1958, AJ 63, 477 *The magnitudes, colors, and motions of stars of spectral class R*
1240	**Slettebak A., Bahner K., Stock J.** 1961, ApJ 134, 195 *Spectra and colors of early-type stars near the north galactic pole*
1241	**Klemola A.R.** 1961, ApJ 134, 130 *The spectrum of the helium star BD +10°2179*
1242	**Preston G.W.** 1961, ApJ 134, 633 *A coarse analysis of three RR Lyrae stars*
1243	**Wallerstein G., Helfer H.L.** 1961, ApJ 133, 562 *Abundances in G Dwarf. V. The metal-rich star 20 Leo Minoris and two comparison stars*
1245	**Gehrels T., Owings D.** 1962, ApJ 135, 906 *Photometric studies of asteroids. IX. Additional light-curves*
1246	**Abt H.A., Jeffers H.M., Gibson J., Sandage A.R.** 1962, ApJ 135, 429 *The visual multiple system containing β Lyrae*
1247	**Wood D.B.** 1958, ApJ 127, 351 *Photoelectric observations of U Coronae Borealis*
1248	**Hardie R.H., Geilker C.D.** 1958, ApJ 127, 606 *Three-color photometry of DY Pegasi*
1249	**Tifft W.G., Smith H.J.** 1958, ApJ 127, 591 *T Sextantis, an RR Lyrae star of type 'c'*
1250	**Hardie R.H.** 1958, ApJ 127, 620 *Light-variation of the spectrum variable HD 124224*

No	Reference
1252	**Lenouvel F., Daguillon J.** 1956, J. Obs. 39, 1 *Observations photoélectriques*
1253	**Deinzer W., Geyer E.** 1959, Z. Astrophys. 47, 211 *The light curve of the eclipsing variable BV143 = BD +35°4496*
1254	**Gascoigne S.C.B.** 1962, MNRAS 124, 201 *NGC 1783, a cluster in the Large Magellanic Cloud*
1255	**Arp H.C.** 1957, AJ 62, 129 *Three-color photometry of cepheids W Virginis, M5 Nos 42 and 84, and M10 Nos. 2 and 3*
1256	**Cousins A.W.J.** 1963, MNASSA 22, 12 *Photometric data for stars in the equatorial zone (Third list)*
1258	**Nekrasova S.V., Nikonov V.B., Polosukhina N.S., Rybka E.** 1962, Izv. Krymsk. Astrofiz. Obs. 27, 228 *Photoelectric magnitudes and colors of photometric standards in Selected Areas*
1260	**Shatzel A.V.** 1954, ApJ 120, 547 *Photometric studies of asteroids. III. The light-curve of 44 Nysa*
1261	**Ahmad I.I.** 1954, ApJ 120, 551 *Photometric studies of asteroids. IV. The light curves of Ceres, Hebe, Flora, and Kalliope*
1263	**Provins S.S.** 1953, ApJ 118, 489 *Light-variations of peculiar A stars*
1264	**Feige J.** 1958, ApJ 128, 267 *A search for underluminous hot stars*
1265	**Przybylski B.** 1962, PASP 74, 230 *The peculiar star ν Indi*
1266	**Krzeminski W.** 1962, PASP 74, 66 *Nova WZ Sagittae, an extremely short-period eclipsing binary*
1267	**Koch R.H.** 1960, AJ 65, 127 *Three-color photometry of AO Cassiopeiae*
1268	**Koch R.H.** 1960, AJ 65, 139 *Photometric photometry of AS Eridani*
1269	**Koch R.H.** 1960, AJ 65, 374 *Photoelectric light curves of XY Leonis*
1270	**Abhyankar K.D.** 1962, Z. Astrophys. 54, 25 *Photometric Elements of YY CMi*
1271	**Rodgers A.W.** 1960, MNRAS 120, 163 *Three-colour photometry in the southern Coalsack*
1272	**Abhyankar K.D.** 1959, ApJ 130, 834 *AD CMi-a new ultrashort-period variable*
1273	**Hogg A.R.** 1957, MNRAS 117, 95 *Variations in the light of σ Scorpii*
1274	**Abt H.A., Hardie R.H.** 1960, ApJ 131, 155 *The 1.3-day variable BL Herculis*
1275	**Popper D.M.** 1956, ApJ 124, 196 *Rediscussion of eclipsing binaries. I. Z Herculis*
1276	**Hardie R.H.** 1955, ApJ 122, 256 *A study of RR Lyrae in three colors*
1277	**Feast M.W., Thackeray A.D., Wesselink A.J.** 1960, MNRAS 121, 337 *The brightest stars in the Magellanic Clouds*
1278	**McNamara D.H., Williams A.D.** 1954, ApJ 121, 51 *The colors of the β Canis Majoris stars*
1279	**Waymann P.A.** 1962, R. Obs. Bull. No 50 *Photoelectric magnitudes and colours of southern double stars*
1280	**Wallerstein G., Spinrad H.** 1960, PASP 72, 486 *A visual binary containing an O-type subdwarf*
1281	**Harris III D.L.** 1956, ApJ 124, 665 *A note on the color-magnitude array of the white dwarfs. I.*
1283	**Herbig G.H.** 1960, ApJS 4, 337 *The spectra of Be- and Ae-type stars associated with nebulosity*
1284	**Gehrels T.** 1956, ApJ 123, 331 *Photometric studies of asteroids. V. The light-curve and phase function of 20 Massalia*
1285	**Blanco V.M., Williams A.D.** 1959, ApJ 130, 482 *A new O-B association with an unusual reddening effect*
1286	**de Vaucouleurs G.** 1959, Planet Space Sci. 2, 26. *Multicolor photometry of Mars in 1958*
1287	**Koch R.H.** 1961, AJ 66, 230 *Departures from the Russell model in TX Ursae Majoris*
1291	**Shao C.Y.** 1964, AJ 69, 858 *Photoelectric observations of ζ Aurigae during the 1963-1964 eclipse*
1292	**Paczynski B.** 1964, AJ 69, 124 *BD +30 2163, a new W UMa variable*
1293	**Koch R.H.** 1963, AJ 68, 785 *AL Geminorum, a possible R Canis Majoris-type binary*
1294	**Herbig G.M., Moorhead J.M.** 1965, ApJ 141, 649 *BD −21°6267A, a new dMe double-line spectroscopic binary*
1295	**Lodén L.O.** 1965, ApJ 141, 668 *On the association Pup I*
1297	**Serkowski K.** 1965, ApJ 141, 1340 *Polarization of galactic clusters M25, NGC 869, 884, 1893, 2422, 6823, 6871, and association VI Cygni*
1298	**Eggen O.J., Sandage A.R.** 1965, ApJ 141, 821 *The colors of some high-latitude blue stars*
1300	**Harris D.L., Morgan W.W., Roman N.G.** 1954, ApJ 119, 622 *Photometric and spectroscopic observations of stars in IC 348*
1301	**Krzeminski W.** 1965, ApJ 142, 1051 *The eclipsing binary U Geminorum*
1306	**Paczynski B.** 1965, Acta Astron. 15, 103 *Three-colour photometry of RR Lyrae stars. I.*
1307	**Paczynski B.** 1965, Acta Astron. 15, 115 *Three-colour photometry of RR Lyrae stars. II*
1308	**Michalowska-Smak A., Smak J.** 1965, Acta Astron. 15, 333 *UBV photometry of eight population II cepheids*
1311	**Przybylski A., Kennedy P.M.** 1965, MNRAS 131, 95 *Radial velocities and three-colour photometry of 166 southern stars*
1312	**Thackeray A.D., Wesselink A.J.** 1965, MNRAS 131, 121 *A photometric and spectroscopic study of the cluster IC 2944*
1313	**Harris D.L.** 1953, ApJ 118, 346 *Note on the cepheid variable ζ Geminorum*
1315	**Grubissich CL.** 1965, Z. Astrophys. 60, 256 *Dreifarben-photometrie von zwei offenen Sternhaufen nahe NGC 7510*
1316	**Popper D.M.** 1964, ApJ 139, 143 *KU Cygni*
1317	**Tifft W.G.** 1964, ApJ 139, 451 *DH Pegasi, an RR Lyrae star of type 'c'*
1318	**Münch G., Münch L.** 1964, ApJ 140, 162 *Radial velocities of distant OB stars*
1319	**Wallerstein G., Greenstein J.L.** 1964, ApJ 139, 1163 *The chemical composition of two CH stars, HD 26 and HD 201626*
1320	**Preston G.W., Paczynski B.** 1964, ApJ 140, 181 *Atmospheric phenomena in the RR Lyrae stars. I. The singly periodic variables*
1321	**Krzeminski W., Kraft R.P.** 1964, ApJ 140, 921 *Binary stars among cataclysmic variables. V. Photoelectric and spectroscopic observations of the ultra-short-period binary Nova WZ Sagittae*
1322	**Larsson-Leander G.** 1961, Ark. Astr. 3, 31 *Photoelectric observations of 31 Cygni and 32 Cygni in spring 1959*
1323	**Popper D.M.** 1955, ApJ 121, 56 *The eclipsing binary V356 Sagittarii*
1327	**Alcaino G.** 1965, Lowell Obs. Bull. 6, 167 No 126 *A photoelectric investigation of the galactic clusters IC 4665 and IC 4756*
1328	**Jerzykiewicz M.** 1965, Lowell Obs. Bull. 6, 175 No 127 *Two-color photoelectric photometry of δ Ceti*
1330	**Fitch W.S.** 1955, ApJ 121, 690 *The cluster-type variable VZ Cancri*
1332	**Kwee K.K., van Genderen A.M.** 1963, BAN 17, 53 *Photo-electric observations of 31 and 32 Cygni during November and December 1961*
1334	**van Genderen A.M.** 1963, BAN 17, 243 *Photo-electric observations in four colours of the ultra-short-period variable SZ Lyncis*
1335	**van Genderen A.M.** 1964, BAN 17, 293 *Nova Herculis 1963*
1336	**Osawa K.** 1963, PASJ 15, 274 *Spectral classification and three-color photometry of A-type stars*
1337	**Osawa K., Nishimura S., Nariai K.** 1963, PASJ 15, 313 *Light variation of HD 30353*
1338	**Smak J.** 1964, Acta Astr. 14, 101 *On the colours of RW Geminorum*

No	Reference
1339	**Paczynski B.** 1964, Acta Astr. 14, 297 *New red variable stars*
1340	**Johnson H.L., Gardiner A.J.** 1955, PASP 67, 74 *The magnitude and color of Mars during the 1954 opposition*
1343	**Connelley M., Sandage A.** 1958, PASP 70, 600 *Photoelectric observations of RS Ophiuchi*
1344	**Spinrad H.** 1959, PASP 71, 53 *Photoelectric observations of the eclipsing system V401 Cygni*
1345	**Wallerstein G.** 1959, PASP 71, 316 *Three-color photometry of U Geminorum during an outburst*
1346	**Stienon F.M.** 1963, PASP 75, 45 *Pre-outburst observations of Nova Herculis 1960*
1347	**Paczynski B.** 1963, PASP 75, 278 *Three-color observations of 2.1937 Ceti*
1348	**Paczynski B.** 1963, PASP 75, 400 *The correct period of TV Leonis*
1349	**Feinstein A.** 1963, PASP 75, 492 *Eta Carinae and the Trumpler 16 cluster*
1350	**Bok B.J., Bok P.F., Graham J.A.** 1963, PASP 75, 514 *The distance of a group of early type stars in Norma*
1351	**Serkowski K.** 1961, Lowell Obs. Bull. 5, 157 No 116 *The Sun as a variable star. II*
1352	**Popper D.M.** 1957, ApJ 126, 53 *Rediscussion of eclipsing binaries. III. Z Vulpeculae*
1355	**Argue A.N.** 1966, MNRAS 133, 475 *UBV photometry of 550 F, G, and K type stars*
1356	**Widing K.G.** 1966, ApJ 143, 121 *The peculiar star 17 Leporis*
1358	**Fernie J.D.** 1966, AJ 71, 119 *Classical cepheids with companions. I. δ Cephei*
1359	**Pyper D.M.** 1966, ApJ 144, 13 *The effective temperatures of Wolf-Rayet stars as derived from their UBV color indices corrected for emission*
1360	**Abell G.O.** 1966, ApJ 144, 259 *Properties of some old planetary nebulae*
1363	**Häggkvist L., Oja T.** 1966, Ark. Astr. 4, 137 *Photoelectric photometry of bright stars*
1364	**Philip A.G.D.** 1966, ApJS 12, 391 *The stellar space distribution in Pegasus at $b^{II} = -29^o$*
1365	**Wildey R.L.** 1966, PASP 78, 132 *Light curve of the cluster variable VZ Cancri*
1366	**Hardie R.H., Heiser A.M.** 1966, PASP 78, 171 *UBV photometry of the Lowell proper motion object G175-34*
1367	**Feinstein A.** 1966, PASP 78, 301 *The open cluster NGC 2451*
1368	**Bok B.J., Kidd C., Routcliffe P.** 1966, PASP 78, 333 *Photometric observations of supergiants in the Large Magellanic Cloud*
1369	**van Altena W.F.** 1966, PASP 78, 345 *The large proper motion white dwarf vA-C216 (= LP9-231)*
1370	**Landolt A.U.** 1966, PASP 78, 516 *Light variation of 4 Canum Venaticorum (= HD 107904)*
1371	**Priser J.B.** 1966, PASP 78, 474 *Photoelectric observations of stars near northern hemisphere Selected Areas*
1375	**Sanders W.L.** 1966, AJ 71, 719 *UBV photometry of 1055 stars*
1376	**Lodén L.O.** 1966, Ark. Astr. 4, 65 *A study of NGC 2467 and the association Pup I*
1377	**Heiser A.M.** 1962, ApJ 135, 78 *Photoelectric photometry of the eclipsing binary V367 Cygni*
1379	**Elvius T., Häggkvist L.** 1966, Ark. Astr. 4, 49 *Photoelectric measurements of stars in Kapteyn's Selected Areas of the +60° zone*
1380	**Oja T.** 1966, Ark. Astr. 4, 199 *A photometric study of a region in Cassiopeia*
1381	**Häggkvist L.** 1966, Ark. Astr. 4, 165 *Photoelectric magnitudes and colours of visual binaries*
1382	**Johnson H.L.** 1960, ApJ 131, 127 *The eclipsing binary Y Leonis*
1383	**Spinrad H.** 1960, ApJ 131, 134 *A study of VZ Cancri*
1385	**Landolt A.U.** 1964, ApJS 8, 329 *The galactic cluster NGC 6087*
1386	**Paczynski B.** 1966, Acta Astron. 16, 97 *Three-colour photometry of RR Lyrae stars. III*
1388	**Preston G.W., Spinrad H., Varsavsky C.M.** 1961, ApJ 133, 484 *The light and radial-velocity variations of TU Ursae Majoris*
1389	**Smak J.** 1966, Acta Astron. 16, 109 *On the spectral classification and reddening determination for long-period variables*
1390	**Grant G.** 1959, ApJ 129, 62 *A photoelectric study of the eclipsing variable RW Tauri*
1391	**Helfer H.L., Wallerstein G., Greenstein J.L.** 1959, ApJ 129, 700 *Abundances in some Population II K giants*
1394	**Osawa K., Nishimura S., Ichimura K.** 1965, PASJ 17, 199 *Light variation of the A-type peculiar star HD 221568*
1395	**Vetesnik M., Perek L.** 1966, BAN 17, 278 *Photoelectric photometry of CV Cygni*
1397	**Fernie J.D.** 1966, AJ 71, 732 *Classical cepheids with companions. II. Polaris*
1399	**Oosterhoff P.Th.** 1960, BAN 15, 199 *Three-colour photometry in the UBV system of 51 northern cepheids*
1400	**Broglia P.** 1965, J. Obs. 48, 124 *Fluctuations dans les courbes de lumière de la variable à éclipses UV Leonis*
1401	**Schild R.E.** 1966, ApJ 146, 142 *Be stars in the region of h and χ Persei*
1402	**Hardie R.H., Tolbert C.R.** 1961, ApJ 134, 581 *Three color photometry of CY Aquarii*
1404	**Krzeminski W., Walker M.F.** 1963, ApJ 138, 146 *Photoelectric observations of UX Ursae Majoris, 1955-1962*
1405	**Hardie R.H., Schroeder N.H.** 1963, ApJ 138, 350 *Three color photometry of 56 Arietis*
1406	**Bok B.F., Bok P.F., Graham J.A.** 1966, MNRAS 131, 247 *A photometric study of the I Scorpii association*
1407	**Thackeray A.D., Wesselink A.F., Harding G.A.** 1962, MNRAS 124, 454 *The cluster NGC 6067*
1408	**Wesselink A.J.** 1962, MNRAS 124, 358 *UBV photometry in and near the Magellanic Clouds*
1409	**Tifft W.G.** 1963, MNRAS 125, 199 *Magellanic Cloud Investigations. I. The region of NGC 121*
1411	**Bell R.A.** 1962, Observatory 82, 68 *Observations of some southern white dwarfs*
1412	**Westerlund B.E., Danziger I.J., Graham J.** 1963, Observatory 83, 74 *Supergiants stars in the wing of the Small Magellanic Cloud*
1413	**Purgathofer A.** 1964, Z. Astrophys. 59, 29 *Lichtelektrische Beobachtungen des Bedeckungsveränderlichen V401 Cygni*
1414	**Knipe G.F.G.** 1966, Republic Obs. Circ. 7, 111 *Photometric observations of binary stars*
1415	**Cousins A.W.J.** 1963, MNASSA 22, 58 *Photometric data for stars in the equatorial zone (Fourth list)*
1417	**Cousins A.W.J.** 1963, MNASSA 22, 130 *Photoelectric data for stars in the equatorial zone (Fifth list)*
1418	**Wallerstein G.** 1966, IAU Inf. Circular No. 1954 *A possible O-subdwarf eclipsing binary*
1419	**Searle L.** 1958, ApJ 128, 61 *A study of three shell stars*
1420	**Jones D.H.P., Lagerwey H.C.** 1966, MNASSA 25, 168 *The delta Scuti variable 1 Monocerotis*
1421	**Williams P.M.** 1966, MNASSA 25, 122 *Photometry of NGC 2451*
1423	**Guetter H.H.** 1964, Publ. David Dunlap Obs. II, 13 *Distances of 97 OB stars*
1460	**Graham J.A.** 1982, PASP 94, 244 *UBVRI standard stars in the E-regions*
1461	**Harris W.E.** 1982, ApJS 50, 573 *Color-magnitude studies of globular clusters. I. The bright stars in NGC 362*

No	Reference

1462 **Henden A.A.** 1980, MNRAS 192, 621 *A search for northern hemisphere double mode cepheids -II. New UBV cepheid photometry*

1463 **Burkhead M.S., Parvey M.I.** 1968, PASP 80, 483 *UBV observations of 3C 273*

1464 **Arp H.** 1959, AJ 64, 1272 *Southern hemisphere photometry VII. The colour-magnitude diagram of NGC 330 and the adjoining region of the Small Magellanic Cloud*

1465 **Martins D.H., Harvel C.A.** 1979, AJ 84, 1725 *UBVR photometric sequences around NGC 5946, 6440, and 6624*

1467 **Raveendran A.V., Mekkaden M.V., Mohin S.** 1982, MNRAS 199, 707 *HD 81417: a new RS CVn binary*

1468 **Evans T.L.** 1983, SAAO Circ. No. 7, 86 *Photoelectric photometry of globular cluster giants*

1469 **Dean J.F.** 1980, MNASSA 39, 13 *UBV observations of two helium weak stars*

1470 **Reed B.C., FitzGerald M.P.** 1983, MNRAS 205, 241 *A photoelectric UBV catalogue of 610 stars in Puppis*

1471 **Vogt N.** 1983, A&AS 53, 21 *Photoelectric UBV photometry of southern and equatorial dwarf novae*

1473 **Meech K.J.** 1983, PASP 95, 662 *Photoelectric sequences for EX Hydrae, V2051 Ophiuchi, and VV Puppis*

1474 **Menzies J.** 1981, MNRAS 195, 67p *Photoelectric photometry of GX304-1 (4U 1258-61)*

1475 **Stone R.P.S.** 1983, Inf. Bull. Var. Stars No. 2380 *UBV observations and a photometric sequence for the highly active T Tauri star VY Tau*

1476 **Walker R.L.** 1983, Inf. Bull. Var. Stars No. 2318 *Six variable stars discovered in binary systems*

1477 **Walker R.L.** 1983, Inf. Bull. Var. Stars No. 2319 *Observations of four variable stars in binary systems*

1478 **Brueck M.T., Hawkins M.R.S.** 1983, A&A 124, 216 *An investigation of faint stars in a region of the Magellanic Stream*

1480 **van den Bergh S., Younger P.F., Brosterhus E.B.F., Alcaino G.** 1983, ApJS 53, 765 *A search for OB associations near southern long-period cepheids. II. CT Carinae, UU Muscae, VZ Puppis, SV Velorum, and EZ Velorum*

1481 **Böhme D.** 1983, Inf. Bull. Var. Stars No. 2442 *Photoelectric observations of R CrB*

1482 **Miller H.R., Mullikin T.L., McGimsey B.Q.** 1983, AJ 88, 1301 *Photoelectric comparison sequences in the fields of four BL Lacertae objects*

1483 **Dean J.F., Bywater R.A., Warren P.R.** 1976, MNASSA 35, 123 *BVI photometry of LMC supergiants*

1484 **Dean J.F.** 1977, MNASSA 36, 3 *B, V, I, photometry of southern cepheids*

1485 **Balona L.A., Martin W.L.** 1978, MNRAS 184, 1 *RS Gruis: a dwarf cepheid in a binary system*

1486 **Davies E.** 1978, MNRAS 185, 573 *The radius of the RRab star UV Octantis*

1487 **Faulkner D.J., Shobbrook R.R.** 1979, ApJ 232, 197 *Cepheid studies. III. Energy changes in the beat cepheid U Trianguli Australis*

1488 **Stobie R.S., Balona L.A.** 1979, MNRAS 189, 641 *Observations of short period cepheids*

1489 **Balona L.A., Stobie R.S.** 1980, MNRAS 192, 625 *On the nature of AI Velorum*

1490 **Gieren W.** 1981, ApJS 47, 315 *A simultaneous photometric and radial velocity study of short-period southern cepheids. II. The photometry*

1491 **Feinstein A.** 1981, PASP 93, 202 *The region of the open cluster Bo 10*

1492 **Graham J.A.** 1981, PASP 93, 291 *Photoelectric UBVRI photometry in two fields near NGC 5128*

1493 **Wizinowich P., Garrison R.F.** 1982, AJ 87, 1390 *UBVRI photometry in the open cluster NGC 3532*

1494 **Reid N.** 1982, MNRAS 201, 51 *New light on faint stars-I. The luminosity function in the solar neighbourhood*

1495 **Reid N., Gilmore G.** 1982, MNRAS 201, 73 *New light on faint stars-II. A photometric study of the low luminosity main sequence*

1496 **Alcaino G., Liller W.** 1982, A&A 114, 213 *The Bok and Tifft UBV sequence in the Large Magellanic Cloud: revised and extended*

1497 **Balona L.A., Stobie R.S.** 1983, south African Astron. Obs. Circ. 7, 19 *Simultaneous photometry and radial velocities of dwarf cepheids and delta Scuti stars*

1498 **Rucinski S.M.** 1983, A&AS 52, 281 *Photometry of the post T Tauri star HD 36705*

1499 **Landolt A.U.** 1983, AJ 88, 853 *UBVRI photometry of stars useful for checking equipment orientation stability*

1500 **Sandage A.** 1983, AJ 88, 1108 *On the distance to M33 determined from magnitude corrections to Hubble's original cepheid photometry*

1501 **Guetter H.H., Hewitt A.V.** 1984, PASP 96, 441 *Photoelectric UBV photometry for 317 PZT and VZT stars*

1502 **Oja T.** 1984, A&AS 57, 357 *UBV photometry of stars whose positions are accurately known. I.*

1503 **Janes K.A.** 1984, PASP 96, 977 *DDO and UBV photometry of red giant stars in NGC 6791*

1504 **Geisler D.P., Smith V.V.** 1984, PASP 96, 871 *Washington photometry of the old open cluster Melotte 66*

1505 **Dachs J., Kaiser D.** 1984, A&AS 58, 411 *UBV photometry of the southern galactic cluster NGC 4755 = κ Crucis*

1506 **Breger M.** 1984, A&AS 57, 217 *Spectral classification and photometry of selected Pleiades stars*

1507 **del Rio G.** 1984, A&AS 56, 289 *UBV photometry of the young open cluster Berkeley 96*

1508 **Stauffer J.R.** 1984, ApJ 280, 189 *Optical and infrared photometry of late-type stars in the Pleiades*

1509 **Cousins A.W.J.** 1984, SAAO Circ. No. 8, 59 *Standardisation of broad band photometry of equatorial standards*

1510 **Ianna P.A., Whitman W.R.** 1984, AJ 89, 568 *Parallaxes and proper motions from the McCormick observatory: list 45*

1511 **Caldwell J.A.R., Coulson I.M.** 1984, SAAO Circ. No. 8, 1 *BVRI photoelectric photometry of cepheids in the Small Magellanic Cloud*

1512 **Holtzman J.A., Nations H.L.** 1984, AJ 89, 391 *UBVRI and Hα photometry of FK Com*

1513 **Nandy K., Morgan D.H., Houziaux L.** 1984, MNRAS 211, 895 *Infrared extinction in the Small Magellanic Cloud*

1514 **Drilling J.S., Landolt A.U., Schönberner D.** 1984, ApJ 279, 748 *Broad-band photometry of extreme helium stars*

1515 **Alcaino G., Liller W.** 1984, ApJS 56, 19 *BVRI main-sequence photometry of the globular cluster M4*

1516 **Kunkel W.E., Liebert J., Boroson T.A.** 1984, PASP 96, 891 *van Biesbroeck 3: a low-luminosity white dwarf, not an M dwarf*

1517 **Petersen J.O., Hansen L.** 1984, A&A 134, 319 *Studies of cepheid type variability I. KZ Centauri: UBVRI photometry of a unique type II field cepheid*

1518 **Caldwell J.A.R., Jones J.H.S., Menzies J.W.** 1984, MNRAS 209, 51 *A photometric study of six nearby common-proper-motion pairs*

1519 **Rydgren A.E., Vrba F.J.** 1984, AJ 89, 399 *The incidence of infrared excesses among G-type stars in the direction of the Orion Ic association*

1520 **Bergmann T.S., Livi S.H.B., Costa R.D.D.** 1984, Ap&SS 100, 341 *B, V, and DDO photometric observations of red variable stars*

1521 **Hodge P.W., Lee S.O.** 1984, ApJ 276, 509 *LMC blue globular clusters containing cepheids. I. NGC 1856*

No	Reference
1522	**McGregor P.J., Hyland A.R.** 1984, ApJ 277, 149 *A photometric comparison of late-type cluster supergiants in the Magellanic Clouds and the Galaxy*
1523	**Olszewski E.W., Canterna R., Harris W.E.** 1984, ApJ 281, 158 *Color-magnitude photometry for the globular cluster NGC 288*
1524	**Turner D.G., Evans N.R.** 1984, ApJ 283, 254 *An investigation of the stellar association containing the 1.95 day cepheid SU Cassiopeiae*
1525	**Olszewski E.W.** 1984, ApJ 284, 108 *Color-magnitude diagram photometry of the LMC red cluster NGC 1978*
1526	**Anthony-Twarog B.J.** 1984, AJ 89, 267 *Further photometric surveys for white dwarfs in Praesepe*
1527	**Oswalt T.D., Peterson B.M., Foltz C.B.** 1984, AJ 89, 421 *MRK 320: a hot DA white dwarf*
1528	**Hall D.S., Cannon III R.O., Rhombs C.G.** 1984, AJ 89, 559 *UBV light curves of the eclipsing binary RS Cephei*
1529	**Rydgren A.E., Zak D.S.** 1984, AJ 89, 1015 *UBVRI monitoring of five late-type pre-main-sequence stars*
1530	**Demers S., Kibblewhite E.J., Irwin M.J., Bunclark P.S., Bridgeland M.T.** 1984, AJ 89, 1160 *Leo A: color-magnitude diagram of its brightest stars*
1531	**Kilkenny D., Malcolm G.** 1984, MNRAS 209, 169 *Four UBVRI sequences near bright Virgo galaxies*
1532	**Robertson T.H.** 1984, AJ 89, 1229 *An objective-prism survey for late M dwarf stars*
1533	**Hesser J.E., McClure R.D., Hawarden T.G., Cannon R.D.** 1984, PASP 96, 406 *A new color-magnitude diagram for the peculiar star cluster E3=C0921-770*
1534	**Turner D.G.** 1984, PASP 96, 422 *The optical companions of the long-period cepheid WZ Sagittarii-remains of an open cluster?*
1535	**Forbes D., Scarfe C.D.** 1984, PASP 96, 737 *UBV photometry of the 1981 eclipse of RZ Ophiuchi*
1536	**Aiad A., Appenzeller I., Bertout C., Isobe S., Shimizu M., Stahl O., Walker M.F., Wolf B.** 1984, A&A 130, 67 *Coordinated spectroscopic observations of YY Orionis stars*
1537	**Neckel T.** 1984, A&A 137, 58 *Polarimetric and new photometric observations in the NGC 6334 / 6357 region*
1538	**Alcaino G.** 1984, A&A 139, 549 *A photoelectric sequence in the region of the globular cluster NGC 5986*
1539	**Heske A., Wendker H.J.** 1984, A&AS 57, 205 *UBV photometry of Tr 24 and its relation to Sco OB1*
1540	**Miller H.R., Wilson J.W., Africano J.L., Quigley R.J.** 1984, A&AS 57, 353 *Photoelectric comparison sequences in the fields of B2 1308+326 and 1418+54*
1541	**Tapia M., Roth M., Costero R., Navarro S.** 1984, Rev. Mex. Astron. Astrofis. 9, 65 *Near-infrared and visual photometry of h and χ Persei*
1542	**Mohan V., Pandey A.K.** 1984, Ap&SS 105, 315 *Photometry of open cluster King 12*
1543	**Coyne G.V., McConnell D.J.** 1983, Vatican Obs. Pub. 2 No 6 *A survey for Hα emission objects in the Milky Way VII. Final zones*
1544	**Platais I.K.** 1984, Sov. Astron. Lett. 10, 84 *New candidate members of the open cluster NGC 7092 (M39)*
1545	**Srivastava J.B., Kandpal C.D.** 1984, Inf. Bull. Var. Stars No. 2574 *HD 126200 - a suspected eclipsing binary ?*
1546	**Bopp B.W., Africano J.L., Goodrich B.D., Henry G.W., Hall D.S., Barksdale W.S.** 1984, Inf. Bull. Var. Stars No. 2604 *HD 91816: a new BY Draconis star*
1547	**Forbes D.** 1984, Inf. Bull. Var. Stars No. 2605 *Additional photometry of the Of-type variable HD 167971*
1548	**Shugarov S.Y.** 1984, Inf. Bull. Var. Stars No. 2612 *Outbursts of the star Case 1*
1549	**Böhme D.** 1984, Inf. Bull. Var. Stars No. 2646 *Photoelectric observations of R CrB*
1550	**Blanco V.M., Blanco B.M.** 1984, PASP 96, 603 *A B,V sequence in Baade's Window*
1551	**Clariá J.J.** 1985, A&AS 59, 195 *Membership and photometric abundances of red evolved stars in open clusters*
1552	**Garcia-Pelayo J.M., Alfaro E.J.** 1984, PASP 96, 444 *UBV photoelectric photometry of faint stars in the clusters NGC 7235 and NGC 1778*
1553	**Janes K.A., Smith G.H.** 1984, AJ 89, 487 *The giant branch of the old open cluster M67*
1554	**Alcaino G., Liller W.** 1984, AJ 89, 814 *Photoelectric UBVRI sequences in the Sculptor group galaxies: NGC 45, NGC 55, NGC 247, NGC 253, NGC 300, and NGC 7793*
1563	**Götz W., Luthardt R.** 1985, Inf. Bull. Var. Stars No 2664 *The behaviour of FG Vul in 1984*
1564	**Gravina R.** 1985, Inf. Bull. Var. Stars No 2834 *Photometry and spectrophotometry of symbiotic stars: CI Cyg, Z And, V1016 Cyg, HM Sge, HBV 475*
1565	**Böhme D.** 1985, Inf. Bull. Var. Stars No 2835 *Photoelectric observations of R CrB*
1566	**Oja T.** 1985, A&AS 59, 461 *UBV photometry of stars whose positions are accurately known. II.*
1567	**Noordanus B., Cottrell P.L.** 1985, MNRAS 212, 931 *HR 5637: The unveiling of a binary*
1568	**Babu G.S.D.** 1985, J. Astrophys. Astron. 6, 61 *A study of the open cluster NGC 2374*
1569	**Oja T.** 1985, A&AS 61, 331 *Photoelectric photometry of stars near the north galactic pole. II.*
1570	**Upgren A.R., Weis E.W.** 1985, AJ 90, 2039 *Photometry of possible members of the Hyades cluster. VI.*
1571	**Carney B.W., Janes K.A., Flower P.J.** 1985, AJ 90, 1196 *The young SMC cluster NGC 330*
1572	**Turner D.G.** 1985, AJ 90, 1231 *A photometric investigation of cluster membership for the cepheid BB Sagittarii*
1573	**Richtler T.** 1985, A&AS 59, 491 *NGC 2112 - a forgotten old open cluster*
1574	**Mateo M., Hodge P.** 1985, PASP 97, 753 *CCD photometry of Large Magellanic Cloud clusters. I. The remote cluster NGC 1777*
1575	**Stauffer J.R., Hartmann L.W., Burnham J.N., Jones B.F.** 1985, ApJ 289, 247 *Evolution of low-mass stars in the α Persei cluster*
1576	**Turner D.G.** 1985, ApJ 292, 148 *The young open cluster Stock 16: an example of star formation in an elephant trunk?*
1577	**van den Bergh S., Younger P.F., Turner D.G.** 1985, ApJS 57, 743 *A search for OB associations near long-period cepheids. III. U Carinae, XZ Carinae, QY Centauri, VX Crucis, and AA Normae*
1578	**Ortolani S., Rosino L., Sandage A.** 1985, AJ 90, 473 *The globular cluster Palomar 13*
1579	**Sandage A., Carlson G.** 1985, AJ 90, 1019 *The brightest star in nearby galaxies. V. Cepheids and the brightest stars in the dwarf galaxy Sextans B compared with those in Sextans A*
1580	**Beers T.** 1985, AJ 90, 2089 *A search for stars of very low metal abundance. I.*
1581	**Mohan V., Sagar R.** 1985, MNRAS 213, 337 *A photometric study of the open cluster M39*
1582	**Alcaino G., Liller W.** 1985, A&A 146, 389 *A UBVRI photoelectric sequence in 47 Tuc*
1583	**Heske A., Wendker H.J.** 1985, A&A 151, 309 *Further photometry and spectroscopy in the young cluster region Tr24 / Sco OB1*
1584	**Stobie R.S., Gilmore G., Reid N.** 1985, A&AS 60, 495 *Galactic structure photoelectric standards*

No	Reference
1585	**Alcaino G., Liller W.** 1985, A&AS 62, 317 *A UBVRI photoelectric sequence in the globular cluster NGC 6584*
1586	**Kozok J.R.** 1985, A&AS 61, 387 *Photometric observations of emission B-stars in the southern Milky Way*
1587	**Coulson I.M., Caldwell J.A.R.** 1985, SAAO Circ. No 9, 5 *Photometry and radial velocities of 27 southern galactic cepheids*
1588	**Kilkenny D., Whittet D.C.B., Davies J.K., Evans A., Bode M.F., Robson E.I., Banfield R.M.** 1985, SAAO Circ. No 9, 55 *Optical and infrared photometry of southern early-type shell stars and pre-main-sequence variables*
1589	**Kilkenny D., Coulson I.M., Laing J.D., Jones J.S., Engelbrecht C.** 1985, SAAO Circ. No 9, 87 *Photometry of R Coronae Borealis and hydrogen-deficient carbon stars, 1982-3*
1591	**Martel M.T., Gravina R.** 1985, Inf. Bull. Var. Stars No. 2750 *UBV observations of symbiotic stars in July and October 1982*
1592	**Walker W.S.G., Marino B.F., Herdman G.** 1985, Inf. Bull. Var. Stars No. 2775 *Photometry of HR 6384 - an 80 day ellipsoidal binary ?*
1593	**Coulson I.M., Caldwell A.R., Gieren W.P.** 1985, ApJS 57, 595 *A photometric and radial velocity study of six southern cepheids. I. The data*
1594	**Norris J., Bessell M.S., Pickles A.J.** 1985, ApJS 58, 463 *Population studies. I. The Bidelman-MacConnell "weak-metal" stars*
1595	**Smith P.S., Balonek T.J., Heckert P.A., Elston R., Schmidt G.D.** 1985, AJ 90, 1184 *UBVRI field comparison stars for selected active quasars and BL Lacertae objects*
1596	**Byrne P.B., Doyle J.G., Menzies J.W.** 1985, MNRAS 214, 119 *Optical photometry and spectroscopy of the flare star Gliese 229 (= HD 42581)*
1597	**Clariá J.J., Lapasset E.** 1985, MNRAS 214, 229 *New photometric data for the red giants in the open cluster NGC 5822: membership, chemical composition and physical properties*
1598	**Thé P.S., Felenbok P., Cuypers H., Tjin A Djie H.R.E.** 1985, A&A 149, 429 *High resolution spectroscopic and photometric study of the possibility that HD 76534 and HD 163296 are Herbig Ae/Be-type stars*
1599	**Heck A., Houziaux L., Manfroid J., Jones D.H.P., Andrews P.J.** 1985, A&AS 61, 375 *Photometric variations of the irregular variable V348 Sgr*
1600	**Sandage A., Carlson G.** 1985, AJ 90, 1464 *The brightest stars in nearby galaxies. VI. Cepheids and the brightest stars in WLM*
1601	**Oja T.** 1986, A&AS *UBV photometry of stars whose positions are accurately known. III.*
1602	**Clariá J.J., Lapasset E.** 1986, AJ 91, 326 *Fundamental parameters of the open cluster NGC 2567*
1603	**McLaughlin S.F.** 1986, Inf. Bull. Var. Stars No. 2857 *Photometry of comparison stars*
1604	**Thé P.S., Wesselius P.R., Tjin A Djie H.R.E., Steenman H.** 1986, A&A 155, 347 *Studies of the Chamaeleon star-forming region II. The pre-main-sequence stars HD 97048 and HD 97300*
1605	**Pandey A.K., Mahra H.S.** 1985, Ap&SS 120, 107 *Photometry of open cluster NGC 1931*
1606	**Herbig G.H., Vrba F.J., Rydgren A.E.** 1986, AJ 91, 575 *A spectroscopic survey of the Taurus-Auriga clouds for pre-main-sequence stars having CaII H,K emission*
1607	**Pound M.W., Janes K.A.** 1986, PASP 98, 210 *The intermediate-age open cluster Mel 71*
1608	**Forbes D.** 1986, PASP 98, 218 *Bright yellow giants in the open cluster Berkeley 82*
1609	**Böhme D.** 1986, Inf. Bull. Var. Stars No. 2893 *Photoelectric observations of γ Cas, X Per and BU Tau*
1610	**Bopp B.W.** 1986, PASP 98, 457 *Chromospheric activity in the Hyades triple system BD +23 635*
1611	**Alcaino G., Liller W.** 1986, A&A 161, 61 *BVRI CCD photometry of the globular cluster NGC 2298*
1612	**Lapasset E., Clariá J.J.** 1986, A&A 161, 264 *Synthetic light-curve method applied the W UMa systems SY Horologii and VY Ceti*
1613	**Tan H.-S., Zhang Z.-S., Zhang Y.-L.** 1986, Chin. A&A 10, 143 *Photoelectric observations of seventeen variable stars reported by Archer*
1614	**Turner D.G.** 1986, AJ 92, 111 *Galactic clusters with associated cepheid variables. I. NGC 6087 and S Normae*
1615	**Harmanec P., Horn J., Koubsk P.** 1986, Inf. Bull. Var. Stars No. 2912 *A new bright Be variable and suspected eclipsing binary*
1616	**Zsoldos E.** 1986, Inf. Bull. Var. Stars No. 2913 *UBV photometry of HR 8752. II*
1617	**Turner D.G., Leonard P.J.T., Madore B.F.** 1986, J. Roy. Astron. Soc. Canada 80, 166 *Photometry of the S Vul cluster*
1618	**Cudworth K.M., Olszewski E.W., Schommer R.A.** 1986, AJ 92, 766 *Proper motions and bright-star photometry in the Ursa Minor dwarf galaxy*
1619	**Upgren A.R., Lu P.K.** 1986, AJ 92, 903 *Photometry of dwarf K and M stars*
1620	**Sandage A., Kowal C.** 1986, AJ 91, 1140 *New subdwarfs. IV. UBV photometry of 1690 high-proper-motion stars*
1621	**Bopp B.W., Africano J., Quigley R.** 1986, AJ 92, 1409 *UBV photometry of ten southern hemisphere active-chromosphere stars*
1623	**Christian C.A.** 1981, ApJ 246, 827 *King 8: A metal-poor disk cluster*
1624	**Pandey A.K.** 1986, Bull. Astron. Soc. India 14, 20 *Photometry of open cluster Stock 17*
1625	**Oja T.** 1987, A&AS 68, 211 *UBV photometry of stars whose positions are accurately known. IV*
1626	**Ianna P.A., Adler D.S., Faudree E.F.** 1987, AJ 92, 347 *Membership in the open cluster NGC 2287*
1627	**Ardeberg A., Lindgren H.** 1987, A&AS 67, 103 *A photoelectric UBV sequence in SA 184*
1628	**Moreno H., Carrasco G.** 1986, A&AS 65, 33 *UBVRI photometry of FK4 and FK4 supplement stars*
1629	**Papousek J.** 1986, Inf. Bull. Var. Stars No. 2965 *Photoelectric photometry of θ CrB in 1984-86*
1631	**Reimann H.G., Jena W.P.** 1987, Astron. Nachr. 308, 111 *Combined UBV and uvby photometry of open clusters. I. NGC 1502*
1632	**Robertson T.H., Hamilton J.E.** 1987, AJ 93, 959 *Photometry of faint red stars*
1633	**Whittet D.C.B., Kirrane T.M., Kilkenny D., Oates A.P., Watson F.G., King D.J.** 1987, MNRAS 224, 497 *A study of the Chamaeleon dark cloud and T-association - I. Extinction, distance and membership*
1634	**Nemec J.M., Harris H.C.** 1987, ApJ 316, 172 *Blue straggler stars in the globular cluster NGC 5466*
1635	**Walker M.F.** 1987, PASP 99, 179 *Photometric zero-point stars in Magellanic Cloud clusters*
1636	**Pettersson B.** 1987, A&AS 70, 69 *An objective prism survey for Hα-emission-line stars of a field in Puppis*
1637	**Stagg C.** 1987, MNRAS 227, 213 *A photometric survey of the bright southern Be stars*
1638	**Alcaino G., Liller W., Alvarado F.** 1987, AJ 93, 1464 *Photoelectric UBVRI sequences in the region of the galactic globular clusters NGC 288, 1851, 1904, 4590, 4833, 5946, 6139, and 7099*

No	Reference

1639 **Rucinski S.M.** 1987, PASP 99, 487 *UBVRI photometry of the open cluster Collinder 359, alleged to contain W Ursae Majoris systems*

1640 **Babu G.S.D.** 1987, J. Astrophys. Astron. 8, 219 *A study of faint young open clusters as tracers of spiral features in our galaxy. Paper 3: Collinder 97 (OCl 506)*

1641 **Evans T.L., Koen M.C.J.** 1987, SAAO Circ. No. 11, 21 *UBV(RI) photometry for CaII emission stars. 1. Observations at Sutherland*

1642 **Cameron A.C.** 1987, SAAO Circ. No. 11, 57 *UBV(RI) photometry for CaII emission stars. 2. Observations at Mt John University observatory and at Mt Stromlo*

1643 **Hawarden T.G., Bingham E.A.E.** 1987, SAAO Circ. 11, 83 *Photoelectric B,V observations of stars in ω Centauri*

1644 **Pedreros M.** 1987, AJ 94, 92 *Photometric study of the southern open cluster IC 2488*

1646 **Hodge P., Flower P.** 1987, PASP 99, 734 *Some studies of Magellanic Cloud clusters*

1647 **Joshi U.C., Sagar R.** 1986, Bull. Astron. Soc. India 14, 95 *UBV photoelectric photometry of the open cluster NGC 2539*

1648 **Oja T.** 1987, A&AS 71, 561 *UBV photometry of stars whose positions are accurately known. V.*

1649 **FitzGerald M.P.** 1987, MNRAS 229, 227 *The distribution of OB stars and dust in a Milky Way field at (l,b)=(335°, 0°)*

1650 **Drilling J.S.** 1983, ApJ 270, L13 *The spectra of 12 new subluminous O stars*

1651 **Perez M.R., Thé P.S., Westerlund B.E.** 1987, PASP 99, 1050 *On the distances to the young open clusters NGC 2244 and NGC 2264*

1652 **Lindgren H., Ardeberg A., Zuiderwijk E.** 1987, A&A 188, 39 *Orbital elements for double stars of population II. The high-velocity system CoD −48°1741*

1653 **Pedreros M.** 1987, AJ 94, 1237 *Photometric study of the southern open clusters NGC 5316 and NGC 6124*

1654 **Zakirov M.M.** 1987, Kinematika Fiz. Nebesn. Tel. 3, 25 *A search for eclipsing variables and the orientation of spectroscopic binaries orbits in the open cluster IC 4665*

1655 **Berdnikov L.N.** 1987, Variable Stars 22, 369 *Photoelectric observations of classical cepheids*

1657 **Carrasco G., Loyola P.** 198 , A&AS *UBVRI photometry of FKSZ stars. I*

1658 **Carney B.W., Latham D.W.** 1987, AJ 92, 116 *A survey of proper-motion stars.I. UBV photometry and radial velocities*

1659 **Evans T.L.** 1987, SAAO Circ. No. 11, 73 *UBVRI photometry of HD 36705 (AB Dor)*

1661 **Suntzeff N.B.** 1988, AJ 95, 91 *Spectroscopy and photometry of giant stars in NGC 2419 and other metal-poor halo clusters*

1662 **van den Bergh S.** 1988, AJ 95, 106 *A sequence of photoelectric standard stars surrounding NGC 6397*

1663 **Dahn C.C., Harrington R.S., Kallarakal V.V., Guetter H.H., Luginbuhl C.B., Riepe B.Y., Walker J.R., Pier J.R., Vrba F.J., Monet D.G., Ables H.D.** 1988, AJ 95, 237 *U.S. Naval Observatory parallaxes of faint stars, list VIII*

1664 **Rucinski S.M.** 1988, Inf. Bull. Var. Stars No. 3146 *TW Hya: A T Tauri star far from any dark cloud. UBVRI Observations in 1986 and 1987*

1665 **Janes K.A., Heasley J.N.** 1988, AJ 95, 762 *NGC 2298-Another perfectly normal globular cluster in the outer halo*

1666 **Sagar R., Cannon R.D., Hawkins M.R.S.** 1988, MNRAS 232, 131 *Photoelectric and electronographic photometry of the globular cluster NGC 1851*

1667 **Jorgensen U.G., Westerlund B.E.** 1988, A&AS 72, 193 *UBVRI observations of stars in the fields of five clusters with nearby carbon stars*

1668 **del Rio G., Huestamendia G.** 1988, A&AS 73, 425 *Photoelectric UBV and photographic RGU photometry of the open clusters NGC 1496 and NGC 1513*

1669 **Alcaino G., Alvarado F.** 1988, AJ 95, 1724 *Photoelectric UBVRI sequences in the Magellanic Cloud clusters Kron 3, NGC 330, NGC 1841 and NGC 2257*

1670 **Crinklaw G., Talbert F.D.** 1988, PASP 100, 693 *UBV photometry of the open cluster NGC 381*

1671 **Clariá J.J., Lapasset E.** 1988, MNRAS 235, 1129 *A UBV and DDO astrophysical study of the open cluster NGC 3532*

1672 **Peterson C.J., FitzGerald M.P.** 1988, MNRAS 235, 1439 *UBV photometry in four southern open clusters with suspected supernova remnants*

1673 **Carrasco G., Loyola P.** 1988, A&AS 76, 1 *UBVRI photometry of FKSZ stars. II.*

1674 **Weis E.W., Hanson B.** 1988, AJ 96, 148 *Photometry of possible members of the Hyades cluster. VII*

1675 **Westerlund B.E., Garnier R., Lundgren K., Pettersson B., Breysacher J.** 1988, A&AS 76, 101 *UBV and uvbyβ photometry of stars in the region of the ζ Sculptoris cluster*

1676 **Alcala J.M., Ferro A.A.** 1988, Rev. Mex. Astron. Astrof. 16, 86 *UBVRI photoelectric photometry of the open cluster NGC 7790*

1677 **Alcaino G., Liller W., Alvarado F.** 1988, AJ 96, 139 *Photoelectric UBVRI sequences in the region of the galactic globular clusters NGC 362, NGC 1261, NGC 6656, and NGC 6809*

1678 **Warner B., Sneden C.** 1988, MNRAS 234, 269 *HD 38451: J. R. Hind's star that changed colour*

1681 **Feinstein A., Vázquez R.A.** 1989, A&AS 77, 321 *New UBVRI photoelectric photometry in the field of the open cluster NGC 2467*

1682 **Forbes D.** 1989, A&AS 77, 439 *Photometry and spectroscopy of stars in northern HII regions*

1683 **Dachs J., Kabus H.** 1989, A&AS 78, 25 *UBV photometry and the structure of the galactic cluster NGC 2516*

1684 **Westerlund B.E., Garnier R.** 1989, A&AS 78, 203 *UBVβ photometry of luminous early-type stars and emission-line stars in the southern Coalsack region*

1685 **Clariá J.J., Lapasset E., Minniti D.** 1989, A&AS 78, 363 *Photometric metal abundances of high luminosity red stars in young and intermediate-age open clusters*

1686 **Maurice E., Bouchet P., Martin N.** 1989, A&AS 78, 445 *BVR photoelectric photometry of late-type stars and a compilation of other data in the Small Magellanic Cloud*

1687 **Hamuy M., Maza J.** 1989, AJ 97, 720 *UBVRI photoelectric photometry in the fields of fifteen active galaxies*

1688 **Seidensticker K.J.** 1989, A&AS 79, 61 *The distance and structure of the Coalsack. I. Photometric data*

1689 **Guetter H.H., Vrba F.J.** 1989, AJ 98, 2 *Reddening and polarimetric studies toward IC 1805*

1690 **Crawford D.L., Perry C.L.** 1989, PASP 101, 601 *Four-color and Hβ photometry near SA 57*

1691 **Dawson P.C., Forbes D.** 1989, PASP 101, 614 *BVRI photometry of 30 proper-motion stars*

1692 **Yoss K.M., Shankar A., Bell D., Neese C., Detweiler H.L.** 1989, PASP 101, 653 *Photoelectric V and (B-V) for Weistrop stars near the north galactic pole*

1693 **Babu G.S.D.** 1989, J. Astrophys. Astron. 10, 295 *Study of young open clusters as tracers of spiral features in our galaxy. Paper 4: Czernik 20 (OCl 427)*

No	Reference

1694 **Stauffer J.R., Hartmann L.W., Jones B.F.** 1989, ApJ 346, 160 *Rotational velocities of newly discovered, low-mass members of the α Persei cluster*

1695 **Clariá J.J., Lapasset E.** 1989, MNRAS 241, 301 *Multicolour photometry of red giants in three southern open clusters*

1696 **Ryan S.G.** 1989, AJ 98, 1693 *Subdwarf studies. I. UBVRI photometry of NLTT stars*

1697 **Castelaz M., Persinger T.** 1989, AJ 98, 1768 *Spectral types and luminosity classification of MAP region stars using UBVRI and DDO photometry*

1698 **Kilkenny D., Menzies J.W.** 1989, SAAO Circ. 13, 25 *UBVRI photometry of southern spectrophotometric standards*

1699 **Jones K., van Wyk F., Jeffery C.S., Marang F., Shenton M., Hill P., Westerhuys J.** 1989, SAAO Circ. 13, 39 *Photometry of hydrogen deficient stars*

1700 **Winkler H.** 1989, SAAO Circ. 13, 63 *Photometric observations of bright stars in the vicinity of Seyfert galaxies (I)*

1701 **Chini R., Wargau W.F.** 1990, A&A 227, 213 *Abnormal extinction and pre-main sequence stars in M16 (NGC 6611)*

1702 **Neese C.L., Yoss K.M.** 1988, AJ 95, 463 *Kinematic and abundance gradients in the galactic disk*

1703 **Thé P.S., de Winter D., Feinstein A., Westerlund B.E.** 1990, A&AS 82, 319 *The extinction law, the distance and the HR diagram of the extremely young open cluster NGC 6611*

1704 **Carrasco G., Loyola P.** 1990, A&AS 82, 553 *UBVRI photometry of FKSZ stars. III.*

1705 **Bessel M.S.** 1990, A&SAS83, 357 *BVRI photometry of the Gliese Catalogue stars*

1706 **Marschall L.A., Comins N.F., Karshner G.B.** 1990, AJ 99, 1536 *Photometry of the young open cluster Trumpler 37*

1707 **Perry C.L., Hill G., Younger P.F., Barnes J.V.** 1990, A&AS 86, 415 *A study of Sco OB1 and NGC 6231. I. New observational data*

1708 **Wegner G., Africano J.L., Goodrich B.** 1990, AJ 99, 1907 *Photoelectric photometry for 106 objects in the Kiso survey*

1709 **Philip A.G.D., Chromey F.R., Dubois P.** 1990, PASP 102, 654 *Photometry of the Rubin Losee blue stars at the anticenter*

1710 **Sanders W.L.** 1990, A&AS 84, 615 *UBV photometry of NGC 6494 and metallicity consideration*

1711 **Landolt A.U., Perry C.L., Levato H.O., Malaroda S.M.** 1990, AJ 100, 695 *Multicolor photometric and spectroscopic observations of the van den Bergh and Hagen cluster No. 99*

1712 **Reed B.C.** 1990, AJ 100, 737 *UBV photometry of OB and field stars at l = 250*

1713 **Grice N.A., Dawson D.W.** 1990, PASP 102, 881 *The open cluster NGC 6716*

1714 **FitzGerald M.P., Harris G.L.H., Reed B.C.** 1990, PASP 102, 865 *The moderately young open cluster NGC 2353*

1715 **Clariá J.J., Lapasset E., Bosio M.A.** 1991, MNRAS 249, 193 *Membership, fundamental parameters, and luminosity function of the open cluster NGC 5662*

1716 **Sato K., Kuji S.** 1990, A&AS 85, 1069 *MK classification and photometry of stars used for time and latitude observations at Mizusawa and Washington*

1717 **Jung J.H., Lee S.W.** 1984, J. Korean Astron. Soc. 17, 37 *Blue stragglers and clump stars in M67*

1718 **Kwon S.M., Lee S.W.** 1983, J. Korean Astron. Soc. 16, 7 *Photoelectric observations of extremely young open clusters*

1719 **Morrison H.L., Flynn C., Freeman K.C.** 1990, AJ 100, 1191 *Where does the disk stop and the halo begin? Kinematics in a rotation field*

1720 **Vázquez R.A., Feinstein A.** 1990, A&AS 86, 209 *The open cluster Tr 18*

1721 **Lahulla J.F.** 1990, Bol. Astron. Obs. Madrid 12, 25 *Photometria UBVRI de la region S298*

1722 **Persinger T., Castelaz M.W.** 1990, AJ 100, 1621 *Spectroscopic parallaxes of MAP region stars from UBVRI, DDO, and uvby Hβ photometry*

1723 **Figueras F., Jordi C., Rosselló G., Torra J.** 1990, A&AS 82, 57 *UBVRI photometry of G, K, M Hipparcos stars*

1724 **Yoss K.M., Bell D.J., Detweiler H.L.** 1991, AJ 102, 975 *Kinematics of Sandage-Fouts stars in three cardinal directions*

1725 **Vázquez R.A., Feinstein A.** 1991, A&AS 87, 383 *The young open cluster NGC 5606*

1726 **Jordi C., Figueras F., Paredes J.M., Rosselló G., Torra J.** 1991, A&AS 87, 229 *UBVRI Photometry of G, K, M Hipparcos stars. II.*

1727 **Kidder K.M., Holberg J.B., Mason P.A.** 1991, AJ 101, 579 *UBV photometry of hot DA white dwarfs*

1728 **Doinidis S.P., Beers T.C.** 1991, PASP 102, 1392 *Photoelectric UBV photometry of northern stars from the HK objective-prism survey*

1729 **Menzies J.W., Marang F., Laing J.D., Coulson I.M., Engelbrecht C.A.** 1991, MNRAS 248, 642 *UBV(RI)_C photometry of equatorial standard stars. A direct comparison between the northern and southern systems*

1730 **Kilkenny D., Laing J.D.** 1990, SAAO Circ. 14, 11 *UBVRI observations of stars in the Guide Star Photometric Catalogue*

1731 **Winkler H., van Wyk F., Glass I.S.** 1990, SAAO Circ. 14, 25 *Photometric observations of bright stars in the vicinity of Seyfert galaxies (II)*

1732 **Menzies J.W., Marang F., Westerhuys J.E.** 1990, SAAO Circ. 14, 33 *Broad-band photometry of selected southern ultraviolet-bright stars*

1733 **Oja T.** 1991, A&AS 89, 415 *UBV photometry of stars whose positions are accurately known. VI.*

1734 **Clariá J.J., Lapasset E.** 1991, PASP 103, 998 *Collinder 223: A moderately young open cluster in the Carina spiral feature*

1735 **Turner D.G.** 1993, A&AS 97, 755 *Star counts and UBV photometry for the open cluster Roslund 3*

1736 **Preston G.W., Shectman S.A., Beers T.C.** 1991, ApJS 76, 1001 *Photoelectric UBV photometry of stars selected in the HK objective-prism survey*

1737 **Drilling J.S.** 1991, ApJS 76, 1033 *UBV photometry of OB+ stars in the southern Milky Way*

1738 **Cutispoto G.** 1991, A&AS 89, 435 *UBVRI observations of southern hemisphere active stars. II. 1987 data*

1739 **Humphreys R.M., Landau R., Ghigo F.D., Zumarch W.** 1991, AJ 102, 395 *BVR photoelectric sequences for selected fields in the Palomar sky survey and the magnitude-diameter relation*

1740 **Vázquez R.A., Feinstein A.** 1991, A&AS 90, 317 *Photometric analysis of the young open cluster Hogg 16*

1741 **Babu G.S.D.** 1991, J. Astrophys. Astr. 12, 187 *Study of faint young open clusters as tracers of spiral features in our Galaxy. Paper 5: NGC 2236 (OCl 501)*

1742 **Mayer P., Hadrava P., Harmanec P., Chochol D.** 1991, Bull. Astron. Inst. Czechosl. 42, 230 *New data on the eclipsing binary V1765 Cyg (HR 7551) and improved orbital and light-curve solutions*

1743 **Vázquez R.A., Feinstein A.** 1991, A&AS 92, 863 *Binary stars: another effect contribution to the abnormal extinction law in NGC 6193?*

1744 **Doinidis S.P., Beers T.C.** 1991, PASP 103, 973 *Photoelectric UBV photometry of northern stars from the HK objective-prism survey. II.*

1745 **Zakirov M.M.** 1990, Kinematica Fys. Nebens. 6, 42 *Open star cluster Dol 42 and the eclipsing variable V478 Cygni*

No	Reference
1746	**Weis E.E** 1991, AJ 102, 1795 *A photometric study of K and M dwarf stars found by Stephenson*
1747	**Reed B.C., Thomas C.G.** 1991, PASP 103, 1094 *UBV photometry of luminous stars at l = 268 and l = 296*
1748	**Castelaz M.W., Persinger T., Stein J.W., Prosser J., Powell H.D.** 1991, AJ 102, 2103 *Photometry of astrometric reference stars*
1749	**Guetter H.H.** 1992, AJ 103, 197 *Photometric studies of stars in the young open cluster NGC 6823*
1750	**Piers R.P.A., The P.S., van Genderen A.M.** 1992, A&AS 92, 609 *A photometric and spectroscopic study of a supposed subclustering of pre-main sequence stars towards the Scorpius OB1 association*
1751	**Forbes D., English D., De Robertis M.M., Dawson P.C.** 1992, AJ 103, 916 *Membership of the WR binary system V444 Cygni in the young open cluster Berkeley 86*
1752	**Martin E.L., Magazzu A., Rebolo R.** 1992, A&A 257, 186 *On the post-T-Tauri nature of late-type visual companions to B-type stars*
1753	**Munari U., Yudin B.F., Taranova O.G., Massone G., Marang F., Roberts G., Winkler H., Whitelock P.A.** 1992, A&AS 93, 383 *UBVRI-JHKL photometric catalogue of symbiotic stars*
1754	**Shobbrook R.R.** 1992, MNRAS 255, 486 *UBV(RI)$_C$ observations for 13 bright cepheids*
1755	**Saurer W., Pfitscher K., Weinberger R., Hartl H.** 1992, A&AS 93, 553 *Photoelectric UBV photometry of stars in four fields near the galactic plane*
1756	**Turner D.G., Forbes D., Pedreros M.** 1992, AJ 104, 1132 *Galactic clusters with associated cepheid variables. II. NGC 129 and DL Cassiopeiae*
1757	**Turner D.G.** 1992, AJ 104, 1865 *Galactic clusters with associated cepheid variables. III. NGC 1647 and SZ Tauri*
1758	**Gonzalez G., Piche F.** 1992, AJ 103, 2048 *Washington photometry of nearby F, G, and K dwarfs*
1759	**Dawson P.C., Forbes D.** 1992, AJ 103, 2063 *Cousins BVRI photometry of high-proper-motion stars*
1760	**Beers T.C., Preston G.W., Shectman A.** 1992, AJ 103, 1987 *A search for stars of very low metal abundance. II.*
1761	**Clariá J.J., Lapasset E.** 1992, Acta Astron. 42, 343 *New color-magnitude diagrams and basic parameters of the open cluster NGC 5749*
1762	**Lahulla J.F., Hilton J.** 1992, A&AS 94, 265 *UBV photometry in the vicinity of L379*
1763	**Covino E., Terranegra L., Franchini M., Chavarra-K C., Stalio R.** 1992, A&AS 94, 273 *UBV(RI) photometric monitoring of Orion population stars in the southern hemisphere*
1764	**Landolt A.U.** 1992, AJ 104, 340 *UBVRI photometric standard stars in the magnitude range 11.5 < V < 16.0 around the celestial equator*
1765	**Landolt A.U.** 1992, AJ 104, 372 *Broadband UBVRI photometry of the Baldwin-Stone southern hemisphere spectrophotometric standards*
1766	**Orsatti A.M.** 1992, AJ 104, 590 *The Puppis region and the last crusade for faint OB stars*
1767	**Matthews J.M., Gieren W.P., Fernie J.D., Dinshaw N.** 1992, AJ 104, 748 *Reexamining the beat cepheid TU Cas: No second overtone, no mode switching*
1768	**Lacy C.H.** 1992, AJ 104, 801 *UBV photometry of selected eclipsing binary stars*
1769	**Rodono M., Cutispoto G.** 1992, A&AS 95, 55 *Long-term monitoring of active stars. I. 1988-89 UBV photometry with the Phoenix APT*
1770	**Slawson R.W., Hill R.J., Landstreet J.D.** 1992, A&AS 82, 117 *A homogeneous catalog of new UBV and Hβ photometry of B-and A-type stars in and around the Scorpius-Centaurus OB association*
1771	**Slawson R.W.** 1992, priv. comm. (Thesis)
1772	**Berdnikov L.N.** 1992, Soviet Astron. Lett. 18, 130 *Photoelectric observations of Cepheids in 1991*
1773	**Ryan S.G.** 1992, AJ 104, 1144 *Halo common proper motion stars: Subdwarf distance scale, halo binary fraction, and UBVRI colors*
1774	**Upgren A.R., Zhao H., Lee J.T.** 1992, AJ 104, 1203 *Parallaxes and proper motions. XIX. The subdwarf sequence*
1775	**Watanabe E., Yutani M., Kondo M.** 1991, Publ. Natl. Astron. Obs Japan 2, 225 *Selection of standard stars for photometric observations with the 91-cm reflector at Okayama*
1776	**Penprase B.E.** 1992, ApJS 83, 273 *Photometric and spectroscopic analysis of high galactic latitude molecular clouds. I. Distances and extinctions of stars towards 25 selected regions*
1777	**Walas O., Dumont M., Remis J.** 1992, Inf. Bull. Var. Stars No. 3758 *B and V photometry and the ephemeris of the W-UMa type star RS Serpentis*
1778	**Zsoldos E.** 1992, Inf. Bull. Var. Stars No. 3761 *What is RX Cephei?*
1779	**Zsoldos E.** 1992, Inf. Bull. Var. Stars No. 3773 *A new supergiant variable: HD 186841 (B1 Ia)*
1780	**Twarog B.A., Twarog B.J.A., McClure R.D.** 1993, PASP 105, 78 *A photometric analysis of the intermediate-age open cluster NGC 5822*
1781	**Clariá J.J., Mermilliod J.-C.** 1992, A&AS 95, 429 *Membership, binarity and metallicity of red giants in the open cluster NGC 6134*
1782	**Stetson P.B., Harris W.E.** 1988, AJ 96, 909 *CCD photometry of the globular cluster M92*
1783	**Oriol R.** 1992, ASS 198, 281 *Analysis of Neckel and Chini standard stars in the UBVRI photometric system*
1784	**Sung H., Lee S.-W.** 1992, J. Korean Astr. 25, 91 *UBV photometry of open cluster M35*
2001	**Arp H.C.** 1958, AJ 63, 118 *Southern hemisphere photometry II. Photoelectric measures of bright stars*
2002	**Arp H.C., van Sant C.T.** 1958, AJ 63, 341 *Southern hemisphere photometry IV. The galactic cluster NGC 4755*
2003	**Braes L.L.E.** 1961, MNASSA 20, 7 *Three-colour photometry of IC 2602*
2004	**Braes L.L.E.** 1962, MNASSA 21, 16 *Photoelectric photometry of the galactic cluster Mel 101*
2005	**Braes L.L.E.** 1962, BAN 16, 297 *The galactic cluster IC 2602*
2006	**Corben P.M.** 1966, MNASSA 25, 44 *Photoelectric magnitudes and colours for bright southern stars*
2007	**Corben P.M.** 1971, MNASSA 30, 37 *Photoelectric magnitudes and colours for bright southern stars*
2008	**Corben P.M.** 1971, MNASSA 30, 79 *Photoelectric magnitudes and colours for bright southern stars*
2009	**Corben P.M., Stoy R.H.** 1968, MNASSA 27, 11 *Photoelectric magnitudes and colours for bright southern stars*
2010	**Cousins A.W.J.** 1961, MNASSA 20, 16 *Standard magnitudes in the Small Magellanic Cloud*
2011	**Cousins A.W.J., Stoy R.H.** 1962, R. Obs. Bull. no 49 *Standard magnitudes in the E regions*
2012	**Cousins A.W.J., Stoy R.H.** 1963, R. Obs. Bull. no 64 *Photoelectric magnitudes and colours of southern stars*
2013	**Cousins A.W.J., Lake R., Stoy R.H.** 1966, ROB no 121 *Photoelectric magnitudes and colours of southern stars, II*
2014	**Cousins A.W.J., Stoy R.H.** 1970, MNASSA 29, 91 *UBV photometry of late B type stars*
2015	**Eggen O.J.** 1963, AJ 68, 483 *Three-color photometry of the components in 228 wide double and multiple systems*

No	Reference
2016	**Eggen O.J.** 1963, AJ 68, 697 *Luminosities, colors, and motions of the brightest A-type stars*
2017	**Eggen O.J.** 1964, R. Obs. Bull. no 84 *A catalogue of high-velocity stars*
2018	**Evans D.S.** 1966, R. Obs. Bull. no 110 *Fundamental data for southern stars*
2019	**Evans D.S., Stoy R.H.** 1959, MNASSA 18, 144 *HD 89499: a high-velocity sub-dwarf*
2020	**Evans D.E., Laing J.D., Menzies A., Stoy R.H.** 1964, R. Obs. Bull. no 85 *Fundamental data for southern stars*
2021	**Feast M.W., Stoy R.H., Thackeray A.D., Wesselink A.J.** 1961, MNRAS 122, 239 *Spectral classification and photometry of southern B stars*
2022	**Fernie J.D.** 1959, MNASSA 18, 57 *Photoelectric magnitudes and colours in NGC 2547*
2023	**Fernie J.D.** 1962, ApJ 135, 298 *WZ Sgr and the association III Sgr*
2024	**Irwin J.B.** 1961, ApJS 6, 253 *Southern cepheid photometry*
2025	**Jankowitz N.E., McCosh C.J.** 1962, MNASSA 22, 18 *Photometric observations of NGC 3114*
2026	**Koelbloed D.** 1959, BAN 14, 265 *Three-colour photometry of the three southern open clusters NGC 3532, 6475 (M7), and 6124*
2027	**Lake R.** 1962, MNASSA 21, 56 *Photoelectric magnitudes and colours for 168 southern stars*
2028	**Lake R.** 1962, MNASSA 21, 191 *Photoelectric magnitudes and colours for 100 southern stars*
2029	**Lake R.** 1963, MNASSA 22, 79 *Photoelectric magnitudes and colours for 242 southern stars*
2030	**Lake R.** 1964, MNASSA 23, 14 *Photoelectric magnitudes and colours for 100 southern stars*
2031	**Lake R.** 1964, MNASSA 23, 136 *Photoelectric magnitudes and colours for 100 southern stars*
2032	**Lake R.** 1965, MNASSA 24, 41 *Photoelectric magnitudes and colours for bright southern stars*
2033	**Stoy R.H.** 1963, MNASSA 22, 157 *Photoelectric three colour magnitudes for 354 southern stars*
2034	**Stoy R.H.** 1965, MNASSA 24, 29 *Photoelectric three colour magnitudes for southern stars*
2035	**Stoy R.H.** 1968, MNASSA 27, 119 *Photoelectric magnitudes and colours for bright southern stars*
2036	**Williams P.M.** 1966, MNASSA 25, 122 *Photometry of NGC 2451*
2037	**Williams P.M.** 1967, MNASSA 26, 30 *Photoelectric photometry of NGC 2451*
2038	**Cousins A.W.J.** 1977, SAAO circ. no 1, 51 *Cape UBV magnitudes and colours of south circumpolar stars*
2039	**Evans D.S., Wild P.A.T.** 1969, MNASSA 89, 15 *An area search for subdwarfs*
2040	**Wild P.A.T.** 1968, MNASSA 27, 47 *The eclipsing binary system, AU Puppis I. Three-colour photoelectric photometry*
2041	**Fernie J.D.** 1979, Private communication
2042	**Becker W., Svolopoulos S.N., Fang C.** 1976, Univ. Basel *Kataloge photographischer und photoelektrischer Helligkeiten von 25 galaktischen Sternhaufen im RGU- und $U_C BV$-system*
3000	**Eggen O.J., Sandage A.R.** 1960, MNRAS 120, 79 *Photometry in the Magellanic Clouds. I. Standard sequences*
3001	**Eggen O.J.** 1967, ApJS 14, 307 *Narrow- and broad-band photometry of red stars. I. Northern giants*
3002	**Eggen O.J.** 1969, PASP 81, 553 *Stellar groups in the old disk population*
3003	**Eggen O.J.** 1968, ApJS 16, 49 *Narrow-and broad-band photometry of red stars. II. Dwarfs*
3004	**Eggen O.J.** 1970, ApJ 161, 159 *A very young cluster with a moderate metal deficiency*
3005	**Eggen O.J., Stokes N.R.** 1970, ApJ 161, 199 *Narrow-band and broad-band photometry of red stars. III. Southern giants*
3006	**Eggen O.J.** 1970, ApJ 159, 945 *Subluminous stars. VI. Photoelectric observations in the red and near-infrared*
3007	**Eggen O.J.** 1970, ApJS 22, 289 *Stellar-population samples at the galactic poles. II. UBVRI photometry of M stars near the south pole*
3008	**Eggen O.J.** 1971, ApJS 22, 389 *Luminosities, temperatures, and kinematics of K-type dwarfs*
3009	**Eggen O.J.** 1971, PASP 83, 251 *The ζ Herculis, σ Puppis, ϵ Indi, and η Cephei groups of old disk population stars*
3010	**Eggen O.J.** 1971, PASP 83, 271 *The Arcturus group*
3011	**Eggen O.J.** 1971, ApJ 166, 87 *The intermediate-age cluster IC 4651*
3012	**Eggen O.J.** 1971, PASP 83, 423 *Some red giants of the old disk population*
3013	**Eggen O.J.** 1971, PASP 83, 741 *Luminosities and motions of the F-type stars. I. Luminosity and metal abundance indices for disk population stars*
3014	**Eggen O.J.** 1971, PASP 83, 762 *The nature of the blue stragglers in the old disk population*
3015	**Eggen O.J.** 1972, ApJ 173, 63 *NGC 2516 and the Pleiades group*
3016	**Eggen O.J.** 1968, R. Obs. Bull. No. 137 *Three-colour photometry of 4000 northern stars*
3017	**Eggen O.J.** 1969, ApJ 155, 439 *The old galactic cluster NGC 3680*
3018	**Eggen O.J.** 1968, ApJ 152, 83 *The intermediate-age cluster NGC 2360*
3019	**Eggen O.J.** 1974, ApJ 188. 59 *NGC 2287 and the Pleiades group*
3020	**Eggen O.J.** 1959, MNRAS 119, 255 *Stellar groups. IV. The Groombridge 1830 group of high velocity stars and its relation to the globular clusters*
3021	**Eggen O.J.** 1961, R. Obs. Bull. No. 27 *Three colour photometry in the southern hemisphere: NGC 6383, NGC 6405 and standard stars*
3022	**Eggen O.J.** 1962, PASP 74, 159 *SZ Lyncis, a new ultrashort-period RR Lyrae variable*
3023	**Eggen O.J.** 1963, AJ 68, 697 *Luminosities, colors, and motions of the brightest A-type stars*
3024	**Eggen O.J.** 1963, AJ 68, 483 *Three-color photometry of the components in 228 wide double and multiple systems*
3025	**Eggen O.J.** 1964, R. Obs. Bull. No. 84 *A catalogue of high-velocity stars*
3026	**Eggen O.J.** 1964, AJ 69, 570 *Colors, luminosities, and motions of the nearer G-type stars*
3027	**Eggen O.J., Greenstein J.L.** 1965, ApJ 141, 83 *Spectra, colors, luminosities, and motions of the white dwarfs*
3028	
3029	**Eggen O.J., Greenstein J.L.** 1965, ApJ 142, 925 *Observations of proper-motion stars. II.*
3030	**Eggen O.J.** 1965, AJ 70, 19 *Masses, luminosities, colors, and space motions of 228 visual binaries*
3031	**Eggen O.J.** 1965, Observatory 85, 191 *The Wolf 630 group*
3032	**Eggen O.J.** 1966, R. Obs. Bull. No. 120 *Three-colour photometry of the components in wide double and multiple systems. II. 298 systems*
3033	**Eggen O.J.** 1967, in The Magnetic and related stars p. 141 *Colors, luminosities, and motions of the peculiar A-type stars*
3034	**Eggen O.J.** 1971, ApJ 163, 313 *Narrow- and broad-band photometry of red stars. V. Luminosities and temperatures for young disk-population red stars of high luminosity*
3035	**Eggen O.J.** 1971, ApJ 165, 317 *Narrow- and broad-band photometry of red stars. VI. Luminosities and temperatures for old disk population red stars of high luminosity*

No	Reference
3036	**Eggen O.J.** 1972, ApJ 172, 639 *Narrow- and broad-band photometry of red stars. VII. Luminosities and temperatures for halo-population red stars of high luminosity*
3037	**Eggen O.J.** 1972, ApJ 175, 787 *Luminosities and motion of the F-type stars. II. Metal-deficient stars*
3038	**Eggen O.J.** 1972, ApJ 174, 45 *The classification of intrinsic variable stars. I. The red variables of type N*
3039	**Eggen O.J.** 1972, ApJ 177, 489 *The classification of intrinsic variable stars. II. The red variables of S and related types*
3040	**Eggen O.J.** 1976, PASP 88. 426 *A sample of old-disk-population red giants*
3041	**Eggen O.J.** 1973, PASP 85, 289 *Luminosity and velocity distribution of high-luminosity red stars near the Sun. I. The very young disk population*
3042	**Eggen O.J.** 1973, Mem. R. Astron. Soc. 77, 159 *Some small amplitude red variables of the disk and halo populations*
3043	**Eggen O.J.** 1974, PASP 86, 960 *High-luminosity red stars in or near galactic clusters. Paper I*
3045	**Eggen O.J.** 1974, PASP 86, 429 *A clump of M stars in Velorum-Carina*
3046	**Eggen O.J.** 1973, ApJ 180, 857 *The classification of intrinsic variable. III. Calibration of the luminosities of small amplitude red variables in the old disk population*
3047	**Eggen O.J.** 1975, PASP 87, 107 *Photometric standards on the UBV and RI system*
3048	**Eggen O.J.** 1975, PASP 87, 111 *Photometry of possible barium stars*
3049	**Eggen O.J.** 1949, PASP 85, 42 *Some variables of spectral type K*
3050	**Eggen O.J.** 1972, PASP 84, 757 *Luminosities and motions of A0 to A2 stars*
3051	**Eggen O.J.** 1972, PASP 84, 406 *The red giants in the Hyades group*
3052	**Eggen O.J.** 1973, PASP 85, 379 *Luminosity and velocity distribution of high-luminosity stars near the Sun. II. The young disk giants*
3053	**Eggen O.J.** 1973, PASP 85, 542 *Luminosity and velocity distribution of high-luminosity red stars. III. Old-disk-population giants*
3054	**Eggen O.J.** 1964, PASP 76, 357 *New double stars of small separation*
3055	**Eggen O.J.** 1973, ApJ 184, 793 *The classification of intrinsic variables. IV. Very-small-amplitude, very-short-period red variables*
3056	**Eggen O.J.** 1968, ApJ 153, 195 *Subluminous late-type stars*
3057	**Eggen O.J.** 1977, ApJ 213, 767 *The classification of intrinsic variables. VII. The medium-amplitude red variables*
3058	**Eggen O.J.** 1976, PASP 88, 402 *Bright metallic-line and pulsating A stars*
3059	**Eggen O.J.** 1971, IAU Symp. No. 42, 8 *Red subluminous stars*
3060	**Eggen O.J.** 1968, ApJS 16, 97 *Luminosities, colors, motions, and distribution of faint blue stars*
3061	**Eggen O.J.** 1976, ApJS 30, 351 *Stellar population samples at the galactic poles. III. UBVRI observations of proper motion stars near the south pole and the luminosity laws for the halo and old disk populations*
3062	**Eggen O.J.** 1969, ApJS 19, 57 *Subluminous stars. V. Photometric UBV photometry of southern proper-motion stars*
3063	**Eggen O.J.** 1969, ApJ 158, L31 *SCO XR-1 a member of the Upper Scorpius complex*
3064	**Eggen O.J.** 1968, ApJ 153, 723 *Stellar-population samples at the galactic poles. I. Proper-motion stars, blue objects, and eclipsing binaries near the south pole*
3065	**Eggen O.J.** 1969, ApJ 157, 287 *Subluminous stars. III. Luminosity calibration for subluminous stars and the space density of the blue subluminous stars south of declination -45°*
3066	**Eggen O.J.** 1968, ApJ 152, 77 *Photometric evidence for the existence of a δ Lyrae cluster*
3067	**Eggen O.J.** 1963, ApJ 138, 356 *Color-luminosity array for NGC 752*
3069	**Eggen O.J.** 1977, ApJ 215, 812 *Intermediate band photometry of late-type stars. II.*
3070	**Eggen O.J.** 1974, PASP 86, 129 *Luminosity and velocity distributions of high-luminosity red stars. IV. The G-type giants*
3071	**Eggen O.J.** 1974, PASP 86, 162 *The oldest disk stars*
3072	**Eggen O.J.** 1974, PASP 86, 697 *The luminosity law for late-type main-sequence stars in the solar neighborhood*
3073	**Eggen O.J.** 1974, PASP 86, 742 *A photometric and spectroscopic survey of large proper-motion stars*
3074	**Eggen O.J.** 1977, ApJS 34, 1 *Classification of intrinsic variables. The long-period cepheids*
3075	**Eggen O.J.** 1977, ApJS 34, 33 *UBVRI observations of Magellanic cepheids*
3076	**Eggen O.J.** 1977, ApJS 34, 233 *UBVRI observations of medium-amplitude red variables*
3077	**Eggen O.J.** 1978, ApJS 37, 251 *Intermediate-band photometry of late-type stars. IV. The catalog*
3078	**Eggen O.J.** 1979, ApJS 39, 89 *Catalogs of proper-motion stars. I. Stars brighter than visual magnitude 15 and with annual proper motion of 1".0 or more*
3079	**Eggen O.J.** 1979, ApJS 43, 457 *Catalogs of proper motion stars. II. Stars brighter than visual magnitude 15 and south of declination $+30^o$ with annual proper motion between 0".7 and 1".0*
3080	**Eggen O.J., Sandage A.R.** 1960, MNRAS 120, 79 *Photometry in the Magellanic Clouds. I. Standard sequences*
4001	**Jennens, P.A., Helfer, H.L.** 1975, MNRAS 172, 667. *A new photometric metal abundance and luminosity calibration for field G and K giants*
4002	**Jennens, P.A., Helfer, H.L.** 1975, MNRAS 172, 681. *Photometric metal abundances for twenty clusters*
4003	**Jennens, P.A.** 1975, MNRAS 172, 695. *Metal abundances for G and K giants and subgiants at the south galactic pole*
5002	**Herbst W.** 1977, AJ 82, 902 *Extinction law in dust clouds and the young southern cluster NGC 6250: Further evidence for high values of R*
5004	**Herbst W., Racine R., Warner J.W.** 1978, ApJ 223, 471 *Optical and infrared properties of the newly formed stars in Canis Major R1*
5005	**Schwartz R.D., Noah P.** 1978, AJ 83, 785 *UBVr photometry of Hα emission stars in southern dark clouds*
5006	**Kunkel W.E., Rydgren A.E.** 1979, AJ 84, 633 *A set of faint equatorial standard stars for BVri photometry*
5007	**Humphreys R.M.** 1979, ApJ 231, 384 *M supergiants and the low metal abundances in the Small Magellanic Cloud*
5008	**Landolt A.U.** 1979, ApJ 231, 468 *Contracting stars in the Pleiades*
5009	**Marraco H.G., Rydgren A.E.** 1981, AJ 86, 62 *On the distance and membership of the R CrA T association*
5011	**Rydgren A.E., Vrba F.J.** 1981, AJ 86, 1069 *Nearly simultaneous optical and infrared photometry of T Tauri stars*
5012	**Herbst W., Miller D.P., Warner J.W., Herzog A.** 1982, AJ 87, 98 *R associations. VI. The reddening law in dust clouds and the nature of early-type emission stars in nebulosity from a study of five associations*
5013	**McElroy D.B., Humphreys R.M.** 1982, PASP 94, 828 *UBVri photoelectric sequences for M83 and NGC 5128*
5014	**Forte J.C., Orsatti A.M.** 1984, ApJS 56, 211 *A deep photometric study of IC 5146*
6001	**Moffett T.J., Barnes III T.G.** 1979, PASP 91, 180 *Secondary standards for BVRI photometry*

No	Reference
6002	**Barnes III T.G., Evans D.S., Moffett T.J.** 1978, MNRAS 183, 285 *Stellar angular diameters and visual surface brightness III. An improved definition of the relationship*
6003	**Slovak M.H., Africano J.** 1978, MNRAS 185, 591 *Flickering of the symbiotic variable CH Cygni during outburst*
6004	**Moffett T.J., Barnes III T.G.** 1979, AJ 84, 627 *Equatorial UBVRI photoelectric sequences*
6005	**Barnes III T.G., Moffett T.J.** 1979, PASP 91, 289 *A UBVRI equatorial extinction star network*
6006	**Carney B.W.** 1980, AJ 85, 38 *Southern metal-poor stars: UBVRI photometry*
6007	**Moffett T.J., Barnes III T.G.** 1980, ApJS 44, 427 *Observational studies of cepheids. I. BVRI photometry of bright cepheids*
6008	**Jacoby G.H.** 1981, ApJ 244, 903 *The peculiar planetary nebula Abell 35*
6009	**Parsons S.B., Montemayor T.J.** 1982, ApJS 49, 175 *Ultraviolet and optical studies of binaries with luminous cool primaries and hot companions. II. BVRI observations*
6011	**Moffett T.J., Barnes III T.G.** 1984, ApJS 55, 389 *Observational studies cepheids. II. BVRI photometry of 112 cepheids*
7002	**Chini R., Elsässer H., Neckel Th.** 1980, A&A 91, 186 *Multicolour UBVRI photometry of stars in M17*
7003	**Chini R.** 1981, A&A 99, 346 *Multicolour photometry of stars in the Ophiuchus dark cloud region*
7005	**Neckel Th., Staude H.J.** 1984, A&A 131, 200 *A survey of bipolar and cometary nebulae. Photographic and photometric observations*
7006	**Chini R., Wink J.E.** 1984, A&A 139, L5 *The galactic rotation outside the solar circle*
7008	**Rosseló G., Calafat R., Figueras F., Jordi C., Nunez J., Paredes J.M., Sala F., Torra J.** 1985, A&AS 59,399 *UBVRI photoelectric photometry of nearby stars*
7009	**Rosseló G., Blanch R., Figueras F., Jordi C., Nunez J., Paredes J.M., Sala F., Torra J.** 1987, A&AS 67,157 *UBVRI photoelectric photometry of nearby stars. II.*
7010	**Rosseló G., Figueras F., Jordi C., Nunez J., Paredes J.M., Sala F., Torra J.** 1988, A&AS 75,21 *UBVRI photoelectric photometry of high proper motion stars*
8001	**Mendoza E.E.** 1963, Tonantzintla Tacubaya 3, 137 *Estudio espectroscopico y fotometrico del cumulo estelar en Coma Berenices*
8002	**Mitchell R.I.** 1964, ApJ 140, 1607 *Nine-color photometry of ε Aurigae*
8003	**Johnson H.L.** 1964, Bol Obs. Tonantzintla Tacubaya 3, 305 *The colors, bolometric corrections and effective temperatures of the bright stars*
8004	**Johnson H.L., Mendoza E.E.** 1964, Tonantzintla Tacubaya 3, 331 *The law of interstellar extinction in Orion*
8005	**Mendoza E.E., Johnson H.L.** 1965, ApJ 141, 161 *Multicolor photometry of carbon stars*
8006	**Johnson H.L.** 1965, ApJ 141, 170 *Infrared photometry of M dwarf stars*
8007	**Johnson H.L.** 1965, ApJ 141, 923 *Interstellar extinction in the Galaxy*
8008	**Johnson H.L., Mendoza E.E., Wisniewski W.Z.** 1965, ApJ 142, 1249 *Observations of infrared stars*
8009	**Mendoza E.E.** 1965, Tonantzintla Tacubaya 4, 3 *La ley de extincion interestelar en las Pleyades*
8010	**Mendoza E.E.** 1965, Tonantzintla Tacubaya 4, 51 *Multicolor photometry of TX Camelopardalis - a Hetzler object.*
8011	**Mendoza E.E., Johnson H.L.** 1965, Tonantzintla Tacubaya 4, 57 *An investigation of late-type stars*
8012	**Johnson H.L.** 1965, Comm. Lun. Plan. Lab. 3, 73 No. 53 *The absolute calibration of the Arizona photometry*
8013	**Mendoza E.E.** 1966, ApJ 143, 1010 *Infrared photometry of T Tauri stars and related objects*
8014	**Johnson H.L.** 1966, ApJ 146, 613 *The bolometric corrections and effective temperatures of two giant stars in the globular cluster M3*
8015	**Johnson H.L., Mitchell R.I., Iriarte B., Wisniewski W.Z.** 1966, Comm. Lun. Plan. Lab. 4, 99 *UBVRIJKL photometry of the bright stars*
8016	**Johnson H.L., Mendoza E.E.** 1966, Ann. Astrophys. 29, 525 *The law of interstellar extinction in Perseus*
8017	**Feinstein A.** 1966, Inf. Bull. South. Hemisphere 8, 29 *Photoelectric observations of southern late-type stars*
8019	**Wisniewski W.Z., Wing R.F., Spinrad H., Johnson H.L.** 1967, ApJ 148, L29 *Additional observations of infrared stars*
8022	**Mendoza E.E.** 1967, Tonantzintla Tacubaya 4, 114 *Multicolor photometry of long period variables*
8023	**Mendoza E.E.** 1967, Tonantzintla Tacubaya 4, 149 *Multicolor photometry of stellar aggregates*
8025	**Johnson H.L., MacArthur J.W., Mitchell R.I.** 1968, ApJ 152, 465 *The spectral-energy curves of subdwarfs*
8027	**Mendoza E.E.** 1968, Obs. Astron. Nacional Cerro Calan no 7, 106 *Some characteristics of infrared colors*
8028	**Feinstein A.** 1968, Z. Astrophys. 68, 29 *A survey of southern Be stars*
8029	**Mendoza E.E.** 1969, Tonantzintla Tacubaya 5, 57 *BVRI photometry of 46 southern bright stars*
8030	**Iriarte E.B.** 1969, Tonantzintla Tacubaya 5, 89 *Photometry in UBVRIJHKL of the early main sequence in the Pleiades down to G0V*
8031	**Serkowski K., Gehrels T., Wisniewski W.** 1969, AJ 74, 85 *Wavelength dependence of polarization. XIII. Interstellar extinction and polarization correlations*
8032	**Lee T.A.** 1970, ApJ 162, 217 *Photometry of high-luminosity M-type stars*
8033	**Mendoza E.E.** 1970, Tonantzintla Tacubaya 5, 269 *BVRI-photometry of the brightest stars in the Magellanic Clouds*
8034	**Helfer H.L., Sturch C.** 1970, AJ 75, 971 *UBVRI photometry of north galactic pole K giants*
8035	**Feinstein A.** 1968, Private communication *Photometric UBVRI observations of Be stars*
8036	**Humphreys R.M., Strecker D.W., Ney E.P.** 1971, ApJ 167, L35 *High-luminosity G supergiants*
8037	**Mendoza E.E.** 1971, Tonantzintla Tacubaya 6, 135 *Infrared photometry of V1057 Cygni*
8038	**Sturch C., Helfer H.L.** 1971, AJ 76, 334 *Upgren's unclassified stars A: a new type of G-giant stars?*
8039	**Iriarte B.E.** 1971, Tonantzintla Tacubaya 6, 143 *Infrared photometry of UV Ceti stars*
8040	**Moreno H.** 1971, A&A 12, 442 *On the transfer of photometric systems*
8041	**Sturch C.R., Helfer H.L.** 1972, AJ 77, 726 *UBVRI photometry of north galactic pole K giants. II*
8042	**Frogel J.A., Kleinmann D.E., Kunkel W., Ney E.P., Strecker D.W.** 1972, PASP 84, 581 *Multicolor photometry of the M dwarf Proxima Centauri*
8043	**Sturch C.** 1972, PASP 84, 666 *A direct photometric comparison of M67 and Hyades stars*
8049	**Wawrukiewicz A.S., Lee T.A.** 1974, PASP 86, 51 *The spectral types of the members of VV Cephei systems from photometric analysis*
8050	**Mendoza E.E., Gonzalez S.F.** 1974, Rev. Mex. Astron. Astrofis. 1, 67 *Multicolor photometry of metallic-line stars I. ν^1 Draconis and ν^2 Draconis*
8051	**Humpheys R.M., Ney E.P.** 1974, ApJ 194, 623 *Visual and infrared observations of late-type supergiants in the southern sky*

No	Reference
8052	**Mendoza E.E.** 1974, Rev. Mex. Astron. Astrofis. 1, 175 *Multicolor photometry of metallic-line stars. III. A photometric catalogue*
8053	**Fernie J.D.** 1976, J. Roy. Astr. Soc. Canada 70, 77 *UBVRI observations of miscellaneous stars*
8054	**Penston M.V., Hunter J.K., O'Neill A.** 1975, MNRAS 171, 219 *Further observations of the Orion nebula cluster*
8055	**Lee T.A.** 1968, ApJ 152, 913 *Interstellar extinction in the Orion association*
8056	**Wisniewski W., Coyne G.V.** 1976, Vatican Obs. Publ. 1, 225 *A survey for Hα emission objects in the Milky Way*
8059	**Fernie J.D.** 1976, PASP 88, 116 *Further photometry of cepheid-like supergiants*
8061	**Wegner G.** 1976, MNRAS 177, 3 *On element abundances in stars belonging to the σ Puppis group*
8064	**Pismis P.** 1977, Rev. Mex. Astron. Astrofis. 2, 59 *On an HII condensation in NGC 2175*
8066	**Humphreys R.M.** 1978, ApJ 219, 445 *Luminous variable stars in M31 and M33*
8067	**Humphreys R.M., Warner J.W.** 1978, ApJ 221, L73 *Infrared detection of luminous stars in M31 and M33*
8070	**Natali G., Fabrianesi R., Messi R.** 1978, A&A 62, L1 *Anomalous light curve of Cyg-X1 during the X-ray increase of April-May 1975*
8071	**Mendoza E.E., Gomez V.T., Gonzalez S.** 1978, AJ 83, 606 *UBVRI photometry of 225 Am stars*
8072	**Warman J., Echevarria J.** 1977, Rev. Mex. Astron. Astrofis. 3, 133 *UBVRI photometry of stars in trapezium-type systems*
8075	**Fernie J.D.** 1979, PASP 91, 67 *Photometry of the classical cepheid SU Cygni*
8076	**Barnes T.G., Feckel F.C., Moffet T.J.** 1977, PASP 89, 658 *A possible eclipse in the triple system 20 Leonis*
8078	**Mendoza E.E., Gomez T.** 1980, MNRAS 190, 623 *UBVRI photometry in the field of NGC 2264*
8079	**Turner D.G., Lyons R.W.** 1980, J. Roy. Astron. Soc. can. 74, 12 *Recent observations of HD 127617-Bidelmann's high latitude Be star*
8080	**Gonzalez S.F., Warman J., Pena J.H.** 1980, AJ 85, 1361 *Observational stability in Am stars*
8082	**Slutski V.E., Stal'bovski O.I., Shevchenko V.S.** 1980, Sov. Astron. Lett. 6, 397 *UBVRI stellar photometry in the field of the Taurus dark clouds*
8083	**Fernie J.D.** 1981, ApJ 243, 576 *89 Herculis : Further misdemeanors*
8084	**Echevarria J., Roth M., Warman J.** 1979, Rev. Mex. Astron. Astrofis. 4, 287 *Photometric study of trapezium-type systems*
8085	**Roth M., Echevarria J., Franco J., Warman J.** 1979, Rev. Mex. Astron. Astrofis. 4, 209 *Visual and infrared observations of trapezium-like objects*
8087	**Burkhead M.S., Hutter D.J.** 1981, AJ 86, 523 *A photometric study of NGC 3623, NGC 3627, and NGC 3628*
8088	**Rucinski S.M.** 1981, Acta Astron. 31, 363 *UBVRI photometry of late-type stars*
8089	**Fernie J.D.** 1982, PASP 94, 172 *R Coronae Borealis near maximum light*
8090	**Rucinski S.M.** 1981, A&A 104, 260 *UBVRI photometry of FK Comae*
8091	**Taranov O.G., Yudin B.F.** 1981, Sov. Astron. 25, 710 *Photometry of symbiotic stars in the UBVRJHKLMN system. 2. Z Andromedae*
8092	**Coleman L.A.** 1982, AJ 87, 369 *Linear radii of 51 red giants in galactic clusters*
8094	**Taranov O.G., Yudin B.F.** 1982, Sov. Astron. 26, 57 *Photometry of symbiotic stars in the UBVRJHKLMN system. 3. AX Per, AG Dra, BF Cyg, V443 Her and YY Her*
8095	**Fernie J.D.** 1982, PASP 94, 537 *UBVRI photometry of HR 7308*
8096	**Cardon de Lichtbuer P.** 1982, Vatican Obs. Publ. 2 no 1 *Photoelectric sequences for three open clusters*
8097	**van Gent R.H.** 1982, Inf. Bull. Var. Stars 2140 *Johnson BVR magnitudes for selected comparison stars*
8098	**Liller W., Alcaino G.** 1982, ApJ 257, L27 *UBVRI photometry of the optical counterparts of X-ray sources in Einstein deep survey fields*
8100	**Fernie J.D.** 1983, ApJS 52, 7 *New UBVRI photometry for 900 supergiants*
8103	**Coyne G.V., Wisniewski W., Corbally C.** 1975, Vatican Obs. Publ. 1, 197 *A survey for Hα emission objects in the Milky Way. II. Cygnus*
8105	**Hartwick F.D.A., Crampton D., Cowley A.P.** 1976, ApJ 208, 776 *Subdwarfs among the old disk population*
8106	**Wisniewski W., Coyne G.V.** 1977, Vatican Obs. Publ. 1, 245 *A survey for Hα emission objects in the Milky Way. IV. Cygnus*
8107	**Coyne G.V., Wisniewski W., Otten L.B.** 1978, Vatican Obs. Publ. 1, 257 *A survey for Hα emission objects in the Milky Way. V. Perseus*
8111	**Rucinski S.M.** 1981, Acta Astron. 31, 37 *The light curve of V471 Tauri in the four-colour uvby photometry*
8112	**Carney B.W.** 1983, AJ 88, 610 *A photometric search for halo binaries. I. New observational data*
8113	**Coyne G.V., Mac Connell D.J.** 1983, Vatican Obs. Publ. 2, 73 *A survey for Hα emission objects in the Milky Way. VII. Final zones*
8114	**Shulov O.S., Kopatskaya E.N., Khudyakova T.N., Khozov G.V.** 1983, Tr. Astron. Obs. Leningrad 38, 76 *UBVRI photometry of stars in the Orion nebula cluster*

Table 4: Additional bibliographic reference keys

Star	References	Star	References
CoD -51 10169	976, 1081	HD 4609	2010, 2013
CoD -47 04551	432, 540, 779, 976	HD 4614	1118, 1197, 1355, 1363, 3016, 8003, 8012, 8015
CoD -45 00450	1657, 1673, 2011	HD 4628	1758, 2024, 3078, 8015
CoD -29 02277	3077, 6006	HD 4656	1363, 1509, 3077, 6002, 8003, 8015
		HD 4727	1077, 1203, 1278, 1363, 8015
BD -18 00359	1705, 1783, 3078, 8039	HD 4813	1197, 2030, 3026, 8015
BD -17 00484	1775, 3037	HD 4817	1379, 8100
BD -15 06290	1705, 1783, 3078, 8039	HD 4976	2001, 8033
BD -12 04523	3078, 8006	HD 5015	6001, 8015, 8040
BD -12 02918	3045, 8003, 8006, 8039	HD 5112	3016, 6001, 8015, 8040
BD -12 02669	2033, 3077, 6006	HD 5133	3077, 4001
BD -3 02525	2017, 2033, 3077	HD 5234	3016, 6001, 8015
BD -1 01792	2034, 3077	HD 5394	1363, 8003, 8015
BD -0 02387	1764, 6004	HD 5395	3035, 4001, 6001, 8015
BD -0 00454	1371, 1729, 6004	HD 5516	1363, 3016, 8015
		HD 5737	1034, 1068, 1075, 2012, 2024, 8015
BD +0 00307	5006, 6004	HD 5848	3016, 8015
BD +1 02447	1783, 3072, 8006, 8039	HD 6116	8015, 8071
BD +1 04774	1355, 1509, 1729, 1775, 2024, 3072, 8039	HD 6172	2010, 3021
BD +2 00348	1783, 3078	HD 6186	3016, 4001, 8015
BD +2 03375	1620, 1658, 3077, 8112	HD 6222	2001, 3021
BD +5 01668	3078, 8006	HD 6482	4001, 8015
BD +8 04887	1632, 3072	HD 6536	2010, 2013, 3021
BD +9 02190	1620, 1658, 3077	HD 6582	1077, 1197, 1355, 1758, 3045, 8015, 8025
BD +11 02576	3078, 8006	HD 6595	8015, 8029
BD +13 03683	1696, 3077	HD 6619	2033, 3050
BD +17 04708	1003, 1064, 1620, 1658, 1696, 3077	HD 6763	1013, 1061, 1509, 8015
BD +18 02647	1298, 3027	HD 6767	1075, 2024, 2038, 3078
BD +25 02534	1298, 1298, 3027	HD 6805	4001, 8003, 8015
BD +26 02606	3077, 8112	HD 6860	8003, 8015, 8032
BD +28 04211	1389, 1783, 8040	HD 6961	1566, 1601, 1625, 1648, 1733, 3023, 8015
BD +30 02431	1298, 1783	HD 7087	6001, 8015
BD +63 00137	1775, 3078	HD 7106	4001, 8015
		HD 7187	2010, 2013, 3021
HD 28	2024, 3016, 8015	HD 7318	8003, 8015, 8100
HD 315	1079, 1509, 1729, 3047	HD 7439	2024, 3077
HD 358	3033, 8003, 8015	HD 7570	1075, 2024, 3026, 8015
HD 432	1118, 1197, 3026, 8003, 8015	HD 7706	977, 1075, 1586, 2011
HD 496	8015, 8029	HD 7795	977, 1075, 1586, 2011, 2024
HD 571	1363, 8015, 8040	HD 7804	3023, 8015
HD 886	1034, 1077, 1117, 1119, 1234, 1278, 1363, 3016, 8003, 8007, 8015	HD 7808	3072, 7008
		HD 7895	3024, 7009
HD 1013	1355, 1389, 1390, 1625, 1648, 1733, 3016, 8003, 8015, 8032, 8040	HD 7983	1658, 1774, 3077
		HD 8362	1704, 2011
HD 1038	3055, 8015	HD 8491	3016, 8015
HD 1280	1077, 1203, 1258, 1355, 1363, 3023, 8015	HD 8501	1704, 2011
HD 1326 AB	1355, 1625, 3078, 8006, 8039	HD 8512	2012, 3016, 4001, 8003, 8015
HD 1368	1775, 3077	HD 8538	1502, 1566, 1569, 1601, 1625, 1648, 1733, 3023, 8003, 8015
HD 1404	1203, 1363, 3023, 8015		
HD 1522	4001, 8003, 8015	HD 8651	1311, 2011, 2038
HD 1581	3078, 8015, 8029	HD 8681	977, 1075, 1460, 1628, 2011, 2038
HD 1967	8022, 8027	HD 9053	3055, 8015, 8029
HD 2151	2038, 3016, 8015, 8029	HD 9138	1363, 1620, 3077, 4001, 8003, 8015
HD 2261	2012, 8015, 8029	HD 9270	1034, 1355, 1363, 1566, 1601, 1625, 1648, 1733, 4001, 8003, 8015
HD 2262	8015, 8029		
HD 2475	2032, 3077	HD 9362	2011, 8015, 8029
HD 2490	3025, 8061	HD 9414	1117, 2011, 2024, 2038
HD 2905	1077, 1363, 3016, 8003, 8007, 8015, 8040, 8100	HD 9826	1355, 1363, 1758, 3053, 6001, 8003, 8012, 8015
HD 3196	2030, 3026	HD 9900	4001, 8015
HD 3360	1363, 8015	HD 9927	4001, 6001, 8003, 8015
HD 3369	1363, 3024, 8015	HD 10126	1775, 3077
HD 3443	2017, 3077	HD 10144	8015, 8029
HD 3546	4001, 8003, 8015	HD 10307	1197, 1286, 1355, 1363, 1758, 3026, 4001, 6001, 8003, 8012, 8015
HD 3567	1097, 1594, 1658, 1696, 3077		
HD 3627	3016, 8003, 8015	HD 10380	1425, 3016, 4001, 6001, 8003, 8015
HD 3651	1197, 1355, 1399, 1758, 3077, 8015	HD 10436	1758, 3072
HD 3689	2001, 2010, 2033	HD 10476	1006, 1077, 1197, 1355, 1758, 3079, 4001, 6001, 8015
HD 3712	8003, 8015		
HD 3719	1425, 2001, 2010, 2013, 2038, 3021	HD 10516	1363, 8003, 8007, 8015
HD 3765	1758, 3008	HD 10553	2011, 2038
HD 3817	4001, 8015	HD 10700	1020, 1034, 1075, 1197, 1425, 1509, 2001, 2012, 3078, 6001, 8003, 8015
HD 4065	2012, 3007		
HD 4128	1034, 1075, 1194, 1425, 2012, 3016, 8003, 8015	HD 10761	3016, 4001, 8003, 8015
HD 4150	2024, 3023, 8029	HD 10780	3008, 4001, 8015
HD 4180	1212, 1223, 1363, 8015	HD 11171	1020, 1075, 1425, 2012, 3077, 8015, 8040
HD 4188	3077, 8015	HD 11353	2012, 3016, 4001, 8003, 8015
HD 4362	6009, 8100	HD 11415	1203, 1223, 8003, 8015
HD 4502	8003, 8015	HD 11443	8003, 8015
HD 4599	2013, 3021	HD 11505	1774, 3062

Table 4 1337

Star	References	Star	References
HD 11559	1425, 3016, 8015	HD 19787	3051, 4001, 8015
HD 11636	1197, 1278, 1363, 3023, 8003, 8015	HD 19820	1118, 1379
HD 11695	3055, 8015, 8029	HD 20010	3026, 8015
HD 11973	3024, 8015	HD 20320	1425, 2012, 3023, 8015
HD 12042	2018, 3037	HD 20336	1203, 1212, 1223, 8015
HD 12274	3016, 8003, 8015	HD 20630	1006, 1020, 1034, 1075, 1197, 1258, 1286, 1355,
HD 12311	1075, 1637, 2012, 3023, 8029		1355, 1363, 1409, 1425, 1509, 1758, 2001,
HD 12447	1363, 1425, 3023, 8015, 8052		3077, 8015
HD 12471	3023, 8015	HD 20720	3051, 8015
HD 12533	8007, 8015	HD 20794	8015, 8029
HD 12929	1006, 1034, 1263, 1351, 1363, 3016, 8003, 8015,	HD 20894	2032, 8015, 8100
	8032, 8040, 8071	HD 21120	1355, 1363, 1425, 1509, 1733, 2001, 4001, 8003,
HD 12953	1110, 1119, 6001, 8007, 8015, 8016, 8100		8003, 8015
HD 12993	1119, 1209	HD 21197	2033, 2034, 3072
HD 13051	1212, 8023	HD 21291	1363, 8003, 8007, 8015, 8100
HD 13136	8016, 8023	HD 21364	3023, 8015
HD 13161	3023, 8003, 8015	HD 21389	1363, 8003, 8007, 8015
HD 13267	1119, 1119, 8015, 8100	HD 21447 AB	1355, 1363, 1390, 1502, 1566, 1601, 1733,
HD 13476	1119, 8015, 8100		3023, 8015
HD 13594	1013, 8015	HD 21543	1774, 2012, 3077, 8025
HD 13611	1363, 1509, 3016, 4001, 6001, 8003, 8015, 8100	HD 21754	3016, 8015
HD 13866	8023, 8100	HD 21770	3037, 8015
HD 13959	1632, 3072	HD 21790	2012, 3023, 8015
HD 13974	1197, 1355, 1363, 1758, 3078	HD 22049	1020, 1034, 1075, 1075, 1194, 1197, 1409,
HD 14055	3050, 6001, 8015		1425, 1509, 3077, 8003, 8006, 8015, 8029,
HD 14142	8016, 8023		8033, 8040, 8071
HD 14228	1075, 2012, 3023	HD 22203	2012, 3023, 8015
HD 14322	8007, 8016, 8023	HD 22484	1355, 1363, 1425, 1769, 2030, 3077, 8003, 8015
HD 14386	8022, 8027	HD 22879	1620, 2012, 3077, 8025
HD 14404	1355, 8007, 8016, 8023	HD 22928	1077, 1203, 1363, 8007, 8015
HD 14489	1363, 8031	HD 22951	1252, 1363, 3016
HD 14818	1083, 1110, 8007, 8015, 8016, 8031	HD 23089	3026, 8015
HD 14826	8007, 8016, 8023	HD 23180 AB	1119, 1203, 1234, 1252, 1363, 1718, 3016, 8015
HD 14872	6001, 8015	HD 23230	1355, 3026, 8003, 8015, 8100
HD 15089	3033, 8015	HD 23249	1075, 1194, 1197, 1409, 1425, 2012, 3045, 4001,
HD 15130	3023, 8015		8003, 8015, 8040
HD 15144	1202, 2012, 3033, 8015	HD 24401	5006, 6004
HD 15318	1034, 1075, 1107, 1203, 1258, 1363, 1425,1509,	HD 23817	2038, 3077
	3023, 8015, 8029, 8033, 8040, 8071	HD 24398 A	8015, 8084
HD 15497	1118, 8023	HD 24398 AB	1009, 1234, 1252, 1363, 8084
HD 15596	3077, 4001	HD 24431	1119, 8031
HD 15798	2012, 3026, 8015	HD 24534 A	1212, 1252, 1609, 1733
HD 16031	2013, 3077	HD 24537	1776, 6004
HD 16046	2012, 3023, 8015	HD 24555	8015, 8100
HD 16160 A	1197, 1256, 1355, 1509, 1620, 3072, 4001,8003,	HD 24640	1203, 1252
	8006, 8015, 8029, 8033, 8040	HD 24706	1075, 2011, 2038, 3035
HD 16161	1256, 1355, 1363, 8015	HD 24760	1009, 1077, 1203, 1223, 1234, 1363, 8007, 8015
HD 16417	2032, 3077	HD 24912	1012, 1077, 1119, 1209, 1234, 1252, 1363, 1569,
HD 16582	1223, 1278, 1363, 1425, 8015		3016, 8003, 8007, 8015, 8100
HD 16895	1758, 3026, 8003, 8015	HD 25025	2012, 3016, 8003, 8015, 8032
HD 16970	1075, 1197, 1203, 1363, 1425, 3023, 8003, 8015	HD 25267	3023, 8015
HD 17081	1425, 2012, 8015, 8040	HD 25291	6001, 8015, 8100
HD 17094	1034, 1075, 1258, 1355, 1363, 1425, 1509,	HD 25329	1758, 1775, 14078
	3026, 8015	HD 25457	2012, 3077
HD 17206	1425, 2012, 3026, 8015	HD 25490	2030, 3023, 8015
HD 17361	1502, 1566, 1601, 1625, 1648, 1733, 3016, 4001,	HD 25604	3016, 8003, 8015
	8003, 8015	HD 25642	1363, 3023, 8015
HD 17378	1110, 1355, 1399, 6001, 8015	HD 25653	1657, 1673, 1704, 2011
HD 17506	1363, 8003, 8007, 8015, 8032	HD 25842 A	1460, 1628, 1657, 1673, 1704, 2011,
HD 17573	3023, 8015, 8084	HD 25843	1657, 1673, 1704, 2011
HD 17584	1355, 1363, 3026, 8015	HD 25940	1048, 1212, 1223, 1363, 8015
HD 17709	8003, 8015	HD 26413	1075, 2011, 2038
HD 17878	8003, 8015	HD 26491	2020, 2038, 3077
HD 17925	1197, 2013, 2030, 3077, 8015, 8025	HD 26574	3026, 8015, 8100
HD 18286	1729, 6004	HD 26612	1020, 1075, 2011, 3026, 8015
HD 18322	2012, 3077, 4001, 8003, 8015	HD 26630 A	8015, 8100
HD 18331	1004, 1006, 1007, 1013, 1020, 1034, 1075, 1254,	HD 26722	3016, 8015
	1263, 1351, 1399, 1409, 1425, 1509, 1566, 1625,	HD 26820	1075, 2011, 2038, 3005
	1648, 1733, 2001, 2012, 3047, 8015, 8029,	HD 26846	3053, 8015
	8033, 8040	HD 26912	1223, 1363, 8015, 8040
HD 18449	1502, 1566, 1569, 1601, 1625, 1648, 1733, 3016,	HD 26913	1738, 1407
	4001, 8015	HD 26965	1013, 1075, 1194, 1197, 1311, 1409, 1425,
HD 18520	1363, 3023, 8015, 8023		1509, 2012, 3078, 8015
HD 18604	1509, 8015, 8040	HD 26967	1075, 2011, 3077, 8015
HD 18884	1355, 1363, 1409, 1425, 1509, 3055, 8003,	HD 27022	4001, 8015
	8015, 8032	HD 27045	8015, 8052
HD 18978	1075, 2012, 3023, 8015	HD 27376	2012, 3023, 8015
HD 19305	2033, 3079	HD 27588	1075, 2011, 2038, 3077
HD 19445	1620, 1658, 1774, 3077, 8025	HD 27778	1211, 3016
HD 19476	4001, 8015	HD 27820	1381, 1415, 3050, 8015
HD 19656	4001, 8015	HD 27861	1469, 2012, 3023, 8015

Star	References	Star	References
HD 28497	8028, 8035	HD 35912	8015, 8055
HD 28749	1509, 2013, 2030, 3047, 8015, 8100	HD 36166	8015, 8055
HD 29085	2012, 3016, 8015	HD 36267	8015, 8055
HD 29139	8015, 8032	HD 36285	3047, 8015, 8055
HD 29140	1084, 1199, 1363, 8015	HD 36351 AB	8015, 8055
HD 29503	2012, 8015	HD 36371	1234, 8015, 8100
HD 29573	2027, 3050, 8052	HD 36389	8007, 8015, 8032
HD 29755	2029, 3055, 8015, 8022	HD 36395	989, 1004, 1006, 1075, 1197, 1197, 1355, 1509,
HD 29763	1211, 1223, 1363, 8015		1729, 1775, 2001, 2033, 3078, 8006
HD 30076	1415, 2029, 8028, 8035	HD 36430	2032, 8015, 8055
HD 30211	1425, 2013, 2030, 8015	HD 36486	1084, 1089, 1119, 1211, 1363, 2012, 8003, 8007,
HD 30495	1197, 2031, 3077, 8015		8015, 8055
HD 30545	1415, 1509, 6005	HD 36512	1020, 1034, 1060, 1075, 1203, 1254, 1390, 1409,
HD 30562	2030, 3077		1425, 1509, 1732, 2001, 2012, 2012, 3000, 8007,
HD 30649	1355, 1620, 3026, 8025		8015, 8029, 8033, 8055, 8071
HD 30652	1006, 1020, 1034, 1075, 1197, 1254, 1291, 1355,	HD 36591	1020, 1060, 1075, 1089, 1425,
	1363, 1390, 1425, 1502, 1509, 1569, 1601, 1625,		1509, 1569, 2012, 8015, 8055
	1648, 1733, 3026, 8003, 8015, 8040	HD 36646	8015, 8027, 8055
HD 30739	1363, 1415, 8015	HD 36673	1034, 1075, 1425, 2012, 3023, 8007, 8015
HD 30834	1291, 1363, 3016, 6007, 8015	HD 36695	8015, 8055
HD 30836	1006, 1020, 1034, 1075, 1203, 1254, 1278, 1291,	HD 36741	8015, 8055
	1363, 1509, 1569, 2031, 8015, 8055	HD 36777	3023, 8015
HD 31109	2012, 3023, 8015, 8040	HD 36779	1732, 8015, 8055
HD 31237	1363, 1425, 2012, 8015	HD 36822	1200, 1203, 1363, 8015, 8055
HD 31278	3023, 8015	HD 36861	1200, 1363, 1425, 1509, 8007, 8015, 8055
HD 31295	1363, 3023, 8015	HD 37055	1415, 8015, 8055
HD 31398	1363, 8003, 8007, 8015	HD 37128	1034, 1060, 1075, 1089, 1107, 2001, 2012, 6002
HD 31647	3050, 8015	HD 37160	1363, 1509, 3077, 4001, 8015
HD 31764	1084, 8084	HD 37202	1363, 8015, 8031
HD 31767	1425, 1509, 4001, 8015, 8100	HD 37394	3072, 8015
HD 31910	8015, 8100	HD 37468	1089, 1206, 1425, 8015, 8055, 8084
HD 31964	8002, 8003, 8015	HD 37479	1206, 8084
HD 32045	2012, 3023, 8015	HD 37490	8015, 8055
HD 32147	1415, 1758, 2013, 2030, 3078, 8015	HD 37507	3023, 8015
HD 32249	2012, 8015	HD 37519	1118, 8015
HD 32309	3023, 8015	HD 37742	1060, 1075, 1089, 1425, 2012, 8007, 8015, 8055
HD 32450	2034, 3072	HD 37744	1415, 1732
HD 32549	3023, 8015	HD 37756	1425, 1732
HD 32630	1077, 1203, 1223, 1234, 1291, 1363, 1390, 8003,	HD 37795	1637, 2012, 8015, 8035
	8007, 8015	HD 37984	1363, 3016, 4001, 8015
HD 32778	2012, 2034, 3062	HD 38089	2006, 3030
HD 32887	8003, 8015	HD 38104	1263, 3050, 8015
HD 33054	1415, 3030	HD 38230	1758, 1774
HD 33111	1007, 1013, 1020, 1034, 1075, 1425, 1509, 2001,	HD 38307	8015, 8022
	2012, 2013, 3023, 8003, 8015	HD 38392	2029, 3077
HD 33276	3026, 8015	HD 38393	1075, 2012, 3077, 8015
HD 33328	1425, 2012, 8015	HD 38656	1080, 1118, 1355, 1363, 1502, 1566, 1569, 1569,
HD 33641	8015, 8052		1601, 1625, 1648, 1733, 4001, 8015
HD 33647	8015, 8055	HD 38678	1020, 1034, 1075, 1075, 1409, 1425, 1509, 2012,
HD 33802	2012, 3023, 8015		2012, 3023, 8015, 8040
HD 33856	1425, 2031, 3016, 8003, 8015	HD 38751	3016, 4001
HD 33904	1425, 2012, 3023, 8015	HD 38771	1034, 1075, 1425, 2001, 6002, 8003, 8007,
HD 33949	2012, 3023, 8015		8015, 8055
HD 34029	3051, 8003, 8015	HD 38899	1079, 1203, 1363, 3023, 8015
HD 34085	2031, 3023, 8003, 8007, 8015	HD 39003 A	1363, 1502, 1566, 1569, 1601, 1625, 1648, 1733,
HD 34317	8015, 8055		3024, 8015
HD 34328	3078, 6006	HD 39045	8015, 8032
HD 34411	1197, 1233, 1291, 1363, 3026, 8012, 8015	HD 39357	3023, 8015
HD 34503	1203, 1425, 2012, 8015	HD 39364	3077, 4001, 8015
HD 34673 AB	2012, 3072	HD 39425	2017, 3035, 8015
HD 34748	8015, 8055	HD 39587	1197, 1355, 1363, 1758, 1784, 3077, 6002, 8015
HD 34816	2012, 8015	HD 39764	2012, 3000, 8015
HD 34959	8015, 8055	HD 39801	8007, 8015, 8032
HD 34989	8015, 8055	HD 39881	2024, 3077
HD 35007	1363, 1415, 8015, 8055	HD 40035	1077, 1080, 1118, 1355, 1363, 1502, 1566, 1569,
HD 35039	1425, 1509, 1732, 8015, 8055		1601, 1625, 1648, 1733, 3016, 4001, 8015
HD 35149 A	1363, 1425, 1732, 2012, 8015, 8055	HD 40111	1784, 8015
HD 35165	2012, 8035	HD 40136	1409, 1425, 2012, 3026, 8015
HD 35299	975, 985, 1004, 1006, 1075, 1089, 1203, 1258,	HD 40183	3023, 8003, 8015, 8023
	1399, 1425, 1509, 1569, 1625, 1648, 1732, 1733,	HD 40239	8015, 8100
	2001, 6001, 8015, 8040, 8055	HD 40312	3033, 8015
HD 35369	3016, 8015	HD 40536	1415, 1420, 2012, 3023
HD 35411	2013, 8007, 8015, 8055	HD 40657	3016, 8015
HD 35439	1425, 1732, 8015, 8055	HD 40967 AB	2012, 8015
HD 35446	2033, 3021	HD 41116 AB	8003, 8015
HD 35468	1034, 1075, 1203, 1234, 1291, 1363, 1425, 6002,	HD 41117	1234, 1258, 1363, 1784, 8003, 8007, 8015, 8055
	8007, 8015	HD 41335	1061, 1212, 8035
HD 35497	1203, 1363, 8007, 8015	HD 41695	3023, 8015
HD 35588	8015, 8055	HD 42087	1207, 1217
HD 35640	1415, 8015, 8055	HD 42088	1209, 1775, 1784, 8064
HD 35715 AB	1732, 8015, 8055	HD 42560	1363, 8015

Table 4 1339

Star	References	Star	References
HD 42581	2012, 3072		8015, 8040
HD 43039	1363, 1724, 1784, 3016, 4001, 6007, 8003, 8015	HD 59612	2012, 2016, 8015, 8100
HD 43232	2012, 3016, 8003, 8015	HD 59717	3026, 8015
HD 43384	1207, 6001, 8015	HD 60179	1363, 8003, 8015
HD 43445	1425, 2012, 3023, 8015	HD 60522	1355, 3035, 4001, 6001, 8015
HD 43544	8028, 8035	HD 60532	1075, 1488, 1705, 2012, 3026, 8015
HD 43834	2038, 3077, 8015	HD 60606	2012, 8035
HD 44402	1637, 2012, 8015	HD 60863	1637, 2027, 3023, 8015, 8084
HD 44458	8028, 8028, 8035	HD 61330	3023, 8015
HD 44478	1194, 3053, 6002, 8003, 8015, 8032	HD 61421	1351, 1363, 1425, 1509, 1705,2012, 3078,
HD 44743	1425, 2012, 8015		8003, 8015
HD 44984	8005, 8015, 8022	HD 61429	3023, 8015
HD 44990	1772, 6007	HD 61935	1425, 2012, 3016, 8015
HD 45067	2012, 3077	HD 62044	4001, 8015
HD 45348	2031, 8015	HD 62345	1077, 1080, 1118, 1351, 1355, 3016, 4001, 8015
HD 45542 AP	1212, 1363, 1784, 8015	HD 62509	1363, 2001, 8003, 8015, 8032
HD 45572	977, 977, 1075, 2011, 2038	HD 62576	8015, 8100
HD 46300	1363, 3023, 6001, 8015, 8100	HD 62623	2012, 2016, 8015, 8100
HD 46407	8005, 8015	HD 62721	3035, 8003, 8015
HD 46652	977, 1066, 1075, 2011	HD 63005	1295, 1376
HD 47105	1107, 1203, 1286, 1363, 3023, 6002, 8003,	HD 63077	2017, 3078
	8015, 8040	HD 63462	1732, 2012, 8015, 8035
HD 47174	1363, 3016, 6001, 8003, 8015, 8100	HD 63700	2012, 8015, 8100
HD 47205	1425, 2012, 3016, 8003, 8015	HD 63975	3023, 8100
HD 47240	1425, 1509, 2012, 8100	HD 64090	1658, 1758, 3078
HD 47442	1075, 1425, 2012, 3000, 8003, 8015, 8100	HD 64096	2012, 3077, 8015
HD 47667	3077, 8015, 8100	HD 64606	2012, 3077
HD 47670	1075, 2011, 3023, 8015	HD 64633	1729, 6004
HD 48069	1673, 2011	HD 64854	5006, 6004
HD 48099	1020, 1075, 1209, 1425, 1509, 6001, 8015	HD 65228	1425, 2012, 2014, 3026, 8015, 8100
HD 48329	1363, 6002, 8003, 8015	HD 65456	2016, 8015
HD 48433	3016, 6001, 8015	HD 65583	3078, 8025
HD 48464	1673, 1704, 2011	HD 65810	1075, 1075, 1425, 2012, 3023, 8015
HD 48855	1704, 2011	HD 65875	1078, 1212, 8035
HD 48915 A	1034, 1705, 2031, 3078, 8003, 8015, 8023	HD 65953	2029, 3016, 8015
HD 49161	1363, 3052, 8041	HD 66141	1080, 1363, 2035, 3001, 8003, 8015
HD 49293	1363, 1425, 1509, 2012, 3016, 8015	HD 66811	1034, 1075, 1209, 1637, 2012, 8003, 8007, 8015
HD 49798	1673, 1704, 1732	HD 67523	3026, 8003, 8015
HD 50013	2012, 8015, 8028, 8035	HD 67594	1084, 1119, 1425, 1509, 2035, 8003, 8015, 8100
HD 50019	1355, 1363, 3023, 8015	HD 67888	1637, 2027, 8035
HD 50196	2011, 2038	HD 68273	2012, 8015, 8029
HD 50209	1729, 6004	HD 68290	3016, 8003, 8015
HD 50223	1075, 2011, 2038, 3026	HD 68553	8015, 8100
HD 50241	1409, 2012, 3023	HD 68601	3023, 8015
HD 50281 A	2033, 3072	HD 68980	1075, 1637, 2012, 8015, 8035
HD 50778	1425, 3077, 8015	HD 69082	1637, 2031
HD 50785	1075, 2011, 2038	HD 69267	985, 1004, 1006, 1007, 1020, 1034, 1075, 1077,
HD 50860	1075, 2011, 2038		1080, 1254, 1263, 1355, 1363, 1390, 1409, 1425,
HD 50973	3023, 8015		1502, 1509, 1566, 1569, 1601, 1625, 1648, 1733,
HD 51754	3077, 6006		2001, 3047, 8003, 8015,8029, 8032, 8033,
HD 51825	2024, 3037		8040, 8071
HD 52362	1075, 2011, 2038	HD 69830	1197, 2029, 3078
HD 52382	1737, 8100	HD 69897	1363, 3037
HD 52437	1637, 1732, 2012, 8028, 8035	HD 70060	2012, 3023, 8015
HD 52918	1415, 1509, 2035, 8015	HD 70272	3016, 8015
HD 53501	2038, 3077	HD 70442	2031, 8015
HD 53623	740, 1732	HD 70958	2012, 3026
HD 53975	1020, 1075, 1203, 1425, 8040	HD 71115	3032, 8015, 8100
HD 54309	1637, 2007, 8035	HD 71155	1034, 1075, 1203, 1330, 1355, 1425, 1509, 2001,
HD 54361	1238, 3038		3023, 8015
HD 54662	1075, 1203, 1209, 1425, 2027	HD 71369	3051, 4001, 8015, 8100
HD 54719	6001, 8003, 8015	HD 72067	1637, 2009, 8035
HD 54810	2012, 3077, 4001, 8003, 8015	HD 72436	2012, 2012
HD 55185	1425, 2035, 3023, 6001, 8015	HD 72905	3026, 8015
HD 56014	1637, 8015, 8028, 8028, 8035	HD 72968	2013, 2027
HD 56096	8022, 8029	HD 73108	4001, 6001, 8015
HD 56139	2012, 8015, 8035	HD 73262	1509, 3023, 8015
HD 56456	3023, 8015	HD 73471	1425, 3016, 8015
HD 56537	1007, 1013, 1022, 1355, 1363, 2001, 3023, 8015	HD 73634	3023, 8015, 8100
HD 56813	2029, 3005	HD 73667	1758, 3077
HD 56986	1363, 3026, 8003, 8015	HD 74000	1775, 2012, 3077, 6006
HD 57146	6009, 8100	HD 74137	1425, 2012, 8015
HD 57150	1732, 2024, 2031, 8015, 8035	HD 74180	1075, 2012, 2031, 3026, 8015, 8100
HD 57219	2012, 8015	HD 74272	3023, 8015
HD 57682	2013, 2028, 3047	HD 74273	2012, 2012
HD 57821	1079, 1425, 2012, 3023, 8015	HD 74280	985, 1004, 1006, 1020, 1034, 1075, 1107, 1203,
HD 58207	1363, 3016, 4001, 8015		1254, 1263, 1363, 1425, 1502, 1509, 1566, 1569,
HD 58367	1355, 1363, 1415, 4001, 8015		1601, 1625, 1648, 1732, 1733, 2001, 3047, 8015,
HD 58526	6009, 8100		8029, 8033, 8040, 8071
HD 58715	1415, 3023, 8015	HD 74371	2012, 8100
HD 58946	1118, 1197, 1330, 1351, 1355, 1620, 3072, 8003,	HD 74395	2012, 6001, 8015, 8100

Star	References
HD 74575	2012, 8015, 8100
HD 74739	1363, 3032, 6001, 8015
HD 74874	1363, 3030, 8015
HD 74918	3016, 8015
HD 74956	3023, 8015
HD 75021	3039, 8005
HD 75063	3023, 8015
HD 75149	1018, 2012, 8100
HD 75276	2012, 8100
HD 75311	2006, 8023, 8035
HD 75530	2017, 3077
HD 75732	1758, 3077
HD 75759	540, 976, 2006
HD 75821	8015, 8100
HD 75914	1728, 1729, 6004
HD 76082	1729, 2012, 6004
HD 76151	1758, 2012, 3077
HD 76294	3016, 4001, 8015
HD 76556	976, 1075, 1737
HD 76566	2012, 2012
HD 76644	1118, 1197, 1355, 3023, 8003, 8015
HD 76728	2012, 3023, 8100
HD 76756	8015, 8052
HD 76838	976, 2012
HD 76943	985, 1077, 1118, 1197, 1351, 1355, 1363, 1502, 1566, 1569, 1601, 1625, 1648, 1733, 3026, 8015
HD 76968	2012, 8100
HD 77140	2012, 3023
HD 77250	3077, 8100
HD 77320	815, 976, 1637, 2012, 8035
HD 77350	1202, 3023, 8015
HD 77581	1034, 1737, 8100
HD 77912	4001, 8015, 8100
HD 78045	2038, 3023, 8052
HD 78154	1355, 3026, 8015
HD 78344	2021, 8100
HD 78541	2012, 3016, 8015
HD 78616	1075, 2011
HD 78647	2011, 8015
HD 78764	2031, 2038, 8035
HD 78791	3026, 8059, 8100
HD 79096	1197, 1351, 1758, 3077, 8041
HD 79186	540, 976, 1018, 1034, 1075, 1509, 1737, 1770, 2011, 8015, 8100
HD 79275	540, 976, 1075, 1732, 1770, 2011, 2038
HD 79351	1075, 2012, 8023
HD 79416	388, 977, 977, 1075, 1637, 2011, 2038
HD 79439	3023, 8015
HD 79447	2012, 8023
HD 79469	1075, 1203, 1355, 1363, 1425, 1509, 2013, 3023, 6001, 8015
HD 79621	1637, 1770, 2011
HD 79694	388, 977, 977, 1075, 1770, 2011, 2038
HD 79735	1075, 1770, 2011, 2038
HD 79900	977, 977, 1075, 1637, 1770, 2011
HD 79910	2029, 3016
HD 79931	2027, 3016
HD 79940	3026, 8015
HD 80007	2038, 3023, 8015
HD 80057	540, 976, 1075, 1770, 2011, 8100
HD 80108	2011, 8100
HD 80404	2012, 3023
HD 80484	1586, 2011
HD 80493	6001, 8003, 8015
HD 80499	3016, 8015
HD 80527	460, 977, 977, 1066, 1075, 1586, 2011
HD 80558	1018, 1034, 1075, 2012, 8100
HD 80586	2012, 3024, 8015
HD 80874	3016, 8015
HD 81136	389, 977, 977, 1075, 2011, 2038
HD 81188	2012, 8015, 8023
HD 81414	1673, 1704, 1770, 2011
HD 81610	1673, 1704, 1770, 2011
HD 81797	1075, 1119, 1194, 1409, 1425, 2012, 3016, 8015
HD 81937	3024, 8015
HD 81997	1075, 1197, 1425, 2033, 3026, 8015
HD 82106	1509, 2033, 3072
HD 82121	1770, 2001, 2011
HD 82152	1311, 2011
HD 82207	2015, 3026
HD 82210	3026, 8015

Star	References
HD 82224	1770, 2011
HD 82241	2015, 3026
HD 82308	8015, 8041
HD 82328	1118, 1119, 3078, 8015
HD 82347	2013, 2029
HD 82434	3026, 8015
HD 82446	1509, 3023, 8015
HD 82455	2011, 6006
HD 82514	2028, 3005
HD 82885	1077, 1080, 1118, 1197, 1351, 1355, 1569, 1758, 3026, 4001, 6001, 8015
HD 83425	1105, 1363, 1425, 4001, 8015
HD 83548	565, 993, 1075, 2011, 2038
HD 83618	1509, 3077, 8015
HD 83754	1425, 2012, 8015
HD 83944	3023, 8023
HD 83953	1212, 1637, 2012, 8015, 8028, 8035
HD 84367	3026, 8015
HD 84441	1194, 1363, 1462, 8003, 8015, 8059, 8100
HD 84737	1197, 1363, 3026, 8012, 8015
HD 84748	8022, 8027
HD 84816	2012, 8023
HD 84937	1097, 1620, 1658, 3077
HD 84971	1729, 6005
HD 84999	1118, 3023, 8015
HD 85444	1425, 2012, 3016, 8015
HD 85990	2012, 6004
HD 86135	2012, 6004
HD 86408	5006, 6004
HD 86440	1075, 2001, 2012, 8023, 8100
HD 86612	2012, 8035
HD 86663	1194, 1355, 1363, 3016, 8015
HD 87152	1075, 2012, 2029
HD 87504	2012, 3023, 8015
HD 87696	1013, 1077, 1118, 1355, 3023, 8015
HD 87737	3023, 8015, 8100
HD 87837	6002, 8015, 8041
HD 87887	1509, 3023, 8015
HD 87901	1020, 1022, 1034, 1079, 1084, 1107, 1203, 1363, 3024, 6002, 8003, 8007, 8015
HD 88230	1625, 1722, 3078, 8006, 8025
HD 88284	2012, 4001, 8015
HD 88661	730, 815, 1637, 2012, 8035, 8035
HD 88725	3026, 6006
HD 88825	2012, 8035
HD 89025	1013, 1118, 1119, 1363, 3026, 8015, 8100
HD 89080	1637, 2038, 3023, 8023
HD 89388	3005, 8017, 8100
HD 89449	1077, 1351, 1355, 1363, 2001, 3026, 8015
HD 89484	8003, 8015, 8032
HD 89668	2033, 3077
HD 89758	8003, 8015
HD 89822	3023, 8015
HD 89890	1637, 2012
HD 90277	3023, 8015
HD 90362	3001, 6001, 8015
HD 90589	2038, 3026
HD 90839	1197, 1351, 1355, 1758, 3072, 6001, 8015
HD 90994	1203, 1363, 1425, 8015
HD 91316	1006, 1020, 1075, 1077, 1119, 1363, 1425, 1509, 1569, 6001, 8003, 8007, 8015, 8100
HD 91480	1363, 3026, 8015
HD 91533	2012, 8100
HD 91612	1363, 1509, 8100
HD 91619	1018, 1034, 2012, 8100
HD 92055	8015, 8022, 8027
HD 92305	2006, 2038, 3055
HD 92449	2012, 3077, 8059, 8100
HD 92839	8022, 8027
HD 92964	815, 1018, 1034, 2001, 2012, 8100
HD 93813	2012, 3016, 6001, 8003, 8015
HD 93843	1034, 3021
HD 94028	3077, 8025
HD 94264	1363, 3035, 4001, 8015
HD 94603	1770, 1775, 1783, 6004
HD 94616	1729, 6004
HD 94683	1075, 2013, 2029, 3005
HD 94808	1729, 6004
HD 95128	1025, 1197, 1363, 3026
HD 95272	1425, 2012, 3077, 4001, 8015
HD 95345	1417, 1509, 3016

Table 4   1341

Star	References	Star	References
HD 95418	1363, 3023, 8015	HD 105937	8015, 8023
HD 95578	1509, 3016, 8015	HD 106038	1620, 1696, 3077, 8112
HD 95689	1080, 1084, 1118, 1363, 3016, 8003, 8015	HD 106068	2012, 2012, 8100
HD 95735	1355, 1569, 1625, 3078, 8003, 8006	HD 106116	1531, 1620, 1658, 2012, 3026
HD 96097	3026, 8015	HD 106321	1075, 2011
HD 96833	1363, 4001, 8003, 8015	HD 106343	1684, 2001, 2012, 2012, 2038, 8100
HD 96917	1737, 8100	HD 106456	2011, 6006
HD 96918	2001, 2012, 8059, 8100	HD 106516	1311, 1417, 1658, 2012, 3078
HD 96919	2012, 2016, 8100	HD 106572	2028, 3077
HD 97277	3023, 8015	HD 106591	1006, 1007, 1013, 1077, 1118, 1197, 1363, 1502,
HD 97503	1729, 3072		1566, 1569, 1601, 1625, 1648, 1733, 3023, 6001,
HD 97534	2012, 3026, 8059, 8100		8003, 8015
HD 97603	1363, 3023, 8003, 8015	HD 106625	1020, 1034, 1075, 1079, 1425, 1509, 2012, 3023,
HD 97633	1363, 3050, 6001, 8003, 8015		8007, 8015, 8097
HD 97707	976, 2021, 8100	HD 106661	1569, 3050
HD 98118	3047, 6001, 8015	HD 106760	3035, 8034
HD 98231	1197, 1355, 3026, 8015	HD 106902	1586, 1770, 2001, 2011, 2024
HD 98262	1355, 3016, 4001, 8015	HD 106983	2012, 8023
HD 98430	2012, 3016, 8015	HD 107259	1425, 1509, 3023, 8015
HD 98664	3023, 8015	HD 107328	1417, 1509, 3026, 4001, 6001, 8015
HD 98718	2012, 8023	HD 107383	6001, 8015
HD 98839	4001, 6001, 8015, 8100	HD 107399	8001, 8023
HD 99028	1363, 1477, 3026, 8015	HD 107468	8023, 8034
HD 99171	2012, 3017	HD 107773	2038, 3077
HD 99211	3023, 8015	HD 107950	6001, 8015
HD 99383	2012, 2039, 3077, 6006	HD 108002	976, 1684
HD 99648	1363, 1417, 3016, 4001, 8015	HD 108177	1658, 3077, 8112
HD 100198	2029, 8100	HD 108483	2012, 8015
HD 100199	1021, 8100	HD 108501	2012, 2018, 2038
HD 100261	2001, 2013	HD 108530	2006, 3077
HD 100262	1075, 2001, 2012	HD 108639	1684, 1688
HD 100363	3077, 6006	HD 108767	1425, 2012, 3024, 6001, 8003, 8007, 8015
HD 100407	3077, 8015	HD 109085	2012, 3026, 8015
HD 100600	985, 1006, 1034, 1502, 1566, 1569, 1586, 1601,	HD 109358	1197, 1351, 1355, 1363, 3026, 8012, 8015
	1625, 1648, 1733, 6001, 8015, 8050, 8071	HD 109379	2012, 8015
HD 100889	1425, 2012, 3023, 8015	HD 109485	3023, 6001, 8015
HD 100920	1425, 2006, 3016, 6002, 8015	HD 109536	8071, 8080
HD 101431	3023, 8015	HD 109799	2012, 3026, 8015
HD 101501	1197, 3026, 4001, 6001, 8012, 8015	HD 109867	1075, 1586, 1684, 1737, 2012, 2038
HD 102070	1425, 2012, 4001, 8015	HD 109914	8005, 8022, 8027
HD 102212	1256, 1355, 1363, 3016, 8015, 8032	HD 110014	3077, 8015
HD 102232	977, 1075, 1770, 2001, 2011, 3023	HD 110066	1240, 3050
HD 102249	2006, 2024, 2038, 3023	HD 110073	3023, 8015
HD 102350	8059, 8100	HD 110304	3023, 8015
HD 102461	2029, 3005	HD 110379	1075, 1075, 1088, 1119, 1197, 1202, 1351, 1363,
HD 102647	1022, 1034, 1363, 2001, 3023, 8003, 8015		1509, 3030, 8003, 8015
HD 102776	1075, 1637, 2012, 8023	HD 110411	6001, 8015
HD 102870	1006, 1007, 1034, 1075, 1197, 1254, 1351, 1363,	HD 110432	1588, 1637, 1684, 1688, 2006
	1409, 1425, 1509, 2001, 3045, 6001, 8003, 8015	HD 110646	3016, 4001, 8015
HD 102964	1020, 1075, 1637, 1637, 2001, 2011, 2024, 8015	HD 110879	2012, 2038, 8023
HD 103079	2038, 8023	HD 110897	1197, 3026, 8015
HD 103095	792, 908, 1006, 1077, 1080, 1118, 1197, 1355,	HD 110956	8015, 8023
	1658, 3078, 6001, 8003, 8015, 8050, 8071	HD 111123	2012, 8015
HD 103287	1007, 1013, 1077, 1118, 1203, 1286, 1363, 1601,	HD 111124	1688, 1737
	1648, 1733, 3023, 8003, 8015	HD 111133	1417, 1509, 1531, 6005
HD 103481	1075, 1586, 2011	HD 111397	1022, 1068, 3050, 8015
HD 103646	5006, 6004	HD 111631	1197, 1509, 1531, 1620, 1758, 2001, 3072, 8006
HD 103729	1704, 1770, 2001, 2011	HD 111812	1080, 1119, 1363, 1690, 3016, 6001, 8015
HD 103746	1075, 1770, 2001, 2011, 2038, 3021	HD 111884	2028, 3077
HD 103779	1684, 8100	HD 112033	3024, 4001
HD 104035	2038, 8100	HD 112078	1075, 1637, 2012, 8023
HD 104304	1417, 1509, 2013, 2029, 3077	HD 112091	1075, 1075, 1637, 2012, 8023
HD 104321	6001, 8015	HD 112092	8015, 8023
HD 104631	1684, 8100	HD 112142	6002, 8015
HD 104671	3023, 8052	HD 112185	1363, 3033, 8003, 8015
HD 104720	1673, 1704, 1770	HD 112300	1363, 1425, 2029, 3025, 8015, 8032
HD 104731	977, 977, 1075, 1311, 1770, 2011, 2038, 3026	HD 112364	1684, 8100
HD 104800	1620, 3077	HD 112412	8003, 8015
HD 104878	2016, 2038	HD 112413	1202, 1240, 1263, 1363, 3023, 8003, 8015
HD 104979	1363, 4001, 8015	HD 112842	1684, 2012
HD 105004	3025, 6006	HD 112943	1705, 3077
HD 105071	1732, 2038	HD 112989	1692, 3041
HD 105313	1628, 1657, 1673, 1704, 1770, 1776, 2011	HD 113012	2021, 8100
HD 105382	2024, 8015, 8023	HD 113120	1637, 1771, 2038
HD 105416	977, 977, 1075, 1637, 1770, 2011, 2038	HD 113139	1007, 1013, 1077, 1118, 1351, 3026, 6001, 8015
HD 105435	1637, 1770, 2006, 8015, 8023	HD 113226	3077, 8003, 8015, 8100
HD 105452	1013, 1075, 1637, 2012, 3026, 8015	HD 113703	8015, 8023
HD 105498	1770, 2001, 2011	HD 113791	8015, 8023
HD 105509	1075, 1770, 2011, 2024	HD 113797	1769, 3023
HD 105707	1075, 2012, 3016, 8015	HD 113865	1690, 3050
HD 105852	977, 1066, 1075, 1586, 2001, 2011, 2038, 3021	HD 113996	1690, 3016, 6001, 8015

Star	References	Star	References
HD 114038	3016, 4001, 8015	HD 125072	2013, 3008
HD 114094	1783, 3077	HD 125241	540, 976
HD 114122	1737, 8100	HD 125288	2014, 8100
HD 114330	1425, 2012, 3023, 8015	HD 125337	1351, 2012, 3023, 6001, 8015
HD 114340	1684, 1737, 8100	HD 125351	6001, 8015
HD 114378	1381, 3030	HD 125451	3026, 8015
HD 114529	2035, 3023	HD 125642	3050, 8015
HD 114606	1658, 1773, 3016	HD 125823	2012, 8015
HD 114642	2007, 3037	HD 125932	3077, 8015
HD 114710	667, 1006, 1019, 1077, 1118, 1197, 1351, 1690,	HD 126053	1197, 1256, 1620, 1658, 3077, 8015
	3078, 6001, 8003, 8012, 8015	HD 126341	8015, 8100
HD 114762	1658, 3026	HD 126354	2001, 2011, 3026, 8015
HD 114873	2013, 2029, 3005, 8100	HD 126504	1770, 2011, 2024, 2038
HD 115043	1119, 1351, 3026, 8023	HD 126660	1197, 1351, 3031, 8015
HD 115114	1684, 1737	HD 126759	2014, 2024
HD 115383	1197, 1256, 1363, 1509, 1531, 3077	HD 126769	2012, 2016
HD 115539	1569, 3077	HD 126868	1509, 6001, 8015
HD 115604	1363, 3030, 6001, 8015	HD 126981	2038, 3023
HD 115617	130, 244, 418, 814, 1006, 1034, 1075, 1351,	HD 127339	2012, 3072
	1425, 1509, 1569, 1586, 2001, 2012, 3078, 8015	HD 127665	1080, 1118, 3016, 4001, 8015
HD 115659	1020, 1075, 1088, 2012, 4001, 8003, 8015	HD 127762	1118, 3023, 6001, 8003, 8015
HD 115842	1075, 1737, 2012, 8100	HD 127864	1770, 2011, 2014
HD 115892	3023, 8015	HD 127972	1075, 1212, 1637, 2011, 8015
HD 116003	1684, 8100	HD 128020	2038, 3077
HD 116064	3077, 6006	HD 128068	1311, 2001, 2011, 2038, 3005
HD 116656	3023, 8015	HD 128167	1197, 1351, 1363, 3037, 6001, 8015
HD 116658	1006, 1020, 1034, 1068, 1075, 1351, 1425, 1509,	HD 128413	863, 977, 977, 1066, 1075, 1586, 2011, 2024
	2012, 8003, 8007, 8015	HD 128429	1003, 1311, 1351, 2012
HD 116842	1013, 1077, 1118, 1502, 1566, 1569, 1601,	HD 128582	1311, 1770, 2011, 2038
	1625, 1648, 1733, 3020, 8015	HD 128726	977, 1066, 1075, 1586, 1770, 2011
HD 116976	2012, 3077, 4001, 8015	HD 128898	2024, 2038, 3023, 8015
HD 117024	1684, 2012	HD 129056	1075, 2001, 2001, 2011, 8015
HD 117176	667, 1006, 1080, 1351, 1363, 1569, 1733, 3077,	HD 129116	2012, 8015
	4001, 6001, 8015	HD 129247	1363, 1381, 3030, 8015
HD 117287	8015, 8022, 8027	HD 129474	1673 , 1704, 1770
HD 117635	1620, 1658, 3077	HD 129502	1197, 2012, 3026, 8015
HD 117675	2012, 3055, 8015	HD 129557	1732, 2012
HD 117876	3077, 8090	HD 129660	1704, 1770, 2011
HD 118022	1013, 1075, 1075, 1088, 1202, 1263, 1363,	HD 129688	1704, 1770, 2011
	3033, 6001, 8015	HD 129712	8022, 8027
HD 118098	1088, 1363, 3023, 8015	HD 129857	1704, 1770, 2011
HD 118216	3026, 6001, 8015	HD 129858	2011, 2038
HD 118246	1775, 6005	HD 129956	1509, 6005
HD 118579	5006, 6004	HD 129989	8003, 8015
HD 118623	6001, 8015	HD 130073	2011, 2038
HD 118716	1075, 2012, 8015, 8023	HD 130109	1006, 1020, 1034, 1075, 1088, 1107, 1203, 1351,
HD 119228	6001, 8015		1363, 1509, 2001, 3023, 8015
HD 119605	3016, 8015, 8059, 8100	HD 130227	2028, 3009, 8061
HD 120136	1197, 1351, 1363, 3026, 8015	HD 130694	8015, 8061
HD 120307	1075, 2012, 2014, 8015	HD 130697	2001, 2011
HD 120315	1077, 1118, 1203, 1234, 8007, 8015	HD 130807	1637, 1770, 2001, 2011, 8015
HD 120323	8003, 8011, 8015, 8017, 8022	HD 130819	1075, 1084, 1088, 1509, 1586, 2012, 3026, 6001,
HD 120324	2032, 8015, 8023		8003, 8015
HD 120477	3016, 8015	HD 130841	1034, 1075, 1084, 1088, 1770, 2012, 3023, 6001,
HD 120818	1118, 8015		8003, 8015, 8052
HD 121370	985, 1006, 1077, 1080, 1118, 1197, 1351, 1363,	HD 130904	2011, 2024
	1502, 1566, 1569, 1601, 1625, 1648, 1733, 3026,	HD 131156	1013, 1197, 1351, 1363, 3077, 8015
	8003, 8015	HD 131657	2011, 2038
HD 121743	2012, 8015	HD 131923	2012, 2020, 3077
HD 121790	2012, 8015	HD 131977	8006, 8015
HD 122196	2012, 3077, 6006	HD 132052	1075, 1088, 3026, 6001, 8015
HD 122223	8015, 8059, 8100	HD 132058	1770, 2011, 8015
HD 122408	1020, 1034, 1075, 1088, 1250, 1363, 1509, 2001,	HD 132142	1705, 3078
	3024, 8015	HD 132200	2001, 2011, 8015
HD 122563	3016, 4001, 8025	HD 132475	2012, 3077
HD 122862	2038, 3077	HD 132813	8022, 8027, 8032
HD 122879	976, 1075, 2014, 2027, 8100	HD 133124	4001, 8015
HD 122980	2012, 8015	HD 133165	1363, 8015
HD 123123	8003, 8015	HD 133200	2001, 2011, 8015
HD 123139	2012, 3077, 8015	HD 133216	1075, 2012, 3055, 4001, 8003, 8011, 8015, 8022
HD 123255	1256, 1351	HD 133242	2011, 8015
HD 123299	1363, 1502, 1566, 1569, 1601, 1648, 1733,	HD 133937	8015, 8023
	3033, 8015	HD 133955	2012, 3030, 8015
HD 123569	2012, 3077	HD 134083	1118, 1351, 3026, 6001, 8015
HD 123657	8015, 8032	HD 134439	1658, 1705, 1773, 2012, 3078, 6006
HD 123999	6001, 8015	HD 134440	1773, 1775, 2012, 3078, 6006
HD 124224	1088, 1363, 3047, 6001, 8015	HD 134481	1770, 3023
HD 124294	4001, 8015	HD 134687	8015, 8023
HD 124367	1075, 1637, 2027	HD 134759	1586, 2012, 3033, 8015
HD 124850	1075, 1088, 2001, 2012, 3077, 8015	HD 135204	1311, 2012, 3078
HD 124897	1351, 1363, 1705, 3078, 8003, 8015, 8032, 8061	HD 135722	1084, 1118, 3024, 4001, 8015

Table 4 1343

Star	References	Star	References
HD 135734	2016, 8015	HD 144608	2012, 8015
HD 135742	1020, 1034, 1075, 1079, 1088, 1203, 1509, 2001, 2012, 3023, 8003, 8007, 8015, 8071, 8097	HD 144969	1737, 8100
HD 136003	976, 8100	HD 145148	1080, 1256, 1620, 3077, 4001
HD 136202	1075, 1088, 1363, 1509, 1586, 3032, 8015	HD 145328	4001, 6001, 8015
HD 136239	1737, 8100	HD 145389	6001, 8015
HD 136352	2032, 3078	HD 145417	2033, 3078, 6006
HD 137006	1586, 3016, 8015	HD 145482	1770, 2012, 8015
HD 137107	1118, 1197, 1381, 3030, 8015	HD 145502	1088, 1770, 2012, 8015
HD 137391	3026, 8015	HD 145570	3023, 6001, 8015, 8052
HD 137613	3035, 8005	HD 146001	1770, 2006
HD 137759	1080, 1118, 1379, 3016, 4001, 6001, 8003, 8015	HD 146051	8003, 8015
HD 137909	1363, 3023, 8015	HD 146143	2001, 2012, 8100
HD 138481	1363, 3016, 8015	HD 146233	1013, 1197, 1256, 2024, 2028, 3077
HD 138485	6001, 8023	HD 146332	1775, 8100
HD 138716	2012, 4001, 8015	HD 146624	3023, 8015
HD 138764	6001, 8015	HD 146686	2012, 3077, 8015
HD 138769	2012, 8015	HD 146791	2012, 4001, 6001, 8003, 8015
HD 139006	1022, 1363, 3023, 8015	HD 147049	1737, 8100
HD 139127	2012, 8015	HD 147084	2024, 8015, 8100
HD 139365	2012, 8015	HD 147165	2012, 3024, 8003, 8007, 8015
HD 139663	6002, 8011, 8015	HD 147318	540, 976
HD 139664	2024, 3026, 8015	HD 147331	432, 540, 976
HD 139997	2012, 8015	HD 147379	3072, 8006
HD 140159	1381, 3050, 8015	HD 147394	985, 1006, 1077, 1118, 1203, 1363, 1399, 1502, 1566, 1569, 1601, 1625, 1648, 1733, 3016, 8007, 8015, 8071
HD 140160	3050, 8015		
HD 140283	1064, 1509, 1594, 1658, 1774, 2033, 3078, 6006, 8003, 8025	HD 147449	1363, 1509, 3026, 6001, 8015
HD 140436	3023, 8015	HD 147513	2032, 3026
HD 140573	985, 1006, 1020, 1034, 1075, 1080, 1088, 1363, 1502, 1509, 1566, 1569, 1601, 1648,1733, 2001, 3077, 8003, 8015, 8032, 8071	HD 147547	1118, 2001, 3024, 8015
		HD 147584	3026, 8015
		HD 147617	540, 976, 1021
HD 140775	3047, 6001, 8015	HD 147677	6001, 8015
HD 140873	3023, 6005	HD 148112	1363, 3033, 6001, 8015
HD 141003	1007, 1013, 1020, 1022, 1028, 1034, 1363, 2001, 3024, 8015	HD 148184	1212, 2012, 8015
		HD 148367	2012, 3023, 6001, 8015
HD 141004	1006, 1020, 1034, 1075, 1088, 1197, 1363, 1399, 1425, 1509, 1586, 2001, 2024, 3045, 6001, 8003, 8012, 8015	HD 148379	1018, 1034, 1586, 1649, 1737, 2001, 2006, 8100
		HD 148387	8003, 8015
		HD 148478	3034, 8003, 8015, 8032
HD 141168	2012, 2030	HD 148546	976, 1770
HD 141477	3016, 8015	HD 148605	2012, 8015
HD 141513	2016, 8015	HD 148688	1075, 1586, 1737, 2001, 2035, 8100
HD 141556	3023, 8015	HD 148703	8015, 8100
HD 141637	1586, 1586, 1637, 2012, 8015	HD 148783	8022, 8027, 8032
HD 141680	1088, 3016, 4001, 8015	HD 148786	2012, 3077, 8015
HD 141702	3037, 6006	HD 148816	1620, 1658, 3078
HD 141795	3023, 8015	HD 148856	3016, 8003, 8015, 8100
HD 141891	2012, 3060, 8015	HD 148898	2012, 3023, 8015
HD 142096	8015, 8023	HD 148937	1586, 1649, 1737
HD 142114	1637, 2012, 8015	HD 148989	1737, 8100
HD 142165	8015, 8023	HD 149161	1363, 3010, 6001, 8015
HD 142184	8015, 8023	HD 149212	6001, 8015
HD 142198	2012, 3005, 8015	HD 149298	1586, 1649
HD 142373	1118, 1197, 3026, 8015	HD 149404	954, 1737, 2009
HD 142378	8015, 8023	HD 149414	1770, 1775, 3077, 7009
HD 142468	403, 540, 567, 976, 1350	HD 149438	1273, 1770, 2012, 8015
HD 142565	976, 1350	HD 149661	1013, 1197, 1256, 1509, 3008, 8015, 2012, 6001, 8003, 8007, 8015
HD 142574	8011, 8015		
HD 142634	540, 976	HD 149933	1729, 6004
HD 142669	1075, 1770, 2012, 8015	HD 150680	1197, 1363, 1569, 3026, 8003, 8015
HD 142780	8015, 8032	HD 150997	4001, 8015
HD 142860	1020, 1034, 1077, 1118, 1197, 2001, 3078, 8015	HD 151515	976, 1406, 1583, 1707, 8100
HD 142983	2012, 8015	HD 151564	1406, 1707
HD 143018	2012, 8015	HD 151613	3026, 6001, 8015
HD 143107	1007, 1077, 1080, 1118, 1399, 1502, 1566, 1569, 1601, 1625, 1648, 1733, 3016, 4001, 6001, 8003, 8015, 8032, 8071	HD 151680	1075, 2012, 8015
		HD 151769	1586, 2012, 3026, 8015
		HD 151804	1707, 2007
		HD 152003	1737, 8100
HD 143275	1075, 2012, 8003, 8007, 8015	HD 152042	1539, 1707
HD 143474	2024, 3030	HD 152107	1363, 3023, 6001, 8015
HD 143699	2012, 8015	HD 152147	1707, 8100
HD 143761	1080, 3026	HD 152235	1586, 1707, 1737, 2007, 8100
HD 143807	1363, 3033, 6001, 8015, 8100	HD 152236	1088, 1586, 1707, 2007, 8100
HD 143894	6001, 8015	HD 152246	1539, 1707
HD 144070	3030, 8015	HD 152391	1013, 1197, 1620, 3078
HD 144206	6001, 8015	HD 152405	1707, 2012, 8100
HD 144217	1509, 1770, 2012, 8007, 8015	HD 152408	1707, 1737, 2032
HD 144284	1118, 1363, 1502, 1566, 1569, 1601, 1625, 1648, 1733, 3026, 8015	HD 152424	1583, 1707
		HD 152614	3023, 8015
HD 144294	2024, 8015	HD 152667	976, 1539, 1586, 1707, 2012, 8100
HD 144470	2012, 8015	HD 152723	1707, 2012
HD 144579	1774, 3026	HD 153210	4001, 8015

Star	References
HD 153597	1620, 3026, 6001, 8015
HD 154090	1586, 2012, 8015
HD 154143	3055, 6001, 8015
HD 154363 A	1034, 1075, 1197, 1509, 3078, 8006, 8040
HD 154363 B	1494, 1586, 3078, 8006, 8039
HD 154368	1018, 1034, 2001, 2012
HD 154811	1737, 8100
HD 155125	2012, 3023, 8015
HD 155203	2011, 2024, 3026, 8015
HD 155603	2001, 8036, 8100
HD 155763	8007, 8015
HD 155826	2020, 3053
HD 155886	2012, 3078, 8015
HD 155985	1770, 2024
HD 156026	1641, 1642, 3078, 8017
HD 156164	1327, 1363, 3023, 6001, 8003, 8015
HD 156236	2011, 2024
HD 156274	1075, 1637, 1770, 2011, 2038, 8015
HD 156283	4001, 8015, 8100
HD 156293	977, 1075, 1770, 2011, 2038
HD 156325	1637, 2006
HD 156384	2030, 3078, 8015
HD 156897	2024, 3026, 8015
HD 156928	3023, 8015
HD 157038	1737, 2001, 2006
HD 157042	1586, 1637, 2011, 2024, 2038
HD 157056	6001, 8003, 8015
HD 157089	3077, 8012, 8025
HD 157214	1077, 1118, 1197, 1258, 1327, 1355, 3078
HD 157243	1075, 1770, 2011
HD 157316	1770, 2011
HD 157477	1066, 1075, 1085, 1460, 1586, 1628, 1657, 1673, 1704, 1770, 2011
HD 157487	1066, 1085, 1460, 1586, 1628, 1657, 1673, 1673, 1704, 1770, 2011
HD 157661	2011, 2038
HD 157697	1704, 1770, 2011
HD 157792	2012, 3023, 8015
HD 157832	1637, 2011
HD 157881	1020, 1034, 1067, 1075, 1197, 1509,1586, 1729, 3078, 8003, 8006
HD 157919	3026, 8015
HD 157950	2012, 3026, 6001, 8015
HD 157999	1363, 1509, 3016, 4001, 8015, 8100
HD 158186	954, 976
HD 158320	954, 976
HD 158427	1637, 2012, 8015
HD 158643	2012, 3023, 8015
HD 158809	1658, 6006
HD 159181	8003, 8015, 8100
HD 159217	2011, 3023, 8015
HD 159402	1586, 2012
HD 159482	1696, 3037, 8112
HD 159532	2011, 3026, 8015
HD 159541	6001, 8015, 8050, 8052, 8071
HD 159560	3023, 6001, 8015, 8050, 8052, 8071
HD 159561	1034, 1197, 1363, 3023, 8015
HD 159707	1770, 2011, 2038, 3021
HD 159868	1586, 1770, 2011, 3026
HD 159876	1509, 2012, 3023, 8015
HD 159975	2012, 3023, 8015
HD 160032	3026, 8015
HD 160263	1770, 2011
HD 160269	1077, 1118, 1197, 1379, 1502, 1566, 1569, 1601, 1648, 1733, 3030
HD 160346	1080, 1197, 1327
HD 160613	3023, 8015
HD 160617	3077, 6006
HD 160691	3077, 8017
HD 160762	1203, 1569, 3016, 8015
HD 160915	2012, 3026, 8015
HD 161096	1006, 1020, 1075, 1080, 1088, 1327, 1355, 1363, 1502, 1509, 1569, 1601, 1648, 1733, 2001, 3016, 8003, 8015, 8032, 8040, 8050, 8071
HD 161471	1075, 2012, 3026
HD 161770	3077, 8112
HD 161797	1197, 1327, 1355, 3045, 4001, 6001, 8003, 8015
HD 162718	1588, 1737
HD 162732	1079, 1733
HD 162978	1012, 2007, 3021
HD 163454	1586, 1737

Star	References
HD 163506	1118, 1327, 1355, 6009, 8015, 8083, 8100
HD 163588	4001, 8015
HD 163758	1586, 1737
HD 163770	3016, 4001, 8015, 8100
HD 163800	1737, 1770, 8007, 8031
HD 163917	1088, 2012, 4001, 6001, 8003, 8015
HD 163955	1075, 1637, 2012, 3023, 8015
HD 163993	3016, 4001, 8015
HD 164058	1502, 1566, 1569, 1601, 1625, 1648, 1733, 3016, 8003, 8015
HD 164136	8015, 8100
HD 164259	1075, 1088, 1258, 1509, 1586, 1586, 2001, 3026, 8015
HD 164349	8015, 8100
HD 164637	1737, 2012
HD 165024	1637, 2012
HD 165040	1075, 2012, 3023
HD 165135	3035, 8015
HD 165222	1705, 3072
HD 165341	1075, 1088, 1197, 1363, 8015
HD 165516	2007, 8100
HD 165634	2012, 6007, 8015
HD 165760	3016, 4001, 6001, 8015
HD 165793	540, 954, 976, 2009, 8100
HD 165908	1118, 1197, 1258, 1355, 3037
HD 166023	1754, 2006, 8100
HD 166620	1118, 1197, 1355, 1399, 3031, 8015
HD 167263	1012, 2007, 8100
HD 167425	2020, 2028, 3077
HD 167618	3055, 8015
HD 168656	1363, 1509, 3024, 4001, 6001, 8015
HD 168720	6001, 8015
HD 168723	1020, 1034, 1075, 1088, 1197, 1509, 2001, 3016, 8015
HD 169022	3023, 8015
HD 169420	2012, 8015
HD 169916	1637, 2012, 8003, 8015
HD 169981	3050, 8015
HD 169985	3016, 8015
HD 170153	1118, 1197, 1363, 1502, 1566, 1569, 1601, 1625, 1648, 1733, 3026, 8015
HD 170235	1586, 1637, 2007
HD 170296	3023, 8015
HD 170493	1729, 3072
HD 171443	1088, 1509, 2012, 3032, 4001, 8015
HD 171635	6001, 8015, 8100
HD 171978	8015, 8071
HD 172167	1197, 1286, 1363,1705, 3023, 8003, 8007, 8015
HD 172555	1075, 1637, 2016, 2038
HD 172748	3026, 8015
HD 172910	2012, 8015, 8023
HD 173009	4001, 8100
HD 173300	3023, 8015
HD 173648	1118, 3023, 8015
HD 173667	1118, 1258, 1355, 1586, 1620, 3026, 8015
HD 173764	1088, 2012, 3016, 4001, 8015
HD 173780	6001, 8015
HD 173948	1088, 1637, 2012, 8028
HD 174115	8071, 8080
HD 174383	1772, 6011
HD 174974	2012, 8015
HD 175179	1696, 3077
HD 175190	8011, 8015
HD 175191	1340, 1637, 2012, 8015
HD 175305	3010, 8112
HD 175329	2007, 2017, 3077, 4001
HD 175518	1620, 3077
HD 175588	8003, 8007, 8015
HD 175638	1327, 1363, 3023
HD 175639	1363, 3023
HD 175751	2012, 4001
HD 175865	8003, 8015, 8022, 8027
HD 176303	1205, 3053, 8015
HD 176437	1063, 1077, 1118, 1203, 1502, 1569, 1601, 1625, 1648, 1733, 3023, 8015
HD 176678	2035, 8015
HD 176687	3030, 8015
HD 177171	2030, 3026
HD 177196	3052, 8015
HD 177565	2024, 4001
HD 177724	1006, 1020, 1063, 1118, 1203, 1363, 3023, 8015

Table 4 1345

Star	References	Star	References
HD 177756	1079, 1088, 3023, 8015	HD 191117	977, 1066, 1075, 1586, 2011
HD 178175	2007, 8028, 8035, 8035	HD 191243	1192, 8100
HD 178524	2012, 3026, 8015	HD 191273	1770, 2011
HD 178596	1502, 1566, 1569, 1569,1601, 1625, 1648,	HD 191349	1770, 2024
	1733, 3053	HD 191408	3078, 8015
HD 179761	1417, 3023	HD 191610	3016, 6001
HD 180163	3016, 6001, 8015	HD 191639	1732, 3047
HD 180262	1355, 3016, 8100	HD 191692	1509, 3023, 6001, 8003, 8015
HD 180316	1601, 1625, 1648, 1733	HD 191849	1628, 1657, 1673, 1704, 2011, 3072
HD 180554	6001, 8015	HD 192310	3078, 8015
HD 180617	1197, 1509, 1620, 1705, 3078	HD 192425	6001, 8015
HD 180777	3026, 8015	HD 192577	8003, 8015
HD 180809	4001, 8007, 8015, 8100	HD 192640	3023, 6001, 8015
HD 180972	8100, 8100	HD 192713	8015, 8040, 8100
HD 181122	1603, 3002, 4001	HD 192836	4001, 8015
HD 181276	4001, 6001, 8003, 8015	HD 192876	2012, 3016, 8003, 8007, 8015, 8040, 8059, 8100
HD 181577	1075, 1088, 2012, 3023, 8015	HD 192886	1075, 1770, 2012, 2024, 2038, 3021
HD 181623	2020, 3026	HD 192907	3023, 8015
HD 181743	1696, 2012, 3077	HD 192909	3016, 8003, 8015
HD 181984	8003, 8015	HD 192934	1070, 1118, 1193, 1196
HD 182255	6001, 8015	HD 192947	2001, 2012, 4001, 8015, 8053
HD 182572	1118, 1197, 1355, 3077	HD 193237	8007, 8015
HD 182640	1363, 1509, 1620, 3026, 6001, 8003, 8015	HD 193322	1118, 1196, 6001, 8015
HD 182835	1088, 1355, 1363, 1509, 2001, 3026, 8015, 8100	HD 193369	1193, 1196
HD 183439	3077, 6001, 8003, 8015	HD 193370	1196, 1218, 3026, 6001, 8015
HD 183912	1363, 3001, 8003, 8015	HD 193432	3023, 8015
HD 184279	1006, 1075, 1212, 1586, 8040	HD 193443	1745, 8031
HD 184406	1417, 4001, 8011, 8015	HD 193495	3026, 6002, 8015
HD 184606	1118, 6001, 8015	HD 193514	1118, 1209
HD 184915	1020, 1075, 1088, 1203, 1258, 1509, 1569,	HD 193571	1770, 2011, 2024, 2038
	3047, 8015	HD 193702	1118, 1193, 1196, 8015
HD 184930	1509, 1586, 8015	HD 193807	1770, 2011, 2038
HD 185025	1764,6004	HD 193901	1658, 1696, 2017, 3078
HD 185144	985, 1077, 1118, 1197, 1355, 1363, 1502, 1566,	HD 194093	1196, 3026, 6001, 8003, 8007, 8015
	1569, 1569, 1601, 1625, 1648, 1733, 3078, 4001,	HD 194433	1075, 2012
	6001, 8003, 8006, 8015	HD 194598	1509, 1620, 1658, 1696, 3077, 8112
HD 185395	1758, 3026, 8015	HD 194943	2012, 3026, 8015
HD 185734	3016, 8015	HD 195050	1196, 3052
HD 185758	3026, 4001, 8015, 8100	HD 195295	1355, 3026, 6001, 8003, 8015, 8100
HD 185859	1733, 8100	HD 195506	1332, 1355, 4001
HD 186219	2020, 2038	HD 195593	1218, 1355, 1399, 3016, 6001, 8007, 8015
HD 186408	6001, 8012, 8015	HD 195810	3016, 8015, 8040
HD 186427	6001, 8012, 8015	HD 196426	1088, 1509, 6005
HD 186791	8003, 8007, 8015, 8100	HD 196574	8003, 8015
HD 187013	1197, 1355, 1758, 3053	HD 196755	1355, 1363, 3032, 4001
HD 187362	6001, 8015	HD 196795	1620, 3072
HD 187459	3016, 8100	HD 196867	1034, 1063, 1079, 1203, 1363, 1569, 1733,
HD 187642	1034, 1075, 1088, 1197, 1363, 1509, 3023,		3023, 8015
	8003, 8015	HD 197051	1075, 2038, 3023
HD 187796	8005, 8022, 8027	HD 197345	3023, 8002, 8003, 8007, 8015, 8037
HD 187811	3016, 6001, 8015	HD 197692	2024, 3026, 8015
HD 187929	1754, 6007	HD 197752	6001, 8015
HD 188031	2011, 3077	HD 197912	3016, 8003, 8015
HD 188119	4001, 8003, 8015	HD 197989	1363, 1502, 1566, 1569, 1601, 1625, 1648, 1733,
HD 188209	1203, 1234, 6001, 8015, 8100		3016, 8003, 8015, 8032
HD 188228	2012, 2038, 3023	HD 198001	1020, 1034, 1075, 1088, 1107, 1509, 2001, 2012,
HD 188293	1084, 1509, 3016		3023, 8015
HD 188294	1084, 1509, 3016	HD 198026	3001, 8003, 8015
HD 188310	3016, 8015	HD 198149	1197, 1355, 1502, 1566, 1601, 1625, 1648, 1733,
HD 188376	4001, 8015		3016, 4001, 8003, 8015
HD 188512	985, 1006, 1020, 1034, 1075, 1080, 1088, 1355,	HD 198183	3016, 8015
	1363, 1502, 1509, 1566, 1569, 1601, 1625, 1648,	HD 198478	1118, 1234, 8015
	1733, 3016, 4001, 8003, 8015, 8040	HD 198542	1024, 1075, 2012, 3005, 8015
HD 188892	3016, 8015	HD 198743	1088, 1509, 2012, 8015
HD 188947	1080, 1118, 1363, 1502, 1566, 1569, 1601, 1625,	HD 198809	4001, 6007, 8015, 8040
	1648, 1733, 3016, 8003, 8015	HD 199288	2012, 3078
HD 189198	2011, 2038	HD 199478	1119, 8015, 8100
HD 189319	3051, 8015	HD 199580	4001, 6011
HD 189502	1770, 2011	HD 200120 A	1119, 1203, 1212, 3024, 8015, 8084
HD 189763	8015, 8022	HD 200310	1196, 1203, 1212, 1615
HD 189933	1704, 1730, 1770, 2011	HD 200499	3023, 8015
HD 190057	2011, 2024	HD 200580	1658, 3037, 8112
HD 190147	1332, 3016, 8100	HD 200761	3023, 8015
HD 190248	2038, 3078, 8015	HD 200905	1218, 1355, 3016, 6001, 8003, 8007, 8015,
HD 190360	1355, 3032		8040, 8100
HD 190406	3077, 6007	HD 200914	2012, 8015
HD 190480	1673, 1673, 1704, 2011	HD 201091	1028, 1067, 1080, 1084, 1355, 1758, 3078, 4001,
HD 190603	1118, 1192, 6001, 8007, 8015, 8100		6001, 8003, 8006, 8015, 8040
HD 190940	8003, 8015	HD 201092	1080, 1084, 1355, 1758, 3078, 4001, 6001, 8003,
HD 191026	1118, 1124, 1355, 3032		8006, 8015, 8040
HD 191046	1124, 3071	HD 201251	3016, 8015, 8100

Star	References
HD 201601	1256, 1263, 1363, 3023, 8015
HD 201891	1658, 1774, 3077
HD 202109	3016, 6001, 8003, 8007, 8015, 8100
HD 202214	1203, 1252
HD 202275	1363, 1381, 2024, 3030, 8015
HD 202444	1363, 1502, 1566, 1601, 1625, 1648, 1733, 3032, 6001, 8003, 8015
HD 202447	1363, 1509, 3016, 8015
HD 202850	1363, 3023, 6001, 8003, 8007, 8015, 8040
HD 203064	1363, 3016
HD 203280	1502, 1566, 1601, 1625, 1648, 1733, 3023, 8003, 8015, 8040
HD 203387	1425, 2012, 8015
HD 203454	6009, 8015
HD 203504	1355, 1363, 1586, 4001, 6001, 8003, 8015
HD 203608	2038, 3045, 8015
HD 204075	2001, 3016, 4001, 8015, 8059, 8100
HD 204139	3005, 8015
HD 204155	1658, 8112
HD 204381	8015, 8053
HD 204867	2001, 2012, 3016, 6001, 8003, 8015, 8040, 8059, 8100
HD 205021	8003, 8015
HD 205435	1363, 3016, 4001, 8003, 8015
HD 205512	1462, 4001, 8015
HD 205637	1149, 1212, 2012, 2032, 8015, 8028, 8035, 8035
HD 205765	3050, 8015
HD 205767	3023, 8015, 8040
HD 206088	2012, 3026, 8015
HD 206165	3016, 8003, 8007, 8015
HD 206570	8022, 8027
HD 206742	3023, 8015
HD 206778	1256, 1363, 1509, 3032, 8003, 8015
HD 206859	1363, 3016, 4001, 8003, 8015, 8059, 8100
HD 206952	4001, 8015
HD 207098	8003, 8015
HD 207198	6001, 8007, 8015
HD 207260	1363, 3016, 8003, 8007, 8015
HD 207330	3016, 8015
HD 207692	2035, 3077
HD 207971	2012, 3023, 8015
HD 209014	1637, 2012, 2024, 8028, 8028, 8035
HD 209100	3078, 8015, 8029
HD 209409	8015, 8028, 8028, 8035, 8053
HD 209481	1077, 1118, 1203
HD 209522	2032, 8028, 8028, 8035
HD 209625	2024, 3023, 8015
HD 209688	3005, 4001, 8015
HD 209747	1355, 1363, 1509, 4001, 6001, 8003, 8015
HD 209750	1088, 1363, 1509, 2001, 3016, 8003, 8007, 8015, 8040, 8059, 8100
HD 209819	1203, 1586, 2012, 3023, 8015
HD 209952	2011, 8015, 8029
HD 209975	1012, 1363, 3016, 6001, 8015
HD 210027	1355, 1502, 1566, 1569, 1601, 1625, 1648, 1733, 3026, 8003, 8015
HD 210049	3023, 8015
HD 210418	1203, 1363, 3023, 8015
HD 210459	3026, 8015, 8100
HD 210702	4001, 8015
HD 210745	1218, 1355, 1363, 3016, 8015
HD 210807	4001, 6001, 8015
HD 210839	1012, 1209, 1234, 1363, 3016, 6001, 8003, 8007, 8015
HD 210934	2024, 3023
HD 211053	1075, 2011, 2038
HD 211073	6001, 8015
HD 211336	3023, 8015
HD 211388	8003, 8015, 8100
HD 211391	2012, 3016, 8015, 8040
HD 211416	1637, 2012, 8029
HD 211838	1509, 2024, 3023
HD 211924	1733, 3016, 8084
HD 211998	2038, 3078
HD 212061	1509, 2024, 3023, 8015
HD 212076	6001, 8015
HD 212097	1363, 3023, 6001, 8015
HD 212120	3024, 8015
HD 212132	2011, 2038
HD 212330	3077, 8017
HD 212496	4001, 8003, 8015, 8040

Star	References
HD 212571	1212, 1363, 1425, 8015
HD 212581	2035, 2038, 3023
HD 212593	3023, 8015
HD 212943	1355, 1363, 1509, 3016, 4001, 6001, 8015
HD 213009	2011, 8015, 8029
HD 213052	1363, 1425, 3026, 8015
HD 213080	2038, 3052, 8015, 8022, 8029
HD 213155	1770, 2011, 2024
HD 213310	8003, 8015
HD 213320	2024, 3023, 6001, 8015
HD 213398	3023, 8015
HD 213420	3016, 8015
HD 213657	2011, 3077
HD 213998	1363, 1425, 1601, 1625, 1648, 1733, 3023, 8015
HD 214168	1212, 8084
HD 214680	1006, 1011, 1077, 1118, 1174, 1203, 1234, 1258, 1278, 1399, 1502, 1566, 1569, 1601, 1625, 1648, 1733, 3016, 8003, 8007, 8015, 8040, 8050, 8071
HD 214748	1079, 1088, 1637, 2012, 3023, 8015, 8028, 8028, 8035
HD 214846	3023, 8029
HD 214923	1203, 1363, 3023, 8015, 8040
HD 214952	8015, 8017, 8029
HD 214953	1075, 2011, 2024, 2038
HD 214994	3023, 6001, 8015
HD 215104	2011, 2024, 3035, 8015
HD 215105	1704, 2011
HD 215167	1425, 2012, 3005, 4001, 8015
HD 215182	3016, 8003, 8015
HD 215257	3077, 8112
HD 215545	2011, 2038
HD 215648	1034, 1118, 1355, 1586, 3077, 8015, 8040
HD 215657	1075, 2011
HD 215665	8015, 8100
HD 215756	1673, 1704
HD 215789	2012, 3023, 8029
HD 215801	2011, 3077
HD 216009	1075, 1586, 1770, 2011
HD 216032	1088, 2012, 3016, 8015
HD 216054	2024, 3026
HD 216131	4001, 8015
HD 216228	4001, 8003, 8015
HD 216291	1075, 2011
HD 216336	2024, 8015
HD 216386	3055, 8015
HD 216406	1066, 1075, 1586, 2011
HD 216494	1079, 1425, 1509, 2012, 2024, 8015
HD 216627	1013, 1075, 1088, 2012, 3023, 8015
HD 216735	1075, 1203, 1355, 1363, 1399, 1425, 1509, 3023, 6001, 8015
HD 216777	2033, 3077
HD 216803	3077, 8015
HD 216899	3078, 8039
HD 216946	1363, 3016, 8015
HD 216956	1075, 3024, 8003, 8015
HD 217014	1355, 3077, 6001, 8003, 8012, 8015, 8100
HD 217364	4001, 8029
HD 217476	8007, 8015
HD 217891	1363, 1425, 8015
HD 217906	8003, 8015, 8032
HD 218031	4001, 8015
HD 218045	1034, 1049, 1119, 1203, 1254, 1363, 3023, 8003, 8015
HD 218227	3026, 8029
HD 218329	1034, 1075, 1258, 1355, 1363, 1425, 1509, 1733, 3016, 8003, 8015
HD 218356	3077, 8015, 8100
HD 218376	1203, 1234, 1363, 6001, 8015, 8040
HD 218594	2012, 8015
HD 218640	3026, 8015
HD 218670	3077, 8015, 8029
HD 219080	3023, 8015
HD 219134	1006, 1077, 1118, 1197, 1355, 1399, 1502, 1566, 1601, 1625, 1648, 1733, 3078, 4001, 6001, 8003, 8006, 8015, 8040, 8050, 8071
HD 219175	1696, 3037
HD 219215	2012, 3071, 6002, 8015
HD 219449	1084, 1425, 3077, 4001, 8003, 8015
HD 219615	1075, 1355, 1363, 1425, 1509, 4001, 8015, 8040
HD 219617	1097, 1509, 1658, 1705, 3078, 8025
HD 219688	1075, 1425, 1586, 2012, 2024, 8015

Table 4 1347

Star	References	Star	References
HD 219734	8003, 8015	NGC 2422 - 9	3016, 8029, 8032
HD 219784	4001, 8015	NGC 2422 - 45	1212, 2012, 2014, 8035
HD 219877	2012, 3026	NGC 2451 - 175	2037, 8015
HD 219916	4001, 8015	NGC 2451 - 187	1637, 2012, 2037, 8035
HD 220061	8015, 8040	NGC 2451 - 251	2013, 2028, 2037
HD 220278	2029, 3030	NGC 2451 - 254	2037, 3041, 8015
HD 220369	1379, 8100	NGC 2467 - 19	1295, 1376, 1681
HD 220657	8003, 8015	NGC 2467 - 77	1681, 1737
HD 220704	2012, 2024, 3016, 4001, 8015, 8040	NGC 2516 - 110	1685, 2012, 3077
HD 220825	1363, 1425, 2024, 3023, 6001, 8015	NGC 2516 - 119	1683, 1685, 3077
HD 220954	1363, 1425, 3016, 4001, 8015	NGC 2516 - 124	2012, 3077
HD 221148	3077, 4001	NGC 2516 - 128	2012, 3077
HD 221253	3024, 8015, 8084	NGC 2516 - 134	1075, 1637, 1683, 2012, 3015, 8035
HD 221507	2012, 2024, 3023, 8015	NGC 2516 - 136	2027, 3015
HD 222095	3023, 8015	NGC 2546 - 682	2012, 8100
HD 222107	8003, 8015	NGC 2632 - 212	8023, 8092
HD 222173	1363, 3023, 8015	NGC 2632 - 253	1245, 1351, 1355, 2001,3077, 4001, 8015, 8092
HD 222368	1004, 1006, 1020, 1034, 1075, 1355, 1363,1425,	NGC 2632 - 265	2001, 8023
	1509, 1566, 1586, 1601, 1625, 1648, 1733, 1758,	NGC 2632 - 283	1569, 3077, 4001, 8015, 8023, 8092
	2024, 3026, 8015, 8040	NGC 2632 - 300	1569, 8015, 8023
HD 222404	1502, 1566, 1569, 1601, 1625, 1648, 1733, 8015	NGC 2632 - 428	8023, 8092
HD 222493	3077, 4001	NGC 2632 - 1133	3077, 8023, 8092
HD 222574	2012, 8015, 8059, 8100	NGC 2682 - 105	8043, 8092
HD 222603	1075, 1355, 1363, 1425, 3023, 8015	NGC 2682 - 131	1717, 8023
HD 222661	2024, 3023, 8015	NGC 2682 - 224	8023, 8092
HD 223047	6009, 8100	NGC 2682 - 244	8023, 8092
HD 223352	2012, 2024, 8015	NGC 3324 - 1	1075, 1732, 2001, 2012, 8100
HD 223719	4001, 8100	NGC 4755 - 1	2001, 8100
HD 224022	2029, 3077	NGC 4755 - 2	2012, 8100
HD 224427	3016, 6001, 8003, 8015	NGC 4755 - 20	1505, 2001, 2012, 3013, 8100
HD 224618	2034, 3078	NGC 6167 - 1155	1081, 1649, 8100
HD 224686	2038, 8029	NGC 6169 - 1	2035, 8100
HD 224889	2038, 8029	NGC 6193 - 1	1732, 1743, 2007
HD 224930	1286, 1758, 3078, 8025	NGC 6193 - 19	1743, 2012, 8100
HD 224935	3047, 8003, 8011, 8015	NGC 6231 - 161	1707, 1737
HD 225132	1075, 1425, 2012, 2024, 8015	NGC 6231 - 293	1707, 8100
HD 225213	2024, 3078	NGC 6383 - 1	2009, 3021
		NGC 6514 - 55	976, 2007, 8100
HDE 260655	3072, 8003, 8006, 8039	NGC 6530 - 7	1161, 1586, 1586, 2009, 3047, 8007, 8015, 8027
HDE 268518	1566, 1569, 1569, 1601, 1625, 1648, 1733, 6007	NGC 6530 - 45	1655, 8027
HDE 269953	3074, 8033	NGC 6530 - 118	1732, 1737
HDE 269992	1277, 8033	NGC 6604 - 1	1547, 1586, 8007
HDE 271182	1483, 3074, 8033, 8100	NGC 6611 - 150	1655, 1703
HDE 283571	8027, 8082	NGC 6611 - 197	1732, 1737
HDE 283668	1783, 8082	NGC 6611 - 205	1703, 1732, 8007
HDE 284248	3077, 8112	NGC 6611 - 246	1703, 1737
HDE 284419	8013, 8027	NGC 6611 - 280	1586, 1703
HDE 285968	1620, 3078	NGC 6611 - 314	1187, 1655, 1703
HDE 290984	5006, 6004	NGC 6611 - 367	1032, 1187, 1703, 1732
HDE 313080	1737, 2023	NGC 6611 - 503	1703, 1737
HDE 326176	1406, 1707	NGC 6823 - 3	1749, 8027
		NGC 6823 - 13	1032, 1749
NGC 104 - 5529	529, 1582	NGC 6871 - 1	1359, 3024, 8084
NGC 457 - 136	1084, 1118, 1182, 1218, 3026, 6001, 8015	NGC 6871 - 4	1124, 1192, 3024, 8100
NGC 869 - 3	1119, 8023	NGC 6871 - 7	1192, 3024
NGC 869 - 16	1119, 8015, 8100	NGC 7160 - 2	1212, 8096
NGC 869 - 1057	1234, 8007, 8016, 8023	NGC 7160 - 3	1211, 8096
NGC 869 - 1162	1119, 8007, 8016, 8023	NGC 7654 - 270	1119, 7006
NGC 884 - 1655	8007, 8016, 8023		
NGC 884 - 1818	8016, 8023	IC 348 - 20	1718, 3024
NGC 884 - 2172	1209, 8023	IC 1805 - 112	1083, 1718
NGC 884 - 2178	8016, 8023, 8031	IC 1805 - 138	1083, 1718
NGC 884 - 2589	8016, 8023	IC 1805 - 148	1012, 1083, 1718, 8031, 8084
NGC 884 - 2691	8007, 8016, 8023	IC 1805 - 160	1012, 1083, 1209, 1718, 8007, 8016, 8023
NGC 884 - 2758	8007, 8016, 8023	IC 1805 - 192	1083, 1718
NGC 2232 - 1	2012, 6001, 8015	IC 1805 - 232	1083, 1718
NGC 2244 - 122	1059, 1075, 1209, 1651, 8007, 8027	IC 1848 - 2	8007, 8016, 8084
NGC 2244 - 180	1075, 8027	IC 2391 - 13	1040, 1075, 2012
NGC 2244 - 203	1012, 1059, 1075, 1209, 1651, 8007, 8027	IC 2391 - 16	1040, 2029
NGC 2244 - 266	1415, 1754, 8015	IC 2391 - 20	1075, 2012, 8015
NGC 2264 - 50	8078, 8114	IC 2391 - 31	2012, 3015
NGC 2264 - 83	8078, 8114	IC 2391 - 34	1040, 2012, 3015, 8023
NGC 2264 - 88	8078, 8114	IC 2395 - 1	1075, 2012
NGC 2264 - 108	8078, 8114	IC 2581 - 1	1018, 1034, 1075, 1088, 2001, 2012, 2041, 3026
NGC 2264 - 131	1075, 1077, 1209, 1234, 1363, 1425, 1509, 1569,	IC 2581 - 2	730, 2012, 2041, 8100
	1625, 1648, 1718, 1732, 1733, 6001, 8007, 8015,	IC 2602 - 27	2012, 2038
	8078, 8084	IC 2602 - 41	3015, 8023
NGC 2287 - 109	1626, 2012	IC 2602 - 48	2038, 3015
NGC 2353 - 1	1732, 2028	IC 2602 - 49	2012, 2016, 2038, 3015
NGC 2362 - 23	1011, 1088, 2001, 2027, 6001, 8015, 8100	IC 2602 - 51	2003, 2005, 2012, 2012, 2038, 3015, 8023
NGC 2362 - 46	976, 1732, 2001	IC 2602 - 54	2005, 3015

Star	References
IC 2944 - 43	976, 1312
IC 2944 - 59	976, 1312
IC 4665 - 23	1327, 1654
IC 4665 - 32	1327, 1509, 1654, 6005
IC 4665 - 50	1389, 1654
IC 4665 - 62	1327, 1509, 1654
IC 4665 - 68	1509, 1654
IC 4725 - 26	1197, 3008
Bl 1 - 42	1068, 1075, 2012, 2024, 3004, 8015
Bo 13 - 1	1730, 1737
Ho 22 - 1	1737, 2012
Pis 12 - 17	863, 977, 1075, 2011
Pis 20 - 8	779, 914, 976, 1672, 1737, 8100
Pis 24 - 1	1537, 1737
St 2 - 31	1119, 8023
St 14 - 4	1417, 1672, 2001, 2012, 8059
St 14 - 13	976, 1672, 2021
St 14 - 14	976, 1672, 2021
St 16 - 113	1684, 2021
Tr 10 - 1	1705, 2031, 8015
Tr 10 - 2	1075, 2012
Tr 16 - 7	1349, 1737
Tr 16 - 180	945, 2042
Tr 37 - 466	3024, 8084
Tr 37 - 1319	8007, 8015, 3016, 8032
Cr 121 - 1	3034, 8100
Cr 121 - 4	2007, 8029
Cr 121 - 22	2012, 8015
Cr 121 - 23	2035, 8028, 8035
Cr 121 - 26	1278, 1409, 2001, 2012, 2014, 3000, 8015
Cr 121 - 28	1088, 2013, 2029, 3034, 8015, 8032, 8100
Cr 121 - 29	2012, 8003, 8007, 8015
Cr 121 - 30	1075, 1425, 2012, 8015, 8100
Cr 121 - 31	1034, 1075, 2012, 3034, 6002, 8015
Cr 121 - 34	1637, 2012, 6002, 8015
Cr 121 - 35	815, 1637, 2007, 8035
Cr 121 - 38	1079, 2009
Cr 121 - 44	2012, 3055, 8015
Cr 132 - 5	1732, 2007
Cr 140 - 1	2029, 3005
Cr 140 - 3	1637, 2027, 8035
Cr 197 - 1	432, 540, 976
Cr 359 - 2	1586, 1639, 3016
Cr 359 - 3	1084, 1149, 1211, 1221, 1363, 1639, 3016, 8015, 8100
Cr 359 - 5	1221, 1256, 1363, 1586, 1639, 3016, 8015
Cr 359 - 10	1639, 3023, 8015
Cr 359 - 13	1006, 1020, 1034, 1075, 1088, 1203, 1286, 1327, 1363, 1509, 1639, 2001, 3023, 8015,8040, 8050, 8071
Alpha Per 44	1197, 1286, 1355, 1363, 1758, 3078, 4001, 8003, 8012, 8015
Alpha Per 605	1355, 3034, 8002, 8003, 8007, 8015
Alpha Per 900	4001, 8015
Alpha Per 1164	1212, 1363, 8015
Coma Ber 40	1298, 3027
Coma Ber 62	8015, 8052
Coma Ber 68	8023, 8052
Coma Ber 79	8001, 8023
Coma Ber 80	8023, 8034
Coma Ber 91	1127, 1363, 3026, 6001, 8001, 8015
Coma Ber 98	8023, 8041
Coma Ber 107	1363, 3023, 8001, 8015
Coma Ber 125	3023, 8001, 8015
Coma Ber 129	1080, 1363, 8015, 8034
Coma Ber 130	1127, 1363, 3023, 8001, 8015
Coma Ber 146	1127, 1202, 3033, 8001, 8015
Coma Ber 160	8001, 8015
Coma Ber 183	8001, 8015
Hyades vB 28	1105, 1127, 1355, 1363, 3077, 4001, 8003, 8015, 8092
Hyades vB 41	1291, 1355, 1363, 3016, 4001, 8003, 8015, 8092
Hyades vB 45	1355, 8015
Hyades vB 54	8015, 8025

Star	References
Hyades vB 56	3050, 8015
Hyades vB 60	3023, 8015
Hyades vB 70	1127, 1291, 1355, 1363, 1390, 4001, 8003, 8015, 8092
Hyades vB 71	1355, 1363, 4001, 8003, 8015, 8092
Hyades vB 72	1084, 1119, 1127, 1286, 1363, 1502, 1566, 1569, 1601, 1625, 1648, 1733, 3023, 8003, 8015, 8025
Hyades vB 74	6001, 8015, 3023
Hyades vB 112	3023, 8015
Hyades vB 117	8023, 8088
Hyades vB 123	3023, 8015
Hyades vB 130	1256, 3023, 8015
Hyades vB 131	8023, 8052
Hyades vB 141	3023, 8015
Hyades vB 169	1199, 1363, 3023, 8015, 8052
Hyades vB 174	8025, 8027
Hyades vB 191	1570, 1619, 3072
Hyades vA 292	1674, 3006
Hyades vA 294	1674, 3072
Hyades vA 490	1727, 3006
Orion B 342	8054, 8055
Orion B 388	1060, 8055
Orion B 442	1060, 8055
Orion B 482	1075, 1084, 8015, 8055
Orion B 493	1075, 1084, 1732, 8015, 8055
Orion B 502	1060, 8055
Orion B 530	1060, 8055
Orion B 587	8015, 8055
Orion B 598	8015, 8055
Orion B 608	1060, 8054, 8055
Orion B 631	1061, 8015, 8055
Orion B 632	1061, 2028, 8015, 8055
Orion B 659	1234, 1425, 1732, 2012, 8015, 8055
Orion B 682	1014, 1060, 1084, 1234, 8007, 8015, 8055
Orion B 714	1060, 1084, 8007, 8055
Orion B 721	542, 1004, 1006, 1011, 1034, 1060, 1075, 1425, 1509, 2001, 2012, 8003, 8007, 8015,8040, 8055
Orion B 722	1078, 2032, 8015, 8055
Orion B 747	1060, 8055
Orion B 760	1060, 8055
Orion B 776	8054, 8055
Orion B 786	1060, 8055
Orion B 806	1415, 3023, 8015, 8114
Orion B 907	1060, 1060, 1212
Orion B 980	1060, 1415, 1732
Orion B 1017	8015, 8055
Orion B 1098	1415, 1425, 1732, 2012, 8015, 8055
Pleiades 447	8015, 8030
Pleiades 468	1212, 1363, 8015, 8030
Pleiades 541	8015, 8030
Pleiades 563	1029, 1203, 1363, 8015, 8030
Pleiades 785	1363, 8015, 8030
Pleiades 817	6001, 8015, 8030
Pleiades 859	8015, 8030
Pleiades 980	1363, 8015, 8030
Pleiades 1084	8030, 8031
Pleiades 1432	1203, 1212, 1363, 8007, 8015, 8030
Pleiades 1823	8015, 8030
Pleiades 2168	1363, 3023, 8015, 8030
Pleiades 2181	1079, 1363, 1609, 3023, 8015, 8030
Cyg OB2 # 35	1167, 1209
Cyg OB2 # 359	1355, 1658, 3008
G 1 - 27	1705, 3078
G 3 - 33	1764, 3078
G 12 - 43	3078, 8088
G 17 - 11	1663, 1705, 3072
G 21 - 15	1764, 3060
G 21 - 22	1696, 3077
G 29 - 33	1632, 1764, 5008
G 64 - 12	3077, 6006
G 65 - 22	1658, 3077
G 93 - 48	1764, 3060
G 102 - 22	3078, 8039
G 140 - 24	1494, 1509, 1697, 3072, 8003, 8006, 8105
HA 92 # 342	1764, 6004

Table 5 1349

Table 5: Additional cross-identifications

Star	Identification	Star	Identification	Star	Identification
CpD -77 00622	T Aps	BD +16 02708	G 138 - 28	BD +48 02138	IDS13338N4839
CpD -72 02640	IDS21402S7234	BD +19 02646	IDS13070N1855	BD +48 03827	IDS22407N4856
CpD -63 02258	IDS12182S6346	BD +19 05116	G 68 - 24	BD +49 03571a	LS III +49 031
CpD -62 02337	LSS 2511		IDS23267N1923	BD +49 03591	IDS21368N5002
CpD -59 04147	IDS12108S5958	BD +20 02243	IDS08509N2014	BD +51 00318	IDS01247N5213
CpD -59 02518	IDS10393S5913	BD +20 02465	G 54 - 23	BD +52 03122	IDS22050N5229
CpD -59 02060	IDS10170S5910	BD +20 03010	IDS14380N1955	BD +53 02911	IDS22287N5316
CpD -58 02721	V426 Car	BD +21 01764	IDS08024N2124	BD +54 02763	IDS22201N5439
CpD -58 02608	IDS10398S5904	BD +21 04747	IDS22200N2204	BD +55 01823	G 202 - 36
CpD -54 06774	variable	BD +21 04923	IDS23178N2135	BD +56 02472	IDS20391N5708
CpD -54 02821	IDS09497S5433	BD +21 04940	IDS23249N2138	BD +57 00579	IDS02230N5714
CpD -52 02315	IDS09213S5241	BD +22 00353	IDS02234N2226	BD +57 02513	IDS22185N5813
CpD -51 09323	LSS 3560	BD +22 02442	IDS12039N2221	BD +57 02525	IDS22214N5720
CpD -76 00652	DI Cha	BD +24 02733	IDS14211N2407	BD +58 00547	IDS02538N5856
CoD -51 13128	LTT 8768	BD +27 00335	IDS02026N2751	BD +58 01772a	T Dra
CoD -48 04314	IDS08569S4838	BD +27 02853	IDS17331N2757		IDS17551N5814
CoD -48 04184	IDS08492S4851	BD +27 02891	IDS17437N2750	BD +59 00333	IDS01442N5931
CoD -48 03636	IX Vel	BD +27 04120	IDS21335N2716	BD +60 00586	IDS02464N6014
CoD -45 11051	IDS16474S4505	BD +27 04445	IDS22523N2728	BD +60 00598	IDS02500N6012
CoD -45 04676	GW Vel	BD +28 00258	IDS01292N2905	BD +60 00639	IDS03041N6034
CoD -45 04447	IDS08400S4547	BD +29 00366	IDS02046N2921	BD +60 02629	UU Cas
CoD -41 11062	IDS16478S4152	BD +29 02250	IDS11582N2935		LS I +60 042
CoD -39 11197	IDS17063S3948	BD +29 03029	IDS17255N2929	BD +60 02664	IDS23589N6017
CoD -39 05049	IDS08564S3910	BD +30 00089	IDS00334N3029	BD +60 02668	IDS00008N6019
CoD -38 11748	AA58,351# 2	BD +30 00206	IDS01158N3049	BD +61 00122	IDS00296N6111
CoD -36 12314	CCS 2543	BD +31 00434	IDS02228N3148	BD +61 00195	G 243 - 52
CoD -35 04415	IDS08127S3520	BD +31 03330	IDS18371N3128	BD +61 00342	IDS01462N6203
CoD -35 04384	IDS08115S3526	BD +33 00099	IDS00382N3319	BD +61 00571	IDS03169N6144
CoD -33 16613	IDS23310S3346	BD +33 01843	IDS08148N3245	BD +61 02350	LS III +62 016
CoD -32 04348	IDS07406S3259	BD +33 02433	IDS14119N3303	BD +61 02352	LS III +62 017
CoD -32 00558	IDS01215S3237	BD +33 02909	IDS17522N3327	BD +61 02353	LS III +62 018
CoD -29 13809	IDS17317S2903	BD +33 04117	IDS20567N3331	BD +61 02357	LS III +62 021
CoD -28 05034	IDS07495S2828	BD +34 03631	IDS19331N3501		IDS22487N6147
CoD -28 00657	IDS02008S2833	BD +35 01169	IDS05276N3545	BD +61 02365	LS III +62 024
CoD -27 04363	IDS07353S2723	BD +35 02436	IDS13149N3540	BD +61 02366	LS III +62 025
CoD -27 04224	IDS07294S2758	BD +35 04077	IDS20174N3518	BD +61 02520	IDS23414N6207
CoD -26 00466	IDS01201S2645	BD +35 04258	IDS20423N3511	BD +61 02526	IDS23428N6130
CoD -24 17814	IDS23378S2440	BD +36 01638	G 87 - 44	BD +62 00258	IDS01233N6304
BD -14 05050a	LSS 5048	BD +36 02322	G 123 - 75	BD +62 00316	IDS01464N6254
BD -21 01074	IDS05026S2143	BD +36 04025	IDS20176N3637	BD +62 00332	IDS01522N6225
BD -21 00311	IDS01442S2123	BD +37 01748	IDS07317N3657	BD +62 00338	IDS01550N6222
BD -20 03283	IDS10468S2043	BD +39 00710	IDS03004N3959	BD +62 02125	LS III +63 013
BD -19 00129	IDS00447S1913	BD +40 01758	IDS06495N4013	BD +62 02127	LS III +63 014
BD -17 06768	IDS23275S1718	BD +40 04212	LS III +41 027	BD +62 02150	LS III +62 033
BD -15 04969	LS IV -15 047		IDS20280N4052	BD +62 02154	LS III +62 017
BD -14 06313	IDS22314S1352		Schulte 3	BD +62 02155	LS III +62 035
BD -14 05043	LS IV -14 058	BD +40 04219	LS III +41 029	BD +62 02166	LS III +63 021
BD -14 05014	IDS18166S1440		Schulte 4	BD +62 02296	IDS23425N6240
BD -13 06464	IDS23432S1333	BD +40 04220	LS III +41 031	BD +62 02332	V375 Cas
BD -8 01666	IDS06558S0844		IDS20288N4058	BD +63 00048	IDS00211N6353
BD -6 01253	IDS05316S0647		Schulte 5	BD +63 00089	IDS00393N6351
BD -4 00938	IDS04416S0448	BD +40 04221	LS III +41 035	BD +63 00537	IDS04410N6309
BD -4 00384	IDS02172S0421		Schulte 6	BD +63 02093	IDS23567N6404
BD -2 04354	V453 Oph	BD +40 04227	IDS20297N4058	BD +64 01677	IDS22280N6457
BD -0 04234	IDS21270S0013		LS III +41 037	BD +65 00671	IDS08472N6521
BD +0 00549	IDS03099N0039		Schulte 8	BD +66 00034	G 243 - 30
BD +0 02058	IDS07386N0011	BD +41 03804	LS III +41 039		IDS00263N6642
BD +1 02570	IDS11239N0138		Schulte 10	BD +66 00717	G 236 - 65
BD +1 03942	IDS19099N0159	BD +41 03807	LS III +41 040	BD +68 00986	IDS18166N6836
BD +1 04381	LS IV +02 013		Schulte 11	BD +68 01345	IDS22552N6830
BD +3 00275	IDS01569N0327	BD +42 02810	IDS17100N4228	BD +70 00035	IDS00344N7028
BD +3 02855	IDS14013N0343	BD +42 03914	IDS20528N4244	BD +71 00851	IDS17401N7122
BD +4 05012	G 29 - 37	BD +42 04182a	Q Cyg	BD +75 00065	IDS01270N7529
BD +5 00043	G 31 - 44	BD +43 01168	IDS04548N4341	BD +75 00403	IDS10017N7537
BD +5 03409	IDS17255N0538	BD +43 04305	G 216 - 16	BD +75 00511	IDS13333N7501
BD +6 02436	IDS11153N0603		IDS22426N4349	BD +77 00361	IDS09063N7740
BD +6 02573	IDS12109N0612	BD +44 00967	IDS04245N4443	BD +79 00022	IDS00501N7935
BD +7 02031	IDS08431N0651	BD +44 01130	IDS05060N4422	BD +80 00245	IDS07550N8012
BD +10 01032	IDS06054N1021	BD +44 01142	IDS05077N4426	HD 3	IDS00000N4440
BD +10 01857	IDS08373N0956	BD +44 02051	G 176 - 11	HD 38	IDS00002N4516
BD +10 04385	IDS20446N1102	BD +44 03571	IDS20178N4430	HD 113	IDS00008N1732
BD +11 02479	NGC 4596 sq1 1	BD +44 03594	IDS20457N4502	HD 123	IDS00010N5753
BD +11 04571	IDS21243N1145	BD +45 03216	IDS20306N4519		G 243 - 13
BD +12 00168	IDS01174N1214	BD +45 03485	IDS21114N4921	HD 142	IDS00011S4938
BD +12 02201	IDS09170N1236	BD +46 02972	IDS20303N4642	HD 166	IDS00014N2829
BD +12 02576	IDS13108N1226	BD +47 00451	IDS01282N4814	HD 319	IDS00027S2304
BD +12 02597	IDS13242N1200	BD +47 01140	IDS05152N4748	HD 358	IDS00032N2832
BD +13 00099	IDS00402N1349	BD +47 02112	IDS13585N4649	HD 431	IDS00038N7910
BD +16 00502	IDS03382N1621	BD +48 01958	IDS11319N4801	HD 432	IDS00038N5836

Star	Identification	Star	Identification	Star	Identification
HD 469	IDS00040S5433	HD 4656	IDS00434N0703	HD 8538	IDS01193N5943
HD 471	IDS00041N2443	HD 4676	IDS00437N1624	HD 8556	IDS01193S0726
HD 489	IDS00043N1834	HD 4730	IDS00444S1406	HD 8589	G 272 - 12
HD 493	IDS00043S2833	HD 4732	IDS00443S2441	HD 8799	IDS01217N4453
HD 560	IDS00049N1035	HD 4757	IDS00445N2710	HD 8801	IDS01216N4035
HD 661	IDS00057S7347	HD 4758	IDS00445N2710	HD 8803	IDS01217N0301
HD 720	IDS00065S2822	HD 4775	B9.5 + G0III-IV	HD 8879	IDS01224S3304
HD 886	IDS00080N1438		IDS00446N6342	HD 8890	IDS01226N8846
HD 895	IDS00082N2626	HD 4967	IDS00467S2327	HD 8921	IDS01228S1125
HD 1014	IDS00093S0820	HD 5005	IDS00470N5605	HD 8949	IDS01231N0727
HD 1061	IDS00098N0816	HD 5015	IDS00470N6034	HD 8997	IDS01236N2112
HD 1070	IDS00099N5913	HD 5042	IDS00472S4415	HD 9132	IDS01248S2209
HD 1083	IDS00100N2644	HD 5098	IDS00478S2433	HD 9138	G 3 - 3
HD 1141	IDS00106N7624	HD 5128	IDS00480N5209		IDS01249N0537
HD 1185	IDS00111N4303	HD 5156	IDS00483S2519	HD 9228	IDS01257S2643
HD 1187	IDS00111S3200	HD 5190	IDS00486S7003	HD 9270	IDS01261N1450
HD 1239	IDS00116N6059	HD 5234	IDS00491N5826	HD 9540	IDS01286S2442
HD 1326	IDS00127N4327	HD 5267	IDS00493N1839	HD 9774	IDS01305N7232
	G 171 - 47	HD 5286	IDS00496N2305	HD 9826	IDS01310N4054
HD 1438	IDS00134N4314	HD 5394	IDS00507N6011	HD 9847	IDS01310S1802
HD 1479	IDS00138N5909	HD 5408	IDS00508N5950	HD 9906	IDS01315S3025
HD 1522	IDS00143S0923	HD 5448	IDS00512N3757	HD 10009	IDS01326S0955
HD 1620	IDS00153S3319	HD 5516	IDS00519N2253	HD 10042	IDS01330S7901
HD 1663	IDS00158N1025	HD 5641	IDS00530N2052	HD 10161	IDS01341S2532
HD 1748	IDS00166S3337	HD 5679	IDS00534N8120	HD 10205	IDS01347N4004
HD 1796	IDS00172N1256	HD 5788	IDS00544N4410	HD 10241	IDS01349S5357
HD 1810	IDS00174N6141	HD 5789	IDS00544N4410	HD 10293	IDS01357N5807
HD 1835	IDS00177S1246	HD 5820	IDS00546N0556	HD 10308	IDS01357N2514
HD 1967	IDS00187N3802	HD 6114	IDS00573N4650	HD 10360	IDS01360S5642
HD 1976	IDS00189N5128	HD 6116	IDS00573N4048	HD 10361	IDS01360S5642
HD 2114	IDS00203N0123	HD 6130	IDS00575N6032	HD 10425	IDS01366N6003
HD 2475	IDS00233S2053	HD 6156	IDS00575S2209	HD 10443	IDS01368N0124
HD 2628	IDS00248N2912	HD 6203	IDS00580S0522	HD 10453	IDS01368S1149
HD 2637	IDS00249S0431	HD 6286	IDS00587N2603	HD 10476	IDS01371N1947
HD 2767	IDS00261N3302	HD 6288	IDS00587N0050	HD 10497	IDS01373N5223
HD 2772	IDS00262N5358	HD 6312	IDS00573N6309	HD 10519	IDS01374S1824
HD 2834	IDS00266S4921	HD 6397	IDS00598N1424	HD 10543	IDS01377N5702
HD 2884	IDS00270S6331	HD 6456	IDS01003N2056	HD 10700	IDS01394S1628
HD 2885	IDS00270S6331	HD 6457	IDS01003N2056	HD 10780	IDS01405N6322
HD 2910	IDS00274N1945	HD 6479	IDS01006N0423	HD 10830	IDS01410S2533
HD 2913	IDS00272N0624	HD 6480	IDS01006N0423	HD 10859	IDS01413S7939
HD 2942	IDS00276N2744	HD 6540	IDS01012N5258	HD 11031	IDS01430N4724
HD 3003	IDS00282S6335	HD 6582	IDS01016N5427	HD 11154	IDS01446N2147
HD 3074	IDS00288S3532	HD 6595	IDS01016S4715	HD 11161	IDS01447N6557
HD 3108	IDS00292S3320	HD 6660	IDS01023N2226	HD 11171	IDS01447S1111
HD 3196	IDS00301N0449	HD 6763	IDS01032N0508	HD 11262	IDS01455S3854
HD 3266	IDS00307N2927	HD 6767	IDS01032S4201	HD 11353	IDS01465S1049
HD 3283	LS I +60 116	HD 6805	IDS01035S1042	HD 11428	IDS01473N4014
HD 3365	IDS00314S3851	HD 6811	IDS01037N4643	HD 11443	IDS01474N2906
HD 3369	IDS00315N3310	HD 6829	IDS01039N6814	HD 11502	IDS01480N1848
HD 3443	IDS00322S2519	HD 6860	IDS01041N3505	HD 11604	IDS01487S8040
HD 3460	IDS00324S3750	HD 6882	IDS01042S5547	HD 11683	IDS01496S1613
HD 3574	IDS00336N4848	HD 6961	IDS01050N5438	HD 11727	IDS01502N3646
HD 3627	IDS00340N3019	HD 7019	IDS01056N5712	HD 11749	IDS01502N3646
HD 3651	IDS00341N2043	HD 7047	IDS01057N0901	HD 11803	IDS01507N0121
HD 3690	IDS00347N2053	HD 7157	IDS01068N6111	HD 11937	IDS01521S5206
HD 3712	IDS00348N5559	HD 7229	IDS01075N2932	HD 11946	IDS01522N6408
HD 3794	IDS00355S1704	HD 7280	IDS01080S2943	HD 11964	IDS01522S1043
HD 3795	G 266 - 152	HD 7318	IDS01083N2403	HD 11973	IDS01524N2306
HD 3807	IDS00356S0454	HD 7344	IDS01085N0703	HD 12051	IDS01533N3244
HD 3883	IDS00363N2405	HD 7345	B9.5 + G0III-IV	HD 12068	IDS01533S2920
HD 3980	IDS00372S5703	HD 7438	IDS01094S0828	HD 12111	IDS01537N7025
HD 4065	IDS00379S3901	HD 7439	IDS01094S0828	HD 12135	IDS01540S3333
HD 4096	IDS00383S0126	HD 7476	IDS01097S0130	HD 12139	IDS01540N2031
HD 4150	IDS00389S5801	HD 7672	IDS01115S0301	HD 12150	IDS01543N5743
HD 4180	IDS00392N4744	HD 7693	IDS01124S6884	HD 12173	IDS01543N7322
HD 4161	IDS00390N7426	HD 7788	IDS01124S6924	HD 12211	IDS01548N2725
HD 4211	G 267 - 161	HD 7853	IDS01131N3652	HD 12230	IDS01551N7648
HD 4222	IDS00396N5440	HD 7895	IDS01135S0123	HD 12292	IDS01555S0900
HD 4294	IDS00402S6303	HD 7916	IDS01136S6656	HD 12301	LS I +64 084
HD 4307	IDS00405S1325	HD 7969	G 268 - 150	HD 12401	IDS01564N5445
HD 4338	IDS00407S1658		IDS01140S2702	HD 12447	IDS01569N0217
HD 4374	IDS00411S0611	HD 7983	IDS01140S0926	HD 12471	IDS01571N3248
HD 4391	IDS00411S4806	HD 8003	IDS01144N6408	HD 12509	IDS01576N6354
HD 4398	G 268 - 44	HD 8036	IDS01147S0101	HD 12533	IDS01578N4151
HD 4502	IDS00421N2344	HD 8272	IDS01170N5737	HD 12534	IDS01578N4151
HD 4597	IDS00427S3728	HD 8350	IDS01177S1936	HD 12558	IDS01580N2527
HD 4614	IDS00430N5717	HD 8389	IDS01180S1329	HD 12641	IDS01587S0049
HD 4628	IDS00430N0446	HD 8491	IDS01189N6736		G5 II-III+G5 V
HD 4631	IDS00431S2537	HD 8511	IDS01190S0832	HD 12727	IDS01598N5635
HD 4635	IDS00433N6954	HD 8512	IDS01190S0842	HD 12873	IDS02010S2451

Table 5

Star	Identification	Star	Identification	Star	Identification
HD 12885	IDS02012N2514	HD 17543	IDS02437N1703	HD 22001	IDS03276S6317
HD 12889	IDS02010S2451	HD 17573	IDS02441N2651	HD 22049	IDS03282S0948
HD 13078	IDS02029N4019	HD 17581	IDS02443N5754	HD 22091	IDS03285N2408
HD 13174	IDS02037N2528	HD 17584	IDS02442N3754	HD 22124	IDS03288N3141
HD 13267	LS I +57 037	HD 17652	IDS02449N3250	HD 22262	IDS03299S3125
	IDS02045N5710	HD 17713	IDS02454S2458	HD 22322	IDS03306S3213
HD 13294	IDS02048N3834	HD 17743	IDS02458N5235	HD 22403	G 6 - 20
HD 13474	G0 II-III+B9 V	HD 17793	IDS02462S3615	HD 22468	IDS03317N0016
	IDS02066N6603	HD 17818	IDS02464N4810	HD 22611	IDS03332N6219
HD 13476	LS I +58 066	HD 17824	IDS02465S2125	HD 22694	IDS03338N1804
HD 13480	IDS02066N2950	HD 17878	IDS02472N5221	HD 22695	IDS03338N1613
HD 13530	IDS02069N5037	HD 17904	IDS02474N3756	HD 22764	IDS03345N5939
HD 13579	IDS02075N6713	HD 17948	IDS02480N6107	HD 22928	IDS03358N4728
HD 13594	IDS02076N4701	HD 17958	IDS02481N6355	HD 22951	IDS03360N3338
HD 13611	IDS02077N0823	HD 18071	IDS01226S2251	HD 23010	IDS03365S1207
HD 13612	IDS02077S0252	HD 18143	IDS02497N2628	HD 23049	IDS03369N4812
HD 13621	IDS02077N5451	HD 18155	IDS02497N4646	HD 23050	IDS03370N4218
HD 13872	IDS02101N2435	HD 18438	IDS02528N7901	HD 23089	IDS03373N6302
HD 13970	IDS02108N5611	HD 18520	IDS02535N2056	HD 23180	IDS03380N3158
HD 13974	IDS02108N3346	HD 18537	IDS02537N5157	HD 23189	IDS03383N6821
HD 13994	IDS02110N5703	HD 18538	IDS02537N5157	HD 23230	IDS03384N4216
HD 14014	IDS02110N5546	HD 18541	IDS02537N3848	HD 23300	IDS03390N4522
HD 14228	IDS02129S5158	HD 18543	IDS02537S0311	HD 23319	IDS03391S3738
HD 14252	IDS02132N2811	HD 18622	IDS02545S4042	HD 23401	IDS03398N7101
HD 14386	IDS02143S0326	HD 18692	IDS02552S2540	HD 23439	IDS03401N4110
HD 14489	IDS02154N5523	HD 18778	IDS02554N8105	HD 23466	IDS03404N0544
HD 14528	IDS02156N5808	HD 18925	IDS02576N5307	HD 23508	IDS03406S4058
HD 14622	IDS02166N4056	HD 18953	IDS02578S0805	HD 23552	IDS03409N5026
HD 14641	IDS02167S5624	HD 19059	IDS02588N3424	HD 23594	IDS03414N5649
HD 14643	IDS02167S7156	HD 19134	IDS02596N2452	HD 23625	IDS03415N3317
HD 14662	IDS02169N5454	HD 19308	IDS03013N3615	HD 23697	IDS03420S5435
HD 14772	IDS02178N1458	HD 19349	IDS03016S0628	HD 23793	IDS03428N1050
HD 14786	IDS02178N1458	HD 19356	IDS03017N4034	HD 23817	IDS03429S6507
HD 14817	IDS02182N6106	HD 19400	IDS03021S7228	HD 23985	IDS03443N2517
HD 14818	LS I +56 048	HD 19476	IDS03027N4429	HD 24072	IDS03449S3756
HD 14872	IDS02189N4950	HD 19656	IDS03048N3914	HD 24131	IDS03455N3403
HD 15089	IDS02208N6657	HD 19836	IDS03063S0411	HD 24263	IDS03466N0615
HD 15144	IDS02212S1547	HD 19926	K1 IIIep+A6 V	HD 24388	IDS03478S0540
HD 15253	IDS02224N5505	HD 19978	IDS03076N7722	HD 24398	IDS03478N3135
HD 15285	IDS02225N0359	HD 19994	IDS03077S0134	HD 24480	IDS03486N6049
HD 15325	IDS02229N5648	HD 20010	IDS03078S2923	HD 24534	IDS03491N3045
HD 15328	IDS02229N0131	HD 20041	LS I +56 085	HD 24546	IDS03492N5024
HD 15453	IDS02242N0907		IDS03081N5646	HD 24550	IDS03492N0454
HD 15497	IDS02247N5715	HD 20104	IDS03088N6517	HD 24554	IDS03493S0315
HD 15524	IDS02248N2448	HD 20121	IDS03089S4448	HD 24555	IDS03493S0315
HD 15548	IDS02249N5613	HD 20193	IDS03096N3229	HD 24640	LS V +34 001
HD 15588	IDS02254S2308	HD 20210	IDS03098N3419	HD 24744	IDS03509S4039
HD 16004	IDS02292N3914	HD 20283	IDS03106N4007	HD 24749	IDS03510N0025
HD 16028	IDS02295N3652	HD 20313	IDS03109S7922	HD 24760	LS V +39 001
HD 16046	IDS02295S2840	HD 20319	IDS03111S0617		IDS03511N3943
HD 16058	IDS02297N3415	HD 20336	IDS03112N6517	HD 24863	IDS03519S5259
HD 16074	IDS02298S0818	HD 20512	IDS03129N1451	HD 24909	IDS03525N4752
HD 16160	G 73 - 71	HD 20547	IDS03133N5848	HD 24982	IDS03530N3832
	IDS02307N0626	HD 20559	IDS03132S0118	HD 25007	IDS03533N8025
HD 16161	IDS02306N0509	HD 20610	IDS03140S2253	HD 25025	IDS03534S1348
HD 16212	IDS02311S0816	HD 20630	IDS03141N0300	HD 25291	LS I +59 196
HD 16232	IDS02312N2413	HD 20631	IDS03141S1855	HD 25330	IDS03563N0943
HD 16234	IDS02312N1200	HD 20720	IDS03151S2207	HD 25555	IDS03583N2350
HD 16246	IDS02312N2413	HD 20727	IDS03152N0840	HD 25604	IDS03588N2148
HD 16327	IDS02321N3718	HD 20756	IDS03155N2047	HD 25680	IDS03594N2144
HD 16429	IDS02331N6051	HD 20766	IDS03160S6253	HD 25833	IDS04005N3311
HD 16467	IDS02334N0301	HD 20798	IDS03160N6114	HD 25887	IDS04009S8534
HD 16523	LS I +56 062	HD 20807	IDS03160S6253	HD 25893	G 39 - 1
HD 16620	IDS02347S1218	HD 20835	IDS03163N4533		IDS04009N374
HD 16628	IDS02348N2638	HD 20898	IDS03171N6008	HD 26038	IDS04023N1704
HD 16735	IDS02359N5306	HD 20995	IDS03182N3311	HD 26413	IDS04055S4608
HD 16739	IDS02359N3946	HD 21018	IDS03184N0432	HD 26553	LS V +57 058
HD 16754	IDS02360S4319	HD 21019	IDS03185S0809	HD 26630	IDS04076N4809
HD 16765	IDS02361S0107	HD 21203	IDS03202N5954	HD 26677	IDS04081N0838
HD 16780	IDS02363N4750	HD 21224	IDS03204N5933	HD 26690	IDS04082N0728
HD 16811	IDS02367N1935	HD 21291	LS I +59 187	HD 26722	IDS04085N0901
HD 16895	IDS02374N4848		IDS03210N5935	HD 26846	IDS04096S1030
HD 16955	IDS02380N2513	HD 21389	LS I +58 125	HD 26906	IDS04101N4558
HD 16970	IDS02381N0249	HD 21427	IDS03221N5901	HD 26913	IDS04102N0556
HD 17098	IDS02395S4057	HD 21447	IDS03224N5506	HD 26923	IDS04102N0556
HD 17240	IDS02409N3508	HD 21448	IDS03224N4442	HD 26965	IDS04108S0749
HD 17326	IDS02417S6708	HD 21467	IDS03226N2227		G 160 - 60
HD 17327	IDS02418N6413	HD 21635	IDS03243S3612	HD 27192	LS V +50 006
HD 17382	IDS02423N2639	HD 21727	IDS03252N4320	HD 27256	IDS04131S6243
HD 17504	IDS02433S6408	HD 21769	IDS03255N5826	HD 27304	IDS04135S6226
HD 17506	IDS02434N5529	HD 21903	IDS03268N5942	HD 27348	IDS04140N3420

Star	Identification	Star	Identification	Star	Identification
HD 27376	IDS04141S3403	HD 31421	IDS04508N1322	HD 34968	IDS05162S2120
HD 27382	IDS04142N2707	HD 31512	IDS04515S0520	HD 34989	LS VI +08 001
HD 27402	IDS04144N5923	HD 31553	IDS04518N2348	HD 35007	IDS05164S0031
HD 27442	IDS04148S5932	HD 31590	IDS04520N7355	HD 35149	IDS05176N0327
HD 27463	IDS04148S6112	HD 31592	IDS04520N2454	HD 35162	IDS05177S2452
HD 27490	IDS04153S3409	HD 31647	IDS04525N3744		G7 II-III+A7IV-V
HD 27588	IDS04161S4430	HD 31662	IDS04526N6056	HD 35165	IDS05177S3427
HD 27604	IDS04162S5306	HD 31739	IDS04532S0222	HD 35186	IDS05179N3718
HD 27611	IDS04163S0020	HD 31761	IDS04534N3915	HD 35281	IDS05185S0831
HD 27638	IDS04165N2524	HD 31764	IDS04533N1423	HD 35295	IDS05186N3446
HD 27639	IDS04165N2035	HD 31910	IDS04545N6018		K1p III-IV+F6 V
HD 27650	IDS04166N4212	HD 31925	IDS04546S1632	HD 35296	IDS05186N1716
HD 27657	IDS04166S6330	HD 31964	LS V +43 023	HD 35317	IDS05188S0058
HD 27710	IDS04174S2558		IDS04548N4341	HD 35328	IDS05189N3005
HD 27778	IDS04180N2404	HD 32039	IDS04553N0328	HD 35343	Sk -69 094
HD 27786	IDS04181N3344	HD 32040	IDS04553N0328	HD 35411	IDS05194S0229
HD 27820	IDS04184N0914	HD 32188	LS V +41 023	HD 35468	IDS05198N0615
HD 27855	IDS04187N5721	HD 32228	Sk -66 028	HD 35497	IDS05200N2831
HD 27941	IDS04195S3547	HD 32273	IDS04568N0128	HD 35520	LS V +34 009
HD 28143	IDS04212S3459	HD 32343	IDS04573N5850	HD 35548	IDS05204S0038
HD 28204	IDS04219N7219	HD 32357	IDS04573N5850	HD 35600	LS V +30 007
HD 28217	IDS04220N1059	HD 32358	IDS04575N4449	HD 35619	IDS05210N3440
HD 28255	IDS04222S5718	HD 32445	IDS04582S5416	HD 35620	IDS05210N3423
HD 28271	IDS04226N3008	HD 32537	IDS04588N5128	HD 35671	IDS05213N1753
HD 28312	IDS04227S2418	HD 32630	LS V +41 027	HD 35708	LS V +21 006
HD 28446	LS V +53 024	HD 32642	IDS04596N1940		IDS05216N2151
	IDS04241N5342	HD 32743	IDS05002S4917	HD 35715	IDS05216N0301
HD 28459	IDS04242N3214	HD 32781	IDS05005N7621	HD 35736	IDS05217S1947
HD 28503	IDS04246N3948	HD 32831	IDS05008S3537	HD 35860	IDS05225S5224
HD 28704	IDS04264N4251	HD 32923	IDS05015N1831	HD 35921	IDS05230N3518
HD 28732	IDS04266S6244	HD 32964	IDS05018S0447	HD 35943	IDS05231N2504
HD 28763	IDS04268S1351	HD 32989	IDS05020N3921	HD 35961	IDS05233N5435
HD 28776	IDS04270S3552	HD 32990	IDS05020N2408	HD 36003	IDS05235S0333
HD 28843	IDS04276S0325	HD 32991	LS V +21 002	HD 36041	IDS05238N3945
HD 28867	IDS04278N1748		IDS05019N2135	HD 36044	IDS05237N2928
HD 28929	IDS04284N2845	HD 33021	IDS05022N0921	HD 36058	IDS05240S0323
HD 29009	IDS04290S0657	HD 33054	IDS05024N0822	HD 36060	IDS05239S4102
HD 29104	IDS04298N1941	HD 33111	IDS05030S0513	HD 36079	IDS05240S2050
HD 29139	IDS04302N1619	HD 33203	LS V +37 007	HD 36090	IDS05241S0446
HD 29140	IDS04302N0957		IDS05036N3710	HD 36167	IDS05240S0110
HD 29173	IDS04305S0957	HD 33224	IDS05036S0848	HD 36267	IDS05254N0552
HD 29180	IDS04306N4430	HD 33276	IDS05039N1529	HD 36313	IDS05256S0027
HD 29227	IDS04310S0349	HD 33366	IDS05046N2502	HD 36351	IDS05260N0313
HD 29248	IDS04313S0332	HD 33412	IDS05050N4146	HD 36371	LS V +32 008
HD 29305	IDS04318S5515	HD 33519	IDS05056S7826	HD 36395	IDS05264S0341
HD 29316	IDS04320N5317	HD 33564	IDS05061N7907	HD 36408	IDS05264N1659
HD 29317	IDS04320N5253	HD 33646	IDS05066N0055	HD 36483	IDS05269N3623
HD 29364	IDS04323N2644	HD 33647	IDS05066N0024	HD 36485	IDS05269S0022
HD 29391	IDS04326S0240	HD 33654	LS V +53 031	HD 36486	IDS05269S0022
HD 29399	IDS04325S6302	HD 33802	IDS05076S1159	HD 36496	IDS05270N6638
HD 29479	IDS04335N1543	HD 33856	IDS05081N0245	HD 36524	IDS05272N1958
HD 29503	IDS04336S1430	HD 33883	IDS05083N0151	HD 36553	IDS05274S4709
HD 29598	IDS04345S8307	HD 33949	IDS05086S1304	HD 36570	IDS05276N6405
HD 29606	IDS04346N5920	HD 33959	IDS05089N3234	HD 36584	IDS05275S6842
HD 29646	IDS04351N2825	HD 34019	IDS05092N5328	HD 36591	Ring Ori # 1
HD 29678	IDS04354N7546	HD 34029	IDS05093N4554		LS VI −01 001
HD 29697	G 8 - 51		G5 IIIe+G0 III		IDS05276S0140
HD 29712	IDS04356S6216	HD 34053	IDS05094N2207	HD 36646	Ring Ori # 2
HD 29721	IDS04358N4947	HD 34078	IDS05097N3412		IDS05281S0147
HD 29755	IDS04361S1952	HD 34085	IDS05097S0819	HD 36673	IDS05283S1754
HD 29763	IDS04362N2246	HD 34334	IDS05116N3316	HD 36705	IDS05284S6532
HD 29833	IDS04370N4214	HD 34335	IDS05115N3002	HD 36777	IDS05290N0342
HD 29875	IDS04373S4203	HD 34411	IDS05121N4001	HD 36779	IDS05290S0106
HD 30003	IDS04386S5908	HD 34467	IDS05125N3541	HD 36822	IDS05293N0925
HD 30020	IDS04388S0859	HD 34493	IDS05127N3802	HD 36861	IDS05296N0952
HD 30021	IDS04388S0859	HD 34499	IDS05128N3353	HD 36862	IDS05296N0952
	G8 III+F4 IIIp	HD 34503	IDS05128S0657	HD 36876	IDS05297S6400
HD 30121	IDS04397N5635	HD 34533	IDS05132N4652	HD 36881	IDS05297N1011
HD 30442	IDS04427N6320	HD 34578	LS V +33 010	HD 37055	IDS05306S0319
HD 30605	IDS04440N1544	HD 34579	IDS05133N2002	HD 37098	IDS05309N2652
HD 30649	IDS04444N4541		G8 II-III+G1IV-V	HD 37128	IDS05311S0116
HD 30652	IDS04444N0647	HD 34656	IDS05139N3720	HD 37226	IDS05318S5458
HD 30958	IDS04469N5506	HD 34673	IDS05141S0311	HD 37232	LS VI +08 002
HD 30985	IDS04470S4130	HD 34721	IDS05141S1814	HD 37269	IDS05322N3026
HD 31093	IDS04478S3504	HD 34760	IDS05148N3317		B9.5 V+ F9 III
HD 31203	IDS04487S5338	HD 34787	IDS05149N5727	HD 37321	IDS05325S0129
HD 31208	IDS04488N0713	HD 34797	IDS05149S1837	HD 37367	LS V +29 006
HD 31278	IDS04493N5336	HD 34798	IDS05149S1837	HD 37394	LS VI −02 002
HD 31283	IDS04492N1116	HD 34842	IDS05153N5324		IDS05332N5326
HD 31295	IDS04494N0959	HD 34880	IDS05155S0528	HD 37468	IDS05337S0239
HD 31327	LS V +36 002	HD 34925	IDS05157N3342	HD 37479	IDS05337S0239

Table 5

1353

Star	Identification	Star	Identification	Star	Identification
HD 37643	IDS05349S1754	HD 42327	IDS06049S1807	HD 47144	IDS06319S3642
HD 37646	IDS05350N2926	HD 42443	IDS06056S2245	HD 47152	IDS06320N2904
HD 37711	IDS05355N1629	HD 42448	IDS06056S4420	HD 47174	IDS06322N4235
HD 37742	IDS05357S0200	HD 42509	IDS06061N1949	HD 47230	IDS06324S3600
HD 37744	LS VI −02 004	HD 42560	IDS06063N1415	HD 47240	LS VI +05 019
HD 37763	IDS05358S7625	HD 42633	IDS06067N6002	HD 47247	IDS06325S2232
HD 37795	IDS05360S3408	HD 42657	IDS06068S0439	HD 47395	IDS06332N2821
HD 37904	IDS05367S0257	HD 42954	IDS06086N1756	HD 47432	LS VI +01 008
HD 38014	IDS05375N0238	HD 42995	IDS06088N2232	HD 47463	IDS06336S3804
HD 38017	IDS05376N3053	HD 43017	IDS06089N3611	HD 47500	IDS06338S3654
HD 38089	IDS05380S0651	HD 43042	IDS06090N1912	HD 47575	IDS06342N1304
HD 38230	IDS05391N3715	HD 43112	LS VI +13 009	HD 47731	IDS06351N2818
HD 38268	Sk −69 243		IDS06095N1353	HD 47827	IDS06354S2336
HD 38284	IDS05396N6246	HD 43157	IDS06097S0432	HD 47973	IDS06360S4808
HD 38309	IDS05398N0358	HD 43232	IDS06100S0615	HD 48099	LS VI +06 009
HD 38392	IDS05403S2229	HD 43247	LS VI +12 003	HD 48189	IDS06369S6127
HD 38393	IDS05403S2229	HD 43319	IDS06106S0453	HD 48250	IDS06374N5933
HD 38495	IDS05411S0418	HD 43358	IDS06107N0112	HD 48279	IDS06375N0149
HD 38527	IDS05414N0929	HD 43362	IDS06107S0900	HD 48329	IDS06378N2514
HD 38563	IDS05416N0002	HD 43384	LS V +23 045	HD 48383	IDS06380S4015
HD 38618	IDS05420N5653	HD 43386	IDS06108N1218	HD 48433	IDS06384N1320
HD 38622	IDS05420N1352	HD 43525	IDS06116N0959	HD 48434	LS VI +03 011
HD 38656	IDS05422N3909	HD 43587	IDS06120N0508	HD 48501	IDS06386S2221
HD 38670	IDS05424N2050	HD 43683	IDS06124N1425	HD 48543	IDS06389S3818
HD 38710	IDS05426N0625	HD 43745	IDS06129S2240	HD 48682	IDS06395N4341
HD 38735	IDS05427S1034	HD 43812	IDS06132N5925	HD 48717	IDS06395N0347
HD 38751	IDS05429N2443	HD 43836	IDS06133N2319	HD 48767	IDS06399N5549
HD 38871	IDS05437S4638	HD 43930	IDS06138N1328	HD 48843	LS VI +12 010
HD 38899	IDS05439N1237	HD 44120	IDS06149S5910	HD 48915	IDS06408S1635
HD 39003	IDS05446N3907	HD 44173	LS VI +11 007	HD 49059	IDS06416N1818
HD 39004	IDS05446N2756	HD 44213	IDS06155N0547	HD 49131	IDS06417S3051
HD 39007	IDS05445N0950	HD 44333	IDS06162N0219	HD 49268	IDS06425S7140
HD 39045	IDS05449N3206	HD 44402	IDS06165S3001	HD 49409	IDS06432N0744
HD 39070	IDS05451S1431	HD 44458	IDS06166S1441	HD 49420	IDS06437N4154
HD 39357	IDS05470N2735	HD 44472	IDS06168N7035	HD 49567	LS VI +01 013
HD 39400	IDS05472N0150	HD 44478	IDS06169N2234	HD 49591	IDS06439S3749
HD 39547	IDS05483S5248	HD 44708	IDS06181N5828	HD 49606	IDS06441N1620
HD 39662	IDS05490N1145	HD 44743	IDS06183S1755	HD 49618	IDS06443N5934
HD 39680	IDS05491N1350	HD 44769	IDS06185N0439	HD 49643	IDS06442S0210
HD 39720	IDS05491S3739	HD 44770	IDS06185N0439	HD 49662	IDS06444S1502
HD 39752	IDS05494S3833	HD 44811	IDS06187N1946	HD 49705	IDS06446S5435
HD 39801	IDS05498N0723	HD 44927	IDS06195N2323	HD 49891	IDS06456S2358
HD 39866	LS V +28 015	HD 44953	IDS06195S1944	HD 49908	IDS06456N2153
HD 39881	IDS05503N1355	HD 44996	IDS06197S1254	HD 49933	IDS06457S0026
HD 39891	IDS05504S2910	HD 45050	IDS06201N0133	HD 50018	IDS06462N3859
HD 39970	LS V +24 006	HD 45088	IDS06203N1849	HD 50019	IDS06462N3405
HD 40035	IDS05512N5416	HD 45145	IDS06206S3639	HD 50037	IDS06464N3834
HD 40111	LS V +25 011	HD 45380	IDS06219S0727	HD 50123	IDS06466S3135
HD 40151	IDS05520S2251	HD 45383	IDS06219S3500	HD 50196	IDS06470S4520
HD 40183	IDS05522N4456	HD 45410	IDS06220N5814	HD 50522	IDS06486N5833
HD 40312	IDS05529N3712	HD 45530	IDS06229N0520	HD 50635	IDS06490N1318
HD 40325	IDS05530N4435	HD 45542	LS V +20 041	HD 50648	IDS06490S2650
HD 40369	IDS05533N1248		IDS06230N2017	HD 50700	IDS06492S0544
HD 40372	IDS05533N2O150	HD 45572	IDS06231S4807	HD 50820	LS VI −01 012
HD 40446	IDS05537N0033	HD 45588	IDS06231S2547	HD 50860	IDS06498S4351
HD 40494	IDS05540S3518	HD 45680	IDS06237S3750	HD 50885	IDS06500N7057
HD 40589	LS V +27 022	HD 45724	IDS06240N0243	HD 51055	IDS06507S2017
	IDS05547N2734	HD 45725	IDS06240S0658	HD 51199	IDS06513S2001
HD 40665	IDS05551S5326	HD 45726	IDS06240S0658	HD 51250	IDS06515S1355
HD 40873	IDS05566N5135	HD 45827	LS VI +09 007	HD 51530	IDS06527N2612
HD 40967	IDS05571S1036	HD 45871	IDS06249S3218	HD 51688	IDS06533N2603
HD 41040	IDS05575N1942	HD 45951	IDS06254N1700	HD 51689	IDS06532N2522
HD 41116	IDS05580N2316	HD 45995	LS VI +11 012	HD 51697	IDS06534S0430
HD 41117	LS V +20 011		IDS06256N1119	HD 51733	IDS06534S2430
HD 41162	IDS05582N3758	HD 46064	IDS06260S1305	HD 51756	IDS06535S0253
HD 41312	IDS05592S2617	HD 46178	IDS06268N1145	HD 51814	IDS06537N0344
HD 41330	IDS05594N3524	HD 46184	IDS06267S1219	HD 51825	IDS06537S3522
HD 41429	IDS06000N2931	HD 46273	IDS06274S5010	HD 51925	IDS06541S2702
HD 41534	IDS06006S3210	HD 46300	LS VI +07 005	HD 52089	IDS06547S2850
HD 41690	IDS06016N2153	HD 46304	IDS06275S0548	HD 52100	IDS06548N3233
HD 41692	IDS06017S0411	HD 46480	IDS06286N6134	HD 52312	IDS06556S0816
HD 41700	IDS06018S4505	HD 46547	IDS06289S3157	HD 52437	IDS06561S2159
HD 41742	IDS06018S4505	HD 46642	IDS06294N0739	HD 52533	IDS06564S0259
HD 41824	IDS06022S4827	HD 46660	IDS06294N1112	HD 52559	LS VI +05 033
HD 41841	IDS06024S2306	HD 46667	IDS06295N3831	HD 52611	IDS06568S0112
HD 42087	LS V +23 032	HD 46769	LS VI +00 009	HD 52721	IDS06571S1110
	IDS06037N2308	HD 46847	IDS06305N0248	HD 52859	IDS06577N5254
HD 42111	IDS06038N0231	HD 46860	IDS06305S5841	HD 52973	IDS06582N2043
HD 42127	IDS06040N4844	HD 46867	IDS06306N0523	HD 53205	IDS06591N0138
HD 42261	IDS06046S0619	HD 47105	IDS06319N1629	HD 53367	IDS06597S1018
HD 42303	IDS06048S4208	HD 47138	IDS06320S1835	HD 53456	IDS07000S1122

Star	Identification	Star	Identification	Star	Identification
HD 53705	IDS07009S4328	HD 60859	IDS07314S1934	HD 68257	IDS08065N1757
HD 53706	IDS07009S4328	HD 60863	IDS07314S2809	HD 68273	IDS08065S4703
HD 53755	IDS07011S1030	HD 60986	IDS07320N3516	HD 68290	IDS08066S1238
HD 53791	IDS07013N2251	HD 61064	IDS07323S0353	HD 68457	IDS08074N6041
HD 53921	IDS07017S5902	HD 61330	IDS07337S3445	HD 68520	IDS08076S6819
HD 53952	IDS07019S3438	HD 61421	IDS07341N0529	HD 68552	IDS08078S3151
HD 53974	IDS07020S1108	HD 61429	IDS07341S2508	HD 68601	IDS08081S4241
HD 54046	IDS07023N1541	HD 61497	IDS07346N5857	HD 68638	IDS08082N5724
HD 54100	IDS07023N1541	HD 61555	IDS07347S2634	HD 68703	IDS08085N1759
HD 54131	IDS07026N1606	HD 61563	IDS07348N0528	HD 68776	IDS08088N1322
HD 54563	G 88 - 9	HD 61606	IDS07350S0322	HD 68895	IDS08093S4558
HD 54719	IDS07048N3025	HD 61709	IDS07354S3104	HD 68951	IDS08097N7243
HD 54764	IDS07051S1604	HD 61774	IDS07358S1926	HD 69002	IDS08098S3316
HD 54810	IDS07052S0405	HD 61931	IDS07366N5040	HD 69080	IDS08102S3150
HD 54912	IDS07056S2504	HD 61966	IDS07366S5303	HD 69081	IDS08102S3601
HD 54958	IDS07057S1831	HD 62044	IDS07371N2907	HD 69082	IDS08102S3601
HD 55130	IDS07066N2724	HD 62153	IDS07374S7403	HD 69142	IDS08105S4003
HD 55185	IDS07068S0020	HD 62264	IDS07380N0026	HD 69144	IDS08105S4641
HD 55187	IDS07068S0358	HD 62345	IDS07384N2438	HD 69267	IDS08111N0930
HD 55383	IDS07076N1620	HD 62509	IDS07392N2816	HD 69302	IDS08112S4532
HD 55458	IDS07078N2511	HD 62555	IDS07395S2516	HD 69445	IDS08119S3037
HD 55579	IDS07083N2453	HD 62556	IDS07395S2810	HD 69863	IDS08138S6236
HD 55621	IDS07086N2504	HD 62864	IDS07409S1427	HD 69882	IDS08139S4213
HD 55718	IDS07089S3623	HD 62898	IDS07411N3340	HD 70003	IDS08145S3704
HD 55775	IDS07092S0344	HD 63208	IDS07426N2323	HD 70013	IDS08146N0415
HD 55856	IDS07096S2244	HD 63295	IDS07430S7222	HD 70267	IDS08159S5851
HD 55865	IDS07096S7020	HD 63302	IDS07431S1544	HD 70302	IDS08161S2237
HD 56003	IDS07102N0001	HD 63323	IDS07431S1544	HD 70340	IDS08163S0117
HD 56014	IDS07102S2611	HD 63336	IDS07433S1157	HD 70556	IDS08176S3610
HD 56096	IDS07105S4429	HD 63382	IDS07435S5629	HD 70647	IDS08179N4220
HD 56196	IDS07111N6536	HD 63462	IDS07439S2541	HD 70734	IDS08184N1057
HD 56239	IDS07111S6301	HD 63589	IDS07446N3329	HD 70761	IDS08186S2602
HD 56456	IDS07119S4806	HD 63700	IDS07451S2437	HD 70930	IDS08195S4810
HD 56537	IDS07124N1643	HD 63754	IDS07454S1957	HD 71046	IDS08201S7112
HD 56577	IDS07124S2308	HD 63799	IDS07456N0332	HD 71066	IDS08201S7112
HD 56820	IDS07135N6005	HD 63804	IDS07455S3305	HD 71093	IDS08204N2813
HD 56855	IDS07136S3655	HD 63922	IDS07462S4607	HD 71115	IDS08205N0753
HD 56986	IDS07142N2210	HD 63926	IDS07462S5613	HD 71150	IDS08207N2716
HD 57150	IDS07147S3633	HD 64042	IDS07468S2416	HD 71152	IDS08207N2452
HD 57219	IDS07147S3633	HD 64067	IDS07470S5610	HD 71153	IDS08207N2452
HD 57264	IDS07154N3657	HD 64090	IDS07472N3055	HD 71176	IDS08207S2343
HD 57423	IDS07160N2038	HD 64096	IDS07471S1338		K5 III+K1 III
HD 57593	IDS07169S2647	HD 64235	IDS07479S0510	HD 71297	IDS08215S0340
HD 57682	LS VI −08 011	HD 64379	IDS07485S3427	HD 71302	IDS08215S4227
HD 57852	IDS07180S5207	HD 64468	IDS07491N1931	HD 71304	IDS08216S4359
	F0/2 IV/V+G0 V	HD 64486	IDS07492N7945	HD 71369	IDS08220N6103
HD 57853	IDS07180S5207	HD 64704	IDS07501N2358	HD 71487	IDS08226S3844
HD 58061	IDS07189S2535	HD 65123	IDS07521N0124	HD 71510	IDS08227S5124
HD 58420	IDS07204S3539	HD 65211	IDS07525S4335	HD 71576	IDS08230S7304
HD 58439	IDS07205S1849	HD 65339	IDS07532N6036	HD 71581	IDS08231S2031
HD 58585	IDS07210S2147	HD 65345	IDS07532N0229	HD 71663	IDS08234S0211
HD 58635	IDS07212S3706	HD 65448	IDS07536N6322	HD 71701	IDS08236S7710
HD 58661	IDS07214N4823	HD 65551	IDS07541S4350	HD 71801	IDS08241S3447
HD 58715	IDS07217N0829	HD 65598	IDS07544S4737	HD 71833	IDS08244S2037
HD 58728	IDS07218N2139	HD 65757	IDS07550N2331	HD 71919	IDS08249S5441
HD 58855	IDS07223N4952	HD 65818	IDS07554S4858	HD 71973	IDS08252N7504
HD 58923	IDS07226N0709	HD 65856	IDS07557N2522	HD 72014	IDS08254S4215
HD 58946	IDS07227N3159	HD 65871	IDS07558N6841	HD 72067	IDS08257S4350
HD 58954	IDS07227S1740	HD 65875	LS VI −02 018	HD 72094	IDS08259N1826
HD 58972	IDS07227N0907	HD 65900	IDS07559N0509	HD 72108	IDS08259S4736
HD 59067	IDS07232S1121	HD 66141	IDS07571N0236	HD 72127	IDS08261S4423
	G8 Ib-II+B8 V	HD 66171	IDS07571N7213	HD 72310	IDS08270S1914
HD 59148	IDS07236N2807	HD 66216	IDS07574N2804	HD 72322	IDS08270S5451
HD 59438	IDS07248S1447	HD 66546	IDS07590S5414	HD 72350	IDS08273S4424
HD 59499	IDS07250S3139	HD 66598	IDS07592S3211	HD 72436	IDS08277S3844
HD 59612	IDS07256S2249	HD 66624	IDS07593S4102	HD 72462	IDS08279S1441
HD 59717	IDS07261S4306	HD 66684	IDS07595N2749	HD 72617	IDS08288N0848
HD 59878	IDS07268N2306	HD 67006	IDS08009N5148	HD 72626	IDS08288S2416
	K0 II-III+F8 V	HD 67159	IDS08016S0857	HD 72737	IDS08293S5252
HD 59984	IDS07273S0840	HD 67243	IDS08019S3317	HD 72945	IDS08305N0658
HD 60107	IDS07279N1603	HD 67523	IDS08033S2401	HD 72946	IDS08305N0658
HD 60168	IDS07281S3540	HD 67536	IDS08033S6233	HD 72954	IDS08305S3215
HD 60179	IDS07282N3166	HD 67594	IDS08036S0242	HD 72993	IDS08307S3716
HD 60284	IDS07286S2739	HD 67751	IDS08043S2004	HD 73262	IDS08324N0603
HD 60312	IDS07287S3545	HD 67767	IDS08044N2549	HD 73476	IDS08335S3324
HD 60318	IDS07288N3111	HD 67880	IDS08049S1557	HD 73495	IDS08336S2554
HD 60335	IDS07289N4315	HD 67921	IDS08051S3002	HD 73603	IDS08342S1923
HD 60357	IDS07291N0335	HD 68146	IDS08059S1330	HD 73667	IDS08344N1154
HD 60369	IDS07290S2807	HD 68242	IDS08064S4221	HD 73752	IDS08348S2219
HD 60522	IDS07298N2707	HD 68243	IDS08065S4703	HD 73882	IDS08355S4004
HD 60584	IDS07301S2315	HD 68256	IDS08065N1757	HD 73887	IDS08355S6230

Table 5 1355

Star	Identification	Star	Identification	Star	Identification
HD 73898	IDS08356S2912	HD 79211	IDS09078N5307	HD 85355	IDS09461S4516
HD 73900	IDS08355S3615	HD 79416	IDS09088S4312	HD 85488	IDS09470N0341
HD 74006	IDS08362S3457	HD 79452	IDS09091N3503	HD 85558	IDS09476S0738
HD 74067	IDS08366S3954	HD 79469	IDS09092N0244	HD 85583	IDS09478N6136
HD 74137	IDS08371S1535	HD 79735	IDS09107S4249	HD 85725	IDS09485S2652
HD 74148	IDS08371S5958	HD 79900	IDS09116S4508	HD 85980	IDS09504S4449
HD 74180	IDS08373S4618	HD 79910	IDS09117S0556	HD 86266	IDS09522S2604
HD 74198	IDS08375N2150	HD 79940	IDS09118S3700	HD 86440	IDS09534S5406
HD 74341	IDS08383S5711	HD 80024	IDS09123N3547	HD 86523	IDS09539S4756
HD 74371	LSS 1124	HD 80057	IDS09124S4429	HD 86629	IDS09546S3525
HD 74375	IDS08384S5924	HD 80064	IDS09124N1155	HD 86661	IDS09549N5605
HD 74377	IDS08386N4203	HD 80081	IDS09126N3714	HD 86728	IDS09553N3225
HD 74395	IDS08388S0652	HD 80290	IDS09138N5141	HD 87030	IDS09572S5628
HD 74405	IDS08387S7002	HD 80383	IDS09143S5226	HD 87076	IDS09576N6010
HD 74442	IDS08390N1831	HD 80441	IDS09147N3837	HD 87318	IDS09591S2450
HD 74688	IDS08403S0214	HD 80479	IDS09148S1525	HD 87344	IDS09593S1737
HD 74738	IDS08406N2908	HD 80499	IDS09150S1133	HD 87543	IDS10005S6124
HD 74739	IDS08406N2908	HD 80558	LSS 1272	HD 87598	IDS10007N6856
HD 74860	IDS08413S1039	HD 80586	IDS09156S0908	HD 87737	IDS10019N1715
HD 74874	IDS08415N0647		G8 III-IV+F5 V	HD 87783	IDS10022S4653
HD 74918	IDS08417S1311	HD 80671	IDS09159S6816	HD 87822	IDS10025N3206
HD 74956	IDS08419S5420	HD 80715	G 116 - 23	HD 87837	IDS10026N1029
HD 75086	IDS08427S5821	HD 80773	IDS09165S3120	HD 87901	IDS10030N1227
HD 75137	IDS08431N0612	HD 80951	IDS09176S7428	HD 88218	IDS10052S3522
HD 75318	IDS08442N0403	HD 81009	IDS09180S0925	HD 88230	IDS10053N4958
HD 75632	IDS08460N7111	HD 81025	IDS09180N5200	HD 88284	IDS10057S1152
HD 75698	IDS08464N3251	HD 81146	IDS09188N2637	HD 88323	IDS10059S6520
HD 75716	IDS08465N2838	HD 81411	IDS09203S3900	HD 88355	IDS10063N1351
HD 75732	G 47 - 9	HD 81688	IDS09221N4603	HD 88437	IDS10067S1434
	IDS08467N2842	HD 81728	IDS09224S0847	HD 88473	IDS10070S6812
HD 75737	IDS08467S0648	HD 81753	IDS09224S2821	HD 88522	IDS10075S2807
HD 75811	IDS08471N0543	HD 81797	IDS09227S0814	HD 88809	IDS10095S3951
HD 75821	LSS 1181	HD 81809	IDS09228S0538	HD 88836	IDS10097S3949
	IDS08472S4609	HD 81830	IDS09230S6131	HD 88842	IDS10096S5116
HD 75860	IDS08473S4323	HD 81858	IDS09231N0930	HD 88849	IDS10098N7134
HD 75916	IDS08478S1251	HD 81873	IDS09232N0837	HD 88987	IDS10108N1814
HD 75959	IDS08482N3057	HD 81919	IDS09235S3434	HD 89010	IDS10110N2356
HD 76072	IDS08487S3610	HD 81937	IDS09236N6330	HD 89025	IDS10110N2356
HD 76143	IDS08492S6625	HD 81980	IDS09240S0049	HD 89055	IDS10113N2622
HD 76230	IDS08496S5145	HD 81997	IDS09240S0220	HD 89125	IDS10118N2336
HD 76360	IDS08505S4708	HD 82087	IDS09247N3406	HD 89169	IDS10120S2010
HD 76370	IDS08506S0735	HD 82205	IDS09255S2609	HD 89263	IDS10126S5924
HD 76376	IDS08506S1752	HD 82242	IDS09258S5237	HD 89268	IDS10128N4716
HD 76483	IDS08512S2718	HD 82328	IDS09262N5208	HD 89388	IDS10137S6050
HD 76535	IDS08516S4702	HD 82381	IDS09266N1009	HD 89455	IDS10144S1201
HD 76566	IDS08518S4440	HD 82384	IDS09265S3127	HD 89484	IDS10145N2021
HD 76572	IDS08519N3037	HD 82434	IDS09268S4002	HD 89497	IDS10145S5900
HD 76595	IDS08522N3611	HD 82514	IDS09274S3516	HD 89714	IDS10160S5654
HD 76644	IDS08524N4826	HD 82523	IDS09274N2849	HD 89715	IDS10160S6410
HD 76728	IDS08528S6016	HD 82536	IDS09275S5755	HD 89828	IDS10169S2201
HD 76756	IDS08530N1215	HD 82543	IDS09275N0218	HD 89890	IDS10172S5532
HD 76757	IDS08530N0156	HD 82685	IDS09284N7332	HD 89995	IDS10180N0612
HD 76805	IDS08533S5220	HD 82694	IDS09284S4012	HD 90125	IDS10190N0253
HD 76813	IDS08534N3248	HD 82780	IDS09291N4024	HD 90470	IDS10216N4207
HD 76943	IDS08542N4211	HD 82785	IDS09291S3441	HD 90508	IDS10219N4918
HD 77002	IDS08545S5851	HD 82885	IDS09297N3616	HD 90537	IDS10221N3713
HD 77104	IDS08553N3239	HD 82901	IDS09297S6221	HD 90569	IDS10224N1016
HD 77140	IDS08555S4651	HD 82984	IDS09302S4834	HD 90839	IDS10243N5589
HD 77175	IDS08558N1540	HD 83023	IDS09304N1450	HD 90972	IDS10250S3006
HD 71973	IDS08252N7504	HD 83104	IDS09309S1908	HD 91106	IDS10260S0707
HD 77190	IDS08558N2818	HD 83261	IDS09321S2415	HD 91130	IDS10262N3254
HD 77250	K1 II-III+F3 IV	HD 83368	IDS09328S4818	HD 91270	IDS10271S6051
HD 77327	IDS08568N4733	HD 83446	IDS09332S4854	HD 91312	IDS10274N4056
HD 77370	IDS08570S5842	HD 83520	IDS09339S5313	HD 91316	IDS10275N0949
HD 77653	IDS08586S5148	HD 83610	IDS09346S3910	HD 91324	IDS10275S5312
HD 77959	IDS09005S4810	HD 83787	IDS09357N3144	HD 91355	IDS10277S4433
HD 78154	IDS09016N6732	HD 83808	IDS09358N1021	HD 91496	IDS10287S7242
HD 78175	IDS09017N2283	HD 83950	IDS09367N5625	HD 91504	IDS10287S4629
HD 78293	IDS09021S5727	HD 83953	IDS09367S2308	HD 91550	IDS10293S2314
HD 78362	IDS09027N6355	HD 84121	IDS09376S5732	HD 91636	IDS10298N0910
HD 78418	IDS09029N2702	HD 84179	IDS09383N6407	HD 91805	IDS10308S4309
HD 78515	IDS09036N2227	HD 84194	IDS09383N1428	HD 91880	IDS10314S1550
HD 78541	IDS09037S2527	HD 84252	IDS09389N1920	HD 91881	IDS10314S2609
HD 78556	IDS09038S0811	HD 84367	IDS09397S2719	HD 91889	IDS10316S1142
HD 78616	IDS09041S4414	HD 84400	IDS09399S5046	HD 92029	IDS10324S8124
HD 78647	IDS09043S4302	HD 84722	IDS09420N1202	HD 92139	IDS10331S4742
HD 78715	IDS09046N2224	HD 84877	IDS09431S5230	HD 92328	IDS10344S4214
HD 78922	IDS09057S2957	HD 84999	IDS09439N5931	HD 92398	IDS10350S5840
HD 79028	IDS09064N6150	HD 85040	IDS09442N2139		HR 4179
HD 79096	IDS09068N1524	HD 85123	IDS09446S6436	HD 92449	IDS10353S5505
HD 79210	IDS09078N5307	HD 85235	IDS09453N5432	HD 92504	IDS10357S5656

Star	Identification	Star	Identification	Star	Identification
HD 92589	IDS10363S3513	HD 98096	IDS11118S4520	HD 104055	IDS11540N0105
HD 92620	IDS10366N3213	HD 98231	IDS11128N3166	HD 104075	IDS11542N3343
HD 92740	LSS 1761	HD 98262	IDS11131N3338	HD 104174	IDS11546S7740
HD 92741	IDS10374S5927	HD 98292	IDS11132S6717	HD 104307	IDS11556S2117
HD 92769	IDS10376N2651	HD 98712	IDS11164S1954	HD 104513	IDS11570N4336
HD 92787	IDS10377N4643	HD 98718	IDS11164S5357	HD 104555	IDS11573S8504
HD 92841	IDS10382N0516	HD 98993	IDS11184S3537	HD 104671	IDS11579S6245
	K3 III+K0 III	HD 99014	IDS11185S7434	HD 104800	IDS11588N0355
HD 92845	IDS10381S3212	HD 99028	IDS11187N1105	HD 104827	IDS11592N2201
HD 93152	IDS10403N3113	HD 99104	IDS11190S6424	HD 104886	IDS11596S0513
HD 93190	IDS10404S5845	HD 99211	IDS11195N1708	HD 104901	IDS11596S6126
HD 93344	IDS10413S7020	HD 99285	IDS11204N1700	HD 105043	IDS12006N6330
HD 93359	IDS10413S7020	HD 99333	IDS11207S3712	HD 105078	IDS12008S3508
HD 93420	IDS10419S5858	HD 99491	IDS11217N0333	HD 105151	IDS12012S6509
HD 93497	IDS10425S4854	HD 99492	IDS11217N0333		G8/K0 III+A3
HD 93502	IDS10424S6005	HD 99556	IDS11221S6034	HD 105211	IDS12017S6403
HD 93526	IDS10427S1443	HD 99564	IDS11221S1148	HD 105219	IDS12018N0622
HD 93742	IDS10442S0330	HD 99574	IDS11221S5237	HD 105304	IDS12023N1336
HD 93779	IDS10443S7956		K0 III+A2/3 V	HD 105382	IDS12032S5010
HD 93873	IDS10450S5855	HD 99582	IDS11223N2435	HD 105383	IDS12032S5010
HD 93903	IDS10453S0822	HD 99648	IDS11227N0326	HD 105435	IDS12032S5010
HD 93943	IDS10454S5848	HD 99651	IDS11228S0109	HD 105639	IDS12045N0227
HD 94132	IDS10467N7023	HD 99787	IDS11237N3953	HD 105686	IDS12049S3409
HD 94180	IDS10471N0133	HD 99803	IDS11238S4207	HD 105852	IDS12058S4452
HD 94220	IDS10474N0532	HD 99872	IDS11242S7155	HD 105943	IDS12065N8216
HD 94363	IDS10483S0143	HD 99898	IDS11244S6223	HD 106022	IDS12070N2906
HD 94370	IDS10484S5813	HD 99922	IDS11247S2355	HD 106116	IDS12074S0232
HD 94388	IDS10486S1935	HD 99946	IDS11248N3031	HD 106257	IDS12084S3314
HD 94402	IDS10486S0135	HD 100180	IDS11266N1455	HD 106321	IDS12088S4510
HD 94465	LSS 1998	HD 100203	IDS11267N6138	HD 106384	IDS12091S0510
HD 94510	IDS10494S5819	HD 100261	IDS11271S5853	HD 106400	IDS12092N1223
HD 94546	LSS 2010	HD 100287	IDS11273S2843	HD 106516	IDS12100S0942
HD 94601	IDS10502N2517	HD 100407	IDS11281S3118	HD 106591	IDS12105N5735
HD 94650	IDS10504S7011	HD 100493	IDS11287S4002	HD 106592	IDS12104N4041
HD 94672	IDS10506N0116	HD 100600	IDS11295N1721	HD 106612	IDS12106S2248
HD 94683	IDS10506S6118	HD 100623	IDS11296S3216	HD 106690	IDS12111N4113
HD 94909	IDS10523S5701	HD 100733	IDS11304S4649	HD 106922	IDS12126S3532
HD 95221	IDS10545S3312	HD 100808	IDS11310N2820	HD 106976	IDS12130S0323
HD 95241	IDS10547N4328	HD 100841	IDS11312S6228	HD 106983	IDS12130S6327
HD 95324	IDS10552S6047	HD 100893	IDS11316S3301	HD 107079	IDS12137S5435
HD 95382	IDS10556N0638	HD 101150	IDS11332N6454	HD 107259	IDS12148S0007
HD 95429	IDS10558S5117	HD 101154	IDS11333S0153	HD 107295	IDS12150S2137
HD 95453	IDS10559N0036	HD 101177	IDS11335N4540	HD 107328	IDS12152N0352
HD 95544	IDS10566N8135	HD 101198	IDS11353N1239	HD 107348	IDS12154S2140
HD 95689	IDS10576N6217	HD 101369	IDS11348S1355	HD 107383	IDS12157N1821
HD 95698	IDS10576S2617	HD 101379	IDS11348S6451	HD 107418	IDS12157S1303
HD 95741	IDS10579S0250	HD 101501	IDS11357N3446	HD 107566	IDS12166S6658
HD 95788	IDS10581S8101	HD 101606	IDS11363N3218	HD 107705	IDS12174N0552
HD 95808	IDS10582S1046	HD 101666	IDS11367S3157	HD 107760	G 237 - 62
HD 95826	IDS10583S5959	HD 101782	IDS11375S8233	HD 107998	IDS12194S4050
HD 95880	IDS10586S5912	HD 101980	IDS11390N2546	HD 108150	IDS12204N6422
HD 95934	IDS10590N3847	HD 102077	IDS11397S4852	HD 108248	IDS12210S6233
HD 96097	IDS10599N0753	HD 102249	IDS11409S6610	HD 108250	IDS12210S6233
HD 96202	IDS11005S2645	HD 102350	IDS11416S6037	HD 108257	IDS12211S5054
HD 96248	LSS 2135	HD 102509	IDS11428N2046	HD 108485	IDS12226S5929
HD 96261	IDS11008S5910		G5 III-IVe+A7 V	HD 108570	IDS12232S5550
HD 96264	IDS11008S6031	HD 102510	IDS11428N0848	HD 108767	IDS12247S1558
HD 96357	IDS11014S6118	HD 102567	LSS 2502	HD 108799	IDS12249S1250
HD 96436	IDS11018N0230	HD 102574	IDS11433S0945	HD 108845	IDS12253N5205
HD 96616	IDS11027S4206	HD 102590	IDS11435N1450	HD 108903	IDS12256S5633
HD 96706	IDS11032S7020	HD 102647	IDS11440N1508	HD 108925	IDS12256S5633
HD 96738	IDS11034N2512	HD 102870	IDS11454N0219	HD 109164	IDS12275S6024
HD 96829	IDS11039S6017	HD 102910	IDS11458N1250	HD 109312	IDS12285S4921
HD 97033	IDS11052N6633	HD 102928	IDS11459S0447	HD 109358	IDS12290N4154
HD 97048	CU Cha	HD 102942	IDS11460N3355	HD 109409	IDS12293S4407
HD 97101	IDS11056N3100	HD 102990	IDS11463S1211	HD 109485	IDS12301N2310
HD 97152	LSS 2198	HD 103079	IDS11470S6439	HD 109510	IDS12301N1856
HD 97334	IDS11071N3621	HD 103095	G 122 - 51	HD 109573	IDS12307S3919
HD 97399	IDS11074S5954	HD 103101	IDS11472S5626	HD 109628	IDS12310N1157
HD 97400	IDS11074S5954	HD 103192	IDS11479S3321	HD 109668	IDS12312S6835
HD 97411	IDS11075S1757	HD 103262	IDS11486N0405	HD 109799	IDS12324S2635
HD 97501	IDS11081N4138	HD 103431	IDS11494N1958	HD 109857	IDS12328S7449
HD 97534	IDS11083S5946	HD 103432	IDS11494N1958	HD 109867	IDS12329S6639
HD 97583	IDS11086S6338	HD 103437	IDS11494S3712	HD 109930	IDS12336S0446
HD 97584	IDS11088N7401	HD 103483	IDS11499N4702	HD 109996	IDS12341N2313
HD 97603	IDS11088N2104	HD 103498	IDS11499N4702	HD 110014	IDS12341S0727
HD 97605	IDS11088N0837	HD 103578	IDS11505N1612	HD 110304	IDS12360S4825
HD 97855	IDS11103N5319	HD 103928	IDS11530N3250	HD 110318	IDS12361S1228
HD 97907	IDS11106N1351	HD 103974	IDS11533S4023	HD 110377	IDS12365N1058
HD 98058	IDS11116S0307	HD 104035	LSS 2555	HD 110379	IDS12366S0054
HD 98088	IDS11119S0635	HD 104039	IDS11538S2521		

Table 5

1357

Star	Identification	Star	Identification	Star	Identification
HD 110423	IDS12369N0721	HD 115533	IDS13125S6122	HD 121325	IDS13497S0734
HD 110532	IDS12377S5821	HD 115617	IDS13132S1745	HD 121336	IDS13498S5338
HD 110678	IDS12387N6142	HD 115659	IDS13135S2238	HD 121370	IDS13499N1854
HD 110786	IDS12394S6140	HD 115704	IDS13137S6129	HD 121384	IDS13499S5412
HD 110829	IDS12398S6026	HD 115762	IDS13141N2508	HD 121518	IDS13507S5712
HD 110863	IDS12401S6000	HD 115778	IDS13142S5915	HD 121557	IDS13510S6519
HD 110879	IDS12402S6733	HD 115810	IDS13149N3540	HD 121932	IDS13534S6547
HD 110956	IDS12406S5556	HD 115967	IDS13154S7137	HD 121995	IDS13539N5229
HD 111028	IDS12413N1006	HD 115995	IDS13156N0328	HD 122064	G 239 - 8
HD 111032	IDS12414S3246	HD 116087	IDS13162S6028	HD 122210	IDS13553S3944
HD 111066	IDS12416N2441	HD 116197	IDS13169S4725	HD 122408	IDS13566N0202
HD 111123	IDS12419S5908	HD 116457	IDS13185S6358	HD 122451	IDS13568S5953
HD 111164	IDS12422N1230	HD 116495	IDS13189N2945	HD 122510	IDS13572S3112
HD 111199	IDS12424S0545	HD 116656	IDS13199N5527	HD 123139	IDS14008S3553
HD 111469	IDS12444N2806	HD 116658	IDS13200S1038	HD 123377	IDS14022S6950
HD 111482	IDS12445S8435		B1 III-IV+B2 V	HD 123445	IDS14027S4300
HD 111515	IDS12446N0145	HD 116781	IDS13208S6208	HD 123515	IDS14030S5102
HD 111597	IDS12453S3327	HD 116795	IDS13208S6121	HD 123569	IDS14033S5258
HD 111720	IDS12462S0948	HD 116836	IDS13211S4838	HD 123782	IDS14046N4956
HD 111790	IDS12466S5317	HD 116873	IDS13214S3939	HD 124224	IDS14072N0253
HD 111862	IDS12472N1737	HD 117176	IDS13236N1419	HD 124314	IDS14077S6114
HD 112014	IDS12484N8357	HD 117187	IDS13236N7255	HD 124367	IDS14080S5637
HD 112028	IDS12484N8357	HD 117200	IDS13237N6515	HD 124471	IDS14088S6607
HD 112033	IDS12484N2147	HD 117201	IDS13237N6515	HD 124601	IDS14094S5927
HD 112091	IDS12487S5638	HD 117287	IDS13242S2246	HD 124674	IDS14099N5215
HD 112092	IDS12487S5638	HD 117360	IDS13246S7703	HD 124675	IDS14099N5175
HD 112185	IDS12496N5630	HD 117376	IDS13252N6027	HD 124677	IDS14099N1629
HD 112244	IDS12501S5618	HD 117408	IDS13250S2308	HD 124679	IDS14099N1034
HD 112300	IDS12506N0357	HD 117436	IDS13252S0557	HD 125040	IDS14119N2035
HD 112412	IDS12514N3851	HD 117440	IDS13252S3854	HD 125161	IDS14126N5150
HD 112413	IDS12514N3851	HD 117460	IDS13253S6231	HD 125184	IDS14127S0704
HD 112486	IDS12519N5438	HD 117470	IDS13255S5115	HD 125276	IDS14133S2522
HD 112573	IDS12527N1914	HD 117555	IDS13261N2445	HD 125383	IDS14139S4236
HD 112661	IDS12532S6145	HD 117661	IDS13266S1813	HD 125451	IDS14145N1328
HD 112846	IDS12545S0316	HD 117789	IDS13275S1451	HD 125455	IDS14144S0441
HD 112989	WEISS 31640	HD 117876	IDS13281N2452	HD 125473	IDS14145S3726
	IDS12555N3119	HD 117919	IDS13284S4746	HD 125628	IDS14154S5800
HD 112992	IDS12554S0250	HD 118054	IDS13294S1242	HD 125721	IDS14161S4752
HD 113014	IDS12556S6139	HD 118156	IDS13300N3918	HD 125810	IDS14167S5019
HD 113022	IDS12557N1855	HD 118261	IDS13304S6111	HD 126053	G 65 - 47
HD 113034	IDS12557S6118	HD 118266	IDS13306N1043	HD 126128	IDS14185N0854
HD 113139	IDS12564N5654	HD 118349	IDS13313S2559	HD 126129	IDS14185N0854
HD 113226	IDS12572N1130	HD 118384	IDS13316S5754	HD 126251	IDS14193S1113
HD 113337	IDS12579N6409	HD 118520	IDS13323S5707	HD 126341	IDS14197S4446
HD 113415	IDS12584S2003	HD 118623	IDS13330N3648	HD 126354	IDS14197S4456
HD 113459	IDS12587S0308	HD 118716	IDS13335S5257	HD 126367	IDS14199S1931
HD 113703	IDS13005S4756	HD 118741	IDS13337N5113	HD 126504	IDS14208S4541
HD 113791	IDS13011S4922		M2 II-III+F3 III	HD 126593	IDS14214S6006
HD 113823	IDS13012S5920	HD 118742	IDS13337N3941	HD 126660	IDS14218N5219
HD 113847	IDS13014N4548	HD 118889	IDS13346N1115	HD 126769	IDS14223S2903
HD 113848	IDS13015N2142	HD 118991	IDS13353S5403	HD 126862	IDS14229S6716
HD 113865	IDS13014N2434	HD 119055	IDS13359N2028	HD 126868	IDS14230S0147
HD 113889	IDS13017N7334	HD 119081	IDS13360N2835	HD 126981	IDS14237S4452
HD 113904	IDS13017S6446	HD 119086	IDS13360S2257	HD 126983	IDS14237S4904
HD 114060	IDS13029N2433	HD 119124	IDS13364N5101	HD 127043	IDS14242N2844
HD 114092	WEISS 52626	HD 119288	G 62 - 63	HD 127067	IDS14242N2844
HD 114330	IDS13048S0500	HD 119361	IDS13377S4134	HD 127243	G 223 - 65
HD 114371	IDS13050S6926	HD 119425	IDS13380N0403	HD 127304	IDS14256N3214
HD 114376	IDS13055N3902	HD 119786	IDS13402S1516	HD 127337	IDS14258N0513
HD 114378	IDS13051N1803	HD 119796	IDS13402S6205	HD 127501	IDS14266S5214
HD 114447	IDS13055N3902	HD 119834	IDS13403S5056	HD 127624	IDS14272S3016
HD 114504	IDS13060N6246	HD 119853	IDS13406S1156	HD 127665	IDS14275N3049
HD 114529	IDS13060S5923	HD 119921	IDS13411S3545	HD 127700	IDS14277N7608
HD 114576	IDS13062S2601	HD 119931	IDS13411N0537	HD 127726	A7IV-V+A7IV-V
HD 114606	IDS13064N0969	HD 120033	IDS13419S0913		IDS14279N2707
HD 114630	IDS13067S5917	HD 120066	IDS13420N0651	HD 127762	IDS14280N3845
HD 114642	IDS13068S1539	HD 120136	IDS13425N1757	HD 127871	IDS14287N0947
HD 114710	IDS13072N2823	HD 120164	IDS13429N3903	HD 127971	IDS14292S4105
HD 114733	IDS13073S5750	HD 120198	IDS13429N5456	HD 128068	IDS14298S4549
HD 114772	IDS13075S5010	HD 120237	IDS13432S3512	HD 128093	IDS14299N3259
HD 114792	IDS13076S6207	HD 120324	IDS13436S4159	HD 128167	IDS14304N3011
HD 114837	IDS13080S5834	HD 120476	IDS13445N2729	HD 128266	IDS14308S4542
HD 114846	IDS13081S1818	HD 120642	IDS13456S5219	HD 128293	IDS14309S6746
HD 114886	IDS13082S6303	HD 120709	IDS13461S3230	HD 128580	IDS14326N1128
HD 114911	IDS13085S6722	HD 120710	IDS13461S3230	HD 128620	IDS14328S6025
HD 115046	IDS13095N1152	HD 120759	IDS13463S3107	HD 128621	IDS14328S6025
HD 115337	IDS13115N8100	HD 120955	IDS13475S3126	HD 128898	IDS14344S6432
HD 115365	IDS13116N2019	HD 120987	IDS13477S3510	HD 129056	IDS14353S4658
HD 115383	IDS13118N0957	HD 120991	IDS13477S4638	HD 129132	IDS14358N2224
HD 115404	IDS13118N1733	HD 121130	IDS13485N6513	HD 129161	IDS14359S3030
HD 115488	IDS13124S0009	HD 121146	IDS13485N6849	HD 129174	IDS14360N1651

Star	Identification	Star	Identification	Star	Identification
HD 129175	IDS14360N1651	HD 136407	IDS15154S1511	HD 142630	IDS15505S3340
HD 129247	IDS14364N1369	HD 136415	IDS15154S5858	HD 142669	IDS15507S2855
HD 129290	IDS14367N1402	HD 136422	IDS15155S3554	HD 142860	IDS15518N1559
HD 129422	IDS14374S6227	HD 136442	G 152 - 10	HD 142908	IDS15521N3814
HD 129557	IDS14380S5511	HD 136479	IDS15158S0528	HD 143018	IDS15528S2550
HD 129580	IDS14382N5823	HD 136504	IDS15159S4420	HD 143107	IDS15534N2710
HD 129798	IDS14396N6141	HD 136512	IDS15160N2959	HD 143118	IDS15535S3807
HD 129893	IDS14400S5158	HD 136514	IDS15160N0105	HD 143248	IDS15541S4009
HD 129926	IDS14402S2501	HD 136672	IDS15168S6757	HD 143275	IDS15544S2220
HD 129932	IDS14402S5147	HD 136901	IDS15181N2559	HD 143333	IDS15547S1614
HD 129954	IDS14403S6610	HD 136933	IDS15182S3921	HD 143454	IDS15553N2612
HD 129978	IDS14404S1502	HD 137015	IDS15186S3749	HD 143474	IDS15554S5730
HD 129980	IDS14405S2045	HD 137058	IDS15189S3823	HD 143619	IDS15565S2851
HD 129989	IDS14406N2730	HD 137107	IDS15191N3039	HD 143761	IDS15573N3337
HD 130458	IDS14432S7247	HD 137387	IDS15206S7303	HD 143928	IDS15580S3735
HD 130529	IDS14435S2350	HD 137391	IDS15207N3742	HD 144046	IDS15588N0516
HD 130559	IDS14438S1344	HD 137392	IDS15207N3742	HD 144061	IDS15588N7110
HD 130603	IDS14440N2447	HD 137432	IDS15209S3625	HD 144070	IDS15589S1106
HD 130669	IDS14444N1038	HD 137465	IDS15212S5115	HD 144217	IDS15596S1932
HD 130807	IDS14451S4310	HD 137597	IDS15218S3701	HD 144218	IDS15596S1932
HD 130819	IDS14453S1537	HD 137759	IDS15228N5919	HD 144362	IDS16004S0601
HD 130841	IDS14453S1537	HD 137763	IDS15228S0900	HD 144390	IDS16007S0553
HD 130917	IDS14457N2902	HD 137778	IDS15228S0900	HD 144415	IDS16007S3629
HD 131040	IDS14463N5147	HD 137909	IDS15237N2927	HD 144515	G 137 - 82
HD 131041	IDS14463N4908	HD 138004	IDS15242N4314		IDS16012N1057
HD 131156	IDS14468N1931	HD 138268	IDS15260S1949	HD 144667	IDS16019S3849
HD 131432	IDS14485S3254	HD 138488	IDS15272S2409	HD 144668	IDS16019S3849
HD 131473	IDS14487N1607	HD 138498	IDS15273S6517	HD 144708	IDS16020S1229
HD 131492	IDS14487S6222	HD 138538	IDS15276S6559	HD 144840	IDS16028S1251
HD 131551	IDS14491S7438	HD 138629	IDS15282N4114	HD 144900	IDS16030S4842
HD 131582	IDS14493N2345	HD 138672	IDS15285S2240	HD 144918	IDS16031S4847
HD 131625	IDS14496S3327	HD 138690	IDS15285S4050	HD 144927	IDS16032S3223
HD 131657	IDS14497S4728	HD 138743	IDS15288S4910	HD 145000	IDS16036N1719
HD 131873	IDS14510N7434	HD 138749	IDS15289N3142	HD 145001	IDS16036N1719
HD 131977	IDS14516S2058	HD 138769	IDS15290S4437	HD 145148	G 16 - 32
HD 132029	IDS14518N3242	HD 138800	IDS15293S7307	HD 145328	IDS16053N3645
HD 132132	IDS14524N0014	HD 138905	IDS15299S1427	HD 145366	IDS16054S7827
HD 132200	IDS14527S4142	HD 138918	IDS15300N1052	HD 145388	IDS16054S7827
HD 132219	IDS14527S2715	HD 138923	IDS15299S3246	HD 145397	IDS16056S5422
HD 132345	IDS14535S1044	HD 139063	IDS15310S2748	HD 145483	IDS16061S2809
HD 132375	IDS14537S0435	HD 139127	IDS15313S4214	HD 145501	IDS16062S1912
HD 132742	IDS14556S0807	HD 139129	IDS15314S5203	HD 145502	IDS16062S1912
HD 132933	IDS14567N0015	HD 139271	IDS15321S3848	HD 145544	IDS16063S6326
HD 132955	IDS14569S3215	HD 139323	IDS15325N4008	HD 145589	IDS16067N0958
HD 133029	IDS14572N4740	HD 139341	IDS15325N4008	HD 145674	IDS16071N5812
HD 133242	IDS14583S4640	HD 139461	IDS15333S0828	HD 145792	IDS16077S2410
HD 133340	IDS14588S4041		F6 V+F6 IV-V	HD 145802	IDS16078N3336
HD 133408	IDS14592N0553	HD 139599	IDS15339S4725	HD 145849	IDS16082N3641
HD 133638	IDS15004S6820	HD 139777	IDS15350N8047	HD 145931	IDS16085N4238
HD 133640	IDS15005N4803	HD 139862	IDS15354N1223	HD 145976	IDS16086N2656
HD 133955	IDS15021S4454	HD 139892	IDS15356N3658	HD 145991	IDS16090N6606
HD 133962	IDS15021N4832	HD 139997	IDS15362S1921	HD 145997	IDS16089S1817
HD 134064	IDS15028N1850	HD 140159	IDS15371N2000	HD 146051	IDS16091S0326
HD 134083	IDS15029N2515	HD 140285	IDS15376S4130	HD 146100	IDS16094N3936
HD 134270	IDS15038S5458	HD 140436	IDS15386N2637	HD 146145	IDS16096S5250
HD 134335	IDS15042N2529	HD 140438	IDS15390N0250	HD 146233	IDS16102S0806
HD 134443	IDS15048S4454	HD 140484	IDS15388S6508	HD 146361	IDS16109N3407
HD 134444	IDS15048S4454	HD 140538	IDS15390N0250	HD 146362	IDS16109N3407
HD 134481	IDS15050S4822	HD 140573	IDS15393N0644	HD 146413	IDS16111N0737
HD 134482	IDS15050S4822	HD 140722	IDS15401S2745	HD 146667	IDS16123S4226
HD 134505	IDS15051S5143	HD 140784	IDS15404S3422	HD 146686	IDS16124S4955
HD 134518	IDS15051S3552	HD 140901	IDS15410S3736	HD 146738	IDS16127N2924
HD 134759	IDS15066S1925	HD 140979	IDS15413S5208	HD 146791	IDS16130S0427
HD 134943	IDS15075N1921	HD 141003	IDS15416N1544	HD 146836	IDS16132S3040
HD 135051	IDS15080S2549	HD 141168	IDS15425S5254	HD 146868	IDS16134N6056
HD 135101	G 136 - 77	HD 141296	IDS15433S4506	HD 146954	IDS16138S3911
	IDS15083N1939	HD 141318	IDS15433S5445	HD 147010	IDS16142S1949
HD 135160	IDS15085S6032	HD 141675	IDS15452N5541	HD 147049	IDS16144S5229
HD 135235	IDS15089S4742	HD 141832	IDS15460S2935	HD 147165	IDS16151S2521
HD 135240	IDS15089S6035	HD 141891	IDS15463S6307	HD 147225	IDS16154S4340
HD 135438	IDS15100N3209	HD 141913	IDS15464S6027	HD 147365	IDS16165N3957
HD 135591	IDS15108S6008	HD 141926	IDS15465S5502	HD 147379	IDS16166N6729
HD 135722	IDS15115N3341	HD 142049	IDS15472S5953	HD 147394	IDS16167N4633
HD 135725	IDS15115S0755	HD 142091	IDS15476N3558	HD 147547	IDS16175N1923
HD 135734	IDS15116S4730	HD 142114	IDS15476S2502	HD 147553	IDS16175S3258
HD 135737	IDS15115S6707	HD 142267	IDS15485N1331	HD 147614	IDS16177S4507
HD 135896	IDS15126S3051	HD 142357	IDS15490N1622	HD 147628	IDS16179S3720
HD 136202	IDS15142N0209	HD 142378	IDS15492S1905	HD 147677	IDS16182N3107
HD 136274	IDS15147N2603	HD 142448	IDS15494S3934	HD 147723	IDS16184S2928
HD 136347	IDS15150S3751	HD 142514	IDS15498S6445	HD 147749	IDS16186N3402
HD 136378	IDS15153N0036	HD 142629	IDS15505S3340	HD 147767	IDS16186N3402

Table 5

1359

Star	Identification	Star	Identification	Star	Identification
HD 147776	IDS16187S1324	HD 152879	IDS16510N1836	HD 156971	IDS17153S1036
HD 147787	IDS16187S6350	HD 152901	IDS16511S3750	HD 157038	IDS17158S3742
HD 147835	IDS16191N3234	HD 152909	IDS16512S1923		LSS 4120
HD 147890	IDS16194S2910	HD 153004	IDS16518S3327	HD 157042	IDS17158S4722
HD 147933	IDS16196S2313	HD 153053	IDS16520S5426	HD 157097	IDS17161S3707
HD 147971	IDS16198S4720	HD 153072	IDS16521S3728	HD 157198	IDS17168N2436
HD 148048	IDS16204N7559	HD 153221	IDS16529S4830	HD 157214	IDS17169N3236
HD 148112	IDS16208N1416	HD 153370	IDS16540S5059	HD 157246	IDS17170S5617
HD 148367	IDS16224S0809	HD 153525	IDS16551N4731	HD 157358	IDS17177N2851
HD 148374	IDS16225N6155	HD 153557	IDS16551N4731	HD 157457	IDS17182S5032
HD 148387	IDS16226N6144	HD 153580	IDS16552S5305	HD 157482	IDS17184N4004
HD 148408	IDS16226S0051	HD 153613	IDS16554S3200	HD 157527	IDS17187S2121
HD 148478	IDS16233S2613	HD 153687	IDS16557S0404	HD 157661	IDS17195S4545
	M1.5Iab-Ib+B4 Ve	HD 153697	IDS16559N6511	HD 157662	IDS17194S5033
HD 148515	IDS16234S0754	HD 153751	IDS16562N8212	HD 157741	IDS17200N1542
HD 148567	IDS16238S4616	HD 153791	IDS16563S4701	HD 157753	IDS17200S5212
HD 148653	IDS16245N1837	HD 153882	IDS16570N1505	HD 157779	IDS17202N3714
HD 148688	IDS16247S4136	HD 153914	IDS16572N0836	HD 157819	IDS17204S5505
HD 148743	LS IV −07 001	HD 154090	IDS16582S3359	HD 157832	IDS17204S4656
HD 148786	IDS16254S1624	HD 154228	IDS16590N1345	HD 157853	IDS17207N3840
HD 148856	IDS16260N2142	HD 154278	IDS16590N1345	HD 157910	IDS17210N3702
HD 148857	IDS16259N0212	HD 154301	IDS16595N1950	HD 157948	IDS17213N3807
HD 148880	IDS16262N5137	HD 154310	IDS16596S3705	HD 157967	IDS17214N1700
HD 148937	IDS16265S4754	HD 154363	IDS16597S0456	HD 158067	IDS17220N2658
HD 149303	IDS16288N4549	HD 154368	IDS16598S3519	HD 158094	IDS17221S6036
HD 149324	IDS16288S7718	HD 154407	IDS17001S3545	HD 158156	IDS17226S3826
HD 149414	IDS16295S0400	HD 154441	IDS17003N1944	HD 158226	IDS17230N3109
HD 149630	IDS16309N4239	HD 154445	LS IV −00 001	HD 158320	IDS17235S3337
HD 149632	IDS16309N1715	HD 154494	IDS17007N1253	HD 158427	IDS17241S4948
HD 149661	IDS16310S0207	HD 154660	IDS17017S0130	HD 158463	IDS17244S0549
HD 149711	IDS16314S4312	HD 154712	IDS17020N5943	HD 158476	IDS17244S4558
HD 149730	IDS16314S5648	HD 154733	IDS17021N2213	HD 158614	IDS17252S0059
HD 149837	IDS16321S6015	HD 154873	IDS17029S4637		G8IV-V+G8IV-V
HD 149886	IDS16324S3701	HD 154895	IDS17031S0057	HD 158661	IDS17254S1704
HD 150100	IDS16339N5308	HD 154903	IDS17031S6704	HD 158837	IDS17264N0248
HD 150117	IDS16339N5308	HD 154906	IDS17033N5436	HD 158926	IDS17268S3702
HD 150275	G 256 - 34	HD 154948	IDS17035S4426	HD 159181	IDS17282N5223
HD 150288	IDS16349S4649	HD 155103	IDS17045N3604	HD 159358	IDS17292S1110
HD 150378	IDS16357N0425		A9III-IV+A9II	HD 159480	IDS17299N0939
HD 150379	IDS16357N0425	HD 155125	IDS17046S1536	HD 159482	IDS17299N0606
HD 150450	IDS16360N4907	HD 155328	IDS17058N5057	HD 159541	IDS17302N5515
HD 150453	IDS16360S1944	HD 155410	IDS17063N4054	HD 159560	IDS17302N5515
HD 150549	IDS16366S6655	HD 155555	IDS17065S3219	HD 159834	IDS17317N2104
HD 150573	IDS16369S4056	HD 155555	IDS17072S6650	HD 159870	IDS17319N5737
HD 150576	IDS16368S5258	HD 155603	IDS17075S3939	HD 159876	IDS17319S1520
HD 150591	IDS16369S4056	HD 155644	IDS17078N1042	HD 159964	IDS17323S7210
HD 150608	IDS16370S3758	HD 155763	IDS17085N6550	HD 159966	IDS17323N6812
HD 150680	IDS16375N3147	HD 155826	IDS17088S3828	HD 160269	IDS17340N6157
HD 150708	IDS16377N6054	HD 155860	IDS17091N4952	HD 160315	IDS17341N0205
HD 150742	IDS16378S4039	HD 155875	IDS17091S6956	HD 160337	IDS17343S3736
HD 150768	IDS16381S2716	HD 155876	IDS17092N4551	HD 160435	IDS17347S5740
HD 150894	IDS16387S2819	HD 155886	IDS17092S2627	HD 160720	IDS17363S5730
HD 150997	IDS16395N3907	HD 155889	IDS17093S3337	HD 160762	IDS17367N4603
HD 151090	IDS16401N0617	HD 155970	IDS17096S1428	HD 160822	IDS17369N3120
HD 151158	IDS16405S4242	HD 155974	IDS17097S3538	HD 160835	IDS17370N2434
HD 151217	IDS16410N0846	HD 156014	IDS17101N1430	HD 160861	IDS17370N6826
HD 151249	IDS16411S5852	HD 156026	IDS17092S2627	HD 160910	IDS17375N1600
HD 151300	IDS16415S4659	HD 156091	IDS17104S5935	HD 160922	IDS17375N6848
HD 151431	IDS16421N0215	HD 156164	IDS17109N2457	HD 160928	IDS17375S4241
HD 151525	IDS16428N0526	HD 156190	IDS17109S7001	HD 161056	LS IV −07 002
HD 151566	IDS16430S4952	HD 156247	IDS17114N0119	HD 161074	IDS17384N2438
	A5 III+F7 III	HD 156266	IDS17114S0020	HD 161208	IDS17391S1837
HD 151613	IDS16434N5658	HD 156274	IDS17115S4632	HD 161270	IDS17396N0237
HD 151771	IDS16442S3720	HD 156322	IDS17119S3216	HD 161289	IDS17396N0237
HD 151862	IDS16450N1326	HD 156324	IDS17119S3218	HD 161321	IDS17397N1427
HD 151890	B1.5 IV+B6.5 V	HD 156325	IDS17118S3227	HD 161390	IDS17403S3804
HD 152082	IDS16461S6306	HD 156327	IDS17118S3457	HD 161471	IDS17406S4005
HD 152107	IDS16463N4609	HD 156331	IDS17118S4958	HD 161797	IDS17425N2747
HD 152125	IDS16465N5519	HD 156349	IDS17119S2411	HD 161814	IDS17426S6008
HD 152127	IDS16464N0123		K0 II/III+F6 IV/V	HD 161832	IDS17427N3922
HD 152217	IDS16469S4105	HD 156350	IDS17119S3417	HD 161833	IDS17427N1744
HD 152303	IDS16475N7741	HD 156384	IDS17121S3453	HD 161892	IDS17431S3701
HD 152408	IDS16480S4059	HD 156398	IDS17122S4407	HD 161897	IDS17430N7227
HD 152424	IDS16480S4156	HD 156633	IDS17136N3312	HD 161912	IDS17432S4003
HD 152569	IDS16490S0127		B1.5 Vp+B5 III	HD 161988	IDS17435S7609
HD 152598	IDS16492N3152	HD 156662	LSS 4091	HD 162003	IDS17437N7212
HD 152723	IDS16500S4021	HD 156717	IDS17140S1739	HD 162004	IDS17437N7212
HD 152751	IDS16501S0809	HD 156768	IDS17143S5755	HD 162102	IDS17443S3341
HD 152781	IDS16503S1639	HD 156846	IDS17147S1914	HD 162189	IDS17445S4045
HD 152849	IDS16508S2259	HD 156897	IDS17150S2100	HD 162220	IDS17448S3032
HD 152863	IDS16510N2554	HD 156928	IDS17152S1244	HD 162756	IDS17476S0753

Star	Identification	Star	Identification	Star	Identification
HD 163151	IDS17496N1109	HD 168656	IDS18159N0320	HD 173949	IDS18431N6057
HD 163181	IDS17497S3228	HD 168723	IDS18161S0255	HD 174069	IDS18435S0835
HD 163217	IDS17500N4002	HD 168905	IDS18170S4410	HD 174080	IDS18438N1039
HD 163302	IDS17503S3444	HD 169022	IDS18175S3427	HD 174116	IDS18437S2026
HD 163336	IDS17506S1548	HD 169226	IDS18186S1215	HD 174160	LS II +23 002
HD 163433	IDS17510S3907	HD 169414	IDS18194N2144	HD 174208	IDS18443S0602
HD 163588	IDS17519N5654	HD 169420	IDS18194S2036	HD 174309	IDS18448S2217
HD 163624	IDS17520N0050	HD 169493	IDS18198S0138	HD 174325	IDS18449S0801
HD 163652	IDS17521S3651		A9 III+F6 III	HD 174366	IDS18451N4858
HD 162734	IDS17475N1521	HD 169646	IDS18206N3841	HD 174464	LS IV −09 031
HD 163755	IDS17527S3015	HD 169689	V2291 Oph	HD 174569	IDS18460N1052
HD 163810	IDS17530S1304		G8 III-IV+A0V	HD 174585	IDS18460N3242
HD 163840	IDS17531N2401	HD 169869	IDS18216S1445	HD 174602	IDS18462N3226
HD 164058	IDS17543N5130		LS IV -14 065	HD 174638	IDS18464N3315
HD 164146	IDS17547S2412	HD 169718	IDS18210N2720		B7 Ve+A8 p
HD 164438	IDS17560S1906	HD 169822	IDS18214N0844	HD 174974	IDS18481S2252
HD 164668	IDS17573N2136	HD 169836	IDS18213S5735	HD 175191	IDS18491S2625
	A5 IIIn+G8 III	HD 169851	IDS18215S2642	HD 175306	IDS18497N5915
HD 164669	IDS17573N2136	HD 169985	IDS18221N0008	HD 175426	IDS18502N3650
HD 164712	IDS17573S7554	HD 170000	IDS18222N7117	HD 175492	IDS18505N2231
HD 164717	IDS17574S2236	HD 170073	IDS18225N5845	HD 175588	IDS18503N3645
HD 164765	IDS17576S0811	HD 170111	IDS18227N2623	HD 175635	IDS18512N3350
HD 164824	IDS17579N3319	HD 170141	IDS18227S2639	HD 175638	IDS18512N0404
HD 164833	IDS17579S2250	HD 170153	IDS18229N7241	HD 175639	IDS18512N0404
HD 164870	IDS17581S3554	HD 170452	IDS18244S1301	HD 175679	IDS18514N0220
HD 164922	IDS17584N2620	HD 170479	IDS18245S3303	HD 175740	IDS18517N4128
HD 164975	IDS17586S2935	HD 170580	LS IV +04 003	HD 175824	IDS18521N4844
HD 165189	IDS17596S4326		IDS18251N0400	HD 175855	IDS18523S3940
HD 165246	IDS18000S2412	HD 170680	IDS18256S1828	HD 175986	IDS18528S6854
HD 165259	IDS17599S7341	HD 170740	LS IV −10 011	HD 176051	IDS18533N3246
HD 165341	IDS17576S0811		IDS18259S1052	HD 176155	IDS18538N1714
HD 165358	IDS18005N4828	HD 170868	IDS18265S3848	HD 176162	IDS18538S1258
HD 165475	IDS18011N1200	HD 171245	IDS18286N2333	HD 176270	IDS18543S3712
HD 165493	IDS18011S4547	HD 171247	IDS18286S0753	HD 176303	IDS18545N1329
HD 165497	IDS18012S5903	HD 171301	IDS18290N3029	HD 176304	LS IV +10 003
HD 165590	IDS18016N2126	HD 171314	G 183 - 44	HD 176411	IDS18551N1456
HD 165645	IDS18019N4156	HD 171348	IDS18293S2210	HD 176437	IDS18552N3233
HD 165777	IDS18026N0933	HD 171369	V4190 Sgr	HD 176502	IDS18555N4033
HD 165814	IDS18027S2529	HD 171653	IDS18310N6523	HD 176560	IDS18558N5805
HD 165908	IDS18032N3033	HD 171745	IDS18314N2331	HD 176664	IDS18562S5110
HD 165910	IDS18032N1303	HD 171746	IDS18314N1654	HD 176668	IDS18563N6216
HD 165955	IDS18033S3453	HD 171759	IDS18314S7131	HD 176687	IDS18562S3001
HD 166006	IDS18035S4732	HD 171779	IDS18317N5216	HD 176795	IDS18569N7539
HD 166014	IDS18036N2845	HD 171834	IDS18318N0636	HD 176884	IDS18572S1923
HD 166023	IDS18036S3045	HD 171957	IDS18324S1405	HD 176984	IDS18576S0351
HD 166045	IDS18038N2605	HD 172044	IDS18330N3323	HD 177017	IDS18577S2251
HD 166046	IDS18038N2605	HD 172088	IDS18332S0317	HD 177109	IDS18581N3329
HD 166063	IDS18038S4558	HD 172167	IDS18336N3841	HD 177196	IDS18586N4648
HD 166107	IDS18040S2347	HD 172171	IDS18336N0845	HD 177230	LS IV −04 033
HD 166182	IDS18045N2048		K1 III+M6 IIIe	HD 177241	IDS18587S2153
HD 166228	IDS18046N4942	HD 172175	LS IV −07 009	HD 177444	IDS18596S0717
HD 166285	IDS18049N0306	HD 172256	LSS 5108	HD 177463	IDS18597S0411
HD 166286	IDS18048S1647	HD 172323	IDS18345N6337	HD 177474	IDS18597S3712
HD 166382	IDS18054N3100	HD 172348	IDS18346N0753	HD 177483	IDS18598N5207
HD 166393	IDS18053S1952	HD 172424	IDS18350N0716	HD 177517	IDS18599S1548
HD 166418	IDS18054S1643	HD 172671	IDS18363N4051	HD 177724	IDS19008N1343
HD 166464	IDS18056S2343	HD 172694	IDS18365S1557	HD 177817	IDS19011S1623
HD 166479	IDS18057N1627	HD 172829	MtW 110 # 53	HD 177863	V4198 Sgr
HD 166540	LSS 4746	HD 172748	IDS18368S0909	HD 177940	IDS19016N0805
HD 166566	IDS18061S1542	HD 172777	IDS18369S3825	HD 177994	IDS19018S3130
HD 166599	IDS18062S6305	HD 172881	IDS18374S7306	HD 178001	IDS19019N5719
HD 166620	G 205 - 5	HD 172910	IDS18376S3544	HD 178125	IDS19023N1055
HD 166841	IDS18073S6816	HD 173009	IDS18381S0822	HD 178129	LS IV +03 011
HD 166865	IDS18076N7959	HD 173084	IDS18385N6702	HD 178208	IDS19026N4946
HD 166866	IDS18076N7959	HD 173087	IDS18385N3439	HD 178428	IDS19035N1642
HD 166937	IDS18078S2105	HD 173383	IDS18400N3912	HD 178449	IDS19036N3221
HD 166988	IDS18081N3325	HD 173460	IDS18403S2230	HD 178475	IDS19037N3557
HD 167050	IDS18084S1906	HD 173495	IDS18406N0524	HD 178524	IDS19038S2111
HD 167224	IDS18091S1858	HD 173524	IDS18407N5526	HD 178555	IDS19039S1958
HD 167263	IDS18093S2025	HD 173582	IDS18410N3934	HD 178845	IDS19051S5039
HD 167356	IDS18096N3841	HD 173607	IDS18410N3934	HD 178911	IDS19054N3426
HD 167425	IDS18100S6355	HD 173638	LS IV −10 024	HD 179366	IDS19072S6650
HD 167618	IDS18109S3647		IDS18412S1014	HD 179484	IDS19077N3837
HD 167647	IDS18110S3408	HD 173648	IDS18413N3730	HD 179588	IDS19081N1641
HD 167771	IDS18116S1830	HD 173649	IDS18413N3730	HD 179761	IDS19087N0207
HD 168021	IDS18128S1840	HD 173654	IDS18413S0104	HD 179950	IDS19094S2526
HD 168092	IDS18129N5633	HD 173667	IDS18414N2027	HD 179957	IDS19095N4940
HD 168323	IDS18140N2315	HD 173739	IDS18418N5927	HD 179958	IDS19080N0338
HD 168339	IDS18140S6132	HD 173740	IDS18418N5927	HD 180163	IDS19104N3858
HD 168387	IDS18143N0713	HD 173880	IDS18426N1804	HD 180262	IDS19107N1455
HD 168454	IDS18146S2952	HD 173948	IDS18430S6218	HD 180553	IDS19119N2717

Table 5

1361

Star	Identification	Star	Identification	Star	Identification
HD 180554	IDS19119N2113	HD 186307	IDS19385N4001	HD 190658	IDS20008N1512
HD 180555	IDS19119N1422	HD 186340	IDS19386N6016	HD 190713	IDS20012N6421
HD 180617	IDS19120N0500	HD 186408	IDS19392N5017	HD 190771	IDS20015N3811
HD 180639	IDS19123N0911	HD 186427	IDS19392N5017	HD 190940	IDS20024N6735
HD 180711	IDS19125N6729	HD 186438	IDS19393N3727	HD 190960	IDS20019N7614
HD 180756	IDS19127N4954	HD 186518	IDS19398N2654	HD 190995	IDS20025N1017
HD 180809	IDS19129N3757	HD 186568	IDS19401N3355	HD 191026	IDS20026N3541
HD 180928	IDS19133S1542	HD 186791	IDS19415N1022	HD 191095	IDS20030S5749
HD 180968	IDS19135N2251	HD 186858	IDS19418N3322	HD 191104	IDS20030N0907
HD 180972	IDS19134N0054	HD 186882	IDS19418N4453	HD 191174	IDS20035N6336
HD 181053	IDS19137N0010	HD 186901	IDS19420N3551	HD 191195	IDS20036N5252
HD 181333	IDS19150N1211	HD 186927	IDS19421N3446	HD 191201	IDS20036N3526
HD 181391	IDS19152S0536		K0 II-III+A2 V	HD 191243	LS II +34 010
HD 181409	LS II +33 002	HD 186957	IDS19423S5927	HD 191396	IDS20047N3750
HD 181454	IDS19154S4439	HD 187013	IDS19426N3330	HD 191408	IDS20046S3621
HD 181558	V4199 Sgr	HD 187038	IDS19428N3239	HD 191570	IDS20055N2037
	IDS19158S1925	HD 187076	IDS19429N1817	HD 191611	IDS20057N3612
HD 181615	B2 IaShell+B2Vpe	HD 187203	LS II +10 009	HD 191639	LS IV −08 030
HD 181828	IDS19169N3500	HD 187235	IDS19439N3810	HD 191692	IDS20061S0107
HD 181987	IDS19175N2523	HD 187259	IDS19440N1134	HD 191738	IDS20065N4735
HD 182286	IDS19188S2930	HD 187362	IDS19445N1853	HD 191785	IDS20066N1553
HD 182308	IDS19190N6412	HD 187420	IDS19447S5514	HD 191862	IDS20068S1254
HD 182369	IDS19192S2442	HD 187421	IDS19447S5514	HD 191877	LS II +21 019
HD 182490	IDS19199N1644	HD 187458	IDS19450N3504	HD 191984	IDS20075N0034
HD 182509	IDS19198S5432	HD 187532	IDS19453S1101	HD 192079	IDS20081N3715
HD 182568	IDS19202N2925	HD 187638	IDS19459N3828	HD 192103	IDS20081N3554
HD 182572	IDS19202N1143	HD 187642	IDS19459N0836	HD 192342	IDS20094N2356
HD 182640	IDS19205N0255	HD 187691	IDS19462N1010	HD 192422	IDS20097N3826
HD 182762	IDS19211N1936	HD 187796	IDS19468N3240	HD 192439	IDS20098N5110
HD 182807	IDS19213N2444	HD 187849	IDS19470N3828	HD 192443	IDS20097N3826
HD 182835	LS IV +00 018	HD 187923	IDS19473N1123	HD 192455	IDS20100N6147
	IDS19214N0008	HD 187961	IDS19475N1006	HD 192472	IDS20099S3646
HD 182955	IDS19221N1942	HD 188001	LS II +18 016	HD 192514	IDS20105N4626
HD 183275	IDS19237S2711	HD 188037	IDS19482N2212	HD 192535	IDS20104N4305
HD 183030	IDS19255N8859	HD 188088	IDS19483S2411	HD 192577	IDS20105N4626
HD 183344	IDS19240S0715	HD 188119	IDS19485N7001	HD 192640	IDS20108N3630
	LS IV −07 035	HD 188154	IDS19487S0850	HD 192641	IDS20108N3622
HD 183439	IDS19245N2427	HD 188209	LS III +46 001	HD 192659	IDS20109N4148
HD 183491	IDS19245N2427	HD 188252	LS III +47 001	HD 192666	IDS20108S1239
HD 183589	IDS19252N0242	HD 188260	IDS19492N2349	HD 192685	IDS20110N2517
HD 183794	IDS19260S0220	HD 188268	IDS19492N0141	HD 192737	IDS20113S2138
HD 183912	IDS19267N2745	HD 188293	IDS19492S0829	HD 192744	IDS20114N3723
HD 183914	IDS19267N2745	HD 188294	IDS19492S0829	HD 192787	IDS20115N3326
HD 183986	IDS19272N3601	HD 188311	IDS19492N0141	HD 192806	IDS20116N2730
HD 184146	IDS19280N8316	HD 188326	IDS19495N3830	HD 192876	IDS20121S1249
HD 184152	IDS19280N0712	HD 188385	IDS19498N0653	HD 192907	IDS20123N7725
HD 184398	K2 II-IIIe +A0V	HD 188405	IDS19500N0700	HD 192909	IDS20124N4725
HD 184406	IDS19292N0710	HD 188512	IDS19504N0609	HD 192947	IDS20125S1251
HD 184467	IDS19295N5824	HD 188651	IDS19511N2956	HD 193077	IDS20133N3707
HD 184586	IDS19300S6655	HD 188753	IDS19516N4136	HD 193092	IDS20134N4003
HD 184606	IDS19302N1933	HD 188793	IDS19518N5927	HD 193094	IDS20134N2850
HD 184700	IDS19306S0027	HD 188891	IDS19523N4007	HD 193150	IDS20136S1926
HD 184707	IDS19306S2506	HD 188947	IDS19526N3449	HD 193216	IDS20140N4959
HD 184759	IDS19309N2915	HD 189037	IDS19530N5210	HD 193281	IDS20143S2931
HD 184884	IDS19314N1056	HD 189118	IDS19534S3458	HD 193322	LS II +40 015
HD 184915	LS IV −07 036	HD 189178	IDS19538N4006		IDS20146N4025
HD 184930	IDS19315S0130	HD 189195	IDS19538S3758	HD 193370	LS II +34 020
HD 184936	IDS19316N5957	HD 189256	IDS19540N4359	HD 193427	IDS20151N3906
HD 185018	IDS19342N1103	HD 189340	IDS19544S1013	HD 193432	IDS20151S1304
HD 185144	IDS19325N6929	HD 189377	IDS19546N4159	HD 193443	IDS20152N3758
HD 185194	IDS19328N1614	HD 189411	IDS19548N1137	HD 193452	IDS20152S1506
HD 185224	IDS19329N3006	HD 189432	IDS19550N3750	HD 193487	IDS20154N3626
HD 185268	IDS19332N2907	HD 189507	IDS19553N1055	HD 193495	IDS20154S1466
HD 185395	IDS19338N4959	HD 189508	IDS19553N1055	HD 193571	IDS20157S4222
HD 185404	IDS19338S2339	HD 189577	IDS19555N1715	HD 193592	IDS20159N5505
HD 185456	IDS19341N4959	HD 189901	IDS19574N3725	HD 193611	IDS20160N3801
HD 185507	IDS19342N0510	HD 190004	IDS19578N2439		LS II +38 044
HD 185622	IDS19349N1621	HD 190056	IDS19580S3220	HD 193680	IDS20165N4734
HD 185644	IDS19350S1631	HD 190147	IDS19585N4950	HD 193702	IDS20166N3905
HD 185657	IDS19352N4903	HD 190211	IDS19588N1813	HD 193721	IDS20166S8118
HD 185734	IDS19354N2955	HD 190306	IDS19592S3317	HD 193793	IDS20171N4332
HD 185758	IDS19356N1747	HD 190323	IDS19592N1442	HD 193807	IDS20171S4245
HD 185762	IDS19356S0051	HD 190360	IDS19594N2936	HD 193911	LS II +24 026
HD 185837	IDS19360N3345	HD 190390	LS IV −11 051	HD 193924	IDS20178S5703
HD 185859	LS II +20 012		IDS19596S1153	HD 193928	IDS20176N3637
HD 185915	IDS19364N2329	HD 190406	IDS19596N1648	HD 193944	IDS20178N5317
HD 185936	IDS19365N1335	HD 190429	IDS19598N3545	HD 193946	IDS20179N3931
HD 186005	IDS19368S1621	HD 190467	IDS20000N3609	HD 194093	LS II +40 023
HD 186155	IDS19378N4517	HD 190544	IDS20004N6432		IDS20186N3956
HD 186185	IDS19378S1542	HD 190603	IDS20007N3156	HD 194220	IDS20195N4240
HD 186203	IDS19379N1135	HD 190606	IDS20007N2021	HD 194244	IDS20195N0045

Star	Identification	Star	Identification	Star	Identification
HD 194298	IDS20198N6340	HD 198390	IDS20449N1210		M1 Ibep+B2pe+B3 V
HD 194433	IDS20204S3744	HD 198404	IDS20450N0510	HD 203358	IDS21166N3202
HD 194636	IDS20216S1832	HD 198478	IDS20455N4545	HD 203504	IDS21175N1923
HD 194649	IDS20218N3954	HD 198513	IDS20457N5132	HD 203562	IDS21179N0623
HD 194882	IDS20230N5916	HD 198571	IDS20461S0600	HD 203585	IDS21180S4126
HD 194883	IDS20229N5421	HD 198625	IDS20465N4617	HD 203705	IDS21187S1318
HD 194939	IDS20232S0341	HD 198639	IDS20466N4341	HD 203784	IDS21197N3655
HD 194943	IDS20232S1809	HD 198700	IDS20470S5850	HD 203803	IDS21194N2351
HD 194951	LS II +34 021	HD 198732	IDS20472S2409	HD 203857	IDS21197N3655
HD 195066	IDS20240N5619	HD 198743	IDS20473S0922	HD 203858	IDS21196N2453
HD 195093	IDS20242S1855	HD 199101	IDS20498N3304	HD 203938	IDS21202N4644
HD 195094	IDS20242S1855	HD 199218	IDS20506N4019	HD 203955	IDS21202S8029
HD 195295	LS II +30 032	HD 199223	IDS20507N0409	HD 204018	IDS21206S4259
HD 195324	LS II +36 080	HD 199254	IDS20509N1211	HD 204075	IDS21210S2251
HD 195325	IDS20255N1034	HD 199305	IDS20513N6148	HD 204131	IDS21214N4853
HD 195330	IDS20255S1523	HD 199345	IDS20515S1004	HD 204172	IDS21217N3614
HD 195434	IDS20262N0452	HD 199442	IDS20521N0005	HD 204585	IDS21244N2145
HD 195483	IDS20264N1055	HD 199478	LS III +47 020	HD 204599	IDS21246N5919
HD 195479	IDS20265N2016	HD 199579	LS III +44 028	HD 204710	IDS21253N4429
HD 195554	IDS20270N5544	HD 199580	IDS20531N4230	HD 204724	IDS21254N2312
HD 195556	IDS20270N4837	HD 199611	IDS20532N5021	HD 204811	IDS21260N4706
HD 195564	IDS20269S1012	HD 199665	IDS20536N1027	HD 204854	IDS21262S3423
HD 195593	LS II +36 082	HD 199766	IDS20541N0355	HD 204867	IDS21263S0601
	IDS20272N3636	HD 199941	IDS20552N1626	HD 205021	IDS21274N7007
HD 195692	IDS20277N2528	HD 199942	IDS20552N0707	HD 205139	LS III +60 002
HD 195774	IDS20282N4853	HD 199951	IDS20552S3239	HD 205471	IDS21304S2637
HD 195873	IDS20288N4132	HD 199955	IDS20553N5004	HD 205541	IDS21309N2400
HD 196067	IDS20298S7542	HD 200011	IDS20556S4323	HD 205637	IDS21315S1955
HD 196093	IDS20300N3454	HD 200044	IDS20559N1856	HD 205688	IDS21319N2936
HD 196124	IDS20302N0546	HD 200077	IDS20561N3952	HD 205765	IDS21324S0050
HD 196171	IDS20305S4738	HD 200120	IDS20564N4708	HD 205767	IDS21324S0818
HD 196178	IDS20306N4621	HD 200245	IDS20572S2807	HD 205811	IDS21328N0610
HD 196379	LS III +51 002	HD 200310	LS III +45 04	HD 205826	IDS21328S1121
HD 196524	IDS20328N1415		IDS20577N4546	HD 205924	IDS21335N0519
HD 196574	IDS20332S0127	HD 200375	IDS20580N0108	HD 206067	IDS21345N0147
HD 196606	IDS20335N3113	HD 200497	IDS20588S0573	HD 206240	IDS21356S8311
HD 196629	IDS20335N3113	HD 200499	IDS20587S2015	HD 206330	IDS21363N4249
HD 196662	IDS20337S1518	HD 200525	IDS20588S7334	HD 206356	IDS21363S2343
HD 196753	K0 II-III+A1 V	HD 200527	V1981 Cyg	HD 206483	IDS21373N5352
HD 196755	IDS20343N0944	HD 200560	IDS20593N4527	HD 206538	IDS21375N4021
HD 196758	IDS20343N0008	HD 200595	IDS20593N4527	HD 206644	IDS21384N4038
HD 196775	IDS20344N1529	HD 200614	IDS20594N5616	HD 206672	LS III +50 029
HD 196787	IDS20345N8105	HD 200644	IDS20596N0508	HD 206697	LS III +43 024
HD 196794	IDS20343N0944	HD 200702	IDS20599S4147	HD 206742	IDS21390S3329
HD 196795	IDS20346N0437	HD 200723	IDS21002N4114	HD 206774	IDS21393N3750
HD 196808	IDS20346N4736	HD 200776	LS III +46 041	HD 206778	IDS21393N0925
HD 196867	IDS20350N1534	HD 200779	IDS21004N0642	HD 206826	IDS21397N2817
HD 196885	IDS20351N1054	HD 200790	IDS21005N0534	HD 206828	IDS21396N2604
HD 196925	IDS20353N8044	HD 200914	IDS21013S2524	HD 206901	IDS21401N2511
HD 196982	IDS20390S3142	HD 201091	IDS21024N3815	HD 207098	IDS21415S1635
HD 197018	IDS20359N4014	HD 201092	IDS21024N3815	HD 207129	IDS21418S4746
HD 197076	IDS20362N1934	HD 201184	IDS21028S2136	HD 207155	IDS21419S3122
HD 197120	IDS20364N2927	HD 201251	IDS21032N4715	HD 207198	LS III +62 002
HD 197177	IDS20370N3157	HD 201416	IDS21043N4827		IDS21422N6200
HD 197226	IDS20373N3844	HD 201433	IDS21044N4827	HD 207260	LS III +60 00
HD 197345	IDS20380N4455	HD 201601	IDS21055N0944	HD 207330	LS III +49 033
HD 197433	G 261 - 28	HD 201616	IDS21055N0944	HD 207491	IDS21442N0515
	IDS20388N7514	HD 201671	IDS21060N2203	HD 207652	IDS21454N1650
HD 197481	IDS20390S3142	HD 201819	LS II +36 086	HD 207673	LS III +40 028
HD 197489	IDS20390N2528		IDS21070N3554	HD 207840	IDS21469N1921
HD 197511	IDS20391N4959	HD 201836	IDS21070N4717	HD 207964	IDS21478S6221
HD 197770	LS III +56 001	HD 202103	IDS21086S5341	HD 207966	IDS21479N4153
HD 197912	IDS20415N3021	HD 202109	IDS21086N2949	HD 208008	IDS21482S1047
HD 197937	IDS20417S4421	HD 202128	IDS21088N1534	HD 208095	IDS21486N5519
HD 197963	IDS20420N1546	HD 202214	LS III +59 003	HD 208132	IDS21491N6517
HD 197964	IDS20420N1546		IDS21093N5935	HD 208177	IDS21494S0347
HD 197989	IDS20422N3336	HD 202240	LS II +36 087	HD 208202	IDS21496N1915
HD 198069	IDS20428N0538	HD 202275	IDS21096N0936	HD 208450	IDS21511S5528
HD 198084	IDS20429N5713	HD 202444	IDS21108N3737	HD 208501	LS III +56 021
HD 198134	IDS20432N3400	HD 202447	IDS21108N0450	HD 208512	IDS21515N5002
HD 198149	IDS20433N6127	HD 202582	IDS21117N6400	HD 208606	LS III +61 004
HD 198151	IDS20433N4609	HD 202730	IDS21127S5352	HD 208682	LS III +65 001
HD 198160	IDS20433S6248	HD 202904	IDS21138N3429		IDS21529N6450
HD 198183	IDS20435N3607	HD 202940	IDS21140S2646	HD 208741	IDS21533S7636
HD 198208	IDS20436S1825	HD 203025	LS III +58 006	HD 208801	IDS21537S0450
HD 198232	IDS20437S3409		IDS21146N5811	HD 208816	IDS21538N6309
HD 198237	IDS20439N4513	HD 203096	IDS21150N4037		M2 Iaep+B8 Ve
HD 198287	IDS20442N3855	HD 203156	IDS21154N3749	HD 208905	IDS21543N6049
HD 198300	IDS20443N5929	HD 203212	IDS21158S7214	HD 208960	IDS21548N6032
HD 198345	IDS20445N4728	HD 203280	IDS21162N6210	HD 209014	IDS21551S2856
HD 198387	IDS20449N5203	HD 203338	IDS21165N5812	HD 209166	IDS21562N1238

Table 5 1363

Star	Identification	Star	Identification	Star	Identification
HD 209278	IDS21570S1727	HD 214953	IDS22367S4743		IDS23038N6306
HD 209339	LS III +62 006	HD 214993	IDS22370N3942	HD 218630	IDS23044S4324
	IDS21576N6201	HD 215104	IDS22377S4156	HD 218634	IDS23045N0808
HD 209369	IDS21578N7242	HD 215114	IDS22378S0850	HD 218640	IDS23046S2300
HD 209515	IDS21589N4410	HD 215182	IDS22383N2942	HD 218658	IDS23047N7451
HD 209693	IDS22002N3228	HD 215242	IDS22388N4638	HD 218738	IDS23053N4725
HD 209750	IDS22006S0048	HD 215359	IDS22396N3856	HD 218753	LS III +59 046
HD 209790	IDS22009N6408	HD 215369	IDS22395S5402		IDS23055N5847
HD 209942	IDS22019N8223	HD 215373	IDS22396N4117	HD 218935	IDS23070N2619
HD 209943	IDS22019N8223	HD 215545	IDS22408S4728	HD 219023	IDS23076S5010
HD 209952	IDS22019S4727	HD 215549	IDS22409N2956	HD 219134	IDS23084N5637
HD 209975	LS III +62 007	HD 215648	IDS22417N1139	HD 219139	IDS23084N1031
	IDS22021N6148	HD 215661	IDS22417N6736	HD 219175	IDS23088S0929
HD 210027	IDS22023N2452	HD 215766	IDS22424S1435	HD 219279	IDS23095S1114
HD 210221	LS III +53 036	HD 215812	IDS22427S0445	HD 219291	IDS23095N2914
HD 210289	IDS22043N4919	HD 216032	IDS22443S1407	HD 219430	IDS23106S0938
HD 210354	IDS22048N3241	HD 216042	IDS22444S3320	HD 219449	IDS23106S0938
HD 210461	IDS22055N1408	HD 216172	IDS22456N6802	HD 219507	IDS23111S4502
HD 210715	IDS22072N5020	HD 216187	IDS22457S6343	HD 219617	IDS23119S1422
HD 210839	LS III +59 023	HD 216259	IDS22464N1326	HD 219623	IDS23121N5241
HD 210855	IDS22081N5620	HD 216336	IDS22470S3324	HD 219657	IDS23125S0204
HD 210884	IDS22084N6938	HD 216380	IDS22474N6109	HD 219734	IDS23131N4828
HD 210939	IDS22087N6016		G8 III-IV+G2 IV	HD 219829	IDS23138N0452
HD 210960	IDS22088S2134	HD 216385	IDS22474N0918	HD 219832	IDS23138S1010
HD 211073	IDS22096N3914	HD 216397	IDS22475N4246	HD 219834	IDS23139S1400
HD 211300	IDS22111N7249	HD 216446	IDS22479N8237	HD 219877	IDS23142S0540
HD 211336	IDS22113N5633	HD 216494	IDS22482S1209	HD 219912	IDS23143S3415
HD 211415	IDS22117S5406	HD 216532	LS III +62 019	HD 219916	IDS23145N6734
HD 211472	IDS22121N5410	HD 216608	IDS22492N4413	HD 219962	IDS23149N4750
HD 211797	IDS22146N3716	HD 216629	IDS22494N5137	HD 220003	IDS23152S5051
HD 211880	IDS22152N6243	HD 216658	IDS22496N6136	HD 220105	IDS23160N4334
HD 211924	IDS22154N0517	HD 216711	LS III +62 023	HD 220117	IDS23161N3739
HD 211998	IDS22160S7244	HD 216718	IDS22500S0532	HD 220140	IDS23163N7827
HD 212061	IDS22166S0153	HD 216763	IDS22504S3304	HD 220222	IDS23170N3116
HD 212087	IDS22166S4627	HD 216777	IDS22506S0821	HD 220278	IDS23174S1536
HD 212097	IDS22167N2750	HD 216831	IDS22511N3550	HD 220363	IDS23180N1146
HD 212120	IDS22169N4602	HD 216863	IDS22514N1656	HD 220392	IDS23182S5421
HD 212132	IDS22170S4626	HD 216898	LS III +62 028	HD 220406	IDS23184S0015
HD 212168	IDS22171S7531	HD 216900	IDS22518N1119	HD 220821	IDS23218N4447
HD 212280	IDS22180N2951	HD 216916	IDS22518N4104	HD 220825	IDS23218N0042
HD 212330	IDS22183S5818	HD 217035	LS III +62 029	HD 220858	IDS23221N0035
HD 212395	IDS22188N2021		IDS22526N6220	HD 221253	IDS23254N5800
HD 212404	IDS22189S0521	HD 217061	LS III +62 030	HD 221503	IDS23276S1723
HD 212581	IDS22202S6529	HD 217086	LS III +62 031	HD 221565	IDS23280S2128
HD 212593	LS III +49 046		IDS22529N6212	HD 221615	IDS23285N2157
HD 212698	IDS22211S1715	HD 217166	IDS22535N0850	HD 221673	IDS23290N3047
HD 212754	IDS22215N0353	HD 217174	IDS22536N6150	HD 221760	IDS23297S4310
HD 212883	IDS22223N3656	HD 217232	IDS22542N1112	HD 221776	IDS23298N3729
HD 212943	IDS22228N0412	HD 217264	IDS22543N0026	HD 222004	IDS23318S3225
HD 212953	IDS22228S3938	HD 217297	LS III +63 016	HD 222093	IDS23325S1337
HD 213009	IDS22233S4400	HD 217312	LS III +62 032	HD 222098	IDS23326N1617
HD 213023	IDS22235N6312	HD 217344	IDS22549S3417	HD 222107	IDS23327N4555
HD 213052	IDS22237S0032	HD 217463	LS III +62 034	HD 222109	IDS23326N4353
HD 213080	IDS22238S4416	HD 217477	IDS22559N3032	HD 222133	IDS23329N1751
HD 213087	LS III +64 003	HD 217490	IDS22560N5905	HD 222143	G 171 - 7
HD 213235	IDS22249N0355	HD 217642	IDS22570S3657	HD 222275	IDS23341N6135
HD 213306	IDS22254N5754	HD 217657	LS III +62 036	HD 222287	IDS23341S4712
HD 213310	IDS22253N4712	HD 217675	IDS22573N4147	HD 222317	G 130 - 1
HD 213398	IDS22258S3252		B6 IIIpe+A2p	HD 222368	IDS23348N0505
HD 213558	IDS22272N4946	HD 217782	IDS22580N4213	HD 222399	IDS23351N3706
HD 213660	IDS22281N3916	HD 217796	IDS22581N6158	HD 222439	IDS23355N4347
HD 213789	IDS22289S0205	HD 217811	IDS22582N4331	HD 222474	IDS23358N1949
HD 213973	IDS22301N6923	HD 217813	IDS22584N5441	HD 222574	IDS23366S1822
HD 214019	IDS22304N6951	HD 217877	G 157 - 8	HD 222661	IDS23375S1505
HD 214122	IDS22310S3211	HD 217906	IDS22589N2732	HD 222670	IDS23376N6358
HD 214150	IDS22311S4106	HD 217919	LS III +63 018	HD 222682	IDS23378N6108
HD 214168	IDS22314N3907	HD 217943	IDS22592N5954	HD 222794	IDS23386N5730
HD 214200	IDS22316N3503	HD 217979	LS III +63 019	HD 222842	IDS23390N2849
HD 214376	IDS22326S0445	HD 218060	IDS23000S0814	HD 222872	IDS23393S2648
HD 214441	IDS22330S5313	HD 218061	IDS22599S1738	HD 223024	IDS23408S1914
HD 214448	IDS22333N0825	HD 218066	LS III +63 020	HD 223047	IDS23411N4552
HD 214479	IDS22333S2108	HD 218195	IDS23010N5742	HD 223128	LS I +66 001
HD 214567	IDS22340N1900	HD 218227	IDS23012S4404	HD 223229	IDS23426N4617
HD 214599	IDS22342S2852	HD 218240	IDS23013S2417	HD 223252	IDS23428S0319
HD 214615	IDS22342S1308	HD 218242	IDS23013S3926	HD 223352	IDS23437S2841
HD 214665	IDS22347N5616	HD 218269	IDS23015S5114	HD 223358	IDS23438N6420
HD 214680	IDS22348N3832	HD 218342	LS III +62 038	HD 223385	IDS23440N6140
HD 214714	IDS22350N3704	HD 218376	LS III +59 043	HD 223466	IDS23446S2553
HD 214810	IDS22356S0404	HD 218395	IDS23027N3217	HD 223498	IDS23449N0219
HD 214850	IDS22359N1401	HD 218452	IDS23031N4550	HD 223515	IDS23450S2958
HD 214923	IDS22365N1019	HD 218537	HR 8808	HD 223552	IDS23454N5104

Star	Identification	Star	Identification	Star	Identification
HD 223778	G 242 - 42	HD 285846	IDS04242N1800	NGC 869 -1586	V359 Per
	IDS23475N7459	HD 286955	IDS04343N0941	NGC 884 -1655	AD Per
HD 223825	IDS23478S0343	HD 292630	IDS06464S0115	NGC 884 -1818	FZ Per
HD 223991	IDS23492S2736	HD 296555	IDS07056S0240	NGC 884 -2088	V506 Per
HD 224085	G 130 - 24	HD 300777	IDS10235S5636	NGC 884 -2165	V507 Per
HD 224427	IDS23527N2436	HD 300814	IDS10285S5614	NGC 884 -2138	V424 Per
HD 224490	IDS23533N5050	HD 303176	IDS10352S5806	NGC 884 -2417	RS Per
HD 224533	IDS23536S0407	HD 303182	IDS10352S5818	NGC 884 -2541	IDS02156N5638
HD 224572	LS I +55 003	HD 305453	IDS10390S5941	NGC 884 -2691	V439 Per
	IDS23539N5512	HD 306145	IDS11070S5923	NGC 884 -2758	V403 Per
HD 224635	IDS23544N3310	HD 306387	IDS11176S6029	NGC 957 - 116	IDS02263N5706
HD 224750	IDS23552S4451	HD 307910	IDS10369S6228	NGC 957 - 213	IDS02263N5706
HD 224758	IDS23552N2622	HD 310376	LSS 2151	NGC 1039 - 166	IDS02348N4215
HD 224855	IDS23561N5948	HD 312603	IDS18007S1713	NGC 1039 - 263	IDS02356N4222
HD 224893	LS I +60 067	HD 316493	IDS17467S2802	NGC 1039 - 266	IDS02356N4222
HD 224927	IDS23567S2621	HD 317690	IDS17248S3118	NGC 1039 - 282	IDS02356N4217
HD 224930	IDS23569N2633	HD 317707	IDS17231S3141	NGC 1039 - 307	IDS02358N4216
HD 225009	IDS23575N6532	HD 317757	IDS17258S3316	NGC 1039 - 328	IDS02360N4220
HD 225160	IDS23590N6144	HD 319703	vdB 86 # b	NGC 1039 - 370	IDS02363N4223
HD 225180	LS I +62 061	HD 320459	IDS17383S3456	NGC 1039 - 374	IDS02363N4223
	IDS23590N6144	HD 323771	vdB 91 # a	NGC 1039 - 503	G 74 - 38
HD 225218	IDS23595N4132	HD 330386	IDS16025S4849	NGC 1444 - 1	IDS03419N5221
HD 225276	IDS23598N2606	HD 330652	IDS16171S4835	NGC 1502 - 1	IDS03590N6204
HD 227247	IDS19586N3502	HD 331161	IDS19425N3147	NGC 1502 - 2	HR 1260, SZ Cam
HD 227452	IDS20005N3940	HD 332407	LS II +29 010		IDS03590N6204
HD 227704	IDS20029N3437	HD 333125	LS II +29 027	NGC 1502 - 10	IDS03590N6204
HD 228199	IDS20077N3612	HD 340730	IDS20314N2502	NGC 1502 - 12	IDS03590N6204
HD 228347	IDS20093N3544	HD 344313	IDS19206N2234	NGC 1502 - 26	IDS03590N6204
HD 228854	IDS20150N3602	HD 344876	IDS19409N2344	NGC 1502 - 36	IDS03590N6204
HD 229159	IDS20193N3853	HD 349425	IDS18457N2036	NGC 1502 - 37	IDS03590N6203
HD 230579	IDS18594N1058			NGC 1502 - 76	IDS03590N6204
HD 231195	IDS19161N1414	NGC 103 - 16	V546 Cas	NGC 1528 - 2	IDS04076N5100
HD 231510	IDS19236N1347	NGC 103 - 39	LS I +61 150	NGC 1545 - 1	HR 1324, B Per
HD 232029	IDS19328N1614	NGC 103 - 43	LS I +61 149	NGC 1545 - 2	HR 1330
HD 232522	IDS01395N5450	NGC 129 - 165	IDS00258N5942	NGC 1545 - 3	IDS04134N5001
HD 232716	IDS02426N5331	NGC 129 - 171	IDS00248N5925	NGC 1545 - 4	IDS04134N5001
HD 233153	IDS05332N5326	NGC 129 - 200	DL Cas	NGC 1545 - 5	IDS04134N5001
HD 233453	IDS07408N5355	NGC 129 - 201	HR 113	NGC 1662 - 1	IDS04429N1045
HD 234677	G 227 - 36		IDS00248N5925	NGC 1662 - 8	IDS04429N1045
HD 236740	IDS01209N5946	NGC 129 - 202	LS I +60 112	NGC 1664 - 578	IDS04438N4319
HD 236970	IDS02262N5553	NGC 188 -2027	ER Cep	NGC 1778 - 31	IDS05013N3655
HD 237051	IDS02541N5827	NGC 457 - 25	V466 Cas	NGC 1778 - 32	IDS05013N3655
HD 237091	IDS03073N5932	NGC 457 - 54	IDS01138N5742	NGC 1778 - 42	IDS05013N3655
HD 237822	IDS09295N5822	NGC 457 - 100	IDS01138N5742	NGC 1778 - 140	IDS05014N3650
HD 238090	IDS12073N5502	NGC 457 - 131	IDS01138N5742	NGC 1778 - 141	IDS05014N3650
HD 239748	LS III +58 013	NGC 457 - 135	IDS01138N5742	NGC 1807 - 12	IDS05050N1622
HD 239960	G 232 - 75	NGC 457 - 136	HR 382	NGC 1807 - 18	G 85 - 47
	IDS22244N5712		IDS01138N5742	NGC 1817 - 247	IDS05061N1637
HD 240764	IDS05014N3016	NGC 457 - 198	IDS01123N5751	NGC 1817 - 248	IDS05061N1637
HD 245310	IDS05304N2107	NGC 559 -1168	LS I +63 146	NGC 1857 - 1	IDS05132N3914
HD 246878	IDS05382N2711	NGC 581 - 70	IDS01266N6010	NGC 1857 - 102	VES 888
HD 250310	IDS05552N2014	NGC 581 - 73	IDS01266N6010	NGC 1857 - 203	IDS05132N3914
HD 250792	IDS05573N1922	NGC 581 - 124	IDS01268N6006	NGC 1893 - 14	IDS05162N3319
HD 250830	IDS05575N2654	NGC 581 - 127	IDS01268N6006	NGC 1912 - 562	HR 1794
HD 253363	IDS06070N2353	NGC 581 - 144	IDS01268N6006		N1907- 251
HD 253683	IDS06082N1900	NGC 581 - 176	IDS01266N6010	NGC 1931 - 19	IDS05248N3412
HD 254477	IDS06114N2228	NGC 581 - 178	IDS01266N6010		LS V +34 046
HD 254657	IDS06121N2153	NGC 581 - 179	IDS01266N6010	NGC 1931 - 26	
HD 255093	IDS06138N2221	NGC 654 - 259	LS I +61 213	NGC 1960 - 8	IDS05298N3404
HD 255312	IDS06145N2322	NGC 654 - 289	LS I +61 217	NGC 1960 - 9	IDS05298N3404
HD 256725	LS V +19 005	NGC 654 - 302	LS I +61 218	NGC 1960 - 138	IDS05300N3400
HD 258728	IDS06254N3436	NGC 654 - 368	LS I +61 229	NGC 2099 - 9	IDS05458N3226
HD 260655	IDS06314N1738	NGC 659 - 111	LS I +60 178	NGC 2099 - 16	IDS05458N3226
HD 263175	IDS06396N3239	NGC 663 - 40	IDS01397N6046	NGC 2099 - 64	IDS05457N3233
HD 267374	LS VI +9 020	NGC 663 - 44	IDS01395N6045	NGC 2099 - 67	IDS05457N3233
HD 268653	Sk -66 005	NGC 663 - 54	IDS01392N6044	NGC 2099 - 69	IDS05457N3233
HD 269154	Sk -67 060	NGC 752 - 60	IDS01502N3716	NGC 2099 - 81	IDS05458N3234
HD 269545	Sk -67 112	NGC 752 - 62	IDS01502N3716	NGC 2099 - 178	IDS05456N3233
HD 269551	IDS05275S6855	NGC 752 - 219	DS And	NGC 2099 - 180	IDS05456N3233
HD 269599	Sk -69 147a	NGC 752 - 309	IDS01529N3712	NGC 2099 - 195	IDS05457N3235
HD 269714	Sk -67 174	NGC 752 - 400	IDS01487N3742	NGC 2099 - 196	IDS05457N3235
HD 269828	Sk -69 209a	NGC 752 - 437	IDS01452N3717	NGC 2099 - 391	IDS05456N3227
HD 269858	Sk -69 220	NGC 869 - 3	IDS02098N5636	NGC 2099 - 437	IDS05454N3235
HD 269927	IDS05398S6932	NGC 869 - 16	HR 654	NGC 2099 - 488	IDS05463N3232
HD 269936	Sk -69 253		IDS02098N5636	NGC 2099 - 489	IDS05463N3232
HD 269964	LB 3459	NGC 869 - 49	V356 Per	NGC 2099 - 687	IDS05451N3228
HD 270046	LSS 294	NGC 869 - 309	V502 Per	NGC 2129 - 7	IDS05550N2320
HD 271191	Sk -65 054	NGC 869 - 612	IDS02114N5633	NGC 2169 - 1	IDS06028N1359
HD 275877	IDS03430N3841	NGC 869 - 899	BU Per	NGC 2169 - 2	V916 Ori
HD 277680	IDS05033N4032	NGC 869 -1057	IDS02120N5640	NGC 2169 - 3	IDS06026N1401
HD 281150	IDS05209N3420	NGC 869 -1162	IDS02122N5642	NGC 2169 - 4	IDS06028N1359

Table 5 1365

Star	Identification	Star	Identification	Star	Identification
NGC 2169 - 5	V917 Ori	NGC 2281 - 117	IDS06430N4107	NGC 2516 - 38	V392 Car
NGC 2169 - 6	IDS06028N1359	NGC 2287 - 1	IDS06420S2039	NGC 2516 - 41	V373 Car
NGC 2169 - 7	IDS06026N1401	NGC 2287 - 23	IDS06418S2041	NGC 2516 - 42	V416 Car
NGC 2169 - 14	V1154 Ori	NGC 2287 - 92	IDS06427S2034	NGC 2516 - 51	V421 Car
NGC 2169 - 31	IDS06028N1359	NGC 2287 - 102	IDS06418S2039	NGC 2516 - 54	V420 Car
NGC 2175 - 1	HR 2344	NGC 2287 - 109	HR 2509, HK CMa	NGC 2516 - 55	V418 Car
	IDS06049N2038	NGC 2287 - 116	IDS06414S2035	NGC 2516 - 59	V417 Car
NGC 2232 - 1	IDS06230S0442	NGC 2287 - 117	IDS06414S2035	NGC 2516 - 60	V419 Car
NGC 2232 - 2	HR 2325	NGC 2287 - 119	IDS06416S2035	NGC 2516 - 63	IDS07584S6030
NGC 2232 - 9	V682 Mon	NGC 2287 - 121	IDS06416S2035	NGC 2516 - 64	IDS07584S6030
NGC 2232 - 16	IDS06201S0423	NGC 2287 - 207	IDS06423S2028	NGC 2516 - 66	IDS07584S6024
NGC 2232 - 21	IDS06234S0424	NGC 2287 - 208	IDS06423S2028	NGC 2516 - 84	V410 Car
NGC 2244 - 80	IDS06262N0455	NGC 2287 - 235	IDS06403S2018	NGC 2516 - 104	IDS07551S6033
NGC 2244 - 83	IDS06260N0454	NGC 2301 - 1	IDS06466N0035	NGC 2516 - 110	HR 3153
NGC 2244 - 84	IDS06260N0454	NGC 2301 - 2	IDS06466N0035	NGC 2516 - 112	IDS07551S6033
NGC 2244 - 121	IDS06266N0461	NGC 2301 - 9	BSD 98-3251	NGC 2516 - 119	HR 3120
NGC 2244 - 122	IDS06466N0501	NGC 2301 - 95	IDS06466N0040	NGC 2516 - 121	IDS07550S6022
NGC 2244 - 123	IDS06266N0461	NGC 2301 - 97	IDS06466N0040	NGC 2516 - 124	HR 3138
NGC 2244 - 125	IDS06265N0460	NGC 2301 - 146	IDS06466N0029		IDS07559S6002
NGC 2244 - 128	IDS06265N0460	NGC 2323 - 1	IDS06582S0812	NGC 2516 - 127	V391 Car
NGC 2244 - 200	V578 Mon	NGC 2343 - 2	IDS07035S1028	NGC 2516 - 130	IDS07567S6035
NGC 2244 - 266	HR 2382	NGC 2353 - 1	HR 2739	NGC 2516 - 133	IDS07572S6022
NGC 2244 - 353	IDS06277N0461	NGC 2353 - 16	IDS07098S1007	NGC 2516 - 134	HR 3147, V374 Car
NGC 2244 - 379	IDS06256N0443	NGC 2353 - 135	IDS07098S1007	NGC 2516 - 136	HR 3152
NGC 2251 - 3	IDS06294N0824	NGC 2354 - 291	IDS07092S2535	NGC 2516 - 206	IDS07572S6022
NGC 2264 - 2	V588 Mon	NGC 2362 - 23	HR 2782	NGC 2516 - 208	IDS07551S6033
NGC 2264 - 7	IDS06337N0944		IDS07146S2446	NGC 2516 - 209	IDS07551S6033
NGC 2264 - 20	V589 Mon	NGC 2362 - 30	IDS07146S2446	NGC 2516 - 210	IDS07567S6035
NGC 2264 - 50	V641 Mon	NGC 2362 - 46	IDS07151S2446	NGC 2516 - 224	HR 3070
	IDS06350N0955	NGC 2367 - 1	IDS07158S2142		IDS07476S6002
NGC 2264 - 66	IDS06350N0955	NGC 2384 - 1	IDS07207S2049	NGC 2516 - 225	HR 3076
NGC 2264 - 67	IDS06350N0955	NGC 2384 - 2	IDS07207S2049	NGC 2533 - 11	IDS08030S2936
NGC 2264 - 74	V684 Mon	NGC 2422 - 1	HR 2897	NGC 2539 - 4	IDS08066S1238
	IDS06350N0955	NGC 2422 - 9	HR 2902, KQ Pup	NGC 2546 - 544	IDS08101S3725
NGC 2264 - 77	NW Mon	NGC 2422 - 21	IDS07303S1357	NGC 2546 - 575	HR 3233
NGC 2264 - 78	LP Mon	NGC 2422 - 45	HR 2921	NGC 2546 - 578	IDS08089S3740
NGC 2264 - 79	NX Mon		IDS07315S1416	NGC 2546 - 682	HR 3219
NGC 2264 - 80	GP Mon	NGC 2422 - 47	IDS07315S1416		IDS08073S3700
NGC 2264 - 89	LQ Mon	NGC 2422 - 78	IDS07320S1416	NGC 2547 - 1	HR 3203
NGC 2264 - 92	V642 Mon	NGC 2422 - 79	IDS07321S1413	NGC 2547 - 8	KW Vel
NGC 2264 - 96	LR Mon	NGC 2422 - 83	IDS07321S1413	NGC 2547 - 14	IDS08067S4854
NGC 2264 - 100	IDS06355N0959	NGC 2422 - 125	HR 2932	NGC 2547 - 45	IDS08067S4901
NGC 2264 - 103	IDS06353N0927		IDS07330S1413	NGC 2547 - 53	IDS08076S4859
NGC 2264 - 104	IDS06355N0959	NGC 2437 - 78	IDS07369S1438	NGC 2548 - 2184	G 113 - 21
NGC 2264 - 105	LT Mon	NGC 2439 - 1	HR 2974, R Pup	NGC 2567 - 54	IDS08144S3020
NGC 2264 - 109	IDS06355N0959	NGC 2439 - 2	IDS07370S3127	NGC 2632 - 45	BR Cnc
NGC 2264 - 115	V419 Mon	NGC 2439 - 69	IDS07367S3128		IDS08320N1952
NGC 2264 - 121	IDS06355N0959	NGC 2451 - 15	IDS07403S3739	NGC 2632 - 114	CY Cnc
NGC 2264 - 126	LU Mon	NGC 2451 - 44	IDS07413S3754	NGC 2632 - 142	IDS08334N1962
NGC 2264 - 131	HR 2456, S Mon	NGC 2451 - 47	IDS07414S3751	NGC 2632 - 150	IDS08334N1962
	IDS06355N0959	NGC 2451 - 57	IDS07417S3757	NGC 2632 - 154	BS Cnc
NGC 2264 - 133	IO Mon	NGC 2451 - 63	IDS07419S3800	NGC 2632 - 155	IDS08334N1962
NGC 2264 - 136	V421 Mon	NGC 2451 - 85	IDS07429S3730	NGC 2632 - 203	IDS08340N2026
NGC 2264 - 137	IDS06355N0960	NGC 2451 - 154	HR 2945	NGC 2632 - 204	BT Cnc
NGC 2264 - 139	IP Mon	NGC 2451 - 163	HR 2955	NGC 2632 - 207	BU Cnc
NGC 2264 - 142	IDS06355N0959	NGC 2451 - 175	HR 2961	NGC 2632 - 212	IDS08342N1954
NGC 2264 - 143	IDS06355N0960		IDS07362S3755	NGC 2632 - 224	IDS08342N1954
NGC 2264 - 151	IDS06357N0959	NGC 2451 - 182	HR 2963	NGC 2632 - 229	IDS08342N1954
NGC 2264 - 153	LW Mon		IDS07362S3755	NGC 2632 - 232	IDS08344N2022
NGC 2264 - 161	LX Mon	NGC 2451 - 184	HR 2964	NGC 2632 - 244	TX Cnc
NGC 2264 - 164	V360 Mon	NGC 2451 - 187	HR 2968	NGC 2632 - 250s	Optical companion
NGC 2264 - 168	V609 Mon	NGC 2451 - 203	HR 2981	NGC 2632 - 253	HR 3427
NGC 2264 - 178	IDS06357N0934	NGC 2451 - 209	HR 2986		IDS08344N2022
NGC 2264 - 179	IDS06357N0959	NGC 2451 - 223	HR 2995	NGC 2632 - 265	IDS08344N2022
NGC 2264 - 181	IDS06357N0959	NGC 2451 - 233	HR 3001	NGC 2632 - 268	IDS08344N2022
NGC 2264 - 184	MM Mon	NGC 2451 - 246	HR 3011	NGC 2632 - 271	IDS08344N2022
NGC 2264 - 197	IQ Mon	NGC 2451 - 251	HR 3016	NGC 2632 - 276	IDS08348N1954
NGC 2264 - 199	V432 Mon	NGC 2451 - 254	HR 3017	NGC 2632 - 279	IDS08346N1961
NGC 2264 - 206	IDS06359N0950	NGC 2451 - 260	IDS07423S3844	NGC 2632 - 283	HR 3428
NGC 2264 - 212	IDS06360N0956	NGC 2451 - 267	HR 3022		IDS08346N2001
NGC 2264 - 215	IDS06360N0956	NGC 2451 - 275	IDS07433S3813	NGC 2632 - 283s	Optical companion
NGC 2264 - 217	MO Mon	NGC 2451 - 277	HR 3032, OX Pup	NGC 2632 - 287	IDS08346N1961
NGC 2264 - 222	IDS06360N0956	NGC 2451 - 283	HR 3035	NGC 2632 - 292	BQ Cnc
NGC 2264 - 233	IDS06364N0918		IDS07439S3816	NGC 2632 - 295	IDS08346N1961
NGC 2264 - 424	V426 Mon	NGC 2451 - 291	IDS07446S3726	NGC 2632 - 300	HR 3429
	HRC 232	NGC 2467 - 40	IDS07486S2601		IDS08348N1954
NGC 2264 - 439	HRC 235	NGC 2467 - 83	OR Pup	NGC 2632 - 318	BV Cnc
NGC 2264 - 491	OY Mon	NGC 2477 - 1044	EG 55	NGC 2632 - 323	BN Cnc
	HRC 240	NGC 2477 - 9001	QT Pup	NGC 2632 - 340	BW Cnc
NGC 2281 - 62	IDS06413N4112	NGC 2516 - 15	V356 Car	NGC 2632 - 348	EP Cnc
NGC 2281 - 64	IDS06413N4112	NGC 2516 - 26	V422 Car	NGC 2632 - 350s	Optical companion

Star	Identification	Star	Identification	Star	Identification
NGC 2632 - 377	IDS08355N2050	NGC 3766 - 240	V855 Cen	NGC 6231 - 253	V945 Sco
NGC 2632 - 385	IDS08356N1937	NGC 3766 - 316	BF Cen	NGC 6231 - 261	V946 Sco
NGC 2632 - 445	BX Cnc	NGC 4103 - 1	AI Cru		IDS16471S4134
NGC 2632 - 449	BY Cnc	NGC 4103 - 4	LSS 2590	NGC 6231 - 290	HR 6260
NGC 2632 - 456s	Optical companion	NGC 4103 - 7	LSS 2592		IDS16470S4138
NGC 2632 - 458	IDS08366N2024	NGC 4103 - 11	LSS 2595	NGC 6231 - 291	IDS16471S4140
NGC 2632 - 552	S Cnc	NGC 4103 - 12	LSS 2594	NGC 6231 - 292	IDS16472S4140
	IDS08382N1924	NGC 4103 - 13	LSS 2593	NGC 6231 - 293	HR 6263
NGC 2632 - 561	EO Cnc	NGC 4103 - 14	LSS 2596	NGC 6231 - 306	IDS16470S4138
NGC 2632 - 1133	HR 3387	NGC 4103 - 330	LSS 2591	NGC 6231 - 338	IDS16474S4140
NGC 2632 - 6391	H 139	NGC 4337 - 17	IDS12194S5734	NGC 6242 - 59	IDS16486S3919
NGC 2632 - 6556	H 262	NGC 4337 - 18	IDS12194S5734	NGC 6242 - 65	IDS16487S3921
NGC 2660 - 9009	GV Vel	NGC 4439 - 2	IDS12230S5932	NGC 6242 - 77	IDS16486S3918
NGC 2682 - 131	ES Cnc	NGC 4609 - 4	IDS12365S6226	NGC 6249 - 2	IDS16504S4438
NGC 2682 - 161	EV Cnc	NGC 4609 - 50	IDS12365S6226	NGC 6250 - 3	IDS16506S4547
NGC 2682 - 184	EW Cnc	NGC 4755 - 1	HR 4887	NGC 6250 - 37	LSS 3887
NGC 2682 - 205	AH Cnc	NGC 4755 - 2	HR 4890	NGC 6281 - 1	HR 6327, V923 Sco
NGC 3033 - 1	HR 3895	NGC 4755 - 3	IDS12480S5948	NGC 6281 - 2	V884 Sco
NGC 3033 - 11	IDS09452S5558	NGC 4755 - 6	BW Cru	NGC 6281 - 3	IDS16580S3745
NGC 3114 - 42	IDS09588S5936	NGC 4755 - 7	BS Cru	NGC 6281 - 6	IDS16580S3749
NGC 3114 - 48	HR 3960	NGC 4755 - 8	IDS12476S5952	NGC 6281 - 9	V974 Sco
NGC 3114 - 125	HR 3966	NGC 4755 - 20	HR 4876	NGC 6281 - 10	IDS16573S3759
NGC 3228 - 1	IDS10160S5104	NGC 4755 - 105	BV Cru	NGC 6281 - 15	V948 Sco
NGC 3228 - 5	IDS10173S5113	NGC 4755 - 106	BU Cru	NGC 6281 - 17	IDS16580S3757
NGC 3293 - 4	IDS10320S5743	NGC 4755 - 115	IDS12476S5948	NGC 6281 - 19	IDS16578S3752
NGC 3293 - 5	V381 Car	NGC 4755 - 223	CC Cru	NGC 6281 - 35	IDS16580S3745
NGC 3293 - 10	V401 Car	NGC 4755 - 305	IDS12479S5950	NGC 6383 - 1	HR 6535
NGC 3293 - 11	V400 Car	NGC 4755 - 418	BT Cru		IDS17282S3231
NGC 3293 - 12	V402 Car	NGC 5168 - 156	IDS13246S6025	NGC 6383 - 14	IDS17281S3231
NGC 3293 - 14	V405 Car	NGC 5168 - 157	IDS13246S6025	NGC 6383 - 27	V486 Sco
NGC 3293 - 15	IDS10319S5741	NGC 5367 - 1	IDS13517S3929	NGC 6383 - 100	V701 Sco
NGC 3293 - 16	V403 Car	NGC 5460 - 1	IDS14011S4750	NGC 6383 - 617	IDS17282S3231
NGC 3293 - 18	V406 Car	NGC 5460 - 31	IDS14012S4754	NGC 6383 - 619	IDS17282S3231
	IDS10322S5741	NGC 5460 - 44	IDS14021S4806	NGC 6383 - 702	V702 Sco
NGC 3293 - 19	IDS10322S5741	NGC 5460 - 54	IDS14010S4757	NGC 6396 - 1	IDS17309S3457
NGC 3293 - 21	V361 Car	NGC 5460 - 62	IDS14019S4740	NGC 6396 - 2	IDS17309S3457
NGC 3293 - 22	IDS10320S5743	NGC 5606 - 1	IDS14204S5911	NGC 6405 - 1	BM Sco
NGC 3293 - 23	V404 Car	NGC 5606 - 2	IDS14204S5911	NGC 6405 - 19	V971 Sco
	IDS10320S5743	NGC 5606 - 5	IDS14204S5911	NGC 6405 - 28	IDS17338S3207
NGC 3293 - 24	V378 Car	NGC 5606 - 12	IDS14204S5911	NGC 6405 - 31	V976 Sco
NGC 3293 - 26	V379 Car	NGC 5606 - 13	IDS14204S5911	NGC 6405 - 32	V862 Sco
NGC 3293 - 27	V380 Car	NGC 5606 - 14	IDS14204S5911	NGC 6405 - 77	V970 Sco
NGC 3293 - 28	LSS 1683	NGC 5606 - 15	IDS14204S5911	NGC 6405 - 100	V994 Sco
NGC 3293 - 65	V412 Car	NGC 5606 - 107	LSS 3238	NGC 6405 - 103	IDS17330S3218
NGC 3324 - 1	HR 4169	NGC 5662 - 104	IDS14284S5608	NGC 6416 - 367	IDS17370S3211
NGC 3324 - 2	IDS10336S5806	NGC 6025 - 12	V350 Nor	NGC 6475 - 26	HR 6647, V957 Sco
NGC 3324 - 5	IDS10336S5806	NGC 6025 - 21	IDS15551S6008	NGC 6475 - 42	HR 6652
NGC 3324 - 11	KU Car	NGC 6031 - 74	HM Nor	NGC 6475 - 48	IDS17465S3443
NGC 3324 - 18	IDS10337S5807	NGC 6067 - 276	IDS16052S5357	NGC 6475 - 55	V958 Sco
NGC 3330 - 66	IDS10346S5337	NGC 6067 - 297	V340 Nor	NGC 6475 - 56	HR 6657
NGC 3496 - 1	IDS10557S5953		IDS16054S5359		IDS17467S3442
NGC 3532 - 27	GV Car	NGC 6067 - 298	IDS16054S5359	NGC 6475 - 58	HR 6658
NGC 3532 - 97	IDS11019S5809	NGC 6067 - 316	IDS16058S5352		IDS17467S3452
NGC 3532 - 98	IDS11019S5809	NGC 6087 - 129	IDS16105S5742	NGC 6475 - 61	IDS17468S3453
NGC 3532 - 155	IDS11005S5810	NGC 6087 - 132	HR 6040	NGC 6475 - 86	HR 6662, V906 Sco
NGC 3532 - 181	IDS11009S5819	NGC 6087 - 155	HR 6062, S Nor		IDS17472S3444
NGC 3532 - 221	HR 4323	NGC 6124 - 36	IDS16191S4037	NGC 6475 - 88	HR 6663
	IDS11022S5808	NGC 6169 - 1	HR 6155, μ Nor	NGC 6475 - 110	HR 6668
NGC 3532 - 248	IDS11012S5804	NGC 6178 - 18	IDS16287S4525	NGC 6475 - 134	HR 6648
NGC 3532 - 252	IDS11013S5805	NGC 6193 - 1	HR 6187	NGC 6475 - 141	HR 6660
NGC 3532 - 253	IDS11013S5805		IDS16338S4834	NGC 6494 - 1	HR 6679
NGC 3532 - 290	IDS10599S5818	NGC 6193 - 2	IDS16338S4834	NGC 6514 - 55	HR 6716
NGC 3532 - 374	IDS11027S5805	NGC 6193 - 3	Ara OB1 # 147		IDS17558S2247
NGC 3532 - 411	IDS11004S5800	NGC 6193 - 5	Ara OB1 # 124	NGC 6514 - 145	IDS17563S2302
NGC 3532 - 421	IDS10595S5815	NGC 6193 - 8	IDS16338S4834	NGC 6514 - 146	IDS17563S2302
NGC 3532 - 447	IDS11030S5810	NGC 6193 - 11	Ara OB1 # 349	NGC 6514 - 149	IDS17563S2302
NGC 3532 - 480	ER Car	NGC 6193 - 15	Ara OB1 # 695	NGC 6530 - 2	IDS17565S2415
NGC 3532 - 522	IDS11000S5756	NGC 6200 - 5	IDS16368S4713	NGC 6530 - 3	IDS17565S2415
NGC 3532 - 612	IDS11010S5832	NGC 6200 - 6	IDS16367S4717	NGC 6530 - 4	HR 6724
NGC 3572 - 72	IDS11062S5944	NGC 6200 - 7	IDS16367S4717	NGC 6530 - 7	HR 6736
NGC 3572 - 124	IDS11063S5936	NGC 6200 - 11	IDS16365S4719	NGC 6530 - 10	SV Sgr
NGC 3603 - 70	LSS 2275	NGC 6200 - 13	IDS16365S4719	NGC 6530 - 43	IDS17581S2415
	IDS11108S6043	NGC 6204 - 1	IDS16388S4651	NGC 6530 - 47	IDS17581S2415
NGC 3766 - 1	V843 Cen	NGC 6231 - 28	V962 Sco	NGC 6530 - 100	IDS17586S2421
NGC 3766 - 7	V844 Cen	NGC 6231 - 92	V870 Sco	NGC 6531 - 1	IDS17582S2230
NGC 3766 - 15	V849 Cen	NGC 6231 - 110	V947 Sco	NGC 6531 - 3	IDS17582S2230
NGC 3766 - 27	IDS11315S6103	NGC 6231 - 150	V920 Sco	NGC 6604 - 1	MY Ser
NGC 3766 - 36	V848 Cen	NGC 6231 - 220	HR 6265	NGC 6604 - 4	IDS18124S1217
NGC 3766 - 63	V846 Cen		IDS16473S4139	NGC 6611 - 409	MZ Ser
NGC 3766 - 67	V847 Cen	NGC 6231 - 224	IDS16473S4140	NGC 6611 - 556	LSS 4885
NGC 3766 - 88	V845 Cen	NGC 6231 - 238	V964 Sco	NGC 6618 - 60	IDS18138S1604

Table 5   1367

Star	Identification	Star	Identification	Star	Identification
NGC 6633 - 10	IDS18200N0711	NGC 7380 - 1	IDS22421N5733	IC 2602 - 78	IDS10316S6337
NGC 6633 - 48	IDS18217N0628	NGC 7380 - 2	DH Cep	IC 2602 - 99	IDS10397S6340
NGC 6633 - 50	IDS18217N0628		IDS22432N5736	IC 2944 - 8	V786 Cen
NGC 6633 - 58	IDS18220N0622	NGC 7380 - 3	IDS22421N5733	IC 2944 - 23	LW Cen
NGC 6633 - 90	IDS18227N0629	NGC 7380 - 34	LS III +57 085	IC 2944 - 43	IDS11335S6239
NGC 6633 - 102	HR 6928	NGC 7510 - 1	LS III +60 049	IC 2944 - 59	IDS11337S6249
NGC 6633 - 136	T Ser		IDS23055N6009	IC 2944 - 96	IDS11345S6246
NGC 6633 - 177	IDS18266N0623	NGC 7510 - 2	IDS23055N6009	IC 2944 - 97	BH Cen
NGC 6649 - 42	IDS18279S1031	NGC 7654 - 133	IDS23148N6058	IC 2944 - 113	IDS11351S6255
NGC 6649 - 64	V367 Sct	NGC 7654 - 265	HR 8886	IC 2944 - 118	IDS11351S6255
NGC 6649 - 117	IDS18279S1031		IDS23162N6139	IC 2944 - 132	IDS11359S6201
NGC 6664 - 80	EV Sct	NGC 7654 - 568	LS III +61 032	IC 2944 - 137	V772 Cen
NGC 6705 - 624	V369 Sct	NGC 7654 - 711	LS III +61 033	IC 4651 - 44	IDS17178S4926
NGC 6709 - 208	IDS18468N1012	NGC 7654 - 759	IDS23198N6103	IC 4651 - 76	IDS17170S4949
NGC 6752 -1013	IDS19018S6012	NGC 7654 - 766	IDS23198N6103	IC 4665 - 23	IDS17390N0553
NGC 6823 - 93	LS II +23 029	NGC 7654 - 769	IDS23198N6103	IC 4665 - 67	IDS17416N0544
NGC 6823 - 34	LS II +23 028	NGC 7654 - 782	IDS23198N6103	IC 4665 - 76	IDS17419N0536
NGC 6823 - 174	LS II +23 020	NGC 7654 - 923	HR 8904	IC 4665 - 82	IDS17421N0544
NGC 6823 - 205	LS II +23 019		IDS23204N6144	IC 4665 - 83	IDS17421N0544
NGC 6823 - 208	LS II +23 017	NGC 7654 -1407	LS I +61 003	IC 4665 - 96	G 140 - 11
NGC 6830 - 24	VES 72	NGC 7654 -1449	LS I +61 004	IC 4665 - 102	IDS17434N0618
NGC 6834 - 67	HE3 1792	NGC 7654 -1736	LS I +60 005	IC 4665 - 108	IDS17440N0460
NGC 6866 - 1	IDS20008N4348	NGC 7686 - 1	HR 8925	IC 4665 - 109	IDS17440N0460
NGC 6871 - 1	V1676 Cyg	NGC 7790 - 384	CF Cas	IC 4756 - 8	IDS18307N0451
	IDS20022N3530	NGC 7790 - 385	CE Cas a+b	IC 4756 - 25	IDS18316N0452
NGC 6871 - 3	IDS20022N3530	NGC 7822 - 7	LS I +66 004	IC 4725 - 50	V3508 Sgr
NGC 6871 - 4	IDS20019N3519	NGC 7822 - 11	IDS23554N6640		IDS18260S1912
NGC 6871 - 5	IDS20023N3529	NGC 7822 - 12	IDS23554N6640	IC 4725 - 51	IDS18260S1912
NGC 6871 - 6	IDS20020N3537			IC 4725 - 91	IDS18260S1912
NGC 6871 - 7	IDS20019N3519	IC 348 - 4	IDS03379N3148	IC 4725 - 93	IDS18260S1912
NGC 6871 - 8	IDS20023N3529	IC 348 - 12	IDS03383N3151	IC 4725 - 96	IDS18260S1912
NGC 6871 - 10	V1821 Cyg	IC 348 - 20	IDS03383N3151	IC 4725 - 97	IDS18260S1912
NGC 6871 - 14	V1820 Cyg	IC 1805 - 125	IDS02248M6122	IC 4725 - 98	IDS18260S1912
NGC 6871 - 31	V453 Cyg	IC 1805 - 148	IDS02251N6101	IC 4725 - 99	IDS18260S1912
NGC 6871 - 35	IDS20022N3530	IC 1848 - 1	IDS02434N6000	IC 4725 - 101	IDS18260S1912
NGC 6871 - 68	IDS20022N3530	IC 1848 - 2	IDS02434N6000	IC 4725 - 102	IDS18260S1912
NGC 6871 - 70	IDS20019N3519	IC 2391 - 2	IDS08336S5241	IC 4725 - 103	HR 6947, U Sgr
NGC 6882 - 1	HR 7718	IC 2391 - 8	HR 3435		IDS18260S1912
NGC 6882 - 2	HR 7711		IDS08359S5244	IC 4725 - 106	IDS18260S1912
NGC 6882 - 3	HR 7719	IC 2391 - 10	IDS08362S5221	IC 4725 - 109	IDS18260S1912
NGC 6882 - 8	IDS20077N2604	IC 2391 - 13	HR 3440, HW Vel	IC 4725 - 110	IDS18260S1912
NGC 6882 - 13	W Vul	IC 2391 - 16	HR 3442	IC 4725 - 111	IDS18260S1912
NGC 6910 - 6	V1973 Cyg		IDS08371S5242	IC 4725 - 153	IDS18266S1910
NGC 6913 - 6	V1322 Cyg	IC 2391 - 18	KR Vel	IC 4725 - 145	IDS18262S1910
	IDS20201N3810	IC 2391 - 19	IDS08373S5330	IC 4725 - 197	IDS18260S1912
NGC 6913 - 7	IDS20201N3810	IC 2391 - 20	HR 3447, o Vel	IC 4756 - 8	FR Ser
NGC 6913 - 10	IDS20201N3810	IC 2391 - 21	HR 3448		IDS18307N0451
NGC 6913 - 111	LS II +38 068	IC 2391 - 28	IDS08392S5224	IC 4756 - 17	IDS18311N0502
NGC 6913 - 225	LS II +38 065	IC 2391 - 31	HR 3466	IC 4756 - 23	IDS18314N0511
NGC 6940 - 120	FG Vul		IDS08396S5245	IC 4756 - 95	IDS18337N0446
NGC 6940 - 147	IDS20307N2812	IC 2391 - 34	HR 3467, HY Vel	IC 4756 - 145	HR 7008
NGC 7039 - 125	G 212 - 28		IDS08396S5245	IC 4996 - 28	IDS20128N3720
NGC 7039 - 170	HR 8103	IC 2391 - 47	IDS08433S5246	IC 4996 - 34	IDS20128N3720
NGC 7039 - 211	IDS21078N4527	IC 2391 - 49	IDS08435S5228	IC 4996 - 39	IDS20128N3720
NGC 7039 - 217	IDS21085N4518	IC 2391 - 50	HR 3503	IC 4996 - 51	V1922 Cyg
NGC 7039 - 219	IDS21086N4524	IC 2391 - 51	IDS08396S5245	IC 5146 - 6	V1578 Cyg
NGC 7039 - 221	IDS21089N4527	IC 2391 - 76	IDS08371S5242	IC 5146 - 42	IDS21496N4648
NGC 7063 - 31	IDS21204N3606	IC 2395 - 1	HR 3462, HX Vel		
NGC 7062 - 262	VES 401		IDS08390S4744	Be 86 - 13	LS II +38 046
NGC 7067 - 1	IDS21210N4735	IC 2395 - 3	IDS08393S4748	Be 87 - 15	V439 Cyg
NGC 7067 - 4	LS III +47 039	IC 2395 - 6	IDS08389S4753	Be 87 - 25	LS II +37 076
	IDS21210N4735	IC 2395 - 8	EP Vel	Be 87 - 78	BC Cyg
NGC 7067 - 10	IDS21210N4735		IDS08394S4743	Biu 2 - 126	IDS20051N3513
NGC 7092 - 38	IDS21272N4801	IC 2581 - 1	HR 4110, V399 Car	Biu 2 - 130	IDS20055N3511
NGC 7092 - 39	IDS21272N4801	IC 2581 - 3	V348 Car	Biu 2 - 131	AT Scl
NGC 7092 - 60	IDS21278N4755	IC 2581 - 78	LSS 1582		IDS20055N3511
NGC 7092 - 65	IDS21279N4750	IC 2602 - 8	IDS10316S6337	Bl 1 - 5	G 267 - 6
NGC 7092 - 100	IDS21291N4802	IC 2602 - 17	V407 Car	Bl 1 - 42	IDS23572S2977
NGC 7128 - 4	V1814 Cyg	IC 2602 - 18	IDS10354S6359	Bl 1 - 59	IDS23583S2933
NGC 7128 - 5	V1481 Cyg	IC 2602 - 25	IDS10361S6434	Bl 1 - 66	G 266 - 33
NGC 7160 - 2	EM Cep	IC 2602 - 27	HR 4185, V364 Car	Bl 1 - 71	PSD 140-194
	IDS21509N6208	IC 2602 - 32	IDS10384S6259	Bl 1 - 109	IDS00008S3053
NGC 7160 - 3	IDS21509N6208	IC 2602 - 33	HR 4196	Bl 1 - 111	IDS00011S2943
NGC 7160 - 4	IDS21506N6207	IC 2602 - 37	HR 4199	Bo 9 - 6	IDS10311S5942
NGC 7209 - 38	IDS22006N4555	IC 2602 - 40	HR 4204	Bo 10 - 10	IDS10374S5838
NGC 7209 - 46	SS Lac	IC 2602 - 41	HR 4205	Bo 10 - 35	LSS 1801
NGC 7209 - 86	HR 8421, HT Lac	IC 2602 - 48	HR 4219	Bo 11 - 2	IDS10433S5934
NGC 7243 - 181	IDS22080N4942	IC 2602 - 49	HR 4220	Cr 121 - 1	HR 2580, o^1 CMa
NGC 7243 - 358	IDS22112N4923		IDS10428S6344	Cr 121 - 2	HR 2578
NGC 7243 - 361	IDS22112N4923	IC 2602 - 51	HR 4222	Cr 121 - 4	HR 2583, EZ CMa
NGC 7243 - 362	IDS22112N4923	IC 2602 - 62	IDS10467S6327	Cr 121 - 17	IDS06508S2408

Star	Identification	Star	Identification	Star	Identification
Cr 121 - 19	IDS06494S2418	Cr 359 - 1	HR 6795	St 7 - 1	IDS02222N6013
Cr 121 - 22	HR 2387, ξ^1 CMa		IDS18046N0359		STF 264 B
	IDS06277S2321	Cr 359 - 2	HR 6873	St 7 - 4	IDS02222N6013
Cr 121 - 23	HR 2492, FT CMa	Cr 359 - 3	HR 6714		STF 264 A
	IDS06407S3058		IDS17556N0256	St 7 - 23	V529 Cas
Cr 121 - 24	HR 2571, EY CMa	Cr 359 - 5	HR 6712, V2048 Oph	St 7 - 27	IDS02222N6013
Cr 121 - 25	HR 2595	Cr 359 - 9	LS IV +6 001		STF 263 B
Cr 121 - 26	HR 2596, ι CMa	Cr 359 - 10	HR 6723	St 7 - 28	V528 Cas
Cr 121 - 28	HR 2646, σ CMa		IDS17567N0118		IDS02222N6013
	IDS06577S2747	Cr 359 - 11	V986 Oph		STF 263 A
Cr 121 - 29	HR 2653	Cr 359 - 13	HR 6629	St 8 - 22	IDS05215N3420
Cr 121 - 30	HR 2657	Cr 463 - 3	IDS01379N7112	St 8 - 23	IDS05215N3420
Cr 121 - 31	HR 2693	HM 1 - 1	LSS 4065	St 8 - 24	IDS05215N3420
	IDS07043S2614	HM 1 - 2	LSS 4067	St 14 - 2	LSS 2472
Cr 121 - 32	HR 2695	HM 1 - 3	LSS 4064	St 14 - 4	HR 4511, V810 Cen
	IDS07046S2353	Ho 15 - 1	BQ Cru	St 14 - 6	LSS 2474
Cr 121 - 33	HR 2734, GY CMa	Ho 15 - 3	LSS 2745, CD Cru	St 14 - 11	LSS 2477
Cr 121 - 34	HR 2827	Ho 22 - 1	IDS16393S4654	St 14 - 12	V346 Cen
	IDS07201S2907	Ki 12 - 19	IDS23482N6124	St 16 - 113	LS 3010
Cr 121 - 35	HR 2855, FY CMa	Ki 12 - 21	LS I +61 083	St 17 - 4	LS I +61 029
Cr 121 - 36	HR 2497		IDS23482N6124	To 1 - 18	IDS06563S2030
	IDS06412S3029	Ki 12 - 31	LS I +61 085	Tr 7 - 6	IDS07232S2344
Cr 121 - 38	HR 2603	Ki 14 - 58	LS I +62 123	Tr 10 - 1	HR 3477
Cr 121 - 39	HR 2611	Ly 2 - 30	IDS14166S6051		IDS08408S4217
Cr 121 - 40	HR 2616	Ly 2 - 90	V807 Cen	Tr 10 - 2	HR 3501
	IDS06545S2517		IDS14178S6052	Tr 10 - 3	IDS08430S4212
Cr 121 - 41	HR 2621	Ly 6 - 75	TW Nor	Tr 10 - 31	IDS08408S4217
	IDS06549S3052	Ly 7 - 1	IDS16026S5505	Tr 14 - 1	IDS10401S5901
Cr 121 - 44	HR 2766	Ma 38 - 1	IDS18094S1861	Tr 14 - 2	IDS10400S5902
Cr 121 - 45	HR 2824	Ma 38 - 2	IDS18094S1861	Tr 14 - 8	IDS10401S5901
Cr 121 - 46	HR 2826	Ma 38 - 3	IDS18094S1861	Tr 15 - 1	IDS10408S5850
Cr 121 - 47	HR 2841	Ma 38 - 5	IDS18094S1861	Tr 15 - 2	IDS10408S5850
Cr 132 - 1	HR 2756	Ma 38 - 11	IDS18094S1861	Tr 15 - 3	IDS10408S5850
Cr 132 - 2	HR 2720	Ma 38 - 30	IDS18094S1861	Tr 15 - 16	RT Car
Cr 132 - 3	HR 2768	Ma 50 - 31	IDS23109N5954	Tr 16 - 1	IDS10412S5910
	IDS07131S3043	Ma 50 - 61	IDS23109N5954	Tr 16 - 7	IDS10412S5910
Cr 132 - 4	HR 2743	Pis 4 - 9	IDS08312S4406	Tr 16 - 34	IDS10413S5914
Cr 132 - 5	HR 2741, GG CMa	Pis 5 - 1	IDS08339S3914	Tr 16 - 64	IDS10412S5910
Cr 132 - 7	IDS07147S3037	Pis 5 - 4	IDS08339S3914	Tr 16 - 65	IDS10412S5910
Cr 132 - 9	FF CMa	Pis 6 - 1	IDS08357S4552	Tr 16 - 175	IDS10403S5903
Cr 132 - 11	HO CMa	Pis 6 - 2	IDS08357S4552	Tr 16 - 176	IDS10403S5903
	IDS07094S3048	Pis 6 - 3	IDS08357S4552	Tr 16 - 178	IDS10407S5913
Cr 132 - 12	IDS07108S3119	Pis 6 - 4	IDS08357S4552	Tr 16 - 179	IDS10407S5913
Cr 132 - 14	IDS07131S3043	Pis 6 - 5	IDS08357S4552	Tr 16 - 182	IDS10413S5914
Cr 132 - 23	IDS07087S3055	Pis 6 - 6	IDS08357S4552	Tr 16 - 183	HR 4210, η Car
Cr 132 - 25	IDS07128S3032	Pis 6 - 7	IDS08357S4552		IDS10412S5910
Cr 132 - 28	IDS07069S3101	Pis 6 - 8	IDS08357S4552	Tr 18 - 13	IDS11073S6006
Cr 132 - 29	IDS07097S3051	Pis 6 - 10	IDS08357S4552	Tr 18 - 36	IDS11075S6009
Cr 140 - 1	HR 2834	Pis 6 - 51	IDS08357S4552	Tr 18 - 133	GH Car
	IDS07209S3137	Pis 11 - 1	LSS 1268	Tr 24 - 4	LSS 3842
Cr 140 - 2	HR 2823	Pis 12 - 17	IDS09166S4445	Tr 24 - 15	LSS 3845
Cr 140 - 3	HR 2819	Pis 16 - 2	IDS09477S5243	Tr 24 - 17	LSS 3847
Cr 140 - 4	HR 2847	Pis 16 - 3	IDS09477S5243	Tr 24 - 92	LSS 3840
Cr 140 - 9	IDS07209S3137	Pis 16 - 4	IDS09477S5243	Tr 27 - 1	IDS17296S3326
Cr 140 - 21	IDS07171S3206	Pis 17 - 1	IDS10572S5918	Tr 27 - 102	V925 Sco
Cr 140 - 26	IDS07212S3202	Pis 20 - 7	IDS15075S5842	Tr 37 - 171	IDS21354N5631
Cr 140 - 33	IDS07222S3209	Pis 20 - 8	IDS15075S5842	Tr 37 - 225	AI Cep
Cr 197 - 1	IDS08410S4055	Pis 20 - 9	IDS15075S5842	Tr 37 - 431	IDS21330N5701
Cr 197 - 2	IDS08410S4055	Pis 20 - 11	IDS15075S5842	Tr 37 - 432	IDS21330N5701
Cr 205 - 1	IDS08572S4837	Pis 24 - 1	IDS17181S3406	Tr 37 - 452	IDS21346N5706
Cr 205 - 2	IDS08572S4836	Pis 24 - 4	AA69,51 # 59	Tr 37 - 455	LS III +57 010
Cr 223 - 2	IDS10260S5942	Pis 24 - 9	AA69,51 # 61	Tr 37 - 466	HR 8281
Cr 223 - 35	LSS 1616	Pis 24 - 16	IDS17181S3406		IDS21358N5702
Cr 228 - 1	IDS10402S5921	Ros 5 - 4	IDS20029N3311	Tr 37 - 477	IDS21364N5702
Cr 228 - 13	IDS10396S5934	Ros 5 - 8	IDS20040N3218	Tr 37 - 686	IDS21284N5727
Cr 228 - 20	IDS10404S5936	Ros 5 - 15	IDS20053N3322	Tr 37 - 778	IDS21430N5719
Cr 228 - 28	DW Car	Ros 5 - 19	IDS20057N3307	Tr 37 - 779	IDS21430N5719
Cr 228 - 33	QZ Car	Ros 5 - 31	IDS20067N3320	TR 37 - 1025	LS III +58 010
	IDS10405S5928	Ros 5 - 33	IDS20067N3320		IDS21367N5747
Cr 228 - 35	IDS10408S5929	Ros 5 - 45	IDS20093N3355	Tr 37 - 1319	HR 8316, μ Cep
Cr 228 - 36	IDS10408S5929	Ros 5 - 46	IDS20093N3355		IDS21404N5819
Cr 228 - 65	IDS10402S5934	Ru 108 - 1	IDS13254S5755	vdB 1 - 45	CV Mon
Cr 228 - 66	IDS10402S5934	Ru 119 - 6	LSS 3612	vdBH 99 - 3	IDS10328S5819
Cr 228 - 67	IDS10402S5935	Ru 127 - 1	IDS17311S3615	vdBH 99 - 19	IDS10350S5840
Cr 228 - 68	IDS10402S5935	St 2 - 22	LS I +59 112		HR 4177
Cr 228 - 69	IDS10400S5934	St 2 - 45	LS I +58 065	vdBH 99 - 20	IDS10350S5840
Cr 228 - 70	IDS10400S5934	St 2 - 66	LS I +59 115		HR 4179
Cr 228 - 92	IDS10433S5934	St 2 - 81	IDS02080N5902	vdBH 99 - 23	IDS10352S5818
Cr 228 - 98	IDS10422S5919	St 2 - 135	LS I +59 119		
Cr 228 - 100	IDS10384S5927	St 2 - 160	IDS02125N5934	Alpha Per 44	HR 937
Cr 258 - 14	IDS12217S6014	St 2 - 161	IDS02125N5934		IDS03018N4914

Table 5 1369

Star	Identification	Star	Identification	Star	Identification
Alpha Per 138	HR 949	Coma Ber 146	HR 4752, AI Com		IDS04249N1558
	IDS03055N4721		IDS12239N2628	Hyades vB 83	HR 1428
Alpha Per 175	HR 956	Coma Ber 149	HR 4753		IDS04249N1529
Alpha Per 220	IDS03073N4813	Coma Ber 150	HR 4766, HZ Com	Hyades vB 84	HR 1430
Alpha Per 225	IDS03070N5000	Coma Ber 160	HR 4766, UU Com		IDS04250N1331
Alpha Per 308	HR 969	Coma Ber 164	IDS12266N2470	Hyades vB 85	IDS04249N1558
Alpha Per 383	HR 987	Coma Ber 183	HR 4780	Hyades vB 89	HR 1432
Alpha Per 401	HR 989	Coma Ber 186	IDS12290N2683	Hyades vB 90	HR 1436
Alpha Per 490	IDS03148N4843			Hyades vB 91	V996 Tau
Alpha Per 497	HR 1001	Hyades vB 1	IDS03121N0717	Hyades vB 92	V997 Tau
	IDS03148N4843	Hyades vB 2	IDS03121N0717	Hyades vB 95	HR 1444, ρ Tau
Alpha Per 501	V459 Per	Hyades vB 3	IDS03247N1946	Hyades vB 97	V938 Tau
Alpha Per 520	V484 Per	Hyades vB 6	HR 1201	Hyades vB 100	HR 1459
Alpha Per 522	IDS03152N5118	Hyades vB 8	HR 1233	Hyades vB 102	V998 Tau
Alpha Per 557	HR 1011	Hyades vB 11	HR 1279	Hyades vB 103	HR 1472
Alpha Per 605	HR 1017		IDS04020N1453		IDS04325N1551
	IDS03171N4930	Hyades vB 14	HR 1292	Hyades vB 104	HR 1473
Alpha Per 606	V461 Per		IDS04060N0516		IDS04326N1218
Alpha Per 665	IDS03187N4640	Hyades vB 20	HR 1319	Hyades vB 107	HR 1480
Alpha Per 675	HR 1029		IDS04101N1509		IDS04336N0741
Alpha Per 774	HR 1034	Hyades vB 21	V984 Tau	Hyades vB 108	HR 1479
Alpha Per 799	IDS03214N4836	Hyades vB 22	V818 Tau		IDS04335N1543
Alpha Per 810	HR 1037	Hyades vB 24	HR 1331	Hyades vB 111	HR 1507
Alpha Per 835	HR 1044		IDS04125N2120		IDS04389N1058
	IDS03222N4910	Hyades vB 25	V985 Tau	Hyades vB 112	HR 1519
Alpha Per 861	HR 1047	Hyades vB 26	V893 Tau		IDS04404N1131
	IDS03225N4636	Hyades vB 28	HR 1346	Hyades vB 117	V808 Tau
Alpha Per 900	HR 1052		IDS04141N1523	Hyades vB 122	IDS04457N1054
Alpha Per 904	HR 1051	Hyades vB 29	IDS04142N1617	Hyades vB 123	HR 1547, V480 Tau
Alpha Per 906	V465 Per	Hyades vB 30	HR 1351, V483 Tau		IDS04455N1840
Alpha Per 934	HR 1056		IDS04143N1348	Hyades vB 124	IDS04462N1329
Alpha Per 953	IDS03245N4731	Hyades vB 31	V986 Tau	Hyades vB 126	HR 1566
Alpha Per 954	HR 1059	Hyades vB 32	HR 1354	Hyades vB 128	IDS04540N1546
Alpha Per 955	IDS03245N4731	Hyades vB 33	HR 1356, V696 Tau	Hyades vB 129	HR 1620
Alpha Per 985	HR 1063, V396 Per	Hyades vB 34	HR 1358	Hyades vB 130	HR 1672
Alpha Per 1082	IDS03272N4817	Hyades vB 38	HR 1368, V775 Tau		IDS05038N0942
Alpha Per 1164	HR 1087, ψ Per		IDS04163N1351	Hyades vB 131	HR 1670
Alpha Per 1225	IDS03306N4747	Hyades vB 40	IDS4171N1449		IDS05035N2754
Alpha Per 1515	V485 Per	Hyades vB 41	HR 1373	Hyades vB 132	IDS05035N2754
Alpha Per 1543	V486 Per		IDS04172N1718	Hyades vB 137	HR 1238
Alpha Per 1556	V487 Per	Hyades vB 43	V988 Tau		IDS03550N1754
Alpha Per 1570	V488 Per	Hyades vB 45	HR 1376	Hyades vB 141	HR 1394, V777 Tau
Alpha Per 1586	V489 Per	Hyades vB 47	HR 1380		IDS04206N1524
			IDS04183N1713	Hyades vB 149	IDS04522N0050
Coma Ber 10	HR 4633	Hyades vB 50	V895 Tau	Hyades vB 152	G 86 - 26
Coma Ber 16	HR 4640	Hyades vB 52	V897 Tau	Hyades vB 154	HR 878
Coma Ber 19	GM Com	Hyades vB 53	HR 1385	Hyades vB 157	HR 656
Coma Ber 39	HR 4667	Hyades vB 54	HR 1387	Hyades vB 160	HR 1254
Coma Ber 47	HR 4673		IDS04194N2204	Hyades vB 164	HR 1517
	IDS12125N2890	Hyades vB 55	HR 1388		NGC 1647 - 1
Coma Ber 59	G 121 - 75		IDS04194N2204	Hyades vB 168	HR 1905
Coma Ber 60	HR 4684, FM Com	Hyades vB 56	HR 1389, V776 Tau	Hyades vB 169	HR 2124
Coma Ber 62	HR 4685		IDS04197N1742		IDS05569N0939
Coma Ber 66	HR 4688	Hyades vB 57	HR 1391	Hyades vB 173	V989 Tau
Coma Ber 70	HR 4694		IDS04199N1543	Hyades vB 174	V990 Tau
Coma Ber 71	HR 4693	Hyades vB 58	IDS04200N1838	Hyades vB 175	V991 Tau
Coma Ber 72	HR 4698	Hyades vB 59	V992 Tau	Hyades vB 181	V995 Tau
	IDS12156N2737	Hyades vB 60	HR 1392, υ Tau	Hyades vB 182	IDS04249N1529
Coma Ber 73	HR 4698		IDS04203N2235	Hyades vB 190	V994 Tau
	IDS12156N2737	Hyades vB 63	V906 Tau	Hyades vA 14	G 7 - 24
Coma Ber 76	IDS12157N2579	Hyades vB 64	V911 Tau	Hyades vA 331	G 7 - 37
Coma Ber 81	G 121 - 80	Hyades vB 65	V993 Tau	Hyades vA 409	LTT 11425
Coma Ber 89	HR 4705	Hyades vB 67	HR 1403	Hyades vA 575	V484 Tau
Coma Ber 91	HR 4707		IDS04221N2124	Hyades J 202	IDS03114N2557
	IDS12175N2584	Hyades vB 68	HR 1408	Hyades CDS 1	HZ 4
Coma Ber 92	IDS12175N2584	Hyades vB 69	V918 Tau	Hyades CDS 3	LP 475-251
Coma Ber 107	HR 4717, GN Com	Hyades vB 70	HR 1409	Hyades CDS 11	LP 357-160
Coma Ber 109	HR 4719		IDS04228N1858	Hyades CDS 13	LP 414-167
	IDS12194N2568	Hyades vB 71	HR 1411	Hyades CDS 16	LP 358-534
Coma Ber 111	IL Com		IDS04228N1545	Hyades CDS 19	LP 358-242
	IDS12194N2568	Hyades vB 72	HR 1412, θ^2 Tau	Hyades CDS 28	LP 415-1574
Coma Ber 113	HR 4725		IDS04228N1545	Hyades CDS 33	LP 415-292
Coma Ber 125	HR 4733	Hyades vB 73	V920 Tau	Hyades CDS 37	LP 415-345
Coma Ber 127	IDS12218N2684	Hyades vB 74	HR 1414	Hyades CDS 45	LP 415-2108
Coma Ber 129	HR 4737	Hyades vB 75	IDS04233N1556	Hyades CDS 46	LP 475-242
Coma Ber 130	HR 4738	Hyades vB 78	IDS04237N1738	Hyades CDS 47	AK 2-188
Coma Ber 133	G 148- 60	Hyades vB 79	V921 Tau	Hyades CDS 48	AK 2-1315
	IDS12222N2735		IDS04237N1738	Hyades CDS 49	V411 Tau
Coma Ber 144	HR 4750	Hyades vB 80	HR 1422	Hyades CDS 52	LP 413-078
Coma Ber 145	HR 4751		IDS04244N1525	Hyades CDS 53	LP 413-018
	IDS12239N2628	Hyades vB 82	HR 1427	Hyades CDS 54	LP 413-019

Star	Identification	Star	Identification	Star	Identification
Hyades CDS 55	LP 356-778		IDS05304S0527	Pleiades 152	V963 Tau
Hyades CDS 56	LP 413-089	Orion B 588	MO Ori	Pleiades 158	V624 Tau
Hyades CDS 57	LP 357-279	Orion B 595	HR 1894, BM Ori	Pleiades 191	MR Tau
Hyades CDS 58	LP 301-069		IDS05304S0527	Pleiades 212	MS Tau
Hyades CDS 59	LP 414-479	Orion B 597	MP Ori	Pleiades 296	V966 Tau
Hyades CDS 60	LP 301-082	Orion B 598	HR 1895	Pleiades 298	IDS03382N2343
Hyades CDS 61	LP 358-724		IDS05304S0527	Pleiades 303	IDS03383N2347
Hyades CDS 63	LP 358-730	Orion B 604	MR Ori	Pleiades 314	V1038 Tau
Hyades CDS 64	LP 475-1699	Orion B 612	IDS05304S0527	Pleiades 324	V632 Tau
Hyades CDS 65	LP 475-1747	Orion B 626	MS Ori	Pleiades 335	MX Tau
Hyades CDS 66	LP 359-243	Orion B 631	HR 1891	Pleiades 347	MY Tau
Hyades CDS 67	LP 359-042		IDS05304S0429	Pleiades 357	NP Tau
Hyades CDS 68	LP 416-565	Orion B 632	HR 1890, V1046 Ori	Pleiades 447	HR 1140
Hyades CDS 69	LP 416-570	Orion B 643	TU Ori	Pleiades 451	V785 Tau
Hyades CDS 70	LP 416-572	Orion B 646	MW Ori	Pleiades 468	HR 1142
Hyades CDS 71	LP 536-063	Orion B 648	CD Ori	Pleiades 541	HR 1144
Hyades CDS 72	LP 416-094	Orion B 653	MX Ori	Pleiades 559	V786 Tau
Hyades HG7- 23	V471 Tau	Orion B 655	V1230 Ori	Pleiades 563	HR 1145
Hyades HG7- 219	IDS04214N0829	Orion B 659	HR 1892		IDS03393N2409
Hyades HG8- 2	V833 Tau		IDS05304S0454	Pleiades 590	OR Tau
	G 8 - 44	Orion B 662	V415 Ori	Pleiades 624	OS Tau
Hyades HG8- 4	G 8 - 40	Orion B 682	HR 1897	Pleiades 625	V811 Tau
Hyades HG8- 14	G 8 - 23		IDS05305S0529	Pleiades 676	OT Tau
Hyades HG8- 18	G 8 - 30	Orion B 690	V358 Ori	Pleiades 686	OU Tau
Hyades HG8- 20	G 8 - 45	Orion B 696	AH Ori	Pleiades 717	IDS03397N2401
Hyades HG8- 21	G 7 - 27	Orion B 698	NP Ori	Pleiades 727	V855 Tau
		Orion B 703	NQ Ori	Pleiades 738	V1041 Tau
Orion B 42	V466 Ori	Orion B 708	AI Ori	Pleiades 739	V969 Tau
Orion B 112	VY Ori	Orion B 714	IDS05305S0529	Pleiades 762	V787 Tau
Orion B 119	HU Ori	Orion B 721	HR 1899	Pleiades 785	HR 1149
Orion B 134	V1006Ori		IDS05305S0559	Pleiades 793	OW Tau
Orion B 136	VZ Ori	Orion B 722	HR 1898	Pleiades 799	V1010 Tau
Orion B 156	SU Ori		IDS05306S0426	Pleiades 801	IDS03399N2250
Orion B 189	WX Ori	Orion B 734	V1073 Ori	Pleiades 817	HR 1151
Orion B 191	V395 Ori		IDS05306S0529	Pleiades 859	HR 1152
Orion B 198	WX Ori	Orion B 737	AL Ori	Pleiades 883	V789 Tau
Orion B 215	BU Ori	Orion B 747	NU Ori	Pleiades 906	OX Tau
Orion B 220	V1044 Ori	Orion B 749	IDS05305S0559	Pleiades 915	V790 Tau
Orion B 223	V744 Ori	Orion B 755	V498 Ori	Pleiades 956	IDS03403N2353
Orion B 224	EZ Ori	Orion B 757	V360 Ori	Pleiades 980	HR 1156, V971 Tau
Orion B 238	IN Ori	Orion B 760	V361 Ori	Pleiades 996	V1045 Tau
Orion B 281	IDS05296S0433		IDS05305S0529	Pleiades 1029	V446 Tau
Orion B 284	V750 Ori	Orion B 761	V359 Ori	Pleiades 1039	V641 Tau
Orion B 312	IU Ori	Orion B 767	NV Ori	Pleiades 1061	PP Tau
Orion B 319	XX Ori	Orion B 785	TV Ori	Pleiades 1100	V642 Tau
Orion B 330	IDS05298S0428	Orion B 786	V566 Ori	Pleiades 1103	PR Tau
Orion B 337	V938 Ori	Orion B 789	AM Ori	Pleiades 1114	V643 Tau
Orion B 339	V398 Ori	Orion B 805	V426 Ori	Pleiades 1124	V814 Tau
Orion B 340	IX Ori	Orion B 806	HR 1901	Pleiades 1207	V1046 Tau
Orion B 341	IY Ori		IDS05307S0455	Pleiades 1266	V534 Tau
Orion B 347	V473 Ori	Orion B 810	BO Ori		IDS03411N2431
Orion B 381	IDS05299S0604	Orion B 824	V363 Ori	Pleiades 1280	V535 Tau
Orion B 388	V372 Ori	Orion B 831	AN Ori	Pleiades 1286	QQ Tau
Orion B 408	YY Ori	Orion B 834	V569 Ori	Pleiades 1305	V644 Tau
Orion B 414	IDS05299S0604	Orion B 850	V500 Ori	Pleiades 1306	FL Tau
Orion B 417	IDS05299S0604	Orion B 853	V389 Ori	Pleiades 1321	V536 Tau
Orion B 422	YZ Ori	Orion B 854	AP Ori	Pleiades 1338	IDS03415N2348
Orion B 424	V776 Ori	Orion B 883	V575 Ori	Pleiades 1355	V646 Tau
Orion B 443	KM Ori	Orion B 884	V815 Ori	Pleiades 1362	V647 Tau
Orion B 466	KS Ori	Orion B 892	V815 Ori		IDS03415N2348
Orion B 467	KR Ori	Orion B 897	OU Ori	Pleiades 1375	IDS03415N2348
Orion B 480	KX Ori	Orion B 901	AR Ori	Pleiades 1397	IDS03415N2336
Orion B 482	HR 1886	Orion B 907	IDS05310S0542	Pleiades 1404	IDS03414N2421
	IDS05301S0605	Orion B 965	V502 Ori	Pleiades 1425	V650 Tau
Orion B 493	HR 1887	Orion B 980	HR 1906	Pleiades 1432	HR 1165
	IDS05301S0605	Orion B 982	AZ Ori		IDS03415N2348
Orion B 509	V481 Ori	Orion B 989	PP Ori	Pleiades 1454	V703 Tau
Orion B 510	LL Ori	Orion B 1012	IDS05317S0608	Pleiades 1485	V540 Tau
Orion B 515	SY Ori	Orion B 1017	HR 1911	Pleiades 1491	V651 Tau
Orion B 517	V376 Ori		IDS05317S0608	Pleiades 1516	V793 Tau
Orion B 518	V355 Ori	Orion B 1032	IDS05319S0532	Pleiades 1531	QX Tau
Orion B 522	V356 Ori	Orion B 1048	PX Ori	Pleiades 1532	V541 Tau
Orion B 530	LP Ori	Orion B 1051	V586 Ori	Pleiades 1553	V652 Tau
Orion B 541	LT Ori	Orion B 1054	V381 Ori	Pleiades 1653	V338 Tau
Orion B 543	AA Ori	Orion B 1098	HR 1918	Pleiades 1726	IDS03422N2350
Orion B 553	V486 Ori			Pleiades 1785	V343 Tau
Orion B 561	V488 Ori	Pleiades 34	V810 Tau	Pleiades 1823	HR 1172
Orion B 563	LX Ori	Pleiades 97	V700 Tau	Pleiades 1827	V346 Tau
Orion B 574	AB Ori	Pleiades 133	V623 Tau	Pleiades 1883	V660 Tau
Orion B 582	LZ Ori	Pleiades 134	LZ Tau	Pleiades 1912	IDS03426N2352
Orion B 587	HR 1893, V1016 Ori	Pleiades 146	MM Tau	Pleiades 2016	V796 Tau

Table 5   1371

Star	Identification	Star	Identification	Star	Identification
Pleiades 2034	V545 Tau	ADS 15262	IDS21398N5711	G 78 - 41A	Comp? to G78 - 41
Pleiades 2082	V354 Tau	AS 373	V1016 Cyg	G 87 - 26	QY Aur
Pleiades 2144	V355 Tau	BC Cha	AAS35,161 # 51	G 97 - 47	V998 Ori
Pleiades 2168	HR 1178	BF Cha	AAS35,161 # 54	G 99 - 47	V1201 Ori
	IDS03432N2345	BK Cha	AAS35,161 # 59	G 106 - 49	V577 Mon
Pleiades 2181	HR 1180, BU Tau	BM Cha	AAS35,161 # 61	G 140 - 24	BD +04 03561a
Pleiades 2193	V357 Tau	UY Dra	IDS17551N5814	G 156 - 31	EZ Aqr
Pleiades 2208	V358 Tau	AO 0235+164	B2 1101+38	G 169 - 34	V749 Her
Pleiades 2220	IDS03434N2405	BPM 25114	V820 Ara	G 170 - 34	V639 Her
Pleiades 2244	V664 Tau	BPM 30551	AX Phe	G 170 - 35	V647 Her
Pleiades 2263	IDS03434N2405	BPM 31594	VY Hor	G 171 - 10	HH And
Pleiades 2368	V667 Tau	BPM 46429	SV Cet	G 174 - 44	V719 Tau
Pleiades 2411	II Tau	BPM 46478	Not CoD -26 00192	G 185 - 32	PY Vul
Pleiades 2425	HR 1183	Cep OB2 # 27	LS III +41 039	G 207 - 9	V470 Lup
Pleiades 2507	IDS03440N2333	Cep OB3 # 16	IDS22496N6136	G 208 - 44	V1581 Cyg
Pleiades 2588	V370 Tau	Cha T # 3	SX Cha	G 210 - 48	V1396 Cyg
Pleiades 2591	V553 Tau	Cha T # 4	SY Cha	G 225 - 67	CM Dra
Pleiades 2601	V371 Tau	Cha T # 6	SZ Cha	G 226 - 29	DN Dra
Pleiades 2602	V372 Tau	Cha T # 7	TW Cha	G 230 - 26	V1513 Cyg
Pleiades 2908	V376 Tau	Cha T # 20	UV Cha	G 237 - 78	DP Dra
Pleiades 2927	V378 Tau	Cha T # 22	UX Cha	G 241 - 76	IDS00076N6846
Pleiades 2940	V707 Tau	Cha T # 23	UY Cha	G 269 - 140	AO Scl
Pleiades 2984	V1051 Tau	Cha T # 24	UZ Cha	G 272 - 61	BL + UV Cet
Pleiades 3019	V381 Tau	Cha T # 27	VV Cha	HBC 244	Cha T # 8
Pleiades 3030	V382 Tau	Cha T # 31	VW Cha	HBC 591	Cha T # 11
Pleiades 3063	V677 Tau	Cha T # 40	VZ Cha	HRC 209	variable
Pleiades 3065	V560 Tau	Cha T # 44	WW Cha	HRC 284	variable
Pleiades 3096	V1054 Tau	Cha T # 45	WX Cha	He3 775	Not CpD -62 02675
Pleiades 3104	V384 Tau	Cha T # 46	WY Cha	IRC +30 381	HY Cyg
Pleiades 3197	V697 Tau	Cha T # 48	WZ Cha	IRC +30 393	ER Cyg
Pleiades 3306	HR 1103	Cha T # 49	XX Cha	IRC +40 442	DG Cyg
Pleiades 3312	IDS03376N2225	Cyg OB2 # 26	Schulte 25	IRC +60 150	TX Cam
Pleiades 3317	IDS03399N2223	Cyg OB2 # 28	Schulte 17	L 68 - 27	IDS12228S7034
Pleiades 3325	HR 1185	Cyg OB2 # 30	Schulte 15	L 68 - 28	IDS12228S7034
Pleiades 9003	LO Tau	Cyg OB2 # 31	Schulte 14	L 221 - 49	JL 230
Pleiades 9010	MZ Tau	Cyg OB2 # 33	Schulte 7	L 222 - 53	JL 247
Pleiades 9018	QU Tau	Cyg OB2 # 34	Schulte 24	L 292 - 41	JL 228
Pleiades 9023	V344 Tau	Cyg OB2 # 35	Schulte 9	LP 731 - 59	LP 791 - 8
Pleiades 9025	V351 Tau	Cyg OB2 # 37	Schulte 19	LP 759 - 82	IDS22120S0917
Pleiades 9035	KY Tau	Cyg OB2 # 38	Schulte 26	LP 783 - 3	IDS07357S1710
Pleiades 9036	LM Tau	Cyg OB2 # 39	Schulte 27	LS I +55 008	V592 Cas
Pleiades 9041	MO Tau	Cyg OB2 # 40	Schulte 22	LS I +56 074	ZZ Per
Pleiades 9061	V399 Tau	Cyg OB2 # 41	Schulte 12	LS I +60 064	QX Cas
Pleiades 9071	MV Tau	Cyg OB2 # 312	Schulte 29	LS I +60 090	IDS00175N6039
Pleiades 9078	OP Tau	Cyg OB2 # 360	Schulte 30	LS I +61 053	IDS23420N6055
Pleiades 9083	V336 Tau	Cyg OB2 # 502	Schulte 21	LS I +57 098	EO Per
Pleiades 9086	V350 Tau	Cyg OB2 # 553	Schulte 13	LS I +61 149	NO Cas
Pleiades 9087	V353 Tau	Cyg OB2 # 560	Schulte 20	LS I +61 303	V615 Cas
Pleiades 9092	V364 Tau	Cyg OB2 # 683	Schulte 23	LS I +62 029	IDS23451N6209
Pleiades 9099	V403 Tau	Cyg OB2 # 705	Schulte 18	LS I +62 114	IDS00244N6217
Pleiades 9105	QV Tau	Cyg OB2 # 1359	G 209 - 35	LS I +62 117	IDS00253N6223
Pleiades 9108	V366 Tau	Cyg OB2 # 1512	LS III +41 025	LS I +62 229	TX Cas
Pleiades 9115	V392 Tau		Schulte 1	LS II +20 019	FG Sge
Pleiades 9127	V402 Tau	Cyg OB2 # 1542	Schulte 2	LS II +39 041	V1972 Cyg
Pleiades 9150	MU Tau	Cyg OB2 # 1621	Schulte 8D	LS III +41 011	V1187 Cyg
Pleiades 9156	V349 Tau	Cyg OB2 # 1623	Schulte 16	LS III +57 026	CX Cep
Pleiades 9197	V444 Tau	Eggen 13138	AP Ser B	LS III +59 040	CR Cas
Pleiades 9201	V475 Tau	Eggen 18672	AK Cyg E	LS V +20 030	IDS06065N2026
Pleiades 9214	V469 Tau	Eggen 18673	AK Cyg W	LS V +23 037	IDS06066N2352
Pleiades 9256	V500 Tau	EG 91	HZ 29	LS V +23 061	IDS06136N2358
Pleiades 9289	V502 Tau	G 3 - 33	TZ Ori	LS VI −04 004	V397 Mon
Pleiades 9308	V488 Tau	G 9 - 38	EI Cnc	LSS 1232	CpD -43 03285
Pleiades 9325	V512 Tau	G 10 - 50	FI Vir	LSS 3013	V864 Cen
Pleiades 9337	V538 Tau	G 12 - 12	V1338 Cyg	LSS 4923	LS IV -15 042
Pleiades 9339	V550 Tau	G 12 - 30	GL Vir	LSS 5040	LS IV -15 046
Pleiades 9369	V681 Tau	G 12 - 43	FL Vir	LTT 5721	V645 Cen, Proxima
Pleiades 9412	V673 Tau	G 24 - 9	V1412 Aql	OI 090 4	0754+101
Pleiades 9477	V739 Tau	G 24 - 16	HU Del	PHL 433	Feige 108
		G 29 - 38	ZZ Psc	PHL 1376	Feige 24
AA45,405 S237 # 4	IDS05248N3412	G 36 - 31	VX Ari	PHL 8054	Feige 16
AA45,405 S237 # 5	IDS05248N3412	G 38 - 29	V468 Per	ROC # 33	V852 Oph
AA45,405 S148 # 1	Per sq 1 # 6	G 44 - 27	RY Sex	Ross 154	V1216 Sgr
AA45,405 S149 # 1	Per sq 1 # 7	G 45 - 20	CN Leo	Ton 245	WD1538+269
AA45,405 S152 # 1	Per sq 1 # 8	G 50 - 4	YZ CMi	Ton 320	WD0823+316
AA50,429 # 1	IDS15538S5352	G 51 - 15	DX Cnc	VES 136	V1173 Cyg
ADS 10841	IDS17464S1532	G 60 - 55	FN Vir	Vyss 111	V1005 Ori
ADS 12100	IDS19054N3714	G 61 - 29	GP Com	Vyss 188	V1285 Aql
ADS 13368	IDS20521N4659	G 63 - 36	VW Com		
ADS 14438	IDS20021N3539	G 64 - 34	HZ Vir		

Table 6: References for the acronyms

Acronym	Reference
A1	Barbieri C., Erculiani L., Rosino L. 1968, Publ. Oss. Astr. Padova no 143
A2	Barbieri C., Rosino L. 1972, Ap&SS 16, 324
AC	Astrographic catalogues
ADS	Aitken R.G. 1932, Carnegie Institution Washington
AG	Astronomische Gesellschaft Katalog
AK	Artyukhina N.M., Kholopov P.N. 1975, Tr. Gos. Astron. Inst. Shternberg 46, 57
	Artyukhina N.M., Kholopov P.N. 1976, Tr. Gos. Astron. Inst. Shternberg 47, 105
AKN 120	Hamuy J., Maza J. 1989, AJ 97, 720
Anti	Neese C.L., Yoss K.M. 1988, AJ 95, 463
AO	Castelaz M.W., Persinger T. 1988, AJ 98, 1768
	Persinger T., Castelaz M.W. 1990, AJ 100, 1621
	Castelaz M.W., Persinger T., Stein J.W., Prosser J., Powell H.D. 1991, AJ 102, 2103
AS	Merrill P.W., Burwell C.G. 1950, ApJ 112, 72
	Miller W.C., Merrill P.W. 1951, ApJ 113, 624
Ara OB1	Herbst W., Havlen R.J. 1977, A&AS 30, 279
AzV	Azzopardi M., Vigneau J. 1975, A&AS 22, 285
an L1159-16	Lippincott S.L., Yang C.Y. 1967, AJ 72, 840 (Star near L1159-16 = Woolley 9066)
BBB	Basinski J.M., Bok B.J., Bok P.F. 1967, MNRAS 137, 5
BPM	Luyten W.J. 1963, Bruce Proper Motion Survey. The General Catalogue. Minneapolis.
BSD	Schwassmann, A., van Rhijn, P.J. 1938, Bergedorfer Spektral Durchmusterung der 115
	Nordlichen Kapteynschen Eichfelder. Hamburger Sternwarte in Bergedorf
Breys	Breysacher J. 1981, A&AS 43, 203
C1947	Turner D.G., Leonard P.J.T., Madore B.F. 1986, JRASC 80, 166
CCS	Stephenson C.B. 1973, Publ. Warner Swasey Obs. 1 no 4
Cha F	Whittet D.C.B., Kirrane T.M., Kilkenny D., et al. 1987, MNRAS 224, 497
Cha T	Whittet D.C.B., Kirrane T.M., Kilkenny D., et al. 1987, MNRAS 224, 497
CMa R1	Herbst W., Racine R., Warner J.W. 1978, ApJ 223, 471
CS	Pier J.R. 1982, AJ 87, 1515
Case *M	Nassau J.J., Blanco V.M., Morgan W.W. 1954, ApJ 120, 478
	Blanco V.M., Nassau J.J. 1957, ApJ 125, 408
Cep OB3	Blaauw A., Hiltner W.A., Johnson H.L. 1959, ApJ 130, 69
CoD	Cordoba Durchmusterung
Comp to F74	Companion to Feige 74, Penston M.J. 1973, MNRAS 164, 133
Compar HZ29	Comparison star for HZ 29 = AM CVn
Cn1-1	Cannon A.J. 1921, Harvard Circ. no 224
CpD	Cape Photographic Durchmusterung
Cyg OB2	Reddish V.C., Lawrence L.C., Pratt N.M. 1966, Publ. R. Obs. Edinburgh 5 no 8
DO	Lee O.J., Baldwin R.J., Hamlin D.W. 1943, Dearborn Annals 5, 1a
Draco	Baade W., Swope H.H. 1961, AJ 66, 300
Dril	Drilling J.S. 1977, AJ 82, 714
E	Harvard STD Regions 1917, Ann. Harv. Coll. Obs. 71, 233
E / Q	Cousins A.W.J., Stoy R.H. 1961, ROB no 49
EG	Eggen O.J., Greenstein G. 1965, ApJ 141, 83; 1965, ApJ 142, 925; 1967, ApJ 150, 927
Eggen	Eggen O.J. 1963, AJ 68, 483
FK X Ray-3	Feigelson E.D., Kriss G.A. 1981, ApJ Lett. 248, l35
Feige	Feige J. 1958, ApJ 128, 267
fld	Saurer W., Pfitscher K., Weinberger R., Hartl H. 1992, A&AS 93, 553
Fld	Cameron Reed B. 1990, AJ 100, 737
Florsch	Florsch A. 1972, Publ. Obs. Astr. Strasbourg 2 Fasc 1
G	Giclas H.L., Burnham Jr R., Thomas N.G. 1971, Lowell Obs. Flagstaff
	Lowell Proper Motion Survey (Northern Hemisphere)
	Giclas H.L., Burnham R. Jr, Thomas N.G. 1978, Lowell Obs. Bull. no 164 (Southern Survey)
GD	Giclas H.L., Burnham R. Jr, Thomas N.G. 1980, Lowell Obs. Bull. no 166
GJ	Gliese W., Jahreiss H. 1979, A&AS 38, 423
GPEC	Morrison H.L., Flynn C., Freeman K.C. 1990, AJ 100, 1191
GR	Giclas H.L., Burnham R. Jr, Thomas N.G. 1980, Lowell Obs. Bull. no 166
H1	Haro G. 1952, Bol. Obs. Tonantzintla Tacubaya 1 no 1 (Table 1)
H2	Haro G. 1952, Bol. Obs. Tonantzintla Tacubaya 1 no 1 (Table 2)
HA	Durchmusterung of Selected Areas, Systematic Plan 1918-1923, Ann. Harv. Coll. Obs. 101, 102, 103
HBC	Herbig G.H., Bell K.R. 1988, Lick Obs. Bull. no 1111
HD	Cannon A.J., Pickering E.C. 1918-1924, Ann. Harv. Coll. Obs. vol. 91-99
HDE	Cannon A.J., Walton Mayall M. 1925-1936, Ann. Harv. Coll. Obs. vol. 100
HIC	Turon C., Crézé M., D. Egret, et al. 1992, The Hipparcos Input Catalogue, ESA SP-1136
HR	Bright Star Catalogue, Hoffleit D., Jaschek J. 1982, Yale Univ. Obs. (New Haven)
HRC	Herbig H.G., Rao N.K. 1972, ApJ 174, 401
HV	Harvard Variables
HZ	Humason M.L., Zwicky F. 1947, ApJ 105, 85

Table 6 1373

Acronym	Reference
Haro 6	Haro G., Iriarte B., Chavira E. 1953, Tonantzintla Tacubaya no 8, 3
He2	Henize K.G. 1967, ApJS 14, 125
He3	Henize K.G. 1976, ApJS 30, 491
Hiltner	Hiltner W.A. 1956, ApJS 2, 389
IDS	Washington Double Star Catalog, Worley C.E., Douglass G.G. 1984, US Naval Obs.
IRC	Neugebauer G.N., Leighton R.B. 1969, Two-micron Sky Survey. California Inst. of Tech., Nasa
Ir-Ch	Iriarte E., Chavira E. 1956, Tonantzintla Tacubaya 14, 31
JL	Jaidee S., Lynga G. 1969, Ark. Astr. 5, 345
Jenkins	Jenkins L.F. General Catalogue of Trigonometric Stellar Parallaxes, 1963, Yale Univ.Obs.
KPS	Kondo M., Watanabe E., Yutani M., Noguchi T. 1982, PASJ 34, 541
KS	Klare G., Szeidl B. 1966, Veröff. Landesst. Heidelberg no 18
KUV	Kondo M., Watanabe E., Yutani M., Noguchi T. 1982, PASJ 34, 541
Klemola	Klemola A. 1961, AJ 67, 740
L	Luyten W.J. 1942, Publ. Astr. Obs. Univ. Minnesota II no 12
	Luyten W.J. 1944, Publ. Astr. Obs. Univ. Minnesota III no 4
LkHα	Herbig G.H. 1957, ApJ 125, 654; 1960, ApJS 4, 337
LB	Luyten W.J. A Search For Faint Blue Stars. University of Minnesota (Minneapolis)
LDS	Luyten W.J. 1941, Publ. Astr. Obs. Univ. Minnesota III no 3
LF 5	McCuskey S.W., Houk N. 1964, AJ 69, 412
LF11	Wooden W.H. II 1970, AJ 75, 324
LF12	Wooden W.H. II 1970, AJ 75, 324
LF13	Wooden W.H. II 1970, AJ 75, 324
LF14	Wooden W.H. II 1970, AJ 75, 324
LHS	Luyten W.J. 1979, LHS catalog (2nd edition) Univ. of Minnesota (Minneapolis)
LMC	Sanduleak N. 1969, Contr. CTIO no 89
LMC *C	Ardeberg A., Brunet J.P., Maurice E., Prévot L. 1972, A&AS 6, 249
	Brunet J.P., Prévot L., Maurice E., Muratorio G. 1973, A&AS 9, 447
LMC *G	Fehrenbach Ch., Duflot M. 1970, A&AS Special no 1
LMC *P	Fehrenbach Ch., Duflot M. 1970, A&AS Special no 1
	Fehrenbach Ch., Duflot M. 1973, A&AS 10, 231; 1981, A&AS 46, 13
LMC *S	Fehrenbach Ch., Duflot M. 1970, A&AS Special no 1
LMC X-1	Jones C.A., Chetin T., Liller W. 1974, ApJ 190,L1
LP	Luyten W.J. 1942, Publ. Astr. Obs. Univ. Minnesota II no 12
	Luyten W.J. 1944, Publ. Astr. Obs. Univ. Minnesota III no 4
LS	Luminous Stars in the Northern Milky Way (Hamburg - Bergedorf)
	Hardorp J., Rohlfs K., Slettebak A., Stock J. 1959, Vol. I
	Stock J., Nassau J.J., Stephenson C.B. 1960, Vol. II
	Hardorp J., Theile I., Voigt H.H. 1964, Vol. III
	Nassau J.J., Stephenson C.B. 1963, Vol. IV
	Hardorp J., Theile I., Voigt H.H. 1965, Vol. V
	Nassau J.J., Stephenson C.B., McConnell D.J. 1965, Vol. VI
LSS	Stephenson C.B., Sanduleak N. 1971, Publ. Warner Swasey Obs. I no 1
LSE	Drilling J.S. 1983, ApJ 270,L13
LTT	Luyten W.J. 1957, A Catalogue of 9867 Stars in the Southern Hemisphere
	with Proper Motion exceeding 0.2 annually. Lund Press, Minneapolis (Minnesota)
	Luyten W.J. 1961, A Catalogue of 7127 Stars in the Northern Hemisphere
	with Proper Motion exceeding 0.2 annually. Lund Press, Minneapolis (Minnesota)
	Luyten W.J. 1962, First Supplement to the LTT Catalogues. Univ. Minnesota
Lambda Ori	Murdin P., Penston M.V. 1977, MNRAS 181, 657
LOD	Lodén L.O. 1967, Ark. Astr. 4, 375; 1968, Ark. Astr. 4, 425
	Lodén L.O. 1969, Ark. Astr. 5, 149; 1969, Ark. Astr. 5, 161
	Lodén L.O., Nordström B. 1969, Ark. Astr. 5, 231
Lod	Lodén L.O. 1973, A&AS 10, 125; 1977, A&AS 29, 31; 1979, A&AS 36, 83
	Lodén L.O. 1979, A&AS 38, 355; 1980, A&AS 41, 173; 1981, A&AS 44, 155
Lucke	Lucke P.B. 1973, Thesis
Ly	Lynga G. 1969, Ark. Astr. 5, 181
MC	McCuskey S.W. 1967, AJ 72, 1199
MCC	Vyssotsky A.N. 1943, ApJ 97, 381; 1946, ApJ 104, 234; 1952, ApJ 116, 117
	Vyssotsky A.N. 1956, AJ 61, 201; 1958, AJ 63, 211
MWC	Merrill P.W., Burwell C.G. 1933, ApJ 78, 87; 1943, ApJ 98, 153; 1949, ApJ 110, 387
Malm	Malmquist K.G. 1927, Medd. Lunds II no 37; 1936, Stockholms Obs. Ann. 12 no 7
	Malmquist K.G. 1960, Uppsala Astr. Obs. Ann. S.IV 17 no 13
Mrk 320	Markarian B.E., Lipovetski V.A. 1971, Afz 7, 511
Met	Metreveli M.D. 1969, Bull. Abastumani Astr. Obs. 38, 93
Mon R1	Herbst W., Miller D.P., Warner J.W., Herzog A. 1982, AJ 87, 98
Mon R2	Herbst W., Racine R. 1976, AJ 81, 840
MtW	Mount Wilson Selected Areas 1930, Carnegie Inst. Publ. no 402
NE	MacConnell D.J. 1968, ApJS 16, 275
NGP	Upgren A.R. 1962, AJ 67, 37

Acronym	Reference
NW	Macconnell D.J. 1968, ApJS 16, 275
Near M8	Herbst W., Miller D.P., Warner J.W., Herzog A. 1982, AJ 87, 98
Neckel	Neckel T., Chini R. 1980, A&AS 39, 411
Nume	Numerova A.B. 1958, Izv. Krym. Astrofiz. Obs. 19, 230
ONRS	Haro G., Chavira E. 1964, ONR Symp. Flagstaff Arizona
Orion P	Parenago P. 1954, Tr. Gos. Astr. Inst. Sternberg Vol 25
Orsatti	Orsatti A.M. 1992, AJ 104, 590
PG	Green R.F., Schmidt M., Liebert J., Green J.A. 1986, ApJS 61, 305
PHI2	Morrison H.L., Flynn C., Freeman K.C. 1990, AJ 100, 1191
PHI4	Morrison H.L., Flynn C., Freeman K.C. 1990, AJ 100, 1191
POSS	Humphreys R.M., Landau R., Ghigo F.D., Zumach W., LaBonte A.E. 1991, ApJ 102, 395
PS	Philip A.G.D., Sanduleak N. 1968, Tonantzintla Tacubaya no 4, 253
PSD	Becker, F., Bruck, H. 1931-1935, Publ. Astrophys. Obs. Potsdam, Vol 27-28,
R	Feast M.W., Thackeray A.D., Wesselink A.J. 1960, MNRAS 121, 337
R CrA T	Vrba F.J., Strom S.E., Strom K.M. 1976, AJ 81, 317
	Marraco H.G., Rydgren A.E. 1981, AJ 86, 62 (no+100)
RL	Rubin V.C., Losee J.M. 1971, AJ 76, 1099
	Rubin V.C., Wespfahl D., Tuve M. 1974, AJ 79, 1406
ROC	Lanz T. 1985, Bull. Inform. CDS 28, 125
Ring Aql	Isserstedt J. 1969, A&A 3, 210
Ring Hydra	Isserstedt J. 1968, Veroeff. Astr. Inst. Univ. Bochum 1, 121
Ring Ori	Isserstedt J. 1968, Veroeff. Astr. Inst. Univ. Bochum 1, 121
Ross	Ross F.E. 1925-1939, AJ 36 to 48
SAO	SAO stars without HD or DM numbers
SB	Slettebak A., Brundage R.K. 1971, AJ 76, 338
SE	MacConnell D.J. 1968, ApJS 16, 275
SOS	Lodén L.O., Lodén K., Nordström B., Sundman A. 1976, A&AS 23, 283
SSII	Slettebak A., Stock J. 1959, Astr. Abh. Hamb. Sternw. Bergedorf V No 5
Steph	Stephsenson C.B. 1986, AJ 91, 144
STO	Stock J., Osborn W., Ibanez M. 1976, A&AS 24, 3
SW	Macconnell D.J. 1968, ApJS 16, 275
Sand	Sanduleak N. 1965, Thesis
Sk	Sanduleak N. 1968, AJ 73, 246
Smethells	Smethells W.G. 1974, Thesis (Case West. Reserve Univ.)
Ton	Iriarte B., Chavira E. 1957, Tonantzintla Tacubaya no 16
	Chavira E. 1959, Tonantzintla Tacubaya no 18, 3
TonS	Chavira E. 1958, Tonantzintla Tacubaya no 17, 3; 1958, Tonantzintla Tacubaya no 17, 5
TonS 180	Hamuy J., Maza J. 1989, AJ 97, 720
Ton 256	Angione R.J. 1971, AJ 76, 412
TPHE	Landolt A.U. 1992, AJ 104, 340
UKST	Stobie R.S., Gilmore G., Reid N. 1985, A&AS 60, 495
UV 1758+36	Giddings J.R., Dworetsky M.M. 1978, MNRAS 183, 265
VES	Coyne G.V., Lee T.A., de Graeve E. 1974, Vatican Obs. Pub. 1 no 5
	Coyne G.V., Wisniewski W., Corbally C. 1975, Vatican Obs. Pub. 1 no 6
	Wisniewski W., Coyne G.V. 1976, Vatican Obs. Pub. 1 no 8
	Wisniewski W., Coyne G.V. 1977, Vatican Obs. Pub. 1 no 11
	Coyne G.V., Wisniewski W., Otten L.B. 1978, Vatican Obs. Pub. 1 no 12
Variable	Hamuy J., Maza J. 1989, AJ 97, 720
Vul R1	Herbst W., Miller D.P., Warner J.W., Herzog A. 1982, AJ 87, 98
Vul R2	Herbst W., Miller D.P., Warner J.W., Herzog A. 1982, AJ 87, 98
van Wijk	Van Wijk U. 1952, Harvard Bull. no 921, 7
vdB	Racine R. 1968, AJ 73, 233
W	Wooden W.H. II 1971, Publ. Warner Swasey Obs. 1 no 2
WEIS	Weistrop D. 1979, PASP 91, 193
WLS	Sanders W.L. 1966, AJ 71, 719
Wray	Wray J.D. (1966) Thesis (in Wackerling, MemRAS 73)
WK X Ray-1	Walter F.M., Kuhi L.V. 1981, ApJ 250, 254
WK X Ray-2	Walter F.M., Kuhi L.V. 1981, ApJ 250, 254
Wolf	Wolff M. 1919-1930, Lists in Astr. Nach.
Wool	Woolley R., Epps E.A., Penston M.J., Pocock S.B. 1970, R. Obs. Ann. no 5
3HLF4	Drilling J.S. 1971, AJ 76, 1072; 1973, AJ 78, 44 (no+1000)
3U 1956+11	Morgon B., Thorstensen J.R., Bowyer S. 1978, ApJ 221, 907
4HLF4	Drilling J.S. 1971, AJ 76, 1072
959	Morrison H.L., Flynn C., Freeman K.C. 1990, AJ 100, 1191
xxx L	Landolt A.U. 1992, AJ 104, 340 (xxx from 98 to 113)

Table 7 1375

Table 7: References for sequences by constellation

Object	Reference
Aql sq 1	Purgathofer A.Th. 1969, Lowell Obs. Bull. 7 no 10
Ara sq 1	Whiteoak J. 1963, MNRAS 125, 105
Aur sq 1	Landolt A.U. 1977, PASP 89, 403
Aur sq 2	Gurzadyan G.A. 1974, Obs. 94, 293
Boo sq 1	Landolt A.U. 1970, PASP 82, 1364
CMa sq 1	Clariá J.J. 1974, AJ 79, 1022
Car sq 1	Lyngå G., Wramdemark S. 1973, A&AS 12, 365
	Wramdemark S. 1976, A&AS 23, 231 (no+200)
	Wramdemark S. 1980, A&AS 40, 81 (*AB, no+100)
Car sq 2	Wramdemark S. 1976, A&AS 23, 231
Cas sq 1	Sarg K., Wramdemark S. 1977, A&AS 27, 403
Cen sq 1	McGruder C.H., Schnur G. 1971, A&A 14, 164
Cen sq 2	McCarthy C.C., Miller E.W. 1973, AJ 78, 33
Cep sq 1	Sarg K., Wramdemark S. 1970, A&AS 2, 251
Cep sq 2	Barbier M., Bernard A., Bigay J.H., Garnier R. 1973, A&A 27, 421
Cet sq 1	Liller W., Alcaino G. 1982, ApJ 257,l27
Cir sq 1	Haug U., Pfleiderer J., Dachs J. 1966, Z. Astrophys. 64, 140
Cru sq 1	Lyngå G., Wramdemark S. 1973, A&AS 12, 365
	Wramdemark S. 1980, A&AS 40, 81 (*CD, no+100)
Cyg sq 1	Stenholm B., Soderhjelm S. 1972, A&AS 7, 385
Cyg sq 2	Stenholm B., Soderhjelm S. 1972, A&AS 7, 385
Cyg sq 3	Loden L.O., Loden K.A. 1964, Ark. Astr. 3, 299
Cyg sq 4	Bregman J., Butler D., et al 1973, Lick Obs. Bull. no 647
Eri sq 1	Liller W., Alcaino G. 1982, ApJ 257,l27
Hya sq 1	Meech K.J. 1983, PASP 95, 662
Nor sq 1	Haug U., Pfleiderer J., Dachs J. 1966, Z. Astrophys 64, 140
Nor sq 2	Drilling J.S. 1972, AJ 77, 463
Nor sq 3	Muzzio J.C., McCarthy C.C. 1973, AJ 78, 924
Oph sq 1	Meech K.J. 1983, PASP 95, 662
	Muzzio J.C. 1974, AJ 79 , 959
Pav sq 1	Liller W., Alcaino G. 1982, ApJ 257,l27
Peg sq 1	Philip A.G.D. 1965, ApJS 12, 391
Per sq 1	Wramdemark S. 1981, A&AS 43, 103
Psa sq 1	Bunclark P.S., Fraser D.W., Dodd R.J. 1980, A&AS 40, 81
Psc sq 1	Landolt A.U. 1970, PASP 82, 1364
Pup sq 1	Lindoff U. 1968, Ark. Astr. 4, 433
Pup sq 2	Munch L. 1954, Tonantzintla Tacubaya 9, 29
Pup sq 3	FitzGerald M.P., Moffat A.F.J. 1975, A&AS 20, 2
Pup sq 4	Hilditch R.W., Dodd R.J. 1977, MNRAS 178, 467
Pup sq 5	Reed B.C., Fitzgerald M.P. 1982, A&AS 49, 521
Pup sq 6	Meech K.J. 1983, PASP 95, 622
Sco sq 1	Roslund C. 1967, Ark. Astr. 4, 341
Sco sq 2	Roslund C. 1964, Ark. Astr. 3, 357
Sct sq 1	Roslund C. 1963, Ark. Astr. 3, 97
Sgr sq 0	Moffat A.F.J., Vogt N. 1975, A&A 41, 413
Sgr sq 1	Janes K.A. 1979, ApJS 39, 130
Tau sq 1	Landolt A.U. 1967, AJ 72, 1012
Vel sq 1	Lyngå G., Wramdemark S. 1973, A&AS 12, 365
Vir sq 1	Harris W.E., Smith M.G. 1981, AJ 86, 90
Fornax sq	Demers S., Kunkel W.E., Hardy E. 1979, ApJ 232, 84
K17 sq	Hodge P.W., Flower P. 1987, PASP 99, 734
Kron 3 sq	Alcaino G., Alvarado F. 1988, AJ 95, 1724
Leo A sq	Demers S., Kibblewhite E.J., et al 1984, AJ 89, 1160
Leo II sq	Demers S., Harris W.E. 1983, AJ 88, 329
Pal 13 sq	Ortolani S., Rosino L., Sandage A.R. 1985, AJ 90, 473
Reticulum sq	Demers S., Kunkel W.E. 1976, ApJ 208, 932
Sculptor sq	Kunkel W.E., Demers S. 1977, ApJ 214, 21
Sextans B sq	Sandage A., Carlson G. 1985, AJ 90, 1019
Sl 4 sq	Hodge P.W., Flower P. 1987, PASP 99, 734
UMi sq	Cudworth K.M., Olszewski E.W., Schommer R.A. 1986, AJ 92, 766
V Cen sq	Turner D.G. 1982, PASP 94, 1003
VLM sq	Sandage A., Carlson G. 1985, AJ 90, 1464

Table 8: References for sequences around NGC and IC objects

Object	Reference
NGC 45 sq1	Alcaino G., Liller W. 1984, AJ 89, 814
NGC 55 sq1	Alcaino G. 1980, A&A 84, 354
NGC 55 sq2	Hanes D.A., Grieve G.R. 1982, MNRAS 198, 193
NGC 104 sq1	Cannon R.D. 1981, MNRAS 195, 1
NGC 121 sq1	Tifft W.G. 1963, MNRAS 125, 199
NGC 152 sq1	Hodge P.W. 1981, ApJ 247, 894
NGC 176 sq1	Hodge P.W., Flower P. 1987, PASP 99, 734
NGC 247 sq1	Hanes D.A., Grieve G.R. 1982, MNRAS 198, 193
NGC 247 sq2	Alcaino G., Liller W. 1984, AJ 89, 814
NGC 253 sq1	Hanes D.A., Grieve G.R. 1982, MNRAS 198, 193
NGC 253 sq2	Alcaino G., Liller W. 1984, AJ 89, 814
NGC 288 sq1	Alcaino G. 1975, A&AS 21, 5
NGC 288 sq2	Alcaino G., Liller W., Alvarado F. 1987, AJ 93, 1464
NGC 300 sq1	Graham J.A. 1981, PASP 93, 29
NGC 300 sq2	Alcaino G., Liller W. 1984, AJ 89, 814
NGC 330 sq1	Arp H.C. 1959, AJ 64, 254
NGC 330 sq2	Carney B.W., Janes K.A., Flower P.J. 1985, AJ 90, 1196
NGC 330 sq3	Alcaino G., Alvarado F. 1988, AJ 95, 1724
NGC 339 sq1	Walker M.F. 1987, PASP 99, 179
NGC 362 sq1	Alcaino G. 1976, A&AS 26, 359
NGC 362 sq2	Harris W.E. 1982, ApJS 50, 573
NGC 362 sq3	Alcaino G., Liller W., Alvarado F. 1988, AJ 96, 139
NGC 458 sq1	Walker M.F. 1987, PASP 99, 179
NGC 609 sq1	Lee V.L., Burkhead M.S. 1971, AJ 76, 467
NGC 1068 sq1	Penston M.J., Penston M.V., Sandage A.R. 1971, PASP 83, 783
NGC 1097 sq1	Hamuy M., Maza J. 1989, AJ 97, 720
NGC 1261 sq1	Alcaino G. 1979, A&AS 38, 61
NGC 1261 sq2	Alcaino G., Liller W., Alvarado F. 1988, AJ 96, 139
NGC 1275 sq1	Penston M.J., Penston M.V., Sandage A.R. 1971, PASP 83, 783
NGC 1399 sq1	Hanes D.A., Grieve G.R. 1982, MNRAS 198, 193
NGC 1466 sq1	Penny A.J. 1981, MNRAS 197, 693
NGC 1499 sq1	Shinohara M., Ishida K. 1976, PASJ 28, 437
NGC 1672 sq1	Hamuy M., Maza J. 1989, AJ 97, 720
NGC 1777 sq1	Mateo M., Hodge P.W. 1985, ApJ 289, 247
NGC 1783 sq1	Walker M.F. 1987, PASP 99, 179
NGC 1841 sq1	Alcaino G., Alvarado F. 1988, AJ 95, 1724
NGC 1851 sq1	Alcaino G. 1976, A&A 50, 299
NGC 1851 sq2	Alcaino G., Liller W., Alvarado F. 1987, AJ 93, 1464
NGC 1856 sq1	Hodge P.W., Lee S.O. 1984, ApJ 276, 509
NGC 1866 sq1	Arp H. 1958, AJ 63, 118
NGC 1868 sq1	Flower P.J., Geisler D., et al 1980, ApJ 235, 769
NGC 1904 sq1	Alcaino G. 1976, A&AS 26, 353
NGC 1904 sq2	Stetson P.B., Harris W.E. 1977, AJ 82, 954
NGC 1904 sq3	Alcaino G., Liller W., Alvarado F. 1987, AJ 93, 1464
NGC 1978 sq1	Olszewiski E.W. 1984, ApJ 284, 108
NGC 1978 sq2	Olszewiski E.W. 1984, ApJ 284, 108
NGC 2004 sq1	Walker M.F. 1987, PASP 99, 179
NGC 2058 sq1	Flower P.J. 1982, PASP 94, 894
NGC 2164 sq1	Hodge P.W., Flower P.J. 1973, ApJ 185, 829
NGC 2164 sq2	Walker M.F. 1987, PASP 99, 179
NGC 2203 sq1	Hodge P.W., Flower P. 1987, PASP 99, 734
NGC 2214 sq1	Penny A.J. 1981, MNRAS 197, 693
NGC 2243 sq1	van den Bergh S. 1977, ApJ 215, 89
NGC 2257 sq1	Penny A.J. 1981, MNRAS 197, 693
NGC 2257 sq2	Alcaino G., Alvarado F. 1988, AJ 95, 1724
NGC 2298 sq1	Alcaino G. 1974, A&AS 13, 55
NGC 2419 sq1	Racine R., Harris W.E. 1975, ApJ 196, 413
NGC 2419 sq2	Suntzeff N.B., Kraft R.P., Kinman T.D. 1988, AJ 95, 91
NGC 2808 sq1	Harris W.E. 1978, PASP 90, 45
NGC 3201 sq1	Alcaino G. 1976, A&AS 26, 251
NGC 3201 sq2	Lee S.W. 1977, A&AS 28, 409
NGC 3227 sq1	Penston M.J., Penston M.V., Sandage A.R. 1971, PASP 83, 783
NGC 3311 sq1	Smith M.G., Weedman D.W. 1976, ApJ 205, 709
NGC 3516 sq1	Penston M.J., Penston M.V., Sandage A.R. 1971, PASP 83, 783
NGC 3783 sq1	Hamuy M., Maza J. 1989, AJ 97, 720
NGC 4051 sq1	Penston M.J., Penston M.V., Sandage A.R. 1971, PASP 83, 783
NGC 4151 sq1	Penston M.J., Penston M.V., Sandage A.R. 1971, PASP 83, 783

Table 8 1377

Object	Reference
NGC 4216 sq1	Hanes D.A., Harris W.E., Madore B.F. 1976, MNRAS 177, 653
NGC 4340 sq1	Hanes D.A., Harris W.E., Madore B.F. 1976, MNRAS 177, 653
NGC 4372 sq1	Alcaino G. 1974, A&AS 13, 345
NGC 4372 sq2	Hartwick F.D.A., Hesser J.E. 1973, ApJ 186, 1171
NGC 4372 sq3	Hartwick F.D.A., Hesser J.E. 1973, ApJ 186, 1171
NGC 4374 sq1	Hanes D.A., Harris W.E., Madore B.F. 1976, MNRAS 177, 653
NGC 4406 sq1	Hanes D.A., Harris W.E., Madore B.F. 1976, MNRAS 177, 653
NGC 4431 sq1	Hanes D.A., Harris W.E., Madore B.F. 1976, MNRAS 177, 653
NGC 4458 sq1	Hanes D.A., Harris W.E., Madore B.F. 1976, MNRAS 177, 653
NGC 4472 sq1	Hanes D.A., Harris W.E., Madore B.F. 1976, MNRAS 177, 653
NGC 4526 sq1	Hanes D.A., Harris W.E., Madore B.F. 1976, MNRAS 177, 653
NGC 4564 sq1	Hanes D.A., Harris W.E., Madore B.F. 1976, MNRAS 177, 653
NGC 4565 sq1	Hanes D.A., Harris W.E., Madore B.F. 1976, MNRAS 177, 653
NGC 4569 sq1	Hanes D.A., Harris W.E., Madore B.F. 1976, MNRAS 177, 653
NGC 4590 sq1	Harris W.E. 1975, ApJS 29, 397
NGC 4590 sq2	Alcaino G. 1977, A&AS 29, 9
NGC 4590 sq3	Alcaino G., Liller W., Alvarado F. 1987, AJ 93, 1464
NGC 4594 sq1	Hanes D.A., Harris W.E., Madore B.F. 1976, MNRAS 177, 653
NGC 4596 sq1	Hanes D.A., Harris W.E., Madore B.F. 1976, MNRAS 177, 653
NGC 4636 sq1	Hanes D.A., Harris W.E., Madore B.F. 1976, MNRAS 177, 653
NGC 4647 sq1	Hanes D.A., Harris W.E., Madore B.F. 1976, MNRAS 177, 653
NGC 4697 sq1	Hanes D.A., Harris W.E., Madore B.F. 1976, MNRAS 177, 653
NGC 4833 sq1	Menzies J. 1972, MNRAS 156, 207
NGC 4833 sq2	Alcaino G., Liller W., Alvarado F. 1987, AJ 93, 1464
NGC 5128 sq1	Roslund C. 1969, Ark. Astr. 5, 249
NGC 5128 sq2	Graham J.A. 1981, PASP 93, 291
	McElroy D., Humphreys R.M. 1982, PASP 94, 828
NGC 5139 sq1	Geyer E.H. 1967, Z. Astrophys. 66, 16
NGC 5139 sq2	Cannon R.D. 1981, MNRAS 195, 1
NGC 5139 sq3	Arp H.C. 1958, AJ 63, 118
NGC 5236 sq1	McElroy D., Humphreys R.M. 1982, PASP 94, 828
NGC 5286 sq1	Alcaino G. 1974, A&AS 18, 9
NGC 5286 sq2	Harris W.E., Racine R., de Roux J. 1976, ApJS 31, 13
NGC 5457 sq1	Sandage A., Tamman G.A. 1974, ApJ 194, 223
NGC 5466 sq1	Cuffey J. 1961, AJ 66, 71
NGC 5466 sq2	Nemec J.M., Harris H.C. 1987, ApJ 316, 172
NGC 5548 sq1	Penston M.J., Penston M.V., Sandage A.R. 1971, PASP 83, 783
NGC 5694 sq1	Harris W.E. 1975, ApJS 29, 397
NGC 5823 sq1	Schnur G.F.D. 1977, Astr. Nach. 298, 167
NGC 5823 sq2	Schnur G.F.D. 1977, Astr. Nach. 298, 167
NGC 5824 sq1	Harris W.E. 1975, ApJS 29, 397
NGC 5897 sq1	Sandage A., Katem B. 1968, ApJ 153, 569
NGC 5927 sq1	White R.E. 1970, AJ 75, 167
NGC 5927 sq3	Alcaino G. 1979, Acta Astr. 29, 281
NGC 5946 sq1	Martins D.H., Harvel C.A. 1979, AJ 84, 1025
NGC 5946 sq2	Alcaino G., Liller W., Alvarado F. 1987, AJ 93, 1464
NGC 5986 sq1	Harris W.E., Racine R., de Roux J. 1976, ApJS 31, 13
NGC 5986 sq2	Alcaino G. 1984, A&A 139, 549
NGC 6093 sq1	Harris W.E., Racine R. 1974, AJ 79, 472
NGC 6101 sq1	White R.E. 1970, AJ 75, 167
NGC 6101 sq2	Alcaino G. 1974, A&AS 18, 9
NGC 6121 sq1	Alcaino G. 1975, A&AS 21, 5
NGC 6139 sq1	Hatzen-Liller M.L. 1984, AJ 89, 1551
NGC 6139 sq2	Alcaino G., Liller W., Alvarado F. 1987, AJ 93, 1464
NGC 6144 sq1	Alcaino G. 1980, A&AS 39, 315
NGC 6171 sq1	Sandage A., Katem B. 1964, ApJ 139, 1088
NGC 6171 sq2	Dickens R.J. 1970, ApJS 22, 249
NGC 6235 sq1	Liller M.H. 1980, AJ 85, 673
NGC 6254 sq1	Harris W.E., Racine R., de Roux J. 1976, ApJS 31, 13
NGC 6256 sq1	Bernard A. 1976, A&AS 25, 281
NGC 6266 sq1	White R.E. 1971, AJ 76, 419
NGC 6266 sq2	Harris W.E. 1975, ApJS 29, 397
NGC 6266 sq3	Alcaino G. 1978, A&AS 32, 379
NGC 6273 sq1	Harris W.E., Racine R., de Roux J. 1976, ApJS 31, 13
NGC 6304 sq1	Bernard A. 1976, A&AS 25, 281
NGC 6316 sq1	White R.E. 1971, AJ 76, 419
NGC 6325 sq1	Harris W.E. 1975, ApJS 29, 397
NGC 6341 sq1	Sandage A., Walker M.F. 1966, ApJ 143, 313
NGC 6352 sq1	Hartwick F.D.A., Hesser J.E. 1972, ApJ 175, 77

Object	Reference
NGC 6388 sq1	White R.E. 1971, AJ 76, 419
NGC 6388 sq2	Alcaino G. 1981, A&AS 44, 33
NGC 6397 sq1	Cannon R.D. 1981, MNRAS 195, 1
NGC 6397 sq2	van den Bergh S. 1988, AJ 95, 106
NGC 6402 sq1	Demers S., Wehlau A. 1971, AJ 76, 916
NGC 6440 sq1	Martins D.H., Harvel C.A. 1979, AJ 84, 1025
NGC 6441 sq2	Hesser J.E., Hartwick F.D.A. 1976, ApJ 203, 97
NGC 6517 sq1	Harris W.E. 1975, ApJS 29, 397
NGC 6522 sq1	van den Bergh S. 1971, AJ 76, 1082
NGC 6522 sq2	Blanco V.M., Blanco B.M. 1984, PASP 96, 603
NGC 6528 sq1	van den Bergh S., Younger F. 1979, AJ 84, 1305
NGC 6535 sq1	Liller M.H. 1980, AJ 85, 1480
NGC 6541 sq1	Alcaino G. 1979, A&AS 35, 233
NGC 6544 sq1	Alcaino G. 1983, A&AS 52, 105
NGC 6584 sq1	Alcaino G., Liller W. 1985, A&AS 62, 317
NGC 6624 sq1	Harvel C.A., Martins D.H. 1977, ApJ 213,l49
NGC 6624 sq2	Liller M.H., Carney B.W. 1978, ApJ 224, 383
NGC 6626 sq1	Alcaino G. 1981, A&AS 44, 191
NGC 6637 sq1	Hartwick F.D.A., Sandage A. 1968, ApJ 153, 715
NGC 6638 sq1	Bernard A. 1976, A&AS 25, 281
NGC 6656 sq2	Alcaino G. 1977, A&AS 29, 383
NGC 6656 sq3	Alcaino G., Liller W., Alvarado F. 1988, AJ 96, 139
NGC 6681 sq1	Harris W.E. 1975, ApJS 29, 397
NGC 6712 sq1	Sandage A., Smith L.L. 1966, ApJ 144, 886
NGC 6715 sq1	Harris W.E. 1975, ApJS 29, 397
NGC 6717 sq1	Goranskii V.P. 1979, SvA 23, 284
NGC 6723 sq1	Menzies J. 1974, MNRAS 168, 177
NGC 6752 sq1	Wesselink A.J. 1974, MNRAS 168, 345
NGC 6752 sq2	Cannon R.D. 1981, MNRAS 195, 1
NGC 6809 sq1	Alcaino G. 1975, A&AS 22, 193
NGC 6809 sq2	Harris W.E. 1975, ApJS 29, 397
NGC 6809 sq3	Lee S.W. 1977, A&AS 29, 1
NGC 6809 sq4	Alcaino G., Liller W., Alvarado F. 1988, AJ 96, 139
NGC 6822 sq1	Kayser S.E. 1967, AJ 72, 134
NGC 6822 sq2	Hodge P.W. 1977, ApJS 33, 69
NGC 6822 sq3	van den Bergh S., Humphreys R.M. 1979, AJ 84, 604
NGC 6838 sq1	Janulis R. 1984, Bull. Vilnius Astr. Obs. 67, 18
NGC 6864 sq1	Harris W.E. 1975, ApJS 29, 397
NGC 6934 sq1	Harris W.E., Racine R. 1973, AJ 78, 242
NGC 7089 sq1	Demers S. 1969, AJ 74, 925
	Harris W.E. 1975, ApJS 29, 397
NGC 7099 sq1	Alcaino G. 1978, A&AS 33, 185
NGC 7099 sq2	Alcaino G., Liller W., Alvarado F. 1987, AJ 93, 1464
NGC 7213 sq1	Hamuy M., Maza J. 1989, AJ 97, 720
NGC 7293 sq1	Dahn S.C., Behall A.L., Christy J.W. 1973, PASP 85, 224
NGC 7469 sq1	Penston M.J., Penston M.V., Sandage A.R. 1971, PASP 83, 783
NGC 7582 sq1	Hamuy M., Maza J. 1989, AJ 97, 720
NGC 7793 sq1	Hanes D.A., Grieve G.R. 1982, MNRAS 198, 193
NGC 7793 sq2	Alcaino G., Liller W. 1984, AJ 89, 814
IC 4329a sq1	Hamuy M., Maza J. 1989, AJ 97, 720
IC 4499 sq1	White R.E. 1971, AJ 76, 419
IC 5146 sq1	Forte J., Orsatti A.M. 1984, ApJS 56, 211

Table 9 1379

Table 9: References for sequences around quasars and active galaxies

Object	Reference
0109+22	Miller H.R., Mullikin T.L., McGimsey B.Q. 1983, AJ 88,1301
0159-11	Angione R.J. 1971, AJ 76,412
0237-23	Angione R.J. 1971, AJ 76,412
0245+18	McGimsey B.Q., Miller H.R. 1977, AJ 82,453
0248+18	McGimsey B.Q., Miller H.R. 1977, AJ 82,453
0350-07	Angione R.J. 1971, AJ 76,412
0405-12	Angione R.J. 1971, AJ 76,412
0957+00	Penston M.J., Penston M.V., Sandage A.R. 1971, PASP 83,783
1156+295	Smith P.S. et al. 1985, AJ 90,1184
1354+19	Angione R.J. 1971, AJ 76,412
1418+54	Miller H.R., Wilson J.W., et al 1984, A&AS 57,353
1Zw 051+12	Penston M.J., Penston M.V., Sandage A.R. 1971, PASP 83,783
2128-12	Angione R.J. 1971, AJ 76,412
2135-14	Angione R.J. 1971, AJ 76,412
2251+11	Angione R.J. 1971, AJ 76,412
2344+09	Angione R.J. 1971, AJ 76,412
2Zw 2130+09	Penston M.J., Penston M.V., Sandage A.R. 1971, PASP 83,783
3C	Angione R.J. 1971, AJ 76,412
3C 2	Penston M.V., Penston M.V., Sandage A.R. 1971, PASP 83,783
3C 9	Penston M.V., Penston M.V., Sandage A.R. 1971, PASP 83,783
3C 47	Penston M.V., Penston M.V., Sandage A.R. 1971, PASP 83,783
3C 48	Penston M.V., Penston M.V., Sandage A.R. 1971, PASP 83,783
3C 57	Penston M.V., Penston M.V., Sandage A.R. 1971, PASP 83,783
3C 68.2W	Angione R.J. 1971, AJ 76,412
3C 93	Penston M.V., Penston M.V., Sandage A.R. 1971, PASP 83,783
3C 120	Angione R.J. 1971, AJ 76,412
3C 147	Penston M.V., Penston M.V., Sandage A.R. 1971, PASP 83,783
3C 175	Penston M.V., Penston M.V., Sandage A.R. 1971, PASP 83,783
3C 190	Penston M.V., Penston M.V., Sandage A.R. 1971, PASP 83,783
3C 196	Penston M.V., Penston M.V., Sandage A.R. 1971, PASP 83,783
3C 208	Penston M.V., Penston M.V., Sandage A.R. 1971, PASP 83,783
3C 216	Penston M.V., Penston M.V., Sandage A.R. 1971, PASP 83,783
3C 217	Penston M.V., Penston M.V., Sandage A.R. 1971, PASP 83,783
3C 232 A	Penston M.V., Penston M.V., Sandage A.R. 1971, PASP 83,783
3C 232 1-7	Angione R.J. 1971, AJ 76,412
3C 245	Penston M.V., Penston M.V., Sandage A.R. 1971, PASP 83,783
3C 263	Angione R.J. 1971, AJ 76,412
3C 273	Penston M.V., Penston M.V., Sandage A.R. 1971, PASP 83,783
3C 286	Penston M.V., Penston M.V., Sandage A.R. 1971, PASP 83,783
3C 298	Penston M.V., Penston M.V., Sandage A.R. 1971, PASP 83,783
3C 323.1	Angione R.J. 1971, AJ 76,412
3C 334	Angione R.J. 1971, AJ 76,412
3C 345	Angione R.J. 1971, AJ 76,412
3C 351	Angione R.J. 1971, AJ 76,412
3C 371	McGimsey B.Q., Miller H.R. 1977, AJ 82,453
3C 390.3	Penston M.V., Penston M.V., Sandage A.R. 1971, PASP 83,783
3C 446	Angione R.J. 1971, AJ 76,412
3C 454.3	Angione R.J. 1971, AJ 76,412
AO 0235+164	McGimsey B.Q., Miller H.R., Williamson R.M. 1976, AJ 81,75
B2 1308+326	Miller H.R., Wilson J.W., et al 1984, A&AS 57,353
BL Lac	McGimsey B.Q., Miller H.R. 1977, AJ 82,453
FAIRALL	Hamuy J., Maza J. 1989, AJ 97,720
II Zw 136	Miller H.R. 1981, AJ 86,87
Ka 102	Vanderriest C., Herpe G. 1980, A&AS 39,395
Lynds 810	Herbst W., Turner D.G. 1976, PASP 88,308
MCG8-11-11	Miller H.R. 1981, AJ 86,87
MCG-5-23-16	Hamuy J., Maza J. 1989, AJ 97,720
Mark 106	Vanderriest Ch., Herpe G. 1980, A&AS 39,395
Mark 304	Vanderriest Ch., Herpe G. 1980, A&AS 39,395
Mark 315	Miller H.R. 1981, 86,87
Mark 478	Vanderriest Ch., Herpe G. 1980, A&AS 39,395
Mark 505	McGimsey B.Q., Miller H.R. 1977, AJ 82,453
Mark 509	Hamuy J., Maza J. 1989, AJ 97,720
Mark 876	Vanderriest Ch., Herpe G. 1980, A&AS 39,395
NAB 0205+02	Vanderriest C., Herpe G. 1980, A&AS 39,395
OE 110	McGimsey B.Q., Miller H.R., Williamson R.M. 1976, AJ 81,75
OF 038	Miller H.R., Mullikin T.L., McGimsey B.Q. 1983, AJ 88,1301

Object	Reference
OI 090.4	Miller H.R., Mullikin T.L., McGimsey B.Q. 1983, AJ 88,1301
OJ 287	Penston M.V., Wing R.F. 1974, Observatory 93,149
OJ-131	Miller H.R., Mullikin T.L., McGimsey B.Q. 1983, AJ 88,1301
OK 290	Wing R.F. 1973, AJ 78,684
ON 231	Wing R.F. 1973, AJ 78,684
ON 254	Miller H.R. 1981, PASP 93,564
ON 325	Wing R.F. 1973, AJ 78,684
OP 471	Miller H.R. 1981, PASP 93,564
OQ 208	Wing R.F. 1973, AJ 78,684
PHL 1377	Penston M.V., et al. 1971,PASP 83,783
PKS 0537-441	Hamuy J., Maza J. 1989, AJ 97,720
PKS 0735+17	Wing R.F. 1973, AJ 78,684
PKS 0736+01	Vanderriest C., Herpe G. 1980, A&AS 39,395
PKS 0846+10	Vanderriest C., Herpe G. 1980, A&AS 39,395
PKS 0934+01	McGimsey B.Q., Miller H.R., Williamson R.M. 1976, AJ 81,75
PKS 2201+04	McGimsey B.Q., Miller H.R. 1977, AJ 82,453
PKS 2155-304	Hamuy J., Maza J. 1989, AJ 97,720
3Zw 008+10	Penston M.J., Penston M.V., Sandage A.R. 1971, PASP 83,783
ZwI 0120+34	Angione R.J. 1971, AJ 76,412

Table 10 1381

Table 10: References for the numbering systems in star clusters

Cluster	Reference
NGC 103	Hardorp J. 1960, Astr. Abh. Hamburg. Sternw. Bergedorf 5 no 7
NGC 104	Lee S.W. 1977, A&AS 27, 381
NGC 129	Arp H.C., Sandage A.R., Stephens C. 1959, ApJ 130, 80
NGC 188	Sandage A.R. 1962, ApJ 135, 333
NGC 225	Hoag A.A., Johnson H.L., et al 1961, Publ. USNO 17, 347
NGC 288	Alcaino G. 1975, A&AS 21, 15
NGC 330	Arp H. 1959, AJ 64, 254
NGC 362	Menzies J.W. 1967, Thesis
NGC 381	Crinklaw G., Talbert F.D. 1988, PASP 100, 693
NGC 457	Boden E. 1946, Uppsala Astr. Obs. Ann. 2, 1
NGC 559	Lindoff V. 1969, Ark. Astr. 5, 221
NGC 581	Steppe H. 1974, A&AS 15, 91
NGC 609	Lee V.L., Burkhead M.S. 1971, AJ 76, 467
NGC 654	Müller E.A. 1955, Z. Astrophys. 38, 110
NGC 659	Steppe H. 1974, A&AS 15,91
NGC 663	Wallenquist A. 1929, Uppsala Astr. Obs. Medd. no 42
NGC 744	Hoag A.A., Johnson H.L., et al 1961, Publ. USNO 17, 347
NGC 752	Heinemann K. 1926, Astr. Nach. 227, 193
NGC 869	Oosterhoff P.Th. 1937, Ann. Sterrw. Leiden 17 part 1
NGC 884	Oosterhoff P.Th. 1937, Ann. Sterrw. Leiden 17 part 1
NGC 957	Gimenez A., Garcia-Pelayo J. 1980, A&AS 41, 9
NGC 1027	Hoag A.A., Johnson H.L., et al 1961, Publ. USNO 17, 347
NGC 1039	Dieckvoss W. 1954, Astr. Nach. 282, 25
NGC 1245	Chincarini G. 1964, Mem. Soc. Astron. Ital. 35, 133
NGC 1252	Bouchet R., Thé P.S. 1983, PASP 95, 474
NGC 1261	Alcaino G., Contreras C. 1971, A&A 11,14
NGC 1342	Hoag A.A., Johnson H.L., et al 1961, Publ. USNO 17, 347
NGC 1444	Hoag A.A., Johnson H.L., et al 1961, Publ. USNO 17, 347
NGC 1496	del Rio G., Huestamendia G. 1988, A&AS 73, 425
NGC 1502	Purgathofer A. 1964, Ann. Univ. Sternw. Wien 26 no 2
NGC 1513	del Rio G., Huestamendia G. 1988, A&AS 73, 425
NGC 1528	Hoag A.A., Johnson H.L., et al 1961, Publ. USNO 17, 347
NGC 1545	Hoag A.A., Johnson H.L., et al 1961, Publ. USNO 17, 347
NGC 1647	Cuffey J. 1937, Ann. Harv. Coll. Obs. 105, 403
NGC 1662	Hoag A.A., Johnson H.L., et al 1961, Publ. USNO 17, 347
NGC 1664	Larsson-Leander G. 1957, Stockholm Obs. Ann. 20 no 2
NGC 1750	Li Hen 1954, Ann. Obs. Astr. Zo-Se 23,1
NGC 1778	Hoag A.A., Johnson H.L., et al 1961, Publ. USNO 17, 347
NGC 1807	Purgathofer A. 1964, Ann. Univ. Sternw. Wien 26 no 2
NGC 1817	Cuffey J. 1938, Ann. Harv. Coll. Obs. 106, 39
NGC 1851	Stetson P.B. 1981, AJ 86, 687
	Alcaino G. 1976, A&A 50,299 (no + 1000); 1969, ApJ 156,853 (no + 1500)
NGC 1857	Cuffey J. 1937, Ann. Harv. Coll. Obs. 105, 433
NGC 1893	Cuffey J. 1937, Ann. Harv. Coll. Obs. 105, 403
NGC 1901	Sanduleak N., Philip A.G.D. 1968, AJ 73, 566
NGC 1904	Stetson P.B., Harris W.E. 1977, AJ 82, 954
NGC 1907	Purgathofer A. 1964, Ann. Univ. Sternw. Wien 26 no 2
NGC 1912	Cuffey J. 1937, Ann. Harv. Coll. Obs. 105, 403
NGC 1931	Pandey A.K., Mahra H.S. 1986, Ap&SS 120, 107
NGC 1960	Boden E. 1951, Uppsala Astr. Obs. Ann. 3 no 4
NGC 1976	Brun A. 1935, Publ. Obs. Lyon 1, Fasc 12, 1
NGC 2024	Johnson H.L., Mendoza E.E. 1963, Tonantzintla Tacubaya 3, 331
NGC 2081	Westerlund B. 1961, Uppsala Astr. Obs. Ann. 5, 58
NGC 2099	van Zeipel H., Lindgren J. 1921, Kungl. Sven. Vet. Handl. 61 no 15
NGC 2112	Richtler T. 1985, A&AS 59, 491
NGC 2129	Cuffey J. 1938, Ann. Harv. Coll. Obs. 106, 39
NGC 2141	Burkhead M.S., Burgess R.D., Harsch B.M. 1972, AJ 77, 661
NGC 2158	Arp H., Cuffey J. 1962, ApJ 136, 51
NGC 2168	Cuffey J. 1938, Ann. Harv. Coll. Obs. 106,39
NGC 2169	Hoag A.A., Johnson H.L., et al 1961, Publ. USNO 17, 347
NGC 2175	Pismis P. 1970, Tonantzintla Tacubaya 5, 219
NGC 2186	Moffat A.F.J., Vogt N. 1975, A&AS 20, 85
NGC 2204	Hawarden T.G. 1976, MNRAS 174, 225
NGC 2215	Becker W., Svolopoulos S.N., Fang C. 1976, Astr. Inst. Univ. Basel' Kataloge photographischer und photoelektrischer Helligkeiten von 25 galaktischen Sternhaufen im RGU- und U_CBV-system
NGC 2232	Clariá J.J. 1972, A&A 19, 303
NGC 2243	Hawarden T.G. 1975, MNRAS 173, 801

Cluster	Reference
NGC 2244	Ogura K., Ishida K. 1981, PASJ 33, 149
NGC 2251	Hoag A.A., Johnson H.L., et al 1961, Publ. USNO 17, 347
NGC 2264	Walker M.F. 1956, ApJS 2, 365
	Vasilevskis S. 1965, AJ 70, 797 (no + 300)
NGC 2269	Moffat A.F.J., Vogt N. 1975, A&AS 20, 85
NGC 2281	Vasilevskis S., Balz A.G.A. 1959, AJ 64, 170
NGC 2286	Li Hen 1954, Ann. Obs. Astr. Zo-Se 23, 1
NGC 2287	Cox A.N. 1954, ApJ 119, 188
NGC 2298	Alcaino G., Liller W. 1974, A&AS 13, 55
NGC 2301	Grubissich C., Purgathofer A. 1962, Z. Astrophys. 54, 41
NGC 2302	Moffat A.F.J., Vogt N. 1975, A&AS 20, 85
NGC 2323	Cuffey J. 1941, ApJ 94, 55
NGC 2324	Cuffey J. 1941, ApJ 94, 55
NGC 2335	Clariá J.J. 1973, A&AS 9, 251
NGC 2343	Clariá J.J. 1972, AJ 77, 868
NGC 2345	Moffat A.F.J. 1974, A&AS 16, 33
NGC 2353	Hoag A.A., Johnson H.L., et al 1961, Publ. USNO 17, 347
NGC 2354	Durbeck W. 1960, Z. Astrophys. 49, 214
NGC 2360	Becker W., Svolopoulos S.N., Fang C. 1976, Astr. Inst. Univ. Basel: Kataloge...
NGC 2362	Johnson H.L. 1950, ApJ 112, 240
NGC 2367	Vogt N., Moffat A.F.J. 1972, A&AS 7, 133
NGC 2374	Fenkart R., Buser R., Ritter H., et al 1972, A&AS 7, 487
NGC 2383	Vogt N., Moffat A.F.J. 1972, A&AS 7, 133
NGC 2384	Vogt N., Moffat A.F.J. 1972, A&AS 7, 133
NGC 2414	Vogt N., Moffat A.F.J. 1972, A&AS 7, 133
NGC 2420	Cannon R.D., Lloyd C. 1970, MNRAS 150, 279
NGC 2421	Moffat A.F.J., Vogt N. 1975, A&AS 20, 85
NGC 2422	van Schewick H. 1966, Veröff. Astr. Inst. Univ. Bonn no 74
NGC 2423	Hassan S.M. 1976, A&AS 26, 13
NGC 2437	Cuffey J. 1941, ApJ 94, 55
NGC 2439	White S.D.M. 1975, ApJ 197, 67
NGC 2447	Becker W., Svolopoulos S.N., Fang C. 1976, Astr. Inst. Univ. Basel, Kataloge...
NGC 2451	Williams P.M. 1967, MNASSA 26, 30
NGC 2453	Moffat A.F.J., FitzGerald M.P. 1974, A&AS 18, 1
NGC 2467	Loden L.O. 1966, Ark. Astr. 4, 65
NGC 2477	Hartwick F.D.A., Hesser J.E., McClure R.D. 1972, ApJ 174, 557
	Hartwick F.D.A., Hesser J.E. 1974, ApJ 192, 391
NGC 2482	Moffat A.F.J., Vogt N. 1975, A&AS 20, 85
NGC 2483	FitzGerald M.P., Moffat A.F.J. 1975, A&A 20, 289
NGC 2506	McClure R.D., Twarog B.A., Forrester W.T. 1981, ApJ 243, 841
NGC 2516	Cox A.N. 1955, ApJ 121, 628
NGC 2527	Lindoff U. 1973, A&AS 9, 229
NGC 2533	Lindoff U. 1968, Ark. Astr. 4, 587
NGC 2539	Pesch P. 1961, ApJ 134, 602
	Joshi U.C., Sagar R. 1986, Bull. Astr. Soc. India 14, 95 (no + 1000)
NGC 2546	Lindoff U. 1968, Ark. Astr. 5, 63
NGC 2547	Clariá J.J. 1982, A&AS 47, 323
NGC 2548	Li Hen 1954, Ann. Obs. Astr. Zo-Sé 23, 1
NGC 2567	Lindoff U. 1968, Ark. Astr. 4,5 87 (1-117)
	Clariá J.J., Lapasset E. 1986, AJ 91, 326 (118-181)
NGC 2571	Lindoff U. 1968, Ark. Astr. 4,5 87
NGC 2632	Klein-Wassink W.J. 1927, Publ. Kapteyn Astr. Lab. 41
	VanderLinden H.L. 1933, Etude de l'amas de Praesepe. J. Duculot, Gembloux (Belgique). (no + 1000)
NGC 2635	Vogt N., Moffat A.F.J. 1972, A&AS 7, 133
NGC 2660	Hartwick F.D.A., Hesser J.E. 1973, ApJ 183, 883
NGC 2669	Vogt N., Moffat A.F.J. 1972, A&AS 9, 97
NGC 2670	Lyngå G. 1959, Ark. Astr. 2, 379
NGC 2682	Fagerholm E. 1906, Uppsala dissertation
NGC 2682	Murray C.A., et al 1965, R. Obs. Bull. no 91, (no + 400)
NGC 2808	Harris W.E. 1975, ApJS 29, 397; Alcaino G. 1969, ApJ 156,853 (no + 500)
NGC 2818	Tifft W.G., Conolly L.P., Webb D.F. 1972, MNRAS 158, 47
NGC 2910	Becker W., Svolopoulos S.N., Fang C. 1976, Astr. Inst. Univ. Basel, Katalogue...
NGC 2972	Vogt N., Moffat A.F.J. 1972, A&AS 9, 97
NGC 3033	Vogt N., Moffat A.F.J. 1972, A&AS 9, 97
NGC 3105	Moffat A.F.J., FitzGerald M.P. 1974, A&AS 16, 2
NGC 3114	Jankowitz N.E., McCosh C.J. 1963, MNASSA 22, 1
NGC 3201	Lee S.W. 1977, A&AS 28, 409
NGC 3228	Hogg A.R. 1963, MNRAS 125, 307
NGC 3255	Moffat A.F.J., Vogt N. 1975, A&AS 20, 125

Table 10 1383

Cluster	Reference
NGC 3293	Turner D.G., Grieve G.R., Herbst W., Harris W.E. 1980, AJ 85, 1193
NGC 3324	Clariá J.J. 1977, A&AS 27, 145
NGC 3330	Becker W., Svolopoulos S.N., Fang C. 1976, Astr. Inst. Univ. Basel, Katalogue...
NGC 3496	Sher D. 1965, MNRAS 129, 237
NGC 3532	Fernandez J.A., Salgado C.W. 1980, A&AS 39, 11
NGC 3572	Steppe H. 1977, A&AS 27, 415
NGC 3590	Steppe H. 1977, A&AS 27, 415
NGC 3603	Sher D. 1965, MNRAS 129, 237
NGC 3680	Eggen O.J. 1969, ApJ 155, 439
NGC 3766	Ahmed F. 1962, Publ. R. Obs. Edinburgh 3, 60
NGC 3960	Janes K.A. 1981, AJ 86, 1210
NGC 4103	Wesselink A.J. 1969, MNRAS 146, 329
NGC 4147	Sandage A.R., Walker M.F. 1955, AJ 60, 230
NGC 4337	Moffat A.F.J., Vogt N. 1973, A&AS 10, 135
NGC 4349	Lohmann W. 1961, Astr. Nach. 286, 105
NGC 4439	Moffat A.F.J., Vogt N. 1973, A&AS 10, 135
NGC 4463	Moffat A.F.J., Vogt N. 1973, A&AS 10, 135
NGC 4609	Feinstein A., Marraco H.G. 1971, PASP 83, 218
NGC 4755	Arp H., van Sant C.T. 1958, AJ 63, 341
NGC 4815	Moffat A.F.J., Vogt N. 1973, A&AS 10, 135
NGC 4833	Alcaino G. 1971, A&A 13,287
NGC 5053	Purgathofer A. 1966, Mitt. Univ. Sternw. Wien 13 no 3
NGC 5138	Lindoff U. 1972, A&AS 7, 231
NGC 5139	Woolley R.R., Epps E., et al 1966, R. Obs. Ann. no 2
NGC 5168	Fenkart R.P., Binggeli B., Good D., et al 1977, A&AS 30, 307
NGC 5272	Johnson H.L., Sandage A.R. 1956, ApJ 124, 379
NGC 5281	Moffat A.F.J., Vogt N. 1973, A&AS 10, 135
NGC 5316	Rahim M.A. 1966, Astr. Nach. 289, 41
NGC 5367	Williams P.M., Brand P.W.J., Longmore A.J., Hawarden T.G. 1977, MNRAS 181, 709
NGC 5460	Clariá J.J. 1971, AJ 76,639
NGC 5606	Moffat A.F.J., Vogt N. 1973, A&AS 10, 135
NGC 5617	Haug U. 1978, A&AS 34, 417
NGC 5662	Haug U. 1978, A&AS 34, 417
	Clariá J.J., et al 1990, MNRAS 249, 193 (no > 300)
NGC 5749	Clariá J.J., Lapasset E. 1992, Acta Astron. 42, 343
NGC 5822	Bozkurt S. 1974, Rev. Mex. Astron. Astrofis. 1, 89
NGC 5823	Brück M.T., Smyth M.J., McLachlan A. 1968, Publ. R. Obs. Edinburgh 6 no 9
NGC 5897	Sandage A., Katem B. 1968, ApJ 153, 569
NGC 5904	Arp H.C. 1962, ApJ 135, 311
NGC 6005	Moffat A.F.J., Vogt N. 1975, A&AS 20, 155
NGC 6025	Feinstein A. 1971, PASP 83, 800
NGC 6031	Lindoff U. 1966, Ark. Astr. 4, 305
NGC 6067	Thackeray A.D., Wesselink A.F., Harding G.A. 1962, MNRAS 124, 445
NGC 6087	Landolt A.J. 1964, ApJS 8, 329
NGC 6121	Greenstein J.L. 1939, ApJ 90, 387
NGC 6124	Koelbloed D. 1959, BAN 14, 265
	Thé P.S. 1965, Contr. Bosscha Obs. no 32 (no > 100)
NGC 6134	Lindoff U. 1972, A&AS 7, 231
NGC 6167	Brück M.T., Smyth M.J. 1966, Publ. R. Obs. Edinburgh 5, 195
NGC 6169	Moffat A.F.J., Vogt N. 1973, A&AS 10, 135
NGC 6171	Sandage A., Katem B. 1964, ApJ 139, 1088
NGC 6178	Moffat A.F.J., Vogt N. 1973, A&AS 10, 135
NGC 6192	Kilambi G.C., FitzGerald M.P. 1983, Bull. Astr. Soc. India 11, 226
NGC 6193	Moffat A.F.J., Vogt N. 1973, A&AS 10, 135
NGC 6200	FitzGerald P.M., Jackson P.D., Moffat A.F.J. 1977, Obs. 97, 129
NGC 6204	Moffat A.F.J., Vogt N. 1973, A&AS 10, 135
NGC 6205	Ludendorf H. 1905, Publ. Astrophys. Obs. Potsdam 15 no 50
NGC 6208	Lindoff U. 1972, A&AS 7, 231
NGC 6218	Racine R. 1971, AJ 76, 331
NGC 6231	Seggewiss W. 1968, Veröff. Astr. Inst. Univ. Bonn no 79
NGC 6242	Seggewiss W. 1967, Thesis (Univ. Münster)
NGC 6249	Moffat A.F.J., Vogt N. 1973, A&AS 10, 135
NGC 6250	Moffat A.F.J., Vogt N. 1975, A&AS 20, 155
NGC 6254	Arp H.C. 1955, AJ 60, 317
NGC 6259	Hawarden F.G. 1974, MNRAS 169, 539
NGC 6281	Feinstein A., Forte J.C. 1974, PASP 86, 284
NGC 6304	Hesser J.E., Hartwick F.D.A. 1976, ApJ 203, 113
NGC 6322	Lohman W., Schnur G. 1962, Astr. Nach 287, 17
NGC 6341	Sandage A.R., Walker M.F. 1966, ApJ 143, 313

Cluster	Reference
NGC 6352	Hartwick F.D.A., Hesser J.E. 1972, ApJ 175, 77
	Alcaino G. 1971, A&A11,7 (no + 1000)
NGC 6362	Alcaino G. 1970, ApJ 159, 325
NGC 6383	Evans T.L. 1978, MNRAS 184, 661
NGC 6396	Moffat A.F.J., Vogt N. 1975, A&AS 20, 155
NGC 6397	Woolley R., et al 1961, R. Obs. Bull. no 43
NGC 6405	Rohlfs K., Schrick K.W., Stock J. 1959, Z. Astrophys. 47, 15
NGC 6416	Thé P.S., Stokes N. 1970, A&A 5, 298
NGC 6425	Thé P.S., Stokes N. 1970, A&A 5, 298
NGC 6441	Hesser J.E., Hartwick F.D.A. 1976, ApJ 203, 97
NGC 6475	Koelbloed D. 1959, BAN 14, 265
NGC 6494	Hoag A.A., Johnson H.L., et al 1961, Publ. USNO 17, 347
	Sanders W.L., Schröder R. 1980, A&A 88, 102 (no > 1000)
NGC 6514	Ogura K., Ishida K. 1975, PASJ 27, 119
NGC 6520	Becker W., Svolopoulos S.N., Fang C. 1976, Astr. Inst. Univ. Basel, Katalogue...
NGC 6530	Kilambi G.C. 1977, MNRAS 178, 423
NGC 6531	Hoag A.A., Johnson H.L., et al 1961, Publ. USNO 17, 347
NGC 6541	Alcaino G. 1971, A&A 13, 399
NGC 6546	Grayzeck E.J., FitzGerald M.P., Luiken M. 1978, PASP 90, 742
NGC 6604	Moffat A.F.J., Vogt N. 1975, A&A 20, 155
NGC 6611	Walker M.F. 1961, ApJ 133, 438
NGC 6613	Lindoff U. 1971, A&A 15, 439
NGC 6618	Ogura K., Ishida K. 1976, PASJ 28, 35
NGC 6626	Alcaino G. 1981, A&AS 44, 191
NGC 6633	Kopff E. 1943, Astr. Nach. 274, 69
NGC 6637	Hartwick F.D.A., Sandage A. 1968, ApJ 153, 715
NGC 6649	Madore B.F., van den Bergh S. 1975, ApJ 197, 55
NGC 6656	Arp H.C., Melbourne W.G. 1959, AJ 64, 28
NGC 6664	Arp H.C. 1958, ApJ 128, 166
NGC 6694	Cuffey J. 1940, ApJ 92, 303
NGC 6704	Forbes D., Dupuy D.L. 1978, AJ 83, 266
NGC 6705	McNamara B.J., Pratt N.M., Sanders W.L. 1977, A&AS 27, 117
NGC 6709	Hakkila J., Sanders W.L., Schroder R. 1983, A&AS 51, 541
NGC 6716	Lindoff U. 1971, A&A 15, 439
NGC 6716	Grice N.A., Dawson D.W. 1990, AJ 102, 881
NGC 6723	Menzies J. 1974, MNRAS 168, 177
NGC 6752	Alcaino G. 1970, ApJ 159, 325
NGC 6755	Hoag A.A., Johnson H.L., et al 1961, Publ. USNO 17, 347
NGC 6791	Kinman T.D. 1965, ApJ 142, 655
NGC 6802	Hoag A.A., Johnson H.L., et al 1961, Publ. USNO 17, 347
NGC 6809	Lee S.W. 1977, A&AS 29, 1
NGC 6811	Lindoff U. 1972, A&A 16, 315
NGC 6819	Auner G. 1974, A&AS 13, 143
NGC 6823	Barkhatova K.A. 1957, Soviet Astr. 1, 822
NGC 6830	Barkhatova K.A. 1957, Soviet Astr. 1, 822
NGC 6834	Funfschilling H. 1967, Z. Astrophys. 66, 440
NGC 6838	Arp H.C., Hartwick F.D.A. 1971, ApJ 167, 499
NGC 6866	Hoag A.A., Johnson H.L., et al 1961, Publ. USNO 17, 347
NGC 6871	Hoag A.A., Johnson H.L., et al 1961, Publ. USNO 17, 347
NGC 6882	Hoag A.A., Johnson H.L., et al 1961, Publ. USNO 17, 347
NGC 6883	Purgathofer A. 1964, Ann. Univ. Sternw. Wien 26 no 2
NGC 6910	Becker W., Stock J. 1948, Astr. Nach. 277, 233
NGC 6913	Becker W., Stock J. 1948, Astron. Nach. 278, 115
NGC 6939	Küstner F. 1923, Veröff. Univ. Sternw. Bonn no 18
NGC 6940	Vasilevskis S., Rach R.A. 1957, AJ 62, 175
NGC 6981	Dickens R.J. 1972, MNRAS 157, 281
NGC 7031	Hoag A.A., Johnson H.L., et al 1961, Publ. USNO 17, 347
NGC 7039	Hassan S.M. 1973, A&AS 9, 261
NGC 7062	Jones B.F., van Altena W.F. 1970, A&A 9, 86
NGC 7063	Hoag A.A., Johnson H.L., et al 1961, Publ. USNO 17, 347
NGC 7067	Becker W. 1965, Mem. Soc. Astron. Ital. 36, 283
NGC 7086	Hassan S.M. 1967, Z. Astrophys. 66, 6
NGC 7089	Arp H.C. 1955, AJ 60, 317
NGC 7092	Mävers F.W. 1940, Astr. Abh. Hamburg. Sternw. Bergedorf 5 no 7
NGC 7099	Dickens R.J. 1972, MNRAS 157, 299
NGC 7128	Hoag A.A., Johnson H.L., et al 1961, Publ. USNO 17, 347
NGC 7142	van den Bergh S., Heeringa R. 1970, A&A 9, 209
NGC 7160	Hoag A.A., Johnson H.L., et al 1961, Publ. USNO 17, 347
NGC 7209	Mävers F.W. 1940, Astr. Abh. Hamburg. Sternw. Bergedorf 5 no 4

Table 10 1385

Cluster	Reference
NGC 7235	Becker W. 1965, Mem. Soc. Astron. Ital. 36, 271
NGC 7243	Lengauer G.G. 1937, Izv. Glav. Astr. Obs. Pulkove 15 no 126
NGC 7261	Fenkart R.P. 1968, Mem. Soc. Astron. Ital. 39, 85
NGC 7380	Hoag A.A., Johnson H.L., et al 1961, Publ. USNO 17, 347
NGC 7419	Handschel G. 1972, Thesis (Univ. Münster)
NGC 7510	Hoag A.A., Johnson H.L., et al 1961, Publ. USNO 17, 347
NGC 7654	Lundby A. 1946, Uppsala Astr. Obs. Ann. 1 no 10
NGC 7686	Hoag A.A., Johnson H.L., et al 1961, Publ. USNO 17, 347
NGC 7762	Chincarini G. 1966, Mem. Soc. Astron. Ital. 37,423
NGC 7788	Becker W. 1965, Mem. Soc. Astron. Ital. 36, 277
NGC 7789	Küstner F. 1923, Veröff. Univ. Sternw. Bonn no 18
NGC 7790	Pedreros M., Madore B.F. 1984, ApJ 286, 563
NGC 7822	Blanco V. 1959, ApJ 130, 482
IC 166	Burkhead M.S. 1969, AJ 74, 1171
IC 348	Gingrich C.H. 1922, ApJ 56, 139
IC 1795	Ogura K., Ishida K. 1976, PASJ 28, 651
IC 1805	Vasilevskis S., Sanders W.L., van Altena W.F. 1965, AJ 70, 806
IC 1848	Hoag A.A., Johnson H.L., et al 1961, Publ. USNO 17, 347
IC 2157	Cuffey J. 1938, Ann. Harv. Coll. Obs. 106, 39
IC 2391	Perry C.L., Hill G. 1969, AJ 74, 899 (no CDS)
IC 2395	Lyngå G. 1959, Ark. Astr. 2, 379
IC 2488	Pedreros D. 1987, AJ 94, 92
IC 2581	Evans L.T. 1969, MNRAS 146, 101
IC 2602	Braes L.L.E. 1962, BAN 16, 297
IC 2714	Becker W. 1960, Z. Astrophys. 51, 49
IC 2944	Ardeberg A., Maurice E. 1977, A&AS 28, 153
IC 4651	Lindoff U. 1972, A&AS 7, 231
IC 4665	Kopff E. 1943, Astr. Nach. 274, 69
IC 4725	Johnson H.L. 1960, ApJ 131, 620
IC 4756	Kopff E. 1943, Astr. Nach. 274, 69
IC 4996	van Schewick H. 1967, Veröff. Astr. Inst. Univ. Bonn 76, 35
IC 5146	Walker M.F. 1960, ApJ 130, 57
Alpha Per	Heckmann O., Doieckvoss W., Kox H. 1956, Astr. Nach. 283, 37
	Stauffer J.R., Hartmann L., et al. 1985, ApJ 289, 247 (no + 1500)
Coma Ber	Trumpler R.J. 1938, Lick Obs. Bull. 18, 167
	DeLuca E.E., Weis E.W. 1981, PASP 93, 32 (no + 1000)
Hyades vB	van Bueren G. 1952, BAN XI, 385
Hyades vA	van Altena W. 1969, AJ 74, 2
Hyades L	Pels G. 1975, A&A 43, 423
Hyades J	Johnson H.L., Mitchell I., Iriarte B. 1962, ApJ 136, 75
Hyades CDS	see cross-identifications in Table 4
Hyades HG7	Giclas H.L. et al 1962, Lowell Obs. Bull. V, 257 no 118
Hyades HG8	Giclas H.L. et al 1962, Lowell Obs. Bull. V, 257 no 118
Pleiades	Hertzsprung E. 1947, Ann. Sterrew. Leiden 19, 3
	Flare stars unified numeration (no PFS + 9000)
Be 11	Jackson P.A., FitzGerald M.P., Moffat A.F.J. 1980, A&AS 41, 211
Be 19	Christian C.A. 1980, AJ 85, 700
Be 21	Christian C.A., Janes K.A. 1979, AJ 84, 204
Be 28	Bijaoui A., Lacoarret M., Mars G. 1984, A&AS 55, 393
Be 62	Forbes D. 1981, PASP 93, 441
Be 82	Forbes D. 1986, PASP 98, 218
Be 86	Forbes D. 1981, PASP 93, 441
Be 87	Turner D.G. 1982, PASP 94, 789
Be 94	Yilmaz F. 1970, A&A 8, 213
Be 96	del Rio G. 1984, A&AS 56, 289
Biu 2	Dupuy D.L., Zukauskas W. 1976, JRAS Canada 70, 169
Bl 1	Westerlund B.E., Garnier R., Lundgren K., et al 1988, A&AS 76, 101
Bo 2	Moffat A.F.J., Vogt N. 1975, A&AS 20, 85
Bo 3	Moffat A.F.J., Vogt N. 1975, A&AS 20, 85
Bo 4	Moffat A.F.J., Vogt N. 1975, A&AS 20, 85
Bo 8	Moffat A.F.J., Vogt N. 1975, A&AS 20, 125
Bo 9	Moffat A.F.J., Vogt N. 1975, A&AS 20, 125
Bo 10	Moffat A.F.J., Vogt N. 1975, A&AS 20, 125
Bo 11	Moffat A.F.J., Vogt N. 1975, A&AS 20, 125
Bo 12	Moffat A.F.J., Vogt N. 1975, A&AS 20, 125
Bo 13	Moffat A.F.J., Vogt N. 1975, A&AS 20, 125
Bo 14	Moffat A.F.J., Vogt N. 1975, A&AS 20, 125
Bo 15	Jackson P.D., FitzGerald M.P., Moffat A.F.J. 1977, A&A 60, 417
Cr 96	Moffat A.F.J., Vogt N. 1975, A&AS 20, 85

Cluster	Reference
Cr 97	Babu G.S.D. 1987, JA&A 8, 219
Cr 121	Feinstein A. 1967, ApJ 149, 107
Cr 132	Clariá J.J. 1977, PASP 89, 803
Cr 135	Clariá J.J., Kepler S.O. 1980, PASP 92, 501
Cr 140	Clariá J.J., Rosenzweig P. 1978, AJ 83, 278
Cr 197	Vogt N., Moffat A.F.J. 1972, A&AS 9, 97
Cr 205	Vogt N., Moffat A.F.J. 1972, A&AS 9, 97
Cr 223	Clariá J.J., Lapasset E. 1991, PASP 103, 998
Cr 228	Feinstein A., Marraco H.G., Forte J.C. 1976, A&AS 24, 389
Cr 258	Moffat A.F.J., Vogt N. 1973, A&AS 10, 135
Cr 268	Moffat A.F.J., Vogt N. 1975, A&AS 20, 155
Cr 271	Moffat A.F.J., Vogt N. 1973, A&AS 10, 135
Cr 307	Moffat A.F.J., Vogt N. 1975, A&AS 20, 155
Cr 347	Moffat A.F.J., Vogt N. 1975, A&AS 20, 155
Cr 359	Rucinski S.M. 1987, PASP 99, 487
Cr 394	Turner D.G., Pedreros M. 1985, AJ 90,1 231
Cr 399	Hall D.S., van Landingham F.G. 1970, PASP 82, 640
Cr 463	Townsed R.E. 1975, PASP 87, 753
Cz 20	Babu G.S.D. 1989, JA&A 10, 295
Dol 24	Moffat A.F.J., FitzGerald M.P., Jackson P.D. 1979, A&AS 38, 197
Dol 25	Moffat A.F.J., Vogt N. 1975, A&AS 20, 85
Dol 42	Zakirov M.M. 1990, Kinematika y Fiz. Nebesn. 6, 42
Haf 8	Fenkart R.P., Buser R., Ritter H., et al 1972, A&AS 7, 487
Haf 14	Jorgensen U.G., Westerlund B.E. 1988, A&AS 72, 193
Haf 15	Vogt N., Moffat A.F.J. 1972, A&AS 7, 133
Haf 16	Vogt N., Moffat A.F.J. 1972, A&AS 7, 133
Haf 18	FitzGerald P.M., Moffat A.F.J. 1974, AJ 79, 873
Haf 19	FitzGerald P.M., Moffat A.F.J. 1974, AJ 79, 873
Ha 20	Turner D.G. 1981, PASP 92, 840
Ho 9	Moffat A.F.J., Vogt N. 1975, A&AS 20, 125
Ho 10	Clariá J.J. 1976, AJ 81, 155
Ho 11	Moffat A.F.J., Vogt N. 1975, A&AS 20, 125
Ho 12	Moffat A.F.J., Vogt N. 1975, A&AS 20, 125
Ho 14	Moffat A.F.J., Vogt N. 1973, A&AS 10, 135
Ho 15	Feinstein A. 1971, PASP 83, 218
Ho 16	Fenkart R.P., Binggeli B., Good D., et al 1977, A&AS 30, 307
Ho 18	Hogg A.R. 1961, Obs. 81, 69
Ho 22	Moffat A.F.J., Vogt N. 1973, A&AS 10, 135
Ki 8	Christian C.A. 1981, ApJ 246, 827
Ki 12	Mohan V., Pandey A.K. 1984, Ap&SS 105, 315
Ki 14	Hardorp J. 1960, Astr. Abh. Hamburg. Sternw. Bergedorf 5 no 7
Ly 1	Peterson C.J., FitzGerald M.P. 1988, MNRAS 235, 1439
Ly 2	Lindoff U. 1968, Ark. Astr. 4, 471
Ly 4	Moffat A.F.J., Vogt N. 1975, A&AS 20, 155
Ly 6	Moffat A.F.J., Vogt N. 1975, A&AS 20, 155
Ly 7	Lyngå G. 1964, Medd. Lunds II, No 139
Ly 14	Moffat A.F.J., Vogt N. 1975, A&AS 20, 155
Ma 38	Grubissich C., Becker W. 1980, A&AS 40, 367
Ma 50	Grubissich C. 1965, Z. Astrophys. 60, 256
Mel 66	Hawarden T.G. 1976, MNRAS 174, 471
Mel 71	Hassan S.M. 1976, A&AS 26, 9
Mel 101	Braes L.L.E. 1962, MNASSA 21, 16
Mel 105	Sher D. 1965, MNRAS 129,2 37
Pis 1	Lindoff U. 1968, Ark. Astr. 5, 63
Pis 4	Moffat A.F.J., Vogt N. 1975, A&AS 20, 85
Pis 5	Vogt N., Moffat A.F.J. 1972, A&AS 9, 97
Pis 6	FitzGerald M.P., Boudreault R., Fich M., et al 1979, A&AS 37, 351
Pis 8	FitzGerald M.P., Boudreault R., Fich M., et al 1979, A&AS 37, 351
Pis 11	Moffat A.F.J., FitzGerald P.M. 1977, A&A 54, 263
Pis 12	Moffat A.F.J., Vogt N. 1975, A&AS 20, 85
Pis 13	Clariá J.J. 1979, Observatory 99, 202
Pis 16	Vogt N., Moffat A.F.J. 1972, A&AS 9, 97
Pis 17	Moffat A.F.J., Vogt N. 1975, A&AS 20, 125
Pis 18	Moffat A.F.J., Vogt N. 1973, A&AS 10, 135
Pis 20	Lyngå G. 1964, Medd. Lunds Astr. Obs. 2 no 139
Pis 21	Moffat A.F.J., FitzGerald P.M. 1977, A&A 54, 263
Pis 22	Westerlund B. 1969, AJ 74, 882
Pis 24	Moffat A.F.J., Vogt N. 1973, A&AS 10, 135
Ros 3	Turner D.G. 1993, A&AS 97, 755

Table 10   1387

Cluster	Reference
Ros 4	Racine R. 1969, AJ 74, 816
Ros 5	Lee P.D., Perry C.L. 1971, AJ 76,464 (no CDS)
Ru 18	Moffat A.F.J., Vogt N. 1975, A&AS 20, 85
Ru 20	Vogt N., Moffat A.F.J. 1972, A&AS 7, 133
Ru 32	Moffat A.F.J., Vogt N. 1975, A&AS 20, 85
Ru 34	Moffat A.F.J., Vogt N. 1975, A&AS 20, 85
Ru 36	Vogt N., Moffat A.F.J. 1972, A&AS 7, 133
Ru 44	Moffat A.F.J., FitzGerald M.P. 1974, A&A 34, 291
Ru 46	Vogt N., Moffat A.F.J. 1972, A&AS 7, 133
Ru 47	Vogt N., Moffat A.F.J. 1972, A&AS 7, 133
Ru 49	Vogt N., Moffat A.F.J. 1972, A&AS 7, 133
Ru 55	Moffat A.F.J., Vogt N. 1975, A&AS 20, 85
Ru 59	Vogt N., Moffat A.F.J. 1972, A&AS 7, 133
Ru 67	Moffat A.F.J., Vogt N. 1975, A&AS 20, 85
Ru 76	Vogt N., Moffat A.F.J. 1972, A&AS 9, 97
Ru 79	Moffat A.F.J., Vogt N. 1975, A&AS 20, 85
Ru 97	Moffat A.F.J., Vogt N. 1975, A&AS 20, 125
Ru 98	Moffat A.F.J., Vogt N. 1973, A&AS 10, 135
Ru 107	Moffat A.F.J., Vogt N. 1975, A&AS 20, 155
Ru 108	Moffat A.F.J., Vogt N. 1973, A&AS 10, 135
Ru 118	Moffat A.F.J., Vogt N. 1973, A&AS 10, 135
Ru 119	Moffat A.F.J., Vogt N. 1973, A&AS 10, 135
Ru 127	Moffat A.F.J., Vogt N. 1975, A&AS 20, 155
Ru 166	Moffat A.F.J., Vogt N. 1973, A&AS 10, 135
Sh 1	Moffat A.F.J., Vogt N. 1975, A&AS 20, 125
Ste 1	Bronkalla W. 1963, Astr. Nach. 287, 249
	Eggen O.J. 1968, ApJ 152, 77 (no + 1000)
St 2	Krzeminski W., Serkowski K. 1967, ApJ 147, 988
St 7	Moffat A.F.J., Vogt N. 1973, A&AS 11, 3
St 8	Mayer P. 1964, Acta Univ. Car. Mat. et Phys. no 1, 25
St 13	Moffat A.F.J., Vogt N. 1975, A&AS 20, 125
St 14	Peterson C.J., FitzGerald M.P. 1988, MNRAS 235, 1439
St 16	Fenkart R.P., Binggeli B., Good D., et al 1977, A&AS 30, 307
St 17	Pandey A.K. 1986, Bull. Astr. Soc. India 14, 20
To 1	Turner D.G. 1983, JRAS Canada 77, 31
Tr 1	Oja T. 1966, Ark. Astr. 4,15 (no - 1000)
Tr 2	Hoag A.A., Johnson H.L., et al 1961, Publ. USNO 17, 347
Tr 7	Vogt N., Moffat A.F.J. 1972, A&AS 7, 133
Tr 9	Pismis P. 1970, Tonantzintla Tacubaya 35,293
Tr 10	Lyngå G. 1959, Ark. Astr. 2, 379
Tr 12	Lyngå G. 1959, Ark. Astr. 2, 379
Tr 14	Feinstein A., Marraco H.G., Muzzio J.C. 1973, A&AS 12, 331
Tr 15	Grubissich C. 1968, Z. Astrophys. 68, 173
Tr 16	Feinstein A., Marraco H.G., Muzzio J.C. 1973, A&AS 12, 331
Tr 17	Steppe H. 1977, A&AS 27, 415
Tr 18	Schmidt H., Diaz Santanilla G. 1964, Ver. Astr. Inst. Univ. Bonn no 7
Tr 21	Moffat A.F.J., Vogt N. 1973, A&AS 10, 135
Tr 24	Seggewiss W. 1967, Thesis (Univ. Münster)
Tr 26	Terzan A., Bernard A. 1981, A&AS 46, 49
Tr 27	Thé P.S., Stokes N. 1970, A&A 5, 298
Tr 28	Thé P.S., Stokes N. 1970, A&A 5, 298
Tr 33	Grubissich C. 1964, Z. Astrophys. 58, 276
Tr 35	Yilmaz F. 1966, Z. Astrophys. 64, 54
Tr 37	Alksnis A. 1961, Latv. Trans. 8, 88
	Marschall L.A., van Altena W.F. 1987, AJ 94, 71 (no > 2000)
Up 1	Rufener F. 1971, A&AS 3, 181
vdB 1	Arp H. 1960, ApJ 131, 322
vdBH 99	Landolt A.U., Perry C.L., Levato O.H., Malaroda S.M. 1990, AJ 100, 695
Wat 3	Moffat A.F.J., FitzGerald M.P., Jackson P.D. 1979, A&AS 38, 197
Wat 6	FitzGerald M.P., Boudreault R., Fich M., et al 1979, A&AS 37, 351
Wat 7	FitzGerald M.P., Moffat A.F.J. 1980, MNRAS 193, 774
Wes 1	Westerlund B. 1961, Uppsala Astr. Obs. Ann. 5, 17
Wes 2	Moffat A.F.J., Vogt N. 1975, A&AS 20, 125